ELLIOTT'S
GUIDE
TO
HOME
ENTERTAINMENT

ELLIOTT'S GUIDE TO HOME ENTERTAINMENT

JOHN ELLIOTT

FOURTH EDITION

AURUM PRESS

First published 1997 by Aurum Press Limited,
25 Bedford Avenue, London WC1B 3AT

A catalogue record for this book is available from the British Library.

ISBN 1 85410 485 3

Cover design by Slatter-Anderson

Typeset by Action Typesetting Ltd, Gloucester

Printed in Finland by WSOY

CONTENTS

PREFACE TO THE FOURTH EDITION   IX

INTRODUCTION   XI

ABBREVIATIONS   XVI

THE FILMS   1

INDEX OF ALTERNATIVE TITLES   915

DISTRIBUTORS LISTING   949

for Jack and Edith

PREFACE TO THE FOURTH EDITION

As its new title implies, the fourth edition of this Guide marks a radical departure from the previous three. Rapid developments in the field of satellite and cable TV are reflected in a wider scope. So while the book still provides extensive video coverage, its broader base means that much material shown on other media is included. This required a major overhaul, and the emphasis is now very firmly directed towards titles produced within the past ten years.

However, although many redundant entries have been dropped to make room for new ones, great care has been taken to ensure that films of major importance have been retained wherever possible. Readers should bear in mind that a book such as this is a "snapshot" that attempts to give a picture of the state of play at the time of writing. The decline of the independent video retailer and the rapid growth of satellite and cable has not increased diversity, and though we do our best to provide accurate information, we cannot guarantee the availability of any individual title. But unlike other guides, we can at least serve as a first point of call, giving viewers the chance to explore the situation further by contacting suppliers if they wish.

Once again, we are indebted to many helpful readers for their valuable comments and suggestions. No book is perfect and we are always grateful to those who take the time and trouble to point out errors and omissions. We hope that this book fulfils both their expectations and ours.

John Elliott, 1997.

INTRODUCTION

This book describes some 10,700 films or TV productions that represent a core selection of home entertainment titles. In general, soap operas, documentaries, filmed artistic or sporting events, instructional tapes and shorts are excluded. On the other hand, a large number of recent cinema releases are included because, in time, they will almost certainly become available on video or some other medium. A very small number of non-fiction works such as documentaries have also been included, wherever we felt there was good reason for doing so.

English is the original language of most of the book's entries but sub-titled or dubbed titles are also included. A note about close-captioned versions is provided. Unfortunately, at present there appear to be no titles offered in an audio-enhanced version, HEAR MY SONG being one of the few that ever appeared in this form.

This book is for all those who enjoy home viewing but are bewildered by the vast choice at their disposal. Our synopses are intended to provide some modest guidance on what is likely to appeal to our readers. We point out, for example, that certain films may lose impact on TV and attempt to mention outstanding scores or the historical context in which a film was made. In judging each entry, like tends to be compared with like, with due regard for budget limitations, genre, year of production etc. We do not expect the special effects in a 1950s SF film to resemble those in INDEPENDENCE DAY or the production values of a low-budget TV action tale to match those of the latest Hollywood blockbuster.

We try to strike a balance between depth and scope, in terms of information, and hope our work will appeal both to the determined film buff and those looking for an evening's entertainment.

Comprehensive details are given in each entry but some gaps are still inevitable, especially in the case of more obscure works. As far as possible, we prefer to include such incomplete entries rather than exclude them. For those hoping to find a particular item on video, details of recording formats and distributors are given. Availability cannot be guaranteed as distributors change and titles are sometimes withdrawn. However, if a video is listed there is a good chance that it is still in circulation.

ABBREVIATIONS

All abbreviations used are listed in the Abbreviations section on page xvi, except for those used for video distributors which are given in the Distributors Listing on page 949.

THE FILM ENTRIES

This is the core of the book, and information is organised as follows:

1. Title

Titles are in alphabetical order with slight variations when this would be more logical. E.g. where films have similarly titled sequels, it is sometimes more appropriate to place them in chronological order. This is also the case with identical titles, which are always put in chronological order.

Definite and indefinite articles (the, a, an) are ignored and placed at the end. Numerals are generally treated as if spelt out in full, and the same applies to abbreviations such as DR, LT, MR, SGT and ST.

If an acronym formed the whole or the first part of the title (e.g. H.E.A.L.T.H.) it is treated as a word in its own right.

American film titles generally use the American spelling, (e.g. HONOR not HONOUR) unless the British spelling is in use in the UK. Foreign films tend to be listed under a well known English alternative, unless there is none or the foreign-language title is in general use. For simplicity, accents and other diacritical marks are omitted and conventions such as the substitution of a double vowel for a German umlaut (as in Maedchen for Mädchen) are not used.

Films intended as a star vehicle (e.g. Elvis Presley), or as part of a series (e.g. the Marx Brothers films) are grouped together under a generic title where this seems best. If a title cannot be found, always check the Index of Alternative Titles.

2. Star Rating
All entries are rated as follows:

no stars	–	abysmal/unwatchable
*	–	poor to bad
**	–	mediocre to fair
***	–	good to very good (or containing sequences of interest)
****	–	excellent
*****	–	outstanding/a masterpiece

3. British Board of Film Classification Certification
See the abbreviations section (page xvi). BBFC certifications are always given for cinema and video releases wherever possible, but only for films shown on other media where it is certain that a BBFC certificated version is being used. If doubts exist as to the BBFC certification, or it is unknown, a suggested certification is generally given in brackets, but there is no guarrantee that this will be the BBFC certification, should one eventually be given.

4. Director(s)
Assumed and real names are given wherever possible, and if an alias was used, the real name (if known) follows in brackets. For film entries comprising separate stories, all made by different directors, the names are given in the order in which their contributions are mentioned in the synopsis section.

5. Country/Countries of Origin
This refers to the country/countries that had overall responsibility for production, rather than simply the location of the studio (e.g. American films are often made in the UK). However, this can be hard to determine exactly and, where doubts exist, the country/countries that provided the principal locations and/or studios tend to share credit with those that provided the bulk of the finance.

6. Year of Production/Release
A single date indicates a film released without delay after production. Where a delay occurred between production and release, this is indicated if it is known.

7. Principal Players
Some actors use assumed names, and we give the real name in brackets if this seems useful. Where actors change the style or spelling of their name, the same rule applies. In general, the credits list players in their order of prominence in the film, rather than according to a conventional pecking order. Where possible, the source of voice-overs for both animations and live-action films (e.g. a narrator) is given.

Sometimes an entry consists of several stories, with each one having a different cast. Here, the actors are given in the order in which the stories are mentioned in the synopsis, and the different casts are separated by a semi-colon.

8. Plot Synopsis
This gives a plot outline plus other relevant details, such as production difficulties, outstanding scores or scripts, sequels, remakes, Academy Awards and any other items felt to be of interest.

Where other titles are noted, they are capitalised if listed in the book, and put in inverted commas if not. Where space permits, the full names of all Academy Award recipients are given.

9. Alternative Title(s)
These are listed alphabetically prefaced by the abbreviation Aka. The same conventions as in the main title entries are used for American and foreign-language titles.

10. Film Category Code

Categories are self-explanatory, are not rigid, and reflect the most essential element in the film.

A	Adult
A/AD	Action/adventure
ANIM	Animation (Cartoon, claymation or puppets. Where there is live-action as well, the synopsis makes this clear.) Wherever possible, the voices credited are those heard.
COM	Comedy
DOC	Documentary (Only included under exceptional circumstances.)
DRA	Drama
FAN	Fantasy and science fiction
HOR	Horror
JUV	Juvenile
MAR	Martial arts
MUS	Musical
THR	Thriller
WAR	War
WES	Western

11. Running Time(s)

These are rounded up or down to the nearest minute. Where useful, alternative running times are given, and distributor codes follow in brackets to distinguish the different versions available where such information would be useful.

If the complete entry occupies several cassettes (e.g. a TV mini-series) this is indicated if known. In such cases, each cassette may be assumed to be of roughly equal duration, unless otherwise stated.

Entries are assumed to be in colour and to have an original English-language soundtrack unless otherwise shown (e.g. silent, sub-titled or dubbed). However, productions from non-English-speaking countries are assumed to be in the language of that country with English sub-titles unless otherwise stated. The running time stated is the one for the medium in which the entry is available. Longer and original running times are also given where known, while information on cuts is given where possible and/or helpful. A film speed of 24-frames-per-second has been assumed for silent films. If the title item is included with others on the same cassette, this is noted, but wherever possible, each item has its own entry (e.g. the various STAR TREK episodes are all listed separately). Information about other titles on the same cassette is restricted to a simple note (e.g. 2-episode cassette) rather than a listing of all the other titles. This is done because such information changes constantly. It would also obscure our main aim, which is to provide information about the chosen entry.

12. Availability/Distribution

The following codes describe the criteria for inclusion and are given in order of decreasing priority:

VIDrel. The entry is available in a video form within certain constraints, and SATrel, CABrel, TVrel and CINrel may also apply.

SATrel. The entry has been shown on a satellite channel and CABrel, TVrel and CINrel may also apply.

CABrel. The entry has been shown on a cable channel and TVrel and CINrel may also apply.

TVrel. The entry has been shown on a terrestrial TV channel and CINrel may also apply.

CINrel. The entry has had a cinema release but, to our knowledge, is at present available nowhere else.

Where the film has been shown on several media, we only list the one that is of the highest priority. For example, STAR WARS is shown as VIDrel. This means that it was definitely available on video and was possibly also shown on satellite, cable, TV and in the cinema. However, availability in other media is not guaranteed. E.g. BEHIND THE WATERFALL is marked as SATrel and is currently not being shown on any other media.

This prioritisation was established to save space and avoid including unnecessary references, our guiding principle being to maximise the value of the information supplied. So if a film shown on satellite, for instance, were to become available on video, the code VIDrel would replace the code SATrel. This means that although, at first glance, the book appears to be primarily a video guide, it nonetheless achieves its aim of covering an enormous amount of material shown elsewhere.

Video takes the top priority, as at present this represents the sole medium on which a chosen title is available to the majority of home viewers. Note that a lower priority code never displaces a higher priority one, unless it is felt that the latter serves no purpose (e.g. if it has been many years since a film was available on video, the VIDrel code may be replaced by another one).

The following notes provide further clarification regarding the various media:

Video Distributors. In almost every case, where our information is less than 18 months old, just one code is given, for the most recent supplier. E.g. VIDrel: WHV means that this film should still be available from Warner Home Video. A full listing (with addresses and other information) of the abbreviations used for video distributors will be found in the Distributors Listing on page 949. No distinction is made between rental and sell-through as the situation changes constantly. In rare cases, more than one distributor is given.

The abbreviation L/A (limited availability) used after the distributor code — e.g. VIDrel: WHV L/A — indicates that our information is more than 18 months old or that we have doubts as to the current availability of the video or the status of the supplier. Always refer to the Distributors Listing to clarify the position. Many readers have suggested that any uncertainty about availability should be indicated in order to avoid disappointment.

Note that we can make no statement as to the legality of any film listed as being supplied by a distributor, nor does the code L/A imply that a given distributor has been handling an illegal version of a film. At all times we have tried to ensure that our information is accurate. The inclusion of a distributor is not an endorsement of the film concerned, nor is the use of the code L/A a criticism.

In some cases the video entry simply has the code VIDrel: L/A and no distributor is given. In these cases, the distributor information is felt to be of no further value, yet the fact the film was shown on video is still considered important. By the same token, if we believe it possible that the film may be released in the future by a new distributor, we use the reference VIDrel: L/A.

Satellite. The relevant satellite channel is given where known as this may prove helpful to the viewer – e.g. SATrel: SKY MOVIES.

Cable. Cable distribution takes two forms. The re-broadcasting of satellite-originated material or direct studio transmission by the cable company. The former is included under satellite, and the small number of cable releases provided relates to the latter, with the company being given where known – e.g. CABrel: HVC.

TV. Where possible, the British terrestrial channel on which the entry was shown is given, but this may not always be the most recent showing, as such information changes constantly — e.g. TVrel: BBC1.

Cinema. This pertains to UK cinema releases and, wherever possible, the cinema running time and BBFC certification are given.

13. Video Format(s)

This section only applies to films marked as video entries (i.e. VIDrel). Where only one distributor is shown, the formats listed refer to that supplier. If more than one distributor is listed and they do not handle identical formats, the position is clarified by giving extra information in brackets – e.g. VIDrel: WHV; PION (LV only).

A full list of codes used for video formats will be found in the Abbreviations section on page xvi. However, this does not generally extend to making distinctions between the various VHS formats available, e.g. stereo, digitally remastered, etc.

14. Additional Information

The abbreviation Boa (Based on a) is used to indicate the literary or other source of a film. Written works are named when films (and remakes) truly derive from them, but only where this is deemed to be appropriate.

The title of the original work is given if it differs substantially from the film title (e.g. GREYSTOKE is based on the novel Tarzan of the Apes). However, this rule does not apply when the title was modified so as to be included as one of several under a generic title (e.g. ELVIS PRESLEY: STAY AWAY, JOE is based on the novel Stay Away, Joe and so the title of the novel is not given).

For a written work to be noted in this section, it should have existed prior to the film being made (e.g. as a novel but not a novelisation of a screenplay), take a different form (e.g. a play or stage musical but not an earlier version of the film) and be the source of both characterisation and plot. Derivation from sources such as songs or poems is more appropriately noted in the synopsis. For TV films, whose sources may not be so well known, such information is provided where available.

INDEX OF ALTERNATIVE TITLES (PAGE 915)

This section lists alternative titles in alphabetical order. However, numerical titles appear at the very beginning of this section, while acronym-based titles are placed at the start of their respective letter groups.

DISTRIBUTORS LISTING (PAGE 949)

This section provides an extensive listing of video distributors, with details of addresses etc. and is arranged alphabetically in the order of the codes used in the body of the book. If a distributor sells wholesale or appears to have become inactive, this is noted if known.

Satellite and cable distributors are not included in this section, but if it is thought useful, a note about them appears after the general abbreviations section.

ABBREVIATIONS

NB. Abbreviations used for video distributors are listed in the Distributors Listing on page 949. Abbreviations relating to Academy Awards are listed separately below.

/a	(after video format code) active play format available	cert	certification(s)
/cav	(after video format code) a version of Laservision that allows perfect freeze-framing and viewing in slow motion without distortion	CINrel	cinema release
		Col	colour
		Col/B/W	colour with B/W sequence(s) or a colour episode followed by a B/W episode (compilation tapes only)
/dm	(after video format code) digitally re-mastered soundtrack available	COM	Comedy (film category)
		comp	compilation
/h	(after video format code) hi-fi soundtrack available	coVer	colourised version available
		DOC	Documentary (film category)
/l	(after video format code) long play format available	DRA	Drama (film category)
		Ex	Exempt from a BBFC certification because of its content
/s	(after video format code) stereo soundtrack available		
		FAN	Fantasy and science fiction (film category)
/sh	(after video format code) stereo and/or hi-fi soundtrack available		
		fps	frames per second
/ss	(after video format code) surround sound	HOR	Horror (film category)
		HVC	Home Video Channel (a dedicated cable channel – see Additional Information below)
12	BBFC rating – only suitable for persons over the age of 12		
15	BBFC rating – only suitable for persons over the age of 15	incl	includes/including
		JUV	Juvenile (film category)
18	BBFC rating – only suitable for persons over the age of 18	(l)	Lyrics
		L/A	Limited availability (noted if at the time of writing the video appears to be unavailable from distributor(s))
A	Adult (film category)		
AD	Action/Adventure (film category)		
AA	Academy Award	LV	Laservision (only the PAL format is covered, as NTSC material has to be imported)
AAN	Academy Award nomination		
Aka	also known as (used with reference to alternative titles)		
		LWT	London Weekend Television (see Additional Information below)
ANIM	Animation (film category)		
asa	also separately available	(m)	music
audE	audio-enhanced version available for the visually impaired (the sound-track generally has additional narration)	MAR	Martial arts (film category)
		mCab	made for cable TV
		min	minute(s)
		MPAA	Motion Picture Association of America
avail	available		
BBC	British Broadcasting Corporation (see Additional Information below)	mPay	made for Pay-TV
		mSat	made for Satellite-TV
BBFC	British Board of Film Classification	mTV	made for TV
Boa	based on a novel, play etc. (only noted if clearly derivative)	MUS	Musical (film category)
		mVid	made for video (in the case of adult films, this often means that the film was shot on video as well)
B/W	black and white		
B/W/Col	mostly B/W with colour sequence(s) or a B/W episode followed by a colour episode (compilation tapes only)		
		NTSC	TV broadcast standard – USA
		orig	original
		ort	original running time
C4	Channel Four (see Additional Information below)	Osca	on the same cassette as (used on rare occasions for tapes with more than one title recorded, where the other titles are named)
C5	Channel Five (see Additional Information below)		
		PAL	TV broadcast standard – UK et al.
CABrel	cable release	PG	BBFC rating – parental guidance
CARLTON	Carlton Television (see Additional Information below)	R18	BBFC rating – only available from licensed sex shops to over 18s
cC	close captioned – this is a form of sub-titling for the hearing impaired. It requires the use of an adaptor (see Additional Information below)	SATrel	satellite release
		sec	second(s)
		SECAM	TV broadcast standard – France et al.

sil	silent(s)
subs	subtitled
subst	substituted
THAMES	Thames Television (see Additional Information below)
THR	Thriller (film category)
TVrel	TV release
U	BBFC rating – suitable for children
Uc	BBFC rating – especially recommended for children
V	VHS format
ver	version(s)
VIDrel	video release
WAR	War (film category)
WES	Western (film category)

wScrn   Wide-screen version available. Films shot in either 70mm or Cinemascope have an aspect ratio that differs markedly from that of a TV screen. Videos and TV transmissions of such films are normally either "panned and scanned" or shown minus their left and right borders. wScrn means that the entire frame is seen on the TV screen, and the film takes the form known colloquially as "letter-box".

WW   World War (the appropriate number following, e.g. WW1)

ADDITIONAL INFORMATION

BRITISH BOARD OF FILM CLASSIFICATION
3 Soho Square, London W1 (Tel: 0171 439 7961)

BRITISH BROADCASTING CORPORATION
BBC Television, Woodlands, 80 Wood Lane, London W12 0TT
(Tel: 0181 576 2000/0181 743 5588)

CARLTON TELEVISION
101 St Martins Lane, London WC2 (Tel: 0171 240 4000)

CHANNEL FOUR
Channel Four Television Corporation, 124 Horseferry Road,
London SW1 (Tel: 0171 396 4444)

CHANNEL FIVE
Channel Five Broadcasting Ltd., 22 Long Acre, London WC2
(Tel: 0171 421 7100/0171 497 5225)

CLOSE-CAPTIONING
European Captioning Institute, Thurston House, 80 Lincoln Road,
Peterborough PE1 2SN

HOME VIDEO CHANNEL
The Home Video Channel Ltd., Aquis House, Station Road, Hayes,
Middlesex UB3 4DX (Tel: 0181 581 7000)

LONDON WEEKEND TELEVISION
London Television Centre , Upper Ground, London SE1
(Tel: 0171 620 1620)

MOVIE CHANNEL see SKY MOVIES

SKY MOVIES
British Sky Broadcasting Ltd., Centaurs Business Park, Isleworth
(Tel: 01506 484777)

THAMES TELEVISION
Thames Studios, Teddington Lock, Broom Road, Teddington
(Tel: 0181 977 3252)

ACADEMY AWARDS

Because of the confusing range of annual Academy Awards, these abbreviations have been listed separately. Over the years the names of the various awards have altered, as new categories have arisen or old ones been replaced or expanded. Where changes have occurred an explanation is given. Note that when a year is given, it refers to the date of the awards ceremony.

Actor	Best Actor.
Actress	Best Actress.
Art	Best Art Direction. From 1940 to 1956/57 separate awards were made for B/W and colour films. These categories were merged in 1957/58, but separated later. Since 1967, B/W and colour films have not been treated separately. Regarding Interior Decoration, this was distinguished from 1941–1946, became Set Decoration in 1947, and has remained a single category ever since. However, the two awards are now usually given together.
Assist Dir	Best Assistant Direction.
Cin	Best Cinematography. Sometimes called Best Photography, this award was split for B/W and colour films from 1939 onwards.
Cost	Best Costume Design. Awarded separately for B/W and colour, this category was created in 1948.
Dance Dir	Best Dance Direction.
Dir	Best Direction.
Doc	Best Documentary. Awarded separately for shorts and feature-length films, from 1941 onwards.
Edit	Best Film Editing.
Effects	Best Special Effects. Originally this category could include both visual and sound effects, but for 1963–1967, sound effects were recognised in their own right. After 1967, only visual effects were awarded in their own right, but not always each year, and were sometimes recognised by a Special Award only. Since 1982, both Special Visual Effects and Special Sound Effects Editing have been recognised, with the former awarded each year and the latter most years. If sound and visual effects are awarded separately, the abbreviations aud and vis are used.
Foreign	Best Foreign Language Film. Before 1956, foreign films were only recognised by a Special or Honorary Award.
Hon Award	Honorary Award. This is sometimes given for an outstanding contribution to a specific film and is sometimes known as a Special Award.
Int	Best Interior Decoration (see note on Best Art Direction)
Make	Best Make-up
Pic	Best Picture
S. Actor	Best Supporting Actor
S. Actress	Best Supporting Actress
Score	Best Score. Over the years this has generally been split to distinguish between musical and non-musical pictures. Up to 1961, recognition was rarely extended to take into account the originality of the score, except for non-musical pictures, and in some years only one award has been made. Generally, this category now recognises Best Scores for musicals, regardless of whether they are original scores or adaptations of another medium.
Score/orig	Best Original Score. Generally used for non- musical pictures.
Score/adapt	Best Score. Only used if the score has clearly been recognised as an adaptation.
Screen	Best Screenplay. Up to 1939 this was the only category for such awards. For 1940–1947, it was given with a Best Original Screenplay award. From 1948 onwards there were a number of different categories, but it was eventually dropped in 1974 in favour of separate awards for original and adapted screenplays.

Screen/orig	Best Original Screenplay written directly for thescreen, or based on material not previously published or produced.
Screen/adapt	Best Screenplay based on material adapted from another medium.
Set	Best Set Decoration. (See Best Art Direction.)
Short	Best Short Film. This is awarded separately for animations and live-action films.
Song	Best Original Song. Where appropriate, the writers of the music and lyrics are individually credited,the abbreviations (m) and (l) being used.
Sound	Best Sound Editing/Recording.
Spec Award	Special Award. This category fell into disuse by1950, and was replaced by an Honorary Award category. In 1972 it was re-instituted to replace the Honorary Award category, but became known as a Special Achievement Award. It is given for an outstanding technical contribution to a specific film.
Story	Best Story/Writing. Originally this was the onlywriting recognition given. For 1930–1934, it wasdropped for categories covering original and adapted writing. For 1935–1956, only the Best Original Story category was retained, after whichseparate awards were made for both Best ScreenplayAdaptation and Best Story/Screenplay written directly for the screen.
Story/orig	Best Story written directly for the screen.
Story/adapt	Best Story based on material from another medium.
Story/Screen	Best Story/Screenplay with both story and screenplay written directly for the screen.

THE FILMS

A

A-TEAM, THE: JUDGEMENT DAY **
PG
David Hemmings USA   1985
George Peppard, Mr T, Dwight Schultz, Dirk Benedict
A dishonest lawyer arranges to have a judge's daughter kidnapped in order to ensure a favourable outcome for a client on trial. The A-Team are called in by the distraught judge and they obligingly help out with their own unique blend of organised mayhem. Based on the popular TV series.
A/AD   90 min   VIDrel: CIC/SONOP   V

A-TEAM, THE: THE COURT MARTIAL **
PG
Tony Mordente/Les Sheldon/Michael O'Herlihy USA   1986
George Peppard, Mr T, Dwight Schultz, Dirk Benedict, Robert Vaughn, Eddie Velez
Based on the popular TV series of the same name, this has the A-Team going to Barcelona in order to rescue a hijacked American plane. However, on board is a man, long believed dead, who can clear the A-Team of a number of serious charges.
Aka: A-TEAM, THE: TRIAL BY FIRE; LAST COURT MARTIAL, THE
A/AD   132 min (ort 140 min)   VIDrel: CIC/SONOP   V/h

ABANDONED AND DECEIVED **
(PG)
Joseph Dougherty USA   1995
Lori Loughlin, Brian Kerwin, Gordon Clapp, Farrah Forke, Eric Lloyd, Bibi Besch, Rosemary Forsyth, Ilene Graff, Robert Hooks, Claudette Nevins, Anthony Tyler Quinn, Amy Steel, Tyler Malinger, Allison Dean, Markus Flanagan
A strong performance from Loughlin sustains this so-so movie about a single mother whose battle with stonewalling bureaucrats becomes a nightmare, when her ex-husband's refusal to pay child support leaves both her and her two sons in poverty. Ultimately, she is obliged to campaign at the national level, becoming something of a champion for others in the same position.
DRA   90 min mTV   SATrel: SKY MOVIES

ABBOTT AND COSTELLO: AFRICA SCREAMS ***
U
Charles Barton USA   1949
Bud Abbott, Lou Costello, Max Baer, Hillary Brooke, Clyde Beatty, Frank Buck, Shemp Howard, Buddy Baer, Joe Besser
Could just as well have been entitled Abbott and Costello on Safari, with our comic duo experiencing various crazy mishaps whilst on safari in Africa, their intention being to locate a hoard of diamonds with a secret map. One of their better efforts with a number of hilarious sequences.
Aka: AFRICA SCREAMS
COM   79 min (ort 87 min) B/W   VIDrel: ORBIT/DISC   V

ABBOTT AND COSTELLO: JACK AND THE BEANSTALK *
U
Jean Yarbrough USA   1952
Bud Abbott, Lou Costello, Dorothy Ford, Barbara Brown, Buddy Baer, David Stollery, William Farnum, Shaye Cogan, James Alexander, Joe Kirk, Johnny Conrad and Dancers, Patrick the Harp
Abbott and Costello version of this tale that begins and ends with black and white sequences. An anaemic version of this famous fairytale with none of verve or funny routines common to their earlier films. Very poor.
Aka: JACK AND THE BEANSTALK
COM   78 min B/W/Col   VIDrel: ORBIT/DISC   V
Boa: short story by Jakob Ludwig Karl Grimm and Wilhelm Karl Grimm.

ABBOTT AND COSTELLO MEET CAPTAIN KIDD **
U
Charles Lamont USA   1952
Bud Abbott, Lou Costello, Charles Laughton, Hillary Brooke, Fran Warren, Bill Shirley, Leif Erickson, Sid Saylor, Rex Lease, Frank Yaconelli, Bobby Barber
A lame pirate spoof with a generous helping of lousy songs which may be of slight interest if only for the presence of Laughton in a rare comic role. Our inept duo are stranded in Tortuga where they get hold of a treasure map. Captain Kidd in the guise of Laughton is however, in hot pursuit.
Aka: MEET CAPTAIN KIDD
COM   70 min   VIDrel: VCC L/A   V

ABDUCTION OF INNOCENCE **
15
James A. Contner USA   1996
Katie Wright, Dirk Benedict, Lucie Arnez, Lochlyn Munro
A wealthy businessman who runs a town as if it is his private domain receives a ransom note when his daughter is kidnapped, but later discovers that it is part of a plot on the part of the girl to extort money from him.
DRA   90 min   VIDrel: ODY/SONOP   V/sh

ABIGAIL'S PARTY **
PG
Mike Leigh UK   1977
Alison Steadman, Tim Stern, John Salthouse
One of Mike Leigh's sharp and barbed character studies in which middle-class mores are put under the microscope in the course of a single night. The first full-length TV drama made in the UK to be partly improvised.
COM   102 min mTV   VIDrel: BBC/TECH   V/h

ABOMINABLE DOCTOR PHIBES, THE ***
15
Robert Fuest UK   1971
Vincent Price, Joseph Cotten, Hugh Griffith, Terry-Thomas, Virginia North, Aubrey Woods, Susan Travers, Alex Scott, Edward Burnham, Peter Gilmore, Peter Jeffrey, Maurice Kaufman, Norman Jones, John Cater, Sean Bury
A man who was horribly disfigured in a car accident that also took the life of his wife, seeks revenge on those whom he considers were responsible. A horror spoof that is a cut above many films of this genre. Followed by the inferior DOCTOR PHIBES RISES AGAIN.
HOR   90 min (Cut at film release – ort 94 min)
VIDrel: VISVID/POLYREC L/A   V

ABOUT LAST NIGHT **
18
Edward Zwick USA   1986
Rob Lowe, James Belushi, Demi Moore, Elizabeth Perkins, George DiCenzo, Michael Alldredge, Robin Thomas, Donna Gibbins, Megan Mollally, Patricia Duff, Rosana De Soto, Sachi Parker, Robert Neches, Joe Greco, Ada Mavis
A teenage love affair comes under threat from peer pressure in this light romantic comedy aimed at the under-twenties. An interesting if rather bland version of a far more engaging one-act play. Let down by the inconsistent direction and poor characterisation, this gave Perkins her screen debut.
COM   102 min (ort 113 min)   VIDrel: VCC/DISC   V/sur
Boa: play Sexual Perversity In Chicago by David Mamet.

ABOVE SUSPICION **
18
Steven Schacter USA   1995
Christopher Reeve, Joe Mantegna, Kim Cattrall, Finola Hughes, Michelle Bonilla, David Byron, Gerald Castillo, Ron Canada, Peter Michael Goetz, Blake Foster, Jonathan Friedman, Ramiro Gonzalez, Tony Mamet, Joy Hooper
After a cop is paralysed by a bullet fired by a drugs runner, his wife and his brother combine forces to care for him. This sparks an erotic attraction between brother and wife, complicated by the fact that the cop has a generous life insurance policy. Being unable to face life in a wheelchair, he devises a scheme for wife and brother to murder him. A dull thriller notable only for being the last film Reeve made before the accident that paralysed him for real.
THR   91 min   VIDrel: MOSAIC/COLUM   V/sur

ABOVE THE LAW **
18
Corey Yuen HONG KONG   1987
Cynthia Rothrock, Yuen Biao, Peter Cunningham, Roy Chiao, Karen Sheperd
A tough female cop and martial arts expert finds herself up against a group of hired killers when she sets out to bring to book a man who has set himself up as a self-appointed vigilante. A standard blend of maximal action and minimal plotting, followed by a sequel.
Aka: RIGHTING WRONGS
A/AD   86 min Cut (11 sec – ort 90 min)
VIDrel: 4-FRONT/POLYREC   V/sur

ABOVE THE LAW 2: THE BLOND FURY **
18
Man Hoi HONG KONG   1987
Cynthia Rothrock, Chin Siu Ho, Man Hoi, Ronnie Yu, Elizabeth Lee, Jeff Falcon
Another action vehicle for martial arts star Rothrock, in this fast-paced tale of a reporter on the Hong Kong Times, who is covering a fraud trial and finds that her investigations lead to a couple of hitmen being sent to silence her for good. Enjoyable if not all that original in design or execution.
Aka: BLOND FURY, THE; BLONDE FURY; LADY REPORTER
A/AD   85 min   VIDrel: 4-FRONT/POLYREC   V/sur

**ABOVE THE RIM ** **   15
Jeff Pollack USA   1994
*Duane Martin, Leon, Tupac Shakur, Marlon Wayans, Tonya Pinkins,
Bernie Mac, David Bailey, Byron Minns, Sherwin David Harris,
Shawn Michael Howard, Henry Simmons II, Iris Little Thomas,
Michael Rispoli, Bill Raftery, Eric Nies*
A young athletic teenager is torn between her ambitions to
improve himself by going to college and the lure of street basket-
ball and its promise of quick money. His predicament is not
made any easier when he gets to know two very different brothers. A failed attempt to say something new about urban and
youth problems. This was Pollack's directorial debut, he also
acted as the movie's co-scriptwriter and co-producer.
DRA   93 min (ort 98 min)   VIDrel: FIRST/SONOP   V/sh

ABOVE US THE WAVES *   U
Ralph Thomas UK   1955
*John Mills, John Gregson, Donald Sinden, James Robertson Justice,
Michael Medwin, James Kenney, O.E. Hasse, Lee Patterson, Lyndon
Brook, William Russell, Thomas Heathcote, Theodore Bikel, Anthony
Newley, Harry Towb*
Realistic, documentary-style film that tells of the relentless
attempts on the part of the British to destroy the German battle-
ship Tirpitz during WW2. A gripping submarine drama held
together by an excellent cast.
WAR   95 min (ort 99 min) B/W   VIDrel: CARL/TECH   V
Boa: book by C.E.T. Warren and J. Benson.

ABRAHAM VALLEY *   PG
Manoel de Oliveira FRANCE/PORTUGAL/
SWITZERLAND   1993
*Mario Barroso, Leonor Silveira, Cecile Sanz de Alba, Luis Miguel
Cintra, Rui de Carvalho, Luis Lima Barreto, Micheline Larpin, Diogo
Doria, Jose Pinto, Joao Perry, Filipe Cochofel, Gloria de Matos,
Antonio Reis, Isabel Ruth, Dino Treno*
A Portuguese variant on Madame Bovary, that examines the
unhappy life of an orphan girl who is brought up with a wealthy
family, whom she leaves for married life with a doctor. But this
marriage brings her little happiness, and she is unfaithful to
him. A melancholy and rather over-intellectualised work, but
one that is always extremely good to look at.
Aka: VALE ABRAAO
DRA   189 min   CINrel
Boa: book by Augustina Bessa-Luis.

**ABRAXAS: GUARDIAN OF THE UNIVERSE **   (15)
Damian Lee USA   1990
*Sven-Ole Thorsen, Jesse Ventura, Marjorie Bransfield, Jim Belushi,
Jerry Levitan, Robert Nasmith, Kris Michaels, Lane Coleman, Francis
Mitchell, Sonja Belliveau plus voices of: Marilyn Lightstone, Moses
Znaimer*
The title character is a powerful intergalactic cop whose job it is
to safeguard the wellbeing of two galaxies. This task becomes
more difficult when he learns that a former partner has become
a dangerous foe who seeks domination of the universe. This
quest has taken the latter to Earth, where he has fathered a child
of strange powers to help him achieve his goal. Fortunately,
Abraxas shows up determined to stop him.
FAN   88 min (ort 90 min)   CABrel: HVC

ABSENCE OF MALICE *   PG
Sydney Pollack USA   1980
*Paul Newman, Sally Field, Bob Balaban, Luther Adler, Melinda
Dillon, Barry Primus, Josef Sommer, Don Hood, John Harkins,
Wilford Brimley, Arnie Ross, Anna Marie Napoles, Shelley Spurlock,
Joe Petrullo, Shawn McAllister*
Absorbing drama telling of reporter Field who is tricked by an
unscrupulous government investigator into printing a story that
harms an innocent man, who then tries to get even. Written by
ex-reporter Kurt Luedtke, this film raises some interesting
points with regard to press freedom versus the rights of the
individual. Filmed in Miami.
DRA   111 min (ort 116 min)   VIDrel: VCC L/A   V

ABSENT-MINDED PROFESSOR, THE *   U
Robert Stevenson USA   1960
*Fred MacMurray, Nancy Olson, Keenan Wynn, Tommy Kirk, Ed
Wynn, Leon Ames, Elliott Reid, Edward Andrews, Wally Brown,
Forrest Lewis, James Westerfield, David Lewis, Belle Montrose, Alan
Carney, Gage Clarke, Alan Hewitt*
Lighthearted Disney story with MacMurray as the title charac-
ter discovering "flubber" or flying rubber; a substance with most
peculiar properties. The only person who believes him is out to

steal it. A bright and breezy comedy with rather good special
effects. Followed by "Son Of Flubber".
COM   97 min B/W coVer   VIDrel: WDV L/A   V
Boa: short story A Situation Of Gravity by Samuel Taylor.

ABSENT WITHOUT LEAVE *   (PG)
John Laing NEW ZEALAND   1993
*Craig McLachlan, Katrina Hobbs, Desmond Kelly, Tina Cleary,
Rebecca Hobbs, Helen Moulder, Joan Foster, Stepehn Lovatt, Margaret
Blay, Robyn Malcolm, David Copeland, Danny Mulheron, Robert
Bennett, Chloe Lang, Peter Yule, Stephen Hale*
True story about a confused and unhappy young soldier who
goes AWOL during WW2 after his wife suffers a miscarriage.
The story is about as simple as they come, but it is really the
acting that makes this film, plus the highly convincing recreation
of the period and attention to detail.
DRA   96 min   SATrel: SKY MOVIES

**ABSOLUTE BEGINNERS **   15
Julian Temple UK   1986
*Eddie O'Connell, Patsy Kensit, James Fox, Lionel Blair, Ray Davies,
David Bowie, Anita Morris, Steven Berkoff, Mandy Rice-Davies, Sade
Adu, Eve Ferret, Graham Fletcher Cook, Bruce Payne, Tenpole Tudor,
Tony Hippolyte*
A lively and stylised musical set in 1950s London and looking
at the beginnings of teenage identity. Disappointing in that the
characters are rather flat, and an examination of the rise of
racism is a contrived attempt to add substance to the film.
Nonetheless, the music of Bowie, Sade and jazz veteran Slim
Gaillard are compensations in an interesting if not entirely satis-
factory experiment.
MUS   107 min   VIDrel: VISVID/POLYREC   V/sur
Boa: novel by Colin MacInnes.

ABSOLUTE POWER *   15
Clint Eastwood USA   1996
*Clint Eastwood, Gene Hackman, Ed Harris, Scott Glen, Judy Davis,
Laura Linney, Dennis Haysbert, E.G. Marshall*
A well-plotted thriller that casts Eastwood as a high-class cat
burglar who steals only from the rich. But out on a job he sees
a murder take place involving the US President and has to
choose between revealing what he has seen and risking imprisonment or remaining silent. Once again, Eastwood
demonstrates his growing assurance behind the camera as well
as in front of it. The screenplay is by William Goldman.
THR   121 min   CINrel
Boa: novel by David Baldacci.

**ABSOLUTION **   18
Anthony Page UK   1978 (released 1981)
*Richard Burton, Dominic Guard, Dai Bradley, Andrew Keir, Billy
Connolly, Willoughby Gray, Hilda Fenemore, Sharon Duce, Hilary
Mason, Robert Addie, Trevor Martin, Robin Soans, Preston Lockwood,
James Ottaway, Brook Williams*
In the cloistered atmosphere of a Catholic boys' school a
humourless priest falls victim to a practical joke played on him
by his best student. Despite a strong performance by Burton, the
film's credibility degenerates as the story progresses. Written by
Anthony Shaffer and filmed in 1978 but not released in the USA
until 1988.
Aka: MURDER BY CONFESSION
DRA   95 min (ort 105 min)   VIDrel: CASPIC L/A   V

ABYSS, THE *   15
James Cameron USA   1989
*Ed Harris, Mary Elizabeth Mastrantonio, Michael Biehn, Leo
Burmester, Todd Graff, John Bedford Lloyd, J.C. Quinn, Kimberly
Scott, Jimmie Ray Weeks, Chris Elliott, George Robert Klek,
Christopher Murphy, Adam Nelson, Richard Warlock*
Underwater adventure written by Cameron, with the US Navy
enlisting the aid of civilian divers working on an experimen-
tal oil-drilling rig when one of their nuclear subma-
rines gets into trouble. Excellent special effects and photography
sustain this sluggish fantasy-adventure. The score is by Alan
Silvestri. In 1992 a special 171-minute edition was released.
See also DEEPSTAR SIX and LORDS OF THE DEEP. AA:
Effects/vis (John Brano/Dennis Muren/Hoyt Yeatman/Dennis
Skotak).
A/AD   133 min Cut (45 sec plus some cuts subst – ort 140
min) wScrn   VIDrel: 20TH/TECH; ENCORE (LV only –
Special Edition)   V/sur LV

ACCATTONE *   15
Pier Paolo Pasolini ITALY   1961
Franco Citti, Franca Pasut, Roberto Scaringella, Adele Cambira, Adriana Asti, Silvana Corsini, Paolo Guidi, Renato Capogna, Mario Cipriani, Piero Morgia, Umberto Bevilacqua, Elsa Morante, Danilo Alleva, Polidor
A tough, unsentimental look at the life of prostitutes, pimps and petty thieves in the slums of Rome, where a young man finds his options strictly limited in his efforts to make an honest living. This engrossing slice of neo-realism was Pasolini's first film and Bernado Bertolucci was one of the assistant directors.
Aka: PROCURER, THE
DRA   112 min (ort 116 min) B/W
VIDrel: CONNO/RTM L/A   V
Boa: novel by Pier Paolo Pasolini.

ACCEPTABLE RISKS *   15
Rick Wallace USA   1986
Brian Dennehy, Cicely Tyson, Kenneth McMillan, Christine Ebersole, Richard Gilliland, Beah Richards, Vic Polizos, Peter Jurasik, Patti Cohoon, Richard McKenzie, Steve Eastin, Tom Simcox, Scott Curtis, Charina Felthous, Edan Gross
The manager of a chemical plant is ordered to cut costs by lowering safety standards but puts his job on the line when he realises that these measures could lead to an ecological disaster for the surrounding community. A solidly crafted but highly predictable drama that never rise above its mTV format.
DRA   88 min (ort 97 min) mTV   VIDrel: ODY/SONOP
V/sh

ACCIDENT *   PG
Joseph Losey UK   1967
Dirk Bogarde, Stanley Baker, Jacqueline Sassard, Delphine Seyrig, Alexander Knox, Michael York, Vivien Merchant, Harold Pinter, Anne Firbank, Brian Phelan, Freddie Jones, Nicholas Mosley, Terence Rigby, Jill Johnson
The script by Harold Pinter sets the tone for a complex story whose main elements – the love of an Oxford professor for a student and the events surrounding a fatal car crash provide a means of examination of the inner thoughts and motives of the characters. A thought-provoking and demanding film in which York appears in his first major film role.
DRA   100 min (ort 105 min)   VIDrel: MGM/WHV   V/h
Boa: novel by Nicholas Mosley.

ACCIDENTAL GOLFER, THE *   (PG)
Lasse Arberg SWEDEN   1991
Lasse Arberg, Jon Skolmen
A golf fanatic makes a bet with a gallery owner that a complete novice cannot beat him after a mere one week's instruction. Having chosen as their novice a roadsweeper, the gallery owner takes him for a week's training in Scotland – her belief being that the instruction there is second to none.
COM   104 min dubbed   SATrel: SKY MOVIES

ACCIDENTAL HERO *   15
Stephen Frears USA   1992
Geena Davis, Dustin Hoffman, Andy Garcia, Joan Cusack, Tom Arnold, Kevin J. O'Connor, Chevy Chase, Maury Chaykin, Stephen Tobolowsky, Janis Paige, Warren Berlinger, Christian Clemenson, Susie Cusack, James Madio, Don Yesso, Lee Wilkof
A petty criminal rescues a number of people from a crashed aircraft, including a woman reporter, and then goes quietly on his way. However, the usual media circus commences when a TV station offers a $1,000,000 reward to this anonymous hero and a down-at-heel hobo comes forward to claim it. Very much a hit-or-miss attempt at a social satire that does not really work, although a set of strong performances are a major asset.
Aka: HERO
COM   113 min (ort 118 min)   VIDrel: COLUM/SONOP L/A
V/sh

ACCIDENTAL MEETING *   15
Michael Zinberg USA   1995
Linda Gray, Linda Pul, Leigh J. McClsokey, Ernie Lively, David Hayward, Kent McCord, Lorna Scott, Nancy Hochman, Bethany Richards, Steve Tom, Duke Stroud, Janet Haley, Yvonne Evans, Patricia North, D.J. Sullivan, Lance Nichols
An auto accident throws two women together by chance and they get to discussing how murder might rid them of some personal encumbrances in the shape of troublesome boyfriends. However, only the psychotic one among them takes this seriously. An updated, female version of STRANGERS ON A

TRAIN, that benefits little from its so-called homage to Hitchcock.
DRA   86 min (ort 91 min) mCab   VIDrel: CIC/SONOP   V

ACCIDENTAL TOURIST, THE *   PG
Lawrence Kasdan USA   1988
William Hurt, Kathleen Turner, Geena Davis, David Ogden Stiers, Amy Wright, Bill Pullman, Robert Gorman, Ed Begley Jr, Bradley Mott, Seth Granger, Paul Williamson, Walter Sparrow, Maureen Kerrigan, Jacob Kasden, Todd Adelman
A man who writes travel guides for a living has become withdrawn following the accidental death of his son in a shooting. This sombre and somewhat wry story follows him as he embarks on an affair with a bubbly and zestful woman after his wife walks out on him, only to find their affair threatened by her sudden return. A perceptive and touching character study. AA: S. Actress (Davis).
DRA   116 min (ort 121 min)   VIDrel: MGM/WHV   V/sur
Boa: novel by Anne Tyler Modarressi, et al.

ACCION MUTANTE *   18
Alex de la Iglesia SPAIN   1994
Antonie Resines, Alex Angluo, Frederique Feder, Juan Viadas, Karra Elejalde, Saturnino Garcia, Fernando Guillen, Jaime Blanch, Ion Garella, Bibi Anderssen, Rosi de Palma, Enrique San Francisco, Feodor Atkine, Felipe Velez, Paco Maestre
Story of a terrorist organisation of the future and their abduction of a millionaire's daughter, whom they hold to ransom, the intention of their deformed leader being to hit back at a society which places too much value on physical perfection. Heavily influenced by punk culture and comics, this is a statement of sorts on the empty values of a consumerist society, but the film lacks both a coherent plot and a clear intention. A pity, as it starts off well.
Aka: MUTANT ACTION
FAN   90 min (ort 94 min) wScrn   VIDrel: IMAG/RTM
V/sur

ACCOMPANIST, THE *   PG
Claude Miller FRANCE   1992
Richard Bohringer, Elena Safonova, Romane Bohringer, Samuel Labarthe, Nelly Borgeaud, Bernard Verley, Niels Dubost, Sacha Briquet, Claude Rich, Nathalie Bach, Gilbert Bahon, Marcel Berbert, Valerie Bettencourt, Gabriel Cattand
In occupied Paris, a woman pianist is engaged as accompanist to a famous singer. When the latter and her manager/husband are suspected of collaboration, all three of them flee to Britain. A slow-paced and mannered drama that is nicely acted and staged but never quite amounts to anything very much.
Aka: L'ACCOMPAGNATRICE
DRA   106 min (ort 111 min) wScrn   VIDrel: TART/20TH
V/sur
Boa: novel L'Accompagnatrice by Nina Berberova.

ACCUSED, THE *   18
Jonathan Kaplan USA   1988
Jodie Foster, Kelly McGillis, Bernie Coulson, Leo Rossi, Ann Hearn, Carmen Argenziano, Steve Antin, Tom O'Brien, Peter Van Norden, Woody Brown, Terry David Milligan, Scott Paulin, Kim Vondrashoff, Stephen E. Miller, Tom Heaton
After a young waitress is brutally gang raped in a bar, a cool and efficient female D.A. bargains with the accused men and the charge against them is reduced to one of reckless endangerment. However, the victim is outraged and in order to make amends, the D.A. decides to press charges against those witnesses who did nothing to prevent the rape. The violent rape scene rather spoils this film's articulate and clear stand. AA: Actress (Foster).
DRA   106 min (ort 110 min)
VIDrel: 4-FRONT/POLYREC/CIC   V/sur

ACE HIGH *   15
Guiseppe Colizzi ITALY   1968
Bud Spencer (Carlo Pedersoli), Terence Hill (Mario Girotti), Eli Wallach, Brock Peters, Kevin McCarthy, Steffen Zacharias, Livio Lorenzon, Tiffany Hoyveld, Remo Capitani, Armando Bandini, Isa Foster, Rick Boyd
Knockabout Western of the spaghetti variety with little of the style of the Sergio Leone epics. One crook steals a fortune from two others who in turn stole it from another... and so on.
Aka: ACE UP MY SLEEVE; ASSO PIGLIA TUTTO; HAVE GUN, WILL TRAVEL; I QUATTRO DELL'AVE MARIA; REVENGE AT EL PASO; REVENGE IN EL PASSO
WES   116 min Cut (2 sec – ort 122 min)
VIDrel: CIC/SONOP   V

ACE VENTURA: PET DETECTIVE *   12
Tom Shadyac USA   1993
Jim Carrey, Sean Young, Courtney Cox, Tone Loc, Dan Marino, John Capodice, Noble Willingham, Raynor Schien, Frank Adonis, Tiny Ron, Troy Evans, Udo Kier, David Margulies, Judy Clayton, Bill Zuckert, Alice Drummond, Rebecca Ferratti
A detective who specialises in locating missing pets is hired to find the kidnapped dolphin mascot of the Miami Dolphins and does so in the nick of time, just before the start of the Super Bowl. A brainless offering that is purely an unashamed vanity vehicle for Carrey and his frenetic brand of slapstick comedy. Although lambasted by the critics, this smash hit gave Carrey his movie breakthrough. A poor and unbearably irritating effort that is barely watchable.
COM   82 min (ort 101 min) cC   VIDrel: WHV   V/sur

ACE VENTURA: WHEN NATURE CALLS *   PG
Steve Oederkerk USA   1995
Jim Carrey, Ian McNeice, Simon Callow, Maynard Eziashi, Bob Gunton, Sophie Okonedo, Tommy Davidson, Adewale, Danny D. Daniels, Sam Motoana Phillips, Damon Standifer, Andrew Steel, Bruce Spence, Thomas Grunke, Arsenio "Sonny" Trinidad
Weak follow-up to the earlier (and better) film that sees Carrey sent off in search of a rare species of bat that is sacred to an obscure tribe. Carrey's incessant mugging becomes more than a little irritating, and is no substitute for a decent script, which this film sorely needs.
COM   89 min (ort 94 min) cC   VIDrel: WHV   V/sur

ACES HIGH ***   PG
Jack Gold UK   1976
Malcolm McDowell, Christopher Plummer, Simon Ward, Peter Firth, Ray Milland, John Gielgud, Trevor Howard, Richard Johnson, David Wood, David Daker, Barry Jackson, Ron Pember, Tim Pigott-Smith, Christopher Blake, Jane Anthony
Competent remake of the early classic film "Journey's End", which in turn was based on the 1929 Sheriff play. McDowell plays a cynical and embittered squadron leader and Firth one of a fresh batch of WW1 pilots whose dislike of his superior turns to admiration as he begins to understand him. Some exciting aerial dogfight sequences highlight this remake.
WAR   109 min (ort 114 min)   VIDrel: WHV   V/h
Boa: play Journey's End by R.C. Sheriff/book Sagittarius Rising by Cecil Lewis.

ACT OF PASSION **   15
Simon Langton USA   1984
Kris Kristofferson, Marlo Thomas, George Dzundza, Jon De Vries, David Rasche, Linda Thorson, Edward Winter, Randy Rocca, Christine Estabrook, Steven Williams, Ron Parady, William Leach, Belinda Bremmer, Karen Cole
A one-night stand with a suspected terrorist ruins a young woman's life; she is persecuted by the newspapers and receives sackfuls of poison-pen letters. The plot is derivative of the 1975 film THE LOST HONOUR OF KATHARINA BLUM, but the adaptation for an American audience (by Loring Mandel) is somewhat uneven.
Aka: LOST HONOR OF KATHRYN BECK, THE
DRA   100 min mTV   VIDrel: 20TH/TECH L/A   V
Boa: novel The Lost Honour of Katharina Blum by Heinrich Boll.

ACT OF PIRACY **   18
John (Bud) Cardos USA   1988
Gary Busey, Belinda Bauer, Ray Sharkey, Nancy Milford, Dennis Casey Park, Ken Gampu, Arnold Vosloo, Anthony Fridjhon, Matthew Stewardson, Nadia Bilchik, Candice Hillebrand, Joe Stewardson, Brian O'Shaughnessy, Gordon Mulholland
Action tale following a group of mercenaries who decide to capture a large motor yacht whose owner was in the process of delivering it to Sydney, Australia.
A/AD   100 min   VIDrel: 20TH/TECH   V

ACT OF VENGEANCE ***   18
Bob Kelljan USA   1974
Jo Ann Harris, Peter Brown, Jennifer Lee, Steve Kanaly, Lada Edmund Jr, Lisa Moore, Connie Strickland, Patrick Estrin, Tony Young, Ross Elliot, Ninette Bravo, John Pickard, Stanley Adams, John McCall, Anneka DeLorrenzo
Rape victims band together to form a vigilante squad for revenge on a vicious sadist who forces his victims to sing while he rapes them. A vigilante tale that's a little more intelligent than is usual for films of this genre. See THE SISTERHOOD for a far more brutal treatment of this theme.
Aka: RAPE SQUAD; VIOLATORS, THE
DRA   82 min Cut (4 min 48 sec – ort 90 min)
VIDrel: RNK L/A   V

ACT OF WILL: PARTS 1 AND 2 **   15
Don Sharp USA   1989
Victoria Tennant, Peter Coyote, Elizabeth Hurley, Kevin McNally, Jean Marsh, Serena Gordon, Lynsey Baxter, Sarah Winman, Judy Parfitt, Simon Merrick, Rachel Robertson, Jeremy Gilley, Jason Savage, Roger Grainger
Long weepie drama following the fortunes of three women, a grandmother, mother and daughter and their trials and tribulations through five decades of love and turmoil. Average.
DRA   200 min (2 cassettes) mTV
VIDrel: 4-FRONT/POLYREC/ODY   V/sh
Boa: novel by Barbara Taylor Bradford.

ACTOR'S REVENGE, AN ***   PG
Kon Ichikawa JAPAN   1963
Kazuo Hasegawa, Fujiko Yamamoto, Ayako Wakao, Eiji Funakoshi, Narutoshi Hayashi, Eijiro Yanagi, Chusha Ichikawa, Ganjiro Nakamura, Saburo Date, Jun Hamamura, Kikue Mori, Masayoshi Kikuno, Raizo Ichikawa, Shintaro Katsu
In 19th century Japan, a Kabuki actor who plays female parts, is aided by a petty gangster in taking a gruesome revenge on the men who murdered his family. In the process, he finds two women falling in love with him. Filmed before in 1935, this is a complex effort that rewards attentive watching.
Aka: REVENGE OF YUKINOJO, THE; YUKINOJO HENGE
DRA   108 min (ort 113 min) wScrn
VIDrel: CONNO/RTM   V
Boa: the "Asahi" newspaper serial by Otokichi Mikami.

ACTS OF LOVE ***   18
Bruno Barreto USA   1995
Dennis Hopper, Amy Irving, Amy Locane, Julie Harris, Gary Busey, Hal Holbrook, Christopher Pettiet, Priscilla Pointer, Gail Cronauer, Alissa Alban, E.J. Morris, Joe Stevens, Connie Cooper, Eleese Lester, Doug Jackson
A rural schoolteacher living on a farm together with his dying mother has a distinct lack of passion in his life, while his established relationship with a local widow also fails to generate any sparks. One day, an attractive seventeen-year-old girl joins his class and he soon finds himself torn between desire and a sense of duty. A solid adaptation of Harrison's novel, brought to life by some first-rate acting.
Aka: CARRIED AWAY
DRA   104 min (ort 109 min)   VIDrel: FIRST/SONOP   V
Boa: novel Farmer by Jim Harrison.

ADAM ***   15
Michael Turner USA   1983
JoBeth Williams, Daniel J. Travanti, Martha Scott, Richard Masur, Paul Regina, Mason Adams, Tony Frank, John M. Jackson, Alex Harvey, John Boston, John Perry Edson, Rick Stokes, Robert Ginnaven, Robin Bradley, Gary Moody
Powerful story of a couple who, after their child is kidnapped and murdered, lobby Congress so that parents of missing children may use the FBI national crime computer to help find them. Scripted by Allan Leicht and followed in 1986 by ADAM: HIS SONG CONTINUES.
DRA   92 min (ort 100 min) mTV   VIDrel: ODY/SONOP   V/h

ADAM AT 6 A.M. ***   15
Robert Scheerer USA   1970
Michael Douglas, Lee Purcell, Joe Don Baker, Grayson Hall, Charles Aidman, Meg Foster, Madge Redmond, Louise Latham, Carolyn Cornwell, Dana Elcar, Anne Gwynne, Richard Derr, Ned Wertheimer, Ed Call, David Sullivan, Jim Lantz
A young Californian college professor becomes disillusioned with his way of life and heads out on the road, eventually winding up in the Midwest where he takes a job as a labourer, getting to know some ordinary people at the same time. A quiet and well-acted drama that makes good use of its locations and players.
DRA   96 min (ort 100 min)   VIDrel: 20TH/TECH   V

ADAM BEDE ***   (PG)
Giles Foster UK   1991
Iain Glen, Patsy Kensit, Susannah Harker, James Wilby, Julia McKenzie, Robert Stephens, Jean Marsh, Freddie Jones, Michael

Percival, Brian Osborne, Edward Jewesbury, Paul Brooke, Tacita Haffenden, Chase Marks, William Holmes
A very handsome adaptation of Eliot's story of the rivalry between two men from very different social backgrounds for the hand of a beautiful woman. Various intrigues and complications flow from this, and though the film is always a treat visually, one often feels a little uninvolved in the plight of the protagonists. A qualified success.
DRA 102 min mTV TVrel: BBC
 Boa: novel by George Eliot.

ADAM: HIS SONG CONTINUES **   15
Robert Markowitz USA   1986
Daniel J. Travanti, JoBeth Williams, Richard Masur, Martha Scott, Paul Regina, Sam McMurray
Sequel to the 1983 film that continues the tragic story of the parents of a small boy whose kidnapping and murder led to such public outcry that the government passed a special act in 1982 to deal with the problem of missing children. Adam's parents have moved to Washington to start again, but their lives are yet again filled with tragedy. Average.
DRA 90 min (ort 120 min) mTV VIDrel:
VISVID/POLYREC L/A V

ADAM'S RIB ***   U
George Cukor USA   1949
Spencer Tracy, Katharine Hepburn, Judy Holliday, Tom Ewell, Jean Hagen, Polly Moran, David Wayne, Hope Emerson, Eve March, Clarence Kolb, Emerson Treacy, Will Wright, Elizabeth Flourmoy, Janna Da Loos, Marvin Kaplan, Sid Dubin
A tough D.A. and his lawyer wife find themselves representing opposing sides in the trial of a woman accused of the attempted murder of a rival for her husband's affection. Great on-screen chemistry between Hepburn and Tracy and a witty script (by Ruth Gordon and Garson Kanin) make this sophisticated examination of the battle of the sexes a sheer comic delight. Holliday stars in the first of her many "scatterbrain" roles. Excellent all round.
COM 97 min (ort 101 min) B/W VIDrel: MGM/WHV V

ADDAMS FAMILY, THE ***   PG
Barry Sonnenfeld USA   1991
Anjelica Huston, Raul Julia, Christopher Lloyd, Dan Hedaya, Elizabeth Wilson, Judith Malina, Carel Struycken, Dana Ivey, Paul Benedict, Christina Ricci, Jimmy Workman, Christopher Hart, John Franklin, Tony Azito, Allegra Kent
For over fifty years the irreverent and sinister cartoon-strip created by Charles Addams amused readers of the New Yorker. Then came an anaemic TV show. Now this very full-blooded and lavish film. Though having little plot (Uncle Fester may or may not be an impostor) it affords ample scope for an excellent cast to do their diabolical best in celebrating the morbid and often tasteless imagination of Addams. A kind of live-action "cartoon". A sequel followed.
COM 95 min (ort 120 min) VIDrel: VCC/DISC/COLUM
V/sur

ADDAMS FAMILY VALUES ***   PG
Barry Sonnenfeld USA   1993
Anjelica Huston, Raul Julia, Christopher Lloyd, John Cusack, Christina Ricci, Carol Kane, Jimmy Workman, Carol Struycken, David Krumholtz, Christopher Hart, Dana Ivey, Peter MacNicol, Christine Baranski, Sam McMurray, Julie Halston
A worthy sequel to the first movie, in which the arrival of a new baby inspires a literally homicidal jealousy in the other Addams children, who plot to rid themselves of this intruder. Meanwhile, the nanny hired to care for the infant, proves to be a gold-digging serial murderess with deadly designs on Uncle Fester. Much black humour and superbly appropriate performances help sustain this movie all the way to the end.
COM 90 min (ort 94 min) cC VIDrel: CIC/SONOP; PION
(LV only) V/sur LV

ADDICTED TO LOVE **   (18)
Paul Ziller USA   1995
Jeff Fahey, Ami Dolenz
Following the murder of his girlfriend, a lonely and desperately unhappy man finds himself still unable to form a new relationship. When his employers ask him to test a virtual reality device, he discovers that he can use it to recreate his dead girlfriend. A nightmarish fantasy whose execution does not really do justice to the excellence of its idea.
FAN 89 min SATrel: MOVIE CHANNEL

ADDICTED TO MURDER **   18
Kevin Lindenmuth USA   1994
Mick McCleery, Laura McLauchlin, Sasha Graham, Nick Staglian, Gordon Linzner
In New York City, a young would-be serial killer acquires a vampire girlfriend, and no matter how many times he kills her, she keeps coming back to him. Crudely made, this low-budget, independent offering is full of daft dialogue and gory effects, and might have worked considerably better had it been made as a black comedy rather than as a straight horror pic. Screenplay is by Lindenmuth and Tom Piccirilli.
HOR 90 min VIDrel: SCEDGE/RTM V

ADDICTION, THE ***   18
Abel Ferrara USA   1995
Christopher Walken, Annabella Sciora, Lili Taylor
When a postgraduate student is bitten by a vampire, she survives the attack but becomes a junkie for human blood, constantly driven by the urge to seek out new victims. Tormented by urges she is unable to control, she faces the inexorable loss of her humanity, until a bloody graduation party that provides her with a redemption of sorts. A memorable and visually powerful film, it attempts an examination of the nature of evil in between the bouts of gore.
HOR 82 min B/W CINrel

ADJUSTER, THE ***   18
Atom Egoyan CANADA   1991
Elias Koteas, Arsinee Khanjian, Maury Chaykin, Gabrielle Rose, Jennifer Dale, David Hemblen, Rose Sarkisyan, Armen Kokorian, Jacqueline Samuda, Gerard Parkes, Patricia Collins, Don McKellar, John Gilbert, Stephen Ouimette
Koteas plays Noah, an insurance adjuster whose work involves dealing with people whose lives have been blighted by house fires. As a man with considerable empathy with his clients, he often indulges in the odd affair, leaving his live-in girlfriend free to pursue her own interests. Soon, the emptiness of their existence is revealed by the intrusion into their lives of a wealthy and self-indulgent couple, in this bleak and frosty drama.
DRA 98 min (ort 102 min) wScrn VIDrel: TART/20TH
V/dm V/sur

ADRIFT **   15
Christian Duguay USA   1993
Kate Jackson, Kenneth Walsh, Bruce Greenwood, Kelly Rowan
In an attempt to save their marriage, a couple take a sailing vacation. They come across a boat that is adrift and unwisely take aboard another couple who prove to be thoroughly nasty psychos. A predictable and blatant clone of DEAD CALM with all the usual excesses but little originality or suspense.
THR 90 min (ort 92 min) mTV VIDrel: MARQ/REFLEC V

ADVENTURE OF SHERLOCK HOLMES' SMARTER
BROTHER, THE **   PG
Gene Wilder UK/USA   1975
Gene Wilder, Madeline Kahn, Marty Feldman, Dom DeLuise, Leo McKern, Roy Kinnear, John Le Mesurier, Thorley Walters, Douglas Wilmer, George Silver, Susan Field, Tommy Godfrey, Joseph Behrmannis, Nicholas Smith, John Hollis
Spoof movie based on Conan Doyle's famous hero with "Sigerson Holmes" as the title character helping damsel in distress Kahn. Look out for DeLuise as a washed up opera star, he provides the liveliest moments in an otherwise weak story. Filmed in England and Wilder's first attempt at acting, writing and directing.
COM 87 min (ort 91 min) VIDrel: 20TH/TECH V

ADVENTURES OF BARON MUNCHAUSEN,
THE ***   PG
Terry Gilliam UK/WEST GERMANY   1988
John Neville, Eric Idle, Sarah Polley, Oliver Reed, Uma Thurman, Jonathan Pryce, Valentina Cortese, Charles McKeown, Winston Dennis, Jack Purvis, Bill Paterson, Peter Jeffrey, Allison Steadman, Ray Cooper, Don Henderson
Long, glossy and very expensively mounted story of this famous baron and his fantastic exploits, which though pure invention on his part, are shown in this colourful production as if they really took place. Unfortunately, despite all the imaginative sets and attention to detail, the film soon grows quite tiresome as one spectacular setpiece follows another. Co-directed by Michele Soavi (uncredited).
FAN 126 min VIDrel: VCC/DISC/COLUM L/A V/sh
Boa: novel by Rudolph Erich Raspe.

ADVENTURES OF BARRY McKENZIE, THE ** 15
Bruce Beresford UK 1972
Barry Crocker, Barry Humphries, Peter Cook, Spike Milligan, Dennis Price, William Rushton, Dick Bentley, Avice Landon, Joan Bakewell
Rumbustious adventures of an Australian tourist in London, that is based on a popular cartoon strip character from the satirical magazine "Private Eye". Although there are a few extremely funny moments, the film is so poorly made that one's attention often wanders. Co-scripted by Beresford and Humphries, who puts in an appearance as his famous alter ego "Dame Edna Everage". Followed by BARRY McKENZIE HOLDS HIS OWN.
COM   102 min   VIDrel: VCC/DISC   V

ADVENTURES OF BREASTMAN, THE * 18
USA 1993
Toni Tedeschi, Heidi Cat, Patricia Kennedy, Mike Horner, Kiss, Marc Wallice, Trinity Loren, Tianna Taylor, Persia, Taylor Wayne, Dominique Simone, Nick East
A man searches for the perfect pair of breasts, and to help him in this search, he carries with him a video camera. A fairly plotless adult movie that is really nothing more than a parade of buxom ladies, some of the footage inevitably lost achieve a BBFC certification.
A   94 min Cut   VIDrel: FALCON/TOTAL   V

ADVENTURES OF CAPTAIN FABIAN * PG
William Marshal FRANCE/USA 1951
Errol Flynn, Vincent Price, Agnes Moorehead, Micheline Presle, Jim Gerald, Victor Francen, Helena Mason, Howard Vernon, Roger Blin, Valentine Camax, Georges Flateau, Zanie Campan, Reggie Nalder, Charles Fawcett, Aubrey Bower
A sea captain in New Orleans gets himself involved with a scheming servant girl whose machinations eventually lead to him being imprisoned for a murder which he did not commit. A failed attempt at a swashbuckler vehicle for Flynn, who was nearing the end of his career.
A/AD   99 min B/W   VIDrel: 4-FRONT/POLYREC   V
Boa: Fabulous Ann Medlock by Robert Shannon.

ADVENTURES OF CAPTAIN ZOOM IN OUTER SPACE, THE ** PG
Max Tash USA
Daniel Riordan, Liz Vassey, Ron Perlman, Duane Davis
A second-rate actor who stars in a kid's TV show is abducted to a distant planet by a group of freedom fighter there, in the hope that he will be able to assist them in opposing the plans of a tyrant. A weak action-comedy.
A/AD   87 min cC   VIDrel: CIC/SONOP   V/dm

ADVENTURES OF DON JUAN, THE *** PG
Vincent Sherman USA 1948
Errol Flynn, Viveca Lindfors, Robert Douglas, Romney Brent, Alan Hale, Ann Rutherford, Robert Warwick, Jerry Austin, Douglas Kennedy, Una O'Connor, Aubrey Mather, Raymond Burr, Jeanne Shepherd, Mary Stuart, Tim Huntley, David Leonard
Flynn is in fine form in this boisterous and slightly tongue-in-cheek tale of the great lover who breaks the hearts of numerous maidens, even directing his blandishments at the Queen as he saves her from the plotting of her minister. Flynn's last big-budget film that contains some interesting attempts at parody.
AA: Cost (Leah Rhodes/Travilla/Marjorie Best).
Aka: NEW ADVENTURES OF DON JUAN, THE
A/AD   111 min B/W   VIDrel: WHV   V

ADVENTURES OF FORD FAIRLANE, THE * 18
Renny Harlin USA 1990
Andrew Dice Clay, Wayne Newton, Priscilla Presley, Morris Day, Lauren Holly, Maddie Corman, Gilbert Gottfried, David Patrick Kelly, Brandon Call, Robert Englund, Ed O'Neill, Vince Neil, Sheila E., Kari Wuhrer, Tone Loc
Stand-up comedian Clay gets his first starring role, in this rather strange melding of an old-fashioned detective yarn with the modern world of rock music, and attempts to act (if that is the right word) the part of a private eye investigating the death of a DJ. A loud and clumsy misfire, based on characters created by Rex Weiner and (as expected) incorporating a good many of Clay's comedy routines.
Aka: FORD FAIRLANE
COM   97 min (ort 104 min)   VIDrel: 20TH/TECH   V/sur

ADVENTURES OF HAMBONE AND HILLIE, THE ** PG
Roy Watts USA 1984
Lillian Gish, Timothy Bottoms, Candy Clark, O.J. Simpson, Robert Walker, Jack Carter, Alan Hale, Anne Lockhart, William Jordan, Paul Koslo, Nancy Morgan, Arnie Moore, Sidney Robin Greenbush, Maureen Quinn, Mark Bentley
An old woman accidentally leaves her dog at an airport over three thousand miles from home. Needless to say, the faithful mutt finds its way home.
Aka: HAMBONE AND HILLIE
COM   86 min (ort 97 min)   VIDrel: MIA/DISC   V/sur

ADVENTURES OF HUCK FINN, THE ** PG
Stephen Sommers USA 1992
Elijah Wood, Courtney B. Vance, Robbie Coltrane, Ron Perlman, Jason Robards Jr, Dana Ivey, Anne Heche, James Gammon, Paxton Whitehead, Laura Bundy, Curtis Armstrong, Tom Aldredge, Mary Louise Wilson, Frances Conroy, Daniel Tamberelli
Competent, Disney-made adaptation of Twain's much filmed novel that provides an enjoyable account of the main events while omitting the racial insults and minstrel dialect. Wood plays the youngster of the title, who befriends a runaway slave who is hiding out along the Mississippi River, and Coltrane and Robards are memorable in cameo roles as a couple of con-men. See also TOM AND HUCK.
DRA   104 min (ort 108 min) cC   VIDrel: WDV/TECH   V/sh
Boa: novel The Adventures of Huckleberry Finn by Mark Twain.

ADVENTURES OF MARCO POLO, THE *** U
Archie Mayo USA 1938
Gary Cooper, Sigrid Guthrie, Basil Rathbone, Ernest Truex, George Barbier, Binnie Barnes, Alan Hale, H.B. Warner, Robert Grieg, Ferdinand Gottschalk, Henry Kolker, Hale Hamilton, Lotus Liu, Stanley Fields, Harold Huber
The exploits of this famed 13th century explorer form the background to what is essentially a lavish adventure, with Polo and his companion surmounting many obstacles on their journey to China. Once arrived at the court of Kublai Khan, they find themseles up to their neck in intrigue, as an evil courtier (Rathbone) plots a takeover. A highly enjoyable adventure spectacle with a few comic touches and zero value as history. Lana Turner appears briefly as a handmaiden.
A/AD   104 min B/W   VIDrel: VCC/DISC   V

ADVENTURES OF MILO AND OTIS, THE *** U
Masanori Hata JAPAN/USA 1986
Dudley Moore (narration only)
A curious kitten and a puppy embark on a series of hazardous adventures when the former hops into a box floating along a small creek. An engaging live-action kid's story that's slightly spoilt by the weak script but is carried along by Moore's amusing narration.
Aka: CHATORAN; KONEKO MONOGATARI; MILO AND OTIS
JUV   72 min Cut (11 sec at UK cinema release – ort 76 min)
VIDrel: VISVID/POLYREC   V/sur
Boa: story by Masanori Hata.

ADVENTURES OF PINOCCHIO, THE *** U
Steve Barron CZECH
REPUBLIC/FRANCE/GERMANY/UK/USA 1996
Martin Landau, Jonathan Taylor Thomas, Genevieve Buchhold, Udo Kier, Bebe Neuwirth, Rob Schneider, Corey Carrier, Marcello Magni, Dawn French, Richard Claxton, Griff Rhys Jones, John Sessions, Jean-Claude Drouot, Erik Averlont
A mixture of live-action and animatronics is used to tell the story of a wooden boy created by a puppet-maker, that is imbued with life by a fairy, but longs to become a real boy. However, he falls into bad company and endures much hardship before his wish comes true. More faithful to the story than the famous Disney animation, it lacks nothing in terms of technical excellence, but is done in a matter-of-fact manner, with no real sense of wonder or charisma.
JUV   90 min (ort 94 min)   VIDrel: POLFIL   V
Boa: novel by Carlo Collodi.

ADVENTURES OF PRISCILLA, QUEEN OF THE DESERT, THE *** 15
Stephan Elliott AUSTRALIA 1994
Terence Stamp, Hugo Weaving, Guy Pearce, Bill Hunter, Sarah Chadwick, Mark Holmes, Rebel Russell, John Casey, June Maria Bennett, Murray Davies, Frank Cornelius, Bob Royce, Leighton Picken, Julia Cortez, Daniel Kellie
A couple of drag artists and their transsexual partner travel around the Australian outbreak performing at a variety of small towns while on their way to gig in the big city. An affectionate

and convincing portrayal of artistic commitment that takes the form of an outrageous and over-the-top comedy. Good performances, especially by Stamp as the transsexual, make it all worthwhile. See also TO WONG FOO, THANKS FOR EVERYTHING! JULIE NEWMAR. AA: Cost.
COM 98 min (ort 104 min) VIDrel: 20VIS/SONOP V/sur

ADVENTURES OF ROBIN HOOD, THE ****   U
Michael Curtiz/William Keighley USA   1938
Errol Flynn, Basil Rathbone, Claude Rains, Olivia De Havilland, Alan Hale, Patric Knowles, Eugene Pallette, Ian Hunter, Melville Cooper, Una O'Connor, Herbert Mundin, Montagu Love, Howard Hill, Leonard Willey, Robert Noble
Splendid, classic version of the Robin Hood story, with Flynn in fine form as Robin foiling the machinations of the Sheriff of Nottingham and Sir Guy of Gisbourne and thereby saving the throne for the absent King Richard. A film that finely balances adventure and comedy, its pace and verve are still unbeaten. Note the early use of three-colour Technicolor. AA: Art (Carl Jules Weyl), Score (Erich Wolfgang Korngold), Edit (Ralph Dawson).
A/AD 102 min VIDrel: WHV V/dm

ADVENTURES OF SHERLOCK HOLMES, THE: A SCANDAL IN BOHEMIA ***   PG
Paul Annett UK   1984
Jeremy Brett, David Burke, Gayle Hunnicutt, Wolf Kahler, Michael Carter, Max Faulkner, Tim Pearce, Rosalie Williams
Holmes is engaged by the King of Bohemia to regain a compromising photograph from a former mistress of the King, in this first episode from the first series of seven atmospheric stories. Written by Alexander Baron.
Aka: SCANDAL IN BOHEMIA, A
DRA 52 min; 105 min (2-film cassette) mTV
VIDrel: HEND/BMGREC L/A V
Boa: short story by Arthur Conan Doyle.

ADVENTURES OF SHERLOCK HOLMES, THE: THE BLUE CARBUNCLE ***   PG
David Carson UK   1984
Jeremy Brett, David Burke, Rosalind Knight, Ros Simmons, Ken Campbell, Desmond McNamara, Amelda Brown, Brian Miller, Rosalie Williams, Frank Mills
Another story from this popular and well mounted series. In this tale Holmes uses his remarkable powers of deduction to return a priceless gem that was stolen from the Countess of Morcar's hotel suite and later turned up in the crop of a Christmas goose. The last tale in the first season of seven stories, it was written by Paul Finney. This story was one of seven on a three-cassette set.
Aka: BLUE CARBUNCLE, THE
DRA 47 min (ort 52 min) mTV VIDrel: CASPIC L/A V
Boa: short story by Arthur Conan Doyle.

ADVENTURES OF SHERLOCK HOLMES, THE: THE COPPER BEACHES ***   PG
Paul Annett UK   1985
Jeremy Brett, David Burke, Joss Ackland, Natasha Richardson, Lottie Ward, Patience Collier, Angela Browne, Peter Jonfield
Another tale in this well made TV series with Holmes investigating strange occurrences at a large country house when a woman who has misgivings about her new post of governess comes to him for advice. This episode was the first in a second season of six tales. Written by Bill Craig, it was also available on a three-cassette box set.
Aka: COPPER BEACHES, THE
DRA 52 min mTV VIDrel: CASPIC L/A V
Boa: short story by Arthur Conan Doyle.

ADVENTURES OF SHERLOCK HOLMES, THE: THE CROOKED MAN ***   PG
Alan Grint UK   1983
Jeremy Brett, David Burke, Norman Jones, Lisa Daniely, Denys Hawthorne, Paul Chapman, Flora Shaw, Shelagh Stephenson, Michael Lumsden, Catherine Rabett, James Wilby, Maggie Holland, Colin Campbell, David Graham Jones
An episode from this excellent Granada TV series in which Holmes is called on to investigate the mystery surrounding a dead colonel that threatens the good name of a regiment. Number five in the first series of seven adventures. Written by Alfred Shaughnessy. Was available on a three-cassette set of seven tales.
Aka: CROOKED MAN, THE
DRA 52 min (ort 55 min) mTV VIDrel: CASPIC L/A V
Boa: short story by Arthur Conan Doyle.

ADVENTURES OF SHERLOCK HOLMES, THE: THE DANCING MEN ***   PG
John Bruce UK   1984
Jeremy Brett, David Burke, Terry Evans, Betsy Palmer, David Ross, Eugene Lupinsky, Lorraine Peters, Wendy Jane Walker
Second episode from the first Granada series with Holmes being called upon to solve the mystery of a series of crudely drawn stick figures that appear to represent a secret message and have terrified a man's new wife. Written by Anthony Skene.
Aka: DANCING MEN, THE
DRA 52 min mTV VIDrel: HEND L/A V
Boa: short story by Arthur Conan Doyle.

ADVENTURES OF SHERLOCK HOLMES, THE: THE ELIGIBLE BACHELOR ***   15
Peter Hammond UK   1992
Jeremy Brett, Edward Hardwicke, Simon Williams, Anna Calder-Marshall, Mary Ellis, Joanna McCullum, Geoffrey Beavers, Paris Jefferson, Phillada Sweell, Elspeth March, Heather Chasen, Bob Sessions, Myles Hoyle, Bruce Myters
Holmes reluctantly agrees to investigate the case of a young heiress who vanished before her wedding, and uncovers a nasty conspiracy. Another excellent, feature-length adaptation in this high-quality series.
Aka: ELIGIBLE BACHELOR, THE; SHERLOCK HOLMES: THE ELIGIBLE BACHELOR
DRA 103 min mTV VIDrel: CASPIC/BMGREC V
Boa: short story The Noble Bachelor by Arthur Conan Doyle.

ADVENTURES OF SHERLOCK HOLMES, THE: THE FINAL PROBLEM ****   PG
Alan Grint UK   1985
Jeremy Brett, David Burke, Eric Porter, Rosalie Williams, Olivier-Pierre, Claude Le Sache, Michael Goldie, Robert Henderson, Paul Sirr, Jim Dunk, Paul Humpoletz, Simon Adams
In this tale Holmes confronts his old enemy Professor Moriarty at the Reichenbach Falls in Switzerland. One of the very best of an excellent series, with a thrilling climax. This episode ended a series of thirteen tales (shown in two seasons) and Holmes did not reappear until 1986 with a new series (starting with THE RETURN OF SHERLOCK HOLMES: THE EMPTY HOUSE). written by John Hawkesworth, this story formed part of a three-cassette set.
Aka: FINAL PROBLEM, THE
DRA 52 min (ort 55 min) mTV VIDrel: CASPIC L/A V
Boa: short story by Arthur Conan Doyle.

ADVENTURES OF SHERLOCK HOLMES, THE: THE GREEK INTERPRETER ***   PG
Derek Marlowe UK   1985
Jeremy Brett, David Burke, Charles Gray, Alkis Kritikos, George Costigan, Nick Field, Anton Alexander, Victoria Harwood, Oliver Maguire, Rita Howard, Peter Mackriel
A further episode from the Granada TV series with Holmes pitting his wits against kidnappers who have abducted a Greek national and forced him to sign important documents against his will. In this episode our sleuth is helped by his brother Mycroft, played with considerable panache by Charles Gray. Number two from the second season of six episodes. Written by Derek Marlowe.
Aka: GREEK INTERPRETER, THE
DRA 52 min mTV VIDrel: HEND L/A V
Boa: short story by Arthur Conan Doyle.

ADVENTURES OF SHERLOCK HOLMES, THE: THE MASTER BLACKMAILER ****   PG
Peter Hammond UK   1991
Jeremy Brett, Edward Hardwicke, Robert Hardy, Norma West, Colin Jeavons, Gwen Ffrangcan-Davies, Nickolas Grace, Serena Gordon, Sarah McVicar, David Mallinson, Brian Mitchell, Hans Meyer, Sophie Thompson, Stephen Simms
This feature-length special introduces Charles Augustus Milverton, one of the most loathsome villains Holmes has ever crossed swords with. A man who has gained great wealth by blackmailing the famous, his operations have created much misery among his victims. Despite the best efforts of Holmes, retribution comes from an unexpected quarter. An excellent adaptation, it was in fact taken from Doyle's "Return Of Sherlock Holmes" collection.
Aka: MASTER BLACKMAILER, THE; SHERLOCK HOLMES: THE MASTER BLACKMAILER
DRA 102 min mTV VIDrel: CASPIC/BMGREC V
Boa: short story Charles Augustus Milverton by Arthur Conan Doyle.

ADVENTURES OF SHERLOCK HOLMES, THE: THE NAVAL TREATY ***
PG
Alan Grint UK 1984
Jeremy Brett, David Burke, David Gwillim, Gareth Thomas, Alison Skilbeck, Ronald Russell, Nicholas Geake, Pamela Pitchford, John Malcolm, David Rodigan, Eve Matheson, Rosalie Williams, Judith Taylor
In this episode Holmes has to locate a missing naval document that deals with a secret treaty between Great Britain and Italy, aware that it could spark off an international crisis were it to fall into the wrong hands. This was number three in the first series of seven tales. Written by Jeremy Paul.
Aka: NAVAL TREATY, THE
DRA 52 min mTV VIDrel: HEND L/A V
Boa: short story by Arthur Conan Doyle.

ADVENTURES OF SHERLOCK HOLMES, THE: THE NORWOOD BUILDER ***
PG
Ken Grieve UK 1985
Jeremy Brett, David Burke, Rosalie Crutchley, Colin Jeavons, Helen Ryan, Jonathan Adams, Matthew Solon, Anthony Langdon, Rosalie Williams, Andy Rasleigh, Ted Carroll
Third tale in the second season of this excellent series, with Holmes being called on to prove the innocence of a man accused of the murder of one Joseph Oldacre. Written by Richard Harris.
Aka: NORWOOD BUILDER, THE
DRA 52 min mTV VIDrel: HEND L/A V
Boa: short story by Arthur Conan Doyle.

ADVENTURES OF SHERLOCK HOLMES, THE: THE RED-HEADED LEAGUE ***
PG
John Bruce UK 1985
Jeremy Brett, David Burke, Roger Hammond, Richard Wilson, Tim McInnerty, Bruce Dukov, John Woodnutt, John Labonowski, Eric Porter, Rog Stuart
In this tale Holmes prevents a serious bank robbery when he is called on to investigate a disbanded organisation. Holmes's evil adversary, Professor Moriarty is behind it all. Number five in the second series of six tales. The script is by John Hawkesworth.
Aka: RED-HEADED LEAGUE, THE
DRA 52 min mTV VIDrel: HEND L/A V
Boa: short story by Arthur Conan Doyle.

ADVENTURES OF SHERLOCK HOLMES, THE: THE RESIDENT PATIENT ***
PG
David Carson UK 1985
Jeremy Brett, David Burke, Nicholas Clay, Patrick Newell, Tim Barlow, Brett Forrest, Charles Cork, John Ringham, David Squire, Rosalie Williams, Norman Mills, Lucy Anne Wilson, Dusty Young
In this tale Holmes is recruited by a young medical man who wishes to help his benefactor, a strange and secretive man who lives in a heavily fortified home. Having refused to confide in the detective, the man is found hanged in mysterious circumstances a short while later. Number four in the second series of six stories. Written by Derek Marlowe.
Aka: RESIDENT PATIENT, THE
DRA 52 min (ort 55 min) mTV VIDrel: HEND L/A V
Boa: short story by Arthur Conan Doyle.

ADVENTURES OF SHERLOCK HOLMES, THE: THE SIGN OF FOUR ***
PG
Peter Hammond UK 1987
Jeremy Brett, Edward Hardwicke, Jenny Seagrove, Ronald Lacey, John Thaw, Alf Joint, Kiran Shah, Rosalie Williams, Emrys James, Derek Deadman, Ishaq Bux, Terence Skelton, Marjorie Suddell, Gordon Gostelow, Lila Kaye, William Ash
Following the strange disappearance of her father, a young woman finds herself receiving a present each year, and consults Holmes in an effort to solve the mystery. Another fine TV adaptation of a Conan Doyle mystery. This tale was also available as part of a three-cassette set of seven episodes.
Aka: SIGN OF FOUR, THE
DRA 92 min mTV VIDrel: HEND/BMGREC V/sur
Boa: story by Arthur Conan Doyle.

ADVENTURES OF SHERLOCK HOLMES, THE: THE SOLITARY CYCLIST ***
PG
Paul Annett UK 1984
Jeremy Brett, David Burke, Barbara Wilshire, John Castle, Michael Siberry, Ellis Dale, Sarah Aitchinson, Simone Bleackley, Penny Gowling, Stafford Gordon, Rosalie Williams
Holmes finds himself working on a case in which a man on a bicycle follows a music teacher for no apparent reason each time she cycles off to give music lessons to the daughter of a widower. Fourth tale in the first series. The script is the work of Alan Plater. This story was available as one of seven on a three-cassette set.
Aka: SOLITARY CYCLIST, THE
DRA 52 min mTV VIDrel: CASPIC L/A V
Boa: short story by Arthur Conan Doyle.

ADVENTURES OF SHERLOCK HOLMES, THE: THE SPECKLED BAND ***
PG
John Bruce UK 1984
Jeremy Brett, David Burke, Jeremy Kemp, Rosalyn Landor, Denise Armon, John Gill, Rosalie Williams
Another adventure based on Conan Doyle's famous sleuth in which he uncovers a vile murder plot that makes use of a snake. Number six in the first set of seven stories. Written by Jeremy Paul.
Aka: SPECKLED BAND, THE
DRA 52 min; 105 min (2-film cassette) mTV
VIDrel: HEND/BMGREC L/A V
Boa: short story by Arthur Conan Doyle.

ADVENTURES OF SHERLOCK HOLMES, THE: THE VAMPIRE OF LAMBERLEY ***
15
Tim Sullivan UK 1992
Jeremy Brett, Edward Hardwicke, Roy Marsden, Keith Barron, Yolanda Vasquez, Maurice Denham, Richard Dempsey, Juliet Aubrey, Jason Hetherington, Peter Geddis, Elizabeth Spriggs, Kate Lansbury, Maria Redmond, Freddie Jones
Holmes and Watson find themselves plunged into a very strange case indeed when they investigate a man whose neighbours are convinced is a vampire. A fine feature-length entry in this memorable series, highly atmospheric and wonderfully acted.
Aka: LAST VAMPIRE, THE; SHERLOCK HOLMES: THE LAST VAMPIRE; SHERLOCK HOLMES: THE VAMPIRE OF LAMBERLEY; VAMPIRE OF LAMBERLEY, THE
DRA 102 min mTV VIDrel: CASPIC/BMGRC V
Boa: short story The Sussex Vampire by Arthur Conan Doyle.

ADVENTURES OF THE FLYING PICKLE, THE *
15
Paul Mazursky USA 1993
Danny Aiello, Dyan Cannon, Shelley Winters, Barry Miller, Jerry Stiller, Christopher Penn, Ally Sheedy, Clothilde Courau
A director on the way down looks to his ex-wives and friends for solace and when he can find none, decides to make a low-budget sex film about a flying pickle. Downright peculiar film that purports to be all about movies made on a shoestring and the problems involved in making them, it was never released in the British cinema, and no wonder. Winters as the director's patient mother injects one of the film's few bright moments. See also LIVING IN OBLIVION.
COM 98 min cC VIDrel: COLUM/SONOP V/sh

ADVENTURES OF TOM SAWYER, THE ***
U
Norman Taurog USA 1938
Tommy Kelly, Ann Gillis, Jackie Moran, May Robson, Walter Brennan, Victor Jory, Spring Byington, Margaret Hamilton, David Holt, Victor Kilian, Nana Bryant, Olin Howland, Donald Meek, Charles Richman, Margaret Hamilton
Excellent David O. Selznick colour adaptation of Twain's classic work which is a little more lighthearted than the original tale but contains a number of entertaining sequences for all that. Originally made with a running time of ninety-three minutes but shortened for general release.
JUV 76 min (ort 93 min) VIDrel: VCC/DISC V
Boa: novel Tom Sawyer by Mark Twain.

ADVISE AND CONSENT ****
U
Otto Preminger USA 1962
Henry Fonda, Walter Pidgeon, Charles Laughton, Don Murray, Burgess Meredith, Meredith, Franchot Tone, Lew Ayres, George Grizzard, Gene Tierney, Will Geer, Peter Lawford, Paul Ford, Betty White, Edward Andrews, Inga Swenson,
A complex and riveting look behind the scenes in Washington. When a new secretary of state has to undergo a thorough grilling by committee before his appointment is confirmed, both blackmail and suicide result from the political manoeuvering set in motion by his appointment. Intelligently made and well acted with great professionalism, this is the kind of film that makes you wonder what happened to Hollywood as a creative force.
DRA 138 min B/W VIDrel: 4-FRONT/POLYREC V
Boa: novel by Allen Drury.

AFFAIR, THE * 15
Paul Seed USA 1995
Courtney B. Vance, Kerry Fox, Leland Gantt, Ned Beatty
Set in England during WW2, this true story tells of the affair a
young married woman has with a black GI, and the inevitable
conflicts this causes, especially when she is forced to make a
heartbreaking choice. Both a look at a doomed love affair and
an examination of the racial prejudices that were commonplace
at the time.
DRA 100 min (ort 105 min) VIDrel: ODY/SONOP V

**AFFAIR TO REMEMBER, AN ** U
Leo McCarey USA 1957
*Cary Grant, Deborah Kerr, Cathleen Nesbitt, Neva Patterson, Richard
Denning, Robert Q. Lewis, Charles Watts, Fortunio Bonanova, Matt
Moore, Nora Marlowe, Louis Mercier, Geraldine Wall, Sarah Selby,
Genevieve Aumont, Jesslyn Fax*
Aboard a luxury liner, a romance gradually develops between
a wealthy bachelor engaged to an heiress and a young woman,
who is also similarly involved. They soon realise that each has
made the wrong choice and agree to meet six months later,
although events conspire to keep them apart. A dull remake of
"Love Affair" (1939) that is an awkward mixture of romantic
comedy and melodramatic tearjerker. See also the 1994 remake
LOVE AFFAIR.
COM 110 min (ort 115 min) VIDrel: 20TH/TECH V/sh

**AFFAIRS WITH DEATH ** 15
Frederick King Keller/Jeffrey Bloom USA 1992
*Laura Robinson, Tony Plana, Christina Pickles, Robert Ruth, Robert
Sutton, Robert Beltran, Dan Hedaya, Jacquelyn Hyde, Ray Stricklyn,
Lois Chiles, Barbara Howard*
The hostess of a popular jazz club hides a secret past that's
bound up with her work in a very different profession, that of
a private investigator.
A/AD 90 min VIDrel: GUILD/SONOP V

**AFRAID OF THE DARK ** 18
Mark Peploe FRANCE/UK 1992
*James Fox, Fanny Ardant, Paul McGann, Ben Keyworth, Clare
Holman, Robert Stephens, Susan Wooldridge, David Thewlis, Struan
Rodger, Rosalind Knight, Niven Boyd, Frances Cuka, Jeremy Flynn,
Star Acri, Sheila Burrell, Hilary Mason*
An offbeat psychological thriller which examines a young boy
whose obsessive fears and daydreams lead to a situation fraught
with danger. A spate of attacks on blind women make him
anxious for his mother, who is blind, while his policeman father
seems unable to make any progress in solving these crimes.
Convoluted in the extreme, this hard-to-follow effort marked
Peploe's directing debut.
THR 87 min (ort 92 min) VIDrel: 20VIS/SONOP V/sur

AFRICAN QUEEN, THE ** U
John Huston USA 1951
*Humphrey Bogart, Katharine Hepburn, Robert Morley, Peter Bull,
Theodore Bikel, Walter Gotell, Peter Swanwick, Richard Marner,
Gerald Onn*
Wonderfully successful combination of Bogart and Hepburn in
the tale of an untidy skipper being persuaded by a prim spin-
ster to take his boat up the Congo during WW1. Together they
face both the elements and hostile Germans and in the process
learn to respect and ultimately, love each other. A classic film
with the two stars in top form. AA: Actor (Bogart).
DRA 100 min (ort 105 min) VIDrel: 20TH/TECH V/dm
Boa: novel by Cecil Scott Forester.

**AFRICAN RUN, THE ** 15
Andrew Sinclair USA
*John Wyman, Carol Royle, Holly Palance, James Coburn Jr, Mike
Samson, John Terry*
A compulsive gambler marries a thief who steals $1,000,000 and
the couple are forced to go on the run, all the while being trailed
by a hired assassin. Adequate action fare of no great conse-
quence.
A/AD 90 min VIDrel: MOPIC/SGSVID V

AFTER DARK, MY SWEET * 18
James Foley USA 1990
*Jason Patric, Rachel Ward, Bruce Dern, George Dickerson, James
Cotton, Rocky Giordani, Corey Carrier, Jeanie Moore, Tom Wagner,
Burke Byrnes, Michael G. Hagerty, James E. Bowen Jr, Vince Mazzella
Jr, Napoleon Walls*
A former boxer absconds from a mental institution and meets

a beautiful but disturbed woman, who together with a friend,
gets him embroiled in a rather clumsy kidnapping plan. An
initially effective adaptation of Thompson's novel that starts
off as a promising example of 1990s film noir but runs out of
ideas at just about the same time the viewer runs out of
patience.
THR 111 min (ort 114 min) VIDrel: VISVID/POLYREC
V/sur
Boa: novel by Jim Thompson.

AFTER DARKNESS * 18
Dominique Othenin-Giraud/Sergio Guerrez SWITZER-
LAND/UK 1985
*John Hurt, Julian Sands, Victoria Abril, Pamela Salem, William
Jacques, Michael Herzog, Philippe Herzog*
A study in schizophrenia in which a man attempts to cure his
young brother himself, after years spent in a mental home have
failed to produce any improvement. This turns out to be a
dangerous undertaking.
Aka: AFTER DARK; NACH DER FINSTERNIS
HOR 104 min (ort 109 min) VIDrel: SPEAR/SONOP V

AFTER HOURS * 15
Martin Scorsese USA 1985
*Griffin Dunne, Rosanna Arquette, Verna Bloom, Thomas Chong, John
Heard, Teri Garr, Linda Fiorentino, Cheech Marin, Catherine O'Hara,
Dick Miller, Bronson Pinchot, Will Patton, Robert Olunket, Rocco
Sisto, Larry Block*
Strange, surrealistic account of a word-processor operator who,
after a date with a strange girl in New York's Soho district, finds
himself effectively marooned there, unable to get away as he
experiences a night of weird and crazy encounters. Only occa-
sionally amusing, this overlong odyssey of oddities has little
going for it plotwise.
COM 93 min (ort 97 min) VIDrel: MGM/WHV V/h

**AFTER THE GLORY ** PG
John Gray USA 1992
*Brad Johnson, Kathleen Quinlan, Tom Sizemore, Josef Sommer, Lisa
Blount, G.W. Bailey, Patricia Clarkson, David Labiosa, John M.
Jackson, Brad Leland, Sean Henningan, Titus Welliver, Shawn
Toovey, Miraida Rios, Jerry Haynes*
Five veterans from WW2 return to their home-town, only
to find it now in the grip of bigots and corrupt cops. They
decide to stand in the local elections in the hope of cleaning
the town up, but this decision exposes them to derision and
enmity.
Aka: AMERICAN STORY, AN
DRA 99 min VIDrel: ODY/SONOP V/sh

AFTER THE PROMISE * PG
David Greene USA 1987
*Mark Harmon, Diana Scarwid, Rosemary Dunsmore, Donnelly
Rhodes, Trey Ames, Mark Hildreth, Richard Billingsley, Chance
Michael Corbitt, David French, Benjamin Turner, Lance Verwoerd,
Gary Verwoerd, Shirley Barclay, Dwight Koss*
Powerful drama based on actual events that tells of the long
struggle of an itinerant carpenter to regain custody of his four
sons, who were taken from him and institutionalised follow-
ing the death of their mother. Set during the Depression
period, this presents an uneasy look at bureaucracy, bungling
and abuse of power.
DRA 89 min (ort 100 min) mTV VIDrel: NWV/SONOP
V/h
Boa: book by Sebastion Milito.

**AFTER THE SILENCE ** 15
Fred Gerber USA
JoBeth Williams, Kelly Martin, Alan Rosenberg
The deaf daughter of a brutal drug dealer is put in hospital by
another attack, but her cause is championed by a feminist.
DRA 94 min VIDrel: ODY/SONOP V/sh

**AFTERMATH ** 15
Glenn Jordan USA 1990
*Richard Chamberlain, Michael Learned, Zeljko Ivanek, Doug Savant,
Darryl Hickman, Denis Heames*
A doctor's wife is killed and his son seriously wounded when
they become innocent victims at the scene of a violent robbery.
As the other children do their best to rally round as the
distraught father, the latter hides his own feelings, devoting his

time to his stricken son. A competent adaptation of Kinder's fact-based bestseller that is hurt by the contrived script and a slight lack of conviction from Chamberlain.

Aka: AFTERMATH: A TEST OF LOVE
DRA   90 min (ort 100 min) mTV   VIDrel: ITC/POLYREC
V/h
Boa: book Victim by Gary Kinder.

AGAINST ALL ODDS **   18
Taylor Hackford USA   1983
Jeff Bridges, James Woods, Rachel Ward, Jane Greer, Alex Karras, Richard Widmark, Dorian Harewood, Swoosie Kurtz, Saul Rubinek, Pat Corley, August Darnell (Kid Creole), Bill McKinney, Allen Williams, Sam Scarber, Ted White
An unemployed American football player in need of cash accepts a job from a former team-mate who is now the owner of a shady nightclub. The job involves finding his girlfriend who has run off to Mexico. He not only finds her but falls in love with her, getting enmeshed in a complex web of corruption and intrigue. This loose remake of OUT OF THE PAST (Greer plays the mother of her character in the original) suffers badly from an over-complex plot.
DRA   116 min (ort 128 min)
VIDrel: VCC/DISC/COLUM L/A   V/sh

AGAINST HER WILL **   15
Delbert Mann USA   1993
Walter Matthau, Susan Blakely, Harry Morgan, Brian Kerwin, Nick Stahl, Stephanie Zimbalist, Bernard Behrens, Lori Hallier, David Nerman, Dennis Strong, James Blendick, Dee McCafferty, Cara Pifko, Eve Crawford, Jeremy Tracz
An easygoing lawyer moves to Baltimore to take up a partnership with a former legal adversary, who feels a pang of guilt for having blighted the career of the former. Bored with his new job, our lawyer jumps at the chance of taking on a most difficult case, that of a woman who committed herself to psychiatric care only to find herself a prisoner in an asylum. Adequate production-line family entertainment that plays like a TV movie.
Aka: AGAINST HER WILL: AN INCIDENT IN BALTIMORE; INCIDENT IN A SMALL TOWN; JUSTIFIABLE HOMICIDE
DRA   86 min (ort 97 min)   VIDrel: MARQ/GUILD   V

AGAINST THE WALL ***   18
John Frankenheimer USA   1993
Kyle MacLachlan, Samuel L. Jackson, Clarence Williams III, Frederic Forrest, Harry Dean Stanton, Philip Bosco, Tom Bower, David Ackroyd, Anne Heche, Carmen Argenziano, Peter Murnik, Steve Harris, Mark Cabus, Bruce Evers, Joey Anderson
A powerful examination of the causes of the 1971 Attica prison riot that cost forty-three lives, a tragedy that to could to a large extent have been avoided had the authorities acted more sensitively and politicians resisted the temptation to make capital from these events. Well directed and acted it may be, but this film breaks no new ground. See also the 1980 TV movie ATTICA.
DRA   110 min (ort 111 min) mCab   VIDrel: MED/DISC
V/sh

AGAINST THEIR WILL ***   (12)
Karen Arthur USA   1994
Judith Light, Stacy Keach, Kay Lenz, Michael Woods, Giuliana Santini, Tonya Pinkins, Chelcie Ross, Tom Butler, Aidan Devine, Eugne Lipinski, Ross Petty, Frank Moore, Theresa Tova, Barbara Eve Harris, Christina Collins, Lili Francks
A female ex-prisoner sues over the inhumane conditions and abuse she suffered at the hands of her guards whilst in prison, the actions of the latter appearing to have received the tacit sanction of the prison management. After many setbacks the lawyer (well played by Keach) is triumphant. A most effective fact-based drama, though given the many prison scandals that have emerged in recent years, one wonders just how realistic its makers allowed it to be.
DRA   91 min   SATrel: SKY MOVIES

AGATHA **   PG
Michael Apted USA   1979
Dustin Hoffman, Vanessa Redgrave, Timothy Dalton, Helen Morse, Tony Britton, Celia Gregory, Paul Brooke, Timothy West, Alan Badel, Carolyn Pickles, Robert Longden, Donald Nithsdale, Yvonne Gilan, David Hargreaves, Sandra Voe
A speculative film examining what happened to mystery writer Agatha Christie when she disappeared for eleven days in 1926.

Redgrave is totally convincing as Christie, though Hoffman is considerably less so as the reporter who tracks her down.
DRA   100 min   VIDrel: WHV   V

AGE OF CONSENT **   15
Michael Powell AUSTRALIA   1968
James Mason, Helen Mirren, Jack MacGowran, Neva Carr-Glyn, Frank Thring, Antonia Katsaros, Michael Boddy, Harold Hopkins, Slim Da Grey, Max Moldrun, Dora Hing, Clarissa Kaye, Judy McGrath, Leonore Katon, Diane Strachan
A jaded painter seeks fresh inspiration in the Great Barrier Reef and meets a young girl only too willing to pose for him. Limp and predictable with the clumsy attempts at comedy an added annoyance. The pretty Australian scenery is however, a slight compensation. This was Mirren's screen debut.
DRA   95 min (ort 103 min) wScrn   VIDrel: TART/20TH   V
Boa: story by Norman Lindsay.

AGE OF INNOCENCE, THE ***   PG
Martin Scorsese USA   1993
Daniel Day-Lewis, Michelle Pfeiffer, Winona Ryder, Richard E. Grant, Alec McCowen, Miriam Margolyes, Sian Phillips, Geraldine Chaplin, Stuart Wilson, Mary Beth Hurt, Michael Gough, Alexis Smith, Jonathan Pryce, Robert Sean Leonard
In upper-class New York society of the 1870s, a man about to be married, falls in love with a beautiful woman, who is surrounded by an aura of scandal, but allows his feelings for her to be stifled by the conventions of the times. Fine attention to period detail is an asset, but over-reverance towards the original Pulitzer Prize-winning novel and severely leaden pacing make it all but impossible to enjoy. A highly competent but stiff epic. AA: Cost.
DRA   132 min (ort 138 min) wScrn cC
VIDrel: VCC/DISC/COLUM   V/sur
Boa: novel by Edith Wharton.

AGE OF TREASON, THE **   (PG)
Kevin Connor USA   1993
Bryan Brown, Matthias Hues, Manda Pays, Art Malik, Patricia Kerrigan, Sophie Okonedo, Richard D. Sharpe, Jamie Glover, Peter Jonfield, William Hootkins, Alan Shearman, Ian McNeice, Shirley Stelfox, Nabil Shaban
A detective movie given a highly unusual setting – Rome 69 A.D., where the central character (Brown) gets the job of locating the missing son of one of Rome's most powerful senators. Not entirely successful in execution, though one feels obliged to applaud the producers for backing such an odd idea.
DRA   90 min   SATrel: SKY MOVIES

AGONY AND THE ECSTASY, THE **   U
Carol Reed USA   1965
Charlton Heston, Rex Harrison, Diane Cilento, Harry Andrews, Adolfo Celi, Alberto Lupo, Venantino Venantini, John Stacy, Fausto Tozzi, Maxine Audley, Thomas Millian, Richard Pearson
Epic account of the life of Michelangelo that examines his conflicts with Pope Julius II and concentrates on his painting of the Sistine Chapel ceiling. A lavish and detailed production that entirely over-shadows the actors, it is preceded by a short documentary on the work of the artist.
DRA   130 min (ort 140 min) wScrn   VIDrel: 20TH/TECH
V/sh
Boa: novel by Irwin Stone.

AGUIRRE, WRATH OF GOD ***   PG
Werner Herzog WEST GERMANY   1974
Klaus Kinski, Ruy Guerra, Del Negro, Helena Rojo, Cecilia Rivera, Peter Berling, Danyel Edes, Edward Roland, Armando Polanah
Beautifully made but empty account of a doomed expedition of Spanish conquistadors in search of seven legendary cities of gold. Led by Aguirre, their deluded and slightly insane leader, they leave Pizarro's expedition and strike out on their own, meeting with death in the remote jungles of the Amazon. A dreamlike, hypnotic and ultimately pointless film. Written by Herzog.
Aka: AGUIRRE, DER ZORN GOTTES; AGUIRRE, THE WRATH OF GOD
DRA   90 min (ort 95 min)
VIDrel: TART/20TH; ENCORE (LV only)   V LV

AI NO BOREI **   18
Nagisa Oshima JAPAN   1978
Kazuko Yoshiyuki, Tatsuya Fuji, Takahiro Tamura, Takuzo Kawatani
A peasant woman has an affair with good-for-nothing and together they plot to murder her husband. However, his ghost

returns to haunt them. A follow-up to AI NO CORRIDA that does little with the interesting subject matter, but offers tedium in large doses.
Aka: EMPIRE OF PASSION; IN THE REALM OF THE PASSIONS; PHANTOM OF LOVE, THE
DRA 100 min (ort 108 min) VIDrel: CONNO/RTM V

AIR AMERICA ** 15
Roger Spottiswoode USA 1990
Mel Gibson, Robert Downey Jr, Nancy Travis, David Marshall Grant, Lane Smith, Ken Jenkins, Burt Kwouk, Art La Fleur, Tim Thomerson
The title refers to an airline that served as cover for the CIA's covert operations in Laos and other parts of Southeast Asia during the Vietnam War, but the film concentrates more on the buddy pairing between a veteran pilot and a young rookie and their wild adventures together, than on the political background. A failed effort at an action film with forced comedy highlights, it shows little return for all the hard work put in by the cast.
A/AD 107 min (ort 113 min)
VIDrel: 4-FRONT/POLYREC/GUILD; PION (LV only)V/sur LV
Boa: book by Christopher Robbins.

AIR FORCE ** PG
Howard Hawks USA 1943
John Garfield, John Ridgely, Gig Young, George Tobias, Arthur Kennedy, James Brown, Charles Drake, Harry Carey, Faye Emerson, Edward S. Brophy, Ward Wood, Ray Montgomery, Stanley Ridges, Willard Robertson, Moroni Olsen
Fairly typical look at a WW2 bomber crew, that covers the period from the attack on Pearl Harbour to a US revenge raid on Tokyo. A dated propaganda piece with banal dialogue that for all its defects, has some moments of excitement and interest.
AA: Edit (George Amy).
WAR 119 min (ort 124 min) B/W VIDrel: WHV V

AIR UP THERE, THE ** PG
Paul Michael Glaser USA 1994
Kevin Bacon, Charles Gitonga Mbugua, Sean McCann, Dennis Patrick, Yolanda Vazquez, Winston Ntshona, Mabutho "Kid" Sithole, Ilo Mutombo, Nigel Miguel, Eric Menyuk, Keith Gibbs, Miriam Owiti, Douglas Leboyare, Ken Gampu, Yusi Kunene
An American basketball coach goes to Kenya to recruit a talented tribesman, who however refuses to play ball and sign a contract, since he is due to assume the leadership of his tribe. However, despite misunderstandings on both sides, our coach eventually prepares a tribal team for a game to resolve a dispute over land ownership. A predictable comedy on the familiar theme of culture clashes in sport that offers some modest entertainment.
COM 103 min (ort 108 min) VIDrel: HOLPIC/TECH L/A V

AIRBORNE ** PG
Rob Bowman USA 1993
Shane McDermott, Chris Conrad, Brittney Powell, Seth Green, Eddie McClurg, Patrick O'Brien
A Californian teenager is forced to exchange the sunny beaches for chilly Cincinnati and finds it hard to gain respect from the locals until he impresses them with how he handles himself on roller skates. The well-staged stunts and rock soundtrack are the only items of interest in this puerile offering.
DRA 86 min (ort 91 min) VIDrel: EIV/SONOP V/sur

AIRHEADS ** 15
Michael Lehmann USA 1994
Brendan Fraser, Adam Sandler, Steve Buscemi, Joe Mantegna, Michael Richards, Michael McKean, Chris Farley, Judd Nelson, Ernie Hudson, Amy Locane, Marshall Bell, Nina Siemaszko, John Melendez, Reg E. Cathey, David Arquette, Allen Covert
Three wannabe musicians are so desperate to get their demo tape played on the airwaves that they resort to taking over a radio station with their toy guns and forcing the resident DJ to oblige them. This leads to unforeseen consequences and a dose of instant media exposure. A mildly amusing comedy whose soundtrack is its main attraction.
COM 88 min cC VIDrel: 20TH/TECH V/sur

AIRPLANE! *** PG
Jim Abrahams/David Zucker/Jerry Zucker USA 1980
Robert Hays, Julie Hagerty, Lloyd Bridges, Peter Graves, Kareem Abdul-Jabbar, Leslie Nielsen, Lorna Patterson, Stephen Stucker, David Zucker, Jerry Zucker, Jimmie Walker, Barbara Billingsley, Ethel Merman

Wildly anarchic spoof on AIRPORT and all those other disaster-in-the-air type movies. A fast-paced series of mainly visual gags revolving around the plight of the passengers and crew aboard a troubled flight, as ground-control battle to bring it safely down. Inventive, witty and very, very funny. The infinitely inferior AIRPLANE 2: THE SEQUEL followed.
COM 84 min (ort 88 min)
VIDrel: 4-FRONT/POLYREC/CIC L/A V/h

AIRPLANE 2: THE SEQUEL ** 15
Ken Finkleman USA 1982
Robert Hays, Julie Hagerty, William Shatner, Peter Graves, Chad Everett, Lloyd Bridges, Sonny Bono, Raymond Burr, Chuck Connors, John Dehner, Rip Torn, Ken McCord, Stephen Stucker, Oliver Robins, James A Watson Jr
A poor sequel to AIRPLANE! with the fun centring around the first passenger space shuttle to the moon. Despite the many cameo appearances and the hard work of all concerned, this one just can't match the pace of the original. The laughs when they do come are few and far between. Written by Finkleman.
COM 80 min (ort 84 min) VIDrel: CIC/SONOP V/h

AIRPORT ** PG
George Seaton USA 1970
Burt Lancaster, Dean Martin, George Kennedy, Helen Hayes, Jean Seberg, Van Heflin, Jacqueline Bisset, Maureen Stapleton, Barry Nelson, Dana Wynter, Lloyd Nolan, Barbara Hale, Gary Collins, John Findlater, Jessie Royce Landis
Top-heavy with a star cast, this boring long film looks at the comings and goings of a busy airport. The inevitable crisis-in-flight goes some way towards enlivening a dull film that spawned three equally boring sequels. AA: S. Actress (Hayes).
DRA 129 min (ort 137 min)
VIDrel: 4-FRONT/POLYREC/CIC V/h
Boa: novel by Arthur Hailey.

AIRPORT '77 ** PG
Jerry Jameson USA 1977
Jack Lemmon, Lee Grant, Brenda Vaccaro, Joseph Cotten, Olivia de Havilland, James Stewart, Darren McGavin, Robert Foxworth, Kathleen Quinlan, Monte Markham, Christopher Lee, James Booth, Robert Hooks, Gil Gerard
Third in the series of AIRPORT dramas. Here a private Boeing 747 crashes into the ocean and sinks, necessitating a daring rescue bid. Lemmon is good as the pilot in an otherwise glossy and fairly vacuous disaster movie. Entertaining if you like that sort of thing.
DRA 108 min (ort 116 min)
VIDrel: 4-FRONT/POLYREC/CIC V/h

AIRWOLF: THE MOVIE ** 18
Donald Bellisario USA 1984
Jan-Michael Vincent, Ernest Borgnine, Alex Cord, David Hemmings, Belinda Bauer, Eugene Roche, Frank Annese, John Calvin, Herbert Jefferson Jr, Tina Chen, Dee Dee Rescher, W.K. Stratten, Deborah Pratt, Philip Bruns, Steven Greenstein
Super-helicopter with massive firepower has to be recovered from Colonel Gaddaffi and a plot to deliver it into the hands of the Russians is to be thwarted. The services of a super-agent are enlisted with this end in mind in this ludicrous and unbelievable yarn. Pilot for a TV series.
A/AD 82 min (ort 98 min) mTV
VIDrel: 4-FRONT/POLYREC/CIC V

AIRWOLF: THE STAVOGRAD INCIDENT * PG
Ken Jubenvill USA 1986
Barry Van Dyke, Michelle Scarabelli, Anthony Sherwood, Geraint Wyn-Davies
After the demise of the original "Airwolf" series, a Canadian company bought the rights to the show and released these cheaper and even more banal clones. In this one a serious nuclear accident in the USSR releases a deadly cloud of radioactive dust and the Russians ask the USA for help in dealing with it. Together with his team, pilot St John Hawke races to the rescue in this badly made and wretchedly acted story.
A/AD 82 min mTV VIDrel: CIC/SONOP V/sh

AKIRA *** 15
Katsuhiro Otomo JAPAN 1988
Voices of: Jimmy Flanders, Drew Thomas, Lewis Lemay, Barbara Larsen, Stanley Gurd Jr
A Japanese comic-book novel forms the basis for this violent story of a gang of teenage bikers who live in a post-apocalyptic

Tokyo. When a gang member becomes imbued with telekinetic powers in the course of a government experiment, he becomes a dangerous menace. Though colourful and technically excellent, the film's muddled script is a handicap. The collector's edition has an interview with Otomo plus a program about the making of the film.
ANIM   119 min dubbed; 173 min (Collector's Edition – subtitled version)   VIDrel: MANGA/SONOP   V/sur

ALADDIN ***   U
Ron Clements/Jon Musker USA   1992
Voices of: Scott Weiniger, Linda Larkin, Robin Williams, Jonathan Freeman, Frank Welker, Gilbet Gottfried, Douglas Seale, Brad Kane, Lea Salonga, Aaron Blaise, Charlie Adler, Jack Angel, Corey Burton, Philip Clarke, Jim Cummings
An outstanding retelling of the story of Aladdin, shown as a street-smart youth who finds a magic lamp inhabited by a genie who grants him three wishes. His new companion proves a useful ally in helping him woo a princess and defeat a dangerous rival for her affections. Will appeal to kids of all ages, but some some scenes may frighten the very young. THE RETURN OF JAFFAR followed. AA: Score/orig (A. Menken), Song ("Whole New World" – A. Menken (m)/Tim Rice (l)).
ANIM   90 min   VIDrel: WDV/TECH L/A   V

ALADDIN **   U
UK   1993
Voices of: Jason Connery, Derek Jacobi, Penelope Keith, Kate O'Mara, Nik Stoter, Edward Woodward
Adequate but certainly not memorable adaptation of this story, made by the same production company that released BEAUTY AND THE BEAST the year before.
ANIM   74 min   VIDrel: 4-FRONT/POLYREC   V

ALADDIN AND THE KING OF THIEVES **   U
USA   1996
Voices of: Robin Williams, Gilbert Gottfried, Jerry Orbach, Scott Weiniger, Linda Larkin, John Rhys Davies
Following on from ALADDIN and THE RETURN OF JAFAR, this is the third and final film in Disney's "Aladdin" trilogy (each adventure being based on one of the three wishes the Genie granted our hero). In this adventure the wedding of Aladdin and the Princess Jasmine is interrupted by Cassim and his Forty Thieves, all of whom are seeking the "Hand of Midas" – a fabled artefact that can turn anything into gold. A tolerable follow-up, though not as good as the first film.
ANIM   78 min cC   VIDrel: WDV/TECH   V/sur

ALAMO, THE ***   PG
John Wayne USA   1960
John Wayne, Richard Widmark, Laurence Harvey, Richard Boone, Frankie Avalon, Carlos Arruza, Patrick Wayne, Linda Cristal, Chill Wills, Joseph Calleia, Joan O'Brien, Ken Curtis, Hank Worden, Denver Pyle, Alissa Wayne, Bill Henry
Wayne directed this over-long account of the heroic stand by 185 Americans who held off 7,000 Mexican troops. A long, rambling and verbose offering that is redeemed by the truly memorable final attack. The excellent score is by Dmitri Tiomkin. AA: Sound (Gordon E. Sawyer/Fred Hynes).
WES   154 min (ort 199 min) wScrn (special edition)
VIDrel: MGM/WHV   V

ALAN AND NAOMI ***   PG
Sterling Van Wagenen USA   1992
Lukas Haas, Michael Gross, Vanessa Zaoui, Amy Aquino, Zohra Lampert, Kevin Connolly, Victoria Christian, Charlie Dow, Randy Williams, Mary McMullan, Stacy Moseley, Richard K. Olsen, Mark Fincannon, Becky Wyatt, Derek Knott, Shaun Ryan
In Brooklyn in 1944, a young Jewish boy learns that there are more important things in life that the fortunes of his stickball teams, when he is asked to be a friend to a catatonic young refugee girl. Having seen her father shot before her eyes by the Nazis, she retreated into a world of her own. A powerful if at times rather slow-paced drama that is supported by some moving and very convincing performances.
DRA   91 min (ort 95 min)   VIDrel: COLUM/SONOP   V/h
Boa: novel by Myron Levoy.

ALASKA ***   PG
Fraser C. Heston USA   1996
Thora Birch, Vincent Kartheiser, Dirk Benedict, Charlton Heston, Duncan Fraser, Gordon Tootoosis, Ben Cardinal, Ryan Kent, Don S.

Davis, Dolly Madsen, Stephen E. Miller, Byron Chief-Moon, Kristin Lehman, Adrien Dorval, Ed Gale
After his wife dies, a man movies with his family from Chicago to a remote town in Alaska. When he is lost in the course of an emergency plane flight, the man's children set off into the wilderness in search of him, but fall foul of a couple of poachers, when they attempt to rescue a bear cub from their clutches. A big budget outdoors adventure, that despite the over-familiar aspects of the storyline, is put together with considerable verve and wit.
A/AD   106 min (ort 109 min)   VIDrel: COLUM/SONOP   V/sur

ALCOVE, THE ***   18
Joe D'Amato ITALY   1982
Laura Gemser, Annie Bell, Al Cliver, Lilli Carati, Robert Caruso
An Italian army officer returns from Africa with a black mistress, whom he treats as his slave. But eventually, he tires of her and passes her over to his bisexual wife, who is only too pleased to have a new playmate, a view not shared by her regular girlfriend. However, all is resolved satisfactorily, as the new girl works her wiles to such an extent that eventually she takes over the household.
Aka: L'ALCOVA; LUST
A   88 min dubbed   VIDrel: JEZ/RTM   V

ALEXA **   18
Sean Delgato USA   1989
Christine Moore, Kirk Bailey, Ruth Corrine Collins, Tom Voth, JOseph P. Giardina, Adam Mechenner, Thomas Walker, Sarah Halley, Trula Hoosier, Lesle Lowe, Fred Cabral, Mary Round, Joseph Haddock, Mark Amato
A hooker falls for a playwright whose latest work deals with her profession, and contemplates retirement and leaving her pimp. A low-budget drama that gives this subject a serious treatment; unfortunately, the film largely fails to bring its characters to life.
Aka: ALEXA: A PROSTITUTE'S OWN STORY
A   77 min (ort 90 min)   VIDrel: MIA/DISC   V

ALEXANDER NEVSKY *****   PG
Sergei Eisenstein USSR   1938
Nikolai Cherkassov, Dmitri N. Orlov, Nikolai P. Okhlopkov, Alexander L. Abrikossov, Vassily K. Novikov, N.N. Arski, V.O. Massalitinova, Vera Ivasbeva, Anna Danilova, V.L. Ersbov, S.K. Blinnikov, I.I. Lagutin
Powerful epic tale of Nevsky and the Russian army repelling an attack by the Teutonic knights in the 13th century. Replete with the required propaganda of the time, this presents a remarkable vision of art in the service of the state. Yet for all that (and the strangely unconvincing battle on the ice), this is a masterwork that rises, together with the brilliant Prokofiev score, far above the stultifying constraints of socialist-realism.
DRA   106 min B/W   VIDrel: HEND L/A   V

ALEXANDER THE GREAT **   U
Robert Rossen USA   1956
Richard Burton, Claire Bloom, Danielle Darrieux, Fredric March, Peter Cushing, Barry Jones, Helmut Dantine, Harry Andrews, Stanley Baker, Niall MacGinnis, Peter Cushing, Michael Hordern, Barry Jones, Marisa De Leza
Remarkably uninspired and dull history lesson that fails to add any interest to its portrayal of Alexander's life and career or give any motivation for his desire for conquest. A beautifully staged and photographed but largely empty spectacle, though Burton has such screen presence that from time to time one forgets the silliness of it all.
DRA   130 min (ort 135 min)   VIDrel: MGM/WHV   V/s

ALEXANDER'S RAGTIME BAND ***   U
Henry King USA   1938
Tyrone Power, Alice Faye, Don Ameche, Ethel Merman, Jack Haley, Jean Hersholt, Helen Westley, John Carradine, Paul Hurst, Wally Vernon, Douglas Fowley, Ruth Terry, Eddie Collins, Joseph Crehan, Robert Gleckler, Joe King
Lavishly and lovingly mounted musical about the rivalry for the love of a young singer that exists between a ragtime band leader and a composer for no less than twenty-three years. Full of great performances and featuring as many as twenty-six compositions by Irving Berlin, including "My Walking Stick" and "Now It Can Be Told". Although corny in many ways, this movie remains hugely enjoyable even today. AA: Score (Alfred Newman).
MUS   109 min B/W cC   VIDrel: 20TH/TECH   V

ALFIE **   15
Lewis Gilbert UK   1966
Michael Caine, Julia Foster, Vivien Merchant, Shelley Winters, Alfie Bass, Millicent Martin, Jane Asher, Shirley Ann Field, Eleanor Bron, Denholm Elliott, Graham Stark, Murray Melvin, Sydney Tafler, Peter Graves
Caine is perfect in this brilliant story of a working-class Romeo who casually enters into relationships with women until a tragedy finally produces in him a sense of responsibility. Fine performances from Winters and the rest of the cast are coupled to a witty and perceptive script. One of the very best British films of the 1960s. Followed by ALFIE DARLING.
DRA   109 min (ort 114 min)
VIDrel: 4-FRONT/POLYREC/CIC L/A   V/h
Boa: play by Bill Naughton.

ALFIE DARLING *   18
Ken Hughes UK   1975
Alan Price, Joan Collins, Jill Townsend, Paul Copley, Annie Ross, Sheila White, Rula Lenska, Hannah Gordon, Roger Lumont, Minah Bird, Derek Smith, Vicki Michelle, Brian Wilde, Robin Parkinson, Jenny Hanley, Sally Bulloch
An attempt to produce a sequel to the smash hit ALFIE from 1966. This time Alfie is a lorry driver pulling the birds along the motorways of Europe. A film having none of the insight or bittersweet humour of the original, with Price an utterly inadequate replacement for Caine. Quite missable.
Aka: OH, ALFIE
COM   98 min (Cut at film release – ort 102 min)
VIDrel: BRAVE/SONOP L/A   V

ALICE *   PG
Jan Svankmajer UK   1988
Kristyna Kohoutova
A mixture of live-action plus puppet animation is used to good effect in this retelling of Lewis Carroll's famous story.
JUV   82 min (ort 86 min) dubbed   VIDrel: VISCOM/RTM V
Boa: novel Alice's Adventures In Wonderland by Lewis Carroll.

ALICE **   15
Woody Allen USA   1990
Mia Farrow, Alec Baldwin, Blythe Danner, Judy Davis, William Hurt, Keye Luke, Joe Mantegna, Bernadette Peters, Cybill Shepherd, Gwen Verdon, Patrick O'Neal, Kim Chan, Julie Kavner, Caroline Aaron, Holland Taylor, Robin Bartlett
A bored and pampered woman becomes tired of her pointless existence and her empty marriage, and embarks on a voyage of self-discovery, helped in no small measure by the mysterious potions supplied by a kindly Asian herbalist (played with great charm by Luke in his last screen role). A gently whimsical tale, it offers undeniable charm if no great insights. As ever, the script is by Allen.
DRA   101 min (ort 106 min)   VIDrel: COLUM/SONOP L/A V/sh

ALICE DOESN'T LIVE HERE ANYMORE **   15
Martin Scorsese USA   1975
Ellen Burstyn, Billy "Green" Bush, Kris Kristofferson, Alfred Lutter, Diane Ladd, Jodie Foster, Harvey Keitel, Vic Tayback, Valerie Curtin, Leila Goldoni, Lane Bradbury, Murray Moston, Harry Northrup, Alfred Lutter, Mia Bendixsen
Excellent and enjoyable account of a young widow coming to terms with her own life after the death of her husband leaves her penniless. Kristofferson as the gentle man who tries to win her love is memorable, as is the fine Robert Getchell screenplay. The TV series "Alice" followed. AA: Actress (Burstyn).
DRA   107 min (ort 113 min) wScrn   VIDrel: WHV   V

ALICE IN THE CITIES *   U
Wim Wenders WEST GERMANY   1974
Rudiger Vogeler, Yella Rottlander, Elizabeth Kreuzer, Edda Kochl, Didi Petrikat, Ernest Bohm, Sam Presti, Lois Moran, Hans Hirschmuller, Sybille Baier, Mirko, Wim Wenders
Another slab of pretentious and heavy Teutonic symbolism. A cynical and world-weary photographer gets saddled with a lovable little girl and gradually loses his negative attitudes in this beautifully photographed but ponderous tale. This was the first film in a trilogy by the director, that continued with THE WRONG MOVE.
Aka: ALICE IN DEN STADTEN
DRA   107 min (ort 110 min) B/W   VIDrel: CONNO/RTM V

ALICE IN WONDERLAND **   U
Clyde Geronimi/Hamilton Luske/Wilfred Jackson USA   1951
Voices of: Kathryn Beaumont, Ed Wynn, Richard Hadyn, Sterling Holloway, Pat O'Malley, Jerry Colonna, Verna Felton, Bill Thompson, Heather Angel, Joseph Kearns, Larry Grey, Queenie Leonard, Dink Trout, Doris Lloyd
Cartoon version containing elements from both "Alice in Wonderland" and "Alice through the Looking Glass". Walt Disney animation at its best, with several entertaining songs and some good characterisations compensating for the uneven script.
ANIM   72 min (ort 75 min) cC   VIDrel: WDV/TECH   V
Boa: novels Alice's Adventures In Wonderland/Through The Looking Glass by Lewis Carroll.

ALICE IN WONDERLAND: PARTS 1 AND 2 *   U
Harry Harris USA   1985
Beau Bridges, Lloyd Bridges, Patrick Duffy, Ringo Starr, Shelley Winters, Telly Savalas, Sammy Davis Jr, Robert Morley, Scott Baio, Red Buttons, Carol Channing, Roddy McDowall, Jonathan Winters, John Stamos, Natalie Gregory
Star-packed version of Lewis Carroll's classic fantasy produced by Irwin Allen – of TOWERING INFERNO and POSEIDON ADVENTURE fame. A colourless and dreary production not helped by some remarkably banal songs.
JUV   180 min (2 cassettes)   VIDrel: WHV   V/h
Boa: novels Alice's Adventures In Wonderland/Through The Looking Glass by Lewis Carroll.

ALICE THROUGH THE LOOKING GLASS *   (U)
Alan Handley USA   1966
Judi Rolin, Ricardo Montalban, Agnes Moorehead, Jack Palance, Tom Smothers, Dick Smothers, Jimmy Durante, Nanette Fabray, Roy Castle, Richard Denning, Robert Coote
Musical adaptation of the classic by Lewis Carroll, in which Alice is assisted in her return to Wonderland and further adventures by a chess piece, that comes to life and shows her the way to get back. An ambitious and colourful kids' adventure, with a witty script and an all-star line-up.
JUV   72 min mTV   SATrel: DISNEY CHANNEL
Boa: novel by Lewis Carroll.

ALICE'S ADVENTURES IN WONDERLAND *   U
William Sterling UK   1972
Peter Sellers, Dudley Moore, Ralph Richardson, Fiona Fullerton, Michael Crawford, Flora Robson, Michael Jayston, Robert Helpmann, Spike Milligan
A poor musical version of the classic children's book with dull music and even duller performances that fails to impart any of the charm or magic of the original.
MUS   95 min   VIDrel: BMGVID   V
Boa: novel by Lewis Carroll.

ALIEN **   18
Ridley Scott USA   1979
Sigourney Weaver, Yaphet Kotto, Tom Skerritt, Ian Holm, Harry Dean Stanton, John Hurt, Veronica Cartwright
State-of-the art special effects were wasted on a routine horror story of a terrifying alien creature that slaughters the crew of a commercial spaceship one by one. Made with infinite care in terms of effects, but poor dialogue and plot development let it down badly. From the opening, it's downhill all the way. Followed by ALIENS which is even more gory. AA: Effects/vis (H.R. Giger/Carlo Rambaldi/Brian Johnson/Nick Allder/Denys Ayling).
HOR   111 min wScrn (ort 117 min)   VIDrel: 20TH/TECH L/A; ENCORE (LV only)   V/sh LV

ALIEN 3 **   18
David Fincher USA   1992
Sigourney Weaver, Charles S. Dutton, Charles Dance, Paul McGann, Brian Glover, Ralph Brown, Danny Webb, Christopher John Fields, Holt McCallany, Lance Henriksen, Chris Fairbank, Carl Chase, Leon Herbert, Paul Brennen, Clive Mantle
In this gore-laden sequel Ripley finds herself being carried to a harsh all-male prison planet where she is forced to shave her head (everybody is so treated) and more or less pretend to be one of the inmates. The prisoners have become the followers of a religious fanatic, there is another monster on the loose, and pretty soon she understands why she has been deliberately spared by it. Brutal, bleak and profoundly depressing.
HOR   110 min (ort 115 min) wScrn   VIDrel: 20TH/TECH L/A; ENCORE (LV only)   V LV

ALIEN ABDUCTION: INTIMATE SECRETS *   *(18)*
Lucian S. Diamonde USA   1995
Darcy Demoss, Pia Reyes, Dumitru Bogomaz, Carmen Iacatus, Alina Chivulescu, Florin Chiviac, Meredyth Holmes, Floriela Grappini, Valentin Lucia, Laura Ilica, Constantin Barbulescu, Alice Balaianu
A group of women relaxing at an indoor swimming baths, start to swap stories of their favourite sexual fantasies. Three of them start recalling strange events from a trip they all took together in Europe when their car broke down on a lonely road. It eventually transpired that what they experienced was due to them being abducted by an alien creature, in this witless blend of soft-core sex film and SF. A very tedious effort, filmed entirely on location in Romania.
A   87 min   SATrel: SKY MOVIES

ALIEN FACTOR, THE **   *18*
Glenn Takakjian USA   1989
Matt Kulis, Patrick Barnes, Tara Leigh, Tony Gigante, Dianna Flaherty, Greg Sullivan, Marcus Powell, Katherine Romaine, George Gerard, Allen Lewis Rickman, Michael D'Andrea, Ralph Ormaldi
A top-secret research lab is conducting experiments on samples of genetic material of extra-terrestrial origin when an accident occurs, and one of the researchers is infected with a virus-like organism. He soon starts to rapidly mutate, turning into an alien flesh-eating monster, a fate that threatens the rest of humanity.
FAN   97 min   VIDrel: POPRO/RTM   V

ALIEN INTRUDER *   *15*
Ricardo Jacques Gale USA   1991
Billy Dee Williams, Tracy Scoggins, Maxwell Caulfield, Gary Roberts, Richard Cody, Stephen Davies, Shano Palovich, Jeff Conaway, Charles Young, Rod Britt, Milto James, Joe Durrenberger, Gwen Somers, Melinda Armstrong, Andrianne Sachs
Four prisoners in the year 2022 are sent on a suicide mission to recover a spacecraft from the most uncharted sector in the universe, where they encounter a beautiful woman who turns out to be a "computer virus" that has infected a virtual reality entertainment system they have onboard ship. An abysmally poor SF offering whose one good idea is totally ruined by a muddled script, lousy production values, unimpressive special effects and uninspired acting.
FAN   90 min   VIDrel: IMPENT   V

ALIEN NATION ***   *18*
Graham Baker USA   1988
James Caan, Mandy Patinkin, Terence Stamp, Kevin Major Howard, Peter Jason, Leslie Bevins, George Jenesky, Jeff Kober, Roger Aaron Brown, Tony Simotes, Tony Perez, Michael David Simms, Ed Krieger, Brian Thompson, Frank McCarthy, Don Hood
Offbeat SF police thriller set at some future period when alien immigrants live and work on Earth. A police detective has his partner killed by some of the aliens in the course of a robbery and is none to pleased to find his new partner is an alien too. Early promise and many clever touches are hampered by a disappointing script. A TV series followed.
FAN   86 min (ort 94 min)   VIDrel: 20TH/TECH L/A   V

ALIEN NATION: BODY AND SOUL **   *12*
Kenneth Johnson USA   1996
Gary Graham, Eric Pierpoint, Michelle Scarabelli
Further adventures in the TV spin-off from the original ALIEN NATION film. In this story we learn that among the alien newcomers were a child and a giant, both of whom being escapees from a secret research lab. Our human and alien detective duo have to expose the secret genetic experiments being conducted that could doom both the human and alien races. Adequate.
FAN   86 min mTV   VIDrel: 20TH/FOXVID   V/sh

ALIEN NATION: DARK HORIZON **   *12*
Kenenth Johnson USA   1995
Scott Patterson, Gary Graham, Eric Pierpoint, Michele Scarabelli, Terri Treas, Nina Foch, Lauren Woodland, Sean Six, Jeff Marcus, Ron Fussler, Michelle Lamar Richards, Jenny Gago, Dana Andersen, Lee Bryant, David Purdham, Diane Cary
A kind of sequel to the 1988 cinema film but starring the cast of the TV series and set some years after the newcomers have begun to adapt to living on Earth. A beacon from the slave ship that brought them here, is located by their former masters who immediately dispatch one of their number to make plans to recover their missing property. A poorly handled and cynical effort to exploit the popularity of the TV series that offers little excitement.
FAN   89 min   VIDrel: 20TH/FOXVID   V

ALIEN NATION: MILLENIUM **   *12*
Kenneth Johnson USA   1996
Gary Graham, Eric Pierpoint, Michelle Scarabelli, Terri Treas, Sean Six, Lauren Woodland
An artefact from the alien's homeworld has appeared, but it poses dangers for those who make use of it, as it traps them in a virtual reality universe. Once again, our crime-busting human/alien duo must expose this danger and save all concerned.
FAN   87 min mTV   VIDrel: 20TH/FOXVID   V/sh

ALIEN SPECIES **   *18*
Dave Payne USA   1996
Maria Ford, Emile Levisetti, Cassandra Leigh, Bob McFarland, Kevin Alber
In this low-budget clone of ALIEN, a group of scientists begin conducting secret genetic experiments five mile underground, but this leads to the creation of a rampaging monster that starts working its way through the assorted boffins. It all takes place in the tunnels and caverns surrounding their laboratories, and though the effects are good enough, what this straight-to-video offering really needed was an injection of fresh ideas.
FAN   90 min   VIDrel: THIRD   V

ALIEN WITHIN *   *18*
Scott Levy USA   1994
Roddy McDowell, Alex Hyde-White, Melanie Shatner, Don Stroud
A research team investigating the use of the sea-bed as a permanent home for humans stumbles across an alien with mind-controlling powers in a film that is derivative of both THE THING and LEVIATHAN. Scripting is weak and direction equally so, and the few good moments are hardly worth the wait.
HOR   83 min   VIDrel: CIC/SONOP   V

ALIENATOR, THE *   *18*
Fred Olen Ray USA   1989
Jan-Michael Vincent, John Phillip Law, Ross Hagen, Teagan Clive, Dyann Ortelli, Jesse Dabson, Dawn Wildsmith, P.J. Soles, Robert Clarke, Richard Wiley, Leo V. Gordon, Robert Quarry
A kind of distant cousin to THE TERMINATOR in which an alien escapes from a prison spaceship and crash-lands on Earth, but is pursued by a murderous android. Forest ranger Law and his companions come across the injured fugitive and set about offering assistance. A film of weak plotting and dialogue; its passable effects and mediocre acting do little to sustain one's interest.
FAN   87 min (ort 93 min)   VIDrel: MOPIC/SGSVID   V

ALIENS ***   *18*
James Cameron USA   1986
Sigourney Weaver, Carrie Helm, Michael Biehn, Paul Reiser, Lance Henriksen, Bill Paxton, Jeanette Goldstein, William Hope, Al Matthews, Mark Rolston, Ricco Ross, Colette Hiller, Daniel Kash, Cynthia Scott, Tip Tipping, Trevor Steedman
A sequel to ALIEN with our heroine arriving at a planet, where she has to do battle with a host of monsters, this time round. The effects are even more gruesome, though the film is as empty as ever. A kind of horror film in space rather than a true science fiction film, the speculative element of this genre completely overwhelmed by gory special effects. AA: Effects/aud (Don Sharpe), Effects/vis (R. Skotak/S. Winston/J. Richardson/S. Benson).
Aka: ALIENS: THIS TIME IT'S WAR
FAN   154 min   VIDrel: 20TH/TECH L/A; ENCORE (LV only)   V/sh LV

ALISON'S BIRTHDAY **   *18*
Ian Coughlan AUSTRALIA   1979
Lou Brown, Joanne Samuel, Bunny Brooke, Vincent Ball, Marion Johns, Belinda Giblin, Lisa Peers, Margie McCrae, Martin Vaughan, Rosalind Speirs, Robyn Gibbes, Lou Brown, Ian Coughlan, Ralph Cotterill, John Bluthal, Eric Oldfield
Horror tale about a girl warned by her dead father, during a seance, to leave her uncle's house before her 19th birthday. She doesn't and discovers that her birthday party is to take the form of a black magic ritual, designed to transfer her personality into the wizened remains of a dead witch in order to make use of her body as the cult's new leader. A standard yarn of poor direction and plotting.
HOR   90 min (ort 99 min)   VIDrel: MOPIC/SGSVID   V

**ALIVE ** **   15
Frank Marshall USA   1992
*Ethan Hawke, Vincent Spano, Josh Hamilton, Bruce Ramsay, John
Haymes Newton, David Kriegel, Kevin Breznahan, Sam Behrens,
Illeana Douglas, Danny Nucci, Christian Meoli, Jack Noseworthy,
John Malkovich, Frank Marshall, Ele Keats*
Recreation of a true incident from 1972, when members of a
rugby team from Uruguay who survived the crash of their plane
in the Andes were forced to turn to cannibalism in order to stay
alive. Fortunately, unlike the 1972 film "Survive!" which also
dealt with this event, this movie concentrates instead on the
broader aspects of their ordeal. The LV disc includes a forty-six
minute documentary on the making of the film plus some interviews with the survivors.
DRA   121 min (ort 127 min) cC   VIDrel: CIC/SONOP; PION
(LV only)   V/sur LV
Boa: book by Piers Paul Reed.

ALIVE AND KICKING **   (PG)
Robert Young UK   1991
*Lenny Henry, Robbie Coltrane, Annabelle Apsion, Imogen Boorman,
Geff Francis, Paul Barber, Jane Horrocks, Cal Macaninch, Beresford
Le Roy, Hakeem Kae-Kazim, Mark A. Newman, Ewen Cummins,
Faith Tingle, Garry Roost, Karl Tessler*
Henry gives a very strong performance as a hardened drug user
and dealer, whose addiction to heroin has blighted his life.
Coltrane plays a drug counsellor (and former addict) who is
determined to help him, using some rather unconventional
methods. Both stars are totally convincing in their demanding
roles, and this strong work is sustained by an incisive script and
a few welcome touches of darkly sardonic humour.
DRA   95 min mTV   TVrel: BBC

ALL ABOUT EVE **   U
Joseph L. Mankiewicz USA   1950
*Bette Davis, George Sanders, Anne Baxter, Celeste Holm, Gary
Merrill, Thelma Ritter, Marilyn Monroe, Hugh Marlowe, Gregory
Ratoff, Barbara Bates, Walter Hampden, Randy Scott, Craig Hill,
Leland Harris, Claude Stroud, Eugene Borden*
Brilliantly acted and scripted story of a young aspiring actress
and her ruthless rise to stardom. Davis is superb as the ageing
star with Merrill playing her young lover. The witty character-
isations are too numerous to mention, but Sanders as a critic and
Marlowe as a playwright stand out. This was the basis for the
Broadway musical "Applause". AA: Pic, Dir, S. Actor (Sanders),
Screen (Mankiewicz), Cost (Edith Head/Charles LeMaire),
Sound (Fox Studios).
DRA   134 min (ort 138 min)   VIDrel: 20TH/TECH   V/sh
Boa: story The Wisdom Of Eve by Mary Orr.

ALL AMERICAN GIRLS *   18
Max Altman USA   1983
*Cassie Blake, Jacqueline Lorians, Stephen Douglas, Jillian Nichols,
Joanna Storm, Ken Star, K.C. Valentine, Laura Lazare, Carrie Sinclair,
Jeff Conrad*
A variant on Chaucer's Canterbury Tales with six sorority sisters
being invited to London by a seventh who just happens to have
married into the aristocracy. The reason for the invitation is to
have them all recount their sexual exploits since graduation,
and as expected, the film takes the form of a sequence of sexual
encounters in various parts of the world. An untidy and rather
inconsequential effort.
A   60 min Cut (8 sec plus some cuts subst – ort 87 min)
VIDrel: SHEP L/A; HAR/GOLD   V

ALL DOGS GO TO HEAVEN **   U
Don Bluth USA   1989
*Voices of: Loni Anderson, Judith Barsi, Dom DeLuise, Melba Moore,
Charles Nelson-Reilly, Burt Reynolds, Vic Tayback, Earleen Carey,
Candy Devine, Rob Fuller, Daryl Gilley, Anna Manahan, Ken Page,
Nigel Peagram*
An orphan is adopted by an ex-con mutt who has returned from
heaven to perform a good deed. The vivid animation work of
Bluth is one of the few strong points in this silly and disorgan-
ised romp of weak plotting and cliched characterisations.
ANIM   81 min (ort 98 min)   VIDrel: WHV   V/sur

ALL I WANT FOR CHRISTMAS **   U
Robert Lieberman USA   1991
*Harley Jane Kozak, Jamey Sheridan, Ethan Randall, Leslie Nielsen,
Lauren Bacall, Kevin Nelson, Thora Birch, Amy Oberer, Andrea
Martin, Michael Alaimo, Patrick Labrecque, Renee Taylor, Joanne
Baron, Darrell Kunitomi, Alan Brooks*

A thirteen-year-old boy and his younger sister of seven contrive
a means of re-uniting their recently divorced parents during the
celebrations for their first Christmas as a divided family. This
sweet and syrupy confection, hampered by over-acting on the
part of the child players, feels like an updated but inferior
version of THE PARENT TRAP.
COM   88 min (ort 92 min)   VIDrel: CIC/SONOP   V/sur

ALL MEN ARE MORTAL **   15
Ate de Jong FRANCE/HOLLAND/UK   1995
*Irene Jacob, Stephen Rea, Colin Salmon, Marianne Sagebracht, Maggie
O'Neill, Steve Nicolson, Chiara Mastroianni, Jango Edwards, Derek
de Lint, John Nettles, Michael Gaunt, Miranda Forbes, Roger Monk,
Fons Elders, Gaia Elders, Jane Wymark*
At the end of WW2, a famous Parisian actress meets an anti-
social misfit who claims to be seven-hundred-years old, but he
is tormented by the memories of all his previous lovers. A
clumsy misfire that takes de Beauvoir's difficult allegorical
novel and creates a film that strains both patience and credibility, not least because of poor casting and a set of singularly
unimpressive flashback sequences.
DRA   90 min   VIDrel: WHV   V
Boa: novel Tous les Hommes Sont Mortels by Simone de
Beauvoir.

ALL NIGHT LONG **   15
Jean-Claude Tramont USA   1981
*Gene Hackman, Barbra Streisand, Diane Ladd, Dennis Quaid, Kevin
Dobson, Ann Doran, William Daniels, Annie Girardot, Hamilton
Camo, Terry Kiser, Charles Siebert, Vernee Watson, Raleigh Bond,
Annie Girardot, Mitzi Hoag*
When a middle-aged executive is demoted by his company,
becoming manager of a downtown all-night drugstore, the
favoured hangout of assorted weirdos; he finds solace by
embarking on an affair with his neighbour's wife. Streisand
replaced Lisa Eichhorn at short notice in this one and she's not
well-cast, yet Hackman performs well in a comedy of considerable charm.
COM   84 min (ort 96 min)   VIDrel: CIC/SONOP L/A   V

ALL OF ME **   15
Carl Reiner USA   1984
*Steve Martin, Lily Tomlin, Victoria Tennant, Madolyn Smith, Dana
Elcar, Richard Libertini, Jason Bernard, Selma Diamond, Eric
Christmas, Gailard Sartain, Neva Patterson, Michael Ensign, Peggy
Feury, Nan Martin, Stu Black*
A lawyer finds that the left side of his body has been taken over
by the soul of an extremely wealthy and eccentric woman.
Martin is superb as the idealistic lawyer driven to distraction by
an often hilarious chain of events, though the script suffers from
unevenness. An ambitious mixture of fantasy and comedy that
actually improves as the story progresses.
COM   88 min (ort 93 min)   VIDrel: WHV   V/h
Boa: novel Me Two by Ed Davis.

ALL QUIET ON THE WESTERN FRONT **   PG
Lewis Milestone USA   1930
*Lew Ayres, Louis Wohlheim, John Wray, Slim Summerville, Russell
Gleason, Ben Alexander, Beryl Mercer, Raymond Griffith, William
Bakewell, Scott Kolk, Walter Browne Rogers, Ben Alexander, Owen
Davis Jr, Harold Goodwin, Pat Collins*
A moving and powerful adaptation of Remarque's great anti-
war novel, tracing the fortunes of a group of German teenagers
who volunteer for service on the Western Front in 1914. Their
passage from naive patriotism to cynical disillusionment is
vividly portrayed with the final sequence being one of the most
haunting on film. Considerably shortened on release, it was
followed in 1937 by "The Road Back", a poor sequel. Remade in
1979. AA: Pic.
WAR   125 min (restored ver incl extra footage – ort 140 min)
B/W   VIDrel: CIC/SONOP   V/h
Boa: novel by Erich Maria Remarque.

ALL QUIET ON THE WESTERN FRONT *   PG
Delbert Mann USA   1979
*Richard Thomas, Ernest Borgnine, Ian Holm, Donald Pleasence,
Patricia Neal, Mark Drewry, Mark Elliott, Dai Bradley, Mathew
Evans, George Winter, Dominic Dephcott, Colin Mayes, Ewan
Stewart, Simon Haywood, Kevin Stoney, Ken Hutchinson*
A remarkably good remake of the 1930 classic anti-war film
telling of the experiences of a group of raw recruits in the
German trenches of WW1. Borgnine gives one of his best perfor-
mances as a world-weary and cynical sergeant whose job is to

mould these recruits into a competent force. The adaptation from Remarque's novel is by Paul Monash.
WAR   123 min (ort 180 min) mTV
VIDrel: 4-FRONT/POLYREC   V
Boa: novel by Erich Maria Remarque.

**ALL SHE EVER WANTED ** **   12
Michael Scott USA   1996
Marcia Cross, James Marshall, C.C.H. Pounder, Leila Kenzle, Bruce Kirby, Tom Nowicki, Richard K. Olsen, Larry Black, Howard Kingkade, Ralph Wilcox, Robby Preddy, Robert Catrini, Nancy McLoughlin, Patricia Clay, Jennifer MacWilliam
A woman has a rare mental disorder tat makes her prone to violent outbursts, but despite this, she is desperate to have a child. When she does fall pregnant, the authorities attempt to force her to have an abortion, and a court battle ensues. A sentimental tale, with the expected happy ending that sees our patient restored to full health and celebrating Mother's Day with her husband and new baby.
Aka: MOTHER'S DAY
DRA   90 min   VIDrel: ODY/SONOP   V

ALL THAT JAZZ *   15
Bob Fosse USA   1979
Roy Scheider, Jessica Lange, Ann Reinking, Leland Palmer, Cliff Gorman, Ben Vereen, Erzsebet Foldi, Sandahl Bergman, John Lithgow, Michael Tolan, William La Messena, Max Wright, Chris Chase, Deborah Geffner, Kathryn Doby, Sue Paul
Surrealistic musical fantasy based on the director/choreographer Fosse's own life, that offers one a strange combination of self-indulgence and gloom. The great opening and splendid dance numbers are enjoyable interludes that compensate to some extent for the film's overall lack of direction or purpose. AA: Art/Set (Philip Rosenberg and Tony Walton/Edward Stewart and Gary Brink), Score (Ralph Burns), Cost (Albert Wolsky), Edit (Alan Heim).
MUS   118 min (ort 123 min)   VIDrel: 20TH/TECH   V/sur

ALL THAT MONEY CAN BUY **   (PG)
William Dieterle USA   1941
James Craig, Walter Huston, Edward Arnold
Superb (if stagebound) retelling of the Faust legend with the Devil (memorably played by Huston) selling poor farmer Craig seven years of good luck in return for his soul. The lively camerawork of Joseph August and creepy score of Herrmann add to the enjoyment of this fantasy, but the highlight has to be a courtroom scene in which lawyer Daniel Webster (Arnold) must battle for the soul of his client. AA: Score (Bernard Herrmann).
Aka: CERTAIN MR SCRATCH, A; DANIEL AND THE DEVIL; DEVIL AND DANIEL WEBSTER, THE; HERE IS A MAN
FAN   112 mins B/W   TVrel: C4
Boa: story The Devil And Daniel Webster by Stephen Vincent Benet.

ALL THAT SEX *   18
Scotty Fox USA   1990
Ashlyn Gere, Madison, Danielle Rogers, Nikki Wilde, Celia Young, Randy Spears, Randy West, Marc Wallice, K.C. Williams
An empty-headed and largely plotless sex film built around the fevered imaginings of a successful movie director who wants to produce the ultimate adult film.
A   51 min (ort 82 min) mVid
VIDrel: GROHOM/MAXSCAN   V

ALL THE KING'S MEN **   U
Robert Rossen USA   1949
Broderick Crawford, Mercedes McCambridge, John Ireland, Joanne Dru, John Derek, Sheppard Strudwick, Anne Seymour, Ralph Dumke, Katherine Warren, Raymond Greenleaf, Walter Burke, Will Wright, Grandon Rhodes, H.C. Miller, Richard Hale
A brilliant account of the rise and fall of a Southern lawyer, who becomes State Governor but is undone by his unbridled lust for power. Crawford is electrifying in the role of an honest man who is all too easily corrupted, and his performance gave his career the boost he richly deserved. His character was clearly inspired by the real-life figure of Huey Pierce Long. See also THE SEDUCTION OF JOE TYNAN. AA: Pic, Actor (Crawford), S. Actress (McCambridge).
DRA   105 min (ort 109 min) B/W   VIDrel: COLUM/SONOP V
Boa: novel by Robert Penn Warren.

ALL THE MORNINGS OF THE WORLD *   15
Alain Corneau FRANCE   1992
Gerard Depardieu, Guillaume Depardieu, Anne Brochet, Jean-Pierre Marielle, Carole Richert, Michel Bouquet, Jean-Claude Dreyfus, Yves Gasc, Jean-Marie Poirier, Yves Lambert, Myriam Boyer, Violaine Lacroix, Nadege Teron
A lavish, long and beautifully staged film biography of seventeenth-century viol musician and composer Monsieur De Sainte Colombe who felt that music was not something that ought to be performed in public. This view, however, was not shared by his far more worldly protege. Father-and-son Depardieu give excellent performances but the unnaturally slow pace of this movie is its greatest drawback.
Aka: TOUS LES MATINS DU MONDE
DRA   110 min (ort 115 min) wScrn
VIDrel: ELPIC/POLYREC   V/sur
Boa: novel Tous Les Matins Du Monde by Pascal Quignard.

ALL THE PRESIDENT'S MEN *   15
Alan J. Pakula USA   1976
Robert Redford, Dustin Hoffman, Martin Balsam, Jason Robards, Hal Holbrook, Jack Warden, Jane Alexander, Stephen Collins, Meredith Baxter, Ned Beatty, Robert Walden, Polly Holliday, F. Murray Abraham, Lindsay Ann Crouse
The film of the book about the Washington Post reporters Bob Woodward and Carl Bernstein, whose investigation into the Watergate break-in ultimately led to the resignation of President Nixon and his political disgrace. A film that skilfully tells a true story without any unnecessary embellishment. AA: S. Actor (Robards), Screen/adapt (William Goldman), Art/Set (George Jenkins/G. Gaines), Sound (A. Plantadosi/L. Fresholtz/D. Alexander/J. Webb).
DRA   132 min (ort 138 min)   VIDrel: WHV   V
Boa: book by Carl Bernstein and Bob Woodward.

ALL THE RIGHT MOVES *   15
Michael Chapman USA   1983
Tom Cruise, Lea Thompson, Craig T. Nelson, Charles Cioffi, Paul Carafotes, Christopher Penn, Gary Graham, Sandy Faison, Paige Price, James A. Baffico, Donald A. Yanessa, Walter Briggs, Leon Robinson
A young boy living in a Pennsylvania steel town sees a football scholarship to university as his only hope of a better life. However, in the process he comes into conflict with his equally ambitious coach. A lighthearted and engaging film very much aimed at the youth audience.
DRA   86 min (ort 91 min)   VIDrel: 20TH/TECH   V/h

ALL THIS AND HEAVEN TOO *   U
Anatole Litvak USA   1940
Bette Davis, Charles Boyer, Jeffrey Lynn, Barbara O'Neil, Virginia Weidler, Charles Hampden, Helen Westley, George Coulouris, Montagu Love, June Lockhart, Anne Todd, Fritz Leiber, Harry Davenport, Sibyl Harris, Janet Beecher
In 19th century France, an aristocrat falls in love with a governess and the pair embark on a doomed love affair that ends in scandal and death. A full-blooded adaptation of the Field novel with a script that provides many opportunities for the assorted stars to give of their best.
DRA   135 min (ort 143 min) B/W   VIDrel: WHV   V
Boa: novel by Rachel Field.

ALL TIED UP *   15
John Mark Robinson USA   1994
Zach Galligan, Teri Hatcher, Lara Harris, Tracy Griffith, Abel Folk, Olivia Brown, Edward Blatchford, Rachel Sweet, Melora Walters, Vivianne Vives, Phyllis Chase, Alvino Johnson
A womanising young man goes over to his girlfriend's home to explain away his behaviour after she breaks off their engagement. However, her friends decide to teach him a lesson by tying him up, hoping that a little bondage will bring him to his senses.
COM   90 min cC   VIDrel: COLUM/SONOP   V/s

ALLAN QUATERMAIN AND THE LOST CITY OF GOLD *   PG
Gary Nelson/Newt Arnold (additional scenes only) USA 1986
Richard Chamberlain, Sharon Stone, James Earl Jones, Henry Silva, Robert Donner, Doghmi Larbi, Aileen Marson, Cassandra Peterson, Martin Rabbett, Rory Kilalea, Alex Heyns, Themsi Times, Philip Borchner, Stuart Goakes
A sequel to KING SOLOMON'S MINES with Quatermain and his fiancee going off in search of his missing brother. He soon

finds himself and his companions menaced by spears, serpents, savages and lions as they trek through uncharted regions in this derivative and lacklustre adventure tale.
A/AD   96 min (ort 110 min)   VIDrel: RNK L/A   V

ALLIGATOR ***   15
Lewis Teague USA   1980
Robert Foster, Michael Gazzo, Robin Riker, Perry Lang, Henry Silva, Jack Carter, Bart Braverman, Dean Jagger, Sidney Lassick, Mike Mazurki, Sue Lyon, Angel Tomkins, James Ingersoll, Robert Doyle, Patti Jerome, Leslie Brown
When a pet alligator gets flushed down the toilet it returns as a twenty-foot monster to terrorise and kill the inhabitants of Chicago. Strange as it may sound, the script (by John Sayles) provides a number of comic moments that nicely balance the expected horror; not least is a wonderful performance by Silva as an egocentric and obsessive Great White Hunter. A sequel followed.
HOR   87 min (ort 94 min)
VIDrel: POLY/POLYREC/BRAVE   V

ALLIGATOR 2: THE MUTATION *   15
Jon Hess USA   1990
Steve Railsback, Joseph Bologna, Dee Wallace Stone, Richard Lynch, Woody Brown, Bill Daily, Holly Gagnier, Brock Peters, Tim Eyster, Julian Reyes, Voyo Goric, Buckley Norris, Deborah White, Bill Anderson, Jack W. Armstrong
This sequel to the 1980 film has chemical pollution leading to the growth of the title beastie to monster size. A property developer tries to hush up evidence of the creature's liking for human flesh, but is eventually forced to call in professional hunters. A flabby and derivative effort of wooden acting and far-from-impressive special effects.
HOR   90 min (ort 92 min)
VIDrel: POLY/POLYREC/BRAVE L/A   V

ALLIGATOR EYES **   15
John Feldman USA   1990
Annabelle Larsen, Roger Kabler, Mary McLain, Allen McCullough
A blind woman hitchhiker is picked up by a three people and gradually begins to involve herself with their private lives and fears, in this far from successful psychological thriller whose initial impetus is soon bogged down in a welter of cliches.
THR   96 min (ort 101 min)   VIDrel: IMAG/RTM   V

ALMOST AN ANGEL **   PG
John Cornell USA   1990
Paul Hogan, Elias Koteas, Linda Kozlowski, Charlton Heston, Doreen Lang, Joe Dallesandro, David Alan Grier, Douglas Seale, Ruth Warshawsky, Larry Miller, Michael Alldredge, Travis Venable, Robert Sutton, Ben Slack, Troy Curvey Jr
A petty thief has a confrontation with the Almighty (nicely played by Heston) in the afterlife, and is sent back to Earth for another chance, becoming a resolute do-gooder in the process. Written by Hogan, the film's executive producer, this syrupy concoction strains for laughs that the plodding script and dialogue just cannot deliver. Music is by Maurice Jarre.
COM   91 min (ort 98 min)   VIDrel: CIC/SONOP   V/sur

ALMOST GOLDEN: THE JESSICA SAVITCH STORY **   15
Peter Werner USA   1995
Sela Ward, Ron Silver, Judith Ivey
The success of a TV network's top news anchorwoman masks her private misery, as she moves from one disastrous relationship to the next. Eventually, she falls victim to drug abuse, putting her career in danger. An overwrought but perfectly watchable drama, based on a true story.
DRA   92 min   VIDrel: ODY/SONOP   V/sh

ALMOST PREGNANT **   18
Michael DeLuise USA   1991
Tanya Roberts, Jeff Conaway, Joan Severance, Dom DeLuise, John Calvin, Christopher Michael Moore, Angela Tiffe, Bruce Lurie, Steve Adell, Eric Amiel, Jennifer Strigler, Carol De Luise, Lezlie Deane, Lisa Comshaw
A woman is so desperate for motherhood that her husband agrees with her plan to take a lover who can get her pregnant. A tasteless and unfunny attempt to wring a few laughs from a serious predicament for many couples, this comedy proceeds along all too familiar lines.
COM   89 min (ort 93 min)   VIDrel: COLUM/SONOP   V

ALMOST YOU *   15
Adam Brooks USA   1984
Brooke Adams, Griffin Dunne, Karen Young, Marty Watt, Christine Estabrook, Josh Mostel, Laura Dean, Miguel Pinero, Joe Silver, Joe Leon, Spalding Gray, Daryl Edwards, Suzzy Roche, Stephen Strimpell, Suzanne Hughes, Wendy Creed
A couple living in New York are obliged to hire a live-in nurse when the wife dislocates her hip. One of those tedious romantic comedies that tries to present some kind of "message" about love, marriage and commitment.
COM   92 min (ort 110 min)   VIDrel: 20TH/TECH   V/h

ALPHABET CITY *   18
Amos Poe USA   1984
Vincent Spano, Kate Vernon, Michael Winslow, Zohra Lampert, Raymond Serra, Jami Gertz, Kenny Marino, Daniel Jordano, Tom Mardirosian, Christina Maire Denihan, Amy Gootenberg, Miguel Pinero, Barry Mitchell, Bob Fuchs, James Cox
A look at the life of an nasty nineteen-year-old gangster on New York City's Lower East Side that is practically devoid of a story-line, relying almost solely on arty direction and surface imagery. A shallow and unsatisfying offering that could easily have been very much better, but not much worse.
DRA   85 min (ort 98 min)   VIDrel: 20TH/TECH   V

ALPHAVILLE **   15
Jean-Luc Goddard FRANCE/ITALY   1965
Eddie Constantine, Anna Karina, Akim Tamiroff, Howard Vernon, Laszlo Szabo, Michel Delahaye, Jean-Andre Fieschi, Jean-Louis Comolli
Confused account of Saris, a computer controlled city of the future where a private detective is sent to rescue a scientist. A tedious and erratic film that suffers badly from the director's failure to impose clarity of vision. Good ideas swim about in there but remain undeveloped. Pity.
Aka: ALPHAVILLE, A STRANGE CASE OF LEMMY CAUTION
FAN   90 min (ort 100 min) B/W   VIDrel: CONNO/RTM
L/A   V

ALTERED STATES **   18
Ken Russell USA   1980
William Hurt, Blair Brown, Bob Balaban, Charles Haid, Thaao Penghilis, Dori Brenner, Miguel Godreau, Drew Barrymore, Peter Brandon, Charles White Eagle, Jack Murdock, Megan Jeffers, Frank McCarthy, Evan Richards
A scientist involved in primal brain research subjects himself to a series of mind-affecting experiments with dire results. Employing some memorable special effects, this excessively talky film had Chayefsky disowning it as an adaptation of his novel (the novel is now credited to Sidney Aaron – his pen-name). This was Hurt's film debut.
FAN   99 min (ort 103 min)   VIDrel: WHV   V/sur
Boa: novel by Paddy Chayefsky.

ALWAYS *   PG
Steven Spielberg USA   1989
Richard Dreyfuss, Holly Hunter, John Goodman, Brad Johnson, Audrey Hepburn, Roberts Blossom, Keith David, Marg Helgenberger, Ed Van Nuys, Dale Dye, Kim Robbilard, Brian Haley, James Lashly, Michael Steve Jones, Jim Sparkman
In this flashy remake of "A Guy Named Joe", a daredevil pilot firefighter is tragically killed at work but returns in spirit form to oversee the welfare of his former girlfriend, who has another chance of happiness with a young pilot. An optimistic and terribly earnest film of little depth but much sentimentality, its device of continual one-way conversations between Dreyfuss and his former girlfriend (she can't hear him and he knows it) soon grows truly tiresome.
COM   117 min (ort 121 min) wScrn   VIDrel: CIC/SONOP
V/sur

ALWAYS REMEMBER I LOVE YOU ***   U
Michael L. Miller USA   1990
Patty Duke, Joan Van Ark, Richard Masur, Stephen Dorff, David Birney, Sam Wanamaker, Jarred Blanchard, Malcolm Steward, Linda Darlow
Having learnt that he was adopted, a teenager also discovers that, unknown to his adoptive parents, he was originally kidnapped. Naturally enough, he sets out to find his real parents. A surprisingly touching tearjerker that avoids the twin pitfalls of predictability and mawkishness.
DRA   93 min (ort 96 min) mTV   VIDrel: ODY/SONOP   V/s

**AMA ** ** (PG)
Kwesi Owusu/Kwate Nee-Owoo UK 1991
Thomas Baptiste, Anima Misa, Roger Griffiths, Nil Oma Hunter, Joy Elias-Rilwan, Georgina Ackerman, Gary Marius, Verona Marshall, Eddie Tagoe, Alexandra Duah, Malcolm Fredericks, Okon Jones, Pauline Bailey, Pitika Ntuli
In London, a young Ghanaian girl encounters a computer and becomes obsessed with the idea that she can see into the future, having obtained a diskette which she reads on the office computer where her mother works as a cleaner. A strange film that is clearly influenced by the African storytelling tradition, it might have worked well, but without a clear narrative it really has nowhere to go. Surprisingly uninteresting, given the promising opening idea.
DRA 100 min CINrel

**AMADEUS ** ** ** PG
Milos Forman USA 1984
Tom Hulce, F. Murray Abraham, Elizabeth Berridge, Simon Callow, Roy Dotrice, Christine Ebersole, Jeffrey Jones, Charles Kay, Kenny Baker, Barbara Bryne, Lisabeth Bartlett, Martin Cavani, Peter DiGesu, Richard Frank, Patrick Hines
A brilliant film version of a play on the life and death of Mozart, filmed in Prague (by Miroslav Ondricek) and boasting one of the most literate film scripts of the 1980s. The story revolves around the unpleasant rivalry between Salieri and Mozart – musical mediocrity and genius respectively. AA: Pic, Dir, Actor (Abraham), Screen/adapt (Schaffer), Sound (Berger et al.), Art/Set (Von Brandenstein/Cerny), Cost (Pistek), Make (LeBlanc/Smith).
DRA 153 min (ort 158 min) wScrn
VIDrel: POLY/POLYREC/BRAVE V/sur
Boa: play by Peter Schaffer.

AMATEUR, THE * 15
Charles Jarrott CANADA 1981
John Savage, Christopher Plummer, Ed Lauter, Marthe Keller, Arthur Hill, Nicholas Campbell, Graham Jarvis, John Marley, Jan Rubes, Chapelle Jaffe, Jan Triska, Miguel Fernandes, Lynne Griffin, Jacques Grodin, George Coe
A CIA computer wizard uses his skills to hunt down the terrorists who seized the American Consulate in Munich and murdered his girlfriend. This foolish and over-complex spy thriller has Savage totally miscast as the computer buff (Plummer would have been a better choice) and suffers badly from the want of a more intelligent script.
THR 107 min (ort 111 min) VIDrel: 20TH/TECH V/sur
Boa: novel by Robert Littell.

AMATEUR * PG
Hal Hartley FRANCE/UK/USA 1994
Isabelle Huppert, Martin Donovan, Elina Lowensohn, Damian Young, Chuck Montgomery, Dave Simonds, Pamela Stewart, Erica Gimpel, Jan Leslie Harding, Terry Alexander, Holt McCallany, Hugh Palmer, Michael Imperioli, Angel Caban
An amnesiac whose ex-wife is a porno star has to run from criminals who want both of them dead and he falls in with a woman writer of pornographic novels, who used to be a nun. A well realised tale with a collection of outlandish characters, thrown together in a plot that is a mystery to all concerned (including the scriptwriter).
DRA 100 min (ort 105 min)
VIDrel: ARTIF/20TH; ENCORE (LV only) V/s LV

AMAZING ADVENTURE, THE * ** U
Alfred Zeisler UK 1936
Cary Grant, Mary Brian, Peter Gawthorne, Henry Kendall, Leon M. Lion, Garry Marsh, Ralph Richardson, John Turnball, Iris Ashley, Moore Marriott, Alf Goddard, Charles Farrell, Arthur Hardy, Frank Stanmore, Beuna Bent
A man inherits millions but fails to find happiness. When a Harley Street doctor diagnoses his problem as underwork, he makes a bet that he can earn his own living for a whole year and never touch his inherited fortune. A naive but enjoyable and ably directed little romp, a remake of a 1920 silent.
Aka: AMAZING QUEST; AMAZING QUEST OF ERNEST BLISS, THE; RICHES AND ROMANCE; ROMANCE AND RICHES
COM 62 min B/W (ort 80 min) VIDrel: CREA/DISC V
Boa: novel The Amazing Quest of Ernest Bliss by E. Phillips Oppenheim.

AMAZING COLOSSAL MAN, THE * PG
Bert I. Gordon USA 1957
Glenn Langan, Cathy Downs, William Hudson, Larry Thor, James Seay, Russ Bender, Lynn Osborn, Diana Darrin, William Hughes, Hank Patterson, Scott Peters, Myron Cook, Jack Kosslyn, Jean Moorehead, Frank Jenks, Bill Cassady
An army officer survives a radioactive explosion but his metabolism is affected and he is unable to stop growing in size. As he grows his mental state degenerates and he becomes a rampaging monster. A competent and quite endearing SF yarn, let down by poor trick photography and a predictable resolution. Followed by WAR OF THE COLOSSAL BEAST.
FAN 80 min (ort 89 min) B/W VIDrel: HEND/BMGREC L/A V

AMAZING HOWARD HUGHES, THE * PG
William A. Graham USA 1977
Tommy Lee Jones, Ed Flanders, James Hampton, Tovah Feldshuh, Lee Purcell, Jim Antonio, Sorrell Brooke, Arthur Franz, Howard Hesseman, Ed Harris, Morgan Brittany, Marty Brill, Marla Carlis, Lee Jones-De Broux, Roy Engel
TV movie covering the career of Howard Hughes, from eccentric millionaire to psychotic recluse, that belies the title by managing to make the weird and faintly sinister career of Hughes seem surprisingly uninteresting. Flanders as Hughes's associate Noah Dietrich, easily steals the scenes from Jones as the millionaire recluse. Feldshuh is interesting too, as the actress Katherine Hepburn. Below average. First shown in two parts.
DRA 118 min (ort 240 min) mTV VIDrel: WHV V/h
Boa: book Howard The Amazing Mr Hughes by Noah Dietrich.

AMAZING PANDA ADVENTURE, THE * PG
Christopher Cain USA 1995
Ryan Slater, Stephen Lang, Yi Ding
A ten-year-old boy travels to China to visit his father, who is involved in a project to save the giant panda from extinction. When the nature reserve is attacked by poachers, he gets separated from his dad and has to trek through the countryside together with a tiny baby panda. An OK adventure tale for kids, the chief stars of which are the beautiful Sichuan locations and the pandas (some of which are special effects creations courtesy of Rick Baker).
JUV 80 min (ort 85 min) cC VIDrel: WHV V/sur

AMAZING STORIES * 15
Steven Spielberg/Bronson Pinchot/
Robert Zemeckis USA 1985
Kevin Costner, Kiefer Sutherland, Bronson Pinchot, Christopher Lloyd, Scott Coffey, Mary Stuart Masterson
A compilation of three short stories, written and presented by Steven Spielberg – "The Mission", "Mummy, Daddy" and "Go To The Head Of The Class". An uneven mixture of horror, comedy and adventure that was followed by several sequels.
FAN 105 min (ort 120 min) VIDrel: CIC/SONOP V

AMAZING STORIES: 2 * 15
Steven Spielberg USA 1985
Robert Blossom, Lukas Haas, Gregory Hines, Danny DeVito, Rhea Perlman
Three more short stories combining elements of fantasy, adventure and comedy, and once more presented by Steven Spielberg. The stories are entitled: "Ghost Train", "The Amazing Falsworth" and "The Wedding Ring". The first story has an old man dismayed to find his son has built his home over the spot where a train nearly killed him years ago. The other tales are similar fantasy mixtures – all are let down by an excess of sentimentality.
FAN 71 min (ort 74 min) VIDrel: CIC/SONOP V

AMAZING STORIES: 3 * PG
Joe Dante/Robert Stephens/Tom Holland USA 1986
Hayley Mills, Stephen Geoffreys, Jon Cryer, Dennis Lipscomb
Director Steven Spielberg once more presents three fantasy tales: "The Greibble", "Moving Day" and "Miscalculation". The first tells of a woman's attempts to get rid of an uninvited guest – her son's lovable monster, the second story examines a boy's plight when he discovers that both he and his parents are extra-terrestrials, and the last tale tells of a photographer with a recipe that can bring photographs to life. Average.
FAN 68 min VIDrel: CIC/SONOP V

AMAZING STORIES: 4 * PG
Brad Bird/Thomas Carter/Matthew Robbins USA 1986
Joe Seneca, Natalie Gregory, Lane Smith

Three more short fantasy tales in this series – "Family Dog", "Dorothy And Ben" and "The Main Attraction". The first tale examines the exploits of a hound after his return from a bizarre obedience school, in the second a man awoken from a forty year coma has in fact been on another world, and in the last tale a meteor shower enhances the attractiveness of a high school romeo. More entertaining fantasy shorts let down by sentimentality.
FAN   71 min   VIDrel: CIC/SONOP V

AMAZING STORIES: 5 ***
PG
Philip Joanou/Mick Garris/Todd Holland USA   1986
John Lithgow, Patrick Swayze, David Carradine, Annie Helm, Rainbow Phoenix, Albert Hague, Sharon Spelman, John Christopher Jones, Hector Elizondo, James T. Callahan, Nicholas Love, T.J. Worzalla, Arnold Johnson, Kyra Sedgwick
Fifth in the series with another three tales of the supernatural. "The Doll" tells of two shy people brought together by the magic of a supernatural toymaker, "Life On Death Row" has Patrick Swayze in the role of a condemned prisoner, and "Thanksgiving" is a humorous episode in which a greedy prospector finds some unearthly carnivores living at the bottom of a well beneath his property. A much better collection with well handled effects.
FAN   70 min (ort 98 min)   VIDrel: CIC/SONOP   V/sh

AMAZING STORIES: 6 ***
U
Martin Scorsese/Paul Michael Glaser/Donald Petrie USA1986
Sam Waterstone, Max Gail, Kate McNeil, Sid Caesar, Dana Gladstone, Valorie Grear, Michael C. Gwynne, Peter Icangelo, Tad Horino, Jay Ingram, Charlie Hawke, Mark Crickson, Larry Gelman, Julius Harris, Tim Herbert
Three more strange stories: "Mirror, Mirror" which has the author of horror tales being chased by a nasty ghoul, "Blue Man Down" in which a cop whose partner was shot dead loses his confidence, and "Mr Magic" which examines the changing fortunes of a washed-up magician who finds a rather special pack of playing cards.
FAN   72 min   VIDrel: CIC/SONOP V

AMAZING STORIES: 7 **
U
Nick Castle/Robert Markowitz/Kevin Reynolds/
Philip Joanou USA   1987
Robert Townsend, Charles Durning, M. Emmet Walsh, Taliesin Jaffe, Craig Richard Nelson, Douglas Seale, Pat Hingle, Gabriel Damon, Marvin J. McIntyre
Four magic stories: "The 12 Inch Sun", "Magic Saturday", "You Gotta Believe It", and "One Amazing Night". The first tale follows the changing fortunes of a struggling scriptwriter, the second is a cliched story of an old man and grandson who swap bodies, the third episode is the predictable story of a man dreaming of a plane crash and the last episode has Santa Claus getting himself arrested.
FAN   91 min   VIDrel: CIC/SONOP V

AMAZING STORIES: 8 **
PG
Norman Reynolds/Leslie Linka Glatter/
Bob Balaban USA   1985/1986
June Lockhart, Milton Berle, Polly Holliday, J.A. Preston, Rich Brinkley, Britt Leach, Billy "Green" Bush, Dianne Hull, Gennie James, Rick Andosca, Matthew Laborteaux, Jimmy Gatherum, Gary Riley, Debbie Carrington
Another set of supernatural tales from this variable series: "The Pumpkin Tale", "Without Diana" and "Fine Tuning". This time round the stories are by Steven Spielberg. Average.
FAN   69 min   VIDrel: CIC/SONOP V

AMAZON WOMEN ON THE MOON **
15
John Landis/Joe Dante/Carl Gottlieb/Peter Horton/
Robert Weiss USA   1987
Rosanna Arquette, Ralph Bellamy, Carrie Fisher, Griffin Dunne, Ed Begley Jr, Steve Guttenberg, Sybil Danning, Steve Forrest, Paul Bartel, B.B. King, Lou Jacobi, Howard Hesseman, Steve Allen, Russ Meyer, Arsenio Hall
A parody of those tacky SF films of the 1950s (especially "Cat Women On The Moon") that tells of three American space voyagers who, with their pet monkey, fall into the clutches of a group of love-starved space maidens. One joke is that the movie is being shown on TV and constant breakdowns in transmission provide the excuse for a series of essentially unrelated skits and parodies, most of which are remarkably unfunny.
COM   82 min (ort 87 min) Col/B/W   VIDrel: CIC/SONOP
L/A   V/h

AMBASSADOR, THE ***
18
J. Lee Thompson USA   1984
Robert Mitchum, Ellen Burstyn, Rock Hudson, Fabio Testi, Donald Pleasence, Michal Bat-Adam, Heli Goldenberg, Ori Levy, Uri Gavriel, Zachi Noy, Joseph Shiloah, Shmulik Kraus, Yossi Virginsky, Iftah Katzur, Yaacov Banai
An American ambassador in the Middle East gets caught up in the Arab-Israeli conflict as he attempts to mediate between the warring groups. Burstyn as the ambassador's adulterous wife is especially good in an intelligent and exciting if somewhat remote adaptation of Leonard's novel (which was originally set in Chicago) that enjoyed a more faithful remake in the 1986 film FIFTY-TWO PICK-UP.
Aka: PEACEMAKER
THR   93 min (ort 95 min)   VIDrel: GUILD/SONOP L/A   V
Boa: novel 52 Pick-up by Elmore Leonard.

AMBULANCE, THE **
15
Larry Cohen USA   1990
Eric Roberts, James Earl Jones, Megan Gallagher, Richard Bright, Janine Turner, Eric Braeden, Red Buttons, Nicholas Chinlund, Laurene Landon, Jill Gatsby, Martin Barter, Jim Dixon, Stan Lee, Matt Norklon, Rudy Jones
In the course of following a girl he is hoping to date, a young man sees her taken ill and whisked off in a battered ambulance. Finding that all attempts to visit her in hospital are stymied, he digs deeper and uncovers a sinister plot on the part of a mad doctor. A muddled thriller that makes use of a well-worn plot. Buttons give a nice cameo performance as an ageing journalist who offers assistance. The script is by Cohen.
THR   91 min (ort 95 min)   VIDrel: 4-FRONT/POLYREC
L/A   V

AMERICA, AMERICA ***
PG
Elia Kazan USA   1963
Stathis Giallelis, Frank Wolff, Harry Davis, Elena Karam, Estelle Hemsley, Gregory Rozakis, Lou Antonio, Salem Ludwig, John Marley, Joanna Frank, Paul Mann, Linda Marsh, Robert H. Harris, Katherine Balfour
In the late 19th century, a Greek immigrant comes to America and experiences many hardships in his efforts to make a life for himself in his new homeland. An excellent and moving example of film-making, based on the memoirs of the director's uncle. AA: Art/Set (Gene Callahan).
Aka: ANATOLIAN SMILE, THE
DRA   161 min (ort 177 min) B/W   VIDrel: CONNO/RTM
V
Boa: book by Elia Kazan.

AMERICAN BLUE NOTE ***
15
Ralph Toporoff USA   1989
Peter MacNichol, Carl Caportoto, Tim Guinee, Jonathan Walker, Charlotte D'Amboise, Bill Christopher-Myers, Trini Alvarado, Zohra Lampert, Phyllis Behar, Louis Guss, Jeff Weiss, Sam Behrens, Margaret Devine, Art Johnson Jr
A comedy-drama that focuses on a month in the life of a 1960s jazz band. An enjoyable film that provides solid entertainment, uncomplicated by any great insights or messages.
Aka: FAKEBOOK
COM   88 min (ort 97 min)   VIDrel: ODY/SONOP V

AMERICAN BUFFALO ***
15
Michael Corrente UK/USA   1995
Dustin Hoffman, Dennis Franz, Sean Nelson
Some men meet regularly for poker sessions at a junk shop, and now set out to rob the home of a coin collector, sending a young assistant out to keep an eye on the man's house. But as their plans for the robbery start to take shape, various petty jealousies and intrigues become apparent, and on the night they are to pull the job off, they all fall out. Scripted by Mamet, this is a simply plotted three-hander, in which the rich dialogue is everything.
DRA   87 min   CINrel
Boa: play by David Mamet.

AMERICAN CYBORG: STEEL WARRIOR *
18
Boaz Davidson USA   1993
Joe Lara, John Ryan, Nicole Hansen, Yosheph Shiloa, Uri Gavriel, Hellen Lesnick, Andrea Litt, Jack Widerker, Kevin Patterson, P.C. Fireberg, Nicole Berger, Allen Nashman, Jack Adalist, David Milton Johnes, Eric Storch
In a devastated post-WW3 world ruled by cyborgs, the last fertile woman is sought by their evil leader but is shielded by a

brave warrior. A derivative and highly unimaginative movie, marred by poor acting, so-so direction and the lack of a decent script.
FAN   90 min   VIDrel: WHV   V/sur

AMERICAN DREAMER **   PG
Rick Rosenthal USA   1984
JoBeth Williams, Tom Conti, Giancarlo Giannini, Coral Browne, James Staley, Huckleberry Fox, C.B. Barnes, Jean Rougerie, Pierre Santini, Leon Zitrone, Alain Frick, Yassan Khan, Christian De Tiliere, Andre Valardy, Robin Coleman
An American housewife writes a novel and wins a trip to Paris where events drive her to the belief that she is in fact the daring heroine of a series of adventure thrillers. A muddled and miscast comedy-adventure that puts one in mind of ROMANCING THE STONE, only don't expect any surprises with this one.
COM   105 min   VIDrel: 20TH/TECH   V

AMERICAN FLYERS **   PG
John Badham USA   1985
Kevin Costner, David Grant, Rae Dawn Chong, Alexandra Paul, Janice Rule, Luca Bercovici, Robert Townsend, John Amos, Jennifer Grey, John Garber, Judy Jordan, Brian Drebber, Tom Lawrence, Jessica Nelson, James Terry, Jan Speck
Two brothers, one terminally ill, take part in a tough professional cycling marathon, in a good-natured if rather glossy look at this gruelling sport. Written by Steve Tesich who also wrote the story for BREAKING AWAY.
DRA   108 min (ort 114 min)   VIDrel: WHV   V/sur

AMERICAN FRIEND, THE ***   PG
Wim Wenders FRANCE/USA/WEST GERMANY   1977
Dennis Hopper, Bruno Ganz, Lisa Kreuzer, Gerald Blain, Jean Eustache, Sam Fuller, Nicholas Ray, Peter Lilienthal, David Schmid, Rudolf Schundler, Sandy Whitelaw, Lou Castel, Andreas Dedecke, David Blue, Heinz Joachim Klein
Vague and ponderous film about the planned assassination of a gangster, that makes a statement of sorts on the Americanisation of European life. The tale has Ganz, as a German picture-framer being hired to carry out the assassination, with film directors Fuller and Ray appearing as thugs.
DRA   120 min (ort 127 min) subs   VIDrel: CONNO/RTM V
Boa: novel Ripley's Game by Patricia Highsmith.

AMERICAN FRIENDS ***   PG
Tristram Powell UK   1991
Michael Palin, Trini Alvarado, Connie Booth, Alfred Molina, Alun Armstrong, Jimmy Jewel
In 1861 a lonely Oxford don on a walking holiday in Switzerland falls madly in love with the eighteen-year-old ward of a wealthy American lady. He sadly returns to his beloved college firmly convinced that the differences between them can never be bridged, but when the two women pay him a surprise visit his hopes are rekindled. Written by Palin and Powell, this well mounted period-piece is based on papers belonging to Palin's great-grandfather.
DRA   91 min   VIDrel: VISVID/POLYREC   V/sur

AMERICAN GIGOLO *   18
Paul Schrader USA   1980
Richard Gere, Lauren Hutton, Nina Van Pallandt, Hector Elizondo, Frances Bergen, K. Callan, Bill Duke, Tom Stewart, Patti Carr, David Cryer, Carole Cook, Carol Bruce, Frances Bergen, Macdonald Carey, William Dozier
A male prostitute is falsely accused of murder and a client of his who can clear him, not unnaturally is reluctant to do so. A poor excuse for a thriller, this look at the seamy side of life in L.A. has little to commend it, not least some remarkably poor acting and dialogue.
THR   112 min (ort 117 min)
VIDrel: 4-FRONT/POLYREC/CIC L/A   V/sh

AMERICAN GRAFFITI ****   PG
George Lucas USA   1973
Richard Dreyfuss, Ronny Howard, Cindy Williams, Paul Le Mat, Charles Martin Smith, Harrison Ford, Wolfman Jack, Candy Clark, Suzanne Somers, Mackenzie Phillips, Bo Hopkins, Kathleen Quinlan, Tim Crowley, Terry McGovern, Jan Wilson
A charming and perceptive coming-of-age tale built around the exploits of four young men about to graduate from a California high school in 1962. A celebration of 1960s nostalgia that is often

very funny and spawned a host of imitations (e.g. SUMMER CITY). Followed by "More American Graffiti".
COM   107 min (ort 112 min)   VIDrel: CIC/SONOP   V/sur

AMERICAN HEART ***   15
Martin Bell USA   1992
Jeff Bridges, Edward Furlong, Lucinda Jenney, Don Harvey, Tracey Kaprisky, John Boylan, Jayne Entwistle, Willie Willaims, Roosevelt Franklin, Don Harvey, Melvyn Howard, Kit McDonough, Wren Walker, Christie McMurdo-Wallis, Loyd Catlett
A recently released convict tries to build bridges to his troubled teenage son while attempting to resist the lure of his former life and prevent the young man from following in his footsteps. Convincing acting by the two leads and realistic locations greatly enhance this moving human drama.
DRA   109 min (ort 114 min)   VIDrel: EIV/SONOP   V/sur

AMERICAN IN PARIS, AN ****   U
Vincente Minnelli USA   1951
Gene Kelly, Leslie Caron, Oscar Levant, Nina Foch, Georges Guetary, Eugene Borden, Martha Bamattre, Mary Jones, Ann Codee, George Davis, Hayden Rorke, Paul Maxey, Dick Wessel
An ex-GI stays in Paris after WW2, and tries to make it as an artist whilst finding himself torn between Caron and Foch. The story has few surprises but this dazzling musical (built around a Gershwin score) hosts an array of fine songs and dances. AA: Pic, Cin (A. Gilks/J. Alton), Story/Screen (Alan Jay Lerner), Score (J. Green/S. Chaplin), Art/Set (C. Gibbons and P. Ames/Edwin B. Willis and K. Gleason), Cost (W. Plunkett/I. Sharaff).
MUS   108 min (ort 113 min)   VIDrel: MGM/WHV   V/dm

AMERICAN KICKBOXER 1 *   15
Franz Nels USA   1990
John Barrett, Brad Morris, Keith Vitali, Terry Norton, Ted Le Plat, Roger Yuan, Michael Huff, Lee Sparrowhawk, Evan Klisser, Gavin Hood, Paddy Lyster, Larry Martin, Gary Chalmers, Frank Notaro, Tootsie Lombard, Jeff Fannell
A world kickboxing champion serves a jail sentence for manslaughter, and emerges from prison to eventually square off against his deadly rival. A cliched action-packed turkey, with mercifully little dialogue spoken by Barrett, in a non-acting role (he also wrote the screenplay). For kickboxing fans only.
MAR   88 min (ort 93 min)   VIDrel: WHV   V/sur

AMERICAN MADNESS **   U
Frank Capra USA   1932
Pat O'Brien, Constance Cummings, Walter Huston, Kay Johnson, Gavin Gordon, Robert Ellis, Jeanne Sorel, Walter Walker, Berton Churchill, Arthur Hoyt, Edward Matindel, Edwin Maxwell, Robert Emmett O'Connor, Anderson Lawler
The president of a bank comes under pressure from his directors for lending money to a businessman who has got into financial difficulties, having done this as an act of faith in the recovery of the American economy. When a robbery leads to a run on the bank, his friends and supporters are hard put to prevent a collapse. An innovative albeit stilted effort, filmed on a vast bank set, that clearly reflects the director's support for Roosevelt's New Deal policies.
DRA   73 min B/W   VIDrel: COLUM/SONOP   V

AMERICAN ME ***   18
Edward James Olmos USA   1991
Edward James Olmos, William Forsythe, Pepe Serna, Evelina Fernandez, Daniel Haro, Cary-Hiroyuki Tagawa, Danny De La Paz, Sal Lopez, Daniel Villarreal, Vira Montez, Daniel A. Haro, Panchito Gomez, Steve Wilcox, Richard Coca
An epic account of thirty years in the life of a Mexican-American family that focuses on the life of its criminal head, who tries too late to go straight. Well acted and directed by Olmos in his directorial debut, this film has a very powerful impact but suffers from a lack of clear structure and an excess of characters. For all that, an extremely watchable effort. See also BLOOD IN BLOOD OUT.
DRA   121 min (ort 125 min)   VIDrel: CIC/SONOP   V/sur

AMERICAN NINJA **   18
Sam Firstenberg USA   1985
Michael Dudikoff, Steve James, Judie Aronson, Tadashi Yamashita, Phil Brock, Guich Koock, John Fujioka, John LaMotta, Tony Carreon, Roi Vinzov, Willie Williams, Jerry Bailey, Christopher Hoss, Joey Galvez, Nick Nicholson
Routine martial-arts tale set at a US Army base in the

Philippines, where Dudikoff, as an American soldier, uses his fighting skills to wipe out drugs dealers, with not a little help from his army pal James. An uninspired stab at the martial arts genre with AMERICAN NINJA 2: THE CONFRONTATION following soon after.

Aka: AMERICAN WARRIOR

MAR 91 min Cut (24 sec – ort 95 min) VIDrel: MGM/WHV V/h

AMERICAN NINJA 2: THE CONFRONTATION *** 18
Sam Firstenberg USA 1987

Michael Dudikoff, Steve James, Larry Poindexter, Gary Conway, Jeff Weston, Michelle Botes, Michael Stone, Len Sparrowhawk, Jonathan Pienaar, Bill Curry, Dennis Folbigge, Elmo Fillis, Ralph Draper, John Pasternack, Gary Ford

Tough GIs Dudikoff and James team up once more as they pit their ninjitsu skills against a psychotic Caribbean drugs baron and his band of warriors, in order to rescue a group of captured US Marines; they are to be trained by the drugs baron for his own evil purposes. A superior sequel to AMERICAN NINJA that benefits from the better script, the work of actors Gary Conway and James Booth. AMERICAN NINJA 3 followed.

Aka: AMERICAN NINJA 2; CONFRONTATION, THE

A/AD 85 min (Cut at film release by 28 sec – ort 105 min) VIDrel: MGM/WHV V/sh

AMERICAN NINJA 3: BLOODHUNT * 18
Cedric Sundstrom USA 1989

David Bradley, Marjoe Gortner, Yehuda Efroni, Steve James, Evan J. Klisser, Michele Chan, Calvin Jung, Adrienne Pearce, Grant Preson, Mike Huff, Alan Swerdlow, Thapelo Mofokeng, Ekard Rabi, Stephen Webber

Following his father's murder by gangsters in the course of a martial arts contest, a young boy is brought up and trained in the fighting arts. Now a young man, he sees his mentor kidnapped by a group of ninja during a martial arts tournament and traces them to a secret research plant. Later, he breaks in only to discover that he has in fact been deliberately enticed there to gauge the effectiveness of a deadly experimental virus. Very disappointing.

Aka: AMERICAN NINJA 3

MAR 84 min (ort 90 min) VIDrel: MGM/WHV V/dm

AMERICAN NINJA 4: THE ANNIHILATION * 18
Cedric Sundstrom USA 1990

Michael Dudikoff, David Bradley, James Booth, Dwayne Alexandre, Robin Stille, Ken Gampu, Franz Doborwosky, Ron Smerczak, Jody Abrahams, Sean Kelly, Anthony Fridjohn, David Sherwood, Jamie Bartlett, Dean Stewardson, David Rees

A bunch of good ninjas set out to rescue commandos being held captive by a bunch of bad ninjas, and also have to prevent terrorists setting off a miniature nuclear weapon a the same time. A very dull actioner with the obligatory martial arts flavour – fourth in a series that appears to have run out of both energy and ideas.

A/AD 94 min Cut (1 sec – ort 99 min) VIDrel: MGM/WHV V/sh

AMERICAN NINJA 5 ** 18
Bobby Gene Leonard USA 1991

David Bradley, Anne Dupont. Lee Reyes, Pat Morita, James Lew, Marc Fiorini, Tadashi Yamashita, Clement Von Franckenstein, Ron Ipale, Norman Burton, Jose Guannchez, Vincente Perez, Jose Salvado, Carlos Ruiz, Tommy Arias, Rufino Perez

A biologist invents a revolutionary pesticide which an evil tycoon wants to use to create a gas bomb. To force the scientist to work on this project, he has his daughter kidnapped and our martial arts hero attempts her rescue, aided by a young fighter. An assembly-line product with no surprises at all.

MAR 97 min (ort 98 min) VIDrel: WHV V/sh

AMERICAN PRESIDENT, THE ** 15
Rob Reiner USA 1995

Michael Douglas, Annette Bening, Martin Sheen, Michael J. Fox, Anna Deavere Smith, Samantha Mathis, Shawna Waldron, David Paymer, Anne Haney, Richard Dreyfuss, Nina Siemaszko, Wendie Malick, Beau Billingslea, Gail Strickland,

The lonely U.S. President hopes to find love in form of Bening, but their romance must take place under the glare of constant publicity and avid media interest. The fact that politically Bening's sympathies are more to the left adds an interesting complication to this pleasant romantic comedy, of appealing

performances and slightly shallow scripting. Both Sheen and Fox are memorable as aides to the President.

COM 109 min (ort 114 min) cC VIDrel: CIC/SONOP; PION (LV only) V LV

AMERICAN SAMURAI, THE ** 18
Sam Firstenberg USA 1992

David Bradley, Mark Dacasacos, John Fujioka, Valarie Trapp, Rex Ryon, Douvey Cohen, Melissa Hellman, Mark Warren, Koby Azarly, Shalom Avitan, Moshe Gal, Arie Muskuna, Barukh Berkin, Michael Morin, John Slater, Moshe Mammon

The adopted American son of a samurai is given a ceremonial sword by his father, which causes great dissension in the family and provides an excuse for ample swordplay, in this workmanlike actioner.

A/AD 86 min (ort 89 min) VIDrel: WHV V/sh

AMERICAN SHAOLIN: KING OF THE KICKBBOXERS 2 ** 18
Lucas Lowe USA 1992

Reese Madigan, Daniel Dae Kim, Billy Chang, Cliff Lenderman, Zhang Zhi Yen, Kim Chan, Alice Zhang Hung, Trent Bushey, Jean Louise Kelly, D.D. Delaney, Sifu Jai, Michael Depasquale Jr, Toki hill, Alan Puttinger, Andrew Shue, Eric Kong

After being defeated in a contest, a teenager goes to China to study the martial arts at first hand from the monks of the Shaolin temple. At first they refuse to admit him but his persistence eventually wins him a place. Filmed on location, this could so easily have been a fascinating insight into a different culture but amounts to no more than a cliched tale of how to win by means of sheer hard work. A disappointing sequel to THE KING OF THE KICKBOXERS.

Aka: AMERICAN SHAOLIN; KING OF THE KICKBOXERS 2

A/AD 101 min (ort 103 min) VIDrel: EIV/SONOP V

AMERICAN SOLDIER, THE ** 15
Rainer Werner Fassbinder WEST GERMANY 1970

Karl Scheydt, Elge Sorbas, Jan George, Hark Bohm, Ingrid Caven, Kurt Raab, Margarethe Von Trotta, Gustl Datz, Marquand Bohm, Katrin Schaale, Rainer Werner Fassbinder, Ulli Lommel, Irm Hermann

After serving in Vietnam, a German-American settles in Munich where he is soon hires himself out as a hitman, becoming deeply embroiled in the local underworld and all its shadowy figures. A kind of absurd spoof on American gangster movies but full of the usual depressive atmosphere for which this director's films are noted.

Aka: DER AMERIKANISCHE SOLDAT

DRA 76 min (ort 80 min) B/W VIDrel: CONNO/RTM V

AMERICAN TAIL, AN ** U
Don Bluth USA 1986

Voices of: Dom DeLuise, Christopher Plummer, Madeline Kahn, Phillip Glasser, Nehemiah Persoff, John Finnegan, Cathianne Blore, Will Ryan

During the late 19th century, a young Russian mouse gets separated from his family as they are about to arrive in America. A glossy and occasionally poignant cartoon feature, the charm of which is seriously hampered by the poorly plotted script. Produced by Steven Spielberg, this was his first attempt at producing a cartoon. Followed by AN AMERICAN TAIL: FIEVEL GOES WEST.

ANIM 77 min (ort 80 min) VIDrel: CIC/SONOP V/sur

AMERICAN TAIL, AN: FIEVEL GOES WEST ** U
Phil Nibbelink/Simon Wells USA 1991

Voices of: James Stewart, Dom DeLuise, John Cleese, Phillip Glasser, Amy Irving, Erica Yohn, Jon Lovitz, Cathy Cavadini, Nehemiah Persoff, Jack Angel, Mickie McGowan, Fausto Bara, Larry Moss, Vanna Bonta, Nigel Pegram, Lisa Raggio

An account of the further adventures of our family of immigrant mice who go West and find no shortage of both danger and opportunity, with Fievel setting his sights on becoming a lawman, while his sister treads the boards as a dance-hall singer.

ANIM 72 min (ort 75 min) VIDrel: CIC/SONOP V/sur

AMERICAN WEREWOLF IN LONDON, AN *** 18
John Landis USA 1981

David Naughton, Griffin Dunne, Jenny Agutter, Brian Glover, John Woodvine, David Schofield, Lila Kaye, Paul Kember, Don McKilliop, Frank Oz, Anne Marie Davies, Paula Jacobs, Gordon Sterne, Mark Fisher, Michele Brisigotti

Two young American lads are attacked by a werewolf whilst out walking on the moors. One dies whilst the other finds himself drawn into an increasingly nightmarish world where dreams and reality intermingle and eventually become indistinguishable. At its very best when horror is mixed with a measure of grisly humour, the ending is a sad anti-climax. Landis scripted this one, but it's the incredible make-up of Baker that one remembers. AA: Make (Rick Baker).

HOR 93 min (ort 98 min) VIDrel: 4-FRONT/POLYREC V

AMERICAN YAKUZA **
Frank Cappelo USA
Viggo Mortensen, Ryo Ishibashi, Michael Nouri, Franklyn Ajaye, Robert Forster, Yuji Okumoto, Christina Lawson, John Fujioka, Nicky Katt, James Katsuyuki Taenaka, Saiko Isshiki, Fritz Mashimo, Jeff Bankert, Rosine Hatem
An FBI agent goes under cover to infiltrate an American-based branch of the Japanese yakuza and as he makes a career for himself within this criminal organisation, he finds himself in the classical dilemma of divided loyalties. Plenty of action papers over the cracks in the plot department, in this competent adventure.
A/AD 95 min VIDrel: POLY/POLYREC/MED V/sh

18
1994

AMERICAN YAKUZA 2: BACK TO BACK ***
Roger Nygard JAPAN/USA
Michael Rooker, Ryo Ishibashi, Danielle Harris, John Laughlin
Two hitmen, an ex-cop (Rooker) and his daughter get involved with each other, with Rooker foiling a robbery, which later leads to a bizarre chain of events that ultimately has him teaming up with a gangster and taking on a Mob boss. A blend of action and black comedy that is in no small measure influenced by PULP FICTION, but has a distinctive style all of its own.
A/AD 87 min VIDrel: MED/20VIS V/sh

18
1996

AMERICA'S MOST WANTED GIRL *
Henri Pachard USA
Carol Cummings, Jerry Butler, Donna Ann, Renee Morgan, Sharon Kane, Charli St Cyr, Eric Edwards, Robert Bullock, Richard Parnes
A society swinger goes into hiding in order to create publicity for her forthcoming memoirs, but her worried sister engages a female private eye to find her. A dreary and limp effort from Pachard (who has produced more imaginative films) that chiefly serves as a vehicle for newcomer Cummings.
Aka: MOST WANTED GIRLS
A 51 min Cut (R18 ver); 85 min (MOON/RIO – 18 ver)
VIDrel: SHEP L/A; MOON/RIO V

R18/18
1989

AMITYVILLE HORROR, THE *
Stuart Rosenberg USA
James Brolin, Rod Steiger, Margot Kidder, Don Stroud, Murray Hamilton, Michael Sacks, Helen Shaver, Natasha Ryan, Meeno Peluce, K.C. Martel, Val Avery, John Larch, Amy Wright, Irene Dailey
Story of a family who move into a haunted Long Island house and endure numerous grisly happenings. A re-run of just about every haunted house cliche going, with Steiger wonderfully hammy as a local priest. Based on a rubbishy pulp fiction novel that was presented as a factual account of real events. A prequel and some sequels followed. Very poor. See also THE HAUNTED, POLTERGEIST and GRAVE SECRETS: THE LEGACY OF HILLTOP DRIVE.
HOR 113 min (ort 118 min)
VIDrel: 4-FRONT/POLYREC/VISVID L/A V
Boa: novel by Jay Anson.

15
1979

AMITYVILLE 2: THE POSSESSION *
Damiano Damiani USA
James Olson, Burt Young, Moses Gunn, Andrew Prine, Rutanya Alda, Jack Magner, Diane Franklin, Leonardo Cimino, Brent Katz, Erica Katz
More of a prequel to THE AMITYVILLE HORROR than a sequel, this purports to be the story of the previous owners of the house, with the older son being eventually compelled by evil forces to murder the rest of the family. A dismal spin-off that's both dull and unpleasant.
HOR 99 min (ort 104 min) VIDrel: GAME/SPEAR V
Boa: novel Murder In Amityville by Hans Holzer.

18
1982

AMITYVILLE 4: THE EVIL ESCAPES **
Sandor Stern USA
Patty Duke, Jane Wyatt, Frederic Lehne, Norman Lloyd, Brandy Gold, Aron Eisenberg, Geri Betzler, Alex Rebar, Jack Nader, Michael Korn, Richard Crystal, John Debello, Dave Elliot, Gary Michael Davies
Fourth outing in the series for another set of unfortunates and

18
1989

this notorious house, where the latest occupants find their daughter falling victim to evil forces. Strong on performances, but no new insights are on offer here.
Aka: AMITYVILLE: THE EVIL ESCAPES; AMITYVILLE HORROR, THE: THE EVIL ESCAPES PART 4
HOR 95 min (ort 100 min) mTV VIDrel: MED/POLY L/A V

AMITYVILLE 1992: IT'S ABOUT TIME *
Tony Randel USA
Stephen Macht, Shawn Weatherley, Megan Ward, Damon Martin, Jonathan Penner, Nita Talbot, Dean Cochran, Terrie Snell, Kevin Bourland, Margarita Franco, Dick Miller, William B. Jackson, Willie C. Carpenter, Alan Berman, Dylan Milo
Yet another film in the AMITYVILLE series that sees an architect bringing home an antique clock resulting in his house becoming possessed by evil spirits that turn his relatives into murderous monsters, Fortunately, his wife is left to deal with the forces of darkness.
HOR 91 min (ort 95 min) VIDrel: 20TH/TECH V/sh

18
1992

AMITYVILLE: A NEW GENERATION **
John Murlowski USA
Ross Partridge, Lala Sloatman, David Naughton, Richard Roundtree, Terry O'Quinn, Julia Nickson-Soul, Jack R. Orend, Barbara Howard, Robert Rusler, Lin Shane, Earl Johnson, Karl Johnson, Ralph Ahn, Tom Wright, Bob Jennings
A further attempt to continue this series of increasingly unimaginative tales of the horror that lurks in a haunted house. This time, an antique mirror plunges a young artist into the dark world of past secrets.
HOR 82 min (ort 92 min) VIDrel: MED/20VIS L/A V/sh

18

AMITYVILLE DOLLHOUSE **
Steve White USA
Robin Thomas, Starr Andreef, Clayton Murray, Allen Cutler
A couple discover an old Victorian dollhouse in their shed and give it to their young daughter. Unfortunately, they have yet to learn that it's a model of the infamous Amityville house and as such, is not really the ideal child's plaything. Starts off with promise, though it is not long before we're back to the old overblown effects (e.g. dolls spurting blood and the like). Passable.
HOR 92 min VIDrel: PROMARK/HIFLI V/h

18
1996

AMONGST FRIENDS **
Rob Weiss USA
Steve Parlavecchio, Joseph Lindsey, Patrick McGaw, Mira Sorvino, Chris Santos, Michael Leb, Christian Thom, David Stepkin, Michael Artura, Brett Lambson, Lou Cantelmo, Jerry Leonard, Greg Bernardi, Lou Bernardi
Three friends from solid Long Island families are attracted to crime as a counterweight to their aimless and boring lives. After one of the them is released from prison, they get together and plan a major drugs deal, which however is threatened by sexual jealousy. An unremarkable reworking of old cliches about disaffected youth that offers no fresh insights. This was Weiss's directorial debut, and he also wrote the screenplay.
DRA 83 min (ort 88 min) VIDrel: SPEAR/SONOP V/s

18
1993

AMORE! **
Lorenzo Doumani USA
Jack Scalia, Kathy Ireland, George Hamilton, Norm Crosby, James Doohan, Elliott Gould, Mother Love, Brenda Epperson, James Doohan, Katherine Helmond, Betsy Russell, Frank Gorshin, Joey DePinto, Allan Rich, Marlon Archey
A bored billionaire decides to change his image and masquerades as an Italian movie star, modelled on a screen idol that he much admires, and is helped in this endeavour by an attractive female screenwriter, for whom he promptly falls.
COM 90 min (ort 93 min) SATrel: SKY MOVIES

(PG)
1993

AMOS AND ANDREW **
E. Max Fyre USA
Nicolas Cage, Samuel L. Jackson, Dabney Coleman, Michael Lerner, Brad Dourif, Giancarlo Esposito, Margaret Colin, Chelcie Ross, Lorettta Devine, Bob Balaban, I.M. Hobson, Jodi Long, Jeff Blumenkrantz, Jordan Lind, Todd Weeks
A famous black playwright moves into a white neighbourhood in a New England resort but the first night in his new home sees him taken for a prowler by his bigoted neighbours who calls the local sheriff. Further misunderstandings follow as the latter tries desperately to cover up this embarrassing mistake.

15
1993

A puerile misfire of a comedy whose attempts at social comment are neither amusing nor effective.
COM 91 min (ort 96 min) VIDrel: POLY/POLYREC
V/sur

AMY JOHNSON STORY, THE ** PG
Nat Crosby UK 1984
Harriet Walter, Clive Francis, Patrick Troughton
The story of this famous female aviator, who on May 5th 1930 became the first woman in aviation history to make a solo flight, going on to make many more until her dramatic disappearance. At its best during the airborne sequences, but considerably less successful when it attempts an examination of her personal life.
Aka: AMY
DRA 82 min (ort 90 min) mTV VIDrel: START/DISC V

ANACONDA * 15
Luis Llosa USA 1996
Jennifer Lopez, Jon Voight, Eric Stoltz
Looking for a lost tribe of Amazonian Indians, anthropologist Stoltz leads a team of television documentary makers up the Amazon. But they have to contend with both a dangerous nutter in the shape of Voight, and the attentions of a predatory forty-foot anaconda. The latter beastie is the only reason to watch this unashamed piece of dross, which bears a lot in common with those 1970s camp exploitation pics like PIRANHA.
A/AD 89 min CINrel

ANASTASIA: PARTS 1 AND 2 ** PG
Marvin J. Chomsky USA 1986
Amy Irving, Olivia De Havilland, Jan Niklas, Nicolas Surovy, Susan Lucci, Elke Sommer, Edward Fox, Claire Bloom, Omar Sharif, Rex Harrison, Jennifer Dundas, Christian Bale, Andrea Bretterbauer, Sydney Bromley, Arnold Diamond
Mini-series about Anna Anderson who was dragged from a Berlin river in 1919 after a failed suicide attempt. After months of amnesia she gradually begins to regain her memory, claiming to be the youngest daughter of Czar Nicholas. However, her failure to offer anything as proof results in her being branded an impostor. This opulent exercise in tedium has a red-eyed Irving hovering on the verge of tears throughout. Watch this and you'll do the same.
Aka: ANASTASIA: THE MYSTERY OF ANNA
DRA 187 min (2 cassettes – ort 200 min) mTV
VIDrel: SCRN/DISC V
Boa: book Anastasia: The Riddle of Anna Anderson by Peter Kurth.

ANATOMY OF A MURDER **** 15
Otto Preminger USA 1959
James Stewart, Lee Remick, Ben Gazzara, Arthur O'Connell, Eve Arden, George C. Scott, Kathryn Grant, Orson Bean, Murrray Hamilton, Joseph N. Welch, Brooks West, Alexander Campbell, Joseph Kearns, Russ Brown, Howard McNear
Exciting courtroom drama about an army officer charged with the murder of the man he alleges raped his wife. An all-star cast, fine acting and the excellent Duke Ellington score combine to make this one a winner. Interestingly, the trial judge is played by Welch, a man who went on to become a real-life judge and defend the U.S. Army against Senator McCarthy in 1954.
DRA 161 min B/W VIDrel: COLUM/SONOP V
Boa: novel by Robert Traver.

ANCHORESS ** 12
Chris Newby BELGIUM/UK 1993
Natalie Morse, Eugene Bervoets, Toyah Wilcox, Peter Postlethwaite, Michael Pas, Christopher Eccleston, Brenda Bertin, Annette Badland, Veronica Quilligan, Julie T. Wallace, Ann Way, Francois Beukelaers, Jan Decleir, David Boyce
In fourteenth century England, a young peasant girl grows obsessed with a statue of the Virgin Mary and after she refuses to enter into an arranged marriage, the Church pronounces her a holy "anchoress" and she is imprisoned in seclusion in a walled up part of the building. She becomes an object of religious interest, and eventually dreams of gaining her freedom. Newby's debut is undeniably powerful but equally annoyingly inconsistent. Filmed in Belgium.
DRA 108 min CINrel

ANCHORS AWEIGH *** U
George Sidney USA 1945
Frank Sinatra, Gene Kelly, Kathryn Grayson, Dean Stockwell, Jose Iturbi, Pamela Britton, Billy Gilbert, Henry O'Neill, Sharon

McManus, Leon Ames, Carlos Ramirez, Rags Ragland, Edgar Kennedy, Henry Armetta, Billy Gilbert, James Burke
Popular musical about a couple of sailors who go on shore leave with 4-day passes and are out for a good time. Meeting up with a Hollywood extra who yearns to be a star, the two buddies decide to help. Features Kelly's memorable dance routine where live action blends with animation as he partners Jerry Mouse. If the story is a little weak the musical numbers more than compensate. AA: Score (Georgie Stoll).
MUS 144 min (ort 146 min) VIDrel: MGM/WHV V/dm

AND A NIGHTINGALE SANG ** PG
Robert Knights UK 1989
Tom Watt, Phyllis Logan, Joan Plowright, John Woodvine, Pippa Hinchley, Stephen Tomkinson, Des Young, Val McLane, Bob Smeaton, Willie Ross, Lyn Douglas, Patrick, Joe Ging, Will Hays, Harry Herring, Donald McBride
Typical British comedy-drama following the affairs of a Geordie family during WW2. Moderately amusing in places but fairly forgettable.
DRA 97 min mTV VIDrel: ODY/SONOP V/h
Boa: play by C.P. Taylor.

AND GOD CREATED WOMAN * 18
Roger Vadim FRANCE 1957
Brigitte Bardot, Curt Jurgens, Jean-Louis Trintignant, Christian Marquand, Georges Poujouly, Jean Tissier, Jane Marken, Paul Faivre, Jacqueline Ventura, Jacques Ciron, Jean Lefevre, Marie Glory, Isabelle Corey, Toscano, Jany Mourey
A dated and dreary vehicle for a display of Bardot's charms in which she plays a young woman with a healthy appetite for men. Filmed on location in Saint Tropez and remade after a fashion in 1987.
Aka: AND WOMAN WAS CREATED; ET DIEU CREA LA FEMME
DRA 87 min (ort 95 min) wScrn VIDrel: ARROW/RTM V

AND GOD CREATED WOMAN ** 18
Roger Vadim USA 1987
Rebecca DeMornay, Vincent Spano, Frank Langella, Donovan Leitch, Judith Chapman, Jaime McEnnan, Benjamin Mouton, David Shelley, Einstein Brown, David Lopez, Thelma Houston, Gail Boggs, Dorian Sanchez, Maria Duval
A woman marries merely to avoid a prison term, but neglects her husband in favour of a rock career and a flirtation with a politician. This watery variation of Vadim's former success has more in common with Goldie Hawn's OVERBOARD than his earlier film. A dull and dreary remake in name alone.
COM 94 min (ort 100 min) VIDrel: VCC L/A V

AND JUSTICE FOR ALL *** 15
Norman Jewison USA 1979
Al Pacino, Jack Warden, John Forsythe, Lee Strasberg, Jeffrey Tambor, Sam Levene, Christine Lahti, Craig T. Nelson, Joe Morton, Robert Christian, Thomas Waites, Larry Bryggman, Dominic Chianese, Victor Arnold, Vincent Beck
Unfunny attempt at a satire on the American system of justice, with a lawyer getting into all kinds of trouble on account of his clients, as he mounts a one-man crusade against the inadequacies of Maryland's legal system. Firing off its shots in all directions (and scoring a few hits along the way), this sometimes sad, sometimes funny movie has the stars performing well despite a basically weak story. Written by Valerie Curtin and Barry Levinson.
DRA 116 min (ort 119 min) VIDrel: CASPIC/BMGREC
L/A V

AND LIFE GOES ON *** (PG)
Abbas Kiarostami IRAN 1992
Ferhed Kherdamend, Hocine Rifahi, Ferhendeh Feydi, Mahrem Feydi, Bahrovz Aydini, Ziya Babai, Mohamed Hocinerouhi, Hocine Khadem, Maassouma Berouana, Mohamed Reda Berouana, Chahrbanov Chefahi, Youssef Branki, Chahine Ayzen
In Iran, a film director drives with his son towards a region devastated by a recent earthquake, where he hopes to meet some of the people he directed in his earlier work – WHERE IS MY FRIEND'S HOUSE? He does indeed meet some of the actors he worked with, but also learns much about human nature and the losses that have so blighted the lives of the people there. An interesting foray into a realm midway between drama and documentary, detailed and carefully constructed.
Aka: ZENDEGI VA DIGAR HICH
DRA 108 min CINrel

AND NOW FOR SOMETHING COMPLETELY DIFFERENT ***

PG

Ian McNaughton UK   1972

John Cleese, Eric Idle, Graham Chapman, Terry Gilliam, Terry Jones, Michael Palin, Connie Booth, Carol Cleveland

Repetition of well known Monty Python sketches from the popular TV show and featuring many examples of their anarchic sense of humour such as "The Dead Parrot", "The Lumberjack Song", "The Upper-Class Twit Of The Year" and "The World's Deadliest Joke". The first entry into films for the Monty Python team and not unnaturally a somewhat uneven collection. Followed by MONTY PYTHON AND THE HOLY GRAIL, the first of several more consistent efforts.

COM   84 min (ort 88 min)   VIDrel: VCC/DISC/COLUM   V

AND NOW THE SCREAMING STARTS **

18

Roy Ward Baker UK   1973

Peter Cushing, Herbert Lom, Stephanie Beacham, Patrick Magee, Ian Ogilvy, Geoffrey Whitehead, Guy Rolfe, Rosalie Crutchley, Janet Key, Gilliam Lind, Sally Harrison, Lloyd Lamble, Norman Mitchell, Frank Forsyth

Bizarre story of horror in a country house, with a newly-wed Beacham finding that she is living in a house that has been under a curse for many years. Better than the usual run of British horror films, this one is helped along by a strong cast.

HOR   86 min (ort 91 min)   VIDrel: VIPCO/SGSVID   V

Boa: novel Fengriffen by David Case.

AND THE BAND PLAYED ON ***

15

Roger Spottiswoode USA   1993

Don Francis, Matthews Modine, Alan Alda, Phil Collins, David Dukes, Richard Gere, Glenn Headly, Anjelica Huston, Swoosie Kurtz, Steve Martin, B.D. Wong, Lily Tomlin, Ian McKellen, Saul Rubinek, Charles Martin Smith, Nathalie Baye

Vast and sprawling drama detailing the early days of the AIDS epidemic, when doctors were puzzled by the appearance of a strange virus that was taking the lives of a large number of young men. Researchers in France and the USA worked tirelessly to try and isolate this organism and understand its workings. Extremely interesting for the most part, although the star cameos are a needless distraction.

DRA   140 min mCab   VIDrel: ITC/POLYREC   V/sur

Boa: book by Randy Shilts.

AND THE SEA WILL TELL ***

15

Tommy Lee Wallace USA   1991

James Brolin, Rachel Ward, Hart Bochner, Richard Crenna, Deidre Hall, John Kapelos, Susan Blakely, Kevin McNulty, Clyde Kusatsu, Marion Gilsenan, Danny Kamekona, Wendy Noel, Paul Jarrett, Garwin Sanford, Bill Dow, Don Thompson

Based on a true story, this bizarre murder mystery revolves around the disappearance of a couple from their luxury yacht and the arrest of another couple, who were found on the missing vessel some time later. Initially only convicted of theft, they were later charged with murder when the body of the missing woman surfaced some six years later (the husband's body was never recovered).

DRA   185 min mTV   VIDrel: COLUM/SONOP   V

Boa: book by Vincent B. Bugliosi with Bruce B. Henderson.

AND THE SHIP SAILS ON **

12

Federico Fellini FRANCE/ITALY   1983

Freddie Jones, Barbara Jefford, Victor Poletti, Peter Cellier, Janet Suzman, Elisa Mai Nardi, Norma West, Paolo Paoloni, Sarah Jane Varley, Fiorenzo Serra, Pina Bausch, Pasquale Zito, Linda Polan, Phillip Locke, Jonathan Cecil

A decidedly bizarre look at a trip undertaken in 1914 by an opera company in order to scatter the ashes of their beloved leading lady at sea, as seen through the eyes of a journalist. There are many delightful touches of fantasy but alas, these are too slight to add anything of real substance to the film. A disappointing souffle of undeniable charm.

Aka: E LA NAVE VA

COM   122 min (ort 138 min)   VIDrel: ARROW/RTM   V

AND THE VIOLINS STOPPED PLAYING ****

15

Alexander Ramati POLAND/USA   1988

Horst Bucholz, Maya Ramati, Piotr Polk, Didi Ramati

The struggle a group of Polish gypsies face to survive during WW2 forms the background to this searing account, in which Ramati provides his own account of the atrocities the Nazis committed in occupied Poland. A powerful film, difficult to

watch without remaining unmoved, yet both rewarding and inspiring.

DRA   129 min   VIDrel: ODY/SONOP   V/sur

Boa: book by Alexander Ramati.

AND THE WALL CAME TUMBLING DOWN **

(PG)

Paul Annett UK   1984

Barbi Benton, Gareth Hunt, Brian Deacon, Barbi Benton, Peter Wyngarde, Pat Hayes, Carol Royle, Ralph Michael, Gary Waldhorn, Robert James, Richard Hampton, Angela Grant, R.W. Armstrong, Peter Baldwin, Iona Jones, Tim Pearce

A strange horror tale involving an old London church due for demolition and a reincarnated Satanist, out for revenge against a coven member who in the year 1649 betrayed others in return for a priest's blessing and thus escaped being burnt at the stake. When a building worker is found dead in the church, a sealed alcove containing a painting and two skeletons is opened, thus releasing our devil-worshipper's evil spirit. An average made-for-TV Hammer offering.

Aka: HAMMER HOUSE OF MYSTERY AND SUSPENSE: AND THE WALL CAME TUMBLING DOWN

HOR   72 min mTV   SATrel: UK GOLD

AND THEN THERE WAS ONE ***

15

David Jones USA   1993

Amy Madigan, Dennis Boutsikaris, Jane Dally, Steven Flynn, Jennifer Hetrick, John Robinson, Martha Henry, Cameron Arnett, Kenneth Walsh, Richard Monette, Dawn Greenhalgh, Henry Ramer, Pam Hyatt, Damir Andrei, Gabe Cohen, Naz Edwards

Fact-based AIDS drama in which a married couple are confronted with the fact that after four years of trying for a baby, the little girl they now have has been infected with AIDS from birth, a fact that comes to their attention when she succumbs to a mystery virus.

DRA   90 min mTV   VIDrel: ODY/SONOP   V/sh

AND THEN YOU DIE ***

18

Francis Mankiewicz CANADA   1987

Kenneth Welsh, Wayne Robson, R.H. Thomson, Tom Harvey, George Bloomfield, Graeme Campbell, Dennis O'Connor, Tom McCamus, Jefferson Mappin, Joran Van Lange, Pierre Chagnon, Guy Thauvette, Maggie Huculak, Donald Davis, David Bolt

Based on a true story, this tells of the last nine days of an Irish drugs baron operating in Montreal. Using Hell's Angels as his muscle, he becomes locked in a power struggle with the Mafia for control of the lucrative cocaine market. Fair action tale.

A/AD   96 min (ort 115 min) mTV   VIDrel: SCRN/DISC   V

ANDERSON TAPES, THE ****

15

Sidney Lumet USA   1971

Sean Connery, Martin Balsam, Dyan Cannon, Ralph Meeker, Alan King, Margaret Hamilton, Christopher Walken, Garrett Morris, Scott Jacoby, Conrad Bain, Dick Williams, Stan Gottleib, Paul Benjamin, Anthony Holland, Judith Scott

An ex-con plans to rob a building, not knowing that his every action is being recorded by surveillance equipment that has been in operation since he left prison. A fast and exciting thriller that has a particularly good climax. The catchy score is by Quincy Jones. This was Walken's film debut.

A/AD   94 min (ort 98 min)   VIDrel: MIA/DISC/COLUM   V

Boa: novel by Lawrence Sanders.

ANDRE ***

U

George Miller USA   1994

Keith Carradine, Tina Majorino, Keith Szarabajka, Chelsea Field, Shane Meier, George Miller, Aidan Pendleton, Joshua Jackson, Shirley Broderick, Andrea Libman, Jay Brazeau, Bill Dow, Joy Coghill, Stephen Dimopoulos, Frank C. Turner

An appealing adaptation of a novel about a baby seal that is adopted by a family in Maine. Having grown to full size, it more or less refuses to return to the wild and be parted from them, although at one point it seems as if an over-zealous animal protection society will separate them. Set in the 1960s, this is a fine piece of family entertainment.

JUV   90 min (ort 94 min)   VIDrel: COLUM/SONOP   V/sur

Boa: novel A Seal Called Andre by Lew Dietz and Harry Goodridge.

ANDREI RUBLEV ****

15

Andrei Tarkovsky USSR   1966

Anatoli Solonitzine, Ivan Lapikov, Nikolai Grinko, Nikolai Sergueiev

A recreation of the turbulent life of a 15th century icon painter, filled with a host of imaginary episodes as he contemplates the role of Art in the service of Life. A splendid and utterly enthralling film that represents one of the director's finest works. Scripted by Tarkovsky and Andrei Milchakov-Konchalovsky, the film was not released by the Soviet authorities until 1971.

DRA   174 min (ort 181 min) wScrn Col/B/W
VIDrel: ARTIF/20TH   V

ANDREW AND FERGIE: BEHIND THE PALACE DOORS **
PG

Michael Switzer
Pippa Hinchley, Sam Miller
It's Royal Family time again as we are graciously granted an allegedly true picture of what actually went on between Prince Andrew and the Duchess of York prior to their separation. People who like this sort of thing will find this the sort of thing they like. See also DIANA: HER TRUE STORY and THE WOMEN OF WINDSOR.

DRA   98 min mTV   VIDrel: COLUM/SONOP   V

ANDROID ***
(12)
1982

Aaron Lipstadt USA
Klaus Kinski, Don Opper, Brie Howard, Norbert Weisser, Crofton Hardester, Kendra Kirchner, Gary Corarito, Darrel Larson, Wayne Springfield, Mary Ann Fisher, Ian Scheibel, Randy Connor, Roger Kelton, Rachel Talalay, Johanne Todd
Stylish and thoroughly enjoyable film about the controller of a strange space station and his android helper Max, who he soon plans to replace with a more up-to-date creation. When three escaped criminals arrive at the space station their quiet routine is interrupted, but the scientist has plans for using one of them. A slightly wacky and offbeat film using sets left over from BATTLE BEYOND THE STARS. Low-budget but good fun. Co-written by Opper.

FAN   77 min (ort 82 min)   SATrel: SKY MOVIES GOLD

ANDROID AFFAIR, THE **
12
1995

Richard Kletter USA
Harley Jane Kozak, Ossie Davis, Saul Rubinek, Peter Outerbridge, Natalie Radford, Griffin Dunne, Chandra Galasso, David Campbell, Ron Hartman, Michelle McFait, Joseph Scorsiani, Heidi Hatashita, Wendy Murphy, Diana Zimmer
In the near future, androids are used as experimental subjects in training doctors in surgical and other techniques. One of these creatures starts to manifest the desire to live a human life and is therefore scheduled to be destroyed. However, the woman doctor to whom he is assigned, gradually begins to develop feelings for him and attempts to save him. Based on the short "Teach 109", this low-key tale benefits from strong performances.

FAN   85 min (ort 90 min) mTV   VIDrel: CIC   V

ANDROMEDA STRAIN, THE **
PG
1970

Robert Wise USA
Arthur Hill, David Wayne, James Olson, Kate Reid, Paula Kelly, Ramon Bieri, George Mitchell, Richard O'Brien, Kermit Murdock, Peter Hobbs, Richard Bull, Walter Brooke, Ivor Barry, Emory Parnell, Eric Christmas, Peter Hell
A team of scientists race against time to isolate a deadly organism brought to Earth by a satellite that has crashed in New Mexico. Seriously overlong, this one was in need of severe editing. However, intelligent scripting helps sustain tension.

FAN   124 min (ort 131 min)   VIDrel: CIC/SONOP   V
Boa: novel by Michael Crichton.

ANDY WARHOL'S DRACULA *
18
1973

Paul Morrissey/Anthony M. Dawson (Antonio Margheriti)
ITALY/FRANCE
Joe Dallesandro, Udo Kier, Arno Juerging, Vittorio De Sica, Maxime McKendry, Roman Polanski, Dominique Darel, Stefania Cassini, Gil Cagne, Milena Vukotic, Silvia Dionisio, Eleonora Zami, Emi Califin
Mildly pornographic spoof on the Dracula story with our vampire discovering that he must have the blood of virgins in order to survive. A lot less gory than his production of Frankenstein, this has some funny moments but the plot remains in serious need of a talent transfusion.

Aka: ANDY WARHOL'S YOUNG DRACULA; BLOOD FOR DRACULA;
DRACULA CERCA SANGUE DI VERGINE E MORI DI SETE; DRACULA
VUOLE VIVERE: CERCA SANGUE DI VERGINA; YOUNG DRACULA
HOR   99 min (ort 106 min)   VIDrel: FIRST/SONOP   V

ANDY WARHOL'S FRANKENSTEIN **
18
1973

Paul Morrissey/Anthony M. Dawson (Antonio Margheriti)
FRANCE/ITALY/WEST GERMANY
Joe Dallesandro, Monique Van Voorien, Udo Kier, Srdjan Zelenovic, Dalia di Lazzaro, Arno Juerging, Lui Bozizio, Carla Mancini, Marco Liofredi, Imelda Marini, Nicoletta Elmi, Christina Gaioni, Fiorella Masselli, Rosita Torosh
Shot in 3-D, this companion piece to ANDY'S WARHOL'S DRACULA is a real camp version, with plenty of blood and severed limbs, as our demented Baron embarks on the creation of a series of beautiful creatures from the assorted pieces of human anatomy. Definitely not one for the squeamish, it is unlikely to ever be available in an uncut form in the UK.

Aka: CARNE PER FRANKENSTEIN; DEVIL AND DR FRANKENSTEIN, THE;
FLESH FOR FRANKENSTEIN; FRANKENSTEIN; FRANKENSTEIN
EXPERIMENT; IL MOSTRO E IN TAUOLA... BARONE FRANKENSTEIN; UP
FRANKENSTEIN; WARHOL'S FRANKENSTEIN
HOR   91 min (ort 95 min)   VIDrel: FIRST/SONOP   V

ANGEL **
15
1982

Neil Jordan EIRE
Stephen Rea, Veronica Quilligan, Alan Devlin, Honor Hefferman, Marie Kean, Peter Caffrey
After a saxophonist witnesses the brutal murder of the band's leader and a deaf-and-dumb girl, he sets out to have his revenge on the culprits, in this formula thriller.

Aka: DANNY BOY
THR   88 min (ort 92 min)   VIDrel: FIRST/SONOP   V/sur

ANGEL **
18
1983

Robert Vincent O'Neil USA
Cliff Gorman, Susan Tyrell, Dick Shawn, Rory Calhoun, John Diehl, Donna Wilkes, Elaine Giftos
A young high school student is a devoted scholar by day and a Hollywood hooker by night. Having said that, little remains in terms of plot as this one goes nowhere. However, do look out for Shawn as a drag queen. AVENGING ANGEL followed.

DRA   92 min (ort 94 min)   VIDrel: L/A   V

ANGEL 3: THE FINAL CHAPTER **
18
1988

Tom De Simone USA
Mitzi Kapture, Maud Adams, Richard Roundtree, Mark Blankfield, Kim Shriner, Emile Beaucard, Barbara Truetellar, Susan Moore, Ann Navarro, Floyd Levine, Kyle Heffner, Dick Miller, Tawny Fere, S.A. Griffin, Cynthia Hoppenfeld
Sequel to AVENGING ANGEL, and this time round our heroine's quest is to rescue her sister from a white slavery ring before she is sold into a fate worse than death. Average.

Aka: ANGEL 3
A/AD   95 min Cut (37 sec – ort 100 min)   VIDrel: NWV
L/A   V

ANGEL AND THE BADMAN ***
U
1947

James Edward Grant USA
John Wayne, Gail Russell, Bruce Cabot, Harry Carey, Irene Rich, Lee Dixon, Stephen Grant, Tom Powers, Paul Hurst, Olin Howlin, John Halloran, Joan Barton, Craig Woods, Marshall Reed, Hank Worden, Pat Flaherty
A gunfighter is won over to an honest life through the love of a young Quaker girl. Despite the simplicity of the plot this fine Western is warm, human and engaging.

WES   94 min (ort 100 min) B/W coVer (VCC)   VIDrel: VCC
L/A; 4-FRONT/POLYREC   V

ANGEL AT MY TABLE, AN ****
15
1990

Jane Campion NEW ZEALAND
Kerry Fox, Alexia Keogh, Karen Ferguson, Iris Churn, K.J. Wilson, Martyn Sanderson, Natalie Ellis, Eddie Hogan, Erin Mills, Virginia Brocklehurst, Glynnis Angell, Sarah Smuts-Kennedy, Andrew Binns, Colin McColl, Francine Clark
A remarkable and moving adaptation of a trio of autobiographical books by Frame, a repressed young girl who was wrongly diagnosed as schizophrenic and spent eight years in a mental hospital, but overcame her experiences to become one of New Zealand's most famous poets and novelists. The various stages of Frame's life are examined with skill and compassion, and the incisive script is by Laura Jones. Originally shown as a 3-part TV series.

DRA   151 min (ort 160 min)   VIDrel: ARTIF/20TH   V
Boa: autobiographies of Janet Frame.

ANGEL BABY *** 15
Michael Rymer AUSTRALIA 1995
Jacqueline McKenzie, John Lynch, Colin Friels, Deborra-Lee Furness, Robyn Nevin
A disturbed young woman who lives in a fantasy world falls in love with an acute schizophrenic, and the couple set up home together. All goes well until they both decide to end their medication and live their lives without the use of sedating drugs. Winner of no less than seven film awards in its native country, this is a mixture of compelling drama and slightly opaque symbolism.
DRA 100 min VIDrel: POLFIL V/sh

ANGEL ENFORCERS ** 18
Godfrey Ho HONG KONG 199-
Sharon Young, Ron Van Lee, Philip Ko
A criminal hatches a plan that involves stealing his own gems, but when one of his accomplices kills an undercover cop, matters take a more serious turn. A general bloodbath ensues, in which a police inspector, his sisters and their father are all murdered. This leaves the way open for three female colleagues of one of the girls to set about having their revenge. Directed with panache if not imagination, this is a non-stop celebration of violent conflicts.
A/AD 87 min wScrn VIDrel: MIA/DISC; ENCORE (LV only) V LV

ANGEL EYES ** 18
Gary Graver USA 1991
John Philip Law, Erik Estrada, Monique Gabrielle, Richard Harrison, Hoke Howell, Robert Quarry, Rachel Vickers, Suzanne Ager, John Coleman, Sherman Scott, Darius Beiderbeck, Martin Nicholas, Sazzy Lee, Gail Carradine
A man's long-lost stepdaughter seems on the surface to be a balanced and attractive young woman, but in reality is a seriously disturbed and potentially lethal killer. As a young child she witnessed the brutal stabbing of her mother and harbours secret feeling of hatred towards all men. A flat and rather mechanically put together thriller whose ample softcore sequences generate neither excitement nor interest.
THR 83 min (ort 90 min) VIDrel: POPRO/RTM V

ANGEL FLIGHT DOWN ** (PG)
Charles Wilkinson USA 1996
Patricia Kalember, David Charvet, Garwin Sanford, Christopher Atkins, Donna Larson, Stephen E. Miller, Paige Magnusson, Gary Graham, Deanna Millian, Donna Belleville, Jesse Moss, Rod Madmos, Paul Coeur, Maureen Thomas, Daniel Busheikin
A mercy flight taking a little girl for urgent medical treatment, turns into tragedy when the plane is downed in a violent snowstorm and those aboard find themselves without any means of communicating with the outside world. A good study of individuals coping with extreme conditions that never resorts to sentimentality and is both dramatic and very watchable. Said to have been based on a true incident. See also ORDEAL IN THE ARCTIC.
A/AD 88 min (ort 90 min) SATrel: MOVIE CHANNEL

ANGEL HEART *** 18
Alan Parker USA 1986
Robert De Niro, Mickey Rourke, Lisa Bonet, Charlotte Rampling, Stocker Fountelieu, Brownie McGhee, Michael Higgins, Charles Gordone, Kathleen Wilhoite, Elizabeth Whitcraft, Elliot Keener, Dawn Florek, George Buck
A tough New York detective pits his wits against a fearsome adversary in a story blending action and the occult and set in the backwoods of New Orleans. The striking visual images are the film's greatest asset, though after a while even they begin to pale in the face of the overly serpentine plot. Parker scripted and directed. A torrid sex scene was cut just prior to release and has not yet been restored.
THR 109 min (Cut at film release by 4 sec – ort 113 min)
VIDrel: POLY/POLYREC L/A V/sh
Boa: novel Falling Angel by William Hjortsberg.

ANGEL IN RED * 18
William Duprey USA 1991
Leslie Bega, Jeffrey Dean Morgan, Pamella D'Pella, Henry Brown, Jason Oliver, Elena Sahagun, Gregory Millar, Sheila Scott-Wilkenson, Wendy J. Cooke, Chelsea Madison-Ciu, Joy Garcia, David Labiosa, Odette Springer, Bob Farnham
When a group of prostitutes learn that their pimp has murdered another one of their number, they set about planning their revenge, in this tough tale of life on the streets. See also ANGEL and AVENGING ANGEL.
Aka: UNCAGED
DRA 75 min Cut (1 sec – ort 78 min)
VIDrel: COLUM/SONOP V

ANGEL OF DESIRE ** 18
Donna Deitch USA 1993
Joan Severance, Anthony John Denison, John Allen Nelson
The son of a US senator gets implicated in the brutal murder of a woman and glamorous cop Severance, who has been assigned to the case, gets involved in a passionate affair with him. An erotic thriller that went straight to video, and apart from a couple of interesting characterisations it has little to recommend it.
THR 92 min VIDrel: HIFLI/SONOP V/h

ANGEL OF FURY * 18
David Worth USA 1992
Cynthia Rothrock, Chris Barnes, Billy Drago, Sam Jones, Peter O'Brian, Greg Stuart, Tanaka
A female security executive with a computer company has to battle the terrorists who are determined to acquire its latest product. Poor and entirely predictable, though Rothrock gives one of her strongest performances in this weak film, and for once shows that she can act decently as well as fight. Screenplay is by Clifford Mohr.
A/AD 90 min VIDrel: COLUM/SONOP V/sh

ANGEL WORE RED, THE ** (PG)
Nunnally Johnson USA 1960
Ava Gardner, Dirk Bogarde, Joseph Cotten, Vittorio De Sica, Aldo Fabrizi, Arnoldo Foa, Finlay Currie, Rossano Rory, Enrico Maria Salerno, Robert Bright, Franco Castellani, Bob Cunningham, Gustavo De Nardo, Nino Castelnuevo, Aldo Pini
During the Spanish Civil War, a former priest who has joined forces on Franco's side, comes across a prostitute whose loyalties lie with the Partisans. He soon falls in love with her but is then captured by government troops who want to recover a religious statue whose whereabouts are unknown to him. Though well staged, this epic period drama fails to exploit its historical background.
DRA 99 min B/W SATrel: TNT MOVIES

ANGELS *** U
William Dear USA 1994
Danny Glover, Tony Danza, Brenda Fricker, Christopher Lloyd, Ben Johnson, Joseph Gordon-Levitt, Jay O. Sanders, Milton Davis Jr, Taylor Negron, Tony Longo, Neal McDonough, Stoney Jackson, Adrien Brody, Tim Conlon, Israel Juarbe
A young child believes that his family will be reunited if the local no-hope baseball team wins the pennant. There seems to be no chance of this but he prays fervently and the team receives some celestial aid in its next game. It is soon on a winning streak that promises to take it to the top, but not until various complications are resolved. A strong remake of the 1951 classic fantasy "Angels In The Outfield" with good performances and solid direction.
Aka: ANGELS IN THE OUTFIELD
FAN 95 min (ort 103 min) VIDrel: WDV/TECH L/A V

ANGELS AND INSECTS *** 15
Philip Haas UK/USA 1995
Mark Rylance, Patsy Kensit, Saskia Wickham, Chris Larkin, Douglas Henshall, Annette Badland, Kristin Scott Thomas, Lindsay Thomas, Michelle Sylvester, Clare Lovell, Jenny Lovell, Anna Massey, Oona Haas, Angus Hodder, Margaret Golder
Having married into a wealthy family, a impoverished naturalist is less than well prepared for the decadence and corruption he discovers within it. Set in the nineteenth century, this offbeat adaptation of Byatt's strange novella works hard to develop tension, even if the nature of the creepy goings-on and eventual outcome are not too hard to guess.
DRA 117 min VIDrel: FILM4/RTM V
Boa: novella Morpho Eugenia by A.S. Byatt.

ANGEL'S BACK! * R18/18
Richard Mailer USA 1988
Angel, Nina Hartley, Rene Morgan, Isabella Rovetti, Mike Horner, Demian Cashmere, Peter North, John Martin, Sasha Gabor
More sexual adventures of the title heroine from ANGEL OF THE ISLAND.
A 57 min Cut (1 min 32 sec); 68 min Cut (18 sec – R18 ver)
VIDrel: SHEP (R18 ver) L/A; FALCON/TOTAL (18 ver) V

ANGELS ONE FIVE ** U
George More O'Ferrall UK   1952
Jack Hawkins, John Gregson, Michael Denison, Dulcie Gray, Cyril Raymond, Humphrey Lestocq, Ronald Adam, Harold Goodwin, Norman Pierce, Geoffrey Keen, Harry Locke, Philip Stainton, Vida Hope, Amy Veness, Richard Dunn
Life at an RAF fighter station in 1940 where a volunteer pilot finds that his C.O. does not quite share his enthusiasm and prefers to do everything by the book. A restrained war film with few heroics in sight and a marked documentary slant, but one that met with strong public approval upon release and achieved great box-office success.
WAR 93 min (ort 98 min) VIDrel: LUMI/SPEAR L/A V

ANGELS OVER BROADWAY *** U
Ben Hecht/Lee Garmes USA   1940
Douglas Fairbanks Jr, Rita Hayworth, Thomas Mitchell, John Qualen, George Watts, Ralph Theodore, Eddie Foster, Jack Roper, Constance Worth, Richard Bon, Frank Conlan, Walter Baldwin, Jack Carr, Al Seymour, Jimmy Conlin, Jack Carr
A rainy New York night brings together a group of losers in a bar. One of them, a hustler, decides to do one good deed in order to redeem himself and enlists the help of the others. An offbeat, black comedy that moves at quite a pace.
DRA 75 min B/W VIDrel: COLUM/SONOP V

ANGELS WITH DIRTY FACES **** PG
Michael Curtiz USA   1938
James Cagney, Pat O'Brien, Humphrey Bogart, Ann Sheridan, George Bancroft, Billy Halop, Leo Gorcey, Huntz Hall, Gabe Dell, Bobby Jordan, Bernard Punsley, Frankie Burke, William Tracy, Marilyn Knowlen, Joe Downing, Earl Dwire
Excellent gangster movie telling of the uneasy friendship between two boyhood friends who go their separate ways, one becoming a priest and the other a gangster who's idolised by some street punks our priest hopes to reform (played by the "Dead End Kids" of the 1937 film DEAD END). Fine entertainment and beautifully acted – the closing "walk to the electric chair" sequence is an all-time great. Scripted by John Wexley and Warren Duff.
DRA 94 min (ort 97 min) B/W VIDrel: MGM/WHV V
Boa: story by Rowland Brown.

ANGI VERA *** 12
Pal Gabor HUNGARY   1978
Veronika Papp, Erszi Pasztor, Tamas Dunai, Eva Szabo, Laszlo Horvath
In Hungary in 1948 a brave young nurse speaks out about the corruption and squalor she sees at the hospital where she works, but for her pains she is packed off to be trained in Party ideology. Finally, corrupted by exposure to this propaganda, she becomes a trusted Party member, and betrays a colleague on her way up. A brutally chilling film that does much to expose the repression so prevalent in Stalinist Hungary just after WW2.
DRA 92 min VIDrel: ARTPRO/RTM V

ANGIE ** 15
Martha Coolidge USA   1993
Geena Davis, Stephen Rea, James Gandolfini, Aida Turturro, Philip Bosco, Jerry O'Hara, Michael Rispoli, Betty Miller, Charlaine Woodard, Bibi Osterwald, Susan Jafee, Jeremy Collins, Robert Conn, Ray Xifo, Rosemary De Angelis
A young woman from Brooklyn finds herself pregnant and leaves the kind but suffocating confines of her Italian-American neighbourhood to embark on a voyage of self-discovery. In many ways this film suffers from a surfeit of melodramatic cliches but Davis' sterling performance helps it rise above what would otherwise have been a banal affair.
DRA 103 min (ort 108 min) cC VIDrel: HOLPIC/TECH V/sur
Boa: novel Angie I Says by Avra Wing.

ANGRY SILENCE, THE ** PG
Guy Green UK   1960
Richard Attenborough, Pier Angeli, Michael Craig, Bernard Lee, Alfred Burke, Geoffrey Keen, Laurence Naismith, Russell Napier, Penelope Horner, Brian Murray, Brian Bedford, Norman Bird, Beckett Bould, Oliver Reed, Edna Petrie
A union agitator comes to a factory and attempts to organise the labour force, eventually fomenting a wildcat strike. However, his efforts fail to earn him the gratitude of the workers who turn on him, since they are unwilling to accept the sacrifice and privation that his actions have caused. A good performance by

Attenborough fails to save this confused drama whose message (if there is one) remains unclear. See the more lighthearted I'M ALL RIGHT JACK.
DRA 90 min (ort 95 min) B/W VIDrel: LUMI/SPEAR L/A V

ANGUS ** 12
Patrick Read Johnson USA   1995
Kathy Bates, George C. Scott, Charlie Talbert, Rita Moreno, Chris Owen, Ariana Richards, James Van Der Beek, Perry Anzilotti, Robert Curtis-Brown, Kevin Connoly, Tony Denman, Yvette Freeman, Salim Grant, Epatha Harris, Steven Hartman
A podgy boy is tormented at school by the class bullies, but wins through in the end and gets the girl for good measure, There is little original in this simple-minded teen-comedy, but a lively performance from newcomer Talbert helps make this an enjoyable if unassuming affair.
COM 86 min (ort 90 min) VIDrel: EIV/SONOP V
Boa: short story by Chris Crutcher.

ANIMAL FARM ** U
John Halas/Joy Bachelor UK   1952/54 (released 1955)
Voices of: Maurice Denham, Gordon Heath (narration only)
Cartoon version of Orwell's cautionary tale, with an ending changed from that of the book to give it a more optimistic flavour. This straightforward adaptation offers a largely diluted set of critical observations on the nature of power and corruption. Not a kiddie film, but only just.
ANIM 75 min VIDrel: BBC L/A V
Boa: novel by George Orwell.

ANIMAL INSTINCTS * 18
Alexander Gregory Hippolyte USA   1992
Maxwell Caulfield, Delia Sheppard, Jan Michael Vincent, Mitch Gaylord, John Saxon, David Carradine, Shannon Whirry, Josh Cruze, Tom Reilly, Erika Mann, Juliet James, Frank Swann, Lynette O'Connell, Robert Johnston, Kismet Salem
Inspired by the real-life tale of a Florida policeman and his insatiable wife, this steamy thriller tells of a couple whose happy marriage hits a rocky patch when the wife becomes a nymphomaniac as a side-effect of taking anti-depressant pills. Their novel solution allows the wife to exercise to the full her amatory instincts, and the husband to satisfy his voyeuristic ones. Followed by several sequels.
THR 97 min VIDrel: POLY/POLYREC V/sh

ANIMAL INSTINCTS 2 ** 18
Alexander Gregory Hippolyte USA   1993
Shannon Whirry, Nick Cassavetes, Don Swayze, Sandahl Begman, Anna Karin, Catherine Parks, Diana Barton, Richard Roundtree
A further instalment in this series of erotic thrillers. A local sexologist gets seduced by one of his patients and is embroiled in a murder. As ever, the mix of eroticism and tension is much the same. Adequate.
THR 96 min VIDrel: MIA/DISC/IMPENT V/s

ANIMAL INSTINCTS 3 ** 18
Alexander Gregory Hippolyte USA   1993
Shannon Whirry, Woody Brown, Elizabeth Sandifer, Al Sapienza, Eric Fleeks, Jennifer Campbell, David Gautreaux, Tom Reilly, Dean Scofield, Alex Walters, Debra Beatty, Charles Boyer, Rob Dorfmann, Douglas T. Jeffery, Hugh Holub
A voyeur enjoys the nightly shows that his sexy neighbour puts on for his benefit but is mightily peeved when he learns that she has got herself a new boyfriend. As he is an electrician, he is able to wire up people's homes in a way that helps him indulge in his hobby, but soon finds that this activity has certain dangers. An assembly-line erotic tale with a few touches of irony.
THR 96 min VIDrel: MIA/DISC/IMPENT V/s

ANIMAL INSTINCTS: THE SEDUCTRESS * 18
Alexander Gregory Hippolyte USA   1995
Wendy Schumacher, James Matthew, John Bates, Marcus Graham
A rock promoter with a taste in voyeurism embarks on a relationship with an exhibitionist. Number four in this series of erotic thrillers offers little new in the way of fresh ideas or interesting characters.
THR 92 min VIDrel: HIFLI/SONOP V/h

ANIMALYMPICS: THE MOVIE ** U
Steven Lisberger USA   1979
Voices of: Gilda Radner, Billy Crystal, Harry Shearer, Michael Fremer
Animated send-up of the Olympics with the human participants

replaced by various animals. OK for kids not yet out of nappies, its surfeit of cuteness will turn others off.
ANIM   78 min (ort 80 min)   VIDrel: MIA/DISC   V/sur

ANNA CHRISTIE ✳✳   PG
Clarence Brown USA   1930
Greta Garbo, Charles Bickford, Marie Dressler, Lee Phelps, George F. Marion, James T. Mack
Garbo's first talkie has her playing a woman with a murky past as a dockside hooker who finds love in the arms of a sailor. A remake of a 1923 silent, it is excessively talky and suffers from all the usual primitive features of early sound films that made them seem so inferior to the silents they ousted. The first film in which Garbo spoke.
DRA   90 min B/W   VIDrel: MGM/WHV   V
Boa: play by Eugene G. O'Neill.

ANNA KARENINA ✳✳✳✳   U
Clarence Brown USA   1935
Greta Garbo, Fredric March, Freddie Bartholomew, Maureen O'Sullivan, May Robson, Basil Rathbone, Reginald Owen, Reginald Denny
An opulent and carefully made romantic tragedy, telling of the wife of a Russian aristocrat who falls hopelessly in love with a handsome cavalry officer. Garbo is excellent in the title role, March gives fine support as her lover as does Rathbone as her government minister husband. Filmed once before as "Love" (also with Garbo) and made several more times since. The fine camerawork is by William Daniels (who understandably became something of a regular for Garbo).
DRA   89 min (ort 96 min) B/W   VIDrel: MGM/WHV   V
Boa: novel by Leo Tolstoy.

ANNA KARENINA ✳✳   PG
Julien Duvivier UK   1947
Vivien Leigh, Ralph Richardson, Kieron Moore, Marie Lohr, Sally Ann Howes, Niall MacGinnis, Michael Gough, Martita Hunt, Hugh Dempster, Mary Kerridge, Heather Thatcher, Helen Haye, Austin Trevor, John Longden, Ruby Miller
Second attempt to film Tolstoy's classic novel of a married woman who falls in love with a handsome Russian officer. The good cast can do little with this dismally stilted and overlong production.
DRA   110 min (ort 139 min) B/W   VIDrel: WHV   V/h
Boa: novel by Leo Nikolai Tolstoy.

ANNA KARENINA ✳✳   PG
Bernard Rose USA   1997
Sophie Marceau, Sean Bean, Alfred Molina, Mia Kershner, James Fox, Fiona Shaw, Danny Huston, Phyllida Law, David Schofield, Saskia Wickham, Jennifer Hall
In 1880s Russia the bored wife of a wealthy but dull aristocrat embarks on an affair with a dashing cavalry officer, and abandons her husband and child to be with him. But eventually he tires of her and she commits suicide by leaping under a train. Tolstoy's great novel is given a glossy and entirely superficial treatment, and though there are some arresting moments, the film remains lifeless and distant.
DRA   110 min   CINrel
Boa: novel by Leo Tolstoy.

ANNABELLE PARTAGEE ✳   18
Francesca Comencini FRANCE   1990
Delphine Zingg, Francois Marthouret, Jean-Claude Adelin, Florence Thomassin, Dominique Regnier, Jeanne Biras, Jean Cherlian, Stefan Elbaum, Pierre Forget, Pascal Gaultier, Gilbert Grosso, Lilliane Liseron, Claudine Taulere
The story of a romantic triangle that opens with a wholly unnecessary and meaningless depiction of ejaculation, possibly to establish the movie's credentials as an art film. The story has a Parisian dance student vacillating between the two men in her life, but the film is so grimly earnest and clinical that one rapidly loses interest. Written by Comencini, who really has nothing of value to say about human relationships.
DRA   76 min (ort 90 min) wScrn   VIDrel: TART/20TH   V

ANNE AND MURIEL ✳✳✳   15
Francois Truffaut FRANCE   1971
Jean-Pierre Leaud, Kika Markham, Stacey Tendeter, Sylvia Marriott, Philippe Leotard, Marie Mansart
A French writer enjoys a holiday in Wales with two sisters and falls in love with both of them. This was the director's second adaptation of a Roche novel (see also JULES AND JIM) and lacks much of that earlier film's sheer sparkle. A mildly diverting

love-triangle of occasional wit, slightly hampered by weak characterisation. The interesting use of low-key colour photography was an attempt by the director to emulate the early two-tone Technicolor effect.
Aka: LES DEUX ANGLAISES ET LE CONTINENT; TWO ENGLISH GIRLS
DRA   124 min (ort 126 min)   VIDrel: ARTIF/20TH   V/h
Boa: novel Les Deux Anglaises et le Continent by Henri-Pierre Roche.

ANNE OF GREEN GABLES ✳✳✳   U
Kevin Sullivan CANADA/USA   1985
Megan Follows, Colleen Dewhurst, Richard Farnsworth, Patricia Hamilton, Marilyn Lightstone, Schuyler Grant, Rosemary Radcliffe, Jonathan Crombie, Charmion King, Jackie Burroughs, Joachim Hansen, Christiane Krueger
A good remake of the 1934 classic film, following the adoption of the title character by a bachelor farmer and his sister through to her blossoming growth into a young woman. Originally shown in two parts, this attractive adaptation of the famous children's classic is always good to look at. Followed by ANNE OF GREEN GABLES: THE SEQUEL
Aka: ANNE OF GREEN GABLES: THE SERIES
JUV   130 min (ort 202 min) mTV
VIDrel: VISVID/POLYREC   V
Boa: novel by Lucy Maud Montgomery.

ANNE OF GREEN GABLES: THE SEQUEL ✳✳✳   U
Kevin Sullivan CANADA/USA   1986
Megan Follows, Colleen Dewhurst, Patricia Hamilton, Jonathan Crombie, Wendy Hiller, Marilyn Lightstone, Schuyler Grant, Rosemary Dunsmore, Kate Lynch, Frank Converse, Genevieve Appleton, Susannah Hoffman, Kathryn Trainor
Lavish sequel to the first film that is nearly as good as its predecessor, and follows the later career of Anne Shirley, who has now become a teacher at the local school. When a teaching post becomes vacant at Kingsport, she must decide whether to leave Avonlea for a new life. A handsome adaptation, whose period atmosphere and acting cannot be faulted, even if the sentimentality is a bit excessive. Followed by a 26-part TV series entitled "Road To Avonlea".
Aka: ANNE OF AVONLEA
JUV   164 min (ort 224 min) mTV
VIDrel: VISVID/POLYREC   V
Boa: novels Anne Of Avonlea, Anne Of The Island and Anne Of Windy Poplars by Lucy Maud Montgomery.

ANNE OF THE THOUSAND DAYS ✳✳   PG
Charles Jarrott UK   1969
Richard Burton, Genevieve Bujold, Anthony Quayle, John Colicos, Irene Papas, Michael Hordern, Katherine Blake, Valerie Gearson, Michael Johnson, Peter Jeffrey, Joseph O'Conor, William Squire, Esmond Knight, Brook Williams
Wooden film version of a boring play about Anne Boleyn, her courtship by Henry VIII, marriage and subsequent execution. Good scenery and costumes plus a nice performance from Bujold are compensations. AA: Cost (Margaret Furse).
Aka: ANNE OF A THOUSAND DAYS
DRA   140 min (ort 145 min)   VIDrel: CIC/SONOP   V/h
Boa: play by Maxwell Anderson.

ANNIE ✳   U
John Huston USA   1982
Albert Finney, Carol Burnett, Aileen Quinn, Ann Reinking, Bernadette Peters, Tim Curry, Geoffrey Holder, Edward Herrmann, Toni Ann Gisondi, Roger Minami, Rosanne Sorrentino, Lara Berk, April Lerman, Lucie Stewart, Robin Ignacio
Film adaptation of the smash Broadway musical about Little Orphan Annie, a famous comic strip character created during the Depression by Harold Gray. Nice performances from Quinn as Annie and Finney as Daddy Warbucks are wasted in a large, ponderous and glitzy effort. Uninspired, unimaginative and unmoving. Written by Carol Sobieski from the play, with music by Charles Strouse and lyrics by Martin Charnin. Followed by ANNIE 2: A ROYAL ADVENTURE.
MUS   122 min (ort 130 min)   VIDrel: COLUM/SONOP   V/sur
Boa: play by Carol Sobieski/book by Thomas Meecham.

ANNIE 2: A ROYAL ADVENTURE ✳✳   U
Ian Toynton   1995
Joan Collins, George Hearn, Ashley Johnson
Sequel to ANNIE that sees both her and Daddy Warbucks back for some more adventures, this time being off to London to see

the Queen. Unfortunately, the wicked Lady Hogsbottom has made plans to spoil their trip.
JUV   89 min   VIDrel: COLUM/SONOP   V/sur

ANNIE HALL ****   15
Woody Allen USA   1977
Woody Allen, Diane Keaton, Tony Roberts, Carol Kane, Paul Simon, Shelley Duvall, Janet Margolin, Colleen Dewhurst, Christopher Walken, Helen Ludlam, Donald Symington, Mordecai Lawner, John Newman, Jonathan Munk, Ruth Volner
Romantic comedy looking at the relationship between a Jewish comedian and a girl from the mid-West. Full of Allen's observations on life, love and fame with warm script and appealing performances, and happily not let down by the unevenness that afflicts so many of his films. One of Allen's best efforts. AA: Pic, Dir, Actress (Keaton), Screen/orig (Woody Allen/Marshall Brickman).
COM   89 min (ort 94 min)   VIDrel: MGM/WHV   V/dm

ANOTHER COUNTRY ***   15
Marek Kaniewska UK   1984
Rupert Everett, Colin Firth, Michael Jenn, Robert Addie, Anna Massey, Betsy Brantley, Rupert Wainwright, Tristan Oliver, Cary Elwes, Geoffrey Bateman, Frederick Alexander, Adrian Ross-Magenty, Philip Dupuy, Jeffrey Wickham
Set in a boys' public school in the 1930s, this film focuses on the oppressive and cloistered atmosphere of corruption and betrayed ideals that influences the lives of some of the pupils there. Based on Mitchell's successful London play, that attempted an examination of the public school system that produced traitors such as Guy Burgess and Donald MacLean. A Cannes Festival Award Winner.
DRA   86 min (ort 90 min)   VIDrel: VISION/DISC   V
Boa: play by Julian Mitchell.

ANOTHER 48 HRS **   18
Walter Hill USA   1990
Eddie Murphy, Nick Nolte, Brion James, Kevin Tighe, Ed O'Ross, David Anthony Marshall, Andrew Divoff, Bernie Casey, Brent Jennings, Tisha Campbell, Cathy Haase, Ted Markland, Felice Orlandi, Edward Walsh, Page Leong, Kelly Goodman
A sequel to 48 HRS that is more like a remake than a new film in its own right, with cynical and world-weary cop Nolte once more turning to Murphy (who has just been sprung from jail) for help in solving a tough case. A totally exploitative rehash if ever there was one, it provides a modicum of entertainment, but is devoid of imagination or sparkle.
A/AD   91 min (ort 98 min)
VIDrel: 4-FRONT/POLYREC/CIC   V/sh

ANOTHER GIRL ANOTHER PLANET **   18
Michael Almereyda USA   1992
Nic Ratner, Elina Lowensohn, Barry Sherman
Ultra low-budget film (it was shot on a Fisher Price PXL 2000 children's camera) that attempts to explore the problems women face in marriage. This might so easily have been another examination of New York angst, but the shoestring budget has forced the director to use all his initiative, and the result is certainly distinctive. Screenplay is by Almereyda.
DRA   56 min B/W   VIDrel: SCEDGE/RTM   V

ANOTHER PAIR OF ACES: THREE OF A KIND **   15
Bill Bixby USA   1991
Kris Kristofferson, Willie Nelson, Joan Severance, Ken Farmer, Dan Kamin, Rip Torn
A sequel to PAIR OF ACES with the same modern Western flavour, in which our Texas Ranger and wily safecracker team up once again to prove the innocence of a friend of the former, who has been wrongly accused of murder. Fair.
WES   89 min (ort 100 min) mTV   VIDrel: 20VIS/SONOP   V

ANOTHER STAKEOUT **   PG
John Badham USA   1993
Richard Dreyfuss, Emilio Estevez, Rosie O'Donnell, Dennis Farina, Marcia Strassman, Cathy Moriarty, John Rubinstein, Miguel Ferrer, Sharon Maughan, Rick Seaman, Christopher Doyle, Sharon Schaeffer, Jan Speek, Gene Ellison, J.R. West
This sequel to STAKEOUT has our two mismatched and far from capable detectives staking out a bad guy who is understandably unwilling to testify against the Mob. This task is made even more difficult by the presence of an assistant D.A. who

insists on bringing her pet dog along. A tired effort that offers little that is new or amusing.
Aka: STAKEOUT 2
A/AD   104 min (ort 109 min) cC   VIDrel: TOUCH/TECH   V/sur

ANOTHER TIME, ANOTHER PLACE ***   15
Michael Radford UK   1983
Phyllis Logan, Giovanni Mauriello, Denise Coffey, Tom Watson, Gian Luca Favilla, Gregor Fisher, Paul Young, Claudio Rosini, Jennifer Piercey, Yvonne Gilan, Carol Ann Crawford, Ray Jeffries, Scott Johnston, Nadio Fortune
A farmer's wife in the Scottish Highlands gradually falls in love with one of a group of Italian POWs billeted on the tiny farming community. A nice, understated WW2 drama, originally made for TV but released to cinemas instead. This was Radford's debut feature.
DRA   90 min (ort 102 min) mTV   VIDrel: GUER/PINN   V
Boa: novel by J. Kesson.

ANOTHER WAY ***   18
Karoly Makk HUNGARY   1982
Jadwiga Jankowska-Cieslak, Grazyna Szapolowska, Jozef Kroner, Adam Szirtes, Gabor Reviczky, Ferenc Bacs, Denes Ujlaki, Eva Igo, Anetta Antal, Gyorgyi Fay, Agnes Kamondy, Janos Kovacs, Vilmos Kun, Sandar Makay
Two female journalists on a Budapest newspaper fall in love, but a tragedy is the result. An interesting European melodrama.
Aka: EGYMASRA NEZRE
DRA   102 min (ort 109 min)   VIDrel: ARTPRO/RTM   V
Boa: novella by Erzsebet Galgoczi.

ANOTHER WOMAN ***   PG
Woody Allen USA   1988
Gena Rowlands, Mia Farrow, Ian Holm, Blythe Danner, Gene Hackman, Betty Buckley, Martha Plimpton, John Houseman, Sandy Dennis, David Ogden Stiers, Philip Bosco, Harris Yulin, Frances Conroy, Fred Melamed, Kenneth Walsh
Another Allen stab at a Bergmanesque study of misery, complete with Sven Nykvist as cameraman. This story focuses on the usual collection of angst-ridden New Yorkers, particularly a woman whose cosy and sheltered lifestyle is shattered by unforeseen events. A mature adult drama of powerful if decidedly narrow appeal.
DRA   77 min (ort 84 min)   VIDrel: VISVID/POLYREC   V

ANOTHER YOU *   15
Maurice Phillips USA   1991
Richard Pryor, Gene Wilder, Mercedes Ruehl, Stephen Lang, Vanessa Williams, Phil Rubinstein, Vincent Schiavelli, Kandis Chappell, Peter Michael Goetz, Craig Richard Nelson, Billy Beck, Jerry Houser, Elsa Raven, Kevin Pollak
Having been confined to a sanatorium because he is a compulsive liar, Wilder finds himself being released into the custody of Pryor, a streetwise trickster who is doing his bit for the community. The pair of them get involved in a clever con-trick in this well meaning but rather shallow effort. Pity to see a couple of first-rate comedians working with such weak material.
COM   90 min (ort 98 min)   VIDrel: VCC/DISC/COLUM   V/sur

ANTAGONISTS, THE **   15
Boris Sagal USA   1981
Peter O'Toole, Peter Strauss, Barbara Carrera, Nigel Davenport, Paul Smith, Alan Feinstein, Giulia Pagano, Anthony Quayle, Denis Quilley, Timothy West, Anthony Valentine, David Warner, Clive Francis, George Innes, David Opatoshu
Cut-down version of the TV film "Masada", telling of the Jewish fortress that defied the might of Rome. There are a few good things in this glossy and overlong TV production, but O'Toole's dreadful over-acting is not one of them.
Aka: MASADA
A/AD   115 min (ort 448 min) mTV   VIDrel: CIC/SONOP   V/h
Boa: novel by Ernest K. Gann.

ANTAGONISTS, THE **   15
Rob Cohen USA   1991
David Andrews, Matt Roth, Lauren Holly, Lisa Jane-Persky, Brett Jennings, Dey Young, Belinda Bauer, Stephen Davies, Harris Laskawy, Erich Anderson, Will Leskin, Geoffrey Rivas, Al Pugliese, Christopher McDonald, Sean Faro

Two lawyers clash in court when they find themselves disagreeing over an apparently clear-cut case.
DRA   85 min (ort 90 min) mTV   VIDrel: CIC/SONOP
V/sh

ANTHONY'S DESIRE *   (18)
Tom Boka USA   1993
Mihaella Stoicov, Douglass DeMarco, Gwen Somers, Tom Boka, Zsuzsa Fontner, Rich Troncone, Kevin Abosch, Michele Jaffe, James Scanlon, Patrick Lim, Mak Stansbury, Miklos Fotos, Steven Burkholder, Annastasia Alexander, Ashlie Rhey
A young man who is uncertain about what to do with his life comes to a small resort whose inhabitants harbour some dark sexual secrets, and once there gets involved with an attractive local woman. A terribly pretentious, arty and hard-to-follow erotic movie, made as one of those irritating films-within-a-film devices, that offers plenty of nudity and softcore couplings, but nothing as unimportant as a plot to maintain interest.
A   90 min   SATrel: MOVIE CHANNEL

ANTONIA'S LINE ***   15
Marleen Gorris BELGIUM/HOLLAND/UK   1995
Willeke van Ammelrooy, Els Dottermans, Dora van der Groen, Veerle van Overloop, Esther Vriesendorp, Carolien Spoor, Thyrza Ravesteijn, Mil Seghers, Jan Decleir, Elsie de Brauw, Reinout Bussemaker, Marina de Graaf, Jan Steen
A portrait of an unconventional and strong-willed woman, who at the end of WW2 raises her daughter by herself, while opening her home to a variety of lost souls. A simple story, much enhanced by the fine acting, the film is strongly feminist in flavour, yet pitched at a gentler level than some of this director's other works, most especially the hateful "A Question Of Silence". The screenplay is by Gorris. AA: Foreign.
Aka: ANTONIA
DRA   104 min   VIDrel: GUILD/20TH/FOXVID   V/sur

ANTONY AND CLEOPATRA ***   U
Jonathan Miller UK   1981
Colin Blakely, Jane Lapotaire, John Paul, Jonathan Adams, Darien Angadi, Janet Key, Howard Goorney, Cassie McFarlane, Emrys James, Kevin Huckstep, Michael Anthony, Mohammad Shamsi, Ian Charleson, Esmond Knight, Harry Waters
Competent BBC production of this famous play, that tells of history's famous love affair. The use of 16th century sets and costumes shows Miller's well known penchant for unconventional interpretations.
DRA   172 min mTV   VIDrel: BBC   V/h
Boa: play by William Shakespeare.

ANY WHICH WAY YOU CAN **   15
Buddy Van Horn USA   1980
Clint Eastwood, Ruth Gordon, Sondra Locke, William Smith, Harry Guardino, Glen Campbell, Anne Ramsey, Logan Ramsey, Geoffrey Lewis, Michael Cavanaugh, Barry Corbin, Roy Jenson, Bill McKinney, William O'Connell, John Quade
Offbeat and mindless follow-up to the successful EVERY WHICH WAY YOU CAN in which bare-fist boxer Eastwood is still slugging it out. This time he is up against a really tough opponent, who is out to prove himself the very best bare-fist boxer around. A mixture of fisticuffs, huumour (some of it really gross) and poor acting (the best actor being our orangutan – Clyde).
COM   110 min Cut (45 sec – ort 116 min)   VIDrel: WHV
L/A   V

ANYONE FOR SEX *   15
Ralph Thomas UK   1973
Hywel Bennett, Nanette Newman, Milo O'Shea, John Cleese
Having had six children, a man's wife decides enough is enough, but when her husband refuses to use any form of birth control, she forces him to sleep in a separate bed. However, the frustration this engenders makes him fantasise about extra-marital relations. A twee and tedious British attempt at a sex comedy.
Aka: IT'S A 2 FOOT 6 INCH ABOVE THE GROUND WORLD; LOVE BAN, THE
COM   97 min   VIDrel: LUMI/SPEAR   V
Boa: play It's a 2 ft 6 inch Above The Ground World by Kevin Laffan.

ANYTHING THAT MOVES **   18
John Leslie USA   1992
Selena Steele, Tracey Wynn, Cassidy, Heidi Nelson, Melody Moore,

Tim Lake, Randy Spears, Steve Drake, Lee Chandler, Nick E., Joel Lawrence, Tody Tedeschi, Nick Santearo, Jake Williams, Kristin Snap, Flame, Howard Kay, Daisy Jefferson
When a low-life is found dead a couple of detectives (a man and a woman) interview his long-suffering stripper girlfriend and her colleagues, and we see just what sort of a person he was and how he treated others. While this one has a stronger plot than most of the sex films being made nowadays, it still remains woefully undeveloped, for the most part being nothing more than a device to tie together a number of sexy dancing sequences etc.
A   79 min (Cut before video submission by 4 min 23 sec – ort 83 min)   VIDrel: GROHOM/MAXSCAN   V

ANZIO **   PG
Edward Dmytryk FRANCE/ITALY/SPAIN   1968
Robert Mitchum, Peter Falk, Robert Ryan, Earl Holliman, Mark Damon, Arthur Kennedy, Reni Santoni, Anthony Steel, Patrick Magee, Joseph Walsh, Thomas Hunter, Giancarlo Giannini, Arthur Franz, Tonio Stewart
Uninspired account of WW2 Allied Anzio landings with plenty of large-scale action and a star-studded cast, but still short on interest.
Aka: BATTLE FOR ANZIO, THE
WAR   112 min (ort 117 min)   VIDrel: VCC/DISC/COLUM
V
Boa: book by Wynford Vaughan-Thomas.

APACHE ***   PG
Robert Aldrich USA   1954
Burt Lancaster, Jean Peters, John McIntire, Charles Buchinsky (Bronson), Ian MacDonald, John Dehner, Walter Sande, Paul Guifoyle, Morris Ankrum, Monte Blue
An Apache Indian continues the fight after the defeat of Geronimo but gradually reverts to his former pacifist views and is reconciled with the enemies of his people. A well-meaning and thoughtful western, sensitively handled, but one whose message is hard to swallow in the cold light of history.
WES   91 min   VIDrel: MGM/WHV   V/h

APARAJITO ****   U
Satyajit Ray INDIA   1956
Pinaki Sen Gupta, Smaran Ghosal, Karuna Banjeri, Kanu Banjeri, Ramani Sen Gupta, Subodh Ganguly, Charu Gosh, Kali Charan Ray, Santi Gupta, K.S. Pandey, Sudipta Ray, Ajay Mitra
Second film in the "Apu" trilogy about the lives of a poor Indian family. Here, the son Apu goes to college in Calcutta and pursues his studies until the sudden death of his mother. A detailed and absorbing film that explores the growing gulf between Apu and his mother and the stark contrast between city and country life. Perhaps not quite as moving as the first film PATHER PANCHALI, but definitely worth seeing. Followed by THE WORLD OF APU.
Aka: UNVANQUISHED, THE
DRA   104 min (ort 127 min) B/W   VIDrel: CONNO/RTM
V
Boa: novel by Bibhutbhusan Bandapaddhay.

APARTMENT, THE ****   PG
Billy Wilder USA   1960
Jack Lemmon, Shirley MacLaine, Fred MacMurray, Ray Walston, Jack Kruschen, Edie Adams, David Lewis, Joan Shawlee, David White, Hope Holiday, Johnny Seven, Naomi Stevens, Frances Weintraub Lax, Joyce Jameson, Willard Waterman
Sharp and witty story of an ambitious executive who lends his apartment to his superiors to curry favour with them, with unforeseen results when he falls in love with his boss's latest girlfriend. Later made into the Broadway musical "Promises, Promises". AA: Pic, Dir, Story/Screen (Wilder/A.L. Diamond), Art/Set (Alexander Trauner/Edward G. Boyle), Edit (Daniel Mandell).
COM   119 min (ort 125 min) B/W   VIDrel: WHV   V

A.P.E.X. **   15
Phillip J. Roth USA   1993
Richard Keats, Mitchell Cox, Marcus Aurelius, Lisa Ann Russell, Adam Lawson, David Jean Thomas, Brian Richard Peck, Ann B. Choi, Kristin Norton, Jay Irwin, Robert Tossberg, Kathy Lambert, Kareem H. Captan, Merle Nicks, Natasha Roth
A researcher from the year 2073 travels back in time to retrieve the title artefact, an advanced killer robot. Upon his return he finds humanity dying from a deadly virus, with the few remaining survivors being hunted down by a robot army. Despite the

low budget, the good special effects help along this unusual exploration of time-travel paradoxes.
Aka: APEX
FAN   98 min (ort 103 min)   VIDrel: MIA/DISC   V

APOCALYPSE NOW ***
18
Francis Ford Coppola USA
1979
Marlon Brando, Martin Sheen, Robert Duvall, Frederic Forrest, Sam Bottoms, Dennis Hopper, Albert Hall, Harrison Ford, Larry Fishburne, G.D. Spradlin, Cyndi Wood, Colleen Camp, Linda Carpenter, Jack Thibeau, Damien Leake, Glen Walken
Hollywood finally discovered the Vietnam War with an epic story of a secret agent sent into the jungle with orders to kill an officer who has set up his own little kingdom in the heart of the Cambodian jungle. Sheen makes a strange, surreal journey upriver, but it is followed by a muddled climax. A film of truly unforgettable imagery and annoying opacity. See also HEART OF DARKNESS. AA: Cin (Vittorio Storaro), Sound (Walter Murch/Mark Berger/Richard Beggs/Nat Boxer).
WAR   147 min (ort 153 min) wScrn   VIDrel: CIC/SONOP; PION (LV only)   V/sur LV
Boa: novella Heart of Darkness by Joseph Conrad.

APOLLO 11 **
(PG)
Norbeto Barba USA
1995
Carmen Argenziano, Tuck Milligan, Dennis Lipscomb, William Mesnik, Jack Conley, Michael Cheiffo, Jeffrey Nordling, James Parks, Barbara Whinnery, Wendie Malick, Jim Metzler, Samantha Dapper, Jane Kaczmarek, Maureen Mueller
Clearly inspired by the success of APOLLO 13, this is a low-budget attempt to relate the events surrounding an earlier moon mission. A competent but not all that exciting space-oriented drama whose appeal lies definitely in its portrayal of the wonders of modern technology.
DRA   90 min mTV   SATrel: MOVIE CHANNEL

APOLLO 13 ***
PG
Ron Howard USA
1995
Tom Hanks, Bill Paxton, Kevin Bacon, Gary Sinise, Ed Harris, Kathleen Quinlan, Mary Kate Schellhardt, Emily Ann Lloyd, Miko Hughes, Clint Howard, Max Elliott Slade, Jean Speegle Howard, Tracy Reiner, David Andrews, Michelle Little
A nailbiting account of this 1970 ill-fated space mission, which was struck by a series of mishaps that could have had fatal consequences. A good study of individuals under pressure, with Hanks and the rest of the cast turning in some very fine performances. The claustrophobic setting aboard the space capsule helps to enhance the movie's dramatic impact. AA: Edit, Sound.
DRA   134 min (ort 140 min) wScrn cC
VIDrel: CIC/SONOP; PION (LV only)   V/sur LV
Boa: book Lost Moon by Jim Lovell and Jeffrey Kluger.

APPALOOSA, THE **
PG
Sidney J. Furie USA
1966
Marlon Brando, Anjanette Comer, John Saxon, Rafael Campos, Frank Silvera, Larry Mann, Alex Montoya, Emilio Fernandez, Miriam Colon, Argentina Brunetti
A cowboy who was planning to start a stud farm goes after the Mexican bandit who has stolen his prize steed. Set in the 1870s, this well-photographed but sluggish yarn serves as little more than a vehicle for Brando. Photography is by Russell Metty.
Aka: SOUTHWEST TO SONORA
WES   94 min (ort 99 min)   VIDrel: CIC/SONOP   V
Boa: novel by Robert MacLeod.

APPLESEED **
15
Kazuyoshi Katayama 1989
JAPAN
Voices of: Larissa Murray, Bill Roberts, David Reynolds, Lorelei King, Vincent Marzello, Julia Brahms
After WW3, human survivors and artificial humanoid creatures known as bionoids coexist peacefully until a terrorist group wrecks havoc upon their common abode, the mega-city Olympus. Another example of Japanese so-called cyberpunk animation that will appeal mainly to fans of this genre.
ANIM   66 min (ort 71 min) dubbed
VIDrel: MANGA/SONOP   V/sh

APPOINTMENT FOR A KILLING **
18
William A. Graham USA
1993
Corbin Bernsen, Markie Post, Don Swayze, Suzanne Barnes, Jeanne Cooper, Laurie O'Brien, Danielle Von Zerneck, Matthew Best, Kelsey Grammer, John Putch, Janet Graham, Geoff Hansen, Anna Maria Sistare, Harry Murphy
Fact-based drama telling of an adulterous dentist who confesses to his wife that he once killed a man. She now begins to recall various other cases of unsolved murders and when people she knows start dying, she confides in a police officer, but finds that escaping from a husband she now fears is not going to be easy.
Aka: APPOINTMENT FOR KILLING
THR   88 min (ort 90 min) mTV
VIDrel: NEWAGE/20VIS L/A   V
Boa: book Appointment For Murder by Susan Crain Bakos.

APPOINTMENT IN LONDON **
U
Philip Leacock USA
1953
Dirk Bogarde, Pier Angeli, Dinah Sheridan, Bill Kerr, Bryan Forbes, William Sylvester, Charles Victor, Anne Leon, Walter Fitzgerald, Anthony Shaw, Carl Jaffe, Ian Hunter, Richard Wattis, Terence Longdon, Sam Kydd, Campbell Singer
An RAF wing commander is grounded after eighty-nine bombing missions but defies orders to take part in one more raid from which he returns to face the expected official disapproval. However, in place of a court martial, he is punished by being posted abroad. Solid acting and some exciting aerial sequences sustain this WW2 drama.
WAR   92 min (ort 98 min)   VIDrel: FABFIL/SPEAR   V

APPRENTICESHIP OF DUDDY KRAVITZ, THE ***
18
Ted Kotcheff CANADA
1974
Richard Dreyfuss, Micheline Lanctot, Jack Warden, Randy Quaid, Joseph Wiseman, Denholm Elliott, Joe Silver, Henry Ramer, Zvee Scooler, Robert Goodier, Alan Rosenthal, Barrie Baldavo, Allan Migicovsky, Barry Pascal, Susan Friedman
Well acted and lively comedy-drama following the rise of an ambitious Jewish boy from Montreal's Jewish quarter who is determined to make good, no matter what it takes. Set in the 1940s and containing many vivid and comical highlights.
DRA   120 min   VIDrel: ARROW/RTM   V
Boa: novel What Makes Duddy Run? by Mordecai Richler.

APRES L'AMOUR **
15
Diane Kurys FRANCE
1992
Isabelle Huppert, Bernard Giraudeau, Hippolyte Girardot, Lio, Yvan Attal, Judith reval, Ingrid Held, Laure Killing, Mehdi Ioossen, Florian Billon, Eva Killing, Ana Girardot, Jean-Claude de Goros, Chrystelle Labaude, Jean Grecault
A novelist loves both the man she lives with and another, and they also have other romantic interests of their own, the object of their affections in turn having yet further involvements. Meanwhile, the girl tries to live her life on her own terms, but her room for manoeuvre is strictly curtailed. A kind of modern-day variant on LA RONDE; a strong cast do what they can in their unappealing roles, but the film is more pretentious than insightful.
DRA   100 min (ort 105 min)   VIDrel: CURZON/20TH   V/sur

APRIL IN PARIS *
U
David Butler USA
1952
Doris Day, Ray Bolger, Claude Dauphin, Eve Miller, George Givot
Dismal star vehicle for Day, with her playing a chorus girl who is invited by mistake to an American arts festival in Paris. Once there she captivates the chief diplomat. The sprinkling of humour (such as various onboard problems for Bolger and Day en route to France) simply cannot hide the sheer lack of ideas in this drab and dull musical.
MUS   96 min (ort 100 min)   VIDrel: WHV   V

ARACHNOPHOBIA **
PG
Frank Marshall USA
1990
Jeff Daniels, Julian Sands, Harley Jane Kozak, John Goodman, Stuart Pankin, Brian McNamara, Mark L. Taylor, Henry Jones, Peter Jason, James Handy, Kathy Kinney, Roy Brocksmith, Mary Carver, Garette Patrick Ratcliff, Marlene Katz
Marshall's first feature film follows the exploits of a deadly tropical spider, that sets up home in a barn on the property of a small-town doctor and causes considerable inconvenience. A desultory rehash of those 1950s bug movies with all the action packed into the last half hour. The last sequence with the doctor's family back in the safety of San Francisco only to suffer the shock of a couple of earth tremors is both cheap and unnecessary.
THR   105 min (ort 109 min)   VIDrel: L/A   V

ARCADE **
15
Albert Pyun USA
1993
Megan Ward, Peter Billingsley, John DeLancie, Seth Green, Sharon

Farrell, Humberto Ortiz, Jonathan Fuller, Norbert Weisser, A.J. Langer, Bryan Dattilo, Don Stark, Dorothy Dells, Todd Starks, Alexandria Byrne, David Sederholm,
After a new arcade opens in town that offers an exiting virtual reality game, a number of local teenagers disappear. A young girl begins to suspect that this new game may be somehow involved and eventually finds herself fighting for life inside a strange and alien world. Computer-generated animation is used to good effect, in this frightening and unusual horror tale.
HOR   81 min (ort 85 min)   VIDrel: EIV/SONOP V

ARCH OF TRIUMPH *** (PG)
Lewis Milestone USA   1948
Ingrid Bergman, Charles Boyer, Charles Laughton, Louis Calhern, Ruth Nelson, Ruth Warrick, Roman Bohnen, Stephen Bekassy, Curt Bois, J. Edward Bromberg, Michael Romanoff, Art Smith, John Laurenz, Leon Lenoir, Franco Corsaro
An ambitious tale of a tragic romance between a cynical refugee and a prostitute in WW2 France, with both being pursued by a Gestapo chief. Slow pacing and the studio-bound sets are serious flaws, though the film is intermittently absorbing and the stars do what they can with the depressing material. An expensive flop, it cost $5,000,000 to make and only grossed $1,500,000. Remade for TV in 1985.
DRA   81 min (ort 131 min) B/W   VIDrel: L/A V
Boa: novel by Erich Maria Remarque.

ARCTIC BLUE ** 15
Peter Masterson USA   1995
Rutger Hauer, Dylan Walsh, Rya Kihlstedt, Richard Bradford, Kevin Cooney, Jan Cuthbert, John Bear Curtis, Stephen E. Miller, Bill Croft, Michael St John Smith, Michael Lawrenchuk, Gunargie O'Sullivan, Ron Chartier, Dan Shea
A marshal and his prisoner find themselves having to rely on each other in the frozen Alaskan wastes where somebody is out to kill them both. An adequate wilderness actioner, enlivened by strong and convincing performances from Hauer and Walsh, who are cast as prisoner and escort respectively.
A/AD   91 min (ort 95 min)   VIDrel: EIV/SONOP V/sur

ARE YOU BEING SERVED? * PG
Bob Kellett UK   1977
Mollie Sugden, John Inman, Frank Thornton, Trevor Bannister, Wendy Richard, Arthur Brough, Nicholas Smith, Arthur English, Harold Bennett, Andrew Sachs, Karan David, Glyn Houston, Penny Irving, Derek Griffiths, Sheila Staefel
One of those feeble TV spin-offs from a popular British sit-com. This one has the staff of Grace Brothers, the department store, going on holiday and enduring a series of catastrophes – the making of this film being one of them.
COM   91 min (ort 95 min)   VIDrel: MGM/WHV V

ARE YOU LONESOME TONIGHT? ** 15
E.W. Swackhamer USA   1991
Jane Seymour, Parker Stevenson, Beth Broderick, Joel Brooks, Robert Pine, Carolyn Kimball, Henry J. Jordan, Jonathan Nichols, Mary Qualls, Ann Tyler, Lael Jackson, Charmame Charles, Rosie Taravelloa, George Chambers
Having discovered that her husband was involved with a phone-sex girl, a wealthy woman hires a private detective to trace him when he vanishes. A competent mTV film of little distinction that offers an hour and a half of undemanding entertainment.
DRA   88 min (ort 91 min) mTV   VIDrel: CIC/SONOP V

ARENA ** 15
Peter Manoogian USA   1988
Paul Satterfield, Claudia Christian, Hamilton Camp, Steve Wang, Marc Alaimo, Shari Shattuck, Armin Shimerman, Brett Porter, Charles Tabansi, Jack Carter, Ken Clark, Michael Deak, William Butler, Grady Clarkson, Dave Thompson, Diana Rose
In the year 4038 the universe is ruled by one government and has a single common language. "Arena" is the name for a popular sport whose contestants compete for cash prizes, but can find themselves up against any kind of creature. This disappointing fantasy tells of a drifter badly in need of some money to return to Earth who decides to try his luck.
FAN   93 min (ort 97 min)   VIDrel: EIV/SONOP V

ARIEL ** (PG)
Aki Kaurismaki FINLAND   1988
Susanna Haavisto, Turo Pajala, Matti Pellonpaa, Etu Hilkamo, Erkki Pajala, Matti Jaaranen, Hanni Viohlainen

A mine worker is made redundant and takes to the road in a white cadillac, but soon loses his money and is forced into casual work, eventually starting a relationship with a young single parent. A surrealistic and quirky film that is well made and acted but will be hard to take for those who are not fans of Kaurismaki. This was the second part of his "Working Class Trilogy", "Shadows In Paradise" and THE MATCH FACTORY GIRL being parts one and three respectively.
DRA   90 min   CINrel

ARISTOCATS, THE *** U
Wolfgang Reithermann USA   1970
Voices: Eva Gabor, Scatman Crothers, Phil Harris, Maurice Chevalier, Lord Tim Hudson, Vito Scotti, Thurl Ravenscroft, Dean Clark, Liz English, Gary Dubin, Stirling Holloway, Nancy Kulp, Pat Buttram, George Lindsay, Monica Evans
A wealthy Frenchwoman dies and bequeaths her fortune to her beloved cat and its three kittens but her butler has other ideas. He drugs them and dumps them in the countryside, believing they will be unable to find their way home. They, however, do just this after meeting up with a tough alley cat and enjoying all kinds of adventures. A top-notch animated tale that took four years to produce and cost $4,000,000, this witty and inventive cartoon can be enjoyed by all.
ANIM   79 min   VIDrel: WDV/TECH V

ARIZONA DREAM *** 15
Emir Kusturica FRANCE/USA   1993
Johnny Depp, Faye Dunaway, Jerry Lewis, Lili Taylor, Paulian Porizkova, Vincent Gallo, Michael J. Pollard, Candyce Mason, Alexa Rayne, Aron Schulman, Polly Noonan, Patricia O'Grady, James R. Wilson, Sal Jenco, Vincent Tocktoo
A young man is lured into coming to Arizona by his larger-than-life uncle but finds himself ill at ease until he falls in love with a rather unstable widow, who shot her husband and dreams about being able to fly. A strange and very moody piece, borne up by its good performances rather then any inherent virtues of its plot.
DRA   135 min (ort 142 min)   VIDrel: POLY/POLYREC V/sh

ARMED AND INNOCENT ** 15
Jack Bender USA   1993
Kate Jackson, Gerald McRaney, Andrew Starnes, James Short, Dru Mouser, Jim Haynie
A young boy defends his home from burglars, shooting two of them. However, the third man is the leader and escapes, and sets about planning his revenge. Competent drama that is based on a true story.
DRA   93 min mTV   VIDrel: ODY/SONOP V/sh

ARMED RESPONSE ** 18
Fred Olen Ray USA   1986
Lee Van Cleef, David Carradine, Michael Berryman, Mako, Lois Hamilton, Ross Hagen, Brent Huff, Laurene Langdon, Dick Miller
Action story set in the Chinatown area of Los Angeles with a retired cop and his eldest Vietnam veteran son hitting the vengeance trail, when the two youngest sons are murdered in a double-cross after they get involved in retrieving a jade statuette for a local Chinese gangster. A fast and brainless story for those who like their action undiluted by a plot.
Aka: JADE JUNGLE, THE
A/AD   82 min Cut (1 sec)   VIDrel: EIV/SONOP V

ARMITAGE 3 – EPISODE 1: ELECTRO BLOOD ** 15
JAPAN   1995
The year is 2179, and on Mars there is a colony of human beings, who live alongside intelligent humanoid robots. In this tale, a cop from the Martian police department investigates the creation of the new type of robot that may pose a threat to humans. He is assisted by his robot partner – "Armitage 3" (hence the title). One in a series of Japanese SF animations.
ANIM   50 min dubbed   VIDrel: PION/RTM V

ARMOUR OF GOD, THE *** 15
Jackie Chan HONG KONG   1986
Jackie Chan, Alan Tam, Lola Forner, Rosamund Kwan, Bozidar Smiljanic, Ken Boyle, John Ladalski, Robert O'Brien, Boris Gregoric, Marcia Chisholm, Linda Denley, Alicia Shonte, Vivian Wickliffe, Stephanie Evans, William Williams
Kung fu actioner with Chan obliged to borrow back some religious artefacts (the "Armour of God") from an antiques dealer to whom he sold them, in order to use the pieces as bait to

retrieve his girlfriend from an evil cult. As ever, the athletic Chan does his own stunts in this fast and furious action tale. A sequel followed.
Aka: LONG XIONG HU DI
MAR   88 min   VIDrel: MIA/DISC/IMPENT   V

ARMOUR OF GOD 2 ***
15
Jackie Chan HONG KONG
Jackie Chan, Carol Cheng, Eva Cobo De Garcia, Shoko Keda, Ken Goodman, Lynn Percival, Alfredo Bael Sanchez, Gregory Tartaglia, Bruce Fontayne, Low Hou Kang, Archer WAyne, Branden Charles, Peter Klemenko, Jonathan Isgar, Mark King
In this enjoyable sequel to the first film, Chan plays an adventurer hired to recover a cache of stolen gold that was hidden in the North African desert by the Nazis at the end of WW2. Naturally, he soon learns that other slightly less scrupulous people are also seeking this treasure.
Aka: OPERATION CONDOR: ARMOUR OF GOD 2
MAR   102 min wScrn   VIDrel: EIV/SONOP   V

ARMY OF DARKNESS **
18
Sam Raimi USA
1992
Bruce Campbell, Embeth Davidtz, Marcus Gilbert, Ian Abercrombie, Richard Grove, Michael Earl Reid, Timothy Patrick Quill, Bridget Fonda, Patricia Tallman, Theodore Raimi, Deke Anderson, Bruce Thomas, Sara Shearer, Ivan Raimi
Third in the Evil Dead series sees a store clerk thrown back in time to the court of King Arthur, where he has a variety of frightening experiences including an encounter with an army of skeletons. However, the impressive special effects are unable to sustain the entire film, which suffers badly from the lack of a strong storyline and is weakened by its ill-advised horror-comedy approach.
Aka: ARMY OF DARKNESS: THE MEDIEVAL DEAD; EVIL DEAD 3
HOR   85 min (ort 109 min)   VIDrel: GUILD/POLYREC
V/sur

AROUND THE WORLD IN 80 DAYS ***
U
Michael Anderson USA
1956
David Niven, Cantinflas, Robert Newton, Shirley MacLaine, Marlene Dietrich, Charles Boyer, Joe E. Brown, Martine Carol, John Carradine, Charles Coburn, Ronald Colman, Melville Cooper, Noel Coward, Finlay Currie, Fernandel
Lavish version of Jules Verne's famous story of a Victorian gentleman and his valet who together win a bet that they can go round the world in eighty days. No less than forty-four star cameos come together in this long, expensively mounted and extremely tiring jaunt. Remade in 1988. AA: Pic, Dir, Cin (Lionel Lindon), Screen/adapt (S.J. Perelman/James Poe/John Farrow), Score (Victor Young), Edit (Gene Ruggiero/Paul Weatherwax).
A/AD   135 min (ort 178 min)   VIDrel: WHV   V/sh
Boa: novel by Jules Verne.

AROUND THE WORLD IN EIGHTY DAYS: PARTS 1 AND 2 **
PG
Buzz Kulik USA
1988
Pierce Brosnan, Eric Idle, Peter Ustinov, Jill St John, Jack Klugman, John Mills, Robert Wagner, Robert Morley, Julia Nickson, Henry Gibson, Lee Remick, John Hillerman, Christopher Lee, Roddy McDowall, Darren McGavin, Stephen Nichols
Overlong and glossy adventure tale loosely based on the Verne classic, with Victorian gentleman Phileas Fogg embarking on his epic journey. By way of a foolish sub-plot, he is pursued by a bumbling detective who suspects him of robbing the Bank of England of ú55,000.
A/AD   275 min (2 cassettes) mTV   VIDrel: GUILD/SONOP
V
Boa: novel by Jules Verne.

ARRANGEMENT, THE **
15
Elia Kazan USA
1969
Kirk Douglas, Faye Dunaway, Richard Boone, Deborah Kerr, Hume Cronyn, John Randolph Jones, Michael Higgins, Carol Rossen, Anne Hegira, William Hansen, E.J. Andre, Charles Drake, Harold Gould, Michael Murphy, Philip Bourneuf, Dianne Hull
The director's own novel forms the basis for this film about an advertising executive who suddenly decides to chuck it all in and re-evaluate his life but has to put up with opposition from those closest to him. A good cast are largely wasted in this flawed and incoherent treatment of a potentially interesting tale.
DRA   120 min (ort 127 min) wScrn   VIDrel: TART/20TH   V
Boa: novel by Elia Kazan.

ARRIVAL, THE ***
18
David Schmoeller USA
1990
John Saxon, Joseph Culp, Robin Frates, Robert Sampson, Michael J. Pollard
Elements of SF are quite cleverly blended with the vampire legend in this movie, which opens with the arrival of an alien object on Earth. a strange force is released, which infects an old man and converts him into a younger and increasingly voracious vampire-like killer. Not far behind him is a determined cop, who is out to stop the carnage. More of a thriller than a shocker, this effective film eschews gore in favour of tension.
HOR   103 min (ort 107 min)   VIDrel: POLY/POLYREC L/A
V

ARRIVAL, THE ***
12
David N. Twohy USA
1996
Charlie Sheen, Teri Polo, Ron Silver, Lindsay Crouse, Richard Schiff, Leon Rippy, Ron Silver, Phyllis Applegate, Alan Coates, Buddy Joe Hooker, Javier Morgan, Tony T. Johnson, Catalian Botello, Georg Lillitisch, Angel De La Pena
A radio astronomer who has taken to monitoring the FM band, picks up a strange signal which he records, giving the tape to his boss. He soon finds unaccountably suspended from his job. allegedly due to government cutbacks. His further investigations take him to Mexico, where he joins forces with a female ecologist, having discovered that aliens have hidden there and are planning to poison the planet prior to a takeover. Quite cleverly plotted SF thriller.
FAN   120 min   VIDrel: EIV/SONOP   V/s

ARROGANT, THE *
(18)
Philippe Blot FRANCE
1987
Sylvia Kristel, Garry Graham, Leigh Wood, Joe Condon, Kimberly Baucum, Brian Storm, Michael Justin, J.D. Zdvorak, Dale Segal, Tony trudnick, David Baxter, Marvin Brody, Teresa Gilmore, Sean Faro, Bill Mullikin, Malika Souri
A man murders his father-in-law with an axe and then takes to the road on his motorcycle, eventually picking up a woman hitchhiker. The rest of the film is a talky exposition of his (and presumably the director's) dull and turgid views. Written and produced by Blot.
DRA   86 min   SATrel: SKY MOVIES

ARSENAL ***
PG
Alexander Dovzhenko USSR
1928
Semyon Svashenko, Nikola Nademsky, Ambrose Buchman
Classic silent film about strikes on the home front during WW1 in the Ukraine and the subsequent struggle to overcome its backwardness. A lyrical blend of many genres that has had a strong influence on documentary film-makers.
Aka: JANUARY UPRISING IN KIEV IN 1918, THE
DRA   90 min (ort 99 min) B/W silent
VIDrel: HEND/BMGREC L/A   V

ARSENIC AND OLD LACE ****
PG
Frank Capra USA
1941 (released 1944)
Cary Grant, Priscilla Lane, Raymond Massey, Peter Lorre, Jack Carson, Jean Adair, Josephine Hull, James Gleason, Grant Mitchell, John Alexander, Edward Everett Horton, Edward McNamara, Garry Owen, John Ridgely, Vaughn Glaser
Extremely funny film version of the famous stage comedy about two old ladies who murder their guests (for the very best of motives). Lorre and Massey are especially good as a couple of unsuspecting murderers who are hiding out in the house.
COM   113 min (ort 118 min) B/W   VIDrel: MGM/WHV   V
Boa: play by Joseph O. Kesselring.

ART OF DYING, THE **
18
Wings Hauser USA
1991 (released 1992)
Wings Hauser, Kathleen Kinmont, Gary Werntz, Mitch Hara, Michael J. Pollard, Sarah Douglas, Sydney Lassick, Angela Rae, T.C. Warner, Jean Levine, Tony Longo, Henry Brown, Pam Dixon, Mary Bon Davis, Michael Easton, Bill Fremer
A psychopath assumes the role of a film director so that he can lure his victims into being filmed before he murders them. An unpleasant little number, gruesome and exploitative, and though well made and acted is really too downbeat to be attractive. Screenplay is by Joseph Merhi, who co-produced with Richard Pepin.
THR   91 min (ort 93 min)   VIDrel: MIA/DISC/IMPENT
V/sh

ART OF LOVE, THE ** **   18
Walerian Borowczyk FRANCE/ITALY   1984
Marina Pierro, Massimo Girotti, Laura Betti, Philippe Lemaire,
Michele Placido, Milena Vukotic, Philippe Tuccini, Antonio Orlando,
Mireille Pame, Simonetta, Pierre Franscesco Aiello
Set in Imperial Rome, this unusual soft-focus effort represents
the director's attempt to provide a very personal interpretation
of the love poems of Ovid. Really little more than a set of erotic
vignettes, that for all its good production values, does little to
maintain one's interest.
Aka: ARS ARMANDI; L'ART D'AIMER
A   93 min wScrn dubbed   VIDrel: JEZ/RTM   V

ARTHUR *   15**
Steve Gordon USA   1981
Dudley Moore, Liza Minnelli, John Gielgud, Geraldine Fitzgerald,
Stephen Elliott, Ted Ross, Barney Martin, Jill Eikenberry, Anne De
Salvo, Thomas Barbour, Marjorie Barnies, Dillon Evans, Maurice
Copeland, Justine Johnston
Some good moments in an uneven comedy about a young
spoiled millionaire who must choose between money and a
planned marriage or true love and possible penury. Often quite
witty with Moore rather endearing as an amiable drunken brat,
but Gielgud steals the show as his caustic butler. Written by
Gordon and followed by a sequel. AA: S.Actor (Gielgud), Song
("Best Thing You Can Do" – Burt Bacharach/Carole Bayer
Sager/Christopher Cross/Peter Allen).
COM   93 min (ort 97 min)   VIDrel: WHV L/A   V/h

ARTHUR 2 *   PG**
Bud Yorkin USA   1988
Dudley Moore, Liza Minelli, John Gielgud, Cynthia Sikes, Stephen
Elliot, Paul Benedict, Geraldine Fitzgerald, Barney Martin, Kathy
Bates, Jack Gilford, Ted Ross, Daniel Greene, Brogan Lane
A disappointing sequel in which Arthur loses all his money and
his wife decides that they ought to adopt a baby. All the charm
and freshness of the original has long since evaporated. Gielgud
pops up briefly – as a ghost.
Aka: ARTHUR 2: ON THE ROCKS
COM   108 min (ort 110 min)   VIDrel: MGM/WHV   V/sur

ARTICLE 99 *   15**
Howard Deutch USA   1992
Ray Liotta, Kiefer Sutherland, Forest Whitaker, John C. McGinley, Lea
Thompson, John Mahoney, Kathy Baker, Eli Wallach, Keith David,
Troy Evans, Noble Willingham, Julie Bovasso, Lynne Thigpen, Jeffrey
Tambor, Rutanya Alda
A Veterans Administration hospital provides the setting for this
witless satire on petty corruption and mismanagement, with
doctors Liotta and Sutherland leading the fight against the
mindless pen-pushers and dealing with various crises, such as
the disgruntled patients staging a take-over. At times it feels
like an extended episode of MASH but without the benefit of an
equally sharp script.
COM   96 min (ort 98 min)   VIDrel: COLUM/SONOP L/A
V/h

AS GOOD AS DEAD *   PG**
Larry Cohen USA   1995
Traci Lords, Crystal Bernard, Judge Reinhold, Carlos Carrasco, George
Dickerson
A woman lets a friend borrow her name so that she can get
medical treatment at a hospital but this act of kindness has fatal
consequences when she is murdered. Convinced that this she
was meant to be the real victim, she is forced to try to find the
murderer before this mistake can be rectified. A moderately
entertaining thriller, strongly directed and quite brisk.
THR   84 min (ort 88 min)   VIDrel: CIC/SONOP   V

AS TEARS GO BY *   18**
Wong Kar Wei HONG KONG   1988
Andy Lau, Maggie Cheung, Jackie Cheung
A saga of gangster life, set in Kowloon, that has two friends
trapped in a lifestyle they want to escape. A kind of Chinese
variant of MEAN STREETS, this was the director's first feature,
and demonstrates his abilities, chiefly by way of a number of
exciting action sequences if not by its characterisations, as in
this latter respect it is little better than countless other Hong
Kong action movies.
A/AD   93 min wScrn   VIDrel: MADE/RTM   V

AS YOU LIKE IT *   U**
Basil Coleman UK   1979
Helen Mirren, Brian Stirner, Richard Pasco, James Bolam, Clive
Francis, Tony Church, Angharad Rees, Richard Easton, John Quentin,
Maynard Williams, Victoria Plucknett, Marilyn Le Conte, Tom
McDonnell, David Lloyd Meredith
A handsome BBC TV production slightly hampered by the
constraints that apply to filmed TV plays but still worth seeing.
This comic tale of love and intrigue that takes place in a myth-
ical forest was made on location at Glamis Castle.
DRA   152 min mTV   VIDrel: BBC   V/h
Boa: play by William Shakespeare.

AS YOU LIKE IT *   U**
Christine Edzard UK   1992
Cyril Cusack, James Fox, Don Henderson, Miriam Margoyles, Emma
Croft, Griff Rhys Jones, Andrew Tiernan, Celia Bannerman, Tony
Armatrading, Ewen Bremner, Valerie Gogan, Roger Hammond,
Murray Melvin, Jonathan Cecil, Cate Fowler
Filmed at London's Docklands, this is a modern version of the
famous play of an exiled king and his changing fortunes. The
text of the comic play is retained, the concession to modernity
being the use of twentieth century costumes. But it is not this
that spoils the work, rather just (with a few exceptions) a distinct
lack of passion on the part of the cast, most of whom seem to go
through their lines as if they are not comfortable with the setting.
DRA   112 min (ort 117 min)   VIDrel: WDV/TECH L/A
V/sh
Boa: play by William Shakespeare.

ASCENT, THE **   PG**
Larisa Shepitko USSR   1976
Boris Plotkinov, Ludmilla Poliakova, Vladimir Gostjuhin, Sergei
Jakovlev, Anatoly Solonitsin
During WW2, a small group of Russian partisans escapes from
the Germans, and makes its way across a frozen landscape,
where they suffer great hardship. Not a trace of the malign influence of Socialist Realism is to be found in this harsh and
uncompromising film, and the death of the director in a car crash
in 1979 (this was only her fourth feature film) was a tragedy for
film-making in general and for Soviet cinema in particular.
Aka: VOSKHOZHDENIE
DRA   104 min B/W   VIDrel: CONNO/RTM   V

ASHES AND DIAMONDS **   (18)**
Andrzej Wajda POLAND   1958
Zbigniew Cybulski, Ewa Krzyzanowska, Adam Pawlikowski, Bogumil
Kobiela, Waclaw Zastrzezynski, Jan Ciecierski, Stanislaw Milski,
Arthur Mlodnicki, Halina Kwiatkoska, Ignacy Machowski, Zbigniew
Skowronski, Barbara Krafft
The last part of Wajda's trilogy about the Polish resistance
(following A GENERATION and KANAL) traces the conflict
between the Communists and the Nationalists at the end of
WW2. Cybulski gives a remarkable performance as a young
Resistance fighter in this harsh and articulate study.
Aka: POPIOL Y DIAMENT
DRA   100 min (ort 104 min) B/W   VIDrel: L/A   V

ASPEN EXTREME *   15**
Patrick Hasburgh USA   1992
Paul Gross, Teri Paolo, Peter Berg, Finola Hughes, William Russ,
Trevor Eve, Martin Kemp, Stewart Finley-McLennan, Tony Griffin,
William McNamara, Andy Mill, Nicolette Scorsese, Julia Royer, Gary
Eimiller, Bill Ferrell
Two friends from Detroit leave that city behind them and head
for a life of pleasure on the ski slopes of Aspen, meeting with
varying degrees of success. A poor excuse of a comedy written
and directed by Hasburgh, a former Aspen ski instructor.
COM   128 min   VIDrel: HOLPIC/TECH L/A   V

ASPHALT JUNGLE, THE **   PG**
John Huston USA   1950
Sterling Hayden, Sam Jaffe, Louis Calhern, Jean Hagen, Marilyn
Monroe, James Whitmore, Marc Lawrence, John McIntire, Barry Kelley
When an ageing crook is released from jail he sets about plan-
ning one last big heist, and puts together a gang for this purpose.
Unfortunately, his carefully made plans are destroyed by some
unforeseen circumstances. A very powerful crime thriller, stark
and realistic, and one of the first to show the events from the
criminal's point of view, and treat its characters with a measure
of sympathy.
THR   107 min (ort 112 min) B/W   VIDrel: MGM/WHV   V
Boa: novel Little Caesar by W.R. Burnett.

ASPHYX, THE *** 15
Peter Newbrook UK   1972
Robert Stevens, Robert Powell, Jane Lapotaire, Fiona Walker, John Lawrence, Alex Scott, Ralph Arliss, Terry Sculler, David Gray, Tony Caunter, Paul Bacon
A 19th century scientist discovers the key to immortality when he finds a way to isolate the spirit of death that hovers over each individual creature at the moment of its death. However, his experiments go horribly wrong and although he gains immortality, he is left to brood for all time on his folly and the tragic death of his daughter. A highly unusual and worthy fantasy tale that is slightly hampered by a wordy script.
Aka: HORROR OF DEATH, THE; SPIRIT OF THE DEAD, THE
HOR   95 min (ort 99 min)   VIDrel: ARTPRO/RTM L/A   V

ASSASSIN, THE ** 18
John Badham USA   1993
Bridget Fonda, Anne Bancroft, Harvey Keitel, Gabriel Byrne, Dermot Mulroney, Miguel Ferrer, Olivia D'Abo, Richard Romanus, Lorraine Toussaint, Geoffrey Lewis, Mic Rodgers, Michael Rapaport, Ray Oriel, Spike McClure, John Capodice
A female drug addict who took part in an armed robbery which ended in murder, gets a reprieve from the death cell when she becomes a secret government assassin. She is relocated and undertakes a number of successful assignments before deciding that it is time for her to quit. A virtual scene-by-scene remake of the superior French hit NIKITA but without much of the earlier film's emotional impact and solid acting. See also BLACK CAT.
Aka: POINT OF NO RETURN
DRA   104 min (ort 109 min)   VIDrel: WHV   V/sur

ASSASSIN OF THE TSAR ** 15
Karen Shakhnazarov RUSSIA/UK   1991
Malcolm McDowell, Oleg Yankovsky, Armen Dzhigarkhanyan, Iurii Sherstnev, Angela Ptashuk, Viktor Seferov, Olga Antonova, Daria Maiorova, Evgeniia Kriukova, Alena Teremezova, Olga Borisova, Aleksei Logunov, Viacheslav Vdovin
In a Moscow psychiatric hospital is a man who claims to be the assassin of the last Tsar of Russia. A doctor decides to assume the identity of the Tsar as an experiment, but fails to foresee the deadly events this will lead to. A high quality production, sustained by strong central performances, but a bit let down by weak scripting.
Aka: TSAREUBIITSA
DRA   103 min (English version)   VIDrel: SPECT/HIFLI V/h

ASSASSIN OF YOUTH (PG)
Elmer Clifton USA   1935
Luanna Walters, Arthur Gardner, Dorothy Short, Earl Dwire, Fern Emmett
A film that purports to describe the hazards and horrors of smoking marihuana, with a bunch of those crazy kids smoking just one too many joints at a wild dope party. Meanwhile, an intrepid reporter investigates. A laughable piece of nonsense that's hugely enjoyable after a few beers, especially for the short film within it – "The Marihuana Menace". See also REEFER MADNESS and MARIHUANA: THE DEVIL'S WEED.
Aka: MARIJUANA
DRA   71 min B/W   VIDrel: L/A   V

ASSASSINATION * 15
Peter Hunt USA   1986
Charles Bronson, Jill Ireland, Stephen Elliott, Jan Gan Boyd, Randy Brooks, Michael Ansara, William Prince, James Staley, Kathryn Leigh Scott, Erik Stern, James Acheson, Robert Axelrod, Lucille Bliss, Robert Dowdell, Stephen Elliott
An agent is hired to protect the President's wife from a vicious assassin whose orders are coming from inside the White House. A wildly ridiculous and over-the-top exercise in explosions and bad directing.
Aka: PRESIDENT'S WIFE, THE
THR   85 min (ort 88 min)   VIDrel: MGM/WHV   V

ASSASSINS * 18
Richard Donner USA   1995
Sylvester Stallone, Antonio Banderas, Julianne Moore, Anatoly Davydov, Muse Watson, Stephen Kahan, Kelly Rowan, Reed Diamond, Kai Wuff, Kerry Skalsky, James Douglas Haskins, David Shark (Fralick), Stephen Liska, John Harms
A professional hit-man finds it is not easy to quit his chosen profession, and learns that he has become the quarry of another. Dimwitted action-thriller that celebrates mindless scripting and occasional over-acting, not least from Banderas who is given the job of eliminating Stallone, his former boss.
A/AD   127 min (ort 133 min) cC   VIDrel: WHV   V/sur

ASSAULT, THE **** PG
Fons Rademakers HOLLAND   1986
Derek De Lint, Marc Van Uchelen, Monique Van De Ven, John Kraaykamp, Huub Van De Lubbe, Elly Weller, Ina Van De Molen, Frans Vorstman, Edda Barends, Wim De Haas, Hiske Van Der Linden, Piet De Wijn, Akkemay, Kees Coolen
A young boy witnesses the brutal murder of his family in the final days of WW2. Haunted by this memory, he finally comes to terms with the truth as he grows to manhood. A long and often harrowing story, suspenseful, intense and without a cliche in sight. The superb performances are matched by Gerard Soeteman's fine screenplay. AA: Foreign.
Aka: DER AANSLAG
DRA   120 min (ort 146 min)   VIDrel: MGM L/A   V
Boa: novel by Harry Mulisch.

ASSAULT AT WEST POINT: THE COURT-MARTIAL OF JOHNSON WHITTAKER *** (12)
Harry Moses USA   1993
Samuel L. Jackson, Sam Waterston, Seth Gilliam, John Glover, Mason Adams, Eddie Bracken, Brad Greenquis, Peter Maloney, Scott Paetty, Ken Garito, Anthony Rapp, John Wehir, Al Freeman Jr, Robert Clohessy, Greg Germann, Ralph Williams
A re-creation of an famous court martial in which the first black cadet at West Point was court-martialled and expelled after he was found tied to his bed after a racist attack. A powerful exposure of the corrupting effect of bigotry and prejudice. Some slowly paced and superficial in its treatment, but still certainly worth seeing.
DRA   90 min (ort 98 min) mCab   SATrel: MOVIE CHANNEL
Boa: book The Court-Martial of Johnson Whittaker by John F. Marszalek.

ASSAULT OF FINAL RIVAL ** PG
HONG KONG   1989
Ko Tin Yan
Like Samson, a martial arts fighter owes his incredible strength to his long, flowing locks. A jealous rival sends a girl to seduce him and cut his hair when he is asleep, thus rendering him powerless. A slightly more original theme enlivens this otherwise fairly conventional offering.
MAR   87 min   VIDrel: IMPENT   V

ASSAULT ON PRECINCT 13 **** 18
John Carpenter USA   1976
Austin Stoker, Darwin Joston, Laurie Zimmer, Martin West, Tony Burton, Nancy Loomis, Kim Richards, Henry Brandon, Peter Bruni, John J. Fox, Gilbert De La Pena, Marc Ross, Charles Cyphers, Alan Koss, Frank Doubleday, Peter Frankland
A nasty gang of psychopaths decide to embark on a series of murders. When one of their members is killed they follow the killer to a nearly totally deserted police station and stage a siege, intending to kill everyone in it. A well made and extremely tense film with minimal dialogue. Written by Carpenter, who also did the score.
THR   91 min   VIDrel: 4-FRONT/POLYREC   V

ASTERIX AND CLEOPATRA *** U
Rene Goscinny/Albert Uderzo BELGIUM/FRANCE   1968
An enjoyable cartoon animation following the adventures of the celebrated comic book character. In this adventure Asterix, Obelix and Getafix go to Egypt with their dog Dogmatic. Here their intention is to help an architect friend build a palace for Queen Cleopatra.
ANIM   69 min   VIDrel: CREMED/LABY   V

ASTERIX AND THE BIG FIGHT ** U
Philippe Grimond FRANCE/WEST GERMANY   1989
Voices of: Bill Oddie, Ron Moody, Brian Blessed, Sheila Hancock, Bernard Bresslaw, Michael Elphick, Andrew Sachs, Tim Brooke-Taylor, Douglas Blackwell
A pleasant cartoon version of two Goscinny-Uderzo tales, one of several adaptations of the popular series of comic book stories, following the affairs of the ancient Gauls and Romans. This story revolves around a custom that requires a would-be chief to defeat another in order to become leader. Getafix has

promised Vitalstatistix a magic potion to help him do this, but the former has an accident and forgets how to make it.
ANIM   76 min   VIDrel: CREMED/LABY   V/h

ASTERIX AND THE TWELVE TASKS ***   U
Rene Goscinny/Albert Uderzo/Matt McCarthy (English version) FRANCE   1975
Voices of: Paul Bacon, George Baker, Sean Barett, Christina Greatex, Ysanne Churchman, Alexander John, Michael Kilgarriff, Barbara Mitchell, Geoffrey Russell, Genni Nevinson, Paddy Turner
Asterix comes to the rescue of his village when it is set twelve impossibly difficult tasks by the Emperor Julius Caesar who wants to take it over.
Aka: LES DOUZE TRAVAUX D'ASTERIX; TWELVE TASKS OF ASTERIX, THE
ANIM   78 min (ort 84 min)   VIDrel: CREMED/LABY   V

ASTERIX CONQUERS AMERICA **   U
Gerhard Hahn GERMANY   1994
Voices of: John Rye, Craig Charles, Howard Lewis, Henry McGee, Geoffrey Bayldon, Christopher Biggins
In 50 B.C. Asterix and Obelix go off in search of their friend, the Druid Getafix, who has been catapulted off beyond the world's end by the Romans. Their search takes them to the New World, where they encounter savage Native Americans and have to deal with a powerful witch doctor. The seventh film inspired by the famous comic-strip by Goscinny and Uderzo, this is slightly spoilt by its crude racial stereotypes and weak plotting, but is adequate enough in its way.
Aka: ASTERIX IN AMERICA; ASTERIX IN AMERIKA
ANIM   82 min (ort 90 min)   VIDrel: 20TH/TECH   V/sur

ASTERIX IN BRITAIN ***   U
Pino Van Lamsweerde DENMARK/FRANCE   1986
Voices of: Robert Barr, Peter Hudson, Patrick Floersheim, Steve Gadler, Judy Rosen-Martinez, Ken Starcevic, Gordon Heath, Jack Beaber, Bill Kearns, Sean O'Neil, Graham Bushnell, Ed Marcus, Herbert Baskind, Jimmy Shuman, Mike Marshall
Just across the English channel an invasion force of Romans are gathering to invade Britain. Only Asterix, Obelix and friends can save the day in this adventure of Britons in 50 BC. A lively and enjoyable animation, based on this popular comic-book character.
Aka: ADVENTUROUS ANTICS OF ASTERIX IN BRITAIN, THE: ONE LITTLE GUY AGAINST ONE BIG EMPIRE; ASTERIX CHEZ LES BRETONS; ASTERIX IN BRITAIN: THE MOVIE
ANIM   75 min (ort 89 min)   VIDrel: CREMED/LABY   V

ASTERIX THE GAUL **   U
Rene Goscinny/Albert Uderzo BELGIUM/FRANCE   1967
Voices of: Paul Bacon, George Baker, Sean Barett, Christina Greatex, Ysanne Churchman, Alexander John, Michael Kilgarriff
Cartoon featuring this popular comic strip character – a little warrior with super-human strength, who in the company of his trusty companions, gets into a variety of scrapes in the course of his battles with the might of Rome.
Aka: ASTERIX LE GAULOIS
ANIM   65 min (ort 90 min)   VIDrel: CREMED/LABY   V

ASTERIX VERSUS CAESAR **   U
Paul Brizi/Gaetan Brizzi BELGIUM/FRANCE   1985
Voices of: Jack Barber, Bill Kearns, Allen Wenger, Gordon Heath, Robert Barr, Patrick Floersheim, Bill Dunn, Steve Gadler, Stuart Seide, Mike Marshall, Bill Doherty, Paul Barrett, Norma Stockle, Peter Semler
Asterix and his overweight buddy Obelisk continue their struggle against the Romans, this time coming to the aid of two lovers who have been captured, one of whom is a prince they must save from being eaten by Caesar's lions. Another pleasant adventure inspired by the popular comic-book characters.
Aka: ASTERIX ET LA SURPRISE DE CESAR
ANIM   74 min (ort 85 min)   VIDrel: CREMED/LABY   V/h

ASYLUM ***   15
Roy Ward Baker UK   1972
Robert Powell, Herbert Lom, Patrick Magee, Geoffrey Bayldon, Barbara Parkins, Richard Todd, Sylvia Sims, Peter Cushing, Barry Morse, Ann Firbank, John Franklyn-Robbins, Britt Ekland, Charlotte Rampling, James Villiers
Quartet of stories recounted by the inmates of an asylum, interwoven with the mystery that surrounds the former director, now an inmate himself. A few chilling moments with segments of Prokofiev's "Pictures at an Exhibition" used to good effect. The shorter running time refers to the film as marketed under

its alternative title. Scripted by Robert Bloch. Stories are: "Frozen Fear", "The Weird Tailor", "Lucy Comes To Stay" and "Mannikins Of Horror".
Aka: HOUSE OF CRAZIES
HOR   88 min (ort 95 min)   VIDrel: VIPCO/SGSVID   V

AT MOTHER'S REQUEST: PARTS 1 AND 2 **   15
Michael Tuchner USA   1987
Stefanie Powers, E.G. Marshall, Doug McKeon, Frances Sternhagen, Corey Parker, Penny Fuller, John Wood, E.G. Marshall, Jenna Von Oy, Dan Luria, Louis Borgenicht, Nancy Borgenicht, Franc Cagney, Walt Filed, John Horton, Gene Pack
A psychological drama based on a true story, with Stefanie Powers as a divorced mother who is constantly pleading for cash from her wealthy father. When he is shot to death there are several suspects. A sluggish and uneven drama of little impact. Based on the notorious Frances Schreuder murder case and originally shown in two parts.
DRA   198 min (2 cassettes) mTV   VIDrel: 20TH/TECH   V
Boa: book by Jonathan Coleman.

AT PLAY IN THE FIELDS OF THE LORD ****   15
Hector Babenco USA   1991
Tom Berenger, John Lithgow, Daryl Hannah, Aidan Quinn, Tom Waits, Kathy Bates, Stenio Garcia, Nelson Xavier, Jose Dumont, Niilo Kivirinta, S. Yriwana Karaja, Jose Renato Lana, Riu Polanah, Carlos Xavante, Ione Machado
The first of several works that focus on the dangers inherent in the endless destruction of the Amazonian rainforest, this film is one of the best. Shot in the Amazon under the most difficult conditions, it explores the unhappy consequences of contact between the forest-dwelling Niaruna Indians and white civilisation, most especially two soldiers-of-fortune and some earnest missionaries. A film of much irony, it does not make easy viewing.
DRA   178 min (ort 190 min)   VIDrel: EIV/SONOP   V/sur
Boa: novel by Peter Matthiessen.

AT THE EARTH'S CORE **   PG
Kevin Connor UK   1976
Doug McClure, Peter Cushing, Caroline Munro, Cy Grant, Godfrey James, Keith Barron, Sean Lynch, Anthony Verner, Michael Crane, Helen Gill, Robert Gillespie, Bobby Parr, Andee Cromarty
A Victorian scientist burrows down to the Earth's core and discovers a land inhabited by subhumans and prehistoric monsters of assorted varieties. A follow-up to THE LAND THAT TIME FORGOT. Occasionally entertaining but generally just rather dull.
FAN   86 min (ort 90 min)   VIDrel: WHV L/A   V
Boa: novel by Edgar Rice Burroughs.

AT THE PORNIES ***   18
Paul Thomas USA
Aja, Shanna McCullough, Nikki Knights, Jamie Gillis, Tom Byron, Ron Jeremy, Don Hart
Built around a spoof of the popular film-review show hosted by Gene Siskel and Roger Ebert, this has Horner and Jeremy playing hosts Sissy and Egbert respectively. As with the originals, they spend most of their time reviewing films (in this case adult ones) and can rarely agree, so much so that Sissy hatches a cunning plan to discredit his co-presenter. An unusual basis for an adult movie, and often quite amusing.
Aka: SISSY AND EGBERT
A   85 min   VIDrel: IMPENT   V

ATLANTIC CITY ****   15
Louis Malle CANADA/FRANCE   1980
Burt Lancaster, Susan Sarandon, Kate Reid, Al Waxman, Robert Goulet, Wallace Shawn, Michel Piccoli, Hollis McLaren, Robert Joy, Moses Znaimer, Sean Sullivan, Angus MacInnes, Harvey Atkin, Eleanor Beecroft, Norma Dell'Agnese
Strange offbeat character study with Lancaster as an ageing small-time crook whose life becomes intertwined with that of some other losers. The film focuses on a series of people who pursue their dreams in a town whose best times have long since gone. Excellent screenplay is by John Guare with Lancaster giving one of his strongest performances.
Aka: ATLANTIC CITY, USA
THR   105 min   VIDrel: ARROW/RTM   V

ATOMIC SUBMARINE, THE **   PG
Spencer Bennet USA   1959
Arthur Franz, Dick Foran, Brett Halsey, Tom Conway, Bob Steele, Joi

Lansing, Victor Varconi, Paul Dubov, Selmer Jackson, Jean Morehead, Jack Mulhall, Sid Melton, Richard Tyler, Ken Becker
Title vessel goes to the North Pole to investigate the disappearance of a number of ships and their passengers and finds a submerged flying saucer. Attempts to recover to this craft disturb its occupant, a cyclops-like space monster, which then proceeds to go on the usual rampage, in this typical 1950s B-movie.
FAN 68 min (ort 80 min) B/W VIDrel: ENCORE/SPEAR V

ATOR, THE FIGHTING EAGLE *
David Hills (Aristide Massaccesi) USA
PG
1982
Miles O'Keeffe, Sabrina Siani, Ritza Brown, Edmund Purdom, Brooke Hart, Laura Gemser, Warren Hillman, Nat Williams, Olivia Goods, Chandra Vazzoler, Jean Lopez
A sword-and-sorcery fantasy in which our young hero battles the Forces of Evil. A ridiculous Italian effort, inspired by those "Conan" films but lacking pace or acting ability from muscle-bound hero O'Keeffe. Followed by ATOR THE INVINCIBLE 2.
Aka: ATOR; ATOR L'INVINCIBLE; ATOR THE INVINCIBLE; HOBGOBLIN, THE
FAN 98 min (ort 100 min) VIDrel: STABL L/A V

ATOR THE INVINCIBLE 2 **
David Hills (Aristide Massaccesi) ITALY/USA
PG
1983
Miles O'Keeffe, Lisa Foster (Lisa Raines), David Cain Houghton, Charles Borromel, Chen Wong, Robert Black, Donald Hodson, Stephan Stoffer, Herschel Curtis, Ned Steinberg, Sandra Carle, Nancy Hall, Linette Ray, Robert Karshin
More sword-and-sorcery adventures on the theme of Good versus Evil. Sequel to ATOR, THE FIGHTING EAGLE with O'Keeffe a prehistoric warrior out to save Earth from the "Geometric Nucleus" – a primitive atomic bomb. An utterly silly effort of far-fetched and unbelievable encounters, but enjoyable in an undemanding way. See also IRON WARRIOR and QUEST FOR THE MIGHTY SWORD.
Aka: ATOR L'INVINCIBLE 2; ATOR, THE RETURN; BLADEMASTER, THE; CAVE DWELLERS
FAN 88 min Cut (27 sec – ort 92 min) VIDrel: L/A V

ATTACK! ***
Robert Aldrich USA
PG
1956
Jack Palance, Eddie Albert, Lee Marvin, Robert Strauss, Richard Jaeckel, Buddy Ebsen, William Smithers, Jon Shepodd, Jimmy Goodwin, Steven Geray, Peter Van Eyck, Strother Martin
Realistic WW2 tale set in Belgium in 1994, where a cowardly captain orders one of his platoons to undertake what proves to be a suicide mission and then abandons them to their fate as they are eventually overwhelmed by superior numbers. A fine comment on the madness of war and the military's gift for political expediency.
WAR 103 min (ort 107 min) B/W VIDrel: MGM/WHV V/h
Boa: play The Fragile Fox by Norman Brooks.

ATTACK FORCE Z ***
Tim Burstall AUSTRALIA/TAIWAN
15
1981
John Phillip Law, Mel Gibson, Sam Neill, Chris Haywood, John Waters, Koo Chuan Hsiung, Sylvia Chang, O Ti, Ku Chun, Lung Shuan, Yi Yuan, Wei Su, Wany Yu, Val Champion, Hsa Li Wen
Commando group sets out to rescue survivors of a plane that's crashed on a Japanese-held island in the Pacific during WW2. A taut and well made little war drama (despite the fact that all the Japanese characters are played by Chinese actors).
Aka: Z MEN
WAR 92 min VIDrel: FIRC/RTM V

ATTACK OF THE 50 FOOT WOMAN **
Christopher Guest USA
12
1994
Daryl Hananh, Daniel Baldwin, William Windom, Frances Fisher, Christi Conaway, Paul Benedict, Lewis Arquette, Xander Berkeley, Hamilton Camp, Richard Edson, Victoria Haas, O'Neal Compton, Berta Waagfjord, Kye Benson, Linda Bisesti
After a close encounter with a UFO, a woman begins to grow to enormous proportions, while her self-esteem increases to a similar degree. This enables her to deal with both her cheating husband and her sneering father. A feminist remake of the 1958 classic SF B-movie that offers reasonable special effects but little else.
FAN 85 min (ort 89 min) VIDrel: EIV/SONOP V/sur

ATTACK OF THE KILLER TOMATOES *
John De Bello USA
PG
1980
David Miller, Sharon Taylor, George Wilson, John DeBello, Jack Riley,
Rock Peace, Eric Christmas, Al Sklar, Ernie Meyers, Jerry Anderson, Ron Shapiro, The San Diego Chicken
The title is the best thing in this low-budget spoof on SF films, dealing with vicious vegetables. Some say it should have been called The Ketchup Killers. Others say it should never have been made. See also SOUR GRAPES, another masterpiece from this director. RETURN OF THE KILLER TOMATOES! followed.
COM 87 min (ort 100 min – director's cut)
VIDrel: FIRST/SONOP V

ATTICA ****
Marvin J. Chomsky USA
PG
1980
Henry Darrow, Charles Durning, Joe Fabiani, Morgan Freeman, George Grizzard, David Harris, Roger E. Mosley, Arlen Dean Snyder, Glynn Turman, Anthony Zerbe, Andrew Duncan, Ron Foster, William Flatley, Noble Lee Lester, Paul Lieber
Gripping and moving account of the Attica prison riot of 1971 which left thirty-nine dead after Governor Nelson Rockefeller sent in the New York state troopers. Despite some network-instigated changes to the script to tone down the criticism of Rockefeller, this powerful film remains a searing and bitter condemnation of needless slaughter. Adapted from Wicker's bestseller by James Henerson. See also AGAINST THE WALL.
DRA 104 min mTV VIDrel: L/A V
Boa: book A Time To Die by Tom Wicker.

AU-PAIR GIRLS *
Val Guest UK
18
1972
Richard O'Sullivan, Gabrielle Drake, Astrid Frank, Me Me Lay, Ferdy Mayne, Nancie Wait, John Le Mesurier, Geoffrey Bayldon, Rosalie Crutchley, Julian Barnes, Lyn Yeldham, Ferdy Mayne, Steve Patterson, John Standing, Johnny Briggs
The comic mishaps of four au-pairs form the basis for this totally forgettable and coy little sex romp, which has not stood the test of time well.
A 80 min (ort 86 min) VIDrel: JEZ/RTM V

AU REVOIR, LES ENFANTS ****
Louis Malle FRANCE
PG
1987
Gaspard Manesse, Raphael Fejto, Philippe Morier-Genoud, Francine Racette, Stanislas Carre De Malberg, Francois Berleand, Francois Negret, Peter Fitz, Pascal Rivet, Benoit Henriet, Richard Leboeuf, Xavier Legrand
During the occupation of France in WW2, three Jewish boys hide out in a Catholic boarding school, where one of them eventually forms a close friendship with another student. Based on an incident in the director's youth, this moving but entirely unsentimental film recalls the course of their relationship and its tragic and brutal ending. A sad but marvellously sensitive tale of childhood and the end of innocence.
Aka: GOODBYE, CHILDREN
DRA 100 min (ort 103 min) VIDrel: CURZON/20TH V/sur

AUGUST ***
Anthony Hopkins UK
PG
1995
Anthony Hopkins, Leslie Phillips, Kate Burton, Gawn Grainger, Rhian Morgan, Menna Trussler, Rhoda Lewis, Hugh Lloyd, Huw Garmon, Rhys Ifans, Susan Flynn, Buddug Morgan, Victoria Pugh, Ioan Meredith, Jams Thomas, Simon Treves
A single weekend of tragi-comic incidents is all it takes to destroy the well ordered calm and efficiency of a Victorian household. This was the directing debut for Hopkins, and is a warmhearted and immensely enjoyable affair, but instead of the impoverished Russian peasants of the original, we are here shown the plight of the workers caught in an explosion at a local quarry.
DRA 90 min (ort 99 min) VIDrel: FILM4/RTM V
Boa: play Uncle Vanya by Anton Chekhov.

AUNT JULIA AND THE SCRIPTWRITER **
Jon Amiel USA
15
1990
Peter Falk, Barbara Hershey, Keanu Reeves, Bill McCutcheon, Dan Hedaya, Patricia Clarkson, Peter Gallagher, Buck Henry, Hope Lange, John Laroquette, Richard Portnow, Elizabeth McGovern, Robert Sedgwick, Henry Gibson, Paul Austin
In 1951 New Orleans, a young man falls in love with his seductive aunt and receives advice on how to woo her from a wildly eccentric radio scriptwriter (Falk). The latter draws on his protege's situation for his material and the former's situation is soon recycled as a radio serial. An oddball movie with flashes of humour that often falls flat (mainly due to a total lack of

screen presence from Reeves). Fortunately, Falk gives an outstanding performance.
Aka: TUNE IN TOMORROW
COM   103 min   VIDrel: EIV/SONOP   V
Boa: novel Tune In Tomorrow by Mario Vargas Llosa.

AURORA **
PG
Maurizio Ponzi ITALY/USA   1984
Sophia Loren, Daniel J. Travanti, Edoardo Ponti, Angela Goodwin, Ricky Tognazzi, Marisa Merlini, Anna Strasberg, Franco Fabrizi, Philippe Noiret, Antonio Alloca, Gianfranco Amoroso, David Caneron, Vittorio Duce
A mother and son track down her former lovers to persuade each of them that they are the boy's father so that they will pay for an operation to restore his sight. Ponti as the boy in question is good in his acting debut, but this tedious and far-fetched tale boasted four writers and even more production companies. A family affair for Loren; Edoardo Ponti is her son and the film was produced by Alex Ponti – her stepson.
Aka: AURORA BY NIGHT; QUALCOSA DI BLONDA; SOMETHING BLONDE
DRA   91 min (ort 100 min) mTV   VIDrel: 20TH/TECH L/A   V

AUTHOR! AUTHOR! ***
PG
Arthur Hiller USA   1982
Al Pacino, Dyan Cannon, Tuesday Weld, Eric Gurry, Bob Dishy, Alan King, Bob Elliott, Ray Goulding, Andre Gregory, Elva Leif, B.J. Barie, Ari Meyers, Ken Sylk, Benjamin H. Carlin, James Tolkan, Tony Munafo, Reuben Singer
Comedy about the marital troubles of a playwright whose wife leaves him with the kids just as his new play is about to open on Broadway. A slight but enjoyable and undemanding tale with some nice performances. The script is by playwright Israel Horovitz.
COM   105 min (ort 110 min)   VIDrel: 20TH/TECH L/A   V

AUTOBUS ***
15
Eric Rochat FRANCE   1991
Yvan Attal, Kristin Scott-Thomas, Marc Berman, Charlotte Gainsbourg, Renan Mazeas, David Burstzein, Dan Herzberg, Luc Lavandier, Francine Olivier, Michele Foucher, Aline Still, Daniel Milgram, Guy Perrot, Francois Lalande
An aimless and argumentative young man hijacks a bus-load of schoolkids on a wild impulse, and though he forces the driver to continue at gunpoint, his intentions are never sinister. The children soon discover this, and he becomes a popular figure with the kids, if not with their teacher. Attal gives a convincing performance as a vulnerable young man who has yet to find his way in life, and the film is a strong blend of tension, drama and irony.
Aka: AUX YEUX DU MONDE
DRA   93 min (ort 98 min) wScrn   VIDrel: ARTIF/20TH   V/h

AUTUMN MOON ***
(PG)
Clara Law CHINA/HONG KONG   1992
Masatoshi Nagase, Li Pui Wai, Choi Siu Wan, Maki Kiuchi, Suen Ching Hung, Sung Lap Yeung, Tsang Yuet Guen, Chu Kit Ming, Ang Ching Yee, Yue Sui Ting, Ngai Wai Man, Lee Yee Ping
An examination of present-day Hong Kong and its uncertain future, that has a minimalist plot taking as its focus the lives and preoccupations of several disparate characters living there. The very thin story follows the fortunes of a Japanese private eye, who strikes up an odd friendship with a young girl who lives with her dying grandmother. A wistful and quite striking film, its loose and rambling structure actually works to its advantage.
Aka: QIUYUE
DRA   108 min   CINrel

AVALANCHE *
15
Corey Allen USA   1978
Rock Hudson, Mia Farrow, Robert Foster, Jeanette Nolan, Steve Franken, Rick Moses, Barry Primus, Steve Franken, Cathey Paine, Peggy Browne, Pat Egan, Joby Baker, Cindy Luedke, John Cathe, Angelo Lamonea
Disaster at a newly opened ski resort where the holidaymakers are at the mercy of a giant avalanche. When it does arrive, the disaster provides a welcome break from a boring script and lacklustre performances.
A/AD   86 min (ort 91 min)   VIDrel: ALLIED/RTM   V

AVALANCHE ***
(15)
Paul Shapiro CANADA   1994
David Hasselhoff, Michael Gross, Deanna Milligan, Miles Ferguson, Don S. Davis, George Josef

Thriller set in Alaska with Hasselhoff playing a nasty murderer who is on the run after committing a diamond robbery, having made his escape from the police by parachuting out of a plane over the Canadian Rockies. When he lands near the cabin of a writer and his family he gets trapped, as his ditched plane explodes on impact, setting off an avalanche. Tense and exciting, with TV's "Baywatch" star Hasselhoff giving a surprisingly strong performance.
THR   90 min mTV   SATrel: MOVIE CHANNEL

AVALON ***
U
Barry Levinson USA   1991
Armin Mueller-Stahl, Elizabeth Perkins, Joan Plowright, Aidan Quinn, Lou Jacobi, Leo Fuchs, Eve Gordon, Kevin Pollak, Israel Rubinek, Elijah Wood, Grant Gelt, Bernard Hiller, Mindy Loren Isenstein, Shifra Lerer, Neil Kirk
Written and directed by Levinson, this is something of a personal story, and provides an affectionate look at the life and times of an immigrant Jewish family in Baltimore, and the changes the years bring. A gentle and often funny and touching portrait that spans four generations, it was the third of Levinson's Baltimore chronicles, following DINER and TIN MEN. Music is by Randy Newman.
DRA   122 min (ort 127 min)   VIDrel: COLUM/SONOP L/A   V/sh

AVANTI! *
15
Billy Wilder USA   1972
Jack Lemmon, Juliet Mills, Clive Revill, Edward Andrews, Gianfranco Barra, Franco Angrisano, Pippo Franco, Franco Acampora, Giselda Castrini, Raffaele Mottola, Lino Coletta, Harry Ray, Guidarino Guidi, Giacomo Rizzo, Janet Agren
An American millionaire goes to Italy to collect the body of his father who died whilst on holiday there. He discovers that he had a mistress who was killed in the same accident. He then finds that the woman's daughter is also at the hotel. From this unpromising beginning develops... nothing, just an endless series of contrived catastrophes and car chases as events pile on top of events in a desperate search for laughs.
COM   139 min (ort 144 min)   VIDrel: MGM/WHV L/A   V
Boa: play by Samuel Taylor.

AVENGING ANGEL *
18
Robert Vincent O'Neill USA   1985
Betsy Russell, Rory Calhoun, Susan Tyrrell, Robert F. Lyons, Ossie Davis, Steven M. Porter, Paul Lambert, Barry Pearl, Estee Chandler, Tim Rossovich, Ross Hagen, Franke Doubleday, Howard Honig, Tracy Robert Austin, Michael A. Andrews
Sequel to ANGEL, in which our heroine graduates from college only to find that the cop who got her off the streets has been murdered. She sets out to avenge his death in this dreadful rubbish. Followed by ANGEL 3: THE FINAL CHAPTER (we hope).
DRA   94 min   VIDrel: RCA L/A   V

AVENGING ANGEL THE **
12
Craig R. Baxley USA   1995
Tom Berenger, James Coburn, Charlton Heston, Fay Masterson, Kevin Tighe, Jeffrey Jones, Tom Bower, Leslie Hope, Daniel Quinn
One of the last few members of the Mormon Church's own militia finds himself thrust into the limelight after a plot to assassinate Brigham Young is revealed and he shoots a would-be assassin. Pompous and self-righteous, the film's anti-gun message is laboriously hammered home, though the film gains a little gravitas from its star-studded cast.
WES   93 min (ort 99 min) mTV   VIDrel: COLUM/SONOP V

AVENGING EAGLE **
15
Sun Chang HONG KONG
Ti Lung, Fu Shong, Ku Feng, Tu Lung
After being rejected by his childhood sweetheart a man takes to crime and becomes a notorious criminal, in this simple martial arts tale.
MAR   80 min   VIDrel: IMPENT   V

AVENGING QUARTET, THE **
18
Stanley Wing Siu (Siu Wing) HONG KONG   1992
Waise Lee, Moon Lee, Michiko Naishiwaki, Yukari Oshima, Cynthia Khan
A man who owns a valuable painting leaves China for Hong Kong when he discovers that it contains the names of Japanese officials who worked there during the occupation. Very soon

two Japanese female fighters attempt to steal it but are resisted by a female cop and her sidekick. Plenty of action papers over the cracks in the confusing and complex plot.
MAR   90 min dubbed   VIDrel: MIA/DISC   V

AVIATOR'S WIFE, THE ***
PG
Eric Rohmer FRANCE   1980
Philippe Marlaud, Marie Riviere, Anne-Laure Meury, Matthieu Carriere, Philippe Caroit, Carolie Clement, Lisa Heredia, Haydee Caillot, Mary Stephen, Neil Chan, Rosett, Fabrice Luchini
Typical Rohmer study of French mores as a young law student spies on his faithless girlfriend and her lover. The first in a series from the director that he called "Comedies and Proverbs".
Aka: LA FEMME DE L'AVIATEUR
DRA   101 min (ort 104 min)   VIDrel: CONNO/RTM   V

AWAKE TO DANGER **
(15)
Michael Tuchner USA   1995
Tori Spelling, Michael Gross, John Getz, Reed Diamond, Laura Johnson, Shae D'Lyn, Michael O'Neill, Holland Taylor, Joanna Lipari, Jennifer Blanc, Chad Cox, David Grieco
A spate of burglaries culminates in tragedy when a man finds that his home broken into in his absence, with his wife lying dead in the bathroom and his teenage daughter knocked unconscious. She survives this brutal attack but spends eighteen months in a coma. Upon regaining consciousness, she at first recalls little of this incident, but her memory starts coming back to her in dreams. Pretty soon, the killer is planning her demise, in this standard thriller.
Aka: AWAKE TO MURDER
THR   86 min mTV   SATrel: MOVIE CHANNEL
Boa: novel Other Side of the Dark by Joan Lowery Nixon.

AWAKENING, THE **
15
Mike Newell UK   1980
Charlton Heston, Susannah York, Jill Townsend, Stephanie Zimbalist, Patrick Drury, Bruce Meyers, Nadim Sawalha, Ian McDiarmid, Ahmed Osman, Miriam Margolyes, Michael Mellinger, Ishia Bennison, Leonard Maguire, Madhav Sharma
A remake of BLOOD FROM THE MUMMY'S TOMB, with an Egyptian Queen reincarnated after 3,800 years, in the body of the daughter of the archaeologist who opens her tomb. Tedious and slow-moving. See also THE TOMB.
HOR   96 min (ort 105 min)   VIDrel: MGM/WHV L/A   V
Boa: novel The Jewel of the Seven Stars by Bram Stoker.

AWAKENINGS ****
15
Penny Marshall USA   1990
Robert De Niro, Robin Williams, John Heard, Julie Kavner, Penelope Ann Miller, Ruth Nelson, Max Von Sydow, Alice Drummond, Anne Meara, Mary Alice, Judith Malina, Barton Heyman, Richard Libertini, Laura Esterman, Dexter Gordon
Based on the early career of neurologist Dr Sacks, this absorbing drama tells of a Bronx doctor who takes over a ward of comatose patients in 1969 (victims of a major sleeping sickness epidemic in the 1920s) and resolves to attempt awakening them with a new experimental drug. De Niro gives a most touching performance as one of the patients, brought back to life after a thirty year coma, and Williams is excellent as the doctor.
DRA   115 min (ort 121 min)   VIDrel: VCC/DISC/COLUM
V/sur
Boa: book by Oliver Sacks.

AWFUL DR ORLOFF, THE *
15
Jess (Jesus) Franco FRANCE/SPAIN   1962
Howard Vernon, Perla Cristal, Ricardo Valle, Diana Lorys, Conrado Sanmartin, Maria Silva, Venancio Muro, Mara Laso, Felix Dafauce, Faustino Cornejo
A mad scientist attempts to restore his daughter's disfigured face by means of skin grafts and blood transfusions that are both taken from the bodies of murdered women. A gory, explicit and openly sadistic offering that tries hard to be impressive with some surrealistic touches, but is merely revolting.
Aka: AWFUL DOCTOR ORLOFF, THE; CRIES IN THE NIGHT; DIABOLICAL DOCTOR SATAN, THE; GRITOS EN LA NOCHE; L'HORRIBLE DOCTEUR ORLOFF
HOR   82 min (ort 90 min) wScrn dubbed
VIDrel: REDEM/RTM   V
Boa: novel by David Kuhne (Jesus Franco).

AWFUL TRUTH, THE ***
U
Leo McCarey USA   1937
Cary Grant, Irene Dunne, Ralph Bellamy, Cecil Cunningham, Mary Forbes, Alexander D'Arcy, Joyce Compton, Molly lamont, Esther Dale, Robert Allen, Robert Warwick, Mary Forbes, Claud Allister, Zita Moulton, Scott Colton, Wyn Cahoon
A couple get divorced almost by mistake, only to discover that they still love each other deeply. However, pride forbids them from coming out and saying so, and they engage instead in a series of hit-and-run wrecking tactics, to prevent one another from getting involved with new partners. A wonderfully screwball comedy, full of crazy scenes, that works a treat. Filmed before in 1925 and 1929 and remade in 1953 as "Let's Do It Again". AA: Dir.
COM   87 min (ort 92 min) B/W   VIDrel: COLUM/SONOP
V
Boa: play by Arthur Richman.

AWFULLY BIG ADVENTURE, AN **
15
Mike Newell UK   1994
Hugh Grant, Alan Rickman, Georgina Cates, Alun Armstrong, Peter Firth, Prunella Scales, Rita Tushingham, Alan Cox, Edward Petherbridge, Nicola Pagett, Carol Drinkwater, Clive Merrison, Gerard McSorley, Ruth McCabe, James Frain
Set in the post-war world of the provincial theatre, this has Cates as a young girl desperate to escape the boredom of her Liverpool existence, joining the local theatre. There, she instantly thinks herself in love with the company's ruthless director. A very typical product of the "life is miserable" school of British comedy that soon reveals itself as both unfunny and in sore need of a better script.
COM   107 min (ort 113 min) cC   VIDrel: 20TH/FOXVID
V/sur
Boa: novel by Beryl Bainbridge.

A.W.O.L. **
18
Sheldon Lettich USA   1990
Jean-Claude Van Damme, Harrison Page, Deborah Rennard, Lisa Pelikan, Ashley Johnson, Brian Thompson, Voyo, George McDaniel, Jason Adams, William Terry Amos, Roz Bosley, Dennis Wayne Rucker, Billy Blanks, Stefanos Miltsakakis
When his brother is badly injured in a dope deal that turned sour and later dies from his injuries, a tough Foreign Legionnaire deserts and takes up bare-knuckle fighting to help the impoverished widow and her daughter. A brutal B-movie that's partly redeemed by the acting, especially Page in the role of Van Damme's manager and mentor. The screenplay is by Lettich and Van Damme.
Aka: A.W.O.L. ABSENT WITHOUT LEAVE; LIONHEART
A/AD   103 min (ort 105 min)
VIDrel: POLY/POLYREC/GUILD; PION (LV only)   V/sur LV

AY! CARMELA ****
15
Carlos Saura ITALY/SPAIN   1990
Carmen Maura, Gabino Diego, Maurizio De Razza, Miguel A. Rellan, Andres Pajares, Edward Zentara, Jose Sancho, Antonio Fuentes, Mario De Candia, Rafael Diaz, Chema Mazo, Mario Martin, Emilio Del Valle, Silvia Casanova, Felipe Velez
A sharply scripted farce that chronicles the plight of a vaudeville entertainer who, having entertained the partisans during the Spanish Civil War, now finds herself, her husband and their mute assistant trapped behind enemy lines. By turns funny and touching, this slightly uneven tale is sustained by a wonderful performance from Maura as the central character.
DRA   98 min (ort 103 min)   VIDrel: ARROW/RTM   V
Boa: play by J. Sanchis Sinistierra.

B

BABAR: THE MOVIE **
U
Alan Bunce CANADA/FRANCE   1989
Voices of: Gordon Pinsent, Elizabeth Hanna, Lisa Yamanaka, Marsha Moreau, Bobby Brecker, Amos Crawley, Gavin Magrath, Sarah Polley, Stephen Oulmette, Ray Landry, Chris Wiggins, John Stocker, Charles Kerr, Stuart Stone, Angela Fusco
Enjoyable but rather bland animation based on the writings of Brunhoff, and following the adventures that befall a baby elephant as he grows to maturity. Compiled from a series of about 24 episodes.
ANIM   73 min (ort 79 min)   VIDrel: ABBEY/POLYREC
V/sh
Boa: short stories by Jean de Brunhoff.

BABE, THE * (PG)
Mark Tinker USA 1991
Stephen Lang, Brian Doyle-Murray, Donald Moffat, Yvonne Sahor, Bruce Weitz, Lisa Zane, William Lucking, Neal McDonough, John Anderson, Peter Rose, William Flatley, Cy Bunyak, Stephen Prutting, Frankie Thorn, Thomas Wagner, Dana Craig
A less than enthralling biopic that concentrates on the life of the title figure off the field, with little attention given to his sporting prowess, and such baseball sequences as are featured lack conviction, mainly die to Lang's weak performance. See the John Goodman film for a much better attempt.
Aka: BABE RUTH
DRA 90 min mTV SATrel: SKY MOVIES
Boa: book Babe Ruth: His Life and Legend by Karl Wagenheim/book Babe: The Legend Comes to Life by Robert W. Creamer.

BABE, THE *** PG
Arthur Hiller USA 1992
John Goodman, Kelly McGillis, Trini Alvarado, Bruce Boxleitner, Peter Donat, Richard Tyson, James Cromwell, Joe Ragno, Bernard Kates, Stephen Caffrey, J.C. Quinn, Michael McGrady, Bob Swan, Ralph Marrero, Gene Ross, Dylan Day
Excellently acted biography of super baseball star Babe Ruth that gives an accurate portrayal of the man, whose vices were also very much larger than life. Goodman is superb in the title role and is well supported by the rest of the cast, but the weak script is a major drawback. Unfortunately, its episodic format never allows Goodman's performance to provide any real psychological insights into the central character.
DRA 110 min (ort 115 min) VIDrel: CIC/SONOP V/sur

BABE **** U
Chris Noonan AUSTRALIA 1995
James Cromwell, Magda Szubanski, Zoe Burton, Paul Goddard, Wade Hayward, Brittany Byrnes plus voices of: Christine Cavanaugh, Miriam Margoyles, Danny Mann, Hugo Weaving, Miriam Flynn, Russie Taylor and Roscoe Lee Brown (narration)
An orphaned piglet with the power of speech turns the life of a farmer upside down with his desire to rise to better things, ultimately finding that his natural charm and intelligence make him an ideal sheep herder. A delightful and magical film, with not a trace of the sugary sentimentality one might have expected. The special effects that make it appear as if the animals are talking cannot be faulted. See also GORDY. AA: Effects/vis (Scott E. Anderson et al.).
Aka: BABE, THE GALLANT PIG
JUV 89 min (ort 93 min) cC VIDrel: CIC/SONOP; PION (LV only) V/sur LV
Boa: novel The Sheep Pig by Dick King-Smith.

BABES IN TOYLAND ** U
Jack Donohue USA 1961
Ray Bolger, Annette Funicello, Tommy Kirk, Gene Sheldon, Henry Calvin, Ed Wynn, Kevin Corcoran, Tommy Sands, Charlotte Henry, Ann Jillian, Brian Corcoran, Marilee Arnold, Melanie Arnold, Jerry Glenn, John Perri, David Pinson
A remake of the 1932 Laurel & Hardy original, with Mary and Tom travelling through the "Forest Of No Return" to Toyland, where they help in making toys for Christmas. But the villainous Mr Barnaby has other ideas. A miscast remake of the earlier film that doesn't lack for special effects but could do with a more talented cast.
MUS 101 min (ort 119 min) VIDrel: WDV L/A V

BABES IN TOYLAND * U
Clive Donner USA 1986
Drew Barrymore, Richard Mulligan, Eileen Brennan, Keanu Reeves, Jill Schoelen, Pat Morita, Googy Gress, Walter Buschoff, Rolf Knie, Gaston Haeni, Pipo Sosman, Shari Weiser, Chad Carlson, Elizabeth Schot, Mona Lee Goss
Sluggish and uninspired remake of the Victor Herbert operetta with only "Toyland" and "March of the Wooden Soldiers" remaining, plus a new and totally forgettable score by Leslie Bricusse. Barrymore is transported to a fantasy village and has to break the domination of the evil Barnaby in this insipid kid's adventure. Written by playwright Paul Zindel.
JUV 91 min (ort 150 min) mTV VIDrel: RCA L/A V/sh

BABETTE'S FEAST **** U
Gabriel Axel DENMARK 1987
Stephane Audran, Jean-Philippe Lafont, Jarl Kulle, Bibi Andersson, Hanne Steensgard, Gudmar Wivesson, Bodil Kjer, Birgitte Federspiel,
Pouel Kern, Vibeke Hastrup, Bendt Rothe, Lisbeth Movin, Prebe Lerdoff Rye, Axel Strobye, Ebbe Rode
Critically acclaimed story set in an austere community in the 1870s, with two spinster sisters living quietly in their small Danish village using religion as a substitute for life. They take in a mysterious Parisian refugee as a cook who later wins a lottery, using the money to prepare a sumptuous banquet for the community. A fresh, evocative and exquisite film. The screenplay is by Axel. AA: Foreign.
Aka: BABETTES GAESTEBUD
DRA 103 min (ort 105 min) dubbed VIDrel: ODY/SONOP V/sh
Boa: short story by Isak Dinesen (Karen Blixen).

BABEWATCH ** 18
Valerie Breiman USA 1993
Julie Strain, Maureen Flaherty, Donna Baltron, Lucky O'Boyle, Becca Rocheford, Rif Coogan, Clayton Halsey, J.C. Palermo
Spoof on the TV show "Bay Watch" with a host of bikini-clad girls getting up to all sorts of shenanigans. Meanwhile the female director (who prior to this project worked on documentaries) begs to be released from her contract and allowed to do anything rather than direct the show. Occasionally amusing, despite that fact that it attempts to spoof a programme that is already an unintentional parody in all but name.
COM 84 min VIDrel: MED/DISC V/sh

BABY BOOM *** PG
Charles Shyer USA 1987
Diane Keaton, Harold Ramis, Sam Wanamaker, Sam Shepard, James Spader, Pat Hingle, Britt Leach, Kristina Kennedy, Michelle Kennedy, Kim Sebastian, Mary Gross, Patricia Estrin, Elizabeth Bennett, Peter Elbling, Shera Danese
A female workaholic yuppie "inherits" a thirteen-month-old baby girl from a relative and finds it impossible to reconcile the demands of motherhood with her career. An amiable comedy that makes a few sharp points in between comic lulls. The script is by Shyer and Nancy Meyers. A TV series followed.
COM 106 min (ort 110 min) VIDrel: MGM/WHV L/A V/sur V/dm

BABY CAT ** 18
Pierre Unia FRANCE 1981
Julie Margo, Felix Marten, Corinne Carson
A wealthy Parisian model hatches a plan to rob her employer and lover by arranging her own "kidnap". However, her plan backfires when she really is kidnapped and she is forced to use all her female wiles to make her escape.
A 77 min (ort 90 min) VIDrel: ELV V

BABY DOLL *** 15
Elia Kazan USA 1956
Carroll Baker, Eli Wallach, Karl Malden, Mildred Dunnock, Lonny Chapman, Rip Torn, Eades Hogue, Noah Williamson
Story of a child bride and her brainless husband and how his business rival tries to manipulate their situation for his own ends. Condemned by the League of Decency and the Catholic Church among others, when it first appeared, it now seems quite tame. However, in some respects it still remains a potent drama.
DRA 109 min (ort 117 min) B/W VIDrel: CONNO/RTM V/sur
Boa: play by Tennessee Williams.

BABY FACE NELSON ** (18)
Scott Levy USA 1995
C. Thomas Howell, F. Murray Abraham, Lisa Zane, Martin Kove
Howell stars as the legendary gangster of the 1920s, whose rise to power eventually brings him into conflict with Al Capone. Adequate, though not quite as good as Don Siegel's 1957 film of the same name.
DRA 85 min SATrel: MOVIE CHANNEL

BABY M *** 15
James Steven Sadwith USA 1984
Jo Beth Williams, John Shea, Bruce Weitz, Robin Strasser, Anne Jackson, Dana Wheeler-Nicholson, Dabney Coleman, Bruce McGill, Jenny Lewis, Annabella Price, Lonny Chapman, Nancy Addison, Tricia O'Neil, Ben Slack
True story that is based on the reports of a child custody battle of the 1908s, when a surrogate mother had second thoughts and attempted to retain custody of the baby she had agreed to give up to a childless couple. An unabashed weepie, literate and well

acted, it does a good job examining the facts surrounding the case of real-life surrogate mum Mary Beth Whitehead.
DRA 188 min (ort 193 min) mTV VIDrel: ODY/SONOP V

BABY OF MACON, THE *** 18
Peter Greenaway FRANCE/GERMANY/HOLLAND/
UK 1993
Julia Ormond, Ralph Fiennes, Philip Stone, Jonathan Lacy, Don Henderson, Celia Gregory, Jeff Nuttall, Jessica Stevenson, Kathryn Hunter, Gabrielle Reidy, Frank Egerton, Phelim McDermott, Leslie Cuss, Tony Vogel, Graham Valentine
Told as if it were a performance before an audience of the 1650s, this is the story of a village stricken with barrenness, where a woman past the age of child-bearing gives birth to baby of unusual beauty. However, the woman's daughter sets out to exploit the situation, by claiming she gave birth to the child by immaculate conception. An intelligent and very fluidly directed film, quite demanding but extremely rewarding if one is patient.
DRA 117 min (ort 122 min) wScrn
VIDrel: ELPIC/POLYREC V/sur

BABY ON BOARD * PG
Francis A. Schaeffer CANADA 1991
Judge Reinhold, Carol Kane, Geza Kovacs, Alex Stapley, Holly Stapley, Errol Slue, Barry Ashley, Conrad Bergschneider, Al Bernado, Peter Blais, Amy Lyle, Jason Blicker, Patty Elsasser, Cecil Halliway, Jerry Levitan, Doris Petrie
When her husband, a mob book-keeper, is accidentally killed in a shooting, a young widow seeks revenge but soon finds herself on the run in New York. Matters become more complex when she is so distracted that she leaves her young child in a cab. Reinhold (as the greatly put upon cabby) was never more bland than in this overlong comedy of errors (the greatest error of which being that this dud ever saw the light of day).
COM 86 min (ort 90 min) VIDrel: MIA/DISC V

BABY SNATCHER ** PG
Joyce Chopra USA 1992
Veronica Hamel, Nancy McKeon, Michael Madsen, David Duchovny, Penny Fuller, John Evans, Roger Bearde, Caitlin Brady, Richard Elmore, Peter Vilkin, Darah DeAngelis, Ernesto Ravetto, Christine P. Bump, Vilma Salva, Pamela Kay Davis
A woman's distress after a miscarriage leads her into the abduction of another woman's child in this reasonably absorbing fact-based drama, which benefits from a strong script written by Hamel.
DRA 89 min mTV VIDrel: ODY/SONOP V/sh

BABYLON 5 ** PG
Richard Compton USA 1993
Michael O'Hare, Tamlyn Tomita, Jerry Doyle, Miru Furlan, Blaire Baron, John Fleck, Paul Hampton, Peter Jurasik, Andreas Katsulas, Johnny Sekka, Patricia Tallman, Steve R. Barnett, William Hayes, Linda Hoffman, Robert Jackson Johnson
Pilot episode for a SF series set on a distant space-station. In this story, the leaders of five different solar systems come together on the station for peace talks, but there is an assassin onboard as well. Adequate TV SF, not exactly groundbreaking in the imagination department, but generally watchable enough. This pilot led to an increasingly tedious series, many of the episodes of which are now available on video.
FAN 80 min (ort 89 min) mTV VIDrel: WHV V/sh

BABY'S DAY OUT ** PG
Patrick Read Johnson USA 1994
Joe Mantegna, Joe Pantoliano, Lara Flynn Boyle, Brian Haley, Cynthia Nixon, Fred Dalton Thompson, John Neville, Eddie Bracken, Matthew Glave, Jacob Joseph Worton, Adam Robert Worton, Brigid Duffy, Guy Hadley, Dan Frick, Jim Foley
The baby son of a wealthy Chicago family evades his would-be kidnappers, who meet with considerable violence, and takes off for a day on the town, in this ridiculous, cynical and calculating recycling of the plot of the HOME ALONE movies.
COM 95 min (ort 99 min) cC VIDrel: 20TH/TECH V/sh

BABYSITTER, THE ** 18
Guy Ferland USA 1994
Alicia Silverstone, Jeremy London, George Segal, J.T. Walsh, Lee Garlington, Nicky Katt, Lois Chiles
An attractive young woman takes up title position with an average suburban family and is soon pursued by several of its members. An unremarkable attempt at an erotic drama that seems tired and uninteresting, though from time to time the

daydreams of the central characters provide a little light relief.
DRA 85 min (ort 90 min) VIDrel: FIRST/SONOP V

BABYSITTERS, THE * PG
John Paragon USA 1994
The Barbarian Brothers (Peter and David Paul), Jared Martin, Barry Dennen, Christian Cousins, Joseph Cousins, Rena Sofer, George Lazenby, Mother Love
Wrestling brothers Peter and David are hired to act as bodyguards to ten-year-old twins, who make their lives hell.
COM 88 min VIDrel: MED/DISC V/sh

BABYSITTER'S SEDUCTION, THE *** (18)
David Burton Morris USA 1996
Kerri Russell, Stephen Collins, Phylicia Rashad, Tobin Bell, John D'Acquino, Linda Kelsey
An eighteen-year-old girl takes a job as a babysitter and finds herself becoming the object of attraction of an older man's passion, which leads to unexpected complications. However, she gradually learns that his pleasant exterior conceals a brutal killer who has murdered his wife and is planning to frame her for this crime. Quite cleverly plotted, one is never quite sure of the eventual outcome.
THR 89 min SATrel: SKY MOVIES

BACHELOR GIRL *** 15
Rivka Hartman AUSTRALIA 1987
Lyn Pierse, Kim Gyngell, Jan Friedl, Bruce Spence
Engaging comedy from Down Under that chronicles the experience of a thirty-two-year-old single girl and would-be writer whose meddlesome aunt and uncle are determined to play matchmaker. When an old university friend comes back into her life, he creates complications that are mirrored in the lives of characters of the TV soap opera she is writing. A light-hearted and entertaining comedy.
COM 79 min VIDrel: GUILD/SONOP V

BACHELOR MOTHER *** U
Garson Kanin USA 1939
Ginger Rogers, David Niven, Charles Coburn, Frank Albertson, Ernest Truex, E.E. Clive, Elbert Coplen Jr, Ferike Boros, Leonard Penn, Paul Stanton, Edna Holland, Frank M. Thomas, Dennie Moore, June Wilkins, Horace MacMahon
A woman accidentally gets left in charge of an abandoned baby in this happy and lively Norman Krasna comedy. Remade in 1956 as "Bundle Of Joy".
COM 81 min B/W VIDrel: VCC L/A V
Boa: story by Felix Jackson.

BACHELOR PARTY ** 18
Neil Israel USA 1984
Tom Hanks, Tawny Kitaen, Adrian Zmed, Bibi Besch, George Grizzard, Robert Prescott, William Tepper, Wendie Jo Sperber, Barry Diamond, Gray Grossman, Michael Dudikoff, Bradford Bancroft, Martina Finch, Deborah Harmon, Tracy Smith
Preparations for a bachelor party get a little out of hand in a comedy that starts off well but rapidly descends into tastelessness in a desperate search for laughs.
COM 101 min (ort 106 min) VIDrel: 20TH/TECH V/h

BACHELOR PARTY 2 * 18
Richard Gabai USA 1993
Linnea Quigley, Michelle Bauer, Burt Ward, Rhonda Shear
Some years ago a high school fraternity was saved from extinction by a wild party, but said party resulted in the loss of two babes "belonging" to another fraternity. Now it is time for revenge. Low-brow comic mayhem, about as funny as watching a drunk fall over.
Aka: BACHELOR JAMBOREE
COM 87 min VIDrel: MARQ/GUILD V

BACK IN ACTION ** 18
Steve Di Marco USA 1993
Billy Blanks, Roddy Piper, Bobbie Phillips, Matt Birman, Nigel Bennett, Damon D'Oliveira, Kai Soremekun, Rob Stefaniuk, Sam Malkin, BBarry Blake, Kelly Fiddick, Gary Robbins, David Ferry, Gerry Quigley, Timm Zemarek, Louis Strauss
When a criminal gang abduct their girlfriends, two martial artists swing into action to deal with these bad guys and much violent action results. A totally unoriginal actioner with no aspirations to anything greater but saved from total mediocrity by the adequate acting.
A/AD 79 min (ort 93 min) VIDrel: GUILD/SONOP V/sh

BACK IN THE USSR ** 15
Deran Sarafian USA 1991
Natalya Negoda, Frank Whaley, Roman Polanski, Dey Young, Andrew Divof, Brian Blessed, Ravil Issyanov, Constantine Gregory, Alexei Yevdokimov, Boris Romanov, Vesvlod Safonov, Yuri Sarntsev, Oleg Anofriev, Nikolai Averiushkin, Igor Klass
A young American on a two-week vacation in the USSR, gets embroiled with a beautiful woman thief who has stolen a rare book from a church. They become lovers but find their lives in danger from the local underworld. A confused and fast-paced actioner that apart from the Moscow locations has few original features. This was the first USA movie to be shot entirely in the USSR.
A/AD 83 min (ort 87 min) VIDrel: WHV V/sur

BACK OF BEYOND ** 15
Michael Robertson AUSTRALIA 1995
Paul Mercurio, Colin Friels, John Polson, Dee Smart, Bob Maza, Terry Serio, Rebekah Elmalogou, Aaron Wilton, Amy Miller-Porter, Glenda Linscott
A man returns home in an attempt to make up for the death of his sister in a motorcycle accident, but gets embroiled in a conflict with three crooks, who stop at the family garage to make repairs to their vehicle. The dry-as-dust Outback locations add a welcome dose of atmosphere to the simple script.
A/AD 85 min (ort 105 min) VIDrel: IMAG/RTM V

BACK ROOM BOY ** U
Herbert Mason UK 1942
Arthur Askey, Googie Withers, Moore Marriott, Graham Moffatt, Vera Frances, Joyce Howard, John Sales, George Merritt
After being banished from the BBC for a minor prank, a young man winds up in charge of a weather station attached to a remote Orkneys lighthouse. He eventually gets a chance to do his bit for the war effort by unmasking a Nazi spy ring and causing the sinking of a battleship. A feeble attempt to create a star vehicle for Askey that fails miserably on account of the weak and unoriginal plot.
COM 82 min B/W VIDrel: PETWAT/WEADIS L/A V

BACK TO SCHOOL *** 15
Alan Metter USA 1986
Rodney Dangerfield, Burt Young, Sally Kellerman, Keith Gordon, Robert Downey Jr, Paxton Whitehead, Terry Farrell, M. Emmett Walsh, Adrienne Barbeau, Ned Beatty, Severn Darden, William Zabka, Edie McClurg, Kurt Vonnegut Jr
A self-made millionaire decides to join his son who is having difficulties fitting in as a college freshman and so his uneducated dad buys his way into the Ivy League college. A simple story that gives ample scope for a winning and hilarious performance from Dangerfield and a funny sequence of sight gags and one-liners.
COM 93 min VIDrel: RCA L/A V/sh

BACK TO THE BEACH * PG
Lyndall Hobbs USA 1987
Frankie Avalon, Annette Funicello, Connie Stevens, Pee-Wee Herman (Paul Reubens), Don Adams, Lori Loughlin, Bob Denver, Tommy Hickley, Demian Slade, Joe Holland, John Calvin, Laura Urstein, Marjorie Gross, Alan Barry, Todd Bryant
A couple take a break and visit their daughter at her Southern California beach apartment. At first bewildered by the swinging lifestyle, they soon adapt, becoming the "beach couple" of the 1980s. The efforts of over six writers achieve little in this thin and uninspired comedy. Contains a host of guest stars from 1950s and 1960s TV sit-coms, but they add little to this one.
COM 88 min (ort 92 min) VIDrel: CIC/SONOP L/A V/sh

BACK TO THE FUTURE *** PG
Robert Zemeckis USA 1985
Michael J. Fox, Crispin Glover, Lea Thompson, Christopher Lloyd, Wendie Jo Sperber, Marc McClure, Claudia Wells, Thomas F. Wilson, James Tolkan, George DiCenzo, Frances Lee McCain, Jeffrey Jay Cohen, Casey Siemaszko, Billy Zane
A clever and lighthearted look at the adventures that befall a teenager who, as a victim of an eccentric scientist, is whisked back into the 1950s where he meets his parents as teenagers, and has the task of persuading them to fall for each other so that he can be born. Features one of John De Lorean's ill-fated gull-winged cars that doubles as the time machine. Followed by a sequel. AA: Effects/aud (Charles L. Campbell/Robert Rutledge).
COM 111 min (ort 116 min) wScrn VIDrel: CIC/SONOP; PION (LV only) V/sur LV

BACK TO THE FUTURE: PART 2 *** PG
Robert Zemeckis USA 1989
Michael J. Fox, Christopher Lloyd, Lea Thompson, Thomas F. Wilson, Harry Waters Jr, Charles Fleischer, Joe Flaherty, Elisabeth Shue, James Tolkan, Casey Siemaszko, Jeffrey Weissman, Billy Zane, J.J. Cohen, Ricky Dean Logan
A frantic, production-line sequel to the first film, with Fox and Lloyd venturing into both the future and the past, as Marty McFly tries to save the future from the effects of his first trip into the past. Incredible plot complications and superb effects mask a film of little warmth, whose ideas inexorably lead to Part 3 for their resolution.
Aka: PARADOX: BACK TO THE FUTURE 2
FAN 103 min (ort 108 min) VIDrel: CIC/SONOP; PION (LV only) V/sur LV

BACK TO THE FUTURE: PART 3 **** PG
Robert Zemeckis USA 1989
Michael J. Fox, Christopher Lloyd, Mary Steenburgen, Thomas F. Wilson, Lea Thompson, Elisabeth Shue, Matt Clark, Richard Dysart, Pat Buttram, Harry Carey Jr, Dub Taylor, James Tolkan, ZZ Top, Christopher Wayne, Mike Watson
This excellent follow-up to Part 2 makes up for the failings of the earlier film, with Fox now back in the Old West of 1885, where he attempts to change the course of history and thereby save the life of his inventor buddy. A noisy comic-book fantasy, imaginative, fast-paced and fun.
FAN 113 min (ort 118 min) wScrn VIDrel: CIC/SONOP; PION (LV only) V/sur LV

BACKBEAT **** 15
Iain Softley GERMANY/UK 1993
Sheryl Lee, Stephen Dorff, Ian Hart, Gary Bakewell, Kai Wiesinger, Chris O'Neill, Scott Williams, Jennifer Ehle, Paul Humpoletz, Bob Spendlove, Charlie Caine, Frieda Kelley, Marcelle Duprey, Fiona Geraghty, John White
Between the years 1960 and 1962 the band that was later to achieve world fame as The Beatles spent a hard and unglamorous time working in the clubs of Hamburg. This excellent and gritty drama revolves around the relationship between John Lennon and Stu Sutcliff, the bass player and so-called "fifth" Beatle, whose departure from the group was precipitated by his romance with a German photographer.
MUS 96 min (ort 100 min) wScrn
VIDrel: VCC/DISC/COLUM V/sur

BACKDRAFT *** 15
Ron Howard USA 1991
Kurt Russell, William Baldwin, Robert De Niro, Scott Glenn, Jennifer Jason Leigh, Rebecca DeMornay, J.T. Walsh, Donald Sutherland, Jason Gedrick, Tony Mockus Sr, Clint Howard, Cedric Young, Junan Ramirez, Kevin M. Casey, Jack McGee
A story of Chicago firemen, in which two brothers working on the same force now spend most of their free time fighting, following the death of their father putting out a blaze. Poor performances hold back a film that in many other ways provides a fascinating glimpse of the work of fire-fighters, though De Niro as an arson investigator and Sutherland as an arsonist both generate a measure of interest.
A/AD 131 min (ort 136 min) VIDrel: CIC/SONOP; PION (LV only) V/sur LV

BACKFIRE ** 18
Gilbert Cates USA 1987
Karen Allen, Keith Carradine, Bernie Casey, Dean Paul Martin, Jeff Fahey, Dinah Manhof, Virginia Capers, Philip Sterling, Frances Flanagan, Antony Holland, Dwight Ross, Gordon McIntosh, Wendy Van Riesen, Eric Schneider
A married woman who is having an affair with an old boyfriend plots to drive her Vietnam veteran husband to suicide, in order to gain control of his property. However, this plan misfires and she finds herself nursing a catatonic invalid in a wheelchair. After her lover leaves, she encounters a mysterious stranger, who soon moves into her home. A conventional thriller of little originality, whose resolution can be guessed far in advance.
THR 89 min VIDrel: VISVID/POLYREC V

BACKSTREET JUSTICE ** 18
Chris McIntyre USA 1994
Linda Kozlowski, Hector Elizondo, John Shea, Paul Sorvino, Tammy Grimes, Viveca Lindfors, William Thunhurst Jr, Keith Randolph Smith, Mark Joy, Bruce Kirkpatrick, Patrick Skeriotis, B.J. Davis. Bernard Canepari, Lou Spencer

A tough woman private eye investigates a series of murders where the victims were all tenants of an apartment block in a poor and rundown area. She concludes that someone is trying to empty the building through scare tactics and discovers a link to the police department. This throws up painful memories of her late father, who had acquired a reputation for corruption. A routine tough thriller, with few surprises and many familiar touches.
THR   87 min (ort 91 min)   VIDrel: PRISM/HIFLI   V/h

BAD AND THE BEAUTIFUL, THE **

PG
1952

Vincente Minnelli USA
Kirk Douglas, Lara Turner, Dick Powell, Gloria Grahame, Bary Sullivan, Leo G. Carroll, Gilbert Roland, Vanessa Brown, Paul Stewart, Sammy White, Walter Pidgeon, Elaine Stewart, Ivan Triesault, Kathleen Freeman, Jonathan Cott
Story of a pushy and obnoxious Hollywood producer who does his best to alienate all those around him. Told in flashbacks by various characters. Overblown and melodramatic but still quite good entertainment. AA: Art (Cedric Gibbons/Edward Carfagno), Set (Edwin B. Willis/F. Keogh Gleason), Cin (Robert Surtees), Screen (Charles Schnee), S. Actress (Grahame).
DRA   113 min (ort 123 min) B/W   VIDrel: WHV   V

BAD ATTITUDE **

18
1992

Bill Cummings USA
Gina Lim, Nathaniel DeVeaux, Susan Finque, Leon
A drugs cop is discharged from the force because of his vigilante tactics and decides to go after a major drug dealer, hoping that his successful capture will get him his job back. Naturally, a couple of his colourful friends are on hand to help him, in this competent thriller. Screenplay is by Crane Webster.
THR   91 min   VIDrel: GUILD/SONOP   V/sh

BAD ATTITUDES **

PG
1991

Alan Myerson USA
Maryedith Burrell, Ethan Randall, Jack Evans, Richard Gilliland, Ellen Blain, Eugene Byrd, Meghann Haldemann, Francis X. McCarthy, Jack Kehler, Raymond Forchion, Phil Proctor, Tony Longo, Patrick Culliton, Alan Shearman
Five high-spirited teenagers escape from summer camp and stow away on an aircraft, little suspecting that a couple of terrorists are planning to abduct the owner. As expected, the kids foil these plans, and learn a thing or two about life in the process. A standard juvenile action yarn, with its moral message coming through loud and clear.
JUV   88 min   VIDrel: 20TH/TECH   V/sh

BAD BEHAVIOUR ***

15
1992

Les Blair UK
Stephen Rea, Sinead Cusack, Philip Jackson, Clare Higgins, Phil Daniels, Mary Jo Randle, Saira Todd, Amanda Boxer, Luke Blair, Joe Coles, Tamlin Howard, Emily Hill, Phillippe Lewison, Ian Flintoff, Kenneth Hadley, Siempre Caliente
A middle-class Irish couple living in London decide to renovate their bathroom and find themselves having to deal with a lot of unresolved issues, in this improvised look-at-life. Fine performances and a good deal of warmth help sustain a thinly plotted affair.
DRA   99 min (ort 104 min) mTV wScrn
VIDrel: FIRST/SONOP   V/sur

BAD BOY BUBBY **

18
1993

Rolf de Heer AUSTRALIA/ITALY
Nicholas Hope, Claire Benito, Ralph Cotterill, Carmel Johnson, Sid Brisbane, Nikki Price, Norman Kaye, Paul Philpot, Peter Monaghan, Natalie Carr, Rachel Huddy, Bridget Walters, Lilli Birme, Aldine Leith, Lucia Mastrantone
Story of a man who has lived within the confines of his tiny home for thirty-five years, until the day he ventures outside and starts off on a wild lifestyle.
DRA   108 min (ort 114 min)   VIDrel: EIV/SONOP   V

BAD BOYS ***

18
1983

Richard Rosenthal USA
Sean Penn, Reni Santoni, Esai Morales, Eric Gurry, Jim Moody, Ally Sheedy, Clancy Brown, Robert Lee Rush, John Zenda, Alan Ruck, Tony Mockus, Laurence Mah, Erik Barefield, Dean Fortunato, Jorge Noa, Ray Caballero, Ray Ramirez
Tough film that examines the treatment of juvenile delinquents at a detention centre and follows the story of a personal vendetta

that reaches its climax within the prison walls. Violent and not always believable, but never less than fully absorbing. This was Sheedy's film debut.
DRA   99 min (ort 123 min)   VIDrel: MGM/WHV L/A   V

BAD BOYS **

18
1995

Michael Bay USA
Martin Lawrence, Will Smith, Tea Leoni, Tcheky Karyo, Theresa Randle, Marg Helgenberger, Nestor Serrano, Julio Oscar Mechoso, Saverio Guerra, Anna Thomson, Kevin Corrigan, Michael Imperioli, Joe Pantoliano, Lisa Boyle, Michael Taliferro
Two Miami cops are assigned the difficult task of locating one million dollars worth of heroin that has disappeared from the police evidence room. This naturally involves them in a succession of both comic and dangerous situations, in this fast-paced action-comedy that thanks to its editing, has something of a mTV appearance about it.
A/AD   124 min wScrn cC   VIDrel: 20VIS/SONOP   V/sur

BAD BOYS 2 **

18
1992

Herb Freed USA
Steve Lieberman, Francine Lepensee, David Winston
Coming-of-age story of tough streetwise youngsters and their eventual trip to prison. Pretty much a re-run of the earlier film in terms of ideas.
A/AD   100 min (ort 109 min)   VIDrel: MARQ/QUANT   V

BAD CHANNELS *

PG
1992

Ted Nicolaou USA
Paul Hipp, Martha Quinn, Aaron Lustig, Ian Patrick Williams, Charlie Spradling, Michael Huddleston, Victor Rogers, Melissa Behr, Ania Sava, Sonny Carl Davis, Daryl Strauss, Roumel Reaux, Rodney Ueno, Robert Factor, Ron Kneel
A wacky radio station attracts an eccentric alien who likes nothing better than collecting women, shrinking his victims in size, and carrying them back to his home-planet in jars. When he attempts to take over the station in a bid to fulfil his plans, he comes up against the resident DJ. A noisy and overblown horror-spoof, unusual in conception, but considerably less effective in execution.
HOR   78 min (ort 90 min)   VIDrel: CIC/SONOP   V

BAD COMPANY **

15
1995

Damian Harris USA
Laurence Fishburne, Ellen Barkin, Frank Langella, Michael Beach, David Ogden Stiers, Spalding Gray, Daniel Hugh Kelly, Gia Carrides
An angry young man (Fishburne) joins an industrial espionage ring and gets involved with an attractive female colleague. However, he is in reality a CIA operative out to smash this organisation. An over-complex action film that concentrates too much on the conflict between the leads and devotes too little attention to a seriously underdeveloped plot. Not released to the cinema in the UK, thanks to its disappointing reception in the States.
A/AD   104 min (ort 108 min) cC   VIDrel: TOUCH/TECH
V

BAD DAY AT BLACK ROCK ****

PG
1954

John Sturges USA
Spencer Tracy, Robert Ryan, Anne Francis, Dean Jagger, Walter Brennan, John Ericson, Ernest Borgnine, Lee Marvin, Russell Collins, Walter Sande
A powerful drama about greed and racism in the small town of the title. A one-armed WW2 arrives there one day on a simple mission to present a dead soldier's father with his son's medal. However, from the very start he meets nothing but a blank wall of hostility that conceals a mystery he must solve at any cost. An excellent example of fine acting and economy of narration rarely equalled.
DRA   78 min (ort 81 min)   VIDrel: MGM/WHV   V
Boa: short story Bad Day at Hondo by Howard Breslin.

BAD DREAMS *

18
1988

Andrew Fleming USA
Jennifer Rubin, Bruce Abbott, Richard Lynch, Harris Yulin, Sy Richardson, Dean Cameron, Susan Barnes, E.G. Daily, Susan Ruttan, Charles Fleischer, John Scott Clough, Sheila Scott Wilson, Damita Jo Freeman, Louis Giambalvo
The sole survivor of a Jim Jones-style mass suicide by members of a cult awakens after 13 years spent in a coma. However, she is haunted by visions of the cult's evil leader and when members of her therapy group start being found murdered, she becomes convinced that the cult leader's spirit is responsible. A visually

unpleasant horror tale whose inadequate script is a major failing.
HOR   81 min Cut (22 sec – ort 84 min)   VIDrel: 20TH/TECH V/sur

BAD GIRLS ** 15
Claude Chabrol FRANCE 1968
Stephane Audran, Jacqueline Sassard, Jean-Louis Trintignant, Nane Germon, Serge Bento, Dominique Zardi, Henri Attal, Claude Chabrol, Henri Frances
The story of an unusual menage-a-trois comprising two lesbians and the young male lover of one of them. The emotional complications are well explored and Audran (then the director's wife) gives a powerful performance, but the characters remain remarkably unengaging and generate neither sympathy nor interest.
Aka: DOES, THE; GIRLFRIENDS, THE; HETEROSEXUALS, THE; LES BICHES
DRA   94 min (ort 97 min) wScrn   VIDrel: ARTPRO/RTM V

BAD GIRLS ** 15
Jonathan Kaplan USA 1993
Madeleine Stowe, Mary Stuart Masterson, Andie MacDowell, Drew Barrymore, Dermot Mulroney, James Russo, Robert Loggia, Jim Beaver, Cooper Huckabee, James LeGros, Nick Chinlund, Neil Summers, Daniel O'Haco, Richard E. Reyes, Alex Kubik
Four hookers hit the road after one kills a man and they get together with a male outlaw. This results in a price on their heads and a posse snapping at their heels. An unsuccessful attempt to create a kind of feminist Western that suffers from poor plotting and an unfortunate tendency to copy ideas from other films of this genre. Despite the star cast, this movie was the box-office flop of 1993. A high-spot is the final gunfight, which is well worth seeing.
WES   96 min (ort 100 min) cC   VIDrel: 20TH/TECH   V/sur

BAD GIRLS 2 ** 18
David I. Frazer USA 1983
Ron Jeremy, David Dukeham, Larry Row, Arthur Knight, Herschel Savage, Honey Wilder, Doug Barris, Paul Barrezi, Michael Warner, Jacqueline Lorians, Blaire Castle, Suzanna Britton
Four photo-models on assignment in a small town turn the place upside down as they drive the men wild.
A   69 min Cut (2 sec – ort 80 min)   VIDrel: HAR/GOLD   V

BAD GIRL'S DORMITORY ** 18
Tim Kincaid USA 1985
Carey Zuris, Theresa Farley, Rick Gianasi, Marita Jennifer DeLora, Natalie O'Connell
The endless saga of exploitative studies of women in prison rolls inexorably on. This one's set in a juvenile detention centre that is privately run. Cut before video submission by 33 sec.
A   92 min   VIDrel: ODY/SONOP   V/sh

BAD GIRLS DOWN UNDER * 18
Aja AUSTRALIA 1990
Sheila Kelly, Kelly Blue, Sunny McKay, Randy Spears, Mel Bourne, Tom Byron
Very little plot is to be has here, just a series of sexual encounters in Australia, with a doctor who specialises in sexual problems taking a group of sex-starved girls to the Outback to relieve their frustrations.
Aka: SEXUAL HEALER
A   76 min (ort 85 min)   VIDrel: GROHOM/MAXSCAN   V

BAD HABITS * 18
John Leslie USA 1994
Deidre Holland, Randy Spears, Tiffany Mynx, Mark Davis, Sasha Strange, Angel Ash, Deborah Wells, Tom Byron, T.T. Boy, Jon Dough, Ted Craig, Dyanna Lauren, Skip, Rick, Joey, Christian, Kelly Nichols
A manipulative shrink sets up a promiscuous female writer for a con-man he has blackmailed into milking her of her money, but they fall in love and turn the tables.
A   85 min (ort 107 min)   VIDrel: PURG/DANTE   V

BAD INFLUENCE ** 18
Curtis Hanson USA 1990
James Spader, Lisa Zane, Rob Lowe, Christian Clemenson, Kathleen Wilhoite, Tony Maggio, Grand L. Bush, John De Lancie, Rosalyn Kandor, Marcia Cross, John Mahon, Palmer Lee Todd, John Verea, Sunny Smith, Susan Lee Hofman, Jeff Kaake
Having befriended a guy who rescued him from a potentially nasty bar-room confrontation, a young man discovers that his

new friend is a disturbed weirdo who is intent upon taking over his life. A heavily derivative film that's reminiscent of both STRANGERS ON A TRAIN and APARTMENT ZERO, it's slick without being memorable, and clever without being original. The script is by David Koepp.
THR   94 min (Cut at UK cinema release by 1 min – ort 99 min)   VIDrel: 4-FRONT/POLYREC/EIV L/A   V/sur

BAD LIEUTENANT *** 18
Abel Ferrara USA 1992
Harvey Keitel, Victor Argo, Paul Calderone, Frankie Thorn, Robin Burrows, Brian McElroy, Frankie Acciario, Peggy Gormley, Stella Keitel, Zoe Lund, Leoanrd Thomas, Daa Dee, Athony Ruggiero, Vincent La Resca, Victoria Bastell
When a police lieutenant learns about a raped nun who refuses to press charges against her attacker, his slumbering Catholic conscience is roused and he tries to leave his corrupt and immoral ways behind him. An over-the-top and very violent urban thriller that is too melodramatic and unsubtle to work effectively as a drama, for all its raw power and gritty realism.
A/AD   90 min Cut (1 min 40 sec – ort 96 min)
VIDrel: POLY/POLYREC/GUILD   V/sh

BAD MEDICINE ** 15
Harvey Miller USA 1985
Steve Guttenberg, Alan Arkin, Julie Hagerty, Bill Macy, Robert Romanus, Curtis Armstrong, Julie Kavner, Joe Grifasi, Taylor Negron, Candi Milo, Arne Gordon, Gilbert Gottfried, Arturo Venegas, Manuel Pereiro Rodriguez
Unfunny medi-comedy about an American student who is forced, by the lack of a place at any US college, to attend a shady medical school somewhere in Central America that is run by Arkin. The latter turns in a highly laudable performance, but this alone cannot breathe life into this wretched dud. Filmed entirely on location in Spain.
COM   96 min   VIDrel: 20TH/TECH   V

BAD SLEEP WELL, THE *** PG
Akira Kurosawa JAPAN 1960
Toshiro Mifune, Masayuki Mori, Takashi Shimura, Chishu Ryu, Takeshi Kato, Akira Nishimura, Kamatari Fujiwara, Gen Shimizu, Kyoko Kagaw, Tatsuya Mihashi, Kyu Sazanka, Seiji Miyaguchi, Nobuo Nakamura, Susumu Fujita
In order to avenge the murder of his father, Mifune gains employment working for a corrupt government official, one of the men he holds responsible for his father's death. Having married the man's daughter, he begins to plan a complex and fitting retribution. Kurosawa's first independent venture is an ambitious attempt to transpose an American novel to Japan, that works surprisingly well, despite some uncharacteristic touches of melodrama.
Aka: ROSE IN THE MUD, THE; WARUI YATSU HODO YOKU NEMURU; WORSE YOU ARE THE BETTER YOU SLEEP, THE
DRA   127 min (ort 152 min) B/W wScrn
VIDrel: CONNO/RTM L/A   V
Boa: novel by Ed McBain.

BAD TASTE ** 18
Peter Jackson NEW ZEALAND 1988
Peter O'Herne, Mike Minett, Terry Potter, Peter Jackson, Craig Smith, Doug Wren, Dean Lawrie, Ken Hammon, Michael Gooch, Laurie Yarrall, Robin Griggs, Shane Yarrall, Cost Botes, Graham Butcher, Bob Halliburton
An insane and overdone celebration of gore, with a bunch of uncontrollable thugs sent on a mission to stop the planet being taken over by a group of aliens, who intend to use the human race as raw material for a string of intergalactic fast-food restaurants. Funny in places, but definitely not one to see having just eaten.
COM   87 min (ort 103 min)   VIDrel: POLY/POLYREC   V

BAD TIMING *** 18
Nicolas Roeg UK 1980
Art Garfunkel, Theresa Russell, Harvey Keitel, Denholm Elliott, Daniel Massey, Dana Gillespie, William Hootkins, Eugene Lipinski, George Roubicek, Stefan Gryff, Sevilla Delofski, Robert Walker, Gertan Klauber, Ania Marson
An American divorcee has an affair with a strange Viennese psychiatrist which eventually leads to tragedy. Despite serious miscasting of Garfunkel, this one has many good things in it, not least being a fine performance from Russell as a woman of self-destructive urges and a good backing score with music from

The Who, Keith Jarrett and Billie Holiday. One of the few Roeg films in which his flashy direction adds to the movie.
Aka: BAD TIMING: A SENSUAL OBSESSION
DRA   118 min Cut (ort 123 min)   VIDrel: VCC L/A   V

BADGE OF BETRAYAL **
Sandor Stern USA
Harry Hamlin, Linda Doucett, Gordon Clapp, Michelle Greene
A female city cop moves to the Northwest to take up a job working with the local sheriff, only to discover that he is not quite the pleasant man she first thought him to be.
THR   91 min   VIDrel: ODY/SONOP   V/sh

15
1996

BADLANDS **
Terence Malick USA
Martin Sheen, Sissy Spacek, Warren Oates, Ramon Bieri, Alan Vint, Gary Littlejohn, John Carter, Bryan Montgomery, Gail Threlkeld, Howard Ragsdale, Charles Fitzpatrick, John Womack Jr, Dona Baldwin, Ben Bravo, Terrence Malick
Two young lovers go on a killing spree in this grim thriller based on an infamous killing spree of the 1950s by Starkweather and Fugate. Now a film with a devoted if minor cult following.
THR   90 min (ort 95 min)   VIDrel: WHV   V/h

18
1973

BAGDAD CAFE ***
Percy Adlon WEST GERMANY
Mariane Sagebrecht, C.C.H. Pounder, Jack Palance, Monica Calhoun, Darron Flagg, Christine Kaufmann, George Aquilar, G. Smokey Campbell, Alan S. Craig, Hans Stadlbauer, Apesanhakwat, Ronald Lee Jarvis, Mark Daneri, Ray Young
A buxom West German woman on holiday in the USA becomes involved with a very strange set of characters who patronise the title establishment, a roadside greasy spoon in the California desert. A largely plotless but endearing comedy from the director of "Sugarbaby", with nice casting of Palance as a former Hollywood set decorator who decides to paint her portrait. A high-spot is the show put on at the end by the staff of the cafe.
Aka: OUT OF ROSENHEIM
COM   87 min   VIDrel: VCC/DISC L/A   V

PG
1988

BAIT, THE **
Bertrand Tavernier FRANCE
Marie Gillain, Olivier Sitruk, Bruno Putzulu, Richard Berry, Philippe Duclos, Marie Ravel, Clotilde Courau, Jean-Louis Richard, Christophe Odent, Jean-Paul Comart, Phillipe Helies, Jacky Nercessian, Alain Sarde, Daniel Russo
An eighteen-year-old Parisian girl and her two friends hatch a plan to use her as seductive bait to obtain the money they need to get to America, luring middle-aged men back to her flat where they can be robbed. Strangely condescending its attitude towards teenagers and their feelings of alienation, this is a weak and awkward affair that never gets beneath the skins of its central characters.
Aka: L'APPAT
DRA   115 min (ort 116 min) wScrn   VIDrel: ARTIF/20TH   V

Boa: book by Morgan Sportes.

18
1995

BAJA **
Kurt Voss USA
Lance Henriksen, Molly Ringwald, Corbin Bernsen, Donal Logue, Michael A. Nickles, Julian Reyes, Wayne Duvall, Karen S. Gregan, Chris Shearer, Nelson Lynch, Rique Renaldo, Thomas G. Romero, Roxanna Michaels, I. Ignacio Alvarez
When a drugs deal goes badly wrong, a couple flee for their lives, arriving at the Mexican town of Baja. But two men are following their trail, a hired assassin and the woman's estranged husband.
THR   88 min (ort 90 min)   VIDrel: HIFLI/SONOP   V/h

18
1994

BALBOA **
James Polakof USA
Tony Curtis, Chuck Connors, Lupita Ferrer, Carol Lynley, Steve Kanaly, Sonny Bono, Catherine Campbell, Cassandra Peterson, Jennifer Chase, Martine Beswicke, Henry Jones
Steamy drama of power and passions among California's idle rich, with Curtis a nasty tycoon in Balboa who's about to pull off a real estate deal whilst everyone else is too busy sleeping around to notice. An episodic and uneven soap opera, edited down from a longer and unsold mini-series.
Aka: BALBOA: MILLIONAIRE'S PARADISE; RICH AND POWERFUL
DRA   90 min (ort 100 min)   VIDrel: MOPIC/SGSVID   V

15
1982 (released 1983)

BALL OF FIRE ***
Howard Hawks USA
Gary Cooper, Barbara Stanwyck, Oscar Homolka, Dana Andrews, Dan Duryea, Gene Krupa, S.Z. Sakall, Richard Hadyn, Henry Travers, Tully Marshall, Leonid Kinsky, Aubrey Mather, Allen Jenkins, Ralph Peters, Kathleen Howard
A group of staid academics writing an encyclopaedia are suddenly brought into contact with rude reality when a burlesque dancer who is to be a witness against her gangster boyfriend hides out with them. This classic screwball comedy is wildly enjoyable and was remade by the same director in 1948 as A SONG IS BORN, with Danny Kaye in the lead role.
COM   111 min B/W   VIDrel: VCC/DISC   V
Boa: short story From A to Z by Thomas Monroe and Billy Wilder.

PG
1941

BALLAD OF CABLE HOGUE, THE ***
Sam Peckinpah USA
Jason Robards, Stella Stevens, L.Q. Jones, Strother Martin, R.G. Armstrong, David Warner, Slim Pickens, Peter Whitney, Gene Evans, William Mims, Kathleen Freeman, Vaughn Taylor, James Anderson, Susan O'Connell, Max Evans
Comedy Western telling of a tough prospector who, after being left to die in the desert by his unscrupulous greedy partners, plots a complex revenge. Stevens is good as a whore who joins him in his quest for the better things in life, in this entertaining feature.
WES   117 min (ort 122 min)   VIDrel: WHV   V/sh

PG
1970

BALLAD OF LITTLE JO, THE ***
Maggie Greenwald USA
Suzy Amis, Bo Hopkins, Ian McKellen, Rene Auberjonois, Carrie Snodgress, Heather Graham, David Chung, Anthony Heald, Ruth Maleczech, Melissa Leo, Sam Robards, Carrie Snodgress, Olinda Turturro, Jeffrey Andrews, Irina Pasmur
After giving birth outside marriage, a woman is disowned by her well-off family and travels west to take part in the 1866 gold rush, having taken the sensible precaution of pretending to be a man. A well-acted but rather ponderous account, based on a true story.
WES   116 min (ort 124 min)   VIDrel: COLUM/SONOP   V/sur

15
1993

BALLAD OF THE SAD CAFE, THE **
Simon Callow UK/USA
Vanessa Redgrave, Cork Hubbert, Keith Carradine, Rod Steiger, Austin Pendleton, Beth Dixon, Lanny Flaherty, Mert Hatfield, Earl Hindman, Anne Pitoniak
In a Depression-era small Southern town, a female tyrant rules the roost until her husband returns, accompanied by a hunchback friend, and publicly humiliates him. A stagebound and overlong excursion into a grotesque and eccentric little world – of interest only for Redgrave's performance, and far too slow-paced to hold the interest.
DRA   96 min (ort 100 min)   VIDrel: CURZON/20TH   V/sh
Boa: novella by Carson McCullers/play by Edward Albee.

15
1989

BALTIMORE BULLET, THE **
Robert Ellis Miller USA
James Coburn, Omar Sharif, Bruce Boxleitner, Ronee Blakly, Calvin Lockhart, Jack O'Halloran, Michael Lerner, Paul Barselou, Cissie Cameron, Jeff Tenkin, Willie Mosconi, Shep Sanders, Jon Ian Jacobs, Ed Bakey, Robert Hughes
A smooth-talking con-man and his young apprentice set out to hustle a suave gambler (Sharif miscast once again) in a high-stakes game of pool. Skip this one and see an infinitely better film – THE HUSTLER.
DRA   98 min   VIDrel: VCC L/A   V

15
1980

BALTO ***
Simon Wells UK/USA
Voices of: Kevin Bacon, Bob Hoskins, Bridget Fonda, Jim Cummings, Phil Collins, Jack Angel, Danny Mann, Robbie Rist, Juliette Brewer, Sandra Searles Dickinson, Donald Sinden, William Roberts, Garrick Hagon, Bill Bailey, Big Al
Rejected by his canine colleagues and the humans of Alaska, half wolf/half husky Balto is unable to win a place on a team of huskies. But an outbreak of diphtheria among the children of an isolated Alaskan town of Nome gives him the chance to prove his worth when he takes part in a 674-mile journey to bring back vital medicine. Loosely based on a true 1925 incident, this is well animated and enjoyable, if slightly let down by a rambling plot and poor characterisations.
ANIM   74 min (ort 77 min) cC   VIDrel: CIC/SONOP   V/sur

U
1995

BAMBI ****
David D. Hand USA
U
1942
Voices of: Bobby Stewart, Peter Behn, Stan Alexander, Cammie King, Donnie Dunagan, Hardy Albright, John Sutherland, Tim Davis, Sam Edwards, Sterling Holloway, Paula Winslowe, Ann Gillis, Mary Lansing, Fred Shields, Bill Wright
This Disney classic about the life of a young fawn in the forest may strike some as excessively sentimental. However, the lavish animation and sheer care taken in its production, make it a timeless masterpiece. Recommended as a hugely enjoyable treat for all, especially for kids exposed to today's far inferior animations.
ANIM 70 min VIDrel: WDV/TECH L/A V
Boa: book by Felix Salten.

BANANAS ***
Woody Allen USA
15
1971
Woody Allen, Louise Lasser, Carlos Montalban, Howard Cosell, Jacobo Morales, Natividad Abascal, Miguel Suarez, David Ortiz, Rene Enriquez, Jack Axelrod, Roger Grimsby, Charlotte Rae, Sylvester Stallone, Stanley Ackerman, Dani Crane
Very uneven comedy set against a guerilla revolution in a backward Latin American dictatorship where Allen winds up as their new president. The funny moments are very funny but there are too few of them, although there is a courtroom scene that is a joy. Look out for Sylvester Stallone, he appears briefly as a hoodlum. The score is by Marvin Hamlisch.
COM 78 min (ort 82 min) VIDrel: MGM/WHV V/dm

BAND OF GOLD **
Charles Beeson UK
15
1995
Cathy Tyson, Geraldine James, Barbara Dickson, Ray Stevenson, Ray Stevenson, David Schofield, Derek Hicks, Judy Braune, Hleta Charnley, Samantha Morton, Sue Cleaver, Alison Bhatti, Mark Addy, Anthony Milner, Rachel Davies, Laura Jones
Raw and highly disagreeable drama about a group of prostitutes that spares no effort to depict their sordid and unhappy lifestyles, bringing all the less agreeable details of their profession into sharp focus, especially the hazards and dangers they face from a variety of punters. Extremely well acted and convincing, but incredibly depressing. The script is by Kay Mellor. Some sequels followed.
DRA 312 min (2 cassettes) mTV VIDrel: VCC/DISC V

BAND WAGON, THE ****
Vincente Minnelli USA
U
1953
Fred Astaire, Cyd Charisse, Jack Buchanan, Oscar Levant, Nanette Fabray, James Mitchell, Robert Grist, Thurston Hall, Ava Gardner, LeRoy Daniels, Dee Turnell, Jack Tessler, Elynne Ray, Peggy Murray, Judy Landon, Jimmie Thompson
Astaire plays a washed-up movie star who goes to Broadway to try his luck in this classic Hollywood musical. If the story is not up to much the numbers are and include highlights such as: "Dancing In The Dark", "Shine On Your Shoes" and "That's Entertainment", all of which are by Howard Dietz and Arthur Schwartz. One of the best numbers is a Mickey Spillane spoof – "The Girl Hunt".
MUS 108 min (ort 112 min) VIDrel: MGM/WHV V/dm

BANDIT 1: BANDIT GOES COUNTRY **
Hal Needham USA
PG
1994
Brian Bloom, Brian Krause, Christopher Atkins, Elizabeth Berkeley, Charles Nelson Reilly, Mel Tillis, Darryl Karolat, Al Wiggins, Heather Lynch, Amy Parrish, Della Basnight, Stanton Barrett, David Barrett, Neil Bonneyy, CK Bibby
A legendary trucker sets off to attend a family reunion and finds that nothing has changed since he left, and that he still has a few old scores to settle. At the same time, he tries to convince his old sweetheart that her intention to marry a local thug is not a wise decision. The first film in a series featuring this character. The SMOKEY AND THE BANDIT series was another very similar set of adventures.
Aka: BANDIT GOES COUNTRY
A/AD 87 min mTV VIDrel: CIC/SONOP V

BANDIT 2: BANDIT BANDIT **
Hal Needham USA
U
1994
Brian Bloom, Brian Krause, Richard Belzer, Gerard Christopher, Ami Dolenz, Gary Collins, Larry Manetti, Mary Ann Mobley, Ronna Reeves, John Schneider, Tom Nowicki, Greg Hohb, Mimi Eisman, Barnaby Carpenter, George Nannarello
In this first sequel, Bandit is out to recover a stolen experimental car, which he was to have driven to its press launch, where

his girlfriend's father (the local governor) is to present it. The loss of the car would endanger the US economy, so when it is stolen by an imposter, he has just three hours to recover it.
Aka: BANDIT BANDIT
A/AD 88 min (ort 90 min) mTV VIDrel: CIC/SONOP V

BANDIT 3: BEAUTY AND THE BANDIT **
Hal Needham USA
15
1994
Brian Bloom, Brian Krause, Henry Cho, Joe Cortese, Tony Curtis, Kathy Ireland, Mark Joy, Rick Warner, Tffany Cara, Norman "Max" Maxwell, Veronica Neal, Paul Nixon, Sal Ruffino, Walalce Merck, Alex Van, Joe Inscoe
Bandit is held up at gunpoint by a woman, and his car is stolen. More high-speed adventures ensue as he gives chase, picking up an Elvis lookalike and some nuns along the way.
Aka: BEAUTY AND THE BANDIT
A/AD 87 min mTV VIDrel: CIC/SONOP V

BANDIT 4: BANDIT'S SILVER ANGEL **
Hal Needham USA
PG
1995
Brian Bloom, Brian Krause, Scott Bloom, Christine Jensen, David Lenthall, Traci Lords, Donald O'Connor, Beth Bruce, Deacon Dawson, Jeffrey Pillards, April Tatro, Lou Criscuolo, Marc Macaulay, Randell Haynes, Fred Ottman, David Dannehl
Bandit's uncle persuades him to join a carnival, but the sheriff is still hot on his trail. More SMOKEY-style semi-comic mayhem.
Aka: BANDIT'S SILVER ANGEL
A/AD 86 min (ort 90 min) mTV VIDrel: CIC/SONOP V

BANDIT QUEEN ***
Shekhar Kapur INDIA/UK
18
1994
Seema Biswas, Nirmal Pandey, Manjoj Bajpai, Rajesh Vivek, Raghuvir Yadav, Govind Namdeo, Saurabh Shukla, Aditya Srivastava, Agesh Markam, Anirudh Agarwal, Anil Sahu, Chotelai Siraswal, Nazim Patel, Nazim Hussain, Pawan Gupta
A low caste child bride embarks on a career as a bandit after she is unable to get justice following her gang rape, becoming so notorious that she earns the admiration of others of her caste. Banned in India, the film is largely based on the diaries Phoolan Devi dictated in prison after her capture by the Indian authorities. Shocking, brutal and graphic, the work is a strong examination of the plight of low caste women in India, but is cumbersome as a piece of cinema.
A/AD 114 min (ort 120 min) VIDrel: MAINPIC/RTM V
Boa: diaries of Phoolan Devi.

BANK ROBBER **
Nick Mead USA
18
1993
Judge Reinhold, Lisa Bonet, Patrick Dempsey, Forest Whitaker, Olivia D'Abo, Paula Kelly, Mariska Hargitay, Michael Jeter, Joe Alaskey, Andy Romano, Warren Munson, Don Perry, Lisa Marie Spikerman, James Farde, Stephen McDonough
A bank robber pulls one last job that goes badly wrong when he is recorded on a surveillance camera and is forced to hide out in a hotel, where his only friend turns out to be a good-hearted hooker. Meanwhile, a couple of bumbling cops are on his trail. A witless and largely unamusing comedy whose attempts at satire seem flat and contrived.
COM 89 min VIDrel: EIV/SONOP V

BANK SHOT ***
Gower Champion USA
PG
1974
George C. Scott, Joanna Cassidy, Sorrell Brooke, G. Wood, James, Bob Balaban, Bibi Osterwald, Frank McRae, Don Calfa, Harvey Evans, Hank Stohl, Liam Dunn, Jack Riley, Pat Zurica, Harvey J. Goldberg, Jamie Reidy
A criminal plans to rob a bank – by moving the entire building! A crazy and entertaining comedy that's a sequel of sorts to THE HOT ROCK.
COM 83 min VIDrel: MGM/WHV L/A V
Boa: novel by Donald E. Westlake.

BAR GIRLS **
Marita Giovanni USA
18
1995
Nancy Allison Wolfe, Liza D'Agostino, Camilla Griggs, Michael Harris
A portrayal of the love-lives of a number of patrons at title lesbian bar in Los Angeles forms the basis for this adaptation of Hoffman's play. Here, the plot revolves around developments in one couple's troubled relationship and the disruptive influence of a third party.
DRA 95 min VIDrel: WILLPRO/RTM V/s
Boa: play by Laura Hoffman.

BARABBAS *   PG
Richard Fleischer ITALY/USA   1962
Anthony Quinn, Jack Palance, Silvana Mangano, Vittorio Gassman,
Valentina Cortese, Katy Jurado, Harry Andrews, Arthur Kennedy,
Michael Gwynn, Ernest Borgnine, Norman Wooland, Valentina
Cortese, Michael Gwynn, Douglas Fowley
Interminable biblical epic that attempts to tell of what happened
to Barabbas who was set free instead of Jesus. Sent to the silver
mines he becomes a Christian and then a gladiator. Well acted
but this one just goes on and on.
DRA   127 min (ort 134 min)   VIDrel: MIA/COLUM   V
Boa: novel by Par Lagerkvist.

BARB WIRE **   18
David Hogan USA   1995
Pamela Anderson Lee, Temuera Morrison, Victoria Rowell, Jack
Noseworthy, Xander Berkeley, Steve Railsback, Udo Kier, Amir
Aboulela, Adriana Alexander, David Andriole, Miles Dougal, Vanessa
Lee Asher, Ron Balicki, Diana Lee Insanto
In a bleak future a tough woman runs a bar that acts as a magnet
for troublemakers. Based on a comic-strip, this is a vanity-
vehicle outing for TV's "Baywatch" star, that demonstrates her
limitations as an actress, though given the central character's
hardbitten and emotionless persona, this works to the film's
advantage. The film was rated 15 in the cinema, but had two
extra minutes added to enhance its rating if not its entertainment
value.
A/AD   110 min cC   VIDrel: POLY/POLYREC   V/sur
Boa: story by Ilene Chaiken.

BARBARELLA ***   15
Roger Vadim FRANCE/ITALY   1967
Jane Fonda, John Phillip Law, David Hemmings, Milo O'Shea, Ugo
Tognazzi, Anita Pallenberg, Marcel Marceau, Claude Dauphin,
Antonio Sabato, Veronique Vendell, Serge Marquand, Talitha Pol,
Nino Muso, Sergio Ferrero, Franco Gula
In the 41st century Barbarella is sent on a mission to investigate
the disappearance of a famous scientist. A highly imaginative
attempt to bring to life a French comic strip character, let down
by a flawed script and poor direction. It now looks very dated,
not least because of its musical score.
Aka: BARBARELLA, QUEEN OF THE GALAXY
FAN   97 min   VIDrel: CIC/SONOP; PION (LV only)   V LV

BARBARIAN AND THE GEISHA, THE *   U
John Huston USA   1958
John Wayne, Eiko Ando, Sam Jaffe, So Yamamura, Norman Thomson,
James Robbins, Morita, Kodaya Ichikawa, Hiroshi Yamoto, Tokujiro
Iketaniuchi, Fuji Kasai, Takeshi Kumagai
The story of the first US diplomat to visit Japan in 1856, who
encounters a good deal of local opposition, but finds solace in
the charms of a beautiful geisha. A flabby and disjointed
costume epic, not helped by a ludicrous piece of casting with
Wayne as the ambassador.
DRA   100 min (ort 105 min) B/W   VIDrel: 20TH/TECH
V/h

BARBARIAN QUEEN 2 *   18
Joe Finley ITALY   1988
Lana Clarkson, Greg Wrangler, Rebecca Wood, Elizabeth Jaegen,
Roger Cuyno, Alejandro Brancho, Cecilia Tijerina
The warlike daughter of a dead king leads a band of female
rebels in an attempt to regain the throne of her father, in this
forgettable costumed romp, full of the obligatory quota of
bulging biceps and ample bosoms.
Aka: BARBARIAN QUEEN 2: THE EMPRESS STRIKES BACK
A/AD   76 min Cut (2 min 48 sec – ort 87 min)
VIDrel: VISVID/POLYREC   V

BARBARIANS AT THE GATE ***   15
Glenn Jordan USA   1992
James Garner, Jonathan Pryce, Peter Riegert, Joanna Cassidy, Fred
Dalton Thompson, Leilani Ferrer, Matt Clark, Jeffrey DeMunn, David
Rasche, Tom Aldredge, Graham Beckel, Peter Dvorsky, Petr Frechette,
Rita Wilson
Dramatisation of one of the nastiest corporate battles of the
avaricious 1980s when the head of RJR-Nabisco tried to arrange
a buyout only to encounter fierce opposition. Fine acting, an
excellent script and capable direction all combine to lend fasci-
nation to an otherwise mundane subject.
DRA   113 min mTV   VIDrel: WHV L/A   V/sh
Boa: book by Bryan Burrough and John Heylar.

BARBARY COAST ***   U
Howard Hawks USA   1935
Joel McCrea, Walter Brennan, Edward G. Robinson, Brian Donlevy,
Frank Craven, Miriam Hopkins, Clyde Cook, Harry Carey, Matt
McHugh, Otto Hoffman, J.M. Kerrigan, Rollo Lloyd, Donald Meek,
Roger Gray, Wong Chung, Russ Powell
Richly melodramatic account of San Francisco in the days of the
Gold Rush, with plenty of excellent acting and a lively plot, as
a dance-hall queen comes into conflict with a local big-shot.
David Niven appears very briefly as an extra.
DRA   86 min (ort 90 min) B/W   VIDrel: VCC/DISC   V

BARCELONA **   12
Whit Stillman USA   1994
Taylor Nichols, Chris Eigeman, Tushka Bergen, Mira Sorvino, Pep
Munne, Nuria Badia, Hellena Schmied, Thomas Gibson, Jack Gilpin,
Pere Ponce, Laura Lopez, Francis Creighton, Edmon Roch, Diana
Sassen, Angels Bassas, Elisenda Bautista
An American naval office with playboy manners and a weak-
ness for a pretty face, visits his sales-rep. cousin who is living
in Barcelona. The two men hit the town but have to cope with
the local anti-American sentiment in their pursuit of female
company. A very talky follow-up to the director's previous film
METROPOLITAN, which it resembles in a number of ways.
COM   97 min (ort 102 min)   VIDrel: COLUM/SONOP
V/sur

BARCHESTER CHRONICLES, THE ***   U
David Giles UK   1982
Donald Pleasence, Nigel Hawthorne, Geraldine McEwan, Susan
Hampshire, Mike Gwilym, Alan Rickman
A very handsome adaptation of two Trollope novels dealing
with the plight of a respected reverend who becomes the victim
of intrigue and dishonesty when the church of which he is a
priest is accused of corruption. A most enjoyable drama, origi-
nally shown in seven 55-minute episodes.
DRA   353 min (2 cassettes) mTV   VIDrel: BBC/TECH   V/h
Boa: novels The Warden and Barchester Towers by Anthony
Trollope.

BARE EXPOSURE **   (18)
Ralph Portillo USA   1993
Tammy Parks, Jack Slater, Ashlie Rhey, Andrea Suzanne, Westey
Scott, Kyle Anderson, Stephanie Carlisle, Leonardo Millan, Michael
Albala, Michael Earl Caza, J.D. Douglas, Lorne B. Green, Vincent
Lemiuex, Rick Scandlin, Matt Duncan
A college kid dreams up wet T-shirt contest to make money to
pay off his debts, and this proves to be a great success. An inno-
cent softcore romp of no great distinction.
A   86 min (ort 90 min)   SATrel: SKY MOVIES

BAREFOOT CONTESSA, THE ***   PG
Joseph Mankiewicz USA   1954
Humphrey Bogart, Ava Gardner, Edmond O'Brien, Marius Goring,
Rossano Brazzi, Elizabeth Sellars, Valentina Cortesa, Warren Stevens,
Franco Interlenghi, Mari Aldon, Bill Fraser, Enzo Staiola, Bessie Love,
Diana Decker, Jim Gerald
A mentor-director propels a beautiful Spanish dancer to
Hollywood stardom but she eventually marries a count.
However, he is unable to father children so she embarks on an
extramarital affair but is eventually shot by him. An absorbing
and well produced melodrama slightly spoilt by an over-wordy
and often obscure script. AA: S. Actor (O'Brien).
DRA   125 min (ort 128 min) B/W   VIDrel: MGM/WHV   V

BAREFOOT IN THE PARK *   PG
Gene Saks USA   1967
Jane Fonda, Robert Redford, Charles Boyer, Mildred Natwick, Herb
Edelman, Mabel Albertson, Fritz Feld, James Stone, Ted Hartley
Plotless and pointless comedy about a couple who live five
storeys up in an apartment block that has no lift. The Neil Simon
script delivers precious few laughs, but the running-joke about
climbing all the stairs begins to grow thin after a while.
COM   101 min (ort 105 min)
VIDrel: 4-FRONT/POLYREC/CIC L/A   V/h
Boa: play by Neil Simon.

BAREFOOT KID, THE ***   15
Johnny To (To Ke-Fung) HONG KONG   1993
Aaron Kwok (Kwok Fu-Shing), Ti Lung, Maggie Cheung, Kent Tsang
Kong
After the death of his father, a young man goes to a small town
in search of work. He is helped by a friend of his father's who

works at a tapestry factory but falls under the control of the local gang when he wins a martial arts contest. Fast-paced, exciting and inevitably light on plotting.

Aka: YOUNG HERO

MAR 86 min (ort 90 min) wScrn VIDrel: MADE/RTM V

**BARGEE, THE ** PG
Duncan Wood UK 1964
Harry H. Corbett, Hugh Griffith, Ronnie Barker, Eric Sykes, Julia Foster, Miriam Karlin, Derek Nimmo, Richard Briers, Eric Barker, Norman Bird, George A. Cooper, Grazina Frame, Brian Wilde, Wally Patch, Godfrey Winn
A barge operator who fancies himself as a bit of a ladies' man gets the daughter of a lock-keeper pregnant and is obliged to marry her. A vulgar comedy that's slightly reminiscent of all those "Carry On" films, and not terribly amusing to boot. Script is by Ray Galton and Alan Simpson, of "Steptoe And Son" fame.

COM 101 min (ort 106 min) VIDrel: LUMI/SPEAR V

BARKLEYS OF BROADWAY, THE * U
Charles Walters USA 1949
Fred Astaire, Ginger Rogers, Oscar Levant, Billie Burke, Gale Robbins, Jacques Francois, George Zucco, Clinton Sundberg, Inez Cooper, Wilson Wood, Carol Brewster, Jean Andren, Laura Treadwell, Margaret Bert, Hans Conreid
After a ten year break Astaire and Rogers teamed up again for this witty Comden-Green story of a showbiz couple who split up but eventually get back together again. Songs include: "You'd Be Hard To Replace" and "They Can't Take That Away From Me".

MUS 104 min (ort 110 min) VIDrel: MGM/WHV V

BARNABO OF THE MOUNTAINS * 12
Mario Brenta FRANCE/ITALY/SWITZERLAND 1994
Marco Pauletti, Duilio Fontana, Carlo Caserotti, Antonio Vecellio, Angelo Chiesura, Alessandra Milan, Elisa Gasperini, Marco Tonin, Francesca Rita Giovannini, Pino Tosca, Alessandro Uccelli, Maria Da Pra, Gianni Bailo
Set in the aftermath of WW1, with the title character a mountain ranger who works in the Dolomites. When he runs into a gang of poachers, he has a crisis of conscience finding that although he has sympathy for them, his duty is to turn them in. Moderately enjoyable fare, with attractive outdoors locations adding immeasurably to the film's atmosphere.

Aka: BARNABO DELLE MONTAGNE

DRA 125 min CINrel
Boa: novel by Dino Buzzati.

BAROCCO * 15
Andre Techine FRANCE 1976
Gerard Depardieu, Isabelle Adjani, Marie-France Pisier, Jean-Claude Brialy, Helene Surgere, Claude Brasseur
A small-time criminal and thug conspires with his girlfriend to blackmail a local councillor, but their plans misfire and result in murder. Meant to be a homage to film noir, this is a slick and gripping thriller. Directed with a few self-indulgent and arty touches, it benefits greatly from some fine set-piece confrontations and strong performances from the two leads.

THR 100 min (ort 120 min) wScrn VIDrel: ARTPRO/RTM V

BARON BLOOD * 15
Mario Bava ITALY/WEST GERMANY 1972
Joseph Cotten, Elke Sommer, Alan Collins (Luciano Pigozzi), Massimo Girotti, Antonio Canatafora, Humi (Umberto) Raho, Nicoletta Elmi, Dieter Tressler, Rada Rassimov
Gruesome horror story telling of the attempts made by the descendants of a vampiric nobleman to deal with him, after a witch's spell bring him back from the grave, intent on having his revenge on the present-day owners of his castle. The standard plot is helped by some highly atmospheric settings and intelligent use made of lighting.

Aka: BLOOD BARON, THE; CHAMBER OF TORTURES; GLI ORRORI DEL CASTELLO DI NORIMBERGA; THIRST OF BARON BLOOD, THE; TORTURE CHAMBER OF BARON BLOOD, THE

HOR 93 min wScrn VIDrel: REDEM/RTM V

**BARRY LYNDON ** PG
Stanley Kubrick UK 1975
Ryan O'Neal, Marisa Berenson, Patrick Magee, Hardy Kruger, Gay Hamilton, Marie Kean, Diana Koerner, Murray Melvin, Frank Middlemass, Andre Morell, Arthur Sullivan, Godfrey Quigley, Leonard Rossiter

Based on a story of a young Irish boy who desires success but is spoilt by it, this is little more than an exquisitely detailed look at 18th century England. An overlong, languid and very beautiful film, but a terribly empty one. Written by Kubrick and with narration by Michael Hordern. AA: Cin (John Alcott), Art/Set (Ken Adam and Roy Walker/Vernon Dixon), Cost (Ulla-Britt Sonderlund/Milena Canonero), Score (Leonard Rosenman).

DRA 185 min VIDrel: MGM/WHV L/A V
Boa: novel The Memoirs Of Barry Lyndon, Esq. by William Makepeace Thackeray.

**BARRY McKENZIE HOLDS HIS OWN * 18
Bruce Beresford AUSTRALIA 1974
Barry Humphries, Barry Crocker, Donald Pleasence, Dick Bentley, Tommy Trinder, Ed Devereaux, Frank Windsor, Deryck Guyler, Arthur English, Roy Kinnear, John Le Mesurier, Desmond Tester, Louis Negin, Paul Humpoletz
A stupid sequel to THE ADVENTURES OF BARRY McKENZIE with both him and his brother plus Dame Edna (who is mistaken for Queen Elizabeth II) being kidnapped and taken to Transylvania. Crocker plays the bill-swilling twit created in the "Private Eye" comic-strip as well as his twin brother. As tiresome as it is witless and a waste of the considerable talents of both Humphries and Crocker.

COM 93 min (ort 95 min) VIDrel: MIA L/A V

**BARTON FINK ** 15
Joel Coen USA 1991
John Goodman, John Turturro, Judy Davis, Michael Lerner, John Mahoney, Jon Polito, Tony Shalhoub, Steve Buscemi, David Warrilow, Richard Portnow, I.M. Hobson, Christopher Murney, Megan Faye, Lance Davis, Harry Bugin, Anthony Gordon
A New York playwright with aspirations that involve making art accessible to the Common Man, is called to Hollywood following the triumph of one of his plays on Broadway. Given the task of writing the screenplay for a wrestling picture, he finds himself assailed both by writer's block and a succession of weirdos, one of whom embroils him in a murder. Well received at Cannes, this is a surreal, nightmarish farce of 1940s Hollywood lacking both depth and impact.

DRA 111 min (ort 116 min) VIDrel: VCC/DISC/COLUM V/sur

**BASED ON AN UNTRUE STORY * 15
Jim Drake USA 1993
Morgan Fairchild, Harvey Korman, Dyan Cannon, Robert Goulet, Dan Hedaya, Victoria Jackson, Dan Hedaya, Ricki Lake, Jeremy Miller, Ellia English, Renee Hambley, Del Hunter White, George McGrath, Raymond Patterson, Joleen Lutz
Unfunny, wildly over-the-top spoof on real-life TV movies that tries hard to get laughs but is just an exercise in bad taste. The mad plot involves a woman creator of perfumes, who discovers that she was adopted, that she must have a life-saving nose operation in seventy-two hours and finally that a serial killer is stalking topless dancers. And whilst she contends with all of this, she still manages to find time to attempt the location of the rest of her family.

COM 89 min (ort 91 min) mTV VIDrel: 20TH/TECH V

BASIC DECEPTION * 18
Ivan Passer USA 1991
Mark Harmon, Mimi Rogers, Paul Gleason, M. Emmet Walsh
A woman's husband walks out on her after seven years of marriage, and she discovers that she never really knew him, and that he had a different identity. She hires a private eye to get at the truth, but as the mystery begins to unravel, evidence is uncovered pointing to her errant spouse being a murderous psychopath. Quite a gripping effort, though the final resolution comes as no surprise. The music is by William Olvis.

Aka: FOURTH STORY

THR 86 min (ort 88 min) VIDrel: VCC/DISC L/A V

**BASIC INSTINCT * 18
Paul Verhoeven USA 1992
Michael Douglas, Sharon Stone, Jean Tripplehorn, George Dzundza, Dennis Arndt, Leilani Sarelle, Bruce A. Young, Chelcie Ross, Dorothy Malone, Wayne Knight, Daniel Von Bergen, Stephen Tobolowsky, Benjamin Mouton, Jack McGee
A police detective assigned to Homicide falls for a cold and over-sexed woman who may in fact be a killer. A nasty and mean-spirited cop thriller that injects a dose of sex in place of a decent plot. A few homophobic touches make it even more disagreeable; a pity, as both Stone and Douglas try to do the best

with their banal and limited dialogue. The LV disc includes a short behind-the-scenes interview with the director and cast.
THR 122 min (ort 150 min) wScrn cC
VIDrel: POLY/POLYREC/GUILD; PION (LV only) V/sur LV

BASIL, THE GREAT MOUSE DETECTIVE *** U
John Musker/Ron Clements/Dave Michener/Burny Mattison
USA 1986
Voices of: Barrie Ingham, Vincent Price, Val Bettin, Alan Young, Susanne Pollatschek, Candy Candido, Eve Brenner, Melissa Machester
Sherlock Holmes gets the Disney treatment in this breathless adventure, which displays Disney's first use of computer animation (the chase sequence where mouse versions of "Holmes" and "Moriarty" battle it out on the face of Big Ben is highlight). The story has our furry sleuth battling the evil Professor Rattigan and locating a missing mouse toymaker. Though nothing like as good as the classic Disney cartoons, this one is well worth a look.
Aka: ADVENTURES OF THE GREAT MOUSE DETECTIVE; GREAT MOUSE DETECTIVE, THE
ANIM 71 min (ort 80 min) cC VIDrel: WDV/TECH
V/sur
Boa: novel Basil Of Baker Street by Eve Titus.

BASKET CASE *** 18
Frank Henenlotter USA 1981
Kevin Van Hentenryck, Terri Susan Smith, Beverly Bonner, Lloyd Pace, Robert Vogel, Diana Browne, Bill Freeman, Joe Clarke, Dorothy Strongin, Ruth Neuman, Richard Pierce, Kerry Ruff
Low-budget horror film about a man who arrives in New York, together with a mysterious padlocked wicker basket that contains his deformed telepathic mutant brother. Very tongue-in-cheek and often quite effective, with a few intriguing animated sequences.
HOR 85 min Cut (35 sec plus film cuts – ort 91 min)
VIDrel: POLY/POLYREC L/A V

BASKET CASE 2 *** 18
Frank Henenlotter USA 1990
Kevin Van Hentenryck, Annie Ross, Heather Rattray, Jason Evers, Kathryn Meisle, Ted Sorel, Judy Grafe, Chad Brown, Beverly Bonner, Leonard Jackson, Alexandra Auder, Brian Fitzpatrick, Gale Van Cott, Kuno Sponholz, Jan Saint
A potent sequel to the earlier film with Duane and his mutant brother Belial arriving at a community of freaks, where they experience a large number of monstrous encounters and the unwelcome attentions of an investigative journalist, whose final comeuppance is clearly borrowed from the 1934 film FREAKS. A gruesome, gory shocker that for all its wealth of horrific images is totally lacking in psychological depth.
HOR 86 min (ort 89 min) VIDrel: MED/POLYREC L/A
V/sh

BASKET CASE 3: THE PROGENY * 18
Frank Henenlotter USA 1991
Kevin Van Hentenryck, Annie Ross, Tina Louise Hilbert, Dan Biggers, Gil Roper, Jim O'Doherty
Dear old Granny Ruth is still as protective towards her charges as ever, and sets about creating a retreat for herself and her family of freaks.
HOR 86 min (ort 90 min)
VIDrel: POLY/POLYREC/BRAVE L/A V

BASKETBALL DIARIES, THE ** 18
Scott Kalvert USA 1995
Leonardo Di Caprio, Lorrraine Bracco, Marilyn Sokol, James Madio, Patrick McGaw, Mark Wahlberg, Roy Cooper, Vinnie Pastore, Bruno Kirby, Jimmy Papiris, Nick Gaetani, Lawrence Barth, Alexander Gaberman, Ben Jorgensen, Josh Mostel
A none too cohesive attempt to recreate the 1960s high-school experiences of the writer Jim Carroll who combined some skill at baseball with an unfortunate indulgence in drugs that soon became an overpowering addiction. Good performances are an asset but they alone cannot give the film the narrative structure it so badly lacks.
DRA 98 min (ort 102 min) VIDrel: FIRST/SONOP V
Boa: novel by Jim Carroll.

BASQUIAT ** 15
Julian Schnabel USA 1996
Jeffrey Wright, Benicio Del Toro, David Bowie, Gary Oldman, Christopher Walken, Dennis Hopper
An attempt to examine the short and unhappy life of Jean-

Michel Basquiat, whose talent for graffiti brought him to the notice of the art world, leading to a brief period of fame and death from a drugs overdose. A flawed and terribly self-indulgent film, that dispenses with a simple narrative structure in favour pretension and an obsession with the superficial. Bowie's performance as the ageing pop icon Andy Warhol is easily the best thing in this pompous movie.
DRA 106 min CINrel

BAT. 21 *** 15
Peter Markle USA 1987
Gene Hackman, Danny Glover, Jerry Reed, Clayton Rohner, David Marshall Grant, Erich Anderson, Joe Dorsey, Michael Ng, Theodore Chan Woei-Shyong, Don Ruffin, Scott Howell, Michael Raden, Timothy Fitzgerald, Stuart Hagen
A lieutenant colonel is shot down behind enemy lines in Vietnam and a rescue mission is mounted, using a pilot to guide him from the air. Having spent his entire career away from ground combat, the officer is shocked at the devastation wrought by American bombing and for the first time in his life sees the horrors of war. A mature film, eschewing RAMBO-style cliches in favour of a look at how both sides have become de-humanised by the conflict.
WAR 100 min (ort 106 min)
VIDrel: 4-FRONT/POLYREC/GUILD V/sur
Boa: book by William C. Anderson.

BATMAN *** 15
Tim Burton USA 1989
Michael Keaton, Jack Nicholson, Kim Basinger, Jack Palance, Billy Dee Williams, Jerry Hall, Pat Hingle, Michael Gough, Robert Wuhl, Tracey Walter, Lee Wallace
The film whose release in the UK prompted a new cinema certification to keep out the tots, this is a dark and moody rendition, bringing the famous D.C. Comics character to the screen, where he is all but swamped by the spectacular sets and Nicholson's wildly overblown portrayal of arch-criminal The Joker. A cruel and harsh story, flawed by weak plotting. Screenplay is by Sam Hamm. BATMAN RETURNS followed. AA: Art (Anton Furst and Peter Young).
A/AD 121 min (ort 126 min) wScrn cC VIDrel: WHV
V/sur

BATMAN RETURNS *** 15
Tim Burton USA 1992
Michael Keaton, Danny DeVito, Michelle Pfeiffer, Christopher Walken, Michael Gough, Michael Murphy, Cristi Conaway, Andrew Bryniarski, Pat Hingle, Steve Witting, Vincent Sciavelli, Jan Hooks, John Strong, Rick Zumwalt, Anna Katarina
It may be Batman who returns in this spectacular adventure, but it is DeVito as the "Penguin" and Pfeiffer as the "Catwoman" who really make their presence felt, as the Caped Crusader's newest adversaries. With a budget reputed to be in the region of $55,000,000, this flashy, exciting and imaginative film has much to enjoy, though a few sado-masochistic references are less welcome, and its severely sombre tone ensures that it is definitely not one for the kids.
A/AD 121 min (ort 126 min) wScrn cC VIDrel: WHV
V/sur

BATMAN FOREVER *** PG
Joel Schumacher USA 1995
Val Kilmer, Tommy Lee Jones, Jim Carrey, Nicole Kidman, Chris O'Donnell, Michael Gough, Pat Hingle, Drew Barrymore, Debi Mazar, Elizabeth Sanders, Rene Auberjonois, Joe Grifasi, Kimberley Scott, Margaret Paul Chan, Jon Favreau
Third outing for this comic character continues the story of his fight against crime. Here he finally acquires his faithful companion Robin and together they take on arch-criminals Two-Face and The Riddler. Kidman plays (very unconvincingly) a woman shrink who gets involved with Batman and his alter ego Bruce Wayne. Somewhat lighter in tone than the first two films but as ever it is the sets and special effects that dominate.
FAN 122 min cC VIDrel: WHV V/sur

BATMAN: MASK OF THE PHANTASM *** PG
Eric Radomski/Bruce W. Timm USA 1993
Voices of: Kevin Conroy, Dana Delany, Hart Bochner, Stacey Keach Jr, Abe Vigoda, Dick Miller, John P. Ryan, Efrem Zimbalist Jr, Robert Costanzo, Bob Hastings, Mark Hamill
A feature-length animation derived from the TV series that features Batman in a struggle with a mysterious figure who is trying to make him unemployed by ridding Gotham City of its

mobsters. Done in a 1940s style of animation, it has a nice dark and brooding feel about it that seems truer to the original comic books.
Aka: BATMAN: THE ANIMATED MOVIE
ANIM 73 min (ort 75 min) VIDrel: WHV V/sur

BATMAN: THE MOVIE ** U
Leslie Martinson USA 1966
Adam West, Burt Ward, Cesar Romero, Frank Gorshin, Burgess Meredith, Lee Meriwether, Alan Napier, Neil Hamilton, Madge Blake, Reginald Denny, Stafford Repp, Milton Frome, Gil Perkins, Dick Crockett, George Sawaya
The only feature film spin-off from the 1960s TV series, in which Batman and his sidekick Robin thwart a united attempt by the four greatest criminal masterminds to hold the entire world to ransom. Kids may enjoy watching the Joker, the Riddler, the Penguin and the Catwoman go about their nefarious business but this one has little appeal for adults.
Aka: BATMAN
COM 105 min VIDrel: 20TH/TECH L/A V

BATTERIES NOT INCLUDED ** PG
Matthew Robbins USA 1987
Hume Cronyn, Jessica Tandy, Frank McRae, Elizabeth Pena, Michael Carmine, Dennis Boutsikaris, Tom Aldredge, Jane Hoffman, John DiSanti, John Pankow, MacIntyre Dixon, Michael Greene
Tenants faced with the demolition of their building get help from an unexpected quarter when small friendly aliens pay them a visit. Another cloyingly sweet sugar-coated Steven Spielberg production.
COM 102 min (ort 107 min)
VIDrel: 4-FRONT/POLYREC/CIC L/A V/dm

BATTLE CRY ** PG
Raoul Walsh USA 1955
Van Heflin, Tab Hunter, Dorothy Malone, Anne Francis, Raymond Massey, Nancy Olson, Mona Freeman, Aldo Ray, James Whitmore, William Campbell, Fess Parker, Justus E. McQueen (L.Q. Jones), Perry Lopez, Carleton Young
Account of US Marines as they train, fight and romance. A disappointingly conventional melodrama that spends more time on their love life than anything else, and totally fails to capture the feeling of the original.
WAR 142 min (ort 149 min) wScrn VIDrel: WHV V/sh
Boa: novel by Leon Uris.

BATTLE FOR THE PLANET OF THE APES ** PG
J. Lee Thompson USA 1973
Roddy McDowall, Claude Akins, John Huston, Natalie Trundy, Severn Darden, Paul Williams, Lew Ayres, Austin Stoker, Noah Keen, Paul Stevens, Francis Nuyen, Bobby Porter, John Landis, Michael Stearns, Richard Eastham
This was the fourth and last sequel to PLANET OF THE APES and has a society in which man and ape live in harmony facing a threat from two opposing factions, one ape and the other mutant-human. A generally sub-standard sequel that uses footage from the earlier films. A dreary TV series followed.
FAN 83 min VIDrel: 20TH/TECH L/A V

BATTLE OF BRITAIN, THE ** PG
Guy Hamilton UK 1969
Laurence Olivier, Robert Shaw, Michael Caine, Susannah York, Christopher Plummer, Kenneth More, Trevor Howard, Ralph Richardson, Patrick Wymark, Curt Jurgens, Michael Redgrave, Nigel Patrick, Robert Flemyng, Edward Fox
Overlong attempt to tell the story of the Battle of Britain that focuses so closely on the details of the conflict that it becomes tedious rather than exciting. Weighed down almost to the point of collapse with star cameos (as is so often the case with British films), the superb flying sequences offer some compensation in the cinema, but will be largely wasted on TV.
WAR 110 min (ort 132 min) VIDrel: MGM/WHV V/h
Boa: book The Narrow Margin by D. Wood and D. Dempster.

BATTLE OF MIDWAY, THE ** PG
John Smight USA 1976
Toshiro Mifune, Charlton Heston, Henry Fonda, Robert Mitchum, Glenn Ford, Edward Albert, James Coburn, Hal Holbrook, Robert Wagner, Robert Webber, Ed Nelson, James Shigeta, Monte Markham, Chris George, Glenn Corbett, Conrad Yama
Account of the 1942 battle when the Japanese attacked the island of Midway during WW2 but found the tide of the war in the

Pacific turning against them. A noisy and confused story that concentrates on too many stars and too little action.
Aka: MIDWAY
WAR 128 min (ort 132 min)
VIDrel: 4-FRONT/POLYREC/CIC V/h

BATTLE OF THE BULGE ** PG
Ken Annakin USA 1965
Henry Fonda, Robert Shaw, Robert Ryan, Charles Bronson, James MacArthur, Dana Andrews, George Montgomery, Ty Hardin, Pier Angeli, Telly Savalas, Barbara Werle, Werner Peters, Hans Christian Blech
Account of the last major German offensive of WW2; it took place in the Ardennes in the middle of the winter of 1944. Apart from some good performances (such as Shaw's), this overlong and dull effort never allows the action to distract one from the banality of the script.
WAR 149 min (ort 163 min) wScrn VIDrel: WHV V/h
Boa: book by R.E. Merriam.

BATTLE OF THE RIVER PLATE, THE ** U
Michael Powell/Emeric Pressburger UK 1956
Peter Finch, John Gregson, Bernard Lee, Anthony Quayle, Michael Goodliffe, William Squire, Andrew Cruickshank, Christopher Lee, Ian Hunter, Jack Gwillim, Lionel Murton, Anthony Bushell, Peter Illing, Patrick Macnee
Rather talky and stilted tale of the British Navy's pursuit and trapping of the German pocket battleship the Graf Spee, in Montevideo harbour in 1939. Done in semi-documentary style with a few good moments, but too many characters vie for attention in this undisciplined effort.
Aka: PURSUIT OF THE GRAF SPEE
WAR 114 min VIDrel: VCC/DISC V
Boa: book Graf Spee by Michael Powell.

BATTLE OF THE SEXES ** U
Charles Crichton UK 1959
Peter Sellers, Robert Morley, Constance Cummings, Jameson Clark, Moultrie Kelsall, Alex Mackenzie, Roddy McMillan, Donald Pleasence, Ernest Thesiger
The uneventful and staid lives of the employees of a stuffy Edinburgh company are upset by the arrival of a pushy female efficiency expert, and one of their number contemplates murder. An interesting attempt at a black comedy that is only intermittently effective.
COM 81 min (ort 84 min) B/W VIDrel: POLY/POLYREC L/A V
Boa: short story The Catbird Seat by James Thurber.

BATTLESHIP POTEMKIN, THE ***** PG
Sergei M. Eisenstein USSR 1925
Alexander Antonov, Grigory Alexandrov, Vladimir Barsky, Mikhail Gomorov, Levshin
A partly fictitious account of the mutiny in 1905, aboard a battleship in Czarist Russia at the port of Odessa, brought vividly to life in a series of powerful images and the brilliant use of montage. Rightfully considered a masterpiece, the scene at the Odessa Steps when innocent civilians are mown down by troops is one of the most famous sequences ever filmed.
Aka: BRONENOSETS POTEMKIN; POTEMKIN
DRA 65 min B/W silent VIDrel: TART/20TH V

BATTLESTAR GALACTICA * PG
Richard A. Colla USA 1978 (re-released 1979)
Lorne Greene, Richard Hatch, Dirk Benedict, Ray Milland, Ed Begley Jr, John Colicos, Patrick Macnee, Lew Ayres, Jane Seymour, Laurette Spang, Terry Carter, Tony Swartz, Rick Springfield, George Murdock, John Fink
Feature film clipped together from the first and fifth episodes of a TV series, and it looks it too. Plenty of special effects in this rather poor story of how the survivors of an attack by evil robots, the Cyclons, escape from their doomed planet in search of a haven on planet Earth, led by a huge spaceship commanded by Greene. The special effects are by John Dykstra. See also MISSION GALACTICA: THE CYCLON ATTACK and CONQUEST OF THE EARTH.
FAN 119 min (ort 125 min) mTV VIDrel: CIC/SONOP V/h

BATTLING FOR BABY * PG
Art Wolff USA 1991
John Terlesky, Doug McClure, Mary Jo Catlett, Leigh Lawson, Courteney Cox, Debbie Reynolds, Suzanne Pleshette, Jeff Olson,

Nancey Silvers, Krista Garfano Capri, Lulee Fisher, Curley Green Jr, Melvin Ward, Robyn Adamson, Betsy Nagel
Two childhood friends find themselves locked into a foolish rivalry when their respective offspring decide to get married. When the wife unexpectedly finds herself pregnant after just two months of marriage, her concert pianist mother recoils in horror at the prospect of becoming a grandmother, while the girl's mother-in-law is absolutely delighted. An anaemic effort that totally fails to ignite and just feels rather silly and stilted.
COM 89 min (ort 93 min) VIDrel: CAPIT/GUILD V/sh

BEACH BABES FROM BEYOND ** 18
Ellen Cabot USA 1992
Joe Estevez, Don Swayze, Joey Travolta, Jacqueline Stallone, Burt Ward, Linnea Quigley, Sara Bellomo, Tamara Landry, Nicole Posey
A sort of EARTH GIRLS ARE EASY with a reversal of gender as three well-endowed female aliens land on Earth and proceed to take part in a bikini beauty contest, in this lightweight tongue-in-cheek spoof.
COM 76 min (ort 78 min) VIDrel: MED/20VIS V/sh

BEACHCOMBER, THE ** U
Muriel Box UK 1954
Glynis Johns, Robert Newton, Michael Hordern, Donald Pleasence, Donald Sinden, Paul Rogers, Ronald Lewis
A remake of the 1938 Charles Laughton film, in which a group of individuals live in retreat from civilisation on the remote tropical island of Baru. For one of these individuals, a drunkard beachcomber, life changes forever with the arrival of a missionary and his sister. This may not be an improvement on the earlier film, but Newton certainly gives a bravura performance in the title role.
COM 77 min B/W VIDrel: VCC/DISC V
Boa: short story Vessel of Wrath by William Somerset Maugham.

BEACHES *** 15
Gary Marshall USA 1988
Bette Midler, Barbara Hershey, John Heard, Spalding Gray, Lainie Kazan, James Read, Grace Johnston, Mayim Bialik, Marcie Leeds, Michael French, Lori Marshall, Frank Campanella, Nicky Blair, Joe Grifasi, Lucinda Crosby
An aspiring singer and a poor little rich girl meet under the boardwalk in Atlantic City and keep alive their friendship over the course of several years by writing letters. Later they become flatmates and the film follows their changing fortunes, with our singer enjoying success but her friend succumbing to a fatal virus. An enjoyable albeit excessively mawkish comedy-drama.
DRA 118 min (ort 123 min) VIDrel: TOUCH/TECH V/sur
Boa: novel by Iris Rainer Dart.

BEANSTALK ** U
Michael Paul Davis USA 1994
J.D. Daniels, Amy-Stock Poynton, Patrick Renna, Richard Moll, Richard Paul, David Naughton, Dominique Adler, Cindy Lou Sorensen, Cathy McAuly, Stuart Pankin, Margot Kidder, Joe Witherell, David O'Callaghan, Kurt A. Roesen
Updated, live-action retelling of "Jack and the Beanstalk" in which a young boy is given some experimental plants by a scientist and of course, these turn out to have unforeseen consequences for him.
JUV 78 min (ort 88 min) VIDrel: CIC/SONOP V/sh

BEAR, THE *** PG
Jean-Jacques Annaud FRANCE 1989
Jack Wallace, Tcheky Karyo, Andre Lacombe
An orphaned bear cub has to survive alone until its "adoption" by a huge Kodiak. However, the two now face danger from a couple of hunters. An absorbing nature story, made over several years and about a million miles removed from the Disney Studios output.
Aka: L'OURS
A/AD 89 min (ort 93 min) VIDrel: COLUM/SONOP L/A V/sh
Boa: novel The Grizzly King by James Oliver Curwood.

BEARSKIN: AN URBAN FAIRYTALE ** (PG)
Ann Guedes/Eduardo Guedes PORTUGAL/UK 1989
Tom Waits, Damon Lowry, Charlotte Coleman, Julia Britton, Bill Paterson, Isabel Ruth, Ian Dury, Alex Norton, Mark Arden, Pip Torrens, David Gant, Karl Collins, Russell Lee, Phil Atkinson, Tom Thompson, Glyn Grimstead, Joe Abdo

A young man on the run from gangsters hits on a novel way of escaping them, by dressing up in a bearskin and hiding out at a Punch and Judy show. An oddball movie, it uses settings in Portugal to stand in for ones in London, and though it has a few arresting moments, is a trifle too self-indulgent to sustain interest.
THR 95 min mTV TVrel: C4

BEAST MUST DIE, THE *** 15
Paul Annett UK 1974
Peter Cushing, Calvin Lockhart, Anton Diffring, Marlene Clark, Charles Gray, Ciaran Madden, Tom Chadbon, Michael Gambon, Sam Mansaray, Andrew Lodge, Carl Bohum, Eric Carte
A millionaire game-hunter invites a group of guests to his mansion for a weekend party, but his real intention is used to sophisticated electronics to unmask one of them as a werewolf. An unusual twist on the TEN LITTLE INDIANS theme that's rather well done, and comes complete with a "guess who?" break towards the end of the film.
Aka: BLACK WEREWOLF
HOR 88 min (ort 93 min) VIDrel: LUMI/SPEAR L/A V
Boa: novel There Shall Be No Darkness by James Blish.

BEAST WITH FIVE FINGERS, THE ** 12
Robert Florey USA 1946
Robert Alda, Andrea King, Peter Lorre, Victor Francen, J. Carrol Naish, Charles Dingle, Belle Mitchell, Pedro De Cordoba, John Alvin, David Hoffman, Patricia White, Barbara Brown, William Edmunds, Ray Walker
A disembodied hand gets up to the usual naughtiness when its owner, a tyrannical, disabled pianist falls down the stairs and dies. Lorre as an ageing servant is of course excellent in an otherwise fairly unremarkable film that is said to have had some of the effects worked on by Luis Bunuel but is hardly any the better for that. See both THE CRAWLING HAND and THE HAND, the latter being an updated version of this tale.
HOR 86 min (ort 90 min) B/W VIDrel: MGM/WHV V/h
Boa: short story by William Fryer Harvey.

BEASTMASTER, THE ** 15
Don Coscarelli USA 1982
Marc Singer, Tanya Roberts, Rip Torn, John Amos, Rod Loomis, Ben Hammer, Ralph Stratt, Josh Milrad, Billy Jacoby, Tony Epper, Janet De May, Chrissy Kellogg, Janet Jones, Vanna Bonta, Kim Tabet, Daniel Zormeier, Jim Driggers
More sword and sorcery as our hero, who can communicate with animals, falls in love with a slave girl and does battle with the forces of evil, in the shape of a wicked priest who had his father killed. Good ideas and excellent cinematography (by John Alcott) are entirely wasted on a film that never rises above the level of a comic-book. Two sequels of similar attributes followed.
FAN 110 min (Cut at film release – ort 118 min) VIDrel: WHV V/sur

BEASTMASTER 2 ** PG
Sylvio Tabet USA 1990
Marc Singer, Sarah Douglas, Wings Hauser, Kari Wuhrer, James Avery, Robert Fieldsteel, Arthur Malet, Robert Z'dar, Michael Berryman, Jim Eagle, David Carrera, Carl Ciarfalio, Larry Dobkin, Steve Donmeyer, John Fifer
Another dreary Good versus Evil saga set in an ancient mythical past, where an evil warrior uses magic to control a cowed populace, only resisted by the brave Beastmaster and his small band of rebels. However, our despotic ruler travels forward in time to present-day L.A (strange that it's always either L.A. or New York) to capture a weapon of great power. A sporadically amusing romp, quite good fun, but woefully unimaginative.
Aka: BEASTMASTER 2: THROUGH THE PORTAL OF TIME
FAN 102 min (ort 107 min) VIDrel: POLY/POLYREC/MED V/sur

BEASTMASTER 3: THE EYE OF BRAXUS ** PG
Gabrielle Beaumont USA 1995
Marc Singer, Tony Todd, Keith Coulouris, Sandra Hess, Casper Van Dien, Patrick Kilpatrick, Leslie-Anne Down, David Warner
A wicked ruler is planning to steal a gem of great power in order to resurrect a demon-god who will then take control of the entire universe. Another helping of Good versus Evil in this sword-and-sorcery series, that proceeds along over-familiar lines and has a very cheap look about it.
FAN 87 min (ort 92 min) mTV cC VIDrel: CIC/SONOP V/dm

BEAT STREET ✶✶ 15
Stan Lathan USA 1984
Rae Dawn Chong, Guy Davis, John Chardiet, Leon W. Grant, Robert Taylor, Saundra Santiago, Mary Alice, Shawn Elliot, Jim Borrello, Dean Elliott, Franc Reyes, Tonya Pinkins, Lee Chamberlin, Antonia Rey, Duane Jones
Lively and noisy film, telling of the efforts of a group of youngsters to escape the poverty of their surroundings through dance and music. Shot entirely on location in New York City. A slick and flashy version of WILD STYLE aimed clearly at the breakdancing set and produced by Harry Belafonte.
MUS 104 min (ort 106 min) VIDrel: L/A V

BEAT THE DEVIL ✶✶ U
John Huston UK 1953
Humphrey Bogart, Jennifer Jones, Peter Lorre, Ivor Bernard, Marco Tulli, Gina Lollabrigida, Robert Morley, Edward Underdown, Marion Perroni, Alex Pochet, Aldo Silviani, Guilio Donnini, Saro Urzi, Juan De Landa, Mimo Peli
Confused satire about con-men in Africa who are intent on double-crossing each other in the search for a uranium mine. Scripted by Huston and Truman Capote but for all that, one of Bogart's less well known films, and rightly so.
COM 96 min (oty 100 min) B/W VIDrel: FABFIL/SPEAR V
Boa: novel by James Helvick.

BEATLES: A HARD DAY'S NIGHT ✶✶✶ U
Richard Lester UK 1964
George Harrison, Paul McCartney, John Lennon, Ringo Starr, Wilfrid Brambell, Norman Rossington, Victor Spinetti, Anna Quayle, John Junkin, Deryck Guyler, Kenneth Haigh, Richard Vernon, Lionel Blair
Wonderfully fresh at the time, this is a zany musical comedy vehicle for The Beatles, that follows them on a train journey from Liverpool to London to take part in a TV show. The first venture for the group into film and containing some of their best songs from that period. The deliberately chaotic style of direction may have dated rather badly, but the music is as good as ever. The tape includes some extra footage not shown in the original release.
Aka: HARD DAY'S NIGHT, A
MUS 108 min (ort 98 min) B/W
VIDrel: VCC/DISC; PION (LV only) V/sh LV

BEATLES: HELP! ✶✶ U
Dick Lester UK 1965
John Lennon, Paul McCartney, George Harrison, Ringo Starr, Leo McKern, Eleanor Bron, Victor Spinetti, Roy Kinnear, Patrick Cargill, John Bluthal, Ronnie Brody, Bob Godfrey, Louis Mansi, Rupert Evans, Andreas Malandrinos
Very much a vehicle film for the Fab Four, this episodic and anarchic attempt to build on the success of A HARD DAY'S NIGHT has The Beatles being chased by a religious sect from the Far East, who wish to regain a sacred ring. Frenetic and tiring, it lacks the freshness of the first film, but is worth hearing if not worth seeing. Songs include "Ticket To Ride" and "You've Got To Hide Your Love Away". The tape includes extra footage not shown before.
Aka: HELP!
MUS 150 min (ort 92 min) VIDrel: VCC/DISC; PION (LV only) V LV

BEATLES: THE MAGICAL MYSTERY TOUR ✶✶ PG
John Lennon/Ringo Starr/Paul McCartney/George Harrison UK 1967
John Lennon, Ringo Starr, Paul McCartney, George Harrison, Jessie Robins, Victor Spinetti
A silly, plotless and self-indulgent vanity vehicle for The Beatles, that follows the fortunes of a coachload of travellers who, having had their day out hijacked by some strange magicians, stagger from one inconsequential adventure to another. Well worth hearing, but too dated and turgid to be worth watching. Songs include "The Fool On The Hill", "Blue Jay Way" and "Your Mother Should Know".
Aka: MAGICAL MYSTERY TOUR, THE
MUS 53 min (ort 60 min) mTV VIDrel: VCC/DISC V/sh

BEATLES: YELLOW SUBMARINE ✶✶✶✶ U
George Dunning UK 1968
Voices of: Paul Angelis, John Clive, Dick Emery, Geoff Hughes, Lance Percival
A charming musical fantasy that offers brilliant animated sequences and Beatles tunes – an unbeatable combination. The

story takes The Beatles on a journey to Pepperland, where they have to free its music-loving folk from the wicked Blue Meanies. A film of rich inventiveness that marked a highpoint for British animation; it has yet to be equalled.
Aka: YELLOW SUBMARINE
ANIM 82 min VIDrel: MGM/WHV L/A V/sh

BEAUMARCHAIS ✶✶ 15
Edouard Molinaro FRANCE 1996
Fabrice Luchini, Manuel Blanc, Sandrine Kiberlain, Michel Serrault, Jacques Weber, Michel Piccoli, Dominique Besnehard, Jean-Francois Balmer, Axelle Laffont, Florence Thomassin, Claire Nebout, Jean-Claude Brialy, Murray Head
Well mounted and opulent costume drama inspired by two famous operas – "The Marriage of Figaro" and "The Barber of Seville". The central character is an eighteenth century wit and firebrand, who divides his time between writing plays and attacking corruption in high places. A very handsome film to look at, yet quite lifeless, and as the camera moving lovingly from one carefully staged set-piece to the next, little attention is given to narrative thrust.
Aka: BEAUMARCHAIS L'INSOLENT
DRA 100 min VIDrel: ARTIF/20TH V
Boa: play (unpublished) by Sacha Guitry.

BEAUTIFUL DREAMERS ✶✶✶ 15
John Kent Harrison CANADA 1990
Colm Feore, Wendel Meldrum, Sheila McCarthy, Rip Torn, Colin Fox, David Gardner, Barbara Gordon, Marsha Moreau, Albert Schultz, Angelo Rozacos, R.D. Reid, Gordon Masten, Gerry Quigley, Roland Hewgil, Roger Clown, Peter Blais
This absorbing human-interest drama, written by Harrison, is set in the Canadian town of London in 1880, where, under the influence of American poet Walt Whitman (nicely played by Torn) the director of an insane asylum agrees to humanise both the treatment and general regime. In adopting a more liberal approach, he naturally has to overcome the prejudices of both staff and townsfolk. A warm and uplifting film, with music by Lawrence Shragge.
DRA 103 min VIDrel: MANGA/SONOP V/sur

BEAUTIFUL GIRLS ✶✶✶ 15
Ted Demme USA 1995
Matt Dillon, Noah Emmerich, Annabeth Gish, Lauren Holly, Timothy Hutton, Rosie O'Donnell, Max Perlich, Martha Plimpton, Natalie Portman, Mira Sorvino, Michael Rapaport, Uma Thurman, Pruitt Taylor Vince, Anne Bobby
An ensemble piece that opens with a group of high school friends getting together for regular drinking sessions, swapping stories and dating the same girls, whilst all the while dreaming of what things might have been like had they plucked up the courage to leave their home-town of Knight's Ridge, Massachusetts. Hutton is the only one who got away, and when he makes a return trip he finds himself falling for a young local girl.
COM 112 min CINrel

BEAUTIFUL MYSTERY: THE LEGEND OF BIGHORN ✶✶✶ 18
Nakamura Genji JAPAN 1987
Nagatomo Tatsuya, Suto Hajime, Shiyuto Kei, Yamashima Kaoru
Reputed to be Japan's first gay porno film, this is a irreverent satire that takes a jaundiced look at some of the recent formative events in Japanese history. From body-building to ritual suicide, the film pays homage to the oddities of Japanese culture, making the central character a Mishima-like figure whose followers are soon to commit suicide for no clear reason. Quite a funny if uneven mixture of slapstick antics and sharp social comment.
DRA 60 min VIDrel: DTK/RTM V

BEAUTIFUL THING ✶✶✶ (15)
Hettie MacDonald UK 1995
Glen Berry, Linda Henry, Scott Neal, Ben Daniels, Tameka Empson, Meera Syal, Martin Walsh, Steven Martin, Andrew Fraser, John Savage, Julie Smith, Jeillo Edwards, Anna Karen, Garry Cooper, Daniel Bowers, Terry Duggan, Sophie Stanton
A working-class comedy-drama set in a South London tenement where two boys discover they are gay, the director happily giving both this realisation and the other trials the boys face something of a comic slant. Debut director MacDonald's film is a trifle over emphatic at times, but also quite touching in examining the growing rapport between the central characters.

Screenplay is by Harvey, with much of the sharp dialogue taken directly from his successful play.
COM   94 min   VIDrel: FILM4   V/sh
Boa: play by Jonathan Harvey.

BEAUTY AND THE BEAST, THE *****   PG
Jean Cocteau FRANCE   1946
Jean Marais, Josette Day, Mila Parely, Michel Auclair, Marcel Andre, Nane Germon, Raoul Marco, Gilles Watteaux, Noel Blin
Cocteau's version of the classic fairy tale Beauty and the Beast represents cinema at its best. One of the greatest films ever made and a work of timeless and magical beauty. The music is by Georges Auric.
Aka: LA BELLE ET LA BETE
FAN   89 min (ort 100 min) B/W   VIDrel: CONNO/RTM   V

BEAUTY AND THE BEAST ***   (U)
Fielder Cook USA   1976
George C. Scott, Trish Van Devere, Virginia McKenna, Bernard Lee, Patricia Quinn, Michael N. Harbour, William Relton
A newer version of the classic fairy tale of the father who is forced to give his youngest daughter to the Beast. Though it cannot compare to Cocteau's masterpiece LA BELLE ET LA BETE, good acting and some striking make-up (by Del Acevedo, John Chambers and Dan Striepke) make it worthwhile.
JUV   86 min mTV   VIDrel: L/A   V

BEAUTY AND THE BEAST ****   U
Gary Trousdale/Kirk Wise USA   1991
Voices of: Paige O'Hara, Robby Benson, Richard White, Jerry Orbach, David Ogden Stiers, Anglea Lansbury, Bradley Michael Pierce, Richard White, Jesse Corti, Rex Everhart, Jo Anne Worley, Kimmy Robertson, Brian Cummings, Tony Jay
A beautiful young woman agrees to live in the gloomy home of a beast and eventually falls in love with him, freeing him from a terrible curse. This outstanding animation took over three years to make, employed some six-hundred animators and comprises more than one million drawings. Disney's thirtieth full-length cartoon is quite an achievement, marking a return to the quality of earlier years. The first cartoon ever to be nominated for a Best Picture Oscar.
ANIM   81 min (ort 85 min)   VIDrel: WDV/TECH L/A   V

BEAUTY AND THE BEAST **   U
UK   1992
Voices of: Jason Connery, Michael Hordern, Christopher Lee, Penelope Keith, Kate O'Mara
A fairly straightforward adaptation of this story. See also ALADDIN (1993), made by the same company and with most of the same cast.
ANIM   67 min   VIDrel: 4-FRONT/POLYREC   V

BEAUTY AND THE BEAST: PART 2 ***   18
USA   1990
Tracy Adams, Victoria Paris, John Leslie, Rachel Ryan, Randy West, Randy Spears, Sabrina Dawn, Jon Dough, Chaz Vincent, Ariel Knight, Dizzy Blonde, Henri Pachard, Ray Victory
The classic fairytale given an adult slant in this rather well conceived sequel, that has our beast now a henpecked husband whose daughter is sleeping with everybody in sight. A blackmail plot develops and this goofy film becomes progressively weirder.
A   83 min (ort 96 min) mVid
VIDrel: GROHOM/MAXSCAN   V

BEAUTY AND THE BEAST: ONCE UPON A TIME IN THE CITY OF NEW YORK **   PG
Richard Franklin/Victor Lobel USA   1988
Linda Hamilton, Ron Perlman, Roy Dotrice, Jay Acovone
Modern reworking of this ancient legend in which a beautiful woman attorney brutally attacked and left for dead is rescued and nursed back to health by the "beast" in his underground lair. However, his appearance is feline and quite striking, not the least frightening, and he displays such noble qualities that she soon falls in love. This was a pilot for a heavily romantic series that attracted a wide female following. Watchable if hardly outstanding.
Aka: BEAUTY AND THE BEAST: ONCE UPON A TIME IN NEW YORK
A/AD   48 min; 100 min (2-episode cassette) mTV
VIDrel: POLY/POLYREC/BRAVE   V

BEAUTY OF THE BARBARIAN **   18
Al Bradley (Alfonso Brescia) ITALY/SPAIN
Lincoln Tate, Lucretia Love, Paola Tedesco, Mirta Miller, Genia Woods, Benito Stefanelli, Solly Stubbing, Robert Vidmark
Fantasy adventure story, in which the Queen of the Barbarians seeks out the secret of the sacred fire, in an effort to conquer The Kingdom, protected by Darma, a supposed god, made immortal by the flames of the sacred fire. Another costumed romp, OK in its way but hardly memorable.
FAN   90 min Cut (31 sec)   VIDrel: MOPIC/SGSVID   V

BEBE'S KIDS **   PG
Bruce Smith USA   1992
Voices of: Bell Carter, Tone Luc
Animated comedy of a man on a first date with an attractive woman who gets conned into taking her neighbour's kids on a trip to an amusement park, where said outing soon turns into a nightmare. A strange, adult-oriented film, based on the popular African-American cartoon characters created by the late comic Robin Harris. Despite its undeniable originality (and the Blues and Hip Hop soundtrack) this unusual piece is rarely amusing.
ANIM   69 min (ort 93 min)   VIDrel: CIC/SONOP   V

BECAUSE MOMMY WORKS ***   (PG)
Robert Markowitz USA   1988
Anne Archer, John Heard, Casey Moses Wurzbach, Ashley Crow, Tom Amandes, Jenny Gago, Jan Triska, Michael Constantine, Page Leong, Ebick Pizzadili, Lance Davis, Trenton Knight, Doug Roberts, Chuck Tamburro, Lee Mary Weilnau
In this fact-based drama a single mother has been given sole custody of her son, who is now six. But she also works as a cardiac nurse, and when her ex-husband gets remarried, he launches a court battle for the child, citing her need to work as a reason for the courts to change the custody order. A well staged courtroom drama, enlivened by solid acting and direction.
DRA   86 min   SATrel: MOVIE CHANNEL

BECKET ****   PG
Peter Glenville UK   1964
Richard Burton, Peter O'Toole, Donald Wolfit, Martita Hunt, Pamela Brown, John Gielgud, Sian Phillips, Paolo Stoppa, Gino Cervi, David Weston, Percy Herbert, Nial McGinnis, Felix Aylmer, John Phillips, Frank Pettingell
A film version of Anouilh's play telling of Becket's career and eventual martyrdom, that moves at a leisurely pace from his early days and friendship with King Henry II to his eventual death at the instigation of the monarch. A powerful and absorbing study, filmed on location in England and strikingly photographed by Geoffrey Unsworth. AA: Screen/adapt (Edward Anhalt).
DRA   142 min (ort 149 min)   VIDrel: L/A   V
Boa: play by Jean Anouilh.

BECOMING COLETTE **   18
Danny Huston FRANCE/GERMANY/UK/USA   1991
Klaus Maria Brandauer, Mathilda May, Paul Rhys, Virginia Madsen, John Van Dreelan, Jean Pierre Aumont, Lucienne Hamon, Georg Tryphon, Frank Demules, Jockel Tschiersch, Vincent Nadal, Maya Thebault, Cecile Bois, Eva Probst
Based on a true story, this details the exploits of a legendary erotic writer, who at the start of her career is a naive country girl in turn-of-the-century France. Seduced and married to a millionaire publisher, she starts a revealing diary, which her husband exploits as a means of clearing his debts, at the same time training her in the ways of the world. A ponderous period drama, with flashes of charm but no great impact.
DRA   90 min (ort 97 min)   VIDrel: MED L/A   V

BED AND BOARD ***   PG
Francois Truffaut FRANCE   1970
Jean-Pierre Leaud, Claude Jade, Hiroko Berghauer, Daniel Ceccaldi, Barbara Laage
A comedy-drama that is loosely autobiographical in structure, and continues the story of Antoine Doinel, a married man who embarks on an affair with a Japanese girl, but eventually returns to his wife when he realises that he has responsibilities that cannot be dismissed lightly. This was the fourth of five films, that began with THE FOUR-HUNDRED BLOWS, and though directed with enormous care, has little of the power of the first film. Followed by LOVE ON THE RUN.
Aka: DOMICILE CONJUGAL
DRA   93 min (ort 97 min)   VIDrel: ARTIF/20TH   V/h

BED AND BREAKFAST ** 15
Robert Ellis Miller USA   1989
Roger Moore, Talia Shire, Colleen Dewhurst, Nina Siemaszko, Ford
Rainey, Stephen Root, Jamie Walters, Cameron Arnett, Bryant
Bradshaw, Victor Slezak, Jake Webber, Cheri Cotton, Frank Dolan,
Harriet Rogers, Marceline Hugot
Badly beaten and half-drowned, a con-man is washed up on a
beach close to a run-down guesthouse that's owned by three
unconventional women. Having been rescued by these ladies,
he sets about helping them make a go of their struggling busi-
ness. Unfortunately, his attackers are still out to finish him off.
Despite the thin plot, this whimsical and engaging comedy
works quite well, mostly thanks to strong performances from
the three female leads.
COM   89 min   VIDrel: EIV/SONOP   V/sur

BED OF ROSES * PG
Michael Goldenberg USA   1995
Christian Slater, Mary Stuart Masterson, Pamela Segall, Josh Brolin,
Brian Tarantina, Debra Monk, Mary Alice, Kenneth Cranham, Ally
Walker, Anne Pitoniak, R.M. Haley, Cass Morgan, Gina Torres, Nick
Tate, Victor Sierra, Michael Mantell
In this romantic comedy, a bank executive on the way up finds
herself the recipient of a succession of bouquets of flowers, and
naturally enough, becomes anxious to learn the identity of her
secret admirer. Daffy, simple and excessively cute, with empty
characters and an anodyne script.
COM   84 min (ort 88 min)   VIDrel: EIV/SONOP   V

BED YOU SLEEP IN, THE ** (18)
Jon Jost USA   1993
Tom Blair, Ellen McLaughlin, Kathryn Sannella, Marshall Gaddis,
Thomas Morris, Brad Shelton, Miki Whittles, Fred Rorie, Cathy
Cadwallader, Rita Roti, Polly Chavarria, Angelika Brooks, Renee
Ballinger, John Murphy, Mark Redpath
Already having to face the possible loss of his business, the
owner of a lumber-mill has to deal with a far more serious
problem, when his daughter makes accusations of sexual abuse
against him. One of those painful family dramas on a subject
that now seems to feature quite often in movies. Fair.
DRA   112 min   CABrel: HVC

BEDAZZLED ** PG
Stanley Donen UK   1967
Peter Cook, Dudley Moore, Eleanor Bron, Raquel Welch, Alba,
Michael Bates, Bernard Spear, Parnell McGarry, Howard Goorney,
Barry Humphries, Daniele Noel, Robert Russell, Michael Trubshawe,
Evelyn Moore, Charles Lloyd Pack
Limp, unfunny and terribly dated attempt to update the Faust
legend, with a short-order chef selling his soul to the Devil in
return for the girl of his dreams. One of the few redeeming
features is Peter Cook's portrayal of the Devil.
COM   101 min (ort 107 min)   VIDrel: 20TH   V/h

BEDFORD INCIDENT, THE *** PG
James B. Harris USA   1965
Richard Widmark, Sidney Poitier, James MacArthur, Martin Balsam,
Wally Cox, Donald Sutherland, Eric Portman, Phil Brown, Michael
Kane, Garry Cockrell, Brian Davies, Warren Stanhope, Colin
Maitland, Edward Bishop, George Roubicek
Cold War story of a US naval destroyer on a routine NATO
patrol near the coast of Greenland, that encounters an uniden-
tified submarine. A battle of wits now takes place, driving all
concerned to near breaking point. A strong cast and tight direc-
tion hold one's attention right to the end.
WAR   98 min (ort 102 min) B/W   VIDrel: ENCORE/SPEAR
V
Boa: book by M. Rascovich.

BEDKNOBS AND BROOMSTICKS *** U
Robert Stevenson USA   1971
Angela Lansbury, David Tomlinson, Sam Jaffe, Roddy McDowall,
John Ericson, Roy Smart, Cindy O'Callaghan, Ian Weighill, Bruce
Forsyth, Tessie O'Shea, Arthur E. Gould-Porter, Reginald Owens,
Ben Wrigley, Rick Traeger, John Orchard
Set in 1940, this film tells the story of three evacuee children
who defeat an invasion attempt against Britain. They are helped
by a kind witch who owns a magic bedstead. Not brilliant but
it has a certain charm. The animated sequences are directed with
considerable panache by Ward Kimball. AA: Effects/vis (Alan
Maley/Eustace Lycett/Danny Lee).
JUV   112 min (ort 117 min) cC   VIDrel: WDV/TECH   V
Boa: novel Bedknob And Broomstick by Mary Norton.

BEDROOM, THE * 18
Sato Hisayasu JAPAN   1992
Alto Kiyomi, Asano Momori, Nakamura Kyoto, Sagawa Issei
Erotic-thriller from Japan's "Pink Cinema" genre, with the girls
who work at a club being plied with a drug that makes them fall
in line with the wishes of the patrons. The fact that the mutilated
corpses of club members keep turning up provides the element
of tension in a repulsive, low-budget exercise in Japanese
fetishism.
Aka: ARIA ON GAZES, AN; SHISENJO NO ARIA
THR   88 min   VIDrel: SCEDGE/RTM   V

BEDROOM WINDOW, THE *** 15
Curtis Hanson USA   1987
Steve Guttenberg, Terry Lambert, Elizabeth McGovern, Isabelle
Huppert, Paul Shenar, Frederick Coffin, Carl Lumbly, Wallace Shawn,
Brad Greenquist, Robert Shenkkan, Maury Chaykin, Sara Carlson,
Mark Margolis, Kate McGregor-Stewart
A man having an affair with his boss's wife, finds himself the
main suspect in a series of brutal killings after reporting an
attack on a girl, seen from his bedroom window by his lover,
who is not prepared to report it herself. A suspenseful and
complex homage to Hitchcock that generally remains engrossing, despite quite a few implausibilities and an unconvincing
performance from Guttenberg.
THR   108 min (ort 113 min)   VIDrel: 20TH/TECH L/A   V
Boa: novel The Witnesses by Anne Holden.

BEETHOVEN ** PG
Brian Levant USA   1992
Charles Grodin, Bonnie Hunt, Dean Jones, Nicholle Tom, Christopher
Castile, Sarah Rose, Karr, Oliver Platt, Stanley Tucci, David
Duchovny, Patricia Heaton, Laurel Cronin, O-Lan Jones, Nancy Fish,
Craig Pinkard, Robi Davidson
Having escaped from the clutches of a couple of dog-nappers,
in the pay of Dr Varnick (an evil animal researcher/vet) a
lovable St Bernard puppy adopts Grodin and family. It moves
into their immaculate home (which rapidly takes on the requi-
site "chewed up" appearance) and captivates all and sundry.
Meanwhile, Varnick is plotting to capture the dog for use in
illegal firearms tests. A cute and predictable comedy that treads
a very well-worn path. A sequel followed.
COM   83 min (ort 87 min) cC   VIDrel: CIC/SONOP   V/sur
V/dm

BEETHOVEN'S 2ND ** U
Rod Daniel USA   1993
Charles Grodin, Bonnie Hunt, Nicholle Thom, Christopher Castile,
Sarah Rose Karr, Debi Mazar, Ashley Hamilton, Danny Masterson,
Maury Chaykin, Catherine Reitman, Heather McComb, Scott Waara,
Jeff Corey, Virginia Capers, Jason Perkins
In this sequel, our lovable canine slob starts a family with the
assistance of a friendly female dog, and his owner's kids decide
to keep the puppies while keeping their existence a secret from
Dad. Complications abound especially as Beethoven's mate is
being used as a pawn in a divorce case by a woman who hates
dogs with a passion. Very much the same mixture as before,
this is undemanding and innocuous entertainment.
COM   85 min (ort 89 min) cC   VIDrel: CIC/SONOP; PION
(LV only)   V/sur LV

BEETHOVEN'S NEPHEW * 15
Paul Morrissey AUSTRIA/FRANCE/WEST GERMANY 1985
Wolfgang Reichmann, Jane Birkin, Nathalie Baye, Ditmar Prinz,
Mathieu Carriere
A glossy and rather irritating drama, following the relationship
between this great composer and his nephew, who was brought
up by Beethoven and lived with him for many years.
Undeniably beautiful to look at, but a superficial study at best.
However, at least the music is worthwhile. See also IMMORTAL
BELOVED.
Aka: LE NEUVEU DE BEETHOVEN
DRA   99 min (ort 103 min)   VIDrel: NWV L/A   V

BEETLEJUICE *** 15
Tim Burton USA   1988
Michael Keaton, Geena Davis, Alec Baldwin, Jeffrey Jones, Sylvia
Sidney, Susan Kellerman, Catherine O'Hara, Winona Ryder, Robert
Goulet, Glenn Shadix, Dick Cavett, Annie McEnroe, Simmy Bow,
Maurice Page, Hugo Stander
A newly deceased couple have a hard time coping with the
obnoxious family who've moved into their home, so they call on
the services of the title spirit. A blend of incredible special effects

and bizarrely comic touches, let down by a weak and rambling script. Written by Michael McDowell and Larry Wilson, with cinematography by Thomas Ackerman and a score by Danny Elfman. AA: Make (Steve Laporte/Robert Short).
COM   88 min (ort 92 min)   VIDrel: WHV   V/sur

BEFORE AND AFTER *   12
Barbet Schroder USA   1996
Meryl Streep, Liam Neeson, Edward Furlong, Julian Weldon, Alfred Molina, Daniel von Bargen, John Heard, Ann Magnuson, Alison Folland, Kaiulani Lee, Larry Pine, Ellen Lancaster, Wesley Addy, Oliver Graney, Bernadette Quigley
A small-town doctor and her husband set out to uncover the killer of a young girl, having found to their dismay that the police regard their son as the prime suspect. A cynical albeit talky look at how the legal process operates in the States, that wisely generates sympathy for the central characters by showing how their lives are blighted by the events they are caught up in.
THR   104 min (ort 108 min) cC   VIDrel: HOLPIC/TECH V/s
Boa: book by Rosellen Brown.

BEFORE SUNRISE *   15
Richard Linklater USA   1995
Ethan Hawke, Julie Delpy, Andrea Eckert, Hanno Poschl, Erni Mangold, Dominik Castell, Karl Bruckschwaiger, Tex Rubinowitz, Haymon Maria Buttinger, Harold Waiglein, Bilge Jeschim, Kurti, Hans Weingartner, Liese Lyon, Peter Ily Huemer
A young American man vacationing in Europe meets a charming French girl on a train and convinces her that the should visit Vienna together. They do so and fall in love in the process. A charmingly old-fashioned romantic comedy that works quite well largely on account of its two charming leads and the atmospheric music.
COM   100 min   VIDrel: COLUM/SONOP   V/sur

BEFORE THE NIGHT **   18
Talia Shire USA   1994
Ally Sheedy, Frederic Forrest, A. Martinez
A love-starved woman is seduced by a handsome stranger and has a one-night stand with him, only to find that when she attempts to locate him there are some unsettling questions about his identity. An adequate directorial debut from Shire, this one went straight to video.
THR   93 min   VIDrel: HIFLI/SONOP   V

BEFORE THE RAIN *   15
Milcho Manchevski FRANCE/MACEDONIA/UK   1995
Katrin Cartlidge, Gregoire Colin, Rade Serbedzija, Labina Mitevska, Jay Villiers, Silvija Stojanovska, Phyllida Law, Josif Josifovski, Kiril Ristoski, Petar Mircevski, Ljupco Bresliski, Igor Madzirov, Ilko Stefanovski
Nominated for the Best Foreign Film Oscar, this tripartite tale switches between London and now independent Macedonia as it tells the story of several individuals caught up in the escalating ethnic conflicts that have convulsed the Balkans, one of whom is a Bosnian photographer who has fled the country to be with his girlfriend in London. A powerful and dramatic movie that also serves as a timely warning.
Aka: PO DEZJU; PRED DOZDOT
DRA   113 min (partly subtitled)   VIDrel: ELPIC/POLYREC V

BEFORE THE REVOLUTION *   PG
Bernardo Bertolucci ITALY   1964
Francesco Barilli, Adriana Asti, Alain Midgette, Molando Morandani, Domenico Alpi
In Parma a young man of bourgeois background finds that he is torn between radical politics and conformity and also between a passionate affair with an attractive aunt and a respectable marriage. Bertolucci's second film (he was only 22 when it was made) is a rich and varied drama that examines issues pertaining to personality, relationships and duty. However, for all its undoubted virtues it remains a work of much unrealised potential.
Aka: PRIMA DELLA RIVOLUZIONE
DRA   112 min (ort 115 min) B/W   VIDrel: CONNO/RTM V
Boa: novel The Charterhouse of Parma by Stendahl.

BEGUILED, THE *   15
Don Siegel USA   1971
Clint Eastwood, Geraldine Page, Elisabeth Hartman, Jo Ann Harris,
Darleen Carr, Pamelyn Ferdin, Mae Mercer, Melody Thomas, Peggy Drier, Pattye Mattick
Brooding atmospheric film about a wounded Yankee soldier who hides out at a Confederate school for young ladies. His presence there leads to jealousy and eventual death. An extremely well made and unusual Eastwood film.
DRA   100 min (ort 109 min)
VIDrel: 4-FRONT/POLYREC/CIC L/A   V
Boa: novel by Thomas Cullinan.

BEHIND CLOSED DOORS *   (18)
Catherine Cyran USA   1994
Barry Bostwick, Lesley-Anne Down, Michael Gross, Teresa Hill, Matt McCoy, Robin Riker, Guy Biyd, Linda Dona, Michael James McDonald, Betsy Lynn George, James Intveld, Michael Buice, Daniel Trent, Lisa Kudrow, Martha Hackett
When a wealthy woman dies after being pushed down the stairs by an intruder, the police put it down as an accident. However, in reality it was a carefully staged murder attempt by her greedy husband and his mistress, her nympho step-daughter by a previous husband. Things start falling apart quite soon afterwards when the will contains some nasty surprises and his lover begins to look for excitement elsewhere. An indifferent thriller of little inventiveness.
THR   93 min mTV   SATrel: MOVIE CHANNEL

BEHIND CONVENT WALLS *   18
Walerian Borowczyk ITALY   1977
Ligia Branice, Marina Pierro, Howard Ross, Gabriella Goaccobe, Loredana Martinez, Olivia Pascal
A classic interpretation of the sexual feelings and fantasies of the women who spend their lives in the cloistered atmosphere of a 19th century convent. A confusing narrative detracts, but this film does not lack for atmosphere.
Aka: BEHIND THE CONVENT WALLS; INTERIOR OF A CONVENT; L'INTERNO DI UN CONVENTO; WITHIN A CLOISTER
DRA   90 min Cut (9 sec plus film cuts) wScrn
VIDrel: REDEM/RTM   V
Boa: story Roman Walks by Stendhal (Marie Henri Beyle).

BEHIND THE WATERFALL **   (U)
Scott Murphy USA   1996
Gary Berghoff, Luke Baird, Alyssa Hansen, Hollis McCarthy, Gena Gale Burghoff, Sam Hennings, Jeff Olson, Michael Scott, Mark McCarthy, Kevin Halladay, Mary Parker Williams, Star Hermann, Carolyn Hucklbart, Joy Miyashoia
After their father dies, a brother and his older sister are sent to live with their aunt. Intrigued by a rumour that a local man who is fond of telling tall stories is really a leprechaun, they attempt to meet him and soon find that they have gained a new friend. An enjoyable little fantasy that works quite well.
FAN   90 min   SATrel: MOVIE CHANNEL

BEING AT HOME WITH CLAUDE **   18
Johanne Boisvert CANADA   1992
Roy Dupuis, Jacques Godin, Jean-Francois Pichette, Gaston Lepage, Hugo Dube, Johanne-Marie Trembley, Nathalie Mallette
A male prostitute kills his lover in a fit of anger but his motives remain a mystery until a policeman starts to interrogate him, in this intense and powerful drama.
DRA   90 min   VIDrel: DTK/TOTAL   V
Boa: play by Rene-Daniel DuBois.

BEING HUMAN *   15
Bill Forsyth UK/USA   1993
Robin Williams, John Turturro, Anna Galiena, Vincent D'Onofrio, Hector Elizondo, Lorraine Bracco, Linsay Crouse, Helen Miller, Charles Miller, William H. Macy, Grace Mahlaba, Dave Jones, Jonathan Hyde, Lizzy McInnerny, Helen Miller
Compilation-style movie, a sort of look at the human condition, set in five different historical epics and dealing with the life of a struggling man, played in each case by Williams. A rather strange effort, well worth seeing for his display of acting skill, but burdened by a most annoying narration, courtesy of Theresa Russell.
DRA   117 min (ort 122 min) cC   VIDrel: WHV   V/sur

BEING THERE *   15
Hal Ashby USA   1979
Peter Sellers, Shirley MacLaine, Jack Warden, Melvyn Douglas, David Clennon, Richard Dysart, Richard Basehart, James Noble, Ruth Attaway, Fran Brill, Denise DuBarry, Oteil Burbridge, Ravenell Keller III, Brian Corrigan, Alfredine Brown

On the death of his employer, a childlike gardener whose entire knowledge of life is drawn from TV, is thrust out into the real world, and an accident brings him into the homes of the powerful, where his simple utterances are mistakenly imbued with great wisdom. A witty but overlong attack on American politics, and very slightly reminiscent of Hans Andersen's "The Emperor's New Clothes". This was the penultimate film for Sellers. AA: S. Actor (Douglas).
COM   121 min (ort 124 min)   VIDrel: L/A   V
Boa: novel by Jerzy Kosinski.

BELIEVED VIOLENT **   15
Georges Lautner FRANCE   1990
Robert Mitchum, Michael Brandon, Sophie Duez, Francis Perrin, Mario Adorf, Marie Laforet, Marc De Jonge, Daniel Ubaud, Jean-Marie Lamaire, Stephanie Bonnet, Steve Klafa, Anne-Marie Kenny, Bill Dunn, Andre Oumansky, Bruce Meyers
When a scientist admitted to a clinic after he killed his wife's lover is abducted by people after top secret information, a special agent is given the job of mounting a rescue. Standard action and dramatics abound in this offering.
Aka: PRESUME DANGEREU
A/AD   100 min Cut (6 sec)   VIDrel: 20VIS/SONOP   V
Boa: novdl by James Hadley Chase.

BELLE DE JOUR ***   18
Luis Bunuel FRANCE   1967
Catherine Deneuve, Jean Sorrel, Michel Piccoli, Genevieve Page, Pierre Clementi, Francisco Rabal, Francois Fabian, Maria Latour, Georges Marchal, Macha Meril, Iska Khan, D. De Roseville, Michel Charrel, Francis Blanche
A bored middle-class housewife, married to a surgeon, is frigid with her husband but full of repressed desires. When she learns that there is a brothel in her neighbourhood, she starts working there during the afternoon. A strange surreal blend of fantasy and reality, heavily atmospheric but equally irritating in its inability to offer any motivation for her actions. The resolution, which reveals it all to be nothing more than a dream is contrived and ineffective.
DRA   95 min (ort 102 min) wScrn
VIDrel: ELPIC/POLYREC   V
Boa: novel by Joseph Kessel.

BELLE EPOQUE **   15
Fernando Trueba SPAIN   1992
Fernando Feman Gomez, Penelope Cruz, Miriam Diaz-Aroca, Michel Galabru, J. Gabano Diego, Jorge Sanz, Ariadna Gil, Maribel Verdu, Agustin Gonzalez, Chus Lampreave, Mary Carmen Ramirez, Juan Jose Olegui, Jesus Bonellia, Maria Galiana
A deserter from the Spanish army comes across a reclusive artist living in the countryside with his four young and very attractive daughters. In next to no time, he has become romantically involved with them, moving from one to the other as each sets out to ensnare him. Handsome to look at, this over-extended farce pokes fun at all its characters, and was a big hit with the critics. AA: Foreign.
DRA   104 min (ort 109 min)   VIDrel: CURZON/20TH   V

BELLES OF ST TRINIANS, THE **   U
Frank Lauder UK   1954
Alastair Sim, Joyce Grenfell, George Cole, Hermione Baddeley, Beryl Reid, Betty Ann Davies, Irene Handl, Mary Merrall, Renee Houston, Joan Sims, Balbina, Guy Middleton, Sidney James, Arthur Howard, Richard Wattis
First in a series of slapstick films centred on an unruly school for girls where more time is spent reading up racing form than on the study of the "three Rs". Based on Ronald Searle's cartoons of a dotty school and its uncontrollable pupils. Occasional sparks of humour are never enough to get this patchy comedy going, though Sim does his best in two roles. BLUE MURDER AT ST TRINIANS followed.
COM   92 min B/W   VIDrel: WHV   V

BELLS OF ST MARY'S, THE **   U
Leo McCarey USA   1945
Ingrid Bergman, Bing Crosby, Henry Travers, William Gargan, Ruth Donnelly, Joan Carroll, Martha Sleeper, Rhys Williams, Dickie Tyler, Una O'Connor, Bobby Frasco, Matt McHugh, Edna Wonacott, Jimmy Crane, Minerva Urecal
A Catholic priest is sent to a poor community where he becomes involved in running a parish school where Bergman is the Sister Superior. A meandering and over-sentimental sequel to "Going

My Way", with Bing introducing the song "Aren't You Glad You're You?". AA: Sound (Stephen Dunn).
COM   121 min (ort 126 min) B/W
VIDrel: 4-FRONT/POLYREC   V

BELLY OF AN ARCHITECT, THE ***   15
Peter Greenaway ITALY/UK   1987
Brian Dennehy, Chloe Webb, Lambert Wilson, Sergio Fantoni, Stefania Casini, Vanni Corbellini, Alfredo Varelli, Geoffrey Copleston, Francesco Carnelutti, Marino Mase, Marne Maitland, Claudio Spadaro, Rate Furlan, Julian Jenkins
A hard-drinking American architect and his beautiful wife travel to Rome to set up an exhibition. As he becomes increasingly obsessed with his failing health and his wife's unfaithfulness the beauty of Rome becomes no more than a backdrop for some symbolic and fascinating observations on art, life and mortality. The striking cinematography is by Sacha Vierny.
DRA   113 min   VIDrel: L/A   V/h

BELOVED ENEMY **   U
H.C. Potter USA   1936
Brian Aherne, Merle Oberon, David Niven, Karen Morley, Donald Crisp, Jerome Cowan, Henry Stephenson, Theodore Von Eltz, John Burton, Leyland Hodgson, David Torrence, Wyndham Standing, Robert Strange, Lionel Pape, P.J. Kelly
Romantic drama set during the 1921 Irish rebellion, when the fiancee of a British officer fells in love with an Irish revolutionary. Superficially enjoyable, but with little basis in the actual events.
DRA   86 min B/W   VIDrel: VCC/DISC   V

BELOW THE RIM *   18
Doug Davenport USA   1994
Bo Summers, Scott Russell, York Powers, Matt Windsor, Clint Benedict, Marc Andrews, Matt Diez, Austin Michaels, Mac Angelo
One guy learns of the joys of homosexuality in this adult movie that is distinguished merely by having all the encounters taking place outdoors.
A   55 min (ort 103 min)   VIDrel: QUANT/TOTAL   V

BELSTONE FOX, THE **   PG
James Hill UK   1973
Eric Porter, Rachel Roberts, Jeremy Kemp, Dennis Waterman, Heather Wright, Bill Travers
Story of an orphaned fox cub reared with the hounds. This leads to tragic consequences in a somewhat mediocre film. Disappointing.
Aka: FREE SPIRIT
DRA   98 min (ort 103 min)   VIDrel: VCC/DISC   V
Boa: novel The Ballad of the Belstone Fox by David Rook.

BELT, THE *   18
Giuliana Gamba ITALY   1989
Eleonora Briglidori, James Russo
An American teacher in Rome is drawn into an intense affair by a woman who enjoys sado-masochistic practices, and she buys him the title item in order to be whipped. Tedious nonsense with unappealing overtones.
DRA   91 min Cut (5 sec)   VIDrel: ARTPRO/RTM   V
Boa: novel by Alberto Moravia.

BEN-HUR ****   PG
William Wyler USA   1959
Charlton Heston, Jack Hawkins, Martha Scott, Stephen Boyd, Hugh Griffith, Sam Jaffe, Finlay Currie, Terence Longdon, Andre Morrell, Haya Harareet, Cathy O'Donnell, John Le Mesurier, Frank Thring, Terence Longden, Jose Greci
A lavish remake of the 1926 silent classic, telling of the conflict between the Jews and Romans at the time of Christ, and of the bitter hatred between two former friends, that is only finally resolved in a spectacular chariot race. AA: Pic, Dir, Actor (Heston), S. Actor (Griffith), Cin (R. Surtees), Art/Set (Horning and Carfagno/Hunt), Edit (Winters/Dunning), Cost (Haffenden), Score (M. Rozsa), Sound (F. Milton), Effects (vis – R. MacDonald/aud – M. Lory).
DRA   222 min wScrn (special edition including a short doc – ort 217 min)   VIDrel: MGM/WHV   V/dm V/s
Boa: novel by Lew Wallace.

BEN-HUR: A TALE OF THE CHRIST ***   PG
Fred Niblo/B. Reeves Eason USA   1926
Ramona Novarro, Francis X. Bushman, May McAvoy, Betty Bronson, Claire McDowell, Carmel Myers, Nigel de Brulier, Myrna Loy

Restored version of this classic silent spectacular with excellent action sequences and featuring a new score by Carl Davis as well as the original tinted and two-strip Technicolor sequences.
A/AD 143 min (ort 148 min) B/W silent
VIDrel: MGM/WHV V/sh
Boa: novel by Lew Wallace.

BENEATH THE PLANET OF THE APES **   15
Ted Post USA   1969
Charlton Heston, James Franciscus, Kim Hunter, James Gregory, Maurice Evans, Linda Harrison, Victor Buono, Paul Richards, Thomas Gomez, Natalie Trundy, Jeff Corey, Don Pedro Colley, Gregory Sierra, Paul Richards, Tod Andrews
This first sequel to PLANET OF THE APES has our astronauts caught up in a battle between the apes and human mutants, who have survived the nuclear holocaust and now live underground. Good ideas are let down by tired and often pompous script. Followed by ESCAPE FROM THE PLANET OF THE APES.
FAN 95 min VIDrel: 20TH/TECH L/A V

BENEFIT OF THE DOUBT *   18
Jonathan Heap GERMANY/USA   1992
Donald Sutherland, Graham Greene, Amy Irving, Christopher McDonald, Theodore Bikel, Ferdiannd Mayne, Gisele Kovach, Rider Strong, Julie Hasel, Patricia Tallman, Ralph McTurk, Shane McCabe, Margaret Johnson, Heinrich James
Having been sent to prison for killing his wife, a crime for which he was convicted largely on his daughter's testimony, a man is released after serving twenty-two years. He pays his daughter and grandson a surprise visit and it is not long before his true psychotic nature becomes apparent. Though well acted, this film is as silly as it is predictable, and lacks both tension and any sense of surprise.
THR 87 min (ort 91 min) cC VIDrel: WHV V/h

BENJI ***   Uc
Joe Camp USA   1974
Peter Breck, Christopher Connelly, Deborah Walley, Edgar Buchanan, Frances Bavier, Patsy Garrett, Allen Fiuzat, Cynthia Smith, Terry Carter, Tom Lester, Mark Slade, Herb Vigran, Larry Swortz, J.D. Young, Erwin Hearne
Lovable mutt saves three children from being kidnapped. An ideal family oriented film, followed by FOR THE LOVE OF BENJI (1977) and then BENJI THE HUNTED (1987).
COM 85 min VIDrel: BEST L/A V

BENJI THE HUNTED **   U
Joe Camp USA   1987
Red Steagall, Frank Inn, Nancy Francis, Mike Francis, Joe Camp, Steve Zanouni, Karen Thorndike, Ben Vaughn
Benji gets lumbered with the task of finding a home for some orphaned cougar cubs in this combination of canine and wilderness adventure films. A mediocre second sequel to BENJI (following on from FOR THE LOVE OF BENJI).
Aka: HUNTED, THE
JUV 86 min (ort 89 min) VIDrel: BUENA L/A V

BENNY AND JOON ***   15
Jeremiah S. Chechik USA   1992
Johnny Depp, Mary Stuart Masterson, Aidan Quinn, Olivier Platt, Julianne Moore, Dan Hedaya, C.C.H. Pounder, Joe Grifasi, William H. Macy, Eileen Ryan, Liane Alexandra Curtis, Don Hamilton, Waldo Larson, Irvin Johnsson
Touching tale of a kind-hearted auto mechanic who takes care of his mentally ill sister but has mixed feelings when she seems to have found an eccentric boyfriend who believes that he is Buster Keaton reincarnated. Fine performances compensate for the thinness of the plot and the superficial approach to mental health.
COM 94 min (ort 99 min) VIDrel: MGM/WHV V/sur

BENNY'S VIDEO **   18
Michael Haneke AUSTRIA/SWITZERLAND   1992
Arno Frisch, Angela Winkler, Ulrich Muhe, Ingrid Stassner, Stephanie Brehme, Stefan Polasek, Christian Pundy, Max Berner, Hanspeter Muller, Shelley Kastner
A young teenage boy with a rich family fails to get much attention from his parents and spends most of his time watching video films, which gradually become a kind of substitute reality. However, he eventually gets himself a girlfriend and takes her home for the weekend while his parents are away, but this action does not prove to have positive consequences.
DRA 105 min CINrel

BERLIN CONSPIRACY, THE **   18
Terence H. Winkless USA   1991
Marc Singer, Mary Crosby, Stephen Davies
Just prior to the collapse of East Germany and the destruction of the Berlin Wall, a CIA agent and an East German security officer fight a dangerous cat-and-mouse game.
DRA 79 min (ort 83 min) VIDrel: COLUM/SONOP V/sh

BERMUDA TRIANGLE *   (PG)
USA   1995
Sam Behrens, Susanna Thompson, Michael Reilly Burke, David Gallagher, Jerry Hardin, Lisa Jakub, Gustavo Laborie, Dennis Neal, Sandra Thigpen, Noami Watts
A family takes a boating trip in the Caribbean but are shipwrecked on an island where they experience some very strange phenomenon, later learning that they are in fact trapped in another dimension. A rather uninventive attempt to squeeze some mileage out of this mythical area.
A/AD 85 min SATrel: MOVIE CHANNEL

BERNARD AND THE GENIE *   PG
Paul Welland UK   1992
Lenny Henry, Rowan Atkinson, John Gabriel, Alan Cumming, Dennis Lill, Kevin Allen, Beau Bryant, Andree Bernad, Melvyn Bragg, Angie Clark, Janet Henrey, Marcia Ashton, Sheila Latimer, Sally Geohegan, Gary Whelan, Trevor Laird
An auction-house employee is sacked from his job at Christmas, and suffers further humiliations before the discovry of a magic lamp that he can use to turn the tables on his adversaries. A Christmas mTV special, done in a modern reworking of the Aladdin fable, but painfully overstretched with a thin script of little wit. Henry's considerable comic talents are wasted in a raucous and overlong tale.
COM 70 min (ort 90 min) mTV VIDrel: THAMES/DISC V

BERSERKER *   18
Jeff Richard USA   1988
Josef Alan Johnson, Valerie Sheldon, Greg Dawson, George Flower
An ancient Viking warrior with a taste for human flesh, comes back to life and threatens six college students on vacation. A new variation on the well-played theme of kids in mortal danger; atrociously made, badly acted and ludicrously plotted.
Aka: BERSERKER: THE NORDIC CURSE
HOR 80 min Cut (1 min 4 sec – ort 85 min)
VIDrel: MOPIC/SGSVID V

BEST INTENTIONS, THE ***   PG
Bille August DENMARK/FINLAND/FRANCE/
GERMANY/SWEDEN   1991
Samuel Froler, Pernilla August, Max Von Sydow, Ghita Norby, Lennart Hjulstrom, Mona Malm, Lena Endre, Keve Hjelm, Bjorn Kjellman, Borje Ahlstedt, Hans Alfredson, Lena T. Hansson, Anita Bjork, Elias Ringquist, Ernst Gunther
A muted account of the difficult courtship and problems that beset Bergman's parents at the start of their marriage. It opens in 1909, with his father studying for the priesthood, tells of his meeting with his future wife, and follows their various tribulations at a remote parish, to which her husband is posted. A complex, sombre and rather disjointed effort, neither accurate as a true autobiographical work nor especially enjoyable as a straight film.
Aka: DEN GODA VILJAN
DRA 173 min (ort 182 min) VIDrel: ARTIF/20TH V/sh
Boa: autobiography Magic Lantern by Ingmar Bergman.

BEST OF THE BEST **   15
Bob Radler USA   1989
Eric Roberts, James Earl Jones, Sally Kirkland, Phillip Rhee, John P. Ryan, John Dye, David Agresta, Tom Everett, Louise Fletcher, Simon Rhee, Christopher Penn, edan Gross, Hee Il Cho, James Lew, Ken Nagayama
Having sworn to have his revenge for the murder of his brother by a Korean karate master, a young man joins the American karate team in order to take part in an international competition. A cliched exercise in fisticuffs that blends together elements of ROCKY and THE KARATE KID. The inevitable sequels soon followed.
MAR 93 min (ort 97 min) VIDrel: EIV/SONOP V

BEST OF THE BEST 2 **   18
Bob Radler USA   1992
Eric Roberts, Christopher Penn, Phillip Rhee, Wayne Newton, Meg Foster, Edan Gross, Ralph Moeller, Sonny Lanham, Meg Foster,

Simon Rhee, Betty Carvalho, Patrick Kilpatrick, Mike Genovese, Claire Stansfield, Hayward Nishioka

Two karate masters set out to avenge the death of their friend, killed in a no-rules match at a shady establishment owned by a champion fighter and his manager. Very similar in scope and theme to the first film, this solidly made sequel breaks no new ground, though the action highlights are directed with considerable flair.

MAR   95 min (ort 101 min)   VIDrel: EIV/SONOP   V/sur

BEST OF THE BEST 3: NO TURNING BACK *   15
Phillip Rhee USA   1995
Phillip Rhee, Gina Gershon, Christopher McDonald, Mark Rolston, Dee Wallace Stone

This time, our intrepid martial artists find themselves facing a ruthless group of neo-Nazis on American soil, when Rhee pays a visit to his sister in a backwater hick town. A dullish entry in this series, neither inspiring nor especially exciting.

MAR   94 min   VIDrel: EIV/SONOP   V

BEST OF TIMES, THE **   15
Roger Spottiswoode USA   1986
Robin Williams, Kurt Russell, Pamela Reed, Holly Palance, Donald Moffat, Margaret Whitton, M. Emmet Walsh, Donovan Scott, R.G. Armstrong, Dub Taylor, Carl Ballantine, Tony Plana, Kirk Cameron, Robyn Lively, Eloy Casados

Williams plays a guy who has never been able to forget the time he muffed a winning pass in a football game at his high school, and decides to arrange a rematch for all concerned twenty years later. A patchy and almost plotless comedy, whose winning performances are chained to a losing script. Written by Ron Shelton.

COM   99 min (ort 105 min)   VIDrel: WHV   V/h

BEST REVENGE, THE **   18
James Becket USA   1996
Robert Pine, Pat Destro, Bruna Lombardi

In 1985 during the civil war in El Salvador a man sees his wife executed on the orders of a government agent. By 1992 he has arrived in L.A. in search of revenge. A straightforward action-thriller, no better or worse than dozens of others.

THR   90 min   VIDrel: GUILD/FOXVID   V

BEST SELLER **   18
John Flynn USA   1987
Brian Dennehy, James Woods, Victoria Tennant, Allison Balson, Paul Shenar, Jenny Gago, George Coe, Mary Carver, Sully Boyer, Kathleen Lloyd, Harold Tyner, E. Brian Dean, Jeffrey Josephson, Edward Blackoff

A policeman/author survives a violent robbery but remembers something of the appearance of one of his masked attackers. Several years later his life is saved by a man of this appearance, who then approaches him to write the story of his life as a hitman to a crooked businessman whom he wants to expose. Two great performances from the leads cannot mask the flaws in the Larry Cohen script.

THR   103 min (ort 110 min)   VIDrel: VCC L/A   V/sh

BEST SHOT ***   PG
David Anspaugh USA   1986
Gene Hackman, Barbara Hershey, Dennis Hopper, Sheb Wooley, Fern Parsons, Brad Boyle, Steve Hollar, Brad Long, David Neidorf, Chelcie Ross, Robert Swan, Michael O'Guinne, Wil Dewitt, John Robert Thompson, Michael Sassone

A former top college coach arrives at a small farming community in the 1950s with plans to make a winning basketball team out of the local school players as a way of redeeming himself. Hackman gives a great performance in this contrived but often entertaining story. The script is by Angelo Pizzo. Hopper received an AAN for his excellent portrayal of an alcoholic father.

Aka: HOOSIERS
DRA   110 min   VIDrel: VCC/RCA L/A   V/sh

BEST YEARS OF OUR LIVES, THE ****   U
William Wyler USA   1946
Dana Andrews, Teresa Wright, Virginia Mayo, Cathy O'Donnell, Fredric March, Myrna Loy, Harold Russell, Hoagy Carmichael, Gladys George, Steve Cochran, Ray Collins, Roman Bohnen, Minna Gonbell, Walter Baldwin, Dorothy Adams

Fine and highly moving drama about three American soldiers returning home to civilian life at the end of WW2 and finding it difficult to adjust. Russell (he lost his hands in WW2) is

outstanding in this great film. The script is by Robert Sherwood. AA: Pic, Dir, Actor (March), S. Actor (Russell), Edit (Daniel Mandell), Screen (Robert E. Sherwood), Score (Hugo Friedhofer), Spec Award (Harold Russell for bringing hope and courage to his fellow veterans).

DRA   170 min (ort 172 min) B/W   VIDrel: VCC/DISC   V
Boa: novel Glory for Me by MacKinlay Kantor.

BETHUNE: THE MAKING OF A HERO **   15
Phillip Borsos CANADA/CHINA/FRANCE   1990
Donald Sutherland, Helen Mirren, Helen Shaver, Harrison Liu, Anouk Aimee, Colim Feore, James Pax, Guo Da, Ronald Pickup, Geoffrey Chater, Tan Zong Yao, Zhang Ke Yaw, Inaki Ayerra, Li Hai Lang, Yvan Ponton, Sophie Faucher

The story of Norman Bethune, an idealistic Canadian doctor who led a crusade for socialised medicine in Depression-era Montreal, treated the wounded during the Spanish Civil and was to be found in China in the 1930s, where he aided Mao Tse-Tung's army during its famous Long March. Sutherland is excellent in a role he first took in a 1977 film, but this expensive effort is both lifeless and disjointed, and is a major disappointment.

Aka: BETHUNE; DOCTOR BETHUNE; HERO OF THE PEOPLE
DRA   111 min (ort 115 min)   VIDrel: COLUM/SONOP   V/sur

BETRAYAL **   15
David Jones UK   1983
Jeremy Irons, Ben Kingsley, Patricia Hodge, Avril Edgar, Ray Marioni, Caspar Norman, Chloe Billington, Hannah Davies, Michael Konig, Alexander McIntosh

This film version of a Pinter play about the relationship between a husband, his wife and her lover, makes use of an unusual device in that the story is told backwards in time. A slack and dull effort despite strong performances from all concerned.

DRA   91 min (ort 95 min)   VIDrel: FABFIL/SPEAR   V/h
Boa: play by Harold Pinter.

BETRAYAL AND REVENGE ***   18
Zheng Kangyu CHINA   1986
An Yaping, Chen Kang, Fang Jian, Pang Lin Tai, Shao Xiao Kui, He Fu Sheng

The son of an executed revolutionary who led an abortive uprising in 1864, takes revenge for his father's death after a period of ten years.

MAR   104 min Cut (11 sec – ort 113 min) dubbed
VIDrel: SCRN/DISC   V

BETRAYAL OF THE DOVE **   15
Strathford Hamilton USA   1992
Helen Slater, Billy Zane, Kelly Le Brock, Alan Thicke, Harvey Korman, Stuart Pankin, David Lander

A divorcee and mother of a seven-year-old goes on a blind-date arranged by her best friend, and finds her partner to be a handsome doctor who soon wins her affections. However, appearances prove to be highly deceptive, and she begins to suspect that both her friend and the doctor have sinister intentions of their own.

THR   90 min (ort 93 min)   VIDrel: EIV/SONOP   V/sur

BETRAYAL OF TRUST **   15
George Kaczender USA   1993
Judith Light, Judd Hrisch, Betty Buckley, Jeff De Munn, Kevin Tighe, Holland Taylor, Nicholas Campbell, Lisa Dar, Amy Aquino, Christine Dunford, John Getz, Colleen Winton, Malcolm Stuart, Gary Chalk, Jon Cuthbert, Matthew Walker

Based on the true story of Barbara Noel, this tells of a female nightclub singer who claims she was drugged and raped by her psychiatrist. Despite her inability to get the police to press charges against him, on learning that he has apparently abused other patients, she resolved to take the matter to his professional body, an action that resulted in a five year suspension. Light's overwrought performance is a weakness in a generally well plotted effort.

DRA   88 min (ort 95 min) mTV   VIDrel: MARQ/QUANT   V

Boa: book You Must be Dreaming by Barbara Noel Wih Kathryn Watterson.

BETRAYED *   18
Constantin Costa-Gavras USA   1988
Tom Berenger, Debra Winger, John Heard, Betsy Blair, John Mahoney, Ted Levine, Jeffrey DeMunn, Albert Hall, David Clennon, Richard Libertini, Robert Swan, Maria Valdez, Brian Bosak, Ralph Moody

A foolish and muddled tale following the exploits of a female FBI agent who is sent to investigate a white supremacist group, but winds up falling in love with her quarry instead, an action which clouds her judgement somewhat. A strangely uninvolving film that explores neither the horrors committed by neo-Nazis nor their twisted motivations.
DRA   121 min (ort 128 min)   VIDrel: MGM/WHV   V/sur

BETRAYED: A STORY OF THREE WOMEN **
(12)
William Graham USA
1995
Meredith Baxter, Swoosie Kurtz, John Terry, John Livingston, Breckin Meyer, Nichholas Pryor, Clare Carey, Bill Brochtrup, Janet Carroll, Janet Graham, Cece Tsou, Robert Kotecki, Janet Cole Notey
Two close women friends learn to their horror that the daughter of one of them is having an affair with the husband of the other. This naturally leads to much anguish and soul-searching, and what initially appears to be the destruction of a lifelong friendship. Although rather overwrought in places one finds oneself absorbed in this painful drama, mostly thanks to the believable way in which the characters are portrayed.
DRA   89 min mTV   SATrel: MOVIE CHANNEL

BETRAYED BY LOVE **
15
John Power USA
1993
Mare Winningham, Steven Weber, Patricia Arquette, Perry Lang, Ned Vaughn, Christopher Curry, Randy Oglesby, Jim Haynie, Dan Bell, Claudette Sutherland, Maynie Lovett, Rachel Duncan, Sean Martin, Robert E. Lee, Pat Harvey
In Kentucky, a bored married woman embarks on an affair with an FBI agent when she starts working as an informant, he role being to provide him with information about a major bank robbery. When the agent brings the affair to an abrupt end, having got what he wanted, she is unable to accept this, vanishes in suspicious circumstances, and is eventually found to have been murdered. A reasonably absorbing drama, based on a true case.
DRA   88 min (ort 90 min) mTV
VIDrel: 4-FRONT/POLYREC/ODY   V/sh
Boa: book The FBI Killer by Aphrodite Jones.

BETTER OFF DEAD **
15
Savage Steve Holland USA
1985
John Cusack, David Ogden Stiers, Kim Darby, Demian Slade, Curtis Armstrong, Scooter Stevens, Diane Franklin, Laura Waterbury, Amanda Wyss
A teen comedy about a young guy who loses his girlfriend to the butch captain of a ski team. To regain his self-respect and win her back, he challenges him to a downhill race. A light-headed comedy that starts off with lots of funny gags and then slides rapidly downhill into cliche. A highlight is the early clay-animation sequence by Jimmy Picker. The feature debut for writer/director Holland.
COM   93 min   VIDrel: 20TH/TECH   V

BETTER TOMORROW, A **
18
John Woo HONG KONG
1986
Chow Yun Fat, Leslie Cheung, Ti Lung, Young Pao-Yi, Emily Chu
A suspense-thriller set in Hong Kong's dangerous underworld, that involves two brothers, one of whom has become a policemen and the other a criminal. However, despite their very different pathways, they both learn the value of family loyalty following a series of violent incidents. A very well made effort that is flawed by the director's excessive reliance on bloodshed (a feature of most of his films). Two sequels followed.
Aka: YINGXIONG BENSE
THR   95 min (ort 195 min) wScrn dubbed
VIDrel: MIA/DISC; ENCORE (LV only)   V LV

BETTER TOMORROW 2, A **
18
John Woo HONG KONG
1987
Chow Yun Fat, Ti Lung, Dean Shek, Leslie Cheung, Kwan Sam, Emily Chu, Kent Tsang, Lung Ming Yun, Regina Kent, Chindy Lau, Paul Francis, Marco Wo, Sammy Lee, Ken Bolye, Wai Sing Chow
Continuation of this bloody gangster saga sees the gang acquiring a new leader who frames one of its members. The latter runs away to New York but returns to take part in a violent showdown during which the body count reaches the hundreds. A gripping and well made film, but Woo's obsession with fountains of blood and explosions is both highly depressing and mighty tedious.
Aka: BETTER TOMORROW, A: PART 2
THR   95 min (ort 100 min) wScrn   VIDrel: MADE/RTM   V

BETTER TOMORROW 3, A ***
18
Tsui Hark HONG KONG
1989
Chow Yun Fat, Anita Mui, Tony Leung
Third part in this trilogy takes the form of a prequel set in Vietnam in 1974 from which one of the characters must escape, together with the woman he loves. A welcome change of pace and much less violence and gore (thanks to a change of director) contribute towards a much more watchable and enjoyable film.
Aka: LOVE AND DEATH IN SAIGON
THR   106 min wScrn   VIDrel: MADE/RTM   V

BETTY BLUE ****
18
Jean-Jacques Beineix FRANCE
1986
Jean-Hugues Anglade, Beatrice Dalle, Consuelo De Haviland, Gerard Darmon, Clemantine Celarie, Jacques Mathou, Claude Confortes, Philippe Laudenbach, Vincent Lindon, Raoul Billeray, Claude Aufaure, Andre Julien, Nataly Dalyan
A disgruntled, tempestuous waitress, embarks on a wild and stormy love affair with a thirty-five-year-old handyman after she discovers that he has written an unpublished novel. Putting his job and emotional stability in jeopardy, they head for Paris and then a provincial town, where finally, tired, drugged and disillusioned, she goes mad. A tragi-comedy of impressive power that won the Grand Prize at the 10th Montreaux Film Festival.
Aka: 37.2° LE MATIN; 37.2 DEGRES LE MATIN; 37.2 DEGREES IN THE MORNING
DRA   117 min (ort 121 min); 176 min (special extended version)   VIDrel: 20TH L/A   V
Boa: novel 37.2° Le Matin by Philippe Dijan.

BETWEEN LOVE AND HONOR *
12
Sam Pillsbury USA
1994
Grant Show, Maria Pitillo, Steve Allie Collura, Joseph Scoren, Howard Jerome, Robert Morelli, Tony De Santis, Barry Flatman, Robert Loggia, Michael Nouri, Cloris Leachman, Harvey Arkin, Geza Kovacs, Nigel Bennet, Vito Rezza
True story of a New York cop given the job of infiltrating a top Mafia family, only to find himself falling in love with an attractive female member. This might so easily have been a gripping account, but sadly is given a thoroughly superficial treatment.
THR   86 min   VIDrel: ODY/SONOP   V/sh

BETWEEN THE LINES ***
15
Joan Micklin Silver USA
1977
John Heard, Lindsay Crouse, Jeff Goldblum, Jill Eikenberry, Bruno Kirby, Gwen Welles, Stephen Collins, Michael J. Pollard, Marilu Henner, Lewis J. Stradlen, Lane Smith, Susan Haskins, Ray Barry, Douglas Kennedy, Jon Korkes
A Boston underground newspaper is threatened with being bought up by a press baron when the owner decides to sell up. Something of an examination of how the innocence of the 1960s was replaced by the cynicism of the 1970s, this is an extremely enjoyable and little known film, which though playing like a sit-com benefits greatly from a strong cast, who at that time were largely unknowns.
COM   100 min   VIDrel: ARROW/RTM   V

BEVERLY HILLBILLIES, THE **
PG
Penelope Spheeris USA
1993
Jim Varney, Cloris Leachman, Dabney Coleman, Lily Tomlin, Rob Schneider, Lea Thompson, Diedrich Bader, Erika Eleniak, Buddy Ebsen, Penny Fuller, Zsa Zsa Gabor, Dolly Parton, Kevin Connolly, Lyman Ward, Leann Hunley, Ernie Lively
Big-screen version of this hit 1960s series, about a mountain family that strike oil and move to Beverly Hills. Once there, they become the target of a cunning con-man and his female companion. A poorly scripted and unfunny comedy that lacks all the innocent charm of the original, but offers at least some convincing portrayals of the various members of the Clampett family.
COM   89 min (ort 93 min) cC   VIDrel: 20TH/TECH   V/sur

BEVERLY HILLS COP ***
15
Martin Brest USA
1984
Eddie Murphy, Judge Reinhold, Lisa Eilbacher, Bronson Pinchot, John Ashton, Ronny Cox, Steven Berkoff, James Russo, Jonathan Banks, Stephen Elliott, Gilbert R. Hill, Art Kimbo, Joel Bailey, Paul Reiser, Michael Champion
After the death of his friend, a Detroit cop takes to the streets of L.A. in search of those responsible. Comedy and drama are blended cleverly with Murphy perfectly cast as a streetwise and

unconventional cop. The script is by Daniel Petrie Jr. Followed by the inevitable sequels.
DRA   101 min (ort 105 min)   VIDrel: CIC/SONOP L/A: PION (LV only)   V/sur LV

BEVERLY HILLS COP 2 ** 15
Tony Scott USA   1987
Eddie Murphy, Ronny Cox, Judge Reihold, John Ashton, Brigitte Nielsen, Paul Reiser, Jurgen Prochnow, Allen Garfield, Dean Stockwell, Paul Reiser, Paul Guilfoyle, Gilbert Hill, Robert Ridgley, Brian O'Connor, Alice Adair
Sound-and-fury sequel that has everything except good acting and a credible, coherent plot. Trite, tiresome and directed with all the sensitivity of a cosh.
A/AD   99 min (ort 103 min)   VIDrel: CIC/SONOP; PION (LV only)   V/sur LV

BEVERLY HILLS COP 3 ** 15
John Landis USA   1993
Eddie Murphy, Judge Reihold, Hector Elizondo, Theresa Ranult, Bronson Pinchot, Timothy Carhart, John Saxon, Alan Young, Stephen McHattie, Al Green, Goerge Lucas, Joe Dante, Martha Coolidge, Arthur Hiller, Ray Harryhausen
Third time round for this fast-talking Detroit cop, sees him investigating an L.A. theme park where the security staff have developed a lucrative sideline in the forgery business. Better than the previous film but still saddled with the same unfunny jokes and an over-emphasis on action stunts at all costs.
A/AD   104 min cC   VIDrel: CIC/SONOP; PION (LV only) V/sur LV

BEVERLY HILLS GIRLS * R18/18
Mike Hall USA   1987
Michelle Bauer, Becky Le Beau, Linnea La Stray, Pamela Mandl, Susan Anton, Julia Jartouer, Trish Adams, Mick Rick, David Allen, Dana Sherwood
An attractive female film producer who has a husband, fame and success, as well as a comfortable mansion, is persuaded by a seductive blonde to attend a wild party in Beverly Hills, where she loses her clothes and her inhibitions. A poorly made adult movie.
A   58 min (18 ver abridged by distributor – ort 73 min)
VIDrel: MIA/DISC/IMPENT   V

BEYOND, THE ** 18
Lucio Fulci ITALY   1981
Katherine McColl, David Warbeck, Sarah Keller, Antoine St John, Veronica Lazar, Anthony Flees, Giovanni de Nava, Michelle Mirabella, Al Cliver (Pier Luigi Conti)
Story of a young woman who inherits a rundown Louisiana hotel that just happens to be built over one of the gateways to Hell. A series of violent deaths take place. Average shocker. See also THE SENTINEL for another treatment of this theme.
Aka: AND YOU'LL LIVE IN TERROR! THE BEYOND; E TU VIVRAI NEL TERRORE! L'ALDILA; L'ALDILA; SEVEN DOORS OF DEATH
HOR   82 min (Cut at film release – ort 89 min)
VIDrel: VIPCO/SGSVID   V/h

BEYOND A REASONABLE DOUBT ** PG
Fritz Lang USA   1956
Dana Andrews, Joan Fontaine, Sidney Blackmer, Philip Bourneuf, Barbara Nichols, Shepperd Strudwick, Dan Seymour, Arthur Franz, Edward Binns, Robin Raymond, William Leicester, Rusty Lane, Joyce Taylor, Carleton Young
A novelist is persuaded to fake evidence that gets him convicted for murder and is then unable to prove his innocence. And why does he do this? To get a first-hand look at the legal system of course. An interesting but utterly unbelievable idea, that just doesn't hold up in this poor effort.
DRA   80 min B/W   VIDrel: ODY/SONOP   V

BEYOND BEDLAM *** 18
Vadim Jean USA   1993
Elizabeth Hurley, Craig Fairbrass, Keith Allen, Anita Dobson, Craig Kelly, Jesse Birdsall, Faith Kent, Samantha Spiro, Stephen Brand, Zoe Heyes, Annette Badland, Natasha Humphrey, Jack McKenzie, Chris Adamson, Shaun Etherton
A convicted serial killer volunteers to undergo a radical new drug treatment that may lead to the suppression of his violent tendencies. Fairbrass plays the detective looking for a link between this controversial treatment and a new killing, his suspicion being that a prison inmate currently taking part in this program has found a way to kill from inside jail. An

extremely violent film that is not recommended for the squeamish.
Aka: NIGHTSCARE
THR   85 min (ort 89 min)   VIDrel: POLY/POLYREC; ENCORE (LV only)   V/sur LV

BEYOND BETRAYAL ** (12)
Carl Schenkel USA   1994
Susan Dey, Richard Dean Anderson, Annie Corley, James Tolkan, Brigita Dau, Michael O'Neill, Dennis Boutsikaris, Ramsin Kelsey, Jerry Wasserman, Ariene Mazevolle, Bryon Lucas, Robin Mossley, B.J. Harrison, D.J. Jackson, Winnie Hung
A wife attempts to escape from her thuggish cop husband, and does this by giving herself a fake identity. She then meets and falls in love with a kind toymaker, but the cop tracks her down and murders the man's estranged wife as an act of revenge. Some further complications come to pass until the final, violent resolution, whose outcome is never seriously in doubt. A standard collection of all the usual psychopath cliches, indistinguishable from numerous similar films.
THR   92 min mTV   SATrel: MOVIE CHANNEL

BEYOND CONTROL: THE AMY FISHER STORY ** 15
Andrew Tennant USA   1993
Drew Barrymore, Anthony John Denison, Harley Jane Kozak, Tom Mason, Laurie Patton, Ken Pogue, Linda Darlow, Gabe Khouth, Garry Darey, Dwight McFee, Ken Angel, Philip Granger, Stephen Cooper, Matthew Walker, Walter Marsh, Terry King
One of three made-for-TV movies devoted to what became known as the Long Island Lolita case, where a young girl attempted to murder the wife of her married lover and was later tried and sent to prison. The films LETHAL LOLITA – AMY FISHER: MY OWN STORY and "Casualties Of Love: The Long Island Lolita Story" also cover this ground, and all three were rushed into production while the trial was in progress, being shown on different networks within six days.
Aka: AMY FISHER STORY, THE
DRA   93 min (ort 96 min) mTV   VIDrel: ODY/SONOP V/sh

BEYOND EVIL ** 15
Herb Freed USA   1980
John Saxon, Lynda Day George, Michael Dante, Mario Milano, Janice Lynde, David Opatoshu, Zitto Kazaan, Anne Marisse, Alan Caillou, Beverly Dixon, Edward Ansara, Jennifer Italiano, Peggy Stewart, Chuck Hicks, Mickey Carouso
A couple move into a house that's inhabited by the hostile spirit of a one-hundred-year-old woman who doesn't take kindly to this intrusion. An adequate horror yarn of no great consequence.
HOR   91 min   VIDrel: VIPCO/SGSVID   V

BEYOND FEAR ** 18
Robert Lyons USA   1993
Mimi Lesseos, Bodo Holst
A woman tourist guide taking a group on a wilderness hike finds herself being chased by two men after one of her charges accidentally videos them committing a murder. Fortunately, our heroine is also a martial arts expert, so the outcome of this routine violent actioner is never really in doubt.
A/AD   83 min   VIDrel: CURB/HIFLI   V

BEYOND FORGIVENESS ** 18
Bob Misiorowski USA   1994
Thomas Ian Griffith, Joanna Trzpiecinska, Rutger Hauer, John Rhys-Davies, Artur Żmijewski, Bozena Szymanska, Jerzy Karaszkiewicz, Cezary Poks, Jacek Poks, Aleksander Wysocki, Ryszard Ronczewski, Jerry Flynn, Leon Niemczyk, Jan Prochwa
After his brother is murdered, a cop from Chicago mounts his own investigation, and learns that this crime is linked to the Polish underworld and its involvement in the illicit trade in human organs for transplantation. Average. See also THE HARVEST.
Aka: BLOOD OF THE INNOCENT
A/AD   89 min (ort 95 min)   VIDrel: 20TH/FOXVID   V/sh

BEYOND OBSESSION ** (15)
David Greene USA   1993
Henry Thomas, Emily Warfield, Victoria Principal, Donnelly Rhodes, Garry Chalk, Joe Reglabuto, Alex Bruhanski, Vince Metcalfe, Linda Darlow, Tom Butler, Forbes Angus, Sarah Clarke, Wally Dalton, Sandra P. Grant, Helen Honeywell
Thomas and Warfield play a teenage couple who are put on

trial for the murder of the girl's mother, their defence being that the woman was an abusive parent. A well acted venture into over-familiar territory, that is (as is often the case with these films) based on a real-life case.
DRA 90 min (ort 92 min) SATrel: SKY MOVIES
Boa: novel by Richard Hammer.

BEYOND RANGOON *   15
John Boorman USA   1995
Patricia Arquette, U Aung Ko, Frances McDormand, Spalding Gray, Tiara Jacquelina, Kuswadinath Bujang, Victor Slezak, Jit Murad, Ye Myint, Cho Cho Myint, Johnny Cheah, Haji Mohd Rajoli, Azmi Hassan, Ahmad Fithi, Adelle Lutz
A doctor is stranded in Rangoon after her passport is stolen, and goes on a tour of Burma whilst waiting for a replacement, but does not realise the dangers of this. Terribly arty and just a bit too self-conscious, Boorman attempts to examine the turbulence of contemporary Burma (as THE KILLING FIELDS did for Cambodia). Though always ravishing to look at, the director's flashy direction cannot mask the emptiness of it all, and Arquette is well out of her depth here.
DRA 100 min VIDrel: COLUM/SONOP V/sur

BEYOND SUSPICION *   (18)
Paul Zillier USA   1993
Jack Scalia, Stepfanie Kramer, Howard G.H. Dell, Francesco Ferrucci, Douglas H. Arthurs, Roger R. Cross, Mark Acheson, Lily Shavick, John Tench, David McKay, Andrew Airlie, Sarah Richardson, Patrick Stevenson, Rick Pierce, Doug Miller
A woman suffering from amnesia is unaware that she is a murder witness and that her life is in danger from the corrupt cop who committed this crime. A standard thriller, well realised but with little originality.
THR 95 min (ort 98 min) mTV SATrel: MOVIE CHANNEL

BEYOND THE CLOUDS *   18
Michelangelo Antonioni FRANCE/GERMANY/ITALY 1995
Sophie Marceau, Vincent Perez, Irene Jacob, Marcello Mastroianni, Fanny Ardant, John Malkovich, Kim Rossi Stuart, Chiara Casselli, Jean Reno, Peter Weller, Jeane Moreau, Ines Sastre, Enrica Antonioni, Carine Angeli
After a break from film-making of ten years, Antonioni returns with this solemn and quite formally composed set of four stories, each one detailing a brief encounter of depth and sadness, and all of them in a sense existing only in the mind of the film's linking character (Malkovich) for whom they are possible movies. Not always the most accessible of directors, yet the intensity of this work is undeniable. The linking sequences are by Wim Wenders.
Aka: AL DI LA DELLE NUVOLE; JENSEITS DER WOLKEN; PAR-DELA LES NUARGES
DRA 104 min (ort 115 min) wScrn VIDrel: ARTIF/20TH V
Boa: short story collection Quel Bowling Sul Tevere by Michelangelo Antonioni.

BEYOND THE POSEIDON ADVENTURE *   15
Irwin Allen USA   1979
Michael Caine, Sally Field, Telly Savalas, Karl Malden, Peter Boyle, Jack Warden, Slim Pickens, Shirley Knight, Shirley Jones, Mark Harmon, Veronica Hamel, Angela Cartwright, Paul Plcerni, Patrick Culliton, Dean Ferrandini
A weak sequel to THE POSEIDON ADVENTURE, telling of attempts by two crews to loot the capsized liner of its valuables before it sinks. The plot in this one is the first thing to go down.
THR 110 min (ort 115 min) VIDrel: MGM/WHV L/A V
Boa: novel by Paul Gallico.

BEYOND THE VALLEY OF THE DOLLS *   18
Russ Meyer USA   1970
Dolly Read, Cynthia Myers, Marcia McBroom, David Gurian, John LaZar, Michael Blodgett, Edy Williams, Erica Gavin
A sequel in name only, that parodies all those trashy soap operas of the time. In this one a female rock band tries to make the big time in L.A. This Hollywood debut for Meyer has many good things in it, not least the utterly over-the-top script by film critic Roger Ebert.
DRA 102 min Cut (53 sec – ort 109 min)
VIDrel: 20TH/TECH V

BEZHIN MEADOW *   PG
Sergei Eisenstein USSR   1937
Vitya Kartashov, Boris Zakhava, Igor Pavlenko, Telesheva

Some fragments of an incomplete Eisenstein film were reconstructed in 1966. Now seen as a series of freeze frames, these few sections show the loss his failure to complete this film represents.
Aka: BEZHIN LUG
DRA 91 min (2 film cassette) B/W VIDrel: HEND L/A V
Boa: story by Ivan Turgenev. Osca: TIME IN THE SUN

BFG, THE *   U
Brian Cosgrove UK   1990
Voices of: David Jason, Amanda Root, Angela Thorne, Ballard Berkley, Michael Knowles, Don Henderson, Mollie Sugden, Jimmy Hibbert, Frank Thornton, Myfanwy Talog
A feature-length adaptation of Dahl's story, telling of a "big friendly giant" whose penchant for mixing up all his words is one of his most endearing features. A few fairly predictable adventures follow, but there are enough dark Roald Dahl touches to keep blandness at bay.
ANIM 88 min (ort 105 min) VIDrel: THAMES/DISC V/sur
Boa: novel by Roald Dahl.

BHAJI ON THE BEACH *   15
Gurinder Chadha UK   1993
Kim Vithana, Lalita Ahmed, Shaheen Khan, Sarita Khajuria, Jimmi Harkishin, Zohra Segal, Peter Cellier, Mo Sessay, Surendra Kochar, Squaad Faress, Tanveer Ghani, Amer Chadha-Patel, Nisha Nayar, Renu Kochar, Surendra Kochar, Dean Gatiss
A group of Indian women living in Britain enjoy a day out in the northern coastal resort of Blackpool, which forms the backdrop against which their various life stories are told. A warm and exotic tale, dealing with the customs and culture of an immigrant group that will be quite unfamiliar to most American viewers.
DRA 96 min (ort 101 min) mTV VIDrel: FIRST/SONOP V/sur

BIBLE, THE *   U
John Huston ITALY/USA   1966
George C. Scott, Peter O'Toole, Ava Gardner, Richard Harris, Ulla Bergryd, Michael Parks, Stephen Boyd, Franco Nero, John Huston, Zoe Sallis, Gabriele Ferzetti, Eleonora Rossi Drago, Robert Rietty, Grazia Maria Spina, Claudie Lange
A dreadful and stilted attempt to film the first 22 chapters of Genesis, following the exploits of Adam and Eve, Cain and Abel, Noah and the Flood and Abraham and his sacrifice. Heavy-handed, ponderous and pompous; the book is a great deal better.
Aka: BIBLE, THE: IN THE BEGINNING, THE; LA BIBLIA
DRA 167 min (ort 174 min) wScrn VIDrel: 20TH/TECH V/h

BICYCLE THIEVES ***   U
Vittoria De Sica ITALY   1948
Lamberto Maggiorani, Lianella Carell, Enzo Staiola, Elena Altieri, Vittorio Antonucci, Gino Saltamerenda, Fausto Guerzoni, Guilio Chiari, Michel Sakara, Carlo Jachino, Nando Bruno, Umberto Spadaro, Massimo Randisi
Wonderfully moving film telling of an Italian workman who, after a long period of unemployment finally gets a job. When the bicycle he needs for the job is stolen he embarks on a long and painful search. A cinema landmark. AA: Spec Award (most outstanding foreign language film released in the USA during 1949).
Aka: BICYCLE THIEF, THE; LADRI DI BICICLETTE
DRA 85 min (ort 90 min) B/W VIDrel: ARTPRO/RTM V
Boa: novel by L. Bartolini.

BIG *   PG
Penny Marshall USA   1988
Tom Hanks, Elizabeth Perkins, Robert Loggia, John Heard, Jared Rushton, Jon Lovitz, David Moscow, Mercedes Ruehl, Josh Clark, Tracy Reiner, Kimberlee M. Davis, Oliver Block, Erica Katz, Allan Wasserman, Mark Ballou, Debra Jo Rupp
With the aid of a wishing machine at a carnival fairground, a youngster gets his chance to become an adult, waking up the next day as a thirty-year-old. However, he finds that life as a grown-up has its complications. A fresh and delightful fantasy that never descends into mawkishness, mainly thanks to a wonderfully artless performance from Hanks and Marshall's skilful direction. The script is by Gary Ross and Anne Spielberg.
COM 100 min (Cut at film release by 2 sec – ort 104 min)
VIDrel: 20TH/TECH V/sur

BIG BANG, THE * 18
Jean-Marc Picha (Jean-Paul Walravens/Boris Szulzinger)
BELGIUM/FRANCE 1987
Voices of: David Lander, Carol Androsky, Marshall Efron, Alice Playten, Marvin Silbersher, Joanna Lehman, Jerry Bledsoe, Josh Daniel, Bob Kaliban, George Osterman, Ray Owens, Deborah Taylor, Ron Vernan, Roberta Wallach
Adult futuristic cartoon comedy set on the eve of WW4, with the shattered USA and USSR forming a single country, the USSR, but finding that they are in conflict with the nation of Vagina, formed by the remaining women who survived WW3. A film that tries hard to be satirical but ends up being merely crude.
Aka: LE BIG BANG
ANIM 73 min Cut (10 sec – ort 90 min)
VIDrel: EIV/SONOP V

BIG BET, THE ** 18
Bert I. Gordon USA 1986
Sylvia Kristel, Kimberley Evenson, Lance Sloane, Ron Thomas, John Smith, Deanna Claire, Kenneth Ian Davies, Stephanie Blake, Elizabeth Blake, Elizabeth Cochrell, Robert Marucci
Trivial teenage fare from a renowned director (of rubbishy SF movies), telling of a nerd who wagers his car against the school bully, that he can bed one of the girls in his class before the week is out. He turns to his beautiful and sophisticated new neighbour for advice.
COM 87 min (ort 90 min) VIDrel: MIA/DISC V

BIG BIRD CAGE, THE *** 18
Jack Hill USA 1972
Pam Grier, Anitra Ford, Sid Haig, Candice Roman, Vic Diaz, Carol Speed, Teda Bracci, Karen McKevic
Amusing spoof on Filipino prison movies with a gang of thieving mercenaries engineering an escape from the outside. A sequel to THE BIG DOLL HOUSE.
Aka: WOMEN'S PENITENTIARY 2
A/AD 86 min Cut (2 min 24 sec – ort 90 min)
VIDrel: SUPVID/RTM V

BIG BLOCKADE, THE ** U
Charles Frend UK 1958
Leslie Banks, Frank Cellier, Alfred Drayton, Will Hay, John Mills, Robert Morley, Michael Redgrave, Quentin Reynolds, Marius Goring, Alfred Drayton, Michael Rennie, Bernard Miles
The British blockade of Germany during WW2 forms the focus for this semi-documentary work, made as a typical flag-waver by Ealing Studios. An uneven work, made with a plethora of stars, some of whom give more convincing portrayals than others; Redgrave as a Russian is especially good.
WAR 71 min (ort 73 min) B/W VIDrel: WHV V

BIG BLUE, THE ** 15
Luc Besson FRANCE 1988
Rosanna Arquette, Jean-Marc Barr, Jean Reno, Paul Shenar, Sergio Castellito, Marc Duret, Griffin Dunne, Andreas Voutsinas, Valentina Vargas, Kimberley Beck, Patrick Fontana, Alessandra Vazzoler, Geoffrey Carey
A colourful action story that traces the changing relationship between a young woman and a man who decides to take up the hazardous sport of free diving. The shallow plot and characterisation are handicaps but the pretty locations will compensate slightly. Filmed in Sicily, Corsica, Paris, New York, the Virgin Islands and the Riviera.
Aka: LE GRAND BLEU
A/AD 114 min (ort 119 min) wScrn; 162 min (long version)
VIDrel: 20TH/TECH V/sur

BIG BULLY ** PG
Steve Miner USA 1996
Rick Moranis, Tom Arnold, Julianne Philips, Carol Kane, Jeffrey Tambor, Don Knotts
Bullied at school by a classmate, a successful author returns to his home-town for a reunion, but learns that his former bully is now none other than his son's teacher. Mostly a succession of sight gags and slapstick shenanigans, with a few sharp barbs to keep it interesting. Screenplay is by Mark Steven Johnson.
COM 86 min (ort 91 min) cC VIDrel: WHV V/sh

BIG BUSINESS ** PG
Jim Abrahams USA 1988
Bette Midler, Lily Tomlin, Fred Ward, Edward Herrmann, Michele Placido, Daniel Gerroll, Barry Primus, Michael Gross, Joe Grifasi, Mary Gross, Deborah Rush, Nicolas Coster, Patricia Gaul, J.C. Quin, Norma MacMillan
Two sets of identical twins, accidentally separated and switched at birth, meet up years later in New York when one set arrives for a showdown with the corporation that's going to erase their little home town, only to find that the other set of girls is in charge of the company. A nice idea for a comedy, but it never really takes off, despite winning performances from Midler and Tomlin. And after a while the trick photography starts to irritate too.
COM 94 min (ort 97 min) VIDrel: TOUCH L/A V

BIG CHILL, THE *** 15
Lawrence Kasdan USA 1983
Tom Berenger, Glenn Close, William Hurt, Kevin Kline, Mary Kay Place, Jeff Goldblum, JoBeth Williams, Meg Tilly, Don Galloway, James Gillis, Ken Place
A look at a group of former college hippies who have now dropped back into society and meet at the funeral of a friend who has just committed suicide. This excellent piece of ensemble acting has the group examining their lives and careers as they come to terms with the suicide. The fine cast and a nice 1960s soundtrack compensate for any deficiencies in the plot. Written by Kasdan and Barbara Benedek. See also PARALLEL LIVES for something similar in conception.
DRA 100 min (ort 108 min) VIDrel: VCC/DISC/COLUM V

BIG CIRCUS, THE *** U
Joseph Newman USA 1959
Victor Mature, Rhonda Fleming, Red Buttons, Kathryn Grant, Peter Lorre, Vincent Price, Gilbert Roland, David Nelson, Howard McNear, Steve Allen, Adele Mara
A lively story that sees the near-bankrupt owner of a circus making an attempt to keep on the road whilst keeping his business rivals at bay at the same time. Despite a loose plot, the film delivers enough thrills to keep it going. Lorre as a clown is one of several highlights.
DRA 103 min (ort 109 min) VIDrel: 20TH/TECH V/h

BIG COMBO, THE *** PG
Joseph H. Lewis USA 1955
Cornel Wilde, Jean Wallace, Brian Donlevy, Earl Holliman, Richard Conte, Lee Van Cleef, Robert Middleton, Helen Walker, Jay Adler, John Hoyt, Ted De Corsia, Helene Stanton, Roy Gordon, Whit Bissell, Philip Van Zandt
Violent film about the struggle between a crime syndicate and the forces of law and order, in the shape of a persistent cop who is out to nail a cunning racketeer. A slick and well directed story.
THR 87 min B/W VIDrel: VISCOM/RTM V

BIG COUNTRY, THE *** PG
William Wyler USA 1958
Gregory Peck, Charlton Heston, Burl Ives, Jean Simmons, Carroll Baker, Chuck Connors, Charles Bickford, Alfonso Bedoya, Chuck Hayward, Buff Brady, Jim Burk, Dorothy Adams, Chuck Roberson, Bob Morgan, John McKee, Jay Slim Talbot
A sea captain goes west to claim his intended, a rancher's daughter, and finds himself involved in a feud over water rights between his prospective father-in-law and a hillbilly clan. Overlong, overblown, overrated Western where even the excellent acting is overshadowed by the title player – the rugged and spectacularly landscape of the Southwest. The fine musical score is the work of Jerome Moross. AA: S. Actor (Ives).
WES 160 min (ort 168 min) wScrn VIDrel: MGM/WHV V/h
Boa: novel by Donald Hamilton.

BIG DOLL HOUSE, THE ** 18
Jack Hill USA 1971
Judy Brown, Pam Grier, Roberta Collins, Pat Woodell, Brooke Mills, Sid Haig, Christiane Schmidtmer, Kathryn Loder, Jerry Franks, Jack Davis, Gina Stuart, Letty Mirasol, Shirley De Las Alas
Women-in-prison exploiter shot in the Philippines and not without touches of humour to leaven the sex and violence, as the inmates learn how to cope with the stock sadistic warden. Gave rise to a sequel THE BIG BIRD CAGE. This was Grier's film debut.
Aka: WOMEN'S PENITENTIARY 1
A/AD 90 min (ort 93 min) VIDrel: SUPVID/RTM V

BIG DREAMS AND BROKEN HEARTS: THE DOTTIE WEST STORY ***
Bill D'Elia USA
PG
1993
Michele Lee, David James Elliot, Larry Gatlin, Chet Atkins, Loretta Lynn, Kris Kristofferson, Willie Nelson, Kenny Rogers, William Russ, Lisa Akey, Dolly Parton, Ben Browder, Tony Higgins, J. Don Ferguson, Rebecca Koon, Tom Nowicki,
Good biopic on the tragic life of Country and Western singer Dottie West who died in 1991. Lee is well cast as the singer, who rose from her impoverished beginnings (she was one of ten children) to the top of her profession, and at the same time brought a touch of much-needed colour to a style of music that prior to this was not noted for its glamorous image. Includes the Grammy-award winning "Here Comes My Baby" which was something of a signature tune for her.
Aka: DOTTIE WEST STORY, THE
DRA   89 min mTV   VIDrel: ODY/SONOP   V/sh

BIG EASY, THE ***
Jim McBride USA
15
1987
Dennis Quaid, Ellen Barkin, Ned Beatty, John Goodman, Ebbe Roe Smith, Lisa Jane Persky, Charles Ludlam, Tom O'Brien, Marc Lawrence, Solomon Burke, Jim Garrison, Grace Zabriskie, Gailard Sartain, Jim Chimento, Robert Lesser
A smart homicide detective finds himself at odds with the new female D.A. whilst investigating a local Mob murder, but the two become romantically involved even though they remain at odds professionally. A vigorous and highly unusual crime drama with a great Cajun music score and some nice New Orleans locations. Written by Daniel Petrie Jr.
DRA   98 min (ort 101 min)   VIDrel: COLUM/VCC L/A V/sh

BIG GAME, THE **
Bob Keen UK
(PG)
1995
Gary Webster, Emma Wray, Chris Jury, Mark McKenna, Mark Williams, Oliver Parker, Doug Bradley, Jean Boht, Imogen Bain, Emma Parish, Sean Connolly, Doug Bradley, Alex Richardson, Nick Forest, Will Petty, Virginia Harrison
A gambling drama that sees a naive factory worker getting a chance to play with the big boys, but in order to be accepted, he gives himself a fake identity as a highly successful stockbroker. However, when he loses his shirt he finds that he has got in well above his head and that his life is now in danger. Quite a good tale, even if the scripting gives it something of a mTV feel about it.
DRA   99 min   SATrel: MOVIE CHANNEL

BIG GREEN, THE **
Holly Goldberg Sloan USA
U
1995
Steve Guttenberg, Olivia D'Abo, Jay O. Sanders, Bug Hall
A bunch of kids in a small town form a soccer team with the help of their unfit sheriff, and train so hard they make it all the way to the championships. Excellent camerawork during the action sequences help nudge the film along, even if the outcome for all concerned is never really in doubt.
COM   97 min (ort 99 min) cC   VIDrel: HOLPIC/TECH V/sur

BIG HAND FOR A LITTLE LADY, A ***
Fielder Cook USA
U
1966
Henry Fonda, Joanne Woodward, Jason Robards, Paul Ford, Charles Bickford, Burgess Meredith, Kevin McCarthy, Robert Middleton, John Qualen, Jin Boles, James Kenny, Allen Collins, Gerald Conklin, Ned Glass, Mae Clarke
When her gambler husband starts to loose heavily at five-card poker, his resourceful wife steps in and tries to reverse this trend and perhaps cure him of this compulsive habit. Originally written as a short TV play, this amusing spoof does seem a trifle overlong, but this is more than compensated for by the competent acting.
Aka: BIG DEAL AT DODGE CITY, A
WES   91 min (ort 95 min) wScrn   VIDrel: WHV   V/sh
Boa: teleplay by Sidney Carroll.

BIG HEAT, THE ***
Fritz Lang USA
15
1953
Glenn Ford, Gloria Grahame, Jocelyn Brando, Alexander Scourby, Lee Marvin, Jeanette Nolan, Carolyn Jones, Peter Whitney
A cop goes all out to nail a vicious gang of crooks after his wife is killed by a car bomb that was meant for him. A harsh and violent film of considerable force, but now chiefly remembered for its nasty "hot coffee in the face" scene.
DRA   86 min (ort 90 min) B/W   VIDrel: VCC/DISC L/A V
Boa: novel by William P. McGivern.

BIG JAKE **
George Sherman USA
15
1971
John Wayne, Richard Boone, Maureen O'Hara, Patrick Wayne, Chris Mitchum, Bobby Vinton, Bruce Cabot, Glenn Corbett, Harry Carey Jr, John Agar, John Doucette, Jim Davis, Gregg Palmer, Robert Warner, Jim Burke, John Ethan Wayne
With the kidnapping of his grandson, an elderly Texan decides to take the law into his own hands and goes after the boy's abductors. A violent and overblown blend of fisticuffs and action, with good production values but an unfortunate element of self-parody.
WES   109 min   VIDrel: 20TH/TECH   V/h

BIG JIM McLAIN **
Edward Ludwig USA
U
1952
John Wyane, James Arness, Nancy Olson, Veda Ann Borg, Hans Conreid, Gayne Whitman, Alan Napier, Hal Baylor, Robert Keys, John Hubbard, Sara Padden, Soo Yong, Dan Liu, Paul Hurst, Vernon McQueen
Cold War tale of a Communist spy ring and their activities on the island of Hawaii, where a hardbitten Federal agent and his partner are sent to put paid to them. A trite and dull stab at an adventure story with exotic locations that fails to ignite.
A/AD   86 min (ort 90 min) B/W   VIDrel: WHV   V/h

BIG JOB, THE **
Gerald Thomas UK
U
1965
Sidney James, Sylvia Syms, Dick Emery, Joan Sims, Jim Dale, Lance Percival
After fifteen years in jail, a bunch of inept crooks are released and return to retrieve their buried loot, but find that a new police station has been built over the field they buried it in.
COM   84 min B/W   VIDrel: WHV   V/h

BIG MAN, THE **
David Leland UK
18
1990
Liam Neeson, Joanne Whalley-Kilmer, Ian Bannen, Billy Connolly, Pat Roach, John Beattie, Amanda Walker, George Rossi, Andrew Meaden, Ashleigh Thoms, Joseph Greig, Sean Scanlan, James Copeland, Micarena Domenguez, Ken Drury
A Scottish miner unemployed since the 1985 strike (during which he was imprisoned) enters the brutal world of bareknuckle fighting that's controlled by a cynical and world-weary criminal. A dour, harsh and uncompromising story, whose plot elements never quite hold together, consequently blurring the impact and message of the original novel. The music is by Ennio Morricone.
Aka: BIG MAN, THE: CROSSING THE LINE
A/AD   111 min   VIDrel: POLY/POLYREC   V/sh
Boa: book by William McIlvanney.

BIG NIGHT ***
Stanley Tucci USA
15
1995
Minnie Driver, Ian Holm, Isabella Rossellini, Tony Shalhoub, Stanley Tucci, Caroline Aaron, Marc Anthony, Allison Janney, Campbell Scott, Larry Block, Andrei Belgrader, Peter McRobbie
Two brothers struggle to make a success of their Italian restaurant in 1950s New Jersey, but while the older brother concentrates on the joys of cooking, the younger one (who runs the business) is more concerned with the trappings of success. In a bid to publicise their venue, they prepare a banquet for a top singing star, but the night in question proves to be full of unexpected encounters and complications. A delightful, bittersweet comedy.
COM   109 min   CINrel

BIG PARADE, THE ***
King Vidor USA
U
1925
John Gilbert, Renee Adoree, Hobart Bosworth, Claire McDowell, Claire Adams, Karl Dane, George K. Arthur
Classic silent WW1 epic about a young man who enlists in the US Army in 1917 and experiences the horrors of war at first hand. A great commercial success, indeed, the biggest grossing silent of all time, this moving anti-war film is now available in a restored version with the original colour toning and a new score by Carl Davis.
WAR   138 min (ort 142 min) B/W silent
VIDrel: MGM/WHV   V/sh

BIG PICTURE, THE ***   15
Christopher Guest USA   1989
*Kevin Bacon, Emily Longstreth, J.T. Walsh, Jennifer Jason Leigh,
Martin Short, Michael McKean, Kim Miyori, Teri Hatcher, Dan
Schneider, Jason Gould, Tracy Brooks Swope, John Cleese, Eddie
Albert, June Lockhart, Elliott Gould*
An award-winning student film-maker gets the chance of a
Hollywood career but soon learns the truth about the movie
world and its predatory inhabitants. A host of amusing cameos
and some clever in-jokes pad out and help maintain interest in
this pleasing satire. Co-written with McKean, this was Guest's
first directing feature.
COM   101 min   VIDrel: 20VIS/SONOP V

BIG RED ONE, THE ***   15
Samuel Fuller USA   1980
*Lee Marvin, Mark Hamill, Robert Carradine, Bobby Di Cicco, Kelly
Ward, Siegfried Rauch, Stephane Audran, Serge Marquand, Charles
Macaulay, Alain Doutey, Maurice Marsac, Joseph Clark, Ken
Campbell, Doug Werner, Perry Lang*
Gritty and realistic WW2 film about a tough infantry sergeant
who, together with his seasoned section, survive while the fresh
recruits die all around them. The title refers to their infantry
badge.
WAR   111 min (ort 113 min)   VIDrel: WHV V/sur

BIG SLEEP, THE ****   PG
Howard Hawks USA   1944 (released 1946)
*Humphrey Bogart, Lauren Bacall, John Ridgely, Charles Waldron,
Martha Vickers, Louis Jean Heydt, Regis Toomey, Peggy Knudsen,
Dorothy Malone, Bob Steele, Elisha Cook Jr, Sonia Darren, Tom
Rafferty, Charles D. Brown*
Classic film noir adapted from Chandler's first book, with
private detective Philip Marlowe being hired by a rich society
lady and getting drawn into a complex web of intrigue and
murder, mostly brought about by the woman's uncontrollable
younger sister. The convoluted plot is wellnigh impossible to
follow (even Chandler couldn't explain one murder), but the
powerful script (by Faulkner, Jules Furthman and Leigh
Brackman) carries it through.
THR   110 min (ort 114 min) B/W   VIDrel: MGM/WHV L/A
V
Boa: novel by Raymond Chandler.

BIG SLEEP, THE ***   15
Michael Winner UK   1978
*Robert Mitchum, Sarah Miles, Richard Boone, Candy Clark, Edward
Fox, John Mills, James Stewart, Joan Collins, Oliver Reed, Harry
Andrews, Richard Todd, Colin Blakely, Diana Quick, James Donald,
John Justin, Simon Turner*
Follow-up to FAREWELL, MY LOVELY that's a fairly faithful
adaptation of Chandler's book, albeit one that's updated and set
in London, thus largely destroying the flavour if not the plot
elements of the novel. As before, private eye Marlowe finds
himself embroiled in murder, blackmail and violence when he is
hired to protect the daughter of a general. Competent rather than
powerful and best not compared to the Howard Hawks classic.
THR   95 min (ort 99 min)   VIDrel: L/A V
Boa: novel by Raymond Chandler.

BIG STAMPEDE, THE **   U
Tenny Wright USA   1932
*John Wayne, Noah Beery, Mae Madison, Luis Alberni, Berton Churchill,
Paul Hurst, Sherwood Bailey, Frank Ellis, Hank Bell, Lafe McKee*
A cattle-rustling rancher believes that he can kill lawmen with
impunity but is overcome by a young deputy who enlists the
help of a bandit. A pleasant and enjoyable early Wayne vehicle
(a remake of "Land Beyond The Law" from 1936, itself a sound
remake of the 1927 silent version).
WES   51 min (ort 63 min) B/W   VIDrel: MGM/WHV V/h

BIG STEAL, THE ***   PG
Nadia Tass AUSTRALIA   1990
*Ben Mendelsohn, Claudia Karvan, Marshall Napier, Steven Bisley,
Tim Robertson, Angelo D'Angelo, Damon Herriman*
Anxious to impress his dream girl, a young man becomes
obsessed with the idea of owning his own Jag. When the young
lady graciously agrees to go out on a date, he hocks the family
car and all he owns to buy a secondhand model from a totally
dishonest car salesman, only to find himself landed with a
barely mobile wreck. A charming and witty comedy, from the
makers of MALCOLM.
COM   96 min   VIDrel: MGM/WHV V/sh

BIG TOP PEE-WEE **   U
Randal Kleiser USA   1988
*Pee-Wee Harman (Paul Reubens), Kris Kristofferson, Penelope Ann
Miller, Susan Tyrell, Valeria Golino. Voices of: Wayne White, Susan
Tyrell, Albert Henderson, Kevin Peter Hall, Kenneth Toby*
A farmer gets a chance to join the circus when a freak storm
dumps a big top on his doorstep. Offbeat and moderately
amusing Pee-Wee comedy vehicle, with a few good moments
but no great laughs. Clever but somewhat insipid. The film was
co-written and produced by Reubens. See also PEE-WEE'S BIG
ADVENTURE.
COM   82 min   VIDrel: CIC/SONOP L/A V

BIG TOWN, THE **   15
Ben Bolt USA   1987
*Matt Dillon, Diane Lane, Tommy Lee Jones, Tom Skerritt, Lee Grant,
Bruce Dern, Suzy Amis, David Marshall Grant, Don Francks, Del
Close, Meg Hogarth, Cherry Jones, Mark Danton, David Elliott, Steve
Yorke, Chris Owens, Kevin Fox*
A small-time gambler comes to Chicago in the 1950s to try his
luck. A flashy and cliched story with Dillon finding the going
tougher than he expected and getting involved with two girls
along the way.
DRA   105 min (ort 109 min)   VIDrel: VCC/DISC L/A V

BIG TRAIL, THE ***   U
Raoul Walsh USA   1930
*John Wayne, Marguerite Churchill, El Brendel, Tully Marshall, David
Rollins, Tyrone Power Sr, Ward Bond, Helen Parrish, Ian Keith*
The story of a wagon train and the hazards faced by its members
as they travel along the Oregon trail. Wayne got his first star-
ring role in this enjoyable early talkie and though he took several
more years to reach stardom, shows considerable promise. The
film was made using an early 70 mm process known as
Grandeur, and is best enjoyed on the wide-screen. A dated but
impressive early Western epic.
WES   116 min (ort 121 min) B/W   VIDrel: 20TH/TECH
V/h

BIG TROUBLE IN LITTLE CHINA ***   15
John Carpenter USA   1986
*Kurt Russell, Kim Catrall, Dennis Dun, James Hong, Kate Burton,
Victor Wong, Suzee Pai, Donald Li, Carter Wong, Peter Kwong, James
Pax, Jeff Imada, Chao Li Chi, Rummel Mor, Craig Ng, June Kim, Noel
Toy, Jade Go, Jerry Hardin*
Russell plays a trucker who finds himself involved in a strange
Chinatown adventure when his friend's fiancee is kidnapped
and he sets out to rescue her, venturing into the underground
domain of a 2,000-year-old wizard in the process. Amusing and
tongue-in-cheek action story of spectacular effects and imagi-
native fantasy elements. The electronic score is by Carpenter.
A/AD   96 min Cut (9 sec)   VIDrel: 20TH/TECH; ENCORE
(LV only) V LV

BIG WEDNESDAY ***   PG
John Milius USA   1978
*Jan-Michael Vincent, William Katt, Gary Busey, Lee Purcell, Barbara
Hale, Patti D'Arbanville, Darell Fetty, Robert Englund, Reb Brown*
Following their stint in Vietnam, three surfing buddies get
together to face the challenge of the waves once more. Starts as
an amusing surfing comedy set in the 1960s, but as it moves
into the 1970s the comic elements give way to drama. Later re-
cut to 104 minutes by Milius and re-titled "Summer Of
Innocence" for pay-TV release.
Aka: SUMMER OF INNOCENCE
DRA   114 min (ort 119 min) wScrn   VIDrel: BLACK/DISC
V/sur

BIGFOOT **   U
Danny Huston USA   1987
*Colleen Dewhurst, James Sloyan, Gracie Harrison, Joseph Maher,
Adam Karl, Candace Cameron, Bernie White*
Familiar family-style adventure about a big furry creature and
its mate who get involved with various humans. Quite similar
in theme to BIGFOOT AND THE HENDERSONS. Pretty routine
stuff though the make-up work of Robert Schiffer stands out.
A/AD   90 min (ort 100 min) mTV   VIDrel: WDV L/A V

BIGFOOT AND THE HENDERSONS ***   PG
William Dear USA   1987
*John Lithgow, Melinda Dillon, Don Ameche, David Suchet, Kevin
Peter Hall, Joshua Rudoy, Lainie Kazan, Margaret Langrick, M.
Emmet Walsh, Bill Ontiverous*

Story of a family who run over a huge beast on their return from a camping outing. Thinking he might sell the carcass, the father straps it to his car roof but later finds that it wasn't killed after all. Whereupon the family decide to adopt the creature. A pleasant, undemanding comedy outing that gave rise to the TV series "Harry And The Hendersons". See also LITTLE BIG FOOT. AA: Make (Rick Baker).
Aka: HARRY AND THE HENDERSONS
COM 106 min (ort 111 min)
VIDrel: 4-FRONT/POLYREC/CIC L/A V/sh

BIGFOOT: THE UNFORGETTABLE ENCOUNTER **
(PG)
Cory Michael Eubanks USA   1995
Matt McCoy, Crystal Chappell, Clint Howard, Rance Howard, David Rasche, Zachary Ty Bryan, Jojo Adams, Nela Mataranzzo, Janice Lynde, Alan Wilder, Dennis Singletary, Clay Lilley, Tohoru Masamune, Ingo Neuhaus
A young boy meets our gentle monster and helps to protect it from both ruthless bounty hunters and the equally ruthless media circus that his encounter with this creature has given rise to.
JUV 83 min (ort 86 min) SATrel: MOVIE CHANNEL

BIGGLES GETS OFF THE GROUND *   PG
John Hough UK   1986
Neil Dickson, Alex Hyde-White, Peter Cushing, James Saxon, Michael Siberry, Marcus Gilbert, Fiona Hutchinson, William Hootkins
But the film stays put. A disjointed account of a young businessman thrown back in time to 1917 where he teams up with our fictional flying ace. The characters are drawn from the "Biggles" WW1 novels of Captain W.E. Johns but the film is firmly rooted in the teen-fantasy adventure genre. Very poor indeed.
Aka: BIGGLES; BIGGLES: ADVENTURES IN TIME
A/AD 89 min (ort 92 min) VIDrel: VISION/DISC V/sur

BIKINI BEACH *   18
Jim Enright USA
T.T. Boy, Randy West, Sierra, Rebecca Bardoux, Lacy Rose, Terry Thomas, Crystal Wilder, Summer Knights, Tiffany Mynx, Jonathan Morgan, Nicholas Rage, Alicia Rio, Nick Knight, Randy West, Will Divide
Virtually plotless tale of a troupe of people who go on vacation south of the border and spend most of their time making love.
A 52 min VIDrel: ONE V

BIKINI CARWASH COMPANY, THE **   18
Ed Hanson USA   1991
Joe Dusic, Neriah Napul, Suzanne Browne, Kristie Ducati
From the same production company that gave us such gems as TAKIN' IT ALL OFF and PARTY FAVORS, comes this story of a bunch of buxom girls who take over a struggling carwash concern, and put it on a more secure footing. What little plot there is in this one serves as nothing more than an excuse to parade a collection of scantily clad beauties. A sequel followed.
Aka: CALIFORNIA HOT WASH
COM 94 min (ort 95 min) VIDrel: MIA/DISC V

BIKINI CARWASH COMPANY 2, THE **   18
Gary Orona USA   1993
Suzanne Browne, Neriah Napaul, Rikki Brando, Greg Raye, Larry De Russy, Kristie Ducati
Sequel to the first film sees the company having become so successful that it has to fight off the inevitable takeover by the stock avaricious businessmen. In order to raise the money the girls need to prevent this from happening, they decide to advertise on a cable TV station. Ample nudity as before but little else of note.
COM 94 min VIDrel: MIA/DISC V

BILL ***   PG
Anthony Page USA   1981
Mickey Rooney, Dennis Quaid, Largo Woodruff, Harry Goz, Anna Maria Horsford, Kathleen Maguire, Jenny Dweir, Tony Turco, Ray Serra, John Towey, Breon Gorman, George Hamlin, Phil Oxnam, Harriet Rogers, Lotta Palfi, Bill Winkler
Sensitive and moving drama about a mentally retarded adult who is struggling to adapt to life outside the institution he has spent forty-six years in. Rooney gives one of the finest performance of his career (for which he won an Emmy), in a film which, but for some contrived padding, would have been quite outstanding. Based on the true story of William Sackter. An

Emmy also went to Corey Blechman for the script. Followed by "Bill: On His Own.
DRA 91 min (ort 100 min) mTV VIDrel: ODY/SONOP V/h
Boa: story by Barry Morrow.

BILL & TED'S BOGUS JOURNEY **   PG
Peter Hewitt USA   1991
Keanu Reeves, Alex Winter, William Sadler, Joss Ackland, George Carlin, Hal Landon Jr, Jeff Miller, David Carrera, Pam Grier, Amy Stock-Poynton, Annette Azcuy, Sarah Trigger, Chelcie Ross, Hal Landon Sr, William Sadler, Robert Noble
A scatterbrained sequel to the first film that has our two dumb nerds setting out on a voyage through space and time and coming up against two malevolent robot duplicates from the future. A lively blend of madcap tomfoolery and frenetic action that begins to pall about halfway through, from which point on it's sustained by clever special effects.
COM 89 min (ort 98 min) VIDrel: VCC/DISC/COLUM V/sur

BILL & TED'S EXCELLENT ADVENTURE **   PG
Stephen Herek USA   1989
Keanu Reeves, Alex Winter, George Carlin, Bernie Casey, Amy Stock-Poynton, Tony Camilieri, Dan Shor, Ted Steedman, Rod Loomis, Al Leong, Robert V. Barron, Clifford David, Jane Wiedlin, Hal Landon Jr, Bernie Casey
Two brainless teenagers must avoid flunking their history exam and their discovery of a time-travelling telephone booth gives them the chance to do so when they meet various characters from the past. An empty-headed jaunt through time that provides a trifle of amusement. A sequel followed.
COM 86 min (ort 90 min) wScrn
VIDrel: BMGVID/BMGREC V/sur

BILL OF DIVORCEMENT, A **   PG
George Cukor USA   1932
John Barrymore, Katharine Hepburn, Billie Burke, David Manners, Henry Stephenson
Drama of a man who returns home after years in a mental home and gets to know his daughter for the first time, despite the fact that his wife is preparing to divorce him. Interesting but rather stilted and now chiefly remembered as Hepburn's screen debut.
Aka: NEVER TO LOVE
DRA 74 min (ort 96 min) B/W
VIDrel: ENTUK/GOLD L/A V
Boa: play by C. Dane.

BILLION DOLLAR BRAIN ***   PG
Ken Russell UK   1967
Michael Caine, Karl Malden, Ed Begley, Oscar Homolka, Francoise Dorleac, Guy Doleman, Vladek Sheybal, Milo Sperber, Mark Elwes, Donald Sutherland, Susan George
This flawed attempt to film Deighton's novel suffers from both stilted dialogue and over-indulgence. However, it's enjoyable for Ed Begley's superb portrayal of a rabid commie-hating billionaire, who plans an attack on the USSR but is foiled by Homolka as a wily KGB colonel. Caine, as agent Harry Palmer is as lifeless as ever, though some see this as his chief strength. Preceded by THE IPCRESS FILE and FUNERAL IN BERLIN.
THR 111 min VIDrel: MGM/WHV L/A V
Boa: novel by Len Deighton.

BILLIONAIRE BOYS' CLUB **   15
Marvin J. Chomsky USA   1987
Judd Nelson, Ron Silver, Fredric Lehne, Brian McNamara, Raphael Sbarge, John Stockwell, Barry Tubb, Stan Shaw, Jill Schoelen, James Sloyan, James Karen, John Dye, Robert Hallak, Robert Krantz, Eric Larson, Allan Miller, Alan Fudge
Based on the true story of convicted killer Joe Hunt, this tells of how a criminal embroiled a group of L.A. youthful millionaires into a dubious money-making scheme, whose eventual outcome was murder. An interesting saga spoilt by stilted dialogue and lack of pace.
Aka: BILLION DOLLAR BOYS' CLUB
DRA 175 min (ort 200 min) (2 cassettes) mTV
VIDrel: VCC L/A V

BILLY BATHGATE ***   15
Robert Benton USA   1991
Dustin Hoffman, Nicole Kidman, Loren Dean, Bruce Willis, Steve Hill, Steve Buscemi, Billy Jaye, John Costelloe, Tim Jerome, Stanley Tucci, Mike Starr, Robert F. Colesberry, Stephen Joyce, Frances Conroy, Moira Kelly, Kevin Corrigan

An account of the decline and fall of Arthur Flegenheimer, also known as "Dutch Schultz", a notorious gangster who met his violent end in 1935. The story is mostly told as seen through the eyes of the newest recruit to his gang. Scripted by Tom Stoppard, the film is neither as sumptuous as THE GODFATHER nor as harsh as MEAN STREETS, but boasts many fine performances, witty dialogue and (as in the novel) a generous dose of brutality.
A/AD   102 min (ort 107 min)   VIDrel: TOUCH/SONOP L/A   V
Boa: novel by E.L. Doctorow.

BILLY LIAR **   PG
John Schlesinger UK   1963
Tom Courtenay, Julie Christie, Wilfred Pickles, Mona Washbourne, Finlay Currie, Ethel Griffies, Gwendolyn Watts, Helen Fraser, Leslie Randall, Rodney Bewes, George Innes, Patrick Barr, Godfrey Winn, Leonard Rossiter
An excellent adaptation of a novel, of a bored undertaker's clerk whose only escape from the dullness of his drab North Country existence is via his own imagination. Imaginative, touching and rather poignant.
COM   94 min (ort 98 min) B/W   VIDrel: WHV   V
Boa: novel by Keith Waterhouse/play by Keith Waterhouse and W. Hall.

BILLY MADISON *   PG
Tamara Davis USA   1995
Adam Sandler, Bradley Whitford, Josh Motel, Bridgette Wilson, Norm MacDonald, Darren McGavin
In order to inherit his father's chain of hotels, a teenager is forced to return to grade school and repeat grades 1-12, otherwise his legacy passes to his late father's assistant. A juvenile comedy of unfunny jokes and other stupidities that is as puerile as it is uninspired.
COM   90 min   CINrel

BILLY THE KID **   15
William A. Graham USA   1988
Val Kilmer, Duncan Regehr, Ned Vaughan, Patrick Massett, Julia Carmen, Nate Esformes, Rene Auberjonois, Albert Salmi, Gore Vidal, Wilford Brimley, John O'Hurley, Andrew Bicknell, Burr Steers, Mike Casper, Jack Dunlap, Bing Blenman
Screenplay is by Gore Vidal, in this sanitised tale of a poor misunderstood lad who just happened to become a killer through events outside his control. An entertaining film that does little to increase our knowledge of the true nature of this legendary Wild West character. An earlier teleplay by Vidal in the 1950s formed the basis for the Paul Newman film, THE LEFT-HANDED GUN. Vidal has a cameo as a graveside minister.
Aka: GORE VIDAL'S BILLY THE KID
WES   93 min (ort 100 min) mCab   VIDrel: L/A   V

BILOXI BLUES **   15
Mike Nichols USA   1988
Matthew Broderick, Christopher Walken, Matt Mulhern, Corey Parker, Markus Flanagan, Casey Siemaszko, Michael Dolan, Penelope Ann Miller, Park Overall, Alan Pottinger, Mark Evan Jacobs, Dave Kienzle, Matthew Kimbrough, Allen Turner
Semi-autobiographical account of Neil Simon's wartime experiences, set in 1943 at an army base in Missouri. This continues the tale of Eugene Jerome that began with BRIGHTON BEACH MEMOIRS and follows him through his ten tough weeks of army basic training. A witty and perceptive look at a young man's journey to manhood. BROADWAY BOUND (which started life as a stage play) is effectively the concluding part of this three-tier autobiography.
COM   102 min (ort 105 min)
VIDrel: 4-FRONT/POLYREC/CIC   V/sur

BINGO **   PG
Mathew Robbins USA   1991
Cindy Williams, David Rasche, Robert J. Steinmiller Jr, David French, Kurt Fuller, Joe Guzaldo, Glenn Shadix, Janet Wright, Wayne Robson, Simon Webb, Suzie Plakson, Tamsin Kelsey, Norman Browning, James Kidnie, Blue Mankuma
A runaway circus dog meets up with a maladjusted young boy and they soon become inseparable. However, when the boy's family decide to move, his parents refuse to allow him to bring his pet. The eventually leads to a journey of about a thousand miles as boy and dog seek to be reunited. A vigorous and worthy successor to all those old "Benji" movies, with our

canine friend giving a most winning performance. Will appeal to kids of all ages.
JUV   86 min (ort 90 min)   VIDrel: COLUM/SONOP L/A V/sh

BIO-DOME **   12
Jason Bloom USA   1995
Stephen Baldwin, Pauly Shore, Kylie Minogue, William Atherton, Joey Adams, Henry Gibson
Two couples drive into the desert and come across a carefully engineered and sealed research environment. When the two boys successfully penetrate this establishment, they get locked in and have to stay there, but while away their time disrupting all the carefully planned experiments, much to the annoyance of various scientists. Initially quite promising, the incessant comic capers and wisecracking become ever more tiresome.
COM   91 min (ort 95 min)   VIDrel: MGM/WHV   V/sh

BIONIC NINJA **   18
Tim Ashby HONG KONG   1986
Kelly Steve, Alan Hemmings, Rick Wilson, Peter Chan, Andy Man, Jack Young, Pauline Chao, Alex Baker
The CIA send their best agent to recover a top secret microfilm, after it has been stolen for the KGB by a criminal Ninja organisation. The mission seems hopeless, our agent is blocked at every turn, but he discovers the power of Ninjitsu, a power so great that our mighty hero is able to smash his way into the heart of this evil empire. High-action nonsense with a low-grade plot.
MAR   86 min Cut (14 sec)   VIDrel: IMPENT   V

BIONIC SHOWDOWN, THE *   PG
Alan Levi USA   1989
Lee Majors, Lindsay Wagner, Richard Anderson, Sandra Bullock, Jeff Yagher, Geraint Wyn Davies, Martin E. Brooks, Robert Lansing, Josef Sommer, Lee Majors II
More bionic shenanigans in this spin-off from "The Six Million Dollar Man", with the pair from the TV series and a couple of similarly equipped youngsters out to foil the work of a bionic spy whose intention is to blight East-West relations. Poorly plotted production-line nonsense.
Aka: BIONIC SHOWDOWN: THE SIX MILLION DOLLAR MAN AND THE BIONIC WOMAN
A/AD   92 min (ort 100 min) mTV   VIDrel: CIC/SONOP   V

BIRD ***   15
Clint Eastwood USA   1988
Forest Whitaker, Diane Venora, Michael Zelniker, Samuel E. Wright, Keith David, Michael McGuire, James Handy, Damon Whitaker, Morgan Nagler, Arlen Dean Snyder, Sam Robards, Hamilton Camp, Anna Levine, Jason Bernard, Penelope Windust
The story of legendary jazz saxophonist Charlie Parker, who dominated the 1940s jazz scene. When it's good it's very good, but the film unwisely focuses on Parker's self-destructive urges, especially his drug addiction. Whitaker's bravura performance and Parker's music (most of the soundtrack features his playing) are considerable compensations. AA: Sound (Les Frescholtz/Dick Alexander/Vern Poore/William D. Burton). See also PETE KELLY'S BLUES.
DRA   154 min (ort 161 min)   VIDrel: WHV   V/sur

BIRD OF PREY **   15
Temistocles Lopez USA   1995
Jennifer Tilly, Lenny Von Dohlen, Richard Chamberlain, Robert Carradine, Lesley Ann Warren, Boyan Milushev
Imprisoned for killing his father's murderer, a young man meets another prisoner who also wants to kill the same gangster. However, when our young man is released from jail, he unwisely falls in love with the daughter of his intended victim. A dark and convoluted thriller, with little in the way of surprises.
THR   96 min   VIDrel: FIRST/SONOP   V

BIRD ON A WIRE *   15
John Badham USA   1990
Mel Gibson, Goldie Hawn, David Carradine, Bill Duke, Stephen Tobolowsky, Joan Severance, Harry Caesar, Jeff Corey, John Piper-Ferguson, Clyde Kusatsu, Jackson Davies, Florence Paterson, Tim Healy, Wes Tritter, Lossen Chambers
A witness who has been given a new identity by the FBI and relocated, finds his life in danger when his cover is blown. After his employer is killed, said witness takes to the road, finding shelter in the arms of a former girlfriend, with whom he endures

a number of close calls before the final confrontation. A messy and unappealing blend of comedy and action that neither excites nor amuses.
COM   106 min (ort 110 min)
VIDrel: 4-FRONT/POLYREC/CIC   V/sur

BIRDCAGE, THE **
15
Mike Nichols USA
1996
Robin Williams, Nathan Lane, Dan Futterman, Dianne Wiest, Gene Hackman, Hank Azaria, Calista Flockhart, Christine Baranski, Tom McGowan, Grant Heslov, Kirby Mitchell, James Lally, Luca Tommassini, Luis Camacho, Andre Fuentes
A remake of LA CAGE AUX FOLLES that stars Williams and Lane as a gay couple who are forced to pretend to be a straight husband and wife when the son of the former gets engaged to the daughter of a right-wing senator. American remakes of French hits are not usually noted for their inventiveness, and this one is no exception, being a virtual scene-for-scene remake of the original. But Lane is outstanding as the drag queen, even if Williams is given precious little to do.
COM   114 min (ort 119 min) wScrn cC   VIDrel: MGM/WHV V/sh

BIRDMAN OF ALCATRAZ, THE ***
PG
John Frankenheimer USA
1961
Burt Lancaster, Karl Malden, Whit Bissell, Edmond O'Brien, Thelma Ritter, Telly Savalas, Neville Brand, Betty Field, Hugh Marlowe, James Westerfield, Crahan Denton, Chris Robinson
A long-term prisoner achieves the astonishing feat of becoming a world-renowned authority on birds by dint of sheer hard work and study. Largely based on the remarkable real-life tale of Robert Stroud, this is a fascinating study despite being confined to a prison setting.
DRA   142 min (ort 148 min) B/W   VIDrel: MGM/WHV V
Boa: book by Thomas E. Gaddis.

BIRDS, THE **
15
Alfred Hitchcock USA
1963
Rod Taylor, Tippi Hedren, Suzanne Pleshette, Jessica Tandy, Veronica Cartwright, Ethel Griffies, Charles McGraw, Ruth Mcdevitt, Joe Mantell, Doodles Weaver, Richard Deacon
One of Hitchcock's weakest films with neither pacing nor tension. Based on an attack by seabirds in 1961 at Santa Cruz, the film has flocks of birds making murderous attacks on the people at a small and isolated California community. The plodding and opaque script is by Evan Hunter. Loved by the critics, but on careful viewing it is really hard to see why. THE BIRDS 2: LAND'S END followed a good many years later. See also DEADLY INVASION: THE KILLER BEE NIGHTMARE.
DRA   113 min (ort 120 min)   VIDrel: CIC/SONOP V
Boa: short story by Daphne Du Maurier.

BIRDS 2, THE: LAND'S END *
15
Alan Smithee USA
1993
Brad Johnson, Chelsea Field, James Naughton, Tippi Hedren, Jan Rubes, Stephanie Millford
After a young family move to a remote part of the East Coast, they become a target for mass attacks by ferocious birds, in a repeat of an event that happened many years before. A pale clone of THE BIRDS, in which Hedren played one of the leads. The use of the "Alan Smithee" pseudonym tells us everything we need to know about the director's feelings about the movie. Hedren (who starred in the earlier film) has a small and pointless cameo as the shopkeeper.
DRA   83 min (ort 87 min) mTV   VIDrel: CIC/SONOP V/sh

BIRDS IN PARADISE: PARTS 1 AND 2 **
(18)
Kent James USA
1985
Jeanine Louise, Sue Morrow, Jennifer Inch, Michael Buschel, Jennifer Whyl, Don Richter, Bruce Mitchell, Eddy Styles, Victoria Lake, John Simmons, Charles Gorson
Two-part erotic story in which a trio of women inherit a luxury yacht, but find that they must make $10,000 if they want to use it. In the second part the girls succeed in setting sail and enjoy various adventures, but at the same time have to contend with a crooked lawyer who is out to separate them from their inheritance.
A   170 min   SATrel: MOVIE CHANNEL

BIRTH OF A NATION, THE ****
15
D.W. Griffith USA
1915
Lillian Gish, Mae Marsh, Miriam Cooper, Robert Harron, Josephine Cromwell, Henry B. Walthall, Wallace Reid, Donald Crisp, Joseph Henabery, Raoul Walsh, Eugene Pallette, Spottiswood Aiken, J.A. Beringer, John French
First and most famous cinema epic depicting the story of the American Civil War and the post-war period, seen through the eyes of two families. A classic and often moving film that is as successful in its depiction of the battle scenes as it is in portraying more intimate moments. The film, originally shown as "The Clansman", is slightly marred by its expression of questionable sentiments. Written by Griffith and Frank E. Woods.
DRA   192 min (ort 190 min) B/W silent
VIDrel: CONNO/RTM V
Boa: novel The Klansman by Thomas Dixon Jr.

BISHOP'S WIFE, THE ***
U
Henry Koster USA
1947
Cary Grant, Loretta Young, David Niven, Monty Wooley, Elsa Lanchester, James Gleason, Gladys Coope, Sara haden, karolyn Grimes, Tito Vuolo, Regis Toomey, Sarah Edwards, Margaret McWade
A sophisticated angel comes to Earth in response to the prayers of a bishop, and helps him and his wife resolve both spiritual and domestic crises and even mend their broken marriage. An appealing fantasy that is made with a light and sure touch, it seems totally believable and works well even though it is slightly overlong. Remade as THE PREACHER'S WIFE. See also IT'S A WONDERFUL LIFE for another tale of angelic intervention. AA: Sound (Gordon Sawyer).
COM   109 min B/W   VIDrel: VCC/DISC V
Boa: novel by Robert Nathan.

BITCH, THE *
18
Gerry O'Hara UK
1979
Joan Collins, Michael Coby, Kenneth Haigh, Ian Hendry, Carolyn Seymour, Sue Lloyd, Mark Burns, John Ratzenberger, Pamela Salem, Anthony Heaton, Maurice O'Connell, Peter Wight, Doug Fisher, George Sweeney, Chris Jagger
In this sequel to THE STUD our heroine, facing divorce from her wealthy husband, will stop at nothing to save her failing disco, and engages in a temporary liaison with a gangster on the run from the Mafia. A weak, noisy disco-style follow-up that unashamedly seeks to cash in on the success of the earlier film.
DRA   90 min (ort 99 min)   VIDrel: ENTUK L/A V
Boa: novel by Jackie Collins.

BITE, THE *
18
Fred Goodwin ITALY/JAPAN
1989
Jill Schoelen, J. Eddie Peck, Jamie Farr, Savina Gersak, Bo Svenson, Sydney Lassick, Marianne Muellereile, Al Fann, Terrence Evans, Sandra Sexton, Bruce Machiaro, Shiri Appleby, Jose Garcia, Tiny Welles, Sommar Betsworth
A young couple travelling through the desert inadvertently come across an abandoned nuclear test site. Unfortunately the man is bitten by one of the mutated snakes that infest the area and eventually becomes a rampaging snake-like creature that must be destroyed. A conventional but rather well made horror film that harks back to the 1950s. Apart from having the same producer, the film bears no relation to THE CURSE.
Aka: CURSE 2: THE BITE
HOR   93 min (ort 98 min)   VIDrel: EIV/SONOP V/sur

BITS AND PIECES **
15
Antonello Grimaldi ITALY
1996
Asia Argento, Enrico Lo Verso, Chiara Mastroianni
An assemblage of thirty different stories, all of which are set in Rome and take place over the same twenty-four hour period. The best and worst of human life is seen, from murders, betrayals and conflicts through to simple encounters and quite a few comic moments. But although the enormous cast (too large to list) struggle gamely with their respective parts, the lack of any over-riding vision is a flaw the film just cannot overcome. See also SHORT CUTS.
DRA   110 min   CINrel

BITTER BLOOD **
15
Jeff Bleckner USA
1994
Kelly McGillis, Keith Carradine, Harry Hamlin, Holland Taylor, Jayne Brook, Ken Jenkins, Elizabeth Wilson, Louise Latham, Wayne Tippit, Tom Aldredge, Anne Pitoniak, Nick Searcy, Tristam Tait, Colleen Flynn, Erik von Detten, Erik Lloyd
Real-life account of a failed marriage and obsessive revenge. Adequate.
Aka: IN THE BEST OF FAMILIES: MARRIAGE, PRIDE AND MADNESS
DRA   178 min (ort 182 min)   VIDrel: ODY/SONOP V/s
Boa: book by Jerry Bledsoe.

BITTER HARVEST ** 18
Duane Clark USA 1992
Patsy Kensit, Stephen Baldwin, Jennifer Rubin, Adam Baldwin, M. Emmet Walsh, James Crittenden, Art Evans, Joanna Jackson, Ed Morgan, David Pawledge, Jim Lovelet, William Hurnley, Bonnie Paul, Kandra Baker, Jeff Yesko, Wally Rose
Having recently inherited a ranch, a young man gets involved in a strange triangular relationship with two willing women that implicates him in robbery and murder. An average erotic-drama on the familiar "femme fatale" theme.
DRA 98 min VIDrel: GUILD/SONOP L/A V/sh

BITTER MOON ** 18
Roman Polanski FRANCE/UK 1992
Peter Coyote, Emmanuelle Seigner, Hugh Grant, Kristin Scott-Thomas, Victor Banerjee, Luca Vellani, Boris Bergman, Sophie Patel, Patrick Albenque, Smilja Mihalovitch, Leo Ekcman, Richard Deux, Danny Garcy, Daniel Dhubert
A young and quite prim British couple take a Mediterranean cruise and meet an American writer now confined to a wheel-chair, and his young and voluptuous French wife. He gradually begins to tell stories from his past that describe in detail the twisted progression of their very strange relationship. An over-the-top exercise in the ludicrous and the grotesque that pretends to be meaningful but is ultimately both empty and futile.
Aka: LUNES DE FIEL
DRA 133 min (ort 139 min) VIDrel: COLUM/SONOP V/sur
Boa: novel Lunes De Fiel by Pascal Bruckner.

BITTER TEA OF GENERAL YEN, THE *** PG
Frank Capra USA 1932
Barbara Stanwyck, Nils Asther, Gavin Gordon, Lucien Littlefield, Richard Loo, Clara Blandick, Walter Connolly, Moy Ming, Robert Wayne, Knute Erikson, Ella Hall, Arthur Millette, Helen Jerome Eddy, Martha Mattox, Jessie Arnold
A woman missionary arrives in 1930s Shanghai to get married but before the wedding can take place, she finds herself captured by a Chinese warlord who becomes deeply enamoured of her. A complex and compelling tale whose theme of inter-racial love is handled with great delicacy, while one of the most memorable passages is a nightmare sequence that reveals Stanwyck's hidden longings. A fascinating film that is well made and highly enjoyable.
DRA 84 min B/W VIDrel: COLUM/SONOP V
Boa: novel by Grace Zaring Stone.

BITTER TEARS OF PETRA VON KANT, THE ** 15
Rainer Werner Fassbinder WEST GERMANY 1972
Margit Carstensen, Irm Hermann, Hanna Schygulla, Eva Mattes, Katrin Schaake, Gisela Fackelday
Ponderous and pretentious film about a lesbian love-triangle, with wealthy fashion designer Carstensen fretting over her unreliable lover Schygulla. Slow, tedious and vastly over-rated. Written by Fassbinder.
Aka: DIE BITTEREN TRANEN DER PETRA VON KANT
DRA 124 min VIDrel: CONNO V

BITTER VENGEANCE ** 15
Stuart Cooper USA 1994
Virginia Madsen, Bruce Greenwood, Kristen Hocking, Eddie Velez, Gordon Jump, Carlos Gomez, Tim Russ, Jack Verrell, Vince Melocchi, Teresa Truesdale, Peter Moore, Mary Ingersoll, Rick Zoerner, Michael White, Virginia Hawkins
An ex-cop gets a job as a security guard but exploits this position to plan and execute an armed robbery, together with his lover. However, he is smart enough to pin both crimes on his estranged wife, who now has to prove her innocence while evading capture by the police. An unremarkable crime thriller of little distinction.
THR 86 min (ort 90 min) mCab VIDrel: CIC/SONOP V

BLACK AND BLUE ** (PG)
David Hayman UK 1992
Christopher John Hall, Linus Roache, Martin Shaw, Iain Glen, Ray Winstone, Don Henderson, Rowena King, Fraser James, Clive Wedderburn, Roger Sewell, David Thewlis, Madhav Sharma, David Morrissey, Patrice Naiambana, Terry Sue Patt
The murder of a local councillor is investigated by a young black policeman, who goes undercover as part of his assignment, but discovers more than he bargained for, for when he finds that there is corruption within the police force at the highest level. A nasty and deeply cynical work, scripted by G.F. Newman, who is

something of a specialist in works on this theme. Quite a gripping tale, even if the distinctly one-sided view of things does get rather tiresome at times.
DRA 85 min mTV TVrel: BBC

BLACK ARROW, THE ** U
Warwick Gilbert/Alex Nicholas/George Stephenson AUSTRALIA 1987
Voices of: Bob Baines, Claire Crowther, Phillip Hinton, Graham Matters, Lloyd Morris
Animated adventure that takes place during the era of the War of the Roses, with a boy and his friends attempting to find the murderer of their father, and throwing in their lot with a band of outlaws.
ANIM 48 min (ort 50 min) VIDrel: CARL/TECH V

BLACK BEAUTY ** U
James Hill UK 1971
Mark Lester, Walter Slezak, Peter Lee Lawrence, Uschi Glas, Patrick Mower, John Nettleton, Maria Rohm, Eddie Golden, Clive Geraghty, Johnny Hoey, Patrick Gardiner, Margaret Lacey, Fernando Bilbao, Vincente Rola, Jose Niero
Film based on the classic story by Anna Sewell about an unhappy horse that is mistreated by a series of cruel owners before finally returning to its original young master. No more than adequate. Screenplay is by Hill and Wolf Mankowicz and the music is by Lionel Bart.
DRA 101 min (ort 109 min) VIDrel: FABFIL/SPEAR V
Boa: novel by Anna Sewell.

BLACK BEAUTY *** U
Caroline Thompson UK 1993
Sean Bean, David Thewlis, Andrew Knott, Jim Carter, Peter Davison, Alun Armstrong, John McEnery, Eleanor Bron, Peter Cook, Adrian Ross-Magenty, Lyndon Davies, Georgina Armstrong, Gemma Paternoster, Alan Cumming (narration only)
A horse in Victorian England tells the story of its life and its treatment at the hands of its various owners. A straightforward rendition of this classic novel that works beautifully thanks to fine acting, good direction and a nicely restrained tone. First-time director Thompson wisely lets the strengths of the novel come to the fore, avoiding the common trap of over-embellish-ment.
DRA 84 min (ort 90 min) cC VIDrel: WHV V/sur
Boa: novel by Anna Sewell.

BLACK BELT ** 18
Charles Philip Moore USA 1992
Don "The Dragon" Wilson, Brad Hefton, Ernest Simmons, Mitch Borrow, Tim Backer, Gerry Blank, Jim Graden, Ian Jacklin, John Graden, Matthias Hughes, Deidre Imershein
A professional kickboxer helps protect a beautiful singer who is being stalked by a mysterious killer, in this conventional martial arts tale. Its threadbare plot serves mainly as an excuse for the usual display of kickboxing skills.
Aka: BLACKBELT
MAR 83 min (ort 85 min) VIDrel: ONE/IMPENT V

BLACK BELT JONES *** 18
Robert Clouse USA 1973
Jim Kelly, Gloria Hendry, Scatman Crothers, Alan Weeks, Eric Laneuville, Nate Esformes, Malik Carter, Mel Novak, Eddie Smith, Alex Brown, Clarence Barnes, Earl Brown, Esther Sutherland, Sid Kaiser, Doug Sides
Black kung fu in this story of a school of self-defence, battling with the Mafia in the Watts district of L.A. A fairly enjoyable action film from the same team that brought us ENTER THE DRAGON. Followed by a sequel.
MAR 82 min (Cut at film release by 1 min 27 sec – ort 87 min) VIDrel: MGM/WHV L/A V

BLACK BELT JONES 2 ** 18
Lee Tso-Nan HONG KONG 198-
Jim Kelly, Norman Wingrove, Bobby Ming, Misaki Nan, Bolo Yeung
A rather pedestrian follow-up to the first film with Kelly making a return as an ex-CIA martial artist. When a priceless diamond known as "The North Star" is stolen from an American courier, our hero is led into the seedy world of night clubs, hookers and strippers, as he fights his way into the heart of a criminal empire. Average.
MAR 90 min VIDrel: MOPIC/SGSVID V

BLACK CANDLE, THE ★★★ 15
Roy Battersby UK 1991
Samantha Bond, Denholm Elliott, Sian Phillips, Nathaniel Parker, Robert Hines, Tara Fitzgerald, James Gaddas, Cathy Sandford, Bob Smeaton, Brian Hogg, Mo Harold, Anne Jameson, Roger Avon, Peggy Shields, Preston Lockwood
Solid, workmanlike adaptation of Cookson's novel of industrial life in Northern England during the last century. A poor but honest working girl is seduced and abandoned, an impoverished aristocrat resorts to a loveless marriage, and an ambitious young foreman is falsely accused of murder. In terms of atmosphere and period detail this work cannot be faulted; if only the plot and characterisation showed a little more depth.
DRA 103 min mTV VIDrel: FOCUS/DISC V
Boa: novel by Catherine Cookson.

BLACK CANDLES ★ 18
Jose Ramon Larraz SPAIN 1981
Martha Belton, Vanessa Ashley, Betty Webster, John McGrat, Jeffrey Healey, Lucille Jameson, Paul Kendall, John Thompson
Produced just after the death of Franco under Spain's newly liberalised laws, this is a curious sex film that mixes eroticism and satanism. It promises much but delivers little, and the scenes of depravity we were so looking forward to fail to make their appearance, leaving us with various odd characters who do their best to look wicked while they conduct a satanic ceremony.
Aka: LOS RITOS SEXUALES DEL DIABLO; NAKED DREAMS; SEX RITES OF THE DEVIL
HOR 82 min VIDrel: REDEM/RTM L/A V

BLACK CAT, THE ★★★ 15
Edgar G. Ulmer USA 1934
Boris Karloff, Bela Lugosi, David Manners, Jacqueline Wells (Julie Bishop), Lucille Lund, Henry Armetta, Egon Brecher, Albert Conti, Anna Duncan, Herman Bing, Andre Cheron, Luis Alberni, Harry Cording, George Davis, Tony Marlow
First pairing of Lugosi and Karloff in this strange tale of a devil-worshipping architect and his doctor adversary. The former lives in a mansion built over the ruins of a WW1 fort where thousands of soldiers met their deaths as a result of his treachery. A strange, surrealistic film that little to do with the Poe story, the character of the architect having been clearly inspired by Aleister Crowley.
Aka: HOUSE OF DOOM; VANISHING BODY, THE
HOR 63 min (ort 70 min) B/W VIDrel: CIC/SONOP L/A V

BLACK CAT, THE ★★ 18
Lucio Fulci ITALY 1981
Patrick Magee, Mimsy Farmer, David Warbeck, Dagmar Lassander, Daniela Doro, Al Cliver (Pier Luigi Conti), Bruno Corazzari, Geoffrey Copleston
Slightly Poe-inspired tale that opens with a strange man who wanders through cemeteries recording messages from the dead. A detective is sent to investigate a series of nasty murders, but it is a female photographer who connects them to the man's sinister cat. A stodgy low-budget affair.
Aka: IL GATTO DI PARK LANE; IL GATTO NERO
HOR 88 min (ort 93 min) VIDrel: REDEM/RTM V

BLACK CAT ★★ 18
Dickson Poon/Stephen Shin HONG KONG 1993
Jade Leung, Simon Yam, Thomas Lam
A woman with a violent and uncontrollable temper murders a cop during an attempted rape, but then agrees to undergo treatment to modify her behaviour. As in THE TERMINAL MAN, this involves the implantation of a microchip (it bears the name of title animal) into her brain. This procedure turns her into a ruthless, cold-blooded assassin. An oriental clone of NIKITA that was followed by a sequel.
MAR 93 min wScrn dubbed VIDrel: MADE/RTM V

BLACK COBRA, THE ★★ 18
Stelvio Massi USA 1987
Fred Williamson, Karl Lundgren, Eva Grimaldi, Riccardo Moni, Vassili Karis
A woman photographer sees a murder being committed and takes a photo of the killer, a vicious gang leader, thereby putting herself at the top of his personal hit list. Williamson plays the cop given the job of protecting her, in this fast-paced and confusingly plotted thriller. A couple of sequels followed.
THR 85 min Cut (1 min 14 sec – ort 95 min)
VIDrel: SUPVID/RTM V

BLACK DEATH ★★★ 15
Sheldon Larry CANADA 1992
Kate Jackson, Al Waxman, Jeffrey Nordling, Chip Zien, Barbara Williams, David Hewlett, Jerry Orbach, Howard Hesseman, Alma Martinez, Luis Guzman, Tom Mardirosian, Kathleen Robinson, Kristina Nicoll, Michael Copeman
A kind of contemporary Typhoid Mary tale, with a young girl infecting New York City with Bubonic Plague, and only the efforts of one determined doctor standing between the inhabitants and total devastation. A compelling TV thriller that is a good deal better than the usual run of such films.
THR 92 min mTV VIDrel: MIA/DISC/IMPENT V

BLACK DETAIL, THE ★ 18
L.S. Talbot/Teri Diver USA 1994
Krista, Sean Michaels, Bill Wilde, Nick East, Nicole London, Damien Zeus, Sahara Sands, Vienna, Chaz Vincent
A porno version of THE LAST DETAIL in which two soldiers escorting a prisoner to the stockade take some time out so that all three can enjoy some sexual adventures. A sequel followed.
A 45 min VIDrel: ONE V

BLACK EAGLE ★★★ 15
Eric Karson USA 1986
Jean-Claude Van Damme, Sho Kosugi, Doran Clark, Bruce French, Vladimir Skomarovsky, William H. Bassett, Jan Triska, Kane Kosugi, Shane Kosugi, Dorota Puzio, Gene Davis, Joe Quattromani, Alfred Mallia, Victor Bartolo
Under the cover of a Japanese marine biologist, a CIA agent undertakes a hazardous mission to recover top-secret laser navigation equipment from a crashed USA fighter. Fast-paced formula actioner with a liberal dose of martial arts combat.
A/AD 89 min (ort 93 min) VIDrel: 4-FRONT/POLYREC V

BLACK EMMANUELLE ★ 18
Albert Thomas (Adalberto Albertini) ITALY 1975
Laura Gemser, Karin Schubert, Angelo Infanti, Don Powell, Isabelle Marchall, Gabriele Tinti, Venantino Venantini
First in an interminable series. Here our heroine is a photographer sent to Nairobi on an assignment. Various sexual adventures follow in this glossy but utterly tedious softcore effort.
Aka: EMMANUELLE IN AFRICA; EMMANUELLE NERA
A 91 min Cut (1 min 28 sec – ort 99 min) dubbed
VIDrel: LUMI/SPEAR V

BLACK EMMANUELLE GOES EAST ★★ 18
Joe D'Amato (Aristide Massacesi) ITALY 1976
Laura Gemser, Gabriele Tinti, Ely Galleani, Ivan Rassimov, Deborah Berger, Kioke Mahoco
This time our heroine is in the Far East. Various sexual adventures follow. Another dreary softcore session enlivened by the exotic locations.
Aka: EMMANUELLE IN BANGKOK; EMMANUELLE NERA: ORIENT REPORTAGE; EMMANUELLE GOES EAST
A 82 min Cut (3 min 25 sec – ort 90 min) dubbed
VIDrel: LUMI/SPEAR V

BLACK FEATHER ★★ (U)
Steve Kroschel USA 1995
Todd Beadle, Amy Wiegart, Roxanne Hagerman, Laura Tyron, Tracy Hinkson, Pius Savage, Alice Welling, Grant Olson, Rob Phillips, David Haynes, Sandra Ramsey, Linda Benson, Steven Tryon, Lara Conway, Lori C. Ostrovsky, Fred Smith
A young girl whose father is totally preoccupied with opening a tourist resort in the mountains, is badly injured when she tries to save a raven's nest in a tree her father had decided to fell. While the doctors remain sceptical as to her likelihood of ever walking again, the raven becomes a treasured pet, much against the wishes of her father. However, he soon has good reason to change his mind. A nicely photographed film, enjoyable and only slightly sentimental.
Aka: ADVENTURES OF BLACK FEATHER, THE
JUV 90 min SATrel: MOVIE CHANNEL

BLACK FOX: BLOOD HORSE ★★ PG
Steven Hilliard Stern USA 1994
Christopher Reeve, Raoul Trujillo, Tony Todd, Janet Bailey, Nancy Sorrel, Chris Wiggins, Chris Benson, Lawrence Dane, Cyndy Preston, Dale Wilson, Don S. Davis, Joel Phage-Wright, Leon Goodstriker, Byron Chief Moon, Buffalo Child
In Texas of 1861 at the outbreak of the Civil War, a Carolina

man runs his smallholding with his bloodbrother, a freed slave. But the Kiowa and Cherokee tribes unite in a move to force out settlers, and kidnap the town's women and children. Both men now face a struggle to survive and safeguard their property. The first story in a trilogy that continued with BLACK FOX: THE PRICE OF PEACE and concluded with BLACK FOX: GOOD MEN AND BAD. Adequate if unremarkable.
Aka: BLOOD HORSE
WES   94 min mTV   VIDrel: MARQ/QUANT   V
Boa: novel by Matt Braun.

BLACK FOX: GOOD MEN AND BAD ** **   (PG)
Steven Hilliard Stern USA   1994
Christopher Reeve, Raul Trujillo, Tony Todd, David Fox, Kim Coates, Janet Bailey, Nancy Sorel, Kelly Rowan, Lawrence Dane, Alan Shearman, Beverly Elliott, Graham McPherson, Elliott, Rainbow Francks, Alan Vansprang, Billy Morton
When racists attack his farm and kill his wife, a man sets about having his revenge and is so obsessed with his quest, that he is in danger of breaking the law. Meanwhile, his black bloodbrother does his best to keep him on the straight and narrow. Last in the trilogy, this concludes a fairly forgettable revisionist Western trilogy, none of which offer anything of note except strong performances from the leads.
Aka: GOOD MEN AND BAD
WES   88 min (ort 90 min) mTV   VIDrel: GUILD/FOXVID V/s

BLACK FOX: THE PRICE OF PEACE ** **   (PG)
Steven H. Stern USA   1994
Christopher Reeve, Raoul Trujillo, Tony Todd, Janet Bailey, Nancy Sorel, Chris Wiggins, Dale Wilson, John Blackwood, Cyndy Preston, Rainbow Francks, Don S. Davis, Michael Rhoades, Luc Corbeil
Having achieved a fragile truce with the local Indians and recovered the woman and children who had been take as hostages, our rancher and his blood-brother are alarmed by the activities of white vigilantes who seem determined to restart the fighting. Second in the trilogy (following on from BLACK FOX: BLOOD HORSE) this film offers more of the same, with capable acting enhancing an over-familiar story. Followed by BLACK FOX: GOOD MEN AND BAD.
Aka: PRICE OF PEACE, THE
WES   87 min (ort 90 min) mTV   SATrel: MOVIE CHANNEL

BLACK HOLE, THE *** **   PG**
Gary Nelson USA   1979
Anthony Perkins, Maximilian Schell, Ernest Borgnine, Yvette Mimieux, Robert Forster, Joseph Bottoms, Tommy McLoughlin. Voices of: Roddy McDowall, Slim Pickens
A film that could have been much better. Poor dialogue and characterisation let down this story of the arrival of an expedition to an almost deserted spaceship, perched precariously on the edge of a black hole, the only inhabitants of which are a crazed scientist and a sinister army of robots. The best features are the effects, and the strange "Heaven or Hell" ending that follows the passage of the ship through the hole.
FAN   91 min (ort 97 min)   VIDrel: WDV/SONOP L/A   V

BLACK JACK ** **   (12)
Kenneth Loach UK   1979
Stephen Hirst, Louise Cooper, Jean Franval, Phil Askham, Pat Wallis, John Young, William Moore, Doreen Mantle, Russell Waters, Brian Hawksley, Packie Byrne, Joyce Smith, Andrew Bennett, David Rappaport, Arthur Davis, Jackie Shinn
In Yorkshire in the 1750s, a wily rogue of a French sailor escapes from a hanging and sets off on the road with a young sidekick, enjoying various adventures, such as rescuing a girl from a private insane asylum. Not one of Loach's strongest works, it is generally competent enough, but the muddled script dissipates tension and interest.
A/AD   110 min   TVrel
Boa: novel by Leon Garfield.

BLACK MAGIC ** **   15
Daniel Taplitz USA   1991
Rachel Ward, Judge Reinhold, Anthony LaPaglia, Brion James, Wendy Makkena, Richard Whitting, Roger Blanck, Tom Mason, Tammy Arnold, John Benens, Lucie Dre McIntyre, Phillip Loch, Mark Jeffrey Miller, Nick Searcy, Jeff Pillars
Haunted by visions of his dead cousin, a man goes to the former's hometown in South Carolina, where he meets the man's ex-girlfriend. He falls in love with her but fails to realise

that she may be a witch. Average mTV fare with but a few thrills.
HOR   89 min (ort 94 min) mCab   VIDrel: CIC/SONOP   V

BLACK MAGIC – M66 ** **   15
Masamune Shirow JAPAN   1987
Another average "anime" offering on the usual good versus evil theme in which a renegade military android sets out to kill the grand-daughter of the man who created it, and a freelance journalist gets caught up in the battle to protect her.
ANIM   47 min (ort 96 min) dubbed
VIDrel: KISEKI/PARADOX   V/sh

BLACK MAGIC WOMAN ** **   18
Deryn Warren USA   1990
Mark Hamill, Apollonia, Amanda Wyss, Marilyn Pitzer, Gwen Wilson, Stella Pacific, Phyllis Flan, Elizabeth Robinson, Jacqueline Coon, Abidan Viera, Larry Hankin, Thomas Meurar, E. Cameron MacRae, Bonnie Ebsen, Alan Toy
A man who has sacrificed everything in pursuit of a glamorous femme fatale, discovers at the end of his affair that both his own life and the lives of those around him are in danger.
THR   87 min (ort 91 min)   VIDrel: CIC/SONOP   V

BLACK MOON RISING *   18**
Harley Cokliss USA   1985
Tommy Lee Jones, Robert Vaughn, Linda Hamilton, Bubba Smith, Keenan Wynn, Richard Jaeckel, Lee Ving, William Sanderson, Nick Cassavetes, Richard Angarola, Dan Shore, Don Opper, William Marquez, David Pressman, Bill Moody
A thief steals a computer tape and hides it in the boot of a high-tech hydrogen-powered supercar. When the car is stolen a chase ensues. Plot minimal, action maximal.
A/AD   95 min (Cut at film release by 40 sec – ort 100 min)
VIDrel: WHV   V/h

BLACK NARCISSUS ** **   U**
Michael Powell/Emeric Pressburger UK   1946
Deborah Kerr, David Farrar, Jean Simmons, Sabu, Shaun Noble, Nancy Roberts, Flora Robson, Ley On, Eddie Whaley Jr, Kathleen Byron, Esmond Knight, Jenny Laird, May Hallatt, Judith Furse
Well filmed story of the problems, emotional and otherwise, that face a group of nuns trying to start a mission in the Himalayas. Dramatic and often moving, with absolutely sumptuous photography that earned it a well-deserved Oscar. A highpoint is the flashback sequence (originally excised from the US prints) in which Kerr recalls her former life. AA: Cin (Jack Cardiff), Art/Set (Alfred Junge).
DRA   96 min (ort 100 min)   VIDrel: VCC/DISC   V
Boa: novel by Rumer Godden.

BLACK ORCHID ** **   18
Michael Ninn USA   1995
Jonathan Morgan, Ariana, Ona Zee, Veronica Hart, Lacy Rose
Having just recently penned a bestseller that contains all his dearest sexual fantasies (the title work) a jaded and burnt-out writer of erotic fiction dreams of discovering love once more. This leads to a series of vignettes, in which he finds that his sexual fantasies have started to come true. A curious film, with high production values and a promising opening, but sadly the latter remains undeveloped.
A   80 min   VIDrel: PURG/DANTE   V

BLACK ORPHEUS *   PG**
Marcel Camus FRANCE/PORTUGAL   1958
Breno Mello, Marpessa Dawn, Lourdes De Oliviera, Lea Garcia, Adhemar Da Silva, Alexandre Constantino, Waldetar De Souza, Jorge Dos Santos, Aurino Cassiano, Maria Alica
Reworking of the Orpheus and Eurydice legend, set in present-day Rio at carnival time where a singing streetcar conductor falls for a country girl fleeing from a man who has sworn to kill her. A fresh and appealing film, full of atmosphere and music, that benefits greatly from its exotic locations. Winner of Best Film award Cannes 1959. AA: Foreign.
Aka: ORFEU NEGRO
DRA   102 min   VIDrel: CONNO/RTM   V
Boa: play Orfeu Da Conceicao by Vincius De Moraes.

BLACK RAIN *   18**
Ridley Scott USA   1989
Michael Douglas, Andy Garcia, Ken Takakura, Kate Capshaw, Yusaki Matsuda, Shigeru Koyama, John Spencer, Guts Ishimatsu, Yuka

Uchida, Tomisaburo Wakayama, Miyuki Ono, Luis Guzman, John A. Costelloe, Stephen Root
A hardbitten New York cop (is there any other kind?) is given the job of delivering a Japanese mobster to the police in Osaka. When the crook escapes, the cop insists on staying in Japan and helping to recapture him, but he now finds himself on unfamiliar ground. A visually exciting rollercoaster of a film with simple plotting and violent action. AA: Sound (Donald O. Mitchell/Kevin O'Connell/Greg P. Russell/Keith A. Webster).
THR   120 min (ort 126 min)
VIDrel: 4-FRONT/POLYREC/CIC; PION (LV only)   V/sur LV

BLACK RAINBOW *** 15
Mike Hodges UK/USA 1989
Rosanna Arquette, Jason Robards, Tom Hulce, Mark Joy, Ron Rosenthal, John Bennes, Linda Pierce, Olek Krupa, Marty Terry, Ed L. Grady, Rick Warner, Jon Thompson, Helen Baldwin, Ed Lillard, Darla N. Warner
A well crafted thriller about a father-and-daughter team who perform clairvoyance tricks at small-town carnivals. When the daughter begins receiving messages apparently warning her about forthcoming murders, the two find their lives in danger. Scripted by Hodges.
THR   98 min (ort 113 min) wScrn   VIDrel: TART/20TH; ENCORE (LV only)   V/sur LV

BLACK ROBE **** 15
Bruce Beresford AUSTRALIA/CANADA 1991
Lothaire Bluteau, August Schellenberg, Aden Young, Sandrine Holt, Tantoo Cardinal, Billy Two Rivers, Lawrence Bayne, Harrison Liu, Wesley Cote, Frank Wilson, Francois Tasse, Jean Brousseau, Yvan Labelle, Raoul Trujillo
A film that examines the often unhappy effects of contact between American Indians and white missionaries whose story opens with a young Jesuit priest being sent to convert the Indians in a remote Canadian wilderness. As his knowledge of them grows, he's forced to abandon the certainties of his faith in favour of mutual respect and compassion. Beautifully filmed in the wilds of Canada, this is a sweeping and epic work. Screenplay is by Moore from his novel.
A/AD   96 min (ort 101 min)   VIDrel: EIV/SONOP   V/sur
Boa: novel by Brian Moore.

BLACK ROSES * 18
John Fasano USA 1988
John Martin, Ken Swofford, Julie Adams, Carla Gerrigno, Sal Viviano, Carmen Appice
Small-town kids turn into monsters after a sleazy hard rock band arrives, and set about murdering their troublesome parents. A gloriously bad teen monster feast, whose promising tongue-in-cheek beginning soon gives way to a disjointed series of over-the-top encounters.
HOR   83 min Cut (31 sec – ort 90 min)   VIDrel: IMPENT   V

BLACK SCORPION * (18)
Jonathan Winfrey USA 1995
Joan Severance, Bruce Abbott, Garrett Morris, Rick Rossovich, Stephen Lee, Terri J. Vaughn, Michael Wiseman, Brad Tatum, Steven Kravitz, Darryl M. Bell, Garrett Morris, Casey Siemaszko, John Sanderford, Paul Tricky, Anita Hart
A female cop goes to the jail to question the local D.A., who is being held there after he inexplicably shot dead her father. But when this leads to her suspension she sets up as a crime-fighting super-hero, complete with sexy outfit and a hi-tech car, and before long uncovers the real culprit – a criminal mastermind who has devised a plan to seize control of the entire city. Severance is barely adequate in this lifeless, cartoon-style, feminist riposte to BATMAN.
A/AD   89 min (ort 92 min)   SATrel: MOVIE CHANNEL

BLACK SHEEP ** 12
Penelope Spheeris USA 1996
Chris Farley, David Spade, Tim Matheson, Christine Ebersole, Gary Busey, Grant Heslov, Timothy Carhart, Bruce McGill, Michael Patrick Carter, Boyd Banks, David St James, Skip O'Brien, Branden R. Morgan, "Gypsy" Spheeris, John Ashker
A politician running for State Governor arranges for a minder to shadow his idiotic younger brother and try to keep him out of embarrassing scrapes. This minder gets him away to a remote cabin in the country for the duration of the election, but this leads to even more catastrophes. This might have been a terrific comedy had all the elements been in place, but the dialogue just

isn't very good, and a succession of sight gags cannot mask its failings.
COM   83 min (ort 87 min) cC   VIDrel: CIC   V/sur

BLACK STALLION, THE *** U
Carroll Ballard USA 1979
Kelly Reno, Mickey Rooney, Teri Garr, Clarence Muse, Hoyt Axton, Michael Higgins, Ed McNamara, Doghmi Larbi, John Burton, John Buchanan, Kristen Vigard, Fausto Tozzi
Adventures of a boy and a magnificent black stallion. They are shipwrecked on an island, and after being rescued go on to win a major horse race. A trifle overlong, but the beautiful photography and a great performance from Rooney as a veteran horse trainer help it along. A rather poor sequel, THE BLACK STALLION RETURNS followed in 1983. AA: Spec Award (Alan Splet for sound editing).
JUV   110 min (ort 120 min)   VIDrel: MGM/WHV   V
Boa: novel by Walter Farley.

BLACK STALLION RETURNS, THE ** U
Robert Dalva USA 1983
Kelly Reno, Vincent Spano, Woody Strode, Allen Goorwitz, Ferdinand Mayne, Jodi Thelan, Teri Garr, Doghmi Larbi
Sequel to THE BLACK STALLION with our boy, now a teenager, searching the Sahara for his Arabian steed. Tired and disappointing sequel with none of the sparkle of the earlier film. The directorial debut for Dalva, who worked on the earlier film as editor.
JUV   99 min (ort 103 min)   VIDrel: MGM/WHV   V
Boa: novel by Walter Farley.

BLACK SUNDAY *** 15
Mario Bava ITALY 1960
Barbara Steele, John Richardson, Ivo Garrani, Andrea Checchi, Arturo Dominici, Enrico Olivieri, Clara Bindi, Antonio Pierfederici, Tino Bianchi, Germana Dominici, Mario Passante
A beautiful witch was put to a cruel death hundreds of years ago. For one day each century she rises from the dead to have her revenge on the descendants of her original tormentors. An atmospheric and intriguing tale, well acted and directed. Remade some years later by Bava's son Lamberto as "Mask Of The Demon."
Aka: DEMON'S MASK, THE; HOUSE OF FRIGHT; LA MASCHERA DEL DEMONIO; MASK OF SATAN; REVENGE OF THE VAMPIRE
HOR   83 min B/W dubbed   VIDrel: REDEM/RTM   V
Boa: short story The Vij by Nikolai Gogol.

BLACK VELVET GOWN, THE *** 15
Norman Stone UK 1991
Janet McTeer, Bob Peck, Geraldine Somerville, Jean Anderson
A woman with four children returns to the fishing village of her childhood only to be greeted with distrust by the locals, and her prospects look pretty bleak until she obtains the position of housekeeper to the owner of an isolated mansion. However, this changes both her life and the lives of her children in ways she could never have imagined. Absorbing drama, set in Northumberland in the 1830s.
DRA   119 min   VIDrel: FOCUS/DISC   V
Boa: novel by Catherine Cookson.

BLACK WIDOW *** 15
Bob Rafelson USA 1986
Debra Winger, Theresa Russell, Dennis Hopper, Nicol Williamson, Sami Frey, Terry O'Quinn, D.W. Moffett, Lois Smith, Mary Woronov, Rutanya Alda, James Hong, Diane Ladd, David Mamet, Leo Rossi
A female investigator at the Justice Department, begins to follow the exploits of a young woman who seduces wealthy men, marrying and then killing them. However, her interest develops into an unhealthy obsession, in this slick and absorbing thriller, full of sexual intensity and often making intriguing comparisons between the two women and their motivations. The photography is by Conrad Hall and the score is by Michael Small.
THR   97 min (ort 101 min)   VIDrel: 20TH/TECH   V/sur

BLACKBOARD JUNGLE *** 12
Richard Brooks USA 1955
Glenn Ford, Anne Francis, Vic Morrow, Louis Calhern, Sidney Poitier, Richard Kiley, Warner Anderson, Margaret Hayes, Emile Meyer, John Hoyt, Rafael Campos, Basil Ruysdael, Paul Mazursky
Fine rendering of a teacher's experiences at the hands of his pupils in a tough New York school, who however, eventually learn to

treat him with a little more respect. A memorable if slightly ambivalent work, it features the famous "Rock Around The Clock" number over its opening credits. Poitier was thirty-one in this film – a mite old perhaps for casting as a high-school student.
DRA   97 min (ort 101 min) B/W   VIDrel: MGM/WHV   V
Boa: novel by Evan Hunter.

BLACKMAIL *   15
Alfred Hitchcock UK   1929
Anny Ondra, John Longden, Donald Calthrop, Cyril Ritchard, Sara Allgood, Charles Paton, Harvey Braban, Phyllis Monkman, Hannah Jones, Percy Parsons, Johnny Butt, Sam Livesey (in the silent version only)
A woman kills a man who tried to rape her. She then finds herself caught between the demands of a blackmailer and her boyfriend, a Scotland Yard detective assigned to the case. Some exciting moments in Hitchcock's (and Britain's) first talkie, which was originally shot as a silent. (Ondra's voice was dubbed over by Joan Barry in the sound version, because of her heavy accent.)
THR   83 min (ort 96 min) B/W
VIDrel: BRAVE/SONOP L/A   V
Boa: play by Charles Bennett.

BLACKMAIL **   15
Ruben D. Preuss USA   1991
Susan Blakely, Dale Midkiff, Beth Toussaint, John Saxon, Mac Davis
A lonely and naive woman succumbs to the attentions of an ardent man, little suspecting that he is setting her up with the help of his girlfriend. A reasonably competent little thriller, slightly hampered by the occasional dull spots.
THR   89 min mCab   VIDrel: CIC/SONOP   V
Boa: short story Passing For Love by Bill Crenshaw.

BLACKSNAKE **   18
Russ Meyer USA   1973
Anouska Hempel, David Warbeck, Percy Herbert, Milton McCollin, Thomas Baston, Dave Prowse
Parody of those slave-owner type movies such as MANDINGO that is set in the Caribbean in 1853, where whip-wielding Hempel runs a plantation, and takes a succession of black slaves as lovers. Eventually, they stage a rebellion. Meyer departs from his usual mindless bosom-fixated offerings and attempts a little social comment, but this ludicrous film is so ineptly plotted and directed it is impossible to take seriously. Its vicious streak makes it no more agreeable.
Aka: SLAVES
A   83 min B/W   VIDrel: ALLIED/RTM   V

BLADE OF FURY **   18
Samo Hung HONG KONG   1993
Cynthia Khan (Yeung Lai-Ching), Ti Lung, Samo Hung Kam-Bo, Ngai Sing, Rosamund Kwan Chi-Lam
Swordplay epic set in China at the turn of the century, where a good part of the countryside is under the domination of the "Black Flag", a nationalist group preaching isolationism. Meanwhile, two brothers find themselves on opposite sides of the fence and ample violence ensues. A stylishly directed if unoriginal effort, it offers no shortage of action.
MAR   105 min wScrn   VIDrel: MADE/RTM   V

BLADE RUNNER **   15
Ridley Scott USA   1983
Harrison Ford, Rutger Hauer, Sean Young, Edward James Olmos, Daryl Hannah, William Sanderson, Joe Turkel, Joanna Cassidy, Brion James, M. Emmet Walsh, Morgan Paull, James Hong, Kevin Thompson, Hy Pyke, John Edward Allen, Kelly Hine
In the 21st century, an ex-cop is forced out of retirement and given the job of hunting down a group of dangerous androids who are hiding out on Earth. Weak dialogue and poor characterisation are flaws, as is bad editing, but for sheer visual impact this film takes some beating. Voice-over is an annoyance, try to ignore it. Or better still see the "Director's Cut", a more tightly directed work, it has a few needless dream sequences but absolutely no voice-over.
FAN   112 min (ort 122 min – Director's Cut) wScrn
VIDrel: WHV   V/sur
Boa: novel Do Androids Dream Of Electric Sheep? by Philip K. Dick.

BLAKE'S 7: THE WAY BACK **   PG
Michael E. Briant UK   1978
Paul Darrow, Michael Keating, Josette Simon, Jan Chappell, Gareth Thomas, Sally Knyvette, Peter Tuddenham, David Jackson

The very first episode of one of the BBC's cheap attempts at a space saga that tells of a group of prisoners who escape and seize a superpowerful starship, thus starting a rebellion against a tyrannical galactic empire. STAR WARS on a flea's budget, with acting to match. All the other entries in the series are now available, but will be of little interest except to devotees.
Aka: BLAKE'S 7: THE BEGINNING
FAN   100 min (2-episode cassette) mTV
VIDrel: BBC/TECH   V/h

BLAME IT ON RIO **   15
Stanley Donen USA   1984
Michael Caine, Joseph Bologna, Valerie Harper, Michelle Johnson, Demi Moore, Jose Lewgoy, Lupe Gigliotti, Michael Menaugh, Tessy Callado, Ana Lucia Lima, Maria Helena Velasco, Zeni Pereira, Eduardo Conde, Betty Von Wien, Nelson Dantes
While on holiday in Rio de Janeiro, Caine has an affair with his best friend's teenage daughter. Even Michael Caine can do little with a lousy script that's based on the French film "One Wild Moment". Written by Charlie Peters and Larry Gelbart.
COM   96 min (ort 110 min)   VIDrel: LUMI/SPEAR L/A
V/h

BLAME IT ON THE BELLBOY **   15
Mark Herman UK/USA   1991
Dudley Moore, Bryan Brown, Richard Griffiths, Patsy Kensit, Bronson Pinchot, Andreas Katsulas, Alison Steadman, Penelope Wilton, Jim Carter, Alex Norton, John Grillo, Andrew Bailey, Ronnie Stevens, Enzo Turrin, Andy Bradord
Produced in Britain (but financed in the States) this thin comedy of mistaken identity attempts to recapture the flavour of those old British Pinewood comedies. In a seedy looking Venice, the lives of a downtrodden executive, an obese banker and a professional hit-man become linked, when the dopey bellboy contrives to mix up all their mail. The expected complications occur, but the pace of the film is so sluggish one soon loses interest.
COM   75 min (ort 79 min)   VIDrel: HOLPIC L/A   V

BLAME IT ON THE NIGHT **   PG
Gene Taft USA   1985
Nick Mancuso, Byron Thomas, Leslie Ackermann, Billy Preston, Merry Clayton, Richard Bakalyan, Merry Clayton, Billy Preston, Ollie E. Brown
A pop star's young son comes to live with him when his mother dies. Brought up in strict private schools, he finds it hard to appreciate his father's way of life. A dull plodder with a script by Taft and Mick Jagger.
DRA   86 min Cut (15 sec)   VIDrel: 20TH/TECH   V/sur

BLANCHE **   PG
Walerian Borowczyk FRANCE   1971
Ligia Branice, Michel Simon, Jacques Perrin, Georges Wilson, Lawrence Trimble, Denise Peronne
Dark tale set in 13th century France, with the beautiful young wife of an ageing baron lives a sexually unfulfilled life at his remote castle, where she is confined virtually a prisoner. However, a rescue is finally engineered by her stepson, assisted by a couple of friends. A visually impressive work that is so flatly directed that at times it becomes most tedious.
DRA   90 min (ort 92 min)   VIDrel: CONNO/RTM   V

BLANCHE FURY **   PG
Marc Allegret UK   1947
Stewart Granger, Valerie Hobson, Walter Fitzgerald, Michael Gough, Maurice Denham, Sybilla Binder, Edward Lexey, Allan Jeayes, Suzanne Gibbs, Ernest Jay, George Woodbridge, Arthur Wontner, Amy Veness. W.E. Clifton-James, Lionel Grose
A woman takes up a post as governess at the request of her uncle and ends up marrying his widowed son, although she really loves the illegitimate scion of the family, who takes care of the stables. Soon her lover murders both men but goes too far when he wants to kill her young charge. A dark and dire melodrama, with an ending worthy of any Greek drama.
DRA   90 min (ort 95 min) B/W   VIDrel: CONNO/RTM L/A
V
Boa: novel by Joseph Shearing.

BLANK CHEQUE **   PG
Rupert Wainwright USA   1993
Brian Bonsall, James Rebhorn, Tone Loc, Karen Duffy, Miguel Ferrer, Jayne Atkinson, Michael Lerner, Michael Faustino, Chris Demetral, Rick Ducommun, Lu Leonard, Debbie Allen, Alex Zuckerman, Alex Allen Morris, Michael Polk

A man who is involved with laundering money, collides with a child on a bicycle and drops a blank check which the child promptly pockets. In need of a bike, he cashes it for $1,000,000 and has quite a time until he realises that happiness is not the same as having money.
Aka: BLANK CHECK
COM   90 min (ort 93 min) cC   VIDrel: WDV/TECH   V/sh

BLANKMAN *   12
Mike Binder USA   1994
Damon Wayans, David Alan Grier, Robin Givens, Jon Polito, Jason Alexander, Christopher Lawford, Lynne Thigpen, Jon Polito
A mild-mannered inventor decides to fight crime in his neighbourhood by donning a home-made costume and using his own inventions instead of the superpowers that he does not have. An unfunny and hysterically played comedy that delivers zero laughs, a great pity given the enormous potential of this idea.
COM   92 min (ort 96 min)   VIDrel: COLUM/SONOP
V/sur

BLAST **   18
Albert Pyun USA   1996
Rutger Hauer, Linden Ashby, Andrew Divoff, Kimberly Warren
Thriller set during the Atlanta Olympics of 1996, when a team of terrorists penetrate the security cordon by taking over a swimming complex.
THR   94 min   VIDrel: BMGREC/BMGVID   V

BLAZE **   18
Ron Shelton USA   1989
Paul Newman, Lolita Davidovitch, Jerry Hardin, Gailard Sartain, Jeffrey De Munn, Garland Bunting, Richard Jenkins, Robert Wuhl
A fictionalised account of Earl K. Long, the larger-than-life governor of Louisiana, whose political career took a nosedive in the 1950s following his affair with Blaze Starr, a stripper of the time. Newman gives a flamboyant performance as the Governor, and Davidovitch is appealing as his girlfriend, but the film remains too distanced from its characters and never really develops. The real-life Starr appears in a cameo as the stripper Lily.
DRA   113 min (ort 119 min)   VIDrel: TOUCH L/A   V
Boa: book Blaze Starr: My Life As Told To Huey Perry by Blaze Starr and Huey Perry.

BLAZING SADDLES **   15
Mel Brooks USA   1970
Cleavon Little, Gene Wilder, Harvey Korman, Madeline Kahn, Slim Pickens, David Huddleston, Alex Karras, John Hillerman, Liam Dunn, Carol Arthur, Dom DeLuise, George Furth, Don Megowan, Burton Gilliam, Count Basie, Tom Steele
Western spoof that was a big hit at the time but now seems very dated. The plot loosely revolves around the arrival in town of the new sheriff, who just happens to be black. Crazy events rapidly pile up until they climax in the breakdown of the entire film – the cast simply rushes off the set and out into the street. Ambitious and high-spirited, but little more than a series of clever one-liners and visual gags – though some are very funny indeed.
COM   89 min (ort 94 min)   VIDrel: WHV   V/h

BLEAK MOMENTS **   PG
Mike Leigh UK   1971
Anne Raitt, Sarah Stephenson, Eric Allen, Liz Smith
In a south London suburb a typist and her retarded sister live out their drab and pointless existence. Scripted by Leigh, this long and difficult film makes a few sharp and wry comments on life in general.
DRA   106 min (ort 111 min)   VIDrel: CONNO/RTM   V

BLESS THIS HOUSE *   U
Gerald Thomas UK   1971
Sid James, Diana Coupland, Terry Scott, June Whitfield, Peter Butterworth, Sally Geeson, Robin Askwith, Carol Hawkins, Janet Brown, George A. Cooper, Patsy Rowlands, Bill Maynard, Wendy Richard, Marianne Stone, Julian Orchard
Another limp British TV spin-off based on a series of the same name. As ever, the material is insufficient to make a full-length film. Stories in the TV series focused on the trials and tribulations of a couple and their two teenage kids – this film offers us more of the same.
COM   85 min (ort 87 min)   VIDrel: VCC/DISC   V

BLIND DATE *   15
Blake Edwards USA   1987
Bruce Willis, Kim Basinger, William Daniels, John Larroquette, Phil Hartman, Alice Hirson, George Coe, Mark Blum, Graham Stark, Stephane Faracy
A yuppie gets fixed up with a blind date but is warned that the girl cannot tolerate alcohol. He takes her along to impress his friends but after giving her champagne she reduces a restaurant and his career to a shambles. Boring and predictable farce with little going for it. This inauspicious beginning was the screen debut for Willis.
COM   91 min   VIDrel: VCC/DISC/COLUM L/A   V/sh

BLIND FURY **   18
Phillip Noyce USA   1989
Rutger Hauer, Terrance O'Quinn, Lisa Blount, Brandon Call, Meg Foster, Nick Cassavetes, Randall "Tex" Cob, Noble Willingham, Rick Overton, Sho Kosugi, Paul James Vasquez, Charles Cooper, Julia Gonzalez, Woody Watson, Alex Morris
In this violent parody of a martial arts movie, a blind swordsman who has been trained in the fighting arts comes to the aid of an ex-army buddy, a chemist whose son has been kidnapped in order to force the father to work for crooks synthesising designer drugs. A kind of westernised adaptation of a popular Japanese character ("Zatoichi") that is played for laughs much of the time.
A/AD   82 min Cut (4 sec – ort 86 min)
VIDrel: COLUM/SONOP   V/sur

BLIND HATE **   (18)
John Korty USA   1990
Corbin Bernsen, Angela Bassett, Jenny Lewis, Tiger Knowles, M. Donald
A campaigning civil rights lawyer comes up against a nasty bunch of white supremacists in the Deep South, when he decides to take them on following the murder of a black youngster. Based on a true story but a no more than adequate foray into this territory. See also INTO THE HOMELAND and IN THE LINE OF DUTY 3: TIME TO KILL.
Aka: IN THE LINE OF FIRE: THE MORRIS DEES STORY
DRA   90 min (ort 96 min) mTV   TVrel

BLIND JUDGEMENT **   15
George Kaczender USA   1991
Lesley Ann Warren, Peter Coyote, Jean Smart, Matt Clark, Don Hood, Freda Ramsey Williams, Marco Perella, Brandon Smith
A fact-based thriller that has a prominent defence lawyer agreeing to take on the case of a woman accused of murdering her husband. Although he is initially all-too-willing to take on the case, he's soon drawn into a web of deceit that endangers his family and threatens to destroy his career.
THR   87 min   VIDrel: NWV/HIFLI   V

BLIND JUSTICE **   15
Rod Holcomb USA   1986
Tim Matheson, Mimi Kuzyk, Tom Atkins, Lisa Eichhorn, Philip Charles MacKenzie, John Kellogg, Marilyn Lightstone, Anne Haney, David Froman, John M. Jackson, Linda Thorson, Jack Blessing, Ann Ryerson, Sam Dalton
A series of unfortunate circumstances lead to the arrest of an innocent man who is accused of a number of armed robberies. He then spends the next 18 months fighting to establish his innocence. Based on a true incident, this is an interesting examination of the forces that operate against a person who is wrongly convicted. Matheson is excellent as the man whose life is blighted by these events.
DRA   90 min mTV   VIDrel: 20TH/TECH   V

BLIND JUSTICE **   15
Richard Spence USA   1994
Armand Assante, Robert Davi, Elisabeth Shue, Adam Baldwin, Ian McElhinney, Danny Nucci, M.C. Gainey, Titus Welliver, Jack Black, Michael O'Neil, Douglas Roberts, Gary Cervantes, Jesse Dabson, Stanton Davis, James Oscar Lee
A blind gunslinger accepts a fee of two-hundred dollars in silver to defend a town from attack by Mexican bandits. An acceptable modern Western that reworks elements from Japanese films such as the "Zatoichi" series as well as SEVEN SAMURAI.
WES   89 min   VIDrel: MIA/DISC   V/sh

BLIND MAN'S BLUFF **   15
James Quinn USA   1991
Robert Urich, Lisa Eilbacher, Patricia Clarkson, Ken Pogue, Ron

Perlman, Brent Stait, Andrea Mann, Sam Nelkin, Yan Michael, Dee Jay Jackson, Samuel Kouth, Joe Maffei, Randi Lynne, Helen Honeywell, Charles Andre, Ken Kramer
A professor who lost his sight in an accident four years ago, has slowly rebuilt his life, only to find that the murder of his next-door neighbour plunges him into a nightmare in which he appears to be the prime suspect. Forced to go on the run in a bid to gain the evidence needed to clear his name, he finds himself being stalked by a killer he cannot see. A most unusual thriller, totally implausible, but undeniably gripping.
THR   83 min (ort 86 min)   VIDrel: CIC/SONOP   V

BLIND SIDE ***

18
Geoff Murphy USA   1992
Rutger Hauer, Rebecca De Mornay, Ron Silver, Jonathan Banks, Mariska Hargitay, Tamara Clatterbuck, Jore Cervera Jr., Josh Cruze, David Labiosa, Bill Dance, Richard L. Duran, Diane Lee Hsu, Geoff Rivas, Joanna Sanchez
A married couple who were involved in a hit-and-run accident while on vacation in Mexico that they did not report, are panic-stricken when a man turns up on their doorstep. He claims that he was also recently in Mexico, and seems to be out to blackmail them. Strong performances by Silver and Hauer enhance this conventional thriller.
DRA   92 min (ort 98 min) mCab   VIDrel: EIV/SONOP
V/sur

BLIND SPOT ***

(PG)
Michael Toshiyuki Uno USA   1993
Joanne Woodward, Laura Linney, Fritz Weaver, Reed Diamond, Cynthia Martels, Patti Yasutake, Patti D'Arbanville, Mark Anthony Wade, Edmond Genest, Sam Turich, Allison Janney, Karrina Arroyave, Sharon Washington, Linda Powell
A strong-willed congresswoman who is running for the Senate is shocked to discover that both her daughter and her son-in-law (who is her assistant) are drug addicts. Events take a tragic turn when the latter dies and the resultant scandal threatens her political aspirations. Woodward's fine performance, ably supported by the rest of the cast, lifts this movie far above any cliched soap opera. A Hallmark Hall of Fame production.
DRA   99 min cC mTV   SATrel: MOVIE CHANNEL

BLIND WITNESS **

15
Richard A. Colla USA   1989
Victoria Principal, Paul Le Mat, Stephen Macht, Matthew Clark, Tim Choate, Marcia Yvette Reider, Jeff Olson, Jesse Bennett, Dennis Saylor, Russ McGinn, Don Re Sampson, Jayne Luke, Will C. Hazlett, Josh Devane, J. Omar Hansen, James Cash
A blind woman is the only "witness" to the brutal murder of her husband by burglars, and the latter set about planning her demise. A surprisingly strong performance from Principal helps redeem the film from its mundane and utterly predictable script.
THR   87 min (ort 96 min) mTV   VIDrel: GENESIS   V

BLINDFOLD: ACTS OF OBSESSION *

18
Lawrence L. Simeone USA   1993
Shannen Doherty, Judd Nelson, Kristian Alfonso, Drew Snyder, Michael Woods, Shell Danielson, Scott Brnadon, Rudolph Williams, Aleksandra Kaniak, Heide Noelle Lenhart, Robert Miano, George Le Porte, Buddy Daniels, Al Sapienza
An attractive woman takes a lover who introduces her to an increasingly bizarre world of sex games and she gradually realises that he is a far from balanced individual. Inevitably, he eventually turns against her and tries to murder her, in this ludicrous psychological thriller. A few anaemic sex scenes add nothing of interest.
Aka: ACTS OF OBSESSION: BLINFOLDED: ACTS OF OBSESSION
THR   85 min (ort 91 min)   VIDrel: POLY/POLYREC   V

BLINDSIDED **

15
Tom Donnelly USA   1992
Jeff Fahey, Mia Sara, Rudy Ramos, Jack Kehler, Brad Hunt, Ben Gazzara, Michael Ornstein, Glenn Ash, Raven, John Kerry, Chris Pedersen, Frank Pesce, Lysa Regina, Julio Medina, Stanley D. Petter, Angela L. Harry, Greg Eagles
An ex-cop turns bad and takes up burglary but is blinded in a drugs raid that goes wrong. While recuperating at an ocean resort he meets a mysterious woman, whom he tries to track down after his sight is restored, only to face more problems. A confused and unimaginative thriller, with little to offer.
THR   89 min (ort 93 min) mCab   VIDrel: CIC/SONOP   V

BLINK **

18
Michael Apted USA   1993
Madeleine Stowe, Aidan Quinn, James Remar, Peter Friedman, Bruce A. Young, Laurie Metcalf, Matt Roth, Paul Dillon, Michael P. Byrne, Anthony Cannata, Greg Noonan, Heather Schwarz, Marilyn Dodds Franks, Michael Stuart Kilpatrick
A woman musician who has been blind for many years undergoes surgery to regain her sight but suffers strange "flashbacks" due to a time lag in her brain's processing of visual information. It soon becomes clear that she has inadvertently witnessed a murder and may be able to identify the killer. Predictably, she soon becomes the next target, in a strongly acted, quite tense but unoriginal rehash of familiar cliches. See also THE EYES OF LAURA MARS.
THR   101 min (ort 106 min)
VIDrel: 4-FRONT/POLYREC/GUILD   V/sur

BLISS *

18
Ray Lawrence AUSTRALIA   1985
Barry Otto, Lynette Curran, Helen Jones, Miles Buchanan, Gia Carides, Tim Robertson, Jeff Truman, Bryan Marshall, Jon Ewing, Paul Chubb, Sara De Teliga, Saski Post, George Whalley, Robert Menzies, Nique Needles, Marco Colombani
A high-powered businessman has a major heart attack and sees himself dying, but returns to life with a major change in his outlook. A good opening to this satire never develops into anything of consequence. Written by Lawrence and Carey from the latter's novel.
COM   106 min (ort 111 min)   VIDrel: VCC/DISC L/A   V/h
Boa: novel by Peter Carey

BLITHE SPIRIT ****

U
David Lean UK   1945
Rex Harrison, Kay Hammond, Margaret Rutherford, Joyce Carey, Constance Cummings, Hugh Wakefield, Jacqueline Clarke
A novelist is troubled by the appearance of the spirit of his first wife at a seance. This playful ghost tries to disturb his second marriage and engineer his swift demise so that they can be reunited in the hereafter. A classic and beautifully crafted comedy-fantasy; scripted by Lean it gave Rutherford a splendid role as the medium Madame Arcati. Photography is by Ronald Neame. AA: Effects (Thomas Howard).
COM   91 min (ort 96 min) B/W   VIDrel: VCC/RTM   V
Boa: play by Noel Coward.

BLOB, THE *

15
Irvin S. Yeaworth Jr USA   1958
Steve McQueen, Aneta Corseaut, Earl Rowe, Olin Howlin, Steven Chase, John Benson, Vince Barbi, Audrey Metcalf, Elinor Hammer, Keith Almoney, Julie Cousins, Robert Fields, James Bonnet, Anthony Granke, Pamela Curran
Some remarkably cheap effects in this story of a meteorite that brings a monster to Earth. It takes the form of a lump of jelly that grows by absorbing whatever gets in its way. A good idea is spoilt by cheap sets and poor direction but somehow the film is endearing for its very ineptness. A comedy sequel – "Son Of Blob" followed. Remade in 1988. The title song was composed by Burt Bacharach. This was McQueen's first starring role.
FAN   82 min   VIDrel: 4-FRONT/POLYREC/BRAVE   V

BLOB, THE ***

18
Chuck Russell USA   1988
Kevin Dillon, Shawnee Smith, Donovan Leitch, Jeffrey DeMunn, Ricky Paull Goldin, Kevin Dillon, Billy Beck, Candy Clark, Joe Seneca, Beau Billingslea, Art La Fleur, Del Close, Michael Kenworthy, Douglas Emerson, Sharon Spelman
Remake of a 1958 dud, with little improvement except in the special effects department (courtesy of Lyle Conway). This $30,000,000 film (the budget was 120 times as much as the original) follows our Blob (in this remake the result of a failed experiment in germ warfare) as he arrives at the small ski resort of Arborville and begins tucking into the residents, abetted by the military authorities who do not want to see the creature destroyed.
FAN   91 min (ort 95 min)   VIDrel: ENTUK L/A   V/sh

BLONDE FIRE **

18
Bob C. Chinn USA   1979
John C. Holmes, Seka (Dorothy Hundley Patton), Jesie St James
Unlikely as it sounds, this is a porno film about a private eye who is hired to bring the Blonde Fire diamond from South Africa to New York, after a client has purchased it for $1,000,000. In the course of this task he beds several women, loses the diamond,

but regains it from a woman who hid it in an unusual place. Cut before video submission by 17 min 1 sec.
A   45 min (ort 96 min)   VIDrel: MOPIC/SGSVID   V

BLONDE FIST *
Frank Clarke UK
15
1991
Margi Clarke, Carroll Baker, Ken Hutchinson, Sharon Power, Angela Clarke, Lewis Bester, Gary Mavers, Jeff Weatherford, Graham Winton, Jane Porter, Lyn Kelly, Genevieve Walsh, David Crean, Julie Aldred, Susan Atkins, Julie Graham
A single mother uses her boxing skills to protect herself and her family. After escaping from prison where she was serving a sentence for assault, she goes to New York in search of her long-lost father, and finds herself entering a prize match to win the money she needs to get herself, her son and her father back to Liverpool for a fresh start. Clarke's directorial debut is a well-acted but relentlessly depressing and patronising view of working-class life.
A/AD   98 min (ort 102 min)   VIDrel: SCRN/TERRY   V

BLONDE IN LOVE, A ****
Milos Forman CZECHOSLOVAKIA
15
1965
Hanna Brejchova, Vladimir Pucholt, Antonin Blazejovsky, Josef Sebanek, Milada Jezkova, Vladimir Mensik, Ivan Kheil, Jiri Hruby, Marie Salacova, Jana Novakova, Jana Crkalova, Zdenka Lorencova, Tana Zelinkova, Jan Vostrcil
A young girl lives in a town and works in the local factory which dominates it, where almost all of the employees are female. When she has a one-night stand with a piano player during a band's stopover in the provinces, she follows him back to his home in Prague, but finds that the man's parents hardly ready to receive her with open arms. Touching and light, this is a bittersweet story of adolescent love, wonderfully well directed – a real delight to watch.
Aka: LASKY JEDNE PLAVOVLASKY; LOVES OF A BLONDE, THE
DRA   88 min B/W   VIDrel: CONNO/RTM   V

BLONDE JUSTICE **
Paul Thomas USA
18
1991
Jannie Lindemulder, Summer Knight, Alex Jordan, Tiffany Minx, Jessica Fox, Lacy Rose, Tony Tedeschi, Nick Rage, Tim Lake, Terry Thomas
A porno version of "Stripped To Kill" that confusingly, was made in two parts. It opens with the strippers doing their stuff at a club, where one of them has been receiving threatening letters from a nutter. But this plot is never properly developed, as the story then moves on to the exploits of an undercover cop, who goes to work there as one of the strippers. A flashy but confusing work that with some more care, could have been considerably better.
Aka: BLOND JUSTICE
A   87 min   VIDrel: VIVID/SCRN   V

BLONDE VENUS ***
Joseph Von Sternberg USA
PG
1932
Marlene Dietrich, Cary Grant, Herbert Marshall, Dickie Moore, Sidney Toler, Hattie McDaniel, Francis Sayles, Robert Emmett O'Connor, Gene Morgan, Rita La Roy, Morgan Wallace, Evelyn Preer, Mildred Washington, Getrude Short
An aspiring young actress who gives up the stage to marry a hopeful scientist, is forced to return to cabaret life after her husband is afflicted with radium poisoning. Episodic and superficial, but it's hard to resist a film with Dietrich at her most engaging.
DRA   95 min (ort 97 min) B/W   VIDrel: CIC/SONOP L/A   V

BLONDES HAVE MORE GUNS! *
George Merriweather USA
18
1995
Richard Neil, Montana, Michael McGaharin, Gloria Lusiak, Elizabeth Key
A smart cop is given the case of a series of murders whose linking factor is a mysterious blonde, and finds the evidence points to a pair of sisters being the culprits. One of those boring Troma duds whose witty title suggests a fun film, an expectation that is far from being fulfilled on this occasion.
COM   89 min   VIDrel: TROMA/RTM   V

BLOOD ALLEY **
William A. Wellman USA
U
1955
John Wayne, Lauren Bacall, Paul Fix, Mike Mazurki, Joy Kim, Barry Kroger, Anita Ekberg, Henry Nakamura, W.T. Chang, George Chan
Simple-minded Cold War thriller with Wayne and company

risking their lives to save Chinese refugees from the communists, all concerned making their way to freedom in Hong Kong. Solid acting (plus a couple of songs) helps compensate for the silliness of the plot.
THR   115 min wScrn   VIDrel: WHV   V/h
Boa: novel by A.S. Fleischmann.

BLOOD AND CONCRETE: A LOVE STORY **
Jeffrey Reiner USA
18
1991
Billy Zane, Jennifer Beals, Darren McGavin, James LeGros, Mark Pellegrino, Nicholas Worth, Marty Shearer, Steve Freedman, William Basiani, Pat Cupo, Pat O'Bryan, Tracy Coles, Ellen Albertini Dow, Lyvinston Holmes
A none-too-successful car thief gets involved with a punk singer who is a drug addict, and is soon drawn into a complex plot that brings him up against a nasty drugs dealer. A flashy and unsuccessful spoof that tries hard for be a stylish film noir but fails miserably to entertain.
A/AD   95 min (ort 97 min)   VIDrel: COLUM/SONOP   V/sh

BLOOD AND ORCHIDS ***
Jerry Thorpe USA
15
1986
Kris Kristofferson, Sean Young, Jose Ferrer, Jane Alexander, Susan Blakely, George Coe, David Clennon, Richard Dysart, Elizabeth Lindsey, Haunani Minn, William Russ, James Saito, Matt Salinger, Madeline Stowe
Story of four native Hawaiians accused in 1937 of having raped the wife of a US Navy officer. A strong examination of racial prejudice and the caste system of the islands, which would have been considerably improved with better casting; Kristofferson as a laid-back detective just does not convince. Based on a script by Katkov, which is in turn based on his book about a real-life case. First shown in two parts.
DRA   175 min (2 cassettes – ort 240 min) mTV
VIDrel: WHV   V/h
Boa: book by Norman Katkov.

BLOOD AND SAND ***
Fred Niblo USA
PG
1922
Rudolph Valentino, Nita Naldi, Lila Lee, George Field, Rose Rosanova, Walter Long, Leo White
A silent classic about the tragic rise and fall of a matador and the women in his life. The dated seduction scenes have not aged well but as a whole the film packs a punch, not least thanks to the star's undeniable charisma. Remade a couple of times since.
DRA   80 min B/W silent   VIDrel: 20TH/TECH L/A   V
Boa: novel by Vicente Blasco Ibanez.

BLOOD AND SAND **
Rouben Mamoulian USA
PG
1941
Tyrone Power, Linda Darnell, Rita Hayworth, Nazimova, Anthony Quinn, J. Carroll Naish, John Carradine, Laird Cregar, Lynn Bari, George Reeves, Monty Banks, Pedro De Cordoba, Vincente Gomez, Fortunio Bonanova, Victor Kilian
A sluggish remake of the 1922 Valentino silent, with a dashing matador finding himself torn between two women. Undeniably handsome to look at, but stilted and lacking in impact. The story was filmed once more in 1989. AA: Cin (Ernest Palmer/Ray Rennahan).
DRA   120 min (ort 123 min)   VIDrel: 20TH/TECH   V/h
Boa: novel by Vicente Blasco Ibanez.

BLOOD AND SAND ***
Javier Elorrieta SPAIN
18
1989
Sharon Stone, Christopher Rydell, Ana Torrent, Guillermo Montesinus, Albert Vidal, Simon Andreu, Antonio Glez Flores, Jose Luis de Villalonga, Tony Fuentes, Margarita Calahorra
The second remake of the 1922 Valentino classic tells of a handsome young bullfighter whose successful career leads to arrogance and the desertion of his faithful wife for the charms of a sexy siren. There are no surprises in store in this careful remake, though the authentic Spanish locations add considerable atmosphere and some torrid sex scenes distinguish it from its predecessors.
DRA   96 min (ort 118 min)   VIDrel: HIFLI/SONOP L/A   V

BLOOD AND WINE *
Bob Rafelson UK/USA
15
1996
Judy Davis, Jack Nicholson, Stephen Dorff, Jennifer Lopez, Harold Perrineau Jr, Michael Caine, Robyn Peterson, Mike Starr, John Seitz, Mark Macaulay, Dan Daily, Marta Velasco, Thom Christopher
Surprisingly untaut thriller about a stolen necklace, a thief and

his sad, betrayed wife, who in the company of her son takes off after having snapped at her husband's behaviour. Unfortunately, she fails to realise that she has accidentally taken the stolen necklace with her, a situation that is bound to lead to tears. Not an edifying film from a director who has done much finer work.
THR   100 min   CINrel

BLOOD BROTHERS **
18
Chang Cheh HONG KONG   1973
Ti Lung, David Chiang, Chen Kwan Tai, Wang Tao, Chan Sin, Chan Hung Lieh, Ching Li
Title heroes mount a violent quest to rid Shanghai of its underworld, in a martial arts actioner given a modern urban setting.
Aka: DYNASTY OF BLOOD
MAR   117 min wScrn dubbed   VIDrel: MADE/RTM   V

BLOOD BROTHERS ***
(PG)
Bruce Pittman USA   1993
Mia Korf, Clark Johnson, Richard Chevolleau, Amir Williams, Ron White, Bill Nunn, Richard Yearwood, Ndehru Roberts, Timothy D. Stickney, Bryon Abalos, Linda Sorensen, Deborah Burgess, Taborah Johnson, Michael Clarke, Rei Miki
In L.A., a black kid sees three black gang members shoot dead a trio of Koreans under his bedroom window, but remains silent as he has seen that one of the killers is his own brother. As the D.A. starts investigations that mostly centre around his home, he finds himself wrestling with a difficult question of conscience. A thought-provoking drama.
AKA: SILENT WITNESS; WHAT A CHILD SAW
DRA   87 min (ort 90 min) mTV   SATrel: SKY MOVIES

BLOOD FROM THE MUMMY'S TOMB **
18
Seth Holt/Michael Carreras UK   1971
Andrew Keir, Valerie Leon, James Villiers, Hugh Burden, George Coulouris, Mark Edwards, Rosalie Crutchley, Aubrey Morris, David Markham, Joan Young, James Cossins, David Jackson, Jonathan Burne, Tamara Ustinov, Penelope Holt
Based on a novel by Bram Stoker, this film tells of the revenge that a long dead Egyptian princess takes, on the members of the team of archaeologists who have disturbed her tomb. Carreras completed the film following the death of Holt. Disjointed and muddled, with Leon being taken over by the spirit of the princess in a silly and contrived reincarnation subplot. Made by Hammer Films and very much a typical example. Remade as THE AWAKENING and THE TOMB.
HOR   90 min (ort 94 min)   VIDrel: LUMI/SPEAR L/A   V
Boa: novel Jewel of the Seven Stars by Bram Stoker.

BLOOD GAMES **
15
Tanya Rosenberg USA   1990
Laura Albert, Luke Shay, Ross Hagen, Gregory Scott Cummins, Shelly Abble II
A female basketball team plays a match in a small town which they are unfortunate enough to win, for the locals respond by killing their coach. This sets in motion a bloody confrontation as the girls seek their revenge. A standard entry, adequately if a little bizarrely plotted.
A/AD   84 min Cut (23 sec – ort 90 min)
VIDrel: COLUM/SONOP   V

BLOOD IN BLOOD OUT **
18
Taylor Hackford USA   1993
Benjamin Bratt, Jesse Borrego, Damian Chapa, Enrique Castillo, Tom Towles, Victor Rivers, Delroy Lindo, Carlos Carrasco, Raymond Cruz, Teddy Wilson, Lanny Flaherty, Valente Rodriguez, Billy Bob Thornton, Karmin Murcelo, Geoffrey Rivas
A powerful and epic tale of three young friends from the Chicano area of east L.A. that follows their lives through the 1970s and 1980s, touching on a good many serious social issues. Unfortunately, the impact of this worthy effort is to some extent marred by its graphic violence and excessive length. See AMERICAN ME for a better film on the subject.
Aka: BLOOD IN, BLOOD OUT: BOUND BY HONOR; BOUND BY HONOR
DRA   176 min (ort 180 min)   VIDrel: HOLPIC/TECH L/A   V

BLOOD LOVE **
18
Thunder Levin USA   1991
Tristan Rogers, Arabella Holzbog, Tyrone Power Jr, Dawn Wells, Paul Bartel, Griffin O'Neal
A pretty college student meets a bestselling writer of horror fiction, but unknown to her he has an evil propensity that he

makes use of in his writing. A weakish vampire movie, offering no new insights and few thrills. Screenplay is by James Hankins.
HOR   79 min   VIDrel: OVER/HIFLI   V/h

BLOOD MONEY **
(18)
John Shepphird USA   1995
James Brolin, Billy Drago, Traci Lords, Dean Tarrolly, Sonny Carl Davis, Alison Moir, Bentley Mitchum, Tony Pierce, Katerine Armstrong, Antony Ponzini
A prisoner denied parole escapes with two others, his intention being to take revenge on the man whose evidence led to his imprisonment. Adequate.
A/AD   88 min   VIDrel: FIRST/SONOP   V

BLOOD OATH **
15
Stephen Wallace AUSTRALIA   1990
Bryan Brown, George Takei, Terry O'Quinn, John Polson, Nicholas Eadie, Toshi Shioya, Deborah Unger, Ray Barrett, Jason Donovan, John Bath, John Clarke, Tetsu Watanabe, Russell Crowe, Sokyu Fujita, Kazuhiro Muruyama, David Argue
A dull and sombre courtroom drama that focuses on the trial of Japanese soldiers involved in war-time atrocities against Australian POWs in 1946 on the Indonesian island of Ambon. Despite the determination of an army lawyer to see that justice is done, political intrigue ensures that the former camp commandant is acquitted. Though based on real-life experiences, this is a plodding albeit undeniably sincere effort.
Aka: PRISONERS OF THE SUN
DRA   104 min   VIDrel: COLUM/SONOP   V/sur

BLOOD OF A POET, THE ***
PG
Jean Cocteau FRANCE   1930
Lee Miller, Enrico Rivero, Jean Desbordes, Feral Benga, Odette Talazac, Jean Cocteau, Feral Benga, Fernand Dichamps, Lucien Jager, Barbette
A young poet muses on life and then passes through a mirror and witnesses a series of strange fantasies, which he sees through the keyholes of the doors in a hotel corridor. A Mexican revolutionary shot and then restored to life, and a boy killed by a snowball, are but two examples of the images seen. Raved over by the critics on release, this is an intriguing experiment in surrealistic imagery – a film of little meaning but much opacity.
Aka: LE SANG D'UN POETE
DRA   49 min (ort 58 min) B/W   VIDrel: TART/20TH   V

BLOOD OF DRAGON PERIL **
18
Rocky Mann HONG KONG   1981
Jerry Chan, Philip Cheung, Marty Chui
A renowned kung fu master in Manchuria is slaughtered when he resists the invading Japanese during WW2. His eldest son is so badly beaten that he suffers permanent brain damage whilst the second son surrenders and joins the army. The youngest son disappears. Later a masked man begins to wage attacks on Japanese army bases and the second son is sent to get him – the inevitable confrontation occurs in a kung fu tale with a well-worn plot but some good action sequences.
Aka: BLOOD OF THE DRAGON PERIL
MAR   83 min   VIDrel: IMPENT   V

BLOOD OF FU MANCHU, THE *
18
Jess (Jesus) Franco UK   1965
Christopher Lee, Richard Greene, Shirley Eaton, Howard Marion Crawford, Gotz George, Tsai Chin, Maria Rohm, Richard Palacios, Frances Kahn
In South America Fu Manchu collects a bevy of beautiful women, his plan is to use them to deal with ten of his enemies, the chosen method being to blind them with poisoned kisses that will slowly kill them. Greene is one of the unfortunates who becomes a target for this devilish scheme. A slackly directed and most unappealing work. See also THE CASTLE OF FU MANCHU and THE BRIDES OF FU MANCHU.
Aka: FU MANCHU AND THE KISS OF DEATH; KISS AND KILL
THR   88 min (ort 90 min)   VIDrel: LUMI/SONOP   V

BLOOD OF THE HUNTER **
15
Gilles Carle USA   1994
Michael Biehn, Alexandra Vandernoot, Gabriel Arcand
A trapper and his wife living in the Canadian wilderness have to cope with an escaped killer who breaks into their cabin while he is away on a hunting trip. Events conspire to have the trapper blamed for a murder, his only solution being to find the real culprit.
A/AD   93 min   VIDrel: TRIM/HIFLI   V/h

BLOOD ON SATAN'S CLAW *** (18)
Piers Haggard UK 1970
Patrick Wymark, Linda Hayden, Barry Andrews, Simon Williams, Tamara Ustinov, Michele Dotrice, Wendy Padbury, Anthony Ainley, James Hayter, Avice Landone, Charlotte Mitchell, Robin Davies, Howard Goorney
The discovery of a devil's claw sparks off a spate of devil worship among a group of farm children in a 17th century English village. A competent horror tale with a gruesome and evocative atmosphere. Hayden is especially good as the possessed girl who is at the centre of all this evil, and if the film tends to concentrate on the more lurid aspects of these events, this proves to be no handicap.
Aka: SATAN'S CLAW; SATAN'S SKIN
HOR 95 min TVrel: C4

BLOOD ON THE SUN *** PG
Frank Lloyd USA 1945
James Cagney, Sylvia Sidney, Robert Armstrong, Wallace Ford, Rosemary De Camp, John Emery, Leonard Strong, Frank Puglia, Jack Halloran, Philip Ahn, Hugh Ho, Joseph Kim, Marvin Miller, Rhys Williams, Porter Hall, James Bell
An American newspaperman in Japan before WW2 discovers plans for world domination. He attempts to smuggle them out whilst being pursued by the agents of a warlord. A fast-paced and suspenseful story. AA: Art/Int (Wiard Ihnen/A. Roland Fields)
DRA 94 min (ort 98 min) B/W VIDrel: VISVID/POLYREC L/A V

BLOOD ORGY OF THE SHE-DEVILS * 18
Ted V. Mikels USA 1972
Linda Zaborin, Tom Pace, William Bagdad, Leslie McRae, Ray Myles, Victor Izay, Paul Wilmoth, Kebrine Kincade, Curt Mason, Linn Henson, John Nicolai, John Ricco, Vincent Barbi, Annik Borell, Sherri Vernon, Erica Campbell
Not quite as exciting as it sounds. A horror film purporting to deal with all aspects of demonology.
Aka: FEMALE PLASMA SUCKERS
HOR 78 min VIDrel: OURVID/SCRN V

BLOOD RED ** 15
Peter Masterson USA 1986
Eric Roberts, Giancarlo Giannini, Dennis Hopper, Burt Young, Carlin Glynn, Lara Harris, Julia Roberts, Elias Koteas, Frank Campanella, Aldo Ray, Susan Anspach, Horton Foote Jr
A turgid and uninteresting tale of family conflicts and intrigues, set among the wine growers of California one hundred years ago. When a group of immigrant farmers learn that plans are afoot to build a railroad right through the heart of their community, they set out to mount a campaign of opposition. Average.
DRA 87 min (ort 91 min) VIDrel: COLUM/SONOP V

BLOOD RIVER ** PG
Mel Damski USA 1990
Wilford Brimley, Rick Schroder, Adrienne Barbeau, John P. Ryan, Dwight McFee, Mills Watson, Henry Beckman, Don S. Davis, Jay Brazeau, Venus Terzo, J.C. "Jim" Roberts, Gordon Tootoosis, Stephen Hair, Maureen Thomas, Jodi Thompson
An aimless young drifter goes on the run after a vengeance killing and teams up with a grouchy mountain man who becomes his companion. Set in the 1880s, this run-of-the-mill Western was written by John Carpenter.
WES 88 min (ort 90 min) mTV VIDrel: 20TH/TECH V

BLOOD RUN ** 18
Boaz Davidson USA 1994
David Bradley, Anna Levine, Lisa Wilson-Deitell, Trevor Short, Ashley Laurence, Jamie Renee Smith, Steven M. Gagnon, Terrance Stone, William Boyett, Stephanie Swinney, Roberto LaSardo, Paul Boyer, Charlie Holliday, Bill Moseley
With the discovery of the body of a young woman, attention focuses on her drug-dealing boyfriend. But he is innocent of the crime. Fortunately, a determined cop begins to suspect that his lover may be concealing the truth. Fair.
THR 91 min VIDrel: 20TH/FOXVID V

BLOOD SABBATH ** 18
Brianna Murphy USA 1972
Anthony Geary, Susan Damante-Shaw, Sam Gilman, Steve Graves, Dyanne Thorne
A GI seeks the sea nymph who saved his life and becomes caught up in a strange world of voodoo.
HOR 85 min VIDrel: SCRN/DISC V

BLOOD SIMPLE *** 18
Joel Cohen/Ethan Cohen USA 1983
John Getz, Frances McDormand, Dan Hedaya, Samm-Art Williams, M. Emmet Walsh, Deborah Neumann, Raquel Gavia, Van Brooks, Senor Ginger, William Creamer, Bob McAdams, Loren Bivens, Shannon Sedwick, Nancy Ginger, William Preston Robertson
A Texan bar-owner hires a private eye to kill his wife and her lover. However, the murders are faked. This leads to complications. A convoluted and flamboyant homage to film noir with good touches of black humour.
THR 98 min VIDrel: ELPIC/POLYREC V

BLOOD SPORT ** PG
Harvey Hart CANADA/EIRE 1989
Ian McShane, Heath Lamberts, Lloyd Bochner, Patrick Macnee, Carolyn Dunn
A millionaire's prize stallion is stolen and a detective hired to locate the beast has to enter the insular and obsessive world of bloodstock. One in a series of Dick Francis mysteries, quite absorbing and enjoyable.
Aka: DICK FRANCIS MYSTERIES: BLOOD SPORT
DRA 88 min (ort 90 min) VIDrel: IMC/DISC V
Boa: novel by Dick Francis.

BLOOD TIDE * (PG)
Richard Jeffries GREECE/UK 1980
Jose Ferrer, James Earl Jones, Lila Kedrova, Deborah Shelton, Martin Kove, Mary Louise Weller, Lydia Cornell
An American artist living on a sinister Greek island, becomes fascinated by the legend of a sinister sea creature, to which virgin girls were said to have been sacrificed. Her brother and sister-in-law visit her, and at the same time a string of brutal murders occurs. Eventually they are traced to our monster, which appears but briefly, just as well as it is a remarkably unconvincing one.
Aka: BLOODTIDE; RED TIDE, THE
HOR 97 min CABrel: HVC

BLOOD VOWS ** (PG)
Paul Wendkos USA 1987
Melissa Gilbert, Joe Penny, Eileen Brennan, Talia Shire, Tony Franciosa, Carmine Caridi, Ron Karabatsos, John Aprea, John Lavachielli, Josh Williams, Mary Ann Pascal, Rhoda Genignani, Santos Morales, William Hubbard Knight
A young girl is mugged soon after arriving in New York and the only help she receives is from a handsome and sophisticated lawyer. Having fallen in love with him she marries, only to find that she has in fact married into a Mafia family. Average mTV effort, quite well acted but with a disappointingly downbeat ending.
Aka: BLOOD VOWS: THE STORY OF A MAFIA WIFE
THR 90 min (ort 95 min) mTV CABrel: HVC

BLOOD WARRIORS ** 18
Norman Benny/Sam Firstenberg USA 1992
David Bradley, Jennifer Campbell, Frans Tumbuan, Frank Zagarino
Two ex-Marines fall out and become deadly enemies when one of them refuses to join a group of mercenaries being organised by the other. An average dose of mindless violence and action.
A/AD 95 min VIDrel: COLUM/SONOP V/sh

BLOOD WEDDING ** U
Carlos Saura SPAIN 1981
Antonio Gades, Christina Hoyos, Juan Antonio Jiminez, Carmen Villena, Pilar Cardenas, El Guito, Elvira Andres, Marisa Nella, Lario Diaz, Azucena Flores, Antonio Quitana, Quico Franco
Flamenco ballet version of the famous play about a bride who runs away with her married lover on her wedding day and is hunted down by her outraged husband.
Aka: BODAS DE SANGRE
DRA 68 min (ort 71 min) VIDrel: PHASE/RTM V
Boa: play by Federico Garcia Lorca.

BLOODFIGHT * 18
Shuji Goto HONG KONG/JAPAN 1989
Yusuaki Kurata, Bolo Yeung, Yam Tat Wah, Cristina Lawson, Lum Ken Ming, John Ladalski, Shinya Ono, Stuart Smita, Ken Boyle, Takaka Nakamura, Chan Fai Ling, Sindy Lim, Richard Foo, Masanari Nasu, Tadashi Sato, Masaru Yamashita
The inevitable revenge theme crops up once more, as a former champion trains his latest protege, only to see him callously murdered in the course of a bout with the reigning champion. A poorly plotted and unconvincingly acted mess, with a silly

sub-plot involving a thuggish former student and lush, scenic-style photography that makes the film resemble a Hong Kong tourist brochure. A sequel followed.
MAR   92 min Cut (4 sec)   VIDrel: MIA/DISC/IMPENT   V

BLOODFIGHT 2: THE DEATHCAGE ** 18
Robert Tai THAILAND 1989
Robin Chou, Joe Lewis, Steve Tagg, Tiger Kim, Annggia Tsai, Vicki Kim, Wayne Archer, Mark Long, Brian Lucey, Tody Russell, Simon Lin, Peter Hsu, John Ladaiski, Master Sken, Nina Burt, Jojo Roe
More of the same in this rather tired sequel, that upgrades the violence in a further set of martial arts bouts.
Aka: DEATH CAGE, THE
MAR   84 min (ort 90 min)   VIDrel: MIA/DISC/IMPENT   V

BLOODFIST ** 18
Terence H. Winkless USA 1989
Don Wilson, Joe Mari Avellana, Michael Shaner, Riley Bowman, Vic Diaz, Rob Kaman, Billy Blanks, Kris Aguilar
The usual revenge plot (this time over a dead brother) is incidental to this enjoyable display of flying fists and feet as various martial arts champs from the World Kickboxing Association go through their paces. A number of sequels have followed, but not all of them are available on video.
A/AD   83 min (ort 86 min)   VIDrel: MGM L/A   V

BLOODFIST 3: FORCED TO FIGHT ** 18
Oley Sassone USA 1991
Don "The Dragon" Wilson, Richard Roundtree, Stan Longidis, Peter Cunningham, Rick Dean, Richard Paul, Tony Di Benedetto, Charles Boswell, Gregory McKinney, John Cardone, Brad Blaisdell, Andre Rosey Brain, J.W. Smith, Laura Stockman
An innocent man is accused of murder and thrown into prison where he finds himself caught between the machinations of rival black and white gangs. A long struggle for survival ensues before the truth is eventually revealed about prison conditions. Adequate action film of little originality but much mayhem.
Aka: FORCED TO FIGHT
A/AD   84 min (ort 90 min)   VIDrel: CIC/SONOP   V

BLOODFIST 4: DIE TRYING ** 18
Paul Ziller USA 1992
Don Wilson, Cat Sassoon, Amanda Wyss, Kale Browne, Liz Torres, Dave Martin, James Tolkan, Dino Holmsey, Gene LeBell, Carolyn Raimondi, Herman Poppe, Stephen James, Carver Gary Daniels, John LMotta, Lenny Citrano
When his daughter is kidnapped, a martial artist takes on all and sundry in his bid for justice and revenge on a gang of terrorists who have stolen a number of detonation devices for nuclear weapons. Average.
A/AD   83 min (ort 100 min)   VIDrel: MIA/DISC   V/sur

BLOODFIST 5: HUMAN TARGET ** 18
Jeff Yonis USA 1993
Don Wilson, Denice Duff, Danny Lopez, Yuji Okumoto, Don Stark, Steve James
A government agent is attacked and left for dead after he is discovered spying on a Korean underworld arms deal. However, he recovers from his injuries and decides to continue his efforts alone now that his cover has been blown. Another rough and tough entry in this series.
Aka: HUMAN TARGET: BLOODFIST 5
A/AD   80 min (ort 90 min)   VIDrel: CIC/SONOP L/A   V

BLOODHOUNDS ** 15
Michael Katleman USA 1996
Corbin Bernsen, Christina Harnos, Markus Flanagan, Adam Tomei, Kirk Baltz, James Pickens
A best-selling author and tough female cop join forces to bring a killer to justice, the latter wants to get revenge for the slaying of her father but the former is just out for a good story. As they bicker their way across the States to Mexico, the inevitable dead bodies start to pile up, but the plot is so very predictable that the film offers little that remains in the memory. This one went straight to video.
DRA   83 min cC   VIDrel: CIC   V/sh

BLOODHOUNDS OF BROADWAY *** PG
Howard Brookner USA 1989
Madonna, Matt Dillon, Jennifer Grey, Julie Hagerty, Rutger Hauer, Esai Morales, Anita Morris, Randy Quaid, Madeleine Potter, Ethan Phillips, Tony Longo, Alan Ruck, David Youse, Louis Zorich, Tony Azito

A loose adaptation of a Runyon story revolving around the exploits of a set of gangsters, gamblers and general hangers-on, all of whom are out to celebrate New Year's Eve 1928. A lightweight but engaging period comedy that captures a good deal of the flavour if not the verve of Runyon's writing. This was documentary-maker Brookner's first work of fiction, unfortunately he died just prior to its release. First filmed in 1952.
COM   88 min (ort 93 min)   VIDrel: VCC L/A   V/sh
Boa: short story by Damon Runyon.

BLOODLUST: SUBSPECIES 3 ** 18
Ted Nicolaou USA 1993
Anders Hove, Denice Duff, Kevin Blair, Melanie Shatner, Pamela Gordon, Michael Dellafemina, Ion Haiduc
Third instalment in this vampire saga of two brothers and their constant struggle to subdue each other (see the earlier SUBSPECIES and BLOODSTONE: SUBSPECIES 2). Here, the evil brother returns once more from the dead and sets about seducing one of two American sisters. Much gore but little sense or style is on offer here, although the revolting special effects are very convincing.
HOR   79 min (ort 83 min)   VIDrel: CIC/SONOP   V

BLOODRAGE ** 18
Joseph Bigwood/Joseph Zito USA 1978
Ian Scott, Lawrence Tierney, Judith-Marie Bergan, James Johnston, Jerry McGee, Jimi Keys, James Johnston, Jerry McGee, Jimi Keys
A vigilante hunts a deranged killer loose on the streets of New York, in this cliched thriller that is somewhat derivative of DEATHWISH, but does little to hold one's interest.
Aka: NEVER PICK UP A STRANGER
THR   78 min (ort 86 min)   VIDrel: VIPCO/SGSVID   V

BLOODSPORT ** 18
Newt Arnold USA 1987
Jean-Claude Van Damme, Leah Ayres, Donald Gibb, Norman Burton, Forest Whitaker, Bolo Yeung
Kung Fu adventure set at a secret martial arts competition held in Hong Kong. Based on the true story of American Frank Dux – the first Westerner ever to win the ultra-tough kumite contest. A violent, low-budget affair that has a few good moments, but not too many of them.
MAR   88 min (ort 92 min)   VIDrel: WHV   V/sh

BLOODSPORT 2 ** 18
Alan Mehrez USA 1995
Daniel Bernhardt, Noriyuki "Pat" Morita, Donald Gibb, James Hong
In-name-alone sequel to the earlier film, that follows the exploits of a tough streetfighter who, having stolen a ceremonial sword from a businessman, is sent to jail. Once there, he is taken under the wing of an older man, and taught a deadly fighting technique known as the "Iron Hand". Adequate.
MAR   90 min   VIDrel: 20VIS/SONOP   V/sur

BLOODSTAINED SHADOW ** 18
Antonio Bido ITALY 1978
Craig Hill, Stefania Casini, Lino Capolicchio
A series of nasty murders have taken place in a small Italian village. A young professor from Venice comes to the village to visit his brother who is the local priest. Having witnessed a murder himself, the priest is now in danger.
HOR   104 min (ort 133 min) wScrn dubbed
VIDrel: REDEM/RTM   V

BLOODSTONE: SUBSPECIES 2 ** 18
Ted Nicolaou USA 1992
Anders Hove, Denice Duff, Kevin Blair, Melanie Shatner, Michael Denish, Pamela Gordon, Ion Haiduc, Tudorel Filimon, Viorel Cominici, Wayne Toth, Viorel Sergonici, Catalina Murgea, Viobea Berbiuc, Nicolae Urs, Vlad Paunescu
A young girl on a visit to Romania is lusted after a vampire, in this assembly-line sequel to SUBSPECIES. Lots of sickening special effects that will doubtless enthral horror fans, but nothing of any wit or interest is in evidence here. Followed by BLOODLUST: SUBSPECIES 3.
Aka: SUBSPECIES 2
HOR   107 min   VIDrel: FULL/HIFLI   V/h

BLOODSTREAM *** 18
Stephen Tolkin USA 1993
Cuba Gooding Jr, Moira Kelly, Alice Drummond, John Seda, Omar Epps, Martha Plimpton, John Cameron Mitchell, David Eigenberg,

Nick Chinlund, Paul Butler, Willie Carson, Mark Boone Jr, Deidre O'Connell, Phil Parolise, Paul Butler
Futuristic fantasy set in a repressive America, where the streets are patrolled by vigilantes and the victims of a new type of plague have been forced to live in quarantined internment camps. A small group of rebels struggle to resist this system as best they can, but are too few in number to achieve their ends. A well conceived effort.
Aka: DAYBREAK
A/AD 90 min mCab VIDrel: NEWAGE/COLUM L/A V
Boa: play Beirut by Alan Bowne.

BLOODSUCKING PHARAOHS IN PITTSBURGH * 18
Alan Smithee (Dean Tschetter) USA 1991
Jake Dengel, Joe Sharkey, Susann Fletcher, Beverly Penberthy, Shawn Elliott, Pat Logan, Jane Esther Hamilton
Silly horror spoof that sees a couple of numbskull detectives being sent to investigate the finding of mutilated bodies in an Egyptian quarter of the city.
COM 88 min VIDrel: COLUM/SONOP V

BLOODY BIRTHDAY ** 18
Edward Hunt USA 1980
Susan Strasberg, Jose Ferrer, Lori Lethin, Melinda Cordell, Julie Brown, Joe Penny, Billy Jacoby, Michael Dudikoff, Andy Freeman, Ellen Geer, Bert Kramer, K.C. Martel, Elizabeth Hoy, Ben Marley, Erica Hope, Cyril O'Reilly
Three women give birth during a solar eclipse. Ten years later a small town is shaken by a series of gruesome murders as three kids born on the day of the eclipse go on a rampage. Despite the unusual basis for the plot, this one just serves up the usual gore.
Aka: BLOODY SUNDAY; CREEPS
HOR 84 min VIDrel: VIPCO/SGSVID V

BLOODY MAMA *** 18
Roger Corman USA 1970
Shelley Winters, Pat Hingle, Don Stroud, Diane Varsi, Robert Walden, Robert De Niro, Pamela Dunlap, Bruce Dern, Clint Kimbrough, Alex Nichol, Michael Fox, Robert Walden, Scatman Crothers, Stacy Harris
Tense and gripping tale of a psychotic mother and her degenerate offspring who together wage war on society. Though violent and sordid it remains one of Corman's better films. Screenplay is by Robert Thom and Don Peters.
DRA 82 min Cut (11 sec – ort 89 min)
VIDrel: VISVID/POLYREC L/A V
Boa: novel by Robert Thom.

BLOODY MOON * 18
Jess (Jesus) Franco SPAIN/WEST GERMANY 1981
Oliva Pascal, Nadja Gerganoff, Jasmin Losensky, Christoph Moosbrugger, Alexander Wachter, Corinna Gillwald, Ann-Beate Engelke, Antonia Garcia, Maria Rubio
A killer preys on the pupils of a language school, in this contrived and woefully unimaginative shocker. Clumsy and quite shoddy, the plot is little more than a framework within which to stage several gory killings, which are realistic for nothing else. A few erotic sequences serve to generate interest when the action flags. Unlikely to ever be available in an uncut form in the UK.
Aka: DIE SAGE DES TODES
HOR 79 min (ort 95 min) VIDrel: VIPCO/SGSVID V/h

BLOSSOMS IN THE DUST ** (U)
Mervyn Le Roy USA 1941
Greer Garson, Walter Pidgeon, Felix Bressart, Marsha Hunt, Fay Holden, Samuel S. Hinds, Kathleen Howard, George Lessey, William Henry, Henry O'Neill, John Eldredge, Clinton Rosemund, Theresa Harris, Charles Arnt
Sentimental but immensely successful star vehicle, the tearjerking story of a woman who founds a Texas orphanage after the deaths of her husband and child. A lavish but essentially hollow tale, based on the life of Edna Gladney. AA: Art/Int (Cedric Gibbons and Uri McCleary/Edwin B. Willis.
DRA 95 min TVrel: C4

BLOW OUT *** 18
Brian De Palma USA 1981
John Travolta, Nancy Allen, John Lithgow, Dennis Franz, Peter Boyden, Curt May, Ernest McClure, Davie Roberts, Maurice Copeland, Claire Carter, John Aquino, John Hoffmeister, Patrick McNamara, Terrence Currier, Tom McCarthy
Interesting variation on the BLOW UP theme. A sound techni-cian accidentally records an accident in which a politician plunges his car into a river, with a girl in the passenger seat; probably suggested by the Chappaquiddick incident that involved US senator Edward Kennedy in controversy. The intriguing story reveals a politically motivated murder, but serious flaws in the plot and irritating camerawork let it down.
THR 108 min VIDrel: VISVID/POLYREC L/A V

BLOW UP ** 15
Michelangelo Antonioni UK 1966
David Hemmings, Sarah Miles, Vanessa Redgrave, Peter Bowles, John Castle, Jane Birkin, Gillian Hills, Verushka, Jill Kennington, Harry Hutchinson, Julian Chagrin, Susan Broderick, Mary Khal, Ronan O'Casey, Tsai Chin, The Yardbirds
How a good idea can be ruined by a lack of plot or characters. A fashion designer taking photos in the park seems to have been an unwitting witness to a murder. After a moment of genuine tension as he develops the film, it's downhill all the way. Written by Antonioni with a score by Herbie Hancock. See also BLOW OUT, which tried a similar idea – but with sound.
THR 106 min (ort 111 min) VIDrel: MGM/WHV V
Boa: short story by Julio Cortazar.

BLOWN AWAY ** 18
Brenton Spencer USA 1992
Corey Haim, Corey Feldman, Nicole Eggert, Gary Farmer, Kathleen Robertson, Jean Leclerc
A young man takes a job at a resort in order to earn money towards his college education and gets involved with a young pretty girl, unaware that she has some deadly plans in mind. Silly erotic thriller with the stars a little too young to give depth to a film of this nature.
THR 92 min VIDrel: 20VIS/SONOP V

BLOWN AWAY ** 15
Stephen Hopkins USA 1994
Jeff Bridges, Tommy Lee Jones, Lloyd Bridges, Forest Whitaker, Suzy Amis, John Finn, Stephi Lineburg, Lloyd Catlett, Caitlin Clarke, Chris De Oni, Ruben Santiago-Hudson, Lucinda Weist, Brendan Burns, Patricia A. Heine, Josh McLaglen
An Irish political activist subjects Boston to a reign of terror with his cleverly designed bombs and directs his aggression forwards a former comrade-in-arms who reformed and is now a member of the city's bomb squad. A noisy and violent thriller, with a hackneyed plot, that is badly hampered by both its unoriginality and the lack of performances that are both strong and convincing.
THR 115 min (ort 121 min) cC VIDrel: MGM/WHV V/sh

BLUE AND THE GRAY, THE *** PG
Andrew V. McLaglen USA 1982
Gregory Peck, Stacy Keach, John Hammond, Lloyd Bridges, Colleen Dewhurst, Julia Duffy, Sterling Hayden, Warren Oates, Geraldine Page, Rip Torn, Diane Baker, Kathleen Beller, Paul Benedict, Rory Calhoun, David Doyle, Dan Shor
Originally an eight-part mini-series, this account of the Civil War from the Harper's Ferry raid to Lincoln's assassination adopts the familiar device of two related families to represent the two sides. (Seen most recently in NORTH AND SOUTH). However, it is distinguished by its realism (it was filmed entirely on location with a huge cast of extras) and occasional moments of heightened tension.
DRA 356 min (2 cassettes – ort 446 min) VIDrel: COLUM V/sh

BLUE ANGEL, THE **** PG
Josef Von Sternberg USA 1930
Emil Jannings, Marlene Dietrich, Kurt Gerron, Hans Albers, Rosa Valetti, Hans Albers, Eduard Von Winterstein, Reinhold Bernt, Hans Roth, Rolf Muller, Robert Klein-Lork, Karl Huszar-Puffy, Wilhelm Diegelmann, Gerhart Bienert
An ageing and straitlaced schoolteacher becomes infatuated with a sleazy nightclub singer and marries her, but she ruthlessly exploits, deceives and humiliates him. The first talkie for Jannings (who gives an endearing if stilted performance) this sombre and grotesque tale did wonders for Dietrich's career and though slow to get started, remains a moving and unforgettable experience. It was also shot in English. Music is by Frederick Hollander.
Aka: DER BLAUE ENGEL
DRA 92 min (ort 103 min) B/W VIDrel: EUREKA/GOLD V
Boa: novel Professor Unrath by Heinrich Mann.

BLUE BLACK PERMANENT ***　　　　　　　　PG
Margaret Tait UK　　　　　　　　　　　　　　1992
Celia Imrie, Jack Shepherd, Gerda Stevenson, James Fleet, Sean Scanlan, Hilary Maclean, Walter Leask, Sheana Marr, Eoin MacDonald, Jimmy Moar, Liz Robertson, Bobby Bews, Keith Hutcheon, Mairi Wallace, Pamela Kelly, Joan Alcorn
A woman returns to her birthplace in the Orkneys, and constantly recalls the love of her mother, whose responsibilities were so very often compromised by a deep passion she felt for the sea. Frequently jumping between the 1930s and the present day, this is a captivating and lyrical film of great beauty, whose convoluted structure and verbose script are but slight drawbacks. This was an impressive feature-film debut from Tait.
DRA   86 min   CINrel

BLUE CHIPS **　　　　　　　　　　　　　　15
William Friedkin USA　　　　　　　　　　　1994
Nick Nolte, Shaquille O'Neal, Malcolm McDowell, Mary McDonnell, Ed O'Neil, J.T. Walsh, Alfre Woodard, Matt Nover, Anfernee "Penny" Hardaway, Anthony C. Hall, Richard Pitino, George Raveling, Bob Cousy, Larry Bird, Bobby Knight
A college basketball coach is forced to disregard his own principles after a losing season and accept the practice of paying money under the table to recruit top-notch players. An interesting look at a corrupt and cynical world, where money seems to be the only consideration.
DRA   103 min (ort 108 min) cC   VIDrel: CIC/SONOP V/sur

BLUE FANTASIES *　　　　　　　　　　　18
Walter Boos WEST GERMANY　　　　　　　1978
Karin Kernke, Elizabeth Welz, Helene Rosenkranz, Nicole Avril
The sex fantasies of a group of college students who are aroused whilst reading a batch of letters. One in a long line of tired and dated German sex films.
Aka: SCHOOLGIRL REPORT NO. 1; SCHULMADCHEN: REPORT 12 TEIL
A   72 min   VIDrel: ELV   V

BLUE FLAME **　　　　　　　　　　　　15
Cassian Elwes USA　　　　　　　　　　　　1995
Brian Wimmer, Jad Magher, Kerri Green, Cecilia Peck, Melissa Behr, Ian Buchanan
A man living in L.A. finds that his mind is being taken over by humanoid aliens from the future, who then abduct his five-year-old daughter. A strange SF fantasy.
FAN   87 min (ort 88 min)   VIDrel: COLUM/SONOP   V

BLUE GRASS: PARTS 1 AND 2 **　　　　　15
Simon Wincer USA　　　　　　　　　　　　1987
Cheryl Ladd, Anthony Andrews, Mickey Rooney, Wayne Rogers, Kieran Mulrony, Brian Kerwin, Diane Ladd, Shawnee Smith
A girl who has grown up with horses on the Kentucky Blue Grass is determined to gain recognition as an equal among the horse breeders of Kentucky but her plans are opposed by a group of breeders who aim to keep their circle exclusive. A pleasant little soap opera tale that will appeal to horse lovers, though the foaling sequence may be a little strong for some.
DRA   177 min mTV   VIDrel: 20TH/TECH   V

BLUE HEAT **　　　　　　　　　　　　15
John Mackenzie USA　　　　　　　　　　　1989
Brian Dennehy, Joe Pantoliano, Jeff Fahey, Bill Paxton, Deborra-Lee Furness, Guy Boyd, Henry Darrow, Lisa Jane Persky, Michael C. Gwynne, Henry Stolow, John Finnegan, J. Kenneth Campbell, Patricia Clipper, Michelle Little
A straight cop and his colleagues discover corruption in high places when they attempt to curtail the activities of drug dealers. Strong performances (especially from Dennehy) are held back by the bizarre script (by Jere Cunningham) which appears to draw some inspiration from the Iran-Contra affair. The music is by Jack Nitzsche.
Aka: LAST OF THE FINEST, THE
A/AD   101 min Cut (5 sec – ort 106 min)
VIDrel: VISVID/POLYREC   V/sh

BLUE ICE **　　　　　　　　　　　　　15
Russell Mulcahy USA　　　　　　　　　　　1992
Michael Caine, Sean Young, Ian Holm, Bobby Short, Alun Armstrong, Sam Kelly, Jack Shepherd, Philip Davis, Patricia Hayes, Alan MacNaughton, Mac Andrews, Todd Boyce, Peter Forbes, Peter Gordon, Oliver Haden, Philip Whitchurch, Bob Hoskins
A retired intelligence agent gets involved with an ambassador's wife, but soon comes to regret his impulsiveness when she involves him in her efforts to get rid of an ardent ex-lover. He soons finds he is suspected of several murders that prove to be linked to a murky plot. A tired thriller filmed in London that is nothing more than a derivative rehash of all the old cliches.
THR   105 min   VIDrel: GUILD/POLYREC L/A   V/s

BLUE IGUANA, THE **　　　　　　　　　15
John Lafia USA　　　　　　　　　　　　　1988
Dylan McDermott, Jessica Harper, James Russo, Pamela Gidley, Dean Stockwell, Tovah Feldshuh, Flea, Michele Seipp, Katia Schkolnik, John Durbin, Eliett, Don Pedro Colley, Pedro Altamirano, Benny Corral, Alejandro Bracho, Siro
A none-too-successful bounty hunter tries to enrich himself by helping the IRS recover $20 million from a small Latin American bank. A spoof on all those "private eye goes south of the border" films, that suffers badly from a messy script and tries too hard for laughs.
COM   86 min (ort 88 min)   VIDrel: 20TH   V/sur

BLUE JEAN COP **　　　　　　　　　　18
James Glickenhaus USA　　　　　　　　　1988
Peter Weller, Patricia Charbonneau, Sam Elliott, Antonio Fargas, Blanche Baker, Richard Brooks, Tom Waits, Kathryn Rossetter, William Prince, Larry Joshua, John McGinley, Augusta Dabney
A highly skilled defence attorney has one last legal aid case to take on before giving up this work for a well paid Wall Street job. This involves defending a drugs dealer who shot dead an undercover cop, and claims it was done in self defence. His investigations uncover a web of corruption within the police force, and a tough but honest cop helps him nail them and clear his client. A competent action movie with good dialogue and an unusual car chase.
Aka: SHAKEDOWN
A/AD   92 min (ort 96 min)   VIDrel: 4-FRONT/POLYREC V

BLUE JUICE ***　　　　　　　　　　　15
Peter Salmi UK　　　　　　　　　　　　　1995
Sean Pertwee, Catherine Zeta Jones, Steven Mackintosh, Ewan McGregor, Peter Gunn, Heathcote Williams, Colette Brown, Michelle Chadwick, Keith Allen, Robin Soans, Jenny Agutter, Guy Leverton, Mark Frost, Paul Reynolds, Edwin Starr
In Cornwall, an obsessive surfboarder has to choose between marrying his longtime girlfriend whose patience is running out or ditching her in favour of a life spent riding the waves. The arrival of three friends from London adds a further complication in this endearing and unexpectedly enjoyable film.
COM   94 min (ort 98 min)   VIDrel: BLACK/DISC   V/sur

BLUE KITE, THE ***　　　　　　　　　PG
Tian Zhuangzhuang CHINA/HONG KONG　　1993
Yi Tian, Zhang Wenyao, Chen Xiaoman, Lu Liping, Pu Quanxin, Li Xuejian, Guo Baochang, Zhong Ping, Chu Quanzhong, Song Xiaoying, Zhang Hong, Liu Yanjin, Li Bin, Lu Zhong, Guo Donglin, Wu Shumin, Zhang Fengyi, Xu Min, Zhang Ju
A fascinating account of the sufferings of a young schoolteacher at the hands of the Communist Party to whom she loses no less than three husbands as well as her own personal freedom. A bleak and harrowing account of the evil inflicted by those infected with fanatical loyalty to an all-pervading ideology. Understandably, the making of this film was subject to political interference and it has now been prohibited from exhibition in its homeland.
Aka: LAN FENGZHENG
DRA   134 min (ort 138 min)   VIDrel: ICAPRO/MANGA V/sur

BLUE LAGOON, THE *　　　　　　　　　15
Randal Kleiser USA　　　　　　　　　　　1980
Brooke Shields, Christopher Atkins, Leo McKern, William Daniels, Glenn Kohan, Elva Josephson, Alan Hopgood, Gus Mercurio, Jeffrey Means, Bradley Pryce, Chad Timmermans, Gert Jacoby, Alex Hamilton, Richard Evanson
Remake of the 1949 film, telling the story of the growing sexual awareness of a young boy and girl, both shipwrecked on an island. Much nudity and adolescent frankness about sex but little else in a thoroughly insipid film. Written by Douglas Day Stewart. The excellent photography is by Nestor Almendros. See also RETURN TO THE BLUE LAGOON.
DRA   100 min (ort 104 min)   VIDrel: MIA/DISC/COLUM V/sh
Boa: novel The Garden Of God by Henry de Vere Stacpoole.

BLUE LAMP, THE *   *PG*
Basil Dearden UK   1949
Jack Warner, Jimmy Hanley, Dirk Bogarde, Robert Flemyng, Bernard Lee, Peggy Evans, Gladys Henson, Tessie O'Shea, Patric Doonan, Bruce Seton, Clive Morton, Frederick Piper, Dora Bryan, Norman Shelley, Campbell Singer, Sam Kydd
A police sergeant on the eve of retirement is shot and killed by a gang robbing a cinema. However, the murderer is brought to justice in a manhunt where even members of the criminal fraternity lend a helping hand as they despise the use of firearms. A very British film in style and attitudes. The dead sergeant later starred in a very long-running TV police series – "Dixon Of Dock Green". A solid and low-key film.
DRA   81 min (ort 84 min) B/W   VIDrel: WHV   V

BLUE MAX, THE *   *PG*
John Guillermin USA   1966
George Peppard, James Mason, Ursula Andress, Jeremy Kemp, Karl Michael Vogler, Anton Diffring, Harry Towb, Peter Woodthorpe, Derek Newark, Derren Nesbitt, Loni Von Friedl, Friedrich Ledebur, Carl Schell, Roger Ostime
Overlong saga of German WW1 combat pilot and his struggle to achieve fame, in his eyes represented by the Blue Max award, even if it means risking the lives of his comrades and seducing the wife of a superior officer. Good photography and aerial sequences compensate for the dullness of the plot.
DRA   137 min (Cut at film release – ort 156 min)
VIDrel: 20TH/TECH   V/sh
Boa: novel by Jack D. Hunter.

BLUE MURDER AT ST TRINIANS *   *U*
Frank Launder UK   1958
Joyce Grenfell, Terry-Thomas, George Cole, Alistair Sim, Lionel Jeffries, Sabrina, Thorley Walters, Eric Barker, Richard Wattis, Lloyd Lamble, Michael Ripper, Judith Furse, Lisa Gastoni, Dilys Laye, Kenneth Griffith
Dreadfully dated and unfunny film, the second in the series, with our anarchic schoolgirls off to Rome, having won a UNESCO prize trip. There they become involved with a jewel thief, who hides out at their school. A film that memorably epitomises all that was and still is wrong with the British film industry; do miss it. Followed by THE PURE HELL OF ST TRINIANS.
COM   83 min (ort 86 min) B/W   VIDrel: WHV   V

BLUE RODEO *   *PG*
Peter Werner USA   1996
Kris Kristofferson, Ann-Margret
A woman and her deaf teenage son movie to Arizona. Kristofferson is her cowboy neighbour who provides the love interest. A pleasant time-filler of no great consequence.
DRA   88 min mTV   VIDrel: WHV   V

BLUE SKY *   *15*
Tony Richardson USA   1990 (released in 1994)
Jessica Lange, Tommy Lee Jones, Powers Boothe, Carrie Snodgress, Amy Locane, Chris O'Donnell, Mitchell Ryan, Dale Dye, Tim Scott, Annie Ross, Anna Klemp, Anna Rene Jones, Jay H. Seidl, David Bradford, Matt Battaglia, John J. Fedak
In the 1960s, a military engineer finds his loyalty coming under strain when he is a witness to a high-level cover-up at a Nevada test site. However, his attempts to do something about this run into difficulties when his superiors use his marital difficulties to call his credibility into question. Not the most carefully developed of plots, but the fine acting of the two leads makes it well worth watching. This was Richardson's last film. AA: Actress (Lange).
DRA   96 min (ort 101 min)   VIDrel: COLUM/SONOP
V/sur

BLUE STEEL *   *U*
Robert N. Bradbury USA   1934
John Wayne, George "Gabby" Hayes, Yakima Canutt, Eleanor Hunt, Ed Peil Sr, George Cleaveland, George Nash, Lafe McKee, Hank Bell, Earl Dwire
Outlaws discover gold deposits beneath the town and resort to murder and extortion until the Duke comes onto the scene. A good, old-fashioned Western with an excellent opening sequence that takes place during a thunderstorm whilst a robbery is in progress.
WES   54 min B/W coVer   VIDrel: ENTUK L/A   V

BLUE STEEL **   *18*
Kathryn Bigelow USA   1989
Jamie Lee Curtis, Ron Silver, Clancy Brown, Elizabeth Pena, Louise Fletcher, Philip Bosco, Richard Jenkins, Kevin Dunn, Markus Flannagan, Mary Mara, Skip Lynch, Mike Hodge, Mike Starr, Chris Walker, Tom Sizemore, David Ilku
A suspense thriller set in New York, where a female rookie cop finds herself being stalked by an obsessive Wall Street broker who also moonlights as a serial killer. Flashy direction and photography cannot hide the shallowness of it all, and this film rapidly becomes as ludicrous (our serial killer appears to be indestructible) as it is boring.
THR   97 min (ort 103 min)   VIDrel: FIRST/SONOP;
ENCORE (LV only)   V/sur LV

BLUE THUNDER: THE MOVIE *   *15*
John Badham USA   1983
Roy Scheider, Warren Oates, Daniel Stern, Malcolm McDowell, Candy Clark, Joe Santos, Paul Roebling, David Sheiner, Ed Bernard, Jason Bernard, Mario Machado, Anthony James, Jim Murtaugh, Pat McNamara, Jack Murdock
Action film that centres around the L.A. police department's use of a super-helicopter, one that is equipped with a formidable arsenal of surveillance devices as well as devastating weaponry. Scheider plays a police officer (and Vietnam veteran) who steals the title craft. A slick and well-crafted film that starts off with promise but gets bogged down in the more ludicrous aspects of the plot. Written by Dan O'Bannon and Don Jakoby.
Aka: BLUE THUNDER
A/AD   105 min (ort 118 min) wScrn
VIDrel: ENCORE/SPEAR/COLUM   V

BLUE TIGER *   *18*
Norberto Barba USA   1994
Virginia Madsen, Toru Nakamura, Ryo Ishibashi, Harry Dean Stanton, Sal Lopez, Dean Hallo, Yuji Okumoto, Brenda Varda, Francois Chau, Henry Mortenson, Chris De Rose, Claudia Templeton, Toshishiro Obata, John Hammil, Chreyy Frey
When her son is tragically killed in the crossfire between two criminal gangs, a woman resolves to track down his killer, an assassin who goes by title name. Calling herself "Red Tiger", she eventually manages to locate him but finds that she has let herself in for a surprise or two. Madsen's convincing portrayal adds weight to this well-made and intelligent thriller.
THR   84 min (ort 88 min)   VIDrel: MED/DISC   V/sh

BLUE TORNADO *   *15*
Tony B. Dobb (Antonio Bido) ITALY   1990
Dirk Benedict, Patsy Kensit, Ted McGinley, David Warner, Chris Ahrens, Tony Allen, John Armstead, Eric Bass, Robert Dawson, Jeff Blynn, Todd Carter, Roger Daniels, Martin Dansky, Pascal Druant, Donald F. Hodson, Jonathan Horn
Two fighter-pilots on a routine mission are blinded by some strange lights in the sky and one of them crashes into a mountainside. His surviving colleague then attempts to probe this mystery and teams up with a woman researcher, but our intrepid couple find many obstacles being placed in their path. A watchable but far from imaginative attempt at an SF thriller that feels overly familiar and predictable.
FAN   87 min (ort 91 min)   VIDrel: 20VIS/SONOP   V/sur

BLUE VELVET *   *18*
David Lynch USA   1986
Isabella Rosselini, Dennis Hopper, Kyle MacLachlan, Laura Dern, Hope Lange, Dean Stockwell, George Dickerson, Brad Dourif, Jack Lance, Priscilla Pointer, Frances Ray, Ken Stovitz, J. Michael Hunter, Dick Green
A weird but original film, telling of a young man and his involvement with a nightclub singer and a sadistic kidnapper, in a seemingly innocent small American town. Hopper gives an outstanding performance as the town's resident psycho, in this bizarre and disturbing study of the seamy side of life. A minor cult film whose use of widescreen will suffer on TV. The script is by Lynch. Rosselini is the daughter of Ingrid Bergman.
DRA   115 min (ort 120 min) wScrn
VIDrel: ELPIC/POLYREC   V/sur

BLUE VILLA, THE *   *(PG)*
Alain Robbe-Grillet/Dimitri de Clerq
BELGIUM/FRANCE/SWITZERLAND   1994
Fred Ward, Arielle Dombasle, Charles Tordjman, Sandrine Le Berre, Dimitri Poulikakos, Christian Maillet, Muriel Jacobs, Michalis

Maniatis, Pandeas Scaramanga, Cai Jiguang, Lee Chong-Lin, Li Lai, Shi Kuifan, Giorgos Grouezas
As a writer approaches a remote island, he works on the screenplay of a film that tells of a vanished sailor, who may have killed his girlfriend, before fleeing to sea. But on the island it is the writer that is accused of the murder, and then further complexities are piled one upon another, for even as we watch the story unfold, it is in the process of undergoing constant revision. A strangely compelling, experimental film, both circular and self-referential.
Aka: UN BRUIT QUI REND FOU
DRA 100 min CINrel

BLUEBERRY HILL ** (12)
Strathford Hamilton USA 1988
Carrie Snodgress, Margaret Avery, Jennifer Rubin, Matt Lattanzi, Ian Patrick Williams, Dendrie Allyn Taylor, Richard Haines, David Shortleff, Cassandra Lee Hamilton, David Shortleff, Hal Havens, Kathryn Attwood, Ira Ingber, Larry Taylor
A young girl growing up in the stifling surroundings of her small hometown has to make the best of her life after the death of her father. But she makes friends with a jazz singer who helps her realise her dreams. The highly music on the soundtrack does much to compensate for the triteness and banality of the plot.
DRA 89 min (ort 93 min) SATrel: MOVIE CHANNEL

BLUES BROTHERS, THE *** 15
John Landis USA 1980
John Belushi, Dan Aykroyd, Cab Calloway, Kathleen Freeman, Chaka Khan, The Blues Brothers Band, Carrie Fisher, James Brown, Henry Gibson, Aretha Franklin, Ben Piazza, Ray Charles, Murphy Dunn, John Candy, Jeff Morris, Lou Marini
Comedy about two brothers, both musical performers, who become involved in a race against time to save their old orphanage – nearly destroying Chicago in the process. A zany and quite enjoyable story that skilfully intercuts some great music, with scenes of wholesale destruction whenever the pace appears to be in danger of slackening. Numerous cameo appearances include Steven Spielberg, Frank Oz, Steve Lawrence, John Lee Hooker and Paul Reubens. Now a minor cult film.
COM 127 min (ort 133 min) VIDrel: CIC/SONOP; PION (LV only) V/sh LV

BOAT, THE *** 15
Wolfgang Petersen WEST GERMANY 1981
Jurgen Prochnow, Herbert Gronemeyer, Klaus Wennemann, Hubertus Bengsch, Martin Semmelrogge, Bernd Tauber, Erwin Leder, Martin May, Heinz Honig, U.A. Ochsen, Claude Olivier Rudolph, Jan Fedder
This account of a German U-boat on a mission during WW2, provides a stark and realistic picture of submarine warfare, spoilt by a terribly contrived ending. A longer version with a running time of 300 minutes was made for TV. Not a film for the claustrophobic as most of the action centres on the behaviour and stresses among the crew, with little respite from the confined space. Written by Petersen.
Aka: DAS BOOT
WAR 123 min (ort 150 min) wScrn dubbed
VIDrel: COLUM/SONOP V/sur
Boa: novel by Lothar-Guenther Buchheim.

BOAT HOUSE ** 15
Aiken Scherberger CANADA 1985
Wendy Crewson, David Ferry, Keith Knight, Robby Benson
An aspiring young reporter is intrigued by the doings of a mysterious woman and her involvement with a boat house on a remote island. He eventually stumbles across a twenty-five-year-old murder mystery. Average.
HOR 92 min VIDrel: MOPIC/SGSVID V

BOB AND CAROL AND TED AND ALICE ** 15
Paul Mazursky USA 1969
Natalie Wood, Robert Culp, Elliot Gould, Dyan Cannon, Horst Ebersberg, Lee Bergiere, Donald F. Muhcich, Noble Lee Holderread Jr, K.T. Stevens, Celeste Yarnall, Carol O'Leeary, Andre Phillipe, Greg Mullavey
A sophisticated and sexually liberated Californian couple try to persuade their less trendy friends to join them in a menage-a-quatre. Some sharp observations on the morality of a vanished era are marred by a stupid ending. Written by Mazursky and Larry Tucker and the directorial debut for the former. Later a short-lived TV series.
COM 101 min (ort 104 min) VIDrel: VCC/RCA L/A V
Boa: novel by P. Welles.

BOB ROBERTS **** 15
Tim Robbins USA 1992
Tim Robbins, Giancarlo Esposito, Alan Rickman, Ray Wise, Brian Murray, Gore Vidal, Rebecca Jenkins, Harry J. Lennix, John Ottavino, Robert Stanton, Kelly Willis, Tom Atkins, Merrilee Dale, David Strathairn, James Spader, Pamela Reed
A razor sharp political satire that examines the politics of image and benefits from a witty and intelligent script from writer-director Robbins. He plays a right-wing politician and wealthy businessman who fakes the radical sentiments of the 1960s in order to dislodge a political opponent. A very clever film, it takes the form of a spoof documentary, and few are spared its barbed humour.
COM 99 min (ort 104 min) VIDrel: COLUM/SONOP L/A V/sh

BODIES IN HEAT: THE SEQUEL ** 18
Henri Pachard USA 1989
Annette Haven, Eric Edwards, Sharon Kane, Jerry Butler, Charli St Cyr, Joey Silvera, Carol Cummings, Herschel Savage, Renee Morgan, Robert Bullock, Billy Dee
A slightly unusual sex film in that it incorporates a murder mystery that is set at a busy brothel, which is the location for a variety of nefarious plots and intrigues.
Aka: BODIES IN HEAT 2
A 53 min (ort 80 min) mVid VIDrel: HAR/GOLD V

BODIES, REST & MOTION ** 15
Michael Steinberg USA 1992
Phoebe Cates, Bridget Fonda, Tim Roth, Eric Stoltz, Alicia Witt, Sandra Lafferty, Sidney Dawson, Jon Proudstar, Scott Johnson, Kerbah Weidner, Peter Fonda, Amaryllis Borrego, Rich Wheeler, Scott Frederick, Warren Burton
A competent character study of four people and their relationships, set in an Arizona desert town, that focuses on their mutual disappointments and frustrations. Very well acted but done in such a stagebound manner that fails to work effectively as a movie.
DRA 94 min VIDrel: ELPIC/POLYREC V/sur
Boa: play by Roger Hedden.

BODILY HARM ** 18
James Lemmo USA 1995
Linda Fiorentino, Daniel Baldwin, Gregg Henry, Troy Evans, Bill Smitrovich, Joe Regalbuto, Millie Perkins, Shannon Kenny, Todd Susaman, William Utay, Ken Lerner, Casey Biggs, Castulo Guerra, Lou Bonacki, Marri Morrow
A female detective and her partner investigate the murder of an erotic dancer, the chief suspect being an ex-cop and former boyfriend, who was sacked when their affair led to the woman's husband committing suicide. As the case progresses more killings occur, the cop, who regrets ever splitting up with him, tries to rekindle their affair. Naturally this complicates matters. A brutal and unpleasant thriller with few sympathetic characters.
THR 87 min (ort 90 min) VIDrel: HIFLI/SONOP V/h

BODY & SOUL **** PG
Robert Rossen USA 1947
John Garfield, Lilli Palmer, Hazel Brooks, Anne Revere, William Conrad, Canada Lee, Joseph Pevney, Lloyd Goff, Art Smith, James Burke, Virginia Gregg, Peter Virgo, Joe Devlin, Mary Currier, Milton Kibbee, Artie Dorrell
Classic boxing film that follows the career of a young man who sets out to reach the top of his profession but is corrupted in the process. He finally regains his self respect when, at the end of his career, he beats a young contender for his title after having previously agreed to take a dive. This tough, final match is one of the best ever put on screen. Written by Abraham Polonsky and remade in 1981. AA: Edit (Francis Lyon/Robert Parrish).
DRA 104 min B/W VIDrel: SECOND/RTM V

BODY AND SOUL * 18
George Bowers USA 1947
Leon Isaac Kennedy, George Bowers, Jayne Kennedy, Peter Lawford, Michael V. Gazzo, Muhammad Ali, Perry Lang, Kim Hamilton
A dreary remake of the 1947 classic in which a welterweight resists the corruption of the fight world (this time round) and goes on to become the champion. Kennedy does his best, but this thin storyline is his toughest opponent.
DRA 95 min VIDrel: RNK L/A V

BODY BAGS *   *(18)*
John Carpenter/Tobe Hooper USA   1992
John Carpenter, Tom Arnold, Tobe Hooper, Robert Carradine, Peter Jason, Molly Cheek, Wes Craven, Sam Raimi, David Naughton, Stacy Keach, David Warner, Sheena Easton, Dan Blom, Deborah Harry, Mark Hamill, Twiggy, John Agar
A trio of horror stories, the first two of which were directed by Carpenter and the last one by Hooper. In "The Gas Station" a nut threatens the young girl who works there at nights, in "Hair" a man is so frightened by the prospect of baldness that he is prepared to try any treatment and in "The Eye" a sportsman who loses his eye has a transplant from a serial killer, with the usual result. Much gore and immature humour but nothing remotely frightening is on offer here.
HOR   91 min (ort 95 min) mCab   SATrel: SKY MOVIES

BODY CHEMISTRY: A DEADLY EXPERIMENT **   18
Kristine Peterson USA   1990
Marc Singer, Mary Crosby, Lisa Pescia, David Kagen, H. Bradley Barneson, Doreen Alderman, Laureen Tuerk, Joseph Campanella
The married director of a research lab investigating sexual behaviour is seduced by a psychopathic female colleague and embarks on a relationship from which he is unable to break free. A blatant clone of FATAL ATTRACTION, well handled but all too familiar. Several sequels followed.
DRA   80 min (ort 87 min)   VIDrel: TRING/COLUM   V

BODY CHEMISTRY 2: VOICE OF A STRANGER **
18
Adam Simon USA   1991
Lisa Pescia, Gregory Harrison, Morton Downey Jr, Robin Riker
A radio phone-in doctor is outwardly well balanced but inwardly psychotic, and when a disturbed former police officer phones her for some advice she sets about using him to live out her fantasies.
THR   80 min (ort 84 min)   VIDrel: 20VIS/SONOP   V

BODY CHEMISTRY 3: POINT OF SEDUCTION **   18
Jim Wynorski USA   1993
Andrew Stevens, Morgan Fairchild, Robert Forser, Stella Stevens
After his friend dies a tragic death, a screenwriter sees a chance of writing a hit film script built around this event. However, he needs the permission of the former's widow and realises that he will have to seduce her and thus betray his own wife.
DRA   83 min (ort 95 min)   VIDrel: HIFLI/SONOP   V/h

BODY CHEMISTRY 4: FULL EXPOSURE **   18
Andrew Stevens USA   1995
Shannon Tweed, Larry Poindexter, Stella Stevens, Chick Vennera, Larry Manetti, Andrew Stevens
A defence attorney accepts the case of a woman psychologist who is accused of murder and unwisely starts an affair with her that could ruin both his marriage and his life.
DRA   89 min (ort 92 min)   VIDrel: HIFLI/SONOP   V/h

BODY DOUBLE **   18
Brian De Palma USA   1984
Craig Wasson, Deborah Shelton, Melanie Griffith, Gregg Henry, Guy Boyd, Dennis Franz, David Haskell, Rebecca Stanley, Al Israel, Douglas Warhit, B.J. Jones, Russ Marin, Lane Davies, Barbara Crampton, Larry Jenkins, MOnte Landis
An unemployed actor spies on a woman who performs a striptease in front of her bedroom window every night. Her sudden murder with a drill is witnessed by him and he soon finds himself drawn into a complex and tangled web of murder. A sleazy film of fetishes and porno movies that has a few good moments even if it does derive its inspiration from Hitchcock's REAR WINDOW. Co-produced and co-written by De Palma. The music is by Frankie Goes To Hollywood.
THR   109 min Cut (5 sec – ort 114 min)
VIDrel: VCC/DISC/COLUM   V/sur

BODY HEAT ***   18
Lawrence Kasdan USA   1980
William Hurt, Kathleen Turner, Richard Crenna, Ted Danson, J.A. Preston, Mickey Rourke, Kim Zimmer, Jane Hallaren, Lanna Saunders, Michael Ryan, Carola McGuinness, Larry Marko, Deborah Luchessi, Lynn Hallowell, Ruth Thom
Somewhat reminiscent of DOUBLE INDEMNITY and THE POSTMAN ALWAYS RINGS TWICE, with a rich married socialite inveigling a Florida lawyer into a plot to kill her husband. Though somewhat over-derivative, it improves as it develops. This was the directorial debut for Kasdan (he also wrote the script) and the screen debut for Turner.
DRA   108 min (ort 113 min)   VIDrel: WHV   V

BODY LANGUAGE **   PG
Arthur Allan Seidelman USA   1992
Heather Locklear, Linda Purl, James Acheson, Edwart Albert, Gary Bissig, Jeff Kizer, Denise Dal Vera, Tim Prulhiere, Russ Fast, Juanita Wyndham, Dennis Bateman, Corey Brunish, Sadie Veraldi, Marianne Doherty, Karen Trumbo
A hard-pressed female executive hires a new secretary who soon proves most adept at her job, but what her boss does not know is that she is a psychotic who has made sinister plans for her own advancement. Another standard variation on the psycho-from-hell theme that ends in the usual bloody way. See also LETHAL CHARM.
THR   88 min (ort 93 min) mCab   VIDrel: CIC   V

BODY LANGUAGE **   18
George Case USA   1995
Tom Berenger, Heidi Schanz, Nancy Travis
A defence lawyer has his hands full representing a Mafia criminal, but on the way home is involved in a car accident with a beautiful woman. This encounter leads to a relationship with the latter, who works as a stripper, and bit by bit he is drawn into a plan she has made to murder her boyfriend. A tangled erotic drama that offers us very little in the way of plotting except a bedroom scene ever twenty minutes or so.
DRA   96 min (ort 100 min) mCab   VIDrel: RYSHER/HIFLI   V

BODY MELT **   18
Philip Brophy AUSTRALIA   1993
Gerard Kenendy, Andrew Daddo, Ian Smith, Vince Gil, Regina Gaigalas, Neil Foely, Anthea Davis, Matt Newton, Lesley Baker, Adrian Wright, Julian Murray, Brett Climon, Lisa McCure, Nick Polites, Maurie Annese, Robert Simper
A scientist invents an experimental drug but finds out too late that it has unfortunate side-effects, when his arrogance leads him to use the inhabitants of a small town as his guinea pigs, only to find that they literally start to melt. A frenzied attack on the drugs industry that relies almost entirely on its revolting special effects (it puts one in mind of STREET TRASH). A poor effort indeed that shows little intelligence or originality.
HOR   79 min (ort 81 min)   VIDrel: FIRST/SONOP   V/sur

BODY OF EVIDENCE **   18
Roy Campanella II USA   1987
Margot Kidder, Barry Bostwick, Tony LoBianco, Caroline Kava, Jennifer Barbour, David Hayward, Debbie Carr, Peter Bibby, Garwin Sanford, Don Davis, George Collins, Blu Mankuma, Don MacKay, Bill Croft, Karen Tilly, Jane McDougal
A pathologist involved in the hunt for a serial killer comes under suspicion from his wife who gradually begins to suspect that he and the murderer could very well be one and the same. Something of a revamp of SUSPICION with a few good moments in a generally poor script. Written by Cynthia Whitcomb and Campanella.
THR   96 min (ort 100 min) mTV   VIDrel: 20TH/TECH   V

BODY OF EVIDENCE *   18
Uli (Ulrich) Edel USA   1992
Madonna, Willem Dafoe, Joe Mantegna, Anne Archer, Julianne Moore, Jurgen Prochnow, Michael Forest, Charles Hallahan, Mark Rolston, Richard Riehle, Frank Langella, Stan Shaw, D. Scott Douglas, Mario DePriest, John DeLay, Frank Roberts
A woman goes on trial accused of having used sex to kill her wealthy, elderly lover and thus benefit from his will. Inevitably, her lawyer finds himself becoming increasingly involved with her. A confused and distasteful mess, poorly acted and unscripted, and with some highly unpleasant sado-masochistic sex scenes that are not really related to the plot, such as it is.
DRA   961 min (ort 141 min)
VIDrel: POLY/POLYREC/GUILD   V/sur

BODY OF INFLUENCE 2 **   18
Brian J. Smith USA   1996
Jodie Fisher, Dan Anderson, Jonathan Goldstein, Pat Brennan
A successful psychiatrist remains haunted by a secret tragedy in his past, but this is revealed when his involvement with a

seductive female patient traps him in a web of lies and intrigue. Quite a clever erotic thriller.
THR   90 min   VIDrel: HIFLI/SONOP   V/h

BODY PARTS ***
18
Eric Red USA
1991
Jeff Fahey, Brad Dourif, Lindsay Duncan, Kim Delaney, Peter Murnik, Paul Benvictor, Sarah Campbell, Lindsay Merrithew, Arlene Duncan, Allan Price, Hal Eisen, Taia Red, James Kidnie, Andy Humphrey, John Walsh
A criminal psychologist loses his arm in a car crash, and is given a replacement one from an unknown donor. However, this grafted limb proves to have a life of its own and embroils the recipient in violence and horror, while he struggles to learn the identity of the donor. This variant on "The Hands Of Orlac" has some excellent if gory effects, but in the main does not benefit from its over-lavish budget. See also THE HAND.
HOR   84 min (ort 88 min)   VIDrel: CIC/SONOP   V/sur
Boa: novel Choice Cuts by Pierre Boileau and Thomas Narcejac.

BODY PUZZLE **
18
Lamberto Bava ITALY
1991
Joanna Pacula, Tomas Arana, Francoise Montagut, Gianni Garko, Erika Blanc
A police commissioner investigates a series of murders in which the bodies of the victims were mutilated and certain organs removed. He soon makes the startling discovery that they were all recipients of organs from the body of the same man, the demented aim of the culprit being to put the dead man back together again.
HOR   90 min (ort 95 min)   VIDrel: ARTPRO/RTM   V

BODY SHOT **
18
Dimitri Logothetis USA
1993
Robert Patrick, Michelle Johnson, Ray Wise, Jonathan Banks, Kim Miyori, William Steis
A fashionable photographer becomes the prime suspect in the murder of a rock star when the police learn that he did a layout for a lookalike and was also obsessed with the dead woman. In order to clear himself, he attempts to track down the killer and makes good use of his professional skills in this quest.
THR   94 min   VIDrel: GUILD/FOXVID   V/sh

BODY SNATCHER, THE ***
PG
Robert Wise USA
1945
Boris Karloff, Henry Daniell, Bela Lugosi, Edith Atwater, Russell Wade, Rita Corday, Shavyn Moffett, Robert Clarke, Mary Gordon, Carl Kent, Donna Lee, Jim Moran, Larry Wheat, Ina Constant, Jack Welch, Bill Williams
Set in Scotland in the 19th century, this tells of a doctor who is forced to employ a bodysnatcher in order to obtain the corpses he requires for his experiments. He eventually finds himself being blackmailed by a villainous coachman, in this atmospheric adaptation of Stevenson's chiller.
HOR   77 min B/W   VIDrel: VCC L/A   V
Boa: story by Robert Louis Stevenson.

BODY SNATCHERS **
15
Abel Ferrara USA
1993
Gabrielle Anwar, Meg Tilly, Forest Whitaker, Billy Wirth, Terry Kinney, Christine Elise, R. Lee Ermey, G. Elvis Phillips, Reilly Murphy, Kathleen Doyle, Stanley Small, Tonea Stewart, Keith Smith, Winston E. Grnat, Phil Neilson
This second remake of INVASION OF THE BODY SNATCHERS is set at a military base where a young girl discovers to her horror that people around are being replaced by emotionless duplicates. Apart from the excellent special effects, this one has little to offer and is oddly lacking in tension, while the violent shootout ending is a major disappointment.
Aka: BODY SNATCHERS: THE INVASION CONTINUES
HOR   83 min (ort 87 min)   VIDrel: WHV   V/sur

BODYGUARD, THE **
15
Mick Jackson USA
1992
Kevin Costner, Whitney Houston, Bill Cobbs, Gary Kemp, Michele Lamar Richards, Ralph Waite, Tomas Arana, Mike Starr, Devaughn Nixon, Christopher Birt, DeVaughn Nixon, Charles Keating, Robert Wuhl, Debbie Reynolds, Ethel Ayler
A former Secret Service agent now working as a private bodyguard accepts an assignment to protect a beautiful singer whose life is being threatened. Against his better judgement (and professional code) he soon sleeps with her, in an overblown, overlong and highly romantic tale. A high point in the film (and

the plot) is Houston singing "I Will Always Love You", but elsewhere the noisy pop score is a distraction and an annoyance. This was Houston's film debut.
DRA   124 min (ort 130 min) wScrn cC   VIDrel: WHV
V/dm V/sur

BOGEY MAN, THE ***
18
Ulli Lommel USA
1980
Suzanna Love, Ron James, John Carradine, Nicholas Love, Raymond Boyden, Felicite Morgan, Bill Rayburn, Llewelyn Thomas, Jay Wright, Natasha Schiano, Howard Grant, Lucinda Ziesing, Jane Pratt, David Swim, Catherine Tambini
A woman is haunted by the spirit of her mother's lover, who was murdered years ago by her now mute brother. Returning to her childhood home in an attempt to purge herself of this trauma, she comes across a mirror which reflects an image of the long-dead lover. She smashes it, but a fragment is imbued with evil power and causes much bloodshed. An intriguing and imaginative fantasy-horror, followed in 1983 by a sequel (banned so far).
Aka: BOGEYMAN, THE; BOOGEY MAN, THE
HOR   79 min Cut (44 sec – ort 91 min)
VIDrel: VIPCO/SGSVID   V

BOILING POINT ***
18
Takeshi Kitano JAPAN
1990
Masahiko Ono, Yuriko Ishida, Takahito Iguchi, Minoru Iizuka, Hitoshi Ozawa, Makoto Ashikawa, Takeshi Kitano, Hisashi Igawa, Bengal, Takahiko Aoki, Hiroshi Suzuki, Kenzo Matsuo, Hiroshi Ide, Tsuneo Serizawa, Johnny Okura
Written and directed by Kitano, this is the story of a young man who works at a gas station and his involvement with gangsters, which comes about when an argument with a customer is resolved with blows. His ex-Yakuza friend attempts to help him, but is beaten up for his pains, and matters conspire to draw the man into an alliance with one of the nastiest gangsters of the lot (Kitano). An odd and memorable blending of violent action and strongly comic elements.
Aka: SAN TAI YON X JUGATSU
A/AD   92 min (ort 96 min)   VIDrel: ICAPRO/MANGA
V/sh

BOILING POINT ***
15
James B. Harris USA
1992
Wesley Snipes, Dennis Hopper, Lolita Davidovich, Dan Hedaya, Valerie Perrine, Viggo Mortensen, Seymour Cassel, Jonathan Banks, Tony LoBianco, James Tolkan, Paul Gleason, Lorraine Evanoff, Stephanie Williams, Christine Elise
A Federal agent goes after the criminal responsible for the death of his partner in a failed drugs bust, in this high-powered action tale that is relentlessly violent and dark in tone. Hopper reprises his standard psycho role and Snipes gives an acceptable performance, but the cliched and contrived plot is a major drawback.
A/AD   88 min (ort 93 min)
VIDrel: 4-FRONT/POLYREC/GUILD   V/sh
Boa: novel Money Men by Gerald Petievich.

BOLERO
18
John Derek USA
1984
Bo Derek, George Kennedy, Andrea Occhipinti, Ana Obregon, Olivia d'Abo, Greg Bensen
A film of mind-numbing stupidity, bad dialogue and worse acting, made by the director as a vehicle with which to show off his pretty wife. Set in the 1920s, it follows a spoiled, rich graduate of an English girls school who, with the help of her chauffeur, is determined to lose her virginity to a handsome Arab sheikh. But he fails her by falling asleep, so she hurries to Spain where a bullfighter eventually helps her out in this laughable effort.
A   100 min   VIDrel: L/A   V

BONANZA: THE RETURN **
(PG)
Jerry Jameson USA
1993
Ben Johnson, Richard Roundtree, Michael Landon Jr., Emily Warfield, Brian Leckner, Alistair MacDougall, Jack Elam, Dirk Blocker, David Sage, Stewart Moss, Dean Stockwell, Linda Gray, John Ingle, Archie Lang, Richard Fullerton
When an unscrupulous land developer tries to gain control of the Ponderosa, the next generation of Cartwrights show that they have not lost the family touch for concerted action. A amiable if none-too-original effort, probably intended as a pilot for a revival of this popular series.
WES   89 min (ort 95 min) mTV   SATrel: SKY MOVIES

**BONDS OF LOVE ** **   PG
Michael Anderson USA   1988
Burt Lancaster, Ben Cross, Olivia Hussey
Romantic drama based on Wojtyla's novel of two couples who
are separated by WW2 and the German invasion of Poland, and
do not meet until many years later, when in New York they find
that their respective children have embarked on a love affair of
their own. Average.
DRA 88 min VIDrel: FUTUR L/A V
Boa: novel The Jeweller's Shop by Karol Wojtyla (Pope John
Paul II).

**BONDS OF LOVE ** **   15
Larry Elikann USA   1992
*Kelly McGillis, Treat Williams, Hal Holbrook, Steve Railsback, Grace
Zabriskie, Kenneth Walsh*
Fact-based drama chronicling the friendship and growing
romantic involvement between a slightly retarded young man
(who lives with his parents) and a former battered wife, who is
attempting to make a new life for herself. The Mel Gibson film
TIM also covered much the same ground.
DRA 90 min VIDrel: 4-FRONT/POLYREC/ODY V/sh

BOUDU SAVED FROM DROWNING * **   PG
Jean Renoir FRANCE   1932
*Michael Simon, Charles Grandval, Jean Daste, Marcelle Hainia,
Severine Lerczinska, Jacques Becker*
A tramp unwillingly rescued from drowning by as bookseller
goes on to seduce the latter's wife and mistress, who unwisely
offer him hospitality. An anarchic and vigorous black comedy,
whose moments of trenchant humour mostly derive from the
stuffy middle-class manners and attitudes of the family.
Aka: BOUDU SAUVE DES EAUX
COM 80 min (ort 82 min) B/W VIDrel: ARTIF/20TH
V/h

**BONFIRE OF THE VANITIES, THE ** **   15
Brian De Palma USA   1990
*Bruce Willis, Melanie Griffith, Tom Hanks, Morgan Freeman, Kim
Cattrall, John Hancock, Saul Rubinek, Kevin Dunn, Clifton James,
Louis Giambalvo, Barton Heyman, Donald Moffat, Alan King, Mary
Alice, Beth Broderick, Kurt Fuller*
With both a wife and an attractive mistress, a cocky young
Wall Street trader enjoys the good things in life. However, his
world collapses when his wife learns of his infidelity and he
injures a destitute black youth in a hit-and-run accident. Hanks
is badly miscast in this clinical and empty adaptation of a
caustic novel whose subtlety and wit are replaced by carica-
ture and flashy camerawork. The disappointing script is by
Michael Cristofer.
COM 120 min (ort 126 min) VIDrel: WHV V/sur
Boa: novel by Tom Wolfe.

BONNIE AND CLYDE ** **   18
Arthur Penn USA   1967
*Warren Beatty, Faye Dunaway, Michael J. Pollard, Denver Pyle, Gene
Hackman, Estelle Parsons, Gene Wilder, Dub Taylor, Evan Evans,
James Stiver, Clyde Howdy, Garry Goddgion, Ken Meyer*
A slick, stylish but heavily glamorised story of Bonnie Parker
and Clyde Barrow – two of the most notorious gangsters to
operate in the 1930s, when they specialised in robbing banks.
Perceptive, funny and violent – this potent mixture is directed
with enormous flair and has great performances from the leads.
The climax is almost unbelievably ferocious. Wilder's screen
debut. AA: S. Actress (Parsons), Cin (Burnett Guffrey).
DRA 106 min VIDrel: WHV L/A V/h

**BONNIE AND CLYDE: OUTLAWS OF LOVE ** **   18
Paul Thomas USA   1992
*Ashlyn Gere, Randy West, Raquel Darian, Derick Lane, Francesca Le,
Nicky Dial, Alex Jordan, Mickey Rae, Nick E. T.T. Boy, Jonathan
Morgan*
Erotic version of the famous tale, with West and Gere playing
the parts that were taken by Beatty and Dunaway in the origi-
nal film. On the run after a bank job, the hole up in a sleepy
town, where they inevitably (this being a sex film) get drawn
into various frolics and orgies with the over-sexed inhabitants.
But when some cops turn up, an obliging hooker keeps them
busy while our robbers get away. Surplus footage from this film
was used to cobble together a "sequel".
Aka: OUTLAWS OF LOVE
A 70 min (ort 94 min) VIDrel: GROHOM/MAXSCAN V

**BONNIE AND CLYDE: THE TRUE STORY ** **   15
Gary Hoffman USA   1992
Dana Ashbrook, Tracey Needham, Michael Bowen, Betty Buckley
Adequate tale of two notorious figures from history, whose
liking for violence and lack of morality resulted in a crime-spree
that ultimately led to their deaths in a police ambush.
A/AD 90 min mTV VIDrel: 20TH V/sh

**BOOK OF LOVE ** **   15
Robert Shaye USA   1990
*Michael McKean, Chris Young, Keith Coogan, Tricia Leigh Fisher,
Aeryk Egan, Josie Bissett, Danny Nucci, John Cameron Mitchell, Beau
Dremann, Jill Jaress, John Achorn, Michael McKean, Michael
Cavalieri, Gary Ellenberg, Brent Fraser*
After his recent divorce, a man looks back to his teenage years
and his first romance. The cast does its best to breathe life into
the scanty story, but the main interest lies in the music, with no
less than 32 oldies being played. Written by Kotzwinkle from his
novel.
COM 84 min (ort 87 min) VIDrel: COLUM/SONOP V/sur
Boa: novel Jack in the Box by William Kotzwinkle.

**BOOMERANG ** **   15
Reginald Hudlin USA   1992
*Eddie Murphy, Robin Givens, Halle Berry, David Alan Grier, Martin
Lawrence, Grace Jones, Geoffrey Holder, Eartha Kitt, Chris Rock,
Tisha Campbell, John Witherspoon, Melvin Van Peebles, Bebe Drake
Massey, John Canada Terrell*
An advertising executive whose attitude towards women is both
shallow and self-centred, tries his charm on his new boss but is
shocked almost beyond belief to discover that she treats men in
a very similar way. A coarse and vulgar sex comedy on the
theme of role reversal that although quite amusing at times,
would have benefited from a more subtle approach.
COM 112 min (ort 118 min) cC
VIDrel: 4-FRONT/POLYREC/CIC V/sur

**BOOST, THE ** **   18
Harold Becker USA   1988
*James Woods, Sean Young, John Kapelos, Steven Hill, Amanda Blake,
Kelle Kerr, John Rothman, Grace Zabriskie, Marc Poppel, Fred
McGarren, Suzanne Kent, Libby Boone, Greg Deason, David Preston,
June Chandler, Edith Fields*
A real-estate developer loses his passport to the good life when
the economy goes into recession, and resorts to cocaine to dull
the pain. A dreary, anti-drug film whose message far outshines
both performance and plot.
DRA 91 min (ort 95 min) VIDrel: COLUM/VCC L/A V/sh
Boa: book Ludes by Benjamin Stein.

**BOOTS AND SADDLES ** **   U
Joseph Kane USA   1937
*Gene Autry, Judith Allen, Smiley Burnette, Gordon (William) Elliott,
John Ward, Chris-Pin Martin, Frankie Marvin, Bud Osborne, Stanley
Blystone, Merrill McCormack*
A ranch foreman helps a British youngster love the outdoors life
and falls for a colonel's daughter in this colourful early Western
with a pleasant mix of comedy, romance and action, culminat-
ing in an exciting horse race finale.
WES 54 min B/W VIDrel: SCRN/DISC V

BOPHA! * **   15
Morgan Freeman USA   1993
*Danny Glover, Alfre Woodard, Malcolm McDowell, Marius Weyers,
Maynard Eziashi, Malick Bowers, Michael Chinyamurinoi
Christopher John Hall, Robin Smith, Grace Mahlaba, Julie Strijdom,
Peter Kampla, Sello Naake Ka-Ncube*
The son of a black township policeman who has a passionate
attachment to the rule of law, rebels against his father and joins
the growing activist movement. The arrest of a prominent black
member of a freedom movement and the machinations of the
South African secret police soon create a situation of fear and
violence that tears father and son apart. A powerful and moving
drama, set in the Apartheid era around 1980. See also CRY
FREEDOM and A DRY WHITE SEASON.
DRA 113 min (ort 120 min) cC VIDrel: CIC/SONOP
V/sur
Boa: play by Percy Mtwa.

**BORDER, THE ** **   18
Tony Richardson USA   1982
*Jack Nicholson, Harvey Keitel, Valerie Perrine, Warren Oates, Elpidia
Carrillo, Shannon Wilcox, Manuel Viescas, Jeff Morris, Mike Gomez,*

Lonny Chapman, Dirk Blocker, Stacey Pickren, Floyd Levine, James Jeter, Alan Fudge

A lukewarm drama about the corruption of an honest border guard who begins accepting bribes to allow in Mexican illegal immigrants. When the child of a poor Mexican girl is kidnapped to be sold on the black market, his sense of decency is put to the test. A film that needed stronger direction and a tighter script to really work.

DRA   107 min   VIDrel: CIC/SONOP L/A   V

BORN FREE **   *U*
James Hill UK   1966
Virginia McKenna, Bill Travers, Geoffrey Keen, Peter Lukoye, Omar Chambati, Bill Godden, Bryan Epsom, Robert Cheetham, Robert Young, Geoffrey Best, Surya Patel

Fair adaptation of Adamson's book based on her experiences working together with her husband, as game wardens in Kenya. Having raised a lion cub in captivity, they are obliged to teach it to fend for itself before releasing it into the wild. A syrupy outdoors adventure saved by the colourful locations and splendid animal shots. LIVING FREE followed. AA: Score/orig (John Barry), Song ("Born Free" – John Barry (m)/Don Blake (l)).

DRA   91 min (ort 96 min)   VIDrel: VCC/DISC/COLUM   V
Boa: book by Joy Adamson.

BORN FREE: A NEW ADVENTURE **   *U*
Tommy Lee Wallace USA   1996
Christopher North, Linda Purl, Jonathan Brandis

In the heart of Africa, Elsa the lioness makes friends with two unhappy teenagers, who are attempting to come to terms with a family move from their downtown Chicago home.

JUV   92 min   VIDrel: ODY/SONOP/COLUM   V

BORN IN EAST L.A. **   *15*
Cheech Marin USA   1987
Cheech Marin, Daniel Stern, Paul Rodriguez, Jan-Michael Vincent, Tony Plana, Kamala Lopez, Alma Martinez, Mike Moroff, Eloy Casados, Terrence Evans, Ted Lin, Sal Lopez, Jason Scott Lee, Jee Teo, Bob McClurg, Diane Bellamy, Damie Valdez

A third-generation American Hispanic who gets caught in an immigration raid without any identification, is taken to be an illegal immigrant and deported to Tijuana. Based on Marin's successful Bruce Springsteen parody record, this follows his crazy antics to get back into the States. A pleasant comedy vehicle whose humour begins to wear thin after the first thirty minutes.

COM   81 min   VIDrel: CIC/SONOP   V/h

BORN KICKING **   *(PG)*
Mandie Fletcher UK   1992
Eve Barker, Denis Lawson, Sheila Ruskin, Julie Hewlett, Carole Hayman, George Irving, John Abineri, Norman Bird, David McAllister, Garfield Morgan, Paul De Bois, James Noble, Robert Perkins, Anthony Warren, John Benfield

A schoolgirl with a great talent for football gets a chance to prove her worth when she becomes something of a media celebrity, but inevitably faces the predictable sexist attitudes towards her sporting prowess. A light and rather superficial film, it tries to make a few valid points but is badly spoilt by the totally contrived ending.

COM   85 min mTV   TVrel: BBC

BORN ON THE FOURTH OF JULY ****   *18*
Oliver Stone USA   1989
Tom Cruise, Willem Dafoe, Raymond J. Barry, Caroline Kava, Kyra Sedgewick, Bryan Larkin, Jerry Levine, Josh Evans, Frank Whaley, Stephen Baldwin, John Getz, Tom Berenger, Abbie Hoffman

The story of anti-war activist Ron Kovic, who joined the Marines as a raw recruit in the 1960s, saw action in Vietnam and came home a cripple, paralysed from the chest down. Based on his memoirs, the film charts his early patriotism, later disillusionment and final ordeal of mental and physical rehabilitation. Cruise is superb in this fine film. The score is by John Williams. Followed by HEAVEN & EARTH. AA: Dir, Edit (David Brenner/Joe Hutshing).

DRA   138 min (ort 144 min) wScrn   VIDrel: CIC/SONOP; PION (LV only)   V/dm LV
Boa: book by Ron Kovic.

BORN TO BE WILD **   *U*
John Gray USA   1995
Wil Horneff, Helen Shaver, John C. McGinley, Peter Boyle, Jean Marie Barnwell, Marvin J. McIntyre, Gregory Itzin, Titus Welliver, Thomas

F. Wilson, Alan Ruck, Janet Carroll, John Procaccino, Obba Babatunde, Keith Swift

A troubled teenager is given a cleaning job at the animal research station where his mother works and gets friendly with a gorilla who is kept there. When he learns that the ape is to sent to a circus, he decides that his new-found friend ought to have his freedom, and this ill-matched pair escape to Canada but a court battle ensues in which the ape pleads her case using sign language. A variant on FREE WILLY, bright and quite likeable, if not especially memorable.
Aka: KATIE

JUV   94 min (ort 99 min)   VIDrel: WHV   V/sur

BORN TO FIGHT **   *18*
Godfrey Ho HONG KONG   1991
Cynthia Luster, Ron Van Lee

A couple of tough cops investigate the disappearance of a number of prostitutes and an apparent link with a drugs ring. Standard fisticuffs and general mayhem.

MAR   88 min   VIDrel: IMPENT   V

BORN TO RUN **   *18*
Albert Magnoli USA   1993
Richard Grieco, Joe Cortese, Jay Acovone, Shelli Lether, Christian Campbell, Tony Romano, Brent Stait, Joe Cortese, Martin Cummins, Wren Roberts, Roger R. Cross, Ken Kerzinger, Tony Romano, Alan C. Peterson, John Novak, James Crescenzo

A Brooklyn street-racer sets out to rescue his brother from the clutches of mobsters and along the way falls for the girlfriend of one of these unsavoury characters. A predictable and trite effort.

THR   93 min (ort 97 min) mTV   VIDrel: 20TH/TECH L/A
V

BORN TOO SOON **   *15*
Noel Nossek USA   1992
Pamela Reed, Michael Moriarty, Terry O'Quinn, Joanna Gleason, Mariangelo Pino, Elizabeth Ruscio

A successful couple looking forward to the birth of their first child have to cope with one that is four months premature, their troubles being compounded by the knowledge that it may not survive. A competent true-life TV drama, quite thoughtful and restrained.

DRA   92 min mTV   VIDrel: 4-FRONT/POLYREC/ODY
V/sh
Boa: book by Elizabeth Mehren.

BORN YESTERDAY ***   *U*
George Cukor USA   1950
Judy Holliday, William Holden, Broderick Crawford, Howard St. John, Frank Otto, Larry Oliver, Barbara Brown, Grandon Rhodes, Claire Carleton, Smoki Whitfield, Helyn Eby Rock, William Mays, David Pardoll, Mike Mahoney

A wealthy junk dealer hires a struggling writer to add a little polish to his rough diamond of a girlfriend, but this simple plan has far-reaching and devastating consequences, as the worm turns with a vengeance. A tour-de-force performance from Holliday, who re-creates her stage role, and an equally good one by Crawford as her bluff paramour, make this a sheer joy. Remade in 1992. AA: Actress (Holliday).

COM   98 min (ort 103 min) B/W   VIDrel: COLUM/SONOP
V
Boa: play by Garson Kanin.

BORN YESTERDAY *   *PG*
Luis Mandoki USA   1992
Melanie Griffith, Don Johnson, John Goodman, Edward Herrmann, Max Perlich, Fred Dalton Thompson, Nora Dunn, Benjamin C. Bradlee, Sally Quinn, Michael Ensign, William Frankfather, Celeste Yarnall, Meg Wittner, William Forward

An ambitious tycoon decides that his dumb bimbo of a mistress is in serious need of a little sophistication and engages a reporter to provide her with the rudiments of an education. However, his little scheme backfires in a way he has failed to predict when she starts getting too smart for his own good. Goodman attacks his role with gusto but his performance alone cannot save this dispirited remake of the 1950 classic.

COM   96 min (ort 100 min)   VIDrel: HOLPIC/TECH L/A   V
Boa: play by Garson Kanin.

BORROWER, THE *   *18*
John McNaughton USA   1991
Rae Dawn Chong, Antonio Fargas, Don Gordon, Tom Towles, Neil Guintoli, Larry Pennell, Pam Gordon, Tony Amendola, Robert Dryer,

Richard Wharton, Bentley Mitchum, Teri Betzler, Tami Clatterback, Tom Allard, Darryl Shelly

A criminal on an alien planet is sentenced to spend the rest of its life in exile, banished in human form to Earth. The limitations of this "genetic devolution" process means that any damage to the creature causes it to revert to its original form. This does indeed happen, and the creature is obliged to keep "borrowing" the heads of various unfortunate humans. A most intriguing idea, that gets lost in a mess of revolting effects, and is hampered by poor acting.

HOR 88 min (ort 97 min) VIDrel: MGM/WHV L/A V/sh

BORSALINO ***
Jacques Deray FRANCE/ITALY
15
1970

Jean-Paul Belmondo, Alain Delon, Michael Bouquet, Catherine Rouvel, Corinne Marchand, Francoise Christophe, Julian Guimoar, Arnoldo Foa, Nicole Calfan, Laura Adani, Christian De Tiliere, Mario David, David Ivernal, Dennis Berry

A stylish and witty film about the growing friendship between two 1930s Marseilles gangsters who decide to join forces. A semi-comic film with the leads giving great performances as two rather likeable hoodlums. The catchy score is by Claude Bolling. A kind of French tribute to American gangster films of the period. The inferior "Borsalino And Co." followed.

DRA 119 min (ort 126 min) VIDrel: CIC/SONOP L/A V
Boa: The Bandits of Marseilles by Eugene Saccomano.

BOSS, THE **
Wesley Emerson USA
18
1994

Tiffany Million, Nicole London, Tiffany Minx, Crystal Wilder, Mike Horner, Randy West, Jonathan Morgan, Terry Thomas

Two aspiring starlets use all the tricks and wiles at their disposal when they find themselves competing for the same part in a new film. Average.

A 48 min (ort 80 min) VIDrel: ONE V

BOSS'S WIFE, THE *
Ziggy Steinberg USA
15
1986

Daniel Stern, Arielle Dombasle, Fisher Stevens, Melanie Mayron, Lou Jacobi, Martin Mull, Christopher Plummer, Thalmus Rasulala, Robert Costanzo

An ambitious young stockbroker, trying to advance his career, has to deal with a demanding boss, a sleazy fellow employee and the oversexed wife of his boss. Disappointing comedy that never really catches fire, despite the best efforts of a strong cast.

COM 80 min (ort 83 min) VIDrel: 20TH/TECH V/sur

BOSTON KICKOUT ***
Paul Hills UK
18
1995

John Simm, Emer McCourt, Marc Warren, Andrew Lincoln, Richard Hanson, Nathan Valente, Derek Martin, Vincent Phillips, Natalie Davies, David Aldous, Sally Grace, Julie Smith, Jeanette Driver, Suzanne Church, Arran Hall, John Rafferty

In the 1980s a youngster moves with his parents from their inner-city home to Stevenage, but on leaving school drifts into drug abuse and crime. The title refers to a nasty game played by the local disaffected youths, who enjoy trampling their way through a series of suburban gardens. Realistic and harsh, this is mostly a study of youthful alienation, but boasts performances of vigour and warmth. A promising debut feature for the director.

DRA 101 min (ort 105 min) VIDrel: FIRST/SONOP V

BOSTON STRANGLER, THE ****
Richard Fleischer USA
18
1968

Tony Curtis, Henry Fonda, George Kennedy, Mike Kelin, Hurd Hatfield, Murray Hamilton, Jeff Corey, Sally Kellerman, William Marshall, George Voskovec, James Brolin, William Hickey, Leora Dana, Carolyn Cornwell, Jeanne Coope

Tony Curtis is superb in this low-key, almost documentary account of the mass murderer who terrorised Boston in the 1960s. A totally absorbing drama, documenting his appearance, career and eventual capture, with a look at his motivations. A utterly chilling film, but the use of complex multi-images will tend to be lost on TV.

DRA 113 min Cut (1 min 5 sec – ort 120 min)
VIDrel: 20TH/TECH V/h
Boa: book by Gerold Frank.

BOSTONIANS, THE ***
James Ivory UK
PG
1984

Christopher Reeve, Vanessa Redgrave, Madeleine Potter, Jessica Tandy, Nancy Marchand, Barbara Bryne, Wesley Addy, Linda Hunt,
Maura Moynihan, Wallace Shawn, Nancy New, Charles McCaughan, John Van Ness Philip, Martha Farrar

Adaptation of the Henry James story of Boston society in the 19th century, with Redgrave well cast as the feminist heroine who tries to interest naive Potter in the women's movement, whilst Southern gent Reeve, vies for her attentions. A slow, literate and somewhat aloof study. The script is by Ruth Prawer Jhabvala.

DRA 117 min (ort 122 min) VIDrel: CARL/TECH V/sur
Boa: novel by Henry James.

BOTTLE ROCKET **
Wes Anderson USA
15
1995

Luke Wilson, Owen C. Wilson, James Caan, Robert Musgrave

Three inept aspiring thieves attempt to embark on a life of crime, but with very little success. Good performances buoy this one up and it moves along quite cheerfully, despite the thin plot.

COM 88 min (ort 91 min) VIDrel: 20VIS/SONOP V/sur

BOULEVARD **
Penelope Buitenhuis CANADA
18
1994

Karin Wuhrer, Lou Diamond Phillips, Rae Dawn Ching, Lance Henriksen, Judith Scott, Joel Bissonnette, Amber Lea Weston, Greg Campbell, Keram Malicki-Sanchez, Katie Griffen, Marcia Bennett, Michael Kramer, Linlyn Lue, James Loxley

An experienced hooker teaches a newcomer how best to survive on the streets and the two of them find themselves involved in a violent conflict between a pimp and a drugs baron. A violent and gritty actioner.

A/AD 96 min VIDrel: MIA/DISC V/h

BOUND ***
Larry and Andy Wachowski USA
18
1996

Jennifer Tilly, Gina Gershon, Joe Pantoliano

Tilly and Gershon are a lesbian couple who steal a chache of money belonging to Tilly's mobster boyfriend and attempt to get away, with the irate ex-lover in pursuit. A dangerous battle of wits ensues, all of which is played out as a black comedy. A funny, irreverent and sometimes irritating movie, of flashy but often impressive direction and complex plotting. This was the Wachowski brothers' debut feature.

COM 104 min (ort 108 min) B/W VIDrel: GUILD/FOXVID
V/sur

BOUND AND GAGGED: A LOVE STORY **
Daniel B. Appleby USA
18
1992

Ginger Lynn Allen, Elizabeth Saltarrelli, Chris Mulkey, Karen Black, Chris Denton, Mary Ella Ross, Abdul Salaam El Razzac, Andrea Scarpa, Gene Larche, Hal Atkinson, Phyllis Wright, Bill Schoppert, Joe Minjares, Peter Williams

Two women search for love and happiness on the backwaters of Minnesota and form a strong attachment to each other. One is a battered housewife, and when she refuses to leave her brutal spouse, the other kidnaps her with the intention of taking her to a centre for abused "brainwashed" women. A bizarre stab at a kinky comedy-thriller that fails to impress.

COM 89 min (ort 96 min) wScrn VIDrel: TART/20TH V

BOUND FOR GLORY ****
Hal Ashby USA
PG
1976

David Carradine, Ronny Cox, Melinda Dillon, Randy Quaid, Gail Strickland, John Lehne, Ji-Tu Cumbuka, Elizabeth Macey, Allan Miller

Biography of Woody Guthrie, the folk balladeer and poet – the eloquent voice of America's downtrodden during the dark years of the Depression and beyond. Carradine is perfectly cast as Guthrie, who travels the country, fighting and singing for the underdog, against the background of Leonard Rosenman's impressive score adaptation. AA: Cin (Haskell Wexler), Score (Leonard Rosenman).

DRA 142 min (ort 149 min) VIDrel: MGM/WHV L/A V
Boa: book by Woody Guthrie.

BOUNTY, THE ***
Roger Donaldson UK
18
1984

Anthony Hopkins, Mel Gibson, Edward Fox, Laurence Olivier, Daniel Day Lewis, Bernard Hill, Philip Davis, Liam Neeson, Wi Kuki Kaa, Tevaite Vernette, Philip Martin Brown, Simon Chandler, Malcolm Terriss, Simon Adams, John Sessions

Third version of the story of the Bounty Mutiny. A well made and lavish study that attempts to paint a truer picture of Captain Bligh and Fletcher Christian, and the events that led up to the famous mutiny. Always good to look at, but directed in a rather

clinical and uninvolved way. The use of wide screen will tend
to be largely lost on TV.
DRA   128 min (ort 133 min)   VIDrel: GAME/SPEAR
V/sur
Boa: book Captain Bligh and Mr Christian by R. Hough.

BOUNTY HUNTERS *   *15*
George Erschbamer USA   1996
Michael Dudikoff, Lisa Howard, Benjamin Ratner, Ashanti Williams
In between bouts of bickering with his ex-wife, a bounty hunter
goes after a car thief who has jumped bail, but finds his quarry
has got involved in a difference of opinion with a nasty Mafia
boss. Abysmal dialogue and pedestrian direction are the most
memorable items here, in a movie that looks just like one of
those forgettable action pics churned out for TV.
A/AD   93 min   VIDrel: EIV/SONOP   V/sh

BOUNTY TRACKER *   *18*
Kurt Anderson USA   1992
*Lorenzo Lamas, Matthias Hues, Cyndi Pass, Paul Regina, Whip
Hubley, Brooks Gardiner, Eugene Robert Glazer, Anthony Peck, Ken
Ober, Eddie Frias, George Perez, Lawrence Lowe, Judd Omen, Steve
Cohen, Leo Lee, Ray Laska, Marty Dudek*
A modern-day bounty hunter goes after the mercenary who
killed his brother, in this standard action tale of much violence
but little entertainment. Lamas gives his usual laid-back tough
guy performance.
A/AD   86 min (ort 90 min)   VIDrel: REFLEC/FIRST L/A   V

BOURNE IDENTITY, THE *   *15*
Roger Young USA   1988
*Richard Chamberlain, Jaclyn Smith, Yorgo Voyagis, Donald Moffat,
Anthony Quayle, Denholm Elliott, Peter Vaughn, Michael Habeck,
Wolf Kahler, Philip Madoc, Bill Wallis, Franciscus Abgottspon,
Frederick Bartman, John Carlin*
Having been washed up on a beach with amnesia and bullet
wounds, a man sets about trying to piece together his true iden-
tity, but finds that he may in fact have been a professional
terrorist and killer. A tense and intriguing thriller.
THR   176 min (ort 192 min) mTV   VIDrel: WHV   V/sh
Boa: novel by Robert Ludlum.

BOX OF DELIGHTS, THE *   *U*
Renny Rye UK   1984
*Patrick Troughton, Devin Stanfield, Robert Stephens, Geoffrey
Lander, Carol Frazer, Jonathan Stephens*
Film adaptation of a six-part serial based on a classic John
Masefield novel. The story tells of a magic box and the adven-
tures it leads a young boy into. A well made and fairly
entertaining kid's fantasy.
JUV   163 min mTV   VIDrel: BBC/TECH   V/h
Boa: novel by John Masefield.

BOX OF MOONLIGHT *   *15*
Tom Dicillo USA   1996
John Turturro, Sam Rockwell, Catherine Keener
Turturro plays an uptight construction boss whose well-ordered
lifestyle is upset by his fears of ageing and the desperate need
for a break from work. An unexpected lay-off gives him the
chance he needs, and he takes off in search of adventure, picking
up an oddball drifter along the way. It is this latter who serves
as a catalyst for his regeneration. A quirky road-movie of
warmth and insight, its lack of any clear direction is quite intentional.
DRA   112 min   CINrel

BOXCAR BERTHA *   *18*
Martin Scorsese USA   1972
*Barbara Hershey, David Carradine, Barry Primus, John Carradine,
Victor Argo, Bernie Casey, David R. Osterhout, Chicken Holleman,
Graham Pratt, Harry Northup, Ann Morell, Marianne Dole, Joe
Reynolds, Gayne Rescher, Martin Scorsese*
Set against the background of the Depression, this is an account
of a small-town Arkansas girl who falls in with a bunch of train
robbers. A kind of BONNIE AND CLYDE without the glamour,
with David Carradine taking on the railroad establishment,
aided by the real-life title character. Scorsese's very first full-
length film shows some of the flair of his later works, but is at
heart superficial and unsatisfying.
DRA   88 min (ort 97 min)   VIDrel: VISVID/POLYREC L/A
V
Boa: book Sister Of The Road by "Boxcar Bertha" Thompson as
told to Ben L. Reitman.

BOXER BLOW *   *18*
Joseph Lai HONG KONG   1989
Jonathan James, Kenneth Woods
Two US agents are dispatched to the Middle East to de-fuse a
delicate situation that could trigger off WW3. They find them-
selves caught up in a clash between a band of fanatics and a
criminal gang. This thin framework provides the minimal plot
for the usual blend of action and martial arts mayhem.
Aka: U.S. CATMAN 2: BOXER BLOW
MAR   85 min Cut (1 min 49 sec)   VIDrel: IMPENT   V

BOXER'S ADVENTURE, THE *   *18*
William Chang HONG KONG
*Pai Ying, Ling Yun, Yeh Han-hsi, Wei Tzu-yun, Tao Liang Tan,
Mong Fei, Show Liang Ko, See Kong Long*
Standard kung fu fare with a boxer fighting criminals at the
time of the Tang Dynasty. Average.
Aka: BOXER ADVENTURE
MAR   89 min   VIDrel: IMPENT   V

BOXING HELENA *   *18*
Jennifer Chambers Lynch USA   1993
*Julian Sands, Sherilyn Fenn, Bill Paxton, Art Garfunkel, Betsy Clark,
Meg Register, Kurtwood Smith, Nicolette Scorsese, Bryan Smith,
Marla Levine, Ted Manson, Lloyd T. Williams, Carl Mazzocone Sr,
Erik Shoaff, Lisa Oz, Amy Levin*
A successful surgeon becomes so totally obsessed with a woman
that he ends up amputating her limbs and keeping her a prisoner in his home. Fortunately, by the end of the film this is
revealed to be a sick fantasy of his, very much like this movie,
which concentrates on the weird and nasty to the exclusion of
everything else. Lynch is the daughter of David Lynch so the
resemblance to his work is no mere accident. Well filmed but
very hard to take in one sitting.
DRA   100 min (ort 105 min)   VIDrel: EIV/SONOP   V

BOY AND HIS DOG, A *   *(15)*
L.Q. Jones USA   1975
*Don Johnson, Susanne Benton, Alvy Moore, Helen Winston, Charles
McGraw, Hal Baylor, Ron Feinberg, Mike Rupert, Don Carter,
Michael Hershman, Tim McIntire (voice only)*
A black comedy set in a post-WW4 world of 2024, with a boy
and his highly intelligent and telepathic dog foraging for food,
eventually being lured by a woman into a bizarre underground
society, where the boy's services are required for procreation
purposes. A faithful adaptation of Ellison's novella, but one that
is severely hampered by a low-budget.
FAN   90 min   TVrel
Boa: novella by Harlan Ellison.

BOY CALLED HATE, A *   *15*
Mitch Marcus USA   1995
Scott Caan, Missy Crider, Elliott Gould, Adam Beach, James Caan
Just out of a juvenile detention centre, a young man foils the rape
of a girl, but in the struggle the would-be rapist is shot dead,
forcing the two youngsters to go on the run. A most promising
directing debut for Marcus, that takes the form of an extended
road-movie in which the script (the work of Marcus) aims for a
realistic feel and believable characterisations.
DRA   95 min (ort 98 min)   VIDrel: IMAG/RTM   V

BOY FROM MERCURY, THE *   *PG*
Martin Duffy IRELAND   1996
James Hickey, Rita Tushingham, Tom Courtenay
Set in the 1960s, this unusual story follows the plight of an
unhappy, eight-year-old who has withdrawn into a fantasy
world ever since his father died. Each night he uses his torch to
"communicate" with the spaceship from Mercury he believes
has sent him to Earth. But his delusions grow ever more bizarre,
and his mother recruits the boy's uncle to help straighten him
out. Writer-director Duffy's film debut is terribly uneven, but
quite charming.
DRA   87 min   CINrel

BOY IN BLUE, THE *   *15*
Charles Jarrot CANADA   1986
*Nicolas Cage, Christopher Plummer, Cynthia Dale, David Naughton,
Sean Sullivan, Melody Anderson, James B. Douglas, Walter Massey,
Austin Willis, Philip Craig, Robert McCormick, Tim Weber, George
E. Zeeman, Geordie Johnson*
Dreary romantic biopic devoted to Ned Hanlan; a Canadian
who was a famed international rowing champion for some ten
years at the start of the century. Filmed in Montreal, with Cage

miscast as this brutish and rather unappealing rower, who comes to Philadelphia at around 1900 to show up the snobs at the first of many regattas.
DRA   93 min (ort 98 min)   VIDrel: 20TH/TECH   V

**BOY MEETS GIRL **   18
Leos Carax FRANCE   1983
Denis Lavant, Mireille Perier, Carroll Brooks, Elie Poicard, Anna Baldaccini
Overstylised, virtually plotless tale of a rootless man's doomed romance, after the break-up of an earlier attachment. A visually striking debut for this young director that won great critical acclaim, despite its essentially derivative nature. The director was only twenty-three when he made this, so perhaps can be forgiven for his self-indulgent direction.
DRA   100 min B/W wScrn   VIDrel: ARTIF/20TH   V

**BOY ON A DOLPHIN **   U
Jean Negulesco USA   1957
Allan Ladd, Clifton Webb, Sophia Loren, Laurence Naismith, Alexis Minotis, Jorge Mistral, Piero Giagnoni, Charles Fawcett, Gertrude Flynn, Charlotte Terrabust, Margaret Stahl, Orestas Rallis
When a Greek girl diver discovers a valuable sunken treasure, the news attracts a ruthless collector and an American archaeologist. An Aegean adventure yarn, in which the beautiful scenery out-acts the sadly miscast male lead. Expect no surprises. This was Loren's US film debut.
A/AD   110 min   VIDrel: 20TH/TECH   V/h
Boa: novel by David Divine.

BOY WHO COULD FLY, THE *   PG
Nick Castle USA   1986
Lucy Deakins, Bonnie Bedelia, Colleen Dewhurst, Jay Underwood, Fred Savage, Louise Fletcher, Fred Gwynne, Mindy Cohen, Janet MacLachlan, Michelle Bardeaux, Aura Pithart, Cam Bancroft, Jason Priestly, Chris Arnold, Sean Kelso, Dan Zale
A young girl moves to a small town after her father dies and discovers a strange autistic boy there, whose only method of coping with reality is to believe he can fly. She gradually gets to know him and in doing so, overcomes her own problems. Though the film's change from drama to fantasy tends to undermine its impact, this sensitive and warm story treats serious ideas with a light touch.
JUV   103 min (ort 114 min)   VIDrel: 20TH/TECH   V/sur

**BOYFRIEND, THE **   U
Ken Russell UK   1971
Twiggy (Lesley Hornby), Christopher Gable, Max Adrian, Barbara Windsor, Bryan Pringle, Murray Melvin, Moyra Fraser, Georgina Hale, Sally Brant, Vladek Sheybal, Tommy Tune, Brian Murphy, Glenda Jackson, Graham Armitage, Caryl Little
Ostensibly an attempt to create a screen version of the Sandy Wilson stage pastiche. The plot (such as it is) follows the attempts of a provincial company to put on a musical. Russell's homage to the genre of the Hollywood musical has a number of good fantasy sequences that mirror the work of Busby Berkley, but the lack of a strong storyline is a drawback. An amiable romp.
Aka: BOY FRIEND, THE
MUS   104 min (ort 125 min)   VIDrel: MGM/WHV   V/sh
Boa: musical by Sandy Dennis.

BOYFRIEND FROM HELL *   15
Alan Smithee (Michael Gottlieb) USA   1990
Cheech Marin, Emma Samms, Vernon Wells, Terence Cooper, Jeanette Cronin, Carole Davis, Bruce Spence, Gary McCormack, Frank Whitten, June Bishop, David Argue, Bruce Allpress, Val Lamond, Claire Glenister, Jonathan Coleman
For you then, Marin leaves the pothead jokes behind, and plays a dumb character who goes to Australia, gets a job working in a Mexican restaurant and becomes involved in a scheme on the part of a wealthy heiress to thwart the marriage plans her father is making for her. An uneven and clumsy work with a few bright moments; the director's use of the ubiquitous pseudonym Alan Smithee on the credits is comment enough.
Aka: SHRIMP ON THE BARBIE
COM   83 min (ort 87 min)   VIDrel: MGM/WHV   V

BOYFRIENDS *   18
Neil Hunter/Tom Hunsinger UK   1996
James Dreyfus, Mark Sands, Andrew Ableson, Michael Urwin, David Coffey, Darren Petrucci, Michael McGrath, Russell Higgs
Shot on 16 mm, this low-budget debut feature explores gay rela-

tionships in the 1990s, the central characters being a bunch of middle-aged men who gather in the country for a weekend. As one couple (Dreyfus and Sands) decide it's about time to break up, another (Urwin and Ableson) are possibly about to put their three-month relationship on a more serious footing. Jealousy, intrigue and petty bickering are on offer here, plus a welcome measure of wit and insight too.
DRA   82 min   VIDrel: DTK/RTM   V

**BOYS **   15
Stacy Cochran USA   1995
Winona Ryder, Lukas Haas, Skeet Ulrich, John C. Reilly, Bill Sage, Matt Malloy, Wiley Wiggins, Russell Young, Marty McDonough, Vivienne Shub, Charlie Hofheimer, Spencer Vrooman, Christopher Pettiet, Andy Davis, David Newsom
About to leave college and get his first job, a young student rescues a wealthy woman when she falls from her horse and they embark on an affair. A rather odd film that has no clear direction and makes no statements of any particular import, though the two main characters are convincing enough.
DRA   86 min   VIDrel: POLFIL   V
Boa: short story Twenty Minutes by James Salter.

**BOYS CLUB, THE **   15
John Fawcett USA   1996
Christopher Penn, Dominic Zamprogna, Stuart Stone, Devon Sawa
Three thirteen-year-old boys find an injured man in their secret hideout, who claims to be an undercover cop on the run from a former partner who has turned against him. They agree to shelter him, but in doing so put their lives at the mercy of his assailant. A vicious thriller, all about the need to grow up quickly when faced with the harsh realities of the real world.
THR   89 min (ort 92 min)   VIDrel: HIFLI/SONOP   V/h

**BOYS FROM BRAZIL, THE **   18
Franklin J. Schaffner USA   1979
Gregory Peck, Laurence Olivier, James Mason, Lilli Palmer, Rosemary Harris, John Dehner, John Rubinstein, Uta Hagen, Steven Guttenberg, Denholm Elliott, Anne Meara, Bruno Ganz, Michael Gough, Prunella Scales, Linda Hayden
Based on a bestseller, this tells of a post-war Nazi plan to use genetic implantation to breed a group of Hitler clones. Sometimes tense, more often nasty, with Olivier totally miscast as a Simon Wiesenthal-type Nazi hunter coming up against Peck's sinister Nazi scientist. The ludicrous insistence on phoney German accents for all and the flat direction make this movie close to parody at times. Very, very disappointing.
THR   118 min (ort 123 min)
VIDrel: 4-FRONT/POLYREC/ITC   V
Boa: novel by Ira Levin.

BOYS FROM THE BLACKSTUFF, THE *   15
Philip Saville UK   1982
Michael Angelis, Bernard Hill, Tom Georgeson, Julie Walters, Alan Igbon, Peter Kerrigan
A long and very downbeat tale of the lives of six former members of a gang of tarmac layers in Liverpool who lose their jobs and have to cope with the callousness of everyday life in Britain. First shown on the BBC as a set of five self-contained plays, all written by Alan Bleasdale. Stories were entitled: "Jobs For The Boys", "Moonlighter", "Shop Thy Neighbour", "Yosser's Story" and "George's Last Ride".
DRA   308 min (2 cassettes) mTV
VIDrel: PARADOX/TOTAL   V/h

BOYS IN THE BAND, THE **   15
William Friedkin USA   1970
Kenneth Nelson, Leonard Frey, Cliff Gorman, Frederik Combs, Laurence Luckinbill, Keith Prentice, Robert LaTourneaux, Reuben Greene, Peter White
Film version of a Broadway play dealing with the lives and loves of a group of men who meet at a birthday party – eight are homosexual, but the ninth insists he isn't. Despite the limitations of a single set, this superbly acted drama is sometimes funny, often sad, but never less than totally absorbing. A badly cut 108 minute version was produced for TV.
DRA   115 min   VIDrel: 20TH/TECH   V/sh
Boa: play by Mart Crowley.

BOYS ON THE SIDE *   15
Herbert Ross USA   1994
Whoopi Goldberg, Drew Barrymore, Mary-Louise Parker, James Remar, Matthew McConaughey, Billy Wirth, Anita Gillette, Dennis

Boutsikaris, Estelle Parsons, Gedde Watanabe, Emily Saliers, Amy Aquino, Stan Egi, Stephen Gevedon, Amy Ray,
A New York nightclub singer travels to Los Angeles in search of better times and shares a ride there with a rather prim young woman. A third woman joins them at Pittsburgh and all three soon form a firm bond as they discuss many of the key issues of the day. Excellent acting makes this film enjoyable to watch even if the plot has a good many shortcomings.
COM   112 min (ort 117 min) cC   VIDrel: WHV   V/sur

BOYS WILL BE BOYS **
William Beaudine UK
_U_
1935
Will Hay, Gordon Harper, Claude Dampier, Jimmy Hanley, Dav Burnaby, Norma Varden, Charles Farrell, Percy Walsh, Peter Gawthorne
A teacher forges his credentials in order to get a job as the head of a private college and is forced to take on as a butler the father of his head boy, who is a jewel thief. The latter is planning to steal a valuable necklace from one of the governors but is eventually foiled by our wily schoolmaster.
COM   73 min (ort 75 min) B/W
VIDrel: CONNO/RTM L/A   V

BOYZ N THE HOOD ***
John Singleton USA
15
1991
Cuba Gooding Jr, Ice Cube, Morris Chestnut, Larry Fishburne, Nia Long, Tyra Ferrell, Angela Bassett, Whitman Mayo, Meta King, Lexie Bigham, Kenneth A. Brown, Nicole Brown, Darneicea Corley, Na'blonka Darden, Dedrick D. Gobert
The tough neighbourhood of South Central L.A. is the setting for this honest and thought-provoking look at the lives of the largely impoverished Blacks who live there, and follows the efforts a divorced father makes to steer his son away from the seemingly inevitable life of crime that beckons. A remarkably strong directing debut from Singleton that sometimes lacks focus but is always absorbing. See also JASON'S LYRIC for something similar (albeit weaker).
Aka: BOYZ N THE HOOD: INCREASE THE PEACE
A/AD   107 min (ort 112 min) cC
VIDrel: VCC/DISC/COLUM   V/sh

BRADY BUNCH MOVIE, THE **
Betty Thomas USA
12
1995
Shelley Long, Michael McKean, Gary Cole, Florence Henderson, Barry Williams, Ann B. Davis, Christopher Knight, Davy Jones, Jean Smart, Henriette Mantel, Christine Taylor, Jennifer Elise Cox, Olivia Hack, Christopher Daniel Barnes
A well-realised but not entirely successful attempt at a parody of this famous 1970s sitcom series about a family who were too wholesome to be true. Here, they find themselves having to raise $20,000 in a hurry in order to save their home. Some very fine performances are in evidence but the many references to the original will baffle those unfamiliar with this cult show.
COM   88 min cC   VIDrel: CIC/SONOP   V/sur

BRAIN, THE **
Freddie Frances UK/WEST GERMANY
12
1962
Anne Heywood, Peter Van Eyck, Cecil Parker, Bernard Lee, Maxine Audley, Jeremy Spenser, Miles Malleson, Jack MacGowran, Frank Forsyth, George A. Cooper, Ann Sears, Allan Cuthbertson, Ellen Schwiers, Irene Richmond
A tycoon has a fatal accident but his brain is kept alive and begins to control the scientist in charge of it, forcing him to execute his evil intentions. A remake of the 1953 film DONOVAN'S BRAIN.
Aka: DEAD MAN SEEKS HIS MURDERER, A; EIN TOTER SUCHT SEINEN MORDER; VENGEANCE
FAN   83 min (ort 87 min) B/W   VIDrel: ENCORE/SPEAR   V
Boa: novel Donovan's Brain by Curt Siodmark.

BRAIN DONORS *
Dennis Dugan USA
PG
1992
John Turturro, Bob Nelson, Mel Smith, Nancy Marchand, John Savident, George De La Pena, Juli Donald, Spike Alexander, Teri Copley, Irene Olga Lopez, Dick Monday, Warren Thomas, Katherine Heard, Micahel Conti, Gary Grossman
A lawyer and his two goofy friends mount an underhand scheme to take over a ballet company, in this misfire comedy which claims to have been suggested by A NIGHT AT THE OPERA, but sadly lacks any real wit, relying instead on contrived and totally unfunny clowning that rapidly becomes

highly tedious. The only thing of interest in this dud are the Claymation opening and closing sequences.
COM   76 min (ort 79 min)   VIDrel: CIC/SONOP   V/sur

BRAIN FIX **
Scott Wallace/Jim Amin USA
18
Charles Copin, Jack Savage
A banned professor believes he is able to cure schizophrenia by implanting parasites in people's brains, and intends to use both his son and a female victim as his next experimental subjects.
HOR   84 min   VIDrel: VIPCO/SGSVID   V/h

BRAIN FROM PLANET AROUS, THE **
Nathan Hertz (Nathan Juran) USA
PG
1957
Robert Fuller, John Agar, Joyce Meadows, Thomas B. Henry, Ken Terrell, Henry Travis, Morris Ankrum, Tom Browne Henry, Tim Graham, E. Leslie Thomas, Bill Giorgio
An evil alien comes to Earth and takes over a scientist as a preliminary to complete domination of the Earth. Meanwhile, a good alien brain has taken up residence in the scientist's dog and proceeds to enlist the help of the scientist's girlfriend in destroying its evil counterpart. A fine example of classically corny 1950s SF.
FAN   68 min B/W   VIDrel: FIRC/RTM   V

BRAIN SMASHER: A LOVE STORY **
Albert Pyun USA
PG
1993
Andrew Dice Clay, Teri Hatcher, Deborah Van Valkenburgh, Tim Thomerson, Yuji Okumoto, Charles Rocket, Brion James, Charles Rocket, Nicholas Guest
Routine adventure yarn in which Clay not only rescues a beautiful model but also ultimately saves the world from falling under the control of a group of Chinese monks. A ridiculous action-comedy with veteran pulp-film director Pyun laying it on thick at every opportunity, thankfully detracting from the terrible acting and dialogue.
Aka: BRAINSMASHER
A/AD   84 min (ort 88 min)   VIDrel: MED/DISC   V/sh

BRAINDEAD **
Peter Jackson NEW ZEALAND
18
1992
Timothy Balme, Diana Penalver, Elizabeth Moody, Ian Watkin, Brenda Kendall, Stuart Devenie, Jed Brophy, Elizabeth Brimilcombe, Stephen Papps, Murray Keane, Glenis Levestam, Lewis Rowe, Elizabeth Mullane, Harry Sinclair, Silvio Fumularo
A man's mother is bitten by a Sumatran Monkey Rat and mutates into a horrific, carnivorous creature. Repulsive and utterly over-the-top horror spoof, full of imaginative ideas of the repellent sort but little in the way of a real plot. However, in terms of sheer shock value this film takes some beating.
HOR   99 min (ort 104 min)   VIDrel: POLY/POLYREC   V

BRAINSCAN **
John Flynn USA
18
1993
Edward Furlong, Frank Langella, T. Ryder Smith, Amy Hargreaves, Jamie Marsh, David Hemblen, Victor Ertmanis, Vlasta Vrana, Dom Fiore, Claire Riley, Tom Fennel, Michele-Barbara Pelletier, Dean Hagopian, Donna Bacalla, Don Jordan
A teenager high-school kid with the usual problems gets hooked on a new computer fantasy game that involves murder, but it soon becomes apparent that these killings may not just be in his imagination. A good idea is soon wasted in an intriguing film that goes rapidly downhill. The movie "Arcade" covered much the same ground.
HOR   91 min (ort 96 min)   VIDrel: GUILD/SONOP   V/sur

BRAINSTORM ***
Douglas Trumball USA
15
1983
Natalie Wood, Christopher Walken, Louise Fletcher, Cliff Robertson, Joe Dorsey, Jordan Christopher, Alan Fudge, Donald Hotton, Stacy Kuhne-Adams, Georgianne Walken, Jason Lively, Lou Walker, John Hugh, David Wood
Two scientists have perfected a sensory device in the form of a headset that can relay images directly into one's consciousness. With potential for both good and evil, the device is sought by some unscrupulous parties. The impressive visual effects (designed for 70 mm stereophonic projection) do not translate well to TV, but hold up an otherwise routine "mad-scientist" type tale. Natalie Wood's last film; she died during production in 1981.
FAN   101 min (ort 106 min)   VIDrel: MGM/WHV   V/sur

BRAMWELL ** 15
David Tucker/Laura Sims UK 1995
Jemma Redgrave, David Calder, Michele Dotrice, Robert Hardy
Story of a female surgeon and her struggle to gain acceptance among her male peers in 1895. Excellent period detail (including ample gore) was the main asset here, as the various intrigues poor Dr Bramwell had to put up with each week (mostly on account of her resentful male colleagues) could make for quite tiresome viewing. This first series ended with her supposedly engaged to be married, but to a doctor who had to leave to take up a distant post.
DRA   360 min (2 cassettes) mTV   VIDrel: CTE/CARL   V

BRANDED TO KILL ** 18
Suzuki Seijun JAPAN 1967
Shishido Jo, Ogawa Mariko
A gunman with weird sexual tastes kills a rival but is then becomes a target of the dead man's lover. An odd film, that blends action and eroticism in a distinctly Japanese way.
A/AD   87 min (ort 90 min) B/W
VIDrel: ICAPRO/MANGA   V

BRANDY AND ALEXANDER * 18
Jack Remy USA 1991
Jeanna Fine, Allison Williams, Jon Dough, Britt Morgan, Allison Sterling, K.C. Williams, Heather Hart, Mickey Ray, Tom Byron, Dorothy De Molay, Edward Penishands
Sex film that tries hard to be a comedy of errors, with a mix-up over some laundry bringing a young couple together.
A   74 min (ort 91 min)   VIDrel: GROHOM/MAXSCAN   V

BRANNIGAN ** 15
Douglas Hickox UK 1975
John Wayne, Richard Attenborough, Judy Geeson, Mel Ferrer, John Vernon, Daniel Pilon, John Stride, James Booth, Del Hanney, Lesley-Ann Down, Barry Dennen, Anthony Booth, Brian Glover, Ralph Meeker, Jack Watson
An Irish-American cop is sent to England to catch a fugitive gangster who has fled to London to avoid extradition. Routine cop thriller with Wayne in an unusual setting.
THR   107 min (ort 111 min)   VIDrel: MGM/WHV L/A   V/h

BRASSED OFF *** 15
Mark Herman UK/USA 1996
Pete Postlethwaite, Tara Fitzgerald, Ewan McGregor, Jim Carter, Ken Colley, Peter Gunn, Mary Healey, Melanie Hill, Philip Jackson, Sue Johnston, Peter Martin, Stephen Moore, Lill Roughley, Stephen Tompkinson, Olga Grahame
Set in 1992 at a time when pits were closing across the country, this unusual drama stars Postlethwaite as a bandleader doing his best to keep brass band music alive in the face of mounting joblessness and despair. In many ways an angry film that wears its heart on its sleeve, but also a film that is hard not to like.
DRA   103 min (ort 107 min) cC   VIDrel: FILM4/RTM   V/s

BRAVADOS, THE *** PG
Henry King USA 1958
Gregory Peck, Joan Collins, Stephen Boyd, Albert Salmi, Henry Silva, Barry Coe, Kathleen Gallant, George Voskovec, Herbert Rudley, Lee Van Cleef, Ken Scott, Andrew Duggan, Gene Evans, Joe Da Rita, Robert Adler, Robert Griffin
An overlong, rambling Western, in which Peck is bent on pursuit of the man who raped and murdered his wife, but realises after a while that he is little better than his quarry. Peck gives a nice, restrained performance and Da Rita an intense one as the "hangman" in this rather austere above-average Western.
WES   93 min (ort 98 min)   VIDrel: 20TH/TECH   V/h

BRAVE LITTLE TOASTER, THE *** U
Jerry Rees USA 1987
Voices of: Jon Lovitz, Tim Stack, Timothy E. Day, Thurl Ravenscroft, Deanna Oliver, Phil Hartman, Joe Ranft, Judy Toll, Wayne Kaatz, Colette Savage, Mindy Stern, Jack Jackman, Randy Cook, Randy Bennett, Joanthan Benair, Louis Conti
Animated feature that follows the adventures of five household items left alone in a holiday home who have faithfully carried out their daily duties despite the fact that no-one lives there anymore. Eventually they tire of this pointless existence and set off in search of a little boy they once knew as "Master". An offbeat but pleasing and well executed effort.
ANIM   90 min   VIDrel: 4-FRONT/POLYREC/ITC   V/sur

BRAVEHEART *** 15
Mel Gibson USA 1995
Mel Gibson, Sophie Marceau, Patrick McGoohan, Catherine McCormack, Brendan Gleeson, James Cosmo, David O'Hara, Angus McFayden, Alun Armstrong, Ian Bannen, Peter Hanly, James Robinson, Sean Lawlor, Sandy Nelson, Sean McGinley
A well made but overlong and historically flawed attempt to turn the 13th century story of Scottish leader William Wallace and his struggle with the king of England into a tale of heroic daring. Filmed on location in Scotland, it is strong on scenic locations and camerawork if not on convincing dramatics. Like DANCES WITH WOLVES, its politically correct "message" was well rewarded at the Oscars ceremony. AA: Pic, Dir, Cin (John Toll), Effects/aud.
A/AD   180 min wScrn cC   VIDrel: 20TH/TECH; ENCORE (LV only)   V/sur LV

BRAVESTARR: THE LEGEND ** U
Tom Tataranowicz USA 1986
Voices of: Charlie Adler, Susan Blu, Pat Fraley, Ed Gilbert, Alan Oppenheimer
Animated space Western about an Indian marshal sent to tame "New Texas", a rough planet whose citizens have pleaded for help in fighting the forces of evil. Fair.
Aka: BRAVESTARR; BRAVESTARR: THE MOVIE; MARSHAL BRAVESTARR
ANIM   87 min (ort 101 min)
VIDrel: 4-FRONT/POLYREC/BRAVE L/A   V

BRAZIL *** 15
Terry Gilliam UK 1985
Jonathan Pryce, Katherine Helmond, Robert De Niro, Ian Holm, Michael Palin, Ian Richardson, Bob Hoskins, Peter Vaughan, Kim Griest, Jim Broadbent, Barbara Hicks, Charles McKeown, Derrick O'Connor, Kathryn Pogson, Bryan Pringle
A chilling and imaginative portrayal of an inefficient but brutal Britain of the not-too-distant future, where a timid clerk clings to his ideals in the face of opposition from the police state but eventually retreats into madness. A cross between 1984 and SLEEPER, this visually impressive film has moments of great power, but its chaotic script and cluttered, dimly lit sets are major flaws. Screenplay is by Gilliam, Tom Stoppard and Charles McKeown.
FAN   137 min (ort 142 min) wScrn   VIDrel: WHV   V/sur

BREACH OF CONDUCT ** 15
Tim Matheson USA 1994
Peter Coyote, Courtney Thorne-Smith, Tom Verica, Beth Toussaint, Keith Amos, Thom Vernon, Tom Mason, Todd McKee, Drew Snyder, Thom McFadden, John Walcutt, Gregg Daniel, Hill Harper, Roger Hewlett, Sharon Mendes, Larry Nash, Tudi Roche
After her Army husband is transferred to a new base, his attractive wife has to cope with the unwelcome attentions of his unbalanced commanding officer, whose growing obsession with her soon pushes him over the edge and leads to her kidnapping.
DRA   89 min (ort 93 min) mCab   VIDrel: CIC/SONOP   V

BREAK, THE ** 12
Lee H. Katzin USA 1995
Vincent Van Patten, Martin Sheen, Rae Dawn Chong, Betsy Russell, Ben Jorgensen, Valerie Perrine
A look at the world of professional tennis with Patten playing an over-the-hill star who is drawn into working as a reluctant coach to a youngster in order to pay off a debt he owes the lad's father, who is the local bookie. Quite a well conceived film that doesn't try to make any especially significant statements, with Patten's wry dialogue and jaded world-weariness adding a little more depth to the film than the plot provides.
DRA   100 min   VIDrel: MED/20VIS   V/sh

BREAKDANCE 2: ELECTRIC BOOGALOO ** PG
Sam Firstenberg USA 1984
Lucinda Dickey, Adolfo (Shabba-Doo) Quinones, Michael (Boogaloo-Shrimp) Chambers, Susie Bono, Harry Caesar, Jo De Winter, John Christy Ewing, Steve (Sugarfoot) Notario, Sabrina Garcia, Lu Leonard, Ken Olfson, Peter MacLean
Happy but empty sequel to the first film, with a rich college-bound kid and some underprivileged youths uniting to save a community centre. A noisy, cheerful stomper for the breakdancing set. Numbers include "Do Your Thang" and "Oye Mamacita".
Aka: BREAKIN' 2 ELECTRIC BOOGALOO; ELECTRIC BOOGALOO, BREAKIN' 2
MUS   89 min (ort 94 min) (Cut at film release)
VIDrel: GUILD/SONOP L/A   V/s

BREAKDANCE: THE MOVIE **
Joel Silberg USA
PG
1984
Lucinda Dickey, Adolfo (Shabba Doo) Quinones, Michael (Boogaloo Shrimp) Chambers, Ben Lokey, Christopher MacDonald, Phineas Newborn III, Bruno (Pop 'N' Taco) Falcon, Timothy (Poppin' Pete) Solomon, Eleanor Zee
The plot is minimal in this story of three breakdancers who try to make it big on Broadway. A breakdancer's answer to FLASHDANCE, with loud music and plenty of breakin'. A sequel followed in 1984.
Aka: BREAKIN'
MUS   83 min (ort 88 min)   VIDrel: L/A   V

BREAKER! BREAKER! *
Don Hulette USA
15
1977
Chuck Norris, George Murdock, Terry O'Connor, Don Gentry, Dan Vandergriff, Michael Augenstein, Ron Cedillos, John Difusco, Douglas Stevenson
A trucker searches for his brother in a town that's run by a corrupt judge, in this stupid comedy-actioner that traded on the CB craze but does little with it.
A/AD   84 min   VIDrel: 4-FRONT/POLYREC   V

BREAKER MORANT ****
Bruce Beresford AUSTRALIA
PG
1979
Edward Woodward, Jack Thompson, John Waters, Charles Tingwell, Bryan Brown, Vincent Ball, Lewis Fitz-Gerald, Frank Wilson, Terence Donovan, Russell Kiefel, Alan Cassell, Judy Dick, Barbara West
An excellent film that's set in the Boer War and tells the story of three Australian soldiers who were court-martialled ostensibly for murdering prisoners, but in reality for political reasons. A powerful drama based on true events and a winner of several Australian Academy Awards.
WAR   106 min   VIDrel: POLY/POLYREC/BRAVE   V
Boa: play by Kenneth G. Ross.

BREAKFAST AT TIFFANY'S ***
Blake Edwards USA
PG
1961
Audrey Hepburn, George Peppard, Patricia Neal, Buddy Ebsen, Mickey Rooney, Martin Balsam, John McGiver, Villalonga, Dorothy Whitney, Stanley Adams, Elvia Allman, Alan Reed Sr, Beverley Hills, Claude Stroud
The bittersweet story of a relationship between a struggling young writer and a zany New York playgirl who comes from a small town. Based on a Truman Capote story, this film has some good moments and a memorable Mancini score. AA: Score (Henry Mancini), Song ("Moon River" – Henry Mancini (m)/Johnny Mercer (l)).
COM   109 min (ort 115 min) cC   VIDrel: CIC/SONOP V/dm
Boa: novel by Truman Capote.

BREAKHEART PASS *
Tom Gries USA
PG
1976
Charles Bronson, Charles Durning, Ben Johnson, Richard Crenna, Ed Lauter, Jill Ireland, David Huddleston, Roy Jenson, Casey Tibbs, Archie Moore, Joe Kapp, Read Morgan, Robert Rothwell, Rayford Barnes, Scott Newman, Eldon Burke
Even the story, based on an Alistair Maclean novel, cannot help in a film that is dull, dull, dull. The confusing plot has Bronson starring as a government undercover agent on the trail of gun-runners. Most of the action or lack of it takes place on an interminable train journey.
WES   95 min (Cut at film release)   VIDrel: MGM/WHV L/A V
Boa: novel by Alistair MacLean.

BREAKING AWAY ***
Peter Yates USA
PG
1979
Dennis Christopher, Dennis Quaid, Daniel Stern, Jackie Earle Haley, Robyn Douglass, Barbara Barrie, Paul Dooley, Hart Bochner, Amy Wright, Peter Maloney, John Ashton, Pamela Jane Soles, Lisa Shure, Jennifer K. Mickel
A film which, despite critical acclaim, never really comes to life. It tells of four friends who, having left school, don't know what to do with their lives. This sincere effort suffers from a lack of direction, though Christopher's worship of his Italian bicycling heroes (he even adopts an Italian lifestyle) and the witty use of Mendelssohn's Italian Symphony are compensations. Later a brief TV series. AA: Screen (Orig) (Steve Tesich).
COM   100 min   VIDrel: 20TH/TECH L/A   V

BREAKING GLASS **
Brian Gibson UK
15
1980
Hazel O'Connor, Phil Daniels, Jon Finch, Jonathan Pryce, Peter Hugo Daly, Mark Wingett, Gary Tibbs, Charles Wegner, Mark Wing-Davey, Hugh Thomas, Derek Thompson, Nigel Humphreys, Ken Campbell, Lowri-Anne Richards
A girl singer forms her own band but cannot cope with the fame that success brings her. A powerful performance from O'Connor as the singer is largely wasted on a predictable script.
DRA   100 min (ort 104 min)
VIDrel: 4-FRONT/POLYREC/ODY   V/sur

BREAKING IN ***
Bill Forsyth USA
15
1988
Burt Reynolds, Casey Siemaszko, Sheila Kelley, Lorraine Toussaint, Albert Salmi, Harry Carey, Maury Chaykin, Steve Tobolowsky, David Frishberg, Tom Laswell, Richard Key Jones, Walter Shane, Frank A. Damiani, John Baldwin
An ageing safebreaker takes on a young protege who has a lot to learn about both life and his profession. A gentle and rather quirky little comedy that affords Reynolds one of his best roles in years. Despite a lack of depth, this is a most pleasing effort. This script is by John Sayles.
COM   90 min Cut (4 sec – ort 95 min)
VIDrel: POLY/POLYREC L/A   V

BREAKING THE SILENCE **
Robert Iscove USA
(12)
1991
Gregory Harrison, Stephanie Zimbalist, Chris Young, Maryann Plunkett, Kelly Rutherford, Fran Bennett, T.C. Warner, David Ackroyd, Nicholas Shields, Jill Jacobson, Hugh Maguire, Arell Blanton, Thomas Wagner, Ernie Lively, Aki Aleong
Courtroom drama that deals with a case of child abuse, but offers neither anything new in the routine script nor especially convincing performances from the leads. Adequate made-for-TV fare with a few overly predictable plot twists.
DRA   94 min mTV   SATrel: MOVIE CHANNEL

BREAKING THE WAVES ***
Lars von Trier
18
DENMARK/FRANCE/HOLLAND/NORWAY/
SWEDEN
1996
Emily Watson, Stellan Skarsgard, Katrin Cartlidge, Jean-Marc Barr, Adrian Rawlins, Jonathan Hackett, Sandra Voe, Udo Kier, Mikkel Gaup, Roef Ragas, Phil McCall, Robert Robertson, Desmond Reilly, Sarah Gudgeon, Finlay Welsh
In the far north of Scotland, a shy, devout and naive local woman marries a Scandinavian oil-rig worker, but her deeply held religious convictions are scant comfort when her husband is paralysed in an accident. He then insists she take lovers as he can no longer satisfy her. Trier's film does much to explore the unhappy woman's inner turmoil, but the intensity of it all becomes just a little overdone at times. A strange, opaque but very powerful movie.
DRA   153 min (ort 159 min)   VIDrel: GUILD/FOXVID V/sur

BREAKOUT ***
Tom Gries USA
15
1975
Charles Bronson, Robert Duvall, John Huston, Alejandro Rey, Jill Ireland, Sheree North, Randy Quaid, Jorge Moreno. Emilio Fernandez, Paul Mantee, Alan Vint, Roy Jenson, Sidney Klute, Chalo Gonzalez
Tough action film in which Bronson plays a lush pilot, who rescues a man from a seedy Mexican jail after he has been framed for a murder. Fast paced and often quite funny, but spoilt by an excessive dependence on violence. Written by Howard B. Krietsek, Marc Norman and Elliott Baker.
A/AD   93 min (ort 96 min)
VIDrel: ENCORE/SPEAR/COLUM   V/sur
Boa: novel Ten Second Jailbreak by C. Asinof.

BREAKTHROUGH **
Andrew V. McLaglen WEST GERMANY/UK
15
1978
Richard Burton, Robert Mitchum, Rod Steiger, Curt Jurgens, Klaus Loewitsch, Helmut Griem, Michael Parks, Veronique Vendell, Joachim Hansen
A disappointing sequel to CROSS OF IRON. Burton plays a German sergeant who becomes enmeshed in a plot to assassinate Hitler in the summer of 1944.
Aka: SERGEANT STEINER
WAR   92 min (ort 111 min)   VIDrel: ENTUK L/A   V

BREATHING FIRE ** **   18
Lou Kennedy/Brandon De Wilde HONG KONG   1990
Jonathan Ke Quan, Jerry Trimble, Addie Saavedra, Ed Neil, Bolo
Yeung, Allen Tackett, Wendell C. Whitaker, Jacqueline Pulliam, Laura
Hamilton, Drake Diamond, Juan Ojeda, Pamela Maxton, Jacqueline
Woolsey, Gary GReen, Annie Rubanoff
A Vietnamese teenager who has grown up in comfort in
California does not realise that his ex-GI father is the leader of
a gang of armed robbers. When a robbery goes wrong the gang
set out to silence the young girl who's the only witness, but the
boy and his brother try to protect her. Somewhat unbelievable
mixture of action and martial arts, that's partially redeemed by
an exciting climax.
A/AD   86 min   VIDrel: MIA/DISC   V

BREATHLESS * **   18**
Jean-Luc Goddard FRANCE   1959
Jean Seberg, Jean-Paul Belmondo, Daniel Boulanger, Jean-Pierre
Melville, Van Doude, Liliane Robin, Henri-Jacques Huet, Claude
Mansard, Michel Fabre, Jean-Luc Goddard, Jean Domarchi, Richard
Balducci, Roger Hanin, Jean-Louis Richard
A young petty criminal steals a car and kills a cop but is even-
tually turned in by his American girlfriend. Goddard's New
Wave offering is long on style and influences but short on
content, with very little in the way of genuine character moti-
vation. Despite its faults, still enjoyable. Remade in 1983.
Aka: A BOUT DE SOUFFLE
DRA   90 min B/W   VIDrel: TART/20TH L/A   V/dm

BREATHLESS ** **   18
Jim McBride USA   1983
Richard Gere, Valerie Kaprisky, Art Metrano, William Tepper, John
P. Ryan, Valerie Kaprisky, Art Metrano, William Tepper, John
Jack Leusing, Waldemar Kalinowski
Lukewarm remake of Godard's 1959 film that lacks any of the
punch of the original. Gere is effective as the young amoral punk
on the run from the police, who embarks on a whirlwind affair
with a beautiful foreign student, but after a while the point-
lessness of the whole exercise begins to tell.
DRA   98 min Cut (24 sec – ort 105 min)
VIDrel: 4-FRONT/POLYREC/VISVID   V/h

BREATHING LESSONS ** **   (U)
John Erman USA   1993
James Garner, Joanne Woodward, Paul Winfeild, Kathryn Erbe, Joyce
Van Patten, Eileen Heckart, Henry Jones, Tim Guinee
A middle-aged couple who are quite devoted despite their
differences in temperament take a ninety-mile car trip to attend
the funeral of a friend. Along the way they quarrel and disagree
as to her impulsive behaviour which so often lands them in the
unexpected. A slow-paced character study that is moderately
enjoyable to watch in spite of the paper-thin plot.
DRA   94 min (ort 98 min) mTV   SATrel: MOVIE
CHANNEL
Boa: novel by Ann Tyler.

BREEDERS ** **   18
Tim Kincaid USA   1986
Theresa Farley, Lance Lewman, Francis Raines, Natalie O'Connell,
Leeanne Baker, Amy Brentano, Matt Mitler, Adrianne Lee, Mae Cerar,
Mark Legan, Dan Geffen, Pat Rizzolino, Derek Dupont, Owen Flynn,
Raheim Grier, Rose Geffen
Nasty mutant spores arrive on Earth and, having assumed
human form, indulge in a rape orgy – the intention being to
reproduce in the form of human duplicates. A chilling idea
that's let down by cliched development.
FAN   75 min (ort 90 min)   VIDrel: EIV/SONOP   V

BREWSTER'S MILLIONS * **   PG
Walter Hill USA   1985
Richard Pryor, John Candy, Lonette McKee, Stephen Collins, Jerry
Orbach, Pat Hingle, Tovah Feldshuh, Joe Grifasi, Hume Cronyn, Peter
Jason, David White, Jerome Dempsey, David Wohl, Ji-Tu Cumbuka,
Milt Kogan, Carmine Caridi
A baseball player is left $30,000,000 by a long-lost uncle, with
the stipulation that in order to inherit he must spend $1,000,000
in 30 days. A brash, noisy and disorganised remake of this old
favourite (first filmed in 1914) that wastes the talents of Pryor
and Candy – good comedians both. This was the seventh outing
for McCutcheon's novel.
COM   98 min (ort 101 min)   VIDrel: CIC/SONOP L/A
V/sh
Boa: novel by George Barr McCutcheon.

BRIAN'S SONG ** **   PG**
Buzz Kulik USA   1970
James Caan, Billy Dee Williams, Jack Warden, Shelley Fabares, Judy
Pace, Bud Furillo, Bernie Casey, David Huddleston, Ron Feinberg,
Jack Concannon, Ji-Tu Cumbuka, Abe Gibron, Ed O'Bradoviich, Dick
Butkus, Mario Machado, Stu Nahan
A true story of the real life friendship between two famous base-
ball stars, one white and the other black, who played for the
Chicago Bears. This film attempts to examine how the team as
a whole was affected when one of them developed cancer and
died at the age of twenty-six. A fine and moving story, scripted
by Edward Blinn and with an outstanding score by Michel
Legrand. See also TRIUMPH OF THE HEART, A: THE RICKY
BELL STORY.
DRA   74 min (ort 76 min) mTV   VIDrel: L/A   V
Boa: story I Am Third by Gale Sayers.

BRIDE, THE ** **   15
Frank Roddam UK   1985
Sting (Gordon Sumner), Jennifer Beals, Clancy Brown, David
Rappaport, Anthony Higgins, Geraldine Page, Alexei Sayle,
Veruschka, Quentin Crisp, Phil Daniels, Andrew De La Tour, Tony
Haygarth, Matthew Guinness, John Sharp
A flawed attempt both to remake THE BRIDE OF FRANKEN-
STEIN and rethink the legend. Here our Baron falls madly in
love with the female he supposedly created for his monster. A
film that succeeds in terms of the look of a gothic horror film,
but fails at any deeper level – mainly due to the poor perfor-
mances of Sting and Beals. Rappaport is however, outstanding
as the midget who befriends our Baron's male monster.
HOR   114 min (ort 119 min)   VIDrel: RCA L/A   V/sh

BRIDE AND THE BEAST, THE * **   PG
Adrian Weiss USA   1958
Charlotte Austin, Lance Fuller, Johnny Roth, Steve Calvert, William
Justine, Jeane Gerson, Gil Frye, Slick Slavin, Jean Anne Lewis,
Bhogwan Singh
An explorer's wife is kidnapped by a gorilla, but it transpires
that she was reincarnated from an ape and went off to mate. Be
warned, the screenplay is by noted director Edward D. Wood
Jr. Need we say more?
Aka: QUEEN OF THE GORILLAS
HOR   73 min (ort 78 min) B/W   VIDrel: CARL/TECH   V

BRIDE CAME C.O.D., THE ** **   U
William Keighley USA   1941
James Cagney, Bette Davis, Harry Davenport, Stuart Erwin, Eugene
Palette, William Frawley, Jack Carson, George Tobias
A charter pilot agrees to kidnap an heiress but gets stuck with
her when they crash in the desert. A limp and empty-headed
comedy that's worth a look, if only to see the stars squeeze a few
laughs out of such unpromising material.
DRA   88 min B/W   VIDrel: MGM/WHV L/A   V

BRIDE OF FRANKENSTEIN, THE ** **   PG**
James Whale USA   1935
Boris Karloff, Colin Clive, Valerie Hobson, Dwight Frye, Elsa
Lanchester, Ernest Thesiger, Una O'Connor, E.E. Clive, Gavin
Gordon, Douglas Walton, O.P. Heggie, John Carradine, Lucien Prival,
Reginald Barlow, Mary Gordon
Classic sequel to FRANKENSTEIN with Frankenstein being
forced by mad doctor Thesiger to make a mate for his monster.
Contains a number of highlights such as the blind hermit scene
and an excellent creation sequence. This shortened version omits
part of the Mary Shelley prologue plus the murder of the mayor.
Scripted by John L. Balterton and William Hurlbut and scored
by Franz Waxman. Followed by "Son Of Frankenstein" and the
remake, THE BRIDE.
HOR   75 min (ort 90 min) B/W   VIDrel: CIC/SONOP L/A
V

BRIDE OF THE MONSTER **   PG
Edward D. Wood Jr. USA   1956
Bela Lugosi, Tor Johnson, Loretta King, Tony McCoy, Harvey B.
Dunn, George Becwar, Don Nagel, Bud Osborne, William Benedict,
Dolores Fuller, Ann Wilner, Eddie Parker, John Warren, Ben
Frommer, Paul Carco
A mad scientist tries to create a race of superbeings and has his
assistant kidnap suitable victims. A worthy candidate for the
worst film of all time, from the creator of PLAN 9 FROM OUTER
SPACE.
Aka: BRIDE OF THE ATOM
HOR   68 min (ort 70 min) B/W   VIDrel: CARL/TECH   V

BRIDES OF FU MANCHU, THE **   U
Don Sharp UK   1966
*Christopher Lee, Douglas Wilmer, Marie Versini, Heinz Drache,
Rupert Davies, Roger Hanin, Howard Marion Crawford, Tsai Chin,
Kenneth Fortescue, Joseph Furst, Carole Grey, Harald Leipnitz*
In his efforts to become ruler of the world, our fiendish villain
kidnaps twelve young women, his intention being to force scien-
tists to develop a powerful disintegration ray for his use. An
enjoyable if slightly weaker sequel to "The Face Of Fu Manchu"
that is based on characters created by Sax Rohmer.
VENGEANCE OF FU MANCHU followed. See also THE
BLOOD OF FU MANCHU and THE CASTLE OF FU
MANCHU.
A/AD   91 min (ort 94 min)   VIDrel: BRAVE/SONOP L/A
V

BRIDESHEAD REVISITED ***   15
Charles Sturridge/Michael Lindsay-Hogg UK   1981
*Jeremy Irons, Anthony Andrews, Diana Quick, Laurence Olivier,
Claire Bloom, Simon Jones, Stephane Audran, Mona Washbourne,
John Le Mesurier, John Gielgud, Jane Asher*
This massive, fanatically faithful adaptation of Waugh's novel
is a slow, ponderous account of how an Oxford student becomes
deeply involved with a wealthy upper-class family. Beautifully
made and acted, this TV drama hosts an array of unpleasant
characters but lacks both pace and impact.
DRA   640 min (3 cassettes – 664 min) mTV
VIDrel: CASPIC/BMGREC   V
Boa: novel by Evelyn Waugh.

BRIDGE, THE ***   15
Sydney MacCartney UK   1990
*Saskia Reeves, David O'Hara, Joss Ackland, Rosemary Harris,
Anthony Higgins, Geraldine James, Tabitha Allen, Dominique Rossi,
Karina Rossi, Anya Phillips, Jo Powell, Michele Wade, Peter Blythe,
Tim Barker, William Job, Ben Daniels*
A woman and her three daughters arrive at a seaside residence
in 1887, where she enjoys the romantic attentions of an artist. A
very lush and glossy romantic drama, of high production values
and strong direction. However, the narrative lacks substance,
and for all the care lavished on the film, there is not really
enough in the story to hold the various strands together. This
was MacCartney's debut feature.
DRA   102 min   VIDrel: COLUM/SONOP   V/sur
Boa: novel by Maggie Hemingway.

BRIDGE AT REMAGEN, THE **   PG
John Guillermin USA   1968
*George Segal, Ben Gazzara, Robert Vaughn, Bradford Dillman, E.G.
Marshall, Peter Van Eyck, Matt Clark, Fritz Ford, Tom Heaton, Bo
Hopkins, Paul Prokop, Robert Logan, Steve Sandor, Frank Webb, Hans
Christian Blech, Joachim Hansen*
Account of a true incident in WW2 when Allied troops had to
hold a vital bridge over the Rhine prior to the final onslaught
on Nazi Germany. Well made but hampered by the utter
predictability of the material – a standard war film.
WAR   112 min (ort 115 min)   VIDrel: MGM/WHV   V/h
Boa: book by K. Hechler.

BRIDGE ON THE RIVER KWAI ****   PG
David Lean UK   1957
*Alec Guinness, Jack Hawkins, Sessue Hayakawa, William Holden,
James Donald, Andre Morell, Geoffrey Horne, Ann Sears, Peter
Williams, John Boxer, Percy Herbert, Harold Goodwin, Henry Okawa,
K. Katsumoto, M.R.B. Chakrabandhu*
During WW2, British soldiers in a Japanese POW camp build a
bridge under the orders of their commander whilst Holden plots
to blow it up. A vigorous but utterly glamorised picture of war
that whitewashes the cruelties of the Japanese, yet for all that,
contains sequences of enormous power. AA: Pic, Dir, Actor
(Guinness), Cin (Jack Hildyard), Edit (Peter Taylor), Score
(Malcolm Arnold), Screen/adapt (Pierre Boulle/Carl
Foreman/Michael Wilson).
WAR   155 min (restored version – ort 161 min) wScrn
VIDrel: COLUM/SONOP   V
Boa: novel by Pierre Boulle.

BRIDGE TO SILENCE **   15
Karen Arthur USA   1988
*Marlee Matlin, Lee Remick, Michael O'Keefe, Josef Sommer, Phyllis
Frelich, Candace Brecker, Allison Silva, Pat Hamilton, Cec Linder, Bob
Hilterman, Anthony Natale, Michael Rhodes, Graham McPherson,
Gerard Parker, Tom Butler*

In her TV acting debut, Matlin plays a woman with a hearing
disability who tries to rebuild her life, shattered in a car crash
that killed her husband. Remick is her cold and heartless
mother, with whom she comes into conflict over the fate of her
daughter. A soap opera style tearjerker which never rises above
the trite script, despite excellent performances all round. See
also CHILDREN OF A LESSER GOD.
DRA   88 min (ort 100 min) mTV   VIDrel: L/A   V
Boa: story by Louisa Burns Bisogno.

BRIDGE TOO FAR, A *   15
Richard Attenborough UK   1977
*Dirk Bogarde, James Caan, Edward Fox, Michael Caine, Sean
Connery, Hardy Kruger, Ryan O'Neal, Robert Redford, Maximilian
Schell, Laurence Olivier, Liv Ullman, Elliott Gould, Gene Hackman,
Anthony Hopkins, Arthur Hill, Siem Vroomrei*
An unsuccessful attempt to tell the tragic story of the failed 1944
Arnhem operation, when Allied commandos dropped behind
the German lines in occupied Holland. A star-studded cast plus
thousands of extras, succeed in producing a film that collapses
under its own weight.
WAR   168 min (ort 175 min) wScrn (special edition)
VIDrel: MGM/WHV   V/sh
Boa: book by Cornelius Ryan.

BRIDGES AT TOKO-RI, THE ***   U
Mark Robson USA   1954
*William Holden, Grace Kelly, Mickey Rooney, Fredric March, Earl
Holliman, Robert Strauss, Charles McGraw, Keiko Awaji, Richard
Shannon, Willis Bouchey, Nadene Ashdown, Cheryl Lynn Calloway,
Teru Shimada*
This drama, based on a bestseller by James Michener, tells the
story of a lawyer recalled to fly jets during the Korean War. A
powerful yet sensitive film showing the ultimate futility of war.
AA: Effects (Paramount Studios).
WAR   99 min (ort 103 min)   VIDrel: CIC/SONOP   V
Boa: novel by James Michener.

BRIDGES OF MADISON COUNTY, THE ****   12
Clint Eastwood USA   1995
*Meryl Streep, Clint Eastwood, Annie Corley, Victor Slezak, Jim
Haynie, Sarah Kathryn Schmitt, Christopher Kroon, Phyllis Lyons,
Debra Monk, Richard Lage, Michelle Benes, Alison Wiegert, Brandon
Bobst, Pearl Faessler, Tania Mishler*
A freelance photographer on an assignment for National
Geographic in rural Iowa, meets a farmer's wife and the two
indulge in a brief four-day affair. A sensitively realised and
touching romantic drama, beautifully acted by Streep and
Eastwood, that is all the better for the wise decision not to tack
on a happy ending. A fine example of how Eastwood has
matured both as an actor and a director.
DRA   129 min (ort 135 min) cC   VIDrel: WHV   V/sur
Boa: novel by Robert James Waller.

BRIEF ENCOUNTER ***   PG
David Lean UK   1945
*Celia Johnson, Trevor Howard, Stanley Holloway, Cyril Raymond,
Joyce Carey, Everley Gregg, Margaret Barton, Dennis Harkin,
Valentine Dyall, Irene Handl, Marjorie Mars, Nuna Davey, Edward
Hodge, Sydney Bromley, Avis Scott, Wally Bosco*
A suburban housewife and a local doctor, both less than content
with their own marriages, meet and enjoy a brief romance. Their
stolen meetings take place at a dismal railway station between
trains. Fine acting and a good score (Rachmaninov's Second
Piano Concerto) serve to point out the serious deficiencies in a
stilted drama, complete with the obligatory stereotyped working-
class characters. Enjoyable but very, very dated. Remade in 1974.
DRA   82 min (ort 86 min) B/W   VIDrel: CARL/TECH   V
Boa: playlet Still Life (from Tonight At 8.30) by Noel Coward.

BRIGADOON **   U
Vincente Minnelli USA   1954
*Gene Kelly, Cyd Charisse, Van Johnson, Elaine Stewart, Jimmy
Thompson, Barry Jones, Hugh Laing, Albert Sharpe, Virginia Bosier,
Tudor Owen, Dee Turnell, Owen McGivney, Dody Heath, Eddie
Quillan*
This adaptation of a Broadway musical, tells the story of a ghost
village in the Scottish Highlands that comes to life once in every
hundred years. Two Americans on holiday in Scotland discover
it. Not all that inventive, this dull musical is helped along by an
enjoyable Alan Jay Lerner score.
MUS   103 min (ort 108 min)   VIDrel: MGM/WHV   V/dm
Boa: musical by Alan Jay Lerner and Frederick Loewe

BRIGHT ANGEL ** **   15
Michael Fields USA   1990
Dermot Mulroney, Sam Shepard, Lili Taylor, Valerie Perrine, Bill Pullman, Burt Young, Mary Kay Place, Benjamin Bratt, Alex Bulltail, Delroy Lindo, Kevin Tighe, Sheila McCarthy, Tom Dixon, Lyle N. Cusson, Myrna Wilken, Tom Connelley
A couple of foolish youngsters find themselves out of their depth when they unwisely get involved with a dangerous con-man.
THR   89 min (ort 94 min)   VIDrel: COLUM/SONOP   V/s

BRIGHT LIGHTS, BIG CITY *   18**
James Bridges USA   1988
Michael J. Fox, Phoebe Cates, Kiefer Sutherland, Jason Robards, Swoosie Kurtz, Frances Sternhagen, Tracy Pollan, John Houseman, Dianne Wiest, David Warrilow, William Hickey, Charlie Schlatter, Alex Mapa, Kelly Lynch, Sam Robards
A young man comes to New York from the Midwest and takes on a job as a fact-checker at a Manhattan magazine, but is soon caught up in a lifestyle that revolves around an endless cycle of drugs and alcohol. Fox is good as a fellow whose frenzied lifestyle threatens to cost him his sanity, but the more complex aspects of the novel are not brought out.
DRA   102 min (ort 110 min)   VIDrel: WHV   V/sur
Boa: novel by Jay McInerney.

BRIGHTER SUMMER DAY, A *   (PG)**
Edward Yang TAIWAN   1991
Zhang Zhen, Lisa Yang, Zhang Guozhu, Elaine Jin, Wang Juan, Zhang Han, Jiang Xiuqiong, Lai Fanyun, Wang Qizan, Ke Yulun, Tan Zhigang, Zhang Mingxin, Rong Junlong, Zhou Huiguo, Liu Qingqi, He Qingxiang, Cai Changda, Li Zhongming
A look at life in Taiwan in the 1960s through the eyes of an idealistic student who refuses to alter his standards despite all that he sees about him. On one level the film tells the simple story of a youngster's involvement with street gangs and violence, but on another, it's a slow, thoughtful but slightly opaque examination of a society in the throes of painful change, well aware of the conflicting demands emanating from America and mainland China.
Aka: BRIGHT SUMMER DAY, A; GULING JIE SHAONIAN SHA REN SHIJIAN
DRA   237 min   CINrel

BRIGHTON BEACH MEMOIRS ** **   15
Gene Saks USA   1986
Blythe Danner, Jonathan Silverman, Bob Dishy, Brian Drillinger, Judith Ivey, Stacey Glick, Lisa Waltz, Steven Hill, Fyvush Finkel, Kathleen Doyle, Alan Weeks, Marilyn Cooper, Jason Alexander, Christian Baskous, Brian Evers
Neil Simon's semi-autobiographical story, that follows the affairs of two families sharing the same house in Brooklyn of 1937. Story is seen through the eyes of the young Simon, whose interests in life are largely confined to baseball and girls. A pleasant but cloying outing, with none of the sharpness of the successful Broadway play. Followed by BILOXI BLUES.
COM   105 min (ort 110 min)   VIDrel: CIC/SONOP   V/h
Boa: play by Neil Simon.

BRIGHTON ROCK *   PG**
John Boulting UK   1947
Richard Attenborough, Hermione Baddeley, William Hartnell, Carol Marsh, Nigel Stock, Wylie Watson, Harcourt Williams, Alan Wheatley, George Carney, Charles Goldner, Virginia Winter, Reginald Purdell, Constance Smith, Lina Barrie
Film version of Graham Greene's novel about 1930s Brighton gangland life. A teenage hoodlum kills a rival but meets his Nemesis in the shape of a music-hall artiste. Attenborough is superb as the vicious gangster who callously seduces a naive, young girl and sets out to use her to escape justice. The script is by Terence Rattigan and Greene. Interestingly, the film's poignant but upbeat ending is quite different to that of the novel, the change being Greene's idea.
Aka: YOUNG SCARFACE
DRA   88 min (ort 92 min) B/W   VIDrel: LUMI/SPEAR L/A V
Boa: novel by Graham Greene.

BRILLIANT DISGUISE, A ** **   18
Nick Vallelonga USA   1994
Lysette Anthony, Anthony John Denison, Corbin Bernsen, Gregory McKinney
A woman with multiple personalities attracts a sports writer who soon becomes quite taken with her. Unfortunately, when

a number of deaths occur, it seems as if she may also be a multiple murderer. An over-the-top-attempt attempt at an erotic thriller with the requisite degree of unnecessary nudity.
THR   93 min   VIDrel: PRISM/HIFLI   V

BRIMSTONE AND TREACLE *   18**
Richard Loncraine UK   1983
Sting (Gordon Sumner), Denholm Elliott, Joan Plowright, Suzanna Hamilton, Mary McLeod, Benjamin Whitrow, Dudley Sutton, Tim Preece
A handsome young man charms his way into the home of a middle-class couple on the pretext of being a friend of their daughter, now in a coma following an accident. A compelling but often unpleasant film with a fine performance from Sting, as the amoral stranger who slowly takes over the household, ultimately raping the daughter. Scripted by Potter from his play (it was originally made as a TV play in 1976, but the BBC banned it). Won the Grand Jury prize at Montreal.
DRA   84 min (ort 87 min) mTV   VIDrel: ARROW/RTM V/sur
Boa: play by Dennis Potter.

BRING ME THE HEAD OF ALFREDO GARCIA *   18
Sam Peckinpah USA   1974
Warren Oates, Isela Vega, Gig Young, Robert Webber, Helmut Dantine, Emilio Fernandez, Kris Kristofferson, Chano Urueta, Jorge Russek, Don Levy, Neri Ruiz, Donnie Fritts, Chalo Gonzalez, Enrique Lucero, Janine Maldonando, Tamara Garina
A sleazy story set in Mexico, with a small-time American piano player getting mixed up with some nasty characters, such as the wealthy Mexican who offers a million dollars for the head of his daughter's seducer. A gruesome and sickening melodrama, well directed but of little purpose or logic.
DRA   108 min (ort 113 min)   VIDrel: MGM/WHV   V/sh

BRING ME THE HEAD OF DOBIE GILLIS ** **   U
Stanley Z. Cherry USA   1988
Dwayne Hickman, Bob Denver, Connie Stevens, Sheila James, Scott Grimes, William Schallert, Tricia Leigh Fisher, Steve Franken
Stars of the American sitcom "The Many Loves Of Dobie Gillis" are reunited after twenty-five years, in this trivial little comedy which has wealthy Thalia Menninger returning to town with the intention of persuading old flame Dobie to dump his wife for her. Many stars from the original series pop up, though Stevens takes over the role originally played by Tuesday Weld.
COM   93 min (ort 100 min) mTV   VIDrel: 20TH/TECH V/h

BRINGING UP BABY **   U**
Howard Hawks USA   1938
Cary Grant, Katharine Hepburn, Charles Ruggles, May Robson, Walter Catlett, Fritz Feld, Jonathan Hale, Barry Fitzgerald, Ward Bond, Leona Roberts, George Irving, Tala Birell, Virginia Walker, John Kelly, Edward Gargan, Buck Mack
Screwball comedy in which a zany heiress with a pet leopard sets her sights on a palaeontology professor and pursues him relentlessly, bringing chaos and havoc into his life. A well-paced comedy that cracks along at breakneck speed, moving from one crazy situation to the next. WHAT'S UP DOC? was an attempt at a remake, but this earlier film is by far the better of the two.
COM   99 min (ort 102 min) B/W   VIDrel: VCC/DISC L/A V

BRITANNIA HOSPITAL *   15
Lindsay Anderson UK   1982
Leonard Rossiter, Brian Pettifer, John Moffatt, Fulton Mackay, Jill Bennett, Vivian Pickles, Barbara Hicks, Graham Crowden, Peter Jeffrey, Marsha Hunt, Mary McLeod, Catherine Wilmer, Joan Plowright, Robin Askwith, Mark Hamill
An old crumbling hospital, threatened by a strike, demonstrations and an impending royal visit, is used as a metaphor for the state of the UK. A ponderous satire of occasional wit and considerable repulsiveness.
COM   111 min (ort 116 min)   VIDrel: WHV   V/h

BROADCAST NEWS *   15**
James L. Brooks USA   1987
William Hurt, Holly Hunter, Albert Brooks, Robert Prosky, Lois Chiles, Joan Cusack, Jack Nicholson, Peter Hackes, Christian Clemenson, Robert Katims, Ed Wheeler, Stephen Mendillo, Kimber Shoop, Dwayne Markee, Gennie James, Amy Brooks
A behind-the-scenes look at the world of TV news broadcasting with a highly strung and neurotic woman producer finding

herself attracted to the handsome anchorman who has joined the
network, despite the fact that he represents everything she
loathes in broadcasting. Meanwhile, her best friend, who is a
first-class reporter, is in love with her. An intelligent and witty
comedy by writer/director/producer Brooks. Set and filmed in
Washington D.C.
COM   127 min (ort 132 min)   VIDrel: 20TH/TECH   V/sur

BROADWAY BILL ***
Frank Capra USA
*Warner Baxter, Myrna Loy, Walter Connolly, Helen Vinson,
Raymond Walburn, Douglass Dumbrille, Lynne Overman, Clarence
Muse, Margaret Hamilton, Paul Harvey, Jason Robards, Charles Lane,
Ward Bond, Claude Gillingwater*
An easygoing horse trainer discovers he has a champion nag in
this lively and zestful romantic comedy whose wisecracking
dialogue is reminiscent of Damon Runyon. Despite the lack of
a strong storyline, Capra's assured direction and a clutch of
good performances result in a film that's more memorable than
it has a right to be.
Aka: STRICTLY CONFIDENTIAL
COM   102 min (director's uncut version) B/W
VIDrel: VCC/DISC   V
Boa: short story by Mark Hellinger.

U
1934

BROADWAY BOUND ***
Paul Bogart USA
*Anne Bancroft, Hume Cronyn, Corey Parker, Jonathan Silverman,
Jerry Orbach, Michele Lee, Marilyn Cooper, Pat McCormick, Jack
Carter*
Written by Neil Simon, this semi-autobiographical work (it
follows on from BRIGHTON BEACH MEMOIRS AND BILOXI
BLUES) has a successful Broadway writer thinking back over his
life to his youth in the 1940s, when he lived at home and worked
in a department store. Three generations occupied these
cramped living quarters, and the film looks at their hopes, fears
and frustrations. A pleasing comedy even if it does cover over-
familiar territory.
Aka: NEIL SIMON'S BROADWAY BOUND
COM   89 min (ort 94 min)   VIDrel: TART/20TH   V/dm
Boa: play by Neil Simon.

PG
1991

BROADWAY DANNY ROSE ***
Woody Allen USA
*Woody Allen, Mia Farrow, Nick Apollo Forte, Howard Storm, Jackie
Gayle, Sandy Baron, Corbett Monica, Monty Gunty, Will Jordan,
Milton Berle, Joe Franklin, Braig Vanderburgh, Herb Reynolds, Paul
Greco, Frank Renzulli*
A Broadway theatrical agent remains ridiculously loyal to his
no-talent acts. He complicates his life even further by falling for
the wife of a Mafia hoodlum. Despite good performances. the
humour becomes a bit strained. Highlights are some of the over-
the-hill untalented performers our agent is trying to push.
Written and directed by Allen.
COM   81 min Cut (ort 86 min) B/W
VIDrel: VISVID/POLYREC   V

PG
1984

BROADWAY MELODY OF 1938 **
Roy Del Ruth USA
*Robert Taylor, Eleanor Powell, Judy Garland, Sophie Tucker, Binnie
Barnes, Buddy Ebsen, Sid Silvers, Billy Gilbert, Raymond Walburn,
George Murphy, Charles Igor Gorin, Robert Benchley, Willie Howard,
Charley Grapewin*
A young girl gets a chance in a troubled show with backstage
problems and soon proves herself to be a star, and when her
racehorse wins a major race, the financial backing for the show
is assured. By no means as good as the other two movies in the
"Broadway Melody" series, but the many song-and-dance
numbers still offer considerable enjoyment.
MUS   106 min (ort 113 min) B/W   VIDrel: MGM/WHV   V/sh

U
1937

BROKEN ARROW **
John Woo USA
*John Travolta, Christian Slater, Samantha Mathis, Delroy Lindo, Bob
Gunton, Frank Whaley, Howie Long, Vondie Curtis-Hall, Jack
Thompson, Vyto Ruginis, Ousaun Elam, Shaun Toub, Casey Biggs,
Jeffrey J. Stephen, Joey Box, Jon W. Kishi*
Slater plays a bomber pilot who has to find a way of stopping his
colleague Travolta from exploding a pair of nuclear warheads. A
spectacular series of set-piece confrontations are barely held
together by the unoriginal and predictable storyline.
THR   104 min wScrn cC   VIDrel: 20TH/TECH; ENCORE
(LV only)   V/sur LV

15
1996

BROKEN BADGES **
Kim Manners USA
*Miguel Ferrer, Eileen Davidson, Jay Johnson, Ismael (East) Carlos,
Tim Neil, Gary Chalk, Leslie Carlson, Gerry Dean, Rosanna Iverson,
Jessica Marlowe, Rob Roy, R. Nelson Brown, Teresa Donahoe, Ernie
Hudson, Ada Maris, Carlos Gomez*
An unconventional cop who doesn't mind breaking the rules
becomes convinced of the innocence of a young man accused of
murdering his parents. Pilot for a TV series, this is an occasion-
ally amusing comedy-thriller, much of the interest being
generated by a parade of oddball characters and wacky encounters.
THR   94 min mTV   VIDrel: COLUM/SONOP   V

15
1990

BROKEN BARS **
Tom Neuwirth USA
*Wings Hauser, Joe Estevez, Donald Gibb, John Cominsky, Benjamin
Kobby*
An FBI agent goes undercover in a tough Federal Penitentiary,
where violence has become a way of life for the inmates. One of
those tiresome "cop in jail" stories, exploring a plot idea that has
been done to death.
A/AD   97 min   VIDrel: GUILD/FOXVID   V

18
1995

BROKEN BLOSSOMS ***
D.W. Griffith USA
*Lillian Gish, Donald Crisp, Richard Barthelmess, Edward Peil, Arthur
Howard, George Beranger*
Silent melodrama set in London's Limehouse district where a
young Chinese falls in love with a street waif and is devastated
when she is killed by her brutal father. He murders the father
and then commits suicide. An often striking but inevitably dated
affair, whose full-blooded Victorian flavour is a little hard to
take seriously today. Remade in the UK in 1936.
Aka: YELLOW MAN AND THE GIRL, THE
DRA   95 min (ort 105 min at 16 fps) B/W (tinted version
available) Silent   VIDrel: THAMES/DISC   V/sh
Boa: short story The Chink And The Child from the book
Limehouse Nights by Thomas Burke.

PG
1919

BROKEN CORD, THE ***
Ken Olin USA
*Jimmy Smits, Kim Delaney, Michael Spears, Frderik Leader-Charge,
Deborah Duchene, Raiul Trujillo, Billy Merasty, Keith Dinicol,
August Schellenberg, Genevieve Appleton, Frank Burning, Chappell
Jaffe, Elizabeth Leigh-Milne*
Absorbing true story of mixed-race professional man who
attempts to adopt a four-year-old native American Indian child
suffering from mental handicap caused during pregnancy by
his mother's alcoholism.
DRA   88 min   VIDrel: CIC/SONOP   V
Boa: book by Michael Dorris.

PG
1991

BROKEN DREAM **
John Korty USA
John Lithgow, Mary Beth Hurt, Linda Kelsey, Ronny Cox
Overwrought melodrama that deals with a couple whose
child is born three months premature, and who find that they
are barely able to pay for the specialised medical care it
requires.
DRA   93 min (ort 95 min)   VIDrel: ITC/POLYREC   V

15
1990

BROKEN LANCE ***
Edward Dmytryk USA
*Spencer Tracy, Richard Widmark, Robert Wagner, Jean Peters, Katy
Jurado, E.G. Marshall, Earl Holliman, Hugh O'Brian, Eduard Franz,
Carl Benton Reid, Philip Ober, Robert Burton, Robert Adler, Robert
Garndlin, Harry Carter*
An ageing land baron remarries and then finds that his empire
is being torn apart by conflict with his three rebellious sons. An
intensely dramatic Western that is virtually a remake of "House
Of Strangers" from 1949. AA: Story (Philip Yordan).
WES   92 min (ort 96 min)   VIDrel: 20TH/TECH   V/sh

U
1954

BROKEN PLEDGES ***
Jorge Montesi USA
*Linda Gray, Leon Russom, David Lipper, Barry Bonds, Jane Galloway,
L. Munro, Malcolm Stewart, Fred Henderson, Chris Marlin, Emily
Perkins, Laurie Grogan, Jed Rees, Wade Anderson, Philip Granger, P.
Lynn Johnson, Jerry Wasserman*
When the son of couple goes to university he dies from alcohol
poisoning in suspicious circumstances during an initiation cere-
mony. His distraught mother attempts to uncover the truth

PG
1994

about what really took place there. A competent fact-based drama.
DRA 88 min VIDrel: NWV/HIFLI V/h

BROKEN PROMISES: TAKING EMILY BACK *** 15
Donald Wrye USA 1993
Cheryl Ladd, Robert Desiderio, Kathleen Wilhoite, Polly Draper, D. David Morin, Ted Levine, Tim Ransom, Patch McKenzie, Megan Braun, Amanda Braun, Robert Donley, Samantha Braun, Buckley Norris, James Harper, Jordan Ladd
A childless couple adopt a baby and find that the real parents (who are homeless) appear to now want their child back and are prepared to resort to blackmail. However, it transpires that this is nothing more than a money-making scheme dreamed up by the husband, which he has used twice before with their other children. A compelling drama, based on a true story of one such couple and their despicable activities.
DRA 89 min VIDrel: MARQ/FIRST V

BROKEN TRUST ** 12
Geoffrey Sax USA 1995
Tom Selleck, Elizabeth McGovern, Marsha Mason, Charles Haid, William Atherton
A district court judge receives a strange request from Federal agents who are planning an undercover operation whose intention is to reveal wrongdoing by one of his colleagues on the bench. However, this "sting" fails to come with the necessary evidence and so he is forced to continue in this role until the matter can be finally resolved.
DRA 92 min (ort 96 min) VIDrel: 20VIS/SONOP V
Boa: novel Court of Honor by William P. Wood.

BRONCO BILLY ** PG
Clint Eastwood USA 1980
Clint Eastwood, Sondra Locke, Geoffrey Lewis, Scatman Crothers, Sam Bottoms, Bill McKinney, Dan Vadis, Sierra Pecheur, Tanya Russell, William Prince, Tessa Richards, Walter Barnes, Beverlee McKinsey, Woodrow Parfrey, Hank Worden
A self-styled cowboy runs a small-time Wild West show which is joined by a spoiled heiress. A screwball comedy that digs deep for laughs but has a modicum of charm if not wit.
Aka: BILLY BRONCO
WES 111 min (ort 119 min) VIDrel: WHV V/h

BRONX TALE, A *** 18
Robert De Niro USA 1993
Robert De Niro, Chazz Palminteri, Francis Capra, Lillo Brancato, Taral Hicks Katherine Narducci, Clem Caserta, Alfred Sauchelli Jr, Frank Pietrangolare, Joe Pesci, Robert D'Andrea, Eddie Montanaro, Fred Rischer, Joseph D'Onofrio
A young boy growing up in an Italian neighbourhood in the Bronx in the 1960s witnesses a murder by the local don but keeps silent. This single act lays the foundation for a friendship and fascination with this gangster, much to the disgust of his honest, hard-working father. But when the boy reaches the age of seventeen he finds himself facing a difficult choice. De Niro's directing debut is a warm and well crafted drama.
DRA 116 min (ort 120 min) VIDrel: POLY/POLYREC L/A V/sur
Boa: play by Chazz Palminteri.

BRONX WARRIORS * 18
Enzo G. Castellari ITALY 1982
Vic Morrow, Christopher Connolly, Fred Williamson, Mark Gregory, Stefania Girolami, John Sinclair, Enio Girolami, George Eastman, Betty Dessy, Rocco Lerro, Massino Vanni, Angelo Ragusa
Set in the Bronx of 1990, this film tells the story of the warfare between Bronx gang leaders and the unsavoury hand of a corporation. Violent, trashy nonsense, filmed in the Bronx and Rome. Followed in 1983 by "Bronx Warriors 2: The Battle Of Manhattan".
Aka: 1990: THE BRONX WARRIORS; I CAVALIERI DEL BRONX
DRA 79 min (ort 84 min) wScrn VIDrel: UNIQUE/RTM L/A V

BROOD, THE *** 18
David Cronenberg CANADA 1979
Samantha Eggar, Oliver Reed, Art Hindle, Cindy Hinds, Nuala Fitzgerald, Susan Hogan, Robert Silverman, Gary McKeehan, Henry Beckman, Michael Magee, Joseph Shaw, Larry Solway, Reiner Schwartz, Felix Silla, John Ferguson
Mental derangement, murderous midgets and buckets of blood – a psychiatrist tries to help his patients by getting them to give birth to physical manifestations of their disturbances. One especially disturbed patient produces a brood of ghastly midget children who attack her perceived enemies, beating them to death with mallets. A sickening but compelling shocker made (as ever) in Cronenberg's inimitable style. Grade 10 on the yuck scale.
HOR 87 min (ort 92 min) VIDrel: ARROW/RTM V

BROTHER FROM ANOTHER PLANET, THE ** 15
John Sayles USA 1984
Joe Morton, Darryl Edwards, Steve James, Leonard Jackson, Maggie Renzi, Tom Wright, Caroline Aaron, Rossette Le Noire, Dee Bridgewater, Ren Woods, John Sayles, Reggie Rock Bythewood, David Strathairn, Josh Mostel
A black mute alien appears in Harlem where he feels at home, and just like Sellers in BEING THERE, impresses everyone he meets, just by letting them do all the talking. He is, however, on the run from his own planet. Made on a minuscule budget, but good dialogue helps it along. A worthy but flawed effort whose foolish drug dealing sub-plot could have been excised to the film's advantage.
FAN 104 min (ort 108 min) VIDrel: ARROW/RTM V

BROTHER FUTURE ** U
Alan Smithee (Roy Campanella II) USA 1991
Frank Converse, Michael Burgess, Carl Lumbly, Phil Lewis, Vonetta McGee, Akosua Busia, Moses Gunn, Bernard Addison, Kenyatta Jackson, Cornell Royal, William Crumby, Michael Flippo, Gene Jones, Johnny Heyward, Frank P. Jarrell
On the run from the Detroit police, a young black lad is involved in an accident and knocked unconscious, but awakens to find himself a slave on a South Carolina plantation in 1822. There he becomes involved in a plot to escape and learns a few useful lessons about the meaning of freedom. Slightly cringe-making if well intentioned, with a few contrived comic moments present to drive the message home. Disowned by the director, hence the pseudonym.
Aka: WONDERWORKS: BROTHER FUTURE
FAN 103 min (ort 110 min) VIDrel: MGM/WHV L/A V/sh

BROTHER SUN, SISTER MOON ** PG
Franco Zeffirelli ITALY/UK 1973
Graham Faulkner, Judi Bowker, Leigh Lawson, Kenneth Cranham, Lee Montague, Valentina Cortese, Alec Guinness, Michael Feast, Nicholas Willatt, John Sharp, Adolfo Celi, Francesco Guerrieri
A flower-power view of the life of St Francis of Assisi, complete with the songs of Donovan. Gentle and well-meaning with excellent photography, but the performances lack dramatic impact, resulting in a lifeless endeavour.
DRA 116 min (ort 122 min) VIDrel: CIC/SONOP V

BROTHERS' DESTINY *** (PG)
Dean Hamilton CANADA 1994
Charles Martin Smith, Dee Wallace Stone, Kris Kristofferson, Mickey Rooney, Danny Aiello, Will Estes, Keegan Macintosh, Robert Prosky
In 1930s New York two orphan boys run away together when they are separated by the adoption of one, their intention being to travel to Nebraska where they hope to find a home at the famed "Boys' Town" orphanage. Along the way they encounter many hazards and are assisted by some kindly hobos. A well made if overly sentimental tale, that won an award at the Chicago International Children's Film Festival.
Aka: LONG ROAD HOME, THE: ROAD HOME, THE
JUV 90 min SATrel: MOVIE CHANNEL

BROTHERS IN TROUBLE ** 15
Udayan Prasad GERMANY/ITALY/UK 1995
Om Puri, Pavan Malhotra, Angeline Ball, Ahsen Bhatti, Bhasker, Pravesh Kumar, Badi Uzzaman, Harmage Singh Kalirai, Kumall Grewal, Omar Salimi, Pal Aron, Shiv Grewal, Avin Shah, Lesley Clare O'Neill, Kulvinder Ghir
In the 1960s, a group of illegal immigrants from Pakistan work in miserable sweatshop conditions to support their families back home. However, the arrival of an idealistic woman gives them cause to hope for an improvement in their lot, but ultimately she proves to be the unwitting cause of their undoing. Deeply sincere in its intentions, this film misfires as a drama chiefly due to its uncertain scripting, and a final resolution that is contrived and awkward.
DRA 98 min (ort 103 min) VIDrel: CONNO/RTM V
Boa: novel Return Journey by Abdullah Hussein.

BROTHERS McMULLEN, THE **   15
Edward Burns USA   1995
Shari Albert, Maxine Bahns, Catharine Bolz, Connie Britton, Edward Burns, Peter Johansen, Jennifer Jostyn, Mike McGlone, Elizabeth P. McKay, Jack Mulcahy
Low-budget comedy (it was shot on 16 mm) that has three brothers finding their strict Catholic upbringing is in conflict with their amatory adventures. The debut film for Burns (who also wrote and produced it), this has a touch of Woody Allen-style angst about it, plus a rather fake Irish charm that eventually becomes just a little bit irritating.
COM   94 min (ort 98 min) cC   VIDrel: 20TH/TECH
V/sur

BROTHERS OF THE FRONTIER ***   (U)
Mark Sobel USA   1995
Joey Lawrence, Jonathan Frakes, Matthew Lawrence, Andrew Lawrence, Mark-Paul Gosselaar, Doug Abrhams, Glen Douglas, Leon S. Goodstriker, Mark Saunders, Richard Sali, Melanie Koskie, David Longworth, Randy Schooley, Theodore Starr
A frontier family in early 19th century America are wrongly accused of deer poaching on land belonging to a family of power-mad settlers who demand several times its value in cash to settle the matter. The father takes his family and runs but their adversaries burn their home to the ground and embark on a vendetta, pursuing them as they move westwards in the hope of capturing them. A nice little Western yarn, with an OK story and good performances.
WES   87 min mTV   SATrel: MOVIE CHANNEL

BROTHER'S PROMISE, A: THE DAN JANSEN
STORY **   (PG)
USA   1996
Matthew Kessler, Jayne Brook, Len Cariou, Christina Cox, Justin Louis, Patricia Gage, David Keely, Scott Speedman, Hugh Thompson, Mark Lutz, Richard Fitzgerald, Nora Sheehan, Scott Wickware, Sharri Hollett, Tom Hankam
A leisurely paced look at the career and life of this speed-skating champion who despite breaking several world records never quite managed to a achieve the success he dreamed of. However, the death of his sister from leukaemia plunged him into a profound despair that was only finally lifted when he won at the Winter Olympics. Very well acted but the lack of dramatic tension offers very little to those viewers who are not devotees of this particular sport.
DRA   90 min mTV   SATrel: MOVIE CHANNEL

BROWNING VERSION, THE ****   U
Anthony Asquith UK   1951
Michael Redgrave, Jean Kent, Nigel Patrick, Bill Travers, Ronald Howard, Wilfrid Hyde-White, Brian Smith, Paul Medland, Ivan Sampson, Peter Jones, Sarah Lawson, Josephine Middleton, Scott Harold, Judith Furse
A classic tale, with Redgrave starring as a once brilliant scholar who's now a middle-aged classics teacher at a boarding school. On the eve of his retirement because of ill health, a single act of kindness gives him the courage to face the future, despite the realisation that he is held in contempt by his adulterous wife, his pupils and fellow teachers. A moving and sensitive adaptation of a fine play. Remade in 1994.
DRA   90 min B/W   VIDrel: L/A   V
Boa: play by Terence Rattigan.

BROWNING VERSION, THE **   15
Mike Figgis UK/USA   1994
Albert Finney, Greta Scacchi, Matthew Modine, Julian Sands, Michael Gambon, Ben Silverstone, Maryma D'Abo, David Lever, Mark Long, Oliver Milburn, Belinda Low, Bruce Meyers, Jeff Nuttall, Heathcote Williams, George Harris, Dinah Stabb
An updated film adaptation of Rattigan's play about a Latin master at a British boarding who is just about to retire and feels unappreciated by all around him, including his younger and unfaithful wife. However, an unexpected act of kindness on the part of one of his pupils helps restore his faith in humanity. Finney is superb, but his efforts alone cannot infuse any life into this curiously flat drama.
DRA   93 min (ort 97 min) cC   VIDrel: CIC/SONOP   V
Boa: play by Terence Rattigan.

BRUBAKER **   15
Stuart Rosenberg USA   1980
Robert Redford, Yaphet Kotto, Jane Alexander, Murray Hamilton, David Keith, Morgan Freeman, Matt Clark, Tim McIntire, Linda
Haynes, M. Emmet Walsh, John McMartin, Richard Ward, Albert Salmi, Wilford Brimley, Everett McGill, Val Avery
An idealistic prison chief faces opposition in his attempts to introduce humane conditions and retain his principles. After a harrowing introduction that effectively depicts the horror and brutality of prison life, the film reverts to a predictable black-and-white tale of how reaction triumphs over reform. Redford plays the title character as an enigmatic loner whose inner feelings and motivations are never brought to the surface. Disappointing.
DRA   125 min (ort 131 min)   VIDrel: 20TH/TECH   V

BRUCE AGAINST IRON HAND **   18
To Lo-Po HONG KONG
Bruce Li (Ho Tsung-Tao), Shao Lung, Bruce Liang
Inspector Li investigates the mysterious deaths of two masters of the martial arts. Another run-of-the-mill kung fu adventure with a Bruce Lee clone.
MAR   85 min (ort 89 min)   VIDrel: IMPENT   V

BRUCE LEE FIGHTS BACK FROM THE GRAVE *   18
Duoo Yong Lee HONG KONG   1976
Bruce K.L. Lea, Deborah Chaplin, Anthony Bronson, Steve Mak, Jack Houston, Charlie Chow, Philip Kenendy, Jimmy Sato, Dan Inosanto, Jun Chong
Turgid martial arts tale with a supernatural flavour as Bruce Lee returns from the dead to wreck havoc on those responsible for his untimely demise. An abysmally poor offering.
MAR   84 min (ort 97 min)   VIDrel: 4-FRONT/POLYREC   V

BRUCE LEE: THE MAN – THE MYTH **   18
Ng See Yuen HONG KONG   1976
Bruce Li (Ho Chung Tao), Donnie Williams, Liang Shao Sung, Chun Yen, Chi Ling Tsao, Cha Chi Song, David Chow, Unicorn Chan, Sham Chien Po, Tsui Chung Shun, Sing Ho, Ke Hung Fung, Ching Wan Fung, Ming Ko, Linda Herst, Tai Ming Lin
Biopic about the life and times of Bruce Lee, from his student days to his involvement in films. Formula martial arts outing trading on the old Bruce magic.
Aka: BRUCE LEE: THE TRUE STORY; LI HSIAO-LUNG CH'UAN-CH'I
MAR   90 min (ort 104 min)   VIDrel: MIA/DISC   V

BRUCE THE KING OF KUNG FU **   18
Bruce Le (Huang Kin Lung) HONG KONG   1984
Bruce Le (Huang Kin Lung), Shih King
OK biopic about the famous kung fu actor whose career was so abruptly cut short. This one traces his life from birth in America to his introduction in Hong Kong to the martial arts.
MAR   87 min   VIDrel: IMPENT   V

BRUCE THE SUPERHERO **   18
Bruce Le (Huang Kin Lung) HONG KONG   1984
Bruce Le (Huang Kin Lung), Lito Lapid, Azenith Briones, Kong To, Yong Sze, Chai Ching Tao
Bruce avenges his father's murder by the Black Dragon Society. If only he would avenge himself on the writers of these interminable Bruce-films.
Aka: BRUCE LEE: SUPERHERO; BRUCE LEE VERSUS THE BLACK DRAGON; BRUCE LE VERSUS NINJA
MAR   84 min Cut (1 min 38 sec – ort 90 min)
VIDrel: IMPENT   V

BRUCE VERSUS BILL **   (PG)
Lam Kwok Cheung HONG KONG   1975
Bruce Le, Bill Louie, Angela Yu Ching, Alexandre, Tien Ching, Liu Yi Fan, Kong Do, Ma Cheung, Fung Ruen Chuen, Cham kou, Lam Kwok Wah
Two Kung Fu fighters clash over buried treasure, a large sum of money hidden in a box, and fight it out. The usual mayhem ensues.
MAR   90 min   SATrel: SKY MOVIES

BUBBLEGUM CRASH: PARTS 1 TO 3 ***   12
Noda Yasuyaki/Kiyotsumum Toshifumi/Michael House
JAPAN   1991
Voices of: Sakakibara Yoshiko, Tachikawa Ryooko, Tomizawa Michie, Hiramatsu Akiko, Furukawa Toshio, Horiuchi Kenyuu, Sogabe Masayoshi, Arakaw Taroo, Suzuki Kiyonobu, Murayama Akira, Yanada Kiyoshi
In this sequel to BUBBLEGUM CRISIS, the Knights Saber are in bad shape and find themselves fighting an old enemy who is planning to sabotage the nuclear power plant on which the city

of MegaTokyo depends. Another helping of this technically innovative but very dystopic version of the future.
ANIM 135 min (3 cassettes – separately available) dubbed
VIDrel: MANGA/SONOP V/sh

BUBBLEGUM CRISIS: VOLS. 1 TO 8 ***
PG/12
Akiyama Katsuhito JAPAN
1987
Unusual cyberpunk saga of a band of high-tech vigilantes called the Knights Saber who come into conflict with the sinister Genom corporation and its evil androids. Set in MegaTokyo in the year 2032, when the city is still recovering from the effects of a devastating earthquake. Followed by the sequel BUBBLEGUM CRASH: VOLS. 1 TO 3.
ANIM 360 min (8 cassettes – certifications vary) dubbed
VIDrel: ANIME/RTM V/sh

BUCCANEERS, THE ***
15
Philip Saville USA
1994
Cherie Lunghi, James Frain, Jenny Agutter, Connie Booth, Michael Kitchen, Mira Sorvino, Rya Khlstedt, Carla Gugino, Alison Elliott, Sheila Hancock
The story of four well-bred American ladies who descend on British society in the 1870s, all intent on catching rich husbands. Their various ploys and intrigues make for fascinating viewing, even if at times the plot (one of them marries a homosexual aristocrat who merely wants to produce an heir) does at times strain credibility.
DRA 288 min (2 cassettes – ort 375 min) mTV
VIDrel: BBC/TECH V/h
Boa: novel by Edith Wharton

BUCK AND THE PREACHER ***
PG
Sidney Poitier USA
1972
Sidney Poitier, Harry Belafonte, Ruby Dee, Cameron Mitchell, Denny Miller, Nita Talbot, John Kelly, Tony Brubaker, James McEachin, Clarence Muse, Ken Menard, Julie Robinson, Bobby Johnson, Lynn Hamilton, Errol John, Fred Waugh
Poitier's directorial debut tells of a scout and con-man preacher helping a band of escaped slaves on the run, the latter having come into conflict with some racist cowboys. Strong characterisations, some sharp and original observations and the music of Benny Carter help make up for the lack of action.
WES 99 min (ort 102 min)
VIDrel: FABFIL/SPEAR/COLUM V

BUCK ROGERS IN THE 25TH CENTURY *
15
Daniel Haller USA
1979
Gil Gerard, Pamela Hensley, Erin Gray, Tim O'Connor, Henry Silva, Joseph Wiseman, Felix Silla, Duke Butler, H.B. Haggerty, Caroline Smith, Kevin Coates, John Dewey-Carter, David Cadiente, Gil Serna, Mel Blanc (voice only)
Pilot for a TV series with Buck Rogers, the legendary space hero, returning in this modern version of the 1930s serial. In this adventure he has to prove that he is not in league with space pirates. A glossy and superficial effort, using much of the hardware and sets left over from shooting another made-for-TV SF outing, BATTLESTAR GALACTICA, which had the same producers. This pilot was released to cinemas first. Mediocre.
Aka: BUCK ROGERS
FAN 89 min mTV VIDrel: CIC/SONOP L/A V

BUCKAROO BANZAI **
15
W.D. Richter USA
1984
Peter Weller, John Lithgow, Ellen Barkin, Clancy Brown, Jeff Goldblum, Christopher Lloyd, Lewis Smith, Rosalind Cash, Robert Ito, Pepe Serna, Ronald Lacey, Matt Clark, William Traylor, Carl Lumbly, Vincent Schiavelli
A multi-talented pulp-fiction hero does battle with aliens from another dimension, who threaten to unleash a nuclear war by taking over the body of a mad scientist. A confused attempt at a credible SF adventure that could have worked well, but was badly let down by the incoherent script. Despite this (or perhaps because of it), the film now has something of a cult following.
Aka: ADVENTURES OF BUCKAROO BANZAI, THE; ADVENTURES OF BUCKAROO BANZAI ACROSS THE EIGHTH DIMENSION
FAN 94 min (ort 103 min) VIDrel: UNIQUE/PARADOX V/sh

BUCKET OF BLOOD, A **
15
Roger Corman USA
1959
Dick Miller, Anthony Carbone, Ed Nelson, Barboura Morris, Julian Burton, Bert Convy, John Brickley, Judy Bamber, Jean Burton, John Shaner, Myrtle Domerel, Bruno VeSota

A coffee-shop waiter longs for fame as a beat poet or artist just like some of his customers. One day, he accidentally kills his landlady's cat and covers its body in clay to hide the evidence. The resulting sculpture is hailed as a masterpiece, his new career is born, but he graduates from felines to humans. A witty, low-budget spoof, with its tongue firmly in its cheek, this is one of the director's most substantial films. Remade (after a fashion) as DARK SECRETS.
HOR 65 min B/W VIDrel: SCREAM/SPEAR V

BUDDHIST FIST **
15
Yuen Woo Ping HONG KONG
Jet Lee
The story of two orphans, both of whom are trained in the title style, but are then separated, only to be reunited when a series of catastrophes bring them into contact once more. Familiar martial arts mayhem, done with considerable panache if not originality.
MAR 92 min wScrn VIDrel: MADE/RTM V

BUDDY BUDDY ***
15
Billy Wilder USA
1981
Jack Lemmon, Walter Matthau, Paula Prentiss, Klaus Kinski, Ed Begley Jr, Dana Elcar, Miles Chapin, Joan Shawlee, Michael Ensign, Fil Formicola, C.J. Hunt, Bette Raya, Ronnie Sperling, Suzie Geller, Bill Manard, Ed Begley Jr
An adaptation of the French film farce, "A Pain In The A—", this black comedy revolves around a hitman and the complications he is caused by a would-be suicide, who's unhappy over the break-up of his marriage. A funny and uncomplicated farce, with the two leads playing their parts to perfection.
COM 92 min (ort 96 min) VIDrel: MGM/WHV V/h

BUDDY HOLLY STORY, THE ****
PG
Steve Rash USA
1978
Gary Busey, Don Stroud, Charles Martin Smith, Maria Richwine, Amy Johnston, Conrad Janis, Dick O'Neill, Bill Jordan, Fred Travalena, Stymie Beard, John F. Goff, Albert Popwell, Jim Beach, M.G. Kelly, Paul Mooney, Freeman King
Musical drama about the life and death of this great rock 'n roll star, with Busey quite outstanding as the singer, in a film that steers well clear of the usual Hollywood glossy excesses and goes for a straight biopic instead. Busey, Smith and Stroud sing and play "live" in this memorable musical docu-drama. In real life Busey is a part-time rock musician. AA: Score (Joe Benzetti).
MUS 108 min (ort 113 min) VIDrel: GUILD/POLYREC V

BUDDY'S SONG **
PG
Claude Whatham UK
1990
Chesney Hawkes, Roger Daltrey, Sharon Duce, Michael Elphick, Liza Walker
A curious film that blends kitchen-sink pathos with nostalgia for those good-old-days of rock 'n' roll, something on the lines of Whatham's first feature THAT'LL BE THE DAY. Daltrey plays an ageing Teddy Boy, who is estranged from his wife, gets on the wrong side of the law and finally goes to jail. His son meanwhile starts writing songs, eventually wins a record contract and lives to see Dad make a fresh start of his own. Screenplay is by Hinton.
DRA 102 min VIDrel: 20TH/TECH V/sur
Boa: novel by Nigel Hinton.

BUFFALO BILL AND THE INDIANS, OR SITTING BULL'S HISTORY LESSON **
PG
Robert Altman USA
1976
Paul Newman, Burt Lancaster, Joel Grey, Kevin McCarthy, Harvey Keitel, Allan Nicholls, Geraldine Chaplin, John Considine, Robert Doqui, Mike Kaplan, Bert Remsen, Bonnie Leaders, Denver Pyle, Will Sampson, Pat McCormick, Shelly Duvall
Buffalo Bill's Wild West Show serves as a vehicle to show how history has been re-written and glamorised. A dull effort that makes a few interesting points regarding the glamorisation of the past, and then has little more to offer. Supposedly a comedy-Western, it does justice to neither genre.
Aka: BUFFALO BILL
WES 100 min (ort 135 min) VIDrel: 4-FRONT/POLYREC L/A V
Boa: play Indians by Arthur Kopit.

BUFFALO GIRLS ***
15
Rod Hardy USA
1994
Angelica Huston, Melanie Grififth, Gabriel Byrne, Peter Coyote, Sam Elliott, Reba McEntire, Jack Palance, Floyd Red Crow Esterman,

Charlayne Woodward, Jhn Diehl, Andrew Bicknell, Paul Lazar, Russell Means
Colourful and larger-than-life made-for-TV miniseries about the life (and loves) of Calamity Jane and her encounters with other famous figures of the Old West, including Buffalo Bill and Annie Oakley. Although hardly accurate in terms of actual history, the well-recreated period atmosphere and detail, and the solid performances, make for quite an enjoyable time.
WES 120 min (ort 180 min) mTV VIDrel: 20TH/FOXVID V/h
Boa: novel by Larry McMurty.

BUFFET FROID *** 15
Betrand Blier FRANCE 1979
Gerard Depardieu, Bernard Blier, Jean Carmet, Genevieve Page, Denise Gence, Carole Bouquet, Jean Benguigui, Michel Serrault
Absurd but entertaining black comedy about a group of incompetent killers, with Depardieu starring as a suspected serial killer, who finds himself matching wits with a crooked police inspector when his switchblade is linked to the murder of a commuter.
Aka: COLD CUTS
COM 89 min (ort 95 min) VIDrel: ARTPRO/RTM V

BUFFY THE VAMPIRE SLAYER ** PG
Fran Rubel Kuzui USA 1992
Kristy Swanson, Luke Perry, Donald Sutherland, Rutger Hauer, Paul Reubens, Michele Abrams, Hilary Swank, Paris Vaughan, David Arquette, Candy Clark, Randall Batinkoff, Natasha Gregson Warner, Mark DeCarlo, Tom Jones, Liz Smith
A Southern Californian girl learns that she is descended from a long line of vampire hunters and must now accept this role and rid L.A. from a plague of these creatures. A mildly amusing spoof that is very good at satirising youth culture but fails to generate any real feeling of suspense, with Hauer's performance a major disappointment.
COM 81 min (ort 94 min) cC VIDrel: 20TH/TECH V/sur

BUFORD'S BEACH BUNNIES * 18
Frderick P. Watkins USA 1992
Jim Hanks, Rikki Brando, Monique Parent, Amy Page, Barrett Cooper, Ina Rogers, Charley Rossman, David Robinson, Adam Wahl, Robyn Webb, Bettina Brancato, Myles Furlow, David Damien, Henry Capana, Aaron Steele, John Callan
A fast food king, whose winning concept involves the use of scantily clad waitresses, wants to leave his business to his son, but is worried by the latter's fear of women, until three of his employees lend a hand. A broad farce with ample nudity but little wit.
COM 101 min VIDrel: IMPENT V

BUGS ** 18
Brian Yuzna USA 1990
Clint Howard, Neith Hunter, Maud Adams, Reggie Bannister, Tommy Hinkley, Aliyce Beasley
A female journalist checks out a case of spontaneous human combustion and her investigations lead her to a weird cult of women who worship a giant insect, their belief being that they can rid themselves of their fear of men by giving birth to a parasitic creature. A bit like the earlier SOCIETY, this is another attempt to employ horror in the service of social satire, only not so effective. Screenplay is by Woody Keith.
Aka: SILENT NIGHT, DEADLY NIGHT 4: THE INITIATION
HOR 86 min VIDrel: GUILD/FOXVID V

BUGSY *** 18
Barry Levinson USA 1991
Warren Beatty, Annette Bening, Harvey Keitel, Ben Kingsley, Joe Mantegna, Elliott Gould, Richard Sarafian, Bebe Neuwirth, Gian-Carlo Scandiuzzi, Wendy Phillips, Stefanie Mason, Kimberly McCullough, Andy Romano, Robert Beltran
The story of Benjamin "Bugsy" Siegel, a shrewd and murderous gangster who made a fortune setting up a gambling empire in Las Vegas, and enjoyed a lifelong love affair with show business. Scripted by James Toback, this is an expensive and lavish portrait of the gangster in his later years, and affords Beatty one of his best roles in years, in a film that displays a grudging admiration for a callous and cold-hearted killer.
DRA 130 min (ort 136 min) wScrn (COLUM/SONOP) cC
VIDrel: VCC/DISC/COLUM; COLUM/SONOP V/sur
Boa: book We Only Kill Each Other: The Life And Bad Times Of Bugsy Siegel by Dean Jennings.

BUGSY MALONE *** U
Alan Parker UK 1976
Scott Baio, Jodie Foster, Florrie Dugger, John Cassisi, Martin Lev, Albin Jenkins, Paul Murphy, Davidson Knight, Sheridan Earl Russell, Paul Christelson, Dexter Fletcher, Vivienne McKonne, Helen Corran, Andrew Paul, Jon Zebrowski
A 1920s gangster musical spoof with all the parts played by kids. The machine-guns fire custard pies and the sedans are driven with pedals. Only the lack of a decent story lets it down. A unique "homage" of sorts, as sweet and messy as the custard pies the kids fire at each other.
MUS 90 min (ort 93 min) VIDrel: VCC/DISC V

BULL DURHAM *** 18
Ron Shelton USA 1988
Kevin Costner, Susan Sarandon, Tim Robbins, Trey Wilson, Robert Wuhl, Jenny Robertson, Max Patkin, William O'Leary, David Neidorf, Danny Gans, Tom Silardi, Lloyd Williams, Rick Marxan, George Buck, Jenny Robertson, Greg Avelone
Story set against the world of major league baseball, where an attractive groupie feels it is her mission in life to pick a new promising youngster each season and teach him some maturity. Costner plays a tough older veteran of the game who has been assigned to one particular player to teach him some discipline. When she chooses this very player a clash of personalities becomes inevitable. A leisurely but charming romantic comedy.
COM 103 min (ort 108 min)
VIDrel: 4-FRONT/POLYREC/GUILD V/sur

BULLET FOR A BADMAN *** PG
R.G. Springsteen USA 1964
Audie Murphy, Darren McGavin, Ruta Lee, Beverly Owen, Skip Homeier, George Tobias, Alan Hale, Berkeley Harris, Edward C. Platt, Kevin Tate, Cece Whitney
A man has to bring in an old friend, who has now become an outlaw. To make matters worse, our hero is now married to his former wife, which hardly adds to his popularity. A strong Murphy vehicle with both him and McGavin in fine acting form.
WES 76 min (ort 80 min)
VIDrel: 4-FRONT/POLYREC/CIC V/sur

BULLET FOR THE GENERAL, A ** 18
Damiano Damiani ITALY 1967
Klaus Kinski, Gian Maria Volante, Lou Castel, Martine Beswick, Bianca Manini, Jaimie Fernandez, Andrea Checci, Jose Manuel Martin, Spartaco Conversi, Joaquin Parra, Aldo Sambrell, Santiago Santos, Valentino Macchi
During the Mexican revolution, a government agent is hired to kill a revolutionary general, and has to win the loyalty of the guerilla leader in order to accomplish this task. Average Italian outing to Mexico.
Aka: EL CUCHO QUIEN SABE; QUIEN SABE
A/AD 112 min (ort 118 min) wScrn
VIDrel: 4-FRONT/POLYREC V

BULLET IN THE HEAD *** 18
John Woo HONG KONG 1990
Tony Leung, Jacky Cheung, Waise Lee, Simon Yam, Bennie Yuen, Yolinda Yam, Kovit Wattanakul, Therdporn Manopaibool, Damrongphandu Sudrak, Suchai Thailuah, Thirasal Sinsoongsud, Somasal Suenguildal
A long, sprawling and ultra-violent tale of three Hong Kong friends who flee from there after a gangland vendetta and unwisely decide to go to Saigon. In Vietnam they fall into the hands of the Viet Cong, who torture them mercilessly, with one of them surviving being shot in the head (hence the title). This injury and the betrayal and disloyalty he experiences eventually combine to drive him insane. A brutal and bloody film, it has strong echoes of THE DEER HUNTER.
A/AD 126 min wScrn VIDrel: MADE/RTM V

BULLET TO BEIJING ** 15
George Mihalka CANADA/RUSSIA/UK 1995
Michael Caine, Jason Connery, Mia Sara, Michael Sarrazin, Michael Gambon, Burt Kwouk, Sue Lloyd
Caine makes a return as Deighton's working class secret agent Harry Palmer, who is brought back out of an enforced early retirement when he is hired to recover a powerful biological weapon that has been stolen in Russia. The expected intrigues and double-crosses are soon taking place as he struggles to locate it and prevent a catastrophe. Refreshing in its low-budget

approach, but not exactly memorable. Followed by MIDNIGHT IN ST PETERSBURG.
Aka: LEN DEIGHTON'S BULLET TO BEIJING
THR   101 min cC   VIDrel: TOUCH/TECH   V/sur
Boa: novel by Len Deighton.

BULLETS OR BALLOTS ***   PG
William Keighley USA   1936
Edward G. Robinson, Joan Blondell, Barton MacLane, Humphrey Bogart, Frank McHugh, Dick Purcell, George E. Stone
A city cop pretends to quit the force, but this is only a ploy in his attempt to infiltrate a mobster's gang. A taut and energetic crime story, one or two rungs short of classic status.
DRA   78 min (ort 81 min) B/W   VIDrel: WHV   V

BULLETS OVER BROADWAY ***   15
Woody Allen USA   1994
John Cusack, Dianne Wiest, Jennifer Tilly, Jack Warden, Chazz Palminteri, Rob Reiner, Mary-Louise Parker, Harvey Fierstein, Jim Broadbent, Tracey Ullman, Tony Sirico, Joe Viterelli, Harvey Broadbent, Stacey Nelkin, Paul Herman
In the 1920s, a New York playwright gets a chance to have his play staged on Broadway but learns that he has to pay a high price. Not only is the money to be provided by a gangster but the latter is only prepared to do so if his mistress gets a leading role. Unfortunately, her acting talent is so small as to be invisible. Excellent period detail, a strong script and much fine acting all combine to make this a most enjoyable offering. AA: S. Actress (Wiest).
COM   95 min (ort 115 min) cC   VIDrel: TOUCH/TECH   V

BULLITT ****   15
Peter Yates USA   1969
Steve McQueen, Robert Vaughn, Jacqueline Bisset, Simon Oakland, Dan Gordon, Robert Duvall, Norman Fell, Georg Stanford Brown, Justin Tarr, Victor Tayback, Carl Reindel, Felice Orlandi, Robert Lipton, Ed Peck, Pat Renella, Paul Genge
Now best remembered for a classic car chase, this film stars McQueen as an unconventional cop whose suspicions are aroused when he's put on an assignment guarding a Mafia squealer. A taut action film with some great San Francisco locations, concise editing and a sharp script. AA: Edit (Frank P. Keller).
THR   109 min (ort 113 min)   VIDrel: WHV   V
Boa: play Mute Witness by Robert L. Pike.

BULLSEYE! *   15
Michael Winner USA   1990
Michael Caine, Roger Moore, Sally Kirkland, John Cleese, Patsy Kensit, Jenny Seagrove
Both Caine and Moore each attempt two roles in this clumsy romp – with them playing an upper-class crook and a cockney con-man who just happen to resemble a pair of scientists. The plot involves the crooks planning the theft of some gems from the scientists. Meanwhile, these latter two have plans of their own to exploit a process for using nuclear fusion to provide cheap energy and hopefully make themselves a fortune. A confused, frenetic and badly scripted dud.
COM   89 min (ort 96 min)   VIDrel: 4-FRONT/POLYREC
L/A   V/sh

BUMP IN THE NIGHT ***   15
Karen Arthur USA   1991
Christopher Reeve, Wings Hauser, Meredith Baxter-Birney, Geraldine Fitzgerald, Shirley Knight, Richard Bradford, Anne Twomey, Terrence Mann, Travis Swords, Corey Carrier, Richard Joseph Paul, Leslie Gail, Audra Blaser
Baxter-Birney plays an alcoholic investigative journalist desperate to rescue her young son who has been abducted by a convicted paedophile whose intention it is to deliver the boy to a porno film-maker. A rather effective thriller that casts its net a little too wide – tackling both alcoholism and paedophilia at once. The script is by Christopher Lofton.
THR   91 min (ort 96 min) mTV   VIDrel: ODY/SONOP
V/sh
Boa: novel by Isabelle Holland.

BURBS, THE **   PG
Joe Dante USA   1989
Tom Hanks, Corey Feldman, Rick Ducommon, Carrie Fisher, Bruce Dern, Courtney Gaines, Wendy Schaal, Gale Gordon, Henry Gibson, Brother Theodore, Robert Picardo, Dick Miller, Cory Danziger, Franklin Ajaye, Rance Howard, Nick Katt
The tale of a fairly ordinary guy, whose nosey neighbours get more than they bargained for when their snooping gets out of hand, especially regarding the eerie Klopeks, whose house looks as though it's haunted. A mildly amusing comedy outing from the director of GREMLINS.
COM   97 min Cut (26 sec – ort 101 min)
VIDrel: CIC/SONOP   V/sur

BURDEN OF PROOF, THE ***   15
Mike Robe USA   1992
Brian Dennehy, Hector Elizando, Mel Harris, Adrienne Barbeau, Anne Bobby, Concetta Tomei, Gail Strickland, Chelcie Ross, Kerri Green, Thomas Anthony Quinn, Neal McDonough, Jeffrey Tambor, Stefanie Powers, Victoria Principal
When he learns of his wife's suicide, a successful attorney is overcome by feelings of grief and guilt. Unable to understand her motives, he begins to dig a little deeper and eventually uncovers a nasty conspiracy that is linked to a case in progress. A long but eminently watchable adaptation of Turlow's novels, with some solid performances, most notably Dennehy who is as convincing as ever. An above-average mini-series, but at three hours best taken in two sittings.
DRA   174 min (2 cassettes – ort 184 min) mTV
VIDrel: ODY/SONOP   V/sh
Boa: novel by Scott Turlow.

BURGLAR *   15
Hugh Wilson USA   1987
Whoopi Goldberg, Bobcat Goldthwait, G.W. Bailey, Lesley Ann Warren, James Handy, Anne De Salvo, John Goodman, Barbara Simpson, Elizabeth Ruscio, Vyto Ruginis, Larry Mintz, Raye Birk, Eric Poppick, Scott Lincoln, Thorn Bray
A cool cat burglar witnesses a murder and then finds that she has been framed as the culprit. With the aid of a crazy friend, she causes general chaos in San Francisco, trying to solve the crime in order to clear herself, whilst both avoiding the police and the killer, who wants to make her his next victim. This clumsy mixture of comedy and suspense is neither funny nor taut, and is well and truly a waste of Goldberg's considerable talents.
COM   98 min (Cut at film release by 11 sec – ort 103 min)
VIDrel: MGM/WHV   V/sur
Boa: novel by Lawrence Block.

BURIED ALIVE *   18
Gerard Kikoine USA   1989
Donald Pleasence, John Carradine, Robert Vaughn, Nia Long, Ginger Allen, Karen Witter
A highly strung female teacher takes a post at a reform school for girls that's run by a charismatic principal. Difficult girls tend to meet nasty ends (such as being buried in concrete) and when our teacher's suspicions are aroused by several disappearances she decides to investigate. A dreary little shocker of neither force nor originality. The alternative title is misleading as there is precious little similarity between this and any Poe story.
Aka: EDGAR ALLAN POE'S BURIED ALIVE
HOR   90 min (ort 97 min)   VIDrel: EIV/SONOP   V

BURNING, THE **   18
Tony Maylam USA   1980
Brian Matthews, Leah Ayres, Brian Backer, Larry Joshua, Shelley Bruce, Lou David, Holly Hunter, Jason Alexander, Ned Eisenberg, Carrick Glenn, Carolyn Houlihan, Fisher Stevens
Five teenagers at summer camp play a nasty trick on a caretaker which goes horribly wrong. Years later, he comes back to claim his revenge. A gory rip-off of an idea that has generally been cultivated by all those FRIDAY THE 13TH films – the gruesome make-up is by Tom Savini. This was Hunter's screen debut. Unlikely to be available in an uncut form in the UK.
HOR   87 min (ort 90 min)   VIDrel: VIPCO/SGSVID   V/h

BURNING BED, THE ***   15
Robert Greenwald USA   1984
Farrah Fawcett, Paul LeMat, Grace Zabriskie, Richard Masur, Christa Denton, Penelope Milford, James Callahan, Gary Grubbs, David Friedman, David Andrews, James Hampton, Virgil Frye, Dixie Wade, Heather Rich, Justin Gocke
A battered working-class wife suffers years of beatings and humiliation at the hands of her violent husband until she snaps and sets him on fire, whilst he is in bed. A harrowing and totally compelling drama with Fawcett giving a fine portrayal of a

woman driven to take an appalling revenge. The script is by Rose Leiman Goldenberg. Based on a true case.
DRA   91 min (ort 100 min) mTV   VIDrel: ODY/SONOP V/h
Boa: novel by Faith McNulty.

BURNING PARADISE ***
Ringo Lam HONG KONG
Lee Tien San, Willie Chi Tian-Sheng, Carmen Lee Yeuk-Tung, Yeung Sing, Maggie Lin Chuan, Yuen Kam-Fai, John Ching Tung, Wong Kam-Kong
18
1994

Two fighters set out to rescue some Shaolin monks who were taken hostage when the evil Red Lotus sect attacked and destroyed their temple, carrying the monks off to their impregnable fortress. A breathless action story, vigorous and full of exciting swordplay, and slightly out of the ordinary thanks to a few supernatural touches.
Aka: RAPE OF THE RED TEMPLE
MAR   104 min wScrn   VIDrel: MADE/RTM   V

BURNING PASSION: THE MARGARET MITCHELL STORY **
Larry Peerce USA
Shannen Doherty, Rue McClanaahan, Dale Midkiff, Stephen Ayres, Robert Treveiler, Matt Mulhern, Ann Wedgeworth, Morgan Weisser, ·John Clark Gable, J. Michael Hunter, Robin Mullens, Beatrice Bush, Shannon Eubanks, Amy Bush
(PG)
1994

Interesting biopic on the title figure, the Pulitzer-prize winning authoress of "Gone With The Wind". The story opens at the turn of the century, when Mitchell first gains an insight into life in the Old South, chiefly thanks to the stories her grandmother regales her with. Various other formative events in her life are then dealt with, such as an early love affair, and the influence these have in shaping her character. A competent and unadorned account.
DRA   87 min (ort 90 min) mTV   SATrel: SKY MOVIES

BURNING SEASON, THE: THE CHICO MENDES STORY ***
John Frankheimer USA
Raul Julia, Sonia Braga, Kamala Dawson, Edward James Olmos, Luis Guzman, Nigel Havers, Tomas Milian, Esai Morales, Tony Plana, Carmen Argenziano, Marco Rodriguez, Carlos Carrasco, Jonathan Carrasco, Jeffrey Licon, Jose Perez
15
1994

Biopic about the life and subsequent martyrdom of Chico Mendes, a Brazilian rubber tapper who became a political activist when he begun to understand the environmental impact of the destruction of rain forests in order to allow cattle ranching. Well acted but the black-and-white presentation of the issues makes for a rather predictable time.
DRA   117 min (ort 123 min) mCab   VIDrel: WHV   V
Boa: novel by Andrew Revkin.

BURNING SECRET ***
Andrew Birkin UK/USA/WEST GERMANY
Faye Dunaway, Klaus Maria Brandauer, Ian Richardson, David Eberts, John Nettleton, Martin Obernigg, Vadav Stekl, Vladimir Pospisil, Ivo Niedrle, K. Karvas-Kratochvil, Jarmila Derkova, Josef Dubicek, Veronika Jenikova
PG
1988

A charming but amoral baron, is recovering from a WW1 bayonet wound at a sanatorium, and embarks on an affair with a woman who has brought her asthmatic son there, initiating a friendship with the son in order to reach the mother. An absorbing drama slightly spoilt by miscasting of Dunaway, though Brandauer as the baron is excellent. Filmed in Prague and Marienbad and made once before in 1933 by Robert Siodmak as "Brennendes Geheimnis".
DRA   103 min   VIDrel: FIRST/SONOP L/A   V
Boa: short story by Stefan Zweig.

BURNT BY THE SUN ****
Nikita Mikhlahov FRANCE/RUSSIA
Nikita Mikhalkov, Oleg Menshikov, Nadia Mikhalkova, Ingeborga Dapkounaite, Andre Umansky, Viacheslav Tikhonov, Svetlana Kriuchkova, Vladimir Ilyin, Alla Kazanskaia, Nina Arkhipova, Avangard Leontiev, Inna Ulianova, Liubov Rudneva
15
1995

In 1936, a man who took part in the 1917 revolution and has been a loyal member of the Party, is unwilling to accept the enormity of Stalin's crimes against the nation. Taking a quiet vacation in the countryside with his younger wife, he is disturbed by the arrival of one of her old boyfriends, who works for the secret police. A beautiful and touching epic,

with touches of fantasy, that offers excellent acting and direction. AA: Foreign.
Aka: SOLEIL TROMPEUR; UTOMLENNYE SOLNTSEM; UTOML'ONY SONTSEM
DRA   130 min (ort 134 min) wScrn   VIDrel: GUILD/20TH V/sh
Boa: story by Nikita Mikhalkov.

BURY ME IN NIAGARA **
Dave Thomas USA
Jean Stapleton, Geraint Wyn Davies, Shae D'Lyn, Denis Akiyama, Ed Sahely, Jayne Eastwood, Zachary Bennett, Bernard Behrens, Stuart Clow, Jeff Pustil, John Thomas, Dennis Sweetung, Jack Duffy, Patrick Patterson, Reg Dreger
(PG)
1992

Quirky comedy with a supernatural flavour in which the main character, who is a doctor in this thirties, retrieves a valuable jewel whilst operating on a Japanese gangster. He passes it on to his mother, but she dies a few days later and is buried with it. A little later she returns as a ghost to help him out of his difficulties as the gangsters want the gem back and in addition, get herself buried at Niagara. Quite good fun, if a bit derivative.
COM   89 min   SATrel: SKY MOVIES

BUS STOP ****
Joshua Logan USA
Marilyn Monroe, Arthur O'Connell, Don Murray, Hope Lange, Betty Field, Robert Bray, Eileen Heckart, Hans Conried, Casey Adams, Henry Slate, Terry Kelman, Linda Brace, Greta Thyssen, Helen Mayon, Lucille Knox
U
1956

Marvellously entertaining comedy about a hick cowhand who falls for a nightclub singer, and sets out to pursue her in the only way he knows how, his intention being to make her his wife. Monroe's first film on her return to Hollywood from the Actors' Studio. A highlight is Monroe singing "That Old Black Magic". Scripted by George Axelrod. Gave rise to a brief TV series.
Aka: WRONG KIND OF GIRL, THE
COM   96 min   VIDrel: 20TH/TECH   V
Boa: play by William Inge.

BUSHFIRE MOON ***
George Miller AUSTRALIA
Dee Wallace Stone, John Waters, Charles Tingwell, Nadine Garner
(U)
1987

Set at the time of Christmas 1891, when a severe drought has struck Southeast Australia, with a resentful eight-year-old not exactly filled with excitement in anticipation of the forthcoming festivities. But when he spots a figure whom he takes to be Santa Claus, his rebellious behaviour ceases and this has a wonderful effect on the rest of his family. A bright and cheerful feel-good movie for the whole family.
DRA   97 min mTV   SATrel: SKY MOVIES

BUSHWACKED **
Greg Beeman USA
Daniel Stern, Jon Polito, Brad Sullivan, Ann Dowd, Anthony Heald, Tom Wood, Blake Bashoff, Corey Carrier, Michael Galeota, Max Goldblatt, Ari Greenberg, Janna Michaels, Natalie West, Michael P. Byrne, Michael O'Neill, Jane Morris
PG
1995

A dumb delivery man who inhabits a world of his own is framed for murder and goes on the run from both the FBI and the underworld. Venturing into the countryside, he meets a group of six Ranger Scouts, who are waiting for the arrival of their scoutmaster, who is to take them on an overnight expedition. This gives our character a fine chance to hide, even though his scouting skills are non-existent. An endearing cast do what they can with this weak material.
COM   86 min (ort 90 min) cC   VIDrel: 20TH/FOXVID V/sur

BUSINESS AFFAIR, A *
Charlotte Brandstrom FRANCE/GERMANY/SPAIN/UK1993
Christopher Walken, Carole Bouquet, Jonathan Pryce, Sheila Hancock, Anna Manahan, Fernando Guillen Cuervo, Tom Wilkinson, Marisa Benlloch, Paul Bentall, Bhasker, Roger Brierly, Allan Corduner, Marian McLoughlin, Miguel de Angel
15

A struggling academic author is lured to work for an American publishing house, and his marriage runs into trouble when the boss of the company starts showing an interest in his very attractive model wife, who has just written her first novel. This one hits the skids very quickly, thanks to inane dialogue and dreary plotting.
COM   97 min (ort 102 min)   VIDrel: EIV/SONOP   V/sur

BUSTED *   R18/18
Gordon Vandermeer USA   1989
Tori Welles, Victoria Paris, Nina De Ponca, Cassandra, Tom Byron,
Peter North, Randy West, Mark Wallace
Title refers to a trashy TV scandal show which stops at nothing
to videotape and then broadcast the sexual peccadillos of noted
celebrities. Very dull.
A   45 min (ort 73 min)
VIDrel: SHEP L/A (R18 ver); HAR/GOLD (18 ver)   V

BUSTER ***   15
David Green UK   1988
Phil Collins, Julie Walters, Larry Lamb, Martin Jarvis, Sheila
Hancock, Stephanie Lawrence, Ellie Beaven, Michael Attwell, Ralph
Brown, Anthony Quayle, Christopher Ellison, Clive Wood, MIchael
Byrne, Harold Innocent, John Benfield
It's homage to train-robbers time as we salute one of the gang
who took part in the Great Train Robbery of August 1963. This
tells of how Buster Edwards fled to Mexico after the raid. A well
made and enjoyable drama that tends to overlook some of the
nastier aspects of this famous raid. (In the course of the actual
robbery, the train driver was viciously beaten – he died not long
afterwards.)
DRA   98 min (ort 102 min)   VIDrel: VCC L/A   V

BUSTER KEATON: COLLEGE **   (U)
James W. Horne USA   1927
Buster Keaton, Ann Cornwall, Snitz Edwards, Flora Bramley, Harold
Goodwin, Grant Withers, Florence Turner, Carl Harbaugh, Sam
Crawford, Lee Barnes, Paul Goldsmith, Morton Kaer, Bud Houser
Keaton's last independently made picture casts him as an inept
bookworm whose girl throws him over for a college athlete.
Determined to win her back, he enrols at the same college and
tries his hand at virtually every athletic event, with a spectacu-
lar lack of success. Some very funny moments enliven an
otherwise trite story.
Aka: COLLEGE
COM   65 min B/W silent   VIDrel: L/A   V

BUSTER KEATON: OUR HOSPITALITY ***   U
Buster Keaton/Jack Blystone USA   1923
Buster Keaton, Natalie Talmadge, Joe Keaton, Buster Keaton Jr, Kitty
Bradbury, Joe Roberts, Leonard Clapham, Craig Ward, Ralph
Bushman, Edward Coxen, Jean Dumas, Monte Collins, James Duffy
A satirical look at the legendary Hatfield-McCoy feud, with
Keaton as the last surviving McCoy, who goes South to claim
his inheritance and falls in love with the daughter of the rival
clan. An inventive silent comedy with a great finale. Keaton
married Talmadge in real life too.
Aka: OUR HOSPITALITY
COM   66 min (ort 75 min) B/W silent
VIDrel: VISVID/POLYREC   V

BUSTER KEATON: SHERLOCK JR ****   U
Buster Keaton/Donald Crisp USA   1924
Buster Keaton, Kathryn McGuire, Ward Crane, Joseph Keaton
In one of his most inventive films, Keaton plays a movie projectionist who is falsely accused of stealing a watch. With his head
filled with fantasies of being a great detective, he dreams himself
into the movies he is showing and merges with the figures and
background on the screen. A joyful, funny romp that ranks as
one of Keaton's best films.
Aka: SHERLOCK JR
COM   46 min B/W silent   VIDrel: VISVID/POLYREC   V

BUSTER KEATON: THE GENERAL ****   U
Buster Keaton/Clyde Bruckman USA   1926
Buster Keaton, Marion Mack, Glen Cavender, Jim Farley, Joseph
Keaton, Frederick Vroom, Charles Smith, Frank Barnes, Mike Donlin,
Tom Nawn
In one of his best silents (based on a true incident) Keaton plays
a Confederate driver whose locomotive is stolen by Union
soldiers. Infiltrating Union lines (along with his girl) he attempts
to get it back, but finds that events go far from smoothly. Not
as fanciful as some of his other films but made with great care
and beautifully photographed by Matthew Brady. The same tale
was done in a more serious vein by Disney in 1956 as THE
GREAT LOCOMOTIVE CHASE.
Aka: GENERAL, THE
COM   77 min (ort 106 min) B/W silent
VIDrel: THAMES/VCC L/A   V/sh

BUSTER KEATON: THE NAVIGATOR ****   U
Buster Keaton/Donald Crisp USA   1924
Buster Keaton, Kathryn McGuire, Frederick Vroom, Noble Johnson,
Clarence Burton, H.M. Clugston
One of Keaton's best comedies, in which he plays a millionaire
who finds himself cast adrift on a deserted liner that's marooned
in the middle of the ocean. With him is his former girlfriend, a
wealthy socialite who had previously rejected his offer of
marriage. Essentially one extended sight gag from start to finish,
a highlight is a crazy chase sequence when they run aground on
an island inhabited by cannibals.
Aka: NAVIGATOR, THE
COM   57 min (ort 80 min) B/W silent
VIDrel: VISVID/POLYREC   V

BUSTIN' LOOSE ***   15
Oz Scott USA   1981
Richard Pryor, Cicely Tyson, Angel Ramirez, George Coe, Bill Quinn,
Roy Jenson, Robert Christian, Alphonso Alexander, Janet Wong, Fred
Carney, Kia Cooper, Edwin DeLeon, Jimmy Hughes, Edwin Kinter,
Tami Luchow, Angel Ramirez
Rather touching film about an ex-con on probation who drives
a busload of handicapped kids and their teacher to a new life in
Seattle. Mainly filmed in 1979 but following an accident to
Pryor, completion was delayed until 1981. The film debut for
Broadway director Scott. Co-written and co-produced by Pryor,
it gave rise to a TV series.
COM   89 min (ort 94 min)   VIDrel: CIC/SONOP   V

BUTCH AND SUNDANCE: THE EARLY DAYS **   PG
Richard Lester USA   1979
William Katt, Tom Berenger, Jeff Corey, Peter Weller, Christopher
Lloyd, John Schuck, Michael C. Gwynne, Brian Dennehy, Jill
Eikenberry, Arthur Hill, Vincent Schiavelli, Paul Plunkett, Wesley
Burgess, Patrick Egan, Lynn Katzman
A good-looking and nicely photographed prequel to BUTCH
CASSIDY AND THE SUNDANCE KID, that's badly let down
by a trite and empty plot. This amiable film meanders along
quite pleasantly, but ultimately the lack of a decent script weighs
heavily against it.
WES   107 min (ort 111 min)   VIDrel: 20TH/TECH L/A   V

BUTCH CASSIDY AND THE SUNDANCE KID ****   PG
George Roy Hill USA   1969
Paul Newman, Robert Redford, Katherine Ross, Strother Martin, Sam
Elliott, Henry Jones, Jeff Corey, George Furth, Cloris Leachman, Ted
Cassidy, Kenneth Mars, Donnelly Rhodes, Jody Gilbert, Timothy
Scott, Don Keefer, Charles Dierkop
A glamorised film about the celebrated outlaws and train
robbers, that's a pure delight to watch and has many memorable
moments – the dynamite episode is a highspot. A semi-comic
tale that follows their exploits and deaths in Bolivia, where they
had fled to avoid capture. AA: Story/Screen (William
Goldman), Cin (Conrad Hall), Score/orig (Burt Bacharach),
Song ("Raindrops Keep Falling On My Head" – Burt Bacharach
(m)/Hal David (l)).
WES   105 min (ort 112 min)   VIDrel: 20TH/TECH L/A;
ENCORE (LV only)   V/h LV

BUTCHER, THE ***   15
Claude Chabrol FRANCE/ITALY   1969
Stephane Audran, Jean Yanne, Antonio Passallia, Mario Beccaria,
Pascal Ferone, William Guerault, Roger Rudel
A series of brutal murders of young girls shocks a small French
town. They are eventually traced to a seemingly inoffensive
local butcher. An extremely tense thriller that explores the
psychology of the murderer and his relationship with a local
schoolteacher.
Aka: LE BOUCHER
THR   90 min (ort 98 min) wScrn   VIDrel: ARTPRO/RTM
V

BUTCHER'S WIFE, THE ***   15
Terry Hughes USA   1991
Demi Moore, Jeff Daniels, George Dzindza, Mary Steenburgen,
Margaret Colin, Frances McDormand, Max Perlich, Miriam
Margoyles, Helen Hanft, Christopher Durang, Luis Avalos, Charles
Pierce, Elizabeth Lawrence, Stephanie Laurence
Moore plays clairvoyant Marina, who lives an empty and unful-
filled life on a remote island, and marries a tubby New York
butcher, having mistakenly come to believe he is the man of her
dreams. In the big city, her gifts are soon exploited by the
lovelorn, but her mystic matchmaking causes a good deal of

chaos, much to the annoyance of a local psychiatrist, many of whose clients are involved. A dopey and rather old-fashioned romantic comedy.
COM 100 min (ort 105 min)
VIDrel: 4-FRONT/POLYREC/CIC V/sur

BUTTERBOX BABIES ***   15
Don McBrearty CANADA   1994
Susan Clark, Peter MacNeill, Michael Riley, Catherine Fitch, Cedric Smith, Nicholas Campbell
The gruesome but true story of Lila and William Young, who ran a maternity home for unmarried mothers on the outskirts of Halifax, Nova Scotia, during the Depression and WW2. The arrival of a health inspector leads to the revealing of the illegal and sometimes murderous disposal of the babies (the title being derived from the method by which the children were disposed of).
DRA 94 min (ort 99 min) mTV VIDrel: ODY/SONOP V/sh

BUTTERCREAM GANG, THE **   (U)
Bruce Neibaur USA   1992
Jason Johnson, Michael D. Weatherred, Brandon Blaser, Jason Glenn, Michael Scott, Stephanie M. Dees, Otto J. Miletti III, Kathryn Little, Miranda Brooke, Chivers, Leo Ware, Ivan A. Crosland, Hazel Cox, Jack Pinette, Ryan Raderbraugh
A group of four kids living in the countryside form a gang that does good deeds and generally provides a helping hand for those who need one. Their companionship, however, is disturbed when one of them moves to Chicago to live with his aunt. There he falls in with a bad crowd, gets expelled from school and returns home much changed. However, his friends eventually manage to get him to reform, in this very sentimental tale. A sequel soon followed.
JUV 94 min SATrel: SKY MOVIES

BUTTERCREAM GANG IN THE SECRET OF
TREASURE MOUNTAIN **   (U)
Scott H. Swofford USA   1993
Brandon Blaser, Jason Glenn, Jason Johnson, Isaac Fugal, Marissa Parritt, Rick Macy, Stephanie Pees, Frank Gerish, John Huntington, Don Shanks, Michael Rudal, Otto Mileti, Steven Anderson, Alexa Pappas, Janet Irick, Ivan Crosland
Sequel to the first film sees our juvenile gang reunited once more and investigating a mystery that provides ample danger and excitement when they comes across an old treasure map.
JUV 84 min (ort 93 min) SATrel: SKY MOVIES

BUTTERFIELD 8 *   15
Daniel Mann USA   1960
Elizabeth Taylor, Laurence Harvey, Eddie Fisher, Dina Merrill, Betty Field, Mildred Dunnock, Jeffrey Lynn, Kay Medford, Susan Olivier, Virginia Downing, George Voskovec, Carmen Matthews, Whitfield Connor, Dan Bergin, Beau Tilden
A high-class call girl falls for a weak-willed married man and contemplates abandoning her chosen profession. A dull drama whose once daring theme now seems trite and dated. AA: Actress (Taylor).
DRA 104 min (ort 110 min) wScrn VIDrel: MGM/WHV V
Boa: novel by John O'Hara.

BUTTERFLIES ARE FREE ***   PG
Milton Katselas USA   1972
Goldie Hawn, Edward Albert Jr, Eileen Heckert, Mike Warren, Michael Glasser
A blind boy falls for the unconventional girl-next-door and comes into conflict with his over-possessive mother. Good performances from Hawn as the daffy aspiring actress, who comes into Albert's life, and Heckert as his mother, for which she won a well-deserved Oscar. Based on the successful Broadway play. AA: S. Actress (Heckert).
COM 105 min VIDrel: RCA L/A V
Boa: play by Leonard Gershe.

BUTTERFLY KISS **   18
Michael Winterbottom UK   1994
Amanda Plummer, Saskia Reeves, Kathy Jamieson, Des McAleer, Lisa Jane Riley, Freda Dowie, Paula Tilbrook, Fine Time Fontayne, Elizabeth McGrath, Joanne Cook, Shirley Vaughan, Paul Brown, Emily Aston, Ricky Tomlinson, Kate Murphy
A rebellious punk girl wanders into a gas-station, and the girl who works there is so taken with her that she invites her home. A seduction follows, but in the morning our punk is gone,

forcing the other to embark on a protracted search, one which draws her into the brutal and murderous world of her lover. A grim, disturbing and anarchic debut feature from Winterbottom with a few touching moments but an excessively negative outlook and no clear message.
THR 84 min (ort 88 min) VIDrel: POLY/POLYREC V/s

BUTTERFLY REVOLUTION, THE **   15
Bert L. Dragin USA   1986
Chuck Connors, Charles Stratton, Adam Carl, Harold P. Pruett, Tom Fridley, Melissa Brennan, Stuart Rogers, Shawn McLemore, Samantha Neward, Rick Fitts, Nancy Calabrese, Michael Cramer, Doug Toby, Shirley Mitchell, Chris Hubbell
Connors stars as the new director of a summer camp, hopelessly out of touch with the youngsters and their teenage counsellors. Stratton stars as one of the counsellors, who challenges his domination, resulting in a coup and the imprisonment of the adults. A contrived and downright peculiar tale, which sees the kids sliding rapidly into barbarism in true LORD OF THE FLIES style.
Aka: SUMMER CAMP NIGHTMARE; SUMMER CAMP MASSACRE
DRA 85 min (ort 89 min) VIDrel: 20TH/TECH V
Boa: novel by William Butler.

BUTTMAN GOES TO RIO *   18
John Stagliano USA   1990
Isabella, Angela, Sylvia, Carolina, Daniela, Denise, Brandy Alexandre, John Stagliano, Vladimir Corea, Aldacir, Menecto, Vanderley Viera, Accaccio Silva, Davit Junior
Another camera-verite offering from this porno director who is completely obsessed with female backsides. In this adventure he decides that carnival time at Rio offers the best chance of pursuing his interest. A long, meandering and incredibly boring effort. Several sequels followed.
A 120 min VIDrel: ONE V

BUY AND CELL **   15
Robert Boris USA   1987
Robert Carradine, Ben Vereen, Michael Winslow, Malcolm McDowell, Randall "Tex" Cobb, Imogene Coca, Fred Travalena, Roddy Piper, Lise Cutter, Michael Goodwin, Tony Plana, Michael Knox, West Buchanan, Tony Carroll, Larry Clark
A stockbroker is framed by his boss and sent to jail, but takes his entrepreneurial talents with him, setting up a business with the help of the inmates. A mediocre comedy, filmed in Italy but set in the USA.
COM 92 min (ort 95 min) VIDrel: EIV/SONOP V

BY DAWN'S EARLY LIGHT ***   15
Jack Sholder USA   1990
Powers Boothe, Rebecca DeMornay, Martin Landau, Rip Torn, Darren McGavin, James Earl Jones, Peter MacNichol, Jeffrey DeMunn, Nicolas Coster, Glenn Withrow, Ronald William Lawrence, Kieran Mulroney, Ken Jenkins,
A nuclear missile is launched from an unknown source in the Middle East and it strikes the Soviet Union. When the Russian leaders become convinced that it represents an American first strike they respond in kind. An effective nuclear thriller that carefully balances tension against plausibility. See also FAIL SAFE and DR STRANGELOVE.
DRA 97 min (ort 100 min) mCab VIDrel: MGM/WHV L/A V/sh
Boa: novel Trinity's Child by William Prochnau.

BY THE LIGHT OF THE SILVERY MOON ***   U
David Butler USA   1953
Doris Day, Gordon MacRae, Leon Ames, Rosemary De Camp, Mary Wickes, Billy Gray, Russell Arms, Maria Palmer, Walter Flannery, Geraldine Wall, John Maxwell, Carol Forman
Sequel to ON MOONLIGHT BAY sees Day waiting for her service sweetheart to return from WW1 and the usual misunderstandings threatening to postpone their wedding for good after he finally arrives. A light and frothy musical entertainment. Songs include the title one and "If You Were The Only Girl In The World".
MUS 97 min VIDrel: WHV V/dm

BY THE SWORD **   15
Jeremy Paul Kagan USA   1991
F. Murray Abraham, Eric Roberts, Mia Sara, Christopher Rydell, Elaine Kagan, Brett Cullen, Doug Wert, Sherry Hursey, Stoney Jackson
Just out of jail, a top fencing instructor returns to the New York

academy he was the master of twenty-five years ago, but finds it now has a dictatorial new chief instructor. With a false name, he takes a job teaching the school's pupils, but incurs the jealousy of the new master, who learns his true identity and expels him. Despite the unusual subject matter, this glossy but entirely predictable effort offers little except some fine swordplay.
A/AD   87 min (ort 91 min)   VIDrel: EIV/SONOP   V/sur

BYE, BYE, BABY *** (PG)
Edward Bennett UK   1992
Ben Chaplin, Jason Flemyng, Colin Tierney, Ewen Bremner, Robert Portal, Nicholas Gleaves, Lyndon Davies, Jonathan Lacey, Mark Alex-Jones, Simon Foy, Timothy Andrews, Chris Chandler, Geoffrey Hutchings, Shaughan Seymour
Written by Jack Rosenthal, this is a semi-autobiographical tale of the 1950s and National Service, with Chaplin looking back over the two years he has spent at an army base in Germany as he celebrates his last day there. Meanwhile, the C.O. is giving a demob speech, but the minds of the lads are elsewhere. An affectionate trip down memory lane, light, detailed and quite charming.
COM   91 min mTV   TVrel: C4

BYE BYE LOVE ** PG
Sam Weisman USA   1995
Paul Reiser, Matthew Modine, Randy Quaid, Janeane Garofalo, Amy Brenneman, Rob Reiner, Eliza Dushku, Ed Flanders, Maria Pitillo, Lindsay Crouse, Johnny Whitworth, Ross Malinger, Wendell Pierce, Mae Whitman, Pamela Dillman, Brad Hall
A worthy but none-too-successful attempt to take a comic look at the effects of divorce, as seen in the lives of three middle-aged fathers who are spending the weekend with their children and are doing their best to come to terms with both their feelings and their situation. Reiner pops up briefly as an obnoxious DJ, injecting one of the few brighter moments.
COM   101 min (ort 107 min) cC   VIDrel: 20TH/FOXVID   V/sh

BYE BYE MONKEY * 18
Marco Ferreri FRANCE/ITALY   1978
Gerard Depardieu, James Coco, Geraldine Fitzgerald, Marcello Mastroianni, Gail Lawrence (Abigail Clayton), Mimsy Farmer, Clarence Muse
Story set in a rundown part of Manhattan where a group of misfits struggle through life, with one of them (Depardieu) finding a baby chimp near the body of a giant ape, and taking it home with him. A strange, bleak comedy, that has a message floating about in there somewhere. Filmed in New York City.
Aka: CIAO MASCHIO
COM   90 min (ort 114 min)   VIDrel: ARTPRO/CASPIC   V

C

CABARET **** 15
Bob Fosse USA   1972
Liza Minnelli, Joel Grey, Michael York, Helmut Griem, Fritz Wepper, Marisa Berenson, Elisabeth Neumann-Viertel, Sigrid Von Richthofen, Helen Vita, Gerd Vespermann, Ralf Wolter, Georg Hartmann, Ricky Renee, Estrongo Nachama
Excellent account of the growing Nazi take-over, seen against and through the setting of a sleazy Berlin cabaret, with the story a loose adaptation of the career of nightclub performer Sally Bowles. The great dance numbers are highlights in a fine musical. AA: Dir, Actress (Minnelli), S. Actor (Grey), Edit (D. Bretherton), Cin (G. Unsworth), Score/adapt (R. Burns), Art/Set (R. Zehetbauer and J. Kiebach/H. Strabel), Sound (R. Knudson/D. Hildyard).
MUS   123 min (ort 128 min)   VIDrel: VCC/DISC; PION (LV only)   V/sur LV
Boa: book Goodbye To Berlin by Christopher Isherwood.

CABIN BOY ** 12
Adam Resnick USA   1993
Chris Elliott, Russ Tamblyn, Ritch Brinkley, James Gammon, Brion James, Brian Doyle-Murray, Melora Walters, Ann Magnusson, Russ Tamblyn, Ricki Lake
A spoiled rich kid mistakes a trawler for the luxury vessel on which he has booked a cruise to Hawaii and finds himself working his passage in the midst of seasoned sailors who do not take kindly to his stuck-up ways. Naturally, this dose of hard work effects a miraculous transformation and helps him gain a more balanced perspective on life. A predictable comedy vehicle

for Elliott, whose comic antics will not be to everyone's taste. CAPTAINS COURAGEOUS it ain't.
COM   77 min (ort 81 min)   VIDrel: TOUCH/TECH L/A   V/sh

CABIN IN THE SKY *** U
Vincente Minnelli USA   1943
Ethel Waters, Eddie Anderson, Lena Horne, Rex Ingram, Louis Armstrong, Duke Ellington, Cab Calloway, John W. Bublett, Kenneth Spencer, John Sablett, Oscar Polk, Mantan Moreland, Willie Best, Fletcher Rivers, Leon James, Bill Bailey
All-black musical in which good and evil forces do battle for the soul of an idle gambler and inconsiderate husband, who is eventually reformed after dreaming of his own death. It now seems more than a little racist in its attitudes and stilted in its production, but its lively musicality compensates for these flaws. Songs include the title one and "Taking A Chance On Love".
MUS   95 min (ort 99 min) B/W   VIDrel: MGM/WHV   V/sh
Boa: play by Lynn Root and John Latouche.

CABINET OF DOCTOR CALIGARI, THE *** U
Robert Wiene GERMANY   1919
Werner Krauss, Conrad Veidt, Lil Dagover, Friedrich Feher, Hans Heinz Von Twardowski, Rudolf Lettinger, Rudolph Klein-Rogge, Ludwig Rex, Elsa Wagner, Henri Peter-Arnolds, Hans Lanser-Ludolff
A landmark Expressionist film in which it would appear that a fairground hypnotist has complete control of a victim he uses for the purpose of murder. Ultimately the sinister acts the victim is forced to carry out are found to be nothing more than the dreams of a disturbed lunatic being treated in a sanatorium. An immensely powerful film of stylised sets and grotesque camera angles; faded but still potent. Remade in 1962.
Aka: DAS CABINET DES DR CALIGARI
HOR   49 min (ort 73 min) B/W silent
VIDrel: REDEM/RTM   V

CABLE GUY, THE * 12
Ben Stiller USA   1996
Jim Carrey, Matthew Broderick, Leslie Mann, Jack Black, George Segal, Diane Baker, Ben Stiller, Eric Roberts, Janeane Garofalo, Andy Dick, Harry O'Reilly, David Cross, Amy Stiller, Owen Wilson, Cameron Starman, Kathy Griffin
Carrey is a lonely cable TV installer who plays a cat-and-mouse game with Broderick after he arrives to install cable TV for the latter and fails to leave, having developed a manic attraction for this unfortunate. There are all the silly jokes and grimaces one expects in a Carrey movie, but the whole thing is so relentlessly grim (it's meant to be a black comedy) that to be effective a more substantial script was needed. Carrey was paid $20,000,000 for this dud.
COM   92 min Cut (1 frame – ort 96 min) cC
VIDrel: COLUM/SONOP; ENCORE (LV only)   V/sh LV

CACTUS FLOWER *** PG
Gene Saks USA   1969
Walter Matthau, Ingrid Bergman, Goldie Hawn, Jack Weston, Rick Lenz, Vito Scotti, Irene Hervey, Eve Bruce, Irwin Charone, Matthew Saks, Eve Bruce, Irwin Charone, Matthew Saks
Wonderfully enjoyable comedy about a dentist who persuades his nurse to pretend to be his wife so that he will not have to marry his girlfriend. The plot is pretty thin, but the fresh and winning performances carry it off. Hawn's first major role, for which she won a well-deserved Oscar. The script is by I.A.L. Diamond from the Broadway play. AA: S. Actress (Hawn).
COM   99 min   VIDrel: VCC L/A   V
Boa: play by Abe Burrows.

CACTUS JACK ** PG
Hal Needham USA   1979
Kirk Douglas, Ann-Margret, Arnold Schwarzenegger, Paul Lynde, Foster Brooks, Ruth Buzzi, Jack Elam, Strother Martin, Robert Tessier
Wildly over-the-top spoof Western that sends up both whites and Indians, telling of an outlaw who pursues a girl and her tough protector. A running gag in the film is that the outlaw constantly finds it necessary to consult a book entitled: "How to be a Badman". (Perhaps the director should have consulted "How to make a comedy Western".)
Aka: VILLAIN, THE
COM   85 min (ort 105 min)   VIDrel: VCC/DISC/COLUM   V

**CADDYSHACK **   15
Harold Ramis USA   1980
Chevy Chase, Rodney Dangerfield, Ted Knight, Michael O'Keefe, Bill Murray, Sarah Holcomb, Scott Colomby, Brian Doyle Murray, Cindy Morgan, Albert Salmi, Dan Resin, Henry Wilcoxon, Ealine Aiken, Ann Ryerson, Lois Kibbee, Scott Powell
Slapstick and zany comedy set in a country club that runs an exclusive golfing establishment. The first opening sequences are very funny, but this effort soon runs out of ideas. Followed by CADDYSHACK 2.
COM   94 min   VIDrel: WHV   V

**CADDYSHACK 2 **   15
Alan Arkush USA   1988
Jackie Mason, Robert Stack, Dan Aykroyd, Chevy Chase, Dyan Cannon, Jonathan Silverman, Dina Merrill, Randy Quaid, Jessica Lundy, Chynna Phillips, Brian McNamara, Paul Bartel, Tony Mockus, Pepe Serna, Bibi Osterwald, Don Draper
This borrows the title of the 1980 film but little else. Mason is good as a self-made construction millionaire, who takes a fitting revenge when his daughter is snubbed by the stuffy members of a local country club. This single idea does not suffice to sustain the film throughout its entire length, and after the first half it soon runs down.
COM   93 min (ort 98 min)   VIDrel: WHV   V/sur

CADFAEL: MONKS HOOD *   (PG)
Graham Theakston UK   1994
Derek Jacobi, Sean Pertwee, Peter Copley, Michael Culver, Julian Firth, Mark Charnock, Aubrey Richards, Ray Llewellyn, Albie Woodington, Sally Baxter, Steven Beard, Freddie Boardley, Thomas Craig, Bernard Gallagher, Huw Garmon
Brother Cadfael is asked to identify the poison that was used to murder a wealthy landowner who cut his stepson out of his will, and is surprised to find that the man's wife is childhood sweetheart. The last in this set of four medieval mysteries featuring our monk with a passion for solving mysteries.
DRA   75 min mTV   VIDrel: CTE/CARL   V
Boa: story by Ellis Peters.

CADFAEL: ONE CORPSE TOO MANY *   15
Graham Theakston UK   1993
Derek Jacobi, Sean Pertwee, Michael Culver, Peter Copley, Julian Firth, Mark Charnock, Maggie O'Neill, Christian Burgess, Michael Grandage, Richard Hender
The first story in what was initially just a set of four detective dramas set in 12th century Shrewsbury, where a monk with a colourful past acts as a medieval sleuth. In this story Brother Cadfael is asked to bury the bodies of some rebel soldiers, and finds the presence of an extra corpse arousing suspicions in his mind, which leads him on a hunt for the killer, in which he is aided by an attractive female fugitive.
DRA   75 min (ort 90 min) mTV   VIDrel: CTE/CARL   V
Boa: novel by Ellis Peters.

CADFAEL: THE LEPER OF ST GILES *   (PG)
Graham Theakston UK   1994
Derek Jacobi, Sean Pertwee, Peter Copley, Michael Culver, Julian Firth, Mark Charnock, Tara Fitzgerald, Susan Fleetwood, Norman Eshley, Sarah Badel, John Bennett, Jonathan Firth, Jamie Glover, Jonathan Hyde, Albie Woodington
Cadfael investigates the disappearance of a rich and powerful baron on the eve of his wedding, his suspicions naturally tending towards foul play. Third entry in this extremely absorbing and carefully made series.
DRA   75 min mTV   VIDrel: CTE/CARL   V
Boa: novel by Ellis Peters.

CADFAEL: THE SANCTUARY SPARROW *   PG
Graham Theakston UK   1994
Derek Jacobi, Sean Pertwee, Roy Barraclough, Steven Beard, Roger Booth, Mark Charnock, Richard Bonneville, Patrick Brennan, Peter Copley, Rosalie Crutchley, Michael Culver, Julian Firth, Fiona Gillies, Toby Jones, Ray LLewellyn
After a goldsmith is robbed and almost killed, a mob chases a young man suspected of this crime into the abbey where he claims sanctuary, and Brother Cadfael is ordered to take charge of his welfare. The second episode in this series, set in 12th century Shrewsbury, which surprisingly (given its strongly realistic look) was created at film studios near Budapest in Hungary.
DRA   75 min mTV   VIDrel: CTE/CARL   V
Boa: novel by Ellis Peters.

**CADILLAC GIRLS **   (PG)
Nicholas Kendall CANADA   1993
Jennifer Dale, Mia Kirshner, Adam Beach, Gregory Harrison, Ann Cameron, Mike Crimp, Morrissey Dunn, Benita Ha, Rachael Clark, Martha Irving, Cameron Diges, Louis Del Grande, Nancy Marshall, Michael Fitzgerald, Ronald Bourgeois
An eighteen-year-old girl and her mother visit their home-town in order to sell their grandfather's house and both of them fall in love. Unfortunately, this causes great problems, since they have both chosen the same man. Average.
DRA   102 min   SATrel: SKY MOVIES

CADILLAC MAN *   15
Roger Donaldson USA   1990
Robin Williams, Tim Robbins, Pamela Reed, Fran Drescher, Anabella Sciorra, Zack Norman, Lori Petty, Paul Guilfoyle, Bill Nelson, Eddie Jones, Judith Hoag, Elaine Stritch, Mimi Cecchini, Lauren Tom, Anthony Powers, Erik King
A pushy car salesman stands to lose his wife, both mistresses, his daughter, job and Mafia protector all at once, when he falls victim to a confused young man who, having crashed into the car showroom, holds everyone hostage at gunpoint. A patchy and often irritating wild rollercoaster of a film that veers from crazy comic to pathos to drama, but is often redeemed by flashes of inspiration and wit. Williams is superb in the title role.
COM   93 min (ort 97 min)   VIDrel: VCC/DISC/COLUM   V/sur

**CAESAR AND CLEOPATRA **   PG
Gabriel Pascal UK   1945
Vivien Leigh, Claude Rains, Stewart Granger, Flora Robson, Francis L. Sullivan, Cecil Parker, Basil Sydney, Ernest Thesiger, Michael Rennie, Anthony Eustrel, Renee Asherson, Raymond Lovell, Olga Edwardes, Esme Percy, Leo Genn
Film version of Shaw's play about the teenage Egyptian queen and her relationship with Caesar. A stilted and utterly tiresome effort with a surprise ending – just when you think it's never going to end it does.
DRA   122 min (ort 138 min)   VIDrel: VCC/DISC   V
Boa: play by George Bernard Shaw.

CAFE FLESH *   18
Rinse Dream/F.X. Pope USA   1982
Andrew Nichols, Paul McGibboney, Pia Snow, Marie Sharp, Kevin Jay, Darcy Nichols, Joey Lennon, Nell Podericki, Robert Dennis
Five years after a catastrophic nuclear war, the population is divided into two groups – the Sex Negatives and the Sex Positives. The former comprise 99% of the population and are human wrecks who abhor physical contact of any kind, whilst the latter are obliged to perform in sex cabarets in a vain attempt to stimulate them back to normality. A nightmarish vision quite unlike any adult film yet made. Cut before video submission by 4 min 16 sec.
A   69 min (ort 90 min)   VIDrel: MIA/DISC/IMPENT   V

**CAGE, THE **   18
Lang Elliot USA   1988
Lou Ferrigno, Reb Brown, Michael Dante, Mike Moroff, Marilyn Tokuda, Al Leong, James Shigeta, Al Ruscio, Branscome Richmond
Standard formula mayhem with Ferrigno a brain-damaged Vietnam veteran who runs a bar with his buddy and is not averse to using his physical prowess to eject troublemakers, usually through the nearest window. Some gangsters witness this and persuade him and his buddy to go in for "cage fighting", a tough and brutal form of combat. People who like Chuck Norris movies will really like this one.
A/AD   96 min Cut (2 sec – ort 101 min)
VIDrel: POLY/POLYREC/BRAVE   V/h

**CAGE 2, THE **   18
Lang Elliot USA   1994
Lou Ferrigno, Reb Brown, James Shigeta, Shannon Lee
A further helping of brutal action in which our muscular hero is kidnapped by evil gangsters and forced to fight for his life in a steel cage, in matches with a variety of opponents. One day, he finds himself facing a former partner who has come looking form and suffered a similar fate.
Aka: CAGE 2: THE ARENA OF DEATH
MAR   94 min   VIDrel: MARQ/QUANT   V/sh

**CAGED FEAR **   18
Fred Olen Ray USA   1997
Jay Richardson, Tim Abell, Ross Hagen, Katherine Victor

After she shoots the gangster who killed her sister, a woman is sent to jail, where she discovers that a contract has been taken out on her life. Average "peril in prison" story.
A/AD   90 min   VIDrel: THIRD   V

CAGED FURY *
18
George Ashwell USA
1993
Isis Nile, Randy West, Crystal Wilder, Sahara, Lana Sands, Krista, Valeria, Peter North, Terry Thomas, Tiffany Minx, Jason Kane, George Rodriguez, Harry Dutchman
One of those women-in-prison movies, that stars Nile as a woman who is wrongly convicted of armed robbery and sent to jail. But fortunately, this establishment is run on very liberal lines, and she has more fun in there than she would probably have enjoyed in the outside world. Low-budget nonsense.
A   58 min (ort 80 min)   VIDrel: ONE   V

CAGED HEARTS **
18
USA
1995
Carrie Genzel, Tane McClure
Two new women prisoners, who were framed for murder, find themselves being forced to work as hookers but eventually turn the tables, in this nonsensical women-in-prison offering.
A/AD   87 min   VIDrel: MARQ   V

CAGED HEAT **
18
Jonathan Demme USA
1974
Juanita Brown, Erica Gavin, Roberta Collins, Rainbeaux Smith, Barbara Steele, Toby Carr-Refelson, Desiree Cousteau, Warren Miller, Lynda Gold, Joe Viola, Mickey Fox, George Armitage, Ann Stockade, Irene Stokes, John Aprea
Tiresome women-in-prison movie set for once in the US and not in the jungle hell, so beloved of this genre. Inevitably, a group of inmates go on the rampage and take their violent revenge. Now has a sizeable cult following. Demme's first feature has a few semi-comic moments, but apart from these it's business as usual.
Aka: CAGED FEMALES; RENEGADE GIRLS
DRA   74 min (ort 92 min)   VIDrel: ODY/SONOP   V

CAGED SEDUCTION ***
15
Karen Arthur USA
1994
Judith Light, Stacy Keach, Kay Lenz
A woman serving a prison sentence embarks on a crusade for justice when she experiences the brutality and sadism of the corrupt prison guards at first hand. With the help of an idealistic lawyer, she sets about trying to change the system. An unusual entry in the women-in-prison genre, that has no gratuitous erotica, instead opting for a more serious story. Presumably, the misleading title was intended to present the movie as an entirely different film.
Aka: CAGED SEDUCTION: THE SHOCKING TRUE STORY
DRA   94 min   VIDrel: ODY/SONOP   V/sh

CAGED WOMEN **
18
Leadro Lucchetti
Pilar Orive, Elena Wiedermann, Christian Lorenz, Isabel Libossart
An American tourist on holiday in South America is framed for drug smuggling by a creep whose advances she rejected, and soon finds herself thrown into a grim all-women penitentiary. She endures various indignities, mostly at the hands of a dominatrix prison guard (see also CHAINED HEAT 2) and learns that worse is to follow. Luckily, she is rescued be her new boyfriend, a man she met just prior to her incarceration. An unashamed piece of titillating nonsense.
DRA   72 min (ort 89 min)   VIDrel: MIA/DISC   V

CAGNEY AND LACEY: THE RETURN **
PG
James Frawley USA
1994
Sharon Gless, Tyne Daly, John Harkkins, James Naughton, David Paymer, Susan Anspach, Anders Hover, Ken Johnston, Molly Orr, Fredd Wayne, Merry Clayton, Robert Hegyes, Paul Mantee, Kelly Jean Peters, Selma Archerd, Rick Pasqualone
Our female duo investigate a weapons heist in this rather tedious feature film based on a TV detective series featuring two female cops. Adequate.
DRA   90 min mTV   VIDrel: IMC/DISC   V

CAGNEY AND LACEY: THE VIEW THROUGH THE GLASS CEILING **
PG
John Patterson USA
1995
Sharon Gless, Tyne Daly, George Coe, Lynne Thigpen, Chip Zien, Sandra Oh, Mark Melymick, Richard Bradford, John Karlen, Molly

Orr, Dwight Bacquie, Glenn Bang, Nigel Bennett, Tyrone Benskin, Kirsten Bishop, Wally Bolland, Matt Cook
Spin-off movie featuring our two female crimebusters, who are now working for the D.A., and find their friendship coming under strain when they investigate the murder of a Chinese male, all the evidence being found to point to a police officer. Matters are not helped by the fact that Lacey has just got divorced and is also being considered for a special award. Thinly plotted, it adds nothing new to what became an increasingly fraught and strident TV series.
DRA   93 min mTV   VIDrel: IMC/DISC   V

CAGNEY AND LACEY: TOGETHER AGAIN **
PG
Reza Badiyi USA
1994
Sharon Gless, Tyne Daly, James Naughton, David Paymer, John Karlen
Another one of those full-length spin-offs from the slightly strident female buddy-buddy detective series. In this story, the duo try to solve the murder of a homeless man. But when they learn that the chief suspect is also homeless and was was a former colleague of Lacey's husband, the investigation takes on a more personal aspect. Average TV fare.
DRA   90 min mTV   VIDrel: IMC/DISC   V

CAGNEY AND LACEY: TRUE CONVICTIONS **
PG
Lynne Littman USA
1995
Tyne Daly, Sharon Gless, Michael Moriarty, Chip Zien, Beau Starr, Darryl Thierese, Sam Coppola, Natalie Bradford, John Karlen, Molly Orr, Kathleen Boyle, Vernon Chapman, Reginald Doressa, Shelley Goldstein, Howard Jerome
The second feature film spin-off from the popular detective series, with one of the girls learning that the girlfriend of a murdered man was in fact her neighbour, and she appears to have died from a drugs overdose. Meanwhile, the investigation of a drive-by shooting leads to romance. A competent rather than outstanding drama, which is in truth nothing more than an extended and updated episode from the former TV series.
DRA   93 min mTV   VIDrel: IMC/DISC   V

CAHILL: U.S. MARSHAL **
15
Andrew V. MacLaglen USA
1972
John Wayne, George Kennedy, Gary Grimes, Neville Brand, Clay O'Brien, Marie Windsor, Morgan Paull, Dan Vadis, Royal Dano, Scott Walker, Denver Pyle, Jackie Coogan, Rayford Barnes, Dan Kemp, Harry Carey Jr, Walter Barnes, Paul Fix
A US marshal finds that his own sons are involved in a robbery that he is investigating. A preachy and earnest Western that follows a well-trodden path – one of Wayne's poorer efforts.
Aka: CAHILL; CAHILL, UNITED STATES MARSHAL
WES   97 min (ort 103 min)   VIDrel: WHV   V/h

CAINE MUTINY, THE ****
U
Edward Dmytryk USA
1954
Humphrey Bogart, Fred MacMurray, Van Johnson, Robert Francis, Jose Ferrer, May Wynn, E.G. Marshall, Lee Marvin, Tom Tully, Claude Akins, Jerry Paris, Whit Bissell, Warner Anderson, Katherine Warren, Steve Brodie, Todd Karns, James Best
Based on the Pulitzer Prize-winning novel, this tells of a ship's captain who hovers on the threshold of breakdown, and receives scant pity from his sullen and contemptuous crew. Bogart as the captain was never better, his courtroom monologue is harrowing. Johnson, Francis and MacMurray play his mutinous chief officers.
DRA   119 min (ort 125 min) B/W
VIDrel: VCC/DISC/COLUM   V
Boa: novel by Herman Wouk.

CAINE MUTINY COURT MARTIAL, THE ***
PG
Robert Altman USA
1987
Eric Bogosian, Brad Davis, Jeff Daniels, Michael Murphy, Peter Gallagher, Kevin O'Connor, Daniel Jenkins
A sequel of sorts to THE CAINE MUTINY, that concentrates on the subsequent court-martial of the officers involved in the mutiny, a highlight being the sequence where Captain Queeg effectively convicts himself. An absorbing and literate study.
DRA   96 min mTV   VIDrel: EIV/SONOP L/A   V

CAIRO ROAD **
U
David McDonald UK
1950
Eric Portman, Laurence Harvey, Maria Mauban, Karel Stepanek, Harold Lang, Camelia, Gregoire Aslan, Oscar Quitak
In his determination to catch drug smugglers, an Egyptian police colonel devises a series of traps. A standard cops and

robbers tale, given a touch more sparkle thanks to the unusual setting.
A/AD 86 min (ort 95 min) B/W VIDrel: WHV V

CAL * 15**
Pat O'Connor UK 1984
Helen Mirren, John Lynch, Donal McCann, John Kavanagh, Ray McAnally, Stevan Rimkus, Catherine Gibson, Louis Rolston, Tom Hickey, Gerald Mannix Flynn, Seamus Ford, Edward Byrne, Audrey Johnson, Brian Munn, Daragh O'Malley, George Shane
A young unemployed man in Northern Ireland is reluctantly drawn into the murder of a policeman and then becomes romantically involved with the dead man's wife. A literate and engrossing study, with a strong performance from Mirren for which she rightly won the Best Actress award at Cannes. Produced by David Puttnam and scripted by MacLaverty.
DRA 99 min (ort 102 min) VIDrel: WHV V/h
Boa: novel by Bernard MacLaverty.

CALAMITY JANE * U**
David Butler USA 1953
Doris Day, Howard Keel, Allyn Ann McLerie, Philip Carey, Gale Robbins, Dick Wesson, Paul Harvey, Chubby Johnson
Bright and bouncy musical Western, with Day in fine form as the tomboyish female gunslinger who changes her ways when she falls in love with Keel. AA: Song ("Secret Love" – Sammy Fain (m)/Paul Francis Webster (l)).
MUS 97 min VIDrel: MGM/WHV V

CALENDAR * (PG)**
Atom Egoyan ARMENIA/CANADA 1993
Atom Egoyan, Arsinee Khanjian, Ashot Adamian, Michelle Bellerose, Natalia Jasen, Susan Hamann, Sveta Kohli, Viva Tsvetnova, Rula Said, Annie Szamosi, Anna Pappas, Amanda Martinez, Diane Kofri
A Canadian couple of Armenian origin take a trip to their homeland where the husband has accepted an assignment to take photographs of old churches for a calendar. While he concentrates on his work, his wife has an affair with their guide at the same time as she develops a deeper attachment to her culture. This rather disorganised account of a crumbling marriage is badly lacking in cohesion and its rambling structure precludes convincing storytelling.
DRA 75 min CINrel

CALENDAR GIRL ** 15
John Whitesell USA 1993
Jason Priestley, Jerry O'Connell, Gabriel Olds, Steve Railsback, Stephen Tobolowsky, Kurt Fuller, Emily Warfield, Maxwell Caulfield, Joe Pantoliano, Michael Quill, Leslie Wing, Elizabeth Quill, Blake McIver Ewing, Sean Fitzgerald
In the summer of 1962, three teenage boys who idolise Marilyn Monroe head for Hollywood with the impossible dream of actually getting a date with this screen goddess. Along the way they have a variety of adventures and learn a thing or two about growing up and adult life. A pleasant, empty-headed but reasonably likeable little film whose cast work hard to turn in some convincing performances (often not succeeding). This was Priestly's feature debut.
COM 87 min (ort 92 min) cC VIDrel: VCC/DISC/COLUM V/sur

CALENDAR GIRL MURDERS, THE ** 18
Willaim A. Graham USA 1985
Tom Skerritt, Sharon Stone, Barbara Bosson, Robert Beltran, Pat Corley, Robert Morse, Alan Thicke, Silvana Gallardo, Michael C. Gwynne, Robert Culp, Barbara Parkins, Wendy Kilbourne, Victoria Tucker, Pamela West
Month by month the pretty girls who grace the centre pages of a men's magazine are being murdered. Having murdered Miss January and Miss February, it would appear that Miss March is next in line. Presumably Miss December is glad she's on the back page. A minor murder mystery.
Aka: INSATIABLE; VICTIMISED
THR 95 min (ort 104 min) mTV
VIDrel: 4-FRONT/POLYREC/CIC V
Boa: story by Gregory S. Dinallo.

CALIFORNIA MAN ** PG
Les Mayfield USA 1992
Sean Astin, Brendan Fraser, Pauly Shore, Megan Ward, Michael De Luise, Derek James, Mariette Hartley, Richard Masur, Patrick Van Horn, Jonathan Quan, Ellen Blain, Esther Scott, Steven Elkins, Wanda Aguna, Furley Lumpkin, Peter Allas
Two cool Californian teenage boys dig up a 10,000-year-old caveman and bring him up to date on the necessities of life in the 20th century. When their guest becomes a popular celebrity, they find they can exploit this for their own benefit. A very juvenile comedy of limited laughs.
Aka: ENCINO MAN
COM 85 min (ort 89 min) VIDrel: HOLPIC/TECH L/A V

CALIFORNIA SUITE * 15**
Herbert Ross USA 1978
Jane Fonda, Alan Alda, Richard Pryor, Bill Cosby, Maggie Smith, Michael Caine, Walter Matthau, Elaine May, Gloria Gifford, Sheila Frazier, Herb Edelman, Denise Galik, David Sheehan, Michael Boyle, Len Lawson, Gino Ardito
Adaptation of a Neil Simon play about four couples who occupy the same suite in a Beverly Hills Hotel. An uneven series of episodes. Smith and Caine as a British couple in town for the Academy Awards are a delight, as are Jewish couple Matthau and May. The divorce drama of Alda and Fonda and the bitchy Pryor and Cosby effort are both duds. AA: S. Actress (Smith).
COM 99 min VIDrel: POLY/POLYREC L/A V
Boa: play by Neil Simon.

CALIGULA * 18
Tinto Brass USA 1979
Malcolm McDowell, Teresa Ann Savoy, Helen Mirren, Peter O'Toole, Guido Mannari, John Gielgud, John Steiner
An attempt to relate the story of this particularly repulsive Roman Emperor that is merely an excuse for sex (six minutes of hardcore in the uncensored version) and explicit violence as we follow this lovable Roman through a seemingly endless series of depravities and decapitations. Written by Gore Vidal and produced by Penthouse, this was the first $15,000,000 porno film. We can't say it was money well spent. Re-issued at 90 minutes.
A 90 min (ort 156 min) VIDrel: NORVID/DISC V

CALL ME ANNA ** 15
Gilbert Cates USA 1990
Patty Duke, Howard Hesseman, Millie Perkins, Deborah May, Ari Meyers, Jenny Robertson, Arthur Taxier, Karl Malden, Timothy Carhart, Matthew L. Perry, David Packer, Dana Gladstone, Woody Eney, Francois Giroday, Lora Staley, Richard Fancy
Patty Duke's autobiography serves as the basis for this honest but somewhat overwrought account of her rise from child actress to successful movie star and her lifelong struggle against manic depression. A fairly absorbing account that unfortunately retains little of the book's impact. Co-produced by Cates and Patty Duke (under the name Anna Duke-Pearce).
DRA 95 min (ort 100 min) mTV VIDrel: GUILD/SONOP V/sh
Boa: autobiography of Patty Duke Astin.

CALL OF THE WILD ** PG
Ken Annakin FRANCE/ITALY/SPAIN/UK/
WEST GERMANY 1972
Charlton Heston, Maria Rohm, Michele Mercier, Raimund Harmstorf, George Eastman (Luigi Montefiore), Friedhelm Lehmann, Horst Heuck, Sancho Garcia, Juan Luis Galiardo, Rik Battaglia, Alf Malland, Alfredo Mayo, Sverre Wilberg
Film version of a classic Jack London story with the beautiful countryside of Finland standing in for the Yukon, as two men and a dog trek across two-hundred miles of frozen wasteland during the 1896 Klondike Gold Rush. A flat and insipid effort that wastes the talents of a good cast, though the beautiful scenery is a compensation.
Aka: IL RICHIAMO DELLA FORESTA
DRA 100 min (ort 105 min) VIDrel: MIA/DISC V
Boa: novel by Jack London.

CALL OF THE WILD * (PG)**
Alan Smithee (Michael Toshiyuki Uno) USA 1992
Rick Schroder, Mia Sara, Gordon Tootoosis, Duncan Fraser, Richard Newman, Brent Stait, Eric McCormack, Kerry Sandomirsky, Tom Heaton, Alan Lysell, Vince Metcalfe, Marie Stillin, Davne Low, Peter Bibby, Bill Croft, Ed Mitchell
Recent adaptation of this much-filmed London classic, detailing the severe hardships suffered by a young man who goes in search of gold in the frozen wastes of the Klondike. A consistent and faithful adaptation of London's novel that focuses on its canine hero. The fine camerawork and beautiful natural locations are major assets and the use of the

"Alan Smithee" alias by the director is both inexplicable and unnecessary.
A/AD   94 min mTV   SATrel: SKY MOVIES
Boa: novel by Jack London.

CAMELOT *
U
1967
Joshua Logan USA
Richard Harris, Vanessa Redgrave, David Hemmings, Franco Nero, Lionel Jeffries, Laurence Naismith, Pierre Olaf, Estelle Winwood, Gary Marshall, Anthony Rogers, Peter Bromilow, Sue Casey, Garry Marsh, Nicolas Beauvy
Film version of the Lerner and Loewe musical that still has the same good songs but is ruined by actors with lousy voices. Excessive use of close-ups is an added annoyance, though on TV this will be less so. A dreary effort. AA: Art/Set (John Truscott and Edward Carrer/John W. Brown), Cost (John Truscott), Score/adapt (Alfred Newman/Ken Darby).
MUS   175 min (ort 181 min)   VIDrel: WHV   V/h
Boa: musical by Lerner and Loewe.

CAMERON'S CLOSET **
18
1987
Armand Mastroianni USA
Cotter Smith, Mel Harris, Chuck McCann, Scott Curtis, Leigh McCloskey, Kim Lankford, Tab Hunter, Gary Hudson, Dort Donald Clark, David Estruardo, Kerry Nakagawa, Wilson Smith, Raymond Patterson, Skip E. Lowe, Bond Bradigan
A ten-year-old boy has had his psychic powers tampered with by his father and a scientist partner. In the course of their experiments the boy is brought to life an evil demon that lives in his bedroom cupboard. He loses control of it and a series of grisly murders takes place. A detective assigned to the case forms a close bond with the boy, and now has the task of rescuing him. A mediocre horror yarn. See also MONSTER IN THE CLOSET.
HOR   83 min (ort 90 min)   VIDrel: MED/POLY L/A   V
Boa: novel by Gary Bradner

CAMILLA **
12
1994
Deepa Mehta CANADA/UK
Jessica Tandy, Bridget Fonda, Hume Cronyn, Elias Koteas, Maury Chaykin, Graham Greene, Ranjit Chowdry, George Harris, Atom Egoyan, Sandi Ross, Gerry Quigley, Devyani Saltzman, Camille Spence, Martha Cronyn, Sheilanne Lindsay
A young woman songwriter rents a room in the home of an elderly retired violinist and the two women strike up a warm relationship. This stands them in good stead, when the latter suddenly decides to drive to Toronto in search of a former lover and tenant joins her for the ride. A thinly plotted female buddy-movie that feels a little contrived but is well worth seeing for Tandy's accomplished performance in her last screen appearance.
DRA   95 min   VIDrel: EIV/SONOP   V

CAMILLE ***
PG
1936
George Cukor USA
Greta Garbo, Robert Taylor, Lionel Barrymore, Elizabeth Allan, Henry Daniell, Leonore Ulric, Laura Hope Crews, Rex O'Malley, Jessie Ralph, E.E. Clive, Russell Hardie, Douglas Walton, Marion Ballou, Joan Brodel
A fine adaptation of the classic Dumas novel, has Garbo at her best as the ailing courtesan who sacrifices everything for the love of a young nobleman. Remade several times since.
DRA   104 min (ort 110 min) B/W   VIDrel: MGM/WHV   V
Boa: novel La Dame aux Camelias (The Lady of the Camelias) by Alexandre Dumas.

CAMILLE CLAUDEL ***
PG
1988
Bruno Nuytten FRANCE
Isabelle Adjani, Katrine Boorman, Alain Cuny, Gerard Depardieu, Laurent Grevill, Daniele Lebrun, Madeleine Robinson, Philippe Clevenot, Maxime Levoux, Roger Planchon, Jean-Pierre Sentier
A lavish, visually impressive but overlong attempt to tell the story of Rodin's pupil, assistant and mistress, whose talent was at least equal to his own. Her passionate relationship with this sculptor, who became virtually a living symbol of the greatness of France, became such a scandal that she was packed off to an asylum where she spent the last decades of her life.
DRA   167 min (ort 174 min) wScrn   VIDrel: ARTPRO/RTM V/sur
Boa: book by Reine-Marie Pavis.

CAMOMILE LAWN, THE ***
15
1992
Peter Hall UK
Felicity Kendal, Paul Eddington, Oliver Cotton, Jennifer Ehle, Rebecca Hall, Tara Fitzgerald, Toby Stephen, Nicholas Le Prevost, Claire Bloom, Rosemary Harris, Richard Johnson, Virginia McKenna, Ben Walden, Jeremy Brook, Joss Brook
Forty years in the life of an English family, from 1939 when as youngsters five cousins gather at their aunt's house, to the present day, when they meet up once more for the funeral of a their violinist friend, a wartime refugee whose life had great impact on them all. A detailed and engrossing study, adapted with enormous care, and originally shown in four parts.
DRA   257 min (2 cassettes) mTV   VIDrel: POLY/POLYREC V
Boa: novel by Mary Wesley.

CAMP NOWHERE **
PG
1994
Jonathan Prince USA
Christopher Lloyd, Jonathan Jackson, Wendy Makkena, M. Emmet Walsh, Andrew Keegan, Marne Patterson, Melody Kay, Kate Mulgrew, Burgess Meredith, Peter Scolari, Romy Walthall, Peter Onorati, Nathan Cavaleri, Brian Wagner,, Kay Baker
A group of kids with assorted asocial habits get together with an unemployed drama teacher and get him to pose as the owner of a summer camp so that they can enjoy themselves without strict supervision. When the parents start to get suspicious, they decide to hold an open day to convince them that everything is as it should be. A formula juvenile comedy that offers a few sparse laughs plus performances of considerable vigour.
COM   82 min (ort 95 min)   VIDrel: HOLPIC/TECH   V/sh

CAMPFIRE TALES **
18
1996
David Semel/Martin Kunert/Matt Cooper USA
Jay R. Ferguson, Christine Taylor, Ron Livingston, Jennifer MacDonald, Alex McKenna, Devon Odessa, Amy Smart, James Marsden, Jacinda Barrett
After they are stranded following a car accident, a bunch of kids pass the time by telling each other scary stories, involving ghosts, maniacs and cannibals. None of the tales are especially strong in plotting, the intention of the makers being to generate a menacing atmosphere and a succession of chills, and in this respect the film largely succeeds.
HOR   83 min   VIDrel: MED/SONOP   V

CAN-CAN **
U
1960
Walter Lang USA
Frank Sinatra, Shirley MacLaine, Maurice Chevalier, Juliet Prowse, Louis Jourdan, Marcel Dalio, Leon Belasco, Nestor Paiva
A Paris nightclub owner faces constant police harassment because of an exciting new dance being performed on his premises. Film version of a dull but tuneful musical set in Paris of the 1890s. Songs include: "C'est Magnifique", "I Love Paris", "Let's Do It" and "Just One Of Those Things".
Aka: CAN CAN
MUS   125 min (ort 131 min) cC   VIDrel: 20TH/TECH   V
Boa: play by Abe Burrows.

CAN I DO IT 'TIL I NEED GLASSES? *
18
1980
T. Robert Levy USA
Roger Behr, Robin Williams, Debra Klose, Jeff Doucette, Victor Dunlop, Moose Carlson, Patrick Wright, Walter Olkewicz, Roger and Roger
The title is the most risque thing in this tired and limp sex comedy. From the same team who produced "If You Don't Stop It You'll Go Blind".
Aka: CAN I DO IT TILL I NEED GLASSES?
COM   67 min (ort 80 min)   VIDrel: MIA/DISC   V

CAN IT BE LOVE? *
18
1992
Peter Maris USA
Richard Beaumont, Mary Ann Mixon, Jennifer Langdon, Blake Pickett, Charles Klausmeyer, Wally "The Wall" Mueller, Karen Trella, Jim Pelish, Jay Derrick, Blake Mitchell, Hesh Rephun, Leesa Christine Moskus, Lesa Stene, Randy Good
During spring break, a boy and his friend cruise the beaches in search of an heiress who bears a distinctive birthmark on her body. A smutty and uninspired teen-comedy of little or no amusement value.
COM   90 min   VIDrel: HIFLI/SONOP   V/h

CAN YOU KEEP IT UP FOR A WEEK? *
18
1974
Tom Atkinson UK
Jeremy Bulloch, Jill Damas, Neil Hallett, Richard O'Sullivan, Sue Longhurst, Jenny Cox, Joy Harrington, Valerie Leon, Olivia Munday, Mark Singleton, Venica Day, Jules Walters, Catherine Howe, Wendy Wax, Mandy Morris

A girl promises to marry her boyfriend if he can keep his hands off other women (and thus keep his job) for a whole week. An utterly dated and dismal sex farce.
COM   90 min (ort 94 min)   VIDrel: JEZ/RTM   V

CANADIAN BACON *   PG
Michael Moore USA   1994
John Candy, Alan Alda, Rhea Perlman, Kevin Pollak, Rip Torn, Bill Nunn, G.D. Spradlin, Kevin J. O'Connor, Steven Wright, James Belushi, Brad Sullivan, Alan Alda, Stanley Anderson, Richard Council, Michael Copeman, Wallace Shawn
A sheriff finds himself being ordered to invade Canada, when the US President declares war on that country as a ploy to boost his falling popularity. A daft and feebly comedy that inevitably recalls THE MOUSE THAT ROARED, this was the last film to be completed by Candy before his death, sadly the film's only point of interest. (In fact Candy died during the shooting of WAGONS EAST!)
COM   91 min (ort 95 min) Col/B/W
VIDrel: COLUM/SONOP   V/sur

CANDIDATE, THE ***   PG
Michael Ritchie USA   1972
Robert Redford, Peter Boyle, Don Porter, Allen Garfield, Karen Carlson, Melvyn Douglas, Quinn Radeker, Michael Lerner, Morgan Upton, Kenneth Tobey, Joe Miksak, Chris Prey, Jenny Sullivan, Tom Dahlgren, Gerald Hiken, Leslie Allen
Political satire about a young Californian lawyer who decides to run for the Senate on a platform of total honesty. A sharp and witty drama that combines a great feel for political campaigning, with a winning performance from Redford. Ritchie and screenwriter Larner were actually involved in some of the campaigns of the 1960s. AA: Story/Screen (Jeremy Larner).
DRA   106 min   VIDrel: MGM/WHV   V

CANDYMAN ***   18
Bernard Rose USA   1992
Virginia Madsen, Tony Todd, Vanessa Williams, Xander Berkley, Kasi Lemmons, DeJuan Guy, Michael Culkin, Gilbert Lewis, Stanley DeSantis, Ted Raimi, Rita Pavio, Vanessa Williams, Marianna Elliott, Mark Daniels, Lisa Anna Poggi
A woman graduate student researching an urban legend about mythical title killer discovers to her horror that it is true and is unable to resist the temptation to summon up this murderous creature. A frightening and highly effective chiller with a good many psychological touches, if a little spoiled by a gory shock ending. A sequel followed.
DRA   94 min (ort 99 min) wScrn
VIDrel: VCC/DISC/COLUM   V/sur
Boa: short story The Forbidden by Clive Barker.

CANDYMAN 2: FAREWELL TO THE FLESH **   18
Bill Condon USA   1995
Kelly Rowan, Tony Todd, Veronica Cartwright, Timothy Carhart, William O'Leary, Fay Hauser, Joshua Gibran Mayweather, Matt Clark, Bill Nunn, David Gianopoulos, Caroline Barclay, Michael Bergeron, Brianna Blanchard
A teacher out to debunk this urban legend performs the requisite ritual and unwittingly raises this murderous demon who then proceeds to stalk her and her family. This time our slasher is the ghost of a plantation slave, murdered by a mob for a crime against his owner. Set in New Orleans at carnival time, this is in many ways a rather lacklustre sequel that proceeds on familiar lines. The colourful locations and Todd's convincing performance are its only assets.
Aka: CANDYMAN: FAREWELL TO THE FLESH
HOR   90 min (ort 95 min) cC   VIDrel: POLY/POLYREC
V/s

CANNIBAL MAN, THE **   18
Eloy de la Iglesias SPAIN   1974
Vincente Parva, Emma Cohen, Vicki Lagos
A man who works in the canning plant of a slaughterhouse kills a taxi driver in self defence. This sparks off a wave of horrific crimes, in this cliched but violent melodrama. Unlikely to be available in an uncut form in the UK.
Aka: APARTMENT ON THE 13TH FLOOR; LA SEMANA DEL ASESINO
DRA   94 min (ort 98 min)   VIDrel: REDEM/RTM   V

CANNONBALL FEVER *   PG
James R. Drake USA   1989
Melody Anderson, Peter Boyle, Donna Dixon, John Candy, Joe Flaherty, Eugene Levy, Tim Matheson, Brooke Shields, Shari Belafonte, Matt Frewer, Mimi Kuzyk, Lee Van Cleef, Smothers Brothers
Yet one more comedy built around the idea of a coast-to-coast auto-race, where a bunch of wealthy folk with nothing better to do decide to hold an illegal cross-country race with no-holds-barred. Despite the new director and cast (Candy replaces Burt Reynolds as the central character), this follow-up to the earlier "Cannonball" films is virtually devoid of new ideas.
COM   92 min   VIDrel: FIRST/SONOP L/A   V

CANNONBALL RUN, THE *   PG
Hal Needham USA   1980
Burt Reynolds, Roger Moore, Farrah Fawcett, Dom DeLuise, Sammy Davis Jr, Edward Asner, Dean Martin, Adrienne Barbeau, Terry Bradshaw, Jack Elam, Bert Convy, Jackie Chan, Molly Picon, Jamie Farr, Bianca Jagger, Mel Tillis
It's no holds barred time when it comes to winning in an illegal car race from New York to California. A noisy and almost unbearable rip-off of THE GUMBALL RALLY. Followed in 1984 by CANNONBALL RUN 2.
COM   96 min   VIDrel: VCC L/A   V

CANNONBALL RUN 2 *   PG
Hal Needham USA   1984
Burt Reynolds, Dom DeLuise, Dean Martin, Sammy Davis Jr, Shirley MacLaine, Jamie Farr, Marilu Henner, Telly Savalas, Susan Anton, Joe Theismann, Frank Sinatra, Sid Caesar, Catherine Bach, Richard Kiel, Tim Conway, Jim Nabors
Sequel to the 1980 film and once more, about a cross-country car race. A sloppy and carelessly thrown together mess, featuring a host of stars in cameo roles.
Aka: CANNONBALL 2
COM   103 min (ort 108 min)   VIDrel: POLY L/A   V

CANTERBURY TALE, A ****   U
Michael Powell/Emeric Pressburger UK   1944
John Sweet, Dennis Price, Eric Portman, Sheila Sim, Esmond Knight, Charles Hawtrey, Hay Petrie, George Merritt, Edward Rigby, Freda Jackson, Betty Jardine, Eliot Makeham, Harvey Golden, Leonard Smith, James Tamsitt
In a small English village a land girl, a sergeant and a GI discover that a self-righteous JP is the mystery assailant who pours glue on the hair of girls during WW2 blackouts. A curious flag-waver that attempts to capture some of the charm and eccentricity of "Old England". The plot has very little to do with Chaucer but the film retains a wonderful period flavour. Quite unlike any other British film of the period, it is as distinctive as it is memorable.
DRA   119 min (ort 124 min) B/W   VIDrel: CARL/TECH   V

CANTERVILLE GHOST, THE ***   U
Paul Bogart USA   1986
John Gielgud, Ted Wass, Andrea Marcovicci, Alyssa Milano, Jeff Harding, Lila Kaye, Harold Innocent, George Baker, Dorothea Phillips, Bill Wallis, Spencer Chandler, Brian Oulton, Deddie Davies, Celia Breckon
A likeable romp through this famous Oscar Wilde tale of a restless 17th century ghost, who will only be released when he can persuade one of his descendants to perform a deed of bravery. Gielgud does justice to the role taken by Laughton in the fondly remembered 1940 version, and the rest of the cast gives good support. Filmed in England.
JUV   100 min mTV   VIDrel: VCC L/A   V
Boa: short story by Oscar Wilde.

CANTON IRON KUNG FU **   15
Li Chao HONG KONG   197-
Liang Jia Ren, Wang Chung
A kung fu movie featuring Canton-style fighting, with one of the ten "tigers" of Quon Tung perfecting his fighting techniques, as a prelude to revenge for a previous defeat.
Aka: CANTONEN IRON KUNG FU; CANTONESE IRON KUNG FU; IRON FIST OF KWANGTUNG
MAR   82 min (ort 90 min)   VIDrel: IMPENT   V

CANVAS **   15
Alain Zaloum USA   1992
Gary Busey, John Rhys-Davies, Vittorio Rossi, Nick Caviola, cary Lawrence, Michael McGill, Jonathan Palis, Mark Comachio, Alexandra Innes, Aron Tager, Mark Bromilow, Tyrone Benskin, Ming Lai Chung, Leni Parker, Tedd Dillon
A ruthless gallery owner recruits a young artist and involves him in a scheme to steal paintings. He proves an adept and quick

learner, but a close brush with murder causes him to reconsider his chosen career. Adequate.
Aka: CANVAS: THE FINE ART OF CRIME
THR 94 min VIDrel: GUILD/SONOP V

CAPE FEAR *** 15
J. Lee Thompson USA 1961
Robert Mitchum, Gregory Peck, Polly Bergen, Martin Balsam, Lori Martin, Jack Kruschen, Telly Savalas, Barrie Chase, Paul Comi, Edward Platt, John McKee, Page Slattery, Ward Ramsey, Will Wright, Joan Staley, Mack Williams
A sadistic ex-con plays a cat-and-mouse game as he stalks both the small-town lawyer responsible for sending him to jail, and the man's wife. As ever, the police are powerless to stop the man's threats and our hero is obliged to take matters into his own hands. Mitchum is excellent as the menacing psycho in a brisk but otherwise unimaginative thriller. Remade in 1991.
THR 101 min (ort 106 min) B/W VIDrel: CIC/SONOP V
Boa: novel The Executioners by John D. MacDonald.

CAPE FEAR *** 18
Martin Scorsese USA 1991
Robert De Niro, Nick Nolte, Jessica Lange, Juliette Lewis, Joe Don Baker, Gregory Peck, Martin Balsam, Ileana Douglas, Robert Mitchum, Fred Dalton Thompson, Zully Montero, Craig Henne, Forest Burton, Edgar Allan Poe IV
Scorsese's remake of this 1960s thriller takes the simple story of an ex-convict who has resolved to terrorise the lawyer he correctly blames for his conviction, jettisons much of the old-fashioned Good versus Evil morality, and injects a dose of 1990s violence. Though De Niro as the ex-con is out to exact a terrible revenge against the family of his former lawyer, one's sympathies are hardly with the latter. A dark and unusually bitter film.
THR 122 min (ort 128 min) VIDrel: CIC/SONOP; PION (LV only) V/sur LV
Boa: novel The Executioners by John D. MacDonald.

CAPONE * 18
Steve Carver USA 1975
Ben Gazzara, Susan Blakely, Harry Guardino, John Cassavetes, Sylvester Stallone, Peter Maloney, Frank Campanella, John D. Chandler, John Orchard, Mario Gallo, Russ Marin, George Chandler, Royal Dano
A tedious attempt to chart the rise and fall of one of America's most notorious gangsters. Gazzara's dialogue is muffled and incomprehensible for most of the time but this is no great loss; the best thing in this movie is the stock footage from THE SAINT VALENTINE'S DAY MASSACRE.
A/AD 97 min (ort 101 min) VIDrel: 20TH/TECH V/sh

CAPRICE * U
Frank Tashlin USA 1967
Doris Day, Richard Harris, Ray Walston, Jack Kruschen, Edward Mulhare, Lilia Skala, Irene Tsu, Michael Romanoff, Michael J. Pollard
When a career woman decides to investigate the murder of her employer she finds herself plunged into the dangerous world of industrial espionage, and attempts to reveal the true nature of a sinister cosmetics empire. A badly dated film that awkwardly blends James Bond-style suspense with an unfunny measure of farce – the result is a glossy, forgettable hotchpotch.
COM 97 min (ort 98 min) VIDrel: 20TH/TECH V/h

CAPRICORN ONE *** PG
Peter Hyams USA 1976
Elliot Gould, James Brolin, Hal Holbrook, Sam Waterston, Karen Black, O.J. Simpson, Telly Savalas, Brenda Vaccaro, Denise Nicholas, David Huddleston, Robert Walden, David Doyle, Lee Bryant, Alan Fudge, Jon Cedar, James Karen
Complications arise when a reporter discovers that the first manned flight to Mars was a fake. The capsule has been reported as burning up on re-entry and the pilots killed, but they are understandably not too willing to help in this hoax. An implausible but entertaining film of intrigues and chases.
A/AD 118 min VIDrel: 4-FRONT/POLYREC V

CAPTAIN AMERICA * PG
Albert Pyun USA 1990
Matt Salinger, Ronny Cox, Ned Beatty, Darren McGavin, Michael Nouri, Melinda Dillon, Kim Gillingham, Scott Paulin, Bill Mumy, Francesca Neri, Massimilo Massimi, Wayne Preston, Norbert Weisser, Garete Ratliff, Bernarda Oman

Based on characters created for Marvel Comics by Joe Simon and Jack Kirby, this sees our title hero being thawed out of the Alaskan ice after fifty years, just in time to deal with his arch enemy the Red Skull, who is once again plotting to take over the world. A crude and unintentionally funny attempt to bring to life a comic-strip hero, the film collapses under the weight of a turgid plot, weak direction and feeble acting.
Aka: CAPTAIN AMERICA: THE MOVIE
A/AD 93 min (ort 103 min) VIDrel: CASPIC/BMGREC L/A V/s

CAPTAIN BLOOD **** PG
Michael Curtiz USA 1935
Errol Flynn, Olivia de Havilland, Basil Rathbone, Lionel Atwill, Ross Alexander, Guy Kibbee, Henry Stephenson, Forrester Harvey, Hobart Cavanaugh, Donald Meek, J. Carrol Naish, Pedro De Cordoba, Leonard Mudie, Jessie Ralph
An unjustly punished surgeon is deported from England and turns to piracy in this first Flynn swashbuckler. Brilliant high adventure, with Flynn at his very best, ably supported by a fine cast. Has also become available in a computer-colourised version for non-purists.
A/AD 94 min (ort 119 min) B/W VIDrel: MGM/WHV V
Boa: novel by Rafael Sabatini.

CAPTAIN COSMOS ** PG
Roy Thomas HONG KONG 1989
Animated space opera set in the not-too-distant future when an emergency signal from Earth reaches the headquarters of a group of galactic warriors. Their help is required to defeat the female head of an evil empire that has been abducting and enslaving space travellers. A fierce battle now ensues as they fight to defeat this empire and free its captives. A watchable film of adequate animation. See also THE COSMOS CONQUEROR and FALCON 7.
ANIM 77 min VIDrel: IMPENT V

CAPTAIN POWER AND THE SOLDIERS OF THE FUTURE ** PG
Otta Hanus/Jorge Montesi CANADA 1989
Tim Dunigan, Peter MacNicol, Sven Thorsen, Maurice Dean Wint, Jessica Steen, David Hemblen, Bruce Gray, Dylan Neal, Paul Humphrey, Peter Snider, Jonathan Wilson, Anthony Dean Rubes, Voices of: Ted Dillion, K. Tedland, John Davies
Fantasy set in the year 2147 A.D., when Earth is in a new dark age and a power-mad tyrant is trying to take over what is left of civilisation by creating a race of bionic fighting machines. Unimpressive special effects abound (despite the use of computer-generated animation) and the flabby plot fails to hold one's interest.
Aka: CAPTAIN POWER: THE SOLDIERS OF THE FUTURE; CAPTAIN POWER AND THE SOLDIERS OF THE FUTURE: THE LEGEND BEGINS
FAN 93 min VIDrel: GENESIS V

CAPTAIN RON * (PG)
Thom Eberhart USA 1992
Kurt Russell, Martin Short, Mary Kay Place, Benjamin Salisbury, Jorge Luis Ramos, Emmanuel Logrono, Meadow Sisto, J.A. Preston, Tanya Soler, John Scott Clough, Raul Estela, Jainardo Batista, Dan Butler, Tom McGowan, Paul Anka
After an inheriting a sailboat, a man and his friends take off for the Caribbean and find themselves all at sea in this water-logged comedy whose idea of amusement is to have someone fall overboard. Very poor.
COM 96 min (ort 99 min) VIDrel: TOUCH/SONOP L/A V

CAPTAINS COURAGEOUS **** U
Victor Fleming USA 1937
Spencer Tracy, Freddie Bartholemew, Lionel Barrymore, Melvyn Douglas, Mickey Rooney, John Carradine, Walter Kingsford, Charley Grapewin, Christian Rub, Leo G. Carroll, Charles Trowbridge, Richard Powell, Jay Ward, Kenneth Wilson
Bartholemew plays a spoilt rich brat who falls off a cruiser and is picked up by a fishing vessel. Tracy gives a remarkable portrayal of the Portuguese fisherman who takes the boy in hand and wins his affection. A splendid film with some great scenes both on land and sea. Tracy won an Oscar in this one, and no wonder. The script is by John Lee Mahin, Marc Connelly and Dale Van Every. Remade as a TV film in 1977.
AA: Actor (Tracy).
DRA 112 min B/W VIDrel: MGM L/A V
Boa: novel by Rudyard Kipling.

CAPTAIN'S TABLE, THE ***
Jack Lee UK   U
1958
John Gregson, Peggy Cummins, Donald Sinden, Nadia Gray, Maurice Denham, Richard Wattis, Reginald Beckwith, Bill Kerr, Nicholas Phipps, John Le Mesurier, Lionel Murton, Joan Sims, Miles Malleson, James Hayter, Nora Nicholson
The captain of a freighter is given trial command of a luxury liner and has to make a number of compromises to get things running smoothly. A lively and spirited adaptation of Gordon's book, written by John Whiting, Bryan Forbes and Nicholas Phipps.
COM   86 min   VIDrel: VCC L/A   V
Boa: novel by Richard Gordon.

CAPTIVE **
Michael Tuchner USA   15
1991
Joanna Kerns, Barry Bostwick, John Stamos, Chad Lowe, Patricia Charbonneau, Teddie Stidder, Florence Paterson, Don S. Davis, Timothy Webber, Jaclyn Hazeldine, Kristy Hazeldine, Douglas Newell, Duncan Fraser, Fred Diehl
Inspired by real events, this unpleasant thriller tells the story of a young married couple who suffer a harrowing ordeal at the hands of two vicious escaped convicts, who take them and their baby daughter hostage.
THR   93 min mTV   VIDrel: CAPIT/GUILD   V

CAPTIVE HEART, THE **
Basil Dearden UK   PG
1946
Michael Redgrave, Basil Radford, Rachel Kempson, Jack Warner, Mervyn Jones, Jimmy Hanley, Gordon Jackson, Ralph Michael, Derek Bond, Karel Stepanek, Gladys Henson, Jack Lambert, Guy Middleton, Meriel Forbes, Robert Wyndham, Jane Barrett
A powerful look at British POWs in WW2 and their life in a German camp, focusing on a Czech officer, who steals the identity papers of a dead British officer and begins to correspond, and later fall in love, with the latter's wife. A fine cast all turn in flawless performances in this moving little drama.
WAR   102 min B/W (ort 98 min)   VIDrel: LUMI/SPEAR L/A   V

CAPTIVE ISLAND **
John Biffar USA   (U)
1994
Jesse Zeigler, Amy Bush, "Banana" George Blair, Bill Cobbs, Arte Johnson, Ernest Borgnine, Bob Hite, Robert Keith, Norma Muller, Kathleen Bryan, Teri Mann, Michael Price, Bull Sharf, Kathleen Bryan, Jonathon Conner, Scott Bennett
A young teenage boy from a family of tycoons, makes an agreement with his father that allows him to spend one week on vacation in Florida before going to summer school in order to prepare for college entrance. Once arrived, he cuts loose and meets up with a young girl and some senior citizens who don't let their age stop them from enjoying life or taking risks. A pleasant little tale with a few serious elements that offers some charming performances.
DRA   95 min (ort 100 min)   SATrel: MOVIE CHANNEL

CAPTIVE OF THE DESERT ***
Raymond Depardon FRANCE   PG
1990
Sandrine Bonnaire, Dobi Kor, Fadi Taha, Dobi Wachink, Badei Barka, Atchi Wahi-li, Daki Kor, Isai Kor, Mohammed Ixa, Brahim Barka, Hadji Azouma, Barkama Hadji, Sidi Hadji Maman plus the inhabitants of Chirfa, Orida and Djaba
The true story of a female French researcher who was captured by rebel soldiers in the Sahara, and held hostage by them for fifteen months. A languid film of few events, most of the story focuses on her slow efforts to gain the trust and respect of her tribesman captors, the success of which finally leads to her release. A demanding and atmospheric film of minimal dialogue, whose director covered the actual story as a photoreporter back in 1975.
Aka: LE CAPTIVE DU DESERT
A/AD   101 min   CINrel

CAPTIVES ***
Angela Pope UK   15
1994
Julia Ormond, Tim Roth, Richard Hawley, Jeff Nuttall, Kenneth Cope, Keith Allen, Bill Moody, Peter Capaldi, Siobhan Redmond, Christina Collingridge, Victoria Scarborough, Aedin Moloney, Tricia Thorns, Nathan Dambuza
Just separated from her husband, an unhappy dentist takes a part-time job in a men prison, and finds herself strongly attracted to one of the inmates. But their deepening relationship leaves both vulnerable to blackmail on the part of another

prisoner. Produced by the BBC, this is an unusually intelligent and well executed story, effective in both conception and resolution. This was Pope's first feature film, but she had already done some good work for TV.
DRA   95 min (ort 100 min)   VIDrel: EIV/SONOP   V

CAR 54, WHERE ARE YOU? *
Bill Fishman USA   15
1991
David Johansen, John C. McGinley, Brian Muldoon, Rosie O'Donnell, Al Lewis, Fran Drescher, Daniel Baldwin, Jeremy Piven, Nipsey Russell, Bobby Collins, Tone Loc, Louis Di Bianco, Barbara Hamilton, Eliza Garrett, Rik Colitti, Sally Cahill
A poor and unfunny attempt to exploit the original TV series of the early 1960s for this feature-length comedy, built around our two bumbling cops and their efforts to catch a gangster while causing general chaos all around them. Some winning performances and elaborate sight gags cannot help to give this one any real humour or cohesive structure.
COM   84 min (ort 89 min)   VIDrel: COLUM/SONOP V/sur

CAR TROUBLE ***
David Green UK   18
1985
Julie Walters, Ian Charleson, Stratford Johns, Vincenzo Ricota, Hazel O'Connor, Dave Hill, Anthony O'Donnell, Vanessa Knox-Mayer, Roger Hume, John Blundell, Veronica Clifford, Laurence Harrington, Jeff Hall, Roy Barraclough
A young wife is forbidden to drive the family car but does so, and in the process becomes somewhat entangled with a young mechanic. A humorous and perceptive farce with Walters and Charleson hilarious as the ill-fated couple.
COM   89 min   VIDrel: LUMI/SPEAR L/A   V/sur

CARAVAGGIO ***
Derek Jarman UK   18
1986
Nigel Terry, Sean Bean, Garry Cooper, Spencer Leigh, Tilda Swinton, Dawn Archibald, Michael Gough, Nigel Davenport, Robbie Coltrane, Noam Almaz, Jack Birkett, Una Brandon-Jones, Imogen Claire, Sadie Corre, Lol Coxhill
A lush and imaginary biopic of the life of the title painter, who died in 1610. As one might expect from the director, it serves both as a statement of the artist's alleged homosexuality and as a framework for images of great beauty, many of which effectively convey the flavour of the artist's best works. Made on a restricted budget, with a share of deliberate jokey anachronisms, it's a self-indulgent but memorable offering.
DRA   89 min (ort 93 min)   VIDrel: CONNO/RTM   V

CARAVAN OF COURAGE: AN EWOK ADVENTURE ***
John Korty USA   U
1984
Eric Walker, Warwick Davis, Fionnuala Flanagan, Guy Boyd, Daniel Frishman, Debbie Carrington, Tony Cox, Kevin Thompson, Maragarita Fernandez, Pam Grizz, Bobby Bell, Aubree Miller, Burl Ives (narration)
A spin-off from the STAR WARS films, with two children who are searching for their parents crash-landing on a strange planet and being rescued by the Ewoks – the benevolent creatures that inhabit it. A wholesome kid's adventure packed with visual marvels. Producer George Lucas's first TV movie, it was followed by EWOKS: THE BATTLE FOR ENDOR.
Aka: CARAVAN OF COURAGE; EWOK ADVENTURE, THE
FAN   95 min (ort 100 min) mTV   VIDrel: 20TH/TECH L/A V
Boa: story by George Lucas.

CARBON COPY **
Michael Schultz USA   15
1981
George Segal, Susan St James, Denzel Washington, Jack Warden, Dick Martin, Paul Winfield, Tom Poston, Vicky Dawson, Macon McCalman, Parley Baer, Ed Call, Vernon Weddle, Edward Marshall, Angelina Estrada
A successful corporate executive who has hidden the fact that he is Jewish from his colleagues, gets a sudden shock when his black seventeen-year-old illegitimate son turns up. A blend of social drama and comedy that scatters a few laughs in several directions, but ultimately runs out of steam. The script is by Stanley Shapiro. See also MADE IN AMERICA and A MODERN AFFAIR.
COM   87 min (ort 92 min)   VIDrel: GUILD/SONOP L/A   V

CARD, THE ***   *U*
Ronald Neame UK   1952
Alec Guinness, Glynis Johns, Valerie Hobson, Petula Clark, Edward Chapman, Veronica Turleigh, George Devine, Joan Hickson, Frank Pettingell, Gibb McLaughlin, Michael Hordern, Alison Leggatt, Wilfrid Hyde-White
In the 1890s a sharp-witted young clerk and son of a washer-woman hits on a number of unorthodox ploys to improve both his social status and his bank balance, eventually becoming the local mayor. A lively and vigorous work, it benefits from a wonderful performance by Guinness in central role, plus Eric Ambler's clever script.
Aka: PROMOTER, THE
COM   87 min (ort 90 min) B/W   VIDrel: CARL/TECH   V
Boa: novel by Arnold Bennett.

CARE BEARS' ADVENTURE IN WONDERLAND, THE **   *U*
Raymond Jafelice CANADA   1987
Voices of: Bob Dermer, Eva Almos, Dan Hennessy, Jim Henshaw, Colin Fox
This insipid and dull cartoon animation has the Care Bears following Alice through the looking glass and embarking on a series of adventures, and assisting the White Rabbit in freeing The Princess of Wonderland from the clutches of an evil wizard. Music is by John Sebastian.
Aka: CARE BEARS IN WONDERLAND
ANIM   72 min (ort 75 min)
VIDrel: 4-FRONT/POLYREC/VISVID   V

CAREFREE ***   *U*
Mark Sandrich USA   1938
Fred Astaire, Ginger Rogers, Ralph Bellamy, Luella Gear, Clarence Kolb, Jack Carson, Franklin Pangborn, Walter Kingsford, Hattie McDaniel, Kay Sutton, Tom Tully, Robert B. Mitchell and his St Brendan's Boys
Typical Astaire and Rogers vehicle with the story of a psychologist falling for one of his wacky female patients sent to him by his best friend. The plot may be paper-thin but this happy and enjoyable musical has some outstanding Irving Berlin numbers such as "Change Partners" and "I Used To Be Color Blind".
MUS   81 min (ort 81 min) B/W   VIDrel: VCC   V

CAREFUL **   *(PG)*
Guy Maddin CANADA   1992
Kyle McCulloch, Gosia Dobrowolska, Sarah Neville, Brent Neale, Paul Cox, Jackie Burroughs, Victor Cowie, Michael O'Sullivan, Vince Rimmer, Katya Gardner, Ross McMillan, Leith Clark, Glen Hubich, Brendan Carruthers, George Toles
In an Alpine village a man undergoes training as a butler while finding the time to chase a pretty girl. Below the picturesque surface, various nasty forces are at work. A black comedy that spoofs a wide variety of targets including film styles and event resorts to the use of title cards between scenes, as in silent movies.
DRA   100 min   CINrel

CARLA'S SONG **   *15*
Ken Loach GERMANY/SPAIN/UK   1996
Robert Carlyle, Oyanka Cabezas, Scott Glenn, Salvador Espinoza, Louise Goodall, Richard Loza, Gary Lewis, Subash Singh Pall, Stewart Preston, Margaret McAdam, Pamela Turner, Greg Friel, Ann-Marie Timoney, Andy Townsley
When a Glaswegian bus-driver falls in love with a refugee from Nicaragua, having stepped in at the right moment to prevent her suicide, he accompanies her on a trip to her home country, where she hopes to lay the ghosts of her unhappy past. Effectively made in two halves, the film never quite captures a sense of conviction or narrative thrust, but is nonetheless a touching love story of worthy sentiments.
DRA   125 min   CINrel

CARLITA'S BACKWAY *   *18*
Stuart Canterbury USA   1995
Leena, Veronica Sage, Angel Bust, Kaitlyn Ashley, Isis Nile, Joey Silvera, Peter North, Mike Horner, Alex Sanders
Sex spoof on CARLITO'S WAY that tells the story of a dancer who is waiting for her lover to be released from jail at the end of his sentence. To help her pass the time she amuses herself with a variety of partners.
A   42 min   VIDrel: ONE   V

CARLITO'S WAY ***   *18*
Brian De Palma USA   1993
Al Pacino, Sean Penn, Penelope Ann Miller, John Leguizamo, Luis Guzman, Ingrid Rogers, James Rebhorn, Viggo Mortensen, Jorge Porcel, Joseph Siravo, Richard Foronjy, Frank Minucci, Adrian Pasdar, John Agustin Ortiz, Angel Salazar
In 1975 a Puerto Rican criminal leaves prison after five years firmly determined to go straight but finds that old habits and loyalties stand in his way, as does his sense of honour. Brilliant performances from Penn and Pacino help enhance this stark gangster story, even if the obligatory doses of violence are as predictable as they are needless.
DRA   138 min (ort 145 min) wScrn cC   VIDrel: CIC/SONOP L/A; PION LV only)   V/sur LV
Boa: novels Carlito's Way and After Hours by Edwin Torres.

CARLTON-BROWNE OF THE F.O. *   *U*
Jeffrey Dell/Roy Boulting UK   1959
Peter Sellers, Terry-Thomas, Ian Bannen, Thorley Walters, Raymond Huntley, Luciana Paoluzzi, Miles Malleson, John Le Mesurier, Marie Lohr, Kynaston Reeves, Ronald Adam, John Van Eyssen, Nicholas Parsons, Irene Handl
Valuable mineral deposits are discovered on a forgotten British island colony in the Pacific and an idiotic British diplomat is dispatched in order to establish good relations. However, his arrival achieves exactly the opposite effect. A trite little comedy replete with embarrassingly poor performances all round.
Aka: MAN IN A COCKED HAT
COM   86 min (ort 88 min) B/W   VIDrel: LUMI/SPEAR   V

CARMEN ***   *PG*
Carlos Saura SPAIN   1983
Antonio Gades, Laura Del Sol, Paco de Lucia, Cristina Hayos, Juan Antonio Jimenez, Sebastian Moreno, Jose Yepes, Pepa Flores
Flamenco interpretation of Bizet's famous opera with the action set among the members of a dance company and following the exploits of a choreographer, who is captivated by the women he is to cast in the title role. Considerably more spirited than many a more conventional interpretation, with superb performances from Gades and Del Sol.
MUS   97 min (ort 103 min) wScrn   VIDrel: PHASE/RTM   V
Boa: opera by Bizet.

CARMEN **   *PG*
Francesco Rosi FRANCE/ITALY   1984
Julia Migenes-Johnson, Placido Domingo, Ruggero Raimnodi, Faith Esham, Jean-Philippe Lafont, Gerard Gardino
This poor film production of the famous opera is stilted and awkward in just about every way except for the singing. On TV the deficiencies are even more apparent, in a film that's best appreciated with one's eyes closed.
PER   148 min (ort 152 min)   VIDrel: COLUM/SONOP V/sh
Boa: opera by Bizet.

CARNAL CRIMES *   *18*
Alexander Hippolyte USA   1991
Martin Hewitt, Linda Carol, Julie Strain, Paula Trickey, Alex Kubik, Rich Carter, Yvette Stefans, Charise Cooper, Prince Hughes, Doug Jones, Andre Rosey Brown, Sergio Saterno, Jasae, Nick Celozzi, Danny Trejo, Robert Cali
A look at the kinky lives of wealthy L.A. residents, that follows a beautiful but unfulfilled Beverly Hills woman who escapes from her unhappy marriage via a string of sordid affairs. When she falls for a handsome photographer, she little suspects that he's a psychopath who likes to blackmail his conquests. Despite high production values, this glossy softcore thriller is both empty and dull. See also SECRET GAMES and MIRROR IMAGES.
THR   94 min Cut (36 sec)   VIDrel: MED/POLYREC L/A V

CARNAL KNOWLEDGE ***   *18*
Mike Nichols USA   1971
Jack Nicholson, Candice Bergen, Art Garfunkel, Ann-Margret, Rita Moreno, Carol Kane, Cynthia O'Neal
This perceptive and thoughtful look at the changing sexual attitudes of two men, from college youthfulness through to middle age, is undeniably absorbing but ultimately a depressing experience. Scripted by Jules Feiffer, with an excellent performance from Ann-Margret as Nicholson's sexy mistress.
DRA   97 min   VIDrel: POLY/POLYREC   V

CARNIVAL OF BLOOD ** 18
Leonard Kirtman USA 1971
Burt Young, Earle Edgerton, Judith Resnick, Martin Barlosky, John Harris (Burt Young), Martin Barolsky, Kaly Mills, Gloria Spivak, Eve Packer, Glenn Kimberley, William Grinnel, Linda Kurtz
A maniac is on the loose in a carnival park, in this gruesome but unmemorable effort.
HOR 85 min VIDrel: IMPENT V

CARNIVAL OF SOULS *** 15
Herk Harvey USA 1962
Candace Hilligoss, Sidney Berger, Frances Feist, Herk Harvey, Art Ellison, Stanley Leavitt, Tom McGinnis, Dan Palmquist, Steve Boozer, Pamela Ballard, Cari Conboy, Larry Sneegas, Karen Pyle, Forbes Caldwell, Bill De Jarnette
A girl is apparently rescued from a river after a car crash, and drives to a new town to take up a post as a church organist. However, she is plagued by bizarre visions of a sinister pursuing figure and finds herself unable to retain a clear hold on reality. Made on a shoestring budget in Lawrence, Kansas, this effective, low-budget chiller conveys a good sense of the supernatural, its B/W photography enhancing the atmosphere considerably.
Aka: CORRIDORS OF EVIL
HOR 75 min (ort 91 min) B/W VIDrel: SCREAM/SPEAR V

CARNOSAUR ** 15
Adam Simon USA 1993
Diane Ladd, Raphael Sbarge, Jennifer Runyon, Harrison Page, Ned Bellamy, Ed Williams, Clint Howard, Frank Novak, Andrew Magarian, Brian Hinkley, Michael Elliott, Myron Simon, Lisa Moncure, Jeff Foster, Martha Hackett, Maud Winchester
A scientist's genetic experiments with chickens go terribly wrong, leading to the creation of the title monsters. JURASSIC PARK meets ALIEN in this cheap-looking work, a quickie that was rushed out from Roger Corman's studios. In many ways it recalls the monster films of the 1950s and has a climax clearly intended to leave the way open for a sequel. (And there was one.)
HOR 83 min (ort 89 min) VIDrel: FIRST/SONOP L/A V

CAROLINA SKELETONS *** 18
John Erman USA 1991
Louis Gossett Jr, Melissa Leo, Paul Robeling, Bruce Dern, Clifton James, Henderson Forsythe, G.D. Spradlin, Rosanna Carter, Trazana Beverly, Marc Macaulay, Kate Bernsohn, Melody Kay, Chris Blackwelder
Set in 1964, this has Gossett Jr returning home a hero from Vietnam, 34 years after his younger brother was wrongfully executed for a double murder. Having promised his dying mother that he will see justice done, he sets out to find the real culprits, but comes up against a wall of racial bigotry and hypocrisy. It may lack the impact of MISSISSIPPI BURNING, but this unusual detective yarn is both absorbing and extremely well acted.
DRA 92 min (ort 94 min) mTV VIDrel: VCC L/A V
Boa: novel by David Stout.

CAROUSEL *** U
Henry King USA 1956
Gordon MacRae, Shirley Jones, Cameron Mitchell, Barbara Ruick, Claramae Turner, Gene Lockhart, Audrey Christie, Susan Luckey, William Le Massena, John Dehner, Jacques D'Amboise, Frank Tweddell
A rough carnival barker tries to improve when he marries, but is killed while committing a robbery to provide for his child. However, he is granted another chance to put his affairs in order. A colourful but stilted film version of the famous musical, based in its turn on a fantasy play. Songs include "If I Love You", "Soliloquy" and "You'll Never Walk Alone". Filmed before as "Liliom" in 1930 and 1934. The use of super-widescreen will be lost on TV.
MUS 128 min wScrn VIDrel: 20TH/TECH L/A V/sh
Boa: play Liliom by Ferenc Molnar/musical by Richard Rodgers and Oscar Hammerstein.

CARRIE ** 18
Brian De Palma USA 1976
Sissy Spacek, Amy Irving, Nancy Allen, William Katt, John Travolta, Piper Laurie, Betty Buckley, P.J. Soles, Priscilla Pointer, Sydney Lassick, Stefan Gierasch, Michael Talbot, Cameron De Palma
A girl with terrifying psychic powers finally turns on the class-

mates who have tormented her, when they play a spiteful trick on her at a school dance. A nasty, depressing film with some gory moments. Screenplay is by Lawrence D. Cohen. The film debut for Irving, Buckley and Soles. Believe it or not, this one formed the basis for a stage musical.
HOR 94 min (ort 98 min) wScrn VIDrel: MGM/WHV V/h
Boa: novel by Stephen King.

CARRINGTON *** 18
Christopher Hampton UK 1995
Emma Thompson, Jonathan Pryce, Steven Waddington, Rufus Sewell, Samuel West, Penelope Wilton, Janet McTeer, Peter Blythe, Jeremy Northham, Alex Kingston, David Ryall, Stephen Boxer, Annabel Mullion, Gary Turner, Georgiana Dacombe
Story of the painter Dora Carrington and her relationship with the writer homosexual writer Lytton Strachey, the couple defying all the conventions of the time to stay together and yet follow their separate lifestyles. Pryce is superb as Strachey (Thompson less so as Carrington) and there is much to enjoy in this handsome and lusty tale, the calibre of the cast being matched in full by the merits of the script. Winner of the Special Jury Prize at Cannes.
DRA 117 min (ort 122 min) cC VIDrel: POLY/POLYREC V/sh
Boa: book Lytton Strachey by Michael Holroyd.

CARRINGTON V.C. *** PG
Anthony Asquith UK 1954
David Niven, Margaret Leighton, Noelle Middleton, Maurice Denham, Geoffrey Keen, Laurence Naismith, Clive Morton, Mark Dignam, Allan Cuthbertson, John Glyn-Jones, Victor Maddern, Newton Blick, Raymond Francis, John Chandos
Courtroom drama about an army major who has embezzled regimental funds and is subsequently court-martialled. As the story unfolds, the reasons behind this become clear, in this solid and absorbing tale.
Aka: COURT MARTIAL
DRA 101 min (ort 106 min) B/W VIDrel: FABFIL/SPEAR V
Boa: play by Dorothy Campbell Christie.

CARRY ON ABROAD ** PG
Gerald Thomas UK 1972
Sidney James, Kenneth Williams, Charles Hawtrey, Joan Sims, Kenneth Connor, Peter Butterworth, Jimmy Logan, Barbara Windsor, June Whitfield, Hattie Jacques, Bernard Besslaw, Derek Francis, Sally Geeson, Carol Hawkins
This time round, a holiday abroad is not quite what the brochures promised, when a bunch of English tourists find themselves spending their holiday at an unfinished hotel in the Mediterranean.
COM 85 min (Cut at film release – ort 89 min)
VIDrel: VCC/DISC V

CARRY ON ADMIRAL ** U
Val Guest UK 1957
David Tomlinson, Peggy Cummins, Brian Reece, Eunice Gayson, A.E. Matthews, Ronald Shiner, Joan Sims, Lionel Murton, Reginald Beckwith, Peter Coke, Desmond Walter-Ellis, Peter Coke, Derek Blomfield, George Moon, Alfie Bass, Tom Gill
When a parliamentary private secretary swaps clothes with a Royal Navy captain in a drunken lark, he finds himself aboard a Naval vessel, where he is taken for an officer. Despite the title, this is not in fact a film in the CARRY ON series, though the silly mishaps and gags are often just as dated. Fortunately, the competence of the strong cast keeps things afloat.
Aka: SHIP WAS LOADED, THE
COM 78 min (ort 85 min) B/W
VIDrel: ETL/POLYREC/BRAVE V
Boa: play Off the Record by Ian Hay and Stephen King-Hall.

CARRY ON AGAIN, DOCTOR ** PG
Gerald Thomas UK 1969
Sidney James, Kenneth Williams, Charles Hawtrey, Jim Dale, Joan Sims, Hattie Jacques, Barbara Windsor, Patsy Rowlands, Peter Butterworth, Patricia Hayes, William Mervyn, Harry Locke, Valerie Leon, Elspeth March, Wilfrid Brambell
More medical antics from the same team who brought you CARRY ON DOCTOR etc. In this one a surgeon sets up a slimming clinic using a potion supplied by an island orderly.
COM 85 min (Cut at film release – ort 88 min)
VIDrel: VCC/DISC V

CARRY ON AT YOUR CONVENIENCE **   PG
Gerald Thomas UK   1971
Sidney James, Kenneth Williams, Charles Hawtrey, Joan Sims, Hattie Jacques, Bernard Bresslaw, Kenneth Cope, Jacki Piper, Richard O'Callaghan, Patsy Rowlands, Bill Maynard, Davy Kaye, Margaret Nolan, Renee Houston, Harry Towb
This farce is set on the premises of Messrs W.C. Boggs and Sons, manufacturers of fine toiletware, and follows the exploits of the factory foreman, whose pet budgie is able to predict race winners. A tired and dated effort.
Aka: CARRY ON ROUND THE BEND.
COM   86 min (Cut at film release – ort 90 min)
VIDrel: VCC/DISC   V

CARRY ON BEHIND **   PG
Gerald Thomas UK   1975
Elke Sommer, Kenneth Williams, Joan Sims, Bernard Bresslaw, Jack Douglas, Windsor Davies, Kenneth Connor, Liz Fraser, Peter Butterworth, Patsy Rowlands, Adrienne Posta, Ian Lavender, Carol Hawkins, Patricia Franklin
A professor of archaeology and his attractive assistant find that they are sharing a caravan site with a lot of buxom beauties. Average.
COM   87 min (Cut at film release – ort 90 min)
VIDrel: VCC/DISC   V

CARRY ON CABBY **   PG
Gerald Thomas UK   1963
Sidney James, Hattie Jacques, Kenneth Connor, Charles Hawtrey, Esma Cannon, Liz Fraser, Bill Owen, Milo O'Shea, Judith Furse, Ambrosine Philpotts, Renee Houston, Jim Dale, Cyril Chamberlain, Norman Chappell, Noel Dyson, Ian Wilson
The neglected wife of a taxi firm owner decides to get her own back by setting up a rival business staffed by buxom girls. Originally entitled "Call Me A Cab" but later retitled to form part of the CARRY ON series.
Aka: CALL ME A CAB
COM   88 min (ort 91 min) B/W   VIDrel: WHV   V/h

CARRY ON CAMPING *   PG
Gerald Thomas UK   1968
Sidney James, Kenneth Williams, Charles Hawtrey, Terry Scott, Joan Sims, Barbara Windsor, Hattie Jacques, Dilys Laye, Bernard Bresslaw, Peter Butterworth, Julian Holloway, Betty Marsden, Trisha Noble
Two men hit on the idea of going on a camping site holiday in Devon as a way of stimulating their unresponsive girlfriends. Later various other groups arrive, including a bunch of hippies intent on holding a pop festival. One of the weakest of the CARRY ON films.
COM   84 min (ort 88 min)   VIDrel: CAR/TECH   V

CARRY ON CLEO **   PG
Gerald Thomas UK   1965
Sidney James, Kenneth Williams, Kenneth Connor, Amanda Barrie, Joan Sims, Charles Hawtrey, Jim Dale, Julie Stevens, Victor Maddern, Sheila Hancock, David Davenport, Jon Pertwee, Tanya Billing, Francis de Wolff, Tom Clegg
Now it's the turn of Cleopatra to get the treatment – in Rome in 50 B.C., British slaves save Caesar from assassination at the hands of an ambitious soldier and the Egyptian Queen.
Aka: CALIGULA'S FUNNIEST HOME VIDEOS
COM   88 min (Cut at film release – ort 92 min)
VIDrel: WHV   V/h

CARRY ON COLUMBUS **   PG
Gerald Thomas UK   1992
Jim Dale, Peter Richardson, Alexei Sayle, Sara Crowe, Bernard Cribbins, Julian Clary, Richard Wilson, Keith Allen, Nigel Planer, Rik Mayall, Andrew Bailey, Burt Kwouk, Tony Slattery, Martin Clunes, Sara Stockbridge, Holly Aird
This, the 30th "Carry On" film, is set in 1492, and follows our adventurer's efforts to put together a motley crew and set sail, a task in which he is constantly hampered by treachery and stupidity. And having eventually arrived at the New World, he finds the local inhabitants just a little bit cleverer than he had expected, which is more than can be said for this film, in which timid dialogue is coupled with a distinct lack of vigour.
COM   87 min (ort 91 min)   VIDrel: WHV L/A   V/sh

CARRY ON CONSTABLE **   PG
Gerald Thomas UK   1959
Sid James, Eric Barker, Kenneth Connor, Charles Hawtrey, Leslie Phillips, Joan Sims, Hattie Jacques, Shirley Eaton, Cyril Chamberlain,

Joan Hickson, Irene Handl, Terence Longdon, Jill Adams, Freddie Mills
This early entry to this long-running series is set in a police training college, and looks at the various mishaps that occur when a new bunch of students arrive. Slightly amusing in a rather episodic way.
COM   83 min (Cut at film release – ort 86 min) B/W
VIDrel: WHV   V/h

CARRY ON COWBOY **   PG
Gerald Thomas UK   1965
Sid James, Kenneth Williams, Jim Dale, Percy Herbert, Joan Sims, Davy Kaye, Bernard Bresslaw, Charles Hawtrey, Peter Butterworth, Angela Douglas, Sydney Bromley, Sally Douglas, Joan Pertwee, Edina Ronay, Peter Gilmore
A former sanitary engineer takes over a town in the Wild West as their new sheriff, and a girl helps him thwart an outlaw who shot her father. An inept period comedy that now looks very dated.
Aka: RUMPO KID, THE
COM   90 min (Cut at film release – ort 95 min)
VIDrel: WHV   V/h

CARRY ON CRUISING **   U
Gerald Thomas UK   1962
Sidney James, Kenneth Williams, Liz Fraser, Kenneth Connor, Dilys Laye, Lance Percival, Jimmy Thompson, Cyril Chamberlain, Esma Cannon, Vincent Ball
The captain of the "Happy Wanderer" has to contend with the fact that, just prior to setting sail on a Mediterranean cruise, the crew of his luxury liner has been replaced with a bunch of inexperienced incompetents. This was the first colour CARRY ON film, and the episodic plot of mishaps and risque jokes remains as unchanged as a seaside postcard.
COM   86 min (ort 89 min)   VIDrel: WHV   V/h

CARRY ON DICK **   PG
Gerald Thomas UK   1974
Sidney James, Barbara Windsor, Kenneth Williams, Hattie Jacques, Bernard Bresslaw, Joan Sims, Kenneth Connor, Peter Butterworth, Jack Douglas, Patsy Rowlands, Bill Maynard, Margaret Nolan, John Clive, David Lodge, George Moon
More broad comedy of the nudge-nudge wink-wink variety based on the legend of Dick Turpin, with our highwayman posing as a village vicar in order to outwit the Bow Street Runners. A pleasant costumed outing.
COM   86 min (ort 91 min)   VIDrel: VCC/DISC   V

CARRY ON DOCTOR *   PG
Gerald Thomas UK   1968
Frankie Howerd, Sidney James, Kenneth Williams, Charles Hawtrey, Jim Dale, Barbara Windsor, Joan Sims, Hattie Jacques, Anita Harris, Bernard Bresslaw, Peter Butterworth, Dilys Laye, Derek Francis, Peter Jones, Dandy Nichols
None of these medical spoofs are much good; this one is no exception. It follows the exploits of a group of patients who revolt against the medical staff, plus the misadventures of a charlatan who thinks he's only got a week to live – but don't pity him, pity us. Very poor.
COM   90 min (ort 94 min)   VIDrel: VCC/DISC   V

CARRY ON DON'T LOSE YOUR HEAD ***   PG
Gerald Thomas UK   1966
Sidney James, Kenneth Williams, Joan Sims, Jim Dale, Charles Hawtrey, Dany Robin, Peter Butterworth, Peter Gilmore, Michael Ward, Leon Greene, Diana Macnamara
One of the best of the CARRY ON films, this is a fairly amusing send-up of the Scarlet Pimpernel story. Here our hero is called The Black Fingernail, and easily outwits the leaders of the French Revolution in his fight to save members of the French nobility from the guillotine.
Aka: DON'T LOSE YOUR HEAD
COM   89 min (Cut at film release – ort 90 min)
VIDrel: VCC/DISC   V

CARRY ON EMMANUELLE **   15
Gerald Thomas UK   1978
Suzanne Danielle, Kenneth Williams, Kenneth Connor, Jack Douglas, Joan Sims, Peter Butterworth, Larry Dann, Beryl Reid, Henry McGee, Howard Nelson, Claire Davenport, Norman Mitchell, Tricia Newby, Robert Dorning, Bruce Boa
Spoof on the well-known sex film with the sexually unsatisfied wife of the French ambassador causing various mishaps.
COM   84 min (ort 88 min)   VIDrel: VCC/DISC   V

CARRY ON ENGLAND **

Gerald Thomas UK PG
1976

Kenneth Connor, Windsor Davies, Patrick Mower, Judy Geeson, Joan Sims, Jack Douglas, Diane Langton, Melvyn Hayes, Peter Joans, David Lodge, Peter Butterworth, Julian Holloway, Johnny Briggs, Brian Osborne, Jeremy Connor

A mixed sex anti-aircraft battery of 1940 is the setting for this wartime CARRY ON comedy. The plot revolves around the attempts of their C.O. to thwart their sexual shenanigans. A weak and foolish story whose risque jokes dealing with sexual and bodily functions, now look remarkably stale.

COM   84 min (Cut at film release – ort 89 min)
VIDrel: VCC/DISC   V

CARRY ON FOLLOW THAT CAMEL **

Gerald Thomas UK PG
1967

Phil Silvers, Kenneth Williams, Jim Dale, Charles Hawtrey, Joan Sims, Angela Douglas, Peter Butterworth, Bernard Bresslaw, Anita Harris, John Bluthal, William Mervyn, Peter Gilmore, Vincent Ball, Larry Taylor, William Hurndell

This time the Foreign Legion gets the CARRY ON treatment, with an English gentleman who has been accused of fraud, joining up and saving the fort from Arab attackers.

Aka: FOLLOW THAT CAMEL
COM   94 min   VIDrel: VCC/DISC   V

CARRY ON GIRLS **

Gerald Thomas UK PG
1973

Sidney James, Joan Sims, Kenneth Connor, Barbara Windsor, Bernard Bresslaw, June Whitfield, Peter Butterworth, Jack Douglas, Patsy Rowlands, Joan Hickson, David Lodge, Valerie Leon, Margaret Nolan, Sally Geeson

When Sid Fidler bulldozes Fircombe Council into holding a beauty contest, the local Women's Lib Action Group opposes the plan. Average CARRY ON capers.

COM   84 min (Cut at film release – ort 88 min)
VIDrel: VCC/DISC   V

CARRY ON HENRY ***

Gerald Thomas UK PG
1971

Sidney James, Kenneth Williams, Joan Sims, Charles Hawtrey, Terry Scott, Barbara Windsor, Kenneth Connor, Julian Holloway, Peter Gilmore, Julian Orchard, Gertan Klauber, David Davenport, William Mervyn, Bill Maynard

Period farce in this comedy series which takes a look at Henry VII and two of his wives, one of whom is a garlic-loving woman who gets pregnant by her lover. One of the better CARRY ON films, with James at his lecherous, leering best.

COM   87 min (ort 89 min)   VIDrel: VCC/DISC   V

CARRY ON JACK **

Gerald Thomas UK PG
1963

Bernard Cribbins, Kenneth Williams, Juliet Mills, Charles Hawtrey, Jim Dale, Peter Gilmore, Donald Huston, Percy Herbert, Ed Devereaux, Ian Wilson, Barry Gosney, George Woodbridge, Cecil Parker, Frank Forsyth, Jimmy Thompson

The mixture as before. This time, the CARRY ON team direct their efforts towards spoofing the British Navy in 1805. This was the first film in the series to use period costumes.

Aka: CARRY ON SAILOR; CARRY ON VENUS
COM   87 min (Cut at film release – ort 91 min)
VIDrel: WHV   V/h

CARRY ON LOVING **

Gerald Thomas UK PG
1970

Sidney James, Kenneth Williams, Charles Hawtrey, Joan Sims, Hattie Jacques, Terry Scott, Richard O'Callaghan, Bernard Bresslaw, Jacki Piper, Imogen Hassall, Julian Holloway, Joan Hickson, Janet Mahoney, Bill Maynard

The action is set around a marriage bureau owned by Sid James and Hattie Jacques, that despite the appearance of being properly run, is in reality a thoroughly disreputable operation. At its premises arrive various individuals in need of marriage partners, who inevitably wind up less than satisfied. Nineteenth in the series is a weakly plotted and repetitive effort, sustained by amusing dialogue and some clever sight gags. Dated but enjoyable.

COM   86 min (Cut at film release – ort 90 min)
VIDrel: VCC/DISC   V

CARRY ON MATRON *

Gerald Thomas UK PG
1972

Sidney James, Kenneth Williams, Hattie Jacques, Charles Hawtrey,

Barbara Windsor, Terry Scott, Joan Sims, Kenneth Cope, Bernard Bresslaw, Kenneth Connor, Bill Maynard, Jacki Piper, Marianne Stone, Derek Francis

Another of the hospital comedies in the CARRY ON series and little better than the 1959 film. This one is set in a maternity hospital with James and a team of crooks intent on stealing a supply of birth control pills. The period farces are generally better.

COM   85 min (Cut at film release – ort 89 min)
VIDrel: VCC/DISC   V

CARRY ON NURSE *

Gerald Thomas UK PG
1960

Shirley Eaton, Kenneth Connor, Charles Hawtrey, Hattie Jacques, Wilfrid Hyde-White, Terence Longdon, Bill Owen, Leslie Phillips, Joan Sims, Susan Stephen, Kenneth Williams, Michael Medwin, Susan Beaumont, Ann Firbank

One of the earliest and feeblest of the CARRY ON films. We are subjected to a display of juvenile antics by patients and staff in a men's hospital ward. A banal plot combines with a paucity of jokes to make this a trial rather than a pleasure to watch. As the series progressed the leading stars developed their personalities – this one is of interest in that it shows a few sparks of undeveloped potential, but little more than that.

COM   84 min (Cut at film release) B/W   VIDrel: WHV
V/h

CARRY ON REGARDLESS **

Gerald Thomas UK PG
1961

Sid James, Charles Hawtrey, Kenneth Williams, Kenneth Connor, Joan Sims, Bill Owen, Liz Fraser, Terence Longdon, Hattie Jacques, Esme Cannon, Sydney Tafler, Julia Arnall, Terence Alexander, Stanley Unwin, Joan Hickson

The owner of an agency set up to provide temporary staff for any occasion, endures one comic mishap after another, from his well-meaning but clumsy staff. As always, a series of blundering episodes are presented for our amusement, culminating in the demolition of a house that the employees merely intended to spruce up.

COM   87 min (ort 93 min) B/W   VIDrel: WHV   V/h

CARRY ON SCREAMING **

Gerald Thomas UK PG
1966

Harry H. Corbett, Kenneth Williams, Fenella Fielding, Charles Hawtrey, Jim Dale, Angela Douglas, Peter Butterworth, Bernard Bresslaw, Jon Pertwee, Tom Clegg, Billy Cornelius, Frank Thornton, Denis Blake

Lightweight spoof on the horror genre, set in a potential Dracula Development Area, with a detective investigating a revived corpse and his vampiric sister, who vitrifies girls for use as shop-window dummies.

Aka: CARRY ON VAMPIRE
COM   92 min (ort 97 min)   VIDrel: WHV   V/h

CARRY ON SERGEANT **

Gerald Thomas UK U
1959

William Hartnell, Bob Monkhouse, Shirley Eaton, Kenneth Connor, Charles Hawtrey, Kenneth Williams, Terence Longdon, Hattie Jacques, Gerald Campion, Cyril Chamberlain, Gorden Tanner, Frank Forsyth, Basil Dignam, John Gatrell

The very first of the CARRY ON films, this feeble comedy centres on the experiences of a bunch of raw recruits at an army training unit. Hartnell plays a sergeant who accepts a bet that the last squad he is to train before retiring, will win the "Star Squad" award. A creaky farce that is slightly redeemed by energetic performances.

COM   81 min B/W   VIDrel: WHV   V/h
Boa: novel The Bull Boys by R.F. Delderfield.

CARRY ON SPYING **

Gerald Thomas UK U
1964

Barbara Windsor, Kenneth Williams, Bernard Cribbins, Charles Hawtrey, Eric Barker, Dilys Laye, Jim Dale, Richard Wattis, Eric Pohlman, Victor Maddern, John Bluthal, Judith Furse, Renee Houston, Gertan Klauber, Frank Forsyth

Windsor plays a trainee spy who joins the rather inept British Secret Service, hoping to put her talents (such as a photographic memory) to good use. In the company of three other trainee spies, she is sent to Vienna and Algiers to prevent a secret formula falling into the hands of a bisexual criminal mastermind.

COM   84 min (ort 87 min)   VIDrel: WHV   V/h

CARRY ON TEACHER **   U
Gerald Thomas UK   1959
Ted Ray, Charles Hawtrey, Leslie Phillips, Joan Sims, Kenneth Wulliams, Hattie Jacques, Rosalind Knight, Cyril Chamberlain, Richard O'Sullivan, Carol White, Paul Cole, Jane White, Larry Dann, Diana Beevers, Roy Hines
The humour centres on a secondary school with headmaster Ted Ray, whose pupils get up to an assortment of pranks when school inspectors visit, their intention being to block their master's transfer to another school. Look out for O'Sullivan in an early appearance as one of the pupils.
COM   83 min (Cut at film release – ort 86 min) B/W
VIDrel: WHV   V/h

CARRY ON UP THE JUNGLE **   PG
Gerald Thomas UK   1970
Sid James, Kenneth Connor, Joan Sims, Frankie Howerd, Charles Hawtrey, Terry Scott, Edward Connor, Bernard Bresslaw, Jacki Piper, Reuben Martin, Valerie Leon, Edwina Carroll
An explorer recalls his amazing adventures in darkest Africa when, whilst searching for a rare bird, he was captured by a tribe of girls. A slight and lacklustre spoof on jungle adventure films, with Scott doing his Tarzan bit as "Jungle Boy".
COM   87 min (ort 89 min)   VIDrel: VCC/DISC   V

CARRY ON UP THE KHYBER ***   PG
Gerald Thomas UK   1968
Sidney James, Kenneth Williams, Charles Hawtrey, Joan Sims, Roy Castle, Terry Scott, Bernard Bresslaw, Peter Butterworth, Angela Douglas, Cardew Robinson, Julian Holloway, Leon Thau, Alexandra Dane, Jeremy Spenser
This is one of the best CARRY ON films and is set in the North West Frontier during the British Raj, when fearless British soldiers faced hostile Afghan tribesmen undaunted. Full of the usual double entendres and with a stronger plot than was usual for this series.
COM   85 min (Cut at film release)   VIDrel: VCC/DISC   V

CARS THAT ATE PARIS, THE **   15
Peter Weir AUSTRALIA   1976
Terry Camilleri, John Meillon, Melissa Jaffa, Kevin Miles, Max Gillies, Peter Armstrong, Edward Howell, Bruce Spence, Derek Barnes, Charlie Metcalfe, Chris Heywood, Tim Robertson, Max Phipps, Frank Saba
The director's second film is a black comedy set in the curiously named Australian town of Paris, whose inhabitants engineer car accidents and then sell the scrap and spare parts from the wrecks. Unfortunately, the initial promise of this comic premise is not maintained, but a fair effort nonetheless.
Aka: CARS THAT ATE PEOPLE, THE
HOR   84 min (ort 91 min)   VIDrel: ARTPRO/RTM   V

CARVE HER NAME WITH PRIDE ***   PG
Lewis Gilbert UK   1958
Virginia McKenna, Paul Scofield, Jack Warner, Denise Grey, Maurice Ronet, Avice Landone, Anne Leon, Nicole Stephane, Billie Whitelaw, William Mervyn, Michael Goodliffe, Bill Owen, Sydney Tafler, Noel Willman
A low-key account of a young and courageous girl – Violette Szabo, who parachuted into Nazi-occupied Europe as a secret agent, only to be betrayed and executed at Ravensbruck. A restrained and well-acted account that is all the more effective for its semi-documentary style. See ODETTE for another film telling of outstanding courage.
WAR   114 min (ort 119 min) B/W   VIDrel: VCC L/A   V
Boa: book by R.J. Minney.

CARVER'S GATE **   15
Sheldon Inkol USA   1995
Michael Pare
Futuristic fantasy set in a devastated world where humanity faces eventual extinction, with the remnants of the world's population being forced to live underground, where the only distraction from their plight is an addictive virtual reality game.
FAN   93 min   VIDrel: HIFLI/SONOP   V/h

CASABLANCA ****   U
Michael Curtiz USA   1943
Humphrey Bogart, Ingrid Bergman, Paul Henreid, Claude Rains, Peter Lorre, Conrad Veidt, Sydney Greenstreet, Dooley Wilson, Marcel Dalio, S.K. Sakall, Madeleine LeBeau, Joy Page, John Qualen, Ludwig Stossel, Leonid Kinskey
By no means a perfect film, but nevertheless one of Hollywood's finest, and used by others as a touchstone for excellence. It tells of the lonely owner of a nightclub in Vichy-controlled Casablanca, and of the brief return of his former lover who now works for the Resistance. A highly entertaining film that deservedly enjoys its reputation as a classic. AA: Pic, Dir, Screen (Julius Epstein/Phillip G. Epstein/Howard Koch).
DRA   132 min (aniversary edition plus documentary – ort 102 min) B/W/Col   VIDrel: MGM/WHV   V/dm V/sh
Boa: play (unstaged) Everybody Comes To Rick's by Murray Burnett and Joan Alison.

CASANOVA **   18
Frederico Fellini ITALY   1976
Donald Sutherland, Cicely Brown, Tina Aumont, John Karlsen, Daniel Emilfork Berenstein, Olimpia Carlisi, Adele Angela Lojodice
Good-looking but essentially flat account of the many adventures of the famous 18th century rover. The music of Nino Rota is a slight compensation in this absurdly stylised and pretentious effort. AA: Cost (Danilo Donati).
Aka: FELLINI'S CASANOVA
A   147 min (ort 163 min)   VIDrel: 20TH/TECH   V

CASANOVA'S BIG NIGHT *   U
Norman Z. McLeod USA   1954
Bob Hope, Joan Fontaine, Basil Rathbone, Vincent Price, Audrey Dalton, Hugh Marlowe, John Carradine, Prino Carrera, Arnold Moss, Lon Chaney Jr, John Hoyt, Hope Emerson, Robert Hutton, Raymond Burr, Frieda Inescort
To escape his creditors, Casanova swaps identities with a tailoring apprentice, thereby giving him the chance to woo a lovely lady, although this lands him right in the middle of a complex court intrigue. A very disappointing big-budget comedy, of few laughs and many yawns.
COM   82 min (ort 86 min)   VIDrel: CIC/SONOP   V/h

CASE CLOSED **   15
Dick Lowry USA   1988
Byron Allen, Charles Durning, Marc Alaimo, James Greene, Eddie Jones, Erica Gimpel, Christopher Neame, Charles Weldon, Erica Gimpel, Jimmy Briscoe, Kimberly McArthur, John Powell, Jon Kohler
A misfire cops versus robbers comedy, with a young police detective teaming up with a retired cop to solve three murders and a jewel theft. A senseless and unfunny attempt to cash in on the success of BEVERLY HILLS COP and those similar 1980s cop comedies. This one's written and co-produced by former stand-up comic Allen.
COM   92 min (ort 100 min) mTV   VIDrel: 20TH/TECH   V

CASE FOR LIFE, A **   12
Eric Laneuville USA
Valerie Bertinelli, Mel Harris, Karl Malden
Two sisters battle it out in the courtroom when one takes it upon herself to attempt to get the other to have an abortion as the pregnancy threatens her life.
DRA   87 min   VIDrel: ODY/SONOP   V/sh

CASE FOR MURDER, A **   15
Duncan Gibbins USA   1993
Jennifer Grey, Peter Berg, Eugene Roche, Belinda Bauer, Samantha Eggar, Justine Arlin, David Hayward, Thomas Kopache, Rosemary Forsyth, Lynne Marta, Robert Do'Qui, Bruce French, Michael Ryan Way, Dick Welsbacher, Nancy McLoughlin
Two lawyers at law firm team up and fall in love whilst investigating the death of a senior partner in a boating accident, but the woman starts to suspect that her colleague is far more involved than he has revealed. Fair.
THR   89 min (ort 94 min)   VIDrel: CIC/SONOP   V

CASEBOOK OF SHERLOCK HOLMES, THE: SHOSCOMBE OLD PLACE ***   PG
Patrick Lau UK   1990
Jeremy Brett, Edward Hardwicke, Robin Ellis, Frank Grimes, Elizabeth Weaver, Denise Black, Michael Bilton, Martin Stone, Michael Wynne, James Coyle, Rosalie Williams, Jude Law, Alan Pattison
This was the third entry in a set of six tales, and has Holmes investigating the case of a man who has staked all he has on a horse he's entering in the Champion Stakes. Added to this is a mystery that revolves around his sister.
Aka: SHOSCOMBE OLD PLACE
DRA   310 min (6-episode set – 2 cassettes) mTV
VIDrel: CASPIC/BMGREC   V
Boa: short story by Arthur Conan Doyle.

CASEBOOK OF SHERLOCK HOLMES, THE: THE BOSCOMBE VALLEY MYSTERY ***

PG
June Howson UK 1990
Jeremy Brett, Edward Hardwicke, Peter Vaughan, Jonathan Barlow, Joanna Roth, Leslie Schofield, James Purefoy, Cliff Howells, Makala Saunders, Mark Jordon, Will Tacey
When an Australian-born farmer is murdered, the son is charged with the crime, but protests his innocence. Holmes is called in to assist, and begins to suspect that the police have not yet found the culprit. Number four in a set of six stories.
Aka: BOSCOMBE VALLEY MYSTERY, THE
DRA   310 min (6-episode set – 2 cassettes) mTV
VIDrel: CASPIC/BMGREC   V
Boa: short story by Arthur Conan Doyle.

CASEBOOK OF SHERLOCK HOLMES, THE: THE CREEPING MAN **

PG
Tim Sullivan UK 1991
Jeremy Brett, Edward Hardwicke, Charles Kay, Adrian Lukis, Sarah Woodward, Anna Mazzotti, James Tomlinson, Peter Guinness, Steve Swinscoe, Colin Jeavons, Anthony Havering, Peter Elliott
The last in this set of six stories is also one of the very weakest, and tells of a sinister, shambling figure a young woman believes she has seen outside her bedroom window. When Holmes learns that she's the daughter of an eminent professor, who is soon to wed a woman many years younger, he begins to draw some very strange conclusions.
Aka: CREEPING MAN, THE
DRA   310 min (6-episode set – 2 cassettes) mTV
VIDrel: CASPIC/BMGREC   V
Boa: short story by Arthur Conan Doyle.

CASEBOOK OF SHERLOCK HOLMES, THE: THE DISAPPEARANCE OF LADY CARFAX ***

PG
John Madden UK 1990
Jeremy Brett, Edward Hardwicke, Cheryl Campbell, Julian Curry, Mary Cunningham, Jack Klaff, Nicholas Fry, Michael Jayston, Anthony Benson, Mary Williams, Anthony Schaeffer, Margo Stanley, Andy Bradford, Elaine Ford
The first of a new set of six Sherlock Holmes adventures, that followed on from the RETURN OF SHERLOCK HOLMES series. This mystery opens in the Lake District, where Watson is on holiday. Events conspire to arouse his fears for the safety of a fellow guest, fears that are realised when she vanishes. Despite the title, this story is taken from the Doyle collection entitled "His Last Bow".
Aka: DISAPPEARANCE OF LADY CARFAX, THE
DRA   310 min (6-episode set – 2 cassettes) mTV
VIDrel: CASPIC/BMGREC   V
Boa: short story by Arthur Conan Doyle.

CASEBOOK OF SHERLOCK HOLMES, THE: THE ILLUSTRIOUS CLIENT ****

PG
Tim Sullivan UK 1991
Jeremy Brett, Edward Hardwicke, Anthony Valentine, Carol Noakes, David Langton, Abigail Cruttenden, Rosalie Williams, John Pickles, Kim Thomson, Roy Holder, Andy Bradford
Number five in this series deals with a totally ruthless Austrian murderer who delights in "collecting" women. When Holmes is asked to prevent his forthcoming marriage to a well-connected young lady, he finds himself pitting his wits against a most dangerous opponent, an enterprise that places his life in danger.
Aka: ILLUSTRIOUS CLIENT, THE
DRA   310 min (6-episode set – 2 cassettes) mTV
VIDrel: CASPIC/BMGREC   V
Boa: short story by Arthur Conan Doyle.

CASEBOOK OF SHERLOCK HOLMES, THE: THE PROBLEM OF THOR BRIDGE ***

PG
Michael A. Simpson UK 1990
Jeremy Brett, Edward Hardwicke, Daniel Massey, Celia Gregory, Catherine Russell, Niven Boyd, Andrew Wilde, Stephen MacDonald, Philip Bretherton, Dean Magri
A wealthy businessman is prepared to do all he can to save a young woman from the gallows, despite the fact that she appears to have shot dead his wife. Holmes conducts a few experiments at the scene of the crime, and forms a completely different opinion. A most effective mystery, number two in this series.
Aka: PROBLEM OF THOR BRIDGE, THE
DRA   310 min (6-episode set – 2 cassettes) mTV
VIDrel: CASPIC/BMGREC   V
Boa: short story Thor Bridge by Arthur Conan Doyle.

CASINO ***

18
Martin Scorsese USA 1995
Robert De Niro, Sharon Stone, Joe Pesci, James Woods, Don Rickles, Alan King, Kevin Pollak, L.Q. Jones, Dick Smothers, Frank Vincent, John Bloom, Pasquale Cajano, Melissa Prophet, Bill Allison, Vinny Vella, Oscar Goodman
Brutal story of greed and corruption, that is set in 1970s Las Vegas, where De Niro runs a casino as the front-man for its mobster owners, with a little help from his psychotic partner, played with gusto by Pileggi. Another chronicle of gangster low-life from Scorsese (see also GOODFELLAS and MEAN STREETS) in which the characters take second place to the way of life depicted, and the constant, underlying sense of menace. Screenplay is by Scorsese and Pileggi.
DRA   171 min (ort 178 min) wScrn cC
VIDrel: CIC/SONOP; PION (LV only)   V/sur LV
Boa: book by Nicholas Pileggi.

CASINO ROYALE *

PG
V. Guest/J. Huston/K. Hughes/R. Parrish/J. McGrath/R. Talmad UK 1967
Peter Sellers, David Niven, Ursula Andress, Joanna Pettet, William Holden, Woody Allen, Daliah Lavi, John Huston, Jacqueline Bissett, Peter O'Toole, Derek Nimmo, Charles Boyer, Deborah Kerr, Orson Welles, George Raft, Terence Cooper
A perfect illustration of the old adage about too many cooks. This unlucky, unhappy spy spoof – nominally about the retired Sir James Bond being asked to resume active service and join the fight against the evil SMERSH – is such a welter of cliches, overblown effects and poor dialogue that it defies description. Never before has so much talent and money been thrown away in the achievement of so little.
COM   126 min (ort 131 min)   VIDrel: VCC L/A   V
Boa: novel by Ian Fleming.

CASPER **

PG
Brad Silberling USA 1995
Christina Ricci, Bill Pullman, Eric Idle, Cathy Moriarty, Amy Brenneman, Ben Stein, Don Novello, Rodney Dangerfield, Ernestine Mercer plus voices of: Malachi Pearson, Joe Nipote, Joe Alaskey, Brad Garrett, John Kassir, Devon Sawa
A psychic researcher and his family move into a house that is haunted by the ghost of a dead boy and the latter falls in love with their adolescent daughter. A failed live-action version of Joseph Oriolo's DC comic-strip character "Casper The Friendly Ghost" that is drenched in special effects, in a vain attempt to mask the lack of a decent script or new ideas. Cameo appearances by a number of Hollywood stars add nothing to this poor and ill-conceived effort.
FAN   96 min (ort 101 min) cC
VIDrel: CIC/SONOP; PION (LV only)   V/sur LV

CAST A DEADLY SPELL **

15
Martin Campbell USA 1991
Fred Ward, Julianne Moore, David Warner, Alexandra Powers, Clancy Brown, Charles Hallahan, Arnetia Walker, Raymond O'Connor, Peter Alias, Lee Tergesen
Set in a parallel world where magic works, this opens in 1940s L.A., where everyone uses the occult, except Lovecraft, a wise-cracking Bogart-style private eye. When a rich recluse hires him to recover the "Necronomicon", a stolen magic book, it's not long before various dangerous weirdos appear on the scene. With a nod towards horror writer H.P. Lovecraft, this clumsy and gory blend of film noir and the supernatural has a few good moments. See also WITCH HUNT.
COM   92 min   VIDrel: MGM/WHV L/A   V

CAST THE FIRST STONE ***

15
John Korty USA 1989
Jill Eikenberry, Richard Masur, Elizabeth Ruscio, Joe Spano, Lew Ayres, Anne Schedeen, Holly Palance, Salome Jens, Dick Anthony Williams, H. Richard Greene, Charles Kimbrough, George McDaniel, Holly Palance, Jeff McCarthy, Richard Riehle
A teacher faces bigotry and intolerance in a small town, where she is fired from her job and suffers persecution because she has chosen to keep the child she gave birth to following a rape. Strong direction and an intelligent and sensitive script are big assets, but this is really Eikenberry's film, and she gives a heart-felt performance.
DRA   95 min mTV   VIDrel: ODY/SONOP   V/sh

CASTAWAY *** 15
Nicolas Roeg UK 1986
*Oliver Reed, Amanda Donohoe, Tony Rickards, Todd Rippon,
Georgina Hale, Len Peihopa, Frances Barber, Virgina Hey, Sarah
Harper, Stephen Jenn, John Sessions, Paul Reynolds, Sean Hamilton,
Arthur Cox, Joseph Blatchley*
Based on the real-life experiences of Lucy Irving, who
responded to an advert from a man seeking a woman to join him
for a year on an isolated desert island, far removed from the
complexities of civilisation. Visually satisfying, but this tale of
conflict and disagreement between two basically extremely
selfish people eventually loses its impact.
DRA 112 min (ort 118 min) VIDrel: MGM/WHV V/sur
Boa: book by Lucy Irvine.

CASTLE FREAK * 18
Stuart Gordon USA 1995
Jeffrey Combs, Barbara Crampton, Jonathan Fuller, Jessica Dollarhide
An American family inherit a castle in Italy but are unaware
that it houses a ghoulish creature whose origins are part of a
long-forgotten family secret. Matters really come to a head when
one of these creatures develops an unreasonable passion for the
family's beautiful but blind daughter. A dismal yarn that is so
idiotic its few striking sequences are too out of place to have
much shock value.
HOR 90 min (ort 93 min) VIDrel: EIV/SONOP V

CASTLE OF ADVENTURE, THE ** U
Terence Marcel UK 1990
*Susan Morgan, Gareth Hunt, Isobel Black, Brian Blessed, Richard
Hanson, Rosie Marcel, Hugo Guthrie, Bethany Greenwood, Eileen
Hawkes*
Five youngsters learn the secrets held within a ruined castle in
this workmanlike adaptation of a Blyton story.
JUV 118 min mTV VIDrel: ABBEY V
Boa: novel by Enid Blyton.

CASTLE OF CAGLIOSTRO, THE ** PG
Hayao Miyazaki JAPAN 1991
Animated tale of a thief who becomes a hero by rescuing a
princess. A violent and action-packed offering.
ANIM 101 min dubbed VIDrel: MANGA/SONOP V

CASTLE OF FU MANCHU, THE ** 18
Jess (Jesus) Franco ITALY/SPAIN/UK/W.GERMANY 1968
*Christopher Lee, Richard Greene, Maria Perschy, Gunther Stoll,
Howard Marion Crawford, Tsai Chin, Rosalba Neri, Jose Manuel
Martin, Werner Aprelat*
Another in the series of films about the evil Chinese criminal
genius and his battle with his arch adversary, Nayland Smith of
the British Home Office. The usual death rays and instruments
of torture abound in this entry, though by way of a variant, the
plot involves using the transplanted heart of an inventor.
Aka: ASSIGNMENT ISTANBUL; DIE FOLTERKAMMER DES DR FU
MANCHU; FU MANCHU'S CASTLE; EL CASTILLO DE FU MANCHU; IL
CASTELLO DI FU MANCHU
THR 92 min VIDrel: LUMI/SONOP V

CASUAL SEX? ** 18
Genevieve Robert USA 1988
*Lea Thompson, Victoria Jackson, Andrew Dice Clay, Mary Gross,
Jerry Levine, Stephen Stellan, Peter Dvorsky, Valeri Breiman, David
Sargent, Cynthia Phillips, Don Woodard, Danny Breen, Bruce Abbott,
Susan Ann Connor, Dan Woren*
Two young women on vacation at a health resort cautiously
look for that elusive meaningful relationship. A slight romantic
comedy for the AIDS age, largely told from the female point of
view.
COM 83 min (ort 87 min) VIDrel: CIC/SONOP V/sur
Boa: play by Wendy Goldman and Judy Toll.

CASUALTIES OF WAR ** 18
Brian De Palma USA 1989
*Michael J. Fox, Sean Penn, Don Harvey, John C. Reilly, John
Leguizamo, Thuy Thu Le, Erik King, Sam Robards*
War drama based on a true incident, when an American patrol
kidnapped, raped and finally murdered a Vietnamese girl. Fox
is the solitary voice of reason in this thought provoking work,
but the jumbled script and interminable moralising dilute the
strong anti-war sentiments.
DRA 108 min (ort 120 min) VIDrel: VCC/DISC/COLUM
V/sur
Boa: article in the New Yorker by Daniel Lang.

CAT, THE *** 15
Pierre Granier-Deferre FRANCE 1971
*Jean Gabin, Simone Signoret, Annie Cordy, Jacques Ruispal, Nicole
Desailly, Harry Max, Andre Rouyer*
After twenty-five years of marriage, a couple no longer speak
to each other. Instead, the husband lavishes all his love on a
stray cat, which makes his wife insanely jealous. A well-acted,
taut little drama, whose flimsy material is imbued with interest
by the inspired acting of the two leads.
Aka: LE CHAT
DRA 79 min (ort 88 min) VIDrel: ARROW/RTM V
Boa: novel Le Chat by Georges Simenon.

CAT AND MOUSE *** 15
Daniel Petrie USA 1974
*Kirk Douglas, Jean Seberg, John Vernon, Bessie Love, Beth Porter, Sam
Wanamaker, James Bradford, Suzanne Lloyd, Stuart Chandler, Valerie
Colgan, Mavis Villiers, Elliott Sullivan, Robert Sherman, James
Berwick*
A mild mannered biology teacher will stop at nothing to have
his revenge on his ex-wife, after learning during their divorce
proceedings that he is not the father of their son. A taut thriller
with a wonderfully sinister performance from Douglas in his TV
movie debut.
Aka: MOUSEY
THR 85 min (ort 87 min) mTV VIDrel: WHV V

CAT BALLOU ** PG
Elliott Silverstein USA 1965
*Jane Fonda, Lee Marvin, Michael Callan, Dwayne Hickman, Tom
Nardini, John Marley, Reginald Denny, Jay C. Flippen, Nat King
Cole, Stubby Kaye, Arthur Hunnicutt, Bruce Cabot, Burt Mustin,
Paul Gilbert, Harvey Clark, Oscar Blank*
Western spoof about a girl who hires a drunken gunfighter to
protect her father and turns outlaw when her dad is killed. Very
uneven and only funny in a few places. The best thing in this
movie is Lee Marvin's two roles – as the drunken gunman and
his twin desperado brother (complete with false tin nose). A
disappointing effort. The script is by Walter Newman and Frank
R. Pierson. AA: Actor (Marvin).
COM 92 min (ort 96 min) VIDrel: VCC L/A V
Boa: novel The Ballad of Cat Ballou by R. Chansler.

CAT CHASER *** 18
Abel Ferrara USA 1989
*Peter Weller, Kelly McGillis, Charles Durning, Frederic Forrest,
Tomas Milian, Juan Fernandez, Phil Leeds, Tony Bolano, Kelly Jo
Minter, Adrienne Sachs, Robert Escobar, Marla M. Ruperto, Vivian
Addison, Brooke Becker*
A kind of modern stab at film noir, this follows the exploits of
an ex-soldier who now runs a hotel in Santo Domingo. His affair
with the wife of the country's head of the secret police plunges
him into a world of intrigue, treachery and murder.
Aka: SHORT RUN
THR 86 min (Cut at film release by 27 sec – ort 93 min)
VIDrel: 4-FRONT/POLYREC L/A V
Boa: novel by Elmore Leonard.

CAT CITY * (PG)
Bela Ternovszky HUNGARY/WEST GERMANY 1986
Set in the year 80 A.M.M. (anno Mickey Mouse), this story tells
of a cat organisation out to rid the world of mice. However, the
rodents engage a scientist to help them fight back.
ANIM 96 min SATrel: MOVIE CHANNEL

CAT ON A HOT TIN ROOF **** 15
Richard Brooks USA 1958
*Elizabeth Taylor, Paul Newman, Burl Ives, Jack Carson, Judith
Anderson, Madeleine Sherwood, Larry Gates, Vaughn Taylor, Patty
Ann Gerrity, Rusty Stevens, Hugh Corcoran, Deborah Miller, Brian
Corcoran, Vince Townsend*
Powerful adaptation of a play by Tennessee Williams about a
Southern family and an autocratic plantation owner dying of
cancer but still dominating his greedy family who are all, with
the exception of one son, competing to win his favour. Adapted
by Brooks and James Poe and remade for TV in 1984.
DRA 103 min (ort 108 min) VIDrel: MGM/WHV V
Boa: play by Tennessee Williams.

CAT O'NINE TAILS * 18
Dario Argento FRANCE/ITALY/WEST GERMANY 1970
*James Franciscus, Karl Malden, Catherine Spaak, Cinzia De Carolis,
Carlo Alighiero*

A former reporter is now blind and spends much of his time doing crossword puzzles. He teams up with an active reporter in order to track down a nasty psycho who has left a gruesome trail of corpses. A brutish murder mystery, loaded with sex and gore and badly dubbed for good measure, though as ever, Argento displays considerable skill during some of the more horrific sequences.
Aka: DIE NEUNSCHWANZIGE KATZE; IL GATTO A NOVE CODE
HOR 107 min dubbed VIDrel: WHV V/h

CAT PEOPLE **
Paul Schrader USA
18
1982
Nastassia Kinski, Malcolm McDowell, John Heard, Annette O'Toole, Ruby Dee, Ed Begley Jr, Scott Paulin, John Larroquette, Frankie Faison, Ron Diamond, Lynn Lowry, Tessa Richarde, Patricia Perkins, Berry Berenson, Fausto Barajas
A not terribly good attempt at a remake of a 1942 film, with a brother and sister having a strange feline ancestry and the power to turn into panthers. This film concentrates more on the sexual aspects of the tale, with Kinski falling in love with a zoo curator whilst a black panther leaves a trail of bloodshed in the community. A good-looking but incoherent story.
HOR 112 min (ort 118 min) VIDrel: CIC/SONOP V/sur

CAT WOMEN OF THE MOON **
Arthur Hilton USA
PG
1953
Sonny Tufts, Marie Windsor, Victor Jory, William Phipps, Douglas Fowley, Susan Morrow, Carol brewster, Suzanne Alexander, Judy Walsh, Betty Allen, Ellye Marshall, Roxann Delman
An American expedition to the moon find it inhabited by a race of ferocious females who rule over an underground empire. Originally made for 3-D cinema, this low-grade SF entry was remade in 1959 as "Missile To The Moon", which was no improvement. However, the sheer silliness of this film has won it an affectionate cult following.
Aka: ROCKET TO THE MOON
FAN 63 min (ort 65 min) B/W VIDrel: FIRC/RTM V

CATACOMBS **
David Schmoeller USA
18
1988
Timothy Van Patten, Laura Schatta, Jeremy West, Vernon Dobtcheff, John Petter, Mapi Golan, Ian Abercrombie, Donald Pleasence, Katrine Michelson, Feodor Chapliani, Brett Porter
For hundreds of years an Italian monastery has been haunted by the legend of a demon that once attacked the previous occupiers and is now imprisoned deep within the catacombs of the ancient building. The arrival of a woman at this formerly male-only community, sparks off a series of events that lead to the release of this creature. Formula creature-released-from-hell horror yarn with some nice settings.
HOR 84 min Cut (12 sec) VIDrel: EIV/SONOP V

CATCH 22 **
Mike Nichols USA
15
1970
Alan Arkin, Martin Balsam, Richard Benjamin, Jon Voight, Art Garfunkel, Bob Newhart, Anthony Perkins, Paula Prentiss, Martin Sheen, Orson Welles, Norman Fell, Bob Balaban, Charles Grodin, Jack Gilford, Buck Henry, Austin Pendleton
Joseph Heller's brilliant anti-war novel contains so much that no film could ever do it justice. This is a worthy but only partially successful attempt to do so and follows the adventures of Yossarian, a bomber pilot who will do anything so as not to have to go on flying.
DRA 116 min (ort 122 min)
VIDrel: 4-FRONT/POLYREC/CIC V
Boa: novel by Joseph Heller.

CATCHFIRE **
Alan Smithee (Dennis Hopper) USA
15
1989
Dennis Hopper, Jodie Foster, Dean Stockwell, Vincent Price, John Turturro, Charlie Sheen, Fred Ward, Julie Adams, G. Anthony Sirico, Sy Richardson, Frank Gio, Joe Pesci, Helen Kallaniotes, John Apicella, Clifford Bartholemew
A hit-man is sent to kill an attractive young artist, who inadvertently saw a gangland slaying. However, he falls in love with his intended victim and is loath to carry out his assignment. A strangely surreal comedy-drama, it suffered extensive re-cutting prior to European distribution, making it almost impossible to follow. Fortunately Hopper (who had his name removed) was allowed to restore it to its original version. Look out for Bob Dylan who appears in a small cameo.
Aka: BACKTRACK
DRA 95 min (ort 105 min) VIDrel: FIRST/SONOP V

CATHOLIC BOYS ***
Michael Dinner USA
15
1985
Donald Sutherland, John Heard, Andrew McCarthy, Mary Stuart Masterson, Kevin Dillon, Malcolm Danare, Jennie Dundas, Kate Reid, Wallace Shawn, Philip Bosco
A look at the lives of a group of boys studying at a Catholic seminary in Brooklyn in 1965. Often very funny, with a good cameo from Shawn as a priest obsessed with preventing lustfulness. Writing debut for Dinner and Charles Purpura.
Aka: HEAVEN HELP US
DRA 99 min VIDrel: WHV V/h

CAT'S VICTIM, THE ***
Antonio Bido ITALY
18
1977
Paola Tedesco, Carrado Pani, Franco Citti, Fernando Cerulli
A female dancer witnesses the first in a string of murders, and becomes a potential victim of the killer, who begins to stalk her. Meanwhile, her boyfriend is working frantically to solve the mystery and save her life. Though hampered by a verbose script, this is nonetheless a superior stalk-and-slash effort, even if the final resolution (the culprit is a Nazi war criminal) is more than a little contrived.
Aka: CAT WITH JADE EYES; IL GATTO DAGLI OCCHI DI GIADA; WATCH ME WHEN I KILL
HOR 91 min (ort 100 min) wScrn dubbed
VIDrel: REDEM/RTM V

CAUGHT IN THE ACT **
Debbie Reinisch USA
PG
1993
Gregory Harrison, Leslie Hope, Patricia Clarkson, Kevin Tighe, Kimberly Scott, Raye Birk, Michael David Lally, Daniel Gerroll, Joe Uria, Scott Smith, Linda Sanders, Kirby Trapper, Zitto Kazann, John Drayman, Ron Orbach
An unemployed actor trying to land a big part in a movie suddenly finds vast amounts of money deposited in his bank account. In no time he falls under suspicion of bank fraud and murder and learns that he has been framed by a new student in the acting class he teaches. However, the rest of the class save him by staging a "play" that gets her to give herself away and reveal her hand. A neat little thriller with a rather unusual plot.
THR 89 min (ort 93 min) mCab VIDrel: CIC/SONOP L/A V

CAUGHT IN THE CROSSFIRE **
Chuck Bowman USA
PG
1994
Dennis Franz, Daniel Roebuck, Alley Mills, Anna Gunn, Ray McKinnon
A news reporter receives a tip about a government cover-up that could make his career, but to research it has to interview a crook with Mafia connections, who is also on a gangster hit-list. The FBI recruits the reporter as part of a clever trap, but their scheme fails to go according to plan. Based on a true story.
THR 85 min mTV VIDrel: NWV/HIFLI V/h

CAVE GIRL *
David Oliver USA
15
1986
Daniel Roebuck, Cindy Ann Thompson, Bill Adams, Larry Gabriel, Jeff Chayette, Valerie Greybe
An anthropology student is thrown back in time whilst exploring a cave, and gets to experience Stone Age living at first hand, in this clumsy and unfunny effort.
COM 85 min VIDrel: 20TH V

CB4 **
Tamra Davis USA
18
1992
Chris Rock, Allen Payne, Deezer D, Phil Hartman, Arthur Evans, Theresa Randle, Chris Elliott, Willard E. Pugh, Charlie Murphy, Khandi Alexander, Stoney Jackson, La Wanda Page, Tyrone Grandson Jones, Victor Wilson, Richard Grant
Confused musical spoof done in the style of THIS IS SPINAL TAP, but targeted this time on gangsta rap, with three friends trying desperately to break into the music business as rappers. When they take over the running of a club owned by a jailed gangster, they name their group after the cell-block where the latter is housed. However, a right-wing politician is out to have them banned. An uneven but fairly entertaining effort, especially for fans of rap.
COM 84 min (ort 89 min) VIDrel: CIC/SONOP V/sur

CELEBRATION FAMILY **
Robert Day USA
U
1987
Stephanie Zimbalist, James Read, Diane Ladd, Ed Begley Jr, Anne Haney, Royce D, Applegate, Sandy Kenyon, Richard Linebakc, Joe

Nesnow, Olivia Burnett, Joel Graves, Johnny Jason Graves, Vaughn Tyree Jeiks, Aaron Lohr, Julie Mannix
A couple who love kids and adopt orphans when the wife is unable to have any more children of her own. This eventually results in a massive family of nearly sixty children, of mixed races and nationalities.
DRA 91 min (ort 95 min) mTV VIDrel: ODY/SONOP V/sh

CELIA ***
15
Ann Turner AUSTRALIA 1988
Rebecca Smart, Nicholas Eadie, Mary-Anne Fahey, Victoria Langley, Margaret Ricketts, Alexander Hutchinson, Adrian Mitchell, Callie Gray, Martin Sharman, Claire Couttie, Alex Menglet, Amelia Frid, Fion Keane
In a dully oppressive 1950s Melbourne suburb, a nine-year-old girl experiences a series of minor disappointments, the effect of which is cumulative. Her growing sense of isolation and mental disturbance leads to an unexpected resolution. A highly unusual study of paranoia, fantasy and childhood fears, it makes several interesting observations despite a slackening of pace towards the climax. Turner's feature debut.
Aka: CELIA: CHILD OF TERROR
DRA 98 min (ort 102 min) VIDrel: FABFIL/SONOP V

CELINE AND JULIE GO BOATING ****
15
Jacques Rivette FRANCE 1974
Juliet Berto, Dominique Labourier, Bulle Ogier, Marie-France Pisier, Barbet Schroeder, Philippe Clevenot, Nathalie Asnar
A girl working as a nightclub magician visits a haunted house in the company of her librarian friend, where the former sometimes works as a governess to a little girl who lives there with her widowed father. When they enter the house they get caught up in a strange, daily ritual enacted by its occupants, which their actions inevitably alter. Shot on 16 mm, this elusive and dreamlike fantasy repays close attention, but one requires patience to sit through it.
Aka: CELINE ET JULIE VONT EN BATEAU
DRA 185 min (ort 187 min) VIDrel: CONNO/RTM V

CELTIC PRIDE **
PG
Tom De Cerchio USA 1996
Damon Wyans, Dan Aykroyd, Daniel Stern, Gail O'Grady, Christopher McDonald, Adam Hendershott, Paul Guilfoyle, Scott Lawrence
Two fanatical supporters of a basketball teams attempt to ensure that their team will win the championship by kidnapping the opposing team's star player the night before the game. This action leads to a catalogue of mishaps and chaotic misadventures.
COM 88 min (ort 124 min) cC VIDrel: HOLPIC/TECH V/sur

CEMENT GARDEN, THE **
18
Andrew Birkin FRANCE/GERMANY/UK 1992
Andrew Robertson, Charlotte Gainsbourg, Sinead Cusack, Alice Coulthard, Ned Birkin, Hanns Zischler, Jochen Horst, Gareth Brown, William Hootkins, Dick Flockhart, Mike Clark
When their mother dies, four children conceal her death by burying her in cellar and covering her body in cement since they are afraid of being placed in care by the authorities and thus separated. Various complications follow, including incest, in this dark and depressing tale.
Aka: SREDNI VASHTAR
DRA 126 min (director's cut – ort 105 min) wScrn
VIDrel: TART/20TH V/sur
Boa: novel by Ian McEwan.

CEMETERY MAN **
18
Michele Soavi ITALY 1994
Rupert Everett, Francois Hadji-Lazaro, Anna Fulchi, Mickey Knox, Clive Riche, Fabiana Formica, Katja Anton, Barbara Cupisti, Pietro Genuardi, Anton Alexander, Patrizia Punzo, Stefano Masciarelli, Vito Passeri, Renato Donis
A cemetery attendant finds his job becoming far too hard when the newly buried dead refuse to remain in their graves, in this tongue-in-cheek horror tale. Problems arise when the attendant thinks he has found himself a beautiful girlfriend, but cannot decide whether she is alive or dead. Special effects are the work of Sergio Stivaletti, who has become one of the leaders for films of this genre.
Aka: DELLAMORTE, DELLAMORE
HOR 99 min (ort 105 min) VIDrel: EIV/SONOP V

CENTURY ***
15
Stephen Poliakoff UK 1993
Charles Dance, Clive Owen, Miranda Richardson, Robert Stevens, Lena Headey, Neil Stuke, Joan Hickson, Liza Walker, Joseph Bennett, Carlton Chance, Graham Loughridge, Alexis Daniel, Ian Shaw, Bruce Alexander, Mark Strong, Dail Sullivan
In turn-of-the-century Britain, a dedicated and idealistic medical researcher joins an institute where his progressive views soon arouse the scorn and distrust of his colleagues. A well-acted study of social attitudes of this period as the Victorian Age was drawing to a close.
DRA 107 min (ort 112 min) wScrn
VIDrel: ELPIC/POLYREC V/sur

CERTAIN FURY *
18
Steven Gyllenhaal USA 1985
Tatum O'Neal, Irene Cara, Nicholas Campbell, Peter Fonda, Moses Gunn, George Murdock
During a courtroom gun battle in which some convicts escape, two juvenile delinquent girls also seize the opportunity to get free, but find that they are in far worse trouble as they are now thought to be accomplices. An inept and rubbishy effort, badly acted and sloppily directed. Made in Canada.
DRA 83 min (ort 88 min) VIDrel: MED/POLYREC L/A V

CHAIN, THE *
PG
Jack Gold UK 1985
Herbert Norville, Denis Lawson, Rita Wolf, Maurice Denham, Nigel Hawthorne, Billie Whitelaw, Judy Parfitt, Leo McKern, Tony Westrope, Bernard Hill, Warren Mitchell, Gary Waldhorn, Ron Pember, Carmen Munroe, David Troughton
Story of the trials of house-moving, revolving around the difficulties faced by seven interlinked households. An utterly predictable and tedious comedy, consisting of a series of dreary and trite vignettes. Screenplay is by Jack Rosenthal.
COM 92 min (ort 100 min) VIDrel: CARL/TECH V

CHAIN, THE **
18
Luca Bercovici USA 1996
Gary Busey, Victor Rivers, Jamie Rose
A tough Boston cop travels to Vera Cruz in search of an old enemy, an international arms dealer. When both men are captured by government forces they are handcuffed together and must find a way to resolve their differences if they are to escape. One of those engaging if terribly derivative actioners, using an idea that betrays something of a lack of originality in the plot department. See also FLED and THE DEFIANT ONES.
A/AD 92 min VIDrel: MOSAIC/COLUM V/sur

CHAIN OF COMMAND **
18
David Worth USA 1993
Michael Dudikoff, Todd Curtis, Keren Tishman, R. Lee Ermey, Jack Adalist, Steve Greenstein, Eil Dankers, Jon J. Heriso, David Menachem, Eli Jaspan, Yaron Levi Sabug, Sami Samir, Johnatan Cherchi, Yosef Shiloa, Shai Schwartz
When a gang of hired mercenaries attempt to overthrow the government of a small country in the Middle East, they encounter stiff resistance from a veteran counter-insurgence operative. An average violent actioner with more than enough gunfire and explosions to satisfy anyone.
A/AD 91 min VIDrel: MGM/WHV V/sh

CHAIN OF DESIRE **
18
Temistocles Lopez USA 1992
Patrick Bauchau, Grace Zabriskie, Linda Fiorentino, Angel Aviles, Malcolm MacDowell, Elias Koteas, Tim Guinee, Assumpta Serna, Jamie Harrold, Dewey Weber, Holly Marie Combs, Seymour Cassel, Kevin Conroy, Suzanne Douglas, Joseph McKenna
Standard erotic drama that offers few surprises as it recounts the sexual adventures of fourteen depraved New Yorkers, all of whom are linked one with the other by way of their sexual liaisons. Lopez's second English language film plays like a version of LA RONDE for the 1990s, and even though a fine cast struggles valiantly with the script, it all gets a bit too repetitive after a while. See also ECLIPSE.
DRA 102 min (ort 107 min) VIDrel: MAINPIC/RTM V

CHAIN REACTION **
12
Andrew Davis USA 1996
Keanu Reeves, Morgan Freeman, Rachel Weisz, Fred Ward, Kevin Dunn, Brian Cox, Joanna Cassidy, Chelcie Ross, Nicholas Rudall, Tzi Ma, Krzysztof Pieczynski, Julie R. Pearl, Godfrey C. Danchimah Jr, Gene Barge, Nathan Davis, James Sie

In Chicago, a team of scientist finally succeeds in developing a new source of energy that is safe, clean and cheap. Reeves plays a gauche, young technician who inadvertently made a discovery that led to the project's success. However, he has to go on the run when he is framed for the murder of the chief scientist. An endless series of chases, attacks and intrigues now occurs, until the final resolution when all is made clear and Reeves is exonerated. A most tedious film.
A/AD 106 min VIDrel: FOXVID V

CHAINED HEAT *   18
Paul Nicolas USA/WEST GERMANY   1983
Linda Blair, John Vernon, Tamara Dobson, Stella Stevens, Sybil Danning, Nita Talbot, Michael Callan, Henry Silva, Louisa Moritz, Edy Williams, Sharon Hughes
Innocent girl in prison finds herself at the mercy of the warden. One of those women-in-prison movies, with plenty of violence and titillation, but nothing of substance. See CAGED WOMEN for more of the same.
DRA 90 min Cut (1 min 38 sec – ort 92 min)
VIDrel: MGM/WHV L/A V/sh

CHAINED HEAT 2 *   18
Lloyd A. Simandl USA   1992
Brigitte Nielsen, Paul Koslo, Kimberley Kates, Kari Whitman, Marek Vasut, David Buonatony, Jana Svandova, Lucie Benes, Marketa Hrubes, John C. Smith, Jiri Popel, Vanda Svarc, Peter Susser, Pan Fiala
A woman framed for possessing drugs in Czechoslovakia faces both the usual sadistic guards and an organised crime syndicate that is forcing the inmates into prostitution and drug smuggling. A sequel of sorts to the first film that is even worse, with Dragnet-style voice-overs at beginning and end. Filmed on location in Prague, this low-budget misfire is amateurish and dull in every way.
DRA 98 min VIDrel: GUILD/POLYREC L/A V/sh

CHAINS *   18
Roger J. Barski USA   1989
Jim Jordan, Michael Dixon, John L. Eves, Howard Friedland, Rengin Altay, Cathleen Zim, John Herriman, Tony Mendez, Hugh Haller, Armando Guzman, Antoine Roshel, Taylor Graye, Don Modona, Tony Smith, Michael Rivero
Two couples out on a double-date find themselves caught up in a gangland war in South Chicago and soon learn just how ruthless they'll need to be in order to stay alive. A violent and exploitative action tale, quite well mounted for all its lack of originality.
Aka: CHAINS, THE: WARRIORS OF CHICAGO
A/AD 93 min VIDrel: IMPENT V

CHAKU MASTER *   18
Luis San Juan HONG KONG
Bruce Ly, Tony Bernard, Rey Malonzo
Villagers hire a martial arts expert to protect them from the depredations of a gang of terrorists. A good demonstration of martial arts prowess is displayed here, but the film is hampered by over-plotting.
MAR 83 min VIDrel: IMPENT V

CHALLENGE OF THE MASTERS *   15
Chang Hsin-Yi HONG KONG   1981
John Lui, Chin Lung
Two rival martial arts schools battle it out after a student is killed in a fight. OK if you know what to expect, just don't look for anything new in the plot department.
MAR 88 min VIDrel: IMPENT V

CHALLENGE THE NINJA *   18
Godfrey Ho HONG KONG   1986
Bruce Baron, Pierre Tremblay, Eric Redner, Alison Ellis, Silvia Rod, Richard Berman, Janet Hansen, Gerry Broad, Billy Lee, Mike Powell, Alex Yang, Michael Wong
A corrupt Ninja gang has spread a reign of terror and murder and a police officer and secret master of Ninjitsu sets out to defeat them. However, a youngster he has befriended who has suffered from the gang, decides to learn the fighting arts for himself and join in the battle to defeat them.
Aka: CHALLENGE OF NINJA; CHALLENGE OF THE NINJA
MAR 82 min Cut (3 min 54 sec – ort 92 min)
VIDrel: IMPENT V

CHAMELEON *   15
Michael Pavone USA   1995
Anthony LaPaglia, Kevin Pollak, Melora Hardin, Wayne Knight, Andy Romano
Alarmed at the rapid growth of the drugs market in Eastern Europe, the US Justice Department sends an agent there undercover, his task being to infiltrate an international drugs ring. LaPaglia as the agent adopts an increasingly bizarre set of disguises, his mission becoming one of retribution, as the crimeboss he is pursuing had his wife and child murdered. Nonsense of a high order, that though totally unbelievable, is fast-paced enough to entertain.
THR 103 min (ort 108 min) VIDrel: MOSAIC/COLUM V/sur

CHAMELEONS *   18
John Leslie USA   1991
Deidre Holland, Ashlyn Gere, P.J. Sparxx, Fawn Miller, Tracy Wynn, Sunset Thomas, Candace Heart, Rocco Siffredi, Jon Dough, Leanna Foxxx, Woody Long
An erotic SF film that has a promising opening premise, in that some of the characters are able to change shape and thus experience sex in the identities of other individuals. However, this interesting idea is soon lost in a welter of sexual encounters, and although the film is undeniably well produced, it soon grows mighty tiresome.
A 74 min (Cut before video submission by 14 min – ort 88 min) VIDrel: GROHOM/MAXSCAN V

CHAMP, THE *   PG
Franco Zeffirelli USA   1979
Jon Voight, Faye Dunaway, Ricky Schroder, Strother Martin, Jack Warden, Arthur Hill, Joan Blondell, Elisha Cook Jr, Mary Jo Catlett, Stefan Gierasch, Allan Miller, Joe Tornatore, Shirlee Kong, Jeff Blum, Dana Elcar
Remake of a 1931 film that deals with a washed-up prize fighter and his son. The boy's mother, having remarried and become a rich and successful fashion writer, turns up and wants to take the boy away. A boring and utterly unbelievable tale, made more so by young Schroder as the boy, who just cries and cries and cries.
DRA 118 min (ort 121 min) VIDrel: MGM/WHV V/sh

CHAMPAGNE CHARLIE *   15
Allan Eastman CANADA/FRANCE   1988
Hugh Grant, Megan Gallagher, Megan Follows, Jean-Claude Descrieres, Stephane Audran, R.H. Thomson,
The romanticised story of Charles Heidsieck, who introduced champagne to America at the time of the Civil War. Arriving in America with his wife, he becomes involved with a Southern girl, is arrested as a spy and experiences various other vicissitudes. An opulent and handsome drama, fatally flawed by excessive length (it was originally broadcast in two parts).
DRA 186 min (ort 200 min) mTV VIDrel: MIA/DISC V

CHAMPION **   U
Mark Robson USA   1949
Kirk Douglas, Marilyn Maxwell, Arthur Kennedy, Ruth Roman, Lola Albright, Paul Stewart, Luis Van Rooten, John Day, Harry Shannon
A ruthless boxer will stop at nothing to get to the top in the fight game. Douglas gives a nice performance as our unlovable pugilist in this loose adaptation of Lardner's story, that has a few faint echoes of BODY AND SOUL. AA: Edit (Harry Gerstad).
DRA 98 min B/W VIDrel: 4-FRONT/POLYREC V
Boa: short story by Ring Lardner.

CHAMPION FIGHTER *   18
Albert Yu
Amior Nissan, Wayne Archer
Two rivals on the underground kickboxing circuit each have their own powerful proteges, whom they send against each other in a major competition. However, one of the fighters has been ordered to throw the match in order to exploit favourable betting odds, a prospect he does not take kindly to. Adequate martial arts mayhem, set against the background of gang warfare.
MAR 85 min Cut (51 sec – ort 90 min) VIDrel: IMPENT V

CHANCES *   15
Buzz Kulik USA   1990
Nicollette Sheridan, Michael Nader, Anne-Marie Johnson, Vincent Irizarry, Eric Braden, Sandra Bullock, Stephanie Beacham, Tim Ryan,

Phil Morris, Luca Bercovici, Alan Rosenberg, David McCallum, Shawnee Smith, Richard Anderson
Following on from LUCKY, this is an overlong and unusually mediocre adaptation of the second Collins novel, the two books spanning fifty years in the lives of the characters. In this sequel, our poor Italian lad has become the powerful and wealthy owner of one of the first casinos in Las Vegas, and his daughter is now grown up. A limp epic that plays like a parody of those GODFATHER films, with endless cliches and wooden acting. Followed by LADY BOSS.
DRA   124 min mTV   VIDrel: 4-FRONT/POLYREC/ODY V/sh
Boa: novel by Jackie Collins.

CHANCES ARE *   *PG*
Emile Ardolino USA   1989
Cybill Sheperd, Ryan O'Neal, Robert Downey Jr, Mary Stuart Masterson, Josef Sommer, Christopher McDonald, Joe Grifasi, Susan Ruttan, Fran Ryan, James Noble, Henderson Forsythe, Lester Lanin, Richard DeAngelis
A college student recalls a previous life as a lawyer who was murdered, and for good measure discovers that the mother of his current girlfriend was in fact his wife in that past incarnation. A rather silly but endearing comedy, its derivative fantasy elements are effective if not all that original.
COM   103 min (ort 108 min)   VIDrel: VCC L/A   V/sh

CHANGELING, THE *   *(18)*
Peter Medak CANADA   1979
George C. Scott, Melvyn Douglas, Trish Van Devere, John Colicos, Jean Marsh, Madeleine Thornton-Sherwood, Barry Morse, James Douglas, Roberta Maxwell, Berrand Behrens, Frances Hyland, Ruth Springford, Helen Burns, Chris Gampel
A recently widowed musician moves into an old house, which is inhabited by the ghost of a crippled child who was murdered by his father about seventy years ago. A patchy and slightly disjointed tale that has some genuinely scary moments, even though it ultimately disappoints. A shorter film would have been far more effective. See also THE CRYING CHILD.
HOR   113 min   VIDrel: L/A   V

CHANGELING 2: THE REVENGE *   18
Lamberto Bava ITALY   1987
Gioia Scola, David Brandon
A clumsy reworking of Mario Bava's SHOCK, in which a woman murders her husband, the owner of a lakeside diner and gets her lover to help dispose of the body. The couple now settle down together, bringing up the child she was unknowingly carrying at the time. But the arrival of a mysterious stranger six years later heralds the start a series of horrifying events. A flabby effort of stilted (and badly dubbed) dialogue and lack-lustre performances.
Aka: UNTIL DEATH
HOR   93 min mTV dubbed   VIDrel: SPEAR/SONOP   V

CHANGES *   *PG*
Charles Jarrott USA   1990
Michael Nouri, Cheryl Ladd, Christopher Gartin, Randee Heller, Charles Frank, James Sloyan, Cynthia Bain, Luis Avalos, Liz Sheridan, Sheila Johns, Christie Clark, Renee O'Connor, Ami Foster, Joseph Gordon-Levitt
A very glossy and handsome adaptation of Steel's soap opera-style novel in which a highly successful but divorced TV anchorwoman meets and falls for an equally successful but widowed surgeon. An undemanding piece of TV entertainment that is devoid of pretensions and depth.
Aka: DANIELLE STEEL'S CHANGES
DRA   92 min (ort 100 min) mTV
VIDrel: MIA/DISC/IMPENT   V
Boa: novel by Danielle Steel.

CHANT OF JIMMIE BLACKSMITH, THE *   18
Fred Schepisi AUSTRALIA   1978
Tommy Lewis, Freddy Reynolds, Ray Barrett, Jack Thompson, Angela Punch, Steve Dodds, Julie Dawson, Tim Robertson, Jane Harders, Peter Carroll, Robyn Nevin, Don Crosby, Ruth Cracknell, Elizabeth Alexander, Rosie Lilley, Rob Steele
A half-caste Aborigine who was brought up by a Methodist minister, exacts a bloody revenge on those who humiliated him. Strong if predictable indictment of racism, set in 1900 and based on some true events.
DRA   112 min (ort 124 min)   VIDrel: ARTPRO/RTM   V
Boa: novel by Thomas Keneally.

CHANTILLY LACE *   *(15)*
Linda Yellen USA   1993
Martha Plimpton, JoBeth Williams, Jill Eikenberry, Lindsay Crouse, Helen Slater, Talia Shire, Ally Sheedy, Matt Battaglia
A woman meets up with six old friends several times during the year and they explore the usual problems caused by the pace of modern living and their own desires and needs. Well acted, with much of the dialogue improvised, this weighty effort takes itself far too seriously and is in great need of a few touches of humour.
DRA   102 min mCab   SATrel: SKY MOVIES

CHAPLIN *   15
Richard Attenborough UK   1992
Robert Downey Jr, Anthony Hopkins, Dan Aykroyd, Kevin Dunn, Geraldine Chaplin, Milla Jovovich, Penelope Ann Miller, Paul Rhys, Kevin Kline, Moira Kelly, Diane Lane, John Thaw, Marisa Tomei, Nancy Travis, James Woods
Carefully crafted but slow-paced account of the life and career of Chaplin that starts with a flashback to his poverty-stricken childhood and follows his career up to his honorary Oscar in 1972. Downey Jr gives a brilliant portrayal in an enjoyable film, which unwisely concentrates on his personal relationships rather than his art. The use of a fictitious, modern-day interview to frame the film is a weakness. The LV disc includes some real behind-the-scenes interviews.
DRA   138 min (ort 145 min) wScrn cC
VIDrel: 4-FRONT/POLYREC/GUILD; PION (LV only)   V/s LV
Boa: book My Autobiography by Charles Chaplin/book Chaplin – His Life And Art by David Robinson.

CHAPPAQUA *   15
Conrad Rooks USA   1966
Conrad Rooks, Jean-Louis Barrault, William S. Burroughs, Allen Ginsburg, Ravi Shankar, Paula Pritchett, Ornette Coleman
The hallucinations of an alcoholic heroin addict during a sleep cure in Paris. A muddled and pretentious piece of nonsense, that attempts surrealism but achieves incoherence. However, this did not prevent it becoming a big hit at the 1966 Venice Film Festival.
DRA   78 min (ort 92 min) B/W/Col   VIDrel: TART/20TH
V

CHARADE *   *PG*
Stanley Donen USA   1963
Cary Grant, Audrey Hepburn, George Kennedy, James Coburn, Walter Matthau, Ned Glass, Jacques Marin, Paul Bonifas, Dominique Minot, Thomas Chelimsky
A woman living in Paris finds her husband murdered when she returns home one day, and then discovers that she has become the prey of crooks looking for the money he stole during WW2. Fortunately however, a mysterious stranger comes to her rescue. A fair comedy-thriller with more than a few nods in the direction of Hitchcock. The script is by Peter Stone and the score is by Henry Mancini.
THR   109 min (ort 113 min)   VIDrel: CIC/SONOP L/A
V/h
Boa: short story The Unsuspecting Wife by Peter Stone and Marc Behm.

CHARGE OF THE LIGHT BRIGADE, THE **   *PG*
Michael Curtiz USA   1936
Errol Flynn, Olivia De Havilland, Patric Knowles, Henry Stephenson, Nigel Bruce, Donald Crisp, David Niven, C. Aubrey Smith, C. Henry Gordon, E.E. Clive, G.P. Huntley Jr, Spring Byington, Lumsden Hare, Robert Barrat
A lavish and stirring action epic, that distorts history to make the fateful charge of the 27th Lancers at Balaclava little more than the work of one over-ambitious officer. That said, its production values are such that, taken as entertainment, it is very good indeed. Allegedly based on the famous poem of Alfred Lord Tennyson and remade in 1968. AA: Assist Dir (Jack Sullivan).
A/AD   111 min (ort 116 min) B/W   VIDrel: WHV   V

CHARGE OF THE LIGHT BRIGADE, THE **   *PG*
Tony Richardson USA   1968
David Hemmings, Vanessa Redgrave, John Gielgud, Harry Andrews, Jill Bennett, Trevor Howard, Mark Burns, Howard Marion-Crawford, Mark Dignam, Alan Dobie, T.P. McKenna, Willoughby Goddard, Corin Redgrave, Norman Rossington, Ben Aris
Opening with some brilliant animated sequences, this straight-forward account of the events leading up to this infamous slaughter, takes a jaundiced and piercing view of the military

and social ethos of the times and its roots in the English class system. Sustained by firm direction and first-rate acting by an elite cast, this ranks among one of the best British films ever but its very honesty and serious intent were probably why it was such a commercial failure.
DRA   130 min (ort 132 min) wScrn   VIDrel: CONNO/RTM V

CHARIOTS OF FIRE ***
Hugh Hudson UK
U
1981
Ben Cross, Ian Charleson, Nigel Havers, Nicholas Farrell, Daniel Gerroll, Cheryl Campbell, Alice Krige, John Gielgud, Lindsay Anderson, Nigel Davenport, Struan Rogers, Ian Holm, Patrick Magee, Dennis Christopher, Brad Davis
True story of two Cambridge students who competed in the 1924 Olympics. Much acclaimed by the critics, it nicely illustrates the British nostalgic yearning for the glories of the past. Nevertheless, a well made and enjoyable if rather ponderous effort. The feature debut for the director. AA: Pic, Screen/orig (Colin Welland), Score/orig (Vangelis), Cost (Milena Canonero).
DRA   118 min (ort 124 min)   VIDrel: 20TH/TECH L/A V/sur
Boa: book Dollar Bottom And Taylor's Finest Hour by J. Kennaway.

CHARLIE BOY **
Robert Young UK
(18)
1984
Leigh Lawson, Angela Bruce, Marius Goring, David Healy, Frances Cuka, Janet Clare Fielding, Jeff Rawle, Michael Culver, Lee Richards, Andrew Parriss, Michael Deeks, Michael Stock, Charles Pemberton
A man buys an African idol which he is told, has the power to kill and he is foolish enough to try it out. Holding a photograph of six people, he plunges a knife into it, and soon enough, deaths start to occur. Desperate to stop the idol, since he and his wife are also on the photo in question, he decides to destroy it, but becomes desperate when the idol is stolen.
Aka: HAMMER HOUSE OF HORROR: CHARLIE BOY
HOR   101 min   SATrel: BRAVO

CHARLIE BROWN: BON VOYAGE **
Bill Melendez USA
U
1980
Voices of: Daniel Anderson, Scott Beach, Casey Carlson, Debbe Muller, Patricia Patts, Arrin Skelley
Celebrating the 30th anniversary of Schulz's "Peanuts" comic strip, this tale takes the Peanuts gang to France for a fortnight as exchange students. A flimsy effort that may appeal to fans of the comic strip. See also SNOOPY, COME HOME!
Aka: BON VOYAGE, CHARLIE BROWN; BON VOYAGE, CHARLIE BROWN (AND DON'T COME BACK!)
ANIM   73 min   VIDrel: 4-FRONT/POLYREC/CIC L/A V

CHARLIE CHAN AND THE CURSE OF THE DRAGON QUEEN *
Clive Donner USA
PG
1981
Peter Ustinov, Lee Grant, Angie Dickinson, Richard Hatch, Brian Keith, Roddy McDowall, Rachel Roberts, Paul Ryan, Johnny Sekka
Attempt to present a slapstick update of the famous detective that quickly degenerates to the level of a bad farce. The story has a curse put on Chan and his descendants by the evil Dragon Queen, but the greatest curse was to be a member of a good cast appearing in rubbish like this.
COM   92 min   VIDrel: MGM/WHV L/A V

CHARLIE CHAPLIN: CITY LIGHTS ****
Charles Chaplin USA
U
1931
Charles Chaplin, Virginia Cherrill, Florence Lee, Harry Myers, Hank Mann, Jean Harlow, Henry Bergman, Albert Austin, Allan Garcia, John Rand, James Donnelly, Robert Parish, Stanhope Wheatcroft
Chaplin at his most inventive in this fine if rather sentimental story of a blind flower girl and a penniless tramp, who somehow finds the money to pay for an operation to restore her sight. Highlights are the sequences dealing with his friendship with a drunken millionaire, who doesn't know him when he's sober, and the finale, when the girl meets up with her benefactor.
Aka: CITY LIGHTS
COM   82 min (ort 87 min) B/W silent
VIDrel: 20TH/TECH   V/dm

CHARLIE CHAPLIN: LIMELIGHT ***
Charles Chaplin USA
U
1952
Charles Chaplin, Claire Bloom, Sydney Chaplin, Nigel Bruce, Norman Lloyd, Andre Edlevsky, Buster Keaton, Melissa Hayden, Charles

Chaplin Jr, Wheeler Dryden, Marjorie Bennett, Geraldine Chaplin, Michael Chaplin, Josephine Chaplin
Rather sentimental story of a washed-up comic from the days of the music-hall who befriends a beautiful young ballerina. Though it suffers badly from stilted dialogue, some flashes of brilliance light it up. It won an Oscar for the music in 1972 when it became eligible for entry (it had not been shown in a Los Angeles theatre until then). AA: Score/orig (Charles Chaplin/Raymond Rasch/Larry Russell).
Aka: LIMELIGHT
DRA   135 min (ort 145 min) B/W   VIDrel: 20TH/TECH V/dm

CHARLIE CHAPLIN: MODERN TIMES ****
Charles Chaplin USA
U
1936
Charles Chaplin, Pauette Goddard, Henry Bergman, Chester Conklin, Stanley "Tiny" Sandford, Hank Mann, Louis Natheaux, Allan Garcia, Stanley Blystone, Sam Stein, Juana Sutton, Jack Low, Walter James, Dick Alexander, Frank Moran
Chaplin's last silent (apart from some contrived sound sequences) is a savagely witty attack on the dehumanising effects of modern industry. Our hero's struggles against the might of big business, his "marriage", a fine ending and Chaplin's own score (including "Smile") make this a classic work.
Aka: MODERN TIMES
COM   83 min (ort 89 min) B/W silent
VIDrel: 20TH/TECH   V/dm

CHARLIE CHAPLIN: MONSIEUR VERDOUX ****
Charles Chaplin USA
PG
1947
Charles Chaplin, Martha Raye, Marilyn Nash, Isobel Elsom, Irving Bacon, William Frawley, Mady Correll, Allison Roddan, Robert Lewis, Fritz Leiber
A sacked bank clerk becomes a murderer, bumping off rich women after marrying them in order to support his family. Overtones of Landru the Bluebeard are treated in a savage and witty attack on the morality of the arms trade. The dialogue is a good deal less stilted than most other Chaplin talkies as ever, Chaplin provides the music.
Aka: MONSIEUR VERDOUX
DRA   118 min B/W   VIDrel: 20TH/TECH   V/dm

CHARLIE CHAPLIN: THE CIRCUS ***
Charles Chaplin USA
U
1928
Charles Chaplin, Merna Kennedy, Allan Garcia, Betty Morissey, Harry Crocker, Stanley J. Sanford, John Rand, George Davis, Henry Bergman, Steve Murphy, Doc Stone
A tramp joins a circus in order to escape the attentions of the law, but finds his attention being drawn to the attractive horse-riding stepdaughter of the circus owner. Though not one of his great films, this comic tale has many hilarious sequences and culminates in a memorable finale. AA: Spec Award (Chaplin).
Aka: CIRCUS, THE
COM   72 min (ort 82 min) B/W silent
VIDrel: 20TH/TECH   V/dm

CHARLIE CHAPLIN: THE GOLD RUSH ****
Charles Chaplin USA
U
1925
Charles Chaplin, Georgia Hale, Mack Swain, Tom Murray, Betty Morrissey, Malcolm Waite, Henry Bergman, Kay Desleys, Joan Lowell, John Rand, Heinie Conklin, Albert Austin, Allan Garcia, Tom Wood, Stanley Sanford, Art Walker
Brilliant Chaplin comedy about a lone Yukon prospector who makes the big time. Still wonderfully funny with some marvellous visual gags, as our hero battles privation and the elements. A highlight is the brilliant sequence where Chaplin and his partner are trapped in their hut as it's about to tumble over a precipice.
Aka: GOLDRUSH, THE
COM   78 min B/W silent   VIDrel: 20TH/TECH   V/dm

CHARLIE CHAPLIN: THE GREAT DICTATOR ***
Charles Chaplin USA
U
1940
Charles Chaplin, Paulette Goddard, Jack Oakie, Reginald Gardiner, Maurice Moscovitch, Billy Gilbert, Henry Daniell, Emma Dunn, Grace Hayle, Carter De Haven, Bernard Gorcey, Paul Weigel, Chester Conklin, Hank Mann, Eddie Dunn
Chaplin said that he would never have been able to make a lighthearted spoof on Hitler had he known the true horrors that would emerge from WW2. Even so, after 45 years this stilted satire on the Nazi dictatorship still retains some interest, even

if most of the humour seems a little strained. The story is one of mistaken identity, with a Jewish barber being mistaken for the dictator of Tomania, Adenoid Hynkel.
Aka: GREAT DICTATOR, THE
COM   126 min (ort 128 min) B/W   VIDrel: 20TH/TECH
V/dm

CHARLIE CHAPLIN: THE KID ***   U
Charles Chaplin USA   1921
Charles Chaplin, Jackie Coogan, Edna Purviance, Lita Grey (Lillita McMurray), Charles Reisner, Carl Miller, Granville Redmond, May White, Tom Wilson, Henry Bergman, Raymond Lee, Edith Wilson, Nellie Bly Baker
Chaplin's first feature is a touching comedy about a tramp who brings up an abandoned baby and has a tough time preventing welfare agencies from taking him into care. A wonderful blend of comedy, tragedy and slapstick. This film launched Coogan as a child star.
Aka: KID, THE
COM   60 min; 82 min (2-film cassette) B/W silent
VIDrel: 20TH/TECH   V/dm

CHARLIE DRAKE: MISTER TEN PER CENT *   U
Peter Graham Scott UK   1966
Charlie Drake, George Baker, Annette Andre, John Le Mesurier
A builder who writes plays as a hobby has one of them accepted by a top London impresario and finds his drama being staged as a successful comedy. A film of little humour that fails to fully exploit the talents of its diminutive star.
Aka: MISTER TEN PER CENT
COM   85 min   VIDrel: WHV   V

CHARLIE DRAKE: PETTICOAT PIRATES **   U
David MacDonald UK   1961
Charlie Drake, Anne Heywood, Cecil Parker, John Turner, Maxine Audley, Thorley Walters
With the seizure of a frigate by a bunch of WRNS, the nervous captive stoker decides it's safer to pose as one of them. Popular comedian Drake does what he can with this contrived comedy vehicle.
Aka: PETTICOAT PIRATES
COM   82 min (ort 87 min)   VIDrel: WHV   V

CHARLIE DRAKE: SANDS OF THE DESERT *   U
John Paddy Carstairs UK   1960
Charlie Drake, Peter Arne, Sarah Branch, Raymond Huntley, Rebecca Dignam, Peter Illing, Harold Kasket
A clumsy and disorganised vehicle for Drake in which he plays a travel agent who investigates sabotage at a desert holiday camp and saves a girl from an Arab sheikh at the same time. The script is by Carstairs.
Aka: SANDS OF THE DESERT
COM   88 min (ort 92 min)   VIDrel: WHV   V

CHARLIE DRAKE: THE CRACKSMAN **   U
Peter Graham Scott UK   1963
Charlie Drake, George Sanders, Dennis Price, Nyree Dawn Porter, Eddie Byrne, Finlay Currie, Percy Herbert
An expert locksmith falls into the clutches of a gang of safebreakers and is duped into assisting them. A cheerfully daft vehicle for this popular comedian that is diluted by overlength.
Aka: CRACKSMAN, THE
COM   108 min (ort 112 min)   VIDrel: WHV   V

CHARLIE'S GHOST: THE SECRET OF
CORONADO **   (PG)
Anthony Edwards USA   1994
Trenton Knight, Cheech Marin, Anthony Edwards, Charles Rocket, J.T. Walsh, Linda Fiorentino, Robert Corman, Veronica Lauren, Dean Cameron, Daphne Zuniga, Bethany Richards, Steve Kearney, Alan Shearman, Leslie Danon, Edwin Smith
When his archaeologist father unearths the grave of a famous Spanish conquistador, a young boy find himself haunted by the latter's ghost. Outraged by this treatment, this angry spirit demands that his mortal remains be re-buried in sacred ground.
Aka: CHARLIE'S GHOST STORY
COM   85 min (ort 91 min)   SATrel: SKY MOVIES
Boa: short story by Mark Twain.

CHARLOTTE'S WEB **   U
Charles A. Nichols/Iwao Takamoto USA   1973
Voices of: Debbie Reynolds, Agnes Moorehead, Paul Lynde, Henry Gibson, Charles Nelson Reilly, Pamelyn Ferdin, Danny Bonaduce,

William B. White, Don Messick, Herb Vigran, Joan Gerber, Robert Holt, John Stephenson, Dave Madden
Cartoon feature based on a children's story about a spider who befriends a piglet and saves his bacon. A fair animation let down by the totally forgettable songs of Richard and Robert Sherman. A Hanna-Barbera production.
ANIM   94 min   VIDrel: MIA/DISC   V
Boa: novel by E.B. White.

CHARLY ***   PG
Ralph Nelson USA   1968
Cliff Robertson, Lilia Skala, Dick Van Patten, Claire Bloom, Leon Janney, William Dwyer, Ed McNally, Barney Martin, Ruth White, Frank Dolan
Based on a Hugo winning short SF story, this is a look at what happens when a mentally retarded adult undergoes experimental neuro-surgery and becomes a genius, but only temporarily. An interesting film that would have been far more convincing minus the touches of sentimentality that tend to spoil it. AA: Actor (Robertson).
DRA   103 min   VIDrel: BRAVE/SONOP L/A   V
Boa: short story Flowers for Algernon by Daniel Keys.

CHARMER, THE ***   15
Alan Gibson UK   1989
Nigel Havers, Rosemary Leach, Bernard Hepton, Fiona Fullerton, Abigail McKern, George Baker, Judy Parfitt
In 1930s Britain, a ruthless con-man uses his wits and his charm to ensnare a variety of more or less willing victims and is even prepared to resort to murder to achieve his ends. A well-acted and firmly directed TV series.
DRA   301 min (6 cassettes – ort 312 min) mTV
VIDrel: CASPIC/BMGREC   V
Boa: novel by Allan Prior.

CHARRO! *   15
Charles Marquis Warren USA   1969
Elvis Presley, Ina Balin, Barbara Werle, Victor French, Lynn Kellogg, Paul Brinegar, Solomon Struges, James Sikking, Harry Landers, Tony Young, James Almanazar, Charles H. Gray, Rodd Redwing, Gary Walbert, Duane Grey
This attempt to give Presley a non-singing role casts him as a reformed outlaw who has to prove his innocence when he is framed for the theft of a cannon. The best efforts of the star in a straight role (he only gets to sing the title song) are as nothing when measured against the dismal production standards of this dud.
WES   98 min   VIDrel: MIA/VCC L/A   V

CHASE, THE **   15
Arthur Penn USA   1966
Marlon Brando, Jane Fonda, Robert Redford, Angie Dickinson, Janice Rule, James Fox, Robert Duvall, E.G. Marshall, Miriam Hopkins, Katherine Walsh, Martha Hyer, Richard Bradford, Diana Hyland, Jocelyn Brando, Henry Hull, Lori Martin
A convict from a small town in Texas escapes, heading in its direction, in a flight that has an effect in one way or another on virtually all its inhabitants. An overlong and violent melodrama that takes a long time to come to its far from exciting climax.
DRA   127 min (ort 135 min)
VIDrel: ENCORE/SPEAR/COLUM   V
Boa: novel by Horton Foote.

CHASE **   15
Rod Holcomb USA   1985
Jennifer O'Neill, Robert S. Woods, Michael Parks, Terence Knox, John Philbin, Kathleen York, J.E. Freeman, Cooper Huckabee, Richard Farnsworth, John Walter Davis, Douglas Newell, Richard Arnold
A female lawyer returns to her small home town and finds herself defending a man accused of murdering an old friend. The case takes on more sinister aspects when she discovers that a group of assassins are out to silence the suspect.
DRA   90 min Cut (25 sec) mTV   VIDrel: 20TH/TECH   V

CHASE, THE **   15
Paul Wendkos USA   1990
Casey Siemaszko, Ben Johnson, Gerry Bamman, Robert Beltran, Barry Corbin, Ricki Lake, Megan Follows
A bank robber who has broken out of jail is trailed by a TV helicopter news team intent on getting a scoop. A reasonably entertaining adventure that is very loosely based on real events. Unfortunately, far too much time is wasted on accounts of the

personal lives of the central characters and too little care devoted to the action sequences.
A/AD 90 min (ort 100 min) mTV VIDrel: CAPIT/GUILD V/s

CHASE, THE ** 15
Adam Rifkin USA 1994
Charlie Sheen, Kristy Swanson, Henry Rollis, Josh Mostel, Ray Wise, Marshall Bell, Wayne Grace, Rocky Carroll, Miles Douglas, Claudia Christian, Alex Allen Morris, Joe Sugal, Marco Perella, Wirt Cain, Bree Walker, Brian Chesney
A man unjustly convicted of a crime escapes from prison and hijacks a young woman in her car as his hostage. Since she is the daughter of the richest man in California, this brings out the media in droves. As their wild odyssey progresses, this mismatched couple fall in love while dodging his pursuers. A noisy and predictable action-comedy, full of sound and fury but little else and quite inferior to the 1966 movie that seems to have served as its inspiration.
COM 84 min (ort 144 min) cC VIDrel: 20TH/TECH V/sur

CHASE MORRAN ** 18
Gilbert Po
Joseph Culp, Raymond Baker, Jocelyn Seagrave
Futuristic fantasy in which a dangerously psychotic criminal breaks out of jail, steals a space-craft and arrives at a small space colony that he sets about taking over, this being the first stage of a master plan he has formulated for taking over an entire planet. Adequate space-opera nonsense.
FAN 90 min VIDrel: FIRST/SONOP V

CHASERS ** 15
Dennis Hopper USA 1993
Tom Berenger, William McNamara, Erika Eleniak, Gary Busey, Marilu Henner, Dean Stockwell, Crispin Glover, Frederic Forest, Dennis Hopper, Matthew Glave, Seymour Cassel, Grand L. Bush, Bitty Schram, Scott Marlowe, Jim Grimshaw
Two Navy shore patrolmen are given an assignment to escort a dangerous prisoner to Charleston and are appalled to learn that the latter is a woman. This single fact serves as the only peg on which to hang all the subsequent antics that occur, in this luke-warm action-comedy. Inevitably it invites comparison with THE LAST DETAIL against which it pales into insignificance.
COM 100 min (ort 101 min) cC VIDrel: WHV V/sur

CHASING THE DEER * PG
Graham Holloway UK 1994
Brian Blessed, Ian Cuthbertson, Matthew Zajac, Fish, Brian Donald, Sandy Welch, Peter Gordon, Carolyn Konrad, Lynn Ferguson, Lewis Rae, Simon Kirk, Andy McCullogh, Callum McDougal, Steven Cooper, Michael Leighton, Dominique Carrara
Set in Scotland in 1745, where Bonnie Prince Charlie tries to drum up support to install his father on the Scottish throne, but father and son become divided by the Civil War. Financed in an odd way by inviting the public to buy shares and appear as extras, this cumbersome film has an amateurish look about it, and for all the atmospheric shots of the Scottish Highlands, its poor production values are very much in evidence.
A/AD 97 min CINrel

CHASING THE DRAGON ** 18
Ian Sander USA 1996
Markie Post, Dennis Boutsikaris, Noah Fleiss
Story of how heroin destroys life of a young single mother after she accepts just one shot of the drug from a supposed friend. An earnest and well meaning movie-with-a-message, that for all its good intentions remains remarkably uninvolving.
DRA 92 min VIDrel: ODY/SONOP V

CHATO'S LAND ** 18
Michael Winner USA 1972
Charles Bronson, Jack Palance, Richard Basehart, James Whitmore, Simon Oakland, Ralph Waite, Richard Jordan, Victor French, William Watson, Roddy McMillan, Paul Young, Lee Paterson, Rudy Ugland, Sonia Rangan, Verna Harvey
An Indian half-breed kills a white man in self-defence and finds that he is pursued by a posse. A competently made film in which the Indian finds that once the posse has entered his land he has the advantage. A violent and often bloody Winner offering (albeit quite well directed) best described as a pointless exercise in nastiness.
WES 95 min Cut (41 sec – ort 100 min)
VIDrel: MGM/WHV L/A V

CHATTAHOOCHEE ** 15
Mick Jackson USA 1990
Gary Oldman, Dennis Hopper, Frances McDormand, Ned Beatty, Pamela Reed, M. Emmet Walsh, William De Acutis, Lee Wilkof, Matt Craven, Gary Klar, Timothy Scott, Richard Portnow, William Newman, Whitey Hughes, Wilbur Fitzgerald
A veteran of the Korean War suffers a mental breakdown but is wrongly placed in a repressive mental institution. A fact-based drama based on some real events of the 1950s that has its heart in the right place but fails to get to grips with its subject matter, for all the excellent work from the cast, and is hurt by clumsy post-production cutting. The comic interlude involving a bank robbery is just an annoying distraction. SHOCK CORRIDOR this ain't.
DRA 93 min (ort 98 min) VIDrel: COLUM/SONOP V/sur

CHECKING OUT * 15
David Leland USA 1988
Jeff Daniels, Melanie Mayron, Michael Tucker, Ann Magnuson, Kathleen York, Allan Harvey, Jo Harvey Allen, Felton Perry, Alan Wolfe, David Byrne
Following the sudden death of his best friend from a heart attack, an executive experiences an acute attack of hypochondria, and spends the rest of the movie wondering if he will experience the same fate. A botched attempt at a black comedy whose amusing moments are few and far between.
COM 90 min (ort 95 min) VIDrel: 20TH V/sur

CHEECH AND CHONG: STILL SMOKIN' * 18
Thomas Chong USA 1983
Cheech Marin, Thomas Chong, Hans Van In't Veld, Carol Van Herwijnen, Susan Hahn, Shireen Strooker, Arjan Ederveen, Kees Prins, Mariette Bout, Fabiola, Carla Van Amstel
Silly pothead comedy with our dope crazy duo, all about a film festival in Amsterdam. As thin as all those Cheech and Chong films and a good deal harder to watch, not least owing to the inclusion of about twenty minutes of concert footage.
Aka: STILL SMOKIN'
COM 87 min (ort 92 min) VIDrel: CIC/SONOP V/h

CHEECH AND CHONG: THINGS ARE TOUGH ALL OVER ** 15
Tom Avildsen USA 1982
Cheech Marin, Thomas Chong, Shelby Fiddis, Rikki Marin, Evelyn Guerero, Rip Taylor, Shelby Chong, Michael Aragon, Toni Attell, Mike Bacarella, Billy Beck, Don Bovingloh, Richard Calhoun, Jennifer Condos, John Corona
Cheech and Chong comedy with two brothers driving a car to California, unaware that it contains $5,000,000. One of their better efforts, with the stars also playing a couple of Arab brothers who cross the path of the first pair. Written by Marin and Chong.
Aka: THINGS ARE TOUGH ALL OVER
COM 86 min (ort 92 min) VIDrel: CASPIC/BMGREC L/A V

CHEECH AND CHONG'S NEXT MOVIE ** 18
Thomas Chong USA 1980
Cheech Marin, Thomas Chong, Evelyn Guerrero, Betty Kennedy, Sy Kramer, Rikki Marin, Bob McClurg, Edie McClurg, Paul Reubens
Unfunny collection of dope-jokes by two of the most famous potheads in the world. Incoherent, uneven and only sporadically amusing.
Aka: HIGH ENCOUNTERS OF THE ULTIMATE KIND
COM 90 min (ort 99 min) VIDrel: CIC/SONOP V/h

CHEECH AND CHONG'S NICE DREAMS ** 18
Thomas Chong USA 1981
Cheech Marin, Thomas Chong, Stacy Keach, Evelyn Guerrero, Timothy Leary, Robert Chong, Pee-Wee Herman (Paul Reubens), Michael Masters, James Faracci, Jeff Pomerantz, Taaffe O'Connell, Suzanne Kent, Michael Winslow, Shirley Prestio
Cheech and Chong decide to use an ice-cream van as a front for selling dope, which naturally leads to complications but not much in the way of real humour.
Aka: NICE DREAMS
COM 84 min (ort 97 min)
VIDrel: FABFIL/SPEAR/COLUM V

CHEECH AND CHONG'S THE CORSICAN BROTHERS * 15
Thomas Chong/Cheech Marin USA 1984
Cheech Marin, Thomas Chong, Roy Dotrice, Shelby Fiddis, Rikki

Marin, Edie McClurg, Rae Dawn Chong, Robbi Chong, Simono, Kay Dotrice, Jennie C. Kos, Martin Pepper, Yvan Chiffre, Dan Schwartz, Jean-Claude Dreyfus, Bernard Szabo
Spoof remake of the 1941 film of the Dumas story of the twins who, though separated at birth, still maintain a psychic link. Their first non-pothead parody that falls flat on its face, despite our boys playing three roles each. Watch it for the nice French locations, but don't expect to laugh.
Aka: CORSICAN BROTHERS, THE
COM   86 min (ort 91 min)   VIDrel: VISVID/POLYREC L/A V

CHEECH AND CHONG'S UP IN SMOKE **
18
Lou Adler USA
1978
Cheech Marin, Thomas Chong, Stacy Keach, Tom Skerritt, Edie Adams, Strother Martin, Zane Busby, Louisa Moritz, Anne Wharton, Mills Watson, Karl Johnson, Rick Beckner, Harold Fong, Richard Novo, Jane Moder, Pam Bille, Ray Vitte
The first in a seemingly endless number of "Cheech and Chong" pothead comedies, based on two dope-smoking, beer-swilling slobs they created in a series of record albums. This one features a vehicle made entirely of pressed dope, and tells of how our two dope-crazy potheads go in search of some good grass. Episodic and largely disjointed, but undeniably funny in parts.
Aka: UP IN SMOKE
COM   82 min (ort 86 min)   VIDrel: CIC/SONOP L/A   V

CHEERLEADER-MURDERING MOM ***
15
Michael Ritchie USA
1992
Holly Hunter, Beau Bridges, Swoosie Kurtz, Matt Frewer, Gregg Henry, Eddie Jones, Frankie Ingrassia, Gary Grubbs
A mother will go to any lengths to ensure her girl has a place on the high school cheerleading team, her actions leading to her being charged with plotting to kill the mother of a rival girl. A comedy-drama similar to films like DEATH OF A CHEER-LEADER and WILLING TO KILL: THE TEXAS CHEERLEADER STORY. However, the last two were simple accounts of a true incident, whereas this one is more of a sharp satire on American family values and the desire to win at all costs.
Aka: POSITIVELY TRUE ADVENTURES OF THE ALLEGED TEXAS CHEER-LEADER-MURDERING MOM, THE
COM   99 min mCab   VIDrel: FIRST/SONOP   V

CHERRY 2000 **
15
Steve De Jarnatt USA
1986
Melanie Griffith, David Andrews, Tim Thomerson, Ben Johnson, Pamela Gidley, Jennifer Balgobin, Marshall Bill, Harry Carey Jr, Larry Fishburne, Michael C. Gwynne, Brion James, Jeff Levine, Jennifer Mayo, Howard Swain
In the world of 2017, sexual partners have been replaced by robot playmates, and one man is forced to embark on a long and arduous search for spare parts when his girlfriend breaks down. An implausible action fantasy let down by poor casting and the lack of a decent plot, though one or two spectacular effects slightly compensate.
FAN   98 min   VIDrel: RCA L/A   V

CHESS PLAYERS, THE ***
PG
Satyajit Ray INDIA
1977
Sanjeev Kumar, Saeed Jaffrey, Amjad Khan, Shabana Azmi, Richard Attenborough
A strange tale set in 1856 at Lucknow, where two wealthy Indian noblemen indulge their passion for chess to the exclusion of everything else, while their Maharajah is in the course of being supplanted by the East India Company. The first film made in Hindi by Ray, this is an ambitious attempt to use the device of a chess game to examine much larger issues, and though not entirely successful, has many telling moments and a deep sense of irony. Music is by Ray.
Aka: SHATRANJ KE KHILARI
DRA   115 min (ort 125 min)   VIDrel: ARTPRO/RTM   V

CHEYENNE WARRIOR ***
(PG)
Mark Griffiths USA
1993
Kelly Preston, Bo Hopkins, Dan Haggerty, Pato Hoffmann, Rick Dean, Charles Powell, Clint Howard, Dan Clark, Winterhawk, Joseph Wolves Kill, Noah Colton, Louise Baker, Frankie Avina, Terrance Fredricks, Mark S. Brien, Mark Costa
After the Civil War, a pregnant woman and her husband trek west but are ill prepared for the hardships they encounter. At a remote trading post they make the acquaintance of the title

figure, to whom the wife begins to feel herself drawn. When the husband is killed by some cutthroats and our warrior badly injured, she nurses him back to health and the two learn to work together to survive and take their revenge. Well-acted, balanced and realistic.
WES   91 min   SATrel: MOVIE CHANNEL

CHICAGO JOE AND THE SHOWGIRL **
18
Bernard Rose UK
1990
Kiefer Sutherland, Emily Lloyd, Patsy Kensit, Alexandra Pigg, Liz Fraser, Harry Fowler, Keith Allen, Angela Morant, John Surman, Janet Dair, Stephen Hancock, Hugh Millais, Harry Jones, John Junkin, Gerard Horan
BONNIE AND CLYDE is transferred to WW2 Britain in this dark tale (based on a real case) of a GI with delusions of grandeur and a liking for the good life, who falls for a British dancer. Together, they embark on a killing spree until finally captured. The sparse and realistic portrayal of wartime Britain is an asset, but the film lacks depth and provides no insight into the motivations of the duo.
DRA   98 min (ort 105 min)   VIDrel: POLY/POLYREC   V/h

CHILD IN THE NIGHT **
15
Mike Robe USA
1990
JoBeth Williams, Tom Skerritt, Season Hubley, Michael Pniewski, Tim Choate, Elijah Wood, Thom Bray, John Procaccino, Darren McGavin, Stephen Godwin, Karen Trumbo, John Aylward, Laura Langwell, Don Hohenstein, Junior Knotts
Having witnessed the slaying of his father, a young boy has withdrawn into a shell. When a child psychologist is assigned to his case in an effort to help him, her attempts to treat him are thwarted by the boy's mother and grandfather. A moderately absorbing TV drama.
DRA   91 min (ort 100 min) mTV   VIDrel: ODY/SONOP V/sh

CHILD IS LOST FOREVER, A **
15
Claudia Weill USA
Beverley D'Angelo, Will Patton
Fact-based drama in which a young woman is forced to give her baby up for adoption and learns a few years later that it died at the age of only three. This knowledge leads her into a struggle to uncover the truth behind her child's death.
DRA   91 min   VIDrel: ODY/SONOP   V/h

CHILD IS MISSING, A **
12
John Power USA
1995
Henry Winkler, Roma Downey, Dale Midkiff, Alberta Watson, Blu Mankuma, Richard Dysart
A backwoods loner becomes the chief suspect when a young boy is abducted, and realises that he has to find the child in order to clear his name.
THR   83 min   VIDrel: NWV/HIFLI   V

CHILD OF DARKNESS, CHILD OF LIGHT **
15
Marina Sargenti USA
1991
Brad Davis, Anthony John Denison, Paxton Whitehead, Claudette Nevins, Sydney Penny, Kristen Dattilo, Viveca Lindfors, Sela Ward, Alan Oppenheimer, Richard McKenzie, Eric Christmas, Joshua Lucas, John DeMitta, Mark Tassoni
An OMEN-inspired supernatural effort, that sees the Church concerning itself with a couple of virgin births, which may in fact represent the coming of both Christ and the Antichrist to the world. An adequate horror yarn, neither especially original nor memorable.
HOR   82 min (ort 100 min) mCab   VIDrel: CIC/SONOP   V
Boa: novel Virgin by James Patterson.

CHILD OF RAGE ***
15
Larry Peerce USA
1993
Mel Harris, Dwight Schultz, Ashley Peldon, Mariette Harley, Rosana Desoto, Sam Gifaldi, Nan Martin, George D. Wallace, Johannah Newmarch, Patricia Gage, Terrence Kelly, Kim Kondrashoff
A young minister and his wife agree to foster two children, a small girl and her younger brother but are not informed that the children have a history of having gone from one foster home to another. As they attempt to build a family, it soon becomes apparent that the girl suffers from a strange and violent personality disorder that makes her literally able to commit murder. A fascinating and well-realised drama that is completely absorbing.
DRA   93 min mTV   VIDrel: ODY/SONOP   V/s

CHILD TOO MANY, A **   PG
Jorge Montesi USA   1993
Michelle Greene, Conor O'Farell, Stephen Macht, Nancy Stafford, Heather Beaty, Alex Doduk, Joel Palmer, Shannon Beaty, Kerry Sandormirsky, Kevin McNulty, Sandra Nelson, Andrew Wheeler, Malcom Stewart, Michele Stewart
A surrogate mother has twins and finds that the adopting couple are only prepared to take one child. A competent TV drama, well acted and quite convincing.
DRA 89 min mTV VIDrel: NWV/HIFLI V/h

CHILDHOOD OF MAXIM GORKY, THE ****   PG
Mark Donskoi USSR   1938
Alexei Lyarsky, Y. Valbert, M. Troyanovski, Valeria Massalitinova
The first part of the Maxim Gorky trilogy, with the orphaned writer being brought up by his grandparents, and learning about life and the tribulations of the Russian people. A rich and detailed film, full of humour and strong characterisations. Followed by MY APPRENTICESHIP and MY UNIVERSITIES.
Aka: DETSTVO GORKOVO
DRA 95 min (ort 110 min) B/W VIDrel: HEND L/A V

CHILDREN OF A LESSER GOD ***   15
Randa Haines USA   1986
William Hurt, Marlee Matlin, Piper Laurie, Philip Bosco, Alison Gompf, John F. Cleary, John Basinger, Philip Holmes, Georgia Ann Cline, William D. Byrd, Frank Carter Jr, John Limnidis, Bob Hiltermann, E. Katherine Kerr
An idealistic teacher of the deaf is intrigued by his discovery of an intelligent deaf woman who works there as a janitor and is isolated owing to her refusal to learn lip-reading. Though less powerful than Medoff's Tony award-winning play, this unusual love story owes much to the convincing and moving performances of its leads. The film debut for both Matlin and Haines. See also BRIDGE TO SILENCE. AA: Actress (Matlin).
DRA 115 min (ort 119 min)
VIDrel: 4-FRONT/POLYREC/CIC V/h
Boa: play by Mark Medoff.

CHILDREN OF THE CORN *   18
Fritz Kiersch USA   1984
Peter Horton, Linda Hamilton, R.G. Armstrong, John Franklin, Courtney Gains, Robby Kiger, Annemarie McEvoy, Julie Maddalena, Jonas Marlowe, Dan Snook, John Philbin, David Cowan, Suzy Southam, D.G. Johnson, Patrick Boylan
A loose adaptation of King's tale of a young couple who get involved with a strange cult, in a small Iowa town in the American corn-belt. Laughable, low-grade nonsense, quite gory in places, and followed by several sequels.
HOR 93 min VIDrel: VCC/DISC V
Boa: short story by Stephen King.

CHILDREN OF THE CORN 2: THE FINAL SACRIFICE **   18
David F. Price USA   1992
Terence Knox, Paul Scherrer, Ryan Bollman, Christie Clark, Rosalind Allen, Ned Romero, Ted Travelstead, Ed Grady, John Bennes, Wallace Merck, Joe Inscoe, Kelly Bennett, Rob Treveiler, Leon Pridgen, Marty Terry, Audry Dollar
Sequel to the first film has an inquisitive reporter working on a story about on the adults who were murdered in a small Iowa town and discovering that this is happening again in a nearby town. A poor effort not directly related to anything written by Stephen King.
HOR 89 min (ort 93 min) VIDrel: VCC/DISC V/sh

CHILDREN OF THE CORN 3 **   18
James R. Hickox USA   1995
Daniel Cerny, Mari Morrow, Ron Melendez, Duke Stroud, Nancy Lee Grahn, Jim Metzler, Jon Clair, Michael Ensign, Rif Hutton, Garvin Funches, Gina St John, Yvette Freeman, Terrence Matthews, James O'Sullivan, Kelly Nelson, Ed Grady
A further instalment in this continuing saga sees two brothers who survived the horrors described in the previous films being taken from the countryside to live in the town with their new adoptive parents. One of them soon starts to show the same evil, murderous tendencies and his brother is forced to try and stop him from staging a further slaughter of adults.
Aka: CHILDREN OF THE CORN 3: URBAN HARVEST
HOR 89 min (ort 91 min) VIDrel: HIFLI/SONOP V/h

CHILDREN OF THE DAMNED **   (15)
Anton M. Leader UK   1964
Ian Hendry, Alan Badel, Barbara Ferris, Alfred Burke, Sheila Allen, Clive Powell, Frank Summerscales, Mahdu Mathen, Gerald Delsol, Roberta Rex, Tom Bowman, Franchesca Lee, Harold Goldblatt, Patrick Whyte, Martin Miller
Sequel to VILLAGE OF THE DAMNED has six super-intelligent children with vast IQs being brought to London, to be studied by UNESCO investigators, while the government ponders their fate, and eventually decides in favour of their destruction. The children, however, have other ideas and soon take control of the proceedings. An adequate low-budget sequel.
FAN 90 min TVrel

CHILDREN OF THE DARK ***   PG
Michael Switzer USA   1994
Peter Horton, Tracy Pollan, Roy Dotrice, Natalija Nogulich, Eric Pierpoint, Sam Horrigan, Lindsey Haun, Annalise Ashdown, Bill Smitrovich, Ramon Bieri, Annie Corley, Mary Pat Gleason, Lenore Kasdorf, Glenn Morshower, Kenenth White
Story of a family and the trauma they suffer because their two children are unable to venture out into the sunlight. A true-life drama that deals with this exceptionally rare medical condition in a restrained way, it allows the characters ample scope to demonstrate their fortitude, and is a touching and surprisingly inspiring work.
DRA 93 min mTV VIDrel: ODY/SONOP V

CHILDREN OF THE NIGHT **   18
Tony Randel USA   1992
Karen Black, Peter De Luise, Mya McLaughlin, Ami Dolenz, Garrett Morris
A small town becomes the prey of a vampire and a teacher takes the unwise decision to investigate, in this standard horror offering.
HOR 87 min (ort 92 min) VIDrel: COLUM/SONOP V/sh

CHILDREN SHOULDN'T PLAY WITH DEAD THINGS *   18
Benjamin (Bob) Clark USA   1973
Alan Ormsby, Anya Ormsby, Valerie Mauches, Jane Daly, Jeffrey Gillen, Paul Cronin, Bruce Solomon, Seth Sklarey, Roy Engleman, Bob Filep, Alecs Baird
A film company goes to an island to make a film and takes over a rural graveyard, unwittingly resurrecting the dead when witchcraft is dabbled in. Unusual but totally unfrightening – the stupid script sees to that.
Aka: ZOMBIE GRAVEYARD
HOR 90 min VIDrel: MOPIC/SGSVID V

CHILD'S CRY FOR HELP, A ***   (PG)
Sandor Stern USA   1994
Veronica Hamel, Pam Dawber, David Hugh Kelly, Lisa Jakub, Cynthia Martells, Daniel Benzali, James Pickens Jr, Jeff Williams, Zachary Charles, Lois Hicks, Tobey Maguire, Regina Krueger, James Gale, Connie Craig, John Ashton
The head of a top children's hospital begins to suspect that a mother of a seriously ill boy is suffering from Munchausen's Syndrome. This condition often causes sufferers to injure themselves or others (for which they fabricate excuses – hence the term "Munchausen") and when she realises that the woman's child may be in danger she has to act to avert a tragedy. An absorbing look at one of psychiatry's most bizarre mental disorders.
THR 90 min mTV SATrel: SKY MOVIES

CHILD'S PLAY ***   15
Tom Holland USA   1988
Catherine Hicks, Brad Dourif, Chris Sarandon, Alex Vincent, Dinah Manoff, Tommy Swerdlow, Jack Colvin, Neil Giuntoli, Alan Wilder, Richard Baird, Raymond Oliver, Aaron Osborne, Tyler Hard, Ted Liss
Horror yarn in which a child is the only one aware of the fact that his new doll is possessed by the spirit of a dead murderer. Eventually he gets the grown-ups to believe him, but not before several murders have taken place. A highlight of this well-made chiller is the excellent special effects in which the doll comes to life. Two sequels followed.
HOR 83 min (ort 87 min) VIDrel: WHV V/sur

CHILD'S PLAY 2 **   15
John Lafia USA   1990
Alex Vincent, Jenny Agutter, Gerrit Graham, Christine Elise, Gregg Germann, Grace Zabriskie, Peter Haskell, Beth Grant, Raymond

Singer, Stuart Mabray, Matt Roe, Charles C. Meshack, Herb Braha, Ed Krieger, Brad Dourif (voice only)
Still struggling to get over the nightmarish events surrounding his first encounter with "Chucky", our fearful youngster has now been taken in by foster parents. Unfortunately, the malevolent doll has made further plans. A weakly plotted sequel that never really holds one's attention until the final fifteen minutes – a celebration of gore that redeems the film as an effective yarn.
HOR   80 min (ort 84 min)   VIDrel: CIC/SONOP L/A   V/sh

CHILD'S PLAY 3 *
Jack Bender USA
Justin Whalin, Perrey Reeves, Jeremy Sylvers, Peter Haskell, Dakin Matthews, Travis Fine, Dean Jacobson, Matthew Walker, Andrew Robinson, Burke Byrnes, Donna Eskra, Terry Wills plus voices of: Edan Gross, Brad Dourif
This further CHILD'S PLAY sequel opens eight years after the climax of the previous tale, with the Chucky doll having been melted down in a toy factory and the "Good Guys" line discontinued. Unfortunately, matters simply cannot be left there, and when they're put back into production (using Chucky's remains in the melting pot) our diabolical doll is soon back in murderous action. A tired and dispirited dud that understandably was released straight to video.
Aka: CHILD'S PLAY 3: LOOK WHO'S STALKING
HOR   86 min (ort 90 min)   VIDrel: CIC/SONOP L/A   V

18
1991

CHILD'S WISH, A **
Waris Hussein USA
Anna Chlumsky, John Ritter, Tess Harper
When a young child is struck down by a serious illness, her parents dedicate their lives to caring for her, and eventually bring about a change in the law to help other similarly stricken families.
DRA   91 min   VIDrel: ODY/SONOP   V/sh

PG
1996

CHIMERA **
Lawrence Gordon Clark UK
John Lynch, Christine Kavanagh, Emer Gillespie, George Costigan, Bhasker, Pip Torrens, David Calder, Gary Mavers, Gillian Barge, Kenneth Cranham, Sebastian Shaw, Debra Gillet, Frank Baker, Dan Mullane, Nicholas Hewetson,
A young nurse accepts a job at a local clinic that specialises in fertility treatment and learns of a strange creature that is slaughtering the staff there. Scripted by Gallagher from his novel, this TV movie has some very chilling moments, even if the whole thing is too implausible to sustain tension.
DRA   104 min (ort 300 min) mTV   VIDrel: FOCUS/DISC   V
Boa: novel by Steven Gallagher.

15
1990

CHINA MOON **
John Bailey USA
Ed Harris, Madeleine Stowe, Benico Del Toro, Charles Dance, Pruitt Taylor Vince, Roger Aaron Brown, Patricia Healey, Tim Powell, Theresa Bean, Allen Prince, Sam Myers, Anson Funderburgh, Jim Milan, Danny Cochran, Gregory Avellone
The familiar story of a private eye who meets an attractive woman who asks for his help and becomes the prime suspect in the murder of her allegedly brutal husband. Set in Florida, this one offers nice scenery, a few sex scenes and good acting that help make up for the paucity of the plot, especially from Harris, who turns in one of his strongest performances.
DRA   96 min (ort 99 min)   VIDrel: COLUM/SONOP   V/sur

18
1993

CHINA O'BRIEN **
Robert Clouse USA
Cynthia Rothrock, Richard Norton, Patrick Adamson, David Blackwell, Keith Cook, Robert Tiller, Steven Kerby, Lainie Watts
A tough female cop and martial arts expert kills in self defence, promptly resigning from the force and returning to her home town. Once there, she finds that she is once more called upon to use her prowess, even though she has sworn never again to use violence.
A/AD   86 min Cut (3 sec)   VIDrel: POLY/POLYREC   V

18
1989

CHINA O'BRIEN 2 **
Robert Clouse USA
Cynthia Rothrock, Richard Norton, Keith Cooke, Frank Magner, Harlow Marks, Tiffany Soter, Tricia Quai, Don Re Sampson, Gary Rogers, Billy Joe Allgood, Jaren Harbrecht, James Horrack, J.R. Glover, Cindy Clark, Michael Anthony
This sequel to the first martial arts action thriller has better production values but essentially provides more of the same. As

18
1991

sheriff of a small town our intrepid heroine is involved with investigations into some gangland killings and soon meets the baddies head-on.
A/AD   82 min (ort 90 min)
VIDrel: 4-FRONT/POLYREC/EIV   V

CHINA SYNDROME, THE ***
James Bridges USA
Jane Fonda, Jack Lemmon, Michael Douglas, Scott Brady, James Hampton, Peter Donat, Wilford Brimley, James Karen, Michael Herd, Daniel Valdez, Stan Bohrman, Michael Alaimo, Donald Hotton, Khalilah Ali, Tom Eure, Nick Pelligrino
Exciting drama about an attempted cover-up of an accident at a nuclear power station in California. The title is the nickname that has been used to describe what could happen in the event of a reactor core meltdown taking place. Tense and quite chilling, with fine performances from Fonda as a TV reporter and Lemmon as a dedicated executive working for the company that runs the plant. Written by Mike Gray, T.S. Cook and James Bridges.
THR   120 min   VIDrel: RCA L/A   V

PG
1979

CHINA WHITE **
Ronny Yu HONG KONG
Billy Drago, Russell Wong, Lisa Schrage, Steven Vincent Leigh, William Ho, Victor Hon, Tommy Wong, Kuk Fung, Ricky Ho, Frank Sheppard, Cahit Olmez, Jules Croisset, Frite Kampinga, Saskia Van Rijswijl, Lex De Regt, Rodney Beddall
A look at the murky world of drug dealing, with a Mafia leader attempting to take over the whole empire, but finding that he has powerful enemies in the form of a rival gangster and a female undercover cop. Fair.
Aka: DEADLY SIN, THE
A/AD   99 min Cut (3 sec)   VIDrel: IMPENT L/A   V

18
1990

CHINATOWN ***
Roman Polanski USA
Jack Nicholson, Faye Dunaway, John Huston, Perry Lopez, John Hillerman, Darrell Zwerling, Diane Ladd, Burt Young, Bruce Glover, Roman Polanski, Dick Bakalyan, Joe Mantell, Nandu Hinds, James Hong, Buelah Quo, Jerry Fujikawa
In 1937 Los Angeles, a private eye is hired by a woman to investigate her husband's love-affairs, and stumbles on a web of greed and corruption. A broodingly atmospheric film rather spoilt by an over-complex plot, though Nicholson as the detective conveys just the right tone of world-weary cynicism. Look out for Polanski who pops up briefly as a knife-wielding nutter. Followed some years later by THE TWO JAKES. AA: Screen/orig (Robert Towne).
THR   125 min (ort 131 min) wScrn   VIDrel: CIC/SONOP   V

15
1974

CHINESE BOXER, THE **
Jimmy Wang Yu HONG KONG
Jimmy Wang-Yu, Wu Szu-Yuan, Yang Chin-Chen, Lo Lieh, Wang Ping, Chen Hsing
One of the first of the kung fu films to reach the West, this one has the highly original plot of a disciple who has to avenge to death of his master.
Aka: HAMMER OF GOD; LUNG HU TOU
MAR   90 min Cut (1 min 3 sec – ort 94 min)
VIDrel: LUMI/SPEAR L/A   V

18
1968

CHINESE FEAST, THE **
Tsui Hark HONG KONG
Leslie Cheung Kwok-Wing, Kenny Bee (Chung Chun-To), Anita Yuen Wing-Yee, Law Kar-Ying, Xiung Xin-Xin (Hung Yan-Yan)
A former master chef now lives as a tramp, but gets a chance to use his skills when he helps a young cook who has entered a competition, with the first prize being 50,000,000 Hong Kong dollars or the ownership of a restaurant.
DRA   107 min wScrn   VIDrel: EAST/DISC   V

12
1995

CHINESE GHOST STORY, A ***
Ching Siu Tung HONG KONG
Leslie Chueng, Wong Tsu Hsien, Wu Ma, Joey Wang, Ching Siu-Tang
A young student who works as a travelling debt collector for a merchant, takes shelter in a temple and meets a beautiful woman who proves to be a trapped spirit. A highly imaginative Gothic tale full of wild special effects, broad humour and a touch of eroticism. Followed by a couple of sequels.
Aka: QIAN NU YOUHAN
FAN   93 min wScrn   VIDrel: MADE/RTM   V

18
1987

CHISUM ** PG
Andrew V. McLaglen USA 1970
*John Wayne, Forrest Tucker, Christopher George, Pamela McMyler,
Geoffrey Deuel, Ben Johnson, Glenn Corbett, Bruce Cabot, Andrew
Prine, Patric Knowles, Richard Jaeckel, Lynda Day (George), John
Agar, Lloyd Battista*
Unbelievable Western in which a cattle baron teams up with
Billy the Kid (!) to fight corrupt officials. Forgettable all the way
through until the climax, which is a laughable, utterly overdone
saloon fight.
WES   107 min (ort 110 min)   VIDrel: WHV   V/h

CHITTY CHITTY BANG BANG * U
Ken Hughes UK 1968
*Dick Van Dyke, Sally Ann Howes, Lionel Jeffries, Gert Frobe, Anna
Quayle, Benny Hill, James Robertson Justice, Robert Helpmann,
Heather Ripley, Adrian Hall, Barbara Windsor, Davy Kaye, Stanley
Unwin, Peter Arne, Victor Maddern*
An inventor rescues a derelict car which he then imbues with
fantastic properties, enabling it to fly. Sloppy special effects and
a lousy score make this kid's musical adventure a most forget-
table experience. Scripted by Roald Dahl and Hughes.
JUV   136 min (ort 145 min)   VIDrel: MGM/WHV   V/sh
Boa: novel by Ian Fleming.

CHOCOLAT *** PG
Claire Denis CAMEROONS/FRANCE/
WEST GERMANY 1988
*Mireille Perrier, Isaach de Bankole, Francois Cluzet, Giulia Boschi,
Cecile Ducasse, Jean-Claude Adelin, Jacques Denis, Emmet Judson
Williamson, Kenneth Cranham*
A white woman travelling in Africa recalls her childhood in
colonial times, when the whites were very clearly in charge and
the natives were restricted to playing a secondary role as the
servants of the former. A very personal and impressively
assured autobiographical effort. Music is by Abdullah Ibrahim.
DRA   99 min (ort 105 min) wScrn
VIDrel: ELPIC/POLYREC   V

**CHOICES OF THE HEART: THE MARGARET
SANGER STORY ***** 15
Paul Shapiro USA 1994
*Dana Delany, Henry Cezerny, Julie Khamer, Tom McCamus, Wayne
Robson, Yank Azman, Jeff Pustil, Kenneth Welsh, Rod Steiger, Ron
Hartman, Catherine Barroll, Nicu Branzea, Patrick Galligan, Sandra
Crijenica, Blake McGrath, Lachlan Murdoch*
Based on the true story of Margaret Sanger, a nurse who led the
fight to legalise abortion back in 1914, but paid a heavy price for
her campaigning when her husband left her and her children
were taken into care.
DRA   92 min mTV   VIDrel: ODY/SONOP   V/s

CHOIRBOYS, THE * 18
Robert Aldrich USA 1977
*Charles Durning, Louis Gossett Jr, Perry King, Clyde Kusatsu,
Stephen Macht, Tim McIntire, Randy Quaid, Chuck Sacci, Don
Stroud, James Woods, Burt Young, Robert Webber, Jeanie Bell, Blair
Brown, Michele Carey, Jim Davis, Joe Kapp*
Black comedy that looks at the lives of the officers in a police
department in Los Angeles, who relieve the pressure of their job
with some riotous forms of entertainment. A ponderous and
vulgar effort, about as witty as a cosh, and only slightly more
articulate. From the novel by Wambaugh, who is said to have
disowned this adaptation – and no wonder.
COM   119 min   VIDrel: 20TH/TECH   V/h
Boa: novel by Joseph Wambaugh.

CHOOSE ME **** 15
Alan Rudolph USA 1984
*Lesley Ann Warren, Keith Carradine, Genevieve Bujold, Rae Dawn
Chong, John Larroquette, Patrick Bauchau, John Considine, Edward
Ruscha, Jodi Buss, Gailard Sartain, Robert Gould, Sandra Will, Mike
E. Caplan, Russell Parr*
Story of lonely people in a big city. A radio counsellor ends up
sharing a flat with one of her confidantes and becomes involved
with her emotional problems. An inventive, witty and percep-
tive look at the roles people play in life, with fine performances
and a great script. Filmed in Los Angeles.
DRA   102 min (ort 114 min)   VIDrel: ARROW/RTM   V

CHOPPER CHICKS IN ZOMBIETOWN ** 18
Dan Hoskins USA 1989
Jamie Rose, Catherine Carlen, Lycia Naff, Vickie Frederick, Kristina

*Loggia, Don Calfa, Martha Quinn, Ed Gale, Gretchen Palmer, Nina
Peterson, Whitney Reis, Martha Quinn, Earl Boen, Billy Bob
Thornton, Dave Adams, Lewis Arquette*
A tough female biker gang come to a small town populated by
zombies created by its resident mad scientist and get involved in
combating this menace to humanity. A tepid horror spoof, full of
typical in-jokes for lovers of this genre, its opening brings to mind
THE WILD ONES and is easily the best thing in the entire film.
A/AD   86 min (ort 90 min)   VIDrel: TROMA/RTM   V

CHOPPING MALL ** 18
Jim Wynorski USA 1986
*Paul Bartel, Mary Woronov, Barbara Crampton, John Terlesky, Kelli
Maroney, Tony O'Dell, Russell Todd, Dick Miller, Karrie Emerson,
Suzee Slater, Nick Segal, Mel Welles, Gerrit Graham, Angela Aames,
Paul Coufos, Arthur Roberts*
Teenagers holding an impromptu party in a store's bed depart-
ment after closing, come under threat from three high-tech
security robots that a thunderstorm has caused to malfunction.
The plot is reminiscent of WESTWORLD as well as countless
teenagers-in-peril movies, and despite a few in-jokes and cameo
roles, quickly lapses into cliche. See also DEATH MACHINE
and HARDWARE.
Aka: KILLBOTS; R.O.B.O.T.
HOR   73 min (ort 77 min)   VIDrel: FIRST/SONOP L/A   V

CHORUS LINE, A ** PG
Richard Attenborough USA 1985
*Michael Douglas, Alyson Reed, Terrence Mann, Audrey Landers,
Blane Savage, Pam Klinger, Charles McGowan, Justin Ross, Gregg
Burge, Nicole Fosse, Janet Jones, Cameron English, Vicki Frederick,
Michael Belvins, Yamil Borges*
This failed attempt to recreate the feeling of the Broadway hit
musical, has performers auditioning for a place on a chorus line
and revealing their hopes and fears as they perform before
cynical director Douglas. A dismal filming of an excellent
musical.
MUS   112 min (ort 118 min)   VIDrel: BMGREC/BMGVID
V/sur
Boa: musical by Nicholas Dante and James Kirkwood.

CHORUS OF DISAPPROVAL, A ** PG
Michael Winner UK 1988
*Jeremy Irons, Anthony Hopkins, Prunella Scales, Sylvia Syms, Gareth
Hunt, Lionel Jeffries, Jenny Seagrove, Richard Briers, Patsy Kensit,
Alexandra Pigg*
Having moved to a small seaside town, a shy widower joins an
amateur theatrical troupe that's dominated as much by its
romantic intrigues as it is by the overbearing director.
Ayckbourn's stage play is clumsily translated to the screen,
where the solid performances of the players can mask neither
the uninspired direction nor the shallowness of the material.
COM   95 min (ort 100 min)   VIDrel: MGM/WHV   V
Boa: play by Alan Ayckbourn.

CHOSEN, THE **** PG
Jeremy Paul Kagan USA 1982
*Robby Benson, Maximilian Schell, Rod Steiger, Barry Miller, Hildy
Brooks, Ron Rifkin, Val Avery, Kaethe Fine, Robert Burke, Lonny
Price, Evan Handler, Douglas Warhit, Stuart Charno, Richard
Lifschutz, Clement Fowler*
Adapted from Potok's novel, this is a wonderfully evocative
recreation of post-war Brooklyn, in which a young Jewish boy
forms a friendship with an orthodox Chasidic boy. Steiger as a
Chasidic rabbi was never more convincing. The film contains
death camp sequences that some may find distressing.
DRA   102 min (ort 112 min)   VIDrel: ARENA   V
Boa: novel by Chaim Potok.

CHRIST STOPPED AT EBOLI **** PG
Francesco Rosi FRANCE/ITALY 1979
*Paolo Bonacelli, Alain Cuny, Lea Massari, Irene Papas, Gian Maria
Volonte, Francois Simon*
A period tale set in 1935, when a doctor and intellectual is exiled
to a backward village in southern Italy because of his opposi-
tion to Fascism. Initially unhappy at his predicament, he grows
accustomed to life in this remote region, and comes to under-
stand a good deal more about human nature. A handsome work,
leisurely in pace, that is part allegory, part political statement
and part simple recounting of Levi's true experiences.
Aka: CRISTO SI E FERMATO A EBOLI; EBOLI
DRA   213 min (2 cassettes)   VIDrel: ARTIF/20TH   V/h
Boa: autobiography of Carlo Levi.

CHRISTINE **   *18*
John Carpenter USA   1982
Keith Gordon, John Stockwell, Alexandra Paul, Robert Prosky, Harry Dean Stanton, Christine Belford, Roberts Blossom, William Ostrander, Steven Tash, David Spielberg, Malcolm Danare, Stuart Charno, Kelly Preston, Marc Poppel
A 1958 Plymouth is a demonic automobile that kills and maims all who come between it and its owner. A gruesome but flimsy horror yarn that starts off with promise but rapidly rolls downhill. Written by Bill Phillips. See THE MANGLER for another film using a Stephen King idea involving the inanimate brought to life by possession.
HOR   105 min (ort 111 min)
VIDrel: VCC/DISC/COLUM L/A   V/sur
Boa: novel by Stephen King.

CHRISTINE'S SECRET ***   *18*
Candida Royalle/R. Lauren Neimi USA   1985
Carol Cross, Taija Rae, Jake West, Chelsea Blake, Marita Ekberg, George Payne, Joey Silvera, Anthony Casino
A lonely woman who spends her summer vacations at the same country hotel finally finds the sexual partner she craves. An adult movie that's filmed in romantic soft focus and though fully explicit still strives to achieve a more sensual note in its portrayal of human sexuality – a typical Royalle hallmark. Unfortunately, despite a few interesting touches, it lacks a real story and its frank photography leaves too little scope to the imagination.
A   60 min (ort 72 min)   VIDrel: MIA/DISC   V

CHRISTMAS CAROL, A ***   *U*
Edwin L. Marin USA   1938
Reginald Owen, Gene Lockhart, Kathleen Lockhart, Terry Kilburn, Leo G. Carroll, Barry MacKay, Lynne Carver, Ann Rutherford, June Lockhart
With the arrival of several ghosts, the life of Scrooge the miser is transformed on the eve of Christmas. An early and carefully made adaptation of the famous Dickens classic, with good period detail. Owen replaced Lionel Barrymore, the latter's lameness preventing him from starring as Scrooge, to which role he gave substance in a number of Christmas radio broadcasts. June Lockhart appears in her screen debut.
FAN   69 min B/W   VIDrel: MGM/WHV   V
Boa: short story by Charles Dickens.

CHRISTMAS CAROL, A ****   *U*
Clive Donner USA   1984
George C. Scott, Nigel Davenport, Frank Finley, Edward Woodward, Lucy Gutteridge, Angela Pleasence, Roger Rees, David Warner, Susannah York, Anthony Walters, Timothy Bateson, Michael Carter, Michael Gough
An excellent version of this famous tale, with Scott in fine form as a memorable Scrooge. A beautifully made film offering much to delight, not least being the film location – Shrewsbury in England. The excellent script is by Roger O. Hirson.
DRA   100 min mTV cC   VIDrel: 20TH/TECH   V/sur
Boa: short story by Charles Dickens.

CHRISTMAS IN CONNECTICUT **   *U*
Arnold Schwarzenegger USA   1992
Dyan Cannon, Kris Kristofferson, Tony Curtis, Richard Roundtree, Kelly Cinnante, Gene Lithgow, Viivan Bonnell, Jimmy Workman, David Arnott, Toni Attell, Jenee Bandler, Bob Braun, Sonny Carl Davis, Judy Forrester, Mary Watson
Updated remake of the 1945 film, with Dyan as a TV hostess and supposedly perfect homemaker who has to cook a traditional Christmas dinner for a Forest Ranger. Unfortunately, this media event threatens to expose the fact that she is completely unable to cook. A bland and lightweight effort that fails to amuse.
COM   93 min mCab   VIDrel: FIRST/SONOP L/A   V

CHRISTMAS MOUSE, THE ***   *U*
Robin Crichton CZECHOSLOVAKIA/UK/USA   1988
Gregor Fisher, John Cairney, Bill McCue, Jack McKenzie, Phil McCall, Richard Blane plus Lynn Redgrave (introduction and narration)
A charming seasonal tale of how a little mouse played a crucial part in the events that led to the creation of the world-famous carol "Silent Night", in Austria in 1887.
Aka: SILENT MOUSE; SILENT NIGHT, HOLY NIGHT
JUV   55 min mTV   VIDrel: RELREC   V

CHRISTMAS ON DIVISION STREET **   *PG*
George Kaczender USA   1991
Fred Savage, Hume Cronyn, Badja Djola, Jim Byrnes, Cloyce Morrow, Casey Ellison, Kenneth Welsh
An upper-class school-kid makes friends with an eccentric eighty-year-old hobo and the latter shows him a side of America he never knew existed, taking him on a tour of poverty, degradation and despair. Loosely based on a true story, the script is by Barry Morrow.
DRA   93 min mTV   VIDrel: ODY/SONOP   V/sh

CHRISTMAS REUNION, A **   *(PG)*
David Hemmings USA   1994
James Coburn, Edward Woodward, Meredith Edwards, Myfanwy Talog, Fraser Caines, Geriant Morgan, Melanie Walters, Gweirydd Gwyndaf, Noel Williams, Nia Elias, Deiniol Wyn Rees, John Davis, Penelope Richards, Andrew Felindre
A young boy recently orphaned is taken to live with his grandparents but feels unwanted and runs away from home. However, when he hears a story told by a man dressed in a Santa costume, both he and his grandfather manage to reach a new understanding. A reasonably well-made seasonal tale.
JUV   92 min   SATrel: SKY MOVIES
Boa: novel by A. Llew Jones.

CHRISTMAS ROMANCE, A **   *U*
Sheldon Larry USA   1995
Olivia Newton-John, Gregory Harrison, Chloe Lattanzi, Stephanie Sawyer, Tom Heaton, Stephen E. Miller, Brent Stait, Susan Astley, Tom McBeath, Anna Ferguson, Teryl Rothery, Melody Ryane
Having just split up with her husband, a mother of two has a visit from the vice-president of the local bank to tell her that her home is to be repossessed, thus making her and her children homeless. Fortunately (and this could only happen in a movie) an accident in a storm forces the bank official to spend the night at her house, thus paving the way for a romance and obligatory happy resolution. One of those feel-good movies that leaves no cliche uncliched.
DRA   91 min mTV   VIDrel: ODY/SONOP   V/sh
Boa: novel by Maggie Davis.

CHRISTMAS STORY, A ***   *PG*
Bob Clark USA   1983
Peter Billingsley, Melinda Dillon, Darren McGavin, Ian Petrella, Scott Schwartz, Tedde Moore, R.D. Robb, Yano Anaya, Zack Ward, Jeff Gillen, Colin Fox, Paul Hubbard, Les Carlson, Jim Hunter, Jean Shephard (narration)
An account of a young boy growing up in the 1950s and of the many devious schemes he invents in order to be given an air rifle for Christmas. A charming and warm-hearted tale, partly based on Shepherd's novel, and narrated by her in the first person. Directed, believe it or not, by the man who gave us PORKY'S, and followed by the sequel MY SUMMER STORY.
COM   89 min (ort 98 min)   VIDrel: MGM/WHV   V
Boa: novel In God We Trust, All Others Pay Cash by Jean Shepherd.

CHRISTOPHER COLUMBUS: THE DISCOVERY **   *PG*
John Glen USA   1992
George Corraface, Tom Selleck, Marlon Brando, Rachel Ward, Robert Davi, Catherine Zeta Jones, Oliver Cotton, Benicio Del Toro, Mathieu Carriere, Manuel de Blas, Glyn Grain, Peter Guinness, Nigel Terry, Nitzan Sharron, Hugo Blick
A film that tries to be all things to all men: in short, an old-fashioned swashbuckling adventure, an accurate historical account and a Politically Correct lecture regarding the peaceful peoples of the Caribbean. Despite all this (and some woefully miscast actors, most especially Brando as Torquemada) there are a few good moments to be had, but they are generally few and far between.
A/AD   115 min (ort 121 min)
VIDrel: POLY/POLYREC/BRAVE L/A   V

CHROME SOLDIERS **   *15*
Thomas J. Wright USA   1992
Gary Busey, Ray Sharkey, William Atheron, Norman Skaggs, Nicholas Guest, Kim Robillard, Yaphet Kotto, D. David Morin, Tony Soper, Shawna Schuh, Joe Ivy, Karen Trumbo, Dylan Taylor, Gary Lee Dansenburg, Robert James White
Five Vietnam veterans come together to take on a group of drug dealers who are terrorising a small town and have murdered one of their friends. A standard by-the-numbers actioner with the requisite violence and shallow characterisations.
A/AD   88 min (ort 92 min)   VIDrel: CIC/SONOP   V

**CHRONICLE OF A LOVE ** **   15
Michelangelo Antonioni ITALY   1950
*Lucia Bose, Massimo Girotti, Ferdinando Sarmi, Gino Rossi, Marika
Rowsky*
The director's first film tells of an adulterous wife who plots
with her lover to do away with her unsuspecting husband. But
an unexpected turn of events casts a different light on their
plans.
Aka: CRONACA DI UN AMORE; STORY OF A LOVE AFFAIR
DRA   96 min B/W   VIDrel: ARROW/RTM   V

**CHRONICLES OF NARNIA, THE:
PRINCE CASPIAN **   U
Alex Kirby UK   1989
*Sophie Wilcox, Sophie Cook, Jonathan Scott, Richard Dempsey,
Barbara Kellerman, Michael Aldridge*
The second part of a competent BBC adaptation of the "Narnia"
books of C.S. Lewis that began with "The Lion, The Witch And
The Wardrobe". In this adventure the four children arrive in
the magical land of Narnia and find that King Miraz is plotting
to take over the kingdom and that Prince Caspian has fled for
his safety.
JUV   56 min mTV   VIDrel: BBC/TECH   V/h
Boa: novel by C.S. Lewis.

**CHRONICLES OF NARNIA, THE: THE LION, THE
WITCH AND THE WARDROBE **   U
Marilyn Fox UK   1988
*Richard Dempsey, Sophie Cook, Jonathan R. Scott, Sophie Wilcox,
Maureen Morris, Michael Aldridge, Jeffrey Perry, Barbara
Kellermann, Big Mick, Kerry Shale, Lesley Nicol, Martin Stone,
Maureen Morris, Ken Kitson*
The first in a series of TV adaptations of the Narnia novels of
C.S. Lewis, this tale follows the exploits of the children as they
first find themselves brought into the magical land of Narnia by
the power of Aslan. A reasonably competent effort, flawed by
weak acting. See also THE LION, THE WITCH AND THE
WARDROBE.
JUV   162 min (2 cassettes – ort 180 min) mTV
VIDrel: BBC/TECH   V/h
Boa: novel by C.S. Lewis.

**CHRONICLES OF NARNIA, THE: THE SILVER
CHAIR **   U
Alex Kirby UK   1990
*David Thwaites, Camilla Power, Warwick Davies, Mike Edmonds,
Richard Henders, Roy Boyd, Barbara Kellerman, Geoffrey Russell,
Tom Baker, Ailsa Berk*
Another entry in this somewhat variable set of TV adaptations,
that has our children enjoying a new Narnian adventure, which
involves them rescuing a Prince from the clutches of a witch
known as the Green Lady. Fortunately, they obtain a little help
from a strange creature known as Puddleglum.
JUV   155 min (2 cassettes – ort 180 min) mTV
VIDrel: BBC/TECH   V/h
Boa: novel by C.S. Lewis.

**CHRONICLES OF NARNIA, THE: THE VOYAGE OF
THE DAWN TREADER **   U
Alex Kirby UK   1989
*Sophie Wilcox, Sophie Cook, Jonathan Scott, Richard Dempsey,
Barbara Kellerman, Michael Aldridge, David Thwaites, Samuel West,
Warwick Davis, Guy Fithen, John Hallam, Neale S. McGrath, Jack
Purvis, Kenny Baker*
The third part of this adequate BBC adaptation. In this adven-
ture three of the children journey to Narnia and sail with Prince
Caspian in search of the seven missing Lords of Narnia.
JUV   108 min (ort 180 min) mTV   VIDrel: BBC/TECH   V/h
Boa: novel by C.S. Lewis.

**C.H.U.D. **   18
Douglas Cheek USA   1984
*John Heard, Daniel Stern, Kim Greist, Brenda Currin, Justin Hall,
Michael O'Hare, Christopher Curry, George Martin, John Ramsey,
Eddie Jones, Graham Beckel, Ruth Maleczech, William Joseph
Raymond, Rocco Siclari, Henry Yuk*
A strange colony of sub-humans live in the sewers of New York
but come up to the surface at night to enjoy some human flesh.
Title is an acronym for Cannibalistic Humanoid Underground
Dwellers. A nasty little film that was followed by a slightly
better sequel.
HOR   83 min (ort 90 min)   VIDrel: MED/POLY L/A   V/h

C.H.U.D. 2 *   15
David Irving USA   1988
*Brian Robbins, Tricia Leigh Fisher, Bill Calvert, Gerrit Graham,
Robert Vaughn, Jack Riley, Norman Fell, Larry Linville, Bianca
Jagger, Larry Cedar, Judd Omen, Sandra Kerns, June Lockhart, Rich
Hall, Robert Symonds*
A group of students steal the body of one of the creatures from
the first movie as a prank. It's soon thawed out and walking the
streets, turning the townsfolk into similar zombie-like creatures.
A crackpot colonel tries to keep the lid on things as our flesh
eating monster starts to get out of control, in this enjoyable but
production-line piece of nonsense.
Aka: C.H.U.D. 2: BUD THE CHUD
HOR   84 min   VIDrel: FIRST/SONOP L/A   V

CHUNGKING EXPRESS *   12
Wong Kar-Wei (Wang Jiawei) HONG KONG   1994
*Brigitte Lin (Lin Qingxia), Takeshi Kaneshiro, Tony Leung (Leung
Chaowei), Faye Wong (Wang Jinwen), Valerie Chow (Shou Jialing)
"Piggy" Chan (Chen Jinquan), Guan Lina, Huang Zhimeng, Liang
Zhen, Zuo Songshen*
Two stories that revolve around cops, drugs and dreams. In the
first a cop meets a woman in a late night bar, but is unaware that
she is a heroin smuggler. In the second a broken-hearted cop fails
to realise that a girl working at the counter of a fast-food joint has
a crush on him. Hard to define, this is as much a sad little character study as it is an urban action film, with an improvised and
natural feel about it that brings to mind movies by Cassavetes.
Aka: CHONGQING SENLIA
THR   100 min   VIDrel: ICAPRO/MANGA   V

**CHURCH, THE **   18
Michele Soavi ITALY   1989
*Asia Argento, Tomas Arana, Hugh Quarshie, Feodor Chaliapin,
Barbara Cupisti*
The renovation of a gothic cathedral that was built on the site
of a mass murder (when the population of an entire village was
accused of witchcraft) leads to an orgy of violence and gore. An
extremely explicit tale, scripted and produced by famous horror
director Dario Argento, whose inventive feel for this genre
exerts an unmistakable influence.
Aka: LA CHIESA
HOR   97 min (ort 110 min)   VIDrel REFLEC/FIRST   V

**C.I.A. – CODENAME ALEXA **   18
Joseph Merhi USA   1992
*Lorenzo Lamas, Kathleen Kinmont, O.J. Simpson, Alex Cord, Michael
Bailey-Smith, Jeff Griggs, Stephen Quadros, Pamela Dixon, Shonna
Cobb, H. Ray Huff, Charles G. Menshack, Dan Tullis Jr., Joe Mehana,
Jim Ishida, Jerry Jacobs*
An agent out to crack an international crime ring decides on a
dangerous strategy when he tries to subvert the loyalty of the
crime boss's beautiful female assistant. A conventional action
thriller with the usual dose of excessive violence. Followed by
CODENAME: ASSASSIN.
Aka: C.I.A. – TARGET ALEXA
A/AD   94 min   VIDrel: 20VIS/SONOP   V/h

**CIDER WITH ROSIE **   PG
Claude Watham UK   1971
Rosemary Leach, Peter Chandler
Account of a childhood spent in a tiny village in the Cotswolds,
this attractive adaptation of Lee's novel covers the usual
coming-of-age ground, including the pain and joy of early
sexual encounters. One of the first film dramas to be produced
by the BBC.
DRA   94 min mTV   VIDrel: PARADOX/TOTAL   V/h
Boa: novel by Laurie Lee.

CINCINNATI KID, THE *   15
Norman Jewison USA   1965
*Steve McQueen, Edward G. Robinson, Karl Malden, Ann-Margret,
Tuesday Weld, Cab Calloway, Rip Torn, Joan Blondell, Jack Weston,
Jeff Corey, Theo Marcuse, Milton Sleezer, Karl Swenson, Emile
Genest, Ron Soble, Irene Tedrow, Midge Ware*
In New Orleans in the 1930s a bunch of roving card-sharks get
together for a big game, with a young challenger taking on an
older champion. Inevitably reminding one of THE HUSTLER,
though as with that film, some contrived romantic interest
detracts from the story. Scripted by Ring Lardner Jr and Terry
Southern, with Jewison replacing Sam Peckinpah as director.
DRA   97 min (ort 101 min)   VIDrel: MGM/WHV   V
Boa: novel by Richard Jessup.

CINDER PATH, THE ***   PG
Simon Langton UK   1994
Lloyd Owen, Catherina Zeta Jones, Maria Miles, Polly Adams, Rosalind Ayres, Tom Bell, Philip Corbitt, Victoria Scarborough, Gary Sefton, Osmond Bullock, Lucy Akhurst, John Warnmby, Rupert Wickham, Allison Jupp, Ralph Ineson
An ordinary soldier becomes an officer and gets married, but does not find happiness easy to achieve. Another handsome adaptation of a Cookson novel, always good to look at if a trifle melodramatic at times.
DRA 151 min (ort 165 min) mTV VIDrel: FOCUS/DISC V
Boa: novel by Catherine Cookson.

CINDERELLA ***   U
Wilfred Jackson/Hamilton Luske/Clyde Geronimi USA 1950
Voices of: Ilene Woods, William Phipps, Eleanor Audley, Rhoda Williams, Lucille Bliss, Verna Felton, James MacDonald, Luis Van Rooten, Don Barclay
A lively adventure that basically offers a simple recounting of this famous fairytale with a few comical overtones and extra characters, notably two mice who set out to assist our heroine, and a villainous cat who does not. Occasionally short on both ideas and action, it's a worthy rather than a brilliant work. Songs include: "A Wish Is A Wish Your Heart Makes" and "Bibbidi Bobbidi Boo".
ANIM 72 min (ort 75 min) VIDrel: WDV/SONOP L/A V
Boa: short story by Charles Perrault.

CINDERELLA LIBERTY ***   15
Mark Rydell USA   1973
James Caan, Marsha Mason, Eli Wallach, Kirk Calloway, Burt Young, Bruce Kirby Jr, Allyn Ann McLerie, Dabney Coleman, Sally Kirkland, Fred Sadoff, Allan Arbus, Jon Korkes, Don Calfa, Ted D'Arms, Diane Schenker, David Proval
Whilst on shore leave, a good-natured but rather simple sailor meets a streetwise prostitute with a black son and falls in love with her. An awkward blend of romance and squalor, held together by sensitive performances from the leads.
DRA 111 min (ort 117 min) VIDrel: 20TH/TECH V
Boa: novel by Darryl Ponicsan.

CINDERELLA 2000 **   18
Al Adamson USA   1977
Catherine Erhardt, Vaughn Armstrong, Jay B. Larson, Erwin Fuller, Bhurni Cowans, Rena Harmon, Adina Ross, Eddie Garetti, Olivia Michelle, Art Cacaro, Sheri Coyle, John Appleton, Steve Puckett, Don Birkley, Harry Kistner
In the year 2047 the Earth is run by computers and robots. Sex is strictly controlled and only allowed for a fortunate few. This forms the background to a space-age revamping of the Cinderella story.
A 89 min VIDrel: IMPENT V

CINEMA PARADISO ****   PG/15
Giuseppe Tornatore FRANCE/ITALY   1989
Philippe Noiret, Jacques Perrin, Antonella Attili, Enzo Cannavale, Salvatore Cascio, Isa Danieli, Leo Gullota, Marco Leonardi, Pupella Maggio, Agnese Nano, Leopoldo Trieste, Leo Gullotta, Tano Cimarosa, Nicola Di Pinto, Roberta Lena
Set in Sicily just after WW2, this loving tribute to the cinema explores the fascination a local cinema exerts on a youngster, where he enjoys a friendship with the kindly projectionist. A warm, poignant and moving film that recaptures some of the magic of the picture palaces of old. First released at 155 minutes but prudently shortened, it won the Special Jury Prize at Cannes in 1989. The script is by Tornatore with music by Ennio Morricone. AA: Foreign.
Aka: NUOVO CINEMA PARADISO
DRA 118 min wScrn; 167 min wScrn (15 cert special edition – ort 175 min) VIDrel: TART/20TH V/dm

CIRCLE OF FRIENDS ***   15
Pat O'Connor EIRE/USA   1995
Minnie Driver, Chris O'Donnell, Saffrom Burrows, Colin Firth, Geraldine O'Rawe, Allan Cumming, Aidan Gillen, Mick Lally, Ciaran Hinds, Seamus Forde, Tom Hickey, Tony Doyle, Britta Smith, John Kavanagh, Ingrid Craigie, Major Lambert
In 1957 three female friends attend college and begin to experience all the problems associated with growing up, especially the conflict between their Catholic upbringing and their own sexual feelings. A candid and extremely well acted comedy-

drama, with enough flashes of bright wit to keep the maudlin sentiments at bay.
DRA 98 min (ort 112 min) VIDrel: POLY/POLYREC V/sh
Boa: novel by Maeve Binchy.

CIRCUITRY MAN **   15
Stephen Lovy USA   1990
Jim Metzler, Dana Wheeler-Nicholson, Vernon Wells, Lu Leonard, Dennis Christopher, Vernon Wells, Andy Goldberg, Paul Willson, Manu Tupou, Garry Goodrown, Jerry Tondo
In the near future after WW3, computer chips that contain artificial fantasies are a valuable item and a woman accepts a mission to take a briefcase full to them to New York. Along the way she runs into a variety of strange creatures, in this imaginative and visually impressive tale that is unfortunately let down by its weak plot.
FAN 88 min VIDrel: 20VIS/SONOP V

CIRCUMSTANCES UNKNOWN **   15
Robert Lewis USA   1995
Judd Nelson, William R. Moses, Isabel Glasser, Rhys Huber
A married couple are unaware that their jeweller friend is murderous psychopath who has targeted them as his next victims, his penchant being to create wedding rings for those he intends to dispatch. However, he meets his match in his latest victim, and a nasty battle of wits ensues as he attempts to silence a witness to his activities.
THR 87 min (ort 91 min) VIDrel: CIC V
Boa: novel by Jonellen Heckler.

CIRCUS BOYS ***   15
Kaizo Hayashi JAPAN   1989
Hiroshi Mikami, Moe Kamura, Xia Jian, Michiru Akiyoshi, Yuki Asayama, Sanshi Katsura, Haruko Wanibuchi, Yoshio Harada, Shiro Sano, Yukio Yamato, Maki Ishikawa, Akira Oizumi, Akaji Maro, Chuck, Masahi Okuda, Baiken Jukkanji
Two brothers grow up in a circus, but when one is injured saving the other from an accident he is crippled and leaves, becoming a dishonest peddler and eventually running off with a gangster's moll, an action that leads to his death. An odd film that fails to find a clear direction and suffers from a lack of realism (especially during the circus sequences) yet has moments of poetry and power.
Aka: NI JU-SEIKI SHONEN DOKUHON
DRA 102 min (ort 106 min) VIDrel: ICAPRO/MANGA V/sh

CISCO KID, THE **   PG
Luis Valdez USA   1993
Jimmy Smits, Cheech Marin, Sadie Frost, Ron Pearlman, Bruce Payne, Tony Amendola
An updated version of this established hero, set in Mexico in the 1860s, where he and his faithful sidekick do their patriotic duty by resisting American gunslingers as well as the French occupying forces of Maximillian. An enjoyable little comedy Western, with a strong emphasis on humour and some engaging performances.
COM 91 min (ort 96 min) mTV VIDrel: FIRST/SONOP V

CITIZEN COHN ***   15
Frank R. Pierson USA   1992
James Woods, Joe Don Baker, Joseph Bolonga, Ed Flanders, Frederic Forrest, Lee Grant, Pat Hingle, Josh McMartin, Josef Sommer, Tovah Feldshuh, Frances Foster, Allen Garfield, Daniel Benzali, John Finn, David Marshall Grant
As a man lies dying from AIDS, he remembers his past and his infamous career as the associate of senator Joseph McCarthy. Although Jewish and homosexual himself, he tried to disguise these facts through an open contempt for both these groups. A fascinating biography of Roy Cohn that is beautifully acted by Woods.
DRA 107 min (ort 112 min) mCab VIDrel: WHV V/sh
Boa: book by Nicholas Von Hoffman.

CITIZEN KANE ****   U
Orson Welles USA   1941
Orson Welles, Joseph Cotten, Agnes Moorehead, Everett Sloane, Dorothy Comingore, Ray Collins, George Coulouris, Ruth Warwick, William Alland, Paul Stewart, Erskine Sanford, Fortunio Bonanova, Gus Schilling, Philip Van Zandt
Outstanding tale of a megalomaniac press baron, told in flashback form with a reporter investigating the circumstances of the

man's death. A powerful film, its thin story is sustained by brilliant direction and Gregg Toland's innovative photography. Newspaper tycoon William Hearst saw the film as a veiled personal attack, but it is his influential columnist Louella Parsons who ensured his papers denied it publicity. AA: Screen/orig (Herman J. Mankiewicz/Orson Welles).
DRA   197 min (includes documentary "Reflections Of Citizen Kane") B/W   VIDrel: 4-FRONT/POLYREC   V

CITIZEN X **   18
Chris Gerolmo USA   1994
Donald Sutherland, Stephen Rea, Max Von Sydow, Jeffrey DeMunn, Joss Ackland, Imelda Staunton, John Wood
Gritty, realistic and entirely convincing account of a Russian detective who spent eight years attempting to catch a shadowy mass murderer who notched up a total of fifty-two victims between 1982 and 1990. This arduous assignment was rendered even more difficult by the politically motivated refusal of some Soviet officials to acknowledge that such a phenomenon could exist in the "workers' paradise". A really first-rate tale, filmed on location in Hungary.
THR   107 min mCab   VIDrel: MOSAIC/COLUM   V/sur
Boa: book The Killer Department by Robert Cullen.

CITY BOY **   (PG)
John Kent Harrison USA   1992
James Brolin, Christian Campbell, Wendel Meldrum, Sarah Chalkie, Christopher Bolton, Gene Heck, Don McKay, Alan C. Peterson, Victor A. Young, Ken Camroux, David Glyn-Jones, Matthew Walker
A young boy travels from Chicago to the Northwest at the turn of the century in search of his father and gets a job guarding a valuable forest. When he makes two new friends who have radically different views on the need to preserve the forests he soon finds himself caught in a conflict of loyalties.
JUV   94 min (ort 120 min)   SATrel: SKY MOVIES
Boa: story Freckles by Gene Stratton Porter.

CITY COPS **   15
Lau Kar Wing (Kar-Wing Lau) HONG KONG   1990
Cynthia Rothrock, Mark Huston, Ken Goodman, Michiko Nishikawa, Suki Kuan, Kiu-Wai Miu, Fu-on Shing, Ken Tong, Fung Wu, Kwong-Chin tsang, Sou-leung O, Wai-Lung Ho, Chuen Luk, Me Lin Yu
An American female cop, skilled in the martial arts, is sent to Hong Kong to extradite an informant who fled there from a Japanese lady assassin who has accepted a contract to kill him. A standard Rothrock vehicle with little plot but plenty of action and more than a little broad humour.
Aka: BEYOND THE LAW
MAR   88 min (ort 92 min) dubbed
VIDrel; MIA/DISC; ENCORE (LV only)   V LV

CITY FOR CONQUEST **   PG
Anatole Litvak USA   1940
James Cagney, Ann Sheridan, Frank Craven, Donald Crisp, Arthur Kennedy, Frank McHugh, George Tobias, Elia Kazan, Anthony Quinn, Jerome Cowan, Lee Patrick, Blanche Yurka, Thurston Hall
A New York truck driver takes up boxing, in the hope of making a better life for himself and his composer brother, but is injured in the course of a fight and loses his sight. Corny and superficial it may be, but both in terms of acting and production it is hard to fault. This was Kennedy's first adult role, and also a rare appearance before the cameras for Kazan as a small-time gangster. See also PARADISE ALLEY, a kind of 1978 update.
DRA   94 min (ort 105 min) B/W   VIDrel: MGM/WHV L/A V
Boa: novel by Aben Kandel.

CITY HALL **   15
Harold Becker USA   1996
Al Pacino, John Cusack, Bridget Fonda, Danny Aiello, Martin Landau, David Paymer, Tony Franciosa, Richard Schiff, Lindsay Duncan, Nestor Serrano, Mel Winkler, Lauren Velez, Chloe Morris, Ian Quinlan, Roberta Peters, Angel David
A naive but idealistic deputy mayor uncovers a web of corruption that spreads across the entire city, and sets about doing what he can to dismantle it. A gripping thriller of strong performances (most especially Pacino's as the town's charismatic mayor) and inventive plotting, that is a little hurt by over-complexity, the script having gone through numerous revisions. See also THE GOOD POLICEMAN for another one of those battle-against-corruption sagas.
THR   107 min (ort 111 min)   VIDrel: COLUM/SONOP V/sur

CITY HEAT **   15
Richard Benjamin USA   1984
Clint Eastwood, Burt Reynolds, Jane Alexander, Madeline Kahn, Irene Cara, Richard Roundtree, Rip Torn, Tony LoBianco, William Sanderson, Nicholas Worth, Robert Davi, Jude Farsee, John Hancock, Tab Thacker, Art La Fleur, Jack Nance
A mean cop teams up with a slick private eye to solve a murder case in Kansas City in the 1930s. Occasionally funny, but most of the time this brutish and simple-minded gangster-versus-cop spoof is just plain nasty. Written by Sam O. Brown (Blake Edwards).
COM   93 min   VIDrel: MGM/WHV   V/sur

CITY HUNTER **   12
Wong Ching HONG KONG   1993
Jackie Chan, Goto Kumiko, Maggie Cheung, Joey Wong, Chingmy Yau, Richard Norton, Gary Daniels
A private eye tracks a runaway girl to a cruise liner that is attacked by a gang of thieves. A caper tale with plenty of heavy-handed humour, including a sequence where our hero assumes the identities of characters from a popular Japanese cartoon and video game. Made in the style of an animated comic-strip, this is often a very funny film, though after thirty minutes or so it does tend to get rather repetitive.
MAR   92 min (ort 102 min)   VIDrel: MIA/DISC/IMPENT V/sur

CITY LIMITS **   15
Aaron Lipstadt USA   1984
Darrell Larson, John Stockwell, Kim Cattrall, Rae Dawn Chong, John Diehl, Don Opper, Robby Benson, James Earl Jones, Danny De La Paz, Norbert Weisser, Pamela Ludwig, Tony Plana, Kelly Stuart, Matt Goulish, Marcia Holley
Futuristic urban adventure movie, in which a plague wipes out the world's adult population, and gangs of young thugs are left to battle it out. A messy and incoherent disaster of a movie, set fifteen years into the future and best left there. Narration is by Jones. See also GAS.
FAN   85 min   VIDrel: POLY L/A   V

CITY OF HOPE **   15
John Sayles USA   1991
Vincent Spano, Joe Morton, Tony Lo Bianco, Barbara Williams, Stephen Mendillo, Chris Cooper, Charlie Yanko, Jace Alexander, Todd Graff, Scott Tiler, John Sayles, Frankie Faison, Gloria Foster, Tom Wright, Angela Bassett
Set in a fictitious New Jersey city of the present-day, the various disparate strands of this complex drama involve an alienated young man, his wealthy businessman father and a black councillor who has come under intense pressure to evict the occupants of a slum tenement block, a decision which leads to a fatal act of arson. Several interlinked stories now unfold simultaneously, in this difficult but compelling and rewarding work. Screenplay is by Sayles.
DRA   124 min (ort 130 min)   VIDrel: VCC/DISC   V/sur

CITY OF JOY **   15
Roland Joffe FRANCE/UK   1992
Patrick Swayze, Om Puri, Pauline Collins, Shabana Azmi, Ayesha Dharker, Santu Chowdhury, Imran Badsah Khan, Art Malik, Nabil Shaban, Debtosh Gupta, Suneeta Sengupta, Mansi Upadhyay, Shyamanand Jalan, Shyamal Sengupta
Lightweight drama set in the slums of Calcutta in which Swayze stars as an idealistic young doctor who chucks up his career and flees to India after he loses a patient. Once there, he wanders about fairly aimlessly until work at a local clinic run by Pauline Collins beckons. Of rather more interest is the parallel story of an impoverished farmer (Puri giving a wonderful performance) who uproots his entire family and brings them to the city.
DRA   129 min (ort 135 min)   VIDrel: WHV   V/sur
Boa: book by Dominique LaPierre.

CITY OF LOST CHILDREN, THE **   15
Jean-Pierre Jeunet/Marc Cano FRANCE/GERMANY/SPAIN 1995
Daniel Emilfork, Ron Perlman, Judith Vittet, Dominique Oinon, Jean-Claude Dreyfus, Genevieve Brunet, Odile Mallet, Mireille Mosse, Serge Merlin, Francois Hadji-Lazaro, Rufus, Ticky Holgado, Jean-Louis Trintignant, Dominique Bettenfeld
Strange, surreal fantasy that tells of a man who ages prematurely because of his inability to dream, so he kidnaps children and takes them to an abandoned oil-rig, where he steals their

dreams. A gothic, adult fairytale of flair and imagination, highly ambitious and not always effective, yet in its way quite unique.
Aka: LA CITE DES ENFANTS PERDUS
FAN   108 min (ort 114 min) wScrn   VIDrel: EIV/SONOP   V

CITY OF SADNESS, A ****   15
Hou Hsiao-Hsien TAIWAN   1989
Li Tine-Lu, Chen Sown-Yun, Kao Jai, Tony Leung
Story of a Taiwanese family, and their fight for survival as the country is torn apart in the wake of the Japanese surrender at the close of WW2. Moving from 1945 when Taiwan became part of China to 1949 when the Nationalists fled there, this is a detailed, painful and demanding work, whose focus is mostly restricted to the domestic difficulties of the central characters. Winner of the Golden Lion at the 1989 Venice Film Festival.
Aka: BEIQING CHENGSI
DRA   155 min   VIDrel: ARTIF/20TH   V

CITY OF SIN **   18
Henri Pachard USA   1992
Jeanna Fine, Deidra Holland, Raven, Johny Nineteen, Ona Zee, John Dough, Tom Chapman, Joey Silvera, Carl Esser, Henri Pachard, Debbie Diamond, T.T. Boy, Trixie Taylor, Jeanna Wells, Charlie W., Chris R., Stacy G., Michele D. Enzo S.
A ruthless mayor stops at nothing to get hold of a list of customers of a high-class brothel, as it includes the names of some of his political rivals who just like him, are regular customers. Very much a standard sex film (and not an especially interesting one) for all the effort made to dress the whole thing up as if it were a satire on political corruption.
A   65 min (ort 80 min)   VIDrel: GROHOM/MAXSCAN   V

CITY OF THE LIVING DEAD **   18
Lucio Fulci ITALY   1980
Christopher George, Janet Agren, Katriona MacColl, Carlo de Mejo, Antonella Interlenghi, Giovanni Lombardo Radice, Daniela Doria, Luca Paismer, Fabrizio Jovine, Lynda Day George, Robert Sampson
A priest commits suicide in a cemetery, and in New York a young girl goes into a trance. This heralds a somewhat serious turn of events, in which the dead rise up in a small Massachusetts town and terrorise the living. A gory and carelessly made shocker, ostensibly set in H.P. Lovecraft's town of Dunwich, but actually shot in Italy.
Aka: FEAR; FEAR IN THE CITY OF THE LIVING DEAD; GATES OF HELL, THE; PAURA NELLA CITTA DEI MORTI VIVENTI; TWILIGHT OF THE DEAD
HOR   86 min (ort 94 min) Cut   VIDrel: VIPCO/SGSVID   V/h

CITY OF WOMEN ***   18
Federio Fellini FRANCE/ITALY   1980
Marcello Mastroianni, Anna Prucnal, Bernice Stegers, Ettore Manni, Donatella Damiani
A womanising businessman is tricked into getting off the train he is travelling on and finds himself in a city populated almost entirely by women. After a series of ever more bizarre encounters, he eventually awakens to discover that the entire episode was a dream. A rather overblown and simplistic exploration of the hero's (and by extension the director's) fears with regard to militant feminism. Disappointing.
Aka: LA CITTA DELLE DONNE
DRA   139 min wScrn   VIDrel: ARTIF/20TH   V/h

CITY ON FIRE **   18
Ringo Lam HONG KONG   1987
Chow Yun Fat, Danny Lee, Roy Cheung, Carrie Ng, Sun Yueh
An undercover cop infiltrates a gang of jewel thieves, in this well-made but rather violent crime actioner.
A/AD   101 min wScrn   VIDrel: MADE/RTM   V

CITY SLICKERS ***   15
Ron Underwood USA   1990
Billy Crystal, Daniel Stern, Bruno Kirby, Patricia Wettig, Helen Slater, Jack Palance, Noble Willingham, Tracey Walter, Josh Mostel, David Paymer, Bill Henderson, Jeffrey Tambor, Phill Lewis, Kyle Secor, Dean Hallo, Jane Alden
Three middle-aged pals who feel their years creeping up on them sign on as hands for a two-week cattle drive as a way of both clearing their minds and regaining their zest for the simple things in life. A comedy that pretends to be a little bit more profound than it really is, but is consistently entertaining, with

nice performances and some very funny lines. AA: S. Actor (Palance). A rather weak sequel followed.
COM   109 min (ort 114 min)   VIDrel: VCC/DISC; PION (LV only)   V/sur LV

CITY SLICKERS 2 **   12
Paul Weiland USA   1993
Billy Crystal, Daniel Stern, Jon Lovitz, Jack Palance, Patricia Wettig, Bill McKinney, Pruitt Taylor Vince, Noble Willingham, David Paymer, Josh Mostel, Beth Grant, Lindsay Crystal, Jayne Meadows, Alan Charop, Kenneth S. Allen, Helen Siff
A weak sequel to the first film sees our three buddies reunited and back out West in order to search from some gold that is also being sought by the twin brother of the trail boss. Once again, the performances are both winning and faultless but the plot is so silly and contrived that this great effort never really amounts to much.
Aka: CITY SLICKERS 2: THE LEGEND OF CURLY'S GOLD
COM   111 min (ort 115 min)   VIDrel: COLUM/SONOP   V/sur

CLAIRE OF THE MOON ***   18
Nicole Conn USA   1992
Trisha Todd, Karen Trumbo, Faith McDevitt, Damon Craig
Two women attending a writers' conference in the Northwest develop a tender and loving relationship while sharing a cabin. A sensitively handled and well-realised drama on the theme of lesbianism.
DRA   108 min   VIDrel: WILLPRO/RTM   V

CLAIRE'S KNEE ***   PG
Eric Rohmer FRANCE   1970
Jean-Claude Brialy, Aurora Cornu, Beatrice Romand, Laurence De Monaghan, Michele Montel, Gerard Falconetti, Fabrice Luchini
The delicate tale of a man who spends the summer enjoying the company of three lovely ladies on the shores of Lake Geneva. Though about to get married, he finds himself becoming obsessed with the knee of a girl towards whom he is otherwise completely indifferent. The fifth in Rohmer's series of "Six Moral Tales", this is a minutely observed drama of manners that does tend to get bogged down in its verbose script. LOVE IN THE AFTERNOON followed.
Aka: LE GENOU DE CLAIRE
DRA   105 min (ort 106 min)   VIDrel: HEND/BMGREC L/A   V

CLARA'S HEART **   15
Robert Mulligan USA   1988
Whoopi Goldberg, Michael Ontkean, Kathleen Quinlan, Spalding Gray, Beverly Todd, Neil Patrick Harris, Hattie Winston, Jason Downs, Maria Broom, Maryce Cooper, Wanda Christine, Angel Harper, Fred Strother, Joseph Muth, Warren Long
A Jamaican housekeeper takes a post with an obnoxious couple hovering on the brink of divorce, and befriends the youngster placed in her care. A mawkish confection of sentimentality and homespun wisdom, that's both badly edited and singularly unmoving. Goldberg's strong screen presence does little for this film. See also CORRINA, CORRINA.
DRA   108 min   VIDrel: WHV   V/sur

CLARENCE, THE CROSS-EYED LION ***   U
Andrew Marton USA   1965
Marshall Thompson, Betsy Drake, Richard Haydn, Alan Caillou, Cheryl Miller, Rockne Tarkington, Maurice Marsac, Bob Do Qui, Albert Amos, Dinny Powell, Mark Allen, Laurence Conroy, Allyson, Daniel, Janee Michele, Naaman Brown
The head of an animal behaviour centre in Africa falls in love with a woman expert on gorillas and together with the man's daughter, they foil a plot to steal the animals. An amiable family movie, with the right blend of fun and drama. Gave rise to the TV series "Daktari" which also starred Clarence, Thompson and Drake.
JUV   88 min (ort 120 min) mTV   VIDrel: MGM/WHV   V/h

CLASH OF THE TITANS **   15
Desmond Davis UK   1981
Harry Hamlin, Judi Bowker, Ursula Andress, Maggie Smith, Laurence Olivier, Burgess Meredith, Claire Bloom, Sian Phillips, Flora Robson, Tim Pigoot-Smith, Neil McCarthy, Susan Fleetwood, Anna Manahan, Jack Gwillim, Vida Taylor
Fantasy adventure based on Greek and Nordic myths. The Gods high up on Olympus, watch Perseus as he battles against various monsters in his march towards everlasting glory. The enjoyable

special effects are by Ray Harryhausen. If only the dialogue were better.
FAN   113 min (ort 118 min)   VIDrel: MGM/WHV   V/sur

CLASS ACT **   15
Randall Miller USA   1991
Kid 'n' Play (Christopher Reid/Christopher Martin), Pauly Shore, Rhea Pearlman, Lamont Jackson, Doug E. Doug, Meshach Taylor, Rick Ducommun, Karyn Parsons, Lamont Jackson
A brilliant student goes to a new high school but finds that his record and identity papers have become switched with those of a young, streetwise thug. As both struggle to cope with this unexpected situation, they learn a few useful lessons. A predictable, role-reversal comedy that relies too heavily on the charm and energy of its two stars, rather than on the merits of a paper-thin plot.
COM   94 min (ort 100 min) cC   VIDrel: WHV   V/sur

CLASS ACTION ***   15
Michael Apted USA   1990
Gene Hackman, Mary Elizabeth Mastrantonio, Joanna Merlin, Colin Friels, Larry Fishburne, Donald Moffat, Jan Rubes, Matt Clark, Fred Dalton Thompson, Dan Hicks, Jonathan Silverman, Joan McMurtrey, Anne Elizabeth Ramsay, David Byron
A lawyer with a well developed social conscience takes legal action against a negligent car company on behalf of a group of plaintiffs. Meanwhile, his daughter has agreed to act for the other side. For all its contrivance and predictable outcome, robust performances from Hackman and Mastrantonio plus an articulate script give this film a sparkle it would otherwise have lacked.
DRA   105 min (ort 109 min)   VIDrel: 20TH/TECH   V/sur

CLASS CRUISE **   PG
Oz Scott USA   1991
Shelley Fabares, Richard Moll, Billy Warlock, Ray Walston, McLean Stevenson
High school shenanigans abound when a couple of rival schools arrange an educational cruise for their pupils. A standard teen-comedy only memorable for its slightly unusual setting.
COM   94 min   VIDrel: COLUM/SONOP   V/sh

CLASS OF '44 **   15
Paul Bogart USA   1973
Gary Grimes, Jerry Houser, Oliver Conant, Deborah Winters, William Atherton, Sam Bottoms, Joe Ponazecki, Murray Westgate, Marion Waldman, Mary Long, Marcia Diamond, Jeffrey Cohen, Susan Marcus, Lamar Criss, Michael A. Hoey
Sequel to SUMMER OF '42 with our hero going to college, growing up and falling in love. Nice period detail but this one has very little to say.
DRA   95 min   VIDrel: MGM/WHV   L/A   V

CLASS OF '61 ***   PG
Gregory Hoblit USA   1994
Christien Anholt, Andre Braugher. Dan Futterman, Joshua Lucas, Sophie Ward, Clive Owen, Sue-Ann Leeds, Laura Linney, Niall O'Brien, Paul Guilfoyle, Dana Ivey, Frederick Rolf, Tim Scott, Beverly Todd, Ed Wiley, Len Cariou
A group of West Point Cadets graduating in the fateful year of 1861 find themselves fighting on different sides as the Civil War engulfs America and divides both the country and individual families. A movng and realistic account of the horrors of this dark epoch in American history.
DRA   90 min   VIDrel: CIC/SONOP   V

CLASS OF 1999 **   18
Mark L. Lester USA   1990
Bradley Gregg, Traci Lin, John P. Ryan, Pam Grier, Joshua Miller, Stacy Keach, Malcolm McDowell, Patrick Kilpatrick, Darren E. Burrows, Jimmy Taggert, Jason Oliver, Sean Sullivan, Jim Gatsby, Sharon Wyatt
THE TERMINATOR joins BLACKBOARD JUNGLE in this gruesome sequel to "Class Of '84", that has a gang of high school thugs meeting their match when they come up against some experimental android teachers. An overblown celebration of violence masquerading as a film with a message. See also THE ULTIMATE TEACHER.
FAN   92 min Cut (15 sec – ort 98 min)
VIDrel: FIRST/SONOP   L/A   V

CLASS OF 1999 2 **   18
Spiro Razaatos CANADA   1993
Sasha Mitchell, Nick Cassavetes, Caitlin Dulany, Jack Knight, Rick

Hill, Gregory West, Diego Serrano, Bernie Pock, Deneny Pierce, Loring Pcikering, John Cothran Jr., Pete Antico, Christopher Brown, Eric Stabenau, Jean Pflueger
A sequel to the previous film that continues the theme of android teachers. Here, a robot solider adopts the guise of one such educator and soon achieves great success in eliminating gang violence, although a shadowy figure is determined to exploit his capabilities for his own ends.
Aka: CLASS OF 1999 PART 2: THE SUBSTITUTE
FAN   87 min (ort 90 min)   VIDrel: REFLEC/FIRST   L/A   V

CLASS OF NUKE 'EM HIGH *   18
Richard W. Haines/Michael Herz USA   1986
Janelle Brady, Gilbert Brenton, Robert Pritchard, R.L. Ryan, James Nugent Vernon, Brad Dunker, Gary Schneider, Gary Rosennblatt, Rick Howard, Theo Cohan, Mary Taylor, Heather McMahan, Anthony Ventola, Seth Oliver Hawkins, Lerae Dean
Nuclear waste turns high school kids into little horrors, in this fairly mindless exercise in low-brow humour, replete with sadistic gags and various overblown episodes.
Aka: NUKE 'EM HIGH
COM   95 min   VIDrel: TROMA/RTM   V

CLASS OF NUKE 'EM HIGH 2: SUBHUMANOID MELTDOWN *   18
Eric Louzil USA   1992
Brick Bronsky, Lisa Gaye, Leesa Rowland
The radioactive waste spewed out by a defective nuclear reactor has created a generation of hideously deformed freaks. When the students of Tromaville Tech learn about this, they join forces with said freaks to put a stop to the activities of the callous plant owners, setting in motion a major conflict.
FAN   90 min (ort 97 min)   VIDrel: TROMA/DISC   V

CLASS OF NUKE 'EM HIGH: THE GOOD, THE BAD AND THE SUBHUMANOID **   18
USA   1995
Brick Bronsky
Third episode in this continuing over-the-top spoof sees a case of mistaken identity leading to the usual complications. Average.
HOR   95 min   VIDrel: TROMA/RTM   V

CLEAN AND SOBER ***   18
Glenn Gordon Caron USA   1988
Michael Keaton, Kathy Baker, Morgan Freeman, M. Emmet Walsh, Brian Benben, Luca Bercovici, Tate Donovan, Henry Judd Baker, Claudia Christian, J. David Krassner, David Matthews, Mary Catherine Martin, Pat Quinn, Ben Piazza
A real-estate salesman who has borrowed a client's money to finance his drug addiction, goes into a treatment centre to stay out of trouble and avoid the police. Once there, he gradually comes to realise that he has a problem and emerges a changed man, if not an entirely reformed character. A gripping drama, despite its length and somewhat deliberate pacing. See also DAYS OF WINE AND ROSES and MY NAME IS KATE.
DRA   119 min (ort 124 min)   VIDrel: WHV   V/sur

CLEAN, SHAVEN **   (18)
Lodge Kerrigan USA   1995
Peter Greene, Molly Castelloe, Jennifer MacDonald, Robert Albert, Megan Owen, Alice Levitt, Jill Chamberlain, Agatha Leclerc, Roget Joly, Rene Beaudin, J. Dixon Byrne, Eliot Rockett, Angela Vibert, Karen MacDonald, Lee Kayman
After being released from a mental home, a schizophrenic man goes in search of his daughter who has been offered fro adoption by her mother. Unknown to him, however, he is being tailed by a detective who is firmly convinced that he is a serial killer. Fair.
DRA   80 min   CINrel

CLEAN SLATE **   12
Mick Jackson USA   1993
Dana Carvey, Valeria Golino, James Earl Jones, Kevin Pollack, Michael Murphy, Michael Gambon, Olivia D'Abo, Tim Scott, Angela Paton, Jayne Brook, Phil Leeds, Gailard Sartain, Michael Monks, Vyto Ruginis, Robert Wisdom
A private eye suffers from a kind of amnesia in which his memory is wiped clean at the start of each new day. This makes for no end of trouble, especially as he is scheduled to testify against a notorious gangster in a case that involves a murder. A sporadically amusing effort that doesn't really make the most of its promising idea.
COM   102 min cC   VIDrel: MGM/WHV   V/sur

CLEAR AND PRESENT DANGER ***
Phillip Noyce USA   12   1994
Harrison Ford, Willem Dafoe, James Earl Jones, Donald Moffat, Harris Yulin, Anne Archer, Joaquim De Almeida, Henry Czerny, Miguel Sandoval, Benjamin Bratt, Raymond Cruz, Dean Jones, Thora Birch, Ann Magnuson, Hope Lange, Tim Grimm
A White House advisor takes over from his sick boss and gets embroiled in a nasty conspiracy involving a USA-backed operation in Columbia to root out drug dealers, but soon finds that those ranged against him include his country's president. This contrived and implausible yarn moves along fast enough to almost make one forget the holes in the plot. Ford is adequate but hardly impressive as the hero while Archer simply reprises her PATRIOT GAMES role as our hero's wife.
THR 135 min (ort 141 min) cC VIDrel: CIC/SONOP; PION (LV only) V/sur LV
Boa: novel by Tom Clancy.

CLEO FROM 5 TO 7 ***
Agnes Varda FRANCE   (18)   1961
Corinne Marchand, Antoine Bourseiller, Dorothee Blanck, Michel Legrand, Jose-Luis de Villalonga, Dominque Davray, Jean-Claude Brialy, Anna Karina (Hanne Karin Blarke Bayer), Eddie Constantine, Sami Frey, Danielle Delorme
A woman pop singer anxious for the results of a medical test for cancer has her fortune read with tarot cards. Told she has the disease, she spends the next two hours wandering through the streets of Paris, with every minor event taking on new meaning as she comes to terms with her own mortality. A fine and sensitive study, if more than a little contrived. See also MY BREAST.
Aka: CLEO DE 5 A 7
DRA 90 min B/W TVrel

CLEOPATRA **
Joseph L. Mankiewicz USA   PG   1963
Elizabeth Taylor, Richard Burton, Rex Harrison, Pamela Brown, George Cole, Hume Cronyn, Cesare Danova, Kenneth Haigh, Andrew Keir, Martin Landau, Roddy McDowall, Robert Stephens, Francesca Annis, Martin Benson, Herbert Berghof
Overlong and overblown retelling of the saga of Cleopatra. Remarkable for its stilted dialogue and cardboard characterisations. An expensive flop that could have been produced far more cheaply without such an emphasis on the overwhelming sets. AA: Cin (Leon Shamroy), Art/Set (DeCuir, Smith, Brown, Blumenthal, Webb, Pelling and Juraga/Scott, Fox and Moyer), Cost (Sharaff, Novarese and Renie), Effects/vis (Emil Kosa Jr).
DRA 248 min (2 cassettes) wScrn VIDrel: 20TH/TECH V/sh
Boa: novel The Life and Times of Cleopatra by C.M. Franzero.

CLEOPATRA JONES *
Jack Starett USA   15   1973
Tamara Dobson, Bernie Casey, Shelley Winters, Brenda Sykes, Antonio Fargas, Bill McKinney, Esther Rolle, Stafford Morgan, Mike Warren, Albert Popwell, Caro Kenyatta, Dan Frazer, Paul Koslo, Joseph A. Tornatore, Keith Hamilton
A black female karate-chopping government agent goes after drug runners. Plenty of action in this violent nonsense. Followed by CLEOPATRA JONES AND THE CASINO OF GOLD, a slightly better film.
THR 84 min (ort 89 min) VIDrel: WHV V/h

CLEOPATRA JONES AND THE CASINO OF GOLD **
15
Chuck Bail HONG KONG/USA   1974
Tamara Dobson, Stella Stevens, Tanny, Norman Kell, Albert Popwell, Caro Kenyatta, Christopher Hunt, Chan Sen, Lin Chen Chi, Liu Loke Hua, Eddy Donno, Bobby Canavarro, Mui Kwok Sing, Join Cheng
Sequel to CLEOPATRA JONES, in which our athletic amazon takes on a Hong Kong drugs ring. A slick, violent and overwrought effort, though Dobson as the title figure has considerable charisma if not an especially good script.
A/AD 92 min (ort 96 min) VIDrel: WHV V/h

CLERKS ***
Kevin Smith USA   18   1994
Brian O'Halloran, Marilyn Ghigliotti, Jeff Anderson, Lisa Spoonauer, Jason Mewes, Kevin Smith, Scott Schiaffo, Scott Mosier, Al Berkowitz, Walt Flanagan, Ed Hapstak, Lee Bendick, David Klein, Pattijean Csik, Ken Clark, Donna Jeanne
Simple tale of a couple of guys who work in a convenience store and a video store that happen to be next to each other and the events of one particular day. An ultra-low budget effort (it cost

a mere $27,000 to make and is filmed on 16 mm stock) it was the director's debut feature and was a surprise hit at Cannes. Hilarious and razor sharp, this is a very vulgar but immensely enjoyable work, that was followed by MALLRATS. See also EMPIRE RECORDS.
Aka: CLERKS, THE
COM 88 min (ort 92 min) B/W wScrn
VIDrel: ARTIF/20TH V/sh

CLIENT, THE ***
Joel Schumacher USA   15   1994
Susan Sarandon, Tommy Lee Jones, Mary-Louise Parker, Brad Renfro, Anthony LaPaglia, Walter Olkewicz, Ossie Davis, Bradley Whitford, Anthony Edwards, J.T. Walsh, Will Patton, Anthony Heald, William H. Macy, Kimberly Scott, Ron Dean
The last person to talk to a Mafia lawyer before his suicide was a boy of only eleven, and he is put under enormous pressure by the police department to reveal what was said. Caught between them and the bad guys, who want him silenced at any cost, he turns in desperation to a tough-as-nails woman lawyer who accepts to act on his behalf. A fine adaptation of Grisham novel, well acted and directed and with a strong sense of suspense.
THR 115 min (ort 121 min) cC VIDrel: WHV V/sur
Boa: novel by John Grisham.

CLIFF RICHARD: FINDERS KEEPERS **
Sidney Hayers UK   U   1967
Cliff Richard, Robert Morley, Peggy Mount, The Shadows, Viviane Ventura, Graham Stark, John Le Mesurier
Cliff and The Shadows head for San Carlos in Spain, where they find the area deserted owing to the presence of an unexploded bomb, dropped in error by the US Air Force. Luckily they find the bomb and save the town, and also have time for some fun as well. An adequate vehicle for the singer and his chums, who really add nothing to the film, which surprisingly would have been better without them.
Aka: FINDERS KEEPERS
MUS 86 min VIDrel: MGM/WHV V

CLIFF RICHARD: SUMMER HOLIDAY **
Peter Yates UK   U   1963
Cliff Richard, Lauri Peters, David Kossoff, Ron Moody, Melvyn Hayes, The Shadows, Lionel Murton, Madge Ryan, Una Stubbs, Teddy Green, Pamela Hart, Nicholas Phipps
Young people have a holiday on the Continent in a borrowed double decker bus. A cute vehicle for Cliff with its appeal largely confined to his large and youthful fan-club. Yates' directorial debut with a script by Peter Myers and Ronnie Cass. Musical direction is by Stanley Black.
Aka: SUMMER HOLIDAY
MUS 103 min (ort 109 min) VIDrel: WHV V

CLIFF RICHARD: TAKE ME HIGH **
David Askey UK   U   1973
Cliff Richard, Hugh Griffith, Debbie Watling, George Cole, Richard Wattis, Anthony Andrews, Madeline Smith, Moyra Fraser, Ronald Hines, Jimmy Gardner, Noel Trevarthen, Graham Armitage, John Franklyn-Robbins, Elisabeth Scott
Cliff Richard vehicle in which he plays a bank employee who is sent to Birmingham to set up a major hamburger restaurant. During his stay there he meets a young girl and falls in love. A tired and empty youth musical, with little of the bounce of Richard's earlier films.
Aka: TAKE ME HIGH
MUS 86 min (ort 90 min) VIDrel: WHV/LUMI V

CLIFF RICHARD: THE YOUNG ONES **
Sidney J. Furie UK   U   1961
Cliff Richard, Robert Morley, Carole Gray, Teddy Green, Richard O'Sullivan, Melvyn Hayes, Annette Robertson, Robertson Hare, Sonya Cordeau, Gerald Harper, The Shadows
Syrupy vehicle for Cliff Richard, who plays a young man fighting to stop his youth club being demolished by his father's company. A badly dated youth musical, made slightly more bearable by a few good songs. The music is by Stanley Black.
Aka: WONDERFUL TO BE YOUNG; YOUNG ONES, THE
MUS 104 min (ort 108 min) VIDrel: WHV V/h

CLIFF RICHARD: WONDERFUL LIFE *
Sidney J. Furie UK   U   1964
Cliff Richard, Susan Hampshire, Walter Slezak, Melvyn Hayes, The Shadows, Richard O'Sullivan, Una Stubbs, Derek Bond, Gerald Harper

Routine Cliff vehicle, with our star getting pressed into service as a stuntman for a film being made on the Canary Islands. So sweet and syrupy it should carry a dental warning; as usual our Cliff combines maximum goodness with minimum personality (a pity, as given the chance he can act). He gets to sing too. Written by Peter Myers and Ronald Cass.
Aka: WONDERFUL LIFE
MUS 108 min (ort 113 min) VIDrel: WHV/LUMI V

CLIFFBANGER ** **   18
Anthony Spinelli USA   1993
Jonathan Morgan, Jake Williams, Saraha, Rebecca Wild, Nikki Sinn, Brigitte Aime, Tina Tedeschi
The title may be borrowed from CLIFFHANGER, but this is just one more sex film, albeit with some nice locations.
A   43 min (ort 80 min)   VIDrel: ONE   V

CLIFFHANGER * **   15**
Renny Harlin FRANCE/USA   1992
Sylvester Stallone, John Lithgow, Janine Turner, Michael Rooker, Rex Linn, Caroline Goodall, Leon, Paul Winfield, Ralph Waite, Craig Fairbrass, Max Pehrlich, Michelle Joyner, Gregory Scott Cummins, Trey Brownell, Zach Grenier
An expert mountain climber who suffers from self-doubt comes up against a bunch of ruthless criminals who are out to recover $100,000,000 and have unwisely kidnapped his girlfriend. A thinly plotted and cliched-ridden thriller that wins out despite these failings thanks to Stallone's screen presence, breathtaking photography and plenty of daredevil stunts. Filmed on location in the Italian Dolomites on a budget of around $70,000,000.
A/AD   106 min (ort 113 min)
VIDrel: 4-FRONT/POLYREC/GUILD; PION (LV only) V/sur LV

CLIFFORD *   PG**
Paul Flaherty USA   1993
Martin Short, Charles Grodin, Mary Steenburgen, Dabney Coleman, Sonia Jackson, Richard Kind, Jennifer Savidge, Brandis Kemp, Ben Savage, Tim Lane, Don Galloway, Susan Varon, Josh Seal, Kevin Mockrin, Timothy Stack, Megan Kloner
A ten-year-old boy comes to visit his unmarried uncle and turns his life upside down in this unoriginal comedy, which seems clearly inspired by DENNIS and other films of that ilk. The diminutive Short has the title role, an inventive piece of casting if ever there was one; a pity this over-extended romp cannot do justice to him.
COM   86 min (ort 90 min)   VIDrel: COLUM/SONOP
V/sur

CLIMAX 2000 *   18**
Michael Zen USA   1994
Kaitlyn Ashley, Vixxen, Tiffany Million, Mike Horner, Tony Tedeschi, Nikki Sinn, Jon Dough, T.T. Boy, Nick East, Leena, Buck Adams
Lust in a cyberpunk world in an adult movie inspired by "The Phantom Of the Opera", that has a shadowy figure living below a film studio that produces sex films. A sequel followed.
Aka: CLIMAX 2000 PART 1: THE PHANTOM'S CURSE
A   65 min   VIDrel: PURG/DANTE   V

CLOCKERS * **   18**
Spike Lee USA   1995
Harvey Keitel, John Turturro, Delroy Lindo, Mekhi Phifer, Isaiah Washington, Keith David, Pee Wee Love, Regina Taylor, Tom Byrd, Sticky Fingaz, Fredro, E.O. Nolasco, Lawrence B. Adisa, Hassan Johnson, Frances Foster, Michael Imperioli
A streetwise drugs-dealer gets involved in a deal with a crime boss that results in the death of a rival dealer, leaving the former at the mercy of the dead man's buddies. From its bloody and brutal opening the film never lets up the pressure, becoming as much a condemnation of a way of life as it is an absorbing character study.
THR   123 min (ort 129 min) cC   VIDrel: CIC/SONOP
V/sur
Boa: novel by Richard Price.

CLOCKWISE *   PG**
Christopher Morahan UK   1985
John Cleese, Penelope Wilton, Alison Steadman, Stephen Moore, Sharon Maiden, Joan Hickson, Michael Aldridge, Benjamin Whitrow, Geoffrey Palmer, Nicholas LePrevost, Sidney Livingstone, John Bardon, Pat Keen
Contrived comedy about a headmaster, whose obsessive need for punctuality is severely tested as he makes his way to an important meeting of headteachers. An unutterably tedious

film, scripted by Michael Frayn, it tries hard to make much of little, but finally collapses for sheer lack of ideas.
COM   92 min (ort 96 min)   VIDrel: WHV   V/h

CLOCKWORK MICE * **   15**
Jean Vadim UK   1994
Ian Hart, Ruaidhri Conroy, Catherine Russell, Art Malik, Claire Skinner, Nigel Planer, John Alderton, James Bolam, Lilly Edwards, Robin Soans, Melissa Simoonds, Jack McKenzie, Carl Proctor, Billy Davey, Hormoz Verahramian
An examination of the relationship that develops between a young boy with a troubled past who is at a school for problem kids, and his idealistic teacher, who believes that he will be able to reach the kids by setting up a running club. Hart as the teacher gives a wonderful performance, sufficient to imbue the film with a potency not to be gained from its uneven direction. The cast of youngsters, most of whom are unknowns, is also to be commended.
DRA   94 min (ort 99 min)   VIDrel: POLY/POLYREC   V/s

CLOCKWORK ORANGE, A * ***   (18)**
Stanley Kubrick USA   1971
Malcolm McDowell, Patrick Magee, Adrienne Corri, Aubrey Morris, James Marcus, Michael Bates, Warren Clarke, Carl Duering, James Farrell, Clive Francis, Michael Glover, Miriam Carlin, Godfrey Quigley, Sheila Raynor
In 1971 what had been merely a savage attack on a future Britain in decline has unfortunately come true in the intervening years. Much of the violence that so appalled the critics now seems tame by comparison with many video offerings. However, the crucial difference is that here it is far from gratuitous, being an essential part of this savage account of a young man's brainwashing at the hands of the state. A brilliant and absorbing film.
FAN   137 min   SATrel: FILMNET 1 (not legally available in the UK)
Boa: novel by Anthony Burgess.

**CLOSE ENCOUNTERS OF THE THIRD KIND:
SPECIAL EDITION *** **   PG**
Steven Spielberg USA   1977 (re-released 1980)
Richard Dreyfuss, Teri Garr, Melinda Dillon, Cary Guffey, Bob Balaban, Francois Truffaut, J. Patrick McNamara, Warren Kemmerling, Roberts Blossom, Phillip Dodds, Shawn Bishop, Adrienne Campbell, Justin Dreyfuss, Jim Mills
Several people are obsessed with strange images linked to imminent alien landings. The government tries at all cost to keep this event secret, but fails to stop a determined few from reaching the contact zone. A film of enormous visual impact, but overlong and rather disjointed. This version was re-edited and has footage not included in the earlier work. AA: Cin (Vilmos Zsigmond), Spec Award (Frank Warner for sound effects editing).
FAN   127 min (ort 152 min) wScrn
VIDrel: COLUM/SONOP   V/sur

CLOSE KUNG FU ENCOUNTER, THE ** **   18
Raymond Chin HONG KONG
Jackie Chang, Roger Mao, Jwang In-Shik, Kim Jung-Sung, Choi Min Kyu, Lee Ye-Min, Hong Jung-Sung, Kim Kee-Joo, King Sung-Nam, Han Myung-Han, Kim Woo-Suk, Lee Suk-Koo, Jang Jung-Kok, Ahn Jin-Soo
A kung fu expert working for the army, attempts to recover some of China's gold from the Japanese. He is captured and forced to fight for his life.
MAR   88 min   VIDrel: IMPENT   V

CLOSE MY EYES * **   18**
Stephen Poliakoff UK   1991
Alan Rickman, Clive Owen, Saskia Reeves, Karl Johnson, Lesley Sharp, Kate Garside, Karen Knight, Niall Buggy, Campbell Morrison, Annie Hayes, Maxwell Hutcheon, Geraldine Somerville, Helen Fitzgerald, Christopher Barr, Jan Winters
A brutally frank examination of incest, in which a womanising brother and his married sister set out to explore their mutual attraction, whilst all the while the woman's husband merely suspects infidelity. A disturbing and incisive treatment of a difficult subject, both well acted and directed.
DRA   104 min (ort 108 min) wScrn   VIDrel: ARTIF/20TH
V/sur

CLOSE TO EDEN ** **   15
Sidney Lumet USA   1992
Melanie Griffith, Eric Thal, John Pankow, Tracy Pollan, Ian Richardson, Mia Sara, Jamey Sheridan, Jake Weber, Ro'ee Levi, David

Rosenbaum, Ruth Vool, David Margulies, Ed Rogers III, Maurice Schell, James Gandolfini, Chris Collins
A tough female cop (well aren't they always?) has to go undercover in search of a killer by entering Brooklyn's tightly-knit and ultra-conservative Chasidic community after a member of it is murdered and robbed of a fortune in gems. An attraction to the rabbi's son adds a further complication. A well acted variant on a theme explored in WITNESS, though Griffith looks far more out of place than Harrison Ford ever did and the plot remains thin. See also WALL OF SILENCE.
Aka: STRANGER AMONG US, A
THR   104 min (ort 109 min)
VIDrel: ENCORE/SPEAR/COLUM   V/sh

CLOSE YOUR EYES AND PRAY *
Skip Schoolnik USA
18
1987
Brittain Frye, Donna Baltron, George Thomas, Annette Sinclair
A bunch of kids set out to spend the night at the furniture store owned by the father of one them, but fall victim to a psychopathic murderer who kills several of them, forcing the others to hunt him down. A standard slasher movie of little merit that's not helped by its daft ending.
Aka: HIDE AND GO SHRIEK
HOR   86 min Cut (50 sec – ort 90 min)
VIDrel: MARQ/QUANT   V

CLOSELY WATCHED TRAINS ****
Jiri Menzel CZECHOSLOVAKIA
15
1966
Vaclav Neckar, Jitka Bendova, Vladimir Valenta, Libuse Havelkova, Josef Somr, Vlastimil Brodsky, Jiri Menzel, Alois Vachek, Jutka Zelenohorska
One of the finest Czech movies ever made. A warm and affectionate look at a young man growing up in WW2 and his fumbling attempts to discover the joys of sex. Set against the background of the German occupation, the irony is so gentle that it is easily overlooked. Won an Oscar as Best Foreign Film and deservedly so. Sit back and enjoy. AA: Foreign.
Aka: CLOSELY OBSERVED TRAINS; OSTRE SLEDOVANE VLAKY
DRA   88 min (ort 92 min) B/W   VIDrel: CONNO/RTM   V
Boa: novel by Bohumil Hrabal.

CLOSER AND CLOSER **
Fred Gerber USA
(18)
1995
Kim Delaney, Scott Kraft, Susan Hogan, Sharon Martin, Bill MacDonald, Mark Melymick, Nicholls King, John Curbt, Nigel Hamer, Sharon Dunne, Gary Crawford, Les Porter, Vidonia Mitchell, Lane White, Henry Alessandroni, Alan Van Sprang
A woman who wrote a bestselling novel about a serial killer three years ago had the misfortune to find that her work inspired a reader to turn killer, and was crippled in an attack the killer meant to be fatal. Having written a sequel, she is about to find history repeating itself, in this ludicrous but undeniably gripping effort.
THR   90 min   SATrel: SKY MOVIES

CLOSET LAND **
Radha Bharadwaj USA
(PG)
1991
Madeleine Stowe, Alan Rickman
In an imaginary totalitarian society, an author of children's books finds herself being interrogated on charges of subversion, in this heavy and preachy drama whose fine performances are not enough to hold the interest.
DRA   95 min   SATrel

CLOTHES IN THE WARDROBE ***
Waris Hussein UK
(PG)
1992
Julie Walters, Joan Plowright, Jeanne Moreau, Lena Headey, David Threlfall, Padraig Casey, Britta Smith, Catherine Schell, Pierre Sioufi, Sherine El Ansari, Annabel Burton, Maggie Steed, John Wood, David Gant, Tommy Duggan, Lamia El Amir
In 1950s Croydon a family makes preparations for the wedding of their young daughter, but she has kept her true feelings to herself, and has nothing but distaste for her fiance, and indeed the whole idea of marriage. When she confides her true feeling to Moreau, who is an old friend of the family, that latter hatches a cunning scheme to so discredit the man that no wedding can take place. Fine ensemble acting sustains this thin little story.
DRA   79 min mTV   TVrel: BBC
Boa: novel by Alice Thomas Ellis.

CLOUD WALTZER **
Gordon Flemyng UK/USA
PG
1986
Kathleen Beller, Francois Eric Gendron, Paul Maxwell, Therese Liotard, Claude Gensac, David Baxt, Dora Doll

A woman journalist ends up falling in love with a millionaire recluse she has been sent to interview. A lush chocolate-box romance, based on a novel in the Harlequin series of romantic fiction. Beautiful photography and competent acting, but the plot is as thin as tissue paper.
Aka: CLOUD WALTZING
DRA   90 min (ort 103 min) mCab   VIDrel: CIC/SONOP   V
Boa: novel by Tony Cates.

CLOWNHOUSE **
Victor Silva USA
15
1989
Nathan Forest Winters, Brian McHugh, Sam Rockwell
The youngest of three brothers has a pathological fear of clowns, but is forced to accompany his eldest brother and girlfriend to the circus, where three escaped lunatics now disguised as clowns, provide the requisite entertainment. A predictable and rather bloodless horror yarn that has a few moments of terror despite the unimaginative direction.
HOR   78 min   VIDrel: EIV/SONOP   V

CLUB, THE **
Kirk Wong HONG KONG
18
1982
Michael Chan Wai Man, Erina Miyai, Hsu Hsiao-Chaing, Mabel Kwong, Kent Cheng, Cheung Kuen, Wilson Tong, Ko Fei
Kung-fu movie set in and around title establishment – a nightclub that seems to operate as a powerful magnet for gangs and assorted troublemakers. As ever, revenge is not far from the minds of those whose story is told here.
MAR   82 min (ort 90 min)   VIDrel: EAST/DISC   V

CLUE **
Jonathan Lynn USA
PG
1985
Eileen Brennan, Tim Curry, Madeline Kahn, Christopher Lloyd, Martin Mull, Michael McKearn, Lesley Anne Warren, Colleen Camp, Lee Ving, Bill Henderson, Howard Hesseman, Don Camp, Jane Wiedlin, Jeffrey Kramer, Kelley Nakahara
Inspired by the board game Cluedo, this is a murder mystery set in one of those Gothic mansions so beloved by murder mystery writers. Reminiscent of, but far inferior to "Ten Little Indians". Released in three versions, each with a different ending, but now all three versions are cobbled together into a single film. A brave attempt, but heavy-going for all concerned. The directorial debut for Lynn.
THR   92 min (ort 97 min)   VIDrel: CIC/SONOP   V

CLUE ACCORDING TO SHERLOCK HOLMES, THE **
Murray Golden UK
U
1980
Keith Michell, Dody Goodman, Keith McConnell, Laurie Main
A Sherlock Holmes-style adventure, in which the mystery of a hidden fortune, reputed to have been left behind by an eccentric millionaire when he died, is eventually solved by a ten-year-old sleuth, with a little help and a good deal of patient endeavour.
Aka: TREASURE OF ALHEUS WINTERBORN, THE
JUV   45 min (ort 50 min) mTV   VIDrel: START/DISC   V

CLUELESS ***
Amy Heckerling USA
12
1995
Alicia Silverstone, Stacey Dash, Brittany Murphy, Paul Rudd, Justin Walker, Donald Faison, Breckin Meyer, Jeremy Sisto, Elisa Donovan, Dan Hedaya, Wallace Shawn, Aida Linares, Twink Caplan, Sebastian Rashidi, Herb Hall, Julie Brown
A much-needed antidote to TV's "Beverly Hills 90210" exposing the vacuous and aimless life style of a group of teenagers whose parents have given them everything they need, except decent values and a sense of purpose. A brash, abrasive and very funny comedy, it benefits greatly from its witty script, the work of the director. A TV series followed.
COM   93 min (ort 97 min) cC
VIDrel: CIC/SONOP; PION (LV only)   V/sur LV

COACH **
Steve Gomer USA
15
1996
Rhea Perlmann, Fredo Starr, Carol Lane
A bunch of tough inner-city kids attending a high school in Brooklyn get a chance to prove their value as a basketball team. An over-familiar story, but made enjoyable largely on account of Perlmann's strong screen presence, although it must be said, the laughs are a bit thin on the ground.
COM   95 min   VIDrel: 20VIS/SONOP   V/sur

COAL MINER'S DAUGHTER **** PG
Michael Apted USA 1980
Sissy Spacek, Tommy Lee Jones, Levon Helm, Phyllis Boyens, Beverly D'Angelo, Bill Anderson Jr, Foister Dickerson, Malla McCown, Pamela McCown, William Sanderson, Kevin Salvilla, Sissy Lucas, Pat Paterson, Brian Warf, Robert Elkins
Rags-to-riches film biography of country singer Loretta Lynn, that is not always all that accurate, but is never less than totally engrossing. Spacek did her own singing, for which she received a well-deserved Oscar. The screenplay was by Tom Rickman.
AA: Actress (Spacek).
DRA 125 min VIDrel: RCA L/A V

COBB *** 18
Ron Shelton USA 1994
Tommy Lee Jones, Robert Wuhl, Lolita Davidovich, J. Kenneth Campbell, Lou Myers, Stephen Mendillo, Scott Burkholder, Allan Malamud, Jeff Fellenzer, Doug Krikorian, Bill Caplan, Ned Bellamy, Gavin Smith, William Utay, Rhoda Griffis
An ageing former baseball player recounts the events of his life and his career to a sports journalist whom he has hired for the purposes of compiling his biography. A fascinating and tour-de-force portrait of the legendary Ty Cobb, who is revealed in this picture as being a highly unpleasant and unattractive human being, whose sole positive quality resided in his sporting prowess.
DRA 123 min (ort 128 min) cC VIDrel: WHV V/sur
Boa: book Cobb: A Biography by Al Stump.

COBRA * 18
George Pan Cosmatos USA 1986
Sylvester Stallone, Brigitte Nielsen, Reni Santoni, Andrew Robinson, Lee Garlington, John Herzfeld, Art La Fleur, Brian Thompson, David Rasche, Val Avery, John Hawk, Nick Angotti, Nina Axelrod, Joe Bonny, John Cahill, Ken Hill
Repellent vigilante drama with Stallone as a cop whose solution to crime is to shoot anything that moves. A DIRTY HARRY clone with our tough cop given the assignment of tracking down serial killers. Some good action sequences, but very little more.
A/AD 83 min Cut (4 sec – ort 87 min) VIDrel: WHV
V/sh
Boa: novel Fair Game by Paula Gosling.

COBRA AGAINST NINJA ** 15
Joseph Lai HONG KONG 1987
Richard Harrison, Stuart Smith, Alan Friss, Paul Branney, Gary Carter, Jimmy Busco, Alfred Pears, Stuart Ting, Shee Wong
Standard adventure fare with the head of the Ninja Empire battling to uphold its honour and integrity.
Aka: COBRA VERSUS NINJA
MAR 86 min Cut (7 sec) VIDrel: IMPENT V

COBRA MISSION * 18
Larry Ludman (Fabrizio De Angelis) ITALY/WEST
GERMANY 1985
John Ethan Wayne, Donald Pleasence, Oliver Tobias, Christopher Connelly, John Steiner, Manfred Lehmann
Another Vietnam POW rescue saga with all the stock characters – sadistic Vietnamese captors, heroic POWs etc. I dare say you've seen it all before.
Aka: DIE RUCKKEHR DER WILDGANSE; OPERATION NAM; RAINBOW PROFESSIONALS
A/AD 90 min VIDrel: EIV/SONOP V/sur

COCA-COLA KID, THE ** 15
Dusan Makavejev AUSTRALIA 1985
Eric Roberts, Bill Kerr, Greta Scacchi, Chris Haywood, Kris McQuade, Max Gilles, Rebecca Smart, Tim Finn, Colleen Clifford, Tony Barry, Paul Chubb, David Slingsby, Esben Storm, Linda Nagle, Julie Nihill, Fiona Hallett, Gia Carides
A Coca-Cola salesman in Australia discovers a small valley where the local bigshot is making his own soft drink and refusing to allow any other to be sold. An offbeat and quirky little film of few laughs.
COM 94 min VIDrel: ARROW/RTM V/sur

COCAINE FIENDS, THE * 18
William A. O'Connor USA 1939
Noel Madison, Lois January, Sheila Manners, Dean Benton, Lois Lindsay, Eddie Phillips
Classic melodramatic warning about the horrors of drug addiction, with a young girl introduced to the pleasures of cocaine (she believes she's taking "headache powder") by a mobster on the run from the law. A tacky and exploitative effort, so bad it's quite funny in places, but not funny enough to rank as a full-blooded comedy. See also REEFER MADNESS and COCAINE: ONE MAN'S POISON.
Aka: PACE THAT KILLS, THE
DRA 74 min VIDrel: CASPIC/TERRY L/A V

COCKLESHELL HEROES, THE ** U
Jose Ferrer UK 1954
Trevor Howard, Jose Ferrer, Anthony Newley, Victor Maddern, David Lodge, Peter Arne, Percy Herbert, Graham Stewart, John Fabian, Judith Furse, John Van Eyssen, Dora Bryan, Walter Fitzgerald, Beatrice Campbell, Karel Stepanek
A WW2 action film that follows the training of ten marines, who are to be sent by canoe to Bordeaux, where their mission is to attach limpet mines to the German vessels in the harbour. The early part of the film deals with their training, and in many ways this is the most interesting section of a restrained, almost documentary-style film, whose predictable stiff-upper-lip script generates neither tension nor surprise.
WAR 93 min VIDrel: VCC/DISC/COLUM V
Boa: book by George Kent.

COCKTAIL * 15
Roger Donaldson USA 1988
Tom Cruise, Bryan Brown, Elisabeth Shue, Lisa Banes, Laurence Luckinbill, Kelly Lynch, Gina Gershon, Ron Dean, Paul Benedict, Robert Donley, Ellen Folley, Andrea Morse, Robert Greenberg, Chris Owens, Justin Lewis, Sandra Will Carradine
A young arrival in New York dreams of making his fortune with a chain of bars, after he become a bartender and falls under the influence of an older barman. A cute, empty and strangely pointless effort. The script is by Gould, from the book he wrote after working for a while as a New York bartender.
DRA 99 min (ort 103 min) VIDrel: TOUCH/TECH V/sur
Boa: book by Heywood Gould.

COCOON *** PG
Ron Howard USA 1985
Don Ameche, A. Wilford Brimley, Hume Cronyn, Steve Guttenberg, Tahnee Welch, Brian Dennehy, Jessica Tandy, Maureen Stapleton, Jack Gilford, Gwen Vernon, Herta Ware, Barnet Oliver, Linda Harrison, Tyrone Power Jr, Clint Howard
A tale of contact between three feisty senior citizens and some benevolent aliens on a mission to Earth, who have hidden a power source with remarkable properties in an unattended swimming pool. It just happens to be the one our trio are in the habit of using. Excessive sentimentality and a slow start are handicaps. A sequel followed. AA: S. Actor (Ameche), Effects/vis (Ken Ralston/Ralph McQuarrie/Scott Farrar/David Berry).
FAN 112 min (ort 117 min) VIDrel: 20TH/TECH L/A;
ENCORE (LV only) V LV
Boa: novel by David Saperstein.

COCOON: THE RETURN ** PG
Daniel Petrie USA 1988
Don Ameche, Jack Gilford, Gwen Verdon, Maureen Stapleton, Hume Cronyn, Steve Guttenberg, Jessica Tandy, Wilford Brimley, Elaine Stritch, Tahnee Welch, Brian Dennehy, Courteney Cox, Barret Oliver, Linda Harrison, Tyrone Power Jr
The elderly folk who were taken to distant stars in the first film, now return to Earth for a visit in this thoroughly disappointing and predictable sequel. None of the sparkle of the first film is here, although Ameche and company do their best with the tired script. An unnecessary sequel if ever there was one.
Aka: COCOON 2
FAN 110 min (Cut at film release by 1 min 5 sec – ort 116 min) VIDrel: 20TH/TECH V/sur

CODE OF HONOUR ** 18
Benny Chan HONG KONG 1996
Chow Yun Fat, Danny Lee, Lam Wei
Story of intrigue and bloodshed among the members of various Triad gangs. Average.
A/AD 90 min dubbed VIDrel: MIA/DISC V

CODE OF SILENCE ** 18
Andrew Davis USA 1985
Chuck Norris, Henry Silva, Bert Remsen, Molly Hagan, Nathan Davis, Joseph Guzaldo, Mike Genovese, Dennis Farina, Allen Hamilton, Ron Henriquez, Ron Dean, Wilbert Bradley, Ralph Froody, Gene Barge, Mario Nieves, Miquel Nino, Joe Kosala
A tough Chicago cop has to fight both an underworld boss and

his own colleagues. Standard action movie of little depth but much mayhem.

A/AD 96 min (ort 101 min) VIDrel: 4-FRONT/POLYREC V/h

CODED HOSTILE ***
David Darlow UK/USA
PG
1989
Ed O'Ross, Gavan Herlihy, George Roth, Mark Burton, Michael Murphy, Debora Weaston, Colin Bruce, Michael Moriarty, Harris Yulin, Shane Rimmer, Chris Sarandon, Torek Bork, Boris Ivarov, Togo Igawa, Takashi Kawahara

An excellent recreation of the behind-the-scenes activities of the US military before and during the shooting down of the Korean passenger plane in 1988. It claims that the US government deliberately ignored the conclusions of its own air force intelligence to the effect that this action was a tragic mistake just in order to score a few political points. A careful, detailed and compelling film. SHOOTDOWN also covered this incident.

Aka: FLIGHT 007; TAILSPIN; TAILSPIN: BEHIND THE KOREAN AIRLINER TRAGEDY

DRA 90 min mCab VIDrel: CASPIC/TERRY L/A V

CODENAME: ASSASSIN *
Lorenzo Lamas USA
18
1993
Lorenzo Lamas, Kathleen Kinmont, Pamela Dixon, Larry Manetti, John P. Ryan, John Savage

When a top secret nuclear missile guidance system is stolen by terrorists, the US government attempts to retrieve it, teaming a CIA agent up with a former operative who was imprisoned on a trumped-up murder charge. The requisite number of explosions and the like are to be found in this painfully cliched effort, that serves as a sequel to C.I.A. – CODENAME ALEXA.

A/AD 87 min VIDrel: HIFLI/SONOP V/h

CODENAME: VENGEANCE **
David Winters ITALY
18
1987
Robert Ginty, James Ryan, Cameron Mitchell, Kevin Brophy, Shannon Tweed, Don Gordon

An ex-CIA agent, who for twelve years was incarcerated in a prison in the Middle East, agrees to undertake a perilous mission and rescue the wife and son of a shah.

A/AD 90 min (ort 97 min) VIDrel: MOPIC/QUANT V

COLD BLOOD **
Ralf Gregan HOLLAND
18
1974
Rutger Hauer, Vera Tschechowa, Horst Frank, Walter Sedlmayr

A pilot who has been undertaking missions for a gang of smugglers attempts to relieve them of a vast sum of money in this formulaic action pic.

Aka: DAS AMULETT DES TODES

A/AD 75 min (ort 90 min) dubbed
VIDrel: 4-FRONT/POLYREC/MIA V/h

COLD BLOODED **
M. Wallace Wolodarsky USA
18
1995
Jason Priestley, Peter Riegert, Kimberly Williams, Michael J. Fox, Robert Loggia, Janeane Garofalo

A young and inexperienced gang member finds himself being trained as a hitman but when he gets new girlfriend, unexpected problems occur as she has no idea of his line of work. An odd black comedy of little merit.

COM 89 min (ort 93 min) VIDrel: POLFIL V

COLD COMFORT ***
Vic Sarin CANADA
15
1989
Maury Chaykin, Margaret Langrick, Paul Gross, Jayne Eastwood, Ted Follows, Richard Flitch, Grant Moll

A travelling salesman crashes his car during a blizzard, but is saved from almost certain death by a passing trucker. However, his rescuer is not quite normal, and our salesman is taken home as a "present" for the man's nubile daughter. Sarin's directorial debut in what is a bizarre thriller of considerable intensity.

THR 90 min (ort 92 min) VIDrel: MOPIC/SGSVID V
Boa: play by James Garrard.

COLD COMFORT FARM ***
John Schlesinger UK
PG
1996
Kate Beckinsale, Joanna Lumley, Rufus Sewell, Ian McKellen, Eileen Atkins, Sheila Burrell, Stephen Fry, Freddie Jones, Miriam Margolyes

Quirky, lightweight send-up of a Jane Austen-style rural saga, that opens with young Flora going to live with her relatives at the title farm after she is orphaned. She finds it a depressing place whose unhappy residents are kept very much in their place by her eccentric aunt, a matter young Flora soon starts changing. This pleasing spoof was first released on TV, but was then seen in American cinemas, where its success led to a film version being made.

COM 104 min CINrel
Boa: novel by Stella Gibbons.

COLD DOG SOUP *
Alan Metter USA
15
1989
Randy Quaid, Frank Whaley, Christine Harnos, Sheree North, Nancy Kwan, Dante Basco, Peter Pan

A man goes out on an innocent date and becomes embroiled in a series of wildly improbable adventures, that begin when he tries to save the life of a dog by giving it mouth-to-mouth resuscitation. A strained and over-extended affair, it might have played well as a thirty-minute sketch.

COM 84 min VIDrel: 20TH V/sur
Boa: book by Stephen Dobyns.

COLD FEVER **
Fridrik Thor Fridriksson GERMANY/ICELAND/SWITZER-LAND/USA
15
1994
Masatoshi Nagase, Lili Taylor, Fisher Stevens, Gisli Halldorsson, Laura Hughes, Seijun Suzuki, Hiromasa Shimada, Masayuki Sasaki, Taizou Mizumura, Ichiko Takashi, Taeko Kato, Toshimori Iwaki, Ari Matthiasson, Magnus Olafsson

A Japanese businessman leaving Tokyo for the cold wastes of Iceland, having agreed to his grandfather's request to lay the souls of his parents to rest, these two having died there seven years ago in a freak accident. A strange, compelling road-movie, full of dream-like images and starkly beautiful vistas, it says nothing of any real importance, but has a curious, austere charm all of its own. The script is by Fridriksson and Jim Stark.

Aka: A KOLDUM KLAKA

DRA 85 min subs VIDrel: POLY/POLYREC V

COLD HEAVEN **
Nicolas Roeg USA
15
1992
Theresa Russell, Mark Harmon, James Russo, Talia Shire, Will Patton, Richard Bradford, Julie Carmen, Diana Douglas, Seymour Cassel, Castulo Guerra, Jeanette Miller, Daniel Addes, Jim Ishida, Martha Milliken, Margarita Cordova, Sal Lopez

A wife sees her husband killed in a speedboat accident before she can tell him that she is leaving him for another man. However, when his body mysteriously disappears from the morgue, she begins to suspect that he may still be alive. To make matters worse, she soon begins to experience religious visions. A confused and irritatingly opaque tale that fails to come over and totally wastes the combined efforts of a very capable cast.

DRA 102 min (ort 105 min) VIDrel: MARQ/QUANT V
Boa: novel by Brian Moore.

COLD JUSTICE **
Terry Green UK
15
1990
Dennis Waterman, Roger Daltrey, Ron Dean, Ralph Foody, Robert Carricart, Bert Rosario, Penelope Milford, Matthew Weartz, Ernest Perry Jr, Joe Greco, Bridget O'Connell

Written by Green, this slow moving thriller is set in Chicago, where an unconventional British Catholic priest comes to a rundown parish in search of his missing brother. He soon achieves considerable popularity with the locals, although his failings, including excessive drinking, begin to take their toll. An undemanding but quite watchable effort. Daltrey gives a nice performance as a former professional boxer now fighting bareknuckle.

THR 102 min VIDrel: 20VIS/SONOP V

COLD LIGHT OF DAY, THE *
Rudolf Van Den Berg UK
18
1996
Richard E. Grant, Simon Cadell, Lynsey Baxter, Perdita Weeks, James Laurenson, Heathcote Williams

A police officer is assigned to the case of a serial killer who has murdered a number of female children, and finds that in his desire to trap the culprit he is developing an unhealthy obsession with the case. Set somewhere in Eastern Europe (it was in fact filmed in the Czech Republic) the film has next to no tension, and for good measure the English accents of the cast do nothing for the film's atmosphere.

THR 96 min VIDrel: POLFIL V/s

COLD RIVER **
Fred G. Sullivan USA
(PG)
1981
Suzanne Weber, Pat Petersen, Richard Jaeckel, Robert Earl Jones, Brad

Sullivan, Elizabeth Hubbard, Augusta Dabney, Adam Petroski, David Thomas, Wade Barnes, Deborah Beck, Trent Gough, Robert Donley, Thomas Kubiak, Muriel Mason
An experienced guide takes his two children on a long trek in the mountains but dies of a heart attack, leaving them to survive on their own in sub-zero temperatures. Nice locations in the Adirondack Mountains, but that's about all.
A/AD   85 min (ort 94 min)   SATrel: SKY MOVIES
Boa: novel by William Judson.

COLD STEEL **   18
Dorothy Ann Puzo USA   1987
Brad Davis, Jonathan Banks, Sharon Stone, Jay Acovone, Adam Ant, Eddie Egan, Sy Richardson, Anne Haney, Ron Karabatos, William Lanteau, Minoy Seeger, Jessie Aragon, Pat Asanti, Robert Cervi, Heidi Korzak, Nick Savage
A tough cop finds that his father has been murdered by a psycho, who holds the cop responsible for an accident that destroyed his vocal chords. He sets out to nail the criminal in a predictable and brutal drama, hampered by poor plotting and dialogue.
DRA   90 min   VIDrel: MED/POLYREC L/A   V/h

COLD SWEAT **   15
Terence Young FRANCE/ITALY   1971
Charles Bronson, James Mason, Liv Ullmann, Michel Constantin, Gabriele Ferzetti, Jill Ireland, Jean Topart, Yannick Delulle, Luigi Pistilli
An American charterboat captain living in France, is forced to run drugs by a sinister crime baron, when his family is taken hostage. A predictable adaptation of Matheson's novel, watchable but lacking in tension.
Aka: DE LA PART DES COPAINS; L'UOMO DALLE DUE OMBRE
THR   93 min   VIDrel: WHV   V
Boa: novel Ride the Nightmare by Richard Matheson.

COLD SWEAT **   18
Gail Harvey USA   1993
Shannon Tweed, Ben Cross, Adam Baldwin, Dave Thomas, Henry Czerny, Lenore Zann, Maria Del Mar
A contract killer quits the business after an innocent person dies accidentally but he is persuaded to undertake one last mission by a man who wants revenge on his business partner who is having an affair with his wife. The usual sexual complications follow, in this murky and sombre erotic thriller.
THR   92 min   VIDrel: OCEAN/FIRST   V

COLDITZ STORY, THE ***   U
Guy Hamilton UK   1954
John Mills, Eric Portman, Frederick Valk, Denis Shaw, Lionel Jeffries, Ian Carmichael, Christopher Rhodes, Richard Wattis, Bryan Forbes, Theodore Bikel, Eugene Deckers, Anton Diffring, Ludwig Lawinski, Carl Duering, Keith Pyott
Nicely balanced and fairly realistic film, dealing with the lives of the POWs held by the Germans in Colditz Castle, in Saxony in 1940. A mixture of humour, tension and factual incidents, set against the background of the boring routine of prison life. Screenplay is by Reid, Ivan Foxwell, William Douglas and Guy Hamilton. A British TV series followed in 1972.
WAR   93 min (ort 97 min) B/W   VIDrel: WHV   V
Boa: books The Colditz Story and The Latter Days by P.R. Reid.

COLLECTABLE **   18
Paul Thomas USA   1991
Randy Spears, John Morgan, Jeanna Fine, Victoria Paris, Tiara West, Sikki Nixx, Tom Chapman, K.C. Williams, Randy West, Tom Byron, Marc Wallice, Nina Hartley, Holly Ryder, T.T. Boy, Trixy Tyler, Buck Adams
A rich eccentric enjoys "collecting" both men and woman as sex objects, but takes on more than he bargained for when he hires a man to assist him in finding some new acquisitions, and comes across a beautiful woman he sets his heart on having. Quite a well plotted and coherent tale, despite the director including some clumsy visual effects in an attempt to impart a weird atmosphere to the movie.
A   52 min (ort 80 min)   VIDrel: GROHOM/MAXSCAN   V

COLLECTOR, THE **   15
William Wyler USA   1965
Terence Stamp, Samantha Eggar, Maurice Dallimore, Mona Washbourne, William Beckley, Gordon Barclay, David Haviland
A remarkably flat attempt to film a fine novel, that deals with a soulless young man who collects butterflies until a football pools

win gives him freedom, whereupon he decides to kidnap a girl to add to his collection.
DRA   114 min (ort 119 min)   VIDrel: CASPIC/BMGREC L/A   V
Boa: novel by John Fowles.

COLLECTOR, THE ***   15
Eric Rohmer FRANCE   1967
Patrick Bauchau, Haydee Politoff, Daniel Pommereulle, Alain Jouffroy, Mijanou Bardot
During a summer vacation, a young man becomes attracted to a young girl, a free spirit who has a new lover every night, However, since he and his friend refuse to becomes part of her "collection", they resolve to try to get her to change her ways. The third of the director's "Six Moral Tales" and the first work to arouse widespread attention. Followed by MY NIGHT WITH MAUD, CLAIRE'S KNEE and finally LOVE IN THE AFTERNOON.
Aka: LA COLLECTIONNEUSE
DRA   82 min (ort 88 min)   VIDrel: CONNO/RTM L/A   V

COLLEGE KICKBOXER **   15
Eric Sherman   1991
Ken Randall Johnson, Matthew Roy Cohens, Mark Williams
James and Mark are two college buddies whose devotion to the martial arts leads to one deciding to enter a major kickboxing tournament. However, before he can compete Mark is badly injured in a gang attack to prevent him from taking part. This leaves his friend with no option but to replace him, an action which brings James face to face with the gang-leader responsible. A formula revenger with a well mounted final reckoning.
MAR   88 min   VIDrel: MIA/DISC/IMPENT   V

COLLISION COURSE **   18
Lewis Teague USA   1988
Pat Morita, Jay Leno, Chris Sarandon, Al Waxman, Ernie Hudson, John Hancock, Soon Teck-Oh, Randall "Tex" Cobb
Two police detectives (one from Japan, the other from Detroit) become rivals in the search for the stolen prototype of a car turbocharger, but eventually join forces in an attempt to find it. An average comedy-thriller, made along similar lines to RED HEAT.
A/AD   96 min (ort 99 min)   VIDrel: 20TH/TECH   V/sur

COLONEL REDL ***   15
Istvan Szabo AUSTRIA/HUNGARY/WEST GERMANY 1985
Klaus Maria Brandauer, Armin Muller-Stahl, Gudrun Landgrebe, Jan Niklas, Hans-Christian Blech, Laszlo Mensaros, Andras Balint, Karoly Eperjes, Dorottya Udvaros, Laszlo Galffi, Gabor Ratonyi, Gabor Svidrony, Eva Szabo, Maria Majlath
An account of an ambitious homosexual career soldier, whose working-class background does not hinder his rapid rise in the Austro-Hungarian army. He is ultimately made head of the secret police, but commits suicide just prior to WW1. Inspired by both historical facts and John Osborne's play "A Patriot For Me", this complex look at power politics has little substance, but much style and a superb performance from Brandauer. Followed by HANUSSEN.
Aka: OBERST REDL; REDL EZREDES
DRA   144 min   VIDrel: L/A   V

COLONY, THE **   PG
Rob Hedden USA   1995
John Ritter, Mary Page Keller, Marshall Teague, Todd Jeffries, Alexandra Picatto, Cody Dorkin, Frank Bonner, Michelle Scarabelli, June Lockhard, Hal Linden, John Welsey, Shirley Spangler, Dan Gilvezan, Douglas Rowe, Steve Kronish
After a tycoon buys the husband's home security system, a couple suffer a car hijacking and move to the former's self-contained estate. Surrounded by high railings and gates, it has its own schools and security force and a thick rule book governs every aspect of life. Soon a more sinister picture emerges of a totally controlled society, ruled over by its paranoid founder. A good idea is thrown away completely in a tepid thriller, badly lacking any sense of suspense.
THR   89 min (ort 93 min) cC   VIDrel: CIC/SONOP   V/dm

COLOR OF MONEY, THE ***   15
Martin Scorsese USA   1986
Paul Newman, Tom Cruise, Mary Elizabeth Mastrantonio, Helen Shaver, John Turturro, Bill Cobbs, Robert Agins, Keith McCready, Carol Messing, Steve Mizerak, Bruce A. Young, Forest Whitaker, Alvin Anastaia, Elizabeth Bracco, Jerry Piller

A sharp and tense sequel to THE HUSTLER, with Newman discovering a younger version of himself in the shape of small-time pool hustler Cruise. They form a partnership and through tutoring him, Newman re-discovers his passion for the game. A flashy and exciting film let down by a poor second half that fails to deliver the expected climax. AA: Actor (Newman).
DRA   115 min (ort 119 min)   VIDrel: TOUCH L/A   V
Boa: novel by Walter Trew.

COLOR OF NIGHT *
Richard Rush USA
Bruce Willis, Jane March, Ruben Blades, Brad Dourif, Scott Bakula, Lesley Ann Warren, Lance Henriksen, Kevin J. O'Connor, Andrew Lowery, Eriq La Salle, Jeff Corey, Kathleen Wilhoite, Shirley Knight, John T. Bower, Avi Korein
18
1994
A psychiatrist takes some time out in Los Angeles after the suicide of one of his patients and agrees to take over a therapy group on behalf of a colleague. This brings into contact with a seductive and mysterious femme fatale who embroils him in murder. A poorly plotted and badly acted mess that was much hyped prior to its release on account of allegedly daring sex scenes, which in truth are as erotic as a cold shower but not so invigorating.
THR   117 min (ort 140 min)   VIDrel: POLY/POLYREC
V/sur

COLOR PURPLE, THE ****
Steven Spielberg USA
15
1985
Whoopi Goldberg, Danny Glover, Rae Dawn Chong, Margaret Avery, Willard Pugh, Oprah Winfrey, Akousa Busia, Desreta Jackson, Adolph Cesar, Dana Ivey, Leonard Jackson, Bennet Guillory, John Patton Jr, Carl Anderson, Susan Beaubian
Sprawling epic tale looking at forty years in the life of a Southern black woman, who is cruelly separated from her sister and endures a lifetime of hardship, until a blues singer enters her life and helps her gain self-respect. A beautifully made and acted tearjerker, that piles on the sentiment all the way. Music is by Quincey Jones, with screenplay by Menno Meyjes and striking cinematography by Allen Daviau. Produced and directed by Spielberg.
DRA   148 min (ort 154 min) wScrn cC   VIDrel: WHV
V/sur
Boa: novel by Alice Walker.

COLORS ***
Dennis Hopper USA
18
1988
Sean Penn, Robert Duvall, Maria Conchita Alonso, Randy Brooks, Grant Bush, Don Cheadle, Rudy Ramos, Trinidad Silva, Gerardo Mejia, Glenn Plummer, Rudy Ramos, Sy Richardson, Charles Walker, Damon Wayans, Fred Asparagus, Virgil Frye
An ageing cop near retirement is assigned a rookie partner, and together they are given the task of cleaning the streets of urban gangs in L.A. A brutal and realistic drama that lacks focus but works surprisingly well, with Penn and Duvall both giving believable and sympathetic portraits. The excellent photography is by Haskell Wexler. See also JUDGEMENT.
A/AD   127 min   VIDrel: 4-FRONT/POLYREC   V/sur

COLOSSUS: THE FORBIN PROJECT ***
Joseph Sargent USA
(15)
1970
Susan Clark, Eric Braeden, Gordon Pinsent, William Schallert, Alex Rodine, Leonid Rostoff, Georg Stanford Brown, William Sage, Martin Brooks, Marion Ross, Byron Morrow, Dolph Sweet, Robert Cornthwaite, James Hong, Lew Brown
A surprisingly well-made and chilling tale of human fallibility. The USA develops a super-computer to control all its nuclear defences, but the machine proves to have a consciousness of its own and by linking up with its Russian counterpart, forcibly ushers in an era of universal peace under an electronic dictatorship. Highly recommended as a fine example of a very intelligent SF movie. See also DEMON SEED.
Aka: FORBIN PROJECT, THE
FAN   100 min   TVrel
Boa: novel Colossus by D.F. Jones.

COLOUR OF LOVE, THE ***
Anthony Drazan USA
15
1992
Michael Rapaport, Ray Starkey, DeShonn Castle, N'Bsuhe Wright, Lois Bradler, Kevin Corrigan, Marsha Florence, Ron Johnson, Glen Dassin, Shirley Benyas, Jon Seda, Bobby Joe Travis, Dan Ziskle, Shula Van Buren, Martin Priest, Liana Pai
A powerful, teenage drama set in a Detroit high school where

two young men find their friendship across the colour barrier coming under a great strain when one of them starts dating the other's cousin. An impressing writing and directing debut for Drazan that benefits from realistic locations and dialogue as well as a fine set of musical numbers.
Aka: ZEBRAHEAD
DRA   98 min (ort 102 min) cC   VIDrel: 20VIS/SONOP
V/sh

COLOUR OF POMEGRANATES, THE ****
Sergo Paradjanov USSR
U
1984
Sofico Chiaureli, M. Aleksanian, V. Galsatian, G. Gregechkori, O. Minassian
Lyrical, almost surrealistic account of various episodes (some fictional) taken from the life of the 18th century poet Arutium Sayadin, who under the name of Sayat Nova rose from humble beginnings to great fame as a court minstrel and later archbishop. A fascinating visual experience that cannot be adequately described in words.
Aka: NRNA GOUYNE; RED POMEGRANATE; SAYAT NOVA; TSVET GRANATA
DRA   73 min   VIDrel: CONNO/RTM L/A   V

COLUMBO: A BIRD IN THE HAND **
Vincent McEveety USA
(15)
1992
Peter Falk, Tyne Daly, Greg Evigan, Frank McRae, Don S. Davis, Leon Singer, Michael Gregory, Steve Forrest, Stephen Liska, G.F. Smith, Carol Swarbrick, Ed McCready, John Petlock, Joel Beeson, Kay Perry, Joanna Sanchez
A man with a gambling habit and very big debts believes that the death of his wealthy uncle would solve this problem nicely. He decides to booby-trap the latter's car but things get complicated when the uncle is killed while out jogging, in what appears to be a hit-and-run accident, and the family gardener is killed while moving the car. This leaves our wily detective with quite a mess to sort out, which becomes more complex when our gambler is shot dead.
DRA   90 min mTV   TVrel: BBC1

COLUMBO: A FRIEND IN DEED **
Ben Gazzara USA
(15)
1974
Peter Falk, Richard Kiley, Michael McGuire, Rosemary Murphy, Val Avery, Eric Christmas, Eleanor Zee, John Finnegan, Arlene Martell, Victor Campos, Joshua Bryant, John Calvin, Byron Morrow, James V. Christy, Alma Beltran
Wealthy Hugh Caldwell murders his wife and then in a panic rings his friend Mark Halperin, who just happens to be the Deputy Police Commissioner. He uses his ingenuity to set up a fake robbery, and by doing this puts his friend in the clear. But he has only done this to blackmail his friend, as he wants to be rid of his wife as well, and it is the second murder that exposes them both. Adequate enough, but this is not an especially memorable entry.
DRA   70 min mTV   TVrel: BBC1

COLUMBO: A MATTER OF HONOR **
Ted Post USA
(15)
1976
Peter Falk, Ricardo Montalban, Pedro Armendariz Jr, A. Martinez, Robert Carricart, Maria Grimm
A legendary retired bullfighter is almost a national institution, revered by thousands for his skill and courage. But his closest friend knows his darkest secret, that he has not been able to face a bull for years, having become prey to a paralysing fear once in the bullring. As is so often the case with these stories, a murder is committed to avoid exposure, and the public humiliation that would result from this. A fair outing for Columbo, but not one of his best.
DRA   70 min mTV   TVrel: BBC1

COLUMBO: A STITCH IN CRIME ***
Hy Averback USA
(15)
1972
Peter Falk, Leonard Nimoy, Anne Francis, Nita Talbot, Will Geer, Jared Martin, Aneta Corsaut, Victor Flores, Kenneth Sansom, Murray MacLeod, Leonard Simon, Ron Stokes, Patsy Garrett
A brilliant heart surgeon is so arrogant and career-minded that he coldly plans to murder his aged colleague so that he can receive sole credit for the development of a drug that suppresses tissue rejection during heart transplants. Unfortunately, a senior nurse discovers his murder plan and is murdered to keep her silent, at which point Columbo enters the scene. A beautifully acted battle of wits follows, with our hero allowed one of his rare outbursts of bad temper.
DRA   74 min mTV   TVrel: BBC1

COLUMBO: AGENDA FOR MURDER ** *(15)*
Patrick McGoohan USA   1989
Peter Falk, Patrick McGoohan, Denis Arndt, Louis Zorich, Penny Fuller, Bruce Kirby, Anne Haney, Stanley Kamel, Steven Ford, Arthur Hill, Michael Goldfinger, Shaun Toub, Annie Stewart, Carol Barbee, Peter Atlas, Kirk Thornton, Eva Charnes
A crafty lawyer carries out a carefully planned murder in order to stop a former associate revealing a past secret that would blight his political aspirations after the next Presidential election. A few minor slip-ups are to be his undoing. Strong on plot, this entry is let down by a lack of tension, and McGoohan's bizarre performance verges at times on caricature. For all the care lavished on this episode, it is one that never sparkles.
DRA   90 min mTV   TVrel: BBC1

COLUMBO: AN EXERCISE IN FATALITY ** *(15)*
Bernard Kowalski USA   1974
Peter Falk, Robert Conrad, Philip Bruns, Pat Harrington, Gretchen Corbett, Collin Wilcox, Jude Farese, Darrell Swerling, Dennis Robertson, Raymond O'Keefe, Victor Izay, Eric Mason, J.R. Clark, Mel Stevens, Manuel Depina, Don Nagel
For years a fitness guru, who owns a string of health clubs, has been skimming off the profits from his business interests, and hiding the money abroad. A business associate has finally uncovered this, and he is murdered for his pains. The fitness expert tries to make it look like an accident that occurred during a workout, by lowering a heavy barbell onto the dead man's neck. Quite good, even if it is an overlong outing for our detective.
DRA   120 min mTV   TVrel: BBC1

COLUMBO: ANY OLD PORT IN A STORM **** *(15)*
Leo Penn USA   1973
Peter Falk, Donald Pleasence, Gary Conway, Julie Harris, Joyce Jillson, Dana Elcar, Vito Scotti, Robert Donner, Robert Ellenstein, Robert Walden, Regus J. Cordic, Reid Smith, John McCann, George Gaynes, Monty Landis, Walker Edmiston
A dedicated winemaker shares responsibility for the business with his brother, to whom the land that the winery is sited on belongs. But the latter, who is a playboy with a passion for scuba diving, announces his intention to sell the site to a company specialising in producing low-grade plonk. Appalled at this decision, the winemaker kills his brother, then dresses him in scuba gear to fake an accidental death. One of the best episodes in this fine series.
DRA   90 min (ort 120 min) mTV   TVrel: BBC1

COLUMBO: BLUEPRINT FOR MURDER ** *(15)*
Peter Falk USA   1972
Peter Falk, Patrick O'Neal, Janis Paige, Pamela Austin, John Fiedler, Forrest Tucker, Bettye Ackerman, John Finnegan, Nick Dennis, Robert Gibbons, Cliff Carnell, Jimmy Joyce
Coming home from a trip abroad, a blustering Texan businessman discovers that an ambitious architect has prevailed upon his wife in his absence to fund a controversial multi-million dollar office development. In a rage he makes it clear he is going to cancel this project, but the architect has found a way out of his difficulties. This was Falk's directing debut, but during this time he was in dispute with the studio bosses, and this shows in unevenness of pacing.
DRA   70 min mTV   TVrel: BBC1

COLUMBO: BUTTERFLY IN SHADES OF GREY ** *(PG)*
Dennis Dugan USA   1993
Peter Falk, William Shatner, Molly Hagan, Jack Laufer, Richard Kline, Yorgo Constantine, Mark Lonow, Beverly Leech, Brian Markinson, Robin Clarke, John C. Anders, Christopher Templeton, Denice Kumagi, Derylo Caitlyn, Glenn Taranto
A radio talk-show host resorts to murder to retain his hold over his adopted daughter, who has aspirations of becoming a novelist that are being encouraged by a researcher who works for him. Hampered by over-complexity, this average entry gave Shatner his second starring role as the villain of the piece, the earlier COLUMBO: FADE IN TO MURDER being the other one.
DRA   89 min mTV   SATrel: SKY MOVIES

COLUMBO: BY DAWN'S EARLY LIGHT *** *(15)*
Harvey Hart USA   1974
Peter Falk, Patrick McGoohan, Tom Simcox, Mark Wheeler, Ray DeBenning, Karen Lamm, Madeleine Thornton-Sherwood, Bruce Kirby Sr, Sidney Armus, Robert Clotworthy, Bruce Kirby Jr
Colonel Rumford is the martinet commander of a strict military

academy, who is shocked to learn that the chairman of the board plans to turn the academy into a co-educational establishment. A strong traditionalist, he sees murder as his only way out, and to do this he sabotages a shell, it being the custom to fire a cannon at the annual Founders Day ceremony. It falls to the chairman to enjoy the dubious distinction of firing the cannon, and his death is the result.
DRA   90 min mTV   TVrel: BBC1

COLUMBO: CANDIDATE FOR A CRIME *** *(15)*
Boris Sagal USA   1973
Peter Falk, Jackie Cooper, Ken Swofford, Joanne Linville, Tisha Sterling, Vito Scotti, Robert Karnes, Jay Verela, Regis Cordic, Sandy Kenyon, Jack Riley, Mario Gallo, Jude Farese, Clete Roberts, Angelo Grisanti, Lew Brown
As election days draws closer, a politician finds it helpful to make use of fake death threats masterminded by his campaign manager, this being a ploy to drum up support for a stand against crime. But the campaign manager has grown too powerful, and now demands that his boss end an adulterous affair. Having murdered his campaign manager to break free of his control, the politician tries to use the phoney threats to put himself in the clear. A competent episode.
DRA   90 min mTV   TVrel: BBC1

COLUMBO: CAUTION, MURDER CAN BE A HAZARDOUS AFFAIR ** *(15)*
Daryl Duke USA
Peter Falk, George Hamilton, Peter Haskell, Penny Johnson, Robert Donner, Steven Gilborn, Rick Najera, Marie Chambers, Dennis Bailey, Patricia Allison, Jack Tate, Paul Ganus, Linda Dona, Michael Russo, Louis Herthum
The head of a hugely popular crime-busting TV show decides to commit a murder to conceal some compromising material, and believes he has found a perfect alibi by pretending to have been working late at his office. However, inconsistencies in his story prompt Columbo to investigate a little more deeply.
DRA   90 min mTV   TVrel: BBC1

COLUMBO: COLUMBO CRIES WOLF ** *(15)*
USA   1989
Peter Falk, Ian Buchana, Deidre Hall, Alan Scarfe, Rebecca Staab
The partner of a publisher goes missing and it is thought she has been murdered, but then she turns up fresh from a tour of Europe. However, this ploy is nothing more than a trick to discredit our sleuth and put him off the scent of a real murder. Some interesting twists are all that's on offer in an entry that outstays its welcome.
DRA   95 min mTV   TVrel: BBC1

COLUMBO: COLUMBO GOES TO COLLEGE ** *(15)*
E.W. Swackhamer USA   1990
Peter Falk, Stephen Caffrey, Gary Hershberger, James Sutorius, William Lucking, Katherine Cannon, Allan Fudge, Maree Cheatham, Bridget Hanley, Jim Antonio, Steven Gilborn, Guy Stockwell, Robert Culp, Les Lannom, Karl Wiedergott
Columbo is asked by his boss to give some classes at a college as part of a lecturer's course and so is on hand when the latter is murdered by two spoiled rich kids. They, like almost everybody else, make the fatal mistake of believing that our detective is not very smart and so they feed him false information but fail to understand that he is not so easily fooled. A standard series entry, with very little suspense.
DRA   90 min mTV   TVrel: BBC1

COLUMBO: COLUMBO GOES TO THE GUILLOTINE *** *(15)*
Leo Penn USA   1989
Peter Falk, Anthony Andrews, Karen Austin, James Greene, Alan Fudge, Dana Andersen, Robert Constanzo, Anthony Zerbe, Michael Bacall, Charles Howerton, Milt Kogan, Tony Amendola, Rob Garrison, Frank Simons, Lenny Hicks, Ben Yudell
One of a number of made-for-TV films that attempted to revive the popular series "Columbo", with our shabby police detective solving ingenious crimes by a process of logical deduction. In this one, a stage magician suffers a beheading that appears to have been self-inflicted, but investigations reveal him to have been involved in an attempt to expose a self-proclaimed psychic as a fraud. Our detective sets up an elaborate charade of his own.
DRA   90 min mTV   VIDrel: CIC/SONOP L/A V

COLUMBO: DAGGER OF THE MIND ** (15)
Richard Quine USA 1972
Peter Falk, Richard Basehart, Wilfrid Hyde White, Bernard Fox, John Williams, Honor Blackman, John Fraser, Richard Pearson, Arthur Malet, Harvey Jason, Ronald Long, Hedley Mattingley, John Orchard, Peter Church
In London, a fading husband-and-wife team sees a new production of Macbeth as a chance of staging a comeback, but it has only been funded by an impresario due to the blandishments of the wife. When he realises he has been duped there is a nasty scuffle and he is killed by accident. The couple now try to cover their tracks by taking the corpse home and leaving it at the foot of the stairs. But their attempts at cleverness are their undoing. A stilted and awkward tale.
DRA 120 min mTV TVrel: BBC1

COLUMBO: DEAD WEIGHT * (15)
Jack Smight USA 1971
Peter Falk, Eddie Albert, Kate Reid, Suzanne Pleshette, John Kerr, Val Avery, Timothy Carey, Clete Roberts, Ron Castro, Glen Vernon, Jimmy Pelham, Jim Halferty
Very boring tale of a retired army colonel who murders a procurement officer who lost his nerve over an investigation into corrupt procurement practices, the latter having accepted bribes to win the former colonel supply contracts for the company he now runs. Having shot this man, he takes his boat out to sea to dump the body, but the shooting was seen by a woman out on the marina with her mother. One of the flattest and least complex of these mysteries.
DRA 70 min mTV TVrel: BBC1

COLUMBO: DEATH HITS THE JACKPOT ** (15)
Vincent McEveety USA 1990
Peter Falk, Rip Torn, Jamie Rose, Gary Kroeger, Betsy Palmer, Warren Berlinger, Antony Ponzini, Penny Santon, Marilyn Tokuda, Britt Lind, Robert Alan Browne, Daniel Trent, Donald Craig, Peter Schreiner, Shane McCabe
A down-at-heel photographer is in the throes of a nasty divorce case when he wins several million dollars on a lottery, and asks his uncle to cash the ticket for him so as to prevent his soon to be ex-wife getting a share of his new fortune. But his uncle decides to keep the lot for himself, and arranges matters to make it appear that his nephew was accidentally drowned. As ever, Columbo's dogged persistence pays off.
DRA 90 min mTV TVrel: BBC1

COLUMBO: DEATH LENDS A HAND **** (15)
Bernard Kowalski USA 1971
Peter Falk, Robert Culp, Patricia Crowley, Ray Milland, Brett Halsey, Eric James, Don Keefer, Len Wayland, Lieux Dressler, Barbara Baldavin
A powerful newspaper magnate has his cheating wife investigated by a private detective, who puts her in the clear to her suspicious spouse, but only in order to blackmail her into feeding him useful titbits of information about her husband's activities. But a meeting with her leads to a scene, and he kills her by accident. One of the very best of the "Columbo" mysteries, tightly scripted and compelling – the clever use of split-screen is very effective.
DRA 80 min mTV TVrel: BBC1

COLUMBO: DOUBLE EXPOSURE *** (15)
Richard Quine USA 1973
Peter Falk, Robert Culp, Robert Middleton, Louise Latham, Arlene Martell, Chuck McCann, Denny Goldman, John Milford, George Wyner, Richard Stahl, Francis Desales, Alma Beltran, Dennis Robertson, Harry Hickox, Ann Driscoll
Dr Kepple is a gifted researcher whose speciality is motivational psychology. But he is averse to using a little blackmail and extortion on the side, by getting some of his male clients to compromise themselves with a pretty girl in Kepple's pay. A client threatens to expose Kepple, so he uses all the expertise at his command (including subliminal advertising) to get the client in the right place for a murder to take place. An ingenious episode indeed.
DRA 70 min mTV TVrel: BBC1

COLUMBO: DOUBLE SHOCK *** (15)
Robert Butler USA 1973
Peter Falk, Martin Landau, Paul Stewart, Julie Newmar, Jeanette Nolan, Tim O'Connor, Dabney Coleman, Kate Hawley, Michael Richardson, Robert Rothwell, Gregory Morton, Tony Cristino
An elderly and wealthy man is soon to marry a much younger woman, the two having been drawn together by a shared interest in physical fitness. But this marriage would inevitably lead to a change in his will, and as a way of putting paid to this the man's nephew electrocutes him in the bath, then drapes him over an exercise bike to suggest a heart attack. Landau plays two parts here, as twin nephews who are both out to gain from their uncle's death.
DRA 70 min mTV TVrel: BBC1

COLUMBO: ETUDE IN BLACK *** (15)
Nicholas Colsanto USA 1972
Peter Falk, John Cassavetes, Blythe Danner, Myrna Loy, Anjanette Comer, James Olson, James McEachin, Don Knight, Pat Morita, Michael Pataki, Michael Fox, Dawn Frame, Charles Macaulay, George Gaynes, Wallace Chadwell
A top orchestra conductor is threatened with exposure by the talented pianist he has been carrying on with, and as his rich mother-in-law chairs the orchestra committee, a divorce could ruin him. So he knocks her out, leaves a typewritten note suggesting suicide and then switches on the kitchen gas oven to finish the job. But Columbo finds too many loose ends here, not least being the fact that the girl had a beloved cockatoo, which was allowed to die as well.
DRA 70 min mTV TVrel: BBC1

COLUMBO: FADE IN TO MURDER *** (15)
Bernard L. Kowalski USA 1976
Peter Falk, William Shatner, Lola Albright, Alan Manson, Bert REmsen, Walter Koenig, Danny Dayton, Timothy Agoglia Carey, J.P. Finnegan, Victor Izay, Shera Danese, Jimmy Joyce, Frank Emmett Baxter, Fred Draper
Shatner plays the part of Ward Fowler, an actor who has the role of top TV detective Inspector Lucerne. His dubious past as a U.S. deserter could ruin him, a secret one of the studio bosses knows, and over the years she has used this to extort money from him. Eventually he can face this no longer, and in disguise shoots her at a diner, his belief being that the police will view this as a botched hold-up. It doesn't take real detective Columbo long to crack this case.
DRA 70 min TVrel: BBC1

COLUMBO: FORGOTTEN LADY *** (15)
Harvey Hart USA 1975
Peter Falk, Janet Leigh, Sam Jaffe, John Payne, Maurice Evans, Ross Elliott, Robert F. Simon, Army Archerd, Linda Scott, Francine York, Jerome Guardino, Danny Wells, Harvey Gold
The gala opening of a Hollywood compilation movie has revived interest in the career of a former screen queen, Grace Wheeler. She has been making plans with her former co-star to appear in a Broadway revival of one of her screen triumphs, but her husband, a top diagnostician, will not hear of it. She kills him but makes it look like suicide, and at the end we learn why her husband opposed her plans and just why Columbo decides to let her escape justice.
DRA 70 min mTV TVrel: BBC1

COLUMBO: GRAND DECEPTIONS ** (PG)
Sam Wanamaker USA 1989
Peter Falk, Robert Foxworth, Andy Romano, Janet Eilber, Stephen Elliott, Michael McManus, James Lashly, Lynn Clark, Bennett Liss, Lee Arenberg, John William Gibson, Stephen Quadros, Milt Kogan, Christopher Titus, Rick Marzan
Another one in this so-so 1980s and 1990s series of detective yarns, based the earlier and much better original "Columbo" TV series. In this tale, our sleuth investigates the murder of a sergeant at a military academy, and soon uncovers the culprit – a instructor who thought he had planned a very clever murder, but inevitably gave himself away by overlooking some seemingly insignificant details.
DRA 91 min mTV TVrel: BBC1

COLUMBO: HOW TO DIAL A MURDER *** (15)
James Frawley USA 1978
Peter Falk, Nicol Williamson, Kim Cattrall, Joel Fabiani, Frank Aletter, Tricia O'Neil, Ed Begley Jr, Fred J. Gordon
A murderous psychologist hatches an incredibly ingenious plan to kill his best friend and colleague, who years ago had an affair with his now deceased wife. Two Doberman pinschers are used here, having been conditioned to attack at the sound of a given word spoken after the sound of a telephone bell. A tense and very atmospheric story, it was together with COLUMBO: MURDER UNDER GLASS, one of the few stories to expose the Columbo to personal danger.
DRA 75 min mTV TVrel: BBC1

COLUMBO: IDENTITY CRISIS **
(15)
Patrick McGoohan USA
1975
Peter Falk, Patrick McGoohan, Leslie Nielsen, Otis Young, Bruce Kirby, Vito Scotti, Val Avery, David White, Barbara Rhoades, William Mims, Carmen Argenziano, Cliff Carnell, Edward Bach, Paul Gleason, Angela May, Betty McGuire
A top government agent has been using his position to work as a double agent, but now finds that an old partner who he thought dead, is not only alive and well but knows too much. Fearing that he may eventually be exposed, he arranges to meet his old buddy, deliberately selecting a spot that is much favoured by muggers. Like is COLUMBO: THE MOST CRUCIAL GAME, this story makes use of a tape recording, and Columbo gains a vital clue from listening to it.
DRA   90 min mTV   TVrel: BBC1

COLUMBO: LADY IN WAITING **
(PG)
Norman Lloyd USA
1971
Peter Falk, Susan Clark, Jessie Royce Landis, Richard Anderson, Leslie Nielsen, Joel Fluellen, Richard Bull, Gary Walberg, Barbara Rhoades, Jon Lormer, Frank Baxter, Susan Barrister
A daffy sister kills her brother to get control of the family business, making it appear as if she shot an intruder, having previously so arranged things that her brother would not be able to get into the house in the normal way. An adequate mystery, but the incessant and intrusive Henry Mancini score is irritatingly smarmy and detracts from the tension in a fairly competent story.
DRA   70 min mTV   TVrel: BBC1

COLUMBO: LAST SALUTE TO THE COMMODORE *
(15)
Patrick McGoohan USA
1976
Peter Falk, Robert Vaughn, Fred Draper, Diane Baker, Wilfrid Hyde White, John Dehner, Dennis Dugan, Bruce Kirby, Joshua Bryant, Susan Foster, Rod McCary, J.P. Finnegan, Joseph Roman, Hanna Hertelendy, Jerry Crews, Fred Porter
Family and friends have gathered at the home of gifted naval architect Commodore Swanson, to celebrate the company's annual party. But Swanson finds it increasingly unbearable to contemplate the operations of his son-in-law Charles, who has turned the company into a large, impersonal corporation. When he tells Charles he has decided to sell the company, he is killed for his pains and taken out to see to fake an accidental drowning. A weak and very tedious story.
DRA   70 min mTV   TVrel: BBC1

COLUMBO: LOVELY BUT LETHAL *
(15)
Jeannot Szwarc USA
1973
Peter Falk, Vera Miles, Martin Sheen, Sian Barbara Allen, Fred Draper, Gino Conforti, Colby Chester, Bruce Kirby, John Finnegan, Dick Stahl, Marc Hannibal, Vincent Price, David Toma, Layne Matthess
The head of a cosmetics company is desperate to prevent an employee and former lover from selling the formula for a beauty cream to her business rival. When he fails to respond to her pleas and rejects an offer of money and a partnership, she is so incensed that she kills him. Later on, she is forced to commit another and more cold-blooded murder. However, she proves no match for our persistent lieutenant who eventually finds a way to prove her guilt.
DRA   74 min mTV   TVrel: BBC1

COLUMBO: MAKE ME A PERFECT MURDER **
(15)
James Frawley USA
1978
Peter Falk, Trish Van Devere, Laurance Luckinbill, Patrick O'Neal, Lainie Kazan, James McEachin, Ron Rifkin, Bruce Kirby, Kenneth Gilman, Milt Kogan, Dee Timberlake, Don Eitner, Morgan Upton, Joe Warfield, George Skaff, Susan Bredhoff
McAndrews is a top network programmer who has just won a major promotion, and his chief assistant feels certain she will get his job, especially as they are lovers. But he just doesn't think she is up to it, so she plans his demise, secure in the knowledge that this action will almost certainly result in her getting his job. A shooting takes place, and she makes use of a film screening to establish her alibi, but a dropped glove is to be her undoing. Adequate.
DRA   90 min (ort 120 min) mTV   TVrel: BBC1

COLUMBO: MIND OVER MAYHEM ***
(15)
Alf Kjellin USA
1974
Peter Falk, Jose Ferrer, Lew Ayres, Robert Walker, Jessica Walter, Lee H. Montgomery, Lou Wagner, Art Batanides, Darrell Zwerling,

Charles Macaulay, John Zaremba, William Bryant, Bert Holland, Ed Fury, Jefferson Kibbee, Diane Turley
A young scientist is soon to win an award for his work on molecular power, despite the fact that he stole much of the original research from a now deceased researcher. But it is the man's tyrannical father who is really to blame, and he resorts to murder to silence a scientist threatening to expose the son. A clever story, its main scriptwriter was Steve Bochco, who also wrote COLUMBO: MURDER BY THE BOOK, and created TV successes such as "Hill Street Blues" and "NYPD Blue".
DRA   70 min mTV   TVrel: BBC1

COLUMBO: MURDER – A SELF-PORTRAIT **
(15)
James Frawley USA
1989
Peter Falk, Patrick Bauchau, Fionnula Flanagan, Shera Danese, Isabel Lorca, Vito Scotti, George Coe
After a woman is murdered, her former husband, an unconventional artist, comes under suspicion. But his apparently water-tight alibi forces Columbo to do some hard thinking. Average.
DRA   86 min mTV   TVrel: BBC1

COLUMBO: MURDER BY THE BOOK ***
(15)
Steven Spielberg USA
1971
Peter Falk, Jack Cassidy, Martin Milner, Rosemary Forsyth, Barbara Colby, Lynette Mettey, Bernie Kuby, Hoke Howell, Marica Wallace, Haven Earle Haley
A mystery writer thinks he has committed a perfect crime when he eliminates his ex-partner, having cleverly got the latter to agree to ring home as if he was still in the office, while these two former friends are in fact somewhere else. Of course, he reckons without our intrepid lieutenant, and the interest of a woman who suspects enough to mount a blackmail attempt. A good episode from the original TV series that was generally very well made and entertaining.
DRA   79 min mTV   TVrel: BBC1

COLUMBO: MURDER IN MALIBU **
(15)
Walter Grauman USA
1990
Peter Falk, Andrew Stevens, Brenda Vaccaro, Floyd Levine, Laurie Walters, Sondra Currie, Janet Margolin, Tom Dresden, James Walker, Ben Slack, Yolanda Lloyd, Efrain Figueroa, Mary Margaret Lewis, Robin Gordon, Bill Zuckert
After a female writer of romantic fiction is found murdered at her Malibu beach house, her fiance eventually confesses to the crime. However, Columbo is forced to release the man when it is discovered that she was in fact already dead when he shot her. Naturally, our detective eventually solves this crime thanks once again to some tiny details. A formula entry in a series that unlike the original stories, was not marked by tightness of plotting.
DRA   90 min mTV   TVrel: BBC1

COLUMBO: MURDER OF A ROCK STAR **
(15)
Alan J. Levi USA
1991
Peter Falk, Dabney Coleman, Cheryl Paris, Julian Stone, Sondra Currie, John Martin, Little Richard, Steven Gilborn, John Finnegan, Grant Heslov, Deborah Rose, Tad Horino, Joseph Chapman, Ann Weldon, B.J. Turner, Steve Tschudy
A lawyer tires of his affair with a rock star because of her infidelity, and decides to have his revenge by killing her and so arranging matters that her lover gets the blame. Average.
Aka: COLUMBO AND THE MURDER OF A ROCK STAR
DRA   90 min mTV   SATrel

COLUMBO: MURDER, SMOKE AND SHADOWS **
(PG)
James Frawley USA
1988
Peter Falk, Fisher Stevens, Molly Hagan, Nan Martin, Jeff Perry, Steven Hill, Jerome Guardino, Elizabeth Ruscio, Al Pugliese, Time Winters, Gayle Harbor, Stewart J. Zully, Avner Garbi, Meg James, Lisa Barnes, Robert Madrid
Our shabby police detective investigates a murder that would appear to have a Hollywood connection, the victim being an old school friend of a young prodigy film-maker and special effects expert. A well-made but rather flashy and terribly contrived episode (the first) from a new series of murder mysteries, largely based on the popular TV series of 1970s.
Aka: MURDER, SMOKE AND SHADOWS
DRA   91 min mTV   TVrel: BBC1

COLUMBO: MURDER UNDER GLASS ***
(15)
Jonathan Demme USA
1978
Peter Falk, Louis Jourdan, Shera Danese, Richard Dysart, Mako,

France Nuyen, Michael V. Gazzo, Larry D. Mann, Antony Alda, Todd Martin, Fred Holliday, Alberto Morin, Jim Murphy, Carolyn Martin, Miyako Kurata, Mieko Kobavashi
The members of the Restaurant Developers Association have for years been paying a proportion of their profits to a powerful food critic, who in return praises their establishments. One of the restauranteurs meets with the critic, and makes it plain that he will pay up no longer, and for good measure will expose this extortion. Poisoned wine is the method used here, and rarely did Columbo seem so pleased to catch the killer.
DRA   70 min mTV   TVrel: BBC1

COLUMBO: NEGATIVE REACTION *** (15)
Alf Kjellin USA   1974
Peter Falk, Dick Van Dyke, Antoinette Bower, Don Gordon, Joanna Cameron, David Sheiner, Larry Storch, Joyce Van Patten, Michael Strong, Vito Scotti, Alice Backes, Harvey Gold, Bill Zuckert, Adrian Ricard, Thom Carney, John Ashton
A top photographer has planned the demise of his domineering wife, but to do this he has to employ the services of an ex-convict, his plan being to make it all look like a kidnapping and ransom demand that went wrong. Everything goes according to plan – the wife is killed and then so is the "kidnapper". The murders seem perfect, but soon this ingenious story reveals all the loose ends that trap the real culprit. Some touches of humour are an added pleasure.
DRA   90 min mTV   TVrel: BBC1

COLUMBO: NOW YOU SEE HIM *** (15)
Harvey Hart USA   1975
Peter Falk, Jack Cassidy, Bob Dishy, Robert Loggia, Nehemiah Persoff, Cynthia Sikes, Patrick Culliton, George Sperdakos, Thayer David, Redmond Gleason, Victor Izay, Robert Gibbons, Michael Payne
The Great Santini is a top stage illusionist whose dark secret is that he was once a guard at a Nazi deathcamp, but has gained a new life with a change of identity. Extortion is his great problem here, in the form of a blackmailing club-owner who is ready to inform U.S. Immigration of the truth should Santini not keep up the payments. Using all his knowledge, Santini eliminates this man, and very nearly gets away with it too, but for one single, insignificant error.
DRA   85 min mTV   TVrel: BBC1

COLUMBO: OLD FASHIONED MURDER * (15)
Robert Douglas USA   1976
Peter Falk, Joyce Van Patten, Celeste Holm, Jeannie Berlin, Tim O'Connor, Jess Osuna, Peter S. Feibleman, Jon Miller, Anthony Holland, Lucy Saroyan, Gary Krawford, Eloise Hardt, Morris Buchanan, Giles Douglas
The Lytton Museum is a family-owned collection of medieval artefacts and works of art and Ruth Lytton is its curator, and looking after the collection is her life's passion. But the museum is making a loss, and her brother plans to initiate moves to shut it down, and as one of the trustees can ensure she is outvoted. She arranges a fake robbery, shoots the "burglar", and then kills her brother with the man's gun. A weak entry, spoilt by far too many comic touches.
DRA   70 min mTV   TVrel: BBC1

COLUMBO: PLAYBACK ** (15)
Bernard Kowalski USA   1975
Peter Falk, Oskar Werner, Martha Scott, Gena Rowlands, Robert Brown, Patricia Barry, Herb Jefferson Jr, Trisha Noble, Bart Burns, Steven Marlo, Joe O'Har
An electronics genius has married into money, and for indulged his passion for gadgetry to the full, but is now soon to be ousted as the boss of the family electronics company. For years he has used the credibility of his invalid wife as a tool against his powerful mother-in-law, but this latter now has a private detective's details of his various infidelities. A sluggish and poorly paced murder mystery, it has a strong final clue, but the abrupt ending is weakness.
DRA   70 min mTV   TVrel: BBC1

COLUMBO: PRESCRIPTION – MURDER *** (15)
Richard Irving USA   1967
Peter Falk, Gene Barry, Katherine Justice, Nina Foch, William Windom, Virginia Gregg, Andrea King, Susanne Benton, Ena Hartman, Sherry Boucher, Anthony James
A seemingly dim-witted and sloppily dressed police officer outwits a smooth-talking doctor who murdered his wife. Made

four years before the series actually began and one of the best stories in it.
Aka: PRESCRIPTION MURDER
DRA   99 min mTV   TVrel: BBC1
Boa: play by Richard Levinson and William Link.

COLUMBO: PUBLISH OR PERISH *** (15)
Robert Butler USA   1974
Peter Falk, Jack Cassidy, Jacques Aubuchon, John Chandler, Mariette Hartley, Mickey Spillane, Gregory Sierra, Alan Fudge, Paul Shenar, Jack Bender, Ted Gehring, Vern Rowe, Lew Palter, George Brenlin, J.S. Johnson, Maurice Marsac
A top thriller writer has decided to quit his publishing house and move to another. But the flamboyant boss is not about to let this happen, and is about to ensure that the writer will never pen another bestseller. To do this he employs the services of a psychopathic ex-army man, who has written a book on home-made explosives, the publisher bribing him into carrying out the murder with false promises to publish it. Complex, concise and extremely clever.
DRA   70 min mTV   TVrel: BBC1

COLUMBO: RANSOM FOR A DEAD MAN *** (15)
Richard Irving USA   1971
Peter Falk, Lee Grant, John Fink, Harold Gould, Patricia Mattick, Paul Carr, Jed Allen, Charles Macaulay, Hank Brandt, Jeane Byron, Richard Roat, Norma Connolly, Harlan Warde, Bill Walker, Timothy Agoglia Carey, Judson Morgan
This was the second pilot for the series (following PRESCRIPTION MURDER) and pits Columbo against a smart female lawyer who has come up with a clever scheme to get rid of her wealthy husband, and make it appear to be a contract killing.
DRA   120 min mTV   TVrel: BBC1

COLUMBO: REQUIEM FOR A FALLING STAR *** (15)
Richard Quine USA   1973
Peter Falk, Anne Baxter, Mel Ferrer, Kevin McCarthy, Frank Converse, Pippa Scott, Edith Head, Sid Miller, William Bryant, John Archer, Jack Griffin, Robert E. Meredith, Bart Burns
A former movie star is reduced to occasional bit parts on TV, and for years has been the victim of a nasty gossip columnist, who in addition to his public displays of bile has been blackmailing the actress over some past financial irregularities. When her private secretary of eighteen years announces that she is to marry this columnist, the actress arranges a fire at his garage, but she apparently misses her intended target and kills the secretary instead.
DRA   70 min mTV   TVrel: BBC1

COLUMBO: REST IN PEACE, MRS COLUMBO ** (15)
Vincent McEveety USA   1989
Peter Falk, Helen Shaver, Tom Isbell, Ian McShane, Ed Winter, Teresa Ganzel, Michael Alldredge, Hugh Gillin, Rosanna Huffman, George Buck, Don Calfa, Roscoe Lee Browne
When her husband finally dies in jail, the man's unbalanced widow looks for those to blame for this, eventually latching on to Detective Columbo, who was instrumental in securing the man's conviction. To get her revenge, she creates a neat plan to poison the detective's wife, but our wily sleuth is one step ahead of her from the very start. The unusual plot device does little for this weak entry.
Aka: REST IN PIECES, MRS COLUMBO
DRA   92 min mTV   TVrel: BBC1

COLUMBO: SEX AND THE MARRIED DETECTIVE ** (15)
James Frawley USA   1989
Peter Falk, Lindsay Crouse, Julia Montgomery, Peter Jurasik, Ken Lerner, Marge Redmond, Stephen Macht, Dave Florek, Harry Johnson, Stewart J. Zully, Pierrino Mascarino, Leeza Vinnichenko, Susan Gibney, Peter Wise, Gary Berner
Another 1980s stab at reviving the popularity of the 1970s police detective series, with the plot revolving around a sex therapist who murders her lover when she discovers that he has been unfaithful. As with the first of these episodes, the story is painfully drawn out, strangely old-fashioned and fairly devoid of tension, though Crouse is more convincing as a therapist than she ever was in HOUSE OF GAMES.
Aka: SEX AND THE MARRIED DETECTIVE
DRA   90 min mTV   TVrel: BBC1

COLUMBO: SHORT FUSE *** (15)
Edward M. Abroms USA   1971
Peter Falk, Roddy McDowall, Anne Francis, Ida Lupino, James Gregory, William Windom, Steve Gravers, Lawrence Cook, Rosalind Miles, Lew Brown, Jason Wingreen, Eddie Quillan, Stuart Nisbet, Annette Molen
A brilliant research scientist is being blackmailed by his devious uncle into co-operating in the sale of his late father's company, but rather than submit to this, he hatches a clever scheme to blow up his uncle's car. But this not only results in the death of the uncle, but also his chauffeur. It doesn't take Columbo too long to reach the correct conclusion, but he has to think up a neat trick to get the killer to incriminate himself.
DRA   70 min mTV   TVrel: BBC1

COLUMBO: SUITABLE FOR FRAMING *** (15)
Hy Averback USA   1971
Peter Falk, Ross Martin, Don Ameche, Kim Hunter, Rosanna Huffman, Vic Tayback, Joan Shawlee, Barney Phillips, Mary Wickes, Sandra Gould, Curt Conway, Claude Johnson, Dennis Rucker
When his millionaire uncle changes his mind about leaving him his priceless art collection, an art critic devises what he thinks is the perfect murder. This involves both a willing accomplice and the use of an electric blanket to keep his uncle's body warm and thus confuse the police as to the time of the murder. However, from the very outset Columbo ferrets out inconsistencies and mistakes that expose the killer. Another excellent episode in this fine series.
DRA   76 min mTV   TVrel: BBC1

COLUMBO: SWAN SONG *** (15)
Nicholas Colasanto USA   1974
Peter Falk, Johnny Cash, Ida Lupino, William McKinney, Sorrell Booke, John Dehner, Bonnie Van Dyke, Janit Baldwin, Vito Scotti, John Randolph, Lucille Meredith, Richard Caine, Donald Mantooth, Jefferson Kibbee, Doug Dirkson
Country singer Tommy Brown enjoys success at the top of his profession, but it is really his tyrannical and fanatically religious wife who controls the finances. She does this by blackmailing her husband, as he used to enjoy a liaison with a sixteen-year-old girl, as he is an ex-convict he cannot afford to risk a charge of statutory rape. A deliberate plane crash is his solution. Cash makes an endearing villain, and Columbo seems genuinely sorry to catch him.
DRA   90 min mTV   TVrel: BBC1

COLUMBO: THE BYE-BYE SKY HIGH I.Q. MURDER CASE ** (15)
Sam Wanamaker USA   1977
Peter Falk, Theodore Bikel, Samantha Eggar, Sorrell Booke, Kenneth Mars, Todd Martin, Basil Hoffman, Howard McGillin, George Sperdakos, Dorrie Thomson, Carol Jones, Jamie Lee Curtis, Carlene Watkins, Fay DeWitt, Kathleen King
Brandt and Hastings are partners in an accounting firm, and also share an interest in the Sigma Society, which just like the real club Mensa, is a group for those with a high I.Q. The pompous Brandt has been embezzling in the funds of wealthy clients, but mostly on account of his wife's lavish tastes, and Hastings has learned of this. An unsatisfactory story, its resolution depends on the murderer incriminating himself, and this is just not believable.
DRA   70 min mTV   TVrel: BBC1

COLUMBO: THE CONSPIRATORS *** (15)
Leo Penn USA   1978
Peter Falk, Clive Revill, Jeanette Nolan, Bernard Behrens, Michael Horton, Albert Paaylsen, L.Q. Jones, Deborah White, Sean McClory, Michael Prince, Donn Whyte, Johnny Silver, Carole Hemingway, Tony Giorgio, John McCann, Doreen Murphy
The public life of an Irish poet and author is one of peace and goodwill to all men, but his private one revolves around shipments of arms to the IRA back home. But his activities have put him at the mercy of an arms dealer, who is demanding more money for his help in supplying weapons for these secret shipments. He is shot with his own pistol, but a scratch on a whisky bottle is the vital clue here. A solid story, this was the last one in the 1970s series.
DRA   120 min mTV   TVrel: BBC1

COLUMBO: THE GREENHOUSE JUNGLE *** (15)
Boris Sagal USA   1972
Peter Falk, Ray Milland, Bradford Dillman, Bob Dishy, Sandra Smith,
Arlene Martel, William Smith, Robert Karnes, Milton Frome, Peggy Mondo, Richard Annis, Larry Watson
A man who sees wealth as a means of winning back his cheating wife agrees to a plan suggested by his uncle to fake his own kidnapping, with a ransom demand being used to unlock money that is held in a trust fund. But the uncle double-crosses his gullible nephew, and kills him with the same pistol they used to set up the earlier kidnapping ruse. A neat little tale, slightly hampered by poor pacing and cutting, but eminently watchable just the same.
DRA   70 min mTV   TVrel: BBC1

COLUMBO: THE MOST CRUCIAL GAME *** (15)
Jeremy Kagen USA   1972
Peter Falk, Robert Culp, Dean Stockwell, Valerie Harper, James Gregory, Dean Jagger, Susan Howard, Val Avery, Los Angeles Lakers, Kathryn Kelly Wiget, Richard Stahl, Don Keefer, Cliff Carnell, Joe Renteria, Ivan Naranjo
The manager of a sports empire, which includes ownership of an L.A. football team, finds it increasingly irksome to work for his boss. This latter is something of a swinger, who cares nothing for his acquisitions, and is seen as a threat to the manager's growing ambitions. Before long the boss is dead, knocked out with a piece of ice and left to drown as if by accident in the pool. But the melted murder weapon leaves traces, and Columbo soon spots their significance.
DRA   70 min mTV   TVrel: BBC1

COLUMBO: THE MOST DANGEROUS MATCH *** (15)
Edward M. Abroms USA   1973
Peter Falk, Laurence Harvey, Kack Kruschen, Lloyd Bochner, Heidi Bruhl, Paul Jenkins, Michael Fox, Oscar Beregi, Mathias Reitz, Drout Miller, Manuel Depina, Stuart Nisbet, Abigail Shelton, John Finnegan
An American grandmaster is out to prove that he is the greatest living chess player, but he can only do this if he can get a top Soviet player, now retired owing to ill health, to agree to a match to settle this matter. These two players meet at a restaurant prior to the match, where using table crockery the Soviet grandmaster demonstrates his superiority. But the American player cannot face the prospect of a public humiliation, and his thoughts turn to murder.
DRA   70 min mTV   TVrel: BBC1

COLUMBO: TROUBLED WATERS ** (15)
Ben Gazzara USA   1975
Peter Falk, Robert Vaughn, Patrick Macnee, Bernard Fox, Poupee Bocar, Dean Stockwell, Jane Greer, Robert Douglas, Susan Damante, Peter Maloney, Curtis Credel
In this story a murder takes place on an ocean liner, with Lieutenant Columbo and his wife onboard (as ever, we don't get to see her) enjoying a cruise down to Mexico. The culprit is an auto executive, who has been having an adulterous affair with a female singer also onboard, but she has now threatened to tell his wife. The woman is shot, her ex-boyfriend (the band's pianist) is framed, and Columbo goes into action. Slack, contrived and quite uninteresting.
DRA   90 min mTV   TVrel: BBC1

COLUMBO: TRY AND CATCH ME *** (15)
James Frawley USA   1977
Peter Falk, Ruth Gordon, Mariette Hartley, Charles Frank, G.D. Spradlin, Mary Jackson, Jerome Guardino, Marie Silva-Alexander
A famous lady mystery writer becomes convinced that her niece did not die in a boating accident but was murdered by her greedy husband. playing on his weakness for money, she lures him into her vault and locks him in, leaving him to suffocate. After the body is discovered some days later, Columbo finds that some very paltry clues but is eventually able to understand their significance and make his case.
DRA   73 min mTV   TVrel: BBC1

COLUMBO: UNDERCOVER *** (PG)
Vincent McEveety USA   1992
Peter Falk, Ed Begley Jr, Burt Young, Harrison Page, Shera Danese, Edward Hibbert, Kristin Bauer, Albie Selznick, Joe Chrest, Robert Donner, Tyne Daly, Hawk Garrett, Denny Santon, Marla Adams, Marianne Muellerleile, Ova Frosh
In this crime mystery, Columbo has to solve a case that has two dead men, who it would appear have killed each other. As ever, matters do not turn out to be quite so straightforward, but a fragment of a photograph clutched in the hand of one of the men plays a crucial part in the story. Columbo goes undercover, following the trail of the $4,000,000 proceeds of a robbery that

took place seven years earlier. A clever and original plot maintains interest.
DRA 90 min mTV SATrel: SKY MOVIES

COLUMBO: UNEASY LIES THE CROWN ** (15)
Alan J. Levi USA 1989
Peter Falk, James Read, Jo Anderson, Nancy Walker, Dick Sargent, Ron Cey, Marshall Teague, Mark Arnott, James A. Watson Jr, Steven Gilborn, Raymond Singer, Victor Bevine, John Roarke, Paul Burke
When a dentist learns that his wife is having an affair with one of his patients, he hatches an ingenious plan to kill this man, making use of a crown that carries some concealed poison. This ingenious idea presents Columbo with a difficult case, which he only solves by making use of a chemistry lesson and a clever bluff to flush out the culprit. Quite a good episode, even if the plot is rather over-extended.
DRA 90 min mTV TVrel: BBC1

COMA *** 15
Michael Crichton USA 1978
Genevieve Bujold, Michael Douglas, Richard Widmark, Elizabeth Ashley, Lois Chiles, Lance LeGault, Harry Rhodes, Rip Torn, Frank Downing, Richard Doyle, Alan Haufrect, Michael MacRae, Betty McGuire, Tom Selleck, Charles Siebert
Patients undergoing routine surgery in a hospital, are being killed and their organs used. A woman doctor defies her male superiors to investigate this. Bujold and Widmark are excellent as opponents in a frightening battle of wits, in which she risks her life to uncover this sinister operation. The script is by Michael Crichton. See also THE HARVEST, BEYOND FORGIVENESS, EXTREME MEASURES and MORTAL FEAR, just four of a number of thrillers given a medical setting.
THR 108 min VIDrel: MGM/WHV V/h
Boa: novel by Robin Cook.

COMANCHE STATION *** PG
Budd Boetticher USA 1960
Randolph Scott, Nancy Gates, Skip Homeier, Richard Rust, Rand Brooks, Dyke Claude Akins, Dyke Johnson, Foster Hood, Joe Molina, Vince St Cyr, John Patrick Noland, P. Holland
A man who is still searching for his wife, kidnapped ten years previously, interrupts his search when he undertakes to rescue a man's wife who has suffered a similar fate. He is successful in this mission but along the way is forced to accept help from a group of outlaws. A concise and sensitively handled tale, it was the last of seven movies the director made with Scott. See also THE SEARCHERS, a film that almost certainly inspired this one.
WES 70 min (ort 74 min)
VIDrel: ENCORE/SPEAR/COLUM V

COMANCHEROS, THE *** PG
Michael Curtiz USA 1961
John Wayne, Stuart Whitman, Ina Balin, Lee Marvin, Bruce Cabot, Nehemiah Persoff, Michael Ansara, Pat Wayne, Joan O'Brien, Jack Elam, Edgar Buchanan, Henry Daniell, Richard Devon, Steve Baylor, John Dierkes, Roger Mobley
A Texas Ranger teams up with a gambler, to rid Texas of renegade whites who sell guns and whisky to the Indians. A likeable and rough Western, with some good action sequences. Curtiz's last film.
WES 103 min (Cut at film release by 11 sec)
VIDrel: 20TH/TECH V/h
Boa: novel by Paul I. Wellman.

COME AND GET IT ** PG
Howard Hawks/William Wyler/Richard Rosson USA 1936
Edward Arnold, Joel McCrea, Frances Farmer, Walter Brennan, Andrea Leeds, Frank Shields, Mady Christians, Mary Nash, Clem Bevans, Edwin Maxwell, Cecil Cunningham, Harry Bradley, Rollo Lloyd, Charles Halton, Phillip Cooper
Sprawling account of the life and loves of a lumber tycoon in 19th century Wisconsin. Both he and his son fall in love with the daughter of a woman whom he rejected so as to concentrate on making his fortune. The fine detail and flawless performances more than compensate for the lack of structure. Farmer gives one of her best performance playing two roles, but it's Brennan who got the Oscar. AA: S. Actor (Brennan).
Aka: ROARING TIMBERS
DRA 99 min B/W VIDrel: VCC/DISC V
Boa: novel by Edna Ferber.

COME DIE WITH ME ** (15)
Armand Mastroianni USA 1994
Rob Estes, Pamela Lee (Anderson), Randi Ingeman, Bert Remsen, Chuck McCann, Joyce Brothers, Jason Schombing, Erica Yohn, Kent Williams, Darlanne Fluegel
Micky Spillane's Mike Hammer looks for man who vanished fourteen years ago, when he is hired by the man's daughter ("Baywatch" star Lee). Predictably, our 'tec is soon in far deeper than he expected to be, especially when he finds himself falling in love with his attractive client. Average entry in this TV movie detective genre.
THR 85 min (ort 88 min) mTV SATrel: SKY MOVIES

COME ON GEORGE * U
James Cruz USA 1938
George Formby, Pat Kirkwood, Joss Ambler, Meriel Forbes, Cyril Raymond, George Hayes, George Carney, Ronald Shiner, Gibb McLaughlin, Hal Gordon, Davy Burnaby, C. Denier Warren, James Hayter, Syd Crossley
Formby plays a young stableboy who by mischance, is on the run from the police. Having taken refuge in a horsebox which holds a remarkably unmanageable beast, he makes friends with it, riding it to victory in a big race. A dated and clumsy Formby comedy.
COM 86 min (ort 88 min) B/W VIDrel: LUMI/SONOP L/A V

COME SEE THE PARADISE *** 15
Alan Parker USA 1990
Dennis Quaid, Tamlyn Tomita, Sab Shimono, Shizuko Hoshi, Stan Egi, Ronald Yamamoto, Akemi Nishino, Naomi Nakano, Brady Tsurutani, Pruitt Taylor Vince
Having made the mistake of falling for and marrying a girl of Japanese parentage in the late 1930s, a union official finds himself separated from his family when the latter are interned following the bombing of Pearl Harbour. A harsh and bleak account of one of the less honourable chapters in American history, that for all its undoubted honesty remains strangely unmoving. Written by Parker and with original music by Randy Edelman.
DRA 127 min (ort 138 min) VIDrel: 20TH/TECH V/sur

COMEBACK, THE ** PG
Jerrold Freedman USA 1988
Robert Urich, Chynna Phillips, Mitchell Anderson, Ronny Cox, Brynn Thayer, Harvey Martin, Allen Hamilton, Sierra Pecheur, Aaron Fletcher, Mark Bradley, Kofi Brewer, Paul Meshejian, Peer Syvertsen, Terry O'Sullivan
A former football star returns to his home town after spending years living as a playboy, and attempts to form a relationship with a son he barely knew. Matters become complicated when the girlfriend of his son falls for him, and he discovers that she is in fact the daughter of a former associate who brought about his early retirement. Standard soap opera tale, with a football flavour.
DRA 90 min VIDrel: 20TH V
Boa: story Eye Of The Beholder by Seymour Epstein.

COMEDY OF ERRORS, THE ** U
Cellan Jones UK 1984
Judi Dench, Francesca Annis, Michael Williams, Wendy Hiller, Cyril Cusack, Suzanne Bertish, Charles Gray, Roger Daltrey, Michael Kitchen, Joanne Pearce, Marsha Fitzalan, David Kelly, Ingrid Pitt, Sam Dastor
Shakespeare's enjoyable comedy of intrigue and mistaken identity is brought to life in this well-mounted production.
COM 130 min VIDrel: POLY/POLYREC V
Boa: play by William Shakespeare.

COMES A HORSEMAN ** 15
Alan J. Pakula USA 1978
Jane Fonda, James Caan, Jason Robards, George Grizzard, Richard Farnsworth, Jim Davis, Mark Harmon, Macon McCalman, Basil Hoffman, James Kline, James Keach, Clifford A. Pellon
Simple Western story set in 1940s Montana, which tells of ranchers trying to save their land from an oilman and an evil cattle baron. A low-key and often ponderous melodrama, sullen and atmospheric, but short on action.
WES 113 min (ort 119 min) VIDrel: MGM/WHV V

COMFORT AND JOY ** PG
Bill Forsyth UK 1984
Bill Patterson, C.P. Grogan, Eleanor David, Alex Norton, Patrick Malahide, Rikki Fulton, Roberto Bernardi, George Rossi, Peter Rossi,

Billy McElhaney, Gilly Gilchrist, Caroline Guthrie, Oma McCracken, Elizabeth Sinclair
The life of a Glasgow D.J. starts to fall apart, when his girlfriend ditches him and he becomes involved in a vendetta between warring Italian ice-cream families. A disappointing film whose comic points fall flat, since the humour is so dry as to be non-existent, and all the mindless violence and stupidity is depressing rather than funny. Written and directed by Forsyth, who gave us the superior GREGORY'S GIRL and LOCAL HERO.
COM   100 min (ort 106 min)   VIDrel: BRAVE/SONOP L/A V

COMFORT OF STRANGERS, THE ***   18
Paul Schrader ITALY/USA   1990
Christopher Walken, Rupert Everett, Natasha Richardson, Helen Mirren, David Ford, Manfredi Aliquo, David Franco, Rossana Cachiari, Fabrizio Castellani, Giancarlo Previati, Antonio Serrano, Mario Catone
Two lovers come to Venice for a holiday, and find themselves being wooed by a creepy married couple whose intentions are less than wholesome. A dark, moody and erotic story that leaves a lasting impression, even if the climax is wholly implausible and the acting variable. Screenplay is by Harold Pinter.
Aka: IN THE COMFORT OF STRANGERS
DRA   100 min (ort 105 min)
VIDrel: ENCORE/SPEAR/COLUM   V
Boa: novel by Ian McEwan.

COMIC STRIP PRESENTS: DIDN'T YOU KILL MY BROTHER? ***   15
Bob Spiers UK   1987
Alexei Sayle, Beryl Reid, Peter Richardson, Graham Crowden, Pauline Melvile
A humorous episode from this uneven comedy series, with the story revolving around the exploits of two brothers – one a demented East End gangster and the other a gentle no-hoper, who has emerged from prison with over four-hundred qualifications and has taken on a job as an "Unstructured Activities Co-ordinator" to a group of juvenile delinquents.
Aka: DIDN'T YOU KILL MY BROTHER?
COM   55 min mTV   VIDrel: POLY/POLYREC   V

COMIC STRIP PRESENTS: MISTER JOLLY LIVES NEXT DOOR **   15
Stephen Frears UK   1987
Peter Cook, Adrian Edmondson, Rick Mayall, Peter Richardson, Nicholas Parsons, Dawn French, Jennifer Saunders
An episode from the "Comic Strip" series with Mayall and Edmondson as a couple of tramps who run an escort agency as a side-line. One day they intercept a package sent to their lunatic hit-man neighbour, and discover a message to "take out" the popular quizmaster Nicholas Parsons. A coarse and vulgar effort, whose occasional moments of humour tended to be swamped in a feast of tastelessness.
Aka: MISTER JOLLY LIVES NEXT DOOR
COM   52 min mTV   VIDrel: POLY/POLYREC   V

COMIC STRIP PRESENTS: MORE BAD NEWS **   15
Adrian Edmondson UK   1986
Adrian Edmondson, Rik Mayall, Nigel Planer, Peter Richardson, Jennifer Saunders, Dawn French
Another tale in the "Comic Strip" series, with a dreadful rock group being brought together five years after they made an unsuccessful tour, the purpose of the reunion being to win a lucrative record contract and play a gig in front of 70,000 fans. As might be expected, nothing turns out quite as planned in this fairly amusing offering.
Aka: MORE BAD NEWS
COM   51 min mTV   VIDrel: POLY/POLYREC   V/sh

COMIC STRIP PRESENTS: THE FUNSEEKERS *   15
Simon Wright UK   1987
Nigel Planer, Keith Allen, Peter Richardson, Kevin Allen, Cathy Burke, Liz Crowther, Katrin Cartlidge, Mark Elliot, Michael Winstanley
Another story from this popular TV series that attempted to present crude parodies of events and genres. In this tale some boozing slobs go on a two-week holiday in Spain and cause general chaos.
Aka: FUNSEEKERS, THE
COM   52 min; 70 min (2-episode cassette) mTV
VIDrel: POLY/POLYREC   V

COMIC STRIP PRESENTS: THE STRIKE ***   15
Peter Richardson UK   1986
Peter Richardson, Nigel Planer, Robbie Coltrane, Keith Allen, Rik Mayall, Adrian Edmondson, Jennifer Saunders, Alexei Sayle, Dawn French
Another episode in this series. This one is a clever parody of how Hollywood sets about making a film of the 1984 British miners' strike, perverting an honest screenplay and destroying the credibility of the writer in the process. Having given Al Pacino and Meryl Streep the roles of Mr and Mrs Scargill respectively, no expense is spared in creating a Hollywood travesty of the truth, in a witty and engaging film-within-a-film.
Aka: STRIKE, THE
COM   59 min; 70 min (2-episode cassette) mTV
VIDrel: POLY/POLYREC   V

COMIC STRIP PRESENTS: THE YOB *   15
Ian Emes UK   1987
Keith Allen, Adrian Edmondson, Peter Richardson, Gary Olson
A parody of films of the 1950s, with a successful director falling victim to a brain transplant experiment and turning into a hooligan.
Aka: YOB, THE
COM   52 min; 92 min (2-episode cassette) mTV
VIDrel: POLY/POLYREC   V/h

COMING, THE ***   18
Bert I. Gordon USA   1980
Susan Swift, Albert Salmi, Guy Stockwell, Tisha Sterling, Beverly Ross, Dana Hardwick, David Rounds, Lauren Dowling, Frank Dolan, John Peters, Judd Dodd, Jennie Babo, Terese Giammarco, Janice Wayne, Harold Jackson
Set at the height of the Salem witch-hunt in 1692, with a woman who has sent 20 innocent people to their deaths now branding a child of five a witch. Though her father has been burnt, his will to save his child carries him 300 years into the future to present day Salem. There he intends to find the reincarnation of this vengeful woman and save his child from death.
HOR   86 min Cut (1 min 11 sec)   VIDrel: IMPENT   V

COMING HOME ***   18
Hal Ashby USA   1978
Jane Fonda, Jon Voight, Bruce Dern, Robert Carradine, Robert Ginty, Penelope Milford, Charles Cyphers, Mary Jackson, Ken Augustine, Tresa Hughes, Ron Amador, Willie Tyler, David Glennon, Olivia Cole, Cornelius H. Austin Jr
A lonely wife whose husband is away fighting in the Vietnam war, falls in love with a paralysed veteran. An unusually mature examination of the Vietnam War and its effects on people's lives, traces of mawkish melodrama do not spoil it. AA: Actor (Voight), Actress (Fonda), Screen/orig (Nancy Dowd/Waldo Salt/Robert C. Jones).
DRA   122 min   VIDrel: MGM/WHV   V
Boa: book by Nancy Dowd.

COMING OF ANGELS, A *   18
Joel Scott USA   1977
Annette Haven, Jamie Gillis, Amber Hunt, Leslie Bovee, Abigail Clayton, Eric Edward, Susan McBain, John Leslie
Inspired in name only by the American TV series "Charlie's Angels", which had three females fighting crime, this film concentrates on the foolish tale of a guy who can only experience orgasms as a voyeur. Forgettable nonsense. A sequel called "Angels" came in 1986.
A   87 min Cut   VIDrel: HAR/GOLD   V

COMING OF CHRISTY, THE **   18
Tina Marie USA
Christy Canyon, Renee Foxxe, Peter North, T.T. Boy, Kristarrah Knight, Mike Horner, Madison, Wayne Summers
The wife of a movie-studio boss tires of his infidelity and embarks on a few escapades of her own. Average.
A   49 min (ort 73 min) mVid   VIDrel: IMPENT   V

COMING OUT **   15
Heiner Carow EAST GERMANY   1989
Mathias Freihof, Dagmar Mazel, Michael Gwisdek, Werner Dissel, Dirk Kummer
A teacher attempts to put his gay past behind him, but without success and is soon back with a young male lover, but this creates a sense of isolation both at work and among his friends.

Grim, depressing and not especially profound, it all takes places against the background of the fall of the Berlin Wall.
DRA   110 min   VIDrel: DTK/TOTAL L/A   V

COMING THROUGH ** **   15
Peter Barber-Fleming UK   1985
Kenneth Branagh, Helen Mirren, Alison Steadman, Philip Martin Brown, Norman Rodway
An examination of D.H. Lawrence's relationship with his eventual wife Frieda von Richthoper, that uses flashbacks to explore the writer's early life, whilst at the same time being rather awkwardly coupled to a present-day tale of a young Lawrence scholar, and her failure to relate to the work of the writer she is studying. A dry lecture of a film, nice to look at but coldly distant.
DRA   78 min (ort 80 min) mTV   VIDrel: CENTV/VCC L/A V

COMING TO AMERICA *   15**
John Landis USA   1988
Eddie Murphy, James Earl Jones, Madge Sinclair, Arsenio Hall, Shari Headley, Paul Bates, Allison Dean, Eriq LaSalle, John Amos, Vanessa Bell, Louie Anderson, Sheila Johnson, Feather, Stephanie Simon, Garcelle Beauvais, Victoria Dillard
The prince of an African royal family decides to get wed, and comes to New York where he believes he will find the girls of his dreams. A sugary romantic comedy in the old style, with numerous in-jokes and a few amusing cameos.
COM   112 min (ort 116 min)
VIDrel: 4-FRONT/POLYREC/CIC L/A   V/dm

COMING TOGETHER ** **   R18/18
Paul G. Vatell USA   1985
Colleen Brennan, Kimberly Carson, Herschel Savage, Greg Ruffner
Tracy is married to Mark and fears that their marriage has grown stale. Despite her husband's insistence that he has stayed faithful, his decision to go duck-hunting with a pal arouses her suspicions, and when she learns that the friend has not seen him, she sets off with an old girlfriend for some adventures of her own.
A   63 min Cut (4 sec); 51 min Cut (2 min 17 sec – 18 ver)
VIDrel: SHEP L/A (R18 ver); SCRN/TERRY L/A; HAR/GOLD   V

COMMANDER, THE ** **   15
Anthony M. Dawson (Antonio Margheriti) ITALY/WEST GERMANY   1988
Lewis Collins, Lee Van Cleef, Donald Pleasence, Manfred Lehmann, Brett Halsey, Chat Silayan, John Steiner, Hans Leutenegger, Christian Bruckner, Frank Glaubrecht, Thomas Danneberg, Anita Lochner, Wolfgang Kuhne
Lightweight and corny actioner, with Collins leading an intrepid band of mercenaries into the steamy jungles of the Far East, in pursuit of a drugs baron who has come into possession of information that may tell him the identity of a mole working for the Drug Enforcement Agency. All the stock cliches are here, in a film with big guns, big trucks, lousy acting and a banal and convoluted script.
A/AD   84 min Cut (7 sec – ort 98 min)
VIDrel: EIV/SONOP   V

COMMANDO *   18**
Mark L. Lester USA   1985
Arnold Schwarzenegger, Dan Hedaya, Rae Dawn Chong, Vernon Wells, David Patrick Kelly, Alyssa Milano, James Olson, Bill Duke, Drew Snyder, Sharon Wyatt, Michael DeLano, Bob Minor, Mike Adams, Carlos Cervantes, Hank Celia, Julie Hayer
A violent, comic-strip adventure, with Schwarzenegger a retired special agent who is forced to don his combat gear once more, when his beloved daughter is kidnapped as part of a plan to restore a Central American dictator to power. Occasional flashes of humour from the star do little to soften the film's mindless noise and violence.
A/AD   86 min (Cut at film release by 12 sec – ort 90 min)
VIDrel: 20TH/TECH L/A; ENCORE (LV only)   V/h LV

COMMANDO LEOPARD ** **   15
Anthony M. Dawson (Antonio Margheriti) ITALY/WEST GERMANY   1985
Lewis Collins, Klaus Kinski, John Steiner, Manfred Lehman, Cristina Donadio, Hans Leutenegger, Alan C. Walker, Francis Derosa, Thomas Danneberg, Michael James, Rene Abadeza, Subas Herrera
An elite commando force is sent to overthrow a corrupt South

American dictatorship, in this competent but uninspired action picture. Filmed in the Philippines.
Aka: KOMMANDO LEOPARD
A/AD   91 min   VIDrel: EIV/SONOP   V/sur

COMMITMENTS, THE *   15**
Alan Parker USA   1991
Robert Arkins, Michael Aherne, Angeline Ball, Maria Doyle, Dave Finnegan, Bronagh Gallagher, Felim Gormley, Glen Hansard, Dick Massey, Johnny Murphy, Kenneth McCluskey, Andrew Strong, Colm Meaney, Anne Kent, Andrea Corr, Ger Ryan
The lives and loves of various eccentric and/or obsessed characters, mostly detailing the efforts of one Jimmy Rabbitte to realise his dream, and bring soul music to Dublin. Having placed an appropriate ad in the local paper, he ruthlessly ploughs through countless hopefuls, eventually forming his own band. As the group grows in stature, various tensions and rivalries appear and this boisterous if uneven comedy ends with the band's inevitable self-destruction.
COM   113 min (ort 118 min)
VIDrel: 20TH/TECH L/A; ENCORE (LV only)   V/sh LV
Boa: novel by Roddy Doyle.

COMMITTED ** **   15
William A. Levey USA   1988
Jennifer O'Neill, Robert Forster, Ron Palillo, Sydney Lassick, William Windom, Richard Alan, Greg Latter, Lynn White, Aletta Bezuidenhout, Dennis Smith, Manfred Seipold, Frank Opperman, John Maytham, Deon Stewardson
Devastated by the suicide of her employer and friend, a nurse accepts a job working at a famous mental hospital, only realising too late that the "routine forms" she signed were in fact her own commitment papers. The expected nightmare of confinement, delusion and murder ensues. Filmed on location in South Africa.
THR   88 min (ort 93 min)   VIDrel: COLUM/SONOP   V

COMMON-LAW CABIN *   18**
Russ Meyer USA   1967
Alaina Capri, Babette Bardot, Adele Rein, Jack Moran, Franklin Bolger, John Furlong, Ken Swofford, Andrew Hagara
At a remote tourist spot, a man runs a game-hunting lodge with his young second wife, and daughter by his first marriage. A doctor and his wife arrive for a day of shooting, but unknown to them an ex-cop turned robber is hiding out there. The doctor drops dead when his wife seduces the ex-cop. More deaths follow, but not before much sex and violence has come to pass in this typical Meyer offering.
Aka: HOW MUCH LOVING DOES A NORMAL COUPLE NEED?
A   69 min   VIDrel: ALLIED/RTM/TROMA   V

COMMUNION *   15**
Philippe Mora USA   1989
Christopher Walken, Lindsay Crouse, Frances Sternhagen, Andreas Katsulas, Terri Hanauer, Joel Carlson, Basil Hoffman, John Dennis Johnston, Dee Dee Rescher, Aileen Fitzpatrick, R.J. Miller, Holly Fields, Paul Shaw, Kate Stern
Screenplay is by Strieber in this account of his alleged real-life experience of alien encounters, when strange creatures habitually "borrowed" both him and his family. Overlong and unlikely to convince, but Walken is well cast as the writer, who believes he requires a psychiatrist, and the film, though an uneven blend of fantasy and psychology, has a few good moments. The main theme was composed and performed by Eric Clapton. See also OFFICIAL DENIAL.
FAN   104 min (ort 107 min)   VIDrel: FIRST/SONOP   V/sur
Boa: book by Whitley Strieber.

COMPANEROS ** **   18
Sergio Corbucci ITALY/SPAIN/WEST GERMANY   1970
Jack Palance, Franco Nero, Fernando Rey, Tomas Milain, Iris Berben, Karin Schubert, Francisco Badalo, Eduardo Fajardo, Luizi Pernice, Alvarado De Luna, Jesus Fernandez, Claudio Scarchilli, Lorenzo Robeldo, Giovanni Petti
A mercenary teams up with a Mexico revolutionary in an attempt to free a professor and his students, who are being held in Texas. However, our hired gun is really intent on locating a cache of gold, whose whereabouts are known to the professor. Along the way, they team up with an American gunslinger with a wooden hand. An overlong spaghetti Western that tries to pretend that it is making some kind of political point, but in so doing becomes quite tiresome.
Aka: VAMOS A MATAR, COMPANEROS!
WES   110 min (ort 118 min)   VIDrel: AKTIV/RTM   V

COMPANION, THE **
Gary Fleder USA
15
1994
Kathryn Harrold, Bruce Greenwood, Talia Balsam, Brion James
A female novelist heads for a remote mountain cabin in order to gain the peace she needs to write her next book, taking along her newly-acquired android companion. Growing lonely, she reprograms her robot to give him a more romantic role and he becomes her lover. But his only conception of what a relationship entails is provided when he reads the book on she is working on. An interesting idea that was handled better in VALERIE 23 – a new OUTER LIMITS episode.
FAN 89 min (ort 94 min) VIDrel: CIC/SONOP V/sh

COMPANY, THE **
Harry Winer USA
PG
1990
Anthony John Dennison, Linda Purl, John Rhys-Davies, Josef Sommer, Colleen Flynn, John Slattery, Bozidar Smiljanic, Raye Birk, Arleen Taylor, Michael Whaley, Dakin Matthews, Milos Kirek, Dawn Comer, Joshua South, Sumer Stamper
A double-agent due to retire from active service, is forced to undertake one last perilous mission behind the Iron Curtain, but when there is a change of management at headquarters, he finds his new masters none to keen to bring him in. A stale and cliched espionage thriller that formed a pilot for a TV series. A sequel followed.
THR 88 min mTV VIDrel: MGM/WHV L/A V/sh

COMPANY 2, THE **
Michael Fresco USA
PG
1991
Anthony John Dennison, Linda Purl, John Rhys-Davies, Josef Sommer, G.W. Bailey, Jesse Borrego, Kassi Lemmons
An experienced FBI agent and two rookies are sent to Kuwait before the Iraqi invasion and get trapped there. The obligatory rescue attempt is mounted. Both this film and an earlier one served as pilots for a TV series entitled "Under Cover".
Aka: COMPANY 2: SACRIFICES
THR 89 min mTV VIDrel: MGM/WHV L/A V/sh

COMPANY BUSINESS ***
Nicholas Meyer USA
15
1991
Gene Hackman, Mikhail Baryshnikov, Kurtwood Smith, Terry O'Quinn, Daniel Von Bargen, Oleg Rudnick, Geraldine Danon, Nadim Sawalha, Michael Tomlinson, Howard McGillin, Louis Eppolito, Toby Eckholdt, Elsa O'Toole, Kate Harper
A post-Cold War espionage thriller that's all about a State Department mole who is released from prison by an ex-CIA agent, who wants to use him (plus a $2,000,000 ransom) as part of a deal to secure the release of a captured US pilot. When this plan goes badly wrong, they make their escape and decide to keep the money, but find themselves pursued by both the CIA and the KGB. Despite the implausible plot, this is an enjoyable buddy-movie offering.
THR 95 min (ort 99 min) VIDrel: MGM/WHV L/A V/sh

COMPANY OF STRANGERS, THE ****
Cynthia Scott CANADA
PG
1990
Alice Diabo, Constance Garneau, Winifred Holden, Cissy Meddings, Mary Meigs, Catherine Roche, Michelle Sweeney, Beth Webber
Scott's first feature film tells of a group of elderly ladies who are stranded in the wilds of Canada, when their coach breaks down and their driver is unable to repair it. As they while away the time awaiting developments, they each do what they can (according to their temperaments) to make their enforced stay easier. A really beautiful film, insightful and profound, that allows the characters to share with us their dreams, hopes and disappointments.
DRA 100 min VIDrel: ELPIC/POLYREC V

COMPANY OF WOLVES, THE ***
Neil Jordan UK
18
1984
David Warner, Angela Lansbury, Graham Crowden, Stephen Rea, Brian Glover, Kathryn Pogson, Tusse Silberg, Micha Bergese, Sarah Patterson, Georgia Slowe, Susan Porrett, Shane Johnstone, Dawn Archibald, Richard Morant
A Freudian variant on the Red Riding Hood tale, with a set of eerie stories told by Granny to her ingenuous granddaughter, who begins to encounter wolves in all manner of guises. A visually powerful fantasy that is spoilt by fragmentary and disjointed development.
HOR 92 min (ort 95 min) VIDrel: VCC L/A V
Boa: short stories by Angela Carter.

COMPLEAT RUTLES, THE ***
Eric Idle/Gary Weiss UK
15
1980
Eric Idle, Neill Innes, John Halsey, Mick Jagger, George Harrison, Ricky Fataar, Michael Palin, Bianca Jagger, John Belushi, Dan Aykroyd, Gilda Radner, Bill Murray, Paul Simon
An ambitious attempt to parody the career of the pop group The Beatles, with an account of a fictitious group – "The Rutles". Worth seeing for the many clever Neill Innes pastiches of Beatles numbers, the careful recreation of Sixties news footage and stadium concerts and a few sharp digs at their more esoteric interests. A carefully detailed spoof of a documentary – it really needs more than just the one idea to give it substance.
Aka: ALL YOU NEED IS CASH; I LOVE THE RUTLES; RUTLES, THE; RUTLES, THE: ALL YOU NEED IS CASH
COM 71 min (ort 78 min) mTV VIDrel: VCC/DISC V

COMPLEX OF FEAR **
Brian Grant USA
18
1992
Hart Bochner, Joe Don Baker, Chelsea Field, Brett Cullen, Farrah Forke, Lisa Darr, Ashley Gardner, Rus Blackwell, Jordan Williams, Ann Bronston, Michelle Little, Kathryn Firago, Terry Beaver, Tom Nowicki, Robert Treveiler, Randi Lane
The women living in a smart apartment block are terrorised by a rapist, and it would appear to be the case that one of the residents is responsible. This one holds together quite well, even if the ideas in it are hardly original.
THR 90 min mTV VIDrel: ODY/SONOP V/sh

COMRADES OF SUMMER **
Tommy Lee Wallace USA
(15)
1993
Joe Mantegna, Michael Lerner, Natalya Negoda, Mark Rolston, John Fleck, Eric Allan Kramer, Ian Tracey, Jay Brazeau, Dwight Koss, David Bener, Mitchel Davies, Roark Critchlow, Todd Duckworth, Grant Forster, Jano Frandsen
A crusty, burnt-out former American baseball player goes to Russia to help put together Russia's very first baseball team in time for the 1992 Olympics. As this may boost his chances of getting back into the sport he agrees. A competent script and good acting by Mantegna help make this sporting story stand out from the crowd.
Aka: COMRADES OF SUMMER, THE
COM 103 min SATrel: MOVIE CHANNEL

CONAGHER ***
Reynaldo Villalobos USA
PG
1991
Sam Elliott, Katherine Ross, Barry Corbin, Billy Green Bush, Ken Curtis, Dub Taylor, James Gammon, Paul Koslo, Gavan O'Herlihy, Daniel Quinn, Cody Braun, Buck Taylor, James Parks, Pepe Serna, Anndi McAfee, John Furlong, Kate Hall
Dedicated to the late Louis L'Amour, this exciting tale opens with Bush taking his new wife and his young kids to his ranch. When he is killed in a fall, she is left to fend for herself in a hostile environment, but when her ranch is used by a new stagecoach company, she finds herself making friends with the driver. Beautifully photographed, this is most enjoyable slice of frontier life. Co-produced by Elliott and Ross.
Aka: LOUIS L'AMOUR'S CONAGHER
WES 112 min VIDrel: FIRST/SONOP V
Boa: novel by Louis L'Amour.

CONAN THE BARBARIAN ***
John Milius USA
15
1982
Arnold Schwarzenegger, Sandahl Bergman, James Earl Jones, Gerry Lopez, Mako, Ben Davidson, Sven Ole Thorsen, Max Von Sydow, Valerie Quennesen, Cassandra Gaviola, William Smith, Luis Barboo, Frank Columbo, Leslie Foldvary, Jorge Sanz
Story of sword-and-sorcery revenge by a warrior in a time of long ago, using characters drawn from the "Conan" novels of Robert E. Howard. In this adventure, our hero seeks revenge on the cult leader who destroyed his village and enslaved him. A spectacular and full-blooded treatment of pulp fiction. Sets are by Ron Cobb, with music by Basil Poulédoris. Followed by CONAN THE DESTROYER.
FAN 121 min (Cut at film release – ort 124 min) wScrn
VIDrel: 20TH/TECH; ENCORE (LV only) V LV

CONAN THE DESTROYER *
Richard Fleischer USA
15
1984
Arnold Schwarzenegger, Grace Jones, Wilt Chamberlain, Mako, Sarah Douglas, Olivia D'Abo, Tracey Walter, Jeff Corey, Sven Ole Thorsen, Pat Roach, Bruce Fleischer, Ferdinand Mayne
Second film based on the "Conan" books of Robert E. Howard,

that tries hard to please but ends up being merely tiresome and ponderous.
Aka: CONAN, KING OF THE THIEVES
FAN 96 min Cut (30 sec) wScrn (COLUM/SONOP) VIDrel: COLUM/SONOP L/A; 4-FRONT/POLYREC V

CONCIERGE, THE **
Barry Sonnenfeld USA
PG
1993
Michael J. Fox, Gabrielle Anwar, Anthony Higgins, Michael Tucker, Udo Kier, Bob Balaban, Dan Hedaya, Isaac Mizrahi, Simon Jones, Dianne Brill, Fyvush Finkel, Bobby Short, Mike G., Daniel green, Paula Laurence, Severio Guerra
A concierge working at a luxury New York hotel is consumed with an overwhelming passion to have his own establishment, and believes that the best way to achieve this is to pander to every whim of his guests. However, just when he seems about to achieve his dream, he becomes romantically involved with a tycoon's mistress. A few good performances are all that's on offer in a laboured comedy devoid of sparkle.
Aka: FOR LOVE OR MONEY
COM 91 min (ort 97 min) cC VIDrel: CIC/SONOP V

CONCORDE AFFAIR, THE **
Roger Deodato (Ruggero Deodato) ITALY
15
1979
James Franciscus, Mimsy Farmer, Van Johnson, Joseph Cotten, Edmund Purdom, Mag Fleming, Mario Maranzana, Venantino Venantini, Francisco Charles, Nando Barberito, Francesco Carnelutti, Ottaviano Dell'acqua, Alessandra Stordy
A Concorde plane crashes near Martinique, and a reporter sent to investigate discovers that sabotage was the cause. Average Euro-thriller.
Aka: CONCORDE AFFAIRE SEVENTY-NINE; S.O.S. CONCORDE
THR 93 min VIDrel: IMPENT V

CONCRETE HELL ***
Eric Till CANADA
18
1985
Nicky Guadagni, Shirley Douglas, Anne Anglin, Jackie Richardson, Bernard Behrens
Account of a framed woman's struggle to survive the horrors of the prison system, as she faces a seven-year stretch for trying to smuggle drugs into Canada. For once a realistic drama that doesn't slide into cliche, with fine performances and an intelligent script. Written by Judith Thompson.
Aka: TURNING TO STONE
DRA 98 min VIDrel: SCRN/DISC L/A V/h
Boa: novel by Judith Thompson.

CONDEMNED *
Wesley Ruggles USA
PG
1930
Ronald Coleman, Ann Harding, Dudley Digges, Louis Wolheim, William Elmer, William Vaughn, Albert Kingsley, Henry Ginsberg, Bud Somers, Baldy Biddle, Stephen Selznick, John George, Arthuro Kobe, Emil Schwartz, John Schwartz
A suave thief condemned to a prison term on Devil's Island, gets a job in the warden's house and eventually falls in love with the warden's wife and inevitably, the couple soon plan a joint escape through the fever-infested swamps and jungle. A predictable romantic drama, interesting only for Colman's performance in this, his second talkie.
DRA 86 min (ort 93 min) B/W VIDrel: VCC/DISC V
Boa: novel Condemned to Devil's Island by Blair Niles.

CONDITION CRITICAL **
Jerrold Freedman USA
(PG)
1992
Kevin Sorbo, Joanna Pacula, Christina Haag, Mark Blum, Suzzanne Douglas, Stephen Eckholdt, Denis Arndt, Anthony Dean Fields, Mike Nussbaum, Richard Grant, Anne Haney, Rudolph Willrich, Scott Lawrence, Ethan Phillips, E.G. Daily
A mysterious and deadly virus plagues the people of L.A., in this fairly gripping TV movie.
Aka: FINAL PULSE
THR 90 min mTV VIDrel: WHV V/h

CONDITION RED **
Mika Kaurismaki USA
18
1995
James Russo, Cynda Williams, Paul Calderone, Victor Argo
A prison officer is seduced by a streetwise female prisoner who gets him to assist in her escape when she tells him she is pregnant.
A/AD 82 min (ort 85 min) VIDrel: MED/COLUM V

CONDUCT UNBECOMING **
Michael Anderson UK
PG
1975
Michael York, Richard Attenborough, Trevor Howard, Stacy Keach,
Susannah York, Christopher Plummer, James Faulkner, Michael Culver, James Donald, Rafiq Anwar, Helen Cherry, Michael Fleming, David Robb, David Purcell
Old fashioned drama that follows the events arising after an officer's wife of the Bengal Lancers is indecently assaulted. A cadet is accused, but in reality the culprit is someone quite different. A stuffy and wooden adaptation of England's play, set in India circa 1878.
DRA 102 min (ort 107 min) VIDrel: WHV V/h
Boa: play by Barry England.

CONEHEADS **
Steve Barron USA
PG
1993
Dan Aykroyd, Jane Curtin, Chris Farley, Laraine Newman, Jason Alexander, Michelle Burke, Chris Farley, Michael Richards, Lisa Jane Persky, Sinbad, Shishir Kurup, Michael McKean, Phil Hartman, David Spade, Dave Thomas
A family of aliens with pointed heads come to Earth and try to live a normal suburban life, in this overlong, one-joke effort that features title characters, who used to feature on the TV show Saturday Night Live. A tedious tale that soon exhausts all the comic potential of its single idea.
COM 83 min (ort 87 min) cC VIDrel: CIC/SONOP V/sur

CONFESSION **
Ken Hughes UK
PG
1955
Sydney Chaplin, Audrey Dalton, John Bentley, Peter Hammond, John Welsh, Jefferson Clifford, Pat McGrath, Robert Raglan, Patrick Allan
In England to launder the money he got from an bank robbery, an American meets a former partner he double-crossed in the past, and a murder results from this. But the killing is seen by a Catholic who tells all to his priest at confession, leading to the latter now becoming a target for the killer. A thin little film, that employs an idea used with considerably more success by Hitchcock in I CONFESS.
Aka: DEADLIEST SIN, THE
THR 86 min (ort 90 min) B/W VIDrel: WHV V
Boa: play by Don Martin.

CONFESSIONAL, THE ***
Robert Lepage CANADA/FRANCE/UK
15
1995
Lothaire Bluteau, Patrick Goyette, Jean-Louis Millette, Kristin Scott Thomas, Ron Burrage, Richard Frechette, Francois Papineau, Marie Gignac, Normand Daneau, Anne-Marie Cadieux, Suzanne Clement, Lynda Lepage-Beaulieu
A remake of Hitchcock's I CONFESS, this tense thriller is set in Quebec, where after three years spent in China, a man comes home and meets up with his adopted brother, and both attempt to uncover the mystery surrounding the true father of the latter. A most impressive and ingenious film from Lepage, whose debut feature this was.
Aka: LE CONFESSIONNAL
THR 97 min (ort 101 min) (some French dialogue) wScrn VIDrel: ARTIF/20TH V/sh

CONFESSIONS FROM THE DAVID GALAXY AFFAIR *
Willy Roe UK
18
1979
Alan Lake, Glynn Edwards, Anthony Booth, Diana Dors, John Moulder-Brown, Milton Reid, Bernie Winters, Kenny Lynch, Mary Millington, Sally Faulkner, Jada Smith, Queenie Watts, Cindy Truman, Vicky Scott, Alec Mango
A girl is not fulfilled sexually until she meets a superstud. And if that wasn't complex enough, he is meanwhile being sought by the police who wish to frame him for a murder.
Aka: SECRETS OF A SEXY GAME; STAR SEX
COM 92 min (ort 96 min) VIDrel: MED/DISC V
Boa: novel by George Evans.

CONFESSIONS OF A CHAUFFEUR ***
J.T. Monroe USA
18
1990
Kelly Royce, Sunny McKay, Buck Adams, Lauren Brice, Eric Edwards, Tony Montana
A chauffeur who worked for a reclusive movie star writes a bestselling book about his job, and this leads to a series of flashbacks about his employer, for although he never had sex with her, he still had to listen to her interminable reminiscences in the course of his work. These reveal her to be a woman whose ambition for fame blighted her first love affair, leading to much loneliness on her part. Unusually, this sex film is sustained by a strong and coherent plot.
A 57 min VIDrel: ONE V

CONFESSIONS OF A SEX MANIAC *

18

Alan Birkinshaw UK 1975
Roger Lloyd-Pack, Vicki Hodge, Derek Royle, Stephanie Marrian, Louise Rush, Candy Baker, Ava Cadell, Cheryl Gilham, Jean Marsden, Carole Hayman, John Aston, Bobbie Sparrow
Forget the title. This one's about the difficulties a young breast-obsessed architect faces, in designing a new building. He finally solves his problem in a novel way, designing a marina that resembles a giant pair of breasts, in this dismal and cheap-looking effort.
Aka: DESIGN FOR LOVE; DESIGN FOR LUST; MAN WHO COULDN'T GET ENOUGH, THE
A 76 min (ort 80 min) VIDrel: JEZ/RTM V

CONFESSIONS OF A SORORITY GIRL **

(12)

Uli Edel USA 1993
Jamie Luner, Betty Rae, Danni Wheeler, Alyssa Milano, Brian Bloom, Lorinne Dills-Vozoff, Sadie Kratzig, Peter Simoon, Kevin Gardner, Judson Mills, David Brisbin, Natalija Nogulich
A remake of "Sorority Girl", a film from 1957, with a girl realising that in order to inherit from her father's will, she will just have to graduate after all, and she starts using every trick in the book to ensure she becomes the top girl. This was one of a number of remakes of 1950s AIP B-movies.
DRA 76 min (ort 90 min) SATrel: MOVIE CHANNEL

CONFESSIONS OF EMMANUELLE ***

18

Joe D'Amato (Aristide Massaccesi) ITALY 1977
Laura Gemser, Karin Schubert, Kristine De Belle, Ivan Rassimov, Don Powell, George Eastman (Luigi Montefiore), Brigitte Petronio
Big budget and colourful sex romp with Gemser travelling across four continents to help a friend who's been kidnapped by sex-slave traders.
Aka: EMMANUELLE: PERCHE VIOLENZA ALLE DONNE?; SHE'S SEVEN-TEEN AND ANXIOUS
A 78 min (Cut at film release by 15 min 45 sec)
VIDrel: IMPENT V

CONFESSIONS: TWO FACES OF EVIL ***

15

Gilbert Cates USA 1993
Jason Bateman, James Wilder, Arye Gross, William Converse Roberts, James Earl Jones, Gloria Reuben, Melinda Dillon, Rex Linn, Ari Meyers, Basil Wallace, Luis Antonio Ramos, Richard Grove, John P. Connolly, Miguel Sandoval
Two men both confess to the slaying of an L.A. cop and the law has to decide who the culprit really is, this difficulty being compounded by the fact that one of the men is a drifter but the other is a college student of apparently good character. A thought-provoking work, that has Jones giving one of his best performances as one of the lawyers involved in the case.
Aka: TWO FACES OF EVIL
DRA 89 min VIDrel: ODY/SONOP V/sh

CONFIDENTIAL REPORT **

PG

Orson Welles UK 1955
Orson Welles, Michael Redgrave, Akim Tamiroff, Katina Paxinou, Mischa Auer, Patricia Medina, Jack Watling, Peter Van Eyck, Paola Mori, Robert Arden, Suzanne Flon, Gregoire Aslan
A long and rambling melodrama in which a wealthy financier pays an American to seek out all the figures from his past who either loved or hated him. The limited budget seriously hampered the director's scope and the film is never more than a loosely handled collection of episodes.
Aka: MISTER ARKADIN
DRA 93 min (ort 100 min)
VIDrel: CONNO/RTM; PION (LV only) V LV

CONFLICT **

PG

Curtis Bernhardt USA 1943 (released 1945)
Humphrey Bogart, Alexis Smith, Sidney Greenstreet, Rose Hobart, Charles Drake, Grant Mitchell, Patrick O'Moore, Ann Shoemaker, Frank Wilcox, James Flavin, Edwin Stanley, Mary Servoss, Doria Caron, Ray Hanson, Billy Wayne
A husband plots to murder his wife in order to marry her younger sister and fakes a car accident in which he pretends to have injured his legs and be unable to walk. However, after disposing of his wife, clues keep appearing that seem to suggest she is still alive. A complex and implausible thriller, made watchable if not believable by the strong cast.
THR 82 min (ort 86 min) B/W VIDrel: MGM/WHV L/A
V

CONFLICT OF INTEREST **

18

Gary Davis USA 1992
Judd Nelson, Alyssa Milano, Christopher McDonald, Dey Young, Gregory Alan Harris
Standard cop thriller with all the usual violence and sleaze as a tough cop plots revenge against the low-life who murdered his wife, framed his son and kidnapped the latter's girlfriend. OK as mindless entertainment but otherwise totally forget-table.
A/AD 84 min (ort 88 min) VIDrel: FIRST/SONOP V

CONFORMIST, THE ***

18

Bernardo Bertolucci FRANCE/ITALY/WEST GERMANY 1971
Jean-Louis Trintignant, Stefania Sandrelli, Dominique Sanda, Pierre Clementi, Gastone Moschin, Enzo Tarascio, Jose Quaglio, Milly, Giuseppe Addobbati, Yvonne Sanson, Fosco Giachetti, Benedetto Benedetti, Marta Lado
Complex story of a man's willingness to conform and compro-mise for the sake of personal advancement. Having repressed his homosexual drives, he joins the Italian Fascist espionage organisation and is entrusted with an assignment to kill a left-wing professor in Paris. Complex and sometimes confused attempt to explain Fascism in psychosexual terms.
Aka: IL CONFORMISTA
DRA 113 min CINrel
Boa: novel by Alberto Moravia.

CONGO **

12

Frank Marshall USA 1995
Dylan Walsh, Laura Linney, Ernie Hudson, Grant Heslov, Tim Curry, Joe Don Baker, Stuart Pankin, Mary Ellen Trainor, James Karen, Lorene Hoh, Misty Rosas, Carolyn Seymour, Romy Rosemont, Bill Pugin, Lawrence T. Wrentz
A lame-brained jungle adventure tale involving an expedition in search of a missing fiance, the return of a tame gorilla to the wild and a fortune in diamonds. As our motley gang make their way through the jungle, the inevitable tensions between them grow until they stumble across an incredible sight. A disap-pointing, poorly plotted and unimpressive tale that harks back to 1930s serials and is hardly any more exciting despite its vastly larger budget.
A/AD 104 min (ort 109 min) cC
VIDrel: CIC/SONOP; PION (LV only) V/sur LV
Boa: novel by Michael Crichton.

CONQUEST OF THE EARTH *

U

Sidney Hayers USA 1980
Lorne Greene, Kent McCord, Barry Van Dyke, Robert Reed, Robyn Douglas, Patrick Stuart, Robbie Rist, John Colicos
A botched attempt to produce a film from various episodes of the TV series "Battlestar Galactica". Haven't we come this way before? This time our heroes must foil a Cyclon attack against Earth. Barely watchable. See also BATTLESTAR GALACTICA.
Aka: GALACTICA 3: CONQUEST OF THE EARTH
FAN 94 min mTV VIDrel: CIC/SONOP L/A V

CONQUEST OF THE PLANET OF THE APES **

15

J. Lee Thompson USA 1972
Roddy McDowall, Don Murray, Ricardo Montalban, Natalie Trundy, Hari Rhodes, Severn Darden, Lou Wagner, John Randolph, Asa Maynor, H.M. Wynant, David Chow, Buck Kartalian, John Dennis, Gordon Jump, Dick Spangler, Joyce Haber
This is the 4th in the series of monkey-movies and tells of how the apes took over from the humans following a revolt against their human masters. A limp and disappointing effort, followed by BATTLE FOR THE PLANET OF THE APES.
FAN 85 min VIDrel: 20TH/TECH L/A V

CONSENTING ADULTS *

15

Alan J. Pakula USA 1992
Kevin Kline, Kevin Spacey, Mary Elizabeth Mastrantonio, Rebecca Miller, E.G. Marshall, Forest Whitaker, Kimberly McCullough, Billie Neal, Lonnie Smith, Joe Muelherin, Benjamin Hendrickson, Rick Hinkle, Artis Edwards Jr, Jerry Campbell
A young suburban couple become corrupted by their next-door neighbours but a little wife-swapping ends in tragedy, in this poorly executed thriller. The usual cataclysmic climax ends a movie that wastes both the talents of a fine cast and our time.
THR 95 min (ort 99 min) cC
VIDrel: HOLPIC/TECH V/sur

CONSPIRACY ** **   18
Christopher Barnard USA   1989
James Wilby, Glyn Houston, Kate Hardie, Steve Pacey, Tony Caunter, Ann Tirard
The US Defence Secretary's partiality for young girls leads to the mother of one such girl committing suicide. A security agent is sent to hush things up and avert a political scandal. A muddled and murky thriller.
THR 87 min VIDrel: 20TH/TECH V

CONSPIRACY OF FEAR *   18**
John Eyres USA   1996
Andrew Lowery, Christopher Plummer, Leslie Hope, Geraint Wyn Davies
A son learns that the explosion that killed his father was deliberate and that his own life is in danger after he escapes from a kidnapping. It all turns out to be because of a mysterious package that was in his dad's possession, and as the youngster starts to dig deeper, he forms an unlikely partnership with an attractive, petty thief. One of those underplotted and totally predictable conspiracy actioners, involving (surprise, surprise) corruption in high places.
A/AD 103 min cC VIDrel: MOSAIC/COLUM V/sh

CONSPIRACY OF SILENCE ** **   15
Francis Mankiewicz CANADA   1991
Michael Mahonen, Michelle St John, Jonathon Potts, Ian Tracey, Carl Marotte, Maury Chaykin, Jonathan Potts, Diego Chambers, Neil Munro, Steve Mousseau, James B. Douglas, Greenhalgh, Catherine Disher, Brooke Johnson, Monique Mojica
In November 1973 a group of drunken louts living in the tightly-knit Canadian town of The Pas, Manitoba, abducted a Cree Indian girl, murdering her when she resisted their advances. Despite the fact that the whole town knew who the culprits were, a wall of silence descended, and sixteen years later the case remained open. Loosely based on real events, this sluggish and unconvincing yarn follows the efforts made by a newly transferred cop to solve the case.
DRA 183 min mTV VIDrel: ODY/SONOP V/s
Boa: book by Lisa Priest.

CONSPIRATORS OF PLEASURE *   (18)**
Jan Svankmajer CZECH REPUBLIC/SWITZERLAND/UK   1996
Petr Meissel, Gabriela Wilhelmova, Barbora Hrzanova, Anna Weltlinska, Jiri Labus, Pavel Novy, Frantisek Polata, Eva Vidimska, Ervin Tomenendal, Josef Chodora, Marie Zemanova, Jan Daniel, Martin Kublak, Eva Vosahlikova
Svankmajer takes a close look at various inhabitants in Prague, who go about their business during the week, most of which seems to involve working on bizarre, fetishistic projects that intend to make use of on the weekend. Without dialogue, we are left with nothing but the pitiless detail of these projects, whose strange purposes eventually become clear. Scripted by Svankmajer, this is a provocative and disturbing examination of perversion and despair.
Aka: SPIKLENCI SLASTI
DRA 75 min CINrel

CONTRA CONSPIRACY ** **   18
Tom DeWeir USA   1988
Michael Williams, Duncan Savage, Tom Maher, Blake Bahner, Vicki Stephenson, Robert Beahl
A Hollywood film crew are attacked when they accidentally discover the secret desert headquarters of a private army, but a woman survivor of the massacre teams up with a mercenary in a desperate escape bid.
A/AD 90 min VIDrel: MOPIC/SGSVID V

CONVERSATION, THE **   15**
Francis Ford Coppola USA   1974
Gene Hackman, John Cazale, Allen Garfield, Frederic Forrest, Robert Duvall, Cindy Williams, Michael Higgins, Elizabeth MacRae, Teri Garr, Harrison Ford, Mark Wheeler, Robert Shields, Phoebe Alexander
A surveillance expert makes the mistake of becoming personally involved with his work, only to find himself caught up in a web of conspiracy and murder. A brilliantly made film that makes a number of intelligent and chilling observations regarding the issue of privacy.
DRA 108 min VIDrel: 4-FRONT/POLYREC/CIC L/A V

CONVICT COWBOY ** **   (18)
Rob Holcomb USA   1995
Jon Voight, Ben Gazzara, Marcia Gay Harden, Kyle Chandler
A rodeo champion commits a murder and is sent to prison but even there he is able to continue in this sport. When another rodeo rider turns up, he is able to take the younger man under his wing and give him the benefit of his experience. A buddy movie with an original setting that works quite well.
DRA 94 min (ort 106 min) VIDrel: MGM/WHV V

CONVICTION OF KITTY DODDS, THE ** **   15
Michael Tuchner USA   1993
Veronica Hamel, Kevin Dobson, Lee Garlington, Mark Rolston, Keith Coulouris, Julie Adams, Mary Tanner
A man who learns that his wife is an escaped Death Row prisoner tries to prove her innocence. Another one of those fact-based stories, so prevalent on US TV. This one offers good performances but little else to make it stand out.
Aka: CONVICTION: THE KITTY DODDS STORY
DRA 92 min mTV VIDrel: ODY/SONOP V

CONVOY *   15**
Sam Peckinpah USA   1978
Kris Kristofferson, Ali MacGraw, Ernest Borgnine, Burt Young, Franklyn Ajaye, Madge Sinclair, Brian Davies, Seymour Cassel, Cassie Yates, Walter Kelly
Ludicrous, semi-comic story of a protest mounted by truckers against police harassment which becomes a mass movement in the southwestern states of the USA. Enjoyable in parts, boring in others and way over-the-top throughout. The script (by B.W.L. Norton) is based on a song by C.W. McCall.
A/AD 106 min (ort 111 min) VIDrel: WHV V/sh

COOK, THE THIEF, HIS WIFE AND HER LOVER, THE *   18**
Peter Greenaway UK   1989
Michael Gambon, Helen Mirren, Richard Bohringer, Alan Howard, Tim Roth, Gary Olsen, Ciaran Hinds, Ewan Stewart, Roger Ashton Griffiths, Ron Cook, Liz Smith, Emer Gillespie, Janet Henfrey, Arnie Breevelt, Tony Alleff, Ian Sears
In an effort to improve his image, a gangster has begun to invest in and patronise a superb French restaurant. His presence there, together with that of his wife and cronies, is reluctantly tolerated by the chef. When one of the regular patrons embarks on an affair with the gangster's wife, the chef does his best to protect them. A flashy and tangled allegory, made with great style, but hampered by irritating camerawork and poor plotting and characterisation.
DRA 118 min (ort 123 min) VIDrel: ELPIC/POLYREC V/sur

COOL AND THE CRAZY *   (12)**
Ralph Bakshi USA   1994
Jennifer Blanc, Matthew Flint, Jared Leto, Alicia Silverstone, Bradford Tatum, Christine Harnos, Tuesday Knight, Christian Frizzell, John Hawkes, John Kapelos, Marianne Bergonzi, Michael Lowry, Richard Singer, Jospeh G. Medalis
Dullish, loose remake of a 1958 movie, in which a woman enjoys a double wedding with her best friend, but doesn't wait long before starting an affair with a local Lothario. A vicious, nasty and exploitative drama, without much in the way of appealing characters or believable situations. One of several remakes of 1950s AIP B-movies.
DRA 82 min (ort 85 min) SATrel: SKY MOVIES

COOL AS ICE *   PG**
David Kellogg USA   1991
Vanilla Ice, Kristin Minter, Michael Gross, Candy Clark, Sydney Lassick, Dody Goodman, Naomi Campbell, Booby Brown, S.A. Griffin, Jack McGee, Victor Dimattia, John Haymes Newton, Allison Dean, kevin Hicks, Deezer D. Ted Swanson
A wandering rebel comes to a conservative small town in the South and causes quite a stir, in this vanity vehicle for rapper Vanilla Ice which gives him an excuse to offer several samples of his musical art. Of interest strictly to rap fans only.
A/AD 87 min (ort 95 min) VIDrel: CIC/SONOP V/sur

COOL BLUE ** **   18
Mark Mullin/Richard Shepard USA   1988
Woody Harrelson, Hank Azaria, Ely Pouget, Sean Penn, John Diehl
In L.A. a struggling artist complicates his life by falling in love with a woman with whom he only enjoyed a single encounter, and spends the rest of the film searching for her. An uneven and

slightly bitter satire that is enlivened by some good performances, especially one from Penn is a rather funny cameo.
DRA 86 min (ort 90 min) VIDrel: EIV/SONOP V

COOL HAND LUKE ***
Stuart Rosenberg USA   15
                                                          1967
Paul Newman, George Kennedy, Dennis Hopper, Strother Martin, J.D. Cannon, Jo Van Fleet, Wayne Rogers, Anthony Zerbe, Ralph Waite, Harry Dean Stanton, Joe Don Baker, Clifton James, Lou Antonio, Morgan Woodward, Luke Askew, Marc Cavell
Splendidly offbeat story of a member of a prison chain gang who becomes a hero to his fellow-convicts for resisting the guards' attempts to break him. Full of memorable moments and great performances but slightly let down by the contrived and downbeat ending. Written by Pearce and Frank R. Pierson and with music by Lalo Schifrin. AA: S. Actor (Kennedy).
DRA 121 min VIDrel: WHV V
Boa: novel by Donn Pearce.

COOL IT CAROL! *
Pete Walker UK   18
                                                          1970
Robert Askwith, Janet Lynn, Peter Elliot, Jess Conrad, Stubby Kaye, Pearl Hackney, Martin Wyldeck, Chris Sandford, Derek Aylward, Peter Murray, Eric Barker
A girl and her boyfriend come to London to make their fortune but find that things do not go as they planned, and they get drawn into the seedier side of London life. The attempt to make this work a film with a message is worthy enough, but adds nothing of entertainment value.
Aka: OH, CAROL
COM 97 min (ort 102 min) VIDrel: JEZ/RTM V

COOL MIKADO, THE **
Michael Winner UK   U
                                                          1962
Frankie Howerd, Stubby Kaye, Tommy Cooper, Dennis Price, Bernie Winters, Mike Winters, Lionel Blair, Kevin Scott, Glenn Mason, Pete Murray, Dermot Walsh, Stubby Kaye, Jacqueline Jones, Jill Mai Meredith, Yvonne Shima, Tsai Chin
Updated version of the Gilbert and Sullivan opera where the soldier son of an American is kidnapped by the boyfriend of the Japanese girl he has fallen in love with.
COM 81 min VIDrel: FABFIL/SPEAR V
Boa: opera The Mikado by W.S. Gilbert and Arthur Sullivan.

COOL RUNNINGS **
Joe Turteltaub USA   PG
                                                          1993
Leon, Doug E. Doug, Rawle D. Lewis, Malik Yoba, John Candy, Paul Coeur, Raymond J. Barry, Peter Outerbridge, Paul Coeur, Larry Gilman, Charles Hyatt, Winston Stona, Bertina Macauley, Pauline Stone Myrie, Kristoffer Cooper
A bob-sled coach accepts the seemingly impossible job of preparing a team to represent Jamaica in the 1988 Winter Olympics in Calgary and finds that his four hopefuls are far removed from the usual dedicated and trained athletes. Given the lack of snow and ice, he has to resort to some unorthodox training methods. A juvenile, likeable but very silly comedy, based on a true story, it's very typical of the Disney approach to family entertainment.
COM 95 min (ort 98 min) cC VIDrel: WDV/TECH V/sur

COOL SHEETS *
Anthony Spinelli USA   18
                                                          1991
Victoria Paris, Tianna, Cheri Taylor, Randy West, Jon Martin
Two friends who have up to now shared everything, get between the sheets for a wild weekend of sexual frolics in this totally forgettable offering.
A 57 min (ort 94 min) VIDrel: FALCON/TOTAL V

COOL SURFACE, THE **
Linda Yellen USA   18
                                                          1993
Robert Patrick, Teri Hatcher, Matt McCoy, Ian Buchanan, Cyril O'Reilly, Steven Tyler, Elizabeth Barondes, Howard Spiegel, Cassian Elwes, Saul Janson, Lisa Marie Kurbikoff, David Niven Jr, Anthony Pena, Rolf Englehart, Paul Timms
A would-be screenwriter comes to Hollywood where he engages in a passionate affair with an aspiring actress. Inspired by this relationship, he produces a script based on their affair and succeeds in selling it. However, problems start to loom when it becomes clear that our actress will do anything to star in the lead role.
THR 88 min (ort 102 min) VIDrel: MED/DISC V/sh

COOL TO THE TOUCH *
Jerry Ross USA   18
                                                          1993
Kim Wilde, Randy West, Jenny Wells, Stacy Nichols, Joey Civera, Rose Hunter, Scott Irish
A women sets out to avenge herself on her two-timing boyfriend, but as ever, the plot is of little consequence in this sex romp.
A 54 min VIDrel: ONE V

COOL WORLD **
Ralph Baski USA   15
                                                          1992
Brad Pitt, Kim Basinger, Gabriel Byrne, Michele Abrams, Deidre O'Connell, Janni-Brenn Lowen, William Frankfather, Greg Collins, Maurice LeMarche, Michael David Lally, Carrie Hamilton, Stephen Worth, Murray Podwal, Jenine Jennings
An underground cartoonist discovers that the bizarre world of his imagination is real and is lured there by a sex-kitten who is possessed by a burning desire to become human. A sort of adult version of WHO FRAMED ROGER RABBIT that is far less explicit than Baski's former offerings and suffers from a lack of humour and an indifferent script. See also VOLERE, VOLARE.
A/AD 97 min (ort 102 min) VIDrel: CIC/SONOP V/sur

COP AND A HALF *
Henry Winkler USA   PG
                                                          1992
Burt Reynolds, Ray Sharkey, Ruby Dee, Norman D. Golden II, Holland Taylor, Sammy Hernandez, Frank Sivero, Rocky Giordani, Marc Macaulay, Tom McLeister, Ralph Wilcox, Tom Kouchalakos, Carmine Genovese, Sean Evan O'Neal, Max Winkler
When a pint-sized kid obsessed with police work witnesses a murder, he forces the cops into giving him a badge in return for his co-operation and is then teamed up with an adult officer. Together, this ill-matched duo set about fighting crime, in a puerile and contrived comedy that can do little to further anyone's career.
COM 88 min (ort 97 min) cC VIDrel: CIC/SONOP V/sur

COP AU VIN ***
Claude Chabrol FRANCE   15
                                                          1984
Jean Poiret, Stephane Audran, Michel Bouquet, Jean Topart, Lucas Belvaux, Pauline Lafont
A young mail boy lives at home with his invalid mother, whilst a trio of nasties are harassing them in order to get them to sell their property. The boy puts sugar in the tank of one of them, causing him to crash his car, and a tough French cop is sent to investigate the murder. A well-paced and entertaining thriller, but one with a disappointing and anti-climactic resolution. Followed by "Inspector Lavardin".
Aka: POULET AU VINAIGRE
THR 109 min VIDrel: L/A V

COP TARGET **
Humphrey Humbert (Umberto Lenzi) ITALY 1990 (released   15
1991)
Robert Ginty, Barbara Bingham, Nina Sue Borrel, Charles Napier, Bradford Devine, Jeff Maldovan, Thomas Bull, Bruce Bartlett, Harry Schreiber, Terri Baer, Alain Marino, Stephen Mignon, Kelly Barnes, Allan Seigel
A cop is entrusted with the task of escorting the widow of a diplomat and her daughter to a Central American country where she is to be presented with an award in honour of her dead husband and his fight against the drugs trade. When the daughter is kidnapped a vast ransom is demanded, and the cop has no choice but to pursue her abductors. A violent, low-budget actioner.
A/AD 89 min VIDrel: MIA/DISC V

COPACABANA **
Waris Hussein USA   U
                                                          1985
Barry Manilow, Annette O'Toole, Joseph Bologna, Ernie Sabella, Estelle Getty, Silvana Gallardo, James T. Callahan, Andra Akers, Cliff Osmond, Dwier Brown, Stanley Brock, Clarence Felder, Hamilton Camp, Hartley Silver
Manilow makes his acting debut in a sort of acknowledgement of 1940s Hollywood musicals, with a plot based on one of his songs, all about a songwriter and his affair with a chorus girl in a nightclub. Sluggish and uninteresting, despite Manilow's engaging personality.
MUS 95 min mTV VIDrel: L/A V

COPPER MOUNTAIN **
David Mitchell CANADA   PG
                                                          1983
Jim Carrey, Alan Thicke, Dick Gauthier, Rod Hebron, Jean Laplac

Carrey plays a lovable prankster who covers up his insecurities with a succession of impersonations, mostly done to hide his fear of women. Meanwhile, a wealthy playboy has arrived at the ski resort where the story is set, his intention being to qualify and enter a major championship event. A pleasing but minor work, clearly released to capitalise on the growing status of Carrey as a comic (if not as an actor).
COM   60 min   VIDrel: SPEAR/SONOP/ARENA   V

COPS & ROBBERSONS **

	PG
Michael Ritchie USA	1994

Chevy Case, Jack Palance, Dianne Wiest, Robert Davi, David Barry Gray, Jason James Richter, Fay Masterson, Mike Hughes, Richard Romanus, Sal Landi, Jack Kehler, Amy Powell, Diano Landon, Jim Howels, Charlie O'Donnell, Preston Hanson
A man addicted to police-action TV shows is delighted when he is approached by a couple of cops who want to use his suburban home for a surveillance operation on his new neighbour, whom they suspect of forgery. Little do they realise that their subject is not above a little murder when it comes to dealing with any threat to his operations. Another very typical Chase comedy, replete with corny dialogue and slapstick antics.
COM   89 min (ort 93 min) cC   VIDrel: COLUM/SONOP
V/sur

COPYCAT ***

	18
Jon Amiel USA	1995

Sigourney Weaver, Holly Hunter, Dermot Mulroney, William McNamara, Harry Connick Jr, J.E. Freeman, Will Patton, John Rothman, Shannon O'Hurley, Bob Greene, Tony Haney, Danny Kovacs, Tahmus Rounds, Scott De Venney, Terry Brown
Another one of those serial-killer-on-the-loose stories with Weaver cast as a criminal psychiatrist being stalked by a psychopath who models his crimes on some classic case studies. The clever plot is often illogical, but it has enough twists to keep one watching, and the direction is mercifully free of unnecessary flourishes.
THR   118 min (ort 123 min) cC   VIDrel: WHV   V/sur

CORMORANT, THE **

	(PG)
Peter Markham UK	1992

Ralph Fiennes, Helen Schlesinger, Thomas Williams, Buddug Morgan, Derek Hutchinson, Karl Francis, Dyfan Roberts, Mici Plwm, Ray Gravell, Stewart Jones, Gwilym Evans
When a writer inherits the home of his uncle in Snowdonia he finds himself growing increasingly disturbed by his late uncle's pet cormorant, a sinister bird whose behaviour is matched by an increasingly dangerous obsession on the part of the writer. A psychological drama that looks as if it is meant to be profound. Unfortunately, the lack of any explanation given for the events examined does not work to the film's advantage.
DRA   88 min mTV   TVrel: BBC
Boa: novel by Stephen Gregory.

CORPSE GRINDERS, THE *

	18
Ted V. Mikels USA	1971

Sean Kenney, Monika Kelly, Sanford Mitchell, J. Bryon Foster, Warren Ball, Ann Noble, Vince Barbi, Harry Lovejoy, Earl Burnam, Zenna Foster, Ray Dennis, Charles Fox, Stephen Lester, William Kirschner, George Bowden, Stephen Lester
Vicious moggies start attacking human beings, in film which could have been entitled "Paws". Suspicion focuses on a new exotic catfood. An endearingly bad exercise in comedy-horror with a few funny moments in amongst the gore.
HOR   74 min (ort 80 min)   VIDrel: OURVID/SCRN   V

CORPSE VANISHES, THE **

	PG
Wallace Fox USA	1942

Bela Lugosi, Luana Walters, Tristam Coffin, Elizabeth Russell, Minerva Urecal, Kenneth Harlan, Vince Barnett, Joan Barclay, Frank Moran, Angelo Rossitto, Gewn Kenyon, George Eldridge
A mad botanist kidnaps young brides in order to provide his elderly wife with the glandular extracts she needs to arrest the ageing process, but a young woman reporter is hot on his trail. Cheaply and quickly made horror programmer that is not without a certain period charm.
HOR   63 min B/W   VIDrel: SCRN/DISC   V

CORRIDORS OF BLOOD *

	15
Robert Day UK	1958 (released 1964)

Boris Karloff, Christopher Lee, Betta St John, Finlay Currie, Adrienne Corri, Francis Matthews, Francis De Wolff, Nigel Green, Charles

Lloyd-Pack, Basil Dignam, Frank Pettingell, Marian Spencer, Carl Bernard
Muddled story of a dedicated surgeon, who undertakes strange experiments in an attempt to find a way of easing the pain of patients undergoing surgery, and falls victim to the effects of his experiments.
Aka: DOCTOR FROM SEVEN DIALS
HOR   82 min (ort 89 min) B/W   VIDrel: ENCORE/SPEAR
V

CORRINA, CORRINA ***

	PG
Jessie Nelson USA	1994

Whoopi Goldberg, Ray Liotta, Tina Majorino, Wendy Crewson, Larry Miller, Erica Yohn, Jenifer Lewis, Joan Cusack, Harold Sylvester, Steven Williams, Don Ameche, Patrika Darbo, Lucy Webb, Noreen Hennessy, June C. Ellis, Mimi Lieber
In the 1950s, a man whose wife has just died of cancer, hires a new maid (Goldberg) and her presence soon makes all the difference to both him and his young daughter, who has lapsed in a non-communicative state and refuses to speak. Excellent performances enhance this heart-warming tale. See also CLARA'S HEART for Goldberg in a similar role.
DRA   116 min   CINrel

CORRUPT JUSTICE **

	18
Ian Barry AUSTRALIA	1993

Jacqueline Bisset, Massaya Kato, Gary Day, John Bach, Gary Sweet, Victoria Langley, Richard Roxburgh, Peter Boxwell, Victoria Campbell, Paul Hanlon, Ralph Cotterill, Peter Hathaway, Justin Lewis, Barry Quin, David Richardson
A female judge has a secret life in which she devotes herself to committing a variety of crimes, including bank robbery. Matters come to a head when her lover tempts her with an audacious but highly dangerous venture.
Aka: CRIMEBROKER
DRA   99 min   VIDrel: COLUM/SONOP   V/sh

COSMIC MAN, THE **

	PG
Herbert Greene USA	1959

John Carradine, Angela Greene, Paul Langston, Bruce Bennett, Lyn Osborn, Scotty Morrow, Walter Maslow, Robert Lytton
An inferior clone of THE DAY THE EARTH STOOD STILL, with an benevolent alien (played by Carradine) paying us a visit to teach us how to live in peace with each other and in harmony with the universe.
FAN   64 min (ort 72 min) B/W   VIDrel: FIRC/RTM   V

COSMIC SLOP **

	(15)
Reginald Hudlin/Warrington Hudlin/Kevin Rodney Sullivan USA	1995

Robert Guillaume, Michele Lamar Richards, Jason Bernard, Edward Edwards, George Wallace, Larry Anderson, Brian Reddy, Nicholas Turturro, Richard Herd, Efrain Figueroa, J. Kenneth Campbell, Paul Jai Parker, Chi McBride, Reno Wilson
Three-part SF film with each story offering a clear and obvious message. In "Space Traders" aliens offer the USA amazing technology in return for all the country's Afro-American citizens. "The First Commandment" tells of a young priest in the South Bronx who gets unexpected problems after a museum buys a holy statue. Finally, "Tang" deals with a bullied black woman and her less than appealing boyfriend. Very disappointing, despite uniformly fine performances.
FAN   87 min (ort 90 min) mCab   SATrel: MOVIE CHANNEL
Boa: short stories by Derrick Bell and Chester Himes.

COSMOS CONQUEROR, THE **

	PG
Johnny T. Howard HONG KONG	1989

Another animated space opera from the same studio as CAPTAIN COSMOS. Here three youngsters with special powers are contacted by the former leader of a conquered planet, who sends them a robot warrior to help defend the Earth from an attack by the evil empire to which his own planet has fallen. A colourful fantasy of good ideas but rather cheap-looking animation techniques. See also FALCON 7.
ANIM   64 min Cut (27 sec – ort 70 min)   VIDrel: IMPENT
V

COTTAGE TO LET **

	U
Anthony Asquith UK	1941

Leslie Banks, Alastair Sim, John Mills, Jeanne De Casalis, Carla Lehmann, George Cole, Michael Wilding, Frank Cellier, Wally Patch, Catherine Lacey, Hay Petrie, Muriel Aked, Muriel George

The inventor of a new bombsight becomes the object of a kidnap plot by Nazi agents but is rescued at the last moment, in this adequate comedy-thriller. Cole makes his screen debut, playing a Cockney kid evacuated to the home of said inventor.
Aka: BOMBSIGHT STOLEN
THR   86 min (ort 90 min) B/W   VIDrel: CARL/TECH   V
Boa: play by Geoffrey Kerr.

COTTON CLUB, THE *** 15
Francis Ford Coppola USA 1984
Richard Gere, Gregory Hines, Bob Hoskins, Lonette McKee, James Remar, Diane Lane, Nicolas Cage, Allen Garfield, Fred Gwynne, Gwen Verdon, Maurice Hines, Joe Dallesandro, Julian Beck, Jennifer Grey, Lisa Jane Persky, Tom Waits, Ed O'Ross
An attempt to evoke the atmosphere of a famous New York jazz club and the gangster-ridden era in which it flourished. Set in Harlem during the time of prohibition, with Gere a rundown trumpet player who saves the life of a powerful gangster, subsequently becoming involved with the New York criminal element. This colourful and episodic tapestry of a film visually overwhelms and has a terrific Duke Ellington soundtrack. All it needs is a stronger story.
DRA   123 min (ort 127 min)   VIDrel: 4-FRONT/POLYREC
V/sur

COUCH TRIP, THE *** 15
Michael Ritchie USA 1988
Dan Aykroyd, Walter Matthau, Charles Grodin, Donna Dixon, Richard Romanus, Mary Gross, David Clennon, Ayre Gross, Victoria Jackson, Michael DeLorenzo, J.E. Freeman, Mickey Jones, David Wohl, Michael Ensign, Carol Mansell
An escaped prisoner takes a job as a replacement for an overworked Beverly Hills psychiatrist, and achieves media notoriety as a radio adviser. Quite a few good laughs are to be had in this uneven tale, which though helped by a strong performance from Aykroyd, is marred by a disappointing ending. The talents of Matthau remain unexploited in this one. See also STRAIGHT TALK, it's similar but a lot lighter in tone.
COM   94 min (ort 98 min)   VIDrel: POLY/POLYREC L/A
V/h

COUNT OF SOLAR, THE **** (PG)
Tristram Powell UK 1991
Tyrone Woolfe, Paul Casey, Jonathan Adams, Peter Needham, Samantha Best, Paul Trussell, Nick Reding, Georgina Hale, Susan Jameson, David Calder, Hermione Norris, Tony Newton, Janet Henfrey, Sally Mates, Charles Simon, Patrick Godfrey
On the eve of the French Revolution a deaf-mute child is found wandering the French countryside, and is taken in by the Abbe de L'Epee, the inventor of one of the first forms of sign-language. Taught to communicate by the Abbe, he eventually reveals himself to be the abandoned child of an aristocratic family, and the holder of a title in his own right. Woolfe (who is profoundly deaf) gives a wonderful performance in the central role in this moving true story.
DRA   77 min mTV   TVrel: BBC
Boa: episode from "When The Mind Hears" by Harlan Lane.

COUNTDOWN *** U
Robert Altman USA 1967
Robert Duvall, James Caan, Charles Aidman, Joanna Moore, Steve Ihnat, Barbara Baxley, Ted Knight, Michael Murphy, Stephen Coit, John Rayner, Charles Irving, Bobby Riha Jr
The story of the first US moonshot and the strains it imposed on the astronauts and their families, with the plot largely following the race to the moon between American and Soviet scientists. An earnest and realistic drama, with the technology of the moonshot the chief star.
DRA   97 min (ort 101 min)   VIDrel: WHV   V
Boa: novel The Pilgrim Project by Hank Searles.

COUNTERFEIT CONTESSA, THE ** U
Ron Lagomarsino USA 1994
Tea Leoni, D.W: Moffett, David Beecroft, Kana Tamburellia, Susan Walters, Moly Roice, William Keane, Nikkie De Boer, Holland Taylor, Lynn Cohen, Louis Guss, Sam Coppola, Jonathan Potts, Pat Mastroianni, Falconer Abraham, Leslie Yeo
A young Italian girl from a working-class families has all the usual girlish dreams of meeting a "Prince Charming". Whilst modelling some dresses for a friend, she impersonates a contessa, and finds herself the object of a wealthy lawyer's attentions. A pleasant romantic comedy.
COM   88 min mTV   VIDrel: 20TH/TECH   V

COUNTESS DRACULA ** 18
Peter Sasdy UK 1971
Ingrid Pitt, Nigel Green, Sandor Eles, Maurice Denham, Patience Collier, Peter Jeffrey, Lesley-Anne Down, Leon Lissek, Jessie Evans, Andrea Lawrence, Nike Arrighi, Charles Farrell, Hulya Babus
Variation on the Dracula legend, the inspiration for which was probably the notorious Countess Bathary. In this tale an ageing countess needs the blood of young virgins to preserve her youth, but the local populace do not take so reasonable a view. Standard Hammer Films offering, with little to distinguish this one from a hundred others.
HOR   89 min (ort 93 min)   VIDrel: VCC/DISC L/A   V
Boa: novel The Bloody Countess by V. Penrose.

COUNTRY DIARY OF AN EDWARDIAN LADY, THE *** U
Dirk Campbell UK 1988
Pippa Guard, James Coombes, Isabelle Amyes, Jill Benedict
A languid and quite pleasing adaptation of Holden's bestseller, built around the life of the title character, a young married woman, who observes the passage of the seasons in rural Warwickshire. Filmed entirely in this part of the country. Originally shown in the UK in twelve 28-minute episodes.
DRA   90 min (ort 336 min) mTV   VIDrel: CASPIC L/A   V
Boa: book by Edith Holden.

COUNTRY LIFE ** 12
Michael Blakemore AUSTRALIA 1994
Sam Neill, Greta Scacchi, John Hargreaves, Kerry Fox, Michael Blakemore, Googie Withers, Patricia Kennedy, Ron Blanchard, Robyn Cruze, Maurie Fields, Bryan Marshall, Tony Barry, Terry Brady, Tom Long, Rob Steele, Ian Bliss
A very loose variant on Chekhov's play "Uncle Vanya" this tells the story of a man who returns to the Australian Outback after spending the past twenty-two years allegedly earning his living as a writer. Having arrived there with his new wife, he receives a warm reception, but it is not long before his tales of success and critical acclaim are exposed as untrue. A comedy-drama of the gentler sort, relaxed and a little lacking in substance.
COM   112 min   VIDrel: TART/20TH   V/s

COUPE DE VILLE ** 15
Joe Roth USA 1990
Patrick Dempsey, Arye Gross, Daniel Stern, Alan Arkin, Annabeth Gish, Rita Taggart, Joseph Bologna, James Gammon, Ray Lykins, Chris Lombardi, Josh Segal, John Considine, Steve Boles, Don Tilley, Terry Loughlin, Reid (Pete) Shook
A trio of feuding brothers have to learn to get along somehow when they drive their dad's car from Detroit to Florida as a present for their mum on her birthday. An OK road-movie that's set in 1963 and makes good use of the music of the period but ultimately says very little of importance.
COM   93 min (ort 99 min)   VIDrel: 20VIS/SONOP   V/sur

COURAGE MOUNTAIN ** PG
Christopher Leitch FRANCE/USA 1989
Charlie Sheen, Leslie Caron, Juliette Caton, Nicola Stapleton, Jan Rubes, Jade Magri, Kathryn Ludlow, Yorgo Voyagis, Laura Betti, Urbano Barberini, Marc Estrada, Massimo Sarchielli, Flora Alberti, David Ogilvie, Joanna Clarke
This bland attempt to update Johanna Spyri's "Heidi" has our heroine being sent off to an Italian boarding school just before the outbreak of WW1, and being rescued by her soldier boyfriend when the army takes it over. Sheen is unconvincing in his role, but as a piece of family entertainment the film is reasonably agreeable.
A/AD   94 min (ort 120 min)   VIDrel: EIV/SONOP   V

COURAGE UNDER FIRE **** 15
Edward Zwick USA 1996
Denzel Washington, Meg Ryan, Lou Diamond Phillips, Michael Moriarty, Matt Damon, Bronson Pinchot, Seth Gilliam, Regina Taylor, Zeljko Ivanek, Scott Glenn, Tim Guinnee, Tim Ransom, Sean Astin, Armand Darrius, Mark Adair-Rios
Washington remains haunted by a decision he took during the Gulf War that led to the loss of life, and as a way of exorcising his feelings of guilt, investigates the posthumous claims of a female pilot to the award of a Medal of Honor. But this proves to be no easy matter, as inconsistencies begin to mount up in the various accounts of her bravery he receives from her former comrades. A deep and thoughtful work, finely acted and cleverly resolved.
DRA   111 min (ort 116 min) cC   VIDrel: 20TH/FOXVID   V/s

COURT JESTER, THE **   U
Norman Panama/Melvin Frank USA   1955
*Danny Kaye, Glynis Johns, Basil Rathbone, Cecil Parker, Mildred
Natwick, Angela Lansbury, Edward Ashley, Robert Middleton,
Michael Pate, Alan Napier, Noel Drayton, Herbery Rudley, John
Carradine, Lewis Martin*
A lively medieval romp in which Kaye gets a wonderful role
(and makes the most of it) as a meek and mild man who, posing
as a jester, finds himself inadvertently embroiled in an attempt
to usurp a despotic king. One of the star's most delightful films,
full of absurd complications, good songs and some unforgettable lines.
JUV   97 min (ort 101 min)   VIDrel: CIC/SONOP   V/h

COURTING JUSTICE **   15
Eric Till USA   1995
Art Hindle, Patty Duke, Linda Dano
A downtrodden wife finally takes her husband to court when he
makes their daughter his sole beneficiary, thus leaving her desti-
tute should he die. A competent drama based on a real case.
DRA   86 min (ort 89 min)   VIDrel: ODY/SONOP   V/sh

COUSIN, COUSINE **   15
Jean-Charles Tacchella FRANCE   1976
*Marie-Christine Barrault, Victor Lanoux, Marie-France Pisier, Guy
Marchand, Ginette Garcin, Sybil Maas, Jean Herbert, Pierre Plesis,
Catherine Verlor, Hubert Gignoux*
A man and a woman who become cousins by marriage meet and
fall in love, although they are both already married. As their
relationship deepens and flourishes (they soon become lovers),
the outrage of their extended bourgeois family knows no
bounds. A charming and witty comedy with a wonderfully light
touch that Hollywood could not resist remaking in 1989 as
COUSINS.
COM   91 min (ort 95 min)   VIDrel: ARROW/RTM   V

COUSINS **   15
Joel Schumacher USA   1989
*Ted Danson, Isabella Rossellini, Sean Young, William L. Petersen,
Norma Aleandro, Lloyd Bridges, George Coe, Keith Coogan, Gina de
Angelis, Katie Murphy, Alex Bruhanski, Stephen E. Miller, Gordon
Currie, Saffron Henderson*
A remake of the French film COUSIN, COUSINE that tells of
two cousins by marriage who refuse to succumb to their mutual
attraction, despite the fact that their respective spouses are
having an affair. Beautifully acted, especially by Danson and
Rossellini as the starcrossed lovers, but the loose and rambling
plot does tends to weaken the impact. The unusual locations
are an attraction.
DRA   108 min (Cut at film release by 2 sec – ort 110 min)
VIDrel: CIC/SONOP   V/sur

COUSINS IN LOVE **   18
David Hamilton FRANCE/WEST GERMANY   1980
*Thierry Tervini, Jean Rougerie, Anja Schute, Hannes Kaetner, Gaelle
Legrand, Anne Fontaine, Carmen Weber, Fanny Meunier, Jean-Louis
Fortuit, Jean-Pierre Rambal, Valerie Dumas, Évelyne Dandry, Elisa
Cervier, Jean-Yves Chatelais*
Soft-focus, softcore story of the sexual education of a 16-year-
old boy, set against the approach of WW2. A typical offering
from this director and not far removed from films such as
BILITIS, but having a somewhat better story.
Aka: TENDRES COUSINES
A   87 min (Cut at film release – ort 90 min)   VIDrel: L/A   V
Boa: novel by Pascal Laine.

COVER GIRL **   U
Charles Vidor USA   1944
*Rita Hayworth, Gene Kelly, Lee Bowman, Phil Silvers, Jinx
Falkenburg, Leslie Brooks, Eve Arden, Otto Krueger, Jess Barker,
Anita Colby, Curt Bois, Ed Brophy, Thurston Hall, Jean Colleran,
Francine Counihan, Helen Mueller, Cecilia Meagher*
A Broadway nightclub singer is tempted into a a career as a
cover girl when she meets a magazine editor who was once
madly in love with her grandmother. Never mind the flimsy,
second-hand plot; this musical offers some brilliant song and
dance numbers, while Silvers is, as ever, a much-appreciated
additional bonus. AA: Score (Morris Stoloff/Carmen Dragon).
MUS   102 min (ort 105 min)   VIDrel: COLUM/SONOP   V

COVER GIRL *   R18/18
Alex De Renzy USA   1981
Cheryl Hanson, John Leslie, David Morris, Joey Civera

A top agent for photo-models is murdered under mysterious
circumstances just before Cheryl arrives for an interview. But
Cheryl has sex with the murderer, believing him to be the
agent. Later she goes to use the shower and finds the dead
agent hidden there. The rest of this utterly tedious film details
her search for the murderer; it would appear that she cannot
remember his face, only the fact that he had a tattoo on his
thigh.
Aka: CHERYL HANSON: COVER GIRL
A   64 min Cut (1 min 21 sec – ort 82 min)   VIDrel: SHEP
L/A (R18 ver); ONE L/A (18 ver)   V

COVER GIRL MODELS **   18
Cirio H. Santiago PHILIPPINES   1974
*John Kramer, Lindsay Bloom, Tara Strohmeimer, Pat Anderson,
Rhonda Leigh Hopkins, Mary Woronov*
Three American fashion models inadvertently get caught up in
an international spy ring when a roll of microfilm gets hidden
in one of their dresses. A dreary little spy thriller with a few nice
outdoors locations.
THR   74 min (ort 85 min)   VIDrel: ALLIED/RTM   V

COVER GIRL MURDERS, THE **   15
James A. Contner USA   1993
*Lee Majors, Jennifer O'Neill, Beverly Johnson, Adrian Paul, Rick
Marotta, Vanessa Angel, Arthur Taxier, Bobbie Phillips, Fawna
MacLaren, Mowana Pryor, Honorato Magaloni, Dick Christie, Jose
Escandon*
Six top models take part in a photo-shoot on a lush tropical
island, their assignment being for a fashion magazine that is in
financial difficulties. They soon learn to their horror that a
murderous killer is lurking in the island's undergrowth, and as
all of them have reason to despise the owner of the magazine,
it's not hard to guess the most likely suspect. Average.
THR   83 min (ort 87 min) mCab   VIDrel: CIC   V

COVER ME **   18
Michael Schroeder USA   1995
*Rick Rossovich, Paul Sorvino, Elliott Gould, Corbin Bernsen,
Courtney Taylor*
In order to trap a murderous psychopath, an agent goes undercover with his detective girlfriend, and they frequent the sleazy
world of strip clubs. However, the girl starts to enjoy this
lifestyle and her undercover work as a stripper, failing to realise
that by embarking on this course she has herself become a potential target. A fairly standard erotic thriller, the most notable
aspect of which is its clever title.
THR   90 min   VIDrel: 20TH/FOXVID   V/sur

COVER UP **   18
Manny Coto USA   1990
*Dolph Lundgren, Louis Gossett Jr, John Finn, Lisa Berkley, Gil Ko Pel,
Ofer Lehavi, Howard Rypp, Zadok Zarum, Oren Neeman, Bruce
Daniel Diker, Danny Friedman, Barry Langford, Sharon Hacohen,
Matt Sevi, Jonathan Dominitz*
Competent Lundgren vehicle in which he plays a journalist
investigating a terrorist raid in Israel, that resulted in the deaths
of American military personnel. He eventually learns that a
cover-up has been mounted to hide the theft of an experimen-
tal and highly lethal new nerve gas. Good performances help
overcome the film's many weaknesses, not least being its indifferent direction.
A/AD   88 min (ort 91 min)
VIDrel: 4-FRONT/POLYREC/GUILD   V/sh

COWBOY AND THE LADY, THE **   U
H.C. Potter USA   1938
*Gary Cooper, Merle Oberon, Patsy Kelly, Walter Brennan, Fuzzy
Knight, Harry Davenport, Emma Dunn, Walter Walker, Berton
Vogeding, Arthur Hoyt, Ernie Adams, Mabel Todd, Henry Kolker,
Charles Richman, Russ Powell, Jack Baxley, Johnny Judd*
The unruly daughter of a politician who is running for President
meets a young rodeo star and falls in love with him. They are
soon married but the difference in their lifestyles threatens to
separate them for good, in a weak comedy that is hampered by
the unsuitable pairing of Cooper and Oberon: AA: Sound
(Thomas T. Moulton).
COM   91 min B/W   VIDrel: VGM/DISC   V

COWBOY WAY, THE **   12
Gregg Champion USA   1993
*Kiefer Sutherland, Woody Harrelson, Dylan Dermott, Ernie Hudson,
Cara Buono, Joaquim Martinez, Marg Helgenberger, Tomas Milian,*

Luis Guzman, Angel Caban, Matthew Cowles, Kristin Baer, Christian Aubert, Emmanuel Xuereb, Francie Swift
Two New Mexico rodeo riders come to New York to find a friend who has gone missing while searching of his daughter, who has also disappeared. Their efforts to locate them, under some very unfamiliar circumstances, bring them up against a nasty gang of immigrant smugglers. A crude comedy-actioner, it offers some modest amusement but little more, though Hudson is a delight as a cop who has more than a passing interest in Westerns.
COM 102 min (ort 107 min) cC VIDrel: CIC/SONOP V

COWBOYS, THE ***
Mark Rydell USA
PG
1971
John Wayne, Roscoe Lee Browne, Bruce Dern, Colleen Dewhurst, Slim Pickens, A. Martinez, Alfred Barker Jr, Nicholas Beauvy, Steve Benedict, Robert Carradine, Norman Howells Jr, Stephen Hudis, Sean Kelly, Clay O'Brien
An ageing rancher is deserted by his drovers and is obliged to hire eleven youngsters to help him drive his herd to market. A handsome film of violence and bloodshed, with an old-fashioned and rather disagreeable eye for an eye message. Later made into a shortlived TV series in 1974.
WES 121 min Cut (1 min 30 sec – ort 128 min)
VIDrel: WHV V/h

COWBOYS DON'T CRY ***
Anne Wheeler CANADA
(PG)
1988
Rebecca Jenkins, Ron White, Joshua Ansley, Michael Hogan, Janet Wright, Val Pearson, Frank Totina, Bairney O'Sullivan, James Defelice, Zachary Ansley, Wendell Smith, Adria Budd, Robert Clinton, Thomas Hauff, Candace Ratcliffe
The wife of an ageing rodeo star is tragically killed in a car accident, and though the man is distraught at this loss, he somehow has to make a new life for himself and his teenage son. A nicely balanced, low-key drama of human hopes and personal relationships that is both well acted and directed.
DRA 96 min SATrel: SKY MOVIES

COWS ***
Julio Medem SPAIN
(18)
1991
Emma Suarez, Carmelo Gomez, Anna Torrent, Karra Elejalde, Txema Blasco, Klara Badiola, Kandido Uranga, Pialr Barden, Miguel Angel Garcia, Ane Sanchez, Magdalena Mikolajczyk, Enara Azkue, Oritz Balda, Elizabeth Ruiz, Ramon Barea
A highly symbolic look at the precariousness of human existence, that opens in 1875, when a Basque woodcutter survives the turmoil of the Second Carlist War by hiding in a cartload of corpses. The years pass, and the focus of the story shifts from him to his illegitimate son, who leaves with his parents when they elope to America, but returns to photograph the Spanish Civil War. Complex and stylised, the film is more a meditation on life than a conventional narrative.
Aka: VACAS
DRA 96 min CINrel

COYOTE RUN **
USA
Michael Pare
18
1996
Pare plays a Vietnam veteran who now works as a deputy sheriff, in which capacity he sets out to catch a Mafia hoodlum who has stolen $15,000,000, this being his chosen way of exorcising the ghosts that still haunt him over his country's ignominious withdrawal from Vietnam. A vicious and crude action pic, devoid of plotting or characterisation.
A/AD 100 min VIDrel: EIV V

CRACKDOWN **
USA
Pamela Dixon, Anthony Gates, Joe Vance, Cynthia Miguel, Lisa Anderson
18
A female cop attempts to rehabilitate two girls she wants to get off the streets of L.A., but after their deaths she sets out to clean up the streets in earnest. The usual violent encounters take place.
A/AD 90 min VIDrel: MOPIC/QUANT V

CRACKDOWN **
Louis Morneau
15
1990
Cliff De Young, Robert Beltran, Jamie Rose, Gerald Anthony
A agent employed by the DEA teams up with a Peruvian cop and together they take on the might of a ruthless and powerful drugs baron.
A/AD 83 min Cut (33 sec plus some cuts subst)
VIDrel: COLUM/SONOP V

CRACKER: ONE DAY A LEMMING WILL FLY **
Simon Cellan Jones UK
15
1993
Robbie Coltrane, Barbara Flynn, Geraldine Somerville, Christopher Eccleston, Lorcan Cranitch, Tim Healy, Frances Tomelty, Christopher Fulford
Criminal psychologist Coltrane investigates the case of a young boy found hanging from a tree. When the boy's former English teacher botches an attempt at suicide, the police decide that this man is a suspect and he is held for questioning. One in a series of absorbing dramas that were occasionally spoilt by too much emphasis on personal conflicts and domestic troubles for the main character, rather than the far more interesting psychological issues raised.
THR 97 min (ort 117 min) mTV
VIDrel: VCC/DISC V/sur

CRACKER: THE BIG CRUNCH **
Julian Jarrold UK
18
1994
Robbie Coltrane, Barbara Flynn, Geraldine Somerville, Lorcan Cranitch, Ricky Tomlinson, Colin Tierney, Wilbert Johnson, Jim Carter, Maureen O'Brien, Cherith Mellor, James Fleet, Samantha Morton, Darren Tighe, Ellie Haddington
Originally shown on TV in three parts, this has our clinical psychologist and occasional sleuth getting caught up in a murder investigation. Meanwhile, his forthcoming separation from his wife threatens to leave him virtually destitute.
THR 148 min mTV VIDrel: VCC/DISC V

CRACKER: THE MAD WOMAN IN THE ATTIC ***
Michael Winterbottom UK
18
1993
Robbie Coltrane, Adrian Dunbar, Barbara Flynn, Christopher Eccleston, Lorcan Cranitch, Geraldine Somerville, Kika Markham, John Grillo, Ian Mercer, Kieran O'Brien, Nicholas Woodeson, Don Henderson
First shown in two parts, this is the first episode from a popular TV series that detailed the exploits of a criminal psychologist, who seemed to be called in to help solve a surprisingly large number of violent murders. This story revolves around the crimes of a serial killer, who stalks lone women travelling by train. When the police find a man suffering from amnesia wandering near a railway line, he becomes their chief suspect.
THR 104 min (ort 117 min) mTV
VIDrel: VCC/DISC V/sur

CRACKER: TO BE A SOMEBODY ***
Tim Flywell UK
15
1994
Robbie Coltrane, Robert Carlyle, Barbara Flynn, Christopher Eccleston, Lorcan Cranitch, Geraldine Somerville, Colin Tierney, Beth Goddard, Tracy Gillman, Glyn Grain, Kieran O'Brien
Another three-part story, with Coltrane getting to grips with the murder of an Asian newsagent, the culprit being ultimately revealed to be a seriously disturbed individual who has never recovered from the death of his father.
THR 148 min mTV VIDrel: VCC/DISC V

CRACKER: TO SAY I LOVE YOU ***
Andy Wilson UK
18
1993
Robbie Coltrane, Barbara Flynn, Christopher Eccleston, Lorcan Cranitch, Geraldine Somerville, David Haig, Andrew Tiernan, Susan Lynch, Ian Mercer, Patti Love, Kieran O'Brien, Keith Ladd
The police ignore the advice of our psychologist, and release a man from custody, only to find that he commits a murder together with his partner. Another tense and well plotted mystery, first shown in three parts.
THR 156 min mTV VIDrel: VCC/DISC V/sur

CRACKERJACK *
Michael Mazo USA
18
1994
Thomas Ian Griffith, Christopher Plummer, Natassia Kinski, George Touliatos, Lisa Bunting, Richard Sali, William Taylor, Frank Cassini, Frank Turner, Dorothy Fehr, Vladimir Kulich, Alex Diajun, Sonny Surowiec, Scott McNeil, Rob Wilton
A cop is so devastated by the death of his family that he becomes almost suicidal but gets a chance to save himself, when a criminal mastermind captures an entire ski resort in a bid to steal some jewels. To make matters worse, our cop's brother is among the hostages. A by-the-numbers, big-scale actioner in the DIE HARD mould, with ample violence to disguise both bad acting and holes in the plot.
THR 96 min VIDrel: MIA/DISC V/sh

**CRADLE OF CONSPIRACY **   15
Gabrielle Beaumont USA   1993
Dee Wallace Stone, Carmen Argenziano, Kurt Deutsch, Danica McKellar, Merle Kennedy, Jeoffrey Thorne, Ellen Crawford, Shannon Fill, Jamie Lunar, Burke Byrnes, Christine Healy, Matthew Faison, Charlie Holliday, Owen Bush
A naive seventeen-year-old is seduced and made pregnant by a criminal drifter whose intention is to corrupt the girl, luring her into a world of vice where she can sell anything, even her unborn child. Another one of those dramas loosely based on a real-life case. Average.
DRA 87 min (ort 90 min) mTV   VIDrel: NWV/HIFLI V/h

**CRAFT, THE **   15
Andrew Fleming USA   1996
Fairuza Balk, Robin Tunney, Neve Campbell, Rachel True, Christine Taylor, Skeet Ulrich, Assumpta Serna, Cliff De Young, Breckin Meyer, Nathaniel Marston, Helen Shaver, Jeanine Jackson, Brenda Strong, Elizabeth Guber, Jennifer Greenhut
Four high school girls start dabbling in the occult, and in no time at all, are calling up satanic creatures to revenge themselves on their enemies. Several deaths later, we find one of the girls (the one with the strongest powers) deciding it's time to quit. Unhappily for her, the others are not ready to let her do so. A special effects rollercoaster with the four leads giving the only portrayals that show some work on the part of the scriptwriters.
HOR 97 min (ort 101 min) cC   VIDrel: COLUM/SONOP V/sur

**CRASH **   18
Charles Wilkinson USA   1995
Michael Biehn, Matt Craven, Leilani Sarelle, Miguel Sandoval, Kim Coates, Ed Lauter
A computer disk that holds all the details of an international drugs money laundering scheme comes into the hands of a female FBI agent and a petty criminal, but they learn that they retain it at their peril, hired assassins having been dispatched to reclaim it. A good action thriller let down by a weak script.
THR 92 min   VIDrel: MED/DISC V/sh

CRASH *   18
David Cronenberg CANADA   1996
James Spader, Holly Hunter, Deborah Kara Unger, Rosanna Arquette, Elias Koteas, Peter MacNeill, Yolande Julian, Cheryl Swarts, Judah Katz, Nicky Guadagni, Ronn Sarosiak
In Toronto, an unhappily married couple keep their marriage going by virtue of a shared obsession with machines and the occasional adulterous affair. When the husband survives a serious car crash, it leads to him finding such incidents a source of sexual arousal, an interest that draws him into a world inhabited by like-minded individuals. A disturbing exploration of an especially sick fetish, it recalls Cronenburg's NAKED LUNCH, but is much harder to watch.
HOR 100 min CINrel
Boa: novel by J.G. Ballard.

**CRASH AND BURN *   15
Charles Band USA   1989
Bill Moseley, Megan Ward, Ralph Waite, Paul Ganus, Eve Larue, Jack McGee, Elizabeth MacEllan, Katherine Armstrong, John Chandler, Kristopher Logan
In the year 2030 a powerful corporation has banned the use of machinery, but is opposed by a group of dissidents. When a girl's grandfather is murdered, she receives help from a mysterious stranger, but the man may not be quite as altruistic as he appears. A totally derivative post-WW3 film that is as dull as it is uninventive.
FAN 81 min (ort 85 min)   VIDrel: EIV/SONOP V

CRASH DIVE *   15
Archie Mayo USA   1943
Tyrone Power, Anne Baxter, Dana Andrews, James Gleason, May Whitty, Henry Morgan, Ben Carter, Charles Tannen, Frank Conroy, Florence Lake, John Archer, George Holmes, Minor Watson, Kathleen Howard, David bacon, Stanley Andrews
A PT boat captain is re-assigned to submarine duty as the executive officer to a captain whose fiancee has aroused his interest and both men find this fact gradually encroaching upon their professional relationship. A rousing flag-waver of a movie with some excellent action scenes and strong acting and direction

that has no need of this trite and tacked-on love story. AA: Effects (vis – Fred Sersen/aud – Roger Herman).
WAR 90 min (ort 105 min)   VIDrel: COLUM/SONOP V/sh

**CRAWLING HAND, THE *   (PG)
Herbert L. Strock USA   1963
Peter Breck, Kent Taylor, Rod Lauren, Arline Judge, Richard Arlen, Alison Hayes, Alan Hale, Ross Elliott, Ed Wermer, Tris Coffin, Syd Saylor, G. Stanley Jones, Ashley Cowan, Jock Putnam, Beverly Lunsford, Andy Andrews
A spacecraft explodes on its return to Earth but an astronaut's hand survives the crash and goes on a killing spree (reminiscent of THE BEAST WITH FIVE FINGERS) because it has become imbued with some strange alien power, in this very poor, almost laughable SF offering.
FAN 89 min B/W   SATrel: BRAVO MOVIES

**CRAZED *   18
Paul Thomas USA   1991
Hyapatia Lee, Christy Canyon, Wayne Summers, Summer Knight, Flame, Chip Knights, Marc Wallice, Melanie Moore, Chuck Martino, Carl Esser, Jonathan Morgan, Tim Lake
Two women are out to have a wild time, in this forgettable porno version of THELMA & LOUISE, that was made in two parts.
A 68 min   VIDrel: VIVID/SCRN V

CRAZIES, THE *   18
George A. Romero USA   1973
Lane Carroll, W.G. McMillan, Harold Wayne Jones, Lloyd Hollar, Lynn Lowry, Richard Liberty, Richard France, Edith Bell, Harry Spillman, Will Disney, Leland Starnes, W.L. Thunhurst Jr, A.C. MacDonald, Robert J. McCully
A plane carrying a sample of a biological agent crashes near a small, remote American town. The virus leaks out into the local water supply causing the inhabitants to go crazy. A gory but very well made tale, that bears more than a passing (and possibly deliberate) resemblance to NIGHT OF THE LIVING DEAD.
Aka: CODE NAME: TRIXIE
HOR 102 min (ort 104 min)   VIDrel: REDEM/RTM V
Boa: play by Paul McCullough.

**CRAZY DESIRES OF A MURDERER *   18
Filippo Walter Ratti ITALY   1977
Isabelle Marchal, Annie Edel
Insane from the time he caught his mother having sex with the gardener, a man finds comfort in murder, embalming his victims after having removed their eyes to add to his collection. One of those gross Italian efforts from the 1970s, recently released by a company that specialises in films of this genre.
HOR 85 min wScrn   VIDrel: REDEM/RTM V

**CRAZY FOR YOU **   15
Harold Becker USA   1985
Matthew Modine, Linda Fiorento, Michael Schoeffling, Ronny Cox, Harold Sylvester, Robert Blossom, Charles Hallahan, Daphne Zuniga, Forest Whitaker, Madonna (Madonna Ciccione), Raphael Sbarge
A young man resorts to extreme action to get a girl to fall in love with him. Standard growing-up type movie. Madonna appears briefly singing, "Crazy for You" and "The Gambler".
Aka: VISION QUEST
DRA 103 min (ort 105 min)   VIDrel: MGM/WHV V/sur
Boa: novel Vision Quest by Terry Davis.

**CRAZY FROM THE HEART **   PG
Thomas Schlamme USA   1991
Christine Lahti, Ruben Blades, William Russ, Louise Latham, Tommy Muniz, Mary Kay Place, Fran Bennett, Vic Trevino, Angela Puton, Nicholas Curtis, Pamela Gordon, Rachel Griffin
The principal of a small town high school in Texas is an attractive but bored woman who has been having a sporadic affair for years with the school's sports coach, but remains unfulfilled. Enter the new janitor, a gentle Mexican farmer who needs a job to make ends meet. Having got her to agree to a date, the couple find the close-knit (and racist) community outraged by her actions. An amiable comedy with a clumsily tacked on social message.
COM 91 min mCab   VIDrel: FIRST/SONOP L/A V

CRAZY IN LOVE *   PG
Martha Coolidge USA   1992
Holly Hunter, Gena Rowlands, Bill Pullman, Julian Sands, Herta Ware, Frances McDormand, Joanne Baron, Michael MacRae, Kit

McDonough, Diane Robin, Krisha Fairchild, Marjorie Nelson, Peter Lohnes, gary Lee Dansenburg, Billy O'Sullivan
Competent and well-acted adaptation of Rice's novel, set in Puget Sound and dealing with the lives and loves of three generations of women.
COM   89 min (ort 93 min) mTV   VIDrel: FIRST/SONOP   V
Boa: novel by Luanne Rice.

CRAZY JUNGLE ADVENTURE *
Harald Reinl USA
Jim Mitchum, Tommy Ohrner, Dawn Chapman, Jenny Jergens, Alexander Hill
The boss of an airline company has made plans to deliberately crash his DC3 to pick up the insurance money, but this plan comes unstuck when his drunken pilot brings the plane down in the remote jungle.
COM   90 min   VIDrel: RAVEN/QUANT   V

PG
1992

CRAZY LOVE ***
Dominique Deruddere BELGIUM/FRANCE
Josse De Pauw, Geert Hunaerts, Amid Chakir, An Van Essche, Florence Beliard
Three very dark stories by Charles Bukowski are linked together as a means of following the growing alienation of one Harry Voss – the film's central character. In 1955, as an innocent twelve-year-old Voss learns of love thanks to a movie. By 1962, he is a troubled, acne-plagued youth whose romantic advances are brutally rebuffed. Finally, in 1976 he is a drug addict and alcoholic living on the fringes of society. Deruddere's debut is a difficult and disturbing work.
DRA   83 min (Dutch dialogue)   VIDrel: MAINPIC/RTM   V

18
1987

CRAZY PEOPLE **
Tony Bill USA
Dudley Moore, Daryl Hannah, Paul Reiser, Mercedes Ruehl, J.T. Walsh, Bill Smitrovich, Alan North, David Paymer, Dick Cusack, Ben Hammer
An overworked advertising executive is committed to a mental institution following a breakdown during which he set about devising a whole series of brutally frank (and often very funny) adverts. However, notwithstanding his incarceration, the ads he has devised are wildly successful. A winning performance from Moore and some witty one-liners help pad out this essentially bland and empty attempt at a screwball business comedy.
COM   87 min (ort 90 min)   VIDrel: CIC/SONOP   V/sur

15
1990

CREATOR **
Ivan Passer USA
Peter O'Toole, Vincent Spano, Mariel Hemingway, Virginia Madsen, David Ogden Stiers, John Dehner, Karen Kopins, Kenneth Tigar, Elsa Raven, Lee Kessler, Rance Howard, Ellen Geer, Jeff Corey, Ian Wolfe, Mike Jolly, Doug Cox
A biologist hopes to recreate his long dead wife from a few of her cells which he has kept. However, to further this end, he requires a human egg which is lovingly donated by his beautiful lab assistant. As his experiments proceed, romantic complications ensue.
COM   103 min (ort 108 min)   VIDrel: EIV/SONOP   V
Boa: novel by Jeremy Leven.

15
1985

CREATURE FROM THE BLACK LAGOON, THE **
Jack Arnold USA
Richard Carlson, Julia Adams, Richard Denning, Antonio Moreno, Nestor Paiva, Whit Bissell, Bernie Gozier, Sidney Mason, Julio Lopez, Rodd Redwing, Ben Chapman, Harry Escalante, Ricou Browning (the creature)
Originally made in 3-D, this tells of an expedition down the Amazon and an encounter with a strange lizard creature. Little in the way of a menacing atmosphere, if anything the creature seems rather likeable. The best thing in this one is the underwater photography. Followed by "Revenge Of The Creature" and "The Creature Walks Among Us".
HOR   79 min B/W   VIDrel: PION   LV

PG
1954

CREATURE OF THE WALKING DEAD *
Frederic Corte (Fernando Cortes)/Jerry Warren
MEXICO/USA
Rock Madison, Ann Wells, Katherine Victor, Bruno VeSota, George Todd, Willard Gross, Fernando Casanova, Sonia Furio
A mad scientist who conducted experiments in immortality is resurrected by his grandson with the predictable dire results. Not helped by its interminable non-action sequences and narra-

(PG)
1960

tion, this awful film consists of a 1960 Mexican release with added new footage.
Aka: LA MARCA DEL MUERTO
HOR   74 min B/W dubbed   SATrel: BRAVO MOVIES

CREATURES THE WORLD FORGOT *
Don Chaffey UK
Julie Ege, Tony Bonner, Robert John, Sue Wilson, Marcia Fox, Rosalie Crutchley, Brian O'Shaughnessy, Don Leonard, Beverly Blake, Doon Baide, Ken Hare, Sue Wilson, Rosita Moulin, Frank Hayden, Gerard Bonthuys, Hans Kiesouw
Another offering from the Hammer stable. This one is a tedious story of the quarrels among rival tribes in the Stone Age. A feeble follow-up to ONE MILLION YEARS B.C. and WHEN DINOSAURS RULED THE EARTH, but this one doesn't even have any dinosaurs to enliven it.
FAN   94 min   VIDrel: RCA L/A   V

18
1971

CREEPERS **
Martin Newlin USA
Jason Saucier, Mary Sellers
Radioactive fallout from a leaking reactor causes the trees of a nearby forest to become carnivorous, an event not exactly welcomed by the local populace.
Aka: CONTAMINATION 7
HOR   90 min   VIDrel: 20VIS/SONOP   V

18

CREEPING FLESH, THE **
Freddie Francis UK
Peter Cushing, Christopher Lee, Lorna Heilbron, George Benson, Kenneth J. Warren, Duncan Lamont, Harry Locke, Hedger Wallace, Michael Ripper, Catherine Finn, Robert Swann, David Bailie, Maurice Bush, Tony Wright
A professor discovers the skeleton of the "Evil One" which he finds will regain its tissues if it comes into contact with water. A series of experiments follows that do not lead to great happiness for him and his family. An absurd horror yarn, but quite chilling in places.
Aka: CRAZE
HOR   87 min (ort 90 min)   VIDrel: ARTPRO/RTM   V

18
1972

CREEPOZOIDS **
David De Coteau USA
Linnea Quigley, Ken Abraham, Michael Aranda, Richard Hawkins, Kim McKamy, Joi Wilson
A bunch of survivors of a nuclear war take refuge in a disused shelter in order to escape from the deadly acid rain. However, they soon find that the shelter is not unoccupied, in this routine chiller that inevitably reminds one of ALIEN.
HOR   69 min (ort 90 min)   VIDrel: ALLIED/RTM   V

18
1987

CREEPSHOW ***
George Romero USA
Hal Holbrook, Adrienne Barbeau, E.G. Marshall, Leslie Nielsen, Fritz Weaver, Viveca Lindfors, Stephen King, Carrie Nye, Ted Danson, Warner Shook, Robert Harper, Elizabeth Regan, Gaylen Ross, Jon Lorner, Don Keefer, Bingo O'Malley
Collection of five separate stories that pay a kind of homage to those pulp comics of the 1950s and from which they draw their inspiration. The direction and acting are fine, but the Stephen King stories are shallow and predictable exercises in horror and grisliness, enlivened by flashes of humour. Each tale has a comic-book type introduction. Followed by CREEPSHOW 2.
HOR   115 min (ort 119 min)   VIDrel: VCC L/A   V
Boa: short stories by Stephen King.

15
1982

CREEPSHOW 2 **
Michael Gornick USA
Lois Chiles, George Kennedy, Dorothy Lamour, Tom Savini, Domenick John, Frank S. Salsedo, Holt McCallany, Dan Kamin, Don Harvey, David Holbrook, Philip Dore, Daniel Bear, Jeremy Green, Page Hannah, Paul Satterfield
Three more Stephen King horror tales are adapted for the screen, but this time the comic-book introduction is dispensed with, though the tales remain just as gruesomely heavy-handed as before (but slightly less stylish). Stories are entitled: Old Chief Wood'nhead", "The Hitch-hiker" and "The Raft".
HOR   85 min (ort 92 min)   VIDrel: VCC/DISC   V/h
Boa: short stories by Stephen King.

18
1987

CRIES AND WHISPERS ****
Ingmar Bergman SWEDEN
Harriet Andersson, Liv Ullmann, Ingrid Thulin, Kari Sylwan, Erland

(15)
1972

Josephson, George Arlin, Henning Moritzen, Anders Ek, Linn Ullmann, Rosana Mariano, Lena Bergman, Monika Priede
Another slab of the old doom and gloom from Bergman. This time we are treated to a story about the lives of a dying woman, her sisters and the servants. As beautifully made as ever but almost unbearably depressing. AA: Cin (Sven Nykvist).
Aka: VISKINGAR OCH ROP
DRA   91 min   VIDrel: L/A   V

CRIME OF THE CENTURY ***
15
Mark Rydell USA   1996
Stephen Rea, Isabella Rossellini, J.T. Walsh, Michael Moriarty
The story of one of America's most grotesque miscarriages of justice, when in the 1930s German immigrant Richard Hauptmann (Rea) was convicted of the kidnap and murder of the baby of aviation hero Charles Lindbergh after he stumbled across the ransom money that had been left out by the family. A very detailed and careful film that avoids unnecessary sentiment and sensationalism, opting instead for strongly drawn characters and a literate script.
DRA   111 min   VIDrel: MOSAIC/COLUM   V/sur
Boa: novel by Ludovic Kennedy.

CRIME STORY **
18
Kirk Wong HONG KONG   1993
Jackie Chan, Kent Chang, Christine Ng, Phua Leng Leng, Au-Yeung Pui Shan, Law Hang Kang
Set against the background of kidnappings now commonplace in Hong Kong, this serious and violent Chan vehicle has him cast a policeman assigned to protect a wealthy business from abduction. Having failed in this task, he has to rescue him and learns to his dismay that his own partner is the mastermind behind this crime. Allegedly based on a true case, this one is full of action and stunts that pad out the simple story.
Aka: CHUNG ON TSOU; POLICE STORY 4: CRIME STORY; SERIOUS CRIMES SQUAD
MAR   105 min   VIDrel: MIA/DISC   V/h

CRIMEBROKER **
(15)
Ian Barry AUSTRALIA   1993
Jacqueline Bisset, Masaya Koto, Gary Day, John Bach, Gary Sweet, Victoria Langley, Richard Roxburgh, Peter Boxwell, Victoria Campbell, Paul Hanlon, Ralph Cotterill, Peter Hathaway, Justin Lewis, Barry Quin, David Richardson
A woman judge is saddled with a secret life that sees her devoting her off-duty hours to committing a variety of crimes, including bank robbery. Things come to a head, however, when her lover tempts her with a proposal for an audacious and highly risky venture.
THR   90 min   SATrel: MOVIE CHANNEL

CRIMES AND MISDEMEANORS ****
15
Woody Allen USA   1989
Martin Landau, Claire Bloom, Anjelica Huston, Woody Allen, Alan Alda, Mia Farrow, Joanna Gleason, Jenny Nichols, Jerry Orbach, Sam Waterston, Caroline Aaron, Christine Aaron, Daryl Hannah, Stephanie Roth, Gregg Edelman, Zina Jasper
Allen takes a good look at envy in this complex film that consists of two distinct stories: in one a wealthy ophthalmologist sets out to silence a troublesome mistress who threatens to expose their affair and in the other a documentary film-maker tries to woo an attractive producer whilst making a film about her obnoxious boss. An ambitious, bittersweet and extremely engaging work; one of Allen's best to date.
COM   100 min (ort 104 min)   VIDrel: VISVID/POLYREC   V

CRIMES OF PASSION **
18
Ken Russell USA   1984
Kathleen Turner, Anthony Perkins, John Laughlin, Annie Potts, Bruce Davison, Norman Burton, James Crittenden, Peggy Feury, Dan Gerrity, Vince McKerwin, Lisa Hayslip, Terry Hoyos, Deanna Oliver, Patricia Stevens, Gordon Hunt, Janice Kent
A married man has an affair with a fashion designer who seems to spend her leisure hours as a high-class prostitute. Less confused than some other work by this director but still heavygoing. A fairly unmemorable effort with Perkins giving his standard nutter performance as he becomes increasingly obsessed with Turner. Several minutes of kinky footage were removed on release.
DRA   102 min (Cut at film release – ort 105 min)
VIDrel: VISVID/POLYREC   V

CRIMES OF SILENCE **
15
James A. Contner USA
Michelle Greene, William R. Moses, Lynda Carter, Joe Penny
A woman is raped by her dentist whilst unconscious under a general anaesthetic, and her torment is added to when she learns that she is pregnant.
DRA   90 min   VIDrel: ODY/SONOP   V

CRIMES OF THE HEART ***
15
Bruce Beresford USA   1987
Diane Keaton, Jessica Lange, Sissy Spacek, Sam Shephard, Tess Harper, David Carpenter, Hurd Hatfield, Beeson Carroll, Jean Willard, Tom Mason, Gregory Travis, Annie McKnight, Eleanor Eagle, Jessica Ezzall, Natalie Anderson
Three Southern sisters meet again in the family home, after the youngest has just completed a prison sentence, for shooting her husband because he beat up her black lover. During their stay assorted family secrets are revealed. A sad, funny and poignant drama based on Henley's Pulitzer Prize-winning play, with screenplay by Henley.
DRA   100 min (ort 105 min)   VIDrel: BMGREC/BMGVID   V
Boa: play by Beth Henley.

CRIMETIME **
18
George Sluizer GERMANY/UK/USA   1996
Stephen Baldwin, Pete Postlethwaite, Sadie Frost, Geraldine Chaplin, Karen Black, James Faulkner, Philip Davis, Marianne Faithfull, Emma Roberts, Anne Lambton, Suzanne Bertish, Stephanie Buttle, Caroline Langrishe, Ron Berglas
After a man kills a woman on a piece of waste ground, he sees the crime re-enacted on a TV show, and becomes obsessed with the actor who plays "him" on TV. Further killings follow, and as the murderer dreams of TV stardom, the real actor slowly slips over the edge, his role in these re-enactments having made him into a celebrity, albeit an unbalanced one. A film of great potential, but one that is exploitative and poorly directed. See also I LOVE A MAN IN UNIFORM.
THR   118 min   CINrel

CRIMINAL, THE ***
PG
Joseph Losey UK   1960
Stanley Baker, Margit Saad, Sam Wanamaker, Patrick Magee, Noel Willman, Gregoire Aslan, Jill Bennett, Kenneth J. Warren, Nigel Green, Patrick Wymark, Laurence Naismith, Edward Judd, Rupert Davies, Murray Melvin
After being sent to prison for a racecourse robbery, a crook is helped to escape by other members of his gang, but attempts a double-cross when he tries to regain some hidden loot. Baker gives a very strong performance in a grim prison film that provokes one's thoughts but never engages one's sympathies.
A/AD   97 min B/W   VIDrel: LUMI/SPEAR   V

CRIMINAL **
18
David Jacobson GERMANY/UK   1994
Ralph Feliciello, Liz Sherman, Sheila York, Eric Reid, Jim Myers, Mikki Moine, Thomas Crouch, Tim Miller
A man steals a large sum of money in the hope of starting a new life with his wife, but takes off alone when he arrives home to find her in bed with her lover. A big success at the 1994 Berlin Film Festival, this was something of a foray into film noir territory, not least thanks to the sombre B/W photography. Quite a downbeat work, oddly plotted and though not great cinema, well worth a look.
DRA   80 min B/W mTV   VIDrel: SCEDGE/RTM   V

CRIMINAL HEARTS ***
(18)
Dave Payne USA   1995
Kevin Dillon, Amy Locane, Morgan Fairchild, M. Emmet Walsh, Michael James McDonald, Cassandra Leigh, Ismael "East" Cario, Michael Todd Curry, Dee Croxton, Julie Araskog, Chelsea Maison-Ciu, Charles Martinez, Bpb McFarland, Don Stroud
Tense road-movie that sees a jilted woman and a hitch-hiker thrown together by force of circumstance. She is trying to locate her missing fiance while he is wanted for robbery and murder and is being hunted by the FBI. What injects a note of interest is that the woman knows nothing of this, and as their journey proceeds, she starts to show a romantic interest in her companion, and they enjoy some breakneck adventures. A fairly standard action-filled road-movie.
A/AD   89 min (ort 92 min)   SATrel: SKY MOVIES

CRIMINAL INTENT **   *18*
Worth Keeter USA   1992
Robert Davi, Jack Scalia, James Russo, Joan Severance, Kent McCord,
Jenilee Harrison, Phil Rubinstein, Michael Monks, Lonnie Burr, Jeff
Carrara, Pamella D'Pella, George Kyle, Talmadge Ragan, Don
Marino, Sondra Currie
When a tough and cynical cop is suspended from duty for using
excessive force, he discovers that his wife has become involved
with the officer put in charge of conducting the internal disci-
plinary investigation.
Aka: ILLICIT BEHAVIOR
DRA   99 min (ort 104 min)   VIDrel: CIC/SONOP   V

CRIMINAL MIND, THE **   *15*
Joseph Vittorie USA   1995
Ben Cross, Frank Rossi, Tahnee Welch, Lance Henriksen, Mark
Davenport, Jeff Austin, Steven Barr, Kiki, Milan Nicksic, Chip Heller,
Philip Tan, Byron Chung, Lynn-Holly Johnson
A D.A. manages to effect a reunion with his brother but his joy
soon evaporates when he discovers that they stand on opposite
sides of the law since the latter has now become a member of
the gangster fraternity.
DRA   92 min   VIDrel: COLUM/SONOP   V/sh

CRIMINAL PASSION **   *15*
Reza J. Badiyi USA
Joanna Cassidy, Jere Burns, Brooke Langton
A student finds that one of her teachers has developed a danger-
ously obsessive interest in her.
THR   90 min   VIDrel: ODY/SONOP   V/sh

CRIMSON TIDE ***   *15*
Tony Scott USA   1994
Gene Hackman, Denzel Washington, Matt Craven, George Dzundza,
Viggo Mortensen, James Gandolfini, Lillo Brancato, Rick Shroder,
Rocky Carroll, Jaime P. Gomez, Scott Burkholder, Michael Milhoan,
Danny Nucci, Vanessa Bell Calloway
After the collapse of the Soviet Union, the world once again
faces the prospect of WW3 when Russian rebels hijack a nuclear
submarine and the USA is forced to send one of its submarines
in order to deal with this threat. However, matters are not
helped by the personal conflict between its veteran commander
and his newly assigned executive officer. A claustrophobic
thriller, of tension and great acting, that works exceptionally
well within a familiar framework.
THR   113 min (ort 115 min) wScrn cC
VIDrel: HOLPIC/TECH; ENCORE/BUENA (LV only)   V/sh LV

CRISIS IN THE KREMLIN: THE LAST DAYS OF
THE SOVIET UNION *   *15*
Jonathan Winfrey USA   1985 (released 1987)
Robert Rusler, Theodore Bikel, Denise Bixler, Doug Wert, Stephan
Danailov, Borris, Loukanov, Jocko, Rositch, George Novakov, Stoycho
Mazgalov, Nikolia Stoilov, Rosen Siromachov, Emil Attnassov, Anna
Vilchanova, Brian Maslo
Political thriller in which a US agent is sent to Moscow to help
prevent the assassination of Gorbachev. An adequate actioner
that incorporates real footage from the attempted coup but holds
little interest, having been overshadowed by events of far
greater interest.
Aka: RED TARGET: THE PLOT TO OVERTHROW THE USSR
THR   84 min   VIDrel: CIC/SONOP L/A   V

CRISS CROSS ***   *(15)*
Robert Siodmark USA   1948
Burt Lancaster, Yvonne De Carlo, Dan Duryea, Stephen McNally,
Richard Long, Tom Pedi, Alan Napier, James (Tony) Curtis, Percy
Helton, Griff Barnett, Meg Randall, Joan Miller, Edna M. Holland,
John Miller, John Doucette, Esy Morales
An armoured-truck guard becomes mixes up in a payroll
robbery because of his unrequited passion for his gold-digging
ex-wife, who has married a notorious gangster. A realistic and
gritty thriller made with the sure Siodmark touch, it marked the
screen debut for Curtis. Remade many years later as THE
UNDERNEATH.
THR   98 min B/W   TVrel

CRISS CROSS **   *15*
Chris Menges USA   1991
Goldie Hawn, Keith Carradine, Arliss Howard, James Gammon, J.C.
Quinn, Steve Buscemi, Paul Calderon, Cathryn DePrume, Nada
Despotovich, Deidre O'Connell, David Anthony Marshall, Anna
Levine Thompson, Neil Guintoli, Cristy Martin

Family drama set in Key West in 1969, where a woman and her
twelve-year-old son struggle to make ends meet after being
abandoned by the husband, a mentally ill Vietnam veteran. A
good human interest story whose potential is largely dissipated
by its slow pace and over-reliance on voice-overs.
Aka: ALONE TOGETHER; CRISSCROSS
DRA   96 min (ort 101 min)   VIDrel: MGM/WHV   V/sur
Boa: novel by Don Tracy.

CRITICAL CONDITION *   *15*
Michael Apted USA   1986
Richard Pryor, Rachel Ticotin, Ruben Blades, Joe Mantegna, Bob
Dishy, Sylvia Miles, Joe Dallesandro, Randall (Tex) Cobb, Garrett
Morris
A con-man takes charge of a prison hospital, in this predictable
and rather thin prison comedy. Pryor gives a good performance,
but the story is too limited for comedy and too disorganised for
drama.
COM   94 min   VIDrel: CIC/SONOP   V/h

CRITTERS *   *15*
Stephen Herek USA   1985 (released 1986)
Dee Wallace Stone, M. Emmet Walsh, Billy Green Bush, Scott Grimes,
Nadine Van Der Velde, Don Opper, Terrence Mann, Billy Zane,
Etham Phillips, Jeremy Lawrence, Lin Shaye, Michael Lee Gogin, Art
Frankel, Douglas Koth, Roger Hampton
Murderous alien criminals escape from their prison planet and
flee to Earth, where they proceed to feast upon the occupants of
an isolated Kansas farm. A rubbishy horror yarn that has a few
echoes of GREMLINS though not its style. Followed by CRIT-
TERS 2.
HOR   82 min (ort 86 min)   VIDrel: VCC/DISC/COLUM   V

CRITTERS 2 **   *15*
Mick Garris USA   1988
Scott Grimes, Liane Curtis, Don Opper, Barry Corbin, Tom Hodges,
Sam Anderson, Lindsay Parker, Herta Ware, Terrence Mann,
Roxanne Kernohan, Doug Rowe, Lin Shaye
Very much a re-run of the first film, set in the same small Kansas
town where, two years on, hatching eggs once more give rise to
the murderous title creatures. Music is by Nicholas Pike.
Aka: CRITTER 2: THE MAIN COURSE
HOR   82 min (ort 87 min)   VIDrel: VCC/DISC/COLUM
V/sh

CRITTERS 3 **   *15*
Kristine Peterson USA   1991
Don Opper, Francis Bay, Leonardo DiCaprio, Aimee Brooks, John
Calvin, Katherine Cortez, Geoffrey Blake, Diana Bellamy, Terrence
Mann, Nina Axelrod, John Cousins, Christian Cousins, William
Dennis Hunt, Jose Luis Valnesuela
A further dose of horror-comedy that sees our title aliens taking
over an apartment block in search of more humans to feed their
voracious appetites. An adequate shocker outing for the undis-
cerning. Followed by CRITTERS 4 (which was filmed at the
same time).
HOR   81 min (ort 86 min) mVid   VIDrel: EIV/SONOP   V

CRITTERS 4 **   *15*
Rupert Harvey USA   1992
Brad Dourif, Terrence Mann, Don Opper, Angela Bassett, Paul
Whitthorne, Eric DaRae, Anders Hove, Anne Elizabeth Ramsay,
Martine Beswick (voice only)
Second half of a back-to-back production with CRITTERS 3, that
sees a mutant strain of these nasty aliens stowing away on a
spaceship (as eggs) and hatching out of their shells to menace
the crew and threaten the entire universe. This continuation of
the horror spoof throws barbs in a few directions, its main target
being ALIEN.
Aka: CRITTERS 4: CRITTERS IN SPACE
HOR   90 min (ort 94 min) mVid   VIDrel: EIV/SONOP   V

CROCODILE DUNDEE ***   *15*
Peter Faiman AUSTRALIA   1986
Paul Hogan, Linda Kozlowski, John Meillon, Mark Blum, Michael
Lombard, David Gulpilil, Irving Metzman, Sue Charlton, Ritchie
Singer, Maggie Blinco, Steve Rackman, Gerry Skilton, Terry Gill,
Peter Turnbull, Christine Totos
A New York reporter is intrigued by stories of an Australian
who tackles crocodiles in the outback, and makes a journey
through the bush in his company. She entices him back to
Manhattan, where muggers, hookers and high society folk fall
victim to his easy-going charm. A humorous and happy old-

fashioned adventure story, co-written by Hogan; who makes this film his screen debut. Followed by CROCODILE DUNDEE 2.
A/AD   93 min Cut (23 sec – ort 98 min)   VIDrel: POLY/POLYREC V/sh

CROCODILE DUNDEE 2 *   *PG*
John Cornell AUSTRALIA/USA   1988
Paul Hogan, Linda Kozlowski, John Meillon, Charles Dutton, Hechter Ubarry, Juan Fernandez, Ernie Dingo, Kenneth Walsh, Gus Mercurio, Gerry Skilton, Jim Holt, Alec Wilson, Stephen Root, Luis Guzman
A pleasant sequel to the earlier film that starts off in Manhattan and winds up in the bush. In between, our laid-back charmer finds himself running foul of an international drugs baron. Pace and tension are all but eliminated in this leisurely journey, but as before, Hogan's easy charm and ready wit carry him through. Written by Hogan and his son Brett.
COM   107 min (Cut at film release by 1 sec – ort 110 min) VIDrel: 4-FRONT/POLYREC/CIC L/A V/dm

CROCODILE SHOES *   *15*
David Richards UK   1995
Jimmy Nail, James Wilby, Roy Pattison, Sammy Johnson, Trevor Fox, Mike Elliott, Barwick Kaler, Harry Jones, Christopher Nichol, Ginny Holder, Zubin Varla, Brian Capron, Angus Wright, Anna Glavin, Anoushka Menzies, Cara Konig
A rock singer/songwriter attempts to solve the mystery surrounding the death of his manager, and his efforts to do so involve him in ever more sinister intrigues, eventually culminating in his own arrest for this crime. Written by Nail, this overly complex tale unfolds at a leisurely pace, and benefits from his strong screen presence.
DRA   364 min (3 cassettes – separately available) mTV
VIDrel: CHRYS/CARL V

CROMWELL *   *PG*
Ken Hughes UK   1970
Richard Harris, Alec Guinness, Robert Morley, Dorothy Tutin, Frank Finlay, Stratford Johns, Timothy Dalton, Patrick Wymark, Patrick Magee, Nigel Stock, Charles Gray, Michael Jayston, Michael Goodliffe, Anna Cropper
A stuffy account of the rise to power of Oliver Cromwell, with Harris badly miscast in the title role. Excellent battle scenes and photography are compensations in this cold and unmoving effort. AA: Cost (Nino Novarese).
DRA   134 min (ort 145 min) wScrn
VIDrel: VCC/DISC/COLUM V

CRONOS *   *18*
Guillermo Del Toro MEXICO/USA   1992
Ron Perlman, Federico Luppi, Claudio Brook, Tamara Shanath, Margarita Isabel, Daniel Gimenez Cacho, Mario Ivan Martinez, Juan Carlos Columbo, Farnesio de Bernal, Luis Rodriguez, Javier Alvarez, Gerardo Moscoso, Eugenio Lobo
An antiques dealer discovers than an artefact from the 14th century possesses the power to confer eternal life but this comes at quite a high price. A highly unusual story that concentrates on the man's struggles to retain his humanity rather than simply exploring the more obvious horror elements, as it could so easily have done. The ingenious and witty script was the work of Del Toro.
Aka: CRONOS: IMMORTAL CURSE
HOR   88 min (ort 92 min) subs   VIDrel: TART/20TH; ENCORE (LV only) V LV

CROOKLYN *   *12*
Spike Lee USA   1994
Alfre Woodard, Delroy Lindo, Zelda Harris, David Patrick Kelly, Carlto Williams, Sharif Rasjed, Tse-Mach Washington, Christopher Knowings, Jose Zuniga, Spike Lee, Frances Foster, Norman Matlock, Joie Susannah Lee, Vondie Curtis-Hall
Long, low-key and episodic tale of a black family living in Brooklyn in the 1970s that is built around the only girl and her relationship with her four brothers and her parents. Written by Lee together with his brother and sister, the biographical elements seem quite obvious, and for all the fine acting, this rambling effort needed a tighter and more disciplined script.
Aka: CROOKLYN: A SPIKE LEE JOINT!
DRA   109 min (ort 114 min) cC   VIDrel: CIC/SONOP V

CROSS MY HEART **   *PG*
Jacques Fansten FRANCE   1990
Sylvain Copans, Nicolas Parodi, Cecilia Rouaud, Olivier Montiege,
Benoit Gautier, Lucie Blossier, Delphine Gouttman, Mathieu Poussin, Kaldi El Hadj, Romauld Jarny, Wilfrid Flandrin, Dominique Lavanant, Jacques Bonnaffe
When a woman dies, a group of children conspire to keep her death a secret so that her son will not be sent to a state-run orphanage. An engaging portrait of childhood, with the warmth and solidarity of the boy's classmates contrasting sharply with the behaviour of the adults. In the end, however, our young hero's fears prove amply justified. Natural and credible performances are firmly in evidence all round from this fine young cast. Warmly recommended.
Aka: LA FRACTURE DU MYOCARDE
DRA   105 min   CINrel

CROSS OF FIRE *   *15*
Paul Wendkos USA   1989
John Heard, Mel Harris, Lloyd Bridges, David Morse, Kim Hunter, George Dzundza, Donald Moffat, Keith Szarabajka, Caroline Kara, Ed Wiley, Dakin Matthews, Douglas Roberts, William Schallert, Dion Anderson, Gilbert Lewis
A first-rate re-enactment of the career of the man responsible for reviving the flagging fortunes of the Ku Klux Klan in Indiana in the 1920s, turning it from an obscure fringe party into a potent political force. Heard gives an outstanding performance in a powerful drama that charts the rise and fall of Klan Grand Dragon D.C. Stephenson. Originally shown in two parts.
DRA   178 min (ort 200 min) mTV
VIDrel: POLY/POLYREC/BRAVE V

CROSS OF IRON *   *18*
Sam Peckinpah UK/WEST GERMANY   1977
James Coburn, Maximilian Schell, James Mason, David Warner, Klaus Lowitsch, Roger Fritz, Vadim Glowna, Fred Stillkraut, Burkhardt Driest, Dieter Schidor, Senta Berger, Veronique Vendell, Arthur Brauss, Slavko Stimac
A brutal, stolid account of the horrors of WW2 as seen through the eyes of a German unit on the Russian front, with competently handled battle sequences but poor characterisation and gratuitous gore. The loosely structured plot tells of a cowardly German officer who is determined to win the Iron Cross. This was the director's first war film. Followed by BREAK-THROUGH.
WAR   127 min (ort 133 min)   VIDrel: WHV V
Boa: novel Das Geduldige Fleisch by Willi Heinrich.

CROSSCUT *   *18*
Paul Raimondi USA   1995
Costas Mandylor, Casey Sander, Allen Cutter, Jay Acovane, Zack Norman, Doug Spinuzza, George Murdock, Greg Collins, Christopher Stanley, Walter Norman
A man who goes to the defence of his friend inadvertently kills the son of a powerful Mafia hoodlum, and has to contend with the enraged father's desire for revenge. The expected nastiness and conflicts are on offer, but unfortunately little that is either fresh or original.
A/AD   94 min   VIDrel: FIRST/SONOP V

CROSSFIRE *   *18*
Rick King USA   1993
Robert Davi, Jeff Fahey, Tia Carrere, Teri Polo, Martin Donnovan
A ruthless female professional assassin has an equally ruthless federal agent for a boyfriend, and the latter is inclined to act beyond the law. The usual double-crosses and violent confrontations ensue.
A/AD   94 min   VIDrel: PROMARK/HIFLI V/h

CROSSING, THE *   *15*
George Ogilvie AUSTRALIA   1990
Russell Crowe, Robert Mammone, Danielle Spencer, Emily Lombers, Rodney Bell, Ben Oxenbould, Myles Collins, Marc Gray, Megan Connolly, John Blair, Gail Lockland, Lea-Ann Tower, Paul Robertson, George Whaley, Jacqy Phillips
A love-triangle drama involving three youngsters whose mutual friendship is compromised by the changing pattern in their relationship. Average in terms of plotting, but the film offers some very convincing performances plus a shattering and unusually downbeat ending.
DRA   90 min (ort 92 min)   VIDrel: 20VIS/SONOP V/sur

CROSSING DELANCEY *   *PG*
Joan Micklin Silver USA   1988
Amy Irving, Peter Riegert, Reizl Bozyk, Jeroen Krabbe, Sylvia Miles,

Suzzy Roche, George Martin, John Bedford Lloyd, Claudia Silver, Rosemary Harris, Amy Wright, Kathleen Wilhoite, Susan Sandler, John Patrick Stanley
A gentle romantic comedy, with Irving as an upwardly mobile Jewish girl being fixed up with a down-to-earth pickle-seller by her doting grandmother. The film has considerable charm but the story stands in sore need of development. The script is by Sandler.
DRA   93 min (ort 97 min)   VIDrel: WHV L/A   V/sh
Boa: play by Susan Sandler.

CROSSING GUARD, THE ***   15
Sean Penn USA   1995
Jack Nicholson, David Morse, Anjelica Huston, Robin Wright, Piper Laurie, Richard Bradford, Priscilla Barnes, David Baerwald, Robbie Robertson, John Savage, Kari Wuhrer, Jennifer Leigh Warren, Kellita Smith, Richard Sarafian
After spending six years in jail for knocking down and killing a little girl as a drunken driver, a man now has to face the wrath of the father whose life he ruined, and who has spent all this time consumed with hatred and the thirst for revenge. Scripted by Penn, this is a thoughtful and often quite unsubtle character study, and if the film does tend to dwell excessively on its issues, the compelling performances of the two leads demand our attention.
A/AD   107 min (ort 114 min) cC
VIDrel: HOLPIC/TECH   V/sur

CROSSING THE BRIDGE ***   15
Mike Binder USA   1992
Stephen Baldwin, Josh Charles, Jason Gedrick, Cheryl Pollack, Richard Edson, Jeffrey Tambor, Rita Taggart, Hy Anzell, Ken Jenkins, David Schwimmer, Abraham Benrubi, Bob Nickman, James Krag, Rana Haugen, Todd Tidgewell, Daniel Hawke
Modest, coming-of-age film about a group of teenagers in Detroit in 1975 and how they pass the time after graduation from high school. On one of their trips to Canada to visit the strip joints, they find themselves being tempted by the opportunity to make some money by smuggling hash back into the USA. The splendid acting is the mainstay of this gripping drama whose principal characters are unfortunately all rather unsympathetic individuals.
DRA   99 min (ort 104 min)   VIDrel: EIV/SONOP   V

CROSSING THE LINE **   15
Gary Graver USA   1989
Rick Hearst, John Saxon, Jon Stafford, Colleen Morris, Cameron Mitchell, Vernon Wells
A portrayal of sibling rivalry in a wealthy family where a badly behaved youth is blamed for the motorcycle accident that left his best friend in a coma. Average.
A/AD   90 min (ort 102 min)   VIDrel: EIV/SONOP   V/sur

CROSSINGS *   12
Karen Arthur USA   1985
Cheryl Ladd, Lee Horsley, Jane Seymour, Christopher Plummer, Stewart Granger, Joan Fontaine, Joanna Pacula, Horst Buchholz, Zach Galligan
Another entry in the Danielle Steel fantasy-factory, this time set during WW2, when the head of a steel company is asked by the President to gather information on Hitler.
Aka: DANIELLE STEEL'S CROSSINGS
DRA   269 min mTV   VIDrel: WHV   V/h
Boa: novel by Danielle Steel.

CROSSROADS **   15
Walter Hill USA   1986
Ralph Macchio, Joe Seneca, Jami Gertz, Joe Morton, Robert Judd, Harry Carey Jr
A young musician traces a legendary bluesman to his hospital bed in Harlem, and agrees to bring him back to Mississippi if he will teach him some long lost songs. This nice idea develops into a tacky morality play, with the bluesman having sold his soul to the devil – Macchio has to play like never before to win it back. The backing score is by Ry Cooder.
A/AD   98 min   VIDrel: CASPIC/BMGREC   V/sur

CROSSWORLDS **   15
Krishna Rao USA   1996
Rutger Hauer, Josh Charles, Stuart Wilson, Andrea Roth
A man acquires a magic sceptre that enables him to travel between different dimensions, but this is sought by a powerful warlord who wants to use it to destroy the dimensional barri-

ers and dominate the entire universe. Adequate SF nonsense, the kind of film to watch on a rainy day.
FAN   87 min   VIDrel: TRIM/HIFLI   V/h

CROW, THE **   18
Alex Proyas USA   1993
Brandon Lee, Ernie Hudson, Michael Wincott, David Patrick Kelly, Angel David, Rochelle Davis, Bai Ling, Tony Todd, Jon Polito, Lawrence Mason, Michael Massee, Marco Rodriguez, Sofis Shinas, Bill Raymond, Anna Thomson, Ron Sykes
A rock musician returns from the dead one year after he and his girlfriend were murdered and stalks those responsible, in this heavily stylised and morbid version of the comic-book by James O'Barr. Very much a triumph of form over content, but eminently watchable for all its faults. Lee (the son of the late Bruce Lee) was accidentally killed during production and innovative digital techniques were used to complete the movie. Followed by THE CROW CITY OF ANGELS.
FAN   101 min   VIDrel: EIV/SONOP   V/sur

CROW CITY OF ANGELS, THE *   12
Tim Pope UK/USA   1996
Vincent Perez, Mia Kirshner, Richard Brooks, Vincent Castellanos, Ian Dury, Tracey Ellis, Thomas Jane, Iggy Pop, Thuy Trang, Eric Acosta, Beverley Mitchell, Aaron Thell Smith, Alan Gelfant, Shelly Desai, Holley Chant, Kerry Rossall
After a father and his son are murdered, the former comes back from the dead to exact his revenge on the gang of thugs responsible. A terribly lame sequel to THE CROW, with no shortage of the same dark and brooding Gothic symbolism, and ample pop music for good measure. However, the film is badly marred by weak direction and unconvincing performances; some originality in the script department would have been most welcome too.
HOR   91 min Cut cC   VIDrel: HOLPIC/TECH   V/sur
Boa: comic strip series created by James O'Barr.

CROWFOOT **   18
James Whitmore Jr USA   1995
Jim Davidson, Tsai Chin, Kate Hodges, Bruce Locke, Larry Manetti, Charles Ka'upu, Mike Genovese, Michael Watson, Erin Gray, Nicolas Surovy, Dann Seki, Kimberly Avila, Sandra Sagisi, Christina Oliver, Melanie S. Cajudoy
Man rescues a woman from attack, but this is only the start of his problems.
DRA   86 min mTV   VIDrel: CIC   V

CRUCIBLE, THE ****   12
Nicholas Hytner USA   1996
Daniel Day-Lewis, Paul Scofield, Winona Ryder, Joan Allen, Bruce Davison, Rob Campbell, Jeffrey Jones, Peter Vaughan, Karron Graves, Charlayne Woodard, Frances Conroy, Elizabeth Lawrence, George Gaynes, Mary Pat Gleason
Miller's powerful play about the 17th century witch trials in Salem, Massachusetts (which cost the lives of nearly a score of innocent people) is such a fine piece of work that that it is hardly remarkable it translates so well to the screen. Both a work of historical fact and a bitter response to the McCarthy with-hunt era of the 1950s, it is given the respect is deserves, in this full-blooded and literate rendition. The script is by Miller.
DRA   124 min   CINrel
Boa: play by Arthur Miller.

CRUDE OASIS, THE **   (15)
Alex Graves USA   1995
Jennifer Taylor, Aaron Shields, Robert Peterson, Mussef Sibay, Lynn Bieler, Roberta Eaton, Anna Roedel, Dean Roedel, Scott McPhail, Grace Kassabaum, Dean Roedel Sr, Lucy Burligame, Kirk Kinsinger, Kelly Lima, Sherry Clymer
A neglected housewife is plagued by nightmares, and spends her days driving about aimlessly while her husband is at work, eventually deciding to commit suicide by using the car exhaust. But she runs out of petrol and in her search for a garage, meets a man who has figured in her dreams, and follows him to a remote bar (the title establishment). Shot in two weeks on a budget of $25,000, this arty and pretentious film is quite atmospheric, but remains undeveloped.
THR   78 min (ort 82 min)   SATrel: MOVIE CHANNEL

CRUEL PASSION **   18
Chris Boger UK   1977
Koo Stark, Lydia Lisle, Martin Potter, Hope Jackman, Katherine Kath, Maggie Petersen, Barry McGinn, Louis Ife, Ann Michele, Jason White,

Alan Rebbeck, David Masterman, Malou Cartwright, Barbara Eatwell, Echo Strade
Dreary erotic drama set in the last century, and following the fortunes of two teenage sisters, who are expelled from a convent and become prostitutes working in a brothel. The films "Justine" and "The Violation Of Justine" were two more attempts to adapt this difficult work.
Aka: JUSTINE; MARQUIS DE SADE'S JUSTINE
A   94 min (ort 97 min) wScrn   VIDrel: JEZ/RTM   V
Boa: novel Justine by the Marquis De Sade.

CRUEL SEA, THE *** PG
Charles Frend UK   1953
Jack Hawkins, Stanley Baker, Denholm Elliott, Moira Lister, Donald Sinden, Virginia McKenna, John Stratton, Liam Redmond, Meredith Edwards, Bruce Seton, June Thorburn, Megs Jenkins, Glyn Houston, Alec McCowen, John Warner, Sam Kydd
Nicely balanced and faithful adaptation of Montsarrat's novel about the crew of a British corvette during WW2. One of those typically British low-key and workmanlike films of the 1950s.
WAR   121 min (ort 126 min) B/W   VIDrel: WHV   V
Boa: novel by Nicholas Monsarrat.

CRUSH, THE ** 15
Alan Shapiro USA   1992
Cary Elwes, Alicia Silverstone, Jennifer Rubin, Kurtwood Smith, Gwynyth Walsh, Amber Benson, Matthew Walker, Deborah Hancock, Beverly Elliott, Andrew Airlie, Sheila Paterson, Brent Chapman, James Kidnie, Betty Phillips
"Lolita" meets FATAL ATTRACTION in this tale of a fourteen-year-old girl who develops a violent passion for a twenty-eight-year-old man who is renting the family guest-house. When her affections are not returned, she soon goes on the expected psychotic rampage. This was Silverstone's film debut.
THR   85 min (ort 89 min) cC   VIDrel: WHV   V/sur

CRUSH ** 18
Alison Maclean NEW ZEALAND   1993
Marcia Gay Harden, William Zappa, Caitlin Bossley, Donough Rees, Pete Smith, Jon Brazier, Geoffrey Southern, Wayne Roberts, Shirley Wilson, Denise Lyness, Trish Howie, Wayne McCoram, Jennifer Karehana, David Stott, Harata Solomon
When her literary critic friend is injured in a car crash that leaves her unconscious, her best friend decides to impersonate her and complete her interview assignment. This brings her in close contact with an author and his teenage daughter. A predictable and arty melodrama that soon proves to be quite tiresome. This was Maclean's first film, and she co-wrote the screenplay with Anne Kennedy.
DRA   92 min (ort 97 min) wScrn   VIDrel: TART/20TH   V/sur

CRUSOE *** 15
Caleb Deschanel USA   1988
Aidann Quinn, Ade Sapara, Warren Clark, Hepburn Grahame, Jimmy Nail, Tim Spall, Michael Higgins, Shane Rimmer, Elvis Payne, Richard Sharp, William Hootkins, Patrick Monkton, Chris Pitt, James Kennedy, Raymond Johnson
Cinematographer-turned-director Deschanel takes a few liberties with Defoe's character, turning him into an arrogant Virginian slave trader in the 19th century, who learns the meaning of human dignity at the hands of a native warrior, who proves to be more than a match for him. The literate script, splendid photography (courtesy of Tom Pinter) and luxurious Seychelles locations add considerably to the film's other virtues.
A/AD   90 min (ort 95 min)   VIDrel: VISVID/POLYREC   V
Boa: novel Robinson Crusoe by Daniel Defoe.

CRY-BABY ** 15
John Waters USA   1990
Johnny Depp, Amy Locane, Susan Tyrrell, Polly Bergen, Iggy Pop, Ricki Lake, Traci Lords, Kim McGuire, Troy Donahue, Mink Stole, Joe Dallesandro, Joey Heatherton, David Nelson, Patricia Hearst, Willem Dafoe, Jonathan Benya
Written and directed by Waters, this polished attempt to remake HAIRSPRAY is set in 1950s Baltimore, where a youngster from the wrong side of the tracks falls for a classy debutante. A kind of parody of 1950s juvenile delinquent films, full of amusing teen-movie cliches and sarcastic insights, it is all performed with great vigour, but remains a flimsy vehicle at best.
DRA   81 min (ort 86 min)   VIDrel: CIC/SONOP L/A   V/sur

CRY FOR HELP, A: THE TRACEY THURMAN STORY *** 18
Robert Markowitz USA   1989
Nancy McKeon, Bruce Weitz, Dale Midkiff, Graham Jarvis, Yvette Heyden, Terri Hanauer, Philip Baker Hall, David Wohl, David Camineho, Priscilla Pointer, Seth Isler, Burton Collins, Paul Carr, Paul Comi, Joe George, Redman Gleason
A harrowing but compelling drama, based on the real-life story of a battered Connecticut housewife, whose casual treatment following a near-fatal attack at the hands of her violent husband, led to a landmark trial and the adoption by the State of a new law. The script is by Beth Miller.
DRA   90 min (ort 100 min) mTV   VIDrel: ODY/SONOP   V/h

CRY FREEDOM *** PG
Richard Attenborough UK/ZIMBABWE   1987
Kevin Kline, Penelope Wilton, Denzel Washington, Kevin McNally, John Thaw, Timothy West, Juanita Waterman, John Hargreaves, Alec McCowen, Zakes Mokae, Ian Richardson, Josette Simon, Miles Anderson, Tommy Buson, Jim Findley
A strong tale of the life of South African activist Steve Biko and his friendship with crusading editor Donald Woods. Following the death of Biko in police custody, Woods mounts a campaign for an inquest, and suffers government harassment as a result. The film runs down in the second half, with too much attention being paid to Woods and his family as they flee persecution. "The Biko Inquest" also dealt with the murder of this civil rights leader.
DRA   151 min   VIDrel: CIC/SONOP   V/sur

CRY IN THE DARK, A **** 15
Fred Schepisi AUSTRALIA/USA   1988
Meryl Streep, Sam Neill, Bruce Myles, Charles Tingwell, Nick Tate, Neil Fitzpatrick, Maurie Fields, Lewis Fitzgerald, Dale Reeves, Michael Wetter, Nicolette Minster, Brian Jams, Dorothy Alsin, Alison O'Connel
The true story of Lindy Chamberlain, who in 1980 was accused of the murder of her baby whilst in the Outback, despite her claim that it was carried off by a dingo. The couple suffered an ordeal by rumour, and fought for several years to clear their name. Written by Schepisi, this is an excellent account of those events. Streep won the Best Actress Award at the Cannes Film Festival. Chamberlain's book "Through My Eyes" gives her side of the case.
DRA   116 min (ort 122 min)   VIDrel: MGM/WHV   V/sur

CRY IN THE WILD *** 15
Charles Correll USA   1991
David Morse, Megan Follows, Dion Anderson, Tom Atkins, Travis Swords, David Soul, Jack Kelher, Taylor Fry, Jim Granna, Mike Girardin, Ronnie Dee Blair, Kathryn Howell, Michelle Leaman, Jason Rojek, Zachary Barton
A dramatic retelling of the kidnapping of young Peggy Ann Bradnick, a teenager who was abducted by a half-crazed mountainman, who had the notion of taking the girl for his wife, and training her to the same level of obedience as his hunting dogs. Needless to say, the abduction led to a huge manhunt, and the girl was rescued after a week-long search.
Aka: CRY IN THE WILD: THE TAKING OF PEGGY ANN
DRA   91 min   VIDrel: CAPIT/GUILD   V

CRY OF THE OWL *** 18
Claude Chabrol FRANCE   1987
Christophe Malavoy, Mathilda May, Virginie Thevenet, Jacques Penot, Jean-Pierre Kalfon, Patrice Kerbrat
A divorced man strikes up a friendship with a woman who misinterprets his intentions and starts to pursue him amorously until rejection sets her mind on thoughts of revenge.
Aka: LE CRI DU HIBOU
DRA   105 min wScrn   VIDrel: LUMI/SPEAR L/A   V
Boa: novel by Patricia Highsmith.

CRY THE BELOVED COUNTRY *** PG
Zoltan Korda UK   1951
Sidney Poitier, Charles Carson, Canada Lee, Joyce Carey, Edric Connor, Geoffrey Keen, Vivien Clinton, Michael Goodliffe, Albertina Temba, Lionel Ngakane, Charles MacRae, Henry Blumenthal, Ribbon Dhlamini, Cyril Kwaza
A black minister goes from his countryside parish to Johannesburg to search for his son, but finds that he is in prison for killing a man during a robbery. Ironically, his victim was the son of a local landowner and a fervent opponent of Apartheid.

A sentimental but still powerful attack on old South Africa's system of racial repression, although lacking the searing impact of Paton's fine novel. Remade in 1974 (as "Lost In The Stars") and once more in 1995.
Aka: AFRICAN FURY
DRA   99 min B/W   VIDrel: LUMI/SPEAR L/A   V
Boa: movel by Alan Paton.

CRYING CHILD, THE ** **   12
Robert Lewis USA   1996
Mariel Hemingway, Finola Hughes, George Del Hoyo, Kim Shriner
Following the death of their child, a couple retire to the family cottage to get over their loss, but the peace of their retreat is broken by the constant crying of an unseen child. All is eventually made clear when a dark secret is uncovered that had long remained hidden. An atmospheric ghost story that has more than a passing similarity to THE CHANGELING.
DRA   88 min cC   VIDrel: CIC   V/sur

CRYING FREEMAN *   18**
Christophe Gans USA   1996
Mark Dacascos, Julie Condra, Rae Dawn Chong, Yoko Shimada
The very first attempt to produce a live-action version of a Japanese Manga cartoon, that follows the plight of a young potter who is kidnapped by a criminal cult organisation and turned into a devastating assassin. A solemn, celebration of the Japanese passion for overblown violence and bloodshed, that does not translate to a live-action format all that well (even though remaining faithful to the novel) but is fast and furious enough to keep one interested.
A/AD   101 min   CINrel
 Boa: graphic novel by Kazuo Koike.

CRYING FREEMAN: VOLS 1 TO 6 **   18
Daisuke Nishio JAPAN   1988/1992
Animated adaptation of a comic-book novel about a young artist forced to be an assassin against his will and his progress through the ranks of a secret criminal society. An odd blending of martial arts action and suspense that is occasionally quite effective. Koike's novel also served as the basis for a live-action film in 1996.
ANIM   300 min (6 cassettes – approx 50 min each) dubbed
VIDrel: MANGA/SONOP L/A   V
Boa: graphic novel by Kazuo Koike.

CRYING GAME, THE *   18**
Neil Jordan UK   1992
Stephen Rea, Miranda Richardson, Jaye Davidson, Forest Whitaker, Adrian Dunbar, Jim Broadbent, Ralph Brown, Breffini McKenna, Joe Savino, Birdie Sweeney, Andre Bernard, Tony Slattery, Jack Carr, Josephine White, Ray De-Haan
An IRA plot goes badly wrong when one of a team who have kidnapped a black British soldier, starts to befriend their prisoner and eventually becomes involved with the latter's girlfriend. An extremely well made and acted drama, drama, which was a great critical success, although perhaps not as deep and meaningful as they would have us believe. AA: Screen/orig (Jordan).
DRA   107 min (ort 112 min)   VIDrel: 4-FRONT/POLYREC V/s

CRYSTAL FIST **   15
Pal Ming/Hwai Hung HONG KONG   1979
Billy Chong, Hau Chiu Sing, Simon Yuen Sui Tin, Kao Chuan, Chia Yu-Kue, Chu Tit Wo, David Woo, Ma Shung Tak, Chaing Tao
After a master of the "Shadow Gate School" of kung fu is killed his only son enrols at another martial arts academy, and works in the kitchen to pay his tuition fees. He learns that the cook is a veteran Shadow Gate fighter and that his father was the man's favourite pupil. Following a violent raid on the school by some old enemies the cook agrees to train the boy, and the latter is able to both avenge his father and kill his mentor's enemies.
Aka: JADE CLAW, THE
MAR   90 min dubbed   VIDrel: CLEAR/DISC/IMPENT   V

CTHULHU MANSION **   18
Juan-Piquer Simon USA   1991
Frank Finlay, Melanie Shatner, Marcia Layton, Brad Fisher, Luis Fernando Alves, Kaethe Cherney, Paul Birchard, Francisco (Frank) Brana
A group of juvenile delinquents invade the estate of a magician and find some deserved supernatural retribution, in this indif-

ferently made and poorly acted film. Allegedly based on stories by H.P. Lovecraft, but the resemblance is hard to spot.
HOR   91 min (ort 95 min)   VIDrel: FIRST/SONOP L/A   V

CUJO *   18**
Lewis Teague USA   1983
Dee Wallace, Danny Pintauro, Daniel Hugh-Kelly, Christopher Stone, Kaiulani Lee, Ed Lauter, Mills Watson, Bill Jacoby, Sandy Ward, Jerry Hardin, Merritt Olsen, Arthur Rosenberg, Terry Donovan-Smith, Robert Elross, Claire Nuno
Nasty tale of a woman and her son who are trapped in their broken-down car by a rabid dog. An unpleasant film with little to recommend it apart from a few chilling moments and a rather horrible climax.
HOR   90 min (ort 97 min)   VIDrel: MIA/DISC/IMPENT   V
Boa: novel by Stephen King.

CUL-DE-SAC *   15**
Roman Polanski UK   1966
Lionel Stander, Donald Pleasence, Francoise Dorleac, Jack McGowran, Ian Quarrier, Renee Houston, Geoffrey Sumner, Robert Dorning, Marie Kean, William Franklyn, Jackie (Jacqueline) Bisset
Two gangsters arrive at the home of a couple with the intention of hiding there. Strange, offbeat film with little plot but a strong undercurrent of menace and a skilful use of location - Holy Island off the Northumberland coast.
DRA   100 min (ort 111 min) B/W   VIDrel: ODY/SONOP   V

CULT RESCUE **   12
Chuck Bowman USA   1994
Joan Van Ark, Stephen Macht, Tom Kurander, Brooke Langton, Daniel Hugh Kelly
Unable to recover after a mystery illness, a young woman tries therapy, but is soon brainwashed and sucked into a sinister cult, forcing her distraught family to mount a rescue attempt. Based on a true story. See also TICKET TO HEAVEN. Average.
DRA   87 min   VIDrel: NWV/HIFLI   V/h

CUP FINAL **   15
Eran Riklis ISRAEL   1990
Moshe Ivgi, Muhamad Bacri, Salim Dau, Basam Zuamut, Yussef Abu Warda, Suheil Haddad, Gasan Abbass, Sharon Alexander, Johnny Arbid, Sami Samir, Meir Swisa, Gadi Fur, Victor Kamar, Rada Ibrahim, Roberto Polak, Shai Aviri, David Sabar
An Israeli soldier captured during the invasion of Lebanon in 1982, finds that he shares a passionate interest in football and the World Cup with one of his captors. Fair.
Aka: ERAN RIKLIS' CUP FINAL; G'MAR GIVIYA; GEMAR GAVIA
DRA   104 min (ort 114 min) wScrn   VIDrel: TART/20TH   V

CUPID *   15**
Doug Campbell USA   1996
Zach Galligan, Ashley Laurence, Mary Crosby, Joseph Kell, Michael Bowen
An obsessive psychopath dreams of finding a perfect love, and when he comes across a single bookstore worker, he decides that she will fit the bill, and proceeds to murder anyone who threatens to come between them. An underplotted stalk-and-slash type movie, totally predictable, unoriginal in conception, and directed with neither flair nor energy.
THR   91 min   VIDrel: FIRST/SONOP   V

CURE, THE **   12
Peter Horton USA   1995
Joseph Mazzello, Brad Renfro, Bruce Davison, Annabella Sciorra, Diana Scarwid, Nicky Katt, Aeryk Egan, Renee Humphrey
An eleven-year-old boy who is new in town make friends with a local boy who is shunned because he has contracted AIDS from a blood transfusion. As their friendship deepens, he takes his new buddy a rafting trip in search of a cure. A sad and moving tale whose references to Huckleberry Finn seem quite out of place.
DRA   98 min   VIDrel: POLFIL   V

CURLY SUE *   PG**
John Hughes USA   1991
James Belushi, Kelly Lynch, Alisan Porter, John Getz, Fred Dalton Thompson, Cameron Thor, Branscombe Richmond, Steven Carell, Gail Boggs, Burke Byrnes, Viveka Davis, Barbara Tarbuck, Edie McClurg, Charles Adams, Lyle Brown
A romantic comedy that sees a gentleman-of-the-road and his young companion (Curly Sue) coming to Chicago and being taken in by a female lawyer. Under her influence, the former

tramp gets a steady job, and after a variety of setbacks, all three come together to form a real family. An unconvincing and terribly manipulative tale, it has no clear delineation of character or motive, and a thoroughly unsatisfying conclusion.
COM   97 min (ort 102 min) cC   VIDrel: WHV   V/sur

CURRY AND PEPPER **
Blackie Ko HONG KONG
18
1990
Stephen Chow Sing Chi, Jacky Cheung Hok Yau, Ann Bridgewater (Park On-Ney), Eric Tsang Chi-Wai, Blacky Ko San-Leung, Bruce Fontain, Michelle Lee, Chow Mei-Yan, Barry Wing Ping-Yiu, Poon Man-Kit, Andrew Lau Wai-Keung
Action-comedy that has two cops going after an arms trader they are determined to bring to justice. Meanwhile, they face the usual bickering and compete for the attentions of a woman news reporter they have both taken a fancy to. A standard blend of comedy and high-paced action.
A/AD   95 min wScrn   VIDrel: MADE/RTM   V

CURSE, THE ***
David Keith USA
18
1986
Wil Wheaton, Claude Akins, Amy Wheaton, Malcolm Danare, Cooper Huckabee, John Schneider, Steve Carlisle, Kathleen Jordan Gregory, Hope North, Steve Davis
Actor David Keith makes a good directing debut in this striking horror yarn telling of a meteorite that brings a nasty parasite to Earth. Crashing near a farm, it begins to poison a family by a strange process that gradually transforms them into inhuman monsters. For good measure the farmhouse is also affected and disintegrates in an enjoyable climax.
Aka: FARM, THE
HOR   83 min (ort 92 min)   VIDrel: EIV/SONOP   V
Boa: short story The Colour Out of Space by H.P. Lovecraft.

CURSE OF FRANKENSTEIN, THE ***
Terence Fisher UK
15
1957
Peter Cushing, Christopher Lee, Robert Urquhart, Hazel Court, Valerie Gaunt, Noel Hood, Marjorie Hulme, Melvyn Hayes, Sally Walsh, Paul Hardtmuth, Fred Johnson, Claude Kingson, Henry Caine, Michael Mulcaster, Patrick Troughton
First of the Hammer horror series is a flashback account of young Victor Frankenstein's experiments, told as he awaits execution for the murder of his wife, who in fact died in the struggle between him and the monster. Stylish if somewhat gory, this film gave rise to an interminable succession of lurid and increasingly cliched horror films. Loosely (and I mean loosely) based on Mary Shelley's novel. REVENGE OF FRANKENSTEIN was the next in line.
HOR   79 min (Cut at film release)   VIDrel: WHV   V

CURSE OF THE CATWOMAN ***
John Leslie USA
18
1991
Selena Steele, Raven, Rocco Siffredy, Patricia Kennedy, Raquel Darrian, Zara Whites, Ashley Nicole, Mark Wallice, Jamie Gillis, Alexandra Quinn, Tom Byron, Randy Spears, T.T. Boy, Derek Lane, Jake Steed, Christian Parker
Two catlike sisters (the plot explains their origins) vie with each other to see who will become supreme Catwoman, a contest that for the most part seems to involve them in proving their sexual athleticism with a variety of willing partners. A dark and moody work, that's as much a bizarre fantasy as a torrid sex film.
A   63 min (ort 75 min) mVid   VIDrel: MIA/DISC   V

CURSE OF THE CRYSTAL EYE, THE *
Joe (Guiseppe) Tornatore PANAMA/USA
(PG)
1988
Jameson Parker, Cynthia Rhodes, Mike Lane, David Sherwood, Johnny Noble, Andre Jacobs, Anton Stoltz, Douglas Bristow, Farcuk Valley Omar, Dawie Maritz, Gamiet Petersen, Paul Ditchfield, Royston Stoffels, Wayne Bowman
A man is given a strange crystal that turns out to be the key to the treasure cave of the legendary Ali Baba. An indifferent adventure tale that does little with its novel premise. Filmed on location in Namibia and Mauritius.
Aka: CRYSTAL EYE
JUV   101 min   SATrel: MOVIE CHANNEL

CURSE OF THE MUMMY'S TOMB **
Michael Carreras UK
15
1964
Terence Morgan, Ronald Howard, Fred Clark, Jeanne Roland, George Pastell, Jack Gwillim, John Paul, Jill Mai Meredith, Michael Ripper, Harold Goodwin, Dickie Owen (the Mummy)
The excavation of a pharaoh's tomb, results in many deaths when the mummy of a dead king comes to life and begins to

stalk the streets of Victorian London, in search of those who broke into the tomb. A well-mounted but predictable yarn, it followed the 1959 film THE MUMMY without being a sequel to it. Made by Hammer Films.
HOR   77 min (ort 81 min)   VIDrel: COLUM/SONOP   V

CURSE OF THE PINK PANTHER, THE *
Blake Edwards USA
PG
1982 (released 1983)
David Niven, Robert Wagner, Herbert Lom, Ted Wass, Joanna Lumley, Capucine, Robert Loggia, Leslie Ash, Harvey Korman, Burt Kwouk, Andre Maranne, Graham Stark, Peter Arne, Patricia Davis, Michael Elphick, Steve Franken, Ed Parker
Niven's last film and what a pointless waste it is. A pathetic and unfunny attempt to keep the PINK PANTHER series going after the death of Sellers, with another bumbling detective selected to replace Inspector Clouseau, who appears to have vanished. Garbage of a very high order. See also THE TRAIL OF THE PINK PANTHER, which was made at the same time as this one. Filmed in the UK.
COM   106 min (ort 110 min)   VIDrel: MGM/WHV   V

CURSE OF THE STARVING CLASS **
Bruce Beresford USA
15
1994
James Wood, Kathy Bates, Randy Quaid, Henry Thomas, Louis Gossett Jr
A portrait of a dysfunctional farming family in the 1930s, who are about to lose their farm, with the husband a hopeless drunk and the wife at her wits' end trying to cope with the situation. The usual relentlessly depressing Shepard offering, dressed up in unconvincing poetic language.
DRA   97 min (ort 102 min) mCab   VIDrel: ARENA/RTM   V
Boa: play by Sam Shepard.

CURSE OF THE VIKING GRAVE ***
Michael Scott CANADA
(PG)
1991
Nicholas Shields, Evan Tlesla Adamas, Michelle St John, Gordon Tootoosis, Wayne Robson, Marianne Jones, Jay Brazeau, Lee J. Campbell, Cedric Smith, David Gillies, Joe Mecredi, Vic Cowie, Michael Meeches, John Bluethner, Jennie Tooto
Set in Canada in 1937, where a young boy stumbles across an ancient spearhead among the ruins of his uncle's old home and is soon approached by an archaeologist who is determined to locate an ancient Viking burial site. But the latter soon reveals himself to be a totally ruthless individual who will stop at nothing to achieve his goal. A nicely realised family adventure of solid performances, attractive locations plus a dash of Indian mysticism.
JUV   96 min   SATrel: SKY MOVIES
Boa: novel by Farley Mowat.

CURSE OF THE WEREWOLF, THE ***
Terence Fisher UK
12
1961
Oliver Reed, Clifford Evans, Yvonne Romain, Catherine Feller, Anthony Dawson, Josephine Llewellyn, Richard Wordsworth, Hira Talfrey, John Gabriel, Warren Mitchell, Anne Blake, George Woodbridge, Michael Ripper, Ewen Solon
A mute servant girl is raped by a beggar and gives birth to a child who is adopted by a kindly professor and his wife. As the child grows to manhood, he falls more and more under the spell of his affliction. Effective, well made and an enthralling recreation of this myth.
HOR   87 min (ort 91 min)   VIDrel: WHV   V/h
Boa: novel The Werewolf of Paris by Guy Endore.

CUSTER OF THE WEST **
Robert Siodmak SPAIN/USA
U
1968
Robert Shaw, Mary Ure, Jeffrey Hunter, Robert Ryan, Ty Hardin, Charles Stalnaker, Robert Hall, Lawrence Tierney
Originally shot in Cinerama, this purports to be the story of General Custer's command of the 9th US Cavalry, who were detailed to keep Cheyenne lands free from incursion after gold was discovered there. However, matters take a more serious turn when the General's own men begin to desert. A melodramatic and dour film, and even the Little Big Horn battle sequence does little to generate interest. See also THEY DIED WITH THEIR BOOTS ON and SON OF THE MORNING STAR.
WES   135 min (ort 140 min)   VIDrel: VCC/DISC   V

CUSTODIAN, THE ***
John Dingwall AUSTRALIA
15
1993
Anthony LaPaglia, Hugo Weaving, Barry Otto, Kelly Dingwall, Bill Hunter, Gosia Dobrowolska, Naomi Watts, Essie Davis, Joy Smithers,

Tim McKenzie, Norman Kaye, Steven Grivers, Bogdan Koca, Russell Newman, Bob Baines, Kee Chan
An honest cop, saddled with an alcoholic wife who has ejected him from their home, decides to root out corruption within the force by the use of his own, highly unorthodox methods. A tough and brutal police drama whose simple story is all too believable, while the fine acting is another major asset.
THR   105 min (ort 110 min)   VIDrel: IMAG/RTM   V/sur

CUTTHROAT ISLAND * PG
Renny Harlin USA 1995
Matthew Modine, Geena Davis, Frank Langella, Maury Chaykin, Patrick Malahide, Stan Shaw, Rex Linn, Paul Dillon, Chris Masterson, Jimmie F. Skaggs, Harris Yulin, Carl Chase, Peter Geeves, Angus Wright, Ken Bones, Mary Pegler
Swashbuckling adventure set on the high seas, where a female pirate and her nasty crew sail through storms and battles to find the treasure secreted on the title island. An expensive dud, whose ambitious sets and attention to detail are scant compensation indeed for the dullness of the plotting and dialogue.
A/AD   117 min Cut (1 min 12 sec – ort 123 min)
VIDrel: GUILD/20TH; PION (LV only)   V/sur LV

CUTTING EDGE, THE * PG
Paul Michael Glaser USA 1992
D.B. Sweeney, Moira Kelly, Roy Dotrice, Terry O'Quinn, Dwier Brown, Chris Benson, Kevin Peeks, Barry Flatman, Rachelle Ottley, Steve Sears, Nahanni Johnstone, Michael Hogan, R.D. Reid, Dick Grant, Melanie Miller, Judy Blumberg
Sweeney is a fallen hockey star and Kelly a world-class figure skater who has seen success pass her by. It is not long before they have teamed up, and are ready to have a stab at the Olympics. Neither a straightforward love story nor a competent account of competitive skating, this simple-minded and tedious tale (complete with noisy rock soundtrack) is as unimaginative as it is unwatchable.
DRA   97 min (ort 102 min)   VIDrel: MGM/WHV   V/sur

CYBER CITY OEDO 808: DATA 1 TO 3 ** 15
JAPAN 1990
Simple stories of a police force of the future, that's set in the year 2808 A.D. In the first story criminals are recruited into the police force, in a bid to contain the ever rising crime rate. One such individual (a former serial killer) is sent to rescue 50,000 people trapped in a skyscraper. The latter tales detail the exploits of a three-man team of cyborg police officers, who operate in Tokyo of the 29th century.
ANIM   127 min (approx 40 min each episode) dubbed
VIDrel: MANGA/SONOP   V/sh

CYBER TRACKER ** 18
Richard Pepin USA 1993
Don "The Dragon" Wilson, Richard Norton, Stacie Foster, John Aprea, Anthony DeLongis, Joseph Ruskin, Jim Maniaci, Tony Burton, John Kassir, Stephen Quadros, Christopher Boyer, Nils Allen Stewart, Stephen Rowe, Steve Bruton, Athena Massey
A government agent narrowly escapes death at the hands of android assassins after being framed for murder and must do battle in order to learn the truth behind the creation of these deadly machines. Despite its SF trappings, this is essentially no more than another violent actioner with a martial arts flavour. A sequel soon followed.
A/AD   91 min   VIDrel: MIA/DISC   V

CYBER TRACKER 2 ** 18
Richard Pepin USA 1995
Don "The Dragon" Wilson, Stacie Foster, Tony Burton, Jim Maniaci, John Kassir, Stephen Rowe, Steve Burton
An agent and his wife must flee for their lives when they learn that android copies of themselves have been created after the theft of this government-controlled technology.
A/AD   95 min (ort 97 min)   VIDrel: MIA/DISC   V/sh

CYBERJACK ** 18
Robert Lee USA 1995
Michael Dudikoff, Suki Kaiser, Brion James
Set in the 21st century, this is the story of a woman and her father who write a computer program that can render the entire Internet immune to computer viruses. Unfortunately, some cyberpunks get to hear about this, and seize the lab in which they are working, their intention being to gain control of the program. Dudikoff plays the caretaker, who as a former cop has

the requisite skills needed to mount a rescue. A dull and derivative DIE HARD clone.
FAN   93 min   VIDrel: COLUM/SONOP   V

CYBORG * 18
Albert Pyun USA 1989
Jean-Claude Van Damme, Vincent Klyn, Deborah Richter, Dayle Haddon, Alex Daniels, Blaise Loong, Rolf Muller,Terrie Batson, Jackson "Rock" Pinckney, Chuck Allen, Janice Graser, Robert Pentz, Sharon K. Tew, Stefanos Miltsakakis
The population of a world of the future has been decimated by evil creatures known as the "Flesh Pirates", who have used plague to consolidate their rule. A female cyborg escapes with vital information that could eradicate the plague, but is recaptured. Luckily, a saviour is on hand to rescue her. A repulsive post-apocalyptic fantasy, with foolish plotting and worse dialogue. Followed by CYBORG 2, a considerably better film in all respects. See also NEMESIS.
FAN   79 min (Cut at film release by 3 min 15 sec – ort 86 min)   VIDrel: MGM/WHV   V/sh

CYBORG 2 ** 18
Michael Schroeder USA 1992
Jack Palance, Elias Koteas, Angelina Jolie, Billy Drago, Karen Shepherd, Ric Young, Allan Garfield, Renee Griifin, Sven Thorsen, Tarcey Walter, Robert Dryer, Jim Youngs, John Durbin, Patrick O'Connell, Sheryl Mary Lewis
Sequel to the earlier CYBORG, that is set in the year 2074 with the Earth a barren and exhausted planet whose future is in the hands of powerful corporations. Having replaced humans with cyborgs at just about every level in society, humanity is now threatened by a new type of cyborg that carries an explosive. Fortunately, a renegade cyborg joins forces with a human being to destroy the plant where these creatures are being made. Adequate SF adventure.
Aka: CYBORG 2: THE GLASS SHADOW
FAN   95 min (ort 99 min)   VIDrel: SPEAR/SONOP   V/s

CYBORG 3: THE RECYCLER ** 18
Michael Schroeder USA 1994
Zach Galligan, Khrystyne Haje, Andrew Bryniarski, Evan Lurie, Michael Bailey Smith, Rebecca Ferratti, Margaret Avery, Ricahrd Lynch, Malcolm McDowell, Bill Quinn, William Katt, Dave McSwain, Barbara Adside, Debra Hall
The third film in the CYBORG series with a new type of female cyborg created that represents yet another deadly threat to humanity, as it has been given the ability to reproduce. More of the same is on offer here, and the lack of fresh ideas is all too obvious.
Aka: CYBORG 3: THE CREATION
FAN   87 min (ort 90 min)   VIDrel: MIA/DISC   V/sh

CYBORG COP ** 18
Sam Firstenberg USA · 1993
David Bradley, Todd Jensen, Alonna Shaw, John Rhys-Davies, Ron Smerzak, Anthony Fridjon, Shalom Kenan, Robert Whitehead, Frank Notaro, Steven Leader, Kurt Egelhof, Billy Mashigo, Ernest Ndolvu, Seldon Ngwena, David Pithar
An ex-cop gets a call for help from his brother who has been captured by a scientist who is intending to turn him into a robot killer programmed for the assassination of a local ruler. Soon our hero is mounting the obligatory rescue mission and happily is eventually able to defeat this evil plan. Very much a by-the-numbers action outing with unimpressive special effects and uninspired acting. Made on location in South Africa.
A/AD   92 min (ort 94 min)   VIDrel: POLY/POLYREC/MED   V/sh

CYBORG COP 2 ** 18
Sam Firstenberg USA 1993
David Bradley, Morgan Hunter, Jill Pierce, Dale Cutts
Sequel to CYBORG COP sees our ex-cop hero battling cyborgs that have been created from the bodies of convicts due to be executed for murder, and have now mutated into super-beings able to regenerate themselves.
Aka: CYBORG SOLDIER
FAN   93 min (ort 97 min)   VIDrel: COLUM/SONOP L/A   V/sh

CYBORG COP 3 * 18
Yossi Wein USA
Frank Zagarino, Brian Genesse, Jennifer Miller, Ian Roberts, Michael Brunner

A couple of FBI agents investigate the disappearance of a group of research students, and predictably stumble on a nasty conspiracy.
A/AD   90 min   VIDrel: 20TH/FOXVID   V/sur

CYCLO ***
Tran Anh Hung FRANCE/VIETNAM
18
1994
Le Van Loc, Tony Leung-Chiu Wai, Tran Nu Yen Khe, Nguyen Nhu Quynh, Nguyen Hoang Phuc, Ngo Vu Quang Hai, Nguyen Tuyet Ngan, Doan Viet Ha, Bjuhoang Huy, Vo Vinh Phuc, Le Dinh Huy, Pham Ngoc Lieu, Le Tuan Anh, Le Cong Tuan Anh
A boy struggles to survive on the streets of Ho Chi Minh City, but when he is mugged he gets subsequently drawn into the world of organised crime. A compelling chronicle of life at the bottom of the heap, and given the way of life of the central characters, the loss of values depicted is as inevitable as it is believable. Won the Critics' Prize at the Venice Film Festival.
Aka: XICH LO
THR   124 min (rt 129 min) wScrn   VIDrel: EIV/SONOP   V

CYCLONE ***
Fred Olen Ray USA
18
1987
Heather Thomas, Jeffrey Combs, Ashley Ferrare, Dar Robinson, Robert Quarry, Martine Beswick, Martin Landau, Huntz Hall, Troy Donahue, Dawn Wildsmith, Tim Conway Jr, Michael Reagan, Bruce Fairburn, Sam Hiona, Russ Tamblyn
When the inventor of "Cyclone", a powerful computer controlled motorcycle is murdered, his girlfriend becomes responsible for his creation. Caught between criminals and government agencies, she finds her life in danger as she attempts to carry out her boyfriend's wishes and deliver the invention to his friend. When he too is killed, she must use her boyfriend's invention to survive. Fast-paced but enjoyable nonsense.
A/AD   83 min   VIDrel: EIV/SONOP   V

CYNARA **
King Vidor USA
PG
1932
Ronald Coleman, Kay Francis, Phyllis Barry, Henry Stephenson, Paul Porcasi
A London barrister has an affair with a working girl while his wife is away, but when he returns to his wife, the girl commits suicide. A competent but dated melodrama, with Colman standing head and shoulders above the shallow and rather obvious plot. Written by Frances Marion and Lynn Starling.
DRA   75 min (ort 78 min) B/W   VIDrel: VGM/DISC   V
Boa: novel An Imperfect Lover by Robert Gore Brown.

CYRANO DE BERGERAC ***
Michael Gordon USA
U
1950
Jose Ferrer, Mala Powers, William Prince, Morris Carnovsky, Ralph Clanton, Virginia Farmer, Lloyd Corrigan, Edgar Barrier, Elena Verdugo, Gil Warren, Don Beddoe, Albert Cavens, Arthur Blake, Percy Hilton, Virginia Christine
A classic seventeenth century tale of love, intrigue and mistaken identity, with Ferrer quite wonderful as the famed wit and soldier of fortune, whose a appearance is such that he despairs of winning the hand of the fair Roxanne, and so writes beautiful love letters to help a friend win her instead. Sadly, this is a cheap looking production that with the notable exception of Ferrer, is generally poorly cast and acted. AA: Actor (Ferrer).
DRA   113 min B/W   VIDrel: FABFIL/SPEAR   V
Boa: play by Edmond Rostand.

CYRANO DE BERGERAC ****
Jean-Paul Rappeneau FRANCE
U
1990
Gerard Depardieu, Jacques Weber, Anne Brochet, Vincent Perez, Roland Bertin, Philippe Morier-Genoud, Josiane Stoleru, Philippe Volter, Pierre Maguelon, Anatole Delande, Jean-Marie Winling, Alan Rimoux, Louis Navarre, Gabriel Monet
A sumptuous and quite remarkable version of Rostand's classic, telling of the gallant 17th century wit and soldier of fortune whose appearance is such that he fails to declare his love for his young cousin until it is too late. Depardieu gives a wonderful and often deeply moving portrayal of the unhappy title character, and the articulate screenplay is by Rappeneau and Jean-Claude Carriere. AA: Cost (Franca Squarciapino).
DRA   132 min (ort 138 min) wScrn
VIDrel: ARTIF/20TH   V/sur
Boa: play by Edmond Rostand.

D

D-DAY THE SIXTH OF JUNE **
Henry Koster USA
PG
1956
Robert Taylor, Richard Todd, Dana Wynter, Edmond O'Brien, John Williams, Jerry Paris, Richard Stapley
A British colonel and an American captain are in love with the same girl, and they reminisce about her whilst on their way to the Normandy invasion of 1944. A flaccid and unremarkable romantic drama with a few excellent action sequences providing some compensation. The usual rose-tinted Hollywood depiction of wartime London prevails.
Aka: SIXTH OF JUNE, THE
WAR   101 min (ort 106 min) wScrn   VIDrel: 20TH/TECH
V/h
Boa: novel by Lionel Shapiro.

D2: THE MIGHTY DUCKS **
Sam Weisman USA
U
1993
Emilio Estevez, Kathryn Erbe, Michael Tucker, Jan Rubes, Carsten Norgaard, Maria Ellingsen, Joshua Jackson, Elden Ryan Ratcliff, Shaun Weiss, Matt Doherty, Brandon Adams, Garette Ratliff Henson, Marguerite Moreau, Vincent A. Larusso
Sequel to THE MIGHTY DUCKS in which Estevez gets the chance to coach a top American hockey team (including members of his former Ducks) after he proved his worth in the first film. The plot has them competing in the Junior Goodwill Games in a contrived and idiotic sequel that pretends to make a few points about commercialism but in truth, is a cynical piece of exploitation if ever there was one. A couple of real-life hockey stars appear in cameo roles.
COM   107 min   VIDrel: WDV/TECH   V

D3: THE MIGHTY DUCKS *
Robert Lieberman USA
PG
1996
Emilio Estevez, Jeffrey Nordling, Heidi Kling, Joss Ackland
More of the same with our boys' ice hockey team starting at a posh prep school, where their tactics on the hockey pitch are not met with universal approval.
COM   104 min cC   VIDrel: WDV/TECH   V/sur

DAD ***
Gary David Goldberg USA
PG
1989
Jack Lemmon, Ted Danson, Olympia Dukakis, Kathy Baker, Kevin Spacey, Ethan Hawke, Zakes Mokae, Chris Lemmon, J.T. Walsh, Peter Michael Goetz, John Apicella, Richard McGonagle, Bill Morey, Marty Fogarty, Art Frankel
A man returns home to take care of his aged and frail father when his mother suffers a heart attack and is hospitalised. Appalled by the degeneration he finds in his father, he attempts to revitalise him and re-kindle their relationship, and eventually does the same with his own son. A moving and warmhearted drama rather spoilt by an excess of sentiment and a lack of strong characterisation. Goldberg's feature directing debut.
DRA   112 min (ort 117 min) VIDrel: CIC/SONOP   V/sh
Boa: novel by William Wharton.

DAD, THE ANGEL AND ME **
Rick Wallace USA
(U)
1995
Judge Reinhold, Carol Kane, Stephi Lineburg, Alan King, Eve Gordon, Shirley Knight, Betsy Brantley, Jane Carr, Caroline Aaron, Kristian Minter, Clare Hoak, Edward Penn, Jeanine Jackson, Allison Mark, Judith Piquet, Del Shores
Since his divorce, a self-centred therapist has lived a hedonistic, bachelor lifestyle, so he is naturally somewhat put out when he learns that his ex-wife has died and that he now has a ten-year-old daughter to care for. As both father and daughter hardly know each other matters do not start out with promise, all the more so when it appears that she already has a guardian angel. An undemanding family drama.
JUV   93 min   SATrel: SKY MOVIES

DADDY LONGLEGS **
Jean Negulesco USA
U
1955
Fred Astaire, Leslie Caron, Terry Moore, Thelma Ritter, Fred Clark, Charlotte Austin, Larry Keating, Kathryn Givney, Kelly Brown, Sara Shane, Numa Lapeye, Ann Codee, Steven Geray, Percival Vivian, Helen Van Tuyl, Joseph Kearns
Musical version of the classic story about an orphaned waif who eventually falls deeply in love with her anonymous benefactor. Unfortunately, this simple storyline cannot really support such a long film, which, although full of music, suffers badly from the

unimaginative choreography. Songs include: "Something's Gotta Give" and "Daddy Longlegs".
MUS   124 min (ort 126 min) cC   VIDrel: 20TH/FOXVID   V
Boa: novel/play by Jean Webster.

**DADDY **   *PG*
Michael Miller USA   1991
Patrick Duffy, Lynda Carter, Kate Mulgrew, John Anderson, Ben Affleck, Jenny Lewis, Matthew Lawrence, Robyn Peterson, Richard McKenzie, Georgia Emelin, Mimi Cozzens, Gloria Dorson, Peter Hansen, Marjorie Lovett, D. Paul Thomas
Having worked hard to give himself and his loved ones a comfortable life, Oliver Watson (Duffy) finds his world beginning to fall apart when his marriage starts to break up and his mother is killed in a car crash. As if this were not enough, his two eldest children reject him, and he's forced to drastically alter his lifestyle in an attempt to rebuild his life. A watchable soap opera (Duffy is unusually good) done in the usual glossy and empty style.
Aka: DANIELLE STEEL'S DADDY
DRA   91 min (ort 95 min) mTV
VIDrel: 4-FRONT/POLYREC   V
Boa: novel by Danielle Steel.

DADDY'S DYIN', WHO'S GOT THE WILL? *   15
Jack Fisk USA   1990
Beau Bridges, Beverly D'Angelo, Tess Harper, Judge Reinhold, Amy Wright, Patrika Darbo, Keith Carradine, Bert Remsen, Molly McClure, Emily Bridges, Carolyn Brooks, Newell Alexander, Schuyler Fisk, Justin Smith, Sandra Will
A weird family of misfits gathers at a Texan homestead to await the death of their father and the reading of his will. However, his long-awaited demise is somewhat delayed, and the rival siblings begin to squabble and re-examine old grievances. A valiant display of ensemble acting from a gifted cast helps enliven a film whose script (by Shores) is a sometimes awkward blend of pathos and wit.
COM   92 min (ort 97 min)   VIDrel: PAL/GUILD L/A   V
Boa: play by Del Shores.

**DADDY'S GIRL **   18
Martin Kitrosser USA
William Katt, Michael Green, Roxana Zal, Gabrielle Boni, Mimi Craven
A ten-year-old girl is adopted by a couple who have no inkling of what they have let themselves in for.
HOR   91 min   VIDrel: FIRST/SONOP   V

DAD'S ARMY *   *U*
Norman Cohen UK   1971
Arthur Lowe, John Le Mesurier, Clive Dunn, Ian Lavender, John Laurie, Arnold Ridley, James Beck, Bill Pertwee, Liz Fraser, Bernard Archard, Derek Newark, Frank Williams, Edward Sinclair, Pat Coombs, Sam Kydd, Fred Griffiths
Inspired by the popular TV series of the same name, this tale is set in a small seaside town in 1939, and follows the attempts of the locals to form a Home Guard unit. Despite the presence of all the stars who made the television series such a hit, this overstretched and under-plotted film has little of the gentle self-mocking humour of the original. Very disappointing.
Aka: DAD'S ARMY: THE MOVIE
COM   91 min (ort 95 min) B/W/Col
VIDrel: VCC/DISC/COLUM   V

DAENS *   15
Stijn Coninx BELGIUM/FRANCE/HOLLAND   1992
Jan Decleir, Gerard Desarthe, Antje De Boeck, Michael Pas, Johan Leysen, Idwig Stephane, Wim Meuwissen, Jappe Claes, Julien Schoenaerts, Karel Baetens, Brit Alen, Brenda Bertin, Alex Wilequet, Rik Hancke, Giovanni Di Benedetto
True story set towards the close of the 19th century, when the booming mill town of Aalst had secured riches for its investors and dreadful working conditions for everyone else. However, the workers are championed by a Catholic priest, who fights for an improvement in their lot, and is eventually elected to Belgium's new parliament. A stirring albeit simplistic film, it won an Oscar nomination for Best Foreign Film of 1992.
DRA   132 min (ort 138 min)   VIDrel: CURZON/20TH
V/sh
Boa: novel Pieter Daens by Louis Paul Boon.

DAGGERS 8 *   *PG*
Cheung Sam/Wilson Tong HONG KONG
Meng Yuen Men, Lily Li, Wilson Tong, Chui Chung Shun, Lee SA Ess
Kung fu adventure in which the main character is forbidden to study the martial arts, so he learns to kill using eight daggers.
MAR   86 min Cut (3 sec)   VIDrel: IMPENT   V

DAISIES *   15
Vera Chytilova CZECHOSLOVAKIA   1966
Jika Cerhova, Ivana Karbanova, Julius Albert
Comedy-drama telling of two bored teenage girls who embark on a series of destructive pranks, mostly out of pique at finding how bad the world is, the focus for their ire being the consumerist society at large. An anarchic and bold film, it makes good use of colour and visual imagery, and so upset the Czech authorities that they banned its release for a year. Written by Chytilova and Ester Krumbachova.
Aka: SEDMIKRASKY
COM   72 min (ort 75 min) B/W   VIDrel: CONNO/RTM   V

DALGLIESH: A TASTE FOR DEATH *   15
John Davies UK   1988
Roy Marsden, Simon Ward, Wendy Hiller, Fiona Fullerton, Penny Downie
An old friend of Dalgliesh, who now has a post as a top government minister, begins to receive a series of threatening anonymous letters. He calls in his friend to help him, but the detective finds himself mounting a murder investigation when one of the man's servants dies under mysterious circumstances. As ever, the typical P.D. James web of lies, deceit, jealousy and hatred is soon revealed.
Aka: TASTE FOR DEATH, A
DRA   292 min (2 cassettes) mTV   VIDrel: L/A   V
Boa: novel by Phyllis Dorothy James.

**DALGLIESH: COVER HER FACE –
PARTS 1 AND 2 ***   *PG*
John Davies UK   1985
Roy Marsden, Phyllis Calvert, Bill Fraser, Mel Martin, Julian Glover, Rupert Frazer, Claire Higgins, Jean Haywood, John Vine, Kim Thompson, Freda Dowle, Barbara Hicks, George Zenios, Matthew Ryan, Charles Morgan, Lucy Hancock
A murder takes place at a country home and our dogged Chief Superintendent faces another difficult investigation.
Aka: COVER HER FACE
DRA   294 min (2 cassettes) mTV   VIDrel: L/A   V
Boa: novel by Phyllis Dorothy James.

**DALGLIESH: DEATH OF AN EXPERT WITNESS –
PARTS 1 AND 2 ***   *PG*
Herbert Wise UK   1983
Roy Marsden, Barry Foster, Geoffrey Palmer, Ray Brooks, Andrew Ray, Cyril Cusack, Meg Davies, Brenda Blethyn, John Vine, Malcolm Terris, Stephen Thorne, Valerie Testa, Rhoda Lewis, Ivor Roberts, Rob Gillion, Annabelle Lanyon
Dalgliesh is called away to Norfolk to investigate the strangling of a young man, but decides that this murder forms no part of another investigation he's currently pursuing into an elusive serial strangler. When a further murder takes place at a forensic science laboratory, he uncovers a nasty plot that involves blackmail and revenge.
Aka: DEATH OF AN EXPERT WITNESS
DRA   300 min (2 cassettes) mTV   VIDrel: L/A   V
Boa: novel by Phyllis Dorothy James.

DALGLIESH: DEVICES AND DESIRES *   15
John Davies UK   1990
Roy Marsden, Betty Marsden, Susannah York, James Faulkner, Gemma Jones, Tony Haygarth, Tom Georgeson, Tom Chadbon, Nicola Cowper, Suzan Crawley, Robert Hines, Lisa Taylor, Harry Burton, Helena Michell, Robert Cotton, Ingrid Statman
Dalgliesh inherits a converted windmill on the Norfolk coast, and takes a break from his duties to make a trip up there. Unfortunately, the peace of his retreat is shattered when he learns of a psychopath who is killing young girls. He becomes involved in local protests about a nearby nuclear power station, and begins to suspect that the murders are linked to it in some way.
Aka: DEVICES AND DESIRES
DRA   296 min (2 cassettes) mTV   VIDrel: L/A   V
Boa: novel by Phyllis Dorothy James.

DALGLIESH: SHROUD FOR A NIGHTINGALE – PARTS 1 AND 2 ***
PG
John Gorrie UK 1984
Roy Marsden, Joss Ackland, Sheila Allen, Margaret Whiting, Liz Fraser, John Vine, Thelma Whitely, Andree Evans, Rosalyn Elvin, John Pennington, David Swift, Natalie Ogle, Judi Maynard, Forbes Collins, Nicholas Coppin, Gabrielle Hamilton
Our Detective Chief Superintendent finds his murder enquiries taking him to a sinister hospital and a bleak adjoining nurses' home. Having interviewed a dying patient, it's not long before he uncovers a complex web of intrigue, jealousy and hatred, but in the process, puts his career at risk.
Aka: SHROUD FOR A NIGHTINGALE
DRA 250 min (2 cassettes) mTV VIDrel: L/A V
Boa: novel by Phyllis Dorothy James.

DALGLIESH: THE BLACK TOWER – PARTS 1 AND 2 ***
PG
Ronald Wilson UK 1987
Roy Marsden, Pauline Collins, Martin Jarvis, Art Malik, Gillian Barge, John Franklyn-Robbins, Carol Gillies, Richard Heffer, Heather James, Harriet Bagnall, Martyn Hesford, Valerie Whittington, Albie Woodington, David Webb
Inspector Dalgliesh investigates the death of an old friend and soon finds himself embroiled in a number of murders that take place at a home for the handicapped. A detailed and sensitive adaptation of the original novel, with a fine atmosphere and well-rounded characters, this provides splendid entertainment right up to its powerful conclusion. One in a series of excellent television adaptations made by Anglia TV.
Aka: BLACK TOWER, THE
DRA 300 min (2 cassettes) mTV VIDrel: L/A V
Boa: novel by Phyllis Dorothy James.

DALLAS ***
PG
Stuart Heisler USA 1950
Gary Cooper, Ruth Roman, Steve Cochran, Raymond Massey, Barbara Payton, Leif Erickson, Antonio Moreno, Jerome Cowan, Reed Haldey, Gil Donaldson, Zon Murray, Will Wright, Monte Blue, Byron Keith, Steve Dunhill, Charles Watts, Gene Evands
A former Confederate soldier fakes his death in s gun duel with Wild Bill Hickok so that he can take over as marshal in the title town and put paid to the criminal activities of three local brothers. When he later learns that they were the thugs who murdered his family, he becomes even more determined that they must not escape justice. A fairly standard plot is much enhanced by fine acting, enjoyable music, firm direction and excellent camerawork.
WES 90 min (ort 94 min) VIDrel: WHV V/h

DALLAS DOLL **
18
Ann Turner AUSTRALIA 1994
Sandra Bernhard, Victoria Longley, Frank Gallacher, Jake Blundell, Rose Byrne, Jonathon Leahy, Douglas Hedge, Melissa Thomas, Elaine Lee, Walter Sullivan, William Usic, Alethea McGrath, Roy Billing, John Frawley, Ken Senga
The well ordered lifestyle of a family is drastically altered by the arrival of a female professional golfer with an outsize ego. However, both the teenage daughter and the family dog remain resistant to her charms. Quite what point the film is trying to make is uncertain, perhaps it's just an offbeat set of observations on the Australian fear of loss of identity. Bernhard is terrific but the film lacks direction.
COM 104 min wScrn VIDrel: TART/20TH V

DAM BUSTERS, THE ***
U
Michael Anderson UK 1954
Michael Redgrave, Richard Todd, Basil Sydney, Ursula Jeans, Derek Farr, John Fraser, Ernest Clark, Nigel Stock, Bill Kerr, George Baker, Robert Shaw, Raymond Huntley, Anthony Doonan, Harold Goodwin, Laurence Naismith, Ewen Solon
Rather a Boy's Own treatment of the development and application of the bouncing bombs of Sir Barnes Wallace, which were used during WW2 in attacks on the Ruhr dams in 1943. Interestingly, security did not permit showing of the real bomb design, but despite its faults, this remains a competent work. The stirring (and eponymous) theme music was written by Eric Coates. See also MOSQUITO SQUADRON.
WAR 120 min (ort 125 min) B/W VIDrel: WHV V
Boa: book Enemy Coast Ahead by Guy Gibson and Paul Brickhill.

DAMAGE **
18
Louis Malle FRANCE/UK 1992
Jeremy Irons, Miranda Richardson, Juliette Binoche, Leslie Caron, Rupert Graves, Peter Stormare, Gemma Clark, Ian Bannen, Julian Fellows, Tony Doyle, Raymond Gravell, Susan Engel, David thewlis, Benjamin Whitrow, Jason Morell
A middle-aged British politician courts disaster when h begins a passionate affair with his son's sexy girlfriend and ends up endangering both his career and his family. Some fine performances help to enhance this well realised film, but Binoche as the mystery woman fails to convince and the ponderous and sometimes pompous direction is a major drawback.
DRA 106 min (ort 112 min) VIDrel: EIV/SONOP V/sur
Boa: novel by Josephine Hart.

DAMIEN: OMEN 2 **
18
Don Taylor USA 1978
William Holden, Lee Grant, Jonathan Scott-Taylor, Lew Ayres, Sylvia Sidney, Leo McKern, Robert Foxworth, Nicholas Pryor, Lance Henriksen, Lucas Donat, Alan Arbus, Meshach Taylor, John J. Newcombe, John Charles Burns, Paul Clark
Sequel to THE OMEN with our Antichrist now fully grown and ready for world domination. An ineffective sequel that has a few good sequences but lacks the force of the earlier film. A disappointing trilogy that concluded with OMEN 3: THE FINAL CONFLICT.
Aka: OMEN 2
HOR 102 min (ort 109 min) VIDrel: 20TH/TECH V/dm

DAMNED, THE ***
12
Joseph Losey UK 1961
Macdonald Carey, Shirley Ann Field, Viveca Lindfors, Alexander Knox, Oliver Reed, Walter Gotell, James Villiers, Thomas Kempinski, Kenneth Cope, Brian Oulton, Barbara Everest, Alan McClelland, James Maxwell, Rachel Clay
An American on holiday at a British seaside resort meets and falls in love with a young woman, whose brother is the brutal leader of a motorcycle gang. He takes a violent objection to their relationship and forces them to take refuge in a cave where they discover a group of strange children, the result of an experiment to breed radiation-resistant humans. A odd hybrid of a film, it never quite finds its genre, and is uncompromisingly bleak right up to the end.
Aka: THESE ARE THE DAMNED
FAN 84 min (ort 96 min) B/W VIDrel: ENCORE/SPEAR V
Boa: novel The Children of Light by Henry Lionel Lawrence.

DAMNED, THE *
18
Luchino Visconti ITALY/WEST GERMANY 1969
Dirk Bogarde, Ingrid Thulin, Helmut Berger, Helmut Griem, Charlotte Rampling, Florinda Bolkan, Renaud Verley, Umberto Orsini, Rene Kolldehoff, Nora Ricci, Albrecht Schoenhals, Wolfgang Hillinger, Bill Vanders
Overblown and melodramatic tale of the break-up of a nasty family of German industrialists, set against the background of the Nazis' rise to power. The running time of domestic versions is 164 minutes.
Aka: GOTTERDAMMERUNG; LA CADUTA DEGLI DEI
DRA 146 min (ort 164 min) wScrn VIDrel: WHV V

DANCE FIRE **
18
John Stagliano USA 1989
Kathleen Jentry, Nina De Ponca, Stephanie Rage, Trinity Loren, Brandy Alexandre, Champagne, Jack Baker, Rick Daniels, Rod Garetto, John Stagliano, Giovanni, Mercedes
Lust and decadence in a dance studio, where the proprietor is not averse to having sex with his students, in this strange and occasionally quite stylish adult movie, that offers a blend of dreams and reality.
A 76 min (ort 105 min) VIDrel: FIFTH/DISC V

DANCE OF THE VAMPIRES **
18
Roman Polanski UK/USA 1967
Jack MacGowran, Roman Polanski, Sharon Tate, Ferdy Mayne, Alfie Bass, Terry Downes, Fiona Lewis, Iain Quarrier, Jessie Robins, Ronald Lacey, Sydney Bromley, Andreas Malandrinos, Otto Di Amant, Matthew Walters
Ponderous and unfunny spoof on vampire films, which nevertheless has some quite good moments. Polanski both directs and stars as the timid assistant to a vampire hunter,

who has come to Transylvania to destroy a nest of these creatures.
Aka: FEARLESS VAMPIRE KILLERS, THE; FEARLESS VAMPIRE KILLERS OR: PARDON ME, BUT YOUR TEETH ARE IN MY NECK, THE
COM   103 min (ort 124 min)   VIDrel: MGM/WHV   V

DANCE WITH A STRANGER ***
15
Mike Newell UK
1984
Miranda Richardson, Rupert Everett, Ian Holm, Tom Chadbon, Jane Bertish, Matthew Carroll, David Troughton, Paul Mooney, Stratford Johns, Joanne Whalley, Susan Kyd, Lesley Manville, Martin Murphy, Michael Jenn, Ian Hurley, Colin Rix
Skilful story of the events leading up to the murder by Ruth Ellis of her lover. As the last woman to be hanged in the UK, her death was instrumental in the eventual abolition of capital punishment. Reminiscent of YIELD TO THE NIGHT, which attempted a similar story. Winner of the Cannes Best Picture Award of 1985.
DRA   98 min (ort 102 min)
VIDrel: POLY/POLYREC/BRAVE L/A   V

DANCE WITH DEATH *
18
Charles Philip Moore USA
1991
Maxwell Caulfield, Barbara Alyn Woods, Martin Mull, Drew Snyder, Catya Sassoon, Tracy Burch, Steven Lloyd-William, Jill Pierce, Shana Arthur, Lisa Kudrow, Aletha Baker, Michael James McDonald, Joe Garcia, Charles Moore
With someone busily bumping off the strippers at a club, it is not long before an attractive female journalist begins to investigate. This involves her going undercover to work as a stripper, a profession she seems to be born for. Meanwhile, a handsome police detective has also been assigned to the same case. A tension-free striptease movie set against a convoluted murder plot, its contrived scripting holds few surprises.
THR   85 min (ort 90 min)   VIDrel: CIC/SONOP   V

DANCES WITH WOLVES ****
15
Kevin Costner USA
1990
Kevin Costner, Mary McDonnell, Graham Greene, Rodney A. Grant, Floyd Red Crow Westerman, Tantoo Cardinal, Robert Pastorelli, Charles Rocket, Maury Chaykin, Larry Joshua, Nathan Lee Chasing His Horse, Jason R. Lone Hill
Costner's remarkable directing debut is the uplifting tale of an idealistic Civil War soldier, whose growing friendship with a band of Sioux leads to an ultimate rejection of his own people for a life among them. A stunning film whose visual impact will somewhat lessened on TV. AA: Pic, Dir, Cin (Dean Semler), Sound (Russell Williams/Jeffrey Perkins/Bill Benton/Greg Watkins), Screen/adapt (Michael Blake), Score/orig (John Barry), Edit (Neil Travis).
WES   173 min (ort 190 min); 223 min wScrn (extended version)   VIDrel: POLY/POLYREC/GUILD; PION (LV only) V/sur LV
Boa: novel by Michael Blake.

DANCING IN THE DARK **
15
Bill Corcoran USA
1995
Victoria Principal, Nicholas Campbell, Dawn Greenhalgh, Robert Vaughn, Geraint Wyn Davies, Kenneth Walsh, Sheila Brand, Maggie Huculak, Marcia Bennett, Anna Louise Ricahardson, Michael Dyson, George Robertson, Jenna Preston,
Based on a true story, this is an account of the harrowing ordeal a woman undergoes (including psychiatric assessment) when her husband refuses to believe her allegations of attempted rape by her father-in-law. None of the cast seem able to breathe life into this dispirited drama, which really should have been more compelling given the nature of the subject matter.
DRA   92 min mTV   VIDrel: ODY/SONOP   V/sh

DANCIN' THRU THE DARK ***
15
Mike Ockrent UK
1989
Claire Hackett, Con O'Neill, Angela Clarke, Julia Deakin, Louise Duprey, Sandy Hendrickse, Mark Womack, Conrad Nelson, Simon O'Brien, Peter Watts, Andrew Naylor, Colin Welland
Written by Willy Russell, and very much a typically bittersweet slice of Liverpudlian life, with a young couple about to be married celebrating separately with their respective friends, but inevitably ending up at the same club. While the bridegroom-to-be lies in a sodden stupor, his future spouse begins to have second thoughts, especially after she meets up with an old flame. Neither insightful nor especially funny, but quite engaging.
COM   91 min (ort 95 min)   VIDrel: PAL/TERRY L/A   V/h
Boa: play Stags And Hens by Willy Russell.

DANCING WITH DANGER **
15
Stuart Cooper USA
1993
Cheryl Ladd, Ed Marinaro, Miguel Sandoval, Pat Skipper, Stanley Kamel
A private eye falls in love with a dance-hall hostess but faces danger when he has to protect her from a mad serial killer who has been stabbing women to death with a pair of scissors. Adequate enough, though without anything that marks it out from similar works.
THR   87 min (ort 90 min) mCab   VIDrel: CIC   V

DANDIN ***
(PG)
Roger Planchon FRANCE
1988
Claude Brasseur, Zabou, Daniel Gelin, Nelly Borgeaud, Jean-Claude Adelin, Evelyne Buyle, Marco Bisson, Vincent Garanger, Martine Merri, Marie Pillet, Philippine Leroy-Beaulieu, Judith Becle, Colette Dompietrini, Anne Guegan
Planchon's directing debut is based on a Moliere's famous farce, and tells of the title figure, who is a rich peasant who marries far above his station. A rumbustious and colourful romp, full of life, vigour and humour.
COM   105 min   TVrel: BBC2
Boa: play by Moliere.

DANGER WITHIN ***
U
Don Chaffey UK
1959
Richard Todd, Bernard Lee, Michael Wilding, Richard Attenborough, William Franklyn, Dennis Price, Donald Houston, Vincent Ball, Peter Arne
A WW2 drama set in as POW camp in 1942, where an informer appears to be at work, putting at risk the escape plans being made by the officers. A strong mixture of comedy and drama (slightly in the style of STALAG 17) whose plot elements are handled with just the right degree of seriousness to maintain tension.
Aka: BREAKOUT
DRA   97 min (ort 101 min) B/W   VIDrel: LUMI/SPEAR L/A   V
Boa: novel Death in Captivity by Michael Gilbert.

DANGEROUS ***
PG
Alfred E. Green USA
1936
Bette Davis, Franchot Tone, Margaret Lindsay, Alison Skipworth, John Eldredge, Dick Foran, Pierre Watkin, Walter Walker, George Irving, William Davidson, Douglas Wood, Richard Carle, Milton Kibbee, George Andre Beranger
A rich man attempts to rehabilitate an alcoholic former star of stage and screen but finds himself falling in love with her. She returns his love but fails to tell that she is separated but still married and so has to resort to desperate measures when her husband refuses to give her a divorce. A passable drama that was later remade as "Singapore Woman". AA: Actress (Davis).
DRA   75 min (ort 78 min) B/W   VIDrel: MGM/WHV   V

DANGEROUS, THE **
18
Maria Dante/Rod Hewitt USA
1995
Robert Davi, Michael Pare, Cary Hiroyuki-Tagawa, Elliott Gould, John Savage, Paula Barbieri, Juna Fernadez, Saemi Nakamura, Joel Grey, June Saruwatari, Marco St. John, Elliott Keener, Layton Martens, Takayo Fisher, Monte Bain, Fred Lewis
Overblown action movie set on the streets of New Orleans, where a gang of Japanese ninja-style assassins are causing havoc among the underworld. Their actions are motivated by revenge for the murder of a girl caught photographing an illicit deal and the only man who knows how to deal with the situation is soon called in. When his opponents make the mistake of kidnapping his girlfriend, the predictably violent rescue attempt soon gets under way.
A/AD   96 min   VIDrel: MED/COLUM   V/sh

DANGEROUS AFFAIR, A *
15
Alan Metzger USA
1994
Connie Sellecca, Gregory Harrison, Christopher Meloni, Rosalind Cash, Jo De Winter, Eileen Seeley, Ryan Todd, Gerald Berns, Brians Evers, Frank Novak, Robin Bartlett, John Marshall Jones, John Bellucci, David Rose, Chip Heller, Jack Hoar
An attractive woman falls for a handsome man whose charming exterior masks his obsessive tendencies, and all too soon he shows himself to be a psychopath given to stalking. A highly conventional psychological thriller with a familiar flashback structure that is devoid of originality and has one of those tiresome twist endings hinting that the nightmare is about to recur.
THR   92 min mTV   VIDrel: ODY/SONOP   V/sh

DANGEROUS DESIRE ** 18
Paul Donovan CANADA 1992
Richard Grieco, Maryam D'Abo, Natalie Redford, Serge Houde, Sean Orr, David McLeod, Benjamin Rather, Christine Lippa, Brenda Crichlow, Cordell Wayne, Roman Podhora
A scientist tries to save the life of her lover by injecting him with hormones derived from a cat, but this turns him into an over-sexed and savage individual who enjoys playing cat-and-mouse games with a string of females. An endearingly silly fantasy-thriller, this one never really finds its genre, and might have worked better as a straight comedy.
Aka: TOMCAT: DANGEROUS DESIRE
THR 91 min VIDrel: 20VIS/SONOP V/sh

DANGEROUS GAME ** 18
Abel Ferrara USA 1993
Harvey Keitel, James Russo, Madonna, Nancy Ferrara, Reilly Murphy, Victor Argo, Leonard Thomas, Kristina Fulton, Heather Bracken, Glenn Plummer, Niki Munroe, Juliette Hohnen, Julie Pop, Lori Eastside, John Snyder, Adina Winston
A film director's latest production plunges him into a state where he is unable to distinguish between dreams and reality as he starts manipulating both crew and cast. A messy and unfocused effort that saves all its fireworks for the last thirty minutes, but this alone is unable to save this sorry effort.
Aka: SNAKE EYES
DRA 104 min (ort 109 min) cC VIDrel: POLY/POLYREC V/sh

DANGEROUS GROUND * 18
Darrell James SOUTH AFRICA/USA 1996
Ice Cube, Elizabeth Hurley, Sechaba Morojele, Eric "Waku" Miyeni, Ving Rhames, Thokazani Nkosi, Ron Smerczak, Wilson Dunster, Peter Kubheka, Roslyn Morapedi
Rap artist Ice Cube is incredibly miscast as an academic specialising in African literature who returns to his native South African village for the first time since leaving for San Francisco as a teenager years ago, having been brought back for his father's funeral. Whilst there he spends time trying to locate his brother, and gets involved with various disagreeable characters. A tedious dud of ludicrous scripting that strains credibility and patience.
A/AD 95 min CINrel

DANGEROUS HEART ** 15
Michael Scott USA 1993
Tim Daly, Lauren Holly, Alice Carter, Joe Pantoliano, Jeffrey Nordling, Bill Nunn, Robert P. Lieb, Michael Ellison, Michael Paul Chan, Michael Keys Hall, Amy Van Norstrand, Brady Ward, Chris Ufland, Doug Kerr, Steve Kehela, Patty Tippo
After her policeman husband is murdered by a drug dealer whom he robbed while working under cover, the latter comes to call on his grieving widow. Having adopted a phoney identity and an equally false charm, he intends to gradually seduce her, in the hope of recovering his money. A poorly acted and realised tale that feels highly implausible and extremely contrived.
THR 89 min (ort 93 min) mCab VIDrel: CIC/SONOP V

DANGEROUS INDISCRETION ** 18
Richard Kletter USA 1994
C. Thomas Howell, Malcolm Dowell, Joan Severance, Sue Matthew
An advertising executive makes the serious error of getting involved with the wife of a ruthless and powerful man who is determined to make him pay a very high price for this indiscretion.
DRA 77 min (ort 81 min) VIDrel: HIFLI/SONOP V/h

DANGEROUS INTENTIONS ** 15
Michael Toshiyuki Olson USA 1995
Donna Mills, Corbin Bernsen, Allison Hossack, Sheila Larken, Ken Pogue, Anna Ferguson, Robin Givens, Alexandra Purvis, Patti Yasutake, Tracey Olson, Roger R. Cross, Thom Cavanagh, Tracey Olson, Ric Reid, Nathaniel Deveaux, Walter Marsh
Another on of those thrillers based on a true story, this one is about a woman whose only escape from her brutal husband is to "disappear" by taking a new identity, having first ensured that her husband was sent to jail for his latest (and most violent) attack. He is soon released and comes looking for her, but in the meantime she has made friends with another battered woman.
THR 92 min mTV VIDrel: ODY/SONOP V/sh

DANGEROUS LIAISONS **** 15
Stephen Frears USA 1988
Glenn Close, John Malkovich, Michelle Pfeiffer, Keanu Reeves, Uma Thurman, Swoosie Kurtz, Mildred Natwick, Peter Capaldi, Joe Sheridan, Valerie Cogan, Laura Benson, Joanna Pavlis, Nicholas Hawtrey, Paulo Abel do Nascimento
An 18th century story of an unscrupulous woman who enjoys manipulating and exploiting those around her for her own amusement, being joined in this pursuit by a count of similar tastes. An excellent and engrossing costume drama set in France, with Close quite outstanding as the woman in question. See also WHEN A WOMAN IS IN LOVE. AA: Art (Stuart Craig/Gerald James), Screen/adapt (Christopher Hampton), Cost (James Acheson).
DRA 115 min (ort 120 min) wScrn VIDrel: WHV V/sur
Boa: novel Les Liaisons Dangereuses by Choderlos de Laclos/play by Christopher Hampton.

DANGEROUS MAN, A: LAWRENCE OF ARABIA **
PG
Christopher Menaul UK 1991
Ralph Fiennes, Dennis Quilley, Paul Freeman, Nicholas Jones, Gillian Barge, Jim Carter, Michael Cochrane, Roger Hammond, Polly Walker, Peter Copley, Robert Arden, Siddig El Fadil, Ray Edwards
Fair drama based on the events that occurred during the 1919 Paris Peace Conference, where T.E. Lawrence attempted to deliver to his war-time Arab allies the land he had promised to win for them – the region that was to form Syria. An interesting fact-based drama that provides some useful insights if not a great deal of dramatic impact.
Aka: DANGEROUS MAN, A: LAWRENCE AFTER ARABIA
DRA 104 min mTV VIDrel: FOCUS/DISC V

DANGEROUS MINDS ** 15
John N. Smith USA 1995
Michelle Pfeiffer, George Dzundza, Courtney B. Vance, Robin Bartlett, John Neville, Beatrice Winde, Lorraine Toussaint, Renoly Santiago, Wade Dominguez, Bruklin Harris, Marcello Thedford, Roberto Alvarez, Richard Grant, Norris Young
A retired woman Marine accepts the challenging task of teaching English to a tough class in a run-down inner-city school and finds ways of achieving a rapport with her teenage pupils and get them to use their minds. Pfeiffer is badly miscast, but the film does have some endearing if corny moments. It may be hard to believe that many successful teachers have been created by Marine Corps training, but this story is based on a real case. See also STAND AND DELIVER.
DRA 95 min (ort 100 min) cC VIDrel: HOLPIC/TECH V
Boa: autobiography My Posse Don't Do Homework by LouAnne Johnson.

DANGEROUS ORPHANS ** 18
John Laing NEW ZEALAND 1987
Peter Stephens, Michael Hurst, Jennifer Ward-Lealand, Ross Girven, Ian Mune, Peter Bland, Zoe Wallace, Grant Tilly, Ann Pacey, Peter Vere-Jones, Michael Haigh, Des Kelly, Tim Lee, Tobby Laing, Kevin Wilson, Marshall Napier
An orphan plots with two friends to avenge his father's murder at the hands of gangsters. A bloody revenger that's well made but pretty standard in terms of its plot.
Aka: VENGEANCE
A/AD 87 min (ort 93 min) Col/B/W VIDrel: SCRN/DISC V

DANGEROUS PASSION *** 15
Michael Miller USA 1990
Carl Weathers, Lonette McKee, Billy Dee Williams, Elpidia Carrillo, Michael Beach, L. Scott Caldwell, Charles Boswell, Tony Di Benedetto, Nancy Fish, Miguel Sandoval, Daniel Ziskie, Rudy RAmos, Shannon Wilcox, Charles Stransky
Having crossed swords with a ruthless mobster, a security expert makes matters worse by falling in love with the man's wife, and learns that he has been framed for a couple of murders. A slow-moving thriller that builds up carefully to a powerful and action-filled climax.
THR 91 min mTV VIDrel: CAPIT/GUILD V/sh

DANGEROUS TOUCH ** 18
Lou Diamond Phillips USA 1993
Lou Diamond Phillips, Kate Vernon, Max Gail, Berlina Tolbert, Mitch Pileggi, Adam Roarke, Robert Prentiss, William Lawrence Allen, Greg Stone, Shanti Khan, Stacie Bourgeois, Karla Montana, Efram Figueroa, Andrew Divoff, Monique Parent

A woman who works as a sex therapist with her own radio show, finds herself becoming irresistibly drawn towards a mysterious stranger and soon becomes his willing slave. This rash behaviour soon embroils her in both blackmail and murder, in this steamy thriller that has the obligatory sex scenes to bolster a thin and poorly conceived plot.
THR 97 min (ort 101 min) VIDrel: POLY/POLYREC L/A
V/s

DANGEROUS WOMAN, A ** 18
Stephen Gyllenhaal USA 1993
Debra Winger, Barbara Hershey, Gabriel Byrne, David Strathairn, Laurie Metcalf, Chloe Webb, John Terry, Jan Hooks, Paul Dooley, Viveka Davis, Richard Riehle, Maggie Gyllenhaal, Jacob Gyllenhaal, Myles Sheridan, Brad Blaisdell
A mentally unbalanced woman who lives in the shadow of her wealthy, widowed sister, finds her life coming apart when she falls for a handyman, who gets involved with both women. Meanwhile, a false charge of theft leads to further complications. An uninspired adaptation that is badly lacking in structure and for good measure, is hampered by a naive and utterly simplistic view of mental illness.
DRA 97 min (ort 102 min) VIDrel: FIRST/SONOP V/sur
Boa: novel by Mary McGarry Morris.

DANNY, THE CHAMPION OF THE WORLD *** U
Gavin Miller UK 1989
Jeremy Irons, Samuel Irons, Robbie Coltrane, Cyril Cusack, Jimmy Nail, Lionel Jeffries, Michael Hordern, Jean Marsh, Ronald Pickup, John Woodvine, William Armstrong, Ceri Jackson, James Walker, Phil Nice, Anthony Collin,
Set in the 1950s, this tells the story of Danny, a nine-year-old living with his father. When their tranquil life in a caravan is threatened by a local developer who has acquired all the surrounding land except their tiny plot, the boy hatches a scheme to teach him a lesson. A nicely made and wholesome kids' film.
Aka: ROALD DAHL'S DANNY, THE CHAMPION OF THE WORLD
JUV 94 min VIDrel: BMGVID/BMGREC V/sur

DANTE'S PEAK *** 12
Roger Donaldson USA 1996
Pierce Brosnan, Linda Hamilton, Jaime Renee Smith, Jeremy Foley, Elizabeth Hoffman, Charles Hallahan, Grant Heslov, Kirk Trutner, Arabella Field
A volcano expert arrives at a small town built close to a long-dormant volcano, having decided to investigate some indications of growing activity. Once there he realises that an eruption is likely, but the townsfolk refuse to believe his warnings. A spectacular disaster movie with no shortage of special effects, which tend to overwhelm all other aspects of the film, including the believable portrayals from the cast. See also VOLCANO: FIRE ON THE MOUNTAIN.
A/AD 109 min CINrel

DANTON *** PG
Andrzej Wajda FRANCE/POLAND 1983
Gerard Depardieu, Wojciech Pszoniak, Patrice Chereau, Roger Planchon, Angela Winkler, Boguslaw Linda, Roland Blanche, Anne Alvaro, Serge Merlin, Andrzej Seweryn, Lucien Meki, Franciszek Starowieyski, Emmanuelle Debever
An overblown, overlong and over-rated account of the struggle between Danton and Robespierre in 1793, when the former fought in vain to curb the excesses of the Terror with its mass executions. Whether or not this film was meant to refer to contemporary events in Poland is irrelevant; what matters is that it is often confusing and hard to follow, especially for non-experts of French history of the period.
DRA 130 min (ort 136 min) VIDrel: ARTIF/20TH V

DANZON *** PG
Maria Novaro MEXICO/SPAIN 1991
Maria Rojo, Carmen Salinas, Tito Vaconcenlos, Victor Carpinteiro, Margarita Isabel, Blanca Guerra, Cheli Godinez, Daniel Rergis, Adyani Chazaro, Caesar Sobreval, Mikhail Kaminin, Rodrigo Gomez, Sergio Colmenares, Luis Gerardo
A telephone operator finds an escape from her monotonous job dancing title dance with her partner at the local dance hall, but when he disappears mysteriously, she sets out to find him. This quest takes her to his home town of Vera Cruz, where many encounters await her. A well-made and quiet human-interest drama, borne up by fine performances.
DRA 102 min (ort 103 min) wScrn VIDrel: TART/20TH
V/dm

DARE TO LOVE ** 12
Armand Mastroianni USA 1995
Josie Bissett, Jill Eikenberry, Chad Lowe, Jason Gedrick
A woman slides into schizophrenia when her brother dies in tragic circumstances, and only emerges after seven years of therapy with the chance of finding love. Based on a true story, but what could have been a truly engrossing and moving film is given the standard soap-opera treatment, doing justice to neither the cast nor the real-life character whose experience led to this movie.
DRA 89 min VIDrel: 20TH/FOXVID V/sh

DARK, THE ** 18
Craig Pryce USA 1993
Cynthia Belliveau, Jaimz Woolvett, Brion James, Stephen McHattie, Neve Campbell, Dennis O'Connor
A strange creature with some remarkable powers lives beneath a churchyard and is pursued by both an FBI agent and a scientist. The former wants to destroy it, while the latter hopes that its abilities can be harnessed to heal the sick.
HOR 88 min (ort 90 min) VIDrel: NORSTAR/HIFLI V/h

DARK ANGEL, THE *** U
Sidney Franklin USA 1935
Merle Oberon, Fredric March, Herbert Marshall, Janet Beecher, John Halliday Henrietta Crosman, Frieda Inescort, George Breakston, Claud Allister
An officer blinded during WW1 tries to get his fiancee to marry another man without her discovering his reasons. Get out the Kleenex - this well made weepie is a three box affair. AA: Art (Richard Day).
DRA 101 min (ort 106 min) B/W VIDrel: VCC/DISC V

DARK ANGEL: THE ASCENT ** 18
Linda Hassani USA 1993
Daniel Markel, Charlotte Stewart, Michael Genovese, Nicholas Worth, Angela Featherstone, Michael C. Mahon, Milton James, Costica Dragensecu, Christina Stoica, Valentina Teodescu, Marius Stanescu, Constantin Costimanis, Liljana Pana
A bored female devil ascends to Earth to seek out evil-doers and punish them in appropriately nasty ways, but her mission is jeopardised when she seems to fall in love. An over-the-top horror tale, with much gore and sick humour, filmed on location in Romania.
FAN 81 min VIDrel: CIC/SONOP V

DARK AVENGER ** 15
Guy Magar USA 1990
Leigh Lawson, Maggie Han, Robert Vaughn, Hector Elizondo
A tough and totally honest young judge gets a little too close in exposing police corruption and an attempt is made on his life. Though badly wounded, he survives and goes undercover in a bid to have revenge. An adequate actioner hampered by implausible plotting.
Aka: I ACCUSE
A/AD 89 min VIDrel: COLUM/SONOP V

DARK BREED * 18
Richard Pepin USA 1995
Jack Scalia, Donna W. Scott, Carlos Carrasco, Jonathan Banks, Robin Curtis, Lance Le Gault
A battered space-shuttle crash-lands back on Earth and is retrieved from a river, but the crew are not found, having all been taken over by a malevolent lifeform that intends to colonise the Earth. What this film really needed was a good dose of the paranoia from THE INVASION OF THE BODY SNATCHERS. However, we are never treated to this and instead have to make do with a daft government conspiracy sub-plot and some contrived nonsense about good aliens.
FAN 92 min VIDrel: MARQ/QUANT V

DARK COMMAND *** U
Raoul Walsh USA 1940
John Wayne, Claire Trevor, Walter Pidgeon, Roy Rogers, George "Gabby" Hayes, Porter Hall, Marjorie Main, Raymond Walburn, Joseph Sawyer, Helen MacKellar, J. Farrell MacDonald, Trevor Bardette, Harry Woods, Glenn Strange
At the end of the American Civil War, a number of private armies ravaged the countryside. Pidgeon plays one such man (loosely based on the 1860s renegade Quantrill), who leads a

band on raids but eventually clashes with the newly-elected marshal. An enjoyable but patchy adventure.
WES   90 min (ort 94 min) B/W
VIDrel: 4-FRONT/POLYREC   V
Boa: novel Dark Command by W.R. Burnett.

DARK EYES *** PG
Nikita Mikhalkov ITALY/USSR   1987
Marcello Mastroianni, Elena Sofonova, Silvana Mangano, Pina Cei, Marthe Keller, Vesvolod Zolothukin, Paolo Baroni, Oleg Tabakov, Youri Boagtirov, Innokenti Smoktunovski, Marthe Keller, Robert Herlitzka, Roberto Herlitzka
On a ship, an ageing Italian playboy meets a friendly Russian and regales him with the story of his life, and all his various romantic peccadilloes. A turn-of-the-century tale, that ably adapts a set of Chekhov short stories ranging in tone from sad and moving to exuberant and farcical. An enjoyable outing for Mastroianni, who easily dominates this unusual comedy-drama.
Aka: BLACK EYES; OCI CIORNIE
DRA   113 min (ort 118 min)   VIDrel: CONNO/RTM   V
Boa: short stories Anna Around The Neck/The Lady With The Little Dog/My Wife/The Name-Day Party by Anton Chekhov.

DARK HABITS *** 18
Pedro Amodovar SPAIN   1983
Cristina S. Pascual, Julieta Serrano, Marisa Paredes, Carmen Maura, Chus Lampreave
After her boyfriend dies from a drugs overdose, a nightclub singer takes to the road and ends up in an eccentric convent. An amusing but wildly uneven black comedy, very typical of this director's output.
Aka: ENTRE TINIEBLAS
DRA   116 min wScrn   VIDrel: TART/20TH   V

DARK HALF, THE ** 18
George A. Romero USA   1991
Timothy Hutton, Amy Madigan, Michael Rooker, Julie Harris, Robert Joy, Kent Broadhurst, Beth Grant, Rutanya Alda, Tom Mardirosian, Chelsea Field, Royal Dano, Patrick Branna, Larry John Meyers, Rohn Thomas, Molly Renfroe, Judy Grafe
A writing instructor who has become a successful writer of horror novels under a pseudonym, announces the death of the latter, thereby causing him to come to life. A not entirely successful adaptation of King's novel with an over-abundance of gore and too few chills to entertain. The variable script is by the director, who would probably have done better to have strayed a bit more from the original novel.
Aka: STEPHEN KING'S THE DARK HALF
HOR   116 min (ort 121 min)   VIDrel: VCC/DISC/COLUM V/sur
Boa: novel by Stephen King.

DARK PASSAGE *** 15
Delmer Daves USA   1947
Humphrey Bogart, Lauren Bacall, Agnes Moorehead, Bruce Bennett, Bob Farber, Tom D'Andrea, Clifton Young, Douglas Kennedy, Rory Mallinson, Richard Walsh, Houseley Stevenson, Clancy Cooper, Pat McVey, Dude Maschemeyer
A man imprisoned in San Quentin for the murder of his wife, escapes and has his face altered by cosmetic surgery, hiding out at Bacall's apartment while his scars heal. She believes in his innocence and he falls in love with her, but his new life is threatened by her best friend, a woman with a guilty past. An uneven loose assembly of the incredible and the unbelievable, but the splendid camerawork and acting are significant compensations.
THR   102 min B/W   VIDrel: MGM/WHV   V
Boa: novel by David Goodis.

DARK RIDER ** 15
Bob Ivy USA   1991
Joe Estevez, Douglas A. Shanklin, Holly Floria, David Shark, Chuck Williams, Alicia Kowalski
Very slightly reminiscent of HIGH PLAINS DRIFTER, this action film sees a leather-clad rider coming to the rescue of a town whose inhabitants live in fear of gangsters.
A/AD   94 min   VIDrel: 20VIS/SONOP   V

DARK SANITY ** 15
Martin Greene/Tim McWhorter/Martha Sudderth USA   1980
Aldo Ray, Chuck Jamison, Kory Clark, Andy Gwyn, Bobby Holt, Harry Carlson, Timothy McCormack, Barry Ray Robinson, Dennis Barnett, Iris Bath, Brenda Bennett, Toni Carleton, Roger Clark, Kevin Downey, Rick Green, Ron Jennings

A rehabilitated woman alcoholic moves into a house that was the scene of a mysterious and unsolved murder, and begins to be haunted by sinister visions of the event. Fair.
Aka: STRAIGHT JACKET
HOR   86 min   VIDrel: VIPCO/SGSVID   V/h

DARK SECRETS * (18)
Michael James McDonald USA   1995
Anthony Michael Hall, Justine Bateman, Sam Lloyd, Alan Sues, Darcy DeMoss, Patrick Bristow, Kin Shriner, Jesse D. Goins, Paul Bartel, Mink Stole, Sheila Traviss, Victor Wilson, Julianna McCarthy, Shadoe Stevens, Jennifer Coolidge
Horror-comedy about a struggling artist who works as a waiter at a cafe frequented by an arty crowd, and who wants to be accepted by them, especially a pretty Italian girl. When he accidentally kills his landlady's cat he hides the body by covering it with plaster, and instantly creates a work of art that gains him recognition. However, events soon conspire to force him to seek further challenges. A very 1990s remake of A BUCKET OF BLOOD, unfunny and rather creepy.
HOR   90 min   SATrel: MOVIE CHANNEL

DARK SECRETS (A TRUE STORY) ** 15
Mimi Leder USA   1991
Pamela Reed, Dwight Schultz, Richard Lineback, Carrie Snodgress, Adam Faraizl, Nick Stahl, James Sloyan, Danielle Von Zerneck, Noble Willingham, Paul Le Mat, Bryan Clark, Olivia Virgil Harper, Robert Schuch, Karen Hensel
A female prisoner with a murky past and a conviction for armed robbery escapes from jail and makes a bid to start a new life. Having married, she tells her husband nothing about her past, but soon learns that the FBI has by no means closed her file.
Aka: WOMAN WITH A PAST
THR   95 min mTV   VIDrel: GENESIS   V

DARK SIDE OF LOVE, THE *** 18
Salvatore Samperi ITALY   1985
Monica Guerritore, Lorenzo Lena
An elegant and slightly disturbing tale of how a sixteen-year-old is seduced and corrupted by his twenty-five-year-old governess, who visits him every night to recount her sexual experiences. These become ever more extreme, and so work on the youngster's imagination that eventually he is drawn into a relationship with both her and another woman. Strongly reminiscent of PRIVATE LESSONS, this is a high-quality softcore effort, lavish and well photographed.
A   89 min wScrn dubbed   VIDrel: JEZ/RTM   V

DARK SKIES ** 12
Tobe Hooper USA   1996
Eric Close, Megan Ward, J.T. Walsh, Conor O'Farrell
Pilot for a far-fetched TV series done in the style of "The X-Files", that attempts to tie in many of the most notorious incidents in American and Western history (e.g. Kennedy's assassination, the U-2 spy-plane incident) with a plot by aliens to take over the world. Naturally, there is a mysterious alien-busting organisation out there working to stop this happening. A ludicrous exercise in paranoia, but quite good fun, even if it does bring to mind TV's "The Invaders".
THR   60 min mTV   VIDrel: COLUM   V

DARK STAR **** PG
John Carpenter USA   1974
Dan O'Bannon, Brian Narelle, Cal Kuniholme, Dre Pahich, Joe Sanders
A highly-acclaimed tongue-in-cheek film all about a sloppy bunch of misfits cruising space in their battered spaceship, blowing up unstable planets prior to colonisation of space. They've been doing it for so long now they hardly remember why. It all comes together in this brilliantly inventive film, Carpenter's first feature, which he expanded from a college short he made with O'Bannon. Written by Carpenter and O'Bannon.
FAN   83 min   VIDrel: HEND L/A   V/s

DARK SUMMER * 12
Charles Teton UK   1994
Steve Ako, Joeline Garnier-Joel, Chris Darwin, Sylvia Amoo, Wayne Ako, Marlene Amoo, Bernie Deasy, Tom Williamson, Marie Higham, Francis Dell, Neil Antony, Dave Rooney, Jimmy Fitz, Dave Murray, Louis Cuddy, Alex Moon, Adam Ryan
Aimless, meandering and totally uninteresting story of a couple of young lovers, their affair starting when the boy begins work

for the girl's father on his construction site. Dad does not approve, possibly because he is black, but more probably because he is an employee, so the girl moves out to set up home with her boyfriend. Teton's first feature (a self-financed project) is a disappointment, and the director's highly stylised approach looks very strained.
DRA 85 min CINrel

DARK TIDE ** **   18
Luca Bercovici USA   1993
Chris Sarandon, Richard Tyson, Brigitte Bako
Three divers on a mission to collect lethal sea snakes for a research laboratory come into conflict with the captain of their boat who has developed an obsessive passion for the female member of their team.
DRA 89 min (ort 92 min) VIDrel: PROMARK/HIFLI V/sur

DARK UNIVERSE ** **   15
Steve Latshoner USA   1993
Blake Pickett, Cherie Scott, Joe Estevez, Bently Tittle, John Maynard, Paul Austin Saunders, Tom Ferguson, Steve Barkett, Patrick Moran
An alien intruder comes to in search of a new food source and finds that humans fit the bill very nicely, in this standard blend of SF and horror genres.
FAN 87 min VIDrel: COLUM/SONOP V

DARK VICTORY *   PG**
Edmund Goulding USA   1939
Bette Davis, George Brent, Ronald Reagan, Humphrey Bogart, Geraldine Fitzgerald, Cora Witherspoon, Henry Travers, Virginia Brissac, Dorothy Peterson, Charles Richman, Herbert Rawlinson, Leonard Mudie, Fay Helm, Diane Bernard
Tear-jerker about a rich and spoiled society girl who finds she is dying from a brain tumour and meets her end with courage. Bogart as an Irishman hits the only jarring note, in this excellent and unashamedly sentimental melodrama. Remade as "The Stolen Hours" and then once again in 1976.
DRA 100 min (ort 106 min) B/W VIDrel: MGM/WHV V
Boa: play by G.E. Brewer and B. Bloch.

DARK WATERS ** **   18
Mariano Baino Carancula RUSSIA/UKRAINE   1993
Louise Salter, Venera Simmons, Maria Kapnist
An Englishwoman tries to uncover the mystery of her father's death and the reason for her loss of memory, and in order to do this she travels to a remote island in the Crimea to delve into the mysteries of a secretive religious sect. Made on a minuscule budget, Carancula's first feature does pretty well given the financial constraints, with ample atmosphere and style. Unfortunately, the film's technical quality is poor, as it the final resolution.
HOR 94 min wScrn dubbed VIDrel: TART/20TH V

DARK WIND, THE *   15**
Errol Morris USA   1991
Lou Diamond Phillips, Gary Farmer, Fred Ward, Guy Boyd, John Karlen, Jane Loranger, Gary Basaraba, Blake Clark, Faye B. Tso, Michelle Thrush, Eugene Sekaquaptewa, Ivory Ocean, James Koots, Arlene Bowman, Neil Kayquoptewa
In the Arizona desert, a young Navajo tribal policeman has been keeping watch on a windmill that has suffered repeated vandalism at night. He hears a light plane crash nearby, and this is only the start of a complex and dangerous drugs-related case that attracts the attentions of a couple of thuggish FBI agents. A most ponderous effort, that sorely tries one's patience.
THR 106 min (ort 111 min)
VIDrel: GUILD/POLYREC L/A V/s
Boa: novel by Tony Hillerman.

DARKMAN *   18**
Sam Raimi USA   1990
Liam Neeson, Frances McDormand, Colin Friels, Larry Drake, Nelson Mashita, Jesse Lawrence Ferguson, Rafael H. Robledo, Danny Hicks, Jenny Agutter, Dan Bell, Theodore Raimi
A brilliant scientist nearing completion of a project that would allow the cloning of human parts is attacked and left for dead by thugs hired by a ruthless businessman. Having been badly disfigured in the attack, he withdraws from society and sets about planning his revenge, using for the purpose a set of holographic synthetic-skin disguises, each one of which only lasts a short while. An entertaining horror yarn done in the style of a comic strip.
HOR 91 min Cut (2 sec – ort 95 min) VIDrel: CIC/SONOP V

DARKMAN 3: DIE, DARKMAN, DIE *   15**
Bradford May USA   1996
Jeff Fahey, Arnold Vosloo, Darlanne Fluegel, Roxann Biggs-Dawson, Alicia Panetta, Nigel Bennett
Vosloo is miscast as Darkman, who is still trying to perfect the synthetic skin with which he hopes to repair his ruined face, and at the same time has to contend with a drug dealer (Fahey) who hopes that he can obtain samples of Darkman's adrenalin to develop as a new designer-drug. This boring dud has very little going for it and Neeson's charisma (he was only in the first film) is badly missed.
FAN 83 min (ort 87 min) cC VIDrel: CIC/SONOP V/sh

DARKNESS BEFORE DAWN *   15**
John Patterson USA   1992
Meredith Baxter, Stephen Lang, Richard Grove, Gwynnyth, L. Scott Caldwell, Chlesea Hertford, Bill Applebaum, Lee Tergesen, Natalie West, Alana Austin, Robert Desiderio, Kirsten Dunst, Tim Farrell, Matthew Linville, James Dean
A nurse is secretly addicted to methadone but is nonetheless still able to fulfil her duties, until her marriage to a former drug-addict threatens to expose her. True-life TV drama, with the ever-dependable Baxter giving the film depth and credibility.
DRA 92 min mTV VIDrel: ODY/SONOP V/h

DARKNESS IN TALLINN **   18**
Ikka Jarvilaturi ESTONIA/FINLAND/SWEDEN/USA   1993
Ivo Uukkivi, Milena Gulbe, Monika Mager, Enn Klooren, Vaino Laes, Peeter Oja, Juri Jarvet, Villem Indrikson, Andreas Raag, Gerardo Contreras, Martin Tulmin, Kadri Kilvet, Salme Poopuu, Ulvi Kreitsmann, Kristel Karner, Tonu Kark
Comedy-thriller that takes as its basis the fact that the Estonians placed $9,000,000 in gold bullion in a Paris bank for safe keeping during the two World Wars. Having achieved statehood, the Estonian government prepares for the safe return of this treasure, but has to contend with the plans of a gang of crooks, who are out to intercept this loot. A very clever movie, full of twists and turns of plotting, that is consistently entertaining.
Aka: TALLINN PIMEDUSES; TALLINN PIMEYS
THR 99 min B/W/Col wScrn subtitles
VIDrel: TART/20TH V/sur

DARKROOM ** **   18
Terence O'Hara USA   1988
Aarin Teich, Jill Pierce, Jeffrey Allen Arbaugh, Sara Lee Wade
A woman returns to her family's farm and learns of a gruesome double murder that took place a few miles away. As the members of her family begin to disappear, she realises that she is meant to be one of his victims. An adequate time-filler of reasonable quality if not originality.
THR 82 min (ort 90 min) VIDrel: GUILD/SONOP V/h

DARKTOWN STRUTTERS ** **   PG
William N. Witney USA   1974
Trina Parks, Edna Richardson, Bettye Sweet, Shirley Washington, Roger E. Mosley, Christopher Joy, Stan Shaw, DeWayne Jesse, Charles Knapp, Dick Miller, Edward Marshall, Milt Kogan, Norman Bartold, Gene Simms, Sam Laws
The varied encounters of an all-black group of female singers as they hit the road form the subject of this hit-and-miss comedy, which claims to be a satire on the contemporary scene. It fails to make any meaningful contribution to race relations on account of its stereotyping of whites as either moronic imbeciles or racists, and is best enjoyed for its energy if not for its wit.
Aka: GET DOWN AND BOOGIE
COM 84 min (ort 93 min) VIDrel: SUPVID/RTM V

DARLING ** **   15
John Schlesinger UK   1965
Julie Christie, Dirk Bogarde, Laurence Harvey, Jose-Luis De Villalong, Alex Scott, Roland Curram, Basil Henson, Helen Lindsay, Tyler Butterworth, Hugo Dyson, Pauline Yates, Peter Bayliss, Ernest Walder, Lucille Soong, Sidonie Bond
A cynical portrait of a young London fashion model who decides to climb the social ladder rather quickly, by way of a few beds. Schlesinger's direction is alternately flashy and perceptive, but Christie is ravishing as the hard-bitten girl who winds up with an Italian nobleman after a few false starts. AA: Actress (Christie), Story/Screen (Frederic Raphael), Cost (Julie Harris).
DRA 122 min (Cut at film release – ort 127 min) B/W
VIDrel: WHV V/h

D'ARTAGNAN'S DAUGHTER ** 15
Bertrand Tavernier FRANCE 1994
Sophie Marceau, Philippe Noiret, Claude Rich, Sami Frey, Jean-Luc Bideau, Raoul Billerey, Charlotte Kady, Nils Tavernier, Jean-Paul Roussillon, Luigi Proietti, Pascale Roberts, Emmanuelle Bataille, Christine Pignet
Adventure-comedy set in the reign of Louis XIV in 1654, when Eloise (whose father is D'Artagnan of Three Musketeers fame) sees the Mother Superior killed by a soldier at the convent where she is staying as a guest. She sets off for Paris, intending to enlist her father's help in avenging this crime. A colourful swash-buckler of gloomy settings and verbose scripting that quickly loses any excitement the story initially generates.
Aka: LA FILLE DE D'ARTAGNAN
A/AD 124 min (ort 130 min) wScrn
VIDrel: ARTIF/20TH; ENCORE (LV only) V LV

DATE WITH AN ANGEL ** PG
Tom McLoughlin USA 1987
Michael E. Knight, Phoebe Cates, Emmanuelle Beart, David Dukes, Phil Brock, Albert Macklin, Pete Kowanko, Vinny Argiro, Bibi Besch, Cheryl A. Pollak, Steven Banks, Charles Lane, J. Don Ferguson, Bert Hogue, O'Clair Alexander
An angel crashes to Earth and is befriended by a musician, about to be wed and absorbed into his father-in-law's cosmetics company. Beart is suitably ethereal as the earthbound spirit who splashes into his pool on the night of his bachelor party, but the oh-so-cute sugary script offers little scope for comedy.
COM 102 min (ort 105 min) VIDrel: 20TH/TECH V/sur

DAUGHTER OF DARKNESS ** 18
Stuart Gordon USA 1990
Mia Sara, Robert Reynolds, Anthony Perkins, Jack Coleman, Dezso Garas, Erika Bodnar, Mari Kiss, Ferenc Nemethy, Istvan Hunyadkurthy, Kati Rak
A young woman journeys to Romania in search of a father she never knew, and in the hope of learning the meaning of her recurrent nightmares. She finds a colony of vampires have been awaiting her arrival, and that she is to be their key to eternal life. An unconvincing horror yarn that gets a touch more impact thanks to a strong performance from Perkins.
HOR 89 min (ort 96 min) mTV VIDrel: MIA/DISC V

DAUGHTER OF THE STREETS ** 18
Ed Sherin USA 1989
Jane Alexander, Roxana Zal, John Stamos, Harris Yulin, Luke Zimmerman, Martha Scott, Brandon Maggart, Peter White, Brynn Horrocks, Randy Brooks, Erika Eleniak, Felton Perry, Lorinne Vozoff, Michael Medeiros, Matt Landers
Teenage prostitution is the social evil on offer here in this predictable TV drama that details the desperate efforts made by a mother to save her daughter. Allegedly based on a true story, this is a preachy melodrama – earnest, intense and self-righteous.
DRA 90 min (ort 100 min) mTV VIDrel: 20TH/TECH V

DAUGHTERS OF DARKNESS *** 18
Harry Kuemel BELGIUM/FRANCE/ITALY/WEST GERMANY 1971
Paul Esser, John Karlen, Delphine Seyrig, Daniele Ouimet, George Jamin, Andrea Rau, Joris Collet, Fons Rademakers
A female vampire and her companion come to a Belgian seaside resort in the off-season, and become erotically involved with a young couple. A mixture of absurd humour and good photog-raphy places this film a notch or two above the average for this genre.
Aka: BLUT AN DEN LIPPEN; CHILDREN OF THE NIGHT; ERZEBETH; LA ROUGE AUX LEVRES; LES LEVRES ROUGE; PROMISE OF RED LIPS; RED LIPS, THE
HOR 95 min VIDrel: TART/20TH V/dm

DAUGHTERS OF THE DUST *** PG
Julie Dash USA 1991
Cora Lee Day, Barbara-O, Alva Rogers, Adisa Anderson, Cheryl Lynn Bruce, Trula Hoosier, Umar Abdurrahamm, Kaycee Moore, Eartha D. Robinson, Bahni Turpin, Tommy Hicks, Malik Farrakhan, Cornell Royal, Vertamae Grosvenor, Sherry Jackson
Slow-paced but fascinating American Playhouse account of the life of a family of West African descent living on an isolated island off the Georgia coast in 1902, where traditional ways and even the local Gullah dialect have been preserved unchanged.
DRA 107 min (ort 113 min) mTV VIDrel: CONNO/RTM V

DAVE *** 15
Ivan Reitman USA 1992
Kevin Kline, Sigourney Weaver, Frank Langella, Kevin Dunn, Ben Kingsley, Charles Grodin, Ving Rhames, Faith Prince, Anna Deavere Smith, Laura Linney, Parley Baer, Stefan Gierasch, Tom Dugan, Oliver Stone, Arnold Schwarzenegger
A presidential lookalike gets called in to impersonate the President when the latter suffers a stroke and incapacitated. However, he excels so well in this role that he is able to thwart a nasty plot and even repair the somewhat frosty relationship with the First Lady. A likeable comedy that inevitably invites comparison with films by Preston Sturges or Capra, but made plausible by some very fine performances, especially Kline.
COM 105 min (ort 110 min) cC VIDrel: WHV V/sur

DAVID *** 15
John Erman UK/USA 1988
Bernadette Peters, John Glover, Matthew Laurance, George Grizzard, Dan Lauria, Christopher Allport, Georgann Johnson, Jordan Charney, Cheryl Anderson, Alexandra Borrie, Frederick Combs, Jack Rader, Lisa Blake Richards
A very determined woman dedicates her life to the welfare of her son, whose mad father tried to burn him to death. An absorbing and well-acted drama that is based, believe it or not, on real-life events as detailed in Rothberg's book. The script is by Stephanie Liss.
DRA 94 min (ort 100 min) mTV VIDrel: ITC/POLYREC V/h
Boa: book by Marie Rothberg and Mel White.

DAVID AND THE MAGIC PEARL ** (U)
Wieslaw Zieba GIBRALTAR/POLAND/SWEDEN 1987
A Chicago kid taken on a jungle vacation by his father, has not only to cope with his new and unfamiliar surroundings, but also gets involved in helping a group of small aliens in their search for a magic pearl which can provide its owner with all the knowledge in the universe. A pleasant kids' adven-ture.
ANIM 68 min (ort 75 min) SATrel: MOVIE CHANNEL

DAVID COPPERFIELD **** U
George Cukor USA 1935
Lionel Barrymore, Maureen O'Sullivan, Madge Evans, Edna May Oliver, W.C. Fields, Lewis Stone, Frank Lawton, Freddie Bartholemew, Elizabeth Allan, Roland Young, Basil Rathbone, Hugh Williams, Elsa Lanchester, Jean Cadell, Jessie Ralph
Despite the worst efforts of a cruel stepfather, a young orphan grows up to become a successful author and marry his child-hood sweetheart. A lavish and literate film that is easily one of the very best Hollywood adaptations of a Dickens novel. Fine performances and unusually good casting (Fields is a memo-rable if not entirely accurate Micawber) are complemented by an excellent script (the work of Howard Estabrook and Hugh Walpole).
DRA 124 min (ort 132 min) B/W VIDrel: MGM/WHV V
Boa: novel by Charles Dickens.

DAVID COPPERFIELD ** U
UK 1993
Voices of: Sheena Easton, Kelly Le Brock, Julian Lennon, Howie Mandel, Michael York
A musical animation of the Dickens tale that replaces the human characters with animals. Here, David is a kitten sent to work in a cheese factory, where he sees brutality and exploitation at first hand.
ANIM 90 min (ort 92 min) mTV VIDrel: ABBEY/VCC V
Boa: novel by Charles Dickens.

DAVID'S MOTHER *** 15
Robert Allan Ackerman USA 1993
Kirstie Alley, Sam Waterston, Stockard Channing, Chris Sarandon, Philicia Rashad, Amanda Blitz, Steven Ivany, Blake Dennis, Renessa Blitz, Nicole Greenspan, Caoline Yeager, Bob Zidel, Nicole Davis, Brandon Roy, Ian Roger
A single mother raises an autistic child and makes this her whole life, but is naturally upset when a special school is suggested as a better alternative. Having alienated everyone with her sarcasm and bitterness, matters do not look promising, but events take a turn for the better when her sister fixes her up with a blind date. A good solid drama, with the three leads working well together to maintain interest.
DRA 92 min mTV VIDrel: ODY/SONOP V/sh

DAVY CROCKETT ★★★ U
Norman Foster USA 1955
Fess Parker, Buddy Ebsen, Basil Ruysdael, Hans Conried, Nick Cravat, William Bakewell, Kenneth Tobey, Pat Hogan, Helene Stanley, Don Megowan, Jim Maddux, Mike Mazurki, Jeff Thompson, Henry Joyner, Robert Booth, Eugene Brindel
Story of the famous Tennessee hunter and Indian scout, who died during the fighting between Texas and Mexico in the Battle of the Alamo. Originally made in three 50-minute parts for a Disney TV show but later released as a feature film. A hugely enjoyable romp with Parker making an admirable Crockett, and enjoying numerous adventures in the company of his buddy George Russel (Ebsen) before meeting his end at the Alamo. "Davy Crockett And The River Pirates" followed.
Aka: DAVID CROCKETT, KING OF THE WILD FRONTIER
JUV 86 min (ort 88 min) mTV cC VIDrel: WDV/TECH V

DAWN RIDER, THE ★★ U
Robert North Bradbury USA 1935
John Wayne, Marion Burns, Yakima Canutt, Reed Howes, Denny Meadows (Dennis Moore), Bert Dillard, Jack Jones, James Sheridan
A man sets out to capture the robber who killed his father, but his mission becomes complicated when he falls for the latter's pretty sister. A very typical "Lone Star" vehicle for Wayne, this early picture has all the flaws and conventions of B-Westerns of this period. Watchable if not exactly outstanding. It was later remade as "Western Trails".
WES 56 min B/W VIDrel: CREMED/LABY V

DAY AFTER, THE ★★★ 15
Nicholas Meyer USA 1983
Jason Robards, JoBeth Williams, Steve Guttenberg, John Lithgow, John Cullum, Amy Madigan, Bibi Besch, Lori Lethin, Jeff East, Georgann Johnson, William Allen Young, Calvin Jung, Lin McCarthy, Dennis Lipscomb, Clayton Day, Doug Scott
The effects of a full-scale nuclear war are seen through the eyes of those living in a small Kansas town. As a statement on the insanity of nuclear war the film was a complete failure, for the effects of the war are so sanitised as to leave no real trace of the true scale of horror. Where the film is at its best is in the opening stages, after which it's downhill all the way. See also TESTAMENT and THE LAST WAR.
DRA 126 min mTV VIDrel: 20TH/TECH L/A V

DAY FOR NIGHT ★★★ PG
Francois Truffaut FRANCE/ITALY 1973
Jacqueline Bisset, Valentina Cortese, Jean-Pierre Leaud, Francois Truffaut, Jean-Pierre Aumont, Alexandra Stewart, Jean Champion, Nathalie Baye, Dani, Bernard Menez, Jean-Francois Stevenin, Nike Arrighi, Gaston Joly
This film, about the making of a romantic film in Nice, cleverly explores all the behind-the-scenes clashes of personality. Truffaut effectively plays himself. AA: Foreign.
Aka: LA NUIT AMERICAINE
DRA 111 min VIDrel: MGM/WHV L/A V

DAY IN OCTOBER, A ★★★ 15
Kenneth Madsen DENMARK/USA 1992
D.B. Sweeney, Kelly Wolf, Daniel Banzali, Ole Lemmeke, Tovah Feldschuh, Kim Romer, Anders Peter Bro, Lars Olof Larsen, Morten Suurballe, Jens Arentzen, Lily Weiding, Arne Hansen, Jorgen Teytaud, Jorgen Bidstrup, Dale Levett
A wounded Danish Resistance fighter is given refuge by a young Jewish woman and her family and later helps them escape to Sweden along with the rest of the country's Jews. A low-key and fact-based history lesson that is solidly made and quite watchable if a little slow. The Danish locations help to maintain a good air of authenticity. Screenplay is by Damien F. Slattery.
DRA 96 min VIDrel: TRANSAT/HIFLI V/sur

DAY IT CAME TO EARTH, THE ★ PG
Harry Z. Thomason USA 1977
Wink Roberts, Roger Manning, Bob Ginnaven, Delight DeBruine, George Gobel, Rita Wilson, Ed Love, Bill Elfstrom, Bill Eubanks, Lyle Armstrong, Joe Barone, Conrad Rothman, Lou Hoffman, LeRoy Slaughter, Bill Boren, J.W. Best
A meteor crashes into a lake containing the body of a gangster in a concrete overcoat, and re-animates his corpse, which then goes on the obligatory revenge rampage. A trite and dull offering.
FAN 83 min (ort 89 min) VIDrel: FIRC/RTM V

DAY OF ATONEMENT ★★ 18
Alexandre Arcady USA 1992
Christopher Walken, Jill Clayburgh, Jennifer Beals, Roger Hanin, Richard Berry
After ten years in prison, a drugs lord comes out of prison and discovers that both his son and his nephew are involved in various aspects of this business. An average crime tale, set in Miami.
A/AD 96 min (ort 127 min) VIDrel: GUILD/SONOP V

DAY OF RECKONING ★★ (12)
Brian Grant USA 1994
Patrick Bauchau, Fred Dryer, Gerard Isamel, Geoffrey Lewis, Julio Oscar Mechoso, Assumpta Serna, Cary-Hiroyuki Tagawa, Meg Wittner, Prachuab Knachalarp. Anne Seekaew, Sumit Schathep, Brad Koerner, Karin Chandrasma, Chee Vimol
A former Special Forces agent has turned away from his old life, become a Buddhist and now works as a tour guide running parties to remote locations in Thailand. But a former adversary blackmails him into setting off on a dangerous mission to locate some ruby mines in Burma, and he finds that his old combat skills are now sorely needed. Adequate.
THR 89 min SATrel: SKY MOVIES

DAY OF THE BEAST, THE ★★★ 18
Alex de la Iglesia ITALY/SPAIN 1995
Alex Angulo, Armando De Razza, Santiago Segura, Terele Pavez, Nathalie Sesena, Jaime Blanch, Maria Grazia Cucinotta, Gianni Ippoliti, El Gran Wyoming, David Pinilla, Antonio Dechent, Ignacio Carreno, Saturnino Garcia, Pololo
After twenty-five years a theology professor finally learns the secret of the Apocalypse of St John, and realises that all the signs point to the imminent birth of the Antichrist. He sets off in search of suitable individuals to help him contact the Devil, his intention being to learn the birthplace of this being and save mankind from destruction. A darkly comic parody of all those horror films dealing with the Antichrist, grotesque, witty and inventive.
Aka: EL DIA DE LA BESTIA
COM 110 min VIDrel: TART/20TH V/sh

DAY OF THE DEAD ★★ 18
George A. Romero USA 1985
Lori Cardile, Terry Alexander, Joseph Pilato, Jarlath Conroy, Antone DiLeo Jr, Richard Liberty, Howard Sherman, G. Howard Klar, John Amplas, Ralph Marrero, Philip G. Kellams, Taso N. Stavrakis, Gregory Nicotero, Jeff Hogan
The second sequel to the original 1968 NIGHT OF THE LIVING DEAD that tells of an Earth dominated by flesh-eating zombie hordes. In this film a female scientist takes shelter with some army personnel in an underground bunker, and tries to study the creatures. A few heart-stopping moments cannot compensate for a sheer lack of ideas in this wordy sequel. See also ZOMBIES: DAWN OF THE DEAD.
HOR 100 min (Cut at film release) VIDrel: ARROW/RTM V

DAY OF THE JACKAL ★★ 15
Fred Zinnemann FRANCE/UK 1973
Edward Fox, Michel Lonsdale, Alan Badel, Eric Porter, Donald Sinden, Timothy West, Jean Martin, Tony Britton, Cyril Cusack, Delphine Seyrig, Derek Jacobi, Ronald Pickup, Adrien Cayla-Legrand, Olga Georges-Picot, Jean Sorel, David Swift
Incredibly wooden rendition of what could have been quite gripping; the story of a lone killer who intends to assassinate President de Gaulle, and the police operation that is mounted to prevent it. Interestingly, the killer is shown getting a forged Danish passport but winds up with a name that could only be Swedish: Per Lundquist. So much for verisimilitude.
THR 137 min (ort 142 min)
VIDrel: 4-FRONT/POLYREC/CIC V/h
Boa: novel by Frederick Forsyth.

DAY OF THE TRIFFIDS, THE ★ 15
Steve Sekely UK 1962
Howard Keel, Janette Scott, Nicole Maurey, Kieron Moore, Mervyn Johns, Ewan Roberts, Alison Leggatt, Janita Faye, Alexander Knox, Geoffrey Matthews, Gilgi Hauser, Carol Ann Ford, Katya Douglas, Victor Brooks, Thomas Gallagher
Plodding workmanlike adaptation of a science fiction tale about what happens when the world, with the bulk of humanity rendered blind, is taken over by semi-intelligent plants.
FAN 94 min wScrn VIDrel: SECOND/RTM V
Boa: novel by John Wyndham.

DAY THE EARTH STOOD STILL, THE ****   U
Robert Wise USA   1951
Michael Rennie, Patricia Neal, Hugh Marlowe, Billy Gray, Sam Jaffe, Frances Bavier, Drew Pearson, Frank Conroy, Carleton Young, Fay Roope, Edith Evanson, Robert Osterloh, Tyler McVey, James Seay, John Brown, Lock Martin (Gort)
A unusual and enjoyable film about an alien who arrives on Earth, to offer mankind the benefit of his race's wisdom, in the form of a powerful robot that can end all wars by punishing aggressors. Shot upon landing, he escapes from custody and hides out in a boarding house, befriending the young son of one of the lodgers. A rarity – an intelligent science fiction film. The script is by Edmund H. North and the effective score is by Bernard Herrmann.
FAN   88 min (ort 92 min) B/W   VIDrel: 20TH/TECH   V/h
Boa: story Farewell to the Master by Harry Bates.

DAY THE SUN TURNED COLD, THE ****   (12)
Yim Ho CHINA   1994
Si Ching Gao Wa, Tao Chuung Wa, Ma Jing Wu, Wu'ai Zi, Shu Zi'ong, Li Hu, Zhao Na Na, Zhang Xue, Wei Pang, Song Yu, Chu Lin, Lu Qi Feng, Ming Yi, Li Wen Fu, Yuan Xiao Jun, Zhao Nai Xun, Wu Gui Lin
In China, a young welder turns his mother in to the police for the murder of his father, an event that took place years ago, after which she remarried. A series of flashbacks now examine the background to this crime, and we learn much about the characters involved, their motivations, petty jealousies and failings. A tightly directed film, its various strands are given space to develop, and finally woven into a memorable whole.
Aka: TIANGUO NIEZI
DRA   100 min   SATrel: MOVIE CHANNEL

DAY THE WORLD ENDED, THE **   PG
Roger Corman USA   1956
Richard Denning, Lori Nelson, Paul Birch, Mike "Touch" Connors, Adele Jergens, Raymond Hatton, Paul Dubov, Jonathan Haze, Paul Blaisdell
A nuclear war ends life as we know it and the radiation turns most of the survivors into horrible mutants. Meanwhile, three unaffected souls are holed up in a cabin in the mountains, but find their sanctuary invaded by a couple of gun-toting thugs. Produced by Corman, this modest SF yarn concentrates on the interplay of the human characters and for all its lack of big budget effects, is surprisingly watchable. Remade as IN THE YEAR 2889.
FAN   79 min (ort 82 min) B/W   VIDrel: HEND/BMGREC L/A   V

DAYBREAKERS, THE ***   U
Robert Totten USA   1979
Glenn Ford, Tom Selleck, Sam Elliot, Jeffrey Osterhage, Ben Johnson, Gilbert Roland, John Vernon, Ruth Roman, Jack Elam, Gene Evans, L.Q. Jones, Paul Koslo, Mercedes McCambridge, Slim Pickens, Pat Buttram, James Gammon
An overlong rambling Western saga, based on two Louis L'Amour novels and telling of three brothers who set out to tame the Wild West just after the Civil War, both seeking their fortunes and avenging a murder out in New Mexico. The first TV adaptation of L'Amour's work and the first half of the complete feature, the other section being known as "The Shadow Riders".
Aka: LOUIS L'AMOUR'S THE SACKETTS; SACKETTS, THE
WES   106 min (ort 120 min); 188 min (2-cassette mini-series version) mTV   VIDrel: WHV   V
Boa: novels The Daybreakers and Sackett by Louis L'Amour.

DAYDREAM BELIEVER **   15
Kathy Mueller AUSTRALIA   1991
Miranda Otto, Martin Kemp, Anne Looby, Alister Smart, Gia Carides, Bruce Venables, Katie Edwards, Russell Kiefel, Kerry Walker, Keith Robinson, Dene Kermond, Adam Cockburn, Jason Meeth, Brian Blain, Howard Vernon, JUlie Godfrey
This bizarre romantic-comedy opens with a young girl being traumatised by her father's sexual attentions in their stables. Years later, she has grown into a disturbed woman who retreats into a "horse" identity whenever under stress. However, an accident with her car sets her up for a romantic encounter, and fate eventually takes a hand in resolving her problems. An awkward blend of comedy and pathos, one is never clear just what the film is trying to say.
COM   82 min (ort 86 min)   VIDrel: COLUM/SONOP   V/sh

DAYLIGHT ***   12
Rob Cohen USA   1996
Sylvester Stallone, Amy Brenneman, Viggo Mortensen, Dan Hedaya, Jay O. Sanders, Karen Young, Claire Bloom, Vanessa Bell Calloway, Barry Newman, Stan Shaw, Renoly Santiago, Colin Fox, Danielle Harris, Trina McGee-Davis
When there is an explosion in a tunnel that runs under the Hudson River, it is fortunate indeed that former emergency services boss (he was forced to resign in disgrace) is on hand, as he quickly comes up with a plan to reach a small band of survivors. As this involves him in moving through ventilation fans and overcoming other hazards to lead them to safety (a bit like in THE POSEIDON ADVENTURE) it offers ample excitement, albeit of the non-cerebral sort.
A/AD   115 min   VIDrel: PION   LV

DAYS, THE ***   (15)
Wang Xiaoshuai CHINA   1993
Yu Hong, Liu Xiaodong, Lou Ye, Wang Xiaoshuai, Chen Jie, Liu Zhongshan, Liu Xiaochun, Yang Jincheng, Liu Baoqin, Zhang Yanchun, Liu Suxian
This sad little film follows the course of a marriage breakdown, seen in retrospect as the husband looks back over the times he shared with his wife. Both are struggling artists/teachers who can barely earn enough to live on, but neither their mutual problems nor lovemaking are sufficient to halt what is in truth, just a growing sense of alienation from each other. Acting is first rate here, which does much to compensate for the slow pace and lack of a strong plot.
Aka: DONG-CHUN DE RIZI
DRA   75 min B/W   CINrel

DAYS OF BEING WILD ***   12
Wong Kar-Wai (Wang Jiawei) HONG KONG   1990/1991
Leslie Cheung (Zhang Guorong), Maggie Cheung (Zhang Manyu), Andy Lau (Liu Dehua), Carina Lau (Liu Jialing), Rebecca Pan (Pan Dihua), Jacky Cheung ((Zhang Xueyou), Tony Leung ((Liang Chaowei), Danilo Antunes, Hung Mei-Mei, Tita Munoz
In Hong Kong in the summer of 1960, a cynical young man lives a life of luxury, being kept by a wealthy courtesan, who indulges his every whim, except his desire to know the identity of his mother. Eventually, he tires of this life, and leaves for the Philippines where he hopes to discover the truth. A film that jumps about from scene to scene, in search of a story that is never made clear, but despite this, there are some moments of extraordinary power.
Aka: AHFEI ZHENJUANG
DRA   90 min (ort 94 min) wScrn   VIDrel: MADE/RTM   V

DAYS OF HEAVEN ***   PG
Terrence Malick USA   1978
Richard Gere, Brooke Adams, Sam Shepard, Linda Manz, Robert Wilke, Jackie Shultis, Stuart Margolin, Tim Scott, Gene Bell, Doug Kershaw, Richard Libertini, Frenchie Lemond, Sahbra Markus, Bob Wilson, Muriel Jolliffe, John Wilkinson
The story of a love triangle involving three young immigrants, from Chicago to the wheatfields of the mid-West. An engrossing and visually satisfying slice of early 20th century life, spoilt by a tendency towards symbolism. Shown originally in 70 mm; unfortunately TV will tend to minimise the fine photography of Alemandros. AA: Cin (Nestor Alemandros).
DRA   89 min (ort 91 min)
VIDrel: 4-FRONT/POLYREC/CIC L/A   V/h

DAYS OF THUNDER *   15
Tony Scott USA   1990
Tom Cruise, Nicole Kidman, Robert Duvall, Randy Quaid, Cary Elwes, Michael Rooker, Fred Dalton Thompson, John C. Reilly, Don Simpson, Caroline Williams, Donna Wilson, Chris Ellis, Peter Appel, Stephen Michael Ayres
From the makers of TOP GUN comes this similar film that tells a noisy, shallow and cliched tale, set in the world of stock car racing, with our obligatory kid from nowhere out to prove that he's the best. Good racing sequences are interspersed with the usual romantic sub-plot. The dreary script is by Cruise and Robert Towne.
A/AD   102 min (ort 108 min)   VIDrel: CIC/SONOP; PION (LV only)   V/sur LV

DAYS OF WINE AND ROSES ****   (15)
Blake Edwards USA   1962
Jack Lemmon, Lee Remick, Charles Bickford, Jack Klugman, Alan Hewitt, Tom Palmer, Debbie Megowan, Maxine Stuart, Katherine

Squire, Jack Albertson, Ken Lynch, Gail Bonney, Mary Benoit, Ella Ethridge, Rita Kenaston, Al Paige
An excellent, stark account of alcoholism and its consequences that pulls no punches, yet is refreshingly free of moralising or sentimentality. Lemmon gives one of his finest performances as the young adman who turns to booze to relieve his stress, with Remick his loving wife who joins him in one drink too many. Easily the director's best film. See also CLEAN AND SOBER and MY NAME IS KATE. AA: Song ("Days of Wine and Roses" – Henry Mancini (m)/Johnny Mercer (l)).
DRA   117 min B/W   TVrel
Boa: TV play by J.P. Miller.

DAZED AND CONFUSED ***
Richard Linklater USA
18
1993
Jason London, Rory Cochrane, Sasha Jenson, Wiley Wiggins, Michelle Burke, Adam Goldberg, Anthony Rapp, Marissa Ribisi, Milla Javovich, Ben Affleck, Joey Lauren Adams, Matthew McConaughey, Shawn Andrews, Jeremy Fox, Deena Martin
A nostalgic look back at life in the 1970s that focuses on a group of Texan teenagers on the occasion of their last day at high school. There's not much by way of a plot but solid acting and an appropriate soundtrack keep the film ticking over well. Thanks to lavish praise from the critics, it achieved an unexpected theatrical release, despite being originally intended to go straight to video.
COM   98 min (ort 103 min) cC   VIDrel: CIC/SONOP V/sur

DAZZLE **
Richard A. Colla USA
15
1994
Lisa Hartman-Black, Cliff Robertson, James Farentino, Dixie Carer, Jeffrey Meek, B.D. Wong, Lisa Eilbacher, Natalija Nogulich, June Chadwick, Kim Johnston Ullrich, Michael Easton, Neil Duncan, Lydie Denier, David Wohl, Joel Polis
A woman fights to retain control of her late father's estate in this rather superficial story, based on the Krantz novel. All the expected intrigues, betrayals and romantic encounters are here, wrapped up in a syrupy script that offers several hours of undemanding entertainment.
Aka: JUDITH KRANTZ'S DAZZLE
DRA   184 min mTV   VIDrel: 4-FRONT/POLYREC/ODY V/sh
Boa: novel by Judith Krantz.

DEACON BRODIE ***
Philip Saville UK
15
1996
Billy Connolly, Catherine McCormanck, Patrick Malahide, Ewen Bremner, Lorcan Cranitch, Siobhan Redmond, Simon Donald, Alex Norton, Peter McNamara, Russell Hunter, Ralph Riach, Sally Dexter, Clive Russell
Period romantic drama set in Edinburgh in the 1780s, and based on the life of a notorious criminal, whose outward appearance as a respected cabinet-maker and local councillor was a front for his more nefarious activities. Eventually he is unmasked and sentenced to death. An entertaining story that in an inspired choice cast stage comic Connolly in the title role. The script is by Simon Donald.
DRA   90 min   VIDrel: POLY/POLYREC   V

DEAD, THE ****
John Huston UK/USA/WEST GERMANY
U
1987
Angelica Huston, Donal McCann, Marie Kean, Donel Donnelly, Helena Carroll, Cathleen Delany, Ingrid Craigie, Rachael Dowling, Dan O'Herlihy, Frank Patterson, Maria McDermottroe, Sean McClory, Katherine O'Toole
John Huston's last film as director is a slow, leisurely and well-crafted look at the lives and hopes of family and friends, gathered at a dinner party to celebrate the Epiphany in Dublin of 1904. The insubstantial nature of the narrative is less important – farewells are exchanged between the guests at the film's end, toasts are drunk to absent friends and this great director makes this film a fitting farewell of his own.
DRA   79 min (ort 83 min)   VIDrel: FIRST/SONOP   V/sur
Boa: short story by James Joyce.

DEAD AGAIN **
Kenneth Branagh USA
15
1991
Kenneth Branagh, Andy Garcia, Emma Thompson, Lois Hall, Richard Easton, Jo Anderson, Patrick Montes, Raymond Cruz, Robin Williams, Wayne Knight, Erik Kilpatrick, Patrick Doyle, Gordana Rashovich, Derek Jacobi, Obba Babatunde
A private detective is hired to help a woman amnesiac found in

the grounds of a school, and resorts to having her hypnotised in a bid to discover her identity. Under hypnosis, she reveals knowledge of a past existence as the murdered wife of an emigre German composer, who was executed for that crime in 1949. An overwrought and artificial film, done in a sub-Hitchcock style, its supernatural framing has little to do with its daft resolution.
THR   103 min (ort 108 min) Col/B/W   VIDrel: CIC/SONOP V/sur

DEAD AHEAD **
Stuart Cooper USA
15
1996
Stephanie Zimbalist, Peter Onorati
Outdoors survival saga that has Zimbalist defending life and limb when a bunch of bank robbers invade her family campsite, after she is left there alone with her two kids, her husband having gone back to the city following a row. Fortunately, she is an expert archer, and puts this skill to good use. Not a million miles away from THE RIVER WILD, this one went straight to video. Fair, if hardly original.
A/AD   88 min mTV   VIDrel: CIC   V

DEAD AIR **
Fred Walton USA
12
1994
Gregory Hines, Debrah Farentino, Beau Starr, Gloria Reuben, Michael Harris, Laura Harrington, Harold Ayer, Eric Boles, W. Earl Brown, Veronica Cartwright, Joe Colligan, John Hawkes, Ron Recasner, Milt Tarver, Steve Jackson Wilde
A night-time DJ finds himself receiving threatening calls from a deranged murderer who has murdered several of his women friends and arranged matters so as to throw suspicion onto him. Fortunately, he makes the acquaintance of a psychology student who helps him in his efforts to track down and capture this killer. A downbeat thriller that offers little that is new or fresh in this familiar genre.
THR   86 min (ort 91 min)   VIDrel: CIC/SONOP L/A   V

DEAD BADGE **
Douglas Barr USA
18
1994
Brian Wimmer, M. Emmet Walsh, Olympia Dukakis, James B. Sikking, Yaphet Kotto, Martha Dubois
A rookie is given the badge of a dead colleague, in a gesture that was intended as an honour, but finds that this has its disadvantages. Curious about the events that lead up to the man's death, he finds himself coming up against evidence of a major corruption scandal among his fellow officers.
DRA   89 min (ort 95 min) mCab   VIDrel: MED/COLUM V/sh

DEAD BEFORE DAWN ***
Charles Correll USA
15
1992
Cheryl Ladd, Jameson Parker, G.W. bailey, Kim Coates, Matt Clark, Keane Young, Stanley Anderson, Hope Lange, Jensen Dagett, Holis McCarthy, Debra Bluford, Ken Boehr, Andrew Gilchrist, Kimberly Horner, Kip Niven, Charles Gordon
The wealthy husband of an abused wife sues her for custody of their two children in retaliation when she tries to leave him and as the situation gets nastier he takes out a contract on her life. However, when she responds by faking her death (done on FBI advice) he takes matters even further by taking out contracts on her parents. A gripping drama that unbelievable as it may seem, was based on a true case. Screenplay is by John Ireland.
DRA   93 min mTV   VIDrel: ODY/SONOP   V/sh

DEAD BOLT **
Douglas Jackson USA
15
1991
Justine Bateman, Adam Baldwin, Chris Mulkey, Michele Scarabelli, Cyndi Press, Colin Fox, Amy Fulco, Ellen Cohen, Grififth Brewer, Isabelle Truchon, Mark Camacho, Anthony Sherwood, Gordon Masten, Shirley Merovitz, Don Jordan
A divorced medical student rents out a room to a young man with whom she becomes emotionally involved but discovers too late that he is a dangerous psychopath who has killed her ex-husband. After he imprisons her in a sealed room, she begins to fear for her life. Another stab at the room-mate from hell syndrome that was explored in SINGLE WHITE FEMALE, this is a film with the same tired and predictable resolution.
Aka: DEADBOLT
DRA   89 min (ort 95 min) mTV   VIDrel: FIRST/SONOP   V

DEAD BY SUNSET ***
Karen Arthur USA
18
1995
Annette O'Toole, Ken Olin, Lindsay Frost, John Terry, Sally Murphy, Titan Crawford, Cody Crawford, Clay Mallensak

A fact-based story telling of an abusive adulterer and his new wife, a female doctor with him he was having an adulterous affair before his first wife was murdered, allegedly by an unknown assailant. But under the pressure of a custody battle over his three sons he grows increasingly irrational, and his new wife begins to suspect that he may have murdered his first spouse. A tense courtroom thriller, highly engrossing and well put together.
THR   171 min mTV   VIDrel: ODY/SONOP   V/sh

DEAD CALM *** 15
Phillip Noyce USA   1989
Sam Neill, Nicole Kidman, Billy Zane, Rod Mullinar, Joshua Tilden, George Shertsov, Michael Long, Lisa Collins, Paula Hudson Brinkley, Shaun Cook, Malinda Butler
A claustrophobic thriller set on the open seas. A couple go sailing in a bid to come to terms with the death of their young son in a road accident. When they spot a man in a dinghy who claims to be the remaining survivor of a bout of food poisoning onboard a nearby schooner, they pick him up and the husband rows out to investigate. Not an especially plausible film, but one that maintains a good sense of tension.
THR   92 min (ort 97 min)   VIDrel: WHV   V/sur
Boa: novel by Charles Williams.

DEAD COLD ** 18
Kurt Anderson USA   1996
Lysette Anthony, Chris Mulkey, Peter Dobson, Alina Thompson, Michael Champion
In an attempt to recover from a carjacking, a couple take a break in the Sierra mountains of Nevada. But when they find and rescue a half-frozen stranger, he recovers only to subject them to a torrent of abuse and violence.
DRA   88 min (ort 91 min)   VIDrel: FIRST/SONOP   V

DEAD END **** PG
William Wyler USA   1937
Sylvia Sidney, Joel McCrea, Humphrey Bogart, Claire Trevor, Wendy Barrie, Marjorie Main, Huntz Hall, Leo Gorcey, Gabriel Dell, Ward Bond, Billy Halop, Bernard Punsley, Allen Jenkins, Bobby Jordan, James Burke, Charles Peck
Excellent study of slum life in New York, with Bogart playing a hoodlum who gets involved with a gang of street kids on a trip back to see his mother and girlfriend. The first film to introduce the "Dead End Kids" who appeared in the original Broadway production and later played in the classic ANGELS WITH DIRTY FACES. Scripted by Lillian Hellman with sets designed by Richard Day.
Aka: DEAD END: CRADLE OF CRIME
DRA   92 min B/W   VIDrel: VCC/DISC   V
Boa: play by Sidney Kingsley.

DEAD END BRATTIGAN ** PG
Gus Trikonis USA   1990
Bruce Greenwood, Jessica Steen, Gregg Henry, Donald Moffat, Aharon Ipale, Harris Laskawy, Robert Ho, Gianni Russo, Paul Guilfoyle, H. Richard Greene, Elizabeth Hoffman, Freddye Chapman, Don Keith Opper, Lloyd Alan, Yomi Perry
A prize-winning journalist soon becomes jaded when he returns to his job on a dull newspaper, which he only got by taking the paper's owner to court. The latter responds by giving the journalist the task of dealing with stories that were never published, but this seemingly pointless job soon involves our reporter in unearthing a major scandal. A strictly mTV assembly-line affair.
A/AD   93 min mTV   VIDrel: COLUM/SONOP   V

DEAD IN THE WATER * 15
William Condon USA   1991
Bryan Brown, Teri Hatcher, Anne De Salvo, Veronica Cartwright, Pruitt Taylor Vance, Seymour Cassel, Ann Thomson, Ron Karabatos, Daniel Reichert, Michael Kaufman, Tim Haldeman, Tom Wright, Ralph Oliver, Eric Christmas, C.H. Evans
A semi-comic thriller with a film noir flavour that has Brown cast as an ambitious and amoral lawyer who has charmed his way to the top of his profession, and now turns to murder to inherit his wife's fortune. As expected, he eventually reaps the full reward of his villainy, but by the time this happens, all interest in this slack and derivative effort has evaporated.
THR   86 min (ort 90 min)   VIDrel: CIC/SONOP   V

DEAD INNOCENT * 18
Sarah Botsford USA   1996
Genevieve Bujold, Graham Greene, Nancy Beatty, Emily Hampshire

A successful defence lawyer is horrified to discover her maid murdered and her daughter missing, and it is not long before she finds that she has become a target too, with her daughter (who has been kidnapped) being held in an attempt to force her to commit suicide. This strained plot might have generated some tension, it really is too bad that it is not developed in a way that gives the actors any scope. Very poor.
THR   92 min   VIDrel: THIRD   V

DEAD MAN ** 18
Jim Jarmusch GERMANY/USA   1995
Johnny Depp, Crispin Glover, Gibby Haines, George Duckworth, Richard Boes, John Hurt, John North, Robert Mitchum, Mili Avital, Peter Schrum, Gabriel Byrne, Lance Henriksen, Michael Wincott, Eugene Byrd, Gary Farmer, Thomas Bettles
After the death of his parents, a man by the name of William Blake travels to the Western town of Machine, where he has a job waiting for him as an accountant in a factory. However, he gets there a month too late and finds that the job has been filled. This is just the start of a series of events that transform him into an outlaw. An infuriatingly opaque Western, replete with New Age overtones that are hard to understand and even harder to sit through.
WES   120 min B/W wScrn   VIDrel: POLY/POLYREC   V

DEAD MAN WALKING **** 15
Tim Robbins USA   1995
Susan Sarandon, Sean Penn, Robert Prosky, Raymond J. Barry, R. Lee Ermey, Celia Weston, Lois Smith, Scott Wilson, Roberta Maxwell, Margo Martindale, Steve Boles, Barton Heyman, Nesbitt Blaisdell, Ray Aranha, Larry Pine
The true story of Sister Helen Prejean and the relationship she entered into with a covicted murderer being held on death-row, the story concentrating on her growing personal involvement in the days leading up to his execution. Scripted by Robbins, this is a remarkably mature and perceptive work. Penn and Sarandon both give outstanding performances. See also A SHORT FILM ABOUT KILLING, LAST LIGHT and KILLER: A JOURNAL OF MURDER. AA: B. Actress (Sarandon).
DRA   117 min (ort 122 min) cC   VIDrel: POLY/POLYREC   V
Boa: book by Sister Helen Prejean.

DEAD MAN'S FOLLY * PG
Clive Donner USA   1985
Peter Ustinov, Jean Stapleton, Tim Pigott-Smith, Constance Cummings, Susan Wooldridge, Jonathan Cecil, Christopher Guard, Nicollette Sheridan, Jeff Yagher
Another Agatha Christie whodunit set (as always) in an English manor house. This one stars Ustinov as the famous sleuth "Hercule Poirot" who gets involved in a series of real murders that take place in the course of a "murder hunt" party being held by an American novelist. As tiresome as it is contrived.
Aka: AGATHA CHRISTIE'S DEAD MAN'S FOLLY
DRA   90 min (ort 100 min) mTV   VIDrel: WHV   V/h
Boa: novel Dead Man's Folly by Agatha Christie.

DEAD MAN'S ISLAND * PG
Peter Hunt USA   1995
Barbara Eden, William Shatner, Morgan Fairchild, Christopher Atkins, Traci Lords, Christopher Cazenove, Roddy McDowall
A successful journalist gets an invitation to spend the weekend at the island home of a wealthy friend she has not seen for twenty-five years, but on arriving learns that he needs her help in catching a hired assassin. Endearingly daft, this weak film has little tension or invention, and the final twist is all too obvious.
THR   85 min   VIDrel: 20TH/FOXVID   V/sh

DEAD MAN'S REVENGE ** PG
Alan J. Levi USA   1993
Bruce Dern, Randy Travis, Michael Ironside, Vondie Curtis-Jackson, Vondie Curtis-Hall, Keith Coulouris, Daphne Ashbrook, Tobin Bell, John M. Jackson, Jack Rader, Melora Walters, Doug McClure, Randy Travis, Ping Wu, Robert Cornthwaite
A homesteader and his family are attacked by a criminal who wants to take over their land and only the former survives. Thrown into jail, he soon escapes and finds himself the prey of a bounty hunter in the pay of our villain. Another modern example of this genre, competently done without being all that exciting, although as ever, Dern is suitably menacing.
WES   87 min (ort 92 min)   VIDrel: CIC/SONOP   V

DEAD MEN DON'T WEAR PLAID *   PG
Carl Reiner USA   1982
*Steve Martin, Rachel Ward, Carl Reiner, Reni Santoni, George
Gaynes, Frank McCarthy, Adrian Ricard, Charles Picerni, Gene
Labell, George Sawaya, Britt Nilsson, Jean Beaudine, John Stuart,
Ronald Spivey, Bob Hevelone, Brad Baird*
A private eye spoof, with a weak story helped slightly by clips
from old 1940s thrillers. A kind of homage to the films of that
period, but after the first five minutes the joke begins to wear
thin. This was the last film for famed costume designer Edith
Head.
COM   84 min (ort 88 min) B/W
VIDrel: 4-FRONT/POLYREC/CIC   V/h

DEAD MOTHER *   18
Juanma Bajo Ulloa SPAIN   1993
*Karra Elejalde, Ana Alvarez, Lio, Silvia Marso, Elena Irureta, Ramon
Barea, Gregoria Mangas, Marisol Saes, Raquel Santamaria, Txarley
Llorente, Super Pake Pekao, Elemen Armengod, Juan Ignacio
Vinuales, Miguel Olmeda, Jose M. Sacristan*
A woman who witnessed the death of her mother at the hands
of a burglar has been rendered mute by this event, which has
so traumatised her that her intelligence is impaired too. Years
later she meets up with the culprit, but only he recognises her.
Nonetheless afraid that she may expose him, he resolves to
remedy this by kidnapping and killing her, but her abduction
results in a bond developing between them. Winner of the Best
Director Award at Montreal 1993.
Aka: LA MADRE MUERTA
THR   107 min wScrn   VIDrel: TART/20TH   V/sh

DEAD NEXT DOOR, THE *   18
J.R. Brookwalter USA   1986
*Peter Ferry, Michael Grossi, Jolie Jackunas, Robert Kokai, Bogdan
Pecic*
A scientist devises a virus that unfortunately proves capable of
re-animating the dead and turning them into flesh-eating
zombies, and the government is forced to set up a special army
squad to deal with this menace. Average.
HOR   85 min   VIDrel: SCEDGE/RTM   V

DEAD OF NIGHT *   PG
Alberto Cavalcanti/Charles Crichton/Basil Dearden/R.
Hamer UK   1945
*Michael Redgrave, Sally Ann Howes, Basil Radford, Naunton Wayne,
Mervyn Johns, Roland Culver, Mary Merrall, Frederick Valk, Renee
Gadd, Anthony Baird, Judy Kelly, Miles Malleson, Michael Allan,
Robert Wyndham, Googie Withers*
A classic collection of five horror tales, linked together by the
device of an architect's visit to a strange country house. On
arrival, he recognises all those present as part of a recurrent
nightmare, after which the various individuals relate their
stories. These are: "The Hearse Driver", "The Christmas Story",
"The Haunted Mirror", "The Golfing Story" and "The
Ventriloquist's Dummy", numbers three and five being the best
stories in the set.
HOR   103 min B/W   VIDrel: LUMI/SPEAR   V
Boa: short stories by H.G. Wells, E.F. Benson, John Baines and
Angus McPhail.

DEAD ON *   18
Ralph Hemecker USA   1993
*Tracy Scoggins, Matt McCoy, Shari Shattuck, David Ackroyd,
Thomas Wagner, Trisha Melynkov, William Anton, Lynn Oddo,
Virginia Watson, Roger Rose, Julie Gregg, Harris Laskaway, Mindy
Seeger, Rochelle Swanson, Christopher Carroll*
A couple meet and initiate a passionate affair and in their desire
to be together they find the best way for them to terminate their
unhappy marriages is to murder each other's spouse. Excessive
concentration on the obligatory sex scenes and lack of attention
to the plot are the main weaknesses of this far from enthralling
offering.
THR   87 min (ort 92 min)   VIDrel: ODY/SONOP   V/sh

DEAD PIT, THE *   18
Brett Leonard USA   1989
*Jeremy Slate, Cheryl Lawson, Steffen Gregory Foster, Danny
Gochnauer, Geha Getz, Joan Betchel, Mara Everett, Randy Fontana,
Michael Jacobs, Jack Sunsari, Frederick Dodge, Neltie Heffner, Luana
Speelman, Skyy Diaz, Steven Strom*
Low-grade tale of a sinister mental home at which strange exper-
iments in a secret chamber were once performed, which reduced
their victims to a zombie-like state. A beautiful female amnesia

victim and other inmates find themselves being stalked by a
deadly killer after this chamber is suddenly exposed during an
earthquake. A dire and derivative effort of little originality and
less merit.
HOR   97 min (ort 102 min)   VIDrel: POPRO/RTM   V

DEAD POETS SOCIETY *   PG
Peter Weir USA   1989
*Robin Williams, Robert Sean Leonard, Ethan Hawke, Josh Charles,
Gale Hansen, Dylan Kussman, Allelon Ruggiero, James Waterston,
Norman Lloyd, Kurtwood Smith, Colin Irving, Laura Flynn Boyle,
Melora Walters, Welker White*
Set at a stuffy New England prep school in the 1950s, where a
newly-arrived and unconventional English teacher inculcates a
love of poetry and intellectual debate among his students, but
not always with the most appropriate results. Melodramatic and
over-wrought, but an impressive performance from Williams
gives this contrived film considerable impact. AA: Screen/orig
(Tom Schulman).
DRA   130 min   VIDrel: TOUCH/TECH   V/sur

DEAD POOL, THE *   18
Buddy Van Horn USA   1988
*Clint Eastwood, Patricia Clarkson, Evan C. Kim, Liam Neeson, David
Hunt, Michael Currie, Michael Goodwin, Darwin Gillett, Anthony
Charnota, John Allen Vick, Christopher Beale, Jeff Richmond, Patrick
Van Horn, James Carrey*
The fifth DIRTY HARRY movie in the series, has our tough
police detective investigating a strange hit-list, with murder
victims apparently linked to a bizarre bet among the members
of a film crew. This well-tried formula puts Harry Callahan
through his paces once more, in an all too predictable if well
made thriller. Fortunately, Eastwood's strong screen presence
makes it watchable.
THR   87 min (Cut at film release by 12 sec – ort 91 min)
VIDrel: WHV   V/sur

DEAD PRESIDENTS *   18
Albert Hughes/Allen Hughes USA   1995
*Larenz Tate, Keith David, Chris Tucker, Freddy Rodriguez, Rose
Jackson, N'Bushe Wright, Alvaletah Guess, James Pickens Jr, Jenifer
Lewis, Clifton Powell, Elizabeth Rodriguez, Terrence Howard, Ryan
Williams, Larry McCoy*
A sequel of sorts to MENACE II SOCIETY that examines the
black American experience through the 1960s and 1970s, taking
as its focus the return of a man from the Vietnam War, where
he has to face up to some of the unresolved problems he left
behind while he was away fighting. A graphically violent film
with a derivative and unsatisfying story that seems full of anger
but has no clear direction. A few well mounted action sequences
provide a small compensation.
A/AD   117 min (ort 119 min) cC   VIDrel: HOLPIC/TECH
V/sur

DEAD RECKONING *   U
John Cromwell USA   1947
*Humphrey Bogart, Lizabeth Scott, Morris Carnovsky, William Prince,
Wallace Ford, Charles Cane, Marvin Miller, James Bell, George
Chandler, William Forrest, Ruby Dandridge, Lillian Wells, Charles
Jordan, Robert Scott*
Two WW2 veterans are on their way to Washington to be deco-
rated when one of them disappears and is found murdered.
Bogart as the tough ex-soldier investigating the death of his
partner is excellent, but this is an example of film noir at its
most opaque; a lighter touch was needed.
DRA   96 min (ort 100 min) B/W   VIDrel: COLUM/SONOP
V

DEAD RINGER *   PG
Paul Henreid USA   1964
*Bette Davis, Karl Malden, Peter Lawford, Philip Carey, Estelle
Winwood, Jean Hagen, George Macready, Cyril Delevanti, George
Chandler, Mario Alcalde, Monika Henreid, Ken Lynch, Bert Remsen,
Charles Watts, Charles Meredith*
Davis plays twin sisters, with the nasty one bearing a long-term
grudge over a man and shooting her other half to settle the
score. The remaining twin now takes over her erstwhile sister's
identity, and hatches a complicated scheme to have her revenge
against the man she feels wronged her. The implausible plot
doesn't matter, this is Davis' film all the way. Remade for TV as
"The Killer In The Mirror".
Aka: DEAD IMAGE
THR   111 min (ort 116 min) B/W   VIDrel: MGM/WHV   V

DEAD ROMANTIC * (PG)
Patrick Lau UK   1992
Janet McTeer, Clive Wood, Jonny Lee Miller, Elspet Gray, Simon Rouse, Robin Weaver, Bernice Stegers, Sarah Burghard, Diana Payan, Rowland Davies, Barbara Keogh, Ralph Arliss, Caroline O'Neill, Rupert Degas, Peter Moreton, Debbie Finch
A feeble murder mystery set in a small town, where a daffy English teacher becomes an object of desire on the part of both a colleague and a pupil. Meanwhile, there is a killer at large in the town, whose speciality is the murder of London prostitutes. Replete with clumsy false leads and contrived clues, one has little reason to watch this dud, as even the undoubted competence of the cast can do little when the script is so flawed.
THR   89 min mTV   TVrel: BBC
  Boa: novel by Simon Brett.

DEAD SILENCE ** 15
Peter O'Fallon USA   1991
Renee Estevez, Lisanne Falk, Carrie Mitchum, Steven Brill, Claudette Nevins, Tim Russ, Beau Starr, Brayn Cranston, Steven M. Gagnon, Al Ruscio, Andrew Zeller, Robert Ackerman, Laura Bastianelli, Bojesse Christopher
A few faint echoes of DIAMOND SKULLS are to be found in this thriller, in which three students celebrating their graduation with a drunken party are involved in a hit-and-run accident. Average.
THR   88 min mTV   VIDrel: 20TH/TECH   V

DEAD SLEEP ** (15)
Alec Mills AUSTRALIA   1991
Linda Blair, Tony Bonner, Andrew Booth, Christine Amor, Sueyan Cox, Suzie MacKenzie, Craige Cronin, Brian Moll, Vassy Cotsopoulos, Peta Downes, Alan Edwards, Pam Byde
A nurse goes to work in a psychiatric hospital whose chief doctor makes use of a controversial treatment that requires heavy sedation and the use of shock treatments. When the woman's former boyfriend apparently commits suicide after a course of this "therapy", she begins to suspect foul play, and learns that the doctor is not quite the dedicated medic he appears to be. A competent thriller, quite watchable but with little originality.
THR   86 min (ort 92 min)   SATrel: MOVIE CHANNEL

DEAD SPACE * 15
Fred Gallo USA   1991
Marc Singer, Laura Tate, Bryan Cranston, Judith Chapman
The commander of a spaceship responds to a desperate SOS sent out from a genetic research station, and upon arrival has to face the danger of a deadly virus which attacks the brain. A derivative SF shocker of limited appeal.
HOR   69 min   VIDrel: COLUM/SONOP   V

DEAD ZONE, THE *** 15
David Cronenberg USA   1983
Christopher Walken, Tom Skerritt, Herbert Lom, Martin Sheen, Brooke Adams, Anthony Zerbe, Colleen Dewhurst, Nicholas Campbell, Jackie Burroughs, Sean Sullivan, Geza Kovacs, Simon Craig, Barry Flatman, Raffi Tchalikian
A man awakens after a five-year coma to discover that he has psychic powers with which he can predict the future. He drifts through life becoming a recluse, but is galvanised into action when he discovers that a man running for the Senate will one day initiate a nuclear war as US President. A film that mixes despair with tension, with Walken giving a performance of anguished sincerity. Abridged before video submission by 13 sec.
THR   103 min   VIDrel: 4-FRONT/POLYREC   V/sur
  Boa: novel by Stephen King.

DEADFALL ** 18
Christopher Coppola USA   1993
Michael Biehn, Sarah Trigger, Nicolas Cage, James Coburn, Peter Fonda, Talia Shire, Charlie Sheen, Michael Constantine, Gigi Rice, Angus Scrimm, J. Kenneth Campbell, Marc Coppola, Micky Dolenz, Brian Donovan, Renee Estevez, Ted Fox
A father-and-son team of con-men plan an elaborate "sting" operation but this goes terribly wrong when the latter accidentally kills the former. He is now forced to team up with his uncle as things go from bad to worse, in this moody and implausible tale. Acted and directed in an excessive and unrestrained manner, this one has little to hold the interest.
DRA   95 min (ort 99 min)   VIDrel: TRIM/HIFLI   V/h

DEADLY ** 15
Ebsen Strom USA   1991
Jerome Ehlers, John Moore, Lydia Miller, Frank Gallacher, Caz Lederman
An investigator is sent to a police station to check the facts behind the apparent suicide by hanging of a man who was being held in custody there, but he uncovers a number of conflicting details.
DRA   101 min   VIDrel: HIFLI/SONOP   V/sur

DEADLY ADVICE ** 15
Mandie Fletcher UK   1994
Brenda Fricker, Jane Horrocks, Imelda Staunton, Jonathan Page, John Mills, Edward Woodward, Billie Whitelaw, Hywel Bennett, Jonathan Hyde, Ian Abbey, Jo Stone-Fewings, Eleanor Bron, Roger Frost, Gareth Gwyn-Jones, Richard Moore
Black comedy detailing the plans two sisters make to kill their domineering mother, helped by a brace of dead British murderers who have all manifested to give advice on this knotty problem, each figure offering suggestions in keeping with the manner of their own crimes. A spiteful and utterly bizarre fantasy, it could have worked beautifully had the script been sharper and done justice to the hard work from all concerned.
COM   86 min (ort 91 min)
VIDrel: CURZON/20TH   V/sur

DEADLY BET ** 18
Richard W. Munchkin USA   1991
Jeff Wincott, Charlene Tilton, Steven Vincent Leigh, Jerry Tiffe, Mike Toney, Michael Delano, Sherrie Rose, Ray Mancini, Eric Lee, Art Camacho, Patty Toy, Carl Butto, Cole McKay, Crete Kara, Tony Williams, Darryl Purpose
A tough gambler puts his life at risk when he places a bet on winning an illegal kickboxing contest, his intention being to start a new life with his fiancee far away from Las Vegas and the underground fighting circuit. When he loses the bet and is unable to pay, he faces a nasty showdown with the mobsters who controls the fighting racket. Another martial arts actioner, sporadically rousing, but generally both predictable and uninspiring.
A/AD   89 min (ort 93 min)   VIDrel: MIA/DISC   V/sh

DEADLY CHINA DOLLS ** 18
Godfrey Ho HONG KONG   199-
Miyamoto Yoko, Sibelle Hu, Maria Jo
This opens with a female assassin killing her gangster boss just after she has made love to him, it appearing to be the case that she is out to consolidate her position as number one. Meanwhile, an equally tough female CIA agent is out to put the damper on these proceedings. The plot gets more than a little complicated when the assassin teams up with another lady, and then finds that there is a third killer out to eliminate them. Fast-paced and mindless.
A/AD   92 min wScrn   VIDrel: EAST/DISC   V

DEADLY DECEPTION * 18
Anthony J. Loma FRANCE/SPAIN   1991
Andrew Stevens, Ivo Pogorelich, Lloyd Bochner, Anthony Eisley, Craig Hill
A private detective and a cop investigate the murder of a local businessman, and find that the trail leads them to a modelling agency and a seductive woman. One of those really dire erotic thrillers that are only worth seeing for making notes on how not to make a movie. There are a few unintentionally hilarious bits of dialogue, scant reward for sitting through the entire movie.
THR   92 min   VIDrel: OVER/HIFLI   V

DEADLY DECEPTIONS *** 15
John Llewellyn Moxey USA   1987
Matt Salinger, Lisa Eilbacher, Bonnie Bartlett, Mildred Natwick, Christopher Allport, James Noble, Robert Harper, Ethan Phillips, Phyllis St James, Jackson Davies, Annie Kidder, Janet Wright, Terrence Kelly, Dale Wilson, Antony Holland
A psychological drama that examines the inability of a man to come to terms with the apparent suicide of his wife, whilst at the same time conducting a frantic search for his missing baby son, presumed dead. Well made and taut, and Natwick gives a splendid performance in a one-scene cameo. Filmed on location in Vancouver, British Columbia.
Aka: DEADLY DECEPTION
DRA   90 min (ort 100 min) mTV   VIDrel: 20TH/TECH   V

DEADLY DREAM WOMAN * 18
Taylor Wong (Wong Ching) HONG KONG 1992
Cheung Man, Chingamy Yau, Jack Cheung, Ken Lo
A young woman becomes a Batman-style vigilante after her
father is harassed by criminals, but problems occur when she
develops amnesia after a fall. A watchable but poorly made
story, set in present-day Hong Kong.
MAR 90 min VIDrel: EAST/DISC V

DEADLY EXPOSURE ** 18
Lawrence Mortorff USA 1992
*Robby Benson, Laura Johnson, Pual Hampton, Andrew Prine, Bentley
Mitchum, Isaac Hayes*
A reporter investigates the murder of his father by right-wing
extremists and finds a trail that leads to the highest levels of
government.
THR 91 min (ort 100 min) VIDrel: HIFLI/SONOP V

DEADLY GAME *** PG
Marshall Brickmann USA 1986
*John Lithgow, Christopher Collett, Cynthia Nixon, Jill Eikenberry,
John Mahoney, Paul Austin, Sully Boyar, Richard Jenkins, Timothy
Carhart, Gregg edelman, John David Cullum, Abe Unger, Richard
Council, Robert Leonard*
A precocious kid steals some plutonium from a top-secret plant
and builds his own nuclear reactor to show how smart he is.
Needless to say, the security authorities do not see it this way
and he soon finds himself in trouble. A slick and well acted
effort, but one that inevitably annoys by its very lack of respon-
sibility, with our irritating kid ready to risk a nuclear
catastrophe just to make a point.
Aka: MANHATTAN PROJECT, THE; MANHATTAN PROJECT, THE: THE
DEADLY GAME
A/AD 107 min (Cut at film release by 12 sec – ort 120 min)
VIDrel: WHV V

DEADLY GAME ** 18
Thomas J. Wright USA 1991
*Jenny Seagrove, Roddy McDowall, Marc Singer, Michael Beck, John
Pleshette, Mitchell Ryan, Brittany Almond, Joseph Arias, Jerry
Basham, Larry Boothby, Richard Duran, Tony Enyart, Russ Fast,
Soren Forrest, Rick Jones, Lisa Sigel*
A horribly scarred and embittered millionaire invites a group
of people to his island to take part in a mysterious game, but his
true intentions are far from benign. A most gory and disagree-
able film, which blends elements of both TEN LITTLE INDIANS
and "The Hounds Of Zaroff" in an unimaginative and
predictable way.
THR 89 min (ort 93 min) mTV VIDrel: CIC/SONOP V

DEADLY HEROES * 18
Menahem Golan USA 1996
Michael Pare, Jan-Michael Vincent, Billy Drago, Claudette Mink
When terrorists take over an airline and demand the release of
their leader, a former Navy SEAL is given the job of handling
this crisis.
A/AD 102 min VIDrel: COLUM/SONOP V

DEADLY INNOCENTS ** 18
John D. Patterson USA 1988
*Mary Crosby, Amanda Wyss, Andrew Stevens, Bonni Hellman, John
Anderson*
A woman with a psychotic personality and a deep-seated hatred
of men makes friends with a young girl and shows how to make
the best of her looks. However, when she acquires a boyfriend,
her true nature begins to make itself known.
DRA 90 min (ort 100 min) VIDrel: BRAVE/SONOP L/A
V

**DEADLY INVASION: THE KILLER BEE
NIGHTMARE ***** (12)
Rockne S. O'Bannon USA 1994
*Robert Hays, Nancy Stafford, Ryan Phillippe, Gina Philips, Gregory
Gordon, Michael A. Nickles, Danielle Von Zerneck, Dennis
Christopher, Whitney Danielle Porter, Jeff Johnson, Mindy Lawson,
Donre Sampson, Carolyn Henensy, Thom Dillon*
Hays and his family try to escape from the title threat, these
beasties having invaded their neighbourhood. Quite a suspense-
ful movie, with writer/director O'Bannon pulling out all the
stops to provide a few good moments. See THE BIRDS for some-
thing similar (albeit a good deal less gripping).
Aka: DEADLY INVASION
THR 83 min (ort 90 min) mTV SATrel: SKY MOVIES

DEADLY MEDICINE * 15
Richard A. Colla USA 1990
*Veronica Hamel, Susan Ruttan, Stephen Tobolowsky, Scott Paulin,
Alan Fudge, Marnie McPhail, Joel Polis, Richard McKenzie, Peter
Vogt, Kim Shriner, Joshua Boyd, J.C. McKenzie, Marina Palmer, Ross
Evans, Nita Whitaker, Dale Harimoto*
A female paediatrician becomes the prime suspect when a series
of patients meet untimely ends at a clinic being run by her.
THR 89 min mTV VIDrel: GUILD/SONOP V/h

DEADLY NIGHTSHADE ** 15
Peter Levin USA 1995
Lindsay Wagner, Piper Laurie, Renee Humphrey
A girl is enticed into the world of drugs and prostitution and
her family makes a determined effort to locate her.
THR 87 min VIDrel: GUILD/FOXVID V

DEADLY RECALL ** 15
Joyce Chopra USA 1993
*Veronica Hamel, Stephen Collins, Dennis Farina, Stan Ivar, Leon
Russom, Bryan Cranston, David Steen, Alyson Reed, Tom Henschel,
James Posolof, Terry Finn, Angela Andrade, Joe Bordeaux, David
Boyce, Raye Brewer, Sean Comey*
A woman suffering from amnesia is recognised as a missing
state lawyer, but memory flashbacks continue to haunt her,
despite being reunited with her husband.
Aka: DISAPPEARANCE OF NORA, THE
THR 90 min (ort 106 min) mTV VIDrel: IMPENT V

DEADLY RELATIONS ** 15
Waris Hussein USA 1991
*Lindsay Wagner, David Dukes, Frances Sternhagen, Maureen
Meuller, Ron Frazier, Ben Savage*
A woman attempts to recover from a two-year coma, brought
about by a violent attack, but she now finds that her life is once
more in danger. A tolerable thriller, sustained by a strong cast.
THR 89 min VIDrel: CAPIT/GUILD V

DEADLY RELATIONS ** 15
Bill Condon USA 1992
*Robert Urich, Shelley Fabares, Gwyneth Paltrow, Anthony Higgins,
Georgia Emelin, Matthew Perry, Ted Marcoux, Roxana Zal, Julian
Boyd, Joy Farmer, Brett Rice, Ed Grady, Nick Searcy, Dan Biggers,
Tom Even, David Dwyer*
True story of a strict father who attempts to bring up his
daughters with his own conception of family values, doing
this by imposing on them a rigid and highly regulated lifestyle.
But as the girls grow, they begin seeking to live their own
lives, a matter our father is not too keen to encourage, even
though he is in reality a secretive (and possibly dangerous)
hypocrite.
THR 89 min mTV VIDrel: CIC/SONOP V
Boa: Deadly Relations: A True Story of Murder in a Suburban
Family by Carol Donahue and Shirley Hall.

DEADLY RIVALS ** 18
James Dodson USA 1992
*Andrew Stevens, Cela Wise, Margot Hemingway, Richard Roundtree,
Joe Bologna, Francesco Quinn, Randi Ingerman, Alan Landers, Jorge
Gil, Robert Mano, Peter Paul DeLeo, Ed Armatrudo, Anthony Giaimo,
Christopher Campbell, Jon Savich*
A laser specialist attending a conference in Miami meets an
attractive woman who is suspected of various criminal activi-
ties and this chance encounter involves him with some very
dangerous people. A routine actioner that offers few surprises.
A/AD 89 min (ort 93 min) VIDrel: 20VIS/SONOP V/sh

DEADLY SINS ** (18)
Michael Robison USA 1995
Alyssa Milano, David Keith, Terry David Mulligan
When a Catholic girls' school becomes the setting for the activ-
ities of a demented killer, a private investigator poses as a new
student in order to catch the person responsible. Adequate
thriller done in the Gothic style.
THR 94 min SATrel: SKY MOVIES

DEADLY SPAWN, THE * 18
Douglas McKeown USA 1983
*Charles George Hildebrandt, Tom de Franco, Jean Taffler, Richard Lee
Porter*
A meteorite brings a germ to Earth which grows into a three-

headed monster and proceeds to gobble up humans. An amateurishly made film, originally shot in 16 mm.
Aka: RETURN OF THE ALIEN'S DEADLY SPAWN
HOR   79 min (ort 90 min)   VIDrel: VIPCO/SGSVID   V/h

DEADLY STRIKE, THE **   18
Huang Lung HONG KONG   1979
Bruce Li (Ho Tsung-Tao), Chen Hsing, Tang Wei, Lung Fei, Choy Hong, Sze Chung Tien, Chu Li, Su Chin Ping, Li Min Liong, Tsang Chiu, Zheng Fu Yiong
Kung fu Western with a new law officer recruiting his men from the local jail, in order to break up a criminal gang. Fair.
Aka: BRUCE LEE: DEADLY STRIKE; YOUNG HERO
MAR   87 min Cut (1 min 23 sec – ort 92 min)
VIDrel: IMPENT   V

DEADLY SURVEILLANCE **   15
Paul Ziller USA   1991
Michael Ironside, Christopher Bondy, Susan Almgren, Vlasta Vrana, David Carradine, Christopher Ozores, Chip Ciuka, George Busa, Doris Malcolm, Roland Nincheri, Michael McGill, Norris Domingue, Irene Kessler, Tyrone Benskin
A variant on those "buddy cop" movies, this one has our police-duo a pair of squabbling misfits who interrupt a burglary only to find the young culprit about to make off with a stash of heroin. Having got him to return the drugs, they stake out the site hoping to catch bigger fry.
A/AD   89 min (ort 92 min)   VIDrel: FIRST/SONOP   V/s

DEADLY TAKEOVER ***   18
Rick Avery USA   1995
Jeff Speakman, Ron Silver, Rochelle Swanson
A plane carrying a group of international scientists is hijacked by a terrorist gang, who kill all the scientists and take their place. This enables them to gain access to a top secret research facility, their plan being to steal a deadly chemical with which to hold US government to ransom. Speakman is the tough ex-marine sergeant they come up against. A strong action film.
A/AD   90 min   VIDrel: 20TH/FOXVID   V/sh

DEADLY VOWS ***   15
Alan Metzger USA   1994
Gerald McRaney, Peggy Lipton, Josie Bissett, Michael MacRae, Venuz Terzo, Ric Reid, P. Lynn Johnson, Larry Musser, Venus Terzo, Roger Allford, Tim Henry, Roger Cross, Nathaniel DeVeaux, Wally Dalton, Eric Keenleyside, John Tench
The fantasies of a married factory-worker lead him into adultery and then a bigamous marriage when he meets a young girl he falls for. His thoughts now turn to murder as he considers which woman he is better off with. A good, solid drama, based on a real case.
DRA   85 min mTV   VIDrel: ODY/SONOP   V/sh

DEADLY WEAPON ***   15
Michael Miner USA   1988
Rodney Eastman, Kim Walker, Gary Frank, Gary Kroeger, Michael Horse, Joe Regalabuto, William Sanderson
A bullied teenager is satisfy his desire for revenge when he comes into possession of a newly-developed and highly experimental secret weapon. A low budget re-working of LASERBLAST, surprisingly effective and inventive.
FAN   86 min (ort 90 min)   VIDrel: 20TH/TECH   V

DEADLY WHISPERS **   15
Bill L. Norton USA   1994
Pamela Reed, Tony Danza, Ving Rhames, Heather Tom, Sean Haberle, Sal Landi, Camryn Manheimn, Ellen Dubin,, Michael P. McCarthy, Amanda Fuller, Kevin Marsh, Richard Gross, Maureen McVerny, Michael Foley, Fred D. Coleman, David Schiro
A strict but kindhearted father learns the terrible truth about his daughter's murder after he gets implicated in her murder investigation. Average.
DRA   90 min mTV   VIDrel: ODY/SONOP   V/sh
Boa: book by Ted Schwarz.

DEADRINGERS ***   18
David Cronenberg CANADA   1988
Jeremy Irons, Genevieve Bujold, Heidi Von Palleske, Barbara Gordon, Shirley Douglas, Stephen Lack, Nick Nichols, Lynn Cormack, Damir Andrei, Miriam Newhouse, David Hughes, Richard Farrell, Warren Davis, Jonathan Haley
A disquieting psychological thriller revolving round twin brothers (Irons in both roles) who work as doctors running a

gynaecological clinic. They share every experience in life until a woman comes between them and disturbs a hitherto cosy set-up. Said to be based on a true story.
THR   110 min (ort 117 min)   VIDrel: 20TH/TECH   V/sur

DEALERS **   15
Colin Bucksey UK   1989
Paul McGann, Rebecca DeMornay, Derrick O'Connor, John Castle, Paul Guilfoyle, Rosalind Bennett, Adrian Dunbar, Nicholas Hewetson, Sara Sugarman, Dikran Tulaine, Douglas Hodge, Annabel Brooks, Simon Slater
A high-flying businessman loses his position in the company, and as he fights to save his career he discovers that he is no more than a pawn in a power struggle. Fair.
A/AD   87 min (ort 110 min)   VIDrel: MGM/WHV   V/sur

DEAR DIARY ***   15
Nanni Moretti FRANCE/ITALY   1994
Nanni Moretti, Giovanna Bozzolo, Sebastiano Nardone, Antonio Petrocelli, Giulio Base, Jennifer Beals, Renato Carpentieri, Raffaella Lebboroni, Marco Paolini, Valerio Magrelli, Sergio Lambiase, Roberto Nobile, Gianni Ferraretto
Three-part study from Moretti in which his observations and travels form the subject matter of this amusing if rather idiosyncratic offering. Blending documentary and fiction, Moretti paints a thoughtful and quirky portrait of modern-day Italy, that stretches from Rome across to the Aeolian Islands, and little escapes his barbed insights. Mostly an effective and joyful work, even if the final section is an overly morbid exploration of his medical encounters.
Aka: CARO DIARIO
COM   100 min wScrn   VIDrel: ARTIF/20TH   V/sh

DEAR MR PROHACK *   U
Thornton Freeland UK   1949
Cecil Parker, Glynis Johns, Hermione Baddeley, Dirk Bogarde, Sheila Sim, Henry Edwards
A civil servant who works in the Treasury comes into a large inheritance, but finds that his skill in handling the financial affairs of the state is not matched by any ability to keep control over his new-found fortune. A mild and not very entertaining comedy, in which the various complications of the plotting generate irritation rather than amusement.
COM   85 min (ort 91 min) B/W   VIDrel: CARL/TECH   V
Boa: play by Edward Knoblock/novel Mr Prohack by Arnold Bennett.

DEAREST LOVE ****   18
Louis Malle FRANCE/ITALY/WEST GERMANY   1971
Lea Massari, Benoit Ferreux, Daniel Gelin, Fabien Ferreux, Marc Winocourt, Michel Lonsdale, Jacqueline Chauveau, Corinne Kersten
An affectionate and satirical examination of the lifestyle of the French upper-class, as a precocious fourteen-year-old boy begins to experience the first onset of sexuality and the possibility of his mother becoming his tutor arises. This delicate comedy of manners manages to say something fresh, witty and perceptive about love, relationships and social conventions.
Aka: LE SOUFFLE AU COEUR; MURMUR OF THE HEART; NOW IS THE TIME
DRA   113 min (ort 118 min)   VIDrel: ELPIC/POLYREC   V

DEATH AND THE MAIDEN ***   18
Roman Polanski FRANCE/ITALY/USA   1995
Sigourney Weaver, Ben Kingsley, Stuart Wilson, Krystia Nova, Jonathan Vega, Rodolphe Vega, Gilberto Cortes, Jorge Cruz, Carlos Morengo, Eduardo Valenzuela, Sergio Ortega Alvarado
After the fall of the regime that had her imprisoned and tortured, a woman spots her former torturer one day and manages to trap him in her home, her intention being to have a fitting revenge. A claustrophobic production with only three characters and one main set, it is undeniably well acted, but is severely hampered by its rigid adherence to the format of the original play. However, it has a strong climax, albeit one that is a trifle contrived and unbelievable.
DRA   99 min (ort 103 min)   VIDrel: ELPIC/POLYREC   V/sur
Boa: play by Ariel Dorfman.

DEATH BECOMES HER **   PG
Robert Zemeckis USA   1992
Meryl Streep, Goldie Hawn, Bruce Willis, Isabella Rossellini, Ian Ogilvy, Adam Storke, Nancy Fish, Alaina Reeh Hall, Michelle Johnson, Mimi Kennedy, Jonathan Silverman, Fabio Lanzoni, Sydney Pollack, Clement Von Frackenstein

An ageing Hollywood actress who will stop at nothing to stay young and beautiful faces problems where her husband's former girlfriend arrives on the scene, and she learns that her youthfulness is due to a magic potion. This wild, over-the-top satire on Hollywood's obsession with outward appearance soon loses its bite in a welter of special effects that are so advanced they overshadow both acting and plot. AA: Effects (vis).
COM   99 min (ort 104 min) cC   VIDrel: CIC/SONOP; PION (LV only)   V/sur V/dm LV

**DEATH BENEFIT **   12
Mark Piznarski USA   1995
Peter Horton, Carrie Snodgress, Wendy Makkena, Nathan Lawrence, Elizabeth Ruscio, Penny Johnson, Belita Moreno, Lee Debroux, Dean Norris, Jack Kehler, James G. McAlpine
A lawyer digs a little too deeply for his own safety when he investigates a fatal accident that proves to be anything but accidental. The victim was a young girl who was believed to have fallen to her death from a cliff, but she may have been killed, as there was an insurance policy on her life.
DRA   85 min (ort 89 min) cC   VIDrel: CIC/SONOP   V/sh
Boa: book by David Heilbroner.

**DEATH DEALERS **   15
Steve Feke   1990
Jane Seymour, Jim Youngs, Denholm Elliott, Omar Sharif, David Warner, Nancy Kwan, Kay Tong Lim, Ric Young
Fair action film built around the efforts made by the Hong Kong Chinese to acquire US passports in preparation for their flight from Communist rule, with a deadly black market trade in said passports being controlled by ruthless gangsters.
Aka: KEYS TO FREEDOM
A/AD   97 min   VIDrel: GENESIS   V

**DEATH DREAMS **   15
Martin Donovan USA   1991
Christopher Reeve, Marg Helgenberger, Fionnula Flanagan, Taylor Fry, George Dickerson, Conor O'Farrell, Cec Verrell, Jim Jarrett, Ian Devereaux, Pat Atkins, Kevin Page, Robert Ward, Harry Johnson, Richard Morrison, Jack Angeles
Devastated by the loss of their young daughter, apparently the victim of an accidental drowning, a married couple become aware of supernatural influences at work. A succession of strange events prompts them to begin an investigation into the truth behind their daughter's death.
THR   94 min   VIDrel: CAPIT/GUILD   V

DEATH IN A FRENCH GARDEN *   18
Michel Deville FRANCE   1985
Michel Piccoli, Nicole Garcia, Anemone, Christophe Malavoy, Richard Bohringer, Anais Jeanneret
A handsome young music teacher arrives at the home of a wealthy industrialist in order to give the daughter guitar lessons. However, the man's attractive and unhappy wife seduces him, and he finds himself enmeshed in a menacing scheme of retribution and murder, in which he is nothing more than an unwitting pawn. A stylish and handsome film, often extremely engrossing, but rather spoilt by a contrived and unsatisfactory conclusion.
Aka: PERIL EN LA DEMEURE
THR   97 min   VIDrel: ARTIF/20TH   V

DEATH IN BRUNSWICK *   15
John Ruane AUSTRALIA   1990
Sam Neill, Zoe Carides, John Clarke, Yvonne Lawley, Nico Lathouris, Nicholas Papademetriou, Boris Brkic, Deborah Kennedy, Doris Younane, Denis Moore, Stephen Hutchinson, Kris Karahisarus, Huriye Balkaya, Orphan Akkus, Daniel Kadamani
A grotesque black comedy that casts Neill as a divorced, leather-jacketed failure who takes a job as a cook at a rock club. There he gets embroiled in a confrontation with the Turkish kitchen-hand, whom he stabs to death. Pursued by the man's friends, he makes a desperate attempt to dispose of the body, yet at the same time contemplates bumping off his shrew of a mother. Ruane's first feature is a violent, mocking, but often very funny affair.
COM   104 min (ort 109 min)   VIDrel: ELPIC/POLYREC   V/sur
Boa: novel by Boyd Oxlade.

**DEATH IN SMALL DOSES **   15
Sondra Locke USA   1993
Richard Thomas, Tess Harper, Glynnis O'Connor, Shawn Elliott, Gary Frank, Matthew Posey, Ann Hearn

Based on a true story, this tells of the death from arsenic poisoning of successful Dallas architect Nancy Lyon, whose gardener husband was accused of the crime. However, when the district attorney begins his investigations, it appears that she had far more to fear from her own brother. A competent routine murder mystery.
THR   92 min mTV   VIDrel: ODY/SONOP   V/sh

**DEATH IN VENICE **   15
Luchino Visconti FRANCE/ITALY   1971
Dirk Bogarde, Mark Burns, Marisa Berenson, Silvana Mangano, Bjorn Andreson, Luigi Battaglia, Romolo Valli, Nora Ricci, Carol Andre, Masha Predit, Leslie French, Franco Fabrizi, Sergio Garfagnoli, Civo Cristofoletti, Luigi Battaglia
Beautifully filmed slow-moving account of the last days of a composer who falls in love with a young boy, and stays in Venice too long to escape the epidemic that is sweeping the city. Movements from Mahler's Third and Fifth Symphonies are high-spots in a lavish but quite vacuous film. Written by Visconti and Nicola Bandalucco.
Aka: MORT A VENISE; MORTE A VENEZIA
DRA   125 min (ort 130 min)   VIDrel: WHV   V
Boa: novella by Thomas Mann.

**DEATH LINE **   18
Gary A. Sherman UK   1972
Donald Pleasence, David Ladd, Sharon Gurney, Christopher Lee, Norman Rossington, Hugh Armstrong, James Cossins, June Turner, Heather Stoney, Hugh Dickson, Clive Swift, Suzanne Winkler, Ron Pember, Jack Woolgar
A far-fetched story of strange cannibalistic being that lives in the bowels of the London Underground system. Quite atmospheric (it was filmed on location at Russell Square Station) and surprisingly sensitive and poignant in places, but gruesomely overdone. The few doses of mocking humour are most welcome.
Aka: RAW MEAT
HOR   83 min (ort 88 min)   VIDrel: CARL/TECH   V

**DEATH MACHINE **   18
Stephen Norrington JAPAN/UK   1995
Brad Dourif, Ely Pouget, William Hootkins, John Sharian, Martin McDougall, Andreas Wisniewski, Richard Brake, Alex Brooks, Stuart St Paul, Anne Marie Zola, Julie Cox, Joe Scott, Matthew Justice, Randall Paul, Ronald Fernee
In 2033 a weapons scientist is sacked by a new plant director and decides to take his revenge by employing his latest project, a highly developed killer robot, and this fearsome creation runs riot on the company property. An average low-budget effort that freely borrows ideas from many other of this genre (see also CHOPPING MALL and HARDWARE), a fact acknowledged by the leading characters being named after noted horror and SF directors.
FAN   111 min   VIDrel: VIPCO/SGSVID   V

DEATH MATCH *   18
Joe Coppoletta USA   1993
Ian Jacklin, Martin Kove, Renee Ammann, Matthias Hues, Steven Leigh, Richard Lynch, Jorge Rivero, Benny "The Jet" Urquidez, Bob Wyatt, Nick Hill, Eric Lee, Carlos Palomino, Tony "The Viking" Halme, Deboarah "Medusa" Micelli, Ian Cook
A former kick-boxing champion agrees to return to the ring in return for a big fee but finds that he is expected to compete in fights "to the death" that are being organised by criminals who are running a gambling operation. A brutal and depressing film, filled with unbelievable violence, but equally devoid of any redeeming qualities.
A/AD   89 min   VIDrel: MIA/DISC   V

**DEATH MERCHANTS, THE **   (15)
Peter Warner/Colin Bucksey/David E. Jackson/
Mark Sobel USA   1991
Jenny Gago, Tom Mason, Byron Keith Minns, Chris Stanley, David Wohl
Action film based on the efforts made by the U.S. Drug Enforcement Administration to combat the global menace of drugs, as seen through the eyes of a group of agents employed by the agency. Fair. A sequel followed.
Aka: MERCHANTS OF MENACE, THE
A/AD   119 min   VIDrel: MGM/WHV L/A   V/sh

DEATH MERCHANTS 2: THE MAFIA CONNECTION * 15
Peter Warner et al. USA 1991
Jenny Gago, Tom Mason, Byron Keith Minns, Chris Stanley, David Wohl
A standard guns-and-action tale of the war between rival drugs barons in Colombia, that follows the career of a woman who is determined enough and ruthless enough to get to the top. Laughable nonsense.
A/AD 115 min VIDrel: MGM/WHV V/sh

DEATH OF A CHEERLEADER ** 15
William A. Graham USA 1994
Kellie Martin, Tori Spelling, James Avery, Eugene Roche, Andy Romano, Terry O'Quinn, Margaret Langrick, Kathryn Morris, Valerie Harper, Christa Miller, Tom O'rourke, Brittney Powell, Marnie Andrews, Jenna J. Leigh, Robyn Bailey
Fact-based drama that explores the pressures behind the murder of a high school cheerleader by another girl, who had plotted a jealousy-driven revenge when she was dropped from the murdered girl's clique. See also WILLING TO KILL: THE TEXAS CHEERLEADER STORY and CHEERLEADER-MURDERING MOM.
DRA 87 min mTV VIDrel: ODY/SONOP V/sh

DEATH OF A NUN * 18
Philip J. Avrech USA
Ellen Barber, Philip English, Sam Gray
Unpleasant shocker in which a lonely ex-convent girl thinks she has met the man of her dreams, only to discover that he has a murderous past.
HOR 86 min VIDrel: VIPCO/SGSVID V/h

DEATH OF A SCHOOLBOY *** PG
Peter Patzak WEST GERMANY/YUGOSLAVIA 1990
Rueben Pillsbury, Christopher Chaplin, Robert Munic, Sinolicka Trpkova, Michele Melega, Alan Cox, Hans Michael Rehberg, Harmut Becker, Alexis Arquette, Phillipe Leotard
Biopic on the life of Gavrilo Princip, the seventeen-year-old who assassinated the Archduke Ferdinand in 1914. A most interesting work, it provides a fascinating glimpse of a vanished era and the pivotal events that shaped history.
Aka: GAVRE PRINCIP – HIMMEL UNTER STEINEN
DRA 89 min wScrn VIDrel: TART/20TH V
Boa: book by Hans Koning.

DEATH ON THE NILE ** PG
John Guillermin UK 1978
Peter Ustinov, Jane Birkin, Lois Chiles, Bette Davis, Mia Farrow, David Niven, Jon Finch, Olivia Hussey, Maggie Smith, Angela Lansbury, George Kennedy, Jack Warden, Simon MacCorkindale, Jane Birkin, Harry Andrews, Sam Wanamaker
One of those typical adaptations of an Agatha Christie novel which totally fails to be anything other than a turgid and static production, in spite of (or possibly because of) a star-studded cast. In this tale, Hercule Poirot is presented with a boatload of suspects after Chiles is murdered on a cruise up the Nile. Visually impressive, but difficult to sit through and Ustinov is badly miscast. The script is by Anthony Shaffer. AA: Cost (Anthony Powell).
DRA 134 min (ort 140 min) VIDrel: WHV V
Boa: novel by Agatha Christie.

DEATH RING ** 18
Robert J. Kizer USA 1992
Billy Drago, Mike Norris, Chad McQueen, Don Swayze, Elizabeth Fong Sung, Isabel Glaser, Branscombe Richmond, Kelly Bennett, Victor Quintero, Donegan Smith, George Kee Cheung, Henry Kingi, Tammy Stones, Judy Peterson, Joe Stoffer
The winner of a gruelling race is chosen as the latest quarry of a group of wealthy hunters who pay large sums for the chance to hunt human prey. A slick, efficient but unimaginative reworking of THE MOST DANGEROUS GAME from 1932, which has since then been remade numerous times in a variety of guises (see the original film for a list of the variants).
MAR 87 min (ort 91 min) VIDrel: TRANSAT/HIFLI V/h

DEATH TRAIN ** 18
Jeff Kwitny ITALY 1989
Mary Kohnert, Sarah Conway Ciminera, William Geiger, Renee Rancourt, Jeremy Sanchez, Alex Vitale, Ron Williams, Victoria Zinny, Savina Gersak, Bo Svenson
A sinister cult of religious fanatics selects a woman as the means of summoning their master, and the unsuspecting victim boards a train on a field trip with some fellow students, little realising that the journey is about to change her life.
Aka: TRAIN, THE
THR 90 min (ort 92 min) VIDrel: 20VIS/SONOP V/sur

DEATH TRAIN ** 15
David S. Jackson CROATIA/SLOVENIA/UK/USA 1992
Pierce Brosnan, Patrick Stewart, Alexandra Paul, Ted Levine, Christopher Lee, John Abineri, Nic D'Avirro, Lorrie Marlow, Clarke Peters, Ron Berglas, Andreas Sportelli, Vili Matula, Terrence Hardiman, Bill Leadbitter
A renegade Soviet general steals plutonium for the building of two A-bombs with which he intends to hold Europe to ransom. These are loaded aboard a hijacked train that travels through Germany in the direction of Switzerland. The United Nations Anti-Crime Organisation responds by sending in a couple of its best agents to stop the train and its deadly cargo. A standard action tale, that feels flat and lifeless for all the efforts of a good cast.
Aka: ALISTAIR MacLEAN'S DEATH TRAIN; DETONATOR
A/AD 94 min mCab VIDrel: EIV/SONOP V
Boa: novel by Alistair McLean.

DEATH TRAP ** 18
Tobe Hooper USA 1976
Neville Brand, Mel Ferrer, Carolyn Jones, Stuart Whitman, Marilyn Burns, Robert Englund, William Finley
Hooper's Hollywood debut is a weird, disagreeable PSYCHO-style shocker, with Brand the psychopathic owner of a seedy swamp-side motel who uses the pet crocodile he keeps in the front yard to deal with his more troublesome guests. A garish and utterly bizarre film, let down by a virtually non-existent plot and an awkward and stylised approach to its material.
Aka: EATEN ALIVE!; HORROR HOTEL; HORROR HOTEL MASSACRE; LEGEND OF THE BAYOU; STARLIGHT SLAUGHTERS
HOR 86 min (ort 96 min) VIDrel: VIPCO/SGSVID V/h

DEATH WARMED UP * 18
David Blyth NEW ZEALAND 1984
Michael Hurst, Margaret Umbers, Bruno Lawrence, David Letch, Norelle Scott, William Upjohn, Gary Day, Geoff Snell, Ian Watkin, David Weatherly, Tina Grenville, Nat Lees, Judy McIntosh, Ken Harris, Karam Haas, Jonathan Hardy
Horror spoof about a mad surgeon and the results of his experiments in neurosurgery. After being treated by the doctor, a young man slaughters his parents. The surgeon now retreats to an island haven, where he sets about creating an army of murderous mutants, but the youngster upon whom he first experimented finds his abode and destroys him. A technically competent but remarkably unappealing work.
HOR 78 min Cut (54 sec – ort 85 min)
VIDrel: VIPCO/SGSVID V/sur

DEATH WARRANT ** 18
Deran Sarafian CANADA 1990
Jean-Claude Van Damme, Robert Guillaume, Cynthia Gibb, George Dickerson, Patrick Kilpatrick, Art Le Fleur, Joshua Miller, Hank Woessner, George Jenesky, Jack Bannon, Abdul Salaam El Razzac, Armin Shimerman, John Lantz, Han Howes
A Royal Canadian Mountie goes undercover as a prisoner to solve a series of murders at a penitentiary, the victims' organs having being sold for transplant. He finally unmasks the culprit, a psychopathic slasher known as the Sandman, and the expected violent confrontation ensues. A disappointing and cliched work in which Van Damme shows his severe limitations as an actor if not as an exponent of the martial arts.
Aka: DUSTED
A/AD 85 min (ort 111 min) VIDrel: MGM/WHV V/sur

DEATH WISH ** (18)
Michael Winner USA 1974
Charles Bronson, Vincent Gardenia, Hope Lange, Stuart Margolin, William Redfield, Steven Keats, Stephen Elliott, Kathleen Tolan, Jack Wallace, Fred Scollay, Chris Gampel, Robert Kya-Hill, Ed Grover, Jeff Goldblum, Floyd Levine
Charles Bronson gives a totally wooden performance as a man who turns into a vigilante after his wife and daughter are so brutally raped that the former dies. An unconvincing and predictable film that has spawned a series of sequels and several imitations. Music is by Herbie Hancock. Look out for Jeff Goldblum in his screen debut as one of the muggers. See also EYE FOR AN EYE.
A/AD 93 min VIDrel: L/A V
Boa: novel by Brian Garfield.

DEATH WISH 2 *   *18*
Michael Winner USA   1982
Charles Bronson, Jill Ireland, Ben Frank, Vincent Gardenia, J.D.
Cannon, Anthony Franciosa, Robin Sherwood, Silvana Gilardo,
Robert F. Lyons, Michael Prince, Drew Snyder, Paul Lambert, Thomas
Duffy, Paul Comi, Hugh Warden
Predictable sequel to DEATH WISH with our lone vigilante back
in action on the streets of L.A., after the gang rape and murder
of his housekeeper and the suicide of his daughter. Poor
Bronson just doesn't seem to have much luck, but he always
makes someone pay in a sequel that's a little more bloody and
a good deal more shallow. Now move on to DEATH WISH 3.
A/AD   85 min (Cut at film release – ort 93 min)
VIDrel: VCC L/A   V
Boa: novel by Brian Garfield.

DEATH WISH 3 *   *18*
Michael Winner USA   1985
Charles Bronson, Ed Lauter, Deborah Raffin, Martin Balsam, Gavin
O'Herlihy, Kirk Taylor, Alex Winter, Tony Spiridakis, Joseph
Gonzalez, Francis Drake, Leo Kharibian, Hana-Maria Pravda, John
Gabriel, Manning Redwood, Joe Cirillo
Third time round and Bronson creaks into action as the lone
vigilante in this yet more stylised, violent and unrealistic
sequel – culminating in a quite remarkable set-piece battle in
which a street gang are blown to pieces with weapons that
include sub-machine guns and a missile-launcher. And who
can blame him, as assorted rapists, muggers and murderers
conspire to attack all his friends? (Perhaps he should move to
an island.)
A/AD   86 min Cut (13 sec – ort 92 min)   VIDrel: VCC L/A
V

DEATH WISH 4: THE CRACKDOWN *   *18*
J. Lee Thompson USA   1987
Charles Bronson, Kay Lenz, John P. Ryan, Perry Lopez, George
Dickerson, Soon-Teck Oh, Dana Barron, Jesse Dabson
A flabby sequel to the earlier film with a different director but
much the same "shoot 'em up" vigilante plot. This time round
Bronson is galvanised into action (if not into acting) by the death
of his girlfriend's daughter from drugs. He now attempts a
wholesale wipe-out for L.A.'s drug dealers.
A/AD   94 min (Cut at film release by 54 sec – ort 98 min)
VIDrel: MGM/WHV   V/h

DEATH WISH 5: THE FACE OF DEATH *   *18*
Allan A. Goldstein USA   1993
Charles Bronson, Lesley Anne Down, Michael Parks, Chuck Shamata,
Kevin Lund, Robert Joy, Saul Rubinke, Miguel Sandoval, Kenneth
Welsh, Lisa Inoye, Erica Lancaster, Jefferson Mappin, Michael
Dunston, Claire Rankin, Elena Kudaba
This entry in an increasingly impoverished series is set in New
York's garment industry, where the mob is attempting to take
over a family business through the use of murder and terror.
Enter our vigilante to redress the balance by dispensing equal
amounts of mayhem. A highly unpleasant and trashy affair that
is just about watchable for those with strong stomachs.
A/AD   91 min (ort 95 min)   VIDrel: GUILD/SONOP   V/s

DEATHRACE 2000 **   *18*
Paul Bartel USA   1975
David Carradine, Simone Griffith, Sylvester Stallone, Louisa Moritz,
Mary Woronov, Don Steele, Joyce Jameson, Fred Grandy, Martin
Kove, Sandy McCallum, Harriett Medin, Carle Bensen, George
Wagner
In the year 2000 the national sport is cross-country racing, with
points scored for knocking down pedestrians. A fast and furious
spoof on all those car-race films, spoilt by too much concentra-
tion on gore. Followed by "Deathsport".
FAN   76 min (ort 80 min)
VIDrel: 4-FRONT/POLYREC/BRAVE   V
Boa: story by Ib Melchior.

DEATHROW GAMESHOW **   *(18)*
Mark Pirro USA   1987
John McCafferty, Robyn Blythe, Beano, Mark Lasky, Darwyn Carlson,
Kent Butler, Debra Lamb, Paul Farbman, Bill Whitehead, esther Elise,
Zach Harris, Paul Brun, John E. Beilin
A misfire of a black comedy in which killers waiting on
deathrow can take part in a new TV quiz show. They stand to
either win their freedom or have their sentence of death carried
out. See also THE PRIZE OF PERIL and THE RUNNING MAN.
COM   80 min   SATrel: BRAVO MOVIES

DEATHSTALKER *   *18*
John Watson ARGENTINA/USA   1983
Richard Hill, Barbi Benton, Richard Brooker, Lana Clarkson, Victor
Bo, Bernard Erhard, Horace Marussi, August Larreta, Lilian Ker,
Marcos Woinsky, Adrian De Piero, George Sorvic, Boy Olmi
Sword-and-sorcery tale with a warrior sent to rescue a princess
from an evil wizard. Like so many of these, the title may change
but the plot remains the same. Filmed in Argentina and followed
by a sequel in 1987.
Aka: DEATH STALKER; EL CAZADOR DE LA MUERTE
FAN   77 min (ort 80 min)   VIDrel: VCC L/A   V

DEATHSTALKER 2 *   *18*
Jim Wynorski USA   1985
John Terlesky, Monique Gabrielle, John La Zar, Toni Naples, Maria
Socas, Marcos Wolinsky, Jake Arnt, Carina Davi, Arch Stanton,
Douglas Mortimer, Leo Nichols, Maria Luisa Carnivani, Nick
Sardansky, Frank Sisty, Queenie Kong
Sequel to the 1983 film with the same elements as before. A
princess in exile is aided in her attempts to regain her throne by
a soldier-of-fortune who must battle the forces of evil and save
a kingdom. A medieval fantasy adventure just like the earlier
film and with just as poor a plot. Filmed on the cheap in
Argentina and released directly to video. DEATHSTALKER 3
soon followed.
Aka: DEATHSTALKER 2: DUEL OF THE TITANS; DUEL OF THE TITANS
FAN   85 min Cut (22 sec)   VIDrel: POLY/POLYREC L/A
V

DEATHSTALKER 3 **   *15*
Alfonso Corona USA   1989
John Allen Nelson, Carla Herd, Thom Christopher, Terri Treas
A further fantasy-adventure in the DEATHSTALKER series,
with a princess who has been entrusted with a valuable jewel
from a lost city, claiming the protection of this warrior. Together
they set out to recover the remaining jewels and have many
adventures, not least being a battle between our hero and an evil
wizard.
Aka: DEATHSTALKER 3: THE WARRIORS FROM HELL; DEATHSTALKER
AND THE WARRIORS FROM HELL
FAN   81 min   VIDrel: FIRST/SONOP L/A   V

DEATHSTALKER 4: MATCH OF TITANS *   *18*
Howard R. Cohen USA   1989
Maria Ford, Michelle Moffett, Brett Clark, Rick Hill
It's the fourth time round for this turgid sword 'n' sorcery series,
and this one is very little different in plotting from its prede-
cessors. Our hero attends a warrior's tournament, which an evil
queen is planning to exploit for her own purposes. The usual
Good versus Evil struggle ensues, in a poor film of unexpected
laughs and all-too-expected bad acting and dialogue.
Aka: DEATHSTALKER 4: CLASH OF THE TITANS
A/AD   75 min (ort 85 min)
VIDrel: POLY/POLYREC/BRAVE   V

DEATHSTONE *   *18*
Andrew Prowse USA   1989
Jan-Michael Vincent, R. Lee Ermey, Nancy Everhard
A woman journalist investigating a wave of murders in Manila
teams up with a former boyfriend and ex-soldier to clear a US
soldier who has been accused. A supernatural sub-plot (the
excavation of an ancient Chinese tomb leads to a series of
deaths) and clashes with the local Triads help pad out this unappealing trifle.
A/AD   90 min   VIDrel: MOPIC/QUANT   V

DECADENCE *   *18*
Steven Berkoff GERMANY/UK   1993
Steven Berkoff, Joan Collins, Christopher Biggins, Michael Winner,
Marc Sinden, Edward Duke, Robert Longdon, David Alder, Tim Dry,
Imogen Bain, Susannah Morley, Ursula Smith, Veronica Lang,
Mathilda Ziegler, Terence Beesley
A wealthy lady of leisure has an affair with a married man in
this offbeat comedy-drama whose intention is as much to put
the mores of the British class system under the microscope as
to tell a story. The rhyming couplet dialogue (taken from
Berkoff's successful stage play) does not translate well to the
big screen and the great charm of the original work is nowhere
in evidence.
DRA   113 min   VIDrel: CURZON/20TH   V/sur
Boa: play by Steven Berkoff.

DECAMERON, THE *** 18

Pier Paolo Pasolini FRANCE/ITALY/WEST GERMANY 1970
Franco Citti, Ninetto Davoli, Angela Luce, Patrizia Capparelli, Jovan Jovanovic, Silvana Mangano, Pier Paolo Pasolini, Nicoletta Machiarelli, Enzo Petriglia, Gianni Rizzo, Silvana Galti, Elisabetta Genovesi, Roberto Simmi
Lavishly made film version of eight tales from Boccaccio's 14th century collection of stories. A colourful and rumbustious film with music by Ennio Morricone and Pasolini. The first work in a trilogy of medieval tales from the director – "The Canterbury Tales" and "The Arabian Nights" followed. Screenplay was by Pasolini.
Aka: IL DECAMERONE
DRA 107 min Cut (22 sec – ort 111 min)
VIDrel: MGM/WHV L/A V
Boa: short stories by Giovanni Boccaccio.

DECEIVED ** 15

Damian Harris USA 1991
Goldie Hawn, John Heard, Ashley Peldon, Robin Bartlett, Tom Irwin, Amy Wright, Jan Rubes, Kate Reid, Francesca Straight, George R. Robertson, Maia Fila, Anais Gronofsky, Heide Von Palleske, Stanley Anderson, Peter Stevens
Hawn plays a successful art-dealer who appears to live an idyllic life with a caring husband and a delightful young daughter. However, when the curator of a museum is murdered in the course of the theft of an ancient Egyptian necklace, she soon learns that her husband is not the man she thinks he is. With little in the way of characterisation and a tendency to give the game away too early, this film begs for a decent script. The director is the son of Richard Harris.
THR 104 min (ort 108 min) cC   VIDrel: TOUCH/TECH
V/sur

DECEIVED BY TRUST ** 12

Chuck Bowman USA 1995
Stepfanie Kramer, Michael Gross, Conor O'Farrell, Shannon Fill, Lisa Gorlitsky, Teryl Rothery, Michelle Goodger, Malcolm Stewart
After a student is sexually assaulted by a high school principal, his victim finds that her assailant is protected by a conspiracy of silence. Finding that no-one will believe her claims, she starts a crusade to expose him, ultimately assisted by a social worker in whom she confides. A fact-based drama with inventive plotting and an effective resolution.
DRA 86 min VIDrel: NWV/HIFLI V/h

DECEIVERS, THE ** 15

Nicholas Meyer INDIA/UK 1988
Pierce Brosnan, Shashi Kapoor, Saeed Jaffrey, Helena Michell, Keith Michell, David Robb, Tariq Yunis, Jalal Agha, Gary Cady, Salim Ghouse, Neena Gupta, H.N. Kalla, Nayeem Hafizka, Bijoya Jena, Kammo, Goga Kapoor, Harish Magon
Period adventure set in 1820s India, where Brosnan plays a British officer given the job of infiltrating the murderous Thuggee cult. Hampered by uneven direction and pacing, this is an expensive looking but undistinguished updating of THE STRANGLERS OF BOMBAY. Despite some nasty authentic touches it remains curiously flat and undeveloped, though it does have its moments.
A/AD 98 min (ort 112 min) VIDrel: MIA/DISC V/sur
Boa: novel by John Masters.

DECEMBER BRIDE ** PG

Thaddeus O'Sullivan UK 1990
Donal McCann, Patrick Malahide, Saskia Reeves, Brenda Bruce, Ciaran Hinds, Michael McKnight, Geoffrey Golden, Frances Lowe, Dervla Kirwan, Peter Capaldi, Cathleen Delaney, Gabrielle Reidy, Catherine Gibson, Karl Hayden
In a remote Northern Ireland rural community at the turn of the century, a young serving girl eventually becomes the mistress of two farming brothers and bears a daughter by one of them. However, her refusal to wed the father incurs the wrath of the local bigots. A film of great visual beauty, ably directed and with sparse dialogue, but for all its virtues strangely distant and unengaging. Won the special jury prize at a 1990 European film festival.
DRA 84 min (ort 90 min) VIDrel: CONNO/RTM V/sur
Boa: novel by Sam Hanna Bell.

DECEPTION **** PG

Irving Rapper USA 1946
Bette Davis, Paul Henreid, Claude Rains, John Abbott, Benson Fong, Richard Walsh, Jane Harker, Suzi Crandall, Dick Erdman, Ross Ford,
Russell Arms, Kenneth Hunter, Einar Nelson, Earl Dewey, Marcelle Corday, boyd Irwin, Bess Flowers
At the end of WW2, a European musician returns to the USA only to find that he has lost his one-time girlfriend to a wealthy man. Davis must now choose between her old flame and rich composer Rains, who has been keeping her in the luxury to which she would like to grow accustomed. A powerful soap opera-style film with Rains quite wonderful as the jealous benefactor and Davis equally good. First filmed as "Jealousy" way back in 1929.
DRA 107 min B/W VIDrel: WHV V
Boa: play Monsieur Lamberthier by Louis Verneuil.

DECEPTIONS * 18

Ruben D. Preuss USA 1989
Harry Hamlin, Nicollette Sheridan, Robert Davi, Marshall Colt, Kevin King
An L.A. cop investigating the murder of a businessman, finds himself falling in love with his widow, who claims to have mistaken her husband for a burglar and shot him, but is now a very wealthy woman. Stealing ideas from a host of muchbetter works (e.g. DOUBLE INDEMNITY) this movie runs through a full range of cliches, but remains a muddled erotic-thriller of little atmosphere.
THR 100 min Cut (6 sec – ort 105 min) mCab
VIDrel: ENCORE/SPEAR/COLUM V
Boa: short story by Ken Dobson.

DECLINE OF THE AMERICAN EMPIRE, THE *** 18

Denys Arcaud CANADA 1986
Dominique Michel, Louise Portal, Dorothee Berryman, Pierre Curzi, Genevieve Rioux, Remy Girard, Yves Jacques, Daniel Briere, Gabriel Arcand
An amusing and perceptive examination of male/female relationships and the differences between the sexes in terms of heir aspirations and desire for happiness. The story focuses on a group of friends who have arranged a dinner party; whilst the women work out at a health club the men prepare the meal. Winner of the Critics Award at the Cannes Film Festival.
Aka: LE DECLIN DE L'EMPIRE AMERICAIN
COM 101 min VIDrel: ARTIF/20TH V/h

DECONSTRUCTING SARAH ** 15

Craig R. Baxley USA 1994
Rachel Ticotin, Sheila Kelly, Rachel Ticotin, David Andrews, Jenifer Lewis, Dwier Brown, Peter Jason, Clyde Kusatsu, John Vicekery, Caroline Williams, Tony Abetemarco, John Andreozzi, James Arone, Camilla Belle, Tony Brubaker
When a professional woman who leads a double life in which she frequented pick-up bars is murdered, her best friend tries to unravel the mystery surrounding her death in order to track down the killer. A slow-paced thriller, slightly erotic in tone, it fails to hold the interest.
DRA 87 min (ort 92 min) mCab VIDrel: CIC/SONOP
V/sh

DECORATION DAY *** PG

Robert Markowitz USA 1990
James Garner, Bill Cobbs, Judith Ivey, Ruby Dee, Larry Fishburne, Norm Skaggs, Jo Anderson, Wallace Wilkinson, Jonathan Peck, Ric Reitz, Jeff Young Lewis, Tracy Perry
A retired US judge who has recently lost his wife hears from an old friend with an unusual proposition. He finds himself involved in researching a WW2 incident that almost cost his friend his life and caused him to refuse to accept the Medal of Honor. A quiet and low-key drama with some winning performances.
Aka: PURPLE HEART
DRA 94 min (ort 99 min) mTV VIDrel: CAPIT/GUILD
V/sh

DECOY ** 18

Victor (Vittorio) Rambaldi USA 1995
Peter Weller, Robert Patrick, Charlotte Lewis, Darlene Vogel
A couple of unstable assassins are hired to protect the daughter of a billionaire, as the latter's life is in danger from a dangerous, female assassin. Various double-crosses and the like take place, with nothing to raise this rather ordinary action thriller above the rest, except perhaps the direction, which is really rather slick.
A/AD 93 min VIDrel: MED/DISC V/sh

DEEP, THE *
15
Peter Yates USA
1977
Robert Shaw, Nick Nolte, Jacqueline Bisset, Eli Wallach, Louis Gossett Jr, Robert Tessier, Earl Maynard, Dick Anthony Williams, Bob Minor, Teddy Tucker, Lee McClain, Peter Benchley, Peter Wallach, Colin Shaw
An overlong and flat version of a novel about a couple who become innocently involved in drug smuggling off Bermuda. A good-looking but almost unbearably tedious effort, with lashings of gratuitous violence that adds nothing whatsoever to the thin and implausible plot. Written by Benchley and Tracy Keenan Wynn.
THR 116 min (ort 126 min) VIDrel: VCC/DISC/COLUM L/A V/sh
Boa: novel by Peter Benchley.

DEEP COVER *
18
Bill Duke USA
1992
Larry Fishburne, Jeff Goldblum, Clarence Williams III, Charles Martin Smith, Victoria Dillard, Gregory Sierra, Robert Guenveur Smith, Cory Curtis, Glynn Thurman, Lira Angel, Rene Assa, Bruce Barbour, Anna Berger, Donald Bishop
A straightlaced cop whose father was killed during a robbery accepts an assignment from the DEA to work under cover and plays his role so well that he finds himself experiencing a conflict of interests. A well realised blend of crime thriller and drama that does justice to both, with good direction and excellent performances. Screenplay is by Henry Bean and Michael Tolkin.
THR 103 min (ort 106 min) cC VIDrel: FIRST/SONOP; ENCORE (LV only) V/sur LV

DEEP DOWN *
18
John Travers USA
1994
Tanya Roberts, George Segal, Kristoffer Tabori, Paul Le Mat, Chris Young
A young man moves from the city into a new apartment. At first, he's quite pleased to learn that he has an attractive woman as one of his neighbours, but soon discovers that she is married to a man who is paranoid in his jealousy.
DRA 86 min (ort 90 min) VIDrel: 20VIS/SONOP V/sh

DEEP IN MY HEART *
U
Stanley Donen USA
1954
Jose Ferrer, Merle Oberon, Henry Traubel, Doe Avedon, Tamara Toumanova, Paul Stewart, Douglas Fowley, Jim Backus, Walter Pidgeon, Paul Henreid, Rosemary Clooney, Gene Kelly, Jane Powell, Vic Damone, Ann Miller, Cyd Charisse
No insights in this very Hollywood account of the life of the famous songwriter Sigmund Romberg with a host of guest stars and several good numbers; highlights being a dance number featuring Cyd Charisse and James Mitchell and a sequence where Ferrer performs an entire show himself.
MUS 126 min (ort 132 min) VIDrel: MGM/WHV V
Boa: book by Elliott Arnold.

DEEP RED *
18
Dario Argento ITALY
1975
David Hemmings, Daria Nicolodi, Gabriele Lavia, Macha Meril, Glauco Mauri, Clara Calamai, Eros Pagni, Giuliana Calandria, Nicoletta Elmi
During a para-psychiatrists' conference in Rome, a psychic gives a demonstration of her powers, and caps this off with an announcement that there is a killer in the audience. She is later murdered, but an English pianist gets a glimpse of the crime, and as he tries to uncover the culprit, further murders take place, and this being an Italian "giallo", all of the killings are marked by much brutality. A bizarre horror-thriller, noisy, gory but very stylish.
Aka: DEEP RED: HATCHET MURDERS; DRIPPING DEEP RED; HATCHET MURDERS, THE; PROFONDO ROSSO; SABRE TOOTH TIGER, THE; SUSPIRIA 2
HOR 121 min (ort 123 min) dubbed VIDrel: REDEM/RTM V

DEEP RED *
15
Craig R. Baxley USA
1993
Michael Biehn, Joanna Pacula, John De Lancie, Tobin Bell, John Kapelos, Steven Williams, Michael Des Barres, Daniel Barringer, John Alder, Lisa Collins, Chayse Dacoda, Kevin Page, Jamie Stern, Hank Cheyn, Jesse Vint, Jack Andreozzi
Imaginative but highly confused SF tale about a group of people whose blood has become infected with alien elements which have the ability to confer immortality. A private detective hired

to find a missing man gets caught up in a mad scientist's hunt for such an infected girl and eventually helps her and her mother to escape. An intriguing idea is absolutely ruined by lack of plot development, incessant flashbacks and an uninspired script.
HOR 81 min (ort 85 min) VIDrel: CIC/SONOP V

DEEP SPACE *
18
Fred Olen Ray USA
1987
Charles Napier, Ann Turkel, Ron Glass, James Booth, Anthony Eisley, Bo Svenson, Julie Newmar, Norman Burton, Jesse Dabson, Elisabeth Brooks, Anthony Eisley, Peter Palmer, Fox Harris, Michael Forest, William Fair
An alien, human-eating monster terrorises a small town and is eventually killed by two police officers, but more horror is in store since it has spawned some offspring.
HOR 87 min VIDrel: EIV/SONOP V

DEEPSTAR SIX *
15
Sean S. Cunningham USA
1988
Greg Evigan, Nancy Everhard, Cindy Pickett, Miguel Ferrer, Taurean Blacque, Marius Weyers, Nia Peeples, Matt McCoy, Elya Baskin, Thom Bray, Ronn Carroll
Set in the near future, this tells of an underwater research base that's threatened by a monster crustacean. A waterlogged and talky yarn with little to recommend it. See also LORDS OF THE DEEP and THE ABYSS.
FAN 94 min (ort 100 min) VIDrel: GUILD/POLYREC L/A; PION (LV only) V LV

DEER HUNTER, THE **
18
Michael Cimino USA
1978
Robert De Niro, John Savage, John Cazale, Meryl Streep, Christopher Walken, George Dzundza, Chuck Aspegren, Shirley Stoler, Rutanya Alda, Pierre Segui, Mady Kaplan, Amy Wright, Mary Ann Haenel, Richard Kuss, Joe Grifasi, Jack Scardino
This story of Pennsylvanian steelworkers before, during and after the Vietnam War is visually quite remarkable and is strongest when concentrating on the emotions and thoughts of the protagonists. Less a direct examination of the war, it is more a study of its effect on their mutual relationships. The haunting score is the work of John Williams. AA: Pic, Dir, S. Actor (Walken), Edit (Peter Zinner), Sound (William McCaughey/Aaron Rochin/Darin Knight).
WAR 175 min (ort 182 min) wScrn (re-mastered collector's edition) VIDrel: WHV V/dm V/sur
Boa: novel by E.M. Corder.

DEFENCE OF THE REALM *
PG
David Drury UK
1985
Gabriel Byrne, Greta Scacchi, Denholm Elliott, Ian Bannen, Bill Paterson, Fulton Mackay, David Calder, Frederick Treves, Robbie Coltrane, Annabel Leventen, Graham Fletcher Cook, Steven Woodcock, Alexei Jawdokimov
An intrepid Fleet Street reporter investigates the relationship between a British MP and a Russian agent, but finds there is more to the case than he first suspected. A competent political thriller, whose many diverse strands are not resolved until the end.
THR 91 min (ort 96 min) VIDrel: CARL/TECH V

DEFENDING YOUR LIFE *
PG
Albert Brooks USA
1991
Albert Brooks, Meryl Streep, Rip Torn, Lee Grant, Michael Durrell, Buck Henry, George D. Wallace, James Eckhouse, Clayton Norcross, Lillian Lehman, Gary Beach, Julie Cobb, Peter Schuck, Marilyn Rockafellow, Roger Behr, Art Frankel
Having just died in a car crash, a yuppie finds himself in "Judgement City", where he is obliged to defend the record of his life on Earth before a tribunal. At the same time, he falls in love with another new arrival, whose record of life on Earth is too good to be true. Written by Brooks, this ambitious comedy-fantasy must have seemed like a great idea on paper, but though slightly amusing it often drags, and fails to capitalise on the unusual opening premise.
COM 106 min (ort 111 min) VIDrel: WHV V/sur

DEFENDING YOUR SEX LIFE *
18
Fank Marino USA
1992
Kelly O'Dell, Sierra, Steve Drake, Woody Long, Tom Chapman, Christina plus voices of: Top Jimmy, C.A.L.
After a woman dies she finds herself in the afterworld, where she is called on to give an account of her sexual transgressions

to an invisible board of review. A silly spoof on DEFENDING YOUR LIFE that is nothing more than a set of loosely strung together vignettes. Dull, unimaginative and badly made.
A   80 min   VIDrel: ONE   V

DEFENSELESS *   18
Martin Campbell USA   1991
Barbara Hershey, Sam Shepard, J.T. Walsh, John Kapelos, Mary Beth Hurt, Sheree North, Randy Brooks, Christopher M. Borwn, Michael Collins, George P. Wilbur, Marabina James, John Achorn, Lisa Darr, Willia Hayes, Glenn Wilson
A female attorney falls in love with a client but becomes the prime suspect when he is murdered. To add to her problems, she finds herself being stalked by a killer, in this well-acted and quite suspenseful thriller that offers some good plot twists.
THR   99 min (ort 106 min)   VIDrel: 20TH   V/sur

DEFIANT ONES, THE *   U
Stanley Kramer USA   1958
Tony Curtis, Sidney Poitier, Theodore Bikel, Lon Chaney Jr, Cara Williams, Charles McGraw, King Donovan, Claude Akins, Lawrence Dobkin, Whit Bissell, Carl Carl Switzer, Kevin Coughlin
Two convicts chained to each other escape from a Southern prison gang. One is white and the other black, and despite racial prejudice and distrust they are gradually forced to co-operate. A simple story beautifully told as well as a powerful comment on race relations that is neither laboured nor contrived. An inferior TV remake appeared in 1985. See also FLED and THE CHAIN. AA: Cin (Sam Leavitt), Screen/Screen (Nathan E. Douglas and Harold Jacob Smith).
DRA   96 min (ort 97 min) B/W   VIDrel: MGM/WHV   V

DEKALOG: THE TEN COMMANDMENTS – PARTS 1 TO 5 *   15
Krzysztof Kieslowski POLAND   1988
Wojciech Klata, Henryk Baranowski, Maja Komorowska; Krystyna Janda, Olgierd Lukaszewicz, Aleksander Bardini; Daniel Olbrychski, Maria Pakulnis; Janusz Gajos, Adrianna Biedrzynska; Mirowslaw Baka, Krzysztof Globisz, Jan Tesarz
Each of these five tales highlights a moral dilemma. In the first a young boy has a doting father but no religious instruction, in the second an unfaithful wife will have an abortion if her dying husband recovers, the third deals with the sins of a man's past, the fourth with the dishonesty of a girl's father. Finally, there's a painfully uncompromising tale of murder, which was expanded into the feature film A SHORT FILM ABOUT KILLING.
Aka: TEN COMMANDMENTS: PARTS 1 TO 5
DRA   287 min (2 cassettes)   VIDrel: ARTIF/20TH   V

DEKALOG: THE TEN COMMANDMENTS – PARTS 6 TO 10 *   15
Krzysztof Kieslowski POLAND   1988
Olaf Lubaszenko, Grazyna Szapolowska, Stefania Twinska; Maja Barelkowska, Anna Polony, Linda Boguslaw; Maria Koscialkowska, Teresa Marczewska, Tadeusz Lomnicki; Piotr Machalica, Ewa Blaszczyk; Jerzy Stuhr, Zbigniew Zamachowski
The final five stories. In the first episode, a shy nineteen-year-old suffers an infatuation with a worldly girl (this led to A SHORT FILM ABOUT LOVE). The next tells of a battle of wills between a mother and daughter over the latter's child, the third is an examination of past sins committed in order to survive, the fourth sees an impotent husband distraught at his wife's unfaithfulness, and finally, two brothers fall out over a valuable legacy.
Aka: TEN COMMANDMENTS: PARTS 6 TO 10
DRA   277 min (2 cassettes)   VIDrel: ARTIF/20TH   V

DELICATESSEN *   15
Marc Caro/Jean-Pierre Jeunet FRANCE   1991
Dominique Pinon, Marie-Laure Dougnac, Jean-Claude Dreyfus, Karin Voard, Rufus, Ticky Holgado, Anne-Marie Pisani, Jacques Mathou, Jean-Francois Perrier, Sylvie Laguna, Chick Ortega, Howard Vernon, Patrick Paroux, Jean-Luc Caron
This bizarre black comedy is set some time in the future, when society has broken down. The story revolves around the crazy eccentric inhabitants of a derelict building, and the cannibalistic "butcher" who rules the roost. Into this strange world stumbles a former circus clown, who falls for the butcher's shortsighted daughter, and though next on the menu, may yet live to see better days. An ingenious blend of surrealism, comedy and horror.
COM   95 min (ort 99 min) wScrn   VIDrel: ELPIC/POLYREC   V/sur

DELINQUENTS, THE *   15
Chris Thompson AUSTRALIA   1989
Kylie Minogue, Charlie Schlatter, Angela Punch-McGregor, Bruno Lawrence, Lyn Treadgold, Todd Boyce, Desiree Smith, Melissa Jaffer, Lynette Curran, Duncan Wass, Rosemary Harris, Yvonne Hooper, Jonathan Hardy, Errol O'Neil, Ove Altman
Soap actress turned pop singer Minogue makes her feature film debut in this lightweight melodrama from Down Under that has a couple of teenage starcrossed lovers fighting to keep their love alive in 1950s Queensland. A remarkably glossy, production-line effort, as lacking in impact as it is in interest.
DRA   100 min   VIDrel: WHV   V/sur
Boa: novel by Criena Rohan.

DELIRIOUS *   PG
Tom Mankiewicz USA   1991
John Candy, Mariel Hemingway, Emma Samms, Raymond Burr, David Rasche, Dylan Baker, Charles Rocket, Jerry Orbach, Renee Taylor, Robert Wagner, Mark Boone Jr, Andrea Thompson, Zach Grenier, Milt Oberman, Tony Steedman, Rita Gomez
Candy plays a writer employed on a downmarket soap opera that's as sordid as it is unimaginative, and features a host of nasty stereotypes. Obsessed with one of the female leads, he awakens after an accident at the studio to find himself living in the very show he has helped create. Finding that he has the power to "write" his own storylines in this world, he sets out to enjoy himself. Candy is wasted in an over-stretched effort that should have been a short sketch.
COM   92 min (ort 96 min)   VIDrel: MGM/WHV   V/sur

DELIVER US FROM EVIL *   18
Clay Borris USA   1991
Nikki De Boer, Alden Kane, Joy Tanner, Alle Ghadban
Seeking sanctuary in the church he attended as a boy, a priest who committed a murder is imprisoned in the crypt. Thirty-three years later he escapes and sets out to slake his thirst for revenge by attacking some youngsters who are out enjoying their school prom. A formulaic stalk-and-slash horror film using a well-worn theme.
HOR   92 min   VIDrel: HIFLI/SONOP   V/h

DELIVERANCE *   18
John Boorman USA   1972
John Voight, Burt Reynolds, Ned Beatty, Ronny Cox, Billy McKinney, Herbert Coward, James Dickey, Ed Ramey, Billy Redden, Seamon Glass, Randall Dean, Lewis Crone, Ken Keener, Johnny Popwell, John Fowler, Kathy Rickman, Louise Coldren
Menacing and concise film about four businessmen whose weekend canoeing trip turns into a nightmare when they are terrorised by two of the more degenerate local inhabitants. Good performances all-round plus the memorable "Guitars Duelling" sequence – one of the more lighthearted sections in a generally nasty film. The film debuts for Beatty and Cox, and scripted by Dickey (who appears briefly as a local sheriff).
DRA   104 min (ort 109 min) wScrn (special edition)
VIDrel: WHV   V/h
Boa: novel by James Dickey.

DELTA FORCE, THE *   18
Menahem Golan USA   1985
Chuck Norris, Lee Marvin, Martin Balsam, Joey Bishop, Robert Forster, Lainie Kazan, George Kennedy, Hanna Schygulla, Susan Strasberg, Bo Svenson, Robert Vaughn, David Menahem, Shelley Winters, Kim Delaney
The story of the creation of a special squad to free the hostages of the real-life 1985 TWA Athens hijack. A simple and unadorned account that works quite well, despite unevenness and poor characterisation. Filmed in Israel.
A/AD   124 min   VIDrel: MGM/WHV   V/sur

DELTA FORCE 2 *   18
Aaron Norris USA   1990
Chuck Norris, John P. Ryan, Paul Perri, Richard Jaeckel, Begonia Plaza, Billy Drago, Mateo Gomez, Hector Mercado, Mark Margolis, Ruth de Sosa, Gerald Castillo, Geof Brewer, Rick Prieto, Sharlene Ross, Michael Heit, Richard Warlock
Sequel to the first film sees our elite force in action again, this time in Latin America, where they are sent in to deal with the operations of a powerful cocaine baron. A violent but quite enjoyable actioner of the usual conventional type, though one wonders why on earth Norris needs to lug around

so much weaponry with the martial arts expertise at his disposal.
Aka: DELTA FORCE 2: OPERATION STRONGHOLD; DELTA FORCE 2: THE COLOMBIAN CONNECTION
A/AD 106 min (ort 110 min) VIDrel: WHV V/sur

DELTA FORCE 3: THE KILLING GAME **
18
Sam Firstenberg USA
1991
Nick Cassavetes, Mike Norris, Matthew Penn, Eric Douglas, John Ryan, Sandy Ward, Hana Azulay-Hasfari, Gregori Tal, Mark Ivanir, Candace Brecker, Dan Turgeman, Jonathan Cherchi, Jonathan Arkin, Kevin Patterson, Baruch Dor
A loosely plotted entry in this series, that's sustained by a succession of exciting action set-pieces, as the crack US commando force of the title joins forces with its Soviet counterpart, to mount an attack on a vicious leader of some Arab terrorists. There's nothing notable in either plot or dialogue, but ample undemanding entertainment for fans of mindless action.
A/AD 94 min (ort 97 min) VIDrel: MGM/WHV V/dm

DELTA FORCE COMMANDO **
18
Frank Valenti ITALY
1987
Brett Clark, Fred Williamson, Bo Svenson, Mark Gregory
Two US fighter pilots fight terrorists in the jungles of Nicaragua, whilst waiting for the crack Delta Force to come to their aid. By way of a wholly unnecessary embellishment, there's also a nuclear bomb ticking away that could do with being defused.
WAR 89 min Cut (52 sec – ort 90 min) VIDrel: MIA/DISC V

DELTA HEAT **
15
Michael Fischa USA
1992
Anthony Edwards, Lance Henriksen, Betsy Russell, Linda Donna, Rod Masterson, John McConnell, Clyde Jones, John "Spud" Campbell, Jack Harris, John Henry Scott, John Feritta, Gregg Brazzel, Harvey Keith, Buddy Carr
A detective and his partner investigate the appearance of a new designer drug in Los Angeles and trace its source to Louisiana. When his partner is tortured and killer by the head of the drugs ring, our cop goes on a one-man revenge rampage to bring in the bad guys.
A/AD 91 min VIDrel: EIV/SONOP V

DELTA OF VENUS *
18
Zalman King USA
1995
Costas Mandylor, Eric Da Silva, Marek Vasut, Zette. Emma Louise Moore, Raven Snow, Rory Campbell, Audie England, Marketka Hrubesova, Daniel Leza, Stephen Hulbert, Dale Waytt, Jiri Dad, Valerie Zawadska, James Donahower, Robert Davi
Very loosely based on the writings on Anais Nin, this tale follows the changing fortunes of a beautiful woman who enjoys the high life when she arrives in Paris, and has to turn to writing erotic literature when her money runs out. A good looking period tale, that is set in the 1930s and has everything going for it except a decent script, the film being essentially just a vehicle for some lukewarm, softcore titillation.
DRA 98 min (ort 101 min) VIDrel: FIRST/SONOP V

DEMENTED DEATH FARM MASSACRE: THE MOVIE *
18
Fred Olen Ray/Donn Davison USA
1986
Ashley Brookes, George Ellis, Trudy Moore, Mike Coolik, John Carradine
A bottom-of-the-barrel horror spoof built around the title event. About as amusing as a cold night out in the rain.
HOR 87 min VIDrel: TROMA/RTM V

DEMENTIA 13 *
15
Francis Ford Coppola USA
1963
William Campbell, Luana Anders, Bart Patton, Mary Mitchell, Patrick Magee, Eithne Dunne, Peter Read, Karl Schanzler, Ron Perry, Derry O'Donovan, Barbara Dowling
Grisly horror movie about a series of bloody axe murders in Ireland whose victims are all members of the same unpleasant family, and thus heirs to a massive inheritance. The plot is indeed rather trite and lightweight, but the film is redeemed from the ranks of the unwatchable largely thanks to Coppola's efforts on this, his first mainstream film.
Aka: HAUNTED AND THE HUNTED, THE
HOR 74 min B/W VIDrel: SCRN/DISC V

DEMETRIUS AND THE GLADIATORS **
PG
Delmer Daves USA
1954
Victor Mature, Susan Hayward, Michael Rennie, Debra Paget, Anne
Bancroft, Richard Egan, Ernest Borgnine, Jay Robinson, William Marshal, Barry Jones, Charles Evans, Everett Glass, Karl Davis, Jeff York, Carmen De Lavallade*
A sequel to THE ROBE with the Emperor Caligula searching for the robe of Christ, which he believes to possess magical powers. Spectacle aplenty is to be had, with some nice scenes where gladiators get to perform in the arena; a shame about Mature's performance out of it.
DRA 97 min (ort 101 min) wScrn VIDrel: 20TH/TECH V/sh

DEMOLITION MAN **
15
Marco Brambilla USA
1993
Sylvester Stallone, Wesley Snipes, Sandra Bullock, Nigel Hawthorne, Benjamin Bratt, Bob Gunton, Glenn Shadix, Denis Leary, Grand L. Bush, Pat Skipper, Steve Kahan, Paul Bollen, Mark Colson, Andre Gregory, John Enos, Troy Evans
After devastating urban wars and an earthquake, Los Angeles is reborn in the 21st century as a peaceful but over-controlled society where even profanity is banned. When a master criminal escapes from prison and commits many murders, the authorities are forced to revive a old-style cop who has been keep in cryogenic suspension for the past three decades. A disappointing futuristic tale that soon degenerates into an irritatingly violent actioner. See also VIRTUOSITY.
A/AD 110 min (ort 115 min) cC VIDrel: WHV V/sur V/dm

DEMOLITIONIST, THE **
18
Robert Kurtzman USA
1995
Nicole Eggert, Bruce Abbott, Richard Grieco, Susan Tyrell, Tom Savini
Another fantasy set in the near future, where law and order has largely broken down, what vestiges of it remaining being maintained by a rather special type of police officer – the "demolitionist" of the title, a bio-regenerated former soldier given full powers to arrest or eliminate criminals. A JUDGE DREDD clone, with the same comic-book feel about it plus the wholesale stealing of ideas from a number of other movies.
A/AD 85 min (ort 96 min) VIDrel: HIFLI/SONOP V/h

DEMON HOUSE **
18
Jimmy Kaufman USA
1996
Amelia Kinkade, Kris Holdenreid, Patricia Rodriguez, Stephanie Bauder, Tara Slone, Gregor Calpakis, Vlastra Vrana
After a Halloween prank turns nasty and leads to a shoot-out at a mini-mart, a bunch of teenagers hide out at a sinister mansion, whose only inhabitant appears to be an attractive woman. However, appearances are deceptive and she turns out to be more than a match for them, using her sinister powers to change her guests into zombies. A very silly horror film, of overblown effects and acting, liberally spiced up with ample teenage sexual shenanigans.
HOR 85 min (ort 107 min) VIDrel: HIFLI/SONOP V/sh

DEMON SEED **
15
Donald Camell USA
1976
Julie Christie, Fritz Weaver, Gerrit Graham, Berry Kroeger, Lisa Lu, John O'Leary, Alfred Dennis, Larry Blake, David Roberts, Patricia Wilson, Michael Glass, Monica MacLean, E. Hampton Beagle plus Robert Vaughn (voice only)
A super-computer takes over the computer controlling the house where the wife of a scientist lives. She is held prisoner, for it has decided to take over the world by cloning a child from her. Some good effects and chilling moments cannot rescue the film from the dullness of the script and its awkward and its ludicrous ending. COLOSSUS: THE FORBIN PROJECT was a lot better.
FAN 91 min wScrn VIDrel: MGM/WHV V/h
Boa: novel by Dean R. Koontz.

DEMONIAC **
18
James P. Johnson (Jesus Franco) BELGIUM/FRANCE/SPAIN
1974
Pierre Taylor, Olivier Mathot, Francoise Goussard, Nadine Pascal, Jesus Franco, Rosa Amiral, Lina Romay, Lynn Monteil, Monica Swinn, France Nicholas, Catherine Lafferiere, Christine Chrieix, David Atta, Sam Maree
Dull horror offering set in Paris where an excommunicated religious fanatic believes that he has been chosen by God to murder degenerates, prostitutes and all those he considers to be fallen

sinners. Originally made as a hard-core movie with footage added to enhance the horror element.
Aka: BLACK MASSES OF EXORCISM, THE; EXORCISME; EXORCISME ET MESSES NOIRES; EL SADICO DE NOTRE DAME; LA SADIQUE DE NOTRE DAME; L'EVENTREUR DE NOTRE DAME; RIPPER OF NOTRE DAME, THE; SADIST OF NOTRE DAME, THE; SEXORCISME
HOR   80 min (ort 96 min)   VIDrel: REDEM/RTM L/A   V

DEMONIC TOYS *
18
Peter Manoogian USA
1991
Tracy Scoggins, Bentley Mitchum, Michael Russo, Jeff Weston, Daniel Cerney, Pete Schrum, Ellen Dunning, Barry Lynch, William Thorne, Robert Stockdale, Larry Cedar, Richard Speight, Jim Mercer, Pat Crawford Brown, June C. Ellis
A policewoman is on a stakeout with her boyfriend when the latter is killed by the villains, and she chases one of them into a toy warehouse. Others follow, and perhaps aroused by the smell of blood, several of the toys come to life, and proceed to stalk our heroine, dispatching the other characters in various messy ways. A gory stalk-and-slash effort that is badly acted and badly shot, and plays like a downmarket version of CHILD'S PLAY or DOLLS.
HOR   79 min (ort 86 min)   VIDrel: EIV/SONOP   V

DEMONS **
18
Lamberto Bava ITALY
1985
Natasha Hovey, Fiore Argento, Urbano Barberini, Paolo Cozza, Bobby Rhodes, Fabiola Toledo, Nicoletta Elmi, Stelio Condelli, Nicole Tessier, Geretta Giancarlo, Bettina Ciampolini
Horror spoof in which the patrons of a Berlin cinema watching a film about zombies become victims of the real thing, as the creatures stream out of the screen and slash the audience to pieces. An utterly over-the-top celebration of gore, well directed but about as illogical as they come. Produced by Dario Argento (hence the film's alternative title) and with special effects by Sergio Stivaletti. A rather weak sequel soon followed.
Aka: DARIO ARGENTO'S DEMONS; DEMONI; DEMONS 1
HOR   84 min Cut (1 min 5 sec – ort 89 min) dubbed
VIDrel: SPEAR/SONOP/CALECO   V

DEMONS 2 **
18
Lamberto Bava ITALY
1987
David Knight, Nancy Brilli, Coralina Catataldi Tassani, Dario Casalini, Bobby Rhodes, Asia Argento, Virginia Bryant, Anita Bartolucci, Antonio Cantafara, Luisa Passega, Davide Marotta, Marco Vivio, Michele Mirabella
Sequel to the earlier film, with our lovable bloodthirsty creatures coming out of TV sets this time and turning a young girl's sixteenth birthday party into a most unpleasant affair. As with the earlier film, this sequel was also produced by Dario Argento.
Aka: DEMONI 2; DEMONI 2: L'INCUBO RITORNA; DEMONS 2: THE NIGHTMARE BEGINS; DEMONS 2: THE NIGHTMARE CONTINUES; DEMONS 2: THE NIGHTMARE RETURNS
HOR   87 min (ort 95 min)   VIDrel: SPEAR/SONOP   V

DEMONS 3 **
18
Lamberto Bava ITALY
1988
Virginia Bryant, David Flosey, Paolo Malco, Sabrina Ferille, Stefania Montorsi
An American writer, staying at a villa with her husband and young son, discovers that a figure out of one of her childhood nightmares is present in reality, and is lurking in the basement of the house. Her husband refuses to believe her until their babysitter is killed and the child taken. A film of effective sequences and rather slow development.
Aka: DEMONS 3: THE OGRE; OGRE, THE
HOR   90 min   VIDrel: MARQ/QUANT   V

DENISE CALLS UP **
15
Harold Salwen USA
1995
Tim Daly, Caroleen Feeney, Dan Gunther, Dana Wheeler Nicholson, Liev Schreiber, Aida Turturro, Alanna Ubach, Sylvia Miles, Mark Blum, Jean Lamarre, Hal Salwen
Salwen's debut film warns of the alienating effects of modern technology in the study of love over a satellite link-up, as a mother attempts to get to know the father of her sperm-bank child. But these calls take place in the context of a host of others, as no-one ever seems to have the time or inclination to actually meet, a point the movie laboriously hammers home. Scripted by Salwen, this is a one-joke affair that runs about sixty-minutes too long.
COM   77 min (ort 80 min) wScrn   VIDrel: ARTIF/20TH
V/h

DENNIS ***
PG
Nick Castle USA
1992
Mason Gamble, Walter Matthau, Christopher Lloyd, Lea Thompson, Joan Plowright, Robert Stanton, Billie Bird, Paul Winfield, Amy Sakasitz, Kellen Hathaway, Arnold Stang, Devin Ratray, Hnak Johnston, Melinda Mullins, Ben Stein
A blonde five-year-old boy is the bane of the neighbourhood and drives his next-door neighbour almost insane with his pranks. An ambitious but not overly successful attempt to transfer the charm of Hank Ketcham's comic-strip character to the screen. Too reminiscent of HOME ALONE and its sequel but saved by its fine cast – especially Matthau, who is perfectly (if a little obviously) cast as the boy's irascible neighbour Mr Wilson. See also CLIFFORD.
Aka: DENNIS THE MENACE
COM   92 min (ort 96 min) cC   VIDrel: WHV   V/dm V/sur

DENTIST, THE **
18
Brian Yuzna USA
1996
Corbin Bernsen, Ken Foree, Linda Hoffman
A millionaire dentist who has an obsession with cleanliness starts to suspect his wife of having an affair. Bernsen gives his best shot as the demented dentist, and Yuzna's sly digs at the seedier side of modern life are very welcome, but this horror-comedy has no dramatic thrust and its impact is very slight.
HOR   90 min   VIDrel: TRIM/HIFLI   V/h

DERBY **
(U)
Bob Clark USA
1995
David Charvet, Joanne Vannicola, Len Cariou, Felton Perry, Dean McDermott, Wayne Robson, Mimi Kozyk, Darren McGavin, Jim McKay, Stuart Parker, Bob Clark, Brenda Radloff, M. Compton James, Carlos Dumont, John Simpson, Ronald Singh
When her mother dies, a fifteen-year-old tomboy leaves her adopted parents' horse ranch to study in Boston, not returning for six years. Shortly afterwards, her father dies and she is forced to shoulder the burden of running his business and realising his dream of winning the title horse race. Competent acting and fine photography are the only items of note in a trite and mundane drama, whose appeal will be restricted mostly to fans of horse-riding and the like.
A/AD   90 min   SATrel: MOVIE CHANNEL

DERSU UZALA ****
U
Akira Kurosawa JAPAN/USSR
1980
Yuri Solomine, Maxim Munzuk, S. Danilchenko, M. Bichkov, V. Khrulev, V. Lastochkin, S. Marin, I. Sikhra, V. Sergiyavov, Ya. Yakobsons, V. Koldin, V. Khlestov, G. Polunik, M. Tettov, S. Sinyavsky, V. Sverba, V. Ignatov
A brilliant evocation of the friendship that develops between a Russian explorer and a native guide-cum-trapper, a wise and gentle man well versed in the art of wilderness survival. A fascinating and beautifully made film that never drags despite its length, its ending is both sad and deeply moving. Shot on 70 mm, some of its grandeur will be lost on the small screen. AA: Foreign.
Aka: DERUSU USARA; HUNTER, THE
DRA   135 min (ort 140 min) wScrn   VIDrel: CONNO/RTM
L/A   V
Boa: novel Dersu the Trapper by V.K. Arseniev.

DESERT FOX, THE ***
PG
Henry Hathaway USA
1951
James Mason, Cedric Hardwicke, Jessica Tandy, Luther Adler, Desmond Young, Leo J. Carroll, Everett Sloane, Richard Boone, George McReady, Eduard Franz, William Reynolds, Charles Evans, Walter Kingsford, John Hoyt, Don De Leo
Film account of the career of Field Marshal Rommel, his military defeat in Africa in the course of WW2, and eventual return to Germany and implication in the plot to assassinate Hitler. A most competent war film with an interesting portrayal of Hitler by Adler. In 1953 Mason played Rommel once again in THE DESERT RATS.
Aka: ROMMEL: THE DESERT FOX
WAR   90 min B/W   VIDrel: 20TH/TECH   V/h
Boa: book by Desmond Young.

DESERT HEARTS **
18
Donna Deitch USA
1985
Helen Shaver, Patricia Charbonneau, Audra Lindley, Andra Akers, Dean Butler, Gwen Welles, James Stanley, Jeffrey Tambor, Katie Labourdette, Sam Minsky, Alex McArthur, Tyler Tyhurst, Denise Crosby, Antony Ponzini, Brenda Beck

A woman in Reno to get a "quickie" divorce, embarks on a passionate lesbian love affair following her pursuit by a young woman who has become enamoured of her. This debut for writer-director Deitch has moments of interest and bags of charm, but is too uneven and corny to convince.
DRA   90 min (ort 96 min)   VIDrel: VISION/DISC   V
Boa: novel Desert of the Heart by Jane Rule.

DESERT LAW ** **   15
Duccio Tessari ITALY   1990
Rutger Hauer, Elliott Gould, Carol Alt, Omar Sharif
A former CIA agent now working as a mercenary takes on a mission to rescue a youngster trapped in the Sahara, and has a tough time evading capture at the hands of two tribes of feuding Arabs, a task made even more difficult by the fact that his former employers are planning to have him killed off.
A/AD   160 min Cut (33 sec)
VIDrel: 4-FRONT/POLYREC/VISVID   V/h

DESERT PASSION *   18**
Carlo G. Gustaff USA   1992
Carrie Janisse, Missy Browning, Tony Bond, Mchael McMillen, Nicole Sassaman, Andy Thomson, Elizabeth Christiansen, Jeri Thompson, Nicole Lyn Marinello, Merle Nicks, Madison Monk, Nichols Hill, Scott McElroy, Nicholas Wright
Two women head for Las Vegas where one of them is to attend an audition but they are kidnapped en route and held prisoner in at a remote installation in the desert. There they are drugged and induced to participate in strange sexual fantasies. Plenty of soft-core, soft-focus sequences of lovemaking which do not really blend well with a decidedly unpleasant and nasty plot.
DRA   71 min (ort 76 min)   VIDrel: 20VIS/SONOP   V

DESERT RATS, THE *   U**
Robert Wise USA   1953
Richard Burton, James Mason, Robert Douglas, Torin Thatcher, Robert Newton, Chips Rafferty, Charles Tingwell, Charles Davis, Ben Wright, James Lilburn, John O'Malley, Ray Harden, John Alderson, Richard Peel, Michael Pate, Frank Pulaski
Made to cash in on the success of THE DESERT FOX, this excellent war film concerns a British captain who takes charge of an Australian company, and has to win their respect if not their affection with his cold, professional manner. Mason adds strong support, repeating his role as Rommel from the earlier film.
WAR   84 min (ort 88 min) B/W   VIDrel: 20TH/TECH   V

DESERT SONG, THE *   U**
Bruce Humberstone USA   1953
Kathryn Grayson, Gordon MacRae, Steve Cochran, Raymond Massey, Dick Wesson, Allyn McLerie, Ray Collins, Paul Picerni, William Conrad
This third version of the Sigmund Romberg operetta set in Africa runs along pretty conventional lines. MacRae as the secret leader of good natives in the battle against nasty ones, still finds time to sing of his love. Filmed before in 1929 and 1944. Highspots are the songs "The Riff Song" and "One Alone". Fair.
MUS   106 min Cut (2 sec)   VIDrel: WHV   V
Boa: play by Otto Harbuch, Lawrence Schwab and Frank Mandel/operetta by Sigmund Romberg.

DESIRE FOR LOVE *   18**
FRANCE   198-
Bernard Hug, Catherine Noel, Karin Adler, Sylvie Dessartre, Maite Lemoine, Johanna Morina, Luc Templar
An aspiring writer, married to a wealthy and very possessive woman, goes to their summer house to work on his novel. There, he meets up with an old girlfriend and her companion, and soon becomes unfaithful. However, the arrival of his wife complicates the situation still further. A poorly plotted softcore romp of very little interest.
A   64 min dubbed   VIDrel: ELV   V

DESIREE *   U**
Henry Koster USA   1954
Marlon Brando, Jean Simmons, Merle Oberon, Michael Rennie, Cameron Mitchell, Elizabeth Sellars, Cathleen Nesbitt, Isobel Elsom
An elaborate but stuffy costume drama presenting a fictionalised account of one of Napoleon's mistresses before he became Emperor. Both Simmons as the mistress and Oberon as the Empress Josephine are suitably ravishing, but Brando looks distinctly out of place in this muddled epic.
DRA   105 min (ort 110 min)   VIDrel: 20TH/TECH   V/sur
Boa: novel by Annemarie Selinko.

DESPERADO *   18**
Robert Rodriguez USA   1994
Antonio Banderas, Steve Buscemi, Salma Hayek, Quentin Tarantino, Cheech Marin, Joaquim De Almeida, Carlos Gomez, Quentin Tarantino, Tito Larriva, Angel Aviles, Danny Trejo, Abraham Verduzco, Carlos Gallardo, Albert Michel Jr
An English-language sequel of sorts to Rodriguez's hit debut EL MARIACHI, in which our tough musician is now a real gunfighter, and in this capacity he proceeds to eliminate his enemies en masse and make them pay for the death of his girlfriend. An insufferably tedious movie, almost plotless and replete with cartoon-style mega violence, clearly intended to cash in on Banderas's sudden popularity with the public and the media.
A/AD   100 min (ort 105 min) wScrn cC
VIDrel: COLUM/SONOP; ENCORE (LV only)   V/sur LV

DESPERATE HOURS *   15**
Michael Cimino USA   1990
Mickey Rourke, Anthony Hopkins, Mimi Rogers, Lindsay Crouse, Kelly Lynch, Elias Koteas, David Morse, Shawnee Smith, Danny Gerard, Gerry Bamman, Matt McGrath, John Christopher Jones, Dean Norris, John Finn, Christopher Curry
An ill-conceived remake of the 1955 classic that examines the plight of a family who are held hostage by a ruthless gang after their home is chosen as a hideout. Nothing like as tense as the earlier film, with poor use of music and some distracting subplots that only serve to weaken the impact of the film. Despite a few good moments the strangely ineffectual script and clumsy ending are serious flaws.
THR   100 min (ort 106 min)   VIDrel: 20TH/TECH   V/sur
Boa: novel by Joseph Hayes.

DESPERATE JOURNEY *   PG**
Raoul Walsh USA   1942
Errol Flynn, Raymond Massey, Ronald Reagan, Nancy Coleman, Alan Hale, Arthur Kennedy, Albert Basserman, Ronald Sinclair, Sig Rumann, Ilka Gruning, Pat O'Moore
An exciting account of the struggles three American pilots face in making their way back to freedom after they escape from a POW camp. One of those wartime efforts in which Flynn wins the war singlehandedly, but despite the dated propaganda this is a rousing and enjoyable adventure.
WAR   103 min (ort 107 min) B/W   VIDrel: MGM/WHV   V

DESPERATE JUSTICE *   PG**
Armand Mastroianni USA   1993
Lesley Ann Warren, Bruce Davison, Shirley Knight, Missy Crider, Allison Mack, David Byron, Annette O'Toole, Joseph Rassulo, Michael Collins, Lance Sollar, Wanda Dittman, William Cook, Barbara Hazlett, Kenneth Bridges
A woman's daughter is raped and beaten senseless, but when her attacker is set free by the court, the girl's mother decides to take the law into her own hands. A provocative revenge drama that raises some important issues regarding the law and retribution without really analysing them. See also EYE FOR AN EYE.
DRA   88 min   VIDrel: ODY/SONOP   V/h

DESPERATE MOTIVE *   18**
Andrew Lane GERMANY/USA   1992
David Keith, Marc Helgenberger, William Katt, Mel Harris, Mary Crosby, Brian Bonsall, Cyndi Pass
A man and his family receive a visit from his distant cousin and the latter's girlfriend. At first sight, they appear to be a charming couple, but our hosts are unfortunately unaware that they have just escaped from a mental asylum.
Aka: DISTANT COUSINS
A/AD   88 min (ort 92 min)   VIDrel: REFLEC/FIRST   V

DESPERATE REMEDIES *   18**
Steward Main/Peter Wells NEW ZEALAND   1993
Jennifer Ward-Lealand, Kevin Smith, Lisa Chappell, Cliff Curtis, Michael Hurst, Kiri Mills, Bridget Armstrong, Timothy Raby, Helen Steemson, Geeling Ching
Melodramatic opera-style 19th century tale of a woman who tries to marry her sister off to a penniless immigrant in order to save the family fortune, yet finds herself growing increasingly fond of him. Overblown and underplotted, this unbelievably contrived work is barely watchable. The daft title refers to the solution required for each sister's dilemma of the heart.
DRA   89 min (ort 93 min)   VIDrel: ELPIC/POLYREC   V/sur

DESPERATE TRAIL, THE **

P.J. Pesce USA
15
1993
Sam Elliot, Craig Sheffer, Linda Fiorentino, Bradley Whitford, Frank Whaley
A marshal becomes obsessed with a female prisoner who was involved in a robbery, and when she escapes he sets off after her. An unusual attempt to introduce some erotic elements into a Western.
WES   89 min (ort 93 min) mVid   VIDrel: COLUM/SONOP V/sur

DESPERATELY SEEKING SUSAN **

Susan Seidelman USA
15
1985
Aidan Quinn, Madonna, Rosanna Arquette, Mark Blum, Robert Joy, Laurie Metcalf, Steven Wright, Richard Hell, Anne Magnuson, Richard Edson, John Laurie, Anne Carlisle, John Turturro, Shirley Stoler, Arto Lindsay, Anna Levine
A bored housewife becomes involved at a distance with a young girl who lives a sort of rootless semi-criminal life in New York. Many twists and turns of the plot ensue before the final climax. Offbeat and intermittently amusing.
COM   119 min   VIDrel: 4-FRONT/POLYREC/VISVID   V

DESTINATION MOON **

Irving Pichel USA
U
1950
John Archer, Warner Anderson, Tom Powers, Dick Wesson, Erin O'Brien-Moore, Ted Warde
Story of a race between the USA and the USSR to reach the Moon. A simple and fairly modest effort, produced by George Pal with Heinlein co-scripting the film from his novel. Enjoyable, but quaint and dated. AA: Effects (George Pal).
FAN   91 min   VIDrel: L/A   V
Boa: novel Rocketship Galileo by Robert A. Heinlein.

DESTINATION TOKYO ***

Delmer Daves USA
U
1944
Cary Grant, John Garfield, Alan Hale, John Ridgely, Robert Hutton, Tom Tully, Whit Bissell, Dane Clark, Warner Anderson, William Prince, Peter Whitney, Faye Emerson, John Forsythe, Warren Douglas, John Alvin, Bill Kennedy
A suspense-filled drama of a US submarine sent into Tokyo Harbour to prepare a mission in which an attack is to be mounted on a Japanese aircraft carrier. Both well made and acted despite the implausible plot and excessive running time. The use of newsreel footage of the time adds a note of authenticity. John Forsythe made his screen debut in this one.
WAR   129 min (ort 135 min) B/W   VIDrel: WHV   V
Boa: novel by S.G. Fisher.

DESTINY TURNS ON THE RADIO *

Jack Baran USA
15
1995
Quentin Tarantino, Dylan McDermott, Nancy Travis, James LeGros, James Belushi, Bobcat Goldthwait, Janet Caroll, David Cross, Richard Edson, Barry "Shabaka" Henley, Lisa Jane Persky, Sarah Trigger, Tracey Walter, Allen Garfield
A bank robber escapes from jail and goes to Las Vegas to collect some loot and go on the run with his girlfriend. However, he and a number of other characters become the playthings of a supernatural being that manipulates their destinies for its own ends. A twee and clumsy attempt at a bizarre comedy that is both pretentious and unfunny, the feeble script and weak direction are additional failings. Very poor indeed.
COM   98 min (ort 102 min)   VIDrel: EIV/SONOP   V

DETECTIVE ***

Jean-Luc Godard FRANCE
(15)
1985
Nathalie Baye, Claude Brasseur, Johnny Hallyday, Laurent Terzieff, Alain Cuny, Jean-Pierre Leaud, Stephane Ferrara
A verbose and stylised homage to the films of John Cassavetes, Clint Eastwood and Edgar G. Ulmer, that is presented in the form of an investigation into a murder that was committed at a Paris hotel a couple of years before. However, the story takes very much second place to the ensemble acting of a strong cast, whose varied antics under the pressure of the investigation form the main item of interest here.
DRA   95 min   VIDrel: ARTIF/20TH   V

DETOUR ***

Edgar G. Ulmer USA
U
1945
Tom Neal, Ann Savage, Claudia Drake, Edmund MacDonald, Tim Ryan, Esther Howard, Roger Clark
A classic Poverty Row B-feature with a cult following, this finely made example of film noir tells of a New York pianist who hitch-

hikes to California to be with his singer girlfriend. Along the way he meets a strange femme fatale and becomes involved in murder. Both intriguing and ironic, this absorbing little film provides an object lesson in how to make the most of a limited budget.
DRA   65 min (ort 69 min) B/W   VIDrel: VISCOM/RTM   V
Boa: novel by Martin Goldsmith.

DEVIL GIRL FROM MARS *

David Macdonald UK
U
1955
Patricia Laffan, Adrienne Corri, Peter Reynolds, Hugh McDermott, Joseph Tomelty, Hazel Court, John Laurie, Sophie Stewart, Anthony Richmond, James Edmond, Stuart Hibberd
A female emissary from Mars comes to Earth to recruit men for reproductive services as their race faces extinction. With her she has a giant robot to convince those who are otherwise not too wild about her proposition. One of the silliest British SF films ever made.
FAN   74 min (ort 76 min) B/W   VIDrel: LUMI/SPEAR L/A V
Boa: play by John C. Mahner and James Eastwood.

DEVIL IN A BLUE DRESS ***

Carl Franklin USA
15
1995
Denzel Washington, Tom Sizemore, Jennifer Beals, Don Cheadle, Maury Chaykin, Terry Kinney, Mel Winkler, Albert Hall, Lisa Nicole Carson, Jernard Burks, David Wolos-Fonteno, John Roselius, Beau Starr, Steven Randazzo, Scott Lincoln
A man agrees to help locate a mysterious woman, but he has little to go on, except the fact that she frequents the seedier side of town. This assignment soon places our sleuth in the obligatory web of blackmail, corruption and danger. An atmospheric film noir outing, set in 1940s L.A., with a great soundtrack, terrific detail and a clever if hard to follow plot.
THR   97 min (ort 101 min)   VIDrel: COLUM/SONOP; ENCORE (LV only)   V/sur LV
Boa: novel by Walter Mosley.

DEVIL IN MISS JONES PART 2 ***

Henri Pachard USA
18
1982
Georgina Spelvin, Samantha Fox, Jacqueline Lorians, Jack Wrangler, Joanna Storm, Anna Ventura, Bobby Astyr, Alan Adrian, Michael Bruce, Ron Jeremy, Joey Civera, Richard Bolla
A sequel to the first film with the Devil returning our heroine to Earth, where she inhabits various bodies as a reward for bringing him to a climax. From body to body and encounter to encounter, Lucifer watches jealously, all the while discussing her with his Advocate. Finally, even he can stand it no longer and abdicates his throne, appearing in her bedroom. A crazy sex-comedy of wild inventiveness and fairly witty dialogue.
A   73 min (Cut at film release – ort 84 min)
VIDrel: EIV/SONOP   V

DEVIL IN MISS JONES PART 3: A NEW BEGINNING ***

Gregory Dark USA
18
1987
Lois Ayres, Jack Baker, Paul Thomas, Vanessa Del Rio, Amber Lynn, Tom Byron
Having apparently slipped and killed herself whilst taking a shower, Ayres gets a guided tour of Hell, in the company of Baker, who describes the various unpleasant and/or erotic vignettes she witnesses with some utterly over-the-top dialogue. A bizarre, disjointed and often tasteless effort, but certainly a memorable one. A further DEVIL IN MISS JONES sequel followed.
Aka: SEX HELL
A   60 min Cut (9 sec – ort 85 min)   VIDrel: QUANT/TOTAL V/sur

DEVIL IN MISS JONES: PART 5 ***

Gregory Dark (Gregory Hippolyte) USA
18
1995
Julie Ashton, Kelly O'Dell, Ariana, Serenity, Amanda Addams
A sexually repressed spinster dies and goes to Hell, where she gets a chance to experience all the pleasures of the flesh she passed up when she was on Earth. A memorable celebration of sleaze and fetish fantasies, full of dwarves, demons and pretty much any visual fantasy the director could think of. Not very strong in the plot department, but in terms of visual impact it's probably a classic of its kind.
Aka: D.M.J.V. – THE INFERNO
A   73 min   VIDrel: PURG/DANTE   V

DEVIL IN THE FLESH *   18
Joe D'Amato USA   1993
Tracy Ray, Wayne Camp, Carmen Di Nietro, Jennifer Loeb, Robert La Brosse
When some soldiers-of-fortune take refuge in a field hospital behind enemy lines, the female doctor and her nurses have to use all their wiles to survive. A tedious erotic thriller of limited appeal, the torrid title being one of its better aspects (but then good titles don't cost any more money to have than bad ones).
THR   78 min (ort 102 min)   VIDrel: RIO/TERRY L/A   V

DEVIL KILLER **   18
Tai Cher HONG KONG
Shan Koon, Wan Lung, Chu Cheng Sin, Ricky Lam
Two powerful fighters arrive in a small town, one to seek his missing older brother and the other, who is a police officer, to carry out some official duties. They're both set up by a local gangster, who sends some assassins to kill them, and it's not long before they unite in a mission to destroy the criminal elements in the town.
MAR   81 min (ort 83 min)   VIDrel: ONE   V

DEVIL MAN: THE BIRTH/DEMON BIRD **   18
JAPAN
Start of a two-part horror tale, with a boy's life being interrupted by the disappearance of his parents. In part two, matters become more serious, as we learn that demons have escaped from beneath the Earth's glacial ice, and now threaten all of humanity.
ANIM   107 min (2 cassettes) dubbed
VIDrel: MANGA/SONOP   V/sh

DEVIL, PROBABLY, THE ****   18
Robert Bresson FRANCE   1977
Antoine Monnier, Tina Irissari, Henri De Maublanc
Despairing at the state of the world, a young man seeks refuge in nihilism and eventually commits suicide. In this clinical study of alienation, Bresson eschews all sentiment and makes the camera an impassive observer of Monnier's journey towards the inevitable, at each stage the attempts to halt this (such as by way of politics or religion) being shown for the shams they are. A remarkable but very painful film. See also WINTER LIGHT.
Aka: LE DIABLE, PROBABLEMENT
DRA   92 min wScrn   VIDrel: ARTIF/20TH   V

DEVIL RIDES OUT, THE **   PG
Terence Fisher UK   1968
Christopher Lee, Charles Gray, Nike Arrighi, Leon Greene, Patrick Mower, Gwen Ffrangcon-Davies, Sarah Lawson, Paul Eddington, Rosalyn Landor, Russell Waters, Monagh Singh, Keith Pyott, John Brown
An aristocrat who is well versed in the occult, sets out to battle a group of Satanists for the soul of a friend who has fallen into their clutches. A laboured adaptation of a rather good Wheatley novel that has a few moments of genuine suspense, several good effects and a lot of talk.
Aka: DEVIL'S BRIDE, THE
HOR   92 min (ort 95 min) wScrn   VIDrel: LUMI/SPEAR   V
Boa: novel by Dennis Wheatley.

DEVIL-SHIP PIRATES, THE **   PG
Don Sharp UK   1964
Christopher Lee, Andrew Keir, John Cairney, Barry Warren, Ernest Clark, Duncan Lamont
When the captain of a Spanish Galleon (Lee) suffers defeat at the hands of Drake, he takes refuge on the coast of Cornwall, where his crew terrorise the locals. This was the only foray into the swashbuckling genre for Hammer Films, and though the direction and acting is certainly competent, there is not enough action to keep the story moving along. A few touches of horror amply demonstrate that the producers were more at home sticking to their familiar territory.
A/AD   86 min wScrn   VIDrel: LUMI/SPEAR   V

DEVIL TO PAY, THE **   U
George Fitzmaurice USA   1930
Ronald Colman, Loretta Young, Florene Britton, Frederick Kerr, David Torrence, Mary Forbes, Paul Cavanagh, Crauford Kent, Myrna Loy
The playboy son of a wealthy Englishman returns home from South Africa almost penniless and proceeds to fall in love with an attractive socialite who is engaged to marry a duke. However, there are many misunderstandings before these two

are united. Colman's screen persona and the witty script help to compensate for the somewhat stiff direction.
COM   72 min B/W   VIDrel: VCC/DISC/COLUM   V
Boa: play by Frederick Lonsdale.

DEVILS, THE **   18
Ken Russell UK   1971
Oliver Reed, Vanessa Redgrave, Dudley Sutton, Max Adrian, Gemma Jones, Murray Melvin, Michael Gothard, Georgina Hale, Brian Murphy, Christopher Logue, Graham Armitage, John Woodvine, Andrew Faulds, Kenneth Colley
Overblown and over-rated account of the alleged demonic possession of a number of nuns in 17th century France. Concentration on the visual aspects of death and horror to the exclusion of everything else, results in a failure to capture any of the psychological aspects of mass-hysteria. Shallow.
DRA   104 min (Cut at film release – ort 111 min)
VIDrel: WHV   V
Boa: book The Devils Of Loudoun by Aldous Huxley/play by John Whiting.

DEVIL'S BED, THE **   (PG)
Sam Pillsbury USA   1994
Nicollette Sheridan, Joe Lando, Adrian Pasdar, Richard Roundtree, Brnadon Smith, Brandon Smith, Eleese Lester, Laura Poe, Fred Andrews, Mary Furse, Donald Sneed, Jerry Haynes, Brady Coleman, Blue Deckert, James Harrell, Steve Flanagin
A turgid erotic-thriller that takes the form of a love triangle, with Sheridan a young woman who returns home and gets involved with the brother of a former boyfriend. But when the latter comes back, fleeing from $60,000 gambling debts and some enforcers out to collect it, their passion is rekindled in secret. Matters come to a head when the brother conceives a desperate plan to pay off his debts. A downbeat drama that is curiously uninvolving.
THR   91 min mTV   SATrel: MOVIE CHANNEL

DEVIL'S EYE, THE **   18
Ingmar Bergman SWEDEN   1960
Bibi Anderssson, Jarl Kulle, Nils Poppe, Sture Lagerwall, Stig Jarrel, Gunnar Bjornstrand, Gertrud Fridh, Georg Funkquist, Gunnar Sjoberg, Allan Edwall, Torsten Winge, Kristina Adolphsson, Ragnar Arvedson, Borje Lundh
The Devil sends Don Juan back to Earth to seduce a virgin bride-to-be whose goodness is more than he can stomach, and is in fact a sty in his eye. The great seducer and his servant arrive at the country vicarage where the girl lives with her parents and set about their task, only to find themselves frustrated at every turn. A disappointing and stilted comedy, from a director capable of much better things even in this field.
Aka: DJAVULENS OGA
COM   83 min (ort 90 min) B/W   VIDrel: TART/20TH   V
Boa: radio play Don Juan Vender Tillbage by Oluf Bang.

DEVILS IN THE CONVENT **   18
Franz Antel (Francois Legrand) ITALY/WEST GERMANY
1973
Galliano Sbarra, Paul Lowinger, Christina Losta, Marika Mindzenthy, Sonja Jeannine, Maja Hoppe, Kurt Grosskurth, Teri Tordai, Margot Hielscher, Hans Terofal, Franz Muxeneder, Alena Penz, Erich Padalewski, Dolores Schmidinger
A sex romp set in a convent, with a randy solicitor working his way through a collection of novice nuns in order to ascertain which of them is the true heir to the fortune left by a dead prostitute.
Aka: DEVIL IN A CONVENT; FRAU WIRTINS TOLLE TOCHTERLEIN; KNICKERS AHOY
A   85 min   VIDrel: MOPIC/SGSVID   V

DEVILS OF DARKNESS *   18
Lance Comfort UK   1965
William Sylvester, Hubert Noel, Tracy Reed, Carole Gray, Diana Decker, Rona Anderson, Peter Iling, Gerard Heinz, Victor Brooks, Avril Angers, Julie Mendes
A vampire disguises himself as a French count (a most original disguise), in order to prey on young girls in this tale of vampirism and black magic. Generally shoddy despite a couple of good scenes.
HOR   85 min (ort 90 min)   VIDrel: ARTPRO/RTM   V

DEVIL'S OWN, THE **   15
Alan J. Pakula USA   1997
Harrison Ford, Brad Pitt, Margaret Colin, Ruben Blades, Treat

Williams, George Hearn, Mitchell Ryan, Natascha McElhone, Paul Ronan, Simon Jones, Julia Stiles, Ashley Carin, Kelly Singer
Ford plays a kindly New York cop who unwisely offers shelter to Pitt, an Irishman who has just arrives in the States. Unfortunately, his new lodger turns out to be a psychopathic terrorist. Directed with efficiency, the cast make the most of their fairly unoriginal material.
THR   111 min   CINrel

DEVIL'S PLAYGROUND, THE ***   15
Fred Schepisi AUSTRALIA   1976
Arthur Dignam, Nikc Tate, Simon Burke, Charles McCallum, John Frawley, Jonathon Hardy, Gerry Duggan, Peter Cox, John Dietrich, Thomas Keneally, Sheila Florance, Gerda Nicholson, John Proper, Anne Phelan, Jillian Archer
Drama set in the cloistered atmosphere of Catholic seminary where students and priests are tempted by carnal desires. Well made and acted but not offering any new insights into this familiar subject.
DRA   95 min (ort 107 min) wScrn   VIDrel: ARTPRO/RTM
V

DIABOLIQUE *   18
Jeremiah S. Chechik USA   1996
Sharon Stone, Isabelle Adjani, Chazz Palminteri, Kathy Bates, Spalding Gray, Shirley Knight, Allen Garfield, Adam Hann-Byrd, Donal Logue, Jeffrey Abrams, Diana Bellamy, Clea Lewis, O'Neal Compton, Bingo O'Malley, Stephen Liska
A dreadful remake of LES DIABOLIQUES with a man so cruel to both his wife and mistress that they get together to plan his demise. Having poisoned him they dump his body in a stagnant pond, only to find that it has vanished when the pool is drained. Bates plays the investigator out to solve the mystery of this disappearance – a serious waste of a top actress.
THR   103 min (ort 108 min) cC   VIDrel: WHV   V/sur

DIAL M FOR MURDER ***   PG
Alfred Hitchcock USA   1954
Ray Milland, Grace Kelly, Robert Cummings, John Williams, Anthony Dawson
An ageing tennis champion plots to kill his wife in the perfect crime. Most of the action takes place in one room and although the cast try hard, the film never really convinces, despite a few moments of genuine tension. Remade for TV in 1981.
DRA   100 min (ort 103 min)   VIDrel: WHV   V
Boa: play by Frederick Knott.

DIAMOND HORSESHOE **   U
George Seaton USA   1945
Betty Grable, Dick Haymes, Phil Silvers, William Gaxton, Beatrice Kay, Carmen Cavallaro, Margaret Dumont
A flashy comedy-musical set in a famous cabaret, where a nightclub singer gives up her career for a her love of a medical student. A colourful vehicle for Grable; efficient rather than appealing. Written and directed by Seaton. Songs include "The More I See You".
Aka: BILLY ROSE'S DIAMOND HORSESHOE
MUS   100 min (ort 104 min)   VIDrel: 20TH/TECH   V/h
Boa: play The Barker by Kenyon Nicholson.

DIAMOND NINJA FORCE **   18
Joseph Lai HONG KONG   1986
Richard Harrison, Melvin Pitcher, Andy Chrorowsky, Clifford Allan, Pierre Trembley, Donald Kong, Maria Francesca
Battling Ninja clans fight a bloody feud in this standard Ninja nonsense.
A/AD   85 min Cut (1 min 22 sec)   VIDrel: IMPENT   V

DIAMOND SKULLS **   18
Nicholas Broomfield UK   1989
Gabriel Byrne, Amanda Donohoe, Michael Hordern, Judy Parfitt, Douglas Hodge, Ian Carmichael, Sadie Frost
Aristocratic Byrne and his drunken soldier cronies are responsible for the death of a chambermaid, when the girl is run over in a car accident and is then callously left to die. Despite the closing of ranks, one of the group is plagued by a guilty conscience and may yet betray the others. Documentary filmmaker Broomfield presents a hollow and clinical study of guilt and decadence. The script is by Tim Rose Price, with music by Hans Zimmer.
Aka: DARK OBSESSION
THR   83 min (ort 87 min)   VIDrel: VIR/RCA L/A   V

DIAMONDS ARE FOREVER **   PG
Guy Hamilton UK   1971
Sean Connery, Jill St John, Charles Gray, Jimmy Dean, Lana Wood, Bruce Cabot, Putter Smith, Bruce Glover, Norman Burton, Joseph Furst, Bernard Lee, Desmond Llewellyn, Leonard Barr, Lois Maxwell, Margaret Lacey, Joe Robinson
James Bond versus the villain, who this time is trying to corner the world's diamond supply. A fast-moving and colourful adventure, set in Las Vegas, but spoilt by some rather unnecessary touches of viciousness. Number seven in this long-running series, and followed by LIVE AND LET DIE, in which Roger Moore took over from Connery.
A/AD   115 min (ort 120 min) wScrn   VIDrel: MGM/WHV
V/dm V/h
Boa: novel by Ian Fleming.

DIANA: HER TRUE STORY *   PG
Kevin Connor UK   1992
Serena Scott Thomas, David Threlfall, Elisabeth Garvie, Donald Douglas, Anne Stallybrass, Jemma Redgrave, Jeremy Child, William Franklyn, Belle Connor, Helen Masters, Aletta Lawson, Cornelia Hayes O'Herlihy, Gabrielle Blunt
A flat rendition of Morton's bestseller that allegedly deals with Princess Diana's unhappy and much-reported marital problems. An overwrought and totally one-sided effort that puts one in mind of TV's "Spitting Image" caricatures, most memorably with Threlfall playing Prince Charles as if half of his face is paralysed. Carelessly thrown together, it invites both ridicule and contempt. See THE WOMEN OF WINDSOR, which attempted a more balanced picture.
DRA   183 min mTV   VIDrel: ENTUK/GOLD L/A   V
Boa: book by Andrew Morton.

DIARY OF A CHAMBERMAID ***   15
Luis Bunuel FRANCE   1963
Jeane Moreau, Michel Piccoli, Georges Geret, Francoise Lugagne, Daniel Ivernel, Jean-Claude Carriere
In Paris in the 1930s a grasping young woman takes a job as a chambermaid, but when a little girl is found murdered, she finds her work bringing her into contact with a sadistic valet. The director skilfully reworks Mirbeau's tale of greed and repression (it was first made by Renoir in 1946) and darkens the tone considerably, using the events described to examine social, religious and sexual conventions in a manner that is pure Bunuel.
Aka: LE JOURNAL D'UNE FEMME DE CHAMBRE
DRA   93 min (ort 98 min) B/W wScrn
VIDrel: ELPIC/POLYREC   V
Boa: novel by Octave Mirbeau.

DIARY OF A HITMAN ***   18
Roy London USA   1991
Forest Whitaker, Sherilyn Fenn, Sharon Stone, James Belushi, Seymour Cassel, John Bedford-Lloyd, Dan Kamin, Wayne Crawford, Laurie Smith, Lois Chiles, Jimmy Butler, Jonny Chrisinger, Winni Flynn, Eva Jackson, Brittany Marsh
A hitman accepts one last job before retiring as he needs the money for a downpayment on an apartment, but begins to have some misgivings when he learns that the victims are a young woman and her baby. Neglecting the code of his calling, the killer makes contact with his intended victims and discovers that his client has not told him the truth. First-rate acting and taut direction make this one stand out. Scripted by Pressman from his play.
DRA   86 min (ort 90 min)   VIDrel: 20TH/TECH   V/sh
Boa: play Insider's Price by Kenneth Pressman.

DIARY OF A LOST GIRL ***   PG
Georg W. Pabst GERMANY   1929
Louise Brooks, Fritz Rasp, Josef Ravensky, Sybille Schmitz, Valeska Gert, Edith Meinhardt, Vera Pawlowa, Andre Roanne, Arnold Korff, Andrews Engelmann, Franziska Kinz
A rich man's daughter suffers a succession of horrible and degrading experiences including rape, imprisonment and working in a brothel. A classic German silent that was often heavily cut by the censors.
Aka: TAGEBUCH EINER VERLORENEN
DRA   74 min (ort 110 min) B/W silent   VIDrel: TART/20TH
V
Boa: novel by Margaret Bohme.

DIARY OF ANNE FRANK, THE ***   U
George Stevens USA   1959
Millie Perkins, Joseph Schildkraut, Shelley Winters, Richard Beymer,

Lou Jacobi, Diane Baker, Ed Wynn, Gusti Huber, Douglas Spencer, Dody Heath, Charles Wagenheim, Frank Tweddell, Delmar Erikson, Robert Boon, Gretchen Goertz
The true story of the fate of a group of Dutch Jews who hid from the Nazis in an Amsterdam attic, based on Frank's diary and a stage play. Despite poor casting of Perkins as the title character, this film makes a brave attempt to bring her story to the screen. Written by Goodrich and Hackett and remade for TV in 1980. AA: S. Actress (Winters), Cin (William C. Mellor), Art/Set (Lyle R. Wheeler and George W. Davis/Walter M. Scott and Stuart A. Reiss).
DRA   VIDrel: 20TH/TECH V
Boa: the diary of Anne Frank/play by Frances Goodrich and Albert Hackett.

DIARY OF LADY M, THE **   18
Alain Tanner BELGIUM/FRANCE/SPAIN/SWITZERLAND 1992
Myriam Mezieres, Juanjo Puigcorbe, Felicite Wouassi, Nanou, Marie Peyrucq-Yamou, Gladys Gambie, Antoine Basler, Makeda, Albert Planes, Roger Mendri, Carlota Soldevila, Olivier Ceyssens
Adult drama in which a painter falls in love with a night-club singer and embarks on a passionate affair with her, until a dropped photograph changes everything. Based on Mezieres's screenplay (which in turn was inspired by her real-life affair with a Spanish painter) this had the potential to become a fascinating example of cinema verite, but is quickly sunk by dull photography and a total lack of atmosphere.
Aka: LE JOURNAL DE LADY M
DRA   107 min (ort 112 min)   VIDrel: MAINPIC/RTM   V

DICK TRACER **   R18/18
John T. Bone USA
Alice Springs, Deidre Holland, Kelly Blue, Misty Regan, Jon Dough, Joey Silvera
The title refers to the speciality of a female private detective, who is asked to locate the missing "Maltese Dick" in this silly sex spoof.
Aka: MISTY UNDERCOVER
A   85 min (R18 ver); 51 min (18 ver)   VIDrel: ELV   V

DICK TRACY ***   PG
Warren Beatty USA   1990
Warren Beatty, Madonna (Madonna Ciccione), Charlie Korsmo, Al Pacino, Glenne Headly, Mandy Patinkin, Charles Durning, Paul Sorvino, William Forsythe, Seymour Cassel, Dustin Hoffman, Dick Van Dyke, Catherine O'Hara, Jim Wilkey
An enjoyable and highly ambitious adaptation of Chester Gould's comic-strip that attempts to retain some of the flavour of the comic by cleverly using a restricted range of colours, as our square-jawed and resolute detective takes on a succession of bizarre villains. Not so much a straight movie as an entertaining parody. AA: Song ("Sooner Or Later (I Always Get My Man)" – S. Sondheim), Make (J. Caglione/D. Drexler), Art (R. Sylbert/R. Simpson).
A/AD   100 min (ort 105 min)   VIDrel: TOUCH/BUENA L/A   V

DIE HARD ***   18
John McTiernan USA   1988
Bruce Willis, Bonnie Bedelia, Paul Gleason, William Atherton, Hart Bochner, Alan Rickman, Alexander Gudonov, Reginald Veljohnson, De'voreaux White, Robert Davi, James Shigeta, Bruno Doyon, Andres Wisniewski, Joey Plewa, Matt Landers
Tough action thriller, about a cop who happens to be attending a Christmas party on the top of a skyscraper, when terrorists burst in and take the guests hostage. Worth seeing for its good action stunts and sequences, though its excessive length is a handicap. Two sequels followed.
A/AD   126 min (ort 132 min) wScrn   VIDrel: 20TH/TECH; ENCORE (LV only)   V/sur LV

DIE HARD 2 ***   15
Renny Harlin USA   1990
Bruce Willis, Bonnie Bedelia, William Atherton, Reginald Veljohnson, Franco Nero, Dennis Franz, William Sadler, John Amos, Art Evans, Sheila McCarthy, Fred Dalton Thomson, Tom Bower, Don Harvey, Tony Ganios, Peter Nelson, Robert Patrick
Formula sequel made to trade on the success of the first film. A gang of terrorists seizes an airport to rescue a drug baron and Detective McClane, there to meet his wife, has to do battle with all and sundry. Meanwhile, his wife's plane is running low on fuel. A film that's very strong on action sequences (its final

budget was $70,000,000) if not on plausibility. DIE HARD WITH A VENGEANCE followed.
Aka: DIE HARD 2: DIE HARDER
A/AD   118 min (ort 124 min) wScrn cC
VIDrel: 20TH/TECH; ENCORE (LV only)   V/sur LV
Boa: novel 58 Minutes by Walter Wager.

DIE HARD WITH A VENGEANCE **   15
John McTiernan USA   1995
Bruce Willis, Samuel L. Jackson, Jeremy Irons, Graham Greene, Colleen Camp, Larry Bryggman, Sam Phillips, Anthony Peck, Nick Wyman, Kevin Chamberlin, Sharon Washington, Stephen Pearlman, Michael Alexander Jackson, Aldis Hodge
Cartoon-style third outing for Willis, this time playing a disgraced cop who becomes the personal target of a mad terrorist who delights in blowing up various buildings in New York until his demands are met. Full of non-stop sound and fury and even more violent than the first two DIE HARD films, but sadly plot and characterisation must take even more of a back seat to the state-of-the-art special effects, which soon grow mighty tiresome.
A/AD   128 min (ort 131 min) wScrn cC
VIDrel: TOUCH/TECH; ENCORE (LV only)   V/sh LV

DIE WATCHING **   18
Charles Davis USA   1992
Christopher Atkins, Tim Thomerson, Vali Ashton, Carlos Palomino, Mike Jacobs Jr.
A number of Hollywood actresses and models get caught up in a director's private fantasies of lust and murder. Average erotic thriller with one or two good moments.
THR   81 min (ort 92 min)   VIDrel: NWV/HIFLI   V

DIFFERENT STORY, A **   15
Paul Aaron USA   1978
Perry King, Meg Foster, Valerie Curtin, Peter Donat, Richard Bull, Barbara Collentine, Guerin Barry, Doug Higgins, Lisa James, Eugene Butler, Linda Carpenter, Allan Hunt, Burke Byrnes, Eddie C. Dyer, Richard Altman
A homosexual and a lesbian marry so as to prevent his deportation, and then proceed to fall in love. Not as bad as it sounds, though as the film progresses it slowly runs out of ideas, despite a really great performance from Foster.
COM   103 min (ort 107 min)   VIDrel: DV8/DISC   V

DIGGER ***   (PG)
Robert Turner CANADA   1994
Adam Hann-Byrd, Joshua Jackson, Barbara Williams, Timothy Bottoms, Olympia Dukakis, Leslie Nielsen, D. Lynn Johnston, Lochlyn Monro, Danielle Fraser, Gabrielle Miller, Andrew B. Parker, Colette Aubin
When his parents' marriage hits the rocks, a young boy is sent to spend some time with his uncle and aunt until things are resolved. Feeling hurt and rejected, his spirits rise when he makes friends with a young boy his own age. Although the latter is physically weak, he manages to teach our hero how to use his imagination and enjoy what life has to offer.
DRA   92 min   SATrel: SKY MOVIES

DIGITAL MAN *   15
Philip J. Roth USA   1994
Ken Olandt, Kristen Dalton, Adam Baldwin, Ed Lauter Paul Gleason, Matthias Hues, Don Swayze, Chase Masterson, Sherman Augustus, Woon Park, Megan Blake, R.J. Bonds, Philip Bruns, Joe Cook, Cliff Emmich, Darren Foreman, Clint Howard
When terrorists gain control of the launch codes for 250 nuclear missiles and demand a vast ransom, an experimental cyborg soldier is dispatched to deal with them. This mission is completely successfully but the robot then proceeds to act on its own and has to be stopped. A laughably silly futuristic adventure tale with poor special effects, a muddled plot and indifferent acting that fails to convince. An undistinguished variation on the "killer cyborg" theme.
FAN   91 min (ort 95 min)   VIDrel: MOSAIC/COLUM V/sh

DIM SUM ***   PG
Wayne Wang USA   1985
Laureen Chew, Kim Chew, Victor Wong, Ida F.O. Chung, Cora Miao, John Nishio, Amy Hill, Keith Choy, Mary Chew, Nora Lee, Joan Chen, Rita Yee, George Wu, Elsa Cruz Pearson, Helen Chew, Jarett Chew
A warm and loving look at three generations of San Francisco Chinese, and especially at the relationships between parents

and children. This gentle and often comical tale focuses on an old-fashioned mother whose modern and free-spirited daughter yearns to leave home for a life of her own. Slow in places, with long leisurely shots of empty rooms, but most definitely worth seeing.
Aka: DIM SUM: A LITTLE BIT OF HEART
DRA   89 min (English and Cantonese with Mandarin subs)
VIDrel: PAL L/A   V

DINER *   15**
Barry Levinson USA   1982
Steve Guttenberg, Daniel Stern, Mickey Rourke, Kevin Bacon, Ellen Barkin, Timothy Daley, Paul Reiser, Michael Tucker, Kathryn Dowling, Jessica James, Colette Blonigan, Kelle Kipp, John Aquino, Richard Pierson, Claudia Cron
An evocation of the problems of growing, up experienced by a bunch of 1950s kids who hang out in a Baltimore diner, looking forward to Christmas and a big ball game, and wondering what life is all about along the way. A warm and quite touching film, it gave Levinson (who also wrote the script) his directing debut. See also SHAKING THE TREE.
COM   105 min (ort 110 min)   VIDrel: MGM/WHV   V

DINNER AT EIGHT **   PG**
George Cukor USA   1933
Jean Harlow, John Barrymore, Marie Dressler, Lionel Barrymore, Lee Tracy, Waalace Beery, Billie Burke, Edmund Lowe, Madge Evans, Jean Hersholt, May Robson, Karen Morley, Phillips Holmes, Louise Closser Hale, Grant Mitchell
Story of the guests at a dinner party who are drawn from various levels of society. Polished, sophisticated and witty – a Hollywood classic. The fine script is by Herman Mankiewicz, Frances Marion and Donald Ogden Stewart. Remade after a fashion in 1989.
COM   107 min (ort 113 min) B/W   VIDrel: MGM L/A   V
Boa: play by Edna Ferber and George S. Kaufman.

DINNER AT EIGHT *   15
Ron Lagomarsino USA   1989
John Mahoney, Marsha Mason, Charles Durning, Lauren Bacall, Ellen Greene, Harry Hamlin, Joel Brooks, Tim Kazurinsky, Stacy Edwards, Kelly Connell, Richard Seff, Jane Alden, Bernadette Birkett, Loyda Ramos, Ralph Bruneau, Julia Sweeney
An updated version of the classic 1930s comedy-drama, in which a variety of characters plan to attend a swanky dinner party, held by the snobbish wife of a minor shipping magnate. A disparate set of sub-plots offers little of interest (though Hamlin's performance as a cocaine-addicted actor at the end of his career is both touching and pathetic). The strong cast do their best with the material, but mumbled dialogue and weak direction are additional handicaps.
COM   94 min (ort 100 min) mCab   VIDrel: L/A   V/sh
Boa: play by Edna Ferber and George S. Kaufman.

DINOSAURS: THE MOVIE **   U
Brett R. Thompson USA   1990
Shawn Hoffman, Tiffanie Poston, Omri Katz, Marc Martorana, Tony Doyle, R.A. Mikhailoff, Don Barnes, Pete Koch, Megan Hughes, Mimi Maynard, Steven Anderson, Jeffrey Asch, Sebastian Mossa, Kimberly Beck, Kevin Thompson, Irwin Keyes
A bunch of kids are projected into their favourite TV program, and have to assist their dinosaur heroes in saving their city from an attack mounted by Neanderthals, under the leadership of a certain Mr Big. An adequate if not especially enthralling kids' fantasy.
Aka: ADVENTURES IN DINOSAUR CITY
JUV   90 min   VIDrel: POLY/POLYREC L/A; PION (LV only)   V/s LV

DIPLOMATIC IMMUNITY **   18
Peter Maris USA   1990
Bruce Boxleitner, Billy Drago, Meg Foster, Tom Breznahan, Christopher Neame, Robert Forster, Fabiana Udenio, Matthias Hues, Sharon L. Case, Lee DeBroux, Ken Foree, Robert Do'Qui, J. Marvin Campbell, Kenneth Kimmins, Rozlyn Sorrell
When his daughter is brutally murdered by a German psychopath whose diplomatic status protects him from arrest and prosecution, a tough marine sergeant travels to Paraguay on a personal mission of vengeance. A well made and ably acted effort that offers few surprises.
A/AD   90 min Cut (29 sec – ort 95 min)   VIDrel: MIA/DISC V
Boa: novel The Stalker by Theodore Taylor.

DIRECT HIT **   18
Joseph Merhi USA   1994
William Forsythe, George Segal, Jo Champa, Richard Norton, John Aprea, Eddi Wilde, Juliet Landau, Steve Garvey, Marc Fiorini, Gary Roberts, Gabrielle Lauren Michel, Mel Novak, Sam Shamshak, Peter Slutsker, David St. James, Sako Kidikian
A CIA assassin agrees to one last job as a condition for his retirement but is angered when he realises that his target is a young and attractive woman who seems innocent of any crime. Soon he finds himself falling in love with her and going up against his former employers. A predictable actioner, reasonably well made.
DRA   85 min (ort 90 min)   VIDrel: COLUM/SONOP   V

DIRTY DANCING *   15**
Emile Ardolino USA   1987
Patrick Swayze, Jennifer Grey, Cynthia Rhodes, Jerry Orbach, Jack Weston, Jane Brucker, Kelly Bishop, Lonny Price, Charles "Honi" Coles, Bruce Morrow, Max Cantor, Wayne Knight, Alvin Myerovitch, Paula Trueman, Miranda Garrison
A naive and gauche seventeen-year-old girl on holiday with her parents in the Catskills in the early 1960s, meets the local dance instructor and develops a crush on him, and he teaches her about love, life and dancing. A charming if superficial story with some great dance numbers. A TV series followed. The LV disc includes some interviews. AA: Song ("(I've Had) The Time Of My Life" – Franke Previte, John Denicola and Donald Markowitz (m)/Frankie Previte (l)).
DRA   100 min (ort 105 min) wScrn   VIDrel: FIRST/SONOP;
PION (LV only)   V/sur LV

DIRTY DOZEN, THE *   15**
Robert Aldrich USA   1967
Lee Marvin, Telly Savalas, Jim Brown, Ernest Borgnine, Robert Ryan, Charles Bronson, Donald Sutherland, John Cassavetes, Clint Walker, Richard Jaeckel, George Kennedy, Ralph Meeker, Trini Lopez, Robert Webber, Tom Busby, Al Mancini
Fairly competent action film with twelve prisoners being recruited for a suicide mission during WW2. Well-acted and highly entertaining, even if the acting is not up to much and the film has dated badly. The inevitable parade of sequels followed. AA: Effects/aud (John Poyner).
WAR   143 min (ort 150 min) wScrn (special edition)
VIDrel: MGM/WHV   V/h
Boa: novel by E.M. Nathanson.

DIRTY DOZEN, THE: THE NEXT MISSION **   PG
Andrew V. McLaglen USA   1985
Lee Marvin, Ernest Borgnine, Ken Wahl, Richard Jaeckel, Larry Wilcox, Sonny Landham, Wolf Kahler, Gavin O'Herlihy, Ricco Ross, Stephen Hattersley, Rolf Saxon, Jay Benedict, Michael John Paliotti, Paul Herzberg, Jeff Harding
We could have done without this flawed attempt to recapture some of the excitement of the first film. Our heroes are coerced into a mission to bump off a German general, who wants to assassinate Adolf and thus prolong the war. A fine example of how the total amounts to less than the sum of the parts – all the necessary action elements are there but the foolish script causes the whole film to sag badly. Followed by two mediocre sequels.
WAR   92 min (104 min) mTV   VIDrel: MGM/WHV   V/h

DIRTY DOZEN, THE: THE DEADLY MISSION **   15
Lee H. Katzin USA   1987
Telly Savalas, Ernest Borgnine, Vince Edwards, Gary Graham, James Van Patten, Vincent Van Patten, Bo Svenson, Randall Cobb, Thom Mathews
Second sequel to THE DIRTY DOZEN, with a bunch of army convicts being assembled once again, this time to go on a dangerous mission behind the German lines. Their task is to destroy a deadly nerve gas factory and rescue the scientists who are being forced to work there. Adequate, but only just.
Aka: DIRTY DOZEN 3: THE DEADLY MISSION; DEADLY MISSION, THE
WAR   90 min (ort 96 min) mTV   VIDrel: MGM L/A   V

DIRTY DOZEN, THE: THE FATAL MISSION **   PG
Lee H. Katzin USA   1988
Telly Savalas, Ernest Borgnine, Erik Estrada, Robert Vaughn, Heather Thomas, Jeff Conaway, Alex Cord, Hunt Block, Ray Mancini
The third TV sequel, with our ever changing group of WW2 misfits out to stop Adolf starting up a Fourth Reich in the Middle East. Set on the Orient Express, where they have to discover which of their members is a spy. Could it be their latest recruit – a woman? Poor fare indeed.
WAR   91 min (ort 96 min) mTV   VIDrel: MGM/WHV   V/h

DIRTY GAMES ** **   15
Gray Hofmeyr SOUTH AFRICA/USA   1986
Jan Michael Vincent, Valentine Vargas, Ronald France, Michael McGovern, Paul Eilers, Frantz Dobrowsky, Lynne White, Colin Sutcliffe, Andrew Buckland, Percy Sieff, John Whitley, Gary Ford, Henry Nairac, Robert Travallyn
A group of terrorists attack a nuclear waste storage plant in Africa that is being visited by a party of scientists, but are eventually overcome after much violence and loss of life. A standard tough actioner with Vincent adequate in the role of the tough hero.
A/AD   96 min   VIDrel: POLY/POLYREC/BRAVE   V

DIRTY HARRY *   18**
Don Siegel USA   1971
Clint Eastwood, Harry Guardino, John Vernon, Reni Santoni, John Larch, Andy Robinson, John Mitchum, Mae Mercer, Lyn Edgington, Ruth Kobart, Woodrow Parfrey, Josef Sommer, William Paterson, Craig G. Kelly, James Nolan, Joe DeWinter
Nice tense thriller in which a tough San Francisco cop breaks all the rules in order to catch a psychotic killer. The dialogue is fairly contrived but the action sequences are nicely handled. Followed by a number of sequels, beginning with MAGNUM FORCE.
THR   98 min (ort 102 min)   VIDrel: WHV   V/h

DIRTY HO ** **   15
Liu Chia Hui HONG KONG   1979
Wong Yue, Liu Chia Hui, Hui Ying Hung, Lo Lieh
When the eleventh son of the Emperor of China is ambushed by assassins, a man known as Dirty Ho comes to his aid, and the two become firm friends. An interesting martial arts movie with the usual revenge-driven themes, but better than average plotting and characterisation.
MAR   99 min wScrn dubbed   VIDrel: MADE/RTM   V

DIRTY LAUNDRY ** **   (15)
William Webb USA   1987
Leigh McCloskey, Jeanne O'Brien, Frankie Valli, Sonny Bono, Nicholas Worth, Robbie Rist, Edy Williams, Carl Lewis, Greg Louganis, John Moschitta Jr, Herta Ware, Johnny B. Frank, Ben Mittleman
The story of an innocent man who becomes involved with gangsters when a satchel full of drugs gets switched with his laundry. The odd cast (including Olympic athletes Lewis and Louganis) add a little sparkle to the basically routine script.
COM   80 min   CABrel: HVC

DIRTY LOVE 2: THE LOVE GAMES *   18**
Joe D'Amato (Aristide Massaccesi) ITALY   1990
Peter Marc, Josie Bisset, Courney Allen, Wanda Murray, Dale Wyatt, Gaby Ford
A naive female music student falls victim to some highly lecherous men, including a professor whose pursuit of her takes place against the background of the Venice Carnival. A dull, glossy travelogue of a sex film.
Aka: DESIRE; DIRTY LOVE 2; LOVE GAMES, THE
A   60 min (ort 96 min)   VIDrel: MOPIC/SGSVID   V

DIRTY ROTTEN SCOUNDRELS *   PG**
Frank Oz USA   1988
Steve Martin, Michael Caine, Glenne Headly, Anton Rodgers, Barbara Harris, Ian McDiarmid, Dana Ivey, Meagan Fay, Frances Conroy, Nicole Calfan, Aina Walle, Cheryl Pay, Nathalie Auffret, Lolly Susi, Rupert Holliday Evans
A successful con artist who makes a career of fleecing wealthy women on the French Riviera, faces competition from a crass new arrival. In an effort to decide who is to remain, they make a wager to see who can be the first to get a foolish soap heiress to part up with $50,000. Something of an improvement on "Bedtime Story" (1964), with both Martin and Caine engaging in countless double-crosses as each one tries to outsmart the other.
COM   105 min (ort 110 min)   VIDrel: 4-FRONT/POLYREC   V/sh

DIRTY TRICKS *   15**
Michael Lindsay-Hogg USA   1992
Diane Keaton, Ed Harris, Ed Begley Jr, Ben Masters, Robert Harper, Brandon Maggart, Russ Tamblyn, Edgar Small, Richard B. Livingston, Stack Keach Sr, Carrie Snow, Roy Soden, Gregg Berger, Flor De Re, Steve Gonzales, Hugh Holub
A successful writer of children's books meets an old flame from

her youth, who is still unmarried and is now running for the Senate. But our writer hates politics and politicians, yet despite the unlikely couple hit it off well. However, a dark secret in the woman's past threatens to ruin her boyfriend's chances of election, when it is uncovered by the press. A pleasing romantic comedy, that takes a nicely cynical view of the press and politics.
Aka: RUNNING MATES
COM   88 min mCab   VIDrel: WHV L/A   V/sh

DIRTY WEEKEND *   18**
Michael Winner UK   1992
Lia Williams, David McCullum, Rufus Sewell, Sylvia Syms, Michael Cole, Mark Burns, Ian Richardson, Christopher Ryan, Sean Pertwee, Nicholas Hewetson, Jack Galloway, Christopher Adamson, Martha Marsh, Shaughan Seymour, Mark Burns
A secretary with a grudge against men in general and her boyfriend in particular sets out to avenge herself on all men, dispatching a variety of unfortunates by bludgeoning, suffocation, stabbing and shooting. Winner has a splendid time turning Zahavi's man-hating, feminist tract into a film of gross effects and graphic violence, shooting it as if it were a set of trailers to advertise the film, but this clever ploy does not make it any more enjoyable.
DRA   97 min (ort 105 min)   VIDrel: POLY/POLYREC   V/sh
Boa: novel by Helen Zahavi.

DIRTY WORK ** **   15
John McPherson USA   1992
Kevin Dobson, John Ashton, Roxann Biggs, Donnelly Rhodes, Jim Byrnes, Mitchell Ryan, Jason Schombirg, Geoorge Josef, Bernie Coulson, Alex Diakun, Dave "Squatch" Ward, Judith Maxie, David Fredericks, Ken Kirzinger, Tom Heaton
Two former cops join forces to run a bail-bond business, friends through a debt of loyalty owed by one to the other for saving his life years ago. But one of them kills a drug dealer and steals his cache of counterfeit money (which was intended for the Mob) and covers his tracks by pinning the blame on the former partner he once saved. The two men now become bitter enemies and the expected violent fight for survival ensues. Adequate.
A/AD   84 min (ort 88 min) mCab   VIDrel: CIC/SONOP   V

DISAPPEARANCE OF CHRISTINA, THE ** **   15
Karen Arthur USA   1993
John Stamos, Kim Delaney, Robert Carradine, CCH Pounder, Claire Yarlett, Dimitra Arlys, Tom Hodges, Joan Hotchkis, Mindy Seeger, Randy Ogelsby, Jack Yates, Rita Zohar, Penelope Crabtree, Wylie Draper, Larry Marks, Fred D.Scott
A man's wife drowns in a boating accident but this tragic event is soon seen in a very different way when it comes to light that she filed divorce papers just a few days before her death. Needless to say, her grieving husband soon becomes the prime suspect, in this acceptable little drama.
DRA   89 min (ort 93 min)   VIDrel: CIC/SONOP   V

DISAPPEARANCE OF NORA, THE ** **   (15)
Joyce Chopra USA   1993
Veronica Hamel, Stephen Collins, Dennis Farina, Stan Ivar, Leon Russom, Bryan Cranston, David Steen, Alyson Reed, Tom Henschel, James Posolof, Terry Finn, Angela Andrade, Joe Bordeaux, David Boyce, Raye Brewer, Sean Comey
A woman wakes up in the middle of the desert suffering from amnesia and with absolutely no idea how she came to be there. Suspicious of all those she encounters and unsure who she can trust, she sets about trying to solve the riddle of her lost memory. A humdrum thriller with a stylish opening sequence, which unfortunately fails to set the tone for the rest of the film.
THR   90 min mTV   SATrel: MOVIE CHANNEL

DISAPPEARANCE OF VONNIE, THE *   (15)**
Graeme Campbell USA   1994
Ann Jillian, Joe Penny, Kim Zimmer, Graham Beckel, Robert Wisden, Alicia Witt, Gary Chalk, Alexander Purvis, Jerry Wassermann, Travis MacDonald, Robert Clothier, Gabrielle Miller, Marlowe Dawn, Aidan Pendleton, Jennifer Meyer
Jillian gives a terrific performance in the story of woman who grows suspicious at the disappearance of her sister, and comes to suspect her brother-in-law, even though there is neither a body nor any evidence of foul play, especially when she learns that he served a prison sentence for murdering two men twenty years earlier. An intense drama based on a real case that led to a long-overdue change in America's legal system.
DRA   90 min   SATrel: MOVIE CHANNEL

DISASTER AT VALDEZ ***   (PG)
Paul Seed UK/USA   1992
Christopher Lloyd, John Heard, Rip Torn, Don S. Davis, Bob Gunton,
Mark Metcalf, Bruce Grey, Wally Dalton, Paul Guilfoyle, Michael
Murphy, Gary Reineke, Jackson Davies, Kenneth Welsh, David Morse,
Timothy Webber, Tamsin Kelsey
Excellent account of the genesis of one of the world's largest
ecological disasters, when an Exxon tanker ran aground off the
coast of Alaska and spilled eleven million gallons of oil into
Prince William Sound. Fine performances all round and a liter-
ate script (based on the research of Michael Baker and Bob
Duffield) make for an above-average TV drama.
Aka: DEAD AHEAD: THE EXXON VALDEZ DISASTER
A/AD   90 min mTV   TVrel: BBC

DISCLOSURE ***   18
Barry Levinson USA   1994
Michael Douglas, Demi Moore, Donald Sutherland, Caroline Goodall,
Dennis Miller, Roma Maffia, Dylan Baker, Rosemary Forsyth, Suzie
Plakson, Nicholas Sadler, Jacqueline Kim, Joe Urla, Michael Chieffo,
Joe Attanasio, Allan Rich
A male executive is put in an invidious position when he
acquires a new boss, a predatory and ruthless woman who for
good measure, was a former lover. Now happily married, he
tries to reject her sexual overtures, eventually finding that he has
no option but to sue her for sexual harassment. However, in
doing so he finds himself caught up in a conspiracy in which he
is being set up as the fall guy. A steamy, over-complex thriller,
saddled with a contrived denouement.
THR   123 min (ort 128 min) cC   VIDrel: WHV   V/sur
Boa: novel by Michael Crichton.

DISCREET CHARM OF THE BOURGEOISIE,
THE ***   15
Luis Bunuel FRANCE/ITALY/SPAIN   1972
Fernando Rey, Sephane Audran, Jean-Pierre Cassel, Delphine Seyrig,
Michel Piccoli, Bulle Ogier, Paul Frankeur, Julien Bertheau, Claude
Pieplu, Muni, Milena Vukotic, George Douking, Benrard Musson,
Francois Mastre, Ellen Bahl
A group of diners are constantly frustrated in their efforts to eat
a meal by a wild succession of bizarre disruptions, both real
and imagined, in this highly confusing surrealist parable. The
usual Bunuel targets are all vehemently attacked with great
savagery, but the work as a whole is too self-indulgent and
contrived to be really effective. AA. Foreign.
Aka: LE CHARME DISCRET DE LA BOURGEOISIE
COM   97 min (ort 105 min) wScrn
VIDrel: ELPIC/POLYREC   V

DISORIENTED: TWO WONGS MAKE A WHITE *   18
Nancy Nemo USA   1992
Anisa, Fantasia Yee Mai Lin, Cassandra Saki, Kitty, Randy West,
Jonathan Morgan
Standard sex romp in which all the girls are American Chinese,
which is the film's only distinguishing feature.
A   60 min   VIDrel: RAVEN/LWV   V

DISTANT JUSTICE **   18
Toru Murakawa JAPAN   1988
Bunta Sugawara, David Carradine, George Kennedy, Yojo Nogiwa,
Eric Lutes, Sakura Sugawara, John Fiore, Peter Korner, Wesley Clark,
Jeremy Goodwin
When his wife is murdered and his daughter kidnapped, a
tough cop sets out to nail those responsible and is unconcerned
about the effect of his actions on the police department, a
number of whose members have more than a few things to hide.
Another violent police actioner whose Japanese origin adds
absolutely no interest whatsoever.
A/AD   87 min (ort 90 min)   VIDrel: OVER/HIFLI   V/h

DISTANT SCREAM, A **   (PG)
John Hough UK   1984
David Carradine, Stephanie Beacham, Stephen Greif, Stephen Chase,
Fanny Carby, Lesley Dunlop, Bernard Horsfall, Ewan Stuart, Edward
Peel
A married woman on holiday in Cornwall with her lover is
haunted by the vision of an old man who proves to be the spirit
of this man, who is dying in prison ten years in the future, after
being convicted of her murder. Somehow able to project himself
back in time, he tries to find out who it was who killed her and
prevent this from happening. Interesting ideas involving tempo-

ral paradoxes are completely ignored in favour of the usual
heavy-handed approach.
Aka: DYING TRUTH; HAMMER HOUSE OF MYSTERY AND SUSPENSE: A
DISTANT SCREAM; PRIMAL SCREAM
HOR   72 min   SATrel: UK GOLD

DISTANT THUNDER, A ****   (PG)
Satyajit Ray INDIA   1973
Soumitra Chatterji, Babita, Sandhya Ray, Gobinda Chakravarty,
Romesh Mukerji
A moving analysis of the Bengal famine of 1942 and the way in
which it affected the lives of high and low alike. This humane
film examines the forces leading to the preventable disaster of
famine which throughout India claimed no less than five million
lives. Voted Best Film at the Berlin Film Festival.
Aka: ASHANTI SANKET
DRA   92 min (ort 100 min)   TVrel

DISTANT THUNDER **   18
Rick Rosenthal CANADA/USA   1988
John Lithgow, Ralph Macchio, Kerrie Keane, Reb Brown, Janet
Margolin, Dennis Arndt, Jamey Sheridan, Tom Bower
The story of a disturbed Vietnam veteran who has taken to
living in the wilds, and whose son decides to visit him, despite
having been abandoned by him years ago. A likeable but rather
contrived drama, saved by good performances, especially
Lithgow's as our unsociable Vietnam vet.
DRA   109 min (ort 114 min)   VIDrel: CIC/SONOP
V/sur

DISTANT VOICES, STILL LIVES ***   15
Terence Davies UK   1988
Freda Dowie, Peter Postlethwaite, Angela Walsh, Dean Williams,
Lorraine Ashbourne, Sally Davies, Nathan Walsh, Susan Flanagan,
Michael Starke, Vincent Maguire, Antonia Mallen, Debi Jones, Chris
Darwin, Marie Jelliman
A musical saga of working class Liverpool life in the years just
after WW2, told almost entirely through songs with little spoken
dialogue. It presents a convincing and depressing picture of
people crippled and restricted by their upbringing and social
circumstances, but lacks a coherent story with well-developed
characters and understandable motivations.
DRA   80 min (ort 87 min)   VIDrel: VISVID/POLYREC
V/sur

DISTINGUISHED GENTLEMAN, THE **   15
Jonathan Lynn USA   1993
Eddie Murphy, Lane Smith, Sheryl Lee Ralph, Joe Don Baker, Victoria
Rowell, Grant Shaud, Kevin McCarthy, Charles S. Dutton, Victor
Rivers, Gary Frank, Noble Willingham, Cynthia Harris, Chi, Sonny
Jim Gaines, Daniel Benzali, James Garner
A black con-man decides that politics is the best possible scam
and manages to get elected to the House of Representatives, but
his initial cynicism gives way to worthy altruism when
confronted with the plight of a sick child. This satire on
American politics has some acid attacks on the usual targets
and if overlong and contrived, remains quite watchable thanks
to Murphy's winning personality.
COM   108 min (ort 112 min)   VIDrel: HOLPIC/TECH L/A
V

DISTURBANCE, THE *   18
Cliff Guest USA   1990
Timothy Greeson, Carole Garlin, Ken Ceresne, Lisa Geoffrion
A disturbed psycho with a king-sized mother complex murders
all the women that he feels attracted to in this dull and unedi-
fying PSYCHO clone. Very poor.
DRA   90 min Cut (1 min 34 sec plus some cuts subst)
VIDrel: 20VIS/SONOP   V

DIVA **   15
Jean-Jacques Beineix FRANCE   1981
Frederic Andrei, Roland Bertin, Richard Bohringer, Wilhelmenia
Wiggins Fernandez, Thuy An Luu, Jacques Fabbri, Dominique Pinon,
Jean-Jacques Moreau, Chantal Dervaz, Patrick Floersheim, Raymond
Aquilon, Eugene Berthier
A young mailboy makes a pirate recording of an opera star who
has never agreed to be recorded, and finds himself drawn into
a nasty web of murder, when his tape gets mixed up with one
made by a prostitute in order to incriminate a gangster. From
this fairly straightforward beginning the film develops into a

flashy, pretentious and only occasionally gripping thriller. The directorial debut for Beineix.
THR 113 min (ort 123 min) wScrn
VIDrel: ELPIC/POLYREC; ENCORE (LV only) V LV
Boa: novel by Delacorta.

DIVE BOMBER ** **   PG
Michael Curtiz USA   1941
Errol Flynn, Ralph Bellamy, Fred MacMurray, Alexis Smith, Robert Armstrong, Regis Toomey, Craig Stevens, Allen Jenkins, Herbert Anderson, Moroni Olsen, Addison Richards, Louis Jean Heydt, Dennie Moore, Cliff Nazarro, Ann Doran
A flight surgeon is stunned by the death of a flyer during pilot training and becomes one himself in order to study the cause of sudden blackouts. His efforts result in the invention of a high-altitude oxygen suit. A well-made WW2 drama with plenty of exciting flying sequences but virtually no romantic interest apart from a brief sequence with Alexis Smith.
DRA 127 min (ort 133 min) VIDrel: WHV V
Boa: story Beyond The Blue Sky by Frank Wead.

DIXIE LANES ** **   (PG)
Don Cato USA   1988
Hoyt Axton, Karen Black, Art Hindle, Ruth Buzzie, John Vernon, Tina Louise, Pamela Springsteen, Moses Gunn
Bizarre comic tale about a small town at the end of WW2 where a woman running a black market operation hits upon a way to keep her young nephew from getting into trouble by involving him in her clandestine business.
COM 90 min CABrel: HVC

DJANGO ** **   18
Sergio Corbucci ITALY/SPAIN   1965
Franco Nero, Loredana Nusciak, Angel Alvarez, Jose Bodalo, Eduardo Fajardo, Jimmy Douglas, Simone Arriaga, Ivan Scratuglia, Erik Schippers, Raphael Albaicin, Jose Canalecas, Eduardo Fajardo
Average Italian Western with plenty of action and violence. The first in the "Django" series, this has a mysterious stranger arriving at a town at the height of a battle between American and Mexican soldiers, and taking off with gold belonging to the Mexican army.
Aka: IL MERCENARIO; MERCENARY, THE; PROFESSIONAL GUN, A
WES 87 min (ort 100 min) VIDrel: 4-FRONT/POLYREC V

DJANGO STRIKES AGAIN ** **   18
Ted Archer ITALY   1987
Franco Nero, Christopher Connolly, Licia Lee Lyon, Donald Pleasence, William Berger, Robert Posse, Alessandro Di Chio, Rodrigo Obregon, Micky, Bill Moore, Consuelo Reina
Django is captured by a renegade Hungarian prince, running a gold mine with slave labour in the Mexican wilderness, but survives to take the usual violent revenge. Another film in the "Django" series with much the same blood-and-bullets plotline, but helped considerably by a strong score and some impressive Colombian locations.
Aka: IL GRANDE RITORNO DI DJANGO
WES 87 min VIDrel: 4-FRONT/POLYREC V

DJANGO THE BASTARD ** **   15
Sergio Garrone ITALY   1969
Anthony Steffen, Paolo Gozlino, Lu Kanante, Tessoro Corra
At the time of the American Civil War, some officers betray a group of their men to the other side, and they are all slaughtered except for the title figure. He now sets off on a vengeance-seeking mission in this adequate albeit very brutal spaghetti Western, one in a series of "Django" films.
WES 95 min dubbed VIDrel: AKTIV/RTM V

DO OR DIE ** **   18
Vincent McEveety USA
Robert Urich, Markie Post, Michael Beck
When a suspicious wife unwisely begins to shadow her husband, she learns something that places her life in danger.
THR 95 min VIDrel: GENESIS V

DO OR DIE * **   18
Andy Sidaris USA   1991
Erik Estrada, Noriyuki "Pat" Morita, Dona Speir, Bruce Penhall, Roberta Vasquez, Carolyn Liu, Stephanie Schick
A team of attractive Federal agents are used to wear out a master criminal, so he hires a bunch of assassins to get rid of them in this ridiculous nonsense.
A/AD 92 min Cut (8 sec – ort 97 min)
VIDrel: 20VIS/SONOP V/s

DO THE RIGHT THING ** **   18
Spike Lee USA   1989
Danny Aiello, Ruby Dee, Ossie Davis, John Turturro, Spike Lee, Richard Edson, Giancarlo Esposito, Bill Nunn, Paul Benjamin, Frankie Faison, Robin Harris, John Lee, Miguel Sandoval, Rick Aiello, John Savage, Sam Jackson
An episodic comedy-drama that follows 24 hours in the life of a pizza delivery-man, who is always ready to calm the frequent disputes that occur between the various ethnic groups that inhabit his neighbourhood. However, some trivial disagreements cause a riot and the film rapidly moves from comedy to sharp social comment. An unusual film in many ways, but rather blighted by the bitter and caustic tone so typical of this director.
DRA 114 min (ort 120 min) VIDrel: CIC/SONOP V/sur

DO YOU KNOW THE MUFFIN MAN? * **   15**
Gilbert Cates USA   1989
Pam Dawber, John Shea, Stephen Dorff, Brian Bonsall, Matthew Laurance, Anthony Geary, Georgann Johnson, Bruce Fairbairn, William Prince, Graham Jarvis, Ruth De Sosa, Natalie West, John Hammond, Janet Eilber, Casey Friel
A well-balanced and sensitively handled examination of child sex abuse, made all the more shocking by the fact that the culprits are several schoolteachers. When a family discover that their young son has been a victim they face an uphill struggle in getting the authorities to believe them and take appropriate action.
DRA 90 min (ort 94 min) VIDrel: ODY/SONOP V

DO YOU REMEMBER LOVE ** **   PG**
Jeff Bleckner USA   1985
Joanne Woodward, Richard Kiley, Geraldine Fitzgerald, Jim Meltzer, Jordan Charney, Marilyn Jones, Carolyn Lagerfelt, Charles Levin, Craig Richard Nelson, Ron Rifkin, Duncan Ross, Rose Gregorio, Susan Ruttan, Sue Giosa
A middle-aged woman teacher of English starts to suffer from strange lapses of memory and encroaching forgetfulness which is eventually diagnosed as Alzheimer's disease. A sensitive and unsentimental treatment of how this complaint affects every aspect of a person's life, especially their personal relationships. Emmys went to Woodward as Outstanding Actress, to Vickie Patik for the script and to the film itself as Outstanding Drama.
DRA 92 min mTV VIDrel: VCC L/A V

D.O.A. * **   PG**
Rudolf Mate USA   1949
Edmond O'Brien, Pamela Britton, Luther Adler, Neville Brand, Henry Hart, Virginia Lee, Beverly Campbell (Garland), Lyn Baggett, William Ching, Laurette Luez, Jess Kirkpatrick, Cary Forrester, Virginia Lee, Michael Ross
A suspense-filled thriller in which O'Brien, as a businessman on vacation in San Francisco, finds that he has been given a slow-acting poison. With only a few days to live unless he finds an antidote he sets out to find out who has done this and why. An effective if excessively convoluted thriller, that strangely never makes it clear exactly what the so-called "luminous toxin" is. Remade in 1969 as "Color Me Dead" and again in 1988.
THR 83 min B/W VIDrel: SECOND/RTM V

D.O.A. ** **   15
Rocky Morton/Annabel Jankel USA   1988
Dennis Quaid, Meg Ryan, Charlotte Rampling, Daniel Stern, Jane Kaczmarek, Christopher Neame, Jay Petterson, Robin Johnson, Rob Knepper, Brion James, Jack Kehoe, Elizabeth Arlen, Karen Radcliffe, William Forward, Lee Gideon
An updated version of the 1949 classic study in paranoia, telling of a college professor of English who searches for the person who has poisoned him in a dispute over a prized fiction manuscript. Aided by a pretty female student, his investigation throws up numerous grim jokes and rather obvious red herrings in a tongue-in-cheek, sardonic remake from the creators of MAX HEADROOM. Written by C.E. Pogue (who co-scripted the remake of THE FLY).
THR 93 min (ort 104 min) VIDrel: TOUCH L/A V

DOC ** **   15
Frank Denny USA   1971
Stacy Keach, Harris Yulin, Faye Dunaway, Mike Witney, Denver John Collins, Dan Greenburg, Penelope Allen, Hedy Sontag, Bruce M. Fisher, James Green, Richard MacKenzie, John Scanlon, Antonia Rey, John Bottoms, Philip Shafer
In this thorough re-assessment of the Earp-Holliday myth, our

good Doctor is constantly incapacitated by spasms of coughing whilst poor Earp gets his face bashed in. A well made anti-Western look at the OK Corral showdown that strives hard for realism but often just delivers plain nastiness. The films GUNFIGHT AT THE O.K. CORRAL, MY DARLING CLEMEN-TINE, TOMBSTONE and WYATT EARP also covered this ground.
WES   92 min Cut (5 sec – ort 96 min)   VIDrel: L/A   V

DOC HOLLYWOOD **
Michael Caton-Jones USA
15
1991
Michael J. Fox, Julie Warner, Barnard Hughes, Woody Harrelson, David Ogden Stiers, Frances Sternhagen, George Hamilton, Bridget Fonda, Mel Winkler, Helen Martin, Roberts Blossom, Tom Lacy, Macon McCalman, Raye Birk, Eyde Byrde
A young surgeon en route for a job interview in Beverly Hills, suffers a minor accident at a small town in South Carolina, and finds himself sentenced to thirty-two hours community service at the local hospital. His initial desire to leave as soon as possible weakens, as he starts to fall in love with both the town and a local beauty. An old-fashioned, slow-moving and rather sugary effort, redeemed by an unusually strong performance from Fox.
COM   99 min (ort 104 min)   VIDrel: MGM/WHV L/A   V
Boa: novel What?... Dead Again by Neil B. Shulman.

DOCTOR, THE **
Randa Haines USA
15
1991
William Hurt, Christine Lahti, Elizabeth Perkins, Mandy Patinkin, Adam Arkin, Charlie Korsmo, Wendy Crewson, Bill Macy, J.E. Freeman, William Marquez, Kyle Secor, Nicole Orth-Pallavicini, Ping Wu, Tony Fields, Brian Markinson
Hurt stars as an arrogant surgeon who thinks he is God's gift to medicine, and cares little for his patients. Matters take an unexpected turn when he develops cancer, and sees quite clearly (by way of an encounter with some of medicine's "rejects") that he is not fated to survive unless he finds a compassionate doctor to undertake his treatment. His own conversion is not long in coming, in an enjoyable but terribly contrived and implausible work.
DRA   123 min   VIDrel: TOUCH/TECH L/A   V
Boa: book A Taste Of My Own Medicine by Ed Rosenbaum.

DOCTOR ALIEN **
Dave De Coteau USA
18
1988
Judy Landers, Billy Jacoby, Linnea Quigley, Olivia Brash, Troy Donahue
A female alien masquerades as a biology professor and experiments with campus freshmen, turning one of the wimps into a gorgeous hunk no-one can resist, including herself. A juvenile comedy, but a sprinkling of one-liners keeps it afloat.
Aka: I WAS A TEENAGE SEX MUTANT
COM   84 min (ort 90 min)   VIDrel: POPRO/RTM   V

DOCTOR AND THE DEVILS, THE *
Freddie Francis UK
15
1985
Timothy Dalton, Jonathan Pryce, Julian Sands, Stephen Rea, Phyllis Logan, Beryl Reid, Nichola McAuliffe, Sian Phillips, Twiggy (Lesley Hornby), Lewis Fiander, Philip Davis, Philip Jackson, Danny Schiller, Bruce Green, Toni Palmer
Based on a 1940s screenplay by Dylan Thomas and inspired by the real-life deeds of the famed grave-robbers Burke and Hare, this Gothic style tale follows the exploits of a doctor, who hires a pair of thugs to supply him with corpses for medical research, being content not to ask too many questions about their sources. Made with a good deal of care but generally unmoving and ineffective.
THR   89 min (Cut at film release by 9 sec – ort 92 min)
VIDrel: 20TH/TECH   V/sur

DOCTOR AT LARGE *
Ralph Thomas UK
U
1957
Dirk Bogarde, Muriel Pavlow, Donald Sinden, James Robertson Justice, Shirley Eaton, Derek Farr, Michael Medwin, Edward Chapman, George Colouris, Ann Heywood, Gladys Henson, Lionel Jeffries, Mervyn Johns, Geoffrey Keen, Dilys Lane
Continuation of a ghastly series. A doctor tries two country practices but returns to medical college. Epitomises everything bad in British comedy films of this period with all the expected stereotypes and stock situations.
COM   94 min (ort 104 min)   VIDrel: VCC/DISC   V
Boa: novel by Richard Gordon.

DOCTOR AT SEA **
Ralph Thomas UK
PG
1955
Dirk Bogarde, Brenda de Nanzie, Brigitte Bardot, James Robertson Justice, Maurice Denham, Michael Hordern, Hubert Gregg, James Kenney, Raymond Huntley, Geoffrey Keen, George Colouris, Noel Purcell, Jill Adams, Joan Sims, Abe Barker
The second in this interminable series of dreadful medical farces. Our doctor now becomes medical officer on a cargo steamer. This one does have a few funny moments however.
COM   89 min (ort 93 min)   VIDrel: VCC/DISC   V
Boa: novel by Richard Gordon.

DOCTOR CRIPPEN ***
Robert Lynn UK
PG
1962
Donald Pleasence, Coral Browne, Samantha Eggar, Donald Wolfit, James Robertson Justice, Geoffrey Toone, Oliver Johnston, Elspeth March, Olga Lindo, Paul Carpenter, John Arnatt, Edward Underdown, Basil Henson
Unsensational reconstruction of a celebrated Edwardian murder case in which a mild-mannered doctor killed his wife and then attempted to escape to the USA with his mistress. Told in a very matter-of-fact and straightforward manner, better suited to a documentary, that greatly lessens any dramatic impact this workmanlike piece might otherwise have achieved.
Aka: DR CRIPPEN
DRA   94 min (ort 98 min) B/W   VIDrel: LUMI/SPEAR L/A
V

DOCTOR DOLITTLE *
Richard Fleischer USA
U
1967
Rex Harrison, Samantha Eggar, Anthony Newley, Richard Attenborough, Peter Bull, Geoffrey Holder, William Dix, Portia Nelson, Norma Varden, Muriel Landers
An unsuccessful attempt to transfer to the screen Hugh Lofting's books, about a 19th century doctor who is an expert in animal languages. It was a bad mistake to make the film as a musical, as this trivialises a fascinating idea. In any case the songs are nearly all dreadful, though some of ambitious sets are quite good fun. AA: Song ("Talk To The Animals" – Leslie Bricusse), Effects/vis (L.B. Abbott).
MUS   138 min (ort 144 min)   VIDrel: 20TH/TECH   V
Boa: novels of Hugh Lofting.

DR GIGGLES **
Manny Coto USA
18
1992
Larry Drake, Holly Marie Combs, Glenn Quinn, Cliff De Young, Richard Bradford, Michelle Johnson, Keith Diamond, John Vickery, Nancy Fish, Sara Melson, Zoe Trilling, Darin Heames, Deborah Tucker, Doug E. Doug, Denise Barnes
A highly intelligent mental patient escapes from his asylum and goes on a murder rampage in a demented bid to avenge the death of his brilliant but seriously unbalanced father. Drake gives an excellent performance in this otherwise totally forgettable tale, whose over-the-top blood and gore is a major drawback, despite being put in to give this unfunny horror-comedy some bite.
HOR   91 min   VIDrel: CIC/SONOP   V/sur

DOCTOR HACKENSTEIN **
Richard Clark USA
18
1989
David Muir, Anne Ramsey, Logan Ramsey, Michael Ensign, Catherine Cahn, Stacy Travis, Dyanne Dirosario, Catherine Davis Cox, John Alexis, Sylvia Lee Baker, William Schreiner, Phyllis Diller, Jeff Rector, Christy Botkin (voice only),
A brainless comedy in the style of YOUNG FRANKENSTEIN with a doctor keeping his dead wife's head alive whilst trying to find her a new body. When three beautiful stranded women turn up at his mansion asking for a bed for the night, he gets his operating theatre ready. A clumsy misfire that mixes comedy and horror genres and comes up with nothing new. See also FRANKENSTEIN GENERAL HOSPITAL.
HOR   90 min   VIDrel: CASPIC L/A   V

DOCTOR IN CLOVER *
Ralph Thomas UK
PG
1966
Leslie Phillips, James Robertson Justice, Shirley Ann Field, John Fraser, Joan Sims, Arthur Haynes, Elizabeth Ercy, Fenella Fielding, Jeremy Lloyd, Noel Purcell, Robert Hutton, Eric Barker, Terry Scott, Alfie Bass
Sixth in this series of dated and geriatric medical farces with our

medic being sent on a refresher course and falling for a French physiotherapist.
Aka: CARNABY, M.D.
COM   97 min (Cut at film release – ort 101 min)
VIDrel: VCC/DISC   V
Boa: novel by Richard Gordon.

DOCTOR IN DISTRESS ** PG
Ralph Thomas UK 1963
Dirk Bogarde, Samantha Eggar, James Robertson Justice, Mylene Demongeot, Donald Houston, Barbara Murray, Dennis Price, Leo McKern, Jessie Evans, Ann Lynn, Fenella Fielding, Jill Adams, Michael Flanders, Frank Finlay, Bill Kerr
Another in the "Doctor" series. Here our hero returns to his old medical college to work under his former professor, who just happens to be in love (for the first time). Slightly more entertaining than the usual run of these comedies with Bogarde giving his last performance as Dr Sparrow.
COM   98 min (Cut at film release – ort 112 min)
VIDrel: VCC/DISC   V

DOCTOR IN LOVE * PG
Ralph Thomas UK 1960
Michael Craig, James Robertson Justice, Virginia Maskell, Carole Lesley, Leslie Phillips, Reginald Beckwith, Joan Sims, Liz Fraser, Nicholas Phipps, Irene Handl, Fenella Fielding, Nicholas Parsons, Moira Redmond, Ronnie Stevens
The amorous adventures of two doctors form the basis for this addition to the "Doctor" series with Craig taking over the role played by Bogarde in the earlier films. A worthy addition to a worthless series.
COM   93 min (ort 97 min)   VIDrel: VCC/DISC   V
Boa: novel by Richard Gordon.

DOCTOR IN THE HOUSE * U
Ralph Thomas UK 1954
Dirk Bogarde, Muriel Pavlow, Kenneth More, Kay Kendall, James Robertson Justice, Donald Houston, Suzanne Cloutier, Geoffrey Keen, George Colouris, Harry Locke, Ann Gudrun, Joan Sims, Shirley Eaton, Nicholas Phipps, Richard Wattis
The first film in a series that spawned several sequels and a TV show. This one purports to tell of the life and pranks of a group of immature medical students. Not funny, merely tiresome, despite an engaging performance from Justice as the pompous Sir Lancelot Spratt (a role he was to repeat throughout the series). Scripted by Gordon, Ronald Wilkinson and actor Nicholas Phipps.
COM   88 min (ort 91 min)   VIDrel: VCC/DISC   V
Boa: novel by Richard Gordon.

DOCTOR IN TROUBLE * PG
Ralph Thomas UK 1970
Leslie Phillips, Harry Secombe, Angela Scoular, Irene Handl, Simon Dee, Robert Morley, Freddie Jones, James Robertson Justice, Joan Sims, John Le Mesurier, Graham Stark, Janet Mahoney, Graham Chapman, Fred Emney, Gerald Sim
Last in the series, with our accident prone doctor stowing away on a cruise liner. Despite good work from the cast this one never rises above the level of a mediocre farce.
COM   87 min (Cut at film release – ort 90 min)
VIDrel: VCC/DISC   V
Boa: novel Doctor on Toast by Richard Gordon.

DOCTOR JEKYLL AND MISTER HYDE *** PG
John S. Robertson USA 1920
John Barrymore, Martha Mansfield, Brandon Hurst, Nita Naldi, Charles Lane, Louis Wolheim, J. Malcolm Dunn, Cecil Clovely
A silent version of the famous tale of split-personality, that has Barrymore giving a great performance as the doctor whose thirst for knowledge drives him to experiment upon himself. Some of Barrymore's transformations to the evil Mr Hyde were done by facial contortion, and though this film is not faithful to the novella (Jekyll is drawn into experimenting by his devious father-in-law) it remains a well-paced and gripping yarn.
HOR   63 min B/W silent   VIDrel: SCREAM/SPEAR   V
Boa: novella The Strange Case of Dr Jekyll and Mr Hyde by Robert Louis Stevenson.

DR JEKYLL AND MR HYDE **** 12
Rouben Mamoulian USA 1931
Fredric March, Miriam Hopkins, Rose Hobart, Holmes Herbert, Halliwell Hobbes, Edgar Norton

In Victorian London an idealistic medical researcher believes that his work will unlock human potential, but when he experiments on himself with a serum he has devised, he changes into a murderous monster. Easily the best and most scary of the numerous adaptations of the famous Stevenson tale, with March bringing a demonic vigour to his demanding role, for which he won an Oscar. Special make-up and filters were used for the transmutation sequences. AA: Actor (March).
HOR   63 min B/W   VIDrel: MGM   V
Boa: novella by Robert Louis Stevenson.

DOCTOR JEKYLL AND MISTER HYDE **** 12
Rouben Mamoulian USA 1931
Fredric March, Miriam Hopkins, Rose Hobart, Edgar Norton, Holmes Herbert, Halliwell Hobbes, Edgar Norton, Arnold Lucy, Tempe Pigott, Eric Wilton, Colonel MacDonnell, Murdock MacQuarrie, Douglas Walton, John Rogers, Murdick MacQuarrie
A dedicated and ambitious doctor creates a potion that he thinks can unlock inner potential, but tries it out on himself and turns into a murderous ghoul. Cinematically, this is the best version of the famous horror story, blessed with good performances and accurate Victorian London settings. The transformation scenes were achieved by March wearing several layers of makeup, each sensitive to different colour filters.
HOR   92 min (ort 98 min) B/W   VIDrel: MGM/WHV   V/h
Boa: novella The Strange Cse of Dr Jekyll and Mister Hyde by Robert Louis Stevenson.

DR JEKYLL AND MS HYDE ** 12
David Price USA 1995
Sean Young, Tim Daly, Lysette Anthony, Stephen Tobolowsky, Harvey Fierstein, Thea Vidale, Jeremy Piven, Polly Bergen, Stephen Shellen, Sheena Larkin, John Franklyn-Robbins, Aron Tager, Jane Connell, Julie Cobb, Kim Morgan Greene
A perfume scientist rediscovers a formula devised by his great-grandfather and experiments with it, transforming himself into an attractive but totally ruthless woman. A bawdy variant on a theme first explored in DOCTOR JEKYLL AND SISTER HYDE it has a few comic moments, but they are nothing more than interludes in a generally foolish romp.
COM   90 min (ort 92 min)   VIDrel: POLFIL   V/sh

DOCTOR JEKYLL AND SISTER HYDE ** 18
Roy Ward Baker UK 1971
Ralph Bates, Martine Beswick, Gerald Sim, Lewis Fiander, Dorothy Alison, Neil Wilson, Ivor Dean, Paul Whitsun-Jones, Philip Madoc, Tony Calvin, Susan Broderick, Dan Meaden, Virginia Wetherell, Irene Bradshaw, Anna Brett
Novel twist to the classic Robert Louis Stevenson story. This time round Dr Jekyll finds that his potion turns him into a beautiful woman, who in her turn kills prostitutes so as to provide corpses for the doctor's experiments. Not in the least bit scary but quite watchable. The film DR JEKYLL AND MS HYDE covers similar ground.
HOR   93 min (ort 97 min)   VIDrel: WHV   V/h

DR M ** 18
Claude Chabrol FRANCE/ITALY/WEST GERMANY 1990
Alan Bates, Jennifer Beals, Jan Niklas, Hans Zischler, Benoit Regent, Peter Fitz, Alexander Radszun, Daniela Poggi, William Berger, Michael Degen, Andrew McCarthy
In a Berlin of the future, the suicide rate has spiralled out of control, and a cop traces this back to the work of a sinister doctor, whose interests include the control of a media empire and the use of indoctrination. A thriller with strong SF elements that appears to have been a homage of sorts to Lang's DR MABUSE: THE GAMBLER and its sequels, but is rapidly sunk by poor casting and possibly the demands inherent in making it a European co-production.
THR   111 min   VIDrel: CURZON/20TH   V/sur
Boa: story by Thomas Bauermeister.

DOCTOR MABUSE: THE GAMBLER *** 18
Fritz Lang GERMANY 1922
Rudolf Klein-Rogge, Alfred Abel, Gertrude Welcker, Lil Dagover, Paul Richter
Celebrated film about a master criminal who uses hypnotism and blackmail in his attempt to rule the world. He is finally cornered and defeated whereupon he reveals himself as a lunatic. Unintentionally funny in parts, fascinating in others, this angst-ridden journey through post-WW1 German cinema takes the form of a Fu Manchu type thriller. Originally made in

two halves (153 minutes and 112 minutes) and followed by several sequels.
Aka: DOCTOR MABUSE: THE FATAL PASSION; DOKTOR MABUSE: DER SPIELER; FATAL PASSION OF DOCTOR MABUSE, THE
DRA 86 min (ort 265 min) B/W silent VIDrel: TART/20TH V
Boa: novel by Norbert Jacques.

DOCTOR MORDRID: MASTER OF THE UNKNOWN **
15
Albert Band/Charles Band USA 1991
Jeffrey Combs, Yvette Nipar, Jay Acovone, Brian Thompson, Keith Colouris, Ritch Brinkley, Pearl Shear, Murray Rubin, Jeff Austin, John Apicella, Julie Michaels, Mark Phelan, Kenn Scott, Scott Roberts, Steen Marca, Jonathan Kruger
A sorcerer from another dimension comes to Earth to save the world from destruction at the hands of an evil colleague who is his sworn enemy. He takes up residence in a New York apartment house, but his task is not made any easier when a female neighbour discovers his true mission. A quite well-made and fairly entertaining movie.
FAN 72 min (ort 102 min) VIDrel: CIC/SONOP V

DOCTOR NO ***
PG
Terence Young UK 1962
Sean Connery, Ursula Andress, Bernard Lee, Joseph Wiseman, Jack Lord, Lois Maxwell, Eunice Gayson, Anthony Dawson, Zena Marshall, John Kitzmiller, Lester Prendergast, Reggie Carter, Peter Burton, William Foster-Davis, Louis Blaazar
Reasonably straight and competent adaptation of one of Ian Fleming's "James Bond" adventures, with our secret agent investigating strange happenings in Jamaica and coming up against a criminal mastermind. This one was the first film in the series and though by no means the best, does not exhibit the annoying reliance on gadgetry that so marred the later entries. Connery's charisma did much to ensure the film's success. Followed by FROM RUSSIA WITH LOVE.
A/AD 105 min (ort 111 min) wScrn cC
VIDrel: MGM/WHV V/dm
Boa: novel Dr No by Ian Fleming.

DOCTOR PETIOT ***
15
Christian De Chalonge FRANCE 1990
Michel Serrault, Pierre Romans, Zbigniew Horoks, Berangere Bonvoisin, Aurore Prieto, Andre Chaumeu, Axel Bogousslavski, Maxime Collion, Andre Julien, Nini Crepon, Nita Klein, Martine Mongermont, Naedege Boscher, Jean Dautremay
During WW2, a French doctor offers to smuggle Jews to safety but murders them by lethal injection instead, incinerating their bodies in his private crematorium. A fact-based film telling of the infamous career of mass-murderer Marcel Petiot, but shot in a strongly expressionist and rather self-indulgent manner, the intention being to fashion the grotesque story into a black comedy and testament to greed and barbarism. A chilling and most effective film.
Aka: DOCTEUR PETIOT
DRA 97 min (ort 102 min) VIDrel: ELPIC/POLYREC V

DOCTOR PHIBES RISES AGAIN ***
15
Robert Fuest UK 1972
Vincent Price, Robert Quarry, Valli Kemp, Fiona Lewis, Peter Cushing, Beryl Reid, Terry-Thomas, Hugh Griffith, Peter Jeffrey, John Cater, Gerald Sim, John Thaw, Keith Buckley, Lewis Fiander, Milton Reid, Caroline Munro
In this sequel to THE ABOMINABLE DOCTOR PHIBES, our dotty doctor is revived from death and travels to Egypt with his assistant, in search of an elixir that will revive his dead wife. He is not alone in his quest but has a rival for the potion and is consequently forced to employ some novel ways of killing off the competition. Price is good in an uneven and decidedly campy sequel.
Aka: DOCTOR PHIBES RISES FROM THE GRAVE
HOR 84 min (Cut at film release – ort 89 min)
VIDrel: VISVID/POLYREC L/A V

DOCTOR STRANGE *
U
Philip DeGuere USA 1978
Peter Hooten, Jessica Walter, John Mills, Clyde Kusatsu, Eddie Benton, Philip Sterling, June Barrett, Sarah Rush, David Hooks, Diana Webster, Blake Marion, Bob Delegall, Frank Catalano, Larry Anderson, Inez Pedroza
A comic strip hero from Marvel comics, Dr Strange is a "Master of the Mystic Arts". He is brought to life in this pilot for a series,

where he becomes the protege of another sorcerer (Mills) in order to foil Walter's plans to enlarge her collection of souls. A pathetic effort that strips a lively and imaginative comic strip of its essential mystic qualities, leaving a series of clumsy battles and endless footage of Hooten at work in his hospital.
FAN 89 min (ort 100 min) mTV VIDrel: CIC/SONOP V

DR STRANGELOVE ****
PG
Stanley Kubrick UK 1963
Peter Sellers, Sterling Hayden, Slim Pickens, George C. Scott, Keenan Wynn, James Earl Jones, Tracy Reed, Peter Bull, Jack Creley, Frank Berry, Glenn Beck, Shane Rimmer, Paul Tamarin, Gordon Tanner, Robert O'Neil, Roy Stephens
A demented American general launches a flight of B-52 bombers at the Soviet Union with the intention of wiping it off the face of the earth. Since he is the only person who knows the code that can recall them the President is faced with a tricky problem. A biting black comedy that dares to think the unthinkable and is more effective than any straight treatment of the subject to date (see also FAIL SAFE and BY DAWN'S EARLY LIGHT). Sellers is quite splendid in three roles.
Aka: DOCTOR STRANGELOVE, OR HOW I LEARNED TO STOP WORRYING AND LOVE THE BOMB
COM 91 min (ort 94 min) B/W
VIDrel: ENCORE/SPEAR/COLUM V
Boa: novel Red Alert (also known as Two Hours To Dream) by Peter George.

DOCTOR SYN **
U
Roy William Neill UK 1937
George Arliss, Margaret Lockwood, John Loder, Roy Emerton, Graham Moffatt, Frederick Burtwell, Meinhart Maur, George Merritt, Athole Stewart, Wally Patch, Muriel George, Wilson Coleman
In 1780 the vicar of Dymchurch is secretly a smuggler and pirate, and also supports of the peasants in their resistance to royal oppression. A rousing adventure movie, remade in 1962.
A/AD 80 min B/W VIDrel: CARL/TECH V
Boa: novel Christopher Syn by William Russell Thorndike.

DOCTOR TERROR'S HOUSE OF HORRORS **
18
Freddie Francis UK/USA 1964
Peter Cushing, Neil McCallum, Urusla Howells, Peter Madden, Alan Freeman, Ann Bell, Bernard Lee, Jeremy Kemp, Roy Castle, Kenny Lynch, Tubby Hayes, Christopher Lee, Michael Gough, Isla Blair, Donald Sutherland, Max Adrian
Horror omnibus of five stories linked by the device of a strange figure (who turns out to death) telling the fortunes of five of his fellow railroad passengers. Unfortunately, none of the stories are really memorable and Francis as a director never showed the same level of skill that he possessed as a cameraman.
Aka: BLOOD SUCKERS, THE
HOR 98 min VIDrel: LARK L/A V

DOCTOR WHO **
12
Geoffrey Sax UK 1996
Paul McGann, Eric Roberts, Daphne Ashbrook, Sylvester McCoy
A feature-length spin-off from the British TV series, with the action now transplanted to San Francisco, where our time traveller is close to death and runs the risk of having his life exchanged for that of an altogether more sinister character by midnight of 1999. Happily, there is an attractive female doctor (well there always is, isn't there) on hand to save both him and the future of mankind.
FAN 90 min mTV cC VIDrel: BBC/TECH V/sh

DOCTOR WHO: AN UNEARTHLY CHILD **
PG
Warris Hussein UK 1963
William Hartnell, Carole Ann Ford, William Russell, Jacqueline Hill, Derek Newark, Alethea Charlton, Jeremy Young, Howard Lang, Eileen Way
First-ever story in this incredibly long-running series sees the Doctor and his granddaughter travelling back in time to the Stone Age, where they get involved in a struggle among various tribes for the secret of fire. Many of the other Dr Who episodes have been compiled into cassettes featuring complete adventures, and can generally be obtained from BBC/TECH, though their availability does fluctuate.
FAN 95 min B/W mTV VIDrel: BBC/TECH L/A V/h

DOCTOR WHO AND THE DALEKS *
U
Gordon Flemyng UK 1965
Peter Cushing, Roy Castle, Jennie Linden, Roberta Tovey, Barrie

Ingham, Michael Coles, Geoffrey Toone, Mark Peterson, John Brown, Yvonne Antrobus

More of the same in this feature film spin-off from a long-running BBC TV series that outlived its ideas. Three children and their grandfather travel in his time machine to a planet ruled by the evil Daleks. A rubbishy British foray in SF, it doesn't even try to do justice to the early (and rather good) "Doctor Who" adventures, on which it is based. Followed by DOCTOR WHO: INVASION EARTH 2150 A.D.
FAN   79 min (ort 83 min) wScrn   VIDrel: WHV   V/dm V/h

DOCTOR WHO: CASTROVALVA **   U
Fiona Cumming UK   1982
Peter Davison, Janet Fielding, Sarah Sutton, Matthew Waterhouse, Anthony Ainley, Derek Waring, Frank Wylie, Michael Sheard, Dallas Cavell
The Doctor's latest regeneration leaves him in a weakened state and he needs to make use of the TARDIS Zero Room. Unfortunately, his old foe the Master prevents the Doctor from gaining access to his recuperative chamber and tricks him into landing on the planet Castrovalva. An adequate adventure compiled from four episodes.
FAN   95 min mTV   VIDrel: BBC/TECH   V/h

DOCTOR WHO: DAY OF THE DALEKS **   U
Paul Bernard UK   1972
Jon Pertwee, Katy Manning, Jean McFarlane, Wilfrid Carter, Tim Condreen, John Scott Martin, Oliver Gilbert, Aubrey Woods, Deborah Brayshaw, Gypsie Kemp, Anna Barry, Jimmy Winston, Scott Fredericks, Valentine Palmer
In this compilation of the TV series our intrepid Doctor and his assistant must fight to save the life of a diplomat and avert a war that will result in the Earth coming under the domination of the evil Daleks and their ferocious slaves – the Ogrons. Originally shown in four parts.
FAN   90 min mTV   VIDrel: BBC L/A; ENCORE (LV only) V/h LV

DOCTOR WHO: DEATH TO THE DALEKS **   U
Michael E. Briant UK   1974
Jon Pertwee, Arnold Yarrow, Roy Heymann, John Abineri, Duncan Lamont, Julia Fox, Joy Harrison, Neil Seiler, Mostyn Evans, Terry Walsh, Steven Ismay, John Scott Martin, Murphy Grumbar, Cy Town
Feature cobbled together from episodes of the popular children's TV series in which a time-travelling adventurer faced a multitude of menaces both human and otherwise. Enjoyable in a clumsy and inept sort of way, this is science fiction for seven-year-olds. In this adventure the good Doctor faces one of his deadliest enemies – those irascible and somewhat hysterical robots known as "The Daleks".
FAN   90 min mTV   VIDrel: BBC/TECH   V

DOCTOR WHO: EARTHSHOCK **   PG
Peter Grimwade UK   1982
Peter Davison, Janet Fielding, Sarah Sutton, Matthew Waterhouse, Clare Clifford, James Warwick, Beryl Reid, June Bland, David Banks, Alec Sabin
In the 25th century the Cybermen make an unwelcome return, and mount an attack against the Earth. Fortunately, the Doctor is on hand to foil their plans, but his friend Adric gets trapped on board a doomed spaceship hurtling back in time to an Earth of the dinosaur age. Originally shown in four parts.
FAN   97 min mTV cC   VIDrel: BBC/TECH   V/h

DOCTOR WHO: INVASION EARTH 2150 A.D. *   U
Gordon Flemyng UK   1966
Peter Cushing, Bernard Cribbins, Ray Brooks, Jill Curzon, Roberta Tovey, Andrew Keir, Godfrey Quigley, Roger Avon, Keith Marsh, Geoffrey Cheshire, Steve Peters, Robert Jewell
Spin-off from a long-running BBC TV series. The Doctor and his friends join the survivors of a Dalek occupation of the Earth in the ruins of a devastated London. A dreary follow-up to DOCTOR WHO AND THE DALEKS, made with neither care nor interest. A pity to see someone as delightful as Cushing in this mess.
Aka: DALEKS: INVASION EARTH 2150 A.D.; INVASION EARTH 2150 A.D.
FAN   81 min (ort 84 min) wScrn   VIDrel: WHV   V/dm

DOCTOR WHO: SILVER NEMESIS **   PG
Chris Clough UK   1989
Sylvester McCoy, Sophie Aldred, Anton Diffring, Fiona Walker, gerard Murphy, Metin Yenal, David Banks
The Doctor and his female companion find the Cybermen plot-

ting with neo-Nazis and other diverse elements to gain control of artefact called Nemesis. This animate metallic statue that now come back to Earth 350 years after it was sent into deep space by the Doctor.
FAN   138 min   VIDrel: BBC/TECH   V/h

DOCTOR WHO: SPEARHEAD FROM SPACE **   U
Derek Martinus UK   1970
Jon Pertwee, Hugh Burden, Neil Wilson, John Breslin, Anthony Webb, Helen Doward, Talfryn Thomas, George Lee, Iain Smith, Tessa Shaw, Ellis Jones, Allan Mitchell, Prentis Hancock, Derek Smee, John Woodnutt, Betty Bowden
Another cobbled up TV feature with Doctor Who and friends out to foil an alien take-over of the planet, the villain this time round being the Nestene, a disreputable creature who has hatched a plan to replace key figures in the government with plastic replicas (will anyone notice?).
FAN   97 min mTV   VIDrel: BBC/TECH   V/h

DOCTOR WHO: THE ARK IN SPACE **   U
Rodney Bennett UK   1974
Tom Baker, Elisabeth Sladen, Ian Marter, Wendy Williams, Kenton Moore, Richardson Morgan, Stuart Fell, Nick Hobbs, Christopher Masters, John Gregg plus voices of: Gladys Spencer, Peter Tuddenham
This four-part story has the Doctor visiting a space station of the future, where a race of insect-like creatures threatens to destroy the human beings held in cryogenic suspension aboard the vessel. Average.
FAN   98 min mTV   VIDrel: BBC/TECH; ENCORE (LV only) V LV

DOCTOR WHO: THE CAVES OF ANDROZANI **   PG
Graeme Harper UK   1984
Peter Davison, Nicola Bryant, Christopher Gable, Maurice Roeves, Robert Glenister, John Normington, Roy Holder, Martin Cochrane, David Neal, Colin Baker
Dr Who and his female companion become embroiled in the politics of the Androzani planets, and while the former does battle with a disreputable gunrunner the latter is captured. Later, when both are infected by a deadly disease the Doctor is obliged to undergo one of his transformations, thus introducing the sixth actor to play him – Colin Baker. First shown on the BBC as a four-part story.
FAN   100 min mTV   VIDrel: BBC/TECH L/A   V/h

DOCTOR WHO: THE CLAWS OF AXOS **   U
Michael Ferguson UK   1971
Jon Pertwee, Katy Manning, Nicholas Courtney, Richard Franklin, John Levene, Roger Delgado, Peter Bathurst, Paul Grist, Donald Hewlett, Bernard Holley, David Savile, Kenneth Benda, Tim Pigott-Smith, Fernanda Marlowe
Axos – a mysterious parasitic creature, arrives on Earth bearing the gift of a powerful substance known as Axonite. The Doctor grows suspicious of the alien's motives when he learns that his old adversary The Master is onboard its vessel, and eventually overcomes it by trapping it in a time-loop. First shown in four parts.
FAN   97 min mTV   VIDrel: BBC/TECH   V/h

DOCTOR WHO: THE CURSE OF FENRIC **   PG
Nicholas Mallett UK   1989
Sylvester McCoy, Bonnie Langford, Sophie Aldred, Dinsdale Landen, Alfred Lynch, Nicholas Parsons, Joanne Kelly, Joanne Bell, Corey Pulman, Marek Anton
This adventure sees the Doctor arriving in England during WW2, where he finds that the Viking graves of an ancient churchyard and an old Norse curse are creating problems for the staff at an isolated research station. Compilation of a four-part story.
FAN   103 min mTV   VIDrel: BBC/TECH   V/h

DOCTOR WHO: THE DAEMONS **   PG
Christopher Barry UK   1971
Jon Pertwee, Katy Manning, Nicholas Courtney, Richard Franlin, John Levene, Roger Delgado, Damaris Hayman, Rollo Gamble, Eric Hillyard, Don McKillop, Jon Croft, Stanley Mason, Stephen Thorne
Compilation of a five-part story in which the Doctor's foe the Master takes advantage of an archeological excavation to a site known as The Devil's Hump, thus releasing (after two-hundred years imprisonment) the last member of an ancient race who possess terrifying powers.
FAN   122 min (full re-constructed colour version) mTV
VIDrel: BBC/TECH L/A   V/h

DOCTOR WHO: THE DALEKS ***   U
Christopher Barry/Richard Martin UK   1963
William Hartnell, Carole Ann Ford, William Russell, Jacqueline Hill,
Alan Wheatley, Philip Bond, John Lee, Marcus Hammond, Gerald
Curtis, Virginia Wetherell, David Graham (Dalek voice only), Peter
Hawkins (Dalek voice only)
Compilation of an early 7-part Doctor Who adventure that intro-
duced the Daleks. When the TARDIS lands on the Planet Skaro
the Doctor finds himself involved in a conflict between the
human-like Thal and the mutated Daleks, both representing the
very different survivors of a neutronic war. Writer Terry
Nation's irascible Daleks proved to be too good an idea to
abandon, and reappeared in several other episodes as well as a
feature film in 1965.
FAN   174 min (2 cassettes) mTV   VIDrel: BBC L/A   V/h

DOCTOR WHO: THE DEADLY ASSASSIN **   PG
David Maloney UK   1976
Tom Baker, Llewellyn Rees, Derek Seaton, George Pravda, Erik Chitty,
Bernard Horsfall, Hugh Walters, Angus Mackay, Peter Pratt,
Llewellyn Rees
Doctor Who returns to Gallifrey where he witnesses the murder
of the President of the Time Lords and learns thas his old enemy
the Master has hatched a plan to destroy the Time Lords, drain-
ing their energy with a black hole. Unfortunately, matters
become somewhat complicated when the Doctor is accused of
the very murder he witnessed.
FAN   95 min mTV   VIDrel: BBC/TECH L/A   V/h

DOCTOR WHO: THE FIVE DOCTORS *   U
Peter Moffatt UK   1983
Jon Pertwee, Peter Davison, Patrick Troughton, Tom Baker, Richard
Hurndall, Janet Fielding, Nicholas Courtney, David Savile, Ray Float,
Lalla Ward, Philip Latham, Elisabeth Sladen, Dinah Sheridan, Carol
Ann Field
A specially designed episode to celebrate the 20th anniversary of
the British TV series, in which the five incarnations of "Doctor
Who" are brought together to face a motley collection of their
enemies. A strictly average entry in a fairly undistinguished series.
FAN   89 min mTV   VIDrel: BBC/TECH   V/sur

DOCTOR WHO: THE KEEPER OF TRAKEN **   PG
John Black UK   1981
Tom Baker, Matthew Waterhouse, Sarah Sutton, Denis Carey, Sheila
Ruskin, Anthony Ainley, John Woodnutt, Margot Van Der Burgh,
Robin Soans, Geoffrey Beevers, Roland Oliver
A once harmonious planet seems to be subjected to a disruptive
influence and the Doctor soon learns that this is the handiwork
of his old enemy the Master, who is using a creature called
Melkur to spread his evil.
FAN   98 min mTV   VIDrel: BBC/TECH   V/h

DOCTOR WHO: THE KROTONS **   U
David Maloney UK   1968/69
Patrick Troughton, Frazer Hines, Wendy Padbury, James Copeland,
Madeleine Mills, Gilbert Wynne, Philip Madoc, Richard Ireson, James
Cairncross plus voices of: Roy Skelton, Patrick Tull
In this compilation of a four-episode tale, Dr Who and friends
land on a barren planet where they get the chance to try out a
device known as the Kroton Teaching Machine. Unfortunately,
their brainwaves awaken the Krotons, a deadly race of crys-
talline beings.
FAN   91 min (ort 121 min) B/W mTV   VIDrel: BBC/TECH
L/A   V/h

DOCTOR WHO: THE PYRAMIDS OF MARS **   U
Paddy Russell UK   1975
Tom Baker, Vik Tablian, Elisabeth Sladen, Bernard Archad, Gabriel
Woolf, Peter Copley, Peter Maycock, Michael Bilton, Michael Sheard,
George Tovey, Nick Burnell, Melvyn Bedford, Kevin Selway
A feature-length compilation of that long-running TV series that
died a long time ago but refused to lay down. Here the Doctor
fights against an ancient Egyptian enemy when he returns to
Earth in the year 1911.
Aka: PYRAMIDS OF MARS, THE
FAN   90 min mTV   VIDrel: BBC/TECH   V

DOCTOR WHO: THE ROBOTS OF DEATH **   U
Michael E. Briant UK   1977
Tom Baker, Louise Jameson, Rob Edwards, Russell Hunter, Pamela
Salem, David Bailie, Brian Croucher, Tariq Yanus, David Collins,
Tania Rogers, Gregory De Polnay, Miles Fothergill, Mark Blackwell
Baker, Mark Cooper, Peter Langtry
Compilation feature from the popular "Doctor Who" series
with Doc and his assistant doing battle with a power-mad
scientist and his array of killer gadgets. First shown in four
parts.
FAN   96 min mTV   VIDrel: BBC/TECH   V

DOCTOR ZHIVAGO **   15
David Lean USA   1965
Omar Sharif, Julie Christie, Rod Steiger, Alec Guinness, Ralph
Richardson, Geraldine Chaplin, Tom Courtenay, Rita Tushingham,
Siobhan McKenna, Noel Willman, Geoffrey Keen, Adrienne Corri,
Jeffrey Rockland, Lucy Westmore
Pasternak's epic tale of the Russian Revolution and its effect
on a group of people, is transformed into a visually outstand-
ing but stilted and lifeless effort. The dialogue is wooden and
Omar Sharif is totally unconvincing as a Russian. Now remem-
bered chiefly for Jarre's fine melodies. AA: Screen/adapt
(Robert Bolt), Cin (Freddie Young), Art/Set (John Box and
Terry Marsh/Dario Simoni), Cost (Phyllis Dalton), Score/orig
(Maurice Jarre).
DRA   200 min wScrn (speial anniversary edition – ort 197
min)   VIDrel: MGM/WHV   V/dm V/sh
Boa: novel by Boris Pasternak.

DODGE CITY ***   PG
Michael Curtiz USA   1939
Errol Flynn, Olivia de Havilland, Ann Sheridan, Bruce Cabot, Frank
McHugh, Alan Hale, John Litel, Victor Jory, Ward Bond, Cora
Witherspoon, Henry Travers, Henry O'Neill, Guinn Williams,
William Lundigan, Gloria Holden
Nothing less than a Western swashbuckler in which a dashing
rancher and soldier of fortune gets elected as sheriff of Dodge
City, and proceeds to take on the bad guys and make the town
fit for decent folks. A simple, value-for-money film with ample
action packed into a meagre plot.
WES   100 min (ort 105 min) B/W   VIDrel: MGM/WHV   V

DODSWORTH ***   PG
William Wyler USA   1936
Walter Huston, David Niven, Paul Lukas, John Payne, Mary Astor,
Geoffrey Gaye, Spring Byington, Geoffrey Gaye, Maria Ouspenskaya,
Odette Myrtil, Harlan Briggs, Beatrice Maude, Gino Corradi, Ines
Palange, Charles Halton
An American businessman and his wife go on a trip of Europe
that radically changes their lives as his wife finally ditches him
in preference for a European baron whom she proposes to marry
me. Fortunately, he meets up with an attractive and loving
window who cares for him. An engrossing and well-made adap-
tation of the original novel. AA: Art (Richard Day).
DRA   101 min B/W   VIDrel: VCC/DISC/COLUM   V
Boa: novel by Sinclair Lewis.

DOG DAY AFTERNOON ***   15
Sidney Lumet USA   1975
Al Pacino, John Cazale, Charles Durning, James Broderick, Sully
Boyar, Penny Allen, Carol Kane, Chris Sarandon, Sandra Kazan, Amy
Levitt, Lance Henriksen, Carmine Foresta, Floyd Levine, Dick
Anthony Williams, Judith Malina
A tense and gritty story about a loser who holds up a bank so
that his homosexual lover will have the money for a sex-change
operation. However, his simple robbery goes badly wrong and
develops into a major incident. An enjoyable film of much
atmosphere and little plot. AA: Screen/orig (Frank Pierson).
DRA   119 min (ort 130 min)   VIDrel: WHV   V

DOG WALKER **   18
John Leslie USA   1994
Christina Angel, Krystu Lynn, Isis Nile, Lana Sands, Maeva, Joey
Silvera, Jamie Gillis, John Dough, Gerry Pike, Tom Byron, Alex
SAnders, Jake Williams, David Pollmen
Three girls tease and tantalise a criminal, and he ends up betray-
ing his gang. A rather dark and complex sex film, not all that
enjoyable, but fairly unusual.
A   71 min (ort 93 min)   VIDrel: PURG/DANTE   V

DOGS IN SPACE *   18
Richard Lowenstein AUSTRALIA   1986
Michael Hutchence, Saskia Post, Nique Needles, Chris Haywood,
Deanna Bond, Tony Helou
A group of young people enjoy the summer of 1978 in the belief
that they will never grow up and join the world of adult squares.
An examination of teen pop culture largely intended as a vehicle
for Hutchence – lead singer with the Aussie group INXS. The

title refers to the epoch-making early space flights, especially those Soviet ones with dogs. Noisy and unbearably tiresome.
COM   108 min   VIDrel: POPRO/RTM   V

DOGS OF WAR, THE ** 15
John Irvin UK 1980
Christopher Walken, Tom Berenger, Colin Blakely, Hugh Millais, Paul Freeman, Robert Urquhart, JoBeth Williams, Jean-Francois Stevenin, Winston Ntshona, Pedro Armendariz Jr, Harlen Carey Pope, Ed O'Neill, Isabel Grandin, Ernest Graves
A sombre adaptation of Forsyth's bestseller, with a cynical and disenchanted mercenary, plotting to overthrow a nasty Amin-like dictator of an impoverished African state, and seize power himself. Lacking in surprises if not in brutality, this tough action film now seems very dated.
A/AD   113 min Cut (5 sec – ort 118 min)
VIDrel: MGM/WHV   V/sur
Boa: novel by Frederick Forsyth.

DOING TIME ON MAPLE DRIVE ** PG
Ken Olin USA 1991
William McNamara, Lori Loughlin, Jim Carrey, James B. Sikking, Bibi Besch, Philip Linton, Bennett Cale, Mark Chaet, Janice Lynde, George Roth, Parker Whitman, Richard Israel, Tom Sawyer, Danielle Michonne, Courtney McWhinney
A family wedding provides the setting for this engaging drama, in which the return of three family members leads to various surprising revelations, including a son's homosexuality, something the boy's father finds very hard to accept.
DRA   88 min (ort 90 min) mTV   VIDrel: 20TH   V/h

DOLL SQUAD, THE * 15
Ted V. Mikels USA 1974
Michael Ansara, Francine York, John Carter, Anthony Eisley, Lisa Todd, Carol Terry, Tura Santana, Rafael Campos, Bert Zeller, Herb Robbins, William Bagdad, Lisa Garrett, Lee Christian, Judy McConnell, Guspave Unger, Bertel Unger
An all-female band battles an ex-CIA agent for world domination, in this unbelievably dire offering.
Aka: HUSTLER SQUAD; SEDUCE AND DESTROY; WILDCATS
A/AD   91 min (ort 101 min)   VIDrel: OURVID/SCRN   V

DOLLMAN ** 18
Albert Pyun USA 1990
Tim Thomerson, Jack Earle Haley, Kamala Lopez, Nicholas Guest, Michael Halsey, Eugene Glazer
Thomerson plays a trigger-happy cop whose beat is to be found on the planet Arturus, where he's involved in a shootout with a disembodied head. This conflict results in both him and his adversary making their way to Earth, where they crash-land in the heart of the Bronx. Unfortunately, by some quirk of space-time, our cop is now only 13 inches tall, but this doesn't stop him from tackling various baddies. An enjoyably daft action-fantasy.
FAN   78 min   VIDrel: EIV/SONOP   V

DOLLY DEAREST * 15
Maria Lease USA 1990
Denise Crosby, Sam Bottoms, Rip Torn, Candy Hutson, Chris Demetral
A ludicrous rip-off of both CHILD'S PLAY films, that's set in Mexico, with a drag version of "Chucky" the demonic doll, going off on its predictable killing spree, after an archaeologist unwittingly releases its evil spirit whilst excavating a tomb belonging to an ancient devil-worshipping sect. A dull and flat dud, with neither shocks nor surprises, and of dubious entertainment value.
HOR   90 min   VIDrel: FIRST/SONOP L/A   V

DOLORES CLAIBORNE *** 18
Taylor Hackford USA 1995
Kathy Bates, Christopher Plummer, Jennifer Jason Leigh, David Strathairn, Judy Parfitt, John C. Reilly, Eric Bogosian, Bob Gunton, Ellen Muth, Roy Cooper, Wayne Robson, Ruth Marshall, Weldon Allen, Tom Gallant, Kelly Burnett
When a Maine housekeeper is accused of the murder of her employer, her daughter returns home after an absence of fifteen years, which brings to the surface a number of unanswered questions about their relationship and her past life. A straightforward adaptation of King's psychological study that is much enhanced by a first-rate and totally convincing performance

from Bates. The strong script and excellent camerawork are additional assets.
DRA   126 min (ort 131 min) cC
VIDrel: COLUM/SONOP; ENCORE (LV only)   V/sur LV
Boa: novel by Stephen King.

DOMINATION * 18
F.J. Lincoln USA 1995
Chasey Lain, Vanessa Chase, Alex Jordan, Barbara Doll, Nick East, Alex Sanders, Tom Byron
A sex-hungry woman gets what she wants by playing the part of a dominatrix in this adult offering with the emphasis firmly on sado-masochism.
A   80 min   VIDrel: PURG/DANTE   V

DOMINICK DUNNE'S 919 FIFTH AVENUE ** (15)
Neal Hager USA 1995
Reed Diamond, Barry Bostwick, Noelle Beck
Adequate soap-opera done in the usual style, telling of an author who undertakes the task of writing a biography of a wealthy and influential Manhattan family, and uncovering a set of scandals the family would have rather kept hidden.
DRA   91 min   SATrel: SKY MOVIES

DOMINION ** 15
Michael Kehoe USA 1994
Brad Johnsn, Brion James, Tim Thomerson, Woody Brown
Six friends take a hunting trip in the wilderness but find themselves the prey of a man who lost his sanity when his young son was accidentally shot dead in a hunting accident.
A/AD   95 min (ort 98 min)   VIDrel: 20TH/SONOP   V/h

DOMINION TANK POLICE: ACTS 1 TO 5 *** 15
Kiochi Mashimo/Takaachi Ishiyama JAPAN 1988/1993
An unusual animation from Japan that takes a look at a future society, where the planet is covered in a permanent belt of smog and the inhabitants are obliged to wear gas masks in a city that is largely under the control of criminals, most especially a couple of sexy but deadly feline sisters. The title police are a special unit formed in a desperate attempt to restore order.
ANIM   180 min (5 cassettes – approx 30 min each)
VIDrel: MANGA/SONOP L/A   V/sh
Boa: story by Masamune Shirow.

DON JUAN DE MARCO *** 15
Jeremy Leven USA 1995
Johnny Depp, Marlon Brando, Faye Dunaway, Rachel Ticotin, Bob Dishy, Talisa Soto, Geraldine Pailhas, Richard Sarafian, Franc Luz, Tresa Hughes, Marita Geraghty, Stephen Singer, Carmen Argenziano, Jo Champa, Esther Scott
A shrink is on the verge of retiring but delays doing so when he acquires an interesting new patient. This young man firmly believes that he is a descendant of the great lover Don Juan and that he has inherited his ancestor's power over women. Some nice performances, most especially from Depp as the disturbed young man, do much to sustain this brittle and whimsical movie, which stands in need of a stronger script and a better effort on the part of Brando.
Aka: DON JUAN AND THE CENTREFOLD
COM   93 min (ort 97 min)   VIDrel: EIV/SONOP   V

DONA FLOR AND HER TWO HUSBANDS ** 18
Bruno Barreto BRAZIL 1978
Sonja Braga, Jose Wilker, Mauro Mendonca, Dinorah Brillanti, Nelson Xavier
A widow who remarries is visited by the ghost of her first husband and finds her loyalty torn between the two. Despite an unquestionably sexy performance from Braga, this one-joke story has too little material to work with and has more in common with softcore films than romantic fantasies. Later remade as "Kiss Me Goodbye".
Aka: DONA FLOR E SEUS DOIS MARIDOS
COM   105 min Cut (34 sec)   VIDrel: RCA L/A   V
Boa: novel by Jorge Amado.

DONA HERLINDA AND HER SON *** 15
Jaime Humberto Hermosillo MEXICO 1986
Guadalpe Del Toro, Arturo Meza, Marco Antonio Trevino, Leticia Lupersio, Angelica Guerrero
A woman tries to get her homosexual son to marry and is relieved when he eventually does so, only to move his

boyfriend in under her roof as well. An amusing satire on attitudes to homosexuality in a macho culture that is often well on target.
Aka: DONA HERLINDA Y SU HIJO
COM   89 min   VIDrel: PRIDE/PARADOX   V

DONATO AND DAUGHTER **   (18)
Rod Holcomb USA   1993
Charles Bronson, Dana Delaney, Xander Berkley
In L.A. a veteran police sergeant Bronson gets embroiled in the hunt for a serial killer, but at the same time the case is being investigated by his police lieutenant daughter, and as we might expect, she is soon to become a potential victim. Unusually gory for a TV movie, this is quite a well plotted thriller, although Bronson gives his usual wooden performance.
THR   90 min mTV   SATrel: SKY MOVIES

DONNIE BRASCO ***   (18)
Mike Newell USA   1996
Johnny Depp, Al Pacino, Anne Heche, Michael Madsen, James Russo, Bruno Kirby, Zeljko Ivanek, Gerry Becker, Zach Grenier, Brian Tarantina
Gangster movie based on a real incident from the 1970s, that sees undercover FBI agent Depp infiltrating the Mafia. His task is made easier by the friendship he strikes up with a small-time hit-man (Pacino) who helps him gain acceptance, but naturally knows nothing about his true identity. As Depp's growing involvement separates him from his family, he finds himself growing ever closer to his mentor. Quite a gruesome film, but consistently fascinating.
THR   120 min   CINrel

DONOR **   18
Larry Shaw USA   1990
Melissa Gilbert-Brinkman, Jack Scalia, Wendy Hughes, Pernell Roberts, Gale Mayron, Gregory Sierra, Marc Lawrence, Wendy Cooke, Michael Boatman, Carol Ann Susi, Hari Rhodes, Virginia Capers, Larry Cedar, Don Alan Croll, David Crowley
Echoes of COMA abound in this gruesome thriller whose plot revolves around the use of a group of unfortunates for experimentation in a big city hospital. Gilbert-Brinkman plays the female medic who uncovers these nefarious goings on. Average offering that takes a nice, cynical view of doctors, but is still constrained by its mTV format.
THR   89 min mTV   VIDrel: 20TH/TECH   V

DONOR UNKNOWN **   15
John Kent Harrison USA   1995
Peter Onorati, Alice Krige, Becky Herbst, Leo Garcia, Sam Robards, John Dorman, Clancy Brown
A man who has been given a new heart learns to his horror that the donor was the victim of a murderous black market trade in human organs and that most of the victims appear to be Mexicans. Suspecting that they may be illegal immigrants who have been murdered, he starts his own investigation, but this is not without danger. A fair thriller, adequately if not especially inventively plotted.
THR   89 min (ort 93 min) cC   VIDrel: CIC/SONOP   V
Boa: novel Corazon by William H. Mooney.

DON'S PARTY ***   18
Bruce Beresford AUSTRALIA   1976
John Hargreaves, Graham Kennedy, Pat Bishop, Ray Barrett, Veronica Lang, Claire Binney, Jeannie Drynan, Candy Raymond, Harold Hopkins, Graeme Blundell
A look at the wife-swapping mores of the middle-class inhabitants of a Sydney suburb, who hold a get-together to watch election returns and have a little fun at the same time. A sharp and biting black comedy, based on a script by Williamson.
COM   86 min (ort 90 min)   VIDrel: ARTPRO/RTM   V
Boa: play by David Williamson.

DON'T BOTHER TO KNOCK ***   PG
Roy Ward Baker USA   1952
Marilyn Monroe, Richard Widmark, Anne Bancroft, Jim Backus, Elisha Cook Jr, Donna Corcoran, Jeanne Cagney, Lurene Tuttle, Gloria Blondell, Verna Felton, Don Beddoe, Willis B. Bouchy, Grace Hayle, Michael Ross, Eda Reis Merin
A young girl taken on as a babysitter proves to be mentally unbalanced in this gripping example of film noir. Monroe is excellent as the girl who's saved from harming both herself and

her charges by the intervention of Widmark, whom she envisions as her dead pilot boyfriend. Bancroft's film debut.
THR   76 min B/W   VIDrel: 20TH/TECH   V
Boa: novel Mischief by Charlotte Armstrong.

DON'T DELIVER US FROM EVIL *   18
Joel Seria FRANCE   1970
Jeanne Goupil, Catherine Wagener, Bernard Dheran, Jean-Pierre Helbert
Two teenage girls who were educated in a convent progress from the torture of animals onto lesbianism, witchcraft and murder.
Aka: MAIS NE NOUS DELIVREZ PAS DU MAL
HOR   102 min   VIDrel: REDEM/RTM   V

DON'T EVER LEAVE ME **   U
Arthur Crabtree UK   1949
Jimmy Hanley, Petula Clark, Hugh Sinclair, Linden Travers, Edward Rigby, Brenda Bruce, Frederick Piper, Barbara Murray, Anthony Newley, Maurice Denham, Sandra Dorne
A naive car salesman gets caught up in the kidnapping of a young girl, and the victim refuses to let him go, as she is a bored, little rich girl who is only too pleased to be involved in something she sees as a great adventure. Clark gives us a couple of songs in this creaking but quite jolly comedy.
COM   81 min B/W   VIDrel: CARL/TECH   V
Boa: novel The Wide Guy by Anthony Armstrong.

DON'T FORGET YOU'RE GOING TO DIE ***   (18)
Xavier Beauvois FRANCE   1995
Xavier Beauvois, Chiara Mastroianni, Roschdy Zem, Bulle Ogier, Jean-Louis Richard, Emmanuel Salinger, Jean Douchet, Pascal Bonitzer, Cedric Kahn, Patrick Chauvel, Stanislas Nordey, Denis Psaltopoulos, Frederic Quiring, Sandra Cheres
An art student tries desperately to avoid military service, but it is not through his efforts that he escapes it, but simply due to a medical test revealing him to be HIV-positive. Having discovered this, he sets out to indulge himself in the pleasures of the senses. A morbid comedy-drama of arresting performances and caustic humour, it has a vigour and humour strangely at odds with the downbeat aspect of the plot.
Aka: N'OUBLIE PAS QUE TU VAS MOURIR
DRA   121 min   CINrel

DON'T GET ME STARTED *   (PG)
Arthur Ellis GERMANY/UK   1994
Trevor Eve, Steven Waddington, Marion Bailey, Ralph Brown, Marcia Warren, Alan David, Patrick O'Connell, Lorna Heilbron, Hannah Taylor-Gordon, Nathan Grower, Hilary Sesta, Barbara Keogh, Stuart Barren, Peter Sproule
A mild-mannered insurance salesman gives up smoking, but the unrelieved tensions he becomes prey to require a new outlet, which he finds when he embarks on a career as a murderer. Extensively re-cut and re-titled after its poor showing at the Cannes Film Festival, this clumsy comedy-thriller still remains a dud, poor dialogue and a woeful lack of inventiveness being its chief defects.
Aka: PSYCHOTHERAPY
THR   76 min   CINrel

DON'T LOOK BACK **   18
Geoff Murphy USA   1996
Eric Stoltz, John Corbett, Josh Hamilton, Annabeth Gish, Amanda Plummer, Dwight Yoakam
In Hollywood, a man has drifted into heroin addiction after his affairs took a turn for the worse, but when he comes across a briefcase containing $200,000 his fortunes appear to be about to improve. But only if he can avoid the wrath of the drug dealers to whom the money belongs. Average.
THR   90 min   VIDrel: EIV/SONOP   V/sh

DON'T LOOK NOW ***   18
Nicolas Roeg UK   1973
Donald Sutherland, Julie Christie, Hilary Mason, Celia Matania, Massimo Serrato, Renato Scarpa, Giorgio Trestini, Leopoldo Trieste, David Tree, Ann Rye, Nicholas Salter, Sharon Williams, Adelina Poerio
A short story serves as the basis for this broodingly atmospheric film about a young couple who go to Venice shortly after one of their children is drowned in a tragic accident. The husband has the gift of second sight and sees things that cannot be explained, but all is resolved in a chilling climax.
DRA   105 min (ort 110 min)   VIDrel: WHV   V/h
Boa: short story by Daphne du Maurier.

DON'T MOVE, DIE AND RISE AGAIN! *** 12
Vitali Kanevsky USSR 1990
Dinara Drukarova, Pavel Nazarov, Elena Popova, Valery Ivchenko, Vyacheslav Bambushek, Vadim Ermolaev, V. Kosobutskaya, N. Mikheev, E. Lipets, A. Rybakov, Iu Rotin, Ryuichi Simidzu, Katia Gromova, F. Turkin, V. Khlusevich, O. Korytin
Two children living in a remote mining town in the distant wastes of Siberia in 1947, survive poverty and immense hardship thanks to the warmth of their friendship and a shared sense of humour. An evocative and beautifully filmed portrayal of childhood that never becomes sentimental or cloying. Scripted by Kanevsky, this impressive first film was his directorial debut.
Aka: DON'T MOVE, DIE AND RESUSCITATE!; FREEZE, DIE, COME TO LIFE!; ZAMRI, UMRI, VOSKRESNI!
DRA 105 min CINrel

DON'T TALK TO STRANGERS ** 12
Robert Lewis USA 1994
Shanna Reed, Terry O'Quinn, Pierce Brosnan, Keegan Macintosh, Michael MacRae, Roger R. Gross, Alan Robertson, Douglas Stewart, David "Squatch" Ward, Colleen Winton, Meredith Woodward, William B. Davis, Peter Hanlon, Tom Picket
A divorced woman remarries and is in the process of moving to California to live with her new husband, when her son is kidnapped. At first, suspicion falls on his cop father, who lost custody of the boy after a court case. However, events soon take an unexpected course. Plotting leaves a lot to be desired in the logic department, but there is no lack of tension.
THR 89 min (ort 94 min) mCab VIDrel: CIC/SONOP V

DON'T TELL HER IT'S ME * 15
Malcolm Mowbray USA 1990
Steve Guttenberg, Shelley Long, Jami Gertz, Kyle MacLachlan, Kevin Scannell, Madchen Amick, Beth Grant, Caroline Lund, Shelley Lund, Perry Anzilotti, Bill Applebaum, Stacy Areheart, O'Neil Compton, Nada Despotovich, Bert Hogue
A nerdish cartoonist is driven mad by his desire for a beautiful woman, but the latter really only cares for the kind of heroic hemen who crop up in romantic fiction. With his appearance terribly blighted by a course of radiation therapy, he enlists the help of his sister to disguise himself as a suitable candidate for her affections. A goofy comedy of mistaken identity that is hard dislike, but equally hard to laugh at, it was simply a bad idea for a comedy.
COM 97 min (ort 102 min) VIDrel: COLUM/SONOP V/s
Boa: novel The Boyfriend School by Sarah Bird.

DON'T TELL MOM THE BABYSITTER'S DEAD ** PG
Stephen Herek USA 1991
Christina Applegate, Joanna Cassidy, John Getz, Keith Coogan, Josh Charles, David Duchovny, Kimmy Robertson, Eda Reiss Merin, Jayne Brook, Concetta Tomei, Robert Hy Gorman, Daniella Harris, Christopher Pettiet, Chris Claridge
When Mom goes on a two-month long vacation to Australia, she leaves the care of her five noisy kids in the hands of a babysitter – an elderly disciplinarian. However, the latter dies and the kids dump her at the local mortuary and start planning for an adult-free summer. Complications abound and the kids show ever more unbelievable examples of self-reliance, in a seriously contrived work that never delivers its earlier promise of black humour.
COM 100 min (ort 143 min) VIDrel: FIRST/SONOP V/sur

DON'T TOUCH THE WHITE WOMAN! ** 18
Marco Ferreri FRANCE/ITALY 1974
Catherine Deneuve, Marcello Mastroianni, Michel Piccolo, Philippe Noiret, Ugo Tognazzi, Alain Cuny, Serge Reggiani, Darry Cowl, Monique Chaumette, Franca Bettoja, Paolo Villagio, Daniele Dublino, Franco Fabrizi
Anarchistic and anachronistic black comedy attacking the expansionist and racist attitudes that lay behind the extermination of the native Americans, with Mastroianni cast as a bombastic and overbearing General Custer, who is seduced by a modest frontierswoman. Some imaginative touches help ram home the message, which is laid on with a trowel. Unfortunately excessive length and a surfeit of bad taste weaken the film's impact considerably.
Aka: DON'T TOUCH WHITE WOMEN; TOUCHE PAS A LA FEMME BLANCHE!
COM 105 min (ort 108 min) VIDrel: ARTPRO/RTM V

DOOM GENERATION, THE * 18
Gregg Araki FRANCE/USA 1995
James Duval, Rose McGowan, Johnathon Schaech, Cress Williams, Skinny Puppy, Dustin Nguyen, Margaret Cho, Lauren Tewes, Christopher Knight, Nicky Katt, Johanna Went, Perry Farrell, Amanda Hearse, Parker Posey, Salvator Xuereb
In this provocative road movie, Araki pulls out all the stops, treating us to a display of fellatio, decapitation and some graphic and outrageous visual gags. The slight story follows the career of a couple of bored L.A. teenagers, who team up with a handsome drifter, and this trio set off across America, enjoying a series of violent encounters with those unlucky enough to come into their path. Written by Araki, this brutal work has little to recommend it.
A/AD 83 min CINrel

DOOMED MEGAPOLIS: VOLS. 1 TO 4 ** 15
Kazuhiko Katayama JAPAN 1992
Four-part "anime" serial about how Tokyo of 1908 is menaced by some strange and rather nasty psychic forces, unleashed when the legendary spirit of Tairo No Masakado (believed to guard the well-being of the city) is disturbed by the activities of a Satanist. Highly imaginative but burdened by the same standardised level of animation.
ANIM 200 min (4 cassettes – approx 40 min each) dubbed
VIDrel: MANGA/SONOP L/A V/sh

DOOMSDAY GUN ** 15
Robert Young USA 1994
Frank Langella, Kevin Spacey, Tony Goldwyn, Alan Arkin, Michael Kitchen, Francesca Annis, James Fox, Aharon Ipale, Zia Mohyeddin, Rupert Graves, Clive Owen, Murray Melvin, Marianne Denicourt, Alexandra Vandernoot, Roger Hammond
An account of what became known in the United Kingdom as the "Supergun" affair in which a brilliant but totally unprincipled weapons scientist was able to build Saddam Hussein the biggest gun in the world. To make matters worse, the materials and equipment for this project were exported by Britain with the tacit approval of its government. A flat, documentary-style account enhanced by a convincing performance by Langella in the lead role.
DRA 106 min VIDrel: GUILD/SONOP L/A V

DOOMWATCH: THE PLASTIC EATERS ** PG
Paul Ciappessoni UK 1970
John Barron, Jennifer Wilson, Michael Hawkins, Kevin Stoney
Story from a so-so SF series with a strong ecological flavour that enjoyed a short period of popularity in the early 1970s. In this tale, the world faces a potential threat in the form of a virus that has the ability to dissolve plastics.
FAN 99 min (2-episode cassette) mTV
VIDrel: PARADOX/TOTAL V/h

DOOMWATCH: TOMORROW, THE RAT ** PG
Terence Dudley UK 1970
Penelope Lee, Eileen Helsby, Robert Sansom, Ray Roberts, John Berryman, Hamilton Dyce, Stephen Dudley
The experiments a genetic scientist has been conducting with her rats get out of hand, and when some of these creatures escape, the people of London are threatened by dangerous mutated creatures that have acquired a taste for human flesh. Another ecological disaster story from the 1970s series.
FAN 99 min (2-episode cassette) mTV
VIDrel: PARADOX/TOTAL V

DOORS, THE *** 18
Oliver Stone USA 1991
Val Kilmer, Meg Ryan, Kevin Dillon, Kyle McLachlan, Frank Whaley, Michael Madsen, Billy Idol, Kathleen Quinlan, Dennis Burkley, Josh Evans, Michael Wincott, John Densmore, Will Jordan, Mimi Rogers, Paul Williams, Billy Vera
An effective documentary-style biopic of Jim Morrison, lead singer with rock group The Doors, that charts their rise to fame in the late 1960s when they became one of the most popular groups in the USA up to Morrison's untimely death in a seedy Paris hotel room. Morrison's self-destructive lifestyle is carefully portrayed if not explained, but the film does convey a wonderful feeling of the 1960s. Kilmer gives a performance of uncanny accuracy.
DRA 134 min (ort 135 min) wScrn
VIDrel: POLY/POLYREC/GUILD; PION (LV only) V/sur LV

DOORWAYS * *15*
Peter Werner USA *1992*
George Newbern, Anne Le Guernac, Robert Knepper, Kurtwood Smith, Carrie-Anne Moss, Signy Coleman, Tisha Putman, Hoyt Axton, Jennifer Rhodes, Jonathan Ward, Max Grodenchik, Rick Dean, Ron Howard George, Patricia Belcher, Wally Crowder
A woman from a parallel world that is under non-human domination lands on Earth where her arrival causes a minor traffic accident. Befriended by the doctor who treats her, this unlikely couple are forced to go on the run when she escapes from custody, and have to stay one step ahead of both the FBI and the humanoids sent to bring her back. An interesting idea is wasted on a tedious chase movie, both inept and unconvincing, and with a tacked-on shock ending.
FAN   83 min mTV   VIDrel: COLUM/SONOP   V

DOPPELGANGER ** *18*
Avi Nesher USA *1992*
Drew Barrymore, George Newbern, Dennis Christopher, Sally Kellerman, Leslie Hope, George Maharis, Luana Anders, Jaid Barrymore, Peter Dobson, Dan Shor, Carl Bressler, Stanley De Santis, Sean Whalen, Thomas Bolack, John Cardone
A young woman claims that she has an evil double who murdered her mother but the police are not so certain. However, she is able to move to L.A. where she rents an apartment with a writer, and confides shares her problems with him. But the writer soon starts to suspect that the woman he is sharing the apartment with may in fact be the killer. A poorly plotted and unimpressive effort, with a just few moments of tension scattered through the film.
Aka: DOPPELGANGER: THE EVIL WITHIN
THR   100 min (ort 105 min)
VIDrel: 4-FRONT/POLYREC/ITC   V/h

DOUBLE CROSS ** *18*
Michael Keusch USA *1993*
Patrick Bergin, Jennifer Tilly, Kelly Preston, Matt Craven, Kevin Tighe, Philip Granger, Jerry Wasserman, Philip Hayes, Deryl Hayes, L. Harvey Gold, Sharlene Martin, Gillian Barner, Michael Iacobucci, James Bell, Monica Lange
A man comes to a small town where his car is involved in a minor collision with that driven by the local femme fatale. Unfortunately, he becomes so enamoured of her that he fails to realise she has implicated him in a nasty murder. Another unsuccessful attempt at an erotic thriller that feels both overwrought and contrived.
THR   87 min (ort 100 min)   VIDrel: MARQ/QUANT   V

DOUBLE, DOUBLE TOIL AND TROUBLE ** *(U)*
Stuart Margolin USA *1995*
Mary-Kate Olsen, Ashley Olsen, Cloris Leachman, Phil Fondacaro, Kelli Fox, Eric McCormack, Wayne Robson, Matthew Walker, Mesach Taylor, Debald Williams, Gary Jones, Babz Chula, Bill Meilen, Nora McLellan, Alex green, Alex Diakun
Twin girls have to cope with a nasty and powerful witch who has imprisoned her good sister in a magic mirror and plans all sorts of evil for a variety of people. A vanity vehicle (one of several) for twin child actresses Olsen whose thespian skills are strictly limited, although this one will undoubtedly appeal to younger children.
JUV   93 min   SATrel: MOVIE CHANNEL

DOUBLE DRAGON ** *12*
James Yukich USA *1994*
Michael Davis, Peter Gould, Robert Patrick, Mark Dacascos, Alyssa Milani, Scott Wolf, Leon Russom, Julia Nickson
In what remains of Los Angeles in the year 2027, a gangster boss wants part of a Chinese amulet that he hopes will bring him great power. A so-so futuristic fantasy, inspired by a video game. A few martial arts flourishes add a slight measure of interest.
FAN   92 min (ort 94 min)   VIDrel: COLUM/SONOP   V/sur

DOUBLE IMPACT * *18*
Sheldon Lettich USA *1991*
Jean-Claude Van Damme, Alan Scarfe, Geoffrey Lewis, Corey Everson, Alonna Shaw, Philip Chan Yan Kin, Bolo Yeung, Sarah-Jane Varley, Wu Fong Lung, Alicia Stevenson, Paul Aylett, Andy Armstrong, Sarah Yuen, Julie Strain, Eugene Choy
Van Damme takes on two roles in this sluggish action film, playing twins Chad and Alex, who were separated at birth when their parents were murdered by a Triad gang in Hong Kong.

Twenty-five years later they are re-united, and after an uneasy start, set off on a mission of vengeance, wiping out various villains on a journey towards an inevitable confrontation with Mr Big. Very poor fare indeed, and Van Damme is surprisingly, just as bad in both roles.
A/AD   105 min (ort 109 min)   VIDrel: COLUM/SONOP   V/sur

DOUBLE INDEMNITY ** *(PG)*
Billy Wilder USA *1944*
Fred MacMurray, Barbara Stanwyck, Edward G. Robinson, Tom Powers, Porter Hall, Jean Heather, Byron Barr, Richard Gaines, Fortunio Bonanova, John Philliber, Bess Flowers, Kernan Cripps, Oscar Smith, Betty Farrington
A brilliantly made example of film noir in this seedy tale of an insurance agent seduced and bamboozled by a glamorous woman into murdering her husband so that they can collect on his policy. Remade as a TV movie in 1973, the terrific script is the work of Wilder and Raymond Chandler.
THR   107 min B/W   TVrel
Boa: novel by James M. Cain.

DOUBLE JEOPARDY ** *18*
Lawrence Schiller USA *1992*
Bruce Boxleitner, Rachel Ward, Sally Kirkland, Sela Ward, Jay Patterson, Tom Everett, Denice Duff, Marjorie Hilton, Rosalind Soulam, Mary Ethel Gregory, Whitney Porter, Louis Schaeffer, Bill Osborn, J. Scott Bronson, Rick Bugg
A man married to a woman lawyer gets re-acquainted with an old girlfriend when she is accused of murdering her boyfriend and his wife is appointed to defend. As she is unaware of their one-time relationship, he tries to keep things this way but encounters a mass of dangerous complications. An unremarkable thriller that boasts some impressive mountain-climbing sequences.
THR   99 min (ort 102 min) mTV   VIDrel: 20TH/TECH   V

DOUBLE LIFE OF VERONIQUE, THE ** *15*
Krzysztof Kieslowski FRANCE/POLAND *1991*
Irene Jacob, Philippe Volter, Sandrine Dumas, Aleksander Bardini, Louis Ducreux, Claude Duneton, Halina Gryglaszewska, Kalina Jedrusik, Wladyslaw Kowalski, Jerzy Gudejko, Guillaume De Tonquedec, Gilles Gaston-Dreyfus
Two women, one French, the other Polish, live parallel lives without ever meeting but share many common things including an ability to sing and a serious heart condition. A very intellectual exercise, atmospheric and impenetrable, but made watchable by Jacob's fine performance and some excellent camerawork. An over-rated film, it won the Best Actress Award at Cannes 1991 and was winner in the Best Film category (International Critics Jury Prize).
Aka: LA DOUBLE VIE DE VERONIQUE
FAN   93 min (ort 110 min) wScrn   VIDrel: TART/20TH   V

DOUBLE O KID, THE ** *15*
Duncan McLachlan USA *1992*
Corey Haim, Brigite Nielsen, Wallace Shawn, John Rhys-Davies, Basil Hoffman, Nicole Eggert, Karen Black, Anne Francis, Leslie Danson, Patrick M. Wright, Bari K. Willerford, Jim Alquist, John Collier, Chick Hicks, Kandra Baker
A seventeen-year-old kid, who is a video games genius, finds himself having to outwit a mad computer scientist and his female companion who are after a package that he has to take to Los Angeles. This eventually involves him in a bid to stop the murder of a prominent scientist. A very juvenile effort, full of special effects but devoid of a solid plot and decent acting.
A/AD   90 min (ort 95 min)   VIDrel: CRYSTAL/HIFLI   V/h

DOUBLE OBSESSION ** *18*
Eduardo Montes USA *1993*
Margaux Hemingway, Maryam D'Abo, Frederic Forrest, Scott Valentine, Beth Fisher, Jamie Horton, Blair Weickgenant, Charles Carroll, Rachel Ward, Mary A. Gaffney, Hank Gaffney, Bea Hurwitz, Don Lambert, Darryl Hogue, Sydney Warner
A young woman at a college is so completely obsessed with her room-mate that she becomes both deranged and homicidal. This becomes especially evident when the object of her affections falls in love. Another tired variant on the "psycho from Hell" theme, with poor scripting and indifferent acting. Allegedly based on a true story. See also SINGLE WHITE FEMALE for something similar.
DRA   84 min (ort 88 min)   VIDrel: COLUM/SONOP   V

**DOUBLE SUSPICION ** **   18
Paul Ziller USA   1993
*Gary Busey, Kim Cattrall, Darlanne Fluegel, Jeff Griggs, Blu
Mankuma, Leam Blackwood, Laurie Briscoe, Michael Tiernan, Doug
Arthurs, Deryl Hayes, Philip Granger, Serge Horde, Stacy Grant,
Frances Ferruci, Sharlene Martin, James Bell*
A former cop must overcome his fear into order to save Seattle
from the grip of a serial murderer who strikes down his victims
with a scalpel and has resumed his bloody handiwork.
Reluctantly he rejoins the police force and takes up pursuit.
Aka: BREAKING POINT
DRA   91 min (ort 95 min)   VIDrel: MED/POLYREC L/A
V/sh

**DOUBLE TROUBLE ** **   18
John Paragon USA   1991
*David Paul, Peter Paul, Roddy McDowall, David Carradine, A.J.
Johnson, James Doohan, Steve Kanaly, Troy Donahue, Corbin
Bernsen, Bill Mumy*
Twin brothers on opposite sides of the law put their differences
aside and team up to tackle a ring of international jewel smug-
glers that is led by a seemingly respectable and power
businessman. A silly vehicle for twin wrestlers David and Peter
Paul (better known as The Barbarian Brothers).
COM   84 min (ort 87 min)   VIDrel: COLUM/SONOP   V

DOUBLE X *   15
Shani S. Grewal UK   1991
*Simon Ward, William Katt, Norman Wisdom, Bernard Hill, Gemma
Craven, Leon Herbert, Chloe Annett, Derren Nesbitt, Vladek Sheybal,
Terry Forrestal, Steve Carlow, Roger Law, Clifford Predgen, Rod
Stenna, Victoria Nairn, Iggy Navarro*
Written and directed by Grewal, this anaemic British thriller
gives Norman Wisdom his first screen role in twenty years – a
pity, he deserves far better. He plays a washed-up safe-breaker
who's on the run from a gangster outfit known as "The
Organisation". He falls in with an American hit-man, but loses
his life just the same. Meanwhile, his erstwhile associate has
fallen for the man's daughter, and sets out to save her life. Very
poor.
THR   94 min (ort 97 min)   VIDrel: DDVID   V/sur
Boa: short story Vengeance by David Fleming.

DOWN AMONG THE Z-MEN *   U
MacLean Rogers UK   1952
*Peter Sellers, Spike Milligan, Michael Bentine, Harry Secombe, Carol
Carr, Clifford Stanton, Graham Stark, Miriam Karlin*
Daft tale of MI5 and atomic research, from the stars of the
popular radio series of the 1950s "The Goons". The plot (if we
may use that term) has a group of criminals visiting a small
town, with the intention of stealing a professor's secret atomic
formula. Sellers does some nice impressions but the whole thing
is so creakingly contrived that it rapidly becomes rather painful
to watch.
Aka: GOON SHOW MOVIE, THE; STAND EASY
COM   68 min (ort 82 min) B/W   VIDrel: MIA/DISC   V

DOWN AND OUT IN BEVERLY HILLS **   15
Paul Mazursky USA   1985
*Nick Nolte, Bette Midler, Richard Dreyfuss, Tracy Nelson, Donald
Penniman, Little Richard, Evan Richards, Valerie Curtin, Elizabeth
Pena, Donald F. Muhich, Paul Mazursky, Barry Primus, Irene Tsu,
Jack Bruskoff, Michael Yama, Ranbir Bhai*
A tramp saved from drowning upsets the neurotic lives of an
over-rich Californian family, exposing the emptiness of their
existence. A slick and often very funny reworking of the 1932
French film BOUDU SAVED FROM DROWNING. Later made
into a TV series.
COM   103 min   VIDrel: TOUCH L/A   V

DOWN, OUT AND DANGEROUS *   15
Noel Nosseck USA   1995
*Richard Thomas, Bruce Davison, Cynthia Ettinger, Steve Hyter,
Christine Cavanagh, Melinda Culea*
A financial analyst takes pity on a tramp and tries to help him,
but his kindness is not returned as the down-and-out sets about
ruining the life of his benefactor. Quite an unappealing offering.
DRA   87 min (ort 90 min)   VIDrel: CIC/SONOP   V/dm

DOWN PERISCOPE **   12
David S. Ward USA   1996
*Kelsey Grammer, Lauren Holly, Rob Schneider, Harry Dean Santon,
Bruce Dern, William H. Macy, Ken Hudson Campbell, Toby Huss,*

*Duane Martin, Jonathan Penner, Brad Tatum, Harland Williams, Rip
Torn, James Martin Jr, Jordan Marder*
Navy larks and shenanigans with Grammer playing an unconventional officer who is assigned a creaking old WW2
submarine (complete with its crew of misfits and no-hopers) by
a spiteful superior. A series of war games ensues and though the
humour is of the simple sort, Grammer's immense charm keeps
the film afloat, as does fine work from everyone else.
COM   89 min (ort 93 min) cC   VIDrel: 20TH/FOXVID
V/sur

**DOWN THE DRAIN ** **   18
Robert C. Hughes USA   1989
*Andrew Stevens, Teri Copley, Don Stroud, John Matuszak, Joseph
Campanella, Jerry Mathers, Stella Stevens, Marco Fiorini, Mickey
Morton, Nick DeMauro, Barry Neikrug, Ken Foree, Sal Lopez, Benny
"The Jet" Urquidez, Dominc Barto*
A smart lawyer with many underworld clients decides to
mastermind a bank robbery with the help of his criminal connec-
tions, but has to run for his life with his girlfriend after he is
double-crossed. Average.
A/AD   101 min (ort 106 min)   VIDrel: COLUM/SONOP   V

**DOWN TO EARTH ** **   U
Alexander Hall USA   1947
*Rita Hayworth, Larry Parks, Marc Platt, Roland Culver, James
Gleason, Adele Jergens, George Macready, William Frawley, Jean
Donahue, Edward Everett Horton, Kathleen O'Malley, William
Haade, James Burke, Fred Sears, Lynn Merrick*
A musical that satirises the nine muses causes great anger
among them and Terpischore, the goddess of classical dance,
comes to Earth to put matters to rights. She gets a part in this
production and attempts to persuade the producer to incorpo-
rate classical Greek dances, but soon sees that jazz is a better
choice. A pleasant and watchable tale that spoofs HERE COMES
MR JORDAN but sunk by its silly plot and poor songs. Remade
as XANADU.
MUS   96 min (ort 101 min)   VIDrel: COLUM/SONOP   V

**DOWNHILL WILLIE ** **   12
David Mitchell USA   1995
*Keith Coogan, Lochlan Monroe, Staci Keanan, Estelle Harris, Fred
Stoller, Lee Reherman*
An accident prone nerd who is a complete failure at everything
has one talent, which is for the ski-slope. In this amiable slap-
stick nonsense he risks life and limb to take part in a dangerous
skiing contest, his hope being to win enough money to save his
friend from bankruptcy and get the girl of his dreams.
COM   85 min (ort 90 min)   VIDrel: MED/COLUM   V/sh

**DOWNPAYMENT ON MURDER ** **   15
Waris Hussein USA   1987
*Ben Gazzara, Connie Sellecca, David Morse, Jonathan Banks, G.W.
Bailey, John Karlen, Sheila Larkin, Miguel Ferrer, Jenny Beck,
Brandon Bluhm, Conrad Bachmann, John Durbin, Kimberly Pistone,
Marta Kober, Penelope Sudrow*
The battered wife of a psychotic real-estate dealer finally plucks
up the courage to leave, taking their kids with her. His efforts
to get the children back involve hiring a hit-man to murder her,
in this conventional tale largely redeemed by Gazzara's chilling
performance.
THR   95 min (ort 100 min) mTV   VIDrel: 20TH   V

DOWNTOWN *   18
Richard Benjamin USA   1990
*Anthony Edwards, Forest Whitaker, Penelope Ann Miller, Joe
Pantoliano, David Clennon, Art Evans, Kimberly Scott, Rick Aiello,
Roger Aaron Brown, Wanda De Jesus, Ron Canada, Frank McCarthy,
ryan McWhorter, Daniel Pipoly*
An inexperienced Philadelphia cop makes some powerful
enemies and finds himself transferred to duty in a rundown
neighbourhood, where his new partner is a solitary and sullen
black officer. With the look of a pilot TV movie about it, this
"odd couple" tale blends elements of comedy, melodrama and
action with neither skill nor originality.
DRA   91 min (ort 96 min)   VIDrel: 20TH/TECH   V/sur

DRACULA **   PG
Tod Browning USA   1931
*Bela Lugosi, Helen Chandler, David Manners, Edward Van Sloan,
Dwight Frye, Herbert Bunston, Frances Dade, Charles Gerrard, Joan
Standing, Moon Carroll, Josephine Velez, Michael Visaroff, Daisy
Belmore*

A classic horror film telling of a Transylvanian vampire count who comes to London and unleashes a wave of terror until he is destroyed. Lugosi's fine performance is rightly considered to be the definitive interpretation, but unfortunately, the strengths of the novel are badly diluted by the weaknesses of the play, from which the film draws much of its stilted dialogue. Followed by DRACULA'S DAUGHTER and countless remakes and spin-offs.
HOR   84 min B/W   VIDrel: CIC/SONOP   V/h
Boa: novel by Bram Stoker/play by John L. Balderston and Hamilton Deane.

DRACULA *   15
Dan Curtis USA   1973
Jack Palance, Simon Ward, Nigel Davenport, Pamela Brown, Fiona Lewis, Murray Brown, Penelope Horner, Sarah Douglas, Virginia Wetherall, Barbara Lindley, George Pravda, Hanna-Maria Pravda, Reg Lye, Fred Stone, Sandra Caron
Another version of the famous legend, with Palance giving an interesting interpretation of the Count, who is now largely shown as a victim of fate. The intelligent script of Richard Matheson and the photography of Oswald Morris put some life into the now over-familiar story.
Aka: BRAM STOKER'S DRACULA
HOR   90 min (ort 97 min) mTV   VIDrel: MIA/DISC   V
Boa: novel by Bram Stoker.

DRACULA *   15
John Badham UK   1979
Frank Langella, Laurence Olivier, Kate Nelligan, Donald Pleasence, Trevor Eve, Jan Francis, Janine Duvitski, Tony Haygarth, Teddy Turner, Sylvester McCoy, Kristine Howarth, Joe Belcher, Gabor Vernon
A more lush and romantic version of the familiar vampire story with good performances all round and some genuinely horrific moments. Let down by rather meandering plot development that weakens any sense of tension, plus Olivier's dreadful over-acting. Worth a look for Langella's remarkable and acclaimed stage characterisation – his vampire is both sinister and sexual, but the trendy and overblown effects swamp him. Filmed in England.
HOR   109 min (ort 112 min)   VIDrel: CIC/SONOP L/A   V
Boa: novel by Bram Stoker/play by John L. Balderston and Hamilton Deane.

DRACULA *   18
Francis Ford Coppola USA   1992
Gary Oldman, Anthony Hopkins, Winona Ryder, Keanu Reeves, Richard E. Grant, Cary Elwes, Bill Campbell, Sadie Frost, Tom Waits, Jay Robinson, I.M. Hobson, Monica Bellucci, Michaela Bercu, Florina Kendrick, Laurie Franks, Maud Winchester
Lavish and highly visual variant on the Dracula legend, replete with wonderful sets and locations, but overlong and lacking both a strong plot and first-rate performances. Oldman plays Dracula as a tragic hero, a medieval warrior who vows to live for ever after the death of his wife. AA: Sound, Cos (Eiko Ishioka), Effects/aud (Tom C. McCarthy and David E. Stone), Make (Greg Cannom, Michele Burke and Matthew W. Mungle).
Aka: BRAM STOKER'S DRACULA
HOR   122 min (ort 128 min)   wScrn cC
VIDrel: COLUM/SONOP   V/sur V/sh

DRACULA A.D. 1972 *   18
Alan Gibson UK   1972
Peter Cushing, Christopher Lee, Stephanie Beacham, Michael Coles, William Ellis, Christopher Neame, Marsha Hunt, Philip Miller, Michael Kitchen, David Andrews, Caroline Munro, Janet Key, Lally Bowers
Dracula comes to swinging London of the 1970s and stays at trendy Chelsea. Forgettable, confusing and unconvincing, in roughly that order. Followed by THE SATANIC RITES OF DRACULA, another film from the Hammer stable.
Aka: DRACULA TODAY
HOR   92 min (ort 100 min)   VIDrel: WHV   V

DRACULA DEAD AND LOVING IT *   PG
Mel Brooks USA   1995
Leslie Nielsen, Peter MacNicol, Steven Weber, Amy Yasbeck, Lysette Anthony, Harvey Korman, Mel Brooks, Mark Blankfield, Megan Cavanagh, Clive Revill, Chuck McCann, Avery Schreiber, Cherie Franklin, Ezio Greggio, Leslie Sachs
The usual plethora of Mel Brooks sight gags adorn this creaking farce and though one is reminded of the earlier YOUNG

FRANKENSTEIN, at least that film had a fairly decent plot. In this story, Nielsen plays the Count, who buys some property in England in the hope of finding fresh victims. Books puts in an appearance as Van Helsing, and would have done better to stick to acting and let someone else tackle the script.
COM   90 min   CINrel

DRACULA HAS RISEN FROM THE GRAVE *   15
Freddie Francis UK   1968
Christopher Lee, Rupert Davies, Veronica Carlson, Barbara Ewing, Barry Andrews, Ewan Hooper, Marion Mathie, Michael Ripper, George A. Cooper, Marion Mathie, Carrie Baker, John D. Collins, Chris Cunningham, Norman Bacon
But the tired old script stays dead and buried. Another tale from Transylvania circa 1905, with our Count pursuing the pretty niece of a small-town priest. Followed by TASTE THE BLOOD OF DRACULA.
HOR   88 min (ort 92 min)   VIDrel: WHV   V/h

DRACULA: PRINCE OF DARKNESS *   15
Terence Fisher UK   1965
Christopher Lee, Andrew Keir, Barbara Shelley, Francis Matthews, Suzan Farmer, Charles Tingwell, Thorley Walters, Walter Brown, Philip Latham, Jack Lambert, George Woodbridge, Philip Ray, John Maxim, Joyce Henson
In 19th century Carpathia, a group of stranded travellers find shelter in the castle of the late Count Dracula, where the latter's evil butler sets about using the blood of one of them to resurrect his master. A muddled film of scant originality, and whose photography is as unmemorable as its dialogue. This dud from Hammer Films is a sequel of sorts to "The Horror Of Dracula" and gains little from the use of clips from their 1958 film – "Dracula".
Aka: BLOODY, THE; DESCIPLINE OF DRACULA; RVENGE OF DRACULA
HOR   86 min (ort 90 min) wScrn   VIDrel: LUMI/SPEAR   V

DRACULA RISING *   18
Fred Gallo USA   1992
Christopher Atkins, Stacey Travis, Doug Wert, Tara McCann, Vessela Karlunovska, Nikolai Sotirov, Zahari Vatahov, Desi Srovanova, Stancho Stanchev, Nelli Vladova
Having become a monk, Dracula meets a women who proves to be the reincarnation of his long-lost love, a witch burnt at the stake many centuries before. Naturally, his dormant feelings are re-awakened and with them his former instincts. Filmed in Bulgaria, this deliriously camp effort (it was produced by Roger Corman) does little to advance the career of either the director or the actors.
HOR   85 min   VIDrel: CIC/SONOP L/A   V

DRACULA SUCKS *   18
Philip Marshak USA   1979
Jamie Gillis, Annette Haven, John C. Holmes, John Leslie, Serena, Seka
Erotic version of the famous legend, that's essentially a series of episodes mixing stilted dialogue, sexual activity and nauseating gore in roughly equal proportions. An unpleasant film that even became available in the USA in an edited softcore version. Cut before video submission by 7 min 11 sec.
Aka: DRACULA'S BRIDE; LUST AT FIRST BITE; THIS VAMPIRE SUCKS
A   72 min Cut (2 min 26 sec – ort 92 min)   VIDrel: ELV L/A; MOPIC/SGSVID   V

DRACULA'S DAUGHTER *   PG
Lambert Hillyer USA   1936
Gloria Holden, Otto Kruger, Edward Van Sloan, Irving Pichel, Hedda Hopper, Marguerite Chuchill, Claud Allister, Nan Grey, E.E. Clive, Gilbert Emery, Billy Bevan, Halliwell Hobbes, Eily Malyon, Christian Rub, Guy Kingsford, Gordon Hart
A logical progression from the 1931 film, with the Count's daughter longing to be free of her inherited vampirism and consulting a psychiatrist in an attempt to be rid of it. But when she falls in love with a man soon to be married, she is not above putting a spell on his fiancee. Meanwhile, Van Helsing languishes in police custody for the murder of Dracula. A restrained low budget film, interesting if not chilling, with Holden most effective in the title role.
HOR   72 min B/W   VIDrel: L/A   V/h
Boa: story Dracula's Guest by Bram Stoker.

DRACULA'S WIDOW *   18
Christopher Coppola USA   1988
Sylvia Kristel, Josef Sommer, Lenny Von Dohlen, Marc Coppola,

Rachel Jones, Stefan Schnable, Traver Burns, Rick Warner, Candice Sims
Countess Dracula finds herself accidentally shipped to the USA from the family castle in Transylvania and upon learning that her husband was killed many years before goes on the rampage in a series of ferocious murders that leave the police baffled. A failed attempt at a film noir that gets nowhere despite its effectively atmospheric photography. The directorial debut of Coppola who still has a lot to learn from his famous uncle.
HOR 86 min VIDrel: POLY/POLYREC L/A V

DRAGNET **
Jack Webb USA
Jack Webb, Ben Alexander, Richard Boone, Ann Robinson, Stacy Harris, Dennis Weaver, Virginia Gregg
An overlong attempt to transfer the format of a very popular TV series to the big screen, that becomes a tedious and drawn-out examination of every fact in the case of an ex-convict murdered by the Mob. The TV series broke new ground in the 1950s with its realistic approach, but tension and brevity are lost in this full-length feature. A TV movie on the same theme (also directed by and featuring Webb) came along in 1969, as did a parody in 1987.
PG
1954
DRA 71 min (ort 89 min) VIDrel: CIC/SONOP L/A V

DRAGNET **
Tom Mankiewicz USA
Dan Aykroyd, Tom Hanks, Harry Morgan, Dabney Coleman, Elizabeth Ashley, Alexandra Paul, Christopher Plummer, Jack O'Halloran, Kathleen Freeman, Joe Altmark, Lenka Peterson, Julia Jennings, Lisa Aliff, Nina Arveson
Aykroyd plays Sergeant Joe Friday's bumbling nephew, the role formerly taken by Jack Webb, in this spoof on the popular TV series of the same name. The plot revolves around an investigation pursued by Aykroyd and his partner into the activities of a TV preacher and a porno magazine king, who are suspected of running a criminal cult. A comedy that runs out of steam very early on and stays like that until the end, although the punchline is pretty funny.
PG
1987
COM 102 min (Cut at film release by 14 sec – ort 106 min)
VIDrel: 4-FRONT/POLYREC/CIC V/sur

DRAGON BALL: THE MAGIC BEGINS **
Joe Chan
Don Wong, Paul Kam, Eddie Chan, Anna Li, Ruby Tse, Peter So, Eagle Kong
Filmed in Thailand, this is a live-action version of a cult status Japanese comic, that tells of an evil king and his quest for the seven magical pearls that will enable him to rule the world.
Aka: DRAGON BALL: THE MOVIE; DRAGON PEARL
12
1992
A/AD 90 min dubbed VIDrel: EAST/DISC V

DRAGON COP **
Alan Roberts USA
Ron Marchini, David Carradine, Carrie Chambers, Michael Bristow, Dana Bentley, Michael M. Foley
In a brutal and lawless nightmare world of the future, a cop who is also a martial arts expert, rescues a lady scientist and returns with her to her fortress home. But this abode is also the site of a device of awesome destructive power, which a ruthless demagogue is rather keen to acquire. Naturally, our cop tries to persuade him otherwise. A low-budget actioner, it treads a familiar path, offering little that is fresh or original.
18
199-
A/AD 90 min VIDrel: NEWAGE/COLUM L/A V

DRAGON FIGHTER **
HONG KONG
Beneath the Castle of Ching is a secret gold mine where people are abducted and forced to work as slaves. One worker revolts and for his pains his hands are broken. But this only fuels his determination to destroy the Castle and its keepers, and to this end he sets about learning the "Heaven Legs" technique. Fair.
18
MAR 92 min Cut (28 sec) VIDrel: IMPENT V

DRAGON FORCE **
Michael King (Key Nam Nam) HONG KONG
Bruce Baron, Mandy Moore, James Barnett, Jovy Couldry, Frances Fong, Olivia Jeng, Randy Channel, Sean Blake, Bruce Li (Ho Tsung-Tao), Kangjo Lee
Kung fu epic set in the 1930s, when Korean fighters use their skills against the Japanese occupation forces.
18
1982
MAR 89 min Cut (28 sec – ort 101 min) VIDrel: IMPENT
V

DRAGON FROM SHAOLIN, THE **
HONG KONG
Richard Kong, Li Ying Ying, Bruce Cheung, Steve Chen, Nam Chi I, Debbie Ling
Martial arts tale set in the Ming Dynasty that sees the winner of a kung fu contest having his family murdered when he refuses to work for the man who organised it. But the man's young daughter escapes and over the next 20 years becomes proficient in the fighting arts, awaiting the day she can have her revenge. Better than average fight sequences enliven the painfully predictable plot.
18
MAR 82 min (ort 90 min) VIDrel: IMPENT V

DRAGON THAT WASN'T... OR WAS HE?, THE **
Bjorn Frank Jensen/Bob Maxfield/Jay Kamen (English ver)
HOLLAND
A boy gets a baby dragon as a pet, but when it grows too big for his home, he is forced to return it to its natural habitat, the land beyond the mountains. On the way there, they face many dangers together.
Aka: ALS JE BEGRIJPT WAT IK BEDOEL; DRAGON THAT WASN'T (OR WAS HE?), THE
U
1983
ANIM 82 min (ort 96 min) VIDrel: CIC/SONOP V

DRAGON: THE BRUCE LEE STORY **
Rob Cohen USA
Jason Scott Lee, Lauren Holly, Robert Wagner, Nancy Kwan, Michael Learned, Kay Tong Lim, Sterling Macer, Ric Young, Sven-Ole Thorson, Luoyang Wang, Ong Soo Han, Eric Bruskotter, Aki Aleong, Chao-Li Chi, Iain M. Parker, Sam Hau
Well-made biopic of this famed martial-arts star who died tragically young in 1973 at only thirty-three. Excellent action sequences blend well with an attempt to show how his determination and ability won him success in his chosen field.
15
1992
A/AD 114 min (ort 120 min) cC VIDrel: CIC/SONOP;
PION (LV only) V/dm LV
Boa: book Bruce Lee: The Man Only I Knew by Linda Lee Caldwell.

DRAGON, THE HERO, THE **
Godfrey Ho HONG KONG
John Liu, Dragon Lee, Tino Wong, Philip Ku, Yang Sze
Two masters of the "Stone Fist" – a powerful fighting style, quarrel, and one of them dies in the fight. Twenty years later their sons meet by chance and Tong, the dead man's son, plots his revenge. His opponent is superior to him and easily defeats him, but when they find that they have the same enemies, they decide to join forces and fight together as their fathers once did.
18
1981
MAR 80 min Cut (2 min 59 sec) VIDrel: IMPENT V

DRAGON, THE YOUNG MASTER, THE **
Godfrey Ho HONG KONG
Dragon Lee, Phoenix Kim, Marty Chui, Ben Lee, Kelvin Chan, Jackie Lee, Tony Min, Steve Lim
Routine kung fu adventure revolving round a hunt for stolen jewellery that has been hidden in the Snow Mountain.
18
MAR 84 min Cut (2 sec) VIDrel: IMPENT V

DRAGONFIRE **
Richard T. Heffron USA
Daniel J. Travanti, Roxanne Hart, Stephen Tobolowsky, William Sadler, Sarah Douglas, Peter Michael Goetz, Lyman Ward, Vince Guastaferro, Noel Harrison, Guy Doleman, Lincoln Kilpatrick, Glenn Morshower, Leon Russom, William Denis
A disabled Vietnam veteran who suffers from bouts of amnesia begins to investigate the nature of the covert operations he took part in, but discovers that others are prepared to have him killed in order to keep the truth hidden. A competent action outing, fast and well edited, and with excellent photography by Billy Dickson.
Aka: TAGGET
15
1990
A/AD 85 min (ort 89 min) mTV VIDrel: CIC/SONOP
V/sh
Boa: novel Tagget by Irving A. Greenfield.

DRAGONHEART ***
Rob Cohen USA
Dennis Quaid, David Thewlis, Peter Postlethwaite, Dina Meyer, Jason Isaacs, Brian Thompson, Lee Oakes, Wolf Christian, Terry O'Neill, Eva Vejmelkova, Peter Hric, Milan Bahul, Sandra Kovacicova plus voices of: Sean Connery, John Gielgud
A dragonslayer and the last dragon in the world team up to put a stop to the depredations of a king, whose life was once saved
PG
1996

by the dragon when the creature gave him half its heart, so if the king dies so will the dragon. A most appealing and imaginative kids' fantasy adventure, enjoyable if not overly taxing on the intellect. Connery provides the voice of the dragon – a task he performs admirably.
FAN   98 min (ort 103 min) cC   VIDrel: CIC; PION (LV only) V/sur LV

DRAGONS FOREVER ***
18
Samo Hung HONG KONG   1988
Jackie Chan, Samo Hung, Yuen Biao, Crystal Kwok, Deannie Yip, Pauline Yeung
The waste from a chemical plant is destroying the local fish farming. When the attractive owner of a farm threatens to go to court, the plant hires Chan, a wily lawyer, to fight her. However, he falls for her star witness and his buddy falls for her. Events take a sinister turn when it transpires that the plant is a disguised drugs plant. The two sides now unite in an effort to smash the drugs ring running it.
A/AD   89 min   VIDrel: POPRO/RTM/IMPENT L/A   V

DRAGONWORLD **
U
Ted Nicolalou USA   1993
Sam MacKenzie, Brittney Powell, John Calvin, Courtland Mead, Andrew Keir, Jim Dunk, John Woodvine, Lila Kaye, Sue Douglas, Janet Henfrey, Stuart Campbell, Marioara Stelian, Valentin Popescu, Cameron Stuart, Claudiu Istodor
A Scottish landowner discovers a dragon on his property but is soon forced to sell his holdings in order to pay his taxes. Unfortunately, this means that this creature falls into the hands of a nasty villain who is planning a theme park in which our dragon will be the main attraction. A so-so fantasy with poor effects that is clearly aimed at very young children.
FAN   82 min (ort 84 min)   VIDrel: CIC/SONOP   V

DRAGSTRIP GIRL **
(12)
Mary Lambert USA   1993
Maria Celedonio, Christopher Crabb, Raymond Cruz, Mark Dascascos, Frderick Coffin, Traci Lords, Richard Portnow, Tracy Wells, Augustus Cesar Sandino, Natasha Gregson Wagner, G. Adam Gifford, Carolyn Mignini, Gary Werntz
One in a series of remakes of 1950s AIP B-movies that are updated to the 1990s. This one hardly resembles its predecessor at all, and is built around the exploits of a young man who takes a job as a car parking valet, but soon gets drawn into involvement with a car-stealing gang, spotting suitable vehicles for their operations. However, matters become fraught when he falls in love with a wealthy young woman.
DRA   79 min   SATrel: SKY MOVIES

DRAUGHTSMAN'S CONTRACT, THE **
15
Peter Greenaway UK   1982
Anthony Higgins, Janet Suzman, Anne Louise Lambert, Neil Cunningham, Hugh Fraser, Dave Hill, David Gant, David Meyer, Tony Meyer, Nicolas Amer, Suzan Crowley, Lynda Marchal, Michael Feast, Alastair Cummings, Steve Ubels, Ben Kirby
Unusual seventeenth century drama about a young draughtsman whose commission includes sexual favours from his female employer. A lavish and complex tale, full of symbols and hidden meanings, but let down by a cold and unmoving script. The music is by Michael Nyman. Made for TV but released theatrically.
DRA   102 min (ort 108 min) wScrn mTV
VIDrel: ARTIF/20TH   V/h

DREAM A LITTLE DREAM *
15
Marc Rocco USA   1989
Jason Robards, Corey Feldman, Piper Laurie, Meredith Salenger, Harry Dean Stanton, Corey Haim, Susan Blakely, William McNamara, Matt Adler, Victoria Jackson, Alex Rocco
Another body-switching comedy as a couple's experiments with transcendental meditation lands them in the bodies of a couple of teenagers. Fairly predictable and most unfunny.
COM   110 min (ort 114 min)   VIDrel: VES L/A   V

DREAM A LITTLE DREAM 2 *
12
James Lemmo USA   1994
Corey Haim, Corey Feldman, Robyn Lively, Stacie Randal, Michael Nicholosi, Lou Bonacki, Bobby Costanzo
This unfunny sequel has an espionage flavour to its story of a hunt for some sunglasses with special powers. Two friends receive a parcel containing said item only to find themselves being pursued by a strange woman. Flat and very disappointing.
COM   87 min (ort 91 min)   VIDrel: FIRST/SONOP   V

DREAM BREAKERS **
15
Stuart Millar USA   1989
Robert Loggia, Kyle MacLachlan, D.W. Moffett, Charles Cioffi, John McIntire, Hal Linden
Drama following the lives of a Chicago building contractor and his two sons, one a dedicated priest and the other a gifted business graduate. The latter has just taken a job with a corrupt and power-hungry land developer, a former partner of his father's. Eventually, all three unite to oppose the man's crooked schemes. Interesting casting is the best feature in this cliched and curiously old-fashioned family saga.
THR   90 min (ort 100 min) mTV   VIDrel: 20TH/TECH   V

DREAM CHASERS ***
(U)
David E. Jackson/Arthur Dubbs USA   1984
Harold Gould, Justin Dana, Wesley Grant, Jeffrey Tambor, Carolyn Carradine, J.J. Lewis
A boy who is dying of cancer, makes one last journey with an elderly friend rather than his father, who has shunned him as he is unable to come to terms with his son's imminent death. The unlikely pair take off and act out a cowboy fantasy together.
DRA   90 min (ort 97 min)   SATrel: SKY MOVIES

DREAM IS A WISH YOUR HEART MAKES, A: THE ANNETTE FUNICELLO STORY **
(PG)
USA   1995
Annette Funicello, Frankie Avalon, Len Cariou, Shelley Fabares, Eva LaRuem Linda Cavin, David Lipper, Austin Basile, Ahnee Boyce, Jay Brazeau, Victoria Brooks, Stephania Ciccone, Damitri Garitsas, Elysa Hogg, Gary Jones
A straightforward biopic that describes the life and career of the title actress, from her early childhood onwards, when she was seen dancing in a school show and was spotted by no less a person than Walt Disney himself, who chose her as one of the "Mouseketeers" for the Mickey Mouse Club. A framing device has the real-life Funicello looking back on her career and also talking about her struggle to cope with the ravages of multiple sclerosis.
Aka: ANNETTE FUNICELLO STORY, THE
DRA   90 min   SATrel: SKY MOVIES

DREAM LOVER **
18
Nicholas Kazan USA   1994
James Spader, Madchen Amick, Frederic Lehne, Larry Miller, Bess Armstrong, Clyde Kusatsu, Kathleen York, Blair Tefkin, Scott Coffey, William Shockley, Michael Milhoan, Robert David Hall, Archie Lang, Janel Moloney, Talya Ferro
A recently divorced architect marries a woman who seems to be completely perfect, but then begins learning of disturbing things that indicate a sinister past, and one that may prove fatal to him. This uneven thriller tries hard to achieve a film noir atmosphere but cannot overcome its undeveloped script in spite of fine work from the cast. The directorial debut for Kazan (the son of Elia Kazan by his first wife).
Aka: SHAME 2: THE SECRET
THR   100 min (ort 103 min)   VIDrel: 20VIS/SONOP   V/sur

DREAM MACHINE **
PG
Lyman Dayton USA   1991
Evan Richards, Susan Seaforth-Hayes, Randall England, Brittney Lewis, Corey Haim, Jeremy Slate, Tracy Fraim, James MacKrell, Suzanne Kent, Don Shanks, Brynja McGrady, Michael Flynn, Sam Shamshak, Julie Simper, Billie Judkins
A youngster gets the gift of a silver Porsche and almost immediately gets embroiled in a series of bizarre adventures and narrow escapes, most of which appear to have arisen from a case of mistaken identity. Directed and produced by Dayton, this is a fairly lightweight blending of action, comedy and thrills.
A/AD   82 min (ort 88 min)   VIDrel: FIRST/SONOP   V

DREAM MAN **
18
David Edwards USA   1994
Michael Kearns
A phone sex host is whatever his callers want him to be, but in reality lives a lonely and isolated life. A low-budget, gay sex film.
A   55 min   VIDrel: MANFOR/GOLD   V

DREAM MAN *
18
Rene Bonniere USA   1995
Patsy Kensit, Bruce Greenwood, Andrew McCarthy, Denise Crosby, Jim Byrnes, Cameron Bancroft, John Cussini, Jay Brazeau, Armin Shimerman, Dawne Pendleburg, L. Harvey Gold, Colleen Rennison, Lisa Bunting, Brian Doctor, Mike Mitchell

A Seattle cop has psychic powers that make her aware of crimes as they are being committed but makes the mistake of falling in love with a dashing playboy, who is the principal suspect in one of her cases. Kensit is totally miscast as the tough cop and this ludicrous piece of casting sinks the film before it even starts. The rest of the cast are slightly better, but are saddled with a daft script.
THR 90 min (ort 94 min) mTV VIDrel: FIRST/SONOP V

DREAM MASTER: THE EROTIC INVADER * (18)
Jackie Garth USA 1996
Cassandra Leigh, Patrick Ahern, Krisitn Knittle, Timothy Di Pri, Jennifer Barnes, Mark Sherman, Patricia Skeriotis, David Ranker, Robbie French
A post-graduate psychiatry student is pursuing some original dream research in when he believes that people can overcome past neuroses by learning to wake up inside their dreams and control the outcome. Ultimately, he also aims to make it possible for two people to share the same dream. However, when he becomes trapped inside his own dream, some students participating in his experiments must rescue him. A soft-core romp that does nothing with some fascinating ideas.
A 86 min SATrel: SKY MOVIES

DREAM ON * 15
Amber Films UK 1991
Maureen Harold, Amber Styles, Anna-Maria Gascoigne, Pat Leavy, Ray stubbs, Brian Hogg, Derek Walmsley, Art Davies, Niall Tobin, Wayne Buck, Mike Christie, Libby Davison, Gwen Doran, Steve Drayton, Mike Elliott, Charlie Hardwick
Ensemble acting is used to good effect in the story of a group of women, all of whom must cope with the many and varied traumas in their lives, which vary from sexual abuse to domestic violence. Not a terribly edifying work, this is a harsh and brutal example of the British penchant for showing life as something to be endured, but there are enough good moments in the film to make it a worthwhile experience. Some scattered moments of comedy are extremely welcome.
DRA 120 min CINrel

DREAM RIDER * U
Bill Brown USA 1992
Matthew Geriak, James Earl Jones, Leigh Taylor-Young, Stacey Bell, Heather D. Haase, E.J. Peaker, Jay Richardson, Murray Rose, Criss Freeman, Lysa Nalin, Victor Gardell, Ginger Erickson, Gloria Steppe, John Gowans, Jack Rule
An American football star learns the true meaning of determination when his playing days are brought to an abrupt end by the loss of a leg. Undaunted, he becomes a motorcycle champion and travels from coast to coast. A true-life drama that both entertains and inspires, not least thanks to the director's strong and unmanipulative script.
Aka: I CAN'T LOSE
DRA 88 min VIDrel: FIRST/SONOP V

DREAM TEAM, THE * 15
Howard Zeiff USA 1989
Michael Keaton, Peter Boyle, Christopher Lloyd, Stephen Furst, Dennis Boutsikaris, Lorraine Bracco, Milo O'Shea, Philip Bosco, James Remar, Jack Gilpin, Macintyre Dixon, Michael Lembeck, Bill Goffi, Jack Duffy, Brad Sullivan
A psychiatrist takes four mental patients on a trip to New York's Yankee Stadium, but the doctor inadvertently witnesses a murder, is injured and hospitalised, leaving his charges to survive as best they can. Their growing sense of co-operation and a strong set of performances, make this chaotic and uneven comedy-thriller a little more enjoyable than it might have been.
COM 108 min (ort 113 min)
VIDrel: 4-FRONT/POLYREC/CIC L/A V/h

DREAMING THE REALITY * 18
Simon Yuen Ching HONG KONG 1991
Moon Lee, Sibelle Hu, Yukari Oshima, Eddie Ko
Some highly-trained girls set out to retrieve the government evidence that could send their gangster boss to prison. A standard explosion of fisticuffs and action.
A/AD 90 min VIDrel: EAST/DISC V

DREAMS * PG
Akira Kurosawa JAPAN 1990
Akira Terao, Chishu Ryu, Martin Scorsese, Mieko Harada, Mie Suzuki, Mitsuko Baisho, Masayuki Yui, Shu Nakajima, Sake Kimura,
Yoshitaka Zushi, Toshihiko Nakano, Mitsunori Isaki, Toshie Negishi, Hirasho Igawa, Chosuke Ikariya
A remarkable effort from writer-director Kurosawa, this is a collection of eight diverse tales that range from charming fairytales to visions of apocalyptic doom. Beautifully photographed and with music that matches each piece's mood, but lacking the coherence of artistic vision that generally typifies the work of this master director.
Aka: AKIRA KUROSAWA'S DREAMS
FAN 114 min (ort 119 min) wScrn VIDrel: WHV V/sur

DREAMS LOST, DREAMS FOUND * PG
Willi Patterson UK/USA 1987
Kathleen Quinlan, David Robb, Betsy Brantley, Colette O'Neil, Charles Gray, Tom Watson, Kay Gallie, Anne Kristen, Fiona Mollison, Tom Mannion, Anne Downie, Raymond Ross, Gary Denis, Louise Breslin
An American widow sells her art gallery and moves to Scotland where she has bought an ancient castle. Newly arrived, she embarks on a tempestuous relationship with the local aristocrat. Another dull entry in the Harlequin Romance series, barely enlivened by attempts to incorporate a supernatural element in the form of a legend of tragic lovers.
DRA 97 min mTV VIDrel: CIC/SONOP V
Boa: novel by Pamela Wallace.

DREAMSCAPE * 15
Joseph Ruben USA 1984
Dennis Quaid, Christopher Plummer, Max Von Sydow, Kate Capshaw, Eddie Albert, George Wendt, David Patrick Kelly, larry Gelman, Redmon Gleason, Peter Jason, Chris Mulkey, Jana Taylor, Madison Mason, Kendall Carly Brown
A man with telepathic ability is hired by an organisation that has built a device enabling people such as him to enter the dreams of others. As he begins work as a therapist, helping disturbed individuals through their dreams, he finds himself caught up in a plot to destroy the US President. A film that's at its best during the dream sequences, which to some extent compensate for the inadequate plot.
FAN 94 min (Cut at film release – ort 99 min)
VIDrel: LUMI/SPEAR L/A; ENCORE (LV only) V/sur LV

DRESS GRAY * 15
Glenn Jordan USA 1986
Hal Holbrook, Alec Baldwin, Eddie Albert, Lloyd Bridges, Alexis Smith, Susan Hess, Patrick Cassidy, Lane Smith, David Harum, Steve Kosko, Patricia Herd, Rick Goldman, Arthur French, Richard Doyle, Alma Martinez
An Eastern military academy is the setting for this taut adaptation of Truscott's novel dealing with the cover-up of the circumstances surrounding the death of a cadet from a well-connected family who proves to have been a homosexual.
DRA 183 min (ort 192 min) mTV VIDrel: WHV V/sh
Boa: novel by Lucian K. Truscott.

DRESSED TO KILL * 18
Brian De Palma USA 1980
Michael Caine, Angie Dickinson, Nancy Allen, Keith Gordon, Dennis Franz, David Marguiles, Brandon Maggart, Fred Weber, Susanna Clemm, Ken Baker, Robert Lee Rush, Bill Randolph, Sean O'Rinn, Mary Davenport
Having told her psychiatrist that her husband is useless sexually, suburban housewife Dickinson finds herself in peril from a sinister figure stalking both her and a streetwise prostitute. When she is killed the hooker teams up with her son to catch the murderer, in this tense and absorbing thriller. Written by De Palma, this film plays on emotion rather than logic, utilising a fine cast and some flashy techniques. The score is by Pino Donaggio.
THR 100 min (ort 104 min)
VIDrel: 4-FRONT/POLYREC/VISVID L/A V

DRESSER, THE * PG
Peter Yates UK 1983
Albert Finney, Tom Courtenay, Edward Fox, Zena Walker, Eileen Atkins, Michael Gough, Cathryn Harrison, Betty Marsden, Sheila Reid, Lockwood West, Donald Eccles, Llewellyn Rees, Guy Manning, Anne Mannion, Kevin Stoney, Ann Way
Acutely observed study of the relationship between an ageing actor and his devoted dresser, who lives his life through the performances of his master. A poignant character study that perfectly captures the atmosphere of wartime England, the

period in which the tale is set, and is largely based on Harwood's own experiences working for Sir Donald Wolfitt.
DRA   113 min (ort 118 min)   VIDrel: VCC L/A   V
Boa: play by Ronald Harwood.

DRESSMAKER, THE *** 15
Jim O'Brien UK 1988
Joan Plowright, Billie Whitelaw, Peter Postlethwaite, Jane Horrocks, Tim Ransom, Rosemary Martin, Pippa Hinchley, Tony Haygarth, Michael James-Reed, Sam Douglas, Bert Parnaby, Margi Clarke, Mandy Humphrey
Two sisters share the same household with their niece in this story set in war-time Liverpool. Their cosy lives are disrupted when the niece brings home an American soldier. The girl is unwilling to go to bed with him so he has a fling with the younger aunt, but their short-lived affair ends in tragedy. The gruesome and wholly unexpected climax may well shock, but in terms of both realism and acting this film is hard to fault.
DRA   88 min (ort 92 min)   VIDrel: MGM/WHV   V/h
Boa: novel by Beryl Bainbridge.

DRIFTER, THE ** 18
Jim Enright USA 1995
Kaylan Nicole, Melissa Monet, Kylie Ireland, Brittany O'Connell, Marc Wallice, Jonathan Morgan, Tom Chapman
A woman seduces a young drifter working as a handyman in her home, in a vain attempt to persuade him to murder her wealthy husband. A sex film whose slightly film noirish plot does very little for it.
A   69 min   VIDrel: ONE   V

DRIFTWOOD ** 18
Ronan O'Leary IRELAND/UK 1996
James Spader, Anne Brochet, Barry McGovern, Anna Massey, Aiden Grenell, Kevin McHugh, John Cleere
A story of obsessive love that opens when a lonely woman (Brochet) discovers an injured, amnesiac man (Spader) on the beach where she has gone to collect driftwood. Taking him back to her cottage, she nurses him and they become lovers. But as his strength begins to return, she grows fearful he will leave, and resorts to ever desperate measures to keep him captive. An atmospheric and oppressive film, but sadly one that gets ever more ludicrous as it develops.
DRA   100 min   CINrel

DRIPPING WITH DESIRE * 18
Eric Edwards USA 1992
Kiss, Sierra, Heather Hart, Celeste, Lilli Xene, Chrissy Ann, Peter North, Jon Dough, T.T. Boy, Kris Newz
A researcher into human sexuality matches up compatible couples in this thoroughly undistinguished adult movie.
A   56 min (ort 80 min)   VIDrel: PASSION/IMC   V

DRIVE-IN MASSACRE * 18
Stuart Segall USA 1976
Jake Barnes, Adam Lawrence, Douglas Gudbye, Newton Naushaus, Norman Sherlock, Valdesta
A crazed killer stalks the audiences at drive-in cinemas in this chilling but unimaginative horror tale.
HOR   71 min (ort 78 min)   VIDrel: VIPCO/SGSVID   V/h

DRIVER, THE *** 15
Walter Hill USA 1978
Ryan O'Neil, Bruce Dern, Isabelle Adjani, Matt Clark, Ronee Blakley, Felice Orlandi, Joseph Walsh, Rudy Ramos, Denny Macko, Frank Bruno, Will Walker, Sandy Brown Wyeth, Tara King, Richard Carey, Fidel Corona, Victor Gilmour
A tough cop and a professional getaway-driver fight a battle of wits in this somewhat convoluted but enjoyable and tense thriller. The car chases are well-handled and Dern gives an intense portrayal of a cop who has carried his desire to uphold the law to the point of obsession. An assured directorial debut for Hill, who also wrote the script.
A/AD   87 min (Cut at film release by 18 sec – ort 90 min)   VIDrel: WHV   V

DRIVING FORCE * 15
Andrew J. Prowse AUSTRALIA 1989
Sam J. Jones, Catherine Bach, Don Swayze
A truck owner faces trouble on two fronts from both his ruthless business rivals and the family of his estranged wife, who are trying to get custody of his young daughter. A formula action story with the usual blend of mayhem and revenge, set

in a future society where law and order have broken down and the roads are haunted by ruthless marauders, very much in the MAD MAX mode.
A/AD   90 min   VIDrel: MED/POLYREC L/A   V

DRIVING ME CRAZY ** PG
Peter Faiman USA 1991
Ed O'Neill, Ethan Randall, JoBeth Williams, Christopher McDonald, L. Scott Caldwell, E.G. Daily, Ari Meyers, Kathleen Freeman, Lisa Figus, Cedering Fox, Shelby Leverington, Kyle Fredericks, David James Alexander, Ross Borden
This amiable road-movie sees a nice working-class guy doing his best to win over his divorcee girlfriend's son, a rich and spoilt brat who is too old to get what he really needs – a good spanking. A war of nerves develops between our ill-matched duo: they lose their car, get robbed but eventually develop a little grudging respect for each other. An undemanding if not exactly inventive time-filler, produced and written by John Hughes, who has done better work.
Aka: DUTCH
COM   103 min Cut (2 sec plus some cuts subst – ort 108 min)   VIDrel: 20TH/TECH   V

DRIVING ME CRAZY ** PG
Jon Turteltaub USA 1991
Billy Dee Williams, James Tolkan, Dom DeLuise, George Wendt, Michelle Johnson, Thomas Gottschalk, Milton Berle, Richard Moll, Steve Kanaly, Morton Downey Jr
An East German inventor takes a trip to America in the hope of finding a market for his ecologically clean car. One or two mild chuckles are to be had here.
COM   84 min (ort 88 min)   VIDrel: COLUM/SONOP   V

DRIVING MISS DAISY *** U
Bruce Beresford USA 1989
Jessica Tandy, Morgan Freeman, Dan Aykroyd, Patti LuPone, Esther Rolle, Joann Havrilla, Alvin M. Sugarman, Clarice F. Geigerman, Muriel Moore, Crystal R. Fox, Sylvia Kaler, Carolyn Gold, Bob Hannall, Ray McKinnon, Ashley Josey
An uneducated but wily black man is hired to work as chauffeur to a peevish old Southern woman, and over twenty-five years the two become inseparable. A film of some charm, it covers the period from the 1940s to the 1960s, says next to nothing about race relations, and yet for all its contrived warmth is undeniably entertaining. AA: Pic, Actress (Tandy), Screen/adapt (Alfred Uhry), Make (Manlio Rocchetti/Lynn Barber/Kevin Haney).
COM   94 min (ort 99 min)   VIDrel: WHV L/A   V/sh
Boa: play by Alfred Uhry.

DROP DEAD FRED * 15
Ate De Jong USA 1991
Rik Mayall, Phoebe Cates, Carrie Fisher, Marsha Mason, Tim Matheson, Bridget Fonda, Keith Charles, Ron Eldard, Daniel Gerroll, Ashely Feldon, Ron Eldard, Sjoukje De Jong Douma, Eleanor Mondale, Bob Reid, Peter Thoemke, Paul Holmes
A withdrawn and unhappy young woman who finds that her life is in turmoil is re-visited by "Drop Dead Fred" – an imaginary childhood playmate with a spitefully mischievous sense of humour. Upon this one interesting idea a depressing and vulgar comedy is built. Give this one a miss, it's just not in the same league as HARVEY.
COM   94 min (ort 99 min)   VIDrel: VCC/DISC/COLUM   V/sur

DROP SQUAD, THE ** 15
David Johnson USA 1994
Eriq Lasalle, Vondie Curtis-Hall, Ving Rhames, Vanessa Williams, Michael Ralph, Kasi Lemmons, Leonard Thomas, Bill Williams, Eric A. Payne, Nicole Powell, Paula Kelly
A group of disenchanted African Americans form a secret society whose object is to restore black pride to various individuals by making use of deprogramming techniques. One of their victims is an executive who has been peddling a series of products for this group that they consider of dubious value. Quite a fascinating idea but one that is fatally flawed by uncertain direction and lack of plot development. Lee also functioned as executive producer on this movie.
DRA   84 min (ort 88 min) cC   VIDrel: CIC/SONOP   V/sur

DROP ZONE *** 15
John Badham USA 1994
Wesley Snipes, Gary Busey, Yancy Butler, Michael Jeter, Corin

Nemec, Kyle Secor, Luca Bercovici, Malcolm-Jamal Warner, Rex Linn, Grace Zabriskie, Claire Stansfield, Robert LaSardo, Sam Hennings, Mickey Jones, Andy Romano, Rick Zieff
Two brothers who are US marshals are assigned to guard a witness but fall victim to an aerial kidnapping that leaves one of them dead. With the witness now gone, the government blames its employees, but the surviving brother mounts his own campaign to track down the culprits. The plot doesn't really hold water, but the spectacular aerial sequences and non-stop action ensure it remains an entertaining and fast-paced movie.
A/AD   101 min (ort 102 min)   VIDrel: CIC/SONOP; PION (LV only)   V LV

DROWNING BY NUMBERS **
Peter Greenaway HOLLAND/UK
18
1987
Bernard Hill, Joan Plowright, Juliet Stevenson, Joely Richardson, Jason Edwards, Bryan Pringle, Trevor Cooper, David Morrissey, John Rogan, Paul Mooney, Jane Gurrett, Kenny Ireland, Michael Percival, Joanna Dickens, Janine Duvitsky
An absurd over-intellectual tale of three generations of women, who all drown their unwanted husbands, eventually coming up against an obsessive coroner. Excellent photography is a compensation for the idiosyncracies (very typical of this director) of this bizarre comedy-thriller.
THR   113 min (ort 121 min)   VIDrel: ELPIC/POLYREC
V/sur

DROWNING POOL, THE **
Stuart Rosenberg USA
12
1975
Paul Newman, Joanne Woodward, Anthony Franciosa, Richard Jaeckel, Murray Hamilton, Melanie Griffith, Gail Strickland, Linda Hayes, Coral Browne, Paul Koslo, Andy Robinson, Richard Derr, Helena Kallianiotes, 'eigh French
Second outing for Newman in his private role of Lew Harper. Here he investigates the murder of a New Orleans businessman and helps an old flame (Woodward) who is being blackmailed. A well made tale but one where the pace is much too slow. This was a sequel to HARPER.
THR   104 min (ort 108 min) cC   VIDrel: WHV   V
Boa: novel by John Ross Macdonald.

DRUGSTORE COWBOY ****
Gus Van Sant Jr USA
18
1989
Matt Dillon, Kelly Lynch, James Remar, James Le Gros, Heather Graham, Beah Richards, Grace Zabriskie, Max Perlich, William S. Burroughs, Eric Hull, Ted D'Arns, John Kelly, George Catalano, Janet Baumhover, Neal Thomas, Ray Monge
A harsh look at the lives of a group of 1970s drug addicts, who support their addiction by robbing drugstores until the day their luck runs out. The pointlessness of their lives is well captured in a film that delivers no sermons or cliches. Dillon and Lynch are totally convincing as a junkie couple. The screenplay is by Van Sant and Daniel Yost.
DRA   97 min (ort 104 min)   VIDrel: VISVID/POLYREC   V
Boa: novel (unpublished) by James Fogle.

DRUM, THE ***
Zoltan Korda UK
U
1938
Sabu, Raymond Massey, Valerie Hobson, Roger Livesey, David Tree, Desmond Tester, Francis L. Sullivan, Archibald Batty, Frederick Culley, Edward Lexy, Amid Taftazani, Roy Emerton, Martin Walker, Charles Oliver, Julien Mitchell
Fairly competent tale set on the Indian Northwest frontier in the days of British rule. Military action by the British against hostile tribesmen forms the backdrop for a story of the friendship between an Indian prince (Sabu) and a drummer (Tester), who shows him how to play the drum. Massey makes a fine villain, in this stirring but stilted adventure. The photography (it was shot on location by George Perinal) has dated a lot less than the crusty script.
Aka: DRUMS
A/AD   104 min   VIDrel: CARL/TECH   V
Boa: novel by A.E.W. Mason.

DRUM *
Steve Carver USA
18
1976
Warren Oates, Ken Norton, Yaphet Kotto, Isela Vega, Pam Grier, John Colicos, Fiona Lewis
Steamy tale of a New Orleans bordello house slave and the sexual harassment he faces. A dull and plodding follow-on to MANDINGO.
DRA   100 min (ort 110 min)   VIDrel: CASPIC L/A   V
Boa: novel by Kyle Onstott.

DRUMS ACROSS THE RIVER **
Nathan Juran USA
PG
1954
Audie Murphy, Lisa Gaye, Lyle Bettger, Walter Brennan, Mara Corday, Hugh O'Brien, Jay Silverheels, Regis Toomey, Morris Ankrum, James Anderson, Emile Meyer, George Wallace, Bob Steele, Lane Bradford, Gregg Barton, Howard McNear
A young man joins a group of gold prospectors whose leader is determined to raid Indian land. When the former begins to have doubts about this endeavour, he finds himself framed for the robbery of a gold shipment. Fortunately, his father's swift intervention saves his life and also stops the beginnings of an Indian war. A conventionally plotted but competently made Western.
WES   75 min (ort 78 min)
VIDrel: 4-FRONT/POLYREC/CIC   V/sur

DRUNKEN ANGEL ***
Akira Kurosawa JAPAN
PG
1948
Takashi Shimura, Toshiro Mifune, Michiyo Kogure, Reizaburo Yamamoto, Chieko Nakakita, Norika Sengoku, Choko Lida
Post-war drama of a dedicated doctor and his work among slum dwellers as he attempts to help them improve the quality of their lives. Mifune plays his moral counterpart, a petty crook and racketeer. An early and little-known film from a master craftsman.
Aka: YOIDORE TENSHI
DRA   93 min (ort 102 min) B/W   VIDrel: CONNO/RTM   V

DRUNKEN MASTER **
Yuen Woo Ping HONG KONG
15
1978
Jackie Chan, Yuan Hsiao Tien, Hwang Jang Lee
A student is sent to learn a special style of combat and comes up against an assassin. Enjoyable but uninspired action. A sequel to TSUI CHUAN.
Aka: TSUI CHUAN
MAR   106 min (ort 113 min) wScrn dubbed
VIDrel: MADE/RTM   V

DRY WHITE SEASON, A ***
Euzhan Palcy USA
15
1989
Donald Sutherland, Janet Suzman, Jurgen Prochnow, Zakes Mokae, Susan Sarandon, Marlon Brando, Winston Ntshona, Thoko Ntshinga, Susannah Harker, Rowan Elmes, Leonard Maguire, Gerard Thoolen, Andrew Whaley, Stella Dickin, John Kani
An absorbing drama written by Colin Welland and Palcy, and telling of a well-meaning but naive white schoolteacher who receives a harsh lesson in the realities of Apartheid, and its implications for his black fellow citizens. Brando is outstanding as a clever lawyer who has resigned himself to the system, and the film is quite memorable, despite contrived plotting and an excess of melodramatics. See also CRY FREEDOM and BOPHA!
DRA   103 min (ort 107 min)   VIDrel: MGM L/A   V
Boa: novel by Andre Brink.

DUDES *
Penelope Spheeris USA
15
1988
Jon Cryer, Daniel Roebuck, Catherine Mary Stewart, Flea, Lee Ving, Billy Ray Sharkey, Glenn Withrow, Michael Melvin, Axxel G. Reese, Marc Rude, Calvin Bartlett, Pete Wilcox, Vance Colvig, Pamela Gidley
Three New York punks head West in search of a new open-air lifestyle, but find themselves running foul of a gang of murderous rednecks in Montana, and get drawn into a revenge-seeking conflict. A repulsive and rather incoherent formula revenger of much violence but little merit.
A/AD   86 min   VIDrel: 20TH   V/h

DUEL ****
Steven Spielberg USA
PG
1971
Dennis Weaver, Tim Herbert, Charles Peel, Eddie Firestone, Shirley O'Hara, Lucille Benson, Alexander Lockwood, Amy Douglass, Gene Dynarski, Cary Loftin
Spielberg's debut was this well made story of a businessman on a trip who begins to realise that the driver of a lorry is out to kill him. The tension is maintained right to the very end in a film that is pared down to the bone. Scripted by Richard Matheson. Additional footage was added for cinema release, hence the longer running time. See also ROADFLOWER.
DRA   86 min (ort 92 min) mTV   VIDrel: CIC/SONOP   V
Boa: story by Richard Matheson.

DUEL AT SILVER CREEK ***
Donald Siegel USA
U
1952
Audie Murphy, Faith Domergue, Stephen McNally, Susan Cabot,

Gerald Mohr, Lee Marvin, Eugene Iglesias, James Anderson, Walter Sande, George Eldredge
A man called the "Silver Kid" arrives in town to enjoy a little gambling but winds up becoming the sheriff's deputy as they tackle a gang of murderous claim-jumpers led by a woman who has kept her identity a secret. A fast-moving and very well made yarn.
WES   73 min (ort 77 min)
VIDrel: 4-FRONT/POLYREC/CIC   V

DUEL IN THE SUN **
King Vidor USA
Jennifer Jones, Joseph Cotten, Gregory Peck, Lionel Barrymore, Walter Huston, Herbert Marshall, Butterfly McQueen, Charles Bickford, Lillian Gish, Joan Tetzel, Tilly Losch, Harry Carey, Otto Kruger, Sidney Blackmer
Sprawling overlong Western with little plot and a strange ending that tells of a half-breed girl torn between two brothers. Lavish, detailed and rather disjointed, despite a number of scenes of undoubted power.
WES   129 min (ort 138 min)
VIDrel: VCC/DISC; PION (LV only)   V LV
Boa: novel by Niven Busch.

PG
1946

DUEL OF HEARTS **
John Haugh UK
Alison Doody, Michael York, Geraldine Chaplin, Benedict Taylor, Beryl Reid, Billie Whitelaw, Virginia McKenna, Richard Johnson, Jeremy Kemp, Tom Adams, Susanna Hamilton, Jolyon Baker, Adalberto Maria Merli, Margaret Mezzantini
A young woman learns of a plot to frame a lord for murder, but finds him curiously indifferent to his fate when she warns him of impending danger. Set in England in 1821, this romantic drama is a fairly typical example of a film based on a Cartland novel; in short - glossy, agreeable and shallow.
DRA   90 min mTV   VIDrel: 4-FRONT/POLYREC   V
Boa: novel by Barbara Cartland.

PG
1990

DUEL OF THE DRAGONS **
Wong Wah Kay
Bruce Li
Two close friends who are both martial artists and acrobatic performers get embroiled in a conflict with bullies from a local martial arts school and give them a drubbing. In time they go their separate ways, but are re-united under their former Master for one final showdown.
Aka: DUEL OF THE DRAGON
MAR   86 min Cut (7 sec)   VIDrel: IMPENT   V

18
1991

DUELLISTS, THE **
Ridley Scott USA
Keith Carradine, Harvey Keitel, Edward Fox, Cristina Raines, Diana Quick, Robert Stephens, Tom Conti, Albert Finney, John McEnery, Alun Armstrong, Maurice Colbourne, Gay Hamilton, Meg Wynn Owen, Jenny Runacre, Alan Webb
Two officers in Napoleon's army fall out over a woman and fight a series of duels stretching over thirty years, in this strangely pointless, interminable but undeniably atmospheric film. Scott's directorial feature debut.
DRA   96 min (ort 101 min)   VIDrel: CIC/SONOP L/A   V
Boa: short story The Point of Honour (also known as The Duel) by Joseph Conrad.

PG
1977

DUET FOR ONE ***
Andrei Konchalovsky USA
Julie Andrews, Alan Bates, Max Von Sydow, Rupert Everett, Cathryn Harrison, Margaret Courtenay, Macha Neill, Liam Neeson
A world famous concert violinist is stricken with multiple sclerosis whilst in her prime. As her illness progresses she is forced to come to terms with her inability to play and her husband's mounting infidelities. Kempinski's acclaimed two-act play (with just virtuoso and psychiatrist) is expanded for the screen and in the process its power is diluted, but it is Andrews's remarkable performance that saves it. Filmed in England.
DRA   102 min (ort 107 min)   VIDrel: GUILD/SONOP L/A   V
Boa: play by Tom Kempinski.

15
1986

DUMB AND DUMBER *
Peter Farrelly USA
Jim Carrey, Jeff Daniels, Lauren Holly, Karen Duffy, Mike Starr, Charles Rocket, Teri Garr, Victoria Rowell, Joe Baker, Cam Neely,

12
1994

Felton Perry, Hank Brandt, Brady Bluhm, Brad Lockerman, Rob Moran, Kathryn Frick, Clint Allen
Two male friends who are borderline mental defectives become unwittingly involved in a kidnapping plot and take a cross-country trip to return a briefcase containing the $100,000 ransom demanded by the kidnappers. Based entirely on crude and vulgar behaviour and Carrey's incessant and rapidly irritating mugging, this unfunny effort is a sad comment on the state of American movie-making. See KINGPIN for another gross exercise in tastelessness.
COM   102 min (ort 107 min) cC   VIDrel: FIRST/SONOP; ENCORE (LV only)   V LV

DUMBO ****
Ben Sharpsteen USA
Voices of: Sterling Holloway, Edward Brophy, Verna Felton, Herman Bing, Cliff Edwards
A marvellous cartoon feature about a baby elephant whose ears are so enormous that he is able to use them to fly. The film contains a remarkable scene, one of Disney's finest, in which Dumbo has a nightmare. AA: Score (Frank Churchill/Oliver Wallace).
ANIM   64 min cC   VIDrel: WDV/TECH; PION (LV only)   V LV
Boa: story by Helen Aberson and Harold Pearl.

Uc
1941

DUNCAN'S WORLD **
John Clayton USA
Larry Tobias, Billy Tobias, Calvin Brown Jr, Ira Shapiro, Jason Roberts, Ben Browder, Vinia Jones, Paul Trimakas, Don Merrill, Marion Mitchell, Rudy Thompson, Ray Berry, Robert Hapenjans, Jeff Daignault, Rusty Russell
Updated Tom Sawyer story where a boy and his friends (plus pet raccoon) learning about fear, courage an friendship when they set about investigating an explosion they suspect the local bullies of having caused.
JUV   93 min   SATrel: MOVIE CHANNEL
Boa: novel by Helen Masson Copeland.

(U)
1980

DUNE ***
David Lynch USA
Kyle MacLachlan, Francesca Annis, Brad Dourif, Jose Ferrer, Linda Hunt, Freddie Jones, Jurgen Prochnow, Richard Jordan, Everett McGill, Kenneth McMillan, Sting (Gordon Sumner), Silvana Mangano, Jack Nance, Max Von Sydow
Though acknowledged to be an expensive failure (it cost $50,000,000) in its attempt to bring Herbert's fantasy to the screen, this film has some superb sets. It tells of a futuristic galactic empire, whose very existence depends on a strange spice found only on one planet. A complex, sluggish and spectacular flop, its greatest failings are the lack of a clear narrative and an assumption of familiarity with the novel. Formerly discarded footage is now included.
FAN   130 min wScrn (incl 50 min extra footage)
VIDrel: POLY/POLYREC   V/sur
Boa: novel by Frank Herbert.

15
1984

DUNE WARRIORS *
Cirio H. Santiago
David Carradine, Rick Hill, Luke Askew, Jilliam McWhirter, Blake Boyd, Val Garay
Futuristic fantasy-adventure that tells of a town ruled by an evil and despotic warlord, and the efforts made by five warriors to win the townsfolk their freedom.
A/AD   75 min   VIDrel: 20VIS/SONOP   V

18
1990

DUNKIRK ***
Leslie Norman UK
John Mills, Richard Attenborough, Robert Urquhart, Bernard Lee, Ray Jackson, Ronald Hines, Sean Barrett, Roland Curram, Meredith Edwards, Patricia Plunkett, Rodney Diak, Michael Shillo, Eddie Byrne, Maxine Audley, Flannagan & Allen
Story of the WW2 evacuation of 350,000 Allied troops from the beaches of Dunkirk in 1940 in the face of an irresistible German onslaught. Made with a good eye for detail, this realistic flag-waver opts for a very low-key approach, with some use made of contemporary newsreel footage. There are no plot devices or unnecessary thrills, just a straightforward story simply told.
WAR   129 min (ort 135 min) B/W   VIDrel: MGM/WHV   V/h
Boa: novel The Big Pickup by Elleston Trevor/book by Ewen Butler and J.S. Bradford.

PG
1958

DUNSTON CHECKS IN ***
Ken Kwapis CANADA
PG
1995
Jason Alexander, Faye Dunaway, Eric Lloyd, Rupert Everett, Graham Sack, Paul Reubens, Glenn Shadix, Nathan Davis, Jennifer Bassey, Judith Scott, Bruce Beatty, Danny Comden, Steven Gilborn, Lois de Banzie, Natalie Core plus "Sam"
Enjoyable farce set at a luxury hotel in LA, where a thief checks in with an orang-utan as a sidekick. However, the son of the hotel manager starts planning to rescue the monkey from its exploitative owner. See also MONKEY TROUBLE.
COM 85 min (ort 88 min) cC VIDrel: 20TH/TECH V/sur

DUPLICATES **
Sandor Stern USA
15
1992
Gregory Harrison, Kim Greist, Cicely Tyson, Lane Smith, William Lucking, Scott Hoxby, Kevin McCarthy, John Delay, Matt Williams, Beth Harper, Don Hibdon, Timi Proulhiere, Erik Alskog, Russ Fast, Shawna Schuh, Barbara Kite
A couple's young son disappears and is later found but seems to have had all his memories erased. As his parents probe this baffling event, the trail leads them to a sinister research facility where some strange experiments are in progress that involve transferring memories to computer. Fair.
THR 87 min (ort 92 min) mCab VIDrel: CIC/SONOP V

DUST DEVIL **
Richard Stanley USA
18
1992
Robert Burke, Chelsea Field, Zakes Mokae, John Matshikiza, Rufus Swart, Luke Cornell, William Hootkins, Madonna, Russell Copley, Andre Odendaal, Phillip Henn, Terri Norton, Robert Stevenson, Marianne Sagenbrecht, Peter Hallr
Two people encounter title figure in the desert and find themselves plunged into a life-and-death struggle, in this average horror movie which offers ample gore but little else. Filmed in the Great Namib desert.
HOR 103 min (ort 105 min) VIDrel: POLY/POLYREC L/A V/s

DWELLING PLACE, THE ***
Gavin Millar UK
PG
1990
Tracy Whitwell, Ray Stevenson, Luke Conway, Fiona Dixon, Lucy Cohu, Edward Rawle-Hicks, James Fox, Leanne Bradford, Andrew McWilliams, Henry Moxon, Rod Culbertson, Mike Elliott, Julie Hesmondhalgh, Helen Lindsay, Sheri Shepstone
In Northumberland of the 1830s, a young girl struggles to keep her family together when the unexpected death of their parents leads to their eviction from their cottage. One of those extremely handsome TV adaptations, always good to watch and highly enjoyable.
DRA 154 min (ort 305 min) mTV VIDrel: FOCUS/DISC V
Boa: novel by Catherine Cookson.

DYING TO LOVE YOU **
Robert Iscove USA
15
1993
Tracy Pollan, Lee Garlington, Tim Matheson, Frances Lee McCain, Jordan Bond, Alan Blumenfeld, Margot Rose, Steve Eastin, Ronald William Lawrence, Lilian Adams, Christine Ebersole, Bradford English, Milt Tarver, Sarah Freeman
When his wife walks out on him, a lonely man puts in an ad in the lonely hearts columns, but gets more than he bargains for when he gets a reply from a murderous, female fugitive who is on the run from the FBI. However, he is not initially aware of the danger he has placed himself in. A tense thriller with an effective if over-familiar plot idea. See also FATAL ATTRACTION and LOVE IN THE STRANGEST WAY.
Aka: LETHAL WHITE FEMALE
THR 90 min mTV VIDrel: 20VIS/SONOP V

DYING TO REMEMBER **
Arthur Allan Seidelman USA
15
1993
Melissa Gilbert, Ted Schackelford, Scott Plank, Christopher Stone, Jay Robinson, Kate Green, Wade Anderson, Sandra Nelson, Babz Chula, James Kidnie, Peter Williams, Alvin Sanders, Brian Mcgugan, Stephen Dimopoulos, Robin Mossley
Plagued by recurrent nightmares, a woman seeks therapy, and under hypnosis recalls a previous life in which she was murdered by being pushed down a lift-shaft. Intrigued by this, she seeks more information, her search taking her to San Francisco where she learns that the building in question is being developed by her husband of the past life, and that he may have been the killer. An unusual thriller whose strong supernatural slant is quite well handled.
THR 85 min mTV VIDrel: CIC/SONOP V

DYING YOUNG *
Joel Schumacher USA
15
1991
Julia Roberts, Campbell Scott, Vincent D'Onofrio, David Selby, Colleen Dewhurst, Ellen Burstyn, Dion Anderson, George Martin, A.J. Johnson, Daniel Beer, Behrooz Afrakhan, Michael Halton, Larry Nash, Alex Trebek, Fran Lucci
When a rather aimless young woman answers a classified ad, she finds herself being hired to act as a nurse and companion for a a wealthy young man who is dying from leukaemia. Predictably enough, the pair fall in love, but this contrived tearjerker has no real substance, though Roberts works hard to generate some interest. Very disappointing and manipulative, it did nothing for the career of Roberts, following her great success in PRETTY WOMAN.
DRA 112 min VIDrel: 20TH/TECH L/A V
Boa: novel by Marti Leimbach.

DYNAMITE SHAOLIN HEROES **
Godfrey Ho HONG KONG
15
Lo Lieh, Sam Kuen, Willie Wong, Tim Ming, Carlo Kim, Roman Lee, Mah Yeung
The prince of a fallen Ming Dynasty is to be used as nothing more than a tool in an ambitious government official's power-struggle. However, the intervention of a mysterious Shaolin master with superb skills thwarts his evil plans.
MAR 78 min VIDrel: IMPENT V

DYNAMO **
Hwai Hung HONG KONG
18
1978
Bruce Li (Ho Tsung-Tao), Man Mali, Ku Feng, Liu Tan, Chaing Tao, Mary Kan, James Griffith, Steve Sandor
Another kung fu movie which makes use of a clone of the late Bruce Lee. A fighter taking part in a tournament must lose in order to save his girlfriend.
MAR 92 min (ort 96 min) VIDrel: IMPENT V

DYNASTY: THE REUNION *
Irving Moore USA
PG
1991
John Forsythe, Linda Evans, Joan Collins, Emma Samms, John James, Heather Locklear, Kathleen Beller, Al Corley, Maxwell Caulfield, Jeroen Krabbe, Robin Sachs, Michael Brandon, Cameron Watson, Alphonsia Emmanuel, Tony Jay
Mini-series entry in the long-running TV soap that is designed to tie up all the loose ends before the final curtain. The daft plot revolves around a conspiracy to take over most of US industry by a sinister consortium, that is headed by a Nazi (Krabbe in a strong performance). As dull and wooden as the TV series, this entry is for fans only.
DRA 181 min (ort 360 min) mTV
VIDrel: MIA/DISC/IMPENT V

E

EACH DAWN I DIE ***
William Keighley USA
PG
1939
James Cagney, George Bancroft, George Raft, Jane Bryan, Maxie Rosenbloom, Thurston Hall, Stanley Ridges, Alan Baxter, Victor Jory, Willard Robertson, Paul Hurst, John Wray, Louis Jean Heydt, Ed Pawley, Joe Downing, Emma Dunn
A reporter is framed for manslaughter and sent to prison where the tough conditions make him into a hardened con. Enjoyable prison drama kept afloat by the three stars though the unbelievable and contrived second half does let the film down considerably.
DRA 88 min (ort 113 min) B/W VIDrel: WHV V
Boa: novel by Jerome Odlum.

EAGLE, THE ***
Clarence Brown USA
U
1925
Rudolph Valentino, Louise Dresser, Vilma Banky, Albert Conti, James Marcus, George Nichols, Carrie Clark Ward, Michael Pleschkoff, Spottiswoode Aitken, Gustav Von Seyffertitz, Mario Carillo, Otto Hoffman, Eric Mayne, Jean De Briac
Valentino plays a Cossack lieutenant who becomes a bandit when his father's lands are annexed and has numerous adventures, including an encounter with Catherine the Great where he spurns her advances. An enjoyable costumed romp with the star at his best.
DRA 70 min (ort 80 min) B/W silent
VIDrel: VISION/SPEAR V
Boa: novel Dubrovsky by Alexander Pushkin.

EAGLE FIST, THE ***   *15*
Lee Tso Nam HONG KONG   1978
Chi Kuan-Chun, Wong Tao, Chang Yi, Hwa Ling, Cheng Kay Ying, Phillip Kao, Leung Kar Yan
Exponents of the Eagle Fist style of kung fu slug it out. A superior martial arts movie with the graceful Chi Kuan-Chun carrying the action with great skill. Well worth a look.
Aka: EAGLE'S CLAW
MAR   90 min (ort 92 min) dubbed wScrn
VIDrel: EAST/DISC   V

EAGLE HAS LANDED, THE ***   *15*
John Sturges UK   1976
Michael Caine, Donald Sutherland, Robert Duvall, Jenny Agutter, Anthony Quayle, Donald Pleasence, Jean Marsh, Sven-Bertil Taube, John Standing, Judy Geeson, Larry Hagman, Maurice Reeves, Treat Williams, Siegfried Rauch
WW2 story with German commandos parachuting into England where they infiltrate a small English village in a bungled attempt to kill Churchill. An imaginative and exciting film with a convoluted and twist-laden plot.
WAR   118 min (ort 135 min)   VIDrel: 4-FRONT/POLYREC V

Boa: novel by Jack Higgins.

EAGLE SHADOW FIST ***   *18*
Hdeng Tsu HONG KONG   1973
Jackie Chan
Martial arts tale set during WW2 with a famous actor of the Chinese theatre leaving the stage to become a resistance fighter against the Japanese.
Aka: IN EAGLE SHADOW FIST; NOT TOO SCARED TO DIE
MAR   85 min Cut (1 min 44 sec)   VIDrel: PARADE/SCRN V

EAGLE VERSUS SILVER FOX **   *18*
Godfrey Ho HONG KONG
Wang Cheng Li, Mario Chan, Richard Kong, Jacky Lee, Sam Yuen, Stan Yuen, Wing Pui Shan, Wu Kam Bo
Set during the time of the Manchu Dynasty with a powerful fighter being chosen by the kung fu schools to lead the battle against a sinister warrior and his followers.
MAR   83 min Cut (5 sec - ort 86 min)   VIDrel: IMPENT   V

EAGLE WARRIORS **   *(PG)*
John Peyser USA   1967
James Drury, Robert Pine, Hank Jones, Tim Nolan, Norman Fell, Jeff Scott, Steve Carlson, Michael Stanwood, Jonathan Daly, Johnny Alladin, Kent McWhirter, Buck Young, George Sawaya, Morgan Jones, Noam Pitlik, Jon Drury, Buck Kartalian
Very loosely based on Matheson's novel (he also did the screenplay), this straightforward WW2 movie looks at how a group of green young soldiers react to the horror and carnage of war. Shot entirely on studio backlots, this low-budget is as stereotyped as they come. Some good camerawork is a slight compensation for the trite plotting and weak direction.
Aka: YOUNG WARRIORS, THE
WAR   89 min (ort 105 min)   SATrel: SKY MOVIES
Boa: novel The Beardless Warriors by Richard Matheson.

EAGLE'S WING *   *PG*
Anthony Harvey UK   1979
Martin Sheen, Sam Waterson, Caroline Langrishe, Harvey Keitel, Stephane Audran, John Castle, Jorge Luke, Jorge Russek, Manuel Ojeda, Pedro Damian, Jose Carlos Ruis, Claudio Brook, Cecilia Camacho, Julio Lucena, Enrique Lucero
Deadly dull but beautifully photographed British-made Western dealing with a contest between an Indian warrior and a trapper for a prize white stallion which the latter has stolen. Derivative of a hundred other films with hardly an original scene of its own. The photography is by Billy Williams with screenplay by John Briley (of GANDHI).
WES   100 min (ort 111 min)   VIDrel: VCC/DISC   V

EARTH ****   *PG*
Alexander Dovzhenko USSR   1930
Semyon Svashenko, Stepan Shkurat, Nikola Nademsky, Yelena Maximova, Pyotr Masokha, V. Mikhailov, Elena Maxinova, Yulia Solntseva, P. Petrik, I. Franko
In a Ukrainian village a landowner resists the drive for collectivisation, but this slim plot merely provides the framework for a beautifully made, poetic hymn to the life of the Russian countryside, indeed so much so that the director suffered severe criticism for his film's lack of socialist realism.
Aka: SOIL; ZEMLYA
DRA   83 min (ort 88 min) B/W silent
VIDrel: HEND/BMGREC L/A   V

EARTH ANGEL **   *PG*
Joe Napolitano USA   1991
Cindy Williams, Cathy Podewell, Rainbow Harvest, Mark Hamill, Erik Estrada, Dustin Nguyen, Roddy McDowell, Alan Young, Brian Krause
Thirty years after she was killed in a prom night car crash, a lovely high school girl returns to Earth, and sets about resolving the problems faced by her former classmates. A remorselessly silly fantasy-comedy that would easily have been more effective as a straight fantasy.
COM   94 min (ort 100 min) mTV   VIDrel: GENESIS   V

EARTH GIRLS ARE EASY ***   *PG*
Julien Temple USA   1989
Geena Davis, Jim Carrey, Damon Wayams, Jeff Goldblum, Julie Brown, Michael McKean, Charles Rocket, Larry Linville, Rick Overton, Angelyne, Damon Wayans, Diane Stilwell, June C. Ellis, Felix Montano, Richard Hurst
Three furry aliens crash-land in a swimming pool belonging to a San Bernado manicurist who's a little short-changed in the brains department. Having shaved the aliens and discovered them to be gorgeous hunks, the girl and her friend take them out for a tour of what L.A. has to offer in the way of fun. A wacky musical-comedy of infectious appeal, its lack of a strong plot and relentless cheerfulness grows just a little tiresome. See also BEACH BABES FROM BEYOND.
COM   100 min   VIDrel: 4-FRONT/POLYREC/BRAVE V/sur

EARTH GIRLS ARE SLEAZY *   *18*
Henri Pachard USA   1993
Kelly Royce, Jerry Butler, Sharon Kane, Debi Diamond, Renee Foxx, Eric Price, Peter North, Tom Byron
A husband and wife spend much of their time trying to contact the dead, but then fall out when the wife accuses him of using this simply as a cover to hide her infidelities with her friends. He shows her a mystical dildo he bought in India, they argue again, and then suddenly swap bodies just before their guests arrive. This interesting idea is thrown away on a the film that concentrates on a virtual mass orgy, with no plot resolution at all. A most amateurish effort.
A   60 min (ort 90 min)   VIDrel: FALCON/TOTAL   V

EARTH VERSUS THE FLYING SAUCERS **   *U*
Fred F. Sears USA   1956
Hugh Marlowe, Joan Taylor, Donald Curtis, Morris Ankrum, Tom Browne Henry, Paul Frees (voice only)
Alien invaders threaten dire consequences if their attempt to take over the Earth is opposed. A low-key and somewhat matter-of-fact story, given an occasional boost by some competent Ray Harryhausen special effects.
FAN   80 min (ort 82 min) B/W   VIDrel: COLUM/SONOP V

Boa: story by Curt Siodmak.

EARTHQUAKE ***   *PG*
Mark Robson USA   1974
Charlton Heston, Richard Roundtree, Lorne Greene, Ava Gardner, George Kennedy, Genevieve Bujold, Lloyd Nolan, Victoria Principal, Marjoe Gortner, Barry Sullivan, Walter Matuschanskayasky (Matthau), Monica Lewis, Gabriel Dell
Disaster epic with a star-studded cast about the destruction of Los Angeles, as the most catastrophic earthquake of all time rips through Southern California, affecting the lives of all who live there. Excellent special effects (which will lose impact on TV) make up for the tedious and cliched plot. AA: Sound (Ronald Pierce/Melvin Metcalf Sr), Spec Award (Frank Brendel/Glen Robinson/Albert Whitlock for visual effects).
A/AD   116 min (ort 129 min)
VIDrel: 4-FRONT/POLYREC/CIC L/A   V/h

EAST L.A. ***   *15*
Gregory Nava USA   1995
Jimmy Smits, Esai Morales, Edward James Olmos, Eduardo Lopez Rojas, Jenny Gago, Elpidia Carrillo, Constance Marie
A detailed but episodic tale that traces three generations of one immigrant family in L.A., that starts from the 1920s when a

young Mexican couple settle in California and raise a family there. Over the next seventy years the various trials, hopes, dreams and changes this family experience are well documented, in this warmhearted if verbose family saga.
Aka: MY FAMILY
DRA   121 min   VIDrel: EIV/SONOP   V

EAST OF EDEN ****   PG
Elia Kazan USA   1955
James Dean, Raymond Massey, Jo Van Fleet, Julie Harris, Burl Ives, Richard Davalos, Albert Dekker, Lois Smith, Harold Gordon, Timothy Carey, Mario Siletti, Lonny Chapman, Nick Dennis
A modern Cain and Abel story with two brothers as rivals for the love of their father. Full of moments of power this is deservedly recognised as a classic. It also gave Dean his first starring role. Remade some years later as a TV mini-series. AA: S. Actress (Van Fleet).
DRA   112 min (ort 115 min) wScrn   VIDrel: WHV   V/s
Boa: novel by John Steinbeck.

EAST OF EDEN ***   15
Harvey Hart USA   1980
Jane Seymour, Timothy Bottoms, Bruce Boxleitner, Soon-Teck Oh, Karen Allen, Hart Bochner, Sam Bottoms, Warren Oates, Howard Duff, Anne Baxter, Richard Masur, Nicholas Pryor, Lloyd Bridges, Nellie Bellflower, M. Emmet Walsh
A competent remake of the earlier classic, that covers far more ground than the 1955 James Dean film, and follows the changing fortunes of the Trask family and the two half-brothers who compete for the affection of their father. The film won Emmy Awards for art direction and set decoration. Originally shown in three parts.
Aka: JOHN STEINBECK'S EAST OF EDEN
DRA   199 min (2 cassettes - ort 640 min) mTV
VIDrel: POLY/POLYREC/BRAVE   V
Boa: novel by John Steinbeck.

EAST SIDE HUSTLE **   18
Frank Vitale USA   1976
Anne Marie Provencher, Man Moyle
Prostitutes decide to quit the game but meet with violent resistance from their pimp in this tough, mean story of violence and revenge.
DRA   87 min Cut (5 sec)   VIDrel: IMPENT   V

EASTER PARADE ****   U
Charles Walters USA   1948
Judy Garland, Fred Astaire, Peter Lawford, Ann Miller, Jules Munshin, Jeni LeGon, Clinton Sundborg, Richard Beavers, Dick Simmons, Jimmy Bates, Bobbie Priest, Dee Turnell, Patricia Jackson, Lola Albright, Joi Lansing, Lynn Romer
Irving Berlin musical about a dancer trying to forget his ex-partner while finding a new one. Bright, breezy and great fun. Songs include "A Couple Of Swells", "Stepping Out With My Baby" and "Shaking The Blues Away". AA: Score (Johnny Green/Roger Edens).
MUS   99 min (ort 103 min)   VIDrel: MGM/WHV   V/dm

EASTERN CONDORS **   18
Samo Hung HONG KONG   1986
Samo Hung, Joyce Godenzi, Yuen Biao, Lam Ching Ying, Haing S. Ngor
Well-made, fast-paced action tale with all the usual elements that is very much an oriental version of THE DIRTY DOZEN. The simple story has a group of hand-picked mercenaries being sent on a suicidal mission to Vietnam.
A/AD   93 min (ort 100 min) wScrn   VIDrel: MADE/RTM   V

EASY KILL **   15
Josh Spencer USA   1989
Frank Stallone, Jane Badler, Cameron Mitchell, Deon Stewardson, Hal Orlandini, Tom Aigner, Shayne Leith, Rick Skidmore, Elliot Frantz, J.Everett Leck, Mark Lancaster, Phillipa Vernon, Jay, The "Bee"
One man takes the law into his own hands in a brutal battle with drugs dealers, in an action tale that is no different from countless others. The plot does try hard however to achieve sophistication, with Badler cast as a treacherous femme fatale.
A/AD   89 min (ort 100 min)   VIDrel: GENESIS   V

EASY RIDER ***   18
Dennis Hopper USA   1969
Peter Fonda, Dennis Hopper, Jack Nicholson, Karen Black, Luke

Askew, Luana Anders, Robert Walker, Phil Spector, Antonio Mendoza, Mac Mashourian, Tita Colorado, Sabrina Scharf, Sandy Wyeth, Robert Walker Jr, Robert Ball
Now a cult film, this was one of the most influential films of the 1960s. It tells of two hippy bikers who go on a long journey of exploration through America. Memorable for an excellent backing of rock songs, it has some good moments but an brutal and shocking ending. The photography is by Laszlo Kovacs. Features music by: Hendrix, The Byrds, Steppenwolf, Roger McGuinn and others. Written by Hopper, Fonda and Terry Southern.
DRA   91 min (ort 94 min) wScrn
VIDrel: VCC/DISC/COLUM   V

EASY WAY, THE *   18
Alex De Renzy USA
Alice Springs, Tianna, Debi Diamond, Samantha Strong, Cheri Taylor, Ashley Dunn, Peter North, Randy Spears, Randy West, Tom Byron
Peter North has just bought his first camcorder, which he uses to record some outdoors action, the film largely taking the form of seven separate vignettes. A disorganised and disjointed mess that exhibits a few sparks of wit.
A   59 min (ort 75 min) mVid   VIDrel: MIA/DISC/IMPENT   V

EASY WAY OUT **   18
John T. Bone USA   1989
Misty Regan, Bob Lowe, Randy West, Victoria Paris, Ray Victory, Tianna, Tina Gordon, Rene Summers, Raven Richards, Sean Michaels
A man wants rid of his frigid wife but our unhappy couple get involved with sex surrogates instead.
A   52 min   VIDrel: GROHOM/MAXSCAN   V

EASY WHEELS *   15
David O'Malley USA   1989
Paul LeMat, Eileen Davidson, Barry Livingston, Marjorie Bransfield, Mark Holton, John Menick, Karen Russell, Jami Richards, Roberta Vasquez, Theresa Randal, Stevie Sterling, Carlos Campean, Mike Leinert
Ultra-low budget, tongue-in-cheek parody of 1960s biker movies that centres around the clash between two gangs. One is led by an ardent feminist, who kidnaps girl babies as part of her plans to rid the world of male domination. The other is headed by an equally macho male. A silly film full of comic-book violence and inane situations that tries desperately to be amusing but without a hint of success.
COM   90 min (ort 94 min)   VIDrel: VISVID/POLYREC   V

EAT A BOWL OF TEA ***   15
Wayne Wang USA   1989
Cora Miao, Russell Wong, Victor Wong, Lau Siu Ming, Eric Tsang Chi Wai, Lee Sau Kee, Yuen Yat Fai, Lau Siu Ming, Hui Fun, Law Lan, Ng Yuen Yee, Wu Ming Yu, Lui Tat, Wong Wai, Philip Chan, Tang Shun Nin, Michael Lee, Woo Wang Tat
Ethnic comedy set in New York's Chinatown in 1949, when the authorities first permitted the immigration of Chinese women into the US (there being a ban in place from 1924 until then). A young Chinese-American war veteran visits China and returns with a Chinese-born wife, but his elders, having been without women for so long, can give him little help as to what marriage involves. A charming little tale, scripted by Judith Rascoe.
COM   99 min (ort 104 min)   VIDrel: CONNO/RTM L/A   V
Boa: novel by Louis Chu.

EAT DRINK MAN WOMAN ***   PG
Ang Lee CHINA   1994
Sihung Lung, Kuei-Mei Yang, Chien-Lien Wu, Yu-Wen Wang, Winston Chao, Ah-Leh Gua, Sylvia Chang, Chao-Jung Chen, Lester Chen, Yu Chen, Ah-Leh Gua, Chi-Der Hong, Gin-Ming Hsu, Huei-Yi Lin, Shih-Jay Lin, Chin-Cheng Lui, Cho-Gin Nei
A bittersweet portrait of a Taipei family whose head is a widowed chef who is gradually losing his sense of taste. To add to his problems, his relationship with his three daughters, all of whom are unmarried, independent and career-minded, is far from harmonious. Sensitively acted and directed, this fine drama is well worth seeing despite the rather mundane nature of its plot. An Oscar-nominated comedy-drama, allegedly based on a true story.
Aka: YINSHI NAN NU
DRA   120 min (ort 104 min)   VIDrel: TOUCH/TECH   V/sh

EAT THE PEACH ***   PG
Peter Ormrod EIRE   1986
Stephen Brennan, Eamon Morrissey, Niall Toibin, Catherine Byrne,
Joe Lynch, Tony Doyle, Takashi Kawahara, Victoria Armstrong,
Barbara Adair, Bernadette O'Neill, Paul Raynor, Martin Dempsey,
Maeliosa Strafford, Jill Doyle
Two friends in rural Ireland lose their jobs, but inspired by
ROUSTABOUT, the film in which Presley rides a "Wall of
Death", they decide to build one near their homes. As their
money runs out they are forced to raise more by dubious means
and encounter some strange characters. An engaging and quirky
tale written by Ormrod and producer John Kelleher and based
on a true story.
COM   91 min (ort 95 min)   VIDrel: CONNO/RTM   V/sur

EATEN ALIVE **   18
Umberto Lenzi ITALY   1980
Robert Kerman, Ivan Rassimov, Mel Ferrer, Janet Agren, Paola
Senatore, Me Me Lai, Meg Fleming, Franco Fantasia, Michele
Schmiegelm, Gianfranco Coduti, Alfred Joseph Berry
A young girl goes to New Guinea in search of a sister who has
disappeared after becoming a member of a strange sect. A
cannibalism story incorporating some elements of the notori-
ous mass-suicide when the followers of demented self-styled
messiah Jim Jones took their lives in the Guyanan jungle. See
also GUYANA: CRIME OF THE CENTURY.
Aka: CANNIBALS; DEFY TO THE LAST PARADISE; DOOMED TO DIE;
EATEN ALIVE BY THE CANNIBALS; EMERALD JUNGLE, THE; MANGIATI
VIVI; MANGIATI VIVI DAI CANNIBALI
HOR   82 min Cut (1 sec plus film cuts)
VIDrel: VIPCO/SGSVID   V/h

EATING RAOUL ***   18
Paul Bartel USA   1982
Paul Bartel, Mary Woronov, Robert Beltran, Susan Suiger, Buck
Henry, Dick Blackburn, Edie McClurg, Ed Begley Jr, John Paragon,
Hamilton Camp, Dan Barrows, Ralph Brannen, Allan Rich, Don
Steele, Billy Curtis, Anna Mathias
Amusing black comedy about a middle-aged couple who hit on
a novel way of financing their restaurant business by luring
wealthy swingers to their apartment, then killing and robbing
them.
COM   87 min   VIDrel: LUMI/SPEAR L/A   V

EBBTIDE **   18
Craig Lahiff USA   1993
Harry Hamlin, Judy McIntosh, John Waters, Susan Lyons
After a young boy dies under mysterious circumstances, a
lawyer is assigned to investigate this case but makes the mistake
of pursuing his attraction to the sexy wife of his prime suspect.
The usual farrago of stock situations and characters follows, in
this cliched and uninspired thriller.
THR   89 min (ort 98 min)   VIDrel: MARQ/QUANT   V/sur

ECHOES IN THE DARKNESS: PARTS 1 AND 2 ***   15
Glenn Jordan USA   1987
Robert Loggia, Stockard Channing, Treat Williams, Peter Coyote,
Cindy Pickett, Peter Boyle, Gary Cole, Zeljko Ivanek, Alex Hyde-
White, Vincent Irizarry, Philip Bosco, Brenda Bazinet, Eugene A.
Clark, Richard Comar
A solid recreation of the bizarre 1979 "Main Line" murder case,
that involved two teachers from an exclusive Philadelphia
school, the killing of a colleague and the disappearance of her
two children. An overlong account but well acted with Coyote
excellent as the devious professor and Loggia equally convinc-
ing as the unconventional principal. The adaptation is by
Wambaugh.
DRA   235 min (2 cassettes)   VIDrel: VCC L/A   V
Boa: book by Joseph Wambaugh.

ECLIPSE, THE ***   PG
Michelangelo Antonioni ITALY   1962
Monica Vitti, Alain Delon, Francisco Rabal, Lilla Brignone, Louis
Seigner, Rossana Rory, Mirella Ricciardi, Cyrus Elias
Depressing but very stylish study of alienation among a group
of affluent people when a young woman breaks with her lover
of four years standing and takes up with his replacement, a
young stockbroker with whom she has little in common. This
bleak film was the best one in the director's trilogy dealing with
society and relationships (following on from L'AVVENTURA
and LA NOTTE). It won the Special Jury Prize at Cannes.
Aka: L'ECLISSE
DRA   123 min B/W   VIDrel: ARTPRO/RTM   V

ECLIPSE **   18
Jeremy Podeswa CANADA/GERMANY   1994
Van Flores, John Gilbert, Pascale Montpetit, Manuel Aranguiz, Maria
Del Mar, Greg Ellwand, Matthew Ferguson, Earl Pastko, Daniel
MacIvor, Kirsten Johnson, Ian Orr, Rosalind Kerr, Dylan Stukator-
McMahon, Tracy Wright, Michael McMurtry
Set against the background of an imminent total eclipse, this
lugubrious effort tells the tale of a group of Toronto men and
women who have sex with each other and then pass on to the
next partner. Unfortunately, this brings them little joy and
merely serves to reinforce the hollowness of their lives. A stylishly photographed but essentially empty movie that fails to
impress, its plot has much in common with LA RONDE and
CHAIN OF DESIRE.
DRA   95 min B/W/Col   CINrel

ECSTASY **   PG
Gustav Machaty CZECHOSLAVIA   1932
Hedy Kiesler (Lamarr), Aribert Mog, Jaromir Rogoz, Leopold Kramer,
Andre Nox, Pierre Nay
A beautiful child bride discovers that her husband is impotent, so
she embarks on an affair with an engineer and her unhappy
spouse eventually commits suicide. Some restrained nude scenes
that show Kiesler cavorting with her lover caused a sensation at
the time and were generally excised (her husband tried to buy up
and destroy every print) but this dated film, though quite stylish,
is in truth a little tiresome and only mildly erotic.
Aka: EXTASE; SYMPHONY OF LOVE
DRA   88 min B/W   VIDrel: L/A   V

ECSTASY GIRLS, THE **   18
Robert McCallum USA   1979
Jamie Gillis, John Leslie, Paul Thorpe, Nancy Sutter, Desiree
Cousteau, Georgina Spelvin, Serena, Leslie Bovee
Sex film about a plot to disinherit five women by filming their
more intimate moments and showing the results to their puri-
tanical father. Not a badly made film but about as contrived as
one of these films can get.
A   59 min (ort 90 min)   VIDrel: HAR/GOLD   V

ED *   PG
Bill Couturie USA   1996
Matt LeBlanc, Doren Ferin, Jayne Brook, Bill Cobbs, Jack Warden
A minor-league baseball pitcher develops an inability to play in
front of large crowds and finds himself only able to play at
home. However, when he befriends the chimp that was brought
in as a mascot for the team, he finds it helping him with his
problem. A very childish comedy that plays strictly to the under-
tens, some of whom may find it quite amusing.
JUV   90 min (ort 95 min)   VIDrel: CIC/SONOP   V

ED McBAIN'S 87TH PRECINCT ***   (15)
Bruce Paltrow USA   1994
Randy Quaid, Alex McArthur
Strong adaptation of one of McBain's many police crime
thrillers, full of sharp dialogue and concise direction, pleasingly
similar in style to the way crime movies were made back in their
1950s heyday. In this story, Quaid and McArthur are on the trail
of a serial killer who only targets female athletes, each one of
whom is left in a sick "victory" pose with a lightning motif
nearby (hence the title of the novel).
THR   85 min mTV   SATrel: SKY MOVIES
Boa: novel Lightning by Ed McBain.

ED McBAIN'S 87TH PRECINCT: ICE **   (15)
Bradford May USA   1995
Dale Midkiff, Joe Pantoliano, Paul Johansson, Andrea Parker, Dean
McDermott, Andrea Ferrell, Diane Douglass, Nigel Bennett, Michael
gross, Lenore Zann, Hugh Thompson, Philip Akin, Tim Koetting,
Judah Katz, Lisa Lacroix, Conrad Dunn
A couple of slayings lead the precinct detectives to the trail on
a vast show-business con scheme in which a number of
diamonds play a key role. An OK police action drama that
moves along familiar lines.
Aka: ICE
DRA   89 min   SATrel: MOVIE CHANNEL

ED WOOD ***   15
Tim Burton USA   1994
Johnny Depp, Martin Landau, Sarah Jessica Parker, Bill Murray,
Patricia Arquette, Jeffrey Jones, G.D. Spradlin, George "The Animal"
Steele, Lisa Marie, Juliet Landau, Max Casella, Vincent D'Onofrio,
Mike Starr, Brent Hinkley

An affectionate tribute to the title figure, a man whose love for the cinema was not matched by any ability as a director, but who has achieved well-deserved fame as the worst director ever, thanks to gems such as PLAN 9 FROM OUTER SPACE. MARS. Depp was an inspired choice as Wood, whilst Landau is equally good as Bela Lugosi, a stalwart of Wood's movies in the twilight of his career. A memorable biopic with fine period atmosphere.
AA: S. Actor (Landau), Make (Rick Baker).
COM   121 min (ort 125 min) B/W cC
VIDrel: TOUCH/TECH   V
Boa: book Nightmare of Ecstasy by Rudolph Grey.

EDDIE **   PG
Steve Rash USA   1996
Whoopi Goldberg, Frank Langella, Dennis Farina, Richard Jenkins, Lisa Ann Walter, John Bnenjamin Hickey, Troy Beyer, John Salley, Rick Fox
A smart and sassy fan of a New York basketball team gets a chance to coach them when she wins a competition designed to gain them some much-needed publicity. This leads to a permanent appointment, and as we might expect, her unconventional ways result in them enjoying a run of good fortune. One of those feel-good comedies that tries hard to please but really has too few ideas to work, and even Goldberg's vigorous performance grows increasingly irksome.
COM   100 min   CINrel

EDDIE AND THE CRUISERS **   PG
Martin Davidson USA   1983
Tom Berenger, Michael Pare, Joe Pantoliano, Matthew Laurance, Ellen Barkin, Helen Schneider, David Wilson, Michael "Tunes" Antunes, Kenny Vance, John Stockwell, Joe Cates, Barry Sand, Vebe Borge
A TV reporter tries to get at the truth behind the demise of a rock band after its lead singer dies in a car crash. Despite the promising start (CITIZEN KANE in a rock 'n' roll genre), the film badly sags after the first 30 minutes. The score is by Kenny Vance. A sequel followed.
DRA   92 min   VIDrel: EIV/SONOP   V/sur

EDDIE AND THE CRUISERS 2: EDDIE LIVES! *   PG
Jean-Claude Lord CANADA   1989
Michael Pare, Marina Orsini, Bernie Coulson, Matthew Laurance, Michael Rhoades, Anthony Sherwood, Paul Markle, Mark Holmes, David Matheson, Vlasta Vrana, Kate Lynch, Larry King, Bo Diddley, Martha Quinn, Merrill Schindler
A sequel to the first film, this has rock star Eddie Wilson alive and well (not having died in a New Jersey car accident as widely believed) and living in Montreal incognito where he earns a living as a construction worker. When interest in Eddie revives, a concept album is mooted, but the master tape is stolen and the singer's old collaborator has to retrieve it. Songs (supplied by John Cafferty) are only so-so, and this dull tale is a big letdown.
DRA   99 min (ort 106 min)   VIDrel: MED/POLYREC L/A V/h

EDEN VALLEY ***   15
Murray Martin/Amber Production Team UK   1994
Brian Hogg, Darren Bell, Mike Elliott, Jimmy Killeen, Wayne Buck, Kevin Buck, John Middleton, Charlie Hardwick, Katja Roberts, Mo Harold, Art Davies, Rose Laidler, Brian Laidler, Rocky Laidler, Cliffy Usher, John Thom, Bill Speed
After being separated for ten years, a rural horse breeder is reunited with his son, now in his teens and a product of the inner city. Despite the differences in their beliefs and values, they soon find each other and set about training a horse with clear champion potential that could win them a good deal of money. A solid slice-of-life drama set in the North-East that gives a convincing account of contemporary realities.
DRA   95 min   CINrel

EDEN: VOLS. 1 TO 6 **   18
Victor Lobl/Kristine Peterson/Costas Tritchonis USA   1992
Barbara Alyn Woods, Darcy De Moss, Steve Chase, Jack Armstrong, Jeff Griggs, Dean Scofield
Sexual soap opera set in a luxurious resort that caters for the personal fantasies of the young and wealthy and focusing on one of its owners, a woman who is having a hard time coming terms with the tragic death of her husband. Other tales include a man planning to kill his heiress wife and a woman who will lose out on a legacy if she fails to find herself a husband. Plenty

of innocuous and glossy soft-core sex but hardly anything to hold the attention.
Aka: INNOCENT OBSESSION
DRA   600 min (six cassettes - availability varies) mCab
VIDrel: SCRN/TOTAL L/A   V

EDGE OF DARKNESS: PARTS 1 AND 2 *   15
Martin Campbell UK   1985
Joe Don Baker, Joanne Whalley, Bob Peck, Jack Watson, Charles Kay, John Woodvine, Kenneth Nelson, Ian McNeice, Tim McInnerny, David Fleeshman, Randal Herley, Bill Stewart, Paul Humpoletz, Michael Meacher
An over-the-edge attempt at a complex thriller, where the justified worries over nuclear power and weapons degenerate into a confused and unbelievable story. Especially annoying is the persistent anti-American tone (more so since the Chernobyl disaster). Originally a BBC TV series, screened in November 1985.
THR   307 min (2 cassettes) mTV
VIDrel: PARADOX/TOTAL   V/h

EDGE OF DOOM *   PG
Mark Robson USA   1950
Dana Andrews, Joan Evans, Farley Granger, Robert Klein, Paul Stewart, Mala Powers, Adele Jergens, Harold Vermilyea, John Ridgely, Douglas Fowley, Mabel Paige, Howland Chamberlain, Houseley Stevenson, Jean Innes, Ellen Corby
When a parish priest is murdered, his successor refuses to believe that a petty crook was the real killer and starts his own investigation. This eventually leads him to an unbalanced youth who has his own reasons for hating the Catholic Church. Suffering badly from post-production editing, this is an oddity that failed to find an audience on release and quickly sank out of sight.
Aka: STRONGER THAN FEAR
THR   97 min B/W   VIDrel: VCC/DISC   V
Boa: novel by Leo Brady.

EDGE OF SANITY **   18
Gerard Kikoine USA   1988
Anthony Perkins, Glynis Barber, David Lodge, Sarah Maurthorp, Ben Cole, Lisa Davis, Kay Jewers, Harry Landis, Briony McRoberts
A doctor experiments with cocaine, bringing out his murderous alter ego in this rather cliched low-budget adaptation of the Jekyll and Hyde story. Not very different from the 1960s Hammer films of this type, Perkins is in good form in a role similar to his demented CRIMES OF PASSION preacher but Barber as his sexually repressed wife is unconvincing. A film of atmosphere let down by lack of imagination, lack of money and an utterly daft ending.
HOR   84 min (ort 90 min)   VIDrel: PAL L/A   V

EDGE OF THE WORLD, THE ***   U
Michael Powell UK   1937
Eric Berry, John Laurie, Belle Chrystall, Finlay Currie, Niall MacGinnis, Grant Sutherland, Kitty Kirwan, Hamish Sutherland, Campbell Robson, Francesca Reidy, George Summers, Michael Powell
A well-acted tale of a remote, inhospitable Scottish island where the peat supply is virtually exhausted and evacuation seems the only choice. This unusual setting provides the many fine locations that compensate for the thin and overly melodramatic plot. Filmed on the North Sea island of Foula, to which Powell and members of his cast and crew returned forty years later to make the documentary "Return To The Edge Of The World".
DRA   71 min (ort 80 min) B/W   VIDrel: CONNO/RTM   V

EDUCATING RITA **   15
Lewis Gilbert UK   1983
Michael Caine, Julie Walters, Michael Williams, Maureen Lipman, Jeananne Crowley, Malcolm Douglas, Godfrey Quigley, Dearbhla Molloy, Pat Daly, Kim Fortune, Philip Hurdwood, Hilary Reynolds, Jack Walsh, Christopher Casson
A young working-class girl wants to broaden her horizons and signs on as a student at the Open University, where she meets her alcoholic tutor. A depressing reworking of the "Pygmalion" theme, though Caine does give one of his best performances. This was Walters's film debut.
DRA   106 min (ort 110 min)   VIDrel: CARL/TECH   V
Boa: play by Willy Russell.

EDUCATION ANGLAISE **   18
Jean-Claude Roy FRANCE   1982
Andre Dupon, Obaya Roberts, Bernard Musson, Caroline Laurence,

Jean-Claude Dreyfus, Catherine Noel, Pierre Risch, Marcelle Barreau, Anne Meriel, Roger Trapp, Veronique Catanzaro, Anne-Marie Le Menu, Christine Magnin, Cornelia Wilms
At a school for girls, punishment takes the form of bare-bottom spankings, administered by the headmistress, who is really a man in drag.
A   87 min (ort 90 min) wScrn dubbed   VIDrel: JEZ/RTM   V

EDVARD MUNCH ***
Peter Watkins NORWAY/SWEDEN
PG
1974
Geir Westby, Gro Fraas, Johan Halsborg, Lotte Teig, Gro Jarto, Rachel Pedersen, Berit Rytter Hasle, Gunnar Skjetne, Kare Stormak, Iselin Bast, Eli Ryg, Alf Kare Strindberg, Eric Allum, Amund Berge, Kerstii Allum, Camilla Falk
A slow-paced recreation of the life of this 19th century Norwegian Expressionist who gradually lost his mind. Told largely with an amateur cast, it devotes great effort to conveying a period atmosphere but this is largely at the expense of dramatic tension.
DRA   211 min   VIDrel: ACAD/RTM   V

EDWARD AND MRS SIMPSON: PARTS 1 AND 2 ***
U
Waris Hussein UK
1979
Edward Fox, Cynthia Harris, Peggy Ashcroft, Maurice Denham, Marius Goring, Nigel Hawthorne, Andrew Ray, Cherie Lunghi, Jessie Matthews, David Waller
A lavish and absorbing account of the events that led up to the abdication of Edward VIII. No expense was spared to make this a realistic and convincing portrayal, and the film captures the atmosphere of the period splendidly. Originally shown as six 50-minute episodes.
DRA   360 min (2 cassettes - ort 364 min) mTV
VIDrel: THAMES/DISC   V

EDWARD II ***
18
Derek Jarman UK
1991
Steven Waddington, Kevin Collins, Andrew Tiernan, John Lynch, Dudley Sutton, Tilda Swinton, Jerome Flynn, Jody Graber, Nigel Terry, Roger Hammond, Allan Corduner, Annie Lennox, Tony Forsyth, Lloyd Newson, Nigel Charnock, Mark Davis
Jarman's version of Marlowe's play filters the story of this British monarch through the lens of his own preconceptions, resulting in a highly theatrical but still polemic account of sexual politics, with a decidedly heterophobic slant. A competent and entertaining rendition, perhaps a touch too idiosyncratic to have wide appeal, but then the same could be said for almost any Jarman film.
DRA   91 min   VIDrel: PAL L/A   V
Boa: play by Christopher Marlowe.

EDWARD SCISSORHANDS ***
PG
Tim Burton USA
1990
Johnny Depp, Winona Ryder, Dianne Wiest, Anthony Michael Hall, Kathy Baker, Robert Oliveri, Conchata Ferrell, Caroline Aaron, Dick Anthony Williams, O-Lan Jones, Vincent Price, Alan Arkin, Susan J. Blommaert, Linda Perry, Biff Yeager
An old lady tells her grandchild a bedtime story of an inventor who created a gentle and naive being, but died before his work was complete, leaving him with scissors-like blades where his hands should be. This strange creature is at first happily adopted by the local townsfolk, but his growing popularity and innocence lead to mistrust and hatred. An utterly bizarre fantasy, ingenious, fascinating and strangely inconsequential.
FAN   100 min (Cut at UK film release - ort 105 min) wScrn
cC   VIDrel: 20TH/TECH; ENCORE (LV only)   V/sur LV

EFFI BRIEST ***
U
Rainer Werner Fassbinder WEST GERMANY
1974
Hanna Schygulla, Wolfgang Schenck, Ulli Lommel, Lilo Pempeit, Hark Bohm, Herbert Steinmetz, Ursula Straetz, Irm Hermann, Karl Scheydt, Barbara Lass, Theo Tecklenburg, Barbara Valentin, Eva Matties, Anndorthe Braker, Peter Gauhe
A young girl is forced by her parents to marry a much older count against her will and suffers a variety of humiliations at his hands, before finally dying. An excellent and well-realised adaptation of this classic novel which has been filmed a number of times, while Schygulla gives a performance of power and depth that provides most of the interest in this austere work.
Aka: FONTANE'S EFFI BRIEST
DRA   134 min B/W   VIDrel: CONNO/RTM   V
Boa: novel by Theodor Fontane.

EGON SCHIELE **
18
Herbert Vesely WEST GERMANY
1980
Mathieu Carriere, Jane Birkin, Christine Kaufmann, Kristina Van Eyck, Nina Finkenstein
Absorbing biopic on the title figure, A German artist who was noted for his self-indulgence and excessive lifestyle.
Aka: EXCESS AND PUNISHMENT
DRA   83 min   VIDrel: REDEM/RTM   V

EIGER SANCTION, THE **
15
Clint Eastwood USA
1975
Clint Eastwood, George Kennedy, Jack Cassidy, Vonetta McGee, Thayer David, Heidi Bruhl, Reiner Schoene, Brenda Venus, Jean-Pierre Bernard, Michael Grimm, Dan Howard, Gregory Walcott, Candice Rialson, Elaine Shore, Jack Kosslyn
Long boring spy story with Eastwood playing a college lecturer who has retired from Intelligence work until forced to return in order to expose a spy. Enjoyable for its action sequences that take place during an ascent of the Eiger.
THR   113 min (ort 125 min)
VIDrel: 4-FRONT/POLYREC/CIC   V/h
Boa: novel by John Trevanian.

EIGHT HUNDRED LEAGUES DOWN THE AMAZON **
(18)
Luis Llosa USA
1993
Daphne Zuniga, Carlos Lopez Moctezuma, Elvia Quintana, Elsa Aguirre, Barry Bostwick, Adam Baldwin, Tom Verica, E.E. Bell, David Camino, Rafael De Lucchi, Toshiro Konishi, David Killerby, Rafael Santa, David Cameron, Gerald Powell
Standard jungle adventure tale (based on a Verne story) about an outlaw returning to his native Brazil in a large river boat and having to cope with a vicious bounty hunter as well as hostile natives and other such hazards.
Aka: 800 LEGUAS POR EL AMAZONAS; JULES VERNE'S EIGHT HUNDRED LEAGUES DOWN THE AMAZON
A/AD   85 min (ort 90 min)   SATrel: SKY MOVIES
Boa: novel by Jules Verne.

EIGHT MEN OUT ***
15
John Sayles. USA
1988
John Cusack, Clifton James, Michael Lerner, Christopher Lloyd, Charlie Sheen, David Strathairn, D.B. Sweeney, Perry Lang, Jane Alexander, John Mahoney, Bill Irwin, Michael Rooker, Studs Terkel, Kevin Tighe, Don Harvey, James Read
A recreation of a true incident from 1919 when members of the Chicago White Sox baseball team were accused of accepting bribes in return for throwing the World Series. The film carefully probes their motivations and the all-pervading atmosphere of corruption that infested the game. Screenplay is by Sayles, who also appears briefly in a cameo role. See FIELD OF DREAMS for a fantasy written around this event.
DRA   114 min (ort 119 min)   VIDrel: VISVID/POLYREC   V
Boa: novel by Eliot Asinof.

EIGHT O'CLOCK WALK **
PG
Lance Comfort UK
1954
Richard Attenborough, Cathy O'Donnell, Derek Farr, Ian Hunter, Maurice Denham, Bruce Seton, Lily Kann, Harry Welchman, Kynaston Reeves, Eithne Dunne, David Hannaford
A taxi-driver is falsely accused of murdering a small girl who is found on a bomb-site. Fortunately, his wife believes him to be innocent, and is able to persuade a determined QC to take up the case. Standard courtroom dramatics are on offer here, and though the cast cannot be faulted, there is nothing here to make it more than just a time-filler.
DRA   83 min (ort 87 min) B/W   VIDrel: LUMI/SONOP   V

EIGHT SECONDS **
PG
John G. Avildsen USA
1993
Luke Perry, Stephen Baldwin, Cynthia Geary, James Rebhorn, Carrie Snodgress, Red Mitchell, Ronnie Claire Edwards, Linden Ashby, Dustin Mayfield, Clyde Frost, Elsie Frost, Gabriel Folse, Joe Stevens, Clint Burkey, John Swasey, Jim Gough
Biopic on the life of Lane Frost, the youngest ever cowboy to achieve fame as a bull rider and win a place in the Rodeo Hall Of Fame. Though both the acting and action sequences are convincing, there is no real examination of the motives for such an interest in this most dangerous of sports. The title was meant to refer to the length of time one has to stay on the bull to score points, it sounded a whole lot better than the real time: twenty-five seconds.
DRA   100 min (ort 105 min)   VIDrel: FIRST/SONOP   V

EIGHTEEN AGAIN! ***
PG
Paul Flaherty USA
1988
George Burns, Tony Roberts, Charlie Schlatter, Anita Morris, Red Buttons, Miriam Flynn, Jennifer Runyon, George DiCenzo, Bernard Fox, Kenneth Tigar, Pauly Shore, Anthony Starke, Emory Bas, Joshua Devane, Benny Baker, Hal Smith
"Vice Versa" once more with an eighty-one-year-old entrepreneur making a wish to be the same age as his grandson when he blows out his birthday candles. On the way home a car crash puts them both in hospital and, guess what, they swap bodies. A fair movie that treads a distinctly well-worn path and contains the expected complications as both individuals re-adjust to their new identities, Schlatter in particular doing a passable impression of Burns.
Aka: 18 AGAIN!
COM   96 min (ort 100 min)   VIDrel: NWV L/A   V

EIGHTH DAY, THE **
PG
Jaco Van Dormael BELGIUM/FRANCE/UK
1996
Daniel Auteuil, Pascal Duquenne, Miou-Miou, Henri Garcin, Isabelle Sadoyan, Michele Maes, Fabienne Loriaux, Alice Van Dormael, Juliette Van Dormael, Sabrina Leurquin, Marie-Pierre Meinzel, Lazlo Harmati, Stephane Keyser, Roland Depauw
A yuppie businessman who has destroyed his family life through insensitivity and overwork, has a road accident that brings him into contact with a retarded man who suffers from Down's Syndrome. The latter has run away from the institution where he lives, and this odd couple eventually built a special relationship of their own. Scripted by former clown Dormael, this is a cloying and mawkish directing debut that lacks both conviction and a light touch.
Aka: LE HUITIEME JOUR
DRA   113 min (ort 117 min)   VIDrel: ELPIC/POLYREC   V

EIGHTY-FOUR CHARING CROSS ROAD ***
U
David Jones USA
1986
Anne Bancroft, Anthony Hopkins, Judi Dench, Jean De Baer, Maurice Denham, Eleanor David, Mercedes Ruehl, Daniel Gerroll, Wendy Morgan, Ian McNeice
Bancroft stars as a New York collector of rare books who begins a 20 year correspondence with the staff of a London bookshop, little dreaming that this will lead to a love-affair conducted by mail but with her unfortunately failing to make the trip over to London before her pen-friend dies. Filmed once before as a BBC TV play, this low-key and detailed film is lovingly made but essentially episodic and lacking in drama.
Aka: 84 CHARING CROSS ROAD
DRA   96 min   VIDrel: RCA/VCC L/A   V
Boa: book by Helen Hanff/play by James Roose-Evans.

EIGHTY-FOUR CHARLIE MOPIC ***
18
Patrick Duncan USA
1989
Jonathan Emerson, Richard Brooks, Christopher Burgard, Nicholas Cascone, Glenn Morshower, Jason Tomlins, Byron Thames, Russ Thurman, Joseph Hieu, Don Schiff
A low-budget and unusual look at life on the front line in Vietnam, where a reconnaissance patrol sets off on a routine mission into the interior, complete with an army motion picture (or MOPIC) cameraman. However, what starts off as a simple mission rapidly develops into a battle to survive and despite a few lulls in the plot, the film (seen entirely through the single hand-held camera) remains totally absorbing. The script is by Duncan.
Aka: 84 CHARLIE MOPIC
WAR   90 min (ort 95 min)   VIDrel: VISVID/POLYREC L/A   V

EL CID ***
U
Anthony Mann SPAIN/USA
1961
Charlton Heston, Sophia Loren, Raf Vallone, Genevieve Page, Herbert Lom, Hurd Hatfield, John Fraser, Gary Raymond, Massimo Serato, Michael Hordern, Frank Thring, Andrew Cruickshank, Douglas Wilmer, Ralph Truman, Carlo Giustini
Highly fictionalised account of the 11th century figure who fought to expel the Moors from Spain. If not true to the past, it remains highly enjoyable for its numerous action sequences and skilful direction. A superior epic of considerable opulence if not veracity. (Look out for the sequence where Heston rides from Valencia to Burgos and back again in a single day.) The score is by Miklos Rozsa.
DRA   172 min (ort 180 min)   VIDrel: 4-FRONT/POLYREC   V/sur

EL DORADO **
PG
Howard Hawks USA
1966
John Wayne, Robert Mitchum, James Caan, Charlene Holt, Ed Asner, Michele Carey, Gayle Hunnicutt, R.G. Armstrong, Paul Fix, Christopher George, Robert Donner, John Gabriel, Jim Davis, Marina Chane, Anne Newman, Johnny Crawford
An ageing gunfighter and a drunken sheriff team up to make peace between a cattle baron and local farmers in a dispute over land rights that threatens to escalate into a range war. A disappointing and overlong Western.
WES   121 min (ort 126 min)
VIDrel: 4-FRONT/POLYREC/CIC   V/h
Boa: novel The Stars in their Courses by Harry Joe Brown.

EL MARIACHI **
15
Robert Rodriguez USA
1992
Carlos Gallardo, Consuelo Gomez, Peter Marquardt, Reinol Martinez, Jaime De Hovos, Ramiro Gomez, Jesus Lopez, Luis Baro, Oscar Fabila, Fernando Martinez, Manual Acosta, Walter Vargas, Roberto Martinez, Jianita Vargas, Yolanda Puga
A "mariachi" musician in a small Mexican town finds himself being mistaken for a notorious killer who has escaped from the local jail, and has to work hard to stay alive. An interesting directorial debut that is reputed to have been made for as little as $7,000. Followed in 1995 by a lavish Hollywood sequel DESPERADO, which was also directed by Rodriguez. Screenplay is by the director.
WES   81 min dubVer (20VIS/SONOP)
VIDrel: 20VIS/SONOP L/A; COLUM/SONOP L/A   V/sh

EL NORTE ***
PG
Gregory Nava USA
1984
Zaide Silvia Gutierrez, David Villalpando, Ernesto Gomez Cruz, Alicia Del Lago, Eraclio Zepeda, Stella Quan, Lupe Ontiveros, Stella Quan, Emilio Del Haro, Rodolfo Alejandre, Rodrigo Puebla, Trinidad Silva, Abel Franco, Mike Gomez
A Guatemalan brother and sister leave their home on a perilous odyssey to the USA (the "North" of the title) in search of a better life and to escape the rampant violence of their homeland. A powerful and compassionate look at the motives behind illegal immigration and the pain of having to accept an inferior status as an illegal alien in an unfamiliar culture.
DRA   139 min (ort 141 min)   VIDrel: ARROW/RTM   V

ELECTRA *
18
Julian Grant USA
1995
Shannon Tweed, Joe Tab, Sten Eirik, Katie Griffin, Lara Daans, Dyanne DiMarco, John Stoneham, Ed Sahely, Rob Wilson, Daniel Levinson, Louise Martin, Ron Sarosiak, Todd Schroeder, Danny Lima, Tig Fong, Peter Schnidelhauer
In this erotic thriller, a man develops a serum that can turn human beings into a superior race, but it can only be passed on by sexual contact. Before he dies he injects his son with this serum and soon this lad is being eagerly sought by both his stepmother and an evil crippled tycoon, who wants this secret for his own ends. A novel idea is soon done to death in a welter of violent effects and bad acting.
THR   87 min   VIDrel: EIV/SONOP   V

ELECTRA GLIDE IN BLUE ****
18
James William Guercio USA
1973
Robert Blake, Billy "Green" Bush, Mitchell Ryan, Jeannine Riley, Elisha Cook Jr, Royal Dano, David Wolinski, Peter Cetera, Terry Kath, Lee Loughnane, Walter Parazaider, Joe Samsil, Jason Clark, Michael Butler, Susan Forristal, Bob Zemko
A slick and incredibly stylish look at the life of an undersized motorcycle cop and his partner. Serious at times, the film has some wry moments as it examines his progress from patrol cop to plainclothes detective working on a strange murder case. Now something of a cult film, and up to the now only work directed by Guercio, a well known figure in the world of rock music.
Aka: ELECTRA
A/AD   108 min (ort 113 min)   VIDrel: MGM/WHV L/A   V/sh

ELECTRIC DREAMS: THE MOVIE **
PG
Steve Barron UK/USA
1984
Lenny Von Dohlen, Virginia Madsen, Maxwell Caulfield, Don Fellows, Alan Polonsky, Wendy Miller, Harry Rabinowitz, Miriam Margoyles, Holly De Jon, Stella Maris, Mary Doran, Diana Choy, Jim Steck, Gary Pettinger, Bud Cort (voice only)
A computer freak has a computer which falls for the girl upstairs

in this bizarre love-triangle tale that starts off well enough but never amounts to anything of substance. Barron's directorial debut is a hurried affair that has the look of the pop videos he used to make, while the heavy reliance on computer graphics and electronic music soon becomes extremely tiresome. This was also the feature film debut for producer Simon Fields. See also HOMEWRECKER.

Aka: ELECTRIC DREAMS
COM   92 min (ort 112 min)
VIDrel: VISVID/POLYREC L/A   V/s

ELECTRIC ESKIMO ** (U)
Frank Godwin UK   1979
Derek Francis, Diana King, Tom Chadbon, David Rowlands, Roger Avon, Victoria Brooks, Charles Pemberton, Norman Mitchell, Kris Emmerson, Ian Sears, Debbie Padbury, Ivor Danvers, Richard Wren, Kenneth Kendall
An Eskimo accidentally becomes charged with electricity, which gives him some strange powers that a criminal gang want to use for their own purposes, so they set out to capture him. A so-so kids' film, produced by the Children's Film Foundation.
A/AD   54 min   SATrel: MOVIE CHANNEL

ELECTRIC HORSEMAN, THE ** PG
Sydney Pollack USA   1979
Robert Redford, Jane Fonda, Valerie Perrine, Willie Nelson, Nicolas Coster, John Saxon, Allan Arbus, Wilford Brimley, Will Hare, Basil Hoffman, Timothy Scott, James B. Sikking, James Kline, Frank Speiser, Quinn Redeker, Louis Areno
A rodeo rider reduced to advertising breakfast cereals suddenly decides that he has had enough of corporate greed and rides off with his $12,000,000 thoroughbred horse out of Las Vegas and into the hills and freedom. A long, meandering journey through some spectacular scenery, with Fonda and Redford displaying their limited acting range; all wrapped up in the slick sentimentality of a happy ending.
DRA   114 min (ort 120 min)   VIDrel: CIC/SONOP   V
Boa: story by Shelly Burton.

ELECTRIC MOON *** 15
Pradip Krishen UK   1991
Roshan Seth, Naseeruddin Shah, Leela Naidu, Gerson Da Cunha, Raghubir Yadav, Alice Spivak, Frances Helm, James Fleet, Francesca Brill, Gareth Forwood, Surendra Rajan, Malcolm Jamieson, Barbara Lott, Vageesh Singh, Mahavir Bhullar
An Indian Maharajah devises a clever and ambitious scheme to make some money, by giving naive tourists what he hopes they will take to be a taste of India, and no effort is too great in his attempts to satisfy the expectations of his guests. A very funny and irreverent look at India, Indians and the gullibility of Westerners who arrive there with their heads packed with foolish stereotypes and preconceptions.
COM   103 min   CINrel

ELEMENT OF DOUBT *** 15
Christopher Morahan UK   1996
Nigel Havers, Gina McKee, Judy Parfitt, Michael Jayston, Mary Woodvine, Denis Lill
The apparently happy life of a married couple masks deep dissatisfactions that eventually come to the surface, in this detailed and quite painful study of a flawed relationship.
DRA   101 min   VIDrel: CARL/TECH   V

ELEMENT OF TRUTH, AN *** 12
Larry Peerce USA   1995
Donna Mills, Peter Riegert, Perrey Reeves, Robin Thomas, Cliff De Young, Brock Peters, Harriet Sansom Harris, Donald V. Allen, Sally Champlin, Caia Coley, Gary Cusano, Jo Ann Dearing, Mark Hutter, Aaron Lustig, Rolanda Mendel
An attractive woman joins a private investment company, and as she learns the ropes she makes plans for her next embezzlement.
DRA   93 min   VIDrel: ODY/SONOP   V/sh

ELENYA ** PG
Steve Gough GERMANY/UK   1992
Margaret John, Pascale Delafouge Jones, Seiriol Tomos, Sue Jones Davies, Klaus Behrendt, Iago Wynn Jones, Lilo Millward, Catrin Llwyd, Edward Elwyn Jones, Ioan Meredith, Eiry Palfrey, Pauline Yates
Set in Wales 1940, where a lonely twelve-year-old girl whose parents are away finds a German pilot, who was badly injured bailing out of his damaged aircraft. A variant on WHISTLE

DOWN THE WIND that examines how the innocence of childhood is necessarily modified by the harsh lessons of life. Filmed in both Welsh and English, though only the latter version is at present available.
Aka: ELENYA: IN KRIEGSZEITEN
DRA   78 min (ort 82 min)   VIDrel: IMAG/RTM   V

ELEPHANT BOY *** U
Robert Flaherty/Zoltan Korda UK   1937
Sabu, Walter Hudd, Walter E. Holloway, Wilfrid Hyde White, Allan Jeayes, Bruce Gordon, D.J. Williams
This film made an international star of Sabu whose portrayal of the title character is outstanding. The story revolves around his efforts to help conservationists and his insistence that he knows of the elephants' burial ground. Dated and rather slow, but the outdoor locations are excellent and the film has some memorable moments.
A/AD   76 min (ort 81 min) B/W   VIDrel: CARL/TECH   V
Boa: novel Toomai of the Elephants by Rudyard Kipling.

ELEPHANT MAN, THE *** PG
David Lynch UK   1980
Anthony Hopkins, John Hurt, John Gielgud, Freddie Jones, Wendy Hiller, Anne Bancroft, Michael Elphick, Hannah Gordon, Helen Ryan, John Standing, Dexter Fletcher, Lesley Dunlop, Phoebe Nicholls, Pat Gorman, Claire Davenport
Beautifully made but rather sentimental story of a hideously deformed man who lived in Victorian England. Rescued from a miserable existence as a circus freak he is taken to live in the London Hospital. (His skeleton was on display there; for those interested he suffered from neurofibromatosis.) Photographed by Freddie Francis and produced by Mel Brooks, who was inspired to choose Lynch as director and ensure he be given full artistic control.
DRA   118 min (ort 125 min) B/W wScrn (re-mastered special edition)   VIDrel: WHV   V/dm V/sur
Boa: book The Elephant Man and Other Curiosities by Sir Frederick Treves/ play by Ashley Montagu.

ELEVEN DAYS ELEVEN NIGHTS ** 18
Joe D'Amato (Aristide Massaccesi) ITALY   1986
Jessica Moore, Mary Sellers, Joshua McDonald, Tom Mojack, Ale Dugas
A man due to marry in eleven days has a chance meeting with a female writer that threatens to upset all his plans. She is near completion of a book detailing her 100 lovers and he is about to become her final conquest for inclusion in "Sarah Asproon And Her One Hundred Men".
Aka: UNDICI GIORNI, UNDICI NOTTI
A   90 min (ort 93 min)   VIDrel: L/A   V

ELEVEN DAYS ELEVEN NIGHTS: PART 2 * 18
Joe D'Amato (Aristide Massaccesi) ITALY   1990
Kristine Rose, Ruth Collins, Frederick Lewis, Maurice Dupre, Alex Dexter, Kristin Cuadraro, Fred Woodruff, James Jackson, Laura Gemser, Garyn Chalet, Russel Pottharst, Michelle McGuire, Debbie Morrant
When the millionaire friend of the woman writer in the previous film dies, he makes her the executor of his will. However, she is given only 11 days in which to interview all his heirs and propose an allocation of the inheritance. The usual softcore encounters transpire.
Aka: HOT CURVES
A   83 min (ort 90 min)   VIDrel: L/A   V

ELEVEN DAYS ELEVEN NIGHTS: PART 3 ** 18
Joe D'Amato (Aristide Massaccesi) ITALY   1988
Valentine Demy, Allen Cort, Robert la Brosse, Carey Salley
A reporter is assigned to a case involving a woman who lost her husband in a strange voodoo ritual. He goes to New Orleans to investigate, but rather unwisely takes his wife along. An erotic thriller.
Aka: AFTERNOON; 11 DAYS 11 NIGHTS: PART 3 - THE FINAL CHAPTER
A   87 min Cut (52 sec - ort 92 min)   VIDrel: MARQ/QUANT V

ELIMINATORS ** 15
Peter Manoogian USA   1986
Andrew Prine, Denise Crosby, Patrick Reynolds, Conan Lee, Roy Dotrice, Peter Schrum, Peggy Mannix, Fausto Barra, Luis Lorenzo, Tad Horino, Pepe Morino, Charly Bravo, Miguel Di Grandi, Gabieno Diego Solis
Clone of THE TERMINATOR in which the android (cast here

as a goodie) and his companions are out to stop his megalomaniacal creator from dominating the world.
FAN 91 min Cut (31 sec - ort 96 min)
VIDrel: NTV/TOTAL/EIV V

ELISA ** 15
Jean Becker FRANCE 1994
Vanessa Paradis, Gerard Depardieu, Clothilde Courau, Dekkou Sall, Florence Thomassin, Werner Schreyer, Michel Bouquet, Philippe Leotard, Catherine Rouvel, Melvil Poupard, Olivier Saladin, Bernard Verley, Reine Barteve, Andre Julien
A teenage hustler is haunted by the memory of her mother's suicide, and sets out to find her father (whom she blames) and take her revenge. Petulant, pouting and angry Paradis is the best thing in this strange Gallic revenge tale, but that isn't saying much, given the abrupt changes of pace and irritating flashbacks we are so often treated to.
DRA 110 min (ort 115 min) wScrn VIDrel: TART/20TH V

ELIZABETH R: PARTS 1 TO 6 *** 12/PG
Roderick Graham/Richard Martin/Claude Whatham/Herbert Wise UK 1971
Glenda Jackson, Ronald Hines, Daphne Slater, Rachel Kempson, Bernard Hepton, Rosalie Crutchley, John Ronante, Peter Jeffrey
A series that deservedly won five Emmys, this richly detailed work covers the life of the title monarch, from her earliest days to her final ones, with Jackson giving a remarkably powerful performance in the title role.
DRA 540 min (6 cassettes) VIDrel: BBC/TECH V/h

ELMER GANTRY **** 15
Richard Brooks USA 1960
Burt Lancaster, Jean Simmons, Dean Jagger, Shirley Jones, Arthur Kennedy, Patti Page, Edward Andrews, Hugh Marlowe, John McIntire, Rex Ingram, Everett Glass, Joe Maross, Michael Whalen, Hugh Marlowe, Philip Ober, Wendell Holmes
An Oscar-winning portrayal of an evangelist in the American mid-West of the 1920s who cynically exploits his talents for commercial gain. Lancaster gives a full-blooded and powerful performance in this fascinating and highly literate film. See also LEAP OF FAITH for an updated treatment of this theme. AA: Actor (Lancaster), S. Actress (Jones), Screen/adapt (Richard Brooks).
DRA 141 min (ort 145 min) VIDrel: MGM/WHV V
Boa: novel by Sinclair Lewis.

ELVIRA MADIGAN *** PG
Bo Widerberg SWEDEN 1967
Pia Degermark, Thommy Berggren, Lennart Malmen, Nina Widerberg, Cleo Jensen
Beautifully photographed, soft-focus recreation of a true story, the tragic, doomed romance between a cavalry officer and a circus tightrope dancer who run away to Denmark, abandoning everything so they can be together. Unfortunately, the Mozart score and exquisite camerawork are all there is in this period tale (set around 1900) and the lack of any real acting is a serious handicap. Not so much a film as a succession of vibrant images. Photography is by Jorgen Persson.
DRA 85 min (ort 90 min) wScrn VIDrel: TART/20TH V/dm

ELVIRA, MISTRESS OF THE DARK ** 15
James Signorelli USA 1988
Elvira (Cassandra Peterson), W. Morgan Sheppard, Daniel Greene, Susan Kellerman, Jeff Conaway, Edie McClurg, Kurt Fuller, William Duell, Pat Crawford Brown, Mario Celario, Lee McLaughlin, Jack Fletcher, Hugh Gillin, Frank Collison
Feature film built around the personality (and impressive bust) of TV horror show presenter "Elvira". The shallow plot, involving a conflict with some small-town conservatives and the hunt for a magic book, is no more than an excuse for a series of risque jokes in this campy and only marginally amusing comedy. A pity, because Peterson has very appealing personality.
COM 91 min (ort 96 min) VIDrel: VCC/DISC/COLUM V/sur

ELVIS: THE MOVIE *** U
John Carpenter USA 1979
Kurt Russell, Shelley Winters, Season Hubley, Pat Hingle, Bing Russell, Ed Begley Jr, Charlie Hodge, Ellen Travolta, Melody Anderson, James Canning, Robert Gray, Charles Cyphers, Peter Hobbs, Les Lannom, Elliott Street, Will Jordan
The story of Presley up to 1969 is retold in a believable way with effective performances, especially from Russell in the title role (though the songs were dubbed by Ronnie McDowell who later worked on ELVIS AND ME). The script is by Anthony Lawrence and though very well put together, is a little bit too reverential for its own good. See also THIS IS ELVIS. Kurt Russell was rightly nominated for an Emmy for his performance in the title role.
Aka: ELVIS
MUS 164 min mTV VIDrel: VCC L/A V

ELVIS AND ME ** 15
Larry Peerce USA 1987
Dale Midkiff, Susan Walters, Billy Green Bush, Linda Miller, Jon Cypher, Anne Haney, Marshall Teague, Ken Gibbel, Hugh Gillin, Mark Thomas Miller, Cynthia Harrison, Linda Dona, Cody Hampton, Jesse Henecke, Wayne Powers
A portrait of Elvis and the women he loved, largely based on his wife's book (she was one of the executive producers). Despite a good performance from Midkiff the film leaves us no wiser about this rock legend. Songs are performed by Ronnie McDowell. See also ELVIS: THE MOVIE and THIS IS ELVIS.
DRA 178 min (ort 240 min) mTV VIDrel: NWV L/A V
Boa: book by Priscilla Presley.

ELVIS & THE COLONEL ** PG
William A. Graham USA 1993
Beau Bridges, Rob Youngblood, Dan Shor, Micole Mercurio, Ben Slack, Scott Wilson, Ron Fussler, Don Stark, Randy Crowder, Ralp Bruneau, Alex Courtney, Patrick O'Connell, James Intveld, Kenla O'Byrne, Bart Braverman, Kirk Scott
Drama that tells of the career of Elvis as it was influenced (or some might say blighted) by the activities of his business manager, the ruthless and predatory "Colonel" Tom Parker. Not a terribly enjoyable film, especially in its portrayal of Elvis as no more than a weak-willed puppet, very much at the mercy of his manager. One strongly suspects that the true state of affairs was a good deal more complex. However, Bridges is outstanding as Parker.
Aka: ELVIS AND THE COLONEL: THE UNTOLD STORY
DRA 88 min (ort 95 min) mTV VIDrel: VCC L/A V

ELVIS PRESLEY: BLUE HAWAII ** PG
Norman Taurog USA 1961
Elvis Presley, Stella Stevens, Angela Lansbury, Joan Blackman, Jenny Maxwell, Roland Winters, John Archer, Howard McNear, Iris Adrian, Gregory Gay, Flora Hayes, Steve Brodie, Darlene Tompkins, Pamela Akert, Jose DeVarga
This story of a returning GI who comes home to Honolulu and becomes a beachcomber, serves as a vehicle for Elvis's star talent. Pleasant and quite entertaining if hardly memorable, but Elvis does get to perform one of his best hits - "Can't Help Falling In Love". The exotic locations are a big asset.
Aka: BLUE HAWAII
MUS 97 min (ort 101 min)
VIDrel: 4-FRONT/POLYREC/BRAVE V

ELVIS PRESLEY: CHANGE OF HABIT ** PG
William A. Graham USA 1969
Elvis Presley, Mary Tyler Moore, Barbara McNair, Edward Asner, Leora Dana, Jane Elliot, Robert Emhardt, Doro Merande, Regis Toomey, Richard Carlson, Ruth McDevitt, Nefti Millet, Laura Figueroa, Lorena Kirk, Virginia Vincent
A slight change from the usual Presley plot gives the singer a storyline in which he plays a doctor in charge of a clinic for the poor. His opinions change after he encounters three nuns who work there, one of whom falls in love with him. Moore is the lucky girl who has to make the difficult choice of a life with the church or a life with the doc. Pleasant if unmemorable fare that gave Presley his last screen role.
Aka: CHANGE OF HABIT
MUS 88 min (ort 97 min)
VIDrel: 4-FRONT/POLYREC/CIC V/h

ELVIS PRESLEY: CLAMBAKE * U
Arthur H. Nadel USA 1967
Elvis Presley, Will Hutchins, Bill Bixby, Shelley Fabares, Gary Merrill, James Gregory, Amanda Harley, Suzie Kaye, Angelique Pettyjohn, Olga Kaya, Arlene Charles, Jack Good, Hal Peary, Sam Riddle, Sue England, Lisa Slagle
Elvis appears as the son of an oil baron, who switches identities with a motorcyclist to work as a water-ski instructor in Miami, and learns all about life. One of the star's weakest vehicles.
Aka: CLAMBAKE
MUS 95 min (ort 98 min) VIDrel: MGM/WHV L/A V/h

**ELVIS PRESLEY: DOUBLE TROUBLE ** ** U
Norman Taurog USA   1967
Elvis Presley, Yvonne Romain, Annette Day, John Williams, Chips Rafferty, Norma Rossington, The Wiere Brothers, Monty Landis, Michael Murphy, Leon Askin, John Alderson, Stanley Adams, Maurice Marsac, Walter Burke
Typical vehicle for Presley's musical talents. This time round he plays a nightclub singer for whom a teenage heiress falls when he performs in England. (An irony not lost on Presley fans is that in real life he never did perform there.) Songs include "Long Legged Girl", one of the star's best numbers.
Aka: DOUBLE TROUBLE
MUS 88 min (ort 90 min)   VIDrel: MGM/WHV L/A V

**ELVIS PRESLEY: EASY COME, EASY GO ** ** U
John Rich USA   1967
Elvis Presley, Dodie Marshal, Pat Priest, Pat Harrington, Skip Ward, Frank McHugh, Elsa Lanchester, Elaine Beckett, Shari Nims, Sandy Kenyon, Mickey Elley, Read Morgan, Diki Lerner, Ed GRiffith, Kay York, Robert Isenberg
This rather aimless musical-comedy casts Presley as a US Navy diver who thinks he's found the location of a cache of sunken treasure off the coast of California. Apart from a couple of good numbers such as "The Love Machine" and "Yoga Is As Yoga Does" this is very forgettable fare indeed.
Aka: EASY COME, EASY GO
MUS 91 min (ort 97 min)
VIDrel: 4-FRONT/POLYREC/CIC V/h

**ELVIS PRESLEY: FOLLOW THAT DREAM ** ** U
Gordon Douglas USA   1962
Elvis Presley, Arthur O'Connell, Anne Helm, Joanna Moore, Jack Kruschen, Simon Oakland, Roland Winters, Alan Hewitt, Frank De Kova, Howard McNear, Gavin Koon, Herbert Rudley, Robin Koon, Robert Carricart, John Duke, Pam Ogles
Easy-going Presley comedy with our star and his family moving to southern Florida with the intention of running a small homestead. No great shakes as a movie but Presley does get to sing "Home Is Where The Heart Is" and "On Top Of Old Smokey".
Aka: FOLLOW THAT DREAM
MUS 106 min (ort 111 min)   VIDrel: MGM/WHV L/A
V/h
Boa: novel Pioneer Go Home by Richard Powell.

**ELVIS PRESLEY: FRANKIE AND JOHNNY ** ** U
Frederick De Cordova USA   1966
Elvis Presley, Donna Douglas, Harry Morgan, Sue Anne Langdon, Nancy Kovack, Audrey Christie, Jerome Cowan, Robert Strauss, Anthony Eisley, Wilda Taylor, Larri Thomas, Dee Jay Mathis, Judy Chapman
An old saloon song is reworked into a suitable vehicle for Presley, with him cast as a singing riverboat gambler down on his luck and afflicted with girl troubles. Now best remembered for songs such as "When The Saints Go Marching In" and "Down By The Riverside". A colourful time-filler.
Aka: FRANKIE AND JOHNNY
MUS 84 min (ort 88 min)   VIDrel: MGM/WHV L/A V/h

**ELVIS PRESLEY: FUN IN ACAPULCO ** ** U
Richard Thorpe USA   1963
Elvis Presley, Ursula Andress, Paul Lukas, Alejandro Rey, Elsa Cardenas, Larry Domasin, Robert Carricart, Teri Hope, Charles Evans, Alberto Morin, Francisco Ortega, Robert De Anda, Linda Rivera, Linda Rand
Another typical Presley vehicle but with superior scenery as our star works as a lifeguard and entertainer in this Mexican resort and finds himself the object of the affections of a lady bullfighter. Song include "No Room To Rhumba In A Sports Car", "Bossa Nova Baby" and "You Can't Say No In Acapulco".
Aka: FUN IN ACAPULCO
MUS 92 min (ort 97 min)
VIDrel: 4-FRONT/POLYREC/BRAVE V

**ELVIS PRESLEY: G.I. BLUES ** ** U
Norman Taurog USA   1960
Elvis Presley, Juliet Prowse, James Douglas, Robert Ivers, Leticia Roman, Ludwig Stossel, Arch Johnson, The Jordanaires
In this, his first film after leaving the army, Elvis plays Tulsa McCauley, a guitar-playing gunner who, in an attempt to win the money he needs to open a nightclub, accepts a bet that he cannot spend the night with a cabaret dancer. However, his plan backfires when he falls in love with the girl in question. Songs

such as "Tonight Is So Right For Love", "Wooden Heart", "Blue Suede Shoes" and the title song compensate for the banal script.
Aka: G.I. BLUES
MUS 99 min (ort 104 min)
VIDrel: 4-FRONT/POLYREC/BRAVE V

**ELVIS PRESLEY: GIRL HAPPY ** * PG
Boris Sagal USA   1965
Elvis Presley, Harold J. Stone, Shelley Fabares, Gary Crosby, Joby Baker, Nita Talbot, Mary Ann Mobley, Chris Noel, Jackie Coogan, Jimmy Hawkins, Fabrizio Mioni, Peter Brooks, John Fiedler, Lyn Edgington, Gale Gilmore
A nightclub entertainer in Fort Lauderdale finds himself acting as chaperone to a group of college girls, one of whom is the daughter of a gangster. A slick and entirely predictable star vehicle, as forgettable as it is innocuous.
Aka: GIRL HAPPY
MUS 93 min (ort 96 min)   VIDrel: MGM/WHV L/A V

**ELVIS PRESLEY: GIRLS! GIRLS! GIRLS! ** * U
Norman Taurog USA   1962
Elvis Presley, Stella Stevens, Laurel Goodwin, Jeremy Slate, Robert Strauss, Guy Lee, Benson Fong, Ginny Tiu, Frank Puglia, Lili Valenty, Beulah Quo, Barbara Beale, Betty Beal, Nestor Paiva, Ann McCrea, Elizabeth Tiu
Presley vehicle with the singer running a fishing boat as a hobby and being chased by a crowd of girls, most of his troubles arising from his inability to decide which one he likes the most. A host of largely forgettable songs follow, but our star does get to sing one classic number, "Return To Sender".
Aka: GIRLS! GIRLS! GIRLS!
MUS 94 min (ort 106 min)
VIDrel: 4-FRONT/POLYREC/BRAVE V

**ELVIS PRESLEY: HAREM HOLIDAY ** * U
Gene Nelson USA   1965
Elvis Presley, Michael Ansara, Mary Ann Mobley, Fran Jeffries, Jay Novello, Theo Marcuse, Philip Reed, Billy Barty, Dirk Harvey, Barbara Werle, Jack Constanza, Larry Chance, Brenda Benet, Gail Gilmore, Joey Russo
Dull Presley vehicle set in the Middle East with a singer/star kidnapped on his way to a premiere of his latest film and finding himself involved in a plot to murder a king. Lots of boring back-lot "desert" locations and a clutch of unremarkable numbers make this one a must to avoid.
Aka: HAREM HOLIDAY; HARUM SCARUM
MUS 81 min (ort 95 min)   VIDrel: MGM/WHV L/A V/h

**ELVIS PRESLEY: IT HAPPENED AT THE WORLD'S FAIR ** ** PG
Norman Taurog USA   1962
Elvis Presley, Joan O'Brien, Gary Lockwood, Yvonne Craig, Vicky Tiu, Kurt Russell, H.M. Wynant, Edith Atwater, Guy Raymond, Dorothy Green, Kam Tong
Presley does his thing at the Seattle World's Fair, in this tale of two crop-dusting pilots who enjoy a little fun and romance. A bright and cheerful musical with songs including "One Broken Heart For Sale", "A World Of Our Own" and "Happy Ending". Look out for young Kurt Russell, who played the star in ELVIS in the 1979 TV movie.
Aka: IT HAPPENED AT THE WORLD'S FAIR
MUS 100 min (ort 105 min)   VIDrel: MGM/WHV L/A
V/h

ELVIS PRESLEY: JAILHOUSE ROCK ** * U/PG
Richard Thorpe USA   1957
Elvis Presley, Judy Tyler, Dean Jones, Vaughn Taylor, Mickey Shaughnessy, Jennifer Holden, Anne Neyland, Hugh Sanders, Mike Stoller, Grandon Rhodes, Katherine Warren, Don Burnett, George Cisar, Glenn Strange, John Indrisano
One of the best of Elvis's films in which he plays a convict sent to jail for accidentally killing a man in a brawl. Here he learns to play the guitar and becomes a star when he gets out, but finds that his ex-cellmate wants to share his success. Has the memorable "Jailhouse Rock" sequence in which Elvis (who choreographed the steps) can be seen at his very best. The score is by Leiber and Stoller. Note that the double cassette pack is rated PG.
Aka: JAILHOUSE ROCK
MUS 92 min (ort 96 min) B/W   VIDrel: MGM/WHV L/A
V

ELVIS PRESLEY: KID GALAHAD ** PG
Phil Karlson USA   1962
Elvis Presley, Lola Albright, Gig Young, Joan Blackman, Charles Bronson, Ned Glass, David Lewis, Robert Emhardt, Michael Dante, Judson Pratt, George Mitchell, Richard Devon, Jeffrey Morris, Liam Redmond
This musical remake of the 1937 film lacks the punch of its predecessor and casts Presley as a peaceable garage mechanic who gets inadvertently drawn into the world of boxing. Presley turns in a rather good performance and sings six unmemorable songs in a film that hardly does him justice. Average.
Aka: KID GALAHAD
MUS   93 min (ort 95 min)   VIDrel: MGM/WHV L/A   V/h
Boa: novel by F. Wallace.

ELVIS PRESLEY: KING CREOLE ** PG
Michael Curtiz USA   1958
Elvis Presley, Carolyn Jones, Walter Matthau, Dolores Hart, Dean Jagger, Vic Morrow, Paul Stewart, Raymond Bailey, Liliane Montevecchi, Jan Shepard, Jack Grinnage, Brian Hutton, Dick Winslow, Raymond Bailey, Ziva Rodann
A failed student gets involved with criminals and a hold-up, but later becomes a big hit when he is forced to start work singing at a gangster's nightclub. A very, very loose adaptation of the Robbins novel which was co-scripted by Michael V. Gazzo. Songs include "Hard Headed Woman", "Trouble" and the title number. Not one of the star's best films, but his zestful performance carries it along.
Aka: KING CREOLE
MUS   111 min (ort 115 min) B/W
VIDrel: 4-FRONT/POLYREC/BRAVE   V
Boa: novel A Stone For Danny Fisher by Harold Robbins.

ELVIS PRESLEY: KISSIN' COUSINS * U/PG
Gene Nelson USA   1964
Elvis Presley, Arthur O'Connell, Glenda Farrell, Pamela Austin, Yvonne Craig, Jack Albertson, Cynthia Pepper, Donald Woods, Tommy Farrell, Beverly Powers, Hortense Petra, Robert Stone
Some of the locals attempt to thwart plans by the US Air Force to build a missile base on Smokey Mountain, and the military man sent to win them over discovers that one of the hillbillies is his double. A feeble effort with the star given a double role and a collection of generally unremarkable songs. Note that only the two-cassette pack is rated PG.
Aka: KISSIN' COUSINS
MUS   93 min (ort 96 min)   VIDrel: MGM/WHV L/A   V

ELVIS PRESLEY: LIVE A LITTLE, LOVE A LITTLE ** PG
Norman Taurog USA   1968
Elvis Presley, Michele Carey, Don Porter, Rudy Vallee, Dick Sargent, Eddie Hodges, Sterling Holloway, Celeste Yarnall, Joan Shawlee, Mary Grover, Emily Banks, Michael Keller, Merri Ashley, Phyllis Davis, Ursula Menzel, Susan Shute
Standard Presley film with our singer playing a pin-up photographer who manages to land two jobs and does both by running back and forth between offices. An inconsequential film that could have done with a few more songs; there are only four.
Aka: LIVE A LITTLE, LOVE A LITLE
MUS   89 min   VIDrel: MGM/WHV L/A   V/dm

ELVIS PRESLEY: LOVE ME TENDER ** U
Robert D. Webb USA   1956
Richard Egan, Debra Paget, Elvis Presley, Robert Middleton, Neville Brand, William Campbell, Mildred Dunnock, Bruce Bennett, James Drury, Russ Conway, Ken Clark, Barry Coe, L.Q. Jones, Paul Burns, Jerry Sheldon, James Stone
Civil War Western which started Presley's film career. The story is one of conflicting politics in a Southern family and of two sons and their mutual love for Paget. Of interest mainly for Presley's singing of several ballads (complete with anachronistic hip swivelling), rather than any strengths of plot or dialogue. "The Brothers Reno" was the title this film was to have had.
Aka: LOVE ME TENDER
MUS   86 min (ort 89 min) B/W   VIDrel: 20TH/TECH   V
Boa: novel The Brothers Reno by Maurice Geraghty.

ELVIS PRESLEY: LOVING YOU ** (U)
Hal Kanter USA   1957
Elvis Presley, Lizabeth Scott, Wendell Corey, Dolores Hart, James Gleason, Paul Smith, Ken Becker, Jana Lund, Ralph Dumke, Yvonne Lime, Skip Young, Vernon Rich, David Cameron, Grace Hayle, Dick Ryan, Steve Pendleton, Sydney Chatton
A publicist discovers hillbilly Elvis working at a petrol station and signs him up to sing with her husband's band in this glossy but vacuous star vehicle. Forget the plot, shut your eyes, and enjoy the music. Songs include "Teddy Bear" and the title number.
Aka: LOVING YOU
MUS   90 min (ort 101 min)   VIDrel: L/A   V

ELVIS PRESLEY: PARADISE, HAWAIIAN STYLE * U
Michael Moore USA   1966
Elvis Presley, Suzanna Leigh, James Shigeta, Donna Butterworth, Irene Tsu, Marianna Hill, Julie Parrish, Philip Ahn, Mary Treen, Linda Wong, Grady Sutton, Jan Shepard, John Doucette, Don Collier, Doris Packer, Mary Treen, Gigi Verone
Another vehicle for the talents of Presley (unfortunately he has little chance to show his true worth) that's essentially a rehash of BLUE HAWAII with our star a pilot who returns to Hawaii to set up a charter helicopter service. Good-looking but empty.
Aka: PARADISE, HAWAIIAN STYLE
MUS   86 min (ort 91 min)
VIDrel: 4-FRONT/POLYREC/BRAVE   V

ELVIS PRESLEY: ROUSTABOUT * U
John Rich USA   1964
Elvis Presley, Barbara Stanwyck, Joan Freeman, Leif Erickson, Sue Ann Langdon, Raquel Welch, Pat Buttram, Joan Staley, Dabbs Greer, Steve Brodie, Norman Grabowski, Jack Albertson, Jane Dulo, Albert Levy, Joel Fluellen
A roving singer joins a carnival, works hard and finds true love in this Presleyscope offering. Despite good work from the support cast, this is a dull and dismal effort. The song "Little Egypt" offers the one highlight.
Aka: ROUSTABOUT
MUS   95 min (ort 101 min)
VIDrel: 4-FRONT/POLYREC/BRAVE   V

ELVIS PRESLEY: SPEEDWAY ** U
Norman Taurog USA   1968
Elvis Presley, Nancy Sinatra, Bill Bixby, Gale Gordon, William Schallert, Carl Ballantine, Ross Hagen, Victoria Meyerinck, Poncie Ponce, Harry Hickox, Christopher West, Beverly Mills, Harper Carter, Bob Harris, Michele Newman
Routine Presley film set in the world of stock car racing with our star a good-natured racing driver and Sinatra the tax inspector he falls for.
Aka: SPEEDWAY
MUS   90 min (ort 100 min)   VIDrel: MGM/WHV L/A
V/dm

ELVIS PRESLEY: SPINOUT ** U
Norman Taurog USA   1966
Elvis Presley, Shelley Fabares, Diane McBain, Deborah Walley, Cecil Kellaway, Una Merkel, Warren Berlinger, Jack Mullaney, Will Hutchins, Carl Betz, Dodie Marshall, Jimmy Hawkins, Frederic Worklock, Dave Barry
A happy-go-lucky singer is persuaded to test-drive an experimental car in a motor race and finds himself having to fend off the usual admirers. A bland Presley musical with nice open air locations and some of his poorer numbers. Songs include "Beach Shack", "Adam And Evil" and "Smorgasbord".
Aka: CALIFORNIA HOLIDAY; SPINOUT
MUS   90 min (ort 93 min)   VIDrel: MGM/WHV L/A   V/dm

ELVIS PRESLEY: STAY AWAY, JOE * U
Peter Tewksbury USA   1968
Elvis Presley, Burgess Meredith, Joan Blondell, Katy Jurado, Thomas Gomez, Henry Jones, L.Q. Jones, Quentin Dean, Anne Seymour, Angus Duncan, Douglas Henderson, Michael Lane, Susan Trustman, Warren Vanders, Buck Kartalian
A half-breed Indian rodeo rider returns to his reservation to help his people set up as cattle ranchers as part of a government rehabilitation scheme. An unusual setting does little for the star in this lacklustre and dreary offering. Written by Michael A. Hoey.
Aka: STAY AWAY, JOE
MUS   97 min (ort 102 min)   VIDrel: MGM/WHV L/A
V/dm
Boa: novel by Dan Cushman.

ELVIS PRESLEY: THAT'S THE WAY IT IS *** U
Denis Sanders USA   1970
Elvis Presley
An absorbing look at Presley that follows him as he prepares for an opening-night performance at a nightclub during his second

Las Vegas season following his return to the stage in 1969. Hardly an in-depth examination of the star, and the interviews with his followers add nothing of interest to the film, but worth a look if only to capture the flavour of his performance.
Aka: THAT'S THE WAY IT IS
DOC 104 min (ort 108 min) VIDrel: MGM/WHV L/A V

ELVIS PRESLEY: THE FLAMING STAR ***   PG
Don Siegel USA   1960
Elvis Presley, Barbara Eden, Steve Forrest, Dolores Del Rio, John McIntire, Rodolfo Acosta, Karl Swenson, Ford Rainey, Richard Jaeckel, Anne Benton, Douglas Dirk, L.Q. Jones, Tom Reese, Roy Jenson, Virginia Christine, Marian Goldina
A half-breed must choose sides when the Indians go on the warpath. There are no songs after the first ten minutes but a surprisingly strong performance from the star and good action sequences make this one of his best films.
Aka: FLAMING STAR, THE
WES 88 min (ort 110 min) VIDrel: 20TH/TECH V
Boa: novel by Clair Huffaker.

ELVIS PRESLEY: THE TROUBLE WITH GIRLS *** U
Peter Tewksbury USA   1969
Elvis Presley, Marlyn Mason, Nicole Jaffe, Sheree North, Edward Andrews, John Carradine, Vincent Price, Anissa Jones, Joyce Van Patten, Pepe Brown, Dabney Coleman, William Zuckert, Pitt Herbert, Anthony Teague, Med Flory
Presley vehicle with him playing the manager of a travelling medicine show in the 1920s and finding himself involved in murder. Written by Arnold and Lois Peyser this is a superior Presley film with a nice feel for the period and a better-than-average script.
Aka: CHAUTAQUA; TROUBLE WITH GIRLS, THE; TROUBLE WITH GIRLS (AND HOW TO GET INTO IT), THE
MUS 95 min (ort 104 min) VIDrel: MGM/WHV L/A V
Boa: novel The Chautauqua by Day Keene and Dwight Babcock.

ELVIS PRESLEY: TICKLE ME **   PG
Norman Taurog USA   1965
Elvis Presley, Julie Adams, Jocelyn Lane, Merry Anders, Jack Mullaney, Connie Gilchrist
An unemployed rodeo star finds some treasure while abetting a girl's escape from a health farm. The threadbare script serves as a vehicle for the usual selection of pleasant Elvis numbers in a film that is largely unmemorable apart from an unusual climax set in a ghost town. The script is by Elwood Ullman and Edward Bernds, who worked with The Three Stooges earlier on in their careers, but there are precious few laughs here.
Aka: TICKLE ME
MUS 86 min (ort 90 min) VIDrel: POLY/POLYREC L/A V

ELVIS PRESLEY: VIVA LAS VEGAS **   U
George Sidney USA   1963
Elvis Presley, Ann-Margret, Cesare Danova, William Demarest, Jack Carter, Nicky Blair, Robert B. Williams, Bob Nash, Roy Engle, Barnaby Hale, Francis Raval, Ford Dunhill, Eddie Quillan, George Cisar, Ivan Triesault, Mike Ragan
Elvis plays a sports car racing driver with the action being set in the gambling resort of Las Vegas. A routine journey for all concerned, though the colourful locations make it bearable.
Aka: LOVE IN LAS VEGAS; VIVA LAS VEGAS
MUS 81 min (ort 86 min) VIDrel: MGM/WHV L/A V

ELVIS PRESLEY: WILD IN THE COUNTRY *   PG
Philip Dunne USA   1961
Elvis Presley, Tuesday Weld, Hope Lange, John Ireland, Gary Lockwood, Millie Perkins, Rafer Johnson, Christina Crawford, Jason Robards Sr, William Mims, Raymond Greenleaf, Robin Raymond, Doreen Lang, Charles Arnt, Ruby Goodwin
A strange vehicle for Presley in which he plays a country boy with ambitions to be a writer who gets involved with various women and is saved from delinquency by the efforts of a female social worker. Earnest, boring and pointless, with a clutch of good performances largely wasted on the cliched script. Written by Clifford Odets.
Aka: WILD IN THE COUNTRY
DRA 110 min (ort 114 min) B/W VIDrel: 20TH/TECH V
Boa: novel The Lost Country by J.R. Salamanca.

ELVIS SLEPT HERE *   18
Scotty Fox USA   1992
Nikki Dial, Christina Applelay, Alyssa Jarreau, Mona Lisa, Jonathan Morgan, Randy West, Randy Spears, Ted Wilson, Meekah

Spears and Meekah are a couple of tabloid journalists who are vying with other to obtain the best story of the week, this being the ploy devised by their female boss to select which one to keep in a job. But this task has to be completed in twenty-four hours, and mostly revolves around interviews of folk who claim they slept with Elvis Presley (though he was in disguise at the time). A foolish and condescending work, that looks as if it was shot in a single day.
A 60 min VIDrel: MOON/TERRY V

EMANON *   (PG)
Stuart Paul USA   1986
Stuart Paul, Cheryl M. Lynne, Jeremy Miller, Patrick Wright, William F. Collard, Joanne Jackson, Kaye Ballard, Yuda Barkan, Paul Bashkin, Al DiNoble, Billie Wallace, David Donham, Morry Flanebaum, Susannah York, Walt Jordan
A clumsy allegorical saga in which the crippled son of a struggling young widow befriends a strange Christ-like tramp, who appears to have the power to bring happiness to all those about him. What must have seemed a promising idea on paper never comes to life on screen.
DRA 98 min CABrel: HVC

EMBRACE OF THE VAMPIRE **   18
Anne Goursand USA   1994
Alyssa Milano, Jennifer Tilly, Martin kemp, Harrison Pruett, Charlotte Lewis
A young woman college student becomes involved in the dark and mysterious world of vampires and eventually has to decide whether or not she wants to become one herself.
HOR 88 minn (ort 92 min) VIDrel: MED/SONOP V/sh

EMBRYO *   15
Ralph Nelson USA   1976
Rock Hudson, Diane Ladd, Roddy McDowall, Barbara Carrera, Anne Schedeen, John Elerick, Vincent Bagetta, Jack Colvin, Dick Winslowe, Lina Raymond, Joyce Spitz, Joyce Brothers
A scientific experiment gets out of hand as a foetus grows into an adult woman in just four and a half weeks. An unpleasant fantasy that borrows heavily from the genre of horror films, especially with regard to the repulsive special effects. See also SPECIES.
Aka: CREATED TO KILL
FAN 104 min (ort 110 min) VIDrel: SCRN/DISC V

EMILIENNE **   18
Guy Casaril FRANCE   1975
Pierre Oudrey, Betty Mars, Nathalie Guerin
The story of a menage-a-trois involving a husband and his wife and mistress in which the man encourages the formation of a lesbian relationship between the two women. An offbeat erotic drama.
Aka: EROTIC MENAGE A TROIS
DRA 89 min (ort 94 min) dubbed VIDrel: ARTPRO/RTM V

EMINENT DOMAIN **   (PG)
John Irvin CANADA/POLAND/UK/USA   1991
Donald Sutherland, Anne Archer, Johdi May, Paul Freeman, Bernard Hepton, Francoise Machaud, Yves Beneyton, Denis Fouqueray, Pip Torrens, Jan Peszek, Tadeusz Bradecki, Marcin Tronski, Joanna Jedryka, Krystyna Froelich
A sluggish and moody thriller set in pre-Solidarity Poland circa 1979, where a high-ranking Politburo official living in Warsaw finds that all his power and privileges have been removed for no apparent reason. Occasional flashes of Kafkaesque paranoia shine through in a film whose flabby and unsatisfying storyline takes very much second place to the atmospheric Warsaw locations.
THR 102 min VIDrel: FIRST L/A V

EMMA ***   U
John Glenister UK   1972
John Carson, Constance Chapman, Donald Eccles, Doran Goodwin
Austen's story of a rich and spoilt young woman who loves meddling in the lives of others is given the usual slightly over-reverent BBC adaptation treatment. Eminently watchable if a touch uninvolving.
DRA 256 min (2 cassettes) mTV VIDrel: BBC/TECH V/h
Boa: novel by Jane Austen.

EMMA ***   U
Diarmuid Lawrence UK   1996
Kate Beckinsale, Samantha Morton, Mark Strong, Bernard Hepton,

Samantha Bond, Dominic Rowan, Prunella Scales, James Hazeldine, Raymond Coulthard

A spoilt young girl lives with her father in a small village, and amuses herself by interfering in the lives of others, not always achieving the results she intended. A full-blooded and vigorous tale of 19th century country life, one of two such films to emerge in the same year. Screenplay is by Andrew Davies, who worked on the BBC's PRIDE AND PREJUDICE.
DRA 103 min (ort 108 min) wScrn VIDrel: CTE/CARL V
Boa: novel by Jane Austen.

EMMA ***   U
Douglas McGrath UK/USA   1996
Gwyneth Paltrow, Toni Collette, Alan Cumming, Ewan McGregor, Jeremy Northam, Greta Scacchi, Juliet Stevenson, Polly Walker, Sophie Thompson, James Cosmo, Denys Hawthorne, Phyllida Law, Kathleen Byron, Edward Woodall, Brett Miley

Tight direction and a most literate script are strong assets in this careful adaptation of Austen's novel of a brash matchmaker, whose penchant for meddling in the lives of her friends is not tempered by any conception of the harm she does. Northam is rather good as Knightly, one of the more dashing figures in the story, but it is really Paltrow as the title character who makes her mark here.
DRA 116 min (ort 121 min) cC VIDrel: TOUCH/TECH V
Boa: novel by Jane Austen.

EMMANUELLE ***   18
Just Jaeckin FRANCE   1974
Sylvia Kristel, Alain Cuny, Marika Green, Daniel Sarky, Jeanne Colletin, Christine Boisson, Samantha, Gaby Brian, Gregory

Supposedly based on Arsan's own experiences, this early soft-core film has had many sequels. The story opens in Bangkok, where Emmanuelle arrives with her diplomat husband after his transfer there. Believing that his wife should experience the full range of sexuality, her husband encourages her to embark on some extramarital affairs. She does so in a beautifully made but utterly empty and loveless study of sexuality and decadence.
A 89 min Cut (1 min 25 sec - ort 92 min)
VIDrel: 4-FRONT/POLYREC/BRAVE V
Boa: book by Emmanuelle Arsan.

EMMANUELLE 2 **   18
Francis Giacobetti FRANCE   1975
Sylvia Kristel, Umberto Orsini, Catherine Rivet, Frederic La Gache, Henry Czarniak, Marion Womble, Tom Clark, Florence Lafuma, Claire Richard, Laura Gemser, Jacqueline May Line, Eva Hampel, Christiane Gibelin, Jean-Pierre Nam

Bangkok is replaced by Hong Kong and Bali in the further adventures of our erotically obsessed heroine. After some encounters on the steamer she gets off at Hong Kong and quickly embarks on a rapid succession of liaisons, mostly with just about anyone who's available. Made with some care, this largely plotless film explores a philosophy of sex totally divorced from notions such as love; the result is a glossy exercise in sleaze.
Aka: EMMANUELLE L'ANTIVIERGE; EMMANUELLE, THE JOYS OF A WOMAN
A 87 min Cut (30 sec - ort 100 min)
VIDrel: 4-FRONT/POLYREC/BRAVE V

EMMANUELLE 3 **   18
 FRANCE
Silvia Castell, Brigette Lahaie, Jean-Marie Pallady

A glamour photographer gets an assignment that takes him to the Orient with two beautiful French models. However, all is not really as it appears to be, for our photographer is in reality a spy. A turgid softcore film not really related to the others in the series.
A 78 min VIDrel: TCX L/A V

EMMANUELLE 4 *   18
Francis Giacobetti FRANCE   1984
Sylvia Kristel, Mia Nygren, Deborah Power, Patrick Bauchau, Sophie Berger, Sonia Martin

One of those endless EMMANUELLE films with our queen of love fleeing from a cruel lover and undergoing cosmetic surgery to emerge with a completely new identity (played by a younger actress). If only she could emerge with a new script. A ludicrous softcore exercise in credibility-stretching.
A 85 min Cut (10 sec plus film cuts - ort 89 min)
VIDrel: L/A V

EMMANUELLE 5 **   18
Steve Barnett/Walerian Borowczyk FRANCE   1986
Monique Gabrielle, Alex Cunningham, Crofton Hardester, Yaseen Khan, Dana Burns Westberg, Harold Kay, Marie Chocolat, Marie Vanille, Isabelle Strawa, Muriel Catan, Jessica Stehl, Peter Lowell, Noelle Fabianni, Francois Clavier

Action erotic thriller set against the backdrop of Cannes and the Far East. After a screening of Emmanuelle's latest film - "Love Express" - she is hounded by the press and runs away with millionaire playboy Charles Foster. Then she is kidnapped by the sinister Rajid, who sets her up as part of his harem. Can Foster rescue her? Now read on...
A 76 min Cut (2 min 41 sec - ort 85 min)
VIDrel: VCC/DISC/COLUM V

EMMANUELLE 6 *   18
Bruno Zincone FRANCE   1988
Natalie Uher, Jean-Rene Gossart, Luis Carlos Mendes, Thomass Obermuller, Gustavo Rodriguez, Hassan Guerrar, Tamira, Dagmar Berger, Edda Kopke, Rania Raja, Ilena D'Arcy, Melissa, Virginie Constantin, Christele Merault

After suffering a variety of hardships in the Amazonian jungle, Emmanuelle receives therapy at a private clinic in a bid to help her regain her memory. This most insubstantial of plots serves as little more than a vehicle for the usual repetitive softcore encounters, all beautifully photographed in a style that resembles nothing so much as a TV commercial.
A 75 min (Cut at film release by 2 sec)
VIDrel: MIA/DISC/IMPENT V

EMMANUELLE FOREVER *   18
Francis Leroi FRANCE   1992
George Lazenby, Sylvia Kristel, Marcella Walerstein, Natala Sevenants, Gerda De Haan, Christine Merlet, Corinne Mafiodo, Evelyne Ruth, Carl Otto Olson, Pham Duc Tu, Huun Daniel Meas, Joel Bui, Jay Hausman, Vibbe Hangaard, Tony Senegal

First in this made-for TV erotic series whose hallmark is its tastefully filmed sex scenes. Our heroine returns to Bangkok after twenty years to find her great love and meets him on the plane but he fails to recognise her. Soon, we learn that a magic perfume given to her in Tibet keeps her young and helps her on her quest as the incarnation of all women. A couple of episodes now follow to show how she uses her magic powers to help the lovelorn. Softcore nonsense.
Aka: ETERNELLE EMMANUELLE
A 86 min (ort 90 min) mCab VIDrel: CREA/DISC V

EMMANUELLE IN AMERICA ***   18
Joe D'Amato (Aristide Massaccesi) ITALY   1979
Laura Gemser, Gabriele Tinti, Roger Browne, Riccardo Salvino, Lars Bloch, Paola Senatore, Maria Piera Begoli, Giulio Bianchi, Efrem Appel, Matilde Dall'Anglio, Carlo Foschi, Maria Renata Franco, Giulio Massimini

In this well made EMMANUELLE sequel Gemser replaces Kristel, playing a photographer and investigative journalist. In her first encounter she uncovers a drug dealer who runs a private harem and later she moves on to Venice, where she meets a count who deals in art forgeries. Other adventures follow, and in each one D'Amato treats us to a strong dose of sexuality and decadence. Cut before video submission by 13 min 3 sec.
Aka: EMANUELLE AROUND THE WORLD; EMANUELLE IN AMERICA; EMMANUELLE NEGRA IN AMERICA
A 85 min Cut (5 min 49 sec) dubbed VIDrel: LUMI/SPEAR V

EMMANUELLE IN SEVENTH HEAVEN *   18
Francis Leroi FRANCE   1992
Sylvia Kristel, Caroline Laurence, Laura Dean, Annie Bellac, Cynthia Van Damme, Julie Jalabert, Roland Waden, Robert Maloe, Joel Bui, Jerome Estienne, Gregoire Wojciechowski, Jean-Philippe Bech, Jean-Marc Vasseur

Good old Emmanuelle introduces her best friend to a man who has enjoyed a most varied sex life, in her new job, which involves her in running a therapy centre at a French chateau. No prizes for guessing how the plot develops, though this one does at least try to keep topical (computers are used at the centre to create virtual reality sex fantasies).
Aka: EMMANUELLE 7; EMMANUELLE SEVEN; EMMANUELLE'S 7TH HEAVEN
A 79 min (ort 90 min) VIDrel: GUILD/SONOP V

EMMANUELLE'S LOVE * 18
Francis Leroi FRANCE 1992
George Lazenby, Sylvia Kristel, Marcella Walerstein, Gilles Brunell, Julie McLaughlin, Luciana Gratival, Babette Esposito, Jean-Marie Picard, Joel Bui, Joy Hausman, Alain Saint-Alix, Pham Duc-Tu, Hunn Daniel Meas, Vibbe Hangaard
Second in the series sets the pattern for the rest of the entries, with Emmanuelle telling her old lover about her adventures. Here, she travels first to Hong Kong to help an old friend seduce her boss and then on to India where the happy couple are taking a honeymoon train trip. There, a friend disfigured in a car accident needs her help to deal with a greedy gigolo. Lastly, in San Francisco her passion for her friend's husband leads her astray.
Aka: L'AMOUR D'EMMANUELLE
A 88 min (ort 90 min) mCab VIDrel: CREA/DISC V

EMMANUELLE'S MAGIC * 18
Francis Leroi FRANCE 1993
Sylvia Kristel, George Lazenby, Krystina Ferentz, Cynthia Van Damme, Pascale Cardan, Didier Wind, Pascal Legrand, Blake Dawson, Corrado, Melanie Moore, Joel Bui, Jay Hausman, Pham Duc Tu, Hunn Daniel Meas, Vibbe Hangaard, Tony Senegal
This adventure is set in Greece where Emmanuelle and a friend are touring the islands where the latter meets up with an old friend, now married to a local man, a sculptor who works with modern materials and has now fallen in love with one of his own creations (just like Pygmalion). Here, some of the soft-core scenes deal with the relationship between these two women friends when they share a room as students. Some nice locations help relieve the tedium.
Aka: LA MAGIE D'EMMANUELLE
A 90 min mCab VIDrel: IMC/DISC V

EMMANUELLE'S PERFUME * 18
Francis Leroi FRANCE 1992
George Lazenby, Sylvia Kristel, Marcella Walerstein, Jean-Pierre Rochette, Cecile Fleury, Dorothee Picard, Julie McLaughlin, Catherine Audray, Steeve, Joel Bui, Pierre Francois Lambert, Emmanuel Marc, Jay Hausman, Pham Duc-Tu .
On her interminable plane journey, Emmanuelle tells her travelling companion about her adventures at the Cannes Film Festival, where she did a favour for a film producer. This involved luring back to civilisation a writer friend who had written the novel, that formed the basis for a film that had been nominated for the Golden Palm award. Armed with a plane ticket and ample cash, Emmanuelle comes to find the writer at his island hideaway.
Aka: LE PARFUM D'EMMANUELLE
A 90 min mCab VIDrel: IMC/DISC V

EMPEROR OF THE BRONX, THE ** 18
Joseph Merhi USA 1988
Alex D'Andrea, Anthony Gioia, William Smith, Licia Sukary, Charlie Gais
A man tries to escape his past as a member of the New York Mafia but is followed to L.A. by his former boss, whom he faces in a final lone showdown.
A/AD 90 min VIDrel: MOPIC/QUANT V

EMPEROR OF THE NORTH *** 15
Robert Aldrich USA 1973
Lee Marvin, Ernest Borgnine, Keith Carradine, Charles Tyner, Harry Caeser, Malcolm Atterbury, Simon Oakland, Matt Clark, Hal Baylor, Elisha Cook Jr, Joe Di Reda, Liam Dunn, Daine Dye, Robert Foulk, James Goodwin, Ray Guth, Sid Haig
A duel to the death between a sadistic train conductor who kills any hobo trying to steal a free ride on his train, and a legendary tramp who claims that he will be the first to succeed and live to tell the tale. Surprisingly effective, with great photography by Joseph Biroc. The script is by Christopher Knopf. Filmed in Oregon.
Aka: EMPEROR OF THE NORTH POLE
A/AD 120 min (ort 131 min) VIDrel: 20TH/TECH V

EMPIRE OF THE SUN ** PG
Steven Spielberg USA 1987
Christian Bale, John Malkovich, Miranda Richardson, Nigel Havers, Leslie Phillips, Joe Pantoliano, Masato Ibu, Emily Richard, Rupert Frazer, Ben Stiller, Robert Stephens, Burt Kwouk, Peter Gale, Paul McGann
A British boy living a comfortable life in Shanghai is separated from his parents following the Japanese invasion of China at the start of WW2 and winds up in an internment camp where

he has to fend for himself. This good opening develops into a confused and manipulative tale full of melodramatic set-pieces and emotion-charged episodes. The over-bearing score is by John Williams in this irritating adaptation of Ballard's autobiographical novel.
DRA 146 min VIDrel: WHV V/sur
Boa: novel by J.G. Ballard.

EMPIRE RECORDS ** 12
Allan Moyle USA 1995
Anthony LaPaglia, Maxwell Caulfield, Debi Mazar, Rory Cochrane, Johnny Whitworth, Robin Tunney, Renee Zellweger, Ethan Randall, Coyote Shivers, Brendan Sexton, Liv Tyler, James "Kimo" Wills, Ben Bode, Gary Bolen, Kimber Monroe
The lives and loves of a bunch of youngsters who work in a record store, and the difficulties the manager has to cope with in the face of a corporate takeover bid. The cast (most of whom are unknowns) is uniformly excellent, as is the soundtrack (which features David Bowie among others). Not at all like CLERKS, this is a safe, chirpy, non-provocative effort, that moves one to smiles if not to chuckles.
COM 86 min (ort 91 min) VIDrel: WHV V/sur

EMPIRE STRIKES BACK, THE **** U
Irvin Kershner USA 1980
Mark Hamill, Harrison Ford, Carrie Fisher, Billy Dee Williams, Dave Prowse, Anthony Daniels, Peter Mayhew, Kenny Baker, Alec Guinness, Jeremy Bulloch, Frank Oz, John Hollis plus voices of: James Earl Jones, Clive Revill
First sequel to STAR WARS but this time the tone is a lot more serious in a film that is both sadder and more mature in its development of this saga, though it does assume one has seen the first film. Followed by RETURN OF THE JEDI. AA: Sound (Bill Varney/Steve Maslow/Gregg Landaker/Peter Sutton), Spec Award (Brian Johnson/Richard Edlund/Dennis Muren/Bruce Nicholson for visual effects).
FAN 119 min (ort 124 min); 126 min (special edition) wScrn VIDrel: 20TH/TECH V/drm

EMPTY CRADLE *** 15
Paul Schneider USA 1993
Kate Jackson, Lori Loughlin, Eriq la Salle, David Lansbury, Jonah Blechman, Karmin Murcello, Peter Rook, Penny Johnson, Michelle Joyner, Ricardo Gutierrez, Walter Addison, Skye Basstee, Camilla Belle, Wendy Bower, Zachary Browne
A mother gives birth to her third child, and is then informed that it died in delivery, but grows suspicious of another woman, whom she becomes convinced has snatched her child (an act she carried out after faking pregnancy in a bid to keep her boyfriend). However, she now has to find a way of proving that her suspicions are well founded. A story that would chill the heart of any parent this was, unbelievable as it might sound, based on a real-life case.
DRA 87 min VIDrel: ODY/SONOP V

ENCHANTED APRIL *** U
Mike Newell UK 1991
Josie Lawrence, Miranda Richardson, Alfred Molina, Jim Broadbent, Michael Kitchen, Joan Plowright, Polly Walker, Stephen Beckett, Matthew Radford, Davide Manuli, Vittorio Duse, Adriana Facchetti, Anna Longhi
Period drama set in the 1920s when four prim and proper Englishwomen spend the winter in the warmer climes of Tuscany, escaping both the weather at home and for two at least, their stifling marriages. A gentle tale of re-awakening, spoilt by poor casting and a script that really fails to capture the depth, honesty and charm of the original novel. Originally produced for the BBC Screen Two series, it was later re-cut for cinema release.
DRA 89 min (ort 99 min) mTV cC VIDrel: WDV/TECH V/sur
Boa: novel by Elizabeth von Arnim.

ENCHANTMENT ** U
Irving Reis USA 1948
David Niven, Teresa Wright, Evelyn Keyes, Farley Granger, Jayne Meadows, Leo G. Carroll, Philip Friend, Shepperd Strudwick, Henry Stephenson, Peter Miles, Colin Keith-Johnston, Gigi Perreau, Sherlee Collier, Warwick Gregson
A man loses his one chance of love when young because of the insane jealousy that his sister feels towards his intended, an orphan who was taken into their household some years before. Fifty years later, in the midst of WW2, he is able to use his expe-

rience to prevent his granddaughter from missing her chance of love. A well-photographed and nicely directed piece of little consequence.
DRA   101 min   VIDrel: VCC/DISC   V
Boa: novel Take Three Tenses by Rumer Godden.

ENCOUNTER AT RAVEN'S GATE *   15
Rolf De Heer/Marc Rosenberg AUSTRALIA   1987
Celine Griffin, Steven Vidler, Ritchie Singer, Vincent Gil, Saturday Rosenberg, Terry Camilleri, Max Cullen, Peter Douglas, Ernie Ellison, Brian O'Connor, Ruth Goble, Max Lorenzin, Phil Bitter
Atmospheric but leaden and confusing account of strange events in the Outback. Power supplies come and go, family pets savage their owners and both murder and madness are rife, but in the end nothing is explained or resolved. A poor effort indeed.
HOR   85 min   VIDrel: COLUM/SONOP   V

ENCOUNTERS OF THE SPOOKY KIND **   18
Samo Hung HONG KONG   199-
Samo Hung, Chung Fat, Chun Lung
A martial arts fighter gets caught up between two powerful wizards in one of those odd hybrid films that blends elements of two genres with moderate success. Followed by SPOOKY ENCOUNTERS, which offered more of the same.
MAR   98 min wScrn   VIDrel: MADE/RTM   V

END OF ST PETERSBURG, THE ***   PG
Vsevolod I. Pudovkin USSR   1927
Ivan Chuvelov, Vera Baranovskaya, A.P. Christiakov, V. Obolenski, Sergei Komarov
A peasant comes to Saint Petersburg in 1914 and becomes a strike-breaker during a dispute, but eventually comes to side with the workers and takes part in the Revolution. A stirring propaganda piece officially commissioned to mark the tenth anniversary of the 1917 October Revolution, it works both as a savage attack on inequality and an effective justification for the sweeping away of Kerensky's provisional government by the Bolsheviks.
Aka: KONYETS SANKT-PETERBURGA
DRA   68 min (ort 122 min) B/W silent wScrn
VIDrel: TART/20TH   V

END OF THE GOLDEN WEATHER, THE ***   PG
Ian Mune NEW ZEALAND   1991
Stephen Fulford, Stephen Paps, Paul Gittins, Gabrielle Hammond, Alexandra Marshall, David Taylor, Ray Henwood, Steven McDowell, Alice Fraser, Alistair Douglas, Bill Johnson, Greg Johnson, Alison Bruce, Francie Gray, Andrea Kelland
A young boy, growing up in the New Zealand countryside in the 1950s, makes friends with a simple-minded recluse who is convinced he will be chosen to represent his country in the forthcoming Olympics. However, this innocent friendship soon lands him in trouble with his rather stern father. A touching coming-of-age tale of strong performances and careful direction.
DRA   98 min (ort 103 min)   VIDrel: SCRN/TERRY   V
Boa: play by Bruce Mason.

END OF THE LINE **   PG
Jay Russell USA   1987
Wilford Brimley, Levon Helm, Mary Steenburgen, Barbara Barrie, Kevin Bacon, Holly Hunter, Bob Balaban, Clint Howard, Rita Jenrette, Howard Morris, Bruce McGill, Trey Wilson
Two buddies lose their jobs on the Southern railroads when their parent company moves over to air freight, so they take off for the corporate headquarters in Chicago in a stolen engine. An offbeat comedy-drama that quickly runs off its own rails, degenerating into little more than a silly farce.
DRA   103 min (ort 105 min)   VIDrel: VIDRI/GROS L/A   V

ENDGAME **   18
Steven Benson (Aristide Massaccesi) ITALY   1983
Joe Spencer, Moira Chen, Jill Elliott, Al Cliver (Pier Luigi Conti), George Eastman (Luigi Montefiore), Jack Davis, Al Yamanouchi, Mario Pedone, Gordon Mitchell, Gus Stone, Al Waterman, Bobby Rhodes, Nat Williams, David Brown
Another post-nuclear holocaust film set in New York where a group of mutants try to escape to join a new civilisation.
Aka: ENDGAME: BRONX LOTTA FINALE; ENDGAME: FINAL BRONX STRUGGLE; ENDGAME, GIOCCIO FINALE; ENDGAMES
DRA   97 min Cut (41 sec)   VIDrel: MOPIC/QUANT   V

ENDLESS NIGHT **   15
Sidney Gilliat UK   1971
Hayley Mills, Hywel Bennett, Britt Ekland, George Sanders, Per Oscarsson, Lois Maxwell, Peter Bowles, Aubrey Richards, Ann Way, Patience Collier, Madge Ryan, Walter Gotell, Helen Horton, David Bauer, Geoffrey Chater
A chauffeur marries a rich American heiress and they move into a stately home only to find it turning into a nightmare mansion. A creaky British thriller of weak plotting and worse characterisation.
Aka: AGATHA CHRISTIE'S ENDLESS NIGHT
THR   95 min (ort 98 min)   VIDrel: WHV L/A   V
Boa: novel by Agatha Christie.

ENEMIES, A LOVE STORY ****   15
Paul Mazursky USA   1989
Anjelica Huston, Ron Silver, Lena Olin, Margaret Sophie Stein, Alan King, Judith Malina, Rita Karlin, Phil Leeds, Elya Baskin, Paul Mazursky, Zypora Spaisman, I.J. Dollinger, Arthur Grosser, Burney Lieberman, Gayle Garfinkle
A Jewish intellectual who escaped the Holocaust lives a strange double life, being married to his wartime protector and yet carrying on a relationship with another woman. The re-appearance of his first wife, whom he thought dead, complicates matters somewhat further. Set in New York in 1949, this strange blend of comedy, pathos and drama is as poignant as it is unusual. The offbeat script is by Mazursky and Roger L. Simon with music by Maurice Jarre.
DRA   114 min (ort 121 min)   VIDrel: 20VIS/SONOP L/A
V
Boa: novel by Isaac Bashevis Singer.

ENEMY BELOW, THE ***   PG
Dick Powell USA   1957
Robert Mitchum, Curt Jurgens, Theodore Bikel, Doug McClure, Russell Collins, Al Hedison, Kurt Kreuger, Frank Albertson, Alan Dexter, Biff Elliott, Jeff Daley, David Blair, Joe Di Reda, Ralph Manza, Robert Boon, Werner Reichow
An American destroyer on an anti-submarine patrol in the South Atlantic plays a cat-and-mouse game with a U-boat commanded by a wily German. An excellent WW2 action film, tense, literate and well directed; and one that portrays the Germans (with one exception) as human beings rather than as caricatures. The underwater effects won a well deserved Oscar. AA: Effects (Walter Rossi).
WAR   93 min (ort 98 min)   VIDrel: 20TH/TECH   V/h
Boa: novel Escort by Commander D.A. Rayner.

ENEMY MINE ***   15
Wolfgang Peterson USA   1985
Dennis Quaid, Louis Gossett Jr, Brion James, Richard Marcus, Lance Kerwin, Carolyn McCormick, Bumper Robinson, Jim Mapp, Scott Kroft, Lou Michaels, Andy Geer, Henry Stolow, Herb Andress, Danmar Wise, Mandy Hausenberger, Emily Woods
An Earthman and his alien enemy are both stranded on a hostile planet, where they are gradually forced to abandon their struggle in order to survive. Highly reminiscent of HELL IN THE PACIFIC with considerable effort being made to portray the alien as an intelligent sentient being rather than a monster. A workmanlike and original film with good special effects its best feature.
FAN   89 min (ort 108 min)
VIDrel: 20TH/TECH; ENCORE (LV only)   V/sur LV

ENEMY OF THE PEOPLE, AN ***   U
George Schaefer USA   1977
Steve McQueen, Charles Durning, Bibi Andersson, Richard Bradford, Eric Christmas, Robin Pearson Rose, Richard A. Dysart, Michael Higgins, Ham Larsen, John Levin, Michael Christofer
A small-town doctor risks his livelihood when he reveals that the water in the local spa has been contaminated by waste from a tannery, a fact the town council is keen to conceal. Adapted from the play by Henry Miller, this sincere and plodding film has a fine performance from McQueen let down by the stilted manner in which it develops. Nevertheless, as an example of one of McQueen's more interesting roles it certainly deserves a look.
DRA   102 min   VIDrel: MGM/WHV L/A   V
Boa: play by Henrik Ibsen.

ENEMY OF THE PEOPLE, AN **   U
Satyajit Ray INDIA   1984
Soumitra Chatterjee, Dhritiman Chatterjee, Ruma Guhathakurta, Mahata Shankar

Written by Ray, the Ibsen tale of a dedicated doctor and his clash with the authorities has been carefully transposed to West Bengal, where our doctor antagonises various vested interests when he reveals that the holy water from the town's temple is contaminated, and has caused an epidemic of jaundice. The director's illness restricted him to a studio setting for this film, which sadly becomes little more than a stilted and wordy rendition.
Aka: GANASHATRU
DRA   95 min   VIDrel: CONNO/RTM   V
Boa: play by Henrik Ibsen.

ENEMY TERRITORY *
Peter Manoogian USA
Gary Frank, Ray Parker Jr, Jan-Michael Vincent, Tony Todd, Peter Teschner, Stacey Dash, Frances Foster, Dean Richmond, Tiger Haynes, Charles Randall, Peter Wise, Robert Lee Rush, Lynnie Godfrey, Teddy Abner, Tad Truesdale
A vicious and fairly mindless look at what happens when a white insurance salesman annoys a nasty black gang on a visit to their ghetto and finds himself trapped in a building by them.
THR   85 min (ort 89 min)   VIDrel: EIV/SONOP   V

18
1976

ENEMY WITHIN, THE **
Jonathan Darby USA
Forest Whitaker, Jason Robards, Sam Waterson, Dana Delany, Josef Sommer, George Dzundza, Isabel Glasser, Dakin Matthews, William O'Leary, Lisa Summerour, Rory Aylward, Greg Brickman, David Combs, Patricia Donaldson, Densie Dowse
Massive cuts in defence spending fuel growing resent in America's armed forces and a general conspires with the vice president and the secretary of defence to engineer a coup. However, they are opposed by a loyal solder who honours his oath to defend the Constitution. A competent remake (of sorts) of SEVEN DAYS IN MAY but lacking very many of that film's superb qualities. See also TWILIGHT'S LAST GLEAMING.
THR   86 min   VIDrel: MOSAIC/COLUM   V/sh

15
1994

ENFORCER, THE ***
Bretaigne Windust USA
Humphrey Bogart, Zero Mostel, Ted De Corsia, Everett Sloane, Roy Roberts, King Donovan, Patricia Joiner, Lawrence Tolan, Bob Steele, Adelaide Klein, Don Beddoe, Tito Vuolo, John Kellogg, Jack Lambert, Susan Cabot, Alan Foster
Bogart plays a crusading D.A. out to crush a criminal gang led by Sloane in this gritty and realistic gangster movie that is without doubt one of the director's best efforts.
Aka: MURDER, INC.
THR   86 min B/W   VIDrel: 4-FRONT/POLYREC   V

PG
1950

ENFORCER, THE **
James Fargo USA
Clint Eastwood, Tyne Daly, Harry Guardino, Bradford Dillman, John Mitchum, DeVeren Bookwalter, John Crawford, Albert Popwell
The second sequel to DIRTY HARRY, with our tough cop coping with a female partner as well as an underground terrorist group. A violent and simplistic celebration of macho cops and big guns, but undeniably slick. Followed by SUDDEN IMPACT.
A/AD   92 min (ort 97 min)   VIDrel: WHV   V/sh

18
1976

ENGLISH PATIENT, THE ***
Anthony Minghella UK
Ralph Fiennes, Kristin Scott Thomas, Juliette Binoche, Naveen Andrews
Badly burnt in a WW2 plane crash, a Hungarian lies in a Tuscany monastery, where whilst recovering from his injuries, he recollects an adulterous affair, and the intrigue and pain this led to. Meanwhile, his Canadian nurse enjoys a love affair of her own. A detailed and opulent film, of pretentious dialogue and beautiful photography, it swept the board at the Academy Awards. AA: Pic, S. Actress (Binoche), Dir, Art, Cin, Sound, Score/orig, Cost, Edit.
DRA   162 min   CINrel
Boa: novel by Michael Ondaatje.

15
1996

ENGLISHMAN WHO WENT UP A HILL, BUT CAME DOWN A MOUNTAIN, THE **
Christopher Monger UK
Hugh Grant, Ian McNeice, Colm Meaney, Tara Fitzgerald, Kenneth Griffith, Ian Hart, Tudor Vaughn, Hugh Vaughn, Robert Pugh, Robert Blythe, Garfield Morgan, Lisa Palfrey, Dafydd Wyn Roberts, Iuean Rhys, Anwen Williams, Fraser Cains
During WW1, two English cartographers come to a Welsh

PG
1995

village on a mapping survey and promptly proclaim that a local landmark is sixteen feet too short to be officially listed as a mountain. With local pride badly dented, the villagers decide to hold the surveyors there until matters can be remedied. Not very funny (for all the media hype that accompanied its release) and as ever Grant is type-cast as a repressed and tongue-tied twit, a part he plays as if born to it.
COM   92 min (ort 99 min) cC   VIDrel: TOUCH/TECH   V/sh

ENIGMA *
Jeannot Szwarc FRANCE/UK
Martin Sheen, Brigitte Fossey, Sam Neill, Derek Jacobi, Michel Lonsdale, Frank Finlay, David Baxt, Kevin McNally, Michael Williams, Warren Clarke, Vernon Dobtcheff
The KGB send five of their top killers to silence five Soviet dissidents and a CIA agent tries to stop them. Never more than mediocre, though an excellent cast do their best. The script is by John Briley.
THR   97 min (ort 122 min)
VIDrel: POLY/POLYREC/BRAVE   V
Boa: novel Enigma Sacrifice by Michael Barak.

15
1982

ENIGMA OF KASPAR HAUSER, THE ****
Werner Herzog WEST GERMANY
Bruno S, Walter Ladengast, Brigitte Mira, Hans Mursaus, Willy Semmelrogge, Michael Kroechner, Henry Van Lyck
In the 1820s a man appeared outside the gates of Nuremberg who stated that he had been kept in confinement from an early age. This mystery has never been solved though it is thought that he may have been of royal birth. This film presents an engrossing look at the world as seen through the eyes of this figure, his years of isolation apparently having given him a dispassionate and unique view of the world. See also KASPAR HAUSER.
Aka: EVERY MAN FOR HIMSELF AND GOD AGAINST ALL; JEDER FUR SICH UND GOTT GEGEN ALLE; KASPAR HAUSER; LEGEND OF KASPAR HAUSER, THE; MYSTERY OF KASPAR HAUSER, THE
DRA   105 min (ort 110 min)   VIDrel: TART/20TH   V

15
1975

ENTANGLED **
Max Fischer FRANCE/USA
Judd Nelson, Pierce Brosnan, Laurence Treil, Roy Dupuis, Max Von Sydow, Lorenzo Caccialanza, Lucie Ganon, Michael McGill, Dorian Joe Clark, Bernadette Li, Christina Chase, Charlie Chagnon, Danielle Bissonette, Alexander Chapman
A struggling writer about to achieve a breakthrough meets a beautiful woman with whom he falls in love and, blinded by passion, discovers too late that he has been framed for murder. A less than enthralling adaptation of a work from a couple of talented thriller writers.
Aka: LES VEUFS
THR   90 min   VIDrel: FIRST/SONOP   V
Boa: novel Les Veufs by Pierre Boileau and Thomas Narcejac.

18
1992

ENTER THE DRAGON ***
Robert Clouse HONG KONG/USA
Bruce Lee, John Saxon, Jim Kelly, Ahna Capri, Yang Tse, Angela Mao, Shih Kien, Bob Wall, Betty Chung, Geoffrey Weeks, Peter Archer
A martial arts fighter infiltrates to be held on an island in an attempt to prevent opium smuggling on behalf of British Intelligence. The foolish plot is best ignored in the face of some excellent martial arts sequences. This was Lee's last complete film role. The music is by Lalo Schifrin.
MAR   94 min Cut (1 min 45 sec); 102 min wScrn (special edition)   VIDrel: WHV   V

18
1973

ENTER THE NINJA **
Menahem Golan USA
Franco Nero, Susan George, Sho Kosugi, Alex Courtney, Will Hare, Zachi Noy, Dale Ishimoto, Christopher George, Jonee Gambroa, Constantin De Goguel, Leo Martinez, Ken Metcalfe, Subas Herrero, Alan Amiel, Bob Jones, Derek Webster
Typical kung fu action film with Nero battling it out with a couple of old enemies. Fairly mediocre. NINJA 2: THE REVENGE OF THE NINJA followed.
MAR   91 min (Cut at film release - ort 94 min)
VIDrel: MIA/DISC/COLUM   V

18
1981

ENTER THE SHOOTFIGHTER **
Richard W. Munchkin USA
Michael Worth, Matyhias Hues, Marshall Teague, Eric Lee, Jenilee Harrison, Connie Llanos, Nick Hill, Bela Lehoczki, Sam Jones
A martial arts expert learns that his best friend was beaten to

18
1994

death in an illegal kickboxing tournament, and he sets out to destroy both the crooked promoter and the fighter he holds responsible.
A/AD   92 min   VIDrel: COLUM/SONOP   V/sur

ENTERTAINER, THE ****
PG
Tony Richardson UK
1960
Laurence Olivier, Joan Plowright, Roger Livesey, Brenda De Banzie, Alan Bates, Albert Finney, Thora Hird, Shirley Ann Field, Daniel Massey, Miriam Karlin, Tony Longridge, Macdonald Hobley, Charles Gray, Geoffrey Toone
Olivier gives a truly remarkable performance in this depressing tale of an obnoxious seaside music-hall entertainer, whose pathetic attempts to stave off the end of his mediocre career are doomed to failure. A stilted but poignant character study, with the star repeating his earlier stage success. The screenplay is by Osborne. Remade for TV (with Jack Lemmon in the title role) in 1975.
DRA   99 min B/W   VIDrel: 4-FRONT L/A   V
Boa: play by John Osborne.

ENTERTAINING MR SLOANE ***
15
Douglas Hickox UK
1969
Beryl Reid, Harry Andrews, Peter McEnery, Alan Webb
Bizarre black comedy about a handsome young layabout who becomes inveigled into an amorous relationship with a sex-starved middle-aged woman whilst at the same time leading on the woman's homosexual brother and contending with hostility from their aged and suspicious father. A patchy adaptation of a witty stage-play with good performances and a few amusing moments, but very little of real substance.
COM   90 min (ort 94 min)   VIDrel: WHV   V
Boa: play by Joe Orton.

EQUINOX **
15
Alan Rudolph USA
1993
Matthew Modine, Marisa Tomei, Lara Flynn Boyle, Lori Singer, Fred Ward, Tyra Ferrell, Kevin J. O'Connor, Tate Donovan, M. Emmet Walsh, Gailard Sartain, Tony Genaro, Angel Aviles, Dirk Blocker, Debra Dusay, Les Podewell, Carlos Sanz
Twin brothers have never met and know nothing of each other, but live in the same city and have strangely parallel lives, despite the fact that one is a gangster while the other earns an honest living as a mechanic. A bizarre mystery that bears many of the director's usual flourishes.
DRA   106 min (ort 110 min) wScrn   VIDrel: TART/20TH
V/sur

EQUUS ***
15
Sidney Lummet UK
1977
Richard Burton, Peter Firth, Colin Blakely, Joan Plowright, Harry Andrews, Jenny Agutter, Eileen Atkins, Kate Reid, John Wyman, Elva Mai Hoover, Ken James, Patrick Brymer, Sheldon Rybowski, Sufi Bikhari, Anita Van Hezewyck, Mark Parr
A psychiatrist investigates the case of a boy who blinded six horses and finds himself questioning his own life and values as he penetrates to the core of his patient's madness. A static and stagebound film, yet not without a certain spellbinding power, rendered thus primarily by the superb acting of the two leads and the poetry of the dialogue.
DRA   132 min (ort 137 min)   VIDrel: MGM/WHV   V
Boa: play by Peter Shaffer.

ERASER **
18
Charles Russell USA
1996
Arnold Schwarzenegger, James Caan, Vanessa Williams, James Coburn, Robert Pastorelli, Andy Romano, James Cromwell, Danny Nucci, Nick Chinlund, Michael Papajohn, Joe Viterelli, Mark Rolston, John Slattery, Robert Miranda
A Federal Marshall specialises in keeping safe those people who have been made part of the Federal Witness Protection scheme, and does this by helping them start new lives with fake identities. He finds his latest assignment a tough one, when he is called upon to protect an attractive woman who is to testify at a trial involving a top-secret super-weapon and illegal arms sales. A noisy, explosive chase adventure, exactly what one might expect from Arnie.
A/AD   114 min Cut (3 min 22 sec) cC
VIDrel: WHV   V/sur

ERASERHEAD ***
18
David Lynch USA
1976
John Nance, Charlotte Stewart, Allen Joseph, Jeanne Bates, Judith

Anna Roberts, Laurel Near, V. Phipps-Wilson, Jack Fisk, Jean Lange, Thomas Coulson, John Monez, Darwin Joston, Neil Moran, Hal Landon Jr, Brad Keeler
An unpleasant surrealistic film that is almost impossible to watch. It consists of a series of nightmarish images loosely tied to the story of an introverted young man whose girlfriend has given birth to a grotesquely repulsive mutant and leaves him to look after it. The film (which took six years to complete) contains images of considerable power, and now enjoys cult status. The bizarre soundtrack is best appreciated in a cinema.
HOR   85 min (ort 100 min) B/W   VIDrel: ELPIC/POLYREC
V/sur

ERIK THE VIKING **
15
Terry Jones USA
1989
Tim Robbins, Gary Cady, Terry Jones, Eartha Kitt, Mickey Rooney, John Cleese, Tsutomu Sekine, Antony Sher, Gordon John Sinclair, Imogen Stubbs, Freddie Jones, Richard Ridings, Samantha Bond, Danny Schiller, Jim Broadbent
Written by ex-Monty Python member Jones, this spoof Viking saga follows the exploits of Erik, a man who abhors rape and pillage, instead choosing to sail off in search of the home of the gods. Cheap looking and dreadfully unfunny to start with, the film eventually delivers a couple of laughs if one is extremely patient.
COM   89 min (ort 104 min)   VIDrel: 20TH/TECH   V/sur

ERMO ****
15
Zhou Xiao Wen CHINA
1994
Ailiya (Alia), Liu Peiqi, Ge Zhijun, Zhang Haiyan, Yan Zhenguo, Yang Xiao, Yang Shenxia, Zhi Yanyan, Shen Enshen, Chen Baochang, Wu Jun, Du Hui, Li Yong'gui, Gao Songhai, Yang Wenming, Yui Guizhi, Ren Fengwu, Wang Wenzhi
A proud female peasant has become obliged by circumstance to adopt the role of breadwinner, supporting her disabled husband and son, but her real ambition is to save up enough money to buy a large TV set, as this would put her on a higher social status than her nasty neighbours. As she toils away at a variety of jobs her savings grow. A wonderfully detailed, vigorous and honest film, made with touching respect for its subject.
DRA   98 min   CINrel
Boa: short story by Xu Baoqi.

ERNEST GOES TO CAMP **
PG
John R. Cherry III USA
1987
Jim Varney, Victoria Racimo, John Vernon, Iron Eyes Cody, Lyle Alzado, Gailard Sartain, Daniel Butler, Hakeem Abdul-Samad, Patrick Day, Scott Menville, Jacob Vargas, Danny Capri, Todd Loyd
Based on a character created in a spate of TV commercials, this tells of a dimwitted and obnoxious fellow who gets his wish to become a summer camp counsellor. He quickly finds his hands full with problems ranging from juvenile delinquency to a takeover by a mining company. A lacklustre and exploitative comedy.
COM   89 min (ort 92 min)   VIDrel: RNK L/A   V

ERNEST GOES TO JAIL ***
PG
John R. Cherry III USA
1990
Jim Varney, Gailard Sartain, Bill Byrge, Barbara Bush, Randall (Tex) Cobb, Barry Scott, Charles Napier, Dan Leegant, Jim Coard, Jackie Welch, Melanie Wheeler, Buck Ford, Daniel Butler, Charlie Lamb, Mac Bennett, Rick Schulman
Having planned the robbery of the bank where our dimwitted hero works, an evil Ernest-lookalike has so arranged things that his innocent double is jailed in his place. Third in the series of "Ernest" films that stretches its material a little further than it might otherwise go, were it not for the terrific zest with which Varney sets about tackling (as expected) both roles.
COM   78 min (ort 82 min)   VIDrel: TOUCH/BUENA L/A
V

ERNEST GOES TO SCHOOL *
PG
Coke Sams USA
1994
Jim Varney, Linda Kash, Bill Byrge, Corrine Koslo, Kevin McNulty, Jason Michas, Sarah Chalk, Gabe Khouth
Below-par fifth entry in a very silly series which sees our hero working as custodian at a high school that is under threat of close due to some underhand machinations. In a bid to keep it open, a special atom-powered device is used to raise Ernest's IQ sufficiently in order to allow him to get an education. A most unappealing and unfunny effort.
COM   85 min (ort 89 min)   VIDrel: FIRST/SONOP   V

ERNEST RIDES AGAIN ** PG
John R. Cherry USA 1994
Jim Varney, Ron K. James, Duke Ernsberger, Jeffrey Pillars, Linda Kash, Tom Butler, Dave "Squatch" Ward, Dee Jay Jackson, Charles Siegel, Alf Humphries, Lillian Carlson, Mitch Kostermann, George Joseph, Tony Morelli, Alan Robertson
Ernest tries his hand with a metal detector and unearths a cannon from the revolutionary wars. Unfortunately, a number of people seem to think that it contains the British Crown Jewels and will stop at nothing to get their hands on it. A tepid and unamusing entry.
COM 90 min (ort 92 min) VIDrel: FIRST/SONOP V

ERNEST SAVES CHRISTMAS ** U
John Cherry USA 1988
Jim Varney, Noelle Parker, Douglas Seale, Oliver Clark, Robert Lesser, Gailard Sartain, Billie Bird, Bill Byrge, Buddy Douglas, Patty Maloney, Beecher Martin, Key Howard, Jack Swanson, Barry Brazel, George Kaplan, Bill Christie
Another Ernest adventure that offers a slight improvement on ERNEST GOES TO CAMP, following the exploits of our irritating hero who this time round has the task of helping Santa find a successor, without which there will be no more Christmases. Mildly amusing in a soppy and over-sentimental way. ERNEST GOES TO JAIL followed.
COM 88 min (ort 91 min) VIDrel: TOUCH L/A V

ERNEST SCARED STUPID * PG
John R. Cherry III USA 1991
Jim Varney, Eartha Kitt, Austin Nagler, Shay Astar, Jonas Moscartolo, John Cadenhead, Bill Byrge, Richard Wolf, Nick Victory, Alec Klapper, Steven Moriyon, Daniel Butler, Esther Huston, Larry Black, Denice Hicks
Another in the series of "Ernest" comedies, this has our inept hero disturbing the ruminations of a malevolent spirit and then having to find a means of saving the townsfolk from peril.
COM 86 min (ort 92 min) VIDrel: TOUCH/SONOP L/A V

EROTIC ADVENTURES OF THE THREE MUSKETEERS, THE * 18
Norman Apstein USA 1993
Larry Paviotti, Martine Helene, Chet Anaszek, Steve Drake, Nina Hartley, Ron Jeremy, Tracy Winn, Britt Morgan, Dino Alba, Scott Gallegos
A witless rendering of this classic that bears little resemblance to this exciting story and concentrates excessively on our trio's sexual adventures. The costumes and scenery create the right period atmosphere but the acting is woefully inadequate. Also filmed in a hard-core version.
A 100 min VIDrel: VCC/DISC L/A V

EROTIC DREAMS OF CLEOPATRA, THE ** 18
Cesar Todd ITALY 1983
Marcella Petrelli, Rita Silva, Jacques Stany, Andrea Coppola
Left alone by Caesar in her villa in Egypt, bored Cleopatra sets out to have some fun. A silly but good-natured Italian sex romp.
Aka: SOGNI EROTICI DI CLEOPATRA
A 85 min wScrn VIDrel: JEZ/RTM V

EROTIKA ** 18
Robert McCallum USA 1995
Samantha Strong, Rbecca Bardoux, Nikki Sinn, Jonathan Morgan, T.T. Boy, E.Z. Ryder, Porsche Lynn, Tanya Storm, Angella Faith, Vince Vouyer, Buck Adams, Rick O'Shea, Randi Hart, Blake Palmer
A woman from a small town sets out to slake her sexual desires in a series of adventures that take her around the world.
A 75 min VIDrel: PURG/DANTE V

EROTIQUE ** 18
Lizzie Borden/Monika Treut/Clara Law GERMANY/HONG KONG/USA 1993
Kamela Lopez-Dawson, Bryan Cranston, Ron Orbach, Ray Oriel, Liane Curtis, Vincent Cook, Kal Clarek, Tanita Tikaram, Marianne Sagenbrecht, Priscilla Barnes, Camile Soeberg, Michael Carr, Tom Lounibos, Hayley Man, Choi Hark-Kin
Three directors examine female sexual fantasies in three short films. "Let's Talk About Sex" deals with the unexpected adventures of a phone-sex operator. In "Taboo Parlour" a pair of lesbians pick up a man and avenge themselves on men in general. Finally, "Wonton Soup" has a young couple exploring their sexuality via a Chinese version of the Kama Sutra. A daft

softcore trilogy, provocative but often just plain pointless. Set in L.A., Hamburg and Hong Kong respectively.
A 90 min VIDrel: HIFLI/SONOP V/h

ESCAPE CLAUSE, THE *** 18
Brian Trenchard-Smith USA 1996
Andrew McCarthy, Paul Sorvino, Connie Britton, Kate McNeil, Stan Egi
McCarthy is a successful insurance executive learns from a hit-man that his wife has taken a contract out on his life, and that he will have to pay $10,000 a day to be allowed to live. However, both wife and assassin are soon killed off and the evidence incriminates McCarthy. Quite a clever thriller of neat twists and turns, if one can forget that ridiculousness of the basic premise, which is not hard given the consistent sense of tension the plot generates.
THR 95 min VIDrel: MGM/WHV V

ESCAPE FROM ALCATRAZ *** 15
Don Siegel USA 1979
Clint Eastwood, Patrick McGoohan, Paul Benjamin, Fred Ward, Roberts Blossom, Jack Thibeau, Danny Glover, Larry Hankin, Bruce M. Fischer, Frank Ronzio, Fred Stuthman, David Cryer, Madison Arnold, Blair Burrows, Bob Balhatchet, Ron Vernan
Retells the true story of a breakout in 1962 from this infamous maximum security prison. A film with hardly any dialogue but held together by a profound sense of tension coupled with some good acting.
THR 112 min VIDrel: 4-FRONT/POLYREC/CIC V
Boa: book by J. Campbell Bruce.

ESCAPE FROM BROTHEL ** 18
Wong Lung Wei HONG KONG 1992
Pauline Chan, Alex Fong, Billy Chow, Rena Murakami
Erotic martial arts tale, with a Hong Kong hooker being sold into prostitution to pay off her husbands's debts.
MAR 89 min (ort 95 min) wScrn VIDrel: EAST/DISC V

ESCAPE FROM JUPITER: THE MOVIE * U
Kate Woods AUSTRALIA/JAPAN 1994
Steve Bisley, Arthur Dignam, Ken Radley, Daniel taylor, Justin Rodniak, Anna Choy, Robyn Mackennzie, Abraham Forsythe, Linden Wilkinson, Ivar Kants, Anne Tenney, Sremi Baba, Kazuhiro Muroyama, Russell Kiefel, Sally Cahill
Kid's action SF adventure, with people escaping from their devastated world in a worn out spaceship that's low on fuel.
FAN 120 min mTV VIDrel: DTK/RTM L/A V

ESCAPE FROM L.A. * 15
John Carpenter USA 1996
Kurt Russell, Stacy Keach, Steve Buscemi, Peter Fonda, George Corraface, Cliff Robertson, Bruce Campbell, Valeria Golino, Pam Grier, A.J. Langer, Jeff Imada, Michelle Forbes, Ina Romero, Peter Jason, Jordan Baker, Caroleen Feeney
Sequel to ESCAPE FROM NEW YORK that is set in 2013 A.D., where Los Angeles has become a independent island region, ruled over by a revolutionary. When the daughter of the US President arrives there with a device that could destroy the entire US satellite system, Russell is dispatched to eliminate her and recover this device. A shameless clone of the earlier film, that replaces the freshness and ideas of the earlier work with the usual parade of big budget effects.
Aka: JOHN CARPENTER'S ESCAPE FROM L.A.
FAN 102 min VIDrel: CIC/SONOP; PION (LV only) V LV

ESCAPE FROM NEW YORK ** 15
John Carpenter USA 1981
Kurt Russell, Lee Van Cleef, Isaac Hayes, Ernest Borgnine, Donald Pleasence, Harry Dean Stanton, Adrienne Barbeau, Season Hubley, Ox Baker, Charles Cyphers, Frank Doubleday, Tommy Atkins, Joe Unger, John Strobel
In 1997 the island of Manhattan has been transformed into a maximum security prison with millions of inmates. The President's plane crashes there and a convicted felon is given the job of going in and rescuing him. A large-scale film of many ideas, but over-complex and excessively grim. Followed by ESCAPE FROM L.A.
A/AD 95 min (ort 106 min; special director's edition — 126 min) wScrn VIDrel: BMGVID/BMGREC V/sur

ESCAPE FROM SOBIBOR *** 15
Jack Gold UK/USA/YUGOSLAVIA 1987
Alan Arkin, Rutger Hauer, Joanna Pacula, Jack Shepherd, Emil Wolk,

Harmut Becker, Kurt Raab, Patti Love, Sara Sugarman, Simon Gregor, Linal Haft, Jason Norman, Robert Gwilyn, Eli Nathenson, Eric P. Caspar, Hugo Bower
Horrifyingly true story of the Nazi's inhumane treatment of Jews at a secluded death camp. Despite the terrible odds, a number of emaciated prisoners successfully organised an escape resulting in the setting loose of over 600 inmates, some 320 actually getting to freedom. Based on two books and Blatt's as yet unpublished manuscript. The script is by Reginald Rose.
DRA 142 min (ort 165 min) mTV VIDrel: CTE/CARL V
Boa: book by Richard Rashke/book Inferno in Sobibor by Stanislaw Szmajner/ book From the Ashes of Sobibor by Thomas Blatt.

ESCAPE FROM TERROR: THE TERESA STAMPER STORY **
15
Michael Scott USA
1994
Adam Storke, Maria Pitillo, Brad Dourif, Tony Becker, Steven Hartley, Cindy Willaims, Phillp A. Lund, Annette Marin, Jan Van Sickle, John Nance, Stanley M. Fisher, Pat Mahoney, Caitlin Flynn
A woman moves from rural Oklahoma to a luxurious mansion when she marries a handsome charmer, but learns to her cost that he is a violently possessive drug-addict. Another unedifying drama, based on a real story.
THR 88 min mTV VIDrel: ODY/SONOP V/sh

ESCAPE FROM THE PLANET OF THE APES **
15
Don Taylor USA
1971
Roddy McDowall, Kim Hunter, Bradford Dillman, Natalie Trundy, Eric Braeden, William Windom, Sal Mineo, Ricardo Montalban, John Randolph, Albert Salmi, Jason Evers, Peter Forster, Steve Roberts, Roy E. Glenn Sr, James Bacon
Third film in the series has three of the apes travelling back in time to present day Los Angeles. Human hospitality turns sour however when it is revealed that one day the world will be ruled by apes. Fairly predictable and hampered by a few clumsy touches of humour. Followed by CONQUEST OF THE PLANET OF THE APES.
FAN 97 min VIDrel: 20TH/TECH L/A V

ESCAPE TO WITCH MOUNTAIN ***
U
John Hough USA
1974
Ray Milland, Eddie Albert, Kim Richards, Ike Eisenmann, Donald Pleasence, Walter Barnes, Reta Shaw, Denver Pyle, Alfred Ryder, Lawrence Montaigne, Don Brodie, Terry Wilson, George Chandler, Dermott Downs, Shepherd Sanders
Two children with strange powers try to discover the secret of their origin while being pursued by a millionaire who wants to use them for his own evil ends. A fine Disney fantasy slightly marred by poor scripting. Followed by RETURN FROM WITCH MOUNTAIN.
JUV 97 min VIDrel: WDV/TECH L/A V
Boa: novel by Alexander Key.

ESCAPE 2000 *
18
Brian Trenchard-Smith AUSTRALIA
1981
Steve Railsback, Olivia Hussey, Michael Craig, Noel Ferrier, Lynda Stoner, Carmen Duncan, Roger Ward, Michael Petrovich, John Ley, Gus Mercurio, Steve Rackman, Bill Young, John Godden, Oriana Panozzo, Marina Finaly
In a future society deviants and political undesirables are put into brutal prison camps as a way of controlling crime, and at one such camp the warden delights in freeing them, but only in order to hunt them to death. An utterly repellent version of THE MOST DANGEROUS GAME, it features an unending stream of whippings, burnings and gruesome decapitations.
Aka: BLOOD CAMP THATCHER; TURKEY SHOOT
FAN 83 min (Cut at film release - ort 88 min) wScrn
VIDrel: VIPCO/SGSVID V

ESCORT GIRLS *
18
Donovan Winter UK
1974
David Dixon, Maria O'Brien, Marika Mann, Gil Barber, Helen Christie, Richard Wren, David Brierly, James Hunter, Teresa Van Ross, Veronica Doran, Brian Jackson, Barbara Wise
A behind-the-scenes look at the escort business following the exploits of several of the girls. Another tired and dated attempt at titillation.
A 102 min VIDrel: JEZ/RTM V

ESKIMO NELL *
18
Martin Campbell UK
1974
Michael Armstrong, Terence Edmond, Christopher Timothy, Roy

Kinnear, Diane Langton, Gordon Tanner, Beth Porter, Richard Caldicott, Prudence Drage, Jeremy Hawke, Rosalind Knight, Katy Manning, Lloyd Lamble, Anna Quayle
Smutty comedy about four different versions of an "Eskimo Nell" film, being made to trick those who have provided the finances for the film. A mediocre offering inspired (if that is the right word) by the bawdy ballad of the same name.
COM 82 min (ort 85 min) VIDrel: MED V

ESTHER WATERS ***
PG
Ian Dalrymple/Peter Proud UK
1947
Dirk Bogarde, Kathleen Ryan, Cyril Cuusack, Ivor Barnard, Mary Clare, Fay Compton, Morland Graham
The dissipated son of a squire seduces a maid and she falls pregnant, but when he fails to marry her she is obliged to bring up their child alone. However, some years later she meets him again, and he now wants to marry her. A fairly entertaining drama of passion and jealousy, set in the 1870s, that is only, given the course of the novel, spoilt by the feeling of tragedy and sadness that hangs over it all.
DRA 106 min (ort 109 min) B/W VIDrel: CARL/TECH V
Boa: novel by George Moore.

E.T. THE EXTRA-TERRESTRIAL ***
U
Steven Spielberg USA
1982
Dee Wallace, Henry Thomas, Peter Coyote, Robert MacNaughton, Drew Barrymore, K.C. Martel, Sean Frye, C. Thomas Howell, David O'Dell, Erika Eleniak, Richard Swingler, Frank Toth, Robert Barton, Michael Darrell
A modern fairytale that tells of a young boy who befriends an alien creature stranded on Earth, who for once is shown as benign in sharp contrast to the cold, clinical forces of the state. A charming if somewhat mawkish fantasy. AA: Score/orig (John Williams), Sound (Robert Knudson/Robert Glass/Don Digirolamo/Gene Cantamessa), Effects/aud (Charles L. Campbell/Ben Burtt), Effects/vis (Carlo Rambaldi/Dennis Muren/Kenneth F. Smith).
FAN 110 min (ort 115 min) VIDrel: CIC/SONOP; PION (LV only) V/sur LV

EUREKA **
18
Nicolas Roeg UK/USA
1983
Gene Hackman, Rutger Hauer, Theresa Russell, Jane Lapotaire, Mickey Rourke, Ed Lauter, Joe Pesci, Helena Kallianiotes, Corin Redgrave, Cavan Kendall, Joe Spinell, Frank Peske, Michael Scott Addis, Norman Beaton, Emrys James
Film based on the true tale of the unsolved murder of a rich gold prospector living on a Caribbean island. The melodramatic plot has personal problems, such as his daughter's affair with an adventurer, flung into the main story in which he watches his thirty-year-old premier collapse as thugs from Miami converge on him. An unusual foray for Roeg, whose flashy style keeps this one going even after one's interest in the characters has evaporated.
THR 124 min (ort 129 min) Cut VIDrel: MGM/WHV V/h
Boa: book Who Killed Sir Harry Oakes? by Marshall Houts.

EUROPA ****
15
Lars Von Trier DENMARK/FRANCE/SWEDEN/WEST GERMANY
1991
Jean-Marc Barr, Barbara Sukowa, Udo Kier, Ernst-Hugo Jaregard, Erik Mork, Jorgen Reenberg, Henning Jensen, Eddie Constantine, Benny Poulsen, Erno Muller, Dietrich Kuhlbrodt, Holger Perfort, Max Von Sydow (narration only)
A young American of German extraction travels to Germany in 1945 in the hope of rediscovering his roots and making some kind of "restitution" to a country he feels his father abandoned. This decision affords him the chance to observe at first-hand the guilt-stained past of its people and their growing defensiveness in the face of Allied war-crimes investigations. This is hardly a film to be enjoyed, but as an experience it is unforgettable.
Aka: ZENTROPA
DRA 107 min (ort 117 min) B/W/Col wScrn (English version) VIDrel: ELPIC/POLYREC V/sur

EUROPA, EUROPA ***
15
Agnieszka Holland FRANCE/WEST GERMANY
1991
Salomon Perel, Marco Hofschneider, Rene Hofschneider, Piotr Kozlowski, Klaus Abramowsky, Michele Gleizer, Marta Sandrowicz, Nathalie Schmidt, Delphine Forest, Andrzej Mastalerz, Wlodzimierz Press, Martin Maria Blau, Klaus Kowatsch
During WW2 a young German Jewish boy manages to survive

because of his supposedly Aryan appearance and winds up as a pupil at a crack Nazi military academy. A shattering tale with some black humour as the youngster attempts to retain his beliefs in the face of all the chaos and inhumanity that surrounds him.
Aka: HITLERJUNGE SALOMON
DRA 107 min (ort 115 min) VIDrel: ARROW/RTM V
Boa: autobiography of Salomon Perel.

EVEN COWGIRLS GET THE BLUES * 15
Gus Van Sant Jr USA 1993
Uma Thurman, Keanu Reeves, Lorraine Bracco, John Hurt, Sean Young, Noriyuki "Pat" Morita, Angie Dickinson, Buck Henry, Roseanne Arnold, Rain Phoenix, Carol Kane, Crispin Glover, Grace Zabriskie, Treva Jeffreys, Allen Arnold, Ken Kesey
A girl endowed with enormous thumbs takes to the road and hitches around the country, ending up on a lesbian dude ranch where she meets a weird bunch of assorted dropouts. An unfunny oddball comedy that is let down by the lack of a coherent script or any characters of real interest. Soundtrack is the work of k.d. lang, but for all its qualities, it can do nothing to rescue this dud and make it any more palatable.
COM 92 min (ort 96 min) VIDrel: ELPIC/POLYREC V
Boa: novel by Tom Robbins.

EVEN HITLER HAD A GIRLFRIEND * 18
Ronnie Cramer 1993
Andren Scott, Monica McFarland, Rebecca Watson, Shannon Strong
A loser who finds himself totally unable to get a girlfriend of any kind is obliged to use up his life savings and resort to call-girls, but once these have gone he decides to sell his apartment and car.
DRA 98 min VIDrel: POPRO/RTM V

EVENING STAR, THE ** 12
Robert Harling USA 1996
Shirley MacLaine, Bill Paxton, Juliette Lewis, Ben Johnson, Miranda Richardson, Jack Nicholson
Sequel to TERMS OF ENDEARMENT that sees MacLaine back as Aurora Greenway, who having raised the three children of her dead daughter, is now trying to find some fulfilment in her own life, but still faces the inevitable parade of difficulties and traumas, such as the jailing of one of her grandchildren. Writer Harling directing debut is an uneven comedy-drama, that is occasionally funny, but is often just as gruelling for us as it is for the characters.
DRA 129 min CINrel

EVENING WITH GARY LINEKAR, AN ** 12
Andy Wilson UK
Clive Owen, Paul Merton, Caroline Quentin, Martin Clunes, Liz McInnery, Gary Linekar
Comedy that revolves around a holiday in Ibiza and the Football World Cup Semi-Final of 1990 between England and Germany.
COM 77 min VIDrel: PICMUS/EMIMUS V/sh

EVERGREEN *** U
Victor Saville UK 1934
Jessie Matthews, Sonnie Hale, Betty Balfour, Barry Mackay, Ivor McLaren, Hartley Power, Patrick Ludlow, Betty Shale, Marjorie Brooks, Richard Murdoch
The daughter of a retired music-hall star is mistaken for her mother and ends up with the lead role in a show, but faces problems when she falls for a man who claims to be her son. An enjoyable musical comedy that was a tour-de-force for its star Matthews, who is in top form in what was to be one of the most popular British films of the decade.
MUS 90 min B/W VIDrel: CONNO/RTM L/A V
Boa: play by Benn W. Levy.

EVERSMILE NEW JERSEY ** PG
Carlos Sorin ARGENTINA/UK 1989
Daniel Day Lewis, Mirjana Jokovic, Gabriela Archer, Julio De Grazia, Igancio Quiros, Miguel Ligero, Ana Maria Giunta, Miguel Ligero, Boy Olmi, Eduardo D'Angelo, Albert Benegas, Roberto Catarineu, Miguel Dedovitch, Jose Maria Rivara
Quirky little satire set in rural Argentina, where an itinerant American dentist travels around in the search for business, while preaching the virtues of modern dentistry. A young woman leaves her boyfriend and takes up with her, eventually overcoming his initial resistance.
COM 87 min VIDrel: COLUM/SONOP V/sh

EVERY BREATH ** 18
Steve Bing USA 1993
Joanna Pacula, Judd Nelson, Patrick Bauchau
A young man gets involved with an attractive woman and her husband and this couple embroil him in some very strange games. Average. See also KNIFE EDGE.
DRA 84 min (ort 89 min) cC VIDrel: COLUM/SONOP V/sh

EVERY WHICH WAY BUT LOOSE ** 15
James Fargo USA 1978
Clint Eastwood, Sondra Locke, Geoffrey Lewis, Ruth Gordon, Beverly D'Angelo, Walter Barnes, George Chandler, Roy Jenson, James McEachin, Bill McKinney, William O'Connell, John Quade, Dan Vardis, Gregory Walcott
Simple-minded knockabout story of a Los Angeles trucker and barefist fighter who wins an orang-otun in a bout, and has a variety of comic adventures with the beast. Intermittently enjoyable, and though Eastwood shows little aptitude for comedy "Clyde" (his ape partner) is wonderful. Followed by ANY WHICH WAY YOU CAN.
COM 110 min (ort 119 min) VIDrel: WHV L/A V
Boa: novel by J.J. Kronberg.

EVERY WOMAN'S DREAM ** 15
Steven Schachter USA 1996
Jeff Fahey, Kim Cattrall, Walter Addison
A real-life movie that casts Fahey as a compulsive liar who gets married to two woman at the same time, and then has the problem of keeping track of the lies this double-life necessitates. The film suffers a slight change of tack towards the end, but for the most part is well acted and absorbing.
DRA 87 min mTV VIDrel: MED V

EVERYBODY WINS * 15
Karel Reisz UK/USA 1989
Debra Winger, Nick Nolte, Will Patton, Judith Ivey, Jack Warden, Kathleen Wilhoite, Frank Converse, Frank Military, Steven Skybell, Mary Louise Wilson, Mert Hatfield, Peter Appel, Sean Weil, Timothy D. Wright, Elizabeth Ann Klein
A private eye hired to prove that a young man did not commit murder, eventually unearths the usual small-town corruption. A dull dud that for all its Arthur Miller screenplay (his first since THE MISFITS) just never comes to life.
DRA 93 min (ort 98 min) VIDrel: VISVID/POLYREC V/sur
Boa: play Some Kind Of Love Story by Arthur Miller.

EVERYBODY'S FINE *** PG
Giuseppe Tornatore FRANCE/ITALY 1990
Marcello Mastroianni, Michele Morgan, Marino Cenna, Roberto Nobile, Valeria Cavali, Norma Martelli, Fabio Iellini, Salvatore Cascio, Matteo Lo Piparo, Gaia Restivo, Mariangela Randazzo, Antonella Attili, Paride Zappala, Nicola Di Pinto
An elderly retired civil servant leaves his native Sicily for the Italian mainland in order to visit each of his five children, all of whom live in different parts of the country. Unfortunately the hopes and dreams he has fondly nurtured for each child are far from sustained by reality. The clumsy and episodic nature of the script tends to deaden the impact of a bittersweet drama of great charm.
Aka: ILS VONT TOUS BIEN; STANNO TUTTI BENE
DRA 121 min (ort 126 min) VIDrel: COLUM/SONOP V

EVERYONE SAYS I LOVE YOU *** 15
Woody Allen USA 1996
Woody Aleen, Alan Alda, Edward Norton, Goldie Hawn, Drew Barrymore, Tim Roth, Lukas Haas, Gaby Hoffman, Julia Roberts
Fluffy, lightweight, romantic comedy-musical, with New York, Paris and Venice settings, and a slight plot that takes a look at the complex and intertwined love lives of a New York family, as revealed by a teenage narrator. What is unusual here is the musical interludes, when the untrained voices of the cast members add a delightfully naive and charming flavour to the proceedings. A joyful confection, free of the pompous pretensions of Allen's earlier works.
MUS 101 min CINrel

EVERYTHING YOU ALWAYS WANTED TO KNOW ABOUT SEX (BUT WERE AFRAID TO ASK) ** 18
Woody Allen USA 1972
Woody Allen, John Carradine, Lou Jacobi, Louise Lasser, Anthony Quayle, Lynn Redgrave, Tony Randall, Burt Reynolds, Gene Wilder,

Meredith MacRae, Geoffrey Holder, Regis Philbin, Jack Barry, Erin Fleming, Elaine Giftos, Toni Holt

A dated and uneven series of sketches whose inspiration was a best-selling sex manual by Dr David Reuben. Each sketch is self-contained and purports to examine a particular aspect of sexuality, though all too often never rising above simple farce. Of the seven, Jacobi's contribution on transvestism and Wilder's on bestiality are the funniest, but that's not saying much.

COM 84 min (ort 88 min) VIDrel: MGM/WHV V/dm

EVIL CLUTCH ** 18
Andreas Marfori ITALY 1991
Coralina C. Tassoni, Diego Ribon, Luciano Crovato, Elena Cantarone, Stefano Molinari

A couple on a European vacation in the Alps encounter some very strange individuals in what proves to be a haunted forest. Good special effects enhance a poorly photographed and badly dubbed horror effort with the unavoidable lashings of gore.

HOR 80 min (ort 88 min) dubbed VIDrel: TROMA/RTM V

EVIL DEAD, THE ** 18
Sam M. Raimi USA 1980 (released 1983)
Bruce Campbell, Ellen Sandweiss, Sarah York, Betsy Baker, Hal Delrich

Five college students on holiday in the backwoods of Tennessee release the spirits of the dead who proceed to take over the bodies of the living, resulting in the students turning into hideous killers. A wild over-the-top horror yarn with lashings of gore, best appreciated by those with very strong stomachs. Followed by EVIL DEAD 2.

Aka: BOOK OF THE DEAD, THE
HOR 79 min Cut (1 min 6 sec plus film cuts - ort 90 min)
VIDrel: POLY/POLYREC V

EVIL DEAD 2 ** 18
Sam M. Raimi USA 1987
Bruce Campbell, Sarah Berry, Dan Hicks, Kassie Wesley, Theodore Raimi, John Peaks, Denise Bixler, Richard Domeier, Lou Hancock

A sequel to THE EVIL DEAD, which is the continuing story of one man's efforts to defeat these gruesome creatures who haunt the cabin in the woods and take possession of those that enter it. Much the same formula as the original film, with no shortage of gore and special effects.

Aka: DEAD BY DAWN; EVIL DEAD 2: DEAD BY DAWN
HOR 81 min (Cut at film release by 2 sec - ort 85 min)
VIDrel: POLY/POLYREC V

EVIL FORCE * 18
Evan Lee USA
Larry Justin, James Habif, Robert Clark, Christopher Lee

A bunch of teenagers attack their teacher and rape his daughter, injuring the former so severely that he almost dies. Having suffered so badly at their hands, the teacher calls up a demon to have his revenge.

HOR 78 min VIDrel: VIPCO/SGSVID V

EVIL HAS A FACE ** 15
Rob Fresco USA 1996
Sean Young, William R. Moses. Joe Guzaldo, Brighton Hertford, Checlie Ross, Jason Zone Fisher, Kate Buddeke, Suzanne Petri, Dick Cusack, Mary Seibel, Morgan McCabe, Jason Wells, Mike Houlihan

A police sketch artist working on a case involving a child kidnapping finds that she has repressed memories of her own that are leading her into danger. A tepid little effort of no great consequence. See also SKETCH ARTIST for something similar.

DRA 89 min (ort 92 min) cC VIDrel: CIC V/sur

EVIL IN CLEAR RIVER ** PG
Karen Arthur USA 1987
Randy Quaid, Lindsay Wagner, Michael Flynn, Thomas Wilson Brown, Stephanie Dees, Gloria Carlin, Carolyn Croft, Spencer Alston, Steven W. Anderson, James Berry, Charles Black, Kimberly Cannaday, Kirk Chambers, Craig Clyde

Fictionalised drama that follows the exploits of a real-life small-town housewife in Canada who discovers that her hockey star son is being taught to hate by an anti-Semitic history teacher he reveres. Quaid is good as the bigoted teacher who is also the hockey coach and town's mayor but this overwrought drama has a cliched and unimaginative script. See also SCANDAL IN A SMALL TOWN.

DRA 94 min mTV VIDrel: SONY L/A V

EVIL SENSES ** 18
Gabriele Lavia ITALY 1986
Gabriele Lavia, Monica Guerritore, Lewis Eduard Ciannelli, Pario Mazzoli, Mimsy Farmer, Gioia Maria Scola, Rene Masrevery, Ragnhild Aslaksen

A professional killer makes a serious mistake when he steals his victim's briefcase and finds it to contain sensitive documents and details of blackmail victims. As of this moment on he finds himself in danger from gangsters who have decided he must be eliminated.

THR 90 min (ort 95 min) dubbed VIDrel: ARTPRO/RTM V

EVIL TEMPTATIONS ** 18
USA 1995
Anna Malle, April, Olivia, Mike Horner, Tony Martino

A woman makes a pact with the Devil to enable her to get the man of her dreams, but she soon realises that he fancied her all along, so she didn't need to sell her soul after all. The way she attempts to get out of her contract forms the basis for the rest of the movie. A cleverly plotted and quite entertaining erotic fantasy.

A 75 min VIDrel: ONE V

EVIL UNDER THE SUN * PG
Guy Hamilton UK 1982
Peter Ustinov, Jane Birkin, Colin Blakely, Roddy McDowall, James Mason, Nicholas Clay, Sylvia Miles, Dennis Quilley, Diana Rigg, Maggie Smith, Emily Hone, John Alderson, Paul Antrim, Cyril Conway, Barbara Hicks, Richard Vernon

Another ponderous adaptation of an Agatha Christie murder mystery, this time set in a resort hotel. Glossy and undeniably well put together but excruciatingly dull. The script is by Anthony Shaffer.

DRA 111 min (ort 117 min) VIDrel: WHV V
Boa: novel by Agatha Christie.

EVITA *** PG
Alan Parker USA 1996
Madonna, Antonio Banderas, Jonathan Pryce, Jimmy Nail, Victoria Sus, Julian Littman, Olga Merediz, Laura Pallas, Julia Worsley, Maria Lujan Hidalgo, Andrea Corr, Servando Villamil, Peter Polycarpou, Gary Brooker, Mayte Yerro

Madonna finally demonstrates that she can act as well as sing. It is all the more of a pity that her vehicle for doing so is a film version of a 1978 stage musical built around the life of the title character, Evita Peron, whose marriage to Argentina's dictator Juan Peron gave her the public platform she so clearly relished. The music is fine, the sets and costumes are magnificent, but within this glossy pageant, the human element is all to often lost completely.

MUS 122 min (ort 134 min) VIDrel: EIV/SONOP V/sh
Boa: musical by Tim Rice and Andrew Lloyd Webber.

EVOLVER ** 15
Mark Rosman USA 1994
Cassidy Rae, Ethan Randall, John De Lancie, Cindy Pickett, Paul Dooley, Tim Griffin, Nassica Nicola, Chance Quinn, Michael Champion

Computer-games players are attracted to a competition at an arcade where the prize is a robot designed to play games and endowed with the ability to learn from its mistakes and thus "evolve". Unfortunately, since its program was originally developed for military use, it has a little difficulty in keeping fantasy and reality apart and soon goes on a killing spree. An average horror flick that seems at times like a small-scale version of the TERMINATOR movies.

HOR 88 min (ort 90 min) VIDrel: MED/DISC V/sur

EWOKS: THE BATTLE FOR ENDOR ** U
Jim Wheat/Ken Wheat USA 1986
A. Wilford Brimley, Warwick Davis, Aubree Miller, Sian Phillips, Carel Struycken, Niki Bothelo, Erick Walker, Marianne Horine, Daniel Frishman, Tony Cox, Pam Grizz, Roger Johnson, Johnny Weissmuller Jr

An expensive, flashy but unsatisfying sequel to CARAVAN OF COURAGE: AN EWOK ADVENTURE, telling of an old hermit who joins a girl and her Ewok pal in a quest to rescue an Ewok family held captive by an evil king. Average.

JUV 93 min (ort 97 min) mTV VIDrel: MGM/WHV L/A V/sh
Boa: story by George Lucas.

EX *** (PG)
Paul Seed UK 1991
Griff Rhys-Jones, Geraldine James, Penny Downie, Jonathan Hackett, Dermot Crowley, Sebastian Knapp, Kate Maberly, Mary Jo Randle, Margery Mason, Bruce Montague, Lindy Alexander, Colin Douglas, Christopher Scoular, Nigel Gregory
One of the writers on a popular TV soap opera finds himself smitten with the show's star but cannot bear to break up with his wife and children. Very much a middle class examination of middle class pre-occupations, though made with a lightness and wit that works to its advantage.
DRA 90 min mTV TVrel: BBC

EX, THE ** 18
Mark L. Lester USA 1996
Yancy Butler, Suzy Amis, Nick Mancuso
Having divorced his first wife, an architect has now settled into domestic bliss with his second, but his happiness is to be short-lived when his vengeful (and psychopathic) former spouse decides to plot his downfall.
THR 88 min VIDrel: HIFLI/SONOP V

EXCALIBUR *** 15
John Boorman UK/USA 1981
Nicol Williamson, Nigel Terry, Helen Mirren, Nicholas Clay, Cherie Lunghi, Corin Redgrave, Paul Geoffrey, Katrina Boorman, Robert Addie, Keith Buckley, Liam Neeson, Gabriel Byrne, Charlie Boorman, Niall O'Brien, Patrick Stewart
Colourful version of the legend of King Arthur that concentrates on visual imagery to the exclusion of nearly everything else. Williamson gives a strange offbeat performance as Merlin that is probably the best thing in the film. Much of the film's sheer visual impact will be lost on the small screen. See also FIRST KNIGHT.
A/AD 135 min (ort 140 min) wScrn VIDrel: WHV V/sh
Boa: poem Le Morte d'Arthur by Thomas Malory.

EXCESSIVE FORCE ** 18
Jon Hess USA 1992
Thomas Ian Griffith, Lance Henriksen, James Earl Jones, Burt Young, W. Earl Brown, Charlotte Lewis, Tony Todd, Paula Anglin, Antoni Corone, Christopher Garbrecht, Danny Goloring, Ian Gomez, Tom Hodges, Richard Mawe, Susan Wood
In Chicago, a drugs dealer revenges himself on the members of a police team who ruined a lucrative deal, and kills them all except for one tough cop. The latter now sets out to have his revenge for the slaying of his buddies, in a fairly competent albeit totally routine actioner. A few martial arts sequences are added to generate a little excitement. A sequel followed.
A/AD 82 min (ort 90 min) VIDrel: EIV/SONOP V

EXCESSIVE FORCE 2 ** 18
Jonathan Winfrey USA 1995
Stacie Randall, Dan Gauthier, Jay Patterson, Bradford Tatum, Dan Luria, Teri J. Vaughn, Michael Wiseman, Henry Brown, John Sanderford, Joe Maruzzo, Tom Wright, Cyril O'Reilly, Joe Mese, Anthony Pennello, Anthony Paul, James Law
Second film in this series is a straightforward action outing that's built around an attractive female cop who has the job of saving L.A. from a lethal hit-squad. An assembly-line product that achieves what it sets out to do, namely offer a dose of mindless (and extremely violent) entertainment.
Aka: EXCESSIVE FORCE 2: FORCE ON FORCE
A/AD 84 min (ort 88 min) VIDrel: FIRST/SONOP V

EXCLUSIVE ** 15
Alan Metzger USA 1992
Suzanne Somers, Michael Nouri, Ed Begley Jr, Joe Cortese, Scott Bryce, Kelly Rowan, Jerry Adler, James Pickens Jr, Eric Doppick, Andrie Neean, Liane Curtis, Lily Mariye, Kevin McClarnon, Joan Stuart Morris, Kevin McDermott
A female reporter gets call with a tip about planned mass murder at a nightclub, and arriving there she finds seven bodies, and initially sees this as nothing more than the scoop she needs to hang onto her job. However, it soon becomes apparent that the killer may well be someone she knows.
THR 87 min (ort 90 min) mTV VIDrel: GUILD/SONOP V/s

EXECUTIVE ACTION * PG
David Miller USA 1973
Burt Lancaster, Robert Ryan, Will Geer, John Anderson, Gilbert Green, Colby Chester, Paul Carr, Ed Lauter, Walter Brooke, Sidney Clute, Deanna Darrin, Lloyd McGough, Richard Hurst, Robert Karnes, James MacColl, Joaquin Martinez
An attempt to explain the events that led up to the assassination of John F. Kennedy with the writers of the book on which the film is based attempting to demonstrate that the president was assassinated by a powerful clique that included high-ranking CIA operatives. Painfully slow and boring, this was Ryan's last film. See also JFK for another conspiracy theory treatment of Kennedy's murder.
DRA 91 min VIDrel: MGM/WHV L/A V
Boa: book Rush To Judgement by Donald Freed and Mark Lane.

EXECUTIVE DECISION *** 15
Stuart Baird USA 1996
Kurt Russell, Steven Seagal, Halle Berry, John Leguizamo, Oliver Platt, Joe Morton, David Suchet, B.D. Wong, Len Cariou, Whip Hubley, Andreas Katsulas, Mary Ellen Trainor, Maria Maples Trump, J.T. Walsh, Ingo Neuhaus, William James Jones
Anti-terrorist political thriller set aboard a hijacked airliner that has been planted with deadly gas bombs. Formulaic it may be, but there are thrills to be had and a clever twist even if our hero's exploits are just a little too contrived to convince.
THR 127 min (ort 133 min) cC VIDrel: WHV V/sur

EXILED ** 18
Paul Leder USA 1990
Maxwell Caulfield, Wings Hauser, Edward Albert Jr, Viveca Lindfors, Kamala Lopez, Stella Stevens
A guerilla leader from Central America flees with his wife and child to the States, where his wife takes a job working as a diner waitress but finds herself falling love with the owner's son. Meanwhile, a team of counter-revolutionaries led by a ruthless CIA man, are sent to eliminate him.
Aka: EXILED IN AMERICA
A/AD 81 min (ort 93 min) VIDrel: 20VIS/SONOP V

EXIT IN RED * 18
Yurek Bogayevicz USA 1996
Mickey Rourke, Annabel Schofield, Anthony Michael Hall, Carre Otis
An unhappy psychiatrist (Rourke) attempts to get his life in order, with the help of an attractive lawyer. However, when he meets a beautiful and sexually liberated stranger, he gets drawn into a murder plot. Yet one more entry in a long line of turgid erotic thrillers, saddled with a derivative script and weak performances, plus the inevitable steamy sex scenes.
THR 92 min VIDrel: MED/20VIS V/sh

EXIT THE DRAGON, ENTER THE TIGER * 18
Lee Tse Nam HONG KONG 1976
David Lee (Shaio Lung), Lung Fei, San Moo, Ma Chi Chiang, An Ping, Chang Sing Yee, Tsao Shao Jung
A martial arts teacher sets out to discover the truth about the death of Bruce Lee, in this boring formula spin-off.
MAR 78 min Cut (1 min 28 sec plus some cuts subst - ort 84 min) VIDrel: ONE/IMPENT V

EXIT TO EDEN * 18
Garry Marshall USA 1994
Rosie O'Donnell, Dan Aykroyd, Dana Delaney, Paul Mercurio, Hector Elizondo, Stuart Wilson, Sean O'Bryan, Stephanie Niznik, Rosemary Forsyth, Donna Dixon, Phil Redrow, Sandra Korn (Sandra Taylor), Julie Hughes, Laurelle Mehus
An island vacation paradise that caters for the sexual fantasies of its wealthy clients, provides the setting for this oddball sex comedy as male and female cops investigate a murder. An unfocused and tedious offering in which feeble scripting is complemented by excessive length.
COM 120 min VIDrel: GUILD/FOXVID V/sh
Boa: novel by Anne Rice.

EXODUS ** PG
Otto Preminger USA 1960
Paul Newman, Eva Marie Saint, Sal Mineo, Ralph Richardson, Lee J. Cobb, Hugh Griffith, Peter Lawford, Jill Haworth, Gregory Ratoff, Felix Aylmer, David Opatoshu, Alexander Stewart, Martin Benson, Martin Miller, Victor Madden
Overlong and strangely lifeless account of the events that led up to the founding of the modern state of Israel. The film rapidly degenerates into an episodic series of clashes, with Newman and Saint utterly miscast and the supporting actors merely providing stereotyped portrayals. AA: Score (Ernest Gold).
DRA 199 min (ort 213 min) VIDrel: WHV L/A V/h
Boa: novel by Leon Uris.

EXORCIST, THE ** *(18)*
William Friedkin USA 1973
Linda Blair, Max Von Sydow, Jason Miller, Ellen Burstyn, Lee J.
Cobb, Kitty Winn, Jack MacGowran, Vasiliki Maliaros, Wallace
Rooney, Titos Vandis plus Mercedes McCambridge (voice only)
Ghastly film that spawned a thousand imitations. A small girl
(Blair) is possessed by the Devil who is most generous with his
displays of physical phenomena. Miller plays the priest who
attempts to exorcise her. Vocal effects are by McCambridge. The
inevitable sequels followed. For some strange reason the BBFC
have consistently refused to certificate this film for video release.
AA: Screen/adapt (William Peter Blatty), Sound (Robert
Knudson/Chris Newman).
HOR 115 min CINrel
Boa: novel by William Peter Blatty.

EXORCIST 2: THE HERETIC * 18
John Boorman USA 1977
Linda Blair, Max Von Sydow, Richard Burton, Louise Fletcher, Paul
Henreid, James Earl Jones, Kitty Winn, Ned Beatty, Belindha Beatty,
Rose Portillo, Barbara Cason, Tiffany Kinney, Joey Green, Fiseha
Dimetros, William O'Malley
Sequel to the 1973 film which falls quite flat despite good special
effects. In this one Burton plays a priest out to discover the
mystery of the demons that still inhabit Blair. Derided by the
critics (and deservedly so) this dreary dud plays like a pompous,
plotless B-movie version of the original film. The constant flash-
backs are as confusing as they are meaningless.
Aka: HERETIC, THE
HOR 113 min (ort 118 min) VIDrel: WHV V/h

EXORCIST 3, THE ** 18
William Peter Blatty USA 1990
George C. Scott, Ed Flanders, Jason Miller, Scott Wilson, Nicol
Williamson, Brad Dourif, Nancy Fish, George DiCenzo, Viveca
Lindfors, Don Gordon, Grand L. Bush, Lee Richardson, Mary Johnson,
Ken Lerner, Tracy Thorne
Touted as the "official" sequel, this further EXORCIST outing
sees a police inspector investigating a series of nasty murders,
and realising that they are the work of a serial killer who was
executed at the same time as the exorcism in the previous film.
His search leads him to an insane asylum and the expected
demonic confrontation takes place. A confused film that begins
well but becomes increasingly absurd; the script is the work of
Blatty.
Aka: EXORCIST 3: LEGION
HOR 105 min (ort 110 min) VIDrel: 20TH/TECH V/sur
Boa: novel Legion by William Peter Blatty.

EXOSQUAD ** *PG*
1993
Voice of: Robby Benson
In the year 2119 genetically engineered super-beings or
Neosapiens have revolted and enslaved their creators. Only the
daring resistance fighters of the Exosquad can save the planet
and the solar system from these creatures.
ANIM 100 min VIDrel: CIC/SONOP V/sur

EXOTICA *** 18
Atom Egoyan CANADA 1995
Bruce Greenwood, Mia Kirschner, Elias Koteas, Arsinee Khanjian,
Mia Kirshner, Don McKellar, Victor Garber, Sarah Polley, David
Hemblen, Calvin Green, Peter Krantz, Damon D'Oliveira, Jack Blum,
Billy Merasty, Ken McDougall
A tangled character study of five very different individuals
whose lives are all linked in some way to the title venue, a strip
club that becomes the catalyst for bringing them together. This
interesting and highly personal work is always good to look at,
even if it is weakened by the contrived ending. Both tragic and
comic, this is an atmospheric, dark and haunting examination
of loneliness and alienation.
DRA 103 min (ort 104 min) VIDrel: ARTIF/20TH V/sh

EXPECT NO MERCY ** 18
Zale Dalen USA 1995
Billy Blanks, Jalal Merhi, Wolf Larson, Laurie Holden, Anthony De
Longis, Michael Blanks
A man sets out to expose a virtual-reality "arts academy" as a
sham that is used to front more sinister activities, its students
being trained in a cyber version of the martial arts to help the
owner in his plans for world domination. The terrific special
effects and animation are a treat, and the film (it often plays like
a computer game) brings to mind both THE LAWNMOWER

MAN and TRON. Sadly, the script is just one more good-versus-
bad action adventure.
A/AD 91 min VIDrel: MIA/DISC V/h

EXPERT, THE ** 18
Rick Avery USA 1994
Jeff Speakman, James Brolin, Elizabeth Gracen, Michael Shaner, Alex
Datcher, Wolfgang Bodison, Norm Woodell, Michelle Nagy, Red West,
William Barry Scott, Jim Varney, Jophery Brown, Robby Robinson,
Ramon Estevez, Dan Chandler
When his sister is brutally murdered and it seems as if her killer
is likely to escape justice, a special operations expert breaks into
the prison where the culprit is held to take matters into his own
hands. Another competently realised but totally unoriginal vigi-
lante revenger.
A/AD 94 min VIDrel: MIA/DISC V/h

EXPERTS, THE * 15
Dave Thomas USA 1987 (released 1989)
John Travolta, Arye Gross, Charles Martin Smith, Kelly Preston,
Deborah Foreman, James Keach, Jan Rubes, Brian Doyle Murray, Rick
Ducommun, Mimi Maynard, Eve Brent, Rick Ducommon, Steve
Levitt, Tony Edwards, Jack Ammon
This tepid comedy, has two cool New Yorkers supposedly being
hired to open a nightclub in a small Nebraska town. In reality,
they have been abducted to the USSR where they arrive at a
Soviet copy of a typical American town, the intention being to
use their "expertise" in the training of spies. A flimsy and senti-
mental mess, as silly as it is contrived. Filmed in Canada.
COM 89 min (ort 94 min) VIDrel: CIC/SONOP V/sur

EXPLOITS AT WEST POLLEY *** *(U)*
Diarmuid Lawrence UK 1985
Anthony Bate, Brenda Fricker, Charlie Condou, Jonathan Jackson,
Jonathan Adams, Noel O'Connell, Frank Mills, Thomas Heathcote,
James Coyle, Jelena Budimir, Diana King, Kelita Groom, Brian Groom,
Sean Bean, George Malpas
In this dramatisation of Hardy's novel, a young boy carrying a
heavy case arrives in a farming village from the city, to stay
with his widowed aunt and his cousin. Striking up an uneasy
relationship the two boys set about exploring a local cave where
they find a vast cavern, and their activities change the course of
an underground river. But this has unexpected consequences for
all the villagers, especially an unpleasant miller. A charming,
little historical tale.
JUV 61 min SATrel: MOVIE CHANNEL
Boa: novel by Thomas Hardy.

EXPOSED * *(15)*
James Toback USA 1983
Nastassia Kinski, Rudolph Nureyev, Harvey Keitel, Ian McShane,
Ron Randell, Bibi Andersson, Pierre Clementi, James Russo, Dov
Gottesfeld, Marion Varella Carrington, Carl Lee, Mariana Magnasco,
Miguel Pinero, Jeff Silverman
A girl from the American Midwest leaves home for New York
where she gets a job as a model, and then becomes involved in
a strange struggle between a terrorist group and their oppo-
nents. A difficult film to follow, it looks like several movies
rolled into one, but none of them are worth watching.
THR 96 min (ort 100 min) SATrel: MOVIE CHANNEL

EXPOSURE ** 18
Walter Salles Jr BRAZIL 1991
Peter Coyote, Amanda Pays, Tcheky Karyo, Raul Cortez, Giulia Gam,
Cassia Kiss, Tonico Pereira, Miguel Angel Fuentes, Eduardo Conde,
Rene Ruiz, Paolo Jose, Iza Do Eirado, Tony Tornadoa, Eduardo
Waddington, Alvaro Freire
An American photographer working in Rio undergoes a consid-
erable personality change as he gets embroiled with the local
underworld and goes in search of the brutal murderer of a
young hooker. After a vicious attack her realises that he must
learn to defend himself and studies to become an expert knife
fighter under the tutelage of a local gangster. A long, dark and
most unrewarding effort that generates much tedium but has
little to hold the interest.
Aka: HIGH ART; KNIFE, THE; KNIFE FIGHTER, THE
THR 100 min (ort 105 min) VIDrel: EIV/SONOP V
Boa: novel High Art by Ruben Fonseca.

EXPRESSO BONGO *** *PG*
Val Guest UK 1959
Laurence Harvey, Sylvia Syms, Cliff Richard, Yolande Donlan, Meier
Tzelniker, Ambrosine Philpotts, Eric Pohlmann, Gilbert Harding,

Hermione Baddeley, Reginald Beckwith, Wilfrid Lawson, Martin Miller, Kenneth Griffith
The story of an over-ambitious talent agent who tries to exploit his latest discovery and make him into a star. Harvey gives a vigorous performance as a Soho hustler and Tzelniker as the agent is also good but an attempt to capture the sleazy side of Soho does not convince. Worth watching anyhow.
DRA 101 min (ort 109 min) B/W
VIDrel: VISVID/POLYREC V
Boa: play by Wolf Mankowitz.

EXQUISITE TENDERNESS * 18
Carl Schenkel GERMANY/USA 1994
Isabel Glasser, James Remar, Sean Haberle, Peter Boyle, Malcolm McDowell, Charles Dance, Beverly Todd, Charles Bailey-Gates, Walter Olkewicz, Mother Love, Gregory West, Juliette Jeffers, Nancy Banks, Kim Robillard, Teryl Rothery
A transplant patient found battered and bleeding leads to suspicions being raised about the activities of a doctor, a brilliant but unbalanced researcher, who as a cripple was sacked three years earlier. However, he has overcome this disability by experimenting with human pituitary gland extract, which he injects into baboons. The results are not pleasant. Seriously weak in terms of plotting, this gory and implausible shocker is a clumsy mixture of tedium and brutality.
Aka: SURGEON, THE
HOR 96 min (ort 100 min) VIDrel: GUILD/FOXVID V/sur

EXTERMINATING ANGEL, THE *** 12
Luis Bunuel MEXICO 1962
Silvia Pinal, Enrique Rambal, Jacqueline Andere, Jose Baviera, Augusto Benedico, Luis Beristein, Antonio Bravo, Claudio Brook, Cesar Del Campo, Lucy Gallardo, Rosa Elena Durgel, Enrico Garcia Alvarez, Ofela Guilmain, Javier Loya
Guests attending a high society function are unable to leave the room where this smart cocktail party is being held, and are forced to stay there for several days. Obscure, surrealistic allegory, allegedly attacking the vacuity of bourgeois life and manners.
Aka: EL ANGEL EXTERMINADOR
COM 89 min (ort 95 min) B/W VIDrel: ELPIC/POLYREC V

EXTERMINATOR, THE *** 18
James Glickenhaus USA 1980
Robert Ginty, Christopher George, Samantha Eggar, Steve James, Tony Di Benedetto, Dick Boccelli, Patrick Farrelly, Michele Harrell, David Lipman, Cindy Wilks, Dennis Boutsikaris, Stan Getz, Judy Licht, Roger Grimsby, Phil Chong
Two Vietnam veterans, one black and one white, work in a warehouse. One is deliberately paralysed in revenge after he foils a robbery attempt by some local thugs. After dealing with the gang his buddy decides to become a lone vigilante and clean up the city. A tough, violent and uncompromising film with a truly stomach-churning opening sequence that will not survive the UK censor. A sequel followed.
A/AD 94 min (Cut at film release by 3 min 38 sec - ort 101 min) VIDrel: 4-FRONT/POLYREC V/sur

EXTRA TERRESTRIAL VISITORS ** 15
Juan Piquer Simon SPAIN 1983
William Anton, Oscar Martin, Ian Sera, Nina Ferrer, Emil Linder, Concha Cuetos, M. Pereiro, Frank Brana, Susi Blasques, Sarah Palmer, Maria Albert
Aliens land in a mountainous region in the USA and kill three hunters whose bodies they take over in this low-budget science fiction yarn.
Aka: LOS NUEVOS EXTRA TERRESTRES
FAN 80 min VIDrel: MOPIC/QUANT V

EXTREME BEHAVIOUR ** 18
Larry Shaw USA 1996
Kristin Davis, Blair Brown, Michael Murphy
A man discovers to his horror that his daughter leads a secret life.
THR 90 min VIDrel: THIRD V

EXTREME JUSTICE ** 18
Mark L. Lester USA 1992
Lou Diamond Phillips, Scott Glenn, Chelsea Field, Yaphey Kotto, Ed Lauter, Richard Grove, Andrew Divoff, William Lucking, L. Scott Cadlwell, Daniel Quinn, Larry Holt, Tom Rosales, Ed Frias, Jay Arlen Jones, G. Adam Gifford
A cop suspended from the L.A. force for his impulsiveness is recruited for service in an elite organisation of police officers who deal with violent criminals in their own way. Based on a true story, this is a competent action-drama, that attempts to examine the problems of police who are tempted to act as judge, jury and executioner in the face of a spiralling level of violent crime. See also THE STAR CHAMBER.
Aka: S.I.S. - EXTREME JUSTICE
A/AD 96 min VIDrel: REFLEC/FIRST L/A V

EXTREME MEASURES *** 15
Michael Apted UK/USA 1996
Hugh Grant, Gene Hackman, Sarah Jessica Parker, David Morse, Bill Nunn, John Toles-Bey, Paul Guilfoyle, Debra Monk, Shaun Austin-Olsen, Andre De Shields, J.K. Simmons, Peter Appel, Diana Zimmer, Nancy Beatty, Gerry Becker
With more than a nod in the direction of COMA, this chilling and tightly directed medical thriller tells of a doctor who believes he may be on the way to curing cancer, but to carry his research forward needs to experiment on a few subjects, and is unconcerned as to the ethics or legality of this. Grant plays a fresh-faced young medic who is on to him, and puts his life at risk in order to expose these activities. Mostly humourless, slightly illogical, but very tense.
THR 113 min (ort 118 min) VIDrel: COLUM/SONOP V
Boa: novel by Michael Palmer.

EXTREME PREJUDICE ** 18
Walter Hill USA 1986
Nick Nolte, Powers Boothe, Maria Conchita Alonso, Michael Ironside, Clancy Brown, Rip Torn, William Forsythe, Matt Mulhern, Larry B. Scott, Dan Tullis Jr
A tough Texas ranger stands on one side of the border and aims to clean up the drugs traffic from Mexico. On the other side is a former boyhood friend who's now a ruthless narcotics dealer. Nolte as the ranger decides to go after him in a film of excesses - excesses of characterisation, of violence and of music. A few tongue-in-cheek touches make it more bearable but the plot - the work of four screenwriters, suffers badly from over-complexity.
A/AD 100 min (ort 104 min)
VIDrel: POLY/POLYREC/GUILD V/sur

EXTREMITIES * 18
Robert M. Young USA 1986
Farrah Fawcett, James Russo, Diana Scarwid, Alfre Woodard, Sandy Martin, Eddie Velez, Tom Everett, Donna Lynn Leavy, Enid Kent
A woman menaced by a rapist manages to turn the tables on him and gain the upper hand when she traps him in her own home. Fawcett is no more than adequate in this clumsy drama. Adapted by Mastrosimone from his play.
DRA 90 min VIDrel: 4-FRONT/POLYREC/EIV L/A V
Boa: play by William Mastrosimone.

EYE FOR AN EYE, AN ** 18
Steve Carver USA 1981
Chuck Norris, Christopher Lee, Richard Roundtree, Mako, Terry Kiser, Matt Clark, Maggie Cooper, Rosalind Chao, Toru Tanaka, Stuart Pankin, Mel Novak, Richard Prieto, Sam Hiona, Dorothy Dells, Dov Gottesfeld, J.E. Freeman
An ex-cop mounts a one-man vendetta against the drug peddlers who killed his partner and his girlfriend, making use of his skill in the martial arts for this purpose. A predictable revenger - short on acting but long on fisticuffs.
A/AD 100 min (ort 106 min)
VIDrel: ETL/4-FRONT/POLYREC V

EYE FOR AN EYE *** 18
John Schlesinger USA 1995
Sally Field, Ed Harris, Olivia Burnette, Alexandra Kyle, Kiefer Sutherland, Joe Mantegna, Beverly D'Angelo, Darrell Larson, Charlayne Woodard, Philip Baker Hall, Keith David, Wanda Acuna, Geoffrey Rivas, Armin Shimerman, Stella Garcia
The mother of a raped and murdered teenager is outraged when the killer escapes conviction on a legal technicality. She embarks on a mission of vengeance and sets out to impose her own death sentence on the culprit. A harrowing and brutal film, it dwells far too long on the girl's murder and this adds little of value to the DEATH WISH-style plot, but the film has tight direction and a uniformly impressive cast. See also DESPERATE JUSTICE.
DRA 97 min (ort 101 min) cC VIDrel: CIC V/sur
Boa: novel by Erika Holzer.

EYE OF THE NEEDLE *** 15
Richard Marquand UK 1981
Donald Sutherland, Kate Nelligan, Ian Bannen, Christopher Cazenove, Philip Martin Brown, George Belpin, Faith Brook, George Lee, Arthur Lovegrove, Colin Rix, Barbara Ewing, Patrick Connor, Rupert Frazer, Alex McCrindle, John Bennett
A German agent on a mission in Britain in 1940, plays a cat-and-mouse game with a lonely woman on a Scottish island in this gripping thriller.
THR   108 min (ort 112 min)   VIDrel: MGM/WHV   V/h
Boa: novel by Ken Follett.

EYE OF THE STRANGER ** 15
David Heavener USA 1993
Martin Landau, Sally Kirkland, Don Swayze, Joe Estevez, Stella Stevens, David Heavener, John Pleshette, Thomas F. Duffy
A town is held in the sway of a ruthless gang, with the sheriff a cowardly alcoholic, who drinks to forget the truth behind an old murder. With the arrival of a mysterious stranger, a few fairly reasonable and rather overdue changes are made. A standard actioner full of violence and the usual cliches. Mindless entertainment of the noisy sort.
A/AD   96 min   VIDrel: MARQ/QUANT   V

EYE OF THE TIGER ** 18
Richard C. Sarafian USA 1986
Gary Busey, Yaphet Kotto, Seymour Cassel, William Smith, Kimberlin, Ann Brown, Bert Ramsen, Denise Galik, Judith Barsi
Busey plays a Vietnam veteran who clashes with a violent motorcycle gang when they invade his town and sadistically murder his wife. In a one-man bid to wipe them out, he attacks their secret desert hide-out at the wheel of a powerful, customised truck. DEATHWISH meets DUEL in this average actioner of blood and gore.
DRA   88 min (ort 90 min)   VIDrel: MED/POLYREC L/A V/sh

EYES OF LAURA MARS, THE ** 15
Irvin Kershner USA 1978
Faye Dunaway, Tommy Lee Jones, Brad Dourif, Rene Auberjonois, Raul Julia, Frank Adonis, Michael Tucker, Lisa Taylor, Darlanne Fluegel, Rose Gregorio, Bill Boggs, Steve Marachuk, Meg Mundy, Marilyn Meyers, Gary Bayer, Jeff Niki
A fashion photographer seems to have a bizarre ability to predict the future and foresees some grisly murders. A detective investigates. What tension there is rapidly becomes dissipated by a series of silly red herrings; the daft ending does little to help. Co-written by John Carpenter. See also BLINK.
THR   99 min   VIDrel: CASPIC/BMGREC L/A   V

EYES WITHOUT A FACE *** 18
Georges Franju FRANCE/ITALY 1959
Pierre Brasseur, Alida Valli, Edith Scob, Francois Guerin
A mad surgeon murders young girls in an attempt to repair the face of his daughter who was mutilated in a car accident. An unpleasant but undeniably powerful horror yarn that over the years has achieved cult status. See also MASSACRE MANSION.
Aka: HORROR CHAMBER OF DR FAUSTUS; LES YEUX SANS VISAGE; TERROR CHAMBER OF DR FAUSTUS, THE
HOR   86 min (ort 90 min)   B/W   VIDrel: CONNO/RTM   V

EYEWITNESS ** 15
John Hough UK 1970
Mark Lester, Lionel Jeffries, Susan George, Tony Bonner, Jeremy Kemp, Peter Vaughan, Peter Bowles, Betty Marsden, Anthony Stamboulish, John Allison, Tom Eytle, Joseph Furst, Robert Russell, Maxine Kalil, David Lodge, Jeremy Young
A young boy on a Mediterranean island is a witness to a political murder but nobody will believe him because of his habit of telling fanciful tales. A kind of sub-Hitchcock thriller with all the usual cliches (and fancy camerawork) but quite well done for all that.
Aka: SUDDEN TERROR
THR   88 min (ort 91 min)   VIDrel: WHV   V/sh
Boa: novel Eye-Witness by Mark Hebden.

F

F FOR FAKE *** PG
Orson Welles FRANCE/IRAN/WEST GERMANY 1973
Orson Welles, Olga Palinkas, Joseph Cotten, Francois Reichenbach
Something of an oddity in the cinema, this purports to be a study of illusion and falsity in all its varied guises, principally using as a core a look at the careers of art forger Elmyr de Hory and writer Clifford Irving, the latter notorious for his forged "autobiography" of billionaire Howard Hughes. Difficult to describe, the film cleverly mixes documentary-style interviews with anecdotes, Welles adding gravitas throughout as a master of ceremonies.
Aka: VERITES ET MENSONGES
DOC   85 min   VIDrel: CONNO   V

FABULOUS BAKER BOYS, THE *** 15
Steve Kloves USA 1989
Jeff Bridges, Michelle Pfeiffer, Beau Bridges, Ellie Raab, Jennifer Tilly, Xander Berkeley, Dakin Matthews, Gregory Itzin, Wendy Girard, David Coburn, Albert Hall, Ken Lerner, Terri Treas, Todd Jeffries, Bradford English
Written by Kloves, this well-crafted tale explores the formation of an unusual love triangle, when two singing brothers hire a cynical hooker to act as vocalist and spice up their act. However, she becomes the catalyst for both their success and jealous rivalry. A polished and confident directing debut for Kloves, this romantic drama is atmospheric, detailed and simplistic, with Pfeiffer ideally cast as the sexy singer.
DRA   109 min (ort 116 min)   VIDrel: MGM/WHV L/A V/dm

FACE, THE ** 15
Jack Bender USA 1996
Yasmine Bleeth, James Wilder, Robin Givens, Richard Beymer, Ricky Paul Goldin, Chandra West, Mitchell Ryan, Mary Ellen Trainor, Ian Abercrombie, Jo De Winter
A disfigured woman who suffered appalling injuries to her face as a child seeks out those she holds responsible.
Aka: FACE TO DIE FOR, A
DRA   90 min mTV   VIDrel: ARENA/RTM   V

FACE OF FU MANCHU, THE ** PG
Don Sharp UK 1965
Christopher Lee, Nigel Green, Tsai Chin, James Robertson Justice, Howard Marion-Crawford, Joachim Fuchsberger, Karin Dor, Walter Rilla, Harry Brogan, Poulet Tu, Peter Mossbacher, Edwin Richfield
Story of a Chinese criminal mastermind who plots to take over the world with the aid of poison gas. An occasionally entertaining spoof set in the 1920s and based on the character created by Sax Rohmer in the 1911 novel. "The Brides Of Fu Manchu" followed.
THR   94 min   VIDrel: BRAVE/SONOP L/A   V

FACE ON THE MILK CARTON, THE ** (PG)
Waris Hussein USA 1995
Kellie Martin, Sharon Lawrence, Edward Herrmann, Richard Masur, Johnny Green, Jill Clayburgh, Kristofer Ryan Winters, Caroline Perreyclear, Paul Dow, Joanna Canton, Lori Lindberg, Leslie Hall, Ellen Lee, Richard K. Olsen
A young girl sees a face of a missing child on a milk carton that seems to be familiar and learns from her parents that she was in fact adopted, and that she has a family history completely unknown to her, including a biological mother who become involved in a sinister cult that she later tried to break free from. A sad little story, of good performances and careful scripting.
DRA   92 min mTV   SATrel: MOVIE CHANNEL
Boa: novels The Face on the Milk Carton and Whatever Happened to Janie by Caroline Gooney.

FACE VALUE ** 15
John Gray USA 1990
Cheryl Pollak, Kirk Baltz, Maria MacDonald, Kari G. peyton, Louis Lotorto, Timothy Shickney, Jane Geesman, Rick Jones, Larry Brooks, Brynn Baron, Mark Vincent, Elizabeth Bradley, J.P. Phillips, Walter Hoppert, Corey Brunush
A pretty girl from a small town agrees to share an apartment with a make-up artist, but doesn't realise this will place her in danger. Based on a true story.
Aka: MARLA HANSON STORY, THE
THR   91 min mTV   VIDrel: GUILD/SONOP   V

FACES **** 15
John Cassavetes USA 1968
John Marley, Gena Rowlands, Lynn Carlin, Seymour Cassel, Fred Draper, Val Avery, Dorothy Gulliver, Joanne Moore Jordan, Darlene Conley, Gene Darfler, O.G. Dunn, Elizabeth Deering, George Sims, Dave Mazzie, Julie Gambol, James Bridges
In L.A. an unhappy executive decides to divorce his wife but

finds that he cannot bear to go through with it. Written and directed by Cassavetes, this intense and demanding study of loneliness, despair and the disintegration of a marriage demands a good deal of patience, but is easily one of the director's most absorbing works.
DRA   124 min (ort 129 min) B/W
VIDrel: ELPIC/POLYREC   V

FACESITTER *
Jim Enright USA
18
1992
Tabatha Cash, Lacy Rose, Leanna Foxx, Kelly O'Dell, Mikala, Sierra, Mark Wallice, Jonathon Morgan, T.T. Boy
Mostly, this has the cast sitting around and discussing how they enjoy the title activity, very little of which is actually seen. The plot has Morgan running a support group that attempts to wean people off this compulsion, leading to the inevitable succession of flashbacks, as each member of the group recounts one of their sexual adventures. A very weak film in terms of plotting, it also has a low-budget look about it. A sequel followed.
A   65 min   VIDrel: MIA/DISC   V

FADE TO BLACK *
Vernon Zimmerman USA
18
1980
Tim Thomerson, Mickey Rourke, Dennis Christopher, Linda Kerridge, Norman Burton, Morgan Paull, Marya Small, Gwynne Gilford, James Luisi, Eve Brent Ashe, John Steadman, Marcie Barkin, Hennen Chambers, Bob Drew, Melinda Free
A film buff is unable to between distinguish film and reality, and takes a terrible revenge on his enemies whilst dressed in the clothes of his favourite cinema villains. A great idea goes nowhere in this violent and sterile film.
HOR   97 min (ort 100 min)   VIDrel: MIA/DISC   V

FADE TO BLACK **
John McPherson USA
15
1992
Timothy Busfield, Heather Locklear, Micahel Beck, Louis Giambalvo, David Byron, Cloris Leachman, Galen B. Schrick, Timi Pulhiere, Bill Birch, John DeLay, Russ Fast, Shawna Schun, Louis A. Lotorto Jr., Curt Hanson, Bunnie Siler
A college professor of psychology given to videotaping the doings of his neighbours for use in his classes, accidentally captures a murder on tape while he is adjusting his video-camera. When he finds himself being framed for this crime, he is forced into mounting his own investigation. A capable tale that benefits from a good performance by Busfield as a with-drawn and rather unlikely hero.
THR   82 min (ort 84 min) mTV   VIDrel: CIC/SONOP L/A   V

FAIL SAFE ***
Sidney Lumet USA
PG
1963
Henry Fonda, Dan O'Herlihy, Walter Matthau, Frank Overton, Edward Binns, Fritz Weaver, Sorrell Brooke, Larry Hagman, Dom DeLuise, Russell Hardie, William Hansen, Russell Collins, Nancy Berg, Stuart Germain, Louise Larabee
A computer system designed to make it impossible for a malfunction to start a nuclear attack malfunctions itself. Because of unforeseen circumstances the unthinkable happens and Moscow is destroyed. The US President sees a way to appease the Soviets and avoid WW3, but only by making an appalling decision. Written by Walter Bernstein, this taut drama makes good use of a literate script and low-key direction. See also DR STRANGELOVE and BY DAWN'S EARLY LIGHT.
DRA   108 min (ort 111 min) B/W
VIDrel: ENCORE/SPEAR/COLUM   V
Boa: novel by Eugene Burdick and Harvey Wheeler.

FAIR GAME **
Mario Orfini ITALY
(15)
1988
Trudie Styler, Gregg Henry, Bill Moseley
A woman whose ex-husband is determined to punish her for rejecting him, traps her in her house with a deadly Black Mamba snake, in this tense but cliched thriller. The music is by Oscar award-winning Giorgio Moroder.
Aka: MAMBA
THR   81 min (Cut at UK cinema release)
SATrel: SKY MOVIES

FAIR GAME **
Andrew Sipes USA
15
1995
William Baldwin, Cindy Crawford, Steven Berkoff, Christopher McDonald, Salma Hayek, Miguel Sandoval, Johann Carlo, John Bedford Lloyd, Olek Krupa, Jenette Goldstein, Marc Macaulay, Sonny Carl Davis, Frank Medrano, Don Yesso

When a group of former KGB operatives come after a tough Florida female lawyer, she is given police protection in the form of an equally tough cop. However, they eventually decide to join forces and go after the bad guys themselves. Very much a production-line actioner, whose only feature of distinction being that it was the vehicle used to start off ex-model Crawford's acting career, but is unlikely to advance it.
A/AD   86 min (ort 91 min) cC   VIDrel: WHV   V/sur
Boa: novel by Paula Gosling.

FALCON 7 **
Roy Thomas HONG KONG
PG
1989
A space opera-style animation from the makers of CAPTAIN COSMOS and THE COSMOS CONQUEROR. The Earth Defence Force must undertake a perilous mission to save a planet and a missing inventor from the ruthless Zoic Empire and its evil Empress. Watchable but not especially well-mounted or scripted.
ANIM   63 min   VIDrel: IMPENT   V

FALL AND RISE OF REGINALD PERRIN, THE **
John Howard Davies/Gareth Gwenlan UK
PG
1976/1980
Leonard Rossiter, Pauline Yates, John Barron, Sue Nicholls, John Horsley, Sally Jane-Spencer, Anne Cunningham, Ken Wynne, Hilary Mason, Charmian May, Roger Brierley, Pamela Manson, John Forbes-Robertson, David Millet, Vi Kane
Compilation of twenty thirty-minute episodes of a TV series, telling the story of a well-paid executive who one day cracks up from boredom and drops out. Rossiter is good in this compilation comedy but the lack of material and the tendency to rely on the same gags lets it down badly. Written by David Nobbs, who has done much better work.
COM   209 min (2 cassettes – ort 360 min) mTV
VIDrel: BBC/TECH   V/h

FALL OF THE HOUSE OF USHER, THE *
Roger Corman USA
15
1960
Vincent Price, Mark Damon, Myrna Fahey, Harry Ellerbe, Bill Borzage, Mike Jordan, Nadajan, Ruth Oklander, George Paul, David Andar, Eleanor LeFaber, Geraldine Paulette, Phil Sylvestre, John Zimeas
A Corman low-budget mangling of a fine Poe story about an ancient house inhabited by a man who is the last of his line. Dull, boring and cheap, all those things that have given Corman cult status as a director. Filmed twice before in 1928 and 1949 and made for TV in 1978.
Aka: HOUSE OF USHER
HOR   76 min (ort 85 min)   VIDrel: VCC L/A   V
Boa: short story by Edgar Allan Poe.

FALL OF THE ROMAN EMPIRE, THE ***
Anthony Mann USA
U
1964
Alec Guinness, James Mason, Sophia Loren, Stephen Boyd, John Ireland, Christopher Plummer, Anthony Quayle, Mel Ferrer, Omar Sharif, Eric Porter, Peter Damon, Douglas Wilmer, Andrew Keir, George Murcell, Lena Von Martens
Long-winded account of the events leading up to the end of the Roman Empire in chaos after the poisoning of Emperor Aurelius by his mad son Commodus. A film that starts off intelligently enough but degenerates into a boring epic, only a parade of spectacular events keeps us watching.
DRA   172 min (ort 182 min)   VIDrel: 4-FRONT/POLYREC   V

FALL TIME *
Paul Warner USA
18
1994
Mickey Rourke, Stephen Baldwin, Sheryl Lee, Jason London, David Arquette, Jonah Blechman, J. Michael Hunter, Richard K. Olsen, Sammy Kershaw, Steve Alden, Michael Edelstein, Jeff Gardner, Tom Hull, Amy Parrish, John Henry Scott
In the 1950s, three high-school chums set out to fake a shooting as a final stunt before graduation, when they attempt to stage a "robbery" at a small bank in Minnesota. However, they unwisely choose a bank at which a real robbery is taking place, and they are taken hostage by a pair of ruthless and psychotic criminals. A pretentious thriller with more than a touch of the preposterous about it, it would really have worked far better as a comedy.
THR   85 min (ort 88 min) mTV   VIDrel: POLY/POLYREC   V

FALLEN ANGELS **
Larry Leahy USA
15
1991
James Remar, Michael Wright, Emily Longstreth
A hitman who is dying steals a fortune from his mobster uncle

and then travels to Las Vegas to avenge his father's death. Travelling by limousine, he passes the time by making calls to those he has wronged in the past.
Aka: CONFESSIONS OF A HITMAN
DRA   92 min   VIDrel: CRYSTAL/HIFLI   V/h

FALLEN ANGELS ✶✶   15
Wong Kar-Wai (Wang Jiawei) HONG KONG   1995
Leon Lai (Li Ming), Michele Reis (Li Jiaxin), Takeshi Kaneshiro (Jin Chengwu), Charlie Young (Yang Caili), Karen Mong (Mo Wenwei), Chan Fai-Hung (Chen Huihong), Chen Wanlei, Toru Saito, Kong To-Hoi, Kwan Lee-Na (Guan Lina)
A professional killer has a business relationship with a woman who acts as his agent and organises his assignments, but this changes when she becomes sexually obsessed with him. Meanwhile, in another part of Hong Kong, a mute man plays at being a shopkeeper by taking over closed shops and turning passers-by into unwilling customers. A riotously noisy, lurid and garish tale, with little dialogue and nervous camerawork. The screenplay is by Kar-Wei.
Aka: DUOLUO TIANSHI
A/AD   96 min   CINrel

FALLING DOWN ✶✶✶   18
Joel Schumacher USA   1993
Michael Douglas, Robert Duvall, Barbara Hershey, Rachel Ticotin, Frederic Forest, Tuesday Weld, Louis Smith, Michael Paul Chan, Raymond J. Barry, D.W. Moffett, Brent Hinkley, Dedee Pfeiffer, Vondie Curtis-Hall, Jack Kehoe
Intense, powerful and extremely violent portrait of an ordinary white-collar worker who snaps under the pressures of city life. Abandoning his car in an L.A. traffic jam, he embarks on a crusade of revenge against society in general and rule-bound officious individuals in particular, having acquired an impressive armoury that includes a bazooka. A highly disturbing film, but one that for all its rage, lacks a clear and unambiguous message.
DRA   108 min (ort 113 min) cC   VIDrel: WHV L/A   V/sur V/dm

FALLING FOR YOU ✶✶   18
Eric Till USA   1995
Jeanie Garth, Billy Dee Williams, Costas Mandylor, Helen Shaver, Currie Graham, Peter Outerbridge, Eugene A. Clarke, Walter Alza, Warren Dexter Beatty, Laura Cataland, Gordon Currie, Lucy Filippone, Kimberly Huie, Paul Molitor
Thriller in which a woman is terrorised by a serial killer whose penchant it is to push his blonde victims out of the windows of tall buildings. An okay thriller with a few original points of interest in an otherwise by-the-numbers plot.
THR   92 min   VIDrel: MED/COLUM   V/sh

FALLING FROM GRACE ✶✶   15
John Mellencamp USA   1991
John Mellencamp, Mariel Hemingway, Claude Akins, Dub Taylor, Kay Lenz, Larry Crane, Kate Noonan, Deidre O'Connell, John Prince, Brent Huff, Tracy Cowles, Joanna Jackson, Melissa Ann Hackman, Mary Tom Crain, Sigmund Balaban
A Country singer returns home with his new wife for his grandfather's eightieth birthday, but once there, finds himself embroiled in all sorts of domestic difficulties and squabbles. An adequate family drama with a competent script by Larry McMurtry. This was singer Mellencamp's acting and directing debut.
DRA   96 min (ort 100 min)   VIDrel: COLUM/SONOP V/sur

FALLING IN LOVE ✶✶   PG
Ulu Grossbard USA   1984
Robert De Niro, Meryl Streep, Jane Kaczmarek, Harvey Keitel, Dianne Wiest, George Martin, David Clennon, Victor Argo, Wiley Earl, Jesse Bradford, Chevi Coltron, Richard Gizza, Frances Conroy, James Ryan, Sonny Abagnale
Two Long Island commuters who are both married, meet and fall in love. A glossy, one-dimensional story set in the comfortable world of professional people who can afford to indulge their emotional whims to the full. Not helped by its totally unsuitable and ultimately irritating theme music (the work of Dave Grusin).
DRA   102 min (ort 106 min)   VIDrel: CIC/SONOP   V

FALLING IN LOVE AGAIN ✶   15
Steven Paul USA   1980
Elliott Gould, Susannah York, Michelle Pfeiffer, Stuart Paul, Kaye

Ballard, Robert Hackman, Steven Paul, Todd Helper, Herb Rudley, Marion McCargo, Bonnie Paul
A husband and wife experience trouble in their marriage when he is unable to stop idolising his youthful days in the Bronx. An inept and contrived feature debut for writer-actor-producer-director Paul, that is told in awkward flashback style, and has a weak and over-sentimental resolution.
Aka: IN LOVE
DRA   95 min (ort 103 min)   VIDrel: ARROW/RTM   V

FALSE ARREST ✶✶   15
Bill L. Norton USA   1991
Donna Mills, Robert Wagner, Steven Bauer, James Handy, Lane Smith, Lewis Van Bergen, Dennis Christopher, Paul Gleason, Penny Fuller, Kiersten Warren, Jason London, Brian Bonsall, Mimi Kuzyk, Warren Frost, Ben Lemmon, Kelly Curtis
A wealthy businesswoman is accused of planning the demise of her husband's business partner and mother-in-law. She embarks on a protracted fight to clear her name and win her freedom, which is achieved only after three years in jail. Based on court reports and the like, this is a true story that would have played better as a documentary; overlength and a TV-movie treatment tend to weaken its impact.
DRA   179 min (ort 182 min) mTV
VIDrel: POLY/POLYREC/BRAVE L/A   V
Boa: book by Joyce Lukezic and Ted Schwarz.

FALSELY ACCUSED ✶✶   PG
Noel Nosseck USA   1993
Lisa Hartman Black, Christopher Meloni, Peter Jurasik, James Staley, Gwynyth Walsh, Martin Kove, David Ogden Stiers, Cloris leachman, Tom McBeath, Babz Chula, Sandra P. Grant, Dons S. Davis. Bill Dow, Eric Keenleyside, Joy Brazeau
A mother is accused of murdering her infant son when he mysteriously dies but is cleared. When the couple have another child that begins to exhibit the same symptoms, it is taken from their care and they are forced to fight through the courts to regain custody. Eventually, a rare genetic condition is shown to be the cause of all their troubles. A solidly made piece of TV fare (based on a true case) it unfortunately occasionally descends into melodrama.
DRA   90 min mTV   VIDrel: 20VIS/SONOP   V

FAME: THE MOVIE ✶✶✶   15
Alan Parker USA   1980
Irene Cara, Lee Curreri, Eddie Barth, Laura Dean, Paul McCrane, Barry Miller, Gene Anthony Ray, Maureen Teefy, Antonio Franceschi, Anna Meara, Albert Hague, Debbie Allen, Joanna Merlin, Tresa Hughes, Steve Inwood
A rather episodic look at the lives and loves of the students at New York's School for the Performing Arts. Despite vigorous performances from a dynamic cast, the lack of a real story does lets one's interest drift. A considerably better TV series followed. Written by Michael Gore. AA: Song ("Fame" – Michael Gore (m)/Dean Pritchard (l)), Score/orig (Michael Gore).
Aka: FAME
MUS   128 min (ort 134 min)   VIDrel: MGM/WHV   V/sur

FAMILY ALBUM ✶✶   PG
Jack Bender USA   1993
Jaclyn Smith, Michael Ontkean, Joe Flanigan, Kristin Minter, Leslie Horan, Tom Mason, Brian Krause, Paul Satterfield, James Curley, Kristen Dalton, Joel Gretsch, Melody Kay, Mary Carver, Dru Mouser, Richard Marion, Ronnie Schelle
After her marriage collapses, a woman fights to bring up her kids alone whilst at the same time rebuilding her Hollywood career. Standard soap opera with all the usual embellishments.
Aka: DANIELLE STEEL'S FAMILY ALBUM
DRA   167 min (ort 200 min) mTV   VIDrel: MIA/DISC V/sh
Boa: novel by Danielle Steel.

FAMILY BUSINESS ✶✶✶   15
Sidney Lumet USA   1989
Sean Connery, Dustin Hoffman, Matthew Broderick, Rosana DeSoto, Janet Carroll, Victoria Jackson, Bill McCutcheon, Deborah Rush, B.D. Wong, Marilyn Cooper, Salem Ludwig, Rex Everhart, James Tolkan, Marilyn Sokol, Wendell Pierce
An unusual drama with three generations of crooks: a professional thief, his reformed son, and his bright and ambitious grandson. When the grandson decides to plan a perfect crime, he enlists the help of his grandfather in order to entice his father, from whom he is estranged, into the robbery. A clever drama,

scripted by Patrick, and if not quite believable, sustained by a set of terrific performances from the leads.
DRA   108 min (ort 115 min)
VIDrel: 4-FRONT/POLYREC/BRAVE   V/sur
Boa: novel by Vincent Patrick.

FAMILY DIVIDED, A ** PG
Scott Rosenfelt USA   1991
Anne Archer, Patti Lupone, Tzvi Ratner-Stauber, Joe Mantegna, Paul Reiser, Allen Garfield, Conchara Ferrell, David Margulies, Shiri Appleby, John Capodice, Julianne Michelle, Ralph Monaco, Keaton Simons, Milt Oberman, Gina Hecht
Portrait of a Jewish family in L.A. in 1969 where a thirteen-year-old boy preparing for his barmitzvah finds himself having to cope with the tension in the family caused by his father's gambling and the intervention of his aunt, who has helped them out financially on several occasions. A watchable but somewhat stereotyped and not entirely convincing coming-of-age tale.
Aka: FAMILY PRAYERS
DRA   104 min (ort 109 min)   VIDrel: COLUM/SONOP
V/sh

FAMILY DIVIDED, A ** 15
Donald Wrye CANADA   1993
Faye Dunaway, Stephen Collins, Cameron Bancroft, Judson Mills, Don S. Davis, Matt Hill, Diane D'Aquila, Aidan Pendleton, Cylk Cozart, Michael Shanks, Andrea Nemeth, Emmanuelle Vaugier, Andy Skely, Suzy Joachim, Sue Mathew, Antony Holland
A woman with a contented family life is shocked to learn that her eldest son took part in the brutal gang rape of a young girl. Matters get more complicated when she finds that her husband is determined to keep their son's involvement in this crime hidden. Dunaway plays her part as the devoted mother with great skill, adding immeasurably to the interest in this somewhat overwrought drama.
DRA   89 min mTV   VIDrel: ODY/SONOP   V/h
Boa: novel Mother Love by Judith Henry Wall.

FAMILY OF COPS * 15
Ted Kotcheff USA   1995
Charles Bronson, Angela Featherstone, Lesley-Anne Down, Daniel Baldwin, Barbara Williams
Not only is a man a top police officer, he also has two sons who are cops plus a defence attorney daughter. When the youngest daughter, who is a bit of a tearaway, wakes up next to a dead businessman, the whole family has to pool their resources in order to prove her innocence. Tired and derivative, this is just about watchable.
DRA   87 min mTV   VIDrel: HIFLI/SONOP   V/h

FAMILY OF STRANGERS ** PG
Sheldon Larry CANADA   1993
Melissa Gilbert, Patty Duke, William Shatner, Ashley Rogers, Martha Gibson, Edie McCormack, John Shaw, Stephen Dinopoulos, Melody Ryane, Michael Jacobucci, Tasha Simms, Dax Belanger, Peter Strebbings, Leslie Hopes, Michell Beuadoin
A woman finds that she must have an operation to remove a blood-clot on her brain, but its chances of success depend on her family medical history, which is not known as she is adopted. A race against time ensues to find her real parents, but the results of her investigation are unexpected. Based on a real case. Fair.
DRA   90 min mTV   VIDrel: BRAVE/SONOP L/A   V
Boa: book Judy by Jerry Hulse.

FAMILY PICTURES *** 15
Philip Saville USA   1993
Anjelica Huston, Sam Neill, Kyra Sedgwick, Dermot Mulroney, Gemma Barry, Tara Charendorf, Torri Higginson, Jamie Harrold, Janet-Laine Green, Alexandra Petrocci, Laura Bertram, Jared Wall, Corey Sevier, Sean Ryan, Jared Cook
A mini-series compilation telling of forty years in the life of a family with an autistic boy whose presence leads to problems for his parents and five siblings. The latter feel resentful and neglected since their mother seems to devote so much time and attention to him, while the father unconsciously blames his wife for their son's condition. Excellent ensemble acting overcomes the thin and episodic script, ensuring us of an absorbing and enjoyable work.
DRA   172 min (ort 240 min) mTV   VIDrel: ODY/SONOP
V/sh
Boa: novel by Sue Miller.

FAMILY PLOT *** 15
Alfred Hitchcock USA   1976
Karen Black, Bruce Dern, William Devane, Barbara Harris, Ed Lauter, Cathleen Nesbitt, Katherine Helmond, Warren J. Kemmerling, Edith Atwater, William Prince, Nicholas Colasanto, Madge Redmond, John Lehne, Charles Tyner, Alexander Lockwood
Hitchcock's last film is not one of his best although it shows a return to the tongue-in-cheek direction of his pre-war films. This is his 54th film and tells of a phoney medium and his girlfriend who are searching for a missing heir but become entangled with a sinister couple planning a kidnapping. A mildly entertaining film, but one in which credibility is stretched rather too far. Scripted by Ernest Lehman.
THR   115 min (ort 120 min)   VIDrel: CIC/SONOP   V
Boa: novel The Rainbird Pattern by Victor Canning.

FAMILY PRAYERS ** (PG)
Scott Rosenfelt USA   1991
Anne Archer, Patti Lupone, Tzvi Ratner-Stauber, Joe Mantegna, Paul Reiser, Allen Garfield, Conchara Ferrell, David Margulies, Shiri Appleby, John Capodice, Julianne Michelle, Ralph Monaco, Keaton Simons, Milt Oberman, Gina Hecht
Portrait of a Jewish family in L.A. in 1969 where a young boy preparing for his bar-mitzvah finds himself having to cope with the tension in the family caused by his father's gambling and the intervention of his aunt, who has helped them out financially on several occasions. A watchable if not entirely convincing tale.
DRA   105 min (ort 109 min)   SATrel: MOVIE CHANNEL

FAMILY RESCUE *** 15
Graeme Campbell USA   1995
George C. Scott, Rachael Leigh Cook, Dom Diamont, Ally Sheedy
A fifteen-year-old is raped by her mother's boyfriend and becomes pregnant, and to add to her torment finds that the courts grant custody of the child to her assailant. She turns to her grandfather for help in getting this insane decision revoked.
DRA   87 min (ort 90 min)   VIDrel: ODY/SONOP   V/h

FAMILY REUNION: A RELATIVE NIGHTMARE ** (U)
Neal Israel USA   1994
Melissa Joan Hart, Jason Marsden, David L. Lander, Romy Walthall, Marcia Strassman, Susan French, Jo Anne Worley, Dody Goodman, Peter Billingsley, Gerrit Graham, Meghann Haldemann, Sara Rue, Joe Flaherty, Alley Mills, Norman Fell
The nutty Dooleys hold an annual "Family Olympics", presided over by their eccentric skateboard-loving mother, who is celebrating her hundredth birthday. As each side of the family attempts to prove its superiority, a youngster who finds the whole thing a frightful chore meets up with a pretty girl who is fleeing her nasty stepsisters. An anarchic, chaotic and fairly mindless comedy.
COM   88 min (ort 90 min) mTV   SATrel: SKY MOVIES

FAMILY THING, A ** 15
Richard Pearce USA   1995
Robert Duvall, James Earl Jones, Michael Beach, Irma P. Hall
Duvall plays a racist, Southern store-keeper in the Mid-west who learns from his dying mother that his birth was the result of his father raping a black maid. He also learns that he has a black brother in Chicago, and sets out to find him. Naturally, they hate each other on sight, but eventually become firm friends. An overly sentimental yarn that wears its heart on its sleeve and was presumably intended to offer a few useful insights regarding racial equality.
DRA   105 min cC   VIDrel: WHV   V

FAMILY WAY, THE ** 15
John Boulting/Roy Boulting UK   1966
Hywel Bennett, Hayley Mills, John Mills, Marjorie Rhodes, Wilfred Pickles, Murray Head, Barry Foster, Liz Fraser, Avril Angers, John Comer, Colin Gordon, Robin Parkinson, Andrew Bradford, Harry Locke, Thorley Walters, Ruth Trouncer
Laboured British comedy-drama about the problems a pair of newlyweds face, not least being the fact that they are obliged to continue living in the boy's parents' house and that Bennett as the husband is suffering from impotence. A conspiracy of circumstances prevents consummation but by the time our hubby does succeed one no longer cares. A film seen as controversial in its day may now be seen for what it really is – dull.
DRA   110 min (ort 114 min)   VIDrel: WHV   V/h
Boa: play Honeymoon Deferred (also known as All In Good Time) by Bill Naughton.

FAN, THE ★★   15
Tony Scott USA   1996
Robert De Niro, Wesley Snipes, Ellen Barkin, John Leguizamo, Benicio del Toro, Chris Mulkey, Patti D'Arbanville-Quinn, Andrew J. Ferchland, Brandon Hammond, Charles Hallahan, Dan Butler, Kurt Fuller, Michael Jace, Frank Medrano
De Niro dons his obsessive psycho persona as a nutty baseball fan whose support of his favourite player (recently dropped from the team) extends to murder, as he sees this as the best way to get the latter reinstated. Snipes proves to be less grateful than was expected, so to make a few more points, he kidnaps the man's son. Illogical and irritating, this is a film that depends too heavily on its stars to remain watchable. Both silly and disappointing.
THR   111 min (ort 114 min)   VIDrel: EIV/SONOP V/sh
Boa: novel by Peter Abrahams.

FANNY AND ALEXANDER ★★★★   15
Ingmar Bergman FRANCE/SWEDEN/WEST GERMANY1982
Gunn Walgren, Ewa Froeling, Jarl Kulle, Erland Josephson, Allan Edwall, Boerje Ahlstedt, Mona Malm, Gunnar Bjornstrand, Jan Malmsjoe, Pernilla Allwin, Bertil Guve, Harriet Andersson
In Uppsala, a well-to-do acting family gather to celebrate Christmas of 1907. When the father dies, the fortunes of his two young children change with his widow's marriage to a sadistic priest. Announced as Bergman's final film, this is a superb and life-affirming evocation of the magic of childhood. The tape is the original longer-running version shown on TV. AA: Foreign, Cin (Sven Nykvist) Art/Set (Anna Asp), Cost (Marik Vos).
DRA   301 min   VIDrel: ARTIF/20TH V

FANTASIA ★★★★   U
Ben Sharpsteen (production supervisor) USA   1940
Narrated by Deems Taylor and with music by Leopold Stokowski and the Philadelphia Orchestra
A brilliant set of cartoons, each of which accompanies a piece of classical music – "The Sorcerer's Apprentice" and "Night On The Bare Mountain" being two highlights. Disney will release this original version once only, as the film is to be digitally restored. Sadly, this slightly sanitised tape has lost one or two fragments, most notably the "black cherub" sequence from the "Pastoral Symphony". AA: Spec Award (Leopold Stokowski and associates).
ANIM   114 min (ort 120 min) Special collector's edition available VIDrel: WDV/SONOP L/A V

FANTASIES ★   15
John Derek USA   1976
Bo Derek, Peter Hooten, Anna Alexiadis, Phaedon Gheorghitis, Therese Bohlin, Nicos Paschalidis, Constantine Beladames, Boucci Simma, Vienneula Koussefhane
The efforts of two men to turn a Greek island into a tourist resort is just a vehicle for displaying Bo Derek plus her minimal acting ability. Another homage by the director to the charms of his wife, who was just sixteen when this amateurish nonsense was made.
Aka: AND ONCE UPON A LOVE; ONCE UPON A LOVE
A   90 min   VIDrel: MOPIC/SGSVID V

FANTASTIC PLANET ★★★   15
Rene Laloux CZECHOSLOVAKIA/FRANCE   1973
Voices of: Barry Bostwick, Marvin Miller, Olan Soule, Cythia Alder, Nora Heflin, Hal Smith, Mark Gruner, Monika Ramirez, Janet Waldo
Imaginative animated SF fantasy about a distant planet people by two distinct races, the Draags and the much smaller Oms, who are descendants of survivors from Earth. The Oms are kept pretty much as pets, but one day a revolt breaks out.
Aka: LA PLANETE SAUVAGE; PLANET OF INCREDIBLE CREATURES
ANIM   72 min   VIDrel: SCREAM/SPEAR V
Boa: novel Oms En Serie by Stefan Wul.

FANTASTIC VOYAGE ★★★   U
Richard Fleischer USA   1966
Stephen Boyd, Edmond O'Brien, Raquel Welch, Donald Pleasence, Arthur O'Connell, William Redfield, Arthur Kennedy, James Brolin, Barry Coe, Jean Del Val, Shelby Grant, Ken Scott, Brendan Fitzgerald
When a famous scientist is shot, a highly experimental technique is used in order to save him. A medical team is placed aboard a submarine, reduced to microscopic size and injected into his bloodstream to remove a blood clot on his brain. An interesting film of good albeit dated special effects (largely wasted on TV) but poor acting and dialogue. Script is by Harry

Kleiner. AA: Art/Set (Smith and Hennesy/Scott and Reiss), Effects/vis (Cruickshank).
Aka: MICROSCOPIA; STRANGE JOURNEY
FAN   96 min (ort 100 min)   VIDrel: 20TH/TECH L/A; ENCORE (LV only) V/h LV
Boa: novel by Otto Klement and Jay Lewis Bixby.

FANTASY ★★   15
Geoffrey Brown/Derek Strahan AUSTRALIA   1990
Colin Borgonon, Clare Chilton, Jane Darley-Jones, Julia Binns, Campbell MacPherson, Elizabeth Stewart, Brendan Strahan
A woman for whom sexual satisfaction is found in a private fantasy world of her own making, meets a doctor who persuades her to take part in an experiment whereby they will each in turn act out the fantasies of the other. A bizarre psychological thriller.
THR   76 min Cut (25 sec – ort 90 min)
VIDrel: COLUM/SONOP V/sh

FAR AND AWAY ★   15
Ron Howard USA   1992
Tom Cruise, Nicole Kidman, Thomas Gibson, Robert Prosky, Barbara Babcock, Cyril Cusack, Eileen Pollock, Colm Meaney, Douglas Gillison, Michelle Johnson, Wayne Grace, Niall Tobin, Barry McGovern, Gary Lee Davis, Jared Harris
A simple, old-fashioned romance, set at the close of the 19th century, when the son of a poor Irish tenant-farmer is obliged by circumstances to sail for America in the company of the daughter of a local landowner. She merely requires his services as a servant, but matters do not proceed to their best advantage, and after many tribulations they discover their love for each other. Shot on 65 mm, this is a beautifully filmed but empty dud.
DRA   134 min (ort 140 min) wScrn cC
VIDrel: CIC/SONOP; PION (LV only) V/dm V/sur LV

FAR FROM HOME: THE ADVENTURES OF YELLOW DOG ★★   U
Phillip Borsos USA   1994
Bruce Davison, Mimi Rogers, Jesse Bradford, Tom Bower, Joel Palmer, Josh Wanamaker, Margot Finley, Matt Bennett, St Clair McColl, Jennifer Weissenborn, Gordon Neave, Karen Kruper, Dean Lockwood, John LeClair plus Dakotah (the dog)
Fourteen-year-old Angus goes on a boating trip with his dog and the pair get trapped in the inhospitable wilds of Canada when their boat capsizes. While they are struggling to survive and reach civilisation the boy's worried parents are doing all they can to locate him. A survival saga that covers familiar ground, and though perfectly adequate viewing, offers no new surprises or insights. Some of the survival sequences may be unsuitable for young kids.
JUV   77 min (ort 100 min) cC   VIDrel: 20TH/TECH V/sur

FAR FROM THE MADDING CROWD ★★★   U
John Schlesinger UK   1967
Julie Christie, Alan Bates, Terence Stamp, Peter Finch, Prunella Ransome, Fiona Walker, Alison Leggatt, Paul Dawkins, Julian Somers, Freddie Jones, Brian Rawlinson, Denise Coffey, Andrew Robinson, John Barrett, Pauline Melville
The story of a beautiful female landowner and her relationship with the three men in her life in Dorset of 1866. Overlong and carefully made, but with little feeling for an England long since gone, perhaps largely because Christie is too much of a modern miss to bring conviction to her part. The excellent photography is by Nicolas Roeg, with a score by Richard Rodney Bennett and a script by Frederic Raphael.
DRA   155 min Cut (12 sec – ort 169 min)
VIDrel: LUMI/SPEAR L/A V
Boa: novel by Thomas Hardy.

FAR NORTH ★★★   15
Sam Shepard USA   1987
Jessica Lange, Charles Durning, Tess Harper, Donald Moffat, Ann Wedgeworth, Patricia Arquette, Nina Draxton, Timothy Hanrahan, Sarah Iverson, Lindsay Hendell, Mary Russell, Sandra Iverson, Sarah Gramse, Paolo Rossi, Pearl Fuller
A restrained drama with a few touches of comedy, that follows the changing fortunes of a Minnesota farming family, where three generations interact and are soon to see the demise of their way of life. When the wayward daughter returns to see her badly injured father, she is given the job of shooting the wild horse that nearly caused his death in a wagon accident. Written by Shepard, this is an impressive directing debut.
DRA   85 min (ort 90 min)   VIDrel: FIRST/SONOP L/A V

FAR OFF PLACE, A **

PG

Mikael Salomon USA   1992

Reese Witherspoon, Ethan Randall, Maximillian Schell, Jack Thompson, Sarel Bok, Robert Burke, Patricia Kalember, Daniel Gerroll, Miles Anderson, Magdalene Damas, Taffy Chihota, Anthony Chunyanya, Brian Cooper, Fidelis Cheza, John Indi

A young American girl growing up together on an African game reserve has to cope with a visiting city boy who is an unpleasant little snob. When her parents are killed by poachers, both teens have to pull together when they are forced to trek through the Kalahari Desert. This would have been a typical Disney effort were it not for the graphic violence.

JUV   116 min   VIDrel: WDV/TECH L/A   V

Boa: books: A Story Like the Wind and A Far Off Place by Laurens van der Post.

FAR PAVILIONS, THE **

PG

Peter Duffell UK   1984

Ben Cross, Amy Irving, Omar Sharif, Christopher Lee, Benedict Taylor, Saeed Jaffrey, John Gielgud, Rossano Brazzi, Robert Hardy, Sneh Gupta, Jennifer Kendal, Felicity Dean, Rupert Everett, Mary Peach, Adam Bareham

A long wallow in nostalgia for the days of the British Raj in this tale of a 19th century officer who is torn between love of his country and his Hindu upbringing. Lavish and incredibly detailed, but the cliched script is a serious flaw. Originally shown in three 96-minute episodes.

Aka: BLADE OF STEEL

DRA   108 min Cut (12 sec – ort 288 min) mTV

VIDrel: BRAVE/SONOP L/A   V

Boa: novel by M.M. Kaye.

FARAWAY, SO CLOSE **

15

Wim Wenders GERMANY   1993

Otto Sander, Horst Buccholz, Nastassja Kinski, Peter Falk, Heinz Ruhmann, Bruno Ganz, Solveig Dommartin, Rudiger Vogler, Lou Reed, Willem Dafoe, Mikhail Gorbachev, Marijam Agischewa, Henri Alekan, Tom Farell, Monika Hansen

This sequel to WINGS OF DESIRE has our angel (Sander) getting his chance to become a mortal, when he saves a girl who falls from a balcony. An altogether less magical work than the earlier film, it has some moments of power, but these are mere islands of intensity, much of the film being given over to various muddled musings on the nature of the world and human purpose. Technically however, the film cannot be faulted.

Aka: IN WEITER FERNE, SO NAH!

FAN   140 min (ort 144 min) Col/B/W

VIDrel: CONNO/RTM   V

FAREWELL, MY CONCUBINE ****

15

Chen Kaige CHINA/HONG KONG/TAIWAN   1993

Gong Li, Leslie Cheung, Zhang Fengyi, Lu Qi, Ying Da, Ge You, Li Chu

Epic, overlong but absolutely captivating tale of a half-century in the lives of two male Peking opera performers and their relationship to each other (one is a clandestine homosexual) and the young prostitute whom the other marries. Their changing fortunes are set against the background of the vast upheavals that China has experienced this century, including the Japanese occupation, the Communist takeover and the Cultural Revolution.

Aka: BA WANG BIE JI

DRA   150 min (ort 170 min) wScrn   VIDrel: ARTIF/20TH V/sh

Boa: novel by Lilian Lee.

FAREWELL, MY LOVELY ***

15

Dick Richards USA   1975

Robert Mitchum, Charlotte Rampling, John Ireland, Anthony Zerbe, Sylvia Miles, Sylvester Stallone, Harry Dean Stanton, Jack O'Halloran, Kate Murtagh, Jim Thompson, John O'Leary, Walter McGinn, Jimmy Archer, Joe Spinell

The third version of a classic detective story with a confusing and incomprehensible plot about a private eye and his search for an ex-con's missing girlfriend. Mitchum is well chosen to play the cynical and world weary Marlowe. Followed by THE BIG SLEEP in 1978.

DRA   91 min (ort 97 min)   VIDrel: POLY L/A   V

Boa: novel by Raymond Chandler.

FAREWELL TO ARMS, A ***

PG

Frank Borzage USA   1932

Gary Cooper, Helen Hayes, Mary Philips, Adolphe Menjou, Jack LaRue, Blanche Frederici

The story of the romance between a wounded American ambulance driver and his nurse in WW1. Avoids all of the deeper issues raised in the novel on which it is based but works perfectly as a simple love story. An inferior remake followed in 1957. AA: Cin (Charles Bryant Lang Jr), Sound (Franklin B. Hansen).

WAR   75 min (ort 78 min) B/W   VIDrel: ORBIT/DISC   V

Boa: novel by Ernest Hemingway.

FAREWELL TO ARMS, A **

15

Charles Vidor USA   1957

Rock Hudson, Jennifer Jones, Vittorio De Sica, Oscar Homolka, Alberto Sordi, Kurt Kasznar, Mercedes McCambridge, Elaine Stritch, Leopoldo Trieste, Franco Interlenghi, Jose Nieto, Georges Brehat, Victor Francen, Joan Shawlee

An inferior remake of the 1932 film of the same name that suffers badly from poor casting and excessive length, treating Hemingway's WW1 love story with a surprising lack of feeling.

WAR   152 min   VIDrel: 20TH/TECH   V/h

Boa: novel by Ernest Hemingway.

FAREWELL TO THE KING ***

PG

John Milius USA   1988

Nick Nolte, Nigel Havers, Frank McRae, Gerry Lopez, James Tokuda, Choy Chang Wing, Aki Aleong, Marius Weyers, William Wise, Wayne Pygram, Richard Morgan, Elan Oberon, James Fox, Michael Nissman, John Bennett Perry

Action wartime adventure set in the exotic jungles of Borneo where a British commando has parachuted into the jungle with the intention of leading a native rebellion against the Japanese. Three years before, a US sergeant turned his back on the war and now lives with a native tribe as their ruler. He is about to find the peace of his haven destroyed in this colourful and exciting adventure, written and directed by Milius.

A/AD   112 min   VIDrel: VCC L/A   V

Boa: novel L'Adieu Au Roi by Pierre Schoendoerffer.

FARGO ***

18

Joel Coen USA   1995

Frances McDormand, Steve Buscemi, Peter Stormare, William H. Macy, Harve Presnell, Kristin Rudrud, Tony Denman, Gary Houston, Sally Wingert, Kurt Schweickhardt, Larissa Kokernot, Melissa Peterman, Steven Reevis, Warren Keith

A black comedy-thriller that tells of how a mid-winter kidnapping goes very badly wrong, when a Minneapolis car salesman arranged to have his wife kidnapped as a desperate means of clearing his debts, his hope being that his wealthy father-in-law will pay the ransom. Quite an entertaining piece, with a few unexpected twists and turns, and an especially good performance from McDormand as the heavily-pregnant police chief attempting to solve a trail of murders.

THR   98 min wScrn   VIDrel: POLFIL   V

FARINELLI IL CASTRATO **

15

Gerard Corbiau BELGIUM/FRANCE/ITALY   1994

Stefano Dionisi, Enrico Lo Verso, Elsa Zylberstein, Caroline Cellier, Marianne Basler, Jacques Boudet, Graham Valentine, Pier Paolo Capponi, Delphine Zentout, Omero Antonutti, Jeroen Krabbe, Renaud de Peloux de Saint Romain

A lovingly recreated 18th century costume drama about the life of the title opera singer, a man so obsessed with fame that he voluntarily had himself castrated for the sake of his career, but with the aid of his brother, still pursued women avidly. A meandering and unfocused affair that fails to throw much light on the man's true character.

Aka: FARINELLI

DRA   106 min (ort 110 min) wScrn

VIDrel: GUILD/FOXVID   V/sur

FARMER'S WIFE, THE *

(U)

Alfred Hitchcock UK   1928

Jameson Thomas, Lillian Hall-Davies, Gordon Harker, Gibb McLaughlin, Maud Gill, Louise Pounds, Olga Slade, Anatonia Brough

A farmer woos a number of high-class women but they all show little interest in his proposal and reject his suit out of hand. Downcast at his failure, he eventually comes to realise that it is his faithful housekeeper that he really loves. An undistinguished and hardly amusing effort, remade with no greater success in 1941.

COM   67 min B/W silent   VIDrel: LUMI/SPEAR   V

Boa: play by Eden Philpotts.

FASCINATION ** 18
Jean Rollin FRANCE 1979
Franca Mai, Brigitte Lahaie, Jean-Marie Lemaire, Fanny Magier, Miriam Watteau, Alain Plumey, Muriel Montosse, Sophie Noel, Evelyne Thomas, Jacques Marboeuf
The story of two women who seek a man they can murder and then devour at a vampiric feast. Complications ensue when one of them falls in love with their intended victim, in this visually fascinating combination of lesbian sex and horror.
HOR 78 min wScrn VIDrel: REDEM/RTM L/A V

FAST COMPANY ** (15)
Gary Nelson USA 1995
Ann Jillian, Tim Matheson, Geoffrey Blake, Dee Elliott Sanders, Tim Ryan, Wendy Phillips, Constance Marie, William Allen Young, Emily Procter, Karla Tamburrelli, Susan Giosa, Mary Kay Adams, Valerie Wildman, Marcy Kaplan
A TV journalist ropes in her policeman husband when she sets out to solve a murder case, the accused being one of her friends, who is suspected of killing the mistress of her cheating husband. Her investigations reveal that the murdered woman had many enemies, having had affairs with a string of married men, and she sets out to find the real killer. Uneven and a trifle frenetic, this crime comedy is not quite the hilarious affair it might have been.
COM 90 min mTV SATrel: MOVIE CHANNEL

FAST FOOD ** PG
Michael A. Simpson USA 1990
Clark Brandon, Randal Patrick, Tracy Griffith, Jim Varney, Michael J. Pollard, Traci Lords, Pamela Springsteen, Kevin McCarthy
A couple of wheeler-dealers on the lookout for a way of making a quick fortune, open a fast-food restaurant that serves hamburgers laced with an aphrodisiac sauce. A very crude teen-oriented comedy, funny in a simple-minded and not overly demanding way.
COM 88 min (ort 91 min) VIDrel: GENESIS V

FAST GETAWAY 2 ** 12
Oley Sassone USA 1993
Corey Haim, Cynthia Rothrock, Sarah G. Buxton, Wally Bujak, Kenny Jacobs, David Alexander Johnston, Ken Lerner, Peter Liapis, Tony Penello, Leo Rossi, Phillip Connery, Stanton Davis, Maggie Grant, Tiffany Grant, John Hrosovsky
Sequel to the first film has our young former bank robber now making an honest living as a security advisor to the banks he used to rob. Unfortunately, somebody else is robbing his customers and he learns from his imprisoned father that this is none other than his former girlfriend. To make matters worse, she is planning to frame him for her misdeeds. An assembly-line sequel of limited merit that was released directly to video.
A/AD 90 min (ort 94 min) VIDrel: FIRST/SONOP L/A V

FAST MONEY ** 18
Alexander Wright USA 1995
Yancy Butler, Matt McCoy, John Ashton, Carole Cook, Trevor Goddard
A journalist strikes up a friendship with a female car thief, but she is on the run after stealing a car that contained a gangster's money, and they are both obliged to flee for the border. But neither the crooks nor the cops are far behind. The usual shootouts and wild chases ensue (plus one really spectacular stunt) before the final resolution, and though this film is routinely plotted, it is fast-moving enough to remain consistently entertaining.
A/AD 88 min VIDrel: MARQ/QUANT V/sur

FASTER PUSSYCAT, KILL... KILL * 18
Russ Meyer USA 1966
Tura Santana, Haji, Lori Williams, Stuart Lancaster, Paul Trinka, Dennis Busch, Ray Barlow, Susan Bernard, Mickey Foxx
Cheap and nasty Meyer offering, with wicked Varla the tough leader of a gang of three women on the lookout for thrills. Varla challenges the boyfriend of a girl to a car race and when he loses she kills him. They tie the girl up and drive off to a remote shack where an old man and his retarded son live. More dollops of gratuitous violence and killing follows but Varla finally meets a fitting end, being left to die in the desert. A mindless and sick anti-film.
Aka: LEATHER GIRLS; MANKILLERS
A 83 min B/W VIDrel: ALLIED/RTM/TROMA V

FATAL ATTRACTION *** 18
Adrian Lyne USA 1987
Michael Douglas, Glenn Close, Anne Archer, Ellen Hamilton Latzen, Stuart Pankin, Ellen Foley, Fred Gwynne, Meg Mundy, Tom Brennan, Lois Smith, Mike Nussbaum, J.J. Johnston, Michael Arkin, Sam J. Coppola, Eunice Prewitt
A lawyer has a quick fling with an attractive woman not knowing she is unbalanced. When he tries to say goodbye she refuses to accept this and pursues him obsessively, turning his family's life into a living hell. A slick and well-acted thriller of considerable tension spoilt by a climax more suited to a Stallone film (the original subtler ending was dropped after an unsuccessful preview). See also LOVE IN THE STRANGEST WAY. THE LV disc includes interviews.
THR 114 min; 149 min (special edition with original ending) wScrn VIDrel: CIC/SONOP; PION (LV only) V/sur LV

FATAL BEAUTY * 18
Tom Holland USA 1987
Whoopi Goldberg, Sam Elliott, Ruben Blades, Jennifer Warren, Harris Yulin, John P. Ryan, Brad Dourif, Mike Jolly, Charles Hallahan, Neil Barry, Richard (Cheech) Marin, Ebbe Roe Smith
A violent and incoherent attempt to cash in on the success of BEVERLY HILLS COP with Goldberg playing an undercover cop out to track down the dealers responsible for pushing poisoned cocaine in L.A. The anti-drugs message comes through loud and clear but this dreadful mess of a film carries such a weight of dross (not least being the awful dialogue) that it fails on all counts.
A/AD 100 min (ort 104 min) VIDrel: MGM/WHV V/sur

FATAL CLAWS AND DEADLY KICKS *** 18
HONG KONG
Hsia Kuang-Li, Peng Kang, Wang Chi-Sheng, Liu Shan, Chang Kuan-Lung, Tai Chi-Hsia, Lu Yi-Lung
Returning home with his wife, a famous escort guard chief is attacked by four bandits who murder him and rape his wife. Rescued by a nun who is skilled in the martial arts, she swears vengeance and spends the next three years training until the time comes for her to track down her assailants and take her revenge. Formula martial arts revenger with some nicely choreographed sequences.
MAR 89 min Cut (1 min 1 sec) VIDrel: IMPENT V

FATAL DECEPTION: MRS LEE HARVEY OSWALD *** (PG)
Robert Dornhelm USA 1993
Helen Bonham Carter, Robert Picardo, Frank Wahley, Bill Bolender, Brandon Smith, Lis Renee Wilson, Deborah Dawn Slaboda, Ingeborga Dapkunaite, Darryl Cox, Vladimir Ilyn, Quenly Bakke, Norman Bennett, Rodger Boyce, Cliff Stephens,
A biopic on the life of the above figure and her quest to uncover the truth about her infamous husband. Carter gives a top-notch performance that imbues her character with great conviction and adds much interest to this mTV offering.
DRA 90 min mTV VIDrel: WHV L/A V/sh

FATAL EXPOSURE ** 18
Peter B. Good USA 1989
Blake Bahner, Ena Henderson, Julie Austin, Dan Schmale, Renee Cline, Gary Wise, Joy Ovington, Susan Welch, Marc Griggs, Jan Riley, Heidi Lawaczek, Tamara Dadd, Beth Cotton
The great-grandson of Jack the Ripper carries on the family tradition but adds his own artistic flourish to his work by taking pictures of his victims. Anxious to maintain the family line, he seeks out a "perfect woman" to bear his son, but is unprepared for her reaction when she learns of his past. Not an edifying experience, the film inevitably reminds one of an earlier work on the same theme: Powell's PEEPING TOM.
Aka: MANGLED ALIVE
THR 77 min (ort 83 min) VIDrel: RIO L/A V

FATAL FRIENDSHIP ** (15)
Bradford May USA 1993
Kevin Dobson, Gerald McRaney, Kate Mulgrew, Patti Yasutake, Will Nye, Michael Faustino, Wanda De Jesus, Stacy Keach Sr, Sharon Mahoney, Jerry Potter, Lindsay Price, Brandi J. Holben, Larry Pine, Michael Paul Chan
Two lifelong buddies have their relationship put in peril when one of them learns that the other has been involved in some very nasty activities, such as murder, and the acquiring of this information is not without risk. An unusual thriller, a little unbelievable, but said to be based on a real-life case.
THR 88 min (ort 90 min) mTV SATrel: SKY MOVIES

**FATAL IMAGE, THE ** *   15*
Thomas J. Wright USA   1990
Michele Lee, Justine Bateman, Francois Dunoyer, Jean-Pierre Casse,
Sonia Petrovna, Francois Guetary, Francoise Delaive, Bernard
Lepinaux, Philippe De Brugada, Michael Davies, Pierre Olivier-
Mornas, Tatum Belkacem, David Jalil
While on holiday in Paris, a divorcee and her daughter inad-
vertently become the quarry of a murderer after the girl
unwittingly videotapes a killing. A series of increasingly sinis-
ter events occur, culminating in the kidnapping of the girl, and
the mother is obliged to turn to others for help. A murky and
uninspired thriller that's a little too gory for its own good.
THR 88 min (ort 100 min) mTV VIDrel: CAPIT/GUILD V

**FATAL INSTINCT ** *   15*
Carl Reiner USA   1993
Armand Assante, Sherilyn Fenn, Kate Nilligan, Sean Young, James
Remar, Tony Randall, Christopher McDonald, Clarence Clemons,
Michael Cumpsty, Blake Clark, John Witherspoon, Edward Blanchard,
David Greenlee, Tim Frisbie, Carl Reiner
A tiresome and very unfunny stab at spoofing all those contem-
porary erotic thrillers such as BASIC INSTINCT. Here, Assante
plays a cop who also works as a defence attorney and is unlucky
enough to get entangled with a female psychotic. Of course he
soon finds that his life is in danger and the gags fly thick and
fast, it's really too bad that most of them misfire, making the film
a real chore to sit through.
COM 86 min (ort 95 min) cC VIDrel: MGM/WHV V/sur

**FATAL LOVE ** *   15*
Tom McLoughlin USA   1992
Molly Ringwald, Lee Grant, Perry King, Roxana Zal, George Coe,
Christopher McLoughlin, Kim Myers, Peter Spears, Robert Bauer, Nancy
McLoughlin, Victor Brandt, Janet MacLachlan, Martin Landau, Jody
Montana, Garn Stephens
Based on a true story, this tells of Alison Gertz, who despite
being far from promiscuous, contracted AIDS after a one-night
stand.
DRA 93 min VIDrel: ODY/SONOP V/sh

**FATAL VOWS ** *   15*
John Power USA   1994
John Stamos, Cynthia Gibb, Ben Gazzara, Sean McCann, Eugene
Clark, David Huband, Patricia Collins, Marc Donato, Tony De Santis,
Gina Wilkinson, Nichole Oliver, Christina Collins, Heidi Von
Palleske, Chandra Muszka, Howard Jerome
A young mother finds that her nice new husband in a danger-
ous serial killer, which soon leads her to fear for her life. A
standard portrait of a psycho on the loose, with all the usual
cliches.
THR 90 min VIDrel: ODY/SONOP V/sh

**FATHER AND SCOUT ** *   U*
Richard Michaels USA   1994
Bob Saget, Brian Bonsall, Heidi Swedberg, Stuart Pankin, David Graf,
Troy Evans
The son of an Eagle Scout has a healthy fear of the great
outdoors but is persuaded to give it a try, and accompanies his
father on a camping trip. But dad proves to be nothing but an
embarrassment, and everything that can go wrong does. A very
contrived effort indeed, though there are a few moments of
endearing, if heavy-handed irony.
JUV 90 min (ort 92 min) VIDrel: EIV/SONOP V

**FATHER AND SON ** *   18*
Georg Stanford Brown USA   1992
Louis Gossett Jr, Blair Underwood, Rae Dawn Chong, Tony Plana,
Clarence Williams III, David Harris, Eddie De Harp, Christopher M.
Brown, Luke Askew, Milt Kogan, Paul Carr, Josie Kim, Edgar Small,
Robert Druer, Ric Mancini
A long-term prisoner in a high-security jail works as a sort of
peace-keeper among the inmates and faces a major dilemma
where a new inmate proves to be the embittered son he abandoned
years before. Naturally, the young man soon sets out to disrupt
his father's efforts in every way possible, a conflict that contin-
ues even after the latter's release.
Aka: DANGEROUS RELATIONS; FATHER AND SON: DANGEROUS RELA-
TIONS; ON THE STREETS OF L.A.
DRA 89 min (ort 93 min) VIDrel: VCC L/A V

**FATHER GOOSE ** *   U*
Ralph Nelson USA   1964
Cary Grant, Leslie Caron, Trevor Howard, Jack Good, Nicole Felsette,
Verina Grennlaw, Pip Sparke, Jennifer Berrington, Stephanie
Berrington, Laurelle Felsette, Sharyl Locke, Simon Scott, John Napier,
Richard Lupino, Alex Finlayson
A layabout on an island in the South Seas works as a coast-
watcher for the Australian Navy, reporting on the movements
of Japanese vessels. However, his lackadaisical days are
numbered when he takes in a young French school-teacher and
her girls who find shelter with him after fleeing the Japanese
during WW2. A trite, modest but fairly enjoyable tale. AA:
Story/Screen (S.H. Barnett/Peter Stone/Frank Tarloff).
COM 112 min (ort 115 min) VIDrel: 4-FRONT/POLYREC
V

Boa: short story A Place of Dragons by S.H. Barnett.

**FATHER HOOD ** *   PG*
Darrell James Roodt USA   1993
Patrick Swayze, Halle Berry, Sabrina LLoyd, Brian Bonsall, Michael
Ironside, Diane Ladd, Bob Gunton, Adrienne Barbeau, Georgann
Johnson, Marvin J. McIntyre, William Bumiller, Vaness Marquez,
Martha Velez-Johnson, Ray De Mathis
A small-time criminal receives an unwelcome visit from his two
kids who have run away from their foster home because of
abuse and force him to evade the police, who are close behind.
To add to his worries, a journalist is also in pursuit, in the hope
of writing a strong human-interest story. A misfire of a comedy
that is frantically unamusing and is unfortunately saddled with
a lame script, an episodic script and uninspired direction.
COM 91 min (ort 94 min) VIDrel: TOUCH/TECH L/A V

FATHER OF THE BRIDE *** *   U*
Vincente Minnelli USA   1950
Spencer Tracy, Elizabeth Taylor, Joan Bennett, Don Taylor, Moroni
Olsen, Billie Burke, Leo G. Carroll, Melville Cooper, Taylor Holmes,
Russ Tamblyn, Paul Harvey, Frank Orth, Tom Irish, Marietta Canty,
Willard Waterman
A man looks back over the trials and tribulations he faced at his
daughter's wedding, and wonders how he ever managed to
survive the chaos this event unleashed on an unsuspecting
household. A charming light comedy that rings very true to life,
with a witty and literate Frances Goodrich/Albert Hackett script
and a host of fine performances (especially from Tracy as the
father) making it a delight to watch. Followed by FATHER'S
LITTLE DIVIDEND and remade in 1991.
COM 89 min (ort 93 min) B/W VIDrel: MGM/WHV V
Boa: novel by Edward Streeter.

**FATHER OF THE BRIDE ** *   PG*
Charles Shyer USA   1991
Steve Martin, Diane Keaton, Kimberly Williams, Kieran Culkin,
Martin Short, George Newbern, B.D. Wong, Peter Michael Goetz,
Kate McGregor Stewart, Carmen Hayward, April Ortiz, Mina
Vasquez, Gibby Brand, Richard Portnow, Barbara Perry
An inferior remake of the 1950 Spencer Tracy hit with Martin
taking the role of a doting dad who is about to lose his darling
daughter in marriage. Paying for the wedding looks set to bank-
rupt him, and as Martin struggles to trim the guest list, the
catastrophes start to pile up. Neither as fresh nor as funny as the
original, it delivers a few chuckles, right up to the syrupy climax.
A sequel followed.
COM 101 min (ort 105 min) VIDrel: TOUCH/TECH L/A
V
Boa: novel by Edward Streeter.

**FATHER OF THE BRIDE 2 ** *   PG*
Charles Shyer USA   1995
Steve Martin, Diane Keaton, Martin Short, Kimberly Williams,
George Newbern, Kieran Culkin, B.D. Wong, Peter Michael Goetz,
Kate McGregor Stewart, Jane Adams, Eugene Levy, Rebecca
Chambers, April Oritz, Dulcy Rogers, Kathy Anthony
A very twee follow-up to the updated FATHER OF THE BRIDE
(the sequel to the Spencer Tracy original being the 1951 film
"Father's Little Dividend") with Martin finding that both his
wife and daughter are pregnant, and that he now has to foot the
bill for a "baby shower" party. By way of a sub-plot, he also
begins to experience a mid-life crisis, though the real crisis in
this sequel is the lack of a decent plot.
COM 102 (ort 106 min) cC VIDrel: TOUCH/TECH V

FATHERLAND * *   15*
Ken Loach FRANCE/UK/WEST GERMANY   1986
Gerulf Pannach, Cristine Rose, Fabienne Babe, Sigfrit Steiner, Robert
Dietl, Heike Schrotter, Stephan Samuel, Thomas Oehlke, Patrick
Gillert

The story of an East German "Liedermacher" whose political songs put him out of favour with the authorities. He reluctantly accepts a one-way ticket to the West, where his "voice of dissent" becomes a marketable commodity. But what he really wants is to find his father, a classical musician who defected many years earlier. A French journalist, who has her own reasons for wanting to find his father, lends assistance. Written by Trevor Griffiths.
DRA   104 min (German dialogue with subs - ort 106 min)
VIDrel: FIRST/SONOP   V

FATHERLAND *   15
Christopher Menaul USA   1994
Rutger Hauer, Miranda Richardson, Peter Vaughan, Michael Kitchen, John Woodvine, John Shrapnel, Clive Russell, Clare Higgins, Jean Marsh, Pavel Andel, Petronella Q. Barker, Sarah Bergeger, Jan Bidlas, Stuart Bunce, Neil Dudgeon
An intriguing thriller set In a post-WW2 Europe of the 1960s where Hitler was not defeated and Germany reigns supreme over a unified continent, with the' exception of the Soviet Union which fights on. Shortly before the signing of a peace treaty with the USA, a German cop is assigned to a murder case and starts to unravel a mystery that seems to have claimed many lives. Hauer is excellent but has to cope with a second-rate script.
THR   106 min mCab   VIDrel: WHV   V/sh
Boa: novel by Robert Harris.

FATHERS AND SONS *   18
Paul Mones USA   1992
Jeff Goldblum, Rosana Arquette, Joie Lee, Rory Cochrane, Mitchell Marchand, Famke Janssen, Natasha Gregson Wagner, Ellen Greene, Samuel L. Jackson, Paul Hipp, Michael Disend, Erika Alexander, Rocky Carrol, Michael Imperioli
After the death of his wife, a movie director tries to sort out his drinking problem and rebuild his relationship with his son, who seems to have inherited his own talent for self-destruction. A warmhearted drama with appealing performances.
DRA   96 min (ort 109 min)   VIDrel: 20VIS/SONOP   V/sur

FATHER'S LITTLE DIVIDEND *   U
Vincente Minnelli USA   1951
Spencer Tracy, Elizabeth Taylor, Joan Bennett, Don Taylor, Billie Burke, Moroni Olsen, Frank Faylen, Marietta Canty, Russ Tamblyn, Tom Irish, Hayden Rorke, Paul Harvey, Beverly Thompson, Dabbs Greer, Frank Sully, Harry Hines
Sequel to FATHER OF THE BRIDE with Tracy bowled over by the news that he is due to become a grandfather. As time proceeds, he finds that this happy event brings with it a whole new series of problems and pleasures. A lavish effort, with great care being taken in every department, but ultimately let down by the meagre plot. Both films were remade forty years later with Steve Martin.
COM   82 min   VIDrel: SCRN/DISC   V

FAUST *   U
F.W. Murnau GERMANY   1926
Emil Jannings, Wilhelm (William) Dieterle, Camilla Horn, Yvetee Guilbert, Gosta Ekman
Acknowledged by many to be the best among the silent versions of the Faust story, this film is a lavish studio production with fine sets, good camerawork and fine performances. Interestingly, it also stars an actor who was to become one of Hollywood's finest directors - the great Dieterle.
DRA   100 min B/W silent   VIDrel: VISION/DISC   V

FAUST *   12
Jan Svankmajer CZECH REPUBLIC/FRANCE   1994
Peter Cepek plus voices of: Jan Kraus, Vladimir Kudla, Antonin Zacpal, Jiri Suchy, Viktorie Knotkova, Jana Mezlova, Miluse Strakova, Josef Fiala, Martin Radimecky, Ervin Tomendal, Frantisek Polata, Josef Chodora, Karel Vidimsky
The Faust legend is transplanted to Prague in the director's second feature, into which he injects a few Kafkaresque elements in the story of a man who sells his soul to the Devil. Using puppetry, live-action and animation, this is an ambitious, multi-layered and over-complex tale, unsettling and distinctive. Though Svankmajer does not quite pull it off as a full-length feature, the film is so unusual and inventive that he can be forgiven its flaws.
Aka: LEKCE FAUST
DRA   97 min   VIDrel: ICAPRO/MANGA   V/sh
Boa: the Faust plays of: Johann Wolfgang Goethe, Christian Dietrich Grabbe and Christopher Marlowe.

FAUSTO *   15
Remy Duchemin FRANCE   1992
Jean Yanne, Ken Higelin, Florence Darel, Francois Hauteserre, Maurice Benichou, Bruce Myers, Marianne Groves, Maite Nahyr, Arthur H, Francois Chattot, Frederique Leyes, Alfred Cohen, Renaud Menager, Arnaud Churin, Georges de Caunes
In 1960s Paris a teenage orphan begins life in a new institution, having got himself expelled from the last one by convincing his fellow inmates he has poisoned them. From there he moves on to an apprenticeship under a dour tailor, which heralds the start of his dreams of fame as a fashion designer. A charming and often very funny film, from first-time director Duchemin, unfortunately ends rather abruptly, just when one is getting really interested in the story.
DRA   78 min (ort 82 min)   VIDrel: CURZON/20TH   V/sh
Boa: novel by Richard Morgieve.

FAVOUR, THE *   15
Donald Petrie USA   1994
Harley Jane Kozak, Elizabeth McGovern, Bill Pullman, Brad Pitt, Ken Wahl, Ginger Orsi, Leigh Ann Orsi, Felicia Robertson, Kenny Twomey, Elaine Mee, Florence Schauffler, John Horn, Wilma Bergheim, Mary Marsh, Marilyn Blechscmidt
A happily married woman with two kids meets an old flame at a high school reunion and, because she feels unable to rekindle the romance herself, encourages her best friend to find out what her old sweetheart is like as a lover. This unusual request is prompted by her obsession with him and her boredom with her marriage. A good idea is eventually ruined as the films slips into a farcical comedy with our dream guy revealed as a stereo-typed macho man.
COM   93 min   VIDrel: COLUM/SONOP   V/sur

FAVOUR, THE WATCH AND THE VERY BIG FISH, THE *   15
Ben Lewin FRANCE/UK   1992
Bob Hoskins, Jeff Goldblum, Natasha Richardson, Michael Blanc, Jean-Pierre Cassel, Angela Pleasence, Jacques Villeret, Samuel Chaimovitch, Jean-Michel Ribes, Sacha Vikouloff, Beth McFadden, Caroline Jacquin, Pamela Goldblum
A photographer who specialises in religious motifs searches for suitable subject for a portrait of Christ and finds it in the shape of a deranged bar pianist who believes he really is Jesus. An unsuccessful attempt at a screwball comedy that fails to raise anything more than a few half-hearted chuckles.
Aka: RUE SAINT-SULPICE
COM   84 min (English version - ort 89 min)   VIDrel: VCC/DISC/COLUM   V/sur
Boa: short story Rue Saint-Sulpice by Marcel Ayme.

FBI STORY, THE *   PG
Mervyn LeRoy USA   1959
James Stewart, Vera Miles, Nick Adams, Murray Hamilton, Larry Pennell, Diane Jergens, Parley Baer, Joyce Taylor, Jean Willes, Victor Millan, Fay Roope, Ed Prentiss, Robert Gist, Buzz Martin, Kenneth Mayer, Paul George, Ann Doran
An FBI agent looks back at his career in this one-sided and totally fictional account of what the FBI was really like under J. Edgar Hoover. Good performances paper over the candyfloss script that tells us nothing of substance.
DRA   144 min (ort 149 min)   VIDrel: L/A   V
Boa: book by D. Whitehead.

FEAR, THE *   18
Vince Robert USA   1994
Eddie Bowz, Anne Turkel, Darin Heames, Antonio Todd, Vince Edwards, Wes Craven, Leland Hayward, Anna Karin, Monique Mannen, Heather Medway
A graduate psychology student takes a group of phobic patients to the mountain college in which he used to spend time as a child. There he played with a large wooden mannequin and this seemingly harmless object brings their worst fears to the surface.
HOR   101 min   VIDrel: 20VIS/SONOP   V

FEAR *   18
James Foley USA   1996
Mark Wahlberg, Reese Witherspoon, William Petersen, Alyssa Milano, Amy Brenneman, Christopher Gray, Tracy Fraim, Gary John Riley, Jason Kristofer, Jed Rees, Todd Caldecott, John Oliver, David Fredericks, Ravinder Toor, Jo Bates
A father's worst nightmare comes true when his sixteen-year-old daughter falls for a young guy in his twenties who cons her into believing that he is sensitive and understanding. All too

soon, however, he reveals himself to be a totally ruthless and brutal psychopath, prepared to kill in order to hang onto her at all costs. A violent and well acted effort but one that offers few surprises all the way to a predictable ending.
THR   97 min   VIDrel: CIC   V

FEAR CITY *
18
Abel Ferrara USA
1984
Tom Berenger, Billy Dee Williams, Melanie Griffith, Rossano Brazzi, Jack Scalia, Rae Dawn Chong, Joe Santos, Michael V. Gazzo, Jan Murray, Ola Ray, Janet Julian, Daniel Faraldo, Maria Chochita, John Foster, Emilia Lesnick
Another one of those psychotic slashers is on the prowl and this time he is attacking showgirls in Manhattan. An ex-boxer and a black detective join forces to catch the culprit. Violent and unappealing in equal measure.
HOR   91 min (Cut at film release – ort 96 min)
VIDrel: LUMI/SPEAR L/A   V/h

FEAR EATS THE SOUL ***
15
Rainer Werner Fassbinder WEST GERMANY
1973
El Hedi Ben Salem, Brigitti Mira, Barbara Valentin, Irm Hermann, Peter Gauhe, Rainer Werner Fassbinder
When a Moroccan guest-worker marries a German woman twice his age, their relationship is marred by the racist attitudes of those around them. A remarkably touching and sensitive study of a doomed love affair.
Aka: ALI; ALI: FEAR EATS THE SOUL; ANGST ESSEN SEELE AUF; ANGST ISS DIE SEELE AUF; DIE ANGST ESSEN SEELE AUF
DRA   89 min (ort 94 min)   VIDrel: CONNO/RTM   V

FEAR IN THE NIGHT **
15
Jimmy Sangster UK
1972
Judy Geeson, Joan Collins, Peter Cushing, Ralph Bates, Gillian Lind, James Cossins, John Brown, Brian Grellis
Horror shocker about a nervous woman who marries a prep schoolmaster and goes to live with him. Strange things begin to happen to her in a school that is as deserted in term time as it is during the school holidays. A competent yarn with several inventive touches and a rather sad resolution.
Aka: DYNASTY OF FEAR; HONEYMOON OF FEAR
HOR   85 min (ort 93 min)   VIDrel: LUMI/SPEAR L/A   V

FEAR INSIDE, THE ***
15
Leon Ichaso USA
1992
Christine Lahti, Dylan McDermott, Jennifer Rubin, David Ackroyd, Thomas Ian Nicholas, Paul Linke, Mike Barger, Gloria McCord, Maria Diaz
A female agoraphobic who makes her living by illustrating books decides to relieve her self-enforced seclusion in her home by inviting a college co-ed to come and live with her. However, when the latter's brother turns up, she soon realises that she has fallen victim to a pair of psychos.
THR   96 min (ort 100 min)   VIDrel: MED/POLYREC L/A   V/sh

FEAR NO EVIL **
18
Frank Laloggia USA
1981
Stefan Arngrim, Elizabeth Hoffman, Kathleen Rowe McAllen, Frank Birney, Daniel Eden, Jack Holland, Barry Cooper, Alice Sachs, Paul Haber, Roslyn Gugino, Richard Jay Silverthorn, Mari Anne Simpson, Joyce Bumpus, Don O'Neil
Classic conflict between good and evil in the shape of two high school students with one of them turning out to be an incarnation of Satan. Fortunately he is opposed by a couple of angels.
HOR   87 min (ort 98 min)   VIDrel: BMGREC/BMGVID   V

FEAR OF A BLACK HAT **
15
Rusty Cundieff USA
1994
Kasi Lemmons, Rusty Cundieff, Larry B. Scott, Mark Christopher Lawrence
Undoubtedly inspired by THIS IS SPINAL TAP, this crude by sometimes amusing parody details a year in the life of a gang of gangsters rappers who go by the name of N.W.H. (Niggaz With Hats). A very tongue-in-cheek offering, but at times the sheer mindlessness of it all makes for an overwhelmingly depressing experience.
MUS   82 min (ort 87 min)   VIDrel: GUILD/SONOP L/A   V

FEARLESS ***
15
Peter Weir USA
1992
Jeff Bridges, Isabella Rossellini, Rosie Perez, Tom Hulce, John Turturro, Benicio Del Toro, Deidre O'Connell, John De Lancie,

Spencer Vrooman, Debra Monk, Daniel Cerny, Eve Roberts, Robin Pearson Rose, Cynthia Mace, Randle Mell
A man survives a plane crash and finds his life is changed by the experience beyond recognition, for he now believes himself to be indestructible, and finds it impossible to continue his everyday life with his family. Instead, he seeks out a fellow survivor – a young woman unable to get over the death of her baby. A long and not terribly successful adaptation (by Yglesias) of a complex novel, it is rendered watchable by a set of first-rate performances from a strong cast.
DRA   117 min (ort 122 min) cC   VIDrel: WHV   V/sur
Boa: novel by Rafael Yglesias.

FEARLESS TIGER **
18
Ron Hulme USA
1993
Jalal Merhi, Bolo Yeung, Monika Schnarre, Jamie Farr
When his brother dies from a drugs overdose, a man swears vengeance on those responsible and travels to Hong Kong, where he immerses himself in the study of the martial arts. After a suitable period of time, he unleashes his wrath on the criminal fraternity in this by-the-numbers effort of much mayhem but little imagination.
MAR   88 min (ort 90 min)   VIDrel: ONE/IMPENT   V

FEAST AT MIDNIGHT, A **
PG
Justin Hardy UK
1994
Freddie Findlay, Aled Roberts, Andrew Lusher, Christopher Lee, Robert Hardy, Samuel West, Carol Macready, Lisa Faulkner, Edward Fox, John Hurley, Sebastian Fernandez-Armesto, Stuart Hawley, Jordan Ruffell, Edward Hassard, Jake Meyer
A boy is terrorised at school by many of the other boys, but most especially a strict Latin master. In response to this, he starts up a club of boys whose main object in life is to mount midnight raids on the school kitchen and indulge in secret gourmet cooking exploits. Quite good fun, with Lee especially good as the odious Latin master.
JUV   102 min (ort 106 min)   VIDrel: EIV/SONOP   V

FEAST OF JULY ***
15
Christopher Menaul UK/USA
1995
Embeth Davidtz, Tom Bell, Graham Jones, James Purefoy, Greg Wise, Kenneth Anderson, Ben Chaplin
A woman abandoned by her lover when she becomes pregnant, is taken in by a farming family and eventually marries one of their three sons. However, their marital happiness is threatened when her ex-lover returns. A powerful and well realised adaptation of the Bates novel.
DRA   112 min (ort 116 min) cC   VIDrel: TOUCH/TECH V/sh
Boa: novel by H.E. Bates.

FEEL THE HEAT *
18
Joel Silberg USA
1986
Tiana Alexandra, David Dukes, Rod Steiger, Brian Thompson, Jorge Martinez, John Hancock, Brian Libby, Toru Tanaka, Jessica Schultz, Russell Clark, Norman Ehrlich, Cecilia Maresca, Jacques Arndy, Miguel Angel Habud, Bill Wood
A drugs baron uses beautiful women couriers to smuggle narcotics inside their bodies, but finally gets his just deserts at the hands of an undercover lady cop who poses as an exotic dancer and for good measure, is a martial arts expert. Fast, violent and predictable.
Aka: CATCH THE HEAT
A/AD   84 min Cut (1 sec – ort 95 min)
VIDrel: EIV/SONOP   V/sur

FEELING MINNESSOTA ***
18
Steven Baigelman USA
1996
Keanu Reeves, Vincent D'Onofrio, Cameron Diaz, Delroy Lindo, Courtney Love, Tuesday Weld, Dan Aykroyd, Levon Helm, Drew DesMarais, Aaron Michael Metchik, Russell Konstans, David Alan Smith, Bill Schoppert, Steve Ghizoni, Jack Walsh
A drug baron's accountant is forced by his employer to marry a sluttish woman as a punishment for having robbed him. As the former hits the bottle, his unbalanced brother comes along and runs off to Vegas with his wife, thus relieving of this problem. A look at the crooked side of life with plenty of unattractive characters and sleazy settings, but still very much of a fantasy as any Hollywood product, although some charming performances add interest.
DRA   94 min (ort 99 min)   VIDrel: EIV/SONOP   V/sh

FELIX THE CAT: THE MOVIE *   U
Tibor Hernardi HUNGARY/USA   1988
Voices of: Chris Phillips, Maureen O'Connell, Peter Neuman, Alice Playton, Susan Montano, Don Oriolo, Christian Schneider, David Kolin
A carefully drawn if very disappointing homage to the famous comic-strip character, that ill-advisedly transposes our feline hero to something akin to a STAR WARS setting, and has him rescuing a princess from the clutches of an evil duke, who has imprisoned her and now rules the kingdom. Felix sets out on his quest, a journey that takes him through other dimensions, where he encounters various fantastical creatures. A sadly derivative effort.
ANIM   79 min (ort 82 min)   VIDrel: COLUM/SONOP
V/sh

FELLINI'S 8½ ***   15
Federico Fellini ITALY   1962
Marcello Mastroianni, Claudia Cardinale, Anouk Aimee, Sandra Milo, Rossella Falk, Barbara Steele, Madelene Lebeau
A noted director is hovering on the verge of a nervous breakdown, completely at a loss as to exactly how he is to get his new film started, and at the same time he has to contend with the constant demands made on his time by his wife, mistress and movie executives. His solution is to retreat ever more deeply into a series of personal fantasies and reminiscences. A visual treat it may be, but this sterile work remains hollow at heart. AA: Foreign, Cost (Piero Gherardi).
Aka: 8½ ; OTTO E MEZZO
DRA   138 min B/W wScrn   VIDrel: CONNO/RTM   V

FELLOW TRAVELLER ***   15
Philip Saville UK/USA   1989
Ron Silver, Imogen Stubbs, Hart Bochner, Daniel J. Travanti, Katherine Borowitz, Jonathan Hyde, Alexander Hanson, John Labanowski, Julian Fellowes, Briony McRoberts, Peter Corey, Richard Wilson, Doreen Mantle, David O'Hara
An interesting political thriller set during the period of the McCarthy witch-hunts. Silver stars as a blacklisted screenwriter who flees to England rather than testify before the House Committee on Un-American Activities, after his best friend, a blacklisted movie star, commits suicide when he implicated him in his HCUA testimony. A highly unusual work that regrettably fails to maintain its momentum. See also THE FRONT and GUILTY BY SUSPICION.
THR   97 min   VIDrel: CONNO/RTM   V/sur

FEMALE PERVERSIONS **   18
Susan Streitfeld GERMANY/USA   1996
Tilda Swinton, Amy Madigan, Karen Sillas, Frances Fisher, Laila Robins, Clancy Brown, Paulina Porizkova, Dale Shuger
According to the novel on which this film is based, women being forced into gender stereotypes is one of the greatest of all perversions, hence the title. (Presumably men suffer no such stereotyping). This film casts Swinton as an L.A. lawyer whose success forces her to adopt an assertiveness she is unhappy with, which latter leads to her falling prey to various anxieties and phobias. An odd art-house movie, occasionally thought-provoking, albeit annoyingly incoherent.
DRA   109 min   CINrel
Boa: novel Female Perversions: The Temptations of Emma Bovary by Louise J. Kaplan.

FEMALE TROUBLE *   18
John Waters USA   1975
Rosie Divine (Glenn Milstead), Mink Stole, Edith Massey, Mary Vivian Pearce, Cookie Mueller, David Lochary, Susan Walsh, Michael POtter, Ed Peranio, Paul Swift, George Figgs, Susan Lowe, George Hulse, Roland Hertz, Betty Woods
An outrageous account of the career of a notorious female transvestite all the way to the electric chair. Another "revolting" John Waters film that has (like so much of his work) achieved cult status. The occasional flashes of wit tend to be swamped in a welter of unpleasantness. See also PINK FLAMINGOES for more of the same.
COM   97 min Cut (5 sec)   VIDrel: CASPIC/BMGREC L/A
V

FEMALE VAMPIRE **   18
Jess Franco WEST GERMANY   1973
Lina Romay, Jack Taylor
Another strong dose of sado-eroticism, in this story of a seductive countess who invariably kills all her male lovers, but

eventually falls in love with one of her intended victims. A dated 1970s blend of sex and eroticism, of some curiosity value.
Aka: BARE BREASTED COUNTESSS
HOR   92 min (ort 94 min) wScrn dubbed
VIDrel: REDEM/RTM   V

FEMME ***   18
Candida Royalle USA   1984
Rhonda Jo Petty, Jerry Butler, Tish Ambrose, Michael Knight, Sharon Caine, Carol Cross, Klaus Multia, George Payne, David Israel-Sander
A most unusual adult film (it led to the Candida Royalle label) that was one of the very first to emphasise the romantic and loving aspects of sexuality. Six plotless vignettes are presented: "Rock Erotica" has two punk rockers making love, "TV Idol" tells of a woman's fantasy, "Gallery" is set in an art gallery, "Photo Session" is exactly that, "Sales Pitch" follows a female bisexual encounter, and finally "Dressing Room" is a self-explanatory title.
A   72 min Cut (1 min 27 sec – ort 75 min)
VIDrel: MIA/DISC   V

FEMME FATALE **   15
Andre Guttfreund USA   1990
Colin Firth, Lisa Zane, Billy Zane, Scott Wilson, Lisa Blount, Carmine Caridi
A newly-married artist comes home only to find that his bride has vanished, leaving behind her a web of mystery and deceit. With the help of a friend he starts a search and eventually finds that he must put his own life in danger if he is to save her.
Aka: FATAL WOMAN
THR   92 min (ort 96 min)   VIDrel: COLUM/SONOP   V

FEMME FATALE **   (18)
Udayan Prasad UK   1992
Sophia Diaz, Donald Pleasence, Simon Callow, Antonella Squadrito, Roberto Nobile, Ricardo Velez, Jacqueline Tong, James Fleet, Colin Welland, Patsy Rowlands, Jason Durr, Margery Mason, Rosalie Crutchley, Anna Mazzotti, Al Ashton
A beautiful woman proves irresistible to more than one man but her boyfriends have a strange habit of turning up dead, in this average erotic thriller that offers all the expected elements.
THR   74 min mTV   TVrel: BBC

FEMME FONTAINE: KILLER BABE FOR THE CIA *
18
Margot Hope USA   1995
Margot Hope, James Hong, Catherine Dao, David Shark, Lynn Paxton, Kevin Fry, Heinz Muller, Ellis Moore, Harry Mok, Brian Fix, Stephanie Shuldner, Margery Page, Betsy Burke, David Garrison, Richard Henry, George Saunders, Mike Bearman
An over-the-top Troma-style spoof about the activities of a female assassin and her various adversaries, most notably the killer of her father, who she attempts to locate with the help of a former CIA agent. As lacking in humour as it is taste and pitched at a moronic level.
DRA   94 min   VIDrel: TROMA/DISC   V

FENCING MASTER, THE **   15
Pedro Olea SPAIN   1992
Omero Antonutti, Assumpta Serna, Joaquim de Almeida, Jose Luis Lopez Vazquez, Miguel Rellan, Alberto Closas, Elisa Matilla, Ramon Goyanes, Juan Jesus Valverde, Francisco Vidal, Tomas Repila, Marcos Tizon, Miguel Angel Salomon
In Madrid of 1868, while Queen Isabel II finds her rule threatened by revolutionaries, a fencing master takes on a variety of students, and breaks his rule barring instructing women to admit a female student. But he falls in love with her and this leads to the inevitable intrigues and complications. Looking much like a fairly cheap TV movie, this is a watchable costume drama, that combines elements of drama, thriller and adventure.
Aka: EL MAESTRO DE ESGRIMA
DRA   84 min (ort 88 min)   VIDrel: CURZON/20TH   V/sh

FERNGULLY: THE LAST RAINFOREST ***   U
Bill Kroyer AUSTRALIA   1992
Voices of: Tim Curry, Samantha Mathis, Christian Slater, Jonathan Ward, Robin Williams, Grace Zabriskie, Geoffrey Blake, Robert Pastorelli, Cheech Marin, Tommy Chong, Tone-Loc, Towsend Coleman, Brian Cummings, Neil Ross
A young fairy who lives deep in the rainforest learns from her mentor of an evil spirit that sought to destroy the forest, but was trapped by magic. In her travels she encounters a young lumber-

jack who has mistakenly become a tool of this evil spirit, but under her influence he comes to respect nature and helps her save the forest from destruction. A slightly ponderous affair with a strong ecological message and excellent computer-generated animation.
Aka: FERNGULLY
ANIM   72 min (ort 76 min) cC   VIDrel: 20TH/TECH V/sur
Boa: short stories by Diana Young.

FEROCIOUS FEMALE FREEDOM FIGHTERS *   18
Jopi Burnama INDONESIA   1982
Eva Arnaz, Barry Prima
A foreign martial movie has been given new dubbed dialogue by the L.A. Connection comedy troupe in the hope no doubt of imitating Woody Allen's success with his similar project, WHAT'S UP TIGER LILY? Unfortunately, this is certainly not the case here and all we are left with is a juvenile and dull effort that is probably worse than the original.
COM   90 min   VIDrel: TROMA/RTM   V

FERRIS BUELLER'S DAY OFF **   15
John Hughes USA   1986
Matthew Broderick, Mia Sara, Alan Ruck, Jennifer Grey, Jeffrey Jones, Cindy Pickett, Lyman Ward, Edie McClurg, Charlie Sheen, Del Close, Ben Stein, Lisa Bellard, Virginia Capers, Richard Edson, Larry Flash Jenkins, Kristy Swanson
The story of a youngster's day of crazy adventures whilst taking the day off school. Starts off brilliantly but suffers from uneven development, sagging badly halfway as the film develops an introspective streak. The script is by John Hughes.
COM   99 min (ort 103 min)
VIDrel: 4-FRONT/POLYREC/CIC   V/sur

FEVER **   18
Larry Elikann USA   1991
Armand Assante, Sam Neill, Marcia Gay Harden, Joe Spano, Greg Henry, Vic Polizos, Tim Ransom, Jonathan Gries, Mark Boone Jr, Gordon Clapp, John Dennis Johnston, John Achorn, Steve Rankin, Jim Pirri, J.D. Callum
Action-packed crime-thriller in which Assante and Neill play a ruthless criminal and a successful lawyer respectively. Both men are rivals for the love of the same woman, and when the crook is released on parole the woman is abducted by a gangster who plans to hold her as hostage, forcing both men to work for him.
THR   93 min (ort 100 min) mTV   VIDrel: VCC L/A   V

FEVER PITCH **   15
David Evans UK   1996
Colin Firth, Ruth Gemmell, Neil Pearson, Lorraine Ashbourne, Mark Strong, Holly Aird, Ken Stott, Stephen Rea, Luke Aikman
A look at the life of a soccer fanatic, or more precisely, a dedicated Arsenal fan, that follows his team through their tremendous 1988/89 season when they moved inexorably towards a final battle with Liverpool for the league championship. As the Gunners head for their first victory in eighteen years, English teacher and obsessive fan Paul finds his personal relationships enduring difficulties of their own. An odd nostalgia film, occasionally quite touching.
DRA   102 min   CINrel
Boa: book by Nick Hornby.

FEVERHOUSE, THE **   15
Howard Walmsley UK   1984
Graham Massey, Joanne Hill, Patrick Nyland
Three characters in a strange establishment share shifting relationships of power and attraction in this unusual tale.
DRA   50 min B/W   VIDrel: IKON/RTM   V

FEW GOOD MEN, A ***   15
Rob Reiner USA   1991
Tom Cruise, Jack Nicholson, Demi Moore, Kevin Bacon, Kevin Pollak, Kiefer Sutherland, James Marshall, J.T. Walsh, Christopher Guest, J.A. Preston, Matt Craven, Wolfgang Bodison, Xander Berkeley, Cuba Gooding Jr, Noah Wyle
A laid-back Naval officer is assigned to defend two soldiers accused of the murder of one of their number at a base in Cuba. Initially he is rather less than concerned about their fate, but after enduring an interview with their commanding officer, he smells a cover-up and decides to investigate. A powerful but contrived effort, sustained by Nicholson's fine performance as

the martinet commander. Scripted by Sorkin who adapted his one-act play for the screen.
DRA   132 min (ort 138 min) wScrn
VIDrel: VCC/DISC/COLUM   V/sur
Boa: play by Aaron Sorkin

FIDDLER ON THE ROOF ****   U
Norman Jewison USA   1971
Chaim Topol, Norma Crane, Molly Picon, Candice Bonstein, Leonard Frey, Paul Mann, Rosalind Harris, Paul Michael Glaser, Michele Marsh, Neva Small, Candy Bonstein, Raymond Lovelock, Elaine Edwards, Shimen Ruskin, Zvee Scooler
A brilliant version of a Broadway musical based on the writings of Sholem Aleichem. Made with enormous care, and though entertaining in its portrayal of life for the Jews in a small village in the Pale, it has a deeper and serious side in examining the sufferings of Jews in Tsarist Russia. Songs are by Sheldon Harnick and Jerry Bock. AA: Cin (Oswald Morris), Score (John Williams), Sound (Gordon K. McCallum/David Hildyard).
MUS   171 min (ort 181 min) wScrn   VIDrel: MGM/WHV V/sur
Boa: short story Tevye And His Daughters by Sholem Aleichem/stage musical by Josef Stein.

FIELD, THE ***   15
Jim Sheridan UK   1990
Richard Harris, John Hurt, Sean Bean, Brenda Fricker, Frances Tomelty, Tom Berenger, John Cowley, Sean McGinley, Jenny Conroy, Ruth McCabe, Malachy McCourt, Ronan Wilmot, Jer O'Leary, Noel O'Donovan, Joan Sheehy, Eamon Keane
In Ireland in the 1930s, a stubborn tenant-farmer who has lovingly nurtured a piece of land, is driven to distraction when he learns that its owner, a widow, has decided to sell it at auction. Harris gives a performance of remarkable power as the tenacious and determined patriarch, but for all the film's undoubted merits, it remains cold and uninvolving. Screenplay is by Sheridan, with music by Elmer Bernstein.
DRA   106 min (ort 113 min)   VIDrel: VCC/DISC   V/sur
Boa: play by John B. Keane.

FIELD OF DREAMS ****   PG
Phil Alden Robinson USA   1988
Kevin Costner, Amy Madigan, Gaby Hoffman, Ray Liotta, Timothy Busfield, James Earl Jones, Burt Lancaster, Frank Whaley, Dwier Brown
An Iowan farmer hears a disembodied voice, and is inspired to build a baseball pitch in the middle of a cornfield. This proves to be what was needed to bring a legendary baseball star (unfairly disgraced in the famous 1919 Chicago White Sox scandal) back to Earth to play one more game. An uplifting and magical fantasy, telling of faith, hope and ultimate redemption. Scripted by Robinson, and with a fine score by James Horner. See also EIGHT MEN OUT.
FAN   101 min (ort 106 min)
VIDrel: 4-FRONT/POLYREC/GUILD; PION (LV only) V/sur LV
Boa: novel Shoeless Joe by W.P. Kinsella.

FIELD OF FIRE **   15
Cirio H. Santiago USA   1991
David Carradine, Eb Lottimer, David Anthony Smith, Scott Utley, Dan Barnes, Henry Strazalkowski, Jim Moss, Tonichi Fructuoso, Joe Mari Avellana, Ruben Ramos, Joe Zucchero, Ken Metcalfe, Joe Towers, Archie Ramirez, Scott Rogers
An army officer who carries in his head plans for a top secret new plane, is lost in the jungles of Vietnam when his plane crashes. Aware that he must not fall into enemy hands, the authorities make a determined effort to mount a rescue. A competently made jungle warfare tale – no better or worse than countless others. See BAT. 21 for something along similar lines, but very much better.
WAR   84 min (ort 96 min)   VIDrel: CIC/SONOP L/A   V

FIELD OF HONOR **   15
USA   1996
Blair Underwood, Delroy Lindo, Mykelti Williamson
Story of the Brooklyn Dodgers, whose owner decided to sign the first black baseball player in the 1940s, an action that does not win him many friends. An interesting slice of recent history without much of a plot, but quite absorbing just the same.
DRA   95 min   VIDrel: THIRD   V

FIERCE CREATURES *

Robert Young UK/USA
12
1997
John Cleese, Kevin Kline, Jamie Lee Curtis, Michael Palin, Robert Lindsay, Ronnie Corbett
An Australian tycoon takes over a zoo and declares that profits are to be increased by 20%, his idea being that this will happen if the zoo only stocks "fierce creatures". As keepers Lindsay and Corbett struggle to find these qualities in their various charges, a catalogue of mishaps and coincidences add to the general confusion. A weak follow-up to A FISH CALLED WANDA with the same cast, very little wit, a plethora of vulgar jokes and a parade of pretty girls.
COM   93 min   CINrel

FIFTEEN STREETS, THE **

David Wheatley UK
15
1989
Owen Teale, Clare Holman, Ian Bannen, Sean Bean, Billie Whitelaw, Frank Windsor, Jane Horrocks, Anny Tobin, Leslie Schofield, Faye Dannell, Mark Mulholland, Christian Rodksa, Margery Bone, Berwick Kaler, Barbara Marten
In turn-of-the-century Newcastle, a docker falls in love with the daughter of a shipbuilder, but the lovers are separated by the social divide to which the title alludes.
DRA   105 min mTV   VIDrel: FOCUS/DISC   V
Boa: novel by Catherine Cookson.

FIFTH MUSKETEER, THE ***

Ken Annakin AUSTRIA/UK
15
1978
Beau Bridges, Sylvia Kristel, Rex Harrison, Ursula Andress, Cornel Wilde, Jose Ferrer, Ian McShane, Lloyd Bridges, Alan Hale Jr, Helmut Dantine, Olivia De Havilland
A virtual remake of "The Man In The Iron Mask" about the struggle between Louis XIII and his twin brother for the throne of France. This one has little new to offer but the Austrian locations and support from a strong cast of veterans are compensations. Written by David Ambrose.
Aka: BEHIND THE IRON MASK
A/AD   115 min (ort 120 min)   VIDrel: CREA/DISC   V
Boa: novel The Man in the Iron Mask by Alexandre Dumas.

FIFTY-FIVE DAYS IN PEKING **

Nicholas Ray SPAIN/USA
U
1962
Charlton Heston, Ava Gardner, David Niven, Flora Robson, Leo Genn, Robert Helpmann, Harry Andrews, Paul Lukas, John Ireland, Elizabeth Sellars, Massimo Serrato, Jacques Sernas, Geoffrey Bayldon, Icchizo Itami, Kurt Kasznar
This overlong star-studded account of the Chinese Boxer uprising of 1900 fails to give a clear insight into the causes of the rebellion, such as the resentment felt by the Chinese at their treatment by the Western powers. A colourful feast of star names, with Robson playing the Dowager Empress who encourages the rebels to take over Peking.
Aka: 55 DAYS AT PEKING
DRA   147 min (Cut at film release – ort 154 min)
VIDrel: 4-FRONT   V
Boa: book 55 Days At Peking by S. Edwards.

FIFTY-TWO PICK-UP **

John Frankenheimer USA
18
1986
Roy Scheider, Ann-Margret, Vanity, John Glover, Robert Trebor, Lonny Chapman, Kelly Preston, Doug McClure, Clarence Williams III
A wealthy industrialist is blackmailed when he has an affair and is forced to re-examine how he feels about his wife, as he struggles to free himself from the grip of these squalid extortionists. A confused and disjointed thriller, heavy on sleazy atmosphere, but without the benefit of a coherent plot. See also THE AMBASSADOR.
Aka: 52 PICK-UP
THR   104 min (Cut at film release by 1 min 36 sec)
VIDrel: RNK L/A   V
Boa: novel 52 Pick-Up by Elmore Leonard.

FIGHT FOR FREEDOM **

John Korty USA
PG
1990
Mark Harmon, Lee Purcell, Leon Russom
Enjoyable drama revolving around the struggles a family face at the time of the Depression.
DRA   91 min (ort 95 min)   VIDrel: GENESIS   V

FIGHT FOR JUSTICE: THE NANCY CONN STORY **   15

Bradford May USA
1995
Marilu Henner, Doug Savant, Ann Wedgeworth, Peri Gilpin, Lisa Jakub, Don Franklin, Sean McCann, Janet Laine-Green, Andrew Keegan, John Novak, Hugh Thompson, John Bourgeois, Sandi Stahlband, Kevin Hickes, Bunty Webb, Rob Wilson
An unbalanced man develops an obsessive interest in a woman, and eventually assaults both her and her cousin, leaving them for dead. He is put in jail following the death of one of the women, but quickly regains his freedom when he is released on parole (this is the USA remember) and starts planning a new attack. An absorbing drama based on a real case, in which the surviving woman started a campaign for the rights of victims in the States.
Aka: NANCY CONN STORY, THE
THR   89 min (ort 90 min) mTV   VIDrel: ODY/SONOP   V/sh

FIGHTING FIST, THE **

Chin Sheng En HONG KONG
15
1977
A martial arts fighter disregards the advice of a fortune-teller and nearly loses his life in an encounter with two underworld figures. Following this defeat, he seeks out a master he believes can teach him enough techniques to make him virtually unbeatable. Average.
MAR   85 min   VIDrel: IMPENT   V

FIGHTING FOR MY DAUGHTER: THE ANNE DION STORY ***

Peter Levin USA
(15)
1994
Renee Humphrey, Lindsay Wagner, Piper Laurie, Chad Lowe, Kirk Blatz, Deidre O'Connell, Paul Lieber, Deanna Milligan, Christopher Gray, Enuka Vanessa Okuma, Jennifer Copping, Lisha Snelgrove, Kevin McNulty, Richard Leacock, Alf Humphreys
A mother and daughter enjoy the closest and most amiable of relationships, but this is put in peril by the girl's latest boyfriend, who turns out to be a pimp and soon tries to exploit his relationship with the girl. A painful and absorbing drama, whose potent script is based on real events.
DRA   86 min mTV   SATrel: MOVIE CHANNEL

FIGHTING KENTUCKIAN, THE **

George Waggner USA
U
1949
John Wayne, Vera Ralston, Philip Dorn, Oliver Hardy, Marie Windsor, John Howard, Hugo Haas, Odette Myrtil, Grant Withers, Paul Fix, Mae Marsh, Jack Pennick, Mickey Simpson, Fred Graham, Mabelle Koening, Shy Waggner
In 1818, a Kentucky riflemen becomes involved in foiling a plot to seize land that had been granted to French veterans of Napoelon's wars. An entertaining, above-average Wayne Western, based on historical fact.
WES   100 min B/W   VIDrel: VCC/DISC/COLUM L/A   V

FIGHTING SEABEES, THE ***

Edward Ludwig USA
U
1944
John Wayne, Susan Hayward, Dennis O'Keefe, William Frawley, Duncan Renaldo, Leonid Kinsley, Grant Withers, Addison Richards, Paul Fix, Ben Welden, J.M. Kerrigan, William Forrest, Addison Richards, Jay Norris, Duncan Renaldo
A story following the work of the C.B.s or Construction Battalion, as tough foreman Wayne and Navy man O'Keefe organise repairs to installations close to the Japanese lines, whilst stationed on an island in the South Pacific. In between they find time to fight for the affections of Hayward, and Wayne goes into battle at one point with a line of construction vehicles. An overblown, melodramatic but rousing war film.
WAR   95 min (ort 100 min) B/W
VIDrel: 4-FRONT/POLYREC   V

FIGHTING 69TH, THE ***

William Keighley USA
PG
1940
James Cagney, George Brent, Pat O'Brien, Alan Hale, Jeffrey Lynn, Frank McHugh, Dennis Morgan, William Lundigan, Dick Foran, Henry O'Neill, Guin "Big Boy" Williams, John Litel, Sammy Cohen, Harvey Stephens, William Hopper
Corny war drama mixing comedy, pathos and action in a tale that follows the exploits of a famous Irish regiment and a new recruit (Cagney) who hides a cowardly heart beneath his brash and cocky exterior. His insubordination leads to a court martial, but he ultimately redeems himself. About as unbelievable as they come, but entertaining nonetheless. A computer coloured version has become available in the States.
WAR   86 min (ort 89 min) B/W   VIDrel: WHV   V

FILIPINA DREAMGIRLS ***

Les Blair UK
(PG)
1991
Bill Maynard, Geoffrey Hutchings, Lee Cornes, David Thewlis,

Charlie Drake, Ray Gravell, Grace Amilbangsa, Angie V. Cantero, Majesty Gaerlan, Max Phipps, Roxanne Silverio, Andrea Arroyo, Sylvia Garde, Juliet Jimenez, Ressie Manuel
Six slightly over-the-hill Welsh bachelors go to Manila to pick up Filipino brides and find that the women to whom they are introduced are neither desperate nor by any means easily fooled. An entertaining TV drama with good performances, it is much enhanced by its interesting locations.
DRA   90 min mTV   TVrel: BBC

FILOFAX **   15
Arthur Hiller USA   1990
James Belushi, Charles Grodin, Anne De Salvo, Loryn Locklin, Stephen Elliott, Hector Elizondo, Veronica Hamel, Mako, Gates McFadden, John De Lancie, Thom Sharp, Ken Foree, J.J., Andre Rosey Brown, Terrence E. McNally, Lenny Hicks
An overstressed Chicago advertising executive leaves his filofax in a phone booth at an L.A. airport and faces the inevitable consequences when an escaped crook (with tickets to the baseball World Series) decides to impersonate him. A remarkably weak comedy of highly improbable plotting that fails to make the most of its thin material. Music is by Stuart Copeland.
Aka: TAKING CARE OF BUSINESS
COM   106 min (ort 108 min)   VIDrel: L/A   V

FILTHY ROTTEN SCOUNDRELS *   18
USA   199-
Kaitlyn Ashley, Barbara Doll, Chanel, Tina Tyler, Marilyn Martin, Montana, Dick Nasty, Ron Jeremy, Mike Horne
A sex spoof on DIRTY ROTTEN SCOUNDRELS set in a resort on the French Riviera where the usual encounters take place. Apart from being shot on location, this has nothing new to offer.
Aka: FILTHY SLEAZY SCOUNDRELS
A   57 min   VIDrel: FALCON/TOTAL   V

FINAL ALLIANCE, THE **   18
Mario Di Leo USA   1990
David Hasselhoff, Bo Hopkins, Jeannie Moore, John Saxon, Ami Artzi, Len Sparrowhawk, Karl Johnson, Frank Notaro, Jeff Fannell, Tom Hoskins, Deon Stewardson, Steed TAylor, Allan Pierce
A tough man fights a lone battle against a bunch of vicious bikers, in a bid to free a town from their domination. Average.
A/AD   89 min Cut (12 sec – ort 94 min)
VIDrel: EIV/SONOP   V

FINAL ANALYSIS **   15
Phil Joanou USA   1992
Richard Gere, Kim Basinger, Uma Thurman, Eric Roberts, Paul Guilfoyle, Keith David, Robert Harper, Agustin Rodriguez, Rita Zohar, George Murdock, Shirley Prestia, Tony Genaro, Katherine Cortez, Wood Moy, Corey Fischer, Lee Anthony
An unbelievable psychological thriller which has Gere playing a suave psychiatrist who gets involved in an affair with the sister of a patient. With the sister's husband a dangerously psychopathic gangster, it's not long before matters resolve themselves into murder, but not before a goodly number of over-complex plot twists have come to pass. And for good measure, there is even a Hitchcock-style climax at the top of a lighthouse.
THR   119 min (ort 125 min) cC   VIDrel: WHV   V/dm   V/sur

FINAL APPEAL **   15
Eric Till CANADA/USA   1993
Brian Dennehy, JoBeth Williams, Tom Mason, Ashley Crow, Betsy Brantley, Lindsay Crouse
A woman shoots her abusive and cheating husband in the presence of his mistress. She claims that it was in self-defence while his bereaved girlfriend says it was murder. An adequate courtroom drama.
DRA   92 min mTV   VIDrel: ODY/SONOP   V/sh

FINAL APPROACH **   15
Eric Steven Stahl USA   1991
James B. Sikking, Hector Elizondo, Madolyn Smith-Osbourne, Kevin McCarthy, Cameo Kheuer, Wayne Duvall, Karen Person, David Bonaiuto, Duffy Rutledge, Colin Vogel, Robert Jay Stahl, Maarten Goslins, Ellen Hilton, James Flynn
A test pilot suffers from amnesia after crashing his plane in the desert and receives treatment from a psychiatrist who tries to uncover the truth behind this incident. The first film to be made with digital sound, this boasts some impressive special effects but suffers badly from a confused and incoherent plot. This movie incorporates much footage from the documentary

"Blackbird The Movie" produced by the Lockheed Corporation on the SR-71 spy plane.
DRA   97 min (ort 100 min)   VIDrel: HIFLI/SONOP   V/sur

FINAL COMBINATION *   18
Nigel Dick USA   1993
Michael Madsen, Lisa Bonet, Gary Stretch, Carmen Argenziano, Paul Leslie Disley, Simon Kenny, Nicholas Kenny, Andrew Shaw, Parker Posey, Johny Clayton Schafer, Connie Blankenship, Jimmy Ortega, Eric Dare, Alex Desir
A vicious serial killer is murdering women after saving them from being raped, and a female journalist, whose sister was one of the victims, teams up with a cop to catch this maniac. To complicate matters, she attaches special conditions to her collaboration with the police. A murky and unpleasant thriller whose basic premise shows a complete lack of originality, and whose lead actors (especially former boxer Stretch) could do with a few acting lessons.
Aka: DEAD CONNECTION
THR   89 min (ort 93 min)   VIDrel: COLUM/SONOP   V/sur

FINAL CUT, THE **   18
Roger Christian USA   1995
Sam Elliott, Charles Martin Smith, Anne Ramsay, Matt Craven, Ray Baker, John Hannah, Amanda Plummer
The people of Seattle are being held to ransom by a demented bomber, who has already destroyed a number of buildings. The devices are booby-trapped in that they are primed to explode in the presence of bomb-disposal teams, and a retired bomb-disposal officer realises that the culprit must in fact be a former colleague. Given the contrived nature of the plot, this action-thriller (it went straight to video) does extremely well, even if the outcome is never in doubt.
A/AD   95 min (ort 99 min)   VIDrel: MED/DISC   V/sh

FINAL DAYS ***   U
Richard Pearce USA   1989
Lane Smith, Richard Kiley, David Ogden Stiers, Ed Flanders, Theodore Bikel, Graham Beckel, James B. Sikking, Gregg Henry
A very clever adaptation of the Woodward/Bernstein book, that follows the latter part of Nixon's political career, from the scandal of Watergate to his resignation as president. Smith gives a sensitive and quite remarkable interpretation, avoiding all caricature in his portrayal of Nixon. Kiley as Buzhardt, the Special White House Counsel, is almost as good.
DRA   144 min (ort 150 min) mTV   VIDrel: COLUM/SONOP   V
Boa: book by Bob Woodward and Carl Bernstein.

FINAL DAYS OF BUTCH AND SUNDANCE **   (PG)
Jack Bender USA   1994
Kenny Rogers, Scott Paulin, Brett Cullen, Mariska Hargotay, Kriss Kamm, Ned Vaughn, Loni Anderson, Richard Riehle, Martin Kove, Darrell Larson, Mark Walters, Kim Walker, Geoffrey Lewis, Stephen Bridgewater, Bruce Boxleitner
Easy-going Western that has Rogers on the trail of his son, whom he believed was living an honest life as a student but has in fact been riding with the title outlaws. Not an especially memorable film, it offers ample entertainment of the undemanding kind.
WES   90 min   SATrel: MOVIE CHANNEL

FINAL EXECUTIONER *   18
Anthony M. Dawson (Antonio Margheriti) ITALY   1984
Woody Strode, Harrison Muller, David Warbeck, William Mang, Marina Costa, Margi Newton, Maria Romano, Tommaso Mesto, Stefano Davanzati, Karl Zinny, Renato Miracco, Luca Giordana, Cinzia Bonfantini, Giovanni Cianfraglia
In a post-nuclear holocaust world the survivors amuse themselves by hunting human prey. A dreary and unappealing effort.
Aka: FINAL EXECUTOR; LAST WARRIOR, THE
FAN   90 min   VIDrel: MOPIC/SGSVID   V

FINAL HEIST, THE **   15
George Mihalka USA   1991
Jan-Michael Vincent, Gabrielle Lazure
A man whose job involves the recovery of stolen paintings and other precious artefacts for an insurance company, becomes the victim of a nasty blackmail attempt when his daughter is abducted. He soon finds himself being forced to plan the theft of a priceless Van Gogh painting in order to save his daughter.
THR   92 min mTV   VIDrel: CAPIT/GUILD   V/sh

FINAL IMPACT **   18
Joseph Merhi/Stephen Smoke USA   1991
Lorenzo Lamas, Jeff Langton, Michael Worth, Kathleen Kinmont, Mike Tony, Mimi Lesseos, Kathrin Lautner
Having received a dreadful pounding at the hands (and feet) of a champion kickboxer, a martial artist who thirst for revenge starts planning his comeback. An adequate formula kickboxing-offering.
MAR   96 min (ort 99 min)   VIDrel: 20VIS/SONOP   V/sh

FINAL JUDGEMENT **   18
Louis Morneau USA   1992
Brad Dourif, Isaac Hayes, Karen Black, Maria Ford, Orson Bean, David Ledingham
A rebellious priest is framed for murder and has to explore the world of L.A.'s sex industry in order to find the real killer, but once there he is forced to confront the dark side of his own nature.
DRA   86 min (ort 90 min)   VIDrel: COLUM/SONOP   V

FINAL MISSION **   (15)
Lee Redmond USA   1993
Billy Wirth, Corbin Bernsen, Elizabeth Gracen, Steve Railsback, Timothy Dale Agee, Tim Moran, Richard Bradford, Frank Zagarino, John Prosky, Beth Tegarden, Patricia Sill, Ezra Gabay, Justin Lord, Hal Havins, Jerry Giles, Jack Eisman
A spate of kamikaze-style deaths among fighter pilots leads to an investigation of their training equipment, which includes an advanced "virtual reality" simulator. This proves to have been sabotaged so that the pilots are being given subliminal instructions to crash their planes. An original, quite fascinating plot idea falls to get suitable treatment in this disappointing thriller.
THR   88 min (ort 91 min)   SATrel: SKY MOVIES

FINAL ROUND *   18
George Erschbamer USA   1993
Lorenzo Lamas, Kathleen Kinmont, Anthony De Longis, Clark Johnson
Three people are kidnapped by a deranged maniac and find themselves being used as human prey as they are stalked by killers armed with high-tech weapons. A low-budget rip-off of THE MOST DANGEROUS GAME, hampered by poor production values, bad lighting, abysmal acting and lousy direction. Lamas gives his usual tough-guy performance, as a professional fighter who has become a victim, but soon turns the tables on his attackers. See also THE NAKED PREY and DEATH RING.
A/AD   79 min (ort 90 min)   VIDrel: OVER/HIFLI   V

FINAL SANCTION, THE **   18
David A. Prior USA   1990
Ted Prior, Robert Z'dar, Renee Cline, David Crawford, William Smith
After a devastating but inconclusive WW3, the USA and the USSR agree to settle the outcome and avoid further destruction, by arranging for single combat to the death between a soldier from each side. A moderately enjoyable blend of action and martial arts. See also ROBOT JOX for something similar.
A/AD   81 min (ort 90 min)   VIDrel: 20VIS/SONOP   V/sh

FINAL SHOT: THE HANK GATHERS STORY **   (PG)
Claude Braverman USA   1992
Victor Love, Duane Davis, Nell Carter, George Kennedy, Baldwin Skies, Ahmad Stoner, Reynaldo Rey, Whitman Mayo, Michole White, Corey Curtis, Ed Arnold, David Netter, Terrence Williams, De'Andre Alfred, Roger Hampton, Ken Foree
Well-made biopic about this fine athlete who devoted himself to his sport despite a serious heart complaint. His determination and drive take him from the ghetto to the top ranks of baseball, until one fateful day when tragedy strikes.
DRA   88 min (ort 90 min) mTV   SATrel: SKY MOVIES

FINAL TABOO, THE ***   R18/18
Henri Pachard USA   1988
John Leslie, Ona Zee, Robert Bullock, Nikki Knights, Shanna McCullough, Alicia Monet, F.M. Bradley, Richard Pacheco, Denise Connors, Veronica Hart
The female head of a call-girl service is hired by two rival TV preachers, each of whom is out to destroy the reputation of the other by spying on each other's sexual activities and filming these moments of passion. The woman accepts both commissions and succeeds in ruining both careers, for it transpires that her father was a preacher who committed suicide after being caught in a similar situation. An intriguing idea that is never developed.
A   VIDrel: SHEP L/A (R18 ver); HAR/GOLD (18 ver)   V

FINAL TEMPTATION **   15
Paco Lara SPAIN/UK   1990
Paul McGann, Sophie Ward, Isla Blair, Frada Dowie, Altana Sanchez-Gijon, Laura Davenport, Suzanne Bertish, Mark Estob, Manuel De Blas, Luis Hostalot, Josefa Sarsa, Fulgencio Saturno, Caspar Cano, Manuel Pereiro
In Madrid in 1743 a monk becomes an object of sexual desire for a young girl who is determined to win him, even if this requires her making a pact with the Devil. Despite having struggled for years against his repressed sexual feelings, our priest feels himself drawn to the girl, but gets involved in a murder in a moment of anger, which leaves him vulnerable to the Inquisition. A steamy drama whose director seems uncertain just where to take the story.
Aka: MONK, THE; SEDUCTION OF A PRIEST
THR   90 min (ort 106 min)   VIDrel: ARROW/RTM   V
Boa: novel by M.G. Lewis

FINAL TEST, THE **   U
Anthony Asquith UK   1953
Jack Warner, Robert Morley, George Relph, Brenda Bruce, Richard Wattis, Adrianne Allen
A champion batsman looking forward to his last test match is dismayed to learn that his son is more interested in being a poet than in seeing him play. A minor comedy-drama, held back by the lack of momentum and unnecessarily padded out with real-life cricketing stars. Scripted by Rattigan.
DRA   87 min (ort 91 min) B/W   VIDrel: ODY/SONOP   V/s
Boa: TV play by Terence Rattigan.

FINAL VENDETTA *   18
Rene Eram USA   1996
Bridgette Wilson, Peter Boyle, Seiko Matsuda, Scott Cohen
Wilson plays a psychopath who is taken into the home of a wealthy middle-class couple and soon makes her indispensable. However, this is all part of a plan she is hatching to take revenge on the person she blames for ruining her life. An over-familiar plot offers little that is fresh or insightful.
THR   92 min   VIDrel: BMGREC/BMGVID   V/sh

FINALLY SUNDAY! **   PG
Francois Truffaut FRANCE   1983
Fanny Ardent, Jean-Louis Trintignant, Phillip Laudenbach, Caroline Sihol, Philippe Morier-Genoud, Xavier Saint Macary, Jean-Pierre Kalfon, Anik Belaubre, Jean-Louis, Richard, Yann Dedet, Nicole Felix, George Kouloris
A wife and her lover are found murdered. Suspicion falls on her estate agent husband whose loyal secretary tries to prove his innocence. A mildly diverting Hitchcock-style thriller with a few touches of humour. The excellent photography is the work of Nestor Almendros.
Aka: CONFIDENTIALLY YOURS; VIVEMENT DIMANCHE
THR   106 min (ort 117 min) B/W   VIDrel: ARTIF/20TH   V/h
Boa: novel The Long Saturday Night by Charles Williams.

FINDERS, KEEPERS, LOVERS, WEEPERS *   (18)
Russ Meyer USA   1968
Paul Lockwood, Joey Duprez, Anne Chapman, Lavelle Roby, Jan Sinclair, Gordon Wescourt, Duncan McLeod, Robert Rudelson, Lavelle Roby, Nick Wolcuff, Pam Collins, Vickie Roberts, John Furlong, Michael Roberts
Another Russ Meyer extravaganza in which a couple of thugs get involved with a strip joint and meet a horde of sex-crazed individuals whose indiscretions they exploit for their own ends. An empty, sick and depressing offering, with a downbeat and violent ending.
A   71 min (ort 90 min)   SATrel: SKY MOVIES

FINE GOLD **   15
Anthony J. Loma (Juan Antonio De Loma) USA   1987
Stewart Granger, Lloyd Bochner, Ted Wass, Jane Badler, Andrew Stevens, Ray Walston
The rivalry between two winegrowing families over the production of a perfect vintage results in much suffering for all concerned, but love later finds a way for them to settle their differences. An unoriginal yarn, adequately handled, but no more than passable.
DRA   96 min   VIDrel: 20TH/TECH   V/sh

FINE THINGS **   PG
Tom Moore USA   1990
D.W. Moffett, Tracy Pollan, Judith Hoag, Cloris Leachman, Noley

Thornton, Darrell Larson, Mia Dillon, Jeanne Hepple, Randy Oglesby, Patricia Gaul, Ralph Meyering Jr, Will MacMillan, Enid Kent, Jeanne Sakata, Troy Davidson

Despite the disapproval of his mother, a man marries the woman of his choice, only to lose her when she gets cancer. During their few years together, he raised her young daughter by a previous marriage as his own, but now finds the woman's drug-dealing former husband using his legal rights to extort money from him. A watchable TV soap opera-style drama, devoid of either plausibility or convincing characterisations.

Aka: DANIELLE STEEL'S FINE THINGS

DRA 137 min (ort 145 min) mTV VIDrel: MIA/DISC V

Boa: novel by Danielle Steel.

FINEST HOUR, THE ** 15
Shimon Dotan USA 1992

Rob Lowe, Gale Hansen, Tracy Griffith, Eb Lottimer, Barcuh Dror, Daniel Dieker, Michael Fountain, Evyazar Lazar, Henry Taejoon Lee, Ari Sorko-Ram, Erim Degn, Nathan Sgan-Cohen, Jon J. Herson, Uri Gavriel, John Phillips

A young man joins a tough unit of Navy divers and immediately comes into conflict with one of its members, a career-minded extrovert with high ambitions. In the course of time, their initial enmity gives way to a deep friendship as both men learn to rely on each other. This stands them in good stead when they are sent on a vital mission on the eve of the Gulf War. An exciting actioner with a surprisingly sober and downbeat ending.

A/AD 100 min (ort 105 min)

VIDrel: 4-FRONT/POLYREC/EIV V/sh

FINGERS * 18
Stuart Canterbury USA 1993

Deborah Wells, P.J. Sparxx, Keisha, Beatrice Valle, China Doll, Jon Dough, Tony Tedeschi, Mike Horner, Joey Silvera

A man's girlfriends are quire enthralled by his use of title objects, in this silly sex film.

A 52 min (ort 80 min) VIDrel: ONE V

FINIAN'S RAINBOW *** U
Francis Ford Coppola USA 1968

Fred Astaire, Petula Clark, Tommy Steele, Don Francks, Keenan Wynn, Barbara Hancock, Al Freeman Jr

A leprechaun does his best to retrieve a crock of gold that has been taken to the American South, in this engaging slice of whimsy. Made some time after the 1947 Broadway success, it now inevitably looks dated, and is weakened by some over-sentimental moments. However, there are still some good things to be enjoyed here, not least the clever Harburg lyrics and a pleasing performance from Astaire as the leprechaun.

MUS 142 min VIDrel: TART L/A V

Boa: play by E.Y. Harburg and Fred Saidy.

FINISHING TOUCH, THE ** 18
Fred Gallo USA 1991

Michael Nader, Shelley Hack, Arnold Vosloo, Art Evans, Clark Johnson, Ted Raimi, John Mariano, Howard Shangraw

A down-at-hell detective takes on a tough case in the hope of saving his career, which involves the murder of two women. His investigations lead to a seedy nightclub where he discovers his ex-wife, a police officer, is working undercover.

THR 78 min (ort 82 min) VIDrel: 20VIS/SONOP V/s

FIORILE *** 12
Paolo Taviani/Vittorio Taviani FRANCE/GERMANY/ITALY 1993

Claudio Bigagli, Galatea Ranzi, Michael Vartan, Lina Capolicchio, Constanze Engelbrecht, Athina Cenci, Giovanni Guidelli, Norma Martelli, Laurent Schilling, Pier Paolo Capponi, Carlo Luca De Ruggieri, fritz Muller Scherz, Elisa Giani

During a car journey, a man relates to his children the origins of a curse that afflicts the family, and how it originated when a woman's soldier lover was executed for a theft of regimental funds committed by her brother. Over the centuries, the wealth that this crime brought the family fails to lead to any luck for the subsequent generations. A complex, multi-layered historical fable, both visually striking and directed with considerable

Aka: WILD FLOWER

DRA 113 min (ort 118 min) VIDrel: ARROW/RTM V

FIRE AND ICE ** PG
Ralph Bakshi USA 1982

Voices of: Randy Norton, Cynthia Leake, Steve Sandor, Sean Hannon, William Ostrander, Leo Gordon, Eileen O'Neil, Elizabeth Lloyd Shaw,

Micky Morton, Susan Tyrell, Maggie Roswell, Stephen Mendel, Tamara Park, Hans Howes

Animated feature about the struggle between Good and Evil with a sword wielding hero and a buxom heroine. Passable but not inspired, this cartoon does not compare well to Bakshi's earlier works such as FRITZ THE CAT or WIZARDS. In this one the animations were traced from live-action footage, and benefited from the design work of illustrator Frank Frazetta. Written by Roy Thomas and Jeff Conway.

ANIM 78 min (ort 81 min) VIDrel: BRAVE/SONOP L/A V

FIRE, ICE AND DYNAMITE ** PG
Willy Bogner WEST GERMANY 1990

Roger Moore, Shari Belafonte, Simon Shepherd, Uwe Ochsenknecht, Geoffrey Moore, Connie De Groot, Celia Gore Booth, Siegfried Rauch, Ursula Karren, Tiziana Stella, John Eaves, Bobby Naish, Stefan Glowacz, Jochen Schweizer

A messy blend of comedy and action that sees a collection of oddball characters competing in the Swiss Alps for a prize of $135,000,000.

A/AD 102 min (ort 106 min) VIDrel: EIV/SONOP V/sur

FIRE IN THE DARK *** PG
David Jones USA 1991

Olympia Dukakis, Lindsay Wagner, Jean Stapleton, Ray Wise, George Hearn, Joan Leslie, Paul Scherrer, Amzie Strickland, Edward Herrmann, Sheila Allen-Jones, Martin Goslins, Michelle Harrell, Denis M. Heames, Carol Bivens

Dukakis plays an ageing widow who has to face up to her growing decrepitude, the death of her lifelong friends and the fact that neither her son nor daughter are all that willing to take responsibility for her care. Despite all this, she endeavours to face life with a measure of courage and wit. A touching albeit contrived comedy-drama that boasts a fine performance from Dukakis. Screenplay is by David J. Hill.

DRA 93 min mTV VIDrel: VCC L/A V

FIRE IN THE SKY, A *** PG
Jerry Jameson USA 1978

Richard Crenna, Elizabeth Ashley, David Dukes, Joanna Miles, Lloyd Bochner, Andrew Duggan, Nicolas Coster, Merlin Olsen, Maggie Wellman, Marj Dusay, John Larch, Kip Niven, William Bogart, Jenny O'Hara, Michael Biehn, Al White

An overlong and vastly expensive disaster movie, in which a comet is seen to be heading for Phoenix, Arizona. Warnings of impending doom from an anxious astronomer are ignored by the town council who dither over evacuation plans until it is almost too late. The tedious multi-character plot is best ignored in favour of the excellent special effects and miniature work, plus the well-handled crowd scenes. See also METEOR.

A/AD 140 min (ort 150 min) mTV VIDrel: RCA L/A V

Boa: story by Paul Gallico.

FIRE IN THE SKY ** 15
Robert Lieberman USA 1992

D.B. Sweeney, Robert Patrick, Craig Sheffer, Peter Berg, James Garner, Henry Thomas, Bradley Gregg, Noble Willingham, Kathleen Wilhoite, George Emelin, Wayne Grace, Scott MacDonald, Kenneth White, Robert Covarrrubias, Bruce Wright

A man whose five-day disappearance in the woods triggered a manhunt claims that he was abducted by aliens, in this film account of an allegedly true event. Naturally, he has more than a little difficulty convincing the authorities of that he is telling the truth. A confused and unfocused tale that is saved by its strong atmosphere and special effects (not least being a fifteen-minute flashback sequence towards the end of the film). See also OFFICIAL DENIAL.

FAN 104 min (ort 111 min) VIDrel: CIC/SONOP V/sur

Boa: book The Walton Experience by Travis Wilson.

FIRE OVER ENGLAND *** 15
William K. Howard UK 1936

Laurence Olivier, Flora Robson, Vivien Leigh, Leslie Banks, Raymond Massey, Tamara Desni, Morton Selten, Lyn Harding, George Thirlwell, James Mason, Henry Oscar, Robert Newton, Donald Calthrop, Charles Carson

A fine cast coupled with good acting make this historical drama dealing with England's struggle against the Spanish Armada a most enjoyable film. Flora Robson's performance as Queen Elizabeth I is considered a gem, and rightly so. Look out for Mason in an early (and unbilled) small part.

DRA 91 min B/W VIDrel: L/A V

Boa: novel by A.E.W. Mason.

FIRE! TRAPPED ON THE 37TH FLOOR *   (15)
Robert Day USA   1991
Lee Majors, Lisa Hartman, Peter Scolari, John Laughlin, Kim Miyori, Michael Beach, David Dunard, Julian Reyes, Ismael Carlo, Paul Linke, Tim Grimm, William Bumiller
A made-for-TV variant on THE TOWERING INFERNO that is based on a real fire that swept through the First Interstate Bank building in L.A. in 1988. The story revolves around the exploits of two bank executives and their struggles to flee the 65-storey building, but there is little plot to be had in this production-line movie, and the tension is contrived and ineffective. A few clumsy touches of humour do not make this effort any more palatable.
A/AD   95 min mTV   TVrel

FIRE WITHIN, THE ***   15
Louis Malle FRANCE/ITALY   1963
Maurice Ronet, Lena Skerla, Yvonne Clech, Hubert Deschamps, Jeanne Moreau, Alexandra Stewart
An alcoholic writer leaves the clinic where he has been drying out, with the intention of committing suicide, but spends the next 24 hours looking up his old friends to see if they can give him a reason to go on living. They don't. Not quite as depressing as it sounds, this harsh study of despair is sustained by a fine central performance from Ronet. The writer Rochelle did in fact kill himself, having collaborated with the Nazis during WW2.
Aka: LE FEU FOLLET; TIME TO LIVE AND A TIME TO DIE, A; WILL O' THE WISP
DRA   103 min (ort 121 min) B/W
VIDrel: ELPIC/POLYREC   V
Boa: novel Le Feu Follet by Pierre Drieu la Rochelle.

FIREFIST OF INCREDIBLE DRAGON **   18
Jimmy Tseng HONG KONG   1980
Jerry Young, Keith Lee, Paul Chan, John Bun, Maple Lin, Dean Chu, Mandy Ding
Another formula martial arts adventure.
MAR   76 min Cut (51 sec – ort 90 min)   VIDrel: IMPENT   V

FIREFOX **   15
Clint Eastwood USA   1982
Clint Eastwood, Freddie Jones, David Huffman, Ronald Lacy, Warren Clarke, Stefan Schnabel, Kenneth Colley, Nigel Hawthorne, Thomas Hill, James Staley, Clive Merison, Kai Wulff, Dimitra Arliss, Austin Willis, Michael Currie
An ex-US pilot is persuaded to undertake a mission to the USSR to steal a supersonic jet fighter that is invisible to radar. Terribly slow and dull plot development mars and ultimately overpowers the exciting flying sequences. Eastwood as the Russian speaking US pilot does little to convince.
THR   119 min (ort 137 min)   VIDrel: WHV   V/sur
Boa: novel by Craig Thomas.

FIREHEAD **   15
Peter Yuval USA   1990
Christopher Plummer, Chris Lemmon, Martin Landau, Gretchen Becker, Brett Porter
A Soviet defector endowed with awesome telekinetic powers after treatment at the hands of American scientists escapes from their custody, triggering a a massive manhunt. By-the-numbers actioner distinguished only by its slightly original plot idea that is submerged all too soon in the usual cliches.
A/AD   84 min (ort 88 min)   VIDrel: 20VIS/SONOP   V/sh

FIREMAN'S BALL, THE **   U
Milos Forman CZECHOSLOVAKIA   1967
Josef Svet, Jan Vostrcil, Josef Sebanek, Josef Kolb, Vaclav Stockel, Maria Jazkova, Karel Valnoha, Frantisek Debelka, Josef Rehorek, Anina Lipoldva, Alena Kvetova, Mila Zelena, Vratislav Cermark, Vaclav Novotny, Jiri Libal
A ball in honour of a retiring fire chief in a small town turns into a total disaster in this rather rambling satire on the Communist party and all its works. Unfortunately, much of the film's bite will be lost on those with no first-hand experience of a communist society.
Aka: HORI, MA PANENKO
COM   69 min (ort 73 min)   VIDrel: TART/20TH   V/dm

FIREPOWER **   18
Richard Pepin USA   1993
Chad McQueen, Gary Daniels, George Murdock, Joseph Ruskin, Jim Hellwig, Alisha Das
In the next century, part of L.A. is a government-sanctioned criminal haven under the control of a ruthless leader. Two cops pursue a wanted felon there and find themselves having to fight for their lives in that area's "Ring Of Death". A derivative and brutal actioner with few futuristic elements but no shortage of violence.
A/AD   91 min (ort 95 min)   VIDrel: MIA/DISC   V/sh

FIRES WITHIN **   15
Gillian Armstrong USA   1991
Greta Scacchi, Jimmy Smits, Vincent D'Onofrio, Luis Avalos, Bertila Damas, Raul Davila, Bri Hathaway, Daniel Fern, Earl Hindman, Kevin Duffis, Victor Rivers, Lazaro Perez, Angelina Estrada, Julia Rodriguez Elliot, Maria Vidal
Scacchi plays the unfaithful wife of a dissident Cuban writer (Smits) who is imprisoned for anti-government activities. Whilst he has been in jail for eight years she has been enjoying a torrid affair, and when he gets released she's unable to decide who she really wants. Some interesting observations regarding Cuban expatriate life and a few titillating moments are all that is on offer here.
DRA   83 min (ort 97 min)   VIDrel: MGM/WHV   V/sur

FIRESTARTER *   15
Mark L. Lester USA   1984
David Keith, Drew Barrymore, Freddie Jones, George C. Scott, Martin Sheen, Heather Locklear, Art Carney, Louise Fletcher, Moses Gunn, Antonio Fargas, Drew Snyder, Curtis Credel, Jeff Ramsey, Keith Colbert, Richard Warlock, Jack Magner
A little girl has strange psychic powers and is able to set fire to anything at will, and the government decides that she could be useful as a weapon. A muddled fantasy yarn with a few good effects but little else of note. Written by Stanley Mann. See also SPECIMEN and WILDER NAPALM.
HOR   109 min (ort 115 min)   VIDrel: GAME/SPEAR   V
Boa: novel by Stephen King.

FIRESTORM *   18
John Shepphird USA   1995
John Savage, Bentley Mitchum, Joseph Culp, Paul Ben-Victor, Paul Williams, Roxana Zal
A fantasy set in the 21st century on the planet Markus 4, where a group of cyborg-humans forced to work as slaves start a revolt. Filmed on the cheap on an industrial estate, this dreadful looking effort does no favours for anyone, not least the unfortunate folk who are unwise enough to watch it.
FAN   88 min   VIDrel: FIRST/SONOP   V

FIREWALKER *   15
J. Lee Thompson USA   1986
Chuck Norris, Louis Gossett Jr, Melody Anderson, Will Sampson, Sonny Landham, John Rhys-Davies, Ian Abercrombie, Richard Lee-Sung, Zaide Silvia Gutierrez, Alvaro Carcano, John Hazelwood, Jose Escandon, Mario Arevalo
After a series of unsuccessful expeditions, two inept adventurers meet a young woman with a map that will lead them to forgotten Aztec treasure buried deep in Guatemala. A dire adventure film done in the style of RAIDERS OF THE LOST ARK but with none of that film's verve.
A/AD   100 min (ort 104 min)   VIDrel: MGM/WHV   V

FIRM, THE ***   18
Alan Clarke UK   1989
Gary Oldman, Lesley Manville, Philip Davis, Andrew Wilde, Charles Lawson, William Vanderpuye
A compelling look at football hooliganism that focuses on the double life of one individual, who away from his gang of thugs, is a happily married family man with a decent job working as an estate agent. Mostly, the story concerns the efforts made by the central character to organise some violence on a trip to Germany for the 1988 European championship. Oldman is chillingly realistic in this disturbing film. See also I.D. and ULTRA for another look at this problem.
DRA   66 min (ort 78 min) mTV   VIDrel: IMAG/RTM; POLY/POLYREC   V

FIRM, THE ***   15
Sydney Pollack USA   1992
Tom Cruise, Jeanne Tripplehorn, Ed Harris, Holly Hunter, David Strathairn, Gary Busey, Hal Holbrook, Gene Hackman, Terry Kinney, Wilford Brimley, Tobin Bell, Steven Hill, Barbara Garrick, Jerry Hardin, Karina Lombard, John Beal
A young law graduate accepts a job offer that seems too good to be true but soon discovers that the prestigious Memphis law

firm he works for is really a front for the Mob. To make matters worse, the FBI blackmails him into assisting their investigation or taking the rap himself. Fast-paced and very watchable if a touch overlong.
DRA 148 min (ort 154 min) cC VIDrel: CIC/SONOP; PION (LV only) V/sur LV
Boa: novel by John Grisham.

FIRST A GIRL ** — U
Victor Saville UK — 1935
Jessie Matthews, Sonnie Hale, Griffith Jones, Anna Lee, Alfred Drayton, Constance Godridge, Martita Hunt, Eddie Gray
Amiable musical vehicle for Matthews in which she plays a messenger girl who poses as a male female-impersonator in a show on the Riviera, having agreed to pose as the regular impersonator (Hale) to help him out. But she is unexpectedly a great success, and some complications flow from this. A lighthearted remake of "Viktor Und Viktoria" that would have been immeasurably better with a stronger plot. See also VICTOR/VICTORIA.
MUS 88 min B/W VIDrel: CONNO/RTM L/A V
Boa: play Viktor und Viktoria by Reinhold Schunzel.

FIRST AMONG EQUALS: PARTS 1 TO 4 ** — 15
John Gorrie/Brian Mills/Sarah Harding UK — 1986
David Robb, Tom Wilkinson, James Faulkner, Jeremy Child, Diana Hardcastle, Anita Carey, Joanna David, Jane Booker
Competent political drama based on Archer's novel, that deals with the careers of four politicians, all of whom have an overwhelming desire to be Prime Minister. Fair.
DRA 484 min (2 cassettes) mTV
VIDrel: 4-FRONT/POLYREC V
Boa: novel by Jeffrey Archer.

FIRST BLOOD * — 15
Ted Kotcheff USA — 1982
Sylvester Stallone, Richard Crenna, Brian Dennehy, David Caruso, Jack Starrett, Michael Talbott, David Crowley, Chris Mulkey, Don Mackay, Chuck Tamburro, Alf Humphreys, John Liam, Bill McKinney, Bruce Barbour, Dan Woznow
An ex-Green Beret gets into trouble with the local police, escapes and uses his military skills to defeat his pursuers. Stallone mumbles and stumbles through this lame-brained actioner, spreading mayhem all about him yet never suffering more than a scratch himself. Followed by RAMBO: FIRST BLOOD, PART 2.
A/AD 89 min (ort 97 min) VIDrel: POLY/POLYREC V/sh
Boa: novel by David Morrell.

FIRST DEADLY SIN, THE *** — 15
Brian G. Hutton USA — 1980
Frank Sinatra, Faye Dunaway, David Dukes, Brenda Vaccaro, Jeffrey De Munn, George Coe, Anthony Zerbe, James Whitmore, Martin Gabel, Joe Spinell, Anna Navar ro, John Devaney, Robert Weil, Hugh Hurd, Jon De Vries, Eddie Jones, Fred Fuster
A New York cop tracks down a vicious homicidal maniac whilst his wife is dying in hospital. Dunaway has little to do except act comatose in this film, but the atmospheric script and a convincing performance from Sinatra do much to make this one of his better movies. The score is by Gordon Jenkins. Bruce Willis has a tiny part as an extra.
DRA 108 min (ort 112 min) VIDrel: L/A V
Boa: novel by Lawrence Sanders.

FIRST DEGREE ** — 18
Jeff Woolnough USA — 1995
Rob Lowe, Leslie Hope, Tom McCamus, Joseph Griffin
A detective investigating the murder of a prominent Manhattan socialite believes that the victim had links with organised crime. Matters soon become complicated when he becomes involved with the man's widow and finds himself caught between both the police and the Mob. Decent enough low-budget thriller, even if neither the plot nor acting are up to much. However, both the soundtrack and clever surprise ending are assets, and deserve a better film.
THR 87 min (ort 98 min) VIDrel: NORSTAR/HIFLI V/h

FIRST DO NO HARM *** — 15
Jim Abrahams USA — 1996
Meryl Streep, Fred Ward, Deth Adkins, Margo Martindale, Tom Butcher, Michael Yarmush
A couple whose son suffers from epilepsy watch him become a

monster under the influence of the drugs that are supposed to control his condition. When she learns of a miracle diet that may offer some hope, she finds it an uphill struggle gaining support from the medical profession, who as ever, are blinkered and narrow-minded. A grim but ultimately inspiring film, that is based on a true story. Some of the cast are former epileptics who are now following this diet.
DRA 90 min VIDrel: ODY/SONOP V/sh

FIRST KNIGHT ** — PG
Jerry Zucker USA — 1995
Richard Gere, Sean Connery, Julia Ormond, Ben Cross, John Gielgud, Stuart Bunce, Liam Cunningham, Christopher Villiers, Valentine Pelka, Colin McCormack, Ralph Ineson, Jane Robbins, Jean Marie Coffey, Paul Kynman, Tom Lucy
A lavish but far from convincing retelling of the legend of King Arthur and the tragic love affair between Queen Guinevere and Lancelot, the noblest knight of the Round Table. Connery as King Arthur is as charismatic as one would expect, and easily outshines both Gere and Ormond and Lancelot and Guinevere respectively. See also EXCALIBUR for a considerably more satisfying adaptation.
A/AD 129 min (ort 134 min) wScrn cC
VIDrel: COLUM/SONOP; ENCORE (LV only) V/sur LV

FIRST LIGHT ** — 15
Bob Misiorowski USA — 1991
Michael Pare, Janis Lee, Amos Lavie, Sasson Gabay, Elkie Jacobs, Jack Widerker, Richard Peterson, Arthur Livingston, Jack Abalist, Ric Roman Waugh, Robert C. Shenk, Irene Handler, Evez Atar, Miki Ben-Harush, Avi Keidar
An expert in the new field of "psychic warfare" is assigned the task of protecting the daughter of the head of the CIA, and finds himself having to make good use of his powers when she is abducted as a hostage by terrorists from the Turkish refugee camp in which she works. The usual mayhem results when a rescue attempt is mounted. A brutal actioner that flirts with its opening ESP premise, then does little to exploit it.
Aka: BLINK OF AN EYE
A/AD 96 min VIDrel: GUILD/FOXVID V/sh

FIRST MAN INTO SPACE ** — PG
Robert Day UK — 1958
Marshall Thompson, Marla Landi, Robert Ayres, Carl Jaffe, Bill Edwards, Bill Nagy, Roger Delgado, Richard Shaw, John McLaren, Bill Nick, Spencer Teakle, John Fabian, Chuck Keyser, Michael Bell, Helen Forrest, Mark Sheldon
An astronaut is exposed to a cloud of radioactive meteor dust and comes back to Earth as a monster that drinks human blood. A modest effort that is distinguished by an attempt to view the man's predicament in a more sympathetic light.
FAN 73 min (ort 78 min) B/W VIDrel: ENCORE/SPEAR V

FIRST MEN IN THE MOON, THE *** — U
Nathan Juran UK — 1964
Edward Judd, Lionel Jeffries, Martha Hyer, Erik Chitty, Marne Maitland, Hugh McDermott, Miles Malleson, Gladys Henson, Gordon Robinson, Sean Kelly, Betty McDowall, John Murray Scott, Lawrence Herder, Paul Carpenter, Peter Finch
A lavish but uneven adaptation of the Wells story of a Victorian inventor who discovers a material impervious to gravity and uses it to construct a spaceship for a voyage to the moon. The addition of some comic elements turns the film into something of a lark but despite this it remains a worthy and enjoyable experience. Special effects are by Ray Harryhausen.
Aka: H.G. WELLS' THE FIRST MEN IN THE MOON
FAN 99 min (ort 103 min) VIDrel: COLUM/SONOP V
Boa: novel by Herbert George Wells.

FIRST NAME: CARMEN ** — 18
Jean-Luc Godard FRANCE/SWITZERLAND — 1983
Maruschka Detmers, Jacques Bonnaffe, Myriem Roussel, Christophe Odent, Jean-Luc Goddard
A self-parodying homage to B-movies with Godard playing a has-been film director of the same name, while the alluring Carmen of the title is a member of a criminal gang using the making of the film as cover for a kidnapping. Set in modern-day Paris and Trouville, much of it revolving around the way Bonnaffe as a cop finds himself embroiled in these events, but the lack of a coherent narrative dissipates tension. A film largely aimed at fans of this director.
Aka: PRENOM CARMEN
DRA 80 min (ort 90 min) VIDrel: ARTIF/20TH V

FIRST OF THE FEW, THE ***
Leslie Howard UK   U
                                                 1942
Leslie Howard, David Niven, Rosamund John, Roland Culver, David Horne, Anne Firth, John H. Roberts, Derrick De Marney, Rosalyn Boulter, Tonie Edgar Bruce, Gordon McLeod, Erik Freund, Filippo Del Guidice, Brefni O'Rorke
Fine biographical drama about R.J. Mitchell who foresaw the advent of WW2 and developed the Spitfire. A restrained account that makes the most of its limited budget. This enjoyable film Howard's last screen appearance. The longer running-time is due to the inclusion of an interview with one of the original Spitfire test pilots – Jeffrey Quill.
Aka: SPITFIRE; SPITFIRE: THE FIRST OF THE FEW
DRA   120 min (special 60th edition – ort 117 min) B/W
VIDrel: 4-FRONT/POLYREC/ODY   V

FIRST REBEL, THE ***
William A. Seiter USA   U
                                                 1939
John Wayne, Claire Trevor, Brian Donlevy, George Sanders, Robert Barrat, John F. Hamilton, Moroni Olsen, Eddie Quillan, Chill Wills, Ian Wolfe, Wallis Clark, Monte Montague, Eddy Waller, Clay Clement, Olaf Hytten
Wayne plays frontiersman James Smith, who leads a band of men against a tyrannical British captain in pre-Revolutionary American colonies as he sets out to smash liquor trafficking with the Indians. A simple and unpretentious tale with a nice performance from Trevor as the girl who is sweet on Wayne.
Aka: ALLEGHENY UPRISING
WES   70 min (ort 81 min) B/W
VIDrel: VCC/DISC/COLUM L/A   V
Boa: novel N.H. Swanson.

FIRST WIVES CLUB, THE **
Hugh Wilson USA   PG
                                                 1996
Goldie Hawn, Bette Midler, Diane Keaton, Maggie Smith, Dan Hedaya, Stockard Channing, Victor Garber, Stephen Collins, Elizabeth Berkley, Marcia Gay Harden, Bronson Pinchot, Jennifer Dundas, Eileen Heckart, Philip Bosco, Rob Reiner
Having all been dumped by their husbands, Midler, Hawn and Keaton put their heads together to plot a fitting retribution and their men. This trio and their comic exploits are pretty much the whole film, with Keaton giving the funniest performance. Highlights are Ivana Trump popping up briefly to proffer advice on revenge and the finale, which takes the form of a song by the three leads. See also SHE-DEVIL.
COM   102 min   CINrel
Boa: novel by Olivia Goldsmith.

FISH CALLED WANDA, A ***
Charles Crichton USA   15
                                                 1988
John Cleese, Michael Palin, Kevin Kline, Jamie Lee Curtis, Maria Aitken, Tom Georgeson, Patricia Hayes, Geoffrey Palmer, Peter Jonfield, Cynthia Caylor, Mark Elwes, Neville Phillips, Ken Campbell, Al Ashton, Roher Hume
A motley gang jewel thieves have just robbed London's Hatton Garden, and only the brains behind the robbery knows where the gems have been hidden, but he is now in jail. However, the gang's sexy member plans to seduce his defence counsel and get the loot for herself. A blend of bad taste jokes and engaging flashes of wit. Written by Cleese and Crichton (the latter being of THE LAVENDER HILL MOB fame) who came out of retirement to direct. AA: S. Actor (Kline).
COM   103 min (ort 109 min)   VIDrel: MGM/WHV   V

FISH THAT SAVED PITTSBURGH, THE *
Gilbert Moses USA   (PG)
                                                 1979
Julius Erving, James Bond III, Stockard Channing, Jonathan Winters, Margaret Avery, Jack Kehoe, Meadowlark Lemon, Nicholas Pryor, Flip Wilson, Michael V. Gazzo, Kareem Abdul-Jabbar, Peter Isacksen, M. Emmet Walsh, Marvin Albert
A basketball team changes its name and uses astrology to win the championship in this simple-minded comedy. The disco soundtrack is an annoyance, but less of one than the sheer lack of laughs.
COM   100 min (ort 104 min)   SATrel: SKY MOVIES

FISHER KING, THE ***
Terry Gilliam USA   15
                                                 1991
Robin Williams, Jeff Bridges, Mercedes Ruehl, Amanda Plummer, Michael Jeter, Adam Bryant, Paul Lombardi, David Pierce, Ted Ross, Lara Harris, Warren Olney, Frazer Smith, Kathy Najimy, Harry Shearer, Melinda Culea, Mark Bowden
A flippant remark made by an arrogant radio D.J. is taken liter-

ally by an unbalanced listener, and a bloody massacre ensues. Racked by guilt, our D.J. becomes an alcoholic and a derelict. He is saved from a mugging by another unhappy soul, a former medieval literature teacher who lost his own wife in the same massacre. Together, they embark on a fantastical "medieval" quest for the Holy Grail, in this imaginative and poetic film. AA: S. Actress (Ruehl).
DRA   132 min (ort 175 min) cC
VIDrel: VCC/DISC/COLUM   V/sur

F.I.S.T. **
Norman Jewison USA   PG
                                                 1978
Sylvester Stallone, Rod Steiger, Peter Boyle, Melinda Dillon, David Huffman, Tony LoBianco, Kevin Conway, Cassie Yates, Henry Wilcoxon, Brian Dennehy, Peter Donat, John Lehne, Richard Herd, Elena Karam, Ken Kercheval, James Karen
The story of the rise and fall of a power-mad union boss loosely based on the life of Jimmy Hoffa. A good script is largely wasted by poor casting, especially with regard to Stallone, who was well and truly out of his depth in this one.
Aka: FIST
DRA   125 min (ort 145 min)   VIDrel: MGM/WHV   V
Boa: novel by Joe Eszterhas.

FIST OF HONOR **
Richard Pepin USA   18
                                                 1992
Joey House, Sam Jones, Bubba Smith, Abe Vigoda, Hary Guardino, Nicholas Worth, Frank Sivero
Violent actioner with all the usual elements as a young boxer seeking revenge for the death of his girlfriend gets caught up in a power struggle between two gangland leaders.
A/AD   95 min (ort 100 min)   VIDrel: IMPENT   V

FIST OF JUSTICE **
Kim Bass USA   18
                                                 1993
Marjean Holden, Corey Everson, Joel Beeson, Sam Jones, Charles Napier, Richard Roundtree
Former women's bodybuilding champ Everson graces the screen in this formula tale of a female cop and martial arts expert who works undercover, but finds herself framed for murder. She enlists the help of her father and brother to clear her name.
A/AD   92 min   VIDrel: GUILD/SONOP   V/sh

FIST OF THE NORTH STAR ***
Toyoo Ashida JAPAN   18
                                                 1992
Voices of: John Vickery, Michael McConohie, Melodee Spyvack, Jeff Corey, Dan Woren, Tony Oliver, Wally Burr, Gregory Shedoff, Holly Sidell, Barbara Goodson, Mike Forest, James Avery, Dave Mallow, Tom Wyner, Doug Stone, Mike Thornton
WW3 has left the planet a bleak radioactive wasteland, with the remnants of humanity surviving as best they can in the shattered remains of the cities. Meanwhile, various outlaws, mutants and petty dictators battle it out to form their own petty kingdoms. Against this backdrop, a powerful warrior is chosen to subjugate these warring factions and bring peace to the planet. A vigorous effort, somewhat crude but often very effective.
Aka: FIST OF NORTH STAR
ANIM   111 min   VIDrel: MANGA/SONOP   V
Boa: graphic novel Hokuto No Sen by Buronson and Tetsuo Hara.

FIST OF THE NORTH STAR *
Tony Randel USA   18
                                                 1995
Gary Daniels, Costas Mandylor, Chris Penn, Malcolm McDowell, Isako Washio, Melvyn Van Peebles, "Downtown" Julie Brown, Leon "Vader" White
The title character is a powerful warrior who wanders the world in search of the man who killed his father and abducted his woman. Fantasy-adventure based on a Japanese Manga comic-book series, in which the characters stagger about in the usual post-Apocalypse wasteland, doing the best they can in a low-budget film whose shortcomings (such as the dire plot) are all too apparent.
FAN   88 min (ort 90 min)   VIDrel: MED/20VIS   V/sh
Boa: novel Hokuto No Ken by Buronson and Tetsuo Hara.

FISTFUL OF DEATH, A **
Gianni Crea ITALY/SPAIN   15
                                                 1972
Lincoln Tate, William Berger, Perry Dell, Dean Stratford, Richard Melvill, Klaus Kinski, Florella Mannoia, Lorenzo Fineschi, Lars Block
A girl and her two brothers investigate the theft of a consignment of gold but get into deep trouble when they discover the truth about it and have to face a notorious gunman who has

been hired to get rid of them. However, the subsequent show-down is not the end of the matter. A well photographed but rather poorly directed and plotted effort.
Aka: ARRIVA! IL CROW; ON THE THIRD DAY ARRIVED THE CROW
WES   84 min   VIDrel: IMPENT   V

FISTFUL OF DOLLARS, A ***
15
Sergio Leone ITALY/SPAIN/WEST GERMANY   1964
Clint Eastwood, John Welles (Gian Maria Volonte), Marianne Koch, Wolfgang Lukschy, Sieghardt Rupp, Antonio Prieto, Pepe Calvo, Benny Reeves, Carol Brown (Bruno Carotenuto), Benito Stefanelli, Richard Stiyvesant (Mario Brega)
One of the first and best of the spaghetti Westerns, loosely based on Kurosawa's YOJIMBO. A mysterious stranger rides into a town that is divided between two warring families and proceeds to play one side off against the other. The excellent score is by Ennio Morricone. Followed by FOR A FEW DOLLARS MORE and THE GOOD, THE BAD AND THE UGLY.
Aka: PER UN PUGNO DI DOLLARI
WES   96 min   VIDrel: MGM/WHV L/A   V

FISTFUL OF FINGERS, A **
15
Edgar Wright UK   1995
Graham Low, Martin Curtis, Oliver Evans, Quentin Green, William Cornes, Edward Scotland, Richard Green, Amy Bowles, Stuart Low, Nicola Stapleton, James Bailey, Ian Crick, Tim Wyatt, Rich Adams, Too Cute To Live Cat, Nick Netsall
A low-budget spoof on the spaghetti Western (it is filmed on 16 mm) that is shot in Somerset and relies heavily on a succession of sight gags to make its points. Made for a mere ú10,000, the slight story follows the career of a bounty hunter whose horse is killed, but this soon gives way to a host of surreal moments and occasionally amusing parodies. Not especially effective, but it has a gusto and irreverence that is quite endearing.
COM   77 min (ort 81 min)   VIDrel: RTM/DISC   V

FISTS OF DRAGONS **
15
Yeh Yung Chu HONG KONG
Huang Cheng Li, Kwo Young Moon, Ou Ti, Huang I Lung
The son of a Chinese boxing master is forced to defend his honour and wipe out his shame. Adequate.
Aka: FIST OF DRAGON
MAR   86 min (ort 92 min)   VIDrel: IMPENT   V

FISTS OF FURY ***
18
Lo Wei HONG KONG   1971
Bruce Lee, Nora Miao, James Tien, Maria Yi, Han Ying Chieh, Tony Liu, Li Hua Sxe, Robert Baker, Miao Ker Hsiu, Malalene, Li Quinn, Chin Shan
A kung fu student returns to Shanghai in 1908 for his master's funeral and takes an oath to refrain from using his skills. He takes a job at a Bangkok ice factory but following the mysterious disappearance of his cousins, discovers it to be no more than a front for a drug smuggling ring. Refusing to be bribed, he sets out to smash the operation. A simple plot is combined with dazzling fight sequences in Bruce Lee's first martial arts film.
Aka: BIG BOSS, THE; CHINESE CONNECTION, THE; FIST OF FURY
MAR   98 min Cut (2 min 51 sec – ort 103 min)
VIDrel: 4-FRONT/POLYREC   V

FISTS OF FURY 2 ***
18
Li Tso-Nan (To Lo Po) HONG KONG   1976
Bruce Li (Ho Tsung-Tao), Lo Lieh, Tien Fong, Ku Feng, Tong Yim Chen, Mgai Ping O, Shum Shim Po, Lee Quinn, Shikamura Yasuyoshi
Kung fu movie set in Shanghai in the 1920s where a man battles a sinister criminal organisation and goes all out for revenge when his mother is murdered.
Aka: CHING-WU MEW SU-TSI; FIST OF FURY 2; FIST OF FURY: PART 2
MAR   94 min Cut (1 min 12 sec – ort 98 min)
VIDrel: 4-FRONT/POLYREC   V

FISTS OF SHAOLIN **
18
Li Hsun HONG KONG   1973
Pai Ying, Han Ying Chieh, Wan Chung Shan
A man hears of a plot to murder his family and rushes home to defend them. He arrives too late finding them massacred and the family's sacred emblem, a Shaolin Fighting Stick, stolen. Honour must be avenged etc.
Aka: FIST OF SHAOLIN
MAR   83 min Cut (5 min 34 sec)   VIDrel: IMPENT   V

FISTS OF VENGEANCE, THE **
18
Cheng Hung Man HONG KONG   1973
Kung Bun, Tung Chi, Shoji Kurata, Lu Pi Chen, Tsao Chien, Lu Ping, Chen Hui Lou, Yong Lung, Pai I Feng, Tang Hsin, Has Su, Chiang Chih Yang, Yu Sung Chao
A Japanese fighter is sent to escort a shipment of red sand, that is used in the making of high-grade steel and has the usual conflicts, in this early and unsophisticated effort.
Aka: FIST OF VENGEANCE; TWO FISTS VERSUS SEVEN SAMURAI
MAR   88 min   VIDrel: BMGVID/BMGREC   V

FIT TO KILL **
18
Andy Sidaris USA   1992
Dona Speir, Julie Strain, Roberta Vasquez, R.J. Moore, Bruce Penhall, Rodrigo Obregon, Cynthia Brimhall, Tony Peck
Two attractive female agents are assigned to guard a priceless diamond and find this assignment is a lot tougher than they anticipated. Another average actioner from this director.
A/AD   90 min (ort 94 min)   VIDrel: TRING/COLUM   V

FITZCARRALDO ***
PG
Werner Herzog WEST GERMANY   1982
Klaus Kinski, Claudia Cardinale, Jose Lewgoy, Paul Hittscher, Miguel Angel Fuentes, Grande Othelo, Huerequeque, Enrique Bohorquez, Peter Berling, David Perez Espinosa, Milton Nascimento, Rui Polanah, Salvador Godinez, Bill Rose
At the turn of the century, an eccentric Irishman is obsessed by the idea of building an opera house in the Peruvian jungle and is forced to drag his boat over the mountains from one river to another. A hauntingly beautiful film; pointless but undeniably hypnotic.
DRA   150 min (ort 160 min)   VIDrel: TART/20TH; ENCORE (LV only)   V LV

FIVE CHILDREN AND IT **
U
Marilyn Fox UK   1992
Simon Godwin, Nicole Mowat, Charlie Richards, Tamzen Audas, Francis Wright, Laura Brattan, Mary Conlon, Ron Welling, Joyce Windsor
A pleasant kid's adventure that follows the exploits of five children who discover a sand fairy, and learn that it has the power to grant their wishes. Originally shown by the BBC in six parts.
Aka: RETURN OF PSAMMEAD, THE; SAND FAIRY, THE
JUV   137 min (2 cassettes – ort 139 min) mTV
VIDrel: BBC/TECH   V/sh
Boa: novel by E. Nesbit.

FIVE DOLLS FOR AN AUGUST MOON *
18
Mario Bava USA   1970
Edwige Fenech, William Berger, Ira Furstenberg, Howard Ross, Helena Ronee
At a beach-front castle, a group of investors gather to meet an inventor, but are killed off one by one. Meanwhile, their wives are enjoying some lesbian encounters or fun with the servants, these erotic interludes serving to sustain interest, as with regard to the killings, very little of this is seen on screen. A trashy, low-budget film, lurid and dated.
HOR   78 min wScrn dubbed   VIDrel: REDEM/RTM   V

FIVE EASY PIECES ***
15
Bob Rafelson USA   1970
Jack Nicholson, Karen Black, Lois Smith, Susan Anspach, Fannie Flagg, Sally (Ann) Struthers, Ralph Waite, Helena Kallianiotes, Billy Bush, Richard Stahl, Toni Basil, Lorna Thayer, William Challee, John Ryan, Irene Dailey
A promising musician working on an oil rig goes home to see his dying father, and ends up throwing over his pregnant girl-friend for a young piano student. A well made and acted character study of a rather unpleasant individual that is sustained by one of the best (and often most moving) performances Nicholson has ever given.
DRA   94 min (ort 98 min)   VIDrel: ITC/POLYREC   V/h

FIVE HEARTBEATS, THE **
15
Robert Townsend USA   1991
Robert Townsend, Michael Wright, Leon, Harry J. Lennix, Tico Wells, Diahann Carroll, Harold Nicholas, John Canada Terrell, Chuck Patterson, Hawthorne James, Tressa Rhomas, Roy Fegan, Troy Beyer, Carla Brothers, O.L. Luke, Paul Benjamin
A terribly corny rags-to-riches showbusiness yarn (based very loosely on the career of The Dells) that charts the rise and fall of a 1960s rhythm and blues quintet. The film takes an inordinate amount of time charting the group's rise, at which point

their drug-induced collapse is almost instantaneous. Despite the over-cliched script (racism is an ever-present backdrop) there are some effective elements, but the forgettable music track is not one of them.
DRA   90 min (ort 121 min)   VIDrel: 20TH/TECH   V/sur

FIVE VENOMS **
18
Chang Cheh HONG KONG
Lu Feng, Lo Meng, Wei Pai, Chiang Sheng, Philip Kwok, Chien Sun, Watsa Mata Yu
As a kung-fu master lies dying, he instructs his last student to check on the activities of five former students to whom he each taught a special and very deadly skill. He is afraid that they will use these gifts for evil and tells his protege to eliminate any of them who have succumbed to this temptation. Unfortunately, this is a tall order as the true identities of the five are known neither to the teacher nor his student.
Aka: FIVE DEADLY VENOMS
MAR   98 min wScrn dubbed   VIDrel: MADE/RTM   V

FIXATION *
18
Alex De Renzy USA
1993
Alicyn Sterling, Flame, Raven, Alexandra Quinn, Kym Wilde, Kay Parker, Jamie Gillis, Buck Adams, Tom Byron
A sex therapist/psychologist explores the darker side of his sexual nature in his free time in a bid to satisfy his obsessions.
A   61 min (ort 65 min)   VIDrel: GROHOM/MAXSCAN   V

FIXING THE SHADOW **
18
Larry Ferguson USA
1992
Charlie Sheen, Linda Fiorentino, Michael Madsen, Courtney B. Vance, Leon Rippy, Dennis Burkley, Lyndsay Riddell, Rino Thunder, Rip Torn, James Oscar Lee, Ed Adams, Hollie Chamberlain, Richard Madsen, Larry Ferguson, Ted Parks
Sheen stars as a young Arizona police officer who goes undercover after he is sacked, and infiltrates a gang of drug-dealer bikers, his hope being that by showing his worth he will get reinstated. However, it is not long before he realises that his own perception of what is right and wrong is beginning to alter in response to his activities and at the same time he is forced to confront some repressed memories, such as his abuse as a child by his father.
Aka: BEYOND THE LAW
DRA   97 min (ort 109 min)   VIDrel: 20VIS/SONOP   V/sur

FLAMBARDS **
U
Lawrence Gordon Clark/Michael Ferguson/Leonard Lewis UK
1980
Christine McKenna, Edward Judd, Streven Grives, Alan Parnaby, Rosalie Williams, Gillian Davey, Sebasitan Abineri, Carol Leader, Gwynne Gray, Olive Pendleton
During WW1 in Britain, a young orphan girl comes to live with tyrannical uncle at his dilapidated estate and has a hard time coping with his brutality and that of his eldest son. A competently made but quite overlong adaptation with good period detail. Originally shown in thirteen 52-minute episodes.
DRA   VIDrel: VCC L/A   V
Boa: novels by K.M. Peyton.

FLAME **
PG
Richard Loncraine UK
1974
Tom Conti, Alan Lake, Dave Hill, Noddy Holder, Don Powell, Johnny Shannon, Kenneth Colley, Sara Clee, Anthony Allen, Tommy Vance, Mike Pasternak, John Dicks, Michael Coles, Nina Thomas, A.J. Brown, Susan Tebbs, John Steel
A look at the ruthless world of the music business, that's set in 1965 and follows the changing fortunes of a group being promoted by a smart advertising executive.
Aka: SLADE IN FLAME
DRA   86 min (ort 91 min)   VIDrel: 4-FRONT/POLYREC L/A   V

FLAME IN MY HEART, A **
18
Alain Tanner FRANCE/SWITZERLAND
1987
Myriam Mezieres, Aziz Kabouche, Benoit Regent, Andre Marcon, Jean-Gabriel Nordman, Biana, Douglas Ireland, Jean-Yves Berteloot
An unhappy and sexually unfulfilled actress tries to break up with her Arab boyfriend and checks into a hotel in order to get away from them. There she meets and seduces a withdrawn journalist with whom she embarks on a live-in relationship. However, when he has to leave her alone in Paris for two weeks while away on an assignment, she begins to lose control of her emotions. A meandering and pretentious effort with elements that recall a soft-core movie.
Aka: UNE FLAMME DANS MON COEUR
DRA   106 min B/W   VIDrel: MAINPIC/RTM   V

FLAME IN THE STREETS ***
PG
Roy Ward Baker UK
1961
John Mills, Sylvia Syms, Brenda De Banzie, Earl Cameron, Wilfred Brambell, Johnny Sekka
An earnest albeit dated attempt to examine the issue of racial prejudice, that has honest union man Mills getting to grips with the problem of an imminent strike by his members, should a black foreman be appointed. Despite his adoption of what he believes to be a liberal position, a problem closer to home provides a far truer test of his tolerance.
DRA   83 min (ort 93 min) B/W   VIDrel: CARL/TECH   V
Boa: play Hot Summer Night by Ted Willis.

FLAME OF THE BARBARY COAST **
U
Joseph Kane USA
1945
John Wayne, Ann Dvorak, Joseph Schildkraut, William Frawley, Virginia Grey, Russell Hicks, Jack Norton, Paul Fix, Manart Kippen, Butterfly McQueen, Eve Lynne, Marc Lawrence, Rex Lease, Hank Bell, Al Murphy, Adele Mara, Emmett Vogan
A saloon singer is pursued by both a rancher and a sophisticate with the former opening his own gambling hall. However, all complications are nicely resolved by a well-timed earthquake. An adequate time-filler in which Wayne played a character named Duke, thus giving rise to his lifelong nickname.
WES   87 min (ort 91 min) B/W
VIDrel: 4-FRONT/POLYREC   V

FLAME TREES OF THIKA, THE: PARTS 1 AND 2 **
PG
Roy Ward Baker UK
1981
Hayley Mills, David Robb, Holly Aird, Ben Cross, Nicholas Jones, Sharon Maughan, Nick Chege, Steve Mwenesi, John Nettleton, Carol MacReady, Robert Rietty, Antony Baird, Paul Onsingo, Mzee Pembe
Story of the fortunes of a family of coffee planters in Kenya in 1914. A competent family drama with a lack of action but some pleasant locations. Originally shown in seven 50-minute episodes. Huxley's novel is to some extent autobiographical.
DRA   350 min (2 cassettes) mTV   VIDrel: VCC L/A   V
Boa: novel by Elspeth Huxley.

FLASH, THE **
PG
Robert Iscove USA
1990
John Wesley Shipp, Amanda Pays, Tim Thomerson, Michael Nader, Lycia Naff, Robert Hooks, M. Emmet Walsh, Priscilla Pointer, Eric DaRae, Paula Marshall
Adequate adaptation of the famous D.C. Comics character, that tells of a forensic chemist who is struck by lightning whilst at work, and finds he has gained the power of super-speed. In a fancy new costume, he becomes the crime-fighting title character. This yarn served as a pilot for a TV series, and as is often the case with super-hero movies, has little to offer apart from its special effects.
FAN   90 min (ort 94 min) mTV   VIDrel: WHV   V/sh

FLASH 2, THE: REVENGE OF THE TRICKSTER **
PG
Danny Bilson USA
1991
John Wesley Shipp, Amanda Pays, Joyce Hyser, Mark Hamill
The Flash is back in action in Central City, righting wrongs and taking on various villains, the principal one in this adventure being a character known as "The Trickster". He has hatched a plan to kill off our hero and take over the town. An adequate caper which is played for laughs rather than thrills.
JUV   88 min mTV   VIDrel: WHV   V/sh

FLASH 3, THE: DEADLY NIGHTSHADE **
PG
Bruce Bilson USA
John Wesley Shipp, Amanda Pays, Richard Belzer
This time round it's a character known as "The Ghost" who must be fought.
JUV   87 min mTV   VIDrel: WHV   V

FLASH GORDON ***
PG
Michael Hodges UK
1980
Sam J. Jones, Melody Anderson, Chaim Topol, Max Von Sydow, Ornella Muti, Timothy Dalton, Brian Blessed, Peter Wyngarde, Mariangela Melato, John Osborne, Richard O'Brien, John Hallam, Philip Stone, Suzanne Danielle, William Hootkins

An updated version of the old serial from Republic. An evil emperor from another planet tries to conquer the Earth but is opposed by Flash and his pals, who have been brought to the planet Mongo. A visually remarkable film with superb sets and costumes, but seriously deficient in characterisation and with Jones making a woefully inadequate hero, especially when up against Von Sydow who is superb as Ming. The dullish rock score soundtrack is by Queen.
FAN   107 min (ort 115 min) wScrn
VIDrel: BMGVID/BMGREC   V

FLASH GORDON CONQUERS THE UNIVERSE **   U
Ford Beebe/Ray Taylor USA   1940
Buster Crabbe, Charles Middleton, Frank Shannon, Anne Gwynne, Carol Hughes, Roland Drew, Shirley Deade, Victor Zimmerman, Don Rowan, Lee Powell, Richard Alexander, Beatrice Roberts, Montague Shaw, Michael Mark, Donald Curtis
Flash and friends join forces with others to tackle the evil Ming, whose plans for conquest involve the use of a strange ray that has caused a plague on Earth. First shown in 12 episodes, this was the third and final "Flash Gordon" adventure from Universal. As ever, a dated yet strangely endearing offering, albeit a little hard to take in one sitting. Note that the second alternative title is generally used for an 87–minute edited version.
Aka: PERILS FROM THE PLANET MONGO; PURPLE DEATH FROM OUTER SPACE; SPACE SOLDIERS CONQUER THE UNIVERSE
JUV   197 min (ort 240 min) B/W   VIDrel: FABFIL/SPEAR V

FLASHDANCE **   15
Adrian Lyne USA   1983
Jennifer Beals, Michael Nouri, Lilia Skala, Sunny Johnson, Kyle T. Heffner, Belinda Bauer, Lee Ving, Ron Ron Karabtsos, Malcolm Danare, Phil Burns, Lucy Lee Flippin, Micole Mercurio, Don Brockett, Cythia Rhodes, Durga McBroom
A young female welder dreams of becoming a ballerina. A sort of overlong video with plenty of music and some high octane dance sequences (most of which were actually performed by uncredited French dancer Marine Jahan). AA: Song ("What A Feeling" – Giorgio Moroder (m)/Keith Forsey and Irene Cara (l)).
MUS   90 min   VIDrel: CIC/SONOP   V/sur

FLASHFIRE **   18
Elliot Silverstein USA   1993
Billy Zane, Louis Gossett Jr, Kristin Minter, Louis Giambalvo, Philip Needs, Tom Mason
A detective who was decorated and promoted when he saved a hostage, is shocked by the murder of his partner. He meet a hooker who witnessed this crime and is prepared to help find the killers, whom she can identify. Their efforts gradually lead them to suspect a conspiracy within the force to cover up a corruption scandal. An adequate if hardly inspired effort on a familiar theme.
THR   84 min (ort 88 min)   VIDrel: 20VIS/SONOP   V

FLASHPOINT **   15
William Tannen USA   1984
Kris Kristofferson, Rip Torn, Treat Williams, Kevin Conway, Miguel Ferrer, Jean Smart, Roberts Blossom, Tess Harper, Kurtwood Smith, Terry Alexander, Guy Boyd, William Frankfather, Nora Heflin, Joaquin Martinez, Mark Slade
Two border patrolmen are in deep trouble when they discover some evidence linked to the assassination of Kennedy. This promising start never develops into anything of note.
A/AD   89 min (ort 95 min)   VIDrel: WHV   V/sh

FLATFOOT *   15
Steno (Stefano Vanzina) ITALY/WEST GERMANY   1978
Bud Spencer (Carlo Pedersoli), Joe Stewardson, Werner Pochat, Enzo Cannavale, Bodo, Dagmar Lassander, Raymond Pellegrin
An over-the-top knockabout cop comedy, detailing the exploits of tough-guy Spencer as he uses his own methods in dealing with drug smugglers. A silly comic strip that uses with live actors, it signals its slapstick humour well in advance.
Aka: FLATFOOT IN AFRICA; KNOCK OUT COP, THE; PIADONE L'AFRICANO; TRNITY: TRACKING FOR TROUBLE
A/AD   103 min Cut (13 sec)   VIDrel: IMPENT   V

FLATLINERS *   15
Joel Schumacher USA   1990
Kiefer Sutherland, Julia Roberts, Kevin Bacon, William Baldwin, Oliver Platt, Kimberly Scott

Five misguided medical students experiment with death, so arranging things that each will undergo a near-death experience (shown by the "flat-line" of their cardiographs) before being resuscitated in the nick of time by the others. Unfortunately, by such meddling they each fall prey to bizarre and frightening supernatural experiences. A most disappointing film, pretentious and pompous, even the effects are incredibly uninventive.
FAN   109 min (ort 111 min)   VIDrel: VCC/DISC/COLUM V/sur

FLAVIA THE HERETIC *   18
Gianfranco Mingozzi FRANCE/ITALY   1974
Florinda Bolkan, Maria Casares, Claudio Cassinelli, Anthony Corlan, Spiros Focas
In 15th century Italy, a young girl is forced to enter a nunnery by her father, and suffers further alienation when she witnesses a brutal rape and various sadistic acts against women. This sets her on the path of a bloody revenge against society in general, seizing her chance to do just that when she joins forces with an invading Moslem army. An unpleasant and deservedly obscure celebration of cruelty, well made but no more agreeable on that account.
Aka: FLAVIA; FLAVIA LA MONACA MUSULMANA; REBEL NUN, THE
HOR   95 min wScrn dubbed   VIDrel: REDEM/RTM   V

FLEA BITES ***   (PG)
Alan Dossor UK   1991
Anthony Hill, Nigel Hawthorne, Dalton Walters, Michelle Fairley, Tim Healy, Ashley Brooks, Charlie Bartle, Anthony Clarke, Thadeus Kaye, Anna Korwin, Crispin Harris, James Hooton, Trevor Peake, Ann Lathlane, Phil Towers
A young man breaks into a home of an elderly Polish exile, and finds the remains of a flea circus there, which he decides to restore. This leads to an unexpected friendship between the two men, which is often touching and quite perceptive. An offbeat character study, moderately absorbing if a little bizarre.
DRA   100 min mTV   TVrel: BBC

FLED *   18
Kevin Hooks USA   1996
Stephen Baldwin, Laurence Fishburne, Will Patton, Robert John Burke, Robert Hooks, Victor Rivers, David Dukes, Ken Jenkins, Michael Nader, Brittney Powell, Salma Hayek, Steve Carlisle, Brett Rice, J. Don Ferguson, Kathy Payne
Mayhem and fisticuffs are the order of the day as fellow convicts Baldwin and Fishburne escape, unhappily still handcuffed together. With one prisoner black and the other white, it comes as no surprise to find the plot calling for rather obvious mutual antagonism. Tired and derivative scripting assures us of a most unrewarding experience. See THE DEFIANT ONES (the original not the remake) for an object lesson in how it should all have been done. See also THE CHAIN.
A/AD   98 min   CINrel

FLESH **   18
Paul Morrissey USA   1969
Joe Dallesandro, Geraldine Smith, Maurice Bradell, Candy Darling, Louis Waldon, Geri Miller, Jackie Curtis, Patti D'Arbanville
A bizarre film about a young hustler in New York (from a graduate of Andy Warhol's film factory), that follows his exploits as he tries to raise the money to pay for an abortion for his wife's girlfriend. A difficult film to follow, but there are a good number of potent observations in there.
COM   85 min (ort 105 min)   VIDrel: FIRST/SONOP   V

FLESH, THE **   18
Marco Ferreri ITALY   1991
Sergio Castellitto, Francesca Dellera, Philippe Leotard, Farid Chopel, Petra Reinhardt, Gudrun Gundlach, Nicoletta Boris, Clelia Piscitello, Pino Tosca, Sonia Topazio, Fulvio Falzarano, Eleonora Cecere, Matteo Ripaldi
A divorced cabaret pianist who has custody of his kids embarks on a passionate affair in this adult comedy.
Aka: LA CARNE
DRA   86 min (ort 90 min)   VIDrel: ARTPRO/RTM   V

FLESH AND BLOOD **   18
Paul Verhoeven HOLLAND/SPAIN/USA   1985
Rutger Hauer, Jennifer Jason Leigh, Tom Burlinson, Jack Thompson, Susan Tyrell, Ronald Lacey, Brion James, Bruno Kirby
A medieval epic with knights in armour, damsels in distress and all the other trappings one expects, shown with a heavy

emphasis on the seamier side of life rather than its romantic aspects, with a young bride-to-be kidnapped and raped but growing to like her captor.
Aka: ROSE AND THE SWORD, THE
A/AD   122 min (Cut at film release – ort 126 min)
VIDrel: 4-FRONT/POLYREC/VISVID   V/sur

FLESH AND BONE ***
Steve Kloves USA
Dennis Quaid, James Caan, Meg Ryan, Christopher Rydell, Scott Wilson, Julia Mueller, Gwyneth Paltrow, Barbara Alyn Woods, Joe Berryman, Ron Kuhlam, James N. Harrell, Ez Perez, Ryan Bohls, Gerardo Johnson, Hector Garcia, Betsy Brantley
A man who lives a rootless life, taking care of vending machines in rural Texas, is haunted by memories of a robbery pulled by his father many years ago, which ended in murder and in which he took part as an unwilling accomplice. When he meets an alcoholic woman he falls in love, little guessing that their lives are strangely linked because of his past. A well acted and low-key drama with striking camerawork and a moody atmosphere.
DRA   121 min (ort 127 min) cC   VIDrel: CIC/SONOP
V/sur
15
1993

FLESH AND THE DEVIL **
Clarence Brown USA
Greta Garbo, John Gilbert, Lars Hanson, Barbara Kent, Marc McDermott
A vamp plays with the affects of various men and brings a swift end to and old friendship. However, she eventually meets with a fitting end when she freezes to death on an ice floe. A melodramatic star vehicle whose huge success may have had something to do with the offstage romance between Garbo and Gilbert.
DRA   113 min B/W silent   VIDrel: MGM/WHV   V/sh
Boa: novel The Undying Past by Hermanin Sudermann.
U
1927

FLESH-EATING MOTHERS *
James Aviles Martin USA
Robert Lee Oliver, Donatella Hecht, Valorie Hubbard, Ken Eaton, Neal Rosen, Terry Hayes
Women who commit adultery become the victims of a strange virus that causes them to eat their offspring in this gruesome and unpleasant tale.
HOR   86 min (ort 90 min)   VIDrel: VIPCO/SGSVID   V
18
1989

FLESH GORDON *
Michael Benveniste/Howard Ziehm USA
Jason Williams, Suzanne Fields, John Hoyt, William Hunt, Joseph Hudgins, Mycle Brandy, Candy Samples, Lance Larson, Nora Witernik, Steven Grummette, Judy Ziehm, Linus Gator, Mark Fore, Donald Harris, Susan Moore, Sally Alt
This semi-porno version of FLASH GORDON has little to recommend it. A sex ray is causing chaos on Earth and Flesh blasts off to the Planet Porno (where the ray originates) in order to do battle. Most of the original graphic sex scenes are cut. A sequel followed.
COM   84 min Cut (1 min 16 sec – ort 90 min)
VIDrel: 4-FRONT/POLYREC L/A   V
18
1974

FLESH GORDON 2 *
Howard Ziehm USA
Vince Murdocco, Robyn Kelly, Tony Travis, Morgan Fox, William Dennis Hunt, Bruce Scott
This dreary sequel to the first film has our hero attempting to save the universe from its worst threat yet – the prospect of galactic impotence. An unutterably silly SF spoof whose vulgarity seems strained, and which does little to either amuse or titillate.
Aka: FLESH GORDON MEETS THE COSMIC CHEERLEADERS
COM   98 min   VIDrel: 4-FRONT/POLYREC/EIV L/A   V
18
1991

FLESH PALACE, THE *
Charles Rothstein USA
J.R. Carrington, Marc Wallice, P.J. Sparxx, Dallas D'Amour, Kimberly Kummings, Roxanne Hall, Blake Hall, Mike Horner, Nick East
Portrait of title sex club where a number of couples engage in the usual encounters.
A   80 min   VIDrel: ONE   V
18
1995

FLESHTONE **
Harry Hurwitz USA
Martin Kemp, Lise Cutter, Tim Thomerson
An artist feels the need for some female company and so
18
1993

---

unwisely phones a sex life, becoming the prime suspect when the woman in question is murdered. A dreary little thriller that runs along over-familiar lines.
THR   86 min (ort 89 min)   VIDrel: PRISM/HIFLI   V

FLETCH **
Michael Ritchie USA
Chevy Chase, Tim Matheson, Dana Wheeler-Nicholson, Joe Don Baker, Richard Libertini, Geena Davis, M. Emmet Walsh, George Wendt, Kenneth Mars, Kareem Abdul-Jabbar, Bill Henderson, William Traylor, George Wyner, Tony Longo
A reporter adopts various disguises in his investigations and becomes involved with a vast drugs-running conspiracy. Okay in small doses but after a while the constant stream of wisecracks begins to irritate. Scripted by Andrew Bergman and followed by FLETCH LIVES.
COM   94 min (ort 98 min)
VIDrel: 4-FRONT/POLYREC/CIC L/A   V/sh
Boa: novel by Gregory McDonald.
PG
1985

FLETCH LIVES *
Michael Ritchie USA
Chevy Chase, Hal Holbrook, Julianne Phillips, Cleavon Little, R. Lee Emery, Richard Libertini, Randall (Tex) Cobb, George Wyner, Patricia Kalember, Geoffrey Lewis, Richard Belzer, Phil Hartman, Titos Vandis, Don Hood
This sequel to FLETCH has our identity-changing reporter now on a trip to Louisiana where he has to deal with the expected stock characters. Even less funny than the earlier film.
COM   91 min (ort 95 min)   VIDrel: CIC/SONOP   V/sur
PG
1989

FLIGHT FROM HELL **
Robet Young USA
Robert Loggia, Scott Bakula
When a pilot who flies a crop-duster gets lost between San Francisco and Sydney he finds his only hope lies in a commercial aircraft, whose pilot tries to guide him to safety.
THR   89 min (ort 96 min)   VIDrel: MARQ/QUANT   V
15
1993

FLIGHT FROM JUSTICE **
Don Kent CANADA/FRANCE
Jean Reno, Carole Laure, Bruce Boxleitner, Vlasta Vrana, David Frances, Jack Langedijk, Jesse Lavendel, Jason Cavalier, Michael Scherer, Yves Langlois, Linda Smith, Peter Colvey, Ken Roberts, Andre Apergis, Chip Cuipka, Michel Perron
In Quebec, a former fighter pilot now works for a dubious businessman who is not averse to a little smuggling. But when he crashes his plane in the wilderness of Canada he faces a tough challenge to survive. An adequate outdoors adventure.
A/AD   92 min   SATrel: MOVIE CHANNEL
(15)
1993

FLIGHT OF THE BLACK ANGEL **
Jonathan Mostow USA
Peter Strauss, William O'Leary, James O'Sullivan, Michele Pawk, Michael Keys Hall, Ben Rawnsley, Marcus Chong, Jerry Bossard, Patricia Sill, Ed Williams, Kim Robillard, James Henriksen, Michael Gregory, Rodney Eastman, Lee Ryan
An unbalanced elite pilot arms his plane with a nuclear-tipped missile and heads off to start WW3, with his courageous superior the only man able to postpone Armageddon.
Aka: FLIGHT OF BLACK ANGEL
A/AD   98 min (ort 115 min) mCab   VIDrel: CIC/SONOP
V/h
15
1991

FLIGHT OF THE DOVE **
Steve Railsback USA
Scott Glenn, Theresa Russell, Lane Smith, Terence Knox, Alex Rocco, Joe Pantoliano
An undercover agent who sells sex in exchange for military secrets tries to leave the secret service, but finds she knows too much to be allowed to do so. One of those intense if clined, cat-and-mouse thrillers, with adequate plotting and characterisation, the sleazy and downbeat opening promising far more than the film is able to deliver.
A/AD   88 min   VIDrel: HIFLI/SONOP   V/h
18
1994

FLIGHT OF THE INNOCENT, THE ***
Carlo Carlei FRANCE/ITALY
Manuel Colao, Francesca Neri, Jacques Perrin, Frederico Paciffici, Sal Borgese, Lucio Zagaria, Giusi Cataldo, Massimo Iodolo, Anita Zagaria, Isabelle Mantero, Nicola Di Pinto, Severino Saltarelli, Gianfranco Pallavicino
When his family are murdered by criminals, a young Sicilian
18
1993

boy is forced to flee for his life. An observant and moving story that has a few sharp things to say about how organised crime blights society as a whole. Co-scripted by Carlei, this was his directing feature.

Aka: LA CORSA DELL'INNOCENTE
THR 105 min CINrel

FLIGHT OF THE INTRUDER, THE *   15
John Milius USA   1990 (released 1991)
Danny Glover, Willem Dafoe, Brad Johnson, Rosanna Arquette, Tom Sizemore, J. Kenneth Campbell, Dann Florek, Madison Mason, Ving Rhames, Christopher Rich, Jared Chandler, Douglas Roberts, Scott Newton Stevens, Justin Williams
A Vietnam war adventure that details the exploits of bomber pilots during that conflict, the stories being set around an aircraft carrier. Plotting is unbelievable and absurd, and the film has more the flavour of a gung-ho 1950s war movie than anything made in the 1990s. Not an edifying experience.
WAR 110 min (ort 119 min) wScrn
VIDrel: 4-FRONT/POLYREC/CIC V/sur
Boa: novel by Stephen Coonts.

FLIGHT OF THE NAVIGATOR *   U
Randall Kleiser USA   1986
Joey Cramer, Veronica Cartwright, Cliff De Young, Howard Hesseman, Sarah Jessica Parker, Robert Small, Albie Whitaker, Jonathan Sanger, Iris Acker, Richard Liberty, Raymond Forchion, Paul Reubens (voice only)
This kid's adventure begins with a twelve-year-old exploring a ravine in the woods and falling in. On waking up he returns home only to find that eight years have passed, his parents have moved but he hasn't aged a single day. He soon discovers a link between his disappearance and a spacecraft when a strange robotic creature summons him to help it steer its craft home. A syrupy and excessively cute fantasy aimed rather too pointedly at kids.
FAN 87 min (ort 89 min) VIDrel: 20TH/TECH V/sur

FLIGHT TO MARS **   U
Lesley Selander USA   1951
Cameron Mitchell, Marguerite Chapman, Virginia Huston, Arthur Franz, John Litel, Morris Ankrum, Richard Gaines, Lucille Barkley, Robert Barratt, Bob Peoples, Edward Earle, William Forrest, Tony Marsh, Tris Coffin, Bill Neff
An expedition to Mars crash-lands on the planet and discovers an advanced civilisation. An early science fiction film that seems very amateurish by today's standards, mainly due to limitations imposed by the restricted budget.
FAN 71 min (ort 75 min) VIDrel: RTM/DISC V

FLINCH **   15
George Erschbamer USA   1993
Judd Nelson, Nick Mancuso, Gina Gershon, Frank Cassini, Marilyn Norry, Veronica Lorenz, Alvin Sanders, Laurence King, Anne Dupont, Andrew Airlie, Suzy Joachim
As a publicity ploy, a store hires two people, a man and a woman, to stand in the window and pretend to be mannequins. Unfortunately, they succeed so well in this task, that they become witnesses to a murder and find themselves facing the killer on their own after the police refuse to believe their story. A fair thriller that treads a familiar path.
THR 88 min VIDrel: HIFLI/SONOP V/h

FLINTSTONES, THE **   U
Brian Levant USA   1993
John Goodman, Rick Moranis, Elizabeth Taylor, Rosie O'Donell, Halle Berry, Elizabeth Perkins, Kyle MacLachlan, Jonathan Winters, Richard Moll, Irwin Keyes, Dann Florek plus voices of: Harvey Korman, Sheryl Le Ralph, Laraine Newman
This over-hyped, live-action version of a popular 1960s animated series offers great energy, inventive effects and inspired casting. It's too bad that it is sunk by the lack of a decent story. Here, Fred is catapulted out of his dead-end job and gets groomed for executive stardom by two crooked employees who need a fall-guy to take the blame for their embezzlement scheme. Along the way Fred dumps his old buddy Barney, but happily lives to see the error of his ways.
JUV 87 min (ort 93 min) cC VIDrel: CIC/SONOP; PION (LV only) V/sur LV

FLINTSTONES, THE: THE HOLLYROCK-A-BYE BABY **   U
USA   1993
The Flintstones journey to Hollyrock (a kind of Stone Age

version of Tinseltown) for the birth of their first grandchild, born to Bamm-Bamm and Pebbles (who were married in an earlier story). However, whilst there Fred gets involved in a case of mistaken identity, and various complications ensue. A competent feature-length animation.
ANIM 92 min VIDrel: FIRST/SONOP V

FLIPPER ***   PG
James Clark USA   1963
Chuck Connors, Luke Halpin, Kathleen Maguire, Connie Scott, Jane Rose, Joe Higgins, Robertson White, George Applewhite
A boy and his dolphin friend enjoy a series of adventures together in this family film which formed the basis for the later TV series. A highly entertaining film, shot in the Florida Keys. Followed in 1964 by FLIPPER'S NEW ADVENTURE.
JUV 86 min VIDrel: MGM/WHV V/h

FLIPPER **   PG
Alan Shapiro USA   1996
Paul Hogan, Elijah Wood, Chelsea Field, Isaac Hayes, Jonathan Banks, Jason Fuchs, Jessica Wesson, Ann Carey, Robert Deacon, Mark Casella, Luke Halpin, Bill Kelley, Lindsay Treco, Mal Jones, Louis Seeger Crume, Bill Nolan, Mary Jo Faraci
A sullen youngster goes to stay with his uncle in Florida, and while he is there he befriends an orphaned dolphin. Based on the 1960s TV series, this is a pleasing revival with a strong 1990s ecological flavour (the baddies in this film are polluting the seas by dumping toxic waste).
JUV 91 min (ort 97 min) cC VIDrel: CIC V/sur

FLIPPER'S NEW ADVENTURE **   PG
Leon Benson USA   1964
Luke Halpin, Pamela Franklin, Helen Cherry, Tom Helmore, Brian Kelly, Francesca Annis, Joe Higgins, Lloyd Battista, Gordon Dilworth, Courtney Brown, William Cooley, Dan Chandler, Robert Baldwin, Ric O'Feldman
Flipper and his owner go to a remote island in the Bahamas where they help to rescue a rich family who have been kidnapped by a gang. A by-the-numbers sequel to FLIPPER, with the dolphin being put through his usual acrobatic paces. Followed by a TV series.
Aka: FLIPPER AND THE PIRATES
JUV 94 min VIDrel: MGM/WHV V/h

FLIRT **   15
Hal Hartley GERMANY/JAPAN/USA   1993
Bill Sage, Martin Donovan, Parker Rosey, Paul Austin, Maria Schrader, Hannah Sullivan, Miho Nikaidoh, Robert Burke, Erica Gimpel
Expanded from a thirty-minute short, this unusual film examines three sets of characters, in three different countries, but all concerned with the same problem, namely how one is to react in a relationship when the other partner demands some kind of permanent commitment. There are some oddly affecting moments, but this is no NIGHT ON EARTH, and the material is too thin to sustain interest.
DRA 75 min CINrel

FLIRTING ***   15
John Duigan AUSTRALIA   1989
Noah Taylor, Thandie Newton, Nicole Kidman, Bartholemew Rose, Felix Nobis, Josh Picker, Kiri Paramore, Marc Gray, Greg Palmer, Joshua Marshall, David Wieland, Craig Black, Leslie Hill, Jeff Truman, Marshall Napier, John Dicks
A sequel to THE YEAR MY VOICE BROKE that is set in rural Australia in 1965, and has the central character from the earlier film now at boarding school. Mocked by his fellow pupils on account of his stammer, he eventually finds friendship in the shape of a sophisticated Ugandan pupil from the nearby girls' school, who is also an outsider because of her colour. A charming film about growing up – fresh, intelligent and well characterised.
COM 95 min (ort 99 min) VIDrel: WHV V/sur

FLIRTING SCHOLAR **   15
Richard Lee (Lok-Chi Lee) HONG KONG   1993
Gong Li, Sing Chi, Stephen Chow Sing-Chi, Chan Bak-Cheung, James Wong Jim, Leung Ka-Yam, Gordan Liu Chic-Hui, Cheng Pei-Pei, Lin Wei, Gabriel "Turtle" Wong Yut-San, Lee Ka-Sing, Peter Lai Bei-Tuk, Ven King-Tan, Ka Tin-Yi, Chu Mai-Mai
Title character employs an unusual blend of martial arts and magical powers to beat his enemies and win the hand of a princess. Having fallen in love with her, he disguises himself as

a servant and seeks employment in her household. A low-brow slapstick piece, set during the Ming Dynasty, and based on the figure of Tong Pak Fu.
MAR   97 min wScrn   VIDrel: MADE/RTM   V

FLIRTING WITH DISASTER ***
David O. Russell USA

15
1996

Ben Stiller, Patricia Arquette, Tea Leoni, Mary Tyler Moore, George Segal, Alan Alda, Lily Tomlin, Richard Jenkins, Josh Brolin, Celia Weston, Glenn Fitzgerald, Beth Ostrosky, Cynthia Lamantagne, David Patrick Kelly
In New York, Stiller learns that he was adopted, and sets off on a the road to find his real parents, supposedly being assisted by a psychologist from the adoption agency, who has the unfortunate knack of trying to match him with a whole succession of outlandish characters. The film is basically a collection of funny or touching encounters, with Stiller lurching from one disastrous meeting to the next. Fortunately, the strong script (by Russell) holds it all together.
COM   92 min   CINrel

FLOOD, THE: WHO WILL SAVE OUR CHILDREN? ***
(PG)
Chris Thomson USA
1993
Joe Spano, David Lascher, Michael Goorjian, Amy Van Nostrand, Norm Skaggs, Renee O'Connor, Scott Michael Campbell, Blayne Weaver, Mark Fairall, Jerome Ehlers, Lisa Rieffel, Kim Kregus, David Franklin, Kelly Rummery, Tim Elston
Fact-based story set in 1987, when a group of kids attending a summer camp organised by the Baptist Church were threatened by an imminent flood. Having gathered all the children together, the camp organisers set off for higher ground, only to have one of their coaches break down while crossing a river, leaving it up to the local townsfolk to attempt a rescue. Twelve children died in this disaster, this compelling drama is an appropriate tribute.
DRA   89 min (ort 93 min) mTV   SATrel: MOVIE CHANNEL

FLOWER OF MY SECRET, THE ***
Pedro Almodovar FRANCE/SPAIN

15
1995

Marisa Paredes, Juan Echanove, Imanol Arias, Carmen Elias, Rossy de Palma, Chus Lampreave, Joaquin Cortes, Manuela Vargas, Kiti Manver, Gloria Munoz, Juan Jose Otegui, Nancho Novo, Jordi Molla, Alicia Agut, Marisol Muriel, Jose Palau
A writer of romantic fiction hovering on the edge of a nervous breakdown finds little support from her family, and retreats into alcoholism instead. The director makes an unusual departure from his usual high-spirited romps, instead presenting this sad character study, with a few bright flashes of irony and strongly drawn characters, the whole presented in such a deadpan way that one doesn't know whether to laugh or cry.
Aka: LA FLEUR DE MON SECRET; LA FLOR DE MI SECRET
DRA   101 min (ort 107 min)   VIDrel: ELPIC/POLYREC   V

FLOWERS IN THE ATTIC **
Jeffrey Bloom USA

15
1987

Louise Fletcher, Victoria Tennant, Jeb Stuart Adams, Kristy Swanson, Lindsay Parker, Ben Granger, Marshall Colt, Nathan Davis, Alex Koba, Brooke Fries, Bruce Neckels, Leonard Mann, Gus Peters, Clare C. Peck (narration only)
Silly, tedious tale adapted from a bestseller that tells of four youngsters kept locked in the attic of the family mansion by their widowed mother as part of a plan to win favour with the wealthy parents that disowned her. The film tones down the novel's sado-masochistic and incestuous aspects and adds little of its own, but good casting generally keeps it going right up to its ludicrous climax. Written by Bloom.
THR   87 min (ort 93 min)   VIDrel: VCC/DISC/COLUM   V
Boa: novel by V.C. Andrews.

FLUKE ***
Carlo Carlei USA

PG
1995

Matthew Modine, Nancy Travis, Eric Stoltz, Jon Polito, Max Pomeranc, Bill Cobbs, Collin Wilcox Paxton, Samuel L. Jackson (voice only)
A businessman dies in a car accident and comes back to Earth inside the body of a title pup, in which form he tries to penetrate the mystery surrounding his death. An intriguing and refreshingly unusual theme is the major asset in this absorbing kid's fantasy, but some ill-advised scenes of animal cruelty make the film unsuitable for very young kids.
JUV   91 min (ort 95 min) cC   VIDrel: MGM/WHV   V/sur
Boa: novel by James Herbert.

FLY, THE **
Kurt Neumann USA

15
1958

Al (David) Hedison, Patricia Owens, Herbert Marshall, Vincent Price, Kathleen Freeman, Betty Lou Gerson, Charles Herbert, Eugene Borden, Torben Meyer, Harry Carter, Charles Tannen, Frank Roehm, Arthur Dulac
A scientist invents a method of transmitting matter and tries it out on himself, but something goes terribly wrong when a fly gets into the transmission chamber. A stilted and wooden film with little but one good idea to hold the interest. A sequel THE RETURN OF THE FLY followed. Remade in 1986.
HOR   91 min (ort 94 min)   VIDrel: 20TH/TECH   V/h
Boa: short story by George Langelaan.

FLY, THE ***
David Cronenberg USA

18
1986

Jeff Goldblum, Geena Davis, John Getz, Joy Boushel, Les Carlson, Gordon Chuvalo, Michael Copeman, David Cronenberg, Carole Lazare, Shawn Hewitt, Brent Meyers, Doron Kernerman, Romuald Vervin
A gory remake of the 1958 film of the same name, telling of a brilliant scientist who develops a matter transporter and falls prey to some hideous changes when he uses. Goldblum is perfectly cast as the eccentric scientist whose gradual evolution into a human fly is handled with some style and wit but leans too heavily on the repulsive effects. Written by Cronenberg and Edward Pogue. AA: Make (Chris Walas/Stephan Dupuis).
HOR   92 min (ort 100 min)   VIDrel: 20TH/TECH L/A;
ENCORE (LV only)   V/h LV

FLY 2, THE **
Chris Walas USA

18
1989

Eric Stoltz, Daphne Zuniga, Lee Richardson, John Getz, Frank Turner, Ann Marie Lee, Gary Chalk, Saffron Henderson, Harley Cross, Matthew Moore
Knowing little of his father, Brundle Jr continues the scientific work of the former, unaware of the reasons behind his rapid growth to adulthood. Meanwhile, the scientists in whose care he has been placed are content to exploit his gifts. A clumsy sequel (the feature debut for the director) that soon jettisons the better human elements of the story in favour of repulsive effects; despite the repulsive and pathetic ending this is a very dull work.
HOR   100 min (ort 105 min)   VIDrel: 20TH/TECH;
ENCORE (LV only)   V/sur LV

FLY AWAY HOME ***
Carroll Ballard USA

U
1996

Jeff Daniels, Anna Paquin, Dana Delaney, Terry Kinney, Holter Graham, Jeremy Ratchford, Deborah Verginella, Michael J. Reynolds, David Hemblen, Ken James, Nora Ballard, Sarena Paton, Carmen Lishman, Christi Hill, Judith Orban
A thirteen-year-old loses her mother in a car crash and goes to live with her divorced dad in Canada, but the pair do not initially get on, until she rescues a clutch of goose eggs, and sets about hatching the chicks. As the pair get closer, they start planning how they can get the adult birds to a distant bird sanctuary, by leading them there in micro-lites. Loosely based on a true story (there was actually no young girl) this is syrupy but enjoyable outing.
JUV   107 min   CINrel
Boa: autobiography of Bill Lishman.

FLY BY NIGHT: RAP ON THE WILD SIDE **
Steve Gomer USA

18
1993

Jeffrey Sams, Ron brice, McLyte, Leo Burmeister, Daryl "Chill" Mitchell, Todd Graff
In New York, young black guys dream of wealth and success through rap but only achieve this goal when they join forces. They eventually realise all their aspirations but soon learn that fame can bring its own problems.
DRA   89 min (ort 93 min)   VIDrel: COLUM/SONOP
V/sur

FLY ME **
Cirio Santiago ITALY

18
1973

Richard Young, Naomi Stevens, Richard Miller, Buzz Albert, Pat Anderson, Lenore Kasdorf, Lyllah Torena
Airline stewardesses and their boyfriends get involved in a series of wild encounters in this adult comedy. A mixed up and fairly mindless saga of sex, mayhem and the occasional kidnapping thrown in for good measure.
COM   71 min   VIDrel: ALLIED/RTM   V

FLYING DOWN TO RIO *   U
Thornton Freeland USA   1933
*Dolores Del Rio, Gene Raymond, Ginger Rogers, Fred Astaire,
Franklin Pangborn, Raul Roulien, Blanche Frederici, Eric Blore,
Walter Walker, Etta Moten, Roy D'Arcy, Maurice Black, Armand
Kaliz, Paul Porcasi, Luis Alberni*
Fun musical memorable mainly for its scenes of girls dancing
on the wings of an aeroplane. The start of a long and fruitful
Astaire-Rogers partnership.
MUS   86 min (ort 89 min) B/W   VIDrel: VCC   V

FLYING LEATHERNECKS **   PG
Nicholas Ray USA   1951
*John Wayne, Robert Ryan, Jay C. Flippen, Janis Carter, Don Taylor,
William Harrigan, James Bell, Carleton Young, Brett King, Maurica
Jara, Steve Flagg, Britt Norton, Adam Williams, Lynn Stalmaster,
Barry Kelly, Sam Edwards*
WW2 drama about a tough Marine flying corps major who's
disliked by his men because of his excessive strictness. All is
forgotten in war however, in this solid but unimaginative
actioner. Highspots are the good aerial battle scenes but the
film's slow pace is a handicap.
WAR   98 min (ort 102 min)   VIDrel: VCC/DISC/COLUM
L/A   V

FLYING SAUCER, THE **   PG
Mikel Conrad USA   1950
*Mikel Conrad, Pat Garrison, Hanz Von Teuffen, Russell Hicks,
Denver Pyle, Roy Engel, Virginia Hewitt, Frank Dairen, Lester
Sharpe, Earle Lyon, Robert Boon, Garry Owen, Phillip Morris, George
Baxter, Lee Langley*
US and Soviet scientists join forces to look for a flying saucer
hidden beneath an Alaskan glacier.
FAN   64 min (ort 70 min) B/W   VIDrel: FIRC/RTM   V

FLYING TIGERS **   PG
David Miller USA   1942
*John Wayne, John Carroll, Anna Lee, Paul Kelly, Mae Clarke, Gordon
Jones, Addison Richards, Edmund MacDonald, Bill Shirley, Tom
Neal, Malcolm McTaggart, David Brice, Chester Gan, James Dodd,
Gregg Barton, John James*
A solid WW2 epic set among an American fighter squadron
based in China. Full of the expected heroics and dogfight
sequences, but quite adequately put together.
WAR   98 min (ort 102 min) B/W
VIDrel: 4-FRONT/POLYREC   V

FLYNN **   18
Frank Howson AUSTRALIA   1996
Guy Pearce, Claudia Karvan, Steven Berkoff, John Savage
A glossy biopic on the early life of Hollywood legend Errol
Flynn, that purports to chart many of the highlights of his inter-
esting and eventful life, such as his trip to the jungles of New
Guinea in search of gold. Quite episodic in structure, it is inter-
mittently absorbing if a little banal, and does no more than hint
at his future status, taking us up to his 1933 debut film: "In The
Wake Of The Bounty".
DRA   95 min   VIDrel: MED/20VIS   V/sh

FOG, THE **   15
John Carpenter USA   1979
*Adrienne Barbeau, Hal Holbrook, John Houseman, Janet Leigh, Jamie
Lee Curtis, Tom Atkins, Nancy Loomis, Charles Cyphers, Ty Mitchell,
George Buck Flower, John Vick, Jim Jacobus, Jim Canning, Regina
Waldon, Darrow Igus*
A small town in California is attacked by the ghosts of mariners
who take a grisly revenge on the descendants of the six people
who lured their ship onto the rocks a hundred years before. A
well-directed film of occasionally scary moments as the remark-
ably substantial ghosts proceed to hack and club their way to
revenge under cover of a dense, sinister fog. A kind of ghostly
but equally gory follow-up to HALLOWEEN.
HOR   86 min (ort 91 min)
VIDrel: ETL/4-FRONT/POLYREC   V

FOLKS! **   15
Ted Kotcheff USA   1992
*Tom Selleck, Don Ameche, Anne Jackson, Christine Ebersole, Wendy
Crewson, Robert Pastorelli, Michael Murphy, Kevin Timothy
Chevalia, Margaret Murphy, T.J. Parish, Joseph Miller, John
McCormack, Peter Burns, Jon Favreau*
A successful stockbroker suffers a series of catastrophes that
brings his world crashing down around his ears. To make

matters worse, his aged parents move in with him, but since they
do not want to be a burden, they volunteer to commit suicide so
that he can collect on the insurance. A tasteless and highly
unfunny black comedy about ageing and bodily decay that is tiresome from beginning to end. Screenplay is by Robert Klane.
COM   108 min (ort 109 min)   VIDrel: MIA/DISC   V/sur

FOLLOW THE FLEET **   U
Mark Sandrich USA   1936
*Fred Astaire, Ginger Rogers, Randolph Scott, Betty Grable, Harriet
Hilliard (Nelson), Astrid Allwyn, Harry Beresford, Russell Hicks,
Brooks Benedict, Ray Mayer, Lucille Ball, Addison (Jack) Randall,
Maxine Jennings, Kay Sutton*
An enjoyable Hollywood musical about two sailors and their
girlfriends. The memorable songs are by Irving Berlin and
feature numbers such as: "Let's Face The Music And Dance",
"Let Yourself Go" and "We Saw The Sea". Borrows heavily from
"Shore Leave (1925) and "Hit The Deck" (1930).
MUS   106 min (ort 110 min) B/W   VIDrel: VCC   V

FOLLOW THE RIVER **   (PG)
Martin Davidson USA   1995
*Sheryl Lee, Ellen Burstyn, Eric Schweig, Tim Guinee, Renee
O'Connor, Tyler Noyes, Andrew Stahl, Gabriel Macht, Tony
Amendola, Sammy D. Miller, Graeme Malcolm, Judson Keith Linn,
Jimmie F. Skaggs*
A woman is captured by hostile Indians who abduct her and
keep her a prisoner in their encampment. There she meets
another female captive and together they plot and finally engi-
neer their escape, facing many hazards as they cross the
Virginian countryside. A competent historical adventure, set in
the colonial era.
A/AD   89 min (ort 93 min)   SATrel: SKY MOVIES

FOLLOW THE SUN: THE BEN HOGAN STORY **   U
Sidney Lanfield USA   1951
*Glenn Ford, Anne Baxter, Dennis O'Keefe, June Havoc, Larry
Keating, Roland Winters, Nana Bryant, Sam Snead, James Demaret,
Cary Middlecoff, Harold Blake, Ann Burr, Harmon Steves, Louise
Lorimer, Harry Antrim, Jeffrey Sayre*
An acceptable film biography of the famous golf player Ben
Hogan that covers most of the key events in his life, especially
the auto crash that almost killed him. However, he was even-
tually able to recover enough to return to the game.
DRA   93 min (ort 96 min) B/W   VIDrel: QUAVID   V
Boa: article by Frederick Hazlitt Brennan.

FOLLOWING HER HEART **   (PG)
Lee Grant USA   1994
*Ann-Margret, George Segal, Brenda Vaccaro, W. Morgan Sheppard,
Kirk Baltz, Scott Marlowe, Greg Mullavey, Tom Key, Edith Fields,
Deanie Gordon, Sandra Reaves, Alexandra Powers, Carla Dorren,
Thom Gossom Jr, Pamela Toll, Ken Schulz*
Story of a Swedish-born woman who marries at sixteen and
goes off to live in America. A pleasant romantic drama of no
great consequence.
DRA   85 min (ort 90 min) mTV   SATrel: SKY MOVIES

FOLLY TO BE WISE **   PG
Frank Launder UK   1952
*Alastair Sim, Roland Culver, Elizabeth Allan, Martita Hunt, Colin
Gordon, Janet Brown, Miles Malleson, Edward Chapman, Peter
Martyn, Robin Bailey, George Cole*
A new Army Captain takes over as Chaplain and Entertainment
Officer at a local Army camp, with disastrous results when he
decides to organise a brains trust that initiates a battle of the
sexes. A film that starts out with vigour and wit, then slowly
runs out of steam. Thankfully, the sparkling performances
(especially Sim's) save it from disaster.
COM   91 min B/W   VIDrel: LUMI/SPEAR   V
Boa: play It Depends on What You Mean by James Bridie.

FOOD OF THE GODS, THE *   18
Bert I. Gordon USA   1976
*Marjoe Gortner, Pamela Franklin, Ida Lupino, Ralph Meeker, John
McLiam, John Cypher, Belinda Belaski, Chuck Courtney, Tom Stovall*
A strange substance oozing out of the ground on a remote island
causes small animals and insects to grow to giant proportions.
A poorly realised attempt to get some mileage out of the classic
Wells story. Filmed eleven years before by Gordon as "Village
Of The Giants".
FAN   84 min   VIDrel: RNK L/A   V
Boa: short story by H.G. Wells.

FOOL FOR LOVE * (15)
Robert Altman USA 1985
Sam Shephard, Kim Basinger, Harry Dean Stanton, Randy Quaid,
Martha Crawford, Louise Egolf, Sura Cox, Jonathan Skinner, April
Russell, Deborah McNaughton, Lon Hill
Story of the complex and self-destructive love-hate relationship
between a pair of lovers, set against the harsh surroundings of
a seedy New Mexico motel. A profoundly depressing experience
with an opaque script (the work of Shepard) and direction that
never allows the film to rise above its stage origins.
DRA 103 min (ort 108 min) TVrel: C4
Boa: play by Sam Shephard.

FOOLS OF FORTUNE *** 15
Pat O'Connor UK 1989
Mary Elizabeth Mastrantonio, Iain Glen, Julie Christie, Michael
Kitchen, Niamh Cusack, Sean T. McClory, Catherine McFadden,
Frankie McCafferty, Ronnie Masterson, Tom Hickey, Hazel Flanagan,
Amy Hastings, Mick Lally, John Kavanagh
A portrait of shattered lives during the aftermath to Irish inde-
pendence in the 1920s. An attack on a peaceful, Protestant family
by the pro-British irregulars, the Black & Tans, leaves the father
and two sisters dead and the only son with a bitter thirst for
revenge. Distinguished by its visual power, this fine drama is
marred by its elaborate flash-forwarding structure that seriously
hampers its narrative flow.
DRA 105 min (ort 110 min) VIDrel: PAL/TERRY L/A V/h
Boa: novel by William Trevor.

FOOTLIGHT PARADE **** U
Lloyd Bacon USA 1933
James Cagney, Joan Blondell, Ruby Keeler, Dick Powell, Guy Kibbee,
Ruth Donnelly, Hugh Herbert, Frank McHugh, Claire Dodd, Herman
Bing, William Granger, Paul Porcasi, Charles C. Wilson, Barbara
Rogers, Billy Taft, Marjean Rogers
Cagney plays a producer of live stage shows that feature between
films, and runs into trouble when he tries to outdo himself staging
some ambitious numbers. A fascinating backstage musical whose
straight parts are every bit as entertaining as the musical
numbers. Ends with three superb Busby Berkeley numbers:
"Honeymoon Hotel", "By A Waterfall" and "Shanghai Lil".
MUS 100 min (ort 104 min) VIDrel: WHV V/h

FOOTLOOSE ** 15
Herbert Ross USA 1984
Kevin Bacon, Lori Singer, John Lithgow, Dianne Wiest, Christopher
Penn, John Laughlin, Sarah Jessica Parker, Elizabeth Gorcey, Frances
Lee McCain, Jim Youngs, Douglas Dirkson, Lynne Marta, Arthur
Rosenberg, Timothy Scott
A big-city kid moves to a small town where dancing has been
banned due to the efforts of a hellfire preacher. He decides to
convert the town to his way of thinking. It's hard to know what
point this film is trying to make but at least the dance sequences
are well handled. The script is by Dean Pritchard, who also
worked on the infinitely superior FAME: THE MOVIE and
SING.
MUS 103 min (ort 107 min) VIDrel: CIC/SONOP V/sur

FOR A FEW DOLLARS MORE *** 15
Sergio Leone ITALY/SPAIN/WEST GERMANY 1965
Clint Eastwood, Lee Van Cleef, Gian Maria Volonte, Jose Egger, Mara
Krup, Rosemarie Dexter, Klaus Kinski, Mario Brega, Aldo Sambrell,
Luigi Pistilli, Robert Camardiel, Benito Stefanelli, Luis Rodriguez,
Panos Papadopulos
An excessively drawn out sequel to A FISTFUL OF DOLLARS.
In this one the "man with no name" teams up with a bounty
hunter to track down an outlaw gang. Vague and ponderous in
places, but enlivened by the occasional flash of humour and a
nice pairing of Eastwood and Van Cleef. The score is by Ennio
Morricone.
Aka: PER QUAICHE DOLLARO IN PIU
WES 127 min (ort 133 min) VIDrel: MGM/WHV V/sh

FOR A LOST SOLDIER *** 18
Roeland Kerbosch HOLAND 1993
Jeroen Krabbe, Maarten Smit, Andrew Kelly
A homosexual man recalls his first sexual encounter which took
place during WW2, when aged only twelve, he became involved
with a Canadian soldier. An extremely frank exploration of
homosexuality, intelligent and touching, but with a sex scene
that some may find shocking.
DRA 89 min (ort 92 min) wScrn VIDrel: DTK/RTM V
Boa: novel by Rudi Van Dantzig.

FOR BETTER OR FOR WORSE ** 15
Gene Quintano USA 1989
Kim Cattrall, Robert Hays, Christopher Lee, Leigh Taylor-Young,
Charles Rocket, Lance Kinsey, Jerry Lazarus, Jonathan Banks, Max
Alexander, Gordon Jump, Doris Roberts, Judy Toll, Kate Benton, Tino
Insano, Jennifer Alin, Laura Cepeda
A slight, romantic-comedy about an ordinary guy who pursues
and marries a glamorous woman after a whirlwind courtship,
blissfully unaware that she has a double life and in reality is a
spy. Complications abound as the couple go on their honey-
moon in this tongue-in-cheek and implausible effort.
COM 89 min Cut (10 sec) VIDrel: EIV/SONOP V

FOR BETTER OR WORSE * 12
Jason Alexander USA 1995
Jason Alexander, James Woods, Lolita Davidovich, Joe Mantenga, Jay
Mohr, Rip Torn
A man who cannot forget the love of his life succumbs to his
brother's new girlfriend, when the latter couple arrive on a visit.
An oddball romantic comedy, full of angst, pathos and
contrivance, that really doesn't work all that well. The self-indul-
gent screenplay is by Alexander.
COM 85 min (ort 95 min) VIDrel: COLUM/SONOP V/sur

FOR LOVE ALONE * PG
Michael Lindsay-Hogg USA 1995
Stephen Collins, Sanna Vraa, Trevor Eve, Brigitte Pacquette, Tom
Rack, Paul Hopkins, Madeline Kahn, Bernard Ranger, Helga Prohoehr
Schmidt, Vlasta Vrana, Denis Mercia, Wendy Fulford, George
Popovich, Francoise Robertson, Cas Anvar
An East European skier meets and marries an American busi-
ness tycoon, in this retelling of the life of Ivana Trump and her
marriage to property tycoon Donald Trump. Totally idea free,
this syrupy dud was possibly of cathartic value for Mrs Trump
following her messy divorce; it certainly has no value as a piece
of entertainment.
Aka: IVANA TRUMP'S FOR LOVE ALONE
DRA 90 min mTV VIDrel: MARQ/20TH V/sur
Boa: novel by Ivana Trump.

FOR LOVE AND GLORY ** PG
Roger Young USA 1992
Daniel Merkel, Tracy Griffith, Robert Foxworth, Victor Love, Kate
Mulgrew, Zack Galligan, Tom O'Brien, Olivia D'Abo, Laura
Carrington, La Chanze, Barton Heyman, Eric Schweig, Grayson
Fricke, Joseph Bias, Walt Goggins, Marc Macaulay
The lives of an aristocratic Southern family are changed forever
by the American Civil War. Fair.
DRA 90 min VIDrel: 20TH/TECH V

FOR LOVE OF IVY * PG
Daniel Mann USA 1968
Sidney Poitier, Abbey Lincoln, Beau Bridges, Leon Bibb, Nan Martin,
Lauri Peters, Carroll O'Connor, Hugh Hurd, Lon Satton, Stanley
Greene, Tony Major, Paul Harris, Clark Morgan, Christopher St John,
Bob Carey, Marlene Clark
A family does not want to lose the services of their black maid
so they find her a boyfriend. A laboured and stale comedy with
precious few jokes and a patronising script. Of slight historical
interest.
COM 102 min VIDrel: VCC/DISC V

FOR PETE'S SAKE ** PG
Peter Yates USA 1974
Barbra Streisand, Michael Sarrazin, Estelle Parsons, Molly Picon,
William Redfield, Louis Zorich, Vivian Bonnell, Richard Ward,
Heywood Hale Broun, Joe Maher, Vincent Schiavelli, Fred Stuthman,
Ed Bakey, Peter Mamakos
Boring attempt at comedy with a cab-driver's devoted wife
borrowing money from the underworld and then finding that
she has some problems. Streisand is great in this one but the
feeble script gives her little to work with. Written by Stanley
Shapiro and Maurice Richlin.
Aka: JULY PORK BELLIES
COM 86 min (ort 90 min) VIDrel: VCC L/A V

FOR QUEEN AND COUNTRY ** 15
Martin Stellman UK/USA 1988
Denzel Washington, Dorian Healy, Amanda Redman, Sean Chapman,
Bruce Payne, George Baker, George McTavish, Geff Francis, Frank
Harper, Craig Fairbrass, Michael Bray, Stella Gonet, Colin Thomas,
Tatitan Strauss, Lisa O'Connor
The story of a black Falklands War hero who returns to his run-

down council housing estate after nine years. Still encountering rejection and hostility, and with few opportunities to make good, he eventually snaps when informed by the immigration authorities that his application to renew his British passport is to be denied. An ambitious but disappointing film that trades on contemporary violence and is let down by a ludicrous final shoot-out.

DRA   100 min (ort 106 min)   VIDrel: SONY L/A   V

FOR THE BOYS ***
Mark Rydell USA

15
1991

Bette Midler, James Caan, George Segal, Patrick O'Neal, Christopher Rydell, Arye Gross, Norman Fell, Rosemary Murphy, Bud Yorkin, Dori Brenner, Jack Sheldon, Karen Martin, Shannon Wilcox, Michael Green, Melissa Manchester

Odd comedy-drama that starts off as a nostalgic account of the days when numerous song-and-dance acts did their bit to entertain the troops and support the war effort. Told in flashback, it casts Midler (in top form) as an all-round entertainer, and follows her through WW2, the Korean War and the Vietnam War. But with each war comes growing disillusionment, and the film's initial cheerful patriotism soon gives way to a far uglier side of American culture.

DRA   139 min (ort 145 min)   VIDrel: 20TH/TECH   V/sur V/dm

FOR THE LOVE OF AARON **
John Kent Harrison USA

(PG)
1994

Meredith Baxter, Keegan MacIntosh, Joanna Gleason, Nick Mancuso, Blu Mankuma, John Kapelos, Malcolm Stewart, Matthew Walker, Tyler Thompson, Michael Rogers, Jannie Woods-Morris, Meredith Woodward, Sheila Paterson, P. Lynn Johnson

A successful author who suffers from schizophrenia is struggling to combat her mental condition whilst bringing up her young son and trying to complete her latest book. When she breaks down under the strain and is taken to hospital, her ex-husband initiates a fight for sole custody of their son. A touching human interest drama (very loosely based on a real case) that strains credibility with its constant parade of disasters. Fortunately, Baxter is quite outstanding.

DRA   89 min mTV   SATrel: MOVIE CHANNEL

FOR THE LOVE OF BENJI ***
Joe Camp USA

U
1977

Cynthia Smith, Patsy Garrett, Ed Nelson, Allen Fiuzat, Peter Bowles, Bridget Armstrong, Art Vasil, Mihalis Lambrinos

Sequel to BENJI in which our canine hero is taken on vacation to Athens but becomes involved in international espionage when he is abducted to be used to smuggle out a secret formula that is stamped onto his paw. He soon gives the bad guys the slip and the chase is on in a likeable and diverting family adventure. Followed by BENJI THE HUNTED. See also OH HEAVENLY DOG! for more canine capers.

A/AD   84 min   VIDrel: BEST   V

FOR THE LOVE OF NANCY **
Paul Schneider USA

PG
1995

Tracey Gold, Jill Clayburgh, Cameron Bancroft, Marl Paul Gosselaar, Michael McRae, Garwin Sanford, William Devane, Deanna Milliaga, Marie Stillin, Kevin McNulty, Teryl Rothery, Henry O. Watson, Colin Cunningham, Glynis Davies

A family realise that their young daughter is dying from anorexia as her desire to slim has now become obsessive. But when she refuses all their offers of help, they are obliged to go to court to ensure that she receives the medical attention she needs. A competent drama based on a real case.

DRA   88 min (ort 90 min) mTV   VIDrel: ODY/SONOP   V/sur

FOR THEIR OWN GOOD ****
Ed Kaplan USA

PG
1994

Elisabeth Perkins, Laura San Giacomo, Charles Haid, C.C.H. Pounder, Gary Basaraba, Kelli Williams, Colleen Camp, Michael O'Neill, David Graf, Glenn Morshower, David Purdham, Hananh Eckstein, Jana Arnold, Trey Ames, Tom Kopache

A factory forces female workers to choose between jobs or sterilisation because it makes use of industrial processes involving toxic chemicals, and the bosses fear possible actions being brought against them by pregnant employees. But the factory is eventually closed down anyway, and a poverty-stricken female worker sues her ex-employers, going all the way to the Supreme Court. A most disturbing film, based on a true events that took place in Texas in 1984.

DRA   91 min   VIDrel: ODY/SONOP   V/sh

FOR WHOM THE BELL TOLLS ***
Sam Wood USA

PG
1943

Gary Cooper, Ingrid Bergman, Akim Tamiroff, Arturo de Cordova, Joseph Calleia, Katina Paxinou, Vladimir Sokoloff, Mikhail Rasumny, Fortunio Bonanova, Eric Feldary, Victor Varconi, Lilo Yarson, Alexander Granach, Adia Kuznetzoff

Romance between an orphan girl and an American schoolteacher set against the background of the Spanish Civil War. Cooper joins the Partisans and falls for Bergman as a refugee before going on a suicide mission. Good photography, but the static and sombre story just doesn't generate any realism. The score is by Victor Young. AA: S. Actress (Paxinou).

DRA   128 min (ort 170 min) B/W   VIDrel: CIC/SONOP L/A   V

Boa: novel by Ernest Hemingway.

FOR YOUR EYES ONLY ***
John Glen UK

PG
1981

Roger Moore, Carole Bouquet, Chaim Topol, Lynn-Holly Johnson, Julian Glover, Cassandra Harris, Jill Bennett, Michael Gothard, John Wyman, Jack Hedley, Lois Maxwell, Desmond Llewellyn, Geoffrey Keen, Walter Gotell, James Villiers

A nuclear submarine activating device has been lost in a sea crash and agent 007 races against time to prevent it falling into the wrong hands. Not one of the best James Bond films, but the plot is quite plausible and Moore gives a nice relaxed performance. This was number twelve in a long-running series that started with DOCTOR NO. The next one in line was OCTO-PUSSY.

A/AD   122 min (ort 128 min) wScrn   VIDrel: MGM/WHV V/sur

FORBIDDEN BEAUTY *
Jim Wynorski USA

(18)
1995

Jennifer Rubin, Doug Wert, Daniel J. Travanti, Maria Ford, Melisa Brasselle, Jay Richardson, Gerrit Graham, Richard Gabai, Johnny Williams, Lenny Juliano, Kimberly Roberts, Fred Olen Ray, Julie Smith, Rod Kerchner, Antonia Dorian

A woman who runs her own successful cosmetics company turns forty and becomes obsessed with the fear that her looks are fading. When a research scientist offers her a revolutionary treatment that can reverse ageing, she welcomes the chance to regain her looks. But the treatment is based on a serum derived from wasp venom, but when she gains access to his lab and over-doses on it, the side-effects are not pleasant. A rubbishy remake of "The Wasp Woman".

Aka: WASP WOMAN

HOR   89 min   SATrel: SKY MOVIES

FORBIDDEN CHOICES **
Jennifer Warren USA

18
1994

Rutger Hauer, Kelly Lynch, Martha Plimpton, Patrick McGaw, Richard Sanders, Michael MacRae, Lance Robinson, Rae Adams, Adriana Lamon-Anderson, Jim Chiros, James Gervasi, Sue Ann Gilfillan, Marjorie Nelson, James Gervasi, Todd Moore

In a small Maine town that goes by the exotic name of Egypt, a member of a large and unruly clan goes to prison, giving his wife a chance to seek comfort elsewhere while his brother does the same with a neighbour. A sprawling and unconcentrated adaptation of Chute's novel.

Aka: BEANS OF EGYPT, MAINE, THE

DRA   95 min (ort 109 min)   VIDrel: GUILD/FOXVID   V

Boa: novel The Beans of Egypt, Maine by Carolyn Chute.

FORBIDDEN MEMORIES ***
Bob Clark USA

(PG)
1995

Mary Tyler Moore, Linda Lavin, Wayne Grace, Nathan Watt, Shirley Knight, Paul Winfield, Allison Mack, Martin Hindley, Christopher Jones, MacKenzie LaCoss, Deborah Knox Meschan, Emily Ciatlin Bivins, Robert C. Johnson

A touching period drama telling of a twelve-year-old who is sent to South Carolina, where he spends the summer with his three maiden aunts. There he forms a close friendship with one of the women, who is slightly retarded, having been traumatised when she witnessed a brutal Ku Klux Klan murder many years before. However, the boy's friendship with this woman marks the start of a long healing process, in which she finally comes to terms with her repressed memories.

DRA   87 min (ort 90 min) mTV   SATrel: MOVIE CHANNEL

FORBIDDEN PLANET ****
Fred McLeod Wilcox USA

U
1956

Walter Pidgeon, Leslie Nielsen, Anne Francis, Warren Stevens, Jack

Kelly, Earl Holliman, Richard Anderson, George Wallace, Robert Dix, Jimmy Thompson, Janes Drury, Morgan Jones, Peter Miller, Harry Harvey Jr, Roger McGee

One of the all-time best science fiction films. A highly intelligent account of a mission to the planet Altair 4 in 2200 A.D. One scientist lives alone with his daughter and Robby, a robot servant, all other members of the settler ship having died. The remains of a powerful non-human civilisation are to be found buried beneath the planet's surface. The sluggish script and poor dialogue are handicaps. Allegedly based on Shakespeare's "The Tempest".
FAN   94 min (ort 98 min) wScrn (special edition)
VIDrel: MGM/WHV   V/sh

FORBIDDEN PLEASURES *   18
Henri Pachard USA   1993
Victoria Paris, Tiffany Minx, Mike Horner, Tom Byron, Mark Wallace, Francesca Leigh, Jaye Lilo
A writer (Horner) becomes obsessed with a voice on a phone-sex line, and sets out to meet the person, but his wife has grown suspicious of him, and sets out to catch him. A thinly plotted effort, in which all is resolved at the end with the obligatory orgy.
A   50 min   VIDrel: GROHOM/MAXSCAN   V

FORCE, THE **   15
Mark Rosman USA   1994
Jason Gedrick, Gary Hudson, Kim Delaney, Cyndi Pass, Lyman Ward, Aki Aleong, Dennis Lipscomb, Chris Kriesa, Gerald Anthony, Susie Singer, Stan Yale, Yasmine Bleeth, Jarrett Lennon, George Saunders, Judd Owen, Steve Wilcox, Frank Novak
A rookie cop joins the force and makes friends with a veteran homicide detective who is notorious for his unconventional methods and lone-wolf approach. When the latter is murdered, his young friend is plagued by nightmares and finds no rest until he resolves to get to the bottom of this mystery, much to the dismay and anger of his superiors. A conventional tale of conspiracy and corruption, sustained by Gedrick's strong performance.
THR   93 min   VIDrel: REFLEC/FIRST   V

FORCE OF EVIL ****   PG
Abraham Polonsky USA   1949
John Garfield, Beatrice Pearson, Thomas Gomez, Roy Roberts, Marie Windsor, Howland Chamberlain, Paul MeVey, Tim Ryan, Sid Tomack, George Backus, Jack Overman, Sheldon Leonard, Jan Dennis, Stanley Prager, Raymond Largay
A powerful study of the corruption of idealism, as seen in the figure of a lawyer who achieves success at the price of working for a gangster, but yearns to break free. An intelligently made and intense crime melodrama, with dialogue that at times is almost poetic. The excellent photography and New York locations add immeasurably to the atmosphere of this classic example of film noir at its best.
THR   78 min (ort 80 min) B/W   VIDrel: SECOND/RTM   V
Boa: book Tucker's People by Ira Wolfert.

FORCE OF ONE, A **   15
Paul Aaron USA   1979
Chuck Norris, Jennifer O'Neill, Clu Gulager, Bill Wallace, Ron O'Neal, James Whitmore Jr, Clint Ritchie, Pepe Serna, Ray Vitte, Taylor Lacher, Kevin Geer, Chu Chu Malave, Eugene Butler, James Hall, Charles Cyphers, Eric Laneuville
A six-times undefeated world karate champion helps a Californian town combat a drugs racket after a number of narcotics agents are murdered. A sound action film built around the never-smiling star with a few good set-piece battles but no deep insights.
MAR   85 min Cut (1 min 2 sec – ort 90 min)
VIDrel: MIA/DISC   V

FORCE OF THE DRAGON **   18
Arthur Wong/Brandy Yuen HONG KONG   1988
Cynthia Khan, Hiroshi Fujioka, Stuart Ong, Nishiwaki Michiko
In Tokyo, an armed couple raid the venue of a jewellery exhibition, shoot dead a police inspector's assistant and escape with the most valuable gems. The inspector trails them to Honk Kong, where he learns that they plan to use the proceeds of their robbery to fund terrorist activities. Formula blending of all-action heroics and martial arts conflicts.
A/AD   80 min   VIDrel: IMPENT   V

FORCE 10 FROM NAVARONE *   15
Guy Hamilton UK   1978
Robert Shaw, Harrison Ford, Edward Fox, Franco Nero, Barbara Bach,

Richard Kiel, Carl Weathers, Alan Badel, Angus MacInnes, Michael Byrne, Philip Latham, Petar Buntic, Michael Sheard, Paul Humpoletz, Dicken Ashworth
An attempt to create a sequel to THE GUNS OF NAVARONE in which the target this time is a bridge and dam in Yugoslavia. A sluggish and implausible tale that has absolutely nothing in common with the earlier film, although the cast give of their best.
WAR   113 min (ort 118 min)   VIDrel: VCC/DISC/COLUM   V
Boa: novel by Alistair MacLean.

FORCED TO KILL **   18
Russell Solberg USA   1993
Corey Michael Eubanks, Michael Ironside, Rance Howard, Don Swayze, Mickey Jones, Clint Howard, Carl Ciarfalio, Cynthia J. Blessington, Allan Wyatt Jr., Brian Avery, Kari Whitman, A.J. Thrasher, Tom Bolger, Cole McKay, Alan Gelfant
A tough young repo-man who's delivering a car gets captured by a degenerate backwoods family who force him to participate in the illegal barefist fights that are organised by the local sheriff, a man who is both brutal and more than a little insane. Another violent actioner, replete with stock characters and situations.
A/AD   88 min   VIDrel: IMPENT   V

FORCED VENGEANCE *   18
James Fargo USA   1982
Chuck Norris, Mary-Louise Weller, Camilla Griggs, Michael Cavanaugh, David Opatoshu, Seiji Sakaguchi, Frank Michael Liu, Bob Minor, Lloyd Kino, Leigh Hamilton, Howard Caine, Robert Emhardt, Roger Behrstock, Jimmy Shaw
A Vietnam veteran who now works as a casino security chief finds himself caught up in a power struggle between two Hong Kong crime syndicates. A formula-ridden martial arts caper which has little going for it.
MAR   86 min Cut (50 sec)   VIDrel: MGM/WHV   V/h

FOREIGN CORRESPONDENT ****   PG
Alfred Hitchcock USA   1940
Joel McCrea, Laraine Day, Herbert Marshall, George Sanders, Edmund Gwenn, Albert Basserman, Eduardo Ciannelli, Barbara Pepper, Eddy Conrad, Robert Benchley, Harry Davenport, Charles Wagenheim, Martin Kosleck
A journalist gets involved in a mysterious espionage plot that unfolds at a furious pace when he is sent to Europe in 1938. Despite a few rambling moments this film builds up to a great climax, with several memorable scenes along the way. One of Hitchcock's best films. Scripted by Charles Bennett and Joan Harrison and with dialogue by James Hilton and Robert Benchley.
THR   115 min (ort 119 min) B/W
VIDrel: 4-FRONT/POLYREC   V/sh
Boa: book Personal History by Vincent Sheean.

FOREIGN FIELD, A ***   (PG)
Charles Sturridge UK   1993
Leo McKern, Alex Guiness, Edward Herrmann, John Randolph, Geraldine Chaplin, Lauren Bacall, Dorothy Grumbar, Jeanne Moreau, Michelle Gheleyns-Hue, Cateline Alteirac
Two WW2 British veterans return to Normandy to visit a fallen comrade's grave on the 50th anniversary of the D-Day landings. There they meet an American ex-soldier who has come together with his son and daughter-in-law, as well as a British woman on a similar mission. A finely acted character study, that is done with quite a light touch despite the underlying seriousness of its theme.
DRA   93 min mTV   TVrel: BBC

FOREIGN STUDENT **   15
Eva Sereny USA   1994
Robin Givens, Marco Hofschneider, Rick Johnson, Charlotte Ross, Hinton Battle, Edward Herrmann, Charles S. Dutton, Jack Coleman, Anthony Herrera
In the 1950s, a male French exchange student comes to a Southern college and commits the cardinal error of falling in love with a black woman who works there. A spotty and episodic effort that falls to justice to the strengths of Labro's autobiographical work.
DRA   91 min (ort 96 min) cC   VIDrel: CIC/SONOP   V
Boa: book by Philippe Labro.

FOREMAN WENT TO FRANCE, THE **   U
Charles Frend UK   1942
Clifford Evans, Constance Cummings, Robert Morley, Tommy

Trinder, Gordon Jackson, Ernest Milton, Paul Bonifas, Charles Victor, John Williams, Anita Palacine, Francois Sully (Francis L. Sullivan), Ronald Adam, Mervyn Jones
During the early days of WW2, a British engineer travels to France to stop some key industrial equipment from falling into the hands of the Nazis. A dated morale-booster comedy thriller of mainly historical interest and dedicated to a real-life hero, Melbourne Jones, whose exploits inspired this film.
Aka: SOMEWHERE IN FRANCE
COM   86 min (ort 88 min) B/W   VIDrel: LUMI/SPEAR L/A V

FOREVER AMBER ** **   *U*
Otto Preminger USA   1947
Linda Darnell, Cornel Wilde, Richard Greene, George Sanders, Richard Hadyn, Jessica Tandy, Anna Revere, John Russell, Jane Ball, Robert Coote, Leo G. Carroll, Natalie Draper, Margaret Wycherly, Alma Kruger, Edmund Breon
In 17th century England, a poor girl uses her female charms to escape a life of poverty but pays a high price in the end, when she becomes one of the mistresses of Charles II. A lavish adaptation of this blockbuster bestseller, but very tame in comparison with the original as the moral climate of the time made it possible or do justice to the book's erotic content.
DRA   132 min (ort 140 min) cC   VIDrel: 20TH/TECH   V
Boa: novel Kathleen Windsor.

FOREVER ENGLAND ** **   *PG*
Walter Forde UK   1935
Betty Balfour, John Mills, Barry MacKay, Jimmy Hanley, Howard Marion Crawford, H.G. Stoker, Percy Walsh, George Merritt, Cyril Smith, Felix Aylmer, Gibb McLaughlin
The illegitimate son of a naval officer proves his worth by single-handedly sinking a German battleship during WW1. Absurdly jingoistic and simplistic, this navy yarn has not stood the test of time well, and given its strongly partisan stance, the cast are not really called upon to show much complexity of thought or feeling.
Aka: BORN FOR GLORY; BROWN ON RESOLUTION; TORPEDO RAIDERS
DRA   68 min B/W   VIDrel: CARL/TECH   V
Boa: novel Brown on Resolution by C.S. Forester.

FOREVER FEMALE * **   *(U)***
Irving Rapper USA   1953
Ginger Rogers, William Holden, Paul Douglas, Pat Crowley, James Gleason, Jesse White, Marjorie Rambeau, George Reeves, King Donovan, Vic Perrin, Russell Gaige, Marion Rose, Richard Shannon, Sally Mansfield, Kathryn Grant
An aspiring playwright agrees to raise the age of his heroine in order to get his play produced but eventually finds a younger actress who can do it justice. A lightweight and quite entertaining comedy, set against a theatre background, and replete with fine performances.
COM   93 min B/W   SATrel: SKY MOVIES
Boa: play Rosalind by James Barrie.

FOREVER MARY * **   *18*
Marco Risi ITALY   1988
Michele Placido, Claudio Amendola, Francesco Benigno, Alessandro Di Sanzo, Tony Sperandeo, Roberto Mariano, Maurizio Prollo, Filippo Genzardi, Giovanni Alamia, Alfredo Di Bassi, Salvatore Termini, Ginaluca Favilla, Matteo Mondello
An idealistic teacher at a boys' reformatory in Palermo encounters an unusual challenge when a transvestite who works as a prostitute is sent to this institution. Meanwhile, for the other inmates the reformatory is little more than a training college for a life of crime. A sensitive and quite moving account of young people in trouble, in a corrupt and uncaring society. A sequel followed.
Aka: MARY FOREVER; MERY PER SEMPERE
DRA   101 min (ort 106 min)   VIDrel: ARTPRO/RTM   V
Boa: novel Mery Per Sempere by Aurello Grimaldi.

**FOREVER YOUNG ** **   *PG*
Steve Miner USA   1992
Mel Gibson, Jamie Lee Curtis, Elijah Wood, Isbael Glasser, George Wendt, Joe Morton, Nicolas Surovy, David Marshall Grant, Robert Hy Gorman, Millie Slavin, Veronica Lauren, Michael Goorjian, Art LaFleur, Eric Pierpoint, Walt Goggins
In 1939, a man's girlfriend is knocked down and put in a coma she appears unlikely to recover from, so in despair he volunteers as a test subject for his friend's cryogenic experiments. However, by accident he is not awakened until 1992, when he finds himself

in a totally unfamiliar and at times baffling world. A gentle fantasy that fails to develop its single intriguing premise, though Gibson gives a performance of great charm and delicacy.
Aka: REST OF DANIEL, THE
FAN   97 min (ort 102 min)   VIDrel: WHV   V/sur

FORGET PARIS * **   *12*
Billy Crystal USA   1995
Billy Crystal, Debra Winger, Joe Mantegna, Cynthia Stevenson, Richard Masur, Julie Kavner, William Hickey, Robert Constanzo, John Spencer, Tom Wright, Cathy Moriarty, Johnny Williams, Marv Albert, Bill Walton, Charles Barkley
A man travels to Paris in order to bury his father but various complications ensure that he is kept there longer than he expected. These include a sudden whirlwind romance with an attractive woman that eventually leads to marriage. A warm and pleasing comedy, whose intelligent script gives the leads a chance to shine. Probably the best of a trio of Paris-inspired romantic comedies (see FRENCH KISS and SABRINA) that came out at about the same time.
COM   98 min (ort 101 min)   VIDrel: COLUM/SONOP; ENCORE (LV only)   V/sur LV

**FORGOTTEN ONE, THE ** **   *18*
Philip Badger USA   1989
Terry O'Quinn, Kristy McNichol, Elisabeth Brooks, Blair Parker, Ed Battle, Michael Osborn, Dwayne Carrington, Bill Alard, Phillip Darlington, Rita Haynes, Roy Yerley, Martin Boyd, Liam Russell, George Psilas, Rex Whitney
A writer moves house in order to gain fresh inspiration, but his new residence turns out to be haunted by the spirit of a dead woman. What is worse, it would seem that she has made plans for him to join her in the Hereafter.
DRA   93 min (ort 98 min)   VIDrel: BRAVE/SONOP L/A V/sh

**FORGOTTEN SINS ** **   *15*
Dick Lowry USA   1996
William Devane, John Shea, Bess Armstrong, Dean Norris, Brian Markinson, Lisa Dean Ryan, Tim Quill, Gary Grubbs
Fact-based story that tells of a devout community where accusations of Devil worship and torture lead to much misery, recrimination and suspicion. Average.
DRA   92 min   VIDrel: ODY/SONOP   V/sh

FORMULA FOR DEATH * **   *15*
Armand Mastroianni USA   1995
Nicollette Sheridan, William Devane, Stephen Caffrey, Dakin Matthews, Kurt Fuller, Barry Corbin, William Atherton, Joe Minjares, Ritch Brinkley, Greg Blanchard, Judith McConnell, Brian Brophy, Jim McMullan, Christopher Kriesa
A doctor's first case as an investigator at a disease control centre is the study of an outbreak of a mysterious and fatal disease in L.A. Having learnt that the disease first occurred twenty years ago, she traces its cause to a high security medical research lab, where some samples of the virus have gone astray. See also OUTBREAK for another film on this theme.
Aka: ROBIN COOK'S FORMULA FOR DEATH; VIRUS
THR   92 min   VIDrel: ODY/SONOP   V/sh
Boa: novel Outbreak by Robin Cook.

FORREST GUMP * **   *12*
Robert Zemeckis USA   1993
Tom Hanks, Sally Field, Robin Wright, Gary Sinise, Mykelti Williamson, Rebecca Williams, Michael Conner Humphreys, Harold Herthum, George Kelly, Rob Penny, John Randall, Sam Anderson, Margo Moorer, Ione M. Telech, John Worsham
A man of low intelligence looks back over his remarkable and impossibly lucky life, when he achieved fame on the sports-field and fought in Vietnam. A mawkish film of clever, computer-enhanced trick photography (like ZELIG, Gump meets some real-life characters) that pays homage to old-fashioned values but lacks bite. The LV disc includes a short documentary. AA: Pic, Actor (Hanks), Dir, Edit (A. Schmidt), Screen/adapt (E. Roth), Effects/vis (K. Ralston et al.).
COM   136 min (ort 142 min) wScrn   VIDrel: CIC/SONOP; PION (LV only)   V LV
Boa: novel by Winston Groom.

**FORREST HUMP * **   *18*
Mitchell Spinelli USA   1995
Mark Davis, Kaitlin Ashley, Bianca Trump, Nikki Sinn, Lynn Le May, Rachel Love, Frank Towers, Blade Baran, Cal Jammer, Yeninas

A poorly made porno parody of FORREST GUMP, in which a man with a very limited I.Q. manages nonetheless to find out all about the birds and the bees.
Aka: FORESKIN GUMP
A 70 min VIDrel: MIA/DISC V

FORSYTE SAGA, THE * PG
David Giles/James Cellan Jones UK 1967
Nyree Dawn Porter, Kenneth More, Eric Porter, Susan Hampshire, Martin Jarvis
Long and rambling adaptation of a series of novels dealing with the fortunes of a single family of London merchants, from the 1870s to the 1920s. A competent albeit rather workmanlike account, that spared no expense and was immensely popular, yet remains curiously distanced from its characters. However, patience is rewarded and this detailed drama becomes progressively more absorbing. First shown on TV as a set of twenty-six 60-minute episodes.
DRA 1,300 min (8 cassettes – ort 1560 min) B/W mTV
VIDrel: BBC/TECH V
Boa: the "Forsyte" novels by John Galsworthy.

FORT APACHE * U
John Ford USA 1948
Henry Fonda, John Wayne, Shirley Temple, Pedro Armendariz, Ward Bond, Victor McLaglen, John Agar, Irene Rich, George O'Brien, Anna Lee, Dick Foran, Jack Pennick, Guy Kibbee, Grant Withers, Miguel Inclan, Mae Marsh, Francis Ford
Story of the conflict between the US Cavalry and the Indians. Fonda plays a martinet who has troubles with his family as well as the Indians. Not up to Ford's usual standard but entertaining with a fine cast and a couple of good action sequences.
WES 127 min B/W VIDrel: 4-FRONT/POLYREC V

FORT SAGANNE * PG
Alain Corneau FRANCE 1984
Gerard Depardieu, Philippe Noiret, Catherine Deneuve, Michel Duchaussoy, Roger Dumas, Sophie Marceau, Jean-Louis Richard
A lavish, expensive and quite empty epic set in North Africa prior to WW1, and dealing with the adventures of a French officer who is posed to a fort in the Sahara desert. Apart from the usual conflicts with the natives, he also has time to conduct a love affair with a journalist. Watchable but seriously overlong and pretentious.
A/AD 180 min (ort 190 min) wScrn VIDrel: LUMI/SPEAR
L/A V

FORTRESS * 15
Arch Wilson AUSTRALIA 1985
Rachel Ward, Sean Garlick, Rebecca Rigg, Robin Mason, Marc Gray, Beth Buchanan, Asher Keddie, Bradley Meehan, Anna Crawford, Richard Terrill, Peter Hehir, Roger Stephen, Vernon Wells, David Bradshaw, Elaine Cusick
A teacher and her class of nine are kidnapped by a gang of four vicious thugs wearing grotesque masks. The victims plot to escape by outwitting their captors. A violent film, but one that tends to hold the interest.
DRA 86 min (ort 90 min) mCab VIDrel: 20TH V

FORTRESS * 18
Stuart Gordon AUSTRALIA/USA 1993
Christopher Lambert, Loryn Locklin, Kurtwood Smith, Lincoln Kilpatrick, Tom Towles, Jeffrey Combs, Clifton Gonzalez Gonzales, Vernon Wells, Alan Zitner, Carolyn Purdy-Gordon, Denni Gordon, Eric Briant Wells, Heidi Stein, Peter Lamb
In a strictly controlled future society, married couples are required by law to limit themselves to a single child. When a woman becomes pregnant with child number two, both she and her husband are sent to a high-security prison whose warden is conducting some strange and painful experiments. A confused and badly plotted film, with good special effects, minimal development and a weak ending; it recalls the equally poor ZERO POPULATION GROWTH. A sequel soon followed.
FAN 91 min (ort 95 min) VIDrel: COLUM/SONOP
V/sur

FORTRESS OF AMERIKKA * 18
Richard W. Haines/Samuel Weil USA 1991
Robert Prichard, R.L. Ryan, Brad Dunker, James Nugent Vernon
The title refers to a band of militant mercenaries who have terrorised the entire nation, and are well on their way to destroying the country. Fortunately, two young lovers are ready to put

an end to their depredations. Action-filled nonsense, watchable if somewhat restricted by the limited budget.
A/AD 120 min VIDrel: TROMA/RTM V

FORTUNATE PILGRIM, THE * 15
Stuart Cooper USA 1987
Sophia Loren, Edward James Olmos, Hal Halbrook, John Turturro, Lucia Angeluzzi, Anna Strasberg, Yorgo Voyagis, Roxann Biggs, Ed Wiley, Mirjana Karanovic, Ron Marquette
A turn-of-the-century story, telling of an Italian immigrant who experiences poverty and the tragic loss of her two husbands, struggling for years to give her five children a better life. A sprawling, melodramatic exercise, strong on atmosphere and detail (despite using Belgrade for New York), but far too long. The opera-style soundtrack is an annoyance. Scripted by John McGreevey and photographed by Reginald Morris. First shown in two parts.
Aka: MARIO PUZO'S THE FORTUNATE PILGRIM
DRA 144 min (ort 250 min) mTV VIDrel: GUILD/SONOP
V
Boa: novel by Mario Puzo.

FORTUNE COOKIE, THE * U
Billy Wilder USA 1966
Jack Lemmon, Walter Matthau, Ron Rich, Cliff Osmond, Judi West, Lurene Tuttle, Noam Pitlik, Ned Glass, Sig Ruman, William Christopher, Harry Holcombe, Les Tremayne, Marge Redmond, Ann Shoemaker, Maryesther Denver, Lauren Gilbert
A TV cameraman is slightly injured filming a football game but falls under the influence of his brother-in-law who is a totally unscrupulous lawyer. His intention is to make these injuries appear to be so serious that he will get a $1,000,000 compensation. Matthau as the lawyer is great, but he's wasted in an over-rated film that lacks both gags and plot development. AA: S. Actor (Matthau).
Aka: MEET WHIPLASH WILLIE
COM 121 min (ort 125 min) B/W VIDrel: WHV V

FORTUNES AND MISFORTUNES OF MOLL FLANDERS, THE * 18
David Attwood UK 1996
Alex Kingston, Daniel Craig, Diana Rigg, Ronald Fraser
In 18th century England the illegitimate daughter of a woman who was transported for theft sets out to secure a place for herself in society, but her journey there involves her in adultery, bigamy and incest. Defoe's bawdy romp is given a carefully plotted treatment, and Kingston makes a ravishing central figure.
Aka: MOLL FLANDERS
DRA 195 min VIDrel: WARVIS/WARUK V/sh
Boa: novel by Daniel Defoe.

FORTUNES OF WAR * 15
James Cellan Jones UK 1987
Kenneth Branagh, Emma Thompson, Ronald Pickup. Charles Kay, James Villiers, Desmond McNamara, Richard Clifford, Mark Drewry, Ronald Fraser, Glyn Grain, Harry Burton, Elena Secota, Magdalena Buznea, Caroline Langrishe
In 1939, a British professor travels to a Balkan country to take up a teaching post and plunges into local anti-fascist politics. Excellent period and local detail and fine performances help to compensate for its excessive length and slow pace. Originally shown in the USA on PBS Masterpiece Theater.
DRA 317 min (2 cassettes – ort 350 min) mTV
VIDrel: BBC/TECH V/h
Boa: novel by Olivia Manning.

FORTUNES OF WAR * 15
Thierry Notz USA 1993
Matt Salinger, Sam Jenkins, Haing S. Ngor, Martin Sheen, Michael Ironside, Michael Nouri, Frankie J.Holden, Vic Diaz, John Getz, Louis Katana, Joonee Gamboa, Ronnie Francisco, Ronnie Lazaro, Adriann Agcaoili, Herbie Go, Bay Drao
Three mercenaries in Southeast Asia join forces and find themselves involved in plans to deal with a local warlord. A very typical jungle adventure of noise, action and violence, with not a new idea in sight.
A/AD 102 min (ort 107 min) VIDrel: 20VIS/SONOP
V/sh

FORTY-EIGHT HRS * 18
Walter Hill USA 1982
Eddie Murphy, Nick Nolte, James Remar, Annette O'Toole, Frank

McRae, David Patrick Kelly, Sonny Landham, Brion James, Kenny Sherman, Jonathan Banks, James Keane, Tara King, Agneta Blackburn, Margot Rose, Denise Crosby

A hard-nosed cop enlists the aid of a jailed robber to help him nab a pair of escaped killers, one of whom being Murphy's former partner. With only two days of freedom for Murphy, this unlikely duo find themselves in a few tricky situations as they comb the streets in their search. Murphy's screen debut is a high-energy performance in this sometimes funny, sometimes noisy, but often downright unpleasant mixture of comedy and action. ANOTHER 48 HRS followed.

A/AD   92 min (ort 97 min) Cut
VIDrel: 4-FRONT/POLYREC   V/h

FORTY GUNS TO APACHE PASS **   PG
William Witney USA   1967
Audie Murphy, Michael Burns, Kenneth Tobey, Laraine Stephens, Michael Keep, Robert Brubaker, Michael Blodgett, Kay Stewart, Kenneth MacDonald, Byron Morrow, Willard Willingham, Ted Gehring, James Beck

Shortly after the Civil War, Cochise leads the Apache in a rebellion but is eventually defeated by a brave cavalry captain who also has to quash a plan to sell guns to the Indians. Standard Murphy vehicle that offers the usual quota of action.

WES   92 min (ort 95 min) VIDrel: VCC/DISC/COLUM   V

FORTY-NINTH PARALLEL, THE ***   U
Michael Powell/Emeric Pressburger UK   1941
Leslie Howard, Raymond Massey, Eric Portman, Laurence Olivier, Anton Walbrook, Glynis Johns, Niall MacGinnis, Finlay Currie, Raymond Lovell, John Chandos, Basil Appleby, Eric Clavering, Charles Victor, Ley On

Five sailors from a sunken U-boat try to escape to the USA from Canada. Generally effective, this episodic propaganda piece gives a fine cast the chance to display their talents. AA: Story/orig (Pressburger).

Aka: INVADERS, THE
DRA   121 min (ort 123 min) B/W
VIDrel: VCC/DISC/COLUM L/A   V

FORTY-SECOND STREET ****   U
Lloyd Bacon USA   1933
Warner Baxter, Ginger Rogers, Ruby Keeler, Bebe Daniels, Dick Powell, George Brent, Una Merkel, Guy Kibbee, Ned Sparks, George E. Stone, Allen Jenkins, Robert McWade, Eddie Nugent, Harry Akst, Clarence Nordstrom, Harry Warren

Problems of an ailing director who has to put on a musical with sensational Busby Berkley numbers. The leading lady is indisposed and a chorus girl takes her role and becomes a star. Performed with great zest with highly enjoyable songs and some great dancing. This definitive Hollywood musical has formed the standard by which later ones are judged. Some colourised versions are available – avoid them if possible. Music is by Al Dubin.

Aka: 42ND STREET
MUS   86 min (ort 89 min) B/W   VIDrel: WHV   V/h
Boa: novel Bradford Ropes by Rian James.

FOUL PLAY *   PG
Colin Higgins USA   1978
Goldie Hawn, Chevy Chase, Burgess Meredith, Dudley Moore, Rachel Roberts, Eugene Roche, Marilyn Sokol, Billy Barty, Marc Lawrence, Brian Dennehy, Chuck McCann, Don Calfa, Bruce Solomon, Cooper Huckabee, Ion Teodorescu

An innocent woman becomes unwittingly embroiled in a murder plot and there are several attempts on her life, but no-one will believe her except a young detective. A tasteless and fairly unamusing comedy with a few Hitchcock touches but little else. Later followed by a brief TV series.

COM   111 min (ort 118 min) VIDrel: CIC/SONOP   V/h

FOUNTAIN, THE ***   (PG)
Yuri Mamin USSR   1988
Asankul Kuttubaev, Sergei Dontsov, Zhanna Kerimtaeva, Viktor Mikhailov, Anatoli Kalmikov, Ljudmila Samokhvalova, Aleksei Zalivalov, Nikolai Trankov, Ivan Krivoruchko, Nina Usatova, Yakov. Stepanov, P. Fetisov, T. Leonova

Absurd comedy set in a rundown St Petersburg apartment block where a chain of disasters is unleashed when an eccentric old man from Central Asia, forced to come and stay with his daughter and son-in law, takes matters into his own hands in order to force the repair of a leaking hot-water pipe in the yard. A humorous allegory on the break-up of the Soviet

system with the director's assured lightness of touch much in evidence.

Aka: FONTAN
COM   101 min   CINrel

FOUR ADVENTURES OF REINETTE AND
MIRABELLE ***   U
Eric Rohmer FRANCE   1986
Jessica Forde, Joelle Miquel, Phillipe Laudenbaum, Marie Riviere, Yasmine Haury, Beatrice Romand, Fabrice Luchini

A charming and witty slice of life revolving around two female students who first meet at the Sorbonne and later become flatmates in Paris. With a limited budget and a largely improvised script (by Rohmer) the contrasting temperaments of reserved country girl Reinette and the more outgoing and sophisticated Mirabelle form the basis for an examination of four separate "adventures" they have with characters from the streets of Paris.

Aka: QUATRE AVENTURES DE REINETTE ET MIRABELLE
DRA   95 min   VIDrel: ARTIF/20TH   V

FOUR EYES AND SIX GUNS **   (PG)
Piers Haggard USA   1992
Judge Reinhold, Patricia Clarkson, Dan Hedaya, Fred Ward, M. Emmet Walsh, Dennis Burkley, John Schuck, Jonathan Gries, Jake Dengel, Shane McCabe, William Duff-Griffin, Austin Pendleton, Neal Thomas, Bill Joe Patton, Bill Getzwiler

A New York optometrist in 1882 who loves to read dime-novels decides to move to Tombstone, Arizona, and arrives just in time to help Wyatt Earp with his failing eyesight. A silly, harmless and moderately amusing effort.

WES   89 min (ort 92 min) mCab   SATrel: SKY MOVIES

FOUR FEATHERS, THE ****   U
Zoltan Korda UK   1939
John Clements, Ralph Richardson, C. Aubrey Smith, June Duprez, Allan Jeayes, Jack Allen, Donald Gray, Frederick Culley, Clive Barker, Robert Rendel, Derek Elphinstone, Archibald Batty, Hal Walters, Norman Pierce, Henry Oscar

Brilliant account of a British stay-at-home who refutes charges of cowardice (he received four feathers after resigning his commission) by going off to help his comrades in the army put down an uprising in the Sudan. A highly atmospheric film that boasts some memorable battle scenes and is a lot less jingoistic than its detractors would have us believe. It was remade for TV in 1977.

A/AD   110 min (ort 130 min) VIDrel: CARL/TECH   V
Boa: novel by A.E.W. Mason.

FOUR-HUNDRED BLOWS, THE ****   PG
Francois Truffaut FRANCE   1959
Jean-Pierre Leaud, Patrick Auffray, Claire Maurier, Albert Remy, Georges Flamant, Guy Decomble, Yvonne Claudie, Robert Beuavais, Claude Mansard, Jacques Monod, Henri Virlojeux, Jeanne Moreau, Jean-Claude Brialy

A brilliant first feature from Truffaut telling the lyrical and potent story of a twelve-year-old schoolboy at odds with his family. His escapades degenerate from truancy to petty crime, and land him in a detention centre for young offenders from which he escapes to the coast. A fine portrayal of a deprived childhood that is allegedly based on the director's own experiences. Followed by: "Antoine And Colette", STOLEN KISSES, BED AND BOARD and LOVE ON THE RUN.

Aka: LES 400 COUPS; LES QUATRE CENT COUPS
DRA   97 min B/W   VIDrel: ARTIF/20TH   V

FOUR MINUTE MILE, THE **   U
Jim Goddard AUSTRALIA/UK   1988
Richard Huw, Nique Needles, Lewis Fitzgerald, John Philbin, Robert Burbage, Adrian Rawlins, Adrian Dunbar, Richard Wilson, Ralph Cotterill, Tracy Mann, Michael York

Story set in 1954, when the athletes of three continents were all trying to break the record for running a mile. Quite a dull film that spends too long concentrating on the training of the athletes involved, but at least the later sequence set at the Empire Games in Vancouver offers a measure of excitement.

DRA   153 min (ort 190 min) mTV   VIDrel: VISION/DISC   V

FOUR MUSKETEERS, THE ***   PG
Richard Lester PANAMA/SPAIN   1975
Oliver Reed, Faye Dunaway, Raquel Welch, Richard Chamberlain, Frank Finlay, Michael York, Christopher Lee, Jean-Pierre Cassel, Geraldine Chaplin, Simon Ward, Charlton Heston, Roy Kinnear, Nicole Calfan

Second half of this version of the Dumas novel is a little too much of a spoof to do justice to it and tends to degenerate into slapstick. Made at the same time as the 1973 version of THE THREE MUSKETEERS, this is in effect the continuation of the first film with the same mixture of visual gags, damsels in distress and acts of bravery. See also THE RETURN OF THE MUSKETEERS.
Aka: FOUR MUSKETEERS, THE: THE REVENGE OF MILADY
A/AD   106 min   VIDrel: ARROW/VCC   V
Boa: novel The Three Musketeers by Alexandre Dumas.

**FOUR ROOMS **   18
Allison Anders/Alexandre Rockwell/R. Rodriguez/Q. Tarantino USA   1995
Tim Roth, Lawrence Bender, Kathy Griffin, Quinn Thomas Hellerman, Marc Lawrence, Unruly Julie McClean, Laura Rush, Paul Skemp, Marisa Tomei, Sammi Davis, Amanda de Cadenet, Valeria Golino, Madonna Ciccione, Jennifer Beals
A bellboy has the wildest night of his life as he visits the various guests in his hotel, each section of the film introducing us to a new room and a new tale (Quentin Tarantino's has an uncredited appearance by Bruce Willis). Stories are entitled: "The Missing Ingredients", "The Wrong Man", "The Misbehavers" and "The Man From Hollywood" and apart from Robert Rodriguez's entry, (two kids are left alone by their parents for an evening) none of them are up to much.
COM   93 min (ort 98 min) cC   VIDrel: TOUCH/TECH
V/sh

**FOUR SEASONS, THE **   15
Alan Alda USA   1981
Alan Alda, Sandy Dennis, Carol Burnett, Rita Moreno, Jack Weston, Len Cariou, Bess Armstrong, Elizabeth Alda, Beatrice Alda, Robert Hitt, Kristi McCarthy, David Stackpole
The various stages or "seasons" in the relations between three married couples are examined in this rather low-key drama which later became a TV series. Mildly amusing in places but generally smug and superficial. Written by Alda, who made his feature directing debut with this film.
DRA   103 min (ort 107 min)   VIDrel: CIC/SONOP L/A
V/h

FOUR TO ONE *   18
USA
Simone, Toni, Brian, Rosie, Gabby
One lucky fellow enjoys the company of four lovely ladies over a weekend in this totally forgettable sex romp.
A   60 min Cut (37 sec)   VIDrel: MOPIC/QUANT   V

**FOUR WEDDINGS AND A FUNERAL **   15
Mike Newell UK   1993
Hugh Grant, Andie McDowell, Krisitn Scott Thomas, Simon Callow, James Fleet, Rowan Atkinson, John Hananh, Charlote Coleman, David Bower, Corin Redgrave, Kenneth Griffiths, Jeremy Kemp, Rosalie Crutchley, Anna Chancellor, Elspet Gray
A tongued-tied, reserved and socially inept Englishman (aptly played by Grant) attends his friends' weddings but despite a fondness for the ladies, cannot sustain a long-term relationship of his own. At one of these events he finds himself smitten by an attractive American woman but, lacking the courage to make his feelings known seems doomed to remain single. A pleasant if over praised outing, with stock figures and cliched incidents, but little sparkle.
COM   112 (ort 117 min) cC   VIDrel: POLY/POLYREC
V/sur

FOURTEEN GOING ON THIRTY *   U
Paul Schneider USA   1988
Steve Eckholdt, Rick Rossovich, Loretta Swit, Patrick Duffy, Daphne Ashbrook, Adam Carl, Gabey Olds, Harry Morgan, Alan Thicke, Dick Van Patten, Rick Rossovich
This Disney reversal of the PEGGY SUE GOT MARRIED comedy has a love-sick teenager getting to be an adult overnight and making a play for the affections of his pretty teacher, thus preventing her from marrying an obnoxious gym instructor. A pleasant, silly, sugary Disney offering.
Aka: 14 GOING ON 30
COM   81 min (ort 100 min) mTV   VIDrel: WDV/TECH L/A
V

**1492: CONQUEST OF PARADISE **   15
Ridley Scott FRANCE/SPAIN/UK/USA   1992
Gerard Depardieu, Sigourney Weaver, Armand Assante, Loren Dean,

Kevin Dunn, Angela Molina, Fernando Rey, Michael Wincott, Tchecky Karyo, Frank Langella, Mark Margolis, Kario Salem, Billy Sullivan, John Heffernan, Arnold Vosloo
An opulently filmed but essentially empty effort to tell the story of Columbus's discovery of the New World and the consequences this had for its native peoples. The historical detail and period sets and costumes have been lovingly and faultlessly recreated, but Depardieu cannot act effectively in English and the lack of a believable performance by the central character is a fatal flaw. One of two films on this subject released in 1992.
A/AD   149 min (ort 155 min) wScrn
VIDrel: 4-FRONT/POLYREC/GUILD; PION (LV only) V/sur LV

FOURTH PROTOCOL, THE *   15
John Mackenzie UK   1986
Michael Caine, Pierce Brosnan, Joanna Cassidy, Ned Beatty, Michael Gough, Julian Glover, Ray McAnally, Ian Richardson, Betsy Brantley, Sean Chapman, Peter Cartwright, Rosy Clayton, Matt Fewer, Jerry Harte, John Horsley
Secret agent Caine has to foil a Soviet plan to detonate one of the nuclear bombs kept by America at an air base in the UK, it being their intention to thus cause the collapse of NATO. A pretty enjoyable if rather predictable spy caper. Brosnan is really good as a Russian agent but Caine gives a wooden and terribly unconvincing performance.
A/AD   114 min (ort 117 min)   VIDrel: VCC/DISC/COLUM
V/sh
Boa: novel by Frederick Forsyth.

**FOURTH WAR, THE **   15
John Frankenheimer USA   1988
Roy Scheider, Jurgen Prochnow, Tim Reid, Lara Harris, Harry Dean Stanton, Dale Dye, Bill MacDonald, Harold Hecht Jr, Alice Pesto
An American colonel cannot adjust to the new peacetime role of the army, and when he finds himself confronted by a like-minded Soviet colonel, the two are able to engage in a private little war of their own. Strong performances are let down by the mediocre script in this unimaginative Cold War tale.
THR   86 min (ort 91 min)   VIDrel: GUILD/POLYREC L/A;
PION (LV only)   V/sh LV
Boa: novel by Stephen Peters.

FOX *   15
Rainer Werner Fassbinder WEST GERMANY   1975
Rainer Werner Fassbinder, Peter Chatel, Karl-Heinz Boehm, Ulla Jacobsen, Adrian Hoven, Ulla Jacobsen, Christiane Maybach, Peter Kern, Hans Zander, Kurt Raab, Irm Herman, Ursula Stratz, Elma Karlowa, Barbara Valentin
The lives and loves of a not terribly endearing group of West German homosexuals are examined in this cruel story that largely focuses on the declining fortunes of an unemployed sideshow performer, whose lottery win marks the starts of his alienation and exploitation. Heavily ironic in tone, the film can be enjoyed on several levels and though memorable and moving, really does go on for too long.
Aka: FAUSTRECT DER FREIHEIT; FIST-RIGHT OF FREEDOM; FOX AND HIS FRIENDS
DRA   118 min (ort 123 min)   VIDrel: CONNO/RTM   V

FOX AND THE HOUND, THE *   U
Art Stevens USA   1981
Voices of: Mickey Rooney, Kurt Russell, Pearl Bailey, Jack Albertson, Sandy Duncan, Jeanette Nolan, Pat Buttram, John Fielder, John McIntire, Dick Bakalyan, Paul Winchell, Keith Mitchell, Corey Feldman
A puppy and a young fox become friends but soon learn that this is something that is both an unnatural friendship and something that does not suit the purposes of human beings. Despite this our two friends manage to avoid harming each other and even save each other's life. A well-made animated tale, far above the usual standard but sadly still inferior to Disney's classic productions.
ANIM   83 min   VIDrel: WDV/TECH   V/sh
Boa: novel by Daniel P. Mannix.

FOXFIRE *   15
Jud Taylor USA   1987
Jessica Tandy, Hume Cronyn, John Denver, Gary Grubbs, Harriet Hall, Collin Wilcox, Paxton, Joshua Bryson, Jenny Winter, Kenny Kosek, Tony Trischka, Roger Mason
A widow living deep in the heart of the Appalachian mountains struggles to hold onto her home and resist the pressure to sell her property to a development corporation. A lyrically

photographed and touching portrait of a vanishing way of life, with some believable supernatural touches. The acting is first rate (Tandy won an Emmy for this role) and the film is convincingly adapted from its stage original.
Aka: FOXFIRE STORY, THE
DRA 96 min (ort 118 min) mTV VIDrel: VCC L/A V
Boa: play by Susan Cooper.

FOXTRAP * 18
Fred Williamson ITALY/USA 1986
Fred Williamson, Christopher Connelly, Beatrice Palme, Donna Owen, Lela Rochon, Arlene Golonka, Cleo Sebastian
A strong-arm private eye is hired by an uncle to find his missing niece but unfortunately he is not told the whole truth by his client and this leads to tragic results. A low-budget offering that is not much more than a vehicle for Williamson.
THR 85 min (ort 89 min) VIDrel: SUPVID/RTM V

FOXY LADY ** 18
George Raminto USA 1992
Steven Bond, Debora Caprioglio, Sharon Twomey
Unmemorable erotic thriller that sees a professional hitman being given the task of killing a drugs baron, but finding himself somewhat distracted by the charms of a mysterious femme fatale.
THR 94 min VIDrel: RIO L/A V

FRAGMENT OF FEAR ** (15)
Richard C. Sarafian UK 1970
David Hemmings, Gayle Hunnicutt, Wilfrid Hyde-White, Flora Robson, Adolfo Celi, Roland Culver, Daniel Massey, Mona Washbourne, Mary Wimbush, Glynn Edwards, Derek Newark, Arthur Lowe, Yootha Joyce, Bernard Archard
A writer and former drug addict attempts to make sense of some bizarre events, such as the murder of an aunt, but is slowly driven to the conclusion that he may be insane. A murky and ponderous thriller that shows great style, but is hampered by a lack of depth. Written by Paul Dehn and photographed by Oswald Morris.
THR 92 min (ort 95 min) SATrel: MOVIE CHANNEL
Boa: novel by John Bingham.

FRAMED ** 15
Geoffrey Sax UK/USA 1992
Timothy Dalton, Timothy West, David Morrissey, Annabelle Apsion, James Findleton, Barry Findleton, Penelope Cruz Sanchez, Rowena King, Jorge Bosso, Glyn Grimstead, Wayne Foskett, Clive Flint, Trevor Cooper, Chris Fairbank
An ambitious policeman spots an ex-con in Spain, where the latter is living a life of ease, and our cop sees his capture as a way of making a name for himself.
A/AD 194 min (ort 240 min) mTV VIDrel: FOCUS/DISC V

FRANCES *** 15
Graeme Clifford USA 1982
Jessica Lange, Sam Shephard, Kim Stanley, Bart Burns, Jeffrey DeMunn, Jordan Charney, Lane Smith, Kevin Costner, Jonathan Banks, Bonnie Bartlett, James Brodhead, J.J. Chaback, Daniel Chodes, Red Colbin, Donald Craig
An aspiring young female film star suffers from self-destructive tendencies and has an eventual nervous breakdown, spending some time in a mental home. A biography of the 1930s star Frances Farmer which chronicles her tragic life in some detail. The film "Will There Really Be A Morning?" covered the same ground.
DRA 113 min (ort 140 min) VIDrel: WHV V/sur

FRANK AND JESSE ** 15
Robert Boris USA 1995
Rob Lowe, Bill Paxton, Randy Travis, Willaim Atherton, Alexis Arquette
At the end of the Civil War, Frank and Jesse James continue the fight by forming a guerilla force and robbing trains and banks but are opposed by Allan Pinkerton, who formed the famous detective agency. An anaemic, revisionist history lesson, it portrays these notorious robbers as victims of society and circumstance, while casting aspersions on the forces of law and order. See also JESSE JAMES.
WES 102 min (ort 105 min) VIDrel: TRIM/HIFLI V/h

FRANKENSTEIN **** PG
James Whale USA 1931
Boris Karloff, Mae Clark, Colin Clive, John Boles, Edward Van Sloan,

Dwight Frye, Frederick Kerr, Lionel Belmore, Michael Mark, Forrester Harvey, Marilyn Harris, Francis Ford, Arletta Duncan, Pauline Moore, Mary Sherman
This is the daddy of them all, the original classic tale which though dated is well worth seeing. A scientist creates a monster out of parts of bodies and brings it to life with electricity. It then goes on a rampage. This film (which bears little resemblance to the novel) has spawned countless imitations and spin-offs. Followed by THE BRIDE OF FRANKENSTEIN.
HOR 69 min (ort 71 min) B/W cC VIDrel: CIC/SONOP V
Boa: novel Frankenstein; Or, The Modern Prometheus by Mary Shelley.

FRANKENSTEIN * 15
James Ormerod UK 1984
Robert Powell, Carrie Fisher, David Warner, John Gielgud, Susan Wooldridge, Edward Judd, Terence Alexander
This straightforward adaptation suffers from having been shot on video, which gives some very muddy photography, miscasting (Fisher) and excessive brevity. All in all, a very uninspired effort.
HOR 90 min mTV VIDrel: VCC/DISC L/A V
Boa: novel Frankenstein; Or, The Modern Prometheus by Mary Shelley.

FRANKENSTEIN *** 15
Kenneth Branagh USA 1994
Robert De Niro, Helena Bonham Carter, Kenneth Branagh, Tom Hulce, Aidan Quinn, Ian Holm, Richard Briers, John Cleese, Robert Hardy, Cherie Lunghi, Celia Imrie, Trevyn McDowell, Gerard Horan, Mark Hadfield, Joanna Roth, Sasha Hanau
A lavish and well-staged attempt to present in full the Shelley's complex tale of a scientist obsessed with the creation of life almost to the point of lunacy, and of how he lives to regret this monstrous conceit. Despite the over-familiar nature of the story, this handsome adaptation offers many points of interest, not least being De Niro's compelling performance as the monster.
Aka: MARY SHELLEY'S FRANKENSTEIN
HOR 118 min (ort 123 min) wScrn cC
VIDrel: COLUM/SONOP L/A V/sur
Boa: novel Frankenstein; Or, The Modern Prometheus by Mary Shelley.

FRANKENSTEIN AND THE MONSTER FROM HELL * 18
Terence Fisher UK 1973
Peter Cushing, Shane Briant, Madeline Smith, Dave Prowse, John Stratton, Charles Lloyd-Pack, Bernard Lee, Patrick Troughton, Sydney Bromley, Janet Hargreaves, Philip Voss
Sixth and last in a series of Frankenstein movies from Hammer Films. Here a young doctor placed in an asylum by the authorities for his experiments on dead bodies finds that this institution is headed by none other than Doctor Frankenstein himself. Together, they create a creature that proves to have an unfortunate taste for human flesh. A turgid and uninspired offering of marginal interest.
HOR 90 min (ort 95 min) VIDrel: WHV V/h

FRANKENSTEIN CREATED WOMAN ** 15
Terence Fisher UK 1967
Peter Cushing, Susan Denberg, Thorley Walters, Robert Morris, Michael Ripper, Duncan Lamont, Peter Blythe, Alan MacNaughton, Peter Madden, Barry Warren, Derek Fowlds, Philip Ray
After "The Evil Of Frankenstein" comes the fourth film in the series from Hammer, and our Baron is now involved in metaphysical experiments. When he attempts to invest the body of a dead woman with the soul of her executed lover, matters do not work out entirely satisfactorily. A disappointing film that is quite ambitious in conception, but is hampered by an excess of gore and crude editing. FRANKENSTEIN MUST BE DESTROYED was next.
Aka: FRANKENSTEIN MADE WOMAN
HOR 87 min VIDrel: LUMI/SPEAR L/A V

FRANKENSTEIN GENERAL HOSPITAL * 15
Deborah Roberts USA 1988
Mark Blankfield, Leslie Jordan, Irwin Keyes, Lou Cutell, Jonathan Farwell, Kathy Shower, Katie Caple, Bobby "Boris" Pickett
At an ineptly-run hospital, an intern is busily employed putting together a monster in the basement. A crude and unappealing comedy, short on both humour and ideas – the title is its best feature. See DOCTOR HACKENSTEIN for something along similar lines.
HOR 87 min (ort 92 min) Col/B/W
VIDrel: VISVID/POLYREC L/A V

FRANKENSTEIN MEETS THE WOLF MAN *** PG
Roy William Neill USA 1943
Lon Chaney Jr, Patric Knowles, Ilona Massey, Bela Lugosi, Lionel Atwill, Maria Ouspenskaya, Dennis Hoey, Rex Evans, Dwight Frye, Don Barclay, Harry STubbs, Martha Vickers, Doris Lloyd, Adia Kuynetzoff, Beatrice Roberts
A sequel to both "The Ghost Of Frankenstein" and THE WOLF MAN, with Chaney discovering that Baron Frankenstein is dead when he comes to him for help. However, he soon finds that the Baron's monster is alive and well. Lugosi is miscast as the monster in his only attempt at the role. Followed by "House Of Frankenstein".
HOR 72 min B/W VIDrel: CIC/SONOP L/A V/h

FRANKENSTEIN MUST BE DESTROYED ** 18
Terence Fisher UK 1969
Peter Cushing, Simon Ward, Veronica Carlson, Maxine Audley, Thorley Walters, Freddie Jones, George Pravda, Geoffrey Bayldon, Colette O'Neill, Peter Copley
But the plot must be kept unchanged. Hammer's fifth outing for the Frankenstein legend has Cushing doing the usual bit with brain transplants and the like, creating the inevitable rampaging monsters. Followed by THE HORROR OF FRANKENSTEIN, but the real horror is the sameness of these films, this undistinguished effort being a case in point.
HOR 96 min (ort 101 min) VIDrel: WHV V/h

FRANKENSTEIN: THE COLLEGE YEARS ** PG
Tom Shadyac USA 1991
William Ragsdale, Christopher Daniel Barnes, Larry Miller, Andrea Elson, De'Voreaux White, Patrick Richwood, Margaret Langrick, Beau Dremann, Robert V. Barron, Vincent Hammond, Richard Clements, Greg Grunberg, Joe Farago
This horror spoof follows the exploits of two young medical students who devise a scheme to revive the title creature, but find the success of their project leads to complications. So-so.
COM 88 min VIDrel: 20TH/TECH V

FRANKENSTEIN: THE TRUE STORY *** (15)
Jack Smight UK/USA 1973
Leonard Whiting, Michael Sarrazin, David McCallum, James Mason, Nicola Pagett, Jane Seymour, John Gielgud, Margaret Leighton, Ralph Richardson, Michael Wilding, Tom Baker, Agnes Moorehead
Vivid, fascinating and highly visual adaptation of Shelley's gothic horror tale that remains faithful to the spirit of the original despite some changes in plot and emphasis. Sarrazin gives a fine performance as the monster, portrayed here (initially at least) as an elegant and handsome dandy and not the familiar lumbering freak of so many other films. An intelligent and quite original approach that pays off, even if it does depart considerably from the novel.
FAN 200 min mTV TVrel: BBC1
Boa: novel Frankenstein; Or, The Modern Prometheus by Mary Shelley.

FRANKENSTEIN UNBOUND ** 18
Roger Corman USA 1990
John Hurt, Raul Julia, Bridget Fonda, Jason Patric, Michael Hutchence, Catherine Rabett, Catherine Corman, Mickey Knox, Cynthia Allison, Matt Cassidy, Myriam Cyr, William Geiger, Nick Brimble (the monster), Terri Treas (voice only)
In the 21st century, a brilliant but eccentric scientist conducts a series of experiments that result in him being catapulted back into the past, to 1817 to be precise. Once there, he meets both Byron and Shelley and learns of a real Dr Frankenstein who has actually created a monster. Corman's first foray into directing in nearly twenty years is a sluggish and verbose work of little merit. The music is by Carl Davis.
Aka: ROGER CORMAN'S FRANKENSTEIN UNBOUND
HOR 82 min (ort 85 min) VIDrel: WHV V/sh
Boa: novel by Brian W. Aldiss.

FRANKIE & JOHNNY *** 15
Garry Marshall USA 1991
Al Pacino, Michelle Pfeiffer, Hector Elizondo, Nathan Lane, Jane Morris, Al Fann, Greg Lewis, Glenn Plummer, Sean O'Bryan, Fernando Lopez, Ele Keats, Phil Leeds, K. Callan, Shannon Wilcox, Kate Nelligan, Tim Hopper, Harvey Miller
Based on a 1978 two-hander stage play, this story (considerably expanded) concerns the struggles of two nobodies to find love, while working in a down-market New York diner. Frankie (Pfeiffer) is the waitress who has withdrawn from life after an abusive and injurious relationship. Johnny (Pacino) is the short-

order cook who, though fresh from a spell in prison, pursues her. A charming, old-fashioned love story. Music is by Marvin Hamlisch.
DRA 113 min (ort 118 min)
VIDrel: 4-FRONT/POLYREC/CIC V/sur
Boa: play Frankie and Johnny in the Clair de Lune by Terrence McNally.

FRANKIE STARLIGHT *** 15
Michael Lindsay-Hogg EIRE/FRANCE/UK 1995
Anne Parillaud, Matt Dillon, Gabriel Byrne, Rudi Davies, Georgina Cates, Corban Walker, Alan Pentony, Niall Toibin, Dearbhla Molloy, Jeane Claude Frissung, Victoria Begeja, Barbara Alyn Woods, John Davies, Amber Hibler
During WW2, a woman stows away on a GI troop-ship and escapes France, but is dumped in Ireland, alone and pregnant. However, she soon becomes an object of interest on the part of three admirers. But this story is actually drawn from the reminiscences of a lonely Irish dwarf, whose book on his family's past becomes a bestseller. Scripted by Raymo and Ronan O'Leary, this charming but slight work was promoted on its video sleeve as an entirely different film.
DRA 96 min (ort 101 min) VIDrel: FILM4/RTM V/s
Boa: novel The Dork of Cork by Chet Raymo.

FRANKIE'S HOUSE ** 15
Peter Fisk AUSTRALIA/UK 1990
Iain Glen, Kevin Dillon, Steven Vidler, Alan David Lee, Stephen Dillane, Alexander Fowler, Caroline Carr, Todd Boyce, Nicholas Hammond, Kay Tong Lime
When a young photographer becomes friends with the son of Errol Flynn, he finds himself caught up in the horrors of the Vietnam War.
Aka: FRANKIE'S WAR
DRA 107 min (ort 180 min) mTV VIDrel: FOCUS/DISC V
Boa: novel Page After Page by Tim Page.

FRANTIC *** 15
Roman Polanski USA 1987
Harrison Ford, Emmanuelle Seigner, Betty Buckley, John Mahoney, Jimmy Ray Weeks, Yorgo Voyagis, David Huddleston, Gerard Klein, Raouf Ben Amor, Boll Boyer, Alexandra Stewart, Robert Barr, Djiby Soumare, Dominique Virton
An American heart surgeon arrives in Paris to attend a convention, only to have his wife suddenly disappear without trace. An intriguing suspense tale that happily never becomes overblown, as so many films of this genre are apt to be.
THR 115 min (ort 120 min) VIDrel: WHV V/sur

FRAUDS ** 15
Stephan Elliott AUSTRALIA 1992
Phil Collins, Hugo Weaving, Josephine Byrnes, Rebel Russell, Peter Mochrie, Mitchell McMahon, Andrew McMahon, Helen O'Connor, Colleen Clifford, Vincent Ball, Ghandi MacIntyre, Christina Ormani, Nicholas Hammond, Kee Chan
An amoral insurance investigator gets wind of a young man who is planning a fraud, by having a friend break into their apartment while the man and his wife are at the opera. A dangerous cat-and-mouse game then ensues. An unusual black comedy that succeeds on most counts, although the script is at times just a little bit disconcerting.
COM 90 min (ort 94 min) VIDrel: FIRST/SONOP V/sur

FREAK SHOW ** 18
Constantino Magnatta USA
Audrey Landers, Peter Read, Will Korbut, Dan Gallagher
A female reporter covers the story of a massacre outside a city theatre, but is later drawn back there, where she discovers a room of freaks.
HOR 91 min VIDrel: COLUM/SONOP V

FREAKED * 15
Tom Stern/Alex Winter USA 1993
Alex Winter, Randy Quaid, William Sadler, Mr. T, Brooke Shields, Megan Ward, Alex Zuckerman, Ray Baker, Morgan Fairchild, Patti Tippo, Lee Arenberg, John Michael Stoyanov, John Hawkes, Derek McGrath, Jeff Kahn, Randy Quaid
A TV actor travels south of the border and goes to visit a circus where some unbelievable freaks are on show. Soon he finds himself captured and being exposed to toxic waste, which is the method the circus owner uses to create his hideous charges. An over-the-top orgy in sick effects, admirable for its technical use of makeup but deficient in every other department and far too

long for its thin material. This horror spoof was Alex Winter's debut feature.
HOR 81 min (ort 96 min)
VIDrel: 4-FRONT/POLYREC/GUILD V/sh

FREAKS * 15
Tod Browning USA 1932
Wallace Ford, Olga Baclanova, Leila Hyams, Roscoe Ates, Harry Earles, Daisy Hilton, Violet Hilton, Daisy Earles, Rose Dione, Edward Brophy, Matt McHugh, Olga Roderick, Johnny Eck, Randian, Schlitzie, Eliva, Jeannie Lee Snow
A fascinating look at the deformed and misshapen performers at a carnival sideshow and the strong bonds of affection between them. When a trapeze artist marries a midget for his money and plans to poison him with the help of her strongman lover, they exact a terrible revenge. The freaks are shown here as feeling and sensitive creatures while the normal members of the circus troupe are made to appear as callous and shallow beings.
Aka: FORBIDDEN LOVE; MONSTER SHOW, THE; NATURE'S MISTAKES
HOR 64 min (ort 90 min) B/W VIDrel: VISCOM/RTM V
Boa: short story Spurs by Todd Robins.

FREAKY FAIRY TALES * 18
Jeffrey S. Delman USA 1985
Nicole Picard, Scott Valentine, Catheryn Le Prume
A trio of horror tales that are all versions of familiar fairy tales, provides the material for this tongue-in-cheek adult horror-comedy.
Aka: DEAD TIME
HOR 88 min Cut (29 sec) VIDrel: EIV/SONOP V

FREAKY FRIDAY * U
Gary Nelson USA 1976
Jodie Foster, Barbara Harris, John Astin, Patsy Kelly, Dick Van Patten, Sorell Brooke, Marie Windsor, Charlene Tilton, Vicki Svreck, Kaye Ballard, Alan Oppenheimer, Marc McClure, Sparky Marcus, Cecil Cabot, Brooke Mills
Inspired by the 1947 British film "Vice Versa" in which a boy and his dad swap minds, this one looks at a mother-and-daughter exchange that takes place due to a magic amulet. A standard Disney comedy somewhat enlivened by winning performances from Foster and Harris. The script is by Mary Rodgers. See also EIGHTEEN AGAIN!
COM 94 min VIDrel: L/A V
Boa: story by Mary Rodgers.

FREDDIE AS F.R.O.7. * U
Jon Acevski UK 1987
Voices of: Ben Kingsley, Jenny Agutter, Brian Blessed, Jenny Funnell, David Ashton, Bruce Purchase, Nigel Hawthorne, Michael Hordern, Edmund Kingsley, Phyllis Logan, Victor Maddern, Jonathan Pryce, Prunella Scales, John Sessions
Unfortunately this ambitious British-made cartoon is neither one thing nor the other, for it tries to be a fairytale, a spy caper and a musical all rolled into one. Freddie is a young prince who's changed into a frog by his wicked aunt in an attempt to cheat him out of his inheritance. He responds by becoming a secret agent and saving the nation. Excellent animation work and some good songs are let down by the muddled and over-stretched plot.
Aka: FREDDIE THE FROG
ANIM 86 min (ort 91 min) VIDrel: 4-FRONT/POLYREC L/A V

FREDDY'S NIGHTMARES: A NIGHTMARE ON ELM STREET * 18
Tobe Hooper/Tom McLoughlin USA 1988
Robert Englund, Gry Park, Hili Park, Ian Williams, William Frankfather, Lori Petty, Yvette Nifar, Lee Kessler, Kane Picoy, Anthony Barton
A spin-off from A NIGHTMARE ON ELM STREET with the unpleasant razor-fingered character of Freddy Krueger giving his account of the Elm Street child murders and his own "death" at the hands of their horrified parents. Now Freddy is back back to have his revenge in this unpleasant and stomach-churning tale. This was the pilot for a TV series.
HOR 90 min mTV VIDrel: BRAVE/SONOP L/A V

FREDDY'S NIGHTMARES: SISTER'S KEEPER/FREDDY'S TRICKS OR TREATS * 18
Ken Wiederhorn USA 1988
Robert Englund, Gry Park, Mili Park, Joshua Cox, Jeff Bennett, Jeff Freilich, Chip Hipkins, Mariska Hargitay, Darren Dalton

Two episodes from the gruesome TV series featuring the devilish character of Freddy Krueger from A NIGHTMARE ON ELM STREET. In "Sister's Keeper" Freddy fights a battle of wits when he returns to Springwood to confront the twins whose father supposedly burnt him to death years ago. In "Freddy's Tricks Or Treats" our lovable rogue has some fun with a sexually repressed girl who is working late at the the the school science labs on the eve of Halloween
HOR 90 min mTV VIDrel: BRAVE/SONOP L/A V

FREDDY'S NIGHTMARES: THE NIGHTMARE BEGINS AGAIN * 18
Tobe Hooper/Tom McLoughlin USA 1988
Robert Englund, William Frankfather, Ian Williams, Lori Petty
Having recounted the story of the Elm Street child murders and his own death at the hands of their enraged parents, that nice Mr Krueger returns once more for another dose of vengeance on the citizens of Springwood. Two episodes from the TV series, entitled "No More Mr Nice Guy" and "Killer Instinct".
HOR 90 min mTV VIDrel: BRAVE/SONOP L/A V

FREE WILLY * U
Simon Wincer USA 1992
Jason James Richter, Jayne Atkinson, Michael Madsen, August Schellenberg, Lori Petty, Michael Ironside, Michael Riehle, Mykelti Williamson, Michael Bacall, Danielle Harris, Isaiah Malone, Betsy Toll, Rob Sample, Mickey Gaines
Simplistic tale of a young boy who befriends a performing whale and decides to free it, much to the annoyance of its evil owners who do their best to stop him. Will appeal mainly to children and animal-lovers and be of little interest to anyone else. (Rumours of cruelty in the treatment of the performing whale used in the movie prompted a changeover to an animated model in the sequel.)
DRA 107 min (ort 112 min) cC VIDrel: WHV V/sur

FREE WILLY 2 * PG
Dwight Little USA 1995
Jason James Richter, Michael Madsen, Jayne Atkinson, Francis Capra, August Schellenberg, Mary Kate Schellhardt, Mykelti Williamson, Elizabeth Peoa, Jon Tenney, Paul Tuerpe, M. Emmet Walsh, John Considine, Steve Kahan, Al Sapienza
Sequel to the first film sees boy and whale reunited two years later. This time Willy has acquired a family, but needs help in finding his way back to the ocean. By way of a sub-plot, an oil spill poses dangers for them all. A wholesome effort, not all that well scripted, but helped along by sturdy action sequences and a good climax. Following adverse publicity regarding the real-life Willy's mistreatment by his owners, this film wisely opted for animatronics.
Aka: FREE WILLY 2: THE ADVENTURE HOME
JUV 93 min (ort 97 min) cC VIDrel: WHV V/sur

FREEBIE AND THE BEAN * 18
Richard Rush USA 1974
Alan Arkin, James Caan, Loretta Swit, Valerie Harper, Jack Kruschen, Mike Kellin, Alex Rocco, Valerie Harper, Linda Marsh, Paul Koslo, Christopher Morley
Two somewhat happy-go-lucky cops almost destroy the city of San Francisco in an attempt to trap a gangster who runs an illegal gambling racket. Caan and Arkin work well together in a frantic, frenetic, noisy failure of a comedy whose highspots are its car crashes rather than its humour. Later the basis for a short-lived TV series.
COM 108 min (Cut at film release by 29 sec – ort 111 min)
VIDrel: MGM/WHV L/A V

FREEFALL * 18
John Irvin USA 1993
Eric Roberts, Jeff Fahey, Pamela Gidley, Rom Smerczak, Anthony Fridjhon, Leslie Fong, Ted Leplat, Warrick Grier, Lucky Shabangu, James Whyle, Terri Norton, Patrick Sands, Todd Jensen, Jenny Steyn
A wildlife photographer is sent to Africa on an assignment to get pictures of a rare bird but finds herself being pursued by several mysterious individuals who act as if she had some ulterior motive for being there. Needless to say, her life is soon in danger. A complex and elaborately plotted adventure of few surprises, it works well by taking no chances. Filmed on location in London, South Africa and Venezuela.
A/AD 96 min VIDrel: POLY/POLYREC/MED V/sur

FREEFALL: FLIGHT 174 **   PG
Jorge Montesi CANADA   1994
William Devane, Scott Hylands, Shelley Hack, Kevin McNulty,
Gwynyth Walsh, Gloria Carlin, Suzy Joachim, Nicholas Turturro,
Winston Rekert, Manette Heartley, Philip Granger, Philip Harris,
David Lewis, Sheelah Megill, John Novak
A ground staff error results in a Boeing 767 running out of fuel
in mid-flight at 41,000 feet, on a journey from Montreal to
Edmonton. With the engines starting to fail, the pilot and his
crew have to attempt to reach the nearest landing strip, which
is in Winnipeg. Quite a decent thriller, based on a true incident
in which disaster was narrowly averted.
THR   93 min mTV   VIDrel: ODY/SONOP   V/sh
Boa: book by William and Marilyn Hoffer.

FREEJACK *   15
Geoff Murphy USA   1991
Emilio Estevez, Mick Jagger, Rene Russo, David Johansen, Anthony
Hopkins, Jonathan Banks, Amanda Plummer, Grand L. Bush, Frankie
Faison, John Shea, Esai Morales, Wilbur Fitzgerald, Jerry Hall, Glen
Trotiner, Jody Waddell, Tom Barnes
Just as he is about to die in a smash-up, racing-driver Estevez
is whisked off into the near future (circa 2009) where he finds
that his body is intended to house the mind of a dying tycoon.
A lucky break gives him a chance to run for his life, but several
bounty hunters are not far behind. This mutilated version of a
fine SF novel is a letdown in most departments. A careless mess
of weak scripting, wooden acting and feeble dialogue.
FAN   105 min (ort 110 min) cC   VIDrel: WHV   V/sur
Boa: novel Immortality Inc. by Robert Sheckley.

FREEWAY **   18
Matthew Bright USA   1996
Kiefer Sutherland, Reese Witherspoon, Amanda Plummer, Brooke
Shields
After her abusive parents are arrested, a youngster sets off to
visit her grandmother, but is picked up by serial killer
Sutherland and tormented, as he too is prone to sexual abuse.
A violent revenge follows as the girl gets even, but this only
lands up her up in prison, where she endures yet more abuse.
A brutal and deeply undefying picture of the seedier side of
modern-day life, its sporadic moments of levity do little to raise
the sombre tone.
A/AD   98 min   VIDrel: HIFLI/SONOP   V/h

FREEZE FRAME *   (PG)
William Bingley USA   1989
Shannen Doherty, Charles Haid, Robyn Douglass, Seth Michaels,
Ryan Lambert, Adam Carl, Steve Russell, Harry Carson, Chi
Blackburn, Mark Krausz, Steve Salge, Angela Marsden, Doug
Johnson, Shelli Halper, Annette Cargioli, Mark Fauser
A young college girl has set her heart on becoming an inves-
tigative journalist rather than a doctor, as her father wishes. To
prove her worth, she sets out with a bunch of fellow students,
and armed with video cameras they expose the nefarious activities of a powerful biotech corporation, that is being assisted by
corrupt politicians. A simplistic juvenile yarn, contrived and
unoriginal.
A/AD   86 min   SATrel: SKY MOVIES

FRENCH CAN-CAN ***   PG
Jean Renoir FRANCE/ITALY   1955
Jean Gabin, Francoise Arnoul, Maria Felix, Jean-Roger Caussimon,
Edith Piaf, Patachou, Gianni Esposito, Philippe Clay, Michel Piccoli,
Jean Paredes, Lydia Johnson, Max Dalban, Jacques Jouanneau, Jean-
Marc Tennberg
An enjoyable fictional account of how the can-can came to be
invented in the early days of the Moulin Rouge nightclub. An
exuberant and joyous vision of 19th century Parisian life that
marked Renoir's return to film-making after a gap of fifteen
years.
Aka: ONLY THE FRENCH CAN
MUS   100 min (ort 106 min)   VIDrel: CONNO/RTM   V

FRENCH CONNECTION, THE ****   18
William Friedkin USA   1971
Gene Hackman, Roy Scheider, Fernando Rey, Tony LoBianco, Marcel
Bozzuffi, Frederic Da Pasquale, Bill Hickman, Ann Rebbot, Harold
Gary, Arlene Farber, Eddie Egan, Andre Ernotte, Sonny Grosso, Pat
McDermott, Alan Weeks
Fast and well-paced story of a hard-headed New York cop who,
together with his buddy is assigned to a serious narcotics case.
Fine action sequences such as an excellent car chase combine

with a sense of real tension as they stake-out their quarry, an
international heroin smuggling king. A sequel, lacking the pace
and tension and pace of the first film followed. AA: Pic, Dir,
Actor (Hackman), Screen/adapt (Ernest Tidyman), Edit (Jerry
Greenberg).
THR   99 min   VIDrel: 20TH/TECH L/A   V
Boa: novel by Robin Moore.

FRENCH CONNECTION 2, THE **   18
John Frankenheimer USA   1975
Gene Hackman, Fernando Rey, Charles Millot, Bernard Fresson, Jean-
Pierre Castaldi, Cathleen Nesbitt, Pierre Collet, Alexandre Fabre,
Philippe Leotard, Jacques Dynam, Raoul Delfosse, Patrick Floersheim
In this sequel Popeye Doyle, the cop from the previous film,
pursues his quarry to Marseilles intent on smashing the drugs
ring. Has none of the sheer impact of the first film and is marred
by a long and tormenting drugs sequence halfway through.
Hackman is as good as ever though.
A/AD   114 min (ort 119 min)   VIDrel: 20TH/TECH   V

FRENCH INVASION, THE *   18
Harold Lime USA   1993
Rebecca Bardoux, Lana Sands, Sharon Kane, Nikki Shane, Julian St.
Jox, T.T. Boy, Scott Turne (Nick E)
A woman's home is invaded by a couple of robbers who raided
a liquor store and took a willing female hostage. Stealing a truck
marked "French cleaners", they decide to indulge in a little
burglary by posing as these cleaners and raiding people's
homes.
A   55 min   VIDrel: REDEEM/TOTAL   V

FRENCH KISS **   15
Lawrence Kasdan USA   1995
Meg Ryan, Kevin Kline, Timothy Hutton, Jean Reno, Francois Cluzet,
Susan Anbeh, Renee Humphrey, Michael Riley, Laurent Spielvogel,
Victor Garrivier, Elizabeth Commelin, Julie Leibowitch, Miquel
Brown, Louise Deschamps
When an engaged woman is told by her boyfriend that he has
decided not to marry her, she hops on a plane to Paris, where
she hopes she'll persuade him to change his mind. But on the
journey she meets a fellow passenger, an amateur jewel thief and
he becomes infatuated with her, a sentiment she fails to return.
A predictable piece of froth, offering little apart from good work
by the leads. See also SABRINA and FORGET PARIS for two
more Paris-inspired comedy romances.
Aka: PARIS MATCH
COM   106 min (ort 111 min) cC   VIDrel: POLY/POLYREC
V/s

FRENCH LIEUTENANT'S WOMAN, THE ***   15
Karel Reisz UK   1981
Jeremy Irons, Meryl Streep, Leo McKern, Lynsey Baxter, Patience
Collier, Hilton McRae, Emily Morgan, Peter Vaughan, Penelope
Wilton, David Warner, Charlotte Mitchell, Jean Faulds, Colin Jeavons,
Liz Smith, John Barrett
The relationship between a gentleman and a jilted mistress in
Victorian times is paralleled by a modern affair between an actor
and an actress playing these roles in a film. After an uncertain
start, this diverting film gets better as it develops, helped in no
small measure by Harold Pinter's literate script. The storyline
does however, still remain painfully thin.
DRA   119 min (ort 124 min)   VIDrel: WHV L/A   V
Boa: novel by John Fowles.

FRENCH MISTRESS, A **   PG
Roy Boulting UK   1960
James Robertson Justice, Cecil Parker, Raymond Huntley, Ian Bannen,
Agnes Laurent, Thorley Walters, Edith Sharpe, Athene Seyler,
Kenneth Griffith
Complications arise with the arrival of an attractive French
teacher at a public boys' school, where the headmaster comes
to believe that the girl is his illegitimate daughter. A haphazard
and unfocused romp, with a competent cast of comedy stalwarts
giving of their best.
COM   94 min (ort 98 min) B/W   VIDrel: WHV   V
Boa: play by Robert Munro (Sonnie Hale).

FRENCH POSTCARDS **   15
Willard Huyuck USA   1979
Valerie Quennessen, David Marshall Grant, Blanche Baker, Debra
Winger, Mandy Patinkin, Marie-France Pisier, Jean Rochefort, Miles
Chapin, Lynn Carlin, George Coe, Christophe Bourseiller, Francoise
Lalande, Anemone, Veronique Jannot

The romantic adventures of a group of American college students spending a year abroad in Paris, form the basis for this amiable but hardly memorable comedy, made as something of a follow-up to AMERICAN GRAFFITI.
COM   90 min (ort 92 min)   VIDrel: CIC/SONOP L/A   V/h

**FRENCH SILK **   (PG)
Noel Nosseck USA   1993
Susan Lucci, Lee Horsley, Shari Belafonte. R. Lee Ermey, Sarah Marshall, Jim Metzler, Bobby Hosea, Joe Warfield, Paul Rosenberg, Taylor Simpson, Victoria Edwards, Monique Viator, Tanya Teague, Michael Bergeron, Robert Florence
The head of title successful lingerie company in New Orleans receives a nasty shock when she learns that the police consider her as their prime suspect in the murder of a television evangelist. Things become even more complicated when she learns that one of her models plays a key role in her efforts to clear her name.
THR   90 min mTV   SATrel: SKY MOVIES
Boa: novel by Sandra Brown.

FRENCH TWIST *   18
Josiane Balasko FRANCE   1995
Victoria Abril, Josiane Balasko, Alain Chabat, Ticky Holgado, Miguel Bose, Catherine Hiegel, Catherine Samie, Michele Bernier, Catherine Lachens, Telsche Boorman, Katrine Boorman, Veronique Barrault, Sylvie Audcoeur, Maureen Diot
A woman gets fed up with her philandering husband and packs her bags, moving into the home of a butch, cigar-smoking lesbian (played with gusto by Balasko). Though he is hardly a model husband, he is most put out by this, but in an attempt to avoid losing his wife entirely, he reluctantly agrees to become part of a strange menage-a-trois. A sexy, comic and very sharp examination of social convention and hypocrisy, with great dialogue and absurd situations.
Aka: GAZON MAUDIT
COM   103 min (ort 107 min) dubbed wScrn
VIDrel: GUILD/FOXVID   V

**FRENCH WOMAN, A **   18
Regis Warnier FRANCE/GERMANY/UK   1994
Emmanuelle Beart, Daniel Auteuil, Gabriel Barylli. Jean-Claude Brialy, Genevieve Casile, Michel Etcheverry, Laurence Masliah, Jean-Noel Broute, Isabelle Guiard, Francois Caron, Maria Fitzi, Samuel Le Bihan, Heinz Bennent
The lives of two sisters, who both marry in 1939, but find that the war blights their happiness. One is married to a soldier who is captured, leaving her free the embark on a couple of affairs, the second of which leads to much bitterness, as it is with the son of a former Nazi. A complex and rambling film that darts about from one setting to another, touching upon a whole range of interesting issues that are never properly explored.
Aka: EINE FRANZOSISCHE FRAU; UNE FEMME FRANCAISE
DRA   99 min   CINrel

**FRENCHMAN'S FARM **   15
Ron Way AUSTRALIA   1986
Ray Barrett, Tracey Tainsh, David Reyne, John Mullion, Norman Kaye, Andrew Blackman, Phil Brock, Tui Bow, Kym Lynch, Andrew Johnston, Lynne Schofield
A woman has a paranormal experience whilst out driving, and finds herself momentarily in the past where she witnesses a murder. She reports her experience to the police who fail to believe her, but eventually the mystery is resolved in this intriguing little thriller.
THR   92 min (ort 96 min)   VIDrel: SPEAR/SONOP   V/sur

**FRENZY **   18
Alfred Hitchcock UK   1972
Jon Finch, Alec McGowen, Barry Foster, Barbara Leigh-Hunt, Anna Massey, Vivien Merchant, Bernard Cribbins, Billie Whitelaw, Michael Bates, Rita Webb, Jimmy Gardner, Clive Swift, Jean Marsh, Madge Ryan, George Tovey
Hitchcock is at his most stilted with a crazed killer who uses neckties to strangle girls, escaping suspicion whilst an innocent man is suspected. Some good camerawork and good casting of Foster as the maniac, but a sense of tension is just not maintained. The script is by Anthony Shaffer.
DRA   110 min Cut (19 sec – ort 116 min)
VIDrel: CIC/SONOP   V/h
Boa: novel Goodbye Piccadilly, Farewell Leicester Square by Arthur La Bern.

FRESH *   18
Boaz Yakin USA   1994
Sean Nelson, Giancarlo Esposito, Samuel L. Jackson, N'Bushe Wright, Ron Brice, Jean LaMarre, Jose Zuniga, Luis N. Lantigua, Yul Vasquez, Cheryl Freeman, Anthony Thomas, Curtis McClarin, Charles Malik Whitfield, Victor Gonzalez
A twelve-year-old chess genius uses his skills to find a way to eliminate the two drug dealers who have infested his neighbourhood, by playing them off against each other. However, this proves to be a dangerous ploy which demands some sacrifices from all concerned. A strong cast give their best in a fine display of ensemble acting, and the film was a big hit with the critics, winning the Special Jury Prize at the Sundance Film Festival.
DRA   108 min (ort 114 min)   VIDrel: EIV/SONOP   V

FRESHMAN, THE *   PG
Andrew Bergman USA   1990
Matthew Broderick, Marlon Brando, Bruno Kirby, Frank Whaley, Penelope Ann Miller, Maximilian Schell, Paul Benedict, B.D. Wong, Jon Polito, Richard Gant, Kenneth Welsh, Pamela Payton-Wright, Bert Parks, Tex Konig
A New York University film student takes a job delivering exotic animals to a pretentious restaurant, and learns that his boss, a Mafia godfather, has taken a shine to him. A delightfully fresh and inventive comedy, whose witty script (the work of Bergman) is nicely complemented by a wonderful rapport between Broderick and Brando, the latter unmercifully spoofing the role he took in THE GODFATHER.
COM   98 min (ort 102 min)   VIDrel: VCC/COLUM L/A
V/sh

FRIDAY *   15
F. Gary Gray USA   1995
Chris Tucker, Bernie Mac, John Witherspoon, Regina King, Nia long, Ice Cube, Tiny "Zeus" Lister Jr, Anna Maria Horsford, Paula Jai Parker, Faizon Love, D.J. Pooh, Angela Means, Vickilyn Reynolds, Ronn Riser, Kathleen Bradley, Tony Cox
Episodic slice of life in the south central Los Angeles ghetto, built around the figures of an unlucky loser who gets fired from his dead-end job and his companion who is unwise enough to smoke the dope he had agreed to sell to a local dealer. One of those movies covering a mere twenty-four hours, it has incident enough, just no overall plot to give it direction and structure.
COM   87 min (ort 91 min)   VIDrel: EIV/SONOP   V

**FRIDAY THE 13TH **   18
Sean S. Cunningham USA   1980
Betsy Palmer, Adrienne King, Harry Crosby, Laurie Bartram, Mark Nelson, Jeannine Taylor, Kevin Bacon, Robbi Morgan, Peter Brouwer, Ari Lehman, Rex Everhart, Ronn Carroll, Ron Millkie, Walt Gorney, Willie Adams
Grisly murders occur again when a summer camp is opened after many years. A film whose only reason for being is to deliver a series of well timed shocks, but it does that very well. The film grossed $17,000,000 and has led to a veritable industry churning out a range of sequels and spin-offs.
HOR   91 min (ort 95 min)   VIDrel: WHV   V/h

FRIDAY THE 13TH, PART 2 *   18
Steve Miner USA   1981
Amy Steel, Adrienne King, John Furey, Betsy Palmer, Kirsten Baker, Stu Charno, Warrington Gillette, Walt Gorney, Marta Kober, Tom McBride, Bill Randolph, Laurie-Marie Taylor, Russell Todd, Betsy Palmer, Cliff Cudney
Sequel in which the sole survivor of the first slaughter is killed and the murders start all over again. This time round the villain is the son of the woman who committed the first lot of killings.
HOR   83 min Cut (42 sec – ort 87 min)
VIDrel: ETL/POLYREC/CIC   V/h

FRIDAY THE 13TH, PART 3 *   18
Steve Miner USA   1982
Dana Kimmell, Paul Kratka, Tracie Savage, Jeffrey Rogers, Catherine Parks, Larry Zerner, Richard Brooker, Rachel Howard, David Katims, Nick Savage, Gloria Charles, Kevin O'Brien, Annie Gaybis, Cheri Maughans, Steve Miner
Some years have passed since Jason gave his first group of happy campers a holiday to remember. Unaware of the latest spate of murders at Crystal Lake, a group of friends decide to spend the weekend at a lakeside cottage. They ignore the warnings of a local and settle in only to find themselves threatened by a motorcycle gang, but receive help rather unexpectedly.

Slightly less gory than its predecessor, but the basic recipe is unchanged.
HOR   91 min Cut (4 sec – ort 96 min)   VIDrel: CIC/SONOP V/h

FRIDAY THE 13TH, PART 4: THE FINAL CHAPTER *

Joseph Zito USA   18
                                                   1984
Kimberly Beck, Corey Feldman, Crispin Glover, Barbara Howard, Ted White, Bruce Mahler, E. Erich Anderson, Joan Freeman, Lawrence Monoson, Thad Geer, Wayne Grace, Alan Hayes, Bonnie Hellman, Frankie Hill, William Irby
Our lovable maniac Jason, dispatches yet more teens in his own inimitable style in this gruesome sequel. And oh yes, at the end he gets his comeuppance, only don't bank on that keeping him from making a sequel.
HOR   88 min Cut (27 sec)   VIDrel: CIC/SONOP   V/h

FRIDAY THE 13TH, PART 5: A NEW BEGINNING *

Danny Steinmann USA   18
                                                   1985
John Shepard, Melanie Kinnaman, Shavar Ross, Richard Young, Carol Lacatell, Marco St John, Juliette Cummins, Vernon Washington, John Robert Dixon, Jerry Pavlon, Caskey Swaim, Mark Venturini, Anthony Barrile, Tiffany Helm
Psychopath Jason Voorhees is dead at last but the killings continue, with inmates of an institute of mental health succumbing this time round. Did I hear there was going to be a new plot? But why change a winning formula?
HOR   87 min Cut (1 min 22 sec plus film cuts – ort 92 min)   VIDrel: CIC/SONOP   V

FRIDAY THE 13TH, PART 6: JASON LIVES *

Tom McLoughlin USA   18
                                                   1986
Thom Mathews, Jennifer Cooke, David Kagan, Renee Jones, Kerry Noonan, C.J. Graham, Tom Fridley, Darcy Demoss, Vincent Gustaferro, Tony Goldwyn, Nancy McLoughlin, Ron Palillo, Alan Blumenfield, Matthew Faison, Bob Larkin
Now young Tommy, as you may recall, finally put paid to psycho-killer Jason Vorhees. However, he begins to suspect that Jason is not really dead and persuades a friend to help him dig up the body. And instead of the rotting corpse they expected to find? Well I won't spoil this one for you.
Aka: JASON LIVES: FRIDAY THE 13TH PART 6
HOR   87 min   VIDrel: CIC/SONOP   V/sh

FRIDAY THE 13TH, PART 7: THE NEW BLOOD *

John Carl Buechler USA   18
                                                   1988
Lar Parc-Lincoln, Kevin Blair, Susan Blu, Terry Kiser, Jennifer Banko, John Otrin, Susan Jennifer Sullivan, Heidi Kozak, Kane Hodder, William Clarke Butler, Staei Grearson, Larry Cox, Jeff Bennett, Diana Barrows, Craig Thomas
A young girl with telekinetic powers accidentally releases Jason from his grave and the killing re-commences. Unfortunately, nothing can levitate this film to the level of the merely mediocre.
HOR   84 min (Cut at film release by 8 sec – ort 90 min)   VIDrel: CIC/SONOP   V

FRIDAY THE 13TH, PART 8: JASON TAKES MANHATTAN **

Rob Hedden USA   18
                                                   1989
Jensen Daggett, Kane Hodder, Peter Mark Richman, Scott Reeves, Barbara Bingham, V.C. Dupree, Sharlene Martin
Yet another slasher film in this series, though this one offers a little imagination, with most of the action taking place on a cruise ship (we only reach Manhattan in the last fifteen minutes) where the usual bunch of unfortunate teens meet grisly ends. Written by Hedden, this overlong shocker is one of the best ones in the series, but that isn't saying much.
HOR   96 min Cut (1 sec – ort 100 min)   VIDrel: CIC/SONOP V

FRIDAY THE 13TH: JASON GOES TO HELL *

Adam Marcus USA   18
                                                   1993
John D. LeMay, Kari Keegan, Allison Smith, Steven Culp, Billy Green Bush, Rusty Schwimmer, Leslie Jordan, Andrew Bloch, Kipp Marcus. Richard Grant, Adam Cranner, Julie Richards, Erin Gray, Kane Hodder, Steven Williams
Allegedly the final instalment in this long-running horror series, this opens with Jason being lured into a police trap where he is unceremoniously blasted to pieces. Unfortunately, his spirit survives to continue the killing until a man learns how he can

be laid to rest. A nasty, extremely violent and derivative movie with a confused plot that lacks any logic or sense.
Aka: JASON GOES TO HELL; JASON GOES TO HELL: THE FINAL FRIDAY
HOR   84 min (ort 91 min – unedited director's cut)
VIDrel: 4-FRONT/POLYREC/GUILD   V/sur

FRIED GREEN TOMATOES AT THE WHISTLE STOP CAFE ****

Jon Avnet USA   PG
                                                   1991
Mart Stuart Masterson, Mary-Louise Parker, Jessica Tandy, Kathy Bates, Cicely Tyson, Nick Searcy, Chris O'Donnel, Stan Shaw, Gailard Sartain, Tim Scott, Gary Basaraba, Lois Smith, Jo Harvey Allen, Macon McCalman, Reid Binion
An account of two lifelong friends and the restaurant they ran in rural Alabama during the Depression, told in flashback form as remembered by an old lady who knew both women, and has now befriended an unhappy soul she met via a feminist support group. Very much an affirmation of feminine power and emancipation (many of the male characters are hateful) this multi-layered and complex story is sharply observant and wryly funny. See also THE SPITFIRE GRILL.
Aka: FRIED GREEN TOMATOES
DRA   124 min (ort 130 min)   VIDrel: VCC/DISC/COLUM V/sur
Boa: novel by Fannie Flagg.

FRIENDS ***

Elaine Proctor FRANCE/SOUTH AFRICA/UK   15
                                                   1993
Kerry Fox, Michele Burgers, Dambisa Kente
During the Apartheid era in South Africa, three women of different races and backgrounds become firm friends. However, their relationship comes under strain when one of them becomes a terrorist. In this role, she plants a bomb that kills two people, an action that faces them with a difficult choice.
DRA   105 min (ort 109 min) wScrn   VIDrel: TART/20TH V/sur

FRIENDS AND LOVERS **

F.J. Lincoln USA   18
                                                   1991
Teri Wiegel, Peter North, Alice Springs, Tracey Winn, April Rayne, Tom Chapman, Joey Murphy, Candice Walker
Romantically inclined sex film about a young couple who decide to get married and their friends and relations. Unfortunately, tragedy strikes when the would-be bride is killed in a road accident. Followed by a sequel.
A   55 min (ort 80 min)   VIDrel: GROHOM/MAXSCAN   V

FRIENDS AND LOVERS: PART 2 **

F.J. Lincoln USA   18
                                                   1991
Teri Wiegel, Peter North, Lois Ayers, Candice Hart, Randy West, Casey Williams, Marc Wallice
After the death of his fiancee, a young man seeks to come to terms with his loss and even consults a medium. However, his grief does not prevent him from seeking solace elsewhere.
Aka: FRIENDS AND LOVERS: THE SEQUEL
A   58 min   VIDrel: GROHOM/MAXSCAN   V

FRIENDS AT LAST **

John Coles USA   PG
                                                   1996
Kathleen Turner, Colm Feore, Julie Khaner, Sarah Paulson, Megan Bouchard, Faith Prince, Krista Marie Bonura, Roger McKeen, Glen Gilbert, Carlo Rota, Ron Gabriel, Arturo Fresolone, Margaret Ozols, Ben Lin, Anne-Marie MacDonald
An ambitious newspaper man is so obsessed with his career that he sacrifices his marriage. Twelve years go by and then he and his former wife finally manage to become good friends. An episodic tale that details both the gradual disintegration of their relationship and how after their divorce they manage to find themselves again as friends. Some fine and touching performances may this one really enjoyable to watch despite the absence of a more solid plot.
DRA   92 min   VIDrel: ODY/SONOP   V/sh

FRIENDSHIPS, SECRETS AND LIES **

Ann Zane Shanks/Marlena Laird USA   12
                                                   1979
Cathryn Damon, Shelley Fabares, Sondra Locke, Tina Louise, Paula Prentiss, Loretta Swit, Stella Stevens, Fran Bennett, Cathee Shirriff, Kyle Richards, Mickey Hartnet, Elizabeth Farley, Krista Kaufman, Sarah Shelby, Estelle Omens
A baby's skeleton is discovered in a house that is being demolished, and the pathologist's report sets that date at 1957, a time when six women spent the summer there. A conventional

murder mystery whose all-female cast and female directors are its most notable features.

DRA 94 min mTV VIDrel: WHV V/h

FRIGHT NIGHT ***
18
Tom Holland USA
1985
William Ragsdale, Chris Sarandon, Roddy McDowall, Amanda Bearse, Stephen Geoffreys, Jonathan Stark, Dorothy Fielding, Art J. Evans, Stewart Stern, Nick Savage, Ernie Holmes, Prince A. Hughes, Heidi Sorensen, Irina Irvine

A young horror-film fan discovers that his new neighbour is a vampire but of course no-one will believe him. Former horror movie star McDowall comes to the rescue, but only after considerable persuasion. An unusually violent vampire film that hovers awkwardly between spoof and horror. The wildly overblown special effects are by Richard Edlund of GHOSTBUSTERS and the script is by Holland. An inferior sequel followed.

HOR 102 min (ort 105 min) VIDrel: VCC/DISC/COLUM V/sur

FRIGHT NIGHT 2 *
18
Tommy Lee Wallace USA
1988
Roddy McDowall, William Ragsdale, Traci Lin, Julie Carmen, Jonathan Gries, Russell Clark, Brian Thompson, Merritt Butrick, Ernie Sabella, Matt Landers, Josh Richman, Karen Anders, Rochelle Ashana, Blair Tefkin, Alexander Folk

A young man meets the girl of his dreams but finds that it is his blood she wants in this slack and disappointing sequel to the first film.

HOR 99 min (ort 108 min) VIDrel: 20TH/TECH L/A V

FRIGHTENED CITY ***
PG
John Lemont UK
1961
Herbert Lom, John Gregson, Sean Connery, Alfred Marks, Yvonne Romain, David Davies, Olive McFarland, Kenneth Griffith, George Pastell, Frederick Piper, Patrick Holt, Bruce Seton, Norrie Paramor

A realistic gangster film about a London dominated by protection rackets, with a master criminal attempting to amalgamate six gangs into a city-wide syndicate. A concise and tightly directed film.

THR 94 min (ort 97 min) B/W VIDrel: WHV V

FRIGHTENERS, THE **
18
Peter Jackson NEW ZEALAND/USA
1996
Michael J. Fox, Trini Alvarado, Peter Dobson, John Astin, Jeffrey Combs, Dee Wallace Stone, Jake Busey, Chi McBride, Jim Fyfe, Troy Evans, Julianna McCarthy, R. Lee Ermey, Elizabeth Hawthorne, Angela Bloomfield, Desmond Kelly, John Leigh

Fox is a ghostbuster who gets help from an unexpected quarter in the shape of three helpful if inept spooks. At the same time there is an evil presence hovering over the town, claiming the lives of a succession of folk, and Fox will have to work hard to prove that he is not involved in these attacks. A frenetic and strained fantasy-comedy, too scary for kids and not really substantial enough for adults, despite some highly competent special effects.

FAN 110 min CINrel

FRIGHTMARE *
18
Pete Walker UK
1974
Rupert Davies, Sheila Keith, Paul Greenwood, Deborah Fairfax, Kim Butcher, Fiona Curzon, Jon Yule, Tricia Mortimer, Pamela Farbrother, Edward Kalinski, Victor Winding, Anthony Hennessey, Noel Johnson, Michael Sharvell-Martin

The story of a couple of psychotic cannibals who go to live in an isolated farmhouse after their release from an asylum. Their young daughter soon joins with them in a number of murderous attacks, disposing of both her half-sister and a psychiatrist who came to investigate. A repulsive parade of gore with a barely discernible plot.

Aka: FRIGHTMARE 2; HORROR STAR; ONCE UPON A FRIGHTMARE

HOR 83 min (ort 87 min) VIDrel: REDEM/RTM V

FRISCO KID, THE **
PG
Robert Aldrich USA
1979
Gene Wilder, Harrison Ford, Ramon Bieri, Val Bisoglio, George Ralph DiCenzo, Leo Fuchs, Penny Peyser, William Smith, Jack Somack, Cliff Pellow, Allan Rich, Beege Barkett, Shay Duffin, Walter Janowitz, Joe Kapp, Clyde Kusatsu

A Polish rabbi travelling across the USA to San Francisco in the 1850s makes friends with an outlaw. An unsuccessful effort that is too episodic to work as a coherent comedy, but is best enjoyed as a series of mildly amusing vignettes instead.

COM 113 min (ort 122 min) VIDrel: WHV V/h

FRITZ THE CAT ****
18
Ralph Bakshi USA
1971
Voices of: Skip Hinnant, Rosetta Le Noire, John McCurry, Judy Engles, Phil Seuling

Inventive and comical story of an alley-cat student and his adventures in New York. All the parts are played by animals with different peoples being represented by different species (for example, the cops are all pigs). A rarity in that the cartoon was X-rated for its graphic sexuality. Largely based on the cartoons of Robert Crumb drawn for an underground magazine. The vastly inferior sequel THE NINE LIVES OF FRITZ THE CAT followed.

ANIM 78 min VIDrel: ARROW/RTM V

FROG PRINCE, THE ***
U
Jackson Hunsicker USA
1987
Aileen Quinn, John Paragon, Clive Revill, Helen Hunt, Seagull Cohen, Jeff Gurner, Eli Gorenstein, Shmuel Atzmon, Aaron Kaplan, Moshe Ish Cassit, Roni Blitz, Ya'acov Booch, Yasha Einstein, Ya'acov Halperin, Nahman Leor

A nice adaptation of a Brothers Grimm tale in which Quinn, as the little Princess Zora, drops a golden ball down a well, encountering a talking frog who promises to retrieve it if she will take him into her home as her companion.

JUV 83 min (ort 86 min) VIDrel: MGM/WHV L/A V/sh

Boa: short story by Jakob Ludwig Karl Grimm and Wilhelm Karl Grimm.

FROM A WHISPER TO A SCREAM **
18
Jeff Burr USA
1986
Vincent Price, Clu Gulager, Terry Kiser, Harry Caesar, Rosalind Cash, Susan Tyrrell, Cameron Mitchell, Martin Beswicke, Angelo Rossitto

Price's last horror movie in which he plays a small-town librarian who tells four tales of horror that took place there. The stories range from the weak to the repulsive and include the tale of a psychopath whose evil lust gives rise to a deadly offspring, and the story of a circus freak who pays a heavy price for his bizarre talent.

Aka: OFFSPRING, THE

HOR 93 min Cut (1 min 45 sec) VIDrel: MED/POLYREC L/A V

FROM BEYOND **
18
Stuart Gordon USA
1986
Jeffrey Combs, Barbara Crompton, Ken Foree, Ted Sorel, Carolyn Purdy-Gordon, Bunny Summers, Bruce McGuire, Del Russel, Dale Wyatt, Karen Christenfeld, Andy Miller, John Leamer, Regina Bleesz

A beautiful psychiatrist is called upon to investigate Dr Pretorious, a mad scientist who has created a device that activates a deadly sixth sense in humans, and is attacked by terrifying creatures from another dimension. The title (if nothing else) is taken from a short story by H.P. Lovecraft. Screenplay is by Dennis Paoli.

HOR 81 min Cut (10 sec – ort 90 min) VIDrel: VCC L/A V

FROM DUSK TILL DAWN *
18
Robert Rodriguez USA
1995
Harvey Keitel, George Clooney, Quentin Tarantino, Juliette Lewis, Cheech Marin, Fred Williamson, Salma Hayek, Marc Lawrence, Michael Parks, Tom Savini, Kelly Preston, John Saxon, Danny Trejo, Ernest Liu, John Hawkes

Armed robbers kidnap Keitel and his daughter and make their way to a bolt-hole in Mexico, where they confront a new danger in the form of vampires. Written by Tarantino (who plays one of the robbers) this is a combined crime film and vampire spoof, with the usual lashings of gore one has come to expect from a Tarantino script. Slickly directed, the over-use of violence quickly loses impact, and the film has no warmth to modify its blood-bath flavour.

A/AD 94 min (ort 108 min) cC VIDrel: HOLPIC/TECH V/sur

Boa: story by Robert Kurtzman.

FROM HERE TO ETERNITY ****
PG
Fred Zinnemann USA
1953
Burt Lancaster, Deborah Kerr, Montgomery Clift, Frank Sinatra, Donna Reed, Ernest Borgnine, Philip Ober, Mickey Shaughnessy, Jack Warden, Claude Akins, George Reeves, John Dennis, Tim Ryan, Barbara Morrison, Kristine Miller

Despite the loss of the novel's harsh criticism of the US Army and its more interesting passages, this remains an excellent

account of life at an army base on Hawaii just before the attack on Pearl Harbour. Brilliantly acted throughout, the film was remade for TV as a dreary six-hour mini-series in 1979. AA: Pic, Dir, S. Actor (Sinatra), S. Actress (Reed), Cin (Burnett Guffey), Screen (Daniel Taradash), Edit (William A. Lyon), Sound (John P. Livadary).
DRA   114 min (ort 118 min) B/W   VIDrel: COLUM/SONOP V
Boa: novel by James Jones.

FROM HERE TO ETERNITY: PARTS 1 AND 2 **
15
Buzz Kulik USA   1979
Natalie Wood, Kim Basinger, William Devane, Steven Railsback, Peter Boyle, Roy Thinnes, Joe Pantoliano, Will Sampson, Rick Hurst, Salome Jens, Andrew Robinson, David Spielberg, Richard Venture, Andy Griffith, Richard Bright
Condensed version of the six-hour mini series that is quite watchable but far inferior to the 1948 classic, with a considerable loss of dramatic impact on account of an almost obsessive desire to cover almost all the areas dealt with in the novel. A brief TV series (of even lower quality) followed.
DRA   280 min (2 cassettes – ort 360 min) mTV
VIDrel: 4-FRONT/POLYREC   V
Boa: novel by James Jones.

FROM NOON TILL THREE *
15
Frank D. Gilroy USA   1976
Charles Bronson, Jill Ireland, Douglas V. Fowley, Stan Haze, Bert Williams, Damon Douglas, Anne Ramsey, Hector Morales, Howard Brunner, William Lanteau, Betty Cole, Davis Roberts, Fred Franklyn, Sonny Jones, Hoke Howell
A strictly second-rate bank robber becomes a legend after he is mistakenly thought to have been gunned down, and a former girlfriend publishes a colourful but highly inaccurate account of his life and times. A strange, lame spoof that never really gets going. The script is by Gilroy.
WES   95 min (ort 99 min)   VIDrel: MGM/WHV L/A   V
Boa: novel by Frank D. Gilroy.

FROM RUSSIA WITH LOVE ***
PG
Terence Young UK   1963
Sean Connery, Robert Shaw, Pedro Armendariz, Daniela Bianchi, Lotte Lenya, Bernard Lee, Eunice Gayson, Walter Gotell, Francis De Wolff, Lois Maxwell, Nadja Regin, George Pastell, Alizia Gur, Martine Beswick, Vladek Sheybal
One of the best of the James Bond adventures, made before the gadgets and special effects had taken over completely. Agent 007 is sent to steal a coding machine from the Russkies and has plenty of narrow escapes. The music is by John Barry. Followed by GOLDFINGER.
A/AD   110 min (Cut at film release – ort 118 min) wScrn cC
VIDrel: MGM/WHV   V/dm
Boa: novel by Ian Fleming.

FROM THE FILES OF JOSEPH WAMBAUGH: A JURY OF ONE **
15
Alan Metzger USA   1992
John Spencer, Charlie Velez, Rachel Ticotin, Sal Lopez, Eugene Clark, Ronald William Lawrence, Red West, Dean Norris, Jon Van Ness, Dan Butten, Margot Rose, Dan Lauria, Egar Small, Harold Canno-Lopez, Glen Morshower, Irma Garcia
In L.A. a washed up detective starts an investigation into a series of gangland killings in this fairly typical detective story.
Aka: JURY OF ONE, A
A/AD   88 min (ort 90 min) mTV   VIDrel: 20VIS/SONOP   V

FROM THE HIP *
15
Bob Clark USA   1986
Judd Nelson, Elizabeth Perkins, John Hurt, Darren McGavin, Ray Walston, Dan Monahan, David Alan Grier, Nancy Marchand, Allan Arbus, Edward Winter, Royce D. Applegate, Richard Zobel, Robert Irvin Elliott, Beatrice Winde, Art Hindle
A young lawyer finds his unconventional approach to defence methods makes him an overnight media sensation, when he undertakes the defence of a cold but brilliant man accused of murder. An irritating film that tries hard to amuse but succeeds in being merely outrageous.
COM   106 min (ort 112 min)   VIDrel: 20TH/TECH L/A V/h

FRONT, THE ****
15
Martin Ritt USA   1976
Woody Allen, Zero Mostel, Michael Murphy, Herschel Bernardi,
Remak Ramsay, Andrea Marcovicci, Joshua Shelley, Lloyd Gough, David Margulies, Norman Rose, Danny Aiello, Scott McKay, Julie Garfield, Charles Kimbrough
During the McCarthy witch-hunt era of the 1950s, blacklisted writers use a bookmaker as a "front" to get their work published, but complications arise in this sharp and witty comedy, which makes a number of salient points. A highlight is a performance from Mostel as a blacklisted comedian fighting to get work. Script is by Walter Bernstein who was blacklisted in reality (as were Ritt, Mostel, Bernardi and Shelley). See also GUILTY BY SUSPICION.
COM   91 min (ort 94 min)   VIDrel: VCC/DISC/COLUM L/A   V

FRONT PAGE, THE **
(PG)
Billy Wilder USA   1974
Jack Lemmon, Walter Matthau, Vincent Gardenia, Susan Sarandon, Allen Garfield, David Wayne, Charles Durning, Austin Pendleton, Carol Burnett, Herbert Edelman, Martin Gabel, Harold Gould, Cliff Osmond, Dick O'Neill, Jon Korkes
A Chicago ace reporter about to get married and quit his much-loathed job is tricked into one last assignment by his calculating and totally unscrupulous editor. Sent to cover the execution of a convicted murderer, he is soon embroiled in endless complications. A strangely flat and disappointing remake of the 1931 film, which makes up in vulgarity what it lacks in real humour. See also HIS GIRL FRIDAY.
COM   105 min   TVrel
Boa: play by Ben Hecht and Charles C. MacArthur.

FRONT PAGE STORY ***
PG
Gordon Parry UK   1953
Jack Hawkins, Elizabeth Allan, Eva Bartok, Derek Farr, Martin Miller, Jenny Jones, Walter Fitzgerald, Joseph Tomelty, Michael Howard, Michael Goodliffe, Patricia Marmont, Helen Haye, Guy Middleton, Henry Mollison, Ronald Adam
The harassed news editor of a London daily tries to keep his private life and the demands of his job separate, but is so involved with his work that he fails to realise his neglected wife has embarked on an affair with a colleague. Various other incidents and human interest stories are examined, the whole giving a dated but entertaining look (perhaps not all that realistic) at Fleet Street in the 1950s. See also THE PAPER.
DRA   99 min B/W   VIDrel: LUMI/SPEAR   V
Boa: novel Final Night by Robert Gaines.

FRONTLINE, THE **
18
Paul Hills UK   1993
Vincent Philipps, Amanda Noar, Geoffrey Leesley, LeRoy Cooper
Bleak story of crime and drug addiction set in Manchester's notorious Moss Side district, where the former girlfriend of a DJ is murdered by an MP with nasty right-wing tendencies. Made on a shoestring budget by the young Hills (he was only twenty when he directed, wrote and produced) this is a film of energy and promise, albeit let down by the preposterous plot and sequences of variable technical quality.
DRA   69 min (ort 77 min)   VIDrel: SCEDGE/RTM   V

FROSTBITER: WRATH OF THE WENDIGO *
18
Tom Chaney/Rick Cioffi/Steve Quick USA   1990
Lori Baker, Patrick Butler, Ron Asheton, Devlin Burton, Tom Franks, Alan Madlane, John Bussard, David Wogh, John "Duke" Mietelka, Bill Siemers, Vicki Howard, Mike Missler, Bret Julyk, Irinia Dvorin, Joel Hale, Kay Davis, Matt Hale
A demonic creature is woken from its sleep by a group of hunters who have unwisely ventured into the depths of a remote island. It sets about killing off these chaps, until only one surviving member of the group remains. Shot on a shoestring on a island in North Michigan, this yawn-inducing effort has very little going for it. The script is by Chaney.
HOR   84 min   VIDrel: TROMA/RTM   V

FROZEN ASSETS *
15
George Miller USA   1993
Shelley Long, Corbin Bernsen, Larry Miller, Dody Goodman, Gerrit Graham, Paul Sand, Teri Copley, Matt Clark, Jeanne Cooper, Gloria Camden, Jennifer Lewis, John Mallory Asher, John Bloom, Sara Ballantine, Gary Basey, Vicky Vose
A young and upwardly mobile corporate executive is put in charge of his employers' latest addition to their commercial empire, a bank. More than a little taken aback to learn that it is a sperm bank, located in a small rural town. Not one to let the grass grow under his feet, he announces a competition for the

most virile donor, with a prize of $1,000,000. A dumb and taste-less comedy that is miserably unfunny.
COM   97 min   VIDrel: 20VIS/SONOP   V/sur

FRUIT MACHINE, THE *   15
Philip Saville UK   1987
Emile Charles, Tony Forsyth, Robert Stephens, Robbie Coltrane, Clare Higgins, Bruce Payne, Carsten Norgaard
Two sixteen-year-old boys – one a streetwise "rent boy" and the other a naive and confused daydreamer – are drawn together by their mutual distrust of society and take to frequenting a local nightclub run by a transvestite. When they witness his murder they flee to Brighton but soon find the killers not far behind. A strange, surreal, but not altogether successful effort, that makes a few interesting observations but suffers from a self-indulgent script.
THR   103 min   VIDrel: FIRST/SONOP   V/sur

F.T.W. *   18
Michael Karbelnikff USA   1994
Mickey Rourke, Lori Singer, Brion James, Rodney A. Grant, Aaron Neville, Charlie Sexton, Peter Berg, Mark Pellegrino
An ex-convict tries to go straight and returns to the rodeo circuit, but gets involved with a young woman who is on the run after taking part in a bank robbery that went wrong. Romantic action film, with a few thrills and a fairly simpleminded plot.
Aka: LAST RIDE, THE; MONTANA
A/AD   97 min (ort 102 min)   VIDrel: MED/COLUM   V/sh

FUDGE-A-MANIA *   U
Bob Clark   1995
Luke Tarsitano, Eve Plumb, Darren McGavin, Florence Henderson
A full-length feature spin-off from a TV series, with Peter and his young brother Fudge going on vacation, but finding that they are to share a house with the Tubman family. Good fun for kids
JUV   90 min mTV   VIDrel: CIC/SONOP   V

FUGITIVE, THE *   15
Andrew Davis USA   1993
Harrison Ford, Tommy Lee Jones, Sela Ward, Jeroen Krabbe, Joe Pantoliano, Andreas Katsulas, Julianne Moore, Daniel Roebuck, L. Scott Caldwell, Tom Wood, Rich Dean, Joseph Kosala, John Drummond, Tony Fosco, Joseph F. Fisher
A doctor, tried and convicted for murdering his wife escapes, through a freak accident and mounts a desperate hunt for the real killer, a man with an artificial arm. A modern version of the classic TV series but done in the style of a 1990s action movie – all special effects but no drama or warmth. Ford as Dr Kimble gives his usual cold-fish performance, while the cops are portrayed as cynical and ruthless. Very disappointing. AA: S. Actor (Jones).
Aka: FUGITIVE, THE: THE MOVIE
THR   125 min (ort 131 min) cC   VIDrel: WHV   V/sur

FUGITIVE AMONG US *   15
Michael Toshiyuki Uno USA   1992
Peter Strauss, Eric Roberts, Elizabeth Pena, Guy Boyd, Lauren Holly, Salvator Xuereb, Kiersten Warren, Annette McCarthy, Dennis Letts, Tyress Allen, Norman Bennett, Aldo Billingslea, Rodger Boyce, Mimi Cochran
Inspired by real events, this taut drama revolves around the obsessive desire of a cop to trap a rapist, and his inability to accept the fact that he may have got the wrong man.
DRA   95 min mTV   VIDrel: ODY/SONOP   V/sh
Boa: chapter "Fugitives" from the book And Deliver Us From Evil by Mike Cochran.

FUGITIVE FAMILY *   (PG)
Paul Krasny USA   1980
Richard Crenna, Diane Baker, Eli Wallach, Don Murray, Ronny Cox, Mel Ferrer, Robin Dearden, K.C. Martel, William Kirby Cullen, Paul Mantee, Felice Orlandi, Sidney Clute, Judy Farrell, Bobby Rolofson, Ken Hill, Burt Marshall
A government witness is forced to go into hiding with his family and take on a new identity after testifying at the trial of a mobster. Predictable and tame crime melodrama.
DRA   96 min (ort 100 min) mTV   SATrel: SKY MOVIES
Boa: story by James G. Hirsch.

FUGITIVE FROM JUSTICE: UNDERGROUND FATHER *   15
Chuck Bowman USA   1996
Stepfanie Kramer, Natalie Cole, Peter MacNicol, Loryn Locklin, Megan Gallagher, Daniel Roebuck, Teryl Rothery

A man who makes the discovery that his estranged wife is abusing their eight-month-old daughter snatches the child and goes into hiding.
THR   93 min mTV   VIDrel: ODY/SONOP   V/sh

FUGITIVE, THE: THE JUDGEMENT: PARTS 1 AND 2 *   PG
Don Medford USA   1966
David Janssen, Barry Morse
The conclusion to a long-running TV series in which Janssen played a man forced to go on the run when he is suspected of the murder of his wife. In this episode all the loose ends are neatly tied up and the "one-armed man" is finally located. Average. All the episodes in this series have become available.
DRA   103 min (2-episode cassette) mTV
VIDrel: SCRN/DISC L/A   V

FULL CONTACT *   18
Rick Jacobson USA   1992
Jerry Trimble, Howard Jackson, Alvin Prouder, Gerry Blanck, Denise Buick, Marcus Aurelius, Raymond Storti, Joe Charles, Manuel luben
A tough country boy from Fresno seeks revenge for his brother's death and goes looking for it in L.A., in the underground world of kickboxing. Average.
MAR   92 min (ort 97 min)   VIDrel: COLUM/SONOP   V

FULL ECLIPSE *   18
Anthony Hickox USA   1993
Mario Van Peebles, Patsy Kensit, Anthony John Denison, Jason Beghe, Paula Marshall, John Verea, Dean Norris, Willie C. Carpenter, Victoria Rowell, Scott Paulin, Mel Winkler, Joseph Culp, Joey De Pinto, Bruce Payne
After his partner is killed, an L.A. cop allows himself to be recruited for service in a special crime-fighting unit. It soon transpires that all its members are injected with a serum that turns them into ferocious werewolves that prey on criminals. An unusual variant on the werewolf theme with one or two original ideas.
FAN   93 min   VIDrel: MED/DISC   V/sh

FULL METAL JACKET *   18
Stanley Kubrick USA   1987
Matthew Modine, Adam Baldwin, Vincent D'Onofrio, Lee Ermey, Dorian Harewood, Arliss Howard, Kevyn Major Howard, Ed O'Ross, John Stafford, John Terry, Kirk Taylor, Ian Tyler, Keiron Jecchinis
A story following a Marine private, from his tough training through to his involvement in the Vietnam War and the heavy fighting of the 1968 Tet offensive. Kubrick's first film for seven years was shot entirely on location in South London and the soft and unmistakably English light coupled with the unreality of the sets does not make for great realism. However, as a stylised statement on the meaninglessness of war it largely succeeds.
WAR   112 min (ort 116 min)   VIDrel: WHV   V
Boa: novel The Short Timers by Gustav Hasford.

FULL MOON IN BLUE WATER *   15
Peter Masterson USA   1988
Gene Hackman, Teri Garr, Burgess Meredith, Elias Koteas, Kevin Cooney, David Doty, Gil Glasgow, Becky Gelke, Marietta Marich, Lexie Masterson, William Larsen, Mitchell Gossett, Mark Walters, Ben Jones, Ed Geldart
The owner of a seedy bar in the South is obsessed with locating his wife, who vanished a year ago. He spends most of his time watching home movies whilst his girlfriend tries to keep the business running. When a crooked estate agent moves in and tries to buy him out, an intriguing mystery slowly develops. The film's leisurely pace and engaging performances are compensations for its weak and sentimental script.
DRA   91 min (ort 96 min)   VIDrel: EIV/SONOP L/A   V

FULL MOON IN PARIS *   15
Eric Rohmer FRANCE   1984
Pascale Ogier, Tcheky Karyo, Fabrice Luchini, Christian Vadim, Virginie Thevenet
In Paris, a young textile designer has a job in the city and an apartment in the suburbs, where she lives with a boyfriend with whom she feels stifled. So she rents an apartment in town and embarks on a couple of affairs. A detailed film that captures much of the essence of everyday life, it is appealing rather than profound. Ogier (the daughter of actress Bulle Ogier) died

prematurely aged 24 in 1984. One of Rohmer's "Comedies and Proverbs" films. See THE GREEN RAY.
Aka: LES NUITS DE LA PLEINE LUNE
DRA   97 min (ort 102 min)   VIDrel: ARTIF/20TH   V

FUN *   18
Rafal Zielinski CANADA/USA   1994
Renee Humphrey, Alicia Witt, William R. Moses, Leslie Hope, Ania Suli, James J. Howard Jr, Frederick D. Adams, Mary Ann Norment, Sabrina Ortega, Patrice F. Battle, Carmina Rubalcava, Malquele Garcia, Cindie Northrup, Denise Fischer
Two teenage girls meet at a bus stop and set off on a bout of criminal activities, such as killing an old lady for kicks. Much of the film is devoted to a discussion of their propensities with the prison psychiatrists (done as B/W footage) interspersed with flashback sequences. However, we learn next to nothing about them, nor develop any interest in doing so. Scripted by Bosley, and given the subject matter, the film is probably destined for cult status.
DRA   105 min wScrn Col/B/W   VIDrel: TART/20TH   V
Boa: play by James Bosley.

FUNERAL, THE ***   18
Abel Ferrara USA   1996
Christopher Walken, Chris Penn, Vincent Gallo, Annabella Sciorra, Benicio Del Toro, Isabella Rossellini, Gretchen Mol, John Ventimiglia, Paul Hipp
Almost a return to the old-style gangster epics of yesterday, this one has various unsavoury fellows attending the funeral of one Johnny Tempio, the youngster of three brothers whose penchant for violence was all but uncontrollable. Having seen via a series of flashbacks just how he met his end, the two remaining brothers set about planning a suitable revenge. Intricate and involved, the movie's great strength is the believability of the characters.
A/AD   99 min   CINrel

FUNERAL IN BERLIN ***   PG
Guy Hamilton USA   1967
Michael Caine, Hugh Burden, Oscar Homolka, Eva Renzi, Rachel Gurney, Guy Doleman, Thomas Holtzman, Gunter Meisner, Heinz Schibert, Wolfgang Volz, Klaus Jepsen, Herbert Fux, Rainer Brandt, Ira Hagen, Marthe Keller, Uschi Heyer
Second outing for cockney secret agent Harry Palmer following THE IPCRESS FILE, with Caine being sent to Berlin to oversee arrangements for the defection of a Russian colonel in charge of Berlin war security. Sluggish and over complex, with a good many superfluous twists to a plot that never really comes to life. Followed by BILLION DOLLAR BRAIN, the last film in the series.
Aka: HARRY PALMER RETURNS
A/AD   100 min (ort 102 min)   VIDrel: CIC/SONOP   V
Boa: novel The Berlin Memorandum by Len Deighton.

FUNNY ABOUT LOVE *   15
Leonard Nimoy USA   1990
Gene Wilder, Christine Lahti, Mary Stuart Masterson, Robert Prosky, Stephen Tobolowsky, Anne Jackson, Susan Ruttan, Jean De Baer, David Margulies, Tara Shannon, Freda Fon Shen, Wendie Maclick, Robert Gorman, Scott Goff
A couple are desperate to have a child, and try everything their fertility doctor can suggest, but to no avail. Events become more complex when Wilder takes up with a female college student. A weakly amusing comedy that delivers minute doses of humour in dribs and drabs, it largely wastes the talents of a strong cast.
COM   100 min (ort 101 min)   VIDrel: CIC/SONOP L/A V/h
Boa: article by Bob Greene.

FUNNY BONES **   15
Peter Chelsom UK/USA   1994
Oliver Platt, Jerry Lewis, Lee Evans, Leslie Caron, Richard Griffiths, Oliver Reed, George Carl, Freddie Davis, Ian McNeice, Christopher Greet, Peter Gunn, Gavin Miller, William Hootkins, Terence Rigby, Ruta Lee, Ticky Holgado
After his comedy act bombs at Las Vegas, the son of a celebrated comedian travels to the British seaside resort of Blackpool in the hope of learning exactly what makes people laugh. A distinctly strange and unstructured movie that never manages to draw its disparate elements into one coherent whole. Evans (as the British comic who offers some much-needed help) is however, a joy.
COM   122 min (ort 128 min) cC   VIDrel: BUENA   V/sh

FUNNY FACE **   U
Stanley Donen USA   1956
Fred Astaire, Audrey Hepburn, Robert Flymyng, Kay Thompson, Michael Auclair, Suzy Parker, Ruta Lee, Dovima, Virginia Gibson, Sunny Harnett, Don Powell, Sue England, Carole Eastman, Alex Gerry, Ipjhigenie Castiglioni, Jean Del Val
A fashion photographer falls in love with a brainy bookshop clerk, or more accurately, her funny face, and turns her into a top Paris model. An enjoyable if hardly top-notch musical; despite the presence of one or two good Gershwin numbers ("He Loves And She Loves", "How Long Has This Been Going On") the film is a little too glossy and stylised for its own good. Astaire's role is loosely based on the life and work of photographer Richard Avedon.
MUS   103 min   VIDrel: CIC/SONOP   V/dm

FUNNY FARM ***   PG
George Roy Hill USA   1988
Chevy Chase, Madolyn Smith, Joseph Maher, Jack Gilpin, Brad Sullivan, MacIntyre Dixon
A New York sportswriter and his wife take up the country life but they find things far from simple in this episodic, uneven but engaging comedy. A film that tends to get better as it goes along, being helped considerably by the photography of Miroslav Ondricek.
COM   97 min Cut (3 sec – ort 101 min)   VIDrel: WHV V/sur

FUNNY GIRL ***   U
William Wyler USA   1968
Barbra Streisand, Omar Sharif, Kay Medford, Anne Francis, Walter Pidgeon, Lee Allen, Gerald Mohr, Frank Faylen, Mae Questel, Gertrude Flynn, Penny Santon, John Harmon, Mittie Lawrence, Thordis Brandt, Bettina Brenna, Virginia Ann Ford,
Streisand is a knockout in this long, lively biopic devoted to the career of comedienne Fanny Brice. The musical numbers are great but the drama tends to drag, especially the account of her troubled first marriage to gambler Nicky Arnstein (Sharif gloriously miscast). Bad as biography, but great as a musical; the score is by Bob Merrill and Jule Styne. Followed by FUNNY LADY. AA: Actress (Streisand – she tied with Hepburn for THE LION IN WINTER).
MUS   141 min (ort 151 min)   VIDrel: VCC/DISC/COLUM V/sh
Boa: play by Isobel Lennart.

FUNNY LADY **   PG
Herbert Ross USA   1974
Barbra Streisand, James Caan, Roddy McDowall, Omar Sharif, Ben Vereen, Larry Gates, Carole Wells, Heidi O'Rourke, Samantha Huffaker, Matt Emery, Joshua Shelley, Corey Fischer, Garrett Lewis, Don Torres, Raymond Guth, Gene Troobnick
This sequel to FUNNY GIRL has Fanny Brice at the height of her career, meeting and marrying ambitious showman Billy Rose. Unfortunately, tiresome bouts of domestic turmoil get in the way of the good moments, these latter being mainly the highly effective musical set-pieces. An uneven, cliched effort that fails to recapture the sparkle of the earlier film.
MUS   132 min (ort 137 min)   VIDrel: VCC/DISC/COLUM V/sh

FUNNY THING HAPPENED ON THE WAY TO THE FORUM, A **   PG
Peter Lester USA   1966
Zero Mostel, Phil Silvers, Jack Gilford, Buster Keaton, Michael Crawford, Michael Hordern, Annette Andre, Patricia Jessel, Leon Greene, Inga Neilsen, Beatrix Lehmann, Myrna White, Pamela Brown, Jennifer Baker, Susan Baker
An over-excited adaptation of a bawdy Broadway musical set in ancient Rome and telling of a cunning slave who connives to win his freedom. Fast and furious but the film tries too hard to be funny, though the slapstick finale is a highlight. Music and lyrics are by Stephen Sondheim. AA: Score/adapt (Ken Thorne).
COM   93 min (ort 99 min)   VIDrel: WHV   V
Boa: musical comedy by Bert Shevelove and Larry Gelbart.

FUNNYMAN **   18
Simon Sprackling UK   1994
Tim James, Benny Young, Matthew Devitt, Pauline Black, Christopher Lee, Ingrid Lacey, Chris Walker, George Morton, Rhona Cameron, Harry Heard, Jamie Heard, Bob Sessions, Ed Bishop, John Chancer, Jana Sheldon, Barnaby North
A self-satisfied record producer wins the ancestral home of a

member of the English aristocracy in a poker game, but fails to realise that it is a shrine to supernatural forces and an evil creature who does not like newcomers. He is soon hard at work killing various individuals in a succession of gruesome ways, and though the film is weak on plotting, it delivers enough chills to satisfy devotees of this genre.
Aka: FUNNY MAN
HOR   89 min (ort 93 min) wScrn   VIDrel: ENCORE/SPEAR;
ENCORE (LV only)   V/sur LV

FURIOUS, THE ** (18)
Joseph Velasco HONG KONG   1981
Bruce Le (Huang Kin Lung), Chan Wei-Min, Lo Lieh, Ku Fung, Huang Ka Tat
A band of freelance martial arts fighters are recruited in order to crush an international drug-smuggling syndicate. A watchable effort with plenty of action if little in the way of a plot.
MAR   90 min   SATrel: SKY MOVIES

FURY, THE * 18
Brian De Palma USA   1978
Kirk Douglas, John Cassavetes, Carrie Snodgress, Amy Irving, Fiona Lewis, Andrew Stevens, Charles Durning, Gordon Jump, Daryl Hannah, Carl Rossen, Joyce Easton, William Finley, Jane Lambert, Sam Laws, Melody Thomas
The head of a government unit tries to prevent his son (who has psychic powers) from being kidnapped by terrorists who want to make use of him. A violent and gory film, quite well made in its way but trashy and painfully contrived. This was Hannah's film debut. The script is by Farris.
A/AD   113 min (ort 118 min)   VIDrel: 20TH/TECH L/A   V
Boa: novel by John Farris.

FURY IN SHAOLIN TEMPLE ** 15
Godfrey Ho HONG KONG   198-
Liu Chia Hui
A kung fu master catches a thief stealing secrets from his temple but takes him on as a student instead of punishing him.
Aka: JAMI SONTEE
MAR   83 min (ort 89 min)   VIDrel: IMPENT   V

FURY OF THE DRAGON ** PG
William Beaudine USA   1966
Bruce Lee, Van Williams
The Green Hornet is a crime-buster by night and a publisher by day. Together with his bodyguards (proof-readers by day?), he fights urban crime. A silly but endearing piece of nonsense, compiled from episodes of a TV series and mainly of interest for showing Bruce Lee in an early role.
Aka: GREEN HORNET, THE
MAR   75 min (ort 93 min) mTV   VIDrel: MERLIN/SPEAR   V

FUTUREKICK ** 18
Damian Klaus USA   1991
Don "The Dragon " Wilson, Meg Foster, Christopher Penn, Eb Lottimer, Linda Dona, Al Ruscio, Jeff Pomerantz, Shaun Phillips, Ryan MacDonald, Dana Lee, Hayden Conner, Joe Mays, William Utay, Fred Scott, Brenda Bolte, Loyda Ramos
Kickboxing actioner set in a dark and uncertain future, with a cyborg bounty hunter battling a gang who are killing people in order to sell their body parts. Despite its SF setting, this is little more than another kickboxing epic.
A/AD   76 min (ort 90 min)   VIDrel: CIC/SONOP L/A   V

FUTUREWORLD *** PG
Richard T. Heffron USA   1976
Peter Fonda, Blythe Danner, Arthur Hill, Yul Brynner, Stuart Margolin, John Ryan, Jim Antonio, Robert Cornthwaite, Angela Greene, Nancy Bell, John Fujioka, Dana Lee, Darrell Larson, Burt Conroy, Dorothy Konrad, Alex Rodine
A sequel to WESTWORLD. The robot complex is repaired but there is a sinister purpose behind it with visitors to the complex being exchanged for robot duplicates in a nasty plan for world domination. A kind of robotic answer to INVASION OF THE BODYSNATCHERS.
FAN   104 min   VIDrel: VCC L/A   V

FUZZ THE HERO ** PG
Craig Clyde USA   1991
Raeanin Simpson, Katherine Willis, Keith Christensen, Reta Patterson, Craig Clyde, Lance Johnson, Jessica Fesh, Bill Green, Carly Fullmer, Peter Hanlon, Kim Stucki, Jana Fillmore, Kay Lewis, Janet Turley, John Klint

Family adventure that sees a young girl finding that her snobbish neighbours do not want to know her as she is poor. However, she finds herself with a loyal friend in the shape of her dog "Fuzz", and together they come to the rescue of a grumpy farmer. A solid and quite pleasing story.
Aka: LITTLE HEROES
JUV   78 min   VIDrel: NEWAGE/COLUM   V

FX: MURDER BY ILLUSION *** 15
Robert Mandel USA   1986
Bryan Brown, Brian Dennehy, Diane Venora, Cliff De Young, Mason Adams, Jerry Orbach, Joe Grifasi, Martha Gehman, Roscoe Orman, Trey Wilson, Tom Noonan, Paul D'Amato, Jossie De Guzman, Jean De Baer, Tim Gallin, Patrick Stack
An expert who creates special effects for films is contacted by the Justice Department to fake the killing of a gangster who has turned squealer. He discovers that he is to be the scapegoat in a murder conspiracy and has to use his specialised skills to save his life. A modest and entertaining thriller whose unusual premise gives rise to some clever moments, but one that is let down by innumerable loose ends and a contrived ending. A sequel followed.
Aka: F/X
THR   104 min (ort 107 min)   VIDrel: COLUM/SONOP
V/sur

FX2: THE DEADLY ART OF ILLUSION ** 15
Richard Franklin USA   1991
Bryan Brown, Brian Dennehy, Rachel Ticotin, Joanna Gleason, Philip Bosco, Kevin J. O'Connor, Tom Mason, Dominic Zamprogna, Josie DeGuzman, John Walsh, James Stacy, Peter Boretski, John Walsh, Lisa Fallon, Lee Broker, Ross Petty
A movie special-effects expert agrees to help the police crack a difficult case, but when things don't go as planned he finds that he is to be the fall-guy in a deadly cover-up. He turns for help to ex-cop and friend Dennehy. A flaccid follow-up to the first film that capitalises on its gimmicks at the expense of the story, which never holds the attention and is flawed by an absurd subplot that adds nothing of value.
Aka: F/X2
THR   103 min (ort 109 min)   VIDrel: VCC/DISC/COLUM
V/sur

G

GABBEH *** (15)
Mohsen Makhmalbaf FRANCE/IRAN   1995
Abbas Sayahi, Shaghayegh Djodat, Hossein Moharami, Rogheih Moharami, Parvaneh Ghalandari, Hassen Kermi, Zineb Kermi, Zahra Kermi, Fatema Kermi, Abreda Kermi, Roustou Kermi, Abdullah Djahaniour, Tahmineh Djahaniour
A poetic and visually overwhelming tale of a young woman, with title name, which is also that of a handwoven carpet. Born into a nomadic tribe of carpet weavers, she meets an old woman washing her carpet in the river and starts to tell her of her life. When her father forbids her to marry the man she loves, her lover refuses to accept this and follow her tribe during its migrations. A colourful and striking movie, with an oblique and non-linear narrative style.
DRA   74 min   CINrel

GABRIELLE AND THE DOODLEMAN ** (U)
Francis Essex UK   1984
Matthew Kelly, Eric Sykes, Windsor Davies, Lynsey De Paul, Gareth Hunt, Josephine Tewson, Bob Todd, Prudence Oliver, Pierre Picton
A little girl who is paraplegic lives with her father, who works as a computer graphics animator. One day a character he created to amuse her comes to life, jumps off the computer screen, and takes the girl on a series of magical adventures. An amiable fantasy, quite inventive and pleasing.
FAN   55 min (ort 58 min)   SATrel: MOVIE CHANNEL

GABY: A TRUE STORY *** 15
Luis Mandoki USA   1987
Rachel Levin, Liv Ullmann, Norma Aleandro, Robert Loggia, Lawrence Monoson, Robert Beltran, Beatriz Sheridan, Tony Goldwin, Danny De La Paz, Paulina Gomez, Enrique Lucero, Eduardo Lopez Rojas, Ana Ofelia Murguia
The story of Gaby Brimmer, a woman born with cerebral palsy to wealthy refugee parents in Mexico. Though almost completely paralysed, she overcame this handicap, eventually becoming a successful author after she was taught to commu-

nicate using her left foot. Levin gives a remarkable performance in an often harrowing but moving story of persistence and courage. Brimmer was the executive producer. See also MY LEFT FOOT.
DRA   110 min   VIDrel: L/A   V

GALAHAD OF EVEREST *   PG
John-Paul Davidson UK   1991
Brian Blessed
Blessed gives a sterling performance as George Mallory, in this effective recreation of the climber's ill-fated ascent of Everest in 1924. A highlight is the Blessed's reading of Mallory's last letters at the summit.
DRA   86 min (ort 90 min) mTV   VIDrel: TART/20TH
V/dm V/sur

GALLANT HOURS, THE *   PG
Robert Montgomery USA   1960
James Cagney, Dennis Weaver, Ward Costello, Richard Jaeckel, Carl Benton Reid
Cagney gives a nice performance in this low-key, documentary-style biopic on Admiral Frederick Halsey Jr, one of the key naval commanders in the Pacific during WW2. A solid if uninspired war film with a few, scattered moments of tension.
WAR   111 min B/W   VIDrel: WHV   V

GALLAVANTS **   U
Art Vitello USA   1983
Voices of: Fred Travelena, Charlie Callas, Barry Gordon, Bob Lydiard, Vic Perrin, Peter Cullen, Joyce Gittlin, Frank Welker, Diane Pershing, Jane Hamilton, B.J. Ward, Wendy Hoffman, Fred McGrath, Ken Sansom
This story of a colony of ants and their daily problems. A rather sugary affair, inspired by a book of the same name, and not helped by the lack of a proper plot and indifferent animations. The interminable and excruciatingly contrived wordplay (involving words incorporating the syllable "ant") has been retained and soon becomes mightily tiresome. Poor fare, indeed. Music is by Stan Wietrzychowski with lyrics by Don Smith.
ANIM   95 min (ort 100 min)   VIDrel: FABFIL/SPEAR
V/sur

GALLIPOLI *   PG
Peter Weir AUSTRALIA   1981
Mark Lee, Mel Gibson, Bill Hunter, Bill Kerr, Robert Grubb, David Argue, Tim McKenzie, Ron Graham, Charles Yunupingu, Heath Harris, Gerda Nicolson, Harold Hopkins, Reg Evans, Jack Giddy, Dane Peterson, Paul Linkson, Jenny Lovell
The story of two young boys who join up to fight in WW1 and become part of the ill-fated Gallipoli landing. An absorbing drama that pays great attention to detail without ever swamping the human element.
WAR   107 min (ort 110 min) wScrn   VIDrel: CIC/SONOP
V/h

GAMBLE, THE *   15
Carlo Vanzina ITALY   1987
Matthew Modine, Faye Dunaway, Jennifer Beals, Corinne Clery, Federica Moro, Ana Obregon, Vernon Wells, Feodor Chaliapin, Gianfranco Bavra, Karina Huff, Cyrus Elias, Marco Stefanelli, Claudia Lawrence, Nazareno Natale, Ian Bannen
A young aristocrat returns home to Venice and finds that his family has gambled away their fortune to a crafty German countess, who proposes one final wager when she meets him, in which he is to be the prize. Competent action comedy set in the 18th century.
Aka: LA PARTITA
COM   100 min (ort 108 min)   VIDrel: MARQ/QUANT   V
Boa: novel by Alberto Ongaro.

GAMBLING MAN, THE *   12
UK   1994
Robson Green, Bernard Hill
A cocky rent-collector is not averse to earning a little spare cash on the side at card games, but comes unstuck when he plays against a trio of thuggish brothers, who are not prepared to allow him to leave with all his winnings. However, even after they have badly injured him, they carry on a vicious campaign of intimidation, culminating in arson and attempted murder.
DRA   149 min (ort 152 min) mTV   VIDrel: FOCUS/DISC   V
Boa: novel by Catherine Cookson.

GAME OF DEATH *   18
Robert Clouse HONG KONG   1979
Bruce Lee, Kim Tai Jong, Gig Young, Hugh O'Brian, Dean Jagger, Chuck Norris, Colleen Camp, Kareem Abdul-Jabbar, Danny Inosanto, Mel Novak, Robert Wall, Billy McGill, Hung Kim Po, Roy Chiao
Released six years after Lee's death and completed using a double, this is a standard fists and feet kung fu movie, in which a young actor fakes his own death to outwit a gang threatening his career. Fairly mundane until the explosive final half hour. See also TRUE GAME OF DEATH.
Aka: BRUCE LEE: GAME OF DEATH
MAR   92 min Cut (2 sec – ort 102 min)
VIDrel: 4-FRONT/POLYREC   V

GAME OF DEATH 2 *   18
Ng See-Yuan HONG KONG   1981
Bruce Lee (Hah), Tang Lung (Kim Tai Ching), Huang Chen-Li, Cassanova Wong, Miranda Austia, Mun Ping, Lung Fei, Kuslai, Sandus
Exploitative martial arts tale built around some out-takes of Bruce Lee from ENTER THE DRAGON. A fighter is suspicious of the death of a friend and at his funeral meets his own death when he tries to prevent a helicopter from carrying off the coffin. These bizarre events send the fighter's brother out to confront those responsible, leading to an exciting climax when he fights an evil drug baron. Good action, poor plot.
Aka: NEW GAME OF DEATH, THE; TOWER OF DEATH
MAR   92 min (ort 96 min)   VIDrel: 4-FRONT/POLYREC   V

GAMES OF DESIRE *   18
Pasquale Fanetti   1988
Malu, Branki Djuric, Lidia Zovkic, Negic Slobadan, Izvdin Bajrovic
An impotent husband is powerless to prevent his sex-starved wife seeking satisfaction elsewhere, and her unbridled passions draw her into a series of bizarre encounters. A glossy but extremely dull effort; a few nods in the direction of Guy de Maupassant give it something of an arty flavour.
A   80 min (ort 95 min)   VIDrel: FABFIL/SPEAR   V
Boa: novel Florentine by Guy de Maupassant.

GANDHI **   PG
Richard Attenborough UK   1982
Ben Kingsley, Candice Bergen, Edward Fox, John Gielgud, Trevor Howard, John Mills, Martin Sheen, Ian Charleson, Athol Fugard, Gunter Maria Halmer, Geraldine James, Amrish Puri, Saeed Jaffrey, Alyque Padamsee, Roshan Seth
Biopic on the life of the man who led India's campaign for independence. Though inclined to sluggishness at times, this excellent recreation is carried along by Kingsley's truly remarkable portrayal. The script is by John Briley. AA: Pic, Dir, Actor (Kingsley), Screen/orig (John Briley), Cin (Billy Williams/Ronnie Taylor), Art/Set (Stuart Craig and Bob Laing/Michael Seirton), Cost (John Mollo/Bhanu Athaiya), Edit (John Bloom).
DRA   182 min (ort 188 min) wScrn (COLUM/SONOP)
VIDrel: VCC/DISC; COLUM/SONOP   V/sur

GANGSTER WARS *   15
Richard C. Sarafian USA   1981
Michael Nouri, Brian Benben, Joe Penny, Markie Post, Richard Castellano, George DiCenzo, Alan Arbus, Madeline Stowe
The first part of a compilation film, condensed from episodes of "The Gangster Chronicles", a saga telling of the rise to power of gangsters Lucky Luciano, Bugsy Siegel and Co. The original TV series was shown in thirteen 50-minute episodes; this detailed account was trimmed considerably from that ponderous original, but shows little improvement.
DRA   116 min (ort 121 min) mTV   VIDrel: CIC/SONOP   V

GANGSTER WARS 2 *   15
Richard C. Sarafian USA   1981
Michael Nouri, Brian Benben, Joe Penny
A continuation of the story of GANGSTER WARS, that's really a condensed version of a much longer TV series. This tale largely revolves around the efforts made by various gangs to control the 1930s trade in illicit alcohol.
DRA   86 min (ort 100 min) mTV   VIDrel: CIC/SONOP   V

GANGWAY *   U
Sonnie Hale UK   1937
Jessie Matthews, Barry Mackay, Olive Blakeney, Liane Ordeyne, Nat Pendleton, Patrick Ludlow, Noel Madison, Alastair Sim, Doris Rogers, Laurence Anderson, Blake Dorn, Graham Moffatt, Peter Gawthorne, Henry Hallatt, Warren Jenkins

A woman reporter suspected of being an international jewel thief inadvertently hides out aboard a liner to New York and gets herself kidnapped by gangsters. Rather less music and dancing than in other Matthews vehicles although the star gives a good comic performance, but hampered by the feeble plot, the film is little more than a set of colourful vignettes.
MUS   88 min (ort 89 min) B/W   VIDrel: CARL/TECH   V

GARAGE GIRLS *   *18*
Robert McCallum (Robert Neeallum) USA   1980
Georgina Spelvin, John Leslie, Lisa De Leeuw, John Seeman, Jon Martin, Dorothy Le May, Chris Cassidy, Brooke West, Susanne Nero, Dewey Alexander
Bawdy sex comedy about four girls who open a garage and provide a few extra services not available from their competitors. A crazy and unashamedly coarse tale, with poolroom encounters, attempts to blow up the garage, a Bonnie and Clyde-style car chase and a hilarious sequence detailing the exploits of a bunch of sex-crazed Youth Campers.
A   50 min (ort 81 min)   VIDrel: MOPIC/SGSVID   V

GARDEN, THE *   *(U)*
Will Dixon CANADA   1990
Jan Rubes, Scott Bremner, Benjamin Woolf, Erin Reesor, Lee Henderson, Jason Ward, Jonah Boyer, Ryan Fullerton, Gerard Lenton-Young, June Mayhew, Alan Bratt, Michael Scholar, Bill Dixon
A group of kids somehow get it into their heads that an old man who lives alone, is really a vampire. They start spying on him, but are pleasantly surprised when they learn the truth and eventually become good friends.
JUV   48 min   SATrel: MOVIE CHANNEL

GARDEN, THE *   *15*
Derek Jarman UK   1990
Kevin Collins, Roger Cook, Jody Graber, Spencer Leigh, Pete Lee-Wilson, Tilda Swinton, Johnny Mills, Jessica Martin, Philip MacDonald, Dawn Archibald, Michael Gough (voice-overs)
A parade of dream-like visual imagery set to music that couples shots of modern Britain with the director's highly personal view of Christianity and his own homosexuality. Totally plotless, this undoubtedly highly powerful and strikingly imaginative film eludes definition and is capable of any number of interpretations.
MUS   88 min (ort 91 min)   VIDrel: ARTIF/20TH L/A
V/sh

GARDEN OF ALLAH, THE *   *U*
Richard Boleslawski USA   1936
Marlene Dietrich, Charles Boyer, Basil Rathbone, Tilly Losch, C. Aubrey Smith, Joseph Schildkraut, Lucille Watson, Henry Kleinbach (Henry Brandon), John Carradine
A Trappist monk and a disillusioned socialite fall in love in the desert, in a film that is nothing more than an interesting star vehicle, albeit a gorgeously photographed one – the cameramen were Howard Greene and Harold Rosson. AA: Spec Award (for colour cinematography).
DRA   78 min (ort 85 min)   VIDrel: VCC/DISC   V
Boa: novel by Robert Hichens.

GARGOYLES: THE MOVIE – THE HEROES AWAKEN *
Kazuo Terada/Saburo Hashimoto/Takamitsu Kawamura   *PG*
JAPAN   1994
Voices of: Keith David, Jonathan Frakes, Salli Richardson, Jeff Bennett, Edward Asner, Bill Fagerbakke, Marina Sirtis
An American buys a Scottish castle and ships it home to be rebuilt on the top of his skyscraper. This relocation breaks a curse on the stone gargoyles on top of the castle (they were immobilised by a spell a thousand years ago) and they are able to come back to life at night, and continue their fight against evil.
ANIM   78 min (ort 80 min) cC   VIDrel: BUENA/TECH
V/sh

GAS *   *18*
Roger Corman USA   1970
Robert Corff, Elaine Giftos, Bud Cort, Talia Coppola (Shire), Ben Vereen, Cindy Williams, Alex Wilson, Lou Oricopio, George Armitage, Jackie Farley, Phil Borneo, David Osterhout, Bruce Karcher, Mike Castle, Country Joe and The Fish
The accidental release of a nerve gas from a defence plant in Alaska, kills everyone over thirty in this meandering, disjointed,

insane but often very funny story. The film CITY LIMITS used the same idea.
Aka: GAS-S-S-S; GAS... OR IT MAY BECOME NECESSARY TO DESTROY THE WORLD IN ORDER TO SAVE IT; GASSSSSSS OR IT BECAME NECESSARY TO DESTROY THE WORLD IN ORDER TO SAVE IT
COM   79 min   VIDrel: CONNO/RTM L/A   V

GAS FOOD LODGING *   *15*
Allison Anders USA   1991
Brooke Adams, Fairuza Balk, Ione Skye, James Brolin, Robert Knepper, David Lansbury, Jacob Vargas, Donovan Leitch, Chris Mulkey, Tiffany Anders, Laurie O'Brien, Julie Condra, Adam Biesk, Leigh Hamilton, Diane Behrens, J. Mascis
A single mother works as a waitress in a small New Mexican town and tries her best to bring up her two daughters, but all three find it hard to cope with their harsh and frustrating lives. A fine character study with some excellent performances that was made on a very slim budget.
Aka: GAS, FOOD AND LODGING
DRA   97 min (ort 101 min)   VIDrel: 20VIS/SONOP   V/sur
Boa: novel Don't Look and It Won't Hurt by Richard Peck.

GASLIGHT *   *PG*
Thorold Dickinson UK   1940
Anton Walbrook, Diana Wynyard, Frank Pettingell, Cathleen Cordell, Robert Newton, Jimmy Hanley, Minnie Rayner, Mary Hinton, Marie Wright, Jack Barty, Angus Morrison, Aubrey Dexter, The Damora Ballet
A wealthy young woman marries an attractive man and they move into a large house that belonged to her family. After a while, the wife begins to suspect she is losing her memory and possibly her sanity. Luckily, a chance social encounter proves to be her salvation. The first and superior film version of this play, that survived no thanks to MGM, who bought up the rights and spitefully tried to destroy every single print, just to boost the popularity of their 1944 remake.
Aka: ANGEL STREET
DRA   80 min (ort 88 min) B/W   VIDrel: MGM/WHV   V/h
Boa: play Angel Street by Patrick Hamilton.

GASLIGHT *   *PG*
George Cukor USA   1944
Ingrid Bergman, Charles Boyer, Joseph Cotten, Dame May Whitty, Barbara Everest, Angela Lansbury, Terry Moore, Eustace Wyatt, Emil Rameau, Edmund Breon, Tom Stevenson, Halliwell Hobbes, Heather Thatcher, Lawrence Harry Adams
Atmospheric period piece with Boyer attempting to drive his wife insane, in order to have a free hand in a search of their attic whilst preserving a guilty secret. This gave Bergman her first Oscar. Lansbury's film debut. Originally filmed in 1939. AA: Actress (Bergman), Art/Int (Cedric Gibbons and William Ferrari/Edwin B. Willis and Paul Huldschinsky).
Aka: MURDER IN THORNTON SQUARE, THE
DRA   110 min (ort 114 min) B/W   VIDrel: MGM/WHV   V
Boa: play Angel Street by Patrick Hamilton.

GATE, THE *   *15*
Tibor Takacs USA   1987
Stephen Dorff, Louis Tripp, Christa Denton, Kelly Rowan, Jennifer Irwin, Deborah Grover, Scott Denton, Ingrid Veninger, Sean Fagan, Linda Goranson, Carl Kraines, Andrew Gunn
A bored young lad is messing about in his backyard with a pal whilst their parents are away, when they inadvertently smash some rocks and open a gateway to Hell. Some good special effects enliven a fairly undemanding horror yarn. A sequel followed.
HOR   84 min (ort 92 min)   VIDrel: MED/POLY L/A   V

GATE 2 *   *15*
Tibor Takacs USA   1989
Louis Tripp, James Villemaire, Pamela Segall, Simon Reynolds, James Kidnie, Neil Munro, Irene Pauzer, Larry O'Brey, Elva Mai Hoover, Gerry Mendicino, Mark Saunders, Todd Waie, Edward Leefe, Layne Coleman, Anita Olawick, Carl Kraines
Having paid a visit to the burnt-out remains of his friend's house, where a demonic battle took place years before, a man summons up a demon from Hell in an effort to solve his problems, but soon finds himself locked in a battle of his own.
Aka: GATE 2: RETURN TO THE NIGHTMARE
HOR   89 min (ort 90 min)
VIDrel: POLY/POLYREC/BRAVE L/A   V

GATOR * *15*
Burt Reynolds USA 1976
Burt Reynolds, Lauren Hutton, Jack Weston, Jerry Reed, Alice Ghostly, Dub Taylor, Mike Douglas, Burton Gilliam, William Engesser, John Steadman, Lori Futch, Stephanie Burchfield, Bob Yeager, Dudley Remus, Alex Hawkins
An ex-con moonshiner is blackmailed by the Justice department into becoming an undercover man in a hoodlum's gang, in an attempt to grapple corruption. A sequel to the 1973 film WHITE LIGHTNING in which action is heavily outweighed by tedium. Reynolds's directing debut.
COM 114 min (ort 116 min) VIDrel: MGM/WHV L/A V

GAUNTLET, THE * *18*
Clint Eastwood USA 1977
Clint Eastwood, Sondra Locke, Pat Hingle, William Prince, Mara Corday, Bill McKinney, Michael Cavanaugh, Carole Cook, Douglas McGrath, Jeff Morris, Samantha Doane, Roy Jenson, Dan Vadis, Carver Barnes
A cop has to escort a prostitute to the trial of a corrupt official where she is a chief witness, and together they run the gauntlet of the Mob who are determined to prevent her testifying. A generally exciting tale if somewhat unbelievable and excessively violent.
THR 104 min (ort 111 min) VIDrel: WHV V

GAY DIVORCEE, THE ** *U*
Mark Sandrich USA 1934
Fred Astaire, Ginger Rogers, Alice Brady, Edward Everett Horton, Eric Blore, Erik Rhodes, Lillian Miles, Betty Grable, Charles Coleman, William Austin, Paul Porcasi, E.E. Clive, George Davis, Alphonse Martell, Charles Hall
Routine 1930s musical with the Astaire-Rogers dance duo. The thin plot has a would-be-divorcee staying in a hotel where she mistakes an author who is in love with her, for a professional co-respondent. Dated but still retaining considerable zest and a number of still comical routines. AA: Song ("The Continental" – Con Conrad (m)/Herb Magidson (l)).
Aka: GAY DIVORCE, THE
MUS 100 min (ort 107 min) B/W VIDrel: VCC V

GENERAL LINE, THE ** *PG*
Sergei Eisenstein/Grigori V. Alexandrov USSR 1929
Marfa Lapkina, Vasya Buzenkov, Kostya Vasiliev, Chukhamarev
Eisenstein's last silent film tells of a peasant woman who is converted to communism and helps a village start up a co-operative. Despite his efforts to avoid trouble over this offering, the director's satirical touches offended the authorities. A mixture of art and propaganda in equal degrees; the famous montage sequence demonstrating the operation of a cream separator is one of the film's highlights.
Aka: GENERALNAYA LINYA; OLD AND NEW; STAROYE I NOVOYE
DRA 97 min (ort 90 min) B/W silent VIDrel: HEND L/A V

GENERATION, A ** *(18)*
Andrzej Wajda POLAND 1954
Tadeusz Lomnicki, Ursula Mordzynska, Roman Polanski, Tadeusz Janczar, Zbigniew Cybulski.
First part of the Wajda trilogy dealing with young Poles and their attitudes towards the German occupation of Poland, particularly exploring the role of the Polish Resistance during WW2, with a young man falling in love with the woman leader of a Resistance group. An absorbing and perceptive study, followed by KANAL and then ASHES AND DIAMONDS.
Aka: POKOLENIE
DRA 86 min B/W VIDrel: L/A V
Boa: novel by Bohdan Czeszko.

GENERATION X * *15*
Jack Sholder USA 1996
Matt Frewer, Finola Hughes, Jeremy Ratchford, Heather McComb, Agustin Rodriguez, Randall Slavin
Spin-off from Marvel's popular "X-Men" comic strip, with a new set of mutant super-heroes getting the chance to put their distinctly odd super powers to the test. A colourful and fairly entertaining romp, it casts Frewer as a demented scientist who has learnt how to invade dreams, giving the X Men the chance to do battle. A slightly tongue-in-cheek affair, its low-budget look works to the film's advantage.
FAN 87 min (ort 89 min) mCab VIDrel: NWV/HIFLI V

GENEVIEVE ** *U*
Henry Cornelius UK 1954
Kenneth More, Kay Kendall, John Gregson, Dinah Sheridan, Geoffrey Keen, Reginald Beckwith, Arthur Wontner, Joyce Grenfell, Michael Medwin, Leslie Mitchell, Michael Balfour, Edie Martin, Harold Siddons
A dose of typical 1950s whimsy about the rivalry between two of the contestants in the London to Brighton veteran car race, who take part in a private wager on the return leg of their journey. An amiable and undemanding frolic, with music composed and played by Larry Adler.
COM 83 min (ort 86 min) VIDrel: CARL/TECH V

GENGHIS COHN ** *(PG)*
Elijah Moshinsky UK 1993
Antony Sher, Robert Lindsay, Diana Rigg, John Wells, Robert Lang, Frances de la Tour, Matthew Marsh, Paul Brooke, Cara Konig, Rowland Davies, Juliette Grassby, Jay Benedict, Peter Penry-Jones, Cheryl Fergison, Daniel Craig
Years after the end of WW2, a former concentration camp commandant living in anonymous prosperity in West Germany is suddenly haunted by the ghost of a Jewish comedian, one of his countless victims. The latter continues to dog him and eventually prepares a fitting retribution, in this dark and biting black comedy.
DRA 80 min mTV TVrel: BBC
Boa: novel The Dance Of Genghis Cohn by Romain Gary.

GENGHIS KHAN * *PG*
Henry Levin UK/USA/WEST GERMANY/YUGOSLAVIA 1965
Omar Sharif, Stephen Boyd, James Mason, Eli Wallach, Francoise Dorleac, Telly Savalas, Robert Morley, Yvonne Mitchell, Woody Strode
Loose, untidy spectacle following the career of this warlord as he grows up to take revenge on the rival chieftain who murdered his father. There is no fine sweep of history here, just poor casting, a pathetic script and a few spectacular moments by way of compensation. Cut before video submission by 1 min 3 sec.
A/AD 119 min (ort 124 min) VIDrel: VCC/DISC/COLUM V

GENO CYBER: PARTS 1 TO 3 * *18*
JAPAN
Three-part manga adventure in which a 21st century medic has plans to resurrect the title creature, an advanced human being with formidable powers of destruction. Episodes are entitled: "A New Lifeform", "Vajranoid Attack" and "Global War". Adequate.
ANIM 93 min dubbed VIDrel: MANGA/SONOP V

GENTLE GUNMAN, THE * *U*
Basil Dearden UK 1952
John Mills, Dirk Bogarde, Elizabeth Sellars, Barbara Mullen, Robert Beatty, Eddie Byrne, Gilbert Harding, Liam Redmond, Jack MacGowran, Joseph Tomelty, James Kenney, Michael Golden, Patric Doonan
Disgusted with the violence of his colleagues, an IRA man sets out to thwart his brother's gang, who have gone to London to mount a bombing campaign. A sincere and moderately effective WW2 drama, set in 1941, but one that is a little spoilt by the weaknesses inherent in the play, which is more than a trifle simplistic.
DRA 82 min (ort 86 min) B/W VIDrel: LUMI/SPEAR L/A V
Boa: play by Roger Macdougall.

GENTLEMAN JIM ** *U*
Raoul Walsh USA 1942
Errol Flynn, Alexis Smith, Jack Carson, Alan Hale, John Loder, William Frawley, Minor Watson, Ward Bond, Arthur Shields, Madeleine LeBeau, Dorothy Vaughn, Rhys Williams, James Flavin, Pat Fleherty, Wallis Clarke, Art Foster
Biopic about Jim Cobbett, an early prizefighter. Flynn as Corbett is good and the fight scenes are handled extremely well. Scripted by Vincent Lawrence and Horace McCoy.
DRA 101 min (ort 104 min) B/W VIDrel: MGM/WHV V
Boa: book The Roar Of The Crowd by James J. Corbett.

GENTLEMAN'S AGREEMENT * *U*
Elia Kazan USA 1947
Gregory Peck, Dorothy McGuire, John Garfield, Celeste Holm, June Havoc, Jane Wyatt, Anne Revere, Dean Stockwell, Albert Dekker,

Nicholas Joy, Sam Jaffe, Harold Vermilyea, Ransom M. Curt Conway, John Newland, Robert Warwick
A journalist pretends to be Jewish in order to investigate the extent and nature of anti-Semitism in the USA. Though considered in its day a daring expose of the corrosive effect of prejudice on human nature, it can now be seen for the ponderous and rather feeble examination it really was. Not so much effective as well-intentioned. AA: Pic, Dir, S. Actress (Holm).
DRA   112 min (ort 118 min) B/W cC   VIDrel: 20TH/TECH V/h
Boa: novel by Laura Z. Hobson.

GENTLEMEN PREFER BLONDES **
Howard Hawks USA
Marilyn Monroe, Jane Russell, Charles Coburn, Tommy Noonan, Elliott Reid, George "Foghorn" Winslow, Norma Varden, Marcel Dalio, Taylor Holmes, Howard Wendell, Steven Geray, Henri Letondal, Leo Mostovoy, Alex Frazer
Two showgirls go to Paris in search of rich husbands, in this musical comedy which is only memorable for the score, and songs such as "Diamonds Are A Girl's Best Friend". Followed by "Gentlemen Marry Brunettes".
MUS   89 min (ort 91 min)   VIDrel: 20TH/TECH   V
Boa: story by Anita Loos.

U
1953

GENUINE RISK **
Kurt Voss USA
Michelle Johnson, Peter Berg, Terence Stamp, M.K. Harris, Teddy Wilson, Sid Haig, John Lavachielli, Hal Shafer, Joe Shea, Steven Brill, Ellen Albertini Dow, Max Perlich, Jeffrey Arbaugh, George Fisher, Tony Gegere, Michael Deluna
A gangster seeks revenge against a down-at-heel gambler he has taken on as a bodyguard, when the latter falls in love with the mistress of his boss. A turgid and thoroughly unconvincing tale that wastes the talents of a strong cast.
DRA   85 min (ort 89 min)   VIDrel: TRANSAT/HIFLI L/A V/h

18
1990

GEORDIE **
Frank Launder UK
Bill Travers, Doris Goddard, Alastair Sim, Molly Urquhart, Jameson Clark, Frances De Wolff, Alex Mackenzie, Raymond Huntley, Brian Reece, Miles Malleson, Stanley Baxter, Jack Radcliffe, Duncan Macrae, Paul Young
A young Scottish boy with a poor physique finally decides to do something about it and a takes a correspondence course that eventually turns him into a fine strapping specimen. He is chosen to throw the hammer for Britain and eventually achieves both fame and romance. A dated but quite charming comedy of slender means that does not take itself too seriously and offers some enjoyable performances.
Aka: WEE GEORDIE
COM   98 min   VIDrel: LUMI/SPEAR   V
Boa: novel by David Walker.

U
1955

GEORGE McKENNA STORY, THE ***
Eric Laneuville USA
Denzel Washington, Lynn Whitfield, Akosua Busia, Richard Masur, Virginia Capers, Ray Buktenica, Barbara Townsend, Israel Juarbe, J.A. Preston, Bill Henderson, Earl Billings, Brent Jennings, Terrance Ellis, Ken Sagoes
A high school principal turns a run-down and gang-ridden school in the poorer part of Los Angeles into a well-run place of learning. A simple and inspiring drama based on the experiences of this real-life principal. The script is by Charles Eric Johnson. Ironically, the film was actually shot on location in Houston, Texas. See also LEAN ON ME for something similar.
DRA   94 min (ort 100 min) mTV
VIDrel: VISVID/POLYREC L/A   V

15
1986

GEORGY GIRL **
Silvio Narizzano UK
Lynn Redgrave, James Mason, Alan Bates, Charlotte Rampling, Bill Owen, Clare Kelly, Rachel Kempson, Denise Coffrey, Dorothy Alison, Peggy Thorp-Bates, Dandy Nichols, Terence Soall, Jolyan Booth
Dated black comedy about a dowdy English girl and her involvement with men, in particular wealthy married man Mason, who wants her for his mistress. Considered in its day to be controversial, this "swinging London" comedy now has the appearance of a film that has been kept in mothballs.
COM   95 min (ort 100 min) B/W   VIDrel: CASPIC/COLUM V
Boa: novel by Margaret Forster.

15
1966

GEREIN' UP **
Michael Craig USA
Ashlyn Gere, Heather Hart, Tianna Taylor, Alicia Rio, Cassidy, Mike Horner, Jon Dough, Scott Irish
A married man accepts a lift when his car breaks down, but learns that the woman who has driven him home has decided to take over his life. This leads to much teasing and eventually some torrid lovemaking that has the man totally captivated by her. Whereupon she dumps him, it transpiring that she was jilted on her wedding day, is now at war with men, and has decided that this is her best way of getting revenge. Quite well made, but also sluggish and uninvolving.
A   61 min (ort 70 min)   VIDrel: GROHOM/MAXSCAN   V

18
1992

GERMANY, YEAR ZERO ***
Roberto Rossellini FRANCE/ITALY
Franz Kruger, Edmund Moschke, Barbara Hintz, Werner Pittschau, Erich Guhne, Alexandra Manys, Baby Reckvell, Ingetraut Hintze, Hans Sange, Hedi Blankner, Count Treiberg, Karl Kauger
This is a bleak look at life in Germany just after WW2, made with a cast of non-professional actors, that traces the efforts of a twelve-year-old boy to feed his family and cope with the burden of a sick father. An uneven but fascinating film, made with both care and compassion.
Aka: GERMANIA, ANNO ZERO
DRA   69 min (ort 87 min) B/W   VIDrel: CONNO/RTM   V

PG
1947

GERMINAL ***
Claude Berri BELGIUM/FRANCE/ITALY
Gerard Depardieu, Miou-Miou, Renaud, Jean Carmet, Judith Henry, Jean-Roger Milo, Laurent Terzieff, Jean-Pierre Bisson, Bernard Fresson, Jacques Ducqmine, Ammy Duperey, Gerard Croce, Frederic Van Den Duchsessne, Amik Alane, Pierre Lafo
Lavish and epic retelling of Zola's novel about the degraded and brutish lives of coalminers and their families in the 19th century, and how they are affected by the actions of an outsider, who stimulates their political consciousness and provokes a fateful strike. As in the novel, the characters serve as mere mouthpieces for the author's message and the film, for all its many virtues, gives the impression of having been commissioned by a committee.
DRA   151 min (ort 160 min)   VIDrel: GUILD/SONOP V/sur
Boa: novel by Emile Zola.

15
1992

GERONIMO! **
Arnold Laven USA
Chuck Connors, Kamala Devi, Pat Conway, Adam West, Enid Jaynes, Larry Dobkin, Denver Pyles, Armando Silvestre, John Anderson, Amanda Amex, Mario Navarro, Eduardo Noriega, Nancy Rodman, Joe Higgins, Robert Huges, James Burk
Account of this famous Indian chief and his reasons for going on the warpath that tries hard to be fair but gives an unrealistic portrayal of the Apaches, toning down their fighting qualities, while the stilted dialogue is another handicap. However, the film is watchable thanks to Connors in the title role and the beautifully photographed locations.
WES   131 min   VIDrel: MGM/WHV   V/h

PG
1962

GERONIMO ***
Roger Young USA
Joseph Runningfox, Nick Ramus, Michelle St John, Michael Greyeyes, August Schellenberg, Jimmy Herman, Ryan Black, Tallinh Forest Flower, Kimberly Norris, Geno Silva, Harrison Lowe
One of two TV films on the famous Indian leader that appeared in 1993, this one covering no less than eighty years in the life of the central character, with three stars sharing the acting honours. Unfortunately, the movies tries so hard to redress the balance in adopting an equitable approach to Native American history that its dramatic impact as a film is seriously weakened. See also GERONIMO: AN AMERICAN LEGEND.
WES   100 min (ort 102 min) mCab   VIDrel: FIRST/SONOP V

15
1993

GERONIMO: AN AMERICAN LEGEND ***
Walter Hill USA
Wes Studi, Jason Patric, Robert Duvall, Gene Hackman, Al Sieber, Matt Damon, Rodney A. Grant, Kevin Tighe, Steve Reevis, Carlos Palomino, Victor Aaron, John Stuart Proud Eagle Grant, Stephen McHattie, Lee De Broux, Rino Thunder, J. Young
Another 1993 biopic of the title leader that concentrates on the efforts of the US government to put paid to the ravages of Geronimo and his band of renegade Apaches. As in DANCES

12
1993

WITH WOLVES, the movie employs the device of a young US Cavalry officer who takes part in the fighting and gradually comes to admire his adversary. Filmed on location in Utah, this offers fine camerawork and first-rate performances from Duvall and Hackman. See also GERONIMO.
Aka: GERONIMO
WES   110 min (ort 115 min) cC   VIDrel: COLUM/SONOP
V/sur

GET CARTER *   18
Mike Hodges UK   1970
Michael Caine, John Osborne, Ian Hendry, Britt Ekland, George Sewell, Geraldine Moffatt, Tony Beckley, Rosemarie Dunham, Dorothy White, Petra Markham, Glynn Edwards, Alun Armstrong, Bryan Mosley
A tough criminal investigates the circumstances surrounding the death of his brother, and his search takes him to Newcastle and a conflict with gangsters dealing in pornography. Strong on atmosphere and quite brutal, the film ends on a very downbeat but all too believable note. Remade in 1972 as "Hit Man".
THR   107 min (ort 112 min)   VIDrel: MGM/WHV   V
Boa: novel Jack's Return Home by Ted Lewis.

GET OUT YOUR HANDKERCHIEFS *   18
Betrand Blier BELGIUM/FRANCE   1978
Gerard Depardieu, Patrick Dewaere, Carol Laure, Riton, Michel Serrault, Eleonore Hirt, Sylvie Joly, Jean Rougerie, Liliane Rovere, Michel Beaune
A man tries everything to cure his sexually frustrated wife's depression, and decides that perhaps the best thing would be to find her a lover, although a young adolescent boy eventually solves her problem. An absurd farces that pokes fun at male attitudes to sex but is far from as funny as it might have been.
AA: Foreign.
Aka: PREPAREZ VOS MOUCHOIRS
COM   108 min   VIDrel: ARROW/RTM   V

GET SHORTY *   15
Barry Sonnenfeld USA   1995
John Travolta, Gene Hackman, Rene Russo, Danny DeVito, Dennis Farian, Delroy Lindo, James Gandolfini, Jon Gries, Renee Props, David Paymer, Martin Ferrero, Miguel Sandoval, Jacob Vargas, Linda Hart, Bobby Slayton, Ron Harabatsos
A loan shark heads for L.A. in search of a man who owes him money, but on his way there falls in with a producer and his actress girlfriend, and these two persuade him to play in their new movie. Travolta is unusually good as the loan shark, as is DeVito as one of the more colourful characters he encounters. A slick outing, colourful and fairly amusing.
COM   101 min (ort 105 min) wScrn cC   VIDrel: MGM/WHV
V/sur
Boa: novel by Elmore Leonard.

GETAWAY, THE *   18
Sam Peckinpah USA   1972
Steve McQueen, Ali MacGraw, Ben Johnson, Slim Pickens, Sally Struthers, Al Lettieri, Richard Bright, Bo Hopkins, Jack Dodson, Dub Taylor, Roy Jensen, John Bryson, Bill Hart, Tom Runyon, Whitney Jones, Raymond Jones, Ivan Thomas
A crook leaves prison to join his wife in planning and carrying out a bank robbery, which does not go as planned. Plenty of car chases and violent action help make up for the deficiencies of the plot. Written by Walter Hill, it was remade in 1993.
THR   117 min (ort 123 min)   VIDrel: WHV   V/h
Boa: novel by Jim Thompson.

GETAWAY, THE *   18
Roger Donaldson USA   1993
Alec Baldwin, Kim Basinger, Jennifer Tilly, James Woods, Michael Madsen, Richard Farnsworth, Burton Gilliam, Philip Hoffman, David Morse, James Stephens, Royce D. Applegate, Daniel Villareal, Scott McKenna, Alex Colon, Justin Williams
A loyal wife manages to spring her husband from jail where he's serving a sentence for robbery, and they pull off one last bank job together, but have a tough time hanging on to the loot. Very much a scene-by-scene remake of the earlier (and better) 1972 film, updated for the 1990s with the expected lashings of sex and violence, but lacking any real depth to the characterisations.
DRA   110 min (ort 115 min)   VIDrel: WHV   V/sur
Boa: novel by Jim Thompson.

GETTING EVEN WITH DAD *   PG
Howard Deutsch USA   1994
Ted Danson, Macaulay Culkin, Glenne Headly, Saul Rubinek, Gailard Sartain, Sam McMurray, Hector Elizondo, Sydney Walker, Kathleen Wulhoite, Dann Florek, Ron Canada, Ralph Peduto, Bert Kinyon, Melvin Thompson, Danny Hunter, Mary Dilts
A petty thief gets lucky when he steals a valuable coin collection but his joy is short-lived when his eleven-year-old son decides to hide it in an effort to force Dad to go straight and starting acting like a proper father. Rather as one might expect, such antics do not go done too well with either Dad or his gang. A contrived and unfunny comedy that could have done with a stronger story, better acting and another child star (Culkin just looks too old to be cute).
COM   104 min (ort 109 min) cC   VIDrel: MGM/WHV
V/sur

GETTING GOTTI *   15
Roger Young USA   1993
Lorraine Bracco, Anthony John Denison, Kathleen Laskey, August Schellenberg, Kenneth Welsh, Jeremy Ratchford, Ron Gabriel, Ellen Burstyn, Lawrence Bayne, Ron Hartman, Jason Blicker, Peter Boretski, John Winston Carroll, Victor Ertmanis
A surprisingly absorbing, effective and well made recreation of the legal efforts that were devoted to putting infamous New York mobster John Gotti behind bars, which was to prove a long and protracted struggle. Firm direction and some excellent performances make this one well worth watching. Bracco is well cast as Diane Giacalone, the assistant D.A. who played a pivotal role in these proceedings. See also GOTTI.
DRA   91 min mTV   VIDrel: ODY/SONOP   V/sh

GETTING OF WISDOM, THE *   PG
Bruce Beresford AUSTRALIA   1977
Susannah Fowle, Hilary Ryan, Alix Longman, Sheila Helpmann, Barry Humphries, Laura Rambotham, Patricia Kennedy, John Waters, Kerry Armstrong, Julia Blake, Dorothy Bradley, Kay Englund, Max Fairchild, Jan Friedl, Diana Greentree
A girl from the outback goes to a snobbish Victorian ladies' college, but refuses to share its values and customs. A curious period drama that is rather spoilt by the bad casting of Fowle, the lack of any clear narrative also being a handicap.
DRA   97 min (ort 100 min)   VIDrel: ARTPRO/RTM   V
Boa: novel by Henry Handel Richardson.

GETTING OUT *   15
John Korty USA   1994
Rebecca De Mornay, Ellen Burstyn, Robert Knepper, Carol Mitchell-Leon, Tandy Cronyn, Richard Jenkins, Norm Skaggs, Sue Bugden, Kevin Dewey, Sean Sweeney, Amy Dott, Jack Swanson, Bruce Evers, Linda Pierce, Suzi Bass, Rosemary Newcott
A woman commits robbery and murder and is sentenced to prison where she serves eight years, During this time she gives birth to a son but the baby is taken into custody and after her release, she is forced to fight a court battle in order to get him back.
DRA   88 min (ort 92 min) mTV   VIDrel: MARQ/QUANT
V
Boa: play by Marsha Norman.

GETTYSBURG *   PG
Jack Bender USA   1990
Jason Robards, Lukas Haas, Campbell Scott, Katherine Helmond, Ed Flanders, Jose Ferrer, Scott Paulin
Set at the time of the American Civil War, this story concerns the efforts made by a man to find his missing brother, and tells of the misery and destruction he encounters in his search.
Aka: PERFECT TRIBUTE
WAR   91 min (ort 94 min)   VIDrel: MARQ/QUANT   V/sur

GETTYSBURG: PARTS 1 AND 2 *   PG
Ronald F. Maxwell USA   1993
Tom Berenger, Jeff Daniels, Martin Sheen, Sam Elliott, Maxwell Caulfield, Kevin Conway, C. Thomas Howell, Richard Jordan, James Lancaster, Stephen Lang, Richard Jordan, Royce D. Applegate, John Diehl, Patrick Gorman, Brian Mallon
Lavish, incredibly expensive and lovingly re-enacted account of the title Civil War battle, that was appropriately enough filmed in the Gettysburg National Park and made use of thousands of extras. Although the battle scenes are exciting and well staged, much-needed relief from the endless scenes of carnage comes in the form of various sub-plots dealing with a handful of the

participants. Producer Ted Turner appears briefly as a Confederate soldier.
Aka: GETTYSBURG
WAR 243 min (2 cassettes – ort 318 min) VIDrel: WHV V/sur
Boa: novel The Killer Angels by Michael Shaara.

GHOST ***
15
Jerry Zucker USA
1990
Patrick Swayze, Demi Moore, Whoopi Goldberg, Tony Goldwyn, Rick Aviles, Vincent Schiavelli, Gail Boggs, Armelia McQueen, Phil Leeds
A New York yuppie is murdered and becomes an earthbound spirit, in this form learning that his girlfriend is also in danger. In a bid to warn her, he uses the mind of a fake medium who actually has psychic powers. A major success of the 1990s, this charming fantasy is effective if not always convincing. Goldberg is memorable as the medium, as is Schiavelli as a demented subway ghost. Scored by Maurice Jarre. AA: S. Actress (Goldberg), Screen/orig (Bruce Joel Rubin).
COM 121 min (ort 127 min) cC
VIDrel: CIC/SONOP; PION (LV only) V/sur LV

GHOST AND MRS MUIR, THE ***
U
Joseph L. Mankiewicz USA
1947
Gene Tierney, Rex Harrison, George Sanders, Edna Best, Robert Coote, Anna Lee, Natalie Wood, Vanessa Brown, Isobel Elsom, Victoria Horne, Brad Slaven, Whitford Kane, William Stelling, Helen Freeman, David Thursby, Heather Wilde
A widow moves into an English cottage and finds it haunted by the ghost of a merchant navy captain who falls deeply in love with her, coming to her rescue when she finds herself in financial difficulties. A sensitively handled fantasy that works beautifully thanks to fine performances by the two leads. Remade in 1955 as "Stranger In The Night" and as a much inferior TV series in 1968.
FAN 100 min (ort 104 min) B/W VIDrel: 20TH/TECH V
Boa: novel by R.A. Dick.

GHOST AND THE DARKNESS, THE ***
15
Stephen Hopkins USA
1996
Michael Douglas, Val Kilmer, Bernard Hill, John Kani, Tom Wilkinson, Brian McCardie, Henry Cele, Om Puri, Emily Mortimer, Kurt Egelhof, Satchu Annamalai, Teddy Reddy, Rakeem Khan, Jack Devnarain, Glen Gabela, Richard Nwamba
A good looking Boy's Own-style outdoors adventure, set in East Africa, where British attempts to build a railway bridge are being seriously hampered by the work of two man-eating lions. Based on true events that occurred in 1896 (the beasts in question claimed 130 lives) it is eventually resolved when a seasoned big-game hunter (Douglas giving a terribly overblown performance) is called in. A bit like JAWS on four legs, with a simple plot but some very tense moments.
A/AD 110 min CINrel

GHOST GOES WEST, THE ***
U
Rene Clair UKA
1935
Robert Donat, Jean Parker, Eugene Pallette, Elsa Lanchester, Ralph Bunker, Patricia Hilliard, Morton Selten, Everley Gregg, Chili Bouchier, Mark Daly, Herbert Lomas, Elliot Mason, Jack Lambert, Colin Leslie, Richard Mackie
A millionaire buys a Scottish castle and has it shipped Stateside, but does not realise that it is haunted by the ghost of a young man who disgraced his ancestors. Despite its age, still a charming and very whimsical comedy that contains some very telling points on the clash between different culture. This was Clair's first English-speaking film.
COM 78 min (ort 82 min) B/W VIDrel: CARL/TECH V
Boa: story Tristram Goes West by Eric Keown.

GHOST IN MONTE CARLO, A **
PG
John Hough UK
1990
Samantha Eggar, Oliver Reed, Sarah Miles, Fiona Fullerton, Christopher Plummer, Lysette Anthony, Joanna Lumley, Lewis Collins, Gareth Hunt, Marcus Gilbert, Ron Moody, Jolyon Baker, Helen Cherry, Elizabeth Sellars
Very much a Barbara Cartland period piece that follows the life of an innocent young woman just out of convent school who accompanies her peevish aunt to Monte Carlo, where love and intrigue await her. A handsome, glossy piece of nonsense, well produced and reasonably absorbing.
DRA 90 min (ort 100 min) mCab
VIDrel: 4-FRONT/POLYREC V
Boa: novel by Barbara Cartland.

GHOST IN THE MACHINE **
18
Rachel Talalay USA
1993
Karen Allen, Chris Mulkey, Ted Marcoux, Wil Horneff, Jessica Walter, Brandon Quintin Adams, Rick Ducommun, Nancy Fish, Jack Lauter, Shevonne Durkin, Richard McKenzie, Mimi Lieber, Mickey Gilbert, Carl Gabriel Yorke, Clayton Landey
A serial killer dies whilst being X-rayed in hospital, and his spirit enters a mainframe computer, from where he is able to reach out through the power grid and kill further victims, making use of their domestic appliances or computers. A ludicrous and seriously unfocused horror movie that never really makes effective use of its one good idea. The film SHOCKER tried something similar with considerably more success.
HOR 91 min (ort 98 min) cC VIDrel: 20TH/TECH V/sur

GHOST IN THE SHELL **
15
Mamoru Oshii JAPAN/UK
1995
Voices of: Richard George, Mimi Woods, William Frederick, Abe Lasser, Christopher Joyce, Mike Sorich, Ben Isaacson, Hank Smith, Steve Davis, Phil Williams
In the year 2029 A.D., a crack secret-service squad is sent after a top criminal known as the "Puppet Master". Best described as a production-line cyberpunk animation, this adventure is well drawn and moves along briskly. But it all becomes rather repetitive after a while, especially the obligatory vistas of bleak urban wastelands. The lack of fresh ideas is clearly apparent.
Aka: KOKAKU KIDOTAI
ANIM 79 min (ort 83 min) dubbed
VIDrel: MANGA/SONOP V/sh
Boa: graphic novel by Shirow (Shiro) Masamune.

GHOST TOWN ***
18
Richard Governor USA
1988
Frank Luz, Jimmie F. Skaggs, Catherine Hickland, Bruce Glover, Michael Aldredge, Penelope Windust, Zitto Kazann, Blake Conway, Laura Schaffer, Ken Kolb, Will Hannah
Hunting for a missing girl, a policeman searches in a dusty ghost town in Arizona that holds a dark secret – a bunch of zombie gunslingers. A stylish and eerie combination of Western and horror genres.
HOR 81 min (ort 85 min) VIDrel: EIV/SONOP V

GHOST TRAIN, THE **
U
Walter Forde UK
1941
Arthur Askey, Richard Murdoch, Kathleen Harrison, Morland Graham, Linden Travers, Peter Murray Hill, Carole Lynn, Herbert Lomas, Raymond Huntley, Betty Jardine, Stuart Latham, D.J. Williams, George Merritt
A detective poses as a passenger in order to catch spies who are using an abandoned track. A so-so remake of the 1931 film, with the leading role rather ill-advisedly split into two characters. See OH! MR PORTER which also made use of the plot from the earlier film.
COM 83 min B/W VIDrel: VCC L/A V
Boa: play by Arnold Ridley.

GHOST WRITER *
PG
Kenneth J. Hall USA
1989
Audrey Landers, Judy Landers, David Doyle, Joey Travolta, John Matuszak, Jeff Conaway, Anthony Franciosa, Dick Miller, Ken Tobey, Nels Van Patten, George "Buck" Flower, Pedro Gonzalez, Jerry Tyminski, Martin Madden
A murdered movie star returns as a ghost to help a young writer expose her killer in this tepid comedy that is a perfect vehicle for the limited acting talent of the Landers twins. A very poor offering indeed.
COM 90 min (ort 94 min) VIDrel: COLUM/SONOP V

GHOSTBUSTERS ***
PG
Ivan Reitman USA
1984
Dan Aykroyd, Bill Murray, Harold Ramis, Sigourney Weaver, Rick Moranis, Annie Potts, Ernie Hudson, William Atherton, David Margulies, Steven Tash, Jennifer Runyon, Slavitza Jovan, Michael Ensign, Alice Drummond
When three scientists lose their research funding they decide to make use of their knowledge of the paranormal, hiring themselves out as a team able to rid buildings of unwanted supernatural phenomena. A brash film that hovers uneasily between comedy and the supernatural, developing eventually into a silly duel between the forces of Good (our three scientists),

and those of Evil. Special effects are by Richard Edlund. Written by Aykroyd and Ramis.
COM   101 min (ort 105 min)   VIDrel: VCC/DISC/COLUM V/sur

GHOSTBUSTERS 2 *   PG
Ivan Reitman USA   1988
Dan Aykroyd, Bill Murray, Harold Ramis, Ernie Hudson, Sigourney Weaver, Rick Moranis, Annie Potts, William Atherton, Peter MacNicol, Harris Yulin, David Marguilies, Kurt Fuller, Janet Margolin, Wilhelm Von Homburg
A likeable sequel to the earlier film that offers little in the way of new ideas, but has our ghostbusting team re-assembled to save New York from an attack by a sea of slime, that's been nourished by the city's famous negative vibes. A loosely plotted comedy, sustained by nice interplay between the leads and some good jokes about the quality of life in the Big Apple. Written by Ramis and Aykroyd.
COM   103 min   VIDrel: VCC/DISC/COLUM   V/sur

GHOSTS FROM THE PAST *   15
Rob Reiner USA   1996
Alec Baldwin, Whoopi Goldberg, James Woods, Craig T. Nelson, Susanna Thompson, Lucas Black, Joseph Tello, Alexa Vega, William H. Macy
In 1963 black civil rights leader Medgar Evers was gunned down outside his Mississippi home by a white racist, who was freed when two all-white juries failed to agree a verdict. This film takes up the story twenty-five years on, when in 1989 a determined assistant D.A. set out to bring the killer to justice. Reiner shows his limitations when tackling a real-life subject, and though most worthy, this flat film does neither its cast nor subject matter justice.
DRA   130 min   CINrel

GHOSTS OF THE CIVIL DEAD *   18
John Hillcoat USA   1988
Dave Field, Mike Bishop, Chris De Rose, Nick Cave, Vincent Gil, Bogdan Koca
This fact-based drama tells of how the inmates of the Central Industrial Prison were kept locked in their cells following a break-out, and a spiralling cycle of violence followed by ever more inhumane restrictions develops when the prisoners find themselves unable to bear their treatment. Harsh and uncompromising, but a film that for all its disturbing images remains completely absorbing and all too believable.
DRA   89 min   VIDrel: ELPIC/POLYREC   V/s

GHOSTWATCH *   (PG)
Lesley Manning UK   1992
Michael Parkinson, Sarah Greene, Mike Smith, Craig Charles, Gillian Bevan, Brid Brennan, Michelle Wesson, Cherise Wesson, Chris Miller, Mike Aiton, Mark Lewis, Linda Broughton, Katherine Stark, Derek Smee, Roger Tebb, Colin Stinton
Made as a one-off for Halloween, this odd supernatural mystery starts with a TV crew descending on a haunted house in Middlesex to make a programme, but finding that their broadcast is effectively sabotaged by poltergeist activity. Done in the form of a fake documentary (one of the more irritating aspects of this work) the film slowly reveals its hand, but both development and resolution are interesting rather then especially memorable.
DRA   95 min mTV   TVrel: BBC

GHOUL, THE *   PG
T. Hayes Hunter UK   1933
Boris Karloff, Ernest Thesiger, Dorothy Hyson, Cedric Hardwicke, Anthony Bushell, Ralph Richardson, Harold Huth, D.A. Clarke-Smith, Kathleen Harrison, Jack Raine
Karloff's first British film since he left in 1909 is a poor attempt to emulate the success of the Universal films. He plays an Egyptologist who returns from the grave to take revenge on those who stole a sacred jewel from the tomb that was intended to assure him of eternal life in the hereafter. The film starts off well but soon deteriorates into a dull farce and was indeed remade as such in 1961 as WHAT A CARVE UP!. One asset: the surprisingly good photography.
HOR   68 min (ort 79 min) B/W   VIDrel: CARL/TECH   V
Boa: novel by Frank King and Leonard Hines.

GHOUL, THE *   18
Freddie Francis UK   1974
Peter Cushing, Alexandra Bastedo, John Hurt, Gwen Watford,

Veronica Carlson, Don Henderson, Ian McCulloch, Stewart Bevan, John D. Collins, Dan Meaden
A group of stranded travellers are attacked by something that lurks in the house of a former clergyman, in this horror yarn set in the 1920s. (I had better not spoil the film for you by revealing that the creature is the cleric's cannibalistic son.)
Aka: THING IN THE ATTIC, THE
HOR   94 min (Cut at film release – ort 88 min)
VIDrel: ARTPRO/RTM   V

GHOULIES *   15
Luca Bercovici USA   1985
Peter Liapis, Lisa Pelikan, Michael Des Barres, Jack Nance, Peter Risch, Tamara de Treux, Scott Thomson, Ralph Seymour, Keith Joe Dick, Mariska Hargitay, David Dayan, Victoria Catlin, Charlene Cathleen
Nasty little monsters start popping up after a boy and his friends mess about with some magic spells. A boring and rather repulsive GREMLINS clone. A number of sequels followed.
COM   77 min (ort 84 min)   VIDrel: NTV/TOTAL   V

GHOULIES 2 *   15
Albert Band USA   1986
Damon Martin, Royal Dano, Phil Fondacaro, J. Downing, Kerry Remsen, Dale Wyatt, Jon Maynard Pennell, Sasha Jensen, Starr Andreeff, William Butler, Donnie Jeffcoat, Christopher Burton, Mickey Knox, Romano Puppo, Ames Morton
Just when Larry is ready to give up running his carnival show – "Satan's Den" – some unexpected visitors arrive and make their home there, proving a big hit with the paying customers. Unfortunately, our demonic imps cause considerable trouble in the process, not to mention a few gruesome murders, in this unsubtle and entirely inevitable sequel.
COM   86 min Cut (55 sec – ort 89 min)
VIDrel: 4-FRONT/POLYREC/EIV L/A   V

GHOULIES 3: GHOULIES GO TO COLLEGE *   15
John Carl Buechler USA   1990
Kevin McCarthy, Evan Mackenzie, Griffin O'Neal, John Johnston, Eva La Rue, Billy Morrissette, Patrick Labyorteaux, Hope Marie Carlton, Stephen Lee, Dan Shor, Marcia Wallace, Jason Scott Lee, Andrew Barach, Sherrie Wills
Second sequel to the original film that sticks strictly to the level of POLICE ACADEMY and the like, with our adorable little nasties being summoned up by a college professor in order to deal with the antics of two rival gangs during the college's annual week of pranks. Apart from a few traces of black humour, this is a very dull effort indeed.
Aka: GHOULIES GO TO COLLEGE
HOR   90 min (ort 94 min)   VIDrel: FIRST/SONOP L/A   V

GHOULIES 4 *   18
Jim Wynorksi USA   1993
Peter Liapis, Barbara Alyn Woods, Stacie Randall, Raquel Krelle, Bobby Di Ciccio
A further entry in this series in which our nasty monsters encounter a female devil-worshipper with a penchant for leather gear and S&M. Another attempt to draw water from a well that has truly run dry.
FAN   80 min (ort 84 min)   VIDrel: WHV L/A   V/h

GIANT *   PG
George Stevens USA   1956
Rock Hudson, Elizabeth Taylor, Mercedes McCambridge, James Dean, Carroll Baker, Chill Wills, Jane Withers, Dennis Hopper, Sal Mineo, Rodney (Rod) Taylor, Earl Holliman, Judith Evelyn, Alexander Scourby, Paul Fix, Robert Nichols
A flat, overlong and boring study, telling of the changing fortunes of several generations on a Texan farm, that's only enlivened by the occasional appearances of Dean. A film that's unaccountably highly regarded by the critics, almost certainly due to the presence of Dean in what was his last movie. It picked up an Oscar but one has to struggle to see why. AA: Dir.
DRA   193 min (ort 201 min)   VIDrel: WHV   V/dm
Boa: novel by Edna Ferber.

GIANT CLAW, THE *   U
Fred F. Sears USA   1957
Jeff Morrow, Mara Corday, Morris Ankrum, Louis D. Merrill, Edgar Barrier, Robert SHayne, Morgan Jones, Clark Howat, Ruell Shayne
Earth is attacked by a giant bird in this laughable SF effort whose attempts at special effects are non-existent (the wires that support this monster being clearly visible). However, a trio of

determined scientists eventually consign the creature to the depths of the ocean.
FAN   84 min B/W   VIDrel: SCREAM/SPEAR   V

GIANT OF THUNDER MOUNTAIN, THE ** (U)
James Robertson USA   1990
Richard Kiel, Jack Elam, Marianne Rogers, Noley Thornton, Chance Michael Corbitt, Ryan Todd, William Sanderson, Foster Brooks, George "Buck" Flower, Ellen Crawford, John Quade, James Hampton plus Cloris Leachman (narration)
A very tall man living alone on a mountain is the subject of various rumours that make him out to be both mad and dangerous. Two kids accept a dare to climb this mountain and end up visiting his cabin. They later return with their sister who becomes his friend, but a subsequent misunderstanding and the greed of some crooks out to steal the giant's gold leads to problems. An adequate if overly sentimental tale that provides acceptable family viewing.
JUV   88 min (ort 90 min)   SATrel: SKY MOVIES

GIANT ROBO: PARTS 1 TO 6 ** PG
JAPAN   1992/1993
A new kind of drive (the "Shizuma Drive") shows great promise as a totally renewable energy source, but the power it can generate is linked to a dangerous secret, known only to a sinister organisation. With the aid of the title creation, an international police force attempt to confront this organisation, and learn the dangers inherent in using the drive.
ANIM   288 min (six cassettes – approx 48 min each) dubbed VIDrel: MANGA/SONOP   V/sh

GIFT OF LOVE, THE ** (PG)
Paul Bogart USA   1994
Andy Griffith, Blair Brown, Penny Fuller, Joyce Van Patten, Richard Herd, Daniel Von Bergen, J.C. Quinn, Olivia Burnette, Will Friedle, Roddy Gray, Randall Haynes, Frank Hoyt Taylor, Tom Parati, John Brasington
A man suffers from a severe cardiac complaint, and when he loses his grandson in an accident he gets the youngster's heart. At the same time he strikes up a friendship with a young girl who has run away from home, and tries to help her sort her life out. A fairly pleasant drama, occasionally quite touching, though the plot does tend to feel both contrived and manipulative. See also SOLOMON'S CHOICE.
DRA   89 min mTV   SATrel: MOVIE CHANNEL
Boa: novel Set For Life by Judith Freeman.

GIGI **** PG
Vincente Minnelli USA   1958
Leslie Caron, Maurice Chevalier, Louis Jourdan, Hermione Gingold, Jacques Bergerac, Eva Gabor, Isabel Jeans, John Abbott, Monique Van Vooren, Lydia Stevens, Edwin Jerome, Dorothy Neumann, Marilyn Sims, Richard Bean
Set in Paris at the turn of the century, this famous musical tells of a young tomboy who grows up into a beautiful woman and is trained to be a courtesan. A colourful joyful romance. AA: Pic, Dir, (J. Ruttenberg), Edit (A. Fazan), Art/Set (W.A. Horning and P. Ames/F.K. Gleason and H. Grace), Screen/adapt (A.J. Lerner), Cost (C. Beaton), Score (A. Previn), Song ("Gigi" – F. Loewe (m)/A.J. Lerner (l)).
MUS   110 min (ort 119 min)   VIDrel: MGM/WHV   V/dm
Boa: novel by Colette.

GIGI AND THE FOUNTAIN OF YOUTH: PARTS 1 AND 2 *** U
Hiroshi Watanabe JAPAN   1984
Voices of: Reva West, Lisa Paulette, Sal Russo, Abe Hurt, Betty Gustafson, Ryan Flanagan, Anita Pia, Sam Jones
Two-part kid's adventure in which Gigi is a princess sent to Earth to learn its ways before she can inherit her parent's kingdom on another world. On Earth she meets Peter, who controls the magical Fountain of Youth and together they embark on a struggle to stop it falling into the wrong hands. A pleasant romp.
Aka; MAGICAL PRINCESS GIGI, THE; MAGICAL WORLD OF GIGI, THE
ANIM   77 min (ort 80 min)   VIDrel: TRING   V

GILDA *** PG
Charles Vidor USA   1946
Rita Hayworth, Glenn Ford, George Macready, Steven Geray, Joseph Calleia, Joe Sawyer, Gerald Mohr, Ludwig Donath, Don Douglas, Lionel Royce, Saul Z. Martel, George J. Lewis, Rosa Rey, Ruth Roman, Ted Hecht, Argentina Brunetti

In South America, a drifter gets taken on as a casino owner's right-hand man and becomes unwillingly involved with the latter's glamorous young wife, having known her before her marriage. An entertaining example of film noir, with good moments of tension but spoilt by a clumsily contrived ending. A highspot is Hayworth singing "Put The Blame On Me".
DRA   105 min B/W   VIDrel: COLUM/SONOP   V

GIMME AN "F" * 15
Paul Justman USA   1984
Steve Shellen, John Karlen, Mark Keyloun, Daphne Ashbrook, Beth Miller, Jennifer C. Cooke, Karen Kelly, Sarah M. Miles, Clyde Kusatsu, Doris Hess, Kathryn Harrison, Tyra Ferrell, Patricia Duff, Leslie Ryan, Audrey Saunders
Undistinguished teen comedy about a cheerleaders' contest with the inevitable high school frolics. Don't they ever do any schoolwork?
Aka: T & A ACADEMY 2
COM   96 min (ort 103 min)   VIDrel: 20TH   V

GINGER AND FRED *** PG
Federico Fellini FRANCE/ITALY/WEST GERMANY   1986
Marcello Mastroianni, Giulietta Masina, Franco Fabrizi, Frederick Von Ledenberg, Martin Maria Blau, Toto Mignone, Augusto Poderosi, Francesco Casale, Frederick Von Thun, Jacques Henri Lartigue, Ezio Marano
Two small-time dancers who once achieved a brief moment of fame as imitators of Rogers and Astaire are reunited for a TV show and rediscover the feelings they once had for each other. However, this very human story is set against a very sharp satire on TV and its inherent nastiness.
COM   122 min (ort 128 min)   VIDrel: ARROW/RTM   V

GIRL CAN'T HELP IT, THE *** U
Frank Tashlin USA   1956
Jayne Mansfield, Edmond O'Brien, Tom Ewell, Henry Jones, John Emery, Julie London, Ray Anthony, Fats Domino, Little Richard, The Platters, Gene Vincent and his Blue Caps, Juanita Moore, The Treniers, Eddie Fontaine
A gangster prevails upon a theatrical agent to groom the former's dumb girlfriend for stardom. A mixture of scatter-brained plotting, good sight gags and an endless series of jokes regarding the star's ample bosom. Worth a look for a set of classic performances from Fats Domino and friends. Songs include "Blue Monday", "You'll Never Know", "BeBop A Lula", "She's Got It" and the title song.
COM   97 min (ort 99 min)   VIDrel: 20TH/TECH   V/h
Boa: short story Do Re Mi by Garson Kanin.

GIRL IN THE CADILLAC ** 15
Lucas Platt USA   1995
Erika Eleniak, William McNmara, Bud Cort, Michael Lerner
A young girl hungry to experience life, leaves her small-town behind her and meets up with an escaped bank robber. Together this odd couple set out on the road.
A/AD   84 min (ort 89 min)   VIDrel: MED/COLUM   V
Boa: novel The Enchanted Isle by James M. Cain.

GIRL MOST LIKELY, THE ** U
Mitchell Leisen USA   1956
Jane Powell, Cliff Robertson, Keith Andes, Kaye Ballard, Tommy Noonan, Una Merkel, Kelly Brown, Judy Nugent, Frank Cady, Nacho Galindo, Chris Essay, Valentin De Vargas, Joseph Kearns, Julia Montoya, Paul Garay, Gloria De Ward
In this musical entertainment a girl has to choose who she wants to marry from among three men, one of whom is a millionaire. A boring remake of TOM, DICK AND HARRY that is partially redeemed by some fair songs.
MUS   94 min (ort 98 min)   VIDrel: VCC L/A   V

GIRL ON A MOTORCYCLE * 18
Jack Cardiff FRANCE/UK   1968
Marianne Faithfull, Alain Delon, Roger Mutton, Marius Goring, Catherine Jourdan, Jean Leduc, Jacques Marin, John G. Heller
One of the dullest and most pointless films ever made. A woman leaves her husband and goes on her motorbike to see her lover. The film consists of an interminable series of her recollections which take place along the journey. The sudden end is meant to shock but instead comes as a blessed relief.
Aka: LA MOTORCYCLETTE; NAKED UNDER LEATHER
DRA   86 min (ort 91 min)   VIDrel: CASPIC/BMGREC L/A   V
Boa: novel La Motocyclette by Pieyre de Mardiargues.

GIRL 6 ***
Spike Lee USA
18
1996
Theresa Randle, Isaiah Washington, Spike Lee, Jennifer Lewis, Debi Mazar, Peter Berg, Michael Imperioli, Dina Pearlman, Maggie Rush, Desi Moreno, Kristen Wilson, K. Funk, Debra Wilson, Naomi Campbell, Gretchen Mol, Shari Freels
A pretty and talented actress is unable to get employment in her field, so she starts work at a sex-line phone service, and so enters into the spirit of her new profession that it's not long before she's running the most popular line in town. With a catchy score by Prince, this is a lightweight and enjoyable affair, marking a radical departure from the director's usual offerings.
COM   104 min (ort 108 min) cC   VIDrel: 20TH/FOXVID
V/sur

GIRL TO KILL FOR, A *
Richard Oliver USA
18
1989
Karen Medak, Sasha Jenson, Karen Austin, Alex Cord, Rod McCary, Sandy Berumen, Adam Chambers, Tony Fasce, Guy Remsen, Eric Stern, Austin Stoker, Irene Tsu, Chopper
Story of a girl called Sue, so voluptuous that men might kill for a date with her. And some do just that, in this low-grade exercise in exploitative nonsense.
THR   85 min (ort 89 min)   VIDrel: EIV/SONOP   V

GIRL WHO SPELLED FREEDOM, THE ***
Simon Wincer USA
U
1986
Wayne Rogers, Mary Kay Place, Kieu Chinh, Kathleen Sisk, Margot Pinvidic, Susan Walden, Blu Mankuma, Jade Chinn, Diana Ung, Linda Wong, Jasmin Tam, Wilson Lo, Raymond Lau, Terry David Mulligan, Tom Heaton, Don Davis, Don Mackay
An American couple sponsor a refugee family for re-settlement in the USA, where they face the painful process of coming to terms with a completely new language and way of life. A surprisingly enjoyable story based on the real-life experiences of a Cambodian refugee who arrived in the USA in 1979, speaking hardly any English, and won a national spelling competition four years later. Scripted by Wincer, Christopher Knopf and David A. Simons.
DRA   91 min (ort 100 min) mTV   VIDrel: WDV/TECH L/A
V

GIRLS IN PRISON **
John McNaughton USA
(18)
1994
Ione Skye, Missy Crider, Diane McGee, Harvey Chao, Bahni Turpin, Ralph Meyening Jr, Letitia Hicks, J. Patrick McCormack, William S. Clark, William Boyett, David Paul Needles, Jon Polito, Annie Heche, Angie Ray McKinney
A remake of a 1950s B-movie, with Crider playing a girl who is sent to jail for her part in an armed robbery. Even though she protests her innocence, she soon learns that her cellmates are determined to learn where she has stashed the loot, and after the jail is wrecked in an earthquake, she is forced at gunpoint to lead some of the others to her place of concealment. Adequate enough, though not much better than the original. See also WHERE'S THE MONEY, NOREEN?
DRA   77 min (ort 85 min) mTV   SATrel: SKY MOVIES

GIRLS IN THE STREET **
Carl Monson USA
18
1972
John Kirkpatric, Frank Bannon, Rosie Stone, Brandy Lyman, Con Covert, Linda York, Tony Scaponi
Two cops have to track down a sadistic rapist and murderer who specialises in killing attractive women in this grimly unpleasant tale, and their investigations take them into the murky world of prostitutes and their pimps.
Aka: SCREAM IN THE STREETS, A; SCREAM STREET
HOR   60 min Cut (14 min 53 sec – ort 90 min)
VIDrel: MOPIC/QUANT   V

GIRLS JUST WANT TO HAVE FUN: THE MOVIE **
Alan Metter USA
PG
1985
Sarah Jessica Parker, Lee Montgomery, Morgan Woodward, Biff Yeager, Shannon Doherty, Jonathan Silverman, Helen Hunt, Holly Gagnier, Ed Lauter, Ian Giatti, Margaret Howell, Terence McGovern, Richard Blade, Kristi Somers
Three girls enter a musical competition for a cable TV station. A fairly cliche-ridden comedy that's saved by good performances from Parker and Hunt.
COM   83 min Cut (33 sec – ort 89 min)
VIDrel: POLY/POLYREC   V/sur

GIRLS' SCHOOL SCREAMERS *
John P. Finegan USA
18
1986
Mollie O'Mara, Sharon Christopher, Mari Butler, Beth O'Malley, Karen Krevitz, Vera Gallagher, Marcia Hinton, Monica Antonucci, Peter C. Cosimano, Charles Braun, Tony Manzo, John Turner, James Finegan Sr, Jeff Menapace
Seven girls spend a weekend in a deserted house to catalogue works of art prior to auction, and become prey to supernatural forces, screaming their heads off in this laughable effort.
Aka: DEATH LEGACY; GIRL SCHOOL SCREAMERS; PORTRAIT, THE
HOR   86 min   VIDrel: TROMA/RTM   V

GIVE ME A BREAK ***
James Lapine USA
PG
1993
Michael J. Fox, Christina Vidal, Nathan Lane, Cyndi Lauper
A former child star who is down on his luck now scratches a living working as a talent agent for kids, but finds his luck taking a turn for the better when he catches an eleven-year-old streetwise kid trying to steal his wallet. Quick to realise that she will be a natural in front of the camera, he sets about grooming her for stardom, but her past is soon to catch up with her. An enjoyable family comedy, inventive and funny.
COM   88 min   VIDrel: BUENA/TECH L/A   V

GIVE MY REGARDS TO BROAD STREET *
Peter Webb UK
PG
1984
Paul McCartney, Bryan Brown, Ringo Starr, Linda McCartney, Ralph Richardson, Barbara Bach, Tracey Ullman, George Martin, Ian Hastings, John Bennett, Luke McMasters, Philip Jackson, Marie Colett, John Harding, Mark Kingston
Musical fantasy revolving around the theft of McCartney's tapes for his latest album. A silly vanity vehicle that is worth listening to if not watching. Songs include "Ballroom Dancing" and "No More Lonely Nights", with the Victorian pastiche the highlight of an otherwise unremarkable film.
MUS   104 min (ort 109 min)   VIDrel: 20TH   V/sdur

GLADIATOR **
Rowdy Herrington USA
15
1992
James Marshall, Cuba Gooding Jr, Brian Dennehy, Robert Loggia, Ossie Davis, Jon Seda, Cara Buono, Lance Slaughter, T.E. Russell, Richard Lexsee, Tab Baker, Dwain A. Perry, Joan Schwenk, Francesca P. Roberts, Emily Marie Hooper
Marshall (in his debut feature) is a well educated teenage boxing champ drawn into Chicago's underground fight circuit in a bid to clear his widowed father's gambling debts. He finds himself trapped in seedy world and fighting a black friend of his, but ultimately wins freedom for both himself and his dad. A ridiculous and unconvincing tale, partly redeemed by Dennehy's performance as an evil promoter specialising in grudge fights between boxers from different races
A/AD   97 min (ort 102 min)   VIDrel: VCC/DISC/COLUM
V/sur

GLASS CAGE, THE **
Michael Schroeder USA
18
1996
Charlotte Lewis, Richard Tyson, Eric Roberts, Stephen Nichols, Joseph Campanella, Richard Moll
A new employee at an erotic strip-club is forced into becoming an assassin for the gangster who owns it.
DRA   92 min (ort 96 min)   VIDrel: 20TH/FOXVID   V/sur

GLASS HOUSE, THE ***
Tom Gries USA
15
1972
Alan Alda, Vic Morrow, Dean Jagger, Billy Dee Williams, Clu Gulager, Scott Hylands, Tony Mancini, Kristoffer Tabori, Luke Askew, Edward Bell, G. Wood, Roy Jenson, Alan Vint
A view of life in a state prison seen from the inside, and following two newcomers, one a college professor who has committed manslaughter and the other an idealistic warder. This harsh and convincing look at prison life follows one prisoner and his brave attempts to get involved in prison reform. The script is by Tracy Keenan Wynn.
Aka: TRUMAN CAPOTE'S THE GLASS HOUSE
DRA   87 min mTV   VIDrel: ODY/SONOP   V
Boa: short story by Truman Capote and Wyatt Cooper.

GLASS MENAGERIE, THE ***
Paul Newman USA
PG
1987
Joanne Woodward, Karen Allen, John Malkovich, James Naughton
Depression-era Southern drama with Woodward as the domineering matriarch whose memories of her happy youth as a Southern belle are her only means of escape from her cramped

existence in a St Louis apartment. Sharing her home with her crippled daughter and rebellious son, she attempts to arrange a date for the girl, but one night the arrival of a "gentleman caller" brings some bitter truths out into the open. A well made and powerfully acted drama.
DRA   129 min   VIDrel: RCA L/A   V
Boa: play by Tennessee Williams.

GLASS SHIELD, THE *** (15)
Charles Burnett USA   1995
Michael Boatman, Lori Petty, Ice Cube, Elliott Gould, Richard Anderson, Michael Ironside, Bernie Casey, Natalja Nogulich, Drew Snyder, M. Emmet Walsh, Gary Wood, Erich Andersen, Richard Anderson, Thomas W. Babson, Monty Bane
A rather aloof account of conditions in the Los Angeles Police Department that takes up the twin problems of racism and corruption (being based on the real-life experiences of John Eddie Johnson) casting Boatman and Petty as two honest cops out to expose these failings. See also SERPICO and THE PRINCE OF THE CITY, two more films dealing with the nasty issue of police corruption.
THR   106 min (ort 110 min)   SATrel: MOVIE CHANNEL

GLEAMING THE CUBE ** PG
Graeme Clifford USA   1988
Christian Slater, Steven Bauer, Ed Lauter, Micole Mercurio, Richard Herd, Charles Cyphers, Le Tuan, Minh Luong, Kieu Chinh, Art Chudabala, Peter Kwong, Max Perlich, Tommy Guerrero, Christian Jacobs, Joe Gosha, Andy Nguyen
A competent murder mystery that's clearly aimed at the teen set, with a rebellious skateboarding youngster out to solve the death of his adopted Vietnamese brother, getting some help along the way from a smart detective. A series of spectacular skateboarding stunts and the pleasing performances keep the film rolling despite the thin storyline.
A/AD   99 min Cut (14 sec plus some cuts substituted – ort 105 min)   VIDrel: MGM/WHV   V/sur

GLEN OR GLENDA? PG
Edward D. Wood Jr USA   1953
Bela Lugosi, Dolores Fuller, Daniel Davis (Edward D. Wood Jr), Lyle Talbot, Timothy Farrell, Tommy Haynes, Charles Crafts, Conrad Brooks, Henry Bederski, George Weiss
Documentary about transvestites which is possibly one of the most inept films ever made. By the same man who brought us the gloriously bad PLAN 9 FROM OUTER SPACE and was no mean cross-dresser himself, this film purports to be a sincere plea for tolerance of transvestism. A wild muddle of crazy dream sequences, stock footage and dire "Bevare!" warnings from Lugosi make this an absolute must for aficionados of bad movies.
Aka: HE OR SHE; I CHANGED MY SEX; I LED TWO LIVES; TRANSVESTITE, THE
DRA   70 min   VIDrel: CARL/TECH   V/dm

GLENGARRY GLEN ROSS *** 15
James Foley USA   1992
Al Pacino, Jack Lemmon, Alec Baldwin, Ed Harris, Alan Arkin, Kevin Spacey, Jonathan Pryce, Bruce Altman, Jude Ciccolella, Paul Butler, Lori Tan Chinn, Neal Jones, Barry Rossen plus voices of: Leigh French, Geore Cheung, Murphy Dunne
Stagebound but excellently acted and directed version of Mamet's bitingly satirical study of forty-eight hours in the lives of a group of real estate men and their tangled relationships with each other and their superiors. Scripted by Mamet from his Tony-award winning play, this is a sharp, funny and touching work, and Lemmon has rarely been better.
DRA   97 min (ort 100 min)   VIDrel: VCC/DISC/COLUM L/A   V/sh
Boa: play by David Mamet.

GLENN MILLER STORY, THE *** U
Anthony Mann USA   1954
James Stewart, June Allyson, Henry Morgan, Charles Drake, George Tobias, Frances Langford, Louis Armstrong, Gene Krupa, Ben Pollack, Marion Ross, Irving Bacon, Kathleen Lockhart, Barton MacLane, Sig Ruman, Phil Garris
An excellent account of the life of this unforgettable band leader, in which the over-sentimental plot takes very much second place to Miller's great music. AA: Sound (Leslie I. Carey).
MUS   108 min (ort 116 min)   VIDrel: CIC/SONOP   V/sur

GLIMMER MAN, THE ** 18
John Gray USA   1996
Steven Seagal, Keenan Ivory Wayans, Bob Gunton, Brian Cox, Michelle Johnson, John M. Jackson, Stephen Tobolowsky, Peter Jason, Ryan Cutrona, Richard Gant, Johnny Strong, Robert Mailhouse, Jesse Stock, Alexa Vega, Nikki Cox, Wendy Robie
When L.A. is struck by a spate of serial killings, a New York detective with special skills in this field is sent for, and is teamed up with a local police officer who reluctantly accepts the role of partner. As this ill-matched couple start their investigations, they are caught up in a complex intrigue. A dullish vehicle for Seagal, who in this movie earns the title nickname for the speed with which he nabs the baddies, but in truth looks lumbering and unconvincing.
A/AD   87 min (ort 91 min) cC   VIDrel: WHV   V/sur

GLORIA *** 15
John Cassavetes USA   1980
Gena Rowlands, Buck Henry, John Adames, Julie Carmen, Lupe Guarnica, Jessica Castillo, Tonu Knesich, Ralph Dolman, Israel Castro, Carlos Castro, Gregory Gleghorne, Philomena Spagnalo, Tom Noonan, Kyle-Scott Jackson, Gary Klarr
An ex-gangster's moll protects a young boy after his parents have been killed by the Mob, but finds that she has to resort to violence. Generally entertaining but seriously overlong. There is a point being made in there but one cannot be sure what it is. Written by Cassavetes.
DRA   116 min (ort 121 min) wScrn   VIDrel: TART/20TH   V

GLORY **** 15
Edward Zwick USA   1989
Matthew Broderick, Denzel Washington, Cary Elwes, Morgan Freeman, Jihmi Kennedy, Andre Braugher, John Finn, Donovan Leitch, John David Cullum, Bob Gunton, Cliff De Young
An anti-war film with a difference, that focuses on the role played by black soldiers in the American Civil War, when a naive Northerner is given the job of training the Union Army's first black unit. Partly based on the letters written by the unit's young commander, this outstanding story is performed with complete conviction and directed with consummate skill. Screenplay is by Kevin Jarre. AA: S. Actor (Washington), Cin (Freddie Francis).
DRA   117 min (Cut at film release by 4 sec – ort 128 min) wScrn   VIDrel: VCC/DISC/COLUM; ENCORE (LV only) V/sur LV

GO-BETWEEN, THE ** PG
Joseph Losey UK   1970
Julie Christie, Alan Bates, Michael Redgrave, Dominic Guard, Edward Fox, Margaret Leighton, Michael Gough, Richard Gibson, Simon Hume-Kendall, Roger Lloyd Pack, Amaryllis Garnett
A twelve-year-old boy is used to carry messages between his friend's sister and a poor tenant farmer. A rather ponderous and over-rated recreation of the life and times of Edwardian gentry. Scripted by Harold Pinter.
DRA   111 min (ort 116 min)   VIDrel: WHV   V/h
Boa: novel by L.P. Hartley.

GO FISH * 18
Rose Troche USA   1994
Guinevere Turner, V.S. Brodie, T. Wendy McMillan, Migdalia Melendez, Anastasia Sharp, Mary Garvey, Daniela Falcon, Tracy Kimme, Jennifer Allen, Walter Youngblood, Arthur C. Stone, Elspeth Kydd, Brooke Webster, Mimi Wadell
A tale of lesbian relationships, set among a group of writers in Chicago, with a couple of women trying to find partners for some of their friends. A poorly made amateur offering with unconvincing acting.
DRA   80 min (ort 83 min) B/W   VIDrel: MAINPIC/RTM   V

GO-KIDS, THE ** PG
Brian Trenchard Smith AUSTRALIA   1985
Henry Thomas, Tony Barry, David Ravensrood, John Ewart, Rachel Friend, Tamsin Wear, Dempsey Knight, John Ewart, Chris Gregory
A young boy investigates an Aborigine legend, and stumbles into an exciting adventure when he attempts to find out what lies at the bottom of the "Donkegin Hole". An enjoyable kids' adventure.
Aka: FROG DREAMING; SPIRIT CHASER
JUV   89 min   VIDrel: SGSVID/GOLD   V

GO TELL THE SPARTANS ***
Ted Post USA   1978
Burt Lancaster, Craig Wasson, Jonathan Goldsmith, Marc Singer, Joe Unger, David Clennon, Dolph Sweet, James Hong, Evan Kim, Dennis Howard, John Megna, Hilly Hicks, Clyde Kusatsu, Denice Kumagai, Tad Horino, Phong Diep
In Vietnam an experienced commander has to rescue a platoon of raw recruits from a Vietcong ambush. A realistic war story that is far better than one would have expected, mainly thanks to Wendell Mayes's literate and often caustic script.
WAR   110 min (ort 114 min)   VIDrel: MGM/WHV   V/h
Boa: novel Incident at Muc Wa by Daniel Ford.

GO TOWARD THE LIGHT **   PG
Mike Robe USA   1989
Linda Hamilton, Piper Laurie, Joshua Harris, Ned Beatty, Gary Bayer, Steve Eckholdt, Rosemary Dunsmore, Brian Bonsall, Mitchell Allen, Brian Lando, Richard Thomas, Ryan McWhorter, Jack Tate, John Wesley, Madison Mason
A sincere but sombre tale of a family who have to live with the certain knowledge that their haemophiliac child has contracted AIDS and is soon to die. The film does what it can with the material, but the constant note of gloom (despite the characters having a strong belief an afterlife – hence the title) makes it something of an ordeal to sit through.
Aka: GO TO THE LIGHT
DRA   89 min (ort 100 min) mTV   VIDrel: BANO/SGSVID   V

GOALKEEPER'S FEAR OF THE PENALTY, THE **   PG
Wim Wenders AUSTRIA/WEST GERMANY   1971
Arthur Brauss, Erika Pluhar, Kai Fischer, Maria Bardischewski, Ligbart Schwartz, Michael Troost, Bert Fortrell, Edda Kochl, Mario Kranz, Ernst Meister, Rosl Dorena, Ruid Schippel, Monika Poschel, Sybile Danzer
After having let in a penalty shot, a goalkeeper meets a cinema cashier and eventually strangles her. Here, sport supposedly serves as some kind of metaphor for modern life and its anxieties, but this bizarre film with its static camera and slow fades between scenes seems as deliberately obscure as it is allusive. The script is by Handke.
Aka: ANXIETY OF THE GOALIE AT THE PENALTY, THE; DIE ANGST DES TORMANNS BEIM ELFMETER; GOALIE'S ANXIETY AT THE PENALTY KICK, THE
DRA   96 min (ort 101 min)   VIDrel: CONNO/RTM   V
Boa: novel by Peter Handke.

GOD OF GAMBLERS, THE **   18
Wong Ching HONG KONG   1989
Chow Yun Fat, Andy Lau, Tony Leung, Chu Chien Lin
The best gambler in the world suffers from amnesia and is taken in by a family who learn about his gift and exploit him in order to make money. Meanwhile, his associates murder this girlfriend as they struggle to take over his business interests. A couple of sequels followed.
A/AD   120 min wScrn   VIDrel: MADE/RTM   V

GOD OF GAMBLERS 3: RETURN TO SHANGHAI **   15
HONG KONG   199-
Gong Li, Chow Sing Chi, Ng Man Tat
A top gambler uses his telekinetic powers to ensure he always wins at cards, but when he gets into a duel with a rival gambler, he suddenly finds himself thrown back in time to 1930s Shanghai. There he not only meets his own grandfather, but finds he is somehow involved in a gangland feud. There is also a rather silly romantic sub-plot too. A ludicrously plotted film that is very hard to follow, and is really best enjoyed with the brain switched off.
A/AD   111 min wScrn   VIDrel: MADE/RTM   V

GODFATHER, THE ****   18
Francis Ford Coppola USA   1971
Marlon Brando, Al Pacino, Robert Duvall, James Caan, Diane Keaton, Sterling Hayden, Richard Castellano, John Cazale, Talia Shire, Richard Conte, John Marley, Al Lettieri, Abe Vigoda, Morgana King, Alex Rocco
A superb evocation of the life of a Mafia family and the problems that arise when one of the sons takes over as head on the death of his father. Not so much a sequel as a closer examination of the story followed soon after. The script is by Francis Ford Coppola and Mario Puzo. AA: Pic, Actor (Brando), Screen/adapt (Puzo/Coppola).
DRA   167 min (ort 175 min)   VIDrel: CIC/SONOP   V
Boa: novel by Mario Puzo.

GODFATHER PART 2, THE ****   15
Francis Ford Coppola USA   1974
Al Pacino, Robert De Niro, Robert Duvall, Diane Keaton, Lee Strasberg, Talia Shire, John Cazale, Michael V. Gazzo, G.D. Spradlin, Morgana King, Mariana Hill, Troy Donahue, Joe Spinell, Abe Vigoda, Fay Spain, Harry Dean Stanton
A sequel that alternates between the early days of the title character and the problems experienced by his son and successor. Somewhat convoluted and hard to follow, but full of memorable moments and thoroughly entertaining. AA: Pic, Dir, S. Actor (De Niro), Screen/adapt (Puzo/Coppola) Art/Set (Dean Tavoularis and Angelo Graham/George R. Nelson), Score/orig (Nino Rota/Carmine Coppola).
DRA   190 min (ort 200 min)   VIDrel: CIC/SONOP   V
Boa: novel by Mario Puzo.

GODFATHER PART 3, THE ****   15
Francis Ford Coppola USA   1990
Al Pacino, Diane Keaton, Talia Shire, Andy Garcia, Eli Wallach, Joe Mantegna, Bridget Fonda, George Hamilton, Sofia Coppola, Raf Vallone, Franc D'Ambrosio, Donal Donnelly, Richard Bright, Helmut Berger, Don Novello
Written by Puzo and Coppola, this GODFATHER sequel works extremely well, and charts the career of Pacino as the new Don, and his ever-frustrated plans to break free of a crime-ridden life. A potent epic of violence and tragedy, with memorable sequences and acting, notably from Garcia and Shire as the Don's nephew and sister. Sadly, the same cannot be said for the abysmal acting of Coppola's daughter, who is badly cast as Don Corleone's daughter.
Aka: GODFATHER 3, THE
DRA   163 min (director's cut – ort 170 min)
VIDrel: CIC/SONOP   V/sur

GODS MUST BE CRAZY, THE ***   PG
Jamie Uys BOTSWANA   1981
N!xau, Marius Weyers, Sandra Prinsloo, Louw Verwey, Jamie Uys, Michael Thys, Nic De Jager, Fanyan Sidumo, Joe Seakatsie, Ken Gampu, Brian O'Shaughnessy, Vera Blacker, Paddy O'Byrne
A Coca Cola bottle dropped from a plane, brings havoc to the life of a tribe of Kalahari bushmen. An unexpectedly engaging comedy that after a slow start parades a sequence of slapstick incidents and catastrophes. Written by Uys and followed by a sequel.
COM   108 min   VIDrel: 20TH/TECH L/A   V

GODS MUST BE CRAZY 2, THE ***   PG
Jamie Uys BOTSWANA/USA   1989 (released 1990)
N!xau, Lena Farugia, Hans Strydom, Eiros, Nadies, Erick Bowen, Pierre Van Pletzen, Treasure Tshabala, Lourens Swanepoel, Richard Loring, Lesley Fox, Simon Sabela, Ken Marshall, Peter Tunstall, Andrew Dibb, Shimane Mpepela
A charming and gently amusing sequel to the first film that follows the exploits of our bushman N!xau, who has to retrieve his two children when they are inadvertently carried off to civilisation in the back of a poacher's truck. The film began production in 1985 but completion and release were delayed for several years.
COM   93 min Cut (1 sec plus some cuts subst – ort 97 min)
VIDrel: 20TH/TECH L/A   V

GODS MUST BE CRAZY 4, THE **   PG
Wellson Chin HONG KONG   1993
N!xau, Carina Lar Kar Ling, Gaco G'oma, Cecilia Yip, Lau Ching Wan, Conrad Janeis, Law Lan, Paul Che, Sze Kai Keung, James Pax, Michael Chow, Kwok Man Ki, Wang Kam Kwong, Stuart Wolfendale, Neil McCarthy, Eliza Marais, Connell Wettan
Whilst filming an American commercial in the jungle, a woman is saved from a lion's attack by a Tarzan-like figure.
Aka: CRAZY HONG KONG
COM   91 min   VIDrel: 20TH/FOXVID   V/h

GODZILLA: DESTROY ALL MONSTERS **   PG
Inoshiro Honda JAPAN   1968
Akira Kubo, Jun Tazaki, Yoshiro Tsuchiya, Kyoko Ai, Yukiko Kobayashi, Kenji Sahara, Nadao Kirino, Haruo Nakayima
Godzilla, Mothra, Rodan and a variety of other stars from the Toho Studio fall under the control of aliens based on the moon and are used to subjugate the Earth. After the obligatory rampages they regain their independence and see off Earth's

aggressors. A delightfully inept monster rally: childish, stilted and unintentionally funny.
Aka: ATTACK OF THE MARCHING MONSTERS; DESTROY ALL MONSTERS; KAIJU SOSHINGEKI; OPERATION MONSTERLAND
FAN   89 min   VIDrel: POLY/POLYREC   V

GODZILLA: EBIRAH – HORROR OF THE DEEP ** PG
Jun Fukuda JAPAN 1966
Akira Takarada, Kumi Mizuno, Akihiko Hirata, Jun Tazaki, Hideo Sunazuka, Toru Watanabe, Haruo Nakayima, Chotaro Togin, Toru Ibuki, Hideko Amamoto, Ikeo Sawamura
An unexpectedly enjoyable monster outing that sees Godzilla in a benign frame of mind (if he has one) and out to see off a gigantic shrimp. By way of a sub-plot, some castaways help a group of captives imprisoned on an island by a nasty bunch of villains. Mothra finds time to put in a brief appearance (and he makes the most of it).
Aka: EBIRAH – HORROR OF THE DEEP; GODZILLA VERSUS THE SEA MONSTER; NANKAI NO DAIKETTO
FAN   87 min   VIDrel: POLY/POLYREC   V

GODZILLA: GODZILLA VERSUS MEGALON * PG
Jun Fukuda JAPAN 1973
Katsuhiko Sasaki, Hiroyuki Kawase, Yutaka Hayashi, Mikita Mori, Kotaro Tomita, Gen Nakajima, Michio Ikeda, Sakei Mikami, Eisuke Nakanishi, Shinji Takaki
Possibly one of the weakest entries in this long-running series, this film has Godzilla battling a brace of monsters as well as a race of underground beings. On hand to aid him in his heroic battles is a jet-propelled super robot. The truly awful special effects and recycled story sink this one beyond redemption.
Aka: GODZILLA VERSUS MEGALON; GOJIRA TAI MEGARO
FAN   82 min wScrn dubbed   VIDrel: POLY/POLYREC   V

GODZILLA: GODZILLA VERSUS MOTHRA * PG
Takao Okawara/Koichi Kawakita JAPAN 1963
Akira Takarada, Yuriko Hoshi, Hiroshi Koizumi, Yu Fujiki, Emi Itoh, Yumi Itoh, Yoshifumi Jajima, Kenji Sahara, Jun Tazaki, Ikio Sawamura, Kenzo Tadake, Susumu Fujita, Yutaka Sada, Yoshio Kosugi, Yutaka Nakayama, Joji Uno
Godzilla battles Mothra the giant moth and kills it, but the latter has laid eggs that hatch into two giant caterpillars that spin a cocoon around our monster hero, thereby rendering him helpless (and ruining his cardigans in the process). It's not long before he is also fighting another creature – the infamous Battra.
Aka: GODZILLA FIGHTS THE GIANT MOTH; GODZILLA VERSUS MOTHRA; GODZILLA VERSUS THE GIANT MOTH; GODZILLA VERSUS THE THING; GOJIRA TAI MOSURA; MOSURA TAI GOJIRA; MOTHRA VERSUS GODZILLA
FAN   102 min dubbed   VIDrel: MANGA/SONOP   V

GODZILLA: GODZILLA VERSUS THE COSMIC MONSTER * (12)
Jun Fukuda JAPAN 1974
Masaaki Daimon, Kazuya Aoyama, Akihiko Hirata, Reiko Tajima, Barbara Lynn, Hiroshi Koizumi, Masao Imafukuu, Mori Kishida, Kenji Sahara
Aliens attempt to invade Earth using a steel replica of Godzilla. The real flesh-and-scales monster comes to the rescue, but finds he needs help from an Okinawan monster spirit in order to vanquish his opponent, who is out to destroy the world. Delightfully insane rubbish, badly made, badly dubbed and badly acted. One of a series.
Aka: GODZILLA VERSUS THE BIONIC MONSTER; GODZILLA VERSUS MECHAGODZILLA; GOJIRA TAI MEKA GOJIRA
FAN   80 min dubbed   VIDrel: L/A   V

GODZILLA: INVASION OF THE ASTRO-MONSTERS * PG
Inoshiro Honda JAPAN 1968
Akira Kubo, Jun Tazaki, Yoshio Tsuchiya, Kyoko Ai, Yukiko Kobayashi, Kenji Sahara, Nadao Kirino, Haruo Nakayima
Aliens are out to borrow Godzilla and Rodan and use them to beat up Ghidrah (who also trades under the name of Monster Zero). But in reality this could all be part of a fiendish plan to grab our monsters and use them against us. A tedious monster movie with few of the unintentional touches of humour that do so much to enliven these epics.
Aka: BATTLE OF THE ASTROS; GODZILLA VERSUS MONSTER ZERO; INVASION OF THE ASTRO-MONSTERS; KAIJU DAI SENSO; MONSTER ZERO
FAN   88 min   VIDrel: POLY/POLYREC   V

GODZILLA: KING KONG VERSUS GODZILLA ** PG
Inoshiro Honda/Thomas Montgomery JAPAN/USA 1963
Michael Keith, James Yagi, Tadao Takashima, Mie Hama, Kenji Sahara, Akihiko Hirata, Ichiro Arishima, Tatsuo Matsumura, Yu Fujiki, Harry Holcombe, Eiko Wakabayshi, Senkichi Omura
Apart from an exciting finale, set atop Mount Fuji where the two title creatures clash, this is a rather slow monster film, short on action and long on talk. The good special effects raise it above the level of the usual run of these films, but cannot compensate for those long boring periods.
Aka: KINGU KONGU TAI GOJIRA; KING KONG VERSUS GODZILLA
FAN   87 min (ort 90 min)   VIDrel: CIC/SONOP L/A   V

GODZILLA: MONSTERS FROM AN UNKNOWN PLANET ** 15
Inoshiro Honda JAPAN 1975
Katsuhiko Sasaki, Tomoko Ai, Akihiko Hirata, Tadao Nakamuru, Katsumasu Uchida, Goro Mutsu, Kenji Sahara, Toru Kawane, Kazunari Mori, Tatsumi Fuyamoto
Aliens are served by an embittered scientist and his daughter, who was repaired as a cyborg by them following a fatal accident. The aliens launch a robot Godzilla and another monster against the Earth, with both of them controlled by the cyborg woman. Fortunately, Godzilla is on hand to see them all off and save the planet, in this agreeably ludicrous nonsense. This was Godzilla's 15th film and marked twenty years in the business for him.
Aka: ESCAPE OF MEGAGODZILLA, THE; MEKAGOJIRA NO GYAKUSHU; MONSTERS FROM AN UNKNOWN PLANET; TERROR OF GODZILLA, THE; TERROR OF MECHAGODZILLA, THE
FAN   80 min (ort 83 min) wScrn dubbed
VIDrel: POLY/POLYREC   V

GODZILLA: SON OF GODZILLA ** 15
Jun Fukuda JAPAN 1967
Tadao Takashima, Akira Kubo, Bibari Maeda, Akihiko Hirata, Yoshio Tsuchiya, Kenji Sahara, Haruo Nakayima, Susumu Kurobe
Godzilla lays an egg which later hatches into Minya, his son, and both our friendly green lizards face assault by giant insects, in this harmless juvenile romp. Average.
Aka: GOJIRA NO MUSUKO; SON OF GODZILLA
FAN   86 min wScrn dubbed   VIDrel: POLY/POLYREC   V

GODZILLA: THE LEGEND IS REBORN * PG
Kohji Hashimodo/R.J. Kizer JAPAN 1985
Raymond Burr, Keiju Kobayashi, Ken Tanaka, Yasuko Sawaguchi, Yusuke Natsuki, Shin Takuma, Eitaro Ozara, Teketoshi Naito, Takeshi Katoh, Nobuo Kaneko, Warren Kemmerling, James Hess, Tesuya Takeda, Travis Swords, Takashi Katoh
Thirty years after wrecking Tokyo in 1956, Godzilla returns, proving that a lack of fresh ideas is not restricted to Hollywood.
Aka: GODZILLA; GODZILLA 1985; GOJIRA
FAN   91 min   VIDrel: NWV L/A   V

GODZILLA: WAR OF THE MONSTERS * PG
Jun Fukada JAPAN 1972
Minoru Takashima, Hiroshi Ichikawa, Tomoko Umeda, Yuriko Hishimi, Kunio Murai, Susumu Fujita, Toshiaki Nishizawa
Aliens from a planet dying from pollution take over a children's amusement park and summon two evil monsters, Ghidorah and Gaigan to attack Earth. As luck would have it, we have Godzilla fighting in our corner so all ends happily. Despite the underlying seriousness of the pollution theme, this is puerile stuff indeed.
Aka: CHIKIYU KOGERI MEIREI; GODZILLA ON MONSTER ISLAND; GODZILLA TAI GAIGAN; GODZILLA VERSUS GIGAN; GODZILLA, WAR OF THE MONSTERS; GOJIRA TAI GAIGAN; WAR OF THE MONSTERS
FAN   89 min wScrn dubbed   VIDrel: POLY/POLYREC   V

GOIN' SOUTH ** PG
Jack Nicholson USA 1978
Jack Nicholson, Mary Steenburgen, Christopher Lloyd, Veronica Cartwright, John Belushi, Richard Bradford, Lucy Lee Flippen, Jeff Morris, Danny DeVito, Tracey Walter, Gerald H. Reynolds, Luana Anders, George W. Smith
An outlaw is saved from hanging by agreeing to marry, but his wife has far from romantic notions about what his obligations towards her are. An offbeat and mildly diverting comedy Western.
WES   104 min (ort 109 min)   VIDrel: CIC/SONOP   V

GOING UNDERGROUND ** 15
David Carson USA 1993
Joanna Kerns, Tim Matheson, LaTanya Richardson, Katherine Cortez,

Elisabeth Franz, Bruce McGill, Justin Isfeld, Khandi Alexander, Corinne Bohrer, Ashley Peldon, Theresa Saldana, Andrew Craig, Charles Walker, Robert Nadir, Ellen Dubin
A woman flees her brutal and abusive husband, but then has to battle to gain custody of her children, having been forced to leave them behind in her desire to escape.
Aka: SHAMEFUL SECRETS
DRA   88 min (ort 92 min)   VIDrel: ODY/SONOP   V

GOLD DIGGERS OF 1933 ****
Mervyn Le Roy USA
Joan Blondell, Dick Powell, Ginger Rogers, Ruby Keeler, Aline MacMahon, Warren Williams, Guy Kibbee, Ned Sparks, Clarence Nordstrom, Sterling Holloway, Robert Agnew, Tammany Young, Ferdinand Gottschalk, Lynn Browning
Flamboyant Busby Berkeley musical built around the attempts of showgirls to help a songwriter save his show. Contains numerous fine song and dance numbers such as: "Remember My Forgotten Man", "We're In The Money" and the "Shadow Waltz". A few slight sub-plots such as the affair between Powell and Keeler are added for good measure, but this one stands or falls by its superb musical sequences.
MUS   96 min B/W   VIDrel: MGM/WHV L/A   V

U
1933

GOLD DIGGERS OF 1935 ****
Busby Berkeley USA
Dick Powell, Adolphe Menjou, Gloria Stuart, Alice Brady, Glenda Farrell, Frank McHugh, Winifred Shaw, Hugh Herbert, Grant Mitchell, Dorothy Dare, Alice Brady, Frank McHugh, Joe Cawthorn, Ramon & Rosita, Matty King
Berkeley's first solo as choreographer and director is a flashy big-scale musical with the usual flimsy plot (a socialite wants to stage a show at her country seat) providing a few laughs in between the superb precision numbers such as: "The Words Are In My Heart" and the Oscar-winning classic that is sung by Shaw. AA: Song ("Lullaby Of Broadway" – Harry Warren (m)/Al Dubin (l)).
MUS   91 min (ort 96 min) B/W   VIDrel: WHV   V

U
1935

GOLD DIGGERS: THE SECRET OF BEAR MOUNTAIN **
Kevin James Dobson USA
Christina Ricci, Anna Chlumsky, Polly Draper, Brian Kerwin, Diana Scarwid, David Keith
A girl from the city comes to the countryside and strikes up a friendship with a local tomboy. They spend their time repairing a boat and take a trip on the river in search of buried treasure. The contrived nature of the script (Keith is given a thankless role as an abusive husband) offers us a story that is filtered through a layer of typical 1990s feminism, spoiling what might well have been an effective adventure.
JUV   89 min (ort 94 min) cC   VIDrel: CIC/SONOP   V

PG
1996

GOLDEN BALLS **
Bigas Luna SPAIN
Javier Bardem, Maria De Medeiros, Maribel Verdu, Elisa Touati, Racquel Bianca, Alessandro Gassmann, Benisio del Toro, Francesco Ma Dominedo, Albert Vidal, Angel de Andres Lopez, Alicia Moro, Enrici Cusi, Pacao Casares
An ambitious construction worker marries a wealthy woman purely for purely selfish reasons, his great dream in life being to have his own skyscraper. Despite being married, he still keeps a mistress. One of those overheated Spanish comedies, often outrageous, occasionally funny and generally quite tiring.
Aka: HUEVOS DE ORO
COM   88 min (ort 95 min) wScrn   VIDrel: TART/20TH   V

18
1993

GOLDEN BRAID ***
Paul Cox AUSTRALIA
Chris Haywood, Gosia Dobrowolska, Norman Kaye, Paul Chubb, Robert Menzies, Jo Kennedy
Inspired by a short Guy de Maupassant story, this film tells of a shy and thoughtful clockmaker and his obsession with a braid of golden hair he finds in an 18th century cabinet. A warmhearted and charming affirmation of life, love and the passage of time.
COM   87 min (ort 88 min)   VIDrel: ARTIF/20TH   V
Boa: short story by Guy de Maupassant.

15
1990

GOLDEN CHILD, THE *
Michael Ritchie USA
Eddie Murphy, Charlotte Lewis, Charles Dance, Randall "Tex" Cobb, Victor Wong, James Hong, J.L. Reate, Shakati, Tau Logo, Tiger Chung Lee, Pons Marr, Peter Kwong

PG
1986

A social worker who specialises in tracing missing kids is given the job of tracking down a "perfect child" who, despite magical powers that an oracle predicts will be used to save the world from evil, has been kidnapped. An insufferable piece of hokum built around Murphy's smart wisecracking flair for comedy and Lewis's non-acting.
COM   89 min (ort 93 min)   VIDrel: 4-FRONT/POLYREC L/A   V/sh

GOLDEN DRAGON, SILVER SNAKE **
Godfrey Ho HONG KONG
Dragon Lee, Johnnie Chan, Merilyn Lee, Patrick Reynolds, Malcolm Levy, Kong Tao, Martin Chong
A fighter known as the "Golden Dragon" teams up with one known as the "Silver Snake" to avenge the murder of his brother.
MAR   85 min   VIDrel: IMPENT   V

18

GOLDEN GATE **
John Madden USA
Matt Dillon, Joan Chen, Bruno Kirby, Teri Polo, Tzi Ma, Stan Egi, Jack Shearer, Peter Murnik, George Guidall, Elizabeth Moreheard, Keone Young, Cully Fredericksen, Leo Downey, Jay Jacobus, Wilbur Jung, Henry Wong, Teresa Huyni
During the McCarthy era an ambitious young FBI agent sets out to frame an innocent Chinese and plays a major role in having him convicted of subversive activities. Some years later, his conscience begins to trouble him when he falls in love with his victim's daughter. A disappointing and unsatisfactory work that somehow manages to do very little with what could have been a fascinating story.
DRA   87 min (ort 91 min)   VIDrel: ENCORE/SPEAR V/sur

15
1993

GOLDEN NINJA WARRIOR **
Joseph Lai HONG KONG
Donald Owen, Queenie Yang, Richard Harrison, Morna Lee, David Chan, Nancy Cheng, Mike Tien
Another assembly-line martial arts caper featuring the familiar figure of the Ninja. In this tale the followers of the Red Ninja Empire are out to pinch a priceless gold statue owned by the Golden Ninja.
Aka: GOLDEN NINJA WARRIORS
MAR   79 min Cut (5 min 58 sec – ort 92 min)
VIDrel: IMPENT   V

18
1986

GOLDEN VOYAGE OF SINBAD, THE ***
Gordon Hessler UK
John Phillip Law, Tom Baker, Caroline Munro, Douglas Wilmer, Gregoire Aslan, Kurt Christian, Takis Emmanuel, John D. Garfield Jr, Martin Shaw, Aldo Sambrelli
Enjoyable 1950s-style rehash of all those earlier Sinbad adventures, with Sinbad and crew encountering one danger after another as they sail to the mysterious island of Lemuria in search of a golden crown. Highlights are the jerky but entertaining effects of Ray Harryhausen, such as a duel with the six-armed statue of a goddess. See also SINBAD AND THE EYE OF THE TIGER.
A/AD   100 min (ort 105 min)   VIDrel: VCC/DISC/COLUM V

U
1973

GOLDEN YEARS, THE **
Kenneth Fink USA
Keith Szarabajka, Felicity Huffman, Frances Sternhagen, R.D. Call, Bill Raymond, John Rothman, Ed Lauter, Mike Mulloy, Adam Redfield, Norman Craig Maxwell, Joe Inscoe, J. Michael Hunter, Richard Whiting, Randall Haynes
The ageing janitor of an agricultural research station who is about to get the push becomes the victim of an accident, which exposes him to chemicals being used in a tissue regeneration experiment. This leads to him getting ever younger and eventually mutating into a being of strange powers, much to the dismay of all around him. An odd fantasy of little merit but much brutality, and spoilt by a cop-out ending. Most disappointing. First shown as seven 50-minute episodes.
Aka: STEPHEN KING'S THE GOLDEN YEARS
FAN   236 min (ort 350 min) mTV
VIDrel: POLY/POLYREC/BRAVE   V
Boa: story by Stephen King.

15
1991

GOLDENEYE **
Don Boyd UK
Charles Dance, Phyllis Logan, Patrick Ryecart, Marsha Fitzalan,

15
1990

Lynsey Baxter, Ed Devereaux, Richard Griffiths, Julian Fellowes, David Forman, Joseph Long, Donald Douglas, David Quilter, Donald Hewlett, Kim Kindersley
Moderately interesting story built around the life and loves of the creator of the Bond novels – the writer Ian Fleming. See also SPYMAKER: THE SECRET LIFE OF IAN FLEMING.
A/AD 100 min VIDrel: CASPIC L/A V

GOLDENEYE **   12
Martin Campbell UK/USA   1995
Pierce Brosnan, Sean Bean, Izabella Scorupco, Famke Janssen, Joe Don Baker, Robbie Coltrane, Tcheky Karyo, Gottfried John, Alan Cumming, Desmond Llewelyn, Samantha Bond, Michael Kitchen, Serena Gordon, Simon Kunz, Pavel Douglas
No relation to the other GOLDENEYE movie, except inasmuch as it is a James Bond movie, with new Bond Brosnan given the task of infiltrating the Russian Mafia in order to locate a secret weapon (the eponymous "Goldeneye") that has fallen into the hands of former secret agent 006, who has now turned traitor. Not an exciting film, the script is dull and weakly plotted, and apart from a couple of good set-piece battles, there is little here to entertain.
A/AD 124 min (ort 130 min) wScrn cC
VIDrel: MGM/WHV V/sur

GOLDFINGER ***   PG
Guy Hamilton UK   1964
Sean Connery, Honor Blackman, Gert Frobe, Shirley Eaton, Bernard Lee, Harold Sakata, Tania Mallett, Martin Benson, Cec Linder, Austin Willis, Lois Maxwell, Bill Nagy, Nadja Regin, Alf Joint, Varley Thomas, Raymond Young, Richard Vernon
The third of the "James Bond" films is a well-plotted and exciting film, about our secret agent's attempt to foil a gold robbery planned by a criminal mastermind. Unfortunately, many of the strengths and interesting sub-plots of the novel are lost in an adaptation that gets bogged down in gadgetry and fisticuffs. Music is by John Barry with the title song performed by Shirley Bassey. THUNDERBALL followed. AA: Effects/aud (Norman Wanstall).
A/AD 105 min (Cut at film release – ort 112 min) wScrn cC
VIDrel: MGM/WHV V/dm V/h
Boa: novel by Ian Fleming.

GOLDWYN FOLLIES, THE **   U
George Marshall USA   1938
Adolphe Menjou, Andrea Leeds, Kenny Baker, The Ritz Brothers, Vera Zorina, Helen Jepson, Bobby Clark, Edgar Bergen and Charlie McCarthy, Phil Baker, Ella Logan, Jerome Cowan, Nydia Westman, Charles Kullmann, Frank Shields
A producer engages a woman to comment on his films from the standpoint of an ordinary person. This flimsy plot is merely an excuse for a repertoire of musical numbers by George and Ira Gershwin. A mixed bag of comedy and song and dance routines, with The Ritz Brothers the main highlight of the film.
MUS 111 min (ort 115 min) VIDrel: VCC/DISC/COLUM V

GOLDY: THE LAST OF THE GOLDEN BEARS **   U
Trevor Black USA   1984
Jeff Richards, Jessica Black, Kate Carlin
Set in California, this tale follows the friendship between a young orphaned girl and a lone prospector, and tells of their contact with a rare golden bear. A pleasant outdoors adventure that was followed by two sequels.
A/AD 91 min VIDrel: L/A V

GOLDY 2: THE SAGA OF THE GOLDEN BEAR *   U
Trevor Black USA   1986
Jessica Black, Jeff Richards, Dan Dalton, John Quinn
This sequel to GOLDY: THE LAST OF THE GOLDEN BEARS is a witless film of flat characterisation and over-cute scripting, that tells of a rescue mission that is mounted to save the title creature from the clutches of a wicked circus-owner.
JUV 91 min VIDrel: FIRST/SONOP L/A V

GOLDY 3: THE MAGIC OF THE GOLDEN BEAR **   U
John Quinn USA   1994
Cheech Marin, Mr T (Lawrence Tureaud), Ronnie Morgan, Danny Woodburn, Jeff Prettyman, John Quinn, Bren McKinley, Hilda Brooks, Kevin Brophy, Mike Moroff, Richard Molinare, Mark Hendrickson, Ken Johnson, Tiffany Quinn, Devon Quinn
A little girl runs away with her pet bear when she learns that a magician is planning to buy it, and the pair take refuge in the

forest. The third instalment in a series of kids' adventures featuring the title creature – a magical performing bear.
Aka: ADVENTURES OF THE GOLDEN BEAR, THE
JUV 101 min (ort 104 min) VIDrel: GUILD/FOXVID V

GOLEM, THE ***   PG
Pael Wegener GERMANY   1920
Paul Wegener, Albert Steinruck, Ernst Deutsch, Lyda Salmanova, Hans Strum, Fritz Feld, Lathar Menthel, Otto Gebuhr
According to film historians, this is the best version of this much-filmed legend of a 16th century rabbi in Prague. He animated a lifeless clay figure to protect the Jews from being expelled from their homes in the Ghetto by Imperial decree. To drive the message home, the rabbi took the monster on a visit to the Emperor, who impressed by its awesome power, rescinded his evil order.
Aka: DER GOLEM, WIE ER IN DIE WELT KAM; GOLEM, THE: HOW HE CAME INTO THE WORLD
HOR 69 min (ort 118 min) B/W silent
VIDrel: SCREAM/SPEAR V

GONE IN THE NIGHT **   15
Bill L. Norton USA
Shannen Doherty, Kevin Dillon, Dixie Carter, Timothy Carhart, Michael Brandon, James Handy, John Finn, Robert Desiderio, Edward Asner
When a couple's young daughter is kidnapped they find the police regard them as the chief suspects, and only with the aid of a cop, a journalist and a lawyer are they finally able to prove their innocence.
THR 171 min VIDrel: ODY/SONOP V/sh

GONE WITH THE WIND ****   PG
Victor Fleming USA   1939
Clark Gable, Vivien Leigh, Olivia De Havilland, Leslie Howard, Thomas Mitchell, Hattie McDaniel, Barbara O'Neil, Victor Jory, Laura Hope Crews, Ona Munson, Harry Davenport, Ann Rutherford, Evelyn Keyes, Carroll Nye
A movie great telling of the turbulent love affair and marriage of a wilful Southern belle during the time of the Civil War. Score is by Max Steiner, with MacDaniel the first black actress to win an Oscar. See SCARLETT: PARTS 1 AND 2. AA: Pic, Dir, Actress (Leigh), S. Actress (McDaniel), Screen (Sidney Howard), Cin (Ernest Haller/Ray Rennahan), Art (Lyle Wheeler), Spec Award (William Cameron Menzies for the use of colour in the enhancement of dramatic mood).
DRA 224 min (restored version – ort 240 min) cC
VIDrel: MGM/WHV V/h
Boa: novel by Margaret Mitchell.

GOOD COP BAD COP **   18
David A. Prior USA   1993
David Keith, Robert Hays, Pamela Anderson, Leo Rossi, Charles Napier, Stacy Keach, Javi Mulero, Bernard Hocke, Ted Prior, April Bogenschutz, Larry McKinley, Marshall Russell, Jeanette Kontomitras, Hal Jeansonne, David Veca, Butch Robbins
A modern-day bounty hunter gets involved in a murder mystery in this odd blending of comedy and action.
A/AD 88 min VIDrel: POLY/POLYREC/MED V

GOOD DIE FIRST FOR A HANDFUL OF SILVER, THE **   15
Giorgio Stegani ITALY   1968
Lee Van Cleef, Anthony (Antonio) Sabato, Bud Spencer (Carlo Pedersoli), Ann Smyrner, Lionel Stander, Graziella Granata, Herbert Fox, Carlo Gaddi, Gordon Mitchell, Enzo Fiermonte, Hans Elwenspoek, Gunther Stoll
A bandit and his comrades rob a stagecoach but later he befriends a man who saves his life and in a career change, becomes the sheriff of a small town. An extremely violent but well made spaghetti Western.
Aka: ABOVE THE LAW; AL DI LA DELLA LEGGE; BEYOND THE LAW; BLOODSILVER; GOOD DIE FIRST, THE
WES 90 min VIDrel: FUNNY/SGSVID V

GOOD EVENING VIETNAM! *   18
USA   1988
Molly O'Brien, Amber Lynn, Liz Randall, Nina Hartley, Pattie Petite, Marc Wallice, John Leslie, D.J. Starr
A sex film rip-off of GOOD MORNING, VIETNAM with a quartet of girls setting up a brothel in 1969 in the jungles of

Vietnam, their intention being to provide aid and comfort to the US personnel stationed in Saigon. A tasteless farce.
Aka: MERCENARIES OF LOVE
A   60 min   VIDrel: RAVEN/QUANT   V

GOOD FATHER, THE *   15
Mike Newell UK   1986
Anthony Hopkins, Jim Broadbent, Harriet Walter, Joanne Whalley, Simon Callow, Fanny Viner, Michael Byrne, Miriam Margolyes, Jeannie Stoller, Johanna Kirby, Stephen Fry, Clifford Rose, Harry Grubb, Tom Jamieson
The story of two recently separated men, and their feelings of rage and resentment at their estrangement from their children. A low-key typically British production.
DRA   86 min (ort 90 min) mTV   VIDrel: FIRST/SONOP   V
Boa: novel by Peter Prince.

GOOD GUYS WEAR BLACK *   15
Ted Post USA   1978
Chuck Norris, Anne Archer, James Franciscus, Dana Andrews, Lloyd Haynes, Jim Backus, Larry Casey, Tony Mannino
A quiet professor uses the skills he learnt as leader of a commando unit in Vietnam, to stay alive when he discovers that he is on a CIA hit list. A fast and furious display of Norris's athletic prowess (too bad about the storyline). Followed by A FORCE OF ONE.
A/AD   90 min (Cut at UK cinema release by 7 sec)
VIDrel: 4-FRONT/POLYREC   V

GOOD LITTLE GIRL *   18
Jean-Claude Roy FRANCE   1972
Beatrice Arnac, Bella Darvi, Jessica Dorn, Michele Girardon, Marie-George Pascal, Cathy Reghin, Sylvie Lafontaine, Pierre Moncorbier, Romain Bouteille, Francois Guerin, Vincent Gauthier, Nicole Isimat, Stephan Lorey, Kristin Genet
Four girls have sexual adventures an isolated chateau, in this early soft-core offering.
Aka: GOOD LITTLE GIRLS; LES PETITES FILLES MODELES
A   85 min   VIDrel: JEZ/RTM   V

GOOD LUCK, MISS WYCKOFF *   18
Marvin J. Chomsky USA   1979
Anne Heywood, Donald Pleasence, Robert Vaughn, Carolyn Jones, Dorothy Malone, Ronee Blakley, John Lafayette, Earl Holliman
A schoolteacher is introduced to sex by being raped, in this uneven and ineffective adaptation of Inge's novel.
Aka: SECRET YEARNINGS; SHAMING, THE; SIN, THE
DRA   90 min Cut (2 min 31 sec of abridged film – ort 105 min)   VIDrel: SCRN/TOTAL L/A   V
Boa: novel by William Inge.

GOOD MAN IN AFRICA, A *   15
Bruce Beresford USA   1993
Sean Connery, Colin Friels, Joanne Whalley-Kilmer, John Lithgow, Diana Rigg, Louis Gossett Jr., Maynard Eziashi, Sarah-Jane Fenton, Jackie Mofokeng, Themba Ndaba, David Phetoe, Patrick Mynhardt, George Lee, Aubrey Molefi, Jenny Mofokeng
A British diplomat in a newly independent African nation demonstrates to everyone's satisfaction that intelligence is not one of the requirements for the job, while the Africans too are not shown in too favourable a light. A silly and totally unamusing mess that wastes the efforts of all concerned. Screenplay is by Boyd.
COM   90 min (ort 94 min)   VIDrel: CIC/SONOP   V
Boa: novel by William Boyd.

GOOD MORNING, BABYLON *   15
Paolo Taviani/Vittorio Taviani FRANCE/ITALY/USA   1986
Vincent Spano, Greta Scacchi, Joaquim De Almeida, Charles Dance, Desiree Becker, Omero Antonutti, David Brandon, Berangere Bonvoisin, Brian Freilino, Margarita Lozano, Massimo Venturiello, Andrea Prodan, Dorotea Ausenda
The story of two Italian brothers who come to America hoping to make their fortune and wind up working on D.W. Griffith's epic film INTOLERANCE. A homage to the early days of film-making, rather simplistic but quite charming in its own way, although the sad ending is something of a surprise.
Aka: GOOD MORNING, BABYLONIA
DRA   112 min (ort 116 min)   VIDrel: ARTIF/20TH   V/h

GOOD MORNING, VIETNAM *   15
Barry Levinson USA   1988
Robin Williams, Forest Whitaker, Tung Tuanh Tran, Chintara

Sukapatana, Bruno Kirby, Robert Wuhl, J.T. Walsh, Noble Willingham, Floyd Vivino, Richard Edson, Richard Portnow, Juney Smith, Cu Ba Nguyen, Dan R. Stanton, DAnny Aiello III
An anarchic, manic anti-establishment DJ blasts the airwaves on the American Forces radio in Saigon in 1965. Williams gets plenty of opportunity to indulge his gift for improvisation in a never-ending stream of monologues, but the thin story fails to make any clear statement about the rights or wrongs of American involvement in the conflict. A comical, insane, fast trip nowhere, based to some extent on the exploits of real-life DJ Adrian Cronauer.
COM   116 min (ort 121 min)   VIDrel: TOUCH/TECH   V/sur

GOOD NEWS *   U
Charles Walters USA   1947
June Allyson, Peter Lawford, Patricia Marshall, Joan McCracken, Mel Torme, Ray McDonald, Robert Strickland, Donald MacBride, Tom Dugan, Clinton Sundberg, Loren Tindall, Connie Gilchrist, Morris Ankrum, Georgia Lee, Jane Green
A college football star faces problems with both his love life and his studies, since he has to pass his French exams or flunk out. As a way out of his dilemma, he agrees to be coached by a bright female student. A lively and lavish remake of the 1930 film, itself a remake of a 1927 movie. Good songs (such as: "Ladies Man", "Pass The Peace Pipe" and "The Best Things In Life Are Free") and a strong performance by Torme in his supporting role make this the best version.
MUS   89 min (ort 93 min)   VIDrel: MGM/WHV   V/dm
Boa: play by Lawrence Schwab, Lew Brown, Frank Mandek, Ray Henderson and Buddy De Sylva.

GOOD OLD BOYS, THE *   PG
Tommy Lee Jones USA   1995
Tommy Lee Jones, Sissy Spacek, Terry Kinney, Frances McDormand, Sam Shepard, Matt Damon, Wilford Brimley, Blayne Weaver, Bruce McGill, Larry Mahan, Richard Jones, Walter Olkewicz, Park Overall
Towards the end of the last century, a Texas cowboy is reluctantly forced to consider settling down and goes to visit his married brother. His presence leads to conflict with his sister-in-law while he also undertakes the uneasy courtship of a local schoolteacher. However, when a crooked banker threatens to steal the family ranch, the two brothers have to put their differences aside to defeat this wicked plan. A modern Western, well done but plotted along purely conventional lines.
WES   113 min (ort 117 min) mCab   VIDrel: 20VIS/SONOP   V/sh
Boa: novel by Elmer Kenton.

GOOD POLICEMAN, THE *   (12)
Peter Werner USA   1993
Ron Silver, Roy Dotrice, Joe Morton, Victor Slezak, Lenny Venito, Joanna Pacula, Blair Brown, Tom Signorellu, Daniel Van Bergen, Tony Darrow, Gene Saks, Jose Zuniga, Juan Cruz, Al Shannon, Judd Henry Baker, Ilene Armstrong, Ron Ryan
In New York, the idealistic Police Commissioner uses unorthodox methods to uphold the law, but this creates conflicts with both the criminal fraternity and the Mayor's Office. Slightly reminiscent of CITY HALL, this is one of those "shouting match" dramas that swiftly degenerates into an all-too-familiar succession of personal conflicts, with the truly interesting aspects of police work left out of the plot. Enjoyable, but only moderately so.
DRA   83 min (ort 90 min) mTV   SATrel: SKY MOVIES
Boa: novel by Jerome Charyn.

GOOD SON, THE *   18
Joseph Rubin USA   1993
Macaulay Culkin, Elijah Wood, Wendy Crewson, David Morse, David Hugh Kelly, Jacqueline Brookes, Quinn Culkin, Ashley Crow, Guy Strauss, Keith Brava, Jeremy Goodwin, Andria Hall, Bobby Huber, Mark Stefanich, Susan Hopper, Rory Culkin
A young boy comes to stay with his cousin and soon shows himself to be a sly and very disturbed psychopath who is able to maintain an air of almost impenetrable innocence. Naturally, everyone steadfastly refuses to believe that our little angel is the culprit. Initially withdrawn from UK cinema release following the Jamie Bulger murder case, but the violence in this film is both contrived and ludicrous.
THR   82 min Cut (ort 87 min) cC   VIDrel: 20TH/TECH   V/sur

GOOD, THE BAD, AND THE UGLY, THE *** 18
Sergio Leone ITALY/SPAIN 1966
Clint Eastwood, Eli Wallach, Lee Van Cleef, Rada Rassimov, Mario Brega, Aldo Giuffre, Chelo Alonso, Silvana Bacci, Luigi Pistilli, Enzo Petito, Al Mulloch, Claudio Scarchilli, Livio Lorenzon, Antonio Casale, Angelo Novi
This is the third and most colourful of the three "man with no name" films of Leone. Set against the background of the Civil War, Eastwood and Wallach form an uneasy partnership in a hunt for the proceeds of a bullion robbery. As before, Ennio Moricone provides an excellent score. Look out for the sequence in which Eastwood obtains his poncho, thus becoming the character of the first two films.
Aka: IL BUONO, IL BRUTO, IL CATTIVO
WES 155 min (ort 180 min) VIDrel: MGM/WHV V/sh

GOOD TO GO ** 15
Blaine Novak USA 1986
Art Garfunkel, Harris Yulin, Robert Doqui, Michael White, Reginald Daughtry, Richard Brooks, Hattie Winston, Paula Davis
A local journalist investigates the rape and murder of a nurse by a drummer, and finds the police less than co-operative, so he is forced to enlist the help of black slum kids. Really a youth musical in disguise, where the black "go-go" sound plays the major role, little else mattering.
Aka: SHORT FUSE
A/AD 86 min (ort 91 min) VIDrel: 20TH/TECH V/sur

GOODBYE BIRD, THE ** (U)
William Clark USA 1993
Cindy Pickett, Christopher Pettiet, Concetta Tomei, Wayne Rogers, Monique Lanier, Jesse Bennett, Marcia Dangerfield, Michael Scott, Tim Shoemaker, Frank Gerrish, Michael Flynn, Jason Tatom, Jeff Olson, Margaret Crowell, Wyn L. Howard
A young boy, accused by his school principal of stealing her talking parrot, meets a kindly vet, who helps him solve this mystery. Average.
JUV 91 min mTV SATrel: DISNEY CHANNEL

GOODBYE CRUEL WORLD * 15
David Irving USA 1982
Dick Shawn, Cynthia Sikes, Chuck Mitchell, Nicholas Niciphor
Before committing suicide, a TV anchorman decides to film his relatives who have driven him to take such a desperate measure. An offbeat comedy misfire.
Aka: UP THE WORLD
COM 90 min VIDrel: MOPIC/SGSVID V

GOODBYE EMMANUELLE * 18
Francois Leterrier FRANCE 1977
Sylvia Kristel, Umberto Orsini, Alexandra Stewart, Jean-Pierre Bouvier, Olga Georges-Picot, Jacques Doniol-Valcroze, Erik Golin, Jack Allen
The setting is the Seychelles this time, and our heroine is now married to an architect in yet another glossy, empty, interminable EMMANUELLE sequel. This was the second such sequel/spin-off in a seemingly endless series.
A 94 min (Cut at film release – ort 100 min)
VIDrel: 4-FRONT/POLYREC/BRAVE V

GOODBYE GIRL, THE *** PG
Herbert Ross USA 1977
Richard Dreyfuss, Marsha Mason, Paul Benedict, Quinn Cummings, Barbara Rhoades, Theresa Merritt, Michael Shawn, Nicol Williamson, Patricia Pearcy, Gene Castle, Daniel Levans, Marilyn Sokol, Anita Dangler, Victor Boothby, Dave Cass
Mason is a dancer whose husband has skipped off, deserting her and their ten-year-old daughter, and at the same time selling their apartment lease to an actor. In the resulting confusion Dreyfuss and Mason find themselves sharing the same apartment, in this warm and funny exploration of their growing, albeit initially reluctant friendship. A succession of funny one-liners sustains the paper-thin plot. Scripted by Simon. AA: Actor (Dreyfuss).
COM 106 min (ort 110 min) VIDrel: WHV V
Boa: play by Neil Simon.

GOODBYE, MISTER CHIPS **** U
Sam Wood UK 1939
Robert Donat, Greer Garson, Paul Von Hernreid (Paul Henreid), Judith Furse, Lyn Harding, Milton Rosmer, Frederick Lesiter, Louise Hampton, David Tree, Austin Trevor, Edmund Breon, Jill Furse, Guy Middleton, Nigel Stock
A sentimental but winning account of a shy schoolmaster and his devotion to his chosen career that focuses on the salient happenings in his life as he recalls the period he spent working at a boys' public school in England, from 1870 to 1928. AA: Actor (Donat).
DRA 110 min (ort 115 min) B/W VIDrel: MGM/WHV L/A V/h
Boa: novel by James Hilton.

GOODBYE, MR CHIPS * PG
Herbert Ross UK 1969
Peter O'Toole, Petula Clark, Michael Redgrave, Michael Bryant, George Baker, Jack Hedley, Sian Phillips, Alison Leggatt, Clinton Greyn, Michael Culver, Barbara Couper, Elspeth March, Clive Morton, Michael Ridgeway
A flashy and superficial musical remake of the classic 1939 study of the life of a shy and kindly schoolmaster. The Leslie Bricusse songs are no more than adequate, direction is uninspired and Clark as the teacher's showgirl sweetheart injects an absurd note into the script. Not so much a film as a rather bad mistake. This was Ross's directorial debut.
MUS 110 min (ort 147 min) VIDrel: MGM/WHV L/A V

GOODBYE PORK PIE ** 18
Geoff Murphy NEW ZEALAND 1980
Tony Barry, Kelly Johnson, Claire Oberman, Shirley Gruar, Jackie Lowitt, Don Selwyn, Shirley Dunn, Paki Cherrington, Christine Lloyd, Maggie Maxwell, John Ferdinand, Clyde Scott, Steven Tozer, Phil Gordon, Bruno Lawrence, Adele Chapman
Three young people are chased by the cops up and down New Zealand's freeways at great speed, and are arrested one by one in this good-looking but rather pointless tale.
A/AD 102 min (ort 105 min) VIDrel: ARTPRO/RTM V

GOODFELLAS **** 18
Martin Scorsese USA 1990
Ray Liotta, Robert De Niro, Paul Sorvino, Joe Pesci, Lorraine Bracco, Frank Sivero, Tony Darrow, Mike Starr, Frank Vincent, Chuck Low, Frank DiLeo, Gina Mastrogiacomo, Christopher Serrone, Henny Youngman, Jerry Vale
Pileggi spent four years interviewing former gangster Henry Hill to produce the book on which this film is based – a study of the life of a youngster growing up in Brooklyn and dreaming of becoming a mobster. An utterly compelling if violent slice of gangster life, rich in detail and atmosphere but perhaps just a touch overlong. Uniformly well acted, but Pesci is truly outstanding. The photography is by Michael Ballhaus. See also CASINO. AA S. Act (Pesci).
A/AD 139 min (ort 146 min) wScrn cC VIDrel: WHV V/sur
Boa: book Wiseguy by Nicholas Pileggi.

GOODNIGHT, SWEET WIFE: A MURDER IN BOSTON *** 15
Jerrold Freedman USA 1990
Ken Olin, Margaret Colin, B.D. Wong, Annabella Price, Anthony Tyler Quinn, James Handy, Michael Gwynne, Carl Anthony Payne II, Bruce McGill, Will Zahrn, Bill Visteen, Shawna Franks, Jim Ortlieb, Larry Brandenburg, Sandy Mashmeyer
A married couple are attacked and shot in their car, on the way home from an ante-natal clinic. Only the husband survives, but his confused account of the crime eventually arouses the suspicions of a local journalist, who mounts her own investigation. A well constructed dramatisation of a real crime, that makes entertaining viewing, mainly thanks to some fine performances and an intelligent and incisive script.
DRA 90 min mTV VIDrel: 20TH V

GOOFY MOVIE, A *** U
Kevin Lima USA 1995
Voices of: Bill Farmer, Jason Marsden, Jim Cummings, Kellie Martin, Rob Paulsen, Wallace Shawn, Jenna Von Oy, Frank Welker, Kevin Lima, Florence Stanley, Jo Anne Worley, Brittany Alyse Smith, Robyn Richards, Julie Brown
That old stalwart of the Disney cartoons, Goofy, lives quietly with his son Max, out somewhere in middle America. At the end of the school-term, father and son take off for a vacation, despite the fact that what the son really longs for is a career as a pop singer. Made by the Walt Disney Company, this unusual animation cleverly explores the tensions of old versus new, or father versus son. A wickedly inventive self-parody, it is only spoilt by the mawkish ending.
ANIM 75 min (ort 78 min) cC VIDrel: WDV/TECH V/sh

GOONIES, THE ** PG
Richard Donner USA 1985
Sean Austin, Josh Brolin, Jeff Cohen, Corey Feldman, Kerri Green, Robert Davi, Martha Plimpton, Ke Huy Quan, John Matuszak, Anne Ramsey, Mary Ellen Trainor, Joe Pantoliano, Lupe Ontiveros, Keith Walker, Curtis Hanson
A group of boys living in a small Oregon town, discover a pirate treasure map and embark on a wild adventure, in an attempt to save their home from land developers. A lively, noisy and badly-articulated film, aimed well and truly at kids, and based on a Steven Spielberg story as well as being produced by him.
JUV 109 min (Cut – sound only – ort 114 min)
VIDrel: WHV V/dm V/sur

GOR ** 15
Fritz Kiersch ITALY 1987
Urbano Barberini, Rebecca Ferratti, Jack Palance, Paul L. Smith, Oliver Reed, Larry Taylor, Graham Clarke, Janine Denison, Donna Denton, Jennifer Oltmann, Martina Brockschmidt, Ann Power, Arnold Vosloo, Chris Du Plessis
Following a car crash, a New England professor is catapulted into a parallel universe by a strange ring, and finds himself on Gor, a primitive planet ruled over by a barbaric tyrant. In order to return home he has to lead the rebels in an attack on the tyrant's domain. A cheap and nasty version of a rather good yarn from Norman's "Gor" series, and one more attempt to revive the fading interest in sword-and-sorcery films. Followed by OUTLAW OF GOR.
FAN 91 min VIDrel: MGM/WHV L/A V
Boa: novel Tarnsman of Gor by John Norman.

GORDY ** U
Mark Lewis USA 1995
Doug Stone, Roy Clark, Mickey Gilley, Kirsty Young, Michael Roescher, Justin Garms, Deborah Hobart, Tom Lester, Ted Manson
A young pig hits the road after his family are sent for slaughter and meets up with two kids who are delighted to learn that he can talk. Together, they enjoy all kinds of adventures, in this OK fantasy. See also BABE.
JUV 87 min (ort 90 min) cC VIDrel: WDV/TECH V/sh

GORGEOUS * 18
F.J. Lincoln USA 1994
Kaylan Nicole, Jewel, Nikole Lac, Alex Jordan, Tom Byron, Joey Silvera, Vince Voyeur
A high society girl tires of her luxurious lifestyle and hankers after a real job, finding her true vocation as the star attraction of a brothel. In this capacity she entertains the patrons by parading about in sexy outfits, and even finds true love when a member of her audience becomes smitten with her. A feeble effort, poorly acted and scripted.
A 65 min (ort 80 min) VIDrel: ONE V

GORGO ** PG
Eugene Lourie UK 1961
Bill Travers, William Sylvester, Vincent Winter, Bruce Seton, Joseph O'Conor, Martin Benson, Christopher Rhodes, Bsil Dignam, Bruce Seton, Maurice Kauffman, Howard Lang, Thomas Duggan, Dervis Ward, Barry Keegan
A prehistoric monster rampages through London on a rescue mission to save her baby from captivity, when the latter is captured off the Irish coast and brought to England to be exhibited. Quite a decent British variant on the GODZILLA theme.
FAN 73 min (ort 78 min) VIDrel: DDVID V

GORGON, THE ** 15
Terence Fisher UK 1964
Peter Cushing, Christopher Lee, Barbara Shelley, Richard Pasco, Michael Goodliffe, Patrick Troughton, Joseph O'Conor, Prudence Hyman, Jack Watson, Redmond Phillips, Jeremy Longhurst, Toni Gilpin, Alister Williamson
A castle ruin near a German village houses a girl whose gaze turns people to stone, for she turns into the title creature at every full moon. Made by Hammer Films, this yarn piles on the Gothic atmosphere but lacks any real suspense, and when we finally catch a glimpse of the creature (briefly – possibly for our own protection) it is disappointing. Direction is workmanlike, as is editing, and a fascinating idea remains unexplored.
HOR 80 min (ort 83 min)
VIDrel: ENCORE/SPEAR/COLUM V

GORILLAS IN THE MIST *** 15
Michael Apted USA 1988
Sigourney Weaver, Bryan Brown, Julie Harris, John Omirah Miluwi,
Iain Cuthbertson, Constantin Alexandrov, Waigwa Wachira, Constantin Alexandrov, Ian Glenn, David Lansbury, Maggie O'Neil, Kong Mbanda, Michael J. Reynolds
A biopic on the life of Dian Fossey, who embarked on a twenty year mission to save the mountain gorilla from extinction, from the time of her arrival in Rwanda up to her death in 1965. The tacked on human interest drama is of little value, but Weaver's committed performance, and the sequences that show her gradually establishing communication with these shy creatures, are easily the film's best points.
DRA 124 min (ort 129 min) VIDrel: WHV V/sur

GORKY PARK *** 15
Michael Apted USA 1983
William Hurt, Lee Marvin, Brian Dennehy, Ian Bannen, Joanna Pacula, Michael Elphick, Richard Griffiths, Rikki Fulton, Alexander Knox, Alexei Sayle, Ian McDiarmid, Niall O'Brian, Henry Woolf, Tusse Silberg, Patrick Field
A Moscow detective investigates the strange case of three bodies buried in Gorky Park. In an attempt to hide their identities their faces and fingerprints have been mutilated. A strongly atmospheric rendition of Smith's novel, that is severely handicapped by the lack of real Moscow locations (once again Helsinki stands in for Moscow) and the basically conventional nature of the plot.
DRA 124 min (ort 128 min) VIDrel: VISVID/POLYREC L/A V
Boa: novel by Martin Cruz Smith.

GOSPEL ACCORDING TO ST MATTHEW, THE *** U
Pier Paolo Pasolini FRANCE/ITALY 1964
Enrique Irazoqui, Susanna Pasolini, Mario Socrate, Margherita Caruso, Marcello Morante, Settimo Di Porto, Otello Sestili, Ferruccio Nuzzo, Giacomo Morante, Alfonso Gatto, Enzo Siciliano, Giorgio Agamben, Guido Cerretani
A low-budget, documentary-style account of the life of Christ, that makes good use of a cast of non-professionals and is generally quite effective in its simplicity, though some typical Pasolini touches now make it look very dated. Additionally, the director's tendency to have Christ launch into noisy tirades at every opportunity is a distinct annoyance. Filmed in Calabria. See also JESUS OF NAZARETH, THE LAST TEMPTATION OF CHRIST and THE GREATEST STORY EVER TOLD.
Aka: IL VANGELO SECONDO MATTEO; L'EVANGILE SELON SAINT MATTHIEU
DRA 129 min (ort 142 min) B/W VIDrel: CONNO/RTM L/A V

GOTHIC *** 18
Ken Russell UK 1986
Gabriel Byrne, Julian Sands, Natasha Richardson, Myriam Cyr, Timothy Spall, Andreas Wisniewski, Alec Mango, Dexter Fletcher, Pascal King, Linda Coggin, Tom Hickey, Kristine Landon-Smith, Chris Chappell, Mark Pickard, Kiran Shah
A recreation of the night in 1816 when Mary Shelley and Dr Polidori were inspired to write their Gothic horror classics, "Frankenstein" and "The Vampyre" respectively. A vivid and enjoyable fantasy full of the usual Russell touches, only this time they add something to the film. The screen debut for Richardson. Remade two years later as HAUNTED SUMMER.
HOR 83 min (ort 90 min) VIDrel: VISVID/POLYREC V/sur

GOTTI ** 18
Robert Harmon USA 1996
Armand Assante, William Forsythe, Anthony Quinn
The story of how one of America's most notorious gangsters became a target for an FBI operation, designed to bring him to justice. See also GETTING GOTTI, which also tells of how the boss of the Gambino family was finally imprisoned.
DRA 117 min VIDrel: RYSHER/HIFLI V/h

GOVERNESS, THE ** 18
Michael Craig USA 1993
Debi Diamond, Teri Diver, Brittany O'Connell, Marc Wallice, Tony Tedeschi, Stephen St Croix, Crystal Wilder, Lacey Rose, Terry Thomas
A woman is hired as governess by a very wealthy couple but she seems to spend most of her time seducing all and sundry, including a number of hanger-on live-in relatives. However, it soon transpires that she is planning to take her revenge on the husband.
A 41 min (ort 80 min) VIDrel: ONE V

GRACE OF MY HEART ** 15
Allison Anders USA 1996
Illeana Douglas, Matt Dillon, Eric Stoltz, Bruce Davison, John Turturro, Patsy Kensit, Jennifer Leigh Warren, Sissy Boyd, Christina Pickles
An occasionally engrossing look at a dozen years in the life of a female singer/songwriter and the success she achieves, mostly through writing songs that are performed by others. With one's eyes shut this often sounds like a great movie (it certainly was a promising idea), unhappily after a good start the story soon degenerates into the usual melodramatic welter of anguish, betrayal and unsatisfying relationships.
DRA   115 min   CINrel

GRADUATE, THE **** 15
Mike Nichols USA 1967
Dustin Hoffman, Anne Bancroft, Katherine Ross, Murray Hamilton, William Daniels, Elizabeth Wilson, Brian Avery, Norman Fell, Marion Lorne, Alice Ghostley, Richard Dreyfuss, Walter Brooke, Elizabeth Fraser, Buck Henry
The film that made Hoffman. A wry look at what happens when a young graduate embarks on a loveless affair with one of his mother's friends but then falls in love with her daughter. The brilliant score is by Simon and Garfunkel. Scripted by Buck Henry (who plays the hotel desk clerk) and Calder Willingham. AA: Dir.
DRA   106 min   VIDrel: ETL/4-FRONT/POLYREC   V
Boa: novel by Charles Webb.

GRAFFITI BRIDGE * 15
Prince USA 1990
Prince, Morris Day, Jerome Benton and The Time, Jill Jones, Mavis Staples, George Clinton, Ingrid Chavez, Robin Power, T.C. Ellis, Tevin Campbell
A kind of dreary sequel to PURPLE RAIN that celebrates the music of Prince, and throws in a few obscure religious references for good measure. Story has Prince the part-owner of a night-club and constantly at odds with his partner over the entertainment they should put on. Unfortunately, there's precious little entertainment to be had in this feeble follow-up, and even the various Prince numbers on offer fail to spark any interest.
MUS   86 min (ort 95 min)   VIDrel: WHV   V/sur

GRAND CANYON *** 15
Lawrence Kasdan USA 1991
Danny Glover, Kevin Kline, Steve Martin, Mary McDonnell, Mary-Louise Parker, Alfre Woodard, Jeremy Sisto, Patrick Malone, Randle Mell, Sarah Trigger, Destinee DeWalt, Candace Mead, Loren Mead, Shaun Baker, K. Todd Freeman
Co-written by Kasdan (together with his wife) this is a film without any clear narrative thrust, that cleverly draws together the separate stories of some ordinary folk, more or less grouped around two families, one black and the other white. A chance meeting between a white lawyer and a black breakdown truck-driver leads to a growing friendship, and the various tales that follow offer much in the way of humour and acute observation.
COM   129 min (ort 134 min) wScrn cC   VIDrel: 20TH/TECH V/sur

GRAND HOTEL *** U
Edmund Goulding USA 1932
Greta Garbo, John Barrymore, Joan Crawford, Wallace Beery, Lionel Barrymore, Lewis Stone, Jean Hersholt, Purnell B. Pratt, Morgan Wallace, Ferdinand Gottschalk, Rafaella Ottiano, Morgan Wallace, Tully Marshall
An ambitious vehicle for a collection of stars, built around the hopes and fears of a group of guests at a plush Berlin hotel, whose lives become intertwined. Faded and certainly stage-bound, but containing several fine performances; John Barrymore as a jewel thief, Garbo as a lonely ballerina and Lionel Barrymore as a dying man are particularly memorable. Later a Broadway musical and used as the plot for several films. AA: Pic.
DRA   107 min (ort 115 min) B/W   VIDrel: MGM/WHV   V
Boa: novel Menschen im Hotel by Vicki Baum.

GRANDVIEW, USA ** 15
Randal Kleiser USA 1984
Jamie Lee Curtis, John Cusack, Michael Winslow, C. Thomas Howell, Patrick Swayze, Troy Donahue, Jennifer Jason Leigh, William Windom, Carole Cook, Ramon Bieri, Elizabeth Gorcey, John Cusack, Camilla Hawk, Melissa Domke
An account of a demolition derby and its participants in a small town, that serves as the basis for a look at life in the Midwest. So-so.
DRA   97 min   VIDrel: 20TH   V/sur

GRAPES OF WRATH, THE **** PG
John Ford USA 1940
Henry Fonda, John Carradine, Jane Darwell, Charley Grapewin, Dorris Bowden, Russell Simpson, Zeffie Tilbury, O.Z. Whitehead, John Qualen, Eddie Quillan, Grant Mitchell, Frank Sully, Frank Darien, Darryl Hickman, Shirley Mills
A superb screen adaptation of Steinbeck's tale of destitute Oklahoma farmers forced to leave their dustbowl-ruined lands during the Depression in the hope of finding a better life in California. Despite the omission of the novel's poignant ending (which would never have passed the censor) this is a splendid film, made with infinite love and care. Scripted by Nunnally Johnson. AA: Dir, S. Actress (Darwell).
DRA   129 min B/W   VIDrel: 20TH/TECH   V/h
Boa: novel by John Steinbeck.

GRASS ARENA, THE *** (PG)
Gillies MacKinnon UK 1991
Mark Rylance, Andrew Dicks, Billy Boyle, Marian McLoughlin, Nick Dawnay, Simon Napper, John Garrett, Brian Hall, Pete Postlethwaite, Gerard Horan, Clive Russell, Louis Mellis, Anna Keaveney, Paddy Joyce, John O'Toole
Having lost his chance to make it as a boxer due to his alcoholism, a man slowly descends ever lower down the social scale, eventually becoming a vagrant who slugs it out among a collection of similar misfits. But a spell in prison leads to his return to normal humanity, when a fellow inmate introduces him to the joys of chess. Based on the true story of John Healy, whose story is as inspiring as it is moving.
DRA   90 min mTV   TVrel: BBC
Boa: autobiographical novel by John Healy.

GRASS IS GREENER, THE ** U
Stanley Donen UK 1960
Cary Grant, Deborah Kerr, Jean Simmons, Robert Mitchum, Moray Watson
An earl and his wife open up their stately home to tourists in order to make ends meet. However, an American millionaire turns up and proceeds to have an affair with the lady of the house. A stilted film version of a lightweight British play that had a good run in the West End play.
COM   100 min (ort 105 min)   VIDrel: 4-FRONT/POLYREC V
Boa: play by Hugh and Margaret Williams.

GRASS ROOTS ** 15
Jerry London USA 1991
Corbin Bernsen, Mel Harris, John Glover, Raymond Burr, Claude Akins, Rod Taylor, James Wilder, Katherine Helmond, Herb Edelman, Christi Conaway, Joanna Cassidy, Henry Jones
A prospective candidate for a seat in the U.S. Senate learns of a nasty world of corruption, intrigue and violence.
DRA   175 min (ort 184 min)   VIDrel: COLUM/SONOP V/s
Boa: novel by Stuart Woods.

GRASSCUTTER, THE ** 15
Ian Mune NEW ZEALAND/UK 1988
Ian McElhinney, Frances Barber, Martin Maguire, Terence Cooper, James Coyle, Ross McKellar, Mitchell Manuel, Wolfgang Khiene, Raymond Hawthorne, Martin Phelan, Judy McIntosh, Hilary Norris, Kate Cribb, Jack Dacy, Marshall Napier
A former member of the Ulster Volunteer Forces tires of the mindless violence and turns Queen's Evidence, leading to the jailing of twenty-three men. In return, he is given a new identity and re-settled in New Zealand, where he leads a tranquil life as a landscape gardener. Unfortunately, his past catches up with him when a hit-squad is sent to kill him and his family.
DRA   104 min   VIDrel: CENTV/VCC L/A   V

GRAVE, THE ** 18
Jonas Pate USA 1996
Craig Sheffer, Gabrielle Anwar, Josh Charles, Donal Logue, John Diehl, Eric Roberts, Anthony Michael Hall
Two car thieves bust out of jail, intending to search for a legendary tomb that they believe is stuffed with treasure. However, one of them dies in the breakout and the other continues the quest alone, but is pursued by a murderous prison guard

and a strange female. An absurdly plotted adventure with a strong dose of black comedy, it never tries to be realistic but at least is generally entertaining.
A/AD 90 min VIDrel: BMGVID/BMGREC V

GRAVE OF THE VAMPIRE **
John Hayes USA
18
1973
William Smith, Michael Pataki, Kitty Vallacher, Lynn Peters, Jay Scott, Jay Adler, Lieux Dressler, Diane Holden, Carmen Argenziano, William Guhl, Abbi Henderson, Inga Neilsen, Margaret Fairchild, Frank Whiteman, Lindus Guiness
A girl raped by a vampire gives birth to a monster which she feeds with her blood in order to sustain it, the son eventually growing up to have his revenge on his father. A convoluted low-budget shocker that starts off with great verve but gradually runs down.
Aka: SEED OF TERROR
HOR 91 min VIDrel: VIPCO/SGSVID V

GRAVE SECRETS: THE LEGACY OF HILLTOP DRIVE **
John Patterson USA
15
1992
Patty Duke, David Selby, Kiersten Warren, Blake Clark, Kelly Rowan, Jonelle Allen, Dakin Matthews, Jon Maynard Oennell, Terry Davis, Kimberly Cullum, David Soul, Maggie Roswell, Rick Fitts, James Lashly, Frances Bay, Jay Brooks
A family buy a house but have scarcely moved in before they experience strange and menacing events that drive them to the brink of desperation. They eventually learn that the house was built over an abandoned graveyard and seek redress through the legal system. A low-key and downbeat recycling of films like POLTERGEIST, but far more restrained in its use of special effects. As ever with these films, the makers claim it to be based on a true story.
HOR 93 min mTV VIDrel: GUILD/SONOP V
Boa: book The Black Hope Horror by Ben Williams, Jean Williams and John Bruce Shoemaker.

GRAVEYARD DISTURBANCE *
Lamberto Bava ITALY
15
1987
Karl Zinny, Beatrice Ring, Leo Martino, Gianmarco Tognazzi, Lino Salemme
A bunch of reckless teenagers on a shoplifting trip make an ill-advised detour into a haunted cemetery, when they are told at a spooky tavern that untold wealth awaits whoever can spend a night in a musty crypt. A largely comical tale of ghoulishness and assorted zombie nastiness, is now unleashed in a film as brainless as its central characters.
Aka: DENTRO IL CIMITERO
HOR 92 min (ort 96 min) VIDrel: SPEAR/SONOP V

GRAVEYARD SHIFT *
Ralph S. Singleton USA
18
1990
David Andrews, Kelly Wolf, Stephen Macht, Brad Dourif, Andrew Divoff, Vic Polizos, Robert Alan Beuth, Ilona Margolis, Jimmy Woodard, Jonathan Emerson, Minor Rootes, Kelly L. Goodman, Susan Lowden, Joe Perham, Dana Packard
There is a sinister link between a long-closed textile mill and the local cemetery, and when the former is re-opened, various nasty creatures erupt from the cellars, to the considerable distress of the workers. A flabby and cliche-ridden horror yarn whose cheap looking effects are its most notable feature.
Aka: STEPHEN KING'S GRAVEYARD SHIFT
HOR 82 min (ort 89 min) VIDrel: VCC/DISC/COLUM V/sur
Boa: short story by Stephen King.

GRAVEYARD STORY ***
B.D. Benedikt CANADA
15
1990
John Ireland, Adrian Paul, Keith Vinsonhaler, Cayle Chernin, Gerry Tucker, Christine Cattell, Courtney Taylor, Maggie MacDonald, Larry Jannison, Victor Altomare, Lawrence Myles, Chris Crumb, Ihor Iomaga, Christopher Calder
Whilst visiting the grave of an old friend, a doctor notices a nearby stone which marks the grave of a little girl who was abducted and murdered. With the help of a former detective he sets about unravelling the mystery surrounding her death.
DRA 91 min VIDrel: 20VIS/SONOP V

GREASE ***
Randal Kleiser USA
PG
1978
John Travolta, Olivia Newton-John, Eve Arden, Stockard Channing, Sid Caesar, Jeff Conaway, Didi Conn, Jamie Donnelly, Dinah Manoff,
Barry Pearl, Kelly Ward, Michael Tucci, Susan Buckner, Eddie Deezen, Lorenzo Lamas
Film version of a Broadway musical looking at the lives and loves of kids in a 1950s high school. An amusing pastiche of 1950s films and manners, in which nothing and nobody is to be taken seriously. The lively dance routines were choreographed by Patricia Birch. Followed by GREASE 2 – one of those sequels that is best forgotten.
MUS 106 min (ort 110 min) wScrn VIDrel: CIC/SONOP V/sur
Boa: musical by Jim Jacobs and Warren Casey.

GREASE 2 *
Patricia Birch USA
PG
1982
Maxwell Caulfield, Michelle Pfeiffer, Adrian Zmed, Lorna Luft, Eve Arden, Didi Conn, Sid Caesar, Dody Goodman, Leif Green, Tab Hunter, Connie Stevens, Dick Patterson, Christopher McDonald, Peter Frechette, Maureen Teefy
An attempt to cash in on the success of the first movie, but this time round the story is one of a British boy who joins the senior high school class. Clumsy, muddled and lacking in good numbers, a dud on all counts. Birch's directorial debut.
MUS 109 min (ort 114 min) wScrn VIDrel: CIC/SONOP V/sur

GREAT BALLS OF FIRE! ***
Jim McBride USA
15
1989
Dennis Quaid, Winona Ryder, Trey Wilson, Lisa Blount, Alec Baldwin, Stephen Tobolowsky, John Doe, Steve Allen, Joshua Sheffield, Mojo Nixon, Lisa Jane Persky, Jimmie Vaughn, David Ferguson, Robert Lesser, Paula Person
Quaid is outstanding in this flashy and stylised account of the rise and fall of rock 'n' roller Jerry Lee Lewis, a thrice married and hard living singer who suffered a temporary setback in his career following his marriage to his third wife, a thirteen-year-old cousin. Lewis's voice is dubbed in, but that's Quaid playing the piano. Entertaining if a trifle clumsily edited. Look out for Baldwin as evangelist Jimmy Swaggart, Lewis's famous cousin.
DRA 102 min (ort 108 min) VIDrel: 4-FRONT/POLYREC L/A V/sur
Boa: book by Myra Lewis.

GREAT CARUSO, THE ***
Richard Thorpe USA
U
1950
Mario Lanza, Ann Blyth, Dorothy Kirsten, Jarmila Novotna, Carl Benton Reid, Eduard Franz, Richard Hageman, Ludwig Donath, Alan Napier, Paul Javor, Carl Milletaire, Shepard Menken, Vincent Renno, Nestor Paiva, Peter Edward Price
The rags-to-riches story of Enrico Caruso, one of the world's greatest opera singers. Largely fictional in content, and fairly well-supplied with the standard Hollywood cliches, but generally enjoyable and skilfully put together. AA: Sound (Douglas Shearer).
MUS 104 min (ort 113 min) B/W VIDrel: MGM/WHV V/dm

GREAT DIAMOND ROBBERY, THE **
Al Waxman CANADA
15
1992
Ben Cross, Kate Nelligan, Brian Dennehy, Tony Rosato, Janet-Lainie Green, Ron Lea, David Huband, Jonathan Welsh, Kurt Reis, John Swindells, Ronnie Pulval, Conrad Bergschneider, Chris Benson, Janet Bailey, Lindsay Leese
An ex-con who has specialised in cracking security systems is hired to guard a diamond worth $5,000,000, but a cop believes that he is using this job as a cover for plans to steal it. Well-acted but let down by a rather undeveloped plot.
Aka: DIAMOND FLEECE, THE
THR 88 min (ort 93 min) mCab VIDrel: 20VIS/SONOP V

GREAT ESCAPE, THE ***
John Sturges/Jud Taylor USA
PG
1963
Steve McQueen, James Garner, Donald Pleasence, David McCallum, James Coburn, Richard Attenborough, Charles Bronson, Gordon Jackson, James Donald, John Leyton, Nigel Stock, Angus Lennie, Jud Taylor, William Russell, Tom Adams
The true story of an Allied breakout from a German prison camp is shown in considerable detail, with the major portion of the film devoted to the preparations. Steve McQueen supplies the comic element as the "Cooler King". Entertaining if overlong, but with an unexpectedly depressing ending. The script is by James Clavell and W.R. Burnett. A TV sequel followed twenty-five years later.
WAR 172 min wScrn VIDrel: MGM/WHV V/h
Boa: book by Paul Brickhill.

**GREAT ESCAPE 2, THE ** ** PG
Paul Wendkos/Jud Taylor USA 1988
Christopher Reeve, Michael Nader, Ian McShane, Donald Pleasence, Judd Hirsch, Anthony Dennison, Charles Haid
Another clear manifestation of the shortage of new ideas in Hollywood, is this ill-conceived attempt to produce a version of the original film. This one is in two parts, with the first half an examination of the planning and execution of this famous WW2 Allied escape, and the second a look at the tragically brutal consequences. Well made, but a film that adds very little to the earlier version of this tale.
Aka: GREAT ESCAPE 2, THE: THE FINAL CHAPTER; GREAT ESCAPE 2, THE: THE UNTOLD STORY – PARTS 1 AND 2
WAR 89 min; 178 min (2 cassettes – ort 200 min) mTV
VIDrel: RCA L/A V

**GREAT EXPECTATIONS ** ** PG
David Lean UK 1946
John Mills, Valerie Hobson, Bernard Miles, Francis L. Sullivan Finlay Currie, Martita Hunt, Anthony Wager, Jean Simmons, Alec Guinness, Ivor Bernard, Freda Jackson, Torin Thatcher, Eileen Erskine, Hay Petrie
The second sound version of the story of a young boy, befriended from afar by an escaped convict he once helped. One of the best films to ever come out of the British cinema; beautifully acted, superbly designed and wonderfully directed. Remade for TV in 1974. AA: Cin (Guy Green), Art/Set (John Bryan/Wilfred Singleton).
DRA 113 min (ort 118 min) B/W VIDrel: CARL/TECH V
Boa: novel by Charles Dickens.

**GREAT EXPECTATIONS ** ** U
Jean Tych AUSTRALIA 1982
Voices of: Bill Kerr, Philip Hinton, Simon Hinton, Barbara Frawley, Robin Stewart, Liz Horne, Marcus Hale, Moya O'Sullivan
Competent animated version of the Dickens story.
ANIM 69 min (ort 84 min) VIDrel: TRING V
Boa: novel by Charles Dickens.

**GREAT GATSBY, THE ** ** PG
Jack Clayton USA 1974
Robert Redford, Mia Farrow, Karen Black, Scott Wilson, Sam Waterston, Lois Chiles, Bruce Dern, Howard da Silva, Edward Herrmann, Patsy Kensit, Roberts Blossom, Elliot Sullivan, Arthur Hughes, Kathryn Leigh Scott, Beth Porter
Second attempt to film the Scott Fitzgerald story, about a mysterious young millionaire who breaks into Long Island society in the 1920s. A vivid recreation of the period that stays faithful to the book but never develops into anything substantial, remaining a frothy jazz-age celebration. The script is by Francis Ford Coppola. AA: Score (Nelson Riddle), Cost (Theoni V. Aldredge).
DRA 135 min (ort 151 min)
VIDrel: 4-FRONT/POLYREC/CIC V/sh
Boa: novel by F. Scott Fitzgerald.

**GREAT LOS ANGELES EARTHQUAKE, THE ** ** 15
Larry Elikann USA 1990
Joanna Kerns, Joe Spano, Lindsay Frost, Richard Masur, Ed Begley Jr, Bonnie Bartlett, Dan Lauria, Brock Peters, Lindsay Frost, Richard Anthony Crenna, Kasi Lemmons, Allan Wasserman, Ross Kettle, Eloy Casados, Jacob Vargas
Having become convinced that L.A. is about to experience a major earthquake, a female seismologist is still debating whether or not to make her fears public when the earthquake begins. Adequate adventure yarn making use of a well-worn theme.
Aka: BIG ONE, THE: THE GREAT LOS ANGELES EARTHQUAKE
A/AD 101 min VIDrel: CAPIT/GUILD V/sh

**GREAT McGONAGALL, THE ** * 15
Joseph McGrath UK 1975
Spike Milligan, Peter Sellers, Julia Foster, Julian Chagrin, John Bluthal, Valentine Dyall, Clifton Jones, Victor Spinetti, Charlie Atom, Janet Adair
Set in Dundee 1890, this fictitious comedy was inspired by the life of one of Britain's worst poets, William McGonagall, who dreams of being made Poet Laureate and saves the Queen from assassination. A feeble effort.
COM 85 min (ort 95 min) VIDrel: FABFIL/SPEAR V

**GREAT MR HANDEL, THE ** ** U
Norman Walker UK 1942
Wilfred Lawson, Elizabeth Allan, Malcolm Keen, Michael Shepley,

Hay Petrie, Morris Harvey, Max Kirby, A.E. Matthews, Frederick Cooper, Robert Atkins, H.F. Maltby, Trefor Jones
A study of the life of this great 18th century composer, who falls from favour when he offends the Prince of Wales, but then writes "The Messiah" and regains the latter's friendship. A well-intentioned and straightforward biopic, competent if not overly inspired. The script is by Gerald Elliott and Victor MacClure. This was one of the very first films to be shot in Technicolor.
DRA 98 min (ort 103 min) VIDrel: CONNO/RTM V
Boa: radio play by D. Du Garde Peach.

**GREAT OUTDOORS, THE ** * PG
Howard Deutch USA 1988
Dan Aykroyd, John Candy, Stephanie Faracy, Annette Bening, Chris Young, Ian Giatti, Hilary Gordon, Rebecca Gordon, Robert Prosky, Zoaunne Leroy
A family's vacation is ruined when the husband's brother-in-law and his uncouth crowd invite themselves along. An unpleasant comedy that fires off its jokes in all directions, without ever engaging one's interest. Scripted by John Hughes.
Aka: BIG COUNTRY
COM 86 min Cut (27 sec plus some cuts subst – ort 91 min)
VIDrel: 4-FRONT/POLYREC/CIC L/A V/dm

**GREAT RACE, THE ** ** U
Blake Edwards USA 1965
Jack Lemmon, Tony Curtis, Natalie Wood, Peter Falk, Keenan Wynn, Larry Storch, Dorothy Provine, Arthur O'Connell, Vivian Vance, Ross Martin, George Macready, Marvin Kaplan, Hal Smith, Denver Pyle, William Bryant, Ken Wales
Overlong and unfunny account of a 1908 car race across three continents from New York to Paris. Over-elaborate and sometimes spectacular, this long and laboured comedy has a few good moments, but most of them occur in the first half. AA: Effects/aud (Tregoweth Brown).
COM 147 min (ort 163 min) VIDrel: VISVID/WHV L/A V

**GREAT ROCK 'N' ROLL SWINDLE, THE ** ** 18
Julian Temple UK 1980
Malcolm McLaren, Sid Vicious, Johnny Rotten, Steve Jones, Paul Cook, Ronald (Ronnie) Biggs, Mary Millington, Irene Handl, Eddie Tudor, Helen of Troy, Sue Catwoman, Liz Fraser, Jess Conrad, Julian Holloway, Dave Dee, Day D'Arcy
A fictionalised account of the career of the punk rock group "The Sex Pistols", that is alternately fascinating and repulsive. See also SID AND NANCY.
Aka: ROCK 'N' ROLL SWINDLE, THE
DRA 100 min (Cut at film release – ort 105 min)
VIDrel: POLY/POLYREC V/sh

**GREAT ST TRINIANS TRAIN ROBBERY, THE ** * U
Frank Lauder/Sidney Gilliat UK 1966
Frankie Howerd, Dora Bryan, George Cole, Reg Varney, Raymond Huntley, Richard Wattis, Portland Mason, Terry Scott, Eric Barker, Godfrey Winn, Desmond Walter-Ellis, Colin Gordon, Barbara Couper, Elspeth Duxbury
Filmed three years after the real Great Train Robbery, this tale follows the exploits of a bunch of criminals who hide their loot on a site that was used to build this notorious girls' school, consequently forcing them to infiltrate the staff in an attempt to regain it. The usual madcap capers and laboured comic moments ensue. Yet another sequel to THE BELLES OF ST TRINIANS, this dismal farce was followed by THE WILDCATS OF ST TRINIANS.
COM 90 min (ort 94 min) VIDrel: WHV V/h

**GREAT SEXPECTATIONS ** ** 18
Henri Pachard USA 1984
Kelly Nichols, Honey Wilder, Joanna Storm, Clelsea Blake, Tanya Lawson, Eric Edwards, John Leslie, R. Bolla
A sex-film director and a producer team up to make a porno movie, and most of the film is taken up with the efforts they make in casting for the leads. While all this is going on, the director finds that he has fallen in love with one of the female stars, and eventually feels impelled to confess his feelings to her. There's a good story in here somewhere, but it never gets developed.
A 30 min (ort 48 min) VIDrel: HAR/GOLD V

**GREAT WALDO PEPPER, THE ** ** PG
George Roy Hill USA 1975
Robert Redford, Bo Svenson, Susan Sarandon, Geoffrey Lewis, Margot

Kidder, Bo Brundin, Edward Herrmann, Scott Newman, Philips Bruns, Roderick Cook, Kelly Jean Peters, James S. Appleby, Patrick W. Henderson Jr, John A. Zee
WW1 flying aces are forced to make a living by giving stunt shows. An interesting film that hovers between drama and comedy, without ever coming to land at either point. However, the flying stunts are superb, as is the photography (by Robert Surtees). The music is by Henry Mancini.
A/AD   102 min (ort 107 min)   VIDrel: CIC/SONOP   V

GREAT WALL, A **
Peter Wang CHINA/USA
PG
1986
Peter Wang, Sharon Iwai, Kelvin Han Yee, Li Qinqin, Hu Xiaoguang, Shen Guanglan, Wang Xiao, Xiu Jian, Ran Zhiluan, Han Tan, Jeanette Pavini, Howard Frieberg, Bill Neilson, Teresa Roberts
A wry comedy about an American computer expert visiting his relatives in Peking, and the resulting clash of cultures that occurs, as this Americanised family encounters life in the People's Republic. Likeable enough, but hardly the stuff of great comedy. Wang's first feature as co-writer, star and director.
Aka: GREAT WALL, THE; GREAT WALL IS A GREAT WALL, THE
COM   100 min   VIDrel: PAL L/A   V

GREAT WHITE HYPE, THE **
Reginald Hudlin USA
15
1996
Samuel L. Jackson, Jeff Goldblum, Peter Berg, Jon Lovitz, Corbin Bernsen, Cheech Marin, John Rhys-Davies, Salli Richardson, Jamie Foxx, Rocky Carroll, Damon Wayans, Albert Hall, Susan Gibney, Michael Jace, Duane Davis
A cynical boxing promoter comes up with a good scheme for making some money when he arranges a re-match between the current heavyweight champ and the only fighter who ever beat him. Meant to be a satire on the corruption that has so plagued this sport, the film's lightweight script is its chief weakness, though there are fine performances all round and some lively and enjoyable moments.
COM   87 min (ort 90 min) cC   VIDrel: 20TH/FOXVID   V/sur

GREATEST, THE **
Tom Gries USA
PG
1977
Muhammad Ali, Ernest Borgnine, John Marley, Lloyd Haynes, Robert Duvall, James Earl Jones, Dina Merrill, Paul Winfield, Roger E. Mosley, Ben Johnson, Malachi Thorne, David Huddleston, Annazette Chase, Mira Waters, Arthur Adama
Ali plays himself in this bland, unfocused and boring mess that attempts to chart the rise of this brilliant boxer from rags to riches, but succeeds only in being ponderous and dull. Ali's acting is adequate, but it's Mosley as former champ Sonny Liston, who engages our attention. Written by Ring Lardner Jr.
DRA   98 min (ort 101 min)   VIDrel: FABFIL/SPEAR   V
Boa: book by Muhammad Ali, Herbert Muhammad and Richard Durham.

GREATEST SHOW ON EARTH, THE ***
Cecil B. De Mille USA
U
1952
Charlton Heston, Betty Hutton, Cornel Wilde, James Stewart, Dorothy Lamour, Gloria Grahame, Henry Wilcoxon, Lawrence Tierney, Lyle Bettger, Emmett Kelly, John Kellogg, John Ringling North
A long epic about life under the big-top, where various dramatic situations are played out. Everything is here: cliches in abundance, spectacular disasters, the excitement of the big-top, and the lives and loves of the people in it. Not so much a film as a parade of stars and situations, but all very impressively done. AA: Pic, Story (Fredrick M. Frank/Theodore St John/Frank Cavett).
DRA   147 min (ort 153 min)   VIDrel: CIC/SONOP   V

GREATEST STORY EVER TOLD, THE **
George Stevens USA
U
1965
Max Von Sydow, Dorothy McGuire, Claude Rains, Jose Ferrer, David McCallum, Charlton Heston, Sidney Poitier, Donald Pleasence, Roddy McDiwall, Gary Raymond, Carroll Baker, Van Heflin, Pat Boone, Sal Mineo, Shelley Winters
A slow, solemn and ponderous epic-length account of the life of Christ, as empty of inner life and real feeling as it is replete with star names, even in bit parts, i.e. John Wayne as a Roman centurion. Need I say more? See also THE GOSPEL ACCORDING TO ST MATTHEW, JESUS OF NAZARETH, THE LAST TEMPTATION OF CHRIST and KING OF KINGS.
DRA   191 min (ort 260 min)   VIDrel: MGM/WHV   V/sur

GREED ***
Erich Von Stroheim USA
PG
1924
Gibson Gowland, ZaSu Pitts, Jean Hersholt, Chester Conklin, Dale Fuller, Tempe Piggot, Gunther von Ritzau, Jimmy Wang, Austin Jewel, Oscar Gotell, Otto Gottell, Joan Standing, Frank Hayes, Fanny Midgeley, Max Tyron
Stroheim's magnificent failure – an attempt to recreate the naturalism of the original novel about a young miner whose life is ruined by his wife's greed and his unemployment. He eventually kills his wife and later, in Death Valley, her lover, even though by then the two men are handcuffed to each other. An intense and totally absorbing adaptation, although inevitably disjointed at times due to the severe cutting by the studio.
DRA   135 min (ort 420 min) B/W silent
VIDrel: MGM/WHV   V/sh
Boa: novel McTeague by Frank Norris.

GREEDY **
Jonathan Lynn USA
12
1993
Kirk Douglas, Michael J. Fox, Nancy Travis, Olivia D'Abo, Phil Hartman, Jere Burns, Ed Begley Jr., Colleen Camp, Bob Balaban, Joyce Hyser, Siobhan Fallon, Mary Ellen Trainor, Kevin McCarthy, Khandi Alexander, Jonathan Lynn, Tom Mason
A bunch of avaricious heirs are faced with a crisis when the patriarch they hope to inherit acquires a new mistress who seems likely to be the one to end up with all his money. To save the situation, they prevail on his favourite but now estranged nephew to get back in his good books again. Allegedly inspired by the plot of Dicken's novel Martin Chuzzlewit, this is a long, meandering and largely unfunny effort that has little to say.
COM   107 min (ort 113 min) cC   VIDrel: CIC/SONOP   V

GREEN BERETS, THE *
John Wayne/Ray Kellogg USA
PG
1968
John Wayne, David Janssen, Ray Kellogg, Jim Hutton, Aldo Ray, Raymond St Jacques, Bruce Cabot, George Takei, Jack Soo, Patrick Wayne, Mike Henry, Irene Tsu, Jason Evers, Luke Askew, Edward Faulkner, Craig Jue, Eddy Donno
Story of the Special Forces in Vietnam. The final scene has the sun setting in the East, a nice comment on the credibility of the whole film.
WAR   136 min (ort 141 min)   VIDrel: WHV   V/h
Boa: novel by Robin Moore.

GREEN CARD **
Peter Weir AUSTRALIA/FRANCE
15
1990
Gerard Depardieu, Andie MacDowell, Bebe Neuwirth, Gregg Edelman, Robert Prosky, Jessie Keosian, Ann Wedgeworth, Ethan Phillips, Mary Louise Wilson, Lois Smith, Simon Jones
A rough-and-ready Frenchman enters a marriage of convenience with a finicky American woman who finds that she can both help him remain in the States and get to rent a smart Manhattan apartment. However, when immigration officials grow suspicious, she's forced to start spending time with him. An agreeable but terribly flimsy romantic comedy with a few amusing touches. This was Depardieu's English language debut. Written, produced and directed by Weir.
COM   102 min (ort 108 min)   VIDrel: TOUCH/SONOP   V/sur

GREEN LEGEND RAN: EPISODES 1 TO 3 **
JAPAN
PG
1992
Futuristic adventure with an orphan named Ran searching for the killers of his mother, in the course of which he tries to join the Hazzard group, who have taken control of all the Earth's water. Joined by a mysterious female, he sets out to destroy this group.
ANIM   150 min (three 50-min cassettes) dubbed
VIDrel: PION/RTM   V/sh

GREEN MAN, THE ***
Robert Day UK
PG
1957
Alastair Sim, George Cole, Terry-Thomas, Jill Adams, Dora Bryan, Raymond Huntley, Avril Angers, Eileen Moore, John Chandos, Colin Gordon, Cyril Chamberlain, Doris Yorke, Vivienne Wood, Arthur Brough, Marie Burke
A seemingly timid clockmaker much prefers his part-time job as paid assassin, and encounters unforeseen obstacles when he takes on an assignment to kill a politician, who's staying at a country inn. A droll and quirky little comedy with Sim quite delightful, in a role that's both wicked and hilarious.
COM   76 min (ort 80 min) B/W   VIDrel: WHV   V/h
Boa: play Meet a Body by Frank Launder and Sidney Gilliat.

GREEN MAN, THE ** 15
Elijah Moshinsky UK 1991
Albert Finney, Sarah Berger, Linda Marlowe, Michael Hordern, Nicky Henson, Natalie Morse, Michael Grandage, Josie Lawrence, Michael Culver, Nickolas Grace, Natalie Moore, Robert Schofield, John Burgess, Peter Treganna
A ghost story set at a country pub, where the owner, who fancies himself as a bit of a ladies' man, comes under the influence of a sinister spirit. A few moments of atmosphere are scattered thinly, in a drawn-out and rather tedious effort. The climax is an exorcism that provokes a remarkably violent response, but this is both overdone and laughable. Disappointing.
FAN 148 min (2 cassettes) mTV VIDrel: BBC/TECH L/A V

GREEN RAY, THE *** PG
Eric Rohmer FRANCE 1986
Marie Riviere, Lisa Heredia, Vincent Gauthier, Beatrice Romand, Carita, Sylvie Richez, Basile Gervaise, Virginie Gervaise, Rene Hernandez
Let down at the last moment by a friend with whom she was to have gone on holiday, a secretary spends the vacation month of August aimlessly wandering the country, until she eventually meets Mr Right. This well made film (the fifth in Rohmer's "Comedies and Proverbs" series) differs from most of its predecessors in that the dialogue is largely improvised. Taken from a Jules Verne novel, the title refers to the last rays of the setting sun. See also PAULINE AT THE BEACH.
Aka: LE RAYON VERT; SUMMER
COM 94 min (ort 98 min) VIDrel: ARTIF/20TH V/h

GREGORY'S GIRL ** PG
Bill Forsyth UK 1980
Gordon John Sinclair, Dee Hepburn, Jake D'Arcy, Clare Grogan, Robert Buchanan, William Greenlees, Alan Love, Caroline Guthrie, Carl Macartney, Chic Murray, Douglas Sannachan, Allison Foster, Alex Norton, John Bett
Over-rated comedy about a young teenager who falls for a girl at his school, but is too bashful to ask her out. After several false starts he finds himself being sought out by a girl who really does fancy him. A mildly engaging comedy with a few nice cameos, but not really enough of a plot to sustain interest. However (unlike the inferior P'TANG, YANG, KIPPERBANG) it does retain a freshness which gives it charm.
COM 89 min (ort 91 min) VIDrel: VCC/DISC/COLUM L/A V

GREMLINS ** 15
Joe Dante USA 1984
Zach Galligan, Phoebe Cates, Hoyt Axton, Frances Lee McCain, Polly Holliday, Judge Reinhold, Corey Feldman, Glynn Turman, Dick Miller, Keye Luke, Scott Brady, Jackie Joseph, Edward Andrews, Chuck Jones, John Louie, Arnie Moore
A boy is given a strange animal as a Christmas present, and ignores the injunction against allowing it to come in contact with water, whereupon it gives rise to a host of murderous creatures that go on a rampage. A spoofy horror yarn, full of film-buff in-jokes, it would have been far better as a straight horror tale. Look out for several cameos, not least one featuring executive producer Steven Spielberg.
HOR 102 min (ort 111 min) wScrn cC VIDrel: WHV V/sur

GREMLINS 2: THE NEW BATCH *** 15
Joe Dante USA 1990
Zach Galligan, Phoebe Cates, John Glover, Robert J. Prosky, Robert Picardo, Haviland Morris, Christopher Lee, Dick Miller, Jackie Joseph, Keye Luke, Gedde Watanabe, Kathleen Freeman, Don Stanton, Dan Stanton, Shawn Nelson
A crazy and somewhat self indulgent sequel to the earlier film, that spends most of its time poking fun at itself, with a collection of in-jokes for film buffs and a host of cameo appearances. Plot centres on a Manhattan office block which the title beasties set about taking over. More akin to AIRPLANE! than the earlier GREMLINS, and often very funny indeed. Music is by Jerry Goldsmith and the script is by Dante and Charlie Haas.
HOR 102 min (ort 107 min) wScrn VIDrel: WHV V/sur

GREMLOIDS *** PG
Todd Durham USA 1986
Alan Marx, Paul Poundstone, Chris Elliott, Robert Bloodworth
A one-joke film, spoofing all those expensive science fiction movies with the title characters being a bunch of midgets, who

scurry about their evil but not terribly powerful leader. Despite a number of funny skits on famous earlier films, this one is too episodic to work as a movie, but as a series of low-budget parodies, it's fine.
Aka: HYPERSPACE
COM 86 min VIDrel: SGSVID/GOLD V

GREY *** 15
JAPAN 1987
Unusual futuristic fantasy set in the year 2588 A.D., which sees society divided into lettered ranks from A to F. Only those in the highest rank of A have full rights of citizenship, and to achieve this ranking they have to become murderers.
ANIM 80 min VIDrel: WESCON/RTM V

GREYHOUNDS ** (PG)
Kim Manners USA 1993
Dennis Weaver, James Coburn, Robert Guillaume, Pat Morita, Roxann Biggs, Kirk Bailey, Ray Young, Orestes Matacena, Robert Weaver, Fabio Alberto, Kathryn Atwood, Diane Dilascio, Nic Da'virro, Michael Scott Greenlee
A set of ageing crimefighters are brought out of retirement by a new female D.A., who is very keen to make a name for herself. Adequate.
COM 93 min mTV SATrel: SKY MOVIES

GREYSTOKE: THE LEGEND OF TARZAN, LORD OF THE APES ** PG
Hugh Hudson UK 1983
Christopher Lambert, Andie MacDowell, Ian Holm, Ralph Richardson, James Fox, Cheryl Campbell, Ian Charleson, Nigel Davenport, John Wells, Richard Griffiths, Paul Geoffrey, Nicholas Farrell, Colin Charles, David Endene, Tristam Jellineck
A flawed attempt to make a more authentic version of the "Tarzan" stories of Edgar Rice Burroughs. Here the boy is discovered by a Belgian explorer, after having been brought up by apes (of an indeterminate species), and taken back to civilisation and his grandfather's estate. A faintly absurd film, that adopts an irritating moral tone. Considered unsuitable, McDowell's Southern drawl was dubbed over by Glenn Close. This was Richardson's last film.
Aka: GREYSTOKE
DRA 131 min wScrn VIDrel: WHV V/sur
Boa: novel Tarzan Of The Apes by Edgar Rice Burroughs.

GRID RUNNERS * 18
Andrew Stevens USA 1995
Don Wilson, Athena Massey, Stella Stevens, Loren Avedon
A blend of martial arts mayhem and virtual reality in a futuristic tale of an agent or "Grid Chaser" who is charged with the task of breaking an armed gang who specialise in smuggling computer games. However, he comes up against some unusual opponents, his enemy being a power-crazed scientist who has found a way of bringing some of the cyber-game characters to life. A very disappointing film that fails to exploit its ideas, being just another standard action movie.
A/AD 87 min VIDrel: EIV/SONOP V

GRIDLOCK *** 15
Sandor Stern GERMANY/USA 1995
David Hasselhoff, Kathy Ireland, Miguel Fernandes, Gotz Otto, Tony Desantis, Allen Scarse, Rofer Dominique, Marjus Parillo, D. Marsman
A bunch of crooks hatch a masterly scheme to facilitate their robbing of the US Federal Reserve: they plant a bomb set to go off during the rush hour. Having achieved a total traffic standstill they now proceed with the rest of their plan, and only a helicopter cop can stop them. See THE ITALIAN JOB for another film that makes use of this idea.
A/AD 88 min (ort 91 min) VIDrel: HIFLI/SONOP V
Boa: novel by C. Bronte.

GRIDLOCK'D ** 18
Vondie Curtis Hall UK/USA 1996
Tim Roth, Tupac Shakur, Thandie Newton, Charles Fleischer, Howard Hesseman, James Pickens Jr, John Sayles, Eric Payne, Tom Toyles, Tom Wright, James Shanta, Jim O'Malley, George Poulos
A tough and sassy urban comedy built around the exploits of two musicians (Shakur and Roth) who are also junkies. When their singer goes into a coma after a drug overdose, they decide to kick the habit, and from this point on, their troubles start. Often sharp and witty (the script is by Hall) this is a pleasing film, even if the plot is so very slight one is hard put to

remember it. The jazz soundtrack is by Stewart Copeland. This was Shakur's last film.
COM 91 min CINrel

GRIEF **
Richard Glatzer USA   18
                                                      1993
Alexis Arquette, Illeana Douglas, Craig Chester, Jackie Beat, Carlton Wilborn, Lucy Gutteridge, Robin Swid, Bill Rotko, Shawn Hoffman, Frank Rehwaldt, Greg Bennett, Mickey Cottrell, Catherine Connella, Mary Woronov, Jeffrey Hilbert
Hectic, fast-paced comedy set behind the scenes of a daytime TV show called "The Love Judge" and focusing on the lives of the production crew and actors. Very uneven, but full of energy and invention, though there is not enough of an overall plot to make the film into anything more than a set of amusing sketches.
COM 87 min VIDrel: DTK/TOTAL V

GRIFTERS, THE ****
Stephen Frears USA   18
                                                      1990
John Cusack, Anjelica Huston, Annette Bening, Pat Hingle, Henry Jones, Michael Laskin, Eddie Jones, J.T. Walsh, Charles Napier, Stephen Tobolowsky, Sandy Baron, Gailard Sartain
A coldly clinical adaptation of Thompson's quirky novel about con artists in L.A. that examines the life of one such trickster, who comes back into the life of her grown-up son after a long absence, only to find that he has taken up with a woman not unlike herself. A totally absorbing film, harsh and brutal, but not without a measure of grim humour and three riveting performances. The gory climax is as shocking as it is unexpected. Music is by Elmer Bernstein.
DRA 105 min (ort 119 min) VIDrel: 4-FRONT/POLYREC V/sur
Boa: novel by Jim Thompson.

GRIM REAPER, THE ***
Bernard Bertolucci ITALY   15
                                                      1962
Francesco Rula, Giancarlo De Rosa, Vincenzo Ciccora, Alvaro D'Ercole, Romano Labate, Marisa Solinas
A derivative and arty debut for Bertolucci, that tells of the investigation into the murder of a prostitute, seen (as in RASHOMON) from the perspective of several different people. The director was only twenty-two when he made this promising work, and the sudden shifts of pace (at one moment the film is a harsh documentary, at another an atmospheric drama) are handled with great skill. Pasolini contributed to the screenplay, and his influence is clearly felt.
Aka: LA COMMARE SECCA
DRA 100 min B/W VIDrel: ARROW/RTM V

GRIZZLY ADAMS: THE LEGEND CONTINUES **
Ken Kennedy USA   U
                                                      1990
Gene Edwards, Link Wyler, Red West, Tony Caruso, Acquanetta, L.Q. Jones
A solitary man and his grizzly-bear companion battle to save a small town from three of the West's meanest desperados. An adequate family-oriented Western that recycles the theme and figures from the original TV series – "The Life And Times Of Grizzly Adams".
JUV 73 min (ort 90 min) VIDrel: POLY/POLYREC L/A V

GRIZZLY ADAMS: TREASURES OF THE BEAR **
(PG)
John Huneck USA   1995
Tom Tayback, Joe Campanella
Grizzly comes to the rescue of a professor friend, who has been kidnapped by a man who is out to locate the sacred burial site of the Indian "Bear People". Another outdoors adventure for this rugged trapper character.
A/AD 88 min SATrel: MOVIE CHANNEL

GROSS MISCONDUCT **
George Miller AUSTRALIA   18
                                                      1993
Jimmy Smits, Naomi Watts, Sarah Chadwick, Tara Judah, Goran Stamenkovic, Leverne McDonnell, Alan Fletcher, Adrian Wright, Brendon Suhr, Susan Ellis, Edwina Exton, Bernadette Walsh, Peter Webb, Paul Murphy, Linda Cable
A university professor is wrongly accused of attacking a female pupil whose advances he rejected and has to endure much suffering before finally being found innocent. A capably made but undistinguished drama that is based on a true case.
DRA 92 min (ort 100 min) VIDrel: SPEAR/SONOP V
Boa: play Assault with a Deadly Weapon by Lance Peters.

GROTESQUE, THE **
John-Paul Davidson UK   18
                                                      1995
Alan Bates, Theresa Russell, Sting, Lena Headey, Jim Carter, Anna Massey, Trudie Styler, Maria Aitken, James Fleet, Steven Mackintosh, John Mills, Chris Barnes, Timothy Kightley, Richard Durden, Nick Lucas, Annette Badland
A bored American woman tires of her husband and looks for sexual excitement with their butler and his wife, but fails to realise that this couple nurture evil ambitions. Something of a black comedy masquerading as a murder mystery, it offers a few laughs. Unfortunately, the episodic nature of the plot (which is little more than a set of funny episodes) weakens the film's overall impact. The script is by McGrath.
COM 98 min VIDrel: IMAG/RTM V
Boa: book by Patrick McGrath.

GROUNDHOG DAY ***
Harold Ramis USA   PG
                                                      1992
Bill Murray, Andie MacDowell, Chris Elliott, Marita Geraghty, Angela Paton, Stephen Tobolowsky, Brian Doyle-Murray, Rick Overton, Rick Du Commun, Rod Sell, Robin Duke, Carol Bivins, Willie Garson, Ken Hudson Campbell, Les Podewll
An obnoxious and incredibly self-centred TV weatherman is sent with two colleagues to cover an annual ceremony in a small town and gets marooned there by blizzards. Worse is to come, when he finds himself living the same day over and over again. A charming and refreshingly original story saved from triteness and mawkishness by excellent performances (Murray is just perfect as the ego-tripped hero) and a strong script. A hugely enjoyable comedy fantasy.
COM 97 min (ort 101 min) cC wScrn (COLUM/SONOP)
VIDrel: COLUM/SONOP; VCC/DISC/COLUM V/sur

GRUMPIER OLD MEN ***
Howard Deutch USA   12
                                                      1995
Jack Lemmon, Walter Matthau, Ann-Margret, Sophia Loren, Daryl Hannah, Kevin Pollak, Burgess Meredith
A surprisingly satisfactory sequel to GRUMPY OLD MEN (unaccountably it had no cinema release in the UK) with old sparring partners Matthau and Lemmon on considerably friendlier terms. This time round they get embroiled in the plans a hot-blooded Italian (Loren) has to turn a local fishing-tackle shop into a classy restaurant. Despite the lack of inventiveness, the film is saved by Lemmon and Matthau, who as ever work wonderfully well together.
COM 97 min (ort 101 min) cC VIDrel: WHV V/sur

GRUMPY OLD MEN **
Donald Petrie USA   12
                                                      1993
Walter Matthau, Jack Lemmon, Ann-Margret, Burgess Meredith, Daryl Hannah, Kevin Pollak, Ossie Davis, Buck Henry, Christopher McDonald, Steve Cochran, Joe Howard, Isabell Monk, Buffy Sedlachek, John Carroll Lynch, Charles Brin
A couple of old men nurture a deep rivalry which becomes more inflamed when an attractive woman moves into their neighbourhood and both men become hopelessly smitten. For all the fine work from the three leads there is little humour to be had in the story, which takes a decidedly mean and jaundiced view of old age. A sequel followed which was considerably lighter in tone.
COM 99 min (ort 103 min) cC VIDrel: WHV V/sur

GUANTANAMERA ***
Tomas Gutierrez Alea/Juan Carlos Tabio   15
CUBA/GERMANY/SPAIN   1995
Carlos Cruz, Mirtha Ibarra, Jorge Perugorria, Raul Eguren, Pedro Fernandez, Luis Alberto Garcia, Conchita Brando, Suset Perez Malberti, Assenech Rodriguez, Luisa Perez Nieto, Idalmis Del Risco, Ikay Romay, Mercedes Arnaez, Alfredo Avila
A world-famous Cuban singer returns after an absence of fifty years and is welcomed to her home town by her niece, a former lecturer in economics, and her husband, an undertaker. Tragedy strikes, however, when our singer suddenly dies after the briefest of reunions with an old love and her body has to be transported back to Havana. A warm and human road-movie that dares to poke fun at the many inadequacies of Cuba's so-called socialist society.
DRA 102 min CINrel

GUARDIAN, THE **
William Friedkin USA   18
                                                      1990
Jenny Seagrove, Dwier Brown, Carey Lowell, Brad Hall, Miguel Ferrer, Natalia Nogulich, Pamela Brull, Gary Swanson, Jack David

Walker, Williy Parsons, Frank Noon, Theresa Randle, Ray Reinhardt, Xavier Berkeley, Jacob Gelman
Friedkin's first horror film since THE EXORCIST has an evil nanny being hired by a couple to look after their child. However, she likes to sacrifice her young charges to a Druidic tree god. A patchy film with tension maintained up to the point where the woman's true nature is revealed, after which the special effects take over and the film goes downhill. Scripted by Greenburg, Friedkin and Stephen Volk. See also MIDNIGHT'S CHILD and THE HAND THAT ROCKS THE CRADLE.
HOR   89 min (ort 92 min)   VIDrel: CIC/SONOP L/A   V
Boa: novel The Nanny by Dan Greenburg.

GUARDIAN ANGEL **

18

Richard W. Munchkin USA   1994
Cynthia Rothrock, Daniel McVicar, Lydie Denier, Marshall Teague, Ken McLeod, John O'Leary, Dale Jacoby, Anna Dalva, Brian Brophy, Robert Miano, Art Camacho, Matthew Walker, Don Doherty, Dale E. Jacoby, Bela Leholczky, Dennis Paladino
A tough female ex-cop accepts an assignment to work as bodyguard to a rich businessman who is being menaced by a deranged killer who murdered her partner and lover. A violent actioner that serves merely as an adequate showcase for Rothrock's fighting skills.
A/AD   91 min (ort 97 min)   VIDrel: MIA/DISC   V/sur

GUARDIAN OF THE ABYSS *

(18)

Don Sharp UK   1984
Ray Lonnen, Rosalyn Landor, John Carson, Paul Darrow, Barbara Ewing, Caroline Langrishe, Sophie Thompson, Sharon Fussey, Barry McDonald
A woman antiques dealer buys an odd lot at auction and finds that this includes a strange mirror that bears an inscription in an unknown language. A male friend takes it to be appraised but gets involved with a young girl who is running away from a Satanist sect. He soon learns that this item is to be used in a ritual intended to incarnate a powerful demon in the body of a sacrificial victim. A flat and confused piece that offers few chills.
Aka: HAMMER HOUSE OF HORROR: GUARDIAN OF THE ABYSSS
HOR   52 min   SATrel: BRAVO

GUARDING TESS **

12

Hugh Wilson USA   1993
Shirley MacLaine, Nicolas Cage, Edward Albert, Austin Pendleton, James Rebhorn, Richard Griffiths, John Roselius, David Graf, Don Yesso, Dale Dye, James Lally, Brant Von Hoffman, Harry J. Lennix, Susan Blommaert, James Handy
A secret service agent is given an assignment that he does everything to avoid since it involves protecting the wife of a former president, who is renowned as a strong-minded and wilful woman. Inevitably, a relationship develops despite the fact that they seem to spend all their time arguing. An unfunny and implausible comedy that could have used the old Tracey-Hepburn magic, though Cage and MacLaine acquit themselves reasonably well.
COM   91 min (ort 99 min) cC   VIDrel: COLUM/SONOP   V/sur

GUESS WHO'S COMING TO DINNER ***

PG

Stanley Kramer USA   1965
Spencer Tracy, Katharine Hepburn, Sidney Poitier, Katherine Houghton, Cecil Kellaway, Beah Richards, Roy E. Glenn Sr, Isabell Sanford, Virginia Christine, Alexandra Hay, Barbara Randolph, Tom Heaton, D'Urville Martin
A glossy, coy and contrived tale of a white, liberal couple, whose daughter announces her intention to marry a man who would be anyone's choice, but for the fact that he is a Negro. This gives rise to serious misgivings on their behalf, which are shared by his parents. A stilted and superficial look at bigotry and mixed marriage, well-acted but filmed like a stage play. This was Tracy's last film. AA: Actress (Hepburn), Story/Screen (William Rose).
DRA   103 min (ort 108 min)   VIDrel: CASPIC L/A   V

GUILTY AS CHARGED **

15

Sam Irvin USA   1992
Rod Steiger, Lauren Hutton, Isaac Hayes, Lauren Hutton, Isaac Hayes, Zelda Rubinstein, Irwin Keyes, Michael Beach
A butcher with an exaggerated and warped sense of justice captures and executes paroled murderers in a home-made electric chair, but runs into problems when an innocent man is framed by a corrupt politician. A very black but not terribly funny comedy-thriller.
COM   90 min (ort 95 min)   VIDrel: COLUM/SONOP   V/sh

GUILTY AS SIN **

15

Sidney Lumet USA   1992
Don Johnson, Rebecca De Mornay, Stephen Lang, Hack Warden, Dana Ivey, Ron White, Norma Dell'Agnese, Sean McCann, Luis Guzman, Robert Kenendy, Tom Butler, Christina Baren, Lynne Cormack, Barbara Eve Harris, Simon Sinn, John Kapelos
A female lawyer is persuaded by an accused womaniser to defend him for the murder of his wife and allows herself to be swayed by his easy charm. However, as she digs into the facts of the case, she learns a few things that put her own life in danger. A sort of re-run of JAGGED EDGE but woefully lacking in suspense and burdened by illogical plotting and a highly contrived ending.
DRA   103 min (ort 120 min)   VIDrel: HOLPIC/TECH L/A   V

GUILTY BY SUSPICION **

15

Irwin Winkler USA   1990
Robert De Niro, Annette Bening, George Wendt, Patricia Wettig, Luke Edwards, Sam Wanamaker, Chris Cooper, Ben Piazza, Barry Primus, Brad Sullivan, Martin Scorsese, Gailard Sartain, Stuart Margolin, Barry Tubb, Robin Gammell
A superficial look at the anti-Communism witch-hunt era of the early 1950s and its effect on Hollywood, with a talented director getting caught up in the hysteria of the time, despite his attempts to remain aloof. This dull history lesson (Winkler's writing and directing debut) neither stimulates nor does justice to its powerful material. Look out for Wanamaker (who spent these years in England) as a crooked lawyer. See also FELLOW TRAVELLER.
DRA   100 min (ort 105 min)   VIDrel: WHV   V/sur

GUINEA PIG, THE **

U

John Boulting/Roy Boulting UK   1948
Richard Attenborough, Sheila Sim, Bernard Miles, Cecil Trouncer, Robert Flemyng, Joan Hickson, Edith Sharpe, Peter Reynolds, Timothy Bateson, Clive Baxter, Basil Cunard, John Forrest, Maureen Glynne, Brenda Hogan, Herbert Lomas
A lower-middle class boy is given the chance of being educated at a public school as an experiment in educational policy, and the initial mutual hostility and contempt he faces from the other pupils leads to a confrontation, but also an eventual degree of acceptance. Attenborough was twenty-five when he was unwisely cast in the central role, which does nothing to help a film that is at best insipid and at worst just rather dated and patronising.
Aka: OUTSIDER, THE
DRA   93 min (ort 97 min) B/W   VIDrel: FABFIL/SPEAR   V
Boa: play by Warren Chetham Strode.

GULLIVER'S TRAVELS *

U

Peter R. Hunt BELGIUM/UK   1976
Richard Harris, Catherine Schell, Norman Shelley, Meredith Edwards.
Voices of: Michael Bates, Denise Bryer, Julian Glover, Stephen Jack, Bessie Love, Murray Melvin, Nancy Nevinson, Robert Rietti, Vladek Sheybal, Graham Stark
A live-action and animation version of this famous tale that is about as ineffective an adaptation as any ever made. Dreary songs are complemented by poor animations.
JUV   80 min   VIDrel: BRITHOM   V
Boa: novel by Jonathan Swift.

GULLIVER'S TRAVELS **

U

Chris Cuddington AUSTRALIA   1979
Voices of: Ross Martin, Hal Smith, John Stephenson, Don Messick, Regis Cordic, Julie Bennet, Janet Waldo
A competent but by no means inspired Hanna-Barbera production, following Gulliver's adventures in Lilliput.
ANIM   48 min   VIDrel: COLUM/SONOP   V/sh
Boa: novel by Jonathan Swift.

GULLIVER'S TRAVELS **

PG

Charles Sturridge UK/USA   1995
Ted Danson, Mary Steenburgen, James Fox, Ned Beatty, Edward Fox, Nicholas Lyndhurst, Phoebe Nicholls, Karyn Parsons, Geraldine Chaplin, John Gielgud, Omar Edward Petherbridge, Robert Hardy, John Standing, John Wells, Alfre Woodard
Incredibly ambitious retelling of Swift's classic adventure, that is weakened both by its tendency to depart from the true narrative and its reliance on an excessive number of flashbacks. For the most part, Gulliver is now being forcibly restrained in an asylum by a devious medic, where he regales the inmates with stories of his adventures. Flashy, irritating and disjointed, with

some alterations to the work that are both unaccountable and ill-advised.
FAN   180 min (ort 270 min) mTV   VIDrel: CH4/RTM   V
Boa: novel by Jonathan Swift.

GUMBALL RALLY, THE ***   PG
Chuck Bail USA   1976
Michael Sarrazin, Normann Burton, Gary Busey, Tim McIntire, Raul Julia, Nicholas Pryor, Susan Flannery, Steven Keats, J. Pat O'Malley, Harvey Jason, Joanne Nail, Vaughn Taylor, Tricia O'Neil, Med Flory
The first of the cross-country road race films, with a variety of vehicles taking part in a race from New York to Long Beach, California. The best of the madcap car race movies so far, but it's the stunts that really score over the humour, though there are a couple of nice cameos. Almost certainly provided the inspiration for the CANNONBALL RUN series.
COM   102 min (ort 107 min)   VIDrel: VISVID/WHV L/A   V

GUN, THE ***   PG
John Badham USA   1974
Stephen Elliott, Jean LeBouvier, Wallace Rooney, David Huffman, Pepe Serna, Edith Diaz, Felipe Turich, Val De Vargas, Ramon Bieri, Michael McGuire, Ron Thompson, John Sylvester White, Richard Bright, Mariclare Costello
Follows the fate of a .38 Police Special handgun, from the day it comes off the production line, through its various owners; never being fired until the end, when it is the instrument of a dreadful tragedy. Inevitably episodic, but utterly absorbing. See also TWENTY BUCKS.
DRA   71 min (ort 78 min) mTV   VIDrel: CIC/SONOP L/A   V
Boa: story by Richard Levinson, William Link and Jay Benson.

GUN CRAZY *   15
Tamra Davis USA   1992
Michael Ironside, Drew Barrymore, Ione Skye, Joe Dallesandro, James Le Gros, Billy Drago, Rodney Harvey, Robert Greenberg, Jeremy Davies, Dan Eisenstein, Willow Tipton, James Oseland, Thomas E. Weaver, Ida Lee, Lawrence Steven Myers
An unhappy teenager who feels trapped in the boredom of her impoverished small-town existence, becomes infatuated with a strange penfriend, an ex-con and murderer, whose obsession with guns she comes to share in a desperate bid to win his love. A very sombre tale, depressing and strangely unmoving.
DRA   92 min (ort 97 min)   VIDrel: MED/POLYREC L/A   V/sh

GUN IN BETTY LOU'S HANDBAG, THE **   (18)
Allan Moyle USA   1992
Penelope Ann Miller, Cathy Moriarty, Julianne Moore, William Forsythe, Eric Thal, Alfre Woodard, Xander Berkeley, Michael O'Neill, Christopher John Fields, Billie Neal, Andy Romano, Ray McKinnon, Gale Myron, Faye Grant, Meat Loaf
A small-town librarian finds a gun that was used in a murder, accidentally discharges it and then decides to confess to this crime, in a bid for publicity and her husband's attention. Her plan however has unforeseen consequences as the real murderer decides that the best course of action would be to eliminate her. A silly and unfunny comedy with an unpleasant streak of violence.
COM   85 min (ort 89 min)   SATrel: SKY MOVIES

GUN LAW **   18
Tonino Valerii ITALY/WEST GERMANY   1969
Lee Van Cleef, Walter Rilla, Giuliano Gemma, Christa Linder, Ennio Balbo, Lukas Ammann, Andrea Bosic, Pepe Calvo, Giorgio Gargiullo, Anna Orso, Benito Stefanelli, Yvonne Sanson
A ruthless gunfighter befriends a young man and teaches him how to handle a gun, but the latter eventually comes to reject a life of violence, and the two of them face a final showdown. A callous and brutal actioner of no great merit.
Aka: DAY OF ANGER; DAY OF WRATH; DAYS OF WRATH; I GIORNI DELL'IRA
WES   81 min (ort 109 min)   VIDrel: MOPIC/SGSVID   V
Boa: novel by Ron Barker.

GUNBUSTER: EPISODES 1 TO 6 ***   PG
Hideaki Anno JAPAN   1988/1989
A young girl embarks on a quest through space in search of her father and becomes involved in defending Earth against alien attack using the fighting machine of the title. This weapons system is so awesome in its power that it becomes the sole means by which our planet can survive. An above-average

animated SF fantasy. Note that episode six is both B/W and in wScrn format.
ANIM   175 min (three cassettes – approx 55 min each)
Col/B/W   VIDrel: KISEKI/PARADOX   V/sh

GUNFIGHT AT THE OK CORRAL ***   PG
John Sturges USA   1957
Burt Lancaster, Kirk Douglas, Jo Van Fleet, Rhonda Fleming, John Ireland, Lyle Bettger, Frank Faylen, Earl Holliman, Ted De Corsia, Dennis Hopper, Martin Milner, Whit Bissell, George Matthews, John Hudson, Olive Carey
Excellent account of the events that led up to a famous gunfight, when Wyatt Earp teamed up with Doc Holliday, and shot it out with the Clanton gang. A little slow getting there, but the final shoot-out is worth waiting for. Written by Leon Uris with Frankie Laine singing the title song. See also MY DARLING CLEMENTINE, WYATT EARP and TOMBSTONE.
WES   117 min (ort 121 min) B/W
VIDrel: 4-FRONT/POLYREC/CIC L/A   V/h

GUNFIGHTER, THE ***   PG
Henry King USA   1950
Gregory Peck, Helen Westcott, Millard Mitchell, Jean Parker, Karl Malden, Ellen Corby, Richard Jaeckel, Skip Homeier, Mae Marsh, Anthony Ross, Verna Felton, Ellen Corby, Alan Hale Jr., John Pickard, Angela Clarke, Cliff Clark
A gunslinger tries to live down his past and effect a reconciliation with his estranged wife and son but is tracked down to the small town where they live, by the brothers of his last victim who are thirsting for revenge. A fine Western with first-rate performances and a good script.
WES   81 min (ort 84 min) B/W   VIDrel: 20TH/TECH   V/h

GUNG HO! ***   12
Ray Enright USA   1943
Randolph Scott, Grace MacDonald, Alan Curtis, Noah Beery Jr, J. Carrol Naish, Robert Mitchum, Milburn Stone, David Bruce, Sam Levene, Peter Coe, Rod Cameron, Richard Lane, Harold Landon, John James, Louis Jean Heydt
Excellent and detailed account of the famous diversionary raid on Makin Atoll in the Gilbert Islands during WW2. We follow a group of Marine Rangers chosen for this task through their initial tough and specialised training and the subsequent bloody and fierce fighting. A well-made and exciting effort.
WAR   88 min B/W   VIDrel: SCRN/DISC   V

GUNG HO! ***   15
Ron Howard USA   1986
Michael Keaton, Gedde Watanabe, George Wendt, Mimi Rogers, John Turturro, Soh Yamamura, Rodney Kageyama, Sab Shimong, Clint Howard, Michelle Johnson Rick Overton, Jihmi Kennedy, Rance Howard, Patti Yasutake
A Japanese company is persuaded by Keaton to re-open a Tennessee auto plant that closed down in his depression-hit home town. The owners of the company are totally unprepared for the ensuing clash of cultures, especially the somewhat happy-go-lucky attitude to work on the part of the local labour force. A lightweight but enjoyable satire making few sharp points on the differences between American and Japanese work attitudes. Later a TV series.
COM   108 min (ort 111 min)   VIDrel: CIC/SONOP L/A   V

GUNGA DIN ****   U
George Stevens USA   1939
Cary Grant, Douglas Fairbanks Jr, Sam Jaffe, Victor McLaglen, Joan Fontaine, Eduardo Ciannelli, Montagu Love, Abner Biberman, Robert Coote, Lumsden Hare, Cecil Kellaway
Exciting adventure set on the North-West Frontier, and revolving around three army veterans who fight the local tribesmen and chase women with equal enthusiasm. Jaffe plays the title character, a water-carrier who has a brief moment of glory as he saves a column of British soldiers from ambush. The script is by Ben Hecht and Charles MacArthur.
A/AD   112 min (ort 117 min) B/W
VIDrel: VCC/DISC/COLUM L/A   V

GUNHED: THE ULTIMATE BATTLE **   15
Masato Harada JAPAN   1989
Masahiro Takashima, Yujin Harada, Kaori Mizushima, Mickey Curtis, Landy Leyes, James B. Thompson, Brenda Bakke, Aya Enjyoji, Yujin Harada, Kaori Mizushima, Yousuke Saito, Doll Nguyen, Jay Kabira, Mickey Curtis
In the year 2038 A.D. a group of adventurers set off in search of

treasure, arriving at a Pacific island that lies in a prohibited zone. Once there, they are unfortunately responsible for re-activating a powerful defence system that nearly eliminated mankind thirteen years before. With their comrades all killed, the last couple of survivors are able to locate and repair a combat robot, which they use to save the world from destruction. Fast-paced and derivative nonsense.
Aka: ROBOT WAR
FAN   96 min (ort 100 min) dubbed
VIDrel: MANGA/SONOP   V/sur

GUNMEN ** 18
Deran Sarafian USA 1993
Christopher Lambert, Denis Leary, Kadeem Hardison, Mario Van Peebles, Sally Kirkland, Patrick Stewart, Brenda Bakke, Denis Leary, Kadeem Hardison, Richard Sarafian, Robert Harper, James Chalke, Humberto Elizondo, Andaluz Russell
A narcotics agent goes after the drug lord who had his father murdered, and in order to achieve his aim, teams up with a slightly unbalanced drug smuggler. Both seek a hidden fortune in narcotics, but for different reasons, and each tries to outwit the other in a violent, over-the-top action movie, that recalls nothing so much as the films of Tarantino.
A/AD   93 min (ort 97 min)   VIDrel: COLUM/SONOP   V/sh

GUNPOINT ** PG
Earl Bellamy USA 1966
Audie Murphy, Joan Staley, Warren Stevens, Edgar Buchanan, Denver Pyle, Royal Dano, Nick Dennis, William Bramley, Kelly Thordsen, David Macklin, Roy Barcroft, Morgan Woodward, Robert Pine, Mike Ragan, John Hoyt, Ford Rainey
When a gang rob a train and kidnap a saloon girl, a serious-minded sheriff gives chase together with his posse and pursues them to New Mexico. A standard oater with a good script and some nice locations.
WES   82 min (ort 86 min)
VIDrel: 4-FRONT/POLYREC/CIC   V/sur

GUNS OF DRAGON ** 18
Tony Leung HONG KONG 1993
Ray Lui, Yvonne Yung, Mark Cheng, Alex Fong, Patrick Lung, Billy Lui, Steven Darrow, Jackson Lou, Chia Sheng Chin, John Sham, Ling Tse, Ming Hung Juan, Tak Wan Eng, Julio Diaz, Ciego Carmen, Kwai Po Chun, Ciara Roberts, Lewis Abernathy
A Hong Kong cop arrives in New York on the run from a criminal gang who are out to assassinate him. Another one of those high-octane action movies, competent enough in its way if not especially remarkable.
Aka: AMERICAN DRAGON
A/AD   88 min   VIDrel: MIA/DISC; ENCORE (LV only)   V LV

GUNS OF HONOUR: REBEL ROUSERS ** 15
Peter Edwards USA 1993
Martin Sheen, Jurgen Prochnow, Corbin Bernsen, Gerard Christopher, Walker Brandt, Todd Jensen, James Van helsen, Ron Smekczak, Jeremy Crulchley, Frank Notaro, Adrian Steed, Robin Smith, Grhama Clark, Bill FLynn, Alan Benatar
Condensed version of mini-series set in the post-Civil War period, when a father and son, both former Confederate officers, are sent to Mexico by the President to carry pardons to rebels who are hiding out there. However, this mission becomes more complicated when our heroes involve themselves with the resistance to the French occupation forces.
WES   95 min   VIDrel: AUDIO/DISC   V
Boa: novels by J.T. Edson.

GUNS OF NAVARONE, THE *** PG
J. Lee Thompson UK/USA 1961
Gregory Peck, David Niven, Stanley Baker, Anthony Quinn, Anthony Quayle, Irene Papas, James Darren, Gia Scala, James Robertson Justice, Richard Harris, Bryan Forbes, Allan Cuthbertson, Albert Lieven, Walter Gotell
Overlong story of a mission to blow up two massive German guns, set atop an all but impregnable fortress on an occupied Greek island in 1943. A nicely balanced mixture of drama, tension and intrigue, and if a trifle slow still highly enjoyable. Written by Carl Foreman and followed by the inferior FORCE 10 FROM NAVARONE. AA: Effects (Bill Warrington/Vivian C. Greenham).
WAR   150 min (ort 157 min) wScrn (COLUM/SONOP)
VIDrel: COLUM/SONOP; VCC/DISC/COLUM; ENCORE (LV only)   V/sur LV
Boa: novel by Alistair MacLean.

GUNSMOKE: ONE MAN'S JUSTICE ** (PG)
Jerry Jameson USA 1993
James Arness, Bruce Boxleitner, Amy Stock-Poynton, Alan Scarfe, Christopher Bradley, Miley Lebeau, Kelly Morgan, Apesanahkwat, Hallie Foote, Ed Adams, Clark Brolly, Wayne Anthony, Bing Blenman, Tom Brinson, Richard Lundin
Another one of those TV spin-off feature films based on the popular TV series "Gunsmoke". In this one, retired marshal Matt Dillon learns that the stagecoach that brought his daughter and son-in-law on a visit was attacked by outlaws, who shot all the occupants. But only one of them, the mother of a young boy, dies from her wounds. Together with the youngster, Dillon sets out to track down those responsible. Fair.
WES   88 min (ort 93 min) mTV   SATrel: SKY MOVIES

GUNSMOKE: RETURN TO DODGE ** 15
Vincent McEveety USA 1987
James Arness, Amanda Blake, Buck Taylor, Fran Ryan, Earl Holliman, Ken Olandt, W. Morgan Sheppard, Patrice Martinez, Tantoo Cardinal, Steve Forrest, Vincent McEveety
Once again, an ancient TV series is relaunched, as part of a spate of films in the late 1980s that attempted to re-establish the Western. The stories in the series "Gunsmoke" followed the career of Marshal Matt Dillon, as he upheld law and order at Dodge City. In this tale the marshal is now retired, but discovers that an old enemy is out to kill him when a friend breaks out of jail to warn him. Average.
WES   90 min (ort 96 min) mTV   VIDrel: 20TH/TECH   V

GUNSMOKE: THE LAST APACHE * PG
Charles Correll USA 1990
James Arness, Richard Kelly, Michael Learned, Amy Stock-Poynton, Geoffrey Lewis, Joe Lara, Joaquin Martinez, Hugh O'Brian, Amanda Blake, Sam Vlahos, Peter Murnik, Robert Covarrubias, Ned Bellamy, Dave Florek, Kevin Sifuentes
This misconceived attempt to revive a once-popular TV series brings former US Marshall, Matt Dillon, out of retirement to rescue a daughter he never knew he had, when he visits an old flame and learns that the girl has just been kidnapped by an Apache brave. A mixture of dull plotting, mawkish sentimentality and unrealistic characterisation.
WES   90 min (ort 96 min) mTV   VIDrel: 20TH   V

GUNSMOKE: THE LONG RIDE ** 15
Jerry Jameson USA 1992
James Arness, James Brolin, Ali McGraw
One of a number of spin-off movies based on the characters from the long-running "Gunsmoke" TV series. In this story Matt Dillon (who has now retired from law enforcement) is enjoying his daughter's wedding day when he is arrested for murder.
WES   95 min (ort 114 min) mTV   VIDrel: 20TH   V

GUNSMOKE: TO THE LAST MAN ** 15
Jerry Jameson USA 1991
James Arness, Pat Hingle, Amy Stock-Poynton, Matt Mulhern, Jason Lively, Joseph Bottoms, Morgan Woodward, James Booth, Amanda Wyss, Jim Beaver, Herman Poppe, Ken Swofford, Don Collier, Ed Adams, Kathleen Erickson, Andy Sherman
When Matt Dillon's cattle are rustled and a couple of men killed, he sets out to catch those responsible. Well acted, competently directed, but never really exciting and with that inevitable assembly-line feeling of most recent made-for-TV westerns.
WES   89 min (ort 94 min) mTV   VIDrel: 20TH   V

GUYS AND DOLLS **** U
Joseph L. Mankiewicz USA 1955
Frank Sinatra, Marlon Brando, Jean Simmons, Vivian Blaine, Stubby Kaye, B.S. Pully, Veda Ann Borg, Sheldon Leonard, Regis Toomey, Johnny Silver, Don Dayton, George E. Stone, Kathryn Givney, Mary Alan Hokanson, Joe McTurk, Kay Kuter
With dialogue drawn straight from Runyon, and some great numbers, this has to be one of the best musicals ever. Set in the underworld of small-time New York petty criminals and gamblers, it tells of one Sky Masterson, who is tricked into a bet that he cannot make a date with a girl from the Salvation Army. Songs (by Frank Loesser) include: "If I Were A Bell", "Luck Be A Lady", "Adelaide's Lament" and "Sit Down, You're Rockin' The Boat".
MUS   143 min (ort 150 min)   VIDrel: VCC/DISC   V
Boa: short stories by Damon Runyon.

GUYVER, THE: DARK HERO **   15
Steve Wang USA   1994
David Hayter, Kathy Christopherson, Bruno Giannotta, Christopher Michael, Stuart Weiss, Billi Lee, Jim O'Donoghue, J.D. Smith, Alisa Merline, Veronica Reed, Wes Deitrick, Stephen Oprychal, Ann George, Marissa Cody, Kristin Calkins
A sequel to "The Guyver" in which various monsters once again come after our hero in an attempt to steal the helmet that gives him his powers. Ample special effects cannot make up for the paucity of ideas that is very much in evidence.
Aka: DARK HERO; GUYVER 2: DARK HERO
FAN   95 min (ort 127 min)   VIDrel: MIA/DISC   V/sh

GUYVER, THE: VOLS. 1 TO 11 **   12/PG
Koichi Ishiguro JAPAN   199-
A student named Sho finds a piece of alien armour that transforms him into a crime-fighting superhero. A collection of episodes that detail his various exploits and adventures, mostly fighting the Chronos organisation that is out to dominate the world, making use of creatures known as Zoanoids (one of whom turns out to be the hero's father).
ANIM   330 min (10 cassettes – approx 30 min each) dubbed
VIDrel: MANGA/SONOP   V/sh

GYPSY ***   U
Emile Ardolino USA   1993
Bette Midler, Cynthia Gibb, Peter Riegert, Ed Asner, Christine Ebersole, Michael Jeter, Andrea Martin, Jennifer Beck, Linda Hart, Rachel Sweet, John La Motta, David Marciano, Keene Curtis, Mike Nussbaum, Lacey Chabert, Sean Sullivan
Another screen version of this famous Broadway stage musical, with Midler in great form as the pushy mother of famous stripper Gypsy Rose Lee who will stop at absolutely nothing to promote her daughter's career. The fine script is by Laurents but this is Midler's film, and her dynamic performance does much to compensate for the weakness of direction and slightly static plot.
DRA   136 min (ort 142 min) mTV   VIDrel: EIV/SONOP
V/sur
Boa: book/musical play by Arthur Laurents.

GYPSY EYES **   18
Vinci Vogue-Anzlovar USA   1993
Jim Metzler, Claire Forlani, Zachary Bogatz, George Di Cenzo, Boris Cavazza, Paul Lightman, Matjaz Lightman, Mirko Derganic, Joans Znidarsic, Ashley Graham, Radko Polic, Simona Pitner, Lojze Scete, Janez Vajevec, Jelena Markovic
When an American embassy official is killed by an assassin, a gypsy girl witnesses the crime and becomes a target for the killer.
THR   83 min   VIDrel: OVER/HIFLI   V/h

H

HACKERS *   12
Iain Softley USA   1995
Jonny Lee Miller, Angelina Jolie, Jesse Bradford, Matthew Lillard, Laurence Mason, Renoly Santiago, Fisher Stevens, Alberta Watson, Darren Lee, Peter Y. Kim, Ethan Browne, Lorraine Bracco, Wendell Pierce, Michael Gaston, Marc Anthony
A young hacker stumbles across a far-flung conspiracy when he breaks into the computer system belonging to a major corporation. This makes him a target for both the FBI and the company's own security expert, and he has to count on the support of his fellow hackers in order to survive. A trivial and simplistic thriller, quite obviously aimed at the teenage market.
THR   101 min (ort 105 min) cC   VIDrel: MGM/WHV   V/sur

HAIR ***   15
Milos Forman USA   1979
Treat Williams, John Savage, Beverly D'Angelo, Annie Golden, Dorsey Wright, Don Dacus, Cheryl Barnes, Nicholas Ray, Charlotte Rae, Richard Bright, Miles Chapin, Fern Tailer, Charles Deney, Herman Meckler, Antonia Rey, Linda Surh
Film version of a 1960s Broadway musical, which was a kind of celebration of the hippy sub-culture. Many memorable scenes and good songs are woven around the story of a rich girl, a group of hippies and a boy waiting to be sent to fight in Vietnam. As a period piece its impact will be minimal, but still enjoyable for the vigour of its performances and songs if not its freshness. The choreography is by Twyla Tharp.
MUS   121 min   VIDrel: MGM/WHV   V/sur
Boa: musical by Galt MacDermot, Gerome Ragni and James Rado.

HAIRDRESSER'S HUSBAND, THE ***   15
Patrice Leconte FRANCE   1990
Jean Rochefort, Anna Galiena, Henry Hocking, Maurice Chevit, Ticky Holgado, Roland Bertin, Philippe Clevenot, Julien Bukowski, Youssef Hamid, Thomas Rochefort, Jacques Mathou, Claude Aufaure, Michele Laroque, Pierre Meyrand
An encounter at the age of twelve with an attractive hairdresser creates a permanent sexual fixation in a young boy. Many years later, he achieves his dream by proposing to and marrying a hairdresser with whom he enjoys a relationship of supreme erotic bliss. A charming and fascinating tale, full of unbelievable characters, it has little to do with logic or real life. The screenplay is by Leconte and Claude Klotz, with music by Michael Nyman.
Aka: LE MARI DE LA COIFFEUSE
DRA   78 min (ort 91 min) wScrn   VIDrel: TART/20TH
V/dm

HAIRSPRAY ***   PG
John Waters USA   1988
Ricki Lake, Divine (Glenn Milstead), Michael St Gerard, Debbie Harry, Jerry Stiller, Pia Zadora, Ruth Brown, Sonny Bono, Colleen Fitzpatrick, Leslie Ann Powers, Shawn Thompson, Ric Ocasek, Mink Stole, John Waters, Clayton Prince
After an absence, Waters marked his return to the screen with this nostalgic and spirited look at the 1960s, by way of a gleeful and sharp parody on all those 1950s and 1960s teen-movies. Contains some wonderfully tacky moments and inspired casting, not least being Lake as hefty Tracy Turnblad, and an appearance by female impersonator Divine in his last role. The slight plot is built around an integrated teen-dance in Baltimore of 1962.
COM   88 min (ort 92 min)   VIDrel: CASPIC/BMGREC
V/sur

HALF A SIXPENCE ***   U
George Sidney UK   1967
Tommy Steele, Julia Foster, Penelope Horner, Cyril Ritchard, Grover Dale, Elaine Taylor, Julia Sutton, Sheila Falconer, Leslie Meadows, Christopher Sandford, Pamela Brown, James Villiers, Gerald Campion, Jeffrey Chandler
An apprentice draper inherits a fortune that enables him to indulge his passion for the good life and social climbing, in this exuberant musical version of Wells' novel. Steele gives an inspired and winning performance, ably assisted by a fine cast who triumph over the cardboard plot and occasional slow patches. The score is by David Heneker and the songs include "Half A Sixpence" and the memorable "Flash, Bang Wallop!".
MUS   139 min (ort 145 min)   VIDrel: 4-FRONT/POLYREC
V
Boa: novel Kipps by H.G. Wells.

HALF HUMAN *   (PG)
Inoshiro Honda/Kenneth G. Crane USA   1958
Akira Takarada, Keni Kasahara, Nobuo Nakamura, Momoko Kochi, Akemi Negishi, John Carradine, Morris Ankrum, Russell Thorson, Robert Karnes
No, the title does not refer to the makers of this Japanese account of the Abominable Snowman, but to the renowned yeti himself and his less famous son. The spliced-in US footage, featuring Carradine and Ankrum, does nothing to make this dull tale any more coherent or watchable.
Aka: JUGIN YUKIOTOKO; MONSTER SNOWMAN
HOR   78 min B/W   SATrel: BRAVO MOVIES

HALF MOON STREET *   18
Bob Swaim USA   1986
Michael Caine, Sigourney Weaver, Keith Buckley, Patrick Kavanagh, Nadim Sawalha, Angus MacInnes, Faith Kent, Ram John Holder, Annie Hanson, Michael Elwyn, Jasper Jacob, Patrick Newman, Niall O'Brian, Vincent Lindon
A woman working on a research fellowship in London, supplements her income by working as a professional escort, and is introduced to a powerful British diplomat in this stuffy and tiresome thriller. Adapted from "Doctor Slaughter", the first half of Theroux's book.
THR   85 min (ort 90 min)   VIDrel: POLY/POLYREC L/A   V
Boa: novella by Paul Theroux.

HALFAOUINE: CHILD OF THE TERRACES **   15
Ferid Boughedir/Mouni Baaziz FRANCE/TUNISIA   1990
Selim Boughedir, Mustapha Adouani, Rabia Ben Abdallah, Mohamed Driss, H. Katzaras, R. Ben Abdallah, F. Ben Saidane. Zahir Ben

Amman, A. Gayess, C. Chelby, A. Stebon, A. Boulabiat, R. Ben Ammar, R. Ben Amor,
A young Arab boy learns about the sexes on a visit to the public baths with his mother.
Aka: HALFAOUINE; HALFAOUINE: L'ENFANT DES TERRASSES; ROOFTOP HOPPER, THE
DRA 95 min (ort 98 min) VIDrel: CONNO/RTM V

HALLELUJAH TRAIL, THE ** U
John Sturges USA 1965
Burt Lancaster, Lee Remick, Martin Landau, Jim Hutton, Brian Keith, Donald Pleasence, Pamela Tiffin, John Anderson, Tom Stern, Robert Wilke, John Dehner (plus narration), Jerry Gatlin, Larry Duran, Jim Burk, Dub Taylor
A temperance leader plans to stop a cavalry-guarded shipment of whisky reaching thirsty Denver miners. Meanwhile, the Indians have their eyes on the shipment too. An easy-going and likeable comedy Western that just goes on for far too long. Originally released in Cinerama.
WES 140 min (ort 165 min) VIDrel: MGM/WHV V
Boa: novel by B. Gullick.

HALLOWEEN *** 18
John Carpenter USA 1978
Donald Pleasence, Jamie Lee Curtis, P.J. Soles, Nancy Loomis, Charles Cyphers, Kyle Richards, Tony Farlow, Brian Andrews, John Michael Graham, Nancy Stephens, Arthur Yablans, Mickey Yablans, Brent LePage, Adam Hollander
A killer escapes from the state mental home in Illinois, and goes on a bloody rampage at Halloween in his home town, reliving the crime that got him put away fifteen years earlier. As a horror film it's excellent, but the end, with its hint of the supernatural, fails to make any real sense. Innumerable sequels and spin-offs have followed.
HOR 87 min; 93 (MIA/DISC) (ort 95 min) wScrn (MIA/DISC) VIDrel: 4-FRONT/POLYREC/MIA; MIA/DISC V

HALLOWEEN 2 ** 18
Rick Rosenthal USA 1981
Jamie Lee Curtis, Donald Pleasence, Charles Cyphers, Jeffrey Kramer, Lance Guest, Pamela Susan Shoop, Hunter Von Leer, Dick Warlock, Leo Rossi, Gloria Gifford, Tawny Moyer, Ana Alicia, Ford Rainey, Cliff Emmich, Nancy Stephens
Sequel to the first film, with our mad killer doing his slash and stab act, as he stalks Curtis in this unpleasant and bloody effort. The script is by John Carpenter and Debra Hill.
HOR 87 min Cut (17 sec) VIDrel: MIA/DISC; GAME/SPEAR V/sur

HALLOWEEN 3: SEASON OF THE WITCH * 15
Tommy Lee Wallace USA 1983
Tom Atkins, Stacey Nelkin, Dan O'Herlihy, Ralph Strait, Michael Currie, Jadeen Barbor, Bradley Schachter, Garn Stephens, Nancy Kyes, Jonathan Terry, Patrick Pankhurst, Al Berry, Wendy Wassberg, Dick Warlock, Norman Merrill
A maniacal toymaker intends to murder millions of children on Halloween as a vast human sacrifice. The title is not related to any of those other HALLOWEEN films but has more in common with THE INVASION OF THE BODYSNATCHERS. Abridged before video submission by 2 min 6 sec.
HOR 92 min (ort 96 min) VIDrel: GAME/SPEAR V

HALLOWEEN 4: THE RETURN OF MICHAEL MYERS ** 18
Dwight H. Little USA 1988
Donald Pleasence, Ellie Cornell, Danielle Harris, George P. Wilbur, Michael Pataki, Beau Starr, Kathleen Kinmont, Sasha Jenson, Gene Ross, Carmen Filpi
Ten years after he went on a bloodthirsty rampage, a demented killer emerges from his coma and sets out for his home town, intending to murder his niece, plus any other folk unfortunate enough to get in his way. A gory sequel that follows on from the first HALLOWEEN film, just as if sequels 1 and 2 had never been made. Technically competent, but totally lacking in surprises or fresh ideas.
HOR 84 min (ort 88 min) VIDrel: POLY/POLYREC/BRAVE V

HALLOWEEN 5: THE REVENGE OF MICHAEL MYERS ** 18
Dominique Othenin-Girard USA 1989
Donald Pleasence, Danielle Harris, Ellie Cornell, Beau Starr, Wendy Kaplan, Donald L. Shanks, Jeffrey Landman, Tamara Glynn

Some more of the same in this further HALLOWEEN sequel, with our old chum Myers slicing his way through a further batch of teenagers, when in reality it is only his young niece that he's after. A ponderous sequence of set-pieces, enlivened by a few very well handled shocks but not much else besides.
HOR 97 min VIDrel: POLY/POLYREC/BRAVE V/sh

HALLOWEEN 6: THE CURSE OF MICHAEL MYERS * (18)
Joe Chappelle USA 1996
Paul Stephen Rudd, Marianne Hagen, Mitchell Ryan, Donald Pleasence, Kim Darby, Bradford English, Keith Bogart, Mariah O'Brien, Leo Geter, J.C. Brandy, Devin Gardner, Susan Swift, George P. Wilbur, Janice Knickrehm, Alan Echeverria
Six years have passed since the death in a fire or mass murderer Michael Myers, or so everyone believes. But one Halloween night sees the start of a fresh set of murders if not a fresh plot. More of the same, this sad, tired, tedious sequel proves that the well has really and truly run dry. However, fans of the first five films will find ample thrills if not originality.
HOR 88 min SATrel: SKY MOVIES

HALLOWEEN TREE, THE ** U
USA 1993
Voices of: Leonard Nimoy, Annie Baker, Lindsay Crouse, Mark Taylor, Alex Greenwald, Edan Gross, Andrew Keegan, Kevin Michaels
Four youngsters are determined to save the soul of a friend from the "Ghost of Halloweens Past", and are obliged to go on a journey discovery in which they learn about the origins of the festival.
ANIM 69 min VIDrel: FIRST/SONOP V
Boa; short story by Ray Bradbury.

HALLS OF MONTEZUMA ** PG
Lewis Milestone USA 1950
Richard Widmark, Walter (Jack) Palance, Robert Wagner, Jack Webb, Karl Malden, Reginald Gardiner, Philip Ahn, Richard Boone, Richard Hylton, Skip Homeier, Neville Brand, Martin Milner, Bert Freed, Don Hicks, Howard Chuman
Although essentially no more than a recruiting film for the US Marine Corps, this is a well-crafted, straightforward war movie with more than its sharp of thrilling (if noisy) action.
WAR 109 min (ort 113 min) VIDrel: 20TH/TECH V

HAMBURGER HILL ** 18
John Irvin USA 1987
Anthony Barrile, Michael Patrick Boatman, Tim Quill, Courtney B. Vance, Don Cheadle, Michael Dolan, M.A. Nickles. Don James, Daniel O'Shea, Steven Weber, Kieu Chinh, Tommy Swerdlow, Tegan West
Clinical recreation of a bloody engagement that took place at the base of Hill 937 in the Ashau Valley in the course of the Vietnam War, when troops of the 101st Airborne Division encountered stiff enemy opposition. Realistic, gory and quite unmoving. Scripted by Jim Carabatsos.
WAR 105 min (ort 110 min) wScrn (ENCORE/SPEAR) VIDrel: VVC/DISC; ENCORE/SPEAR V/sur

HAMLET *** U
Laurence Olivier UK 1948
Laurence Olivier, Eileen Herlie, Basil Sydney, Jean Simmons, Norman Wooland, Felix Aylmer, Terence Morgan, Stanley Holloway, John Laurie, Esmond Knight, Anthony Quayle, Niall McGinnis, Harcourt Williams, Peter Cushing
Famous version of Shakespeare's play of the tragic Danish prince, that's beautifully photographed in Denmark. Only Olivier's tendency to over-act mars this one, but that is more a reflection of his stage origins (where a different style of acting is called for) than a criticism of his failings as an actor. AA: Pic, Actor (Olivier), Art/Set (Roger K. Furse/Carmen Dillon), Cost (Roger K. Furse).
Aka: OLIVIER'S HAMLET
DRA 155 min B/W VIDrel: CARL/TECH V
Boa: play by William Shakespeare.

HAMLET *** PG
Grigori Kozintsev USSR 1964
Innokenti Smoktunovsky, Mikhail Nazvanov, Elza Radzin-Skolonis, Yuri Tolubeyev, Anastasia Vertinskaya, S. Oleksenko
Shakespeare's famous tragedy was filmed in the Soviet Union to mark the 400th anniversary of the playwright's birth, and is a bold and rich work, both visually powerful and carefully crafted. There are few ambiguities to be found in the text

(considerably shortened) of this stirring rendition, and the music of Shostakovich is used to good effect.
DRA   142 min (ort 150 min) B/W   VIDrel: HEND L/A   V
Boa: play by William Shakespeare.

HAMLET *** PG
Franco Zeffirelli USA 1990
Mel Gibson, Glenn Close, Alan Bates, Paul Scofield, Ian Holm, Helena Bonham-Carter, Stephen Dillane, Nathaniel Parker, Sean Murray, John McEnery, Michael Maloney, Trevor Peacock, Richard Warwick, Christien Anholt, Dave Duffy
Gibson gives a strong performance in this handsome production, though this owes more to his sheer physical presence than to any great insights into the poetry. Close is miscast as his mother, though their similarity of ages is to the film's advantage when their near-incestuous relationship is examined. Filmed on location in Scotland, and well photographed and staged. Scofield is memorable as the ghost of Hamlet's dad. Music is by Ennio Morricone.
DRA   129 min (ort 135 min)   VIDrel: COLUM/SONOP
V/sur
Boa: play by William Shakespeare.

HAMLET **** PG
Kenneth Branagh UK/USA 1996
Kenneth Branagh, Julie Christie, Billy Crystal, Gerard Depardieu, Charlton Heston, Derek Jacobi, Jack Lemmon, Rufus Sewell, Robin Williams, Riz Abbasi, Richard Attenborough, David Blair, Brian Blessed, Richard Briers, Judi Dench
In Denmark, the king succumbs to poison administered by his brother, who then marries the man's widow. It falls to Hamlet to take revenge for this deed, after he receives a visit from the ghost of his murdered father. Branagh shows that when it comes to bringing the Bard to the screen, there are few directors who are his equal. A richly detailed, full-blooded and vigorous rendition, so faithful to the text that one can almost see the pages turning.
DRA   242 min   CINrel
Boa: play by William Shakespeare.

HAMLET, PRINCE OF DENMARK **** U
Rodney Bennett UK 1979
Derek Jacobi, Claire Bloom, Patrick Stewart, Eric Porter, Lalla Ward, David Robb, Patrick Allen, Robert Swann, Jonathan Hyde, Geoffrey Bateman, Emrys James, Jason Kemp, Geoffrey Reevers, Bill Homewood, Peter Richards
An outstanding TV adaptation of this famous play, with a brilliant central performance, strong support and excellent staging.
Aka: HAMLET
DRA   214 min (2 cassettes) mTV   VIDrel: BBC L/A   V/h
Boa: play by William Shakespeare.

HANCOCK *** (PG)
Tony Smith UK 1991
Alfred Molina, Frances Barber, Mel Martin, Nick Burnell, Malcolm Sinclair, Jim Carter, Clive Russell, Ken Kitson, Stephen Bill, Danny Schiller, Fred Darlington, Salih Ozdermirciler, Jane Kaeser, Harry Ditson, Terry Diab
A very carefully crafted biopic charting the rise and eventual fall of one of television's most gifted (but unhappily also most insecure) comedians. Sharply focused and often very witty, it's sustained by Molina's uncanny ability to really get under the skin of his character, portraying the self-destructive star with sensitivity and utter conviction.
DRA   116 min mTV   TVrel: BBC

HANCOCK: THE PUNCH AND JUDY MAN *** U
Jeremy Summers UK 1962
Tony Hancock, Sylvia Sims, Ronald Fraser, Barbara Murray, John Le Mesurier, Hugh Lloyd, Mario Fabrizi, Pauline Jameson, Norman Bird, Walter Hudd, Eddie Byrne, Peter Myers, Hattie Jacques, John Dunbar, Brian Bedford, Russell Waters
A nicely observed, bittersweet comedy, with Hancock playing a seaside Punch and Judy man who tries, and fails, to become part of the snobby social set. Written by Philip Oakes and Tony Hancock.
Aka: PUNCH AND JUDY MAN, THE
COM   88 min (ort 96 min) B/W   VIDrel: WHV   V/h

HANCOCK: THE REBEL *** U
Robert Day UK 1960
Tony Hancock, George Sanders, Irene Handl, Paul Massie, Margit Saad, Dennis Price, Gregoire Aslan (Krikor Aslanian), Mervyn Johns,

John Le Mesurier, Liz Fraser, Nanette Newman, Peter Bull, John Wood
A clerk with delusions of artistic grandeur, goes to Paris in order to achieve fame as a painter, and eventually does so, but thanks only to the work of someone else. Enjoyable feature length spin-off from the popular TV comedy series "Hancock's Half Hour" and written by the same team; Ray Galton and Alan Simpson.
Aka: CALL ME GENIUS; REBEL, THE
COM   101 min (ort 105 min)   VIDrel: WHV   V/h

HAND, THE * 18
Oliver Stone USA 1981
Michael Caine, Andrea Marcovicci, Annie McEnroe, Bruce McGill, Viveca Lindfors, Rosemary Murphy, Nicholas Hormann, Charles Fleischer, Ed Marshall, John Stinson, Richard Altman, Sparky Watt
A successful cartoonist has his hand severed in a car accident, it then disappears, taking on a life of its own and killing his enemies. This feeble imitation of THE BEAST WITH FIVE FINGERS has Caine well and truly miscast as the cartoonist. A flabby script doesn't help much either.
HOR   105 min   VIDrel: MGM/WHV L/A   V
Boa: novel The Lizard's Tail by Marc Brandel.

HAND OF DEATH, THE ** 15
Wu Sum-Yum (John Woo) HONG KONG 1976
Cheng Long (Jackie Chan), Samo Hung, Tam Tao Liang, James Tien, Dorian Tan, Chu Ching, Chen Yuan-Lung
A young Shaolin temple student is ordered by the monks to find and kill a renegade monk who has murdered his father and put the Chung Dynasty in danger by joining the Manchu forces and becoming a feared commander. A standard revenge tale of little distinction but much ire.
MAR   92 min dubbed   VIDrel: MIA/DISC   V

HAND THAT ROCKS THE CRADLE, THE *** 15
Curtis Hanson USA 1992
Annabella Sciorra, Rebecca De Mornay, Matt McCoy, Ernie Hudson, Julianne Moore, Madeline Zima, John de Lancie, Kevin Skousen, Mitchell Laurance, Justin Zaremby, Eric Melander, Jennifer Melander, Ashley Melander, Mary Anne Owen
An independent-minded woman of the 1990s makes the mistake of hiring as a nanny a woman who is both cold-blooded and vengeful, and soon both women are pitted against each other in a nasty battle of wills. For all the hype that accompanied the film (it just isn't the FATAL ATTRACTION powerhouse its makers claimed it to be) this competent effort has several chilling moments. Hudson as a slightly retarded handyman is quite outstanding. See also THE GUARDIAN.
THR   106 min (ort 110 min) cC   VIDrel: HOLPIC/TECH
V/sur

HANDFUL OF DUST, A **** PG
Charles Sturridge UK 1988
James Wilby, Kristen Scott Thomas, Rupert Graves, Anjelica Huston, Judi Dench, Alec Guinness, Pip Torrens, Cathryn Harrison, Richard Beale, Jackson Kyle, Norman Lumsden, Jeanne Watts, Kate Percival, Richard Leech, Roger Milner
A handsome adaptation of Waugh's 1930s novel looking at the life of an upper-class couple whose marriage is on the rocks, and at the inevitable break-up of their relationship, brought on by the death of their son. This detailed and sharp examination of the stifling social conventions of 1930s Britain, has some beautiful performances and an excellent script by Tim Sullivan, Derek Granger and Sturridge.
DRA   113 min (ort 118 min)   VIDrel: ARROW/RTM   V/sur
Boa: novel by Evelyn Waugh.

HANDGUN ** 18
Tony Garnett USA 1982
Karen Young, Clayton Day, Ben Jones, Suzie Humphreys, Helena Humaun
After her rape, a young girl joins a gun club in order to have her revenge on the man responsible. A dull exploitation movie, masquerading as one of those "films with a message".
Aka: DEEP IN THE HEART
THR   95 min   VIDrel: WHV   V/h

HANDGUN *** 18
Whitney Ransick USA 1994
Seymour Cassel, Treat Williams, Paul Schulze, Anna Thompson
Having stolen $500,000 a man goes on the run, being the only surviving member of the gang that pulled off the robbery. He is finally obliged to hide out with his con-man brother, but finds

that his own sons have joined forces with others to track him down. Writer/director Ransick's debut feature is an offbeat crime-thriller, quite compelling and often unexpectedly funny, even if the plot is not up to much.

A/AD 86 min VIDrel: 20TH/FOXVID V/sh

HANDMAID'S TALE, THE ** 18
Volker Schlondorff USA 1990
Natasha Richardson, Faye Dunaway, Aidan Quinn, Elizabeth McGovern, Victoria Tennant, Robert Duvall, Blanche Baker, Traci Lind, David Dukes, Zoey Wilson, Kathryn Doby, Reiner Schoene, Karma Ibsen Riley, Lucile McIntire
In a despotic America of the near future, only a few fertile women remain, and a military elite exploits them as "baby-makers" to selected, infertile couples. Richardson as one such "handmaid" has to cope with both her master's advances and the hostility of his sterile wife, in this faithful, earnest yet strangely unmoving adaptation of Atwood's visionary novel. Screenplay is by Pinter, with music by Ryuichi Sakamoto.
FAN 104 min (ort 118 min) VIDrel: VISVID/POLYREC V/sur
Boa: novel by Margaret Atwood.

HANDS OF A STRANGER: PARTS 1 AND 2 *** 15
Larry Elikann USA 1987
Armand Assante, Blair Brown, Beverly D'Angelo, Michael Lerner, Phillip Casnoff, Arliss Howard, Patricia Richardson, Sam McHoward, Ben Affleck
An ambitious narcotics detective obsessed with finding the man who raped his wife, becomes romantically involved with a female assistant D.A. A film that promises much excitement but very quickly runs out of steam. Adapted from Daley's novel by playwright Arthur Kopit and originally shown in two parts.
DRA 187 min (2 cassettes – ort 240 min) mTV
VIDrel: 20TH/TECH V/h
Boa: novel by Robert Daley.

HANG 'EM HIGH *** 18
Ted Post USA 1967
Clint Eastwood, Inger Stevens, Ed Begley, Pat Hingle, Arlene Golonka, James MacArthur, Ben Johnson, Bruce Dern, Charlie McGraw, L.Q. Jones, Ruth White, Bob Steele, Alan Hale Jr, Dennis Hopper, Jack Ging, Bert Freed, Michael O'Sullivan
A cowboy survives a lynching and plans his revenge on the nine men responsible in this brutal but well made Hollywood stab at a spaghetti Western.
WES 110 min (ort 114 min) VIDrel: MGM/WHV V/h

HANGAR 18 * PG
James L. Conway USA 1980
Robert Vaughn, Gary Collins, Philip Abbott, James Hampton, Darren McGavin, Joseph Campanella, William Schallert, Cliff Osmond, Tom Hallick, Steven Keats, Pamela Bellwood, Andrew Bloch, H.M. Wynant, Bill Zuckert, Stuart Pankin
A UFO crashes and the American government tries to keep it a secret by keeping the spacecraft and the dead bodies of its crew hidden away in a hangar. A flashy but implausible piece of nonsense, re-released for TV as "Invasion Force" with a new ending, but this did nothing to improve matters, and the film still looks like an episode of TV's "The X-Files" on a bad day.
Aka: INVASION FORCE
FAN 97 min (ort 104 min) VIDrel: VISION/DISC V
Boa: novel by Charles E. Sellier and Robert Weverka.

HANGFIRE ** 18
Peter Maris USA 1990
Brad Davis, Yaphet Kotto, Jan-Michael Vincent, Lee De Broux, James Tolkan, George Kennedy, Kim Delaney
A group of convicts escape from prison in New Mexico and take control of a small town but soon have to face a battle with the National Guard. The guns and stunts act beautifully in this exciting but totally cliched adventure yarn.
A/AD 85 min (ort 91 min) VIDrel: 20VIS/SONOP V/sh

HANGING GALE, THE *** 15
Diarmuid Lawrence EIRE/UK 1995
Joe McGann, Mark McGann, Paul McGann, Stephen McGann, Michael Kitchen, Fiona Victory, Tina Kellegher, Sean McGinley, Gerald McSorley, Joe Pilkington, Claire Fitzgerald, Ciara Marley, Dyland O'Connell, Dave Duffy, Birdy Sweeney, Mal Whyte
Melodramatic tale set in Ireland in the 1840s at the time of the Great Famine, when the failure of the potato harvest led to famine and widespread starvation. As a family of tenant farmers

struggles to avoid eviction, the story details the great inhumanity the English landowners showed to their impoverished tenant farmers. Made with great care it has the unusual feature of the McGanns playing four direct ancestors in a work partly based on their family history.
DRA 198 min (2 cassettes) mTV VIDrel: BBC/TECH V/h

HANGING TREE, THE ** PG
Delmer Daves USA 1958
Gary Cooper, Maria Schell, Karl Malden, George C. Scott, Ben Piazza, Karl Swenson, Virginia Gregg, John Dierkes, King Donovan, Slim Talbot, Guy Wilkerson, Bud Osborne, Annette Claudier, Clarence Straight
A doctor whose past conceals a dark secret lives in a rough gold-mining community and falls in love with a woman patient. A colourful portrayal of frontier life.
WES 103 min (ort 106 min) VIDrel: WHV V/h
Boa: novel by Dorothy M. Johnson.

HANGIN' WITH THE HOMEBOYS *** 15
Joseph B. Vasquez USA 1991
Doug E. Doug, Mario Joyner, John Leguizamo, Nestor Serrano, Kimberly Russell, Mary B. Ward, Reggie Montgomery, Christine Claravall, Victor Mack, Rosemary Jackson, Steven Randazzo, LaTanya Richardson, Cheryl Freeman
A sensitive examination of the problems of four young men from the south Bronx, two Blacks and two Puerto Ricans, on their weekly night out to the very different world of Manhattan. As the evening progresses and they encounter a variety of adventures and setbacks, we learn much more about these four very different characters on their journey of self-discovery. A refreshing and candid film that offers no easy answers or patent solutions.
COM 86 min (ort 90 min) cC VIDrel: 20VIS/SONOP V/sur

HANKY PANKY ** 15
Sidney Poitier USA 1982
Gene Wilder, Gilda Radner, Kathleen Quinlan, Richard Widmark, Robert Prosky, Josef Sommer, Johnny Sekka, Jay O. Sanders, Sam Gray, Johnny Brown, Larry Bryggman, Pat Corley, Bill Beutel, Madison Arnold, Nat Habib, James Tolkan
A female spy involves an architect in her mission and together they are chased by spies, cops etc. in this predictable spoof of NORTH BY NORTHWEST. Written by Henry Rosenbaum and David Taylor and originally intended to be a follow-up to STIR CRAZY, but Pryor's part was rewritten for Radner.
COM 103 min (ort 110 min) VIDrel: RCA L/A V

HANNAH AND HER SISTERS **** 15
Woody Allen USA 1986
Woody Allen, Michael Caine, Mia Farrow, Carrie Fisher, Barbara Hershey, Lloyd Nolan, Maureen O'Sullivan, Daniel Stern, Max Von Sydow, Dianne Wiest, Tony Roberts, Sam Waterston, Julie Kavner, Bobby Short, Joanna Gleason
This examination of three sisters whose lives intertwine is a sharp, witty and affectionate look at their loyalties, conflicts and neurotic tendencies. The thin plot of an affair between Hannah's husband and the youngest sister is nothing more than a vehicle to support a perceptive and often funny look at life, with Allen himself in top form as Farrow's hypochondriac ex-husband. AA: S. Actor (Caine), S. Actress (Wiest), Screen/orig (Allen).
COM 102 min VIDrel: VISVID/POLYREC V

HANNIE CAULDER ** 15
Burt Kennedy UK 1971
Raquel Welch, Robert Culp, Ernest Borgnine, Jack Elam, Strother Martin, Christopher Lee, Diana Dors, Stephen Boyd
A woman takes revenge on the outlaw who raped her and killed her husband, after getting shooting lessons at the hands of a bounty hunter, in this gory and offbeat tale.
WES 81 min (ort 85 min) VIDrel: UNIQUE/PARADOX V

HANS AND THE SILVER SKATES ** (U)
Geoff Collins/Peter Jennings/Richard Slapczynski
AUSTRALIA 1994
Voices of: Ric Herbert, Juliet Jordan, Robert Mag, Joanna Moore, Lee Perry
A young boy desperately needs to help his family raise the money that they need to pay for his father's operation, so he enters an ice-skating race in which the first prize is a pair of silver skates. A cartoon version of a traditional story filmed before as a feature and a TV movie.
ANIM 50 min SATrel: MOVIE CHANNEL

HANS CHRISTIAN ANDERSEN ** U
Charles Vidor USA 1952
Danny Kaye, Farley Granger, Zizi Jeanmaire, Roland Petit, John Qualen, Joey Walsh, Philip Tonge, Erik Bruhn, John Brown, Jeanne Lafayette, Fred Kelsey, Robert Malcolm, George Chandler, Gil Perkins, Peter Votrian, Betty Uitti
This biography has little to do with the life of this great writer, but has some fine songs by Frank Loesser and Richard Day, that compensate for the emptiness of the script.
MUS 108 min (ort 120 min) VIDrel: VCC/DISC/COLUM V

Boa: book by Myles Connolly.

HANSEL AND GRETEL *** U
Len Talan ISRAEL/UK 1987
Cloris Leachman, David Warner, Nicola Stapleton, Hugh Pollard, Emily Richard, Susie Miller, Eugene Kline, Warren H. Feigin, Josh Buland, Lutuf Nouasser, Beatrice Shimshoni
A newer version of this famous tale that is often dull in plot but never less than striking visually. The music of Humperdinck is replaced by a bland and forgettable score, but despite this and other defects, this is one of the best versions yet, and one with a sense of humour.
Aka: CANNON MOVIE TALES: HANSEL AND GRETEL
MUS 81 min (ort 84 min) VIDrel: WHV V/sur
Boa: short story by Jakob Ludwig Karl Grimm and Wilhelm Karl Grimm.

HANUSSEN *** 15
Istvan Szabo HUNGARY/WEST GERMANY 1988
Erland Josephson, Klaus Maria Brandauer, Ildiko Bansagi, Karoly Eperjes, Adriana Biedrzynska, Grazyna Szapolowska, Peter Bobai, Walter Schmidinger, Colette Pilz-Warren, Gyorgy Cserhalmi, Lajos Kovacs
Near to the end of WW1, an Austrian corporal is wounded in the head, and at a military hospital comes under the care of a Jewish doctor, with whose nurse he enjoys an affair. Our corporal now finds that he can predict the future and after the war gives public displays of his gift until his arrest by the Nazis. A sombre and atmospheric parable on life and fate that completes a most unusual trilogy: MEPHISTO and COLONEL REDL being the two earlier works.
DRA 112 min (ort 117 min) VIDrel: CONNO/RTM L/A V

HAPPIEST DAYS OF YOUR LIFE, THE ** U
Frank Launder UK 1950
Alastair Sim, Margaret Rutherford, Joyce Grenfell, Edward Rigby, Bernadette O'Farrell, Guy Middleton, John Betley, Muriel Aked, John Turnbull, Richard Wattis, Arthur Howard, Millicent Wolf, Myrette Morven, Russell Waters
During the London Blitz, a girls' school is evacuated from London to the country and assigned to a boys' school due to a bureaucratic foul-up. The resulting situation causes great distress to both head teachers, who are soon at each others' throats. A highly talented cast works hard to inject life into this ancient farce, but although there some funny moments, the appeal of this British comedy has clearly lessened with the years.
COM 78 min (ort 81 min) B/W VIDrel: LUMI/SPEAR L/A V

Boa: play by John Dighton.

HAPPY GILMORE ** 12
Dennis Dugan USA 1996
Adam Sandler, Christopher McDonald, Julie Bowen, Frances Bay, Carl Weathers, Alan Covert, Robert Smigel, Bob Barker, Richard Kiel, Dennis Dugan, Joe Flaherty, Lee Trevino, Kevin Nealon, Verne Lundquist, Jared Van Snellenberg
A hockey player dreams of making the big time as a top professional, but gives all this up when he discovers that his abilities are far more suited to the golfing green. A slapstick comedy that pokes fun at the world of golf, the snobbery of its players and the petty rule-bound mentality of its organisers. Not quite the dud one might have feared, this is often a sharp and occasionally very caustic affair, well served by "Saturday Night Live" comedian Sandler.
COM 88 min (ort 92 min) cC VIDrel: CIC/SONOP V

HAPPY NEW YEAR ** PG
Claude LeLouch FRANCE/ITALY 1973
Francoise Fabran, Lino Ventura, Charles Gerard, Andre Falcon
A thief is released on parole and goes looking for his former partner, his plan being to rob a jewellery store. However, when one of the accomplices falls for the lady owner of the antiques shop next door to the store they intend to rob, their plans begun to go wrong. A bright and breezy comedy-thriller, remade in the USA in 1987.
Aka: HAPPY NEW YEAR CAPER, THE; LA BONNE ANNEE
COM 110 min (ort 115 min) VIDrel: ARROW/RTM V

HAPPY TOGETHER * 15
Mel Damski USA 1989
Helen Slater, Patrick Dempsey, Dan Schneider, Marius Weyers, Barbara Babcock
A shy would-be novelist and a scatterbrained drama student enrol at the same college and a computer glitch throws them together as room-mates; the expected romantic consequences ensue. A bland comedy that would have been better as a thirty-minute TV sit-com. Our lovers split up and get back together innumerable times, and though they work hard in their roles, there just isn't enough going on to maintain interest. THE SURE THING is a superior variant on this idea.
COM 99 min (ort 102 min) VIDrel: GUILD/SONOP L/A V

HARD-BOILED ** 18
John Woo HONG KONG 1992
Chow Yun-Fat, Tony Leung, Philip Chan, Anthony Wong, Teresa Mo, Bowie Lam, Bobby Au-Yuen, Ng Shui-Ting, Kwan Hoi-Shan, Tung Wai, Y. Yonemura, John Woo, Philip Kwok, Tung Wai, Johnson Law, Lau Kong, Lam Wai-Sun, Benny Lam, Kenny Lam
A complex and unusually violent police actioner in which a police inspector joins up with an undercover cop masquerading as a mob hitman in order to capture the gang of smugglers who killed the former's partner.
Aka: LASHOU SHENTAN
A/AD 122 min subs (dubbed version available – ort 126 min) wScrn VIDrel: TART/20TH; ENCORE (LV only) V/s LV

HARD-BOILED 2: THE LAST BLOOD ** 18
Wong Jing HONG KONG 1991
Andy Lau, Alan Tam, Eric Tsang, Leung Kar Yan
An assassination attempt on the life of the Daka Lama causes the severe injury of a woman, it transpiring that both transfusions of the same rate blood group in order to survive. Unfortunately, the terrorists start killing all suitable blood donors. An action-filled sequel to the first film that strains credibility.
Aka: LAST BLOOD, THE
A/AD 90 min wScrn VIDrel: EAST/DISC V

HARD DRIVE ** 18
James Merendino USA 1993
Matt McCoy, Christina Fulton, Edward Albert Jr, Leo Damian, Belinda Waymouth, Stella Stevens
A former child star has become a virtual recluse and very much of an Internet junkie and meets a mysterious woman through the network with whom he indulges in a fantasy affair. They eventually decide to meet but when she turns up dead, our hero finds himself cast as the prime suspect. An adequate thriller, enlivened with a few erotic overtones.
THR 89 min (ort 92 min) VIDrel: FIRST/SONOP L/A V

HARD EVIDENCE ** 15
Jan Egleson USA 1994
Kate Jackson, John Shea, Terry O'Quinn, Beth Broderick, Jennifer Cuthrie, Rand Courtney, Megan Gallagher, Gustave Johnson, Dean Stockwell, Robert Pentz, Nello Tare, Marc McCaulay, Ed Lillard, James Martin, Robert Raiford, Phil Loch
A woman discovers that she has the misfortune of working for a new boss who though outwardly a well respected local official, is in fact a leading figure in the criminal underworld, and runs an empire built on drugs and prostitution. With two children to support following her recent divorce, our heroine needs the financial security the job brings, but realises that her duty is to expose the activities of her boss. A reasonably diverting tale, based on a real case.
DRA 90 min VIDrel: ODY/SONOP V/sh

HARD EVIDENCE ** 18
Michael Kennedy USA 1995
Gregory Harrison, Joan Severance, Cali Timmins, Andrew Airlie, Nathaniel Deveaux, Colin Cunningham, Jon Cuthbert, James Crescenzo, Paul Jarrett, Nick James, Maxine Guess, Syliva Mitchell, Janet Craig, Tanya Dargel, Jason Smith
A married architect has an affair but is unlucky enough to get involved with a scheming woman who is not above a little drug-

dealing and murder as she ensnares her lover's wife in her nefarious plans. This however is all part of a sinister plot by a close associate who is desperate to clear his gambling debts. A standard thriller with erotic elements that offers some undemanding viewing while making no pretentions as to originality or imagination. Fair.

THR   90 min (ort 100 min)   VIDrel: THIRD   V

HARD HUNTED **   15
Andy Sidaris USA   1992
Dona Speir, Roberta Vasquez, Cynthia Brimhall, Bruce Oenhall, R.J. Moore, Tony Peck, Rodrigo Obregon, Al Leong, Mciahel J. Shane
Another assembly-line actioner from this director that relies heavily on the presence of pretty girls in minimal swimwear (many of whom are former Playboy models) to bolster its flimsy plot about three undercover agents out to thwart a gang trying to steal nuclear weapons.

A/AD   93 min (ort 97 min)   VIDrel: MIA/DISC   V/sh

HARD JUSTICE **   18
Greg Yaitanes USA   1995
David Bradley, Charles Napier, Yuji Okumoto, Vernon Wells
After his partner is murdered, a government agent goes under cover and goes to prison in order to find his killer. Unfortunately, he soon discovers that the latter is involved with the warden in running a gun-smuggling scheme.

A/AD   87 min (ort 95 min)   VIDrel: 20TH/FOXVID   V/sur

HARD MEN *   18
J.K. Amalou FRANCE/UK   1996
Vincent Regan, Ross Boatman, Lee Ross, Ken Campbell, "Mad" Frankie Fraser, Mirella D'Angelo, Irene Ng, Robyn Lewis, Andrew Weatherall, Roger Griffiths, Stuart Jason Cole, Finola Geraghty, Nadio Fortune, Corin Mellinger, Vic Tablian
So clearly inspired by both PULP FICTION and RESERVOIR DOGS that one is hard put to find any originality of its own, this film follows the exploits of three vicious East End villains, who make their way through the underworld, dishing out rough justice to anyone unwise enough to cross them. When one of the three decides to quit this line of work, his former colleagues come after him. A gratuitously violent exercise in tedium and nastiness.

THR   87 min   CINrel

HARD PROMISES **   PG
Martin Davison USA   1992
Sissy Spacek, William Petersen, Brian Kerwin, Mare Winningham, Jeff Perry, Peter MacNicol, Ann Wedgeworth, Amy Wright, Lois Smith, Rip Torn, Olivia Burnette
An errant husband returns home when he learns that he has been divorced and he is invited to his ex-wife's wedding. When he learns that her intended is a former high-school rival, he devises a plan to win her back. An unfunny and uninspired comedy with a nasty and unpleasant streak.

COM   91 min (ort 95 min)   VIDrel: 20TH/TECH   V/sur

HARD TARGET **   18
John Woo USA   1992
Jean-Claude Van Damme, Lance Henriksen, Yancy Butler, Wilford Brimley, Chuck Pfarrer, Arnold Vosloo, Bob Apisa, Douglas Forsyth Rye, Willie Carpenter, Dave Efron, Jules Sylvester, Barbara Tasker, Kasi Lemmons, Randy Cheramie, Tom Lupo
A brave sailor comes into conflict with a group of wealthy, sadistic hunters whose penchant is going after human prey, and in the process assists a group of veterans escape a nasty fate. The US debut for Hong Kong action director John Woo bears all the hallmarks of his brand of mindless action and violence with precious little by way of plot development or characterisations.

A/AD   92 min (ort 97 min)   cC   VIDrel: CIC/SONOP; PION (LV only)   V/sur LV

HARD TICKET TO HAWAII **   18
Andy Sidaris USA   1987
Dona Speir, Hope Marie Carlton, Ronn Moss, Harold Diamond, Rodrigo Oberon, Cynthia Brimhall, Wolf Larson, Kiram Hi Lam, Rustam Branaman, Peter Browmilow
A sequel to MALIBU EXPRESS, with two female undercover Federal agents using the front of an inter-island cargo service for their investigations. When they stumble on a cache of diamonds, they soon find themselves menaced by a drug dealer's contract killer. A picturesque but forgettable trip round some of Hawaii's more colourful islands.

A/AD   91 min Cut (1 min 13 sec)   VIDrel: RCA L/A   V

HARD TIMES ***   U
Peter Barnes UK   1994
Alan Bates, Bob Peck, Bill Paterson, Emma Lewis, Harriet Walter, Richard E. Grant, Miriam Margoyles
A solid and well-acted adaptation of the Dickens novel dealing with the harshness of 19th century life in industrial England.

DRA   102 min (2 cassettes – ort 120 min) mTV
VIDrel: BBC/TECH   V/h
Boa: novel by Charles Dickens.

HARD TIMES: PARTS 1 AND 2 ***   PG
John Irvin UK   1977
Patrick Allen, Timothy West, Alan Dobie, Edward Fox, Michelle Dibnah, Ursula Howells, Jacqueline Tong, Harry Markham
In Victorian England an abandoned circus girl is taken in by a stuffy northern industrialist and made into his ward. A strong cast and good attention to detail make this rather talkative Dickens adaptation quite entertaining. It was originally shown as four approximately 50-minute episodes.

DRA   220 min (2 cassettes – ort 240 min) mTV
VIDrel: CASPIC/BMGREC   V
Boa: novel by Charles Dickens.

HARD TO KILL **   18
Bruce Malmuth USA   1990
Steven Seagal, Kelly LeBrock, Bill Sadler, Frederick Coffin, Bonnie Burroughs, Zachary Rosencrantz, Dean Norris
A martial arts-trained police detective uncovers political corruption and is shot and left for dead, but survives in a seven-year coma. Awakening from this he sets out to have his revenge. Tight direction and well handled chase sequences enliven this otherwise fairly standard action-and-revenge tale.

A/AD   91 min (ort 95 min)   VIDrel: WHV   V/sur

HARD TRUTH, THE **   18
Kristine Peterson USA   1993
Eric Roberts, Michael Rooker, Lysette Anthony, Ray Baker, Don Yesso, Jason Schombing, Katherine Cortez, Yvonne Farrow, Michael Cascone, Heather Cummings, Michael Rose, Brian Markison, Lee Wessof, Charlie Brewer, Reg E. Cathey
A world-weary cop decides to opt out and devises a plan to steal a fortune in Mafia money from a safe in the office of his girlfriend's boss. However, in order to pull off this job, he has to blackmail an electronics expert into taking part. A standard crime thriller in which an alluring woman proves as deadly and as ruthless as she is irresistible. The film noir atmosphere adds some marginal interest to an otherwise unremarkable work.

THR   96 min (ort 100 min)   VIDrel: PROMARK/HIFLI   V/h

HARD WAY, THE *   15
John Badham USA   1990
Michael J. Fox, James Woods, Stephen Lang, Annabella Sciorra, Delroy Lindo, Luis Guzman, Mary Mera, LL Cool J, John Capodice, Christina Ricci, Conrad Roberts, John Costelloe, Bill Cobbs, Penny Marshall, George Cheung, Frank Geraci
A spoilt brat of a movie star is to play a street cop in his next film, so by way of a little homework, he takes some lessons from a hardbitten, reckless New York patrolman. A messy blend of pointless action and dumb humour, the film rapidly slides out of control, with nothing much to show for all the energy expended. The Hitchcock-style ending is as contrived as it is inappropriate.

A/AD   106 min (ort 111 min)   VIDrel: CIC/SONOP   V/sur

HARDCOVER **   18
Tibor Takacs USA   1988
Jenny Wright, Clayton Rohner, Randall William Cook, Stephanie Hodge, Vance Valencia, Michelle Jordan, Mary Baldwin, Rafael Nazario, Bob Frank, Bruce Wagner, Kevin Best, Steven Hemel, Vincent Lucchesi, Murray Rubin, Tom Badal
A young actress who enjoys reading horror stories, becomes the victim of a fictional demon that crosses over into the real world. An intriguing idea, but an unsatisfactory development.
Aka: I, MADMAN

HOR   85 min   VIDrel: EIV/SONOP   V

HARDER THEY COME, THE ***   15
Perry Henzell JAMAICA   1973
Jimmy Cliff, Janet Barkley, Ras Daniel Hartman, Carl Bradshaw, Bobby Charlton, Basil Keame, Winston Stona
Jamaica's first feature film is a story of a country boy who comes to Kingston to become a singer, but has to resort to crime in

order to become a success. A simple tale of considerable vigour, now something of a cult film.
DRA   103 min (ort 105 min)   VIDrel: ARTPRO/RTM   V

HARDER THEY FALL, THE ****   15
Mark Robson USA   1956
Humphrey Bogart, Rod Steiger, Jan Sterling, Mike Lane, Max Baer, Edward Andrews, Harold J. Stone
In his last performance, Bogart plays a cynical press agent and former sportswriter who sets out to expose the ruthless exploitation of prizefighters by their managers. A powerful and often brutally honest expose of the less endearing aspects of the fight game. Scripted by Philip Yordan and photographed by Burnett Guffey (for which he obtained an AAN).
DRA   105 min B/W   VIDrel: COLUM/SONOP   V
Boa: novel by Budd Schulberg.

HARDWARE ***   18
Richard Stanley UK   1989
Dylan McDermott, Stacey Travis, John Lynch, William Hootkins, Carl McCoy, Iggy Pop, Mark Northover, Lemmy, Mac MacDonald, Chris McHallem, Barbara Yu Ling, Oscar James, Arnold Lee, Susie Ng, Fred Leewn, Mimi Chinn
Writer-director Stanley makes his directorial debut with this futuristic fantasy, in which the remains of a scrapped killer-robot is foolishly given to a female sculptress as a present. When the device re-activates itself her apartment is turned into a combat zone as the robot sets out to kill her and any other living things in its path. A flashy and gruesome blend of ALIEN and THE TERMINATOR. See also DEATH MACHINE and CHOPPING MALL for two similar films.
FAN   89 min (ort 95 min)   VIDrel: 4-FRONT/POLYREC   V/sur

HAREM, THE **   15
Marco Ferreri ITALY   1967
Carroll Baker, Gastone Moschin, John Phillip Law, Ugo Tognazzi, Renato Salvatori, MIchel Le Royer
A young female architect breaks off her engagement to a wealthy businessman and takes a holiday, getting involved with a number of men. A trite and fairly unremarkable comedy-drama with a strong erotic element.
DRA   91 min (ort 100 min)   VIDrel: WESCON/RTM   V

HARLEM NIGHTS **   18
Eddie Murphy USA   1989
Eddie Murphy, Richard Pryor, Redd Foxx, Danny Aiello, Arsenio Hall, Berlinda Tolbert, Della Reese, Jasmine Guy, Michael Lerner, Stan Shaw, Vic Polizos, Lela Rochon, David Marciano, Thomas Mikal Ford, Michael Goldfinger, Joe Pecorraro
Written and produced by Murphy in his directorial debut, this is a fairly enjoyable and somewhat old-fashioned gangster tale in which a black 1930s Harlem club-owner takes a stand against a white mobster who is out for a slice of their business. Short on plotting and long on profanity, this film never quite develops the expected impact. Music is by Herbie Hancock.
DRA   111 min (ort 118 min)
VIDrel: 4-FRONT/POLYREC/CIC   V/sur

HARLEY DAVIDSON AND THE MALBORO MAN **   18
Simon Wincer USA   1991
Mickey Rourke, Don Johnson, Chelsea Field, Daniel Baldwin, Giancarlo Esposito, Vanessa Williams, Julius Harris, Tia Carrere, Julius Harris, Eloy Casados, Big John Studd, Tom Sizemore, Mitzi Martin, Kelly Hu, James Nardini
This lavishly mounted buddy-movie casts Rourke as a leather-clad biker and Johnson as his cowboy dude partner. Both hustlers, they hang out at their favourite bar, which is up for sale now the lease has expired. With a few others, they plan a "rescue" which involves robbing a security van, but find not money but a shipment of drugs. Various baddies now come after them in a fairly mindless action caper.
A/AD   95 min (Cut at UK cinema release by 2 sec – ort 98 min)   VIDrel: MGM/WHV L/A   V

HARMFUL INTENT **   15
John D. Patterson USA   1992
Tim Matheson, Emma Samms, Robert Pastorelli, Kurt Fuller, John Walcutt, Tom Isbell, Heather Fairfield, Alex Rocco, Mathew Faison, Leo Geter, Romy Rosemont, Joyce Cohen, Ja'nelle Dixon, Jan Gardner, Jonathan Goldberg, Susan Dolan
When one of his patients dies in suspicious circumstances, a

doctor is obliged to go on the run, his hope being to elude arrest until he can clear his name. Fair.
Aka: ROBIN COOK'S HARMFUL INTENT
THR   89 min (ort 93 min) mTV   VIDrel: ODY/SONOP V/sh
Boa: novel by Robin Cook.

HARNESSING PEACOCKS **   15
James Cellan Jones UK   1992
Serena-Scott Thomas, Peter Davison, John Mills, Nicholas Le Prevost, Renee Asherson, Jeremy Child, Brenda Bruce, Tom Beasley, Richard Huw, Tom Beard, Abigail McKern, David Harewood, Dilys Hamlett, Marsha Fitzalan, Delia Lindsay
A lover from a woman's colourful past re-appears on the scene, but she is now a single mother, whose struggles to keep her son at a private school require her to work as a high-class hooker. A fairly sterile outing, that says and does nothing of especial note.
DRA   100 min (ort 108 min) mTV   VIDrel: ODY/SONOP   V
Boa: novel by Mary Wesley.

HAROLD AND MAUDE ****   15
Hal Ashby USA   1971
Ruth Gordon, Bud Cort, Vivian Pickles, Cyril Cusack, Charles Tyner, Ellen Geer, Eric Christmas, G. Wood, Judy Engles, Shari Summers, M. Borman, Ray Goman, Gordon Devol, Harvey Brumfield, Henry Dieckoff, Philip Schultz
Brilliant black comedy following the partnership of a twenty-year-old youth obsessed with death and a vital and lively eighty-year-old woman. A film that has achieved cult status. Highlights are Cort's phony suicide attempts. The music is by Cat Stevens.
COM   90 min   VIDrel: CIC/SONOP L/A   V
Boa: story by Colin Higgins.

HAROLD LLOYD: FEET FIRST ***   U
Clyde Bruckman USA   1930
Harold Lloyd, Robert McWade, Lillianle Leighton, Barbara Kent, Alec Francis, Noah Young, Henry Hall, Arthur Housman, Sleep 'n' Eat (Willie Best), Noah Beery, Buster Phelps, Leo Willis, Nick Copeland, James Finlayson
A shoe salesman gets himself hopelessly entangled in a series of strange encounters. Not only does he have to cope with dangling from a skyscraper, but he must also prevent the girl he loves from discovering that he is not a rich businessman, as she has been led to believe. Lloyd's second talkie is a solidly made and quite amusing affair, though not as funny as one might have hoped.
Aka: FEET FIRST
COM   69 min (ort 93 min(B/W   VIDrel: CONNO/RTM   V

HAROLD LLOYD: GIRL SHY ***   U
Fred Newmeyer/Sam Taylor USA   1924
Harold Lloyd, Jobyna Ralston, Richard Daniels, Carlton Griffin
A painfully shy tailor's apprentice who has nevertheless written a book on courtship, comes to the city in search of a publisher, and falls for a girl he meets on the train. A fine Lloyd comedy, with some hilarious sequences that feature several of his comic fantasies involving women. A witty and inventive vehicle, with a strong climax as our hero races the clock to save his beloved from marrying a bigamist.
Aka: GIRL SHY
COM   89 min (ort 65 min) B/W silent
VIDrel: CONNO/RTM   V

HAROLD LLOYD: MOVIE CRAZY ***   U
Clyde Bruckman USA   1932
Harold Lloyd, Constance Cummings, Kenneth Thomson, Sydney Jarvis, Eddie Feherston, Robert McWade, Louise Closser Hale, Spencer Charters, Mary Doran, Harold Goodwin, Lucy Beaumont, DeWitt Jennings, Noah Young, Grady Sutton
A young Kansas boy is totally obsessed with the movies and sends a letter to a Hollywood studio offering his services. He encloses a photo but by mistake this is replaced by that of a much better-looking lad and he is soon called for a screen test. Upon arrival in Hollywood, however, his bumbling ways caused a great many complications. A more than slightly auto-biographical tale that offers a fascinating look at the studio system.
Aka: MOVIE CRAZY
COM   79 min B/W   VIDrel: CONNO/RTM   V

HAROLD LLOYD: SAFETY LAST ****   U
Fred Newmeyer/Sam Taylor USA   1923
Harold Lloyd, Mildred Davis, Bill Strother, Noah Young, Westcott B. Clarke, Mickey Daniels, Anna Townsend, Charles Stevens, Gus Leonard, Helen Gilmore
Classic Lloyd comedy about a young man's efforts to make good in the big city, in order not to lose his girlfriend. However, the only job he can find is that of humble clerk in a department store, but his letters to her give a very different impression. The classic Lloyd film, par excellence, with the famous scene of him hanging from a clock; the star performing most of the stunts himself. Inventive, witty and very enjoyable.
Aka: SAFETY LAST
COM   71 min (ort 82 min) B/W silent
VIDrel: CONNO/RTM   V

HAROLD LLOYD: SPEEDY ****   U
Ted Wilde USA   1928
Harold Lloyd, Ann Christy, Bert Woodruff, Brooks Benedict, Geroge Herman, Babe Ruth, Dan Wolheim, Byron Douglas, Hank Knight, Walter Hiers, Herbert Evans, Ernie S. Adams, Gus Leonard, Bobby Dunn, James Dime, Jack Hill
A young man falls in love and gets deeply involved in helping to save the horse-drawn trolley car business, operated by his girlfriend's grandfather, from attempts to close it. A well-paced and brilliantly crafted film, with all the expected Lloyd touches, including a hair-raising climax. The title was Lloyd's nickname and recalls a running gag from THE FRESHMAN. This was the famed comedian's final silent movie.
Aka: SPEEDY
COM   86 min B/W silent   VIDrel: CONNO/RTM   V

HAROLD LLOYD: THE CAT'S PAW, THE **   PG
Sam Taylor USA   1934
Harold Lloyd, Una Merkel, George Barbier, Nat Pendleton, Grace Bradley, Alan Dinehart, Grant Mitchell, Fred Warren, Warren Hymer, J. Farrell MacDonald, James Donlan, Edwin Maxwell, Frank Sheridan, David Jack Holt, Vince Barnett
A man comes home after twenty years spent in missionary work in China with his father and is set up by corrupt politicians, who persuade him to run in the mayoral race for his home town. Unexpectedly elected in a landslide victory, he starts to root out wrong-doing while spouting Chinese proverbs. A strange and barely amusing effort that marked Lloyd's return to the screen after an absence of two years.
Aka: CAT'S PAW, THE
COM   97 min (ort 100 min) B/W   VIDrel: CONNO/RTM   V

HAROLD LLOYD: THE FRESHMAN ****   U
Sam Taylor/Fred Newmeyer USA   1925
Harold Lloyd, Jobyna Ralston, Brooks Benedict, James Anderson, Hazel Keener
Lloyd plays a college guy who'll do anything to win popularity, blissfully unaware that his best efforts are giving rise to much mirth behind his back. The football game finale is one of several great highlights.
Aka: FRESHMAN, THE
COM   75 min B/W silent   VIDrel: VISVID/POLYREC L/A   V

HAROLD LLOYD: THE KID BROTHER ****   U
Terry Wilde/J.A. Howe USA   1927
Harold Lloyd, Jobyna Ralston, Leo Willis, Olin Francis, Walter James, Eddie Boland, Constantine Romanoff, Ralph Yearsley, Gus Leonard
Lloyd plays a timid Cinderella-type kid brother in a tough all-male family, whose dad is the local sheriff. However, he eventually gets to show his mettle in a hilarious finale in which he subdues the town bully and gets the girl. One of Lloyd's best films, this is a strongly plotted blend of comedy, romance and clever sight gags.
Aka: KID BROTHER, THE
COM   82 min (ort 84 min) B/W silent
VIDrel: CONNO/RTM   V

HARPER ***   12
Jack Smight USA   1966
Paul Newman, Lauren Bacall, Julie Harris, Shelley Winters, Robert Wagner, Janet Leigh, Arthur Hill, Pameal Tiffin, Robert Webber, Strother Martin, Harold Gould, Roy Jenson, Martin West, Jacqueline De Wit, Eugene Iglesias
A private eye is hired by a woman to find her missing husband, and comes across some really strange individuals in his search. Engrossing adaptation of a detective novel with excellent dialogue and acting. A much less effective sequel THE DROWNING POOL appeared in 1975.
Aka: MOVING TARGET, THE
THR   115 min (ort 121 min) wScrn   VIDrel: WHV   V
Boa: novel The Moving Target by Ross MacDonald.

HARRIET THE SPY **   PG
Bronwen Hughes USA   1996
Michelle Trachtenberg, Vanessa Lee Chester, Gregory Smith, Rosie O'Donnell, J. Smith-Cameron, Robert Joy, Eartha Kitt, Charlotte Sullivan, Teisha Kim
This is the first film venture from Nickelodeon, the kid's TV channel, and has a feisty eleven-year-old who dreams of becoming a writer spying on her neighbours in order to put together a private notebook, this being her chosen means of gaining relevant experience. But when her classmates find her notes, all hell breaks lose. Drawing fine performances from its young cast, the film promises much, but is weakened by poor development and its pop video approach.
JUV   101 min   CINrel
Boa: novel by Louise Fitzhugh.

HARRISON BERGERON **   (12)
Bruce Pittman CANADA   1995
Sean Aston, Christopher Plummer, Miranda DePencier, Buck Henry, Eugene Levy, Howie Mandell, Andrea Martin, Peter Boretski, David Calderisi, Roger Dunn, Emmanuelle Chrioli, Hayden Christensen, Cindy Cook, Jayne Eastwood, Hale Eisen
A bleak version of the future in which the young man of the title becomes part of the ruling elite in an America where averageness is the supreme goal and the populace are forced to wear headbands that scramble thought and impair mental processes. Kidnapped and taken to a secret centre, he eventually rebels against the system in a novel and unexpected way. A valiant but unsuccessful attempt to convey the feeling and depth of Vonnegut's story.
Aka: KURT VONNEGUT'S HARRISON BERGERON
FAN   95 min (ort 99 min) mTV   TVrel: BBC2
Boa: short story by Kurt Vonnegut.

HARRISON: CRY OF THE CITY **   (15)
James Frawley USA   1995
Edward Woodward, Elizabeth Hurley, Cynthia Harris, Robert Montano, Jeffrey Nordling, Jude Ciccolella, Felicity Huffman, Elva Mai Hoover
A British former cop goes to New York for his daughter's wedding, and while there gives some assistance to a lawyer, who is defending a man accused of killing a popular police officer who was responsible for rooting out corruption in the drugs squad. Adequate, this was one of several such TV movies, and the competence of the cast is not matched by much merit in the script. Woodward more or less reprises the role he had in the TV series "The Equalizer".
DRA   94 min mTV   SATrel: SKY MOVIES

HART TO HART: SECRETS OF THE HART **   (PG)
Kevin Connor USA   1995
Robert Wagner, Stepfanie Powers, Lionel Stander, Jason Bateman, Wendie Malick, Michael Parks, Edward Mulhare, John Beck, Taylor Negron, James Warwick, Mickey Cottrell, John Pinette, Ross Hagen, Pat Morita, Natasha Gregson Wagner
In San Francisco to attend a charity auction to raise money for the orphanage where he grew up, Jonathan Hart is intrigued when he finds a locket containing a photo of himself as a young boy. In a bid to learn more about his origins, he soon finds himself claimed by a long-lost sister and her nephew. Meanwhile a couple of ruthless thieves are planning to steal a fortune in gold coins on display at the auction. An assembly-line series spin-off.
Aka: SECRETS OF THE HART
DRA   86 min mTV   SATrel: MOVIE CHANNEL

HARVEST, THE **   18
David Marconi USA   1993
Miguel Ferrer, Leilani Sarelle Ferrer, Harvey Fierstein, Anthony John Denison, Matt Clark, Henry Silva, Mike Vendrell, Randy Walker, Mario Ivan Martinez, Angelica Aragon, Juan Antonio Llanes, Jose Lavat, Jorge Zepeda
A screenwriter in search of material goes to Mexico to investigate a series of murders and awakens one morning in his hotel after having been drugged and kidnapped. He soon learns that one of his kidneys has been removed and thus gets unwillingly involved with a group of ruthless black-market dealers in

human organs. This frightening idea is spoilt by poor treatment, resulting in a film that brings to mind COMA, but has none of the tension of that earlier work.
THR   97 min   VIDrel: FEATFIL/RTM   V/sh

HARVEY ***
Henry Koster USA   U
                                                                   1950
James Stewart, Josephine Hull, Victoria Horne, Peggy Dow, Cecil Kellaway, Ida Moore, Charles Drake, Jesse White, Wallace Ford, Nana Bryant, William Lynn, Clem Bevans, Richard Wessel, Pat Flaherty, Norman Leavitt, Ed Max, Anna O'Neal
An amiable drinker has an imaginary companion: a six-foot tall invisible rabbit called Harvey. Despite the fact that this apparent delusion is totally harmless, the man's sister still attempts to have him certified. A surprisingly witty and engaging comedy. The script was adapted by Mary Chase and Oscar Brodney from the former's Pulitzer Prize-winning play. This was White's film debut. AA: S. Actress (Hull).
COM   107 min B/W   VIDrel: CIC/SONOP   V/h
Boa: play by Mary Chase.

HARVEY GIRLS, THE ***
George Sidney USA   U
                                                                   1946
Judy Garland, Ray Bolger, John Hodiak, Preston Foster, Virginia O'Brien, Angela Lansbury, Marjorie Main, Chill Wills, Kenny Baker, Selena Royale, Cyd Charisse, Ruth Brady, Catherine McLeod, Jack Lambert, Edward Earle
At the turn of the century, a railway restaurant chain recruits young girls to work out West as waitresses in their elegant establishments. This true but rather lightweight story forms the basis for an entertaining musical comedy that offers excellent acting (especially from Garland) and some wonderful music. AA: Song ("On The Atchison, Topeka And The Santa Fe" – Harry Warren (m)/Johnny Mercer (l)).
MUS   97 min (ort 102 min)   VIDrel: MGM/WHV   V/dm

HATCHET MAN **
Victor Salva USA   18
                                                                   199-
Lance Henriksen, Eric Roberts, Brion James
Story of a nasty battle of wits that ensues between two men, one being a travelling salesman who has just stolen $1,000,000 from a Las Vegas Casino, while the other is a serial killer dubbed the "Hatchet Man". At a roadside diner they get to meet, and the killer soon realises that his reluctant companion has something of great value in his briefcase. This is a bit like DUEL, but updated to the 1990s, and given a slightly homo-erotic flavour for good measure.
THR   87 min   VIDrel: COLUM/SONOP   V

HAUNTED, THE ***
Robert Mandel USA   15
                                                                   1991
Sally Kirkland, Jeffrey DeMunn, Louise Latham, Joyce Van Patten, Stephen Markle, Diane Baker, George D. Wallace
An AMITYVILLE clone featuring the allegedly true experiences of a family that moved into a haunted house, where they were hard put to it, battling a series of malevolent entities over a ten-year period. Not exactly original, but highly watchable, not least thanks to some very well handled special effects. See also POLTERGEIST and GRAVE SECRETS: THE LEGACY OF HILLTOP DRIVE.
HOR   87 min (ort 100 min) mTV   VIDrel: 20TH/TECH   V
Boa: book The Haunted: One Family's Nightmare by Jack and Janet Smurl, Ed and Lorraine Warren and Robert Curran.

HAUNTED **
Lewis Gilbert UK/USA   15
                                                                   1995
Aidan Quinn, Kate Beckinsale, Anthony Andrews, Anna Massey, John Gielgud, Alex Lowe, Geraldine Somerville, Victoria Shalet, Peter England, Liz Smith, Linda Bassett, Alice Douglas, Hilary Mason, Edmund Moriarty, Emily Hamilton
A para-psychology professor arrives at an isolated country house to investigate a reports of a haunting, promptly falls for the attractive owner and finds his initial scepticism turning to belief as the macabre history of the building is revealed by a series of inexplicable events. An enjoyable period ghost story, generally well executed (apart from some unimpressive special effects) and atmospheric.
FAN   100 min (ort 107 min)   VIDrel: EIV/SONOP   V
Boa: novel by James Herbert.

HAUNTED COTTAGE *
Hugh DeWitt UK   18
                                                                   199-
Amber Nectar, Anna Cooper, Crystal D'Canter, Kelly Marsh

Two friends rent a cottage for a short break, but a ghost in the kitchen sets the scene for some torrid encounters. Lesbian sex film.
A   55 min (ort 80 min)   VIDrel: MIST/FALCON   V

HAUNTED GOLD **
Mack Wright USA   U
                                                                   1932
John Wayne, Sheila Terry, Erville Anderson, Harry Woods, Martha Mattox, Edgar "Blue" Washington, Slim Whitaker, Jim Corey, Ben Corbett, Bud Osborne
Wayne's first Western for Warner is a remake of the 1928 silent The Phantom City footage of which has been included here. The plot relates to a spooky gold mine in the desert that a group of bandits are attempting to take over.
WES   55 min (ort 57 min) B/W   VIDrel: MGM/WHV   V/h

HAUNTED HEART, THE *
Frank LaLoggia USA   15
                                                                   1994
Diane Ladd, Olympia Dukakis, Morgan Weisser, Ele Keats, Matt Clark, Scott Wilson, Lucy Lee Flippin
Ladd is the overly protective mother who is quite prepared to kill anyone she sees as a threat to her son, in this remarkably unpleasant variant on PSYCHO (the film even includes a shower sequence). Not exactly original, and not even especially well made.
THR   90 min   VIDrel: FOXVID   V

HAUNTED SCHOOL, THE **
Frank Arnold AUSTRALIA/FRANCE/UK/WEST   (U)
GERMANY   1986
Carol Drinkwater, James Laurie, Michael Becher, Emil Minty, Grant Navin, Beth Buchanan, Leigh Nicholls, Mouche Phillips, Lynne Porteous, Brett Levy, Mervyn Drake, Harry Lawrence, Duncan Wass, Jennifer West, Elizabeth Gentle
Pleasing period tale set in the 19th century, telling of a schoolteacher who leaves England for Australia, her hope being to obtain a better job there and a more secure future. However, upon arrival she finds that life is just as precarious and cannot obtain work. Instead, she explores the option of starting up a school at a remote spot in the Blue Mountains, using for this purpose a derelict building said to be haunted.
JUV   90 min mTV   SATrel: MOVIE CHANNEL

HAUNTED SUMMER **
Ivan Passer USA   18
                                                                   1988
Eric Stoltz, Laura Dern, Philip Anglim, Alice Krige, Alex Winter, Peter Berling, Don Hodson, Giusto Lo Pipero, Antoinette McLain, Terry Richards
Another look at the events of the summer of 1816, when Byron and Shelley and their lovers and friends, stayed at the Villa Deodati in Italy. Extremely well made and pleasing to look at, but stilted and unconvincing. See also GOTHIC for Ken Russell's attempt.
DRA   101 min (ort 106 min)   VIDrel: PATHE L/A   V
Boa: novel by Anne Edwards.

HAUNTING, THE ***
Robert Wise UK   12
                                                                   1963
Julie Harris, Claire Bloom, Russ Tamblyn, Richard Johnson, Lois Maxwell, Valentine Dyall, Fay Compton, Rosalie Crutchley, Diane Clare, Ronald Adam, Freda Knorr, Janet Mansell, Pamela Buckley, Howard Lang, Mavis Villiers
An assorted group of people investigate an allegedly haunted house and experience a sequence of terrifying events. The lack of any visible ghosts is a great advantage in building up a genuine atmosphere of unseen menace. Effective despite the slightly overblown camerawork. See also THE LEGEND OF HELL HOUSE.
HOR   107 min (ort 112 min) B/W wScrn
VIDrel: MGM/WHV   V/h
Boa: novel The Haunting of Hill House by Shirley Jackson.

HAUNTING OF HARRINGTON HOUSE, THE **
Murray Golden USA   PG
                                                                   1982
Dominique Dunne, Edie Adams, Roscoe Lee Browne, Phil Leeds, Vitto Scotti, Rena Craig, Marry Betten, Dolores Albin, Dean Dittman, Marie Earle, Michael Goldfinger, Al Secunda
A young girl decides to uncover the mystery surrounding a strange house that is reputed to be haunted. A neatly made and quite competent story, clearly aimed at a juvenile audience.
A/AD   45 min (ort 50 min)   VIDrel: START/DISC   V

HAUNTING OF SEACLIFF INN, THE ** 15
Walter Klenhard USA 1995
Ally Sheedy, William R. Moses. Louise Fletcher, Lucinda West, Tom McCleister, Maxine Stuart, Shannon Cochran, Jay W. MacIntosh, James Horan, Frederick Dal
In an attempt to save their failing marriage, a couple leave San Francisco and move to a small town where they buy a bed-and-breakfast establishment. But it soon proves to be haunted by the spirit of the former owner, and various nasty happenings (unexplained fires etc.) are soon taking place. Don't expect any surprises in the plot, though for all its lack of originality the film works hard to generate atmosphere and a strong sense of foreboding.
HOR 89 min (ort 94 min) mCab VIDrel: CIC/SONOP L/A V

HAVANA ** 15
Sydney Pollack USA 1990
Robert Redford, Lena Olin, Alan Arkin, Tomas Milian, Daniel Davis, Tony Plana, Betsy Brantley, Lise Cutter, Richard Farnsworth, Mark Rydell, Raul Julia, Vasek Simek, Fred Asparagus, Richard Portnow, Dion Anderson, James Medina
Redford is an aging card shark trying his luck in Cuba during the closing day's of the Batista regime, where a romantic encounter with a married Swedish woman (Olin giving a surprisingly poor performance) leads to him being drawn into supporting the communists and compromising his neutrality. A film that offers a few faint echoes of CASABLANCA in terms of plotting, but is hampered by weak acting and dialogue, and a general lack of vigour.
DRA 138 min (ort 145 min)
VIDrel: 4-FRONT/POLYREC/CIC V/sur

HAWK, THE *** 15
David Hayman UK 1993
Helen Mirren, George Costigan, Rosemary Leach, Owen Teale, Marie Hamer, Melanie Hill, Clive Russell, Caroline Paterson, David Harewood, Christopher Madin, Pooky Quensel, Helen Ryan, John Duttine, Jayne Mackenzie, Nadim Sawalha
When a series of nasty murders are committed by a serial killer (who in the process acquires the title name) a married woman slowly comes to fear that this sinister figure may in fact be her husband. A dark and very disturbing tale of suspicion and guilt, with a capable cast turning in some credible performances.
DRA 83 min (ort 86 min) mTV VIDrel: POLY/POLYREC L/A V
Boa: novel by Peter Ransley.

HAWKEYE ** 15
Leo Fong USA 1988
George Chung, Troy Donahue, Chuck Jeffreys, Hidy Ochiai, Stan Wertlieb, Elizabeth frieje, Michelle McCormick, Juan Chapa, Frank Parrish, Jerry Wilson, Ronnie Lott, Kathleen Brady Overton, Joe Lynum, John Armado, Steve Sweeters
After the title character's best friend is murdered by gangsters in a sordid deal over drugs, our hero, a tough Texan cop, gets a new partner and heads for L.A. to have his revenge. Simple action tale using a tried and tested formula.
A/AD 87 min (ort 90 min) VIDrel: IMPENT V

HAWKS *** 15
Robert Ellis Miller UK 1988
Timothy Dalton, Anthony Edwards, Janet McTeer, Jill Bennett, Sheila Hancock, Connie Booth, Camille Coduri, Robert Lang, Bruce Boa, Pat Strarr, Julie T. Wallace, Caroline Langrishe, Geoffrey Palmer
The story of two terminally ill young men; a cynical but brave British lawyer and an American football player. They meet in hospital where the former faces his imminent death with wry humour whilst the latter is more than a little fearful. The lawyer persuades the ex-footballer to fight back and when remission comes they steal an ambulance and have several amusing encounters. A poignant, sad, funny but not altogether successful film.
COM 105 min (Cut at film release) VIDrel: MGM/WHV L/A V/sh

HAWKS AND SPARROWS *** PG
Pier Paolo Pasolini ITALY 1966
Toto, Ninetto Davoli, Rossana Di Rocco, Renato Capogna, Pietro Davoli
Comedy-drama that tells of two vagabonds, a father and son, who attempt to follow in the footsteps of St Francis of Assissi, accompanied on their journey by a talking bird with a nice line in Marxism. As the bird regales his companions with various left-wing parables, a debate takes place about the true nature of belief, and the conflict that exists between the Catholic Church and Communist ideology. A witty intellectual exercise of great charm.
Aka: HAWKS AND THE SPARROWS, THE; UCCELLACCI E UCCELLINI
DRA 85 min B/W VIDrel: CONNO/RTM L/A V

HAWK'S VENGEANCE ** 18
Marc Voizard USA 1996
Gary Daniels, Jayne Heitmeyer, Cass Magda, George Chiang
Hawk Kelly sets off for the States, his intention being to find the killers of his step-brother. However, he learns that the gang responsible are in the employ of a martial arts master. Adequate action tale, with little that is new or surprising; the revelation that there is a sinister plot involving the operation of a human organ smuggling ring is hardly the most original of ideas.
A/AD 92 min VIDrel: EIV/SONOP V/sh

HAXAN ** 15
Benjamin Christensen SWEDEN 1922
Maren Pedersen, Clara Pontoppidian, Elith Pio, Oscar Stribolt, Tora Teje, Johs Andersen, Benjamin Christensen, Ellen La Cour, Emmy Schonfeld, Karen Fabian, Karen Winther, William S. Burroughs (narration)
A docu-drama history of witchcraft with acted sequences that is now seems hopelessly antiquated and slightly ridiculous. Not helped either by the complete absence of any humour. This seven-part study has a few interesting ritual sequences (which have given it its undeserved reputation) but hardly enough to make it worth seeing, except perhaps for its curiosity value. The version available in the USA has added narration by William Burroughs.
Aka: WITCHCRAFT THROUGH THE AGES; WITCHES, THE
DRA 87 min (ort 113 min) B/W silent
VIDrel: REDEM/RTM V

HAZARD OF HEARTS, A * PG
John Hough UK/WEST GERMANY 1987
Helena Bonham-Carter, Edward Fox, Fiona Fullerton, Diana Rigg, Neil Dickson, Stewart Granger, Anna Massey, Gareth Hunt, Marcus Gilbert, Eileen Atkins, Christopher Plummer
A glossy and totally superficial romantic drama telling of a spoilt young virgin who is lost in a game of cards by her wealthy father to a lecherous nobleman, but falls under the spell of a mysterious marquis. A big-budgeted Lew Grade production of considerable splendour and vast tedium.
DRA 89 min (ort 92 min) VIDrel: 4-FRONT/POLYREC V
Boa: novel by Barbara Cartland.

HE SAID, SHE SAID * 15
Ken Kwapis/Marisa Silver USA 1991
Kevin Bacon, Elizabeth Perkins, Sharon Stone, Nathan Lane, Anthony LaPaglia, Stanley Anderson, Charlaine Woodard, Danton Stone, Phil Leeds, Rita Karin, Erika Alexander, Paul Butler, Ashley Gardner, M.K. Harris, Damien Leake
An irritating comedy that focuses on the dissolution of a romance between a pair of yuppie TV talk show hosts, with the break-up being shown first from his point of view (Kwapis directs) and then hers (Silver directs). A clumsy and gimmick-laden effort that's rendered more so by a few fantasy sequences of no great import. RASHOMON it is not.
COM 111 min (ort 115 min) VIDrel: CIC/SONOP V/sur

HE WALKED BY NIGHT *** PG
Alfred L. Werker/Anthony Mann USA 1948
Richard Basehart, Scott Brady, Roy Roberts, Whit Bissell, Jimmy Cardwell, Jack Webb, Bob Brice, Reed Hadley, Chief Bradley, John McGuire, Lyle Latell, Jack Bailey, Mike Dugan, Garret Craig, Bert Moorhouse, Gaylord Pendleton
Gritty and realistic police drama about the hunt for a psychotic cop killer, done in a semi-documentary style and distinguished by fine performances and excellent direction.
DRA 79 min (ort 80 min) B/W VIDrel: SECOND/RTMC V

HEAD OF THE FAMILY ** 18
Robert Talbot USA 1996
Blake Bailey, Jacqueline Lovell, J.W. Pera, Bob Schott, James Jones, Diane Colazzo
A low-budget horror pic with a vicious streak, that has a man discovering that the head in question belongs to a wheelchair-bound creature, whose three siblings are almost equally strange. When he prevails upon the head to have his girlfriend's husband kidnapped, matters do not end there, as the creature decides to

have the girl abducted too. Not unlike the kind of films made by Troma, this one is a mixture of black humour and nastiness.
HOR 78 min (ort 82 min) VIDrel: EIV/SONOP V/sh

HEAD OFFICE ** 15
Ken Finkleman USA 1986
Judge Reinhold, Jane Seymour, Danny DeVito, Eddie Albert, Rick Moranis, Don Novello, Michael O'Donaghue, Richard Masur, Don King, Wallace Shawn, Lori-Nan Engler, Merritt Butrick, George Coe, Ron Frazier, Ron James
The corruption of innocence is the theme of this lame comedy about a wide-eyed college graduate, who lands a job with a powerful multi-national (thanks to his father's efforts behind the scenes) and soon learns how to succeed in business. Some funny moments in an episodic and disjointed tale, but Reinhold is far too bland an actor to do justice to the script.
COM 87 min (ort 90 min) VIDrel: WHV V/sur

H.E.A.L.T.H. * (12)
Robert Altman USA 1979
Glenda Jackson, James Garner, Carol Burnett, James Garner, Lauren Bacall, Diane Stilwell, Henry Gibson, Donald Moffatt, Paul Dooley, Dick Cavett, Alfre Woodard
An association of health-food cranks hold a convention in a Florida hotel to elect a new president and various parties conspire behind the scenes, eventually reducing the place of a shambles. Altman's unfunny satire on politics if full of caricatures in place of characters and very contrived situations that are completely unamusing. A remarkably untalented work that is very hard to sit through and was not much seen when finally released two years after completion.
COM 96 min (ort 102 min) SATrel: SKY MOVIES

HEAR MY SONG *** 15
Peter Chelsom UK 1991
Ned Beatty, Adrian Dunbar, Tara Fitzgerald, Shirley Anne Field, William Hootkins, Harold Berens, David McCallum, John Dair, Stephen Marcus, Britta Smith, Gladys Sheehan, Gina Moxley, James Nesbitt, Andrew Sachs (narration only)
Unusually for Britain, this film is neither a kitchen-sink saga nor a look at the glories of the past, and is very loosely based on the life of tenor Josef Locke, who left Ireland in the 1950s for America and fortune. The story has some delightfully funny twists, the music is wonderful, and the work can be enjoyed as simple entertainment at its best. Co-written by Chelsom in his directing debut, it was the first film available in the UK in an audio-enhanced form.
DRA 101 min (director's version)
VIDrel: 4-FRONT/POLYREC V/sur

HEAR NO EVIL ** 15
Robert Greenwald USA 1992
Marlee Matlin, D.B. Sweeney, Martin Sheen, John C. McGinley, Christina Carlisi, Greg Elam, Charley Lang, Marge Redmond, Billie Worley, George Rankins, Karen Trumbo, Candice Kingrey, Mary Marsh, Ron Graybeal, Bill Pugin, Pat Codekas
A deaf woman (ably played by Matlin) finds herself being stalked in a terrifying game of cat-and-mouse, by a crooked cop who is trying to recover a valuable stolen coin that he hid in her beeper. A simple-minded version of WAIT UNTIL DARK that is badly lacking in suspense.
HOR 92 min (ort 98 min) cC VIDrel: EIV/SONOP V

HEART AND SOULS ** PG
Ron Underwood USA 1993
Robert Downey Jr, Charles Grodin, Alfred Woodard, Kyra Sedgwick, Elizabeth Shue, Tom Sizemore, David Paymer, Bill Cavert, Lisa Lucas, Shannon Orrock, Will Nye, Chasiti Hampton, Michael Zebulon, Wanya Green, Janet MacLachlan
A young man grows up, surrounded by four guardian spirits who died the same night he was born. In return for their help and guidance, he tries to complete the things they did not have time to accomplish. A good cast helps to compensate for a rather sentimental plot.
FAN 99 min (ort 103 min) cC VIDrel: CIC/SONOP V/sur

HEART CONDITION ** 15
James D. Parriott USA 1990
Bob Hoskins, Denzel Washington, Chloe Webb, Ray Baker, Jeffrey Meek, Kieran Mulroney, Roger E. Mosley, Alan Rachins, Ja'net Dubois, Eva Larue, Frank R. Robert Apisa, Lisa Stahl, Clayton Landey, Julie Silverman, Phyllis Hamlin
A racist L.A. cop (Hoskins) suffers a stroke and is given the heart of his favourite hate object, a recently murdered black lawyer whom he suspected of all kinds of crimes. The latter now appears as a ghost and torments him into going after his killers. Strong performances from the two leads enhance this very formula comedy which has nothing useful to say on the subject of bigotry and prejudice. Written and directed by Parriott.
COM 96 min VIDrel: VCC/DISC/COLUM V/sur

HEART IN WINTER, A *** 15
Claude Sautet FRANCE 1992
Emmanuelle Beart, Daniel Auteuil, Andre Dussolier, Elisabeth Bourgine, Luben Yordanoff, Myriam Boyer, Brigitte Catillon, Maurice Garrel, Jean-Luc Bideau, Jean-Claude Bouillard, Stanislas Carre De Malberg, Dominique De Williencourt
A master violin maker and repairer falls in love with a convert violinist but their relationship runs into problems straight away as both reserve all their passion for their respective professions. A low-key and highly restrained romantic tale.
Aka: UN COEUR EN HIVER
DRA 100 min (ort 104 min) wScrn VIDrel: ARTIF/20TH V/h

HEART IS A LONELY HUNTER, THE *** PG
Robert Ellis Miller USA 1968
Alan Arkin, Stacy Keach, Sondra Locke, Laurinda Barrett, Cicely Tyson, Chuck McCann, Biff McGuire, Percy Rodriguez, Jackie Marlowe, Johnny Popwell, Wayne Smith, Peter Mamakos, John O'Leary, Hubert Harper, Anna Lee Carroll
The moving story of a lonely deaf mute living in a small Southern town. Arkin gives one of the best performances of his career, and if the story is somewhat over-sentimental, it's held together by fine performances from all concerned. The film debuts of Locke and Keach.
DRA 123 min VIDrel: MGM/WHV L/A V
Boa: novel by Carson McCuller.

HEART OF A CHAMPION: THE RAY MANCINI STORY ** PG
Richard Michaels USA 1985
Doug McKeon, Robert Blake, Mariclare Costello, Curtis Conway, Tony Burton, Ray Buktenica, James Callahan, Dick Bakalyan, Luisa Leschin, Norman Alden, Ben Frank, Carl Steven, James Arone, Marty Denkin, Richard Doyle
This TV movie focuses on the challenge by Mancini for the lightweight boxing title, his quest being given added impetus by the certain knowledge that his father would have won it had not WW2 cut short his career. The film's executive producer was Sylvester Stallone, who choreographed the boxing sequences, but these do little to enliven a shallow offering, despite the fact that McKeon is convincing in the title role.
DRA 90 min (ort 100 min) mTV VIDrel: 20TH/TECH V

HEART OF A CHILD ** PG
Sandor Stern USA 1993
Ann Jillian, Michele Green, Bruce Greenwood, Malcolm Stewart, Rip Torn, John Procaccino, Matt Walker, Andrew Wheeler, Jane MacDougal, Joel Palmer, William B. Davis, Sheila Moore
A mother learns that her baby girl is incurably ill with a brain disorder and at the same time learns of another woman who has an equally ill child, but one whose life can be saved by a heart transplant from a suitable donor. As expected, she eventually agrees to the use of her child's heart in a bid to save the other child's life. Based on a true case, this is a proficient but slightly unmoving story, the two leads giving capable if not memorable performances.
DRA 88 min (ort 90 min) mTV VIDrel: ODY/SONOP V

HEART OF DARKNESS *** 15
Nicolas Roeg USA 1993
John Malkovich, Tim Roth, James Fox, Isaach De Bankole, Morten Faldaas, Jan Triska, Patrick Ryecart, Michael Fitzgerald, Geoffrey Hutchings, Peter Vaughn, Phoebe Nicholls, Allan Corduner, Alan Scarfe, Michael Cronin, Timothy Bateson
In 19th century Africa, a steamboat captain is asked to locate the missing representative of an ivory trading company who may have lost his sanity. A long and sometimes confused adaptation of Conrad's work that is very good to look at and undeniably well acted but lacking a tighter and more cohesive script. The film APOCALYPSE NOW was also based on this work, albeit a good deal more loosely.
DRA 101 min (ort 105 min) mTV VIDrel: FIRST/SONOP V/sur
Boa: novella by Joseph Conrad.

HEART OF DIXIE * 15
Martin Davidson USA 1989
Ally Sheedy, Virginia Madsen, Phoebe Cates, Treat Williams, Don Michael Paul, Kyle Secor, Francesca Roberts, Kurtwood Smith, Richard Bradford, Peter Borg, Barbara Babcock, Ashly Gardner, Hazen Gifford, Jenny Robertson
A period drama set around a 1950s Alabama college sorority, with a young girl becoming aware of the civil rights movement and the plight of blacks in the South, and using her post as a college newspaper reporter to write on this issue. Meanwhile, other Southerners enjoy their way of life while it lasts. An intensely self-righteous and irritating film, that never gets to grips with the problems it purports to examine.
DRA 91 min (ort 96 min) VIDrel: VISVID/POLYREC V/sur

HEART OF GLASS * 15
Werner Herzog WEST GERMANY 1976
Stefan Guttler, Sepp Muller, Josef Bierbichler, Clemens Scheitz, Volker Prechtel, Sonia Skiba
An intense, bizarre allegory that's set in a 19th century Bavarian village, where a wandering shepherd supplies a near-bankrupt glass factory with the lost formula for a precious ruby glass and foretells the coming of the industrial age. A film that attempts to convey a sense of mysticism and the apocalyptic, it is by turns arresting, profound, meaningless and opaque. Herzog had most of the cast perform under hypnosis throughout.
Aka: HERZ AUS GLAS
DRA 88 min (ort 94 min) VIDrel: TART/20TH V

HEART OF MIDNIGHT * 18
Matthew Chapman USA 1988
Jennifer Jason Leigh, Peter Coyote, Frank Stallone, Brenda Vaccaro, Jack Hallett, Nick Love, James Rebhorn, Tito Wells, Sam Schact, Nina Lova, Steve Buscemi, Denise Dummont, James Geallis, Ken Moser, Nicholas Cimino
Bizarre tale of a young woman hovering on the edge of a mental breakdown, who inherits a dilapidated former nightclub on the death of her uncle. She moves into it and workmen begin the process of renovation. Her troubles start when she learns that it was used by her uncle to satisfy his perversions. Later, she is raped by one of the workmen, but on regaining herself drawn into a murderous web of intrigue. Scripted by Chapman.
THR 106 min (Cut at film release by 8 sec – some cuts subst) VIDrel: MIA/DISC V/sur

HEART OF THE LIE, THE * 15
Jerry London USA 1992
Timothy Busfield, Lindsay Frost, John Karlen, Linda Blair, John Pleshette
Fact-based drama that details the tribulations of an honest lady cop who is kicked out of the police false on fabricated charges. As she battles to clear her name, she finds her efforts being blocked by the real villains of the piece.
DRA 86 min
VIDrel: ODY/SONOP V/sh Boa: book by John Geenya.

HEARTBEAT * PG
Michael Miller USA 1992
John Ritter, Polly Draper, Kevin Kilner, Michael Lembeck, Nancy Morgan, Christian Cousins, Victor DiMattia, Suzanne Suciu, Seth Isler, Steven Gilborn, David Selburg, Jeanine Jackson, Joseph Svezia, Katsy Chappell, Danny Hart
Ritter plays a successful (well they always are in Danielle Steel novels) author who makes his living penning TV soap operas. In the throes of a messy and acrimonious divorce, he meets a woman whose own marriage is breaking up, and against the odds love blossoms once more. A lightweight romantic trifle, offering a pleasant enough way to kill ninety minutes.
Aka: DANIELLE STEEL'S HEARTBEAT
DRA 89 min mTV VIDrel: MIA/DISC V
Boa: novel by Danielle Steel.

HEARTBREAK KID, THE * PG
Elaine May USA 1972
Charles Grodin, Cybill Shepherd, Jeannie Berlin, Eddie Albert, William Prince, Audra Lindley, Mitchell Jason, Augusta Dabney, Doris Roberts, Marilyn Putnam, Jack Hausman, Erik Lee Preminger, Art Metrano, Tim Browne, Jean Scoppa
A Jewish boy regrets his marriage during his honeymoon and plans to drop his wife in favour of a cool blonde. A modern social comedy somewhat reminiscent of THE GRADUATE, but lacking in warmth. Screenplay is by Neil Simon.
COM 102 min VIDrel: MIA/DISC V/sur
Boa: story A Change of Plan by Bruce Jay Friedman.

HEARTBREAK KID, THE * (15)
Michael Jennings AUSTRALIA 1993
Gloria Karvan, Alex Dimitriades, Steve Bastoni, Nico Lathouris, George Vidalis, Louise Mandylor, Doris Younane, William McInnes, Jasper Bagg, Fonda Goniadis, Vikash Prasad, Buo Quach, Kathy Holliday, Denny Stamatopoulos
A full-length feature spin-off from the Australian high-school TV drama series "Heartbreak High". Here, problem student Nick falls for his young teacher Christina, works with her to set up the school's football team, but comes into conflict with the racist sports teacher (who hates both of them because they are of Greek extraction). Adequate high-school nonsense, reasonably watchable if at times painfully trite.
DRA 92 min SATrel: MOVIE CHANNEL
Boa: play by Richard Barrett.

HEARTBREAK RIDGE * 15
Clint Eastwood USA 1986
Clint Eastwood, Marsha Mason, Everett McGill, Eileen Heckart, Moses Gunn, Bo Svenson, Mario Van Peebles, Arlen Dean Snyder, Roman Franco, Vincent Irizany, Tom Villard, Mike Gomez, Rodney Hill, Peter Koch, Richard Venture
One of those formula films in which a tough marine sergeant whips a bunch of raw recruits into shape, before sending them off to battle. Overlong and predictable, but rescued by Eastwood's vigorous performance.
A/AD 125 min Cut (15 sec – ort 130 min) VIDrel: WHV V/sur

HEARTBURN * 15
Mike Nichols USA 1986
Jack Nicholson, Meryl Streep, Maureen Stapleton, Jeff Daniels, Stockard Channing, Richard Masur, Catherine O'Hara, Steven Hill, Milos Forman, Karen Akers, Anna Maria Horsford, Mercedes Ruehl, Joanna Gleason, Yakov Smirnoff
Bittersweet comedy with Streep a divorced cookery writer, meeting Nicholson, a Washington columnist at a friend's wedding. There follows a marriage, a baby, an affair, a separation and finally, a reconciliation. Though this poignant look at modern marriage lacks the cutting edge of Ephron's autobiographical best-seller, fine performances hold it together. Director Foreman makes his acting debut.
COM 105 min (ort 108 min) VIDrel: CIC/SONOP V/h
Boa: novel by Nora Ephron.

HEAT * 18
Paul Morrissey USA 1972
Sylvia Miles, Joe Dallesandro, Andrea Feldman, Pat Ast, Ray Vestal, Eric Emerson, Lester Persky
An unemployed actor who has come to Hollywood to become a star, involves himself with the usual outcasts and takes up with a faded star, who lives with a former child actor in a seedy motel. An intense, moody and offbeat film, highly watchable, despite its improvised feel and rambling plot, which in many ways has clearly been lifted straight out of SUNSET BOULEVARD. A few comic touches help lighten the tone a little.
Aka: ANDY WARHOL'S HEAT
DRA 95 min Cut (36 sec – ort 100 min)
VIDrel: FIRST/SONOP V

HEAT ** 18
Michael Mann USA 1995
Al Pacino, Robert De Niro, Val Kilmer, Jon Voight, Tom Sizemore, Diane Venora, Amy Brenneman, Ashley Judd, Mykelti Williamson, wes Studi, Ted Levine, Dennis Haysbert, William Fichter, Natalie Portman, Tom Noonan, Kevin Gage
A career criminal comes up against the obsessive LAPD Homicide cop who has sworn to get him, and a nasty battle-of-wits (though not without some mutual respect on both sides) takes place. A tough, brutal and highly effective crime thriller with the two stars in top form and working together for the first time. Despite the banal storyline and flashy direction, their truly fine performances turn a potential dud into a really terrific movie. See also L.A. TAKEDOWN.
THR 164 min (ort 171 min) wScrn cC VIDrel: WHV V/h

HEATED VENGEANCE * 18
Edward Murphy USA 1984
Richard Hatch, Michael J. Pollard, Dennis Patrick, Mills Watson,
Cameron Dye, Ron Max, Jolina Mitchell-Collins, Robert Walker
A man returns to Vietnam looking for the woman he left behind
thirteen years earlier, but soon discovers he's in trouble with a
nasty bunch of deserters.
A/AD   87 min (ort 91 min)   VIDrel: 20TH   V

HEATHERS ** 18
Michael Lehmann USA 1989
Winona Ryder, Christian Slater, Shannon Doherty, Kim Walker,
Lisanne Falk, Penelope Milford, Glenn Shadix, Lance Fenton, Patrick
Labyorteaux, Jeremy Applegate, Jon Matthews, Carrie Lynn
A sharp black comedy that examines a variety of issues of inter-
est to the high school set, most notably peer group pressure and
sexuality. Largely built around the tale of a girl who finds herself
unable to fit in with the reigning college clique. Witty, irrever-
ent and often quite tasteless, the lack of a strong plot weakens
its impact. The feature debut for Lehmann.
Aka: LETHAL ATTRACTION
COM   98 min (ort 102 min)   VIDrel: COLUM/SONOP L/A
V

HEATSEEKER ** 18
Albert Pyun USA 1995
Norbert Weisser, Keith H. Cooke, Gary E. Daniels, Thom Mathews,
Tina Cote, Tim Thomerson
Futuristic story with a kickboxing android posing a major threat
to everyone, until it is opposed by a human kickboxer. A reasonably satisfactory blending of martial arts and SF elements, with
the former clearly predominating. The effects are cheap looking,
the action scenes are handled adequately enough, and Pyun
shows that for all his competence as a director, his production-
line work schedule makes the creation of a quality work unlikely.
FAN   88 min (ort 91 min)   VIDrel: TRIM/HIFLI   V/h

HEATWAVE *** 18
Phillip Noyce AUSTRALIA 1982
Judy Davis, Richard Moir, Bill Hunter, John Mellon, John Gregg,
Chris Haywood, Anna Jemison, Dennis Miller, Peter Hehir, Carole
Skinner, Gillian Jones, Frank Gallacher, Tui Bow, Don Crosby,
Lynette Curan
A property development company resorts to violence during
the Christmas heatwave in Sydney, to push through the
construction of a new apartment block. A kind of Australian
answer to CHINATOWN, polished but melodramatic.
DRA   90 min (ort 99 min)   VIDrel: ARTPRO/RTM   V

HEAVEN & EARTH ** 15
Oliver Stone USA 1993
Tommy Lee Jones, Joan Chen, Haing S. Ngor, Hiep Thi Le, Debbie
Reynonds, Dale Dye, Dustin Nguyen, Thuan K. Nguyen, Bussaro
Sanruck, Lan Nguyen Calderon, Thuan Le, Mai Le Ho, Vinh Dang,
Khiem Thai, Michael Lee, Lung Nguyen, Vivian Wu
A Vietnamese woman endures various tragedies in her life,
including war and colonial occupation, before marrying an
American sergeant and moving to the USA. Following
PLATOON and BORN ON THE FOURTH OF JULY, this is a
personal view of the Vietnam conflict and how the USA height-
ened the sufferings of the Vietnamese, that exhibits Stone's
penchant for over-statement, with the characters reduced to
mere pawns. A fine cast struggle valiantly, but cannot make this
film shine.
DRA   134 min (ort 140 min) cC   VIDrel: WHV   V/sur
Boa: books When Heaven and Earth Changed Places/Child of
War, Women of Peace by Le Ly Hayslip (with Jay Wurts and
James Hayslip).

HEAVEN CAN WAIT ** PG
Warren Beatty/Buck Henry USA 1978
Warren Beatty, Julie Christie, Dyan Cannon, Jack Warden, Charles
Grodin, James Mason, Buck Henry, Vincent Gardenia, Joseph Maher,
Hamilton Camp, Arthur Arthur Malet, Stephane Faragy, Jeannie
Unero, Harry D.K. Wong, George J. Manos
A football player is called to Heaven too soon and returns to
Earth in another man's body. Remake of the 1941 film "Here
Comes Mr Jordan" that despite its talented cast, feels lifeless
and uninspired, with a total lack of wit and painfully listless,
uninspired direction. AA: Art/Set (Paul Sylbert and Edwin
O'Donovan/George Gaines).
COM   96 min (ort 100 min)
VIDrel: 4-FRONT/POLYREC/CIC   V/h

HEAVEN IS A PLAYGROUND *** 15
Randall Fried USA 1991
Michael Warren, D.B. Sweeney, Richard Jordan, Victor Love, Hakeem
Olajuwon, Bo Kimble
Compelling drama set on Chicago's South Side where a young
and idealistic lawyer joins forces with a basketball coach to set
in motion a scheme to help high school students improve their
chances of keeping away from drugs and crime by applying for
college athletics scholarships.
DRA   102 min (ort 104 min)   VIDrel: 20VIS/SONOP   V

HEAVENLY CREATURES *** 18
Peter Jackson NEW ZEALAND 1994
Melanie Lynskey, Kate Winslet, Sarah Peirse, Diana Kent, Clive
Merrison, Jed Brophy, Simon O'Connor, Peter Elliott, Gilbert Goldie,
Geoffrey Heath, Kirsti Ferry, Danon Takle, Ben Skjellererup, Elizabeth
Moody, Liz Mullane, Moreen Eason
In New Zealand in the 1950s, a pair of mismatched teenage girls
develop an obsessively close relationship that is based around
a shared fantasy world. When the mother of one of the girls
threatens to split them up, they resort to murder. This stylised
recreation of a true incident works pretty well for the most part,
thanks chiefly to immaculate performances from Lynskey and
Winslet as the murderous duo (who were both only detained for
five years for this crime).
DRA   97 min cC   VIDrel: TOUCH/TECH   V

HEAVENLY PURSUITS * 15
Charles Gormley UK 1986
Tom Conti, Helen Mirren, David Hayman, Brian Pettifer, Jennifer
Black, Dave Anderson, Tom Busby, Sam Graham, Kara Williams,
Robert Paterson, James Gibb, John Mitchell, Ewen Bremner, Philip
Maxwell, Grace Kirby, Juliet Cadzow
A Glaswegian schoolteacher seems to gain miraculous powers
after saving a pupil from a fall. A waste of the talents of a good
cast.
Aka: GOSPEL ACCORDING TO VIC, THE; JUST ANOTHER MIRACLE
COM   86 min (ort 92 min)   VIDrel: 20TH   V

HEAVENS ABOVE! ** PG
John Boulting/Roy Boulting UK 1963
Peter Sellers, Ian Carmichael, Irene Handl, Cecil Parker, Isabel Jeans,
Eric Sykes, Bernard Miles, Brock Peters, Roy Kinnear, Miriam Karlin,
Joan Miller, Miles Malleson, Eric Barker, William Hartnell, Joan
Hickson, Mark Eden
A down-to earth parson is appointed to a snobby parish and
manages to turn the entire area upside down. A typical Boulting
Brothers satire, which waits until the end, and then pulls its
punches.
COM   113 min (ort 118 min) B/W   VIDrel: WHV   V/h

HEAVEN'S GATE ** 18
Michael Cimino USA 1980
Kris Kristofferson, Isabelle Huppert, Christopher Walken, Jeff Bridges,
John Hurt, Brad Dourif, Richard Masur, Joseph Cotten, Sam
Waterston, Geoffrey Lewis, Terry O'Quinn, Mickey Rourke, Roseanne
Vela, Ronnie Hawkins, Paul Koslo
A vast overblown Western that cost $44,000,000 to make. It is
supposed to recreate the period of the Johnson County War,
when cattlemen tried to force out new settlers. Wonderfully
photographed with magnificent locations and sets, but the
absence of a clear story utterly destroys the film (which in its
turn bankrupted the studio – United Artists).
WES   207 min (Cut at film release by 1 min 14 sec – ort 220
min)   VIDrel: WHV   V/sur

HEAVEN'S PRISONERS * 15
Phil Joanou USA 1996
Alec Baldwin, Kelly Lynch, Mary Stuart Masterson, Eric Roberts,
Teri Hatcher, Vondie Curtis-Hall, Badja Djola, Samantha Lagpacan,
Joe Viterelli, Truck Milligan, Hawthorne James, Don Stark, Carl A.
McGee, Paul Guilfoyle
An ex-cop from New Orleans is obliged to go back onto the
streets on his family's behalf, when their lives are threatened by
criminals. Numerous tedious sub-plots (drink problems, little
girls rescued from aeroplane crashes etc.) quickly dissipate
tension and slow down the action in what might have been quite
a good movie, had the script undergone some judicious pruning.
DRA   132 min   VIDrel: POLFIL   V
Boa: novel by James Lee Burke.

HEAVY ***
James Mangold USA
15
1995
Pruitt Taylor Vince, Shelley Winters, Liv Tyler, Deborah Harry, Joe Grifasi, Evan Dando, David Patrick Kelly, Marian Quinn, Meg Hartig, Peter Ortel, George Alvarez, Cordis Heard, J.C. Mackenzie, Allan D'Arcangelo
A sad misfit lives with his domineering mother and spends much of his time pandering to her, but attempts to gain some freedom when he meets a pretty college drop-out. A mournful independent offering, with splendid ensemble acting and an earnest, literate and incisive script. The film was a success at the Sundance Film Festival, where it received a special jury award.
DRA   103 min wScrn   VIDrel: ARTIF/20TH   V/sh

HEAVY METAL **
Gerald Potterton USA
18
1981
Voices of: Richard Romanus, John Candy, Joe Flaherty, Don Franks, Eugene Levy, Harold Ramis, John Vernon, Roger Bumpass, Jackie Burroughs, Martin Lavut, Marilyn Lightstone, Alice Playten, Susan Roman, August Schellenberg
Experimental SF-oriented animation which is really a compilation of different vignettes, some of them with a certain soft-core content, whilst others have a strongly sadistic flavour. As the stories unfold, a floating green globe informs us that it is the essence of all evil. Worth seeing for the high quality animations if not the stories, which are carelessly put together, the influence of underground comic-books being very noticeable.
ANIM   90 min   VIDrel: COLUM   V

HEAVY TRAFFIC **
Ralph Bakshi USA
18
1973
Voices of: Joseph Kaufman, Beverly Hope Atkinson, Frank De Kova, Terri Haven, Mary Dean Lauria, Jacqueline Mills, Lilian Adams, Jim Bates, Jamie Farr, Robert Easton, Charles Grodone, Michael Brandon, Morton Lewis
Animated X-rated satire about the sexual and other adventures of a young New Yorker. Bakshi made this one on the strength of his success with FRITZ THE CAT. Intercut with a number of pointless live-action sequences and let down by lack of story development. See also WIZARDS, HEY GOOD-LOOKIN' and FIRE AND ICE; three other films by this director.
ANIM   73 min (ort 76 min)   VIDrel: ARROW/RTM L/A   V

HEAVYWEIGHTS **
Steven Brill USA
PG
1995
Tom McGowan, Aaron Schwartz, Shaun Weiss, Tom Hodges, Leah Lail, Paul Feig, Jeffrey Tambor, Ben Stiller, Kenan Thompson, Jerry Stiller, Anne Meara
A group of overweight boys attending a summer health camp are dismayed to find themselves placed on a strict fitness program by the new owner, who is planning to film their activities for commercial reasons. Happily, the kids eventually rebel, in this predictable Disney comedy.
COM   94 min (ort 97 min)   VIDrel: WDV/TECH   V

HECK'S WAY HOME **
Michael Scott USA
(U)
1995
Alan Arkin, Chad Krowchiuk, Michael Riley, Shannon Lawson, Don Francks, Gabe Khouth
When his owners move 2,000 miles in order to prepare to travel to Australia, the family dog refuses to left behind and goes after them, in this well-made and very predictable effort. A little nonsense with the local dog-catcher injects the obligatory note of tension. Fair. See also HOMEWARD BOUND: THE INCREDIBLE JOURNEY.
JUV   88 min (ort 92 min)   SATrel: SKY MOVIES

HEDD WYN ***
Paul Turner UK
PG
1992
Huw Garmon, Sue Roderick, Catrin Fychan, Judith Humphreys, Phil Reid, Gwyn Vaughan, Ceri Cunnington, Llio Silyn, Grey Evans, Gwen Ellis, Emma Kelly, Sioned Jones Williams, Llyr Joshua, Emlyn Gomer, Angharad Roberts, Geraint Roberts
The story of WW1 poet Ellis Evans (who used the pseudonym Hedd Wynn). Having left school at fourteen, he is sent off to fight in France during the Great War, but was killed at Passchendale. However, in lulls during the fighting he worked on his fine poem "The Hero" (Yr Arwr) which was sent home by post, winning him the posthumous literary recognition he had always dreamt of. This seriously underrated and little known film received a Best Foreign Film Oscar nomination.
DRA   110 min (Welsh dialogue)   VIDrel: S4C   V

HEIDI ***
Allan Dwan USA
U
1937
Shirley Temple, Jean Hersholt, Arthur Treacher, Helen Westley, Pauline Moore, Mary Nash, Thomas Beck, Sidney Blackmer, Mady Christians, Sig Rumann, Marcia Mae Jones, Christian Rub, Delmar Watson, Egon Brecher, George Humbert
Film version of a famous children's classic about a young orphan sent to live with his gruff grandfather in the Swiss Alps. A classic children's story that forms a useful vehicle for the talents of Temple. Written by Walter Ferris and Julian Josephson.
JUV   84 min (ort 88 min) B/W   VIDrel: 20TH/TECH L/A   V
Boa: novel by Johanna Spyri.

HEIDI **
Toshiyuki H. Takashi USA
U
1994
Story of a young girl sent to stay with her grandfather in the mountains, where she soon settles down to a new life surrounded by animals. However, her aunt soon reveals that she is making other plans for her.
ANIM   50 min   VIDrel: ABBEY/POLYREC   V
Boa: novel by Johanna Spyri.

HEIRESS, THE *
Paul Thomas USA
18
1988
Jeff Stryker, Samantha Strong, Stephanie Rage, Ona Zee, Siobhan Hunter, Damien Cashmere, Randy West, Nick Ferrar, Louis Paul
A biker saves a rich girl from being attacked by some hooligans and gets involved with her and her stepmother, who gets into bed before allowing him out with her daughter.
A   60 min   VIDrel: VISION/DISC   V

HELICOPTER SPIES, THE **
Boris Sagal USA
(PG)
1967
Robert Vaughn, David McCallum, Bradford Dillman, Carol Lynley, Leo G. Carroll, Barbara Moore, Julie London, Lola Albright, John Carradine, Sid Haig
Compilation from the TV series "The Man From U.N.C.L.E." that was released theatrically, and has special agents Vaughn and McCallum travelling to Greece to foil the fiendish schemes of a megalomaniac, who plans to control the world by means of a secret weapon. Dated fast-moving hokum, full of cliches and silly chases, but retaining some of the charm that made the TV series so popular in its day. Carradine has an interesting cameo as a wise and ancient mystic.
THR   91 min mTV   SATrel: SKY MOVIES

HELL HATH NO FURY **
Thomas J. Wright USA
15
1990
Loretta Swit, Barbara Eden, David Ackroyd, Amanda Peterson, Kim Zimmer, Jim Haynie, William Lucking, Conor O'Farrell, Stephen Iff, Natalie Core, Don Craig, Robert Rockwell, John Marshall Jones, Vernee Watson-Johnson, Clifford Dalton
A happily married housewife is shattered by the murder of her husband. However, worse is to follow when the female psychotic responsible sets out to frame her. A flat adaptation of Battin's bestseller that is badly lacking in tension, despite a convincing performance by Swit.
THR   88 min (ort 100 min) mTV   VIDrel: GUILD/SONOP   V
Boa: novel Smithereens by B.W. Battin.

HELL HATH NO FURY... **
Nicholas Medina USA
18
1996
Shawna O'Brien, Paul Michael Robinson, Jeff Rector, Jenna Bodnar
A married businessman has a one-night stand with a woman he meets whilst on holiday, and returns home to find this individual now firmly ensconced in his household as a nanny. Standard psychological warfare ensues.
DRA   85 min   VIDrel: THIRD   V

HELL IN THE PACIFIC ***
John Boorman USA
PG
1969
Lee Marvin, Toshiro Mifune
During WW2, an American pilot and a Japanese naval officer find themselves alone on a Pacific island and confront each other. From an initial desire to murder each other, their relationship goes through a number of phases as they realise the value of co-operation. An engrossing character study spoilt by the contrived ending. See ENEMY MINE for an SF variant on this theme.
WAR   101 min   VIDrel: VCC/DISC   V

HELL IS A CITY ***
Val Guest UK
PG
1959
Stanley Baker, John Crawford, Donald Pleasence, Maxine Audley, Billie Whitelaw, Joseph Tomelty, George A. Cooper, Vanda Godsell, Geoffrey Frederick, Charles Houston, Sarah Branch, Russell Napier
A police inspector heads a campaign to recapture a jewel thief who has escaped from prison in Manchester, after beating a warder to death. The presence of the mineral malachite on some stolen banknotes helps bring him closer to his quarry. A grim and realistic police thriller that moves along at a rapid pace and (surprisingly for films of this period and genre) is filmed in a very fluid style. The use of Manchester locations is a major asset.
DRA   92 min (ort 98 min) B/W   VIDrel: WHV   V/h
Boa: novel by Maurice Proctor.

HELL NIGHT *
Tom De Simone USA
18
1981
Linda Blair, Vincent Van Patten, Peter Barton, Jenny Neumann, Kevin Brophy, Suki Goodwin, Jimmy Sturtevant, Hal Ralston, Cary Fox, Ronald Gans, Gloria Hellman
A group of young college students have to spend a night in a haunted house, as part of their initiation into the college fraternity. There is however, no ghost, just a crazed killer. Dull.
HOR   101 min Cut (1 sec – ort 101 min)   VIDrel: MIA/DISC
V

HELL TOWN **
Charles Barton USA
U
1938
John Wayne, Johnny Mack Brown, Marsha Hunt, John Patterson, Syd Saylor, Monte Blue, Lucien Littlefield, Nick Lukats, James Craig, Jack Kennedy, Vester Pegg, Earl Dwire, Jim Thorpe, Jennie Boyle, Alan Ladd, Jack Daley
Two cowboy drifters get work at a ranch, but one of them finds himself being framed when he starts sweet-talking the girlfriend of his boss in this modestly entertaining effort.
Aka: BORN TO THE WEST
WES   55 min (ort 60 min) B/W   VIDrel: SCRN/DISC   V
Boa: novel by Zane Grey.

HELLBOUND *
Aaron Norris CANADA/ISRAEL/USA
18
1993
Chuck Norris, Calvin Levels, Sheree J. Wilson, Christopher Neame, David Robb, Cherie Franklin, Jack Adalist, Erez Atar, Jack Messinger, Elki Jacobs, Niko Nitai, Shabtai Konorty, Albert Ilouz, Eli Dor Ham, Ovi Levy, Zoe Trilling
A couple of Arab grave-robbers inadvertently release a powerful demon that was walled up in the 12th century by Richard The Lionheart. In no time at all it has embarked on the obligatory rampage and after it commits a murder in Chicago, a hardbitten cop and his partner travel to Israel, where they confront this menace. A poorly realised affair of wooden acting, which neither entertains nor frightens, and totally fails to develop any real tension.
HOR   90 min (ort 95 min)   VIDrel: WHV L/A   V/sur

HELLBOUND: HELLRAISER 2 **
Tony Randel UK
18
1988
Ashley Laurence, Kenneth Cranham, Clare Higgins, Imogen Boorman, Sean Chapman, Doug Bradley, William Hope, Oliver Smith, Barbie Wolde, Nicholas Vince, Simona Bamford, Angus McInnes, Deborah Joel, James Tilitt, Ron Travis
A film that begins where the earlier HELLRAISER left off, with our young heroine from the first film being forced to do battle with the psychiatrist in charge of the mental hospital to which she has been confined, and taking a trip to Hell in the hope of rescuing her father. As ever, the nasty Cenobites are out to cause misery, in a thinly-plotted film that's all but overwhelmed by its gruesome effects.
HOR   91 min Cut (7 sec – ort 99 min) wScrn
VIDrel: VCC/DISC/COLUM   V

HELLCATS OF THE NAVY ***
Nathan Juran USA
U
1957
Ronald Reagan, Nancy Davis (Reagan), Arthur Franz, Robert Arthur, William Leslie, William Phillips, Harry Lauter, Selmer Jackson, Joseph Turkel, Michael Garth, Maurice Manson
Set during WW2 aboard a US submarine, this story tells of the ship's captain who is ordered to undertake a dangerous mission, which takes him to the heavily mined waters off the Asiatic mainland. A standard war film with some excellent action sequences.
WAR   78 min (ort 82 min) B/W   VIDrel: COLUM/SONOP   V
Boa: novel Hellcats Of The Sea by C.A. Lockwood and H.C. Adamson.

HELLFIGHTERS **
Andrew V. McLaglen USA
PG
1968
John Wayne, Jim Hutton, Katherine Ross, Vera Miles, Jay C. Flippen, Bruce Cabot, Barbara Stuart, Edward Faulkner, Edmund Hashim, Valentin De Vargas, Frances Fong, Alberto Morin, Alan Caillou, Laraine Stephens, John Alderson
Adventure tale of the men who have the dangerous job of fighting oil-well fires. The action sequences and special effects are fine in themselves, and the cast play their parts well, but the thin plot cannot sustain the dramatic situations or plausibly handle an examination of their personal relationships or family lives.
A/AD   115 min (ort 120 min)
VIDrel: 4-FRONT/POLYREC/CIC   V/sur

HELLO AGAIN **
Frank Perry USA
PG
1987
Shelley Long, Judith Ivey, Gabriel Byrne, Corbin Bernsen, Madeleine Potter, Sela Ward, Austin Pendleton, Carrie Nye, Robert Lewis, Thor Fields, Tony Sivic, John Cunningham, Mary Fogarty, Elkan Abramowitz, Shirley Rich
A Long Island woman chokes to death, but is resurrected a year later by her witch sister. She soon finds that she's unable to pick up her life again where she left off – her husband has remarried and is none too thrilled at her return. A contemporary social comedy that does little with the basic premise, except scatter a few laughs in different directions.
COM   92 min   VIDrel: TOUCH L/A   V

HELLO, DOLLY! ***
Gene Kelly USA
U
1969
Barbra Streisand, Walter Matthau, Michael Crawford, Tommy Tune, E.J. Peaker, Louis Armstrong, Marianne MacAndrew, David Hurst, Danny Lockin, Joyce Ames, Judy Knaiz, Fritz Feld, Richard Collier, J. Pat O'Malley
Overlong and somewhat tedious film version of a smash hit musical about a celebrated matchmaker who decides it's time to get hitched herself. Some good songs, a vigorous performance from Streisand and wonderful costumes make it all the more bearable. AA: Art/Set (John DeCuir, Jack Martin Smith and Herman Blumenthal/Walter M. Scott, George Hopkins and Raphael Bretton), Score/adapt (Lennie Hayton/Lionel Newman), Sound (Jack Solomon/Murray Spivack).
MUS   139 min (ort 146 min) wScrn   VIDrel: 20TH/TECH   V
Boa: musical by Jerry Herman and Michael Stewart/play The Matchmaker by Thornton Wilder.

HELLO, HEMINGWAY ***
Fernando Perez CUBA
(PG)
1990
Laura de la Uz, Raul Paz, Herminia Sanchez, Caridad Hernandez, Enrique Molina, Maria Isabel Diaz, Marta Del Rio, Micheline Calvert, Jose Antonio Rodriguez, Ana Gloria Buduen, Yanara Moreno, Wendy Guerra, Carlos Manuel Barco
In 1950s Cuba, outside Ernest Hemingway's grand estate a sixteen-year-old schoolgirl dreams of getting a college education, but the poverty of her family makes her application for a scholarship to the USA a source of much bitterness. However, a friend's reading of Hemingway's short story "The Old Man and the Sea" parallels her determination. A complex film of considerable irony, with an apt central metaphor and a fine performance from de la Uz as the main character.
DRA   88 min (ort 90 min)   CINrel

HELLRAISER ***
Clive Barker UK/USA
18
1987
Andrew Robinson, Clare Higgins, Ashley Lawrence, Sean Chapman, Oliver Smith, Robert Hines, Anthony Allen, Leon Davis, Michael Cassidy, Frank Baker, Doug Bradley, Kenneth Nelson, Nicholas Vance, Simon Bamford, Grace Kirkby
Larry and Julia move into Larry's childhood home where the remains of Frank, his brother and Julia's ex-lover, lie upstairs in the attic. The victim of some macabre experiments with demons from another world, he is accidentally brought back to life when Larry cuts himself. Julia now agrees to lure men into the home and murder them in order to provide the blood he requires. An imaginative albeit grisly debut for writer Barker. HELLBOUND: HELLRAISER 2 followed.
HOR   89 min Cut (4 sec – ort 93 min) wScrn
VIDrel: VCC/DISC   V/sh
Boa: novella The Hellbound Heart by Clive Barker.

HELLRAISER 3: HELL ON EARTH **   *18*
Anthony Hickox USA   1992
Terry Farrell, Paula Marshall, Kevin Bernhardt, Peter Boynton, Doug Bradley, Ashley Laurence, Lawrence Mortorff, Ken Carpenter, Sharon Hill, David Young, Brent Bolthouse, Peter Atkins, Paul Vincent Coleman, George Lee, Aimee Leigh
Second sequel to HELLRAISER continues the story of Pinhead who is brought to our world when the owner of a New York nightclub obtains a very strange statuette. Meanwhile, a woman TV journalist finds herself become deeply involved with some very weird events. Despite the tongue-in-cheek approach, this film wallows in a welter of gore and other sickening special efforts, which cannot mask the lack of a strong storyline.
HOR   89 min (ort 97 min)   VIDrel: VCC/DISC   V/h

HELLRAISER: BLOODLINE *   *18*
Alan Smithee USA   1996
Bruce Ramway, Valentina Vargas, Charlotte Chatton, Adam Scott, Kim Myers, Mickey Cottrell, Christine Harnos
A further tale in the series set onboard a space-station of the future (albeit with some flashbacks to 18th century France and present-day New York). Dear old "Pinhead" is still hard at it, bumping of various unfortunates in a variety of gruesome ways, and there is also a dog from Hell that puts in an appearance. A truly dire effort, carelessly plotted, gory and inept. Not a film a director would be proud of, as the use of the Alan Smithee pseudonym attests.
HOR   81 min cC   VIDrel: HOLPIC/TECH   V/sur

HELL'S ANGELS ON WHEELS **   *18*
Richard Rush USA   1967
Adam Roarke, Jack Nicholson, Sabrina Scharf, Jana Taylor, John Garwood, Sonny Barger, Richard Anders, Mimi Machu, James Oliver, Jack Starrett, Gary Littlejohn, Bruno Ve Sota, Robert Kelljan, Kathryn Harrow
Exploits of a bunch of Hell's Angels on a violent trip across America. A tough but trashy film, watchable chiefly thanks to the excellent work of photographer Laszlo Kovacs.
Aka: ANGEL WARRIORS 2; LEADER OF THE PACK, THE
A/AD   80 min Cut (10 min 59 sec – ort 95 min)
VIDrel: 4-FRONT/POLYREC/MIA   V

HELLZAPOPPIN' **   *U*
H.C. Potter USA   1941
Ole Olsen, Chich Johnson, Robert Paige, Jane Frazee, Lewis Howard, Martha Raye, Clarence Kolb, Nella Walker, Mischa Auer, Richard Lane, Elisha Cook Jr, Hugh Herbert, Olive Hatch, Shemp Howard, Jody Gilbert, Geroge Davis, Harry Monti
Hollywood adaptation of this Broadway screwball comedy, which here starts off with Olson And Johnson in Hell and trying to make a movie without any real plot. Despite plenty of sight gags that sometimes work (and sometimes don't) the film is weakened by an all too conventional romantic subplot.
COM   80 min (ort 84 min) B/W
VIDrel: 4-FRONT/POLYREC   V
Boa: play by Nat Perrin.

HENRY & JUNE ***   *18*
Philip Kaufman USA   1990
Fred Ward, Uma Thurman, Maria de Medeiros, Richard E. Grant, Kevin Spacey, Jean-Philippe Ecoffey, Bruce Myers, Jean-Louis Bunuel, Feodor Atkine, Sylvie Huguel, Artus De Penguern, Pierre Eldernac, Gaetan Bloom
Well-photographed but doomed attempt to breathe life into the triangular relationship between the writer Henry Miller, his wife June, and the French diarist Anais Nin. Set against the picturesque background of 1931 Paris, the film devotes ample time to their erotic encounters but fails to explore both literary aspirations and character motivations. A leaden-paced and overlong effort that fails to hold the interest. See also THE ROOM OF WORDS.
DRA   130 min (ort 140 min)
VIDrel: 4-FRONT/POLYREC/CIC   V/sur
Boa: diaries of Anais Nin.

HENRY IV: PARTS 1 AND 2 ***   *U*
David Giles UK   1979
Jon Finch, Anthony Quayle, David Gwillim, Michele Dotrice, Robert Eddison, Brenda Bruce, Frances Cuka, Rob Edwards, Martin Neil, Roger Davenport, Bruce Purchase, David Neal, Michael Miller, Richard Bebb, John Humphry
Competent and well-acted TV version of this famous play, originally shown in two separate parts. Part 1 examines the conflict

between Prince Hal, the dissolute heir apparent, and his father Henry IV. Part 2 deals with the Prince's growing maturity and his final transition to ruler.
DRA   301 min (2 cassettes) mTV   VIDrel: BBC   V/h
Boa: play by William Shakespeare.

HENRY: PORTRAIT OF A SERIAL KILLER ***   *18*
John McNaughton USA   1986
Michael Rooker, Tom Towles, Tracy Arnold, Mary Demas, Anne Bartoletti, Denise Sullivan, Elizabeth Kaden, Ted Kaden, Anita Ores, Megan Ores, Cheri Jones, Monica Anne O'Malley, Bruce Quist, Erzsebet Szilky, David Katz, Flo Spink
Harsh, low-budget examination of the mind and actions of serial killer Henry Lee Lucas, that has the murderer revealing his predilection to a former cell-mate and the latter's sister. A compelling but intensely chilling study of a cold-as-ice sociopath for whom the taking of a human life is no more significant than swatting a fly. Though repulsive in content, the film provides a fascinating illustration of the banality of evil. See also TO CATCH A KILLER.
DRA   89 min (Cut at UK cinema release – ort 83 min)
VIDrel: ELPIC/POLYREC; ENCORE (LV only)   V LV

HENRY V ****   *U*
Laurence Olivier UK   1944
Laurence Olivier, Robert Newton, Leslie Banks, Renee Asherson, Esmond Knight, Leo Genn, George Robey, Ernest Thesiger, Ivy St Helier, Ralph Truman, Harcourt Williams, Max Adrian, Valentine Dyall, Felix Aylmer
Excellent film version of the famous play, with superb acting and some unforgettable battle scenes. An outstanding film in many ways, stirring, stylised and utterly absorbing. The film opens and closes with a typical performance at the Globe Theatre, set in the year 1603. The splendid score is by William Walton. AA: Spec Award (Laurence Olivier for outstanding achievement as actor, producer and director).
DRA   137 min   VIDrel: CARL/TECH   V
Boa: play by William Shakespeare.

HENRY V ***   *U*
David Giles UK   1979
David Gwillim, Alec McCowen, Jocelyne Boisseau, Martin Smith, Rob Edwards, Roger Davenport, Clifford Parrish, Derek Hollis, Robert Ashby, David Buck, Rod Beacham, Trevor Baxter, John Abineri, Michele Dotrice, Frances Cuka
One in a series of BBC adaptations, with Gwillim effective in the title role and ably supported by a strong cast.
DRA   163 min (ort 165 min) mTV   VIDrel: BBC   V/h
Boa: play by William Shakespeare.

HENRY V ****   *PG*
Kenneth Branagh UK   1989
Kenneth Branagh, Derek Jacobi, Paul Scofield, Judi Dench, Emma Thompson, Alec McCowen, Ian Holm, Robbie Coltrane, Brian Blessed, Richard Briers, Robert Stephens, Christian Bale, Michael Maloney, Geraldine McEwan
In this remarkable directorial debut, Branagh has succeeded in creating a stunning and highly original rendition of Shakespeare's famous play of the warrior-king and his war against France. Quite different in feel from the famous Olivier film, this interpretation is no less visually impressive, and yet offers one an intimacy and warmth absent from that earlier classic. AA: Cost (Phyllis Dalton).
DRA   131 min (ort 137 min) wScrn
VIDrel: VCC/DISC/COLUM   V/sur
Boa: play by William Shakespeare.

HENRY VIII AND HIS SIX WIVES ***   *PG*
Waris Hussein UK   1972
Keith Michell, Frances Cuka, Charlotte Rampling, Donald Pleasence, Jane Asher, Lynne Frederick, Barbara Leigh-Hunt
The story of one of England's most degenerate and bloodthirsty monarchs, seen as a series of flashbacks on his deathbed. A generally well made drama, adapted from a BBC TV series, with Michell giving one of his best performances as the monarch. Music is performed by The Early Music Consort of London with David Munrow, who also handled arrangement. See also THE SIX WIVES OF HENRY VIII – a TV production.
DRA   120 min   VIDrel: BRAVE/SONOP L/A   V

HER ALIBI **   *PG*
Bruce Beresford USA   1988
Tom Selleck, Paulina Porizkova, William Daniels, James Farentino,

Hurd Hatfield, Patrick Wayne, Tess Harper, Joan Copeland, Ronald Guttman, Victor Argo, Bill Smitrovich, Bobo Lewis, Jane Welch, W. Benson Terry, Alan Mixon
A comedy thriller, in which a successful writer of mysteries hopes to cure himself of a temporary slump in his writing, by providing a false alibi for an accused murderess and taking her back home to observe her. Neither tense nor particularly funny, this strange film is a dud on both counts. Likeable performances partially rescue it.
THR   90 min (ort 94 min)   VIDrel: MGM/WHV   V/sur

HER HIDDEN TRUTH **   15
Daniel Lerner USA   1995
Kellie Martin, Antonio Sabato Jr, Ken Howard, Reed Diamond, Bruce Weitz, Condy Pickett, Gordon Clapp, Mary Donnelly Haskell, Lisa Thornhill, Red West, Harold Surratt, Allison Potter, Dennis Letts, Diana Taylor, David Dwyer
A girl of ten was branded an arsonist and placed in detention, when a fire was started in her home and led to the death of her mother and baby sister. At the age of eighteen she escapes with the intention of finding the real arsonist and clearing her name. Fortunately, she gets a little help from a sympathetic detective. A fairly good time-filler, but equally devoid of any real surprises or unexpected twists.
THR   90 min mTV   VIDrel: ODY/SONOP   V/sh

HER LAST CHANCE **   15
Richard A. Colla USA   1995
Kellie Martine, Patti Lupone, Jonathan Brandis, Tony Lucca
The mother of a young girl struggles to save her daughter from ruining her life with drink and drugs, in a competent TV drama.
DRA   90 min mTV   VIDrel: ODY/SONOP   V/sh

HER LIFE AS A MAN ***   PG
Robert Ellis Miller USA   1983
Robyn Douglass, Marc Singer, Robert Culp, Laraine Newman, Miriam Flynn, Joan Collins, Anthony Holland, Patricia Barry, Malcolm Campbell, Steve Fogel, Debbie Gilbert, Dino Gigante, Loretta Greenwood, Bobby Hosea, Suze Lanier, Liz Torres
A female reporter has to disguise herself as a man, in order to land a job as a sportswriter. A likeable and amusing addition to films such as TOOTSIE and VICTOR/VICTORIA. The adaptation (based on a true story) is by Joanna Crawford. See also JUST ONE OF THE GUYS.
COM   92 min (ort 100 min) mTV   VIDrel: 20TH/TECH L/A V
Boa: magazine article in "The Village Voice" by Carol Lynn Mithers.

HERBIE GOES BANANAS *   U
Vincent McEveety USA   1980
Cloris Leachman, Charles Martin-Smith, John Vernon, Steven W. Burns, Harvey Korman, Elyssa Davalos, Fritz Feld, Richard Jaeckel, Joaquin Garay III, Alex Rocco, Vito Scotti, Jose Gonzalez, Rubin Moreno, Tina Melard, Jorge Moreno
The fourth in the series of THE LOVE BUG comedies, about a Volkswagen with a life of its own. This time it takes its owners on a motoring holiday in South America. Thankfully this was the last one in an increasingly dreary chain of sequels. However, a TV series soon followed in the States.
JUV   88 min (ort 100 min) cC   VIDrel: WDV/TECH   V

HERBIE GOES TO MONTE CARLO **   U
Vincent McEveety USA   1977
Dean Jones, Don Knotts, Julie Sommers, Roy Kinnear, Jacques Marin, Bernard Fox, Eric Braeden, Xavier Saint Macary, Francois Lalande, Alan Caillou, Laurie Main, Mike Kulcsar, Stanley Brock, Jeard Jugnot, Johnny Haymer
One of an endless series of sequels to THE LOVE BUG, with our car entering the Monte Carlo Rally where a spy ring hides diamonds in its petrol tank. Followed by HERBIE GOES BANANAS.
JUV   100 min (ort 104 min) cC   VIDrel: WDV/TECH   V

HERBIE RIDES AGAIN ***   U
Robert Stevenson USA   1974
Helen Hayes, Ken Berry, Stefanie Powers, Keenan Wynn, John McIntire, Huntz Hall, Ivor Barry, Dan Tobin, Vito Scotti, Raymond Bailey, Liam Dunn, Elaine Devry, Chuck McCann, Richard X. Slattery, Hank Jones, Rod McCary
First sequel to THE LOVE BUG. The Volkswagan Beetle with a mind of its own, saves its owner from the attentions of a greedy property developer. Fairly typical Disney slapstick, good fun in

a mindless sort of way. Followed by HERBIE GOES TO MONTE CARLO.
JUV   85 min (ort 88 min) cC   VIDrel: WDV/TECH   V

HERCULES AND THE AMAZON WOMEN **   PG
Bill L. Norton USA   1993
Kevin Sorbo, Anthony Quinn, Roma Downey, Michael Hurst, Lloyd Scott, Lucy Lawless, Christopher Brougham, Tim Lee, Kim Michalis, Maggie Tarver, John Steemson, Rose McIver, Jennifer Ludlam, Nick Kemplen, Heidi Anderson, Jill Sayre
On his way to attend a friend's wedding, Hercules finds that a group of sinister women warriors have been ravaging the area and is soon called upon to save a village from their attacks. Another film in this fantasy series that offers some simple-minded entertainment.
A/AD   87 min (ort 90 min) mTV   VIDrel: CIC/SONOP V/sh

HERCULES AND THE CIRCLE OF FIRE **   12
Doug Lefler USA   1994
Kevin Sorbo, Anthony Quinn, Tawny Kitaen, Kevin Atkinson, Stephanie Barrett, Christopher Brougham, Joseph Greer, Kerry Gallagher, Leonard Twins, John Watson, Sharon Tyrell, Mark Ferguson
With mankind facing a frozen future, it falls to Hercules to confront the Goddess Hera, who has removed the Eternal Torch, thus creating a world in which no fires are able to burn. Another entry in this generally rather weak series of tales; Sorbo as Hercules acquits himself well enough, though the script is not exactly demanding.
A/AD   87 min   VIDrel: CIC/SONOP   V/dm

HERCULES AND THE LOST KINGDOM **   PG
Harley Cokliss USA   1993
Kevin Sorbo, Anthony Quinn, Renee O'Connor, Robert Trebor, Eric Close, Elizabeth Hawthrone, Nathaniel Lees, Jay Saussey, John Sumner, Lee-Jane Foreman, Onno Boelee, Chic Littlewood, Alex Beasley, Te Whatanui Skipwith, Daniel Warren
One in a series of films built around title hero, the mortal son of Zeus, the ruler of the gods in Greek mythology. Here, he embarks on a mission to find the lost city of Troy, and enjoys many hazardous adventures along the way together with the enmity of his wicked stepmother the Goddess Hera. A watchable fantasy tale with unremarkable special effects and so-so acting. A TV series was also made.
A/AD   86 min (ort 90 min) mTV   VIDrel: CIC/SONOP V/sh

HERCULES IN NEW YORK *   PG
Audrey Wiseberg/Arthur Allan Seidelman USA   1970
Arnold Strong (Arnold Schwarzenegger), Arnold Stang, Deborah Loomis, Taina Elg, James Karen, Ernest Graves, Tanny McDonald, Michael Lipton, Howard Blustein, Merwin Goldsmith, George Bartenieff, Erica Fitz, Diane Goble
Hercules is bored with his life among the Gods on Mount Olympus and ends up on a visit to New York. A weak attempt at spoofing the many films based on this heroic character.
Aka: HERCULES GOES BANANAS; HERCULES THE MOVIE
COM   78 min (ort 95 min)
VIDrel: ETL/4-FRONT/POLYREC   V

HERCULES IN THE MAZE OF THE MINOTAUR **   15
Josh Bender USA   1994
Kevin Sorbo, Anthony Quinn, Tawny Kitaen, Michael Hurst, Ray Anthony Parker, Nic Fay, Andrew Turtell, Paul McIver, Simon Lewthwaite, Rose McIver, Katrina Hobbs, Warren Carl, Maya Dalziel, Sydney Jackson, Marise Wipani, John Mello
Now a married farmer with three kids, Hercules has been having a quiet time recently but is soon called upon the face the Minotaur, which has been inadvertently released from imprisonment in its cave, by two boys looking for some hidden treasure. A poor effort, heavily padded with footage from the other films in this series and replete with bad acting and some abysmal dialogue.
Aka: HERCULES AND THE MAZE OF THE MINOTAUR
A/AD   87 min (ort 90 min) mTV   VIDrel: CIC/SONOP V/dm

HERCULES IN THE UNDERWORLD **   12
Bill L. Norton USA   1994
Kevin Sorbo, Anthony Quinn, Marlee Shelton, Tawny Kitaen, Clif Curtis, Jorge Gonzales, Timothy Balme
The ground has opened up beneath a village, revealing an

entrance to Hades. A beautiful girl from the village asks Hercules for his assistance and he agrees to help, but as ever, has to contend with the machinations of his evil stepmother Hera.
A/AD   87 min   VIDrel: CIC/SONOP   V/sur

HERCULES RETURNS ***   15
David Parker AUSTRALIA   1993
David Argue, Michael Carman, Mary Coustas, Bruce Spence, Brendon Suhr, Nick Polites, Lance Anderson, Laurie Dobson, Richard Moss, Burt Cooper, Tom Coltraine plus voices of: Des Mangan, Sally Patience, Matthew King, Bruce Spence
A film buff loses his job with a cinema chain and sets up his own cinema, intending to show classic films. But his former employers resort to a catalogue of dirty tricks to sabotage his operation. A funny spoof that takes its inspiration from those tacky 1960s Italian sword-and-sandal epics, one of which has to be dubbed when it arrives in its original unsubtitled Italian. Screenplay is by Des Mangan, developed from his live show "Double Take Meets Hercules".
COM   77 min (ort 80 min)   VIDrel: TART/20TH   V

HERE COMES MR JORDAN ****   U
Alexander Hall USA   1941
Robert Montgomery, Claude Rains, James Gleason, Evelyn Keyes, Rita Johnson, Edward Everett Horton, John Emery, Donald MacBride, Halliwell Hobbes, Don Costello, Benny Rubin, Bert Young, Ken Chrsity, Joseph Crehan, Billy Newell
A prizefighter killed in an air crash he was supposed to survive, is sent back to Earth. However, he is forced to find another body and eventually takes over one belong to a murdered multi-millionaire in a sudden resurrection that causes considerable consternation all round. A charming comedy fantasy, much imitated (remade as HEAVEN CAN WAIT for example) but never surpassed. AA: Story/orig (Harry Segall), Screen (Sidney Buchman/Seton I. Miller).
COM   90 min (ort 94 min) B/W   VIDrel: COLUM/SONOP V/dm
Boa: play Halfway to Heaven by Harry Segall.

HERE COMES THE SON: A TRUE STORY **   12
Paul Schneider USA   1996
Scott Bakula, Chelsea Field, Dan Luria, Cynthia Martells, Pamelaa Brull, Belita Moreno, Brian Smiar, Nancy Linari, Jeanne Mori, Brian Reddy, Jonathan Selstad, Thomas Selstad, Bill Hayes, Barry Neikrug, Judith Scarpone, Ken Kerman
A product reviewer meets his baby son for the first time when a woman arrives one night and dumps him on his doorstep. Claiming she is unable to take care of him, his eager but unprepared dad steps into the breach despite the opposition of the welfare authorities. However, when mum changes her mind and wants him back, our proud father decides to mount a petition for custody of his child.
Aka: BACHELOR'S BABY, THE
DRA   95 min mTV   VIDrel: ODY/SONOP   V/sh

HERO AND THE TERROR **   15
William Tannen USA   1988
Chuck Norris, Steve James, Brynn Thayer, Jack O'Halloran, Jeff Kramer, Billy Drago, Ron O'Neal, Tony Di Benedetto, Joe Guzaldo, Murphy Dunne, Peter Miller, Heather Blodgett, Karen Witter, Lorry Goldman, Christine Wagner
A cop tracks down a demented killer he once caught years ago. A film that marks a change for Norris, from fairly non-stop martial arts action, to some attempt at characterisation, a change which is not entirely successful.
A/AD   92 min (ort 97 min)   VIDrel: WHV   V/sh

HERO IN THE FAMILY **   (U)
Mel Damski USA   1986
Christopher Collett, Cliff De Young, Annabeth Gish, Darleen Carr, Keith Dorman, David Wohl, M. Emmet Walsh, Jay Brazen, Bernard Cuffling, Don Davis, Bill Down, Deryl Hayes, Max Margolin, Alicia Michelle, Stephen E. Miller
An astronaut is sent on a space mission together with a chimp, but they discover a crystal there that results in an exchange of personalities. A standard Disney "swap-over" comedy.
JUV   87 min (ort 100 min) mTV   SATrel: DISNEY CHANNEL

HEROES ***   PG
Donald Crombie AUSTRALIA   1989
Paul Rhys, John Bach, John Hargreaves, Bill Kerr, Jason Donovan,
Cameron Daddo, Christopher Morley, Timothy Lyn, David Wenham, Jeff Truman, Gerry Skilton, Mark McAskill, Wayne Scott Kermond, Don Halbert, Tom Considine
Story of a WW2 suicide-mission by a small group of men, who use a fishing boat to mount an attack on the Japanese fleet in Singapore after the fall of that island in 1942. Based on true events.
WAR   208 min mTV   VIDrel: 4-FRONT/POLYREC   V
Boa: book by Ronald Neame.

HEROES 2: THE RETURN **   PG
David Crombie AUSTRALIA   1990
Nathaniel Parker, Craig McLachlan, Christopher Morsley, John Bach, Simon Burke, Ken Teraizumi, Mark Lewis Jones, Ian Bolt, Anne Louise Lambert, Troy Willats, Miranda Otto, Wayne Scott Kermond, Kelly Dingwall, John O'Hare
Following the success of the first raid on the Japanese Singapore fleet in HEROES, another mission is planned, but the consequences are less satisfactory.
WAR   208 min mTV   VIDrel: 4-FRONT/POLYREC   V

HEROES OF TELEMARK, THE ***   U
Anthony Mann UK   1965
Kirk Douglas, Richard Harris, Ulla Jacobsson, Michael Redgrave, Mervyn Johns, Anton Diffring, David Weston, Roy Dotrice, Eric Porter, Jennifer Hilary, Barry Jones, Ralph Michael, Geoffrey Keen, Maurice Denham
Low-key and realistic tale of Norwegian resistance to the German occupation in WW2, and how they helped destroy a heavy water plant that was to be used in the development of an atomic bomb. Good action sequences but not a very exciting film as a whole.
WAR   124 min (ort 131 min)   VIDrel: VCC L/A   V
Boa: books But For These Men by J.D. Drummond/Skies Against The Atom by Knut Haukelid.

HEROES SHED NO TEARS **   18
John Woo HONG KONG   1986
Edy Ko, Lam Ching-Ying, Chen Yue-San, Lee Hio Seu, Lau Chau Sang, Jang Soo Hee, Philip Loffredo, Cecile Le Bailly, Lee Hoi Suk
An ex-soldier is assigned by the Thai government to assemble and train a mercenary troop to take on drug dealers in the Golden Triangle and capture the head of the gang. They in turn abduct his girlfriend and her son, forcing him to mount a successful rescue mission, after which they must all flee for their lives. Very typically a John Woo movie experience: noisy, fast-paced and full of gunfire and fury, though this adventure is far from one of his best.
MAR   94 min wScrn dubbed   VIDrel: MIA/DISC   V

HEROIC LEGEND OF ARISLAN, PARTS 1 TO 4 **   PG
JAPAN   199-
A warrior falls into disfavour with his ruler and as a punishment is assigned the task of guarding his son: Prince Arislan. The latter proves to be a most resourceful and enterprising young man, who soon involves his followers in many colourful adventures. Set in the mythical kingdom of Pulse in 320 A.D.
ANIM   240 min (4 cassettes) dubbed
VIDrel: MANGA/SONOP   V

HEROIC TRIO, THE ***   18
Johnny To (Du Qifeng) HONG KONG   1992
Anita Mui (Mei Yanfang), Maggie Cheung (Zhang Manyu), Michelle Yeoh (Yang Ziqiong), Damian Lee (Liu Songren), Paul Chin (Qin Pei), James Pak (Bai Shiqian), Ren Shiguan, Anthony Wong (Huang Qiusheng), Zhu Mimi, Jiang Haowen
Three super-heroines unite to stop a demon who is planning to dominate the world by stealing babies to be trained as an invincible army, with one of them destined to be a puppet ruler over China. An imaginative celebration of Hong Kong's popular pulp-fantasy action films, it proved to be a major box-office hit, and was soon followed by a sequel. Not a film of any depth, but as a perfect example of the genre it takes some beating.
Aka: DONGFANG SAN XIA
A/AD   84 min (ort 104 min) wScrn   VIDrel: MADE/RTM V

HEROIC TRIO 2, THE: EXECUTIONERS ***   18
Johnny To/Ching Siu Tung HONG KONG   199-
Anita Mui, Michelle Yeoh, Maggie Cheung
Sequel to the earlier film that has our three vixens battling the "Black Knight", an unsavoury rogue who is out to take over the

world. Set in a post-nuclear holocaust Hong Kong, the complex plot is well served by an array of impressive effects and excellent combat sequences, making it even more enjoyable than the first film.
A/AD   105 min wScrn   VIDrel: MADE/RTM   V

HESTER STREET ***
PG
Joan Micklin Silver USA
1975
Carol Kane, Doris Roberts, Steven Keats, Mel Howard, Dorrie Kavanaugh, Stephen Strimpell, Lauren Frost, Paul Freedman, Martin Garner, Leib Lensky, Zane Lensky, Zvee Scooler, Eda Resiis Merin, Robert Lesser, Joanna Merlin
A rapidly Americanised Jewish tailor living on New York's Lower East Side at the turn of the century sends for his wife from the old country. While trying to please him, however, she refuses to accept his rejection of traditional rites and customs. A beautifully made and acted film that brings the past to life in a wonderfully convincing way.
DRA   190 min B/W   VIDrel: ARROW/RTM   V
Boa: short story Yekl by Abraham Cahan.

HEXED *
15
Alan Spencer USA
1992
Arye Gross, Claudia Christian Rolls, Adrienne Shelly, Norman Fell, Michael Knight, R. Lee Ermey, Robin Curtis, Brandis Kemp, Roy Baker, Norman Fell, Pamela Roylance, Billy Jones, John Davies, Fred Mata, Marilyn Staley, Julio Cedillo
A bellboy with an active fantasy life gets involved with a former fashion model whose past conceals a deadly secret, in this over-the-top spoof on erotic thrillers such as BASIC INSTINCT etc. Unfunny and extremely tedious.
Aka: ALL SHOOK UP!
COM   89 min (ort 93 min)   VIDrel: 20VIS/SONOP   V/sh

HEY GOOD-LOOKIN' **
18
Ralph Bakshi USA
1975 (re-edited for release in 1982)
Voices of: Richard Romanus, David Proval, Jesse Welles, Tina Bowman, Angelo Grisanti, Danny Wells, Bennie Massa, Gelsa Palao, Paul Roman, Larry Bishop, Tabi Cooper, Juno Dawson, Shirley Jo Finney, Martin Garner, Terry Haven
Film animation set in the 1950s and based on the director's own youth in Brooklyn, largely telling of the exploits of two young womanisers. An ambitious stab at a youth culture film via animation; interesting but not all that effective. Written by Bakshi, whose other films include FRITZ THE CAT, WIZARDS, FIRE AND ICE and HEAVY TRAFFIC.
ANIM   74 min (ort 87 min)   VIDrel: L/A   V

HI HONEY, I'M DEAD **
U
Alan Myerson USA
1990
Curtis Armstrong, Kevin Conroy, Catherine Hicks, Joseph Gordon-Levitt, Paul Rodriguez, Robert Briscoe Evans, Harvey Jason, Ernest Harada, Richard Stahl, Carol Androsky, Andre Rosey Brown, Betty Carvalho, Wendy Cutler, Ron Leath
An arrogant property developer suffers poetic justice when he dies in a lift accident caused by his skimping on safety standards to save money. He discovers to his horror that he has been reincarnated in the body of a vagrant and that it is in this guise that he must win back the love and affection of his neglected family. Another sentimental comedy with a ghostly theme that runs out of steam all too soon. A watchable trifle.
Aka: HULLO HONEY, I'M DEAD
COM   88 min (ort 100 min) mTV   VIDrel: 20TH/TECH   V

HIDDEN, THE *
18
Jack Sholder USA
1987
Michael Nouri, Kyle McLachlan, Ed O'Ross, Clu Gulager, Claudia Christian, Clarence Felder, William Boyett, Richard Brooks, Catherine Cannon, Larry Cedar, John McCann, Chris Mulkey, Lin Shaye, James Luisi, Frank Renzulli
An alien creature moves from the body of one victim to the next, causing each host to commit acts of violence. Nouri plays a Los Angeles cop who is given the task of investigating a bizarre wave of violence, fortunately he receives a little help from one of the good aliens in the shape of McLachlan. A flawed marriage of fantasy and urban violence genres that does justice to neither.
HOR   93 min (ort 96 min)   VIDrel: POLY/POLYREC L/A   V

HIDDEN 2, THE **
18
Seth Pinkser USA
1993
Rapahel Sbarge, Kate Hodge, Jovin Montanaro, Christopher Henry, Michael Weldon, Michael A. Nickles, Tony Di Benedetto, Tom Tayback, Dennis Bertsch, Cate Caplin, Edith Varon, Peter Gregory, Bobby Foxworth, Honey Lauren, Danny McBride
Sequel to the first film is very much more of the same with our nasty alien having survived and taking up residence first in a dog and then in a variety of human hosts. In order to destroy it, a good alien is forced to enlist the help the daughter of a dead cop, who was also an alien. Incessant gunfire and some truly sickening special effects do nothing to add any interest or suspense to this by now familiar tale.
HOR   90 min (ort 91 min)   VIDrel: POLFIL   V

HIDDEN AGENDA ***
15
Ken Loach UK
1990
Frances McDormand, Brad Dourif, Mai Zetterling, John Benfield, Des McAleer, Brian Cox, Jim Norton, Maurice Roeves, Bernard Archard, Bernard Bloch, Michele Farley, Patrick Kavanagh, Ian McElhinney
When an American civil rights lawyer is shot dead at a police roadblock in Northern Ireland, a senior mainland police officer is sent there to investigate. Undaunted in his task, he soon detects a cover-up that extends right up to the top levels of government, and the existence of a secret conspiracy. A well-assembled and essentially paranoid view of "The Troubles" that's both realistic and well acted but ultimately too propagandist to work.
THR   104 min (ort 110 min)   VIDrel: 20TH/TECH   V

HIDDEN FORTRESS, THE ***
PG
Akira Kurosawa JAPAN
1958
Toshiro Mifune, Misa Uehara, Minoru Chiaki, Takashi Shimura, Susumu Fujita, Takashi Shimura, Kamatari Fujiwara, Eiko Miyoshi, Toshiko Higuchi, Kichijiro Ueda
A classic film about a samurai in the service of a princess who must escape to safety in another province after her father is defeated and killed. Can be described as almost a western, and elements of this film seem to have found their way into STAR WARS, no less. Highly recommended.
Aka: KAKUSHI TORIDE NO SAN-AKUNIN; THREE BAD MEN IN THE HIDDEN FORTRESS; THREE RASCALS IN THE HIDDEN FORTRESS
A/AD   138 min B/W wScrn   VIDrel: CONNO/RTM   V

HIDDEN OBSESSION **
18
John Stewart USA
1992
Jan Michael Vincent, Heather Thomas, Nick Celozzi, Davi Glasser, Linda Krus, Zita Stone, Everett Smith, John Lisbon Wood, Joe Celozzi, Chick, Richard Baird, Bob Raitblat, Turk Muller, Joel Tatom, Marge Royce, Tony McGuide
A news anchorwoman retreats to cabin in lonely woods for a much-needed vacation, but her tranquillity is shared when a serial killer escapes from a nearby prison. Despite the presence of a her police-officer neighbour, our killer soon comes after her in this highly predictable offering.
Aka: HIDDEN RAGE
THR   91 min mTV   VIDrel: GUILD/SONOP   V/s

HIDE IN PLAIN SIGHT **
PG
James Caan USA
1980
James Caan, Jill Eikenberry, Robert Viharo, Joe Grifasi, Barbra Rae, Thomas Hill, Kenneth McMillan, Josef Sommer, Danny Aiello, Chuck Hicks, Andrew Gordon Fenwick, Heather Bicknell, David Marguiles, David Clennon, Ken Sylk, Anne Helm
A divorced man discovers that his children have disappeared, because their new father is a government witness who has been given a new identity. A cliched and irritating melodrama whose unusual plot premise offers something by way of compensation. See also THE WHEREABOUTS OF JENNY.
DRA   98 min   VIDrel: L/A   V
Boa: novel by Leslie Waller.

HIDEAWAY **
18
Brett Leonard USA
1995
Jeff Goldblum, Christine Lahti, Alicia Silverstone, Jeremy Sisto, Alfred Molina
A man is returned to life after being clinically dead for two hours and discovers that he now has a psychic link with a crazed serial killer who is planning to murder his daughter. A clumsy chiller with moderately good special effects, it lacks pace and is seriously hampered by the unbelievable serial killer plot, which is pure contrivance and rapidly sinks what might have been a perfectly good movie.
THR   106 min   VIDrel: COLUM/SONOP   V/sur
Boa: novel by Dean R. Koontz.

HIDEOUS SUN DEMON, THE * PG
Robert Clarke USA 1959
Robert Clarke, Patricia Manning, Nan Peterson, Patrick Whyte, Fred La Porta, Bill Hampton, Robert Garry, Pearl Driggs, Richard Cassarino, Helen Joseph, Bill Currie, Fran Leighton, Chuck Newell, John Murphy, Daryl Westerbrook
A scientist accidentally exposed to radiation finds that sunlight turns him into a scaly lizard-like monster. Low-budget, low-brained 1950s SF at its very worst.
Aka: HIDEOUS SUN DEMONS
FAN 71 min (ort 74 min) B/W VIDrel: FIRC/RTM V

HIDING OUT * 15
Bob Giraldi USA 1987
Jon Cryer, Keith Coogan, Annabeth Gish, Oliver Cotton, Claude Brooks, Ned Eisenberg, Tim Quill, Tony Soper, Marta Gerahty, Steven Small, John Walker, John Spencer, Gretchen Cryer, Anne Pioniak, Lou Walker, Beth Ehlers, Nancy Fish
A silly, contrived comedy with Cryer playing a stockbroker who is due to testify against organised crime. When his FBI bodyguard is killed, he goes on the run from the mobsters responsible, and hides out by assuming the role of a high school senior.
COM 99 min VIDrel: 20TH/TECH V/sur

HIGH ANXIETY ** 15
Mel Brooks USA 1977
Mel Brooks, Madeline Kahn, Cloris Leachman, Harvey Korman, Ron Carey, Howard Morris, Dick Van Patten, Murphy Dunne, Jack Riley, Charlie Callas, Rudy De Luca, Ron Clark, Barry Levinson, Lee Delano, Richard Stahle, Al Hopson
A spoof of Hitchcock films about a psychiatrist who gets into trouble when he takes charge of a sanatorium. An affectionate pastiche of a hundred other films, with several isolated moments of brilliance strung together in a haphazard fashion.
COM 94 min VIDrel: 20TH/TECH V/sur

HIGH ART ** (18)
Walter Salles Jr BRAZIL 1992
Peter Coyote, Tcheky Karyo, Amanda Pays, Raul Cortez, Giulia Gam, Cassia Kiss, Tonico Pereira, Miguel Angel Fuentes, Eduardi Conde, Rene Ruiz, Paolo Jose, Iza Do Eirado, Tony Tornadoa, Eduardo Waddington, Alvaro Freire
Coyote plays a photographer embroiled in a nasty underworld conspiracy when he comes across a stolen computer disk, the finding of which leads to a brutal attack which leaves him injured and his girlfriend dead. Driven by an overwhelming desire for revenge, he seeks out a master to train him in the martial arts, before planning his next moves. Perfectly capable actioner, with few surprises and the requisite dose of thrills.
A/AD 101 min SATrel: MOVIE CHANNEL
Boa: novel by Ruben Fonseca.

HIGH BOOT BENNY *** 15
Joe Comerford EIRE 1993
Frances Tomelty, Alan Devlin, Marc O'Shea, Fiona Nicholas
A Catholic delinquent flees from the IRA to the Republic where he hides out at a school on the border there. But he is soon involved in murder when the school's caretaker is found dead, and finds himself suffering various conflicts of interest. A morbid exercise in symbolism, that uses the story as an allegory for the "troubles", and for the most part succeeds in making its points with potency if not wit. The sombre mood of the whole film gives one little comfort.
A/AD 82 min VIDrel: IMAG/RTM V

HIGH BRIGHT SUN, THE ** PG
Ralph Thomas UK 1965
Dirk Bogarde, George Chakiris, Susan Strasberg, Denholm Elliott, Gregoire Aslan, Colin Campbell, Joseph Fuerst, Nigel Stock, Katherine Kath, George Pastell, Paul Stassino
Drama set in Cyprus in 1957, when the struggle for independence was in full swing, and an officer in British Intelligence finds that his actions have put the life of his archaeologist girlfriend in danger. In the hands of a better director this might well have been a top-notch adventure, but weak direction and acting holds it back, as does the confused attempt to turn what should have been a straightforward action-thriller into a romantic drama.
Aka: DATE WITH DEATH, A; McGUIRE GO HOME!
A/AD 109 min (ort 114 min) VIDrel: CARL/TECH V
Boa: novel by Ian Stuart Black.

HIGH FINANCE WOMAN * 18
Joe D'Amato ITALY 1989
Tara Buckman, Charlie Edwards, Paul Van Gent, Dan Smith, Julia Howard, Louie Elias
A glossy and vacuous film that resembles a TV advert, telling of a clever and unscrupulous stock market dealer who is always ready to use a little sex when there's useful information to be had. Already involved with an older man, she finds herself falling in love with his son, and the father begins plotting his revenge.
Aka: LOVES OF A WALL STREET WOMAN, THE; WALL STREET WOMAN
DRA 89 min VIDrel: MED/POLYREC/COLUM L/A V/h

HIGH HEELS *** 18
Pedro Almodovar SPAIN 1991
Victoria Abril, Marissa Paredes, Miguel Bose, Pedro Diez del Corral, Feodor Atkine, Ana Lizaran, Rocio Munoz, Maraita O'Wisiedo, Miriam Diaz Aroca, Bibi Andersson, Cristina Marcos, Nacho Martinez, Placido Guimaraes, Eva Siva
A complex melodrama about the relationship between a stubborn singer and the girl she deserted as a child. Though most of the film concentrates on the bitterness and rivalry that exists between mother and now grown-up daughter, there are quite enough of the director's touches (dancing lesbians, drag artists etc.) to make the film resemble something of a circus performance. A kind of black semi-farce, this is not one of Almodovar's best films.
Aka: TACONES LEJANOS
DRA 109 min (ort 114 min) VIDrel: VCC/DISC/COLUM V/sur

HIGH HOPES *** 15
Mike Leigh UK 1988
Philip Davies, Ruth Sheen, Edna Dore, Philip Jackson, Heather Tobias, Lesley Manville, David Bamber, Jason Watkins, Cheryl Prime, Judith Scott, Linda Beckett, Diane-Louise Jordan
Mike Leigh's first cinema feature since BLEAK MOMENTS seventeen years ago, is a semi-farcical tale of a good-natured working-class couple's involvement with a suburban moron, a near-senile mother, nouveau-riche relations and some appalling, over-the-top yuppie neighbours. Scripted by Leigh.
COM 107 min (ort 112 min) VIDrel: IMAG/RTM V

HIGH LONESOME ** 12
Jeff Bleckner USA 1994
Louis Gossett Jr, Joseph Mazello, James Greene, Don Swayze, David hart, William Lucking, Jack Kehler, Mark Cabus, Patrick Labyortuex, William Fichtner, Evan Rachel Wood, John Bellucci, Courtenay McWhitney, Chuck Butto, Mark Phelan
The title refers to a town tucked away in the Ozarks, where the only black man left there faces the hatred of the villainous and racist townsfolk, whilst trying to scratch a living on his farm. He employs a white man to help run the farm, but when the latter takes off he is left with the man's little boy to look after, not an easy task as the child has inherited some of his dad's prejudices. Rather overdone and stereotyped, yet the film does have its moments.
DRA 89 min VIDrel: ODY/SONOP V/sh

HIGH NOON **** U
Fred Zinnemann USA 1952
Gary Cooper, Grace Kelly, Lloyd Bridges, Thomas Mitchell, Katy Jurado, Otto Kruger, Lon Chaney Jr, Henry Morgan, Ian MacDonald, Eve McVeagh, Harry Shannon, Lee Van Cleef, Robert Wilke, Sheb Wooley, Tom London, Ted Stanhope
A classic Western in which an ageing sheriff has to save the townsfolk from a gang of ruthless outlaws, who have been freed from jail and are out for revenge. An all-time great. AA: Actor (Cooper), Edit (Elmo Williams/Harry Gerstad – note how clever editing makes the film "appear" to unfold in real-time), Song ("Do Not Forsake Me, Oh My Darlin'" – Dmitri Tiomkin (m)/Ned Washington (l)) and Score (Tiomkin).
WES 81 min (ort 84 min) B/W VIDrel: SECOND/RTM V
Boa: short story The Tin Star by John W. Cunningham.

HIGH PLAINS DRIFTER *** 18
Clint Eastwood USA 1973
Clint Eastwood, Verna Bloom, Marianna Hill, Mitchell Ryan, Jack Ging, Stefan Gierasch, Billy Curtis, Ted Hartley, Geoffrey Lewis, Walter Barnes, Paul Brinegar, Dan Vadis, Jack Kosslyn, Belle Mitchell, John Mitchum, Pedro Regas
A town hires a mysterious stranger, to protect them from

outlaws who have just been released from prison. Skilful use of intercut sequences show us how the original sheriff met his death, whilst the townsfolk looked on. His replacement bears an uncanny resemblance in all but features to the murdered man. A compelling film, both under-stated and atmospheric, that demonstrated Eastwood's abilities as a director. The film PALE RIDER is a remake of sorts.
WES   100 min (ort 105 min)
VIDrel: 4-FRONT/POLYREC/CIC   V

HIGH SCHOOL HIGH *
Hart Bochner USA
12
1996
Jon Lovitz, Tia Carrere, Mehki Phifer, Louise Fletcher, Malinda Williams
A parody of those rough, tough, inner-city high school movies (especially DANGEROUS MINDS) that sees Lovitz cast as the son of a man who runs a rich kid's private school, where the former also teaches. The son runs off to find himself, winding up at a school that would challenge even Schwarzenegger, where he is soon locked in a battle of wills with both the kids and the principal. There may be numerous gags, it's just such a pity they almost all misfire.
COM   89 min   CINrel

HIGH SEASON **
Clare Peploe UK
15
1987
Jacqueline Bisset, James Fox, Irene Papas, Sebastian Shaw, Kenneth Branagh, Lesley Manville, Robert Stephens, Paris Tselios, Ruby Baker, Geoffrey Rose, Mark Williams, Shelley Laurent, George Diakoyorgio, Father Bassili
A look at the effects of tourism on a small Greek island, with Bisset a female photographer whose peaceful life on Rhodes, is shattered by the arrival of a pair of obnoxious tourists, a spy and her self-centred husband. A fluffy, romantic comedy-drama of little substance and pretty locations (rather beautifully photographed by Chris Menges).
COM   90 min (ort 101 min)   VIDrel: ARROW/RTM   V

HIGH SIERRA ***
Raoul Walsh USA
PG
1941
Humphrey Bogart, Ida Lupino, Joan Leslie, Alan Curtis, Arthur Kennedy, Henry Hull, Henry Travers, Barton Maclane, Jerome Cowan, Cornel Wilde, Donald McBride, Minna Gombell, Paul Harvey, John Eldredge, Isabel Jewell, Willie Best
Bogart stars as a killer on the run from the police, who befriends a lame girl and pays for her to have an operation. Eventually, his luck runs out and he is shot and wounded, hiding in the mountains for a final showdown. A sombre gangster movie that gave the star his first good role (George Raft unwisely turned the part down). Scripted by John Huston and W.R. Burnett and remade as "I Died A Thousand Times" and as the Western "Colorado Territory".
A/AD   96 min (ort 100 min) B/W   VIDrel: WHV   V
Boa: novel by W.R. Burnett.

HIGH SOCIETY **
Charles Walters USA
U
1956
Bing Crosby, Frank Sinatra, Grace Kelly, Celeste Holm, Louis Calhern, Louis Armstrong, Sidney Blackmer, Margalo Gillmore, Lydia Reed, John Lund, Gordon Richards, Richard Garrick, Richard Keen, Ruth Lee, Helen Spring, Paul Keast
A lightweight and clinical remake of the 1940 film THE PHILADELPHIA STORY in which a rich socialite is about to re-marry when her former husband arrives on the scene. Songs include: "True Love", "Did You Evah?", "You're Sensational" and "Now You Has Jazz". The last acting role for Kelly.
MUS   103 min (ort 107 min)   VIDrel: MGM/WHV   V/dm
Boa: play The Philadelphia Story by Philip Barry.

HIGH SPIRITS ***
Neil Jordan USA
15
1988
Daryl Hannah, Peter O'Toole, Steve Guttenberg, Beverly D'Angelo, Jennifer Tilly, Liam Neeson, Donal McCann, Mary Coughlan, Liz Smith, Tom Hickey, Preston Lockwood, Ray McAnally, Aimee Delamain, Ruby Buchanan, Tony Rohr, Isolde Cazelet
The soused debt-ridden owner of an Irish castle decides to open it as a haunted bed-and-breakfast, and gets his staff to dress up in sheets in order to scare the guests. However, events take an unexpected turn when the former but now deceased residents, decide to join in the fun.
COM   94 min   VIDrel: POLY/POLYREC L/A   V/sh

HIGHER LEARNING ***
John Singleton USA
15
1994
Jennifer Connelly, Ice Cube, Omar Epps, Michael Rapaport, Tyra Banks, Kristy Swanson, Laurence Fishburne, Jason Wiles, Cole Hauser, Bradford English, Regina King, Busta Rhymez, Jay Ferguson, Andrew Bryniarski, Trevor St John
A black athlete goes to college and is helped to deal with his anger and confusion by one of the professors working there. An overlong look at campus life in the 1990s that addresses a number of contemporary issues but all too easily resorts to stereotyping in its characterisations.
DRA   123 min (ort 128 min)   VIDrel: COLUM/SONOP   V/sur

HIGHLANDER *
Russell Mulcahy UK/USA
15
1986
Christopher Lambert, Sean Connery, Clancy Brown, Alan North, Beatie Edney, Sheila Gish, Roxanne Hart, John Polito, Hugh Quarshie, Christopher Malcolm, Peter Diamond, Billy Hartman, James Cosmo, Celia Emrie, Alistair Findley
Two immortal beings fight it out across time and space, as a 16th century Scotsman discovers that his old enemy has pursued him to 20th century Manhattan. An interesting idea is thrown away by flashy direction and ridiculous plotting. As much of a pain to listen to as it is to watch (the director started out by making rock videos). Several sequels and numerous episodes of a truly dire TV series have followed.
FAN   111 min (ort 116 min) wScrn   VIDrel: WHV   V/sur

HIGHLANDER 2: THE QUICKENING **
Russell Mulcahy USA
15
1991
Christopher Lambert, Sean Connery, Virginia Madsen, Michael Ironside, Allan Rich, John C. McGinley, Phil Brock, Rusty Schwimmer, Ed Trucco, Stephen Grives, Jimmy Murray, Pete Antico, Peter Bucossi, Peter Bromilow, Jeff Altman
Sequel set in the 21st century, with the Earth now having a shield to replace its destroyed ozone layer, that's operated by a corrupt corporation. MacLeod, the original "Highlander" from the first film, and the shield's inventor, an alien, return to the latter's home-world to combat its evil ruler. Filmed in Argentina, this muddled mess offers little but good effects and the usual fine performance from Connery. Music is by Stewart Copeland.
FAN   86 min (Cut at UK cinema release – ort 106 min)
VIDrel: EIV/SONOP   V/sur

HIGHLANDER 3: THE GATHERING *
Thomas J. Wright USA
15
1992
Christopher Lambert, Adrian Paul, Richard Moll, Vanity, Alexandra Vandernoot, Stan Kirsh
Compiled from episodes of the TV series, this is a poor effort indeed with McLeod, the "Highlander" of the first film taking on a rather nasty fellow Immortal. Flat, unbelievable and burdened with acres of swordplay but little else. Despite the title, this mTV effort is a poor relation to the other HIGLANDER films.
FAN   90 min (ort 98 min) mTV   VIDrel: EIV/SONOP   V

HIGHLANDER 3: THE SORCERER *
Andy Morahan CANADA/FRANCE/UK
15
1995
Christopher Lambert, Mario Van Peebles, Deborah Unger, Mako, Raoul Trujillo, Jean-Pierre Perusse, Martin Neufeld, Frederick Y. Okimura, Daniel Do, Gabriel Kakon, Louis Bertignac, Michael Jayston, Zhenhu Han, Akira Inoue, David Francis
Another outing for our heroic fighter, who must now faces the wrath of a powerful wizard, who is intent on world domination. Lambert is still swinging his sword as the hero (even if Connery wisely dropped out of this one) and the battle takes place in present-day New York. Tedious dross; badly directed, badly acted and badly in need of a decent script.
FAN   93 min (ort 98 min)   VIDrel: EIV/SONOP   V/sur

HIGHTIDE ***
Gilliam Armstrong AUSTRALIA
15
1987
Judy Davis, Jan Adele, Claudia Karvan, Colin Friels, John Clayton, Frankie J. Holden, Toni Scanlon, Monica Trapaga, Barry Rugless, Mark Hembrow, "Cowboy" Bob Purtell, Emily Stocker, Marc Gray, Sarah Oord, Jane Boreham, May Howlett
An intelligent and perceptive film that looks at the life of a self-absorbed woman, who having lost her job as one of the back-up singers working with an Elvis Presley impersonator, finds herself stranded with a broken down car at an Australian beach resort. There she meets a young girl who turns out to be her daughter, whom she abandoned to her mother-in-law when her

husband died. Often bleak and rather preachy, the film's strong script is its greatest asset.
DRA   101 min (ort 104 min)   VIDrel: CONNO/RTM   V

HIGHWAY PATROLMAN ***
15
Alex Cox JAPAN/MEXICO/USA   1992
Roberto Sosa, Bruno Bichir, Vanessa Bauche, Zaide Silvia Gutierrez, Pedro Armendariz Jr, Ernesto Gomez Cruz, Jorge Russek, Karl Braun, Guillermo Rios, Farnesio De Bernal, Eduardo Lopez Rojas, Malena Doria, Damian Alcazar
The story of drugs, police corruption and prostitution in Mexico, that takes the form of an offbeat road-movie, its central character being a corrupt cop who believes that everyone he encounters has something to hide, and is prepared to offer a bribe to get on the right side of him. An independent production, it amply demonstrates that given free rein, the director is capable of telling a strongly plotted and absorbing story, albeit a typically offbeat one.
Aka: EL PATRULLERO
DRA   100 min (Spanish dialogue – ort 104 min) subs wScrn
VIDrel: TART/20TH   V

HIGHWAY TO HELL *
18
Mark Griffiths USA   1983
Eric Stoltz, Monica Carrico, Stuart Margolin, Virgil Frye, Richard Bradford
A prisoner under sentence of death escapes with the help of his female pen-friend, but has to kill a guard in the process. The couple are then obliged to go on the run.
THR   90 min   VIDrel: MOPIC/SGSVID   V

HIGHWAY TO HELL **
15
Ate De Jong USA   1991
Patrick Bergin, Chad Lowe, Kristy Swanson, Adam Storke, Pamela Gidley, C.J. Graham, Jarrett Lennon, Richard Farnsworth, Lita Ford, Gilbert Gottfried, Anne Meara, Amy Stiller, Ben Stiller, Jerry Stiller, Be Deckard, Troy Tempest
A couple of newlyweds taking a trip through the desert to Las Vegas encounter a cop from Hell who kidnaps the wife on behalf of his master, the ruler of the Netherworld. Naturally, the husband follows in a bid to rescue her. A spoofy horror comedy with some nice effects, clever sight gags and sharp dialogue.
COM   90 min (ort 94 min)
VIDrel: ENCORE/SPEAR/COLUM   V/sur

HIJACKING OF THE ACHILLE LAURO, THE ***
PG
Robert Collins USA   1988
Karl Malden, Lee Grant, E.G. Marshall, Christina Pickles
Adequate recreation of the hijacking of this liner, off the coast of Egypt in 1985, when PLO terrorists boarded the ship, threatening the lives of the 380 passengers and murdering wheelchair-bound Leon Klinghoffer. See also VOYAGE OF TERROR: THE ACHILLE LAURO AFFAIR.
Aka: SEA OF TERROR; TERROR SQUAD
THR   91 min   VIDrel: BRAVE/SONOP L/A   V

HILLS HAVE EYES, THE ***
18
Wes Craven USA   1977
Susan Lanier, Robert Houston, Martin Speer, Virginia Vincent, Russ Grieve, Michael Berryman, Dee Wallace, James Whitworth, John Steadman, Janus Blythe, Lance Gordon, Brenda Marinoff, Cordy Clark, Flora Stricker, Arthur King
A family crossing the Californian desert is attacked by a ghastly family of cannibalistic mutants. A gory low-grade shocker that is undeniably well made and has several complex plot twists and unpleasant surprises. Now a cult film. A trashy sequel followed in 1983.
HOR   89 min Cut (2 sec plus film cuts)
VIDrel: POLY/POLYREC L/A   V

HILLS HAVE EYES, THE: PART 2 *
18
Wes Craven UK/USA   1983 (released 1985)
Michael Berryman, John Bloom, John Laughlin, Tamara Stafford, Janus Blythe, Kevin Blair, Peter Frechette, Robert Houston, David Nichols, Willard Pugh, Colleen Riley, Penny Johnson, Brenda Marinoff, Edith Fellows, Lance Gordon
A sequel to the previous film. This time a group of teenagers run out of petrol in the desert and are attacked by the local mutant family. Gruesome, unadulterated rubbish, with numerous flashback sequences that borrow footage from the earlier (and superior) effort, including an unintentionally funny episode that is meant to represent the recollections of a dog.
HOR   86 min   VIDrel: WHV   V/h

HINDENBURG, THE **
PG
Robert Wise USA   1975
George C. Scott, Anne Bancroft, Burgess Meredith, William Atherton, Roy Thinnes, Gig Young, Charles Durning, Robert Clary, Rene Auberjonois, Richard A. Dysart, Katherine Helmond, Joanna Moore, Stephen Elliott, Joyce Davis
Advances the theory that the crash of the Hindenburg airship in New York in 1937, was due to sabotage by anti-Nazis. A drawn out and lightweight tale with some ludicrous characterisations and an anti-climactic ending. Some use of original newsreel footage adds a little interest, to this over-long and stilted yarn. AA: Spec Award (Peter Berkos for sound effects), Spec Award (Glen Robinson/Albert Whitlock for visual effects).
DRA   110 min (ort 125 min)   VIDrel: L/A   V
Boa: novel by Michael M. Mooney.

HIRED HAND, THE ***
15
Peter Fonda USA   1971
Warren Oates, Peter Fonda, Robert Pratt, Verna Bloom, Severn Darden, Ted Markland, Rita Rogers, Megen Denver
A man runs out on his wife to go gold prospecting in California with two companions. When he returns seven years later, she will only take him back as a paid worker. However, he is soon obliged to leave once more, when a buddy is taken prisoner by bandits. A quirky and uneven story, aided by excellent acting and photography.
WES   86 min (ort 98 min)   VIDrel: CIC/SONOP L/A   V

HIROSHIMA ***
15
Peter Werner USA   1990
Brian Dennehy, Michael Tucker, Max Von Sydow, Judd Nelson, Noriyuki "Pat" Morita, Ben Wright, Tamlyn Tomita, Brady Tsuratani, Sab Shimono, Kim Miyori, Hal Holbrook, Shizuko Hoshi, Natsuko Ohama, Elizabeth Sung, Rodney Kageyama, Ping Wu
A TV film that traces the development of the A-bomb and, thanks to its length, gives a more detailed account of this complex process as well as the tangled chain of events that led to its use on Hiroshima and Nagasaki during WW2. See SHADOW MAKERS for another look at this subject. John McGreevey's rather preachy script is very loosely based on Hachiya's book.
Aka: HIROSHIMA: OUT OF THE ASHES; OUT OF THE ASHES
DRA   120 min (ort 180 min) mTV   VIDrel: MOPIC/SGSVID   V
Boa: book Hiroshima Diary by Michihiko Hachiya.

HIROSHIMA MON AMOUR ****
PG
Alain Resnais FRANCE/JAPAN   1959
Emmanuelle Riva, Eiji Okada, Bernard Fresson, Stella Dassas, Pierre Barbaud
Whilst working on location in Hiroshima in 1950, a French actress has a brief fling with a Japanese architect, and tries to overcome her sadness when recalling both the destruction of the city during WW2 and an earlier affair she had with a German soldier. Having spent eleven years making shorts, Resnais' feature debut is a beautifully evocative and highly individual study of personal grief and public tragedy – complex, thoughtful and remarkably innovative.
DRA   86 min (ort 91 min) B/W   VIDrel: CONNO/RTM   V

HIS GIRL FRIDAY ***
U
Howard Hawks USA   1940
Rosalind Russell, Cary Grant, Ralph Bellamy, Gene Lockhart, Porter Hall, Ernest Truex, Cliff Edward, Clarence Kolb, Roscoe, Karns, Frank Jenks, Abner Biberman, Frank Orth, John Qualen, Helen Mack, Billy Gilbert, Alma Krueger
The most successful version of the classic stage comedy "The Front Page", also filmed in 1931 and 1974. A female newspaper reporter covers a scoop on an escaped criminal, this being part of a cunning plan by her ex-husband to stop her leaving his paper to get married. A frantic zany comedy with breakneck pace and machine-gun dialogue.
COM   91 min B/W   VIDrel: SCRN/DISC   V
Boa: play The Front Page by Charles MacArthur and Ben Hecht.

HISTORY OF THE WORLD: PART 1 **
15
Mel Brooks USA   1981
Mel Brooks, Madeline Kahn, Harvey Korman, Gregory Hines, Pamela Stephenson, Dom DeLuise, Shecky Greene, Jack Carter, Jan Murray, Sid Caesar, Jackie Mason, Cloris Leachman, Ron Carey, Spike Milligan, Orson Welles (narration)
Episodic comedy ranging from the Stone Age to the French Revolution. Despite the best efforts of a talented cast, this frenzied, frenetic scattershot comedy fails to sustain the momentum

of its opening, and soon degenerates into a series of one-line gags. The few funny moments (such as Caesar's role as a caveman in the opening) could have been compressed into a 30-minute short.
COM   92 min   VIDrel: L/A   V

HIT LIST, THE **
18
William Webb USA
1992
Jeff Fahey, Yancy Butler, James Coburn, Michael Beach, Randy Oglesby, Jeff Kober, Michael Harris, Chris Pederson, Tony Amendola, Charles Lanyer, Sherman Howard, Steven Marcus, La Joy Farr, Carl Ciarfalio, David Brisbyn
A contract killer gets a beautiful woman client and then finds himself having to save her, an action that plunges him into a murky plot that threatens his own life. Average.
THR   93 min (ort 97 min)   VIDrel: EIV/SONOP   V

HITCHER, THE **
18
Robert Harmon USA
1986
Rutger Hauer, C. Thomas Howell, Jennifer Jason Leigh, Jeffrey DeMunn, John Jackson, Bill Greenbush, Jack Thibeau, Armin Shimerman, Eugene Davis, Jon Van Ness, Henry Darrow, Tony Epper, Tom Spratley, Colin Campbell
Unbelievable tale of mass-murder, with the story of a psychotic hitch-hiker who doggedly keeps reappearing in a man's life. Occasionally intriguing but generally disagreeable.
DRA   93 min (ort 97 min) wScrn   VIDrel: WHV   V/sur

HITLER, A FILM FROM GERMANY ****
15
Hans Jurgen Syberberg WEST GERMANY
1977
Heinz Schubert, Andre Heller, Hellmuth Lange, Amelie Syberberg, Harry Baer, Peter Kern, Peter Luhr, Rainer von Artenfels, Peter Moland, Martin Sperr, Johannes Buzalski, Alfred Edel
Syberberg attempts to understand the rise of Fascism in his native country by putting his country's culture and people under the microscope. Made in four parts, the sections being entitled: "The Grail", "A German Dream", "The End Of The Winter's Tale" and "We, The Children Of Hell". An unflinching and brave work, which though employing many artifices to make its points (such as the use of choruses) is never less than totally absorbing. See also THE NASTY GIRL.
Aka: HITLER, EIN FILM AUS DEUTSCHLAND
DOC   400 min   VIDrel: ACAD/RTM   V

HITMAN, THE **
18
Aaron Norris USA
1991
Chuck Norris, Michael Parks, Al Waxman, Alberta Watson, Salim Grant, Ken Pogue, Marcel Sabourin, Bruno Gerussi, Frank Ferucci, James Purcell, Candus Churchill, Alan Peterson, Paris Mileos, Alex Buhanski, Stephen Dimopulos
Set up by his crooked partner, an honest New York cop is shot and left for dead. However, he recovers and some time later takes a job with the DEA, only to learn that his erstwhile partner is now running drugs for an Iranian criminal ring.
A/AD   90 min (ort 94 min)   VIDrel: WHV   V/sh

HITMAN IN THE HAND OF BUDDHA **
15
Hwang Jan Lee HONG KONG
Hwang Jan Lee, Eddie Ko, Tino Wong, Fan Mei San
Two brothers who work in a rice trading company are taken to task by some business rivals who are upset at the crafty trading methods of the former. A tough martial arts fighter is hired to chastise them.
MAR   90 min   VIDrel: EAST/DISC   V

HOBSON'S CHOICE ****
U
David Lean UK
1953
Charles Laughton, John Mills, Brenda De Banzie, Daphne Anderson, Prunella Scales, Richard Wattis, Derek Blomfield, Helen Haye, Julien Mitchell, Joseph Tomelty, Gibb McLaughlin, Dorothy Gordon, John Laurie, Raymond Huntley
A tyrannical Lancashire bootmaker is brought to heel by his eldest daughter when she marries his boothand in the face of his opposition, and sets about starting up their own business. Laughton is a delight to watch, as the tyrant in this fine period comedy. Scripted by Norman Spencer from the play and remade for TV in 1983.
COM   102 min (ort 107 min) B/W   VIDrel: WHV   V/h
Boa: play by Harold Brighouse.

HOCUS POCUS **
PG
Kenny Ortega USA
1993
Bette Midler, Kathy Najimy, Sarah Jessica Parker, Thora Birch, Doug

Jones, Omri Katz, Vinessa Shaw, Penny Marshall, Garry Marshall, Amanda Shepherd, Larry Bagby III, Tobias Jelinke, Stephanie Faracy, Charlie Rocket, Karyn Malchus
Three 17th century witches executed in Salem somehow travel 300 years in time to the present day where they proceed to get their revenge in some rather nasty ways. Some good performances by the leads help to pad out the paper-thin plot but even with the good special effects, this disappointing effort does not really yield much amusement.
COM   92 min (ort 97 min) cC   VIDrel: WDV/TECH   V/sur

HOFFA ***
15
Danny DeVito USA
1992
Jack Nicholson, Danny DeVito, Armand Assante, J.T. Walsh, Frank Whaley, John P. Ryan, John C. Reilly, Kevin Anderson, Robert Prosky, Natalija Nogulich, Karen Young, Nicholas Pryor, Cliff Gorman, Paul Guilfoyle, Joanne Neer, Joe V. Greco
Long, flashback style biopic on the career of Teamster union boss Hoffa that paints an episodic and favourable portrait of this controversial figure while avoiding any searching exploration of his inner drives. Nicholson gives a fine portrayal that is totally convincing, and is ably assisted by DeVito (proving he is more than just a comic actor) as a union aide. The powerful if chaotic script is by David Mamet. See also TEAMSTER BOSS: THE JACKIE PRESSER STORY.
DRA   134 min (ort 140 min)
VIDrel: 4-FRONT/POLYREC/GUILD   V/sur

HOFFMAN **
U
Alvin Rakoff UK
1970
Peter Sellers, Sinead Cusack, Ruth Dunning, Jeremy Bulloch, David Lodge, Ron Taylor, Kay Hall, Karen Murtagh, Cindy Burrows, Elizabeth Bayley
A middle-aged creep blackmails a typist about to be married, into spending a dirty weekend with him. A bizarre character study that's occasionally funny but more often just tiresome.
COM   107 min (ort 116 min)   VIDrel: BRAVE/SONOP   V
Boa: novel/play Shall I Eat You Now? by Ernest Gebler.

HOLD ME THRILL ME KISS ME **
18
Joel Hershman USA
1993
Adrienne Shelly, Max Parrish, Andrea Naschak, Timothy Leary, Bela Lehoczky, Ania Suli, Vic Trevino, Allan Warnick, Joseph Anthony Richards, Sean Young, Diana Ladd, Martha Shaw, Frank Noon, Mary Lanier, Bruce E. Morow, Jo Farkas
After accidentally shooting his girlfriend after a quarrel at their marriage ceremony, a man runs for his life and takes refuge with a screwball stripper who lives in a trailer park full of equally eccentric characters. However, his fiancee survives and soon comes after both of them. A crude, over-the-top black comedy that is mildly amusing at best and seems more than a touch inspired by the films of Russ Meyer.
Aka: HOLD ME, THRILL ME, KISS ME
COM   95 min   VIDrel: ARTPRO/RTM   V/h

HOLD THE DREAM: PARTS 1 AND 2 **
15
Don Sharp UK
1986
Jenny Seagrove, Stephen Collins, Deborah Kerr, James Brolin, Claire Bloom, Fiona Fullerton, Paul Daneman, Suzanna Hamilton, Nigel Havers, John Mills, Liam Neeson, Pauline Yates, Valentina Pelka, Sarah-Jane Varley
A sequel to A WOMAN OF SUBSTANCE, in which our elderly heroine struggles to retain control of her commercial empire, although she ultimately intends to hand it over to her favourite granddaughter. A glossy and inferior sequel, scripted by Barbara Taylor Bradford. Followed by TO BE THE BEST.
Aka: HOLD THAT DREAM
DRA   196 min (2 cassettes) mTV
VIDrel: 4-FRONT/POLYREC/ODY   V
Boa: novel A Woman Of Substance by Barbara Taylor Bradford.

HOLIDAY ***
U
George Cukor USA
1938
Cary Grant, Katharine Hepburn, Lew Ayres, Edward Everett Horton, Doris Nolan, Ruth Donnelly, Henry Kolker, Binnie Barnes, Henry Daniell, Jean Dixon, Charles Trowbridge, George Paucefort, Marion Ballou, Howard Hickman
A struggling young nonconformist lawyer ditches his snobby fiancee in favour of her free-minded sister. Sparkling and witty comedy, a remake of the 1930 version, with excellent perfor-

mances, great dialogue and real panache. A delight in all departments.
Aka: FREE TO LIVE; UNCONVENTIONAL LINDA
COM 91 min (ort 94 min) B/W VIDrel: COLUM/SONOP V/dm
Boa: play by Philip Barry.

**HOLIDAY AFFAIR ** **   PG
Alan Myerson USA
David James Elliott, Cynthia Gibb, Curtis Blanck
On a trip to Manhattan, a lonely widow and her son meet a sales clerk and a romance ensues. Fair.
DRA 90 min VIDrel: ODY/SONOP V/sh

HOLIDAY INN *   U
Mark Sandrich USA   1942
Bing Crosby, Fred Astaire, Marjorie Reynolds, Walter Abel, Virginia Dale, Louise Beavers, Irving Bacon, James Bell, John Gallaudet, Shelby Bacon, Leon Belasco, Harry Barris, Judith Gibson, Katherine Booth, Joan Arnold
The two owners of a country inn that's only open on holidays, fall for the same girl, in this flimsy romantic triangle plot that serves simply as a vehicle for some great Irving Berlin songs, including the famous "White Christmas". A breezy fun-filled musical, superior to that partial remake WHITE CHRISTMAS.
AA: Song ("White Christmas" – Irving Berlin).
MUS 100 min (ort 104 min) B/W VIDrel: CIC/SONOP V

HOLIDAY ON THE BUSES *   PG
Bryan Izzard UK   1973
Reg Varney, Doris Hare, Stephen Lewis, Anna Karen, Michael Robbins, Bob Grant, Wilfrid Brambell, Kate Williams, Arthur Mullard, Queenie Watts, Henry McGee, Adam Rhodes, Michael Sheard, Franco Derosa, Gigi Gatti, Eunice Black
Third in this series of spin-offs from a TV series about the adventures of the crew of a bus company. This time round the family goes on holiday, but ends up driving for a holiday camp. Another one of those dismal and dated British comedies inspired by the somewhat better series "On The Buses". See also ON THE BUSES and MUTINY ON THE BUSES.
COM 82 min (ort 85 min) VIDrel: MGM/WHV V

**HOLLISTER ** **   15
Vern Gillum USA   1991
Brian Bloom, Jamie Rose, Jorge Cervera Jr, David Carradine, James Remar, Mark Ballou, Deborah Falconer, Shawn Levy, Phil Fondacaro, Shannon Sturges, Joan Bahan, Kim Blacklock, Curtis Brown, Kenneth V. Dickerson, Jerry Gardner
Thanks to his skill with a gun, a veteran of the American Civil War has become a living legend. Whilst travelling across country with his brother, the latter is murdered by bandits, and our hero goes after them. A standard revenger, no better or worse than countless others.
WES 86 min VIDrel: CIC/SONOP V

**HOLLOW REED ** **   15
Angela Pope GERMANY/UK   1995
Sam Bould, Martin Donovan, Ian Hart, Joely Richardson, Jason Flemyng, Kelly Hunter, Shaheen Khan, Tim Crouch, Jane Hill, Glen Hammond, Simon Chandler, Dilys Hamlett, Andy Rashleigh, David Calder, Maeve Murphy, Victoria Scarborough
After he reveals his homosexuality, a family doctor's marriage breaks down and custody of his son is awarded to his ex-wife, who subsequently remarries. When his son comes to visit and shows signs of having been beaten, he claims that he was attacked in a park, but his father soon comes to suspect that his stepfather is to blame. With the support of his lover, he now engages in a protracted and tortuous legal battle to regain custody of his son.
DRA 104 min CINrel
Boa: story by Neville Bolt.

**HOLLY DOES HOLLYWOOD 4 ** **   18
Jon Stallion USA   1990
Christy Canyon, Carol Cummings, Michelle Monroe, Kim Kane, Jon Stallion, Ron Jeremy, Tony Montana, Randy West, Marc Wallice, Wayne Summers, Hans Mueller, Dick James, Paul Pounder
Newly arrived in Hollywood, a young hopeful begins her slow climb to the top in the porno movie industry, in the course of which her director (and mentor) attempts to get the studio bosses to use her in place of an established star. Despite being a sex film this one is quite well plotted, bearing more than a passing similarity to A STAR IS BORN. Stallion appears in it as

Holly's director. Parts 1, 2 and 3 of this movie appear to have never been released.
A 61 min (ort 88 min) mVid VIDrel: MIA/DISC/IMPENT V

HOLLYWOOD CHAINSAW HOOKERS *   18
Fred Olen Ray USA   1988
Linnea Quigley, Gunnar Hansen, Dawn Wildsmith, Michelle Bauer, Michelle Bauer, Michelle McLellan, Esther Alise, Tricia Burns, Jerry Fox, Jimmy Williams, Dukey Flyswatter (Michael Sonye), Dennis Monney, Jerry Miller
A nasty piece of sleaze, with a private eye hired to locate a missing teenage girl, discovering that she has been kidnapped by a strange sect of hookers who worship chainsaws, their intention being to make her a virgin sacrifice to their leader. A grisly black comedy, inspired by the repulsive low-budget 1960s films of Herschell Gordon Lewis and equally full of gore and mutilation. Hansen was "Leatherface" in THE TEXAS CHAINSAW MASSACRE.
Aka: CHAINSAW HOOKERS; HOLLYWOOD HOOKERS
HOR 70 min Cut (1 min 6 sec – ort 90 min)
VIDrel: POPRO/RTM V

HOLLYWOOD COP *   18
Amir Shervan USA   1988
Jim Mitchum, Cameron Mitchell, David Goss, Julie Schonhofer, Lincoln Fitzgerald, Troy Donahue, Aldo Ray, Larry Frio, Brandon Angle
When a woman's former husband steals several million dollars from the Mafia, her young son is kidnapped, but a tough cop and a bunch of Hell's Angels set out to rescue him. A fairly mindless action tale, written by Shervan.
A/AD 100 min Cut (1 min 3 sec) VIDrel: IMPENT V

**HOLLYWOOD DREAMING ** **   15
Jim Marshall USA   1986
Ben Glass, Orson Bean, Natasha Kautsky, Anthony Alda, Kerry Remsen, Orson Bean, Lucinda Crosby, David Hedison, Bill Henderson, Zsa Zsa Gabor
A chronicle of the frustrations facing an eccentric aspiring filmmaker, as he attempts to find the money to finance his ambition. A dullish and rather stupid comedy.
Aka: MOVIE MAKERS, THE; SMART ALEC
COM 87 min VIDrel: 20TH/TECH V

**HOLLYWOOD DREAMS ** **   (18)
Ralph Portillo USA   1993
Kelly Cook, Danny Smith, Debra Beatty, Rick Scandlin, Jonathon Murray, Brian Palermo, Anthony Holuin, Natasha Pulanova, Stephanie Carlisle, Kathy Lewis, Tony Belline, Kathy Passmore, Melvin Coots, Jacqueline St. Claire
A young girl leaves the Midwest behind, when she arrives in Hollywood and finds that fame and fortune seem to depend on how she uses her attractive face and figure. A silly soft-core effort that is reasonably well made but impossible to take all that seriously.
A 83 min (ort 90 min) SATrel: SKY MOVIES

**HOLLYWOOD MADAM ** **   18
Fred Gallo USA   1994
Michael Nouri, Robert Constanzo, Shannon Whirry, Crystal Chappell, Karen Kopins, Meg Foster, Charles Grant, Paul Linke, Taylor Locke, William Devane, Sandra Korn, Dan Blom, Paula MacLaren, David Young, Jack L. Harrell, Hank Galia
When an L.A. detective embarks on an investigation into a series of murders of high-class hookers, he finds himself getting involved with a beautiful woman.
DRA 80 min (ort 90 min) mTV VIDrel: ODY/SONOP V/sh

HOLLYWOOD SHUFFLE *   15
Robert Townsend USA   1987
Robert Townsend, Anne-Marie Johnson, Starletta Dupois, Helen Margin, Craigus R. Johnson, Keenan Ivory Wayans, Domenick Irrera, John Witherspoon, Paul Mooney, Lisa Mende, Robert Shafer, Ludie Washington
A look at the struggles of a black aspiring actor to get parts. As he goes after yet another stereotyped "cool cat" or "butler" role, he dreams of fame and a real part in a Hollywood film. He does eventually get a part, but only because the directors were looking for an Eddie Murphy lookalike. A most enjoyable and intelligent story, that's by turns gleeful, affectionate and perceptive. Written by Townsend and Wayans.
COM 78 min (ort 81 min) VIDrel: VISVID/POLYREC L/A V

HOLLYWOOD SUPERSTAR ** 18
Gene Nash USA 1970
Andy Davis, Vitra Videt, Joe Taylor, Reid Smith, Ray Foster, Matt Bennett, Jeremy Stockwell
A male actor dresses in drag in order to create a career and has to continue playing his female role. Mildly diverting, fact-based drama that doubtless provided some of the inspiration for the later film TOOTSIE.
Aka: DINAH EAST
DRA 89 min (ort 96 min) VIDrel: L/A V

HOLLYWOOD WIVES * 15
Robert Day USA 1985
Anthony Hopkins, Angie Dickinson, Candice Bergen, Rod Steiger, Mary Crosby, Robert Stack, Steve Forrest, Stefanie Powers, Joanna Cassidy, Suzanne Somers, Roddy McDowall
The story of the bed-hopping adventures and other activities of some of the women in celluloid city. Dross of a high order, based on a trashy novel of similar merits.
DRA 270 min (2 cassettes – ort 360 min) VIDrel: WHV V/h
Boa: novel by Jackie Collins.

HOLY MATRIMONY ** PG
Leonard Nimoy USA 1993
Patricia Arquette, Joseph Gordon-Levitt, Tate Donovan, Armin Mueller-Stahl, Lois Smith, John Schuck, Courtney B. Vance, Jeffrey Nordling, Richard Riehle, Mary Pat Gleason, Alaine Byrne, Dan Cossolini, Lori Alan, Franz Novak, Ted Hoff
A female robber hits the road after pulling a job and eventually finds herself hiding out among the members of an old-fashioned religious community. A surprise soon awaits her however, when she learns that by their customs she is required to marry a twelve-year-old boy. A distinctly unimpressive comedy that is too contrived to be effective.
COM 89 min (ort 93 min) VIDrel: 20VIS/SONOP V/sur

HOLY VIRGIN VERSUS THE EVIL DEAD ** 18
Choy Fat/Wang Zhen-Yi HONG KONG 1990
Donnie Yen, Sibelle Hu, Yang Bao-Ling, Cathy Chow, Lin Guo-Bin
A Cambodian princess who is armed with a magic sword must confront a terrifying creature called the "Moon Monster" that feeds on human flesh. A melding of Chinese erotic fantasy with supernatural horror.
FAN 88 min (ort 90 min) wScrn VIDrel: EAST/DISC V

HOMBRE *** PG
Martin Ritt USA 1967
Paul Newman, Fredric March, Martin Balsam, Richard Boone, Diane Cilento, Cameron Mitchell, Barbara Rush, Margaret Blye, Peter Lazer, Skip Ward, Frank Silvera, Val Avery, David Canary, Larry Ward, Pete Hernandez, Linda Cordova
A strange silent man, raised by the Indians, joins the passengers of an Arizona stagecoach in the 1880s. They soon come to despise him, but when they are trapped by outlaws, find that it is their own actions and thoughts that are worthy of contempt. An entertaining Western, with Newman as the half-breed, playing a dangerous cat-and-mouse game with Boone, as the sinister leader of the outlaws.
WES 105 min (ort 111 min) VIDrel: 20TH/TECH V/h
Boa: novel by Elmore Leonard.

HOME ALONE *** PG
Chris Columbus USA 1990
Macaulay Culkin, Joe Pesci, Daniel Stern, John Heard, Catherine O'Hara, John Candy, Roberts Blossom, Billie Bird, Angela Goethals, Devin Ratray, Gerry Bamman, Hillary Wolf, Larry Hankin, Michael C. Maronna, Kristin Minter
In the chaos surrounding their departure for a Christmas holiday in Paris, the eight-year-old son in a family of fifteen is inadvertently left behind at home. Relieved at having the house all to himself for some time, he is soon disturbed by the two most inept burglars ever. They try repeatedly to break in but are repulsed at every turn in a series of hilarious encounters. An amusing but excessively violent comic cartoon, with music by John Williams.
COM 98 min (ort 110 min) VIDrel: 20TH/TECH L/A V/h

HOME ALONE 2: LOST IN NEW YORK ** PG
Chris Columbus USA 1992
Macaulay Culkin, Joe Pesci, Daniel Stern, John Heard, Catherine O'Hara, Tim Curry, Daniel Stern, Devin Ratray, Brenda Fricker,

Eddie Bracken, Dana Ivey, Kieran Culkin, Rob Schneider, Hilawy Wolf, Maureen Elizabeth Shay, Donald Trump
Sequel to the first film is very much of a carbon copy in terms of plot and characters. As Kevin's family depart for a holiday in Florida, he gets left behind and boards the wrong plane, ending up in New York. There he soon encounters our two incompetent burglars who have broken out of prison. Many of the gags are lifted straight from the first film and the level of cartoon-like violence is even greater, although the film still retains a little humour.
COM 115 min (ort 120 min) VIDrel: 20TH/TECH L/A V

HOME AT SEVEN *** U
Ralph Richardson UK 1953
Ralph Richardson, Margaret Leighton, Jack Hawkins, Meriel Forbes, Campbell Singer, Frederick Piper, Diana Beaumont, Michael Shepley, Margaret Withers, Gerald Case
Failing to return home at seven for the first time in many years, a bank clerk is shocked to learn that the previous twenty-four hours are a blank. Worse is to come when he discovers that the safe at his sports club has been rifled and a steward murdered. After the police break a phoney alibi he gave them, he becomes convinced of his own guilt. A remarkably tense story, produced on a shoestring in only fifteen days and Richardson's only directorial effort.
Aka: MURDER ON MONDAY
THR 82 min (ort 85 min) B/W VIDrel: WHV V/h
Boa: play by R.C. Sheriff.

HOME FOR CHRISTMAS *** U
Peter McCubbin USA 1990
Mickey Rooney, Simon Richards, Lesley Kelly, Noah Plener, Chantellese Kent, Joel Kaiser, Susan Hamann, Ken McKenzie, Hersh Kalles, Karen Inwood, Ken Innes, Peter Ferri, Paul Babiak, John Tench, Albert Davidson, Austin Schatz
A six-year-old girl asks Santa Claus to bring her a grandfather for Christmas, and her wish is strangely fulfilled when a lonely old tramp comes into her life. Rooney as the destitute down-and-out gives a heartwarming performance in this unashamedly sentimental tearjerker.
JUV 96 min VIDrel: DDVID V

HOME FOR THE HOLIDAYS *** 15
Jodie Foster USA 1995
Holly Hunter, Anne Bancroft, Robert Downey Jr, Charles Durning, Steve Guttenberg, Dylan McDermott, Cynthia Stevenson, Claire Danes, Geraldine Chaplin, Austin Pendleton, David Strathairn, Zachary Duhame, Emily Ann Lloyd
A single mother returns home for Thanksgiving, but thanks to the antics of her relatives, she has a most trying time. The various members of her family are mostly eccentrics and incompetents, and as she has just lost her job, she has troubles enough of her own, and is ill equipped to deal with the conflicts and recriminations that occur. A film that relies on its oddball characters to generate its laughs, it takes a while to get going but is worth sticking with.
COM 98 min (ort 102 min) VIDrel: POLFIL V/sh
Boa: short story by Chris Radant.

HOME OF OUR OWN *** PG
Tony Bill USA 1993
Kathy Bates, Edward Furlong, Soon Teck-Oh, Clarrisa Lassig, Miles Feulner, T.J. Lowther, Tony Campisi, Dave Jensen, H.E.D. Redford, Melvin Ward, Michael Flynn, Don Re Sampson, Frank Gerrish, Rosalind Sullivan, Joseph Schaefer
In the early 1960s, a woman from L.A. relocates to rural Idaho with her six kids, and despite a lack of funds sets about putting a roof over their heads. Through sheer persistence and the goodwill of her neighbours, she finally achieves her goal. An old-fashioned film that tends towards sentimentality, but the sheer charm and dynamism of Bates keeps it alive.
DRA 100 min (ort 104 min) VIDrel: COLUM/SONOP V/sur

HOMEBOY ** 15
Michael Seresin USA 1988
Mickey Rourke, Christopher Walken, Debra Feuer, Thomas Quinn, Kevin Conway, Anthony Alda, Jon Polito, Bill Slayton, David Taylor, Joseph Ragno, Mathew Lewis, Will Devill, Reuben Blades, Sam Gray, Dondre Whitfield, Teddy Abner
A young boxer who dreams of the big time finds that his

unscrupulous manager has made a crooked deal that could ruin his career.
DRA   110 min (ort 158 min)
VIDrel: 4-FRONT/POLYREC/BRAVE   V/sur

HOMER & EDDIE **
Andrei Konchalovsky USA
James Belushi, Whoopi Goldberg, Karen Black, Nancy Parsons, Anne Ramsey, Ernestine McClendon, Beah Richards, Vincent Schiavelli, Robert Glaudini, John Waters, Angelyne, Tad Horino, James Thiel, Jeffrey Thiel, Andy Jarrel
15
1990
A mentally retarded man robbed of his life savings while on the road teams up with an escaped mental patient, who is dying of a brain tumour, and the two go in pursuit of the robbers. Unknown to him, however, his companion is subject to violent behaviour and is not above a little murder or robbery to satisfy her needs. A more bitter than sweet road-movie, well acted by Belushi and Goldberg, but too downbeat to work, even as a black comedy.
COM   95 min (ort 99 min)   VIDrel: VISVID/POLYREC   V/s

HOMETOWN STORY **
Arthur Pierson USA
Donald Crisp, Alan Hale Jr, Jeffrey Lynn, Marjorie Reynolds, Marilyn Monroe, Melinda Plowman, Glenn Tyron
U
1951
A recently defeated politician decides that a local businessman was responsible for his plight and schemes to get even. A faded drama of little interest apart from the presence of Monroe in one of the supporting parts.
DRA   61 min (ort 75 min) B/W   VIDrel: SCRN/DISC   V

HOMEWARD BOUND ***
Richard Michaels USA
David Soul, Moosie Drier, Barnard Hughes, Judith Penrod, Jeff Corey, Carmen Zapata, Michelle Downey, Jim Haynie, Lynne Marta, Carol Vogel, Jim Antonio, Robert Elross, Titos Vandis
(PG)
1980
A divorced father and his terminally ill son spend a last summer at the grandfather's vineyard. A remarkably moving and unmanipulative drama, with a literate script by Burt Pretutsky. The score is by Fred Karlin.
DRA   96 min (ort 100 min) mTV   SATrel: SKY MOVIES

HOMEWARD BOUND: THE INCREDIBLE STORY ***
Duwayne Durham USA
Robert Hays, Kim Greist, Jean Smart, Veronica Lauren, Kevin Chevalia, Benj Thall, William Edward Phipps, Mark I. Taylor, Don Alder, David McIntye, Mariah Milner, Jane Jones plus voices of: Michael J. Fox, Sally Field, Don Ameche
U
1992
Quality remake of Disney's THE INCREDIBLE JOURNEY telling the story of two dogs and a cat who travel some 250 miles across the wilderness to rejoin their owners. An enjoyable family affair, with the human actors inevitably playing second fiddle to their animal counterparts. Unfortunately, the unwise use of human voices for the animals and the all-pervasive sentimentality are annoying drawbacks. A sequel followed. See also HECK'S WAY HOME.
JUV   81 min (ort 85 min) cC   VIDrel: WDV/TECH   V/sh
Boa: book The Incredible Journey by Sheila Burnford.

HOMEWARD BOUND 2: LOST IN SAN FRANCISCO **
USA
Voices of: Michael J. Fox, Sally Field, Ralph Waite, Sinbad
U
1995
Chance the bulldog, Shadow the golden retriever and Sassy the cat have further adventures together, this time on the streets of San Francisco, when they take off in search of their owners.
JUV   88 min cC   VIDrel: WDV/TECH   V/sh

HOMEWORK *
James Beshears USA
Joan Collins, Michael Morgan, Shell Kepler, Lanny Horn, Lee Purcell, Carrie Snodgress, Wings Hauser, Mel Welles, Beverly Todd, Betty Thomas, Erin Donovan, Renee Harris, John Romano, Joy Michael, Dee Dee Downs, Bill Knight
18
1979
A sixteen-year-old boy can't make it with girls his own age, so he tries an older woman. Despite some clumsy attempts to simulate nude scenes with Joan Collins, this dismal dud (promoted as a comedy) is best avoided. See PRIVATE LESSONS or MY TUTOR, that at least attempted to inject some style into this theme.
Aka: GROWING PAINS
DRA   88 min (ort 90 min)   VIDrel: CIC/SONOP L/A   V/h

HOMEWORK ***
Jaime Humberto Hermosillo MEXICO
Maria Rojo, Jose Alonso, Xanic Zepeda, Christopher
18
1990
A film student tries to video herself and her ex-husband making love without the knowledge of the latter (who keeps declaring that he feels someone is watching him). Sometimes a little claustrophobic (there are mostly just the two characters) but the script is so witty that one can easily overlook this. Made in just four days on the one set, it brings to mind the technique Hitchcock used rather less effectively in ROPE. The script is by Hermosillo.
Aka: LA TAREA
COM   85 min wScrn   VIDrel: TART/20TH L/A   V/dm

HOMEWRECKER **
Fred Walton USA
Robby Benson, Sydney Walsh, Sarah Rose Karr, Kate Jackson
PG
1992
A scientist invents a domestic robot with feminine characteristics to serve as both a housekeeper and companion. As time passes his creation takes over more aspects of his life, revealing jealousy and murderous tendencies when his estranged wife pays him a visit. See also ELECTRIC DREAMS: THE MOVIE for something rather similar.
HOR   83 min (ort 88 min)   VIDrel: CIC/SONOP L/A   V

HOMICIDAL **
William Castle USA
Glenn Corbett, Patricia Breslin, Jean Arless, Eugenie Leontovich, James Westerfield, Anna Bunce
15
1961
A pretty nurse cares for a wheelchair-bound stroke victim after attending a murderous wedding ceremony in this offbeat, low-budget examination of a female psychopath and her increasingly bizarre behaviour. The plot doesn't hold water, the acting is good if not great, but the film does retain a certain dated, B-movie distinctiveness.
HOR   88 min B/W   VIDrel: ENCORE/SPEAR   V

HOMICIDE ***
David Mamet USA
Joe Mantegna, William H. Macy, Natalija Nogulich, Ving Rhames, Vincent Guastaferro, Rebecca Pidgeon, J.J. Johnston, Jack Wallace, Lionel Mark Smith, Adolph Mall, Paul Butler, Colin Stinton, Roberta Custer, Charles Stransky
15
1991
Something of a study of anti-Semitism in the New York Police Force, this has Mantegna as an alienated Jewish cop who has always striven to play down his background, assigned to a case involving the murder of a Jew. At first resentful at this, he becomes increasingly drawn into the ramifications of the investigation, uncovers some very nasty things, but loses his own self-respect in the process. A perplexing and highly provocative film.
DRA   97 min (ort 102 min)   VIDrel: FIRST/SONOP; PION (LV only)   V/sur LV

HONDO ***
John Farrow USA
John Wayne, Geraldine Page, Ward Bond, Michael Pate, James Arness, Rodolfo Acosta, Leo Gordon, Tom Irish, Lee Aker, Paul Fix, Rayford Barnes
U
1953
A former gunman comes across an isolated ranch run by an abandoned wife and her young sun. He tries to persuade them to leave because of the danger of raids by the Apache but they refuse. He leaves but after surviving several dangers returns and takes them with him to California. A fine adaptation of the original story, with some nicely rounded characterisations, including those of the Indians.
WES   80 min (ort 84 min)   VIDrel: VCC/DISC/COLUM   V/sh
Boa: story The Gift of Cochise by Louis L'Amour.

HONEY, I BLEW UP THE KID *
Randal Kleiser USA
Rick Moranis, Lloyd Bridges, Marcia Strassman, Robert Oliveri, Amy O'Neill, Daniel Shalikar, Joshua Shalikar, John Shea, Keri Russell, Gregory Sierra, Julia Sweeney, Ken Tobey, Linda Carlson, Leslie Neale, Ron Canada, Lisa Mende
U
1992
Sequel of sorts to HONEY, I SHRUNK THE KIDS with our crazy inventor this time somehow managing to make his two-year-old grow to the dizzy height of 112 feet before it proceeds to wander the streets of Las Vegas. Full of amazingly poor special effects, tedious and repetitive, and in many ways a throwback to those duff SF films of the 1950s. Kleiser does not impress as a director.
COM   85 min (ort 89 min) cC   VIDrel: WDV/TECH   V/sur

HONEY, I SHRUNK THE KIDS ***
Joe Johnston USA   U 1989
Rick Moranis, Matt Frewer, Marcia Strassman, Amy O'Neill, Jared Rushton, Kristine Sutherland, Thomas Brown, Robert Oliveri, Carl Steven, Mark L. Taylor, Lou Cutell, Laura Waterbury, Trevor Galtress, Janet Sunderland
An old-fashioned Disney fantasy-comedy done in the style of THE ABSENT-MINDED PROFESSOR, that sees a scientist inventing a shrinking machine and then being forced to mount a frantic search when it accidentally causes his own kids and those of a neighbour to shrink to the size of ants. An entertaining family film, somewhat bland, but enlivened by the excellent special effects. A sequel followed.
COM 89 min (ort 101 min) VIDrel: WDV/TECH V/sur

HONEYMOON IN VEGAS **
Andrew Bergman USA   15 1992
James Caan, Sarah Jessica Parker, Nicolas Cage, Pat Morita, Anne Bancroft, Johnny Williams, John Capodice, Robert Coztanzo, Peter Boyle, Burton Gilliam, Seymour Cassel, Jerry Tarkanian, Tony Shalhoub, Brad Blumenthal, J.J. Bostick
A woman finally convinces her reluctant boyfriend to marry her and they go to Las Vegas to tie the knot but no sooner have they arrived then he manages to lose her in a game of poker with a gangster. The latter whisks her off to Hawaii for a spot of wooing, while hubby mounts a desperate search. An over-the-top farce, full of Elvis imitators of every shape and size, but too contrived to be really amusing.
COM 92 min (ort 96 min)
VIDrel: FIRST/SONOP; PION (LV only) V/sur LV

HONEYMOON KILLERS, THE ***
Leonard Kastle USA   18 1970
Shirley Stoler, Tony LoBianco, Mary Jane Highby, Doris Roberts, Dortha Duckworth, Kip McArdle, Marilyn Chris, Barbara Cason, Ann Harris, Guy Sorel, Mary Breen, Elsa Raven, Mary Engel, Mike Haley, Diane Asselin, William Adams
A ruthless couple murder lonely old women for their money. Based on the true story of multiple murderers Martha Beck and Raymond Fernandez who were executed in 1951. The chilling script is by Kastle.
Aka: LONELY HEARTS KILLERS, THE
DRA 103 min (ort 108 min) B/W wScrn
VIDrel: TART/20TH V

HONEYSUCKLE ROSE ***
Jerry Schatzberg USA   15 1980
Willie Nelson, Dyan Cannon, Slim Pickens, Amy Irving, Joey Floyd, Charles Levin, Mickey Rooney Jr, Priscilla Pointer, Diana Scarwid
A Country music star finds that his life is falling to pieces and that he can no longer cope with the pressures of life on the road, when he embarks on an affair with the young daughter of one of his colleagues. A restrained account (said by some to be based on INTERMEZZO, but I have my doubts) that has lively performances and fine music.
Aka: ON THE ROAD AGAIN
DRA 114 min (ort 119 min) wScrn VIDrel: TART/20TH V/sur

HONG KONG '97 **
Albert Pyun USA   18 1994
Robert Patrick, Ming-Na Wen, Brion James, Tim Thomerson, Michael Lee, Andrew Divoff, Selena Mangh, Steve Day, Joey Leung, Nonong Talbo, Terri Conn, Chad Stahelski, Jody McAndrew, Joseph Pe, Yvonne Victa, Augusto Victa
On the eve of the hand-over of power in Hong Kong to Communist China, a contract killer is ordered to assassinate a general but learns to his cost that he has been set up. Soon he is forced to seek out some unlikely allies in order to survive, in this standard actioner.
A/AD 87 min (ort 91 min) VIDrel: TRIM/HIFLI V/h

HONKY TONK FREEWAY **
John Schlesinger USA   15 1981
William Devane, Beverly D'Angelo, Beau Bridges, Jessica Tandy, Hume Cronyn, Geraldine Page, George Dzundza, Teri Garr, Joe Grifasi, Deborah Rush, Howard Hesseman, Paul Jabara, Frances Lee McCain, Alice Beardsley, Daniel Stern
A satire on the American love affair with the car, with assorted car freaks assembling, in a tiny town determined to keep its tourist trade despite being bypassed by a motorway. An expen-sive and flashy farce that has some nice cameos but no real story to get one's teeth into.
COM 102 min (ort 107 min) VIDrel: WHV V/h

HONKYTONK MAN **
Clint Eastwood USA   15 1982
Clint Eastwood, Kyle Eastwood, John McIntire, Alexa Kenin, Verna Bloom, Matt Clark, Barry Corbin, Jerry Hardin, Tim Thomerson, Macon McCalman, Rebecca Clemens, Joe REgalbuto, Garry Grubbs, John Gimble, Linda Hopkins, Bette Ford
Eastwood and his son team up, in this story of a Depression-era Country singer dying of cancer, and his relationship with his young nephew, who has tagged along with him on one last singing trip. A touching but overlong and contrived story, with Eastwood coming across well as the singer, but having too little material to work with. Features appearances by some of the greats of Country music, such as Marty Robbins, who died just before film release.
DRA 118 min (ort 123 min) VIDrel: WHV V/h
Boa: novel by C. Carlile.

HONOR AND GLORY **
Godfrey Hall USA   15 1992
Cynthia Rothrock, Donna Jason, Chuck Jeffreys, Gerald Klein, Robin Shou, Richard Yuen, Hing Yip Yam
An FBI agent who is a martial arts expert teams up with her Interpol counterpart to stop a ruthless bank from stealing a cache of nuclear weapons. Another assembly-line effort that serves mainly as a vehicle for Rothrock.
MAR 86 min (ort 89 min) VIDrel: MIA/DISC V

HONOR BOUND **
Jeannot Szwarc USA   15 1989
John Philbin, Tom Skerritt, Gabrielle Lazure, George Dzundza, Lawrence Pressman, Gene Davis, Eric Douglas, Edward Meeks, Hana Baczynska, Relja Basic, Tibor Belisza, Ranko Zidaric, Ijarko Janes, Slavko Brankov, Bob Delegall
When a vital US surveillance satellite appears to suffer sabotage, a soldier trained in anti-terrorism is given the job of driving a surveillance car into East Germany, in order to find out what the Soviets are up to. But his partner is killed and he is captured by the KGB, interrogated and returned to the US authorities. However, he absconds from the US base and goes to uncover the truth. A murky and cheap looking spy thriller, that now looks very dated indeed.
THR 102 min CINrel
Boa: book Recovery by Steven L. Thompson.

HONOR THY FATHER AND MOTHER: THE MENENDEZ KILLINGS **
Paul Schneider USA   15 1993
James Farentino, Jill Clayburgh, Billy Warlock, David Beron, Susan Blakely, Erin Gray, Elaine Joyce, John David Conti, Stanley Kamel, Meg Witther, Sean Graham, Marcos Ferraez, Jeffrey Lee Broadhurst, Charles Lucia, Bobbie Phillips
Dramatised reconstruction of a real-life murder case where two sons of a wealthy Beverly Hills family were charged with the murder of their parents and subsequently stood trial. Well acted and competently realised but the nature of the subject makes inevitably for a rather downbeat and restrained effort. See also HONOR THY MOTHER for something rather similar.
Aka: TRUE STORY OF THE MENENDEZ MURDERS, THE
DRA 90 min (ort 97 min) nTV VIDrel: ODY/SONOP V/sh
Boa: book Blood Brothers by John Johnston and Ronald L. Soble.

HONOR THY MOTHER **
David Greene USA   15 1992
Sharon Gless, William McNamara, Paul Scherrer, Christian Hoff, Brian Wimmer, Jonathan Ward, Suzanne Ventulett, Dion Anderson, Lee Garlington, Mathew Faison, Gary Grubbs, Wayne Tippit, Will Nye, Jack Rasder, Ryan Reid
Based on a true case, this tells of a wealthy couple who are badly injured in the vicious attack, which leaves the husband dead. As the police mount their investigation, evidence starts to accumulate pointing to the couple's nineteen-year-old son being involved in this crime. An absorbing story, well acted and scripted.
DRA 90 min mTV VIDrel: CIC/SONOP L/A V
Boa: novel Blood Games by Jerry Bledsoe.

HOOK **
Steven Spielberg USA   U 1991
Robin Williams, Dustin Hoffman, Julia Roberts, Bob Hoskins, Maggie

Smith, Caroline Goodall, Charlie Korsmo, Amber Scott, Laurel Cronin, Phil Collins, Arthur Malet, Isaiah Robinson, Jasen Fisher, Dante Basco, Raushan Hammond

Peter Pan has become Peter Banning, a forty-year-old lawyer whose work exerts a stronger pull than either his past or his family. However, on a trip to London he meets his old playmate Wendy (now a grandmother) and begins to remember the past. When Captain Hook kidnaps his children, he is forced to return to Neverland and confront his old adversary. A messy and syrupy extravaganza, under-plotted and at $70,000,000, most definitely over-budgeted.

FAN   136 min (ort 142 min) wScrn
VIDrel: VCC/DISC/COLUM   V/sur
Boa: play Peter Pan plus short stories by J.M. Barrie.

HOPE AND GLORY *** 15
John Boorman UK 1986
Sarah Miles, David Hayman, Sebastian Rice-Edwards, Ian Bannen, Derrick O'Connor, Susan Wooldridge, Sammi Davis, Geraldine Muir, Jean-Marc Barr, Annie Leon, Jill Baker, Amelda Brown, Katrine Boorman, Colin Higgins

Story drawn from the director's experiences as a child, living through the wartime London Blitz, and focusing on the experiences that befall a nine-year-old boy, for whom the war was a time of great excitement. An affectionate and carefully drawn tale, which though lacking a sense of harsh realism, can be enjoyed for its rich performances and striking period detail.

DRA   108 min (ort 112 min)   VIDrel: VCC/COLUM L/A V/sh

HOPPITY GOES TO TOWN *** Uc
Dave Fleischer USA 1941
Voices of: Kenny Gardner, Gwen Williams, Jack Mercer, Ted Pierce, Mike Meyer, Stan Freed, Pauline Loth

Good-looking cartoon, telling of a village of bugs threatened by the humans and their building work. Grasshopper "Hoppity" takes them on a long Exodus to safety, atop a new apartment block. The thin storyline and forgettable songs (courtesy of Frank Loesser and Hoagy Carmichael) are serious flaws in an otherwise imaginative and lively animation.

Aka: MISTER BUG GOES TO TOWN
ANIM   74 min   VIDrel: BBC L/A   V

HOPSCOTCH *** 15
Ronald Neame USA 1980
Walter Matthau, Glenda Jackson, Ned Beatty, Sam Waterston, Herbert Lom, Douglas Dirkson, David Matthau, George Baker, Ivor Roberts, Jacquelyn Hyde, George Pravda, Lucy Saroyan, Mike Gwilym, Allan Cuthbertson, Terry Driscoll

A seasoned CIA man teaches his idiotic superior a lesson by publishing some highly revealing memoirs. Despite a somewhat thin plot, this is an enjoyable and well made comedy. Co-written by Garfield from his novel.

COM   100 min (ort 104 min)   VIDrel: LUMI/SPEAR L/A V
Boa: novel by Brian Garfield.

HORROR EXPRESS ** 15
Eugenio Martin SPAIN/UK 1972
Christopher Lee, Peter Cushing, Telly Savalas, Silvia Tortosa, Jorge Rigaud, Helga Line, Jorge Rigaud, Alberyo De Mendoza, Angel De Pozo, Julio Pena, Jose Jaspe

A turn-of-the-century horror yarn, with a long-frozen monster being discovered by an anthropology professor who thinks he has found the missing link. He loads his discovery onto a train to take it from Asia to the West, thus giving it a chance to thaw out and make use of its ability to turn humans into zombies, in this standard horror tale.

Aka: PANIC IN THE TRANS-SIBERIAN TRAIN; PANIC ON THE TRANS-SIBERIAN TRAIN; PANICO EN EL TRANSSIBERIANO
HOR   83 min (ort 88 min)   VIDrel: FABFIL/SPEAR   V

HORROR HOSPITAL ** 18
Anthony Balch UK 1975
Michael Gough, Robin Askwith, Vanessa Shaw, Skip Martin, Ellen Pollock, Dennis Price, Kurt Christian, Kenneth Benda, Barbara Wendy, George Herbert

Gough plays a demented surgeon, who runs the title establishment, where he carries out brain surgery of the folk unlucky enough to enter the hospital, his latest victim being a young girl on whom he intends to perform an experimental lobotomy. Fortunately, a visiting songwriter saves her and puts an end to all this. An overblown horror spoof saddled with a weak script,

but worth seeing for its outlandish cast, which includes the inevitable prowling monster.

Aka: COMPUTER KILLERS; DOCTOR BLOODBATH
HOR   89 min (ort 91 min)   VIDrel: VIPCO/SGSVID   V/h

HORROR OF DRACULA, THE *** 15
Terence Fisher UK 1958
Peter Cushing, Christopher Lee, Melissa Stribling, Carol Marsh, Michael Gough, John Van Eyssen, Valerie Gaunt, Miles Malleson, Charles Lloyd-Pack, Janina Faye, Olga Dickie, George Woodbridge, Barbara Archer, George Benson

Well-made reworking of the 1930 film with a good use of colour and an imaginative treatment of the sexual themes of Vampirism. In many ways, the high point of the Hammer horror output.

Aka: DRACULA
HOR   81 min   VIDrel: WHV   V

HORROR OF FRANKENSTEIN, THE * 18
Jimmy Sangster UK 1970
Ralph Bates, Kate O'Mara, Veronica Carlson, Dennis Price, Joan Rice, Bernard Archard, Graham James, Dave Prowse

Frankenstein's son takes over on the death of his father. When not making monsters, he chases women instead. One of the last in the Hammer series and an attempt at black comedy that is certainly laughable if nothing else. This gem was followed by "Frankenstein And The Monster From Hell".

HOR   93 min (ort 95 min)   VIDrel: LUMI/SPEAR L/A   V

HORRORS OF THE BLACK MUSEUM ** 15
Arthur Crabtree UK/USA 1959
Michael Gough, June Cunningham, Graham Curnow, Shirley Ann Field, Geoffrey Keen, Gerald Anderson, John Warwick, Beatrice Varley, Austin Trevor, Malou Pantera, Howard Greene, Dorinda Stevens, Stuart Saunders, Hilda Barry

A crippled journalist, who would like to be a writer of mystery novels, uses drugs and hypnosis to get his assistant to kill in a variety of grisly ways and then report so that he can incorporate the material in his writings. A nasty, little film with the emphasis on gore and sadism.

DRA   78 min (ort 80 min)   VIDrel: LUMI/SPEAR L/A   V

HORS LA VIE *** 15
Maroun Bagdadi BELGIUM/FRANCE/ITALY 1991
Hippolyte Girardot, Rafic Ali Ahmad, Hussein Sbeity, Habib Hammond, Magdi Machmouchi, Hassan Farhat, Hassan Zbib, Nabala Zeitouni, Hamzah Nasrullah, Sami Hawat, Sabrina Leurquin, Roger Assaf, Nidal El Achkar, Fady Abou Khalil

A freelance photographer is taken hostage in Beirut and held for nine months, the intention of his captors being to force the release of a political prisoner. Based on a real-life case, the director wisely opts for a semi-documentary style, and this plus Girardot's fine performance in the central role, do much to make this a most compelling piece. However, the scenes of emotional torment he undergoes as a hostage do not make for easy viewing.

DRA   97 min B/W/Col   VIDrel: VCC/DISC/COLUM L/A V
Boa: book by Roger Auque with Patrick Forestier.

HORSE FOR DANNY, A ** (U)
Dick Lowry USA 1993
Leelee Sobieski, Robert Urich, Ron Brice, Gary Basaraba, Erik Jensen, Ed Bruce, Karen Carlson, Brian Michael, Ed Grady, Randal Patrick, Stuart Greer

A young girl with a burning interest in horses and her trainer uncle fall for a horse in which they see a potential champion. This, however, involves them in a serious confrontation with some powerful and ruthless individuals. Engaging performances and a pleasing script are the strengths of this enjoyable story.

JUV   92 min   SATrel: DISNEY CHANNEL

HORSE SOLDIERS, THE ** PG
John Ford USA 1959
John Wayne, William Holden, Constance Towers, Althea Gibson, Hoot Gibson, Anna Lee, Russell Simpson, Stan Lee, Carleton Young, Basil Ruysdael, Willis Bouchey, Ken Curtis, O.Z. Whitehead, Judson Pratt, Denver Pyle, Hank Worden

Big-budget Civil War Western, about the daring exploits of a troop of Union soldiers led by a pair of bickering officers, who penetrate 300 miles behind Confederate lines to destroy a railroad junction, and thus disrupt the enemy's communication and

transport lines. Based on some true incidents and Ford's only film set in the Civil War period, but this is little more than a sprawling and overlong affair.
WES 115 min (ort 119 min) VIDrel: MGM/WHV V
Boa: novel by Harold Sinclair.

HORSEMAN ON THE ROOF, THE *** 15
Jean-Paul Rappeneau FRANCE 1995
Olivier Martinez, Juliette Binoche, Laura Marioni, Paul Chevillard, Patrick Medioni, Philippe Guegan, Jean-Francois Pages, Richard Sammel, Claudio Amendola, Elizabeth Margoni, Carlos Moreno, Jean-Claude Dumas, Jean-Paul Journot
In 1834, during a French cholera epidemic, a young woman is trapped by quarantine regulations in a small town, where she meets and falls in love with a handsome hussar, an Italian nationalist who is trying to escape the Austrian assassins sent to eliminate him. Sadly, these two never get the chance to consummate their love, even though their feelings are never in doubt. A splendid costume drama, slightly detached and unmoving, but always good to look at.
Aka: LE HUSSARD SUR LE TOIT
DRA 131 min (ort 136 min) wScrn VIDrel: ARTIF/20TH; ENCORE (LV only) V LV
Boa: novel by Jean Giono.

HOSPITAL, THE * 15
Arthur Hiller USA 1971
George C. Scott, Diana Rigg, Barnard Hughes, Nancy Marchard, Richard Dysart, Stephen Elliott, Robert Blossom, Robert Walden, Lenny Baker, Frances Sternhagen, Stockard Channing, Andrew Duncan, Donald Harron, Tresa Hughes
Black medical farce about mishaps at a hospital, that are really the work of a deranged killer, with Scott giving an unappealing performance as a cynical doctor battling chaos and apathy. Supposedly a madcap comedy, but the note of bitterness injected by the central character makes the whole affair rather unrewarding. AA: Story/Screen (Paddy Chayefsky).
COM 103 min VIDrel: MGM/WHV L/A V

HOSTAGE, THE ** 18
Robert Young ARGENTINA/UK 1992
Sam Neill, Talisa Soto, James Fox, Michael Kitchen, Art Malik, Cristina Higueras, Nigel Hastings, Jean Pierre Reguerraz, Elizabeth Garvie, Trevor Banister, Leone Mellinger, Vando Villame, Joyce Baza De Candia, Bill Stewart
Having completed his mission in Argentina, a British agent wants out but finds that his superiors are not willing to let him go. In order to outwit the assassins they have sent to eliminate him, he is obliged to come up with a desperate plan. Quite an aimless and cumbersome affair, with a little tacked on love interest in the shape of Soto, that does nothing for the story.
A/AD 96 min (ort 100 min) mTV
VIDrel: POLY/POLYREC V/sh
Boa: novel No Place To Hide by Ted Albeury.

HOSTAGE FOR A DAY ** PG
John Candy CANADA/USA 1993
John Candy, George Wendt, John Vernon, Robin Duke, Peter Tovokrei, Kathleen Laskey, Don Lake, Frank Moore, Currie Graham, John Hemphill, Walter Alza, Monika Deol, Christopher Templeton, Rick Bennett, Vic Cummings, Dan Gallagher
A man with a shrewish wife meets an old flame and runs away, but when his wife empties a secret bank account, he resorts to some desperate measure to get the money back, generating ample confusion all round. A good cast do their best with some rather thin material.
COM 86 min (ort 96 min) mTV VIDrel: MARQ/QUANT V

HOSTILE ADVANCES *** 15
Allan Kroeker USA 1996
Rena Sofer, Don Francks, Karl Furner, Richard Fitzpatrick, Sherry Miller, Bernard Behrens, Luynne Cormack, Leon Pownall, Ron Hartman, Geoffrey Bowes, Ray Paisley, Briar Boacke, Edward Jaunz, Robert Bidman, Maureen Cassidy
A woman's complaint of sexual harassment is dismissed by the courts and she has to fight a lone battle for justice, her campaign eventually leading to a change in American law regarding workplace sexual harassment. A depressing and cogent tale, the actors giving substance to what is in fact a true story.
Aka: HOSTILE ADVANCES: THE KERRY ELLISON STORY
DRA 89 min mTV VIDrel: ODY/SONOP V/sh

HOSTILE FORCE ** (18)
Michael Kennedy GERMANY/USA 1996
Andrew McCarthy, Cynthia Geary, Wolf Larson, Cali Timmins, Hannes Jaenicke, Brent Stait, Peter Hanlon, Sean Milliken, Aaron Smolinski, Andrew Wheeler, Colin Cunningham, Ken Camroux, Jason Gray-Stanford, Nick James, Reese McBeth
After her policeman fiance is killed during a supermarket robbery, a woman police officer leaves the force and takes a job with a hi-tech security firm. However, a quiet weekend shift becomes deadly serious when she and her colleagues are held hostage by a gang intending to rob one of their clients. This tense thriller starts off extremely well, but its action-style ending and cliched portrayal of the criminals as unbalanced psychos strike false notes.
THR 95 min SATrel: SKY MOVIES

HOSTILE HOSTAGES ** 15
Ted Demme USA 1994
Denis Leary, Judy Davis, Kevin Spacey, Robert J. Steinmiller Jr, Glynis Johns, Raymond J. Barry, Richard Bright, Christine Baranski, Adam LeFevre, Phillip Nicoll, Ellie Raab, Bill Raymond, John Scurti, Jim Turner, Ratanya Alda
When a robbery goes wrong, a cat burglar finds himself holding a family hostage on the eve of Christmas Eve. But husband and wife are less concerned with this then their forthcoming divorce, and as their incessant bickering gets out of hand the burglar realises he has taken on more than he can handle. Quite an amusing black comedy for the first half hour or so, though it does all get rather repetitive. Screenplay is by Richard La Gravenese and Marie Weiss.
Aka: REF, THE
COM 92 min (ort 97 min) VIDrel: TOUCH/BUENA L/A V

HOSTILE INTENTIONS ** 18
Catherine Cyran USA 1995
Tia Carrere, Lisa Dean Ryan, Tricia Leigh Fisher, Carlos Gomez, Roman Cisneros, Luis Antonio Ramos, Ramon Franco
Three American girls cross the border into Mexico for a weekend party there but on arrival, they find many of the other guests distinctly unfriendly and a good deal worse the wear from drugs. Soon their true nature becomes evident, and the death of two police is the result, placing our girls in a most dangerous situation as they attempt to flee back to the States.
A/AD 86 min (ort 90 min) VIDrel: MED/DISC V/sh

HOT IN THE CITY ** 18
Tina Marie USA 1991
Christy Canyon, Sasha Strange, Stacy Lords, Arcie Miller, Andrea, Billie Dee, Peter North, Tom Byron, Johnny Ace
Canyon plays the wife of a farmer who is attending to the needs of just about every local woman, whilst ignoring his wife's. She leaves him, and hitches a lift with a biker for the big city, where she has decided to look up an old friend. Having had a good taste of freedom, she phones her husband and he begs her to return. But she can hear that he has a girl with him and understandably, decides to stay in town.
A 45 min Cut (9 min 7 sec) VIDrel: MIA/DISC/IMPENT V

HOT PURSUIT ** 15
Steven Lisberger USA 1987
John Cusack, Robert Loggia, Wendy Gazelle, Jerry Stiller, Monte Markham, Shelley Fabares, Terrence Coope, Dah-Ve Chodan, Keith David, Paul Bates, Miguel angel Fuentes, Ursaline Bryant, Carlos Horcasitas, Martin Lasalle, Ted White
Owing to his bad grades, a student is prevented from going on vacation with his girlfriend and her wealthy family. A last minute reprieve from one of his teachers enables him to set off in pursuit. Once he arrives at their tropical island holiday resort, he encounters some desperate hijackers and faces other perils, in this foolish formula comedy.
COM 90 min (ort 93 min) VIDrel: FIRST/SONOP L/A V

HOT ROCK, THE *** PG
Peter Yates USA 1972
Robert Redford, George Segal, Zero Mostel, Ron Leibman, Paul Sand, Moses Gunn, William Redfield, Charlotte Rae, Topo Swope, Graham P. Jarvis, Harry Belalver, Seth Allen, Robert Levine, Lee Wallace, Robert Weil, Lynne Gordon
Four inept crooks plan to steal a valuable diamond from a museum, but commit a series of blunders every step of the way in this engaging crime caper. Most of the film is taken up, not

with the robbery, but with their bungling attempts to get back the gem after having lost it. Screenplay is by William Goldman. Followed by BANK SHOT.
Aka: HOW TO STEAL A DIAMOND IN FOUR UNEASY LESSONS
COM   97 min (ort 105 min)   VIDrel: 20TH/TECH L/A   V
Boa: novel by Donald Westlake.

HOT SHOTS! *
Jim Abrahams USA
Charlie Sheen, Lloyd Bridges, Cary Elwes, Valeria Golino, Jon Cryer, Kevin Dunn, William O'Leary, Kristy Swanson, Efrem Zimbalist Jr, Bill Irwin, Heidi Swedberg, Brue A. Young, Rino Thunder, Mark Arnott, Ryan Cutrona, Don Lake
A naval pilot who wrecked a $30,000,000 aircraft and subsequently went to live on an Indian reservation is recruited to join a special force training for an attack on Iraq's nuclear facilities. A wildly unfunny spoof on TOP GUN and other films of that kind that parodies almost everything in sight. The humour is so strained and the gags so contrived that very few laughs result, despite being made by the team that gave us AIRPLANE! and THE NAKED GUN. A sequel followed.
COM   81 min (ort 85 min)   VIDrel: 20TH/TECH L/A   V

PG
1991

HOT SHOTS! PART DEUX *
Jim Abrahams USA
Charlie Sheen, Lloyd Bridges, Valeria Golino, Brenda Bakke, Richard Crenna, Miguel Ferrer, Rowan Atkinson, Jerry Haleva, Ryan Stiles, Michael Colyar, Bob Vila, Gregory Sierra, Andrew Katsuals, Clyde Kusatsu, David Wohl, Ryan Stils
Sequel to the first film sees Admiral Benson being elected President and sending Charlie Sheen to mount a Rambo style raid on Saddam Hussein. Very much the same style as before with wild spoofs on the "Rambo" films and APOCALYPSE NOW that offer very few laughs.
COM   83 min (ort 90 min)   VIDrel: 20TH/TECH L/A   V

PG
1993

HOT SPOT, THE ***
Dennis Hopper USA
Don Johnson, Virginia Madsen, Jennifer Connelly, Charles Martin Smith, William Sadler, Jerry Hardin, Barry Corbin, Leon Rippy, Jack Nance, Virgil Frye, John Hawker, Margaret Bowman, Debra Cole, Karen Culley, Cody Haynes
A womanising drifter comes to a small Texas town and lands himself a job as a car salesman. His arrival is greeted by the locals who are at a loss to understand why anyone should choose to settle there. Meanwhile, while his true plans are gradually revealed, he settles for a little romance with both the wife of his boss and his secretary. A languid but atmospheric film with a complex plot of blackmail, murder and robbery. Music is by Jack Nitzsche.
THR   124 min (Cut at UK cinema release by 4 sec – ort 130 min)   VIDrel: COLUM/SONOP L/A   V
Boa: novel Hell Hath No Fury by Charles Williams.

18
1990

HOT, THE COOL AND THE VICIOUS, THE **
Lee Tso Nam HONG KONG
Wang Tao, Tommy Lee, Delon Tan Tao-Liang, Sun Chia-Lin
Three martial arts fighters meet an inn and decide to join forces in a fight against corruption. There's nothing new to be found here, in a tired rehash of stock figures and situations.
MAR   89 min wScrn   VIDrel: EAST/DISC   V

15
197-

HOTEL DU LAC ***
Giles Foster UK
Anna Massey, Denholm Elliott, Julia McKenzie, Googie Withers, Patricia Hodge, Barry Foster, Irene Handl
After narrowly avoiding an unsuitable marriage, a woman novelist retreats to an elegant Swiss hotel in the off-season, where she ponders her life whilst carefully observing her fellow guests. A nicely paced and detailed adaptation of a Booker Prize-winning novel, filmed amidst the lakes and mountains of Switzerland.
DRA   75 min mTV   VIDrel: BBC L/A   V/h
Boa: novel by Anita Brookner.

PG
1986

HOTEL NEW HAMPSHIRE, THE **
Tony Richardson USA
Jodie Foster, Beau Bridges, Rob Lowe, Nastassja Kinski, Wilford Brimley, Paul McCrane, Jennie Dundas, Dorsey Wright, Matthew Modine, Amanda Plummer, Wallace Shawn, Anita Morris, Lisa Barnes, Wall Aspell, Joely Richardson
An American moves to Bavaria to manage a hotel, and becomes involved in a series of bizarre adventures. A long and disor-

18
1984

ganised adaptation of Irving's semi-comic work, that makes a few interesting observations on life, and boasts a number of good performances. Unfortunately it all wears rather thin about halfway through.
DRA   104 min (ort 110 min)   VIDrel: WHV; IMAG   V/sur
Boa: novel by John Irving.

HOTEL ROOM **
David Lynch/James Signorelli USA
Harry Dean Stanton, Glenne Headly, Griffin Dunne, Chelsea Field, Crispin Glover, Deborah Unger, Mariska Hargitay, Grifin Dunne, Freddie Jones, Alicia Witt, Clark Heathcliffe Brolly, Camilla Overbye Roos
A room in a dingy New York hotel is the setting for this trilogy of stories set in different years, i.e., 1936, 1969 and 1992, respectively. In "Blackout" a husband tries to cope with his wife's insanity. "Tricks" describes a strange encounter between a hooker, her john and her talkative husband, while "Getting Rid Of Robert" sees three girlfriends discussing how one of them should go about dumping her boyfriend. A passable comedy-drama but by no means memorable.
DRA   100 min mCab   VIDrel: SCRN/TERRY L/A   V

18
1992

HOTEL SORRENTO **
Richard Franklin AUSTRALIA
Caroline Goodall, Caroline Gillmer, Tara Morice, Joan Plowright, Nicholas Bell, Ray Barrett, Ben Thomas, John Hargreaves, Dave Barnett, Peter O'Callaghan, Jane Edmanson, Bill Howie, Sam Newman, Shane Healy, Phillip Lee
An Australian author living in London has her latest novel on the Booker Prize shortlist, and while she awaits the outcome, takes a trip back home for a family reunion, and while there enjoys much discussion with a local journalist over the nature of the Australian psyche and its cultural values. A detailed and overly verbose film, it soon betrays its stage origins, and though replete with sharp observations, never really comes to life as a movie.
DRA   107 min (ort 112 min)   VIDrel: POLFIL   V
Boa: play by Hannie Rayson.

15
1994

HOTHOUSE ROSE **
Henri Pachard USA
Jeanna Fine, Victoria Paris, John Dough, Nina Hartley, Casey Williams, Tiarra, Sikki Nixx, T.T. Boy, Randy West, Mark Wallice, F.J. Lincoln, Henri Pachard, Trixie Tyler, Holly Ryder, Tom Chapman, E.Z. Ryder, Wayne Summers
This one opens with a woman finding herself unable to satisfy her lover, whose fascination with the eponymous "Hothouse Rose" results in the woman developing an obsession of her own about her. Despite visits to a psychiatrist, this fixation develops to the point where she tries to take over the woman's identity, leading to a sequence of sexual episodes in which she imagines herself to be Rose. A bizarre sex film, it ends with her arrest. A sequel followed.
A   54 min (Cut before video submission by 16 min 18 sec)
VIDrel: GROHOM/MAXSCAN   V

18
1992

HOTHOUSE ROSE 2 **
Henri Pachard USA
Jeanna Fine, Victoria Paris, John Dough, Nina Hartley, Buck Adams, Trixy Tyler, Candace Heart, Tom Byron, Jordon Smith, Holly Ryder
This sequel to the first film opens with Fine telling her shrink all about her previous sexual encounters, which are then seen as flashbacks. The slight plot seems to revolve around the attempts she has been making to get into porno movies, while all the while her idol Rose is trying to get out of them. An odd, albeit well acted film, saddled with a plot that is inconsistent and almost impossible to follow.
A   60 min (ort 70 min)   VIDrel: GROHOM/MAXSCAN   V

18
1992

HOUND OF THE BASKERVILLES, THE ***
Terence Fisher UK
Peter Cushing, Christopher Lee, Andre Morell, Maria Landi, Miles Malleson, David Oxley, Francis De Wolff, Ewen Solon, John Le Mesurier, Sam Kydd, Helen Goss, Judi Moyens, Dave Birks, Michael Mulcaster, Michael Hawkins, Ian Hewitson
Hammer-style remake of the 1939 film version of the Sherlock Holmes tale, telling of a supernatural dog said to bring death to an aristocratic family. A fairly decent remake of this much-filmed classic, spoilt by the usual clumsy touches that generally find their way into Hammer films. Despite this, Cushing is well cast and the film doesn't lack for atmosphere.
HOR   86 min   VIDrel: L/A   V
Boa: novella by Arthur Conan Doyle.

PG
1959

HOUR OF THE PIG, THE *** 15
Leslie Megahey FRANCE/UK 1993
Colin Firth, Ian Holm, Donald Pleasence, Amina Annabi, Nicol Williamson, Michael Gough, Harriett Walker, Jim Carter, Lysette Anthony, Justin Chadwick, Sophie Dix, Michael Cronin, Elizabeth Spriggs, Emil Wolk, Vincent Grass
In a remote part of medieval France, a pig is accused of the murder of a young Jewish boy, and an idealistic lawyer sets out to prove that the animal is innocent and at the same time expose the real culprits, who have engineered this travesty to escape retribution. Inspired by the real trials that took place at that time, the film attempts to examine the nature of hypocrisy and falsehood, a task it sets about with wit and skill. The script is by Megahey.
DRA 107 min (ort 117 min) VIDrel: CURZON/20TH
V/sur

HOURGLASS ** 18
C. Thomas Howell USA 1995
C. Thomas Howell, Sofia Shinas, Kiefer Sutherland, Ed Begley Jr, Terry Kaiser, Timothy Bottoms, Anthony Clark
A major figure in the fashion industry falls head over heels for a very sexy woman but finds that this encounter may cost him both his livelihood and his life. A standard erotic thriller that was released directly to video, it has the feel of a sleazy exploitation pic, plus an irritating voice-over for good measure. A few comic touches in the dialogue make one wonder whether or not it was meant to be a spoof.
THR 88 min (ort 91 min) VIDrel: 20VIS/SONOP V

HOURS AND TIMES, THE *** 18
Christopher Munch USA 1991
David Angus, Ian Hart, Stephanie Pack, Robin McDonald, Sergio Martino, Unity Greenwood
Fictionalised drama built around the idea of what might have occurred in the Spring of 1963 when Brian Epstein took Beatles member John Lennon to Barcelona for a four-day break. A piece of speculative film-making that entertains thanks to its witty script, much of which consists of a dialogue between the two men, who in terms of background and character could hardly be more different.
Aka: HOURS AND THE TIMES, THE
DRA 54 min (ort 60 min) VIDrel: ICAPRO/MANGA V

HOUSE * 15
Steve Miner USA 1985
William Katt, Kay Lenz, George Wendt, Richard Moll, Mary Stavin, Susan French, Erik Silver, Michael Ensign, Mark Silver, Alan Autry, Michael Ensign, Steven Williams, Ronn Carroll, Jim Calvert, Mindy Sterling, Jayson Kane
A horror film with comic touches. After his divorce a writer moves to the house where his aunt killed herself and finds himself plagued by nightmarish fantasies. A messy and disjointed film that works as neither comedy nor horror. Followed by the inevitable sequels.
HOR 93 min VIDrel: POLY/POLYREC L/A V

HOUSE 2: THE SECOND STORY ** 15
Ethan Wiley USA 1987
Arye Gross, Lar Park Lincoln, Jonathan Stark, Royal Dano, John Ratzenburger, Devin Devasquez, Jayne Modean, Ronn Carroll, Dean Cleverdon, Doug McHugh, Bill Maher, Amy Yasbeck, Gregory Walcott, Dwier Brown, Lenora May
As much a comedy as a horror film, this bizarre sequel has a man moving into the house where his parents were murdered years ago. He digs up the coffin of his mischievous 170-year-old great-grandfather, and together they embark on a series of rather strange adventures. Followed by another film in the HOUSE series.
HOR 84 min (ort 94 min) VIDrel: 4-FRONT/POLYREC L/A V

HOUSE 3: THE HORROR SHOW ** 18
Jim Isaac USA 1989
Lance Henriksen, Rita Taggart, Brion James, DeeDee Pfeiffer, Aron Eisenberg
Another weird in-name-only sequel to HOUSE, with a mass murderer finally being sent to the electric chair after being hunted down by an intrepid detective. However, his death catapults him into an unearthly plane of existence where he still has scope for more evil. Occasionally chilling but generally just rather gruesome.
HOR 91 min VIDrel: POLY/POLYREC/BRAVE L/A V

HOUSE 4 * 15
Lewis Abernathy USA 1991
Terri Treas, Ned Romero, Scott Burkholder, Melissa Clayton, William Katt, Danny Dillon, Dabbs Greer, Ned Bellamy, John Santucci, Mark Gash, Paul Keith, Kevin Goetz, Judith Jordan, Steve Vinovich, Carolyn Mignini, Rebecca Rocheford
A crippled girl and her mother move into a derelict old house and soon enough, find the inevitable sequence of nasty happenings taking place. The fourth film in this series is a terribly weak horror-comedy, with minimal gore and little in the way of special effects. A unutterably dull work of very limited appeal.
Aka: HOUSE 4: HOME DEADLY HOME; HOUSE 4: THE REPOSSESSION
HOR 90 min Cut (7 sec – ort 94 min)
VIDrel: MED/POLYREC L/A V/sh

HOUSE BY THE CEMETERY, THE ** 18
Lucio Fulci ITALY 1981
Katherine McColl, Paolo Malco, Giovanni Frezza, Giovanni De Nava, Dagmar Lassander, Anja Pierani, Silvie Collatina, Daniele Doria, Carlo De Mejo
A family moves into a house once owned by a certain Dr Freudstein, and finds that he is alive and well and living in the cellar, even though he is by now about 150 years old. Buckets of blood and other horrors abound in this incredibly over-the-top Italian offering. Cut before video submission by 34 seconds. Unlikely to be available in an uncut form in the UK.
Aka: HOUSE OUTSIDE THE CEMETERY, THE; QUELLA VILLA ACCANTO AL CIMITERO
HOR 75 min Cut (4 min 11 sec plus film cuts – ort 89 min)
VIDrel: VIPCO/SGSVID V/h

HOUSE IN NIGHTMARE PARK, THE *** 15
Peter Sykes UK 1973
Frankie Howerd, Ray Milland, Kenneth Griffith, Hugh Burden, Rosalie Crutchley, John Bennett, Ruth Dunning, Elizabeth MacLennan, Aimee Delamain, Peter Munt
In 1907, a family of greedy relatives are out to steal a fortune, and invite a ham actor to perform at the title venue. But he is in fact the dead man's long lost son and therefore has a claim to the man's fortune, a fact that gives his invitation a good deal more significance. Something of a homage to "The Cat And The Canary", replete with eerie atmosphere, sinister goings-on and sufficient gags to keep us watching.
COM 92 min (ort 95 min) VIDrel: LUMI/SPEAR L/A V

HOUSE IN THE HILLS, A *** 18
Ken Wiederhorn USA 1992
Michael Madsen, Helen Slater, James Laurenson, Elyssa Davalos, Jeffrey Tambor, Taylor Lee, Toni Barry, James Noellenhoff, Margaret Parke, LaRue M. Mall, Vincent Eaton, Sally Daykin, Pieter Riemens, Brenda Colom, Ron La Paz
An aspiring young actress agrees to house-sit and gets herself inadvertently entangled with an ex-con who is seeking revenge against the owner who had his brother killed and him framed. A well-acted cat-and-mouse psychological thriller with an inventive plot and plenty of twists, while Slater is both appealing and highly convincing as our would-be performer.
THR 86 min (ort 89 min) VIDrel: WHV V

HOUSE OF ANGELS *** 15
Colin Nutley SWEDEN 1992
Helena Bergstrom, Rikard Wolff, Sven Wollter, Ernst Gunther, Viveka Seldahl, Reine Brynolfsson, Per Oscarsson, Ing-Marie Carlsson, Peter Andersson, Johannes Brost, Jan Mybrand, Jakob Eklund, Tord Petterson, Gorel Crona, Gabriella Boris
A woman inherits a country house when one of her relatives dies, and moves in together with a motley group of assorted friends. But her liberated manners and unconventional companions prove quite a culture shock for the conservative locals, but both sides eventually reach some kind of understanding. A massive hit in Sweden, this is a long and episodic tale, but succeeds largely thanks to the fine work of an accomplished cast. A sequel followed.
Aka: ANGLAGARD
DRA 126 min (ort 119 min) VIDrel: CURZON/20TH
V/sh

HOUSE OF CARDS *** 15
Paul Seed UK 1991
Ian Richardson, Susannah Harker, Diane Fletcher, David Lyon, Isabelle Amyes, Miles Anderson, Alphonsia Emmanuel, Nicholas Selby, Malcolm Tierney, Damien Thomas, Christopher Owen, Kenneth Gilbert, James Villiers, Kenny Ireland

A cynical and very wry look at British party politics and infighting in the near future, with a ruthlessly manipulative Conservative politician doing his level best to bring about the downfall of the new Prime Minister, in pursuit of his own ambitions. An intricately plotted and totally absorbing drama, with a tour-de-force performance from Richardson. Followed by TO PLAY THE KING.
DRA   181 min (2 cassettes) mTV   VIDrel: BBC/TECH V/sh
Boa: novel by Michael Dobbs.

HOUSE OF CARDS **
Michael Lessaic USA
15
1992
Kathleen Turner, Tommy Lee Jones, Asha Menina, Shiloh Strong, Ester Rolle, Park Overall, Michael Horse, Anne Pitoniak, Joaquim Martinez, Jacqueline Cassel, John Henderson, Craig Fuller, Rick Marshall, Reuben Valiquette Murray
After her father is killed in accident at a Mayan archaeological site, a six-year girl stops speaking. Upon their return to the USA, her mother consults a psychiatrist for advice (which she does not always take). Meanwhile, her daughter amuses herself by constructing an elaborate house of cards. A long and quite pretentious effort that never overcomes the weaknesses of its plot.
DRA   109 min   VIDrel: MIA/DISC   V/sur

HOUSE OF ELLIOT, THE: VOLS. 1 TO 6 ***
Rodney Bennett UK
12/PG/U
1991
Stella Gonet, Louise Lombard, Barbara Jefford, Aden Gillett, Peter Birch, Jeremy Brudnell, Colin Jeavons, Cathy Murray, Francesca Folan, Kelly Hunter, Robert Daws, Burt Kwouk, Jill Melford, Anne Lambton, Clare Bryan-Shaw
After their father dies, two sisters find themselves completely impoverished but improve their fortunes by founding a small dressmaking business that eventually grows to be an influential fashion house. Set in London of the 1920s, this well mounted series benefits from strong performances and a painstaking attention to detail. A few contrived touches of melodrama are no more than one would expect to find in any extended TV drama. Originally shown in twelve parts.
DRA   948 min (6 cassettes – certifications vary) mTV
VIDrel: BBC/TECH   V/h

HOUSE OF GAMES ***
David Mamet USA
15
1987
Lindsay Crouse, Joe Mantegna, Lilia Skala, Steve Goldstein, Mike Nussbaum, J.T. Walsh, Willo Hauseman, Karen Kolhaas, Jack Wallace, Ricky Jay, G. Roy Levin, Bob Lumbra, Andy Potok, Allen Soule, Ben Blakeman, Scott Zigler
Playwright David Mamet makes his directorial debut in this tale of a successful psychiatrist who, in the course of treating a gambler, meets a con artist, and finds herself so fascinated by his world of deception that she is drawn into it far more deeply than was her intention. Mantegna gives one of his best performances in an elegant and absorbing study, that is only slightly let down by a flat climax and Crouse's lack of conviction.
THR   98 min (ort 102 min)   VIDrel: CONNO/RTM L/A V/s

HOUSE OF GOD, THE **
Donald Wrye USA
(12)
1979
Tim Matheson, Bess Armstrong, Charles Haid, Michael Sacks, Lisa Pilikan, George Coe, Ossie Davis, Howard Rollins Jr, James Cromwell, Sandra Bernhard
Black comedy detailing the exploits of some young intern in hospital and their various attempts to deal with the normal pressures of hospital life. Scripted by Wrye from the novel, this weak comedy misfire has one or two mildly diverting moments but in general does nothing with a fine cast, most of whom deservedly went on to better things. The film was never released theatrically.
COM   104 min (ort 108 min)   SATrel: SKY MOVIES
Boa: novel by Samuel Shem.

HOUSE OF SECRETS **
Mimi Leder USA
15
1994
Bruce Boxleitner, Melissa Gilbert, Kate Vernon, Michael Boatman, Cicely Tyson, John Henry Scott, Tayor Simpson, Sally Birdsong, Silas Cooper, Joseph Chrest, Michael Audley, Kris Shaw, Graham Timbes, Stocker Fontelleu, Elaine West
Based on same book as LES DIABOLIQUES, this tells of two women who have both suffered at the hands of the same man plotting his demise. One is the man's wife, who has inherited a

sanatorium, the other is the physiotherapist who works there, and has been blackmailed into having an affair in exchange for silence over her criminal past. A diverting little work, nothing like as good as the French classic, but perfectly watchable.
THR   88 min   VIDrel: ODY/SONOP   V/sh
Boa: novel Celle Qui N'Etait Plus by Pierre Boileau and Thomas Narcejac.

HOUSE OF SLEEPING BEAUTIES **
Paul Thomas USA
18
1992
Jamie Summers, Madison, Savannah, Flame, Francesca Le, Marc Wallace, Tom Byron, Steve Drake, Tom Chapman
A painter lives in the mountains in a house together with his wife and a group of sundry hangers-on. One day he seems to discover title building in the woods where he spies a woman with whom he gradually becomes obsessed. An oblique and pretentious work, with many dream-like sequences, that was released in the American hard-core version in two parts.
A   68 min   VIDrel: VIVID/SCRN   V

HOUSE OF SPIRITS, THE **
Bille August DENMARK/GERMANY/PORTUGAL/USA1993
15
Meryl Streep, Glenn Close, Jeremy Irons, Winona Ryder, Antonio Banderas, Armin Mueller-Stahl, Vanessa Redgrave, Maria Conchita Alonso, Vincent Gallo, Teri Polo, Miriam Colon, Oscacar A. Colon, Franco Diogene, Jane Gray, Sasha Hanan
Badly flawed, ponderous and seriously overlong adaptation of Allende's novel about fifty years in the lives of a powerful South American family and the fate of its various members. The elements of "magic realism" present in the original novel are badly realised and the whole effort is hardly improved by miscasting for some of the key roles and a badly stilted script.
Aka: HOUSE OF THE SPIRITS, THE
DRA   132 min (ort 145 min)   VIDrel: EIV/SONOP   V
Boa: novel by Isabel Allende.

HOUSE OF USHER, THE **
Alan Birkinshaw USA
18
1988
Oliver Reed, Donald Pleasence, Rufus Swart, Romy Windsor, Norman Coombes, Anne Stradi, Carole Farquhar, Philip Godewa, Leonorah Ince, Jonathan Fairbirn
An updated version of Poe's classic tale that's set in England in modern times, where Ryan Usher and his fiancee have been invited to stay at the family home by an uncle. However, the uncle has designs on the girl, for as the last of his line he badly wants to produce an heir. A low-budget film, well photographed but devoid of the atmosphere so necessary to give substance to the work of Poe. Filmed in South Africa.
HOR   87 min (ort 98 min)   VIDrel: ENTUK L/A   V
Boa: short story The Fall of the House Of Usher by Edgar Allan Poe.

HOUSE OF WAX **
Andre De Toth USA
PG
1953
Vincent Price, Frank Lovejoy, Phyllis Kirk, Carolyn Jones, Paul Cavanagh, Charles Buchinski (Bronson), Paul Picerni, Roy Roberts, Dabbs Greer, Angela Clarke, Reggie Rymal, Philip Tonge, Darwin Greenfield, Jack Kenney
A sculptor disfigured in a fire at his wax museum takes a grisly revenge on those around him, rebuilding his damaged showplace with the bodies of his victims. A remake of "The Mystery Of The Wax Museum", that's packed with sudden shocks (mainly because it was an early 3-D film) but is hampered by poor narrative development.
HOR   84 min (ort 88 min) wScrn   VIDrel: WHV   V/sh

HOUSE OF WHIPCORD **
Pete Walker UK
18
1974
Barbara Markham, Patrick Barr, Ray Brooks, Penny Irving, Anne Michelle, Sheila Keith, Dorothy Gordon, Robert Tayman, Ivor Salter, Judy Robinson, Karen David, Jane Howard, Celia Quicke, David McGillivray, Tony Sympson
A senile and vicious old judge is so appalled by the decline in moral standards that he sets up a private house of correction for young girls. The tale follows the ordeal of one such girl who is imprisoned after accepting an invitation on the part of her boyfriend to meet his parents. A polished sexploitation film that dissipates its power with gratuitous titillation and an implausible climax.
HOR   98 min (ort 100 min)   VIDrel: REDEM/RTM   V

**HOUSE ON CARROLL STREET ** **   PG
Peter Yates USA   1987
*Kelly McGillis, Jeff Daniels, Mandy Patinkin, Jessica Tandy,
Christopher Rhode, Jonathan Hogan, Remak Ramsey, Ken Walsh,
Randal Mell, Michael Flanagan, Paul Sparer, Brian Davies, Mary
Diveny, Bill Moor, Cliff Cudney*
A woman editor on Life magazine, sacked when she refuses to
take part in the anti-Communist witch-hunt of the 1950s, inad-
vertently stumbles across a sinister government-inspired
conspiracy to smuggle ex-Nazis into the USA. A detailed film
whose plot improbabilities weigh heavily against it.
A/AD   97 min (ort 111 min)   VIDrel: VISVID/POLYREC
V

HOUSE PARTY *   15
Reginald Hudlin USA   1990
*Christopher Reid, Robin Harris, Tisha Campbell, Martin Lawrence,
A.J. Johnson, Christopher Martin, Paul Anthony, John Witherspoon,
Barry Diamond, Michael Pniewski, George Clinton, Kelly Jo Minter*
A zestful comedy-musical built around a party held in a house
in a black neighbourhood that will be enjoyed by fans of rap
music and for its keen sense of fun. Music is by Marcus Miller
and Lenny White and features: Full Force, Kid 'n' Play
(Christopher Reid and Christopher Martin), L.L. Cool J, Flavor
Full and others. Writer-director Hudlin's directorial debut.
COM   99 min (ort 101 min)   VIDrel: COLUM/SONOP L/A
V/sh

HOUSE PARTY 2 *   15
George Jackson/Doug McHenry USA   1991
*Christopher Reid, Christopher Martin, Martin Lawrence, Tisha
Campbell, Queen Latifah, Iman, Georg Stanford Brown, Kamron, B-
Fine, Helen Martin, William Schallert, Tont Burton, Louie Louie, D.
Christopher Judge, Daryl M. Mitchell*
Inferior sequel with Kid going to college but finding himself in
trouble when an administrative slip confuses him with another
student and his tuition fee gets lost in the process. An episodic
effort that never really comes together despite the efforts of its
young and talented cast.
Aka: HOUSE PARTY 2: THE PAJAMA JAM
MUS   90 min (ort 94 min)   Vidrel: FIRST/SONOP   V/sur

HOUSE PARTY 3 *   15
Eric Meza USA   1993
*Kid 'N' Play, Bernie Mac, Gilbert Gottfried, Angela Means, TKC,
Immature, Free Luve, Khandi Alexander, Ketty Lester, David
Edwards, Anthony Johnson, Michael Colyar, Tisha Campbell, Ron
Fegan, Daniel Gardner, Eddie Griffin*
Kid finally decides to tie the knot but finds his attention also
taken up with various other problems include attempts to sign
an up-and-coming rap group and conflicts with a promoter.
Fully of wild and undirected energy but rather unconcentrated
and unappealing.
MUS   93 min   VIDrel: EIV/SONOP   V/sur

HOUSE THAT MARY BOUGHT, THE *   (15)
Simon MacCorkindale FRANCE/USA   1995
*Susan George, Ben Cross, Maurice Thorgood, Charlotte Valandrey,
Jean-Paul Muel, Vernon Dobtcheff, Chantal Grisset, Francois Trehard,
Michele Kerhoas, Alan Francoise, David Valentine, Anne Marie
Berny, Sophie Langevin, Serge Ruest*
Soon after they move into their new home in Brittany, strange
events start to occur, which make the wife suspect her husband
of having an affair. Meanwhile, he thinks his wife is having a
breakdown, but all is made clear when they learn that the place
is haunted.
COM   102 min   SATrel: MOVIE CHANNEL
Boa: novel Odd's End by Tim Wynne Jones.

HOUSE WHERE DEATH LIVES, THE *   15
Alan Beattie USA   1980
*Joseph Cotten, Patricia Pearcy, David Hayward, John Dukakis, Leon
Charles, Alice Nunn, Patrick Pankhurst, Simone Griffeth, Louis Basill,
Abraham Alvarez, James Purcell, Shelby Leverington*
A young woman takes up a post as nurse to an elderly invalid,
but becomes emotionally involved with his sixteen-year-old
grandson, and shortly after her arrival a series of murders begins.
Aka: DELUSION
HOR   79 min (ort 93 min)   VIDrel: VIPCO/SGSVID   V

HOUSEBOAT *   U
Melville Shavelson USA   1958
Cary Grant, Sophia Loren, Martha Hyer, Harry Guardino, Eduardo

*Ciannelli, Murray Hamilton, Mimi Gibson, Paul Petersen, Charles
Herbert, Madge Kennedy, John Litel, Werner Klemperer, Peggy
Connelly, Kathleen Freeman, Helen Brown*
A socialite becomes housekeeper to a widow with three children, and takes both him and the kids in hand. An amiable
romantic comedy that makes up in easy charm what it lacks in
inventiveness. Guardino has a very funny role as the houseboat
handyman. Written by Shavelson and Jack Rose.
COM   105 min (ort 110 min)   VIDrel: CIC/SONOP   V/h

HOUSEGUEST *   PG
Randall Miller USA   1995
*Sinbad, Phil Hartman, Jeffrey Jones, Kim Griest, Stan Shaw, Tony
Longo, Paul Ben-Victor, Mason Adams, Kim Murphy, Chauncey
Leonard, Leopardi, Talia Seider, Ron Glass*
In trouble with some loan sharks working for organised crime,
a black guy takes refuge with an uptight white family by
pretending to be the childhood buddy of the father. He gradually talks his way into their hearts and helps them loosen up and
enjoy life. Trite and predictable, this dull misfire takes a long
time to go nowhere and is burdened by unfunny situations and
poor editing, though US comedian Sinbad works hard to keep
the film afloat.
COM   105 min (ort 109 min)   VIDrel: HOLPIC/TECH   V

HOUSESITTER *   PG
Frank Oz USA   1992
*Steve Martin, Goldie Hawn, Dana Delaney, Julie Harris, Donald
Moffat, Peter MacNichol, Richard B. Shull, Lauren Cronin, Roy
Cooper, Christopher Durang, Heywood Hale Broun, Cherry Jones,
Vasek Simek, Suzanne Whang, Mary Klug*
In this very uneven comedy Martin plays an architect who
builds a dream home for his childhood sweetheart (it's a
strangely insipid dwelling) but is dumped by her. Enter Hawn
who, giving her usual scatty performance as a compulsive liar,
moves into the house, masquerades as Martin's wife, and turns
his life upside down. But when this prompts fresh interest on
the part of his true love, he decides to play along. Lightweight
froth with some funny moments.
COM   97 min (ort 102 min) cC   VIDrel: CIC/SONOP; PION
(LV only)   V/sur LV

HOW GREEN WAS MY VALLEY *   U
John Ford USA   1941
*Walter Pidgeon, Maureen O'Hara, Roddy McDowall, Donald Crisp,
Sara Allgood, Anna Lee, Barry Fitzgerald, Patric Knowles, Arthur
Shields, John Loder, Morton Lowry, Frederic Worlock, Ann Todd,
Richard Fraser, Evan S. Evans*
A flawed but moving account of the lives of Welsh coalminers
that focuses on the members of a family with six sons, of whom
only the youngest seems destined for a better life. Set around the
turn of the century, when trade unions were far from being
accepted by employers, this is a stilted and sentimental work,
redeemed by splendid performances. AA: Pic, Dir, S. Actor
(Crisp), Cin (Arthur C. Miller), Art/Int (Richard Day/Nathan
Juran and Thomas Little).
DRA   118 min B/W cC   VIDrel: 20TH/TECH   V/sh
Boa: novel by Richard Llewellyn.

HOW I GOT INTO COLLEGE *   15
Savage Steve Holland USA   1989
*Anthony Edwards, Corey Parker, Christopher Rydell, Lara Flynn
Boyle, Finn Carter, Brian Doyle-Murray, Nina Saatchi, Phil
Hartman, Nora Dunn, Gary Owens, Tichina Arnold, Bill Raymond,
Phillip Baker Hall, Nicolas Coster, Charles Rocket*
This juvenile comedy starts off with a promising premise: a
look at the antics ambitious kids use to get into top colleges
and the tricks the colleges employ in turn to get the students
they want. The flimsy plot follows the exploits of a student
who applies to one such college, but only because a girl he is
after is doing so. A few fantasy sequences pep up an otherwise dull effort.
COM   83 min (ort 89 min)   VIDrel: 20TH/TECH   V/sur

HOW I WON THE WAR *   12
Richard Lester UK   1967
*Michael Crawford, John Lennon, Roy Kinnear, Lee Montague, Jack
McGowran, Michael Hordern, Jack Hedley, Karl Micheal Vogler,
Ronald Lacey, Aleaxander Knox, James Cossins, Robert Hardy, Sheila
Hancock, Charles Dyer, Bill Dysart*
A young and inexperienced WW2 officer finds himself in charge
of a bunch of sundry failures in this tiresome and unsuccessful
attempt an anti-war black comedy that is too over-the-top and

undisciplined to work. What humour there is belongs the British "life-is-misery" school and feels forced and unnatural.
COM   106 min (ort 110 min)   VIDrel: MGM/WHV V/h
Boa: novel by Patrick Ryan.

HOW THE WEST WAS WON ***
John Ford/Henry Hathaway/George Marshall USA   1962
Henry Fonda, Gregory Peck, James Stewart, George Peppard, Debbie Reynolds, John Wayne, Carroll Baker, Lee J. Cobb, Spencer Tracy, Karl Malden, Eli Wallach, Carolyn Jones, Richard Widmark, Walter Brennan, Raymond Massey, Agnes Moorehead
An epic story that follows a family as it moves West, encountering many dangers along the way. The film loses much of its impact on TV (it was shown in Cinerama) but still retains considerable force, even if there are tedious moments in between the spectacular set-pieces. The fine score is by Alfred Newman. AA: Story/Screen (James R. Webb), Sound (Franklin E. Milton), Edit (Harold F. Kress).
WES   157 min Cut (2 sec – ort 165 min) wScrn
VIDrel: MGM/WHV V/sh

HOW TO BE A WOMAN AND NOT DIE IN THE ATTEMPT ***
Ana Belen SPAIN   1991
Carmen Maura, Antonio Resines, Juanjo Puigcorve, Carmen Conesa, Tina Sainz, Pasca Casares, Victor Garcia, Olalla Aguirre, Juan Diego Botto, Luis Perez Agua, Asuncion Balaguer, Enriqueta Carballeira, Miguel Rellan, Mercedes Lezcano
Screenplay is by Rico-Godoy, in this story of a woman who has to balance the demands of her job as a journalist with her other responsibilities as both a mother and the wife of an alcoholic. Some very typical anti-male stereotyping (largely drawn from the novel) tends to diminish the impact of this work.
Aka: COMO SER MUJER Y NO MORIR EN EL INTENTO
DRA   85 min (ort 89 min)   VIDrel: CURZON/20TH V/sh
Boa: novel by Carmen Rico-Godoy.

HOW TO GET AHEAD IN ADVERTISING **
Bruce Robinson USA   1988
Richard E. Grant, Rachel Ward, Richard Wilson, Jacqueline Tong, John Shrapnel, Susan Wooldridge, Mick Ford
A bizarre parody on advertising, this sees a young advertising executive beset with doubts regarding his profession during a campaign for a new pimple cream. To add to his troubles a boil erupts on his neck, and develops into a talking head that spews forth a constant stream of abrasive advertising slogans. Written by Robinson, this ponderous satire on marketing has its funny moments, but is generally as tasteless as it is disagreeable.
COM   90 min (ort 94 min)   VIDrel: 20TH/TECH V/sur

HOW TO MAKE AN AMERICAN QUILT ***
Jocelyn Moorhouse USA   1995
Winona Ryder, Anne Bancroft, Ellen Burstyn, Kate Nelligan, Alfre Woodard, Kaelyn Craddick, Sara Craddick, Kate Capshaw, Adam Baldwin, Dermot Mulroney, Maya Angelou, Lois Smith, Jean Simmons, Denis Arnot, Rip Torn, Derrick O'Connor
Members of a rural quilting group tells tales that highlight their experiences from life in this amiable but aimless celebration of womanhood. Tales of passion and tragedy abound and though having little by way of a plot, the movie has a generally pleasing feel and benefits greatly from the care with which it was made.
DRA   112 min (ort 117 min)   VIDrel: CIC/SONOP V
Boa: novel by Whitney Otto.

HOW TO MARRY A MILLIONAIRE **
Jean Negulesco USA   1953
Marilyn Monroe, Betty Grable, Lauren Bacall, William Powell, Rory Calhoun, David Wayne, Alex D'Arcy, Fred Clark, Cameron Mitchell, George Dunn, Harry Carter, Robert Adler, Tudor Owen, Percy Helton, Maurice Marsac, Abney Mott
Three girls rent an expensive apartment in New York and go hunting for millionaires to marry. Monroe does her best in a dullish film that has a few bright moments, but not enough of them. A remake of a 1932 film "The Greeks Had A Word For Them".
COM   92 min (ort 96 min)   VIDrel: 20TH/TECH V

HOW TO MURDER YOUR WIFE ***
Richard Quine USA   1964
Jack Lemmon, Virna Lisi, Terry-Thomas, Eddie Mayehoff, Claire Trevor, Sidney Blackmer, Max Showalter, Jack Albertson, Mary Wickes

A cartoonist gets drunk and wakes up next morning to find himself married to a beautiful Italian girl. As he's a confirmed bachelor he feels honour bound to get rid of her. A delightfully wacky comedy that generally holds up pretty well, though towards the end it runs out of steam. A highlight is the courtroom scene.
COM   114 min (ort 118 min)   VIDrel: WHV V

HOW TO STEAL A MILLION ***
William Wyler USA   1966
Audrey Hepburn, Peter O'Toole, Charles Boyer, Eli Wallach, Hugh Griffith, Fernand Gravey, Marcel Dalio, Jacques Moustache, Roger Treville, Eddie Malin, Bert Bertram, Louise Chevalier, Remy Longa, Gil Delamare
A sophisticated comedy about a million dollar art theft at a Paris museum, with Hepburn and O'Toole working well together and generally overcoming the sluggishness of the script.
COM   118 min (ort 127 min)   VIDrel: 20TH V

HOW TO STUFF A WILD BIKINI *
William Asher USA   1965
Annette Funicello, Dwayne Hickman, Brian Donlevy, Buster Keaton, Mickey Rooney, Harvey Lembeck, Beverly Adams, Jody McCrea, John Ashley, Frankie Avalon
Another one of those dreary "Beach Party" comedies, with an advertising man out to find a typical "girl next door". Followed by THE GHOST IN THE INVISIBLE BIKINI.
Aka: HOW TO FILL A WILD BIKINI
COM   97 min   VIDrel: RCA L/A V

HOW TO SUCCEED IN BUSINESS WITHOUT REALLY TRYING ***
David Swift USA   1967
Robert Morse, Rudy Vallee, Michele Lee, Anthony Teague, Maureen Arthur, Sammy Smith, Murray Matheson, Kay Reynolds, John Myhers, Jeff DeBenning, Ruth Kobart, George Fenneman, Anne Seymour, Erin O'Brien-Moore, Joey Faye
A window cleaner advances to the top of a New York company with the aid of a helpful handbook. An enjoyable and brash musical that moves along at a good pace and has several well-staged musical numbers. Songs (by Frank Loesser) include: "I Believe In You" and "Brotherhood Of Man".
MUS   121 min   VIDrel: MGM/WHV L/A V
Boa: novel by Stephen Mead/musical by Abe Burrows and Frank Loesser.

HOWARD: A NEW BREED OF HERO *
William Huyck USA   1986
Lea Thompson, Tim Robbins, Jeffrey Jones, Paul Guilfoyle, Liz Sagal, Holly Robinson, Dominique Davalos, Tommy Swerdlow, Richard Edson and as Howard: E. Gale/C. Zien/T. Rose/S. Sleap/P. Baird/M. Wells/L. Sturtz/J. Prentice
A duck from another planet mysteriously arrives on Earth and is far from impressed by what he sees, but becomes instrumental in saving the planet from invasion by malevolent creatures from another dimension. An unutterably tedious film of wildly overblown special effects and little else. This noisy and misguided attempt to bring a Steve Gerber's Marvel comic-strip character to the screen had no clear target audience and bored adults and children alike.
Aka: HOWARD THE DUCK
COM   105 min (Cut at UK cinema release – ort 111 min)
VIDrel: 4-FRONT/POLYREC/CIC L/A V/dm

HOWARDS END ***
James Ivory UK   1991
Anthony Hopkins, Emma Thompson, Helena Bonham Carter, Vanessa Redgrave, James Wilby, Samuel West, Prunella Scales, Joseph Bennett, Adrian Ross Magenty, Jo Kendall, Jemma Redgrave, Ian Latimer, Mary Nash, Susie Lindeman, Mark Tandy
From the makers of A PASSAGE TO INDIA comes this story of two very different upper-middle class Edwardian families, and the disagreements that arise when the liberal and open-minded Schlegel sisters become embroiled with the stuffy and self-seeking Wilcoxes. The dying Ruth Wilcox decides to leave her home to her friend Margaret Schlegel, but a conflict of interest results. A fascinating look at a vanished era. AA: Actress (Thompson), Art, Screen/adapt (R.P. Jhabvala).
DRA   136 min (ort 142 min)   VIDrel: CURZON/20TH
V/sur
Boa: novel by E.M. Forster.

HOWARDS OF VIRGINIA, THE ***   U
Frank Lloyd USA   1940
Cary Grant, Martha Scott, Cedric Hardwicke, Alan Marshal, Richard Carlson, Paul Kelly, Irving Bacon, Elizabeth Risdon, Anne Revere, Tom Drake, Phil Taylor, Rita Quigley, Libby Taylor, Richard Gaines, George Houston, Sam McDaniel
A backwoods man is catapulted into Virginia society when he marries a young aristocratic lady. The couple set up home in the wilderness, but the American Revolution shatters their happiness and they find themselves supporting opposing sides in the conflict. An overlong but patriotic and often surprisingly rousing melodrama, slightly spoilt by poor casting (though Houston is worth a look for his portrayal of George Washington).
Aka: TREE OF LIBERTY, THE
DRA   111 min (ort 117 min) B/W   VIDrel: COLUM/SONOP V
Boa: novel The Tree of Liberty by Elizabeth Page.

HOWLING, THE ***   18
Joe Dante USA   1980
Dee Wallace, Patrick Macnee, Dennis Dugan, Christopher Stone, Belinda Balaski, Kevin McCarthy, John Carradine, Slim Pickens, Elisabeth Brooks, Robert Picardo, Dick Miller, Margie Impert, Noble Willingham, James Murtaugh
A strange Californian encounter group community is in reality a den of werewolves, in this spoof full of film-buff jokes. The superb special effects are the work of Rob Bottin. Followed by several "sequels" that are, in reality, totally unrelated to this film.
HOR   87 min (ort 90 min)   VIDrel: BMGVID/BMGREC   V

HOWLING 2, THE *   18
Phillipe Mora USA   1984
Christopher Lee, Reb Brown, Annie McEnroe, Marsha A. Hunt, Ferdy Mayne, Sybil Danning (Sybelle Danninger), Judd Omen, Jimmy Nail, Steven Bronowski, Patrick Field, James M. Crawford, Jiri Krytmar, Ladislav Kregmer, Jan Kraus
More werewolf nonsense in darkest Transylvania, as an expert leads an expedition to destroy a werewolf queen. Unrelated to the 1981 film, this clumsy attempt at parody is one big dud.
Aka: HOWLING 2, THE: YOUR SISTER IS A WEREWOLF
HOR   87 min (ort 90 min)   VIDrel: VCC L/A   V

HOWLING 3 **   18
Phillippe Mora AUSTRALIA   1987
Barry Otto, Imogen Annersley, Leigh Biolos, Max Fairchild, Dasha Blahova, Frank Thring, Michael Pate, Ralph Cotterill, Barry Humphries, Brian Adams, Carole Skinner, Bill Collins, Christopher Pate, Jenny Vuletic, Alan Penney
Learning that werewolves are to be found in the outback, a professor goes off to find them and discovers a marsupial variety. He later casts his lot in with them in the face of government plans for their eradication. A gentle but uneven spoof on this horror genre, with a few amusing touches such as Thring as a Hitchcock-like director and Annersley's pretty female werewolf who carries her child in a pouch on her belly. Third film in the series.
Aka: HOWLING 3: THE MARSUPIALS
HOR   94 min   VIDrel: 20TH   V/sur

HOWLING 4: THE ORIGINAL NIGHTMARE ***   18
John Hough UK   1988
Romy Windsor, Michael T. Weiss, Suzanne Severeid, Anthony Hamilton, Lamya Derval, Norman Anstey, Kate Edwards, Clive Turner, Dennis Folbigge, Anthony James, Dale Cutts, Megan Kruskal, Dennis Smith, Gregg Latter, Maxine John
A female writer is plagued by a series of disturbing visions and decides to take a much-needed holiday in the countryside, staying at a remote cottage her husband has rented. However, her visions worsen and she is soon drawn into a sinister and frightening mystery. Despite being slow to develop, this effective chiller is well worth a look.
HOR   87 min (ort 94 min)   VIDrel: 20TH   V

HOWLING 5: THE RE-BIRTH *   15
Neal Sundstrom USA   1989
Philip Davis, Victoria Catlin, Ben Cole, Elizabeth She, Mary Stavin, Mark Silvertsen, William Shockley, Stephanie Faulkner, Clive Turner, Nigel Triffitt, Jill Pearson, Joszef Madaras, Renata Szatler
Only the title bears any relation to the earlier films, with this in-name-only sequel following the adventures of a group of European travellers who take refuge at a spooky castle that has

been shut up for 500 years. The werewolf doesn't appear until near the end, by which time one no longer cares.
HOR   92 min (ort 99 min)   VIDrel: VCC L/A   V

HOWLING 6: THE FREAKS *   18
Hope Perello USA   1990
Brendan Hughes, Michele Matheson, Antonio Fargas, Sean Gregory Sullivan, Carol Lynley, Jered Barclay, Bruce Martin Payne
Another entry in this long-running series. Unrelated to any previous film, it revolves instead around a sinister freak-show and the exploits of its evil owner. By way of a sub-plot, we have a werewolf seeking revenge for the destruction of his family, but neither this nor the passable special effects enliven an anaemic tale of poor plotting and weak direction.
HOR   96 min   VIDrel: L/A   V/sh

HUCKSTERS, THE **   U
Jack Conway USA   1947
Clark Gable, Deborah Kerr, Sydney Greenstreet, Adolphe Menjou, Ava Gardner, Keenan Wynn, Edward Arnold, Aubrey Mather, Richard Gaines, Frank Albertson, Clinton Sundberg, Douglas Fowley, Gloria Holden, Connie Gilchrist, Vera Marshe
Having returned from the rigors of WW2, an advertising man finds himself fighting almost equally hard to hang onto his principles and his job, as he faces the problems of dealing with his clients. Quite an entertaining adaptation by Wakeman, although it does have a tendency to soften the pungency of the original novel.
DRA   115 min B/W   VIDrel: MGM/WHV   V
Boa: novel by Frederic Wakeman.

HUD ****   12
Martin Ritt USA   1963
Paul Newman, Melvyn Douglas, Patricia Neal, Brandon DeWilde, John Ashley, Crahan Denton, Val Avery, Sheldon Allman, Pitt Herbert, Peter Brooks, Curt Conway, Yvette Vickers, George Petrie, David Kent, Frank Kilmond, N. Candido
Brilliant story of the no-good son of a Texan rancher, his transitory involvement with the housekeeper, and his ruthless and self-seeking ways. A fascinating character study that is both well acted and directed. A joy to watch despite the depressing nature of the storyline. AA: Actress (Neal), S. Actor (Douglas), Cin (James Wong Howe).
DRA   111 min B/W   VIDrel: 4-FRONT/POLYREC/CIC   V
Boa: novel Horseman, Pass By by Larry McMurty.

HUDSON HAWK *   15
Michael Lehmann USA   1991
Bruce Willis, Richard E. Grant, Danny Aiello, Andie MacDowell, Sandra Bernhard, Donald Burton, James Coburn, Don Harvey, David Caruso, Andrew Bryniarski, Lorraine Toussaint, Frank Stallone, Leonardo Cimino, Frank Page
A cat burglar just released from Sing Sing is coerced into undertaking a perilous robbery at a New York auction house. He does so, only to find that this is merely the beginning of a series of wild adventures involving the theft of a valuable Leonardo da Vinci manuscript (hence the flashy opening). An incredibly incoherent and flaccid spoof whose well-staged action sequences do not make up for the irritating personality of Willis or the lack of a cohesive plot.
A/AD   95 min (Cut at UK cinema release by 3 sec – ort 100 min)   VIDrel: VCC/DISC/COLUM   V/sur

HUDSUCKER PROXY, THE **   PG
Joel Cohen USA   1993
Tim Robbins, Paul Newman, Jennifer Jason Leigh, Charles Durning, Jim True, John Mahoney, Bill Cobbs, Bruce Campbell, Harry Bugin, John Seitz, Joe Grifasi, Roy Bucksmith, J.M. Hobson, John Scanlon, Jerome Dempsy, John Wyle, Gary Allen
A young man who has just taken on a job in the mail department of a vast corporation, is summoned to the boardroom and told that he has just become the next company president, following the death of the founder. Naturally, this turns out to be nothing more than an elaborate scheme for which he is to be the fall guy. This over-the-top affair tries too hard to combine special effects with a savage indictment of modern industrial society, doing justice to neither.
COM   106 min (ort 112 min)   VIDrel: COLUM/SONOP V/sur

HUE AND CRY ***   U
Charles Crichton UK   1946
Harry Fowler, Alastair Sim, Jack Warner, Joan Dowling, Jack

Lambert, Valerie White, Frederick Piper, Vida Hope, Gerald Fox, Grace Arnold, Douglas Barr, Stanley Escane, Ian Dawson, Paul Demel, Joey Carr, Robin Hughes, Joey Carr

Crooks use a boys' comic-paper to pass information for robberies, but fail to reckon with the resourcefulness of the youngsters. An excellent film with a period charm of its own. A highlight is the appearance by Sim, as the slightly dotty writer whose stories are used by the crooks. The script is by T.E.B. Clarke.

Aka: SHOOT TO KILL
COM   78 min (ort 82 min) B/W   VIDrel: LUMI/SPEAR L/A V

HUMAN ERROR ***   PG
Clyde Ware USA   1989
Susanne Wouk, Vincent Cobb, Rob Garrison, Joe Estevez, Rod McCary, Pepe Serna, Loraine Venture, Lynn Turner, Papi Mandeville, Betty Rae, Eric Show, Cara Mia Show, Barry Donovan, Ted Rae, David Hearty, Ray Hickman

A revenge movie with an intriguing and original premise. A man designs a nuclear-proof shelter and agrees to test it with his family, but is unaware that they are all the subjects of an experiment involving controlled exposure to radiation. This eventually kills his wife and children, and our designer sets out to have his revenge.

HOR   103 min   VIDrel: IMPENT   V

HUMAN HIGHWAY *   (12)
Dean Stockwell/Bernard Shakey (Neil Young) USA   1982
Neil Young, Dennis Hopper, Russ Tamblyn, Dean Stockwell, Charlotte Stewart, Sally Kirkland, Geraldine Baron, Devo's Nuclear Garbagemen

One of the attendants at a gas station located near a nuclear power station, receives a blow to the head and starts to have strange fantasies involving the "new wave" rock band Devo. A feeble movie that took four years to complete and was hardly worth the effort. Despite all that, it still enjoys a definite cult status, chiefly thanks to the music of Neil Young if not that of Devo.

Aka: HUMAN HIGHWAY: A FILM BY NEIL YOUNG
COM   83 min   VIDrel: WARMUS/WARUK   V

HUMAN MONSTER, THE ***   PG
Walter Summers UK   1939
Bela Lugosi, Hugh Williams, Greta Gynt, Edmond Ryan, Wilfrid Walter, Alexander Field, Arthur E. Owen, Julie Suedo, Gerald Pring, Bryan Herbert, May Haliatt, Charles Penrose

Well-made atmospheric tale about the director of a home for the blind who is in reality the head of an insurance company who gets residents of the home to make out their policies in his favour. When several of them are later found drowned in the Thames, suspicions of foul play are aroused and the police start an investigation. Effectively underplayed by Lugosi with a chilling atmosphere. The first film in the States to get an H (for Horror) certificate.

Aka: DARK EYES OF LONDON, THE
HOR   73 min (ort 78 min) B/W   VIDrel: LUMI/SPEAR L/A V

Boa: novel Dark Eyes of London by Edgar Wallace.

HUMAN SHIELD, THE **   15
Ted Post USA   1991
Michael Dudikoff, Steve Inwood, Tommy Hinkley, Hana Azulay-Hasfari, Geula Levy, Uri Gavriel, Avi Keidar, Gil Dagon, Michael Shillo, Roberto Pollak, Albert Ilouz, Gilles Ben-David, Irving Kaplan, Nagd Tarabshi, Gitan Londner

Though he has been forbidden to enter Iraq, a tough ex-marine disobeys orders and makes a rescue attempt, when he learns that his brother has been kidnapped by the Iraqis following their invasion of Kuwait. A most unremarkable albeit competent action film, shot on location in Israel.

A/AD   87 min   VIDrel: WHV   V/sh

HUMAN TIMEBOMB: A MAN WITH A MISSION **   18
Marc Roper USA   1995
Bryan Genesse, Joe Lara, J. Cynthia Brooks, Shelley Andrews, Jo DaSilva, Franz Dobrowsky, Bill Flynn, Leslie Jong, Anthony Fridjohn, Mike Joyce, Paul Brandt, Gerrie Buader, Crispin De Nuys, Lisa De Villiers, Kurt Egelhof

The FBI are led on an undercover mission by a tough special agent, and their investigations result in the discovery of a top-secret micro-chip that can be used to control human behaviour through implantation in the brain.

A/AD   95 min   VIDrel: 20TH/FOXVID   V/sh

HUMMINGBIRD TREE, THE ***   (PG)
Noella Smith UK   1992
Patrick Bergin, Susan Wooldridge, Tom Beasley, Desha Penco, Sunil Y Ramjitsingh, Valerie Laurent Stevens, Clive Wood, Rebecca Aldred, Bret Kenny, Kirk Collins, Amy Lazzari, Damien Farah, Charles Applewaite, David Sammy

In 1940s Trinidad an upper-class white boy of twelve falls in love with the eleven-year-old daughter of the East Indian couple who work as servants for his parents. Predictably, this young love affair does not meet with universal approval. A charming, coming-of-age tale, loosely based on McDonald's own experiences and recollections. The pleasing location photography adds much to the atmosphere.

DRA   81 min mTV   TVrel: BBC
Boa: novel by Ian McDonald.

HUNCHBACK HAIRBALL OF L.A., THE *   15
Jeremy Paul Kagan USA   1990
Allan Katz, Corey Parker, Cindy Williams, Melora Hardin, Jessica Harper, Tom Skerritt, John Finnegan, Bill Morey, Andrew Bloch, Armin Shimerman, Sam Assaid, Steven Barr, Catherine Bernstrom, Nat Bernsleid, Anna Cannold

Campus-comedy about a strange Quasimodo-like character who lives in a clock tower. After attempting the rescue of a pretty girl student he is eventually entrusted to the care of a none-too-bright psychology student who teaches him to speak. A juvenile and tasteless affair that takes its single idea and drives it into the ground.

Aka: BIG MAN ON CAMPUS
COM   105 min   VIDrel: VES L/A   V

HUNCHBACK OF NOTRE DAME, THE ***   PG
Wallace Worsley USA   1923
Lon Chaney, Patsy Ruth Miller, Norman Kerry, Ernest Torrance, Tully Marshall, Gladys Brockwell, Kate Lester, Brandon Hurst

An early version of this classic story, of the deformed bellringer of Notre Dame Cathedral and his unrequited love for a gypsy girl he rescues from death. Though silent, the lavish sets and incredible make-up of Chaney ensure that this film stands the test of time.

DRA   93 min B/W silent   VIDrel: EUREKA   V/s
Boa: novel by Victor Hugo.

HUNCHBACK OF NOTRE DAME, THE ****   PG
William Dieterle USA   1939
Charles Laughton, Cedric Hardwicke, Maureen O'Hara, Thomas Mitchell, Edmond O'Brien, Walter Hampden, George Zucco, Harry Davenport, Alan Marshal, Katherine Alexander, Fritz Leiber, Rod La Roque, Etienne Girardot, Spencer Charters

Laughton gives a most moving performance in this first sound version of the classic tale, as the lame and deformed bellringer of the Cathedral who is obliged to do his master's bidding when the latter (Hardwicke) falls in love with a gypsy girl. A lavish recreation of Paris of the Middle Ages that spares no expense – the assault on the Cathedral being a high-spot.

DRA   111 min (ort 117 min) B/W
VIDrel: 4-FRONT/POLYREC   V
Boa: novel by Victor Hugo.

HUNCHBACK OF NOTRE DAME, THE ***   PG
Michael Tuchner USA   1982
Anthony Hopkins, Derek Jacobi, Lesley-Anne Down, Robert Powell, John Gielgud, David Suchet, Gerry Sundquist, Tim Piggott-Smith, Alan Webb, Roland Culver, Nigel Hawthorne, Rosalie Crutchley, Joseph Blatchley, Dave Hill

A spirited version of the Hugo classic that remains faithful to the novel, and boasts Hopkins's excellent characterisation as Quasimodo, and John Stoll's detailed replica of Notre Dame Cathedral, built at Pinewood Studios, England. The adaptation of the novel is by John Gay.

DRA   102 min (ort 150 min) mTV
VIDrel: VCC/DISC/COLUM L/A   V
Boa: novel by Victor Hugo.

HUNCHBACK OF NOTRE DAME, THE **   U
Gary Trousdale/Kirk Wise USA   1996
Voices of: Jason Alexander, Mary Kay Bergman, Corey Burton, Jim Cummings, Tom Hulce, Demi Moore, Bill Fagerbakke, Tony Jay, Paul Kandel, Charles Kimbrough, Kevin Kline, Heidi Mollenhauer, Patrick Pinney, Gary Trousdale

Victor Hugo's classic story is given the Disney treatment, and though slighter darker than the usual Disney offerings, there are the expected changes, such as the friendly gargoyles on hand to give shy, lonely and unloved Quasimodo every encourage-

ment to win the heart of Esmeralda. Really quite nonsensical in conception, there are one or two good moments, but on the whole this is a badly drawn and executed effort, with weak and rather foolish songs. Disappointing.
ANIM 87 min (ort 91 min) cC VIDrel: WDV/TECH V
Boa: novel Notre Dame de Paris by Victor Hugo.

HUNGER, THE ** 18
Tony Scott USA 1983
Catherine Deneuve, Susan Sarandon, David Bowie, Cliff De Young, Beth Ehlers, Dan Hedaya, Ann Magnuson, Willem Dafoe, Rufus Collins, Suzanne Bertish, James Aubrey, John Stephen Hill, Shane Rimmer, Douglas Lambert, Bessie Love, Oke Wambu
Two vampires live together in New York. One of them suddenly starts to age and the other is forced to try and find some way of helping him. The ageing scene is the film's highlight in an otherwise bland and unmemorable effort. See also NADJA, a considerably more stylish work.
HOR 92 min (ort 100 min) wScrn VIDrel: MGM/WHV V/h
Boa: novel by Whitley Strieber.

HUNGRY HILL ** PG
Brian Desmond Hurst UK 1946
Margaret Lockwood, Dennis Price, Cecil Parker, Michael Denison, Jean Simmons, Siobhan McKenna, F.J. McCormick, Dermot Walsh, Eilenn Crowe, Peter Murray, Eilenn Herlie, Arthur Sinclair, Barbara Waring, Dan O'Herlihy
Story of two Irish copper-mining families over three generations, detailing their loves, feuds and problems, that starts off with Lockwood playing a feisty beauty whose fortunes undergo many declines over the years. An over-extended melodrama, quite trite and rambling, despite the fact that du Maurier co-wrote the script (and was sorely disappointed with the finished product).
DRA 92 min (ort 109 min) B/W VIDrel: CARL/TECH V
Boa: novel by Daphne du Maurier.

HUNK ** 15
Lawrence Bassoff USA 1986
John Allen Nelson, Steve Levitt, Deborah Shelton, Rebeccah Bush, James Coco, Robert Morse, Avery Schreiber
A computer wizard without any social graces, sells his soul to the Devil to become an irresistible lady's man. After the transformation, he schemes to find a way out of the contract. A foolish and muddled piece of nonsense, carried off by a good cast. Coco plays the Devil in an inspired piece of casting. See also WITCH ACADEMY.
COM 98 min (ort 102 min) VIDrel: FIRST/SONOP L/A V

HUNT FOR RED OCTOBER, THE **** PG
John McTiernan USA 1990
Sean Connery, Alec Baldwin, Sam Neill, Tim Curry, Peter Firth, James Earl Jones, Scott Glenn, Joss Ackland, Richard Jordan, Courtney B. Vance, Jeffrey Jones, Stellan Skarsgard, Larry Ferguson, Timothy Carhart
Connery plays the captain of a Soviet nuclear sub who may be planning to use the vessel to defect to the West during its maiden voyage, or may be about to launch an attack. Meanwhile, Western intelligence agencies attempt to predict his every move. A tense and complex Cold War thriller, with a superb cast and terrific direction. A flawed attempt at having the actors speak Russian is but a minor distraction. AA: Effect/aud (Cecilia Hall/George Watters).
THR 129 min (ort 135 min) wScrn VIDrel: CIC/SONOP; PION (LV only) V/sur LV
Boa: novel by Tom Clancy.

HUNTED *** PG
Charles Crichton UK 1952
Dirk Bogarde, Jon Whiteley, Elizabeth Sellars, Kay Walsh, Frderick Piper, Julian Somers, Jane Aird, Jack Stewart, Geoffrey Keen, Joe Linnane, Leonard White, Gerald Anderson, Denis Webb, Gerald Case, Katherine Blake, Alec Finter
After killing his wife's boyfriend, a man goes on the run and meets up with a six-year-old boy who has run away after inadvertently setting his home on fire. As they travel around, their relationship begins to grow and deepen so that when the boy falls seriously ill, the killer gives up his chance to evade capture and makes sure he gets urgent medical care. A competent drama, well made if a little contrived. The excellent photography is by Eric Cross.
Aka: STRANGER IN BETWEEN, THE
DRA 81 min (ort 84 min) B/W VIDrel: CARL/TECH V

HUNTED ** PG
Rob Iscove USA 1993
Victoria Principal, Peter Onorati, Sean Murray, David Beecroft, Gary Grubbs, Ari Meyers, Lee Garlington, Dirk Blocker, Raymond Baker, Sal Lopez, Fran Ryan, Stan Ivar, John Fleck, Valorie Armstrong, Larisa Oleynik, Tom Noga
A woman goes with her new boyfriend on a trip to the Rio Grande, but they become the target of a psychotic killer.
Aka: MURDER ON THE RIO GRANDE
THR 88 min VIDrel: IMPENT L/A V

HUNTED, THE ** 18
J.F. Lawton USA 1995
Christopher Lambert, John Lone, Joan Chen, Yoshizo Harada
A businessman learns that his death has been ordered by a Japanese gangster and teams up with one of the latter's enemies, in this standard actioner. The noisy drumbeat soundtrack is the movie's most distinctive feature, though the action sequences (especially a bloody conflict aboard an express train) are handled with considerable skill.
A/AD 105 min (ort 111 min) cC VIDrel: CIC/SONOP V/sur

HUNTER, THE ** 15
Buzz Kulik USA 1980
Steve McQueen, Eli Wallach, LeVar Burton, Ben Johnson, Kathryn Harrold, Tracey Walter, Richard Venture, Tom Rosales, Theodore Wilson, Ray Bickel, Bobby Bass, Karl Scueneman, Margaret O'Hara, James Spinks, Frank Delfino, Murray Rubin
The true story of a modern-day bounty hunter who goes after people who have skipped bail. McQueen's last film and not a particularly well made or inspiring one, though it does have a few humorous moments. Based on the life of bounty hunter Ralph (Pappy) Thorson, this slack and unrewarding affair is a poor tribute to the late actor.
A/AD 98 min (ort 117 min)
VIDrel: 4-FRONT/POLYREC/CIC V/h
Boa: book by Christopher Keane.

HURRICANE, THE **** PG
John Ford USA 1937
Dorothy Lamour, John Hall, C. Aubrey Smith, Mary Astor, Raymond Massey, Thomas Mitchell, John Carradine, Jerome Cowan, Al Kikume, Kuulei DeClercq, Layne Tom Jr, Mamo Clark, Movita Castenada, Reri, Francis Kaai, Mary Shaw
The peaceful life of a tropical island comes to an abrupt halt with the arrival of a new and evil-minded governor. Contains some classic hurricane scenes. Remade in 1979 at vast cost by the Swedish director Jan Troell. The fine score is by Alfred Newman and James Basevi takes the credit for the outstanding special effects. AA: Sound (Thomas T. Moulton).
A/AD 99 min (ort 110 min) B/W
VIDrel: VCC/DISC/COLUM V
Boa: novel by Charles Nordhof and James Norman Hall.

HURRICANE EXPRESS ** U
Armand Schaefer/J.P. McGowan USA 1932
John Wayne, Shirley Grey, Tully Marshall, Conway Tearle, Joseph Girard, J. Farrell McDonald
Feature version of a longer twelve-part serial, about a young transport pilot who tries to catch a saboteur responsible for the train crash in which his father died. Average.
DRA 77 min (ort 223 min) B/W VIDrel: SCRN/DISC V

HUSBANDS AND WIVES *** 15
Woody Allen USA 1992
Blythe Danner, Liam Neeson, Juliette Lewis, Lysette Anthony, Woody Allen Mia Farrow, Judy Davis, Sydney Pollack, Cristi Conaway, Timothy Jerome, Ron Rifkin, Jerry Zaks, Bruce Jay Friedman, Benno Schmidt, Caroline Aaron
A long-married couple decide to split up, a decision that comes as a surprise to their closest friends, who, however, soon begin to follow in their footsteps. Allen's humourless look at the problems that beset modern relationships is well acted and scripted, but suffers from an annoying use of hand-held cameras. On release it was inevitably overshadowed by the real-life events surrounding the director's life, but can now be judged on its own merits.
DRA 103 min (ort 108 min) cC
VIDrel: VCC/DISC/COLUM V/sur

HUSH-A-BYE BABY ***   15
Margo Harkin UK   1989
Emer McCourt, Michael Liebmann, Cathy Casey, Julie Marie Reynolds, Sinead O'Connor
At a convent school in Derry, a group of girls occupy much of their time in talking about their abiding obsession – boys. When one of them meets a local lad at evening classes, it's not long before she becomes pregnant. Unfortunately, this coincides with the arrest of her boyfriend. Harkin's first film is a gritty and realistic look at life in Northern Ireland, that rises far above the constraints of its limited budget. Music is by O'Connor.
DRA   76 min (ort 80 min)   VIDrel: CONNO/RTM   V

HUSH, HUSH SWEET CHARLOTTE ***   15
Robert Aldrich USA   1964
Bette Davis, Olivia De Havilland, Joseph Cotten, Cecil Kellaway, Agnes Moorehead, Victor Buono, William Marshall, Mary Astor, Bruce Dern, Wesley Addy, William Campbell, Frank Ferguson, George Kennedy, Dave Willock, John Megna
The story of a southern belle who lives for thirty-seven years in her lonely mansion tormented by her belief that she killed her fiance. However, some strange events begin to convince her otherwise. A moody and generally absorbing tale, supported in large measure by the star cast.
DRA   132 min (ort 134 min) B/W   VIDrel: 20TH/TECH V/h
Boa: story by Henry Farrell.

HUSH LITTLE BABY **   (18)
Jorge Montesi CANADA   1993
Diane Ladd, Wendel Meldrum, Geraint Wyn Davies, Ilya Woloshyn, Ingrid Veninger, Norma Edwards, Dave Nichols, Paul Soles, Andrew Aguanno, Patrick Aguanno, Bradley Sewell, Brendan Sewell, James Howardm David Ferry, J.W. Carroll
After being adopted as a child, a grown woman meets up with her biological mother. However, she is blissfully unaware of her murderous tendencies and that she tried to kill her when she was very young. A shrill and overwrought thriller.
THR   90 min mCab   SATrel: SKY MOVIES

HUSTLER, THE ****   15
Robert Rossen USA   1961
Paul Newman, Jackie Gleason, George C. Scott, Piper Laurie, Myron McCormick, Murray Hamilton, Michael Constantine, Stefan Gierasch, Jake La Motta, Gordon B. Clarke, Alexander Rise, Carolyn Coates, Carl York, Vincent Gardenia, Tom Ahearne
A poolroom hustler hits rock bottom after a severe beating, but regains his self-respect when he wins a match against a legendary champ. The seedy and unpleasant world of poolroom hustlers and drifters is realistically brought to life in this stylish and memorable film. Followed 25 years later by an excellent sequel – THE COLOR OF MONEY. AA: Cin (Eugen Shuftan), Art/Set (Harry Horner/Gene Callahan).
DRA   129 min (ort 134 min) B/W   VIDrel: 20TH/TECH L/A V
Boa: novel by Walter S. Tevis.

HUSTLER, THE **   R18/18
John T. Bone AUSTRALIA   1989
Kelly Blue, Deidre Holland, Sheila Kelly, Alicia Springs, Randy West, Joey Silvera, John Dough
One in a series of Australian sex films, this story opens with West inheriting what he takes to be a horse-ranch in Australia, but learning that in reality it's a whorehouse. He soon finds out the truth, and makes the most of his new acquisition, but strange to relate, eventually decides to turn the place into a real horse-ranch.
Aka: BUSH WACKERS
A   89 min (R18 ver); 71 min Cut (15 min 40 sec – 18 ver)
VIDrel: ELV   V

HYPER SAPIEN: PEOPLE FROM ANOTHER PLANET **   U
Peter Hunt CANADA/USA   1986
Ricky Paull Goldin, Sydney Penny, Rosie Marcel, Keenan Wynn, Dennis Holahan, Gail Strickland, Chuck Shamata
Children of an alien species stowaway on a trip to Earth, and make friends with a couple of kids living in a small town in the Midwest. They all get on extremely well, but trouble soon looms when adults on both sides start to interfere. An earnest, over-sentimental plea for universal tolerance and understanding, that's aimed fair and square at the juvenile market.
Aka: HYPER SAPIEN: PEOPLE FROM ANOTHER STAR
JUV   89 min (ort 96 min) mTV   VIDrel: WHV   V/dm

I

I AM A CAMERA **   (PG)
Henry Cornelius UK   1955
Julie Harris, Laurence Harvey, Shelley Winters, Ron Randell, Lea Seidl, Anton Diffring, Jean Gargoet, Frederick Valk, Tutte Lemkow, Patrick McGoohan, Stanley Maxted, Julia Arnall, Zoe Newton, Stan Bernard Trio
A young writer observes Berlin life in the 1930s and develops a platonic relationship with a rather reckless English girl. A stolid adaptation of the work of Isherwood and Van Druten that is enlivened by a good performance from Harris. Remade with a lot more style (if with less verisimilitude) as CABARET.
DRA   95 min (ort 98 min) B/W   VIDrel: L/A   V
Boa: short stories by Christopher Isherwood/play by John Van Druten.

I AM A FUGITIVE FROM A CHAIN GANG ****   (PG)
Mervyn LeRoy USA   1932
Paul Muni, Glenda Farrell, Helen Vinson, Preston Foster, Allen Jenkins, Edward J. Macnamara, Berton Churchill
An out-of-work man gets unwillingly embroiled in a botched robbery and is jailed, but he escapes and goes on the run, eventually achieving success and stability under a false name. But the law is not far behind him, and soon he is forced to flee once more, becoming a criminal in order to survive. Based on the hard-hitting autobiography of Burns (also told in "The Man Who Broke 1,000 Chains") this is one of the greatest pleas for penal reform ever made.
Aka: I AM A FUGITIVE
DRA   90 min B/W   TVrel
Boa: book by R.E. Burns.

I AM CURIOUS – YELLOW **   18
Vilgot Sjoman SWEDEN   1967
Lena Nyman, Borje Ahlstedt, Peter Lindgren, Chris Wahlstrom, Magnus Nilsson, Marie Goranzon, Ulla Lyttkens, Holger Lowenadler, Hans Hellberg, Bim Warne
Seized by the US Customs as obscene, this was one of the first X-rated foreign films to be shown in the USA. As such, it is an intelligent account of a young woman's exploration of sex and politics in her homeland and her attempts at making sense of both.
Aka: I AM CURIOUS (YELLOW); JAG AR NYFIKEN – GUL
DRA   116 min (ort 121 min) B/W   VIDrel: JEZ/RTM   V

I BOUGHT A VAMPIRE MOTORCYCLE ***   18
Dirk Campbell USA   1989
Neil Morrisey, Amanda Noar, Michael Elphick, David Daker, Anthony Daniels, Andrew Powell, George Rossi, Midge Taylor, Daniel Peacock, Paula-Ann Bland, Burt Kwouk, Brendan Donnison, Graham Pudden, Tedie Thompson, Terence Budd
A hairy slob of a biker purchases a brand new Norton, and said bike turns out to be from Hell. Stuck in its garage during the hours of daylight, at night is comes to life, replenishing its fuel-tank by recourse to the blood of traffic wardens, hookers and the like. Our biker turns to the local priest for help. An unusual and effective horror spoof, not exactly mature comedy, but sharp and very entertaining. Campbell's directorial debut.
HOR   101 min   VIDrel: POLY/POLYREC   V/sur

I CLAUDIUS: PARTS 1 AND 2 ***   15
Herbert Wise UK   1975
Derek Jacobi, Sian Phillips, Brian Blessed, John Hurt, David Robb, Patrick Stewart, Fiona Walker, George Baker, Stratford Johns, John Paul, Christopher Guard, Carleton Hobbs, Angela Morant, Sheila Ruskin, Renu Setna, Guy Siner
A most enjoyable if a trifle uneven attempt to portray the life and times of one of Rome's more eccentric emperors, that if not terribly convincing at times is nonetheless sustained by some outstanding acting. The occasionally ludicrous script is by Jack Pulman.
DRA   648 min (4 cassettes – ort 652 min) mTV   VIDrel: BBC L/A   V/h
Boa: novels I, Claudius and Claudius The God by Robert Graves.

I CONFESS ***   PG
Alfred Hitchcock USA   1953
Montgomery Clift, Anne Baxter, Karl Malden, Brian Aherne, Dolly Haas, O.E. Hasse, Roger Dann, Charles Andre, Judson Pratt, Ovila Legare, Nan Boardman, Henry Corden, Carmen Ginras, Renee Hudson, Albert Goderis

A priest hears a confession and refuses to reveal what he has heard to the police, even though he is suspected himself. Despite the limitations of the plot, Hitchcock manages to imbue this one with a fair degree of tension, even if the outcome is never in doubt. Set in Quebec, with good use made of this location. Remade a good many years later as THE CONFESSIONAL. See also CONFESSION.
DRA 90 min (ort 93 min) B/W VIDrel: WHV V
Boa: play Our Two Consciences by Paul Anthelme.

I COULD GO ON SINGING ** U
Ronald Neame USA 1963
Judy Garland, Dirk Bogarde, Aline MacMahon, Jack Klugman, Gregory Phillips, Pauline Jameson, Jeremy Burnham, Russell Waters, Gerald Sim, Leon Cortez
A singer from the US comes to Britain to visit a former boyfriend and claim custody of their son, who, however, wants nothing to do with her when he learns of his true parentage. Although seasoned with a few good songs, this mediocre and downbeat effort has little that is memorable. This was Garland's last film and a rather low note on which to end such a splendid career.
MUS 95 min (ort 99 min) VIDrel: MGM/WHV V/h

I DO PART 3: THE OTHER WOMAN * 18
Paul Thomas USA 1991
Hyapatia Lee, P.J. Sparks, Kim Wild, John Dough, Scott Irish
Sex film revolving around the love life of a wealthy woman who is content to fool around with married men who are not totally satisfied by their wives. A dull film that has no connection with the two earlier films except the presence of Lee.
Aka: I DO 3
A 45 min (ort 80 min) VIDrel: VIVID/SCRN V

I DON'T KISS *** 18
Andre Techine FRANCE 1991
Philippe Noiret, Emmanuelle Beart, Manuel Blanc, Helene Vincent, Ivan Desny, Christophe Bernard, Roschdy Zem, Rapha line Goupilleau, Michele Moretti, Philippe Adam, Jean-Christophe Bouvet, Paulette Bouvet, Kleber Bouzzone
Adult drama in which a would-be actor from a remote village in the Pyrenees travels to Paris in the hope of achieving success, but instead finds himself forced into male prostitution to make a living. However, he eventually falls in love with a hooker. An acutely observed drama, quite poignant and believable, though the episodic nature of the script and the director's tendency to labour the point are slight annoyances.
Aka: J'EMBRASSE PAS
DRA 111 min (ort 116 min) wScrn VIDrel: TART/20TH V/dm

I KNOW MY FIRST NAME IS STEVEN *** 15
Larry Elikann USA 1989
Cindy Pickett, John Ashton, Corin Nemec, Arliss Howard, Luke Edwards, Pruitt Taylor Vance, Brenda Tarbuck, Scott Curtis, Stephanie Walski, Sumer Stamper, Cassy Friel, June C. Ellis, Billy O'Sullivan, Jason Preston, Stephen Durff
Based on a true story, this tells of a seven-year-old boy who vanished in 1972 while on his way to school. Seven years later he turns up, but can remember little of his past life. A highly competent and rather harrowing drama.
DRA 183 min VIDrel: ODY/SONOP V/h

I KNOW WHERE I'M GOING! *** U
Michael Powell/Emeric Pressburger UK 1945
Wendy Hiller, Roger Livesey, Pamela Brown, John Laurie, Finlay Curry, Nancy Price, Valentine Dyall, Petula Clark, George Carney, Walter Hudd, Murdo Morrison, Jean Cadell, Margot Fitzsimmons, Norman Shelley, Catherine Lacey
A headstrong young lady has decided to marry for money, and sets off for the Hebrides to meet her spouse – a wealthy old man. She is stranded on Mull during a storm, and gradually falls in love with a young naval officer. An uneasy blend of romance, comedy and drama that has little plot, but is sustained by a good measure of wit and some beautiful scenery. Photography is by Erwin Hiller.
DRA 91 min B/W VIDrel: CARL/TECH V

I LIKE IT LIKE THAT ** 15
Darnell Martin USA 1994
Lauren Velez, Jon Seda, Tomas Melly, Desiree Casado, Isaiah Garcia, Jesse Borrego, Lisa Vidal, Griffin Dunne, Rita Moreno, Vincent Laresca, E.O. Nolasco, Sammy Melandez, Jose Soto, Gloria Irizarry, Emilio Del Pozo, Donald Jackson

A bittersweet look at a hard-pressed Bronx family where the mother is forced to become the breadwinner after her husband is jailed for looting. Naturally, she has more than her fair share of problems, in this over-extended ethnic tale.
COM 102 min (ort 106 min) VIDrel: COLUM/SONOP V/sur

I LIKE YOU, I LIKE YOU VERY MUCH *** 18
Oki Hiroyuki JAPAN 1994
Oki Hiroyuki, Shibuya Kazunori, Kitakaze Hisanori, Nishimoto Kazufumi
Gay sex film making much use of hand-held cameras that tells a modern-day love story in Kochi, where a young man finds his affection for a stranger he approaches at a crowded tram station is not reciprocated, and spends the rest of the movie trying to come to terms with his sense of rejection. Largely plotless, this is a strangely absorbing and poignant little tale, as much about alienation and loneliness as it is about homosexuality.
A 60 min VIDrel: DTK/RTM V

I LIVE WITH ME DAD *** 15
Paul Moloney AUSTRALIA 1987
Peter Hehir, Haydon Samuels, Rebecca Gibney, Tony Hawkins, Gus Mercurio, Dennis Miller, Robyn Gibbes, Esben Storm
Sentimental tearjerker based on a real story, with a young boy living with his down-and-out loafer of a father, much against the wishes of the child welfare authorities. They present the boy's father with an ultimatum; get a job or lose the kid and the pair are split up despite the father's attempts to curtail his drinking. All ends happily however, but not before there have been a series of incidents in this competent but often harsh drama.
DRA 91 min VIDrel: 20TH V
Boa: short story by Derry Moran.

I LOVE A MAN IN UNIFORM ** 18
David Wellington CANADA 1993
Tom McCamus, Brigitte Bako, Kevin Tighe, David Hemblen, Alex Karzas, Graham McPherson, Daniel McIvor, Wendy Hopkins, Kiersten Lieferle, Paulina Gillis, Dana Broooks, Steve Ambrose, Michael Hogan, Mark Melymick, Cynthia Gillespie
An actor working as banker gets the chance of a part in a cop series and uses method techniques to prepare for this role. However, he becomes dangerously obsessed and is no longer able to distinguish between fact and fiction, and takes to patrolling the streets in a police uniform. A chilling and rather unpleasant tale, with enough unexpected nastiness to keep one watching if not exactly edified. See also CRIMETIME.
Aka: MAN IN UNIFORM, A
DRA 94 min (ort 97 min) wScrn VIDrel: TART/20TH V

I LOVE TROUBLE ** PG
Charles Shyer USA 1993
Nick Nolte, Julia Roberts, Saul Rubinek, Robert Loggia, James Rebhorn, Kelly Rutherford, Olympia Dukakis, Marsha Mason, Eugene Levy, Charles Martin Smith, Dan Butler, Paul Gleason, Nora Dunn, Keith Gordon, Dorothy Lyman, Stuart Pankin
An ageing news-hound and a young and far more attractive female rival go after the same story that centres around a train derailment that may have been caused by sabotage. Predictably, events conspire to bring them together and they eventually fall in love whilst fighting hard to stay alive. This overlong and awkward blending of comedy and tension deliberately imitates many a film of the 1930s, but poor casting of the leads spoils it.
COM 123 min (ort 125 min) cC VIDrel: TOUCH/TECH V

I LOVE YOU ** 18
Marco Ferreri FRANCE/ITALY 1986
Christophe Lambert, Eddy Mitchell, Flora Barillano, Agnes Soral, Anemone, Marc Berman, Patrick Bertrand, Paula Dehelly, Mayrizio Donadoni, Fabrice Dumeur, Carole Fredericks, Laurence Leguellan, Olivia Link, Laura Manszky
A man is unable to find a woman who will devote herself totally to him, but gains solace when he falls in love with a lip-shaped electronic key-ring which says "I love you" whenever he whistles. This quirky black comedy purports to say something profound about the dehumanisation of relationships and the tendency for women to be portrayed as mere sexual objects, but without a proper narrative to give it structure and meaning, one rapidly finds one's attention wandering.
COM 96 min wScrn VIDrel: ARTPRO/RTM V

I LOVE YOU NO MORE *   18
Serge Gainsbourg FRANCE   1975
Jane Birkin, Serge Gainsbourg, Joe Dallesandro, Hugues Quester,
Gerard Depardieu
The anatomy of a brief love affair, with a second man complet-
ing the triangle as a homosexual lover. With minimal dialogue
and heavy symbolism, writer/director Gainsbourg attempts to
say something profound about love and human relationships,
but this bitter film is so clumsy and pretentious that whatever
message was in there gets lost amongst all the self pity.
Aka: I LOVE YOU, I DON'T; JE T'AIME, MOI NON PLUS
DRA   84 min (ort 90 min)   VIDrel: WESCON/RTM L/A   V

I LOVE YOU TO DEATH **   15
Lawrence Kasdan USA   1990
Kevin Kline, Tracey Ullman, River Phoenix, Joan Plowright, William
Hurt, Keanu Reeves, James Gammon, Victoria Jackson, Miriam
Margoyles, Heather Graham, Jon Kasdan, Michelle Joyner, John
Kostmayer, Kathleen York
Allegedly based on a true-life story, this tells of the happily
married wife of a pizza baker who is so shattered by the reve-
lation of his many infidelities, that she resolves to murder him.
Fortunately for the intended victim, her attempts are so ludi-
crously inept that he lives to tell the story. A failed black comedy
that soon loses its sharpness and rapidly descends into slapstick. Disappointing.
COM   93 min (ort 96 min)   VIDrel: RCA/VCC L/A   V

I MARRIED A VAMPIRE **   15
Jay Raskin USA   1983
Rachel Golden, Brendan Hickey, Ted Zalewski, Deborah Carroll,
Temple Aaron, David Dunton, Kathryn Kames
A thoroughly campy Troma-produced horror spoof, in which a
naive young girl from a small town comes to the big city, where
nothing but lies and deceit await her, gets romanced by a dashing
suitor, and eventually finds herself married to a vampire.
HOR   89 min   VIDrel: 20VIS/SONOP L/A   V

I, MONSTER ***   15
Stephen Weeks UK   1971
Christopher Lee, Peter Cushing, Mike Raven, Richard Hurndall,
Kenneth J. Warren, George Merritt, Susan Jameson, Marjie Lawrence,
Aimee Delamain
A fairly competent reworking of the Jekyll and Hyde theme,
with Lee quite convincing as the doctor whose thirst for knowl-
edge forces him to experiment on himself. Despite a slow start
and some awkward melodramatic flourishes, the film builds up
to a satisfactory climax.
HOR   75 min   VIDrel: LUMI/SPEAR L/A   V
Boa: novella The Strange Case of Dr Jekyll and Mr Hyde by
Robert Louis Stevenson.

I NEVER PROMISED YOU A ROSE GARDEN **   18
Anthony Page USA   1977
Bibi Andersson, Kathleen Quinlan, Sylvia Sidney, Reni Santoni, Signe
Hasso, Diane Varsi, Susan Tyrrell, Dennis Quaid, Ben Piazza,
Lorraine Gary, Norman Alden, Martine Bartlett, Robert Viharo, Jeff
Conaway, Dick Herd, June C. Ellis
The story of a teenage girl undergoing treatment for schizo-
phrenia which concentrates on the relationship between her and
her psychiatrist. A ponderous account that benefits from its
graphic approach, but never sufficiently to make it much better
than average. See also STRANGE VOICES.
DRA   88 min (ort 96 min)   VIDrel: L/A   V
Boa: novel by Hannah Green.

I POSED FOR PLAYBOY *   15
Stephen Stafford USA   1991
Lynda Carter, Michelle Greene, Amanda Petersen, Brittany York,
John Finn, Josis Bassett, Lee Garlington, Don S. Davis, David Megey,
Lochlyn Munro, Dale Wilson, Malcolm Stewart, Jonathan Bruce, Josh
Murray, David Newman
Based on three allegedly true stories of women who posed nude
for Playboy magazine, and the effect this had on their lives.
Dull, very dull.
Aka: I POSED FOR PLAYBOY BEHIND THE SCENES; POSING; POSING:
INSPIRED BY THREE REAL LIFE STORIES
DRA   96 min   VIDrel: GUILD/SONOP   V

I SAW WHAT YOU DID ***   18
Fred Walton USA   1987
Tammy Lauren, Shawnee Smith, Robert Carradine, David Carradine,
Candace Cameron
Remake of the 1965 film, with two teenage girls playing a silly
phone prank on a man who has just committed a murder, and
thus making themselves prime candidates for his next crime.
THR   90 min   VIDrel: CIC/SONOP L/A   V/sh
Boa: novel by Ursula Walter.

I SEE A DARK STRANGER ***   U
Frank Launder UK   1946
Deborah Kerr, Trevor Howard, Raymond Huntley, Norman Shelley,
Brenda Bruce, Michael Howard, Liam Redmond, Brefni O'Rorke,
James Harcourt, Garry Marsh, W.G. O'Gorman, George Woodbridge,
Tom Macauley, Olga Lindo, Kathleen Harison
WW2 thriller about a young Irish girl whose support of the IRA
leads her to fall easy prey to a Nazi agent, who involves her in
aiding the escape from prison on the Isle of Man of a German
spy. However, her role in this venture brings her into contact
with a young British officer, with whom she soon falls in love.
A rather contrived comedy-thriller, not exactly brilliantly
plotted, but made with some care and well supported by the
cast.
Aka: ADVENTURESS, THE
THR   107 min (ort 112 min) B/W   VIDrel: CONNO/RTM
L/A   V

I SEE ICE ***   U
Anthony Kimmins UK   1938
George Formby, Cyril Richard, Kay Walsh, Betty Stockfeld, Garry
Marsh, Frederick Burtwell, Ernest Sefton, Gavin Gordon, Gordon
McLeod, Archibald Batty, Frank Leighton, Roddy McDowall, Ernest
Jay
An accident-prone photographer's assistant in an ice-ballet
company invents a bow-tie camera, with which he inadvertently
takes a snap of some crooks, landing him a job on a newspaper
as a photographer. Formby is at his clumsy, blundering best in
this silly but quite endearing knockabout comedy, made to cash
in on the ice-skating craze that was briefly in vogue in the 1930s.
Songs include: "In My Little Snapshot Album".
COM   83 min B/W   VIDrel: LUMI/SPEAR L/A   V

I SHOT ANDY WARHOL *   18
Mary Harron UK/USA   1995
Lili Taylor, Jared Harris, Lothaire Blutheau, Martha Plimpton,
Stephen Dorff, Danny Morgentstern, Anna Thompson, Tahnee Welch,
Michael Imperioli, Reg Rogers, Coco McPherson, Donovan Leitch,
Craig Chester, James Lyons, Myrian Cyr
Story of the bizarre events surrounding the shooting of Warhol,
his attacker being one Valerie Solanas, a seriously disturbed
individual whose hatred of men eventually found an outlet in
violence. Screenplay is by Harron and Daniel Minahan, and one
comes away with the impression that they are not entirely out
of sympathy with this attack. An extremely unpleasant and
barren movie.
DRA   99 min (ort 103 min)   VIDrel: BMGREC/BMGVID   V

I SPY RETURNS ***   15
Jerry London AUSTRIA/USA   1993
Bill Cosby, Robert Culp, George Newburn, Salli Richardson, Nikolaus
Paryla, Lynsey Baxter, Jonathan Hyde, Sheila Wills, Brent Huff, Greg
Blanchard, Kateryn Lucius, Ilene Kreshka, Paul Krehska, Christoph
Hohlfeld, Johannes Krisch
Spin-off of the 1960s TV series that re-unites its stars, who now
have to help their kids with their first assignment, the latter
having becomes Secret Service agents just like their parents once
were. Filmed on location in Vienna, this is a handsome and
enjoyable blending of action and comedy, retaining much of the
flavour of the earlier series.
A/AD   92 min mTV   VIDrel: FIRST/SONOP   V

I, THE WORST OF ALL ***   15
Maria Luisa Bemberg ARGENTINA   1990
Assumpta Serna, Dominqiue Sanda, Hector Alterio, Lautaro Murua,
Alberto Segado, Franklin Cacideo, Graciela Araujo
A long, detailed and very well-made recreation of the life of a
17th century Mexican woman writer who eventually entered a
convent but continued to practice her craft until her archbishop
attempted to stop her. Told partly in flashback, this film explores
the political and moral climate of the times and is well served
by Serna's excellent acting.
Aka: YO, LA PEOR DE TODAS
DRA   105 min (ort 107 min)   CINrel
Boa: The Traps of Faith from the book Sor Juana: Her Life and
Her World by Octavio Paz.

I VITELLONI *** U
Federico Fellini ITALY   1953
Franco Fabrizi, Franco Interlenghi, Eleonora Ruffo, Alberto Sordi, Leopoldo Trieste, Riccardo Fellini, Leonora Ruffo, Lida Baarowa, Arlette Sauvage, Maja Nipora, Jean Brochard, Claude Farere, Carlo Romano, Enrico Viariso
Acutely observed, nostalgic tale of five restless young men and their lives in a small Adriatic town, where they pass their days in pursuit of aimless pleasures. The film bears traces of the neo-Realist style that Fellini later abandoned. The title stands for "The Young Calves" – the nickname by which the group of friends are known.
Aka: SPIVS; YOUNG AND THE PASSIONATE, THE
DRA 103 min (ort 109 min) B/W VIDrel: FABFIL/SPEAR V

I WAKE UP SCREAMING *** PG
Bruce Humberstone USA   1941
Betty Grable, Carole Landis, Victor Mature, Laird Cregar, William Gargan, Alan Mowbray, Allyn Joslyn, Elisha Cook Jr
An absorbing mystery tale in which the sister of a murdered model teams up with the chief suspect (Mature) in order to catch the real culprit, whilst all the while being pursued by a determined cop. Sharp, concise and highly atmospheric; the twist ending is as effective as it is unexpected. Remade as "Vicki".
Aka: HOT SPOT
DRA 78 min (ort 82 min) B/W VIDrel: 20TH/TECH V
Boa: novel by Steve Fisher.

I WANT YOU * U
Mark Robson USA   1951
Dana Andrews, Farley Granger, Dorothy McGuire, Mildred Dunnock, Peggy Dow, Robert Keith, Ray Collins, Martin Milner, Jim Backus
Story of two men who are called up to fight in the Korean War, and the impact this has on the loved ones they leave behind. Producer Sam Goldwyn thought that this hollow flag-waver would achieve the same kind of success THE BEST YEARS OF OUR LIVES had. It didn't.
DRA 101 min B/W VIDrel: VCC/DISC V

I WAS A MALE WAR BRIDE ** U
Howard Hawks USA   1949
Cary Grant, Ann Sheridan, William Neff, Marion Marshall, Randy Stuart, Eugene Gericke, Ruben Wendorf, John Whitney, Ken Tobey, Joe Haworth, John Zilly, William Pullen, William Self, Bill Murphy, Robert Stevenson, Harry Lauter
A French officer in WW2 marries a WAC lieutenant and finds himself confronted with a situation not foreseen by military regulations. This forces him to spend much time in drag and masquerading as a war bride, the only way he and his wife can get into the country.
Aka: YOU CAN'T SLEEP HERE
COM 101 min (ort 105 min) B/W VIDrel: 20TH/TECH V
Boa: novel by Henri Rochard.

I WAS A TEENAGE FRANKENSTEIN * 15
Herbert L. Strock USA   1957
Whit Bissell, Phyllis Coates, Robert Burton, Gary Conway, Marshall Bradford, George Lynn, John Cliff, Joy Stoner, Claudia Bryar, Russ Whiteman, Angela Blake, Charles Seel, Gretchen Thomas, Paul Keast, Pat Miller, Larry Carr
A mad scientist "rescues" a youngster from a car crash and sets about using him, grafting his face onto a monstrous creature he has fashioned out of the corpses he has salvaged from similar accidents. The monster eventually rebels, destroys his maker and is electrocuted on a power board. Produced by Herman Cohen as a follow-up to his I WAS A TEENAGE WEREWOLF, this gruesome shocker is but fitfully entertaining, and now looks very dated indeed.
Aka: TEENAGE FRANKENSTEIN
FAN 74 min B/W (plus one Colour sequence)
VIDrel: HEND/BMGREC L/A V

I WAS A TEENAGE TV TERRORIST * PG
Stanford Singer USA   1987
Adam Nathan, Julie Hanlon, John MacKay, Walt Willey, Saul Alpiner
Two teenagers working at a cable TV station stage a bomb scare to get even with their sadistic boss but this plus a catalogue of other pranks fails to raise a laugh, in this highly amateurish offering.
Aka: AMATEUR HOUR
COM 85 min VIDrel: TROMA/RTM V

I WAS A TEENAGE WEREWOLF ** 15
Gene Fowler Jr USA   1957
Michael Landon, Yvonne Lime, Whit Bissell, Vladimir Sokoloff, Guy Williams, Malcolm Atterbury, Eddie Marr, Louise Lewis, S. John Launer, Tony Marshall, Ken Miller, Dawn Richards, Barney Phillips, Joseph Mell, Cindy Robbins
A professor attempts to treat a teenager for hyper-sensitivity, but his experimental techniques cause a process of regression that transforms his subject into a ravenous werewolf. The title is the best-remembered aspect of this blend of horror and juvenile delinquency themes, which is surprisingly well handled for a film with such an established cult status. Followed by the inferior I WAS A TEENAGE FRANKENSTEIN.
HOR 75 min (ort 76 min) B/W VIDrel: HEND/BMGREC L/A V

I WAS MONTY'S DOUBLE *** U
John Guillermin UK   1958
John Mills, Cecil Parker, Michael Hordern, Leslie Phillips, Bryan Forbes, Marius Goring, Patrick Allen, M.E. Clifton-James, James Hayter, Sidney James, Vera Day, Victor Maddern, Marne Maitland, Alfie Bass, Duncan Lamont
A well made and fact-based story of an actor being used to confuse the Germans, by impersonating Montgomery in the North African campaign during WW2.
Aka: HELL, HEAVEN AND HOBOKEN
WAR 96 min (ort 100 min) B/W VIDrel: WHV V
Boa: book by M.E. Clifton-James.

I WAS ON MARS *** 15
Dani Levy GERMANY/SWITZERLAND/USA   1991
Maria Schrader, Dani Levy, Mario Giacolone, Antoni Rey, Penny Arcade (Susan Ventura), Luis Caballero, Rafael Clements, Stuart Duckworth, Pasquale Gaeta, Ben Berman, Lisa Langford, Ronnie Shades, Arndt Wiegering, Cyndi Coyne, Bob Albert
A Polish girl comes to the New York but spends most of her money and what little remains is stolen. Days later, destitute and starving she locates the man she believes to be the one who robbed her, and follows him incessantly, eventually having an affair with both him and his brother. A touching, unnerving and occasionally funny look at the Great American Dream, very much done from an outsider's point of view. The script is by Dani Levy.
DRA 83 min (ort 87 min) wScrn VIDrel: TART/20TH V/dm

I WILL IF YOU WILL *** 18
Nello Rossati ITALY   1975
Ursula Andress, Luciana Paluzzi, Jack Palance, Lino Toffolo, Duilio Del Prete, Mario Pisu, Carla Romanelli
A wine merchant's relatives engage a sexy nurse, in the hope of finishing the old man off, but this plan backfires when she seduces every other male in the house to satisfy her appetite. Fair Italian sex film with comic overtones.
Aka: L'INFERMIERA; NURSE, THE; SECRETS OF A SENSUOUS NURSE; SENSUOUS NURSE, THE
A 77 min (ort 85 min) VIDrel: FUNNY/SGSVID V

ICE ** 18
Brook Yeaton USA   1993
Traci Lords, Phillip Troy, Zach Galligan, Jorge Rivero, Jaime Alba, Jean Pflieger, Michael Bailey Smith, Eddi Wilde, Marc Siegler
After a mob boss robs a number of jewellery stores and steals diamonds to a value of $60,000,000, the insurance company calls in a married couple who are both professional burglars. Their task is to raid our villain's home and steal back the gems. A moderately exciting action tale of no great distinction.
A/AD 87 min (ort 91 min) VIDrel: IMPENT V

ICE COLD IN ALEX *** PG
J. Lee Thompson UK   1960
John Mills, Sylvia Sims, Anthony Quayle, Harry Andrews, Liam Redmond, Peter Arne, Diane Clare, Richard Leech, Allan Cuthbertson, David Lodge
A British ambulance driver in Libya in 1942, successfully charts his vehicle through German minefields and other dangers. An overlong but extremely engaging war film, that benefits from good direction and solid performances.
Aka: DESERT ATTACK
WAR 125 min (ort 132 min) B/W VIDrel: WHV V
Boa: novel by Christopher Landon.

ICE RUNNER, THE ** 15
Barry Alan Samson USA 1993
Edward Albert, Victor Wong, Olga Kabo, Eugene Lazarev, Alexander Kuznitzov, Basil Hoffman
An American spy agent in Russia is betrayed to the Soviet secret police who make sure that he is sent to a labour camp in the frozen Siberian wastes. However, his spirit is far from broken and he plots to escape.
DRA 122 min VIDrel: GUILD/FOXVID V

ICE STATION ZEBRA ** U
John Sturges USA 1968
Rock Hudson, Ernest Borgnine, Patrick McGoohan, Jim Brown, Tony Bill, Lloyd Nolan, Gerald S. O'Loughlin, Alf Kjellin, Ted Hartley, Ron Masak, Murray Rose, Lee Stanley, Sherwood Price, Joseph Bernard, John Orchard, Jim Goodwin
Cold War tale of murder and espionage at a remote Arctic station, with Hudson as the commander of a submarine sailing to the North Pole, where he must obtain the hidden film from an American spy satellite before it falls into Russian hands. A mixture of tension and tedium, as the various MacLean plot convolutions unravel. Written by Douglas Heyes and Harry Julian Fink.
A/AD 139 min (ort 150 min) VIDrel: MGM/WHV V
Boa: novel by Alistair MacLean.

ICEMAN COMETH, THE ** 18
Clarence Fok Yiu Leung HONG KONG 1991
Yuen Bao, Yuen Wah, Maggie Cheung
During the Ming dynasty, an Imperial policeman is on the track of a notorious murderer and rapist when both men become trapped in a glacier in the Himalayas. A thousand years later, their bodies are exhumed and taken to Hong Kong where they are thawed out and come back to life, thereby resuming their struggle. Fortunately, a young hooker is on hand to help our policeman find his way in an unfamiliar world.
A/AD 114 min VIDrel: MADE/RTM V

ICICLE THIEF, THE ** PG
Maurizio Nichetti ITALY 1990
Maurizio Nichetti, Caterina Sylos Labini, Claudio G. Fava, Heidi Komarek, Renata Scarpa
A black-and-white drama being shown on TV is interrupted by so many commercials that the two become hopelessly intermingled and the film's author is forced to enter his own film to put matters right. A contrived satire on commercialism that has some clever ideas but fails to place them in a coherent setting. Written by Nichetti.
Aka: LADRI DI SAPONETTE
DRA 81 min (ort 84 min) B/W/Col wScrn
VIDrel: TART/20TH L/A V

I.D. *** 18
Philip Davis GERMANY/UK 1994
Reece Dinsdale, Richard Graham, Perry Fenwick, Philip Glenister, Warren Clarke, Claire Skinner, Saskia Reeves, Sean Pertwee, Charles De'Ath, Lee Ross, Terry Cole, Steve Sweeney, Nicholas Bailey, Nick Bartlett, David Schaal
Four undercover police officers are ordered to infiltrate a gang of football hooligans in a bid to halt an ever increasing level of violence. However, one of them finds his personality being corrupted by the strong sense of fellowship he begins to feel for the various gang members he gets to know. An impressive debut feature from Davis, very carefully put together, but hampered by a low budget and an excessive tendency towards the polemical. See also THE FIRM and ULTRA.
DRA 103 min (ort 107 min) VIDrel: POLY/POLYREC V/sh

IDENTIFICATION OF A WOMAN *** 18
Michelangelo Antonioni FRANCE/ITALY 1982
Tomas Milian, Daniela Silverio, Christine Boisson, Marcel Bozzufi, Lara Wendel, Veronca Lazar
Whilst searching for the right woman to star in his next movie, a divorced, middle-aged director starts and finishes two relationships. Despite the undoubted care with which the film is made, and the arresting beauty of many of its images, this is not so much an examination of the difficulties that beset human relationships in modern times as an intellectual experiment of but limited success. A clear direction for the narrative to go would have helped greatly.
Aka: IDENTIFICAZIONE DI UNA DONNA
DRA 123 min (ort 130 min) VIDrel: CONNO/RTM V

IF... ** 15
Lindsay Anderson UK 1968
Malcolm McDowell, David Wood, Richard Warwick, Robert Swann, Peter Jeffrey, Christine Noonan, Arthur Lowe, Mona Washbourne, Graham Crowden, Anthony Nicholls, Geoffrey Chater, Hugh Thomas, Rupert Webster, Mary McLeod, Ben Aris
A strange, allegorical, somewhat empty film, set in a boys' public school and full of rather meaningless images that you can take any way you like. Is it an anti-establishment satire? You tell me. The film O LUCKY MAN! serves in some ways as a companion piece.
DRA 107 min (ort 111 min) Col/B/W VIDrel: CIC/SONOP V

IF DON JUAN WERE A WOMAN ** 18
Roger Vadim FRANCE 1973
Brigitte Bardot, Jane Birkin, Maurice Ronet, Michelle Sand, Robert Hossein, Robert Walker Jr
A torrid tale of love, passion and murder with Bardot cast as a female Don Juan who delights in conquering men and then ruining them. A good vehicle for Bardot, but not really memorable on any other count.
Aka: DON JUAN; DON JUAN 73; DON JUAN 73 OR IF DON JUAN WERE A WOMAN; DON JUAN 73 OU SI DON JUAN ETAIT UNE FEMME; MS. DON JUAN; SI DON JUAN ETAIT UNE FEMME
DRA 89 min (ort 95 min) wScrn VIDrel: ARROW/RTM V

IF EVER I SEE YOU AGAIN * PG
Joe Brooks USA 1978
Joe Brooks, Shelley Hack, Jerry Keller, Jimmy Breslin, George Plimpton, Danielle Brisebois, Kenny Karen, Michael Decker, Julie Ann Gordon, Branch Emerson, Shannon Bolin, Caroline Mignini, Joe Leon, Ed Kovins, Bob Kaliban
A couple find that the pressures of their jobs and family commitments threaten to put paid to their love affair. A dreary and soppy follow-up to YOU LIGHT UP MY LIFE, this gave Hack her starring debut but failed to make good use of her talents. Written and directed by Brooks.
DRA 100 min (ort 105 min) VIDrel: L/A V

IF LOOKS COULD KILL ** 15
Sheldon Larry USA 1996
Brad Dourif, Antonio Sabato Jr, David Keith, Maura Chaykin
A con-man fakes a traffic accident and uses the insurance money to finance his activities involving the setting up of a chain of stores. But when he runs out of money he has to start planning a new accident, and this one involves faking his own death.
DRA 86 min VIDrel: 20TH/FOXVID V/sh

IF LUCY FELL * 15
Eric Schaeffer USA 1996
Sarah Jessica Parker, Eric Schaeffer, Ben Stiller, Elle MacPherson, James Rebhorn, Robert John Burke, David Thornton, William Sage, Dominic Chianese, Scarlett Johanssen, Michael Storms, Jason Myers, Emily Hart, Paul Greco
Two twenty-year-old friends make a suicide pact that they must both jump off the Brooklyn bridge if either of them have failed to meet the man of their dreams within twenty-eight days, by which time they will both be thirty. A vulgar and poorly realised tale with repellent characters and uninspired dialogue.
COM 93 min VIDrel: COLUM/SONOP V/sur

IF SOMEONE HAD KNOWN *** 15
Eric Laneuville USA 1995
Kellie Martin, Kevin Dobson, Linda Kelsey, Ivan Sergei, Ann Gunn, Kristin Dattilo-Hayward, Alan Fudge, Jennifer Savidge, Tom Amandes, James Harper, Ramsay Midwood, Noah Emmerich, Ted Hayden, Jennifer Griffith, Rodney Rowland, Greg Gaul
A woman brought up to obey her husband endures a life of beatings at the hands of her spouse, but outwardly ensures that she appears to enjoy a happy marriage. Eventually, both she and her ex-cop father are driven to take drastic measures to end her years of torment. A fact-based drama that pulls no punches, offering us compelling performances and a harrowing script.
DRA 86 min (ort 92 min) VIDrel: ODY/SONOP V/sh

IF THESE WALLS COULD TALK *** 15
Nancy Savoca/Cher USA 1996
Demi Moore, Sissy Spacek, Cher, Anne Heche, Hedy Burress
Three separate stories, set in 1952, 1974 and 1996, but at the same house, where in each case a woman has to deal with an unwanted pregnancy. In the first story (easily the best) Moore seeks a backstreet abortion, in the second tale Spacek has to

contend with her feminist daughter, while the final story (the weakest one) has a pregnant Heche attending an abortion clinic, where pro-life demonstrators are staging a protest. This was Cher's directing debut.
DRA   93 min Cut (3 sec)   VIDrel: MED   V/sh

I'LL DO ANYTHING **
James L. Brooks USA
12
1993
Nick Nolte, Albert Brooks, Julie Kavner, Joely Richardson, Tracey Ullman, Whittni Wright, Jeb Brown, Joely Fisher, Ian McKellan, Suzanne Douglas, Robert Joy, Harry Shearer, Rosie O'Donnell, Ken Page, Woody Harrelson, Patrick Cassidy
A Hollywood actor whose career has seen better days, suddenly finds himself having to play the new role of father, when he is unexpectedly saddled with the welfare of his six-year-old daughter. Naturally, it proves quite difficult for him to reconcile the conflicting demands that fatherhood and acting make on his time. An uneven comedy that fails to blend its opposing elements of satire and drama into an effective whole.
COM   111 min (ort 116 min)   VIDrel: 20VIS/SONOP V/sur

I'LL SEE YOU IN MY DREAMS **
Michael Curtiz USA
U
1951
Danny Thomas, Doris Day, Frank Lovejoy, Patrick Wymore, James Gleason, Mary Wickes, Jim Backus, Julie Oshins, Minna Gombell, Harry Antrim, Bunny Lewbel, William Forrest, Robert Lyden, Mimim Gibson, Christy Olson, Dick Simmons
Reasonably well made musical biography of the songwriter Gus Kahn tracing his career from his beginnings in Chicago to his death in 1941. Packed full of songs, all of whose lyrics were penned by Kahn. Songs include "Ain't We Got Fun", "It Had To Be You" and "I'M Through With Love."
MUS   105 min (ort 110 min) B/W   VIDrel: WHV   V

I'LL TAKE MANHATTAN: PARTS 1 AND 2 **
Douglas Hickox/Richard Michaels USA
15
1986/1987
Valerie Bertinelli, Perry King, Francesca Annis, Barry Bostwick, Staci Love, Jane Kaczmarek, Jack Scalia, Paul Hecht, Timothy Daly, Donald Trump, Katherine Houghton, John Colicos, Louis Guss, Robert Milli, Corinne Bohrer
Following the mysterious death of her father, a spoilt young girl takes over his publishing empire and has to deal with the underhand schemes of a wicked uncle. A competent adaptation of Krantz's soap opera tale of love, deceit and intrigue, all wrapped up in a glossy and colourful package.
DRA   375 min (2 cassettes – ort 480 min) mTV
VIDrel: BRAVE/SONOP L/A   V
Boa: novel by Judith Krantz.

ILLEGAL ENTRY **
Henri Charr USA
18
1992
Barbara Lee Alexander, Gregory Vignolle, Arthur Roberts, Carol Hoyt, Sabryn Genet, David Milbern, Leigh Stevens, Robert Arentz, Mia Leigh, Dan Koko, Paul Brewster, Ron Althoff, Josh Levine, Rita Stephens, Alan Silverblatt, Nola Jones
A young girl's life turns into a nightmare when her parents are murdered and she and her boyfriend are forced to flee for their lives. When further killings occur, they also find themselves wanted by the police. A standard, cliched thriller with indifferent acting, poor plotting and a contrived ending.
Aka: ILLEGAL ENTRY: FORMULA FOR FEAR
THR   84 min (ort 88 min)   VIDrel: IMPENT   V

ILLEGAL ENTRY **
Richard W. Munchkin USA
18
1995
Corey Feldman, Kimberly Stevens, Brion James, Mark Derwin, Una Damon, Nicole Durant, Bela Lehoczky, Michael Phenici, Lorelei Leslie, Stacey Randall, Sy Richardson, Kimberly Stevens
A top model is pursued by an admirer whose attentions become increasingly unwelcome, all the more so when she begins to suspect him of being a killer who may be responsible for the deaths of no less than twelve of her colleagues. When the threat he poses becomes too much, she hires a private detective for protection.
Aka: EVIL OBSESSION
THR   93 min   VIDrel: GUILD/FOXVID   V

ILLEGAL IN BLUE **
Stu Stegall USA
18
1995
Dan Gauthier, Stacey Dash, Louis Giambalvo, Trevor Goddard, Michael Durrel, David Groh
A cop meets a blues singer who works in a nightclub but when

he initiates a passionate affair with her, he finds himself deeply embroiled in murder, as it transpires that she has possibly shot her millionaire husband and now needs a fall guy. A run-of-the-mill offering with pretensions in the erotic thriller department.
THR   90 min   VIDrel: MARQ   V

ILLICIT DREAMS **
Andrew Stevens USA
18
1993
Andrew Stevens, Shannon Tweed, Joe Cortese, Stella Stevens, Michelle Johnson, Brad Blaisdell, Rochelle Swanson, Dave Carlton, Jennifer Bassey, Bryan Goodwin, Carrie Dobro, Elizabeth Sandler, Roger Toussaint, Errol O. Coughlen
A frustrated housewife indulges in steamy erotic fantasies but finds that her dreams are coming alive and may prove deadly. An unimpressive effort in which former Playmate Tweed gives an adequate performance but there is little she or the rest of the cast can do to give this one any real interest.
DRA   91 min (ort 93 min)   VIDrel: 20TH/FOXVID   V/sur

ILLUSIONS **
Victor Kulle USA
15
1991
Robert Carradine, Heather Locklear, Emma Samms, Ned Beatty
A wife trying to save her marriage and recover her sanity, begins to suspect that her visiting sister-in-law may have more than sisterly feelings for her husband. Meanwhile, she confides her fears to her landlord only to find this does little to help her fragile state of mind.
DRA   98 min   VIDrel: CIC   V
Boa: play I'll Be Back Before Midnight by Peter Colley.

ILLUSTRATED MAN, THE **
Jack Smight USA
15
1969
Rod Steiger, Claire Bloom, Robert Drivas, Don Dubbins, Jason Evers, Tom Weldon, Christie Matchett
A brave but failed attempt to bring to the screen Bradbury's collection of fantasy tales, linked by the device of a tattooed man, each of whose pictures tells a sinister story. Atmospheric in parts, but generally too disjointed and opaque to be really effective.
FAN   98 min (ort 103 min) wScrn (special edition)
VIDrel: WHV   V/h
Boa: short stories: The Long Rain, The Veldt and The Last Night of the World by Ray Bradbury.

I'M ALL RIGHT JACK ***
John Boulting/Roy Boulting UK
U
1959
Ian Carmichael, Terry-Thomas, Peter Sellers, Richard Attenborough, Dennis Price, Irene Handl, Margaret Rutherford, Miles Malleson, Victor Maddern, Liz Fraser, John Le Mesurier, Marne Maitland, Kenneth Griffith, Terry Scott
An idealistic graduate takes a job at his uncle's factory, starting at the bottom. However, his well-meaning and conscientious ways eventually provoke a strike. Sellers is memorable as a gruff and bloody-minded union leader. Initially a brilliant satire on labour relations in 1950s Britain, the film progressively degenerates into a contrived web of intrigue, culminating in an ending that pulls all of its punches. See also THE ANGRY SILENCE.
Aka: I'M ALRIGHT JACK
COM   100 min (ort 104 min) B/W   VIDrel: WHV   V/h
Boa: novel Private Life by Alan Hackney.

I'M GONNA GIT YOU, SUCKA! ***
Keenen Ivory Wayans USA
15
1988
Keenan Ivory Wayans, Bernie Casey, Antonio Fargas, Isaac Hayes, Jim Brown, Steve James, Ja'net DuBois, Dawnn Lewis, John Vernon, Clu Gulager, Kadeem Hardison, Damon Wayans, George James, Marc Figueroa
Written by Wayans, this is a sharp parody of 1970s blaxploitation films that sees one Jack Spade out to avenge the death of his brother and clean up the ghetto at the same time. Disjointed and episodic it may be, but the gags fly thick and fast, and most of them hit their targets. Followed by a sequel of sorts – A LOW DOWN DIRTY SHAME.
COM   84 min (ort 89 min)   VIDrel: MGM/WHV L/A V/sh

IMAGE, THE **
Peter Werner USA
15
1990
Albert Finney, Kathy Baker, John Mahoney, Marsha Mason, Swoosie Kurtz, Spalding Gray, Wendie Jo Sperber, David Clennon, Brett Cullen, Jim Haynie, Robert Schenkkan, Nicholas Cascone, Beth Grant, Banks Harper, Brad Pitt
A ratings-hungry TV news anchorman begins to believe his own

reviews, and his ruthless thirst for good audience figures creates conflict and ultimate disillusionment. A variant on a theme explored in NETWORK, this provocative look at tele-journalism is certainly well acted, but ultimately says nothing of much importance.
DRA 89 min (ort 110 min) mCab VIDrel: MGM/WHV L/A V/sh

IMAGE OF BRUCE LEE, THE * 18
Yeung Kuen HONG KONG 1978
Bruce Li (Ho Tsung-Tao), Chang Wu Lang, Chang Lei
Our martial arts hero fights to expose a ring of international forgers in this totally forgettable tale.
MAR 88 min Cut (2 sec) VIDrel: IMPENT V

IMAGES OF DESIRE * 18
John T. Bone USA 1989
Victoria Paris, Deidre Holland, Jeanna Fine, Chessie Moore, Kelly Blue, Viper, Jamie Leigh, Raven, Madison, Shelly Kelly, Candice, Delta Force, Leslie, Kitty Reynolds
Lavish plotless sex film filmed over two continents but merely a parade of sex scenes.
A 68 min (ort 90 min) VIDrel: GROHOM/MAXSCAN V

IMAGINARY CRIMES *** PG
Anthony Drazan USA 1994
Harvey Keitel, Fairuza Balk, Kelly Lynch, Vincent D'Onofrio, Seymour Cassel, Chris Penn, Elisabeth Moss, Diane Baker, Amber Benson, Richard Venture, Tori Paul, Melissa Bernstein, Annette O'Toole, Bill Geisslinger, William Shilling
After his wife dies, a man tries to raise his two teenage daughters to the best of his ability, but has to face the fact that his career as a con-man does nothing to help matters. Excellent performances all round (especially from the ever-dependable Keitel) help to enhance this fine drama.
DRA 101 min (ort 106 min) cC VIDrel: WHV V/sur
Boa: book by Sheila Ballantyne.

IMAGINARY TALE, AN ** (PG)
Andre Forcier FRANCE 1990
Jean Lapointe, Louise Marleau, Charlotte Laurier, Marc Messier, France Castel, Jean-Francois Pichette, Tony Nardi, Marc Gelinas, Louis De Santis, Warren "Slim" Williams, Donald Pilon, Leo Munger, Louise Gagnon, Angelo Cadet
In Montreal a Jazz trumpeter falls is love with an actress playing in a local production of Othello, but his involvement with her draws him ever more deeply into the various intrigues, petty jealousies and squabbles of the theatre folk she works with. A lighthearted romantic drama with a few touches of comedy, its exuberance does much to entertain, but the meandering plot eventually becomes a handicap it cannot overcome.
Aka: UNE HISTOIRE INVENTEE
DRA 100 min CINrel

IMAGINE * 18
Michael Craig USA 1991
Ashlyn Gere, Randy West, Brittany Morgan, Erica Boyer, Raven, Missy Warner, Ashley Nicole, Dusty, Mike Horner, John Dough, T.T. Boy
A female executive, unsatisfied in her married sex life, has an over-active imagination and begins to think that she sees couples indulging themselves everywhere she looks.
A 62 min (ort 71 min) VIDrel: FIFTH/DISC V

IMMACULATE CONCEPTION ** 18
Jamil Dehlavi UK 1991
James Wilby, Melissa Leo, Shabana Azmi, Zia Mohyeddin, James Cossins, Shreeram Lagoo, Ronny Jhutti, Tim Choate, Bhaskar, Bill Bailey, Zafar Hameed, Zahoor Ahmad, Yaqoob Zakaria, Nasir Anwar Khan, Sultan Khan, Intezar Hussain
Together with her husband, the daughter of a top US senator goes on a journey to a shrine in India, her hope being that by making this pilgrimage they will be blessed with a child. A strange, quirky and over-emotional drama, that offers a number of interesting observations without presenting any clear message or useful insights. This was Dehlavi's feature debut.
DRA 118 min (ort 120 min) VIDrel: VCC L/A V

IMMORAL MR TEAS, THE ** 18
Russ Meyer USA 1958
Bill Teas, Ann Peters, Marilyn Wesley, Dawn Denelle, Michele Roberts
The first of many films produced by a director obsessed with big-breasted women. This one tells the story of a sexually frustrated bachelor, who satisfies his desires by ogling women and imagining them naked. As time goes on he gets better at this and the viewer is treated to much nudity, if not action (poor Mr Teas never does meet anyone). An innocuous and innocent film that now looks very dated, but despite this, retains a certain naive charm.
A 63 min VIDrel: L/A V

IMMORAL TALES * 18
Walerian Borowczyk FRANCE 1974
Lise Danvers, Fabrice Luchini, Charlotte Alexandra, Paloma Picasso, Pascal Christophe, Florence Bellamy, Jacopo Berinizi
A collection of four mediocre erotic tales that are supposedly an essay on life, love and women, with each episodes set in a different time and place. In "The Tide" a young man teaches his cousin the joys of sex, "Therese, The Philosopher" tells of a girl who finds a book of erotic drawings and learns to masturbate, in "Erzsebet Bathory" a countess rounds up the local virgins and in "Lucrezia Borgia" a female Borgia makes love to Pope Alexander VI.
Aka: COMTES IMMORALS; CONTES IMMORAUX
A 98 min (ort 100 min) VIDrel: CONNO/RTM V

IMMORTAL BELOVED *** 15
Bernard Rose USA 1994
Gary Oldman, Jeroen Krabbe, Isabella Rosselini, Johanna Ter Steege, Valeria Golino, Marco Hofschneider, Miriam Margoyles, Barry Humphries, Gerard Horan, Luigi Dilberti, Matthew North, Leo Faulkner, Christopher Fulford, Alexander Pigg
After the death of Beethoven, his devoted assistant and admirer faces an uphill task in trying to fulfil the terms of his will, which left the bulk of his estate to a mysterious woman known only by the title pseudonym. A sumptuous, if heavily fictionalised account, that benefits enormously from its location shooting in Prague and of course, its wonderful soundtrack. See also BEETHOVEN'S NEPHEW for another look (albeit also fictionalised) at the life of this genius.
DRA 115 min (ort 121 min) VIDrel: EIV/SONOP V/sur

IMMORTAL SINS ** 15
Herve Hachuel SPAIN/USA 1991
Cliff De Young, Maryam D'Abo, Shari Shattuck, Tony Isbert, Paloma Lorena, Miguel De Grandy, Jose Yepes, Manuel Pereiro, Paul Soldevilla, Oswaldo Delgado Sanchez, Natalia Diaz, Isidro Espino, Francisco Moreno, Jesus Garcia
A couple inherit a castle in Spain, little suspecting that the property carries a curse, and when the husband becomes involved with a woman he believes to be a distant cousin, he finds himself trapped in a dangerous relationship. A low-grade horror film with no lack of special effects but little else, and with the standard fiery climax seen in countless films of this ilk.
HOR 89 min (ort 100 min) VIDrel: 20VIS/SONOP V

IMMORTAL STORY, THE *** 15
Orson Welles FRANCE 1969
Orson Welles, Jeanne Moreau, Rogger Coggio, Norman Eshley, Fernando Rey
A fabulously wealthy merchant in 19th century Macao gets his clerk to hire a beautiful woman to play his wife, and a sailor to make love to her, this being his way of bringing to life an old legend about an aged miser who gets a sailor to make love to his wife. With none of the flashy camerawork of early Welles movies, this is a most unusual and well-staged adaptation of Dinesen's novella, and it has a strange, magical atmosphere all of its own.
Aka: UNE HISTOIRE IMMORTELLE
DRA 55 min (ort 58 min) mTV VIDrel: CONNO/RTM V
Boa: novella Skibsdrengens Fortaeling by Isak Dinesen (Karen Blixen).

IMMORTALIZER, THE * 18
Joel Bender USA 1990
Ron Ray, Chris Crone, Melody Patterson, Clarke Lindsley, Steve Jamieson, Heg Joujon-Roche, Bekki Armstrong, Cynthia Chase, Bob Verne, Terry Miller, Brian Savid Zola, Tommy Lamparski, Raye Hollitt, Bo Byers, Elmarie Wendel, Karen Moore
A brilliant but demented doctor discovers a way of putting old brains into new bodies, and starts offering his services at a price, but is obliged to resort to kidnapping in order to obtain the young donors he needs. A gory, tedious and totally over-the-top shocker, badly acted and hardly helped along by some woefully unimpressive special effects that just seem to go on for ever.
HOR 92 min (ort 96 min) VIDrel: 20VIS/SONOP V

IMMORTALS, THE ***
Brian Grant USA   18
                                                1995
Eric Roberts, Tia Carrere, Joe Pantoliano, Tony Curtis, Chris Rock, Kevin Bernhardt, Kieran Mulroney
Roberts manages a rundown nightclub owned by his Mafia boss Curtis, and recruits four groups of strangers to take part in a complicated robbery. His plan relies on the fact that all the men recruited have a fatal illness that will eventually kill them anyway, but the reason he has chosen them only becomes clear as the story develops. An unusual and clever story, a bit let down by its occasional forays into slapstick comedy.
A/AD   94 min   VIDrel: 20TH/FOXVID   V/sh

IMP, THE **
David De Cocteau USA   18
                                                1987
Linnea Quigley, Michelle Bauer, Andras Jones, Robin Rochelle, Brinke Stevens, Kathi Orbrecht, Buck Flower, John Stuart, Hal Havner, Carla Baron, John Wildman, Michelle McKlintock, Craig Caton
A crazy and overblown fantasy horror tale in which a nasty little imp is released from a bowling trophy, and wreaks havoc among the college kids who unwittingly released him.
Aka: SORORITY BABES IN THE SLIME BOWL-A-RAMA
HOR   75 min Cut (1 min 9 sec – ort 85 min)
VIDrel: ALLIED/RTM   V

IMPORTANCE OF BEING EARNEST, THE ****
Anthony Asquith UK   U
                                                1952
Michael Redgrave, Michael Denison, Edith Evans, Richard Wattis, Margaret Rutherford, Joan Greenwood, Dorothy Tutin, Miles Malleson, Walter Hudd, Ivor Barnard, Aubrey Mather
Wilde's famous comedy of manners, about two bachelors and their troubled love lives. Set in Victorian England and boasting an impeccable cast who play their parts to perfection.
COM   91 min (ort 95 min)   VIDrel: CARL/TECH   V
Boa: play by Oscar Wilde.

IMPROMPTU ***
James Lapine FRANCE/UK/USA   15
                                                1990
Judy Davis, Hugh Grant, Mandy Patinkin, Bernadette Peters, Julian Sands, Ralph Brown, George Corraface, Anton Rodgers, Emma Thompson, Anna Massey, David Birkin, Nimer Rashee, Fiona Vincente, John Savident, Elisabeth Spriggs
A witty and light-hearted account of George Sand's determined wooing of Frederic Chopin between 1836 and 1838, and of the jealousies and rivalry she had to overcome after an initial setback to her plans. The period atmosphere is lovingly recreated while some fine performances bring these historical figures to life as fully rounded individuals. A most enjoyable film, all the more remarkable for being Lapine's directorial debut.
DRA   103 min (ort 109 min)
VIDrel: ENCORE/SPEAR/COLUM   V

IMPROPER CONDUCT **
Jag Mundhra USA   18
                                                1994
Steve Bauer, Lee Anne Beaman, Tahnee Welch, John Laughlin, Kathy Shower, Nia Peeples, Adrian Zmed, Patsy Pease, Stuart Whitman
When her sister suffers sexual harassment at work from her boss, she brings an action against her tormentor, but loses it and suffers public humiliation. When this leads to her committing suicide, her feminist sister sets out to plan a fitting revenge.
THR   94 min   VIDrel: NEWAGE/TECH L/A   V

IMPUDENT GIRL, AN ***
Claude Miller FRANCE   15
                                                1985
Charlotte Gainsbourg, Bernadette La Font, Jean-Claude Brialy, Raoul Billerey, Clothilde Baudon, Julie Glenn, Jean-Philippe Ecoffey
A young French girl lives an isolated life with her widowed father and older brother, but this changes when a noted pianist arrives in town, who is in fact a child prodigy of her own age. A devoted fan of the latter, she pursues he with great determination, her dearest wish being to become her manager. A sensitive and delightful film, it explores the follies and traumas of childhood with style and wit.
Aka: L'EFFRONTEE
DRA   93 min (ort 97 min) wScrn   VIDrel: ARTIF/20TH   V
Boa: book The Member of the Wedding by Carson McMullers.

IMPULSE **
Graham Baker USA   18
                                                1985
Meg Tilly, Tim Matheson, Hume Cronyn, John Karlen, Bill Paxton, Claude Earl Jones, Amy Stryker, Robert Wightman, Lorinne Vozoff, Peter Jason, Mary Celio, Abigail Booraem, Jack T. Collis, Christian Crane, Chuck Dorsett, Anne Haney
The inhabitants of a small town suddenly and inexplicably begin to behave in a violent and irrational way. A woman goes home to investigate her mother's attempted suicide and gets caught up in this mystery. A modest little fantasy-thriller that offers no explanation and has no real resolution.
FAN   87 min (ort 95 min)   VIDrel: POLY/POLYREC L/A   V

IN A CHILD'S NAME **
Tom McLoughlin USA   15
                                                1991
Michael Ontkean, Valerie Bertinelli, Louise Fletcher, Timothy Carhart, David Huddleston, John Karlen, Caroline Kava, Joanna Merlin, Mitchell Ryan, Karla Tamburelli, Vincent Guastaferro, Christopher Meloni, Andy Hirsch
A fact-based drama in which a man with a history of violence is charged with his wife's murder, and immediately commences an action to ensure he gains custody of his five-month-old son.
DRA   156 min (ort 174 min) mTV   VIDrel: NWV/SONOP V/h
Boa: book by Peter Maas.

IN A LONELY PLACE ***
Nicholas Ray USA   PG
                                                1950
Humphrey Bogart, Gloria Grahame, Frank Lovejoy, Robert Warwick, Carl Benton Reid, Art Smith, Jeff Donnell, Martha Stewart
A sour and arrogant screenwriter who suffers from self-destructive urges and a violent temper, becomes the chief suspect when a woman he briefly met is found murdered. Though innocent, the charge poisons and eventually destroys the only relationship that might have given him lasting happiness. An unusual and harsh melodrama of little warmth.
DRA   89 min (ort 93 min) B/W   VIDrel: COLUM/SONOP V
Boa: novel by Dorothy B. Hughes.

IN A STRANGER'S HANDS **
David Greene USA   15
                                                1991
Robert Urich, Megan Gallagher, Isabella Hofmann, Brett Cullen, Maria O'Brien, Dakin Matthews, Alan Rosenberg, Christine Dunford, Erica Dill, Vondi Curtis-Hall, Janel Mahoney, Walter Addison, Richard Hardacre, Russell Curry
A wealthy man sets out to help a woman whose young daughter has been abducted, and together they reveal the existence of a nasty gang of professional child-snatchers.
THR   89 min (ort 93 min) mTV   VIDrel: ODY/SONOP V/sh
Boa: novel The Last Stop by Matt Benjamin.

IN A YEAR OF 13 MOONS **
Rainer Werner Fassbinder WEST GERMANY   18
                                                1978
Volker Spengler, Ingrid Caven, Gottlieb John, Elizabeth Trissenaar, Eva Mattes, Gunther Kauffmann
A man has a sex-change operation out of love for his business partner but finds himself abandoned and left to indulge in a number of self-destructive relationships. A ponderous, pessimistic and pretentious piece, full of the director's favourite obsessions, but having very little of worth to say about the human condition.
Aka: IN A YEAR WITH 13 MOONS; IN EINEM JAHR MIT 13 MONDEN
DRA   119 min   VIDrel: CONNO   V

IN BETWEEN **
Thomas Constantinides USA   (PG)
                                                1991
Wings Hauser, Robin Mattson, Robert Forster, Alexander Paul
Three people awake to find themselves in a strange house and learn that they have died and have gone to a kind of halfway house. Here they must decide between reincarnation on Earth or passing on to some form of afterlife.
COM   92 min   CABrel: HVC

IN BROAD DAYLIGHT **
James Steven Sadwith USA   15
                                                1990
Marcia Gay Harden, Brian Dennehy, Cloris Leachman, John Anderson, Robert Schenkkan, Ken Jenkins, Chris Cooper, David Neidorf, Bill Thurman, Brandon Smith, Colleen Keegan, Tony Frank, Bill Bolender, Greta Muller, Debra London
A small town finally revolts against the rule of the local armed psychopath who swears vengeance on all and sundry after his daughter is arrested for shoplifting. A nasty tale of gratuitous evil and violence, with Dennehy in fine form in a highly unsympathetic role. Based on a true story.
DRA   91 min (ort 100 min) mTV   VIDrel: NWV/HIFLI   V

**IN COUNTRY ** 15
Norman Jewison USA 1989
Bruce Willis, Emily Lloyd, Joan Allen, Kevin Anderson, John Terry, Peggy Rhea, Judith Ivey, Richard Hamilton, Patricia Richardson, Jim Beaver
A shell-shocked Vietnam veteran and a young girl who lost her father in the conflict, each try to come to terms with the war in their own way. Willis is excellent as the girl's uncle, but the flawed script is as uneven as it is well-intentioned, though a final sequence set at the Veterans' Memorial in Washington D.C. is surprisingly moving.
DRA   110 min (ort 120 min)   VIDrel: WHV   V/sur
Boa: novel by Bobbie Anne Mason.

**N CUSTODY ** U
Ismail Merchant INDIA 1994
Shashi Kapoor, Shabana Azmi, Om Puri, Neena Gupta, Rupinder Kaur, Shahid Masood, Tiblu Khan, Afzal Khan, Manzoor Ahtesham, Alakh Nandan, Tinnu Anand, Amit Goswani, Maza Bi, Pragya Mekherjee, Sushma Seth, Umer Sahib, Gopal Janum
A university teacher and would-be poet is engaged by a magazine to interview his idol, the greatest Urdu-language poet in India, but is shocked to find that the latter has fallen on hard times, and is reduced to taking rooms in a brothel. An ambitious but failed attempt to produce a satire on life in modern India that does not really work.
Aka: HIFAZAAT
DRA   120 min (ort 126 min)   VIDrel: CURZON/20TH
V/sur
Boa: novel by Anita Desai.

IN DEFENSE OF SAVANNAH * 18
Paul Thomas USA 1992
Savannah, Christy Canyon, Britt Morgan, Jeanna Fine, Jon Dough, Peter North, Scott Irish, T.T. Boy, Mickey Ray, Carl Esser, Henri Pachard
A porno director's latest film is so provocative that he finds himself in court defending his work against obscenity charges. All the film's stars are present at the hearing, and it's decided that the film will have to be shown for the charges to be tried. The film-within-a-film structure adds little of interest to this one.
A   81 min (Cut before video submission by 30 min 2 sec)
VIDrel: GROHOM/MAXSCAN   V

**IN EXCESS ** 18
Mauro Bolognini ITALY 1992
Julian Sands, Joanna Pacula, Tcheky Karyo, Lara Wendel, Marco Di Stefano, Jeanne Valerie, Veronica Del Chiappa, Inez Nobili, Sonia Topazio
A young couple decide on open marriage with total honesty and when the wife takes a lover, the husband insists on hearing all the details. However, as things become more intense, their relationship comes under increasing strain, in this confused and unappealing adaptation.
Aka: HUSBANDS AND LOVERS
DRA   91 min (ort 94 min)   VIDrel: VCC/DISC L/A   V
Boa: novel by Albert Moravia.

**IN GOLD WE TRUST ** 15
P. Chalong THAILAND 1990
Jan-Michael Vincent, Sam Jones, James Phillips, Sherrie Rose, Michi McGee, Robert Cespedes, James Phillips, Nappon Gomarchun, Christoph Kluppel, Senior Gomarchun, Herb "Superb" Jones, Alan Chedester, Dean Alexander
A mindless but exciting actioner, set in the now over-familiar jungles of Southeast Asia, where a group of former US troops and Japanese soldiers left behind in 1945, fight it out over a vast fortune in gold. A standard entry in this genre, that offers no surprises, and whose minimal plotting hardly matters in the face of an endless welter of gunfire and explosions.
A/AD   87 min (ort 89 min)   VIDrel: FIRST/SONOP   V

**IN HARM'S WAY ** PG
Otto Preminger USA 1965
John Wayne, Kirk Douglas, Henry Fonda, Patricia Neal, Carroll O'Connor, Tom Tryon, Paula Prentiss, Dana Andrews, Brandon de Wilde, George Kennedy, Slim Pickens, Larry Hagman, Bruce Cabot, Jill Haworth, Franchot Tone
Overlong and highly tedious account of one navy man's part in the struggle against Japan after the attack on Pearl Harbour. A

few action sequences fail to provide more than a momentary distraction.
WAR   160 min (ort 165 min) B/W
VIDrel: 4-FRONT/POLYREC/CIC   V/h
Boa: novel by James Bassett.

**IN LOVE AND WAR ** 15
Richard Attenborough USA 1996
Sandra Bullock, Chris O'Donnell, Mackenzie Astin
In the last year of WW1, a young Ernest Hemingway is working for the American Red Cross in Italy, but finds himself bored with his duties. In an attempt to get a closer look at the fighting, he sets out for the trenches, but is wounded. He then embarks on a passionate affair with his nurse (Agnes Von Kurowsky) but eventually loses out to a rival, a wealthy doctor. Partially based on fact, this stilted tale is a curious mixture of romance and tedium.
DRA   115 min   CINrel

IN LOVE WITH AN OLDER WOMAN * PG
Jack Bender USA 1982
John Ritter, Karen Carlson, Jamie Ross, Robert Mandan, Jeff Altman, George Murdock, Robin Curtis, Wendall Wright, Deborah Tilton, Blaine Novak, Robert Townsend, Jo Anne Astrow, Mary-Alan Hockanson, Michael Cummings, Sandy Ward
A younger man and an older woman face scorn and prejudice because of their relationship. A generally agreeable comedy-drama that is better than one might have expected, mainly thanks to good work from Ritter and Carlson. The script is by Michael Norell. See also WHITE PALACE.
DRA   100 min mTV   VIDrel: L/A   V
Boa: novel Six Months With An Older Woman by David Kaufelt.

**IN OLD CALIFORNIA ** U
William McGann USA 1942
John Wayne, Binnie Barnes, Albert Dekker, Helen Parrish, Patsy Kelly, Edgar Kennedy, Dick Purcell, Harry Shannon, Charles Halton, Emmett Lynn, Milton Kibbee, Bob McKenzie, Paul Sutton, Anne O'Neal, Frank McGlynn, Jack O'Shea
A pharmacist comes to Sacramento and has various adventures in the midst of the gold rush in a rather anaemic film that is not one of Wayne's better efforts.
WES   89 min B/W   VIDrel: 4-FRONT/POLYREC   V/h

IN OLD CHICAGO * U
Harry King USA 1938
Tyrone Power, Alice Faye, Don Ameche, Alice Brady, Andy Devine, Tom Brown, Brian Donlevy, Phyllis Brooks, Sidney Blackmer, Berton Churchill, June Storey, Paul Hurst, Tyler Brooke, J. Anthony Hughes, Gene Reynolds
Sprawling, colourful account of the background to the great Chicago fire of 1871 that is built around the lives of an Irish widow and her three sons as they grow up and become involved with various aspects of the city's life. One of the most expensive productions of its time, costing well over $1,800,000, it has a twenty-minute sequence of the fire that audiences found highly impressive. AA: S. Actress (Brady).
DRA   91 min (ort 115 min) B/W   VIDrel: 20TH/TECH   V/h

IN PRAISE OF OLDER WOMEN * 18
George Kaczender CANADA 1978
Tom Berenger, Karen Black, Susan Strasberg, Helen Shaver, Alexandra Stewart, Marilyn Lightstone, Marianne McIsaac, Alberta Watson, Louise Marleau, Monique Lepage, Ian Tracey, Jill Frappier, Mignon Elkins, Joan Stuart
The story of a Hungarian emigre and his numerous sexual encounters, seen as a series of tiresome flashback sequences, in which semi-clothed women are paraded across the screen for our edification. But for all that the film remains a boring dud.
DRA   105 min Cut (Cut at film release by 8 sec – ort 110 min)
VIDrel: MGM/WHV L/A   V
Boa: novel by Stephen Vizinczey.

IN PURSUIT OF HONOR * 15
Ken Olin USA 1995
Don Johnson, Craig Sheffer, Rod Steiger, Gabrielle Anwar, Bob Gunton, James B. Sikking, John Dennis Johnston
After the US cavalry switches to tanks, two soldiers are ordered to destroy the remaining five-hundred horses. However, as cavalry men they find it impossible to carry out their orders and mutiny instead, heading for Canada along with their four-legged charges.
DRA   110 min   VIDrel: MOSAIC/COLUM   V/sh

IN SEARCH OF THE CASTAWAYS ***
Robert Stevenson UK/USA
Hayley Mills, Maurice Chevalier, Wilfrid Hyde White, George Sanders, Wilfrid Brambell, Michael Anderson Jr, Antonio Cifariello, Keith Hamshere, Jack Gwillim, Ronald Fraser, Inia Te Waita, Norman Bird, Michael Wayne, Milo Sperber
Three children and a professor go to South America in search of a missing sea captain, encountering various disasters and hazards along the way. A colourful Disney adventure tale, that suffers rather badly from an uneven script, but is generally good clean fun.
A/AD   94 min (ort 100 min)   VIDrel: WDV/TECH   V
Boa: novel Captain Grant's Children by Jules Verne.

U
1961

IN THE ARMY NOW *
Daniel Petrie Jr USA
Pauly Shore, Lori Petty, David Alan Grier, Andy Dick, Esai Morales, Lynn Whitfield, Art LeFleur, Tom Villard, Keith Coogan, Paul Mooney, Fabriana Udenio, Beau Billingslea
A meek electronics-store clerk with pacifist leanings, joins the US Army Reserves only to find himself shipped off overseas. As expected, he makes an absolute mess of military life, proving that there are no ideas on offer as far as this kind of comedy is concerned. Amazingly, this unfunny dud was the work of no less than five writers: Ken Kaufman, Stu Kriger, Daniel Petrie Jr, Fax Bahr, Adam Small.
COM   92 min   VIDrel: HOLPIC/TECH   V

PG
1994

IN THE BEST INTEREST OF THE CHILD **
David Greene USA
Meg Tilly, Ed Begley Jr, Michael O'Keefe, Michele Greene, Marta Woodward, David Wohl, James Eckhouse, Jim Byrnes, Angela Bassett, Peter Hansen, Tom Irwin, Rosana Huffman, Cynthia Mace, Maria O'Brien, Robin Pearson Rose, Kenneth Tigar
A divorced mother is forced into flouting the law by refusing to allow her ex-husband custody of their young daughter in order to stop him from continuing to abuse her sexually. An effective indictment of the failings of the American legal system.
DRA   90 min (ort 100 min) mTV   VIDrel: CIC/SONOP L/A V

15
1990

IN THE BEST INTEREST OF THE CHILDREN ***
Michael Ray Rhodes USA
Sarah Jessica Parker, Elizabeth Ashley, Sally Struthers, Lexi Randall
Based on true events, this harrowing story tells of the attempts made by a mother to regain custody of her children, after they were placed with foster parents following her treatment at hospital for manic depression.
DRA   90 min   VIDrel: ODY/SONOP   V/s

PG

IN THE BLEAK MIDWINTER ***
Kenneth Branagh UK
Michael Maloney, Richard Briers, Celia Imrie, Hetta Charnley, Joan Collins, Nicholas Farrell, Mark Hadfield, Gerard Horan, Jennifer Saunders, Julia Sawalha, John Sessions, Ann Davies, James D. White, Robert Hines, Allie Byrne
A young actor struggles to put on a Christmas production of Hamlet in an abandoned village church, but without any money finds he attracts nothing but misfits. This behind-the-scenes look at the antics, jealousies and intrigues of the characters has some very funny cameos and if not overly amusing, still remains a gentle and pleasing work. See also NOISES OFF!
COM   95 min (ort 98 min) B/W   VIDrel: COLUM/SONOP V/sur

15
1995

IN THE BLINK OF AN EYE **
Micki Dickoff USA
Veronica Hamel, Mimi Rogers, Carlos Gomez, Jeffrey Dean Morgan, Brian Markinson, Mary Mara, Judith-Marie Bergan, Polly Bergen
A film-maker fights a lone battle to secure the release of a childhood friend, the latter having been imprisoned for a crime she did not commit. A fact-based drama.
DRA   86 min   VIDrel: ODY/SONOP   V/sh

12
1995

IN THE COLD OF THE NIGHT **
Nico Mastorakis USA
Jeff Lester, Marc Singer, Shannon Tweed, Adrienne Sachs, David Soul, John Beck, Tippi Hedren
A photographer has a strange dream in which he murders a woman. Later he is absolutely astonished to meet her in the flesh when she turns up on his doorstep.
THR   107 min (ort 112 min)   VIDrel: CIC   V

18
1990

IN THE COMPANY OF DARKNESS **
David Anspaugh USA
Helen Hunt, Steven Weber, Jeff Fahey, Juan Ramirez, Margaret Travolta, Don Conway, Annabel Armour, Michael Barcella, Julian Brams, Marilyn Dodds Franks, Danny Goldring, Irma Hall, Don James, Kim Klutznik, Tony Mockus Jr
A rookie cop is assigned to the case of a series of child killings in a small Wisconsin town and goes there undercover to unmask the culprit, who it would appear is a highly intelligent man who enjoys playing cat-and-mouse games with the authorities. Strong characterisations, especially of the lady cop (who has a traumatic past of her own) helps make this one stand out, even if there are some sequences both disturbing and unnecessary.
THR   91 min mTV   VIDrel: CIC/SONOP   V

(18)
1992

IN THE CUSTODY OF STRANGERS ***
Robert Greenwald USA
Martin Sheen, Jane Alexander, Emilio Esterez, Kenneth McMillan, Ed Lauter, Matt Clark, John Hancock, Virginia Kiser, Jon Van Ness, Judyann Elder, Deborah Foreman, Susan Peretz, Peter Jurasik, Pat McNamara, Ramon Estevez
Explores the troubled relationship between a father and his wild son, when the latter is put in jail on charges of drunkenness and assault, and the parents are unable to get him out. A thought-provoking melodrama with a good script by Jennifer Miller.
DRA   91 min (ort 100 min) mTV   VIDrel: MIA/DISC   V

15
1982

IN THE DEEP WOODS **
Charles Correll USA
Rosanna Arquette, Anthony Perkins, Will Paton, D.W. Moffett. Chris Rydell, Amy Ryan, Beth Broderick, Harold Sylvester, Kimberly Beck, Paul Perri, Greg Kean, Ned Bellamy, Mary Gregory, David Grant Wright, Ava Lazar, Donalee Wood
The search for a serial killer leads the police to a young woman and children's author. She may know more about the criminal's identity than she realises, having lost a childhood friend to this killer, who appears to target successful career women. Under suspicion herself, she forms an alliance with a private detective who is also trying to solve this case. A tense and rather unpleasant psychological thriller, sustained by the good work of the two leads.
THR   91 min (ort 94 min)   VIDrel: COLUM/SONOP   V
Boa: novel by Nicholas Conde.

18
1992

IN THE EYES OF A STRANGER ***
Michael Tashiyuki Uno CANADA
Richard Dean Anderson, Justine Bateman, Gordon Pinsent, Cynthia Dale, Denis Akiyana, Geza Kovacs, Colin Fox
A woman on a subway witnesses a professional assassination, and the fatally wounded man (who is a robber) reveals the location of $2,000,000 in stolen money to her. Fearing for her life, she obtains police protection, and it is not long before this couple start making plans to retrieve the money for themselves, an action not without danger. A cleverly plotted caper movie, with enough plot twists to keep one guessing.
THR   92 min (ort 94 min)   VIDrel: GUILD/SONOP   V

15
1992

IN THE FRAME **
Wigbert Wicker UK
Ian McShane, Barbara Rudnik, Lyman Ward, Peter Sattman, Amadeus August, Patrick Cauderlier, Rainer Grenkowitz, Cedric Smith, Joseph Ziegler, Laura Dickson, Hans-Peter Korff, Liliane Clune, Karin Rasenack
A Jockey Club investigator has to deal with a bombing, a murder and a couple of burglaries, the victims all turning out to be race-horse owners. See also TWICE SHY, another film in this series.
Aka: DICK FRANCIS MYSTERIES: IN THE FRAME
DRA   95 min   VIDrel: IMC/DISC   V
Boa: novel by Dick Francis.

PG
1989

IN THE HEAT OF PASSION **
Rodman Flender USA
Sally Kirkland, Nick Corri, Michael Greene, Jack Carter, Gloria LeRoy, Carl Franklin
Low-budget attempt at a erotic thriller with a young mechanic taking a fancy to a married woman psychiatrist but facing problems from her irate husband. Well acted by marred by low production values and an over-ambitious script.
DRA   82 min (ort 86 min)   VIDrel: COLUM/SONOP   V

18
1991

IN THE HEAT OF THE NIGHT ****
Norman Jewison USA
Sidney Poitier, Rod Steiger, Warren Oates, Lee Grant, Scott Wilson,

15
1967

Larry Gates, Quentin Dean, James Patterson, Anthony James, William Schallert, Jack Teter, Matt Clark, Kermit Murdock, Khalil Bezaleel, Beah Richards, Clegg Hoyt
A superb drama about a black Philadelphia homicide cop who is reluctantly roped into helping a Southern sheriff in a murder case, thus exposing himself to local bigotry and hatred. Poitier and Steiger complement each other brilliantly and the music of Ray Charles is used to great effect. The vastly inferior "They Call Me Mr Tibbs" followed. AA: Pic, Actor (Steiger), Edit (Hal Ashby), Screen/adapt (Stirling Silliphant), Sound (Goldwyn Studios).
DRA   105 min (ort 109 min)   VIDrel: MGM/WHV   V
Boa: novel by J. Ball.

IN THE LAKE OF THE WOODS **
Carl Schenkel USA
Peter Strauss, Kathleen Quinlan, Peter Boyle
A politician running for the Senate finds his dubious past in the Vietnam era catching up with him, and retreats with his wife to escape the pressures of a bunch of journalists keen to get at the truth.
THR   87 min   VIDrel: 20TH/FOXVID   V/h

15
1995

IN THE LINE OF DUTY ***
Yuen Wo Ping HONG KONG
Michelle Khan (Yeoh Chu Kheng), Michael Wong, Donnie Yen
When an agent from the Anti-Narcotics Bureau is murdered whilst investigating drug-running links between criminals and American Intelligence, a tough female martial arts expert is given the task of cracking the case. The start of a series of such films, offering what was to become a standard blending of martial arts and action within the format of a police story.
A/AD   90 min   VIDrel: ONE/IMPENT   V

18
1989

IN THE LINE OF DUTY 2 **
Cha Chun Yee (Cha Fu-Yi) HONG KONG
Cynthia Khan (Yang Li Chiang)
More high-kicking action as our lady heroine goes after the baddies. This film was re-released as "In The Line Of Duty 5" and to confuse matters still further, that title is also sometimes used for a film completely unrelated to this extensive series.
Aka: MIDDLE MAN
A/AD   91 min Cut (18 sec)   VIDrel: IMPENT L/A   V

18
1987

IN THE LINE OF DUTY 2: BLOOD BROTHERS **
Dick Lowry USA
Charles Haid, James Farentino, Susan Walters, Stephen Weber, Harold Sylvester
A cops-in-action film revolving around the efforts made by a city's police force to fight crime and keep morale up in the face of the murder of one of their colleagues. Followed by IN THE LINE OF DUTY 3: TIME TO KILL.
A/AD   90 min   VIDrel: GENESIS   V

15
1990

IN THE LINE OF DUTY 3: TIME TO KILL **
Dick Lowry USA
Rod Steiger, Michael Gross, Gary Basaraba, Christopher Rich, Brad Sullivan, Amy Wright, Beth Fowler, Henderson Forsythe, Tony Higgins, Daniel Ziskie, David Hart, Michael Dolan, Madison Arnold, Kristin Griffith, Randall Patrick
Third in this series of films has an intrepid FBI agent tracking down a mad killer who has murdered two Federal marshals. He soon learns that his quarry heads a shadowy, sinister white supremacist organisation whose supporters are prepared to do anything to protect their leader. A competent exercise in contemporary nastiness, but hardly outstanding cinema. See INTO THE HOMELAND and BLIND HATE for two more films on this theme.
Aka: IN THE LINE OF DUTY: THE TWILIGHT MURDERS; TIME TO KILL: IN THE LINE OF DUTY 3; TWILIGHT MURDERS, THE
A/AD   91 min (ort 95 min)   VIDrel: GENESIS   V

15
1990

IN THE LINE OF DUTY: AMBUSH AT WACO *
Dick Lowry USA
Tim Daly, William O'Leary, Neal McDonough, Lewis Smith, Marlee Shelton, Jeri Lynn Ryan, Stephen Mailer, Debra Jo Rupp, Clu Gulager, Gordon Clapp, Richard McGonagle, Susanna Thompson, Heather McAdam, Kris Kamm, Dan Lauria
Hurriedly made docu-drama on this incident that was actually in the process of being shot while these tragic events were unfolding. A dull and tasteless effort of little merit.
Aka: AMBUSH AT WACO: IN THE LINE OF DUTY
DRA   88 min (ort 93 min)   VIDrel: ODY/SONOP   V/sh

18
1993

IN THE LINE OF DUTY: HUNT FOR JUSTICE **
Dick Lowry USA
Nicholas Turturro, Adama Arkin, Dan Lauria, Adam Arkin
An FBI agent and a New Jersey cop join forces to catch the terrorists who have been carrying out a series of armed robberies, in the course of which they killed a police officer. One in a series of American cop films (it has no connection with the similarly titled martial arts series). Fair.
A/AD   87 min (ort 90 min)   VIDrel: ODY/SONOP   V/sh

12
1995

IN THE LINE OF DUTY: KIDNAPPED **
Bobby Roth USA
Timothy Busfield, Lauren Tom, Dabney Coleman, Tracey Walter, Henry Sanders, Barbara Williams, Carmen Argenziano, Richard Zavaglia, David Purdhamn, Christine Eastabrook, Erik Von Detten, Martin Davidson, Susie Singer, Justin Garms
An FBI agent reacts wildly when his son is kidnapped by a criminal whose speciality is targeting the children of the super-rich in L.A. With the help of his partner he sets out to track down the criminal responsible, who turns out to be a corrupt employee of the Internal Revenue – hence his ability to select the wealthy as his victims. Another well written and enjoyable story in the series, though it is claimed that this one is based on fact.
A/AD   87 min (ort 90 min) mTV   VIDrel: ODY/SONOP   V/sh

PG
1994

IN THE LINE OF DUTY: STREET WARS **
Dick Lowry USA
Mario Van Peebles, Ray Sharkey, Peter Boyle, Courney B. Vance, Michael Boatman, Morris Chesnut, Laurie Morrison, Merlin Santana, Trazana Beverly, Kenny Leon, Damon Pooser, Troy Hogan, C. Harrison Avery Jr, Jaime Tirelli
Two cops are firm buddies and partners who patrol the streets of New York City. When one of them is killed by a local drug baron the other sets out to avenge his death. Based on a true story, this is another entry in the series, that despite being well acted, has very much the look of a TV movie about it.
Aka: STREET WAR
A/AD   88 min (ort 90 min) mTV   VIDrel: POLY/POLYREC   V
Boa: article in New York Daily New Magazine by Mark Kriegel.

15
1992

IN THE LINE OF DUTY: THE F.B.I. MURDERS ***
Dick Lowry USA
David Soul, Michael Gross, Ronny Cox, Doug Sheehan, Bruce Greenwood, Teri Copley, Ronald G. Joseph, Peter McRobbie, Anne Lange, Kathleen Layman, Deborah May, Becky Gelke, Randal Patrick, Ashton Wise, Jaime Tirelli, Geoffrey Deuel
A fact-based crime drama telling of the six-month investigation by the FBI into a wave of bank robberies that took place in Miami in 1986. The crimes were the work of two outwardly normal individuals, who in reality were dangerous psychopaths. Their true natures ultimately drive them to plan a robbery, in the course of which they commit a particularly callous murder. One in a set of such films, it bears no relation-ship to the similarly titled Hong Kong series.
Aka: F.B.I. MURDERS, THE
A/AD   92 min mTV   VIDrel: CIC/SONOP L/A   V

18
1988

IN THE LINE OF DUTY: THE PRICE OF VENGEANCE **
Dick Lowry USA
Dean Stockwell, Michael Gross, Brent Jennings, Bruce A. Young, Tina Lifford, Brian Markinson, Kathleen Robertson, David Harris, Shaun Baker, Coutney Gains, Enrique Castillo, Richard Zavaglia, Susanna Thompson, Mary Kay Place
When a detective assigned to an investigation into the activities of a local gangster is killed, an L.A. cop is put on the case and secures a conviction that gets the gangster jailed on a petty robbery charge. Our cop now begins the task of amassing enough evidence to send the gangster to death-row, as he blames him for the murder of his colleague. Another entry in this series, but this one is based on a real case.
Aka: PRICE OF VENGEANCE, THE
A/AD   88 min (ort 90 min) mTV   VIDrel: ODY/SONOP   V/sh

15
1993

IN THE LINE OF DUTY: WISE GUYS **
Peter Markle USA
Tony Danza, Samuel Jackson, Ted Levine
A fact-based drama telling of the murder of a DEA officer who

15
1991

was engaged in an undercover mission, and the strenuous efforts made by an assistant D.A. to bring his killers to justice.
Aka: WISE GUYS: IN THE LINE OF DUTY
DRA 86 min VIDrel: GENESIS V

IN THE LINE OF FIRE *** 15
Wolfgang Petersen USA 1992
Clint Eastwood, Rene Russo, Dylan McDermott, John Malkovich, Gary Cole, Fred Dalton Thompson, John Mahoney, Jim Curley, Clyde Kusatsu, Patrika Darbo, John Heard, Greg Alan-Williams, Sally Haynes, Tobin Bell, Bob Schott, Juan A. Riojas
A grizzled Secret Service agent who failed to stop Kennedy's assassination receives a call from a psychopath who taunts him over his past failure and then calmly announces his intention to assassinate the incumbent president. This sets the scene for a complex cat-and-mouse, with Eastwood as the agent doing all he can to stop him. A taut and very exciting thriller, marred only by some unnecessarily violent scenes. However, Eastwood's performance is superb.
THR 123 min (ort 129 min) wScrn
VIDrel: COLUM/SONOP V/sur

IN THE MOUTH OF MADNESS *** 18
John Carpenter USA 1994
Sam Neill, Julie Carmen, Jurgen Prochnow, John Glover, Charlton Heston, David Warner, Bernie Casey, Peter Jason, Frances Bay, Wilhem von Homburg, Kevin Rushton, Gene Mack, Conrad Bergschneider, Marvin Scott, Katherine Ashby
A private eye is given an unusual assignment to locate a secretive horror writer whose latest work seems to make its readers lose their sanity. He tracks his quarry to a small town, but learns to his cost that it harbours a strange and terrifying secret. Extremely well scripted and tense, but this horror yarn starts to runs out of ideas as it approaches its distinctly lame climax. A pity, as in many respects this is one of the director's most coherent works.
Aka: JOHN CARPENTER'S IN THE MOUTH OF MADNESS
HOR 91 min (ort 95 min) VIDrel: EIV/SONOP V

IN THE NAME OF LOVE: A TEXAS TRAGEDY ** 12
Bill D'Elia USA 1995
Laura Leighton, Michael Hayden, Bonnie Bartlett, Richard Crenna
A wealthy bachelor who was crippled in a motor accident stands to inherit a fortune from his grandfather. However, the latter begins to consider changing his will when his grand-son starts a relationship with an obvious gold-digger.
DRA 88 min VIDrel: 20TH/FOXVID V/sur

IN THE NAME OF THE FATHER *** 15
Jim Sheridan EIRE/USA 1993
Daniel Day-Lewis, Emma Thompson, Pete Postlethwaite, John Lynch, Corin Redgrave, Mark Sheppard, Beatie Edney, Marie Jones, Britta Smith, John Benfield, Paterson Joseph, Marie Jones, Gerard McSorley, Frank Harper, Don Baker
An understandably partial but extremely dramatic reconstruction of one of the worst ever miscarriages of British justice, when a group of innocent Irish people were framed by the police and sent to prison after the horrific IRA bombing of a pub in Guilford, that was popular with members of the British Army. All the nastiest aspects of the British establishment are revealed, while the fifteen years it took to clear their names is the worst aspect of this scandal.
DRA 127 min (ort 133 min) cC VIDrel: CIC/SONOP; PION (LV only) V/sur LV
Boa: book Proved Innocent by Jerry Conlon.

IN THE NICK OF TIME *** (U)
George Miller USA 1991
Lloyd Bridges, Michael Tucker, Allison LaPlaca, A Martinez, Cleavon Little, Jessica DiCiccio, Jenny Parsons, Wayne Robson, Adurey Webb, Michael Lamport, Lucy Filippone, Ken James, Thomas Hauff, Corad Bergschneider, Matt Birman
Santa learns to his dismay that his elves has made a mistake in their calculations as to how long he has left to reign, it being the case that every three-hundred years a new mortal must take over the job. With only seven days to go, he has to find a replacement, or the magic of Christmas will end forever. A sprightly and enjoyable fantasy, with an idea not a million miles away from the one explored in THE SANTA CLAUSE.
JUV 88 min SATrel: DISNEY CHANNEL

IN THE SHADOW OF A KILLER *** PG
Alan Metzger USA 1992
Scott Bakula, Lindsay Frsot, Miguel Ferrer, J.T. Walsh, Tony LoBianco, Diane Houghton, Robert Closhessy
True story of a police officer given the job of breaking a local protection racket who is responsible for the apprehension of a notorious cop killer, but cannot bring himself to give the testimony that will result in this man's execution.
DRA 90 min (ort 93 min) VIDrel: ODY/SONOP V/sh

IN THE SHADOW OF EVIL ** 15
Daniel Sackheim USA 1995
Treat Williams, Margaret Colin, Joe Morton, William H. Macey, Timothy Busfield, Brad Greenquist, Art Cazares, Craig Clyde, D.J. Faulk, Michael Flynn, Frank Gerrish, Leo Geter, Dave Jensen, Joey Miyashima, Bill Osborn, Tom Proctor
After a car accident, a cop who is chasing a serial killer loses his memory, and the psychiatrist who tries to help him regain it puts herself in danger by doing this. A fairly run-of-the-mill thriller with the requisite number of thrills; only the slightly unusual plot premise makes it stand out from the rest.
THR 89 min mTV VIDrel: 20TH V/sh

IN THE SOUP ** 15
Alexandre Rockwell USA 1992
Steve Buscemi, Seymour Cassel, Jennifer Beals, Will Patton, Pat Moya, Jim Jarmusch, Stanley Tucci, Carol Kane, Sully Boyer, Steven Randazzo, Francesco Messina, Rockets Redglare, Elizabeth Bracco, Debi Mazer, Sam Rockwell
A writer who has produced a brief screenplay advertises it for sale and comes into contact with a con-man, who is masquerading as a producer. The latter agrees to produce it and sets about raising the money, in this haphazard and rather absurd comedy.
COM 95 min (ort 96 min) VIDrel: TART/20TH V/s

IN THE SPIRIT ** 18
Sandra Seacat USA 1990
Elaine May, Marlo Thomas, Jeannie Berlin, Peter Falk, Olympia Dukakis, Michael Emil, Melanie Griffith, Christopher Durang, Rockets Redglare, Agda Antonio, Brian Hickey, Laurie Jones, Phil Harper, Steve Powers, Hope Cameron
After her neighbour is murdered, a gutsy housewife and a new acquaintance come into possession of a notebook containing evidence that can identify her killer. Needless to say, the murderer now turns his attention to them in this incoherent black comedy, which starts out as a satire on New Age philosophy but soon proceeds to its obvious conclusion. The small parts by Falk, Dukakis and help enhance an otherwise forgettable offering.
DRA 90 min (ort 94 min) VIDrel: MGM/WHV L/A V

IN THE YEAR 2889 * 18
Larry Buchanan USA 1966
Paul Petersen, Quinn O'Hara, Charl Doherty, Neil Fletcher, Bill Thurman, Hugh Feagin, Max Anderson
After WW3, the blasted surface of the Earth is inhabited by telepathic cannibal mutants who stalk the few remaining human survivors. An uncredited remake of THE DAY THE WORLD ENDED.
Aka: YEAR 2889
FAN 80 min VIDrel: SCRN/DISC V

IN TOO DEEP ** (18)
Colin South/John Tatoulis USA 1990
Samantha Press, Hugo Race, Ebekah Elmaloglou, John Flaus, Dominic Sweeney
A female singer at a jazz bar finds herself attracted to a singer with a rock band, a heartless petty criminal who is only too happy to exploit her feelings. In fact, the worse he treats her, the more obsessed she becomes, in this competent but not especially engaging adult movie.
A 90 min (ort 106 min) CABrel: HVC V

IN TOO DEEP ** 18
Bob Misiorowski USA 1993
Michael Pare, Michael Ironside, Barbara Carrera, Lehua Reid, Henry Cele, Ian Yule, Michael McGovern, Angus Douglas, Ketan Larhani, Tony Caprari, Lara Logan, Colleen Nicholas Ordman, Nik Rujevic, Shane Mooieya, Robin Napier, Steve Fatarr
When his partner is killed during a customs operation, a Miami cop is accused of complicity in this crime and forced to resign from the force. He soon finds new employment as bodyguard to a drug lord's beautiful mistress, which gives him a chance to

investigate police corruption and eventually clear his name. A predictably violent actioner that moves along at a fast pace and offers the usual thrills.

Aka: POINT OF IMPACT; SPANISH ROSE

A/AD   94 min (ort 100 min)   VIDrel: MED/20VIS   V/sur

IN WHICH WE SERVE ****   U
Noel Coward/David Lean UK   1942
Noel Coward, Bernard Miles, John Mills, Celia Johnson, Kay Walsh, Michael Wilding, Joyce Carey, Penelope Dudley Ward, Philip Friend, Frederick Piper, Derek Elphinstone, Geoffrey Hibbert, Richard Attenborough
WW2 morale-booster written by Coward, that describes the life of a destroyer and those who serve on her. Told in flashbacks as the survivors of the torpedoed ship recall their lives at war and on leave. A low-key affair, but in its way excellent. AA: Spec Award (Noel Coward for outstanding production achievement).

WAR   109 min (ort 114 min) B/W   VIDrel: CARL/TECH   V

INCIDENT AT DARK RIVER **   (PG)
Michael Pressman USA   1989
Mike Farrell, Tess Harper, Helen Hunt, Arthur Rosenberg, Philip Baker Hall, K. Callan, Nicolas Coster, Jonjiorgi Enos, Daniel Beecher, Becky Hubricj, Jim Platt, Steve Anderson, Michaela Nelligan, Hester Schell, Steve Wigdahl
A poorly educated factory worker attempts to uncover the truth about the death of his young daughter, who died from poisoning after regularly playing near a river that carried toxic waste discharged from the town's battery factory (and major employer). Naturally, he comes up against many obstacles. Co-scripted by Farrell, this is a well acted if preachy drama, with little depth but a strong ecological message.

DRA   95 min mCab   TVrel: BBC

INCIDENT AT DECEPTION RIDGE **   15
John McPherson USA   1994
Michael O'Keefe, Linda Purl, Ed Begley Jr, Miguel Ferrer, Colleen Flynn, Ian Tracy, Jesse Moss, Tom Glass, Alan C. Petersen, D. Neil Clark, Tom Heaton, Morgan Brayton, Lucinda Nielsen, Frank C. Turner, Roman Podhora, Austin Schatz
A man just out of prison gets caught up in a deadly kidnapping drama, when he gets onto a bus where one of the passengers is a man carrying the ransom with which to obtain his wife's freedom. When he steals the cash and runs off, the irate kidnappers begin a vicious chase, forcing him to react in kind.

DRA   89 min (ort 94 min) mTV   VIDrel: CIC/SONOP   V

INCONVENIENT WOMAN, AN ***   15
Larry Elikann USA   1991
Jason Robards, Jill Eikenberry, Rebecca DeMornay, Chelsea Field, Peter Gallagher, Joseph Bologna, Chad Lowe, Roddy McDowall, Roy Thinnes, Paxton Whitehead, Elaine Stritch, Alex Rocco, Warren Frost, David Marguiles
Robards plays a wealthy powerbroker who adopts an actress as a mistress when he tires of his wife. With the murder of a close family friend, he mounts an elaborate cover-up, for he cannot afford a scandal. However, an outsider is on hand to mount his own investigation. Though cliched in terms of both characterisation and scripting, this well made drama (it's reminiscent of the Arthur Bloomingdale/Vikki Morgan affair) is surprisingly effective.

DRA   182 min (2 cassettes) mTV   VIDrel: ODY/SONOP   V/sh

Boa: novel by Dominick Dunne.

INCREDIBLE HULK, THE *   PG
Kenneth Johnson USA   1977
Bill Bixby, Susan Sullivan, Jack Colvin, Lou Ferrigno, Susan Batson, Charles Siebert, Mario Gallo, Eric Server, Eric Deon, Jake Mitchell, Lara Parker, Olivia Barash, William Larsen, George Brenlin, June Whitley Taylor
Pilot for a TV series based on a character created by the Marvel Comic group. Doctor David Banner is investigating strange cases of superhuman strength, and subjects himself to an experiment which results in him becoming an irritable green giant when angered. A pretty laughable affair, very much made for the small screen. An interminable TV series followed.

FAN   94 min (ort 100 min) mTV   VIDrel: CIC/SONOP L/A   V

INCREDIBLE HULK RETURNS, THE ***   PG
Nicholas Corea USA   1988
Bill Bixby, Lou Ferrigno, Jack Calvin, Tim Thomerson, Lee Purcell,

Eric Kramer, William Riley, Tom Finnegan, Donald Willis, Carl Nick Ciafalio, Bobby Travis McLaughlin, Burke Denis, Nick Costa, Peisha McPhee
The most recent in a series of feature spin-offs from the popular TV series based on a Marvel Comics character. In this tale Dr David Banner develops a device he hopes will rid him of his dangerous alter-ego, but the forces of evil put an end to his scheme, and as the Hulk he is forced to team up with Thor the Thunder God to fight them. Enjoyable hokum.

Aka: RETURN OF THE INCREDIBLE HULK

A/AD   93 min (ort 100 min) mTV   VIDrel: NWV L/A   V/h

INCREDIBLE JOURNEY, THE ***   Uc
Fletcher Markle CANADA   1963
Emile Genest, John Drainie, Tommy Tweed, Sandra Scott, Syme Jago
Two dogs and a Siamese cat travel 250 miles across Canada to rejoin their owners. Wholesome Disney fare with our trio encountering various hazards in the form of unfriendly animals but arriving safe and sound in the end. A good film for the family. Remade in 1993 as HOMEWARD BOUND: THE INCREDIBLE JOURNEY.

JUV   80 min   VIDrel: WDV L/A   V

Boa: book by Sheila Burnford.

INCREDIBLE KUNG FU MISSION **   15
Chang Hsin-Yi HONG KONG   1982
John Liu (Liu Chung-Liang), Shang Kuan Lung, Chen Lung, Hso Chung-Hsu, Ting Hwa-Choong, Cheng Ching
An expert martial artist recruits five mercenaries of varying ability to rescue a rebel who is being held prisoner at a fortress. Not all that well plotted, but this film has some of the best displays of martial arts prowess to be seen, and on that account is certainly worth a look.

Aka: KUNG FU COMMANDOS

MAR   85 min (ort 91 min) dubbed   VIDrel: IMPENT   V

INCREDIBLE MELTING MAN, THE *   18
William Sachs USA   1977
Alex Rebar, Burr DeBenning, Myron Healey, Michael Aldredge, Ann Sweeney, Cheryl Rainbeaux Smith, Lisle Wilson, Julie Drazen, Leigh Mitchell, Janus Blythe, Dorothy Love, Edwin Max, Bonnie Inch, Sam Gelfam, Chris Whitney
The survivor of a space mission develops a taste for human flesh as his own body begins to melt and fall apart, in this repulsive low-budget nonsense. The remarkable work of make-up expert Rick Baker partially redeems it.

FAN   85 min (ort 90 min)   VIDrel: VIPCO/SGSVID   V/h

INCREDIBLY STRANGE CREATURES WHO STOPPED LIVING AND BECAME MIXED UP ZOMBIES, THE **
15
Ray Dennis Steckler USA   1963
Cash Flagg (Ray Dennis Steckler), Brett O'Hara, Carolyn Brandt, Atlas King, Sharon Walsh, Madison Clarke, Erina Enyo, Jack Brady, Toni Camel, Neil Stillman, Son Hooker
Now a camp classic, this is probably the world's first (and last?) horror musical, about a gypsy fortune-teller who turns her customers into zombies. Badly acted, with dreadful dialogue, but well-filmed and surprisingly atmospheric. Followed by THE THRILL KILLERS.

Aka: TEENAGE PSYCHO MEETS BLOODY MARY, THE

HOR   81 min (ort 90 min)   VIDrel: RTM/DISC   V

INCREDIBLY TRUE ADVENTUES OF TWO GIRLS IN LOVE, THE ***
15
Maria Maggenti USA   1995
Laurel Holloman, Nicole Parker, Kate Stafford, Sabrina Artel, Toby Poser, Nelson Rodriguez, Dale Dickey, Nicole Parker, Andrew Wright, Katlin Tyler, Anna Padgett, Chelsea Catthouse, Stephanie Berry, Babs Davy, John Elson
Two high-school teenage girls from radically different home backgrounds meet and become friends. But as their relationship progresses they both find that they are sexually drawn to each other and become lovers. Thankfully, their lesbianism is never sensationalised and they are shown as just as caring and supportive as any heterosexual pair. Only the rushed and somewhat farcical ending spoils the gentle and sensitive tale.

DRA   94 min   CINrel

INDECENCY **   15
Marisa Silver USA   1992
Jennifer Beals, Barbara Williams, James Remar, Sammi Davis-Voss, Christopher John Fields, Ray McKinnon, John Fleck, Anna Gunn,

Terry Hoyos, Scott Allan Campbell, Gian-Carlo Scandiuzzi, Timothy D. Stickney, Steve Kehela
A woman recovering from a nervous breakdown takes a job at an agency that is running a campaign to promote the title perfume. But once there, she gets drawn into an affair with the ex-husband of the agency's owner, and events take a more sinister turn when her boss appears to commit suicide. A fairly watchable thriller, even if the mounting catalogue of "incidents" eventually strains both credibility and patience.
HOR 82 min (ort 88 min) mTV VIDrel: CIC/SONOP V

INDECENT BEHAVIOUR ** 18
Lawrence Lanoff USA 1992
Shannon Tweed, Gary Hudson, Jan-Michael Vincent, Michelle Moffett, Lawrence Hilton-Jacobs, Brandy Sanders, George Shannon, Penny Peyser, Brenda Swanson, Robert Sampson, Ken Steadman, Joy Davison, Angela Black, Stephen Fiachi
When a man dies from an overdose of a designer drug that enhances sexual pleasure, his therapist becomes the prime suspect in a murder investigation and soon gets involved with the detective in charge. Another undistinguished attempt at an erotic thriller, it was followed by a couple of sequels.
THR 94 min VIDrel: PROMARK/HIFLI V/h

INDECENT BEHAVIOUR 2 ** 18
Carlos G. Gustaff USA 1994
Shannon Tweed, James Brolin, Cynthia Steele, Rochelle Swanson, Elizabeth Sandifer, Craig Stepp, Chad McQueen, Laura Rogers, Stephen Polk, Irena Maximova, Nikki Fritz, Juan Carlo Vasquez
Sequel to the first film sees our sex therapist in trouble when one of her patients is involved in a scandal and murdered, and she becomes drawn into a complex intrigue, forcing her to request help from a colleague.
THR 90 min (ort 96 min) VIDrel: PROMARK/HIFLI V/h

INDECENT BEHAVIOUR 3 ** 18
Kelley Cauthen USA 1995
Shannon Tweed, Sam Hennings, Colleen Coffey, Doug Jeffrey, Beau Billingslea, Laura Rogers,
In this second sequel, our sex therapist takes on a client who proves to have some very nasty urges, of which she is initially unaware. However, all becomes clear when a renegade cop reveals that he is hunting a serial killer who has already claimed four lives. A preposterous erotic-thriller offering the same kind of softcore encounters as the earlier films, the serial killer subplot really being nothing more than a device to give the film some direction.
THR 92 min VIDrel: MED/DISC V/sh

INDECENT PROPOSAL ** 15
Adrian Lyne USA 1993
Robert Redford, Demi Moore, Woody Harrelson, Seymour Cassel, Oliver Platt, Billy Bob Thornton, Rip Taylor, Billy Connolly, Joel Brooks, Sheena Easton, Herbie Hancock, Danny Zorn, Kevin West, Tommy Bush, Mariclare Costello
A couple in financial difficulties go to Vegas to try their luck and meet a millionaire who is immediately taken with the wife. In fact, his desire is so great that he offers them one million dollars for a single night with her. They decide to go ahead with this strange deal, but are unprepared for the flood of negative emotions this unleashes. Well directed and acted, this dumb and rather ludicrous film offers no great insights into either marriage or fidelity.
DRA 112 min (ort 119 min) wScrn cC
VIDrel: CIC/SONOP; PION (LV only) V/sur LV
Boa: novel by Jack Engelhard.

INDECENT SEDUCTION ** 15
Alan Metzger USA 199-
Gary Cole, Nicholle Tom, Mary Kay Place, Mac Davis
Fact based drama telling of how a high school girl develops a friendship with a married football coach at her school, but is raped by him when she unwisely agrees to spend the night at his house.
DRA 90 min VIDrel: ODY/SONOP V/sh

INDECENT WOMAN, THE ** 18
Ben Verbong HOLLAND 1991
Josie Way, Coen Van Vrijberghe De Coningh, Huub Stapel, Marieke Van Leeuwen, Lydia Van Nergena, Theo De Groot, Peter Bolhuis, Niels Wolf, Peter Smits, Aga De Wit, Regina General, Roos General, Aukje Jetten, Earl Van Es, Jack Wouterse
A happily married woman embarks on a dangerous game in which she acts out her secret fantasies, helped along by a mysterious stranger has just met.
Aka: DE ONFATSOENLIJKE VROUW
DRA 92 min (ort 93 min) VIDrel: IMAG/RTM V

INDEPENDENCE DAY *** 12
Roland Emmerich USA 1996
Will Smith, Bill Pullman, Jeff Goldblum, Mary McDonnell, Judd Hirsch, Randy Quaid, Margaret Colin, Robert Loggia, James Rebhorn, Harvey Fierstein, Adam Baldwin, Brent Spiner, James Duval, Viveca A. Fox, Lisa Jakub, Ross Bagley
Despite the millions lavished on some truly staggering special effects, this film amply justifies the title of the most expensive B-movie in cinema history. A fleet of mile-long spaceships surround planet Earth, and pretty soon reveal their hostile intentions when they proceed to destroy the world's major cities. Just as it appears to be the end of life as we know it, some brave Americans save the day. Highly watchable, despite being noisy, cliched and poorly plotted.
FAN 139 min (ort 145 min) wScrn
VIDrel: 20TH/TECH; ENCORE (LV only) V/sur LV

INDIAN FIGHTER, THE *** PG
Andre De Toth USA 1955
Kirk Douglas, Walter Matthau, Elsa Martinelli, Walter Abel, Lon Chaney Jr, Diana Douglas, Eduard Franz, Alan Hale Jr, Elisha Cook, Michael Winkelman, Harry Landers, William Phipps, Buzz Henry, Ray Teal, Frank Cady, Hank Worden
A scout taking a wagon train through Sioux territory tries to obtain safe passage from the Indians by getting them to sign a peace treaty at the local fort. His efforts, however, are impeded by two white villains who are set on the Indians' gold. A vigorous and well-acted tale that gives a sympathetic and worthy portrayal of Indian life and customs, with ample action to make up for a rather thin story.
WES 88 min VIDrel: MGM/WHV V

INDIAN IN THE CUPBOARD, THE *** PG
Frank Oz USA 1995
Hal Scardino, Litefoot, Lindsay Crouse, Richard Jenkins, Rishi Bhat, Steve Coogan, David Keith, Sakina Jaffrey, Vincent Kartheiser, Nestor Serrano, Ryan Olson, Leon Tejwani, Lucas Tejwani, Christopher Conte, Cassandra Brown
A boy receives a toy cupboard and a little Indian figure for his birthday. One night he leaves the figure in his cupboard, and finds to his amazement that it has come alive (for the cupboard has the magical power of bestowing life on anything left inside it). This marks the start of an enjoyable adventure, in this fairly well realised if slightly moralistic children's fantasy.
JUV 92 min (ort 96 min) cC VIDrel: CIC/SONOP V/sur
Boa: novel by Lynn Reid Banks.

INDIAN RUNNER, THE **** 15
Sean Penn USA 1991
David Morse, Viggo Mortensen, Patricia Arquette, Charles Bronson, Sandy Dennis, Dennis Hopper, Jordan Rhodes, Enzo Rossi, Harry Crews, Eileen Ryan, Trevor Endicott, Brandon Fleck, Kathy Jensen, Jim Devney, Annie Pearson
Said to have been inspired by Bruce Springsteen's song "Highway Patrolman", this evocative drama takes its title from a Plains Indians legend – the "hunter" being a semi-religious motif representing stealth, courage and strength. Set in Nebraska in the late 1960s, the film has a small-town cop embarking on a long and difficult journey, his intention being to "hunt", capture and ultimately rescue his younger brother, a disturbed and unhappy Vietnam veteran.
DRA 121 min (ort 126 min) VIDrel: 20VIS/SONOP V/sur

INDIAN SUMMER ** 18
Paul Thomas USA 1992
Hyapatia Lee, Savannah, Tianna, Madison, Paula Price, Biff Malibu, Randy West, Peter North, Eric Price, Bud Lee, Wayne Summers
The best friends of a male and a female workaholic hatch a plot to bring a little love into the lives of their respective friends. This story was released in two versions, a "soft" one-part version for the UK and a "hard" and more coherent two-parter for the USA. Consequently, how the characters get along with the matchmaking schemes of their friends is only covered in "Indian Summer: Part 2".
A 73 min (ort 90 min) VIDrel: GROHOM/MAXSCAN V

INDIAN SUMMER *** (15)
Mike Binder USA 1992
Alan Arkin, Elizabeth Perkins, Vincent Spano, Diana Lane, Bill Paxton, Matt Craven, Kevin Pollak, Sam Raimi, Julie Warner, Kimberly Williams, Anne Holloway, Richard Chevolleau, Robert Feldman, Cliff Woolner, Emily Creed, Barth Deutch
A group of friends in their early thirties gather for a nostalgic reunion at the summer camp they enjoyed so much when they were in their teens. They talk, learn something about themselves and eventually save it from closure. A pleasant little film, whose fine cast do their best with the admittedly thin material. Based on the recollections of the director, it was filmed at the real-life camp he attended as a youngster.
COM 94 min (ort 98 min) SATrel: SKY MOVIES

INDIANA JONES AND THE LAST CRUSADE *** PG
Steven Spielberg USA 1988
Harrison Ford, Sean Connery, Alison Doody, John Rhys-Davies, Denholm Elliott, Julian Glover, River Phoenix, Michael Byrne, Alex Hyde-White
This big-budget sequel to INDIANA JONES AND THE TEMPLE OF DOOM sees Indy joining his father on a quest for the Holy Grail, when the latter vanishes whilst on that mission. Father and son are soon battling assorted Nazis in a story that is almost a clone of RAIDERS OF THE LOST ARK, but lacks that film's panache. Nevertheless, this stylish sequel has many virtues, and the two leads are great together. AA: Effects/aud (Ben Burtt/Richard Hymns).
A/AD 121 min (ort 127 min) VIDrel: CIC/SONOP; PION (LV only) V/sh LV

INDIANA JONES AND THE TEMPLE OF DOOM **
PG
Steven Spielberg USA 1984
Harrison Ford, Kate Capshaw, Ke Huy Quan, Amrish Puri, Roshan Seth, Philip Stone, Dan Aykroyd, David Yip, Ric Young, Chua Kah Joo, Philip Tan, Akio Mitamura, Michael Yama, D.R. Nanayakkara, Dharmadasa Kuruppu, Ruby DeMiel
A sequel to the highly enjoyable RAIDERS OF THE LOST ARK, but somewhat less successful in terms of story and entertainment, with the film whizzing along at breathtaking speed. In this tale Ford has to regain a sacred jewel whose loss has plunged a village into despair. A noisy, headache-inducing affair. Followed by INDIANA JONES AND THE LAST CRUSADE. AA: Effects/vis (Dennis Muren/Michael McAlister/Lorne Peterson/George Gibbs).
A/AD 112 min (Cut at film release – ort 118 min) VIDrel: CIC/SONOP; PION (LV only) V/sh LV

INDICTMENT: THE McMARTIN TRIAL **** 15
Mick Jackson USA 1995
James Woods, Mercedes Ruehl, Lolita Davidovich, Sanda Thompson, Shirley Knight, Henry Thomas, Robert Bassin, Mark Blum, Gabrielle Boni, Bob Clenedin, Betsy Brockhurst, Patsy chappel, Richelle Churchill, Leigh Curran
A well-staged recreation of the trial of title family on charges of child abuse and their stout defence of their name by the attorney Danny Davis (played here by Woods). This witch-hunt of a trial lasted seven years, cost about $16,000,000, and resulted in one of the accused spending five years in jail until the case collapsed. A frightening example of how hysteria and ignorance can effectively nullify the safeguards of a legal system. See also LIAR LIAR.
Aka: McMARTIN TRIAL, THE
DRA 130 min (ort 132 min) mTV
VIDrel: MOSAIC/COLUM V

INDIO * 15
Anthony M. Dawson USA 1989
Francesco Quinn, Brian Dennehy, Marvin Hagler
An American Indian and former marine returns to his tribal homelands only to find that his jungle-dwelling people are threatened by the activities of a construction company. Having seen that the company is prepared to use murder to prevent opposition, and having had his own father killed, he wages a lone battle against the company and its ruthless boss. A foolish action film of poor acting and ludicrous plotting.
A/AD 89 min VIDrel: 20TH V/sur

INDIO 2: THE REVOLT ** 15
Anthony M. Dawson (Antonio Margheriti) ITALY 1992
Marvin Hagler, Frank Cuervo, Dirk Galuba, Jacqueline Carol, Charles Napier, Tetchie Agbayani, Maurizio Fardo

A sergeant in the U.S. Marines is disturbed to see how the lives of native Indians have become blighted by the construction of a motorway through their land, and decides to help them fight for their rights. A carbon copy of the first film, and just as uninspired.
A/AD 97 min (ort 100 min) VIDrel: 20TH/TECH V/sur

INDISCREET *** PG
Stanley Donen UK 1958
Cary Grant, Ingrid Bergman, Phyllis Calvert, Cecil Parker, David Kossoff, Megs Jenkins, Oliver Johnston, Michael Anthony, Middleton Woods, Frank Hawkins, Richard Vernon, Eric Francis, Diane Clare
A NATO officer meets a beautiful actress and falls in love with her, but protects himself by saying his is married. Despite the thin plot, both stars are in top form and carry off this charming comedy of manners.
COM 96 min (ort 100 min) VIDrel: 4-FRONT/POLYREC V
Boa: play Kind Sir by Norman Krasna.

INDOCHINE *** 15
Regis Wargnier FRANCE 1991
Catherine Denueve, Vincent Perez, Linh Dan Pham, Jean Yanne, Henri Marteau, Dominique Blanc, Carlo Brandt, Gerard Lartigau, Hubert Saint Macary, Mai Chou, Andrzej Seweryn, Alain Froamger, Chu Hung, jean-Baptiste Huynh, Eric Nguyen
Well-staged, lavish and totally empty melodrama set in French Indo-China in the 1930s, where a rich French colonist finds that his native ward is now beginning to think for herself. Beautifully photographed and competently, this overlong effort offers little that is memorable. AA: Foreign.
DRA 151 min (ort 155 min) wScrn
VIDrel: ELPIC/POLYREC V/sur

INFERNO ** 18
Dario Argento ITALY 1978
Leigh McCloskey, Irene Miracle, Sacha Pitoeff, Eleanora Giorgi, Alida Valli, Daria Nicolodi, Feodor Chaliapin, Veronica Lazar, Gabriele Lavia, Leopoldo Mastelloni
An American returns from his studies in Rome in order to investigate the brutal murder of his sister, and discovers that it was committed by a group of Satanists. A disjointed and largely unsuccessful shocker with a few moments of surreal power.
Aka: INFERNO '80
HOR 101 min Cut (28 sec – ort 107 min) wScrn
VIDrel: 20TH/TECH; ENCORE (LV only) V LV

INFERNO IN SAFEHAVEN * 18
Brian Thomas Jones/James McCalmont USA 1988
Rick Gianasi, John Wittenbauer, Roy MacArthur, William Beckwith, Sammi Gavich, Mollie O'Hara, Marcus Powell, Sharon Shahinian, Ric Siver, Damon Clarke, Tere Malson, John Skalar, Diamond Geary, Andrea Black, Robert Celli
In a brutalised post-WW3 future, a family are allocated living-quarters in "Safehaven"; they discover that this allegedly safe refuge is dominated by ruthless thugs and they are forced to fight to survive. Another poorly mounted apocalyptic vision whose paltry ideas are drowned in a welter of bad acting and cheap visual effects.
Aka: ESCAPE FROM SAFEHAVEN
FAN 83 min Cut (10 sec – ort 87 min) VIDrel: MED L/A V

INFIDELITY ** PG
David Lowell Rich USA 1987
Kirstie Alley, Lee Horsley, Laurie O'Brien, Robert Englund, Lindsay Parker, Courtney Thorne-Smith, Michael Carren, Vera Lockwood, Alfred Dennis, Ben Wright, Lynne Charney, Jeb Ellis-Brown, Terence Cooper, John Hamelin
A yuppie couple attempt to combine full-time careers with family life but when the wife miscarries after an accident for which she blames her husband, they begin to drift apart. They soon each find some one else and agree an amicable divorce but eventually re-kindle their love. A soap-opera treatment of some serious themes that never rises above its TV format.
DRA 94 min (ort 104 min) mTV VIDrel: COLUM/SONOP V/s

INFILTRATOR, THE *** 18
John MacKenzie USA 1995
Oliver Platt, Arliss Howard, Tony Haygarth, Peter Riegert, Alan King, Julian Glover, Michael Byrne, George Jackos, Alex Kingston
Fifty years after the end of WW2, an Israeli journalist discovers

a group of neo-Nazis consisting of old Nazis still in hiding. He takes on a suitable Aryan identity and proceeds to infiltrate them. A dramatic reconstruction of a true story.
DRA   91 min (ort 102 min)   VIDrel: ODY/SONOP   V/h
Boa: book In Hitler's Shadow by Yaron Svoray and Nick Taylor.

INHERIT THE WIND ****
Stanley Kramer USA
Spencer Tracy, Gene Kelly, Fredric March, Dick York, Donna Anderson, Claude Akins, Florence Eldridge, Harry Morgan, Elliott Reid, Philip Coolidge, Paul Hartman, Jimmy Boyd, Noah Beery Jr, Gordon Polk, Ray Teal, Norman Fell
The famous Scopes "monkey trial" of 1925 provided the subject for this classic account of a schoolteacher brought to trial for teaching Darwin's theory of evolution in a small Southern town. Although its stage origins are always in evidence, this long but compelling courtroom film has splendid performances and a lively, articulate script. Remade for TV in 1988.
DRA   123 min (ort 127 min) B/W   VIDrel: WHV   V
Boa: play by Jerome Lawrence and Robert E. Lee.

U
1960

INHERIT THE WIND **
David Greene USA
Kirk Douglas, Jason Robards, Darren McGavin, Jean Simmons, Megan Follows, Kyle Secor, John Harkins, Michael Ensign, Don Hood, Josh Clark, Scotch Byerley, Ebbe Roe Smith, Douglas Dirkson, Richard Lineback, Thom McCleister
A remake of the 1960 Spencer Tracy film, with our proponent of the theory of evolution and his opponent, battling it out in a courtroom. A feeble version indeed, that makes bad use of the stars and weakens the story, by removing several important supporting characters. Written by John Gay.
DRA   96 min (ort 100 min) mTV   VIDrel: MGM L/A   V
Boa: play by Jerome Lawrence and Robert E. Lee.

PG
1988

INHUMANOIDS: THE EVIL THAT LIES WITHIN ***
(PG)
Jay Bacal/Wally Burr/Jim Graziano USA
Voices of: Michael Bell, William Callaway, Fred Collins, Brad Crandel, Dick Gautier, Ed Gilbert, Chris Latta, Neil Ross, Stanley Ralph Ross, Richard Sanders, Susan Silo, John Stephenson
The planet is threatened by the spread of the "Inhumanoid" terror from creatures that live below the surface of the planet and are intent on world domination. To resist them and their evil human collaborator, a defence force known as the "Earth Corps" is formed, and travels to the centre of the Earth to do battle. A lively kiddie's adventure based on some Hasbro toy characters.
Aka: INHUMANOIDS, THE
ANIM   85 min   SATrel: MOVIE CHANNEL

1986

INN OF THE SIXTH HAPPINESS, THE ***
Mark Robson USA
Ingrid Bergman, Curt Jurgens, Robert Donat, Ronald Squire, Athene Seyler, Richard Wattis, Moultrie Kelsall, Noel Hood, Joan Young, Edith Sharpe, Peter Chong, Athene Seyler, Michael David, Zed Zakari, Burt Kwouk
The story of Gladys Aylward, who became a missionary in China and endured great hardship, leading 100 children to safety (singing "Knick, Knack, Paddywack") following the Japanese invasion of WW2. Overlong, but fine performances from Bergman and Donat as a Chinese mandarin (his last role) work to its advantage. The score is by Malcolm Arnold.
DRA   152 min (ort 158 min) wScrn   VIDrel: 20TH/TECH   V
Boa: novel The Small Woman by Alan Burgess.

PG
1958

INNER CIRCLE, THE ***
Andrei Konchalovsky ITALY/USSR
Tom Hulce, Lolita Davidovich, Bob Hoskins, Aleksandr Zbruev, Bess Meyer, Feodor Chaliapin Jr, Marla Baranova, Irina Kuptchenko, Valdimir Khulishov, Vsevolod Larionov, Aleksander Filippenko, Evdokia Germanova, Aleksandr Garin
A brave attempt to examine Stalin's cult of personality, as seen through the eyes of his private projectionist. Based on the memories of projectionist Alexander Ganshin (who remained pro-Stalin throughout his life) it opens with the projectionist's marriage. His skill with the camera having led to his acceptance into Stalin's "inner circle", he sees much, but comprehends little. A flawed film, it would have fared better as a documentary.
Aka: IL PROIEZIONISTA
DRA   132 min (English version – ort 137 min) cC
VIDrel: VCC/DISC/COLUM   V/sh

15
1991

INNER SANCTUM **
Fred Olen Ray USA
Tanya Roberts, Margaux Hemingway, Joseph Bottoms, Valerie Wildman, William Butler, Brett Clark, Baxter Reed, Suzanne Ager, Jay Richardson, Ted Newson
Deeply involved with two women, a man makes plans to do away with his rich invalid wife, hiring a new nurse for this purpose. But the intended victim gets an inkling of what's in store for her when she learns that the nurse is a murder suspect. A quickly made erotic drama of limited appeal. Followed by NATURAL COLD KILLER.
THR   86 min (ort 90 min)   VIDrel: TRING/COLUM   V

18
1991

INNERSPACE ***
Joe Dante USA
Dennis Quaid, Martin Short, Meg Ryan, Kevin McCarthy, Fiona Lewis, Vernon Wells, Robert Picard, Wendy Schaal, Harold Sylvester, William Schallert, Ken Tobey, Henry Gibson, Orson Bean, Kevin Hooks, Kathleen Freeman, Dick Miller
Comedy tale in which a Navy test pilot, miniaturised as part of a secret experiment, is accidentally injected into a hypochondriac supermarket clerk instead of a laboratory rabbit as was intended. A boisterous comedy with some good effects, and a nice performance from Short as the unhappy, comical hero. AA: Effects/vis (Dennis Muren/William George/Harley Jessup/Kenneth Smith).
COM   115 min (ort 120 min) wScrn   VIDrel: WHV   V/sur

PG
1987

INNOCENCE UNPROTECTED ***
Dusan Makavejev YUGOSLAVIA
Dragoljub Aleksic, Ana Milosavljevic, Vera Jovanovic, Ivan Zivkovic, Bratoljub Gligorijevic, Pera Molsavljevic
Interesting and whimsical example of cinema montage in which the director took a forgotten 1942 melodrama produced, directed by and starring Aleksic (the story is set in Nazi-occupied Belgrade where an acrobat saves an orphan from a terrible fate) and intercut it with documentary and newsreel footage, as well as some new sequences featuring interviews with the surviving members of the original cast. A fascinating and highly unusual movie.
Aka: NEVINOST BEZ ZASTITE
DRA   78 min B/W/Col   VIDrel: CONNO/RTM   V

U
1968

INNOCENT, THE **
Luchino Visconti FRANCE/ITALY
Giancarlo Giannini, Laura Antonelli, Jennifer O'Neil, Rian Morelli, Massimo Girotti, Didier Haudepin, Marie Dubois, Roberts Paladini, Claude Mann, Marc Porel
A free-thinker has both wife and mistress and believes himself immune to the power of emotions like jealousy, but when his other half decides to pay him back in kind, his downfall is not far off. Visconti's last work, and far from his best.
Aka: INTRUDER, THE; L'INNOCENTE
DRA   123 min (ort 125 min) wScrn   VIDrel: FABFIL/SPEAR
V
Boa: novel by Gabriele D'Annunzio.

15
1976

INNOCENT, THE **
John Schlesinger GERMANY/UK
Anthony Hopkins, Campbell Scott, Ronald Nitschke, Isabella Rossellini, Hart Bochner, James Grant, Jeremy Sinden
In 1955, a British engineer is sent to Berlin to take part in a joint USA-UK intelligence project to monitor communications between East Germany and the USSR. This soon leads to tensions with his American opposite number, who professional paranoia extends to the German woman with whom our engineer starts an affair. A rather disappointing affair that fails to generate the necessary atmosphere or suspense.
Aka: UND DER HIMMEL STEHT STILL
THR   114 min   VIDrel: EIV/SONOP   V/sur
Boa: novel by Ian McEwan.

15
1992

INNOCENT, THE **
Mimi Leder USA
Kesley Grammer, Polly Draper, Jeff Kober, Gary Werntz, Carlos Gomez, Amy Steel, Keegan Macintosh, Dean Stockwell, Baron Kelly, Bob McCracken, Ellia English, Jack Black, Esther Scott, Nancy Kandal, Iqbal Thieba, Richard Whiten
A police lieutenant is reminded of his own dead son by a witness he has to interview, an autistic nine-year-old who saw a brutal robbery and murder. As he attempts to communicate with the boy, his finds his thwarted paternal instincts coming into play.

(15)
1993

An unusual psychological drama that raises some interesting issues.
DRA 88 min (ort 90 min) mTV SATrel: SKY MOVIES

INNOCENT BLOOD * 15
John Landis USA 1992
Anne Parillaud, Robert Loggia, Anthony LaPaglia, David Proval, Don Rickles, Rocco Sisto, Kim Coates, Chazz Palminteri, Elaine Kagan, Frank Oz, Luis Guzman, Tony Lip, Rohn Thomas, Angela Bassett, Leo Burmester, Tony Sirico, Sam Raimi
Gory and violent vampire tale in which a gangster meets up with a real femme fatale but finds that being one of the Undead is not really any handicap in his chosen profession. Meanwhile, an undercover cop who has infiltrated a Mafia family is out to stop this gangster turning the town into a safe haven for the undead. A silly and overlong spoof that is most unamusing. A bit sad to see Landis, who made AN AMERICAN WEREWOLF IN LONDON, wasting his time on this dud.
COM 113 min VIDrel: WHV L/A V/sur

INNOCENT LIES ** 18
Patrick Dewolf USA 1995
Stephen Dorff, Gabrielle Anwar, Adrian Dunbar, Joanna Lumley, Florence Hoath, Sophie Aubry, Alexis Denisof, Marianne Denicourt, Melvil Poupaud, Bernard Haller, Rosalind Bennett, Keira Knightley, Tobias Saunders, Robin Saunders
A detective investigating the murder of an old friend meets the sister of his prime suspect. He soon learns that she may be involved in an incestuous relationship with her brother, which complicates both his case and his growing attraction to her. Another highly derivative erotic-thriller with all the stock ingredients.
THR 84 min (ort 88 min) VIDrel: POLY/POLYREC V/sh

INNOCENT MAN, AN * 18
Peter Yates USA 1989
Tom Selleck, F. Murray Abraham, Laila Robins, David Rasche, Richard Young, Badja Djola, Todd Graff, M.C. Gainey, Peter Von Norden, Bruce A. Young, James T. Morris, Terry Golden, Dennis Burkley, Charles Landry, Vito Petersen
An innocent man's life becomes a nightmare when he is framed by two corrupt cops and sent to jail, their intention being to cover up the fact that they made a mistake. In prison he learns to survive whilst nurturing plans for his eventual revenge. An implausible tale that (apart from Abraham who is excellent) has weak performances to match.
Aka: HARD RAIN
THR 109 min (ort 113 min) VIDrel: TOUCH/SONOP L/A V

INNOCENT MOVES *** PG
Steve Zaillian USA 1994
Max Pomeranc, Joe Mantegna, Joan Allen, Ben Kingsley, Laurence Fishburne, Robert Stephens, David Paymer, Hal Scardino, Vasek Simek, William H. Macy, Dan Heday, Laura Linney, Anthony Heald, Steven Randazzo, Chelsea Moore, Josh Mostel
A seven-year-old chess-playing genius presents his parents with a dilemma as they try to do their best for him. At the same time they must contend with the pressure of two coaches, one of whom wants him to take a conventional route to the top while the other wishes to groom him for the more demanding speed-chess bouts. Surprisingly absorbing and accessible (even for non-chess players) this endearing family film (Zaillian's directing debut) has much to commend it.
Aka: SEARCHING FOR BOBBY FISCHER
DRA 105 min (ort 110 min) cC VIDrel: CIC V
Boa: book by Fred Waitzkin.

INNOCENT SLEEP, THE *** 15
Scott Michell UK 1995
Rupert Graves, Oliver Cotton, Tony Bluto, Paul Brightwell, Michael Gambon, Campbell Morrison, Graham Crowden, Franco Nero, Hilary Crowson, Kieran Smith, Annabella Sciorra, Sean Gilder, Brian Lipson, Dermot Kerrigan, Dermot Keanby
A young man sleeping rough on the streets of London sees a gangland killing at close hand, but later finds that the report of this crime is not consistent with the facts. Soon enough, he discovers that there is a price on his head, but fortunately, he finds help in the form of a hardbitten American reporter. A tense and imaginative foray into the crime-thriller genre, with good use made of the London locations. Gambon as a corrupt police officer is cliched but good.
THR 95 min (ort 99 min) VIDrel: MOSAIC/20VIS V/sur

INNOCENT VICTIM *** (15)
Michael Winner USA 1976
Yvette Mimieux, Tommy Lee Jones, Robert Carradine, Frederic Cook, Severn Darden, Howard Hesseman, Mary Woronov, Betty Thomas, John Lawlor, Britt Leach, Patrice Rohmer, Nancy Noble, Lisa Copeland, Clifford Emmich, Michael Ashe
A successful female advertising executive packs in her job in L.A. and takes off on her travels, but is thrown in jail on a minor offence, where she is raped by a psychotic warder. Eventually, she manages to break out of jail with another fellow inmate, and the inevitable chase and final confrontation takes place. A fast-paced story that has a minor following. The film was later remade for TV as "Outside Chance".
Aka: JACKSON COUNTY JAIL
DRA 84 min SATrel: SKY MOVIES

INNOCENT VICTIMS ** 15
Gilbert Cates USA
Hal Holbrook, Ricky Schroder, Rue McClanahan, John Corbett, Tom Irwin
A US Army Sergeant is wrongfully convicted of rape and murder and awaits the death penalty whilst in prison. Meanwhile, his parents and lawyers fight for him.
DRA 180 min VIDrel: ODY/SONOP V/sh

INNOCENT WAR, AN ** PG
Gabe Torres USA 1991
Wil Wheaton, Paul Balthazar Getty, Brian Krause, Jason London, Chris Young, Robert Miller, Ann Hartfield, Soren Bailey
On the eve of WW2, five students have to come to terms with the likelihood of having to fight as America gears up for war following the Pearl Harbour attack. A verbose film that has most of the action (such as it is) set in the student dorms, and quickly dissipates what little interest one initially had in the proceedings.
Aka: DECEMBER
DRA 87 min VIDrel: COLUM/SONOP V/sh

INNOCENT YOUNG FEMALE * 18
Bobby Houston USA 1991
David Keith, Deborah May, Kristen Cloke, Ray Sharkey, Loretta Devine, Karen Black
A criminal who robbed jewellery store pretends to be phoney warden to save his girlfriend from the torments of a women's prison staffed with all the usual lesbian and sadistic guards
Aka: HOTEL OKLAHOMA
DRA 99 min VIDrel: NEWAGE L/A V

INNOCENTS WITH DIRTY HANDS *** 15
Claude Chabrol FRANCE/ITALY/WEST GERMANY 1975
Romy Schneider, Rod Steiger, Paolo Giusti, Jean Rochefort, Pierre Santini, Hans Christian Blech, Dominique Zardi, Henri Attal
A married woman finds satisfaction in the arms of a friend of the family, but is soon looking for a more permanent solution to her marital discontent. and starts plotting to be rid of her uncouth husband. Chabrol's stylish thriller generates an intense and often uneasy atmosphere, and benefits from sharply drawn characters if not from the plot, which in common with much of this director's work, is rather too convoluted for its own good.
Aka: LES INNOCENTS AUX MAINS SALES
THR 120 min wScrn VIDrel: ARTPRO/RTM V

INSEMINOID * 18
Norman J. Warren UK 1980
Robin Clarke, Jennifer Ashley, Stephanie Beacham, Steven Grives, Judy Geeson, Barry Houghton, Rosalind Lloyd, Victoria Tennant, Trevor Thomas, Heather Wright, David Baxt, Dominic Jephcott, John Segal, Kevin O'Shea
An alien creature takes over the body of a female member of a team of space archaeologists, and she gives birth to twin monsters. A repulsive ALIEN clone with none of the sheer style of that earlier film.
Aka: HORROR PLANET
FAN 89 min (ort 93 min) wScrn VIDrel: VIPCO/SGSVID V

INSERTS * 18
John Byrum UK 1975
Richard Dreyfuss, Jessica Harper, Veronica Cartwright, Bob Hoskins, Stephen Davies
A once-famous director in Hollywood of the 1930s, is reduced to making sex films for a living. A boring and pretentious affair,

in which the five characters stumble through the interminable script in search of a few good lines.
DRA 114 min (ort 117 min) VIDrel: MGM/WHV L/A V

INSIDE DAISY CLOVER **
Robert Mulligan USA
15
1965
Natalie Wood, Robert Redford, Ruth Gordon, Christopher Plummer, Roddy MacDowall, Katherine Bard
In Hollywood of the 1930s, an adolescent actress enjoys a rise to fame, but not without the inevitable anguish. A slightly amusing, yet often gloomy look at life through the distorting mirror of Hollywood stardom, that has a couple of good moments, but is hampered by contrived plotting, superficial characterisation and coy dialogue. A STAR IS BORN (1937) did it much, much better.
DRA 128 min wScrn VIDrel: TART/20TH V/h
Boa: novel by Gavin Lambert.

INSIDE EDGE **
Warren Clark USA
18
1992
Richard Lynch, Michael Madsen, Tony Peck, Rosie Vela, George Jenesky, Branscombe Richmond
A tough guy suffers in a variety of ways after he infiltrates a drugs gang and makes the mistake of falling for the nightclub singer who is the girlfriend of its psychotic leader. A standard tough actioner.
DRA 84 min (ort 90 min) VIDrel: MARQ/QUANT V

INSIDE MAN, THE *
Tom Clegg SWEDEN/UK
15
1985
Dennis Hopper, Gosta Ekman, Hardy Kruger, David Wilson, Cory Molder, Celia Gregory, Lena Endre, Per Mattson, Torsten Wahlund, Janos Hersko, Leif Ahrle, Lill Lindfors, Charlie Elvegard, Jan Hermfelt, Carl-Ivar Nilsson
Dreary spy thriller with Stockholm locations, and the novel plot of a laser invention for hunting submarines, but the usual collection of spies who are after it.
Aka: SLAGSKAMPEN
THR 89 min (ort 100 min) VIDrel: ARROW/RTM L/A V
Boa: novel The Fighter by Harry Kullman.

INSIGNIFICANCE *
Nicolas Roeg UK
15
1984
Tony Curtis, Theresa Russell, Gary Busey, Michael Emil, Will Sampson, Lou Hirsch, Ray Charleson, Patrick Kilpatrick, Jan O'Connell, George Holmes, Richard Davidson, Mitchell Greenberg, Raynor Scheine, Jude Ciceolella
Story of a fictitious meeting at a hotel between Einstein, McCarthy, Joe DiMaggio and Marilyn Monroe. An ambitious but meandering contemplation of fame and its implications, coupled with a look at the perils of atomic war. Set in New York during 1953, the four characters are "unnamed" but readily identifiable. Despite some sharp dialogue, this sluggish effort is both self-indulgent and pompous, and does not make for easy viewing.
DRA 105 min (ort 110 min) VIDrel: L/A V
Boa: play Relatively Speaking by Terry Johnson.

INSPECTOR CALLS, AN ***
Guy Hamilton UK
PG
1954
Alastair Sim, Arthur Young, Brian Worth, Eileen Moore, Bryan Forbes, Jane Wenham, Olga Lindo, George Woodbridge, Barbara Everest, John Welsh, Norman Bird, Pat Neal, George Cole, Jenny Jones, Amy Green, Catherine Wilmer
A mysterious policeman calls on a wealthy family in Yorkshire in 1912, after a young girl has taken her own life with poison. As he questions them, a series of flashbacks show how each member of the family bears some of the blame for her death. A nice adaptation of Priestley's play, with a neat twist by way of a finale.
DRA 76 min (ort 79 min) B/W VIDrel: LUMI V
Boa: play by John Boynton Priestley.

INSPECTOR CLOUSEAU *
Bud Yorkin UK
U
1968
Alan Arkin, Frank Finlay, Patrick Cargill, Beryl Reid, Barry Foster, Delia Boccardo, Clive Francis, Richard Pearson, Michael Ripper, Tutte Lemkow, Anthony Ainley, Wallas Eaton, Eric Pohlmann
Inspector Clouseau is called in by Scotland Yard, to deal with the threat of a series of serious robberies. A lame attempt to cash in on the success of THE PINK PANTHER films, that very quickly runs out of steam.
COM 92 min (ort 105 min) VIDrel: MGM/WHV V

INSPECTOR MORSE: ABSOLUTE CONVICTION **
Antonia Bird UK
PG
1992
John Thaw, Kevin Whately, Diana Quick, Sean Bean, Jim Broadbent, Suzanna Hamilton, Richard Wilson
When a convict at an open prison is found dead, Morse begins a murder investigation, eventually uncovering a trail that leads to the affairs of some businessmen jailed for fraud. One of the weakest entries in a generally excellent series, with a meandering plot and obvious resolution.
Aka: ABSOLUTE CONVICTION
DRA 105 min mTV VIDrel: CTE/CARL V
Boa: novel by Colin Dexter.

INSPECTOR MORSE: CHERUBIM AND SERAPHIM **
Danny Boyle UK
PG
1992
John Thaw, Kevin Whately, Lisa Walker, Jason Isaacs, John Junkin, Charlotte Chatton, Charlie Caine, Paul Brightwell, Anna Chancellor, Freddie Brooks, Sorcha Cusack, Edwina Day, Celia Blaker, Mathew Terdre, Glean Mead
While he carries out an investigation into the death of a young girl, Morse finds himself having to contend with various domestic crises, forcing him to leave much of the footwork to his assistant Sergeant Lewis. Not a terribly good entry, it lacks both believability and pace.
Aka: CHERUBIM AND SERAPHIM
DRA 105 min mTV VIDrel: CTE/CARL V
Boa: novel by Colin Dexter.

INSPECTOR MORSE: DAUGHTERS OF CAIN **
UK
15
1996
John Thaw, Kevin Whately
A case that at first appeared to be straightforward gets ever more involved as Morse and Lewis delve deeper, uncovering the use of drugs at college, the sacking of a college employee and the reasons behind the suicide of a student. Matters take a more sinister turn when a ceremonial knife is stolen from a museum and another murder takes place.
DRA 103 min mTV VIDrel: CARL/TECH V

INSPECTOR MORSE: DEAD ON TIME ***
John Madden UK
PG
John Thaw, Kevin Whately, Joanna David, Samantha Bond, David Haig, Adrian Dunbar, James Grout, Richard Pasco, Susan-Jane Tanner, James Walker, Martyn Waites, Dominic Keating, Greer Gaffney
Morse finds himself haunted by his past when he investigates the tragic suicide of a brilliant Oxford don, who apparently took his life in the face of a terminal illness. The man's wife turns out to be the woman Morse had hoped to marry, and as the case progresses, Lewis becomes increasingly concerned for his superior. An unusual and rather implausible outing for Morse, that has our character losing his normal clinical detachment.
Aka: DEAD ON TIME
DRA 105 min mTV VIDrel: CTE/CARL V

INSPECTOR MORSE: DEADLY SLUMBER ****
Stuart Orme UK
PG
1993
John Thaw, Kevin Whately, Brian Cox, Janet Suzman, James Grout, Carol Starks, Jason Durr, Richard Owens, Jestyn Philips, Adam Maxwell, Su Eliot, Ian MacNeice, Robert Swann, David Goudge, Lou Wakefield, Patrick Godfrey
The plight of a young girl, who lies on a hospital bed in an irreversible coma, is linked to the murder of the doctor who operated on her. Morse investigates, his prime suspect being the father of the girl, who is found to have been sending death threats to the doctor responsible. However, the real facts are only slowly discovered, in a complex and totally absorbing story, one of the very best ones in this generally excellent series.
Aka: DEADLY SLUMBER
DRA 105 min mTV VIDrel: CTE/CARL V

INSPECTOR MORSE: DECEIVED BY FLIGHT ***
Anthony Simmons UK
PG
1991
John Thaw, Kevin Whately, Amanda Hillwood, Norman Rodway, Sharon Maughan, Daniel Massey, Jane Booker, Nicky Henson, Bryan Pringle, Geoffrey Beevers, Nat Parker, Stephen Moore, Peter Amory
Sergeant Lewis finds that he is to have a chance to show off his cricketing prowess when the Claret XI old boys' cricket team gathers in Oxford for its annual match. Unfortunately, one of the members of the team meets a rather sudden end just before the

match starts, giving Morse a chance to show off his investigative skills.

Aka: DECEIVED BY FLIGHT
DRA 103 min mTV VIDrel: CTE/CARL V

INSPECTOR MORSE: DRIVEN TO DISTRACTION ***
15
Sandy Johnson UK 1988
John Thaw, Kevin Whately, Patrick Malahide, David Ryall, Christopher Fulford, Mary Jo Randle, James Grout, Julia Lane, Tariq Yunus, Tessa Wojtczak, Richard Huw, Carolyn Choa, Ken Mazarin, Al Ashton
The apparently motiveless murder of a young woman gives Morse and Lewis a baffling case to investigate. Initially, it would seem that a local used car salesman may be the culprit, especially as he has been putting undue pressure on the dead girl's friend to conceal his relationship with the victim. A tense and very carefully worked out story, whose final resolution is both unexpected and rather chilling. Unusually, it's an entry in which Morse faces personal danger.

Aka: DRIVEN TO DISTRACTION
DRA 105 min mTV VIDrel: CTE/CARL V

INSPECTOR MORSE: FAT CHANCE **
PG
Roy Battersby UK 1989
John Thaw, Kevin Whately, Maurice Denham, David Gant, Julian Gartside, Caroline Ryder, Zoe Wanamaker, Maggie O'Neill, Sarah Carpenter, Arbel Jones, James Grout, Una Brandon-Jones, Nicholas Selby, Eileen Dunwoodie, Kenneth Haigh
Morse investigates the suspicious death of a female deacon against the background of the appointment of the city's first female chaplain, an event not altogether welcomed by the church's more conservative members. It's in this unlikely setting that our indefatigable inspector finds a little romance. An implausible and rather ineffective story.

Aka: FAT CHANCE
DRA 105 min mTV VIDrel: CTE/CARL V
Boa: novel by Colin Dexter.

INSPECTOR MORSE: GREEKS BEARING GIFTS ***
15
Adrian Shergold UK 1989
John Thaw, Kevin Whately, James Hazeldine, Mike Kremastoules, Jan Harvey, Andreas Markos, Eileen Way, Maureen Bennett, Andrew Kazamia, James Faulkner, Eve Adam, James Grout, Martin Jarvis, Richard Pearson, Johnny Lee Miller, Ian Sharp
Morse and Lewis investigate a murder among Oxford's Greek community, and discover a link between this crime and the work of a group of scholars of ancient Greece at the university.

Aka: GREEKS BEARING GIFTS
DRA 105 min mTV VIDrel: CTE/CARL V
Boa: novel by Colin Dexter.

INSPECTOR MORSE: HAPPY FAMILIES ***
PG
Adrian Shergold UK
John Thaw, Kevin Whately, Anna Massey, Gwen Taylor, Alun Armstrong, Rupert Graves, Charlotte Coleman, Andrew Ray, Sukie Smith, Martin Clunes, Jonathan Coy, George Raistrick, Mark Draper, Jamie Foreman, Tony Guilfoyle, Sophie Heyman
A well-known business tycoon is found murdered in his large country house, and suspicion falls on his unstable wife and his two feuding sons. However, the mystery of the killer's identity deepens when further murders take place.

Aka: HAPPY FAMILIES
DRA 105 min mTV VIDrel: CTE/CARL V

INSPECTOR MORSE: LAST BUS TO WOODSTOCK ***
PG
Peter Duffell UK 1988
John Thaw, Kevin Whately, Anthony Bate, Terrence Hardiman, Fabia Drake, Jill Baker, Holly Aird, Paul Geoffrey, Peter Woodthorpe
When the body of a young secretary is found in a pub car-park, the investigations lead to a local insurance office, where Morse uncovers a nasty web of intrigue. Another clever entry in this generally very well made detective series.

Aka: LAST BUS TO WOODSTOCK
DRA 105 min mTV VIDrel: CTE/CARL V
Boa: novel by Colin Dexter.

INSPECTOR MORSE: LAST SEEN WEARING ***
PG
Edward Bennett UK 1987
John Thaw, Kevin Whately, Peter McEnery, Glyn Houston, Frances Tomelty

The schoolgirl daughter of a wealthy family has been missing for six months, and routine investigations appear to have achieved nothing. Morse is called in, and soon becomes convinced that the girl has been murdered.

Aka: LAST SEEN WEARING
DRA 103 min mTV VIDrel: CTE/CARL V

INSPECTOR MORSE: MASONIC MYSTERIES ***
PG
Danny Boyle UK 1988
John Thaw, Kevin Whately
When a girlfriend of Morse is murdered during a rehearsal for a local production of The Magic Flute, all the clues seem to point to him as being her killer.

Aka: MASONIC MYSTERIES
DRA 105 min mTV VIDrel: CTE/CARL V

INSPECTOR MORSE: PROMISED LAND **
15
John Madden UK 1989
John Thaw, Kevin Whately, Con O'neill, Philip Anthony, James Grout, Kevin Leslie, Bill Young, Vanessa Patterson, Rhondda Findleton, Marie Armstrong, John Jarratt, Maureen Green, Noah Taylor, Paul Hunt, Max Phipps, Bill Young
Morse and his ever-loyal assistant Lewis, undertake a search for a former informer that takes them halfway round the world to Australia, where they suffer a bad case of culture shock, before finding their feet and getting the investigation under way. Another well acted story in this series, that unfortunately suffers badly from over-complexity.

Aka: PROMISED LAND
DRA 105 min mTV VIDrel: CTE/CARL V
Boa: novel by Colin Dexter.

INSPECTOR MORSE: SECOND TIME AROUND ***
PG
Adrian Shergold UK 1989
John Thaw, Kevin Whately, Jenny Laird, Oliver Ford Davies, Kenneth Colley, James Grout, Maurice Bush, Ann Bell, Pat Heywood, Sam Kelly, Mark Draper, Simon Adams, Christopher Eccleston, Peter Waddingtyon, Liz Kettle, David Baucham
With the mysterious death of a former deputy commissioner, Morse finds himself being brought into conflict with an old colleague. When a link is discovered between this crime and the unsolved murder of a young girl that took place eighteen years ago, it becomes apparent that the earlier case must be solved before any further progress can be made.

Aka: SECOND TIME AROUND
DRA 105 min mTV VIDrel: CTE/CARL V

INSPECTOR MORSE: SERVICE OF ALL THE DEAD ***
PG
Peter Hammond UK 1991
John Thaw, Kevin Whately, Angela Morant, John Normington, Maurice O'Connell, Judy Campbell, Michael Hordern, Peter Woodthorpe
A murder takes place at a quiet country church, and both the vicar and his loyal band of voluntary helpers come under suspicion. Morse and Lewis set out to solve this macabre mystery.

Aka: SERVICE OF ALL THE DEAD
DRA 102 min mTV VIDrel: CTE/CARL V

INSPECTOR MORSE: THE DAY OF THE DEVIL ***
PG
Stephen Whittaker UK 1993
John Thaw, Kevin Whately, Harriet Walker, Richard Griffiths, Keith Allen, James Grout, Gilly Coman, Patrick O'Connell, Anthony Hunt, Lloyd McGuire, John Bleasdale, Aran Bell,Richard Grahama, Susan Ellen Flynn, Patrick Drury
A mental patient escapes from a high security hospital to which he has been confined in order to serve life imprisonment for a murder committed as part of his Satanist beliefs. Soon this unusually resourceful and intelligent psychopath forces a woman therapist to indulge in a very dangerous game when she agrees to be exchanged for a woman hostage he is holding. A powerful and taut effort that benefits from a very satisfying twist ending.

Aka: DAY OF THE DEVIL, THE
DRA 105 min mTV VIDrel: CTE/CARL V

INSPECTOR MORSE: THE DEAD OF JERICHO ***
15
Alistair Reid UK 1989
John Thaw, Kevin Whately, Gemma Jones, Patrick Troughton, Norman Jones, Richard Durden, James Laurenson, Peter Woodthorpe
The very first episode from a popular TV police detective series that gave Thaw one of his best-ever roles as a

world-weary and highly unorthodox police officer who saw his cases as intellectual challenges as much as crimes. When the body of an attractive woman who apparently committed suicide is discovered, Morse has good reason to believe she was murdered.
Aka: DEAD OF JERICHO, THE
DRA   105 min mTV   VIDrel: CTE/CARL   V
Boa: novel by Colin Dexter.

INSPECTOR MORSE: THE DEATH OF THE SELF **
PG
Adrian Shergold UK   1992
John Thaw, Kevin Whately, Frances Barber, Michael Kitchen, George Corraface, James Grout, Chris Hunter, Julia Goodman, Kate Harper, Alan Rowe, Jane Snowden, Jane Wenham, Peter Blythe, Cesare Landricina, Jolyon Baker
Morse and Lewis investigate the strange death of an Englishwoman in Italy, but while on their assignment there, Morse finds himself coming under the influence of an attractive opera singer.
Aka: DEATH OF THE SELF, THE
DRA   105 min mTV   VIDrel: CTE/CARL   V

INSPECTOR MORSE: THE GHOST IN THE MACHINE ***
15
Herbert Wise UK   1988
John Thaw, Kevin Whately, Amanda Hillwood, Patricia Hodge, Clifford Rose, Bernard Lloyd, Michael Godley, Michael Thomas, Patsy Byrne, Irina Brook, Lill Roughley, Robert Oates, John Elmes, Rainbow Dench
Some valuable erotic paintings are stolen from a stately home, and the owner disappears under mysterious circumstances soon after. As ever, Morse is on hand to unravel the complexities of the case.
Aka: GHOST IN THE MACHINE, THE
DRA   104 min mTV   VIDrel: CTE/CARL   V

INSPECTOR MORSE: THE INFERNAL SERPENT ***
PG
John Madden UK   1988
John Thaw, Kevin Whately, Geoffrey Palmer, David Neal, Barbara Leigh-Hunt, Pearce Quigley, Michael Attwell, Cheryl Campbell, John Joyce, Ian Brimble, Tom Wilkinson, Irene Richard, George Costagan, Denys Hawthorne, Sydnee Blake
A child abuse case that involves the death of a prominent environmentalist leads Morse and Lewis to start questioning an academic family, whose outwardly respectable exterior hides a cupboard full of skeletons. A moody and gripping mystery that presents the viewer with a set of baffling events, until all is made clear in the rather disturbing resolution.
Aka: INFERNAL SERPENT, THE
DRA   105 min mTV   VIDrel: CTE/CARL   V

INSPECTOR MORSE: THE LAST ENEMY ***
PG
James Scott UK   1991
John Thaw, Kevin Whately, Amanda Hillwood, Barry Foster, Michael Aldridge, Tenniel Evans, Beatie Edney, Sian Thomas, James Grout, Lana Morris, Bert Parnaby, Mark Tandy, Pauline Munro, Albert Welling, Kevin McMonagle
A body is recovered from a canal in Oxford, and the only clue of any value points to a link with one of the university's colleges. An adequate entry in this well mounted series.
Aka: LAST ENEMY, THE
DRA   102 min mTV   VIDrel: CTE/CARL   V

INSPECTOR MORSE: THE SECRET OF BAY 5B ***
PG
Jim Goddard UK   1988
John Thaw, Kevin Whately
A playboy is found strangled in his car, with a car park ticket and diary the only possible clues to this mystery. Morse investigates and finds the case giving him little time to cultivate his budding romance with a woman pathologist.
Aka: SECRET OF BAY 5B, THE
DRA   105 min mTV   VIDrel: CTE/CARL   V

INSPECTOR MORSE: THE SETTLING OF THE SUN ***
15
Peter Hammond UK   1988
John Thaw, Kevin Whately, Anna Calder-Marshall, Robert Stephens, Derek Fowlds, Robert Lang, Avis Bunnage, Amanda Burton, Peter Woodthorpe, Philip Middlemiss, Eiji Kusuhara, Tim Barker, Llewellyn Rees, Blue Macaskill

Inspector Morse and his colleague Lewis set out to solve the mysterious death of a Japanese student in Oxford in this absorbing detective mystery.
Aka: SETTLING OF THE SUN, THE
DRA   105 min mTV   VIDrel: CTE/CARL   V

INSPECTOR MORSE: THE SILENT WORLD OF NICHOLAS QUINN ***
15
Brian Parker UK   1986
John Thaw, Kevin Whately, Michael Gough, Frederick Treves, Roger LLoyd Pack, Clive Swift, Amanda Hillwood, Barbara Flynn, Elspet Gray, Peter Woodthorpe, Phil Nice, Anthony Smee, Julie Neubert, Arthur Cox, Philip Voss, Gabrielle Blunt
The second tale from this excellently-scripted crime series, in which Detective Inspector Morse could always be relied upon to use his highly intellectual and unconventional approach to solve his cases. In this episode a deaf man working for Oxford University's examinations board appears to have come into possession of a piece of information that cost him his life.
Aka: SILENT WORLD OF NICHOLAS QUINN, THE
DRA   101 min mTV   VIDrel: CTE/CARL   V
Boa: novel by Colin Dexter.

INSPECTOR MORSE: THE SINS OF THE FATHERS ***
15
Peter Hammond UK   1988
John Thaw, Kevin Whately
An old-fashioned brewery faced with a takeover bid becomes the background to a series of baffling murders, in this complex and well-plotted entry in a long-running series.
Aka: SINS OF THE FATHERS, THE
DRA   105 min mTV   VIDrel: CTE/CARL   V

INSPECTOR MORSE: THE WAY THROUGH THE WOODS ***
15
John Madden UK   1995
John Thaw, Kevin Whately, James Grout, Nicholas Le Prevost, Malcolm Storry, Neil Dudgeon, Michelle Fairley
The notorious "Lover's Lane Killer" is murdered in a prison fight just before he is due to go on trial, but as far as Morse is concerned the case is far from closed. Our intrepid inspector looks for clues to a murder at a lake by returning to Wytham Woods.
Aka: WAY THROUGH THE WOODS, THE
DRA   104 min mTV   VIDrel: CTE/CARL   V

INSPECTOR MORSE: THE WOLVERCOTE TONGUE ***
PG
Alastair Reed UK   1987
John Thaw, Kevin Whately, Simon Callow, Kenneth Cranham, Roberta Taylor, Peter Woodthorpe, Robert Arden, Christine Norden, Bill Reimbold, Shirley Brown, John Bloomfield, Mildred Shay, Jane Bertish, Christine Kavanagh
A wealthy American, one of a party visiting Oxford, is found dead in her hotel room, apparently from a heart attack. The suspicions of our Inspector are aroused, especially when he learns of the strange disappearance of the "Wolvercote Tongue", and foul play is soon being offered as an explanation.
Aka: WOLVERCOTE TONGUE, THE
DRA   104 min mTV   VIDrel: CTE/CARL   V

INSPECTOR MORSE: TWILIGHT OF THE GODS **
PG
Herbert Wise UK   1993
John Thaw, Kevin Whately, John Gielgud, Sheila Gish, Robert Hardy, Jean Anderson
Morse falls for a celebrated Welsh opera who is shot dead during an Oxford University procession. She was due to honoured by an award and Morse begins to suspect that her death may be linked to the murder of a journalist investigating the background of one of those who was due to honoured together with her. As he digs ever deeper, he comes to the realisation that the public images of the rich and famous conceal a multitude of failings and even greater sins. Adequate.
Aka: TWILIGHT OF THE GODS
DRA   105 min mTV   VIDrel: CTE/CARL   V

INSPECTOR MORSE: WHO KILLED HARRY FIELD? ***
PG
Colin Gregg UK   1989
John Thaw, Kevin Whately, Trevor Byfield, John Castle, Geraldine James, Andy Mulligan, Nicola Cowper, Steven Payne, Maureen Bennett, Philip Locke, Veronica Lang, James Grout, Ronald Pickup, Freddie Jones, Vania Villiers, Jeremy Clyde

Morse and Lewis investigate the murder of a painter who, at first glance might be described as a lovable rogue. However, his paintings provide the key to a side of his character that's a good deal less savoury.
Aka: WHO KILLED HARRY FIELD?
DRA   105 min mTV   VIDrel: CTE/CARL   V
Boa: novel by Colin Dexter.

INSPECTOR WEXFORD: A GUILTY THING SURPRISED ***
PG
Mary McMurray UK   1988
George Baker, Christopher Ravenscroft, Michael Jayston, Nigel Terry, Padraig O'Loinsigh, Ellis Van Maarseveen, Catherine Neilson, Louie Ramsay, Ann Penfold, Kenm Kitson, David Swift
A woman is found battered to death in the woods near a local manor house, and Wexford finds himself having to cope with far too many potential suspects. A superior murder mystery, with an intelligent plot and some convincing performances.
Aka: GUILTY THING SURPRISED, A
DRA   145 min (ort 150 min) mTV   VIDrel: IMC/DISC   V
Boa: novel by Ruth Rendell.

INSPECTOR WEXFORD: A NEW LEASE OF DEATH ***
PG
Herbert Wise UK   1991
George Baker, Christopher Ravenscroft, Diane Keen, Ken Kitson, Dave Hill, Dorothy Tutin, Peter Egan, Sharon Maughan, Kate Lansbury, Tina Marian, Trevor Sellers, Eilenn Davies, Ruth Trouncer, John Evitts
A thirty-year-old murder case is re-opened, and Wexford is forced to re-examine his investigation, which he now fears may have resulted in sending an innocent man to the gallows. An engrossing murder mystery, first shown in two parts.
Aka: NEW LEASE OF DEATH, A
DRA   156 min mTV   VIDrel: VGM/VCC L/A   V
Boa: novel by Ruth Rendell.

INSPECTOR WEXFORD: A SLEEPING LIFE ***
PG
Bill Hays UK   1989
George Baker, Christopher Ravenscroft, Sylvia Syms, Dave Hill, Imelda Staunton, William Simons, Mamta Kaash
In this feature-length murder mystery, Wexford investigates the murder of a woman whose body was found on a tow-path, the only clue being a brand new but empty wallet. As ever, Wexford wends his slow and deliberate way in search of the culprit.
Aka: SLEEPING LIFE, A
DRA   144 min (ort 150 min) mTV   VIDrel: IMC/DISC   V
Boa: novel by Ruth Rendell.

INSPECTOR WEXFORD: ACHILLES HEEL ***
PG
Sandy Johnson UK   1992
George Baker, Christopher Ravenscroft, Diane Keen, Louie Ramsay, Robert Hands, Stephen Dillane, Kim Barclay, Julia Lane, Charlotte Howard, Norman Eshley, Saira Todd, Kate O'Connell, Preston Lockwood, Ann Way, Richard Strange
Whilst on holiday in Corsica Wexford gets to know a happily married couple, and on his return to England receives a call a short while later from the distraught husband, who claims that his wife has been kidnapped. But this is far from being the true state of affairs, and the detective learns that the realities of the case are far more sinister.
Aka: ACHILLES HEEL
DRA   104 min mTV   VIDrel: VCC L/A   V
Boa: novel by Ruth Rendell.

INSPECTOR WEXFORD: AN UNKINDNESS OF RAVENS ***
PG
John Gorrie UK   1991
George Baker, Christopher Ravenscroft, James Snell, Norma West, Karen Westwood, Louie Ramsay, Sidney Livingstone, Natascha Taylor, Emma Smith, Diane Keen, Brian Smith, Trevor Cooper, Charon Bourke, Deborah Poplett
Chief Inspector Wexford finds his involvement in the strange disappearance of a parent and the effect this has had on a school taking on a significance that's more than purely professional.
Aka: UNKINDNESS OF RAVENS, AN
DRA   100 min mTV   VIDrel: IMC/DISC   V
Boa: novel by Ruth Rendell.

INSPECTOR WEXFORD: AN UNWANTED WOMAN **
PG
Jenny Wilkes UK   1992
George Baker, Christopher Ravenscroft, Louie Ramsay, Ann Penfold,

Ken Kitson, Deborah Oplett, Dave Hill, Emma Smith, Diane Keen, Sean Pertwee, John Burgess, Robert Hands, Marjorie Sommerville, Anna Mottram, John Barrard
Wexford has to investigate a puzzling murder that took place in the tiny village of Kingmarkham, where it would appear that a woman has disposed of her own mother. But as ever, nothing is quite as it seems. A fair entry that was first shown in two parts. Adapted from the novel by Rosemary Anne Sisson.
Aka: UNWANTED WOMAN, AN
DRA   104 min mTV   VIDrel: VCC L/A   V
Boa: novel by Ruth Rendell.

INSPECTOR WEXFORD: FROM DOON WITH DEATH ***
PG
Mary McMurray UK   1991
George Baker, Christopher Ravenscroft, Diane Keen, Louie Ramsay, Ken Kitson, Dave Hill, Colin Campbell, Amanda Redman, Julia Chambers, Elizabeth Bennett, Geoffrey Bateman, John Salthouse, Malcolm Storry, Christopher Fulford
In this three-part mystery Wexford investigates the strange disappearance of a man's wife. Though foul play is not suspected, the worried husband insists that his wife has never done anything so out of character before.
Aka: FROM DOON WITH DEATH
DRA   108 min mTV   VIDrel: VCC L/A   V
Boa: novel by Ruth Rendell.

INSPECTOR WEXFORD: KISSING THE GUNNER'S DAUGHTER ***
PG
Mary McMurray UK   1992
George Baker, Christopher Ravenscroft, Sean Pertwee, Diane Keen, Louie Ramsay, Jacqueline Tong, Jacqueline Defferary, Anne Reid, Deborah Poplett, John Warnaby, Charles Collingwood, Angela Chow, David Jackson, James Ryland
Wexford has the unenviable task of investigating the death of a colleague, who was killed in the course of a bank robbery. At the same time, a massacre at a country mansion demands his attention. A standard entry in the series, that is absorbing and cleverly plotted, even if the addition of some domestic difficulties for our detective (he doesn't approve of his daughter's new boyfriend) seems more than a little contrived. First show in four parts.
Aka: KISSING THE GUNNER'S DAUGHTER
DRA   156 min mTV   VIDrel: VCC L/A   V
Boa: novel by Ruth Rendell.

INSPECTOR WEXFORD: MEANS OF EVIL ***
PG
Sarah Hellings UK   1991
George Baker, Christopher Ravenscroft, Louie Ramsay, Diane Keen, Ken Kitson, Patrick Malahide, Wendy Morgan, Michael Hadley, Fay Masterson, Cheryl Campbell, Colin Campbell, Gerard Horan, John Burgess, John Humphrey
A two-part mystery that revolves around a newlywed couple whose happiness is blighted by a dark secret. Wexford investigates.
Aka: MEANS OF EVIL
DRA   108 min mTV   VIDrel: VCC L/A   V
Boa: novel by Ruth Rendell.

INSPECTOR WEXFORD: MURDER BEING ONCE DONE ***
PG
John Gorrie UK   1991
George Baker, Christopher Ravenscroft, Diane Keen, Louie Ramsay, Michael Elwyn, Polly Adams, Lizzy McInnerny, Philip Glenister, David Cheesman, John Forgeham, Charlotte Attenborough, George Sweeney, Barbara Keogh, Derek Hicks
A three-part mystery, which has Burden being transferred to the Metropolitan Police whilst his colleague Wexford recovers from surgery. With Wexford meant to take it easy, both men enjoy a short holiday with their wives. Yet our intrepid inspector still finds himself taking an interest in police work, and is soon back at work solving the murder of a young girl.
Aka: MURDER BEING ONCE DONE
DRA   156 min mTV   VIDrel: VCC L/A   V
Boa: novel by Ruth Rendell.

INSPECTOR WEXFORD: NO CRYING HE MAKES ***
PG
Mary McMurray UK   1988
George Baker, Christopher Ravenscroft, Jane Horrocks, Clive Wood, Christine Kavanagh, Louie Ramsay, Ann Penfold, Charon Bourke, Emma Smith, Noah Huntley, Jonathan Lacey, Alison Rose, Dorothy Vernon, Natasha Williams

Just before Christmas, a mother discovers that her 4-month-old child has been abducted, and a similar child left in its place. As concern mounts for the safety of the missing child, Wexford begins unravelling the clues.
Aka: NO CRYING HE MAKES
DRA   77 min; 178 min (2-film cassette) mTV   VIDrel: L/A V
Boa: novel by Ruth Rendell. Osca: INSPECTOR WEXFORD: THE VEILED ONE

INSPECTOR WEXFORD: NO MORE DYING THEN ***
Jan Sargent UK   PG
                                                  1989
George Baker, Christopher Ravenscroft, Celia Gregory, Simon Shepherd, Nick Edmett, Dave Hill, Sam Moore
The disappearance of two children, separated by a period of 12 months, leads Wexford to conclude that the two incidents are linked, and he begins to fear worse is to follow. Another complex crime mystery that manages to be both intriguing and entertaining at the same time.
Aka: NO MORE DYING THEN
DRA   145 min (ort 150 min) mTV   VIDrel: IMC/DISC   V
Boa: novel by Ruth Rendell.

INSPECTOR WEXFORD: PUT ON BY CUNNING ***
Sandy Johnson UK   PG
                                                  1991
George Baker, Christopher Ravenscroft, Rossano Brazzi, Janet Maw, Diane Keen, Cherie Lunghi, Louie Ramsay, Charon Bourke, Beryl Reid, Sally Home, Helena McCarthy, Michael Bilton, Malcolm Tierney, Amanda Boxer, Amy Werba
A world-renowned flautist is found drowned just prior to his marriage to a woman many years younger. Though apparently the victim of a tragic accident, Wexford grows suspicious when he hears that the man's daughter is an impostor. Our dogged detective travels to France in search of the truth.
Aka: PUT ON BY CUNNING
DRA   99 min mTV   VIDrel: IMC/DISC   V
Boa: novel by Ruth Rendell.

INSPECTOR WEXFORD: SHAKE HANDS FOREVER ***
Don Leaver UK   PG
                                                  1988
George Baker, Christopher Ravenscroft, Margery Mason, Tom Wilkinson, Louie Ramsay, June Ritchie, Bernard Holley, Ken Kitson, Patrick Drury
When a woman is found murdered at her country cottage, Wexford becomes convinced that her husband was responsible, but finds that this suspect appears to have a cast-iron alibi. Undaunted, he persists in his efforts, to the inevitable and unpleasant conclusion.
Aka: SHAKE HANDS FOREVER
DRA   146 min (ort 150 min) mTV   VIDrel: IMC/DISC   V
Boa: novel by Ruth Rendell.

INSPECTOR WEXFORD: SOME LIE AND SOME DIE ***
Sandy Johnson UK   PG
                                                  1990
George Baker, Christopher Ravenscroft, Louie Ramsay, Diane Keen, Noah Huntley, Emma Smith, Ken Kitson, John Burgess, Marian McLoughlin, Dave Hill, Ann Penfold, Donald Sumpter, Gemma Jones, Mary McLeod, Annette Keer
A rock concert provides the setting for this mystery, when the body of a woman is discovered close by. As ever, Wexford finds that there's more to this murder than meets the eye.
Aka: SOME LIE AND SOME DIE
DRA   143 min (ort 150 min) mTV   VIDrel: IMC/DISC   V
Boa: novel by Ruth Rendell.

INSPECTOR WEXFORD: THE BEST MAN TO DIE ***
Herbert Wise UK   PG
                                                  1990
George Baker, Christopher Ravenscroft, Louie Ramsay, Julia Ormond, Diane Keen, Barbara Leigh-Hunt, Phoebe Burridge, Emma Smith, Adrian McLoughlin, John Burgess, Ken Kitson, Stephen Boxer, Emma Dewhurst, Richard Graham
In this story Wexford, for the first time ever, discovers a body of his own. The victim is one Charlie Hatton, a man who was to have attended a friend's wedding as best man that very afternoon.
Aka: BEST MAN TO DIE, THE
DRA   144 min (ort 150 min) mTV   VIDrel: IMC/DISC   V
Boa: novel by Ruth Rendell.

INSPECTOR WEXFORD: THE MOUSE IN THE CORNER **
Rob Walker UK   PG
                                                  1992
George Baker, Christopher Ravenscroft, Louie Ramsay, Diane Keen, Emma Smith, Katrina Levon, Che Walker, Victoria Fairbrother, Pip Donaghy, Lynn Farleigh, George Innes, Elizabeth Spriggs, Belinda Sinclair, Michael Irving, Ken Kitson
When the body of a popular local farmer is found murdered in his own kitchen, Wexford is called in to investigate but finds that the family are unwilling to assist him. As he delves into the victim's background, he discovers facts that paint the dead man in a most unsavoury light. First shown in two parts.
Aka: MOUSE IN THE CORNER, THE
DRA   97 min (ort 104 min) mTV   VIDrel: VCC L/A   V
Boa: short story by Ruth Rendell.

INSPECTOR WEXFORD: THE SPEAKER OF MANDARIN ***
Rob Walker UK   PG
                                                  1992
George Baker, Christopher Ravenscroft, Norman Rodway, Annette Crosbie, Virginia McKenna, Louie Ramsay, Marjorie Yates, Frances Cuka, Maggie Steed, Michael Carter, Mary Jo Randle
Wexford visits China on official business, but whilst there is plagued by disturbing dreams, and later on his return to England he begins to suspect that the key to a murder investigation is to be found in that country. An unusual setting adds much to this intriguing story, which sees Wexford obliged to investigate everyone who accompanied him on the trip. Originally shown in three parts, this story was scripted by Trevor Preston.
Aka: SPEAKER OF MANDARIN, THE
DRA   156 min mTV   VIDrel: VCC L/A   V
Boa: novel by Ruth Rendell.

INSPECTOR WEXFORD: THE VEILED ONE ***
Mary McMurray UK   PG
                                                  1989
George Baker, Christopher Ravenscroft, Paola Dionisotti, Deborah Poplett, Louie Ramsay, Ian Fitzgibbon, Camille Couri, Hugh Lloyd, Simon Chandler, Tony Vogel, David Fleeshman, Arthur Hewlett, Philip Bretherton, Paula Jacobs
A murdered woman is found, her body being discovered in a car park. This marks the start of a complex case for Wexford, but our police inspector is almost killed himself when he's injured in a bomb explosion. An atmospheric entry in this popular series (replete with dream sequences that are not found in the other stories) it's well worth seeing, despite the fact that the killer's identity is never made clear at the end.
Aka: VEILED ONE, THE
DRA   101 min mTV   VIDrel: IMC/DISC   V
Boa: novel by Ruth Rendell.

INSPECTOR WEXFORD: WOLF TO THE SLAUGHTER ***
John Davies UK   12
                                                  1987
George Baker, Christopher Ravenscroft, Kim Thomson, Carmel McSharry, Russell Hunter, Robert Reynolds, Donald Hewlett
Chief Inspector Wexford is baffled when he receives an anonymous note about a murdered girl called Ann, for he's unable to find any evidence that such a murder ever took place. An intriguing mystery tale of considerable complexity, that moves along smoothly to its final resolution.
Aka: WOLF TO THE SLAUGHTER
DRA   190 min (2 cassettes) mTV   VIDrel: IMC/DISC   V
Boa: novel by Ruth Rendell.

INSTITUTE BENJAMENITA, OR THIS DREAM PEOPLE CALL LIFE ***
Brothers Quay UK   PG
                                                  1995
Mark Rylance, Alice Krige, Gottfried John, Daniel Smith, Joseph Alessi, Jonathan Stone, Cesar Sarachu, Peter Lovstrom, Uri Roodner, Peter Whitfield
A boarding school for training servants provides the focus for this truly bizarre exploration of human aspirations and frustrations, with a man who has enrolled at the title institute learning that the lessons are exercises in repetition and monotony but may (or possibly, may not) serve a useful purpose. A film that is opaque and demanding, yet always good to look at, and is best looked on as a puzzle without an answer.
DRA   104 min B/W   VIDrel: ICAPRO/MANGA   V
Boa: novella Jakob von Gunten (and other works) by Robert Walser.

INTENT TO KILL ** 18
Charles T. Kanganis USA 1992
Traci Lords, Scott Patterson, Angelo Tiffe, Kevin Benton, Elena Sahagun, Sabrina Ferrand, Michael Foley, Luis Perez, Yaphet Kotto
A female vice cop comes up against a gang of drug dealers in this standard actioner that has more than the usual quota of sex and violence.
A/AD 95 min VIDrel: 20VIS/SONOP V

INTERCEPTOR ** 15
Michael Cohn USA 1992
Andrew Divoff, Elizabeth Morehead, Jurgen Prochnow, Jon Cedar, J. Kenneth Campbell, David Namath, Woodford Croft, Lawrence Cook, Michael Buice, Rick Marzan, Thom Adcox, John Prosky, Tim Moran, Billy Bates, Dennis Madalowe
A stealth bomber becomes the object of a terrorist hijack plot, with a pilot the only one on hand to foil their plans. A standard adventure yarn, enhanced by action sequences that are partially based on computer-generated graphics.
A/AD 89 min (ort 92 min) VIDrel: CIC/SONOP V

INTERIORS *** 15
Woody Allen USA 1978
E.G. Marshall, Geraldine Page, Diane Keaton, Maureen Stapleton, Kristin Griffith, Mary Beth Hurt, Richard Jordan, Sam Waterston
Woody Allen attempts a kind of Bergmanesque study of well-off but unhappy people, marking a change in direction from his earlier work that so often parodied films such as these. A film that is too lightweight to be really tragic and too anguished to be funny, but is undeniably well put together.
DRA 91 min (ort 93 min) VIDrel: MGM/WHV L/A V

INTERMEZZO *** PG
Gregory Ratoff USA 1939
Ingrid Bergman, Leslie Howard, John Halliday, Edna Best, Cecil Kellaway, Ann Todd, Enid Bennett, Douglas Scott, Eleanor Wesselhoeft, Marie Flynn
A world-famous violinist has an affair with his musical protege, but she finds the courage to leave him, thus freeing him to return to his wife and children. An overly sentimental remake of the Swedish original of 1936, which is the better film of the two. Its touching moments (and there are some) are badly diluted by an intrusive instrumental backing that is rarely silenced. This was Bergman's first English-speaking film.
Aka: ESCAPE TO HAPPINESS; INTERMEZZO: A LOVE STORY
DRA 69 min B/W VIDrel: VCC/DISC V

INTERNAL AFFAIRS ** 18
Mike Figgis USA 1990
Richard Gere, Andy Garcia, Nancy Travis, Laurie Metcalf, William Baldwin, Michael Beach, Richard Bradford, Katherine Borowitz, Faye Grant, Xander Berkeley, John Kapelos, John Capodice, Victoria Dillard, Pamela D'Apella
A young cop joins the Internal Affairs department of the L.A. police and becomes obsessed with busting a corrupt cop, so much so that his determined pursuit of justice becomes a personal vendetta. Gere gives a convincing portrayal of the dishonest cop, in a film that for all its promise fails to develop its ideas. An oblique and strangely lifeless work, both over-stylised and as remoselessly violent as any Shakespearian tragedy.
DRA 110 min (ort 117 min)
VIDrel: 4-FRONT/POLYREC/CIC V/sur

INTERNATIONAL AFFAIRS ** 18
Stuart Canterbury USA 1995
Asia Carrera, Misty Rain, Julia Channel, Barbara Doll, Isis Nile, Sabrina, Mike Horner, Buck Adams, Don Fernando, Jack Slater
An international dating agency forms the background to the usual encounters in this standard soft-core sex film.
A 58 min (ort 81 min) VIDrel: FALCON/TOTAL V

INTERNATIONAL VELVET ** PG
Bryan Forbes UK 1978
Tatum O'Neal, Christopher Plummer, Anthony Hopkins, Nanette Newman, Peter Barkworth, Dinsdale Landen, Sarah Bullen, Jeffrey Byron, Richard Warwick, Daniel Albineri, Jason White, Martin Neil, Douglas Reith, Norman Wooland
An attempt to produce a sequel to the 1944 hit film "National Velvet", that told of a champion horse. This story is one of an orphan girl who develops into an Olympic horsewoman, coached by her Aunt Velvet (the grown-up main character from the first film). An agreeable and undemanding tale, quite unde-

serving of the opprobrium heaped on it by the critics, despite the insipid dialogue.
DRA 89 min (ort 132 min) VIDrel: L/A V

INTERSECTION ** 15
Mark Rydell USA 1993
Richard Gere, Sharon Stone, Lolita Davidovich, Martin Landau, David Selby, Jenny Morrison, Ron White, Matthew Walker, Scott Bellis, Patricia Harris, Alan C. Paterson, Sandra P. Grant, Robyn Stevan, David Hurtabise, Gary Jones
A fatal road accident suffered by an architect sets in motion a series of flashbacks in which he remembers the events of his life, including his failed marriage to a frigid wife, his affair with another woman and the effect this had on their teenage daughter. An uninspired and untalented remake of the superior 1969 French film THE THINGS OF LIFE, although Stone as the unhappy wife gives an unusually strong performance.
DRA 94 min (ort 99 min) cC VIDrel: CIC/SONOP; PION (LV only) V/sur LV

INTERVIEW WITH THE VAMPIRE ** 18
Neil Jordan USA 1994
Brad Pitt, Christian Slater, Tom Cruise, Stephen Rea, Antonio Banderas, Kirsten Dunst, Virginia McCollam, John McConnell, Mike Seelig, Bellina Logan, Thandie Newton, Lyla Hay Owen, Lee Emery, Indra Ove, Helen McCrory
A two-hundred-year-old vampire in modern-day Los Angeles recounts the events of his life to a writer. This flashy and totally superficial adaptation of Rice's fine novel fails to bring to the fore any of the strengths of the story, relying so very predictably on visual effects, ample gore and the photogenic qualities of the lead actors.
Aka: INTERVIEW WITH THE VAMPIRE: THE VAMPIRE CHRONICLES
HOR 117 min (ort 122 min) wScrn cC VIDrel: WHV V/sur
Boa: novel by Anne Rice.

INTERVISTA **** 18
Federico Fellini ITALY 1987
Marcello Mastroianni, Anita Ekberg, Sergio Rubini, Federico Fellini, Tonino Delli Colli, Maurizio Mein, Antonio Cantafora, Lara Vendel
Hard at work on his next film at the Cinecitta studios, the director finds himself being interviewed by a Japanese TV crew. This leads to a journey of remembrance, in which Fellini takes a gentle, self-indulgent and very slightly mocking look at his life, his movies, his wife and his stars. An affectionate and absorbing self portrait, offering a fascinating blend of fantasy and documentary style sequences. Music is by Nino Rota.
DRA 106 min VIDrel: WESCON/RTM L/A V

INTERZONE ** 18
Deran Sarafian 1987
Bruce Abbott, Tegan Clive
Another post-nuclear holocaust tale, in which the Earth is a dead wasteland except for one region, which is protected from incursions by a powerful force field. However, two rival groups find a way of gaining access.
FAN 89 min Cut (9 sec) VIDrel: EIV/SONOP V

INTIMATE BETRAYAL *** (PG)
Robert M. Lewis USA 1987
James Brolin, Melody Anderson, Morgan Stevens, Pamela Bellwood, Joe Spano, Mona Abiad, Drew Borland, Beverly Elliott, Jill Diane Filion, Merrilyn Gann, Michele Goodyer, Don Granberry, Lee Jeffrey, Campbell Lane, Walter Marsh
A man hides a secret from his wife, and when his picture appears in a magazine a chain of events is set in motion that leads to him faking his own death. With the apparent loss of her husband, his wife begins to piece together details of his secret life. A complex and effective thriller.
Aka: DEEP DARK SECRETS
THR 96 min CABrel: HVC

INTIMATE OBSESSION ** 18
Lawrence Unger USA 1992
Jodie Fisher, James Quarter, Richard Abbott Booth
Bored with her lawyer husband and her philandering ways, a rich woman seeks thrills and satisfaction elsewhere and thus gets involved in a highly dangerous game. Meanwhile, her husband is making some plans of his own with his mistress. Average erotic-thriller of little distinction.
THR 90 min VIDrel: MARQ/QUANT V

INTIMATE PROPOSAL ** 18
Toby Phillips USA
Sarah Suzanne Brown, Diane Hurley, Dan Frank, Michael Artura
A married couple whose penchant it is to play sexual games, find their cosy set-up put in jeopardy by a handsome stranger.
DRA 87 min VIDrel: COLUM/SONOP V/sur

INTIMATE STRANGER ** 18
Allan Holzman USA 1991
Debbie Harry, James Russo, Tim Thomerson, Paige French, Grace Zabriskie, Mel Johnson Jr, Neal Israel, Lee Wallace, Billy Vera, Ed Bernard, Tia Carrere, Amy Hill, Mark Daneri, E.J. Castillo, Jan Monroe, Robert Glaudini
Angel (Harry) is a telephone sex-line hostess who makes a living by talking to assorted weirdos, one of whom calls her up to let her listen to the screams of a woman he has taken hostage and is now about to murder. When she takes her story to the police, only a young and ambitious cop is ready to believe her, and together they try to track down the culprit, who has promised to call again. A moderately gripping thriller.
THR 90 min Cut (2 min 23 sec) mTV
VIDrel: MED/POLYREC L/A V

INTIMATE STRANGERS ** 15
Robert Ellis Miller USA 1986
Stacy Keach, Teri Garr, Cathy Lee Crosby, Priscilla Lopez, Justin Deas, Max Gail, Justin Deas, Max Barabas, Tresa Hughes, Bob E. Hannah, Robert Goodman, Ray Forchion, Manny Bronz, Ernest Aruba, Carol Gun, Christine Page
A married couple worked as doctor and nurse during the Vietnam War and became separated after the fall of Saigon. After ten years in captivity the wife returns home, but is still haunted and scarred by her experiences. Meanwhile, her husband has taken up with another woman and is now forced to choose between them. An average soap opera-style melodrama of weak acting and scripting.
DRA 90 min (ort 100 min) mTV VIDrel: GUILD/SONOP L/A V

INTIMATE TERROR: ANGEL OF DEATH ** 15
Bill L. Norton USA 1990
Jane Seymour, Gregory Harrison, Brian Bonsall, Peggy Rea, Ray Walston, Chris Mulkey, Terence Knox, John De Lancie, Susan Hess, Grand L. Bush, Sonny Carl Davis, Dana Gladstone, Frank Birney, Tommy Hinkley, Bernard Kates, Janni Brenn
An attractive female teacher living alone with her son finds herself the object of a man's attentions, but the latter is not quite the charmer he at first sight appears. Expect the usual histrionics in a standard secret-psychopath thriller.
THR 90 min mTV VIDrel: CAPIT/GUILD V/sh

INTIMATE WITH A STRANGER * 18
Mel Roberts UK 1994
Roderick Mangin-Turner, Daphne Nayer, Amy Tolsky, Lorelei King, Ellenor Wilkinson, Janis Lee, Darcey Ferrer, Kaethe Cherney, Colleen Passard, Sara Mason, Francesca Wilde, Tamsin Hollo, Gifty Garton, John Guerrasio
In L.A., a disillusioned student switches from philosophy to becoming a gigolo, and makes his living by satisfying the requirements of women who are as deeply unfulfilled as he is. Poorly acted and pretentious, this boring and talky film was the directing debut for Roberts, whose attempt to give a simple erotic story an intellectual gloss does not meet with success.
DRA 94 min CINrel

INTIMIDATOR, THE ** 15
Noel Black USA 1988
Bruce Boxleitner, David Graf, Pat Hingle
A brutal thug returns to his home town after a couple of years in jail and proceeds to terrorise the inhabitants, who eventually conspire to murder him. However, the young public prosecutor is outraged by this and sets about gathering sufficient evidence to prosecute those responsible. A muddled "film with a message" that is too ambiguous to be effective.
DRA 90 min (ort 97 min) VIDrel: ODY/SONOP V

INTO THE BADLANDS *** 15
Sam Pillsbury USA 1991
Bruce Dern, Mariel Hemingway, Helen Hunt, Lisa Pelikan, Andrew Robinson, Michael Metzger, Dylan McDermont, Adam Sanchez, Jerry Gardner, Glen Burns, Oryan Walsky, Loren Haynes, Reynaldo Canti, Steven Schwartz-Hartley
Three tales of the Old West of the 1870s, with Dern playing a

character whose relentless quest for a killer with a price on his head serves as a linking device for stories of bounty hunters, hired killers, lawmen and other colourful figures.
WES 85 min (ort 90 min) VIDrel: CIC/SONOP V/h
Boa: short stories Streets Of Laredo by Will Henry, Time Of The Wolves by Marcia Muller and The Last Pelt by Bryce Walton.

INTO THE HOMELAND ** 15
Lesli Linka Glatter USA 1987
Powers Boothe, Paul Le Mat, C. Thomas Howell, Cindy Pickett, David Caruso, Shelby Leverington, Emily Longstreth
A dissipated former policeman goes after his missing daughter, and gets embroiled in a violent clash with a nasty bunch of white supremacists. A formula action yarn of no great consequence. Scripted by Anna Hamilton Phelan, this was the feature debut for Glatter. See also IN THE LINE OF DUTY 3: TIME TO KILL and BLIND HATE for something rather similar.
A/AD 90 min (ort 115 min) mCab VIDrel: L/A V/sh

INTO THE NIGHT * 15
John Landis USA 1985
Jeff Goldblum, Michelle Pfeiffer, Richard Farnsworth, Irene Papas, Kathryn Harrold, Paul Mazursky, Roger Vadim, Dan Aykroyd, David Bowie, Bruce McGill, Vera Miles, Clu Gulager, John Landis, Hadji Sadjadi, Michael Zand
Comedy-thriller about a man who becomes involved with a beautiful girl on the run from Iranian assassins. A film made for film buffs, with a number of funny cameos by directors (Don Siegel, Jonathan Demme and David Cronenberg among them) that is far too self-centred and contrived to work as a movie for the general public, as well as being an ill-matched blend of gratuitous violence and clumsy slapstick that does little to entertain.
COM 109 min (ort 115 min) VIDrel: CIC/SONOP L/A V

INTO THE WEST ** PG
Mike Newell EIRE/UK 1992
Ellen Barkin, Gabriel Byrne, David Kelly, Colm Meaney, Ciaran Fitzgerald, Johnny Murphy, John Kavanagh, Brenham Gleeson, Jim Norton, Anita Reeves, Ray McBride, Dave Duffy, Stuart Dannell, Becca Hollinshead, Bianca Hollinshead
Scripted by Jim Sheridan of MY LEFT FOOT fame, this is an engaging fantasy telling of two Dublin boys who, inspired by the legend of a magical white stallion, befriend a horse they take to be this very creature, and rescue it from the clutches of a cruel farmer. Over-sentimental at times, but the acting cannot be faulted and the ending is touching and believable.
Aka: INTO THE WEST: WHERE MYTH AND MAGIC WALK THE EARTH
A/AD 96 min (ort 102 min) VIDrel: EIV/SONOP V/sur
Boa: short story by Michael Pearce.

INTOLERANCE **** PG
D.W. Griffith USA 1916
Lillian Gish, Mae Marsh, Constance Talmadge, Robert Harron, Elmo Lincoln, Eugene Pallette, Bessie Love, Elmer Clifton, Seena Owen, Alfred Paget
A classic of the early cinema, with four parallel stories linked and intercut to demonstrate man's inhumanity to man. Despite the quaint little title cards and the over-ripe performances, this powerful work is an all-time great, with many spectacular scenes of undeniable power. Years ahead of its time, it presents one with incredible crowd scenes whose brilliance remains undimmed. Some prints have a running time of 208 minutes.
DRA 115 min (ort 123 min) B/W (tinted version) silent
VIDrel: VISION/DISC V

INTRUDER, THE ** U
Guy Hamilton UK 1953
Jack Hawkins, George Cole, Dennis Price, Michael Medwin, Susan Shaw, Dora Bryan, Edward Chapman, Nicholas Phipps, Hugh Williams, Duncan Lamont, Arthur Howard, George Baker, Richard Wattis, Gene Anderson, Patrick Barr
A retired colonel challenges a burglar, and recognises him as one of the tank corps troopers he had in his command during the war. He sets about delving into the man's past to see why he has become a thief. A series of cameo dramas unfolds competently but rather artificially, no doubt intended to remind one of the problems returning soldiers faced in post-war in Britain.
DRA 81 min (ort 84 min) B/W VIDrel: BRAVE/SONOP L/A V
Boa: novel Line On Ginger by Robin Maugham.

INTRUDER, THE *** 15
Roger Corman USA 1961
William Shatner, Frank Maxwell, Beverly Lunsford, Robert Emhardt, Jeanne Cooper, Leo Gordon, Charles Beaumont
A racist travels from one Southern town to the next, doing his best to incite the townsfolk against enforced racial integration in the local schools. One of Corman's better films which, though cheaply made, neither made any money nor was a quickie exercise in exploitation. The literate script is by Beaumont.
Aka: I HATE YOUR GUTS!; SHAME; STRANGER, THE
DRA 79 min (ort 84 min) B/W VIDrel: CONNO/RTM V
Boa: novel by Charles Beaumont.

INTRUDERS *** 15
Dan Curtis USA 1992
Richard Crenna, Mare Winningham, Susan Blakely, Ben Vereen, Daphne Ashbrook, Steve Berkoff, Alan Autry, G.D. Spradlin, Jason Beghe, Joseph Cousins, Lorry Goldman, Christian Cousins, Roslind Chao, Robert Mandan, Warren Frost, Ron Masak
Unusual account of a level-headed psychologist who becomes involved with a number of people who suffer strange experiences. Under hypnosis they seem to reveal evidence of alien abduction that gradually convinces our sceptical shrink that this is indeed the case. A well made and gripping tale that establishes a a convincing atmosphere of menace, although the theme of a government cover-up conspiracy does seem more than a little contrived.
FAN 162 min mTV VIDrel: 20TH/TECH V
Boa: book by Budd Hopkins.

INVADERS, THE ** 15
Paul Shapiro USA 1995
Scott Bakula, Elizabeth Pena, Richard Thomas, Delane Matthews, Richard Belzer, Roy Thinnes
A pilot uncovers a terrifying plan by aliens to destroy our environment in order to pave the wave for colonization. An updated feature based on the cult TV series. Competently made but essentially unoriginal, with our hero now given the obligatory attractive female companion who aids him in his efforts to alert the authorities to this menace.
FAN 169 min (ort 180 min) mTV VIDrel: POLFIL V/sh

INVADERS FROM MARS *** PG
William Cameron Menzies USA 1953
Helena Carter, Arthur Franz, Jimmy Hunt, Leif Erickson, Hillary Brooke, Bert Freed, Morris Ankrum, Max Wagner, Milburn Stone, William Phipps, Walter Sande, Douglas Kennedy, Bert Freed, Robert Shayne, Janine Perreau, John Eldredge
Well made story of a young boy, who witnesses an alien ship landing and the gradual takeover of the town's inhabitants, his own parents included. The first half of the film is by far the best. Once the aliens are exposed (all too easily, it seemed), the film develops (via the standard stock footage of tanks rolling) into a "we can shoot 'em and beat 'em" story. The twist ending is all too predictable. Originally shot in 3-D and remade in 1986.
FAN 70 min (ort 82 min) VIDrel: RTM/DISC V

INVADERS FROM MARS ** PG
Tobe Hooper USA 1986
Karen Black, Timothy Bottoms, Hunter Carson, Laraine Newman, James Karen, Louise Fletcher, Bud Cort, Jimmy Hunt, Eric Pierpoint, Christopher Allport, Donald Hotton, Kenneth Kimmins, Charlie Dell, William Bassett, Chris Hebert
A flashy but inferior remake of the atmospheric 1953 film, about a small boy who witnesses the landing of a spacecraft near his small town and sees its inhabitants taken over one by one. There are some good moments to be had and it is all done with great panache, but the film slowly and surely unravels as it develops.
FAN 94 min (Cut at film release – ort 102 min)
VIDrel: RNK L/A V

INVASION FORCE * 15
David A. Prior USA 1990
Renee Cline, Wally Cox, David Shark, Richard Lynch, Douglas Harter, Graham Timbs, Angie Synodis, Charlie Steadman, Rebecca McGowin
Hundreds of terrorists are parachuted into the States, their mission being to wreak as much havoc as possible. When their secret camp is discovered by a film crew doing a spot of location work, the latter make a desperate attempt to bluff them into surrender, armed only with fake weapons and other film props. An implausible and weakly plotted actioner.
A/AD 88 min VIDrel: 20VIS/SONOP V

INVASION OF PRIVACY ** 18
Kevin Meyer USA 1992
Jennifer O'Neill, Lydie Denier, Robby Benson
Standard nutter-on-the-loose shenanigans with an attractive female journalist beings stalked by a dangerous psychopath who has been released from prison on parole.
THR 92 min VIDrel: HIFLI/SONOP V/h

INVASION OF THE ANIMAL PEOPLE * (PG)
Virgil Vogel/Jerry Warren SWEDEN/USA 1960
Jerry Warren, Robert Burton, Barbara Wilson, Sten Gester, Bengt Bomgren, John Carradine, Ake Gronberg, Brita Borg, Jack Hefner
Poor SF nonsense about a monster on the rampage in Lappland (Northern Sweden) where it attacks everyone in sight, including the scientist who was investigating reports of a meteor crash.
Aka: HORROR IN THE MIDNIGHT SUN; RYMDINVASION I LAPPLAND; SPACE INVASION FROM LAPPLAND; SPACE INVASION OF LAPLAND; TERROR IN THE MIDNIGHT SUN
FAN 73 min B/W SATrel: BRAVO MOVIES

INVASION OF THE BEE GIRLS * 18
Denis Sanders USA 1973
Victoria Vetri, William Smith, Anitra Ford, Cliff Osmond, Ben Hammer, Wright King, Anna Aries, Katie Saylor, Andre Philippe, Sid Kaiser, Beverly Powers, Tom Pittman, Willaim Keller, Cliff Emmich
Despite its title, there is little sting in this pedestrian tale of women who are transformed by a strange force into title creatures and then proceed to love their men to death. The Troma film TEENAGE CATGIRLS IN HEAT uses much the same idea.
Aka: GRAVEYARD TRAMPS
FAN 85 min VIDrel: SCRN/DISC V

INVASION OF THE BODY SNATCHERS, THE *** PG
Don Siegel USA 1956 (re-issued 1979)
Kevin McCarthy, Dana Wynter, Larry Gates, King Donovan, Carolyn Jones, Jean Willes, Ralph Dumke, Virginia Christine, Tom Fadden, Sam Peckinpah, Bobby Clarke, Beatrice Maude, Everett Glass, Richard Deacon, Whit Bissell, Guy Way
Tense and atmospheric tale of alien duplicates, who gradually hatch from pods and take over the inhabitants of a small town. (Towards the end the storyline falters, as the hero's girlfriend is taken over even though no pod is present.) An effective demonstration of what is possible with a limited budget. The script is by Daniel Mainwaring. The clumsy prologue and epilogue tacked onto the film by the studio were removed on re-issue. Remade in 1978.
FAN 77 min (ort 80 min) B/W VIDrel: VCC L/A V
Boa: Collier's Magazine serial The Body Snatchers by Jack Finney.

INVASION OF THE BODY SNATCHERS, THE ** 15
Philip Kaufman USA 1978
Donald Sutherland, Brooke Adams, Veronica Cartwright, Leonard Nimoy, Jeff Goldblum, Kevin McCarthy, Don Siegel, Robert Duvall, Art Hindle, Leila Goldoni, Stan Ritchie, Tom Luddy, David Fisher, Gary Goodrow, Tom Dahlgren
A remake of the 1956 classic now updated to the 1970s and transplanted to San Francisco. Some superb effects and genuinely tense moments fail to offer much improvement on the original. Watch out for Kevin McCarthy and director Don Siegel, both of whom have cameo roles. The score is by Denny Zeitlin. See also BODY SNATCHERS, the second remake of the original and THE PUPPET MASTERS, another movie using this type of idea.
FAN 114 min wScrn VIDrel: MGM/WHV V/sur

INVASION, USA * 18
Joseph Zito USA 1985
Chuck Norris, Richard Lynch, Melissa Prophet, Alex Colon, Alexander Zale, Billy Drago, Eddie Johns, John DeVries, James O'Sullivan, Jaime Sanchez, Dehl Berti, Shane McCamey, Stephen Markle, Martin Shakar
A former CIA operative takes on the combined might of a Cuban/Soviet invasion force and single-handedly routs them, emerging unscathed from this encounter. Violent, bloody and sterile.
A/AD 96 min Cut (14 sec – ort 107 min)
VIDrel: MGM/WHV V/sur

INVESTIGATION OF A CITIZEN ABOVE SUSPICION *** 18
Elio Petri ITALY 1970
Gian Maria Volonte, Florinda Bolkan, Salvo Randone, Gianni Santuccio

Taut drama about a police chief who murders his mistress, and then proceeds to investigate the crime, having implicated another of her lovers. The score is by Ennio Morricone. AA: Foreign.
Aka: INDAGINE SU UN CITTADINO AL DI SOPRA DI OGNI SOSPETTO; INVESTIGATION INTO A CITIZEN ABOVE SUSPICION; INVESTIGATION OF A PRIVATE CITIZEN; STORY OF A CITIZEN ABOVE ALL SUSPICION
DRA   115 min   VIDrel: RCA L/A   V

INVINCIBLE ARMOUR, THE **   18
Ng See Yeun HONG KONG   1981
Huang Cheng Li, John Liu, Lee Lau Fer, Lar Kin Tee, Lu Chung Liang, Wang Chiang, Tino Wong, Phillip Kao
A general is framed for the murder of his superior and has to find the real culprit in this standard martial arts film.
MAR   100 min Cut (10 sec) wScrn dubbed
VIDrel: EAST/DISC   V

INVINCIBLE OBSESSED FIGHTER **   18
John King HONG KONG   198-
Elton Chong, Mike Wong
Two powerful warriors fight to possess a stolen treasure.
Aka: INVINCIBLE FIGHTER
MAR   85 min Cut (26 sec)   VIDrel: IMPENT   V

INVINCIBLE SHAOLIN **   15
Chang Cheh HONG KONG   1978
Kuo Chui, Lu Feng, Chiang Sheng
A Manchu warlord dreams up a scheme to destroy the influence of the Shaolin monks by getting the various sects to fight one another. He does this by arranging false bouts in which Shaolin fighters take on novices, the resultant deaths then being blamed on the Shaolin. But eventually the monks realise they have been duped, and set about putting matters right. An incredibly gory kung fu movie, of much mutilation and bloodshed.
MAR   101 min (ort 107 min) wScrn dubbed
VIDrel: MADE/RTM   V

INVINCIBLE SHAOLIN KUNG FU **   15
Ko Shih Hao HONG KONG
Hsiao Po Le, Li Yi Min, Ho Chun, Chen Wai Lan, Yue Chung Chiu, Sen Wing Je, Ying Kwok Chung
Three men avenge the murder of a great kung fu fighter in this routine tale.
Aka: SECRET SHAOLIN KUNG FU, THE
MAR   86 min Cut (15 sec – ort 91 min)   VIDrel: IMPENT   V

INVISIBLE GHOST, THE *   PG
Joseph H. Lewis USA   1941
Bela Lugosi, Polly Ann Young, John McGuire, Betty Compson, Clarence Muse, Terry Walker, Ernie Adams, George Pembroke, Fred Kelsey, Jack Mulhall
A kindly doctor is heartbroken at the death of his wife in an auto accident, but is unaware that she survived but suffered amnesia and is being cared for by the family gardener. However, for some strange reason her gaze turns her husband into a raging murderer. A ludicrous shocker that is a disappointing waste of a talented cast who do their best with this silly story.
HOR   100 min B/W   VIDrel: SCRN/DISC   V

INVISIBLE MAN, THE ***   PG
James Whale USA   1933
Claude Rains, Gloria Stuart, Dudley Digges, William Harrigan, Una O'Connor, E.E. Clive, Dwight Fyre, Henry Travers, Forrester Harvey, Holmes Herbert, John Carradine, Harry Stubbs, Donald Stuart, John Merivale, Merle Tottenham
Despite its venerable age, a brilliant rendering of Wells's story of how a scientist discovers the secret of invisibility and uses himself as a guinea-pig. Unable to cope with his condition, he descends slowly into megalomania. Though the film does not stick too closely to the book, a good feel for the spirit of the work is captured if not its substance.
FAN   68 min (ort 71 min) B/W   VIDrel: CIC/SONOP   V
Boa: novel by H.G. Wells.

INVISIBLE MOM **   (PG)
Fred Olen Ray USA   1995
Dee Wallace Stone, Barry Livingston, Trenton Knight, Russ Tamblyn, Phillip Van Dyke, Christopher Stone, Brinke Stevens, Stella Stevens, Joey Andrews, Vanessa Koman, Beth Ulrich, Tripp Reed, William C. Martell, Steve Barrett
A nerdish inventor has a nasty superior who bullies him incessantly, and takes all the credit for his inventions. So when he devises an invisibility formula he keeps it a secret and tries it out on the family dog. But to make matters worse his wife drinks some before he can devise and antidote. Meanwhile, his boss plans to have him railroaded into a mental hospital. A Disney-style comedy of stock characters and over-familiar complications.
COM   88 min   SATrel: MOVIE CHANNEL

INVISIBLE: THE CHRONICLES OF BENJAMIN KNIGHT *   18
Jack Ersgard USA   1993
Brian Cousins, Jennifer Nash, Michael Dellafemina, Curt Lowens, David Kaufman, Alan Oppenheimer, Aharon Ipale, Jake McKinnon, Dana Magdici, Daniela Nane, Valentin Popescu, Marian Hudae, Constantin Cotimanis, Geo Dobre, Doru Ana
Sequel to MANDROID in which our nasty villain from that film has taken to having one girls abducted for his own sadistic pursuits. Naturally, our crippled hero and his companions get involved in stopping him with the aid of their indestructible robot, but find their task complicated by an experiment that renders one of them invisible. A perfectly dreadful effort that recalls many a cheap 1930s serial episode.
Aka: CHRONICLES OF BENJAMIN KNIGHT, THE
FAN   76 min (ort 80 min)   VIDrel: CIC/SONOP   V

INVITATION TO THE WEDDING *   PG
Joseph Brooks UK   1983
Ralph Richardson, John Gielgud, Paul Nicholas, Elizabeth Sheperd, John Standing, Edward Duke, Susan Brooks, Ronald Lacey, Janet Burnell, Jeremy Clyde, Allan Cuthbertson, Aimee Delamain, Leslie French, Kate Harper
During a wedding rehearsal everything that can possibly go wrong does, with an American who is standing in for the groom finding himself married to the daughter of an impoverished earl. Bad casting and a flabby script ruin an idea that might have worked well enough as a thirty-minute short.
COM   85 min (ort 90 min)
VIDrel: SPEAR/SONOP/CALECO   V/sur

IP5 ***   15
Jean-Jacques Beineix FRANCE   1992
Yves Montand, Olivier Martinez, Sekou Sall, Geraldine Pailhas, Colette Renard, Stogui Kougate, Georges Staquet, Arlette Didier, Kleber Bouzonne, Bernard Lepinaux, Laurent Duequesnoy, Samir Geismi, Carole Pichert, Jane Hugon
A couple of Parisian street kids hitch a ride to Grenoble where they steal a car, unaware that an elderly man is asleep in the back seat. They eventually learn that he is some sort of eccentric, but he teaches them a few useful lessons about life before suffering a fatal heart attack. Worth seeing for Montand in his last role (like the character he also died from heart failure after completing this work) but the film is both oblique and self-indulgent.
Aka: IP5: L'ILE AUX PACHYDERMES; IP5: THE ISLAND OF PACHYDERMS
DRA   114 min (ort 119 min) wScrn   VIDrel: ARTIF/20TH   V/sur

IPCRESS FILE, THE **   PG
Sidney J. Furie UK   1965
Michael Caine, Nigel Green, Guy Doleman, Sue Lloyd, Gordon Jackson, Frank Gatliff, Aubrey Richards, Freda Bamford, Thomas Baptiste, Peter Ashmore, Oliver MacGreevy, Pauline Winter, Anthony Blackshaw, Barry Raymond, David Glover
Introduces Harry Palmer, the cockney secret agent. In this film he has to trace a missing scientist and discovers that his superior is a spy. A convoluted and twisty tale, that is never less than impossible to follow but has a nice understated performance from Caine. Unfortunately it all now looks rather dated. The score is by John Barry. Followed by FUNERAL IN BERLIN and BILLION DOLLAR BRAIN.
THR   103 min (ort 108 min)   VIDrel: CARL/TECH   V
Boa: novel by Len Deighton.

I.Q. **   U
Fred Schepisi USA   1994
Meg Ryan, Tim Robbins, Walter Matthau, Stephen Fry, Gene Saks, Lou Jacobi, Joseph Maher, Tony Shalhoub, Frank Whaley, Keene Curtis, Charles Durning, Alice Playten, Danny Zorn, Helen Hanft, Roger Berlind, Lewis J. Stradlen, Jeff Brooks
Too much in awe of her uncle, Einstein's niece is an intellectual snob who believes that intelligence (as defined by I.Q. tests) is more important than anything else. She thus rejects the advances of a garage mechanic who has fallen for her, but her uncle steps

in to teach her a much-needed lesson. An intriguing and original plot that sadly is not all that well realised.
COM 91 min (ort 95 min) cC VIDrel: CIC/SONOP V/sh

IRMA VEP ***
Olivier Assayas FRANCE 1996
Maggie Cheung (Zhang Manyu), Jean-Pierre Leaud, Nathalie Richard, Antoine Basler, Nathalie Boutefeu, Alex Descas, Dominique Faysse, Arsinee Khanjian, Bernard Nissile, Olivier Torres, Bulle Ogier, Lou Castel
A kind of satire on the pomposity and pretensions of French cinema, especially the New Wave directors, that is full of clever in-jokes and the like. Title refers to a character in a 1915 serial "Les Vampire", which a washed-up director decides to remake, casting Cheung as the title figure, but only because he fancies her. As the feuds and problems mount up, it becomes clear the film is never going to get made. Uneven, but great fun, the script is by Assayas.
COM 98 min CINrel

IRON ANGELS **
18
Raymond Leung/Teresa Woo HONG KONG 1986
Hideki Saiju, Moon Lee, Yukari Oshima
After the police destroy some opium plantations in Thailand, the villains set out to have their revenge and begin to kill off the officers who took part in the operation. Eventually a group of tough crime-busters are sent for in order to destroy the female mastermind who is behind the shipping of drugs into Hong Kong. Mindless nonsense, but well handled and with an explosive climax. Two sequels followed.
Aka: ANGEL; ANGELS
MAR 86 min (ort 93 min) wScrn dubbed
VIDrel: MIA/DISC; ENCORE (LV only) V LV

IRON ANGELS 2 **
18
Raymond Leung/Theresa Woo HONG KONG 1988
Moon Lee, Elaine Lui, Alex Fong, Karinna Andrews, Nathan Chan
Violent and very fast sequel to the first film that sees an all-girl martial arts trio battling (and battering) a renegade cop and his gang of gun-runners whilst on holiday in Malaysia. However, one of the trio succumbs to the charm of the chief villain – a former friend who has strong fascist leanings.
Aka: ANGEL 2; ANGELS 2
MAR 90 min wScrn dubbed
VIDrel: MIA/DISC; ENCORE (LV only) V LV

IRON DRAGON STRIKES BACK, THE **
18
Siu Kwai HONG KONG
Bruce Li (Ho Tsung-Tao), Philip Ko, Hau Kwok Choi, Wei Liet
Martial arts adventure involving gold smuggling, deep sea diving and a quest for revenge.
MAR 88 min VIDrel: ONE/IMPENT V

IRON EAGLE *
15
Sidney J. Furie USA 1986
Louis Gossett Jr, Jason Gedrick, David Suchet, Tim Thomerson, Larry G. Scott, Caroline Lagerfelt, Jerry Levine, Michael Bowen, Robbie Rist, Bobby Jacoby, Melora Hardin, David Greenlee, Michael Alldredge, Tom Fridley
And you'll need an iron posterior to sit through this adolescent and turgid nonsense. A teenage boy steals an F-16 fighter plane to rescue his father, who is being held prisoner somewhere in the Middle East. A sequel of similar merits followed in 1988.
A/AD 112 min (ort 119 min) VIDrel: VCC/DISC/COLUM V/sur

IRON EAGLE 2 *
PG
Sidney J. Furie CANADA/ISRAEL 1988
Louis Gossett Jr, Stuart Margolin, Mark Humphrey, Sharon H. Brandon, Alan Scarfe, Maury Chaykin, Colm Feore, Jason Blicker, Mark Ivanir, Douglas Sheldon, Uri Gavriel, Neil Munro, Jesse Collins
A foolish sequel to the first film that sees Gossett back again as Charles "Chappy" Sinclair, whose new mission is to assemble a group of Soviet and American pilots for a raid on a secret nuclear missile site somewhere in the Middle East. As silly as the first film, and slightly more boring.
Aka: IRON EAGLE 2: BATTLE BEYOND THE FLAG
A/AD 95 min (ort 105 min)
VIDrel: GUILD/POLYREC L/A V/sh

IRON EAGLE 3 **
15
Jogn Glen USA 1991
Rachel McLish, Paul Freeman, Christopher Cazenove, Louis Gossett Jr, Horst Buchholz, Paul Freeman, Fred Dalton Thompson, Mitchell Ryan
Another tale in the IRON EAGLE series, with Colonel Sinclair back in action, this time out to nail Nazi drug runners and avenge the murder of one of his buddies. Adequate action film, quite silly but reasonably watchable nonetheless.
Aka: ACES: IRON EAGLE 3
A/AD 95 min (ort 98 min) VIDrel: POLY/POLYREC; PION (LV only) V/sur LV

IRON EAGLE 4 *
15
Sidney J. Furie USA 1996
Lou Gossett Jr, Jason Cadieux, Joanne Vannicola, Max Piersig, Ross Hull, Karen Gayles, Rachel Blanchard, Dominic Zamprogna, Sean McCann, Victoria Snow, Jason Blicker, Al Waxman, Jack Nicholson, Dean McDermott, Aidan Devine
Having retired from the US Air Force, "Chappy" Sinclair starts up a school for young offenders, but his proteges learn that some maverick USAF officers have stockpiled supplies of toxic chemicals, having hatched a crackpot scheme to bomb Cuba. Fortunately, Sinclair's youngsters set out to take appropriate action of their own and avert a major crisis. Made with little care, this dreary sequel cheats on the action scenes by using footage from earlier films.
A/AD 91 min (ort 96 min) VIDrel: GUILD/FOXVID V/sur

IRON FIST BOXER **
15
Chang-Tse-Tsou HONG KONG 1989
Two young boys become apprentices to a master of "Back Kung Fu" in order to have revenge for a savage beating they suffered at the hands of a brothel guard. An undistinguished effort – average in all departments.
MAR 86 min Cut (2 min 48 sec) VIDrel: IMPENT V

IRON MAZE **
15
Hiroaki Yoshida JAPAN/USA 1991
Bridget Fonda, Jeff Fahey, J.T. Walsh, Gabriel Damon, Hiroaki Murakami, John Randolph, Peter Allas, Carmen Filpi, Francis John Thornton, Jeffrey J. Stephen, Mark Lowenthal, Goh Misawa, J. Michael Hunter, Lenora Nemetz, Steve Aronson
A reworking of the RASHOMON theme, set in a derelict Pennsylvania steel town, where the local steelmill has been bought up by a Japanese tycoon, who plans to turn it into an amusement park. When the son of the tycoon is found injured, various people come forward with information about the attack. A laboured effort indeed, whose attempt to pass comment on American-Japanese relationships is both trite and superficial.
DRA 98 min (ort 106 min) VIDrel: MIA/DISC V/sur
Boa: short story In a Grove by Ryunosuke Akutagawa.

IRON MONKEY *
15
Chen Kwan Tai HONG KONG 1996
Chen Kwan Tai, Kam Kong, Chi Kuan Chun, Wuilson Tong, Leung Kar Yan
The title character is a young gambler who sets out to have his revenge for the murder of his family. Another tedious revenge-driven martial arts saga.
MAR 90 min wScrn dubbed VIDrel: EAST/DISC V

IRON PETTICOAT, THE **
U
Ralph Thomas UK 1956
Bob Hope, Katharine Hepburn, James Robertson Justice, Robert Helpmann, David Kossof, Alan Gifford, Paul Carpenter, Noelle Middleton, Nicholas Phipps, Sidney James, Alexander Gauge, Doris Goddard, Tutte Lemkow, Sandra Dome, Richard Wattis
A US Air Force captain forces down a Russian MIG fighter, only to find that the pilot is an attractive woman. He spends the rest of the film doing his best to woo her away from Communism. A NINOTCHKA-style comedy that strains for every one of its laughs, and though its pacing is fast enough for any comedy, there is too little substance in the story and the stars just fail to convince. Writer Ben Hecht took his name off the credits when he saw the finished result.
COM 90 min (ort 96 min) VIDrel: FABFIL/SPEAR V

IRON WARRIOR *
15
Al Bradley (Alfonso Brescia) ITALY 1985
Miles O'Keeffe, Savina Gersak, Tim Lane, Frank Daddi, Elizabeth Kaza, Iris Peynado, Malcolm Borg, Conrad Borg, Tiziana Altieri, Josie Coppini, Jon Rosser
A foolish sword-and-sorcery offering, with a muscle-bound hero out to rescue a fair damsel, from a witch who has the usual plans

for world domination. A kind of slack follow-up to ATOR, THE FIGHTING EAGLE and ATOR THE INVINCIBLE 2, both of which starred O'Keeffe.
Aka: ECHOES OF WIZARDRY
FAN   84 min   VIDrel: RNK L/A   V

IRON WILL *   *U*
Charles Haid USA   1993
MacKenzie Astin, Kevin Spacey, David Ogden Stiers, August Schellenberg, John Terry, Brian Cox, George Gerdes, Penelope Windust, Jeffrey Allen Chandler, James Cada, Michael Laskin, Rex Linn, Allan "RJ" Joseph, Alvin William "Dutch" Lunak
Disney adventure based on true events and set in 1917, with a determined young man attempting to win the $10,000 prize on offer to the victor of a tough 522-mile dog-sleigh race from Canada to Minnesota. Winning will save the family from financial ruin and the loss of their farm following the death of the father. Spacey has a good role as the journalist who covers the race, but it is really Astin (as the youngster) to whom the acting honours go.
JUV   104 min (ort 109 min)   VIDrel: WDV/TECH   V

IRONCLADS *   *PG*
Delbert Mann USA   1991
Alex Hyde-White, Virginia Madsen, Reed Edward Diamond, Philip Casnoff, E.G. Marshall, Fritz Weaver, Leon B. Stevens, Kevin O'Rourke, Joanne Dorian, Burt Edwards, Beatrice Bush, Conrad McLaren, James Getty, Phil Whiteway, Marty Terry
A competent reconstruction of the first naval battle of the American Civil War between two ironclad men o' war, the Confederate Merrimac and the Union ship The Monitor. However, the cloak-and-dagger antics that precede this clash, which involve a beautiful Southern lady spy, take up far too much of the film and are told in such an uninspired fashion that ones interest very rapidly evaporates.
WAR   91 min (ort 100 min) mCab   VIDrel: FIRST/SONOP L/A   V

IRONHEART *   *18*
Robert Clouse USA   1992
Bolo Yeung, Richard Gordon, Britton Lee, Richard Nortn, Karman Krushke
When his former partner is killed by nasty white slavers, a cop uses his martial arts skills to get his revenge. Another assembly-line actioner of the standard violent variety.
MAR   88 min (ort 90 min)   VIDrel: MIA/DISC   V/sh

IRONWEED *   *15*
Hector Babenco USA   1987
Jack Nicholson, Meryl Streep, Carroll Baker, Michael O'Keefe, Tom Waits, Fred Gwynne, Diane Venora, Margaret Whitton, Jake Dengel, Joe Grifasi, Nathan Lane, James Gammon, Will Zahrn, Laura Esterman, Hy Anzell, Bethel Leslie
Depression-era tale set in Albany, New York, where a vagrant seeks a way back into the life he left behind years before. Streep plays his companion in misery who, like him, is a confirmed alcoholic. Both stars give fine performances in an atmospheric but almost unbearably depressing work, Babenco's first film in America. Scripted by Kennedy from his Pulitzer Prize-winning novel.
DRA   137 min (ort 143 min)
VIDrel: 4-FRONT/POLYREC/BRAVE   V/sur
Boa: novel by William Kennedy.

IRRESISTIBLE FORCE *   *15*
Kevin Hooks USA   1994
Stacy Keach, Cynthia Rothrock, Christopher Neame, Kathleen Garrett, Michael Bacall, Nicholas hammond, Jerome Ehlers, Paul Winfield, Penne Hackforth-Jones, Kim Knucky, Peter Kent, Len Kaserman, Philip Hinton, Martin Sacks, Malcolm Corke
A Police sergeant, who is about to retire, is partnered with a tough, young and very keen lady cop, who also just happens to be a martial arts expert. With his chance of a peaceful few weeks before retirement gone forever, this unlikely duo proceed to bring in the bad guys. A by-the-numbers buddy movie with enough high-kicking action to keep one watching – just.
A/AD   74 min mTV   VIDrel: 20TH/TECH   V

ISHI: THE LAST OF HIS TRIBE *   *(PG)*
Robert Ellis Miller USA   1978
Dennis Weaver, Eloy Phil Casados, Devon Ericson, Joseph Running Fox, Lois Red-Elk, Joaquin Martinez, Geno Silva, Gregory Norman

Cruz, Michael Medina, Arliene Nofchissey Williams, Patricia Ganera, Eddy Marquez, Dennis Dimster
Fact-based story of the last of the Yahi Indians, who is befriended by an anthropologist who takes him through the formative events in his life, which are seen in flashback. A moving and poignant tale, made with care and a good deal of respect for its subject, who spent his last years at a museum in San Francisco, dying in 1917. The late Dalton Trumbo's script was completed by his son Christopher. Remade as THE LAST OF HIS TRIBE. See also LAST OF THE DOGMEN.
DRA   150 min mTV   TVrel
Boa: novel Ishi In Two Worlds by Theodora Kroeber Quinn.

ISHTAR *   *PG*
Elaine May USA   1987
Warren Beatty, Dustin Hoffman, Isabelle Adjani, Charles Grodin, Jack Weston, Tess Harper, Carol Kane, Aharon Ipale, David Margulies
Two untalented songwriters are advised by their agent to go as far away from him as possible, and so they head for Morocco where they become involved in a budding revolution, and wind up on opposite sides. An expensive flop that tries awfully hard to be funny without ever finding a direction to go in. The deliberately bad songs are by Paul Williams and the film is scripted by May.
COM   103 min (Cut at film release by 8 sec – ort 107 min)
VIDrel: VCC L/A   V

ISLAND, THE *   *18*
Michael Ritchie USA   1980
Michael Caine, David Warner, Angela Punch McGregor, Frank Middlemass, Don Henderson, Jeffrey Frank, Christopher F. Bean, Zakes Mokee, Dudley Sutton, Clyde Jeavons, Brad Sullivan, John O'Leary, George McLaughlin, Jimmy Casino
A reporter investigating strange happening in the Caribbean, is captured by a colony of throwbacks from an 18th century pirate community. A ludicrous and sometimes unintentionally funny effort, scripted by Benchley from his novel.
A/AD   108 min (ort 114 min)   VIDrel: CIC/SONOP L/A   V
Boa: novel by Peter Benchley.

ISLAND CITY *   *15*
Jorge Montesi USA   1994
Kevin Conroy, Brenda Strong, Eric McCormack, Pete Koch, Constance Marie, Veanne Cox, Rick Porter, Joe Marchman, Jerry Haynes, Angie Bolling, Alex Allen Morris, Cynthia Dorn, Paul Pender, Caroline Summers, Shea Fowler, Clint Freeman
SF fantasy set in the year 2035 A.D., when the world's population enjoys a disease-free existence ten years after the introduction of a revolutionary vaccine that has wiped out most disease. Unfortunately, humanity is now threatened by the appearance of a new type of mutation.
FAN   87 min   VIDrel: WHV   V/h

ISLAND OF DOCTOR MOREAU, THE *   *15*
Don Taylor USA   1977
Burt Lancaster, Michael York, Nigel Davenport, Barbara Carrera, Richard Basehart, Nick Cravat, The Great John "L", Bob Ozman, Fumio Demura, Gary Baxley, John Gillespie, David Cass
Flat, useless attempt to film the classic story of a man on a lonely Pacific island, who spends his time conducting experiments on animals, in order to give them human form. A remake of ISLAND OF LOST SOULS. York is especially memorable for the woodenness of his acting, the others are little better.
DRA   98 min (ort 104 min)   VIDrel: L/A   V
Boa: novel by Herbert George Wells.

ISLAND OF DOCTOR MOREAU, THE *   *12*
John Frankenheimer USA   1996
Marlon Brando, Val Kilmer, David Thewlis, Fairuza Balk, Ron Perlman, Marco Hofschneider, Temuera Morrison, William Hootkins, Daniel Rigney, Nelson De la Rosa, Peter Elliott, Mark Dacascos, Miguel Lopez, Neil Young, David Hudson
In a Pacific island laboratory, a demented vivisectionist has succeeded in conferring human attributes onto animals. Kilmer plays the sullen and aimless young drifter who arrives at this hell-hole, Brando plays (with characteristic over-the-top gusto) the title character. The splendid special effects will cause a few shudders, but Kilmer is out of his depth here, as were the scriptwriters.
FAN   91 min (ort 96 min)   VIDrel: EIV/SONOP   V/sh
Boa: novel by Herbert George Wells.

ISLAND OF LOST SOULS *   12
Erle C. Kenton USA   1932
Charles Laughton, Bela Lugosi, Richard Arlen, Leila Hyams, Kathleen Burke, Hans Steinke, Alan Ladd, Harry Ekezian, Rosemary Grimes, Paul Hurst, George Irving, Joe Bonomo
Charles Laughton, Bela Lugosi, Richard Arlen, Leila Hyams, Kathleen Burke, Robert Kortman, Tetsi Komai, Stanley Fields, Hans Steinke, Alan Ladd, Harry Ekezian, Rosemary Grimes, Paul Hurst, George Irving, Joe Bonomo
A shipwrecked man is washed up on a small island and soon learns that its seemingly pleasant owner is in reality a scientist who is totally obsessed with conducting strange experiments to transform animals into human beings. A highly effective adaptation of Wells's novel that is very atmospheric and leaves much scope for the imagination. Remade in 1959 as "Terror Is A Man" and twice again as THE ISLAND OF DOCTOR MOREAU.
FAN   68 min cC B/W   VIDrel: VISCOM/RTM   V
Boa: novel The Island of Dr Moreau by H.G. Wells.

ISTANBUL *   15
Mats Arehn SWEDEN/TURKEY   1989
Timothy Bottoms, Twiggy (Lesley Hornby), Emma Kilberg, Robert Morley, Lena Endre, Sverre Anker Ousdal, David Gartenkraut, Pierre Stahre, Merden Taner, Engin Inal, Celal Khosrowshahi, Zeki Goker, Nurit Ozdogru, Sarl Sahbaz
An American journalist travels to Istanbul with his daughter when he receives a mysterious videotape sent by the father of his stepson, but when his daughter is kidnapped he is drawn into a murky and complex intrigue. A dully opaque thriller, as uninteresting as it is impenetrable.
Aka: ISTANBUL: KEEP YOUR EYES OPEN
THR   87 min   VIDrel: ODY/SONOP   V

IT *   15
Tommy Lee Wallace USA   1990
Harry Anderson, Dennis Christopher, Tim Curry, Olivia Hussey, Richard Masur, Annette O'Toole, Tim Reid, John Ritter, Richard Thomas, Jonathan Brandis, Adam Faraizl, Brandon Crane, Seth Green, Ben Heller, Emuly Perkins
Seven childhood friends meet again after thirty years when their small town is plagued by a series of murders of children, lured to their death by an evil force that takes the shape of a circus clown. A competent but overlong film (it was compiled from a mini-series) that tends to diminish the impact of King's novel. However, there are some genuinely creepy moments, and Curry is truly memorable as the evil clown "Pennywise".
Aka: STEPHEN KING'S IT
HOR   187 min (ort 193 min) mTV   VIDrel: WHV   V/sh
Boa: novel by Stephen King.

IT CAME FROM BENEATH THE SEA *   PG
Robert Gordon USA   1955
Kenneth Tobey, Faith Domergue, Ian Keith, Donald Curtis, Harry Lauter, Dean Maddox Jr, Richard W. Peterson, Del Courtney, Ed Fisher, Rudy Puteska, Jules Irving, Charles Griffith, Tol Avery, Ray Storey, Jack Littlefield
An H-bomb explosion disturbs a giant squidlike creature that was sleeping peacefully on the sea-bed and it vents its spleen on the fair city of San Francisco for having been so rudely awakened. Entertaining nonsense with special effects (courtesy of Ray Harryhausen) that were good for the time.
FAN   80 min B/W   VIDrel: COLUM/SONOP   V

IT CAME FROM OUTER SPACE *   PG
Jack Arnold USA   1953
Richard Carlson, Barbara Rush, Charles Drake, Kathleen Hughes, Russell Johnson, Joe Sawyer, Alan Dexter, George Eldredge, Brad Jackson, Warren MacGregor, George Selk, Edgar Dearing, Kathleen Hughes, William Pullen
An alien ship crashes in the desert, and its occupants assume the identities of some of the local townsfolk while they undertake repairs. Originally shown in 3-D on a wide screen, this fairly competent yarn (one of the first to use the theme of impersonation) will lose a good deal of its impact on TV.
FAN   77 min (ort 81 min) B/W   VIDrel: CIC/SONOP L/A   V
Boa: short story by Ray Bradbury.

IT CAME FROM OUTER SPACE 2 *   PG
Roger Duchowny USA   1995
Brian Kerwin, Elizbeth Pena, Jonathan Carrasco, Adrian Saprks, Bill McKinney, Dean Norris, Dawn Zeek, Lauren Tewes, Mickey Jones, Iilana B'tiste, Jerry Giles, Howard Morris, Michael Ray Miller, Clement Baker, Thom Adcox
Remake of the original SF film, with an isolated desert commu-

nity taken over by invaders from another planet after their spacecraft crashes and they are forced to make repairs.
FAN   84 min   VIDrel: CIC/SONOP   V/sh
Boa: short story by Ray Bradbury.

IT CONQUERED THE WORLD *   PG
Roger Corman USA   1956
Peter Graves, Beverly Garland, Lee Van Cleef, Sally Fraser, Russ Bender, Charles B. Griffith, Dick Miller, Jonathan Haze, Karen Kadler, Paul Harbor, Taggart Casey, Tom Jackson, Marshall Bradford, Paul Blaisdell (the monster)
A vegetable-creature arrives from Venus, is foolishly sheltered by a naive scientist, and becomes a menace to mankind. One of Roger Corman's early low-budget quickies, this enjoyable piece of nonsense is surprisingly effective despite its clumsily-plotted script. Remade as "Zontar, The Thing From Venus".
HOR   68 min (ort 70 min) B/W   VIDrel: HEND/BMGREC L/A   V

IT COULD HAPPEN TO YOU *   PG
Andrew Bergman USA   1994
Nicolas Cage, Bridget Fonda, Rosie Perez, Isaac Hayes, Seymour Cassel, J.E. Freeman, Stanley Tucci, Richard Jenkins, Ann Dowd, Wendell Pierce, Victor Rojas, Red Buttons, Robert Dorfman, Charles Busch, Beatrice Winde, Ginny Yang
When a cop finds himself without money to tip a waitress, he gallantly offers to split his lottery ticket with her. He is later astounded to learn that it has won him $4,000,000 and being a man of principle, he tries to let her have half. However, his spitfire of a wife is outraged, and soon starts to make both their lives hell. Allegedly based on a true incident, this Capra-style comedy relies for its appeal almost entirely on the fine work of the three leads.
COM   98 min (ort 101 min) cC   VIDrel: COLUM/SONOP V/sur

IT HAPPENED HERE *   PG
Kevin Brownlow/Andrew Mollo UK   1963
Sebastian Shaw, Pauline Murray, Nicolette Bernard, Bart Allison, Stella Kemball, Fiona Leland, Frank Bennett, John Herrington
Restrained account of what might have happened in Britain had Germany won WW2 and mounted an invasion, with much of the story detailing the experiences of a Welsh nurse, who joins the Nazi party in London. A semi-professional production that is slightly hampered by its tight budget, yet for all that remains surprisingly effective.
DRA   96 min B/W   VIDrel: CONNO/RTM   V

IT HAPPENED ONE NIGHT **   U
Frank Capra USA   1934
Clark Gable, Claudette Colbert, Walter Connolly, Roscoe Karns, Alan Hale, Ward Bond, Claire McDowell, Arthur Hoyt, Blanche Frederici, Jameson Thomas, Wallis Clark, Hal Price, Eddy Chandler, Kay Robinson, Frank Holliday
A film that established Columbia's reputation and marked their change from a studio making quickies. This tells of a runaway heiress who falls in love with a reporter on a bus trip. A delightful comedy, the first film to win all five top Oscars. Not until ONE FLEW OVER THE CUCKOO'S NEST 41 years later, was this done again. AA: Pic, Dir, Actor (Gable), Actress (Colbert), Story/adapt (Robert Riskin).
COM   101 min (ort 105 min) B/W
VIDrel: COLUM/SONOP   V
Boa: story Night Bus by S.H. Adams.

IT TAKES TWO *   PG
Andy Tennant USA   1995
Kirstie Alley, Steve Guttenberg, Mary-Kate Olsen, Ashley Olsen, Phillip Bosco, Jane Sibbett, Michelle Grisom, Desmond Roberts, Tiny Mills, Shanelle Henry, Anthony Aiello, La Tonya Borsay
Just on the point of marrying a cynical gold-digger, a billionaire widower (Guttenberg) gets involved with a woman who works in an orphanage, and who just happens to be looking after a child who is identical to his own nine-year-old daughter. The obvious mix-ups and complications ensue, until our tycoon learns just who he would be better off marrying. A clumsy farce, laboured, pretentious and really rather dull.
COM   100 min   CINrel

ITALIAN JOB, THE *   PG
Peter Collinson UK   1969
Michael Caine, Noel Coward, Benny Hill, Maggie Blye, Tony Beckley, Raf Vallone, Rossano Brazzi, Irene Handl, John Le Mesurier, Fred

Emney, Graham Payn, Robert Rietty, Simon Dee, Henry McGee, Robert Powell
The rather neat story of a criminal mastermind (played by Coward) who plans a brilliant bullion robbery from inside prison, by causing the biggest traffic jam in the history of Turin. An enjoyable and ingenious film sadly let down by poor characterisation and dialogue, though Coward's cameo is a joy. See also GRIDLOCK.
Aka: MASTERMIND, THE
COM   95 min (ort 100 min)   VIDrel: 4-FRONT/POLYREC/CIC; PION/CIC (LV only)   V LV

ITALIAN STALLION, THE **   18
Morton Lewis USA   1970
Sylvester Stallone, Henrietta Holm, Jodi Van Prang, Frank Micelli, Nicholas Warren, Barbara Strom
Said to be his first screen role, for which he is reputed to have been paid $200, Stallone stars as Stud, a loving guy frustrated by his intense sex drive. Even though his girlfriend makes love to him whenever she can, he still cannot keep his thoughts off other women. A film of minimal plot that ends with a general orgy and one that displays a naive innocence common to early sex films. Cut before video submission by 7 min 49 sec.
Aka: PARTY AT KITTY AND STUDS
A   57 min Cut (15 sec – ort 72 min)   VIDrel: MIA/DISC   V

IT'S A MAD, MAD, MAD, MAD WORLD *   U
Stanley Kramer USA   1963
Spencer Tracy, Phil Silvers, Terry-Thomas, Ethel Merman, Mickey Rooney, Dick Shawn, Jimmy Durante, Milton Bearle, Jonathan Winters, Buddy Hackett, Dorothy Provine, Sid Caesar, Edie Adams, Buster Keaton, Jack Benny, Jerry Lewis
And it's one long, long, long, long and very unfunny film, as a group of ill-assorted fortune-seekers attempt to recover the loot hidden by a dying gangster who reveals its approximate whereabouts. Endless frantic chases do not a comedy make. The film "Money Mania" tried something very similar. AA: Effects/aud (Walter G. Elliott).
COM   148 min (remastered version – ort 192 min)
VIDrel: MGM/WHV   V/h

IT'S A WONDERFUL LIFE ****   U
Frank Capra USA   1946
James Stewart, Donna Reed, Lionel Barrymore, Henry Travers, Thomas Mitchell, Ward Bond, Gloria Grahame, Frank Faylen, Beulah Bondi, H.B. Warner, Todd Karns, Frank Albertson, Samuel S. Hinds, Mary Treen, Sheldon Leonard
An angel shows a despairing man that his life has not been a failure despite what he thinks, by allowing him to see how those around him would have fared had he never been born. A wonderfully charming classic that shows film-making at its best. Remade for TV as "It Happened One Christmas". (The 143 min running time given includes a fifteen minute tribute to Capra, hosted by his son.) See also MR DESTINY and THE BISHOP'S WIFE.
DRA   125 min (ort 143 min) B/W coVer (THAMES)
VIDrel: THAMES L/A; 4-FRONT/POLYREC   V

IT'S ALIVE! ***   18
Larry Cohen USA   1974
John Ryan, Sharon Farrell, Andrew Duggan, Guy Stockwell, James Dixon, Michael Ansara, Michael Emhardt, William Wellman Jr, Daniel Holzman, Shamus Locke, Mary Nancy Burnett, Diana Hale, Patrick MacAllister, Gerald York
A woman's newborn babe turns out to be a terrifying monster. An unusual and chilling shocker that wore out its welcome with two further instalments. The score is by Bernard Herrmann. See also THE UNBORN.
HOR   87 min (ort 91 min)   VIDrel: MGM/WHV L/A   V

IT'S ALIVE 2 *   15
Larry Cohen USA   1978
Frederic Forrest, Kathleen Lloyd, Andrew Duggan, John P. Ryan, John Marley, Eddie Constantine, James Dixon
Sequel to IT'S ALIVE! with our mutant baby and two others, escaping from a research centre and going on the obligatory rampage. Followed by a third outing for our murderous babes.
Aka: IT LIVES AGAIN
HOR   91 min   VIDrel: MGM/WHV L/A   V

IT'S ALIVE 3: ISLAND OF THE ALIVE **   18
Larry Cohen USA   1986
Michael Moriarty, Karen Black, Laurene Landon, Gerrit Graham, Neal

Israel, James Dixon, Art Lund, Ann Dane, Macdonald Carey, William Watson, Patch MacKenzie, C.L. Sussex, Rick Garia, Carlos Palomino, Tony Abatemarco
The third instalment in this series has the father sending his "monster" son to a remote island to be with others of its kind. Years later, these creatures return to threaten civilisation as we know it. A wildly overblown sequel with some touches of black humour, but still not all that good.
Aka: ISLAND OF THE ALIVE
HOR   90 min   VIDrel: MGM/WHV L/A   V/sh

IT'S ALL HAPPENING **   U
Don Sharp UK   1963
Tommy Steele, Michael Medwin, Angela Douglas, Jean Harvey, Danny Williams, Bernard Bresslaw, Walter Hudd, Richard Goolden, Dick Kallman, John Barry
A record company talent scout stages a show for an orphanage, saving it from closure and making himself into a star in the process. A bland but fairly endearing musical comedy.
Aka: DREAM MAKER, THE
MUS   96 min (ort 101 min)   VIDrel: WHV   V

IT'S CALLED MURDER BABY **   18
Sam Weston USA   1981
John Leslie, Cameron Mitchell, Lisa Trego, Lisa De Leeuw, Seka
An ex-cop private eye is hired to stop a blackmail scheme and finds that he has bitten off more than he can chew, in this cutdown version of a hardcore film that was originally entitled "Dixie Ray: Hollywood Star".
Aka: DIXIE RAY: HOLLYWOOD STAR
THR   89 min (ort 129 min)   VIDrel: MOPIC/SGSVID   V

IT'S IN THE AIR **   U
Anthony Kimmins UK   1938
George Formby, Polly Ward, Garry Marsh, Jack Hobbs, Hal Gordon, Julien Mitchell, C. Denier Warren, Jack Melford, Michael Shepley, Frank Leighton, Ilena Sylva, O.B. Clarence, Esma Cannon
When motorcycle-mad Formby poses as his sister's fiance, who is a dispatch rider with the RAF, the usual parade of mishaps is inevitable. Not one of the best of the star's comedies, though the songs are pretty good and there are some well handled stunts.
Aka: GEORGE TAKES THE AIR
COM   87 min B/W   VIDrel: LUMI/SONOP L/A   V

IT'S MY PARTY ***   15
Randal Kleiser USA   1996
Eric Roberts, Bruce Davison, Lee Grant, Devon Gummersall, Gregory Harrison, Marlee Matlin, Roddy McDowall, Olivia Newton-John, Bronson Pinchot, Paul Regina, George Segal, Margaret Cho, Steven Antin, Dimitra Arlys, Christopher Atkins
Family and friends gather for the expected death of an architect, who is dying of AIDS and has decided to throw a farewell party before committing suicide. Roberts as the central character gives a remarkably restrained and thoughtful portrayal, and despite the subject matter, this is a gentle film of depth and warmth, though at times it does becomes just a little too gruelling. See also THE LAST BEST YEAR, THE LAST SUPPER and WHEN THE TIME COMES.
DRA   106 min (ort 110 min) cC   VIDrel: MGM/WHV   V

IT'S NOTHING PERSONAL **   15
Bradford May USA   1992
Amanda Donohoe, Bruce Dern, Yaphet Kotto, Veronica Cartwright, Xander Berkley, S. Epatha Merkerson, Elizabeth Franz, Miguel Sandoval, Dean Norris, Tom Hodges, Natalija Nogulich, Joe Urta, Claire Bloom, Eileen Ryan, Jeanne Mori
When the brother of a female cop is murdered, the former finds herself unable to shake off the feeling that she is to blame, and when the police decide to close the file on this case, she teams up with a bounty hunter to catch the killer. A moderately diverting action tale, with little original added to this wellworn idea.
A/AD   89 min (ort 96 min) mTV   VIDrel: WHV   V/h

IT'S PAT: THE MOVIE *   (12)
Adam Bernstein USA   1994
Julia Sweeney, David Foley, Charles Rocket, Larry Hankin, Tom Meadows, Arleen Sorkin, David Foley, Julie Hayden, Timothy Stark, Mary Scheer, Beverly Leech, Cathy Najimy, Jerry Tongo, Philip McNoven, Michael Yama, Nyoko Yamaguchi
Two androgynous individuals meet and fall in love. One is gentle and sweet, the other, a perfect slob and so there relationship differs in no way from those of those whose gender identity is more clear-cut. While their sexual ambiguity fasci-

nates their friends and neighbours, this fact alone cannot breathe any life into this moribund dud.
COM   78 min   SATrel: SKY MOVIES

IVAN THE TERRIBLE *****   PG
Sergei Eisenstein USSR   1944 and 1946
Nikolai Cherkassov, Serafima Birman, Ludmila Tselikovskaya, Eric Pyriev, Mikhail Nazvanov, Pavel Kodochnikov, Andrei Abrikosov, Vsevolod Pudovkin, Mikhail Zharov, A. Ngebrov
A magnificent and utterly enthralling account of the life and times of one of Russia's greatest (and most ruthless) rulers, tracing his life from early childhood up to cynical and embittered middle-age. A difficult, stylised and rewarding masterpiece. Unfortunately only a fragment of Part 3 exists as Stalin banned the film (he took offence to Eisenstein's depiction of the Tsar's secret police.) The score is by Prokofiev and the last reel has one colour sequence.
Aka: IVAN GROZNYI; IVAN THE TERRIBLE: PART 1 (asa); BOYAR'S PLOT, THE: IVAN THE TERRIBLE – PART 2 (asa)
DRA   177 min (2 cassettes – ort 185 min) B/W/Col
VIDrel: TART/20TH   V

IVANHOE **
Stuart Orme UK   1996
Stephen Waddington, Ciaran Hinds, Susan Lynch, Victoria Smurfit, James Cosmo, Christopher Lee
Story of the title figure and his exploits in defending King Richard's throne from the devious plans for usurpation of his brother John. Quite an ambitious and well mounted affair, even if the constant scenes of fighting and bloodshed do become rather repetitive after a time.
A/AD   100 min (2 cassettes) mTV   VIDrel: BBC/TECH
V/sh

Boa: novel by Sir Walter Rayleigh.

IVAN'S CHILDHOOD ***   PG
Andrei Tarkovsky USSR   1962
Kolya Burlyaev, Valentin Zubkov, E. Zharikov, I. Tarkovskaya, Nikolai Grinko, V. Malyavina, S Krylov, D. Miliutenko
After his family is massacred by the Nazis, a twelve-year-old Russian boy joins the Partisans, undertaking a variety of dangerous spying missions. Tarkovsky, in his film debut, eschews the realistic or propagandistic approach in favour of a quiet lyricism that captures the eerie beauty of the forest landscape through which Ivan moves. Slightly spoilt by a romantic sub-plot, but still a mightily impressive first work.
Aka: MY NAME IS IVAN; YOUNGEST SPY, THE
DRA   90 min B/W (ort 97 min)   VIDrel: ARTIF/20TH   V/h

I'VE HEARD THE MERMAIDS SINGING ***   15
Patricia Rozema CANADA   1987
Sheila McCarthy, Paule Baillargeon, Ann-Marie MacDonald, John Evans, Brenda Kamino, Richard Monette
A young and rather naive woman takes a job in an art gallery run by another woman, finds both her and the art world strange and intimidating, but quickly develops an enormous crush on her attractive and self-assured boss. A touching, quirky, low-budget comedy of manners, winner of the "Prix De La Jeunesse" at the 1987 Cannes Film Festival.
Aka: LE CHANT DES SIRENES
COM   82 min (ort 85 min)   VIDrel: DTK/RTM   V/s

J

JABBERWOCKY **   PG
Terry Gilliam UK   1977
Michael Palin, Max Wall, Deborah Fallender, Neil Innes, Dave Prowse, Harry H. Corbett, John Le Mesurier, Annette Badland, Warren Mitchell, Rodney Bewes, Bernard Bresslaw, Derek Francis, Alexandra Dane, Frank Williams
Medieval satire from the "Monty Python" team, in which a cooper's apprentice finds himself having to slay a dragon after having been mistaken for a prince, but is rewarded by the hand of a princess. Patchy and largely unfunny, with the emphasis firmly on the more unpleasant aspects of life in that period. Written by Charles Alverson and Terry Gilliam.
COM   100 min (ort 104 min)   VIDrel: 20TH/TECH   V

JACK *   PG
Francis Ford Coppola USA   1996
Robin Williams, Diane Lane, Jennifer Lopez, Brian Kerwin, Fran Drescher, Bill Cosby, Michael McKean, Don Novello, Allan Rich,

Adam Zolotin, Todd Bosley, Seth Smith, Mario Yedidia, Jeremy Lelliott, Rickey D'Shon Collins, Dani Faith
Williams plays a rapidly ageing boy who has to cope with the problems he encounters as a youngster trapped in a man's body. The star is given free rein to go into over-the-top mode as we witness him enduring a variety of supposedly embarrassing situations. Progeria really is no laughing matter and this ill-advised effort scores few points for sensitivity. Fortunately, a sequel is wellnigh impossible.
COM   109 min (ort 113 min) cC   VIDrel: HOLPIC/TECH
V/sur

JACK & SARAH **   15
Tim Sullivan UK   1995
Richard E. Grant, Samantha Mathis, Judi Dench, Iam McKellen, Cherie Lunghi, Eileen Atkins, Imogen Stubbs, David Swift, Laurent Grevill, Kate Hardie, Bianca Lee, Sophia Lee, Sophia Sullivan, Niven Boyd, Tracy Thorne, Lorraine Ashbourne
A highly successful lawyer loses his wife in childbirth, and his efforts to cope with his young child are constantly thwarted by the wishes of his family to take over. Fortunately, he finds help in the form of an attractive waitress whom he decides to take on as a nanny. Dench and Atkins are memorable as battling grandmothers, and they have the best lines in this bittersweet romantic comedy, which suffers from uneven scripting if not from a lack of good characterisation.
COM   105 min (ort 110 min) cC   VIDrel: POLY/POLYREC
V/sh

JACK BE NIMBLE **   18
Garth Maxwell NEW ZEALAND   1994
Alexis Arquette, Sarah Smuts-Kennedy, Bruno Lawrence, Tony Barry, Elizabeth Hawthorne, Brenda Simmons, Gilbert Goldie, Tricia Phillips, Paul Minifie, Sam Smith, Hannah Jessop, Nicholas Antwis, Olivia Jessop, Ricky Plester, Rohan Stace
A brother and sister abandoned by their parents are separated when they are adopted but find each other again thanks to their psychic powers. These also prove useful when the brother sets out to get even with his abusive adoptive family. Stylish and often surprising, Maxwell's debut feature has great energy and invention, and though at times the script is uneven and derivative, this does not detract too much from the whole.
THR   91 min (ort 95 min) wScrn   VIDrel: TART/20TH
V/sur

JACK REED: A SEARCH FOR JUSTICE **   12
Brian Dennehy USA   1994
Brian Dennehy, Charles S. Dutton, Susan Ruttan, Joe Grifasi, Rex Linn, Miguel Ferrer, Charles Hallahn, Michael Talbott, Amber Neson, Allison Mackie, Marjorie, Monaghan, Justin Burnette, Michael C. Gwynne, Elizabeth Dennehy
Police Sergeant Reed is put on the case of the murder of a night-club waitress, and is teamed up with a black colleague who got the promotion he was supposed to have. As they investigate, they find the clues pointing to her employer, who has powerful friends at City Hall. Dennehy is terrific in an otherwise undistinguished tale, and the over-emphasis on corruption in high places (a common feature of other stories in the series) grows a bit tiresome.
DRA   91 min mTV   VIDrel: ODY/SONOP   V/sh

JACK REED: BADGE OF HONOR **   15
Kevin Connor USA   1994
Brian Dennehy, Susan Rutan, Alice Krige, R.D. Call, Amy Aquino, Jo Anderson, Bryan Keith Minns, Neal McDonough, Justin Burnette, Joey Zimmerman, Michele Lamar Richards, William Sadler, Michael Talbott, Udo Kier, Amber Benson
Sergeant Reed sets out to catch the killer of a young mother in this fairly ordinary entry in the series.
DRA   89 min mTV   VIDrel: ODY/SONOP   V/sh

JACK REED: ONE OF OUR OWN **   15
Brian Dennehy USA   1995
Brian Dennehy, Charles S. Dutton, Susan Ruttan, Kevin Dunn, Suki Kaiser, Michael Talbott, Bernie Coulson, Amber Benson, Cusse Mankuma, Justin Burnette, Megan Leitch, Justin Louis, CCH Pounder, Terence Kelly, Lorena Gale, Gavin Buhr
A tough cop is framed for the murder of his partner by corrupt officials in City Hall, who want to discredit him as his investigations into the activities of a hit-man are getting too close for comfort, especially as he has contacted a former girlfriend of one of them. Co-written by Dennehy, this is a watchable if unmemorable entry in the series. Said to be based on a true

case, but this corruption issue is a common feature of the other stories too.
DRA   89 min mTV   VIDrel: ODY/SONOP   V/sh

JACK THE BEAR **   15
Marshall Herskovitz USA   1992
Danny DeVito, Robert J. Steinmiller Jr, Miko Hughes, Gary Sinise, Stefan Gierasch, Art LaFleur, Erica Yohn, Julia Louis-Dreyfus, Reese Witherspoon, Bert Remsen, Andrea Marcovicci, Carl Gabriel Yorke, Lee Garlington, Lorinne Vozoff
A widowed man who hosts a TV horror show has a hard time keeping his family together and takes to drink, which places a heavy burden on his oldest son who is just twelve. Meanwhile, a Nazi neighbour also threatens the survival of his family. Set in 1972, this confused, overlong and depressing comedy-drama fails to amuse or impress despite the hard work of a fine cast.
DRA   94 min (ort 110 min)   VIDrel: 20TH/TECH   V/sh
Boa: novel by Dan McCall.

JACK THE RIPPER ***   15
David Wickes UK   1988
Michael Caine, Lewis Collins, Jane Seymour, Susan George, Lysette Anthony, Armand Assante, Ray McAnally, Ken Bones, Jonathan Moore, Kelly Cryer, Michael Gorhard, Harry Andrews, Ann Castle, Trevor Baxter, Angela Crow
The latest retelling of this gory tale is a lavish, well-made and well-acted film, that retraces the events in the Ripper's career in considerable detail, but adds some fictional elements, with Caine ultimately unmasking the murderer. First shown in two parts.
DRA   183 min (ort 200 min) mTV   VIDrel: THAMES/DISC
V

JACKALS **   18
Gary Grillo USA   1985
Jack Lucarelli, A. Wilford Brimley, Gerald McRaney, Jameson Parker, Jeannie Wilson
The title refers to the local police who prey on illegal Mexican immigrants, robbing and murdering them as they try to cross the border. An uncorrupted ex-city cop mounts a bloody campaign to reveal the truth, but the virtue of this interesting idea is marred by excessive gore.
Aka: AMERICAN JUSTICE
A/AD   92 min   VIDrel: 20TH/TECH   V

JACKNIFE ***   15
David Jones USA   1989
Robert De Niro, Ed Harris, Kathy Baker, Charles Dutton, Loudon Wainwright III
A Vietnam veteran visits an old army pal in the hope of getting him to face up to his unhappy memories of the conflict and the loss of a mutual friend, and his arrival brings a little romance into the life of the man's shy sister. A very fine drama, superbly acted and intelligently scripted by Metcalf from his stage play.
DRA   98 min (ort 102 min)   VIDrel: FIRST/SONOP   V/sur
Boa: play Strange Snow by Stephen Metcalf.

JACK'S BACK **   18
Rowdy Herrington USA   1987
James Spader, Cynthia Gibb, Rod Loomis, Jim Haynie, Wendell Wright, Robert Picardo, Chris Mulkey, Danitza Kingsley
The story of Jack the Ripper is updated and transplanted to modern-day L.A., where a series of grisly murders of prostitutes takes place 100 years after the crimes of the legendary Victorian figure. When the chief suspect is murdered, the police are content to close their files, but his twin brother arrives to clear his name. A sluggish tale of preposterous twists and turns, with a few thrills near the end as a partial compensation.
THR   92 min (ort 97 min)   VIDrel: 20TH/TECH L/A   V/h

JACKSONS, THE: AN AMERICAN DREAM ***   PG
Karen Arthur USA   1992
Lawrence Hilton-Jacobs, Angela Bassett, Holly Robinson, Margaret Avery, Alex Burall, Jermaine Jackson II, Bumper Robinson, Shakiem Jamar Evan, Floyd Roger Meyers Jr, Monica Calhoun, Jason Weaver, Angela Vargas, Jacene Wilkerson
Overlong, glitzy and superficial account of the genesis of the Jackson 5 that traces their career from their first beginnings in Gary, Indiana, to their 1984 farewell tour. Quite enjoyable for its music and acting, but do not expect any deep insights into character or motivation. Very much the standard Hollywood showbiz biography.
DRA   215 min (ort 300 min) mTV   VIDrel: POLY/POLYREC
V

JACOB'S LADDER ***   18
Adrian Lyne USA   1990
Tim Robbins, Elizabeth Pena, Danny Aiello, Matt Craven, Pruitt Taylor Vince, Jason Alexander, Patricia Kalember, Eriq LaSalle, Ving Rhames, Macaulay Culkin, Brian Tarantina, Anthony Alessandro, Brent Hinkley, Suzanne Shepherd
A Vietnam veteran working as a New York postman begins to experience strange nightmares rooted in his battlefield experiences. These visions increase in intensity until he begins to see demons everywhere. After suffering a mental breakdown, he eventually learns that other members of his former unit are showing the same symptoms. A visually brilliant but seriously flawed film that fails to do justice to its complex and imaginative script.
HOR   108 min (ort 116 min) cC
VIDrel: 4-FRONT/POLYREC/GUILD   V/sur

JACQUELINE BOUVIER KENNEDY *   PG
Steven Gethers USA   1981
Jaclyn Smith, Rod Taylor, Stephen Elliott, James Franciscus, Claudette Nevins, Donald Moffat, Joseph Chapman, Heather Hobbs, Dolph Sweet, James F. Kelly, Maurice Marsac, Eve Roberts, Robert Easton, Lauree Berger, Ned Wilson
Dull biopic about Jackie Kennedy's rise to fame from age five onwards, that proves as interesting as its subject matter.
DRA   143 min (ort 149 min) mTV
VIDrel: POLY/POLYREC/BRAVE L/A   V

JACQUOT DE NANTES ***   PG
Agnes Varda FRANCE   1991
Philippe Maron, Edouard Joubeard, Laurent Monnier, Brigitte de Villepoix, Daniel Dublet, Clement delaroche, Rody Averty, Helene Pors, Marie-Sidonie Benoist, Julien Mitard, Jeremie Bader, Jeremie Bernard, Cedric Michaud
True story of film-maker Jacques Demy and his love affair with the cinema, made by his wife Agnes Varda, with whom he lived from 1958 until his death in 1990. A true labour of love for his widow, featuring sequences from some of his best films, intercut with many interesting scenes from his youth, when despite his father's opposition, he decided on the career he would follow. A highlight has Demy, now close to death, looking back over his fascinating life.
DOC   114 min (ort 118 min) B/W/Col wScrn
VIDrel: TART/20TH   V

JADE *   18
William Friedkin USA   1995
David Caruso, Linda Fiorentino, Chazz Palmientieri, Michael Biehn, Richard Crenna, Donna Murphy, Ken King, Holt McCallany, David Hunt, Angie Everhart, Jay Jacobus, Kevin Tighe, Victoria Smith, Drew Snyder, Bud Bostwick, Darryl Chan
When a rich female socialite is found murdered, the D.A. in charge of the case finds himself considering a former lover as one of his suspects. As he gradually uncovers her secret double life, he becomes drawn into a web of intrigue and deceit. Scripted by Joe Eszterhas, this overheated erotic thriller is memorable for its daft dialogue, having neither tension nor eroticism.
THR   125 min cC   VIDrel: CIC/SONOP; PION (LV only)
V/sur LV

JAGGED EDGE **   18
Richard Marquand USA   1985
Jeff Bridges, Glenn Close, Peter Coyote, Robert Loggia, John Dehner, Leigh Taylor-Young, Karin Austin, Lance Henricksen, James Karen, Maria Matenzet, Dave Austin, Richard Partlow, William Allen Young, Ben Hammer, Sanford Jensen
A woman lawyer defends a publisher on a charge of murder and falls in love with him. However, he may not be as innocent as she thinks. The title refers to the murder weapon: a hunting knife. Implausible nonsense, but quite well done.
THR   105 min (ort 108 min)   VIDrel: COLUM/VCC L/A
V/sh

JAIL OF NO RETURN **   18
Ng Doy Yung HONG KONG
Wong Chi Yeung, Ku Feng, Richard Grosse
Fact-based tale of a penal colony near Singapore, whose inmates are only released on their sixtieth birthday, but whose regime is so tough that many do not make it to their release date.
DRA   90 min   VIDrel: EAST/DISC   V

JAILBAIT *   *PG*
Edward D. Woods Jr USA   1954
Lyle Talbot, Steve Reeves, Herbert Rawlinson, Dolores Fuller, Clancey
Malone, Theodora Thurman, Mona McKinnon
A crime melodrama from the director of PLAN 9 FROM OUTER
SPACE in which the son of a plastic surgeon turns to crime
under the influence of a delinquent companion who eventually
murders him. However, our father exacts a fitting revenge.
Rawlinson's last film as he died the day after its completion and
really of curiosity value only.
DRA   68 min (ort 80 min) B/W   VIDrel: CARL/TECH
V/dm

JAILBREAKERS *   *(18)*
William Friedkin USA   1994
Shannen Doherty, Adrienne Barbeau, Adrien Brody, Vince Edwards,
George Gerges, Daia Barron, Chris Conrad, Sean Whalen, Taobert
Morton, Charles Napier, Julie Ariola
A remake of a cult 1960s movie that was a big hit at the drive-
ins, this is set in the 1950s, and has Doherty playing a pretty
cheerleader who gets involved with a biker. Having been egged
on to steal a necklace for her, the young man is sent to prison
for six years, finds he cannot live without her and breaks out of
jail. Re-united with his love once more, the pair are forced to go
on the run. Hardly better than the earlier film, this has little
originality or interest.
DRA   73 min mCab   SATrel: SKY MOVIES

JAMES AND THE GIANT PEACH ***   *U*
Henry Selick UK/USA   1996
Paul Terry, Joanna Lumley, Miriam Margoyles, Pete Postlethwaite,
Steven Culp, Girocco Dunlap, Michael Giradin, J. Stephen Coyle,
Tony Haney plus voices of: Paul Terry, Simon Callow, Richard
Dreyfuss, Jane Leeves, Susan Sarandon
An unhappy orphan gets away from his spiteful aunts by hiding
inside a giant magical peach that hangs from a barren tree and
is carried off by it on some wonderful adventures. A witty and
entertaining film by the same director who did THE NIGHT-
MARE BEFORE CHRISTMAS, this is part live-action and part
animation. As with the earlier film, it skilfully blends pathos,
humour and frights, but will almost certainly be too scary for
very small children.
ANIM   78 min   VIDrel: GUILD/FOXVID   V/sur
Boa: book by Roald Dahl.

JAMON, JAMON **   *18*
Bigas Luna ITALY/SPAIN   1992
Penelope Cruz, Javier Bardem, Anna Galiena, Stefania Sandrelli, Jordi
Molla, Juan Diego, Tomas Penco, Armando Del Rio, Diana Sassen,
Chema Mazo, Isabel De Castro Oros, Marianne Hermitte, Maria
Renio, Nadia Godoy, Susana Koska
When her daughter announces her intention to marry the son
of the town whore, a woman decides that the best way to stop
this is to hire a sexy young man to act as his rival. Matters
develop according to plan until the mother is herself unable to
resist the charms of her hired Lothario and beds him, but this
event leads to an ultimate tragedy. An uneven and spiteful black
comedy that makes a few sharp points about morals and
honesty. See THE TIT AND THE MOON.
COM   90 min (ort 95 min) wScrn   VIDrel: TART/20TH
V/sur

JANE EYRE ***   *PG*
Robert Stevenson USA   1943
Joan Fontaine, Orson Welles, Margaret O'Brien, John Sutton, Henry
Daniell, Agnes Moorehead, Elizabeth Taylor, Peggy Ann Garner, Sara
Allgood, Aubrey Mather, Hillary Brooke, Edith Barrett, Ethel Griffies,
Barbara Everest
A languid but careful adaptation of Bronte's classic tale, in
which a poor orphan girl obtains the post of governess at a
mansion, but finds that her master has a sullen disposition and
a mysterious past. An atmospheric and often effective offer-
ing; Welles is at his menacing best as a brooding Rochester.
Scripted by Stevenson, Aldous Huxley and John Houseman.
The music is by Bernard Herrmann and photography by
George Barnes.
DRA   96 min B/W   VIDrel: 20TH/TECH   V/h
Boa: novel by Charlotte Bronte.

JANE EYRE ***   *PG*
Franco Zeffirelli FRANCE/ITALY/UK/USA   1995
William Hurt, Charlotte Gainsbourg, Joan Plowright, Anna Paquin,
Geraldine Chaplin, Billie Whitelaw, Maria Schneider, Fiona Shaw,
Elle Macpherson, John Wood, Amanda Root, Samuel West, Josephine
Serre, Leanna Rowe, Nic Knight
A very dark and brooding adaptation telling of the young
governess who starts work for the mysterious Mr Rochester,
and slowly learns of her master's unhappy past. The bleak
English countryside of the region is put to good use in devel-
oping atmosphere, but the cast generally lack charisma, and
though Hurt as Rochester does his best to look suitably glow-
ering his efforts are not well matched by those of Gainsbourg,
whose odd English accent sounds strained indeed.
DRA   109 min (ort 113 min)   VIDrel: GUILD/FOXVID
V/sur
Boa: novel by Charlotte Bronte.

JANE EYRE: PARTS 1 AND 2 **   *U*
Julian Amyes UK   1983
Zelah Clarke, Timothy Dalton, Judy Cornwell, Robert James, Kate
David, Sally Osborn, Christine Labsalome, Sian Paeterda
A lively, well-made TV version of this famous classic, telling of
the stormy relationship that develops between a governess and
her brooding and melancholy employer.
DRA   238 min (2 cassettes) mTV   VIDrel: BBC/TECH   V
Boa: novel by Charlotte Bronte.

JANE'S HOUSE ***   *(PG)*
Glenn Jordan USA   1993
James Wood, Anne Archer, Missy Crider, Graham Beckel, Diane
D'Aquila, Keegan Macintosh, Barry Bonds, Jeff Irvine, Carrie Cain-
Sparks, Eric Keenleyside, Terrence Kelly, Fred Henderson, Austin
Basile, Debbie Podowski, Dona Yamamoto
A recently widowed father-of-two meets a successful career
woman and the couple marry, but his new wife not only finds
it hard to win over his children but also to adapt to life in a
house that bears so many traces of the former wife. A charming
romantic drama, well performed and believable.
DRA   85 min (ort 90 min) mTV   SATrel: SKY MOVIES
Boa: novel by Robert Kimmel Smith.

JANITOR, THE **   *15*
Peter Yates USA   1981
William Hurt, Sigourney Weaver, Christopher Plummer, James
Woods, Irene Steven Hill, Kenneth McMillan, Pamela Reed, Albert
Paulsen, Sharon Goldman, Morgan Freeman, Alice Drummond, Chao-
Li Chi, Keone Young, Dennis Sakamoto
A janitor in a New York office building is infatuated with a
woman TV news reporter. When a Vietnamese businessman is
murdered there, he pretends to know something about the crime
in order to get to know her. The characters are well delineated
except for Weaver, who fails to convince as a New York Jewess.
In addition, the plot device (a scheme to get Jews out of the
USSR) seems strained and contrived. A flawed work. Written by
Steve Tesich.
Aka: EYEWITNESS
DRA   98 min (ort 108 min)   VIDrel: 20TH   V

JASON AND THE ARGONAUTS ***   *U*
Don Chaffey UK   1963
Todd Armstrong, Niall MacGinnis, Gary Raymond, Nancy Kovack,
Honor Blackman, Laurence Naismith, Nigel Green, Michael Gwynn,
Douglas Wilmer, Jack Gwillim, Andrew Faulds, Patrick Troughton,
John Cairney, Gernando Poggi
Jason's voyage in search of the Golden Fleece, and his adven-
tures, as he and his men face many hardships and dangers, are
brought to life in a film with fine special effects by Ray
Harryhausen. The score is by Bernard Herrmann.
A/AD   99 min (ort 104 min)   VIDrel: VCC/DISC/COLUM
V

JASON'S LYRIC **   *18*
Doug McHenry USA   1994
Allen Payne, Jada Pinkett, Bokeem Woodbine, Forest Whitaker,
Suzzanne Douglas, Anthony "Treach" Criss, Lisa Carson, Eddie
Griffin, Lahmard Tate, Clarence Whitmore, Asheamu Earl Randle,
Rushion McDonald, Bebe Drake
A depressing tale of Afro-American life that centres around an
honest man who works in an electronics store, and his drug-
addict brother, who steals to support his habit. When the former
meets and falls in love with an attractive waitress, this soon
leads to problems. Another flawed attempt to cover the same
familiar territory as several other recent films (such as BOYZ N
THE HOOD). Weak plotting and poor characterisations detract
from the film's impact.
DRA   116 min (ort 119 min)   VIDrel: COLUM/SONOP   V/sh

JASSY ★★   U
Bernard Knowles UK   1947
Margaret Lockwood, Patricia Roc, Dennis Price, Basil Sydney, Dermot
Walsh, Nora Swinburne, Linden Travers, Ernest Thesiger, Cathleen
Nesbitt, Jean Cadell, Esma Cannon, John Laurie, Clive Morton, Grey
Blake, Torin Thatcher
Story of a gypsy girl who is gifted with second sight, and
becomes the wife of a wealthy man she despises, but only in the
hope of gaining his estate for the man she loves. When her
boorish husband is badly injured in a riding accident, she nurses
him devotedly, but is put on trial for murder when he is
poisoned. Set in the 1830s, this period drama is full of melodramatics and quite dated, yet is colourful and atmospheric
enough to maintain interest.
DRA   102 min   VIDrel: CARL/TECH   V
Boa: novel by Nora Lofts.

JAWS ★★★★   PG
Steven Spielberg USA   1975
Roy Scheider, Robert Shaw, Richard Dreyfuss, Lorraine Gary, Jeffrey
Kramer, Murray Hamilton, Susan Backlinie, Carl Gottlieb, Jonathan
Filley, Jay Mello, Ted Grossman, Chris Rebello, Lee Fierro, Jeffrey
Voorhees, Craig Kingsbury
A brilliant look at what happens when a monstrous shark
attacks a small coastal island. The film really examines the rela-
tionship that develops between the three men who set out to
hunt it. The script is by Benchley who has a small cameo as a
TV reporter. A superb film that has been followed by a series of
utterly dreadful sequels. AA: Sound (Robert L. Hoyt/Roger
Herman/Earl Madery/John Carter), Score/orig (John
Williams), Edit (Verna Fields).
A/AD   119 min (ort 124 min) wScrn
VIDrel: CIC/SONOP; PION (LV only)   V/dm LV
Boa: novel by Peter Benchley.

JAWS 2 ★   PG
Jeannot Szwarc USA   1978
Roy Scheider, Lorraine Gary, Murray Hamilton, Joseph Mascolo,
Collin Wilcox, Jeffrey Kramer, Ann Dusenberry, Mark Gruner, Barry
Coe, Susan French, Gary Springer, Donna Wilkes, Gary Dubin, John
Dukakis, G. Thomas Dunlop, Marc Gilpin
A feeble formula spin-off attempting to cash in on the success
of its predecessor. Waterlogged dialogue is coupled with an
interminable final sequence in which the shark attacks a bunch
of kids and is finished off by Scheider. Pity the director didn't
meet up with it prior to filming.
A/AD   111 min (ort 116 min)   VIDrel: CIC/SONOP   V

JAWS 3 ★   15
Joe Alves USA   1983
Dennis Quaid, Bess Armstrong, Simon MacCorkindale, Louis Gossett
Jr, John Putch, Lea Thompson, P.H. Moriarty, Dan Blasko, Liz Morris,
Lisa Maurer, Harry Grant, Andy Hansen, P.T. Horn, John Edson Jr,
Kaye Stevens, Steve Mellor
This time the shark is on the prowl in Florida's Sea World, where
the two grown-up sons of the police chief from the first two
films are now living. When a baby Great White shark is
captured, its mother charges to the rescue, flattening everything
in its path. A dull, silly second sequel, with the useless gimmick
of 3-D. Followed by JAWS 4.
Aka: JAWS 3-D
A/AD   94 min (ort 97 min)   VIDrel: CIC/SONOP   V/sh

JAWS 4 ★   15
Joseph Sargent USA   1987
Lance Guest, Lorraine Gary, Mario Van Peebles, Karen Young,
Michael Caine, Judith Barsi, Lynn Whitfield, Mitchell Anderson, Jay
Mello, Cedric Scott, Charles Bowleg, Melvin Van Peebles, Mary
Smith, Edna Billotto, Lee Fierro
The widow of the sheriff from the first two films, re-lives past
horrors in a nightmare that starts when her son is killed by a
shark, and she goes to the Bahamas to visit her other son. There,
she comes to the startling conclusion that members of her family
are being deliberately attacked by a Great White shark, in a
personal vendetta for the deaths of its relatives. A truly ludicrous premise, not helped by shoddy handling and an inept
ending.
Aka: JAWS: THE REVENGE
A/AD   86 min (ort 100 min)   VIDrel: CIC/SONOP   V/sur

JAYNE MANSFIELD STORY, THE ★   PG
Dick Lowry USA   1980
Loni Anderson, Arnold Schwarzenegger, Raymond Buktenica,

Kathleen Lloyd, G.D. Spradlin, Dave Shelley, Laura Jacoby, Whitney
Rydbeck, John Medici, Lewis Arquette, James Jeter, Janice Kent, Lynne
Seibel, Gwen Van Dam
Biopic on this platinum blonde sex symbol of the 1950s, who met
an untimely end in a car crash. A film that is neither interesting
nor accurate.
Aka: JAYNE MANSFIELD: A SYMBOL OF THE 50s
DRA   92 min (ort 100 min) mTV   VIDrel: ODY/SONOP   V/h

JAZZ SINGER, THE ★★★   U
Alan Crosland USA   1927
Al Jolson, May McAvoy, Warner Oland, Eugenie Besserer, Otto
Lederer, William Demarest, Roscoe Karns
Despite its corny story of how the son of a Jewish cantor defied
his father and made the big time in show business, this film is
pure cinema history. Known as the first talkie, although it is
actually a silent whose added sound sequences caused a sensa-
tion at the time. Now notable primarily on this account, but
Jolson's fine performance spelt certain commercial success.
Remade in 1953 and 1980. AA: Spec Award (Warner Brothers).
MUS   84 min silent (with sound sequences – ort 90 min) B/W
VIDrel: WHV   V/h
Boa: play by Samson Raphaelson.

JAZZ SINGER, THE ★★   PG
Richard Fleischer USA   1980
Neil Diamond, Laurence Olivier, Lucie Arnaz, Catlin Adams, Paul
Nicholas, Franklyn Ajaye, Sully Boyar, Mike Kellin, James Booth,
Luther Walters, Rod Gist, Oren Waters, Walter Joanowitz, Janet
Brandt, John Witherspoon
An updated remake of the earlier classic, in which Diamond
plays a rock singer at odds with his Orthodox Jewish father
(played by Olivier with appropriate hammy cliches of "I hef no
son!"). Tepid, dull and mawkish, despite a really outstanding
performance from Diamond.
MUS   110 min (ort 115 min)   VIDrel: WHV   V/sur
Boa: novel by Samson Raphaelson.

JEAN DE FLORETTE ★★★   PG
Claude Berri FRANCE   1986
Gerard Depardieu, Yves Montand, Daniel Auteuil, Elisabeth
Depardieu, Marcel Champel, Ernestine Mazurowa, Marc Betton,
Bertino Benedetto, Armand Meffre, Margarita Lozano, Pierre Jean
Rippert, Andre Dupon, Pierre Nougaro
Lavish first part of Pagnol's tale of a simple man who leaves the
city to live on an inherited piece of land, but is ruined when he
fails to find water, the location of an underground spring having
been kept secret by two scheming villagers. The conclusion to
this story comes in the final part: MANON DES SOURCES.
DRA   122 min wScrn   VIDrel: ELPIC/POLYREC   V
Boa: novel L'eau des Collines by Marcel Pagnol.

JEFFERSON IN PARIS ★★   12
James Ivory UK/USA   1995
Nick Nolte, Greta Scacchi, Thandie Newton, Jean-Pierre Aumont,
Todd Boyce, Estelle Eonnet, Gwyneth Paltrow, Seth Gilliam, Simon
Callow, James Earl Jones, Nigel Whitney, Michael Lonsdale, Nancy
Marchand, Daniel Mesguich, Lambert Wilson
During Thomas Jefferson's period of service as American
ambassador to France in the 1780s, he met and fell in love with
an attractive Englishwoman while at the same time, feeling
drawn to a young, black slave. Both these affairs arouse the
opposition of his daughter, a strong-willed woman with a mind
of her own. A well acted and attractive looking historical biopic
that sadly, is both lifeless and leaden-paced.
DRA   137 min (ort 139 min)   VIDrel: TOUCH/TECH   V

JEFFREY ★★★   15
Christopher Ashley USA   1995
Steven Weber, Michael T. Weiss, Patrick Stewart, Bryan Batt,
Sigourney Weaver, Kathy Najimi, Christine Baranski, Nathan Lane,
Olympia Dukakis, Peter Jacobson, Tom Cayler, David Thornton, Lee
Mark Nelson, John Ganun, Joseph Dain
A young gay man adopts celibacy to save himself from getting
AIDS but is forced to reconsider his stand when he meets a
potential partner who is HIV-positive. Confused and bewil-
dered, he turns to his friends for advice and guidance. A open
and warmhearted gay comedy with a few mawkish moments
but nice vigorous performances, especially from Stewart as one
of Jeffrey's friends. The film never escapes its stage origins, the
screenplay being by Rudnick.
COM   90 min (ort 94 min)   VIDrel: FILM4/RTM   V/sh
Boa: play by Paul Rudnick.

JEM: THE MOVIE * (U)
Jay Bacal/Wally Burr/Bill Dubay USA 1985
Voices of: Charlie Adler, Pat Albrecht, Marlene Aragon, Ellen Bernfeld, Bobbi Block, Cathianne Blore, Susan Blu, Anne Bryant, Kim Carlson, T.K. Carter, Walker Edmiston, Diva Grey, Michael Horton, Ford Kinder, Cathy Marcuccio
Feature-length animation about the adventures of the title, all-girl band. Poor workmanship combines with even worse pop music to produce a work of great tedium.
ANIM 91 min mTV SATrel: MOVIE CHANNEL

JENNIFER 8 * 15
Bruce Robinson USA 1992
Andy Garcia, Uma Thurman, Lance Henriksen, Kathy Baker, Graham Beckel, Kevin Conway, John Malkovich, Perry Lang, Lenny Von Dohlen, Bob Gunton, Paul Bates, Perry Long, Bryan Larkin, Nicholas Love, Michael O'Neil
A burnt-out cop puts L.A. behind him and relocates to a small town in North California that is being stalked by a serial killer. After finding a hand belonging to his latest victim (hence the title), he meets a woman witness and becomes involved with her. A fine, moody atmosphere and great performances are major strengths but a stronger and more consistent plot was sorely needed.
THR 120 min (ort 127 min) VIDrel: CIC/SONOP V/dm V/sur

JENNY'S WAR: PARTS 1 AND 2 * U
Steven Gethers USA 1985
Dyan Cannon, Robert Hardy, Nigel Hawthorne, Elke Sommer, Patrick Ryecart, Richard Todd, Harmut Becker, Christopher Cazenove, Sion Tudor-Owen, John Moulder-Brown, Denis Lill, Garfield Morgan, Michael Elphick, Hugh Grant
An American woman goes behind German lines in occupied Europe, disguised as a man in order to find her POW son, an RAF pilot shot down over Germany. A disappointing and unworthy treatment of the real-life story and exploits of Jenny Baines.
DRA 208 min (2 cassettes) mTV VIDrel: VCC L/A V
Boa: novel by Jack Stoneley.

JEREMIAH JOHNSON * PG
Sydney Pollack USA 1972
Robert Redford, Will Geer, Allyn Ann McLerie, Stefan Gierasch, Charles Tyner, Josh Albee, Paul Benedict, Matt Clark, Joaquin Martinez, Jack Colvin, Richard Angarola, Delle Bolton
Nice low-key story of an ex-soldier in the 1850s who becomes a mountain trapper to get away from civilisation but finds himself inadvertently involved in conflict with the local Indians. Overlong, but good fun. The script is by John Milius and Edward Anhalt.
WES 106 min VIDrel: WHV V
Boa: book Mountain Man by Vardis Fisher./short story Crow Killer by Raymond W. Thorp and Robert Bunker.

JERICHO FEVER * PG
Sandor Stern USA 1993
Stephanie Zimbalist, Perry King, Branscombe Richmond, Alan Scare, Ari Barak, Elyssa Davalos, Kario Salem, Don Harvey, Andrew Brye, Michael Yama, Jamie Stern, Nicholas Glaser, Kathleen Erickson, Peter Grano, Connie Perez, Ramon Chavez
After assassinating a number of negotiators at Israeli-Arab peace talks in Mexico, terrorists cross the border into the USA. Unfortunately, they have been exposed to a deadly infectious virus and a medical team is formed to deal with this threat, but nothing can be done until the terrorists have been captured. A standard, race-against-the-clock thriller – in no way original but quite gripping and generally well handled.
THR 85 min (ort 88 min) mTV VIDrel: CIC/SONOP V

JERICHO MILE, THE * 12
Michael Mann USA 1979
Peter Strauss, Roger E. Mosley, Brian Dennehy, Billy Green Bush, Ed Lauter, Beverly Todd, Richard Lawson, William Prince, Miguel Pinero, Geoffrey Lewis, Richard Moll, Edmund Penney, Burton Gilliam, Ji-Tu Cumbuka, Wilmore Thomas
A prisoner sentenced to life, is given a chance to run in the Olympics. An unusual tale with a thought-provoking script by Mann and Nolan.
DRA 97 min (ort 100 min) mTV VIDrel: VCC/DISC V
Boa: story by Patrick J. Nolan.

JERK, THE * 15
Carl Reiner USA 1979
Steve Martin, Bernadette Peters, Catlin Adams, Mabel King, Richard Ward, Dick Anthony Williams, Bill Macy, Jackie Mason, M. Emmet Walsh, Maurice Evans, Dick O'Neill, Helena Carrol, Rene Wood, Pepe Serna, Sonny Terry, Brownie McGee
The white adopted son of black sharecroppers makes a fortune, only to lose it again. Stand-up comic Martin's first starring film role, and later remade as a TV film which he produced. Some very funny clowning moments cannot hide the sheer lack of substance. Reiner makes a brief appearance in an amusing cameo.
COM 90 min (ort 94 min)
VIDrel: 4-FRONT/POLYREC/CIC V/h

JERK TOO, THE * (12)
Michael Schultz USA 1983
Mark Blankfield, Ray Walston, Stacey Nelkin, Thalmus Rasulala, Barrie Ingham, Mabel King, Pat McCormick, Gwen Verdon, Jimmie Walker, Martin Mull, Lainie Kazan, Robert Sampson, Patricia Barry, John LeClerc, Bill Saluga
Not so much a TV sequel to THE JERK as a remake, this reprises the story of a white youngster brought up by black sharecroppers, who goes in search of his true love. Some bright comic moments enliven an otherwise dull film, that was intended as a TV pilot for a prospective series.
COM 88 min (ort 100 min) mTV VIDrel: CIC/SONOP L/A V

JERKER * 18
Hugh Harrison USA 1990
Tom Wagner, Joseph Stachura
A series of late-night phone calls bring together two homosexuals who share their fantasies, in this adaptation of a controversial hit play.
DRA 96 min VIDrel: MANFOR/GOLD V

JERRY LEWIS: AT WAR WITH THE ARMY * U
Mal Walker USA 1950
Jerry Lewis, Dean Martin, Polly Bergen, Angela Greene, Mike Kellin, Jimmie Dundee, Dick Stabile, Tommy Farrell, Frank Hyers, Dan Dayton, William Mendrek, Kenneth Forbes, Paul Livermore, Ty Perry, Jean Ruth, Angela Greene
Dated comedy about a private helping his sergeant who has girl trouble. A high-spot is a soda machine gag in this strained comedy – the first starring feature for Lewis and Martin.
Aka: AT WAR WITH THE ARMY
COM 90 min (ort 93 min) B/W VIDrel: ORBIT/DISC V
Boa: play by James Allardice.

JERRY LEWIS: CINDERFELLA * U
Frank Tashlin USA 1960
Jerry Lewis, Ed Wynne, Judith Anderson, Anna Maria Alberghetti, Henry Silva, Robert Hutton, Count Basie
A garish and overblown adaptation of the famous fairytale suitably adjusted to serve as a vehicle for the comic talents (such as they are) of Lewis. Talky and ineffectual, this exercise in pretentiousness works neither as a comedy nor as a fantasy. A dull and dismal dud.
Aka: CINDERFELLA
COM 91 min VIDrel: L/A V

JERRY LEWIS: SMORGASBORD * PG
Jerry Lewis USA 1983
Jerry Lewis, Herb Edelman, Zane Buzby. Guest stars: Dick Butkus, Milton Berle, Sammy Davis Jr, Foster Brooks, Buddy Lester, Francine York, Bill Richmond, Robin Bach, Paul Davidson
A sketch-type film in which Lewis tries extremely hard to amuse, this time as a near suicidal patient who is being treated by a psychiatrist. Shelved (understandably) for some time after production.
Aka: CRACKING UP
COM 85 min VIDrel: 20TH/TECH L/A V

JERRY LEWIS: THE BELLBOY * U
Jerry Lewis USA 1960
Jerry Lewis, Alex Corry, Bob Clayton, Sonny Sands, Milton Berle, Walter Winchell, Eddie Shaeffer, Herkie Styles, David Landfield, Bill Richmond, Larry Best, Jimmy Gerrard, Tilly Gerrard, Duke Art Jr, Eddie Barton
The comic tale of a blundering bellboy working at the Fontainbleu Hotel in Miami Beach. No plot, just a series of madcap slapstick adventures as Lewis clumsily staggers from

one mishap to another. A very dated comedy that gave Lewis his first opportunity to direct.
Aka: BELLBOY, THE
COM   69 min (ort 72 min) B/W   VIDrel: L/A   V

JERRY LEWIS: THE BIG MOUTH *
U
Jerry Lewis USA   1967
Jerry Lewis, Harold J. Stone, Buddy Lester, Susan Bay, Del Moore, Paul Lambert, Jeanninie Riley, Charlie Callas, Leonard Stone, Frank DeVol, Vern Rowe, Dave Lipp
A timid bank clerk out fishing gets embroiled in murder and a madcap chase, when he hooks a frogman who hands him a map giving the location of stolen loot, just before being killed by gangsters. A frantic hunt now ensues, with Lewis donning disguise to stay one step ahead of his pursuers, in an unfunny and interminable spoof on IT'S A MAD, MAD, MAD, MAD WORLD that (unfortunately) give the star ample scope for his tiresome clowning.
Aka: BIG MOUTH, THE
COM   103 min (ort 107 min)   VIDrel: RCA L/A   V

JERRY LEWIS: THE ERRAND BOY **
U
Jerry Lewis USA   1961
Jerry Lewis, Brian Dunlevy, Fritz Feld, Howard McNear, Sig Rumann, Iris Adrian, Kathleen Freeman, Joe Besser, Mike Mazurki, Dick Wesson, Felicia Atkins, Mary Ritts, Paul Ritts, Isobel Elsom, Fritz Feld, Stanley Adams
Frenetically paced account of a bumbling idiot who masquerades as an errand boy but has in reality been hired by a studio mogul to uncover waste and inefficiency. Has a few funny moments but most gags misfire.
Aka: ERRAND BOY, THE
COM   92 min (ort 95 min) B/W   VIDrel: L/A   V

JERRY LEWIS: THE NUTTY PROFESSOR **
PG
Jerry Lewis USA   1963
Jerry Lewis, Stella Stevens, Del Moore, Kathleen Freeman, Med Flory, Howard Morris, Elvia Allman, Henry Gibson, Skip Ward, Norman Alden, Milton Frome, Buddy Lester, Marvin Kaplan, David Landfield, Celeste Yarnall, Francine York
A college professor undergoes a Jekyll and Hyde transformation, in this very unfunny Lewis comedy, which like so many of his other efforts demonstrates clearly the limits of his talent. One of the comic's best vehicles (but that's not saying much) it was scripted by Lewis and Bill Richmond. Remade (under the same title) in 1996 with Eddie Murphy.
Aka: NUTTY PROFESSOR, THE
COM   103 min (ort 107 min)   VIDrel: CIC/SONOP L/A   V

JERRY LEWIS: THE PATSY **
U
Jerry Lewis USA   1964
Jerry Lewis, Everett Sloane, Ed Wynne, Phil Harris, Keenan Wynn, Peter Lorre, John Carradine, Hans Conreid, Phil Foster, Richard Deacon, Scatman Crothers, Nancy Culp, Ed Wynn, Ed Sullivan, Mel Torme, Hedda Hopper
A bellboy is groomed for stardom after a popular entertainer dies in a plane crash. Sad to relate, this unfunny turkey was Lorre's last film.
Aka: PATSY, THE
COM   101 min   VIDrel: L/A   V

JERRY MAGUIRE ***
15
Cameron Crowe USA   1996
Tom Cruise, Renee Zellweger, Cuba Gooding Jr, Jonathan Lipnicki, Kelly Preston, Jerry O'Connell, Jay Mohr, Regina King, Bonnie Hunt, Jonathan Lipnicki
Cruise plays a sports agent who is unwise enough to tell his ruthless employers that they should take more pains with their clients, and is promptly sacked for his pains. However, by way of proving he is not finished, he takes over the managing of a sporting star, with predictably satisfying results. There is not really enough pace and direction to make this a really strong film, despite a stirring speech at the end. However, Cruise was rarely better.
DRA   136 min   CINrel

JERSEY GIRL ***
15
David Burton Morris USA   1992
Jami Gertz, Dylan McDermott, Molly Price, Aida Turturro, Star Jasper, Sheryl Lee, Joseph Bologna, Joseph Mazzello, Philip Casnoff, Pat Collins, Regina Taylor, Amy Johanna Sakasitz, Mary Beth Peil, Jordan Dean, Richard Maldone
Gertz is a sassy, streetwise Italian-American girl whose dreams

of finding Mr Right come true when she crashes her car into the brand new Mercedes being driven by McDermott. Despite their initial mutual attraction, there are difficulties to be resolved, for they have different backgrounds and there is a jealous ex-girlfriend to contend with. Quite a likeable film, that has nice performances and smart dialogue, if not exactly a riveting plot.
COM   91 min (ort 95 min)   VIDrel: EIV/SONOP   V

JESSE JAMES ***
U
Henry King USA   1938
Henry Fonda, Tyrone Power, Randolph Scott, Nancy Kelly, Henry Hull, Brian Donlevy, Slim Summerville, J. Edward Bromberg, John Carradine, John Russell, Donald Meek, Jane Darwell, Charles Tannen, Claire Du Brey, Willard Robertson
A highly inaccurate but entertaining biography of outlaws Jesse and Frank James, that relates how they took to a life of crime after their mother was murdered at the end of the Civil War. A vigorous, full-blooded and enjoyable knockabout Western. A sequel "The Return Of Frank James" followed in 1940 and the film was remade as "The True Story Of Jesse James" in 1957. See also FRANK AND JESSE.
WES   105 min Cut (13 sec)   VIDrel: 20TH/TECH   V/h

JESSE JAMES MEETS FRANKENSTEIN'S DAUGHTER *
(PG)
William Beaudine USA   1966
John Lupton, Estelita, Cal Bolder, Steven Geray, Jim Davis, Narda Onyx, Felipe Turich, Rosa Turich, Raymond Barnes, William Fawcett, Nestor Paiva, Dan White, Page Slattery, Roger Creed, Fred Stromsoe, Mark Norton
Mixture of Western and horror genres, with Jessie James battling to save the world from Frankenstein's grand-daughter, who has turned a member of his gang into a killer monster called Igor. Ludicrous, low-budget shenanigans.
HOR   63 min (ort 95 min)   SATrel: BRAVO MOVIES

JESUS CHRIST, SUPERSTAR ***
PG
Norman Jewison USA   1973
Ted Neely, Carl Anderson, Yvonne Elliman, Barry Dennen, Joshua Mostel, Bob Bingham, Larry T. Marshall, Kurt Yahgjian, Philip Toubus, Pi Douglass, David Devir, Jonathan Wynne, Richard Molinare, Jeffrey Hyslop, Robert LuPone
Film version of the rock opera based on the life of Christ, that combines fine music with a somewhat stagebound approach. Highly unusual but definitely showing its age.
MUS   102 min (ort 108 min) wScrn
VIDrel: 4-FRONT/POLYREC/CIC   V/sur
Boa: rock opera by Andrew Lloyd Webber and Tim Rice.

JESUS OF MONTREAL ****
18
Denys Arcand CANADA/FRANCE   1989
Lothaire Bluteau, Catherine Wilkening, Johanne-Marie Tremblay, Remy Girard, Robert Lepage, Gilles Pelletier, Marie-Christine Barrault, Yves Jacques, Denys Arcand
Written by Arcand, this unusual film offers the premise that a group of actors who have assembled to mount an unconventional production of the Passion Play may in fact be in the presence of Jesus. Bluteau gives a superb performance in the demanding title role, and this insightful work loses no opportunity to examine religious bigotry and various other issues.
Aka: JESUS DE MONTREAL
DRA   114 min (ort 120 min) subs   VIDrel: ARTIF/20TH
V/sur

JESUS OF NAZARETH **
PG
Franco Zeffirelli ITALY/UK   1977
Robert Powell, Anne Bancroft, Laurence Olivier, Ralph Richardson, James Mason, Anthony Quinn, Peter Ustinov, Rod Steiger, Christopher Plummer, Ernest Borgnine, Claudia Cardinale, Valentina Cortese, James Farentino, Ian McShane
A long and detailed account of the life of Christ, filmed in Morocco and Tunisia, best described as sincere but rather dull. Originally a four-part TV production, it has moments of genuine power and atmosphere, but suffers from an over-reverential approach to its subject matter. A host of stars have cameo roles, doing nothing to enliven a film that should inspire rather than bore. See also THE GOSPEL ACCORDING TO ST MATTHEW and THE GREATEST STORY EVER TOLD.
DRA   383 min (4 cassettes) mTV
VIDrel: POLY/POLYREC   V

JETSONS: THE MOVIE *　　　　　　　　　　U
William Hanna/Joseph Barbera USA　　　　　1990
*Voices of: George O'Hanlan, Penny Singleton, Mel Blanc, Tiffany,
Don Messick, Patric Zimmerman, Jean Vander Pyl, Ronnie Schell,
Patti Deutsch, Dana Hill, Russi Taylor, Paul Kreppel, Rick Dees,
Michael Bell, Jeff Bergman*
Feature-length tale of this 21st century family whose adventures
were first shown on TV in the 1960s. Here, the head of the family
is given promotion (albeit with an ulterior motive) by his boss,
and placed in charge of an ore extraction plant on a distant aster-
oid. An assembly-line product that suffers from a thin storyline
and a tacked-on environmental message – and the animation as
abysmal as ever.
ANIM　78 min (ort 82 min)
VIDrel: 4-FRONT/POLYREC/CIC L/A　V/sh

JEWEL IN THE CROWN, THE *　　　　　　15
Jim O'Brien/Christopher Morahan UK　　　　1984
*Peggy Ashcroft, Charles Dance, Saeed Jaffrey, Geraldine James, Rachel
Kempson, Rosemary Leach, Art Malik, Tim Pigott-Smith, Zia
Mohyeddin, Wendy Morgan, Judy Parfitt, Eric Porter, Susan
Wooldridge, Anna Cropper*
An account of the closing years of the British Raj. Massive in
scope but unfortunately overladen with all the stock British
"pukka" characters, and Indians who rarely rise above the level
of cardboard characters.
DRA　752 min (3 cassettes) mTV
VIDrel: CASPIC/BMGREC　V
Boa: the Raj Quartet novels of Paul Scott.

JEWEL OF THE NILE, THE *　　　　　　　PG
Lewis Teague USA　　　　　　　　　　　1985
*Michael Douglas, Kathleen Turner, Danny DeVito, Spiros Focas, Paul
David Magid, Avner Eisenberg, Howard Jay Patterson, Randall
Edwin Nelson, The Flying Karamazov Brothers*
A sequel to ROMANCING THE STONE, with our madcap duo
getting involved in a hunt for a mysterious gem in the deserts
of North Africa, and a race to outwit an arch-villain. The same
mixture of violent action, romance and languors as before.
A/AD　101 min (ort 104 min)　VIDrel: ENCORE　LV

JEWELS *　　　　　　　　　　　　　DRA
Roger Young UK/USA　　　　　　　　　1993
*Annette O'Toole, Anthony Andrew, Jurgen Prochnow, Corinne
Touzet, Sheila Gish, Simon Gates, Robert Wagner, Bradley Cole,
Ursula Howells, Geoffrey Whitehead, Arthur Cox, Josianne Peiffer,
Pen Turner, Nicholas Klein*
The lives and loves of an American woman, from marriage into
the nobility to her ownership of a high class jewellery company.
Very much a standard soap opera no different from hundreds
of others.
Aka: DANIELLE STEEL'S JEWELS
DRA　242 min mTV　VIDrel: 4-FRONT/POLYREC　V/sh
Boa: novel by Danielle Steel.

JEZEBEL *　　　　　　　　　　　　U
William Wyler USA　　　　　　　　　　1938
*Bette Davis, Henry Fonda, George Brent, Margaret Lindsay, Donald
Crisp, Fay Bainter, Spring Byington, Richard Cromwell, Henry
O'Neill, John Litel, Eddie Anderson, Gordon Oliver, Irving Pichel,
Georges Renavent, Fred Lawrence*
A temperamental and self-centred Southern beauty outrages all
and sundry by her behaviour, but finally gets a chance to
redeem herself when yellow fever strikes, and she volunteers to
act as nurse to her sick fiance and accompany him into quaran-
tine. AA: Actress (Davis), S. Actress (Bainter).
DRA　100 min (ort 103 min) B/W　VIDrel: WHV　V
Boa: play by Owen Davis Sr.

JFK *　　　　　　　　　　　　　　15
Oliver Stone USA　　　　　　　　　　　1991
*Kevin Costner, Tommy Lee Jones, Kevin Bacon, Laurie Metcalf, Gary
Oldman, Michael Rooker, Jay O. Sanders, Sissy Spacek, Jack Lemmon,
Sally Kirkland, Brian Doyle Murray, Edward Asner, Joe Pesci, Walter
Matthau, John Candy, Jodi Farber*
A highly ambitious and detailed attempt to retell the events that
followed the assassination of President John F. Kennedy, that
created enormous interest (and controversy) upon release by its
recourse to a compelling (if unproven) conspiracy theory.
Costner plays a determined D.A. who suspects a government
cover-up, and fights a lone battle to get at the truth. See also

RUBY, EXECUTIVE ACTION, RUBY AND OSWALD and THE
TRIAL OF LEE HARVEY OSWALD.
DRA　197 min (director's cut – ort 189 min) wScrn cC
VIDrel: WHV　V/sur
Boa: book Trail of the Assassins by Jim Garrison/book Crossfire:
The Plot That Killed Kennedy by Jim Marrs.

JIGSAW MAN, THE *　　　　　　　　　15
Terence Young UK　　　　　　　　　　1983
*Michael Caine, Laurence Olivier, Susan George, Robert Powell,
Charles Gray, Michael Medwin, Vladek Sheybal, Anthony Shaw,
Maureen Bennett, Patrick Dawson, Juliet Nissen, David Kelly, Peter
Burton, Maggie Rennie*
A defector undergoes plastic surgery and is then sent back to the
UK to recover secret documents. A verbose and muddled
thriller.
THR　90 min (ort 95 min)　VIDrel: ARENA　V/h
Boa: novel by Dorothea Bennett.

JIMMY HOLLYWOOD *　　　　　　　15
Barry Levinson USA　　　　　　　　　1994
*Joe Pesci, Christian Slater, Victoria Abril, Jason Beghe, John Cothran
Jr, Robert La Sardo, Richard Hind, Marcus Giamatti, Ralph Tabakin,
Blanche Rubin, Lopez, Cynthia Steele, Helen Brown, James Pickens
Jr, Earl Billings, Rob Weiss*
A man with an obsessive desire to act has never had a chance
to realise his dreams, but finally gets one when he becomes a
sort of community crime-fighter and is immortalised on video-
tapes by a friend. A predictable affair, built around a figure
both desperate and unattractive, that takes a succession of
easy pot-shots at Hollywood at all its attendant myths. Pesci
is first-rate but his undeniable screen presence cannot save this
dud.
COM　112 min (ort 118 min) cC　VIDrel: CIC/SONOP
V/sur

JIM'S GIFT *　　　　　　　　　　　(U)
Bob Keen UK　　　　　　　　　　　　1994
*Robert Llewellyn, Chris Jury, Jennifer Calvert, Ann Gosling, Jean
Boht, Doug Bradley, Luciano Romanor, Greg Rose, David Schaal,
Freddy White, Sean Connolly, Leon Petit, Danny Green, Daniel
Burke, Amy Griffin, Ryan Smith, Joe Garner*
A young boy who lost his dog, attends a car-boot sale where he
meets a strange stall-holder who sells him a video recorder for
just two pounds but warns him never to touch the fast-forward
button. Once at home, he gets it working and discovers that this
machine allows him to see past events, with the assistance of the
stall-holder who pops up as a sort of guide and mentor. A wry
kid's fantasy with some strained attempts at comedy and silly
sibling rivalry.
JUV　95 min　SATrel: MOVIE CHANNEL
Boa; novel by Sylvia Wickham.

JINGLE ALL THE WAY *　　　　　　　(12)
Brian Levant USA　　　　　　　　　　1996
*Arnold Schwarzenegger, Phil Hartman, Sinbad, Rita Wilson, Robert
Conrad, Martin Mull, Jake Lloyd, James Belushi, E.J. De La Pena,
Laraine Newman, Justin Chapman, Harvey Korman, Richard Moll,
Daniel Riordan, Jeff L. Deist*
Rough and brutal comedy of noise and mayhem, that sees big
Arnie promising his son he'll have a "Turbo Man" toy in time
for Christmas, having by an oversight completely forgotten to
buy it in time. Naturally enough, in his quest to get same he
faces competition from other equally determined dads. For
those who like seeing department stores torn to shreds, this is
a must.
COM　86 min (ort 89 min) Cut　VIDrel: 20TH/FOXVID
V/sh

JIT *　　　　　　　　　　　　　　(PG)
Michael Raeburn ZIMBABWE　　　　　　1993
*Dominic Makuvachuma, Sibongile Nene, Farai Sevenzo, Oliver
M'tukudzi, Winnie Mdemera, Lawrence Simbarashe, Jackie Eeson,
Cecil Zilla Mamanzi, Zanape Fazilahmed, Taffy Marichidza, Jones
Muguse, Fidelis Cheza, Emmanuel Boro*
A country boy falls for a girl from the city and sets out to win
her heart although this presents all sorts of problems, which he
solves with a little help from a guardian spirit. A fresh and
charming comedy, strongly acted and made with an appealing
directness and simplicity. The first major feature film to be
produced in Zimbabwe.
COM　92 min　CINrel

**JITTERS, THE ** ** 15
John Fasano JAPAN/USA 1989
Sal Viviano, Marilyn Tokuda, James Hong, Frank Dietz
Dead Chinatown shopkeepers become zombies and take their
revenge on the gang who murdered them. A routine horror
offering with all the standard ingredients, despite it having been
adapted from a series of Chinese films that feature such creatures.
HOR 86 min VIDrel: POPRO/RTM V

**JOAN OF ARC ** ** PG
Victor Fleming USA 1948
*Ingrid Bergman, Jose Ferrer, George Coulouris, Francis L. Sullivan,
J. Carroll Naish, Gene Lockhart, Ward Bond, John Ireland, Leif
Erickson, William Conrad, Hurd Hatfield, George Zucco*
Bergman stars as Joan of Arc, the young French peasant girl
who led the French army against the English. Betrayed to the
enemy, she was forced to confess to heresy and was burnt at the
stake as a witch. A competent rather than exciting adaptation
of the original play, let down by a distinct lack of spectacle. AA:
Cin (Joseph Valentine/William V. Skall/Winton C. Hoch), Cost
(Dorothy Jeakins/Karinska), Spec Award (Walter Wanger).
DRA 96 min (Abridged by distributor – ort 145 min)
VIDrel: L/A V
Boa: play by Maxwell Anderson.

**JOAN OF ARC ** ** PG
Jacques Rivette FRANCE 1994
*Sandrine Bonnaire, Andre Marcon, Jean-Louis Richard, Marcel
Bozonnet, Patrick Le Mauff, Didier Sauvegrain, Jean-Pierre Lorit,
Bruno Wolkowitch, Pierre Baillot, Mathias Jung, Quentin Ogier,
Stephane Boucher, Xavier Maly*
A minimalist biopic on the title character, replete with stilted
acting and low-budget battle scenes that would have more
wisely been left out. However, the attempts to show the human
face of Joan benefit much from Bonnaire's powerful presence.
Sadly the rest of the cast do little to make an impact. Incredible
to think that this four-hour epic (it is in two parts – "Les
Bataillesor" and "Les Prisons") was in fact trimmed down from
an even longer work.
Aka: JEANNE LA PUCELLE
DRA 240 min (ort 360 min) CINrel

JOCK OF THE BUSHVELD ** * PG
Gray Hofmeyr SOUTH AFRICA 1986
*Jonathan Rands, Gordon Mulholland, Jocelyn Broderick, Wilson
Dunster, Michael Brunner, Oliver Ngwena, Marloe Scott-Wilson,
Mfubu*
A classic South African novel about a boy's bush childhood
forms the basis for this enjoyable adventure, which concerns
the exploits of our hero who, accompanied by his dog, searches
for riches in the South African Gold Rush of the 1890s.
DRA 93 min (ort 101 min) VIDrel: FABFIL/POLYREC L/A
V
Boa: novel by Percy Fitzpatrick.

**JOE KIDD ** ** 15
John Sturges USA 1972
*Clint Eastwood, Robert Duvall, John Saxon, Don Stroud, Stella
Garcia, James Wainwright, Paul Koslo, John Carter, Gregory Walcott,
Pepe Hern, Chuck Hayward, Dick Van Patten, Buddy Van Horn,
Lynne Marta, Joaquin Martinez, Ron Soble*
A gunfighter is hired by a landowner to hunt down some
Mexican squatters. Another one of those features that helped
nudge Eastwood towards stardom, but this one is little better
than mediocre.
WES 84 min (ort 88 min)
VIDrel: 4-FRONT/POLYREC/CIC L/A V/h

JOE VERSUS THE VOLCANO ** * PG
John Patrick Shanley USA 1990
*Tom Hanks, Meg Ryan, Lloyd Bridges, Robert Stack, Abe Vigoda,
Dan Hedaya, Barry McGovern, Ossie Davis, Amanda Plummer, Jayne
Haynes, David Burton, Jon Pocharan, Jim Hudson, Antoni Gatti,
Darrell Zwerling, Jim Ryan, Karl Rumburg*
A bullied wage-slave is diagnosed as suffering from a mysterious
terminal disease, but is offered a fabulous remaining six months
by a millionaire if he accepts a challenge to throw himself into a
live volcano on a Polynesian island, for the natives require a hero
to appease their gods and the millionaire has worked out a
method of getting control of the island's mineral deposits. A
chaotic and disorganised screwball romantic-comedy.
COM 98 min (ort 102 min) VIDrel: WHV V/sh

JOE'S APARTMENT ** * 12
John Payson USA 1996
Jerry O'Connell, Megan Ward, Robert Vaughn,
A youngster newly arrived in New York goes in search of some
cheap lodgings, which he obtains when a woman drops dead in
front of him and he grabs her keys and takes over her apartment.
The trouble is that it is infested with singing cockroaches, but
this is just one of his problems, as he is more interested in
winning the girl of his dreams. An offbeat, occasionally hilarious work, with a wacky, anarchic style that recalls film's such
as WAYNE'S WORLD.
COM 82 min mTV VIDrel: WHV

**JOEY BOY ** * PG
Frank Launder UK 1965
*Harry H. Corbett, Stanley Baxter, Bill Fraser, Percy Herbert, Lance
Percival, Reg Varney, Moira Lister, Derek Nimmo, Thorley Walters,
John Arnatt, Eric Pohlmann, John Phillips, Lloyd Lamble, Edward
Chapman, Norman Rossington*
War-time comedy set in 1941, that follows the exploits of a
bunch of black marketeers, who having been forced into the
army, make the best of it by running various rackets. A cheap
looking army comedy, it wastes the talents of a fine cast on weak
gags and foolish shenanigans.
COM 90 min (ort 91 min) B/W VIDrel: LUMI/SPEAR V
Boa: novel by Eddie Chapman.

**JOHNNY AND THE DEAD ** ** U
Gerald Fox UK 1995
George Baker, Brian Blessed, Jane Lapotaire, Geoffrey Whitehead
On his way to school one day a youngster takes a short-cut
trough the local cemetery, where he finds that he is able to see
the dead. They explain their need for help in a venturing into
the outside world, and with the boy's help they succeed.
COM 94 min VIDrel: WARVIS/WARUK V/sh

**JOHNNY BE GOOD ** * 15
Bud Smith USA 1987
*Anthony Michael Hall, Robert Downey Jr, Paul Gleason, Uma
Thurman, Steve James, Seymour Cassel, Michael Greene, Robert
Downey Sr, Jim McMahon, Howard Cosell, Jennifer Tilly, Marshall
Bell, Deborah May, Michael Alldredge*
A dull comedy about the illegal recruitment of high school
athletes by US colleges, concentrating on one athlete who has
promised his girlfriend they would attend the local university
together. Full of nudity and crude jokes, to make up for the lack
of plot and competent acting.
COM 87 min (ort 98 min) VIDrel: VISVID/POLYREC
V/sur

**JOHNNY COME LATELY ** ** PG
William K. Howard USA 1943
*James Cagney, Grace George, Marjorie Lord, Marjorie Main, Hattie
McDaniel, Edward McNamara, Bill Henry, Bill Barrat, George
Cleveland, Margaret Hamilton, Lucien Littlefield, Irving Bacon,
Norman Willis, Edwin Stanley*
A wandering journalist stops at a small town where he helps an
elderly editor expose the corruptness of the local politicians.
Made by Cagney as an independent production, this mildly
amusing piece of turn-of-the-century whimsy has a few enjoy-
able moments scattered here and there.
Aka: JOHNNY VAGABOND
DRA 97 min B/W VIDrel: VISVID/POLYREC L/A V
Boa: novel McLeod's Folly by Louis Bromfield.

**JOHNNY DANGEROUSLY ** * 15
Amy Heckerling USA 1984
*Michael Keaton, Joe Piscopo, Marilu Henner, Peter Boyle, Maureen
Stapleton, Griffin Dunne, Richard Dimitri, Glynnis O'Connor, Byron
Thames, Danny DeVito, Dom DeLuise, Ray Walston, Sudie Bond,
Dick Butkus, Ron Carey, Alan Hale*
A wild spoof on American gangster films, that takes a look at
the career of the title character. A scattering of feeble jokes serve
merely to highlight the atrociousness of the whole effort.
COM 86 min (ort 90 min) VIDrel: 20TH/TECH L/A V/h

JOHNNY GUITAR ** * PG
Nicholas Ray USA 1953
*Sterling Hayden, Joan Crawford, Scott Brady, Mercedes
McCambridge, Ward Bond, Ernest Borgnine, Ben Cooper, Royal
Dano, John Carradine, Paul Fix, Frank Ferguson, Rhys Williams, Ian
MacDonald, Will Wright, John Maxwell*
An old flame comes to work at a saloon run by a woman whose

relationship with the local townsfolk is none too friendly. Hayden as the reluctant gunslinger, and Crawford as the saloon-keeper with a penchant for outlaws, strike sparks off each other in this one. McCambridge was never more sinister. The script is by Philip Yordan.
WES   110 min   VIDrel: VCC L/A   V
Boa: novel by R. Chanslor.

JOHNNY HANDSOME ***   15
Walter Hill USA   1989
Mickey Rourke, Ellen Barkin, Charles Roven, Elizabeth McGovern, Forest Whitaker, Scott Wilson, Lance Henriksen, Morgan Freeman, Yvonne Bryceland, Jeff Meek, David Schramm, Peter Jason, J.W. Smoth, Allan Graf, Ed Zang
A badly disfigured criminal is ditched by his comrades during a robbery, captured and sent to prison. However, once there he's given the chance to have cosmetic surgery as part of a reha-bilitation programme, and on being eventually released sets out to have his revenge on his former colleagues. A moody, violent and offbeat yarn, not without its virtues despite the lack of a strong plot. Music is by Ry Cooder.
DRA   89 min (ort 94 min)   VIDrel: 4-FRONT/POLYREC L/A; PION (LV only)   V/sh LV
Boa: novel The Three Worlds of Johnny Handsome by John Godey.

JOHNNY MNEMONIC **   15
Robert Longo CANADA   1995
Keanu Reeves, Dolph Lundgren, Ice T, Henry Rollins, Dina Meyer, Takeshi Kitano, Udo Kier, Denis Akiyama, Tracy Tweed, Don Francks, Barbara Sukowa, Falconer Abraham, Diego Chambers, Sherry Miller, Arthur Eng, Von Flores
In the near future, human beings are able to transport vast amounts of computer data in their heads, such couriers being considered a safe and reliable means of transporting this commodity around the world. One such individual undertakes a final journey and is menaced by villains in the pay of a company that will stop at nothing to steal the data our hero is carrying. What starts off as an intriguing tale soon degenerates into a standard violent actioner.
FAN   93 min (ort 113 min) cC   VIDrel: 20TH/TECH   V/sur
Boa: novella by William Gibson.

JOHNNY SUEDE *   15
Tom DiCillo FRANCE/SWITZERLAND/USA   1991
Brad Pitt, Richard Boes, Cheryl Costa, Michael Luciano, Calvin Levels, Nick Cave, Peter McRobbie, Ashley Gardner, Dennis Parlato, Ron Vawter, Tina Louise, Ralph Marrero, Wilfredo Giovanni Clark, Alison Moir, Wayne Maugans, Joseph Barry
Writer-director DiCillo's debut feature tells of an aspiring New York musician with a penchant for strange encounters, who despite his lack of knowledge of the fairer sex, falls for a girl who has a psycho for a boyfriend. Various adventures follow, some comical, and some considerably less so, as this weakly-scripted film wends its weary way to the obligatory happy ending. More of an exercise in empty style than a movie in its own right.
COM   93 min (ort 97 min)   VIDrel: ARTIF/20TH   V/h

JOHNNY'S GIRL **   (PG)
John Kent Harrison USA   1995
Treat William, Mia Kirschner, Dianne Debassige, Janne Mortil, Frank Cassini, Dave "Squatch" Ward, Jay Brazeau, Barry Pepper, Lalannia Lindbjerg, Michael Tayles, Franck C. Turner, Alex Diakun, John Destry, Josh Holmes, Kelly Sheridan
A petty thief wanted by the police takes off for Alaska, leaving his wife in hold her for herself. After her death, their daughter goes to join him but finds their relationship quite a troubled one, largely on account of his operation of a gaming joint and less than honest ways. However, when she is placed by the welfare authorities in the custody of a local cop, our reluctant dad has to shape up if he is not to lose her for good. A competent but shrill family drama.
DRA   88 min (ort 92 min)   SATrel: MOVIE CHANNEL
Boa: novel by Kim Rich.

JOHNS *   18
Scott Silver USA   1995
David Arquette, Lukas Haas, Wilson Cruz, Keith David, Christopher Gartin, Elliott Gould, Terrence Dashon Howard, Richard Timothy Jones, Nicky Katt, Richard Kind, Kurtis Kunzler, John C. McGinley
A look at the life of a male prostitute, who in a short period of time is robbed, stabbed, beaten up and eventually murdered by a client. A seriously unpleasant character study, that unlike films

by directors such as Bruce LaBruce is so very lacking in warmth and humour that it becomes quite gruelling to watch. Not so much a film as a painful diatribe.
DRA   96 min   CINrel

JOLSON SINGS AGAIN **   U
Henry Levin USA   1949
Larry Parks, Barbara Hale, William Demarest, Ludwig Donath, Bill Goodwin, Tamara Shayne, Myron McCormick, Eric Wilton, Robert Emmett Keane, Frank McLure, Jock Mahoney, Betty Hill, Margie Stapp, Nelson Leigh, Dick Cogan
A flaccid attempt to build on the success of THE JOLSON STORY, which takes us through the singer's second marriage and later career, in which he gained a new lease of life as an entertainer of the troops during WW2. As before, Parks mimes and Jolson sings, but apart from the songs (which include "Baby Face", "Sonny Boy" and "Back In Your Own Yard") this is a singularly uninspiring affair.
MUS   91 min (ort 96 min)   VIDrel: COLUM/SONOP V/sur

JOLSON STORY, THE ***   U
Alfred E. Green USA   1946
Larry Parks, Evelyn Keyes, William Demarest, Bill Goodwin, Ludwig Donath, Tamara Shayne, Scotty Beckett, John Alexander, Jo-Carroll Dennison, Ernest Cossart, William Forrest, Ann E. Todd, Edwin Maxwell, Emmett Vogan
An excellent biopic on the famous singer, with Jolson dubbing his own songs, that is done very much in the traditional Hollywood style but eminently enjoyable nonetheless. Followed in 1949 by the sequel JOLSON SINGS AGAIN. Songs include: "April Showers", "Avalon", "You Made Me Love You" and "My Mammy". AA: Score (Morris Stoloff), Sound (John Livardy).
MUS   124 min (ort 129 min)   VIDrel: COLUM/SONOP V/sur

JONATHAN LIVINGSTON SEAGULL **   U
Hal Bartlett USA   1973
Voices of: James Franciscus, Juliet Mills, Hal Holbrook, Philip Ahn, David Ladd, Dorothy McGuire, Richard Crenna
An account of a seagull who wants to fly higher and faster than any other in the world, that offers excellent photography, jarring music and a totally silly voice-over, all elements that fail in their attempt to convey the mysticism of the original novel. The intru-sive score is by Neil Diamond.
DRA   94 min (ort 120 min)   VIDrel: CIC/SONOP   V/sh
Boa: novel by Richard Bach.

JONATHAN: THE BOY NOBODY WANTED ***   PG
George Kaczender USA   1992
JoBeth Williams, Chris Burke, Jeffret DeMunn, Tom Mason, Robert Cicchini, Dana barron, Mason Adams, Paul Linke, H. Richard Greene, Madge Sinclair, Alley Mills, Lorraine Morin-Torre. Brandon Bauer, K.C. Clarizio, Linden Chiles
A Down's Syndrome boy is placed in a home by his parents and a volunteer worker there attempts to give the child some love, despite the wishes of the parents.
DRA   93 min mTV   VIDrel: ODY/SONOP   V/sh

JOSEPHINE BAKER STORY, THE ***   15
Brian Gibson UK/USA   1991
Lynn Whitfield, Ruben Blades, David Dukes, Craig T. Nelson, Louis Gossett Jr, Kene Holliday, Vivian Bonnell, Mayah McCoy, Ainsue Currie, Luis Reyes, Pierre Magny, Franco liriti, George Faison, Robert Lesser, Eartha Robinson
A fast-paced account of the colourful life of this multi-talented performer that follows her rise to stardom in the late 1920s in Paris, where her skin colour proved no obstacle to her success. Whitfield is ravishing in the title role and is ably supported by a strong cast, but the movie is far too short to do justice to all the events it attempts to describe. Filmed, strangely enough, on location in Budapest.
DRA   129 min mCab   VIDrel: FOCUS/DISC   V

JOSH AND S.A.M. *   12
Billy Weber USA   1993
Joan Allen, Stephen Tobolowsky, Matha Plimpton, Jacob Tierney, Noah Fleiss, Chris Penn, Ronald Guttman, Maury Chaykin, Udo Kier, Sean Baca, Jake Gyllenhaal, Anne Lange, Ann Hearn, Christian Clemenson, Allan Arbus, Kayla Allen, Amy Wright
Two brothers take off on their own in a car when their parents split up, but neither one is old enough to hold a driver's licence. A juvenile road-movie without much of a plot or anything else,

except one rather funny idea: the older brother Josh has convinced Sam that he is a "Strategically Altered Mutant" (hence his name, and the title of this trite film).
COM   93 min (ort 98 min)   VIDrel: POLY / POLYREC L / A   V

JOSH KIRBY: TIME WARRIOR! – EGGS FROM 70 MILLION B.C. * (U)
Mark Manos USA
Corbin Allred, Jennifer Burns, Derek Webster, Steve Wilder, Gary Kaspar, Barrie Ingram
Some prehistoric eggs gets attached to our hero's time pod and hatch into enormous worms that prove to have an appetite to match. Another adequate adventure for the title character, the fourth one in the series.
Aka: EGGS FROM 70 MILLION B.C.
JUV   87 min (ort 93 min)   SATrel: MOVIE CHANNEL

JOSH KIRBY: TIME WARRIOR! – JOURNEY TO THE MAGIC CAVERN * (U)
Ernest Farino USA   1995
Corbin Allred, Jennifer Burns, Derek Webster, Matt Winston, Nick De Gruccio, Cindy L. Sorensen, Barrie Ingram, Michael Hagiwara, Lomax Study, Mihan Niculescu
In this adventure, Josh and his companions land in a cavern that is home to mushroom-like people that fear a monster known as "The Muncher". Events force them to go in search of one of its victims and the resultant encounter leads to some unexpected developments. The fifth story in the series.
Aka: JOURNEY TO THE MAGIC CAVERN
JUV   89 min   SATrel: MOVIE CHANNEL

JOSH KIRBY: TIME WARRIOR! – LAST BATTLE FOR THE UNIVERSE * (U)
Frank Arnold USA   1995
Corbin Allred, Jennifer Burns, Derek Webster, Jonathan Charles Kaplan, Barrie Ingham
After being trapped in a cavern by Erwin 1138 (who proved to be the villain of the piece) Josh, Azebeth and Dr Zoetrope go on a trip through time, ending up on Earth in 1980. Here they part company, as only Josh can travel forward to the 25th century for a final showdown. A weak and padded-out concluding episode to this six-part series, with a minimal plot, poor effects and acting to match.
Aka: LAST BATTLE FOR THE UNIVERSE
JUV   89 min   SATrel: MOVIE CHANNEL

JOSH KIRBY: TIME WARRIOR! – PLANET OF THE DINO-KNIGHTS * (U)
Ernest Farino USA   1995
Corbin Allred, Jennifer Burns, Derek Webster, Barrie Ingham, John DeMita, Spencer Rochfort, Sandra Guibord, Jonathan Charles Kaplan, Time Winters, Robert Louis Kempf, Michael Mahon, Helen Biff, Charisma Carpenter, Johnny Green
First story in this series sees our young hero accidentally taken aboard a time travel vehicle that takes him back to a strange time where knights rode dinosaurs and not horses. He and his companions are drawn into a struggle between two brothers of royal blood, one good and the other bad. A terribly poor SF adventure, followed by five episodes of similar merits.
Aka: PLANET OF THE DINO-KNIGHTS
JUV   87 min (ort 90 min)   SATrel: MOVIE CHANNEL

JOSH KIRBY: TIME WARRIOR! – THE HUMAN PETS * (U)
Frank Arnold USA   1995
Corbin Allred, Jennifer Burns, Derek Webster, Barrie Ingham, John DeMila, Richard Lineback, Spencer Rochfait, Sandra Giobord, Jonathan Charles Kaplan, Henrich James, Dimitri Bogomas, Time Winters, Robert Louis Kempf, Mary-Pat Green
Second tale in this anaemic juvenile time travel series (following on from the PLANET OF THE DINO-KNIGHTS story) that has Josh and his friends moving forward in time some 70,000 years. At this point, mankind has evolved into giant creatures, one of whom captures our time-travellers, holding them as pets for its own amusement. Followed by TRAPPED ON TOY WORLD.
Aka: HUMAN PETS, THE
JUV   87 min (ort 90 min)   SATrel: MOVIE CHANNEL

JOSH KIRBY: TIME WARRIOR! – TRAPPED ON TOY WORLD * (U)
Frank Arnold USA   1995
Corbin Allred, Jennifer Burns, Derek Webster, Sharon Lee Jones, Buck Kartalian, J.P. Hubbell, Barrie Ingham

Third episode in this juvenile SF series finds our young hero once again in a tricky situation when he is trapped on a world inhabited by toys, and finds himself locked in a battle with his old enemy, Dr Zoetrope. With the latter having taken over the minds of the robot toys there, Josh flees into the forest with the inventor of the toys, but both get back in the nick of time to defeat Zoetrope's dastardly plans. EGGS FROM 70 MILLION B.C. followed.
Aka: TRAPPED ON TOY WORLD
JUV   87 min (ort 90 min)   SATrel: MOVIE CHANNEL

JOSHUA TREE ** 18
Vic Armstrong USA   1993
Dolph Lundgren, George Segal, Kristian Alfonson, Geoffrey Lewis, Michael Paul Chan, Bert Remsen, Beau Starr, Ken Foree, Michelle Philips, Nicholas Chinlund, Khandi Alexander, Marcus Brown, Matt Battaglia, Edward Stone
A man comes to a small town in the Southwest after having escaped from prison, where he was sent after his wrongful conviction. In his thirst for revenge against the man who had him framed, he sets about taking on all and sundry. Very much an action vehicle for Lundgren, who demonstrates his martial arts skills in a film that treads a well worn path. Entertainment of an undemanding kind.
Aka: ARMY OF ONE
A / AD   96 min (ort 102 min)   VIDrel: EIV / SONOP   V / sur

JOUR DE FETE **** U
Jacques Tati FRANCE   1947
Jacques Tati, Guy Decombie, Paul Frankeur, Santa Relli, Maine Vallee, Roger Rafal, Beauvais, Delcassan, The Inhabitants of St Severe-Sur-Indre
A village postman tries to improve the service with hilarious results, after seeing a documentary about the efficiency of the US Mail. Tati's full-length directorial debut (expanded from his earlier work – a short entitled "L'Ecole Des Facteurs") was shot in both colour (using a new process) and B/W. But colour quality was so poor that only a B/W print (sometimes partially hand-coloured) was released, but recent technical advances have made a colour print available.
Aka: HOLIDAY; VILLAGE FAIR, THE
COM   77 min (ort 87 min – originally B/W)
VIDrel: CONNO / RTM; 4-FRONT / POLYREC   V

JOURNEY INTO DARKNESS: THE BRUCE CURTIS STORY *** 15
Graeme Campbell CANADA   1991
Simon Reynolds, Jaimz Woolvett, Claire Coulter, Richard Donat, Nicola Lipman, John Kapelos, Kenneth Walsh, Raquel Duffy, Lisa O'Brien, Benny Fong, Trevor McCarthy, John Fulton, David Renton, Tara Wilde, Barrie Dunn
Grisly drama based on the true events that surrounded the deaths of a New Jersey couple in 1982 and the trial of their son Scott and his eighteen-year-old friend Bruce for the crime. Invited by Scott to spend the weekend with his mother and stepfather, Bruce becomes embroiled in a double murder when the stepfather (who has an uncontrollable temper) threatens them with a shot-gun. An unpleasant and often quite shocking tale.
Aka: BRUCE CURTIS STORY, THE: JOURNEY INTO DARKNESS
DRA   93 min mTV   VIDrel: FABFIL / POLYREC L / A   V

JOURNEY OF HOPE **** PG
Xavier Koller TURKEY / SWITZERLAND   1990
Necmettin Cobanoglu, Nur Srer, Emin Sivas, Erdinc Akbas, Yaman Okay, Yasar Gner, Hseyin Mete, Yaman Tarcan, Mathias Gnadinger, Dietmar Schonherr, Fritz Denoth, Theo Marti, Herbert Leiser, Andrea Zogg, Hansjorg Schneider
A simple shepherd in Southern Turkey receives a postcard from an unemployed relative in Switzerland and is so impressed by his description of that country as paradise that he resolves to follow suit. He sells all he owns and buys tickets and false documents for the long journey together with his wife and six-year-old son. A moving account of the universal desire to seek a better future and the heavy price such dreams can exact. AA: Foreign.
Aka: REISE DER HOFFNUNG
DRA   105 min (ort 110 min)   VIDrel: VCC L / A   V

JOURNEY TO THE CENTER OF THE EARTH *** U
Henry Levin USA   1959
James Mason, Pat Boone, Arlene Dahl, Diane Baker, Thayer David, Peter Ronson, Alan Napier, Alan Caillou, Ben Wright, Alex Finlayson, Frederick Halliday, Mary Brady, Robert Adler, Ivan Triesault, John Epper, Peter Wright

A vigorous evocation of Verne's classic tale, of a scientific expedition that descends into the bowels of the earth, where various incredible sights await. Rather poorly remade in 1978 as "Where Time Began".
FAN 126 min (ort 132 min) VIDrel: 20TH/TECH L/A V/h
Boa: novel by Jules Verne.

JOURNEY TO THE SOUTH *** 18
Juan Bautista Stagnaro ARGENTINA 1988
Adrian Ghio, Mira Jokovic, Zarko Lausevic, Osvaldo Santoro
Story of a man from South America who returns to his native village on the pretext of choosing a wife, but in reality has arranged for a number of women (seven in all) to be tricked into sailing for Argentina, where they are forced into prostitution. A sad tale, made more poignant with the knowledge that such dreadful abuses really did occur, yet also a story that is not without hope.
DRA 91 min VIDrel: WESCON/RTM V

JOY LUCK CLUB, THE *** 15
Wayne Wang USA 1993
Kieu Chinh, Tsai Chin, France Nuyen, Rosalind Chao, Tamlyn Tomita, Lisa Lu, Lauren Tom, Ming-Na Wen, Michael Paul Chan, Andrew McCarthy, Christopher Rich, Russell Wong, Victor Wong, Vivian Wu, Jack Ford, Diane Baker, Chao-Li Chi
Sweeping, evocative and deeply moving account of four Chinese women living in the USA, their relationships with their daughters and the sadness and tragedy that lies in their past. Every week they meet to play Mah Jong and gossip (this explains the film's title) and these weekly get-togethers serve as an effective linking device for the various flashback episodes.
DRA 133 min (ort 139 min) cC VIDrel: HOLPIC/TECH V/sur
Boa: novel by Amy Tan.

JU DOU *** 15
Zhang Yimou CHINA/JAPAN 1987
Gong Li, Li Baotian, Li Wei, Zhang Yi, Zheng Jian, Niu Xingli, Jia Zhaoji, Wu Fa, Ma Chong, Cong Zhijun, Jin Jia, Yang Quanbin, Jian Wen
Visually striking, moving tale, set in China in the 1920s where an ageing and embittered factory owner takes a third wife in the hope of obtaining a male heir. Tragically, his young bride begins an affair with his nephew that has devastating results. Subsequently banned in China, allegedly because of its concentration on the fate of individuals, this was the first Chinese film to be nominated for the Academy Awards.
DRA 91 min (ort 98 min) VIDrel: ICAPRO/SONOP V

JUAREZ *** U
William Dieterle USA 1939
Paul Muni, Bette Davis, John Garfield, Brian Aherne, Claude Rains, Donald Crisp, Gale Sondergaard, Gilbert Roland, Louis Calhern, Pedro De Cordoba, Henry O'Neill, Joseph Calleia, Montagu Love, Harry Davenport, Grant Mitchell
Muni gives a fine performance as this famous Mexican leader, in a colourful film full of splendid sets and locations, that describes the fight to oust the Emperor Maximillian from the Mexican throne, on which he had been placed by the troops of France's ruler Napoleon III. A film that for all its lavish sets and fine acting remains somewhat aloof and clinical.
DRA 106 min (ort 132 min) B/W VIDrel: L/A V
Boa: book Phantom Crown by B. Harding.

JUBILEE * 18
Derek Jarman UK 1978
Toyah Wilcox, Jordan, Little Nell, Jenny Runacre, Hermione Demoriane, Ian Charleson, Karl Johnson, Linda Spurrier, Neil Kennedy, Orlando, Wayne County, Richard O'Brien, David Haughton, Adam Ant, Claire Davenport
Punk-rock film with Queen Elizabeth I being transported to the present day by her astrologer, and witnessing the virtual collapse of civilisation. A typical Jarman extravaganza of shock effects and pointless nastiness. The film has something valid to say, but it never has a chance to say it.
DRA 100 min (ort 103 min) VIDrel: TART/20TH V

JUDE *** 15
Michael Winterbottom UK 1996
Christopher Eccleston, Kate Winslet, Liam Cunningham, Rachel Griffiths, June Whitfield, Ross Colvin Turnbull, James Daley, Berwick Kaler, Sean McKenzie, Richard Albrecht, Caitlin Bossley, Emma Turner, Lorraine Hilton, James Nesbitt

Eccleston is extraordinarily good as the eponymous Jude, a 19th century farmhand who dreams of bettering himself, in Hardy's difficult and depressing novel. Winslet is the liberal-minded cousin under whose spell he falls, but unhappily neither their liaison nor his dreams of an education are to be crowned with success.
DRA 117 min (ort 122 min) VIDrel: POLFIL V/s
Boa: novel Jude the Obscure by Thomas Hardy.

JUDGE AND THE ASSASSIN, THE *** 15
Betrand Tavernier FRANCE 1975
Phillippe Noiret, Michel Galabru, Isabelle Huppert, Jean Claude-Brialy, Yves Robert, Renee Faure, Cecile Vassort, Jean-Roger Caussimon, Monique Chaumette, Jean Bretonniere, Francois Dyreck, Liza Braconnier, Jean Amos
In 19th century France, a judge trying the case of a child murderer has to decide on the man's sanity. His efforts to come to a just decision involve him in a relationship with the man, and he eventually learn of the forces that shaped his life and lead him to commit this crime. A well realised and complex tale whose ultimate meaning is hard to unravel.
Aka: LE JUGE ET L'ASSASSIN
DRA 122 min (ort 130 min) VIDrel: ARROW/RTM V

JUDGE DREDD ** 15
Danny Cannon USA 1995
Sylvester Stallone, Armand Assante, Rob Schneider, Diane Lane, Joan Chen, Jurgen Prochnow, Max Von Sydow, Balthazar Getty, Joanna Miles, Maurice Roeves, Ian Dury, Chris Adamson, Ewen Bremner, Peter Marinker, Angus MacInnes
Bleak fantasy set in the year 2139 AD, on an over-populated and ravaged Earth, whose people live in vast high-rise mega-cities. The rule of law is maintained by heavily armed "judges", who dispense justice with rigour (they can even execute criminals on the spot). When Dredd is framed on false charges, he sets out to clear his name. John Wagner and Carlos Ezquerra created a memorable character in the comic-strip 2000 AD; this messy film does not do them justice.
FAN 92 min (ort 96 min) wScrn VIDrel: GUILD/20TH V

JUDGE PRIEST ** U
John Ford USA 1934
Will Rogers, Henry B. Walthall, Tom Brown, Anita Louise, Rochelle Hudson, David Landau, Berton Churchill, Stepin Fetchit, Hattie McDaniel, Frank Melton, Roger Imhof, Charley Grapewin, Francis Ford, Paul McAllister, Matt McHugh
The story of a judge in the old South and his professional and domestic life, set in 1890 when memories of the Confederacy were still very much alive. alive. A star vehicle for Rogers but marred today by the crude and rather obvious stereotyping of the black characters.
DRA 71 min (ort 80 min) B/W VIDrel: SCRN/DISC V
Boa: stories by Irvin S. Cobb.

JUDGEMENT ** 18
William Sachs USA 1989
Elliott Gould, Karen Black, Emilia Crow, Thalmus Rasulala, Cuba Gooding Jr, Ed Lauter, Sydney Lassick, Jimmy Solera, Francesco Quinn, Richard Coca
When one member of an L.A. gang kills two rivals, he sets in motion a series of bloody event as the court system struggles hard to cope with these violent juvenile criminals. A melodramatic and uneven tale. See also COLORS for another look at L.A. street gangs.
Aka: HITZ
A/AD 87 min (ort 96 min) VIDrel: PROM/HIFLI V

JUDGEMENT *** 15
Tom Topor USA 1990
Keith Carradine, Blythe Danner, David Strathairn, Jack Warden, Michael Faustino, Mitchell Ryan, Robert Joy, Bob Gunton, Steve Hofvendahl, Mary Joy, Brad Sullivan, Dylan Baker, Bob Barnes, Deborah Borane, Douglas Brush
A couple find that their local priest, whom everybody in their small Louisiana parish treats with great reverence, is actually a child molester whose victims include their son. The Church, however, is predictably more concerned to hush matters up rather than take action, and they are forced to resort take legal proceedings. A well made and nicely acted drama, based on an actual court case.
DRA 86 min (ort 96 min) mCab VIDrel: MGM/WHV L/A V

JUDGEMENT AT NUREMBERG ***　　　　PG
Stanley Kramer USA　　　　　　　　　　1961
Spencer Tracy, Burt Lancaster, Richard Widmark, Marlene Dietrich,
Judy Garland, Maximilian Schell, Montgomery Clift, William
Shatner, Kenneth MacKenna, Edward Binns, Werner Klemperer,
Torben Meyer, Alan Baxter
An overlong film version of what started life as an entry in a
1950s TV series, "Playhouse 90" in the USA. Here the fictional
story of a Nazi judge on trial for war crimes, serves as a means
of examining the atrocities committed or sanctioned by the offi-
cials of that hideous regime. AA: Actor (Schell), Screen/adapt
(Abby Mann).
DRA　178 min (ort 190 min) B/W　VIDrel: MGM/WHV　V
Boa: TV play by Abby Mann.

JUDGEMENT DAY: THE JOHN LIST STORY **　　15
Bobby Roth USA　　　　　　　　　　　1992
Robert Blake, Beverly D'Angelo, David Caruso, Melinda Dillon, Alice
Krige, David Purdham, Micole Mercurio, Carroll Baker, Tom Butler,
Gary Chalk, Gina Gallego, Roger R. Cross
The childhood of an apparently normal and law-abiding man
was blighted by parental abuse, the scars of which have long
since been hidden. However, one day he cracks and slaughters
his entire family, vanishing after leaving a signed confession to
this crime. Despite the best efforts of the FBI, he eludes capture
and a local policeman attempts to track him down. A competent,
fact-based drama.
DRA　94 min mTV　VIDrel: ODY/SONOP　V/sh

JUDGEMENT IN BERLIN ***　　　　　PG
Leo Penn USA　　　　　　　　　　　　1988
Martin Sheen, Sam Wanamaker, Max Gail, Heinz Hoenig, Carl
Lumbly, Harris Yulin, Nora Chmiel, Sean Penn, Max Volkert
Martens, Juerger Hemrich
An East German couple have to stand trial for using a toy gun
to hijack a Polish airliner to West Berlin, and the German author-
ities decide to ask an American judge to preside. A gripping
courtroom drama that offers some splendid performances from
its leading players. The script is by Joshua Sinclair and Leo Penn,
and though world events have overtaken the film, it remains a
most worthy effort.
Aka: ESCAPE TO FREEDOM
DRA　92 min　VIDrel: MGM/WHV L/A　V
Boa: book by Herbert J. Stern.

JUDGEMENT IN STONE, A **　　　　15
Ousama Rawi CANADA　　　　　　　　　1986
Rita Tushingham, Tom Kneebone, Ross Petty, Shelley Peterson, Jessica
Stern, Jonathan Crombie, Jackie Burroughs
A woman housekeeper who is dyslexic, is driven mad by the
pressure of trying to conceal her disability from her kindly
employers and eventually kills them. A clumsy adaptation of
Rendell's sinister story, that takes an inordinately long time to
develop into anything, and when it does so, the acting is so poor
that the film plays more like a black comedy than the psycho-
logical thriller its makers intended. Rendell's novel was adapted
with far more success in 1995.
Aka: HOUSEKEEPER, THE
HOR　94 min (ort 102 min)　VIDrel: VISVID/POLYREC L/A
V
Boa: novel by Ruth Rendell.

JUDGEMENT IN STONE, A ***　　　　15
Claude Chabrol FRANCE　　　　　　　　1995
Sandrine Bonnaire, Isabelle Huppert, Jean-Pierre Cassel, Jacqueline
Bisset, Virginie Ledoyen, Valentin Merlet, Julien Rocquefort,
Dominique Frot, Philippe Le Coq, Jean-Francois Perrier, Yves
Verhoeven, Ludovic Brillant, Claire Chiron
The story of the wealthy, upper-class Lelievre family and their
ill-educated housekeeper, who is kept in her place and eventu-
ally sacked by the former. Both this woman and her friend the
local postmistress hide secrets, and both have ample reason to
hate the Lelievres, who are not quite as nice as they at first
appear. Chabrol takes a coldly dispassionate look at the characters in this intense drama, which is both compelling and
slightly unsettling.
Aka: LA CEREMONIE
DRA　108 min (ort 112 min)　VIDrel: TART/20TH　V/sh
Boa: novel by Ruth Rendell.

JUDGEMENT NIGHT **　　　　　　18
Stephen Hopkins USA　　　　　　　　　1993
Emilio Estevez, Cuba Gooding Jr, Denis Leary, Stephen Dorff, Jeremy
Piven, Erik Schrody, Peter Greene, Michael DeLorenzo, Christine
Harnos, Galyn Gorg, Angela Alvarado, Michael Wiseman, Relioues
Webb, Will Zahrn, Lauren Robinson
A group of young men go out for a night on the town but take
a wrong turning that takes them to a dangerous neighbourhood.
There they unfortunately witness a a gangland murder by a
drugs dealer and are forced to flee for their lives as he comes
after them. A grim and violent saga of urban survival.
A/AD　105 min (ort 110 min) cC　VIDrel: CIC/SONOP
V/sur

JUDICIAL CONSENT **　　　　　　18
William Bindley USA　　　　　　　　　1994
Bonnie Bedelia, Dabney Coleman, Billy Wirth, Lisa Blount, Will
Patton
Crime drama set in the legal world where a woman judge
presiding over a murder trial where the victim was a much
unloved lawyer, finds herself being gradually framed for this
crime. This danger may in turn be related to the sexual relationship she is having with her law library clerk. Convincing
acting helps to overcome some of the more silly aspects of the
script, which only holds water for about the first thirty
minutes.
DRA　95 min (ort 100 min) mTV　VIDrel: GUILD/FOXVID
V/sh

JUICE ***　　　　　　　　　　　15
Ernest R. Dickerson USA　　　　　　　　1992
Tupac Shakur, Omar Epps, Jermaine Hokins, Khalil Kain, Queen
Latifah, Cindy Hatton, Vincenta, Latesca, Samuel L. Jackson, George
O. Gore, Grace Garland, Idina Harris, Victor Campos, Eric Payne,
Sharon Cook, Darien Berry, Maggie Rush
Four young black youths in Harlem try to earn some respect and
plan to rob a local store, but one of them takes the fatal step of
taking along a gun. A powerful and moving account that is
highly dramatic, even if it does not have any real answers to
offer. A promising directorial debut from Dickerson (who acted
as Spike Lee's cinematographer on DO THE RIGHT THING and
JUNGLE FEVER) with an effective music soundtrack.
DRA　91 min　VIDrel: 20TH/TECH　V/sur

JULES AND JIM ***　　　　　　　PG
Francois Truffaut FRANCE　　　　　　　1961
Oskar Werner, Henri Serre, Jeanne Moreau, Marie Dubois, Vanna
Urbino, Boris Bassiak, Sabine Haudepin
Story of an amicable love triangle involving a temperamental
woman and two men, one French and the other Austrian, span-
ning some twenty years. An exuberant and often joyful work,
but also one with depth and pathos. The novel that formed the
basis for this film was itself inspired by Lucien Rebatet's "Les
Deux Etendards". See also ANNE AND MURIEL.
Aka: JULES ET JIM
DRA　101 min (ort 105 min) B/W wScrn
VIDrel: ARTIF/20TH　V/h
Boa: novel by Henri-Pierre Roche.

JULIA **　　　　　　　　　　　PG
Fred Zinnemann USA　　　　　　　　　1977
Jane Fonda, Vanessa Redgrave, Jason Robards, Maximilian Schell,
Rosemary Murphy, Hal Holbrook, Meryl Streep, John Glover, Lisa
Pelikan, Cathleen Nesbitt, Maurice Denham, Susan Jones, Gerard
Buhr, Stefan Gryff, Dora Doll, Hans Verner
The story of two women and their friendship, told against the
background of the resistance to the Nazis prior to WW2.
Wooden direction and acting, combine in a beautifully
photographed but sterile film, that is utterly lacking in solid
and credible characterisations. This was Meryl Streep's film
debut. AA: S. Actor (Robards), S. Actress (Redgrave),
Screen/adapt (Alvin Sargent).
DRA　113 min (ort 118 min)　VIDrel: 20TH/TECH L/A　V
Boa: book Pentimento by Lillian Hellman.

JULIA AND JULIA *　　　　　　　18
Peter Del Monte ITALY　　　　　　　　1987
Kathleen Turner, Sting (Gordon Sumner), Gabriel Byrne, Gabriele
Ferzetti, Angela Goodwin, Lidia Broccolino, Alexander Van Wyk,
Renato Scarpa, Norman Mozzato, Yorgo Voyagis, Mirella Falco,
Francesca Muio, John Steiner
A woman whose husband died on their wedding day, comes
home one day and finds that he is still alive, they have a six-
year-old son, and she has a lover. A weird, pseudo-SF tale,
ineptly handled and with an irritatingly inconclusive ending.
Notable only because it is the first feature to be shot on high-

definition video (with 1125 instead of 625 lines) and then transferred to 35 mm film.
Aka: GIULIA E GIULIA
THR 97 min VIDrel: 20TH V/sur

JULIA HAS TWO LOVERS ** 15
Bashar Shbib CANADA 1990
Daphna Kastner, David Duchovny, David Charles, Tim Ray, Clare Bancroft, Martin Donovan, Anita Olanick, Al Samuels, Julie Roswal, C.h. Lehenhof, Lauren Fitch
A children's authoress lives with her publisher but is undecided about his proposal of marriage. A surprise phone call from a man claiming to be an old acquaintance of hers soon gets her deeply involved with a mysterious stranger. A sort of sub-par SEX, LIES AND VIDEOTAPE contrasting predictable domesticity with sexual obsession, but hampered by poor dialogue and a shoestring budget, the entire film having been shot in just two weeks.
COM 81 min (ort 91 min) VIDrel: VCC/DISC/COLUM V

JULIET OF THE SPIRITS *** 15
Federico Fellini FRANCE/ITALY/WEST GERMANY 1965
Giuletta Masina, Sandra Milo, Mario Pisu, Valentina Cortese, Lou Gilbert, Sylva Koscina, Caterina Boratto, Luisa Della Noce, Sabrina Gugli, Rosella Di Sepio, Silvana Jachino, Elemea Fondra, Jose-Luis De Vilallonga
Complex fantasy involving a woman who fears that her husband is unfaithful and is persuaded to attend a seance at which she finds that she can conjure up the spirit of her grandfather as well as others. An inventive and highly imaginative movie that is visually hugely enjoyable, with a host of fascinating scenes. Like most of Fellini's work, this film appeals far more to the emotions than the intellect and is best appreciated this way.
Aka: GIULIETTA DEGLI SPIRITI; JULIA UND DIE GEISTER; JULIETTE DES ESPRITS;
FAN 121 min (ort 148 min) wScrn VIDrel: FABFIL/SPEAR V

JULIUS CAESAR **** U
Joseph L. Mankiewicz USA 1953
John Gielgud, Marlon Brando, James Mason, Deborah Kerr, Greer Garson, Louis Calhern, Edmond O'Brien, George Macready, Michael Pate, John Hoyt, Alan Napier, Richard Hale, William Cottrell, Ian Wolfe, Douglass Dumbrille, Edmund Purdom
Led by Cassius and Brutus, a gang of conspirators murder Caesar, but Mark Antony sets about avenging their cowardly attack. Uninspired direction and awkward editing detract but little from this lavish and literate adaptation of Shakespeare's play (albeit using a truncated text) with fine acting (Gielgud as Cassius stands out) and memorable sets. Screenplay is by Mankiewicz. AA: Art (Cedric Gibbons and Edward Carfagno), Set (Edwin B. Willis and Hugh Hunt).
DRA 116 min (ort 121 min) B/W VIDrel: MGM/WHV V
Boa: play by William Shakespeare.

JULIUS CAESAR ** PG
Stuart Burge UK 1969
John Gielgud, Charlton Heston, Jason Robards, Richard Johnson, Diana Rigg, Robert Vaughn, Richard Chamberlain, Jill Bennett, Christopher Lee, Andrew Crawford, Alan Browning, Norman Bowler, David Dodimead, Michael Gough
Third film version of Shakespeare's play offers a hardly exciting view of this drama. Adequate but far from inspiring, with Robards's dismal portrayal of Brutus a serious handicap.
DRA 117 min VIDrel: 4-FRONT/POLYREC V
Boa: play by William Shakespeare.

JULIUS CAESAR ** U
Herbert Wise UK 1979
Richard Pasco, Charles Gray, Keith Michell, David Collings, Virginia McKenna, Elizabeth Spriggs, Sam Dastor, John Sterland, Brian Coburn, Garrick Hagon, Leonard Preston, Alex Davion, Darien Angadi
An efficient TV production of this famous play. One of a series of BBC Shakespeare adaptations which were characterised by the faithfulness of their interpretation and a pleasing lack of gimmickry, if not always by their inspiration.
DRA 160 min (ort 162 min) mTV VIDrel: BBC V/h
Boa: play by William Shakespeare.

JUMANJI *** PG
Joe Johnston USA 1995
Robin Williams, Jonathan Hyde, Kirsten Dunst, Bradley Pierce,

Bonnie Hunt, Bebe Neuwirth, David Alan Grier, Patricia Clarkson, Adam Hann-Byrd, Laura Bell Bundy, James Handy, Gillian Barber, Brandon Obray, Cyrus Thiedeke, Leonard Zola
A couple of orphans find a magical board game that they start playing, only to find it causes the appearance of a strange unkempt figure who reveals that he was trapped for twenty years in another dimension as a result of playing the game. Unfortunately, it must now be played to its conclusion, so they have to continue with it, and this leads to ever more bizarre events. The incredible special effects are memorable, but the film lacks all the charm of the novel.
FAN 100 min (ort 117 min) cC VIDrel: COLUM/SONOP; ENCORE (LV only) V/sur LV
Boa: books of Chris Van Allsburg.

JUMPIN' AT THE BONEYARD ** 15
Jeff Stanzler USA 1992
Tim Roth, Alexis Arquette, Danitra Vance, Kathleen Chalfant, Samuel L. Jackson, Luis Guzman, Elizabeth Bracco, Jeffrey Wright, Richard Morris, Agustin Agustin Rodriguez, Ginny Young, Steve Catala, James Williams, Traci Kindell
Well-acted but extremely depressing tale of a young, out-of-work man who one day finds his drug addict brother turning over his apartment, together with a friend. Despite his initial anger at this action, he gradually becomes obsessed with the idea of getting him a place on a rehabilitation program. A grim study of addiction, it takes place over twenty-four hours and ends in tragedy.
DRA 102 min (ort 107 min) VIDrel: 20TH V/sur

JUMPING FOR JOY ** U
John Paddy Carstairs UK 1956
Frankie Howerd, Stanley Holloway, A.E. Matthews, Tony Wright, Joan Hickson, Lionel Jeffries, Susan Beaumont, Bill Fraser, Colin Gordon
Howerd stars as a sacked sweeper/handyman who frequents the local greyhound track, where he does his best to outwit bookies, track officials and several less savoury characters. Tricked into buying a duff greyhound, he sets out to train it for a big race. At its best this is no more than watchable, a distinct lack of decent dialogue being its chief handicap, which no amount of leering and grimacing on the part of Howerd can replace.
COM 86 min (ort 88 min) B/W VIDrel: CARL/TECH V

JUMPIN' JACK FLASH * 15
Penny Marshall USA 1986
Whoopi Goldberg, Stephen Collins, Carol Kane, Lawrence Gordon, John Wood, Joel Silver, Annie Potts, Roscoe Lee Browne, Sara Botsford, Jeroen Krabbe, Peter Michael Goetz, Jonathan Pryce, Tracy Reiner, Jim Belushi, Tony Hendra
A computer operator is drawn into the complex world of international espionage when a British agent sends a message for help to her terminal. A confused and disjointed comedy-thriller, not improved by the fact that the original director, Howard Zieff, was replaced during shooting. The talents of Goldberg are wasted.
COM 101 min VIDrel: 20TH/TECH L/A V/sh

JUNGLE ASSAULT ** 18
David A. Prior USA 1989
William Smith, Ted Prior, William Zipp, Mario Rosado, David Mariott, Jeannie Moore
A general's daughter falls in with left-wing rebels in Central America (who brainwash her into assisting them), but is eventually returned safe and sound to the USA when her father dons uniform once more and mounts his own rescue mission. Another violent actioner from this director.
A/AD 82 min Cut (26 sec – ort 90 min) VIDrel: POLY/POLYREC/BRAVE V

JUNGLE BOOK, THE *** U
Zoltan Korda USA 1942
Sabu, Joseph Calleia, John Qualen, Frank Puglia, Rosemary De Camp, Patricia O'Rourke, Ralph Byrd, John Mather, Faith Brook, Noble Jackson
A wonderful and lively fantasy of a boy brought up by wolves, who is at home in the jungle and at one with all (or nearly all) the animals. He dreams of killing Shere-Khan the tiger. A film of enormous visual impact with some remarkably beautiful sets (it was filmed entirely in-studio). Sabu is memorable (even if the fight to the death with a stiff-jointed model tiger is less so). The

score is by Miklos Rozsa. Remade by Disney as a cartoon animation in 1967.
Aka: RUDYARD KIPLING'S JUNGLE BOOK
JUV 101 min (ort 109 min) B/W   VIDrel: IMAG/RTM   V
Boa: novel by Rudyard Kipling.

JUNGLE BOOK, THE ****   *U*
Wolfgang Reithermann USA   1967
Voices of: George Sanders, Phil Harris, Louis Prima, Sebastian Cabot,
Clint Howard, Sterling Holloway, J. Pat O'Malley, Bruce Reitherman,
Verna Felton, Chad Stuart, Lord Tim Hudson, Digby Wolfe, John
Abbott, Ben Wright
Superb animated adaptation of Kipling's classic tale of a young
boy who was abandoned at birth and subsequently raised by
wolves. After some ten years in the jungle he is due to be
returned to live among humans, but falls in instead with a
happy-go-lucky bear called Baloo, with whom he enjoys many
adventures. Excellent voice characterisations and good animation blend well with the music to make Walt Disney's last project
a wonderful success.
ANIM 75 min (ort 78 min)   VIDrel: WDV/TECH L/A   V
Boa: novel by Rudyard Kipling.

JUNGLE BOOK, THE **   *PG*
Stephen Sommers USA   1994
Jason Scott Lee, Sam Neill, Adam Beach, Cary Elwes, Lena Headey,
John Cleese, Jason Flemyng, Stefan Kalipha, Ron Donachie, Anirudh
Agrawal, Faran Tahir, Sean Naegeli, Joanna Wolff, Liza Walker,
Rachel Robertson, Natalie Morse
Lavish, live-action version of Kipling's classic tale of a young
man raised in the jungle by wolves, who possesses a remarkable
rapport with all the animals except the fearsome tiger Shere-
Khan. Though very well made, this is a bland affair that
seriously lacks both conviction and pace. The idea of having
Mowgli brought back to civilisation (having been re-united with
the daughter of a British major he knew as a child) is an interesting if slightly contrived touch.
Aka: RUDYARD KIPLING'S THE JUNGLE BOOK
JUV 108 min (ort 111 min)   VIDrel: WDV/TECH   V
Boa: novel by Rudyard Kipling.

JUNGLE FEVER ***   *18*
Spike Lee USA   1991
Wesley Snipes, Annabella Sciorra, Spike Lee, Ossie Davis, Ruby Dee,
Samuel L. Jackson, Lonette McKee, John Turturro, Frank Vincent,
Anthony Quinn, Tyra Ferrell, Halle Berry, Veronica Webb, Veronica
Timbers, Tim Robbins, Brad Dourif
A successful black architect is drawn to his white secretary, and
the couple eventually embark on an affair that ends unhappily.
A noisy and crowded examination of the prejudices attached to
such relationships and the pain they can cause, coupled with a
timely warning on the perils of crack addiction. The entire cast
give first-rate performances, but too many characters and the
lack of any overall structure make this film hard to watch at
times.
DRA 126 min (ort 132 min)   VIDrel: CIC/SONOP   V/sur

JUNGLE KING, THE **   *U*
Kamoon Song   1993
Average animated tale built around title character, a lion. When
he is kidnapped by a hyena, his twin brother is forced to step
into his shoes and must learn to overcome his own timidity in
the process.
ANIM 49 min   VIDrel: COLUM/SONOP   V

JUNGLE LAW *   *18*
Damian Lee CANADA   1994
Jeff Wincott, Paco Christan Prieto, Richard Yearwood, Christina Cox
A lawyer is haunted by his criminal past, when he escaped after
a robbery went wrong, leaving behind his partner who was
caught and imprisoned. The latter is now a powerful gangster,
and as our lawyer is both deeply in debt and handy with his
fists, he agrees to work for his ex-buddy as a bare-knuckle
fighter. Natty Wincott enjoys several changes of dress in this
badly staged actioner, pity the script (by Lee) is not as inventive.
A/AD 87 min   VIDrel: EIV   V

JUNGLE 2 JUNGLE *   *PG*
John Pasquin USA   1997
Tim Allen, Sam Huntingdon, JoBeth Williams, Lolita Davidovich,
Martin Short, Valerie Mahaffey, LeeLee Sobieski, Frankie Galasso,
Luis Avalos, Bob Dishy, Rondi Reed, Oni Faida Lampley

The French hit "Un Indien Dans Le Ville" (or Little Indian, Big
City) gets the remake treatment, and has Allen cast as a wealthy
New York stockbroker who sets off for a remote South American
island to finalise his divorce. But there, he learns he has a son,
whom his wife has raised among the native Indians for the past
twelve years. Father and son return to the Big Apple, where the
latter soon makes it obvious he is a stranger to city life. A
clumsy, irritating dud.
COM 105 min   CINrel

JUNIOR **   *PG*
Ivan Reitman USA   1994
Arnold Schwarzenegger, Danny DeVito, Emma Thompson, Frank
Langella, Pamela Reed, Judy Collins, James Eckhouse, Aida Turturro,
Welker White, Megan Cavanagh, Merle Kennedy, Mindy Seeger,
Christopher Meloni, Antoinette Peragine
A scientist is persuaded by a ruthless and unscrupulous
colleague to take part in an experiment that involves him being
implanted with a fertilised human egg. It eventually gives rise
to a viable embryo and he gives birth. This disappointing second
pairing of Schwarzenegger and DeVito has none of the magic
of TWINS, and though they work very well together, this is a
one-joke film. A few mildly amusing moments are scattered
about a predictable script.
COM 105 min (ort 110 min) cC   VIDrel: CIC/SONOP;
PION (LV only)   V LV

JUNIOR BONNER ***   *PG*
Sam Peckinpah USA   1972
Steve McQueen, Ida Lupino, Robert Preston, Ben Johnson, Joe Don
Baker, Barbara Leigh, Mary Murphy, Bill McKinney, Sandra Deel,
Donald Barry, Dub Taylor, Rita Garrison, Charles Gray, Matthew
Peckinpah, Sundown Spencer
An over-the-hill rodeo rider returns to the bosom of his family
but finds his troubles are by no means over in this pleasant
rodeo comedy-drama, that is sustained by fine acting and a good
feel for the subject. One of the few Peckinpah films in which the
characters don't end up full of bullet holes.
DRA 100 min (ort 105 min)
VIDrel: 4-FRONT/POLYREC/BRAVE L/A; PION (LV only) V LV

JURASSIC PARK **   *PG*
Steven Spielberg USA   1993
Sam Neill, Laura Dern, Richard Attenborough, Jeff Goldblum, Bob
Peck, Samuel L. Jackson, Martin Ferrero, Joseph Mazzello, Ariana
Richards, Wayne Knight, B.D. Wong, Jerry Molen, Miguel Sandoval,
Cameron Thor, Christian John Fields
An eccentric billionaire uses his wealth to genetically clone
dinosaurs from a sample of prehistoric DNA and builds a theme
park for his new pets. He invites a palaeontologist couple and
a mathematician to visit and view the creatures but things go
badly wrong and they escape (the creatures, that is). Superb
special effects make the dinosaurs seem alive, but the film is
both verbose and boring. Co-scripted by Crichton from his
novel. AA: Sound, Effects/aud, Effects/vis.
FAN 121 min (ort 127 min) wScrn cC   VIDrel: CIC/SONOP;
PION (LV only)   V/sur LV/cav
Boa: novel by Michael Crichton.

JUROR, THE *   *18*
Brian Gibson USA   1996
Alec Baldwin, Demi Moore, Joseph Gordon-Levitt, James Gandolfini,
Lindsay Crouse, Tony Lo Bianco, Anne Heche, Michael Constantine,
Matt Craven, Todd Susman, Michael Rispoli, Julie Halston, Frank
Adonis, Matthew Cowles
A single mother who is serving as a juror in the trial of a Mafia
boss, finds herself being alternately seduced and threatened by
an attractive young man, who wants to force her into persuading the other members of the jury to bring in a not guilty verdict.
A weakly plotted and unconvincing film, it starts off slowly and
never gets any better, and though Moore is always good to look
at, she'll be winning no Oscars for this film. See also TRIAL BY
JURY.
THR 113 min (ort 118 min) cC   VIDrel: 20VIS/SONOP;
ENCORE/COLUM (LV only)   V/sur LV
Boa: novel by George Dawes Green.

JURY DUTY *   *12*
John Fortenberry USA   1995
Pauly Shore, Tia Carrere, Shelley Winters, Brian Doyle-Murray,
Stanley Tucci, Abe Vigoda, Charles Napier, Richard Edson, Richard
Riehle, Alex Datcher, Dick Vitale, Andrew "Dice" Clay
A nerdish guy is called for jury service and manages to exploit

this civic duty for his own ends, An uninspired spoof on America's obsession with criminal trials that tries hard to raise laughs but falls flat on its face, while its ill-chosen references to the O.J. Simpson case are both tasteless and unfunny.
COM   86 min (ort 88 min)   VIDrel: COLUM/SONOP V/sur

JUST A GIGOLO *   15
David Hemmings WEST GERMANY   1978
David Bowie, Sydne Rome, Kim Novak, David Hemmings, Maria Schell, Curt Jurgens, Marlene Dietrich, Erika Pluhar, Rudolf Schundler, Hilde Weissner, Werner Pochath, Bela Erny, Friedhelm Lehmann, Rainer Hunold, Evelyn Kunneke
A Prussian war veteran tries to make a living in Berlin of the 1920s and finds his niche before the inevitable tragic ending. A severely edited melodrama that makes little sense. Don't watch it to see Dietrich – her part is tiny.
Aka: SCHONER GIGOLO, ARMER GIGOLO
DRA   89 min (ort 147 min)   VIDrel: VISVID/POLYREC L/A V

JUST ANOTHER GIRL ON THE I.R.T. **   15
Leslie Harris USA   1992
Ariyan Johnson, Kevin Thigpen, Ebony Jerido, Chequita Jackson, William Badget, Jerard Washington, Karen Robinson, Tony Wilkes, Johnny Roses, Shawn King, Kisha Richardson, Monet Dunham, Wendall Moore, Laura Ross, Rashmella
A bright seventeen-year-old black girl from Brooklyn sees her hopes of a college education and medical school go up in smoke when she finds herself pregnant. Having scraped together the money for an abortion, she blows it instead on a shopping spree. A low-budget tale, filmed in just seventeen days for $170,000, that is full of rage but rather empty of meaning or purpose. The usual stereotyping of all the white characters is an added annoyance.
DRA   93 min (ort 97 min) wScrn   VIDrel: TART/20TH V

JUST ASK FOR DIAMOND ***   U
Stephen Bayly UK   1988
Peter Eyre, Susannah York, Rene Ruiz, Nickolas Grace, Patricia Hodge, Saeed Jaffrey, Dursley McLinden, Colin Dale, Michael Medwin, Roy Kinnear, Jimmy Nail, Bill Paterson, Michael Robbins, Donald Standen, Jim McManus, Forbes Collins
A sharp thirteen-year-old boy and his incompetent elder brother are paid ú200 to look after a box of confectionery. Soon, they find they are threatened by nasty characters who will stop at nothing to get the box. A witty and neat re-working of the private eye genre, set in contemporary London. The script is by Horowitz from his novel.
Aka: DIAMOND'S EDGE; FALCON'S MALTESER, THE
JUV   89 min (ort 94 min)   VIDrel: 20TH/TECH V
Boa: novel The Falcon's Malteser by Anthony Horowitz.

JUST BETWEEN FRIENDS ***   15
Allan Burns USA   1986
Mary Tyler Moore, Ted Danson, Christine Lahti, Sam Waterston, Susan Rinell, Salome Jens, Jane Greer, James MacKrell, Timothy Gibbs, Mark Blum, Castulo Guerra, Chet Collins, Terri Hanauer, Helene Winston, Gary Riley, Leda Siskind
Two women meet and develop a firm friendship, yet they remain unaware that they are both involved with the same man, as wife and lover respectively. The directing debut for Burns, and though the film suffers badly from the contrived plot, fine performances redeem it.
DRA   106 min (ort 120 min)   VIDrel: RCA L/A V

JUST CAUSE ***   18
Arne Glimcher USA   1994
Sean Connery, Laurence Fishburne, Kate Capshaw, Blair Underwood, Ed Harris, Christopher Murray, Ruby Dee, Scarlett Johansson, Daniel J. Travanti, Ned Beatty, Liz Torres, Lynne Thigoen, Kevin McCarthy, Hope Lange, Chris Sarandon
A black man condemned to death for the rape and murder of a little girl continues to strenuously protest his innocence even after his conviction, and manages to interest a Harvard law professor in his case. As the latter starts to investigate, he soon uncovers a nasty conspiracy in which the condemned man has been made the fall-guy. Conventionally plotted this thriller may be, but the fine work of its leads keeps one watching.
THR   98 min (ort 102 min) cC   VIDrel: WHV V/sur
Boa: novel by John Katzenbach.

JUST DENNIS: THE MOVIE **   U
Doug Rogers USA   1988
Victor Di Mattia, William Windom
An adaptation of a long-running comic strip by Hank Ketchum, this tale follows the exploits of our angelic-looking title character, after he discovers a dinosaur bone in his back garden. A cute little kid's film, made very much in the same style as all those Disney adventures.
JUV   92 min (ort 97 min)   VIDrel: ODY/SONOP V/h

JUST HEROES **   18
John Woo HONG KONG   1989
David Chiang, Chen Tuan-Kai, Kelly Chu, Danny Lee, Chow Sing Chi, Lee San San Yin, Woo Ma
After the murder of a powerful gangster boss, a bloody struggle erupts among his potential successors, in this empty actioner that is full of gunfire and melodramatics but lamentably devoid of any real interest. Woo directs the bloodshed with all his usual efficiency, and though the movie appears to carry an anti-violence message, one is not convinced the director is sincere.
A/AD   93 min wScrn   VIDrel: MIA/DISC; ENCORE (LV only)   V LV

JUST LIKE A WOMAN ***   15
Christopher Monger UK   1992
Julie Walters, Adrian Pasdar, Paul Freeman, Susan Wooldridge, Gordon Kennedy, Ian Redford, Shelley Thompson, Togo Igawa, Jill Spurrier, Corey Cowper, Mark Hadfield, Joseph Bennett, Brooke, Eve Bland, Jeff Nuttall, Sayo Inaba
Bitter-sweet comedy with a young man finding he is only really at ease when he dresses as a woman and takes on an alternate personality. Fortunately, his landlady proves to be both broad-minded and supportive, and when he loses his job she assist him in his scheme to revenge himself on his employer. Walters as his kindly landlady gives a performance of delicacy and charm, something of a first for her.
COM   101 min (ort 106 min) cC
VIDrel: VCC/DISC/COLUM V/sur
Boa: book Geraldine, For the Love of a Transvestite by Monica Jay.

JUST LIKE DAD **   (U)
Blair Treu 1996   USA
Wallace Shawn, Ben Diskin, Nick Cassavetes, Jarrett Lennon, Laura Innes, Michael Tucci, George Fisher, Frank Gerrish, Michael Flynn, Duane Stephens, Brittney Lewis, Christy Summerhays, Elizabeth Lund, Nick Murdock, Sarah Scharaub
A young boy, who lives alone with his father since his mother abandoned them many years before. He is rather ashamed of him since he never stands up for his rights and advises him to pay money to the school bullies instead of fighting. When the school's fathers-and-sons athletics event comes around, he hires an athletic man to impersonate his dad but this fellow's murky past soon gets them into hot water. A sentimental and predictable offering, aimed at kids.
JUV   90 min   SATrel: DISNEY CHANNEL

JUST MY IMAGINATION **   (PG)
Jonathan Sanger USA   1993
Jean Smart, Tom Wopat, Richard Gilland, Cristine Rose, Orson Bean, Pat Carroll, Audra Lindley, Mary Kay Place, Gilbert Lewis, Lynn Milgrim, Ernie Sabela, Peter Slusker, Jonathan Sanger, Jack Betts, Karen Morrow
In a sleepy town in North Carolina the life of a bored teacher receives a jolt when a rock singer who was born in the town releases his latest hit, a song that describes a passionate relationship they once enjoyed. Despite the fact that the song is pure fiction, its repercussions are considerable. Fair.
DRA   90 min mTV   SATrel: MOVIE CHANNEL
Boa: novel Bobby Rex's Greatest Hit by Gingher.

JUST ONE OF THE GIRLS *   15
Michael Keusch USA   1993
Corey Haim, Nicole Eggert, Cameron Bancroft, Gabe Khouth, Kevin McNulty, Wendy Van Riessen, Johannah Newmarch, Lochlyn Munro, Rachel Hayward, Molly Parker, Shane Kelly, Matt Bennett, Janet Craig, R. Nelson Brown, Kathy Tong
A young man finds himself menaced by the school bully and decides to disguise himself off as a girl in order to avoid a beating. As ever, the expected romantic complications ensue (he rather unwisely falls for the bully's sister) and to make matters worse, said bully falls for his female persona. A dumb high

school comedy with faint echoes of TOOTSIE. See also JUST
LIKE A WOMAN.
Aka: ANYTHING FOR LOVE
COM   90 min (ort 94 min)   VIDrel: 20VIS/SONOP   V/sh

JUST ONE OF THE GUYS *** 15
Lisa Gottlieb USA 1985
Joyce Hyser, Billy Jacoby, Clayton Rohner, Toni Hudson, Sherilyn
Fenn, Leigh McCloskey, William Zabka, Emily Ragside, Michael
Guinn, Stacy Blythe, Linda Kelly, Anthony Galde, Jim Norwitz, Kim
Studer, Joseph Finsterwald
A teenage girl disguises herself as a boy and changes her school
in order to win a journalism contest. The usual transsexual
complications a la TOOTSIE arise, in a film that is a lot funnier
than one might have expected. See also HER LIFE AS A MAN
and VICTOR/VICTORIA.
COM   96 min (ort 100 min)   VIDrel: RCA L/A   V

K

K-9 ** 15
Rod Daniel USA 1989
James Belushi, Mel Harris, Kevin Tighe, Ed O'Neill, Jerry Lee, James
Handy, Cotter Smith, Daniel Davis, John Snyder, Pruitt Taylor
Vance, Sherman Howard, Jeff Allin, David Haskell, Alan Blumenfeld,
Bill Sadler, Marjorie Bransfield
Belushi gives a great performance that's the best thing in this
tale of an unconventional cop, whose obsessive desire to nail a
bunch of drug dealers makes him so difficult to work with that
he's given a dog as a partner. However, the mutt proves to be
more than a match for him, but the film, though both slick and
well paced, is really far too derivative for its own good.
Disappointing. See also TURNER & HOOCH, K9000 and TOP
DOG for more cop/dog "partnerships".
JUV   97 min (ort 111 min)
VIDrel: 4-FRONT/POLYREC/CIC   V/sur

K9000 * 15
Kim Manners USA 1989
Catherine Oxenberg, Chris Mulkey, Dennis Haysbert, Dana
Gladstone, Judson Scott, Anne Haney, Thom McFadden, Ted Barba,
David Renan, Ivan E. Roth, Rick Aiello, Jim Burk, Jason Corbett,
Kenny Endoso, Jerry Houser (voice of K9000)
Hard on the heels of K-9, comes this blatant rip-off. A dog that
has had a miniature computer implanted in its brain is stolen
by a ruthless maniac. A cop tracks them to an abandoned ware-
house, but in the ensuing rescue is accidentally implanted with
a receiver that allows him to communicate with the dog tele-
pathically, thereby making him a valuable ally in the fight
against crime. A poorly-realised, woodenly-acted, idiotic film.
See also TURNER & HOOCH.
A/AD   93 min (ort 96 min)   VIDrel: MOPIC/SGSVID   V

K2 *** 15
Franc Roddam USA 1991
Michael Biehn, Matt Craven, Raymond J. Barry, Hiroshi Fujioka,
Patricia Charbonneau, Luca Bercovici, Julia Nickson-Soul, Jamal
Shah, Annie Grindlay, Elena Stiteler, Blu Mankuma, Charles
Oberman, Christopher Brown, Leslie Carlson
The title refers to a mountain in the Karagoram range, which lies
on the border between China and Pakistan. At 28,250 feet above
sea level, it is the second highest peak in the world, and is recog-
nised by many as being tougher than Everest. Based both on a
stage play and the exploits of Wickwire and Reichardt (who
surmounted it in 1978) the film works best as a simple outdoors
adventure; a few moralising touches are considerably less
welcome.
A/AD   105 min (ort 111 min)   VIDrel: EIV/SONOP   V/sur
Boa: play by Patrick Meyers.

KABLOONAK *** 15
Claude Massot CANADA/FRANCE/USA 1994
Charles Dance, Adamie Quasiak Inukpuk, Seporah Q. Ungalaq, Natar
Ungalaq, Bernard Bloch, Peter Hudson, Matthew Saviakjuk-jaw,
Georges Claisse, Pauloosie quiliqtalik, Tattigat Arnatsiaq, Aleega
Ragee Killiktee, Tony Vogel
Opening in 1922, this speculative tale is built around the exploits
of Irish-American film-maker Robert Flaherty. Looking back
over his life, he recalls the hardship he endured in the Arctic,
when in 1919 he made "Nanook Of The North", a fascinating
documentary on Inuit traditions and culture. A visually impres-

sive film that dispenses with an examination of just how true to
life Flaherty's film was, focusing on his character instead.
Aka: NANOOK
A/AD   103 min   VIDrel: POLFIL   V

KABUTO: THE GOLDEN EYE MONSTER ** 15
JAPAN 1992
Animated tale of a martial artist combining traditional Japanese
mythology with the usual cyberpunk features of anime. In this
story the hero roams the land fighting evil, eventually returning to his place of birth to find that the rightful ruler has been
deposed by a tyrant.
ANIM   45 min dubbed   VIDrel: MANGA/SONOP   V

KADAICHA ** (18)
James Bogie AUSTRALIA 1988
Zoe Carides, Tom Jennings, Eric Oldfield, Fiona Gauntlett, Kerry
McKay, Sara Dakin, Bruce Hughes, Deborah Kennedy, Harry Cripps,
Terry Markwell, Nicholas Flannegan, Nicholas Ryan, Steve Dodd,
John Paramor
After experiencing a horrible nightmare, a young girl wakes up
with a strange stone in her hand, which her teacher informs her
is an Aboriginal "death stone". After she is found cut to shreds,
the stone moves on to other kids who also meet violent deaths,
in this Australian clone of FRIDAY THE 13TH.
Aka: KADAICHA: THE DEATH STONE
HOR   86 min   SATrel: SKY MOVIES

KAFKA ** 15
Steven Soderbergh FRANCE/USA 1991
Jeremy Irons, Theresa Russell, Joel Grey, Ian Holm, Jeroen Krabbe,
Armin Mueller-Stahl, Alec Guinness, Brian Glover, Keith Allen,
Simon McBurney, Robert Flemyng, Matyelok Gibbs, Ion Caramitru,
Hilde Van Meighem, Jan Namejovsky
A mild-mannered insurance clerk becomes intrigued at the
disappearance of a colleague from his office and starts to inves-
tigate, eventually unearthing a vast and terrifying conspiracy.
A highly stylised exercise in conveying the atmosphere, if not
the content, of Kafka's stories that starts off well enough but
finishes as just one more effects-laden horror movie. The cinematography (in both black-and-white and colour) is a major
asset.
FAN   94 min (ort 98 min) B/W/Col
VIDrel: GUILD/SONOP   V/sur

KAGEMUSHA *** PG
Akira Kurosawa JAPAN 1980
Tatsuya Nakadai, Tsutomo Yamazaki, Kenichi Hagiwara, Jinpachi
Nezu, Shuji Otaki, Daisuke Ryu, Masayuki Yui, Kaori Momori,
Mitsuko Baisho, Hideo Morata, Koji Shimizu, Sen Yamamoto,
Takayuki Shino, Noburo Shimizu, Shuhei Sugimori
"Kagemusha" means "The Double", and refers to the practice of
Japanese warlords of employing others to impersonate them-
selves during battles. In this film, a thief about to be executed is
reprieved and given the task of impersonating a one such man.
When the latter dies the impersonation continues until he is
unmasked, whereupon the story moves on to explore the end of
a warrior clan. An impressive, lavish, but strangely lifeless work.
Aka: KAGEMUSHA THE SHADOW WARRIOR
WAR   153 min (ort 181 min)   VIDrel: L/A   V/h

KALEIDOSCOPE ** PG/15
Jud Taylor USA 1990
Jaclyn Smith, Perry King, Patricia Kalember, Claudia Christian,
Colleen Dewhurst, Donald Moffat, Terry O'Quinn, Bruce Abbot,
Kim Thomson, Penny Johnson, Mary Jo Keene, Ben Lemon, Erika
Flores, Becky Herbst, Tasia Scutt
A standard Danielle Steele saga of passion, jealousy and
intrigue, with three orphaned sisters sharing a mysterious past
that brings them together after thirty years apart, having been
fostered with different families when their parents died. When
a detective, acting out the wishes of a dying client, attempts to
reunite them, the scene is set for some typical Danielle Steel
complications.
Aka: DANIELLE STEEL'S KALEIDOSCOPE
DRA   89 min (15 ver); 91 min (PG ver) mTV
VIDrel: IMPENT L/A (15 ver); MIA/VCC (PG ver)   V
Boa: novel by Danielle Steel.

KALIFORNIA *** 18
Dominic Sena USA 1992
Brad Pitt, Juliette Lewis, David Duchovny, Michelle Forbes, John
Zarchen, J. Michael McDougal, Patricia Sill, Gregory Mars Martin,

Sierra Pecheur, Catherine Larson, John Dullaghan, David Milford, Tommy Chappelle, Judson Vaughn
A writer and his photographer girlfriend (who are collaborating on a book about serial killers) plan a trip across country by car and advertise for someone to share their costs. This brings them into contact with a psychopath who has just murdered his landlord, and his equally dippy female companion. A tour-de-force performance from all four leads stands out in this almost unbearably violent tale of a descent into hell. Both fascinating and disturbing.
DRA   113 min (ort 118 min)   VIDrel: VCC/DISC/COLUM
V/sur

KANAL ****   (18)
Andrzej Wajda POLAND   1956
Teresa Izewska, Tadeusz Janczar, Vladek (Wladyslaw) Sheybal, Wienczylaw Glinski, Emil Kariewicz, Tadeusz Gwiazdowski, Stanislaw Mikulski, Teresa Berezowska, Emil Karewiez, Teresa Berezowska, Zofia Lindorf
The second film in Wajda's excellent WW2 trilogy. This one looks at a group of partisans who take part in the 1944 Warsaw uprising and fight their way through the sewers. Preceded by A GENERATION and followed by ASHES AND DIAMONDS.
Aka: THEY LOVED LIFE
WAR   93 min (ort 96 min) B/W   VIDrel: L/A   V
Boa: novel Kloakerne by Jerzy Stawinski.

KANDYLAND **   18
Robert Schnitzer USA   1987
Sandahl Bergman, Kim Evenson, Charles Laulette, Irwin Keyes, Bruce Baum, Cole Stevens, Alan Toy, Steve Kravitz, Catlyn Day, Ja-Net Hintzen, Chrissy Ratay, Brenda Winston, Israel Jurabe, Beth Peters, Ken Olfson, Richard Neil
A young girl takes a job at a sleazy nightclub as a stripper, learning the ropes from former stripper Bergman. She is soon drawn into a sordid world of vice and drugs, in this inconsequential sex-thriller that inevitably has a few similarities to STRIPPED TO KILL, but generally manages to sustain its story right up to the tragic ending, despite the excessive footage devoted to actual stripping.
THR   79 min (ort 94 min)   VIDrel: NWV L/A   V/h

KANE AND ABEL: VOLS. 1, 2 AND 3 **   PG
Buzz Kulik USA   1985
Sam Neill, Peter Strauss, Veronica Hamel, David Dukes, Fred Gwynne, Tom Roberts Byrd, Alberta Watson, Reed Birney, Vyto Ruginus, Jill Eikenberry, Richard Anderson, Kate McNeil, Lisa Banes, Christopher Cazenove, Jan Rubes
Two men, born on the same day and destined to become powerful tycoons, become implacable enemies and bitter rivals largely by chance, rather than due o any personal animosity. Strauss imparts real life to the figure of Abel Rosnovski, the Polish immigrant of a hotel chain, but Neill gives his usual flat performance. A watchable if not excessively entertaining adaptation of a readable if not excessively entertaining book.
DRA   310 min (2 cassettes – ort 320 min) mTV
VIDrel: 4-FRONT/POLYREC   V
Boa: novel by Jeffrey Archer.

KANSAS *   15
David Stevens USA   1988
Matt Dillon, Andrew McCarthy, Kyra Sedgwick, Leslie Hope, Arlen Dean Snyder, Brent Jennings, Harry Northrup, Gale Myron, Alan Toy, Andy Romano, Brynn Thayer, Clint Allen, Linda Dawson, James Lovelett, Louis Giambalvo, James Lee Raupp
A young man becomes innocently embroiled in a bank robbery, after his car and possessions are destroyed on the road. A fast journey nowhere.
A/AD   106 min (ort 111 min)   VIDrel: EIV/SONOP   V

KANSAS: A JOURNEY OF THE HEART ***   PG
Robert Mandel USA   199-
Patricia Wettig, Matt Craven, Bethel Leslie, Deidre O'Connell
When her father dies, leaving her the family farm, a mother of two has to make a difficult choice, as she can either run the farm with her sister or stay in her successful career. A slow moving drama that gets better and considerably more involving as the story develops.
DRA   93 min   VIDrel: ODY/SONOP   V/sh

KANSAS CITY ***   15
Robert Altman FRANCE/USA   1996
Jennifer Jason Leigh, Miranda Richardson, Harry Belafonte, Dermot Mulroney, Michael Murphy, Steve Buscemi, Brooke Smith, Jane Adams, Jeff Feringa, A.C. Smith, Martin Martin, Albert J. Burnes, Ajia Mignon Johnson, Tawanna Benbow
In Kansas City in the 1930s, a woman whose petty thief of a husband is being held prisoner by a ruthless gangster, is forced to take desperate measures to free him. She kidnaps the unloved wife of a local Democrat politician and holds her hostage, thereby blackmailing him into using his influence to help her. A simple tale, embroidered with much fine jazz and a mass of colourful characters. Based on Altman's own recollections and scripted by him and Frank Barhydt.
DRA   115 min   CINrel

KANSAS CITY CONFIDENTIAL ***   12
Phil Karlson USA   1952
Preston Foster, Neville Brand, Lee Van Cleef, Jack Elam, Neville Brand, John Payne, Coleen Gray, Howard Negley, Mario Siletti, Dona Drake, Helen Kleeb, Ted Ryan, George Wallace, Vivi Janiss, Archie Twitchell, House Peters Jr.
An ex-police captain masterminds a major bank robbery with the aid of three gunman, all the participants in the raid wearing masks to conceal their identities from each other. The police arrest an ex-con as their prime suspect and to clear his name, he is forced to go after the gang who have fled to Guatemala. A tautly directed and tense thriller, spoilt only by a disappointing climax. Scripted by George Bruce and Harry Essex.
Aka: SECRET FOUR, THE
THR   96 min (ort 98 min) B/W   VIDrel: SCRN/DISC   V
Boa: story by Harold Green and Rowland Brown.

KARATE COP *   18
Kurt Anderson USA   1991
Cynthia Rothrock, Jeff Wincott, Paul Johansson, Evan Lurie
Wincott is a police detective who teaches the martial arts to his comrades. Rothrock is an officer of similar capabilities, whom he left behind in the course of a transfer. When a cop buddy of Wincott's is killed in a car crash (apparently having been drunk at the time) he grows suspicious, and our two fighting cops set out to investigate. They uncover the drug-dealing activities of a murderous club owner, in a lightweight and superficial dud.
A/AD   87 min   VIDrel: 4-FRONT/POLYREC/EIV   V

KARATE KID, THE ***   15
John G. Avildsen USA   1984
Ralph Macchio, Noriyuki (Pat) Morita, Martin Kove, Elisabeth Shue, Randee Heller, William Zabka, Chad McQueen, Tony O'Dell, Larry Drake, Ron Thomas, Rob Garrison, Israel Juarbe, William H. Bassett, Larry B. Scott, Juli Fields
A sugar-sweet confection all about a teenager who is taught karate by the local janitor of the apartment block where he lives, so that he can face up to bullies. Excellent performances from Macchio and Morita are coupled with a solid script. Followed by three progressively inferior sequels.
DRA   122 min (ort 126 min) subH
VIDrel: COLUM/SONOP   V/sur

KARATE KID 2, THE *   PG
John G. Avildsen USA   1986
Ralph Macchio, Noriyuki (Pat) Morita, Nobu McCarthy, Danny Kamekona, Yuji Okumoto, Tamlyn Tomita, Martin Kove, Pat E. Johnson, Bruce Malmuth, Eddie Smith, Garth Johnson, Brett Johnson, William Zabka, Chad McQueen, Tony O'Dell
Pupil and mentor from THE KARATE KID now travel to Japan where they must face the latter's great rival and enemy in a series of adventures. Strictly kiddie-fare and inferior to the first film. Now move on to number three in the series.
DRA   120 min   VIDrel: COLUM/SONOP   V/sur

KARATE KID 3, THE *   PG
John G. Avildsen USA   1989
Ralph Macchio, Noriyuki (Pat) Morita, Sean Kanan, Robyn Elaine Lively, Thomas Ian Griffith, Martin Kove, Jonathan Avildsen
Young Daniel (a hormone imbalance formerly prevented twenty-seven-year-old Macchio from looking his age) now has to fight to retain a karate championship title, but finds that his old mentor is unwilling to train him. Eventually he acquires a new trainer who proves to be secretly in league with an old enemy of the Kid. An over-sentimental dud, of cliched homilies and a predictable outcome. Followed by THE NEXT KARATE KID.
Aka: KARATE KID PART 3, THE
A/AD   108 min (ort 111 min)   VIDrel: COLUM/SONOP
V/sur

KARATE WARRIOR ** 15
Larry Ludman (Fabrizio de Angelis) ITALY 1988
Jared Martin, Janet Agren, Kim Stuart, Ken Watanabe, Jannelle Barretto, Enrio Torralba, Jonny Tuazon, Rudy Meyer, Enrico Orbita, Arnulfo C. Quiwa,Cyrus Bautista, Rey Solo, Julio Garcia
A martial arts master trains a boy who was brutally attacked by a former pupil, and who has now turned to crime, in this fairly standard action tale.
Aka: FIST OF POWER; IL RAGAZZO DAL KIMONO D'ORO
A/AD 85 min Cut (25 sec – ort 90 min) VIDrel: IMPENT V

KASPAR HAUSER **** 18
Peter Sehr GERMANY 1993
Andre Eisermann, Udo Samel, Katharina Thalbach, Jeremy Clyde, Cecile Paoli, Hermann Beyer, Hansa Czypionka, Dieter Mann, Johannes Silberschneider, Peter Lohmeyer, Tilo Nest, Dieter Laser, Uwe Ochsenknecht, Anja Schiller
A story based on the real-life mystery surrounding the identity of the 19th century title figure, who may have been a prince brought up in total isolation for twelve years and robbed of his birthright, then finally released as an alienated and pathetic figure. A deeply moving film, it can be enjoyed both as an account of one of history's strangest puzzles and as a tale of deprivation and its effect on the human spirit. See also THE ENIGMA OF KASPAR HAUSER.
Aka: VERBRECHEN AM SEELENLEBEN EINES MENSCHENS
THR 137 min (ort 180 min) VIDrel: ARROW/RTM V

KATIA ISMAILOVA *** 18
Valerii Todorovsky FRANCE/RUSSIA 1994
Vladimir Nashkov, Ingeborga Dapkunaite, Aleksandr Feklistov, Alisa Freindlikh, Natalia Shchukina, Iurii Kuznetsov, Avangard Leontiev, M. Openkina, E. Vakhovskaia, S. Razguliaeva, K. Charmadov
Married to the neglectful son of a top female novelist, Katia types out her mother-in-laws manuscripts with resignation, until her seduction by the family handyman. But when this affair is discovered by the mother-in-law, the woman's long suppressed passion finds a violent outlet. Told with enormous simplicity, this atmospheric tale unfolds like a carefully worked out puzzle, while beneath the deceptively calm exterior can be sensed fury and emotion.
Aka: MOSCOW NIGHTS; PODMOSKOVNYE VECHERA
DRA 95 min CINrel
Boa: novel by Leskov.

KATIE'S PASSION: A GIRL CALLED KATIE TIPPEL ** 18
Paul Verhoeven HOLLAND 1976
Rutger Hauer, Monique Van De Ven
In Holland in 1881, poverty forces a strong-willed girl into prostitution, but she uses it as a means of obtaining the social status she craves. Average.
Aka: CATHY TIPPEL; HOT SWEAT; KEETJE TIPPEL
DRA 102 min (ort 107 min) dubbed VIDrel: MIA/DISC V
Boa: novels by Neel Doff.

KAVANAGH Q.C. – A FAMILY AFFAIR *** 15
Penny Rye UK 1996
John Thaw, Lisa Harrow, Anna Chancellor, Daisy Bates, Tom Brodie, Oliver Ford Davies, Nicholas Jones, Cliff Parisi, Holly Aird, George Costigan, Phyllis Logan, Rita Wolf, Toyah Wilcox
Kavanagh appears in court on behalf of a businessman accused of snatching his own son from a school as he believes the child is suffering abuse.
DRA 76 min mTV VIDrel: CARL/TECH V

KAVANAGH Q.C. – HEARTLAND *** 12
Colin Gregg UK 1996
John Thaw, Anna Chancellor, Lisa Harrow, Daisy Bates, Tom Brodie, Oliver Ford Davies, Nicholas Jones, Cliff Parisi, Phoebe Nicholls, Robert Glenister
Kavanagh is called on to defend a former policeman who drove a car the knocked down a teenager, the case being complicated by the mother's belief that this was no accident.
DRA 75 min mTV VIDrel: CTE/CARL V

KAVANAGH Q.C. – NOTHING BUT THE TRUTH *** 15
Colin Gregg UK 1995
John Thaw, Lisa Harrow, Anna Chancellor, Daisy Bates, Nicholas Jones, Geraldine James, Alison Steadman, Ewan McGregor, Pip Torrens
Kavanagh defends a student accused of raping a middle-aged housewife. Another episode in this generally well written and absorbing series, that deals with the career and cases of a top defence barrister.
DRA 101 min mTV VIDrel: CTE/CARL V

KAVANAGH Q.C. – THE SWEETEST THING *** 15
Paul Greengrass UK 1995
John Thaw, Lisa Harrow, Anna Chancellor, Daisy Bates, Tom Brodie, Oliver Ford Davies, Nicholas Jones, Cliff Parisi, Jenny Jules, Anastasia Hille, Jesse Birdsall
A prostitute is accused of the murder of a wealthy client, and Kavanagh sets out to prove her innocence.
DRA 76 min VIDrel: CARL/TECH V

KAZAAM ** 12
Paul Michael Glaser USA 1996
Shaquille O'Neal, Francis Capra, Ally Walker, Marshall Manesh, James Acheson, Fawn Reed, John Costello, Jo Anne Hart, Bob Clendenin, Fawn Reed
A youngster chased by a gang into a warehouse, accidentally knocks over a box and so released a genie who has been trapped for a very long time and gives him three wishes. However, the boy really misses his father so they both set out to find him, enjoying many adventures along the way. More than a little inspired by Disney's ALADDIN and produced by the same company, this is a tired assembly-line effort that is built around the appeal of basketball star O'Neal.
COM 89 min (ort 93 min) VIDrel: POLFIL V

KEEP THE CHANGE ** (PG)
Andy Tennant USA 1992
William L. Petersen, Lolita Davidovich, Rachel Ticotin, Buck Henry, Jack Palance, Fred Dalton Thompson, Jeff Kober, Lois Smith, Angela Paton, Frank Collison, James Ellis, William Frankfather, Ron Ray, Charlie Carpenter
When his inspiration dries up, a painter now living in California returns to his family's Montana ranch, which is now being run by his uncle and aunt. There he finds that the father of a former girlfriend is now greedily eyeing this property. A leisurely paced adaptation, quite atmospheric and with some good characterisations but lacking a good, solid plot.
DRA 95 min mCab SATrel: TNT MOVIES
Boa: novel by Thomas McGuane.

KEEPER OF THE CITY ** 15
Bobby Roth USA 1991
Louis Gossett Jr, Anthony LaPaglia, Peter Coyote, Renee Soutendijk, Aeryk Egan, Tony Todd
The son of a Mafia member decides to make amends to society by acting as a lone vigilante and executing a number of gangsters. An experienced detective is assigned to stop him and finds that a crusading reporter is also very interested. Set in Chicago, this tough tale makes little of its intriguing premise. Scripted by DiPego from his novel.
DRA 91 min (ort 96 min) VIDrel: 20TH/TECH V/sur
Boa: novel by Gerald DiPego.

KEEPER OF THE FLAME ** U
George Cukor USA 1942
Spencer Tracy, Katherine Hepburn, Richard Whorf, Margaret Wycherly, Forrest Tucker, Frank Craven, Stephen McNally, Audrey Christie, Frank Craven, Percy Kilbride, Howard Da Silva, Darryl Hickman, William Newell, Rex Evans
A war correspondent is commissioned to write the biography of a recently deceased American patriot and manages to draw close to his reclusive widow. As their friendship grows, he learns that the man in question was far from the democrat that he pretended to be. A well-acted but essentially muddled tale that bombed at the box office.
DRA 96 min (ort 100 min) B/W VIDrel: MGM/WHV V
Boa: novel by I.A.R. Wylie.

KEEPING SECRETS *** PG
John Korty USA 1991
Suzanne Somers, Ken Kercheval, Miuchael Learned, James Sutorious, John Scott Clough, Kim Zimmer, Michael Horton, Susan Krebs, Rae Allen, Lynn Llewelyn, David Birney, Jon Steuer, Norbert Weisser, John Christy Ewing
Somers plays herself in this sad tale of her unhappy life, that takes in her upbringing within a family of whom three members were alcoholics, a failed marriage, affairs, abortions and various other unedifying episodes. Not so much a film as a piece of

therapy, but honest and often quite moving, despite being rather hard to take in a single sitting.
Aka: KEEPING SECRETS: SUZANNE SOMERS IN HER OWN TRUE STORY
DRA 96 min mTV VIDrel: CIC/SONOP V
Boa: book by Suzanne Somers.

KEEPING TRACK ** 15
Robin Spry CANADA 1985
Margot Kidder, Michael Sarrazin, Alan Scarfe, Ken Pogue, Vlasta Vrana, John Boylan, Daniel Pilon, James D. Morris, Shawn Lawrence, Pierre Zimmer, Louis Negin, Terry Haig, Patricia Phillips, Renee Girard, Leo Ilial
A female bank executive and a work-obsessed TV journalist witness a strange robbery and find themselves involved in a deadly game. Yet another one of those tiring espionage comedies, where innocent bystanders find themselves up to their necks in diverse and complex conspiracies.
THR 100 min mTV VIDrel: MED L/A V

KELLY'S HEROES ** PG
Brian G. Hutton USA/YUGOSLAVIA 1970
Clint Eastwood, Donald Sutherland, Telly Savalas, Don Rickles, Carroll O'Connor, Gavin MacLeod, Stuart Margolin, (Harry) Dean Stanton, Hal Buckley, Jeff Morris, Richard Davalos, Perry Lopez, Tom Troupe, Dick Balduzzi, Len Lesser
Overlong and unbelievable story of a plan to steal a hoard of Nazi gold, from a bank behind the German lines in the middle of a bloody battle of WW2. The strong cast do their best and there are one or two bright spots, but this one was sorely in need of a more intelligent script.
WAR 137 min (ort 145 min) VIDrel: MGM/WHV V/sh

KENNEDY *** PG
Jim Goddard UK/USA 1983
Martin Sheen, John Shea, E.G. Marshall, Vincent Gardenia, Geraldine Fitzgerald, Blair Brown, Kevin Conroy, Charles Brown, Nesbitt Blaisdell, Peter Boyden, Kent Broadhurst, James Burge, William Cann, Joanne Camp
Originally a 7-hour mini-series, this long saga covers the presidency of John F. Kennedy, and was produced to mark the 20th anniversary of his death. Starting with his triumph over Nixon, it's essentially made in three distinct parts, each of which examines the most important events of his presidency, and closes with his assassination in Dallas. Despite being somewhat overdramatic at times, this is a detailed and absorbing work.
DRA 313 min (2 cassettes – ort 420 min) mTV
VIDrel: CENTV/VCC L/A V

KENTUCKIAN, THE ** PG
Burt Lancaster USA 1955
Burt Lancaster, Diana Lynn, Walter Matthau, Dianne Foster, John McIntire, Una Merkel, John Carradine, Donald MacDonald, John Litel, Rhys Williams, Edward Norris, Lee Erickson, Clem Bevans, Lisa Ferraday, Douglas Spencer
A rugged frontiersman and his son travel to Texas in search of a place where they can start a new life. Matthau makes his film debut in this minor piece that has a few good moments, but nothing of great impact. Nowadays, it is chiefly remembered for being the film that gave Matthau his screen debut.
WES 99 min (ort 104 min) VIDrel: MGM/WHV V
Boa: novel Gabriel Horn by F. Holt.

KENTUCKY FRIED MOVIE, THE *** 18
John Landis USA 1977
Evan Kim, Bill Bixby, George Lazenby, Henry Gibson, Donald Sutherland, Tony Dow, Richard A. Baker, Master Bong Soo Han, Boni Enten, Marilyn Joi, Saul Kahan, Marcy Goldman, Joe Medalis, Barry Dennem, Rich Gates, Tara Strohmeir
A series of skits and sketches satirising many themes from American TV and film; several spin-offs followed. The idea originated with a theatre troupe known as the "Kentucky Fried Theatre" which numbered Jim Abrahams, David Zucker and Jerry Zucker among its members, these three going on to make AIRPLANE! and AMAZON WOMEN ON THE MOON. Often vulgar and brash, but some of the gags (especially the long ENTER THE DRAGON spoof) are hilarious.
COM 80 min Cut (9 sec – ort 85 min)
VIDrel: POLY/POLYREC L/A V

KEOMA ** 15/18
Enzo G. Castellari ITALY 1979
Franco Nero, William Berger, Olga Karlatos, Orso Maria Guerrini, Gabriella Giacobbi, Antonio Marsina, John Offredo, Donald O'Brien,
Leon Lenoir, Woody Strode, Wolfango Soldat, Victoria Zinni, Alfio Caltabiani
A half-breed Indian fights against both an outbreak of the plague and his own half-brothers when he returns to his hometown at the end of the Civil War.
Aka: DJANGO RIDES AGAIN; DJANGO'S GREAT RETURN; KEOMA, THE VIOLENT BREED; VIOLENT BREED, THE
WES 96 min wScrn dubbed (AKTIV/RTM)
VIDrel: AKTIV/RTM L/A (18 ver); 4-FRONT/POLYREC (15 ver) V

KES ** PG
Ken Loach UK 1969
David Bradley, Lynne Perrie, Colin Welland, Freddie Fletcher, Brian Glover, Bob Bowes, Robert Naylor, Trevor Hesketh, Geoffrey Banks, Eric Bolderson, Joey Kaye
A boy's harsh life in a Northern town is temporarily enlivened when he acquires a pet kestrel. A film of almost unbearable grimness where the beauty of a creature of the wild clashes harshly with the brutalised nature of the human inhabitants.
DRA 106 min (ort 113 min) VIDrel: MGM/WHV V
Boa: novel A Kestrel For A Knave by Barry Hines.

KEY, THE ** 18
Tinto Brass ITALY 1984
Frank Finlay, Stefania Sandrelli, Franco Branciaroli, Barbara Cupisti, Maria Grazia Bon, Armando Maura, Gino Cavalieri, Piero Bortoluzzi, Irma Veithen, Eolo Capritti, Maria Pia Colonnelo, Milly Corinaldi, Edgardo Fugagnoli
Story of the erotic relationship between an elderly English hotelier living in Venice in 1940 and his much younger wife, with whom he enjoys sexual fantasy games. A beautifully photographed but empty film.
A 104 min (Cut at film release) VIDrel: WESCON/RTM V
Boa: novel by Junichiro Tanizaki.

KEY LARGO **** PG
John Huston USA 1948
Humphrey Bogart, Edward G. Robinson, Lauren Bacall, Lionel Barrymore, Claire Trevor, Thomas Gomez, Marc Lawrence, Jay Silverheels, Monte Blue, Alberto Morin, Rodric Redwing, Harry Lewis, John Rodney, William Haade, Joe P. Smith
Excellent thriller about a tough hoodlum holed up in a run-down hotel during a storm, and holding its occupants captive until he can make his getaway, with the unwilling help of boat-owner Bogart. Scripted by Huston and Richard Brooks. The final shoot-out onboard Bogart's boat, plus a terrific performance from Trevor as Robinson's alcoholic moll, are highlights. AA: S. Actress (Trevor).
THR 97 min (ort 101 min) B/W VIDrel: MGM/WHV V
Boa: play by Maxwell Anderson.

KEYS, THE ** PG
Richard Compton USA 1992
Geoffrey Blake, Brian Bloom, Scott Bloom, Ben Masters
Two brothers sent to live with their estranged father on one of those islands off the coast of Florida find themselves drawn into defending his interests, when they come up against a corrupt businessman who is out to make a killing in real estate. A narcotics sub-plot adds a little more interest to this adequate film, whose attractive location is its chief strength.
A/AD 92 min mTV VIDrel: CIC/SONOP V/h

KEYS ** 15
John Sacret Young USA 1995
Marg Helgenberger, Gary Dourdan, Brett Cullen, Vondie Curtis Hall, Neil Giuntoli, Don McManus, Robert Schenkkan, Richard Masur, Ralph Waite, Bobby Lee Grissom, Riza Hernandez, Mark Macaulay, Mayte Vilar, Ed Amatrudo, Steve Small
A freelance woman pathologist who lives on one of the Florida Keys (hence the title) and flies her own helicopter, is called in to help investigate a savage case of murder and child abduction. Tortured by memories of her own failed life, she begins to probe deeper and eventually discovers that the house where the attack took place, holds a dark secret. A mishmash of styles that do not make up for the lack of a solid script. Watchable but hardly engrossing.
THR 89 min mTV VIDrel: ODY/SONOP V/sh

KHARTOUM ** PG
Basil Dearden UK 1966
Charlton Heston, Laurence Olivier, Ralph Richardson, Richard

Johnson, Nigel Green, Michael Hordern, Alexander Knox, Johnny Sekka, Zia Mohyeddin, Marne Maitland, Hugh Williams, Ralph Michael, Douglas Wilmer, Edward Underdown
An overlong, boring and historically inaccurate account of the conflict in Sudan of the 1880s, between the British and the charismatic leader of the Dervish – the Mahdi. Khartoum falls after a ten-month siege when British forces fail to relieve it. Olivier is excellent as the Mahdi, but no meeting between him and Gordon ever took place (though it made pretty good cinema). Well acted throughout, but the film drags until the final attack on the city.
DRA   123 min Cut (26 sec – ort 136 min) wScrn
VIDrel: MGM/WHV   V/sh
Boa: book by A. Caillou.

KICK FIGHTER ** 18
Anthony Maharaj USA   1988
Richard Norton, Benny "The Jet" Urquidez, Steve Rackman, Glenn Ruehland, Franco Guerrero, Erica Von Wagener
Two youngsters become orphans when their parents are killed by rebels in Thailand, and are brought up by the family's servant. The boy grows up to become a streetfighting champ, and enters backstreet kickboxing matches to earn money, but one day his sister is kidnapped to force him to throw a fight. Adequate martial arts capers, directed and produced by Maharaj.
Aka: FIGHTER, THE
MAR   92 min Cut (31 sec)   VIDrel: MIA/DISC/IMPENT   V

KICKBOXER ** 18
David Worth USA   1989
Jean-Claude Van Damme, Dennis Alexio, Eric Sloane, Haskell V. Anderson, Dennis Chan, Rochelle Ashana, Tung Po, Steve Lee, Richard Foo, Ricky Liu, Sin Ho Ying, Tony Chan, Brad Kevner, Dean Harrington
When a man's brother is deliberately paralysed in a martial arts contest, the former vows to have his revenge and undertakes a period of training in order to equip him for this task. A formula revenger with much mindless mayhem but little plotting.
Aka: KICK BOXER
MAR   98 min Cut (1 sec plus film cuts)
VIDrel: EIV/SONOP   V

KICKBOXER 2 ** 18
Albert Pyun USA   1990
Sasha Mitchell, Peter Boyle, Dennis Chan, John Diehl, Michael Qissi, Vincent Murdocco, Cary-Hiroyuki Tagawa, Heather McComb, Matthias Hues, Humberto Ortiz, Emmanuel Kervyn, Joe Restivo, Vincent Klyn, Brian Green, Brent Kelly
Weak attempt at a sequel to the 1989 film but without the presence of Jean-Claude Van Damme, this invents a third Sloan brother, David, who runs a kickboxing gym in L.A. and devotes himself to the street kids who use it. Forced by financial necessity to a make a deal with a crooked promoter, he eventually finds himself fighting for his life in the ring. Competent but marred by far too much emphasis on blood and gore.
Aka: KICKBOXER 2: THE ROAD BACK; KICKBOXER 2: THE ROAD HOME
MAR   86 min (Cut at UK cinema release – ort 90 min)
VIDrel: EIV/SONOP   V

KICKBOXER 3: THE ART OF WAR ** 18
Rick King USA   1992
Sasha Mitchell, Dennis Chan, Richard Comar, Noah Verduzco, Alethea Miranda, Milton Goncalves, Ricardo Petraglia, Gracindo Junior, Miguel Orniga, Kate Lyra, Leonor Gottlieb, Renato Coutinho, Ian Jacklin, Manitu Felipe, Shuki Ron
A kickboxer travels to Latin America in search of action but gets more than he expected when he is tricked into a fight to the death with a hired killer. The outcome of this fight will determine the fate of a young girl, who has been kidnapped by the local kickfighting sponsor, a ruthless drugs baron. An actionful reworking of familiar elements, with the usual excessive emphasis on violence, blood and sudden death.
MAR   88 min (ort 92 min)   VIDrel: 20VIS/SONOP   V/sh

KICKBOXER FROM HELL * 18
Eric Tsui HONG KONG   1991
Mark Houghton, Sooni Shroff, Richard Edwards, Kieran Hanlon, Richard Ryrko, Roger Bingham, Steve Brettingham, Wagne Archer, William R. Webb
An American kickboxing champ is in Hong Kong for a major contest when he meets a nun who has learnt that the Devil has been incarnated as the master of an evil team of kickboxers. Having fought off an attack by these Satanic disciples, he understandably agrees to help her defeat the Devil after his brother is murdered by them. A ludicrously-plotted melding of two very different genres that doesn't make for an effective film.
MAR   86 min (ort 93 min)   VIDrel: IMPENT   V

KICKBOXER KING ** 18
Alton Cheung HONG KONG   1991
Kenneth Goodman, Bruce Fontaine, Nick Brandon, Steve Brettingham
Ordinary martial arts mayhem story, set against the seedy world of drugs, criminal gangs and police corruption.
Aka: KICKBOXING KING
MAR   85 min   VIDrel: IMPENT   V

KICKBOXER THE CHAMPION ** 18
Alton Cheung HONG KONG   1991
Donald Murray, Vince Par, Errol Baxter, Wayne Archer, Carter Hwong
Between the wars, a ruthless crook in Shanghai sets out to take control of all the shipping routes in order to corner the region's opium trade. Two amateur kickboxers who oppose his activities are inveigled into entering a contest, unaware that their enemy has a secret weapon in the form of a deadly martial artist. Fair.
MAR   88 min Cut (3 sec – ort 90 min)
VIDrel: ONE/IMPENT   V

KICKING AND SCREAMING *** 15
Noah Baumbach USA   1995
Eric Stoltz, Olivia D'Abo, Chris Eigeman, Josh Hamilton, Parker Possy, Cara Buono, Elliott Gould
A group of graduates living in New York find it difficult to cope with the fact that jobs are very hard to get and turning to each other for support, they talk interminably about their hopes, their prospects and their feelings. This sarcastic conversation piece is initially depressing, but the characters are so very well drawn that bit by bit one finds oneself relating to them, at which point the film's humour becomes more evident. Screenplay is by Noah Baumbach.
COM   92 min (ort 96 min)   VIDrel: EIV   V

KID * 18
John Mark Robinson USA   1990
C. Thomas Howell, Sarah Trigger, Brian Austin Green, Dale Dye, Michael Bowen, Damon Martin, Lenore Kasdorf, Michael Cavanaugh, Tony Epper, Don Collier, R. Lee Ermey
A determined but puerile attempt to produce a teen-market version of HIGH PLAINS DRIFTER, in which a young man returns to a small town to avenge the murder of his hippy parents by local rednecks, one of whom was the sheriff. An abysmal film, of sparse dialogue (a point in its favour) and predictable plotting, that's mediocre at best, and is just about passable as entertainment.
A/AD   88 min (ort 94 min)   VIDrel: EIV/SONOP   V

KID DIVINE ** 15
Giorgio Rossi UK   1992
Jesse Birdsall, Michael Elphick, Neil Morrissey, Martin Clunes, Ben Cole, Vincenzo Nicoli, David Sternberg, Sue Graham
Oddball Peckinpah pastiche in the form of a spaghetti Western with a couple of Irish bounty hunters on the trail of the title character, an outlaw with a $25,000 reward on his head.
Aka: BALLAD OF KID DIVINE, THE: THE COCKNEY COWBOY
WES   53 min   VIDrel: FABFIL/SPEAR   V

KID FROM BROOKLYN, THE *** U
Norman McLeod USA   1946
Danny Kaye, Virginia Mayo, Eve Arden, Vera-Ellen, Steve Cochran, Walter Abel, Lionel Stander, Fay Bainter, Clarence Kolb, Victor Cutler, Charles Cane, Jerome Cowan, Don Wilson, Knox Manning, Kay Thompson, Johnny Downs
A Brooklyn milkman who is hardly the bravest of men, inadvertently takes up a boxing career in this remake of the Harold Lloyd comedy "The Milky Way". A contrived and often overblown effort, but Kaye's considerable comic gifts carry it off.
COM   109 min 9ort 113 min)   VIDrel: VCC/DISC/COLUM   V
Boa: play The Milky Way by Lynn Root and Harry Clark.

KID FROM SPAIN, THE *** U
Leo McCarey USA   1932
Eddie Cantor, Lyda Roberti, Robert Young, Ruth Hall, John Milijan, J, Carrol Naish, Noah Beery, Robert Emmett O'Connor, Stanley Fields, Paul Porcasi, Sidney Franklin, Julian Rivero, Theresa Maxwell Conover, Walter Walker

Two college roommates are expelled after being found in a girls' dorm and decide to south to Mexico, the homeland of one of them. However, on the way there they get involved with a bank robbery and later with bullfighting. A bright and breezy musical comedy, with some nice Busby Berkeley dance routines and a few songs. These include "In The Moonlight" and "What A Perfect Combination".
MUS 96 min (ort 118 min) B/W
VIDrel: VCC/DISC/COLUM V

KID GALAHAD *   PG**
Michael Curtiz USA   1937
Edward G. Robinson, Bette Davis, Humphrey Bogart, Wayne Morris, Jane Bryan, Harry Carey, Veda Ann Borg, William Haade, Soledd Jiminez, Joe Cunnigham, Ben Welden, Joseph Crehan, Frank Faylen, Harland Tucker, Bob Evans
A sharp boxing promoter takes a naive bellhop and makes him into a boxing star, but jealousy blights his relationship with the latter and he loses his girlfriend to him for good measure. An enjoyable and fast-paced boxing melodrama. Remade as "The Wagons Roll At Night" and after a fashion as a Presley musical.
Aka: BATTLING BELLHOP, THE
DRA 98 min (ort 101 min) B/W VIDrel: WHV V
Boa: novel by Francis Wallace.

KID MILLIONS *   PG**
Roy Del Ruth USA   1934
Eddie Cantor, Ethel Merman, Ann Sothern, George Murphy, Warren Hymer, Jesse Block, Eve Sully, Berton Churchill, Paul harvey, Otto Hoffman, Doris Davenport, Edgar Kennedy, Stanley Fields, Jack Kennedy, John Kelly, Guy Usher, Henry Kolker
A dull guy from Brooklyn inherits a fortune and has the time of his life spending it, in this dated but enjoyable musical comedy that offers some wonderful numbers and a show-stopping performance by Merman. Look out for Lucille Ball as one of the Goldwyn Girls in the final dance number.
MUS 90 min B/W/Col VIDrel: VCC/DISC/COLUM V

KIDNAPPED **   U
Robert Stevenson UK   1959
Peter Finch, James MacArthur, Bernard Lee, Niall McGinnis, John Laurie, Finlay Currie, Miles Malleson, Duncan Macrae, Peter O'Toole, Andrew Cruickshank, Alex Mackenzie, Oliver Johnston, Norman MacOwan, Eileen Way, Jack Stewart
A faithful adaptation of this adventure yarn telling of a young boy's adventures in Scotland at the time of the Jacobite Rebellion. A strong cast and colourful locations compensate for the dullness of the script.
A/AD 90 min (ort 97 min) VIDrel: WDV/TECH V
Boa: novel by Robert Louis Stevenson.

KIDNAPPED **   U
Delbert Mann UK   1971
Michael Caine, Trevor Howard, Jack Hawkins, Donald Pleasence, Gordon Jackson, Vivien Heilbron, Lawrence Douglas, Freddie Jones, Jack Watson, Andrew McCulloch, Eric Woodburn, Roger Booth, Russell Waters, John Hughes, Jack Watson
A young boy is kidnapped and sent to sea during the time of the Jacobite rebellion due to the machinations of a wicked uncle, but is eventually restored to his rightful position in this third bash at the classic yarn. A rather stilted rendition, but Caine's performance as Alan Breck is pleasing.
A/AD 103 min (ort 107 min) VIDrel: VCC L/A V
Boa: novels Kidnapped and Catriona by Robert Louis Stevenson.

KIDNAPPED *   18
Howard Avedis USA   1986
Barbara Crampton, David Naughton, Charles Napier, Lance LeGault, Michelle Rossi, Kim Evinson, Chick Vennera, Jimmie Walker, Kim Shriner, Robert Dryer, Gary Wood, Etan Boritzer, Cosie Costa, Michael Pappas, Joe Portaro, Violet Manes
A young girl who dreams of a career in films, is offered a part by an apparently respectable film-maker who has a secret side-line making porno films. The girl is kidnapped whilst on holiday with her sister, but the latter teams up with an undercover cop in order to mount a rescue. A cliched and unimaginative effort.
A/AD 94 min Cut (2 min 8 sec – ort 100 min)
VIDrel: VISVID/POLYREC V

KIDNAPPED **   15
John Pasquin USA   1991
Victoria Principal, Paul Sorvino, Jonathan Banks, Danielle Harris, Gregg Henry, Christopher Wynne

When a schoolgirl is abducted and then almost immediately released, a police officer finds that matters are a little too complex to be resolved by a straightforward prosecution. Average.
Aka: DON'T TOUCH MY DAUGHTER; NIGHTMARE
DRA 90 min (ort 95 min) VIDrel: GENESIS V

KIDS *   18**
Larry Clark USA   1995
Chloe Sevigny, Leo Fitzpatrick, Justin Pierce, Sajan Bhagat, Yakira Peguero, Sarah Henderson, Rosario Dawson
A bleak and uncompromising look at the meaningless existence of a group of young but hard-as-nails street kids who steal, take drugs, have sex and generally indulge in all forms of anti-social and self-destructive behaviour.
DRA 90 min Cut (59 sec) CINrel

KIDS IN THE HALL: BRAIN CANDY **   15
Kelly Makin CANADA   1996
Dave Foley, Scott Thompson, Mark McKinney, Bruce McCulloch, Kevin McDonald, Kathryn Greenwood, Amy Smith, Lachlan Murdoch, Nicole deBoer, Krista Bridges, Christopher Redman, Erica Lancaster, Jackie Harris, Jonathan Wilson, Tony Ning
A pharmaceutical company on the verge of bankruptcy, rushes into production a new anti-depressant drug against the wishes of one of its employees, a scientist who believes that more testing is required. Soon his worst fears are realised as people start behaving in bizarre ways. A feature film spin-off from a Canadian TV comedy group, in which they play a confusing number of multiple roles. Their hard work generates much irony but very few real laughs.
COM 89 min CINrel

**KID'S RETURN **   **
Takeshi Kitano JAPAN   1996
Masanobu Ando, Ken Kaneko, Leo Morimoto, Hatsuo Yamaya, Mitsuko Oka, Ryo Ishibashi, Susumu Terajima, Koichi Shigehisa, Michisuke Kashiwaya, Yuko Daike, Atsuki Ueda, Moro Morooka
After a near fatal motorcycle accident, Kitano makes an interesting return with this offbeat, action tale of two high school misfits, who are not averse to mugging the younger kids in their school. Eventually, they go their separate ways, one into training as a boxer and the other into a Yakuza gang, but neither achieve any lasting success. With little of the raw energy of the earlier VIOLENT COP or BOILING POINT, this odd film has the feel of an unfinished experiment.
A/AD 107 min CINrel

KIDZ IN THE WOOD **   (PG)
Neal Israel USA   1994
Dave Thomas, Julai Duff, Tatyana M. Ali, Candace Cameron, David Lascher, Darius McCrany, Alfonso Ribeiro, Byron Chief-Moon, Sam Khouth, Ryan Brown, Andrews Mark Berman, Don S. Davis, Gary Chalk, Ed Hong-Louie, Gaetana Kobbin
A bunch of underprivileged inner-city kids are taken on a field trip into the Great Outdoors, and have to fend for themselves when they accidentally get separated from their guides. A pleasant if unremarkable comedy, the title (if nothing else) clearly derived from the drama BOYZ N THE HOOD.
COM 90 min SATrel: MOVIE CHANNEL

KIKA **   18
Pedro Almodovar SPAIN   1994
Veronica Forque, Victoria Abril, Alex Casanova, Peter Coyote, Rossy de Palma, Santiago La Justicia, Anabel Alonson, Bibi Andersen, Jesus Bonilla, Karra Elejalde, Mauel Bandera, Charo Lopez, Francisca Caballero, Monica Bardem
A liberated woman is involved with at least two men, one of whom may be a serial killer. Meanwhile, a female TV reporter scours the city in the hope of witnessing crimes she can record on videotape. Another over-the-top Almodovar offering that purports to be an attack on the media's glorification of violent criminals, but in truth is just another excuse to parade the same old crazy characters and their repulsive lifestyles.
COM 109 min (ort 114 min) wScrn
VIDrel: ELPIC/POLYREC V/sur

KILL AND KILL AGAIN **   15
Ivan Hall USA   1981
James Ryan, Anneline Kriel, Ken Gampu, Norman Robinson, Michael Meyer, Bill Flynn, Marloe Scott-Wilson, Stan Schmidt, John Ramsbottom, Eddie Dorie, Mervyn Johns
A martial arts champion rescues a researcher who has won the

Nobel Prize, from a mad scientist who wants to take over the world. A fairly lively martial arts outing that followed KILL OR BE KILLED. Filmed in South Africa.
MAR   95 min Cut (1 min 13 sec – ort 100 min)   VIDrel: L/A
V

KILL CRAZY *
David Heavener USA
18
1988
David Heavener, Danielle Brisebois, Burt Ward, Bruce Glover, Rachelle Carson, Lawrence Hilton-Jacobs, Gary Owens
A Vietnam veteran goes on a camping trip with some friends as a way of slowly forgetting the horrors of war, but the group are attacked by a bunch of fanatical white supremacists. A typical Heavener survival tale, written and directed by him.
A/AD   90 min Cut (13 sec)   VIDrel: IMPENT   V

KILL ME AGAIN **
John R. Dahl USA
18
1990
Val Kilmer, Joanne Whalley-Kilmer, Michael Madsen, Jonathan Gries, Michael Greene, Pat Mulligan, Nick Dimitri, Bibi Besch, Joseph Carberry, Dominic Dinino, Robert Schuch, Duane Tucker, Molly Flanegin, Daniel Dorse, Jim Boeke
A woman double-crosses her boyfriend and runs off with the money they stole from the Mob. With her lover in hot pursuit, she approaches a down-at-heel private eye and persuades him to fake her death. A highly derivative and excessively studied attempt at a modern film noir atmosphere with a slightly tongue-in-cheek flavour. The plot, however, is far too convoluted and contrived to be ever convincing. Scripted by Dahl and David Warfield.
THR   92 min (ort 94 min)   VIDrel: ITC/POLYREC L/A
V/h

KILL OR BE KILLED **
Ivan Hall USA
15
1980
James Ryan, Norman Combes, Charlotte Michelle, Danie DuPlessis, Stan Schmidt, Norman Robinson
A martial arts master is lured into a trap disguised in the form of a martial arts tournament, and all to slake the thirst for revenge of a former Nazi coach who was beaten by his Japanese counterpart in a WW2 contest. Not bad for this type of film, with a better-than-average script. Followed by KILL AND KILL AGAIN.
Aka: KARATE OLYMPIA; KARATE KILL
MAR   90 min   VIDrel: MOPIC/SGSVID   V

KILLER, THE ***
John Woo HONG KONG
18
1989
Chow-Yun Fat, Sally Yeh, Danny Lee, Kenneth Tsang, Chu Kong, Lam Chung
A contract killer accidentally blinds a singer and is forced into doing one more job to pay for an operation to restore her sight. This plot nonetheless holds up well in a fast-paced tale of considerable length and complexity. A highlight is the final decisive gunfight, a sequence that marks this violent action tale as one of the best films of its type.
Aka: DIE XUE SHUANG XIONG
A/AD   106 min (ort 135 min) wScrn dubbed
VIDrel: MADE/RTM   V/h

KILLER ***
Mark Malone USA
18
1994
Anthony LaPaglia, Mimi Rogers, Matt Craven, Peter Boyle, Monika Schnarre, Joseph Maher, Mark Acheson, Philip Hayes, Christopher Mark Pinhey, Claudio De Victor, Justine Priestly
A top assassin is hired to eliminate an attractive woman, but unwisely develops a romantic interest in her, as he is intrigued by her ready acceptance of her fate. An unusual thriller that is offbeat enough to keep one interested if not on the edge of the seat.
Aka: BULLETPROOF HEART
THR   93 min (ort 98 min)   VIDrel: FIRST/SONOP   V

KILLER: A JOURNAL OF MURDER ****
Tim Metcalfe USA
18
1995
James Woods, Robert Sean Leonard, Ellen Greene, Cara Buono, Robert John Burke, Richard Riehle, John Bedford Lloyd, Jeffrey DeMunn, Conrad McLaren, Steve Forrest
A strongly scripted prison drama (it's also based on a true story) telling of Carl Panzram, America's first known serial killer, who awaits his fate on Death Row. There, a kindly prison guard takes it upon himself to befriend him, smuggling a notebook into the man's cell so he can write his life story. In his directing debut,

Metcalfe achieves a potent examination of violence and the morality of capital punishment, that in its own way, equals DEAD MAN WALKING.
DRA   92 min   CINrel

KILLER AMONG US, A **
Peter Levin USA
15
1990
Jasmine Guy, Dwight Schultz, Anna Maria Horsford, Mykel T. Williamson, Lisa Barnes, Neil Gray Giuntoli, Yvonne Coll, Anita Dangler, Richard Reihle, Dave Florek, Earl Billings, Burton Collins, Mike Genovese, Fitzhugh G. Houston
A juror at a murder trial gradually comes to the opinion that they have convicted the wrong man and attempts to redress this wrong by tracking down the true culprit. Average thriller full of the usual stock characters and situations, but sustained by a genuine sense of tension and firm direction.
THR   88 min (ort 90 min) mTV   VIDrel: GUILD/SONOP
V/sh

KILLER ELITE, THE *
Sam Peckinpah USA
18
1975
James Caan, Robert Duvall, Arthur Hill, Bo Hopkins, Mako, Burt Young, Gig Young, Tom Clancy, Tiana, Kate Heflin, James Wing Woo, George Kee Cheung, Simon Tam, Rick Alemany, Hank Hamilton, Walter Kelley, Billy J. Scott
A veritable feast of gore with Caan out for revenge against Duvall, who double-crossed him whilst they were working as mercenaries. A standard offering from Peckinpah, with a trashy plot that drags in the CIA. Some good action sequences make it slightly more bearable.
THR   118 min (ort 124 min)   VIDrel: WHV   V
Boa: novel by Robert Rostand.

KILLER IMAGE ***
David Winning USA
18
1991
M. Emmet Walsh, John Pyper-Ferguson, Krista Errickson, Michael Ironside, Al Duerr, Paul Austin, Chantelle Jenkins, Kristie Baker, Barbara Gajewskia, Joel Stewart, Jack Ackroyd, Danny Shuttleworth, Brian Martell, Dwayne Pearce
Visiting the apartment of his recently-deceased brother, a photographer finds that it has been burgled, and searching through his brother's possessions, finds a roll of film that shows a prominent senator in a most incriminating situation. With the senator's murderous brother out to retrieve the film, he finds himself suddenly framed for a double murder. A tense and convoluted if rather unconvincing thriller.
THR   94 min (ort 97 min)   VIDrel: FIRST/SONOP L/A   V

KILLER INSTINCT **
David Tausik USA
18
1993
Scott Valentine, Charles Napier, Vanessa Angel, Talia Balsam, Michael Traeger, Brian Cousins, Mary-Ellen Dunbar, Kevin West, David Weiss, Ivy Bethune, Brigitta Stenberg, Tisa Chess, Chuck Bulot, Brendon W. Dawson
A bright lawyer has no stomach for a political career, and is passed over for a complicated case by an ambitious D.A. But a young woman who claims she is the D.A.'s niece comes to his rescue and promises to blackmail her uncle into reconsidering his decision. However, she eventually murders him but manages to arrange things so that the lawyer thinks he is guilty of this crime and is thus motivated to get rid of the body.
Aka: HOMICIDAL IMPULSES
DRA   83 min   VIDrel: 20VIS/SONOP   V/sur

KILLER KLOWNS FROM OUTER SPACE *
Stephen Chiodo USA
15
1987
Grant Cramer, Suzanne Snyder, John Allen Nelson, Royal Dano, John Vernon, Michael Siegel, Peter Licassi
Another variation on the CRITTERS theme, our aliens with a taste for human flesh this time taking the form of sadistic circus clowns who stalk their prey with a formidable arsenal of bizarre weapons, after landing their circus-tent/spaceship at a small secluded town. A gruesome comedy that leans rather too heavily on its repulsive ideas (e.g. our "klowns" like to coat their victims in candyfloss before drinking their blood).
HOR   83 min (ort 88 min)   VIDrel: EIV/SONOP   V/sur

KILLER LOOKS **
Toby Philips USA
(18)
1994
Sara Suzanne Brown, Michael Artura, Len Donato, Diane Hurley (Dyanna Lauren), Dan Frank, Janine M. Lindemulder, Lene Hefner, John P. Hubbell, Christina Johns, P.J. O'Connor, Paul Baumgartner, Stas Connely

A married couple like to play sex games in which they get outsiders to watch them making love. However, their latest recruit turns out to be a bit more dangerous than they anticipated, as he has an agenda of his own, and this one involves blackmail. A steamy erotic drama.
DRA   97 min   SATrel: MOVIE CHANNEL

KILLER NUN *   18
Giulio Berruti ITALY   1978
Anita Ekberg, Joe Dallesandro, Alida Valli, Massimo Serato, Lou Castel, Laura Nucci, Paola Morra, Lee De Barriault
A man-hating nun who works in an asylum, slaughters the male patients in her care whilst diverting attention to another nun who happens to be a drug addict. To avoid a scandal, she is poisoned by the Mother Superior. A crude and nasty anti-clerical sexploiter.
Aka: SUOR OMICIDI
HOR   80 min (ort 90 min) Cut wScrn dubbed
VIDrel: REDEM/RTM   V

KILLER PARTY *   (18)
William Fruet USA   1986
Martin Hewitt, Ralph Seymour, Elaine Wilkes, Paul Bartel, Alicia Fleer, Sherry Willis-Burch, Woody Brown, Joanna Johnson, Terri Hawkes, Deborah Hancock, Laura Sherman, Jeff Pustil, Pam Hyatt, Howard Busgang, Jason Warren
A variation of the slasher in the dorm movies – the spirit of a murdered student possesses a young girl in order to gain its revenge. An ineffective mixture of chills and puerile high school humour.
Aka: FOOL'S NIGHT
HOR   88 min (ort 92 min)   SATrel: TNT MOVIES

KILLER SHREWS, THE *   12
Ray Kellogg USA   1959
James Best, Ingrid Goude, Ken Curtis, Gordon McLendon, Baruch Lumet, J.H. Dupree, Alfredo DeSoto
A charmingly inept tale of title rodents that go on the rampage after a mad scientist's experiments cause them to grow into 100-pound monsters. A dismal and dull offering of little action but much interminable dialogue. The very badness of this low-budget dud makes it an inevitable candidate for cult status.
HOR   69 min B/W   VIDrel: SCREAM/SPEAR   V

KILLERFISH *   15
Anthony M. Dawson (Antonio Margheriti) BRAZIL/ITALY 1978
Lee Majors, Karen Black, James Franciscus, Margaux Hemingway, Marisa Berenson, Gary Collins, Dan Pastorini, Roy Brocksmith, Charlie Guardino, Frank Pesce, Anthony Steffen, Fabio Sabag, Chico Arago, Sonia Citicica
A gang of emerald thieves stash their loot in a reservoir stocked with piranha fish. Attempts by various parties to recover the loot form the basis for this tired outing – the fish are deadly, and so is this story. Filmed in Brazil. See also PIRANHA.
Aka: DEADLY TREASURE OF THE PIRANHA; KILLER FISH
THR   97 min (ort 101 min)   VIDrel: L/A   V

KILLERS, THE ***   18
Don Siegel USA   1964
Lee Marvin, John Cassavetes, Angie Dickinson, Clu Gulager, Ronald Reagan, Claude Akins, Norman Fell, Virginia Christine, Don Haggerty, Jimmy joyce, Robert Phillips, Kathleen O'Malley, Ted Jacques, Irvin Mosley, Davis Roberts
Two hoodlums kill the man they were hired to murder, but are so intrigued by the way in which he accepted his death, that they decide to piece together his past. An absorbing if loose adaptation of Hemingway's story, originally filmed for TV but rejected as being too violent and released to cinemas instead. This was Reagan's last film role.
Aka: ERNEST HEMINGWAY'S THE KILLERS
DRA   90 min (ort 95 min) mTV   VIDrel: L/A   V
Boa: short story by Ernest Hemingway.

KILLER'S EDGE, THE **   18
Joseph Merhi USA   1990
Wings Hauser, Robert Z'dar, Karen Black, Joe Palese, Elaine Pelino, Gino Dinocente, Robert Figaro, Wendy MacDoanld, Robert Gallo, Andres Carranza, Walter Cox, Dallas Cole, Talbot Simmons, Delores Nascar, Charlie Ganis, Robert Axelrod
A cop is given a tough assignment that proves extra difficult

when he learns that the man he has to bring in once saved his life back in Vietnam.
Aka: BLOOD MONEY
A/AD   91 min   VIDrel: IMPENT   V/sh

KILLERS INVINCIBLE **   18
HONG KONG
Alexander Lou, Yau Jin Thomas, Lor Mei
A police officer attached to a special drug unit is also a skilled Ninja fighter. He has to use his skills to the full when the murder of a scientist who was developing a cure for drug addiction takes him to Hong Kong where he has to confront a criminal Ninja gang.
MAR   88 min   VIDrel: IMPENT   V

KILLER'S KISS ***   (PG)
Stanley Kubrick USA   1955
Frank Silvera, Jamie Smith, Irene, Jerry Jarret, Mike Dana, Felice Orlandi, Ralph Roberts, Phil Stevenson, Julius Adelman, David Vaughan, Alec Rubin, Ruth Sobotka
A boxer saves a young dancer who lives in his apartment house from being raped by the manger of the club in which she works. They decide to leave town together and make a new start in Seattle, but have to cope with murder and abduction. A realistic, gritty and downbeat second feature, made on a tiny budget but with many interesting and innovative touches.
DRA   65 min (ort 67 min) B/W   SATrel: MOVIE CHANNEL

KILLER'S ROMANCE *   18
Philip Ko/X.R. Tu HONG KONG   1990
Simon Yam, Joey Wang
The boss of a Japanese criminal empire is murdered by a Chinese gang, the man's son leaves his homeland and travels to Hong Kong on a mission of vengeance. A fast paced and extremely violent action movie, very loosely based on "Crying Freeman", a Japanese comic-book.
A/AD   89 min (ort 95 min) dubbed   VIDrel: POPRO/RTM
V

KILLING AT HELL'S GATE, THE **   PG
Jerry Jameson USA   1981
Robert Urich, Deborah Raffin, Lee Purcell, Joel Higgins, George DiCenzo, Paul Burke, Mitch Carter, Brion James, John Randolph, Maya Braddock, Vicci Cooke, William D. Cottrell, Bob Griggs, Curtis Hanson, Kenny Kinsner
Similar in plot to DELIVERANCE, this tells of how a canoe trip becomes a nightmare when the group is attacked by snipers. The inclusion of two women in the group does little to improve the cliched plot. Average. The initial title of the film was "Hell And High Water".
HOR   100 min mTV   VIDrel: 20TH/TECH L/A   V

KILLING BOX, THE ***   18
George Hickenlooper USA   1993
Martin Sheen, Corbin Bernsen, Adrian Pasdar, Ray Wise, Cynda Williams, Roger Wilson, Dean Cameron, Matt LeBlanc
Assigned to investigate the slaughter of hundreds of men during the American Civil War, a Confederate colonel finds out that the enemy is not human. This proves to be a voodoo entity that was disturbed by some slave traders and is now busy re-animating corpses to create its own zombie army. A dark and atmospheric tale, with an unusual setting.
Aka: GREY NIGHT; GHOST BRIGADE; GHOST BRIGADE: THE KILLING BOX
THR   78 min (ort 86 min)   VIDrel: 20VIS/SONOP   V/sh

KILLING DAD **   15
Michael Austin UK   1989
Julie Walters, Richard E. Grant, Denholm Elliott, Anna Massey, Ann Way, Laura de Sol, Jonathan Phillips, Kevin Williams, Tom Radcliffe, Emma Longfellow, Ronnie Stevens, Pearce Quigley, Peter Geddis, Anthony Higginson
A black comedy in which a man sets out to get to know his father, a cad who walked out on the family when he was a child. In order to do this he changes his name so he can meet the man incognito. However, his father turns out to be a person so disgusting that he decides to murder him. A good cast do their best in a film that tries very hard to be zany, and merely ends up being chaotic. The screenplay is by Austin.
COM   89 min (ort 93 min)   VIDrel: PAL/TERRY L/A   V/h

KILLING FIELDS, THE ***   15
Roland Joffe UK   1984
Sam Waterston, Haing S. Ngor, John Malkovich, Julian Sands, Athol

Fugard, Craig T. Nelson, Bill Paterson, Spalding Gray, Graham Kennedy, Katherine Kragum Chey, Oliver Pierpaoli, Edward Entero Chey, Tom Bird, Ira Wheeler
A harrowing account of the reign of terror imposed by the Khmer Rouge, and of an American journalist who tries to get his Cambodian guide and interpreter out of the country. The depiction of a country being torn apart by war and savagery cannot be faulted. This was documentary-maker Joffe's first feature. AA: S. Actor (Ngor – who lived through the turmoil only to be deliberately gunned down in 1996), Cin (Chris Menges), Edit (Jim Clark).
DRA 136 min (ort 142 min)
VIDrel: 4-FRONT/POLYREC/BRAVE V/sur
Boa: memoirs of New York Times reporter Sidney Schanberg.

KILLING GAME, THE *** 15
Alain Jessua FRANCE 1967
Jean-Pierre Cassel, Claudine Auger, Michel Duchaussoy, Anna Gaylor, Eleanore Hirt, Guy Saint-Jean, Nancy Holloway, Regine, Oyo, Nora, Ysmane My, Roger Curel, Jean Dewever
A married couple who write and draw a mystery comic-strip meet a playboy who is a great fan of their work and is obsessed with acting out the stories they tell. When they start a new strip about a killer, they find themselves in mortal danger. A neat and inventive thriller with a non-violent resolution.
Aka: ALL WEEKEND LOVERS; JEU DE MASSACRE
THR 95 min VIDrel: ARTPRO/RTM L/A V

KILLING GAME, THE ** 18
Joseph Merhi USA 1987
Robert Zdar, Chard Hayward, Cynthia Killion, Geoffrey Sadwith, Bette Rae, Julie Noble, Monique Monet, Brigitte Burdine, Janet Jimmi Parker, Ron Gilchrist, Leia Luahiwa
Made-for-video thriller set on the Californian "Gold Coast" where a hit-man becomes a blackmail victim after carrying out a contract killing, and the prime suspect appears to be a Las Vegas drugs baron.
A/AD 90 min Cut (7 sec) mVid VIDrel: MARQ/QUANT V

KILLING IN A SMALL TOWN, A *** 18
Stephen Gyllenhaal USA 1990
Barbara Hershey, Brian Dennehy, John Terry, Richard Gilliland, Hal Holbrook, Lee Garlington, Matthew Posey, James Black, Dennis Letts, Marco Perella, Jerry Haynes, Norman Bennett, Rodger Boyce, Daphne Eckler, Jan DeWitt, Blue Deckert
A powerful dramatisation of the true-life case of a God-fearing Texas housewife who brutally murdered her best friend with an axe and stood trial for this crime. Hershey won a well-deserved Emmy for her searing performance which is the best thing by far in this mTV drama.
Aka: EVIDENCE OF LOVE
DRA 91 min (ort 100 min) mTV VIDrel: ODY/SONOP V/sh
Boa: book Evidence of Love by John Bloom and Jim Atkinson.

KILLING JAR, THE * 18
Evan Cooke USA 1996
Brett Cullen, Tamlyn Tomita, Wes Studi, M. Emmet Walsh, Xander Berkeley, Tom Bower, Frank McRae, John Philbin
Cullen in the chief suspect after a series of murders take place, but all becomes clear when his traumatic childhood is revealed in an extended flashback sequence. One of those overwrought and underplotted psychological thrillers, treading a path that has by now truly been done to death. Tomita is good as the man's frightened wife, who has to live with someone she hardly recognises any longer.
THR 97 min VIDrel: GUILD/FOXVID V

KILLING MACHINE, THE ** 18
David Mitchell CANADA 1996
Michael Ironside, Jeff Wincott, Terri Hawkes, David Campbell, Callista Carradine, Richard Fitzpatrick, Jeff Pustil, Michael Copeman, David Bolt, Mark Duffus, Tyrone Benskin, Doug O'Keefe, Neil Crone, Rupert Harvey
After he is injured in a gun battle with the police, a contract killer is given both a new identity and a chance to stay out of jail by working for a secret government agency. A derivative little film that brings to mind NIKITA, THE ASSASSIN and several other such works.
Aka: KILLING MAN, THE
A/AD 92 min (ort 95 min) VIDrel: MARQ/QUANT V/sh

KILLING MIND, THE ** 15
Michael Ray Rhodes USA 1991
Stephanie Zimbalist, Tony Bill, Daniel Roebuck, K. Todd Freeman, Candy Ann Brown, Lee Tergesen, Stan Ivar, John Durbin, Keith MacKechnie, Cameron Thor, Billy Beck, Gordon Currie, Tim De Zarn, Danielle Harris, Darlene Kardon
A former FBI agent now working as a police sergeant in Los Angeles attempts to solve a twenty-year-old murder case by using psychiatric techniques on suspects. He is also assisted by a journalist who sees in his efforts some good material for a story. A diverting if distinctly unoriginal yarn.
THR 94 min (ort 100 min) mCab VIDrel: 20VIS/SONOP V/sh

KILLING OF A CHINESE BOOKIE, THE ** 15
John Cassavetes USA 1976
Ben Gazzara, Timothy Agoglia Carey, Seymour Cassel, Azizi Johari, Meade Roberts, Morgan Woodward, Alice Friedland, Donna Gordon, Robert Phillips, John Red Kullers, Virginia Carrington, Al Rubin, Soto Joe Hugh, Haji
Written and directed by Cassavetes, this moody and offbeat film tells of an L.A. nightclub owner who is heavily in debt to gangsters, and is told he can murder the title character as a way of clearing his debt. A most pretentious film, shot through with opaque and meaningless sequences, yet often remarkably atmospheric. Not surprisingly, in some quarters it has achieved a minor cult following.
DRA 104 min (ort 108 min) VIDrel: ELPIC/POLYREC V

KILLING OF SISTER GEORGE, THE *** 18
Robert Aldrich USA 1969
Beryl Reid, Susannah York, Coral Browne, Ronald Fraser, Patricia Medina, Hugh Paddick, Cyril Delevanti, Sivi Aberg
An engrossing tragi-comedy with Reid in one of her finest roles, as an ageing lesbian who loses both her job and her lover, when she is axed from a long-running TV farmyard soap opera. Not quite retaining the light touch of the play but undeniably effective. Scripted by Lukas Heller and filmed in England.
DRA 133 min (ort 139 min) VIDrel: VCC/DISC V
Boa: play by Frank Marcus.

KILLING STREETS ** 18
Stephen Cornwell USA 1991
Michael Pare, Lorenzo Lamas, Jennifer Runyon, Michael Downs, Hashem Yassin, Rahely Chimeyan, Yaacov Gvir-Cohen, Alon Abootboul, Itzik Atzmon, Menahem Eini, Shachar Cohen, Travis Ducsay, Ofer Levi, Noa El-Rom, Mati Seri, Rafi Milo
A captain in the U.S. Marines is out with his girlfriend when the latter is murdered and he is kidnapped. Despite hearing that he is dead, the man's twin brother sets out on a dangerous mission to get at the truth and rescue a brother he believes is still alive.
A/AD 102 min (ort 109 min) VIDrel: MIA/DISC V

KILLING TIME, THE ** 15
Rick King USA 1987
Kiefer Sutherland, Beau Bridges, Wayne Rogers, Joe Don Baker, Camelia Kath, Janet Carroll, Gracie Harrison, Harvey Vernon, Michael Madsen, Harriet Medin, Shiri Appleby, Jeb Ellis-Brown, Richard Bolik, Paul McKenna
An unfaithful wife and her sheriff boyfriend, hatch a plot to kill the husband of the former and pin the murder on Sutherland, but things do not turn out as expected, in this muddled thriller.
THR 89 min (ort 94 min) VIDrel: NWV L/A V/h

KILLING ZOE ** 18
Roger Avary FRANCE/USA 1994
Eric Stoltz, Jean-Hugues Angalde, Julie Delpy, Gary Kemp, Bruce Ramsay, Tai Thai, Kario Salem, Salvator Xuereb, Gian Carlo Scanduzzi, Cecilia Peck, Elise Renee, Martin Raymond, Eric Pascal Chaltiel, Ron Jeremy Hyatt, Gerard Brown
An American safe-cracker comes to Paris to take part in a bank heist that is scheduled for Bastille Day (July 14) but is shocked to find that his associates are more concerned with drug-taking than the job in hand. When the robbery goes wrong, a series of bloody clashes ensue. An incredibly violent and over-stylised movie that is more than a little reminiscent of PULP FICTION, a work Avary co-scripted. He both directed and wrote the script of this one.
THR 96 min VIDrel: POLY/POLYREC V/sh

KILLING ZONE, THE ** 18
Addison Randall USA 1990
Deron McBee, James Dalesandro, Melissa Moore, Armando Silvester,

Felicia Mercado, Augustine Beral, Sydne Squire, Quentin Gutierrez, Michael Easton, Vic Laroux, Raymond Martino, Charles K. Sullivan, Wally K. Berns, Gerald Thomas
The leader of a powerful Mexican drug cartel vows revenge when his jailed brother is callously murdered in prison. Exiled from the USA, he returns and begins a campaign of slaughter stretching right across L.A. In desperation, the cops call on the services of an old enemy, a former DEA officer, who, in the company of his nephew, hunts down the gangster for a final, bloody confrontation.
A/AD   83 min (ort 90 min)   VIDrel: IMPENT   V/sh

KIM ***
John Howard Davies USA
PG
1984
Peter O'Toole, Bryan Brown, John Rhys-Davies, Ravi Sheth, Julian Glover, Lee Montague, Alfred Burke, Mick Ford, Bill Leadbitter, Sneh Gupta, Roger Booth, Peter Childs, Noel Coleman, Nadira, Lavlin, Jalal Agha, Sean Scanlon
A colourful remake of the 1950 tale of colonial India, telling of a resourceful British boy who is brought up by an Indian mystic and becomes a spy for the British Secret Service. Sheth makes his debut as the title character, with O'Toole giving a splendidly over-the-top performance as a Buddhist monk. The script is by James Brabazon.
A/AD   136 min (ort 150 min) mTV
VIDrel: MANAGE/TOTAL   V
Boa: novel by Rudyard Kipling.

KIND HEARTS AND CORONETS ****
Robert Hamer UK
U
1949
Dennis Price, Alec Guinness, Joan Greenwood, Valerie Hobson, Miles Malleson, Hugh Griffith, Jeremy Spenser, Arthur Lowe, Audrey Fildes, Clive Morton, Lyn Evans, Cecil Ramage, John Penrose, John Salew, Eric Messiter, Anne Valery
Superb black comedy about a man who ruthlessly eliminates eight of his relatives (all played by Guinness) in order to inherit a title and stately home, part of his motivation being as an act of revenge in memory of his disowned mother, and part just plain greed. Price is perfectly cast as the charming but ruthless villain of the piece, in this beautifully crafted comedy.
COM   101 min (ort 106 min) B/W   VIDrel: WHV   V/h
Boa: novel Israel Rank (Noblesse Oblige) by Roy Horniman.

KIND OF LOVING, A ****
John Schlesinger UK
15
1962
Alan Bates, June Ritchie, Thora Hird, Bert Palmer, Gwen Nelson, Malcolm Patton, Pat Keen, James Bolam, Jack Smethurst, John Ronane, David Mahlowe, Patsy Rowlands, Michael Deacon, Jerry Desmonde, Leonard Rossiter
A young draughtsman in a Northern industrial town, is forced into an early marriage when he gets a girl at work pregnant. A fine realistic drama that examines his relationship with those around him, not least a hostile mother-in-law. All the cast are uniformly excellent but Bates as the young man and Hird as his noxious mother-in-law are outstanding.
DRA   107 min (ort 112 min) B/W   VIDrel: WHV   V
Boa: novel by Stan Barstow.

KINDERGARTEN COP **
Ivan Reitman USA
15
1990
Arnold Schwarzenegger, Penelope Ann Miller, Pamela Reed, Linda Hunt, Richard Tyson, Carroll Baker, Cathy Moriarty, Park Overall, Richard Portnow, Jayne Brook, Joseph Cousins, Christian Cousins, Tom Kurlander, Alix Koromzay
A tough Los Angeles cop is forced to masquerade as a kindergarten teacher in order to locate the estranged wife and son of a ruthless drugs dealer. His young charges prove quite a handful but he manages to tame them in due course before going on to get the bad guys. A shapeless blend of comedy and action that does justice to neither and is well laced with sentimental platitudes about kids. Offers a few easy laughs but little else.
COM   106 min (ort 111 min)   VIDrel: CIC/SONOP   V

KING AND COUNTRY ***
Joseph Losey UK
PG
1964
Dirk Bogarde, Tom Courtenay, Leo McKern, Barry Foster, James Villiers, Peter Copley, Barry Justice, Vivan Matalon, Jeremy Spenser, James Hunter, David Cook, Larry Taylor, Jonah Seymour, Keith Buckley, Richard Arthure, Derek Patridge
After three years of war, a young soldier walks away from the trenches of Passchendale and is court-martialled. An army captain is assigned to defend him, but the man is sentenced to

death. An earnest and intelligent adaptation of a fine anti-war play, much of whose intensity is ditched in favour of excessive moralising. Nevertheless, some beautiful performances make a depressing film more bearable. The score is the work of Larry Adler.
DRA   82 min (ort 88 min) B/W   VIDrel: CONNO/RTM   V
Boa: play Hamp by John Wilson/novel Return to the Wood by James Lansdale Hodson.

KING AND I, THE ****
Walter Lang USA
U
1956
Deborah Kerr, Yul Brynner, Rita Moreno, Martin Benson, Terry Saunders, Alan Mowbray, Rex Thompson, Carlos Rivas, Patrick Adiarte, Geoffrey Toone, Robert Banas, Yuriko, Marion Jim, Dusty Worrall, Gemze De Lappe, Charles Irwin
Musical version of "Anna And The King Of Siam" telling the story of the British governess who went to Siam in 1862 and spent six years teaching the King's many children at the Royal Palace of Bangkok. A lively and appealing film tailor-made for Brynner. AA: Actor (Brynner), Art/Set (Lyle R. Wheeler and John DeCuir/Walter M. Scott and Paul S. Fox), Sound (Carl Faulkner), Score (Alfred Newman/Ken Darby), Cost (Irene Sharaff).
MUS   128 min (ort 133 min) wScrn   VIDrel: 20TH/TECH
V/dm
Boa: musical Anna and the King of Siam/book The English Governess at the Siamese Court by Anna Leonowens.

KING ARTHUR WAS A GENTLEMAN **
Marcel Varnel UK
U
1942
Arthur Askey, Evelyn Dall, Anne Shelton, Max Bacon, Jack Train, Al Burnett, Peter Graves, Vera Frances, Brefni O'Rorke, Ronald Shiner, Freddie Crump, Ernie (Victor) Feldman, John Wynn, Veronica Turleigh, Elizabeth Flateau
A man called Arthur King is inducted into the Army and as his is obsessive with the story of King Arthur, his friends present him with a sword that he some how believes to be the famous Excalibur. This inspires him to acts of great bravery in the North African desert but the revelation of the truth comes as quite shock. An overlong tale that is only mildly amusing at best.
COM   99 min B/W   VIDrel: PETWAS/WEADIS L/A   V

KING BOXER **
Liu Chia Liang HONG KONG
18
1973
Liu Chia-Hui, Hui Ya-Mung, Wang Lung-Wei
A young martial arts devotee comes up against rivals, and has his hands broken as a warning. However, he makes a remarkable recovery and learns a secret "iron-fist" technique with which he has his revenge. An early martial arts offering with several good sequences, but some ludicrous dubbing and a pointless romantic sub-plot.
Aka: FIVE FINGERS OF DEATH; INVINCIBLE BOXER
MAR   101 min Cut (7 sec) dubbed   VIDrel: L/A   V

KING BOXER 2 **
Joseph Kong HONG KONG
18
Bruce Le (Huang Kin Lung), Chang Lee, Lita Vasquez
A follow-up to the first film, with a formula tale of honour and revenge.
MAR   90 min   VIDrel: MOPIC/SGSVID   V

KING DAVID **
Bruce Beresford USA
PG
1985
Richard Gere, Edward Woodward, Denis Quilley, Alice Krige, Niall Buggy, Jack Klaff, Cherie Lunghi, Hurd Hatfield, Jim Castle, Tim Woodward, David Keysear, Luigi Montefiori, Ian Sears, Simon Dutton, Jean-Marc Barr, Arthur Whybrow
A Hollywood biblical epic that fails, largely due to poor casting of Gere in the title role. However, despite a panning from the critics there are some good things in it, such as a fine performance from Woodward as Saul and the excellent photography of Donald McAlpine. The script is less memorable and gradually unravels as the story progresses. The score is by Carl Davis.
Aka: STORY OF DAVID, THE
DRA   109 min (ort 114 min)   VIDrel: CIC/SONOP   V/sur

KING FRAT *
Ken Wiederhorn USA
18
1979
Dan Chandler, Dan Fitzgerald, Mike Grabow, Ray Mann, John Di Santi, Charles Pitt
A routine teenage college comedy, with some pretty weird types enrolled at Yellowstream University, who indulge in the usual predictable antics and vulgar jests.
COM   84 min (ort 90 min)   VIDrel: VIPCO/SGSVID   V

KING KONG ***
Merian C. Cooper/Ernest B. Schoedsack USA   PG   1933
Fay Wray, Bruce Cabot, Robert Armstrong, Frank Reicher, Noble Johnson, James Flavin, Sam Hardy, Victor Wong, Steve Clemento, Ethan Laidlaw, Charlie Sullivan, Vera Lewis, Leroy Mason, Dick Curtis, Lynton Brent, Frank Mills, Jim Thorpe
A rather wooden tale of the capture of a giant ape, and its exhibition when brought back to civilisation, where it escapes and runs amok. Though terribly dated now, the film must be seen in the context of the technical limitations of the 1930s, and on that basis alone can be regarded as something of a milestone in cinema. It was remade in 1976. See also THE SON OF KONG and MIGHTY JOE YOUNG. The tape includes the documentary entitled "It Was Beauty Killed The Beast".
Aka: KING KONG: THE EIGHTH WONDER OF THE WORLD
A/AD   96 min (director's cut plus documentary – ort 105 min) B/W   VIDrel: 4-FRONT/POLYREC   V
Boa: novel by Edgar Wallace and Merian C. Cooper.

KING KONG ***
John Guillermin USA   PG   1976
Jeff Bridges, Jessica Lange, John Randolph, Charles Grodin, Julius Harris, Rene Auberjonois, Ed Lauter, Jack O'Halloran, Mario Gallo, Jorge Moreno, Sid Conrad, John Agar, John Lone, Gary Walberg, George Whitman, Rick Baker
Excellent remake of the rather cardboard 1933 film that handles both human and animal characters with considerable intelligence. After it captures the girl, both human and ape develop a touching and rather ambivalent relationship that is well shown. The special effects in this one are superb. AA: Spec Award (Carlo Rambaldi/Glen Robinson/Frank Van Der Meer for visual effects).
A/AD   128 min (ort 135 min) wScrn
VIDrel: BMGVID/BMGREC   V
Boa: novel by Edgar Wallace and Merian C. Cooper.

KING KONG LIVES *
John Guillermin USA   PG   1986
Brian Kerwin, Linda Hamilton, John Ashton, Peter Michael Goetz, Frank Maraden, Peter Elliott, George Yiasomi, Alan Sader, Lou Criscuolo, Marc Clement, Richard Rhodes, Larry Souder, Ted Prichard, Jayne Lindsay-Gray
A sequel to the 1976 remake of the original, in which Kong is brought out of the coma caused by his fall, and restored to full health and vigour by a heart transplant. To make his happiness complete, he even gets a mate. A film that's as idiotic as it is unconvincing.
A/AD   101 min (ort 105 min)   VIDrel: 20TH/TECH   V/sur

KING LEAR **
Grigori Kozintev USSR   PG   1971
Yuri Yarvet, Elsa Radzinya, Galina Volchek, Valentina Shendrikova, Karl Sebris, Oleg Dal
Ambitious film version of Shakespeare's play, with some fine visual imagery but nowhere near as firmly directed as the director's version of Hamlet. Yarvet is good if somewhat one-dimensional in the title role, but is not helped by a weak supporting cast. Giving the story a political slant is not well advised, as it weakens the dramatic elements. Worth seeing if only to fully appreciate the Shostakovich score. The script is adapted from Boris Paternak's translation.
Aka: KOROL LIR
DRA   137 min (ort 140 min) B/W   VIDrel: TART/20TH   V
Boa: play by William Shakespeare.

KING LEAR ***
Peter Brook DENMARK/UK   PG   1971
Paul Scofield, Irene Worth, Alan Webb, Cyril Cusack, Tom Fleming, Jack MacGowran, Susan Engel, Patrick Magee, Robert Lloyd, Soeren Elung-Jensen, Ian Hogg, Annelise Gabold, Barry Stanton, Soren Elung Jensen
An effective version of Shakespeare's play (filmed in Jutland and other parts of Denmark) that benefits from a strong cast and a wonderful feel for the material. This restrained version is a little hard to follow but rewards patience.
DRA   132 min (ort 137 min) B/W   VIDrel: RCA/VCC L/A V
Boa: play by William Shakespeare.

KING LEAR ****
Jonathan Miller UK   PG   1982
Michael Hordern, Frank Middlemass, John Shrapnel, Norman Rodway, Michael Kitchen, Gillian Barge, Brenda Blethyn, Penelope Wilton, John Bird, Julian Curry, David Weston, Harry Waters, Anton Lesser, John Grillo, Iain Armstrong
An outstanding television production, easily one of the best in this large and ambitious series of the complete set of Shakespeare's plays. Acting and direction are absolutely first rate.
DRA   185 min mTV   VIDrel: BBC   V/sur
Boa: play by William Shakespeare.

KING LEAR ***
Michael Elliott UK   15   1983
Laurence Olivier, Colin Blakely, Anna Calder-Marshall, John Hurt, Jeremy Kemp, Robert Lang, Robert Lindsay, Leo McKern, Diana Rigg, David Threlfall, Dorothy Tutin
This well-staged adaptation represented one of Olivier's last demanding roles, and though it fails to rival the power of earlier works (such as the 1971 Peter Brook production), it's certainly made with enough care to merit attention.
Aka: OLIVIER'S KING LEAR
DRA   158 min (ort 180 min) mTV
VIDrel: CASPIC/BMGREC   V/sur
Boa: play by William Shakespeare.

KING OF COMEDY, THE ***
Martin Scorsese USA   PG   1981 (released 1982)
Robert De Niro, Jerry Lewis, Ed Herlihy, Diahnne Abbott, Sandra Bernhard, Tony Randall, Shelley Hack, Fred de Cordova, Leslie Levinson, Margo Winkler, Lou Brown, Whitey Ryan, Doc Lawless, Marta Heflin, Tony Boschetti, Richard Dioguardi
De Niro gives a bravura performance as an obsessive would-be-comic, who kidnaps a comedian with his own TV show in order to be given the chance of making a guest appearance. Lewis is suitably insincere as the successful comic, De Niro is wonderfully obnoxious as the aspiring one. A patchy and only slightly amusing comedy, scripted by Paul D. Zimmerman. Scorsese has a small part as a TV director.
DRA   104 min (ort 109 min)   VIDrel: WHV   V

KING OF KINGS **
Nicholas Ray USA   U   1961
Jeffrey Hunter, Robert Ryan, Sibohan McKenna, Frank Thring, Hurd Hatfield, Rip Torn, Harry Guardino, Viveca Lindfors, Rita Gam, Carmen Savillia, Brigid bazlen, Guy Rolfe, Maurice Marsac, Gregoire Aslan, Orson Welles (narration)
A devout but dullish account of the life of Christ that is admittedly well made but somewhat lacking in dramatic impact. Despite the efforts of all concerned, this prime example of Hollywood biblical epic never really comes to life. See also the following: THE LAST TEMPTATION OF CHRIST, JESUS OF NAZARETH, THE GREATEST STORY EVER TOLD and THE GOSPEL ACCORDING TO ST MATTHEW.
DRA   163 min   VIDrel: MGM/WHV   V/s

KING OF NEW YORK **
Abel Ferrara ITALY/USA   18   1989
Christopher Walken, David Caruso, Larry Fishburne, Victor Argo, Wesley Snipes, Janet Julian, Joey Chin, Giancarlo Esposito, Paul Calderon, Steve Buscemi, Theresa Randle, Leonard Lee Thomas, Roger Smith, Carrie Nygren
Stylish but insubstantial tale of a newly released New York Mafia don who plans to reassert his control over the drug trade by eliminating his rivals. To show he has a human side, he also toys with the idea of opening a closed hospital in his old South Bronx neighbourhood. Walken gives a chilling and charismatic performance in the title role, but this simply cannot compensate for the vastly muddled plotline and the incessant violence.
A/AD   99 min (ort 106 min)   VIDrel: VCC/DISC/COLUM V/sur

KING OF THE HILL ***
Steven Soderbergh USA   15   1993
Jesse Bradford, Karen Allen, Jeroen Krabbe, Lisa Eichhorn, Joseph Chrest, Elizabeth McGovern, Spalding Gray, Adrien Brody, Cameron Boyd, Ron Vawter, Chris Samples, Katherine Heigl, Amber Benson, John McConnell, John Durbin, Ron Yerxa
In St Louis during the Depression years, a twelve-year-old boy has to cope as best he can in the absence of his parents and younger brother. Forced to live in a run-down hotel, he makes friends both there and at school, and they provide help and support. An atmospheric and very well presented adaptation of Hochner's autobiographical work, which avoids sentimentality

in preference for attention to detail and respect for its subject matter.
DRA   98 min (ort 103 min)   VIDrel: CIC/SONOP   V/sur
Boa: memoirs of A.E. Hotchner

KING OF THE KICKBOXERS, THE **
18
Lucas Lowe USA
1989
Loren Avedon, Keith Cooke, Billy Blanks, Richard Jaeckel, Don Stroud, Jerry Trimble, Sherrie Rose, William Long, David Michael Sterling, Ong, John Kay, Bruce Richard Fontaine, Patrick Schuck, Michael DePasquale Jr, Vincent Lin
The murder of a man's brother by a martial arts champ leaves him filled with bitterness, but when the culprit sets about recruiting people to star in his new kickboxing movie the opportunity arises to have revenge. Followed by AMERICAN SHAOLIN: KING OF THE KICKBOXERS 2.
MAR   91 min Cut (46 sec – ort 93 min)
VIDrel: 4-FRONT/POLYREC/EIV L/A   V

KING OF THE WIND ***
U
Peter Duffell UK
1989
Frank Finlay, Jenny Agutter, Nigel Hawthorne, Navin Chowdhry, Glenda Jackson, Richard Harris, Ralph Bates, Neil Dickson, Barry Foster, Anthony Quayle, Ian Richardson, Norman Rodway, Peter Vaughn, Jill Gascoine
A mute Arab boy accompanies a stallion on a perilous journey from North Africa to France and then Britain. They suffer many hardships and misfortunes and have a number of adventures before the inevitable happy ending. An enjoyable children's film that, despite its stereotyped characterisation and over-burdened plotting, is sustained by fine photography and performances.
DRA   102 min mTV   VIDrel: FIRST/SONOP L/A   V
Boa: novel by Marguerite Henry.

KING RALPH **
PG
David S. Ward USA
1990
John Goodman, Peter O'Toole, John Hurt, Richard Griffiths, Leslie Phillips, Camille Coduri, James Villiers, Joely Richardson, Niall O'Brien, Julian Glover, Judy Parfitt, Ed Stobart, Rudolph Walker, Gedren Heller, Ann Beach, Tim Seely
When a freak accident wipes out the entire extensive British Royal Family, a desperate search is mounted for any distant surviving relative. The only heir is found to be a Las Vegas entertainer best described as a decent, lovable but uncouth slob. Goodman gives an amiable performance in the title role but his talent alone is not enough to save this overlong misfire, whose single comic idea soon grows wearying in the extreme.
COM   92 min (ort 105 min)
VIDrel: 4-FRONT/POLYREC/CIC   V/sur
Boa: novel Headlong by Emlyn Williams.

KING RAT ***
12
Bryan Forbes USA
1965
George Segal, Tom Courtenay, James Fox, Patrick O'Neal, Denholm Elliott, John Mills, Leonard Rossiter, Gerald Sim, Todd Armstrong, James Donald, Alan Webb, Geoffrey Bayldon, Sammy Reese, William Fawcett, Joseph Turkel
Set in a Japanese prison camp in Singapore towards the end of WW2, where the title character, a U.S. corporal, lives by his wits and manages to survive, but often at the expense of his fellow prisoners. However, the camp's Provost Marshal has vowed to make this fellow pay dearly for his collaboration with their captors. A grim study of human behaviour under captivity that lacks pace but has an intense and claustrophobic atmosphere. The script is by Forbes.
WAR   129 min (ort 134 min) B/W
VIDrel: ENCORE/SPEAR/COLUM   V
Boa: novel by James Clavell.

KING SOLOMON'S MINES *
PG
J. Lee Thompson USA
1985
Richard Chamberlain, Sharon Stone, John Rhys-Davies, Herbert Lom, Ken Gampu, June Buthelezi, Sam Williams, Shai K. Ophir, Fidelis Chea, Mick Lesley, Bob Greer, Vincent Van Der Byl, Oliver Tengende, Neville Thomas, Isiah Murett
A rather bad version of this classic adventure tale, somewhat influenced by RAIDERS OF THE LOST ARK etc. Cardboard characters combine with wooden acting to produce a film lacking in credibility. Never fear, there is a sequel: ALLAN QUARTERMAIN AND THE LOST CITY OF GOLD.
A/AD   96 min (ort 100 min)   VIDrel: L/A   V
Boa: novel by H. Rider Haggard.

KINGDOM, THE: PARTS 1 TO 3 ***
15
Lars Von Trier DENMARK
1994
Ernst-Hugo Jaregard, Kirsten Rolffes, Ghita Norby, Udo Kier, Holger Juul Hansen, Soren Pilmark, Jens Okking, Otto Brnadenburg, Annevi Schelde Ebbe, Baard Owe, Birgitte Raaberg, Peter Mygind, Vita Jensen, Morten Rotne Leffers
At Copenhagen's Kingdom Hospital, the position of a tyrannical and hateful Swedish surgeon comes under threat when his attempts to cover up his negligence begin to unravel. Clearly inspired by the commercial success of TWIN PEAKS, Von Truer presents his own potent blend of hauntings, phantom ambulances and other such paraphernalia instead of a decent plot. However, the performances (especially Jaregard's) make it worthwhile. Only Part 1 is at present available.
Aka: RIGET
FAN   51 min (ort 279 min)   VIDrel: ICAPRO/MANGA
V/sh

KINGFISH: A STORY OF HUEY P. LONG **
12
Thomas Schlamme USA
1994
John Goodman, Matt Craven, Anne Heche, Ann Dowd, Jeff Perry, Bob Gunton, Bill Cobbs, Hoyt Axton
A look at the career and assassination of this controversial politician, who in the early 1900s became a senator and ran for President, but was tainted by corruption charges. Goodman is outstanding in his portrayal of this memorable character, but his efforts are rather let down by an uninspired script and poor direction. There was another TV movie on this subject in 1977 – "The Life And Assassination Of The Kingfish".
Aka: KINGFISH
DRA   93 min (ort 97 min) mCab   VIDrel: 20VIS/SONOP
V/sh

KINGPIN **
12
Peter Farrelly/Bobby Farrelly USA
1996
Woody Harrelson, Randy Quaid, Vanessa Angel, Bill Murray, Chris Elliott, William Jordan, Richard Tyson, Lin Shaye, Zen Gesner, Prudence Wright Holmes, Rob Moran, Danny Green, Will Rothhaar, Jill Lytle, Willie Beauchene
Having been double-crossed by a con-man, a former champion bowler finds himself in dire straits, but comes up with a scheme to get a naive Amishman to enter a $1,000,000 prize bowling contest, his hope being that his protege will beat the bad guy who was responsible for hacking off his hand for hustling and thus ending his bowling career. A cruel slapstick comedy of spite, tastelessness and occasional funny antics, from the makers of DUMB AND DUMBER.
COM   121 min   VIDrel: EIV/SONOP   V/sh

KINGS OF THE ROAD ***
18
Wim Wenders WEST GERMANY
1975
Rudiger Vogler, Hanns Zischler, Liza Kreuzer, Rudolf Schundler, Marquard Bohm
A New Wave German film that provides a slow, insightful and absorbing examination of a host of issues, all wrapped up in the form of a road-movie. A travelling projectionist/repairman teams up with a depressed hitch-hiker to explore a dreary, forgotten border region between East and West Germany. A film about fast cars, rock 'n' roll and industrial decline, that is well complemented by its fine rock score. See also ALICE IN THE CITIES and THE WRONG MOVE.
Aka: IM LAUF DER ZEIT; IN THE COURSE OF TIME
DRA   168 min (ort 176 min) B/W   VIDrel: CONNO/RTM
V

KING'S WHORE, THE **
15
Axel Corti FRANCE/ITALY/UK
1990
Valeria Golino, Robin Renucci, Paul Crauchet, Amy Werba, Francesca Reggiani, Feodor Chaliapin, Stephane Freiss, Dominique Marcas, Timothy Dalton, Eleanor David, Margaret Tyzack, Elizabeth Kaza, Lea Padovani, Anna Bonaiuto
A Parisian woman and her ambitious husband come to the court of the kingdom of Piedmont in Northern Italy but their honeymoon is cut short when the country's ruler conceives a violent lust for her. Encouraged by her husband, she gives way to this fierce passion and becomes his mistress. A visually striking, well staged and convincingly acted historical tale that for all its undoubted merits comes across as curiously lifeless and uninvolving.
Aka: JEANNE; PUTAIN DU ROI
DRA   89 min (ort 132 min)   VIDrel: 20VIS/SONOP   V
Boa: novel Jeanne de Luynes, Comtesse de Verue by Jacques Tournier.

KINJITE: FORBIDDEN SUBJECTS **   18
J. Lee Thompson USA   1989
Charles Bronson, Juan Fernandez, James Pax, Kumiko Hayakawa, Perry Lopez, Peggy Lipton, Amy Hathaway, Bill McKinney, Sy Richardson, Alex Hyde-White, Richard Egan Jr
A long-serving vice cop sets out to smash a child prostitution ring, run by a ruthless pimp who has kidnapped the daughter of a Japanese businessman recently arrived in Los Angeles. Standard hard-boiled Bronson thriller dressed up with a few new angles.
Aka: KINJITE
A/AD 93 min (ort 97 min) VIDrel: MGM/WHV V/sh

KISMET ***   U
Vincente Minnelli USA   1955
Howard Keel, Ann Blyth, Dolores Gray, Monty Woolley, Sebastian Cabot, Vic Damone, Jay C. Flippen, Mike Mazurki, Jack Elam, Ted De Corsia, Patricia Dunn, Reiko Sato, Wonci Lui, Julie Robinson
An entertaining Arabian Nights fantasy, with some great songs (and some dreadful ones too) about a beggar and his daughter and their adventures in Baghdad. The best songs, "Stranger In Paradise" and "Baubles, Bangles and Beads", are based on music by Borodin.
MUS 108 min (ort 113 min) VIDrel: MGM/WHV V/dm
Boa: musical by Edward Knoblock.

KISS, THE **   18
Pen Densham USA   1988
Meredith Salenger, Joanna Pacula, Mimi Kuzyk, Nicholas Kilbertus, Jan Rubes, Sabrina Boudot, Shawn Levy, Celine Lomez, Dorian Joe Clark, Richard Dumont, Priscilla Mouzakiotis, Talya Rubin, Philip Pretten, Johanne Herelle
An African voodoo priestess is subject to a curse that is handed down from one generation to the next by means of a woman-to-woman kiss. In a search for an heiress she enters the lives of the family of her dead sister, where the young girl she has chosen begins to have doubts about the wholesomeness of her newly-arrived aunt. A cumbersome yarn with several scary moments but a general failure to explore the premise's intriguing sexual element.
HOR 94 min (ort 101 min) VIDrel: VCC L/A V/sh

KISS BEFORE DYING, A **   18
James Dearden USA   1991
Matt Dillon, Sean Young, Max Von Sydow, Diane Ladd, James Russo, Martha Gehman, Ben Browder, Joy Lee, Adam Horovitz, Sam Coppola, Jim Fyfe, Elzbieta Czyzewska, Freddy Koehler, Leslie Lyles, Shane Rimmer, Sarah Keller
A psychopathic killer uses his charm to ingratiate himself with the wealthy family of a local copper magnate (Sydow in a brief but excellent performance). He decides that marriage to one of the family's twin daughters is his best route to the top but is forced to change his plans when she becomes pregnant. A studio-bound, over-contrived and poorly realised remake of the 1956 film, with Dearden also responsible for the screenplay.
THR 89 min (ort 96 min) VIDrel: CIC/SONOP V/sh
Boa: novel by Ira Levin.

KISS GOODNIGHT, A **   18
Daniel Raskov USA   1993
Paula Trickey, Mark Moses, Al Corley, Lawrence Tierney, Brett Cullen, James Karen, Robert Wuhl, Lisa Engelman, Charles J. Filer, Sydney Walsh, Shelly Desai, Philip O'Brian, Lori Miller, Mark Ginther, Scott Parker, Kevin Schon
Tired of her uncommitted boyfriend, a woman unwisely has a fling with as stranger she meets one night, and then ends the affair wen he boyfriend begs her for another chanced. However, the new man in her life is unwilling to let her go.
THR 84 min (ort 88 min) VIDrel: HIFLI/SONOP V/h

KISS ME A KILLER **   15
Marcus De Leon USA   1991
Julie Carmen, Robert Beltran, Guy Boyd, Ramon Franco, Charles Boswell
A musician working at a club in East L.A. conspires with the owner's unhappy wife to murder the husband. A low-budget but adequate reworking of THE POSTMAN ALWAYS RINGS TWICE with a Hispanic flavour.
THR 81 min (ort 92 min) VIDrel: COLUM/SONOP V

KISS ME DEADLY ***   PG
Robert Aldrich USA   1955
Ralph Meeker, Albert Dekker, Paul Stewart, Maxine Cooper, Gaby Rodgers, Wesley Addy, Juano Hernandez, Nick Dennis, Cloris Leachman, Marian Carr, Jack Lambert, Jack Elam, Jerry Zinneman, Percy Helton, Fortunio Bonvanova
Moody, atmospheric and very confused adaptation of Spillane's Mike Hammer novel in which our detective hero soon finds himself embroiled with a gang of crooks out to steal a box containing a radioactive isotope. The plot is so very convoluted it can hardly be followed, but in terms of atmosphere and menace the film takes some beating, even if the cataclysmic resolution is a little unbelievable (the film cleverly played on the nuclear war fears of the time).
THR 106 min B/W VIDrel: MGM/WHV V
Boa: novel by Mickey Spillane.

KISS ME KATE ***   U
George Sidney USA   1953
Howard Keel, Kathryn Grayson, Ann Miller, Bobby Van, Keenan Wynn, James Whitmore, Bob Fosse, Tommy Rall, Kurt Kaszner, Ron Randell, Willard Parker, Dave O'Brien, Claud Allister, Ann Codee, Carol Haney, Jeanne Coyne, Hermes
The lead couple in a musical version of Shakespeare's "The Taming Of The Shrew" have a married life that in many respects resembles their on-stage roles. Many fine tunes grace this highly enjoyable adaptation of the smash hit Broadway musical and include numbers such as: "So In Love", "Always True To You In My Fashion", "Brush Up Your Shakespeare" and "From This Moment On".
MUS 105 min (ort 111 min) VIDrel: MGM/WHV V/dm
Boa: play by Samuel and Bella Spewack/musical by Cole Porter.

KISS ME MONSTER *   18
Jess Franco WEST GERMANY   1967
Janine Reynaud, Rossana Yani, Adrian Hoven, Michel Lemoire, Chris Howland, Carlos Mendi
Sequel to SADISTEROTICA, with a sexy female detective duo suffering various indignities at the hands of mutant body-builders and sexually voracious lesbians. Their task appears to involve a hunt for a missing scientist, but in truth it may equally have been a hunt for a missing scriptwriter. There are some amusing moments, but mostly they are unintentional.
Aka: CASTLE OF THE DOOMED
HOR 75 min wScrn dubbed VIDrel: REDEM/RTM V

KISS ME, STUPID *   PG
Billy Wilder USA   1964
Dean Martin, Ray Walston, Kim Novak, Felicia Farr, Cliff Osmond, Barbara Pepper, Doro Merande, Henry Gibson, John Fiedler, Mel Blanc, Howard McNear, Alan Dexter, Tommy Nolan, Alice Pearce, John Fiedler, Arlen Stuart, Cliff Norton
A lewd comedy vehicle for Martin, in which he plays a womanising singer who feigns interest in the work of an unsuccessful songwriter as a ploy to bed the latter's wife. Despite good casting of Martin and a scattering of amusing gags, the passage of time has not been kind to this sluggish and inept affair, which remains as tasteless as it is tiresome. Following a heart attack, Peter Sellers dropped out of the role taken over by Walston.
COM 121 min (ort 126 min) VIDrel: WHV V
Boa: play L'Oro Della Fantasia by Anna Bonacci.

KISS OF A KILLER **   15
Larry Elikann USA   1992
Annette O'Toole, Eva Marie Saint, Brian Wimmer, Gregg Henry, Vic Polizos, Amy Stock-Poynton, Cassy Friel, Jim Haynie, Lee Garlington, Gordon Clapp, Marnie McPhail, Brian Smiar, Francesca Buller, Amy Morton, Stephen Burks
A woman badly abused as a child by her mother, develops a second personality that allows her to cope and function, especially in her relations with men. However, this gets her into grave danger when she meets up with a lethal rapist.
DRA 91 min (ort 93 min) mTV VIDrel: GUILD/SONOP V
Boa: novel The Point Of Murder by Margaret Yorke.

KISS OF DEATH **   18
Barbet Schroeder USA   1994
David Caruso, Nicolas Cage, Kathryn Erbe, Helen Hunt, Samuel L. Jackson, Michael Rapaport, Ving Rhames, Stanley Tucci, Philip Baker Hall, Anthony Heald, Angel David, John Costelloe, Lindsay J. Wrinn, Megan L. Wrinn, Katie Sagona
A loose remake of the 1955 film about a small-time thief whose efforts to go straight are stymied by a psychotic associate, with the result being that he is ultimately forced into a collaboration with the cops. This undistinguished effort lacks both the realism

of the original and the care with which it was made. Cage's usual over-the-top performance is ill suited to the film and is a major drawback.
THR   96 min (ort 101 min) cC   VIDrel: 20TH/FOXVID V/sur

KISS OF THE SPIDER WOMAN ***   15
Hector Babenco BRAZIL/USA   1985
William Hurt, Raul Julia, Milton Goncalves, Sonia Braga, Jose Lewgoy, Nuno Leal Maia, Antonio Petrim, Denise Dummont, Miriam Pires, Fernando Torres, Patricio Bisso, Herson Captri, Nildo Parente, Antonio Petrin, Wilson Grey
A much-acclaimed film about two prisoners who share a cell in a Latin American prison. One is a revolutionary, the other a homosexual and a strange multi-faceted relationship develops between them, as the latter entertains the former with his memories of trashy Hollywood movies. The engrossing script is by Leonard Schrader, but this is really Hurt's film, and he gives a towering performance. AA: Actor (Hurt).
Aka: BEIJO DA A MUHER ARANNHA
DRA   116 min (ort 120 min) Col/B/W
VIDrel: ELPIC/POLYREC   V
Boa: novel by Manuel Puig.

KISS AND BE KILLED **   18
Tom E. Milo USA   1991
Crystal Carson, Tom Reilly, Jimmy Baio, Ken Norton, Caroline Ludvik, William Smith, Chip Hall
A woman seeks vengeance when a knife-wielding Vietnam veteran nutter murders her husband on their wedding night.
DRA   82 min (ort 89 min)   VIDrel: CIC/SONOP   V

KISS SHOT **   PG
Jerry London USA   1989
Whoopi Goldberg, Dennis Franz, Tasha Scott, Dorian Harewood, David Marciano, Teddy Wilson, Adilah Barnes, Charles Branklyn, Phyliss Coates, Bob Ernst, Richard Dupell, Eva Gholson, Michael Halton, Chuck L. Hibert, Kathryn Keats
A single mother faces eviction from her home after losing her job, unless she can find several thousand dollars to pay off her mortgage. When her estranged father refuses her a loan, she resorts to becoming a pool hustler. She achieves some measure of success but her plans nearly come unstuck when she falls for one of her opponents. A conventional and remarkably lightweight romantic comedy for Goldberg in her TV-movie debut.
COM   93 min (ort 96 min) mTV   VIDrel: VCC/DISC L/A V/s

KISSING PLACE, THE **   15
Tony Wharmby USA   1989
Meredith Baxter-Birney, David Ogden Stiers, Nathaniel Moreau, Victoria Snow, Michael Kirby, Ted Dykstra, Chris Benson, George Buza, Patricia Caroll Brown, Nikki De Boer, Eugene Clark, Tom Douglas, Martin Karon, Justin Louis
A ten-year-old boy discovers his parents' dark secret and the reason why the family is never able to stay in one place for long. An interesting if flawed drama whose restrained direction is singularly at odds with Baxter-Birney's overwrought performance.
DRA   84 min (ort 96 min)   VIDrel: CIC/SONOP L/A   V

KITCHEN TOTO, THE ***   15
Harry Hook UK   1987
Bob Peck, Phyllis Logan, Edwin Mahinda, Robert Urquhart, Kirsten Hughes, Edward Judd, Nicholas Charles, Nathan Dambuza Mdledle, Ann Wanjugu, Job Seda, Ronald Pirie
The powerful story of a young Kikuyu boy sent to work for the British police chief, in the White Highlands of Kenya in 1950, the year when the support for the Mau-Mau rebels was rapidly growing among the Kikuyu tribe. Scripted by Hook in his directorial debut. The film was a prizewinner at the Tokyo Film Festival.
DRA   92 min (ort 95 min)   VIDrel: MGM/WHV L/A   V

KLUTE ***   18
Alan J. Pakula USA   1971
Donald Sutherland, Jane Fonda, Roy Scheider, Charles Cioffi, Dorothy Tristan, Rita Gam, Jean Stapleton, Vivian Nathan, Nathan George, Morris Strassberg, Barry Snider, Anthony Holland, Richard Shull, Betty Murray
An ex-cop investigating a disappearance, becomes involved with a call-girl in this excellent drama, which offers plenty of

suspense despite its downbeat and low-key approach. AA: Actress (Fonda).
THR   108 min (ort 114 min)   VIDrel: WHV   V
Boa: novel by William Johnston.

KNIFE EDGE **   15
Kurt Voss USA   1991
Brad Dourif, Sammi Davis, Max Perlich, Vic Tayback, M.K. Harris
A young couple who pose as a brother and sister plan the seduction of a stranger in this adequate psychological thriller. See also EVERY BREATH for a film using a similar idea.
THR   85 min   VIDrel: HIFLI/SONOP   V/h

KNIFE IN THE WATER ***   PG
Roman Polanski POLAND   1962
Leon Niemczyk, Jolanta Umelka, Zygmunt Malanowicz
A tense and gripping tale of a young couple who pick up a hitchhiker and invite him onto their yacht, where a strange game begins. The director's only feature made in his native country, it demonstrates a sure hand in a sparse and compelling style. Malanowicz's voice was dubbed over with Polanski's.
Aka: NOZ W WODZIE
DRA   94 min (ort 98 min) B/W   VIDrel: CONNO/RTM;
PION (LV only)   V LV

KNIGHT MOVES **   18
Carl Schenkel GERMANY/USA   1992
Christopher Lambert, Diane Lane, Tom Skeritt, Daniel Baldwin, Katherine Isobel, Sam Malkin, Codie Luca Wilbee, Josh Murray, Franck C. Turner, Megan Leitch, Don Thompson, Alex Diakun, Ferdinand Mayne, Mark Wilson, Blu Mankuma
A series of grisly murders occur during a chess tournament and the evidence seems to point to one of the participating champions. However, the killer taunts him by sending cryptic messages that he has to decipher for himself. A dark, depressing and quite violent thriller with few surprises. See also INNOCENT MOVES for another tale built around chess, albeit one that is considerably lighter in tone.
THR   111 min (ort 116 min)   VIDrel: VCC/DISC/COLUM V/sur

KNIGHT RIDER: THE ORIGINAL TV MOVIE **   PG
Daniel Haller USA   1982
David Hasselhoff, Edward Mulhare, Phyllis Davis, Richard Anderson, Vince Edwards, Pamela Susan Shoop, Lance LeGault, Noel Conlon, Michael D. Roberts, Bert Rosario, Richard Basehart, Edmund Gilbert, Shawn Southwick
A cop injured in a shoot-out is rebuilt with the aid of plastic surgery, and becomes a one-man police force by being given a radio link with his partner, an indestructible computer-controlled car. Pilot for a TV series. In this episode he hunts down a group of industrial saboteurs.
A/AD   91 min mTV   VIDrel: CIC/SONOP L/A   V/h

KNIGHT RIDER 2: NIGHT OF THE JUGGERNAUT **   PG
Georg Fenady USA   1985
David Hasselhoff, Edward Mulhare, Patricia McPhenan
Pilot for the second "Knight Rider" TV series that started in 1985 (the first one ran from 1982). As before, episodes are built around the exploits of a cop who has been equipped with an "intelligent" patrol car that has its own built-in computer. Average.
A/AD   91 min mTV   VIDrel: CIC/SONOP L/A   V

KNIGHT RIDER 2000 **   PG
Alan J. Levi USA   1991
David Hasselhoff, Edward Mulhare, Susan Norman, James Doohan, Eugene Clark, Carmen Argenziano, Megan Butler, Mitch Plleggi, Christine Healy, Lou Beatty Jr, Francis Hulnan, John Cannon Nichols, Chris Bonno, Phillip Hafer
Set in the year 2000, this action tale sees our rebuilt cop and his computer-controlled car being brought out of retirement and sent back into action, the mission this time being to clean up the city and destroy a web of corruption. A further spin-off of the popular TV series "Knight Rider".
A/AD   90 min mTV   VIDrel: CIC/SONOP   V/h

KNIGHT RIDER 2010 *   PG
Sam Pilsbury USA   1997
Richard Joseph Paul, Heidi Leick, Michael Beach, Don McManus, Nicky Katt, Badja Djola, Mark Pellegrino, Una Damon, Kimberly

Moris, Brion James, Jim Cody Williams, Betty Matwick, Joseph Redondo, Wanda Dittman, Ramon Chavez
In title year, a man makes a dangerous living smuggling illegal immigrants into the USA via the desert areas which have reverted to a form of barbarism. His exploits interest the head of a powerful corporation who has him bailed out of jail ostensibly for the purposes of producing a cyberspace game but his real intentions are far more sinister. A low-grade futuristic actioner of little imagination and poor acting that seems like a pilot for a prospective series.
FAN   87 min mTV   VIDrel: CIC   V

KNIGHT WITHOUT ARMOUR *** U
Jacques Feyder UK   1937
Robert Donat, Marlene Dietrich, Irene Vanbrugh, John Clements, Austin Trevor, Herbert Lomas, Miles Malleson, Basil Gill, David Tree, Lawrence Hanray, Hay Petrie, Lyn Harding, Frederick Culley, Lisa D'Esterre
A British agent is sent to Russia where he is captured and sent to Siberia, but after the 1917 revolution he is set free and appointed as commissar for a small town. Ordered to escort a beautiful countess to Petrograd for trial, he chooses to flee with her to the West. A fine adventure film with just the right blend of romance and action. The score is by Miklos Rozsa.
DRA   103 min (ort 108 min) B/W   VIDrel: CARL/TECH   V
Boa: novel Without Armour by James Hilton.

KNIGHTS ** 15
Albert Pyun USA   1992
Kris Kristofferson, Lance Henriksen, Kathy Long, Scott Paulin, Gary Daniels, Nicholas Guest, Vince Klyn, Ben McCrery, Bobby Borwn, Jon Epstein, Edmond Tyler Wrenn, Burton Richardson, Nancy Thurston, Brad Langeberg, Borovnisa Blervaque
In a post-WW3 future, the Earth is dominated by mutant cyborgs who use human blood as a fuel and enforce their rule through the use of a human army. A woman whose parents died in a cyborg raid lives only for revenge and gets her chance when she encounters a cyborg that has been programmed to hunt down and terminate the mutants. A highly derivative effort that suffers badly from an unimaginative plot and indifferent acting. The unresolved climax hints at a sequel to come.
FAN   90 min   VIDrel: 20TH/TECH   V/sur

KNIGHTS AND EMERALDS ** PG
Ian Emes UK   1986
Christopher Wild, Warren Mitchell, Rachel Davies, Tony Milner, Tracie Bennett, Beverly Hills, Bill Leadbitter, Nadim Sawalha, Patrick Field, Maurice Dee, David Keys, Andrew Goodman, Annette Badland, Rodney Litchfield
A talented drummer in a Birmingham marching band has to contend with both a racist father and a cash shortage, the latter forcing a merger with an all-female band. After their defeat in the first round of a contest, he joins an all-black group, an action that sparks off a major family crisis. An earnest if stilted and clumsy attempt to say something of social significance within the context of a comedy.
COM   87 min (ort 94 min)   VIDrel: MGM/WHV L/A V/sh

KNOWLEDGE, THE * PG
Bob Brooks UK   1979
Nigel Hawthorne, Michael Elphick, Maureen Lipman, Mick Ford, Kim Taylforth, Jonathan Lynn, Jonathan Ryall, Lesley Joseph, June Watson, Philippa Howell, Gary Holton, Natalie Ann King, Nigel Humphreys, Tim Stern, James Duggan, Hugh Quarshi
Stupid and tiresome attempt to compress the ordeal London cabbies have to go through in order to gain their badge, into a lightweight and empty film. The script is by Jack Rosenthal.
COM   90 min (ort 100 min) mTV   VIDrel: THAMES/DISC   V

KOJAK: THE BELARUS FILE ** PG
Robert Markowitz USA   1985
Telly Savalas, Suzanne Pleshette, Max Von Sydow, Herbert Berghof, Betsy Aidem, Alan Rosenberg, Charles Brown, George Savalas, David Leary, Harry Davis, Rita Karin, Mark B. Russell, Vince Conti, Margaret Thomson
A fair pilot for a new TV series about a canny bald-headed cop. Russians who survived the concentration camps are being murdered forty years on, as our detective faces one of his most baffling cases.
Aka: BELARUS FILE, THE
DRA   93 min (ort 100 min) mTV   VIDrel: L/A   V
Boa: novel The Belarus Secret by John Loftus.

KOJAK: THE PRICE OF JUSTICE ** 15
Alan Metzger USA   1987
Telly Savalas, Kate Nelligan, Jack Thompson, Pat Hingle, Brian Murray, Tony di Benedetto, John Bedford-Lloyd, Jeffrey DeMunn
Kojak is assigned to a case involving the murder of two small boys who were found dumped in the Harlem river. Their mother is accused of the murder, but a sudden break in this tortuous case seems to point in the direction of her magnate lover. A further episode in this new series of tales that were based (somewhat unwisely) on original stories rather than specially written screenplays. Average.
Aka: PRICE OF JUSTICE, THE
THR   92 min (ort 100 min) mTV   VIDrel: L/A   V
Boa: novel The Investigation by Dorothy Uhnak.

KOLYA **** 12
Jan Sverak CZECH REPUBLIC/FRANCE/UK   1996
Zdenek Sverak, Andrej Chalimon, Libuse Safrankova, Ondrej Vetchy, Stella Zazvorkova, Ladislav Smoljak, Irina Livanova, Sylvia Suvadova, Lilian Malkina, Karel Hermanek, Petra Spalkova
Set in Prague in 1988, just before the Russians pulled out of Czechoslovakia, where a middle-aged musician gets saddled with a Russian woman's small son after he goes through a marriage of convenience with her. The growing bond that develops between man and child is examined with clarity, wit and honesty, in a film that may be light on plotting, but is utterly charming for all that. AA: Foreign.
DRA   105 min   CINrel

KORCZAK **** (18)
Andrzej Wajda FRANCE/POLAND/UK/
WEST GERMANY   1990
Wojtek Pszoniak, Ewa Dalkowska, Teresa Budzisz-Krzyzanowska, Jan Peszak, Marzena Trybala, Piotr Kozlowski, Zbigniew Zamachowski, Aleksander Bardini, Marek Bargielowski, Maria Chwalibog, Andrzej Kopiczynski, Michal Staszczak
Written Agnieszka Holland, this is a stark, simple and totally absorbing tale of a doctor who worked at a Jewish orphanage in Warsaw. After the German invasion of 1939, he stayed with his charges and even accompanied them to the Treblinka death camp where they all perished. Intensely moving, it doubtless inspired SCHINDLER'S LIST, with which it invites comparison. The unusual double ending will not be to everyone's taste and weakens the film's overall impact.
DRA   118 min B/W   CINrel

KOTCH *** PG
Jack Lemmon USA   1971
Walter Matthau, Deborah Winters, Felicia Farr, Charles Aidman, Ellen Greer, Darrell Larson, Paul Picerni, Lucy Saroyan, Jane Connell, Jesica Rains, Dean Kowalski, James E. Brodhead, Lawrence Linville, Donald Kowalski, Biff Elliot
A lovable old widower refuses to be put into a home, moves away, buys his own place and takes in a young unmarried mother. Matthau gives an excellent characterisation of a man who refuses to be deprived of his dignity and sense of worth in a highly original comedy-drama. Lemmon's directorial debut.
DRA   93 min (ort 114 min)   VIDrel: VCC/DISC/COLUM V
Boa: novel by Katherine Topkins.

KRAMER VERSUS KRAMER *** PG
Robert Benton USA   1979
Dustin Hoffman, Meryl Streep, Justin Henry, Jane Alexander, Howard Duff, George Coe, JoBeth Williams, Bill Moor, Howland Chamberlain, Jack Ramage, Jess Osuna, Nicholas Hormann, Ellen Parker, Shelby Brammer, Carol Nadell
An adman's wife gets a divorce and walks out leaving her ex-husband to take care of their son. Glossy and skilfully made production that's quite moving in parts, though a red-eyed Streep doing her best to cry her eyes out is an irritation. AA: Pic, Dir, Actor (Hoffman), S. Actress (Streep), Screen/adapt (Robert Benton).
DRA   100 min (ort 105 min)   VIDrel: VCC/DISC/COLUM V
Boa: novel by Avery Corman.

KRAYS, THE *** 18
Peter Medak UK   1989
Gary Kemp, Martin Kemp, Billie Whitelaw, Susan Fleetwood, Charlotte Cornwell, Alfred Lynch, Steven Berkoff, Tom Bell, Kate Hardie, Patti Love, Barbara Ferris, Victor Spinetti, John McEnery, Philip Blommfield, Roger Monk

A stylised portrayal of two of Britain's most notorious and vicious gangsters, that neglects many aspects of their early life and subsequent career in favour of a one-sided concentration on their relationships with women, most notably their mother. The Kemp brothers (of pop group Spandau Ballet) are excellent, but the script fails to probe. The allegation that the film rights cost ú255,000 caused an understandable furore at the time.
DRA   115 min (ort 119 min)   VIDrel: ETL/POLYREC
V/sur

KREUTZER SONATA, THE ***   (18)
Mikhail Schweitzer USSR   1987
Oleg Yankovsky, Alexander Trofimov, Irinia Selnyova, Dmitri Pokrovsky
Two travellers strike up a conversation and one of them gradually reveals how his irrational jealousy eventually led to him murdering his wife. An overlong and excessively reverent adaptation of a novel that is ill-suited to the demands of the cinema, yet for all these failings, Yankovsky's performance as the husband is very powerful that one remains involved in the proceedings.
Aka: KREITZEROVA SONATA
DRA   135 min   TVrel: C4
Boa: novel by Leo Tolstoy

KRULL **   PG
Peter Yates UK   1983
Ken Marshall, Lysette Anthony, Francesca Annis, Freddie Jones, Liam Neeson, Alun Armstrong, David Battley, Bernard Bresslaw, John Welsh, Graham McGrath, Tony Church, Bernard Archard, Belinda Mayne, Dicken Ashworth, Todd Carty
Sword-and-sorcery SF epic with the usual theme of good versus evil, as seen in a perilous quest to rescue a beautiful maiden from captivity, at the hands of an evil creature. An abundance of special effects and elaborate settings are hampered by the slow development of the story.
FAN   116 min (ort 121 min)   VIDrel: VCC L/A   V/sh

KUFFS *   15
Bruce A. Evans USA   1991
Christian Slater, Milla Jovovich, Tony Goldwyn, Bruce Boxleitner, Troy Evans, George De La Pena, Leon Rippy, Ric Waugh, Steve Holladay, Clarke Coleman, Chad Randall, Craig Benton, Ashley Judd, Joshua Cadman, Aki Aleong, Gary Munch
The basic premise of this film is that the San Francisco police force operates a bizarre "franchise" system, with the city divided up into areas that each have their own "patrol specials", who operate alongside the regular police. Kuffs is a high school drop-out who overnight takes on the enormous responsibility of police chief, when his brother is gunned down by villains. Much mayhem results in a daft comedy-thriller that defies belief.
THR   97 min (ort 102 min)   VIDrel: EIV/SONOP   V/sur

KUNG FU EXECUTIONERS **   15
Paul D. Robinson
Gordon Mitchell, Shirley Corrigan, Bruce Liang
A Mafia big-shot who runs an international drug smuggling ring has successfully eliminated the Interpol agents sent to destroy his operation, but in Hong Kong comes up against a superb martial artist who frustrates his plans. Having lured him to Rome with the promise of a film contract, the scene is set for a final showdown.
MAR   93 min   VIDrel: IMPENT   V

KUNG FU GENIUS **   15
Wilson Tong HONG KONG
Cliff Lok, Wilson Tong, Siu Hau, Allan Eu, Annie Liu
The title fighter is the only one courageous enough to stand up against a bunch of thugs who are trying to take over his hometown. Fortunately, having studied the martial arts diligently, he has evolved his own powerful style.
MAR   84 min   VIDrel: EAST/DISC   V

KUNG FU KIDS **   15
Yu Chih Ping HONG KONG   1984
Wang Ye Lung, Ma Chang, Weng Shao Fu, Ou Ti, Ter Re Zee
An orphan and two petty thieves survive as best they can by living on their wits and giving displays of kung fu. However, they clash with a murderous criminal overlord who controls all the local rackets, and a difference of opinion leads to a vigorous exchange of ideas.
MAR   84 min   VIDrel: ONE/IMPENT   V

KUNG FU MASTER *   PG
HONG KONG   1979
Chang Wu Lang, Yuen Siu Ting, Kao King
A man tricked into smuggling diamonds has to fight to the finish.
Aka: KUNG FU MASTER NAMED DRUNK CAT
MAR   84 min Cut (1 sec)   VIDrel: IMPENT   V

KUNG FU THE HEAD CRUSHER **   (18)
Chiang Hung HONG KONG   1970
Chen Hsing, Henry Yu Young, Chang Lie, Linda Ling, Charlie Chiang
Two underground policemen infiltrate a criminal gang and find themselves in deadly peril, in this typically violent martial arts actioner. The title is derived from a deadly hold the hero is able to administer to those unwise enough to oppose him. A well staged film, but one without much originality. At present the movie is only available on a tape entitled "Eastern Heroes - The Video Magazine: Vol 2".
Aka: TOUGH GUY
MAR   90 min   VIDrel: EAST/DISC   V

KUNG FU WARRIOR **   18
HONG KONG
Chang Lei, Kuan Hai Shan
A young kung fu student encounters unforeseen setbacks, when he moves to Hong Kong to find work and perfect his martial skills.
MAR   82 min (ort 91 min)   VIDrel: ONE/IMPENT   V

KUNG FU ZOMBIE **   18
HONG KONG   1980
Billy Chong, Ching Tao, Cheng Kay Ying, Kwon Young Moon
A meddlesome priest causes a man's enemies to come back from beyond the grave, in yet another martial arts film with a supernatural slant. The film "Kung Fu From Beyond The Grave" also dealt with this idea.
MAR   90 min wScrn dubbed   VIDrel: EAST/DISC   V

KURONEKO ***   15
Kaneto Shindo JAPAN   1968
Kichiemon Nakamura, Nobuko Otowa, Kiwako Taichi, Kei Sato, Hideo Kanze, Rokko Toura, Taiji Tonoyama
Two women raped and murdered by a samurai, return to terrorise the area in this fantasy set in medieval Japan. A celebrated warrior is dispatched to deal with their ravages, and discovers to his horror that they are the avenging spirits of his wife and mother. Scenes of imaginative power alternate with ones of clumsy horror, in a remarkable film that in many ways gives the impression of having been made by two directors.
Aka: BLACK CAT, THE; YABU NO NAKA KURONEKO
HOR   95 min (ort 99 min) B/W wScrn   VIDrel: TART/20TH
V

KWAIDAN ***   (15)
Masaki Kobayashi JAPAN   1964
Rentaro Mikuni, Ganemon Nakamura, Katso Nakamura, Michiyo Aratama, Keiko Kishi, Tatsuya Nakadai, Takashi Shimura, Misako Watanabe, Yoichi Hayashi, Tetsuo Tamba, Noboru Nakaya, Ganjiro Nakamura
Four tales of the supernatural involving samurai, monks, spirits etc. that are all put together with great skill but are best enjoyed in two sittings. The superb use of wide-screen format will be lost on TV, but fortunately the imaginative use of colour and setting can still be enjoyed.
Aka: GHOST STORIES; WEIRD TALES
FAN   154 min (ort 164 min) wScrn   VIDrel: TART/20TH   V
Boa: short stories by Lafcadio Hearn.

L

L-SHAPED ROOM, THE **   15
Bryan Forbes UK   1962
Leslie Caron, Brock Peters, Tom Bell, Patricia Phoenix, Cicely Courtnidge, Emlyn Williams, Avis Bunnage, Nanette Newman, Harry Locke, Ellen Dryden, Jenny White, Anthony Booth, Gerry Duggan, Joan Ingram, Mark Eden, Gerald Sim
A London suburban lodging house is inhabited by a variety of characters including an author and a pregnant French girl with whom he falls in love. A mild little drama with watchable performances but nothing of substance.
DRA   120 min (ort 142 min) B/W   VIDrel: WHV   V/h
Boa: novel by Lynne Reid Banks.

L.627 ***
15

Betrand Tavernier FRANCE
1991

Didier Bezace, Jean-Paul Comart. Charlotte Kady, Jean-Roger Miilo, Philippe Torreton, Nils Tavernier, Lara Guirao, Cecile Garcia-Fogel, Claude Brosset, Fabrice Roux, Jean-Luc Abel, Martial, Jacques Pratoussy, Didier Castello

A look at the work of the Paris Drugs Squad that adopts a low-key approach, with the story focusing on a decent cop who is transferred to the squad. Seeing at first hand the brutalising nature of the job, it soon becomes apparent just how hopeless the nature of the task is. Overlong and lacking pace, it remains consistently absorbing thanks to its eye for detail and an outstanding cast. The title refers to the legislation used to take drug offenders into custody.

Aka: LAW 627

DRA   140 min wScrn   VIDrel: ARTIF/20TH   V/s

LA AMIGA ***
(PG)

Jeanine Meerapfel ARGENTINA/WEST GERMANY   1988

Liv Ullmann, Cipe Lincovsky, Federico Luppi, Victor Laplace, Harry Baer, Lito Cruz, Greger Hansen, Nicolas Frei, Cristina Murta, Amancay Espindola, Chela Cardala, Gonzalo Arguimbau, Fernan Miras, Victoria Solarz, Maria Carla Bustos

Story built around the mothers of the Plaza de Mayo who wait forlornly for their sons and husbands, who were recruited into the Argentine army but never returned. The film opens with two girls of growing up in Buenos Aires in the 1940s and 1950s, following very different lives and yet meeting again in adulthood, when one has become a leader of the "mothers" after her eldest son's abduction. A moving tale that hints at themes it has little scope to examine.

Aka: GIRLFRIEND, THE

DRA   108 min   CINrel

LA BALANCE ****
18

Bob Swain FRANCE
1982

Nathalie Baye, Philippe Leotard, Richard Berry, Maurice Ronet, Christophe Maavoy, Jean-Paul Connart, Geoffrey Carey, Michel Amphoux, Gerard Beaune, Pierre-Marie Escourou, Claude Villiers, Mostefa Zerguine, Patrick Guillaumes

A one-time Parisian criminal is forced by an unscrupulous cop, to inveigle a former associate into a robbery in order to betray him to the police, who are desperate to catch him. A tough and compelling thriller that won several awards in its homeland, it remains one of country's best examples of a film of this genre.

THR   98 min (ort 104 min)   VIDrel: ARROW/RTM   V/s

LA BAMBA ***
15

Luis Valdez USA
1986

Lou Diamond Phillips, Esai Morales, Rosana De Soto, Elizabeth Pena, Joe Pantoliano, Danielle Von Zerneck, Rick Dees, Marshall Crenshaw, Brian Setzer

Biopic on the tragically brief career of Ritchie Valens, who achieved almost overnight fame but died in the same plane crash that killed Buddy Holly. The excellence of the lead performance and the music, help make up for the occasional cliched treatment. The music is performed on the soundtrack by Los Lobos (who appear as the Tijuana Band) and the score was written by Carlos Santana and Miles Goodman.

DRA   104 min   VIDrel: VCC/COLUM L/A   V/sh

LA BELLE NOISEUSE ***
PG/15

Jacques Rivette FRANCE
1991

Michel Piccoli, Jane Birkin, Emmanuelle Beart, David Bursztein, Marianne Denicourt, Gilles Arbona, Bernard Dufour

An ageing painter who has produced nothing of importance since he scrapped a painting of his wife ten years ago, is inspired to start again by a beautiful young woman he takes on as a model. As he makes preparations to undertake work on a possible "masterpiece", various tensions and petty jealousies are revealed in those around him. Very loosely based on the Balzac story, this long (a shorter version is available) difficult film will not be to all tastes.

Aka: DIVERTIMENTO; LA BELLE NOISEUSE: DIVERTIMENTO (shorter PG version)

DRA   125 min (PG version); 228 min (2 cassettes – ort 239 min)   VIDrel: ARTIF/20TH   V/h

Boa: short story Le Chef D'Oeuvre Inconnu (The Unknown Masterpiece) by Honore de Balzac.

LA BETE HUMAINE ***
PG

Jean Renoir FRANCE
1938

Jean Gabin, Simone Simon, Julien Carette, Fernard Ledoux, Jean Renoir, Jenny Helia, Gerard Landy, Blanchette Brunoy, Colette Regis, Jacques Berlioz, Leon Larive, Georges Spanelly, Emile Genevois, Jacques B. Brunius, Marcel Perez

A train engineer is the son of a drunkard and inherits his father's unstable temperament, but is able to restrain his psychopathic urges until he falls madly in love with a station master's wife and gets involved with her murder plans. A finely observed and beautifully made classic French melodrama.

Aka: HUMAN BEAST, THE; JUDAS WAS A WOMAN

DRA   98 min B/W   VIDrel: ELPIC/POLYREC   V

Boa: novel by Emile Zola.

LA CAGE AUX FOLLES ***
15

Edouardo Molinaro FRANCE/ITALY
1978

Ugo Tognazzi, Michel Serrault, Michel Galabru, Claire Maurier, Remy Laurent, Benny Luke, Carmen Scarpitta, Luisa Manieri

Two ageing homosexual lovers, one a nightclub owner and the other his star "lady", pretend to be a normal married couple for the sake of the son of the former, who wishes to marry a respectable girl. An amusing adaptation of a French stage farce that soon runs out of steam, being at its best during the opening nightclub sequences. Two sequels and a hit Broadway musical followed, as well as an Americanised version (THE BIRDCAGE) in 1996.

Aka: BIRDS OF A FEATHER

COM   87 min (ort 93 min)   VIDrel: MGM/WHV   V

Boa: play by Jean Poiret.

LA CAGE AUX FOLLES 3 **
15

Georges Lautner FRANCE/ITALY
1985

Michel Serrault, Ugo Tognazzi, Michel Galabru, Benny Luke, Stephane Audran, Antonella Interlenghi, Saverio Vallone, Gianluca Favilla, Umberto Raho, Pier Francesco Aiello, Flora Carabella Mastroianni, Piero Di Carlo, Reanto De Montis

The third attempt to extract some laughs from the rather tired theme of a homosexual couple, one of whom is a transvestite performer at the nightclub owned by the other. Here, the "female" half is in line for a vast inheritance if he marries and has a child within eighteen months. An unfunny second sequel.

Aka: LA CAGE AUX FOLLES 3: ELLES SE MARIENT; LA CAGE AUX FOLLES 3: THE WEDDING

COM   87 min   VIDrel: RCA L/A   V

L.A. CONNECTION ***
18

Philip Ko HONG KONG
1989

Brent Gilbert, Mike Abbott

A special agent is ordered to stop a pair of drug runners operating a route from Hong Kong to L.A. However, this involves the agent in taking on the drugs producer, a vicious killer known as the "Black Dragon". A tough, fast-paced martial arts police adventure.

A/AD   85 min   VIDrel: IMPENT   V

L.A. CRIMEWAVE *
15

Michael Mann USA
1989

Scott Plank, Michael Rooker, Daniel Baldwin, Xander Berkeley, Ely Pouget, Ed De Fusco, R.D. Call, Clarence Gilyard Jr, Donals Grant, Laura Harrington, Mimi Lieber, Gil Parra, Alex McArthur, Richard Chaves, Victor Rivero, John Santucci

A disappointingly conventional crime drama built around the three-way relationship of a dedicated cop who heads L.A.'s robbery and murder unit, the criminal mastermind who is behind a series of murders and burglaries and the girlfriend of the former. The film fails to blossom, mostly due to poor scripting and acting. Devised by Mann, who has enjoyed considerably more success with an earlier creation, the TV series "Starsky And Hutch".

Aka: CRIMEWAVE; MADE IN L.A.

A/AD   92 min mTV   VIDrel: MOPIC/SGSVID   V

LA CRISE ***
15

Coline Serreau FRANCE
1992

Vincent Lindon, Patrick Timsit, Annik Alane, Valerie Alane, Gilles Privat, Nanou Garcia, Christian Benedetti, Didier Flamand, Zabou, Maria Pacome, Yves Robert, Michele Laroque, Clothilde Mollet, Nicolas Serreau, Yves Llobregat

A man who works as a successful corporate lawyer is made redundant and looks to family and friends for emotional support, but finds they are so involved with their own troubles they can spare no time to listen to his. Eventually, he does find a sympathetic ear, but only from a slow-witted misfit. A melancholy black farce of uncertain direction, sustained by a terrific central performance and a few sharp observations.

DRA   91 min (ort 95 min)   VIDrel: ELPIC/POLYREC   V

LA DOLCE VITA *** 18
Federico Fellini FRANCE/ITALY 1961
Marcello Mastroianni, Anouk Aimee, Claudia Cardinale, Anita Ekberg, Alain Cuny, Yvonne Furneaux, Magali Noel, Nadia Gray, Lex Barker, Annibale Ninchi, Walter Santesso, Jacques Sernas, Valeria Ciangottini, Alan Dijon, Harriet White
Rambling, overlong and episodic look at the decadent lifestyle of Roman high society, seen through the eyes of a gossip columnist who aspires to straight journalism, but is unable to break away from his debauched life. Some powerful scenes inevitably stick in the memory, but the self-indulgent world shown here inhibits any sympathy one might feel for the characters. AA: Cost (Piero Gherardi).
Aka: SWEET LIFE, THE
DRA 167 min (ort 174 min) B/W wScrn
VIDrel: ELPIC/POLYREC V

LA FILLE DE L'AIR *** 15
Maroun Bagdadi FRANCE 1992
Beatrice Dalle, Thierry Fortineau, Hippolyte Girardot, Roland Bertin, Jean-Claude Dreyfus, Jean-Paul Roussillon, Catherine Jacob, Liliane Rovere, Louise-Laure Mariani, Arnaud Chevrier, Elisabeth Macocco, Isabelle Chandelier
Inspired by novel rather than closely based on it, this tells the story of Nadine Vaujour who helped her husband break out of jail in 1986, after he was sentenced to thirty-six years for armed robbery, the law stating that as his crime was committed whilst he was on the run, his sentences on each count had to be served consecutively. A worthy account of a notorious event that gripped the public's imagination, it lambastes the French penal system for good measure.
DRA 102 min (ort 108 min) wScrn VIDrel: TART/20TH V/sur
Boa: autobiography of Nadine Vaujour.

LA FRONTERA *** 18
Ricardo Larrain CHILE/SPAIN 1991
Patricio Contreras, Gloria Laso, Alonso Venegas, Hector Noguera, Aldo Bernales, Sergio Schmied, Patricio Bunster, Elsa Poblete, Grisela Nunez, Anibel Reyna, Sergio Hernandez, Eugenio Morales, Sergio Madrid, Joaquin Velasco
Following his arrest during a political rally, a man is effectively exiled by the Chilean military to a remote village, where his initial loneliness is eventually tempered by his friendship with a local beauty. A funny and moving debut feature from Larrain, that is ostensibly about Pinochet's Chile, but in truth is as much about human relationships as it is about the contemporary politics of that country.
DRA 115 min wScrn VIDrel: TART/20TH V

LA GRANDE BOUFFE ** 18
Marco Ferreri FRANCE 1973
Michel Picoli, Ugo Tognazzi, Marcello Mastroianni, Philippe Noiret, Andrea Ferreol
Wearing with the burden of living, four male friends decide to end it all by means of an orgy in self-indulgence in sex and food and retire to a quiet villa to do just that. Some amusing performances add a great deal of interest, but this seriously overlong satire seems very uncertain of its target and eventually degenerates into an empty exercise in bad taste for its own sake.
Aka: BIG FEAST, THE; BLOW-OUT; GREAT FEED, THE
DRA 124 min VIDrel: ARTPRO/RTM V

LA GRANDE ILLUSION ***** U
Jean Renoir FRANCE 1937
Pierre Fresnay, Jean Gabin, Erich Von Stroheim, Marcel Dalio, Dita Parlo, Julien Carette, Gaston Modot, Jean Daste
During WW1, a couple of French Army officers from different backgrounds escape from the prison camp they are being held in, and go on the run, eventually finding shelter with a German widow. Based on a true story that was told to the director, this is one of the best anti-war films that has ever been made – sensitive, concise and profound, the only side it is on is the side of humanity, and that is the mark of its greatness.
Aka: GRAND ILLUSION
DRA 108 min (ort 117 min) B/W VIDrel: ARTPRO/RTM V

LA HAINE **** 15
Mathieu Kassovitz FRANCE 1995
Vincent Cassel, Hubert Kounde, Said Taghmaoui, Karim Blehadra, Edouard Montoute, Francois Levantal, Solo, Marc Duret, Heloise Rauth, Rywka Wajsbrot, Tadek Lokcinski, Choukri Gabteni, Nabil Ben Mhamed, Felicite Wouassi
A savage attack on slum life that centres on the world of the inhabitants of a run-down Paris housing estate and the ensuing race riots that take place there, all seen through the eyes of three friends. A sharp and immensely potent film, it is nowhere near as depressing as it might have been, thanks to its energy and caustic humour. The film won awards for Best Director (Cannes Film Festival 1995) and Best Editing (Cesar Awards 1996).
Aka: HATRED
THR 97 min B/W wScrn VIDrel: TART/20TH V/sh

LA JETEE ***** PG
Chris Marker FRANCE 1962
Quite unique in the history of the cinema is this experimental film, that employs narration and a succession of stills to create a photo-montage story of immense power. The world of the future is bleak and hopeless, most of mankind having been destroyed by WW3. One man's strong memory of a pretty woman he saw as a child at an airport provides the key for a trip back in time, and perhaps a chance for the survivors to escape their fate. See also TWELVE MONKEYS.
Aka: PIER, THE
FAN 26 min (ort 29 min) B/W VIDrel: ACAD/RTM V

L.A. LAW ** 15
Gregory Holbit USA 1986
Richard Dysart, Harry Hamlin, Corbin Bernsen, Jill Eikenberry, Jimmy Smits, Michael Tucker, Alan Rachins, Juanin Clay, Rob Knepper, Michele Greene, Shannon Wilcox, Susan Dey, Susan Ruttan
Pilot for the popular TV series about a busy L.A. legal firm. Slick and well acted with good characterisation but lacking in real bite as the storyline is painfully thin and contrived.
DRA 90 min (ort 104 min) mTV VIDrel: 20TH/TECH L/A V

LA NOTTE ** (15)
Michaelangelo Antonioni ITALY 1961
Marcello Mastroianni, Jeanne Moreau, Monica Vitti, Bernhard Wicki, Maria Pia Luizi, Rosy Mazzacurati, Guido A. Marsin, Gitt Magrini, Vincenzo Corbella, Giorgio Negro, Roberta Speroni, Ugo Fortunati, Vittorio Bertolini
Examination of twenty-our hours in the life of an alienated middle-class couple who are unable to communicate. After visiting a friend in hospital who is dying of cancer, they go to a party but quarrel on the way there, with the wife going off on her own. Having arrived at the party, the husband is then free to chase a young girl. A contrived and depressing work, with a heavy-handed and quite hackneyed message. Followed by THE ECLIPSE, the final film in the trilogy.
Aka: NIGHT, THE
DRA 122 min CINrel

LA PASSIONE * 15
John B. Hobbs UK 1996
Thomas Orange, Paul Shane, Sean Gallagher, Shirley Bassey, Jan Ravens, Carmen Silvera, Keith Barron, Benedick Blythe, Anna Pernicci, Belinda Stewart-Wilson, Ruth Harford, Gavin Abbott, Raymond Brody
In the north of England, the son of an Italian immigrant has a passion for racing cars, and dreams of becoming part of the Ferrari racing team. His work helping out in his father's cafe grows increasingly irksome, but he finds a way to escape when he realises that his dad's secret ice-cream vanilla flavouring would make a great after shave. A sugary and annoyingly banal confection of a movie, that plays like a misconceived pop video.
DRA 91 min CINrel

LA REGLE DU JEU **** PG
Jean Renoir FRANCE 1939
Marcel Dalio, Nora Gregor, Jean Renoir, Mila Parely, Gaston Modot, Julien Carette, Roland Toutain, Paulette Dubost, Odette Talazac, Claire Gerard, Anne Mayen, Lise Elina, Roland Toutain, Pierre Magnier, Eddy Debray, Pierre Nay
A wealthy land-owning couple stage a weekend party at their country home, during which time various sexual intrigues and tensions among servants and masters alike come to the surface. A complex, multi-layered and sharply observed picture of this type of society, often held to be a masterpiece of world cinema, but only restored to its original length as late as 1956. Sad, funny and inventive; it is one of the director's best works. Screenplay is by Renoir.
Aka: RULES OF THE GAME, THE
DRA 110 min (ort 113 min) B/W VIDrel: CONNO/RTM V

LA REINE MARGOT ★★★   18
Patrice Chereau FRANCE/GERMANY/ITALY   1993
Isabelle Adjani, Daniel Auteuil, Jean-Hugues Anglade, Vincent Perez,
Virna Lisi, Dominique Blanc, Pascal Greggory, Claudio Amendola,
Miguel Bose, Asia Argento, Julien Rassam, Thomas Kretschmann,
Jean-Claude Brialy, Albano Guaetta
The Catholic sister of King Charles IX is forced to marry the
Protestant Henri de Navarre in an effort to stem the rising tide
of sectarian bloodshed, but this marriage has little impact, and
the actions of the unbalanced king result in the 1572 St
Bartholemew's Day massacre of Protestants. Confusing, opulent
and brutal, the film is more a historic document than a piece of
entertainment, but its production values cannot be faulted.
DRA   137 min (GUILD/FOXVID); 158 min (ort 162 min)
wScrn   VIDrel: GUILD/FOXVID; GUILD/20TH   V/sur
Boa: novel by Alexandre Dumas.

LA RONDE ★★★   PG
Max Ophuls FRANCE   1950
Anton Walbrook, Simone Signoret, Serge Reggiani, Simone Simon,
Daniel Gelin, Danielle Darieux, Fernand Gravet, Jean Louis Barrault,
Odette Joyeux, Isa Miranda, Gerard Philipe, Robert Vattier
Assorted amoral goings-on in Vienna of 1900, built around the
device of a circular series of linked love affairs. Considered
greatly daring at the time, this amazingly stilted work retains
some charm, not least thanks to Walbrook's role as a cynical
narrator. The music is by Oscar Straus. Remade in 1964 by Roger
Vadim. See also LOVE CIRCLES, CHAIN OF DESIRE and
ECLIPSE.
Aka: MERRY-GO-ROUND, THE
DRA   89 min (ort 97 min) B/W   VIDrel: CONNO/RTM;
PION (LV only)   V LV
Boa: play Reigen (Merry Go Round) by Arthur Schnitzler.

LA ROUTE DE CORINTHE ★★★   12
Claude Chabrol FRANCE   1967
Jean Seberg, Maurice Ronet, Christian Marquand, Michel Bouquet,
Saro Urzi, Antonio Passalia, Claude Chabrol
When a security officer who works for NATO is murdered
whilst investigating the jamming of US radar installations in
Greece, the dead man's wife is framed for the murder and
kicked out of the country. But she gives her minders the slip and
sets out to uncover the real culprits. A stylish thriller of intelli-
gence and wit, that despite the subject matter is directed in a
fluid and surprisingly lighthearted manner. The terrific photography is the work of Jean Rabier.
Aka: ROAD TO CORINTH, THE
THR   86 min wScrn   VIDrel: ARTPRO/RTM   V

LA RUPTURE ★★★   18
Claude Chabrol BELGIUM/FRANCE/ITALY   1970
Stephane Audran, Michel Bouquet, Jean-Pierre Cassel, Annie Cordy,
Michel Duchaussoy, Marguerite Cassan, Jean-Claude Drouot, Mario
David, Catherine Rouvel, Domonique Zardi, Margo Lion, Jean Carmet
A man takes LSD and then attacks his wife and child but she
strikes back and injures him so badly that he has to be hospi-
talised. However, her in-laws use this incident as part of a
campaign to gain custody of their grandchild, in this potent
examination of family evil.
Aka: BREAKUP, THE
DRA   119 min (ort 124 min) wScrn   VIDrel: ARTPRO/RTM
V
Boa: novel The Balloon Man by Charlotte Armstrong.

LA SCORTA ★★★   15
Ricky Tognazzi ITALY   1994
Carlo Checchi, Claudio Amendola, Enrico Lo Verso, Tony Sperandeo,
Lorenza Indovina, Ugo Conti, Rita Sagnone, Francesca D'Aloja, Ricky
Memphis, Angelo Infanti, Leo Guillota, Giovanni Palalvicino,
Benedetto Raneli, Claudia Bonivento
A prosecuting attorney assigned to deal with Mafia activities in
Sicily is given four young policemen as his bodyguards, and
this close-knit group soon finds itself at the centre of a complex
conspiracy. A taut and well structured picture of the pervasive
nature of political corruption at every level of Italian govern-
ment.
THR   91 min wScrn   VIDrel: TART/20TH; ENCORE (LV
only)   V/sur LV

LA SEPARATION ★★★   PG
Christian Vincent FRANCE   1994
Isabelle Huppert, Daniel Auteuil, Jerome Deschamps, Karin Viard,
Laurence Lerel, Louis Vincent, Nina Morato, Estelle Larrivaz, Gerard

Jumel, Jean-Jacques Vanier, Frederic Gelard, Christian Benedetti,
Claudine Challier
A couple live together quite happily with their young son until
the day the woman tells her boyfriend she has fallen out of love
with him, a confession that is ultimately to have bitter consequences for them all. Not so much a film with a strong story as
an examination (and a very detailed one at that) of the death of
a relationship, and though the film has its flaws and cliches, it
is the enormous conviction of the actors one remembers best.
Screenplay is by Franck.
DRA   84 min (ort 88 min) wScrn   VIDrel: GUILD/20TH
V/sh
Boa: novel by Dan Franck.

L.A. STORY ★★   15
Mick Jackson USA   1991
Steve Martin, Victoria Tennant, Richard E. Grant, Marilu Henner,
Sarah Jessica Parker, Kevin Pollak, Sam McMurray, Patrick Stewart,
Iman, Susan Forristal, Andrew Amador, Gail Grate, Eddie DeHaro,
M.C. Shan, Larry Miller
A very weak and patchy portrait of the superficial life of a group
of rich L.A. residents, principally wacky TV weatherman
Martin, who finds true love in the shape of a visiting Times jour-
nalist (his wife Tennant giving a remarkably frosty and
unconvincing performance). In his clumsy attempts to woo her,
he finds an ally in a helpful road-traffic indicator. Despite
Martin's funny clowning, this unusually insipid effort lacks both
structure and depth.
COM   90 min (ort 98 min)
VIDrel: 4-FRONT/POLYREC/GUILD; PION (LV only)
V/sur LV

LA STRADA ★★★★   PG
Federico Fellini ITALY   1954
Anthony Quinn, Giulietta Masina, Richard Basehart, Aldo Silvana,
Marcella Marcella Rovere, Livia Venturini
A brutal travelling strongman in need of an assistant buys a
slow-witted young peasant girl from her destitute mother. He
treats her abysmally but she stays with him as they go on the
road, acting as his servant and eventually his mistress and learn-
ing routines for his act. A moving and haunting film, with
hypnotic theme music, it is poetic and realistic in equal measure.
AA: Foreign (the first in this category).
Aka: ROAD, THE
DRA   103 min (ort 107 min) B/W   VIDrel: CONNO/RTM
V

L.A. TAKEDOWN ★★★   15
Michael Mann USA   1989
Scott Plank, Michael Rocker, Alex McArthur, Vincent Gustaferro,
Ely Pouget
The director described this film as a kind of dress rehearsal for
HEAT (it uses the same script) that has a determined cop going
after an equally determined (and highly successful) bank robber.
A tough little film, it was shot in just over two weeks, has much
of the dialogue and incident of the latter movie, and if not its
sheer power, works very well indeed in its own right as a
simple, low-budget thriller.
THR   95 min mTV   VIDrel: MIA/DISC   V

LA TERRA TREMA ★★★   U
Luchino Visconti ITALY   1948
Inhabitants of the village of Aci Trezza in Sicily
A moving and lovingly made study of the harsh lives of Sicilian
fisherman, telling of their exploitation at the hands of ruthless
middlemen. Intended as the first part of a trilogy that was never
made. A commentary spoken by Visconti in standard Italian
was added after the film bombed, probably since the amateur
cast all use Sicilian dialect.
Aka: EARTH MOVED, THE; EARTH TREMBLES, THE; EARTH WILL
TREMBLE, THE; EPISODA DEL MARE
DRA   153 min (ort 161 min) B/W   VIDrel: CONNO/RTM
V
Boa: novel by Verga.

LA VIE DU CHATEAU ★★★   12
Jean-Paul Rappeneau FRANCE   1965
Catherine Deneuve, Philippe Noiret, Pierre Brasseur, Mary Marquet,
Henri Garcin, Carlos Thompson
Rappeneau's directorial debut is an easygoing and often charm-
ing tale of a bored housewife living in Normandy in WW2. A
mysterious stranger discovered at her chateau declares his love
for her, but only to protect his true identity as a Free French

agent on a secret mission. Various other complications follow in this contrived but enjoyable blend of drama and farce.
Aka: GOOD LIFE, THE; MATTER OF RESISTANCE
DRA   95 min B/W   VIDrel: ARROW/RTM   V

LA VIE EST BELLE ***
PG
Bernard Lamy/Ngangura Mweze
BELGIUM/FRANCE/ZAIRE 1987
Papa Wemba, Bibi Krubwa, Landu Nzunzimbu Matshia, Kanku Kasongo, Lokinda Meji Feza
An impoverished traditional musician whose music has fallen out of fashion comes to the sprawling city of Kinshasa in order to become a music star. After various diversions, he obtains employment and falls in love, but is dismayed to learn that his wealthy boss is also after his girl. A wonderfully exuberant and charming comedy of African urban life, full of music and a sheer joy to watch. Warmly recommended.
Aka: LIFE IS ROSY
COM   85 min   VIDrel: L/A   V

LABOR OF LOVE: THE ARLETTE SCHWEITZER STORY **
PG
Jerry London USA 1993
Ann Jillian, Tracey Gold, Diana Scarwid, Frances Sternhagen, Donal Logue, Bill Smitrovich, Robert Curtis-Brown, Andy Stahl, Helen Baldwin, Alan Sader, Joe Dorsey, Rob Treveiler, Grady Bowman, Wells Struble, Brett Kelly
A woman who is unable to have children finds a surrogate in the shape of her own mother, in this fact-based drama.
Aka: LABOUR OF LOVE
DRA   89 min   VIDrel: 20TH   V

LABYRINTH ***
U
Jim Henson USA 1986
David Bowie, Jennifer Connelly, Toby Freud, Shelley Thompson, Christopher Malcolm, Natalie Finland, Shari Weiser, Brian Henson, Ron Mueck, Rob Mills, Dave Goetz, David Barclay, David Shaughnessy, Karen Prell, Timothy Bateson
Elaborate fantasy tale of adventure as a young girl tries to rescue her baby stepbrother, who has been kidnapped by the king of the goblins. Enjoyable most of the time, though there are one or two sluggish moments. Written by Terry Jones.
FAN   97 min (ort 101 min)   VIDrel: POLY L/A   V/sh

LABYRINTH OF PASSION **
18
Pedro Almodovar SPAIN 1982
Antonio Banderas, Imanol Arias, Cecillia Roth, Helga Line, Angel Alcazar, Marta Fernandez-Muro, Agustin Almodovar, Ana Pegamoide, Eduardo Pegamoide, Nacho Pegamoide, Bernardo Bonezzi, Cristina S. Pascual, Fabio McNamara
Screwball comedy from this tiresome and much overrated director, whose flimsy plot revolves around young female rock singer who joins forces with a laundry-maid, in a bid to deal with their rather annoying parents. Full of vastly overblown characters and much frantic action but equally empty of any depth or real attempt at humour. Screenplay is by Almodovar, who reveals his shortcomings as a writer/director.
Aka: LABERINTO DE PASIONES
COM   94 min (ort 100 min) wScrn   VIDrel: TART/20TH   V

LACE *
12
William Hale USA 1984
Brooke Adams, Bess Armstrong, Phoebe Cates, Angela Lansbury, Anthony Higgins, Herbert Lom, Anthony Quayle, Honor Blackman, Arielle Dombrasle, Nickolas Grace, Leigh Lawson, Simon Chandler, Trevor Eve, Francois Guetary
An international sex symbol is determined to trace her mother, having decided to ruin her life because she abandoned her as a baby. A glossy but extremely boring and empty Hollywood production soap. A sequel followed all too soon.
DRA   116 min (2 cassettes – ort 240 min) mTV
VIDrel: WHV   V
Boa: novel by Shirley Conran.

LACE 2 *
15
William Hale USA 1985
Brooke Adams, Deborah Raffin, Arielle Dombrasle, Phoebe Cates, Anthony Higgins, Christopher Cazenove, James Read, Patrick Ryecart, Michael Gough, Francois Guetary, Michael Fitzpatrick, Walter Gotell, Paul Shelley
More of the same with Lili having to find the identity of her father in order to raise a ransom of $1,000,000 for her mother. A sequel to the earlier dross, not based on any novel this time.

Very much another celebration of empty glitz, and if possible, slightly worse than its predecessor.
DRA   180 min mTV   VIDrel: L/A   V
Boa: novel by Shirley Conran.

LACEMAKER, THE ****
15
Claude Goretta FRANCE/SWITZERLAND/WEST GERMANY 1977
Isabelle Huppert, Yves Beneyton, Florence Giorgetti, Anne Marie Duringer, Renata Schroefer, Jean Obe, Michael De Re, Monique Chaumette, Odile Poisson
The relationship between a young Parisian girl and a student she meets while on a holiday, fails because of differences of class and education. A rather beautiful little film, made with enormous care and attention to detail. The script is by Laine and Claude Goretta.
Aka: LA DENTELLIERE
DRA   102 min (ort 107 min)   VIDrel: ARROW/RTM   V
Boa: novel La Dentelliere by Pascal Laine.

LADDER OF SWORDS **
15
Norman Hull UK 1988
Martin Shaw, Juliet Stevenson, Eleanor David, Bob Peck
A second-rate circus act is stranded in a remote lay-by, awaiting a phone call that may result in an offer of work. When the bored wife of the troupe leader exposes herself to a passing motorist a bizarre chain of events unfolds. A very strange mixture of drama, intrigue and death, made more so by a few touches of black humour.
DRA   94 min   VIDrel: ODY/SONOP   V/sur

LADIES OF THE BOIS DE BOULOGNE ***
PG
Robert Bresson FRANCE 1944
Maria Casares, Paul Bernard, Elina Labourdette, Lucienne Boagaert
A woman abandoned by her lover engineers a spiteful revenge by placing in his path a fallen women who appears to be just her opposite, yet has carefully maintained an aura of respectability. Impressed by her beauty, charm and intelligence they marry, only to find their happiness and obvious affection for each other put at risk when the man's former lover reveals the truth. Stylish and poignant, the film's apt resolution is deeply touching.
Aka: LADIES OF THE PARK; LES DAMES DU BOIS DE BOLOGNE
DRA   82 min B/W   VIDrel: CONNO/RTM   V
Boa: story by Denis Diderot.

LADIES WHO DO **
U
C.M. Pennington-Richards UK 1964
Peggy Mount, Robert Morley, Harry H. Corbett, Miriam Karlin, Avril Elgar, Dandy Nichols, Jon Pertwee, Nigel Davenport, Graham Stark, Ron Moody, Cardew Robinson, John Laurie, Arthur Howard, Margaret Boyd, Joan Benham, Harry Fowler
A trio of cleaning-ladies salvage tips from the waste-bins of a property speculator, and form their own company with the assistance of a retired army officer. Mount, Karlin and Nichols are the three ladies in question, and it is thanks to their acting talents rather than any inherent merits in the script (by Michael Pertwee) that the film works as well as it does.
COM   81 min (ort 85 min) B/W   VIDrel: FABFIL/SPEAR   V

LADY AND THE HIGHWAYMAN, THE ***
PG
John Hough UK 1988
Emma Samms, Oliver Reed, Claire Bloom, Christopher Cazenove, Lysette Antony, Hugh Grant, Michael York, Gordon Jackson, Gareth Hunt, John Mills, Robert Morley, Bernard Miles, Stephanie Pitt, Floyd Bevan, Wayne Michaels
Standard British swashbuckler, set in the Civil War period and with all the stock characters. Despite being fairly predictable an excellent cast and good production values sustain it.
A/AD   89 min (ort 100 min) mTV
VIDrel: 4-FRONT/POLYREC   V
Boa: novel Cupid Rides Pillion by Barbara Cartland.

LADY AND THE TRAMP, THE ****
U
Hamilton Luske/Clyde Geronimi/Wilfred Jackson USA 1955
Voices of: Peggy Lee, Barbara Luddy, Larry Roberts, Stan Freberg, Verna Felton, Alan Reed, George Givot, Dallas McKennon, Lee Millar, Bill Baucon, Bill Thompson, The Mello Men
One of Disney's most delightful cartoons, in which a pedigree dog runs away from home after the arrival of a baby makes her feel unwanted, and meets up with a stray who lives by his wits. The two dogs survive various hazards and win through in the end, when they prove their worth by rescuing the baby. The

first Disney film in Cinemascope, and a skilful (if anthropomorphic) blend of comedy, drama and pathos. Songs are by Peggy Lee and Sonny Burke.
ANIM   77 min   VIDrel: WDV/TECH L/A   V/sh
Boa: short story Happy Dan, The Whistling Dog by Ward Greene.

LADY BLUE **   15
Gary Nelson USA   1985
Jamie Rose, Danny Aiello, Katy Jurado, Kate Mahoney, Babi Besch, Jim Brown, Tony LoBianco, Marco Rodriguez, Ajay Naidu, Ron Dean, Zaid Farid, Henry Godinez, Ricardo Guiterrez, John Mahoney, Steven Memel, Kathryn Jodsten
A hardboiled female cop uses unconventional methods to trap a murderer who wiped out an entire family. Pilot for a violent and short-lived TV series. A kind of feminist answer to DIRTY HARRY, with all the appropriate trimmings.
DRA   91 min (ort 104 min) mTV   VIDrel: MGM L/A   V

LADY BOSS *   15
Charles Jarrott USA   1992
Kim Delaney, Jack Scalia, Alan Rachins, Anthony John Dennison, Joan Rivers, David Selby, Phil Morris, Yvette Mimieux, Beth Toussaint, Vanity, Daniel Quinn, Scott Valentine, Robin Strasser, John Randolph, Joseph Cortese, Jeff Kaake
Sequel to the two Jackie Collins films LUCKY and CHANCES that continues the story of the daughter of a 1930s gangster, who has taken over her father's legitimate businesses and naturally (this being Joan Collins territory) has to deal with an endless succession of intrigues, betrayals and romantic encounters.
DRA   175 min   VIDrel: 4-FRONT/POLYREC/ODY   V/sh

LADY CAROLINE LAMB *   15
Robert Bolt ITALY/UK   1972
Sarah Miles, Jon Finch, Richard Chamberlain, Laurence Olivier, John Mills, Margaret Leighton, Ralph Richardson, Pamela Brown, Silvia Monti, Peter Bull, Charles Carson, Sonia Dresdel, Nicholas Field, Robert Harris, Felicity Gibson
The tempestuous affair between a young married aristocratic lady and the famous poet Lord Byron, makes for a remarkably dull film that follows in laborious detail the tortuous sequence of passion and humiliation that outraged contemporary society. The ridiculous script is by Bolt who also directed (if that is not too strong a word).
DRA   118 min (ort 123 min)   VIDrel: WHV   V

LADY CHATTERLEY **   18
Ken Russell UK   1993
Sean Bean, Joely Richardson, Shirley Ann Field
Fairly unremarkable adaptation of the Lawrence novel, from a director who has never really regained the exuberance and imagination of his earlier years. However, Bean is well cast as the gamekeeper Mellors, and brings to the role a vigour often lacking in adaptations of the story.
DRA   205 min (ort 230 min) mTV   VIDrel: CARL/TECH V/sur
Boa: novel Lady Chatterley's Lover by D.H. Lawrence.

LADY CHATTERLEY'S LOVER *   18
Just Jaeckin FRANCE/UK   1981
Sylvia Kristel, Nicholas Clay, Shane Briant, Ann Mitchell, Elizabeth Spriggs, Pascale Ridault, Anthony Head, Frank Morey, Peter Bennett, Bessie Love, Michael Huston, Mark Colleano, John Tynan, Fran Hunter, Ryan Michael
Soft-focus porno film with literary pretensions that takes a vastly over-rated novel and turns it into a mediocre romp, in which the wife of an impotent mine-owner embarks on a wild affair with a gamekeeper. Of little value except perhaps as a chance to admire Kristel's beauty if not her acting ability.
Aka: L'AMANT DE LADY CHATTERLEY
DRA   99 min (ort 105 min)   VIDrel: VCC/DISC/COLUM V
Boa: novel by D.H. Lawrence.

LADY DANCER **   18
Michael Paul Girard USA   1996
April Breneman, Kim Dawson, John McCafferty, Cara Vanlandingham
A couple specialise in performing erotic dances at private clubs, but their cosy set-up is threatened by the interest of a wealthy seductress.
DRA   85 min   VIDrel: THIRD   V

LADY DRAGON **   18
David Worth USA   1992
Cynthia Rothrock, Richard Norton, Robert Ginty, Bella Esperence, Henry Tornado
Martial arts actioner whose thin plot has Rothrock looking for revenge after her husband was callously gunned down on their wedding day by thugs in the pay of Norton. She goes undercover as a prostitute in order to get at her adversary, is raped and severely beaten, but makes a comeback after being nursed back to health. There are one or two decent combat sequences plus an entertaining climax, but little of the verve of Rothrock's earlier films.
A/AD   97 min   VIDrel: VCC/DISC   V

LADY FOR A DAY ****   U
Frank Capra USA   1933
May Robson, Guy Kibbee, Warren William, Glenda Farrell, Jean Parker, Walter Connolly, Ned Sparks, Barry Norton, Robert Emmett O'Connor, Wallis Clark, Hobart Bosworth, Nat Pendleton, Blind Dad Mills, Shorty, Halliwell Hobbes
A tough gangster with a soft spot for an impoverished apple vendor, helps the latter pose as a wealthy society lady when her daughter comes to town on a visit. An exceptionally fine blend of comedy and unabashed sentiment, the excellent Robert Riskin script is one of the best adaptations of Runyon for the screen. Followed by a sequel entitled "Lady By Choice", and remade by Capra many years later as "Pocketful Of Miracles".
COM   96 min (director's archive original) B/W
VIDrel: VCC/DISC   V
Boa: short story Madame La Gimp by Damon Runyon.

LADY FROM THE SHANGHAI CINEMA, THE **   15
Guilherme De Almeida Prado BRAZIL   1988
Maite Proenca, Antonio Fagundes, Jose Lewgoy, Jorge Doria, Jose Mayer, Paulo Villaca, Miguel Falabella, Sergio Mamberti, Helena Imara Reis, John Doo, Julio Calasso Jr, Carlos Takeshi plus voices of: Susana de Moraes, Maristela Morena
During a heatwave an estate agent with time on his hands wanders into a cinema, meets a married woman there and embarks on an affair with her. Subsequently, he sells an apartment to her husband, and the pair use it to carry on their liaison. However, their affair embroils them in violence and danger. A complex homage to film noir, that is just too convoluted for its own good, and is not exactly helped by the unappealing natures of the main characters.
Aka: A DAMA DO CINE SHANGHAI
THR   117 min   CINrel

LADY IN THE DARK ***   (U)
Mitchell Leisen USA   1942 (released 1944)
Ginger Rogers, Warner Baxter, Ray Milland, Jon Hall, Mischa Auer, Mary Philips, Barry Sullivan
When the editor of a fashion magazine is unable to choose between three suitors, she resorts to the services of a psychoanalyst, who is able to help her interpret the strange dreams she has been having. A slick and very attractive film version of a hit stage musical, which though lacking many of the songs that were in the original, still remains a good example of the high quality romantic musicals that were so much a feature of the 1940s.
MUS   100 min   SATrel: SKY MOVIES GOLD
Boa: play by Moss Hart.

LADY IN WHITE ***   15
Frank LaLoggia USA   1988
Lukas Haas, Katherine Helmond, Len Cariou, Alex Rocco, Jason Presson, Jared Rushton, Renata Vanni, Angelo Bertolini, Gregory Levinson, Joelle Jacobi, Lucy Lee Flippen, Tom Bower, Sydney Lassick, Rita Zohar, Hal Bokar, Rose Weaver
A youngster locked in a school cloakroom is the terrified witness to the spectral re-enactment of the murder of a little girl that took place three years before. This event leads to his involvement with an unsolved series of child murders. Set in the 1960s, this effective chiller doesn't lack for atmosphere, even if the final outcome is rather too easy to guess.
HOR   109 min (ort 112 min)   VIDrel: VISVID/POLYREC L/A   V

LADY JANE **   PG
Trevor Nunn UK   1985
Jane Lapotaire, John Wood, Patrick Stewart, Sara Kestelman, Helena Bonham Carter, Cary Elwes, Michael Hordern, Jill Bennett, Joss Ackland, Richard Vernon, Warren Saire, Ian Hogg, Richard Johnson, Lee Montague, David Waller, Pip Torrens

Historical drama about Lady Jane Grey, who was Queen of England for a mere seven days in 1553. A solid account that is rather too slow to really hold the attention. This story was filmed once before as "Tudor Rose" (1936).

DRA 136 min (ort 142 min) VIDrel: CIC/SONOP V/sur

LADY KILLER ** (18)
Steven Schachter USA 1995
Judith Light, Jack Wagner, Ben masters, Tracey Gold, Elizabeth Lennie, Diana LeBlanc, Patricia Carroll Brown, J.R. Zimmerman, Pixie Bigelow, Wendy Hopkins, Hadley Sandiford, Mishu Vellani, Raymond O'Neill, Sheila Brand
A married woman has a brief affair but decides to end it, which provokes her psychotic lover to rape her and then start dating her daughter out of pure spite. This places her in a difficult position as she is unable to defend her family without revealing her own indiscretion. An unedifying and overwrought psychological thriller.

THR 86 min (ort 90 min) mTV SATrel: MOVIE CHANNEL

LADY OF THE CARMELIAS, THE *** 15
Mauro Bolognini FRANCE/ITALY/WEST GERMANY 1981
Isabelle Huppert, Gian Maria Volonte, Bruno Ganz, Fabrizio Bentivoglio, Clio Goldsmith, Mario Maranzana, Jann Banilee, Carla Fracci, Fernando Rey, Cecile Vassort, David Jalil, Piero Vida, Fabio Traversa, Remo Remotti, Mattia Sbragia
In 19th century France, a courtesan moves to Paris and lives among the aristocracy, where she is accompanied by her drug-addict father, her husband and even a patron. This film is not so much a chronicle of the Dumas character as an examination of the life of the real woman who inspired his novel, who is portrayed as greedy, ruthless and quite deserving of her eventual fall from grace. An opulent and sharply focused period drama.
Aka: DIE KAMELIENDAME; LA DAME AUX CAMELIAS; LA DAMA DELLE CAMELIE; TRUE STORY OF CAMILLE, THE
DRA 121 min VIDrel: ARROW/RTM V
Boa: novel by Alexandre Dumas.

LADY OF THE NIGHT * 18
Piero Schivazappa ITALY 1986
Serena Grandi, Fabio Sartor, Francesca Topi
A man is so involved with his work that his neglected wife sets out to provoke him into showing her more interest, and does this by flirting with all and sundry and accepting their sexual advances. However, this ploy is not without an element of danger. Poorly dubbed (and the daft dialogue is often unintentionally funny) and directed in haphazard fashion, this is a dreary softcore drama as often incomprehensible as it is lifeless.

DRA 90 min wScrn dubbed VIDrel: ARTPRO/RTM V

LADY VANISHES, THE **** U
Alfred Hitchcock UK 1938
Margaret Lockwood, Michael Redgrave, Paul Lukas, Dame May Whitty, Cecil Parker, Linden Travers, Mary Clare, Naunton Wayne, Basil Radford, Emile Boreo, Googie Withers, Philip Leaver, Catherine Lacey, Charles Oliver
Enjoyable film in which two people investigate the disappearance of an old woman from a train, and discover a plot to pretend she was never on the train in the first place. A lively comedy-drama that, though somewhat over-rated, has many good things in it, not least being the witty script (by Frank Launder and Sidney Gilliat) and some splendid performances. Remade in 1979.
DRA 91 min (ort 95 min) B/W
VIDrel: VCC/DISC/COLUM L/A V
Boa: novel The Wheel Spins by Ethel Lina White.

LADY VANISHES, THE * PG
Anthony Page UK 1979
Elliott Gould, Cybill Shepherd, Angela Lansbury, Herbert Lom, Arthur Lowe, Ian Carmichael, Jenny Runacre, Gerald Harper, Jean Anderson, Vladek Sheybal, Medlena Nedeva, Wolf Kahler, Madge Ryan, Rosalind Knight, Jonathan Hackett
A dismally flawed remake of the 1938 original which, although fairly faithful to the story (lady vanishes en route back to England from Switzerland) is fatally ruined by incompetent casting. The script is by George Axelrod.
DRA 96 min (ort 99 min) VIDrel: VCC L/A V
Boa: novel The Wheel Spins by Ethel Lina White.

LADY WITH A LAMP, THE *** U
Herbert Wilcox UK 1951
Anna Neagle, Michael Wilding, Gladys Young, Felix Aylmer, Julia D'Albie, Arthur Young, Helena Pickard, Peter Graves, Sybil Thorndike, Monckton Hoffe, Charles Carson, Edwin Styles, Barbara Couper, Helen Shingler, Nigel Stock
A workmanlike account of Florence Nightingale's pioneering efforts in the field of nursing, well acted if somewhat less than accurate in terms of historical detail. Neagle is particularly well cast.
DRA 99 min (ort 110 min) B/W VIDrel: WHV L/A V/s
Boa: play by Reginald Berkeley.

LADYBIRD LADYBIRD * 15
Ken Loach UK 1994
Crissy Rock, Vladimir Vega, Sandie Lavelle, Mauricio Venegas, Ray Winstone, Clare Perkins, Jason Stracey, Luke Brown, Lily Farrell, Scottie Moore, Linda Ross, Kim Hartley, Jimmy Batten, Sue Sawyer, Pamela Hunt, Alan Gold, Sue Gifford
A single mother who has four children by different fathers battles the welfare authorities who have taken them into care. Having finally met a man who does not abuse her, she becomes anxious to see her family re-united. An angry and tendentious attack on 1990s Britain with Loach persisting in his rigid and stereotypical view of the working classes as ill-educated, brutalised and socially incompetent. In defence, one notes that the film is based on fact.
DRA 97 min (ort 101 min) VIDrel: ELPIC/POLYREC V/sur

LADYBUGS ** PG
Sidney J. Furie USA 1992
Rodney Dangerfield, Jackee, Jonathan Brandis, Ilene Graff, Vinessa Shaw, Tom Parks, Jeanette Arnetta, Nancy Parsons, Blake Clark, Tommy Lasorda, Vanessa Monique Roussel, Jennifer Francis Lee, Lacrystal Cooke, Johna Stewart, Valentino
A junior executive agrees to coach a girls' soccer team sponsored by his boss, in the hope that this will help him win promotion. However, the girls know very little about the game, he persuades his girlfriend's son to don drag and join the team in order to give them some much-need instruction.
COM 85 min (ort 91 min) VIDrel: WHV L/A V

LADYHAWKE *** PG
Richard Donner USA 1985
Rutger Hauer, Michelle Pfeiffer, Leo McKern, Matthew Broderick, John Wood, Richard Donner, Lauren Schuler, Ken Hutchison, Alfred Molina, Giancarlo Prete, Loris Loddi, Alessandro Serra, Charles Borromel, Massimo Sarchielli
Medieval fantasy in which two young lovers change at night into a wolf and a hawk because of a spell cast by an evil bishop. Overlong but generally quite well sustained, despite bad casting of Broderick and the intrusive score by Andrew Powell.
FAN 118 min (ort 124 min) VIDrel: 20TH/TECH L/A; ENCORE (LV only) V/sh LV

LADYKILLER ** 15
Michael Scott USA 1992
Mimi Rodgers, John Shea, Tom Irwin, Alice Krige, Bob Gunton, Bert Remsen, Art Kimbro, Robert Gossett, Elizabeth Keifer, Geoff Rivas, Courtney Barilla, Erick Weiss, Jim Holmes, Thomas Knickerbocker, Jean Kauffman, Emile Kuroda
A burnt-out policewoman goes to a video dating agency and finds a suitable professional to whom she takes a fancy. She soon begins to believe that he is the one for her, but her curiosity about his past proves to have fateful consequences. The clever script is the work of Shelley Evans.
THR 87 min (ort 97 min) mTV VIDrel: CIC/SONOP V

LADYKILLERS, THE **** U
Alexander Mackendrick UK 1955
Alec Guinness, Katie Johnson, Peter Sellers, Cecil Parker, Herbert Lom, Danny Green, Frankie Howerd, Jack Warner, Philip Stainton, Fred Griffiths, Edie Martin, Kenneth Connor, Jack Melford, Ewan Roberts, Harold Goodwin
Amusing black comedy about a little old lady who lets a room to a man who is regularly visited by his friends. They pretend to be musicians but are really planning a robbery. A minor classic with a wonderful performance from Guinness. The script is by William Rose.
COM 87 min (ort 97 min) VIDrel: WHV V/h

LAGUNA HEAT ** 18
Simon Langton USA 1987
Harry Hamlin, Jason Robards, Rip Torn, Catherine Hicks, Anne Francis, James Gammon

An L.A. cop goes back to his home town but finds its former tranquillity shattered by a series of murders that seem to be linked in some way to people associated with his father. Generally atmospheric but a little stolid in pacing.
THR   105 min (ort 110 min) mCab   VIDrel: GUILD/SONOP
L/A   V
Boa: novel by T. Jefferson Parker.

LAID IN HEAVEN **
Jake Craig USA
Ashlyn Gere, Raven Richards, Kandi Valentine, Alexandria Quinn, Marc Wallice, T.T. Boy, Wayne Summers, Mike Reynolds
Having slipped whilst getting out of the bathtub, a virgin bride strikes her head and gets killed. She finds herself in heaven where it's made clear that she has arrived before time, and is to be given 24 hours on Earth to learn all she can about sexuality. A series of adventures follow, but ultimately all is revealed to have been nothing more than the fantasies of a young bride with an over-active imagination.
A   49 min   VIDrel: IMPENT   V

18
1991

LAIR OF THE WHITE WORM, THE ***
Ken Russell UK
Amanda Donohue, Hugh Grant, Catherine Oxenberg, Peter Capaldi, Sammi Davis, Stratford Johns, Paul Brooke, Imogen Claire, Chris Pitt, Gina McKee, Lloyd Peters, Christopher Gable, Mirand Coe, Linzi Drew, Caron Anne Kelly
The screenplay is by Russell in this overblown (as ever) and over-ripe adaptation of Stoker's novel, telling of an archaeologist's investigations into an estate where a huge wormlike skull was discovered, and his discovery that some of these bizarre creatures are alive and well. Generally far too self-indulgent to work as a horror yarn, but as a farce it is effective. See also THE REPTILE.
HOR   89 min (ort 93 min)   VIDrel: FIRST/SONOP L/A   V
Boa: novel by Bram Stoker.

18
1988

LAKE CONSEQUENCE **
Rafael Eisenman USA
Billy Zane, Joan Severance, May Karasun, Whip Hubley, Courtland Mead, Dan Reed, Christi Allen, Andria Litto, Ron Howard George, Allan Graf
A housewife is accidentally locked in her camper and finds herself being abducted by a handsome criminal and his female companion. She finds herself attracted to her captors and cannot decide whether to escape or engage in a menage-a-trois. A curious erotic tale, produced by Zalman King.
Aka: LAKE CONSEQUENCE: A MAN AND TWO WOMEN
DRA   86 min (ort 90 min)   VIDrel: COLUM/SONOP   V/sh

18
1992

LAMB ***
Colin Gregg UK
Liam Neeson, Harry Towb, Hugh O'Conor, Frances Tomelty, Ian Bannen, Ronan Wilmot, Denis Carey, Eileen Kennally, David Gorry, Andrew Pickering, Stuart O'Connor, Ian McElhinney, Bernadette McKenna, Jessica Saunders, Roger Booth
A young priest at a Catholic school for delinquent youths forms a close relationship with a disturbed fourteen-young boy who suffers from epilepsy. Shattered by the death of his father, both he and his young charge, who is in much need of a father figure, take off together on a journey of self-discovery. A touching and downbeat tale, much enhanced by quality acting.
DRA   105 min (ort 110 min)   VIDrel: FIRST/SONOP
V/sur
Boa: novel by Bernard MacLaverty.

15
1986

LAMBADA: THE FORBIDDEN DANCE *
Greydon Clark USA
Laura Herring, Jeff James, Sid Haig, Richard Lynch, Barbara Brighton, Kid Creole, Miranda Garrison, Angela Moya, Shannon Farnon, Linden Chiles, Ruben Moreno, Steven Lloyd Williams, Pilar Del Rey, Tom Alexander, Sabrina Mance
A rainforest princess goes to the USA to protest about the destruction of her habitat and teaches her native dance to a rich man with whom she falls in love. They enter and win a dance contest, giving her a chance to appear on TV and appeal for help in stopping a powerful logging corporation who are cutting down the trees. The dancing is the best thing in this cynically exploitative film, whose ecological sub-plot is entirely out of place.
Aka: FORBIDDEN DANCE, THE
MUS   97 min   VIDrel: RCA L/A   V

15
1989

L'AMORE MOLESTO ***
Mario Martone ITALY
Anna Bonaiuto, Angela Luce, Gianni Cajafa, Peppe Lanzetta, Licia Maglietta, Anna Calato, Italo Celoro, Carmela Pecoraro, Giovanni Viglietti, Lina Polito, Enzo Decaro, Francesco Paolantoni, Piero Tassitano, Marita D'Elia, Alda Massa
When her mother disappears on the train from Naples to Rome and her body is later found on a beach, her daughter becomes obsessed with the idea of finding her killer. She begins to recall details of her childhood and eventually learns things about her mother that come a complete surprise. Writer-director Martone has created a compelling tale in the style of those old film noir murder mysteries, and uses it to examine some interesting wider issues.
DRA   103 min   CINrel
Boa: novel by Elena Ferrante.

15
1995

LAMP, THE ***
Tom Daly USA
Deborah Winters, James Huston, Scott Bankston, Andra St Ivanyi, Danny Daniels, Mark Mitchell, Andre Chimene, Damon Merrill, Barry Coffing, Tracye Walker, Raan Lewis, Hank Amigo, Brian Floores, Michelle Watkins, Coy Sevier
A girl discovers a lamp at her father's museum, and unwittingly releases a nasty demon when she rubs it (the lamp that is). Under the spell of said demon, she invites some of her chums back to the museum to spend Halloween there, and our demon arranges a series of gruesome surprises. Quite a predictable film, but not without a good number of chilling effects.
Aka: OUTING, THE
HOR   85 min (Cut at film release)   VIDrel: POPRO/RTM
V

18
1986

LANCELOT DU LAC ***
Robert Bresson FRANCE/ITALY
Luc Simon, Laura Duke Condominas, Humbert Balsan, Vladimir Antolek-Oresek, Patrick Bernard, Arthur De Montalembert
A personal and very different presentation of the Arthurian legend that concentrates on the period after the Knights returned from their quest for the Holy Grail. Their disillusionment and bitterness soon lead to internal divisions and even fighting among the main characters, who are certainly not portrayed in a romantic light. Filmed in a visually striking style, with restrained performances and a prominent soundtrack.
Aka: GRAIL, THE; LANCELOT OF THE LAKE; LE GRAAL
DRA   84 min   VIDrel: ARTIF/20TH L/A   V/h

PG
1974

LAND AND FREEDOM ***
Ken Loach GERMANY/SPAIN/UK
Ian Hart, Rosana Pastor, Iciar Bollain, Tom Gilroy, Marc Martinez, Frederic Pierrot, Andres Aladren, Sergi Calleja, Raffaele Cantatore, Pascal Demolon, Paul Laverty, Josep Magem, Eoin McCarthy, Jurgen Muller, Roca, Emili Samper
Story of the 1936 Spanish uprising, with an unemployed Liverpool man leaving his fiancee behind as he travels to Spain as a volunteer for the Partisans. However, once there he meets a beautiful Spanish girl who is already involved in the struggle and at the same time finds his eyes opened by the petty squabbling of those around him. An immensely well crafted tale that is perhaps a little hampered by the director's usual moralising, but remains a most worthy effort.
DRA   109 min (partly subtitled – ort 110 min) wScrn
VIDrel: ARTIF/20TH; ENCORE (LV only)   V/sur LV

15
1995

LAND BEFORE TIME, THE ***
Don Bluth USA
Voices of: Pat Hingle, Gabriel Damon, Helen Shaver, Candice Houston, Judith Barsi, Will Ryan, Burke Barnes, Will Ryan
When a young dinosaur is orphaned, he sets off with several of his fellows to find a secret valley where they can find safety from a plague that is ravaging the world. On the way they experience danger but also have several amusing encounters. Excellent animation work helps sustain this sluggish yarn, which could so easily have been a great deal better (and certainly more colourful). A couple of sequels followed.
ANIM   66 min (ort 69 min)   VIDrel: CIC/SONOP   V/sur

U
1988

LAND BEFORE TIME 3, THE: THE TIME OF GREAT GIVING **
Roy Allen Smith USA
When a meteorite cuts off the water supply to the valley tension mounts among the adult dinosaurs, so our brave young saurians mount a search for a new supply. Along the way they brave

U
1995

many dangers and have quite a few adventures. This second sequel to the original dinosaur film was released straight to video.
ANIM   68 min (ort 71 min)   VIDrel: CIC/SONOP   V/dm

LAND OF FARAWAY, THE **
Vladimir Grammatikov NORWAY/USSR   1987
Timothy Bottoms, Susannah York, Christian Bale, Christopher Lee, Nicholas Pickard
A young boy releases a spirit from a bottle and is whisked off to a magical kingdom to which all the children have been abducted by a wicked knight.
Aka: MIO, MIN MIO; MIO IN THE LAND OF FARAWAY
FAN   95 min   VIDrel: MIA/DISC   V
Boa: novel Mio, My Son by Astrid Lindgren.

LAND OF SILENCE AND DARKNESS ****   U
Werner Herzog WEST GERMANY   1971
Fini Straubinger
A truly fascinating documentary about a fifty-six-year-old deaf and blind woman, who by virtue of her determination and powerful personality, overcame her severe handicaps to live a fulfilled and rewarding life. A memorable, inspiring and deeply moving work.
DOC   81 min (ort 90 min)   VIDrel: TART/20TH   V

LAND THAT TIME FORGOT, THE ***   PG
Kevin Connor UK   1974
Doug McClure, John McEnery, Susan Penhaligon, Keith Barron, Anthony Ainley, Declan Mulholland, Godfrey James, Bobby Farr, Ben Howard, Colin Farrell, Roy Holder, Andrew McCulloch, Grahame Mallard, Brian Hall, Peter Sproule
Germans and Americans on board a submarine discover an island of prehistoric monsters and savages in this engaging fantasy yarn that is let down by rather poor special effects. Followed by THE PEOPLE THAT TIME FORGOT.
A/AD   87 min (ort 91 min)   VIDrel: WHV   V
Boa: novel by Edgar Rice Burroughs.

LANGOLIERS, THE **   15
Tom Holland USA   1995
Mark Lindsay Chapman, Kate Maberly, Julie Arnold Lisnet, David Morse, Michael Louden, Patricia Wettig, Kymberly Dakin, Bronson Pinchot, Chris Hendrie, Jennifer Nichole Porter, Christoper Collet, Frankie R. Faison, Dean Stockwell
Ten passengers aboard a plane find themselves having some very strange experiences in which time seems to be playing tricks on them, when they wake up after a sleep to find that all the other passengers and crew have vanished. An average adaptation of King's novella.
Aka: STEPHEN KING'S THE LANGOLIERS
HOR   180 min   VIDrel: POLY/POLYREC   V/sh
Boa: novella Four Past Midnight by Stephen King.

LAP DANCER *   18
Arthur Egeli USA   1995
Robert Emmett, Elizabeth Wagner, Steven Kesmodel, Jennifer Wolf
By night a woman is studying for a business degree, and by day she earns just enough to get by on, refusing offers of help from her rich boyfriend. But as the bills start to pile up, she realises she can clear her debts by working for a month or so as a lap dancer. However, she finds it considerably easier to get into this scene then to get out of it. A rather boring erotic drama, doubtless made to cash in on the expected success of SHOWGIRLS.
DRA   82 min   VIDrel: MARQ/QUANT   V

LAPSE OF MEMORY ***   PG
Patrick Dewolff CANADA/FRANCE   1991
John Hurt, Marthe Keller, Matthew Mackay, Kathleen Robertson, Marion Peterson, Serge Dupire, Robert Watson Barr, Terry Haig, Gordon Masten, Michael McGill, Jude Berry, John Lambert
A gripping psychological thriller about a fifteen-year-old who tries to find out his true background after learning that he has a false identity, that his parents may not be his real ones and that he supposedly died two years ago. But it was the work of a Federal Witness re-location scheme that created this mystery, and having become a target for assassins, the boy has to go on the run. The director's debut film, this is a loose remake of "I Am The Cheese" (1983).
Aka: MEMOIRE TRAQUEE
THR   81 min (ort 85 min)   VIDrel: VCC/DISC/COLUM   L/A   V/sh
Boa: novel I Am the Cheese by Robert Cormier.

L'ARGENT ***   PG
Robert Bresson FRANCE/SWITZERLAND   1983
Christian Patey, Michel Briguet, Caroline Lang, Sylvie Van Den Elsen, Jeanne Aptekman
An innocent young man is the victim of a schoolboy prank involving a forged 500-franc note, and because of this single action eventually becomes a thief and murderer. A masterfully restrained and austere work, made to illustrate the director's religious beliefs, but flawed by a tendency to manipulate the characters like puppets in order to tell an updated version of Tolstoy's parable.
DRA   80 min (ort 82 min)   VIDrel: ARTIF/20TH   V
Boa: short story by Leo Tolstoy.

LARGER THAN LIFE *   PG
Howard Franklin USA   1996
Bill Murray, Janeane Garofalo, Matthew McConaughey, Keith David, Pat Hingle, Jeremy Piven, Lois Smith, Anita Gillette, Maureen Mueller, Harve Presnell, Tracey Walter, Jerry Adler
A suave salesman learns that he has acquired a circus elephant by way an inheritance when his father passes away. The small amounts of humour to be found here mostly derive from Murray's problems regarding the disposal of this beast, which he can either sell to a shady animal trainer or send to Sri Lanka as part of a breeding program. Meanwhile, it turns his life upside down in a series of totally predictable and most unfunny escapades. A waste of Murray's comic gifts.
COM   93 min   CINrel

LASSIE ***   U
Daniel Petrie USA   1994
Thomas Guiry, Helen Slater, Frederic Forrest, Brittany Boyd, Jon Tenney, Michelle Williams, Charlie Hofheimer, Clayton Barclay, Earnest Poole Jr, Yvonne Brisendine, Jeffrey H. Gray, Joe Inscoe, David Bridgewater
After a city family suffers financial disaster, they move to the countryside where their son meets up with a very smart dog that decides to befriend them. This meeting proves quite beneficial as soon everybody is rallying around in an attempt to come to terms with all their problems. A competent updating of this perennial favourite, with the canine role played by a descendent of the original.
JUV   92 min   VIDrel: CIC/SONOP L/A   V

LASSITER **   18
Roger Young USA   1984
Tom Selleck, Lauren Hutton, Jane Seymour, Bob Hoskins, Joe Regalbuto, Warren Clarke, Ed Lauter, Edward Peel, Paul Antrim, Christopher Malcolm, Barrie Houghton, Peter Skellern, Harry Towb, Belinda Mayne, Morgan Sheppard
An ace jewel thief in London in 1936 uses his skills to steal documents from the German embassy, having been blackmailed by Scotland Yard into accepting this dangerous assignment. Quite carefully made, but the pedestrian pacing spoils it.
DRA   96 min (ort 100 min)   VIDrel: MGM/WHV L/A   V

LAST ACTION HERO **   15
John McTiernan USA   1992
Arnold Schwarzenegger, Anthony Quinn, Mercedes Ruehl, F. Murray Abraham, Art Carney, Charles Dance, Austin O'Brien, Robert Prosky, Tommy Noonan, M.C. Hammer, Brigitte Wilson, Sharon Stone, Chevy Chase, Jean-Claude Van Damme
Critically lambasted, mega-budget action spoof that failed completely to find an audience. The plot, such as it is, has a young teenage boy going to the cinema and entering a film starring his tough action idol Jack Slater. However, when the movie villains escape into the real world, this unlikely duo give chase. Full of star cameos and Hollywood in-jokes that add nothing at all to a noisy and ultimately very tiresome effort.
A/AD   125 min (ort 131 min) wScrn
VIDrel: VCC/DISC/COLUM   V/sur

LAST AMERICAN HERO, THE ***   PG
Lamont Johnson USA   1973
Jeff Bridges, Valerie Perrine, Geraldine Fitzgerald, Art Lund, Gary Busey, Ed Lauter, Ned Beatty, William Smith III, Gregory Walcott, Tom Ligon, Ernie Orsatti, Erica Hagen, James Murphy, Lane Smith
A backwoods moonshiner and racing fanatic pits his wits against the established authorities. Extremely convincing characterisations are the mainstay of this witty and offbeat hillbilly

comedy-drama (based to some extent on the early career of racing driver Junior Johnson).
Aka: HARD DRIVER
DRA   91 min (ort 95 min)   VIDrel: 20TH/TECH L/A   V

LAST BEST YEAR, THE *** (15)
John Erman USA   1990
Mary Tyler Moore, Bernadette Peters, Brian Redford, Dorothy McGuire, Kate Reid, Carmen Mathews, Kenneth Welsh, Erika Alexander, Lawrence Dane, Albert Schultz, Michael Hogan, Batsheba Garnett, Michael J. Reynolds, Chris Walker
A female cancer patient is finally told by her doctors that there is nothing more that can be done for her, and is helped to deal with her impending fate by a lonely therapist. Strong performances by Peters and Moore stand out in this otherwise unremarkable offering. See also WHEN THE TIME COMES and IT'S MY PARTY.
Aka: HER LAST BEST YEAR
DRA   90 min (ort 94 min) mTV   VIDrel: L/A   V

LAST BOY SCOUT, THE ** 18
Tony Scott USA   1991
Bruce Willis, Damon Wayans, Chelsea Field, Noble Willingham, Taylor Negron, Danielle Harris, Halle Berry, Bruce McGill, Badja Djola, Kim Coates, Chelcie Ross, Joe Santos, Clarence Felder, Tony Longo, Frank Collinson, Bill Medley
A foul-mouthed comedy-actioner that sees Willis in action as an untidy private detective and former secret service agent who once saved the life of a president, but lost his job just the same. Despite the fact that his wife was having an affair with his best friend, he still resolves to catch the killers when said friend is blown up by a car bomb. Further explosions, shootouts and car chases follow, but the good guys win through in the end.
A/AD   101 min (ort 105 min) cC   VIDrel: WHV   V/sur

LAST CHANCE ** (18)
Dan Golden USA   1995
James Brolin, Bryan Genesse, Kehli O'Byrne, Elena Sahagun, Susan Africa, Nikki Fritz, Richard Lynch, Craig JUdd, Bon Vibar, Nick Nicholson, Bobby Greenwood, Zarina Torres, Cristina Villiegars, Tom Tauss, Lorne Greenwood
In a bleak and devastated post-WW3 world most of humanity is dead, and men and women have split into two warring factions, sexual contact being unknown between them as it leads to the transfer of deadly man-made plague. Meanwhile, an isolated outpost of civilisation has perfected a serum to overcome this, but it is attacked and destroyed by a local warlord. A lone survivor attempts to save the world. A weakly and unconvincing effort that spoils its one good idea.
FAN   89 min   SATrel: MOVIE CHANNEL

LAST DANCE, THE ** 18
Anthony Markes USA   1991
Cynthia Bassinet, Elaine Hendrix, Kurt T. Williams, Allison Rhea, Erica Ringstrom
Five contestants for a beauty contest find themselves the targets of a crazed killer, in this standard slasher tale.
THR   80 min (ort 94 min)   VIDrel: COLUM/SONOP   V

LAST DANCE ** 18
Bruce Beresford USA   1995
Sharon Stone, Rob Morrow, Randy Quaid, Peter Gallagher, Jack Thompson, Don Harvey, Pamela Tyson, Jayne Brook, Skeet Ulrich, Diane Sellers, Patricia French, Jeffery Ford, Dave Hager, Christine Cattell, Peg Allen, Peggy Walton Walker
A young man gets a job through his brother, who works in the state governor's office, and he finds himself saddled trying to obtain a pardon for a female double murderer who is due to be executed. Another debate on the theme of capital punishment, with Stone suitably unappealing in a difficult and quite unsympathetic role as a cold-blooded killer.
THR   100 min (ort 103 min) cC   VIDrel: TOUCH/TECH   V/sur

LAST DAYS OF CHEZ NOUS, THE ** 15
Gillian Armstrong AUSTRLAIA   1992
Lisa Harrow, Bruno Ganz, Kerry Fox, Miranda Otto, Bill Hunter, Lex Marinos, Kiri Paramore, Mickey Camilleri, Lynne Murphy, Claire Haywood, Leanne Bundy, Wilson Alcorn, Tom Weaver, Bill Brady, Eva Di Cesare, Danny Caretti, Steve Cox
Unengaging study of a homesick Frenchman (Ganz) married to a novelist who finds the arrival of his sister-in-law having a

complicating effect on his tangled and unsatisfying relationships. A well-composed film of insight and wit, but burdened with an excess of unsympathetic characters.
DRA   93 min (ort 97 min) wScrn   VIDrel: TART/20TH   V/sur

LAST DAYS OF FRANKIE THE FLY, THE * 18
Peter Markle USA   1996
Dennis Hopper, Kiefer Sutherland, Daryl Hannah, Michael Madsen
A man who works as a menial for a Hollywood gangster specialising in making porno films decides to strike out on his own, and make his own movies.
A/AD   92 min   VIDrel: GULD/FOXVID   V/sur

LAST DAYS OF PARADISE, THE *** 15
Floyd Mutrux USA   1994
Dermot Mulroney, Ricky Schroder, Kelli Williams, Noah Wyle, Jill Schoelen, Kristin Minter, Lucy Deakins, Kenny Ransom, Seymour Cassel, Paul Gleason, Janet MacLachlan, Frederick Coffin, Andrew Robinson
Drama set in Los Angeles 1965 against the background of the Watts riots and the anti-Vietnam marches. Soundtrack features many of the excellent songs from the period, and if the plot shows a lack of work, the film makes up for it with its depiction of a vanished era.
DRA   94 min   VIDrel: COLUM/SONOP   V/sur

LAST DAYS OF PATTON, THE ** PG
Delbert Mann USA   1985
George C. Scott, Eva Marie Saint, Richard Dysart, Lee Patterson, Ed Lauter, Murray, Hamilton, Kathryn Leigh Scott, Horst Janson, Daniel Benzali, Ron Berglas, Don Fellows, Errol John, Alan MacNaughton, Paul Maxwell
Scott reprises the role he created in PATTON in this account of the last six months in the life of this colourful WW2 general. No more than adequate with a ludicrous and tiresome deathbed scene that just goes on and on. The script is by William Luce.
DRA   140 min (ort 180 min) mTV   VIDrel: 20TH/TECH L/A   V/h
Boa: book by Ladislas Farago.

LAST DETAIL, THE *** 18
Hal Ashby USA   1973
Jack Nicholson, Otis Young, Randy Quaid, Clifton James, Carol Kane, Michael Moriarty, Nancy Allen, Launa Anders, Kathleen Miller, Gerry Salsberg, Don McGovern, Pat Hamilton, Michael Chapman, Jim Henshaw, Derek McGrath, Jim Horn
Two sailors escorting a prisoner, to serve a long sentence in a naval prison for a foolish and trivial theft, try to give him a memorable time before he goes inside. A complex movie that has strong characterisations, with a downbeat but appropriate ending. Written by Robert Towne.
DRA   100 min (ort 105 min)   VIDrel: CASPIC/BMGREC   L/A   V
Boa: novel by Darryl Ponicsan.

LAST DRAGON, THE * 15
Michael Schultz USA   1985
Taimak, Julius J. Carey III, Chris Murney, Leo O'Brien, Thomas Ikeda, Faith Prince, Vanity, Mike Starr, Jim Moody, Glen Eaton, Ernie Reyes Jr, Roger Campbell, Esther Marrow, Keshia Knight Pullam, Jamal Mason
Story of a kung fu quest by a disciple of the martial arts. On his way, he has to rescue a young damsel and fight the self-styled Shogun of Harlem. Ponderous overblown nonsense, very clumsy and disorganised.
Aka: BERRY GORDON'S THE LAST DRAGON; LAST DRAGONS, THE
MAR   107 min Cut (1 min 59 sec)   VIDrel: 20TH/TECH   V/sur

LAST EMBRACE, THE *** 18
Jonathan Demme USA   1979
Roy Scheider, Janet Margolin, John Glover, Sam Levene, Christopher Walken, Charles Napier, Jacqueline Brooks, David Margulies, Andrew Duncan, Marcia Rodd, Gary Goetzman, Lou Gilbert, Mandy Patinkin, Max Wright, Sandy McLeod
An agent's wife is killed in an attack meant for him and he is forced to go underground. A tight and punchy thriller that builds up to an exciting climax at Niagara Falls. The score is by Miklos Rozsa.
THR   98 min (ort 101 min)   VIDrel: MGM/WHV L/A   V
Boa: novel The 13th Man by Murray Teigh Bloom.

LAST EMPEROR, THE ****   15
Bernardo Bertolucci CHINA/ITALY/UK   1987
John Lone, Joan Chen, Peter O'Toole, Ying Ruocheng, Victor Wong, Dennis Dun, Ryuichi Sakamoto, Maggie Han, Ric Young, Wu Vuu, Tijger Tsou, Wu Tao, Fan Guang, Wu Jun Mei, Cary Hiroyuki Tagawa, Jade Go, Fumihiko Ikeda
Vast, sprawling epic account of China's last emperor that spares no expense in following the career of Pu Yi, from his cloistered upbringing to his enforced exile after being deposed. A cold but lavish odyssey. AA: Pic, Dir, Screen/adapt (M. Peploe/B. Bertolucci), Cin (V. Storaro), Art/Set (F. Scarfiotti/B. Cesari and O. Desideri), Cost (J. Acheson), Sound (B. Rowe/I. Sharrock), Edit (G. Cristiani), Score/orig (R. Sakamoto/D. Byrne/C. Su).
DRA   156 min (ort 164 min)   VIDrel: VCC/DISC/COLUM V/sur
Boa: book by Pu Yi.

LAST EXIT TO BROOKLYN ***   18
Uli Edel WEST GERMANY   1989
Stephen Lang, Jennifer Jason Leigh, Burt Young, Peter Dobson, Jerry Orbach, Stephen Baldwin, Ricki Lake, John Costelloe
Written by Desmond Nakano, this is a harsh and gloomy adaptation of Selby's cult novel, which paints a brutal and uncompromising picture of life around a 1950s Brooklyn waterfront, where poverty, crime and labour unrest are the most influential features of life. Despite being memorable for its detail and strong characterisation, the film suffers from a lack of focus and a serious shortage of sympathetic characters.
Aka: LETZE AUSFAHRT BROOKLYN
DRA   98 min (ort 102 min)   VIDrel: ARROW/RTM   V/sur
Boa: novel by Hubert Selby Jr.

LAST FLIGHT OUT **   PG
Larry Elikann USA   1990
Richard Crenna, James Earl Jones, Haing S. Ngor, Eric Bogosian, Rosalind Chao, Arliss Howard, Elizabeth Lindsay, Barry Corbin, James Hong, Stephen Tobolowsky, Soon-Teck Oh, James Morrison, Kieu Chinh, Bradford English, Alice Lo
A dullish TV drama that attempts to recreate the chaos of those few last days in Saigon during 1975, just before the fall of the South Vietnamese regime, when Pan-Am was the last commercial airline operating from there. We follow the fortunes of a number of its employees as they desperately struggle to bring family and friends to safety.
DRA   94 min (ort 96 min) mTV   VIDrel: ODY/SONOP   V

LAST GASP **   18
Scott McGinnis USA   1995
Robert Patrick, Joanna Pacula, Vyto Ruginis, Mimi Craven, Alexander Enberg, Shashawanee
Patrick plays a ruthless property developer who has an Indian tribe wiped out when they interfere with his plans to develop a piece of land. He even kills the Indian chief, but these are Totec Indians and have magical powers, so the dead chief takes him over, forcing him to kill and devour a succession of victims. Pacula is a woman in search of her missing husband, and she proves to be his undoing. An illogical and gory tale, but Patrick is quite charismatic.
HOR   88 min (ort 90 min)   VIDrel: MED/DISC   V/sh

LAST GREAT WARRIOR, THE **   PG
Xavier Koller USA   1994
Adam Beach, Michael Gambon, Mandy Patinkin, Nathaniel Parker, Irene Bedard, Sheldon Peters Wolfchild, Eric Schweig, Leroy Peltier, Alex Norton, Bray Poor, Stuart Pankin, Mark Margolis, Julian Richings, Donal Donnelly, Paul Klemtowicz
Period adventure that casts Beach as Squanto, a legendary American Indian warrior, who is kidnapped by British explorers and taken to England as a slave. As expected, he struggles to gain his freedom and return to his homeland.
Aka: SQUINTO: A WARRIOR'S TALE
A/AD   98 min (ort 102 min)   VIDrel: WDV/TECH   V

LAST HERO IN CHINA, THE **   12
Wong Ching HONG KONG   1993
Jet Lee (Li Lian-Ji), Gordon Lui (Liu Chia-Hui), Leung Kar Fei
The owner of a martial arts school is forced to vacate when he can no longer pay the rent and relocates to another town where he takes premises next to a brothel. This soon puts him in conflict with the leader of a rival school and a corrupt police chief, both of whom are involved in white slavery. An adequate

effort, built around the character of Wong Fei Hong from the "Once Upon A Time In China" series.
Aka: IRON ROOSTER VERSUS THE CENTIPEDE
MAR   104 min wScrn   VIDrel: MADE/RTM   V

LAST HIT, THE **   15
Jane Egleson USA   1992
Bryan Brown, Brooke Adams, Harris Yulin, Daniel Von Bargen, Sally Kemp, Paul Blott, John David Garfield, Lawrence Parke, Angelina Torres
A CIA assassin finds himself no longer able to kill, and retires from the service, buying a house in New Mexico, and for good measure getting romantically involved with the former owner, an attractive widow. However, when he learns that the last person he was meant to kill happens to be her father, he realises that another assassin will soon appear to do the job, and that his quiet retirement will be no easy matter. Fair.
THR   88 min (ort 93 min)   VIDrel: CIC/SONOP   V

LAST HOUR, THE **   18
William Sachs USA   1991
Michael Pare, Shannon Tweed, Robert Pucci, Bobby Di Ciccio, George Kyle, Danny Trejo, Robert Miano, Raye Hollitt, Anthony Gioia, Thomas Nelson Webb
A woman is held hostage by a gang of ruthless criminals, and two men start planning a rescue mission.
Aka: CONCRETE WAR
A/AD   81 min (ort 85 min)   VIDrel: PROMARK/HIFLI   V

LAST HOUSE ON THE LEFT PART 2, THE **   18
Mario Bava ITALY   1971
Claudine Auger, Chris Avram, Luigi Pistilli, Claudio Volonte, Anna Maria Rosati, Laura Betti, Brigitte Skay, Isa Miranda, Leopoldo Triste, Paola Rubens
Teenagers in an apparently deserted resort find themselves at the mercy of a silent killer, as a series of murders occur, all appearing to be linked in some way to an attempt to possess a valuable piece of real estate. A fair chiller that bears no relation to the similarly titled "The Last House On The Left".
Aka: ANTEFATTO; BAY OF BLOOD; BEFORE THE FACT; BLOODBATH BAY OF DEATH; CARNAGE; ECOLOGIA DEL DELITTO; ECOLOGY OF A CRIME, THE; REAZIONE A CATENA; TWITCH OF THE DEATH NERVE
HOR   80 min (ort 87 min) wScrn dubbed
VIDrel: REDEM/RTM   V

LAST HURRAH FOR CHIVALRY **   (18)
John V.S. Woo HONG KONG   1978
Wei Pai, Liu Sung Sen, Lu Chiang, Wei Chiu Hua, Peng Ke An, Yi kar, Lee Hoi Sang, Chen Lei, Chang Hung Chang, Wang Kuan Yi, Liu Kuo Ping, Chien Yu Sheng, Chang Chin Po, Wen hsiu, Li yung Lung, Feng Ke Feng, Feng Jen Chien
One of those innumerable vengeance tales from Hong Kong. In this story a warrior is attacked on his wedding night by a sworn enemy and starts making plans to have his revenge, recruiting for the purpose a powerful but impoverished fighter. But unknown to this latter fellow, his death is also part of the revenge plan. Adequate.
MAR   101 min   SATrel: SKY MOVIES

LAST IMAGES OF THE SHIPWRECK **   15
Eliseo Subiela ARGENTINA   1989
Lorenzo Quinteros, Noemi Frenkel, Hugo Soto, Pablo Brichta, Sara Benitez, Andres Tiengo, Alicia Aller, Alfredo Stuart
An insurance salesman who's a frustrated writer rescues a young woman from an apparent suicide attempt, finds that she did this only to "acquire" a rescuer, and meets the rest of her strange family, all of whom have now fallen on hard times after their father's desertion. They begin to see him as bringing meaning into their empty lives, but are ultimately disappointed. A strange blend of powerful imagery and sterile, inconclusive dialogue.
Aka: ULTIMAS IMAGENES DEL NAUFRAGIO
DRA   127 min (ort 129 min)   VIDrel: PAL/TERRY L/A   V

LAST INNOCENT MAN, THE ***   18
Roger Spottiswoode USA   1986
Ed Harris, Roxanne Hart, David Suchet, Darrell Larson, Bruce McGill, Rose Gregorio, Clarence Williams III, Robert Lesser, Joe Mays, Meshach Taylor, Michael Durrell, Frank Koppola, Charles Lampkin, Robert Biheller, Lance Rosen
A disenchanted lawyer decides to leave his profession but changes his mind after beginning a passionate affair with a woman, who persuades him to defend her husband against a

charge of murder. An unusual drama with arresting performances and a literate script by Dan Bronson.
DRA   109 min (ort 114 min) mCab   VIDrel: GUILD L/A   V
Boa: novel by Phillip M. Margolin.

LAST LIGHT ***   18
Kiefer Sutherland USA   1993
Kiefer Sutherland, Forest Whitaker, Amanda Plummer, Lynne Moody, Kathleen Quinlan, Clancy Brown, Cameron Dye, Tony T. Johnson, Lydell Cheshier, Eddie Bunker, Mike Gomez, Danny Trejo, Robert Eisele, Valeri Ross, Matt Galle
A prisoner on Death Row strikes up an unlikely relationship with one of the guards who works there and the two men begin to realise that they have certain things in common. However, their friendship comes to an end when our prisoner is executed. An impassioned plea for the abolition of the death penalty, well acted and competently directed by Sutherland in his directorial debut. See also DEAD MAN WALKING.
DRA   100 min (ort 104 min) mCab   VIDrel: TRIM/HIFLI   V/h

LAST MAN STANDING **   18
Joseph Merhi USA   1995
Jeff Wincott, Jillian McWhirter, Steve Eastin, Jonathan Fuller, Jonathan Banks
A cop seeking the killers of his partner, shot while tracking a bank robber, learns that corrupt colleagues were responsible for this outrage. Unable to tell who is friend or foe, he is forced to wage a solitary struggle in his bid for justice.
A/AD   97 min   VIDrel: 20VIS/SONOP   V

LAST MAN STANDING ***   18
Walter Hill USA   1996
Bruce Willis, Christopher Walken, Alexandra Powers, David Patrick Kelly, William Sanderson, Karina Lombard, Ned Eisenberg, Michael Imperioli, Ken Jenkins, R.D. Call, Leslie Mann, Bruce Dern, Ted Markland, Patrick Kilpatrick
Set in a small Texas town in the 1930s, where gang rule has replaced law and order, until the arrival of a mysterious stranger who sets about playing one gang off against the other, his intention being to deal with the survivor himself. An interesting and mostly effective blending of several genres, helped along by the atmospheric music of Ry Cooder and the director's strong script, who doubtless looked to YOJIMBO and A FISTFUL OF DOLLARS for inspiration.
THR   123 min wScrn   VIDrel: EIV/SONOP   V/sh

LAST METRO, THE **   PG
Francois Truffaut FRANCE   1980
Catherine Deneuve, Gerard Depardieu, Jean Poiret, Heinz Bennent, Andrea Ferreol, Paulette Dubost, Sabine Haudepin, Jean-Louis Richard, Maurice Risch, Marcel Berbert, Richard Bohringer, Jean-Pierre Klein, Jean-Jose Richer
A Jewish theatre-owner in occupied Paris hides in the cellars of his establishment while his wife struggles to keep things going and prevent the Germans from finding him. A good idea wasted in a well-made but flatly acted film, not helped by Deneuve's characteristically icy performance and an abrupt ending. Disappointing.
Aka: LE DERNIER METRO
DRA   125 min (ort 135 min)   VIDrel: ARTIF/20TH   V/h

LAST OF HIS TRIBE, THE ***   15
Harry Hook USA   1991
Jon Voight, Graham Greene, David Ogden Stiers, Jack Blessing, Ann Archer, Daniel Bezali, Christianne Hauber, Charles A. Martinet, Carl D. Parker, Angela Paton, Benne B. Adler, Marie Bain, Loryn Barlese, Gilbert Bear
A remake of the TV film ISHI: THE LAST OF HIS TRIBE, that traces the life of the last Yahi Indian, from his childhood to his death in 1917. The story opens with his discovery in 1911, when a museum curator and his wife learn of his existence, and having befriended him, decide to make a record of the customs of his people; a quest that takes on an added urgency in the face of the tribe's imminent extinction. A sad and touching movie. See also LAST OF THE DOGMEN.
WES   89 min   VIDrel: VCC/DISC L/A   V
Boa: novel Ishi In Two Worlds by Theodora Kroeber Quinn.

LAST OF THE DOGMEN **   PG
Tab Murphy USA   1995
Tom Berenger, Barbara Hershey, Kurtwood Smith, Steve Reevis, Andrew Miller, Gregory Scott Cummins, Mark Boone Jr, Helen
Calahasen, Eugene Blackbear, Dawn Lavand, Sidel Standing Elk, Hunter Bodine, Parley Baer, Georgie Collins
Three escaped prisoners are pursued into the Rockies by a bounty hunter who makes an unbelievable discovery when he comes across a group of Native Americans belonging to a tribe thought to have become extinct a century ago. Helped by an anthropologist, he must take action to save them from the ravages of the modern world. The fine landscapes of Montana are the best players in this film, done in a variety of styles by writer-director Murphy. See also THE LAST OF HIS TRIBE.
A/AD   113 min (ort 118 min)   VIDrel: GUILD/FOXVID   V/sur

LAST OF THE HIGH KINGS, THE ***   15
David Keating DENMARK/EIRE/UK   1996
Jared Leto, Christina Ricci, Gabriel Byrne, Catherine O'Hara, Stephen Rea, Lorraine Pilkington, Jason Barry, Emily Mortimer, Karl Hayden, Colm Meaney, Des Braden, Ciaran Fitzgerald, Darren Monks, Peter Keating, Renee Weldon
Gentle, coming-of-age saga set in Ireland, set among a family of oddball characters, such as a father (Byrne) who is mostly absent and a mother (O'Hara) of very set opinions. Set in the summer of 1977, it mostly deals with the exploits of a youngster, who cannot wait for school to end so that he can get on with his real interests: watching girls and messing about. Covering familiar ground, this likeable film says nothing new, but its charm keeps one involved.
DRA   100 min (ort 104 min)   VIDrel: FIRST/SONOP   V
Boa: novel by Ferdia MacAnna.

LAST OF THE MOHICANS ***   PG
James L. Conway USA   1977
Steve Forrest, Ned Romero, Don Shanks, Andrew Prine, Robert Tessier, Jane Actman, Michele Marsh, Robert Easton, Whit Bissell, Beverly Rowland, Dehl Berti, John G. Bishop, Coleman Creel, Rosalyn Mike, Reid Sorenson
A "Classics Illustrated" version of this much-filmed Western yarn about a white hunter and his Indian blood brothers, who together help a British officer escort two women through Indian country, at the time of the French and Indian War. Colourful and spirited.
WES   95 min (ort 100 min) mTV   VIDrel: CASPIC L/A   V
Boa: novel by James Fenimore Cooper.

LAST OF THE MOHICANS, THE ***   15
Michael Mann USA   1992
Daniel Day-Lewis, Madeleine Stowe, Russell Means, Wes Studi, Eric Schweig, Johdi May, Steven Waddington, Maurice Roeves, Patrick Chereau, Colm Meaney, Edward Blatchford, Terry Kinney, Tracey Ellis, Justin M. Rice, Dennis J. Banks
In 1757, a white frontiersman raised by the Mohican Indians, tries to remain neutral in the colonial wars between the British and the French. However, when he rescues the daughter of a British officer, he and his Indian companions find themselves drawn into this conflict. A visually striking and lavishly mounted adaptation of this much-filmed novel, but with little dialogue or depth of characterisation. AA: Sound.
A/AD   107 min (ort 122 min) wScrn cC   VIDrel: WHV   V/dm
Boa: novel by James Fenimore Cooper.

LAST OF THE SUMMER WINE: UNCLE OF THE BRIDE **   PG
Alan J.W. Bell UK   1985
Bill Owen, Peter Sallis, Michael Aldridge, Thora Hird, Jane Freeman, Joe Gladwin, Kathy Staff
Spin-off of a popular British TV series dealing with the escapades of a trio of eccentric senior citizens in a small Northern town. Various strange individuals meet at a wedding in this feature.
Aka: UNCLE OF THE BRIDE
COM   84 min mTV   VIDrel: BBC/TECH   V/h

LAST OUTLAW, THE **   15
Geoff Murphy USA   1993
Mickey Rourke, Dermot Mulroney, Ted Levine, John C. McGinley. Keith David, Steve Buscemi, Daniel Quinn, Gavan O'Herlihy, Richard Fancy, Tom Connor, Paul Ben-Victor, Greg Doty, John David Garfield, Sid Klinge, Phil Mead, Jake Walker
Left for dead by his own men who could no longer stand his blood lust, a renegade ex-Confederate officer turned outlaw survives and sets out on a bloody revenge. A bleak and unedi-

fying oater that recalls many a spaghetti Western of earlier years, but without any style or humour.
WES   89 min mCab   VIDrel: POLY/POLYREC   V/sur

LAST PICTURE SHOW, THE ****   15
Peter Bogdanovich USA   1971
Cybill Shepherd, Jeff Bridges, Timothy Bottoms, Cloris Leachman, Ellen Burstyn, Ben Johnson, Eileen Brennan, Randy Quaid, Clu Gulager, Sharon Taggart, Joe Heathcock, Bill Thurman, Barc Doyle, Jessie Lee Fulton, Gary Brockette
A penetrating study of a small and dusty Texas town in 1951 and of its inhabitants and their relationships. Despite the mundane nature of the script, a clutch of effortless performances makes for a fascinating film, which winds up to a strong and fitting climax with the closing of the town's local cinema. The beautiful B/W photography is by Robert Surtees. Followed by TEXASVILLE a good few years later. AA: S. Actor (Johnson), S. Actress (Leachman).
DRA   115 min (ort 148 min) B/W   VIDrel: VCC L/A   V
Boa: novel by Larry McMurty.

LAST PLATOON *   18
Paul D. Robinson ITALY   1988
Richard Hatch, Max Laurel, Anthony Sawyer, Vassili Karis, Maricar, Donald Pleasence, Mike Monti, David Light, Mylene Thy-Sanh
A bunch of raw recruits and their tough sergeant are sent on a mission into the jungles of Vietnam, but find themselves trapped and at the mercy of the enemy. A standard offering.
WAR   95 min   VIDrel: MIA/DISC/VPD   V

LAST PROSTITUTE, THE **   15
Lou Antonio USA   1991
Sonia Braga, Wil Wheaton, Cotter Smith, David Kaufman, Dennis Letts, Woody Watson, Dru Mouser, Babs George, Richard Pillard, Brad Leland, Henry Ellerman
Two teenage boys set off on a cross-country journey, their sole aim being to lose their virginity. However, events conspire to turn their trip into a journey of self-discovery. A sharply focused coming-of-age tale that (despite the rather silly title) is not in the least bit titillating, but instead offers some perceptive and touching insights into the problems of early adulthood.
DRA   89 min (ort 95 min) mCab   VIDrel: CIC/SONOP   V/h

LAST ROMANTICS, THE ***   (PG)
Jack Gold UK   1991
Ian Holm, Sara Kestelman, Leo McKern, Alan Cumming, Rufus Sewell, John Lloyd Fillingham, Helga Brindle, Miranda Forbes, Charles Dale, Alice Douglas, Sean O'Callaghan
In the quiet, isolated and peaceful world of Cambridge University, a tutor has to contend with the arrival of a difficult and disruptive student, the activities of whom forces him to re-examine his own life, and the manner in which he betrayed a university tutor when he was himself a student. A brooding and thoughtful character study, that does much to bring to life the slightly claustrophobic atmosphere of the setting. The script is by Nigel Williams.
DRA   90 min mTV   TVrel: BBC

LAST SEDUCTION, THE ***   18
John Dahl USA   1995
Linda Fiorentino, Bill Pullman, J.T. Walsh, Peter Berg, Bill Nunn, Herb Mitchell, Dean Norris, Brian Varady, Donna Wilson, Mik Sriba, Mike Liscenco, Serena, Micahel Raysses, Zack Phifer, Erik-Anders Nilsson, Patrick K. Caprio
A wife steals the money her husband made in a drugs deal and takes to the road, where she meets up with a number of men who all suffer as a result of an encounter with this deadly woman. Fiorentino give a chilling portrait of a totally manipulative and utterly ruthless creature, who is as cunning as she is beautiful. A fascinating and very dark piece, that maintains its cynical atmosphere right up to the climax.
DRA   105 min (ort 110 min)   VIDrel: ITC/POLYREC   V/h

LAST STAND AT LANG MEI **   18
Cirio H. Santiago   1990
Steve Kanaly, Ken Wright, Peter Nelson, John Vargas, Carl Franklin
Standard Vietnam heroics that tells of the fate of a bunch of GIs deserted by their country and left to face the Vietcong alone. Average.
WAR   88 min   VIDrel: TRING/COLUM   V

LAST SUPPER, THE ***   (15)
Cynthia Roberts CANADA   1995
Daniel MacIvor, Jack Nicholson, Ken McDougall
A dancer who has been infected with AIDS decides to cut short his suffering by taking his own life and organises a farewell ceremony attended by his lover and his doctor. A harrowing and deeply moving film, that is part fiction and part documentary, it was made as a tribute to the actor McDougall, who died in 1994, just four days after shooting was completed. See also IT'S MY PARTY, which attempted to explore the same theme.
DRA   96 min   CINrel
Boa: play by Hillar Liitoja.

LAST SUPPER, THE ***   15
Stacy Title USA   1995
Cameron Diaz, Ron Eldard, Annabeth Gish, Jonathan Penner, Courtney B. Vance, Jason Alexander, Nora Dunn, Charles Durning, Bryn Erin, Mark Harmon, Dan Rosen, Bill Paxton, Ron Perlman, Amber Taylor, Matt Cooper, Gil Segel, Rachel Chagall
Five young yuppies run a discussion group, and each week they have a guest speaker to air his or her views on life and the world, as the guests are invariably right-wing bigots, it sometimes happens that they decide to poison a guest speaker if the views expressed are too extreme for their tastes. A wickedly sharp black comedy, it might have worked better as a thirty-minute revue, having not quite enough substance for a full-length feature.
COM   87 min (ort 91 min)   VIDrel: 20VIS/SONOP   V/sh

LAST TANGO IN PARIS ***   18
Bernardo Bertolucci FRANCE/ITALY/USA   1972
Marlon Brando, Maria Schneider, Jean-Pierre Leaud, Darling Legitimus, Catherine Sola, Mauro Marchetti
A middle-aged American meets a young French girl when they both go to view the same apartment, and initiates a strange sexual relationship with her. Massively over-rated when released, this is a rather pretentious and fairly meaningless film, though Brando does give a wonderful performance.
Aka: ULTIMO TANGO A PARIGI
DRA   124 min (ort 129 min)   VIDrel: MGM/WHV   V

LAST TEMPTATION OF CHRIST, THE ***   18
Martin Scorsese USA   1988
Willem Dafoe, Harvey Keitel, Barbara Hershey, Harry Dean Stanton, David Bowie, Paul Greco, Steven Shill, Andre Gregor, Verna Bloom, Roberts Blossom, Barry Miller, Gara Basaraba, Irvin Kerschner, Victor Argo, Michael Been
Despite the wave of accusations of blasphemy unjustifiably heaped upon this film, this portrayal of Christ in terms of his human feelings is neither shocking nor sensational, but merely unexciting and unworthy of the novel on which it is based. There are some genuinely moving moments; if only the film had been directed less self-indulgently. The score is by Peter Gabriel. See also JESUS OF NAZARETH, THE GOSPEL ACCORDING TO ST MATTHEW and KING OF KINGS.
DRA   156 min (ort 163 min)   VIDrel: CIC/SONOP   V/sur
Boa: novel by Nikos Kazantzakis.

LAST TO GO, THE **   15
John Erman USA   1990
Tyne Daly, Terry O'Quinn, Annabeth Gish, Tim Ransom, Julianne Moore, Sarah Trigger, Matthew Labyroteaux, Justic Gocke, Lauren Woodland, Scott Nell, Hartley Haventy, Curt Lowens, Tuck Milligan, Noelle Parker, John Petlock
A married woman is shattered when her husband abandons her for a younger woman, and faces a painful re-adjustment to a situation that she never envisaged. A fine performance by Daly lends interest to this simple tale, which is not helped by the unoriginal script. See also AN UNMARRIED WOMAN and LOVE AND BETRAYAL.
DRA   91 min (ort 100 min) mTV   VIDrel: ITC/HIFLI L/A   V/h

LAST TRAIN FROM GUN HILL, THE ***   15
John Sturges USA   1959
Kirk Douglas, Anthony Quinn, Carolyn Jones, Earl Holliman, Brad Dexter, Ziva Rodann, Brian Hutton, Val Avery, Walter Sande, Lars Henderson, John P. Anderson, Lee Hendry, William Newell, Sid Tomack, Charles Stevens, Jack Lomas
A marshal's efforts to bring the killer of his wife to justice, are complicated by the fact that the latter is the son of an old friend. A taut and well-paced Western that poses the moral

question as to whether turning a blind eye is the best way to resist evil.
WES   90 min (ort 98 min)
VIDrel: 4-FRONT/POLYREC/BRAVE   V

LAST TYCOON, THE ***   15
Elia Kazan USA   1976
Robert De Niro, Tony Curtis, Robert Mitchum, Jeanne Moreau, Jack Nicholson, Donald Pleasence, Ray Milland, Ingrid Boulting, Dana Andrews, Peter Strauss, John Carradine, Theresa Russell, Jeff Corey, Angelica Huston
Film adaptation of an unfinished novel about Hollywood and the film moguls of the late 1920s and 1930s. De Niro is cast as an Irving Thalberg-type movie boss, who finds himself falling for Boulting as she reminds him of his late wife. A restrained adaptation by Harold Pinter with De Niro giving one of his best performances. Music is by Maurice Jarre.
DRA   118 min (ort 124 min)   VIDrel: L/A   V
Boa: novel by F. Scott Fitzgerald.

LAST UNICORN, THE *   U
Jules Bass/Arthur Rankin Jr USA   1982
Voices of: Alan Arkin, Jeff Bridges, Tammy Grimes, Angela Lansbury, Mia Farrow, Robert Klein, Christopher Lee, Keenan Wynn, Paul Frees, Rene Auberjonois,
Animated story of a unicorn who goes in search of others of her kind and after various adventures, frees all the other unicorns from captivity. A slack and dispirited affair.
ANIM   84 min (ort 95 min)   VIDrel: 4-FRONT/POLYREC V/sur
Boa: novel by Peter S. Beagle.

LAST WAR, THE **   (15)
Shue Matsubayashi JAPAN   1961
Yuriko Hoshi, Frankie Sakai, Akira Takarada, Nabuko Otawa, Yumi Shirakawa
During a period of high international tension, a misunderstanding between the superpowers leads to a nuclear exchange and Earth is destroyed. Supposedly set in the year 2015 (but firmly rooted in the 1960s) this rather cheap-looking and dated film is as ponderous as it is earnest. See also THE DAY AFTER and TESTAMENT.
Aka: FINAL WAR, THE; SEKAI DAISENSO
FAN   76 min (ort 79 min)   VIDrel: L/A   V

LAST WARRIOR, THE **   18
Martin Wragge AUSTRALIA   1989
Gary Graham, Maria Holvoe, Cary Hiroyuki-Tagawa, John Carson, Steven Ito, Al Karaki, Peggy Champion, Victoria Cooper, Ingrid Ernsley, Pippa Duffy, Eva Davis, Sheryl Burton
A clone of HELL IN THE PACIFIC that has a Japanese and an American soldier battling it out on a remote island during WW2. Quite a good looking yarn, despite the pronounced lack of originality.
Aka: COASTWATCHER
A/AD   90 min (ort 94 min)   VIDrel: 20VIS/SONOP   V

LAST YEAR AT MARIENBAD ****   U
Alain Resnais FRANCE/ITALY   1961
Delphine Seyrig, Giorgio Albertazzi, Sacha Pitoeff, Francoise Bertin, Luce Garcia-Ville, Helena Kornel, Francois Spira, Karin Toeche-Mittler, Pierre Barbaud, Wilhem Von Deek, Jean Danier, Gerard Lorin, Davide Montemuri
In an enormous baroque hotel, a young man attempts to persuade a female guest that they had an affair the year before, and that she should leave her partner and run away with him. Lacking a clear narrative and chronological structure, this elegant, erotic and dreamlike film blends reality with fantasy and past with present. Not so much a film as a stylised and haunting puzzle without a solution. This was Seyrig's first feature film.
Aka: L'ANNEE DERNIERE A MARIENBAD; L'ANNO SCORSO A MARIENBAD
DRA   90 min (ort 100 min) B/W wScrn
VIDrel: CONNO/RTM   V

L'ATALANTE ****   PG
Jean Vigo FRANCE   1934
Dita Parlo, Jean Daste, Michel Simon, Gilles Margarites, Louis Lefevre, Diligent Raya, Maurice Gilles, Fanny Clar
An acknowledged classic of the French cinema from a director who died at twenty-nine, leaving this behind as his only full-length work. Here, reality and dreams are mixed in a study of

a barge captain and his young wife as they begin their married life sailing down the Seine. The pace is leisurely, but there is much to enjoy for those willing to accept this film on its own terms. See also ZERO DE CONDUITE, Vigo's only other available work.
Aka: LE CHALAND QUI PASSE
DRA   86 min (ort 89 min) B/W   VIDrel: ARTIF/20TH   V

LATE FOR DINNER **   PG
W.D. Richter USA   1991
Brian Wimmer, Peter Berg, Marcia Gay Harden, Colleen Flynn, Kyle Secor, Michael Beach, Peter Gallagher, Cassy Friel, Ross Malinger, John Prosky, Steven Schwartz-Hartley, Bo Brundin, Donald Hotton, Billy Vera, Jeremy Roberts
Two young men on the run meet a character called Dr Chilblains, and he freezes them as part of his cryogenic research. When they thaw out twenty-nine years later, they pay a visit to their somewhat older families. Fortunately, true love is enough to conquer all, in this quirky celebration of home values and the simple things in life.
COM   89 min (ort 93 min)   VIDrel: FIRST/SONOP   V/sur

LATE SHOW, THE ***   15
Robert Benton USA   1977
Art Carney, Lily Tomlin, Bill Macy, Eugene Roche, Joanna Cassidy, John Considine, Howard Duff, Ruth Nelson
An ex-private eye comes out of retirement when his old partner is killed and investigates the case. Carney and Tomlin work well together, with the latter giving a convincing performance as an aimless young woman who decides to assist him. Complex, amiable and quite funny. Later gave rise to a brief TV series. The script is by Benton.
COM   94 min   VIDrel: MGM/WHV L/A   V

LATEX ***   18
Michael Ninn USA   1995
Sunset Thomas, Julia Ashton, Jon Dough, Tyffany Million, Sam Cooper
In a dehumanised and bleak world of the future, a man finds he has the ability to bring to life people's secret fantasies just by touching them. As these fantasies are often bound up with feelings of shame or guilt, the effect of their release is cathartic and his services are much in demand. Eventually, the authorities catch up with him, and he is tried and imprisoned in an insane asylum. A strongly plotted and most unusual offering.
A   71 min   VIDrel: PURG/DANTE   V

LATINO ***   15
Haskell Wexler USA   1985
Robert Beltran, Annette Cardona, Tony Plana, Ricardo Lopez, Gavin McFadden, Marta Tenorio, Michael Goodwin, Luis Torrentes, Juan Carlos Ortiz, Julio Medina, Mayra Juarro, Walter Marin, James Karen
Set in Nicaragua, this purports to be an examination of the undercover wars of the US. Secretly sent into combat minus identification, a Green Beret finds himself fighting a war that is not his, and one that includes innocent civilians. An irritating film with a pronounced left-wing bias, but remarkably well directed and acted. Filmed in Nicaragua.
A/AD   105 min (ort 108 min)   VIDrel: 20TH/TECH   V/sur

LAUGHING DEAD, THE *   (18)
S.P. Somtow USA   1989
Matthew De Merritt, Premika Eaton, Ryan Effner, Larry Kagen, Krista Kem, Raymond Ridenour, Patrick Roskowick, Wendy Weil, George Salazar, Billy Silver, S.P. Somtow, Vanna Sucharitkul, Tim Sullivan, Maritz Tamara, Joey Acedo
A novelist who becomes possessed by a Mayan demon after attending the festival of the dead in Mexico in a town where a Mayan site is being studied by archaeologists and much mayhem and gore ensue. A typical low-budget horror offering with little in the way of decent plot or acting.
HOR   92 min   CABrel: HVC

LAUGHTER IN PARADISE **   U
Mario Zampi UK   1951
Alastair Sim, Fay Compton, Beatrice Campbell, Veronica Hurst, Guy Middleton, George Cole, A.E. Matthews, Joyce Grenfell, Anthony Steel, Ronald Adam, John Laurie, Eleanor Summerfield, Leslie Dwyer, Ernest Thesiger, Hugh Griffith
A man with a penchant for practical jokes dies and leaves a large sum to each of four relatives, but with strange conditions attached, A female snob must work as a maid, a playboy has to marry the first girl he sees, a timid bank clerk must carry out a

hold-up and a crime novelist must spend twenty-eight days in jail. Well acted and with some amusing sequences, but too patchy to succeed. Remade in 1972 as SOME WILL, SOME WON'T.
COM   92 min (ort 94 min) B/W   VIDrel: LUMI/SPEAR   V

LAURA ***
U
Otto Preminger USA   1944
Gene Tierney, Dana Andrews, Clifton Webb, Vincent Price, Judith Anderson, Dorothy Adams, Grant Mitchell, Clyde Fillmore, Ralph Dunn, Kathleen Howard, James Flavin, Lane Chandler, Harold Schlickenmayer, Harry Strang, Frank La Rue
A detective investigating the alleged murder of a beautiful woman finds himself falling madly in love with her. A classic example of film noir with a perfect cast and a witty script by Jay Dratler, Samuel Hoffenstein and Betty Reinhardt. The theme song is by David Raskin. Rouben Mamoulian started directing this one but Preminger took over. The 88 minute running time was reduced owing to a dispute over music rights. AA: Cin (Joseph LaShell).
THR   88 min B/W   VIDrel: 20TH/TECH   V/h
Boa: novel by Vera Caspary.

LAUREL AND HARDY: A CHUMP AT OXFORD **
U
Alfred Goulding USA   1939
Stan Laurel, Oliver Hardy, Forrester Harvey, Wilfrid Lucas, Forbes Murray, James Finlayson, Anita Garvin, Peter Cushing, Frank Baker, Eddie Borden, Gerald Fielding, Gerald Rogers, Victor Kendall, Rex Lease, Stanley Blystone
Our two stars play a couple of street-cleaners whose efforts in foiling a bank robbery win them a college education as a reward. A highly uneven and only occasionally amusing romp, whose effectiveness is seriously diluted by unrelated comic episodes. However, with their arrival at Oxford (patience is needed) matters improve somewhat. Laurel's transmutation into the fearsome Lord Paddington is a highlight.
Aka: CHUMP AT OXFORD, A
COM   61 min (ort 63 min) B/W   VIDrel: VISVID   V

LAUREL AND HARDY: A-HAUNTING WE WILL GO **
U
Alfred Werker USA   1942
Stan Laurel, Oliver Hardy, Harry A. Jansen, Sheila Ryan, John Shelton, Don Costello, Elisha Cook Jr, Dante the Magician, Edward Gargan, Addison Richards, George Lynn, James Bush, Lou Lubin, Robert Emmett Keane, Willie Best
A criminal gang trick our heroes into carrying a wanted gangster across the state line in a coffin. However, it gets switched with one used in an act by a stage magician who takes our duo on as inept if eager assistants. Laurel and Hardy look uncomfortable and out of place in a film whose faster 1940s tempo suits neither their personalities nor their timing. A ragbag of semi-comic incidents, only a few of which work.
Aka: A-HAUNTING WE WILL GO
COM   61 min (ort 69 min) B/W   VIDrel: 20TH/TECH   V/h

LAUREL AND HARDY: AIR RAID WARDENS *
U
Edward Sedgwick USA   1943
Stan Laurel, Oliver Hardy, Edgar Kennedy, Jacqueline White, Horace (Stephen) McNally, Nella Walker, Donald Meek, Russell Hicks, Howard Freeman, Henry O'Neill, Paul Stanton, Robert Emmett O'Connor, Lee Phelps, Martin Cichy
Having been rejected as unfit for military service, two idiotic air raid wardens capture a bunch of Nazi spies entirely by accident. A weak and flawed comedy that had an appeal on release that has now largely faded. Very disappointing.
Aka: AIR RAID WARDENS
COM   68 min B/W   VIDrel: MGM/WHV   V

LAUREL AND HARDY: BABES IN TOYLAND ****
PG
Gus Meins/Charles R. Rogers USA   1934
Stan Laurel, Oliver Hardy, Charlotte Henry, Henry Brandon, Felix Knight, Jean Darling, Florence Roberts, Johnny Downs, Gus Leonard, Ferdinand Munier, Marie Wilson, William Burress, Virginia Karns, Johnny Downs, Frank Austin
Film version of a Victor Herbert operetta which incorporates many fairy tale figures. An exuberant comedy-fantasy with our two comedians giving one of their best performances, easily making the most of the thin and uneven script.
Aka: BABES IN TOYLAND; LAUREL AND HARDY IN TOYLAND; MARCH OF THE WOODEN SOLDIERS; WOODEN SOLDIERS
MUS   70 min (ort 79 min) B/W   VIDrel: EUREKA/GOLD V
Boa: story by Glen MacDonough/operetta by Victor Herbert.

LAUREL AND HARDY: BLOCKHEADS ***
U
John G. Blystone USA   1938
Stan Laurel, Oliver Hardy, Billy Gilbert, Patricia Ellis, James Finlayson, Minna Gombell
Stan becomes a hero having guarded a trench for twenty years since WW1. But he finds it a little difficult to adjust to civilian life when he is taken home by Ollie. A pleasant outing that is a little too noisy and repetitive for its own good, but several classic gags are a compensation.
Aka: BLOCKHEADS
COM   67 min B/W coVer   VIDrel: VISVID/POLYREC L/A V

LAUREL AND HARDY: BOGUS BANDITS ***
(U)
Hal Roach/Charles R. Rogers USA   1933
Oliver Hardy, Stan Laurel, Dennis King, James Finlayson, Thelma Todd, Henry Armetta, Lucille Browne, Arthur Pierson, Matt McHugh, Lane Chandler, Nena Quartero, Wilfred Lucas, James C. Morton, Carl Harbaugh, George Miller
Laurel and Hardy travel to the Italian Alps by mule, getting involved in some hilarious situations. Based on a romantic operetta.
Aka: BOGUS BANDITS; DEVIL'S BROTHER, THE; FRA DIAVOLO; VIRTUOUS TRAMPS, THE
COM   74 min (ort 88 min) B/W   VIDrel: L/A   V

LAUREL AND HARDY: GREAT GUNS **
U
Montague Banks USA   1941
Stan Laurel, Oliver Hardy, Sheila Ryan, Dick Nelson, Edmund MacDonald, Allan Webb, Charles Trowbridge, Ludwig Stossel, Mae Marsh, Kane Richmond, Ethel Griffies, Paul Harvey, Charles Arnt, Pierre Watkin, Russell Hicks
In this feeble comedy, Stan and Ollie join the Texas Cavalry to be with their millionaire employer, to whom they are entirely devoted. Quite watchable but far off the standard set by their best work. Watch out for Alan Ladd who puts in a brief appearance.
Aka: GREAT GUNS
COM   70 min (ort 74 min) B/W   VIDrel: 20TH/TECH   V/h

LAUREL AND HARDY: HEROES OF THE REGIMENT ***
U
James Horne USA   1935
Stan Laurel, Oliver Hardy, June Lang, James Finlayson, William Janney, Anne Grey, Vernon Steele, David Torrence, James Mark, Mary Gordon, Maurice Black, Daphne Pollard, Lionel Belmore, James May, Kathryn Sheldon, Minerva Urecal
Having arrived in Scotland to claim a non-existent inheritance, Stan and Ollie inadvertently join a Scottish regiment and find themselves in India. A parody on "Lives Of A Bengal Lancer" in which the paper-thin plot is a minor distraction from a number of very funny sequences, of which a clean-up detail is particularly memorable.
Aka: BONNIE SCOTLAND; HEROES OF THE REGIMENT
COM   77 min (ort 81 min) B/W   VIDrel: MGM/WHV   V

LAUREL AND HARDY: NOTHING BUT TROUBLE *
U
Sam Taylor USA   1944
Stan Laurel, Oliver Hardy, Mary Boland, Philip Merivale, David Leland, Henry O'Neill, John Warburton, Matthew Boulton, Connie Gilchrist, Paul Porcasi, Jean De Briac, Joe Yule Sr, Eddie Dunn, Ray Teal, Howard Mitchell
A minor Laurel and Hardy vehicle, in which they are hired to work as servants in a household where they inadvertently foil an attempt to poison a young king. A weak and shallow effort, the last comedy the pair made for MGM.
Aka: NOTHING BUT TROUBLE
COM   71 min (ort 90 min) B/W   VIDrel: MGM/WHV   V

LAUREL AND HARDY: OUR RELATIONS **
U
Harry Lachman USA   1936
Stan Laurel, Oliver Hardy, Sidney Toler, Alan Hale, James Finlayson, Daphne Pollard, Betty Healy, Iris Adrian, Noel Madison, Ralf Harolde, Arthur Housman, Jim Kilganon, Charlie Hall, Harry Bernard, Harry Arras, John Kelly
Our comic duo gets into all kinds of trouble on account of their long-lost twin brothers. A patchy and uneven work that starts off with great promise, but dissipates much of its energy on overlong chase sequences without developing the inherent potential of the plot.
Aka: OUR RELATIONS
COM   70 min (ort 72 min) B/W coVer
VIDrel: VISVID/POLYREC   V
Boa: short story The Money Box by W.W. Jacobs

LAUREL AND HARDY: PACK UP YOUR TROUBLES **   U
George Marshall/Raymond McCarey USA   1931
Stan Laurel, Oliver Hardy, Charles Middleton, George Marshall, Mary Carr, James Finlayson, Donald Dillaway, Dick Cramer, Tom Kennedy, Billy Gilbert, Grady Sutton, Jacquie Lyn, C. Montague Shaw, Muriel Evans, Bill O'Brien
When her father dies, a young girl is entrusted to the care of his two pals, a couple of former WW1 combatants. Good fun but disjointed and generally aimless.
Aka: PACK UP YOUR TROUBLES
COM   65 min (ort 68 min) B/W coVer
VIDrel: VISVID/POLYREC   V

LAUREL AND HARDY: PARDON US **   U
James Parrott USA   1931
Stan Laurel, Oliver Hardy, Wilfred Lucas, Walter Long, James Finlayson, June Marlowe, Charlie Hall, Sam Lufkin, Silas D. Wilcox, George Miller, Frank Holliday, Harry Bernard, Stanley J. (Tiny) Sanford, Frank Austin, Otto Fries
First feature from our comic duo is a rather dated prison spoof that has some humorous moments, most of which revolve around some difficulties Ollie has with a loose tooth. Patchy and muddled but quite likeable. A highlight is the unusual song-and-dance routine about halfway through.
Aka: JAILBIRDS; PARDON US
COM   52 min (ort 55 min) B/W coVer
VIDrel: VISVID/POLYREC   V

LAUREL AND HARDY: SAPS AT SEA ***   U
Gordon Douglas USA   1940
Stan Laurel, Oliver Hardy, James Finlayson, Richard Cramer, Ben Turpin, Harry Bernard, Eddie Conrad, Harry Hayden, Charlie Hall, Patsy Moran, Gene Morgan, Charles A. Bachman, Bud Geary, Jack Greene, Eddie Bordon
Laurel and Hardy's last film for Hal Roach has Ollie recuperating after suffering a breakdown brought on by his job as a horn factor. But his fishing vacation with Stan soon has them both at sea. A solid work that's a little short on gags.
Aka: SAPS AT SEA
COM   56 min (ort 60 min) B/W coVer
VIDrel: VISVID/POLYREC   V

LAUREL AND HARDY: SONS OF THE DESERT ****   U
William A. Seiter USA   1933
Stan Laurel, Oliver Hardy, Charley Chase, Dorothy Christy, Mae Busch, Lucien Littlefield, John Elliott, Charley Young, John Merton, William Gillespie, Charles McAvoy, Robert Burns, Al Thompson, Eddie Baker
Having given their oath to attend a fraternal convention, Stan and Ollie discover that their wives have other ideas and are forced to employ a complex ruse that involves Ollie contracting a disease requiring a long sea voyage. One of the very best of their comedies; complex, vigorous and hilarious.
Aka: FRATERNALLY YOURS; SONS OF THE DESERT
COM   64 min (ort 69 min) B/W coVer
VIDrel: VISVID/POLYREC L/A   V

LAUREL AND HARDY: SWISS MISS **   U
John G. Blystone USA   1938
Stan Laurel, Oliver Hardy, Della Lind, Walter Woolf King, Eric Blore, Adia Kuznetzof, Charles Judels, Eddie Kane, Anita Garvin, Franz Hug, Sam Lufkin, Ludovico Tomarchio, Tex Driscoll, George Sorel, Harry Semels, Otto Jehle
Two mousetrap salesmen in Switzerland become involved with an actress, who is scheming to arouse her husband's jealousy. The romantic setting doesn't give our comic duo much chance to shine, although there is a nice moment when Ollie gets to serenade his beloved.
Aka: SWISS MISS
COM   67 min (ort 73 min) B/W coVer   VIDrel: VISDVID   V

LAUREL AND HARDY: THE BOHEMIAN GIRL **   U
James Horne/Charles R. Rogers USA   1936
Stan Laurel, Oliver Hardy, Mae Busch, Darla Hood, Jacquelle Wells (Julie Bishop), Antonio Moreno, James Finlayson, Thelma Todd, William P. Carleton, Zeffie Tilbury, Harry Bowden, James C. Morton, Mitchell Lewis, Eddie Borden
Laurel and Hardy in their last comic opera as the guardians of a young girl who is a kidnapped princess, although nobody

seems to realise this. The singing unfortunately tends to get in the way of the gags in this one, which all too often misfire.
Aka: BOHEMIAN GIRL, THE
COM   67 min (ort 74 min) B/W coVer
VIDrel: VISVID/POLYREC   V

LAUREL AND HARDY: THE BULLFIGHTERS **   U
Mal St Clair USA   1945
Stan Laurel, Oliver Hardy, Margo Woode, Richard Lane, Carol Andrews, Diosa Costello
Laurel and Hardy play two detectives, who pursue a female criminal over the border into Mexico, where Laurel is mistaken for a famous matador, and winds up having his moment of glory in the bullring. This last American feature shows a few flashes of humour but is largely a poor vehicle for their talents.
Aka: BULLFIGHTERS, THE
COM   61 min B/W   VIDrel: 20TH/TECH   V/h

LAUREL AND HARDY: THE DANCING MASTERS **   U
Mal St Clair USA   1943
Stan Laurel, Oliver Hardy, Trudy Marshall, Bob Bailey, Margaret Dumont, Matt Briggs, Robert Mitchum
Laurel and Hardy are cast here as the owners of a ballet school who find themselves involved with gangsters. Our duo work well together but lack material, and the clumsily back-projected runaway bus sequence fails to amuse.
Aka: DANCING MASTERS, THE
COM   64 min B/W   VIDrel: 20TH/TECH   V/h

LAUREL AND HARDY: THE FLYING DEUCES **   U
A. Edward Sutherland USA   1939
Stan Laurel, Oliver Hardy, Jean Parker, James Finlayson, Reginald Gardiner, Charles Middleton, Jean Del Val, Clem Wilenchick, Crane Whitley, Rychard Cramer, Michael Visaroff, Monica Bannister, Bonnie Bannon, Mary Jane Carey
Stan and Ollie join the Foreign Legion so that Stanley can forget an unhappy romance – the usual mixture of mayhem and complications follow. An over-extended farce that has the charming musical number – "Shine On, Harvest Moon", an interminable chase sequence, a plane crash and finally Stan seeing Ollie reincarnated as a horse.
Aka: FLYING ACES; FLYING DEUCES, THE
COM   61 min (ort 65 min) B/W
VIDrel: 4-FRONT/POLYREC/ODY; ORBIT/DISC   V

LAUREL AND HARDY: UTOPIA *   U
Leo Joannon/John Berry FRANCE   1950
Stan Laurel, Oliver Hardy, Suzy Delair, Max Elloy
An island paradise in the South Pacific is threatened by the discovery of uranium and the plans of a crooked lawyer. Our duo's last film, and not one of their best.
Aka: ATOLL K; ROBINSON CRUSOELAND; UTOPIA
COM   82 min B/W   VIDrel: ORBIT/DISC   V

LAUREL AND HARDY: WAY OUT WEST ***   U
James W. Horne USA   1937
Stan Laurel, Oliver Hardy, Sharon Lynn, James Finlayson, Rosina Lawrence, Stanley Fields, Jim Mason, James C. Morton, Frank Mills, Dave Pepper, Vivien Oakland, Harry Bernard, Mary Gordon, May Wallace, The Avalon Boys, Jack Hill
Well-paced and funny Laurel and Hardy comedy, with our inept pair being tricked into handing over the deeds to a mine, to a girl who is impersonating the rightful heiress. Easily one of their best films with some great routines and a couple of charming musical interludes such as "Shoe Shuffle" and "The Trail Of The Lonesome Pine".
Aka: WAY OUT WEST
COM   60 min (ort 66 min) B/W   VIDrel: VISVID/POLYREC   V

LAURIN **   18
Robert Sigl HUNGARY/WEST GERMANY   1990
Dora Szimetar, Brigitte Karner, Zoltan Gera, Karoly Eperjes, Hedi Temessy, Barnabas Toth, Katalin Sir, Zoltan Gera, Endre Katay, Janos Dersi, Ildiko Hamori, Ottilia Marschek, Gabriella Marschek, Attila Hajdu, Gabor Nemeth
A woman who is troubled by recurrent nightmares and hallucinations puts her life at risk when she attempts to learn their cause.
HOR   80 min dubbed   VIDrel: REDEM/RTM   V

LAVENDER HILL MOB, THE ****
Charles Crichton UK   U
                                                      1951
Alec Guinness, Stanley Holloway, Sidney James, Alfie Bass, John Gregson, Marjorie Fielding, Clive Morton, Ronald Adams, Sydney Tafler, Jacquis Brunius, Meredith Edwards, Edie Martin, Patrick Barr, Marie Burke, John Salew, Peter Bull
Sparkling Ealing comedy about an underpaid bank clerk who plans and executes a brilliant gold robbery, but his colleague's botched attempt to hide the loot leads to some bizarre complications. Look out for Audrey Hepburn, who appears very briefly. AA: Story/Screen (T.E.B. Clarke).
COM 77 min (ort 78 min) B/W VIDrel: WHV V/h

L'AVVENTURA **
Michelangelo Antonioni FRANCE/ITALY   PG
                                                      1960
Monica Vitti, Gabriele Ferzetti, Lea Massari, Dominique Blanchar, James Addams, Lelio Luttazi, Dorothy De Poliolo, Giovanni Petrucci, Franco Cimino, Enrico Bologna, Giovanni Danesi, Rita Mole, Renato Pincicoli, Vincenzo Trachina
Slow, ponderous examination of the relationships among a group on a vacationers, when one of their number, a woman, disappears while they are visiting a volcanic island. Seemingly pointless and overlong, the work offers little to hold the attention. This was the first film in the director's trilogy on alienation and was in its day considered daring and innovative; it still retains flashes of the director's undoubted vision. Followed by LA NOTTE.
DRA 136 min (ort 145 min) B/W VIDrel: CONNO/RTM V

LAW LORD, THE **
Jim Goddard UK   (PG)
                                                      1991
Anthony Andrews, Bernard Hill, Tom Baker, John Rowe, Kate Lynn-Evans, T.P. McKenna, Tim Preece, George Harris, Roger Brierley, John Stratton, David Rebb, John Southworth, Leonard Maguire, John Boswall, Duncan Bell
A sluggish political drama set at a time when there is a doctrinaire government in place, which attempts to usurp the independence of the judiciary, but its efforts are opposed by Andrews as the title figure. This should have been an intriguing story, but the lack of pace is a handicap it cannot overcome.
DRA 90 min mTV TVrel: BBC

LAW OF DESIRE **
Pedro Almodovar SPAIN   18
                                                      1986
Carmen Maura, Eusebio Poncela, Antonio Banderas, Helga Line, German Cobos, Fernando Guillen Geurvo, Maria Fernandez Muro, Lupo Barrado, Maruchi Leon, Alfonso Vallejo, Jose Manuel Bello, Agustin Almodovar, Rosy Von Donna
A man's obsessive passion leads him and others into a tangled web of complications, in this rather typical offering, the director's seventh film, which takes a fast-moving and comic look at a director of homo-erotic movies. Plenty to interest and offend, according to taste, but certain to appeal mostly to Almodovar fans.
Aka: LA LEY DEL DESEO
COM 97 min (ort 100 min) wScrn VIDrel: TART/20TH V

LAWLESS FRONTIER **
Robert N. Bradbury USA   U
                                                      1934
John Wayne, Sheila Terry, George "Gabby" Hayes, Earl Dwire, Yakima Canutt, Jack Rockwell, Gordon D. Woods, Lloyd Whitlock, Eddie Parker, Artie Ortego, Buffalo Bill Jr
Having lost his parents to a murderous Mexican bandit, a young man joins forces with an old man and his daughter on a mission of revenge. One of Wayne's early Lone Star vehicles, this creaky film shows its age, but has a few good moments.
WES 52 min B/W coVer VIDrel: ENTUK L/A V

LAWLESS RANGE **
Robert North Bradbury USA   U
                                                      1935
John Wayne, Sheila Mannors, Yakima Canutt, Jack Curtis, Frank McGlynn Jr, Glenn Strange, Earl Dwire, Wally Howe, Jack Kirk, Fred Burns, Slim Whitaker, Julia Griffin
An undercover agent sent to investigate cattle rustling is captured by the gang responsible and finds himself in considerable danger. A well-acted and directed standard Western. The TRING tape comprises both the title film and a tribute to Wayne – hence the longer running time.
WES 105 min (ort 56 min) B/W VIDrel: TRING V

LAWMAN **
Michael Winner USA   PG
                                                      1971
Burt Lancaster, Robert Ryan, Lee J. Cobb, Robert Duvall, Sheree

North, Albert Salmi, Joseph Wiseman, J.D. Cannon, Richard Jordan, John McGiver, Ralph Waite, John Beck, John Hillerman, Robert Emhardt, Richard Bull
A marshal comes to a small town to arrest six ranch hands for the accidental killing of a drunk and finds himself blocked at every turn by the locals, including the town sheriff. As he persists in his efforts to see justice done, he finds himself caught up in a spiral of violence that culminates in a bloody showdown. A dour and depressing offering that has the cynical feel of a spaghetti Western.
WES 94 min VIDrel: MGM/WHV V

LAWNMOWER MAN, THE ***
Brett Leonard UK/USA   15
                                                      1992
Jeff Fahey, Pierce Brosnan, Jenny Wright, Mark Bringleson, Geoffrey Lewis, Jeremy Slate, Dean Norris, Colleen Coffey, Troy Evans, Rosalee Mayeux, Joe Hart, Austin O'Brien, Michael Gregory, John Laughlin, Ray Lykins, Jim Landis
Touted as the first "virtual reality" movie, this flawed but highly unusual effort combines the basics of King's story with "CyberGod", the director's own project. Fahey plays a computer scientist who is researching computer simulated realities. When he uses his mentally retarded gardener as an experimental subject, a being of awesome power is created. A most uneven work, visually strong (albeit reminiscent of TRON) but let down by script inconsistencies.
FAN 103 min wScrn; 141 min (director's cut FIRST/SONOP – ort 108 min) VIDrel: VCC/DISC; FIRST/SONOP V
Boa: short story by Stephen King.

LAWNMOWER MAN 2: BEYOND CYBERSPACE **
Farhad Mann USA   12
                                                      1995
Patrick Bergin, Matt Frewer, Austin O'Brien, Ely Pouget, Camille Cooper, Patrick la Brecque, Crystal Celeste Grant, Sean Parhm, Mathew Valencia, Kevin Conway, Trevor O'Brien, Richard Fancy, Ellis Williams, Castulo Guerra
An inferior sequel to the first film that sees our former retarded gardener now mutated into a powerful creature that inhabits computer cyberspace, but is attempting to gain control of the real world. He is opposed by a computer scientist who has assembled a band of hackers to assist him. Watchable this may be, but it has neither imagination nor originality.
Aka: LAWNMOWER MAN 2: JOBE'S WAR
FAN 89 min (ort 92 min) VIDrel: FIRST/SONOP V

LAWRENCE OF ARABIA ***
David Lean UK   PG
                                      1962 (restored 1989)
Peter O'Toole, Omar Sharif, Arthur Kennedy, Anthony Quinn, Anthony Quayle, Jack Hawkins, Claude Rains, Jose Ferrer, Alec Guinness, Michel Ray, I.S. Johar, Zia Mohyeddin, John Dimech, Donald Wolfit, Howard Marion Crawford, Jack Gwillim
An exciting but highly romanticised look at the life of T.E. Lawrence, which established O'Toole as a star but unfortunately provides little insight into the inner drives of the title character. A film of immense visual beauty whose 70 mm wide-screen camerawork will be largely lost on TV. AA: Pic, Dir, Cin (Fred A. Young), Art/Set (John Box and John Stoll/Dario Simoni), Score/orig (Maurice Jarre), Sound (John Cox), Edit (Ann Coates).
DRA 217 min (restored director's cut – ort 221 min) wScrn VIDrel: COLUM/SONOP V
Boa: book Seven Pillars of Wisdom by T.E. Lawrence.

LAWS OF GRAVITY **
Nick Gomez USA   18
                                                      1992
Adam Trese, Peter Greene, Edie Falco, Arabella Field, Peter Schulze, Saul Stein, James McCauley, Anibel Lierras, Miguel Sierra, Larry Meistrich, Rick Greol, David Troup, Patricia Sullivan, Tony Fernandez, John Gallagher
A bleak view of street life and violence in Brooklyn as seen through the eyes of a couple of petty thieves as they ply their trade. Made on a tiny budget, with much use made of a hand-held camera. Sometimes funny, sometimes not, this is a film about nothing much, and though the basic plot elements are hardly original, the director (in his film debut) does well with his limited resources.
DRA 98 min CINrel

LE BAL **
Ettore Scola ALGERIA/FRANCE/ITALY   PG
                                                      1982
Christophe Allwright, Marc Berman, Regis Bouquet, Chantal Capron, Nani Noel, Danielle Rochard, Jean-Claude Penchenat, Etienne Guichard, Jean-Francois Perrier, Liliane Delval, Monica Scattini

Set in a Parisian ballroom, this is a survey of life and love between 1936 and 1983, all told in musical form without dialogue. Highly original and certainly stylish, but it all becomes something of a strain after a while. Those who like films without dialogue will like this one a lot. Written by Scola, Ruggero Maccari, Furio Scarpelli and Jean-Claude Penchenat, and based on a stage production by the Theatre du Campagnol.
MUS   112 min   VIDrel: MGM/WHV L/A   V

LE BEAU MARIAGE ***
Eric Rohmer FRANCE

PG
1982

Beatrice Romand, Arielle Dombasle, Andre Dussollier, Huguette Faget, Feodor Atkine, Thamila Mezbah, Sophie Renoir, Herve Duhamel, Pascal Greggory, Ann Mercier, Virginie Thevenet, Denise Baily, Vincent Gauthier, Catherine Rethi
A young woman tires of her single life and decides to acquire a husband, and it seems that just about any man will do. However, when she finally makes her choice, her would-be mate has other ideas. Second in the "Comedies and Proverbs" series is a charming study of a character with some rather contradictory ideas. As ever the acting and direction are faultless.
Aka: GOOD MARRIAGE, A; WELL-MADE MARRIAGE, THE
COM   95 min (ort 100 min)   VIDrel: CONNO/RTM   V

LE BONHEUR EST DANS LE PRE ***
Etienne Chatiliez FRANCE

15
1995

Michel Serrault, Eddy Mitchell, Sabine Azema, Carmen Maura, Francois Morel, Guilaine Londez, Virgine Darmon, Alexandra London, Christophe Kourotchkine, Jean Bousquet, Eric Cantona, Joel Cantona, Catherine Jacob, Daniel Russo
Serrault plays an unhappy middle-aged factor owner with a nasty wife and a hateful daughter who is slowly recovering from a serious heart attack. When he sees an attractive woman and her two daughters making a TV appeal for their missing husband to come back, he sets out for their home to see if he can fill this role, and once there finds all the things that have been so lacking in his domestic life. A charming comedy-drama, funny and quite touching.
DRA   101 min (ort 106 min)   VIDrel: GUILD/FOXVID; GUILD/20TH   V/sur

LE CHANT DU MONDE ***
Marcel Camus FRANCE/ITALY

PG
1965

Catherine Deneuve, Hardy Kruger, Charles Vanel, Marilu Tolo, Andre Lawrence, Reinhard Kolldehoff, Saro Urzi, Michel Vitold, Georgette Anys, Nane German, Maria Rosa Rodriguez, Ginette Leclerc, Christian Marin, Serge Marquand
A Romeo and Juliet-style tale of the passionate love affair between two members of rival clans that are forever locked in conflict. An enjoyable drama, set in the French department of Haute-Provence.
Aka: WORLD SONG, THE
DRA   95 min   VIDrel: CASPIC L/A   V
Boa: novel by Jean Giono.

LE COLONEL CHABERT **
Yves Angelo FRANCE

12
1994

Gerard Depardieu, Fanny Ardant, Fabrice Luchini, Andre Dussollier, Daniel Prevost, Olivier Saladin, Maxime Leroux, Eric Elmosnino, Guillaume Romain, Patrick Bordier, Claude Rich, Jean Cosmos, Jacky Nercessian, Albert Delpy
In Paris of 1817, a man appears in the office of a lawyer, laying claim to the fortune and rank of a supposedly deceased colonel, a heroic figure who it is believed died at the battle of Eylau. A leisurely and reverent adaptation of Balzac's novel, it tracks endlessly across battlefields and drawing rooms, but this wealth of visual detail overcomes and stifles the film's narrative. Echoes of THE RETURN OF MARTIN GUERRE abound, but with none of that film's intensity.
DRA   106 min (ort 111 min) wScrn   VIDrel: GUILD/20TH V/s
Boa: novel by Honore de Balzac.

LE COP ***
Claude Zidi FRANCE

(18)
1985

Philippe Noiret, Thierry Lhermitte, Regine, Grace De Capitani, Julien Guiomar
A cynical and hardbitten cop accepts bribes to turn a blind eye to crimes, and is perfectly happy with this state of affairs. When he gets a new partner he has to give him lessons in corruption. A funny and lighthearted film that despite the downbeat nature of the subject matter, never takes itself seriously enough to be

anything more than an enjoyable romp. Noiret's great charm as the veteran cop is one of the film's chief assets.
Aka: LES RIPOUX
COM   106 min   CINrel

LE COP 2 **
Claude Zidi FRANCE

PG
1989

Philippe Noiret, Thierry Lhermitte, Guy Marchand, Line Renaud, Grace De Capitani, Michel Aumont, Jean-Pierre Castaldi, Jean-Claude Brialy, Jean Benguigui, Christian Bouillette, Roger Jendly, Georges Montillier, Ren Morard
Our two lovable rogue cops are back in action, with Francois the idealistic younger one trying to go straight in the face of the rank dishonesty of his older partner, Rene. However, when a corruption investigation results in their suspension they retreat to the country. Meanwhile, their replacements prove to be even more dishonest than they were, but fortunately vindication is not too long in coming in this rather disagreeable sequel.
Aka: RIPOUX CONTRE RIPOUX
THR   102 min (ort 108 min) wScrn   VIDrel: TART/20TH   V

LE CRIME DE MONSIEUR LANGE ****
Jean Renoir FRANCE

PG
1936

Rene Lefevre, Jules Berry, Florelle, Jean Daste, Maurice Baquet
A sparkling comedy-thriller set in pre-war France during the rise to prominence of the Popular Front. Told in flashback form, it is largely concerned with the fortunes of an author, who is exploited by the unscrupulous boss of a publishing house but eventually sees the latter get his well-deserved comeuppance. The political references may be dated, but the humour is as fresh as ever. The script is by Jacques Prevert.
COM   80 min B/W   VIDrel: CONNO/RTM   V

LE JEUNE WERTHER ***
Jacques Doillon FRANCE

15
1993

Ismael Joie-Menebhi, Thomas Bremond, Simon Claviere, Pierre Mezerette, Faye Anastasia, Miren Capello, Sunny Lebrati, Mirabelle Rousseau, Jessica Tharaud, Pierre Encreve, Margot Abascal, Herve Duhamel, Marie de Laubier, Eve Guillou
The story of a group of thirteen-year-old school-kids and how they struggle to come to terms with the suicide of a classmate and find a reason for this act. Eventually they come to believe that it has two causes: the nastiness of a teacher and the boy's unrequited love for a pretty girl, and they set out to plan appropriate revenges. Wonderfully well acted, this fine film is by no means as morbid as it sounds, but instead is a work of warmth and perception.
DRA   90 min (ort 95 min) wScrn   VIDrel: TART/20TH   V
Boa: novel The Sorrows of Young Werther by Goethe.

LE JOUR SE LEVE ****
Marcel Carne FRANCE

(PG)
1939

Jean Gabin, Jules Berry, Arletty, Jacqueline Laurent, Rene Genin, Mady Berry, Bernard Blier, Marcel Peres, Jacques Baumer, Rene Bergeron, Gabrielle Fontan, Arthur Devere, George Douking, Germaine Lix
A murderer trapped in an attic by a police siege recalls the events that led to his present predicament. A marvellously poetic example of classic French film noir. RKO later remade it as "The Long Night", and having bought up the film rights the studio bosses tried to have all the prints of the original destroyed. Happily, this example of cultural barbarism was not successful.
Aka: DAYBREAK
DRA   93 min (ort 95 min) B/W   TVrel

LE MIRACULE ***
Jean-Pierre Mocky FRANCE

18
1987

Michel Serrault, Jean Poiret, Jeanne Moreau
A man who is in the process of attempting to swindle an insurance company by faking an injury gets invited to Lourdes by a well meaning friend. However, their trip is being closely monitored by the boss of the insurance company, who is determined to expose the fraud. An oddball comedy-thriller, quite witty and inventive.
COM   84 min   VIDrel: LUMI/SPEAR   V

LE PARFUM D'YVONNE ***
Patrice Leconte FRANCE

18
1994

Jean-Pierre Marielle, Hippolyte Girardot, Sandra Majani, Richard Bohringer, Paul Guers, Corinne Marchand, Philippe Magnan, Claude Dereppe, Claude Aufaure, Isabelle Tinard, Luc Palun, Didier Lafaye, Louis-Marie Audubert, Marie Cosnay

A would-be writer remembers the summer of 1958, which he spent hiding from military service at a Lake Geneva hotel during the Algerian War. There he came to know a beautiful woman named Yvonne and her enigmatic, wealthy male companion. An ambiguous study of doomed relationships, the passage of time and the inevitability of loss, this is a languid, plotless meditation on life, visually impressive if not completely successful as a drama.
DRA   85 min (ort 89 min) wScrn   VIDrel: ARTIF/20TH V/h
Boa: novel Villa Triste by Patrick Modiano.

LE PETIT PRINCE A DIT ***
Christine Pascal FRANCE/SWITZERLAND
Richard Berry, Anemone, Marie Kleiber, Lucie Phan, Mista Prechac, Claude Muret, Jean Cuenoud, John Gutwirth, Baptiste Adatte, Huguette Bonfils, Carlo Boso, Sergio Colella, Barbara DeRosa, Bass Dhem, Christine Youilloz
While her divorced doctor father is away on a trip, the mother of a ten-year-old girl comes to visit, and starts to suspect that her daughter has a health problem, as she has begun to suffer unexplained blackouts. A brainscan shows the presence of a fatal tumour, but the girl's father cannot face the loss of his daughter, and takes her away for a holiday in a forlorn attempt to hide the truth from her. A touching and acutely observed drama.
Aka: AND THE LITTLE PRINCE SAID
DRA   106 min   CINrel
PG
1991

LE SAMOURAI ***
Jean-Pierre Melville FRANCE/ITALY
Alain Delon, Nathalie Delon, Cathy Rosier, Francois Perier, Jacques Leroy, Jean-Pierre Posier, Catherine Jourdan, Michel Boisrond, Robert Favart, Andre Salgues, Roger Fradet, Carlo Nell, Robert Rondo, Andre Thorent, Pierre Vaudier
A hired assassin arrested who is always very careful to have a strong alibi falls in love with a girl who inadvertently betrays him, and she becomes one of the chief witnesses in his trial. A detailed and very restrained film of sparse dialogue, the director's low-key approach to his subject matter is quite memorable, as is Delon's coldly effective performance in the central role.
Aka: SAMURAI, THE
DRA   100 min (ort 105 min) wScrn   VIDrel: ARTIF/20TH; ENCORE (LV only)   V/h LV
Boa: novel The Ronin by Joan McLeod.
PG
1967

LEADER OF THE BAND ***
Nessa Hyams USA
Steve Landesberg, Mercedes Ruehl, Gaillard Sartain, James Martines, Calvert DeForest (Larry 'Bud' Melman)
The story of a musician who dreams of leading a marching band and gets his opportunity when a local band comes off worst in an argument in a bus. An amusing little tale that makes the most of its material.
COM   87 min (ort 90 min)   VIDrel: 20TH   V
15
1987

LEAGUE OF GENTLEMEN, THE ***
Basil Dearden UK
Jack Hawkins, Nigel Patrick, Richard Attenborough, Roger Livesey, Bryan Forbes, Kieron Moore, Terence Alexander, Norman Bird, Robert Coote, Melissa Stibling, Nanette Newman, Gerald Harper, Patrick Wymark, David Lodge
Nice stylish film about a military man who assembles a bunch of ex-officers who left the service for a variety of reasons, and trains them to carry out a bank robbery with military precision. The manner in which they get caught is painfully contrived, as one feels almost inclined to believe they should have got away with it. However, as a well made crime caper the film can stand alongside most others. The script is by Bryan Forbes.
DRA   108 min (ort 115 min) B/W   VIDrel: VCC/RTM   V
Boa: novel by John Boland.
PG
1960

LEAGUE OF THEIR MOANS, A **
Mitch Spinelli USA
Taylor Wayne, Sierra, Toni Tedeschi, Jake Williams, Trinity Loren, Alicia Rio, Cal Jammer
The story of a female baseball team and their off-field activities forms the basis for this sex spoof on A LEAGUE OF THEIR OWN. However, these exploits prove to be the reminiscences of their boozy old former coach, who is thinking back to the fun he had during WW2, when all the men were away fighting. Not a terribly well plotted affair, this one consists mostly of a series of disconnected encounters.
A   62 min (ort 94 min)   VIDrel: FALCON/TOTAL   V
18
1992

LEAGUE OF THEIR OWN, A **
Penny Marshall USA
Geena Davis, Tom Hanks, Jon Lovitz, Madonna (Madonna Ciccione), Lori Petty, Rosie O'Donnell, Bitty Schramm, Ann Cusack, Anne Elizabeth Ramsay, Megan Cavanagh, Tracy Reiner, Freddie Simpson, Renee Coleman, Robin Knight
The story of an all-female baseball team that's established during WW2, when all the men are away fighting. Really no more than a set of vignettes built around the personalities of several of the players, it doesn't provide very much of a story, and what laughs there are come mainly from the aggressive behaviour of the truculent male coach and various other male stereotypes. A kind of soft-centred feminist movie that never finds its focus.
COM   122 min (ort 128 min) cC
VIDrel: VCC/DISC/COLUM   V/sur
Boa: short story by Kim Wilson and Kelly Candaele.
PG
1992

LEAN ON ME **
John G. Avildsen USA
Morgan Freeman, Robert Guillaume, Beverly Todd, Alan North, Lynne Thigpen, Robin Bartlett, Michael Beach, Ethan Phillips
A discipline-minded principal is given just one school year to turn a hopeless hellhole into a model seat of learning, and succeeds in doing so by dint of some very dubious methods. A fine performance from Freeman is the best thing in this slack and superficial tale, loosely based on the exploits of Crazy Joe Clark, a real-life New Jersey principal. See also THE PRINCIPAL and THE GEORGE McKENNA STORY for something not too dissimilar.
DRA   104 min (ort 109 min)   VIDrel: MGM/WHV   V/sur
15
1989

LEAP OF FAITH **
Stephen Gyllenhaal USA
Anne Archer, Sam Neill, Frances Lee McCain, Louis Giambalvo, James Tolkan, Elizabeth Ruscio, C.C.H. Pounder, Michael Constantine, Norman Parker, James Hong, Kent Williams, Kelly Wolf, Aaron Lustig, Tony Abatemarco, Dean Morris
A woman learns that she has terminal cancer but refuses to accept the verdict of her doctors and turns instead to alternative medicine in her search for a cure. A sincere story, told with conviction; its obvious TV origins tend to dilute its impact.
Aka: QUESTION OF FAITH
DRA   94 min (ort 100 min) mTV
VIDrel: POLY/POLYREC L/A   V
PG
1988

LEAP OF FAITH **
Richard Pearce USA
Steve Martin, Debra Winger, Lolita Davidovich, Liam Neeson, Lukas Hass, La Chanze, Meat Loaf, Philip Seymour Hoffman, M.C. Gainey, Delores Hall, Ricky Dillard, John Toles-Bey, Albertina Walker, Vince Davis, Troy Evans, Mark Walters
A phoney evangelist becomes stranded in a small town in a poor farming community and sets up his tent. However, both he and his assistant are changed by unexpected events – for Winger it's a love affair but for Martin it's the witnessing of a real miracle. Martin works hard in a very demanding role but ultimately fails to convince, and though it has good detail and atmosphere, the film's weak storyline is a major flaw. ELMER GANTRY did it very much better.
COM   103 min (ort 110 min) cC
VIDrel: 4-FRONT/POLYREC/CIC   V/dm V/sur
PG
1992

LEAPIN' LEPRECHAUNS! ***
Ted Nicolaou USA
John Bluthal, Grant Cramer, Sharon Lee Jones, Gregory Edward Smith, Godfrey James, Tina Martin, Erica Nicole Kess, Sylvester McCoy, James Ellis, Andrew Smith, Ray Bright, Mike Higgins, James Ellis, Jon Haiduc, Mihal Niculescu
A Irishman owns a castle and land where leprechauns and fairies live. He is good friends with them but they become alarmed when his son, who lives in the USA, has the site surveyed, allegedly by mistake. When he goes to the USA for a family visit, the fairy queen and three leprechauns, including their king, hide in his suitcase. It soon becomes clear that the son is planning to develop his father's property for an amusement park. An engaging family fantasy.
JUV   80 min (ort 84 min) cC   VIDrel: CIC/SONOP   V/dm
U
1994

LEAPIN' LEPRECHAUNS 2 **
Ted Nicolaou USA
John Bluthal, Madeleine Potter, Gregory Edward Smith, Godfrey James, Tina Martin, Sylvester McCoy, James Ellis
U
1996

In this sequel to the earlier film, a group of leprechauns get some welcome assistance when they start making plans to defeat a sinister witch.
JUV   81 min   VIDrel: CIC   V/sh

LEATHER BOYS, THE **   15
Sidney J. Furie UK   1963 (released in USA 1966)
Rita Tushingham, Colin Campbell, Dudley Sutton, Gladys Henson, Avice Landone, Lockwood West, Betty Marsden, Martin Matthews, Johnny Briggs, James Chase, Geoffrey Dunn, Sandra Caron, Dandy Nichols, Valerie Varnam, Sylvia Kay
A sleazy study of a teenage marriage and its gradual break-up. Well observed but profoundly gloomy, and with a few daring (for the time) observations on homosexuality.
DRA   90 min (ort 108 min) B/W   VIDrel: MIA/VCC L/A V
Boa: novel by Eliott George (Gillian Freeman).

LEATHER JACKETS **   18
Lee Drysdale USA   1990
Bridget Fonda, Cary Elwes, D.B. Sweeney, Judi Taylor, Neil Giuntoli, James LeGros, Jon Pochron, Phil Chong, Jeff Imada, Al Goto, April Tan, Christopher Penn, Jon Polito, Joseph Lewis, Heather Haase, Mary Ella Ross, Marshall Bell
The grim story of three characters screwed up by poverty and circumstance: a bored young girl who is very free with her favours, an aimless dreamer who is one of her admirers and finally a tough delinquent who has graduated to becoming a professional criminal and whose murder of a Vietnamese gang member involves all three.
A/AD   90 min   VIDrel: MED/POLYREC L/A   V/sh

LEATHERNECKS **   18
Paul D. Robinson ITALY   1988
Richard Hatch, James Mitchum, Tony Marsina, Robert Marius, Anthony Sawyer, Vassili Karis, Tania Gomez
Standard war heroics tale that's set in Vietnam, where a US Marines sergeant is ordered to bring in two deserters and learns that one of them is his much-decorated close friend. The latter is subsequently captured by the Vietcong and the sergeant and his unit find themselves fighting a desperate battle against a numerically superior force. A predictable low-budget affair with ample action but little else.
WAR   90 min Cut (1 min 17 sec)   VIDrel: IMPENT L/A   V

LEAVE HER TO HEAVEN **   U
John M. Stahl USA   1945
Gene Tierney, Cornel Wilde, Jeanne Crain, Vincent Price, Mary Phillips, Ray Collins, Gene Lockhart, Reed Hadley, Darryl Hickman, Chill Wills, Paul Everton, Olive Blakeney, Addison Richards, Harry Depp, Grant Mitchell, Milton Parsons
Lush, melodramatic tale of a woman who is so pathologically obsessed with the need to be loved that she even resorts to murder and deliberate miscarriage. However, when her foster sister comes into her life, this provokes a crisis that has unforeseen consequences for all concerned. Tierney is excellent as our deranged heroine but is not matched by Wilde or Crain whose performances lack strength. Remade for TV as TOO GOOD TO BE TRUE. AA: Cin (Leon Shamroy).
DRA   105 min (ort 110 min) cC   VIDrel: 20TH/TECH   V/h
Boa: novel by Ben Ames Williams.

LEAVE OF ABSENCE ***   15
Tom McLoughlin USA   1994
Brian Denenhy, Jacqueline Bisset, Blythe Danner, Polly Bergen, Noelle Parker, Tim Lounibos, Jessica Walter, Grayce Spence, Michael Burgess, Tonea Stewart, Tom Atwater, Elizabeth Grillo Barbeau, James Burgess, Angelo Pope
A married architect puts his marriage in jeopardy by falling for a colleague at new firm and insisting on taking time to be with her when she learns that she is terminally ill. Dennehy's sensitive and touching performance makes all the difference in this well-handled human drama.
DRA   88 min mTV   VIDrel: ODY/SONOP L/A   V/sh

LEAVING LAS VEGAS ***   18
Mike Figgis USA   1995
Nicolas Cage, Elisabeth Shue, Julian Sands, Richard Lewis, Steven Weber, Kim Adams, Emily Proctor, Stuart Regen, Valeria Golino, Al Henderson, Shashi Bhatia, Carey Lowell, Anne Lange, Thomas Kopache, Vincent Ward, French Stewart
An alcoholic no-hoper comes to Las Vegas with the intention of drinking himself to death, but once there falls in love with

an equally unhappy individual – a woman who works as a hooker. A touching, funny, sharp and witty character study, made on a shoestring (it was shot on Super-16) and boasting a really wonderful performance from Cage as the loser in question. The director appears very briefly as a mobster. AA: B. Actor (Cage)
DRA   102 min (ort 112 min) wScrn   VIDrel: EIV/SONOP   V
Boa: novel by John O'Brien.

LEAVING LENIN ***   12
Endaf Emlyn UK   1993
Sharon Morgan, Wyn Bowen Harries, Ifan Huw Dafydd, Steffan Trevor, Catrin Mai, Ivan Shvedov, Richard Harrington, Shelley Rees, Anna Vronskaya, Nerys Thomas, Helen Louise Davies, Geraint Francis, Mikhail Maizel
Comedy-drama revolving around the exploits of a group of Welsh sixth-form students who arrive in St Petersburg on a cultural trip, but initially have far less elevated pursuits in mind. However, as time passes they find themselves becoming captivated both by the city and its people. Insightful and deftly directed, it draws believable and touching performances out of its young cast.
Aka: GADAD LENIN; GADAEL LENIN
DRA   89 min (ort 90 min) (Welsh dialogue) subs
VIDrel: POLY/POLYREC L/A   V

LEAVING NORMAL ***   PG
Edward Zwick USA   1992
Christine Lahti, Meg Tilly, Lenny Von Dohlen, James Gammon, Maury Chaykin, Patrika Darbo, Eve Gordon, James Eckhouse, Brett Cullen, Lachlan Murdoch, Robyn Simons, Ken Angel, Darrell Dennis, Barbara Russell, Ahnee Boyce
Having just walked out on her failed marriage, naive Marianne meets Darly, a cynical waitress who has just chucked her boring job. These two team up and head for Alaska, where Darly has inherited some property, but on the way have to deal with crazy encounters, romantic interludes and strange predicaments. An enjoyable blend of comedy and drama, this is a female buddy-movie that inevitably recalls THELMA & LOUISE but has a more upbeat tone.
DRA   105 min (ort 110 min)   VIDrel: CIC/SONOP   V/sur

LEFT-HANDED GUN, THE ***   PG
Arthur Penn USA   1958
Paul Newman, John Dehner, Lita Milan, Hurd Hatfield, James Congdon, James Best, Colin Keith-Johnston, John Dierkes, Bob Anderson, Wally Brown, Ainslie Pryor, Martin Garralaga, Denver Pyle, Nestor Paiva, Robert Foulk, Paul Smith
Penn's debut is a psychological study of Billy the Kid and his quest for revenge on the four men who murdered his friend. Harsh and realistic, but not always successful in its aspirations to explain the workings of a pathological mind. See also BILLY THE KID.
WES   99 min (ort 102 min) B/W   VIDrel: WHV   V
Boa: TV play by Gore Vidal.

LEGACY, THE *   18
Richard Marquand UK   1978
Katherine Ross, Sam Elliott, Ian Hogg, John Standing, Margaret Tyzack, Lee Montague, Charles Gray, Roger Daltrey, Hildegard Neil, Marianne Broome, William Abney, Patsy Smart, Reg Harding, Mathias Kilroy
All of the guests at a country house have been inveigled there, in order to determine which one is to inherit a satanic legacy. The selection proceeds with the progressive murder of each guest who is deemed not to qualify. A streak of viciousness mars this tale, and the expected revelation has all the horror of a bad joke.
Aka: LEGACY OF MAGGIE WALSH, THE
HOR   94 min (ort 102 min)   VIDrel: ODY/SONOP   V/sur

LEGACY OF LIES **   15
Bradford May USA   1992
Michael Ontkean, Martin Landau, Eli Wallach, Michael Nicolosi, Danny Goldring, Lee Lai Demoz, Juan Ramirez, Amy Carlson, Rick Snyder, David Darlow, Gretcher Becker, David Razowsky, Dick Sollenberger, Jim Aquino, Allen A. Secher
An honest cop gets involved in a politically motivated murder and it is up to his dad, a crooked Chicago cop to see if he can clear his son's name.
Aka: LEGACY OF LIES: THREE GENERATIONS ON TWO SIDES OF THE LAW
DRA   90 min mCab   VIDrel: CIC/SONOP   V

LEGACY OF LOVE, THE ** 18
Jim Enright USA 1992
Tom Byron, Nikki Dial, Tiffany Million, Jon Dough, Kelly O'Dell, Lacy Rose
An insurance investigator dealing with a million dollar claim by a wealthy man's widow, goes to investigate her in an effort to learn more about the circumstances surrounding his client's death.
A 48 min VIDrel: ONE V

LEGACY OF RAGE ** 18
Ronnie Yu HONG KONG 1986
Brandon Lee, Michael Wong, Regina Kent, Onno Boelee, Tanya George, Wai-Man Chan, Randy Mang, Meng Hoi
A man is framed for murder by his best friend but gets out of prison just in time to have his revenge, in this predictably violent martial arts actioner, the only film Lee ever made in Hong Kong.
MAR 82 min (ort 90 min) dubbed
VIDrel: 4-FRONT/POLYREC V

LEGACY OF SIN ** PG
Steven Schachter USA 1995
Neil Patrick Harris, Bonnie Bedelia, Meredith Salenger, John Pennell, Brante DeMorneau, Bruce Gray, J.J. Johnston, Terry Kiser, Ernie Lively, Michael MacRae, Arthur Taxier, Steven Williams
Loosely based on real events, this is an account of a young man whose loyalty to his evil mother is severely tested when he attempts to uncover the truth regarding the murder of his father. A chilling portrayal of a ruthless sociopath.
Aka: LEGACY OF SIN: THE WILLIAM COIT STORY
DRA 92 min VIDrel: ODY/SONOP V/sh
Boa: book by Stephen Singular.

LEGAL EAGLES ** PG
Ivan Reitman USA 1986
Robert Redford, Debra Winger, Daryl Hannah, Brian Dennehy, Terence Stamp, Steven Hill, David Clennon, John McMartin, Jennie Dundas, Roscoe Lee Browne, Christine Baranski, Sara Botsford, David Hart, James Hurdle, Gary Klar
An assistant D.A. becomes involved with a woman defence attorney when they are brought together in a fraud case. A frantic attempt to recreate the witty partnership of Tracy and Hepburn in their best comedies, but this one fails to come anywhere near. The music is by Elmer Bernstein.
COM 115 min VIDrel: 4-FRONT/POLYREC/CIC V/sur

LEGAL TENDER *** 18
Buck Adams USA
Victoria Paris, Tracy Adams, Aja, Mandi Wine, Jerry Butler, Buck Adams, Peter North, Raven Richards, Ray Victory
Butler plays a newly released crook who is back to have his revenge on some former "buddies". Meanwhile a couple of cops are on his trail. A blend of "Miami Vice"-style action and sexual escapades, quite well put together.
A 57 min (ort 100 min) mVid VIDrel: IMPENT V

LEGEND *** PG
Ridley Scott UK 1985
Tim Curry, Tom Cruise, Mia Sara, David Bennent, Alice Playton, Billy Barty, Cork Hubbert, Peter O'Farrell, Kiran Shah, Annabelle Lanyon, Robert Picardo, Tina Martin, Ian Longmuir, Mike Crane, Lis Gilbert, Eddie Powell
Fairytale about a princess who inadvertently causes the death of a unicorn, thus unleashing the powers of darkness against her forest domain. Tim Curry is outstanding as the demon who desires her as his bride, in an otherwise lavishly mounted but shallow offering.
FAN 90 min (ort 95 min) VIDrel: WHV V/sur

LEGEND OF BLACK THUNDER MOUNTAIN, THE ** U
Tom Breemer USA 1979
Holly Beeman, Steve Beeman, F.A. Milovich, Ron Brown, Keith Sexson, John Sexson, Vance Cleveland, Dick Albertson, Glen Porter, Tim Stabb
Two children lost in the mountains are adopted by a grizzly bear in this lighthearted kids' adventure.
JUV 90 min VIDrel: SGSVID/GOLD V

LEGEND OF EMMANUELLE, THE ** 18
Francis Leroi FRANCE 1992
Sylvia Kristel, Marcella Walerstein
One of a number of made-for-cable softcore films built around the exploits of the title figure, most of which begin with the orig-

inal Kristel recounting her earlier sexual adventures to anyone interested. Having gained the power to enter the bodies of other women, she uses it in this story to help a lonely and frustrated widow regain her sexual confidence. THE SECRET WORLD OF EMMANUELLE and EMMANUELLE FOREVER are another two titles in this extensive series.
A 88 min (ort 90 min) mCab VIDrel: CREA/DISC V

LEGEND OF HELL HOUSE, THE ** PG
John Hough UK 1973
Pamela Franklin, Roddy McDowall, Clive Revill, Gayle Hunnicutt, Roland Culver, Peter Bowles, Michael Gough
A scientist, his wife and a medium are hired by a millionaire who is obsessed with knowing whether there is life beyond the grave, and sent to investigate a haunted house, where previous hauntings have often led to the death of the psychics who attempted to probe its secret. There are certainly chilling moments but the expected revelation is hardly the stuff of which nightmares are made. Scripted by Matheson. See also THE HAUNTING.
HOR 94 min VIDrel: L/A V
Boa: novel Hell House by Richard Matheson.

LEGEND OF SURAM FORTRESS, THE **** U
Dodo Abashidze/Sergei Paradjanov USSR 1984
Levan Outchanechvili, Zourab Kipchidze, Lela Alibegashvili, Dodo Abashidze, Veriko Andzhaparidze
Soon after its completion, a medieval fortress starts to disintegrate under the influence of a mysterious force. A local soothsayer is called in to solve this problem, and demands that the son of the lover who rejected her must be walled up alive inside the building. Based on a Gregorian legend, this poetic celebration of the colourful history of the region eschews simple patriotic fervour in favour of images of striking vigour and beauty.
Aka: LEGEND OF THE SURAM FORTRESS, THE; LEGENDA SURAMSKOI KREPOSTI
DRA 90 min VIDrel: CONNO/RTM V

LEGEND OF THE GOLDEN PEARL, THE *** PG
Teddy Robin Kwan HONG KONG 1987
Ti Lung, Bruce Baron, Teddy Robin Kwan, Wong Joe Yin, Heidi Makinen, Lo Ta Yu, Peter Pau
An explorer well versed in the martial arts, is asked by a man to find the title object, which is reputed to have been left behind by a flying dragon and to possess great power. He refuses to be drawn into the quest until a boyhood friend takes him to a monastery in Kathmandu, where he finds the pearl and learns a little more about its properties. A lively action-adventure with a strong mystical flavour.
A/AD 77 min (ort 80 min) VIDrel: EIV/SONOP V

LEGEND OF THE HOLY DRINKER, THE *** PG
Ermanno Olmi ITALY 1988
Rutger Hauer, Anthony Quayle, Sandrine Dumas, Dominique Pinon, Sophie Segalen
A Parisian tramp with a criminal past is asked by a stranger to take 200 francs to a holy shrine but succumbs to the temptation to spend the money instead on his own pleasures. A long and leisurely paced adaptation of Roth's novella of religious faith, brought to the screen with great sensitivity and boasting a fine performance by Hauer in the title role.
Aka: LA LEGGENDA DEL SANTO BEVITORE
DRA 122 min (ort 125 min) VIDrel: ARTIF/20TH V
Boa: novella by Joseph Roth.

LEGEND OF THE LONE RANGER, THE ** PG
William A. Fraker USA 1981
Klinton Spilsbury, Jason Robards, Matt Clark, Christopher Lloyd, Michael Horse, Juanin Clay, John Bennett Perry, David Hayward, John Hart, Richard Farnsworth, Lincoln Tate, Ted Flicker, Marc Gilpin, Patrick Montoya
An attempt to tell the story of the origins of our masked avenger and his first contact with Tonto, his Indian friend and subsequent sidekick. Episodes of humour alternate with ones in a more serious vein and the film starts off with promise, unfortunately bad casting (plus the clumsy dubbing of Spilsbury) and the awkward narrative style eventually pull the film down.
WES 92 min (ort 98 min) VIDrel: L/A V

LEGEND OF THE LOST ** U
Henry Hathaway ITALY/PANAMA/USA 1957
John Wayne, Sophia Loren, Rossano Brazzi, Kurt Kasznar, Sonia Moser, Angela Portaluri, Ibrahim El Hadish

Two men roam the Sahara searching for a lost city. What they find however is a young slave girl who presence has a disturbing influence on them both. A dull desert adventure yarn that gets nowhere at a very leisurely pace.
A/AD   104 min (ort 109 min) wScrn   VIDrel: MGM/WHV
V/h

LEGEND OF THE SEVEN GOLDEN VAMPIRES, THE **
18
Roy Ward Baker HONG KONG/UK   1973
Peter Cushing, Julie Ege, John Forbes-Robertson, David Chiang, Robin Stewart, Shih Szu, Chan Shen, Robert Hanna, James Ma, Liu Chia Yung, Feng Ko An, Wong Han Chan
In 1904 Professor Van Helsing finds Dracula behind a cult of Chinese vampires. An uneasy mix of horror and martial arts genres that does nothing for either.
Aka: DRACULA AND THE SEVEN GOLDEN VAMPIRES; SEVEN BROTHERS MEET DRACULA, THE
HOR   85 min Cut (12 sec – ort 89 min)   VIDrel: WHV   V/h

LEGEND OF THE WEREWOLF, THE *
18
Freddie Francis UK   1974
David Rintoul, Peter Cushing, Ron Moody, Hugh Griffith, Roy Castle, Lynn Dalby, Stefan Gryff, Renee Houston, Marjorie Yates, Norman Mitchell, Patrick Holt, John Harvey, Mark Weavers, David Bailie, Michael Ripper, Pamela Green
Poorly realised account of a werewolf who grows up with a travelling circus but finds his true vocation after a few false starts. Cushing investigates.
HOR   86 min (ort 90 min)   VIDrel: ARTPRO/RTM   V

LEGEND OF THE WHITE HORSE, THE ***
PG
Jerzy Domaradzki/Janusz Morgenstern POLAND   1985
Christopher Lloyd, Dee Wallace Stone, Soon-Teck Oh, Christopher Stone, Luke Askew, Allison Balson
Enjoyable action tale in which an environmentalist and his young son travel to a distant country, where the father is carrying out a study of a mining operation. Together, father and son have some strange encounters with a dragon-like creature. Fair.
A/AD   87 min (ort 91 min)   VIDrel: 20TH/TECH   V

LEGEND OF VALENTINO, THE **
15
Melville Shavelson USA   1975
Franco Nero, Suzanne Pleshette, Judd Hirsch, Lesley Ann Warren, Milton Berle, Yvette Mimieux, Harold J. Stone, Alicia Bond, Michael Thoma, Brenda Venus, Constance Forslund, Ruben Moreno, Penny Stanton, Jane Alice Brandon
A so-so biopic on the romantic heart-throb of the 1920s that is about as incisive as a magazine interview and just as glossy. Nero is adequate as this silent-screen idol, but certainly displays nothing one could label charisma. The 1977 film VALENTINO is somewhat better.
DRA   92 min (ort 120 min) mTV   VIDrel: L/A   V

LEGENDS OF THE FALL **
15
Edward Zwick USA   1994
Brad Pitt, Aidan Quinn, Anthony Hopkins, Julia Ormond, Henry Thomas, Karina Lombard, Tantoo Cardinal, Gordon Tootoosis, Paul Desmond, Christina Pickles, Robert Wisden, John Novak, Kenneth Welsh, Bill Dow, Sam Sarkar, Nigel Bennett
A man turns his back on fighting against the Indians and settles down in Montana, where he raises his three sons. But when one of the boys returns home with his new bride, this sets in motion a process that eventually leads to the disintegration of the family. A long, meandering but well photographed saga, let down rather badly by its lack of pace and structure and the superficiality of most of the characterisations. AA: Cin (John Toll).
DRA   127 min (ort 134 min) wScrn cC
VIDrel: COLUM/SONOP   V/sur
Boa: novella by Jim Harrison.

LEGENDS OF THE NORTH **
PG
Rene Manzor USA   1994
Randy Quaid, Georges Corraface, Sandrine Holt, Randy Quaid, John Dunhill
Three diverse characters join forces to hunt for a Yukon lake that they believes to be rich in gold deposits, in this capable adaptation of London's short story.
A/AD   95 min   VIDrel: TRIM/HIFLI   V/h
Boa: short story by Jack London.

LEGION OF IRON *
18
Yakov Bentsvi USA   1989
Kevin T. Walsh, Erica Nann, Reggie De Morton, Camille Carrigan, Nelson Anderson
A young couple are mysteriously abducted to a strange desert region, where they find themselves fighting to stay alive and under threat from an evil warlord.
A/AD   86 min Cut (51 sec – ort 90 min)
VIDrel: EIV/SONOP   V

LEMON SISTERS, THE **
15
Joyce Chopra USA   1990
Diane Keaton, Carol Kane, Kathryn Grody, Elliott Gould, Ruben Blades, Aidan Quinn, Estelle Parsons, Richard Libetini, Sully Boyar, Bill Boggs, Emily A. Rose, Ashly Peldon, Nicky bronson, Francine Fargo, Jo Milazzo, Neil Miller
The somewhat unusual title refers to a child act which three women decide to revive and take one the road when the husband of one of them loses the family fortune in a bad business deal. A good cast struggle bravely but their efforts are just not enough to inject any life or interest into this failed drama.
DRA   88 min (ort 93 min)   VIDrel: EIV/SONOP   V/sur

LENA: MY HUNDRED CHILDREN ***
U
Ed Sherin USA   1988
Linda Lavin, Torquil Campbel, Lenore Harris, George Touliatos, Cynthia Wilde, Suzannah Hoffman, John Evans, Sam Mackin, Victoria Wauchophe, Megan Fahlenbock
A Jewess who survived WW2 by pretending to be Aryan, tries to purge the guilt she feels as a survivor, by taking care of a large group of Jewish orphans in Poland. An above-average and well-handled drama, based on the life of Lena Kuchler-Silberman, who died in 1987.
Aka: LENA: MY 100 CHILDREN
DRA   95 min (ort 100 min) mTV   VIDrel: SONY L/A   V
Boa: book by Lena Kuchler-Silberman.

LENINGRAD COWBOYS GO AMERICA **
15
Aki Kaurismaki FINLAND   1989
Matti Pellonpaa, Kari Vaananen, Sakke Jarrenpaa, Heikki Keskinen, Sakari Kuosmanen, Pimme Korhonen, Puka Oinonen, Silu Seppala, Mauri Sumen, Pekka Virtanen, Mato Valtonen, Nicky Tesco, Olli Tuominen, Kari Lamn, Jim Jarmusch
A no-talent Finnish pop group unable to secure any gigs in their own country come to the USA where, as their manager claims, any rubbish will do in the rural backwaters they frequent. An absurd comedy with innumerable surreal touches that is both very silly and very clever, although never quite as amusing as one might have hoped. Destined, however, for undisputed cult status.
COM   75 min (ort 80 min) wScrn   VIDrel: ARTIF/20TH
V/sur

LENNY ***
18
Bob Fosse USA   1974
Dustin Hoffman, Valerie Perrine, Jan Miner, Stanley Beck, Gary Morton
Film biography of stand-up comic Lenny Bruce and his controversial brand of humour, which made him both admired and derided in the 1950s. Hoffman is well cast, but the film owes much to the fine camerawork of Bruce Surtees. Scripted by Julian Barry.
DRA   107 min (ort 111 min) B/W   VIDrel: MGM/WHV   V
Boa: play by Julian Barry.

LENSMAN: THE POWER OF THE LENS **
PG
Tom Weiner/Kazuyuki Hirokawa/Yoshiaki Kawajiri
JAPAN/USA   1985
Voices of: Ben Gibson, Mickey Godzilla, Ray Michaels, Ryan O'Flannigan, Greg Snow, Colin Phillips, Leonard Pike, Jeffrey Platt, Doug Stone, Drew Thomas, Philboyd Studge, Kerrigan Mahan, Tom Wyner, Greg Snegoff, Michael McConnohie
Set in 2500 A.D. and with more than a nod in the direction of STAR WARS, this has young boy becoming the "Lensman" and battling a variety of hazards to get information back to Galaxy Patrol that will help neutralise the fearful weapons of the Boskone Empire. A routine Good versus Evil space opera which, despite some impressive computer-generated graphics, never amounts to more than another assembly-line piece of second-rate animation.
Aka: LENSMAN
ANIM   107   VIDrel: MANGA/SONOP   V
Boa: "Lensman" novels by E.E. "Doc" Smith.

LEOLO *** 18
Jean-Claude Lauzon CANADA 1992
Maxime Collin, Ginette Reno, Julien Guiomar, Pierre Bourgault, Giuditta del Vecchio, Andree Lachapelle, Denys Arcand, Germain Houde, Yves Montmarquette, Lorne Brass, Roland Blouin, Genevieve Samson, Marie-Helene Montpetit
Evocative, original and often weird tale of a young boy growing up in Montreal who seeks refuge from the pressures of his strange family through a vivid fantasy life. Visually striking but often very bleak.
DRA 102 min (ort 107 min) wScrn VIDrel: TART/20TH
V/sur V/dm

LEON *** 18
Luc Besson USA 1994
Jean Reno, Gary Oldman, Danny Aiello, Nathalie Portman, Peter Appel, Michael Badalucco, Ellen Greene, Elizabeth Regen, Carl J. Matusovich, Randolph Scott, Keith A. Glascoe, Frank Senger, Lucius Wyatt "Cherokee", Luc Bernard
In New York City a twelve-year-old girl is orphaned when her family are massacred by a corrupt cop, but fortunately she is taken under the wing of the title figure, a tough loner who teaches her the art of killing so she can have her revenge. In return, she teaches her mentor much about human values and compassion for others. An explosive and noisy thriller, whose well handled action sequences are nicely balanced by quieter moments.
THR 105 min (ort 110 min) wScrn cC
VIDrel: TOUCH/TECH V

LEON THE PIG FARMER ** 15
Vadim Jean/Gary Sinyor UK 1992
Mark Frankel, Maryam D'Abo, Janet Suzman, Brian Glover, Connie Booth, David De Keyser, Gina Bellman, Annette Crosbie, Bernard Bresslaw, John Woodvine, Vincenzo Ricotta, Jean Anderson, Annette Crosbie, Stephen Greif, Burt Kwouk
A young Jewish man's life is badly disturbed when he learns that he was the result of experiments in artificial insemination and that his biological father is a pig father in the North of England. A confused and rather silly one-joke comedy whose brand of absurd, strained humour fails to amuse. This was Bresslaw's last screen appearance.
COM 99 min (ort 102 min) VIDrel: ELPIC/POLYREC;
ENCORE (LV only) V/sur LV

LEONA HELMSLEY: THE QUEEN OF MEAN ** PG
Richard Michaels USA 1990
Suzanne Pleshette, Lloyd Bridges, Joe Regalbuto, Raymond Singer, Bruce Weitz
A somewhat different biopic, devoted to the real-life wife of a hotel-chain owner whose pride and meanness led to her eventual conviction on charges of tax evasion. This is almost a classical tale of hubris, but with a central character so loathsome and uninteresting that little dramatic tension is developed. Pleshette gives an outstanding performance as the title figure.
Aka: QUEEN OF MEAN, THE
DRA 94 min (ort 100 min) mTV
VIDrel: FABFIL/POLYREC L/A V
Boa: book The Queen of Mean by Ransdell Pierson.

LEOPARD FIST NINJA, THE *** 18
Godfrey Ho HONG KONG
Jack Lam, Willie Freeman, James Exshaw, Dick Hunt, Chung Wok, Peter Sho, Chuck Horry
Story of a loner who has travelled across the country, honing his martial arts skills to perfection with the sole aim of avenging the death of his parents. However, his adversary awaits his arrival with the support of a band of Ninja warriors in this enjoyable adventure.
MAR 83 min VIDrel: IMPENT V

LEPRECHAUN ** 15
Mark Jones USA 1992
Warwick Davis, Jennifer Aniston, Ken Olandt, Mark Holton, Robert Gorman, John Sanderford, Shay Duffin, John Voldstad, Pamela Maint, William Newman, David Permenter, Raymond Turner, Heather Kennedy, Timothy Garrick
A man goes to Ireland to attend his mother's funeral and discovers the location of a cache of gold belonging to a leprechaun. He steals this but accidentally takes the creature back with him to the USA. Naturally, its bloody revenge is not slow in coming in this confused and poorly made horror tale. A sequel followed.
HOR 88 min (ort 91 min) VIDrel: REFLEC/FIRST L/A V

LES AMANTS DU PONT-NEUF *** 18
Leos Carax FRANCE 1991
Juliette Binoche, Denis Lavant, Klaus-Michael Gruber, Daniel Buain, Marion Statens, Chrichan Larson, Paulette Berthonnier, Roger Berthonnier, Edith Scob, Georges Apperighis, Michel Vandestien, Georges Casterp, Alain Dahan
One of the most expensive French films of all time, this tells of the love that develops between two dropouts in Paris, one of whom is an artist who is going blind and the other a shaven-headed professional fire-eater. Their growing love is interrupted when the girl learns that her wealthy father is trying to trace her (he wants to pay for an operation to save her sight) and they part, but later enjoy a reunion. A mellow and strangely moving film.
Aka: LOVERS ON THE BRIDGE
DRA 120 min (ort 126 min) wScrn VIDrel: ARTIF/20TH
V/sur

LES APPRENTIS ** 15
Pierre Salvadori FRANCE 1995
Guillaume Depardieu, Francois Cluzet, Judith Henry, Claire Laroche, Philippe Girard, Bernard Yerles, Marie Trintignant, Jean-Pol Brissart, Jean-Michel Julliard, Blandine Pelissier, Maryvonne Schiltz, Claude Aufaure, Helene Roussel
An introverted, middle-class writer and a feckless, easy-going layabout share a flat, and finding all their efforts to improve their lifestyle defeated, eventually turn to crime and carry out an inept robbery. A clumsy and meandering comedy, it touches on problems such as loneliness and despair, makes a few funny observations and generally avoids the more obvious cliches. Scripted by the director, this is an offbeat and not entirely successful work.
COM 98 min VIDrel: TART V

LES BIJOUTIERS DU CLAIR DE LUNE ** 18
Roger Vadim FRANCE 1957
Brigitte Bardot, Alida Valli, Stephen Boyd
A young convent girl elopes to Spain with her lover, who murdered her uncle and seduced her aunt. Their passion intensifies as time runs out and they are eventually cornered in the mountains. Well-made nonsense that serves as a good showcase for Bardot, but is certainly one of her least memorable films.
Aka: HEAVEN FELL THAT NIGHT; NIGHT HEAVEN FELL, THE
DRA 95 min VIDrel: CASPIC L/A V

LES DIABOLIQUES *** PG
Henri-Georges Clouzet FRANCE 1955
Simone Signoret, Vera Clouzet, Paul Meurisse, Charles Vanel, Jean Brochard, Noel Roquevert, Therese Dorny, Pierre Larquey, Michel Serrault, Yves-Marie Maurin, George Poujouly, Jean Themerson, Georges Chamarat, Jacques Varennes
The wife and mistress of the tyrannical headmaster of a rural boarding school conspire to murder him. They gave him poison and dump the body in the swimming pool but are terrified when it disappears after the pool is drained. A classic suspenser that keeps us guessing right to the surprise ending. A much imitated film that was remade in 1976 as the TV movie "Reflections Of Murder", again as the movie HOUSE OF SECRETS and once more in 1996 as DIABOLIQUE.
Aka: DIABOLIQUE; FIENDS, THE
THR 116 min (ort 117 min) B/W VIDrel: ARROW/RTM
V
Boa: novel Celle Qui N'Etait Plus by Pierre Boileau and Thomas Narcejac.

LES ENFANTS DU PARADIS ***** PG
Marcel Carne FRANCE 1945
Arletty, Jean-Louis Barrault, Pierre Brasseur, Marcel Herrand, Louis Salou, Maria Casares, Pierre Renoir, Gastin Modot, Jeanne Marken, Jacques Castelot, Jean Gold, Guy Faviere, Paul Frankeur, Albert Remay, Leon Larive, Pierre Palau
An exploration of the lives and loves of a group of characters, whose fates are all linked with their involvement with a seductive courtesan (Arletty), in the theatre world of 1840s Paris. This magnificent tapestry of a bygone age is too detailed and rich for any description to do it justice, but is without doubt one of the most enthralling and evocative films ever made.
Aka: CHILDREN OF PARADISE
DRA 181 min (restored version – ort 195 min) B/W
VIDrel: ARTIF/20TH V/h

LES GIRLS ***
George Cukor USA   *U*
                                                                   1957
Gene Kelly, Kay Kendall, Mitzi Gaynor, Taina Elg, Jacques Bergerac, Leslie Phillips, Henry Daniell, Patrick Macnee, Stephen Vercoe, Philip Tonge, Francis Ravel, Owen McGiveney, Adrienne D'Ambricourt, Maurice Marsac, Cyril Delevanti
Three dancers fall out when one of the girls publishes her memoirs in a film that, though hampered by a narrative that makes much use of flashbacks, remains a thoroughly lively affair, if a little short on wit. The use of Cole Porter's music is a major asset. AA: Cost (Orry-Kelly).
MUS   113 min   VIDrel: MGM/WHV   V
Boa: novel by Vera Caspary.

LES GRANDES MANOEUVRES ***
Rene Clair FRANCE   *PG*
                                                                   1955
Brigitte Bardot, Michele Morgan, Gerard Philipe, Yves Robert, Pierre Dux, Jean Desailly, Jacques Francois, Lise Delamere, Jacqueline Maillan, Magalie Noel, Simone Valere, Catherine Anouilh, Jacques Fabbri, Raymond Cordy
Clair's first colour film is a witty and ironic romantic-comedy set in a pre-WW1 garrison town. A dragoon with a reputation as a ladies' man accepts a bet that he can seduce an icy divorcee. He sets to this challenge with gusto only to fall in love with her, but his hopes are dashed by the notoriety his many affairs have achieved. A film of both charm and insight.
Aka: GRAND MANEUVER, THE; SUMMER MANOEUVRES
DRA   106 min   VIDrel: CASPIC L/A   V

LES LIAISONS DANGEREUSES ***
Roger Vadim FRANCE   *18*
                                                                   1959
Gerard Philipe, Jeanne Moreau, Annette Vadim, Jeanne Valerie, Jean-Louis Trintignant
An updating of the 18th century novel that has Valmont (Philipe) and his wife Juliette (Moreau) encouraging each other in their sexual conquests, for they delight in nothing so much as comparing notes. When they plan the seduction of an innocent and sweet-natured young girl, they fall into a trap created by their actions and reap a fitting reward for their callousness. An absorbing little moral fable, remade as DANGEROUS LIAISONS and "Valmont".
Aka: DANGEROUS LIAISONS 1960; DANGEROUS LOVE AFFAIRS
DRA   105 min B/W   VIDrel: CASPIC L/A   V
Boa: novel by Choderlos de Laclos.

LES MISERABLES ****
Richard Boleslawski USA   *(12)*
                                                                   1935
Frederic March, Charles Laughton, Cedric Hardwicke, Rochelle Hudson, Frances Drake, John Beal, Jessie Ralph, Florence Eldridge, Marilyn Knowlden, Vernon Downing, John Beal, Ferdinand Gottschalk, Jane Kerr, Eily Malyon, John Bleifer
An outstanding production of the classic novel about a man sentenced to the galleys in France in 1796 and his eventual redemption that is brilliant in all departments, although the sheer power of Laughton's performance is probably the film's most memorable feature. Filmed numerous times before, as well as being made for TV in 1978 and 1982.
DRA   108 min B/W   TVrel
Boa: novel by Victor Hugo.

LES MISERABLES ***
Glenn Jordan USA   *PG*
                                                                   1978
Richard Jordan, Anthony Perkins, Cyril Cusack, Claude Dauphin, John Gielgud, Flora Robson, Celia Johnson, Joyce Redman, Christopher Guard, Ian Holm, Caroline Langrishe, Angela Pleasence, John Moreno, Roy Evans, David Swift
The story of a man sentenced to serve in the galleys in France of 1796, his escape and ultimate redemption. A lavish and effective remake that, though lacking the sheer power of the 1935 Charles Laughton classic, nevertheless remains quite a respectable effort, despite the sluggish and padded out second half. This was Dauphin's last film.
DRA   131 min (ort 180 min) mTV
VIDrel: ETL/POLYREC/ITC   V
Boa: novel by Victor Hugo.

LES MISERABLES ****
Claude LeLouch FRANCE   *12*
                                                                   1995
Jean-Paul Belmondo, Michel Boujenah, Alessandra Martines, Annie Giradot, Clementine Celarie, Philippe Leotard, Rufus, Jean Marais, Micheline Presle, Salome Lelouch, Philippe Khorsand, Ticky Holgado, Nicolle Croisille, Jean Marais
Inspired by rather than closely based on the Victor Hugo novel,

this updating of his work is set in the days just prior to the Nazi invasion of France, and tells of a family's flight to freedom, and of the man who leads them. Winner of the Golden Globe for Best Foreign Film, this is an intense, literate and hugely enjoyable drama in its own right, and Belmondo has never been better.
Aka: LES MISERABLES DU VINGTIEME SIECLE
DRA   167 min (ort 175 min) cC   VIDrel: WHV   V/h
Boa: novel by Victor Hugo.

LES NOCES ROUGES **
Claude Chabrol FRANCE/ITALY   *15*
                                                                   1972
Stephane Audran, Michel Piccoli, Claude Pieplu, Eliana De Santis, Clothilde Joano, Francois Robert
An assistant to the mayor in a small town is having an affair with the wife of his boss and at the same time is speculating in a crooked land deal. But his clever schemes start to come apart when his wife is found murdered. Not one of the director's strongest works, it lacks both a clear identity and direction, and though based on a real-life murder case, is no more interesting for all that. It was briefly banned on release, as the villain was clearly a Gaullist.
Aka: BLOOD WEDDING; RED WEDDING; WEDDING IN BLOOD
THR   92 min wScrn   VIDrel: ARTPRO/RTM   V

LES ROSEAUX SAUVAGES ****
Andre Techine FRANCE   *15*
                                                                   1993
Elodie Bouchez, Frederic Gorny, Gaol Morel, Michele Moretti, Nathalie Vignes, Jacques Nolot, Eric Kreikenmayer, Michel Ruhl, Fatia Maite, Claudine Taulere, Elodie Soulinhac, Paul Simonet, Charles Picot, Christophe Maitre
A coming-of-age drama that explores the sexual awakening of four teenagers at a boarding school in France in the 1960s. A masterful work, of depth and perception, with some of the scenes so carefully composed that they look like paintings, but the feeling of intensity and realism is never lost. The music of the period is used to good effect. Winner of Best Director, Best Screenplay, Best French Film and Best Young Actress at the Cesar Awards 1994.
Aka: WILD REEDS, THE
DRA   109 min (ort 113 min) wScrn   VIDrel: TART/20TH   V

LES VALSEUSES **
Bertrand Blier FRANCE   *18*
                                                                   1974
Gerard Depardieu, Miou-Miou, Patrick Dewaere, Jeane Moreau, Jacques Chailleux, Brigitte Fossey, Isabelle Huppert, Jacques Rispal
Two amoral malcontents take out their frustrations when they hit the open road, and set off on an orgy of petty crime, abduction and debauchery. As often disagreeable as it is funny, this irreverent film makes no moral judgements, has no particular message and goes in no clear direction. A flawed and uneven film of limited appeal, its title is French slang for testicles, an apt choice if ever there was one.
Aka: GOING PLACES; MAKING IT
DRA   113 min (ort 118 min) wScrn   VIDrel: TART/20TH   V/dm
Boa: novel by Betrand Blier.

LES VISITEURS ***
Jean-Marie Poire FRANCE   *15*
                                                                   1993
Christian Clavier, Jean Reno, Valerie Lemercier, Marie-Anne Chazel, Isabelle Nanty, Christian Bujeau, Didier Pain, Gerard Sety, Jean-Paul Muel, Pierre Vial, Ariel Semenoff, Michel Peyrelon, Francois Lalande, Didier Benureau, Anna Gaylor
Two characters from the 12th century are transported into the 20th century, with comic results in this knockabout farce that achieved an enormous success in its native country. While they cause no end of havoc in their attempts to get to grips with the modern world, they continue to try to find a way back to their own time. Extremely good in terms of detail and effects, this is a broad, rather childish comedy, not exactly thoughtful, but quite good fun.
COM   102 min (ort 107 min)   VIDrel: ARROW/RTM   V

LESS THAN ZERO *
Marek Kanievska USA   *18*
                                                                   1987
Andrew McCarthy, Robert Downey Jr, James Spader, Nicholas Pryor, Donna Mitchell, Jami Gertz, Tony Bill, Michael Bowen
An attempt to examine the superficial lifestyle of the Beverly Hills crowd that results in a film that's generally as unappealing as the characters in it.
DRA   94 min (ort 98 min)   VIDrel: 20TH/TECH   V/sur
Boa: novel by Bret Easton Ellis.

"LET HIM HAVE IT" ★★★ 15
Peter Medak UK 1991
Christopher Eccleston, Paul Reynolds, Tom Courtenay, Tom Bell, Eileen Atkins, Clare Holman, Mark McGann, Michael Gough, Ian Deam, Steve Nicholson, Bert Tyler-Moore, Niven Boyd, Serena Scott-Thomas, Ronald Fraser, Peter Eyre
The story of Derek Bentley, who was hanged for saying the title words when, in the course of a robbery that went wrong, he claimed he was calling upon his armed colleague to surrender. Set in 1951, it carefully details the events that led up to Craig shooting dead a policeman. Yet it was Bentley who was hanged, for Craig was only sixteen and was therefore too young to be sent to the gallows. An absorbing story that should really have been made as a documentary.
DRA   110 min (ort 115 min)   VIDrel: FIRST/SONOP; PION (LV only)   V/sur LV

LET IT RIDE ★ 15
Joe Pytka USA 1989
Richard Dreyfuss, Teri Garr, David Johansen (Buster Poindexter), Jennifer Tilly, Allen Garfield, Ed Walsh, Michelle Phillips, Mary Woronov, Robbie Coltrane, Richard Edson, Cynthia Nixon
A compulsive gambler whose wife is on the verge of leaving him decides to make one last bet at a Florida racetrack on the strength of some inside information. However, he hits a winning streak, and cannot stop. Pytka's directorial debut attempts a Damon Runyon-style fast-paced comedy, but is as uneven as it is overstretched.
COM   86 min   VIDrel: CIC/SONOP L/A   V

LETHAL CHARM ★★ PG
Richard Michaels UK/USA 1990
Heather Locklear, Barbara Eden, Stuart Wilson, David James Elliott, Jed Allan, Allan Miller, Julie Fulton, Tom Klunis, Jordan Charney, Gloria Henry, Bill Morey, Maurce Bernard, Jim Caldwell, Eric Kohner, Rose Malek-Yonan
An attractive young broadcaster who is obsessed with her career, moves to Washington where she is given a post under a top reporter. Anxious to displace her superior, she embarks on a devious scheme to discredit her, but this results in a deadly power struggle that threatens the life of another journalist. A soap opera-style variant on ALL ABOUT EVE that is reasonably well acted, but never rises above its mTV format. See also BODY LANGUAGE.
Aka: HER WICKED WAYS
DRA   92 min mTV   VIDrel: ITC/HIFLI L/A   V/h

LETHAL LOLITA – AMY FISHER: MY OWN STORY ★★ 15
Bradford May USA 1992
Noelle Parker, Ed Marinaro, Boyd Kestner, Pierrette Grace, Lawrence Dane, Kate Lynch, Kathleen Laskey, Rino Romano, Gemma Barry, Typer Ross, Kirsten Kieferle, Evan Sabba, Henriette Ivanans, Marianna Pascal, Angelo Pedari
One of three made-for-TV versions of the relationship between teenager Amy Fisher and thirty-eight-year-old Joey Buttafuoco that led to the shooting by Amy of his wife. This one was made with the assistance of Fisher and reflects her view that she was manipulated into committing this dreadful crime. The films BEYOND CONTROL: THE AMY FISHER STORY and "Casualties Of Love: The Long Island Lolita Story" cover the same ground.
DRA   90 min (ort 92 min)   VIDrel: ODY/SONOP   V
Boa: book by Maria Eftimiades.

LETHAL OBSESSION ★ 18
Dean Hamilton USA 1994
Robert Eastwick, Michelle Lamothe, Margie Peterson, Clark Katz, Michelle Garrin, Veronica Saners, Eric Paul, Shelby ane, Jeanine Antoine, Sindy Tenens, Thomas Prisco, Christina Gancevitch, Gregory Vlahakis, Aaron "Rocky" Hamilton
A glamour photographer has a cop boyfriend who is investigating a robbery while she is doing a shoot in desert. Meanwhile, a serial killer is out stalking them. Lots of posing models do little to enliven a dull and poorly acted erotic thriller.
Aka: STRIKE A POSE
THR   86 min (ort 90 min)   VIDrel: MARQ/QUANT   V

LETHAL PANTHER ★★ 18
Philip Ko HONG KONG 1993
Yukari Oshima, Philip Ko
Standard martial arts action tale with a female protagonist, but

apart from this, the same tale of revenge and bloodletting applies.
MAR   84 min (ort 90 min)   VIDrel: EAST/DISC   V

LETHAL PURSUIT ★★ 18
Don Jones USA 1987
Blake Bahner, Mitzi Kapture, John Stuart Wildman, Bill Kerr
A rock star and her boyfriend visit her home town after a ten year absence, their intention being to take a relaxing break there. However, her arrival rekindles the passion of the girl's former high school sweetheart, and he becomes a ruthless killer, forcing them to flee the town for their lives.
A/AD   89 min Cut (19 sec)   VIDrel: MED/POLYREC   V/h

LETHAL TENDER ★★ 15
John Bradshaw CANADA 1996
Jeff Fahey, Kim Coates, Carrie-Anne Moss, Gary Busey, Dennis Akayama, David Mucci, David Fraser, Martin Roach, Jonathan Potts, Karen Dwyer, Chris Pickles, Linda Barnett, Chris Gillett, Bill Lake, Emily Roberts, Ian Downie, Eileen Sword
The employees at a water treatment plant are taken hostage by terrorists who threaten to poison the city's water supplies. An unconventional cop teams up with the plant supervisor to rescue them. However, both men are unaware they our terrorists are in reality criminals who have mounted this attack as a diversion to hide the fact that a robbery is in progress elsewhere.
A/AD   89 min   VIDrel: HIFLI/SONOP   V

LETHAL WEAPON ★★ 18
Richard Donner USA 1987
Mel Gibson, Danny Glover, Gary Busey, Gustav Vintas, Mitchell Ryan, Tom Atkins, Darlene Love, Traci Wolfe, Jackie Swanson, Damon Hines, Ebonie Smith
Two cops team up to tackle a drugs ring – one carries a gun but the other doesn't, as he is a lethal weapon in his own right. A fast-moving saga of loud bangs and nasty villains, who all too soon get what's coming to them. A sequel followed in 1989.
A/AD   105 min (ort 110 min) wScrn   VIDrel: WHV   V/sur

LETHAL WEAPON 2 ★★★ 18
Richard Donner USA 1989
Mel Gibson, Danny Glover, Joe Pesci, Joss Ackland, Derrick O'Connor, Patsy Kensit, Darlene Love, Traci Wolfe, Steve Kahan
This flashy and violent sequel sees the return of our maverick police duo, this time out to bust a drug smuggler who does not scruple to hide behind diplomatic immunity. The mixture of humour and non-stop action is as before, in a stylised cartoon-like film that will certainly appeal to fans of the first one.
A/AD   109 min Cut (32 sec – ort 113 min) wScrn VIDrel: WHV   V/sur

LETHAL WEAPON 3 ★★★ 15
Richard Donner USA 1992
Mel Gibson, Danny Glover, Joe Pesci, Rene Russo, Stuart Wilson, Steve Kahan, Darlene Love, Traci Wolfe, Damon Hines, Ebonie Smith, Gregory Millar, Nick Chinlund, Jason Meshover-Iorg, Alan Scarfe, Delores Hall, Mary Ellen Trainor
Some more of the same with Gibson and Glover back in a new set of adventures. Having escaped death in a massive explosion (they failed to defuse the bomb) the action moves to their efforts in tracking down a bunch of crooks who specialise in bumping off cops. Russo as a tough, martial arts-trained officer, offers ample assistance. Mindless, quickfire mayhem, enlivened by some good banter and a dash of romance. Pure, noisy escapism.
A/AD   113 min (ort 118 min) wScrn cC   VIDrel: WHV V/sur

LET'S MAKE LOVE ★★★ U
George Cukor USA 1960
Marilyn Monroe, Yves Montand, Tony Randall, Wilfrid Hyde-White, David Burns, Frankie Vaughan, Bing Crosby, Gene Kelly, Milton Berle, Michael David, Mara Lynn, Dennis King Jr, Joe Besser, Madge Kennedy, Ray Foster, Mike Mason
A millionaire learns that he is to be lampooned in an off-Broadway show and tries to have it stopped, but changes his mind after meeting Monroe, falls for the female lead and is eventually hired to play himself. A lightweight musical, inspired by "On The Avenue", that's redeemed by excellent interaction between the two leads.
MUS   115 min (ort 118 min)   VIDrel: 20TH/TECH   V

LETTER, THE **** PG
William Wyler USA 1940
Bette Davis, Herbert Marshall, James Stephenson, Sen Yung, Frieda Inecort, Gale Sondergaard, Bruce Lester, Tetsu Komai
A plantation owner's wife in the former British colony of Malaya claims to have killed a man in self-defence, but a letter that proves it was a crime of passion is used to blackmail her. A remake of the 1929 film (which also starred Marshall), with Davis excellent in a difficult and unsympathetic role. Remade several times since.
DRA 91 min (ort 95 min) B/W VIDrel: MGM/WHV L/A V
Boa: play/story by William Somerset Maugham.

LETTER TO BREZHNEV ** 15
Chris Bernard UK 1985
Margi Clarke, Peter Firth, Alfred Molina, Alexandra Pigg, Neil Cunningham, Ken Campbell, Angela Clarke, Tracy Lea, Susan Dempsey, Ted Wood, Carl Chase, Sharon Power, Robbie Dee, Eddie Ross, Syn Newman, Gerry White, Pat Riley
Two bored unemployed Liverpool lasses out for an evening's fun, meet up with a couple of Russians visiting Britain as part of a public relations exercise. Love blossoms between the young couples who have only twenty-four hours to spend together before parting, perhaps for ever. A charming idea ruined by the stereotyped portrayal of the working class as lewd, crude and vulgar, and of critics of the Soviet Union as ignorant bigots.
DRA 91 min (ort 100 min) VIDrel: 4-FRONT/POLYREC V/sur

LETTER TO MY KILLER ** PG
Janet Meyers USA 1995
Rip Torn, Mare Winningham, Nick Chinlund, Josef Sommer, Eddie Jones, James Murtagh, Dey young
A construction worker finds a thirty-year-old letter from a woman who was murdered and he and his wife use the information it contains to blackmail those responsible for this crime.
THR 88 min (ort 92 min) VIDrel: CIC V

LETTER TO THREE WIVES, A ** U
Joseph L. Mankiewicz USA 1949
Jeanne Crain, Linda Darnell, Ann Sothern, Kirk Douglas, Paul Douglas, Jeffrey Lynn, Thelma Ritter, Barbara Lawrence, Connie Gilchrist, Florence Bates, Hobart Cavanaugh, Patti Brady, Ruth Vivian, Celeste Holm (voice only)
Details the story of three wives who each receive a letter from the same woman, claiming that she has run off with their husbands. An absorbing comedy-drama that is full of sharp dialogue but little development. Holm narrates in the role of the writer of the letters. Remade for TV in 1985. AA: Dir, Screen (Mankiewicz).
DRA 98 min (ort 103 min) B/W VIDrel: L/A V
Boa: novel by John Klempner.

LETTERS FROM THE EAST ** (PG)
Andrew Grieve GERMANY/FINLAND/SWEDEN/UK 1995
Ewa Froling, Mark Womack, Ingeborga Dapkunaite, Marta Laurent, Rein Oja, Mikk Mikiver, Gertrud Talvik, Gerli Nurmsalu, Laura Kunnapas, Nicholas Le Prevost, Terence Longdon, Hanns Zischler, Juri Krjukov, Artur Talvik, Anu Lamp
In London an Estonian cellist lives in exile, and still tries to come to terms with her painful WW2 memories. When her father dies, she goes through his papers, and learns that he had lied for years about her mother, whom he had claimed had died during the War. Despite her misgivings, she sets off for Estonia, in the hope of finding her mother. Scripted by Grieve, the film is both an overly simplistic history lesson and an absorbing human drama.
DRA 107 min SATrel: SKY MOVIES

LETTERS OF LOVE ** R18/18
Paul G. Vartell USA 1985
Bridgette Monet, Cyndee Summers, Ginger Lynn, Kelly Howe, David Cannon, Tom Byron, Debbie Northup, Pam Nimmo, Greg Ruffner, Greg Rome, Herschel Savage
Monet applies to work on the letters column of a magazine to which people write with their sexual problems. This provides the linking device for a set of episodes which are meant to dramatise the letters sent, and the responses they elicit. By way of a twist, the final letter is from the boss of the magazine (Cannon, Monet's real-life husband) who writes in to reveal that

he's in love with her. A satisfactory effort, quite restrained and romantic.
A 58 min Cut (40 sec plus some cuts subst – R18 ver); 42 min VIDrel: HAR/GOLD V/sur

LEVIATHAN ** 18
George P. Cosmatos USA 1989
Peter Weller, Richard Crenna, Amanda Pays, Daniel Stern, Ernie Hudson, Lisa Eilbacher, Michael Carmine, Hector Elizondo, Meg Foster, Eugene Lipinsky, Larry Dolgin, Pascal Druant, Steve Pelot
An underwater clone of ALIEN, in which the crew of an undersea mining platform discover a sunken Soviet ship that was deliberately scuttled to prevent an aborted and potentially disastrous genetic experiment from endangering the world. An unoriginal and derivative film with little to sustain interest except the expected gruesome effects (which are very well done). Music is by Jerry Goldsmith. See also ALIEN WITHIN.
THR 93 min (ort 98 min) VIDrel: 20TH/TECH L/A V

LIAR LIAR ** 15
Jorge Montesi CANADA 1992
Rosemary Dunsmore, Susan Hogan, Art Hindle, Michelle St John, Janne Mortil, Vanessa King, Kate Nelligan, Roman Podhora, Wendy Van Riesen, Philip Akin, Kevin McNulty, Joel Palmer, Ashley Rogers, Kaj-Erik Eriksen, Ken Kramer
An eleven-year-old girl accuses her father of sexual abuse and the case goes to court, where inconsistencies in her story are revealed, it soon becoming apparent that her accusations may be simply a way of getting back at her father for being disciplined. Based on a real-life case that made the headlines in Canada, this is a mature and careful look at the problem of sexual abuse and false accusation. See also INDICTMENT: THE McMARTIN TRIAL and UNSPEAKABLE ACTS.
DRA 90 min mTV VIDrel: 4-FRONT/POLYREC/ODY V/sh

LIAR LIAR ** 12
Tom Shaydac USA 1996
Jim Carrey, Maura Tierney, Justin Cooper, Cary Elwes, Anne Haney, Jennifer Tily, Amanda Donohoe, Jason Bernard, Swoosie Kurtz, Mitchell Ryan, Chip Mayer, Eric Pierpoint, Randall "Tex" Cobb, Cheri Oteri
Carrey play a lecherous lawyer whose complex lifestyle is founded on a mass of lies, much to the annoyance of his ex-wife, who never finds him arriving as promised to visit his young son, Max. But this all changes when Max makes a birthday wish that his dad will be unable to lie for twenty-four hours. Carrey has a wonderful time, getting bogged down in an increasingly hectic series of revelations, and those who like his manic style will like this movie a lot.
COM 87 min CINrel

LIAR'S CLUB, THE ** (18)
Jeffrey Porter USA 1992
Wil Wheaton, Brian Krause, Soleil Moon Frye, Michael Cudlitz, Jennifer Burns, Bruce Weitz, Aron Eisenberg, Shevonne Durkin, Kim Loughran, Alan Fudge, Robin Groves, Tom Urich, Jim Hudson, Arthur Duffy, David Delus
A member of a football team is sexually attracted to a girl he meets and pursues her relentlessly but eventually goes too far. When she accuses him of rape, his friend and a fellow member of their gang does all he can to force her to retract this accusation. An unedifying if fairly well made thriller whose attempts to examine the dangers of obsessive loyalty put one in mind of the equally unpleasant DIAMOND SKULLS.
THR 98 min (ort 100 min) SATrel: SKY MOVIES

LIAR'S EDGE ** 18
Ron Oliver CANADA 1992
Joseph Bottoms, Shannon Tweed, David Keith, Nicholas Shields, Joanna Parica, Kathleen Robertson, Stephen Hunter, Christoppher Plummer, Tom Melisses, Joy Tanner, Brock Simpson, David Strattten, Elena Kudoba, Glen Ottaway
A teenage youth suffers from violent nightmares involving his dead father and finds that his sanity is not improved when his mother remarries and he acquires a step-uncle with some murderous tendencies. A poorly realised attempt at a psychological thriller that is violent but far from effective.
THR 98 min VIDrel: HIFLI/SONOP V

LIAR'S MOON **
15
David Fisher USA
1981
Matt Dillon, Cindy Fisher, Christopher Connelly, Hoyt Axton, Maggie Blye, Susan Tyrrell, Yvonne De Carlo, Broderick Crawford
A poor man and a rich woman fall in love and embark on a secret relationship despite their different backgrounds. A trite little yarn, made with two different endings. The script is by Fisher.
DRA   101 min (ort 105 min)   VIDrel: ARROW/RTM   V

LIBERACE: THE UNTOLD STORY ***
PG
David Greene USA
1987
Victor Garber, Saul Rubinek, Michael Dolan, Maureen Stapleton, Shawn Levy
A portrait of this pianist and showman that attempts to strip away the myths and examine the man behind the public persona, resulting in a far more interesting film than the "approved" Billy Hale version. Garber attempts to give his character some substance rather than merely attempting mimicry, and Stapleton is quite memorable as the performer's domineering mother.
Aka: LIBERACE: BEHIND THE MUSIC
DRA   90 min (ort 100 min) mTV   VIDrel: L/A   V

LICENCE TO KILL ***
15
John Glen UK
1989
Timothy Dalton, Carey Lowell, Robert Davi, Talisa Soto, Anthony Zerbe, Frank MacRae, Wayne Newton, Everett McGill, Benecio Del Toro, Desmond Llewelyn, David Hedison, Priscilla Barnes, Robert Brown, Caroline Bliss, Anthony Starke
The second Bond outing for Dalton (following his debut in THE LIVING DAYLIGHTS) in which he goes after a drugs mobster in order to avenge the maiming of his best friend and murder of the man's bride. Despite the lack of Connery's suave sophistication, Dalton brings a toughness and intensity to his role that makes this violent adventure entertaining, if sometimes a little disagreeable. The spectacular stunts are an added attraction. See GOLDENEYE.
A/AD   127 min (ort 133 min) wScrn (special edition)
VIDrel: MGM/WHV   V/sur V/dm

LICENSE TO DRIVE **
PG
Greg Beeman USA
1988
Corey Haim, Corey Feldman, Carol Kane, Richard Masur, Heather Graham, Harvey Miller, Michael Manasseri, Grant Goodeve, Parley Baer, M.A. Nickles, Nina Siemaszko, James Avery, Grant Heslov, Michael Ensign, Kimberly Hope
A sixteen-year-old boy fails his driving test, but this doesn't stop him from borrowing his grandfather's car for a date, that turns into an endless catalogue of mishaps. A likeable comedy whose lack of a coherent narrative dilutes the strong performances from Haim and Masur.
COM   86 min Cut (7 sec – ort 88 min)   VIDrel: 20TH/TECH
V/sur

LIE, THE ***
15
Francois Margolin FRANCE
1991
Nathalie Baye, Didier Sandre, Helene Lapiower, Marc Citti, Adrien Beau, Christophe Bourseiller, Evelyne Ker, Domonique Besnehard, Louis Ducreaux, Bruno Todeschini, Josiane Stoleru, Francis Girod, Andree Tainsy, Jean-Claude Lecas
A couple of journalists enjoy a happily married life together, but when the wife falls pregnant with her second child she learns to their shock that she is HIV positive. Having never cheated on her husband in ten years of marriage, she can only conclude that the same is not true of him. An engrossing and quite painful drama that explores a contemporary issue with competence and insight.
Aka: MENSONGE
DRA   85 min (ort 89 min) wScrn   VIDrel: TART/20TH
V/dm

LIEBELEI **
U
Max Ophuls AUSTRIA
1932
Magda Schneider, Wolfgang Liebneiner, Luise Ullrich, Willy Eichberger, Gustaf Grundgens, Olga Tschechowa
Romantic tale of a violinist's daughter and her tragic love affair with a young officer whose death in a duel over a married woman provokes her own suicide. Set in Vienna 1910, this is a well conceived effort whose dialogue plays very much second fiddle to the music and camerawork and additionally, lacks many of the ironic overtones of the original.
DRA   82 min (ort 88 min) B/W   VIDrel: ARTPRO/RTM   V
Boa: play by Artur Schnitzler.

LIEBESTRAUM ****
18
Mike Figgis USA
1991
Kevin Anderson, Pamela Gidley, Bill Pullman, Kim Novak, Graham Beckel, Zach Grenier, Thomas Kopache, Anne Lange, Harper Harris, Karen Silas, Tracy Thorne, Max Perlich, Hugh Hurd, Catherine Hicks, Taina Elg, Joe Aufiery
Nick Kaminsky is an architectural writer who arrives at the fictitious "Elderstown" to visit his dying mother. There he meets an old college buddy who now works as a contractor, and is about to tear down a building of architectural interest. Nick decides to write a feature on it, but a crime of passion that took place there many years ago begins to exert a strange influence on his life. Scripted by Figgis, this is a most intriguing and unusual film.
Aka: LOVE DREAM
DRA   108 min (ort 113 min)   VIDrel: MGM/WHV   V/sur

LIES BEFORE KISSES **
15
Lou Antonio USA
1991
Jaclyn Smith, Ben Gazzara, Nick Mancuso, Greg Evigan, Penny Fuller, James Karen, William Allen Young, Jean Hale, Jim Antonio, Clyde Kasatsu, Sara Rose Johnson, Lisa Rinna, Laura Dobbin, Alina Cenel, Milt Kogan, Herb Mitchell
The wife of a wealthy publisher learns that her husband is being blackmailed for the alleged murder of a hooker. A glossy soap opera-style thriller, light on plotting but heavy on romantic interest. Adequate.
THR   93 min (ort 100 min) mTV   VIDrel: GUILD/SONOP
V/sh

LIES BOYS TELL, THE ***
PG
Tom McLoughlin USA
1994
Kirk Douglas, Craig T. Nelson, Bess Armstrong, Bonnie Bartlett, Richard Galliland, Ernest Thompson, Nancy McLaughlin, Lee Garlington, Anne Haney, Jason Herrey, Eileen Brennan, Martey Shelton, Glenn Walker Harris, Geraldine Leer
A terminally ill man resolves to visit the past loves of his life, make his peace with his estranged son and if possible bring about a reconciliation between the son and his wife.
DRA   87 min mTV   VIDrel: ODY/SONOP   V/sh
Boa: novel by Lamar Herrin.

LIES FROM THE TWINS **
15
Tim Hunter USA
1991
Isabella Rossellini, Aidann Quinn, Iman, Claudia Christian, John Pleshette, Hurd Hatfield, Angela Paton, Barbara Eda-Young, Brenda Varda, Iris Peynard, Jeanine Bisignano, Leonard Termon, Richard Harrison, Tracey Ross, Sav Farrow
A beautiful young woman finds herself becoming deeply involved with a pair of identical twins, one of whom is good natured and honest, whilst the other is charismatic but evil. A tightly knit and well acted psychological thriller that maintains a strong sense of tension. See also DEADRINGERS for another film on this theme.
Aka: LIES OF THE TWINS
THR   88 min (ort 95 min) mCab   VIDrel: CIC/SONOP   V
Boa: novel Lies of the Twins by Rosamond Smith (Joyce Carol Oates).

LIES OF THE HEART: THE STORY OF LAURIE KELLOGG **
(18)
Michael Toshiyuki Uno USA
1994
Jennie Garth, Gregory Harrison, Steven Keats, Franicis Guinand, T.C. Warner, Robin Frates, Alex Arquette, Sharon Spelman, Jeff Doucette, Virginya Keehne, Gina Philips, Phil Buckman, William Wellma Jr, Robert Cavanaugh, Robert Factor
Yet another in a seemingly endless series of TV films detailing the events surrounding a real-life crime. In this unedifying tale, a woman is accused of involving her four teenage sons in a plot to murder her husband, as ever her excuse being that her was an abusive husband. What tension there is revolves around the decision the court has to make as to her culpability. Adequate.
DRA   90 min mTV   SATrel: SKY MOVIES

LIFE AND DEATH OF COLONEL BLIMP, THE ***
U
Michael Powell/Emeric Pressburger UK
1943
Roger Livesey, Anton Walbrook, Deborah Kerr, John Laurie, James McKechnie, Albert Lieven, Roland Culver, Arthur Wontner, David Hutcheson, Ursula Jeans, Harry Welchman, Reginald Tate, Carl Jaffe, Valentine Dyall, Patrick Macnee
The title character bears no relation to the famous character created by cartoonist David Low. This is a story of a British soldier's career through three wars and for the most part is a touching and memorable film. The less appealing latter portion

offers an ill-advised change of pace and some comic interludes during a military exercise, but fortunately doesn't detract from the film as a whole.
Aka: COLONEL BLIMP
COM 157 min (ort 163 min) VIDrel: VCC/DISC/COLUM V

LIFE AND EXTRAORDINARY ADVENTURES OF PRIVATE IVAN CHONKIN, THE *** 15
Jiri Menzel CZECH
REPUBLIC/FRANCE/ITALY/RUSSIA/UK 1994
Gennadiy Nazarov, Zoya Buryak, Vladimir Ilyin, Valeriy Dubrovin, Alexei Zharkov, Yuriy Dubrovin, Sergei Stepanchenko, Sergei Garmash, Zinovii Gerdt, Marian Labuda, Maria Vinogradova, Tatiana Gerbachevskaia, Liubov Rudneva
During WW2, a Red Army private cannot believe his good fortune when he finds himself left behind in a remote Ukrainian village to guard a downed military plane, while his comrades go off to fight. Confident that his superiors have forgotten his very existence, he settles down to a life of relative ease and pursues a romance with a local woman. However, the jealousy of a neighbour proves to be an unexpected hazard. A biting black comedy of wit and invention.
COM 111 min CINrel

LIFE AND NOTHING BUT **** PG
Betrand Tavernier FRANCE 1989
Philippe Noiret, Sabine Azema, Pascale Vignal, Maurice Barrier, Francois Perrot
Noiret plays a French Army officer who has become obsessed with the toll of WW1 casualties, and two years after the end of the war, is still hard at work on his assigned task of tabulating and identifying French dead; a task which brings him into contact with a wealthy woman who is desperately searching for her missing husband. Despite the unfocused script, this remarkable film is redeemed by some outstanding images and performances.
DRA 130 min (ort 135 min) wScrn VIDrel: ARTIF/20TH V

LIFE AND TIMES OF JUDGE ROY BEAN, THE *** 15
John Huston USA 1972
Paul Newman, Ava Gardner, Stacy Keach, Victoria Principal, John Huston, Anthony Perkins, Ned Beatty, Jim Burk, Tab Hunter, Roddy McDowall, Anthony Zerbe, Jacqueline Bisset, Richard Farnsworth, Roy Jenson, LeRoy Johnson
A self-appointed judge rules over a vast territory, and dispenses rough justice to all and sundry, in a colourful and slightly tongue-in-cheek tale, whose appeal largely resides in a series of engaging vignettes. Scripted by John Milius.
WES 118 min (ort 124 min) VIDrel: WHV V

LIFE IN EMERGENCY WARD 10 ** U
Robert Day UK 1959
Michael Craig, Wilfrid Hyde White, Dorothy Allison, Glyn Owen, Rosemary Miller, Charles Tingwell, Frederick Bartman, Joan Sims, Rupert Davies, Sheila Sweet, David Lodge, Dorothy Gordon, Christopher Witty, Douglas Ives
Full-length feature from the TV long-running TV series "Emergency Ward 10", that followed the day-to-day work and traumas of the medical staff and patients at a hospital. There is little plot to be had here except the usual intrigues and squabbles, though one interesting strand has a devoted medic experimenting with a heart-lung machine that he uses to save the life of a young boy. Sadly, the acting and dialogue is so terribly stilted that one soon loses interest.
DRA 82 min (ort 86 min) B/W mTV VIDrel: ODY/SONOP V

LIFE IS A LONG QUIET RIVER ** 15
Etienne Chatiliez FRANCE 1987
Benoit Magimel, Valerie Lalande, Tara Romer, Jerome Floch, Sylvie Cubertafon, Emmanuel Cendrier, Helene Vincent, Andre Wilms, Christine Pignet, Maurice Mons, Catherine Jacob, Patrick Bouchitey, Claire Prevost
A malicious nurse reveals to a well-to-do family that twelve years ago she swapped their newborn child with one from a poor, semi-criminal family, an act of malice designed to revenge herself on her doctor lover, who had refused to marry her. This sets in motion a chain of events that corrupts the first family and benefits the second. A spiteful comedy-drama, sluggish in pacing and far too unpleasant to amuse.
Aka: LA VIE EST UN LONG FLEUVE TRANQUILLE
DRA 90 min (ort 95 min) VIDrel: L/A V

LIFE IS CHEAP... BUT TOILET PAPER IS EXPENSIVE * (18)
Wayne Wang USA 1990
Chan Kim Wan, Spencer Nakasako, Victor Wong, Cheng Kwan Min, Cora Miao, Lam Chung, Allen Fong, John K. Chan, Bonnie Ngai, Lo Wai, Cinda Hui, Gary Kong, Rocky Ho, Yu Chien, Wu Kin Man, Lo Lieh, Mr and Mrs Kai-Bong Chau
Initially this was meant to be a documentary on Hong Kong, but somehow the director changed directed, and the result is a peculiar mishmash of a film that starts with a courier coming to Hong Kong with a mysterious package in a briefcase that's chained to his wrist. But on arrival he finds he cannot deliver the package so he takes in the sights and sounds of the city instead. A bizarre, opaque and quite unpleasant film, with a few scenes of bloodshed thrown in.
DRA 88 min CINrel

LIFE IS SWEET *** 15
Mike Leigh UK 1990
Alison Steadman, Jim Broadbent, Timothy Spall, Claire Skinner, Jane Horrocks, David Thewlis, Moya Brady, Stephen Rea, David Neilson, Harriet Thorpe, Paul Trussel, Jack Thorpe Baker
A warm, life-affirming, but not really convincing portrait of a larger-than-life working-class family, where the father suddenly decides that a fast-food caravan will be a sure-fire recipe for instant wealth. Meanwhile, his two very different daughters face problems of their own. Both rambling yet skilful, this enormously likeable film is a good deal more mellow in tone than previous Leigh films, such as HIGH HOPES.
COM 99 min (ort 102 min) VIDrel: IMAG/RTM V/sur

LIFE OF NINJA, A ** 18
HONG KONG
The police are powerless in the face of a Ninja mob who appear able to carry out their assassinations with impunity. However, a man steeped in their traditions and skills is inadvertently drawn into a conflict with them and challenges their dominance. A set of well staged fight sequences are on offer in this utterly predictable martial arts movie.
Aka: LIFE OF A NINJA, A
MAR 84 min Cut (23 sec) VIDrel: IMPENT V

LIFE OF OHARU, THE **** 18
Kenji Mizoguchi JAPAN 1952
Kinuyo Tanaka, Toshiro Mifune, Hisako Yamane, Yuriko Hamada, Ichiro Sugai, Tsuki Matsura, Toshiaki Konoe, Masao Shinizu, Eitaro Shindo, Jukichi Uno, Akira Akira Oizumi, Masao Shimizu, Daisuke Kato, Toranosuke Ogawa, Eijiro Yanagi
A beautifully made account of a woman's life of degradation in 17th century Japan, when her love for a page brings disgrace on her family. After he is executed and she fails to kill herself, she becomes a prince's concubine and bears him a son. Sent away by him, she falls into a life of prostitution. A deeply moving and beautifully composed work from this master director and enjoyable despite its formal style.
Aka: DIARY OF OHARU; SAIKAKU ICHIDAI ONNA
DRA 130 min (ort 136 min) B/W VIDrel: TART/20TH V
Boa: novel Koshukuo Ichidal Onna by Ibara Saikaku.

LIFE ON A STRING *** PG
Chen Kaige CHINA/GERMANY/UK 1992
Liu Zhongyuan, Huang Lei, Xu Qing, Zhang Zhengyuan, Ma Ling, Zhang Jinzhan, Zhong Ling, Yao Erga
A young man searches for some way to cure his blindness and learns of a myth that may enable him to do so, but in return must devote his life entirely to the playing of music. Having become an itinerant singer, he travels the countryside in the company of an old man, who has become his mentor and companion. A fascinating tale from China's distant past, that benefits from excellent camerawork and a fine soundtrack.
Aka: BIAN ZOU BIAN CHANG
DRA 102 min (ort 105 min) VIDrel: ICAPRO/MANGA V/sur
Boa: short story by Shi Tiesheng.

LIFE STINKS ** PG
Mel Brooks USA 1991
Mel Brooks, Lesley Ann Warren, Jeffrey Tambor, Stuart Pankin, Howard Morris, Rudy De Luca, Teddy Wilson, Michael Ensign, Matthew Faison, Billy Barty, Brian Thompson, Raymond O'Connor, Carmine Caridi, Sammy Shore, Frank Roman, John Welsh
More coherent than usual, this film tells of a wealthy and callous developer who takes a bet that he won't survive for one month

as a derelict on the streets of L.A. The usual messy clutch of Brooks gags is found here, and though there are some extremely funny moments, the film's attempt to say something meaningful about homelessness is not very convincing. Lesley Ann Warren makes an utterly ravishing (if not believable) bag-lady. COM 91 min (Cut at UK cinema release by 18 sec – ort 95 min) VIDrel: 20TH/TECH V/sur

LIFEBOAT **** PG
Alfred Hitchcock USA 1944
William Bendix, Tallulah Bankhead, Walter Slezak, Mary Anderson, John Hodiak, Henry Hull, Heather Angel, Hume Cronyn, Canada Lee
The survivors of a torpedoed liner are adrift in a lifeboat. Having picked up the U-boat commander responsible for their plight, they are forced to rely on him to steer the vessel. A gripping propaganda film, with unusual casting and a literate script (by Jo Swerling). The director uses the constraints imposed by the cramped set to his advantage. Bankhead, playing a spoilt rich girl in her one major role, was never better.
DRA 96 min B/W VIDrel: 20TH/TECH V
Boa: story by John Steinbeck.

LIFEFORCE ** 18
Tobe Hooper USA 1985
Frank Finlay, Steve Railsback, Mathilda May, Peter Firth, Patrick Stewart, Nicholas Ball, Michael Gothard, Aubrey Morris, Nancy Paul, John Hallam, John Keegan, Paul Cooper, Christopher Jagger, Bill Malin, Jerome Willis
An alien spacecraft is brought back to Earth where its occupants escape and wreak havoc, by draining people they contact of their life-force, after which the victims become short-lived sex-crazed zombies. A blend of intriguing SF elements and ridiculous supernatural touches, with the latter eventually gaining precedence.
FAN 97 min (ort 116 min) VIDrel: VCC L/A V
Boa: novel The Space Vampires by Colin Wilson.

LIFEFORCE EXPERIMENT, THE ** (PG)
Piers Haggard CANADA/UK 1993
Donald Sutherland, Mimi Kuzyk, Vlasta Vrana, Corin Nemec, Hayley Reynolds, Miguel Fernandez, Michael Rudder, Michael Reynolds, Bronwen Martel, Peter Colzey, Richard Zeman, Ann Page, Harrison Walcott, Philip Pretten, Michael Caloz
A scientist finds way to capture the life energy of dying people and arouses the interest of the CIA who promptly dispatch one of their agents to investigate his activities. She is shocked by what she learns and recommends that his experiments be terminated, but her superiors have a secret agenda of their own. A failed attempt to adapt Du Maurier's story for the screen that degenerates all too soon into the usual government conspiracy cliches.
Aka: BREAKTHROUGH, THE
FAN 88 min (ort 90 min) mCab SATrel: SKY MOVIES
Boa: short story by Daphne Du Maurier.

LIFEPOD ** 15
Ron Silver USA 1992
Robert Loggia, Stan Shaw, Ron Silver, C.C.H. Pounder, Adam Storke, Jessica Tuck, Kelli Williams, Ed Gate, Lisa Waltz, Sam Whipple, Cork Hubbert, John Mahon, Pat Destro
In the year 2168, a spaceship taking hundreds of passengers to Earth meets with disaster due to sabotage. A group of nine people take refuge in a damaged lifepod but soon learn that their lives are in danger from one of their number, the terrorist responsible for this outrage. A very loose variation on the 1944 Hitchcock movie LIFEBOAT, with adequate sets and effects.
FAN 85 min (ort 100 min) mTV VIDrel: ITC/POLYREC L/A V/h

LIFESPAN ** 18
Alexander (Sandy) Whitelaw HOLLAND/USA 1975
Klaus Kinski, Hiram Keller, Tina Aumont, Fons Rademakers, Eric Schneirer, Franz Mulder, Lyde Polak, Joan Remmelts, Andre Van Den Heuvel, Onno Molenkamp, Dick Schefer, Albert Van Doorn, Adrian Brine, Helen Van Meurs
A wealthy industrialist fights a young scientist for control of a drug that promises to be an elixir of life, conferring immortality on its recipients. An erotic thriller with a strong SF flavour.
FAN 77 min Cut (1 min 14 sec – ort 85 min) VIDrel: ARTPRO/RTM V

LIFT, THE *** 15
Dick Maas HOLLAND 1983
Huub Stapel, Willeke Van Ammelrooij, Josine Van Dalsun, Hans

Verman, Hans Dagelet, Ab Abspoel, Frederick DeGroot, Onno Molenkamp, Henri Serge Valcke, Liz Snijdijink, Wiske Sterringa, Huib Broos, Pieter Lutz, Dick Scheffer, Piet Romer
A lift seems to have the ability to cause a series of nasty killings in a new apartment block, in this flashy and taut chiller that unfortunately ends without any clear resolution or explanation.
Aka: DE LIFT; GOING UP
HOR 94 min (ort 99 min) dubbed VIDrel: MGM/WHV L/A V

LIFT TO THE SCAFFOLD *** PG
Louis Malle FRANCE 1957
Jeanne Moreau, Maurice Ronet, Lino Ventura, Georges Poujouly, Yori Bertin
An executive plots the murder of his boss with the connivance of the man's wife, but is trapped in the building overnight. Meanwhile, his car is stolen by a thief who also commits a murder, and this leads to his arrest. Malle's first non-documentary feature as director is a coldly clinical work of some complexity; it starts out with great promise but ultimately disappoints. The strong jazz score was improvised at a screening by Miles Davis and others.
Aka: ASCENSEUR POUR L'ECHAFAUD; ELEVATOR TO THE GALLOWS; FRANTIC
THR 87 min (ort 90 min) B/W VIDrel: ELPIC/POLYREC V

LIGHT IN THE JUNGLE, THE ** (15)
Gray Hofmeyr SOUTH AFRICA 1990
Malcolm McDowell, Susan Strasberg, Andrew Davis, Patrick Shai, John Carson, Helen Jessop, Henry Cele, Michael Huff, Stuart Parker, Barbara Nielsen, Michael Khumalo, Masbatha Jaffa, Dominique Moser, Roy Dilamine, Martin Adamiel
A dullish biopic on the life of Dr Albert Schweitzer, who received the Nobel Peace Prize for his work in Africa, where he established a hospital and sought to improve the health of the natives. A strangely flat and mediocre effort, slightly redeemed by its African locations.
Aka: OUT OF DARKNESS: TRIUMPH OF COURAGE
DRA 93 min SATrel: SKY MOVIES

LIGHT SLEEPER ** 15
Paul Schrader USA 1991
Willem Dafoe, Dana Delaney, Susan Sarandon, David Clennon, Mary Beth Hurt, Victor Garber, Jane Adams, Paul Jabara, Robert Cicchini, Sam Rockwell, Rene Rivera, David Spade, Steven Posen, Ken Ladd, Brian Judge, Vinny Capone
Writer-director Schrader takes some of the themes touched on in earlier films (such as TAXI DRIVER) and has as his central character a drug-dealer who finds that he is getting just a little tired of his trade. Much of the story revolves around his relationship with another dealer, a glamorous ex-girlfriend and former addict, and though the seedy side of New York adds considerable atmosphere, the film never really develops any momentum.
THR 99 min (ort 103 min) VIDrel: GUILD/POLYREC L/A; PION (LV only) V LV

LIGHTHORSEMEN, THE ** PG
Simon Wincer AUSTRALIA 1987
Jon Blake, Peter Phelps, Anthony Andrews, Tony Bonner, Sigrid Thornton, Bill Kerr, John Walton, Tim McKenzie, Ralph Cotterill, Grant Piro, Serge Lazareff, Gary Sweet, Jon Blake, Anthony Hawkins, Gerard Kennedy, Shane Briant
A simple-minded WW1 actioner, dealing with the exploits of a small cavalry contingent in Palestine. Beautifully filmed, but spoilt by one-dimensional characterisations and excessive length. The original 128 minute version (or director's cut) included a commentary from Wincer, a theatrical trailer and a few scenes from the 1940 film "40,000 Horsemen".
WAR 110 min Cut (6 sec – ort 128 min)
VIDrel: 4-FRONT/POLYREC V/sur

LIGHTNING INCIDENT, THE ** 15
Michael Switzer USA 1991
Nancy McKeon, Tantoo Cardinal, Miriam Colon, Tim Ryan, Polly Bergen, Elpida Carrillo, Joaquin Martinez, Gary Clarke, George Salazar, Sheree Spargo, Dave Adams, George Pompa, Kathleen Erickson, Barbara Glover, Lillie Richardson
A woman goes to a psychologist for help in dealing with her nightmare, but under hypnosis reveals that she is being persecuted by a satanic cult who have chosen her son to be a sacrifice.
HOR 86 min (ort 90 min) mTV VIDrel: CIC/SONOP V

LIGHTNING JACK **
Simon Wincer USA 1993
PG

*Paul Hogan, Cuba Gooding Jr, Beverly D'Angelo, Kamala Dawson,
Pat Hingle, Richard Riehle, Frank McRae, Roger Daltrey, L.Q. Jones,
Max Cullen, Mark Miles, Roy Brocksmith, Douglas Stewart, Kevin
O'Morrison, Cliff Stokes, Bob Sorenson*
A short-sighted outlaw in the old West wants to be a bank robber
but his poor eyesight forces him to team up with a sharp-eyed
mute. Together, this unlikely couple plunder one bank after
another in a weird comedy of errors that is something of a variation on SEE NO EVIL, HEAR NO EVIL. Fans of Hogan will like
this one a lot, others might find it just a bit over-extended.
COM 93 min (ort 114 min) VIDrel: TOUCH/TECH L/A V

LIGHTS OF VARIETY ***
Federico Fellini/Alberto Lattuada ITALY 1950
PG

*Peppino De Felippo, Carla Del Poggio, Giulietta Masina, John
Kitzmiller*
Story of a third-rate troupe of travelling musicians, which is
joined by a stage-struck young girl, who stays with them for a
while until leaving for better things. In his youth Fellini was
involved in the music-hall, and draws on his experiences to
recreate the seediness, warmth and humour of it all, in this
charming film.
Aka: LUCI DEL VARIETA; VARIETY LIGHTS
DRA 92 min (ort 94 min) B/W VIDrel: CONNO/RTM L/A
V

LIGHTSHIP, THE **
Jerzy Skolimowski USA 1985
15

*Robert Duvall, Klaus Maria Brandauer, Tom Bower, Robert Costanzo,
Badja Djola, William Forsythe, Arliss Howard, Michael Lyndon*
Three criminals are rescued by the crew of a lightship whom
they harass and terrorise. A stylish film full of rather opaque
symbolism – some critics saw the lightship as representing the
Weimar Republic.
DRA 84 min (ort 89 min) VIDrel: 20TH/TECH V
Boa: novel Das Feuerschiff by Siegfried Lenz.

LIKE FATHER, LIKE SON *
Rod Daniel USA 1987
15

*Kirk Cameron, Dudley Moore, Sean Astin, Margaret Colin, Catherine
Hicks, Patrick O'Neal, Cami Cooper, Micah Grant, Bill Morrison,
Skeeter Vaughan, Larry Sellers, Tami David, Maxine Stuart, David
Wohl, Michael Horton, Christine Healy*
A father and son accidentally change bodies with predictable
complications all round. A limp and unfunny comedy built
around this "Vice Versa" idea, followed by several films that
handled this perennial favourite with a good deal more style.
See also EIGHTEEN AGAIN!
COM 95 min (ort 101 min) VIDrel: RCA L/A V

LIKE GRAINS OF SAND **
Ryosuke Hashiguchi JAPAN 1995
(PG)

*Yoshinori Okada, Kota Kusano, Ayumi Hamazaki, Koji Yamaguchi,
Kumi Takada, Shizuka Isami, Cho Bang-Ho, Yoshihiko Hakamada,
Miyako Yamaguchi, Yoshie Negishi, Kunio Murai*
Though his fellow classmate stands by him when he is taunted
for being gay, a youngster nonetheless finds that his affectionate overtures are not reciprocated by this individual, who unlike
his friend is completely heterosexual. An honest attempt to
explore sexuality in a youthful setting, the film's undoubted
strengths are not well served by its lack of a clear narrative.
Aka: NAGISA NO SINDBAD
DRA 129 min CINrel

LIKE WATER FOR CHOCOLATE ***
Alfonso Arau MEXICO 1992
15

*Lumi Cavazos, Marco Leonardi, Regina Torne, Mario Ivan Martinez,
Yareli Arizmendi, Ada Carrasco, Claudette Maille, Pilar Aranda,
Farnesio De Bernal, Joaquin Garrido, Rodolfo Arias, Margarita Isabel,
Sandra Arau, Andres Garcia Jr*
In the Mexico of the early 1900s, the youngest daughter of a
widow grows up in a strange household whose housekeeper
has a magical way with food. Refused permission to marry her
by her mother who insists on the family tradition being carried
until she dies, our heroine puts all her feelings into her cooking.
A touching and fascinating tale, very much in the magical
realism tradition of Latin American literature.
Aka: COMO AGUA PARA CHOCOLATE
DRA 109 min (ort 114 min) VIDrel; ELPIC/POLYREC
V/sur
Boa: novel by Laura Esquivel.

LIKELY LADS, THE *
Michael Tuchner UK 1976
PG

*Rodney Bewes, James Bolam, Brigit Forsyth, Mary Tamm, Sheila
Fearn, Zena Walker, Anulka Dubinska, Judy Brixton, Alun
Armstrong, Vicki Michelle, Penny Irving, Michelle Newell, Susan
Tracy, Gordon Griffin, Edward Wilson*
Two Geordie friends and their girlfriends go on a caravan
holiday together. A spin-off from a long-running TV series but
with none of the latter's sharply observed social comment.
Written by Dick Clement and Ian La Frenais.
COM 86 min (ort 90 min) B/W VIDrel: WHV L/A V

LILI MARLEEN **
Rainer Werner Fassbinder WEST GERMANY 1981
15

*Hanna Schygulla, Giancarlo Giannini, Mel Ferrer, Christine
Kaufman, Karl Heinz, Udo Kier, Hark Bohm, Karin Baal, Erik
Schumann*
The story of a female singer who is catapulted to fame by
singing an old song during WW2 in Nazi Germany. A strange
film that never seems able to make up its mind whether to
become a satire or a straight drama, and ultimately does justice
to neither genre.
DRA 116 min (ort 120 min) VIDrel: MIA/DISC V
Boa: novel The Sky Has Many Colours by Lale Anderson.

LILY IN WINTER **
Delbert Mann USA 1994
PG

*Natalie Cole, Brian Bonsall, Marla Gibbs, Cecil Hoffman, Dwier
Brown, Monte Russell*
Set in New York in 1957 just on the eve of Christmas, with a
black woman who works as a maid to a wealthy family being
forced to flee, when her brother robs her employer's home. To
make matters worse, the couple's young son comes along with
her, which leads to her being suspected of kidnapping. Back in
her home state of Alabama, she searches desperately for a way
out of her predicament. A bit overdone and heavy-going, this
gave singer Cole's her first movie role.
DRA 89 min (ort 94 min) VIDrel: CIC/SONOP L/A V/sh

LILY WAS HERE **
Ben Verbong HOLLAND 1989
18

Marion Van Thijn, Thom Hoffman, Monique Van De Ven
When her black American soldier boyfriend is killed by a bunch
of racist thugs, a young and pregnant Dutch woman leaves the
home of her unsympathetic parents and falls into a life of
squalor. A cold and clinical piece of European kitchen-sink
melodrama, that really fails to engage the emotions, and seems
even more remote because of its inappropriate dubbing with
American voices.
DRA 105 min (ort 110 min) dubbed VIDrel: 20VIS/SONOP
V/sh

LIMBIC REGION, THE ***
Michael Pattinson USA 1996
18

*Edward James Olmos, George Dzundza, Gwyneth Walsh, Don Davis,
Tom McBeath*
An intense and moody psychological thriller, mostly told as a
series of flashbacks, from which we learn all about a cop and his
obsession with catching a serial killer. As the cop is obliged to
play an increasingly nasty cat-and-mouse game with the killer,
the plot unfolds in a fairly predictable way, and the final resolution is never in doubt. However, it is all handled with enough
panache to keep one involved.
THR 92 min cC VIDrel: MGM/WHV V/sh

LIMIT UP **
Richard Martini USA 1989
PG

*Nancy Allen, Dean Stockwell, Brad Hall, Danitra Vance, Rance
Howard, Luana Anders, Ray Charles, Sally Kellerman, Sandra Bogan,
William J. Woff, Robbie Robbie Martini, Ava Fabian, Teresa Ovetta
Burrell, Winifred Freedman*
A modernised version of the Faust legend, in which a female
would-be commodities broker does a deal with the Devil, who
acquires her soul in return for allowing her to gain control of the
world's trade in soybeans. Not quite the success it might have
been, as the fantasy elements do not blend well with the
comedy, but casting Ray Charles as God is a nice touch. Look
out for Kellerman, who has a cameo as a nightclub singer.
COM 85 min Cut (4 sec – ort 88 min) VIDrel: MED L/A V

LINDA ***
Nathaniel Gutman USA 1993
15

Virginia Madsen, Richard Thomas, Ted McGinley, Laura Harrington,

J.E Russell, J. Don Ferguson, Ricahrd K. Olsen, David Dwyer, Michael Goodwin, Maria Howell, Vito Mirabella, Mary Lynn Riner, John Keena, Winston Hemingway
A couple make friends with their new neighbours and the wife is so attracted to the other woman's husband that she devises a cunning plan to get rid of her own spouse by murdering her lover's wife and framing him. A well-acted but dark tale of human evil, with strong performances and enough twists and turns in the plot to keep the suspense going.
THR 84 min (ort 88 min) mTV VIDrel: CIC/SONOP
V/sur
Boa: novella by John D. MacDonald

LINGUINI INCIDENT, THE *   15
Richard Shepard USA   1991
Rosanna Arquette, David Bowie, Marlee Matlin, Buck Henry, Andre Gregory, Eszter Balint, Viveca Lindfors, Eloy Casados, Michael Bonnabel, Lenaa Hall, Susan Mechsner, Maura Tierney, Kristina Loggia, Pat Dubroff, Roxanne Beckford
A waitress at a trendy New York restaurant who is obsessed with Houdini entertains plans to rob this joint and gets involved with a variety of other way-out characters. A messy, unfunny and contrived comedy that is full of shouting and over-the-top acting, but has little entertainment value.
COM 104 min VIDrel: VCC/DISC/COLUM L/A V/sh

LION IN WINTER, THE ***   15
Anthony Harvey UK   1968
Peter O'Toole, Katharine Hepburn, Jane Merrow, Timothy Dalton, Anthony Hopkins, Nigel Terry, Nigel Stock, O.Z. Whitehead, Kenneth Griffith, John Castle, Kenneth Ives, Henry Wolff, Karol Hager, Mark Griffith
King Henry II and his Queen meet at Christmas to consider the question of a successor to the throne. Despite the literate script, the film meanders between farce and high drama, but has many moments of brilliance among the dross. Hepburn tied with Barbra Streisand (in FUNNY GIRL) at the Academy Awards. AA: Actress (Hepburn), Score/orig (John Barry), Screen/adapt (James Goldman).
DRA 128 min (ort 134 min) VIDrel: POLY L/A V
Boa: play by James Goldman.

LION KING, THE ***   PG
Roger Allers/Rob Minkoff USA   1993
Voices of: Whoopi Goldberg, Cheech Marin, James Earl Jones, Jeremy Irons, Matthew Broderick, Madge Sinclair, Robert Guillaume, Jonathan Taylor Thomas, Rowan Atkinson, Nathan lane, Ernie Sabella, Niketa Calame, Moira Kelly
After his father is betrayed an killed by his uncle, a young lion cub is forced into exile but eventually learns that his rightful place has been usurped and resolves to return and claim it. This extremely well crafted animation was a massive success but is possibly a touch too frightening for very young children. AA: Score/orig (Hans Zimmer), Song ("Can You Feel The Love Tonight?" – Elton John (m)/Tim Rice (l)).
ANIM 84 min (ort 88 min) wScrn cC VIDrel: WDV/TECH
V

LION OF THE DESERT **   15
Moustapha Akkad LIBYA/UK/USA   1979 (released 1980)
Anthony Quinn, Oliver Reed, Irene Papas, Rod Steiger, John Gielgud, Raf Vallone, Gastone Moschin, Andrew Keir, Takis Emmanuel, Stefano Patrizi, Sky Dumont, Robert Brown, Eleonora Stathopoulou, Adolfo Lastretti
Quinn plays Omar Mukhtar, a Libyan guerilla leader who fought Italy's occupation of the country from 1911 to 1931. A glamorised account of some spectacle but little verisimilitude. Steiger gives a good performance as Mussolini.
Aka: OMAR MUKHTAR
DRA 156 min (ort 163 min)
VIDrel: POLY/POLYREC/BRAVE V/sur

LION STRIKE **   18
Rik Jacobson USA   1995
Don "The Dragon" Wilson, Bobby Phillips, Morgan Honier
A network of international terrorists starts dealing in nuclear weapons, and a doctor finds himself unexpectedly embroiled in their affairs, forcing him to make use of his kickboxing skills in order to survive. Wilson has little ability as an actor, which is fortunate, as this noisy and explosive picture has a barely discernible plot.
A/AD 91 min VIDrel: MIA/DISC V/sh

LION, THE WITCH AND THE WARDROBE, THE *   Uc
Bill Melendez UK/USA   1978
Voices of: Rachel Warren, Susan Sokol, Reg Williams, Simon Adams, Victor Spinetti, Dick Vosburgh, Don Parker, Liz Proud, Stephen Thorne, Beth Porter
Fantasy animation about a country kept in eternal winter by the power of an evil witch and telling of the three children who journey there. A rather insipid adaptation that has none of the poetry or magic of Lewis's novel, though surprisingly, it did win an Emmy in the category reserved for children's features. Written by Melendez and David Connell. See also THE CHRONICLES OF NARNIA series, a set of adaptations produced by the BBC.
ANIM 90 min (ort 96 min) mTV VIDrel: L/A V
Boa: novel by C.S. Lewis.

LIONHEART **   PG
Franklin J. Schaffner USA   1987
Eric Stoltz, Gabriel Byrne, Nicola Cowper, Dexter Fletcher, Deborah Barrymore, Nicholas Clay, Bruce Purchase, Neil Dixon, Penny Downie, Chris Pitt, Nadim Sawalha, John Franklyn-Robbins, Matthew Sim, Paul Rhys, Sammi Davis
A young knight riding off to join King Richard in the Crusades to the Holy Land is diverted from this quest when he saves a band of homeless children from the clutches of an evil knight who planned to sell them into slavery. A very poor effort, quite clearly aimed at children, but unlikely to have much appeal in that or any other quarter.
A/AD 100 min (ort 104 min) VIDrel: MGM/WHV L/A V

LIPSTICK *   18
Lamont Johnson USA   1976
Margaux Hemingway, Anne Bancroft, Robin Gammell, Chris Sarandon, Perry King, Mariel Hemingway, Francesco, Bill Burns, Meg Wylie, Inga Swenson, John Bennett Perry
A raped model takes revenge when the courts fail her in this nasty little exploiter that treads the same well-worn path as DEATHWISH. This was Mariel Hemingway's film debut. See also THE SISTERHOOD.
DRA 83 min Cut (2 min 54 sec – ort 90 min) VIDrel: L/A
V

LIPSTICK CAMERA **   18
Mike Bonifer USA   1994
Brian Wimmer, Ele Keats, Terry O'Quinn, Sandahl Bergman, Charlotte Lewis, Richard Portnow, Corey Feldman
A woman determined to break into video journalism sets up her equipment at the home of a friend and accidentally films a vicious sex murder. But this enables her to fulfil her ambition, which it to work alongside her idol – a top reporter. However, one of the assignments she is sent on proves to be a good deal less straightforward than it initially appeared to be. A pretentious erotic thriller, it misses the chance to say anything substantial about the media.
THR 86 min (ort 93 min) VIDrel: 20TH/FOXVID V/h

LIPSTICK ON YOUR COLLAR **   15
Renny Rye UK   1993
Peter Jeffrey, Louise Germaine, Giles Thomas, Ewan McGregor, Kymberley Huffman, Clive Francis, Douglas Henshall, Nicholas Farrell, Nicholas Jones, Roy Hudd, Maggie Steed, Bernard Hill
Written by Dennis Potter and originally shown in six parts, this drama follows the lives of four people in Britain, from the Suez Crisis through to the 1950s. A moderately enjoyable trip down nostalgia lane.
DRA 356 min (2 cassettes – ort 390 min) mTV
VIDrel: POLY/POLYREC V

LIQUID SKY **   18
Slava Tsukerman USA   1983
Anne Carlisle, Paula E. Sheppard, Susan Doukas, Otto Von Wernherr, Bob Brady, Elaine Grove, Stanley Knap, Jack Adalist, Lloyd Ziff, Roy MacArthur, Harry Lum, Sara Carlisle, Nina V. Kerova, Alan Preston, Christine Hatfield
A lesbian punk in Manhattan becomes involved with a UFO and other strange happenings, in this utterly bizarre low-budget effort that has some good moments but not much of a plot.
DRA 108 min (ort 112 min) VIDrel: TART/20TH V/dm

LISA AND THE DEVIL *   18
Mickey Lion (Mario Bava) ITALY/SPAIN/
WEST GERMANY   1975
Telly Savalas, Elke Sommer, Sylva Koscina, Robert Alda, Alida Valli,

Gabriele Tinti, Alessio Orano, Eduardo Fajardo, Carmen Silva, Franz Von Treuberg, Espartaco Santoni
A tourist in Rome sees a church fresco of the Devil and becomes drawn into a strange world of Satanism. An incoherent and muddled film that exists in two versions, this original and another called (among other titles) "The House Of Exorcism", that incorporates new footage featuring Robert Alda as a priest. Both however, do not spare us the graphic effects of possession.
Aka: DEVIL AND THE DEAD, THE; DEVIL IN THE HOUSE OF EXORCISM; EL DIABOLO SE LLEVA A LOS MUERTOS; HOUSE OF EXORCISM, THE; IL DIAVOLO E I MORTI; IL DIAVOLO E IL MORTO; LA CASA DELL'EXORCISMO; LISA E IL DIAVOLO
HOR   91 min (ort 100 min) dubbed   VIDrel: REDEM/RTM
V

LISZTOMANIA *   18
Ken Russell UK   1975
Roger Daltrey, Sara Kestelmann, Paul Nicholas, Fiona Lewis, Ringo Starr, John Justin, Veronica Quilligan, Nell Campbell, Andrew Reilly, Anulka Dziubinska, Imogen Claire, Rick Wakeman, Rikki Howard, Felicity Devonshire
Ken Russell's usual uncontrolled exercise in self-indulgence is this time applied to the life of Liszt, seen in terms of pop performers. A visual and aural inundation.
MUS   99 min (ort 105 min)   VIDrel: MGM/WHV L/A   V

LITTLE BIG FOOT **   PG
Art Camacho USA   1996
Ross Malinger, Matt McCoy, Kenneth Tigar, P.J. Soles, Kelly Packard, Don Stroud
A family on holiday at Cedar Creek discover the title creature in this amiable outdoors adventure, and rescue both it and its wounded mother from hunters. A corny rip-off of BIGFOOT AND THE HENDERSONS, but pleasant enough in its way to offer some mild diversion to young kids.
JUV   92 min   VIDrel: MARQ/QUANT   V

LITTLE BIG LEAGUE **   PG
Andrew Scheinman USA   1993
Luke Edwards, Timothy Busfield, John Ashton, Ashley Crow, Kevin Dunn, Billy L. Sullivan, Miles Feulner, Jonathan Silverman, Dennis Farina, Jason Robards, Wolfgang Bodison, Duane Davis, Leon "Bull" Durham, Kevin Elster, Joseph Latimore
On the death of his grandfather, a twelve-year-old boy finds himself in the strange position of having inherited a basketball team. Unable to procure the services of a manager, he is obliged to assume this position himself. An over-extended and unsuccessful attempt at a family comedy that is too unfocused to work. A number of real-life basketball players appear in cameo roles. This was Scheinman's directing debut.
COM   116 min (ort 119 min) cC   VIDrel: COLUM/SONOP
V/sur

LITTLE BIG MAN ***   15
Arthur Penn USA   1970
Dustin Hoffman, Faye Dunaway, Chief Dan George, Martin Balsam, Richard Mulligan, Jeff Corey, Amy Eccles, Jean Peters, Carole Androsky, Cal Bellini, Robert Little Star, Thayer David, James Anderson, Jesse Vint, Jack Bannon
Long, rambling and episodic account of the Old West seen through the eyes of a 121-year-old man, who was brought up by the Indians and was present at the Battle of the Little Big Horn. The best incidents are quite excellent, but there are not enough of them. The script is by Calder Willingham.
WES   133 min (ort 150 min)   VIDrel: 20TH/TECH; ENCORE
(LV only)   V/h LV
Boa: novel by Thomas Berger.

LITTLE BOY LOST, A **   PG
Alan Spires AUSTRALIA   1978
Tony Barry, Lorna Lesley, John Hargreaves, John Jarratt, James Elliott, Les Foxcroft, Nathan Dawes, Robert Quilter, Don Crosby, John Nash, Brian Anderson, Bernadette Hughson, Redmond Phillips, Ray Marshall, Julie Dawson, Max Osbiston
A recreation of some true events of the 1960s, when a massive search took place to discover the whereabouts of a small boy. Quite well handled but otherwise unmemorable.
DRA   90 min (ort 92 min)   VIDrel: MOPIC/SGSVID   V

LITTLE BUDDHA **   PG
Bertolo Bertolucci FRANCE/UK   1993
Keanu Reeves, Ying Ruocheng, Chris Isaak, Alex Wiesendanger, Raju Lal, Greishma Makar Singh, Sogyal Rinpoche, Khyongla Rato

Rinpoche, Bridget Fonda, Jo Champa, Jighe Kunsang, Thubtem Jampa, Surekha Sirki, Ruchaprasad Sengupta
A Seattle family with a ten-year-old son are astonished when a Tibetan monk arrives at their home and announces that the boy may be the reincarnation of a famous Buddhist master. Soon the entire family are on their way to Tibet and during the long journey, are regaled with the tales of Prince Siddartha, who founded Buddhism 2,500 years ago. A visually striking but badly structured epic, hampered by indifferent acting and a general lack of conviction.
DRA   118 min (ort 140 min)   VIDrel: TOUCH/TECH L/A
V

LITTLE CAESAR ****   PG
Mervyn Le Roy USA   1930
Edward G. Robinson, Douglas Fairbanks Jr, Glenda Farrell, Stanley Fields, Sidney Blackmer, George E. Stone, Thomas Jackson, Armand Kaliz, Lucille La Verne, Landers Stevens, Maurice Black, Noel Madison, Nick Bela, George Daly
Robinson is excellent as the central character in this tale of the rise and fall of a small-time crook, based on the career of Al Capone. The script is by Francis Faragoh and Robert E. Lee.
THR   75 min (ort 80 min) B/W   VIDrel: WHV   V
Boa: novel by W.R. Burnett.

LITTLE DEATH, THE **   18
Jan Verheyen USA   1995
Dwight Yoakam, Pamela Gidley, J.T. Walsh, Brent Fraser, Richard Beymer
A photographer meets a beautiful married woman with whom he becomes obsessed. When he accidentally kills her husband, he becomes implicated in a cunning inheritance scheme, in which the dead man's son is being manipulated by his cunning stepmother. One of those cliche-ridden erotic thrillers, in which the cast work hard to breathe life into the hackneyed script, whose final twist comes as more of a relief than a surprise.
THR   87 min (ort 91 min)   VIDrel: POLFIL   V/s

LITTLE DEVILS: THE BIRTH **   15
George Pavlov USA   1993
Russ Tamblyn, Marc Price, Nancy Valen, Wayne McNamara, Stella Stevens, Jerry Levitan, Donald Saunders, David Campbell, Henry Roth, Tania Leil, Gregory Dunn, Sean P. Carroll, Cliff Makinson, Vanessa Walton Bern, Elliott Stein
A demonic creature emerges from a long imprisonment in a deserted tomb, and takes over the mind of a curious scientist who just happened to be there, forcing him to sculpt a host of miniature devils that will take over the planet. Fortunately, his neighbour learns all about this plan, and with the help of a couple of others, takes on these creatures. Diabolical nonsense, this is a horror-comedy somewhat better in conception than in execution.
HOR   100 min   VIDrel: NWV/HIFLI   V

LITTLE DORRIT: PARTS 1 AND 2 ***   U
Christine Edzard UK   1990
Derek Jacobi, Alec Guinness, Sarah Pickering, Robert Morley, Joan Greenwood, Cyril Cusack, Max Wall, Patricia Hayes, Miriam Margolyes, Roshan Seth, Bill Fraser, Luke Duckett, Eleanor Bron, Michael Elphick, Sophie Ward, Liz Smith
A masterly adaptation of Dickens's dark and teeming novel of 1820s England, that tells of a young man who is thrown into a debtor's prison where he meets a man who has been imprisoned there for 25 years. This absorbing drama is told in two overlapping segments seen from different perspectives, and boasts a collection of fine performances.
DRA   282 min (2 cassettes)   VIDrel: MGM/WHV   V/sur
Boa: novel by Charles Dickens.

LITTLE DRUMMER GIRL, THE ***   15
George Roy Hill USA   1984
Diane Keaton, Klaus Kinski, Sami Frey, Yorgo Voyagis, Anna Massey, Thorley Walters, Michael Cristofer, David Suchet, Eli Danker, Kerstin De Ahna, Dana Wheeler-Nicholson, Robert Pereno, Moti Shirin, Ben Levine, Jonathan Sagalle
A young American actress gets herself involved in the Arab-Israeli conflict and is forced to take sides. A ponderous and pompous thriller that remains worth seeing for an electric performance from Keaton.
DRA   125 min (ort 130 min)   VIDrel: MGM/WHV L/A
V/h
Boa: novel by John Le Carre.

LITTLE FOXES, THE ****   PG
William Wyler USA   1941
Bette Davis, Herbert Marshall, Teresa Wright, Dan Duryea, Richard
Carlson, Patricia Collinge, Charles Dingle, Carl Benton Reid, Jessie
Grayson, Jessica Grayson, John Marriott, Russell Hicks, Lucien
Littlefield, Virginia Brissac
Film version of a play about a mean and scheming family in the
post-Civil War period, with Davis quite outstanding and the
other stars almost as good. The film debuts for Collinge, Duryea,
Reid and Wright, with the first three actors recreating the roles
they had in the Broadway version of the play.
DRA   116 min B/W   VIDrel: VCC/DISC   V
Boa: play by Lillian Hellman.

LITTLE GANG, THE ***   U
Michel Deville FRANCE   1983
Andrew Chandler, Helene Dassule, Nicole Palmer, Hamish
Scrimgeour, Katherine Scrimgeour, Nicolas Sireau, Remi Usquin,
Valerie Gauthier, Yveline Ailhaud, Michel Amphoux, Roland
Amstutz, Pierre Ascaride, Jean-Pierre Bagot, Jacques Blot
A group of English schoolchildren stowaway on a cross-Channel
ferry and disembark in France, where they enjoy a whole series
of adventures, hitch-hiking in cars, getting chased by villains
and escaping in a balloon. A rollicking and surreal comedy,
extraordinarily imaginative and highly entertaining. The film
ends with the kids shipwrecked on a desert island, presumably
leaving the way open for a sequel.
Aka: LA PETITE BANDE
JUV   91 min   CINrel

LITTLE GIANTS **   PG
Duwayne Dunham USA   1994
Rick Moranis, Ed O'Neill, Shawna Waldron, John Madden, Susanna
Thompson, Brian Haley, Devon Sawa, Michael Zwiener, Triy
Simmons, Sam Horrigan, Harry Shearer, Dabbs Greer, Todd Bosley,
Danny Pritchett, Marcus Troji, Joe Bays
A man launches a rival kids' football team after his daughter is
turned down for a place on the team run by his elder brother, a
swollen-headed player and coach. Pretty soon, their sibling
rivalry has escalated into an all-out war as they seek to outdo
each other on the football field. Tolerably well made, this
vacuous assembly-line effort will appeal mainly to kids (and
possibly football fans). Some real-life players appear in cameo
roles to inject a note of realism.
COM   101 min (ort 106 min) cC   VIDrel: WHV   V/sh

LITTLE GIRL LOST **   15
Sharron Miller USA   1988
Tess Harper, Frederic Forrest, Patricia Kalember, Lawrence Pressman,
Marie Martin, Christopher McDonald
The story of a couple and their struggle to keep their foster child,
in the knowledge that she has been abused by her real father. A
competent but terribly depressing drama that is to some extent
based on a real case.
DRA   90 min (ort 96 min) mTV   VIDrel: BANO/SGSVID   V

LITTLE HERO OF THE SHAOLIN TEMPLE, THE *   18
HONG KONG   198-
Cheng-Taid Syh
The heir to the throne of China is forced to hide in a Shaolin
temple, after the Emperor orders his assassination by a band of
Ninja killers.
Aka: LITTLE HEROES OF SHAOLIN TEMPLE
MAR   82 min Cut (1 min 1 sec – ort 90 min)
VIDrel: IMPENT   V

LITTLE LORD FAUNTLEROY **   U
Andrew Morgan UK   1994
George Baker, Betsy Brantley, Michael Benz, Bernice Stegers, John
Castle, Martin Ball, Helen Lindsay, Jacki Webb, Bernadette Short,
Christopher Bowen, David Healy, Truan Munro, Andrew Robertson,
Adrian Cairns, Virginia Beare
A young New York boy has to adapt to being the new Earl of
Dorincourt in a fairly competent adaptation of the Burnett novel.
DRA   158 min (2 cassettes) mTV   VIDrel: BBC/TECH   V/s
Boa: novel by Frances Hodgson Burnett.

LITTLE MAN TATE ***   PG
Jodie Foster USA   1991
Jodie Foster, Dianne Wiest, Adam Hann-Byrd, Harry Connick Jr,
David Pierce, Debi Mazar, P.J. Ochlan, Alex Lee, Michael Shulman,
Nathan Lee, Celia Weston, Danitra Vance, Richard Fredette, George
Plimpton, Jennifer Trier, John Bell

Foster's first film as director is the story of a child prodigy, a shy
and sensitive genius who finds himself torn between the
demands of his mother that he live a normal life, and the ambitions of a cold and ruthless expert in gifted children. The
youngster's clumsy attempts to make friends, the sterile
American cult of success, and the simple and unchanging truths
of life are all strong elements in this unusual and rewarding
story.
DRA   95 min (ort 99 min)   VIDrel: VCC/DISC/COLUM
V/sur

LITTLE MATCH GIRL, THE **   U
Michael Lindsay-Hogg USA   1987
William Daniels, John Rhys-Davies, Keshia Knight Pulliam, Rue
McClanahan, Jim Metzler, William Youmans, Hallie Foote, Maryedith
Burrell, Robyn Steven, Stephen Dimopoulos, Charles Andre, Norma
Macmillan, Nikki Sharp, Blu Mankuma
A version of this famous tale set in the 1920s and altered some-
what to have our poor waif taken in by a wealthy family. Quite
good to look at, but lacking sufficient material to develop into
a full-length feature. The script is by Maryedith Burrell.
JUV   94 min (ort 100 min) mTV   VIDrel: MIA/DISC   V
Boa: short story by Hans Christian Andersen.

LITTLE MERMAID, THE ***   U
John Musker/Ron Clemente USA   1989
Voices of: Jodi Benson, Pat Carroll, Samuel E. Wright, Kenneth Mars,
Buddy Hackett, Christopher Daniel Barnes, Rene Auberjonois, Ben
Wright
A very loose adaptation of the Andersen tale of a young
mermaid who wishes to become human and experience life
above the waves, this highly ambitious animation jettisons much
of the flavour of the original story, replacing it with typical
Disney adventures and characterisations (plus a happy ending).
Nevertheless, both animation and music are memorable. AA:
Score/orig (Alan Menken), Song ("Under The Sea" – Menken
(m)/Howard Ashaman (l)).
ANIM   79 min (ort 82 min)   VIDrel: WDV/TECH L/A   V
Boa: short story by Hans Christian Andersen.

LITTLE MISS MARKER *   U
Walter Bernstein USA   1980
Walter Matthau, Julie Andrews, Tony Curtis, Bob Newhart, Lee
Grant, Sara Stimson, Brian Dennehy, Kenneth McMillan, Andrew
Rubin, Joshua Shelley, Tom Pedi, Randy Herman, Nedra Volz,
Jacquelyn Hyde, Jessica Rains, Henry Slate
A little girl is left with a bookmaker as an I.O.U. and reforms
the entire gambling fraternity. A flat and uninspired remake of
the charming Shirley Temple film of 1934, and one that marked
the directorial debut for screenwriter Bernstein.
COM   99 min (ort 103 min)   VIDrel: CIC/SONOP L/A   V
Boa: short story by Damon Runyon.

LITTLE MISS MILLIONS **   (PG)
Jim Wynorski USA   1993
Howard Hesseman, Steve Landesberg, Anita Morris, Robert Fieldsteel,
Love Howitt, James Avery, Terri Treas, Paul Hertzberg, Queen Kong,
Ace Mask, Lenny Juliano, Michelle Failey, Patrick Statham, Rick
Dean, Tony Naples, Pat Brady
When a nine-year-old heiress to vast fortune runs away from her
avaricious stepmother, the latter engages a private eye to find
her. He does so but when this resourceful girl accuses him of
kidnapping her, both of them find themselves on the run.
COM   88 min (ort 90 min)   SATrel: MOVIE CHANNEL

LITTLE MONSTERS **   PG
Richard Alan Greenberg USA   1989
Howie Mandel, Fred Savage, Daniel Stern, Margaret Whitton, Ben
Savage, Frank Whaley, Rick Ducommun, Amber Barretto
A fantasy comedy with a strong BEETLEJUICE flavour, in which
a youngster finds a mischievous gremlin under his bed, and
their resultant friendship and adventures create havoc for all
concerned. The fascinating premise is let down by a lack of
inventiveness, bad editing and a decidedly spiteful outlook.
COM   97 min (ort 100 min)   VIDrel: VCC L/A   V

LITTLE NEMO: ADVENTURES IN
SLUMBERLAND **   (U)
Masami Hara/William T. Hurtz JAPAN   1992
Voices of: Gabriel Damon, Mickey Rooney, Rene Auberjonois, Daniel
Mann, Laura Mooney, Bernard Erhard, William E. Martin
A young boy travels in his dreams to the land of Slumberland
and unwittingly releases a monster that abducts its ruler. To

make up for this, he mounts a rescue together with the latter's daughter and a variety of companions. An okay animated tale with passable animation.
ANIM   84 min   SATrel: MOVIE CHANNEL

LITTLE NIGHT MUSIC, A ** 12
Harold Prince AUSTRIA/WEST GERMANY   1978
Elizabeth Taylor, Diana Rigg, Lesley-Anne Down, Len Cariou, Hermione Gingold, Lawrence Guittard, Christopher Guard, Chloe Franks, Heinz Maracek, Lesley Dunlop, Jonathan Tunick, Hubert Tscheppe, Rudolf Schrympf, Jean Sincere
Film version of a Stephen Sondheim stage musical which is in turn based on an Ingmar Bergman film, about a weekend party at a country estate – SMILES OF A SUMMER NIGHT, and tells of a middle-aged lawyer who rekindles an affair with an old flame in Vienna at the turn of the century. Pleasant enough but hardly the stuff of greatness, and suffering badly from a stilted script. Filmed in Austria. AA: Score/adapt (Jonathan Tunick).
MUS   119 min (ort 125 min)   VIDrel: CREA/DISC   V
Boa: musical by Stephen Sondheim and Hugh Wheeler.

LITTLE NIKITA ** 15
Richard Benjamin USA   1988
Sidney Poitier, River Phoenix, Richard Bradford, Richard Lynch, Caroline Kava, Loretta Devine, Lucy Deakins, Jerry Hardin, Albert Fortell, Ronald Guttman, Jacob Vargas, Roberto Jimenez, Robb Madrid, Chez Lister, Tom Zak
A teenager's parents are revealed as Soviet sleeper agents when their teenage son applies to the Air Force Academy, and he is caught in an agonising conflict of loyalties, when they are eventually instructed to undertake an espionage mission. This fascinating idea makes for a rather disappointing film that does little with its central idea, though Poitier is good as an FBI agent.
DRA   93 min (ort 98 min)   VIDrel: RCA L/A   V

LITTLE NINJA DRAGON, THE ** 15
Joseph Merhi USA
Stephen Furst, Joseph Campanella, Billy Hutsey, Ted Jan Roberts, Shonda Whipple
Together with his sister, a twelve-year-old martial arts experts tries to get his uncle out of trouble. Average.
JUV   86 min (ort 89 min)   VIDrel: POLY/POLYREC/BRAVE V

LITTLE NOISES *** 15
Jane Spencer USA   1991
Crispin Glover, Steven Schub, Tatum O'Neal, Rik Mayall, Tate Donovan, John C. McGinley
A failed writer steals the work of a dumb poet and passes it off as his own, but is stricken by his conscience when he learns that his victim has become homeless. Spencer's directorial debut is an incisive and mocking work, whose articulate script has a few useful things to say about human nature and the creative process.
DRA   87 min (ort 110 min)   VIDrel: COLUM/SONOP   V

LITTLE ODESSA *** 15
James Gray USA   1994
Tim Roth, Maximillan Schell, Edward Furlong, Vanessa Redgrave, Moira Kelly, Paul Guilfoyle, Natasha Andreichenko, David Vadim, Mina Bern, Boris McGiver, Mohammed Ghaffari, Michael Khumrov, Dmitry Preyers, David Ross, Ron Brice
A Russian-Jewish hitman for the Russian Mafia comes back to Brooklyn to do a job and attempts to re-establish his relationship with his family and say his farewells to his dying mother. However, his presence has a disturbing effect on them, most especially his younger brother. A dark, brooding and slightly boring film, enhanced by strong acting and direction. The title refers to that part of Brooklyn Bridge which has become home to many Russian immigrants.
DRA   94 min (ort 98 min)   VIDrel: FIRST/SONOP; ENCORE (LV only)   V LV

LITTLE PRINCESS, A *** U
Alfonso Cuaron USA   1995
Liesel Matthews, Eleanor Bron, Liam Cunningham, Rusty Schwimmer, Arthur Malet, Vanessa Lee Chester, Errol Sitahal, Heather DeLoach, Taylor Fry, Darcie Bradford, Rachel Bella, Alexandra Rea-Baum, Camilla Belle, Lauren Blumenfeld
When he father goes off to war, a young girl is sent to a strict boarding school in New York, where her grace and charm succeeds in eventually winning over both staff and inmates. A

lightweight little tale, but one of enormous appeal; newcomer Mathews is first-rate as the central character.
JUV   93 min (ort 97 min) cC   VIDrel: WHV   V/sur
Boa: novel Little Lord Fauntleroy by Frances Hodgson Burnett.

LITTLE RASCALS, THE ** U
Penelope Spheeris USA   1994
Travis Tedford, Bug Hall, Britanny Ashton Holmes, Kevin Jamal Woods, Zachary Mabry, Ross Elliot Bagley, Sam Saletta, Blake Jeremy Collins, Blake McIver Ewing, Jordan Warkol, Courtland Mead, Juliette Brewer, Heather Karasek, Petey
Another example of the desperate recycling of old ideas as the juvenile heroes of the 1930s Hal Roach comedies return to the screen in this far from impressive remake. Such plot as there is revolves around attempts by Spany and his fellow misogynists to sabotage the romance between Darla and Alfalfa. Aimed squarely at kids, who will probably find the film's old-fashioned antics fairly unamusing.
COM   79 min (ort 83 min) cC   VIDrel: CIC/SONOP   V

LITTLE ROMANCE, A *** PG
George Roy Hill FRANCE/USA   1979
Laurence Olivier, Dane Lane, Thelonius Bernard, Arthur Hill, Sally Kellerman, Broderick Crawford, David Dukes, Andrew Duncan, Claude Brosset, Claudette Sutherland, Graham Fletcher-Cook, Ashby Semple, Anna Massey
A pair of thirteen-year-old lovers run away to Venice where an ageing Frenchman (played by Olivier) acts as their guide and mentor. A film loaded with sentiment that nevertheless remains rather appealing, mainly due to winning performances from the leads. AA: Score/orig (Georges Delerue).
COM   108 min   VIDrel: MGM/WHV L/A   V
Boa: novel E = MC² Mon Amour by Patrick Cauvin.

LITTLE SHOP OF HORRORS, THE *** 15
Roger Corman USA   1960 (re-released 1987)
Dick Miller, Jonathan Haze, Jackie Joseph, Mel Welles, Jack Nicholson, Myrtle Vail, Leola Wendorff, Charles B. Griffith (voice of Audrey)
A high camp horror comedy all about Seymour, a shy errand boy for a run-down flower shop, who creates a hybrid plant that develops a taste for blood. As the plant grows it begins to develop a voracious appetite which becomes ever harder to satisfy. Originally a B/W offering with Nicholson in his first film performance, which has been colourised and re-released. Written by Charles B. Griffith. Later made as a musical both on the stage and on film.
COM   71 min B/W coVer   VIDrel: L/A   V

LITTLE SHOP OF HORRORS *** PG
Frank Oz USA   1986
Rick Moranis, Ellen Greene, Steve Martin, Vincent Gardenia, John Candy, Bill Murray, Jim Belushi, Christopher Guest, Tichinia Arnold, Tisha Campbell, Michelle Weeks, Levi Stubbs (voice of Audrey II)
Film version of the off-Broadway musical that was itself based on the 1960 Roger Corman film, about a young man and his attachment to a man-eating plant. A wonderfully tacky black comedy with our hero thinking the plant will make him his fortune, only things don't quite turn out as expected. Unfortunately, the film has a vicious streak which tends to spoil the fun.
MUS   91 min   VIDrel: WHV   V/sh
Boa: musical by Howard Ashman and Alan Menken.

LITTLE VEGAS ** 15
Perry Lang USA   1990
Anthony John Denison, Catherine O'Hara, Anne Francis, Michael Nouri, Jerry Stiler, John Sayles, Jay Thomas, Bruce McGill, Bob Goldthwait
A man inherits a considerable fortune from his girlfriend and finds that her family are very hostile to him and his plans. They live in a small desert town full of very strange characters, while in the background the Mafia have their own plans for the area. A quirky little comedy-drama.
COM   88 min (ort 91 min)   VIDrel: 20VIS/SONOP   V/h

LITTLE WITCHES ** 18
Jane Simpson USA   1995
Mimi Reichmeister, Sheeri Rappaport, Jennifer Rubin, Jack Nance, Zelda Rubinstein, Tommy Stork, Eric Pierpoint, Zoe Alexander, Melissa Taub, Clea Duvall, Landon Hall, Lalaneya Hamilton, Constance Crossen, Erica Doering
A group of girls at a repressive Catholic School find a book of satanic spells and a secret place of worship. The usual nastiness

takes place in a film that offers few surprises in development or outcome.
HOR   87 min   VIDrel: HIFLI/SONOP   V/h

LITTLE WOMEN ****   U
George Cukor USA   1933
Katharine Hepburn, Paul Lukas, Joan Bennett, Frances Dee, Jean Parker, Spring Byington, Edna May Oliver, Henry Stephenson, Douglass Montgomery, John Davis Lodge, Samuel Hinds, Mabel Colcord, Marion Ballou, Nydia Westman
A faithful adaptation of the book telling of the growing up of four sisters in America during the period just before and after the Civil War. A fine cast perform well in a film of considerable visual beauty. Remade several times since. AA: Story/adapt (Sarah Y. Mason/Victor Heerman).
DRA   111 min (ort 115 min) B/W   VIDrel: MGM L/A   V
Boa: novel by Louisa May Alcott.

LITTLE WOMEN ***   U
Gillian Armstrong USA   1994
Winona Ryder, Gabriel Byrne, Kirsten Dunst, Eric Stoltz, Susan Sarandon, Trini Alvarado, Samantha Mathis, Claire Danes, Mary Wickes, Christina Bale, John Neville, Mary Wickes, Florence Paterson, Robin Collins, Corrie Clark
A spirited and convincing adaptation of Alcott's story of four sisters growing up in a prosperous home in the years prior to the civil War. It tells of their fears and expectations as they embark on womanhood and the prospect of independent lives of their own. Very well acted by the entire cast, with Wickes quite brilliant as a very grumpy relative.
DRA   113 min (ort 118 min) wScrn cC
VIDrel: COLUM/SONOP   V/sur
Boa: novel by Louisa May Alcott.

LITTLEST REBEL, THE ****   PG
David Butler USA   1935
Shirley Temple, John Boles, Jack Holt, Karen Morley, Bill "Bojangles" Robinson, Guinn Williams, Willie Best, Frank McGlynn Sr, Bessie Lyle, James Flavin, Hannah Washington
A small Southern girl visits President Lincoln to persuade him to release her Confederate father from imprisonment in the North, where he is due to be executed as a spy, after he tried to cross the front line to visit his sick wife. A classic Temple vehicle, done with enormous style and wit. Songs include "Polly Wolly Doodle".
MUS   75 min B/W   VIDrel: 20TH/TECH L/A   V
Boa: play by Edward Peple.

LIVE AND LET DIE **   PG
Guy Hamilton UK   1973
Roger Moore, Jane Seymour, Yaphet Kotto, Clifton James, Bernard Lee, Lois Maxwell, David Hedison, Julius W. Harris, Geoffrey Holder, Gloria Hendry, Tommy Lane, Earl Jolly Brown, Roy Syewart, Lon Satton, Arnold Williams, Ruth Kemp
James Bond adventure about a drug smuggling racket in the Caribbean run by a black mastermind who isn't averse to using voodoo to control and terrify the locals. A plethora of stunts and high-speed action masks a very thin plot. This was number eight in the series, and gave Moore his first chance as 007. THE MAN WITH THE GOLDEN GUN followed.
A/AD   116 min (ort 121 min) wScrn   VIDrel: MGM/WHV
V/dm
Boa: novel by Ian Fleming.

LIVE BY THE FIST **   18
Cirio H. Santiago USA   1992
Jerry Trimble, George Takei, Ted Markland, Laura Albert, Vic Diaz, Romy Diaz, Roland Nates, Mike Nicholson, Steve Rogers, Bert Labra, John Crank, Ramon D'Salva, Zernan Manahan, Jim Moss, Ned Hourani, Ronald Asinas, Archie Adamos
A former member of a Navy SEAL team has to abandon his pacifist principles after he is framed on a murder charge and sent to prison. There he has to fight to stay alive in this derivative and wholly routine effort.
A/AD   73 min (ort 90 min)   VIDrel: ONE/IMPENT   V

LIVE FROM DEATH ROW **   18
Patrick Duncan USA   1991
Bruce Davison, Joanna Cassidy, Art LaFleur, Calvin Levels, Jason Tomlins, Julio Oscar Mechoso, Michael D. Roberts, Martha Velez-Johnson, Kathleen Wilhoite, David Bowe, Keith MacKechie, Lee Arenberg, Gene Butler, Lisa Niemi
A murderer awaiting execution on death row, takes a film crew

hostage who were in the process of filming a report for prime-time TV. He sets about using them as a bargaining tool in a dangerous battle of wits with the authorities. This might well have been a most tense film, but writer/director Duncan shows his failings as a scriptwriter, and the cast have little to get their teeth into.
Aka: LIVE! FROM DEATH ROW
THR   92 min (ort 94 min) mTV   VIDrel: CAPIT/GUILD   V

LIVE NUDE GIRLS *   18
Julianna Lavin USA   1995
Dana Delany, Kim Cattrall, Cynthia Stevenson, Laila Robins, Lora Zane, Olivia D'Abo, Glenn Quinn, Tim Choate, Jeremy Jordan, V.C. Davis, Simon Templeman, Brian Markinson, Paul Perri, Joshua Beckewt, Jerry Spicer
Six women who have are old girlhood friends hold a bachelorette party for one of their number, who is getting married for the third time. As the event progresses they spend most of their time bitching about men and reviving past memories, which inevitably leads to the surfacing of hidden tensions and resentments. A superficial and highly unamusing comedy that is additionally burdened by the definite lack of sympathetic characters.
COM   90 min (ort 92 min)   VIDrel: POLFIL   V

LIVE WIRE **   15
Christian Duguay USA   1992
Pierce Brosnan, Ron Silver, Ben Cross, Lisa Eilbacher, Al Waxman, Michael St Gerard, Philip Baker Hall, Brent Jennings, Tony Plana, Ivan Roth, Clement Von Franckenstein, Ivan Roth, Selma Acherd, Rick Cicetti, Norman Burton
An FBI bomb expert is assigned to protect a US senator from terrorists who have already killed a number of senators, using a new liquid explosive that is virtually undetectable. To add to his worries, his wife is having an affair with the man that he has to protect. An adequate action film that offers undemanding entertainment, despite the plot eventually becoming almost impossible to follow.
A/AD   81 min (ort 87 min) mCab   VIDrel: EIV/SONOP
V/sur

LIVING *****   PG
Akira Kurosawa JAPAN   1952
Takashi Shimura, Nobuo Kaneko, Kyoko Seki, Miki Odagiri, Yunosuke Ito, Kamatari Fujiwara, Makoto Koburi, Nobuo Nakamura, Minosuke Yamada, Haruo Tanak, Bokuzen Hidari, Monoru Chiaki, Shinichi Himori, Kazao Abe, Ko Kimura
As colleagues gather at the funeral of a minor civil servant who was dying of stomach cancer, we are shown his life in a series of flashbacks. Rejected by an embittered son, disliked by his subordinates and with no-one to confide in, we watch him sink into despair until the discovery of a task that can give his life purpose and value. This is a perfect film; its appeal is universal, its message timeless.
Aka: DOOMED; IKIRU
DRA   141 min (ort 143 min) B/W   VIDrel: CONNO/RTM
V

LIVING A LIE **   15
Larry Shaw USA   1991
Jill Eikenberry, Peter Coyote, Roxanne Hart, Jarred Blancard, Claudette Sutherland, Michael Waltman, Castulo Guerra, Allison Mack, David Andrews, Don Collier, Francesca Jarvis, Nick Sean Gomez, Manny Simo-Maceo, Babs Bram
Reluctantly driven to the conclusion that her husband is a dangerous arsonist, a woman has to decide whether or not to inform the police. However, she eventually learns that he took part in the burning down of a Mexican church in which two youngsters died, whilst under the influence of a white supremacist friend. A lame study of racism in which as ever, our racists are both stupid and brutal. Average.
DRA   88 min mTV   VIDrel: ODY/SONOP   V/sh

LIVING DAYLIGHTS, THE ***   PG
John Glen UK   1987
Timothy Dalton, Maryam D'Abo, Jeroen Krabbe, Joe Don Baker, Art Malik, John Rhys-Davies, Geoffrey Keen, Desmond Llewelyn, Andreas Wiesniewski, Robert Brown, Thomas Wheatley, Caroline Bliss, Walter Gotell
Dalton gives a good demonstration of his acting ability in his debut as James Bond, making him a credible figure at last. Still as action-packed as ever (for this outing Bond has to combat a double-dealing Russian general) and not without some wry

humour, it makes a welcome relief from the increasingly irritating parodies that Roger Moore starred in (these ended with A VIEW TO A KILL). Bliss makes her debut as Miss Moneypenny. Followed by LICENCE TO KILL.
A/AD 125 min (ort 130 min) wScrn (special edition)
VIDrel: MGM/WHV V/sur

LIVING DEAD GIRL, THE *** 18
Jean Rollin FRANCE 1982
Marina Pierro, Francoise Blanchard, Mike Marshall, Carina Barone, Fanny Magier, Jean-Pierre Bouyxou, Dominique Treillou, Jean Cherlian
When a young girl dies, she is brought back to a semblance of life when an earthquake disturbs toxic waste and causes the release of gases from it. In order to survive, she needs constant supplies of human blood, which she gets from innocent strangers with the connivance of her best friend, who has remained devoted to her. But as memory returns, so does an awareness of her condition. A surreal and disturbing film, rather excessively gory but well made nonetheless.
Aka: LA MORTE VIVANTE
HOR 86 min wScrn VIDrel: REDEM/RTM V

LIVING END, THE *** 18
Gregg Araki USA 1992
Craig Gilmore, Mike Dytri, Darcy Marta, Scott Goetz, Joanna Went, Nicole Dillenberg, Mary Woronov, Mark Finch, Paul Bartel, Bretton Veil, Maggie Song, Stephen Holman, Peter Lanigan, Jon Gerrens, Jack Kofman, Michael Haynes
Two young HIV-positive men set out a journey of discovery, in this very different road-movie that was made on a minuscule budget. Along with the anger, there is also a good measure of black humour and a set of very fine performances. The screenplay is by Araki.
A 36 min (ort 85 min) VIDrel: PRIDE/PARADOX V

LIVING FREE ** U
Jack Couffer UK 1972
Susan Hampshire, Nigel Davenport, Geoffrey Keen, Edward Judd, Peter Lukoye, Shane De Louvre, Robert Beaumont, Nobby Noble, Charles Hayes, John Hayes, James Kamau, Aludin Quershi
Sequel to BORN FREE taking up the story of Elsa and her three cubs. A film that ambles along very pleasantly, but can be enjoyed just as much with eyes half closed. Hampshire does her best in an undemanding role, with neither this film nor the earlier one providing anything more than a highly sanitised and anodyne portrait of Joy Adamson, a woman whose irascibility was as legendary as her devotion to her lions.
DRA 87 min (ort 92 min) VIDrel: VCC/DISC/COLUM V
Boa: book by Joy Adamson.

LIVING IN OBLIVION *** 15
Tom DiCillo USA 1995
Steve Buscemi, Catherine Keener, Dermot Mulroney, Danielle Von Zernack, James Legros, Peter Dinklage, Hilary Gilford, Michael Griffiths, Matthew Grace, Robert Wightman, Kevin Corrigan, Tom Jarmusch, Ryan Bowker, Francesca DiMauro
A satire on the tribulations of film-making as the creator of a low-budget movie experiences one nightmarish day in which everything that possibly can go wrong does. An unusual and endearing little film: often funny, generally fascinating and always very well acted. See also THE ADVENTURES OF THE FLYING PICKLE, though nowhere near as good, this is another film about low-budget movie-making.
COM 94 min VIDrel: EIV/SONOP V

LIZ: THE ELIZABETH TAYLOR STORY ** 12
Kevin Connor USA 1994
Sherilyn Fenn, Ray Wise, Eric Gustavson, John Saxon, Troy Harris, Nigel Havers, Katherine Helmond, Angus MacFayden, Kevin McCarthy, Corey Parker, Dan McVicar, Christine Healy, Charls Frank, Victor Raider-Wexler, Michael Cavanaugh
This unauthorised biopic on the life of the title actress briskly takes us from one marriage to the next, giving us occasional glimpses of the more well known aspects of her career, but offering absolutely no insights into the star's beliefs or inner drives. A very superficial piece indeed, though there are a couple of good moments, not least the well-charted conflicts between Taylor and Richard Burton during their two stormy marriages.
DRA 171 min mTV VIDrel: ODY/SONOP V/sh
Boa: biography by C. David Heymann.

LOADED * 18
Anna Campion NEW ZEALAND/UK 1994
Oliver Milburn, Dearbhla Molloy, Danny Cunningham, Catherine McCormack, Thandie Newton, Nick Patrick, Biddy Hodson, Matthew Eggleton, Caleb Lloyd, Joe Gecks, Bridget Brammall, Tom Welsh
A group of seven students, all worthy representatives of the lost generation, go to a remote country house to shoot a low-budget film that one of them has scripted. Thrown together for an entire weekend, their underlying tensions and fears soon surface. However, after they decide to take LSD, events get out of hand and a fatal accident occurs, which provides them with a moral dilemma. An unimpressive debut from writer/director Campion, the sister of Jane.
DRA 96 min (ort 108 min) CINrel

LOBSTER MAN FROM MARS ** PG
Stanley Sheff USA 1989
Tony Curtis, Deborah Foreman, Patrick Macnee, Billy Barty, Anthony Hickox, Tommy Sledge
Silly but enjoyable film-within-a-film spoof that borrows its central idea from THE PRODUCERS. A movie mogul desperately in need of a tax loss meets a young man who has penned the title epic, and we then get to see the film being made in all its atrocious glory. A few laughs are on offer but like most parodies, once the central idea has been uncovered, it's slowly downhill from that point on.
COM 78 min VIDrel: EIV/SONOP V

LOCAL HERO ** PG
Bill Forsyth UK 1983
Burt Lancaster, Peter Riegart, Denis Lawson, Fulton MacKay, Peter Capaldi, Jenny Seagrove, Christopher Rozycki, Jennifer Black, Christopher Asante, Rikki Fulton, Alex Norton, Norman Chancer, David Anderson, Sandra Voe, Alan Clark
An oil company representative tries to buy up a Scottish village for use as a refinery site, but encounters unexpected difficulties. A quirky little comedy that is very occasionally slightly amusing rather than hilarious, but for the most part is uneventful and a little tedious.
COM 107 min (ort 111 min)
VIDrel: 4-FRONT/POLYREC/BRAVE V/h

LOCH NESS ** PG
John Henderson UK 1994
Ted Danson, Joely Richardson, Ian Holm, Harris Yulin, James Frain, Keith Allen, Nick Brimble, Kirsty Graham, Harry Jones, Phillip O'Brien, Joseph Greig, John Dair, John Verea, Deborah Weston, Wolf Kahler, Julian Curry, Roger Sloman
A boffin is sent to Scotland to put paid once and for all to the myths surrounding the title lake, but his growing involvement with the locals, especially an attractive new girlfriend, finally convinces him otherwise. Danson as the investigator is suitably cynical and scornful, and though the facts are for the most part the usual stereotyped characters, the light blend of action and comedy is handled with care. A mildly witty affair.
COM 100 min cC VIDrel: POLY/POLYREC V/s

LOCK UP ** 18
John Flynn USA 1989
Sylvester Stallone, Donald Sutherland, John Amos, Darlanne Fluegel, Frank McRae, Sonny Landham, Tom Sizemore, William Allen Young, Larry Romano, Dean Duval, Jordan Lund, Jerry Strivelli, David Anthony Marshall, Kurek Ashley
A model prisoner with only six months left to serve is transferred to a prison hell-hole at the behest of its sadistic governor, who aims to exact a cruel revenge for a past humiliation, and hopes to provoke Stallone into a breach of the rules. Much brutality now follows, but our prisoner escapes to have his revenge and this simple-minded tale delivers the expected happy resolution. Music is by Bill Conti.
A/AD 104 min (ort 115 min)
VIDrel: 4-FRONT/POLYREC/GUILD V/sur

LOCK UP YOUR DAUGHTERS! * 15
Peter Coe UK 1969
Christopher Plummer, Jim Dale, Susannah York, Glynis Johns, Ian Bannen, Kathleen Harrison, Roy Kinnear, Richard Wordsworth, Roy Dotrice, Vanessa Howard, Fenella Fielding, Peter Bayliss, Georgia Brown, Fred Emney
A dull farce set in 18th century London, and following the exploits of an aristocrat and three sailors, all of whom set off in search of female companionship and suffer the consequences of mistaken identity. A brash and vulgar effort, that tries hard to

achieve some of the verve of TOM JONES, but is held back by its crudity and lack of wit.
COM 96 min Cut (27 sec – ort 103 min) VIDrel: L/A V
Boa: plays by Bernard Miles, Laurie Johnson and Lionel Bart/Rape Upon Rape by Henry Fielding/The Relapse by John Vanbrugh.

LOCKED UP: A MOTHER'S RAGE ** 15
Bethany Rooney USA 1991
Cheryl Ladd, Jean Smart, Dean Norris, Ariana Richards, Angela Bassett, Kimberly Scott, Vanessa Marquez, Diana Muldaur
A fact-based tale that tells of the anguish of an innocent woman wrongfully given a life sentence, and follows the battle she fights with the authorities for the rights of her children and the chance of a new trial.
DRA 90 min VIDrel: GUILD/SONOP V/sh

LOGAN'S RUN * PG
Michael Anderson USA 1976
Michael York, Jenny Agutter, Richard Jordan, Farrah Fawcett, Roscoe Lee Browne, Michael Anderson Jr, Randolph Roberts, Lara Lindsay, Peter Ustinov, Gary Morgan, Michelle Stacy, Denny Arnold, Bob Neill, Glen Wilder, Randolph Roberts
In a city of the future no-one is permitted to live beyond thirty. A member of the elite death squad responsible for enforcing this rule escapes from the city and discovers the truth about his society. A hammy and irksome display of bad acting and worse direction with a flashy first half and an utterly dreary second. AA: Spec Award (L.B. Abbott/Glen Robinson/Matthew Yuricich for visual effects).
FAN 113 min (ort 120 min) wScrn VIDrel: MGM/WHV V/sur
Boa: novel by William F. Nolan and George Clayton Johnson.

LOLA ** PG
Jacques Demy FRANCE/ITALY 1960
Anouk Aimee, Marc Michel, Jacques Harden, Elina Labourdette, Alan Scott
A Nantes nightclub singer and dancer finds herself having to chose from among no less than three suitors, of whom two are sailors. An enjoyable but very frothy affair that marked the director's debut, with a plot clearly lifted from ON THE TOWN and more than a nod in the direction of Max Ophuls.
DRA 84 min (ort 91 min) B/W wScrn
VIDrel: ELPIC/POLYREC V

LOLA *** 15
Rainer Werner Fassbinder WEST GERMANY 1981
Barbara Sukowa, Armin Muller-Stahl, Mario Adorf, Matthias Fuchs, Ivan Desny
Set in 1950s Bavaria, this variant on THE BLUE ANGEL follows the career of a nightclub hostess and her seduction of a pillar of society who falls madly in love with her. But she is in truth mistress to the grasping owner of the venue, and happily exploits the situation to her own ends. Fassbinder uses the setting and story to make his own observations on his native country, which he does not paint in an especially appealing light.
DRA 109 min (ort 115 min) VIDrel: CONNO/RTM V

LONDON KILLS ME ** 18
Hanif Kureshi UK 1991
Justin Chadwick, Steven Mackintosh, Emer McCourt, Roshan Seth, Brad Dourif, Fiona Shaw, Tony Haygarth, Stevan Rimkus, Eleanor David, Alun Armstrong, Nick Dunning, Naveen Andrews, Garry Cooper, Gordon Warnecke, Evelyn Doggart
Everyone is a "victim" in this film, in which writer-director Kureshi (in his directing debut) examines the lives of various shiftless and unlovable drug addicts and dealers. Central to the story is the luckless Clint, who having lost his girl and suffered a beating, resolves to go straight. An intriguing film of opaque symbolism (Clint searches endlessly for a decent pair of shoes) and a meandering and unfocused narrative.
DRA 102 min (ort 107 min) VIDrel: COLUM/SONOP V/sur

LONDON'S BURNING: THE MOVIE ** 15
Les Blair UK 1986
Gary McDonald, Mark Arden, James Hazeldine, James Marcus, Sean Blowers, Rupert Baker, Richard Walsh, Gerard Horan, Jerome Flynn, Katherine Rogers, Eric Deacon, Yvonne Edgell, Hetty Baynes, Corinne Skinner-Carter, Jason Rose
The feature-length pilot that launched a popular TV series

dealing with the exploits of the London Fire Brigade. Blue Watch B25 are joined by a female recruit who has to prove herself in their almost exclusively male environment. Meanwhile, there are fires to be put out. Not exactly brilliant drama, but nicely handled and quite watchable.
DRA 105 min mTV VIDrel: CLEAR/DISC V

LONE RUNNER * 15
Roger (Ruggero) Deodato ITALY 1988
Miles O'Keeffe, Ronald Lacey, Savina Gersak, John Steiner, Michael J. Aronin, Donal Hodson
Desert warrior O'Keeffe sets out to rescue a kidnapped princess in this banal and lacklustre effort.
A/AD 83 min Cut (29 sec) VIDrel: EIV/SONOP V/sur

LONE STAR *** 15
John Sayles USA 1995
Chris Cooper, Elizabeth Pena, Joe Morton, Matthew McConaughey, Kris Kristofferson, Stephen Mendillo, Stephen Lang, Oni Faida Lampley, Eleese Lester, Joe Stevens, Gonzalo Castillo, Richard Coca, Clifton James, Tony Frank
A rather long but intriguing "modern" Western, built around the discovery of a corpse outside a Texas town. The evidence points to it being that of a feared and hated sheriff (Kristofferson) who was supposedly run out of town in 1957, but it is up to the present sheriff (Copper) to solve the mystery. Intricately plotted (the script is by Sayles) and happily free from cliche or contrivance.
WES 130 min (ort 135 min) VIDrel: 20VIS/SONOP V/sur

LONE WOLF McQUADE *** 18
Steve Carver USA 1983
Chuck Norris, David Carradine, Barbara Carrera, Leon Isaac Kennedy, Robert Beltran, L.Q. Jones, Dana Kimmell, R.G. Armstrong, Jorge Cervera, Sharon Farrell, Daniel Frishman, William Sanderson, John Anderson, Robert Arenas
A martial arts Texas Ranger takes on a gun-running operation in this fast and eventful tale that gives Norris a good chance to show off his skill in the fisticuffs department (if not as a Thespian). Shallow but entertaining.
A/AD 105 min (ort 107 min) VIDrel: 4-FRONT/POLYREC V

LONELINESS OF THE LONG DISTANCE RUNNER, THE **** 15
Tony Richardson UK 1962
Tom Courtenay, Michael Redgrave, James Bolam, John Thaw, Alec McCowen, Avis Bunnage, Peter Madden, James Fox, Julia Foster, Joe Robinson, Dervis Ward, Topsy Jane, James Cairncross, Philip Martin, Arthur Mullard, Ray Austin, Anthony Sayer
An engrossing character study examining the life of a rebellious and dishonest youngster, who is sent to reform school following a robbery. There his prowess as a runner leads to his selection in an inter-school competition, and as he runs he thinks back over his life's pointless and empty life. Screenplay is by Sillitoe.
Aka: REBEL WITH A CAUSE
DRA 100 min (ort 104 min) B/W
VIDrel: 4-FRONT/POLYREC V
Boa: short story by Alan Sillitoe.

LONELY GUY, THE ** 15
Arthur Hiller USA 1983
Steve Martin, Charles Grodin, Judith Ivey, Robyn Douglass, Steve Lawrence, Merv Griffin, Joyce Brothers, Candi Brough, Randi Brough, Julie Payne, Madison Arnold, Roger Robinson, Joan Sweeney, Daniel P. Hannafin, Leon Jones
A man is thrown out by his girlfriend and discovers an entire society of lonely guys in New York, being introduced to said society by Grodin, who is quite wonderful in this otherwise low-key and sombre comedy. Adapted from Friedman's book by Neil Simon and scripted by Ed Weinberger and Stan Daniels.
COM 87 min (ort 90 min) VIDrel: CIC/SONOP L/A V/h
Boa: book The Lonely Guy's Book Of Life by Bruce Jay Friedman.

LONELY HEARTS *** 15
Paul Cox AUSTRALIA 1981
Norman Kaye, Wendy Hughes, Jon Finlayson, Julia Blake, Jonathan Hardy, Vic Gordon, Irene Inescourt, Ted Grove-Rogers, Ronald Falk, Chris Haywood, Diana Greentree, Margaret Steven, Kris McQuade, Maurie Fields, Laurie Dobson
Story of an unlikely romance between a middle-aged piano tuner and a shy office worker who come together courtesy of a

dating agency. An uneven and offbeat tale, done with considerable warmth. Screenplay is by Cox.
DRA   91 min (ort 95 min)   VIDrel: ARTPRO/RTM   V

LONELY HEARTS ** 15
Andrew Lane USA   1991
Joanna Cassidy, Eric Roberts, Beverly D'Angelo, Herta Ware, Bibi Besch, Robert Ginty, Rebecca Street, Sharon Farrell, Miriam Flynn, Sandy Baron, Marlyn Mason, Charles Napier, Simone Study, Jack Jozefson, Ellen Geer
A handsome but ruthless man is able to have any woman he wants, and exploits his charm by moving from one lady to another, leaving them the moment he has obtained what he wants, which is generally their money. Eventually he encounters a woman who sees through him, but despite this she cannot give him up, even though the relationship may result in her death.
THR   105 min (ort 109 min)   VIDrel: FIRST/SONOP L/A V

LONELY IN AMERICA *** (U)
Barry Alexander Brown USA   1990
Ranjit Chowdry, Adelaide Miller, Robert Kessler, Melissa Christopher, David Toney, Franke Hughes, Anila Singh, R. Ganesh, Tiriok Malik, R. Ganesh, Anila Singh, Ken Forman, Richard Raphael, Matt Midler, Louis Farber, Cee-Cee Rider
A young Indian immigrants leaves the sub-continent to come and work for his relatives in New York, experiencing the initial joys and disappointments of any newcomer to the Big Apple. However, the time comes when he realises that in true American fashion he must find his own way and reject his family's choice of his wife and career. A charming and well-acted romantic comedy with a light, deft touch.
DRA   96 min   TVrel

LONELY PASSION OF JUDITH HEARNE, THE *** 15
Jack Clayton UK   1988
Maggie Smith, Bob Hoskins, Wendy Hiller, Marie Kean, Ian McNeice, Prunella Scales, Alan Devlin, Rudi Davies, Sheila Reid, Niall Buggy, Aini Ni Mhuiri, Kate Binchy, Martina Stanley, Frank Egerton, Kevin Flood, Catherine Cusack
An Irish spinster finds love at long last, but discovers that her young lover is only interested in her money. Sparkling performances save this flawed adaptation that tends to wallow in melodrama at the expense of its characters.
DRA   111 min (ort 120 min)   VIDrel: MGM/WHV L/A   V
Boa: novel by Briane Moore.

LONESOME COWBOYS ** 18
Paul Morrissey USA   1968
Viva, Taylor Mead, Tom Hompertz, Louis Waldon, Joe Dallesandro, Eric Emerson, Julian Burroughs, Francis Francine
Produced by Andy Warhol, this is as good as any a look at his world of misfits, and is built around a kind of sex Western, in which affection between the cowboys depicted forms the central theme for the work. Not terribly well acted and with no discernible plot, this high camp exercise is by turns comic, bizarre, poignant and boring.
DRA   105 min (ort 110 min)   VIDrel: VISVID/POLYREC L/A   V

LONESOME DOVE: PARTS 1, 2 AND 3 *** 15
Simon Wincer USA   1988
Robert Duvall, Anjelica Huston, Tommy Lee-Jones, Danny Glover, Robert Urich, Ricky Schroder, Diane Lane, Frederic Forrest, D.B. Sweeney, Barry Corbin, Chris Cooper, Tom Scott, Glenne Headly, William Sanderson, Dave Cubitt
The story of two ageing Texas Rangers and their cattle drive from the title town in Texas to Montana. Both an enjoyable old-fashioned Western and an absorbing character study though definitely not, as some of the advance publicity would have us believe, the greatest Western ever made (even if McMurtry's novel did win the Pulitzer Prize). Followed by RETURN TO LONESOME DOVE: PARTS 1, 2 AND 3.
WES   360 min (3 cassettes – separately available) mTV
VIDrel: MARQ/QUANT   V/sur
Boa: novel by Larry McMurtry.

LONG AND THE SHORT AND THE TALL, THE *** PG
Leslie Norman UK   1960
Richard Todd, Laurence Harvey, Richard Harris, David McCallum, Ronald Fraser, John Meillon, John Rees, Kenji Takaki
WW2 drama set in Malaysia. A British patrol captures a

Japanese scout and agonise over whether to shoot him or not. A powerful character study that has not aged too well but still retains impact.
Aka: JUNGLE FIGHTERS
WAR   101 min (ort 110 min) B/W   VIDrel: WHV   V
Boa: play by Willis Hall.

LONG DAY CLOSES, THE *** PG
Terence Davies UK   1992
Marjorie Yates, Leigh McCormack, Anthony Watson, Nicholas Lamont, Ayse Owens, Tina Malone, Jimmy Wilde, Robin Polley, Peter Ivatts, Joy Blakeman, Denise Thomas, Patricia Morrison, Gavin Mawdsley, Kim McLaughlin, Marcus Heath
Eleven-year-old Bud is a quiet and withdrawn boy who, though surrounded by a warm and protective family, still finds himself alone. This being so, he takes delight in closely (if unobtrusively) observing all those around him, and the film, though devoid of a conventional narrative, makes good use of his memories to paint a detailed, lyrical and absorbing portrait of childhood and a vanished age. An undeniably fragmentary but rewarding film. Screenplay is by Davies.
DRA   81 min (ort 85 min)   VIDrel: CURZON/20TH   V/sur

LONG DAY'S JOURNEY INTO NIGHT *** U
Sidney Lumet USA   1961
Ralph Richardson, Katharine Hepburn, Dean Stockwell, Jason Robards Jr, Jean Barr
The anguish of an American family as it tears itself apart. Set in 1910, this acutely-observed melodrama pulls out all the stops in an effort to tug at our hearts, with the wife a drug addict, one son an alcoholic, and the other son dying of TB. Overblown (it largely exhibits the faults of the play) but still powerful. The script is by O'Neill, whose play was based to some extent on his early life.
DRA   135 min (ort 174 min) B/W
VIDrel: 4-FRONT/POLYREC   V
Boa: play by Eugene O'Neill.

LONG GOOD FRIDAY, THE *** 18
John Mackenzie UK   1980
Bob Hoskins, Helen Mirren, Eddie Constantine, Dave King, Brian Hall, Bryan Marshall, Derek Thompson, Stephen Davis, George Coulouris, P.H. Moriarty, Paul Freeman, Charles Cork, Paul Barber, Patti Love, Ruby Head, Loe Dolan, Patti Love
Rival gangsters in London battle it out in an East End setting unaware that a far more powerful criminal organisation has become involved. A brutish and unpleasant saga, redeemed by a strong performance from Hoskins and a number of sharp observations.
THR   109 min (ort 114 min)   VIDrel: CIC/SONOP; PION (LV only)   V/sur LV

LONG GOODBYE, THE ** 18
Robert Altman USA   1973
Elliott Gould, Nina Van Pallandt, Mark Rydell, Sterling Hayden, Henry Gibson, Jim Bouton, David Arkin, Warren Berlinger, Jo Ann Brody, Tammy Shaw, Jack Knight, Vince Pamieri, Arnold Strong, Ken Sansom, Danny Goldman, Steve Colt
Private detective Philip Marlowe gets involved south of the border helping out a friend suspected of murdering his wife. A less than completely serious treatment, that adds comedy at the expense of the tension and feeling of paranoia that this genre demands. The score is by John Williams. Arnold Schwarzenegger pops up briefly, but I leave you to guess his role.
DRA   111 min   VIDrel: MGM/WHV L/A   V
Boa: novel by Raymond Chandler.

LONG HOT SUMMER, THE: PARTS 1 AND 2 *** PG
Stuart Cooper USA   1985
Don Johnson, Jason Robards, Cybill Shepherd, Judith Ivey, Ava Gardner, Wings Hauser, William Russ, Alexandra Johnson, Stephen Davies, Charlotte Stanton, Albert Hall, William Forsythe, James Gammon, Rance Howard, Bill Thurman
An excellent remake of the 1958 film that tells of a tyrannical Southerner and his conflict with a tenant farmer in a small Mississippi town. The screenplay is by Dennis Turner and makes use of the one for the earlier film. Originally shown in two parts.
DRA   185 min mTV   VIDrel: 20TH/TECH   V
Boa: short stories/novel The Hamlet by William Faulkner.

LONG KISS GOODNIGHT, THE * 18
Renny Harlin USA 1996
Geena Davis, Samuel L. Jackson, Patrick Malahide, Craig Bierko, Brian Cox, David Morse, G.D. Spradlin, Tom Amandes, Yvonne Zima, Melina Kanakaredes, Alan North, Joseph McKenna, Dan Warry-Smith, Kristen Bone, Jennifer Pisana, Rex Linn
This is as over-the-top as they come, and casts Davis as a female government assassin (and apparently top martial artist as well) who for the past eight years has lived a most peaceful life, having come to believe her fake identity is the real thing following her near termination and resultant amnesia. Enter private eye Jackson, who uncovers her past. An unplotted and unfocused series of killings, explosions and perils now ensues. Entertaining, if a little mindless.
A/AD 120 min VIDrel: EIV/SONOP V/sh

LONG RIDERS, THE * 18
Walter Hill USA 1980
Stacy Keach, James Keach, Pamela Reed, David Carradine, Keith Carradine, Robert Carradine, Dennis Quaid, Randy Quaid, Kevin Brophy, Harry Carey Jr, Christopher Guest, Nicholas Guest, Shelby Leverington, Felice Orlandi
The story of various outlaw gangs in the Old West told in violent detail with little else in the way of fresh ideas, except perhaps the novelty of using a bunch of brothers to play almost all the leading parts. The score is by Ry Cooder.
WES 95 min Cut (1 min 35 sec – ort 99 min)
VIDrel: MGM/WHV V

LONG ROADS, THE * (PG)
Tristam Powell UK 1992
Edith Macarthur, Robert Urquhart, John Buick, Paul Morrow, Paul Sykes, Sam Davies, Bill Riddoch, Ralph Riach, John McGlynn, Anne Marie Timoney, Rory John Mackay, Iain James Mackay, Jimmy Chisholm, Matthew Costello, David Meldrum
An elderly couple leave their home in Skye and embark on an extended visit to all their children, who are scattered about the country, from Glasgow to London. At each stopover a drama unfolds and we are regaled with the events that have placed each particular family in the circumstances it now faces. Very slightly reminiscent of TOKYO STORY, this is touching, perceptive and carefully made, even if the plot device does at times feel just a little too contrived.
DRA 87 min mTV TVrel: BBC

LONG VOYAGE HOME, THE * U
John Ford USA 1940
John Wayne, Thomas Mitchell, Ian Hunter, Ward Bond, Barry Fitzgerald, John Qualen, Wilfrid Lawson, Mildred Natwick, Arthur Shields, Joseph Sawyer, Ward Bond, J.M. Kerrigan, Rafaela Ottiano, David Hughes, Billy Bevan, Jack Pennick
A close study of the crew of a tramp steamer during WW2 which is carrying a cargo of explosives across the Atlantic. Undeniably studiobound in its approach, but still worth seeing for the fine acting and excellent photography. The script is by Dudley Nichols.
DRA 100 min B/W VIDrel: 4-FRONT/POLYREC V/h
Boa: four one-act plays by Eugene Gladstone O'Neill.

LONG WALK HOME, THE ** PG
Richard Pearce USA 1990
Sissy Spacek, Whoopi Goldberg, Dwight Schultz, Ving Rhames, Dylan Baker, Erika Alexander, Lexi Faith Randall, Richard Habersham, Jason Weaver, Dan E. Butler, Crystal Robbins, Chelcie Ross, Mary Steenburgen (narration only)
A look at the changing pattern of life in the Deep South in the mid-1950s, when segregation was both a way of life and a creed. The story is built around the boycott that took place in Montgomery, Alabama, when blacks finally refused to submit to various indignities, such as sitting at the back of the bus. Spacek is a Southern belle who grows in social awareness, and Goldberg is her hardworking housekeeper. A detailed and potent film.
DRA 91 min (ort 98 min) VIDrel: 20TH V/sur

LONG WEEKEND, THE * 15
Colin Eggleston AUSTRALIA 1978
John Hargreaves, Briony Behets, Mike McEwen, Michael Aitkins, Roy Day, Sue Kiss Von Soly
An unhappy married couple spend a long weekend by the sea and rather thoughtlessly kill a few animals. However, their short break turns into a nightmare as they fall victim to Nature's revenge on Man for his abuse of the flora and fauna. There is a message in here somewhere.
HOR 92 min (ort 97 min) VIDrel: ARTPRO/RTM V

LONGEST DAY, THE * PG
Ken Annakin/Andrew Martin/Bernard Wicki USA 1962
John Wayne, Rod Steiger, Robert Ryan, Robert Mitchum, Henry Fonda, Robert Wagner, Mel Ferrer, Paul Anka, Fabian, Tommy Sands, Richard Beymer, Jeffrey Hunter, Sal Mineo, Roddy McDowall, Stuart Whitman, Richard Burton, Eddie Albert
Vast sprawling epic of a film that attempts to recreate the Allied landing on Normandy during WW2. Brilliant handling of spectacle is hampered by lack of focus, making it rather difficult to follow and a little tiring to watch. AA: Cin (Jean Borgoin/Walter Wottitz), Effects (Robert MacDonald/Jacques Maumont).
WAR 168 min (ort 180 min) B/W wScrn
VIDrel: 20TH/TECH; ENCORE (LV only) V/sh LV
Boa: novel by Cornelius Ryan.

LONGTIME COMPANION * 15
Norman Rene USA 1990
Stephen Caffrey, Patrick Cassidy, Brian Cousins, Bruce Davison, Mark Lamos, John Dossett, Dermot Mulroney, Mary-Louise Parker, Michael Schoeffling, Campbell Scott, Robert Joy
A skilfully wrought PBS production that closely follows the growth of AIDS through the 1980s, and its impact on the homosexual community in New York City, as seen through the eyes of two close friends. Excellent performances and a faultless script that is never cloying or mawkish put this film among the very best of those that have attempted to deal with the ravages of this disease.
DRA 95 min (ort 100 min) VIDrel: VISION/DISC V

LOOK BACK IN ANGER * PG
Tony Richardson UK 1959
Richard Burton, Claire Bloom, Edith Evans, Mary Ure, Gary Raymond, Donald Pleasence, Glen Byam-Shaw, Phyllis Neilson-Terry, Stanley Van Beers, John Dearth, Jordan Lawrence, Michael Balfour, George Devine, Anne Dickins
A film adaptation of one of the first of the influential British dramas dealing with anti-establishment sentiments. Burton is rather striking as our irascible young man who has decided to rebel against life and family, but the play on which this film was based is now looking very dated. Scripted by Nigel Kneale.
DRA 95 min (ort 101 min) B/W VIDrel: 4-FRONT V
Boa: play by John Osborne.

LOOK WHO'S TALKING * 15
Amy Heckerling USA 1990
John Travolta, Kirstie Alley, Olympia Dukakis, George Segal, Abe Vigoda, Twink Caplan, Jason Schaller, Jaryd Waterhouse, Jacob Haines, Christopher Aydon, Joy Boushel, Dan S. Davis, Andrea Mann plus Bruce Willis (voice only)
An unmarried mother is ditched by her boyfriend and sets out to find a perfect surrogate father for the child. Travolta becomes the baby's somewhat unconventional sitter and most of the humour is derived from the baby's sharp observations, which from conception to the age of one are relayed courtesy of the Bruce Willis voice-over.
Aka: DADDY WANTED
COM 92 min (ort 96 min) VIDrel: VCC/DISC/COLUM V/sur

LOOK WHO'S TALKING TOO * 15
Amy Heckerling USA 1990
John Travolta, Kirstie Alley, Olympia Dukakis, Elias Koteas, Twink Kaplan, Gilbert Gottfried, Lorne Sussman, Megan Milner, Georgia Keithley, Nikki Graham plus voices of: Bruce Willis, Roseanne Barr, Damon Wayans, Mel Brooks
An ill-conceived sequel that adds a dollop of tastelessness to the feeble plot, and dispenses with any of the charm of the original. This time round, New York cabby Travolta is married to accountant Alley, and her son now has a baby sister. Two voiceovers compete for attention in a noisy and vulgar mess, which has a few bright moments, but a climactic apartment fire is most definitely not one of them.
COM 77 min (ort 85 min) VIDrel: VCC/DISC/COLUM V/sur

LOOK WHO'S TALKING NOW * 12
Tom Ropelowski USA 1993
John Travolta, Kirstie Alley, David Gallagher, Tabitha Lupien, Olympia Dukakis, Danny De Vito, Diane Keaton, Lysette Anthony, George Segal, Charles Hartley, John Stocker, Elizabeth Leslie, Caroline Elliott, Vanessa Morley
Second sequel to the original now switches its attention from the family's children to its dogs, who put their heads together to

save their master's marriage when a slinky seductress appears on the scene. Very silly indeed, and the voice-overs are more than a little annoying, all the more so as the family pets now have them. A movie devoid of charm or entertainment value.
COM   92 min (ort 95 min) cC   VIDrel: COLUM/SONOP
V/sur

LOOKALIKE, THE ***
Gary Nelson USA
Melissa Gilbert-Brinkman, Diane Ladd, Thaao Penghlis, Bo Brinkman, Frances Lee McCain, Jason Scott Lee, April Stevens, C.K. Bibby, Irene Ziegler, Dave Hager, Janelle Cochrane, Carole Shoemaker, Lew Gallo, Sandy Shackleford
A woman is recovering from the death of her young daughter, and thinks she has seen an exact double of her child. A rather good thriller, well handled and neatly scripted.
THR   88 min (ort 100 min) mCab   VIDrel: CIC/SONOP
V/sh
Boa: short story by Kate Wilhelm.
15
1990

LOOKING FOR MR GOODBAR *
Richard Brooks USA
Diane Keaton, Tuesday Weld, Richard Kiley, Richard Gere, William Atherton, Tom Berenger, LeVar Burton, Alan Feinstein, Priscilla Pointer, Laura Prange, Joel Fabiani, Julius Harris, Richard Bright, Marilyn Coleman, Carole Mallory
A woman teacher of deaf children haunts singles bars by night for one-night stands. A film that starts off in an engrossing manner and then degenerates into a tiresome and pretentious display of sordidness. Later gave rise to the TV movie "Trackdown: Finding The Goodbar Killer".
DRA   130 min (ort 136 min)   VIDrel: CIC/SONOP L/A   V
Boa: novel by Judith Rossner.
18
1977

LOOKING FOR MR RIGHT **
Joseph L. Scanlan USA
Donna Mills, Brian Wimmer
In despair at the prospect of ever finding her soul-mate and getting married, a highly successful lawyer finds herself becoming more hopeful when she starts looking for a new secretary.
COM   90 min   VIDrel: CAPIT/GUILD   V/s
PG
1990

LOOKING FOR RICHARD ****
Al Pacino USA
Al Pacino, Harris Yulin, Penelope Allen, Alec Baldwin, Kevin Spacey, Estelle Parsons, Winona Ryder, Aisdan Quinn, Gordon MacDonald, Madison Arnold, Vincent Angell, Timmy Prairie, Landon Prairie, Kevin Conway, Larry Bryggman, Ira Lewis
Under the probing eye of Pacino, the true nature of one of Shakespeare's most demanding characters, Richard III, is revealed, by way of a series of rehearsals, insights and explanations, as a host of characters make ready for a performance of the play. Part documentary, part performance and part interview, this was a project Pacino (in his directing debut) worked on for years. The finished result is clearly worthy of the care lavished on it.
DOC   113 min   CINrel
12
1996

LOOKING GLASS WAR, THE *
Frank R. Pierson UK
Christopher Jones, Pia Degermark, Ralph Richardson, Anthony Hopkins, Paul Rogers, Susan George, Anna Massey, Ray McAnally, Maxine Audley, Cyril Shaps, Robert Urquhart, Frederick Jaegar, Paul Maxwell, Vivien Pickles, Peter Swanwick
Story of a spy mission to photograph a missile site in East Germany. A dull and incoherent spy thriller that fails to do justice to the original novel. Jones just does not convince as the young Pole sent over to do British Intelligence's dirty work for them. Written by Pierson.
DRA   103 min (ort 108 min)   VIDrel: RCA L/A   V
Boa: novel by John Le Carre.
15
1969

LOOSE CANNONS *
Bob Clark USA
Gene Hackman, Dan Aykroyd, Dom DeLuise, Ronny Cox, Nancy Travis, Paul Koslo, Robert Prosky, Dick O'Neill, Jan Triska, Leon Rippy, Robert Elliott, Robert Dickman, Herb Armstrong, David Alan Grier, Alex Hyde-White, Reg E. Cathey
A pair of mismatched police detectives solve a serial murder case that involves an incriminating video tape of Hitler and his henchmen that could ruin the career of a German politician. A clumsy farce that throws in just about every ingredient on offer, but the whole is very definitely less than the sum of the parts.
COM   90 min (ort 94 min)   VIDrel: RCA/VCC L/A   V
15
1990

LOOT ***
Silvio Narizzano UK
Hywel Bennett, Richard Attenborough, Lee Remick, Milo O'Shea, Dick Emery, Joe Lynch, Roy Holder, John Cater, Aubrey Woods, Robert Raglan, Jean Marlow, Enid Lowe, Andonia Katsaros, Harold Innocent, Kevin Brennan, Robert Raglan
A funny but uneven black comedy about bank robbers who make frantic efforts to retrieve the proceeds of a robbery, having hidden it in the coffin of the mother of one of them. However, they do not bargain for Attenborough's Inspector Truscott who fancies himself as something of a sleuth. Bad casting of Remick is a handicap, as is the weak plotting.
Aka: LOOT... GIVE ME MONEY, HONEY!
COM   97 min (ort 101 min)   VIDrel: WHV   V/h
Boa: play by Joe Orton.
15
1970

LORD JIM **
Richard Brooks UK
Peter O'Toole, James Mason, Eli Wallach, Curt Jurgens, Jack Hawkins, Paul Lukas, Akim Tamiroff, Daliah Lavi, Ichizo Itami, Christian Marquand, Andrew Keir, Jack MacGowran, Noel Purcell, Walter Gotell, Noel Chester, Serge Reggiani
A poor and shallow adaptation of Conrad's novel telling of a young sailor who commits an act of cowardice, and then spends the rest of his life struggling to find some way to atone and regain his self-respect. An outstanding cast do their best. Scripted by Brooks.
A/AD   148 min (ort 154 min)   VIDrel: ENCORE/SPEAR   V
Boa: novel Joseph Conrad.
PG
1964

LORD OF ILLUSIONS **
Clive Barker USA
Scott Bakula, Kevin J. O'Connor, Famke Janssen, Daniel Von Bargen
A private detective is engaged to track down a powerful supernatural being who has been posing as a stage magician, but whose powers are so great that it could destroy the entire world. This search plunges our detective into a dark and very macabre world, but unfortunately not a very exciting one; the soggy script tending to see to that. A few strongly directed and suitably repulsive scenes are scant compensation.
Aka: CLIVE BARKER'S LORD OF ILLUSIONS
HOR   116 min (ort 120 min) cC   VIDrel: MGM/WHV
V/sur
18
1995

LORD OF THE FLIES ***
Peter Brook UK
James Aubrey, Tom Chapin, Roger Elwin, Hugh Edwards, Tom Gaman, Sam Surtees, Eric Surtees, Roger Allen, David Brunjes, Peter Davy, Kent Fletcher, Alan Heeps, Jonathan Heaps, Nicholas Hammond, Christopher Harris, Richard Hope
Having been stranded on a remote tropical island by a plane crash, a group of English schoolboys slowly descend into savagery and tribalism. A flawed attempt to adapt Golding's book, it has some powerful moments but is generally weighed down by the novel's symbolism rather than assisted by it. A semi-professional production that is sometimes a little too ambitious for its own good but has some powerful moments. Screenplay is by Brook. Remade in 1990.
DRA   86 min (ort 91 min) B/W   VIDrel: LUMI/SPEAR L/A
V
Boa: novel by William Golding.
PG
1963

LORD OF THE FLIES **
Harry Hook USA
Paul Balthazar Getty, Chris Furrh, Danuel Pipoly, Michael Green, Badgett Dale, Edward Taft, Andrew Taft, Gary Rule, Terry Wells, Braden McDonald, Angus Burgin, Martin Zentz, Brian Jacobs, Vincent Amabile, David Weinstein
A lush updating of Golding's classic apocalyptic fable that turns our British schoolboys into American kids, adding a certain contemporary feel to the story, but at the expense of the novel's sub-Freudian insights. The excellent score is by Phillipe Sarde, and a fine cast (largely of unknowns) play their parts well, but the depth and compelling (if sporadic) force of the earlier Peter Brook film is entirely absent.
DRA   86 min (ort 95 min)   VIDrel: 4-FRONT/POLYREC
V/h
Boa: novel by William Golding
15
1990

LORD OF THE RINGS, THE **
Ralph Bakshi USA
Voices of: Christopher Guard, William Squire, John Hurt, Michael Sholes, Dominic Guard
15
1978

Animated version of this fantasy saga dealing with the battle between good and evil in the realm of "Middle Earth", and a struggle for possession of a magical ring which can confer great power on its owner. An ambitious but flawed adaptation of the Tolkien saga that ends rather abruptly, having only got halfway through the trilogy. One interesting aspect is the way in which the cartoon was made, using film of live actors as a basis.
ANIM   60 min Cut (2 sec – ort 133 min)   VIDrel: VES L/A V
Boa: trilogy by J.R. Tolkien.

LORDS OF DISCIPLINE, THE **
Franc Roddam USA
15
1982
David Keith, Mark Breland, Robert Prosky, G.D. Spradlin, Rick Rossovitch, Mitchell Lichtenstein, Barbara Babcock, Michael Biehn, Judge Reinhold, Bill Paxton, John Lavachielli, Mark Breland, Malcolm Danare, Gregg Webb, Ed Bishop
An ugly story of racism at a military academy in South Carolina in 1964, where the first black student is being systematically assaulted by a secret society. Good performances hold up the murky and unappealing script. Filmed in England.
DRA   98 min (ort 103 min)   VIDrel: CIC/SONOP L/A   V
Boa: novel by Pat Conroy.

LORDS OF THE DEEP **
Mary Ann Fisher USA
PG
1988
Bradford Dillman, Priscilla Barnes, Daryl Haney, Ed Lottimer, Melody Ryane, Stephen Davies, Gregory Sobeck
The crew of an underwater research station become trapped by an earthquake but discover an alien colony beneath the seabed. Like THE ABYSS and DEEPSTAR SIX, this film explores a late 1980s interest in underwater fantasies.
FAN   75 min   VIDrel: MGM L/A   V

LORELEI'S GRASP, THE *
Armando de Ossorio SPAIN
18
1972
Tony Kendall (Luciano Stella), Helga Line, Silvia Tortosa, Loretta Tovar, Luis Induin, Jose Theman, Angel Menendez, Luis Barboo, Josefina Jartin
A low-budget quickie, very loosely based on the German legend of a siren who ate the hearts of young people to keep her beauty. This tale combines elements of the "Nibelungen" myth and has our creature living in a cave, from where she ventures forth as a monster to attack young women and tear out their hearts. A warrior is hired to destroy her and eventually does so, with a blade forged from the sword of Siegfried.
Aka: LAS GARRAS DE LORELEI; LORELEY'S GRASP, THE; WHEN THE SCREAMING STOPS
HOR   81 min Cut (37 sec – ort 102 min)
VIDrel: IMPENT L/A   V

LORENZO'S OIL ***
George Miller USA
15
1992
Susan Sarandon, Nick Nolte, Peter Ustinov, Kathleen Wilhoite, Zack O'Malley Greenburg, Gerry Bamamn, Margo Martindale, James Rebhorn, Ann Hearn, Colin Ward, Maduka Steady, Noah Banks, E.G. Daly, Michael Hader, Bill Amman, Sandy Gore
When a young couple discover that their child's abnormal behaviour is due to a rare and incurable disease, they soon learn that the doctors offer little hope and less comfort. However, they refuse to accept that there is no cure and move heaven and earth in their efforts to find any kind of treatment that may offer a chance. Based on a true case, this is a sensitive and often profoundly moving tale, of humanity and warmth.
DRA   129 min (ort 136 min) cC   VIDrel: CIC/SONOP V/sur

LORNA DOONE ***
Basil Dean UK
U
1934
Victoria Hopper, John Loder, Margaret Lockwood, Roy Emerton, Frank Cellier, George Curzon, Herbert Lomas, Roger Livesey, Mary Clare, Edward Rigby, D.A. Clarke-Smith, Lawrence Hanray, Amy Veness, Eliot Makeham, Wyndham Goldie
Set on Exmoor in 1625, this classic romance tells of a farmer who falls in love with the daughter of an outlaw, who in reality is an heiress stolen from her family as a child. A well mounted and polished piece of work, it makes good use of some lovely exterior locations, but is somewhat lacking in dramatic impact.
DRA   83 min (ort 90 min) B/W   VIDrel: LUMI/SPEAR L/A V
Boa: novel by R.D. Blackmore.

LORNA DOONE ***
Andrew Grieve UK
PG
1990
Polly Walker, Sean Bean, Clive Owen, Michael Mackenzie, Andrew Ferguson, Claire Maddeen, Paul Young, Jane Gurnett, Billie Whitelaw, Miles Anderson, Kenneth Haigh, Evan Grant MacLachlan, Robert Stephens, Martin Heller
In 17th century England, a man swears vengeance on the outlaw clan that killed his family when he was a small child but finds himself falling in love with a young innocent girl who is a member of that family. A well mounted adaptation of this romantic drama, filmed several times before.
DRA   86 min (ort 90 min) mTV   VIDrel: ARENA/SPEAR V
Boa: novel by R.D. Blackmore.

LOSING ISAIAH **
Stephen Gyllenhaal USA
15
1995
Jessica Lange, Halle Berry, Samuel L. Jackson, La Tanya Richardson
A black former drug addict has to battle to regain custody of her baby boy, who has been adopted by a white couple, and in the end has to mount an action through the courts. Lange is highly effective as the white doctor who has gained custody of the child and Berry as the addict does what she can with her unpromising and stereotyped role. This is a film that struggles for realism but instead delivers manipulation, and the happy ending is weak and unconvincing.
DRA   102 min cC   VIDrel: CIC/SONOP   V

LOSING TRACK ***
Jim Lee UK
(PG)
1992
Alan Bates, Geraldine James, Ben Holden, Michael Culver, Sue Roderick, James Copnall, Brinley Jenkins, Lynette Edwards, Kirsten Jones, Caroline Stubbs, Jack Walters, Judith Humphreys, Giles Thomas
A former colonial civil servant returns from India to England for his wife's funeral, and gets to meet the twelve-year-old son he has not seen for five years. Bates gives a wonderful performance as a pompous man who finds it easy to shrug off his responsibilities, but far from easy to communicate his feelings. A leisurely and fairly uneventful tale, set in the 1950s, that benefits from strong period detail if not from its rather contrived ending.
COM   75 min mTV   TVrel: BBC

LOST BOYS, THE **
Joel Schumacher USA
15
1987
Corey Feldman, Jami Gertz, Corey Haim, Jason Patric, Kiefer Sutherland, Edward Herrmann, Barnard Hughes, Dianne Wiest, Jamison Newlander, Brooke McCarter, Billy Wirth, Alexander Winter, Chance Michael Corbitt
Two brothers move to a new town and learn that the local biker gang are members of the Undead. The younger one now has his work cut out when his older brother falls for the one female in this bunch of motorised vampires. A flashy but predictable chiller with some humorous touches. The climax is easily the best thing in the film.
HOR   93 min (ort 98 min) wScrn   VIDrel: WHV   V/sur V/sh

LOST CAPONE, THE ***
John Gray USA
15
1990
Adrian Pasdar, Ally Sheedy, Eric Roberts, Titus Welliver, Jimmie F. Skaggs, Maria Pitillo, Anthony Crivello, Dominic Chianese, Andrew Palmacci, William Andrews, Barton Heyman, Max Maxwell, Ed Grady, Bill Luhrs, Karma Ibsen
The story of Al Capone's younger brother Jimmy, whose desire to distance himself from his gangster sibling led to a change of name and the job of marshal in Homer, Nebraska, where he did his best to ruin his brother's bootlegging operations. An absorbing morality play, with Roberts giving a performance of barely controlled fury as Al, that works well against Pasdar's more restrained role as Jimmy.
DRA   89 min (ort 100 min) mCab   VIDrel: NWV/SONOP V/h

LOST HONOUR OF KATHARINA BLUM, THE ***
Volker Scholondorff/Margarethe von Trotta WEST
GERMANY
15
1975
Angela Winkler, Mario Adorf, Dieter Laser, Heinz Bennent, Jurgen Prochnow, Hannelore Hoger, Rolf Becker, Harald Kuhlmann, Herbert Fux, Regine Lutz, Werner Eichhorn, Karl Heinz Vosgerau, Angelika Hillebrecht, Horatius Haeberle
A young woman fleetingly involved with an alleged terrorist is effectively crushed between the millstones of media persecu-

tion and police suspicions, in this highly atmospheric drama that deals with issues of political repression and individual freedom. Remade as the TV movie ACT OF PASSION.
Aka: DIE VERLORENE EHRE DER KATHARINA BLUM
DRA   106 min   VIDrel: CONNO/RTM   V
Boa: novel by Heinrich Boll.

LOST HORIZON ****
Frank Capra USA
Ronald Colman, Jane Wyatt, H.B. Warner, Sam Jaffe, Thomas Mitchell, Edward Everett Horton, Isabel Jewell, John Howard, Margo, Hugh Buckler, Lawrence Grant, John Burton, John Miltern, John T. Murray, Max Rabinowitz, Willie Fung
A brilliant evocation of our longing for a land beyond time and human imperfection finds expression in this exquisite tale of five travellers who are abducted and brought to the remote Himalayan community of Shangri La, the intention being to select a new leader. Edited from 130 to 118 minutes on release. Unfortunately portions of the original uncut film are now lost. Remade in 1972. AA: Art (Stephen Goosen), Edit (Gene Havlick/Gene Milford).
A/AD   132 min B/W   VIDrel: COLUM/SONOP   V
Boa: novel by James Hilton.

U
1937

LOST IN SIBERIA ***
Alexander Mitta UK/USSR
Anthony Andrews, Ira Mikhalyova, Yelena Mayorova, Vladimir Ilyin, Yevgeni Mironov, Alexei Zharkov, Valentin Gaft, Alexander Bureyev, Vladimir Prozorov, Hark Bohm, Nikolai Pastukhov, Yuri Sherstnyov, Nicolas Chagrin, Elena Secota
Very loosely based on a true story, this details the exploits of a British archaeologist (in reality the central character was an American soldier) who is kidnapped by the KGB whilst working in Persia during WW2. Despite torture and imprisonment in a Siberian labour camp, he learns to survive both the harsh conditions and the brutality of his fellow prisoners. A bleak and relentlessly depressing film, similar in tone to "Coming Out Of The Ice" and "Gulag".
Aka: ZATERYANI V SIBIRIY
DRA   103 min (ort 108 min) (English version)
VIDrel: HIFLI/SONOP   V/h

15
1991

LOST IN THE BARRENS **
Michael Scott USA
Nicolas Shields, Evan Adams, Lee J. Campbell, Marianne Jones, Victor Cowie, Graham Greene, Paul Gray, Harry Nelken, Ken Babb, Jeff Madden, Fred Robinson, Brian Richardson, Eric Robinson, Adam Beach, Louie Camerone, Joe Onyuk
Lost in the Canadian wilderness, a youngster from the big city and an Indian boy learn the meaning of co-operation in their struggle to survive.
JUV   92 min (ort 95 min)   VIDrel: CAPIT/GUILD   V
Boa: novel by Farley Mowat.

PG
1990

LOST IN TIME ***
Anthony Hickox USA
Zach Galligan, Monika Schnarre, Martin Kemp, Alexander Godunov, Bruce Campbell, Sophie Ward, Marina Sirtis, Michael Des Barres, Juliet Mills, David Carradine, Patrick Macnee, Billy Kane, Joe Baker, John Ireland, Jack Eiseman
This sequel to WAXWORK is an enjoyable horror-spoof that sees a young couple discovering a key that can open a door in time. However, their explorations bring them into contact with aliens and other unsavoury beings, and they face a struggle in finding their way back home. Music is by Steve Schiff.
Aka: WAXWORK 2: LOST IN TIME
HOR   100 min   VIDrel: EIV/SONOP   V

15
1992

LOST IN YONKERS ***
Martha Collidge USA
Richard Dreyfuss, Mercedes Ruehl, Irene Worth, David Strathairn, Brad Stoll, Mike Damus, Robert Guy Miranda, Jack Laufer, Susan Merson, Illya Haase, Calvin Stillwell, Dick Hagerman, Jesse Vincent, Howard Newstate, Peter Gannon
When their mother dies and their father goes off to look for work, two teenage brothers are packed off to live with his mother, in a household that also includes his married sister. The grandmother is a domineering and iron-hard woman, easily able to repressing her dreamy, other-worldly daughter. A sensitively handled adaptation of Simon's play that won him the

PG
1992

Pulitzer Prize and very watchable if slightly stagebound in places.
Aka: NEIL SIMON'S LOST IN YONKERS
COM   109 min (ort 114 min) cC   VIDrel: COLUM/SONOP   V/sur
Boa: play by Neil Simon.

LOST LANGUAGE OF CRANES, THE ***
Nigel Finch UK
Brian Cox, Eileen Atkins, Angus MacFayden, Corey Parker, Rene Auberjonois, John Schlesinger, Cathy Tyson, Richard Warwick, Nicholas Le Prevost, Ben Daniels, Frank Middlemass, Nigel Whitney, Edmund Kente, Paul Cottingham
Portrait of a family and the crisis that develops as a number of skeletons in its cupboards are gradually revealed. These dark secrets include both marital infidelity and homosexuality. The strange title refers to a thesis on a dysfunctional family written by a social worker.
DRA   90 min mTV   TVrel: BBC
Boa: novel by David Leavitt.

(18)
1991

LOST WORLD, THE ***
Harry Hoyt USA
Wallace Beery, Louis Stone, Bessie Love, Lloyd Hughes, Arthur Hoyt, George Bunny, Bull Montana, Margaret McWade, Jules Cowles, Alma Bennett, Charles Wellesley, Finch Smiles, Virginia Brown
A scientist leads an expedition to South America and encounters a prehistoric world complete with living dinosaurs and other monsters. The special effects are excellent and much ahead of their time. An inferior remake appeared in 1960 and again in 1991.
FAN   60 min (ort 77 min) B/W silent
VIDrel: VISION/DISC   V
Boa: novel by Arthur Conan Doyle.

U
1925

LOST WORLD, THE *
Timothy Bond CANADA
John Rhys-Davies, Tamara Gorski, Nathania Stanford, David Warner, Darren Peter Mercer, Eric McCormack, Sala Cane, Fideliz Chea, John Chinosiyani, Geza Kovacs, Innocent Choga, Brian Cooper, Charles David, Kate Egan, Mike Grey
Although reasonably faithful to Doyle's exciting novel, there is little enjoyment to be had in this film version, as it recounts the story of a Victorian scientist who leads an expedition to a plateau in Latin America where time has stood still since the prehistoric era. Poor dialogue, a feeble plot and acting to match make this one hard to sit through. Followed by RETURN TO THE LOST WORLD.
A/AD   99 min (ort 180 min) mTV   VIDrel: AUDIO/VCC   V
Boa: novel by Arthur Conan Doyle.

PG
1991

LOULOU ***
Maurice Pialat FRANCE
Gerard Depardieu, Isabelle Huppert, Guy Marchand, Humbert Balsan, Bernard Tronczyk, Christian Boucher, Frederique Cerbonnet, Jacquelien Dufranne, Willy Safar, Agnes Rosier, Patricia Coulet, Jean-Claude Meillard
A smart businesswoman leaves her respectable lover to live with a boozy layabout, to whom she is drawn by lust alone. Restrained direction and detailed camerawork sustain this curious character study; its pessimistic barbs are as often directed at 1970s consumerism as they are at the couple's drunken and aimless excesses.
DRA   100 min (ort 110 min)   VIDrel: ARTIF/20TH   V

18
1980

LOVE AFFAIR **
Glenn Gordon Caron USA
Warren Beatty, Annette Bening, Garry Shandling, Chloe Webb, Kate Capshaw, Pierce Brosnan, Katherine Hepburn, Paul Mazursky, Brenda Vaccaro, Glenn Shadix, Barry Miller, Harold Ramis, Elya Baskin, Taylor Dayne, Carey Lowell, Ray Charles
A womanising TV sports journalist meets an attractive female aboard a plane and sets about romancing her during the journey. An uninspired remake of a 1939 film of the same name (it was remade once before in 1959 as AN AFFAIR TO REMEMBER). The attractive camerawork offers a partial compensation for the weak plotting in a drama that seems both flat and undeveloped.
DRA   103 min cC   VIDrel: WHV   V/sur

12
1994

LOVE AMONG THIEVES ** PG
Roger Young USA 1987
Audrey Hepburn, Robert Wagner, Jerry Orbach, Patrick Bauchau, Brion James, Samantha Eggar, Christopher Neame
A synthetic blend of romantic comedy and action adventure set in Mexico, where a female jewel thief goes to ransom her kidnapped fiancee with the booty from her latest robbery. This weak effort was Hepburn's TV movie debut, and this talented actress does her best to make the most of her role.
COM 98 min (ort 100 min) mTV VIDrel: GUILD/SONOP
L/A V

LOVE AND A .45 ** 18
C.M. Talkington USA 1994
Gil Bellows, Renee Zellweger, Rory Cochrane, Jeffrey Combs, Jace Alexander, Michael Bowen, Ann Wedgeworth, Peter Fonda
A feckless young couple get involved in a robbery that goes badly wrong when a colleague turns it into a massacre. Armed with a handgun they go on the run, and indulge in an orgy of violent crime, all the while being pursued by various nutters. Despite being atmospheric and intense, this low-budget film (it recalls NATURAL BORN KILLERS) is an unpleasant and unoriginal piece, whose director seems to be of the opinion that gore is a suitable substitute for talent.
THR 98 min (ort 102 min) VIDrel: TRIM/HIFLI V/h

LOVE AND BETRAYAL ** 15
Richard Michaels USA 1989
Stefanie Powers, David Birney, Fran Descher, Amanda Peterson, Martha Scott, Josh C. Williams, Reiner Schoene, Lisa Aliff, Ann Ryerson, Pamella D'Pella, Mary Gordon Murray, Cathy McAuley
Powers plays a happily married housewife with two kids and a husband who makes his living from writing self-help guides. Having sacrificed a life of her own to look after her family, she's unable to cope when her husband has an affair (it's prompted by his mid-life crisis) and decides to leave her. She never really recovers from this shock, nor us from this painful divorce saga, which is bleak, intense and not especially endearing. See AN UNMARRIED WOMAN.
Aka: THROWAWAY WIVES
DRA 94 min (ort 100 min) mTV VIDrel: ITC/HIFLI L/A
V

LOVE AND BETRAYAL: THE MIA FARROW
STORY ** PG
Karen Arthur USA 1994
Patsy Kensit, Dennis Boutsikaris, Richard Muenz, Robert Lupone, Gina Wilkinson, Frances Helm, Taryn Davies, Grace Una, Kristi Goteke, Nigel Bennett, Heidi Von Palleske, Natalie Miller, Caley Wilson, Christine Andreas, Tina Su
Dramatised account of the life of the title actress, from her childhood through to her stormy and highly public split with her long-time partner, Woody Allen.
DRA 158 min (ort 200 min) mTV VIDrel: 20TH/FOXVID
V
Boa: books Mia & Woody: Love and Betrayal by Kristi Goteke with Marjorie Rosen/Mia: The Life of Mia Farrow by O. Epstein and J. Morella.

LOVE AND CURSES... AND ALL THAT JAZZ ** (PG)
Gerald McRaney USA 1991
Delta Burke, Gerald McRaney, Elizabeth Ashley, Derria Brown, Tony Todd, Rachael Fontenot, Wendy Gardner, Brett Ashby, Randy Cheramie, John Wilmot, Joe Warfield, Scott Murphy, Arthel Neville, Leslie Nicholson, Victoria Pevels
A young female psychologist who is making plans for her forthcoming wedding gets a most intriguing case – that of a catatonic woman who was mistakenly pronounced dead the month before.
THR 90 min mTV SATrel: UK LIVING

LOVE AND DEATH ** PG
Woody Allen USA 1975
Woody Allen, Diane Keaton, Harold Gould, Alfred Lutter, Olga Georges-Picot, Jessica Harper, Zvee Scooler, Despo, Frank Adu, James Tolkan, Henry Czarniak, Georges Adet, Edmond Ardisson, Feodor Atkine, Albert Augier, Lloyd Battista
A very uneven Woody Allen look at Russian literature and its obsession with the meaning of life, made in the form of a spoof on foreign films. The one-line jokes are the best things in it.
COM 81 min (ort 85 min) VIDrel: WHV V

LOVE AND HATE *** 15
Francis Mankiewicz CANADA 1990
Kate Nelligan, Kenneth Walsh, John Colicos, R.H. Thomson, Duncan Ollrenshaw, Leon Pownall, Brent Carver, Cedric Smith, Victoria Snow, Eugene Lipinski, Doris Petrie, Vicki Wauchope, Noam Zylberman, Bernard Behrens, Robert Benson
Courtroom drama based on the true case of the murder of Jo-Ann Thatcher, whose body was discovered in a small Canadian town in 1983. As the ex-wife of Colin Thatcher, the millionaire cabinet minister and son of the Premier, the case received massive publicity, especially when the woman's former husband was arrested, tried and convicted of her murder (a charge he denies to this day). An overlong but compelling tale.
DRA 184 min mTV VIDrel: ODY/SONOP V/h
Boa: book A Canadian Tragedy by Maggie Siggins.

LOVE AND HUMAN REMAINS ** 18
Denys Arcand CANADA 1993
Thomas Gibson, Ruth Marshall, Mia Kirschner, Cameron Bancroft, Rick Roberts, Joanne Vannicola, Matthew Ferguson, Aidan Devine, Robert Higden, Sylvain Morin, Ben Watt, Karen Young, Serge Houde, Alex Wylding, Polly Shannon, Annie Juneau
A group of thirty-year-olds go looking for meaningful relationships in the big city and encounter a host of obstacles. Meanwhile, a serial killer of women is at large, and this inevitably casts a gloomy shadow over their quest. A pretentious and rather pompous little affair that offers nothing of insight.
Aka: LOVE & HUMAN REMAINS
DRA 99 min VIDrel: ELPIC/POLYREC V
Boa: play Unidentified Human Remains and the True Nature of Love by Brad Fraser.

LOVE AND OTHER CATASTROPHES *** 15
Emma-Kate Croghan AUSTRALIA 1996
Frances O'Connor, Alice Garner, Matthew Dyktynski, Radha Mitchell, Matt Day
The lives of a group of film students at university provide the focus here, principally two girls who move into a new flat and start looking for a third flatmate, but at the same time have to get to grips with their university course and their complex personal relationships. A touching, funny, offbeat and multi-stranded film, most of the story taking place over a single day, with our two girls planning a house-warming party. This was Croghan's feature film debut.
DRA 79 min CINrel

LOVE AND WAR *** 15
Paul Aaron USA 1987
James Woods, Haing S. Ngor, Jane Alexander, Concetta Tomei, Jon Cedar, James Pax, Richard McKenzie, Sally Klein, Lillian Lehman, Stephen Dorff, Lou Fant, Leo Geter, Raymond Ma, Steven Vincent Leigh, James Lashly, George Milan
Another film with a Vietnam background, dealing with a couple, and telling of how their love survived the husband's captivity at the hands of the Viet Cong. Despite enduring years of torture by his captors, the husband did not crack and was even able to secrete coded messages in his letters home to his wife, alerting her as to the conditions he was enduring. A harrowing and well-made tale.
Aka: IN LOVE AND WAR
DRA 92 min VIDrel: SONY L/A V
Boa: book In Love And War by Jim and Sybil Stockdale.

LOVE AT FIRST BITE *** 15
Stan Dragoti USA 1979
George Hamilton, Susan St James, Richard Benjamin, Dick Shawn, Arte Johnson, Sherman Hemsley, Isabel Sandford, Eric Laneuville, Barry Gordon, Michael Pataki, Robert Ellenstein, Rolfe Sedan, Bob Basso, Hazel Shermet
Count Dracula is evicted from his home in Transylvania and comes to New York where he falls in love with a model. Hamilton is quite outstanding in a spirited and entertaining comedy. The script is by Robert Kaufman.
COM 96 min VIDrel: MIA/DISC V

LOVE BITES ** 18
Marvin Jones USA 1988
Kevin Glover, Tom Wagner
A vampire-hunter falls in love with his quarry, a male vampire who is 347 years old. An unusual variant on the vampire legend with a gay perspective.
A 69 min VIDrel: PRIDE/PARADOX V

LOVE BITES ★★   15
Malcolm Marmorstein USA   1992
Adam Ant, Kimberly Foster, Roger Rose, Michelel Forbes, Philip Bruns, Judy Tenuta, Jacqueline Schuhltz, Rhonda Lee Doroton, Eric Lawrence, Julie Strain, Ava Fabian, Peggy Patrick
A spoof on vampires that has one of their number getting more than he bargains for when he falls in love, and finding that the only way out of his predicament is to become a mortal once again.
COM   94 min (ort 95 min)   VIDrel: REFLEC/FIRST L/A   V

LOVE BUG, THE ★★★   U
Robert Stevenson USA   1968
Dean Jones, Michele Lee, Buddy Hackett, David Tomlinson, Joe Flynn, Benson Fong, Iris Adrian, Joe E. Ross, Robert Reed, Bert Convy, Hope Lange, Barry Kelley, Andy Granatelli, Dale Van Sickel, Regina Paton, Bob Drake, Hal Brock
A Volkswagen has a mind of its own and gets up to all sorts of mischief in this fast and furious celebration of slapstick and mayhem. Followed by three rather less inventive "Herbie" sequels, the first one being HERBIE RIDES AGAIN.
COM   104 min (ort 110 min) cC   VIDrel: WDV/TECH L/A   V

LOVE CAN BE MURDER ★★   (15)
Jack Bender USA   1992
Jaclyn Smith, Colin Berensen, Cliff De Young, Tom Bower, Nicholas Pryor, Susan Brown, Elaine Kagan, John Carter, Anne Francis, Bruce Vilanch, Pamela Roberts, Doug Hale, Scott Stevens, Kimberly La Marque, Cameron Watson
Comedy-thriller that casts Smith as a former lawyer who sets up alone as a private detective when her job with an L.A. law firm comes to an end. But when another private eye asks for her help in solving a murder case from 1948, she finds herself getting in over her head. Quite watchable nonsense.
COM   90 min   SATrel: MOVIE CHANNEL

LOVE, CHEAT AND STEAL ★★   18
William Curran USA   1993
John Lithgow, Eric Roberts, Madchen Amick, Richard Edison, Donald Moffat, David Ackroyd, Dan O'Herlihy, Jason Workman, Claude Earl Jones, Jack Axelrod, Bill McKinney, John Pyper-Ferguson, Tom Kindle, Mary Fanaro, Chuck Zito
A man is happily married to an attractive woman, who seems perfect in every respect. However, he is blissfully unaware of her past, in which she framed a former boyfriend for murder and was responsible for him being sent to jail for seven years. Now released, he is determined to take his revenge and make their lives hell. A standard psycho-on-the-loose tale, with all the obligatory elements, well enacted by a capable cast who deserve better material.
THR   91 min (ort 96 min)   VIDrel: ITC/HIFLI   V/s

LOVE CIRCLES ★★   18
Gerard Kikoine UK   1984
John Sibbit, Marie France, Michel Siu, Josephine Jacqueline Jones, Pierre Burton, Sophie Berger, Philippe Baronne, John Allen, Lisa Allison, Timothy Wood
A young man travels the world in search of love and sexual experience. An international and updated version of LA RONDE, with a packet of cigarettes being passed from one person to the next in a series of erotic encounters. Trite and uninteresting. See also CHAIN OF DESIRE.
A   94 min (ort 97 min)   VIDrel: MGM L/A   V

LOVE CRIMES ★★   18
Lizzie Borden USA   1991
Sean Young, Patrick Bergin, Arnetia Walker, James Read, Ron Orbach, Fern Dorsey, Tina Hightower, Donna Biscoe, Danielle Shuman, David Shuman, Rebecca Wackler, Jill Jane Clementis, Roe Sabordo, Sarah Bork, Dianne Butler
An assistant female DA goes under ground to trap a photographer who has been preying on women by exploiting their sexual fantasies. A confused and unfocused thriller, at least in the R-rated version. A director's cut of 97 minutes also exists and is much more coherent than this baffling thriller, whose touches of sado-eroticism really add nothing of value to the plot.
THR   86 min (ort 97 min)
VIDrel: ENCORE/SPEAR/COLUM   V/sh

LOVE ETERNAL ★★   U
Jean Delannoy FRANCE   1943
Jean Marais, Madeleine Sologne, Jean Murat, Yvonne De Bray, Roland Toutain, Pieral
A new twist is given to the "Tristan and Isolde" legend, which tells the unhappy story of two lovers who find peace only in death. In this variant Marais brings Sologne to his widowed friend's chateau, his intention being to effect a marriage. But they imbibe poison administered by an evil dwarf and fall madly in love with each other, an event that leads to an inevitable tragic conclusion. Made during the German occupation, this is an ambitious but deeply flawed work.
Aka: ETERNAL RETURN, THE; L'ETERNEL RETOUR
DRA   107 min (ort 111 min) B/W   VIDrel: ARTPRO/RTM   V

LOVE FIELD ★★★   15
Jonathan Kaplan USA   1992
Michelle Pfeiffer, Dennis Haysbert, Brian Kerwin, Stephanie McFadden, Louise Latham, Peggy Rea, Beth Grant, Johnny Ray McGhee, Cooper Huckabee, Bob Gill, Mark Miller, Pearl Jones, Janell McLeod, Bob Minor, Ron Shelly, Rhoda Griffis
A Dallas beautician who is fixated on the Kennedys walks out on her husband immediately after the assassination and heads for Washington D.C. On the way there she meets and falls for a enigmatic black man who is travelling together with his young daughter. A contrived road-movie that offers some outstanding performances but little else of note. Pfeiffer achieved a Best Actress Oscar nomination, but in the UK this film went straight onto video.
DRA   100 min (ort 104 min)   VIDrel: VCC/DISC/COLUM   V/sur

LOVE, HONOR AND OBEY: THE LAST MAFIA MARRIAGE ★★   15
John Patterson USA   1993
Eric Roberts, Nancy McKeon, Ben Gazzara, Alex Rocco, Phyllys Lyons, Tomas Milian, Mike Nussbaum, Peter Jurasik, Joanna Merlin, Dylan Baker, Kimber Riddle, Joe Petruzzi, John Harkins, H. Richard Greene, Kane Picoy, Sanna Vraa
Fact-based gangster saga based on the story of Bill Bonanno and Rosalie Proface, who by marriage brought two powerful Mafia families together. Average.
Aka: LAST MAFIA MARRIAGE, THE
A/AD   177 min (ort 200 min) mTV   VIDrel: 20TH/TECH   V

LOVE HURTS – 1ST SERIES ★★★   PG
Carol Wiseman/Guy Slater/Roger Bamford UK   1992
Zoe Wanamaker, Adam Faith, Jane Lapotaire, Stephen Moore, Richard Cordery, Tony Selby, Hilary Mason, Richard Pearson, John Flanagan, Robin Weaver, Garrick Hagon, Thomas Baptiste, Richard Whitmore, Frank Mills, Shireen Shah, Sandra Voe
The story of a romance between a career-woman in her early forties who has given up her job to work for a Third World charity and a former plumber who has become a self-made millionaire. Something of a celebration of mid-life angst, this pleasing drama was sustained by good performances and dialogue. A second series followed, but it never quite recaptured the freshness and charm of this one.
DRA   245 min mTV   VIDrel: VCC/DISC/COLUM   V

LOVE HURTS – 2ND SERIES ★★   PG
Carol Wiseman/Guy Slater/Roger Bamford UK   1992
Zoe Wanamaker, Adam Faith, Jane Lapotaire
Having finally taken the plunge and set up home together, our middle-aged couple Tessa and Frank now find the long-term relationship fraught with difficulties. A rather contrived continuation of the TV drama, whose final outcome is never really in doubt.
DRA   245 min mTV   VIDrel: VCC/DISC/COLUM   V

LOVE IN THE AFTERNOON ★★★   15
Eric Rohmer FRANCE   1972
Zouzou, Bernard Verley, Francoise Verley, Daniel Ceccaldi, Malvina Penne, Babette Ferrier, Frederique Hender, Claude-Jean Philippe, Sylvaine Charlet, Daniele Malat, Suze Randall, Tina Michelino, Jean Louis Livi, Pierre Nunzi
A married man becomes innocently involved with the former mistress of a friend in this poignant drama, and their furtive meetings take place during his lunch break – hence the title. But when it comes to it, he decides to go back to his wife. This was the last of Rohmer's "Six Moral Tales", that began with "The Baker Of Monceau", "Suzanne's Career" and THEN THE COLLECTOR (but the first two films were never released theatrically).
Aka: CHLOE IN THE AFTERNOON; L'AMOUR L'APRES-MIDI
DRA   95 min (ort 98 min)   VIDrel: HEND/BMGREC L/A   V

LOVE IN THE STRANGEST WAY **
15
Christopher Frank FRANCE
1994
Thierry Lhermitte, Maruschka Detmers, Nadia Fares, Patrick Timsit, Umberto Orsini, Johann Martel, Vincent Planchais, Bernard Freyd, Marie-Christine Adam, Patrick Floersheim, Marina Rodriguez Tome, Alain Frerot, Georges Siatidis
Strongly reminiscent of FATAL ATTRACTION in its plotting, this film examines the actions of a scorned woman and her desire for revenge against the married man who has spurned her after a brief fling. Despite his attempts to discourage her, she is soon installed in his home as the new child-minder, from which position she starts to plot his destruction. A tired foray into over-familiar territory. See also DYING TO LOVE YOU and THE HAND THAT ROCKS THE CRADLE.
Aka: ELLES N'OUBLIENT PAS
THR   107 min wScrn   VIDrel: TART/20TH   V

LOVE IS A GUN ***
18
David Hartwell USA
1993
Eric Roberts, Kelly Preston, Eliza Roberts, R. Lee Ermey, Joe Sirola, John Toles-Bay, Jack Kelhler, Harvey Vernon, Michael Krawic, Peter Pit, Darryl Fong, Marshall Bell, Grant Cramer, Homsell Joy, Frank Di Paolo, Chuck Zito
A police photographer becomes seriously infatuated with a pretty model but when she is found dead, he finds himself the only suspect in a murder investigation. Excellent performances enhance this atmospheric thriller and the occasional touches of black comedy and the supernatural give the film an interesting feel all of its own.
THR   103 min   VIDrel: HIFLI/SONOP   V/h

LOVE IS A MANY-SPLENDORED THING, A **
U
Henry King USA
1955
William Holden, Jennifer Jones, Isobel Elsom, Jorja Curtwright, Virginia Gregg, Torin Thatcher, Richard Loo, Murray Matheson, Soo Young, Philip Ahn, Donna Martell, Candace Lee, Kam Tong, James Hong, Herbert Heyes, Angela Loo
A Eurasian woman doctor falls in love with an American war correspondent at the time of the Korean War. Mediocre plot beautifully realised in a film that is chiefly remembered for its tragic ending and Oscar-winning theme song. AA: Score (Alfred Newman), Song ("Love Is A Many Splendored Thing" – Sammy Fain (m)/Paul Francis Webster (l)), Cost (Charles LeMaire).
Aka: MANY-SPLENDORED THING, A
DRA   97 min (ort 102 min) wScrn cC   VIDrel: 20TH/TECH   V
Boa: novel A Many Splendoured Thing by Han Suyin.

LOVE IS NEVER SILENT ****
PG
Joseph Sargent USA
1985
Mare Winningham, Phyllis Frelich, Ed Waterstreet, Fredric Lehne, Cloris Leachman, Sid Caesar, Susan Ann Curtis, Mark Hildreth, Lou Fant, Julianna Fjeld, Jeremy Christall, Stephen E. Miller, Alex Diakun, Gregory Hayes
A girl with normal hearing has to function as the ears of her deaf parents and is torn between this role and a desire for a life of her own. The sensitive script is by Darlene Craviotto. A splendid drama with fine performances. The won Emmy Awards for Outstanding Director and Outstanding Drama Special.
Aka: SHATTERED SILENCE
DRA   97 min (ort 100 min) mTV   VIDrel: VCC L/A   V
Boa: novel In The Sign by Joanne Greenberg.

LOVE KILLS **
15
Brian Grant USA
1991
Virginia Madsen, Jim Metzler, Lenny Von Dohlen, Erich Anderson, Kate Hodge, Michael Flyn, David Jensen, Alison Harris, Thor Nielsen, Frank Magner, Michael Ruud, Michael Alvarez
An unhappy wife embarks on an affair with a handsome young man, but learns to her horror that her lover is in fact a paid assassin in the employ of her husband.
THR   84 min (ort 92 min) mTV   VIDrel: CIC/SONOP   V/sur

LOVE LESSONS ***
15
Bo Widerberg DENMARK/SWEDEN
1995
Johan Widerberg, Mariuka Lagercrantz, Tomas von Bromssen, Karin Huldt, Bjorn Kjellman, Nina Gunke, Kenneth Milldoff, Frida Lindholm, Magnus Andersson, Linus Ericsson, Peter Nilsson
One of those teenage rites-of-passage films, that is set in 1943 at a school in Malmo, where Stig (played by the director's son) finds the attractive new teacher awakening yearnings he was formerly largely unaware of. His attentions are not rebuffed,

and as their relationship flourishes, he finds himself clearly out of his depth, but obliged to mature owing to growing outside influences. A most likeable film, slightly stilted, but made with honesty and care.
Aka: LAERERINDEN; LUST OCH FAGRING STOR
DRA   128 min   CINrel

LOVE, LIES AND MURDER ***
18
Robert Markowitz USA
1990
Clancy Brown, John Ashton, Sheryl Lee, Moira Kelly, Ramon Bieri, Kenneth Walsh, Tom Bower, John M. Jackson, Cynthia Nixon, Nestor Serrano, Shelley Morrison, Cathryn De Prume, Caitlin Clarke, McGregor Stewart, L. Scott Caldwell
Rather overlong study of a nasty murder that took place in 1985, when a teenager was convicted of killing her stepmother. None of the sordid details of the youngster's miserable home life are spared, from her devious and disagreeable dad to her creepy stepmother, who also happens to be the younger sister of the girl's dead mother. Not well received by the critics it is, despite its unhappy subject matter, both absorbing and convincing.
DRA   180 min (ort 200 min) mTV   VIDrel: ODY/SONOP   V/sh

LOVE MISTRESS, THE *
18
USA
Hyapatia Lee, Barbara Dare, Rachel Ryan, Champagne, Marc Wallice
The adventures of a group of high-class call girls are amply described here, with one of them even managing to save a client's marriage.
A   60 min (ort 90 min)   VIDrel: MOPIC/SGSVID   V

LOVE ON THE DOLE ***
PG
John Baxter UK
1941
Deborah Kerr, Clifford Evans, George Carney, Joyce Howard, Frank Cellier, Geoffrey Hibbert, Mary Merrall, Maire O'Neill, Marjorie Rhodes, A. Bromley Davenport, Marie Ault, Iris Vandeleur, Kenneth Griffith, B. John Slater
Well-made account of the Depression-era story set in Northern England, where Lancashire cotton workers struggle to make ends meet. A film that eschews a happy ending in favour of a far more convincing and realistic approach. Quite a rare piece of social comment for the time, in many ways it foreshadowed aspects of British film-making in the 1960s.
DRA   94 min (ort 100 min) B/W   VIDrel: FABFIL/SPEAR   V
Boa: novel by Walter Greenwood/play by Ronald Gow.

LOVE ON THE RUN **
15
Francois Truffaut FRANCE
1978
Jean-Pierre Leaud, Marie-France Pisier, Claude Jade, Dani, Rosy Varte, Julien Bertheau, Daniel Mesguich, Dorothee
Final appearance of the character Antoine Doinel who was first seen in THE FOUR-HUNDRED BLOWS, illustrated with shots of all the previous films. Once again, we are treated to a rather tiresome exposition of the central figure's inability to sustain an adult relationship with a woman.
Aka: L'AMOUR EN FUITE
DRA   95 min B/W/Col   VIDrel: ARTIF/20TH   V

LOVE OR MONEY? *
15
Todd Hallowell USA
1990
Kevin McCarthy, Haviland Morris, David Doyle, Timothy Daly, Shelley Fabares, Michael Garin, Allan Harvey, Tisha Roth
A yuppie faces a hard choice when he discovers that he can either get the girl of his dreams, or make a fortune on a real estate deal, but cannot have both at the same time. Simple, trite, and only marginally appealing.
COM   87 min (ort 92 min)   VIDrel: COLUM/SONOP   V

LOVE POTION NO. 9 *
15
Dale Launer USA
1992
Tate Donovan, Sandra Bullock, Dale Midkiff, May Mara, Hilary Bailey Smith, Anne Bancroft, Dylan Baker, Blake Clark, Bruce McCarty, Rebecca Staab, Ric Reitz, Adrian Paul, Jordan Baker, Ken Strong, Gary Watkins, Scott Higgs
A couple of misfit researchers come across a secret gypsy love potion that is said to have remarkable properties and test it out on some of the lab's chimpanzees. Having been greatly encouraged by the results of this, their next step is to test it on some humans volunteers, and they choose themselves for this purpose. A poorly made and dire sex comedy of few laughs.
Aka: LOVE POTION #9
COM   93 min (ort 97 min)   VIDrel: FIRST/SONOP L/A   V

LOVE STORY *** U
Leslie Arliss UK 1944
Margaret Lockwood, Stewart Granger, Patricia Roc, Tom Walls, Reginald Purdell, Moira Lister
WW2 romance set in Cornwall, where a concert pianist with a weak heart falls for a mining engineer who is losing his sight. Casting the female lead as a pianist does seem a little contrived (this romantic device has been used in several other works) but given that, within the context of the film it works well, and there is ample scope for the two leads to show their feelings. Dated and corny it may be, but the film has charm and a certain touching poignancy.
Aka: LADY SURRENDERS, A
DRA 108 min (ort 112 min) B/W VIDrel: CARL/TECH V

LOVE STORY * PG
Arthur Hiller USA 1970
Ryan O'Neal, Ali MacGraw, Ray Milland, John Marley, Katherine Balfour, Tom (Tommy) Lee Jones, Russell Nype, Sydney Walker, Robert Modica, Walker Daniels, John Merensky, Andrew Duncan, Bob O'Connell, Sudie Bond, Milo Boulton
Synthetic weepie that was a huge box-office success in which boy meets girl, boy and girl fall in love, girl dies suddenly of something quite painless that is never mentioned by name. High-grade romantic slush that was followed by a sequel, OLIVER'S STORY in 1978. The script is by Segal. AA: Score/orig (Francis Lai).
DRA 96 min (ort 100 min)
VIDrel: 4-FRONT/POLYREC/CIC V/h
Boa: novelette by Erich Segal.

LOVE STREAMS ** 15
John Cassavetes USA 1984
John Cassavetes, Gena Rowlands, Seymour Castel, Diahnne Abbott, Risa Martha Blewitt, Margaret Abbott, Jakob Shaw
A typical Cassavetes-style talky drama about the trials and tribulations of a couple's relationship at a difficult period in their lives. Not everyone's idea of a fun film but it has won many European film awards. The script is by Cassavetes and Ted Allan.
DRA 135 min (ort 141 min) VIDrel: GUILD/SONOP L/A V
Boa: play by Ted Allan.

LOVE THY NEIGHBOUR * PG
Stuart Allen UK 1973
Jack Smethurst, Kate Williams, Rudolph Walker, Nina Baden-Semper, Bill Fraser, Charles Hyatt, Keith Marsh, Tommy Godfrey, Patricia Hayes, Melvyn Hayes, Azad Ali, Arthur English, Clifford Mollison, Lincoln Webb, Andria Lawrence
Spin-off film from a dire TV series, that followed the exploits of two next-door neighbours, one black and the other white. Comedy (such as it was) arose from the bigoted and racist views of the white man (which wife didn't share). The series worked neither as a comedy nor as a piece of social comment, and this film shares those faults, and adds a few more of its own.
COM 85 min VIDrel: THAMES/DISC L/A V

LOVE WITH A PERFECT STRANGER ** PG
Desmond Davis (uncredited) UK/USA 1986
Marilu Henner, Daniel Massey, Sky Dumont, Stephen Greif, Delia Paton, Tim Stern, Shirin Taylor, Robert Rietty, Phillip O'Brien, Gay Barnes, Tim Price, Rita Lester, Ray Marioni, Bruce Lidington, Anthony Wingate
A beautiful, successful and wealthy American businesswoman travelling by train to Italy, finds love in the shape of a dashing Irishman. First in a series of TV films from Harlequin – purveyors of romantic novels. Written by Pamela Wallace.
DRA 90 min (ort 98 min) mCab VIDrel: CIC/SONOP V
Boa: novel by Pamela Wallace.

LOVEJOY: HIGHLAND FLING *** PG
Francis Megahy UK 1991
Ian McShane, Phyllis Logan, Dudley Sutton, Chris Jury, Simon Ward, Eleanor David, Bill Traver
Lovejoy and Lady Jane travel to Scotland to help an old friend raise money to repair his house. An episode from an amusing comedy drama series that cast McShane as a wily antiques dealer, who was forever in search of the elusive find that would make his fortune. A number of the other episodes in this series have become available.
DRA 98 min mTV VIDrel: VCC/DISC V

LOVELINES * 15
Rod Amateau USA 1985
Greg Bradford, Mary Beth Evans, Michael Winslow, Tammy Taylor, Joanna Lee, Stacey Toten
Rival groups of teenagers behave as only they can in this witless comedy that follows the exploits of rival bands.
COM 93 min VIDrel: 20TH/TECH V/sur

LOVER, THE ** 18
Jean-Jacques Annaud FRANCE/UK 1992
Jane March, Tony Leung, Frederique Meininger, Arnaud Giovaninetti, Melvil Poupard, Lisa Faulkner, Xiem Mang, Raymonde Heudeline, Philippe Le Dem, Jeanne Moreau, Ann Schaufuss, Quach Van An, Tania Torrens, Yvonne Wingerter
The story of an illicit love affair between a poor French girl of fifteen and a wealthy Chinese of thirty-two, that is set in pre-Independence Vietnam of the 1920s. The novel is highly regarded in France (it won the Prix Goncourt in 1984) and on that account, the film has much of its prose, often in the form of a voice-over from Moreau, who plays the central character recollecting the affair many years later. A blend of shallow cliche and sensuous imagery.
Aka: L'AMANT
A 101 min (English version – ort 115 min) cC
VIDrel: 4-FRONT/POLYREC; PION (LV only) V/s LV
Boa: novel L'Amant by Marguerite Duras.

LOVERS, THE *** 15
Louis Malle FRANCE 1958
Jeanne Moreau, Jean-Marc Bory, Alain Cuny, Jose-Luis Villalonga, Judith Magre, Gaston Modot, Patricia Garcin, Claude Mansard, Georgette Lobbe
Bored with both her rich spouse and their dull provincial life, a sexy woman follows a secret life in Paris, but finds a far more meaningful relationship with an overnight guest of her husband's. An elegant and erotic satire, made with much style but little depth, it caused a sensation in its day, and catapulted Moreau to international stardom.
Aka: LES AMANTS
DRA 86 min (ort 90 min) B/W VIDrel: ELPIC/POLYREC V

LOVERS ** 18
Vicente Aranda SPAIN 1991
Victoria Abril, Jorge Sanz, Maribel Verdu
A story of passionate and violent love, set in Madrid during the 1950s. A hardworking young man who is engaged to be married, unwittingly embarks on an affair when he rents a room from a young widow who proves to be a con artist. However, his fiancee reacts to this relationship in a totally unexpected and emotional way.
Aka: AMANTES; LOVERS: A TRUE STORY
DRA 104 min VIDrel: MAINPIC/RTM V

LOVERS AND OTHER STRANGERS **** 15
Cy Howard USA 1969
Gig Young, Bonnie Bedelia, Anne Jackson, Harry Guardino, Michael Brandon, Beatrice Arthur, Richard Castellano, Cloris Leachman, Robert Dishy, Diane Keaton, Anne Meara, Marian Hailey
After having lived together, a young couple get married only to discover that there are still many problems to be faced. A witty, charming and splendidly directed farce, full of memorable moments and great performances. This was Diane Keaton's film debut. AA: Song ("For All We Know") – Fred Karlin (m)/Robb Royer and James Griffin (l)).
COM 91 min (ort 106 min) VIDrel: L/A V
Boa: play by Renee Taylor and Joseph Bologna.

LOW DOWN DIRTY SHAME, A * 18
Keenen Ivory Wayans USA 1994
Keenen Ivory Wayans, Jada Pinkett, Salli Richardson, Charles S. Dutton, Gary Cervantes, Andrew Divoff, Corwin Hawkins, Gregory Sierra, Kim Wayans, Andrew Shaifer, Christopher Spencer, Devin Devasquez, John Capodice, Craig Ryan Ng
In an effort to locate over $20,000,000 in drugs money, the DEA takes the unusual step of hiring the services of a black private eye, who soon learns that he will find it hard to stay alive long enough to fulfil his task. A coarse and noisy action-comedy (and a kind of sequel to I'M GONNA GIT YOU, SUCKA!) that relies almost wholly on stock figures and violent special effects to

sustain its hackneyed plot, which was written by Wayans. The title aptly sums up the film.
Aka: MISTER COOL
COM   97 min (ort 100 min)   VIDrel: HOLPIC/TECH   V

LOWER DEPTHS, THE **   PG
Akira Kurosawa JAPAN   1957
Toshiro Mifune, Isuzu Yamada, Kyoko Kagawa, Bokuzen Hidari, Ganjiro Nakamura, Kyoko Kagawa, Minoru Chiaki, Kamatari Fujiwara, Eijiro Tono, Eiko Miyoshi, Akemi Negishi, Koji Mitsui, Nijiko Kiyokawa, Haruo Tanaka, Yu Fujiki
A flawed attempt to transpose Gorky's powerful treatise on human suffering to Japan resulted in this tale of a peddler on the run from the police who hides out with a group of derelicts. Relentlessly gloomy, this artificial work (which takes place almost exclusively on a single set) fails to fascinate and is sometimes very hard to take.
Aka: DONZOKO
DRA   119 min (ort 125 min) B/W   VIDrel: CONNO/RTM   V
Boa: play by Maxim Gorky.

LOWER LEVEL **   15
Kristine Peterson USA   1991
David Bradley, Elizabeth Gracen, Jeff Yagher, Shari Shattuck, David Sterry
Working late at her office, an attractive young architect learns that she has been stood up by her boyfriend, and strikes up a conversation with the security guard. Having accepted his invitation to join him in the bowels of the building, she finds herself trapped by a psychopathic loner with whom she is forced to play a dangerous cat-and-mouse game.
THR   84 min (ort 88 min)   VIDrel: CIC/SONOP   V

LUCAS ***   15
David Seltzer USA   1986
Corey Haim, Kerri Green, Charlie Sheen, Courtney Thorne-Smith, Winona Ryder, Thomas E. Hodges, Guy Bond, Ciro Poppiti, Jeremy Riven, Kevin Gerard Wixted, Emily Seltzer, Erika Leigh, Annie Ryan, Jason Robert Alderman, Tom Mackie
A tale of teenage love between a fourteen-year-old boy and an older girl of sixteen and how their affair affects those around them. A delightful little drama that has engaging performances and sharply defined but believable characters. The directorial debut for Seltzer.
DRA   96 min (ort 100 min)   VIDrel: 20TH/TECH   V/sur

LUCKY **   15
Buzz Kulik USA   1990
Nicollette Sheridan, Michael Nader, Anne-Marie Johnson, Irizarry, Stephanie Beacham, David McCallum, Eric Braeden, Sandra Bullock, Phil Morris, Tim Ryan, Lenni Golden
The usual Jackie Collins sex-lies-intrigue-revenge concoction, that has a former 1930s booze racketeer moving into the lucrative hotel and gambling business, falling in love and fathering a daughter he names Lucky. His young girl proves to be something of a handful, and various complications and plots unfold against a variety of luxurious settings. A glossy and vacuous soap opera, followed by CHANCES.
DRA   144 min (ort 180 min) mTV
VIDrel: 4-FRONT/POLYREC/ODY   V/sh
Boa: novel by Jackie Collins.

LUCKY JIM **   U
John Boulting UK   1957
Ian Carmichael, Hugh Griffith, Terry-Thomas, Sharon Acker, Jean Anderson, Clive Morton, Maureen Connell, Kenneth Griffith, John Welsh, Jeremy Hawk, Reginald Beckwith, Harry Fowler, John Cairney, Charles Lamb, Ian Wilson
A junior history lecturer at a provincial university staggers from one mishap to another in this mildly enjoyable farce that (unlike the novel) relies on broad humour rather than subtlety.
COM   91 min (ort 95 min) B/W   VIDrel: BRAVE/SONOP   L/A   V
Boa: novel by Kingsley Amis.

LUCKY LUKE *   U
Terence Hill ITALY   1992
Terence Hill, Nancy Morgan, Ron Carey, Dominic Barto, Bo Gray, Fritz Sparberg, Arsenio 'Sonny' Trinidad, Neil Summers, Mark Hardwick, Buff Douthy, Sky Fabin, Marc Mouchet, Robin Westphal, Deborah Mansy, Paula Baz
Title cowboy hero draws faster than his own shadow, a skill

that proves highly useful in outwitting the succession of bad guys with whom he has to contend. A live-action version of the doings of this animation figure that fails to amuse, largely on account of its coarse humour and poor acting. First in a series.
COM   86 min (ort 91 min)   VIDrel: FIRST/SONOP   V

LUCKY TEXAN, THE **   U
Robert North Bradbury USA   1934
John Wayne, Barbara Sheldon, Yakima Canutt, George "Gabby" Hayes, Lloyd Whitlock, Gordon D. (Demain) Woods, Eddie Parker, Earl Dwire, Jack Rockwell, Artie Ortego, Tex Palmer, Tex Phelps, George Morrell
A young man comes West to work his late father's share of a gold claim with the latter's partner but has to contend with claim jumpers. A fairly undemanding but enjoyable effort, one in a series of John Wayne "Lone Star" Westerns, produced by Paul Malvern.
WES   55 min (ort 61 min) B/W coVer   VIDrel: ENTUK L/A   V

LULLABY OF BROADWAY **   U
David Butler USA   1951
Doris Day, Gene Nelson, Billy De Wolfe, Gladys George, Florence Bates, S.Z. Sakall, Anne Triola, Hanley Stafford, Sheldon Jett, Murray Alper, Edith Leslie, Hans Herbert, Herschel Dougherty, Elizabeth Flournoy, Donald Kerr
A woman achieves stardom in a musical comedy but is unaware of the fact that her mother is in urgent need of a helping hand. A competent but uninspired musical, lively but hardly memorable. A few good songs keep it going.
MUS   88 min (ort 92 min)   VIDrel: WHV   V

LUNAR COP **   18
Boaz Davidson   1994
Michael Pare, Billy Drago
Another one of those innumerable futuristic cop fantasies. This one is set in the year 2050 A.D., with planet Earth a blighted wasteland, the wealthy having left for the more congenial surroundings of a lunar colony. When a chemical that could regenerate the Earth is stolen from the colony, a lunar cop is given the task of retrieving it.
FAN   88 min   VIDrel: 20TH/FOXVID   V

LUNATIC, THE ***   15
Lol Creme USA   1990
Paul Campbell, Julie T. Wallace, Carl Bradshaw, Reggie Carter, Winston Stona, Linda Gambrill, Rosemary Murray, Lloyd Reckford, Abe Dabdoub, Tony Hendricks, Michael London, Roy Thomas, Raymond Mair, Ruth Ho Shing
Formerly one half of the rock band 10 CC, Creme makes his feature debut with this amusing tale, set in Jamaica, that follows the adventures of the half-crazed Aloysius, after his encounter with a sex-starved German tourist named Inga. Inga brings some joy into his life, establishing a menage-a-trois with both him and a local butcher, and persuades them to take part in a foolhardy burglary. Scripted by Winkler, this has music by Wally Badarou.
COM   94 min   VIDrel: MANGA/SONOP   V
Boa: novel by Anthony C. Winkler.

LUNATICS ***   15
Josh Becker USA   1991
Theodore Raimi, Deborah Foreman, Bruce Campbell, Brian McCree, Eddie Rosmaya, Michelle Stacey
Two individuals who are both victims of their private nightmares are brought together when one of them makes a phone call that misroutes. One is a former mental patient and poet who fears the world outside his apartment, whilst the other is a girl abandoned by her boyfriend and obsessed by the belief that she causes the death of all those she loves. A touching story of two lonely misfits who find strength and companionship in each other.
Aka: LUNATICS: A LOVE STORY
DRA   83 min (ort 87 min)   VIDrel: 20VIS/SONOP   V

LUPIN 3: THE FUMA CONSPIRACY **   PG
Seijun Suzuki JAPAN   1987
The wife of a samurai is kidnapped by a rival clan, their intention being to use her as a bargaining point to obtain the family treasure of the wronged samurai. One in a series of animations featuring this character. Average.
ANIM   73 min   VIDrel: WESCON/RTM   V

LURKING FEAR ** 18
C. Courtney Joyner USA 1993
Jon Finch, Blake Bailey, Ansley Lauren, Jeffrey Combs, Allison Mackie, Paul Mantee, Joe Leavengood, Vincent Schiavelli
Very loosely based on the Lovecraft tale, this tells of an ex-con who discovers that his dad (who was also a crook) secreted a stash of loot inside a corpse that is buried at the local cemetery. Having gone to the cemetery he gets trapped in the church, as others are also after the loot. Soon, everyone learns that a race of grotesque creature live below the cemetery, and come above ground every now and now. Imaginative and entertaining, but limited by its low budget.
HOR 74 min (ort 78 min) VIDrel: CIC/SONOP V
Boa: short story by H.P. Lovecraft.

LUSH LIFE *** 15
Michael Elias USA 1994
Jeff Goldbaum, Forest Whitaker, Kathy Baker, Tracey Needham, Lois Chiles, Zack Norman, Don Cheadle, Alex Desert, Tom LaGrua, Ron Taylor, Ernie Andrews, Perry Moore, 'Nita Whitaker, Patti Yasutake, Jack Sheldon, Buddy Arnold
Two male musicians share a passion for jazz that eventually proves stronger than anything else in their lives. A moving and detailed exploration of the demands of such devotion to one's art. A trifle overlong, but full of great music and some very winning performances.
DRA 105 min (ort 106 min) VIDrel: ARENA/SPEAR V/sh

LUST *** 18
Joe D'Amato ITALY 1985
Laura Gemser, Martin Philips, Lilli Carati, Al Cliver (Pier Luigi Conti), Annie Bell, Noemie Chelkoff, Ursula Foti, Roberto Caruso
A white hunter returns from Africa with a native girl whom he promptly installs in his home. His wife and servants are none too pleased at first, but soon learn that the girl is very free with her favours – our hunter now finding that he has to get to the back of the queue. Not a bad effort from D'Amato.
Aka: LUSSURIA
A 84 min Cut (1 min 6 sec plus film cuts – ort 88 min) wScrn dubbed VIDrel: JEZ/RTM V

LUST FOR A VAMPIRE ** 18
Jimmy Sangster UK 1970
Ralph Bates, Suzanna Leigh, Michael Johnson, Barbara Jefford, Mike Raven, Yutte Stensgaard, Helen Christie, Pippa Steel, David Healy, Michael Brennan, Jack Melford, Erik Chitty, Luan Peters, Christopher Cunningham
A pupil at a European finishing school for girls is a vampire and wreaks havoc on the pupils. A typical period Hammer film (circa 1830) with the appropriate wooden performances.
Aka: TO LOVE A VAMPIRE
HOR 91 min (ort 95 min) VIDrel: WHV V/h
Boa: novel Carmilla by J. Sheridan Le Fanu.

LUST FOR LIFE **** PG
Vincente Minnelli. USA 1956
Kirk Douglas, Anthony Quinn, James Donald, Pamela Brown, Everett Sloane, Niall MacGinnis, Noel Purcell, Henry Daniel, Lionel Jeffries, Madge Kennedy, Jill Bennett, Laurance Naismith, Eric Portman, Jeanette Sterke, Toni Gerry
A splendid filmisation of Stone's account of the life of Van Gogh, eschewing naturalism in favour of a dramatised approach to plot and character that works beautifully. Quinn gives a strong if unsubtle portrayal of the painter's friend Gaugin, and the love/hate relationship between these two men is well captured. Unsurpassed, despite two films (VINCENT & THEO and VAN GOGH) made in celebration of the 200th anniversary of the artist's birth. AA: S. Actor (Quinn).
DRA 117 min (ort 123 min) VIDrel: WHV V
Boa: novel by Irving Stone.

LUST IN THE DUST ** 15
Paul Bartel USA 1984
Tab Hunter, Divine (Glenn Milstead), Lainie Kazan, Geoffrey Lewis, Henry Silva, Cesar Romero, Woody Strode, Gina Gallegro, Pedro Gonzales-Gonzales, Nedra Volz, Courtney Gains, Daniel Fishman, Al Cantu, Ernie Shinagawa
Spoof Western about hidden gold and sundry fortune seekers, with Divine trying to locate the treasure in order to achieve "her" ambition of becoming a saloon singer. A high-camp celebration of bad taste and general nonsense, but enjoyable in small doses.
COM 81 min (ort 87 min) VIDrel: RCA L/A V/h

LUST IN THE WOODS ** 18
Paul Thomas USA 1989
Hyapatia Lee, Mike Horner, Megan Lee, Joey Silvera, Lola, Tanja de Vries
A loving couple find the romantic weekend they've planned in the woods not turning out as expected, but make the most of it with a couple of unexpected visitors. Average.
Aka: BODY FIRE
A 60 min Cut (13 sec plus some cuts subst – ort 75 min) mVid VIDrel: RAVEN/QUANT V

LUST ON THE ORIENT XPRESS *** 18
Tim McDonald USA 1986
Gina Carrera, John Leslie, Tracy Adams, Eric Edwards, Jamie Gillis, Paul Thomas, Pat Manning, Keli Richards, Andrew Herbert, Jim Woodman, Francois Papillon, Raymond Roberts, F. Powers Gireau, Mike Kypros
A husband and wife team write sexy mystery novels and are being harassed by a publisher for their next book. They take a trip on the Orient Express to recharge their batteries, but in the course of the journey a murder is committed and a priceless diamond goes missing. They set out to probe each suspect in a well made sex film with more than a few twists to the plot.
A 60 min Cut (3 min 51 sec – ort 77 min)
VIDrel: ELV/DISC V

LUSTFUL FEELINGS * 18
Kemal Horulu USA 1973
Leslie Bovel, Jamie Gillis
Deeply in debt, a man persuades his girlfriend to appear in some pornographic home-movies as a way of making some money. An average adult movie, though now looking rather dated in comparison to what is being produced nowadays.
A 61 min (ort 90 min) VIDrel: FIFTH/DISC V

M

M **** PG
Fritz Lang GERMANY 1931
Peter Lorre, Otto Wernicke, Gustav Grundgens, Theodore Loos, Ellen Widmann, Georg John, Inge Landgut, Ernst Stahl-Manchbaur, Paul Kemp, Franz Stein, Rudolf Blumner, Karl Platen
A psychopathic child-murderer is at large in Berlin, and panic ensues until he is caught by the city's criminals, who are repelled by his activities. Lang's first sound film is replete with the expected Expressionist touches, and works both as a classic melodrama and a disturbing social satire. Lorre gives a remarkable performance as the killer, and was rewarded in Hollywood by forever being cast in petty criminal roles. A dated but truly unforgettable work.
DRA 100 min (ort 118 min) B/W VIDrel: REDEM/RTM V

MA SAISON PREFEREE *** 15
Andre Techine FRANCE 1993
Catherine Deneuve, Daniel Auteuil, Marthe Villalonga, Jean-Pierre Bouvier, Chiara Mastroianni, Carmen Chaplin, Anthony Prada, Michele Moretti, Jacques Nolot, Bruno Todeschini, Jean Bousquet, Roschady Zem, Ingrid Caven
A brother and sister find themselves having to care for their elderly and ailing mother, a task which puts them under a great deal of strain. As the seasons pass, the changing nature of the relationship of all three main characters is explored with wit and insight.
DRA 121 min (ort 127 min) VIDrel: ARROW/RTM V/sur

MABOROSI *** (PG)
Hirokazu Koreeda JAPAN 1995
Makiko Esumi, Takeshi Naito, Tadanobu Asano, Goki Kashiyama, Naomi Watanabe, Midori Kiuchi, Akira Emoto, Mursuko Sakura, Hidekazu Akai, Hiromi Ichida, Minori Terada, Ren Osugi, Kikuko Hashimoto, Shuichi Harada, Takashi Inoue
A debut feature from documentary film-maker Koreeda, that tells of a troubled young woman who is convinced that a curse ensures she will always eventually cause the death of those she loves. Having lost her first husband to suicide, she finds a little happiness with another man, but all the while her mind is subject to fear and worry. A most absorbing character study, devoid of both unnecessary sentimentality and plot contrivances.
Aka: MABOROSHI NO HIKARI
DRA 109 min CINrel
Boa: short story by Teru Miyamoto.

MAC ***   18
John Turturro USA   1992
*John Turturro, Ellen Barkin, Katherine Borowitz, John Amos, Carl
Capotorto, Michael Badalucco, Olek Krupa, Joe Paparone, Matthew
Sussman, Steven Randazzo, Nicholas Turturro, Dennis Farina,
Kaluhani Lee, Richard Spare, James Modio*
Beautifully crafted tale, set in the 1950s, as the three sons of an
Italian immigrant struggle to make their own world after their
father's death. One of them is a skilled carpenter who refuses to
cut corners and leaves his boss to start a house construction
company of his own. An impressive directorial debut that is
largely based on autobiographical material.
DRA   113 min (ort 118 min)   VIDrel: EIV/SONOP   V/sur

MACABRE **   18
Lamberto Bava ITALY   1980
*Bernice Stegers, Stanko Molnar, Veronica Zinny, Roberto Posse,
Ferdinand Orlandi, Fernando Pannullo, Elisa Kadiga Bove*
A woman who has lost both her small son and her lover, keeps
the severed head of the latter, which she takes to bed with her
at night. A nosy neighbour hears strange noises coming from her
apartment and decides to investigate.
Aka: FROZEN TERROR; MACABRO
HOR   86 min (ort 90 min)   VIDrel: ARTPRO/RTM   V

MACAO **   PG
Josef Von Sternberg USA   1952
*Robert Mitchum, Jane Russell, William Bendix, Gloria Grahame,
Thomas Gomez, Philip Ahn, Brad Dexter, Edward Ashley, Vladimir
Sokoloff, Don Zelaya, Emory Parnell, Nacho Galindo, George Chan,
Seldon Jett, Genevieve Bell, Tommy Lee*
A singer has a tempestuous affair with a wandering adventurer,
against a background of Oriental intrigue and suspense set in
the port of Macao, and together they set out to capture a gang-
ster wanted in the USA. A disappointing yarn, with stylish
direction but an unimaginative script.
DRA   78 min (ort 81 min) B/W   VIDrel: VCC L/A   V

MacARTHUR **   PG
Joseph Sargent USA   1977
*Gregory Peck, Ed Flanders, Dan O'Herlihy, Sandy Kenyon, Dick
O'Neil, Marj Dusay, Ivan Bonar, Ward Costello, Art Fleming, Russell
D. Johnson, Addison Powell, Robert Mandan, Allan Miller, Dick
O'Neill, Tom Rosqui, G.D. Spradlin*
Biopic tracing the career of this famous general during both
WW2 and the Korean War. Sincere and well crafted but lacking
in dramatic impact.
Aka: MacARTHUR THE REBEL GENERAL
DRA   124 min (ort 144 min)
VIDrel: 4-FRONT/POLYREC/CIC   V/h

MACBETH ***   PG
Orson Welles USA   1948
*Orson Welles, Jeanette Nolan, Dan O'Herlihy, Edgar Barrier, Roddy
McDowall, Robert Coote, Erskine Sanford, Alan Napier, John Dierkes,
Peggy Webber, Lionel Branham, Gus Schilling, Archie Heugly,
Christopher Welles, Morgan Farley*
Welles's version of the famous play is sorely hampered by lack
of money, but is well worth seeing for its sheer inventiveness
and exuberance, as well as its excellent use of interiors to create
a heightened mood, entirely in keeping with the sombre spirit
of the play. The cheap-looking sets and an ill-advised attempt
to present an authentic Scottish accent are unfortunately major
failings, but the movement of the forest is well-handled. An
ambitious, atmospheric failure.
DRA   107 min (ort 111 min) B/W   VIDrel: SECOND/RTM
V
Boa: play by William Shakespeare.

MACBETH ****   15
Roman Polanski UK   1971
*Jon Finch, Francesca Annis, Martin Shaw, Nicholas Selby, John
Stride, Stephan Chase, Paul Shelley, Terence Bayler, Andrew
Laurence, Frank Wylie, Bernard Arcgard, Brian Purchase, Keith
Chegwin, Noel Davis, Elsie Taylor*
Ranks almost equally with Kurosawa's "Throne Of Blood", as
one of the most powerful adaptations of Shakespeare's famous
tragedy, telling of lust for power, murder, treachery and a final
reckoning. Brilliantly acted, this full-blooded rendition has some
graphic scenes of violence that may disturb those used to a more
restrained presentation.
DRA   134 min (ort 140 min)   VIDrel: COLUM/SONOP   V
Boa: play by William Shakespeare.

MACBETH **   PG
Jack Gold UK   1983
*Nicol Williamson, Jane Lapotaire, Tony Doyle, Ian Hogg, Brenda
Bruce, Eileen Way, Anne Dyson, Mark Dignam, James Hazeldine,
Christopher Ellison, John Rowe, Gawn Grainger, David Lyon, Gordon
Kane*
A fairly efficient TV adaptation, a little spoilt by a bland perfor-
mance from Williamson in the title role, that lacks both vigour
and conviction. One of a number of BBC adaptations of
Shakespeare's many plays.
DRA   148 min mTV   VIDrel: BBC   V/h
Boa: play by William Shakespeare.

MACBETH **   12
Jeremy Freeston UK   1997
*Jason Connery, Helen Baxendale, Graham McTavish, Kenny Bryans,
Kern Falconer, Hildegard Neil, Chris Gormlie, Jean Trend, Phillipa
Peak, Iain Stuart Robertson*
With his future foretold by the three witches, Macbeth uses
duplicity to ensure that he be crowned king. But this brings him
no joy, and with his treachery revealed, he does battle with
Macduff and is slain. A terribly bland rendition of Shakespeare's
masterpiece, Connery and Baxendale are so lacking in screen
presence that the majesty of the piece is never realised, and they
are consistently overshadowed by the supporting cast.
DRA   129 min   CINrel
Boa: play by William Shakespeare.

MACHINE, THE *   18
Francois Dupeyron FRANCE/GERMANY   1993
Gerard Depardieu, Nathalie Baye, Didier Bourdon
A psychiatrist who develops a means of recording and trans-
ferring brain patterns, records the brainwaves of a killer, but
during this test he finds his consciousness is transferred into
the body of this man, and the latter walks free inside his body.
Later, the psychiatrist's son undergoes the same experience.
Despite the promise of the idea, having Depardieu act as three
different people renders the film ludicrous, dubbing renders it
unwatchable.
Aka: LA MACHINE
FAN   83 min dubbed   VIDrel: POLFIL   V/s

MACK THE KNIFE *   15
Menahem Golan USA   1989
*Raul Julia, Rachel Robertson, Julie Walters, Richard Harris, Roger
Daltrey, Julia Migenes-Johnson, Clive Revill, Bill Nighy, Erin
Donovan, Elizabeth Seal, Julie T. Wallace, Louise Plowright, Chrissie
Kendall, Miranda Garrison*
Brecht's musical satire on the evils of 19th century capitalism
suffers a mutilation from which it never recovers. Its heavily
stylised format (it is more a filmed review than a proper movie)
and a revamped script (the work of the director) ensure that
none of the flavour of the original remains. Both overblown and
anaemic, its dreadful lyrics and artificial sets stick in the mind,
but are remembered without fondness.
MUS   118 min (ort 122 min)   VIDrel: 20VIS/SONOP L/A
V
Boa: play The Threepenny Opera by Bertold Brecht and Kurt
Weil.

MacKENNA'S GOLD **   15
J. Lee Thompson USA   1969
*Gregory Peck, Omar Sharif, Telly Savalas, Camilla Sparv, Keenan
Wynn, Lee J. Cobb, Julie Newmar, Raymond Massey, Eli Wallach,
Edward G. Robinson, Anthony Quayle, Burgess Meredith, Eduardo
Ciannelli, Rudy Diaz, Ted Cassidy, Duke Hobbie*
The story of a long and difficult hunt for a lost treasure of gold,
confuses both the audience and the searchers, in this dull epic
that was not improved by pre-release cutting. Written by Carl
Foreman with narration by Victor Jory. The score is by Quincy
Jones.
WES   123 min (ort 135 min)   VIDrel: VCC L/A   V
Boa: novel by Will Henry.

MACKINTOSH MAN, THE **   15
John Huston UK   1973
*Paul Newman, James Mason, Harry Andrews, Dominique Sanda, Ian
Bannen, Nigel Patrick, Michael Hordern, Peter Vaughan, Roland
Culver, Percy Herbert, Leo Genn, Robert Lang, Jenny Runacre, John
Bindon, Hugh Manning, Wolfe Morris, Eric Mason*
A complex spy thriller about a government agent deliberately
framed and sent to prison, so that he can eventually expose a
criminal gang led by a powerful politician. A strictly run-of-the-

mill affair that wastes both time and talent. The murky script is by Walter Hill.
DRA   95 min (ort 100 min)   VIDrel: MGM/WHV L/A   V
Boa: novel The Freedom Trap by Desmond Bagley.

MACON COUNTY LINE ** 18
Richard Compton USA   1973
Alan Vint, Cheryl Waters, Max Baer Jr, Jesse Vint, Joan Blackman, Geoffrey Lewis, James Gammon, Leif Garret, Doodles Weaver, Sam Gilman, Timothy Scott, Emile Meyer
Three strangers are chased by the local sheriff in the mistaken belief that they murdered his wife. Allegedly based on a true incident that took place in Georgia in 1954, but really just another re-run for that tired old theme of corrupt power-crazy Southern lawmen doing their own thing. Followed by RETURN TO MACON COUNTY. The script is by Max Baer Jr.
Aka: KILLING TIME
DRA   84 min (ort 90 min)   VIDrel: L/A   V

MacSHAYNE: THE FINAL ROLE OF THE DICE ** (PG)
E.W. Swackhamer USA   1993
Kenny Rogers, Michael Mckean, Daniel Hugh Kelly, Maria Conchita Alonso, Alan Baxter, Bayler Brunmeier, Gwen Gastaldi, Scott Henry, Kelly Junkermann, Lincoln Lageson, Ron Kusiak, Jacqueline Lear, Nick Mazzola, William Novack, Zane Weber
Another outing for Rogers as the Las Vegas casino detective and general troubleshooter.
Aka: FINAL ROLL OF THE DICE, THE
DRA   88 min (ort 92 min) mTV   SATrel: SKY MOVIES

MacSHAYNE: WINNER TAKES ALL ** (PG)
E.W. Swackhamer USA   1993
Kenny Rogers, Terry O'Quinn, Ann Jillian, John Karlen, Wendy Phillips, J.A. Preston, Richard McGonagle, Robert Guy Miranda, Jeff Allin, Debra Jo Rupp, Barry Newman, Stephen Wesley Birdgewater, Ray Buktenica, William Collins, Roy Conrad
Rogers is as plastic as ever in his role as a detective who works at a Las Vegas nightclub, having just been released from prison. As he attempts to locate his ex-wife and son he finds himself being pressured by a retired police chief into helping the latter plan a casino scam. One of several films in the series. Adequate.
Aka: WINNER TAKES ALL
DRA   87 min mTV   SATrel: SKY MOVIES

MAD DOG AND GLORY *** 15
John McNaughton USA   1992
Robert De Niro, Uma Thurman, Bill Murray, Kathy Baker, David Caruso, Mike Starr, Tom Towles, Derek Anunciation, J.J. Johnston,, Richard Belzer, Anthony Cannaya, Doug Hara, Guy Van Searingen, Jack Walalce, Clem Caserta, Fred Squillo
While off duty, a police photographer helps to save the life of a small-time gangster and is rewarded with a pretty girl who is his for one week. Trouble ensues, however, when these two fall for each other and want to stay together. An interesting and offbeat drama, with Murray cast against type as our low-life criminal.
COM   92 min (ort 97 min) cC   VIDrel: CIC/SONOP; PION (LV only)   V/dm V/sur LV

MAD DOGS AND ENGLISHMEN ** 18
Henry Cole UK/USA   1994
Elizabeth Hurley, C. Thomas Howell, Joss Ackland, Claire Bloom, Frederick Treves, Andrew Connolly, Jeremy Brett, Louise Delamere, Chris Adamson, Marcus Bentley, Russ Cane, Cheryl Doll, Nicola Duffett, Alan Freeman, Paula Hamilton
A motorcycle messenger gets involved with a beautiful, upper-class woman who has a severe drug addiction, and through her is drawn into this sleazy world. Meanwhile, a policeman whose daughter is also an addict is hunting the aristocratic drug pusher who is their common source of supply. A nasty and surprisingly repulsive drama, with reasonably competent performances, good direction, but a woeful lack of fresh ideas (even the title is a tired cliche).
THR   93 min (ort 115 min)   VIDrel: EIV/SONOP   V

MAD LOVE ** 15
Antonia Bird USA   1995
Chris O'Donnell, Drew Barrymore, Matthew Lillard, Richard Chaim, Robert Nadir, Joan Allen, Jude Ciccolella, Kevin Dunn, Liev Schreiber, Amy Sakasitz, Matthew Lillard, T.J. Lowther, Elaine Miles, Sharon Collard, Selene H. Virgil
A high-school student in his senior year falls for an attractive girl but the opposition of her parents forces them to elope and

she is later sent to a mental home because of this escapade. He gallantly engineers her escape and they take to road but he learns to his cost that his love is in reality a mentally unstable maniac-depressive. A lightweight road-movie of good performances and weak plotting.
DRA   93 min (ort 96 min) cC   VIDrel: TOUCH/TECH   V

MAD MAX *** 18
George Miller AUSTRALIA   1979
Mel Gibson, Joanne Samuel, Steve Bisley, Hugh Keays-Byrne, Tim Burns, Roger Ward, Geoff Parry, Paul Johnstone, John Ley, Jonathon Hardy, Vince Gil, Sheila Florence, Reg Evans, Stephen Clark, Howard Eynon, John Farndale, Max Fairchild
The weird story of a desolate post-WW3 world, in which the police have their hands full combating roving motorcycle gangs, and one cop seeks revenge for the slaughter of his family. A visually impressive cult favourite that has spawned several sequels and a clutch of imitations.
FAN   89 min Cut (48 sec – ort 100 min) wScrn
VIDrel: WHV   V/h

MAD MAX 2 *** 18
George Miller AUSTRALIA   1981
Mel Gibson, Bruce Spence, Emil Minty, Vernon Wells, Mike Preston, Virginia Hay, Kjell Nilsson, Syd Heylen, Moira Claux, David Slingsby, Max Phipps, Steve J. Spears, Arkie Whitely, William Zappa, Jimmy Brown, David Downer, Guy Norris
Sequel to MAD MAX, with our hero helping a small oil-producing community defend itself against gangs desperate for fuel, in this post-WW3 world. Plenty of incredible car stunts make this a must, even if the script has a good deal less imagination. Followed by MAD MAX 3: BEYOND THE THUNDERDOME.
Aka: ROAD WARRIOR, THE
FAN   91 min (ort 96 min) wScrn   VIDrel: WHV   V/dm V/sur

MAD MAX 3: BEYOND THE THUNDERDOME *** 15
George Miller/George Ogilvie AUSTRALIA   1985
Mel Gibson, Tina Turner, Angelo Rossitto, Helen Buday, Rod Zuanic, Frank Thring, Angry Anderson, Bruce Spence, Robert Grubb, George Spartels, Adam Cockburn, Paul Larrson, Mark Spain, Mark Kounnas, Tom Jennings, Adam Willits
Third in the MAD MAX series, and as ever set in a post WW3 world, where civilisation has broken down, and scattered enclaves struggle for survival. Here the action takes place in Bartertown, where Max arrives in search of some stolen property and has to fight a duel in the Thunderdome arena. With the same formula as the two earlier films, but showing a definite lack of fresh ideas, though Turner does make a most appealing villainess.
FAN   102 min wScrn   VIDrel: WHV   V/dm V/sur

MADAGASCAR SKIN ** (PG)
Chris Newby UK   1995
Bernard Hill, John Hannah, Mark Anthony, Mark Petit, Danny Earl, Robin Neath, Simon Bennett, Matthew Davies, Alex Hooper, Alex Symons-Sutcliffe, Virginia Davies, Susan Harries, George Thomas, Sarah Thomas, William Burke
A young gay man inhibited by the birthmarks on his skin (hence the title) gets drawn into a relationship with a petty crook when he saves the life of the latter. The romance that develops from this serves as an examination of their sexuality, but the limitations of the director's script provide little scope for either drama or comedy. However, as a slightly offbeat character study it does make a few interesting observations if nothing else.
DRA   93 min   CINrel

MADAME BOVARY ** (12)
Vincente Minnelli USA   1949
Jennifer Jones, Van Heflin, Louis Jourdan, James Mason, Christopher Kent (Alf Kjellin), Gene Lockhart, Gladys Cooper, George Zucco, John Abbott, Frank Alleby, Henry Morgan, George Zucco, Ellen Corby, Eduard Franz
Fourth film version of the classic novel has some fine moments but fails to encapsulate all the feeling of the original story of a young woman trapped in a loveless marriage to a dull provincial doctor. Filmed before in 1932, 1934 and 1937 and lavishly remade in 1991. Average.
DRA   114 min B/W   TVrel: C4
Boa: novel by Gustave Flaubert.

MADAME BOVARY *** PG
Claude Chabrol FRANCE   1991
Isabelle Huppert, Jean-Francois Balmer, Christopher Malavoy, Jean

Yanne, Jean-Louis Maury, Florent Gibassier, Jean-Claide Bouillard, Sabeline Campo, Yves Verhoeven, Marie Mergey, Francois Maistre, Thomas Chabrol, Phillipe Abitbal
A young woman marries a dull and unsuccessful country doctor and becomes so bored with provincial life and her lifeless marriage that she begins a series of affairs. But these leave her feeling as unsatisfied as ever, and her mounting extravagance soon brings about a tragic reckoning. A long, complex and very well realised adaptation of Flaubert's novel, which has been filmed several times before.
DRA 136 min VIDrel: ARROW/RTM V/s
Boa: novel by Gustave Flaubert.

M. BUTTERFLY ** 15
David Cronenberg CANADA 1993
Jeremy Irons, John Lone, Barbara Sukowa, Ian Richardson, Vernon Dobtcheff, Annabel Leventon, Shizuko Hoshi, Richard McMillan, Vernon Dobtcheff, David Hemblen, Damir Andrei, Antony Parr, Margaret Ma, Tristram Jellinek, David Neal
In French Indo-China a diplomat has a long affair with a beautiful female opera star that stretches over twenty years, but fails to discover that his lover is not only a spy, but a man as well. Based on a true story, this bizarre tale was a great hit on the stage, where it provided a fascinating look at the nature of sexual obsession. Sadly, it fails to work as a film, due to flat direction and the casting of Lone as the star, who looks far too masculine.
Aka: MADAME BUTTERFLY
DRA 96 min (ort 101 min) cC VIDrel: WHV V/sur
Boa: play by David Henry Hwang.

MADAME SIN *** PG
David Greene USA 1971
Bette Davis, Robert Wagner, Roy Kinnear, Paul Maxwell, Denholm Elliott, Gordon Jackson, Dudley Stratton, Pik-Sen Lim, David Healy, Alan Dobie, Al Mancini, Charles Lloyd Pack, Arnold Diamond, Frank Middlemass, Burt Kwouk
An evil female oriental genius plots to take over the world, using an ex-CIA agent in order to gain control of a Polaris submarine. An entertaining pilot for a proposed series, well worth seeing for Davis's performance.
THR 86 min mTV VIDrel: L/A V

MADAME SOUSATZKA *** 15
John Schlesinger UK 1988
Shirley MacLaine, Peggy Ashcroft, Navin Chowdray, Leigh Lawson, Twiggy (Lesley Hornby), Shabana Azmi, Geoffrey Bayldon, Lee Montague, Robert Rietty, Sam Howard, Jeremy Sinden, Christopher Adey, Barry Douglas
The story of an eccentric, ageing Russian piano teacher, who takes on a fifteen-year-old Indian boy as a pupil, and is drawn into an obsessive desire to both teach him music and instruct him in life. Extremely well made, but MacLaine's role, calling for her to wear a ludicrous amount of make-up and jewellery, has an air of absurdity about it.
DRA 116 min (ort 121 min) VIDrel: VISVID/POLYREC V/sur

MADDENING, THE ** 18
Danny Huston USA 1996
Burt Reynolds, Angie Dickinson, Mia Sara, Briam Wimmer, Josh Mostel, William Hickey
A man who owns a gas station picks up a mother and daughter after they are stranded when their car breaks down, but what they do not realise is that this individual is seriously disturbed, and is prepared to kill in order to preserve his family's dark secrets. An unappetising tale, which was released directly to video.
HOR 92 min (ort 97 min) VIDrel: EIV/SONOP V

MADE FOR EACH OTHER *** PG
John Cromwell USA 1938
James Stewart, Carole Lombard, Lucile Watson, Charles Coburn, Alma Kruger, Ward Bond, Harry Davenport, Eddie Quillan, Esther Dale, Louise Beavers, Alma Kruger, Fred Fuller, Edwin Maxwell, Harry Depp, Michael Rentschler
A touching and skilful romantic examination of the married life of a young couple, and the pitfalls that lie ahead as they struggle with babies, poverty and interfering in-laws. Written by Jo Swerling.
DRA 90 min (ort 95 min) B/W VIDrel: SECOND/RTM V

MADE IN AMERICA ** 15
Richard Benjamin USA 1992
Whoopi Goldberg, Ted Danson, Nia Long, Will Smith, Jennifer Tilly,

Peggy Rea, Clyde Kusatsu, Lu Leonard, Paul Rodriguez, David Bowe, Jeffrey Joseph, Fred Mancuso, Rawley Valverde, Charlene Fernetz, Shaun Lew, Joe Lerer, Janice Edwards
The daughter of a black single mother learns not only that she is the result of artificial insemination but also that her father is apparently a brash car salesman who happens to be white. Goldberg and Danson (her former boyfriend) work well together, but this daft, one-joke movie wastes their comic talents. The revelation that our man did not in fact father a black child is just the kind of cop-out ending one would expect from Hollywood. See also CARBON COPY.
COM 106 min (ort 111 min) cC VIDrel: WHV V/sur

MADE IN BRITAIN ** 18
Alan Clarke UK 1982
Tim Roth, Terry Richards, Bill Stewart, Eric Richard, Geoffrey Hutchings, Sean Chapman, John Bleasedale, Noel Diamond, Maurice Quick, Sharon Courtney, Kim Benson, Catherine Clarke, Jean Marlowe, Jim Dunk, Vass Anderson
Story of an anti-social skinhead and the efforts made by social services to rehabilitate him. Roth's disturbing portrait of a Nazi-clad malcontent is the best thing in this rather simplistic attempt to analyse the beliefs and motivations of a racist skinhead. This was one of four TV plays written by Leland (the others were "Birth Of A Nation", "Flying Into The Wind" and "Rhino") each of which explored the problems of education in Britain.
DRA 72 min (ort 78 min) mTV VIDrel: IMAG/RTM V
Boa: play by David Leland.

MADE IN HEAVEN ** PG
Alan Rudolph USA 1987
Timothy Hutton, Kelly McGillis, Maureen Stapleton, Don Murray, Anne Wedgeworth, Mare Winningham, Amanda Plummer, Ellen Barkin, Debra Winger, Maej Dusay, Ray Gideon, Timothy Daly, Neil Young, Tom Petty, Ric Ocasek
After a man has accidentally cut short his life, he and a woman "newly created" fall madly in love. When she is sent to Earth to live her first life, the man follows, but has just thirty years in which to find her. A sugary romantic comedy spoilt by too many gags and a reliance on funny cameos instead of a funny script.
COM 98 min (ort 103 min) VIDrel: GUILD/SONOP L/A V

MADHOUSE ** 18
James Clark UK 1974
Vincent Price, Peter Cushing, Adrienne Corri, Robert Quarry, Linda Hayden, Natasha Pyne, Michael Parkinson, Barry Dennen, Ellis Dayle, Catherine Wilmer, John Garrie, Ian Thompson, Jenny Lee Wright, Julie Crosthwaite, Peter Halliday
A mad horror film star comes out of retirement to take revenge for the slaying of his girlfriend, and is soon implicated in a series of grisly murders, in this entertaining but uneven yarn. A few touches of grim humour help it along somewhat.
Aka: REVENGE OF DOCTOR DEATH, THE
HOR 88 min (ort 92 min) VIDrel: RCA L/A V
Boa: novel Devilday by Angus Hall.

MADHOUSE * 15
Tom Ropelewski USA 1990
John Larroquette, Kirstie Alley, Alison La Placa, John Diehl, Jessica Lundy, Bradley Gregg, Dennis Miller, Robert Ginty, Wayne Tippit, Paul Eiding, Aeryk Egan, Deborah Otto, Mark Bringelson, Karen Krownwell, Heather McNair
Yuppie couple Larroquette and Alley are unable to enjoy the pleasures of their new home, which is suddenly invaded by dreadful unexpected guests, grumbling relatives, and some very peculiar neighbours, none of whom want to leave. The gags take forever to reach their conclusion, the characters are simplistic and the plot (the work of the director) is almost non-existent. This is essentially a one-joke film.
COM 86 min (ort 99 min) VIDrel: COLUM/SONOP L/A V

MADNESS OF KING GEORGE, THE *** PG
Nicholas Hytner UK/USA 1994
Nigel Hawthorne, Helen Mirren, Ian Holm, Amanda Donohoe, Rupert Graves, Rupert Everett, Julian Rhind-Tutt, Julian Wadham, Jim Carter, Geoffrey Palmer, Charlotte Curley, Anthony Calf, Matthew Lloyd Davies, Adrian Scarborough
Towards the latter part of his reign, King George III began to suffer bouts of insanity (possibly due to the blood disorder poryphia) which affected his personality and incapacitated his mind. The resultant power vacuum led to much political

intrigue in which the Prince of Wales played no small part. Scripted by Bennett, this vivid and moving drama says much about medical ignorance and court life in 18th century England. AA: Art (Ken Adam).

Aka: MADNESS OF GEORGE III, THE

DRA   105 min (ort 110 min) wScrn cC

VIDrel: COLUM/SONOP; ENCORE (LV only)   V/sur LV

Boa: play The Madness of George III by Alan Bennett.

MADNESS OF THE HEART *
Charles Bennett UK
U
1949

Margaret Lockwood, Paul Dupuis, Kathleen Byron, Maxwell Reed, Thora Hird, Raymond Lovell, Maurice Denham, David Hutcheson, Cathleen Nesbitt, Peter Illing, John McNaughton, Pamela Stirling, Marie Burke, Marie Ault, Kynaston Reeves

A blind girl weds the son of a French aristocrat and has to contend with the hatred and machinations of a bitter rival. A rather thinly plotted variant on a theme done much better in REBECCA. This unconvincing dud did nothing to revive Lockwood's flagging film career.

DRA   105 min B/W   VIDrel: CARL/TECH   V

Boa: novel by Flora Sandstrom.

MADONNA: INNOCENCE LOST **
Bradford May USA
15
1995

Terumi Matthews, Wendie Malick, Jeff Yagher, Diana LeBlanc, Dean Stockwell, Nigel Bennett, Dominique Briand, Don Francks, Tom Melissis, Christian Vidosa, Rod Wilson, Kenner Ames, Dino Bellesario, Leo Burns, Mischke Butler, Gil Filar

A by-the-numbers biopic about Madonna's struggle to achieve fame and fortune that follows her career from her days as an impoverished performer in New York to the release of her first album. Of real interest only to hard-core fans, this bland and tedious film will tell us nothing of interest (assuming that we want to know). However, Matthews is perfectly cast as the singer.

DRA   87 min (ort 90 min) mTV   VIDrel: 20TH/FOXVID V/sh

Boa: book Madonna: Unauthorized by Christopher Andersen.

MADONNA OF THE SEVEN MOONS **
Arthur Crabtree UK
PG
1944

Stewart Granger, Phyllis Calvert, Patrica Roc, Peter Glenville, John Stuart, Jean Kent, Nancy Price, Peter Murray Hill, Reginald Tate, Dulcie Gray, Amy Veness, Hilda Bayley, Alan Haines, Evelyn Darvell, Danny Green, Eliot Markeham

Raped in her youth by a gypsy, the wife of a wealthy Italian wine merchant develops a split personality. This shows itself in long absences during which she is only aware of herself as the mistress of a thief. An overblown, romantic melodrama that never rises above the mediocre.

DRA   88 min (ort 110 min) B/W   VIDrel: CONNO/RTM L/A   V

Boa: novel by Margery Lawrence

MAGEE AND THE LADY **
Gene Levitt AUSTRALIA/USA
(12)
1978

Tony LoBianco, Sally Kellerman, Ann Semler, Rod Mullinnar, Kevin Leslie

A man fights off a takeover bid for his freighter, by kidnapping the spoilt daughter of a businessman so as to force a meeting. Their mutual hostility slowly turns to affection as they fight off various perils together. A standard romantic adventure, with contrived dialogue and a wholly predictable outcome. Now turn to THE AFRICAN QUEEN to see how it should have been done.

Aka: MAGEE; SHE'LL BE SWEET

A/AD   100 min mTV   TVrel

MAGENTA *
USA
18
1996

Julian McMahon, Crystal Adams

A successful doctor is lured into the seedy world of drug taking and sexual shenanigans by his envious friends. A forgettable erotic drama.

DRA   90 min   VIDrel: MARQ   V

MAGIC ***
Richard Attenborough USA
15
1978

Anthony Hopkins, Ann-Margret, Burgess Meredith, Ed Lauter, Jerry Houser, E.J. Andre, David Ogden Stiers, Lillian Randolph, Joe Lowry, I.W. Klein, Beverly Sanders, Stephen Hart, Patrick McCullough, Bob Hackman, Mary Munday

Excellent and gripping tale of a ventriloquist whose attachment to his dummy becomes an unhealthy dependence, to the point where he is driven "by the dummy" to murder. Hopkins gives a truly remarkable performance (for which he had to learn the skills of a ventriloquist). Screenplay is by Goldman.

DRA   98 min (ort 106 min)   VIDrel: 4-FRONT/POLYREC L/A   V

Boa: novel by William Goldman.

MAGIC ADVENTURE, THE **
Cruz Delgado SPAIN
(U)
1973

A lively animated fantasy inspired by some of the tales of Hans Christian Andersen, with two children being taken to a magical land and enjoying a series of adventures. Fair.

Aka: HANS CHRISTIAN ANDERSEN'S MAGIC ADVENTURE; MAGICA AVENTURA

ANIM   67 min (ort 90 min)   SATrel: MOVIE CHANNEL

MAGIC BOX, THE ***
John Boulting UK
U
1951

Robert Donat, Maria Schell, Margaret Johnston, John Howard Davies, Renee Asherson, Richard Attenborough, Robert Beatty, Michael Denison, Leo Genn, Joyce Grenfell, Marius Goring, Robertson Hare, Kathleen Harrison

Made for the Festival of Britain, and with a host of guest stars, this is devoted to the life and career of William Friese-Greene, a British portrait photographer who helped pioneer the development of the cine camera but died forgotten and in poverty. A careful and expensive biopic, in which sincerity replaces inventiveness and a series of vignettes replace the narrative.

DRA   103 min (ort 118 min)   VIDrel: WHV   V

Boa: book Friese-Greene by Ray Allister.

MAGIC BUBBLE, THE ***
Alfredo Ringel/Deborah Taper Ringel USA
(PG)
1993

Diane Salinger, John Calvin, Colleen Camp, Priscilla Pointer, George Clooney, Shera Danese, Nicholas Guest, Anthony Peck, Don Diamont, Michael Greene, Dayle Haddon, Lyndsay Riddell, Loyda Ramos, Michael Boatman, Bill Erwin

On her fortieth birthday, a woman is depressed by the prospect of growing old but a seemingly magic potion restores her lust for life, in this unusual and whimsical offering.

COM   88 min (ort 90 min)   SATrel: MOVIE CHANNEL

MAGIC COP **
David Lain (Li Da-Wei) HONG KONG
PG
1989

Lam Ching Ying, Wilson Lam, Billy Chow, Michiko Nishiwaki

An ex-cop rejoins the force and discovers that an international drugs syndicate is linked to supernatural manifestations involving ghosts and vampires.

Aka: QU MO JING CHAN

MAR   90 min   VIDrel: EAST/DISC   V

MAGIC CRYSTAL, THE **
Jacques Rissi/Richard Norton HONG KONG
15
1988

Ricky Chan, Cindy (Cynthia) Rothrock, Ivan, Richard Norton

Our hero is employed by the Hong Kong police, and is asked to go to Greece by a friend who has learnt of an ancient gem with magical powers (it is in fact an alien lifeform). When the friend vanishes, a KGB officer and a couple of Interpol agents appear to be involved. The inevitable violent resolution takes place.

A/AD   91 min (Cut at UK cinema release by 16 sec) dubbed VIDrel: ONE/IMPENT   V

MAGIC IN THE MIRROR: FOWL PLAY **
Ted Nicolaou USA
U
1995

Jaime Renee Smith, Kevin Wixted, Saxon Trainor, David Brooks, Godfrey James, Eileen T'Kaye

Mirrors are doorways to another world, but they are guarded by "Mirror Minders". However, when one of these creatures falls asleep, a little girl inadvertently enters this magical other world, her doorway being an antique mirror bequeathed to her by her granny. Once there, she experiences encounters with some bizarre, unfriendly talking ducks. A flat "Alice Through The Looking Glass" variant, nothing like as exciting or imaginative as it might have been.

FAN   83 min   VIDrel: CIC/SONOP   V

MAGIC IN THE WATER, THE **
Rick Stevenson USA
PG
1995

Rick Harmon, Harley Jane Kozak, Joshua Jackson, Sarah Lisette Wayne, David Rasche, Sotonoma Salsedo, Willie Mark-Om

An overworked psychiatrist, totally unaware of how neglected his family feels, takes them on a vacation to a small community.

There they learn that a legendary monster is said to inhabit a local lake and their gradual involvement with this creature helps to turn their domestic situation around. A sentimental, mushy and generally ill-conceived attempt at wholesome family entertainment.
Aka: GLENORKY
DRA   96 min (ort 101 min)   VIDrel: COLUM/SONOP   V

MAGIC KID 2, THE **
Stephen Furst USA
(PG)
1992
Stephen Furst, Ted Jan Roberts, Dana Barron, Don Gibb, Hugo Napier, Jennifer Savidge, Susan Angelo, Sebastian White, Jeff Rector, Michael Mitz, Wil Shriner, William Daniels, Howie Mandel, David Morse, Allyce Beasley, Shevonne Durkin
A young martial arts fighter is all set for Hollywood stardom after appearing in his own movie until he gets disillusioned and decides to quit the business. But the promoter of the show in which he appears is determined to hang onto him at all costs. A slightly amusing satire on Hollywood and its phoney values, perfectly watchable but not very original.
Aka: LITTLE NINJA DRAGON, THE
COM   86 min (ort 90 min)   SATrel: SKY MOVIES

MAGIC RIDDLE, THE *
Yoran Gross/Junko Aoyama/Sue Beak/Nobuko Burnfield
AUSTRALIA
U
1991
Voices of: Robyn Moore, Keith Scott, Rod Hay
An awkward blend of CINDERELLA, PINOCCHIO, "Snow White And The Seven Dwarfs" and Little Red Hiding Hood, in which our chief character, Cindy, is the maid-of-all-work to her ugly half-sisters and wicked step-mother. Though her step-mother is determined to do her out of her rightful inheritance, she gains several allies and overcomes various trials to win through, and even gets to marry a handsome prince. A disorganised and mind-numbing mess.
ANIM   90 min (ort 94 min)   VIDrel: VCC/DISC   V/sur
Boa: short stories by Hans Christian Andersen, Carlo Lorenzini and The Brothers Grimm.

MAGICIAN, THE ***
Ingmar Bergman SWEDEN
18
1958
Max Von Sydow, Ingrid Thulin, Gunnar Bjornstrand, Naima Wifstrand, Bibi Andersson, Lars Ekborg, Gertrud Fridh, Toivo Pawlo, Erland Josephson, Oscar Ljung, Sif Ruud, Ake Fridell, Ulla Sjoblom, Axel Duberg, Brigitta Pettersson
A well-made exercise in typical Bergman introspection, with an illusionist and mesmerist taking a grim revenge when he is apparently unmasked as a fake. Though enjoyable and mystifying, the film is spoilt by the use of a clumsy deus ex machina resolution which has our magician being summoned to Stockholm where he is to give a show for the Swedish royal court. However, for all its faults it still won the Special Jury Prize at the Venice Film Festival 1959.
Aka: ANSIKTET; FACE, THE
DRA   100 min (ort 103 min) B/W   VIDrel: TART/20TH   V/dm

MAGICIAN OF LUBLIN, THE *
Menahem Golan ISRAEL/WEST GERMANY
15
1979
Alan Arkin, Louise Fletcher, Lou Jacobi, Valerie Perrine, Shelley Winters, Warren Berlinger, Shai K. Ophir, Lisa Whelchel, Maia Danziger, Linda Bernstein, Lachi Nov, Friedrich Schonfeder, Ophelia Stral, Buddy Elias
Film version of a story about a magician with a strange power over women, that he uses to reach the top of Warsaw society at the turn of the century. A slack and ponderous rendition of Singer's novel, with a few clever moments but a general air of apathy.
DRA   101 min (ort 114 min)   VIDrel: RNK L/A   V
Boa: novel by Isaac Bashevis Singer.

MAGNET, THE **
Charles Frend UK
U
1950
William (James) Fox, Kay Walsh, Stephen Murray, Meredith Edwards, Thora Hird, Gladys Henson, Wylie Watson, Julien Mitchell, Michael Brooke Jr, Keith Robinson, Thomas Johnson, Bryan Michie, James Robertson Justice, Joss Ambler
An eleven-year-old boy steals a magnet and by a set of accidental circumstances becomes a hero. An odd little kids' comedy, quite endearing in an undemanding way. Shot on Merseyside, a now much-changed location that adds a measure of interest to the thin story.
COM   74 min (ort 79 min) B/W   VIDrel: WHV   V/h

MAGNIFICENT AMBERSONS, THE ****
Orson Welles USA
U
1942
Tim Holt, Joseph Cotten, Dolores Costello, Anne Baxter, Agnes Moorehead, Ray Collins, Richard Bennett, Erskine Sanford, Tim Holt, J. Louis Johnson, Charles Phipps, Don Dillaway, Dorothy Vaughn, Elmer Jerome, John Elliott
This film might have been Welles' masterpiece but was cut to pieces by the studio. It tells of a family unable to come to terms with a changing world, and of a mother and son conflict over her lover. Many of the Welles touches are still there but re-cutting and re-shooting by the studio all but ruined it. In the States a laserdisc version with running commentary by Robert Carringer became available.
DRA   88 min   VIDrel: VCC L/A   V
Boa: novel by Booth Tarkington.

MAGNIFICENT SEVEN, THE ***
John Sturges USA
PG
1960
Yul Brynner, Steve McQueen, James Coburn, Eli Wallach, Horst Buchholtz, Charles Bronson, Robert Vaughn, Brad Dexter, Vladimir Sokoloff, Rosenda Monteros, Jorge Martinez de Hoyos, Whit Bissell, Val Avery, Bing Russell
Seven gunfighters come together, when their services are required by a poor Mexican village subject to periodic raids by bandits. A direct steal from Kurosawa's brilliant film THE SEVEN SAMURAI, yet quite enjoyable for all that, and with a memorable score by Elmer Bernstein. Followed by a spate of sequels, beginning with "Return Of The Seven", then "Guns Of The Magnificent Seven" and finally THE MAGNIFICENT SEVEN RIDE!
WES   123 min (ort 138 min) wScrn   VIDrel: MGM/WHV   V/h

MAGNIFICENT SEVEN DEADLY SINS, THE *
Graham Stark UK
PG
1971
Bruce Forsyth, Joan Sims, Roy Hudd, Harry Secombe, June Whitfield, Julie Ege, Leslie Phillips, Harry H. Corbett, Cheryl Kennedy, Ian Carmichael, Alfie Bass, Spike Milligan, Ronald Fraser, Arthur Howard, Stephen Lewis
A compilation of comedy sketches of variable quality, not one of which is memorable, but all of which attempt to highlight a legendary human failing. The sins examined are: avarice, envy, gluttony, lust, pride, sloth and wrath.
COM   103 min (ort 107 min)   VIDrel: FABFIL/SPEAR   V

MAGNIFICENT SEVEN RIDE!, THE **
George McGowan USA
PG
1972
Lee Van Cleef, Stefanie Powers, Michael Callan, Pedro Armendariz Jr, Luke Askew, Mariette Hartley
Third and final sequel to THE MAGNIFICENT SEVEN, that has Lee Van Cleef now working as a small-town marshal. When his wife is kidnapped, he releases a bunch of convicts from jail in return for their help in tracking down the kidnapper. A low-budget follow-up to a classic Western, with a tired script and a distinct lack of vigour. Van Cleef is a poor substitute for Brynner.
WES   96 min (ort 100 min)   VIDrel: MGM/WHV   V/h

MAGNUM FORCE **
Ted Post USA
18
1973
Clint Eastwood, David Soul, Mitchell Ryan, Hal Holbrook, Felton Perry, Tim Matheson, Robert Urich, Kip Niven, Christine White, Adele Yoshioka, Albert Popwell, Margaret Avery
A police inspector in San Francisco has to find out who is responsible for a wave of gangster killings. Second in the DIRTY HARRY series with Eastwood as mean and tough as ever but the script showing a distinct lack of imagination. Written by John Milius and Michael Cimino. Followed by THE ENFORCER.
DRA   116 min (ort 124 min)   VIDrel: WHV L/A   V

MAHABHARATA, THE ***
Peter Brook UK
U
1990
Robert Langton-Lloyd, Bruce Myers, Vittorio Mezzogiorno, Antonin Stahly-Vishwandam, Andrzej Seweryn, Mamdou Dioume, Jean-Paul Denizon, Mahmoud Tabrizi-Zadeh
Lavishly staged and visually impressive staging of an ancient Hindu epic detailing the conflict between powerful clans that results in a disastrous war. Unfortunately, the use of different non-native English-speaking actors tends to diminish the dramatic impact of the script. The three cassettes are entitled: "The Game Of Dice", "Exile In The Forest" and "The War".
DRA   313 min (3 cassettes – ort 360 min)
VIDrel: CONNO/RTM   V
Boa: epic poem by Vyasa.

MAHLER ** 15
Ken Russell UK 1974
Robert Powell, Georgina Hale, Richard Morant, Lee Montague, Rosalie Crutchley, Antonia Ellis, Benny Lee, David Collings, Ronald Pickup, Ken Colley, Arnold Yarrow, Dana Gillespie, Elaine Delmar, Michael Southgate, Otto Diamant
Here we go, it's Ken Russell time and we're in for another wild, overblown extravaganza in which the music of a great composer once more serves as a vehicle for Russell's fantasies. Looks pretty dated now (it looked pretty dated at the time).
DRA 111 min (ort 115 min) VIDrel: ODY/SONOP V/sh

MAID, THE ** PG
Ian Toynton FRANCE 1990
Martin Sheen, Jacqueline Bisset, Jean-Pierre Cassel, Victoria Shalet, James Faulkner, Isabelle Guiard, Dominic Gould, Catherine Lachens, Carina Barone, Joe Cosgrave, Jerry Di Giacomo, Jean Martin, Philippe Dehesdin, Bela Grushka
Unconvincing and implausible comedy about a successful Wall Street financier who is posted to Paris, where he becomes smitten by a beautiful woman, and spends a month between jobs working as her maid, as this offers a useful way of getting to know her. Despite the weakness of the script, the film offers a few moments of charm and some amusing complications.
COM 87 min (ort 91 min) VIDrel: BUENA/SONOP L/A V

MAID TO ORDER ** 15
Amy Jones USA 1987
Ally Sheedy, Beverly D'Angelo, Michael Ontkean, Valerie Perrine, Dick Shawn, Tom Skerritt, Merry Clayton, Begona Plaza, Rainbow Phoenix
A spoilt rich girl is transformed into a penniless brat, when her father wishes that he did not have a daughter. What's worse, she forced to work as a maid for a couple just as rich and insufferable as she used to be. A cute fairytale, with lively performances and a couple of nice songs too.
COM 89 min (ort 96 min) mTV VIDrel: 20TH/TECH V

MAIGRET ** 15
Paul Lynch UK 1988
Richard Harris, Patrick O'Neal, Victoria Tennant, Barbara Shelley, Don Henderson, Ian Ogilvy, Eric Deacon, Caroline Munro, Andrew McCulloch, Dominique Barnes, Richard Durden, Annette Andre, Mark Audley, Eve Ferret
A Parisian policeman searches for the killer of his best friend, who worked as a private eye. This pilot for a prospective series based on Georges Simenon's famous creation fails to convince (despite its locations) largely on account of the inability of the cast to breathe life into the characters they portray or give them realistic accents.
DRA 95 min (ort 100 min) mTV VIDrel: VCC L/A V

MAIN EVENT, THE * 15
Howard Zieff USA 1979
Barbra Streisand, Ryan O'Neal, Paul Sand, Whitman Mayo, James Gregory, Richard Lawson, Patti D'Arbanville, Chu Chu Malave, Richard Altman, Lindsay Bloom, Joe Amsler, Seth Banks, Earl Boen, Roger Bowen, Badja Medu Djola
A businesswoman finds herself almost totally bankrupt, with no assets but a broken-down boxer about to retire. To save her perfume business, she becomes his manager and tries to force him to make a comeback. Poorly scripted and indifferently acted, this comedy is an unworthy vehicle for the talents of all concerned.
COM 105 min (ort 112 min) VIDrel: WHV V

MAJOR DUNDEE ** PG
Sam Peckinpah USA 1965
Charlton Heston, Richard Harris, Jim Hutton, James Coburn, Michael Anderson Jr, Senta Berger, Slim Pickens, Mario Adorf, Brock Peters, Warren Oates, Ben Johnson, R.G. Armstrong, L.Q. Jones, Karl Swenson, Michael Pate
An assorted bunch of Confederate soldiers fight Apaches and antagonise the French in Mexico. Peckinpah disowned the film which was re-cut by others, but it is certainly better than one might have expected, and the excellent cast and spectacular battles more than compensate for the lack of warmth or coherence. Written by Harry Julian Fink, Oscar Saul and Peckinpah.
WES 117 min (ort 134 min) VIDrel: RCA L/A V

MAJOR LEAGUE ** 15
David S. Ward USA 1989
Tom Berenger, Charlie Sheen, Corbin Bernsen, Margaret Whitton,
James Gammon, Rene Russo, Bob Uecker, Wesley Snipes, Charles Cyphers, Chelcie Ross, Dennis Haysbert, Andy Romano, Steve Yeager, Peter Vuckovich, Bill Leff
A low-brow comedy about a no-hope baseball team of misfits and oddball characters, that is groomed for stardom by their new lady boss, who has inherited the team. Strident and rather crude, but the attractive cast make the most of their thin material.
COM 102 min (ort 107 min)
VIDrel: 4-FRONT/POLYREC/BRAVE V/sur

MAJOR LEAGUE 2 * PG
David S. Ward USA 1994
Charlie Sheen, Tom Berenger, Corbin Berensen, Omar Epps, Dennis Haysbert, David Keith, Bob Uecker, Randy Quaid, James Gammon, Alison Doody, Michelle Burke, Margaret Whitton, Eric Bruskotter, Takaaki Ishibashi, Skip Griparis
Very much a standard sequel with some of the same actors and much the same antics and episodic tale (single-season baseball champs "The Cleveland Indians" are back for an encore as one of the worst teams in baseball history). It's all a lot less funny the second time around.
COM 100 min (ort 105 min) cC VIDrel: WHV V/sur

MAJOR PAYNE ** 12
Nick Castle USA 1995
Damon Wayans, Karyn Parsons, Michael Ironside, William Hickey, Albert Hall
A former US Marine takes a job at a military academy where he has to take a bunch of teenage misfits and turn them into a crack cadet unit that will win some imminent military games. A familiar story told in a familiar way, this is nothing more than a clumsy remake of the 1955 comedy "The Private War Of Major Payne". The childish slapstick antics grow increasingly tiresome, but fortunately there are one or two lighter moments.
COM 93 min (ort 98 min) cC VIDrel: CIC/SONOP V

MAJORITY RULE ** PG
Gwen Arner USA 1993
Blair Brown, John Getz, John Glover, Jensen Daggett, Robin Gammell, James Handy, Richard Herd, Nicholas Pryor, Mitchell Ryan, Fran Bennett, Tim Conlon, Paul Gleason, Rene Levant, Miguel Sandoval, Donald Moffat, Michael Gregory
A female general embarks on a political career, her aim being to become the USA's first female President, but does not reckon with the opposition and deceit she will encounter along the way. Brown is well cast in the lead role, and this movie provides a few fascinating glimpses of the American political scene, even if the script is really too lightweight to do the intriguing premise justice.
DRA 90 min mTV VIDrel: NEWAGE/20VIS L/A V

MAKING A CASE FOR MURDER: THE HOWARD BEACH STORY * 15
Dick Lowry USA 1989
Daniel J. Travanti, William Daniels, Joe Morton, Cliff Gorman, Bruce Young, Dan Lauria, Regina Taylor, Bill Cwikowski, Gerry Becker, Kurt Naebig, Johnny O'Donnel, Michael Nathaniael Robinson, Anthony Russel Jr, Fred Stone
A state prosecutor faces an uphill struggle when he tries to solve the murder of a black man in a white working-class district. A straightforward account of the notorious murder that took place at Queens, New York City in 1986. Fair.
Aka: HOWARD BEACH: MAKING A CASE FOR MURDER
DRA 94 min (ort 100 min) mTV VIDrel: MOPIC/SGSVID V

MAKING MR RIGHT * 15
Susan Seidelman USA 1987
John Malkovich, Ann Magnuson, Glenne Headly, Ben Masters, Laurie Metcalf, Polly Bergen, Harsh Nayyar, Hart Bochner, Susan Berman, Polly Drapes, Sidney Arbus, Christiane Clemenson, Sid Raymond, Merwin Goldsmiwth, Robert Trebor
A woman public relations expert is hired to promote a scientist's android creation. An interesting comic premise fails to develop its full potential. Filmed in Miami Beach, and set there too.
COM 95 min VIDrel: RCA L/A V/sh

MAKING OF A HOLLYWOOD MADAM, THE: THE HEIDI FLEISS STORY * (18)
Michael Switzer USA 1996
Michael Gross, Cindy Pickett, Tricia Leigh Fisher, Jennifer Crystal, Lois Nettleton, George Segal, Michael Pataki, Cathy Lind Hayes, Kim

Oja, Christopher John Fields, Linda Dangoli, Thomas Bellin, Greg Longstreet, Bonnie Beck
A woman finds fame and fortune in Hollywood in a novel manner by providing members of the acting community with call-girls willing to cater for all their sexual needs. A reasonably straightforward portrait of the sordid affairs of this real-life character, no more than mildly diverting, its only point of interest being that Hollywood idols are as flawed as normal mortals.
Aka: HEIDI FLEISS STORY, THE
DRA   88 min   SATrel: MOVIE CHANNEL

MAKING UP ***
15
Katja von Garnier GERMANY   1992
Katja Riemann, Nina Kronjager, Gedeon Burkhard, Max Tidof, Daniela Lunkewitz, Peter Sattmann, Jochen Nickel, Carola Hohne, Stefan von Moers, Jophi Ries, Sybille Hamm
Two young women, a cartoonist and a nurse, are both on the look out for some romance in their lives, and when the former arranges a double date, the latter finds herself falling for the other woman's partner. A quirky little comedy without much of a plot, it takes a sharp look at female friendships in the 1990s, makes a few wry observations, but says nothing terribly profound.
Aka: ABGESCHMINKT
COM   54 min; 68 min (2-film cassette)
VIDrel: ELPIC/POLYREC   V/s
Osca: MOST BEAUTIFUL BREASTS IN THE WORLD, THE

MAKYU SENJO: VOLS 1 AND 2 **
15
JAPAN   1991
Adventures of a programming genius who is about to be awarded the Nobel Prize. When he learns that he was created by a sinister research foundation as part of its scheme for world domination he sets out to thwart these evil plans. An average Manga animation.
ANIM   99 min (2 cassettes) dubbed
VIDrel: MANGA/SONOP   V

MALCOLM X ***
15
Spike Lee USA   1992
Denzel Washington, Spike Lee, Angela Basset, Al Freeman Jr, Delroy Lindo, Albert Hall, Theresa Randle, Kate Vernon, Lonette McKee, Ernest Thompson, Tommy Randle, James McDaniel, Jean LaMarre, Tommy Hollis, Giancarlo Esposito
Long, detailed and finely made film biography of this controversial black leader, charting his career from this early days as a street hustler to a position of prominence within the Black Muslim movement and his eventual assassination. As ever the director exercises his penchant for over-statement, but despite that Washington is really terrific in the title role.
DRA   193 min (ort 201 min)
VIDrel: 4-FRONT/POLYREC/GUILD   V/sur
Boa: book The Autobiography of Malcolm X as told to Alex Haley.

MALEDICTION *
18
Bert I. Gordon USA   1989
Robert Forster, Lydie Denier, Caren Kaye, Phillip Glasser, M.K. Harris, Jack Carter, Ellen Geer, Henry Brown, Marlena Giovi, Leslie Huntley, Al Pugliese, Nick Angotti, Trent Dolan, Rena Riffel, RCB, J.P. Bumstead, Daryl Anderson
A former cop-turned-private-eye is hired to find a missing girl and gets involved with the female owner of a modelling agency. He sees a strange medieval portrait that hangs in her office, but its significance is not revealed until later, in a mediocre and gory film that's as devoid of logic as it is of conviction on the part of its actors.
Aka: SATAN'S PRINCESS
HOR   87 min Cut (57 sec – ort 90 min)   VIDrel: CIC/SONOP
L/A   V

MALIBU BEACH VAMPIRES, THE **
(18)
Francis Creighton USA   1991
Angelyne, Becky LeBeau, Joan Rudelstein, Marcus A. Frishman, Rod Sweitzer, Francis Creighton, Anet Antelle, Yvette Buchanin, Cherie Romaors, Kelley Galindo
A congressman and two other corrupt individuals have installed their mistresses in houses on Malibu Beach. Unfortunately for them, the girls are really vampires who, to make matters, worst have been injected with a truth serum.
COM   90 min   CABrel: HVC

MALIBU EXPRESS *
18
Andy Sidaris USA   1984
Darby Hinton, Sybil Danning, Art Metrano, Niki Dantine, Lynda Wiesmeier, Shelly Taylor, Lori Sutton, Barbara Edwards, Kimberly McArthur
A private eye becomes involved in a plot to steal high-tech secrets and give them to the Russians, but this corny story takes second place in a film that serves merely as a vehicle to display the talents of a number of Playboy models. A very loose remake of "Stacey", an earlier film by this director. Followed by a similar film (though not a sequel) entitled PICASSO TRIGGER.
THR   97 min (ort 101 min)   VIDrel: CIC/SONOP L/A   V

MALIBU HOT SUMMER *
18
Richard Brander USA   1974 (released 1986)
Terry Congie, Leslie Brander, Roselyn Royce, Robert Acey, Kevin Costner, Larry DeCraw, James Pascucci, Justin Scott
Three young women go to Los Angeles to fulfil their ambitions in various fields. Low-grade nonsense. This poor offering gave Costner his first screen role.
Aka: SIZZLE BEACH; SIZZLE BEACH, USA
DRA   80 min (ort 93 min)   VIDrel: TROMA/RTM   V

MALIBU NIGHTS *
18
Alex De Renzy USA   1991
Jeanna Fine, K.C. Williams, Ashlyn Gere, Angela Summers, Moana Pozzi, Miss Pomadoro, Jon Dough, T.T. Boy, Sean Michaels, Marc Wallice, Joey Silvera, Randy West, Jack Baker, Delta Force, Dave Rock, Rip Hymen, Jake Steed
Shot in a gym, this film has a large cast and by way of a plot, has Jon Dough donating his Malibu Fitness Center to charity for tax reasons. As ever, this slim opening premise leads into an endless series of erotic encounters.
Aka: MALIBU SPICE
A   54 min (ort 80 min)   VIDrel: GROHOM/MAXSCAN   V

MALICE **
15
Harold Becker USA   1993
Alec Baldwin, Nicole Kidman, Bill Pullman, Peter Gallagher, Josef Sommer, Bebe Neuwirth, Anne Bancroft, George C. Scott, Debrah Farentino, Tobin Bell, William Duff-Griffin, Gwyneth Paltrow, Davie Bowe, Diana Bellamy, Michael Hatt
An experienced surgeon comes to a small town and rents a room from a college professor and his wife, thus setting in motion a chain of events, including a suit for malicious malpractice and much else besides. A hardly believable tale with the obligatory erotic overtones and a slight echo of PACIFIC HEIGHTS.
THR   102 min (ort 107 min) wScrn
VIDrel: 4-FRONT/POLYREC   V/sur

MALICIOUS **
18
Ian Corson USA   1995
Molly Ringwald, John Vernon, Patrick McGaw, Mimi Kuzyk, Sarah Lassez, Rick Henrickson, Jennifer Copping, Stephen E. Miller, Joe Maffei, Ryan Michael, Jerry Wasserman, Jay Brazeau, Marlene Worrall, Paul McLean, Judith Maxie, Philip Hayes
An up-and-coming college athlete meets a mysterious female stranger, and discovers too late that she is intent on ruining his life. Standard psychological drama.
DRA   88 min (ort 92 min)   VIDrel: MED/DISC   V/sh

MALLENS, THE: PART 1 – THE MALLEN STREAK **
PG
Richard Martin/Ronald Wilson/Brian Mills UK   1978
John Duttine
The story of a wealthy squire who squanders his money and fathers a number of illegitimate children, each of whom is recognisably his child, thanks to a streak of white hair. Compiled from "The Mallens", this rather tedious drama wended its weary way, but improved somewhat by the time the second series began dealing with the fortunes of the squire's many offspring.
DRA   160 min (ort 171 min) mTV   VIDrel: FOCUS/DISC V
Boa: The Mallen trilogy by Catherine Cookson.

MALLENS, THE: PART 2 – THE MALLEN GIRLS **
PG
Richard Martin/Ronald Wilson/Brian Mills UK   1978
John Duttine
The second part of the first TV series of "The Mallens". The once wealthy squire has squandered his money and is forced to vacate High Banks Hall and move to a small cottage with his two

wards. His two illegitimate sons become regular visitors there and one night Barbara is savagely raped.
DRA 177 min mTV VIDrel: FOCUS/DISC V
Boa: The Mallen trilogy by Catherine Cookson.

MALLENS, THE: PART 3 – THE MALLEN SECRET **
PG
Mary McMurray/Roy Roberts UK 1980
Juliet Stevenson, Gerry Sundquist, Caroline Blakiston
Based on the second series of tales in "The Mallens", this follows the career of Barbara, one of the late Squire's many illegitimate children, who has been kept from the truth of her origins by her governess. However, matters become complicated when she falls in love with her cousin Michael, who has also had the truth of his illegitimacy kept from him. A fair drama of no great weight, but absorbing enough in its way.
DRA 140 min mTV VIDrel: FOCUS/DISC V
Boa: The Mallen trilogy by Catherine Cookson.

MALLENS, THE: PART 4 – THE MALLEN CURSE **
PG
Mary McMurray/Brian Mills UK 1980
Juliet Stevenson, Gerry Sundquist, Caroline Blakiston
The second series of tales based on Cookson's "The Mallens" tells of Barbara who, rejected by her cousin, enters a loveless marriage with the son of the Benshams, who now own the High Banks Hall since the bankruptcy of her late father. Meanwhile, the widowed Mr Bensham has proposed to Anna Brigmore, who has ambitions of her own regarding the former home of the Mallens. A glossy soap opera-style yarn, but somewhat better than the first series.
DRA 147 min mTV VIDrel: FOCUS/DISC V
Boa: The Mallen trilogy by Catherine Cookson.

MALLRATS **
18
Kevin Smith USA 1995
Shannen Doherty, Jeremy London, Jason Lee, Claire Forlani, Priscilla Barnes, Michael Rooker, Jason Mewes, Kevin Smoth, Ben Affleck, Lauren Adams, Renee Humphrey, Art James
The story of two loafers who spend almost all their time at a shopping mall, forms the basis for this teen comedy. A kind of follow-up to the much brighter and funnier CLERKS, this occasionally amusing work exhibits the deadening influence of major studio funding (in this case Universal Studios) and amply proves that nothing concentrates the mind so much as a lack of money. However, despite the film's more conventional feel, it is still reasonably entertaining.
COM 92 min (ort 96 min) cC VIDrel: CIC/SONOP V/sur

MALONE **
18
Harley Cokliss USA 1987
Burt Reynolds, Cliff Robertson, Scott Wilson, Kenneth McMillan, Lauren Hutton, Cynthia Gibb, Philip Anglim, Dennis Burkley, Alex Diakun, Mike Kirton, Brooks Gardner, Duncan Fraser, Janne Mortil, Christopher Lane, Jack Weston
An ex-CIA agent, disillusioned with the workings of the intelligence agencies, is poised to leave his job, when he suddenly unearths a strange conspiracy in a small town, that poses a major threat to national security. Amiable high-action dross that is best enjoyed with the mind disengaged.
THR 88 min Cut (1 sec – ort 92 min)
VIDrel: VISVID/POLYREC L/A V
Boa: novel Shotgun by William Wingate.

MALPERTUIS ***
18
Harry Kumel BELGIUM/FRANCE/WEST GERMANY 1972
Orson Welles, Susan Hampshire, Michel Bouquet, Mathieu Carriere, Walter Rilla, Daniel Pilon, Jean-Pierre Cassel, Dora Van Der Groen, Sylvia Vartan
An old sailor discovers the gods of ancient Greece, who are exhausted by years of neglect. He captures them and sews them into human skins with the aid of a taxidermist, and they are brought to a house where they live fairly drab lives, but occasionally reveal their true natures. When a young sailor strays into their abode he is turned into a statue by Medusa. An opulent and surreal tale, quite unlike any other horror film.
Aka: MALPERTUIS: HISTOIRE D'UNE MAISON MAUDITE; MAUDITE: THE LEGEND OF DOOM HOUSE
HOR 89 min VIDrel: L/A V
Boa: novel by Jean Ray.

MALTA STORY, THE ***
U
Brian Desmond Hurst UK 1953
Alec Guinness, Jack Hawkins, Anthony Steel, Flora Robson, Muriel Pavlow, Renee Asherson, Hugh Burden, Nigel Stock, Reginald Tate, Ralph Truman, Rosalie Crutchley, Michael Medwin, Ronald Adam, Stuart Burge, Jerry Desmonde
Thrilling account of the RAF action during WW2, in a battle to prevent the island fortress of Malta from falling into German hands. Low-key it may be, but this only serves to highlight the merits of this well-crafted flag-waver.
WAR 99 min (ort 103 min) B/W VIDrel: VCC L/A V

MALTESE FALCON, THE ***
PG
John Huston USA 1941
Humphrey Bogart, Sydney Greenstreet, Peter Lorre, Mary Astor, Gladys George, Ward Bond, Elisha Cook Jr, Jerome Cowan, Lee Patrick, Barton MacLane, James Burke, Walter Huston, Murray Alper, John Hamilton, Emory Parnell, Robert Homas
Vastly over-rated mystery about the hunt for a statuette, so valuable that murders are committed to gain it. Doors fly open, doors fly shut, bodies turn up here, there, everywhere. Yet for all its contrived plotting, the film remains highly memorable, chiefly thanks to a fine cast and some brilliant dialogue. This was Huston's directorial debut and Greenstreet's first screen role. Filmed twice before, in 1931 by Roy Del Ruth and in 1936 by Dieterle.
DRA 99 min (ort 101 min) B/W VIDrel: MGM/WHV V
Boa: novel by Dashiell Hammett.

MAMBO KINGS, THE **
15
Arne Glimcher USA 1992
Armand Assante, Antonio Banderas, Cathy Moriarty, Maruschka Detmers, Pablo Calogero, Scott Cohen, Mario Grillo, Ralph Irizarry, Pete MacNamara, Jimmy Medina, Marcos Quintanilla, J.T. Taylor, William Thomas Jr, Yul Vazquez
The bleak tale of a one-hit-wonder mambo group, who enjoy a brief moment of success in New York during the 1950s. Cesar and Nestor are two brothers in the band, and they engage in an endless series of confrontations and petty squabbles. The Pulitzer Prize-winning novel had a few profound things to say about life, but this is lost through over-simplification and an irritating soap opera-style treatment. The music however, is always worth hearing.
DRA 99 min (ort 104 min) cC VIDrel: WHV V/sur
Boa: novel The Mambo Kings Play Songs of Love by Oscar Hijuelos.

MAME *
PG
Gene Saks USA 1974
Lucille Ball, Robert Preston, Beatrice Arthur, Jane Connell, Bruce Davison, Kirby Furlong, Joyce Van Patten, John McGiver, Don Porter, Audrey Christie, Doria Cook, Bobbi Jordan, George Chiang, Roger Price
After his father dies, a nine-year-old boy goes off to live with his aunt and discovers that she is a strange and flamboyant person. This film version of the musical is much inferior to the straight 1958 "Auntie Mame" in which Rosalind Russell gave an outstanding performance in the title role. Songs include: "If He Walked Into My Life", "Open A New Window" and "We Need A Little Christmas".
MUS 126 min (ort 131 min) VIDrel: WHV V
Boa: novel by Patrick Dennis/play by Jerome Lawrence and Robert E. Lee.

MAMMA DRACULA *
18
Boris Szulzinger BELGIUM/FRANCE 1988
Louise Fletcher, Maria Schneider, Marc-Henri Wajnberg, Alexander Wajnberg, Jess Hahn
A female vampire who enjoys bathing in the blood of virgins, is forced by the increasing rarity of this commodity to employ a scientist who makes artificial blood. However, the mass disappearances have focused attention on her, and her birthday party to which she has invited a bevy of girls, is joined by an undercover female cop. A slack-witted and disappointing horror spoof.
COM 89 min VIDrel: EIV/SONOP V

MAN ABOUT THE HOUSE *
PG
John Robins UK 1974
Paula Wilcox, Sally Thomsett, Richard O'Sullivan, Yootha Joyce, Brian Murphy, Peter Cellier, Patrick Newell, Spike Milligan, Arthur Lowe, Doug Fisher, Aimi MacDonald, Jack Smethurst, Melvyn Hayes, Michael Ward

Another dreadful TV spin-off from an even worse TV series, in which a man and the two girls he shares the flat with, join forces with their disreputable landlord to fight off redevelopment of the area. The script could have used some redevelopment.
COM   86 min (ort 90 min)   VIDrel: WHV   V/h

MAN AND A WOMAN, A **
Claude LeLouch FRANCE
PG
1966
Anouk Aimee, Jean-Louis Trintignant, Valerie LaGrange, Pierre Barouk, Simone Paris, Antoine Sire, Souad Amidou, Yane Barry, Paul Le Person, Henri Chemin, Gerard Sire
The ultimate romantic experience: a hyper-glossy account of a racing driver and a script girl, both widowed with small children, meeting and falling in love. Followed by a sequel twenty years later. AA: Foreign, Story/screen (Claude Lelouch/Pierre Uytterhoeven).
Aka: UN HOMME ET UNE FEMME
DRA   103 min   VIDrel: WHV   V

MAN AND A WOMAN, A: TWENTY YEARS LATER *
Claude LeLouch FRANCE
15
1986
Anouk Aimee, Jean-Louis Tritignant, Richard Berry, Evelyne Bouix, Charles Gerard, Marie-Sophie Pochat, Antoine Sire, Andre Engel, Patrick Poivre D'Arvor, Yane Barry, Alain Berry, Jean-Philippe Chatrier, Maurice Illouz
Twenty years may have gone by, but they've made very little difference in terms of either plot or conception to this delayed remake, with a film producer seeking out his old love in order to immortalise their romance on celluloid. Music is by Francis Lai.
Aka: UN HOMME ET UNE FEMME: VINGT ANS DEJA
DRA   108 min (ort 120 min)   VIDrel: MGM/WHV L/A V/sh

MAC AND ME *
Stewart Raffill USA
U
1988
Christine Ebersole, Jonathan Ward, Tina Caspary, Lauren Stanley, Vinnie Torrente, Jade Gregory, Martin West, Ivan Jorge Rado, Danny Cooksey, Barbara Allyne Bennet, Laura Waterbury, Jack Eiseman, Richard Bravo, Gary Brockette
A family of fairly grotesque but quite harmless aliens are accidentally brought to Earth by a returning space probe, and rapidly make themselves at home in this deliberate clone of E.T. THE EXTRA-TERRESTRIAL. See also ALF.
JUV   95 min (ort 99 min)   VIDrel: POLY/POLYREC L/A V/sh

MAN AT THE GATE, THE **
Norman Walker UK
U
1935
Harold Simpson, Wilfrid Lawson, Mary Jerrold, William Freshman, Kathleen O'Regan
During WW2, the wife of a fisherman loses her faith when her sailor son is reported drowned. A competent little drama that attempts to examine the effects of the war at home, and is both restrained and convincing. The use of a speech made by George VI at the time is an effective highlight.
Aka: MEN OF THE SEA
DRA   48 min; 126 min (2-film cassette) B/W
VIDrel: CONNO/RTM L/A   V
Boa: poem by Louise Haskin. Osca: TURN OF THE TIDE

MAN AT THE TOP **
Mike Vardy UK
15
1973
Kenneth Haigh, Nanette Newman, Harry Andrews, John Quentin, Mary Maude, Danny Sewell, Paul Williamson, Margaret Heald, Angela Bruce, Charlie Williams, Anne Cunningham, William Lucas, John Collin, Norma West
An attempt to cash in on the success of a popular TV series of the same name, that was inspired by John Braine's novel "Room At The Top", and followed the adventures of Joe Lampton, a character who first saw the light of day in the 1958 film ROOM AT THE TOP (followed by the 1965 sequel "Life At The Top"). In this predictable spin-off, Lampton is now a successful executive, and faces a nasty dilemma over the use of a new, untested drug.
DRA   91 min (ort 95 min)   VIDrel: BRAVE/SONOP L/A   V

MAN BETWEEN, THE ***
Carol Reed USA
U
1953
James Mason, Claire Bloom, Hildegard Neff, Geoffrey Toone, Aribert Waescher, Ernst Schroeder, Dieter Krauser, Hilde Sessak, Karl John
A young British woman visits her brother and his German-born

wife in West Berlin and is soon plunged into a murky intrigue involving a black-marketeer who specialises in hunting down fugitives wanted by the East German regime. Our heroine soon falls for this world-weary anti-hero but events take a complicated course when she is kidnapped by mistake. A highly atmospheric but very downbeat tale.
THR   97 min (ort 101 min) B/W   VIDrel: WHV   V
Boa: novel Susanne In Berlin by Walter Ebert.

MAN BITES DOG **
Benoit Poelvoorde/Remy Belvaux/Andre Bonzel BELGIUM
18
1992
Benoit Poelvoorde, Jacqueline Poelvoorde-Pappaert, Nelly Pappaert, Jenny Drye, Malou Madou, Willy Vandenbroeck, Rachel Deman, Andre Laime, Edith Lemerdy, Sylvaine Gode, Zoltan Tobolik, Valerie Parent, Alexandra Fandango, Oliver Cotica
A serial killer hires a film crew to document his campaign of murder and robbery, committing the latter in order to provide the necessary funds for this enterprise. A sick pseudo-documentary that spoofs TV and film violence. The 124 min tape includes Bonzel's short film "Pas De C4 Pour Daniel".
Aka: C'EST ARRIVE PRES DE CHEZ VOUZ
DRA   96 min B/W; 124 min wScrn (special edition)
VIDrel: TART/20TH   V/dm

MAN CALLED HORSE, A ***
Elliot Silverstein USA
15
1970
Richard Harris, Judith Anderson, Jean Gascon, Manu Tupou, Corinna Tsopei, Dub Taylor, William Jordan, James Gammon, Eddie Little Sky, Manuel Padilla, Iron Eyes Cody, Lina Marin, Tamara Garina, Michael Baseleon, Tom Tyon
In 1825 an English aristocrat is captured by the Sioux, but eventually proves his worth as a warrior by undergoing torture, and becoming their leader in the process. An intense and often bloody film, with a fine Leonard Rosenman score. Followed by RETURN OF A MAN CALLED HORSE and TRIUMPHS OF A MAN CALLED HORSE. The film "Run Of The Arrow" handled a similar theme.
WES   109 min   VIDrel: 20TH/TECH   V
Boa: short story by Dorothy M. Johnson.

MAN ESCAPED, A ***
Robert Bresson FRANCE
U
1956
Francois Leterrier, Charles Le Clainche, Roland Monot, Maurice Beerblock, Jacques Ertaud, Roger Treherne, Jean-Paul Delhumeau, Jean-Philippe Delamare, Jacques Oerlemans, Klaus Detlef Grevenhorst, Leonard Schmidt
Spartan recreation of how a figure in the French Resistance engineered his escape from imprisonment in the fortress at Montluc in 1943, accompanied by his cell-mate (who functioned as technical adviser on this film). An involving and fascinating account and a fine illustration of Bresson's directorial style and skill. Winner of the Best Director Award at Cannes 1957.
Aka: MAN ESCAPED, OR THE WIND BLOWETH WHERE IT LISTETH, A; OU LE VENT SOUFFLE OU IL VENT; UN CONDAMNE A MORT S'EST ECHAPPE
DRA   95 min (ort 102 min) B/W   VIDrel: ARTIF/20TH   V

MAN FOR ALL SEASONS, A **
Fred Zinnemann UK
U
1966
Paul Scofield, Wendy Hiller, Susannah York, Leo McKern, Robert Shaw, Orson Welles, John Hurt, Nigel Davenport, Corin Redgrave, Colin Blakely, Vanessa Redgrave, Cyril Luckham, Jack Gwyllim, Yootha Joyce, Anthony Nichols, Eira Heath
Mechanical version of a play about how Sir Thomas More's opposition to Henry VIII's divorce inevitably led to his execution. Despite its excellent photography and costumes, this stiff piece is weighed down by leaden prose and wooden acting, though this did not prevent it winning several Oscars. Remade for cable in 1988. AA: Pic, Dir, Actor (Scofield), Cin (Ted Moore), Cost (Elizabeth Haffenden/Joan Bridge), Screen/adapt (Robert Bolt).
DRA   116 min (ort 120 min)   VIDrel: VCC/DISC/COLUM V
Boa: play by Robert Bolt.

MAN FROM ATLANTIS, THE *
Lee H. Katzin USA
PG
1977
Patrick Duffy, Belinda J. Montgomery, Victor Buono, Art Lund, Dean Santoro, Lawrence Pressman, Mark Jenkins, Allen Case, Joshua Bryant, Steve Franken, Virginia Gregg, Curt Lowens, Charles Davis, Lilyan Chauvin, Alex Rodine
Pilot for a mercifully short-lived TV series, about a strange man whose ability to breathe underwater plus other aquatic skills,

makes him a valuable recruit to the US Navy, after he is unable to find his way back to his underwater homeland. Written by Mayo Smith.
FAN   105 min mTV   VIDrel: CASPIC L/A   V

MAN FROM BEYOND, THE ** PG
Burton King USA 1922
Harry Houdini, Arthur Maude, Nita Naldi, Frank Montgomery
A man frozen alive is revived after a hundred years and meets a girl who seems the reincarnation of his lost love. Written and produced by Houdini, the stuntman, illusionist and escape artist, to demonstrate his belief in life after death. A full-blooded melodrama with plenty of stunts and chases.
DRA   80 min (ort 91 min) B/W silent
VIDrel: VISION/DISC   V

MAN FROM LARAMIE, THE *** U
Anthony Mann USA 1955
James Stewart, Arthur Kennedy, Donald Crisp, Cathy O'Donnell, Alex Nicol, Wallace Ford, Jack Elam, Aline MacMahon, John War Cloud, James Millican, Gregg Barton, Boyd Stockman, Frank De Kova, Frosty Royse, Eddy Waller
A man sets out to find the gunrunners responsible for selling rifles to the Apaches, as these were used in the murder of his brother. A well-handled and suspenseful revenge yarn. Well photographed in Cinemascope, though this additional asset will be lost on the TV screen.
WES   98 min (ort 104 min)   VIDrel: ENTUK/GOLD L/A   V

MAN FROM LEFT FIELD, THE ** PG
Burt Reynolds USA 1993
Burt Reynolds, Reba McEntire, Bradley Barfield, Jonatahn Bouck, Suzanne Dodd, Clarence Goss, Danny green, Biran Iannitti, Ken kay, Tom Kouchalakos, Richard Lepore, Marc Macaulay, Michael O'Smith, Avery Sommers, David Benardi
A man takes a job coaching a Little League team and manages to inspire a turnaround in its fortunes that leads to final victory. A pleasant enough time-filler that moves along familiar lines and relies rather too heavily on the easy-going charm of Reynolds.
JUV   96 min mTV   VIDrel: MARQ/20VIS L/A   V

MAN FROM MONTEREY, THE ** U
Mack V. Wright USA 1933
John Wayne, Ruth Hall, Luis Alberni, Francis Ford, Nina Quartaro, Lafe McKee, Donald Reed, Lillian Leighton, Claude Whitaker, Jim Corey
An Army captain is sent to Monterey to make sure that Mexican landowners register properties and comes up against a villain who has been operating a land swindle. An okay Wayne vehicle that marked his last Western for Warner Brothers.
WES   55 min (ort 59 min) B/W   VIDrel: MGM/WHV   V/h

MAN FROM UTAH, THE ** U
Robert North Bradbury USA 1934
John Wayne, Polly Ann Young, George "Gabby" Hayes, Yakima Canutt, Ed Piel Sr, Anita Campillo, Lafe McKee, George Cleveland, Earl Dwire, Artie Ortego
A cowboy sets out to bring to justice a gang of crooks who are killing rodeo riders in this fair actioner. The rather clumsy use of stock rodeo footage is an annoyance in an otherwise well-made tale.
WES   52 min (ort 55 min)   VIDrel: NTV/TERRY   V

MAN IN GREY, THE *** PG
Leslie Arliss UK 1945
James Mason, Phyllis Calvert, Margaret Lockwood, Stewart Granger, Ryamond Lovell, Nora Swinburne, Helen Haye, Martita Hunt, Amy Veness, Diana King, Beatrice Varley, Roy Enerton, A.E. Matthews, Ann Wilton, Drusilla Wills
A saga of love and passion among the English aristocracy in the Regency period. A woman trapped in a loveless marriage takes on a friend as governess to her child, but her friendship is repaid by treachery and deceit. It made stars of Lockwood and Mason and started the Gainsborough series of "wicked lady" films.
DRA   88 min (ort 116 min) B/W   VIDrel: VCC/RTM   V
Boa: novel by Lady Eleanor Smith.

MAN IN LOVE, A * 18
Diane Kurys FRANCE/ITALY/USA 1987
Peter Coyote, Jamie Lee Curtis, Greta Scacchi, Peter Riegert, John Barry, Claudia Cardinale, John Barry, Vincent Lindon, Jean Pigozzi, Elia Katz, Jean Claude de Goros, Constantin Alexandrov, Michele Melega, Iole Silvani

An actor playing an important role embarks on a torrid romance with a fellow player. Impressive sex scenes but little else, make for a very slight drama. Written by Kurys and Israel Horovitz.
Aka: UN HOMME AMOUREUX
DRA   106 min (ort 110 min)   VIDrel: VISVID/POLYREC L/A   V

MAN IN MY LIFE, THE *** 15
Jean Charles Tacchella CANADA/FRANCE 1992
Maria de Medeiros, Thierry Fortineau, Jean-Pierre Bacri, Anne Letourneau, Ginette Garcin, Ginette Mathieu, Alain Doutey, Alix de Konopka, Carmela Valente, Betrand Lacy, Frederique Lazarini, Emmanuelle Oriheul, Samuel Sogno
This slight comedy is about a young girl who is determined to find herself a rich husband after she grows tired of moving from job to job, and announces to her foster mother that she'll be married by Christmas. Her first conquest is a bookseller whose impending poverty puts her off, but further conquests are even less satisfactory and she eventually realises that love is more important than riches. A lightweight film of much charm, hampered by a weak plot.
Aka: L'HOMME DE MA VIE
COM   98 min (ort 104 min) wScrn   VIDrel: TART/20TH   V

MAN IN THE ATTIC, THE ** 15
Graeme Campbell USA 1994
Anne Archer, Neil Patrick Harris, Len Cariou, Alex Carter, Rock Roberts, Deborah Drakeford, Martha Cronyn, Toby proctor, Nahanni Johnstone, Richard Liss, Judith Orban, Bruce Vavrina, James Mainprize, Pixie Bigelow, Sam Nelkin
A married woman begins an affair with a teenage boy who eventually moves into her home. While they can be together during the day, at night he must hide away in the attic and remain quite still so as not to arouse her husband's suspicions.
DRA   93 min (ort 97 min) mCab   VIDrel: 20TH/TECH V/sh

MAN IN THE BROWN SUIT, THE ** PG
Alan Grint USA 1988
Stephanie Zimbalist, Rue McClanahan, Edward Woodward, Tony Randall, Ken Howard, Nickolas Grace, Simon Dutton
A resourceful young woman sees a man's fatal stabbing at the airport in Cairo, and becomes intrigued by the circumstances of his death. A piece of paper clutched in his hand and a few other clues lead her to a Nile cruise ship and a cat-and-mouse game with a sinister gang over some missing gems. A poor adaptation that attempts to blend comedy, romance and thrills on the lines of ROMANCING THE STONE, but succeeds in being merely tedious.
THR   91 min   VIDrel: MGM/WHV L/A   V
Boa: novel by Agatha Christie.

MAN IN THE IRON MASK, THE *** PG
Mike Newell USA 1977
Richard Chamberlain, Patrick McGoohan, Jenny Agutter, Ian Holm, Louis Jourdan, Ralph Richardson, Vivien Merchant, Brenda Bruce, Esmond Knight, Godfrey Quigley, Emrys James, Denis Lawson, Anne Zelda, Stacy Davis
Film version of a classic novel about the strange fate of a prisoner forced to wear an iron mask so as to hide his identity. This lively remake of the 1939 film has Chamberlain well cast and ably supported by a fine cast. The adaptation is by William Bast.
DRA   101 min mTV   VIDrel: POLY/POLYREC L/A   V
Boa: novel by Alexandre Dumas.

MAN IN THE MOON, THE *** PG
Robert Mulligan USA 1991
Sam Waterston, Tess Harper, Reese Witherspoon, Gail Strickland, Jason London, Emily Warfield, Bentley Mitchum, Ernie Lively, Dennis Letts, Earleen Bergeron, Anna Chappell, Brandi Smith, Sandi Smith, Derek Ball, Spencer Ball
In 1950s rural Louisiana, two sisters (aged fourteen and seventeen) fall for the same handsome cousin, and despite the fact that the older girl really has no shortage of boyfriends, she and her younger sibling become reluctant rivals. Ultimately, a tragedy brings them together again, but not before many sharply observed moments have come to pass, in this endearing rites-of-passage film.
DRA   95 min (ort 99 min) cC   VIDrel: MGM/WHV   V/sur

MAN IN THE WHITE SUIT, THE *** U
Alexander McKendrick UK 1951
Alec Guinness, Joan Greenwood, Cecil Parker, Michael Gough, Ernest

Thesiger, Miles Malleson, Vida Hope, Howard Marion Crawford, John Rudling, Patric Doonan, Duncan Lamont, Harold Goodwin, Colin Gordon, Joan Harben, Arthur Howard

A mere laboratory assistant invents a cloth that is virtually indestructible and so causes a panic in the textile industry, which sees his invention as a deadly threat. Unions and management unite in a bid to stop the march of progress in a witty farce that unfortunately, pulls its punches at the end, neatly restoring the status quo by having the fabric ultimately prove to be worthless. Written by Macdougall, John Dighton and McKendrick.

COM   81 min (ort 85 min) B/W   VIDrel: WHV   V
Boa: play by Roger MacDougall.

MAN INSIDE, THE **   15
Bobby Roth FRANCE/USA   1990
Jurgen Prochnow, Peter Coyote, Nathalie Baye, Monique Van De Ven, Dieter Laser, Phillip Anglim, Henry G. Sanders

In West Germany, an investigative reporter infiltrates a sordid newspaper, his intention being to expose their corrupt journalistic practices. Based on the real-life exploits of Gunter Wallraff, the film starts off with great promise, but simply cannot sustain interest all the way through, despite the merits of a strong cast and skilful direction. In any case, rapidly changing world events have overtaken the story, which now looks pointless and dated.

DRA   94 min (ort 101 min)   VIDrel: VIR/RCA L/A   V

MAN NEXT DOOR, THE ***   15
Lamont Johnson USA   1995
Michael Ontkean, Pamela Reed, Annette O'Toole, Susan Baskin

A serial rapist is freed on parole from a seven-year sentence and goes to live in a small town, where the only person who knows about his past is his parole officer (who for good measure just happens to be one of his neighbours as well). Of course, when another woman is raped, he comes under suspicion and has to decide whether to run for it or attempt to prove his innocence. A nicely plotted thriller which holds together well.

THR   89 min   VIDrel: ODY/SONOP   V/h

MAN OF FLOWERS *   18
Paul Cox AUSTRALIA   1984
Norman Kaye, Alyson Best, Chris Haywood, Sarah Walker, Julia Blake, Bob Ellis, Barry Dickins, Patricia Cook, Victoria Eagger, Werner Herzog, Hilary Kelly, James Stratford, Eileen Joyce, Marianne Baillieu, Lirit Bilu

Pretentious story of a mother-fixated rich man unable to have a relationship with a woman, who hires a young girl to strip for him and conceives a suitable artistic revenge on her brute of a boyfriend. An empty, depressing and sterile work, predictably raved over by the critics, telling us more about their taste than it does about the film.

DRA   86 min (ort 91 min)   VIDrel: ARTPRO/RTM   V

MAN OF LA MANCHA *   PG
Arthur Miller USA   1972
Peter O'Toole, Sophia Loren, James Coco, Harry Andrews, John Castle, Brian Blessed, Ian Richardson, Julie Gregg, Rosalie Crutchley, Gino Conforti, Marne Maitland, Dorothy Sinclair, Miriam Acevedo, Dominic Barto, Fred Evans

Ponderous attempt to make a film of the Broadway musical, which has Cervantes telling the story of Don Quixote to his fellow inmates, after having been cast into prison by the Inquisition. A flashy and hollow work.

MUS   124 min (ort 132 min)   VIDrel: WHV   V
Boa: play by Dale Wasserman/musical by Mitch Leigh and Joe Darion.

MAN OF NO IMPORTANCE, A ***   15
Suri Krishnamma EIRE/UK   1994
Albert Finney, Brenda Fricker, Michael Gambon, Tara Fitzgerald, Rufus Sewell, Patrick Malahide, Anna Manahan, Joe Pilkington, Brendan Conroy, John Killalea, Pat Killalea, Joan O'Hara, Eileen Conroy, Eileen Reid, David Kelly

A homosexual bus conductor in 1960s Dublin charms his passengers with his knowledge of literature and his fascination with Oscar Wilde, but none of those around him seem aware of his true nature. One day, he is inspired to stage Wilde's play "Salome", casting some of his passengers in the roles, a decision that arouses some disquiet. A whimsical parable that does not ring quite true although Finney's superlative performance makes for a very enjoyable time.

DRA   94 min (ort 98 min)   VIDrel: FIRC/RTM   V

MAN OF THE HOUSE *   U
James Orr USA   1994
Chevy Chase, Farrah Fawcett, Jonathan Taylor Thomas, George Wendt, David Shiner, Art LaFleur, Richard Portnow, Richard Foronjy, Peter Appel, Chief Leonard George, George Greif, Ron Canada, Chris Miranda, Zachary Browne

Having moved into the home of his fiancee and her young son, a man is less than well prepared for the jealous son's campaign to drive him out, his star ploy being to get him signed up as a woodland guide. Chase demonstrates once more his reliable penchant for appearing in duds.

COM   97 min   VIDrel: BUENA/TECH   V

MAN OF THE YEAR ***   15
Dirk Shafer USA   1995
Dirk Shafer, Vivian Paxton, Deidra Shafer, Michael Ornstein, Claudette Sutherland, Calvin Bartlett, Beth Broderick, Fabio, Mary Stein, Cynthia Szigeti, Dennis Bailey, Charles Sloane, Patricia Domiano, Dawn Christie, Phyllis Franklin

A spoof documentary starring, scripted and directed by "Playgirl" magazine centrefold of the year Shafer, who became a pin-up following an innocent photo-shoot. He worked for the magazine throughout 1992, dispensing advice on what women really want in a man (all the while hiding his homosexuality). The clumsily contrived attempt to make the film look like a real documentary doesn't work, but the star's insights are both entertaining and revealing.

COM   86 min   VIDrel: DTK/RTM   V

MAN ON FIRE *   18
Elie Chouraqui FRANCE/ITALY   1987
Brooke Adams, Scott Glenn, Danny Aiello, Joe Pesci, Paul Shenar, Jonathan Pryce, Jade Malle, Laura Morante, Giancarlo Prati, Inigo Lezzi, Alessandro Haber, Franco Trevisi, Lou Castel, Lorenzo Piani, Giuseppe Cederna

A disillusioned former CIA agent goes on a rampage when the twelve-year-old girl he was hired to protect is kidnapped. A disagreeable mixture of violence and implausibility, wasting a good cast.

Aka: PERICOLO IN AGGUATO; WITHOUT MERCY
A/AD   88 min Cut (31 sec – ort 92 min)   VIDrel: MIA/DISC
V/sur
Boa: short story by A.J. Quinnell.

MAN TO MAN ***   (PG)
John Maybury UK   1992
Tilda Swinton

With a mass of digital video techniques, this interesting film casts Swinton (who repeats her much admired stage portrayal) as a German woman whose survival in Nazi Germany came about when she impersonated her husband, a ruse she continued for another forty years. Hampered by the clumsy use of disturbing visual images that interrupt the narrative flow rather than complement it, yet the film is so very unusual it is certainly worth a look.

DRA   74 min mTV   TVrel: BBC
Boa: screenplay Jacke Wie Hose by Manfred Karge (translated by Anthony Vivis).

MAN TROUBLE *   15
Bob Rafelson USA   1992
Ellen Barkin, Jack Nicholson, Harry Dean Stanton, Beverly D'Angelo, Michael McKean, Saul Rubinek, Veronica Cartwright, Viveka Davis, David Clennon, John Capelos, Paul Mazursky, Lauren Tom, Betty Carvalho, Mark J. Goodman, Robin Greer

After a frightening break-in, an opera singer hires a security man who specialises in attack dogs. The two become involved but she is unaware that he has also been hired to steal the manuscript of a book written by her sister. A terribly disappointing attempt at a zany comedy that falls flat on its face.

COM   95 min (ort 100 min)   VIDrel: FIRST/SONOP   V/sur

MAN UPSTAIRS, THE ***   PG
Don Chaffey UK   1958
Richard Attenborough, Bernard Lee, Donald Houston, Dorothy Alison, Virginia Maskell, Kenneth Griffith, Patricia Jessel, Charles Houston, Maureen Connell, Walter Hudd, Edward Judd, Alfred Burke, David Griffith, Polly Clark, Raymond Ray

Attenborough gives a splendid performance as a neurotic scientist who after going berserk, injures a policeman and barricades himself in his room. While the forces of law and order attempt to reason with him, he contemplates both the events that led to his breakdown and the option of suicide. An absorbing

character study that is slightly reminiscent of LE JOUR SE LEVE, but minus the same degree of tension.
DRA 84 min (ort 88 min) B/W VIDrel: LUMI/SPEAR V

MAN WHO COULD WORK MIRACLES, THE *** U
Lothar Mendes UK 1937
Roland Young, Joan Gardner, Ralph Richardson, Ernest Thesiger, Edward Chapman, Sophie Stewart, Robert Cochrane, George Zucco, Lawrence Hanray, George Sanders, Wallace Lupino, Lady Tree, Joan Hickson, Wally Patch
A mild-mannered drapery store assistant is given unlimited power to do anything, as a result of a bet made by three heavenly figures (one of the best sequences in the film). However, he is unable to rise above his own pettiness and limited vision, and almost causes the destruction of the world in his clumsy attempts to improve the lot of mankind. Despite some floundering about midway through, this is a charming fantasy. Scripted by Wells from his own story.
FAN 79 min (ort 82 min) B/W VIDrel: CARL/TECH V
Boa: short story by H.G. Wells.

MAN WHO FELL TO EARTH, THE *** 18
Nicolas Roeg UK 1976
David Bowie, Rip Torn, Candy Clark, Buck Henry, Bernie Casey, Jackson D. Kane, Tony Mascia, Rick Ricardo, Linda Hutton, Adrienne Larussa, Hillary Holland, Peter Prouse, Richard Breeding, Lilybell Crawford, James Lovell
Confusing story of an alien who comes to Earth to save his arid planet from dying, but is prevented from leaving when he is in a position to complete his mission. Excellent camerawork and inspired casting of Bowie as the alien fail to rescue an incoherent plot, and Roeg shows an irritating penchant for self-indulgence. Written by Paul Mayersburg and remade for TV in 1987.
FAN 132 min (ort 138 min) VIDrel: WHV L/A V
Boa: novel by Walter Tevis.

MAN WHO FELL TO EARTH, THE ** PG
Robert J. Roth USA 1986
Lewis Smith, Beverly D'Angelo, James Laurenson, Wil Wheaton, Bruce McGill, Robert Picardo, Annie Potts, Henry Sanders, Bobbi Jo Lathan, Carmen Argenziano, Chris De Rose, Ritch Shydner, Rob Neilson, Steve Natole
An alien from a dying world lands on Earth, where he tries to find a means of saving his home planet. Although this represents a somewhat unnecessary remake when one considers the 1976 British production starring David Bowie, it does at least attempt a more straightforward version of the original story. The script is by Richard Kletter. Pilot for a prospective series.
FAN 93 min (ort 100 min) mTV VIDrel: MGM L/A V
Boa: novel by Walter Tevis.

MAN WHO HAD POWER OVER WOMEN, THE * 15
John Krish UK 1970
Rod Taylor, Carol White, James Booth, Penelope Horner, Charles Korvin, Clive Francis, Alexandra Stewart, Keith Barron, Marie-France Boyer, Magali Noel, Jimmy Jewel, Geraldine Moffat, Wendy Hamilton, Ellis Dale, Philip Stone, Sara Booth
Black comedy about a public relations man and the ethical problems he faces after he comes to hate his job. An overblown exercise in breast-beating with few endearing aspects, despite the fine work of the cast. Written by Andrew Meredith.
COM 89 min VIDrel: 20TH/TECH L/A V
Boa: novel by Gordon Williams.

MAN WHO KNEW TOO MUCH, THE ** PG
Alfred Hitchcock USA 1956
James Stewart, Doris Day, Bernard Miles, Brenda De Banzie, Ralph Truman, Alan Mowbray, Daniel Gelin, Christopher Olsen, Mogens Wieth, Hilary Brooke, Noel Willman, Reggie Nalder, Richard Wattis, Alix Talton, Yves Brainville
Hitchcock's remake of his own 1934 film about an American couple who become involved in an international conspiracy to assassinate an important figure. A thoroughly disappointing film that shows none of the lightness of touch the director exhibited in the earlier film. AA: Song ("Whatever Will Be, Will Be" (Que Sera, Sera) – Jay Livingston/Ray Evans).
THR 115 min (ort 120 min) VIDrel: CIC/SONOP V/h

MAN WHO LOVED CAT DANCING, THE ** 18
Richard C. Sarafian USA 1973
Burt Reynolds, Sarah Miles, Lee J. Cobb, George Hamilton, Jack Warden, Bo Hopkins, Robert Donner, Sandy Kevin, Nancy Malone,

Jay Silverheels, Jay Varela, Owen Bush, Larry Littlebird, Sutero Garcia Jr, Larry Finley
An unhappily married woman is taken hostage in the course of a robbery and finally falls for one of the outlaws. Fairly slow-moving but moderately diverting, and somewhat inspired by "No Orchids For Miss Blandish". The script is by Eleanor Perry.
WES 118 min (ort 127 min) VIDrel: MGM/WHV V
Boa: novel by Marilyn Durham.

MAN WHO LOVED WOMEN, THE * 15
Blake Edwards USA 1983
Burt Reynolds, Julie Andrews, Kim Basinger, Marilu Henner, Barry Corbin, Cynthia Sikes, Jennifer Edwards, Tracy Vaccaro, Sela Ward, Ellen Bauer, Denise Crosby, Ben Powers, Jill Carroll, Schweitzer Tanney, Regis Philbin, Roger Rose
An American remake of a 1977 Truffaut film, about a bachelor obsessed with women to the point of eventually seeking help from a female psychologist. A verbose and ponderous affair that lacks the panache of the earlier film, despite a winning performance from Reynolds. Written by Milton Wexler, Geoffrey Edwards and Blake Edwards, who gives a good demonstration of his shortcomings as a director.
COM 106 min (ort 110 min) VIDrel: RCA L/A V

MAN WHO SHOT LIBERTY VALANCE, THE *** U
John Ford USA 1962
James Stewart, John Wayne, Vera Miles, Lee Marvin, Edmond O'Brien, Andy Devine, John Carradine, Jeanette Nolan, John Qualen, Woody Strode, Denver Pyle, Strother Martin, Lee Van Cleef, Ken Murray, Willis Bouchey, Paul Birch
A lawyer looks back on his political career and recalls that much of his success was due to the fact that he was credited with shooting a notorious outlaw, although the fatal shot was actually fired by his best friend. A classic Ford Western that combines fine performances with a literate script.
WES 118 min VIDrel: CIC/SONOP V/h

MAN WHO STAYED AT THE RITZ, THE *** 15
Desmond Davies FRANCE/UK/USA 1988
Perry King, Leslie Caron, Cherie Lunghi, David McCullum, David Robb, Joss Ackland, Patachou, Mylene Demongeot, Sophie Barjac, Terry Taplin, Brigitte Kahn, Barrie Houghton, John Grillo, Nathalie Gerda, Wolf Kahler
During the early days of WW2, an American expatriate living at the Ritz in Paris, realises that he cannot remain neutral in the face of the German occupation, and is gradually drawn into collaborating with the Resistance.
Aka: MAN WHO LIVED AT THE RITZ, THE
DRA 185 min (o2 cassettes – 360 min) mTV
VIDrel: OURVID/SCRN V
Boa: novel by A.E. Hotchner.

MAN WHO WOULD BE KING, THE *** PG
John Huston USA 1975
Sean Connery, Michael Caine, Saeed Jaffrey, Christopher Plummer, Jack May, Doghmi Larbi, Karroom Ben Bouih, Mohammad Shamsi, Albert Moses, Paul Antrim, Graham Acres, Shakira Caine, Kimat Singh, Gurmuks Singh, Yvonne Ocampo
Caine and Connery star as two engaging soldiers of fortune who ply their less-than-respectable trade in 19th century India. They contrive and carry out a daring plan to become rulers in the small isolated land of Kafiristan. A splendid, full-blooded adventure yarn of the old-fashioned kind; highly entertaining and well-made, but lacking the genuine Indian locations (it was filmed in Morocco) that would have added more to the atmosphere.
A/AD 123 min (ort 129 min) VIDrel: VCC/DISC/COLUM V
Boa: short story by Rudyard Kipling.

MAN WHO WOULDN'T DIE, THE * (PG)
Bill Condon USA 1994
Roger Moore, Malcolm McDowell, Nancy Allen, Jackson Davies, Eric McCormack, Kevin McNulty, Mina E. Mina, Don MacKay, Bernard Cuffling, Jessica Va Der Veen, Sheila Paterson, Teryl Rothery, Wendy Van Riesen, Scott Bellis, Roger R. Cross
A writer of mystery-thrillers learns that one of the plots of his latest novel has provided the framework for the activities of a real-life killer. In a bid to track down this individual, he calls on the services of a psychic. With such a flawed, uninvolving and derivative story, the real mystery about this movie is how it ever came to be made.
DRA 88 min SATrel: SKY MOVIES

**MAN WITH A GUN ** ** 18
David Wyles USA 1994
Michael Madsen, Jennifer Tilly, Gary Busey, Robert Loggia, Ian tracey, Bill Cobbs
A Mafia hitman having an affair with his boss's wife is devastated when he is ordered to kill her. When he tells her of his predicament, she suggests that he murder her twin sister instead, a course of action that he finds equally abhorrent.
DRA 91 min (ort 96 min) VIDrel: 20TH/FOXVID V/sh
Boa: novel The Shroud Society by Hugh C. Rae.

**MAN WITH THE DEADLY LENS, THE ** ** 15
Richard Brooks USA 1982
Sean Connery, George Grizzard, Katherine Ross, Robert Conrad, Henry Silva, Leslie Nielsen, John Saxon, G.D. Spradlin, Robert Webber, Dean Stockwell, Hardy Kruger, Rosalind Cash, Ron Moody, Jennifer Jason Leigh
An ace reporter discovers a US government plot to assassinate an Islamic leader, in this unusual satire that cracks along at a fair pace but tends to miss most of its targets. The script is by Brooks.
Aka: WRONG IS RIGHT
A/AD 113 min Cut (2 sec – ort 117 min)
VIDrel: MIA/DISC V
Boa: novel The Better Angels by Charles McCarry.

MAN WITH THE GOLDEN ARM, THE * ** 15
Otto Preminger USA 1955
Frank Sinatra, Kim Novak, Darren McGavin, Eleanor Parker, Arnold Stang, Doro Merande, Robert Strauss, John Conte, George E. Stone, Emil Meyer, Shorty Rogers, Shelly Manne, Leonid Kinskey, Frank Richards, Ralph Neff, Joe McTurk
A gambler tries to kick his addiction to drugs and eventually succeeds in this harrowing drama that is enlivened by a fine performance from Sinatra and a memorable jazz score by Elmer Bernstein. The muddled narrative and some bad casting are but minor faults.
DRA 119 min (ort 120 min) B/W
VIDrel: 4-FRONT/POLYREC V
Boa: novel by Nelson Algren.

**MAN WITH THE GOLDEN GUN, THE ** ** PG
Guy Hamilton UK 1974
Roger Moore, Christopher Lee, Britt Ekland, Maud Adams, Marc Lawrence, Herve Herve Villechaize, Bernard Lee, Clifton James, Lois Maxwell, Desmond Llewellyn, Richard Loo, Soon-Teck Oh, James Cossins, Chan Yiu Lam, Carmen Sautoy
Another espionage adventure with James Bond getting involved in a plan to liquidate a professional assassin in the East. Good car stunts and exotic locations are chiefly all this sluggish film has to offer, despite Lee's immense charisma as the villain and a set-piece encounter in a hall of mirrors. Ekland's presence adds absolutely nothing to this film. Number nine in the series, it was followed by THE SPY WHO LOVED ME.
A/AD 119 min (ort 125 min) wScrn VIDrel: MGM/WHV
V/dm
Boa: novel by Ian Fleming.

**MAN WITH THREE ARMS, THE ** ** 15
Adam Rifkin USA 1991
Judd Nelson, Wayne Newton, Lara Flynn Boyle, Bill Paxton, James Caan, Rob Lowe, Claudia Christian
A hopeless stand-up comic who cannot raise any laughs gets some unexplained help, when he starts to grow a third arm in the middle of his back. A strange B-movie offering, set in a depressed and broken-down world, that tries hard for cult status.
Aka: DARK BACKWARD, THE
COM 99 min VIDrel: 20VIS/SONOP V

**MAN WITH TWO BRAINS, THE ** ** 15
Carl Reiner USA 1983
Steve Martin, David Warner, Kathleen Turner, Paul Benedict, James Cromwell, Richard Brestoff, James Cromwell, George Furth, Randi Brooks, Peter Hobbs, Earl Boen, Bernie Hern, Frank McCarthy, William Taylor, Sissy Spacek (voice only)
A brilliant brain surgeon learns to his dismay that he has married a heartless gold-digger, but this discovery coincides with his growing affection for a female brain being kept alive by a crazy scientist. This crazy and disorganised spoof is reminiscent of AIRPLANE! but is let down by a general failure to exploit its many varied ideas.
COM 86 min (ort 93 min) VIDrel: WHV V/h

**MAN WITH X-RAY EYES, THE ** ** PG
Roger Corman USA 1963
Ray Milland, Diana Van Der Vlis, Harold J. Stone, John Hoyt, Don Rickles, John Dierkes, Kathryn Hart, Lorie Summers, Vicki Lee, Carol Irey, Dick Miller, Barboura Morris, Morris Ankrum, Jonathan Haze
A surgeon discovers a means of giving himself X-ray vision and is at first intoxicated by his discovery, but eventually becomes an outcast and finally succumbs to madness. A moody and compelling little story that starts off with great promise but eventually gets bogged down in cheap effects and a muddled script. The unpleasant climax ends the film abruptly without resolving anything.
Aka: X: THE MAN WITH X-RAY EYES, THE
FAN 76 min (ort 79 min) VIDrel: L/A V

MAN WITHOUT A FACE, THE * ** 15
Mel Gibson USA 1993
Mel Gibson, Margaret Whitton, Fay Masterson, Geoffrey Lewis, Richard Masur, Nick Stahl, Gaby Hoffmann, Michael Deluise, Jean De Baer, Jack De Mave, Viva, Ethan Phillips, Justin Janew, Sean Kellman, Chris Linburg, Kelly Wood
A young boy in Maine who has no father, is helped in his loneliness by a man with a badly scarred face and body, injuries which he received in a horrifying road accident. Initially at first only motivated by morbid curiosity, he approaches the man but the two eventually develop a warm and mutually supportive relationship. A moving and warm-hearted tale of some depth. Ably directed by Gibson in his directorial debut.
DRA 109 min (ort 115 min) VIDrel: EIV/SONOP V/sur
Boa: novel by Isabelle Holland.

MAN, WOMAN AND CHILD * ** PG
Dick Richards USA 1982
Martin Sheen, Blythe Danner, Sebastian Dungan, Craig T. Nelson, Arlene McIntyre, David Hemmings, Nathalie Nell, Maureen Anderman, Sebastian Dungan, Missy Francis, Billy Jacoby, Ruth Silveira, Jacques Francois
A happily married man learns that a brief affair ten years before produced a son who has just become an orphan. Scripted by Erich Segal and David Selag and despite the expected lashings of sentimentality, surprisingly effective.
DRA 96 min (ort 100 min) VIDrel: RCA L/A V
Boa: novel by Erich Segal.

MANCHURIAN CANDIDATE, THE ** ** 15
John Frankenheimer USA 1962
Frank Sinatra, Laurence Harvey, Janet Leigh, James Gregory, Angela Lansbury, Henry Silva, John McGiver, Leslie Parrish, Khigh Deigh
Gripping thriller about a Korean war hero who comes back home after having been brainwashed and programmed to assassinate a liberal politician. The feeling of tension and political paranoia never lets up throughout and the film builds up to a shattering climax. A first rate work, with Lansbury giving a remarkably creepy performance as the man's mother and secret controller. The daft comedy GOING BERSERK was inspired by this film.
THR 121 min (ort 126 min) B/W VIDrel: WHV V
Boa: novel by Richard Condon.

MANDELA * ** PG
Philip Saville USA 1987
Danny Glover, Alfre Woodard, John Matshikiza, Warren Clarke, Allan Corduner, Julian Glover, John Indi
An absorbing examination of the life of Nelson and Winnie Mandela and the former's struggle against the South African apartheid regime. Despite a tendency to glamorise the subject and an unwillingness to examine the more questionable activities of the ANC, the film certainly has many effective moments. The script is by Ronald Harwood. Filmed on location in Zimbabwe.
DRA 144 min mCab VIDrel: IMC/DISC V

**MANDINGO ** * 18
Richard Fleischer USA 1975
James Mason, Perry King, Susan George, Richard Ward, Ken Norton, Brenda Sykes, Lillian Hayman, Paul Benedict, Ji-Tu Cumbuka, Ben Masters, Louis Turenne, Ray Spruell, Debbie Morgan, Irene Tedrwo, Reda Wyatt, Simon McQueen, John Barber
Passions run high on a slave-breeding plantation in the steamy South, with Mason playing a bigoted plantation owner to perfection, one of the few good things in this trashy and overblown

melodrama. Followed by the sequel DRUM. See also BLACK-SNAKE.
DRA 120 min Cut (47 sec – ort 127 min)
VIDrel: 4-FRONT/POLYREC/CIC L/A V/sh
Boa: novel by Kyle Onstott/play by Jack Kirkland.

**MANDROID ** ** 15
Jack Ersgaard USA 1992
Brian Cousins, Jane Caldwell, Michael Dellafemina, Curt Lowens, Patrick Ersgaard, Costel Constantin, Robert Symnonds, Ion Haiduc, Mircea Albulescu, Adrian Pintea, Radu Minculescu
Having discovered a powerful new element, a father-and-daughter team of scientists build title robot to handle it, but their powerful creation is stolen by a mad, disfigured scientist called Dr Drago. A standard tale, badly told and filmed on location in Romania, which is burdened by its low budget, indifferent acting and direction, and poor special effects. However, this did not stop the making of equally bad sequel: INVISIBLE: THE CHRONICLES OF BENJAMIN KNIGHT.
FAN 78 min (ort 81 min) VIDrel: FULL/HIFLI L/A V/h

**MANDY ** ** PG
Alexander Mackendrick UK 1952
Phyllis Calvert, Terence Morgan, Mandy Miller, Jack Hawkins, Godfrey Tearle, Marjorie Fielding, Patricia Plunkett, Dorothy Alison, Nancy Price, Jane Asher, Edward Chapman, Eleanor Summerfield, Colin Gordon, Julian Amyes, Gabrielle Brune
A couple's marriage almost founders when the wife defies her husband's wishes and takes their dumb-and-deaf daughter to a special school, where she can receive a suitable education. Rumours of his wife's infidelity reach him and he goes to the school to investigate but has a change of heart when he sees how well his daughter is doing there. A capable, small-scale human drama.
Aka: CRASH OF SILENCE
DRA 93 min (ort 100 min) B/W VIDrel: LUMI/SPEAR V
Boa: novel The Day is Ours by Hilda Lewis.

**MANGLER, THE ** ** 18
Tobe Hooper USA 1995
Robert Englund, Ted Levine, Daniel Matmor, Jeremy Crutchley, Vanessa Pike, Demetre Phillips, Lisa Morris, Vera Blacker, Ashley Hayden, Danny Keogh, Gerrit Schonhoven, Ted Leplat, Todd Jensen, Nan Hamilton, Adrian Waldron
An enormous combined mangler and folding machine at a vast industrial laundry is possessed by evil spirits that causes it to maim and injure the employees and a cop sets out on a mission of exorcism. A visually impressive but overblown piece that is much in need of a stronger plot and some believable characters. The earlier film CHRISTINE and MAXIMUM OVERDRIVE (both based on King stories) explored much the same idea as did the 1974 TV film "Killdozer".
HOR 101 min (ort 105 min) VIDrel: GUILD/FOXVID V/sh
Boa: short story by Stephen King.

**MANHATTAN ** ** ** 15
Woody Allen USA 1979
Woody Allen, Diane Keaton, Mariel Hemingway, Michael Murphy, Meryl Streep, Anne Byrne, Karen Ludwig, Michael O'Donoghue, Gary Weis, Kenny Vance, Damion Sheller, Tisa Farrow, Wallace Shawn, Helen Hanft, Bella Abzug, Victor Truro
A wry film about the life and loves of a New York writer and his friends, built around his love affair/obsession with New York and everything about it. A rich, funny, sharp and poignant comedy, and one that sums up the director's idiosyncratic view of life better than any of his earlier works. The music of Gershwin is used to great effect and the wonderful photography is by Gordon Willis. Look out for Karen Allen, who has a small cameo.
COM 92 min (ort 96 min) B/W wScrn VIDrel: MGM/WHV V

**MANHATTAN MURDER MYSTERY ** ** ** PG
Woody Allen USA 1993
Woody Allen, Diane Keaton, Angelica Huston, Alan Alda, Jerry Adler, Melanie Norris, Jon Rifkin, Joy Behar, Lynn Cohen, William Addy, John Doumanian, Sylvia Kanders, Marge Redmond, Zach Braff, George Manos, Linda Taylor, Aida Turturro
Allen and Keaton are a middle-aged couple who live in an upmarket apartment block, where they are on nodding terms with their neighbours, an elderly couple. When the wife apparently dies from a heart attack, Keaton becomes increasingly

suspicious, eventually stumbling upon evidence that the husband may have killed her. An amazingly clever piece, witty and well conceived but not especially funny. However, there are numerous good moments and a strong sense of tension.
COM 103 min (ort 107 min) cC
VIDrel: VCC/DISC/COLUM V/sur

**MANHUNTER ** ** ** 18
Michael Mann USA 1986
William L. Petersen, Joan Allen, Stephen Lang, Brian Cox, Tom Noonan, Dennis Farina, Kim Griest, David Seaman, Benjamin Hendrickson, Michael Talbott, Michele Shay, Dan E. Butler, Robin Moseley, Paul Perri, Patricia Charbonneau, Alex Neil
An FBI agent who has received special psychological training and possesses the ability to read the minds of killers and understand their thinking, is recalled from retirement to deal with a very difficult case; a serial killer who chooses his victims with great care. See also THE SILENCE OF THE LAMBS, a considerably more ambitious adaptation of a Harris novel.
A/AD 115 min (ort 124 min) VIDrel: 20TH/TECH V/sur
Boa: novel Red Dragon by Thomas Harris.

**MANIAC COP ** ** 18
William Lustig USA 1988
Tom Atkins, Bruce Campbell, Laurene Landon, Richard Roundtree, Sheree North, William Smith, Robert Z'dar, Nina Aversen, Nick Barbaro, Lou Bonacki, Barry Brenner, Victoria Catlin, Jim Dixon, Corey Eubanks, Jill Gatsby, Jon Green
A series of brutal murders in New York prove to be the handiwork of a deranged cop, causing a situation that culminates in clashes between the police and members of the public. Some clumsy attempts at black humour and flat direction kill this one off before it even gets started. Followed by some inferior sequels.
HOR 81 min Cut (5 sec – ort 92 min)
VIDrel: 4-FRONT/POLYREC/MED V

**MANIAC COP 2 ** * 18
William Lustig USA 1990
Robert Davi, Claudia Christian, Michael Lerner, Bruce Campbell, Laurene Landon, Robert Z'dar, Clarence Williams III, Leo Rossi, Lou Bonacki, Charles Napier, Paul Trickey, Santos Morales, Robert Earl-Jones, Andrew Hill Newman
In this low-budget sequel to the earlier film, our murdered and apparently unstoppable cop returns once more to embark on yet another round of blood-letting, this time joining forces with a serial killer. A trashy and gore-laden film, pitched at about the same level as all those NIGHTMARE ON ELM STREET sequels, only a good deal less witty.
HOR 83 min Cut (4 sec – ort 90 min)
VIDrel: POLY/POLYREC/MED V/sur

**MANIAC COP 3: BADGE OF SILENCE ** * 18
William Lustig USA 1992
Robert Davi, Robert Z'dar, Caitlin Dulany, Gretchen Becker, Paul Gleason, Jackie Earle Haley, Julius Harris, Grand Bush, Doug Savant, Robert Forster, Bobby Di Ciceo, Frank Pesce, Lou Diaz, Vanessa Marquez, Denney Pierce
Second sequel continues the story of our murderous cop who continues to practice his handiwork on all and sundry, resulting in the usual torrent of gore. An assembly-line effort in a series that threatens to be as long-running as all those FRIDAY THE 13TH films.
HOR 81 min (ort 85 min) VIDrel: POLY/POLYREC/MED V/sh

**MANITOU, THE ** * 18
William Girdler USA 1978
Tony Curtis, Susan Strasberg, Michael Ansara, Ann Sothern, Burgess Meredith, Stella Stevens, Jon Cedar, Paul Mantee, Jeanette Nolan, Lurene Tuttle, Joe Glieb, Ann Mantee, Hugh Corcoran, Tenaya, Jan Heininger, Carole Hemingway
An Indian medicine man is resurrected in a growth on a girl's neck and bursts upon the world in a welter of gory special effects. A graphic foray in territory explored by "The Exorcist" with a silly plot but rather good special effects.
HOR 99 min (ort 105 min) VIDrel: POLY L/A V
Boa: novel by Graham Masterton.

**MANNEQUIN ** * PG
Michael Gottlieb USA 1986
Andrew McCarthy, Kim Cattrall, Meshach Taylor, G.W. Bailey, James Spader, Estelle Getty, Carole Davis, Stephen Vinovich, Christopher

Maher, Jake Jundeff, Phyllis Newman, Phil Rubenstein, Jeffrey Lampert, Kenneth Lloyd, Harvey Levine
An unemployed sculptor get a job as a window-dresser in a large department store, where he finds that a mannequin he once created has magically come to life. The expected love affair transpires, in a clumsy slapstick film that is spoilt by its incessant pop background. Taylor as gay window-dresser "Hollywood" is easily the best thing in this dreary effort (he reprises this role in MANNEQUIN ON THE MOVE). See also ONE TOUCH OF VENUS and A MOM FOR CHRISTMAS.
COM   86 min (ort 90 min)   VIDrel: WHV   V/sur

MANNEQUIN ON THE MOVE *
Stewart Raffill USA
PG
1991
Kristy Swanson, William Ragsdale, Meshach Taylor, Terry Kiser, Stuart Pankin, Cynthia Harris, Andrew Hill Newman, Julie Foreman, John Edmondson, Phil Latella, Mark Gray, Erick Weiss, Jackye Roberts, John Casino, Laurie Wing
Less a sequel than a reworking of the earlier film MANNEQUIN, with the story slightly altered, making our enchanted beauty a stone statue of a peasant girl, who was frozen a thousand years ago by an evil count. The latter's descendant now arrives on the scene to claim her for himself. As flat and mechanical a re-run as one could imagine (despite good work from Taylor, largely repeating the role he had in MANNEQUIN) with dialogue and acting to match.
Aka: MANNEQUIN 2: ON THE MOVE
COM   91 min (ort 98 min)   VIDrel: MGM/WHV L/A   V/s

MANON DES SOURCES ***
Claude Berni FRANCE
PG
1987
Yves Montand, Daniel Auteuil, Emmanuelle Beart, Hippolyte Girardot, Andre Dupon, Margarite Lozano, Elisabeth Depardieu, Gabriel Bacquier, Armand Meffre, Pierre Nougaro, Jean Maurel, Roger Souza, Didier Pain
Second and concluding part of the story that began with JEAN DE FLORETTE, that deals with the revenge inflicted by Jean's daughter on those responsible for his death. An absorbing tale that makes its leisurely way to an unexpected climax.
Aka: MANON OF THE SPRING
DRA   115 min (ort 120 min) wScrn
VIDrel: ELPIC/POLYREC   V
Boa: novel by Marcel Pagnol.

MAN'S BEST FRIEND **
John Lafia USA
15
1993
Ally Sheedy, Lance Henriksen, Robert Costanzo Fredric Lehne, John Cassini, J.D Daniels, William Sanderson, Trula M. Marcus, Robin Frates, Rick Barker, Bradley Pierce, Robert Arentz, Cameron Arnett, Adam Carl, Tom Rosales Jr
An ambitious journalist investigating a story breaks into a top-secret research facility and comes face to face with a genetically engineered guard dog, and gradually forms a relationship with this aggressive and wily beast. A standard monster story, with plot development and logic sacrificed in favour of ample gore.
HOR   90 min   VIDrel: GUILD/SONOP L/A   V/sh

MANSFIELD PARK: PARTS 1 AND 2 **
David Giles UK
U
1986
Anna Massey, Bernard Hepton, Angela Pleasence, Nicholas Farrell, Sylvestra le Touzel, Samantha Bond
Television adaptation of a story about a poor girl of good family snubbed by her arrogant rich relatives in 18th century England. Fair.
DRA   260 min (2 cassettes) mTV   VIDrel: BBC/TECH   V/h
Boa: novel by Jane Austen.

M.A.N.T.I.S. **
Eric Laneuville USA
PG
1994
Carl Lumbly, Gina Torres, Bobby Hosea, Christopher, M. Brown, Yvonne Farrow, Steve James, Obba Battunda, Marcia Cross, Wendy Racquel Robinson, Philip Baker Hall, Yvonne Farrow, Francis X. McCarthy, Alan Fudge, Grant Heslov, Luis Ramos
A bio-physicist becomes paraplegic after being shot in riot but invents an exo-skeleton suit that gives him super strength and agility, and allows him to become a crime-fighter. The script for this work was originally intended for Sam Raimi, and there is more than a passing similarity in places to DARKMAN. But given that, it's still a perfectly watchable effort, quite fast-paced and well conceived.
FAN   84 min mTV   VIDrel: CIC/SONOP   V

MAP OF THE HUMAN HEART ***
Vincent Ward AUSTRALIA/CANADA/FRANCE/UK
15
1992
Jason Scott Lee, Patrick Bergin, Anne Parillaud, Robert Joamie, Annie Galipeau, John Cusack, Jeanne Moreau, Ben Mendelsohn, Clotilde Courau, Jerry Snell, Jayko Pitseolak, Matt Holland, Rebecca Vevee, Josape Kopalee, Harry Hill
An Inuit boy is taken from his original home and raised as white by a Montreal mapmaker in the 1930s, but is scarred by the racism he experiences, especially the reaction to his friendship with a half-cast Mestis girl he meets in hospital. Years later, he meets up with his lost love in Dresden. A beautifully filmed tale, with excellent scenery, but slightly marred by the rambling plot. However, in terms of sheer visual impact, it cannot be faulted.
DRA   104 min (ort 109 min)   VIDrel: VCC/DISC/COLUM   V/sh

MAPANTSULA ***
Oliver Schmitz SOUTH AFRICA
15
1988
Thomas Mogotlane, Marcel Van Heerden, Thembi Mtshali, Dolly Rathebe, Peter Sephuma, Gabriel Dichabe, Brad Morris, Darlington Michaels, Eugene Majola, Duana Ngembe, Polite Dlamini, Jerry Mokgoko, Similio Makahambi
In Soweto, a happy-go-lucky crook attempts to gain access to the white world in furtherance of his criminal ambitions, but is soon caught and sent to jail. Once there, influenced by his brutal police mistreatment and his more radical cell-mates, he slowly grows in political awareness. A sharp and vigorous piece of social commentary, made in the face of South African censorship restrictions by the ploy of using a script different to the one submitted.
Aka: HUSTLER
DRA   99 min (ort 109 min)   VIDrel: CONNO/RTM   V

MARAT/SADE ***
Peter Brook UK
(15)
1966
Patrick Magee, Ian Richardson, Glenda Jackson, Michael Williams, Clifford Rose, Freddie Jones, Hugh Sullivan, Morgan Sheppard, Jonathan Burn, John Steiner, Jeanette Landis, Robert Lloyd
Excellent adaptation of the stage play by Weiss that attempts to "recreate" the story of Marat's death, as performed by the inmates of a French lunatic asylum. A powerful work that makes for compelling viewing, despite the fact that it demands patience and concentration.
Aka: PERSECUTION AND ASSASSINATION OF JEAN-PAUL MARAT AS PERFORMED BY THE INMATES OF THE ASYLUM OF CHARENTON UNDER THE DIRECTION OF THE MARQUIS DE SADE, THE
DRA   115 min   SATrel: SKY MOVIES
Boa: play by Peter Weiss.

MARATHON MAN ***
John Schlesinger USA
18
1976
Dustin Hoffman, Laurence Olivier, Roy Scheider, William Devane, Marthe Keller, Fritz Weaver, Marc Lawrence, Richard Bright, Allen Joseph, Ben Dova, Tito Goya, Lou Gilbert, Jacques Marin, James Wing Woo, Nicole Deslauriers
A marathon runner gets drawn into an intricate web of intrigue and finds himself the quarry of a Nazi war criminal, who has returned from Uruguay to the USA in order to obtain some diamonds that were being kept for him by his now dead brother. An over-complex film that appears to be about to make some valid observations, when it changes direction and becomes a vicious-minded thriller. It is however, all very well done. The script is by Goldman.
THR   119 min (ort 126 min)
VIDrel: 4-FRONT/POLYREC/CIC   V
Boa: novel by William Goldman.

MARGARET'S MUSEUM ***
Mort Ransen CANADA/UK
15
1995
Helen Bohham-Carter, Kate Nelligan, Clive Russell, Craig Olejnik, Andrea Morris, Peter Boretski, Kenneth Welsh, Barry Dunn, Norma Dell'Agnese, Glenn Wadman, Elizabeth Richardson, Ida Donovan, Gordon Joe, Wayne Reynolds
In 1940s Nova Scotia, a spirited young woman dreams of marriage to a man who is not obliged to work in the coal-mines that so dominate the area, as she has already lost both her father and older brother. She does indeed meet and marry one such man, but her happiness is shortlived, for when he loses his job he too enters the mines and is eventually killed. The death unhinges her mind, and she responds with a gruesome act of mourning. A tragic, bizarre love story.
110 min   CINrel
Boa: novel The Glace Bay Miner's Museum by Sheldon Currie.

MARIA'S CHILD ** *(PG)*
Malcolm McKay UK 1992
Yolanda Vazquez, David O'Hara, Jemmais Keval-Baxter, Sophie Okonedo, Linda Davidson, Rudi Davis, Gary Mavers, Nick Woodeson, Bob Goody, Ricardo Aliaga, Kathryn Hunter, Seeta Indrani, Ray Marioni, Robert Purvis, Alec McCowen
An Italian dancer falls pregnant, a complication to her life she does not view with pleasure, but matters take a distinctly bizarre turn when the unborn child starts communicating with her. A weird psychological drama that at times verges on farce, with heavy symbolism and a message swimming about in there somewhere. Some interesting issues are raised, but then discarded without examination.
DRA 93 min (ort 95 min) mTV TVrel: BBC

MARIA'S LOVERS *** 15
Andrei Konchalovsky USA 1984
Nastassja Kinski, John Savage, Robert Mitchum, Keith Carradine, Anita Morris, Bud Cort, Tracy Nelson, Vincent Spano, Karen Young, Tracy Nelson, John Goodman, Danton Stone, Lela Ivey, Elena Koreneva, Anton Sipos, Larry John Meyer
An American GI returns home from a Japanese POW camp to marry his childhood sweetheart, but is unable to consummate their marriage because of the harrowing wartime experiences he endured. A detailed and absorbing tale that would have been far better if the general air of gloom had been a good deal less marked. The opening sequence makes good use of footage from a John Huston documentary: "Let There Be Light".
DRA 104 min (ort 110 min) VIDrel: GUILD/SONOP L/A V

MARIHUANA: THE DEVIL'S WEED *(12)*
Dwain Esper USA 1936
Harley Wood, Hugh McArthur, Pat Carlyle, Paul Ellis, Dorothy Dehn, Richard Erskine
A 1930s tale that cautions on the dangers of smoking dope. Wonderfully inept and overblown, and a fine companion to REEFER MADNESS and ASSASSIN OF YOUTH.
Aka: MARIHUANA; MARIHUANA: DEVIL'S WEED WITH ROOTS IN HELL
DRA 54 min (ort 56 min) B/W VIDrel: L/A V

MARINE ISSUE ** 18
Christopher Bentley (Denis Amar) GIBRALTAR 1986
Michael Pare, Tawny Kitaen, Peter Crook, Charles Napier, Eddie Avoth, G. Scott Del Amo, Lynda Bridges, Maurice E. Aronow, Lionel A. Ephraim, Aldo San Brell, Peter Boulter, Thomas Abbott, Scott Miller, Manuel De Blas
When his sister is brutally murdered by drug smugglers, a young marine soldier feels compelled to resign in order to track down and punish the culprits. Good action sequences partially redeem a hackneyed plot.
Aka: INSTANT JUSTICE
A/AD 97 min (ort 101 min) VIDrel: WHV V/sh

MARIO AND THE MOB ** *PG*
Virgil Vogel USA 1990
Robert Conrad, Ann Jillian, Byrne Pivan, Bo Kaprall, Mitch Rouse, La Velda Fann, Mark Morettini, Michael Nicolosi, Jay Leggett, Steve Cory, Kyle Cory, Christopher Noga, Tammy Brady, Jesse Erwin, Jim Ortlieb, John Finn, James Krag
A gangster with a kind heart finds himself obliged to adopt five kids in this fairly anodyne effort.
COM 92 min mTV VIDrel: CAPIT/GUILD V

MARK OF THE DEVIL ** 18
Michael Armstrong UK/WEST GERMANY 1969
Herbert Lom, Olivera Vuco, Udo Kier, Reginald Nalder, Herbert Fuchs, Michael Maien, Ingeborg Schoener, Gaby Fuchs, Dorothea Carrera, Adrian Hoven, Doris Von Danwitz, Marlies Peterson, Gunther Clemens, Johannes Buzalski
Story of a cult of devil worshippers who bring death and destruction to the countryside during the 18th century. An interminable parade of amputations, torturings and sexually depraved torments, most of which are carried out at the behest of the local witchfinder (Lom). Clearly made to cash in on the success of the superior WITCHFINDER GENERAL, this film makes up in gore what it lacks in imagination. Followed by a sequel in 1972. Average.
Aka: AUSTRIA 1700; BRENN HEXE BRENN; BURN WITCH BURN; HEXEN BIS AUFS BLUT GEQUAELT; MARK OF THE WITCH; SATAN
HOR 88 min (ort 98 min) VIDrel: REDEM/RTM V

MARK OF ZORRO, THE **** *U*
Rouben Mamoulian USA 1940
Tyrone Power, Linda Darnell, Basil Rathbone, Gale Sondergaard, Eugene Pallette, Gale Sondegaard, J. Edward Bromberg, Montagu Love, Janet Beecher, Robert Lowery, Chris-Pin Martin, George Regas, Belle Mitchell, John Bleifer
Unsurpassed film version of this novel, set in early 19th California during the Spanish period. A young man returns home from his studies in Spain to find his father having been replaced as governor, and the region in the grip of the new governor's tax collectors. Outwardly, a vain and foppish fool, our hero dons a mask and raises the flag of revolt. Fine performances, excellent action sequences and a strong script make this one unforgettable.
A/AD 90 min (ort 93 min) B/W cC VIDrel: 20TH/TECH V/h
Boa: novel The Curse Of Capistrano by Johnston McCulley.

MARK TWAIN AND ME *** *U*
Daniel Petrie CANADA 1991
Jason Robards, Talia Shire, R. H. Thompson, Fiona Reid, Chris Wiggins, Amy Stewart, Annie Ferguson, Chapelle Jaffe, Colin Fox, Bunty Webb, Brian Paul, Jenny Turner, Corinne Condey, Richard Poley, James Mainprice, Lee MacDougall
A young girl travelling ship is delighted to discover that her favourite author, Mark Twain, is also one of the passengers. Despite his gruff and severe exterior and manners, the two soon strike up quite a cordial relationship, in this sensitive and well-handled tale.
JUV 90 min (ort 93 min) mTV VIDrel: WDV/TECH L/A V
Boa: novel Enchantment: A Little Girl's Friendship With Mark Twain by Dorothy Quick.

MARKED WOMAN ** *PG*
Lloyd Bacon USA 1937
Bette Davis, Humphrey Bogart, Lola Lane, Isabel Jewell, Jane Bryan, Eduardo Ciannelli, Allen Jenkins, Mayo Methot, Ben Welden, Henry O'Neill, Rosalind Marquis, John Litel, Damian O'Flynn, Robert Strange, James Robbins, Sam Wren
A tough D.A. has to persuade a clip-joint hostess and her four girlfriends to give evidence against their gangland boss. A rather average gangster drama, with a female slant and the usual strong performance by Davis.
DRA 91 min (ort 97 min) B/W VIDrel: MGM/WHV V/h

MARKED FOR DEATH ** 18
Dwight H. Little USA 1990
Steven Seagal, Joanna Pacula, Keith David, Basil Wallace, Tom Wright, Bette Ford, Elizabeth Gracen, Danielle Harris, Al Israel, Arlen Dean Snyder, Victor Romero Evans, Michael Ralph, Jeffrey Anderson-Gunter, Tony Di Benedetto
A retired narcotics agent returns to his hometown to live a quiet life, but finds it in the grip of a drugs baron. When the agent kills one of his henchmen, the latter swears revenge in a fight to the death. Another undistinguished and violent tale that gives Seagal ample scope to do what he does best.
A/AD 89 min Cut (22 sec – ort 98 min) VIDrel: 20TH/TECH V

MARKED FOR MURDER ** 15
Mimi Leder USA 1992
Powers Boothe, Laura Johnson, Michael Ironside, Jay Acovone, Lou Liberatore, Gary Werntz, Billy Dee Williams, Richard Grove, Sean O'Bryan, Badja Djola, Tom Dahlgren, Stogie Kenyatta, Michael Shamus Wiles, Bill Henderson, Robert Z'dar
A convict is released from jail and promised a fresh start if he agrees to take part in a dangerous scheme to uncover a group of crooked cops in one of the city's precincts.
DRA 90 min mTV VIDrel: IMPENT L/A V

MARKED MAN ** 18
Marc Voizard USA 1996
Roddy Piper, Jane Wheeler, Miles O'Keeffe, Chris Bolton, Tyrone Benskin, Alina Thompson, Dennis O'Connor, Richard Zeman, Jason Cavalier, Chip Chuipka, Claire Sims, David Nichols, Desmond Campbell, Minor Mustain, Johnny Gore
After a man's fiancee dies after a hit-and-run accident, he kills the driver in revenge and is sent to a minimum security prison for three years. There he witnesses two corrupt guards murdering an inmate and finds himself framed for this crime. In order to have a chance to clear his name, he is forced to escape and go on the run.
A/AD 90 min VIDrel: FIRST/SONOP V

MARNIE ** 15
Alfred Hitchcock USA 1964
Tippi Hedren, Sean Connery, Martin Gabel, Diane Baker, Louise Latham, Alan Napier, Bruce Dern, Henry Beckman, Bob Sweeney, Milton Selzer, Edith Evanson, Mariette Hartley, S. John Launer, Meg Wyllie, Louise Lorimer
A curiously flat drama about a frigid kleptomaniac who robs a series of employers but is forced into marrying the latest one who is wise to her wiles. Marred by an unsatisfactory resolution and the use of very obvious backcloths and studio sets, that one really wonders whether the director cared a jot about creating the illusion of reality.
DRA 124 min (ort 130 min) VIDrel: CIC/SONOP V/h
Boa: novel by Winston Graham.

MAROONED ** U
John Sturges USA 1969
Gregory Peck, Richard Crenna, David Janssen, James Franciscus, Gene Hackman, Lee Grant, Nancy Kovak, Mariette Hartley, Scott Brady, George Gaynes, Walter Brooke, Mauritz Hugo, Craig Huebling, John Carter, Frank Marth, Duke Hobbie
A rescue mission is mounted to bring back three astronauts stranded in space when their retro-rockets fail to ignite and they are left with only a 42-hour supply of oxygen. Despite the flashy effects and sets, the film moves along at such a pedestrian pace that no tension is ever developed. Written by Mayo Simon. AA: Effects/vis (Robbie Robertson).
FAN 124 min (ort 134 min) VIDrel: RCA L/A V
Boa: novel by Martin Caidin.

MARQUIS ** 18
Henri Xhonneux BELGIUM/FRANCE 1989
Voices of: Bien De Moor, Philippe Bizot, Gabrielle Van Damme, Francois Marthoreut, Valerie Kling, Michel Robin, Isabel Cane-Wolfe, Vicky Messica, Rene Lebrun, Nathalie Juvet, Bob Morel, Roger Crouzet, Willem Holtrop, Erica De Saria
An intellectual satire on the French Revolution and much else, loosely based on the writings of the Marquis De Sade, in which all the actors wear animal masks and Claymation figures are also used.
DRA 79 min (ort 88 min) VIDrel: ICAPRO/MANGA V

MARQUISE OF O, THE ** PG
Eric Rohmer FRANCE/WEST GERMANY 1976
Edith Clever, Bruno Ganz, Peter Luhr, Edda Seippel, Otto Sander
A young widowed noblewoman is saved from raped but finds herself pregnant some months later and is totally perplexed at her condition, as she was unconscious when her rescuer took advantage of her. Slavishly faithful to the literary original, but related as a series of lifeless tableaux vivants, and accompanied by a tiresome narration. A well made drama, but one that lacks both conviction and life.
Aka: DIE MARQUISE VON O; LA MARQUISE D'O
DRA 98 min (ort 102 min) VIDrel: CONNO/RTM V
Boa: novella by Heinrich von Kleist.

MARRIAGE ITALIAN STYLE ** 15
Vittorio De Sica FRANCE/ITALY 1964
Sophia Loren, Marcello Mastroianni, Aldo Puglisi, Pia Lindstrom, Vito Moriconi, Giovanni Ridolfi, Generoso Cortini, Tecla Scarano, Marilu Tolo, Raffaello Rossi Bussola, Vincenza Di Capua, Vincenzo Aita
A prostitute has to use all her wits to get a client of long-standing to take the plunge and marry her when he announces his forthcoming marriage to a young girl. Having taken to her bed, she sends him word that she's dying, and he rushes to her side to propose, whereupon she makes a miraculous recovery (and announces the existence of three grown-up sons. A likeable comedy that makes a few sharp points about Italy's cumbersome divorce laws.
Aka: MATRIMONIO ALL'ITALIANA
COM 96 min (ort 102 min) dubbed VIDrel: ARROW/RTM V
Boa: play Filomena Marturano by Eduardo de Filippo.

MARRIAGE OF MARIA BRAUN, THE *** 15
Rainer Werner Fassbinder WEST GERMANY 1978
Hanna Schygulla, Klaus Lowitsch, Ivan Desny, Gottfried John, Gisela Uhlen, Gunter Lamprecht, Mark Bohm, George Byrd, Elizabeth Trissenaar, Liselotte Eder, Rainer werner Fassbinder, Isolde Barth, Peter Berling, Sonja Neudorfer
A bride in the post-war period builds an industrial empire after her husband is imprisoned for murdering her black GI boyfriend. A heavy mixture of irony and drama, but never less than fully absorbing. The first of three post-WW2 accounts by the director, with "Lola" and then "Veronika Voss" following.
Aka: DIE EHE DER MARIA BRAUN
DRA 115 min (ort 120 min) VIDrel: CONNO/RTM V

MARRIED PEOPLE, SINGLE SEX ** 18
Mike Sedan USA 1993
Chase Masterson, Josef Pilato, Darla Slavens, Wendi Westbrook, Shelley Michelle, Robert Zachar, Samuel Mongiello. Teri Thompson
Three couples decide to enlarge their range of sexual experience in a variety of ways, in this overlong erotic drama.
DRA 110 min VIDrel: MARQ/QUANT V

MARRIED PEOPLE, SINGLE SEX 2: FOR BETTER OR WORSE ** 18
Mike Sedan USA 1994
Kathy Shower, Monique Parent, Craig Stepp, Doug Jeffrey, Liza Smith, Sam Schueler
Sequel to the first film with another three couples having to contend with a variety of marital problems, in this so-so offering.
DRA 99 min VIDrel: MARQ/QUANT V

MARRIED TO IT ** 15
Arthur Hiller USA 1991
Beau Bridges, Stockard Channing, Robert Sean Leonard, Mary Stuart Masterson, Cybill Shepherd, Ron Silver, Don Francks, Donna Vivino, Jimmy Shea, Ed Koch, Nathaniel Moreau, Gerry Bamman, Chris Wiggins, Larry Reynolds, Djanet Sears
A children's planned school play brings together three very different New York couples who gradually become friends, in this cliche-ridden drama that does very little with the efforts of a capable cast.
DRA 107 min VIDrel: VCC/DISC/COLUM V/sur

MARRIED TO THE MOB ** 15
Jonathan Demme USA 1989
Michelle Pfeiffer, Matthew Modine, Dean Stockwell, Mercedes Ruehl, Alec Baldwin, Joan Cusack, Trey Wilson, Charles Napier, Tracey Walter, Al Lewis, Paul Lazar, Marlene Willoughby, Frank Acquilino, Charles Napier, Jason Allen
When a woman's gangster husband is killed she goes on the run, intending to make a new life for herself. However, she is pursued by both an FBI agent, who wants information, and the Mafia boss who had her husband killed and who is besotted with her. A screwball comedy that has several engaging cameos and some witty dialogue, but is a little too chaotic for its own good. The score is by David Byrne.
COM 99 min (ort 104 min)
VIDrel: 4-FRONT/POLYREC/VISVID V/sur

MARS ATTACKS! ** 12
Tim Burton USA 1997
Jack Nicholson, Glenn Close, Annette Bening, Pierce Brosnan, Danny DeVito, Martion Short, Sarah Jessica Parker, Michael J. Fox, Rod Steiger, Tom Jones, Lukas Haas
After a fleet of Martian spacecraft are revealed to be on their way to Earth, the American President prepares to welcome them, having been convinced that their intentions are friendly. But this proves to be just the opposite, as these repulsive looking creatures embark on a wholesale slaughter, but all the while espousing messages of friendship. A cruel and unfunny spoof on all those 1950s SF B-movies, liberally supplied with special effects in true Burton style.
COM 106 min CINrel

MARSHAL LAW * (18)
Stephen Cornwell HONG KONG 1996
Jimmy Smits, James Le Gros, Vonte Sweet, Scott Plank, Channon Roe, Micahel Cavalieri, Rodney Rowland, Tai Thai, Kristy Swanson
Smits has to take on a gang that is led by an experienced escaped con, and the usual variety of violent clashes, thrills and spills ensue, in this highly formulaic action-thriller.
A/AD 96 min VIDrel: 20VIS V

MARTIAL ARTS MASTER WONG FEI HUNG ** 12
Lee Hang/Lee Chiu HONG KONG 1992
Lam Ching Ying, Chin Kar Lok, Wu Chine Lien (Ng Sin Lin)
Story of the early life of the title figure, who has to deal with the murder of his father and his attempt to take over his position. As ever, the simple plot forms no more than a framework within

which to demonstrate the usual plethora of martial arts fisticuffs etc.
MAR   90 min   VIDrel: EAST/DISC   V

MARTIAL LAW **   18
S.E. Cohen USA   1990
Cynthia Rothrock, Chad McQueen, David Carradine, Philip Tan, Tony Longo, Andy McCutcheon, V.C. Dupree, Jim Malinda, Rick Walters, Patricia J. Wilson, Lars Lundgreen, Toru Tanaka, John Fujioka, Ethan Bortizer, Marty Dudek
McQueen (he's the son of Steve McQueen) plays an L.A. police officer and karate master whose girlfriend (Rothrock) is also a police officer and martial arts expert. However, the brother of the former is secretly in the pay of a ruthless gangster and martial artist, who is responsible for a series of underworld murders.
A/AD   84 min Cut (1 min 9 sec – ort 90 min)
VIDrel: 4-FRONT/POLYREC/EIV   V

MARTIAL MONKS OF SHAOLIN **   18
Godfrey Ho HONG KONG   1983
Dragon Lee, Wong Chen Li, Petty Suh, Jose Wong
All-action martial arts tale as a villain tries to seize control of a monastery. As ever, the plot is no more than a vehicle for some exciting sequences that display a variety of fighting styles.
Aka: MARTIAL MONKS OF SHAOLIN TEMPLE
MAR   85 min Cut (56 sec – ort 90 min)   VIDrel: IMPENT   V

MARTIAL OUTLAW **   18
Kurt Anderson GERMANY/USA   1993
Jeff Wincott, Gary Hudson, Richard Jaeckel, Krista Errickson, Vladimir Skomarovsky, Liliana Komorowska, Gary Wood, Natasha Pavlova, Avi Barak, Thomas Rietz, Christopher Kriesa, Stefanos Miltsakakis, Ed Wilde, Anan Karin, Ed Moore
A DEA agent tracks a Russian drug dealer and former KGB man to Los Angeles and meets up with his older brother who begs for a chance to take part in his investigation. However, the younger brother soon begins to suspect that his sibling is playing both sides against the middle. Average.
A/AD   85 min (ort 89 min)   VIDrel: FIRST/SONOP   V

MARTIAN CHRONICLES, THE *   PG
Michael Anderson UK/USA   1980
Rock Hudson, Gayle Hunnicutt, Darren McGavin, Roddy McDowall, Maria Schell, Bernadette Peters, Fritz Weaver, Jon Finch, Bernie Casey, Christopher Connelly, Joyce Van Patten, Nicholas Hammond, Linda Lou Allen, Michael Anderson Jr
A film version of a group of stories telling of the early colonisation of Mars by settlers from Earth. Rather flat with none of the poetic strength of the original work and seriously hampered by the writer's dated vision of the future (the stories were written about thirty years ago). Originally shown in three 75-minute episodes.
Aka: MARTIAN CHRONICLES PART 1, THE: THE EXPEDITIONS; MARTIAN CHRONICLES PART 2, THE: THE SETTLERS; MARTIAN CHRONICLES PART 3, THE: THE MARTIANS; EXPEDITIONS, THE; SETTLERS, THE; MARTIANS, THE
FAN   194 min (ort 300 min) mTV   VIDrel: CASPIC L/A   V
Boa: book The Silver Locusts by Ray Bradbury (several stories only).

MARTIN ***   18
George A. Romero USA   1977
John Amplas, Lincoln Maazel, Sarah Venable, Christine Forrest, Elayne Nadeau, Tom Savini, Fran Middleton, George A. Romero, Al Levitsky, James Roy, J. Clifford Forrest Jr, Robert Ogden, Donaldo Soviero, Donna Siegal
A seventeen-year-old boy thinks he is a vampire and carries out a number of attacks using razor blades and syringes instead of the fangs he lacks. Eventually, his notoriety comes to the attention of a local radio station and he becomes a celebrity, phoning in to discuss his problem. A chilling little piece of work, bleak and disturbing but undeniably well done.
HOR   94 min (ort 97 min)   VIDrel: REDEM/RTM   V
Boa: story by George A. Romero.

MARTIN CHUZZLEWIT ***   U
James Pedr UK/USA   1994
Paul Scofield, Tom Wilkinson, Emma Chambers, John Mills, Julia Sawalha, Philip Franks, Pete Postlethwaite, Maggie Steed
Long, richly detailed and finely acted adaptation of the classic

novel, in which a wealthy man becomes the focus of a greedy family.
DRA   337 min (2 cassettes) mTV   VIDrel: BBC/TECH V/sh
Boa: novel by Charles Dickens.

MARTY ****   U
Delbert Mann USA   1955
Ernest Borgnine, Betsy Blair, Joe Mantell, Joe De Santis, Esther Minciotti, Jerry Paris, Karen Steele, Frank Sutton, Walter Kelley, Robin Morse, Augusta Ciolli
A touching, warm and wholesome tale of a lonely Bronx butcher who lives with his mother and dreams forlornly of finding love, but happily eventually does so. An utterly charming film, as fresh and vibrant as when it was first made. AA: Pic, Dir, Actor (Borgnine), Screen (Chayevsky).
DRA   89 min (ort 91 min) B/W   VIDrel: MGM/WHV   V
Boa: play by Paddy Chayevsky.

MARX BROTHERS: A DAY AT THE RACES ***   U
Sam Wood USA   1937
Groucho Marx, Chico Marx, Harpo Marx, Allan Jones, Maureen O'Sullivan, Sig Rumann, Margaret Dumont, Douglass Dumbrille, Esther Muir, Leonard Ceeley, Robert Middlemass, Vivien Fay, Charles Trowbridge, Frank Dawson, Max Lucke, Si Jenks
The Marx Brothers team up with a girl who owns both a sanatorium and a racehorse, but they create havoc at the sanatorium when they pay it a visit. Contains several of their funniest sequences, but is let down by a feeble storyline that holds back their zany humour, and some truly forgettable songs. Often cited as the first film to show signs of their decline, but the comic elements (especially the racecourse climax) remain potent.
Aka: DAY AT THE RACES, A
COM   109 min (ort 111 min) B/W   VIDrel: MGM/WHV   V

MARX BROTHERS: A NIGHT AT THE OPERA ****   U
Sam Wood USA   1935
Groucho Marx, Harpo Marx, Chico Marx, Kitty Carlisle, Allan Jones, Margaret Dumont, Sig Ruman, Walter Woolf King, Edward Keane, Robert Emmet O'Connor, Gino Corrado, Purnell Pratt, Frank Yaconelli, Billy Gilbert, Sam Marx
One of the greatest of the Marx Brothers' zany comedies with a fast-paced plot that almost defies description. A few good numbers sung by Carlisle and Jones blend well with Marx Brothers lunacy, as Groucho and company become involved in the production of a Sig Ruman opera, but ultimately destroy it. The script is by George S. Kaufman and Morrie Ryskind. From this film onwards, Zeppo no longer appeared.
Aka: NIGHT AT THE OPERA, A
COM   87 min (ort 96 min) B/W   VIDrel: MGM/WHV   V

MARX BROTHERS: A NIGHT IN CASABLANCA ***   U
Archie Mayo USA   1946
Groucho Marx, Harpo Marx, Chico Marx, Sig Ruman
Groucho is the owner of the exotic Hotel Casablanca and both he and his brothers become involved with Nazi villains, who are out to retrieve some hidden treasure. Having officially "retired" from the screen in 1941, this film was meant to be the Marx Brothers' big comeback, but does not open with promise. However, patience is rewarded and after a while it develops the usual frenetic pacing. A few very funny sequences add to the fun.
Aka: NIGHT IN CASABLANCA, A
COM   85 min B/W   VIDrel: CONNO/RTM   V

MARX BROTHERS: ANIMAL CRACKERS ****   U
Victor Heerman USA   1930
Groucho, Harpo, Chico and Zeppo Marx, Margaret Dumont, Lilian Roth, Louis Sorin, Hal Thompson, Robert Greig
Though this, their second film, suffers from a stage-bound approach and is somewhat patchy, the comic exchanges are still extremely funny after all these years. Adapted from a successful Broadway run, the story, such as it is, deals with a stolen painting that turns up at an elegant party.
Aka: ANIMAL CRACKERS
COM   93 min (ort 98 min) B/W   VIDrel: CIC/SONOP   V

MARX BROTHERS: DUCK SOUP ***   U
Leo McCarey USA   1933
Groucho Marx, Harpo Marx, Chico Marx, Zeppo Marx, Margaret Dumont, Louis Calhern, Raquel Torres, Edgar Kennedy, Edmund Breese, Verna Hillie, Leonid Kinskey, Edwin Maxwell
A crazy but very uneven Marx Brothers comedy set in a ficti-

tious country that declares war on one of its neighbours. The jokes and comic sequences (especially the mirror one) are quite wonderful; if only the plot had been sharper and the ending less abrupt. But still worth seeing for all that. This was the last film in which Zeppo appeared.
Aka: DUCK SOUP
COM 70 min (ort 72 min) B/W VIDrel: CIC/SONOP L/A V

MARX BROTHERS: GO WEST ** U
Edward Buzzell USA 1940
Groucho Marx, Chico Marx, Harpo Marx, John Carroll, Diana Lewis, Walter Woolf King, Robert Barrat, June MacCloy, George Lessey, Mitchell Lewis, Tully Marshall, Clem Bevans, Joe Yule, Arthur Houseman
Humdrum story of three zany characters and their adventures out in the Wild West. Memorable for a funny opening sequence at a ticket office and a wonderful finale in which they dismantle a moving train. Unfortunately, what comes between these two highlights is some of their weakest material, and for the most part revolves around two feuding families and a land deed. Very disappointing.
Aka: GO WEST; MARX BROTHERS GO WEST, THE
COM 80 min (ort 82 min) B/W VIDrel: MGM/WHV V

MARX BROTHERS: LOVE HAPPY ** U
David Miller USA 1949
Groucho Marx, Harpo Marx, Chico Marx, Ilona Massey, Vera-Ellen, Marion Hutton, Raymond Burr, Eric Blore, Bruce Gordon, Melville Cooper, Leon Belasco, Paul Valentine, Marilyn Monroe
Not the funniest of the Marx Brothers' films but still quite watchable, this story is built around the exploits of a bunch of impoverished actors who come into possession of the Romanov diamonds. It all looks a trifle strained now and the comic gags of earlier films have given way to doses of mawkishness. Look out for Marilyn Monroe who appears very briefly. Apart from a few guest appearances in other films, this was the last "Marx Brothers" outing.
Aka: LOVE HAPPY
COM 81 min (ort 91 min) B/W
VIDrel: 4-FRONT/POLYREC V

MARX BROTHERS: MONKEY BUSINESS **** U
Norman Z. McLeod USA 1931
Chico Marx, Groucho Marx, Harpo Marx, Zeppo Marx, Thelma Todd, Rockcliffe Fellowes, Ruth Hall, Harry Woods, Ben Taggart, Otto Fries, Evelyn Pierce, Maxine Castle
Four ship's stowaways have a tough time evading capture, gatecrash a society party and catch a bunch of crooks at the same time. A vintage Marx Brothers comedy, of clever gags, improbable situations and inspired lunacy.
Aka: MONKEY BUSINESS
COM 74 min (ort 81 min) B/W VIDrel: CIC/SONOP V

MARX BROTHERS: ROOM SERVICE **** U
William A. Seiter USA 1938
Groucho Marx, Harpo Marx, Chico Marx, Lucille Ball, Ann Miller, Frank Albertson, Donald MacBride, Cliff Dunstan, Philip Loeb, Philip Wood, Charles Halton, Alexander Asro
The Marx Brothers play penniless playwrights who have to find a way to avoid eviction from their hotel room, while waiting for a Broadway backer to turn up. Adapted in the inimitable Marx style from a Broadway farce and written by Morris Ryskind. Later remade as the musical STEP LIVELY.
Aka: ROOM SERVICE
COM 75 min (ort 78 min) B/W
VIDrel: 4-FRONT/POLYREC V
Boa: play by John Murray and Alan Boretz.

MARX BROTHERS: THE BIG STORE *** U
Charles Riesner USA 1941
Groucho Marx, Chico Marx, Harpo Marx, Tony Martin, Virginia Grey, Margaret Dumont, Douglass Dumbrille, Henry Armetta
Groucho plays an unconventional detective who's hired to investigate the running of a department store. When not creating the usual mayhem with his brothers, he manages to save it from crooks. One of the most conventionally plotted of their comedies, this one has some excellent comic moments separated by dull interludes. Often cited as the weakest comedy they made for MGM, this was their last feature for that studio.
Aka: BIG STORE, THE
COM 84 min B/W VIDrel: MGM/WHV V

MARY POPPINS *** U
Robert Stevenson USA 1964
Julie Andrews, Dick Van Dyke, Glynis Johns, David Tomlinson, Ed Wynn, Karen Dotrice, Hermione Baddeley, Matthew Garber, Arthur Treacher, Reginald Owen, Reta Shaw, Elsa Lanchester, Jane Darwell, Arthur Malet, Cyril Delevanti
The story of a magical nanny whose arrival at the home of a staid and stuffy bank employee heralds some magical adventures for his two children. Some good song and dance routines plus lively animated sequences make this a charming film. Andrews's screen debut. AA: Actress (Andrews), Score/orig (Richard M. Sherman/Robert B. Sherman), Edit (Cotton Warburton), Song ("Chim Chim Cheree" – Sherman/Sherman), Effects/vis (Ellenshaw et al.).
MUS 133 min (ort 140 min) cC VIDrel: WDV/TECH V/sur
Boa: novel by P.L. Travers.

MARY REILLY ** 15
Stephen Frears USA 1996
Julia Roberts, John Malkovich, Glenn Close, George Cole, Michael Gambon, Kathy Staff, Michael Sheen, Bronagh Gallagher, Linda Bassett, Henry Goodman, Tim Barlow, Ciaran Hinds, Sasha Hanau, Moya Brady, Emma Griffiths Malin, David Ross
An interesting variant on the Jekyll-and-Hyde theme that retells this familiar story from the point of view of the title character, the good doctor's maid. Happy at having a secure position with a kind employer, she finds herself falling in love with both him and his alter ego, Mr Hyde. Slow and ponderous, it offers bags of atmosphere and a few unexpected laughs, mostly on account of Roberts and her rather dreadful Irish accent.
DRA 104 min (ort 109 min) VIDrel: COLUM/SONOP V/sur
Boa: novel by Valerie Martins.

MASALA *** 18
Srinivas Krishna CANADA/INDIA 1991
Srinivas Krishna, Sakina Jaffrey, Zohra Segal, Saeed Jaffrey, Heri Johal, Madhuri Bhatia, Ronica Sajnani, Les Porter, Ishwarial M. Mooljee, Raju Ahsan, Jennifer Armstrong, Wayne Bowman, Don Callaghan, Paul Persofsky, Ram Ghoman
An Indian drug addict living in Canada tries to overcome the death of his family in a plane crash and turns to Lord Krishna for help, in this strange and unusual comedy, made in a variety of styles and with both musical and erotic sequences.
COM 106 min CINrel

MASCULINE/FEMININE ** 15
Jean-Pierre Luc Goddard FRANCE/SWEDEN 1965
Jean-Pierre Leaud, Chantal Goya, Marlene Jobert, Michael Deborb, Catherine-Isabelle Duport, Eva-Britt Strandberg, Birger Malmstein, Elsa Leroy, Francoise Hardy, Chantal Darget, Brigitte Bardot, Antoine Bourseiller
Godard's look at young people in the Sixties, "the children of Marx and Coca-Cola", with the emphasis squarely on the former. It tells in his usual meandering and episodic style of a young man just released from military service who meets a girl and gets a job on the magazine where she works, but then neglects her in favour of politics and his two female roommates.
Aka: MASCULIN/FEMININ
DRA 100 min (ort 103 min) B/W VIDrel: CONNO/RTM V

M*A*S*H ** 15
Robert Altman USA 1970
Donald Sutherland, Elliott Gould, Sally Kellerman, Tom Skerritt, Robert Duvall, Gary Burghoff, Jo Ann Pflug, Rene Auberjonois, Roger Bowen, John Schuck, G. Wood, Fred Williamson, Bud Cort, Danny Goldman, Dawne Damon, Carl Gottlieb
Wildly episodic black comedy about the antics of a couple of crazy surgeons posted to a field hospital during the Korean War. A disappointing effort whose emphasis is firmly focused on the crazy antics and sexual exploits of all concerned, with little or no attention paid to the absurd horror of patching up combat personnel in order to send them back into action. The film gave rise to a superior TV series that just ran and ran. Screen/adapt (Ring Lardner Jr).
COM 111 min (116 min) wScrn VIDrel: 20TH/TECH; ENCORE (LV only) V/sh LV
Boa: novel by Richard Hooker.

M*A*S*H – GOODBYE, FAREWELL AND AMEN * PG
Alan Alda USA   1983
Alan Alda, Loretta Swit, Mike Farrell, Harry Morgan, Jamie Farr,
William Christopher, David Ogden Stiers
Final full-length episode from a long-running TV series that
should have been put out of its misery years earlier. The medics
who patch up US soldiers fighting in the Korean War have to
pack up and go home when peace finally breaks out. A feature-
length wake for a show that outstayed its welcome, lasting no
less than eight years more than the duration of the Korean War.
Sloppy, silly and sentimental.
WAR   115 min (ort 120 min) mTV   VIDrel: 20TH/TECH
L/A   V

MASK ****   15
Peter Bogdanovich USA   1985
Eric Stoltz, Cher, Sam Elliott, Estelle Getty, Richard Dysart, Harry
Carey Jr, Laura Dern, Nicole Mercurie, Harry Carey Jr, Dennis
Burhley, Laurence Monoson, Ben Piazza, Alexandra Powers, L. Craig
King, Todd Allen, Joe Unger
Based on a true life account, this is the study of a teenager
suffering from a disorder that causes his cranial bones to
continue growing, giving him a grotesque mask-like appear-
ance and leading to his eventual death. His relationship with
his mother and others is sympathetically dealt with in a film
that carefully eschews sentimentality. Stolz and Cher give
really outstanding performances. AA: Make (Michael
Westmore/Zoltan Elek).
DRA   115 min (ort 120 min) wScrn
VIDrel: 4-FRONT/POLYREC/CIC   V/h

MASK **   18
Paul Thomas USA   1993
Lene Hefner, Mark Davis, Jon Dough, Sierra, Brittany O'Connell,
Lacy Rose, Steven St. Croix, Nick E., Paul Thomas
A scientist disfigured in a laboratory accident, is abandoned
by his fiancee who takes one look at the remains of his face
and runs away. However, when his face is artificially restored
by the use of a mask, he begins to pursue her again, while also
getting involved with some hookers. When he finally gets her
into bed, he wisely decides that this time he ought to be the
one to do the running. Standard sex film, enlivened with an
unusual plot.
A   45 min   VIDrel: VIVID/SCRN   V

MASK, THE **   PG
Charles Russell USA   1994
Jim Carrey, Cameron Diaz, Peter Riegert, Peter Greene, Amy Yasbeck,
Richard Jeni, Orestes Matacena, Timothy Bagley, Nancy Fish, Johnny
Williams, Reginald E. Cathey, Jim Doughan, Denis Forest, Joseph
Alfieri, Robert Keith, Catherine Berge
A quiet and withdrawn young man finds a strange wooden
mask and tries it on, thereby transforming himself into an
outlandish superhero with formidable powers and an array of
weapons to match. A frenetic Carrey vanity-vehicle that relies
almost exclusively on its violent, computer-generated special
effects rather than the star's irritating rubber-faced persona. For
all the appeal inherent in its opening premise, this film is most
definitely not one for younger kids.
COM   97 min (ort 101 min) wScrn   VIDrel: EIV/SONOP   V

MASK OF DEATH **   18
David Mitchell CANADA   1995
Lorenzo Lamas, Rae Dawn Chong, Jerry Wasserman, Billy Dee
Williams
An off-duty cop is shot in the face and undergoes extensive
cosmetic surgery, but his face is deliberately rebuilt to resemble
that of the crook who shot him, thus enabling him to go undercover to smash and drug dealing ring. A preposterously plotted
identity-switch movie that stretches both our credibility and the
far-from-impressive acting skills of the star. experience.
DRA   88 min   VIDrel: EIV/SONOP   V

MASK OF DIMITRIOS, THE ***   (PG)
Jean Negulesco USA   1994
Peter Lorre, Sidney Greenstreet, Zachary Scott, Faye Emerson, Victor
Francen, Steven Geray
Lorre plays a timid Dutch novelist who with some help from
Greenstreet, is drawn into a hunt for a mysterious East
European villain, whose international intrigues and machina-
tions leave a trail of crimes that demands investigation. Scott is
suitably sinister in his screen debut, and this complex film (it
makes much use of flashbacks) has no shortage of atmosphere,

even if the story does not quite fulfil its initial promise of great-
ness.
THR   95 min B/W   SATrel: TNT MOVIES
Boa: novel A Coffin for Dimitrios by Eric Ambler.

MASQUE OF THE RED DEATH, THE ***   15
Roger Corman USA   1964
Vincent Price, Jane Asher, Hazel Court, David Weston, Patrick
Magee, Skip Martin, John Westbrook, Nigel Green, Gaye Brown,
Julian Burton, Doreen Dawn, Paul Whitsun-Jones, Jean Lodge, Verina
Greenlaw, Brian Hewlett, Harvey Hall
Probably the most successful of Corman's attempts to adapt an
Edgar Allan Poe story to the screen, thanks largely to the
photography of Nicholas Roeg. While plague devastates an
Italian province in the 12th century, its ruler, a sadist and
Satanist, holds a masked ball to which Death comes as an unin-
vited guest. The tale "Hopfrog" is presented by way of a
sub-plot.
HOR   85 min (Cut at film release – ort 89 min)
VIDrel: BRAVE/SONOP L/A   V/sh
Boa: short stories The Masque of the Red Death/Hopfrog or the
Eight Chained Orang-outangs by Edgar Allan Poe.

MASQUE OF THE RED DEATH, THE **   18
Larry Brand USA   1989
Adrian Paul, Patrick Macnee, Tracy Reiner, Claire Hoak, Jeff
Osterhage, Kelly Ann Satsso, Maria Ford
Produced by Roger Corman, this is effectively a remake of one
of his better works, that adds a smattering of sex and violence
to Poe's tale of paranoia and death. The debauched character of
Prince Prospero is carefully delineated, and the pleasing sets
and costumes help sustain a film whose lack of momentum and
limited budget are serious handicaps.
HOR   80 min (ort 83 min)   VIDrel: MGM L/A   V
Boa: short story by Edgar Allan Poe.

MASQUERADE **   18
Bob Swaim USA   1988
Rob Lowe, Meg Tilly, Kim Cattrall, Doug Savant, John Glover, Dana
Delany, Erik Holland, Brian Davies, Barton Heyman, Bernie
McInerney, Bill Lopatto, Pirile MacDonald, Maeve McGuire, Ira
Wheeler, Timothy Mandyfield, Edwin Bordo
A wealthy woman whose mother has recently died falls in
love with a charming man, unaware that there is a sinister plan
to kill her for her money. A loose remake of the Hitchcock
classic SUSPICION (1941) that might have been very good
indeed without Lowe, who is far too bland to be the least bit
sinister.
THR   87 min (ort 91 min)   VIDrel: MGM/WHV L/A   V

MASQUERADE ***   18
Silvio Bandinelli ITALY   199-
Nellie Marie Vickers, Teresa Weigel, Joseph Nassivera (Joey Silvera)
Glossy but reasonably watchable erotic-thriller in which a
female investigator is sent to Florence to find out why some of
Italy's top international art dealers have been murdered. The
acting and settings in this steamy thriller are uniformly good,
and the storyline ensures it all moves along at a decent pace. It's
a pity the dubbing is so poor.
THR   89 min (ort 93 min) dubbed   VIDrel: MIA/DISC   V

MASQUERADER, THE **   PG
Richard Wallace USA   1933
Ronald Colman, Elissa Landi, Halliwell Hobbes, Juliette Compton,
David Torrence
A drug-addicted M.P. swaps has an identical cousin, a London
policeman, and when the two swap roles a complicated romantic intrigue develops. Colman does well in two roles here, and
though dated and unconvincing (Hollywood could never quite
capture the flavour of London) there is enough excitement here
to keep the story going. Hobbes has a nice role as the politician's patient and long-suffering butler.
DRA   77 min B/W   VIDrel: VCC/DISC   V
Boa: novel by Katherine Cecil Thurston.

MASSACRE AT CENTRAL HIGH **   18
Renee Daalder USA   1976
Andrew Stevens, Robert Carradine, Derrel Maury, Kimberly Beck,
Steve Bond, Roy Underwood, Steve Sikes, Lani O'Grady, Damon
Douglas, Rainbeaux Smith, Dennis Court, Jeffrey Winner, Thomas
Logan
Nine students are brutally murdered at a Californian high
school when a newcomer decides on a novel way to combat

bullying and eliminates an entire gang. A brutal and offbeat tale that gives a new twist to an old formula.
Aka: BLACKBOARD MASSACRE
HOR 84 min VIDrel: VIPCO/SGSVID V/h

MASSACRE IN ROME *** 15
George Pan Cosmatos FRANCE/ITALY 1973
Richard Burton, Leo McKern, Peter Vaughan, John Steiner, Anthony Steel, Marcello Mastroianni, Delia Boccardo, Renzo Montagnani, Giancarlo Prete, Robert Harris
Burton plays a German colonel who must execute 330 Roman hostages as a reprisal for the death of 33 German soldiers. Mastroianni plays the priest who opposes him in a desperate attempt to save them. Based on a true WW2 incident and on the whole effectively handled. The most moving and stark moment comes at the end, when the names of all 330 victims roll across the screen. A fine tribute.
Aka: RAPPRESAGLIA; SS REPRESSAILLES
WAR 97 min (ort 104 min) VIDrel: MIA/DISC V
Boa: book Death In Rome by Robert Katz.

MASSACRE MANSION * 18
Michael Pataki USA 1975
Richard Basehart, Trish Stewart, Gloria Grahame, Lance Henriksen, Vic Tayback, Arthur Space, Libbie Chase, Al Ferrara
A doctor kidnaps people, and uses their eyes to restore the sight of his daughter, blinded in a crash that he caused. None of the transplantations he attempts are successful, and he ends up with a collection of sightless victims whom he imprisons. They eventually escape and give their tormentor a taste of his own medicine. Low-grade and barely watchable rubbish, this is a poor reworking of George Franju's EYES WITHOUT A FACE.
Aka: MANSION OF THE DOOMED; TERROR OF DR CHANEY, THE
HOR 82 min (ort 90 min) VIDrel: VIPCO/SGSVID V

MASSEUSE, THE *** 18
Paul Thomas USA 1990
Hyapatia Lee, Randy Spears, Viper, Porsche Lynn, Paul Thomas, Danielle Rogers
A twenty-eight-year-old virgin who works in a library, rejects the advances of a fellow worker and patronises a massage parlour instead. There he gradually gets involved with the title figure, eventually having sex with her in his apartment. He falls for her, but she is already married and has good reasons for not getting more involved. An above-average offering, this one has a decent plot for a change plus an unusually strong cast.
A 80 min VIDrel: HAR/GOLD V

MASSEUSE ** 18
Peter Daniels USA 1995
Griffin Drew, Tim Abbel, Monique Perent
Tired of her boyfriend's infidelities, a woman profits from his absence on a business trip to turn her home into a massage parlour, which soon brings her a good many satisfied customers. A plot-free erotic drama, no better or worse than dozens of others.
Aka: AMERICAN MASSEUSE
A 85 min (ort 90 min) VIDrel: THIRD V

MASTERS OF MENACE * 15
Daniel Raskov USA 1990
David Rasche, Catherine Bach, Lance Kinsey, David L. Lander, James Belushi, Dan Aykroyd
Outrageous and excessive spoof on biker movies, that is built around the efforts made by a gang to give one of their members the funeral he deserves; in Las Vegas no less. To do this, however, they have to violate the terms of their probation, which brings their old adversary the local D.A. after them. A tedious chase yarn whose crude attempts at humour are more embarrassing than amusing.
COM 93 min (ort 97 min) VIDrel: FIRST/SONOP L/A V

MASTERS OF THE UNIVERSE ** PG
Gary Goddard USA 1987
Dolph Lundgren, Frank Langella, Courtney Cox, Cristina Pickles, Billy Barty, James Tolkan, Meg Foster, Jon Cypher, Chelsea Field, Tony Carroll, Pons Mar, Anthony DeLongis, Robert Towers, Barry Livingston
A live-action feature based on the incredibly popular but banal children's animated series, in which the heroic He-Man battles the evil Skeletor on the war-devastated planet Eternia, in his bid to defeat the forces of darkness. Langella plays Skeletor with considerable panache, and almost redeems a film whose fantasy

element is seriously weakened by the inclusion of a long sequence that takes place on Earth.
Aka: MASTERS OF THE UNIVERSE: THE MOTION PICTURE
FAN 101 min (ort 106 min) VIDrel: WHV L/A V/sur

MATA HARI * 18
Curtis Harrington UK 1985
Sylvia Kristel, Oliver Tobias, Christopher Cazenove, Gaye Brown, Gottfried John, William Fox, Michael Anthony, Vernon Dobtcheff, Anthony Newlands, Brian Badcoe, Tutte Lemkow, Taylor Ryan, Tobias Rolt, Victor Langley
Film version of the life and death of the famous female spy whose activities on behalf of the Germans during WW1 attracted more attention on account of fact that she was exotic, than as any reflection of the use she was to the Germans. A plodding effort in which plotting is abandoned in favour of glimpses of an unclothed Kristel.
DRA 103 min (ort 108 min) VIDrel: VCC L/A V

MATADOR * 18
Pedro Almodovar SPAIN 1986
Carmen Maura, Assumpta Serna, Antonio Banderas, Nacho Martinez, Eva Cobo, Julieta Serrano, Chus Lampreave, Eusebio Poncela, Bibi Andersen, Luis Ciges, Eva Siva, Veronica Forque, Pepa Merino, Lola Peno, Marisa Tejada, Milton Diaz
A messy and unclear tale of love and death with a former bullfighter encountering a woman who shares his passion for killing their partner as a means of achieving sexual pleasure. Full of the director's typical over-the-top effects and oddball characters, and weighed down by leaden and pretentious dialogue.
DRA 92 min (ort 106 min); 96 min wScrn (special edition)
VIDrel: TART/20TH; ENCORE (LV only) V/dm LV

MATCH FACTORY GIRL, THE *** (15)
Aki Kaurismaki FINLAND 1990
Kati Outinen, Elina Salo, Esko Nikkari, Vesa Vierikko, Silu Seppala
A young girl living a drab life with a job in a match factory and a dull-witted mother and stepfather, finds a way to transform her existence when she buys a bright party dress. This enables to find a boyfriend but her lover soon abandons her after getting her pregnant. Part three of the director's "Working Class Trilogy" (following on from ARIEL), this is a compelling but profoundly depressing work.
Aka: TULITIKKUTEHTAAN TYTTO
DRA 70 min B/W TVrel

MATERNAL INSTINCTS * 15
George Kaczender USA 1996
Delta Burke, Beth Broderick, Garwin Sanford, Sandra Nelson, Gillian Barber, Tom Butler, Tom Mason
A woman becomes demented after she is operated on by a doctor who performs a hysterectomy, and sets out to have revenge both on her husband and the female medic responsible. But the doctor is herself pregnant, so this aggrieved woman concocts a plan to have the ultimate revenge by abducting the woman's child. About as corny and formulaic as they come, this dire thriller is written without a shred of conviction or originality.
THR 88 min VIDrel: CIC V/sh

MATEWAN **** 15
John Sayles USA 1987
Chris Cooper, Will Oldham, Mary McDonnell, Bob Gunton, Ken Jenkins, James Earl Jones, Kevin Tighe, Gordon Clapp, David Straithairn, Josh Motel, Joe Grifasi, Maggie Renzi, Jace Alexander, Nancy Mette, Jo Henderson, Nancy Mette
A realistic account of the coal-miners' struggles for trade union rights in the 1920s, and their fight against the gangster methods employed by the mine-owners. A splendid period piece without a single discordant note. The script is by Sayles (who even wrote some labour songs) and the fine photography is by Haskell Wexler.
DRA 126 min (ort 130 min) VIDrel: MGM L/A V
Boa: novel Union Dues by John Sayles.

MATILDA *** PG
Danny De Vito USA 1996
Mara Wilson, Rhea Perlman, Danny De Vito, Embeth Davidtz, Pam Ferris, Paul Reubens, Tracey Walter, Brian Levinson, Jean Speegle Howard, Sara Magdalin, R.D. Robb, Goliath Gregory, Fred Parnes, Kiami Davael, Leor Livneh Hackel
In a small town, Matilda grows up with a passion for books and practical jokes on those who annoy her. Eventually, she is sent

to a school run by the monstrous Agatha Trunchbull, where she makes friends with her kindly teacher. As Matilda has developed telekinetic powers, she uses these to her own advantage and to help her teacher, who was cheated of her inheritance by Trunchbull. A most spirited adaptation, capturing the very essence of Dahl's grotesque novel.
Aka: ROALD DAHL'S MATILDA
COM   94 min (ort 98 min) cC   VIDrel: COLUM/SONOP V/sur
Boa: novel by Roald Dahl.

MATINEE ★★   PG
Joe Dante USA   1993
John Goodman, Cathy Moriarty, Simon Fenton, Omri Katz, Kellie Martin, Lisa Jakub, Jesse Lee, Lucinda Jenney, James Villemaire, Robert Picardo, Dick Miller, John Sayles, Mark McCracken, Jesse White, David Clennon, Lucy Butler
On the eve of the Cuban missile crisis, a B-movie king (aptly played by Goodman) decides to preview his latest creation in Key West and gets involved with the lives of some of its inhabitants. A nostalgic look at the 1950s, with the usual quota of in-jokes for movie buffs, it builds up to a wild climax set during a showing of his movie, that threatens to demolish the entire theatre. An intermittently enjoyable romp, with some nice performances all round.
COM   94 min (ort 99 min)   VIDrel: GUILD/POLYREC L/A V/s

MATRIARCH, THE ★   18
Luca Bercovici USA   1995
Stella Stevens, Shannon Whirry, Luca Bercovici
A wealthy old woman comes into possession of an elixir that is said to grant eternal youth, but in her eagerness to be rejuvenated, drinks too much and dies. However, when her family gather to celebrate and share out her fortune, she returns from the grave to chastise them. This starts out as if it is going to be a straightforward horror film, then abruptly changes into a crazy and gore-laden black comedy, and not a very funny one at that.
Aka: MATRIARCH, THE: MOTHER OF THE DEAD
HOR   85 min   VIDrel: HIFLI/SONOP   V/h

MATTER OF LIFE AND DEATH, A ★★★★   PG
Michael Powell/Emeric Pressburger UK   1946
David Niven, Roger Livesey, Marius Goring, Kim Hunter, Abraham Sofaar, Raymond Massey, Robert Coote, Joan Maude, Kathleen Byron, Bonar Colleano Jr, Richard Attenborough, Robert Atkins, Edwin Max
An RAF pilot bails out of his plane without a parachute and should have been killed on hitting the ground, but by an oversight is "missed". A messenger is dispatched from the Afterlife in an attempt to remedy this, but finds himself unable to prevail on the pilot to accompany him, as the latter has now fallen in love. A marvellously poetic fantasy, that cleverly offers a rational explanation for the various supernatural events portrayed. Bold, inventive and moving.
Aka: STAIRWAY TO HEAVEN
FAN   103 min   VIDrel: CARL/TECH   V/sur

MATTERS OF THE HEART ★★★   PG
Michael Ray Rhodes USA   1990
Jane Seymour, Christopher Gartin, James Stacy, Geoffrey Lewis, Nan Martin, Allen Rich, Katherine Cannon, Clifford Davis, Jon Maynard Pennell, Lisa Picotte, Mai-Lis, Kuniholm, James Johnston, Derek Basco, Wendy Rhodes
A terminally-ill pianist, of world renown, buries herself in a remote mansion and wallows in drunken misery. However, a chance meeting draws her into a tempestuous relationship with an eighteen-year-old, who is himself a talented pianist. Eventually she dies, but not before having instilled in him the intense passion for playing she once had. An unusually good performance from Seymour raises this film above its essentially soap opera script.
DRA   90 min (ort 100 min) mCab   VIDrel: CIC/SONOP L/A V/h
Boa: novel The Country Of The Heart by Barbara Wersba.

MAURICE ★★★   15
James Ivory UK   1987
James Wilby, Hugh Grant, Rupert Graves, Denholm Elliot, Simon Callow, Billie Whitelaw, Barry Foster, Judy Parfitt, Phoebe Nicholls, Ben Kingsley, Patrick Godfrey, Helena Bonham-Carter, Mark Tandy, Kitty Aldridge
Story of a boy's growing awareness of his own sexuality, that's set in pre-WW1 England, where two young men studying at

Cambridge meet and fall in love, but find their relationship fraught with dangers and difficulties owing to the repressive attitudes prevalent at the time. Overlong but quite effective. Co-scripted by Ivory and Kit Hesketh-Harvey.
DRA   134 min (ort 140 min)   VIDrel: L/A   V
Boa: story by E.M. Forster.

MAVERICK ★★   PG
Richard Donner USA   1994
Mel Gison, Jodie Foster, James Garner, Graham Greene, Alfred Molina, James Coburn, Dub Taylor, Geoffrey Lewis, Paul L. Smith, Dan Hedaya, Dennis Fimple, Denver Pyle, Clint Black, Max Perlich, Art La Fleur, Leo V. Gordon, Paul Tuerpe
Inspired by the TV Western series of the same name, this long and leisurely affair sees our gambler on his way to an important game of poker, and enjoying a variety of encounters into the bargain. Pleasant performances and an amiable atmosphere sustain this self-conscious movie, which is hardly improved by a set of cameos courtesy of former bit players. Garner (who had the lead in the TV series) is cast in the role of a US marshal.
WES   121 min (ort 127 min) cC   VIDrel: WHV   V/sur

MAX HEADROOM FILM, THE ★★★   PG
Annabel Jankel/Rocky Morton UK   1985
Matt Frewer, Nickolas Grace, Amanda Pays, Hilary Tindall, Morgan Shepherd, Paul Spurrier, Hilton McRae, George Rossi, Roger Sloman, Anthony Dutton, Constantine Gregory, Lloyd McGuire, Elizabeth Richardson, Gary Hope
A fascinating and highly original look at the near future, with a story of an investigative journalist's search for the truth behind "blipverts" i.e. compressed adverts, used by powerful TV corporations as they fight ratings battles. Actually a pilot for a pop music show compered by the title character, a computer-simulated talking head. Several "Max Headroom" episodes followed, in an attempt to cash in on the success of this one.
Aka: MAX HEADROOM: THE ORIGINAL STORY; MAX HEADROOM STORY, THE
FAN   60 min mTV   VIDrel: VISVID/POLYREC L/A   V/sh

MAXIE ★★   PG
Paul Aaron USA   1985
Glenn Close, Mandy Patinkin, Ruth Gordon, Barnard Hughes, Michael Ensign, Michael Laskin, Valerie Curtin, Googy Gess, Harry Hamlin, Lou Cutell, Nelson Welch, Leeza Gibbons, Evan White, Harry Wong, Charles Douglas Laird, David Sosna
A "flapper" from the 1920s is reincarnated in the body of the private secretary to the Bishop of San Francisco in this light comedy very much in the vein of ALL OF ME. Gordon's last film is a rather weak affair, and does little to exploit the range of an actress of Close's stature.
Aka: FREE SPIRIT; I'LL MEET YOU IN HEAVEN
COM   94 min (ort 98 min)   VIDrel: RNK L/A   V
Boa: novel Marion's Wall by Jack Finney.

MAXIMUM BREAKOUT ★   18
Tracy Lynch Britton USA   1991
Sydney Cole Phillips, Tom Blanton, Steve Rally, Caine Murray, Bobby Johnston
Action film that revolves around the nasty activities of a "baby auction" ring, who kidnap a beautiful woman, leaving her boyfriend for dead. Not an edifying experience and entirely predictable in outcome (said boyfriend recovers from his wounds and mounts a rescue mission).
A/AD   92 min   VIDrel: 20VIS/SONOP   V

MAXIMUM FORCE ★★   18
Joseph Merhi USA   1991
Sam Jones, John Saxon, Sherrie Rose, Richard Lynch, Jeff Langton, Sonny Landham, Jason Lively, Mickey Rooney
In L.A., three maverick cops decide to take on police corruption and the underworld in their own way by infiltrating a criminal gang. An average actioner with much violence and few surprises.
A/AD   90 min   VIDrel: MIA/DISC   V/sh

MAXIMUM OVERDRIVE ★   18
Stephen King USA   1986
Emilio Estevez, Pat Hingle, Laura Harrington, Christopher Murvey, John Short, Yeardley Smith, Ellen McElduff, J.C. Quinn, Holter Graham, Frankie Faison, Pat Miller, Jack Canon, Barry Bell, John Brasington, J. Don Ferguson
King's directorial debut tells of a rogue comet that passes close to Earth causing mechanical devices to initially malfunction and

then take on a life of their own. At first the problem is one of only minor inconveniences but the situation eventually deteriorates when trucks come to life and start attacking the population. A poor affair. See also THE MANGLER and CHRISTINE.

HOR   97 min   VIDrel: 20TH/TECH L/A   V

MAXIMUM SECURITY *   18
Fred Olen Ray USA   1996
Paul Michael Robinson
A former CIA agent is framed for a murder and sent to deathrow, but gains a chance to prove his worth when the prison is attacked by government agents intent on subversion. A noisy, crashing bore of a movie.
A/AD   90 min   VIDrel: GUILD/FOXVID   V

MAY THE BEST MAN WIN *   15
Michael McCarthy USA   1989
Shawn Weatherly, Michael Nouri, Lee Van CLeef, Liz Torres, Craig Gardner, Russel Savadier, John Hussey, Toni Caprari, Joe Ribiero, Nadia Bilchik, Coaudia Udy, Danie Voges, Graham Weir, Joe Maytham, Kenneth Hendel
Before he dies, a wealthy Argentinian rancher appoints a former employee to decide which one of his two relatives is to inherit his vast fortune. One of them is a disgraced stockbroker, the other a young girl called Peter who has been raised as a boy by her father. An unfunny spoof-style comedy, replete with caricatures and flat jokes, despite which, Nouri gives a good performance, though Van Cleef's brief appearance is rather less impressive.
Aka: CHAMELEON; POWER PLAY; THIEVES OF FORTUNE
COM   95 min   VIDrel: EIV/SONOP   V

MAYBE BABY *   15
John G. Avildsen USA   1988
Molly Ringwald, Randall Batinkoff, Kenneth Mars, Miriam Flynn, Conchata Ferrell, Sean Frye, Allison Roth, Trevor Edmond, Hailey Ellen Agnew, Jaclyn Bernstein, Michelle Downey, Janet MacLachlan, Steve Eckholdt, Robin Morse
Two high school students in love find their world turned upside down by an unexpected pregnancy and are forced to make some tough decisions at a tender age. A well acted and assembled teenage bubblegum comedy.
Aka: FOR KEEPS
COM   94 min (ort 98 min)   VIDrel: RCA L/A   V

MAYBE BABY... AGAIN? *   PG
Tom Moore USA   1988
Dabney Coleman, Jane Curtin, Julia Duffy, Florence Stanley, David Doyle, Peter Michael Goetz
A happily-married businesswoman of 39, sets her heart on having a baby before it's too late, but just about everyone else including her husband try to dissuade her. A mild little blend of comedy and pathos that's not quite one thing or the other. Fair.
Aka: MAYBE BABY
COM   90 min (ort 100 min) mTV   VIDrel: CASPIC L/A   V

MAYERLING *   PG
Terence Young FRANCE/UK   1968
Omar Sharif, Catherine Deneuve, James Mason, James Robertson Justice, Ava Gardner, Genevieve Page, Ivan Desny, Andrea Parisy, Fabienne Dali, Moustache, Maurice Teynac, Bernard Lajarrige, Veronique Vendell, Charles Millot
A dull and overlong remake of the 1936 French film, detailing the doomed love affair between Crown Prince Rudolf of the Austro-Hungarian Empire and his mistress, a seventeen-year-old girl. The film concentrates on the events that led to their suicide pact and though lavishly staged and not too badly acted, it is ruined by a poor script, excessive length and miscasting of Sharif (for the umpteenth time). This one sank without a trace, and no wonder.
DRA   135 min (ort 141 min)   VIDrel: LUMI/SPEAR L/A   V
Boa: novels Idyll's End by Claude Anet and The Archduke by Michael Arnold.

MAYOR OF CASTERBRIDGE, THE *   U
David Giles UK/USA   1978
Alan Bates, Anna Massey, Anne Stallybrass, Jack Galloway, Janet Maw, Ronald Lacey, Freddie Jones, Jeffrey Holland, Gilly Brown, Patricia Fincham
Period drama that revolves around the misdeeds of a man who sold his wife years ago and later went on to become mayor of a

small town, never imagining that his wife would return to expose him.
DRA   349 min (2 cassettes) mTV   VIDrel: BBC/TECH   V/h
Boa: novel by Thomas Hardy.

McBAIN *   15
James Glickenhaus USA   1991
Christopher Walken, Maria Conchita Alonso, Michael Ironside, Steve James, Jay Patterson, Chick Vennera
This film opens with a flashback that takes us to Vietnam where Walken is rescued from a Vietcong POW camp and certain death. Ten years later he sets about repaying the debt he owes his rescuer, an endeavour that involves him in a trip to Columbia and involvement with revolutionaries out to topple the corrupt government. A straightforward action film, simple in scope and devoid of unnecessary complications.
A/AD   103 min   VIDrel: MGM/WHV L/A   V

McCABE AND MRS MILLER *   15
Robert Altman USA   1971
Warren Beatty, Julie Christie, Rene Auberjonois, John Schuck, Bert Remsen, Hugh Naughton, Keith Carradine, William Devane, Shelley Duvall, Michael Murphy, Corey Fisher, Anthony Holland, Elizabeth Murphy, Hugh Millais, Jack Riley
At the turn of the century, a gambling gunslinger comes to a mining town that is enjoying a boom, joins forces with a prostitute, and opens a lavish brothel. Altman's attempt to de-glamorise the Hollywood image of the Wild West is generally successful, but for all the care taken in terms of detail, this obscure effort does not make especially good entertainment, and is hampered by a most unsatisfactory ending. The ponderous script is by Altman and Brian McKay.
WES   116 min (ort 121 min) wScrn   VIDrel: TART/20TH V/h
Boa: novel McCabe by Edmund Naughton.

McGYVER: THE LOST TREASURE OF ATLANTIS *   (PG)
Michael Vejar USA   1993
Richard Dean Anderson, Brian Blessed, Christian Burgess, Sophie Ward, Oliver Ford Davies, Tim Woodward, Kevork Malikyan, Geoffrey Beevers, Hugh Quarshie, George Jackos, Andreas Markos, Barry McCormick
A movie spin-off from the TV series "McGyver" that casts Anderson as a former Special Forces agent who now has an academic post working for the Phoenix Foundation. In this adventure he sets out for a volcano that he hopes will enable him to prove that the lost island city of Atlantis really did exist. As ever, our prof has to contend with the usual array of villains. Adequate.
A/AD   90 min mTV   SATrel: MOVIE CHANNEL

McLINTOCK! *   U
Andrew V. McLaglen USA   1963
John Wayne, Maureen O'Hara, Patrick Wayne, Stefanie Powers, Yvonne De Carlo, Jack Kruschen, Chill Wills, Jerry Van Dyke, Edgar Buchanan, Bruce Cabot, Perry Lopez, Michael Pate, Strother Martin, Gordon Jones, Robert Lowery
A tough cattle baron hires a widow as his housekeeper and she brings along her two children, the elder of whom falls in love with her employer's daughter. However, trouble soon arrives in the shape of our rancher's estranged wife who wants custody of the girl along with a divorce, and a battle royal erupts between these two protagonists. A loud and overblown comedy Western full of bluster and broad humour.
WES   90 min; 127 min (VCC/DISC – producer's cut)
VIDrel: SIMIT L/A; VCC/DISC   V

McQ *   15
John Sturges USA   1974
John Wayne, Eddie Albert, Diana Muldaur, Colleen Dewhurst, Clu Gulager, David Huddleston, Julie Adams, Al Lettieri, Roger E. Mosley, William Bryant, Joe Tornatore, Ken Sanford, Richard Kelton, Richard Eastham, Dick Friel, Fred Waugh
A cop leaves the force and stops at nothing to get the crooks who killed a couple of his buddies. A number of good action sequences are about all there is to find in this embarrassing effort, in which an ageing Wayne stumbles through a muddled and rambling mess.
A/AD   106 min (ort 116 min)   VIDrel: WHV   V/h

McVICAR *   18
Tom Clegg UK   1980
Roger Daltrey, Adam Faith, Cheryl Campbell, Georgina Hale, Steven

Berkoff, Brian Hall, Peter Jonfield, Matthew Scurfield, Leonard Gregory, Joe Turner, Jeremy Blake, Anthony Trent, Terence Stuart, Charlie Cork, Ronald Herdman
Film based on the true story of a violent criminal, his escape from Durham Prison, his capture, reform and eventual rehabilitation. Exploitative and dreary, though Daltrey is certainly convincing. Scripted by McVicar from his book.
DRA   107 min (ort 112 min)   VIDrel: 4-FRONT/POLYREC V/sur
Boa: book McVicar By Himself by John McVicar.

M.D. GEIST **   15
JAPAN   198-
A bio-engineered fighting machine comes back to the world from which it was exiled and must choose to help former adversaries stop a doomsday weapon that is threatening to destroy their planet.
ANIM   50 min dubbed   VIDrel: KISEKI/PARADOX   V

ME AND THE COLONEL **   U
Peter Glenville USA   1958
Danny Kaye, Curt Jurgens, Nicole Maurey, Francoise Rosay, Akim Tamiroff, Martita Hunt, Alexander Scourby, Liliane Montevecchi, Ludwig Stossel, Gerard Buhr, Franz Roehn, Celia Lovsky, Clement Harari, Alain Bouvette, Albert Godderis
A Polish Jew manages to get to Paris ahead of the Nazis and escapes from their by car, together with a Polish colonel with virulently anti-Semitic views. After they collect the latter's girlfriend, several precarious situations ensue from which they manage to extricate themselves. Despite good performances by Kaye and Jurgens, this episodic comedy is decidedly lacking in substance.
COM   105 min (ort 110 min) B/W   VIDrel: RCA L/A   V
Boa: play Jacobowsky And The Colonel by Franz Werfel.

ME AND THE KID ***   (PG)
Dan Curtis USA   1993
Danny Aiello, Alex Zuckerman, Joe Pantoliano, Cathy Moriarty, David Dukes, Anita Morris, Ric Aiello, Desmond Wilson, Ben Stein, Robin Thomas, Alaina ReedHall, Abe Vigoda, Todd Bryant, Mowava Pryor, Joseph Pecoraro, Eric Wylie
A couple of small-time crooks are forced to kidnap a young boy who was a witness to their crime and one of them decides to demand a ransom from his wealthy father. The other soon realises that the boy's life is in danger and resolves to protect him at any cost. Strong and believable performances by Aiello and Zuckerman make this unoriginal tale work quite well despite the limitations of the script.
COM   93 min (ort 97 min)   SATrel: SKY MOVIES

MEAN MACHINE, THE **   15
Robert Aldrich USA   1974
Burt Reynolds, Eddie Albert, Ed Lauter, Michael Conrad, Bernadette Peters, Jim Hampton, Charles Tyner, Mike Henry, Harry Caesar, Richard Kiel, Robert Tessier, Malcolm Atterbury, Pervis Arkins, Tony Cacciotti, Anitra Ford
A football star imprisoned for drunken driving, is blackmailed into forming a team to play a match against the prison guards. A brutal and disjointed comedy that fires off lots of gags in every direction without the benefit a clear narrative drive would have supplied. Scripted by Tracy Keenan Wynn.
Aka: LONGEST YARD, THE
COM   116 min (ort 121 min)   VIDrel: CIC/SONOP   V

MEAN SEASON, THE **   15
Phillip Borsos USA   1985
Kurt Russell, Mariel Hemingway, Richard Jordon, Richard Masur, Andy Garcia, Joe Pantoliano, Richard Bradford, Rose Portrillo, William Smith, John Palmer, Lee Sandman, Dan Fitzgerald, Cynthia Caquelin, Fred Ornstein
A crime reporter finds himself the unwilling confidant of a crazy killer who supplies him with exclusive details of each murder he commits, until the reporter's growing celebrity status causes the murderer to kidnap the reporter's girlfriend in a fit of jealousy. The intriguing basic premise that underpins the film is wasted in the second half. The music is by Lalo Schifrin.
THR   99 min (ort 106 min)   VIDrel: RNK L/A   V
Boa: novel In The Heat Of The Summer by John Katzenbach.

MEAN STREETS **   18
Martin Scorsese USA   1973
Robert De Niro, Harvey Keitel, David Proval, Amy Robinson, Richard Romanus, Cesare Danova, George Memmoli, Robert Carradine, David

Carradine, Jeannie Bell, Lenny Scaletta, Victor Argo, Murray Mosten, Lois Walden, Harry Northup
Realistic account of the lives of four young Italian-American men and their involvement in the world of the Mafia and small-time criminals. The film takes a harsh and uncompromising look at their world but ultimately has nothing to say worth hearing. Screenplay is by Scorsese and Mardik Martin. See also GOODFELLAS.
DRA   103 min (ort 112 min)   VIDrel: ELPIC/POLYREC   V

MEANTIME **   15
Mike Leigh UK   1983
Tim Roth, Jeff Robert, Alfred Molina, Gary Oldman, Pam Ferris, Phil Daniels, Peter Wright, Marion Bailey, Eileen Davies, Herbert Norville, Biran Hoskin
The story of an East London family and the various problems they face, such as unemployment etc. A typically negative Mike Leigh portrait of the working classes from a director whose undoubted eye for realistic detail is so often hampered by his relentlessly bleak view of life.
DRA   102 min (ort 104 min)   VIDrel: IMAG/RTM   V

MEASURE FOR MEASURE ***   PG
Desmond Davis UK   1979
Tim Pigott-Smith, Kenneth Colley, Kate Nelligan, Christopher Strauli, John McEnery, Jacqueline Pearce, Frank Middlemass, Alun Armstrong, Adrienne Corri, Ellis Jones, John Clegg, William Sleigh, Neil McCarthy, Yolande Palfrey
A competent TV production of the Shakespeare play. Casting is strong and as ever in this ambitious BBC series, the text of the play is rendered in full.
DRA   145 min (ort 147 min) mTV   VIDrel: BBC   V/h
Boa: play by William Shakespeare.

MEATBALLS *   15
Ivan Reitman CANADA   1979
Bill Murray, Harvey Atkin, Kate Lynch, Russ Banham, Kristine DeBell, Sarah Torgov, Chris Makepeace, Jake Blum, Keith Knight, Cindy Girling, Todd Hoffman, Margot Pinvidic, Matt Craven, Norma Dell'Agnese, Michael Kirby, Greg Swangson
A low-brow look at a summer camp and the strange characters who inhabit it, not least being a camp counsellor who has a grudge against a rival camp. Followed by several similarly unfunny sequels.
COM   89 min (ort 92 min)
VIDrel: 4-FRONT/POLYREC/CIC L/A   V/h

MEATBALLS 2 *   15
Ken Wiederhorn USA   1984
Richard Mulligan, John Mengatti, Hamilton Camp, Kim Richards, Tammy Taylor, John Laroquette, Archie Hahn, Misty Rowe, Pee-Wee Herman (Paul Reubens), Felix Silla, Vic Dunlop, Elayne Boosler, Nancy Glass, Patti Kirkpatrick, Paul Stout
Not really a sequel, more a desperate attempt to wring some laughs out of a number of disparate elements such as a juvenile alien with amazing powers, sex-mad camp counsellors, a manic coach driver etc. As before the action is set at a summer camp. Pretty feeble stuff.
Aka: MEATBALLS: PART 2
COM   84 min (ort 90 min)   VIDrel: EIV/SONOP   V

MEATBALLS 3 *   18
George Mendeluk USA   1987
Sally Kellerman, Patrick Dempsey, Al Waxman, Isabelle Mejias, Shannon Tweed, Ian Taylor, George Buza, Maury Chaykin, Ronnie Hawkins, Mark Blutman
A dead porno queen has to help a young lad lose his virginity before she can enter heaven. Another outing for a series that began badly and never got any better. Followed by "Meatballs 4".
Aka: MEATBALLS 3: SUMMER JOB
COM   90 min (ort 95 min)   VIDrel: L/A   V

MECHANIC, THE ***   15
Michael Winner USA   1972
Charles Bronson, Keenan Wynn, Jan-Michael Vincent, Jill Ireland, Linda Ridgeway, Frank De Kova, Lindsay H. Crosby, Takayuki Kubota, Martin Gordon, James Davidson, Steve Cory, Patrick O'Moore, Kevin O'Neal, Linda Grant
A professional hit-man takes on an apprentice but finds that he has made a serious error of judgement. Harsh and brutal, but

definitely engrossing, with a neat twist at the end. Scripted by Lewis John Carlino. One of the director's better films.
Aka: KILLER OF KILLERS
DRA 96 min Cut (7 sec – ort 100 min) VIDrel: MGM/WHV L/A V

MEDEA * PG
Pier Paolo Pasolini FRANCE/ITALY/WEST GERMANY 1969
Maria Callas, Guiseppe Gentile, Massimo Girotti, Laurent Terzieff, Margareth Clementi, Anna Maria Chio, Paul Jabor, Luigi Urbini, Gerard Weiss, Giorgio Trombetti, Franco Jacobbi, Gian Paolo Durgar
Following his successful quest, Jason returns to Corinth with the Golden Fleece and Medea, the High Priestess who helped him. Having lived together for several years with their children, he grows tired of her and she is driven to seek vengeance. Despite the unusual casting of Callas in the title role, this sluggish and uninspiring adaptation is hampered by poor direction and does little with the fiery (and dubbed) star.
DRA 105 min (ort 118 min) VIDrel: CONNO/RTM V
Boa: play by Euripides.

MEDICINE MAN * PG
John McTiernan USA 1991
Sean Connery, Lorraine Bracco, Jose Wilker, Rodolfo de Alexandre, Francisco Tsirene Tsere Rereme, Elias Monteiro da Silva, Edinei Maria Serrio Dos Santos, Bec-Kana-Re Dos Santos Kaiapo, Angelo Barra Moreira, Jose Lavat
Connery plays a pony-tailed eccentric botanist, who having gone native in the rainforests of Brazil, believes he has found a cure for cancer. Having requested assistance from a research institute to develop it, he begins to have serious misgivings about the true environmental cost of developing the drug. For once, Connery is well and truly miscast in this rambling, ambiguous and annoying dud, whose sentiments are sound, even if the script isn't.
A/AD 100 min (ort 105 min) cC
VIDrel: 4-FRONT/POLYREC/GUILD V/sh

MEDICINE RIVER *** (PG)
Stuart Margolin CANADA 1993
Graham Greene, Janet-Laine Green, Sheila Tousey, Byron Chief Moon, Jimmie Herman, Maggie Black Kettle, Tina Louise Bomberry, Ben Cardinal, Michael C. Lawrencechuck, Raoul Trujillo, Dakota Horse, Shirley Cheechoo, Cherilyn Cardinal
After twenty years working as a photojournalist, an Indian returns to Toronto from a dangerous foreign assignment to learn that his mother died in his absence and she is to buried in the next few days. This forces him to travel to the reservation where his brother and relatives still and rediscover his roots, eventually falling in love with a woman very different from his current lover. A small-scale and gentle drama, low-key and quite heartwarming. Very watchable.
DRA 97 min SATrel: SKY MOVIES
Boa: novel by Thomas King.

MEDITERRANEO *** 15
Gabriele Salvatores ITALY 1991
Diego Abatantuono, Claudio Bigagli, Giuseppe Cederna, Claudio Bisio, Vanna Barba, Luigi Alberti, Ugo Conti, Memo Dini, Vasco Mirandola, Luigi Montini, Irene Grazioli, Antonio Catania
During WW2, eight totally incompetent Italian soldiers are sent to take charge of a small Greek island but manage to make a total mess of this assignment. Their radio set is destroyed as is the battleship that was to evacuate. Effectively out of the war, they proceed to make friends with the locals who only now emerge from hiding. A languid and rather idyllic antiwar comedy; not at all profound but quite watchable. AA: Foreign.
COM 86 min (ort 90 min) VIDrel: CURZON/20TH V

MEET JOHN DOE *** U
Frank Capra USA 1941
Gary Cooper, Barbara Stanwyck, Edward Arnold, James Gleason, Walter Brennan, Spring Byington, Gene Lockhart, Irving Bacon, Regis Toomey, Ann Doran, Rod La Rocque, Warren Hymer, Andrew Tombes, Aldrich Bowker, Sterling Holloway
A naive tramp is suckered into representing the common man in a false goodwill drive that's designed to benefit a corrupt politician. However, he eventually exposes the true intentions of latter, but only at the cost of branding himself a fake. A wordy populist epic that's overlong but always interesting, with the usual Capra themes and an unsatisfactory conclusion that

detracts from the central core of the film. Music is by Dmitri Tiomkin.
DRA 130 min (director's archive film) B/W
VIDrel: VCC L/A V

MEET ME IN ST LOUIS **** U
Vincente Minnelli USA 1944
Judy Garland, Tom Drake, Mary Astor, Margaret O'Brien, Leon Ames, Marjorie Main, Lucille Bremer, June Lockhart, Harry Davenport, John Carroll, Hugh Marlowe, Robert Sully, Chill Wills, Hank Daniels, Donald Curtis, Ken Wilson
Musical about the life and times of a family in St Louis during the 1903 World Fair. Judy sings several numbers that became classics including: "The Boy Next Door", "Have Yourself A Merry Little Christmas" and "The Trolley Song". An uneven but captivating tale. Songs are by Ralph Blane and Hugh Martin and the script is by Irving Brecher and Fred F. Finkelhoffe. AA: Spec Award (Margaret O'Brien as outstanding child actress).
MUS 110 min (ort 113 min) VIDrel: MGM/WHV V/dm V/h
Boa: novel by Sally Benson.

MEET THE APPLEGATES ** 15
Michael Lehmann USA 1990
Ed Begley Jr, Stockard Channing, Dabney Coleman, Bobby Jacoby, Gami Cooper, Glen Shadix, Susan Barnes, Adam Biesk, Savannah Smith Boucher, Roger Aaron Brown, Lee Garlington, Philip Arthur Ross, Steven Robert Ross, Chuck Lafont
The Applegates are a bunch of man-eating Brazilian beetles who have adopted the guise of an all-American middle-class family, their mission being to destroy a nuclear reactor and thus make the world safer for insects. Despite their benign appearance, they still retain a liking for human flesh. A grotesque and patchy "ecological" comedy, that starts off with an intriguing idea, but is hurt by a lack of pace and the weak ending.
Aka: APPLEGATES, THE
COM 86 min (ort 90 min) VIDrel: COLUM/SONOP L/A V

MEET THE FEEBLES * 18
Peter Jackson NEW ZEALAND 1989
Danny Mulheron plus voices of: Donna Akersten, Stuart Devenie, Mark Hadlow, Ross Jolly, Brian Sargent, Peter Vere Jones, Mark Wright
A kind of wild, overblown bad-taste movie that features Muppet-like creatures indulging in a variety of disgusting and/or violent activities in an attempt to generate some laughs. What plot there is (and there isn't much) has a puppet troupe getting ready for their live TV show, and encountering innumerable problems along the way. An ill-conceived romp that plays like a puppet version of FRITZ THE CAT, but minus the wit.
ANIM 92 min (ort 97 min) VIDrel: ISLPIC/SONOP L/A V

MEETING VENUS *** 15
Istvan Szabo FRANCE/HUNGARY/UK 1991
Glenn Close, Niels Arestrup, Moscu Alcalay, Macha Meril, Johanna Ter Steege, Maite Nahyr, Victor Poletti, Marian Labuda, Jay O. Sanders, Dieter Later, Maria de Medeiros, Ildiko Bansagi, Dorottya Udvaros, Roberta Pollak, Andre Chaumeau
A well-known Hungarian musician comes to the West to conduct a performance of Wagner's Tannhauser in Paris, but upon arrival is immediately assailed by a catalogue of increasingly ludicrous difficulties. As if this were not enough, he is soon locked in a passionate affair with his leading diva (Close with Kiri Te Kanawa's singing dubbed-in). Despite some crude attempts to make the film parallel the opera itself, this remains a refreshing comedy.
COM 115 min (ort 120 min) VIDrel: WHV V/sur

MEETINGS WITH REMARKABLE MEN ** U
Peter Brook UK 1979
Dragan Maksimovic, Mikica Dimitrijevic, Terence Stamp, Athol Fugard, Gerry Sundquist, Warren Mitchell, Bruce Myers, Donald Sumpter, Natasha Perry, Tom Fleming, Fahro Konjhodzix, David Markham, Fabijan Sovagovic, Bruce Purchase
Biopic on the Russian philosopher and mystic G. I. Gurdjieff, hailed by some as a modern-day saint and vilified by others as a charlatan, which depicts his wanderings through Asia and the Middle East in search of spiritual enlightenment. Filmed in the mountains of Afghanistan, it was initially withheld from video and TV release by the executors of the Gurdjieff estate.
DRA 102 min (ort 107 min) VIDrel: CURZON/20TH V

MEGAVILLE * 15
Peter Lehner USA 1989
Daniel J. Travanti, Billy Zane, Grace Zabriskie, J.C. Quinn, Kirsten Cloke, Stefan Gierasch
Stylish but failed futuristic thriller hampered by its confusing plot. In the corrupt title city of the future, a law enforcement employee is persuaded to impersonate a vanished racketeer, but is unaware that a monitoring device has been implanted in his skull and is responsible for his paralysing headaches. A disjointed film that never realises more than a tiny fraction of its potential.
FAN 89 min Cut (7 sec – ort 96 min)
VIDrel: POLY/POLYREC/BRAVE V

MELANCHOLIA * 15
Andi Engel UK 1989
Jeroen Krabbe, Susannah York, Jane Gurnett, Kate Hardie, Ulrich Wildgruber, John Sparkes, Saul Reichlin, Mochetee Van Helsdingen, John Joyce
The first feature for former film critic Engel tells of a retired German left-wing activist living in London, who now works as an art critic. Yet his apparently successful life is in fact quite empty, and when a voice from the past calls upon him to assassinate a visiting Chilean doctor and former torturer, he experiences a difficult conflict of loyalties. A dark and moody film, bleakly cynical and surprisingly tense.
THR 84 min (ort 87 min) VIDrel: ARTIF/20TH V

MELODY IN LOVE * 18
George Morton (Hubert Frank) WEST GERMANY 1978
Britta Glatzeder, Sascha Hehn, Claudine Bird, Melody O'Brien, Wolf Goldan, Scarlett Gunden
A woman visits her female cousin in Mauritius and discovers that she and her husband have various lovers. She subsequently has a number of relationships with various men and women. A first-class yawn-inducer.
A 91 min Cut (11 sec – ort 94 min) VIDrel: IMPENT V

MEMOIRS OF A SURVIVOR * 18
David Gladwell UK 1981
Julie Christie, Leonie Mellinger, Christopher Guard, Nigel Hawthorne, Debbie Hutchings, Pat Keen, Georgina Griffiths, Christopher Tsangarides, Mark Dignam, Alison Dowling, John Franklyn-Robbins, Rowena Cooper, Adrienne Byrne
In the near future civilisation is collapsing, and a housewife finds that she is able to escape from the unpleasantness of reality by retreating into a strange dreamworld. A flat and uninspired attempt to deal with a fascinating idea.
DRA 110 min (ort 115 min) VIDrel: MGM/WHV L/A V
Boa: novel by Doris Lessing.

MEMOIRS OF AN INVISIBLE MAN * PG
John Carpenter USA 1992
Chevy Chase, Daryl Hannah, Sam Neill, Michael McKean, Stephen Tobolowsky, Jim Norton, Pat Skipper, Paul Perri, Richard Epcar, Steven Barr, Gregory Paul Martin, Patricia Heaton, Barry Kivel, Donald Li, Rosalind Chao, Jay Gerber
A stock-market analyst is sent to a conference at a research lab, but turns up ill-prepared to cover the conference (he has a hangover) and sneaks off as soon as he can, settling down in an empty sauna to have a nap. He awakens to find himself invisible, the victim it would appear of some serious accident at the lab. Some spectacular special effects and very clever sight gags add to the enjoyment of this unusual if annoyingly muddled comedy.
COM 95 min (ort 99 min) cC VIDrel: WHV V/sur
Boa: novel by H.F. Saint.

MEMORIES OF ME * 15
Henry Winkler USA 1988
Billy Crystal, Alan King, JoBeth Williams, Janet Carroll, David Ackroyd, Phil Fondacaro, Robert Pastorelli, Sidney Miller, Mark L. Taylor, Peter Elbling, Larry Cedar, Sheryl Bernstein, Joe Shea, Jay "Flash" Riley
A New York surgeon recovering from a mild heart attack visits Los Angeles to effect a reconciliation with his estranged bit-actor father whom he finds terminally ill. A messy and contrived affair that is too superficial to be anything more than mediocre.
COM 99 min (ort 105 min) VIDrel: MGM L/A V

MEMORIES OF MIDNIGHT: VOLS 1 AND 2 * 15
Gary Nelson USA 1991
Jane Seymour, Omar Sharif, Taro Meyer, Thaao Penghlis, Stephen Macht, Theodore Bikel, Ken Howard, Bozidar Alic, Peter Cetera, Julia Lane, Kim Weeks, Nada Arbus, Boris Bakal, Relja Basic, Vedran Derek, Brank Balce, Nadio Fortune
Seymour plays an amnesiac woman who becomes the protege of an unscrupulous millionaire (Sharif). She eventually learns that she lost her memory in an accident that swept her out to sea and that the authorities, having become convinced that she had been murdered, tried and executed both her husband and his mistress. A typical Sheldon soap opera, glossy, superficial and nice to look at (it's set in Greece) if nothing more.
Aka: SIDNEY SHELDON'S MEMORIES OF MIDNIGHT
DRA 186 min (2 cassettes) mTV
VIDrel: 4-FRONT/POLYREC V
Boa: novel by Sidney Sheldon.

MEMORIES OF MURDER * (18)
Robert Lewis USA 1990
Nancy Allen, Robin Thomas, Vanity, Olivia Brown, Don Davis, Linda Darlowe
A woman suffering from amnesia finds herself the victim of a mysterious killer and wonders if this is somehow linked to visions of a murder that keep flashing into her mind. A confused and unimpressive thriller with little to recommend it. The ludicrous climax (clearly inspired by FATAL ATTRACTION) is both derivative and ineffective.
THR 104 min mCab TVrel: BBC1

MEMORY RUN * 18
Allan A. Goldstein CANADA 1993
Karen Duffy, Saul Rubinek, Matt McCoy, Lynn Cormack, Torri Higginson, Chris Makepeace, Barry Morse, Nigel Bennett, Natalie Radford, Robbie Fox, Eric Murphy, John Stoneham Sr, Diana Rowland, Corey Serier, Lazar Rockwood, Doug O'Keefe
Futuristic fantasy set in the year 2015 A.D., where a corrupt legal system has had the brain of an innocent man put into the body of a dead woman. When this creation fails to function properly, it is sent for termination but escapes into the underworld. Freely borrowing ideas from other films (such as FREEJACK) this cliched effort just gets increasingly confusing as the story progresses. Clearly a film whose budget did not run to a decent script.
FAN 89 min VIDrel: HIFLI/SONOP V/h
Boa: novel Season of the Witch by Hank Stine.

MEMPHIS * 15
Yves Simoneau USA 1991
Cybill Shepherd, John Laughlin, Martin C. Gardner, J.E. Freeman, Richard Brooks, Moses Gunn
This slow and coldly efficient thriller is played out against the growing civil rights movement in the Deep South, where Shepherd and two men plan the kidnapping of the son of a wealthy black family. Believing that the police will do little, they are certain the family will have to pay a $60,000 ransom. Matters do not quite work out thus, and the kid's grandfather hatches a rescue plan. Not an endearing effort.
THR 88 min VIDrel: FIRST/SONOP V
Boa: novel by Shelby Foote.

MEMPHIS BELLE * PG
Michael Caton-Jones UK 1990
Matthew Modine, Eric Stoltz, Sean Astin, Harry Connick Jr, John Lithgow, Reed Edward Diamond, Tate Donovan, D.B. Sweeney, Billy Zane, Courtney Gains, Neil Giuntoli, David Strathairn, Jane Horrocks, Mac Macdonald, Jodie Wilson
This fictional retelling of the last WW2 bombing run of the B-17s (inspired by William Wyler's 1944 documentary of the same name) follows the exploits of a clean-cut crew who, having completed 24 bombing missions, will be going home if they can successfully manage just one more. Unconvincing acting and a feast of WW2 cliches spoil the film, but when the action starts, there are some exciting sequences to be had.
DRA 102 min Cut (5 sec – ort 107 min) VIDrel: WHV V/sur

MEN, THE * PG
Fred Zinnermann USA 1950
Marlon Brando, Teresa Wright, Jack Webb, Everett Sloane, Howard St John, Nita Hunter, Patricia Joiner, John Miller, Cliff Clark, Ray Teal, Marguerite Martin, Obie Parker, Ray Mitchell, Pete Simon, Paul Peltz, Tom Gillick
An ex-GI tries to readjust to civilian life, in this story of paraplegics who face more difficulties than other veterans. A restrained but powerful drama that was shocking at the time for

its honest examination of the sexual frustrations such people suffered. The script is by Carl Foreman. This was Brando's film debut.
Aka: BATTLE STRIPE
DRA   85 min B/W   VIDrel: VCC L/A   V

MEN AT WORK *
Emilio Estevez USA
15
1990
Charlie Sheen, Emilio Estevez, Leslie Hope, Keith David, Darrell Larson, Dean Cameron, John Getz, Cameron Dye, John Putch, Hawk Wolinski, Geoffrey Blake, John Lavichielli, Tommy Hinkley, Sy Richardson, Kari Whitman, Troy Evans
Estevez and Sheen (his real-life brother) star as a couple of care-free Californian garbagemen, whose carefree lives are blighted by the discovery of a corpse on their route. This embroils them in a murky and muddled intrigue, which ultimately centres on the illegal dumping of toxic waste. Screenplay is by Estevez, and some dumb clowning makes the film just about bearable. The music is by Miles Copeland.
COM   94 min Cut (5 sec – ort 99 min)
VIDrel: 4-FRONT/POLYREC/EIV L/A   V

MEN DON'T LEAVE **
Paul Brickman USA
15
1990
Jessica Lange, Arliss Howard, Joan Cusack, Kathy Bates, Charlie Korsmo, Chris O'Donnell, Tom Mason, Jim Haynie, Belita Moreno, Corey Carrier, Kevin Corrigan, Shannon Moffett, David Cale, Constance Shulman, Mark Hardwick
Recently widowed Lange, left to struggle with two sons, moves her family from Maryland to Baltimore, gets a job in a deli-catessen, and eventually meets Mr Right in the shape of an eccentric musician. An undemanding comedy-drama that's quite appealing, despite a tendency to go for pathos more often than for laughs. Based on the French film "La Vie Continue", it's scripted by Brickman and Barbara Benedek.
DRA   110 min (ort 115 min)   VIDrel: MGM/WHV L/A
V/sh

MEN DON'T TELL ***
Harry Winer USA
(12)
1993
Peter Strauss, Judith Light, James Gammon, Noble Willingham, Stephen Lee, Mary Kane, Richard Gant, Reni Santoni, Carroll Baker, Ashley Johnson, Michael Rand, Cliff Bemis, Nick Angotti, Guy Killum, Katherine Cortex, David Correia
The story of a husband who is too ashamed to reveal that he lives a life of torment, being physically abused by his wife, as his spouse suffered an abusive childhood that has left her emotionally scarred. Often harrowing, this disturbing drama makes for compelling viewing, being one of the very few number of films that attempt to examine this hidden problem.
DRA   91 min   SATrel: SKY MOVIES

MEN OF RESPECT **
William Reilly USA
18
1990
Rod Steiger, John Turturro, Katherine Borowitz, Peter Boyle, Dennis Farina, Stanley Tucci, Stephen Wright, Julie Garfield, Lilia Skalia, Carl Capotorto, Michael Balducco, Robert Modica, David Thornton, Michael Sergio, Tony Gigante
A peculiar gangster/Shakespeare hybrid, that updates Macbeth, presenting the story as a contemporary morality play, with Turturro playing a mobster who rises in the criminal hierarchy by bumping off his rivals, all the while being encour-aged by his wife. Not an especially effective film, it's over-ambitious and a little ludicrous, never finds an identity of its own, and largely wastes a set of strong performances. Screenplay is by Reilly.
DRA   108 min (ort 113 min)   VIDrel: VCC/DISC/COLUM
V/h
Boa: play The Tragedy Of Macbeth by Wiliam Shakespeare/ short story by William Reilly.

MEN OF WAR **
Perry Lang USA
18
1995
Dolph Lundgren, Charlotte Lewis, B.D: Wong, Anthony John Denison, Tim Guinee, Don Harvey, Tiny "Zeus" Lister, Tom Wright, Catherine Bell, Kevin Tighe, Trevor Goddard, Thomas Gibson, Perry Lang, Aldo Sambrell, Juan Pedro Tudela
When a ruthless mining corporation encounters opposition to its plans from the inhabitants of a small tropical island, they decide to recruit a group of mercenaries to bully the natives into submission. However, this scheme misfires and our fighting heroes change sides and spearhead a popular revolt. A highly predictable jungle adventure yarn, shot on location in Thailand,

whose beautiful countryside is the only real attraction in this assembly-line offering.
A/AD   98 min (ort 102 min)   VIDrel: EIV/SONOP   V

MENACE II SOCIETY ***
Albert Hughes/Allen Hughes USA
18
1993
Tyrin Turner, Larenz Tate, Jada Pinkett, Toshi Toda, Samuel L. Jackson, Arnold Johnson, Bill Duke, Charles S. Dutton, Vonte Sweet, Glenn Plummer, MC Eight, Marilyn Coleman, Clifton Powell, Pooh Man, Ryan Williams, Toshi Toda
Powerful, violent and very bloody slice of life in L.A.'s Watts district, where a teenager is unable to break away from his back-ground and seems destined to come to a bad end. Acted in a very convincing way, this searing movie is impressive directorial debut and paints a very convincing if depressingly bleak picture of ghetto life. The script is by Tyger Williams and the directors.
DRA   92 min (ort 97 min)   VIDrel: FIRST/SONOP L/A;
ENCORE (LV only – NTSC version)   V/sur LV

MEN'S CLUB, THE *
Peter Medak USA
18
1986
Roy Scheider, David Dukes, Richard Jordan, Harvey Keitel, Frank Langella, Craig Wasson, Treat Williams, Stockard Channing, Gina Gallegos, Cindy Pickett, Gwen Welles, Jennifer Jason Leigh, Ann Wedgeworth, Ann Dusenberry, Marilyn Jones
Seven men meet and talk mainly about woman and their rela-tionships with them. Later on five of them visit a brothel. Despite its curious theme, this was a major hit in the cinemas.
DRA   101 min   VIDrel: 4-FRONT/POLYREC/EIV L/A   V
Boa: novel by Leonard Michaels.

MENU FOR MURDER **
Larry Pearce USA
PG
1990
Ed Marinaro, Julia Duffy, Douglas Barr, Morgan Fairchild, Joan Van Ark, Cindy Williams
When the female president of a school PTA is poisoned during a dinner, the investigating police detective discovers that the lives of all the murder suspects are now in danger in this adequate comedy-thriller.
Aka: MURDER AT THE P.T.A. LUNCHEON
THR   89 min   VIDrel: CAPIT/GUILD   V

MEPHISTO ***
Istvan Szabo HUNGARY
15
1981
Klaus Maria Brandauer, Krystyna Janda, Rolf Hoppe, Ildiko Bansagi, Karin Boyd, Christine Harbot, Gyorgy Cserhalmi, Martin Hellberg, Christiane Graskoff, Peter Andorai, Ildiko Kishonti, Tamas Major, Maria Bisztrai
An actor remains in Germany after 1933 and becomes the willing tool of the Nazis in order to further his career. A confusing but powerful examination of how ideals are betrayed by the desire for fame. The first film in a trilogy by the director that was followed by COLONEL REDL and HANUSSEN. AA: Foreign.
DRA   138 min (ort 144 min)   VIDrel: ARTPRO/RTM   V
Boa: novel by Klaus Mann.

MERCENARY *
Avi Nesher ISRAEL/USA
15
1996
Oliver Gruner, John Ritter, Robert Culp, Ed Lauter, Martin Kove
Yet another one of those action tales that are written to a formula, in this case dealing with a bitter executive (Ritter) who dreams of getting revenge for the murder of his wife by a terror-ist, and to this end, hires a mercenary (Gruner) to help him do this. There are the requisite explosions, gun battles and confla-grations, but absolutely no unexpected plot twists or surprises.
A/AD   99 min cC   VIDrel: HOLPIC/TECH   V/sur

MERCI LA VIE **
Betrand Blier FRANCE
18
1991
Charlotte Gainsbourg, Anouk Grinberg, Gerard Depardieu, Michel Blanc, Jean Carmet, Catherine Jacob, Thierry Fremont, Francois Perrot, Didier Benureau, Jean-Louis Trintignant, Jurgen Mash, Christianne Jean, Annie Girardot
This imponderable, chaotic and self-indulgent film-within-a-film almost defies description, and starts with two rebellious girls becoming firm friends after various mishaps. The film then switches to a sexual mystery involving a highly contagious AIDS-like disease, then as suddenly to an unpleasant Nazi-era exercise, and then back again. This is an unsatisfactory and irri-tating demonstration of intellectualism without self-discipline.
Aka: THANK YOU, LIFE
COM   113 min (ort 118 min) wScrn Col/B/W
VIDrel: ARTIF/20TH   V/sh

MERCY MISSION: THE RESCUE OF FLIGHT 771 **
(PG)
Roger Young USA   1993
Robert Loggia, Scott Bakula, Rebecca Riff, Alan Fletcher, Michael Bishop, Kit Taylor, Scott Mackenzie, Steven Tandy, Robert Benedetti, Sarah Kemp, Ingrid Mckillop, Bob Reynolds, Peter Kent, Sean Surgess, Eddy Sentosa, Lafe Charlton
An over-sure pilot takes his small plane out over the Pacific where he runs into trouble and is forced to call for help. Unbelievably, his call is heard by the pilot of a New Zealand airliner who devises a daring rescue plan that implies risk to his own plane and passengers. Loggia and Bakula are suitably square-jawed and determined, and the whole effort is carried off with a general air of conviction.
THR   89 min mTV   SATrel: MOVIE CHANNEL

MERMAIDS ***
15
Richard Benjamin USA   1990
Cher, Bob Hoskins, Winona Ryder, Michael Schoeffling, Christina Ricci, Jan Miner, Caroline McWilliams, Betsy Townsend, Richard McElvain, William Paul Steele, Paula Plum, Dossy Peabody, Rex Trailer, Pete Kovner, Carol Moss
This quirky comedy is set in the early 1960s, with Cher playing an eccentric woman who constantly moves home with her two daughters whenever she gets too involved with a man. A severe embarrassment to her older teenage girl, she finally falls for a shoe salesman in a small New England town, but only after her daughter has given her a much-needed push. A whimsical blend of comedy and drama that both amuses and occasionally irritates. Music is by Jack Nitzsche.
COM   105 min (ort 111 min)   VIDrel: 4-FRONT/POLYREC V/sur
Boa: novel by Patty Dann.

MERRY CHRISTMAS, MR LAWRENCE ***
15
Nagisa Oshima JAPAN/UK   1983
David Bowie, Tom Conti, Ryuichi Sakamoto, Jack Thompson, Takeshi, Johnny Okura, Alistair Browning, James Malcolm, Chris Broun, Yuya Uchida, Ryunosuke Kaneda, Takashi Naito, Tamio Ishikura, Rokko Toura, Kan Mikami, Yuji Honma
A strange drama set in a Japanese POW camp with an attempt at exploring the clash of cultures with a battle of wills taking place between camp commander Sakamoto and British major Bowie. The haunting score is also by Sakamoto. This was Oshima's first English language film and is a difficult but compelling piece of cinema. The script is by Oshima and Paul Mayersberg.
DRA   118 min (ort 123 min)   VIDrel: L/A   V
Boa: book The Seed And The Sower by Laurens Van Der Post.

MERRY WIVES OF WINDSOR, THE **
U
David Jones UK   1983
Richard Griffiths, Prunella Scales, Alan Bennett, Ben Kingsley, Elizabeth Spriggs, Bryan Marshall, Judy Davis, Richard O'Callaghan, Tenniel Evans, Gordon Gostelow, Nigel Terry, Michael Robbins, Miranda Foster, Ron Cook
One in a series of BBC adaptations of Shakespeare's plays, most of which are competent rather than inspired. This one is no exception.
DRA   167 min (ort 169 min) mTV   VIDrel: BBC   V/h
Boa: play by William Shakespeare.

MESA OF LOST WOMEN, THE *
15
Ron Ormond/Herbert Tevos USA   1953
Jackie Coogan, Richard Travis, Mary Hill, Allan Nixon, Robert Knapp, Tandra Quinn, Harmon Stevens, Samuel Wu, George Barrows, Chris-Pin Martin, John Martin, Angelo Rossitto, Lyle Talbot (narration)
Grade-Z junk about a mad scientist and his attempts to create a super-race by combining the best bits of beautiful women with deadly spiders.
Aka: LOST WOMEN
FAN   67 min (ort 69 min) B/W   VIDrel: FIRC/RTM   V

MESSAGE, THE **
PG
Moustapha Akkad LEBANON/LIBYA/
SAUDI ARABIA   1976
Anthony Quinn, Irene Papas, Michael Ansara, Johnny Sekka, Michael Forrest, Neville Jason, Andre Morell, Martin Benson
The story of the career and life of Mohammed, with the Prophet never appearing in the film so as to avoid offending Moslem sensibilities. The film was made in both an English and an Arabic version. Visually impressive it may be, but this tedious film is made even more so by over-reverential direction, and is fatally flawed by the decision never to show Mohammed.
Aka: MOHAMMED, MESSENGER OF GOD
DRA   182 min   VIDrel: POLY/POLYREC   V

MESSAGE FROM HOLLY, A **
PG
Rod Holcomb USA   1992
Lindsay Wagner, Shelly Long, Molly Orr, Tony Colitti, P.L. Wilson
A dying businesswoman wants her best friend to become a surrogate mother to her six-year-old daughter in this fact-based drama.
DRA   90 min mTV   VIDrel: 4-FRONT/POLYREC/ODY V/h

MESSAGE FROM NAM **
PG
Paul Wendkos USA   1993
Jenny Robertson, Nick Mancuso, Ed Landers, Ted Marcoux, Hope Lange, Steven Eckholdt, Christopher Allport, Tracy Griffith, Billy Dee Williams, Esther Rolle, Rue McClanahan, Ken Marshall, Kieu Chinh, Russell Curry, Nick Angotti
Story of an idealistic female reporter sent on assignment to Vietnam, whose honest reports from Saigon acquire an enormous following in the States. But this being a Danielle Steel story, there's room in the plot for a passionate affair with an army captain, the ramifications of which lead to a re-appraisal of our heroine's beliefs and values. Lightweight and terribly superficial, this is not a film to provide any insights into this tragic period of history.
Aka: DANIELLE STEEL'S MESSAGE FROM NAM
DRA   172 min (ort 200 min) mTV
VIDrel: MIA/DISC/IMPENT   V
Boa: novel by Danielle Steel.

MESSENGER OF DEATH **
18
Fred Williamson ITALY/USA   1987
Fred Williamson, Christopher Connelly, Joe Spinelli, Cameron Mitchell, Val Avery, Jasmine Maimone, Sandy Cummings
A man just released from prison seeks revenge for the murder of his wife, and begins to pick off his targets from a hit-list of underworld figures. He soon finds himself in danger from her killers in this tough B-movie actioner, a fairly typical Williamson product.
Aka: IL MESSAGERO; MESSENGER, THE
A/AD   93 min Cut (26 sec – ort 95 min)
VIDrel: SUPVID/RTM   V

MESSENGER OF DEATH **
18
J. Lee Thompson USA   1988
Charles Bronson, Trish Van Devere, Laurence Luckinbill, John Ireland, Daniel Benzali, Marilyn Hassett, Charles Dierkop, Jeff Corey, Penny Peyser, Gene Davis, John Solari, Jon Cedar, Tom Everett, Duncan Gamble
Bronson plays a reporter who has the task of trying to unravel a murder case that involved two hostile Mormon sects. A production-line affair.
THR   87 min (ort 91 min)   VIDrel: MGM/WHV   V/sh
Boa: novel The Avenging Angel by Rex Burns.

METEOR *
PG
Ronald Neame USA   1979
Sean Connery, Natalie Wood, Karl Malden, Brian Keith, Henry Fonda, Trevor Howard, Martin Landau, Joseph Campanella, Richard Dysart, Bo Brundin, Katherine DeHetre, Trevor Howard, Joseph Campanella, James G. Richardson, Michael Zaslow
Disaster movie in which an enormous meteor threatens the Earth and all the parties involved in averting this danger immediately fall out and start bickering. See A FIRE IN THE SKY for a far better treatment of this theme.
FAN   103 min (ort 107 min)   VIDrel: MGM/WHV L/A   V

METEOR MAN, THE **
PG
Robert Townsend USA   1993
Robert Townsend, Marla Gibbs, Eddie Griffin, Robert Guillaume, James Earl Jones, Bill Cosby, Roy Fegan, Cynthin Belgrave, Martin Coleman, Frank Gorshin, Bobby McGee, Luther Vandross, Sinbad, Big Daddy Kane, LaWanda Page, Tiny Lister
After being hit by a meteor, a timid schoolteacher acquires a set of modest super-powers and reluctantly takes on the role of a super hero, beginning his career by cleaning up his own neighbourhood. A moderately amusing spoof, it is endearingly tongue-in-check but stands in need of a stronger story and sharper direction.
COM   95 min (ort 100 min) cC   VIDrel: MGM/WHV V/sur V/dm

METRO *   18
Thomas Carter USA   1997
Eddie Murphy, Michael Rappaport, Kim Miyori, Art Evans, James Carpenter, Donal Logue, Jeni Chua, Dick Bright, David Michael Silverman, Denis Arndt, Frank Somerville, Malou Nubla, Carmen Ejogo
A carelessly thrown together police story, clearly made to cash in on the surge of popularity that followed Murphy in the wake of his performance in THE NUTTY PROFESSOR. Here, he is a smart-talking San Francisco cop who specialises in dealing with hostage incidents, and is given the job of chasing after a dangerous psychopath who decides to make the battle of wits personal. A glossy, noisy, totally unoriginal action-thriller that insults our intelligence.
A/AD   117 min   CINrel

METROPOLIS ****   U
Fritz Lang GERMANY   1926
Brigitte Helm, Alfred Abel, Gustav Frohlich, Rudolf Klein-Rogge, Fritz Rasp, Theodor Loos, Erwin Biswanger, Heinrich George, Olaf Storm, Hanns Leo Reich, Georg John, Margaretta Lanner, Heinrich Gotho, Max Dietze, Walter Kohle
Classic early SF story of a dehumanised future society in which the workers toil below ground for the benefit of an elite who live above. Fascinating if dated sequences are woven into the sentimental story of the romance between the son of the city's ruler and a saintly girl from below ground. (Most tapes now available are re-edited and tinted versions, with some restored scenes, but often also with an irritating modern-music soundtrack – this is one such tape.)
FAN   138 min (restored version) B/W silent
VIDrel: EUREKA/GOLD   V/sh
Boa: novel by Thea von Harbou.

METROPOLITAN ***   15
Whit Stillman USA   1990
Carolyn Farina, Edward Clements, Taylor Nichols, Christopher Eigeman, Dylan Hundley, Allison Parisi, Isabel Gillies, Bryan Leder, Will Kempe, Elizabeth Thompson, Roger W. Kirby, Linda Gillies, John Lynch, Tom Voth, Frank Creighton
Articulate and witty look at the lives of a group of privileged upper-crust New York college kids and debutantes, who gather one Christmas break and talk over their hopes and fears. A vigorous display of ensemble acting from a lively young cast, with Clements as an "outsider" especially good. The directing debut for Stillman, who also wrote and produced the film.
COM   94 min (ort 98 min)   VIDrel: MAINPIC/RTM   V

MIAMI BLUES ***   18
George Armitage USA   1989
Fred Ward, Alec Baldwin, Jennifer Jason Leigh, Charles Napier, Paul Gleason, Nora Dunn, Jose Perez, Obba Babatunde, Martine Beswicke
Baldwin plays a cold-hearted crook and killer, who arrives in Miami fresh out of prison, and takes up with a young hooker who cannot see his failings. Later he steals a cop's badge and I.D., and a nasty battle of wits ensues. Written and directed by Armitage, this is a gripping albeit violent thriller, chiefly memorable for some black comic overtones and three excellent performances from the leads.
THR   92 min (ort 99 min)   VIDrel: VISVID/POLYREC   V/sur
Boa: novel by Charles Willeford.

MIAMI RHAPSODY **   15
David Frankel USA   1995
Sarah Jessica Parker, Antonio Banderas, Mia Farrow, Paul Mazursky, Gil Bellows, Kevin Pollak, Barbara Garrick, Carla Cugino, Bo Eason, Naomi Campbell, Jeremy Piven, Kelly Bishop, Mark Blum, Norman Steinberg, Ben Stein, Donal Logue
A woman accepts her finance's marriage proposal but soon gets cold feet once she starts observing the failed relationships of her parents and other family members. This verbose and insubstantial stab at a romantic comedy suffers badly from the lack of a solid plot, though the occasional Woody Allen-style observations on marriage and fidelity are a partial compensation.
COM   95 min   VIDrel: HOLPIC/TECH L/A   V

MIAMI SPICE **   18
Svetlana USA   1987
Amber Lynn, Barbara Dare, Sheri St Clair, Danielle, Randy West, Eric Edwards, Danielle, Janette Littledove, Mike de Marco, Robert Bullock, Candie Evans, Blondi, Porsche Lynn, Tony Martino, Rick Poins
Agents go undercover at a porno pool party to catch a drug

dealer in this sex spoof on TV's "Miami Vice", that is sustained by quite high production values (in fact all the outdoor shots were filmed in Miami). A sequel followed.
A   50 min   VIDrel: FIFTH/DISC   V

MIAMI SPICE 2 **   18
Svetlana USA   1988
Amber Lynn, Sheri St Clair, Stacey Donnovan, John Leslie, Joey Silvera, Randy West, Danielle, Candie Evans, Mike de Marco, Paul Baresi, Cindy Arhh, Francois, Chuck Gee
In this sequel to the earlier film, Leslie replaces Eric Edwards as the baddie, who as with the earlier film, is a major drug dealer who hides his activities under a philanthropic front. That said, the plot gets so very confusing it becomes almost impossible to follow, but all ends satisfactorily enough, even if most of the action revolves around various sexual encounters. As with the first film, high production values mark this one out from the rest.
A   49 min   VIDrel: FIFTH/DISC   V

MIAMI VICE ***   15
Thomas Carter USA   1984
Don Johnson, Philip Michael Thomas, Michael Talbot, Saundra Santiago, John Diehl, Gregory Sierra, Bill Smitrovich, Jimmy Smits, Belinda Montgomery, Michael Santoro, Martin Ferrero, Jossie DeGuzman, Harold Bergman
Pilot episode for a popular TV series showing the sleazy side of Miami. A cop seeks revenge on a drug dealer who murdered another cop, his brother. Quite a stylish opener for a series that got progressively more disagreeable as it developed.
Aka: MIAMI VICE: THE MOVIE
A/AD   90 min (ort 99 min) mTV
VIDrel: 4-FRONT/POLYREC L/A   V/sur

MIAMI VICE: DOWN FOR THE COUNT **   15
Richard Compton USA   1987
Don Johnson, Philip Michael Thomas, Saundra Santiago, Michael Talbot, Pepe Serna, Olivia Brown, Edward James Olmos, Dan King, Mark Breland
Based on the popular police series, this has detectives Crockett and Tubbs on the trail of a major narcotics importer. The crook has a passion for boxing and gambling, and the two detectives pose as cable TV promoters looking for a deal, in order to trap him. A glossy and shallow tale that's redeemed by the colourful locations and cheerfully mindless action.
A/AD   88 min (ort 93 min) mTV   VIDrel: CIC/SONOP   V/h

MIAMI VICE: THE GOLDEN TRIANGLE *   15
Georg Stanford Brown/David Aspaugh USA   1985
Don Johnson, Philip Michael Thomas, Edward James Olmos, Keye Luke, Joan Chen, Saundra Santiago, Michael Talbott, John Diehl
Feature spin-off from a popular TV series cop series. Our two detectives are sent to work under cover in Thailand in order to crack a drug-smuggling ring, but there's little in terms of plot or action, and the same minimalist style of acting, plus musical montages for the boring bits, very much in the style of a pop video.
A/AD   92 min Cut (5 sec) mTV   VIDrel: CIC/SONOP   V

MIAMI VICE: THE PRODIGAL SON *   15
Paul Michael Glaser USA   1985
Don Johnson, Philip Michael Thomas, Saundra Santiago, Edward James Olmos, Michael Talbot, Penn Jilette, John Diehl, Olivia Brown, Pam Grier
Pilot for the second series of TV cop films, with our two heroes tracking down a drugs ring, and discovering a link between a trafficker and a New York bank, in the jungles of Latin America.
A/AD   93 min (ort 99 min) mTV   VIDrel: CIC/SONOP   V/sh

MICHAEL **   PG
Nora Ephron USA   1996
John Travolta, Andie MacDowell, William Hurt, Robert Pastorelli, Bob Hoskins, Jean Stapleton, Teri Garr, Wallace Langham, Joey Lauren Adams, Carla Cugino, Tom Hodges, Catherine Lloyd Burns
Travolta is the eponymous Michael, an angel who has descended from Heaven to help the people of Earth, but has to contend with a couple of hotshot reporters who are out to get the scoop of the century. The main joke in this movie (which was a huge success in the States) is that Michael is a chain-smoking, uncouth slob of an angel, whose powers are severely limited, as

is the range of ideas in this painfully over-stretched, one-joke film.
COM   105 min   CINrel

MICHAEL COLLINS ***   15
Neil Jordan USA   1996
Liam Neeson, Aidan Quinn, Stephen Rea, Alan Rickman, Julia Roberts, Ian Hart, Richard Ingram, John Kenny, Roman McCairbe, Ger O'Leary, Michael Dwyer, Martin Murphy, Sean McGinley, Gary Whelan, Frank O'Sullivan, Frank Laverty
After the 1916 Irish Rebellion, Collins surrenders to the British and is interned. However, this does nothing to dampen his Republican fervour, and upon his release he dedicates his life to Sinn Fein and the elimination of a British presence from Irish soil, eventually becoming head of the Irish Free State that was set up by the treaty of 1921. Scripted by Jordan, this is a powerful but simplistic work, and its decidedly partisan stance does it little credit.
DRA   134 min   VIDrel: WHV   V

MICKI AND MAUDE ***   PG
Blake Edwards USA   1984
Dudley Moore, Anne Reinking, Amy Irving, Richard Mulligan, George Gaynes, Wallace Shawn, Lu Leonard, Priscilla Pointer, Andre the Giant, John Pleshette, H.B. Haggerty, Robert Symonds, George Coe, Gustav Vintas, Ken Olfson
A man's career-minded wife refuses to have children so he finds another woman who will. When she becomes pregnant, he decides to divorce his wife and marry her, but before he can do this she triumphantly announces that she is pregnant too. Bigamy soon seems to be a small price to pay in order to keep both women happy. A lighthearted farce that quickly runs out of ideas and energy, but for the most part remains extremely effective.
COM   113 min (ort 118 min)   VIDrel: RCA L/A   V

MIDDLEMARCH ***   U
Anthony Page UK/USA   1994
Juliet Aubrey, Simon Chandler, Ian Driver, Pam Ferris, Jonathan Firth, Peter Jeffrey, Robert Hardy, Caroline Harker, Douglas Hodge, Michael Hordern, Patrick Malahide, Trevyn McDowell, Stephen Moore, Rachel Power, Rufus Sewell
An excellent and lovingly made adaptation of Elliot's novel of life in a small town in Victorian England where a young widow falls in love with a cousin of her dead husband. However, she faces the prospect of losing her inheritance should she decide to re-marry.
DRA   357 min (2 cassettes – 390 min) mTV
VIDrel: BBC/TECH   V/h
Boa: novel by George Elliot.

MIDNIGHT **   18
John Russo USA   1981
Lawrence Tierney, Melanie Verlin, John Amplas, John Hall, Charles Jackson, Doris Hackney, Robin Walsh, David Marchick, Greg Besnak
A young woman leaves home and hitch-hikes to see a friend but gets caught up with a group of Satanists. A pretty fair low-budget offering, adapted from the novel by Russo.
Aka: BACKWOODS MASSACRE
HOR   90 min (ort 94 min)   VIDrel: VIPCO/SGSVID   V/h
Boa: novel by John Russo.

MIDNIGHT *   15
Norman Thaddeus Vane USA   1988
Lynn Redgrave, Tony Curtis, Steven Parrish, Rita Gam, Gustav Vintas, Frank Gorshin, Karen Witter, Wolfman Jack
The hostess of a late-night horror movie show finds herself involved in a deadly conflict over the rights to the show, with everyone around her starting to die in unusual circumstances. Redgrave does what she can with her role (doubtlessly inspired by TV's horror-hostess "Elvira") but the film is both sluggish and insipid; a longer "director's cut" was shown theatrically after video release, but was not well received.
HOR   82 min (ort 90 min)   VIDrel: BRAVE/SONOP L/A   V

MIDNIGHT ANGEL **   18
Yee Chik-Ki HONG KONG   1988
Yukari Oshima, May Lo Mei-Mei, Angile Leung Wan-Yu, Min Kiu-Wai, Mark Cheung No-Ham, Shih Kien, Melvin Wong Kam-Sum, Cho Tat-Wah, Lily Chung Suk Wai, Kam Seung-Yuk, Yim Chau-Wah, Hon Chun, Yu Ming, Christine Duhler
Following in the footsteps of their grandfather, three girls take turns to become a caped crime-fighter after the boyfriend of one

of them is murdered by drug-dealers. A confused spoof that has no shortage of action but is badly let down by its undeveloped script.
Aka: JUSTICE WOMEN, THE
MAR   90 min   VIDrel: EAST/DISC   V

MIDNIGHT BLUE *   18
Skott Snider USA   1996
Annabel Schofield, Damian Chaps, Dean Stockwell, Steve Kanaly, Harry Dean Stanton
A businessman who has been carrying on with a hooker loses track of her, only to find that when he moves to L.A. to start a new job, his boss appears to be married to the woman in question. Unmemorable erotic drama.
DRA   90 min   VIDrel: 20TH/FOXVID   V/sur

MIDNIGHT CLEAR, A ***   15
Keith Gordon USA   1992
Peter Berg, Kevin Dillon, Arye Gross, Ethan Hawke, Gary Sinise, Frank Whaley, John C. McGinley, Larry Joshua, Curt Lowens, David Jensen, Rachel Griffin, Tim Shoemaker, Kelly Gately, Bill Osborn, Andre Lamal
At Christmas of 1944, a squad of young American soldiers who have lost half their number, are sent on a dangerous mission to capture an abandoned house in the Ardennes Forest. A solidly made anti-war drama, filmed in a languid and dreamlike style, but full of moving performances.
WAR   103 min (ort 108 min)   VIDrel: VCC/DISC/COLUM V/sur
Boa: novel by William Wharton.

MIDNIGHT CONFESSIONS **   (18)
Allan Shustak USA   1993
Carol Hoyt, David Milbern, Julie Strain, Ricahrd Lynch, Christina Rich, Cori Hansen, Callie Michael, Steve Michael, John Dagnen, Derek Mitchell, De Ann Ponco, Doug Demarco, Lisa Comshaw, Raven Alexander, Diana Cuevas, Tate Sprenger
Callers to a midnight sex talk show are being stalked and murdered by a deranged killer in this dark and unpleasant film, whose attempts at eroticism sit ill with its violent theme.
DRA   82 min (ort 90 min)   SATrel: SKY MOVIES

MIDNIGHT COWBOY ****   18
John Schlesinger USA   1969
Dustin Hoffman, Jon Voight, Sylvia Miles, John McGiver, Brenda Vaccaro, Bob Balaban, Barnard Hughes, Viva, Taylor Mead, Paul Morrissey, Jennifer Salt, Ruth White, Gil Rankin, Gary Owens, T. Tom Marlowe, George Eppersen, Linda Davis
Sharp and offbeat story of a simple-minded "cowboy" who arrives in New York determined to make his fortune as a stud to rich ladies. Hoffman is superb as the crippled loser he teams up with on the strength of a promise to find him some work in this line. The fine score is by Harry Nillson. AA: Pic, Dir, Screen/adapt (Waldo Salt).
DRA   108 min (ort 113 min)   VIDrel: MGM/WHV   V/sur
Boa: novel by James Leo Herlihy.

MIDNIGHT CROSSING *   18
Roger Holzberg USA   1988
Faye Dunaway, Daniel J. Travanti, Kim Cattrall, John Laughlin, Ned Beatty, Pedro De Pool, Doug Weiser, Vincent Fall, Michael Thompson, Chick Bernhardt, Janet Constable, Mara Goodman, Armando Gonzales, Pat Selts, Rhonda Johnson
A businessman arranges a trip to the Caribbean on board a luxury yacht as a treat for his blind wife. But, unknown to her, his real reason for the present trip is to recover a hoard of stolen money from a tiny island. The journey soon turns into a nightmare of deceit, violence and treachery. A better title for this boring mess would have been "Mayhem At Midnight".
THR   92 min (ort 104 min)   VIDrel: VES L/A   V

MIDNIGHT DANCERS ***   18
Mel Chionglo PHILIPPINES   1995
Ryan Aristorenas, Perla Bautist, Richard Cassity, Grandong Cervantes, Luis Cortez, Lawrence David, Alex Del Rosario, Leonard Manalanson, John Mendoza, Danny Ramos, Grandong Cervantes Jr.
A searing account of sex and prostitution in Manila, as three brothers resort to working in a gay bar in their efforts to survive. Banned by the authorities in its homeland.
Aka: SIBAK
DRA   122 min   VIDrel: DTK/RTM   V

MIDNIGHT EXPRESS ★★★   18
Alan Parker UK/USA   1978
Brad Davis, Irene Miracle, Bo Hopkins, Randy Quaid, John Hurt,
Mike Kellin, Paul Smith, Paolo Bonacelli, Norbert Weisser, Franco
Diogene, Michael Ensign, Gigi Ballista, Kevork Malikyan, Peter
Jeffrey, Ahmed El Shenawi
Traces a young American's descent into the hell of a Turkish
prison after being convicted of drug-smuggling and facing the
ghastly prospect of years of imprisonment. A powerful but
brutal drama that strays rather too often from the real-life events
on which it was based. See also PRISON HEAT. AA: Screen/
adapt (Oliver Stone), Score/orig (Giorgio Moroder).
DRA   116 min (ort 120 min) wScrn
VIDrel: COLUM/SONOP   V/sur
Boa: book by Billy Hayes and William Hoffer.

MIDNIGHT FEAR ★★   18
Bill Crain USA   1990
Craig Wasson, David Carradine, Page Fletcher, Evan Richards,
August Ward, Mark Carlton, Charlie Schmidt, Jill Worley, Rob
Wheeler, Ross Wilburn, Jack Green, Tammy Schneider, Kurt Stamps,
Jay Komosenski, Pat Atkins, Joe Turner
When the body of a young girl who was brutally murdered is
discovered in a small town, a burnt-out cop embarks on an
obsessive hunt for the killer, the success of which may resurrect
his career. As he closes in on the killer, he finds himself trapped
in a nasty web of deceit, and is eventually drawn to a deserted
house for a violent confrontation.
THR   90 min   VIDrel: NWV/HIFLI L/A   V

MIDNIGHT HEAT ★★   18
Allan A. Goldstein USA
Brian Bosworth, Brad Dourif, Claire Yarlett, Marta Dubois
Having suffered a head injury that has left him with amnesia, a
man starts getting flashes of his past memories, and these
encourage him to attempt to find out about his former life. But
this attempt is fraught with peril. An interesting if rather derivative psychological thriller.
THR   94 min   VIDrel: EIV/SONOP   V

MIDNIGHT HEAT ★★   18
Richard Mahler USA   1983
Jamie Gillis, Howard Feline, Cheri Champagne, Sharon Mitchell,
Susan Nero, Joey Carson, Tish Ambrose, Lucretia Love, Michelle
Perelo
A contract killer hiding out from a gangster recalls his career,
in this blend of sex and gangster genres.
A   75 min (ort 78 min)   VIDrel: ELV   V

MIDNIGHT HEAT ★★   18
John Nicolella USA   1992
Dennis Hopper, Michael Pare, Adam Ant, Little Richard, Charlie
Schlatter, Tracy Tweed, Daphne Ashbrook, Luca Bercovic, Stephanos
Miltsakakis, Cheryl Freeman, Tony tood, Joe Lara, Michael Talbott,
Cindy Valentine, Julie Strain
A former drug dealer becomes a photographer and returns
home to L.A. where he stays with a friend who is still working
for the psychopath who used to be his boss. When the latter is
robbed, he blames our hero and sets out to get even, in this
action-filled if derivative thriller.
Aka: SUNSET HEAT
THR   94 min (ort 96 min)   VIDrel: FIRST/SONOP L/A   V

MIDNIGHT HEAT ★★   (18)
Harvey Frost USA   1994
Tim Matheson, Stephen Mendel, Mimi Craven
A football player is foolish enough to indulge in an affair with
the wife of the team's owner. When the latter is murdered, he
finds himself cast in the role of prime suspect. Watchable, but
no more than that.
DRA   94 min (ort 97 min)   SATrel: MOVIE CHANNEL

MIDNIGHT HOUR ★★   18
Rodney McDonald USA   1995
Paula Barbieri, Jeff Trachta, Andrew Stevens
A police officer arrives at the home of an attractive psychiatrist
following a call for assistance, and stays the night to allay her
fears. As a relationship develops between the two, our patrol-
man begins to realise that his affair with her has certain dangers.
THR   97 min   VIDrel: HIFLI/SONOP   V/h

MIDNIGHT IN ST PETERSBURG ★★   15
Douglas Jackson UK/CANAD/RUSSIA   1995
Michael Caine, Jason Connery, Michelle Rene, Thomas Michael
Gambon, Michael Sarrazin, Tanya Jackson
Sequel to BULLET TO BEIJING that once more casts Caine as
retired agent Harry Palmer, who is brought back into service to
recover some stolen plutonium. Another low-key, low-budget
outing, that gives Caine a chance to show his mettle, despite
being hampered (once more) with Connery as the most wooden
sidekick imaginable. The weak dialogue and stale ideas are not
much help either. The dreadful score is the work of Rick
Wakeman, who has done much better work.
THR   86 min cC   VIDrel: TOUCH/TECH   V/sur
Boa: novel by Len Deighton.

MIDNIGHT RIDE ★★   18
Bob Bralver USA   1993
Michael Dudikoff, Robert Mitchum, Mark Hamill, Savina Gersak,
Pamela Ludwig, Timothy Brown, Lezlie Dean, Steven Ingrassia,
Cynthia Szigeti, Dee Dee Rescher, Mary Peters, Regina Krueger, Judy
Bartram, Suzanne Celeste, R.A. Rondell
A woman who has decided to leave her police officer husband
hitches a lift, but learns that her new companion is a serial killer.
An adequate if rather derivative thriller. The director appears
briefly in a tiny cameo.
THR   89 min   VIDrel: WHV   V/sh

MIDNIGHT RUN ★★★   18
Martin Brest USA   1988
Robert De Niro, Charles Grodin, John Ashton, Dennis Farina, Yaphet
Kotto, Joe Pantoliano, Wendy Phillips, Richard Foronjy, Robert
Miranda, Jack Kehoe, Danielle Duclos, Philip Baker Hall, Thom
McCleister, John Toles-Bey, Mary Gillis
An ex-cop bounty hunter goes after a bail-jumper wanted by
both the FBI (for his testimony) and the Mafia (for embezzling
$15,000,000). The strongly focused and witty script is nicely
complemented by good work from the leads. Written by
George Gallo, and with an effective score by Danny Elfman.
Followed by MIDNIGHT RUN 2: ANOTHER MIDNIGHT
RUN.
COM   121 min Cut (15 sec – ort 125 min)
VIDrel: 4-FRONT/POLYREC/CIC   V/sur

MIDNIGHT RUN 1: MIDNIGHT RUNAROUND ★★   PG
Frank De Palma USA   1995
Christopher McDonald, Kyle Secor, Rebecca Cross, Ed O'Ross, Dan
Hedaya, John Fleck, Leon Russom, Gary Grubbs, Beverly Leech, Don
Gibb, Dick Miller, Jack R. Grend, Sarah Lassez, Tom McLesiter, Jeff
Doucette, Tom De Zarn, Scott Burkholder
TV spin-off from the movie MIDNIGHT RUN, that follows the
exploits of a Los Angeles bounty-hunter, who in this adventure
has to go to Oklahoma to pick up a young crook who is very
popular with the opposite sex. Followed by MIDNIGHT RUN
2: ANOTHER MIDNIGHT RUN.
Aka: MIDNIGHT RUNAROUND
A/AD   85 min (ort 90 min) mTV   VIDrel: CIC/SONOP   V

MIDNIGHT RUN 2: ANOTHER MIDNIGHT RUN ★★   PG
James Frawley USA   1995
Christopher McDonald, Cathy Moriarty, Jeffrey Tambor, Ed O'Ross,
Dan Hedaya, John Fleck, Sam Shamshack, Julie Lott, Frank Pesce,
Bryan Clark, Suzanne Kent, Virginia Watson, John William Young,
Jack Nance, John P. Connolly, Daniel Leslie
Action-comedy that continues the story from the earlier film,
with a bounty-hunter in L.A. where he hopes to catch two top
con-artists, but faces stiff competition from a rival who is in the
same line of work. McDonald takes over from De Niro in the
central role, and does what he can with a script that offers little
in the way of surprises.
Aka: ANOTHER MIDNIGHT RUN
A/AD   88 min (ort 90 min) mTV   VIDrel: CIC   V

MIDNIGHT RUN 3: MIDNIGHT RUN FOR YOUR
LIFE ★★   PG
Frank Sackheim USA   1995
Christopher McDonald, Ed O'Ross, Melora Walters, John Fleck, Dan
Hedaya, Joe Bradley, Richard Herkert, Jennifer Manasseri, Millie
Perkins, Stephen Lee, Louis Mustillo, Vincent Gustaferro, Stephen
Hynter, Richard Bradford, Gabriel Dell Jr
Last in this set of three comedy-action films, this time dealing
with a woman who narrowly escapes death and flees to Mexico.
Aka: MIDNIGHT RUN FOR YOUR LIFE
A/AD   85 min (ort 90 min) mTV   VIDrel: CIC/SONOP   V

MIDNIGHT STING *** 15
Michael Ritchie USA 1992
James Woods, Oliver Platt, Orestes Matacena, Louis Gossett Jr, Bruce Dern, Randall "Tex" Cobb, Thomas Wilson Brown, Duane Davis, Willie Green, George D. Wallace, Marshall Bell, Heather Graham, Kim Robillard, John Short, Jeff Benson
A con-man leaves prison and dreams up a scam in which he plans to skin a nasty individual who arranges unregulated boxing matches in title locale. Hoping to call on the services of a former prizefighter (ably played by Gossett), he makes a bet that over twenty-four hours his boxer can beat any ten put up by the other side.
Aka: DIGGSTOWN
COM 93 min (ort 98 min) cC VIDrel: MGM/WHV V/dm
Boa: novel The Diggstowm Ringers by Leonard Wise.

MIDNIGHT WITNESS ** 18
Peter Foloy USA 1993
Paul Johansson, Maxwell Caufield, Karen Moncrieff, Jan-Michael Vincent
A young couple videotape a routine traffic incident but are shocked when the police involved beat and murder the vehicle's driver. When they realise that this crime has been filmed, the culprits give chase in a bid to eliminate both them and the evidence.
A/AD 86 min (ort 90 min) VIDrel: 20VIS/SONOP V

MIDNIGHT'S CHILD ** 15
Colin Bucksey USA 1992
Olivia D'Abo, Marcy Walker, Cotter Smith, Elissabeth Moss, Jim Norton, Judy Parfitt, Roxan Biggs, Mary Larkin
A housewife hires a Swedish au pair to act as nanny for their young girl, but do not realise that she is a Satanist who wants the child for a sacrifice. Another one of those "nanny from Hell films" (see THE GUARDIAN) that treads an over-familiar path, but carries off the enterprise with conviction and skill if not originality.
HOR 85 min (ort 89 min) mTV VIDrel: GUILD/SONOP V

MIDSUMMER NIGHT'S DREAM, A *** U
Max Reinhardt/William Dieterle USA 1935
James Cagney, Olivia De Havilland, Mickey Rooney, Dick Powell, Joe E. Brown, Jean Muir, Ross Alexander, Hugh Herbert, Arthur Treacher, Frank McHugh, Otis Harlan, Dewey Robinson, Victor Jory, Verree Teasdale, Anita Louise, Fred Sale
A star-packed, visually impressive version of this famous fantasy telling of two lovers who sort out their problems in a magical wood in Athens with a little help from the fairies. Imaginative casting gave Cagney the role of "Bottom", a part he makes the most of, whereas Rooney as "Puck" becomes truly tiresome. However, the music of Mendelssohn is used to good effect. This was the film debut of De Havilland. AA: Cin (Hal Mohr), Edit (Ralph Dawson).
MUS 112 min (ort 133 min) B/W VIDrel: MGM/WHV L/A V
Boa: play by William Shakespeare.

MIDSUMMER NIGHT'S DREAM, A ** U
Elijah Moshinsky UK 1981
Helen Mirren, Peter McEnery, Nigel Davenport, Estelle Kohler, Hugh Quarshie, Geoffrey Lumsden, Pippa Guard, Nicky Henson, Robert Lindsay, Cherith Mellor, Geoffrey Palmer, Brian Glover, John Fowler
Another BBC television adaptation, generally well cast and agreeable despite the constraints of the studio setting.
COM 112 min mTV VIDrel: BBC V/h
Boa: play by William Shakespeare.

MIDSUMMER NIGHT'S DREAM, A ** PG
Adrian Noble UK 1996
Lindsay Duncan, Alex Jennings, Desmond Barrit, Barry Lynch, Monica Dolan, Kevin Doyle, Daniel Evans, Emily Raymond, Alfred Burke, Howard Crossley, Robin Gillespie, John Kane, Mark Letheren, Kenn Sabberton, Ann Hasson, Osheen Jones
In ancient Athens, a youngster dreams of the Duke of Athens, Theseus, and his intention to marry the Amazon queen Hippolyta, whom he defeated in battle. Meanwhile, two young lovers run off to hide in the nearby forest, for the girl has been betrothed to another. Shakespeare's magical fantasy on love is given an overly clever play-within-a-play treatment, but this inventive approach does not work on film, and is best kept on the stage, where it originated in 1994.
FAN 105 min VIDrel: VCC/DISC V
Boa: play by William Shakespeare.

MIDSUMMER NIGHT'S SEX COMEDY, A ** 15
Woody Allen USA 1982
Woody Allen, Mia Farrow, Jose Ferrer, Mary Steenburgen, Tony Roberts, Julie Hagerty, Tony Farentino, Adam Redfield, Moishe Rosenfeld, Timothy Jenkins, Sol Frieder, Michael Higgins, Boris Zoubok, Thomas Barbour, J. David Copeland
Allen's version of Bergman's 1955 film "Smiles Of A Summer Night" about the sexual relations among three couples spending a summer weekend on a country estate around 1900. Mildly diverting but rather slack, though the splendid photography of Gordon Willis is an asset.
COM 85 min (ort 88 min) VIDrel: WHV V

MIDWEST OBSESSION ** 15
William A. Graham USA 1995
Courtney Thorne-Smith, Kyle Secor, Stephen Fanning, Tracey Gold, Ewan "Sudsy" Clark, Joely Collins, Don S. Davis, Bruce Harwood, Duane Keogh, David Lewis, Kevin McNulty, Chris Martin, Sheelah Megill, Jenny Mitchell, Larry Musser
Based on a true story, this tells of a local beauty whose desire to find a "perfect" husband masks her psychopathic tendencies.
Aka: BEAUTY'S REVENGE
DRA 87 min (ort 92 min) mTV VIDrel: ODY/SONOP V/sh

MIGHTY APHRODITE *** 15
Woody Allen USA 1995
Woody Allen, Helena Bonham Carter, F. Murray Abraham, J. Smith Cameron, Steven Randazzo, David Ogden Stiers, Olympia Dukakis, Jeffrey Kurland, Tucker Robin, Peter Weller, Donald Symington, Claire Bloom, Nolan Tuffey, Jimmy McQuaid
Some years after a New York couple adopted a baby boy, the father becomes curious as to the child's origins, as he appears to possess a genius-level of intelligence. However, he is astounded to discover that the boy's natural mother is a bright and good-natured woman who works as a hooker. The two become firm friends and he resolves to save her from this life and find her a suitable mate. A breezy and witty comedy, with first-rate performances. AA: Actress (Sorvino).
COM 92 min (ort 945 min) cC VIDrel: TOUCH/TECH V/sur

MIGHTY DUCKS, THE ** PG
Stephen Herek USA 1992
Emilio Estevez, Lane Smith, Joss Ackland, Heidi Kling, Josef Sommer, Joshua Jackson, Elden Ratliff, Shaun Weiss, Matt Doherty. M.C. Gainey, Brandon Adams, J.D. Daniels, Aaron Schwartz, Garette Ratliff Henson, Marguerite Moreau
A young, self-centred lawyer is sentenced to community service for drunk driving and finds himself cast in the role of coach to an inner-city hockey team. The usual misunderstandings occur on both sides before he manages to win their respect and inspire them with the necessary self-confidence. A preachy and predictable affair that provides some light and undemanding entertainment. Followed by D2: THE MIGHTY DUCKS.
Aka: CHAMPIONS; MIGHTY DUCKS ARE THE CHAMPIONS, THE
COM 99 min (ort 103 min) cC VIDrel: WDV/TECH V

MIGHTY JOE YOUNG *** (12)
Ernest B. Schoedsack USA 1949
Terry Moore, Ben Johnson, Robert Armstrong, Frank McHugh, Douglas Fowley, Regis Toomey, Dennis Green, Paul Guilfoyle, Nestor Paiva, Dale Van Sickel, Primo Carnera, Wee Willie Davis, Henry Kulky, Sammy Stein, Karl David, Ivan Rasputin
An updated version of KING KONG with the giant ape being captured to form the main attraction at a night club, running amok when taunted, but later redeeming himself when he saves some kids from a fire. A tongue-in-cheek effort that is surprisingly well crafted. The splendid effects are by Willis O'Brien and Ray Harryhausen. AA: Effects (Cooper/R.K.O. Radio).
DRA 90 min (ort 94 min) B/W VIDrel: L/A V

MIGHTY MORPHIN POWER RANGERS: THE MOVIE * PG
Bryan Spicer USA 1995
Karan Ashley, Johnny Yong Bosch, Steve Cardenas, Jason David Frank, Amy Jo Johnson, David Yost, Paul Freeman, Paul Schreier, Jason Narvy, Nicholas Bell, Gabrielle Fitzpatrick, Peta-Maree Rixon, Jean Paul Bell, Kerry Casey
Feature-length Rangers inspired by the TV series that places our heroic teens on familiar ground as they combat their mortal enemy in a succession of set-piece encounters. Having lost their powers in attempting to rescue their leader, they are taught the

martial arts by a female fighter. There is little to distinguish this from any episode of the cult juvenile series, except its length. For devoted fans only, others will find it unbearably trite and tedious.
Aka: POWER RANGERS: THE MOVIE
JUV 95 min CINrel

MIKE'S MURDER *
James Bridges USA   1982 (released 1984)   18
Debra Winger, Mark Keyloun, Darrell Larson, Paul Winfield, Brooke Alderson, William Ostrander, Dan Shor, Robert Crosson, Gregory Hormel, John Michael Stewart, Victor Perez, Mark High, Ken Y. Nmaba, Ruth Winger, April Ferry
A woman bank-teller's sometime boyfriend is killed in a drugs deal and she decides to go looking for clues to his murder, aided by the man's addict friend who is next on the hit-list. After less than ecstatic reviews, the film was withdrawn for extensive re-cutting and received very limited release in 1984. An interminable film that fails in all departments.
DRA 97 min (ort 105 min) VIDrel: MGM/WHV L/A V

MIKEY & NICKY *
Elaine May USA   15   1976
Peter Falk, John Cassavetes, Ned Beatty, Sandford Meisner, Rose Arrick, Joyce Van Patten, William Hickey, M. Emmet Walsh, Carol Grace, Sy Travers, Peter Scoppa, Sanford Meisner, Virginia Smith, Jean Shevlin, Danny Klein
The story of two small-time hoods who were childhood friends. One of them is wanted by the Mob and the other may be setting him up to be hit by them. An opaque and over-sentimental mess that is made barely watchable by excellent performances from the leads.
DRA 101 min (ort 119 min) VIDrel: CONNO/RTM V

MILAGRO BEANFIELD WAR, THE ***
Robert Redford USA   15   1987
Sonia Braga, Chick Vennera, Ruben Blades, Christopher Walken, Daniel Stern, John Heard, Melanie Griffith, Carlos Riquelme, Freddy Fender, Tony Genaro, Jerry Hardin, Ronald G. Joseph, Mario Arrambide, Alberto Morin, Julie Carmen
A poor bean farmer in New Mexico illegally diverts water in a desperate attempt to save his parched fields. His action rouses and unites his poor, downtrodden community, but at the same time sets it on a collision course with local big business interests. A vivid and entertaining film, scripted by Nichols and David Ward. AA: Score/orig (Dave Grusin).
COM 113 min (ort 117 min) VIDrel: CIC/SONOP V/sur
Boa: novel by John Nichols.

MILDRED PIERCE ***
Micheal Curtiz USA   PG   1945
Joan Crawford, Ann Blyth, Jack Carson, Zachary Scott, Eve Arden, Bruce Bennett, George Tobias, Lee Patrick, Moroni Olsen, Jo Ann Marlowe, Barbara Brown, Charles Trowbridge, John Compton, Butterfly McQueen, Garry Owen
A woman leaves her husband and by dint of sheer hard work builds up a chain of restaurants, but pays a high price for this success, including a murder case and a troubled relationship with her daughter. An excellent and highly enjoyable melo-drama, with Crawford turning in a fine performance. AA: Actress (Crawford).
DRA 106 min (ort 113 min) B/W VIDrel: MGM/WHV V/h
Boa: novel by James M. Cain.

MILES FROM HOME ***
Gary Sinise USA   15   1988
Richard Gere, Kevin Anderson, Penelope Ann Miller, Laurie Metcalf, John Malkovich, Brian Dennehy, Judith Ivey, Helen Hunt, Terry Kinney, Francis Guinan, Randy Arney, Jason Campbell, Austin Baumgarner, Helen Hunt, Robert Otis
An Iowa farm goes from success in the 1950s to bankruptcy in the 1980s, so Gere burns it down rather than see a bank take it over. He then takes off with his brother on a crime spree, the pair becoming folk heroes in the process. A kind of distant cousin to BONNIE AND CLYDE but a good deal more perceptive, despite the miscasting of Gere (who just doesn't convince as a country-boy).
A/AD 103 min (ort 114 min) VIDrel: EMPIRE/TOTAL L/A V/dm

MILES FROM NOWHERE ***
Buzz Kulik USA   PG   1991
Ricky Schroder, James Farentino, Shawn Phelan, Melora Hardin, Marlyn Mason, Tom Schanley, Kaj-Erik Eriksen, Johanna Newmarch,

Tom McBeth, Linda Sorensen, Sheila Moore, Frank C. Turner, Simon Webb, Miriam Sirois, Tom Heaton
A youngster who appears to have everything life can offer is devastated when his brother is seriously injured in a car accident and rushed to hospital in a deep coma. Having faced up to the unpleasant consequences of his own former lack of responsibility, he endeavours to redeem himself by devoting every free moment to the care of his brother. An absorbing family drama.
DRA 87 min VIDrel: NWV/HIFLI V/h

MILES TO GO **
David Greene USA   15   1986
Jill Clayburgh, Tom Skeritt, Mimi Kuzyk, Rosemary Dunsmore, Cyndy Preston, Andrew Bednarski, Peter Dvorsky, Caroline Arnold, Sheena Larkin, Catherine Golvey, Lorena Gale, Linda Smith, Danette Mackay, Donna Faron, Ian Finlay
A close-knit loving family have to come to terms with the fact that their mother is dying of cancer. A real weepie in the manner of LOVE STORY that is sustained by a dignified performance from Clayburgh.
Aka: LEAVING HOME
DRA 94 min (ort 100 min) mTV VIDrel: ODY/SONOP L/A V
Boa: story by Beverly Levitt.

MILK MONEY ***
Richard Benjamin USA   15   1994
Ed Harris, Melanie Griffith, Michael Patrick Carter, Malcolm McDowell, Anne Heche, Casey Siemaszko, Philip Bosco, Brian Christopher, Adam LaVorgna, Kevin Scannell, Jessica Wesson, Amanda Sharkey, Margaret Nagle, Kati Powell, Tom Coop
Three boys pool the title source of funds and go to the city in the hope of being able to see naked women, but instead they meet a hooker with a heart of gold. She gives them a ride back home, where she gets involved with the widowed father of one of the boys. Hardly an edifying plot, though this silly and sugary romantic-comedy does at least give Griffith a chance to shine.
COM 104 min (ort 109 min) cC VIDrel: CIC/SONOP V

MILKY WAY, THE **
Luis Bunuel FRANCE/ITALY   15   1968
Laurent Terzieff, Paul Frankeur, Delphine Seyrig, Edith Scon, Alain Cluny, Bernard Verley, Michel Piccoli, Pierre Clementi, Georges Marchal, Jean Piat, Denis Manuel, Daniel Pilon, Claudio Brook, Julien Guiomar, Marcel Peres
Two tramps on a pilgrimage from Paris to Spain enjoy various adventures en route, making detours in both time and space. An obtuse examination of Catholicism, hard to take and even harder to appreciate.
Aka: LA VOIE LACTEE; LA VIA LATTEA
DRA 97 min (ort 102 min) wScrn
VIDrel: ELPIC/POLYREC V

MILL ON THE FLOSS, THE ***
Graham Theakston UK   (PG)   1996
Emily Watson, Cheryl Campbell, James Frain, Bernard Hill
A Victorian tragic drama, telling of a brother and sister who grow up together, forging a strong bond despite their very different temperaments, the girl being something of a free spirit. An initial romance ensues with the son of the lawyer who ruined their father, but this is opposed by her brother, and a later romance fares little better, leading to much suffering when she returns home. A well mounted adaptation of one of Eliot's most depressing tales.
DRA 120 min mTV cC VIDrel: BBC/TECH V/sh
Boa: novel by George Eliot.

MILLENNIUM *
Michael Anderson CANADA/USA   PG   1989
Kris Kristofferson, Cheryl Ladd, Robert Joy, Daniel J. Travanti, Brent Carver, Maury Chaykin, David McIlwraith, Al Waxman, Lloyd Bochner, Lawrence Dane, Thomas Huff, Peter Dvorsky, Philip Akin, David Calderisi, Gary Reineke
A mysterious woman from another time takes people off aircraft just as they are about to crash, her intention being to repopulate a disease-ridden world of the future. Kristofferson is the suspicious crash investigator who becomes involved with time-travelling Ladd, who is their agent from the future. The fascinating opening premise is not done justice by the script, which decays into cliched tedium. The script is by Varley.
FAN 101 min (ort 108 min) VIDrel: POLY/POLYREC V/sur
Boa: short story Air Raid by John Varley.

MILLER'S CROSSING ** 18
Joel Coen USA 1990
Gabriel Byrne, Marcia Gay Harden, John Turturro, Jon Polito, Albert Finney, J.E. Freeman, Mike Starr, Al Mancini, Michael Jeter, Frances McDormand, Richard Woods, Thomas Toner, Steve Buscemi, Mario Todisco, Olek Krupa
One of those flashy Prohibition Era gangster yarns, that expends most of its energies on form rather than substance, with Byrne playing a murderous Irish mobster who operates by his own obscure code of ethics. A violent comic-book exercise, dark, sombre and convoluted; the best thing in it is Barry Sonnenfeld's excellent photography. Written by the Coen brothers, with Ethan producing and Joel directing.
A/AD 110 min (ort 115 min) VIDrel: 20TH/TECH; ENCORE (LV only) V/sur LV

MILLION DOLLAR DUCK ** U
Vincent McEveety USA 1971
Dean Jones, Sandy Dennis, Joe Flynn, Tony Roberts, James Gregory, Virginia Vincent, Jack Kruschen, Edward Andrews, Arthur Hunnicutt, Sammy Jackson, Lee Harcourt Montgomery, Billy Bowles, Jack Bender, Frank Wilcox, Ted Jordan
After accidental exposure to radiation a duck starts to lay golden eggs. Both gangsters and government are interested in getting hold of the bird. A mild Disney comedy inspired by "Mister Drake's Duck".
Aka: $1,000,000 DUCK
COM 92 min VIDrel: WDV/TECH L/A V

MILLION DOLLAR SCREW, THE * 18
Hal Freeman USA
Barbara Dare, Gail Force, Peter North, Buck Adams, Don Fernando
After she receives title sum in an insurance claim, a woman finds that she is attracting a lot of men who are only interested in her money. Very poor, even for films of this genre.
A 60 min VIDrel: MOPIC/QUANT V

MILLION POUND NOTE, THE ** U
Ronald Neame UK 1954
Gregory Peck, Jane Griffith, Joyce Grenfell, Ronald Squire, A.E. Matthews, Wilfrid Hyde White, Maurice Denham, Reginald Beckwith, Brian Oulton, John Slater, Hartley Power, George Devine, Bryan Forbes, Ann Gudrun, Ronald Adam
A man is given a million pound note and will be given any job that can be found for him if he can return it unspent after a month. Though this amiable comedy is hard to fault in terms of production values, it wears out its one joke very quickly.
Aka: MAN WITH A MILLION; ONE MILLION POUND NOTE, THE
COM 85 min (ort 91 min) VIDrel: VCC L/A V
Boa: short story by Mark Twain.

MILLION TO ONE, A ** (PG)
Paul Rodriguez USA 1993
Paul Rodriguez, Ruben Blades, Polly Draper, Gerardo, Cheech Marin, Edward James Olmos, Pepe Serna, Paul Willaims, Tony Plana, Bert Rosario, Jonathan Hernandez, Larry Linville, Victor Rivers, Maria Rangel, Leslie Danon
A poor Puerto Rican immigrant in Los Angeles, who is widowed and has a ten-year-old son to support, mysteriously receives a cheque for $1,000,000. With it come instructions that this money is to be "used" but not spent. An updated film adaptation of Twain's short story that is well intentioned but terribly unfocused in its efforts to combine social satire with a touching melodrama.
Aka: MILLION TO JUAN, A
COM 93 min (ort 97 min) SATrel: SKY MOVIES
Boa: short story The Million Pound Note by Mark Twain.

MILLIONAIRESS, THE ** U
Anthony Asquith UK 1960
Sophia Loren, Peter Sellers, Alastair Sim, Vittorio De Sica, Dennis Price, Gary Raymond, Alfie Bass, Miriam Karlin, Noel Purcell, Virginia Vernon, Basil Hoskins, Diana Coupland, Willoughby Goddard, Pauline Jameson
Unimpressive film version of Shaw's play that discards much of its wit and almost all of its ideas in favour of farce. A millionairess with a father complex decides that an impoverished physician is just the man for her and sets about pursuing him with all the means at her disposal. Good performances alone fail to inject enough interest into this sorry effort. Loren looks ravishing, but Sellers looks bored (and so are we).
COM 90 min VIDrel: ARROW/RTM V
Boa: play by George Bernard Shaw.

MILLIONS ** 18
Carlo Vazina ITALY 1990
Billy Zane, Lauren Hutton, Carol Alt, Alexandra Paul, Jean Sorel, Donald Pleasence, Catherine Hickland
The heir to a family fortune will stop at nothing to get his way even if he has to seduce his sister-in-law and his cousin.
DRA 95 min VIDrel: MARQ/QUANT V

MILOU IN MAY *** 15
Louis Malle FRANCE/ITALY 1989
Michel Piccoli, Miou Miou, Michel Duchaussoy, Dominique Blanc, Harriet Walter, Bruno Carette, Paulette Dubost, Renaud Danner, Herry Le Clerc
It is May 1968, and the members of a large family descend on a provincial country house for the funeral of Milou's mother. As the various guests strive to sort out the complexities of the estate, they face growing difficulties in the face of the apparent collapse of the country into strikes and fuel shortages. Eventually, overcome by paranoia, they adopt the wisest course and flee to the hills. A bleakly mocking film, redeemed by a measure of wit.
Aka: MAY FOOLS; MILOU EN MAI
COM 114 min VIDrel: WHV V

MINA TANNENBAUM **** 12
Martine Dugowson FRANCE 1993
Romane Bohringer, Elsa Zylberstein, Florence Thomassin, Nils Tavernier, Stephane Slima, Chantal Krief, Jany Gastaldi, Dimitri Furdui, Eric Defosse, Jean-Philippe Ecoffey, Harry Cleven, Alexandre von Sivers, Artus de Penguern
Two young Jewish girls remain friends throughout their teenage years and on into adulthood, until they are parted when one falls in love. Though quite dissimilar in temperament, they are drawn together by their shared insecurities and interests, and over the years the relationship alters in tune with their changing personalities. An absorbing and honest work, of grace and insight, it gains much from totally convincing performances from the two leads.
DRA 123 min (ort 128 min) VIDrel: CURZON/20TH V/sur

MIND RIPPER ** 18
Jonathan Craven USA 1995
Lance Henriksen, Claire Stansfield, Natasha Wagner, John Diehl, Dan Blom, Giovanni Ribisi, Gregory Sporleder, John Apicella, Peter Shepard, Adam Solomon
A group of scientists at a remote, desert-based research lab become prey to a murderous mutant when they become trapped after a genetics experiment goes wrong. This low-budget variant on ALIEN relies heavily on gory shock effects, but is woefully deficient in both plotting and acting.
Aka: OUTPOST, THE; WES CRAVEN PRESENTS: MIND RIPPER; WES CRAVEN'S THE MIND RIPPER
HOR 91 min VIDrel: MED/DISC V/sh

MINDER ON THE ORIENT EXPRESS ** PG
Francis Megahy UK 1985
Dennis Waterman, George Cole, Patrick Malahide, Glynn Edwards, Honor Blackman, Adam Faith, Ronald Lacey, Robert Beatty, Maurice Denham, Ralph Bates, Linda Hayden, Amanda Pays, Peter Childs
Based on the TV series "Minder", here our seedy heroes our involved in an adventure on the world's most famous train. Other countries make films, the UK makes TV spin-offs. What plot there is involves the daughter of a dead crook, the hidden number of his Swiss bank account, and a collection of greedy characters all out to get the money. This one does at least have good excellent dialogue and characterisation.
COM 106 min mTV VIDrel: VCC/DISC L/A V

MINDWARP ** 18
Steve Bennett USA 1990
Bruce Campbell, Angus Scrimm, Elizabeth Kent, Marta Alicia, Mary Becker, Wendy Sandow, Brian Brill, Bekki Vallin, Matt Hensley, Keith Rodenberger, Gene McCarr, Roger Perkovich, Gereald Shidell, Tyrone Yonash, Troy Siegmeier
In a post-WW3 world, most of humanity live in a sterile underground city where they spend most of their time connected to machines that create dreamlike illusions. When a girl anxious to find out about her missing father, kills her mother accidentally she is cast out to live on the barren surface, populated by mutant predators. A film that borrows heavily from MAD MAX and many similar works in this genre. Produced by the Fangoria publishing company.
FAN 91 min VIDrel: 20VIS/SONOP V

MINES OF KILIMANJARO, THE ***   18
Mino Guerrini ITALY   1986
Christopher Connelly, Gordon Mitchell, Elena Pompei, Tobias Hoesl, Francesca Ferre, Peter Berling, Matteo Corsini, Josette Martial, Franco Diogene, Al Cliver (Pier Luigi Conti), Tino Castaldi, Kit Dickinson, Luca Giordana, Huera Kiro
In 1934 the Third Reich is re-arming, and when an exiled German professor is murdered, his student sets off for Africa in search of his killers, stumbling on the mines of Kilimanjaro and a secret Nazi undertaking. A lively but totally derivative clone of RAIDERS OF THE LOST ARK, complete with indestructible hat-wearing hero, but done as a very self-conscious spoof. Substandard effects and weak acting do nothing to assist.
A/AD   93 min   VIDrel: IMPENT L/A   V

MIRACLE AT CHRISTMAS: EBBIE'S STORY ***   PG
George Kaczender CANADA/USA   1995
Susan Lucci, Wendy Crewson, Ron Lea, Molly Parker, Lorena Gale, Jennifer Clement, Nicole Parker, Susan Hogan, Kevin McNulty, Tarah Noah Smith, Jeffrey DeMunn, Adrienne Carter, Bill Croft, Elan Ross Gibson, Laura Harris, Gary Jones
Yet another adaptation of the Charles Dickens story, this time updated, gender-changed and transplanted to the States, where a top executive in a toy store has forgotten about all the important things that once had meaning in her life. See also SCROOGED for another such contemporary rendition.
Aka: EBBIE
DRA   90 min mTV   VIDrel: ODY/SONOP   V/sh
Boa: novella A Christmas Carol by Charles Dickens.

MIRACLE BEACH **   15
Skott Snider USA   1992
Ami Dolenz, Dean Cameron, Noriyuki "Pat" Morita, Felicity Waterman, Martin Mull, Alexis Arquette, Vincent Schiavelli, Dean Cain, Pamela Pond
A genie is sent to help a man who has just lost his job, his girlfriend and his home. Amiable romantic-comedy which tries hard to be scatty, but lacks enough inventiveness to really succeed.
COM   84 min (ort 88 min)   VIDrel: COLUM/SONOP   V/sh

MIRACLE IN MILAN ***   U
Vittorio de Sica ITALY   1951
Francesco Golisano, Brunella Boyo, Emma Gramatica, Paolo Stoppa, Guglielmo Barnabo, Alba Arnova, Virgilio Riento, Ermino Spall, Flora Cambi, Arturo Bragaglia, Riccardo Bertazzolo, Angelo Prioli, Francesco Rissone
An abandoned child taken in by a kindly old woman joins a group of squatters after her death and brings light into their lives, eventually helping them to fly away to a better land. A charming post-WW2 fable highlighting the fate of the poor and dispossessed.
Aka: MIRACOLO A MILANO
DRA   92 min (ort 101 min) B/W   VIDrel: ARTPRO/RTM   V
Boa: story Toto Il Buono by Cesare Zavattini

MIRACLE IN THE WILDERNESS **   PG
Kevin Dobson USA   1991
Kris Kristofferson, Kim Cattrall, John Dennis Johnston, Rino Thunder, David Oliver, Sheldon Peters Wolfchild, Steve Reevis, Peter Alan Morris
A mawkish tale set in the wilds of the American West, where an ex-cavalry man has settled down to a simple farming life, with his wife and young child. This idyllic existence is not to last, for when an Indian Blackfeet chief sets his heart on having revenge against his old enemy, he has the entire family taken hostage. But the Indians are soon captivated by the story of the Nativity in this crass, clumsy but surprisingly well photographed adventure.
WES   85 min (ort 88 min) mTV   VIDrel: FIRST/SONOP   V
Boa: novella by Paul Gallico.

MIRACLE LANDING **   PG
Dick Lowry USA   1989
Connie Sellecca, Wayne Rogers, Michelle Honda, Mimi Thompkins, Robert Schornsteimer, Ann-Alicia, Nancy Kwan, James Cromwell, Jay Thomas, Herta Ware, Armin Shimerman, Jeff Allin, Kathleen Bailey, James Grant Benton
Adequate drama that follows the fate of Flight 243, which took off from Hawaii in 1988 on a routine flight and miraculously escaped disaster when a section of the roof gave way at 24,000 feet.
DRA   84 min mTV   VIDrel: 20TH/TECH   V/h

MIRACLE MILE **   15
Steve De Jarnatt USA   1988
Anthony Edwards, Mare Winningham, John Agar, Denise Crosby, Lou Hancock, Mykel T. Williamson, Kelly Minter, Kurt Fuller, Robert Doqui, Danny De La Paz, O. Lan Jones, Claude Earl Jones, Earl Boen, Diane Delanoe, Joe Mercado
A musician answers a ringing payphone, and learns that the US has launched a missile attack and that he has 70 minutes to go before WW3 erupts. He reacts by trying to get his newly-found waitress girlfriend to safety before the bombs start falling. A fascinating idea is taken absolutely nowhere in this confused and illogical thriller. A few flashes of black humour add nothing to the plot. Written by Jarnatt, and with music by Tangerine Dream.
THR   84 min (ort 87 min)   VIDrel: COLUM/SONOP   V

MIRACLE ON 34TH STREET ****   U
George Seaton USA   1947
Edmund Gwenn, Maureen O'Hara, John Payne, Gene Lockhart, Natalie Wood, James Seay, Porter Hall, Philip Tonge, Harry Antrim, Mary Field, William Frawley, Thelma Ritter, Percy Helton, Jerome Cowan, Alvin Greenman, Anne Staunton
A man who works as a Santa Claus in a department store has to convince a sceptical child that he is the real McCoy. Despite some ill-advised romantic interest, this fresh and energetic fantasy is now rightly regarded as a classic. This was Ritter's screen debut. Remade in 1973 and 1994. AA: S. Actor (Gwenn), Story/orig (Valentine Davies), Screen (George Seaton).
Aka: BIG HEART, THE
COM   92 min (ort 96 min) B/W   VIDrel: 20TH/TECH   V/sur
Boa: story by Valentine Davies.

MIRACLE ON 34TH STREET ***   U
Les Mayfield USA   1994
Richard Attenborough, Elizabeth Perkins, Dylan McDermott, J.T. Walsh, Mara Wilson, James Remar, Janes Leeves, Simon Jones, William Windom, Robert Prosky, Jack McGee, Joe Pentangelo, Mark Damiano II, Casey Moses Wurzbach, Lisa Sparman
Updated remake of the charming 1937 fantasy in which a man called Kris Kringle causes consternation by claiming to be the real Santa Claus, and setting out to prove this to a disbelieving little girl. Attenborough is excellent in the lead role, but the movie suffers from both excessive length and an unwise decision to make some changes to the original story. See also THE SANTA CLAUSE.
FAN   109 min (ort 114 min) cC   VIDrel: 20TH/TECH   V/sur

MIRACLE ON INTERSTATE 880 ***   PG
Robert Iscove USA   1993
Ruben Blades, David Morese, Len Carriou, Sandy Duncan, Ada Morris, Scott Hylands, Don S. Davis, Ian Tracey, Blu Mankuma, John Pyper Ferguson, Ric Reid, Jerry Wasserman, Sabrina Wiener, Jeffrey Licon, Suki Kaiser, Emily Perkins
Drama telling of the events that took place on October 1987, when the San Francisco earthquake caused widespread damage, trapping many thousands under tons of rubble.
DRA   89 min mTV   VIDrel: COLUM/SONOP   V

MIRACLE WORKER, THE ***   PG
Arthur Penn USA   1962
Anne Bancroft, Patty Duke, Victor Jory, Inga Swenson, Andrew Prine, Beah Richards, Kathleen Comegys, Jack Hollander, Peggy Burke, Mindy Sherwood, Grant Code, Michael Darden, Dale Ellen Bethea, Walter Wright Jr, Donna Bryan
Harrowing but fascinating account of how Annie Sullivan taught the deaf and blind Helen Keller, the two stars recreating the Broadway performances they gave to general acclaim. A powerful film that engages one's emotions, yet never descends into pathos. Made as a TV movie in 1979. AA: Actress (Bancroft), S. Actress (Duke).
DRA   102 min B/W cC   VIDrel: MGM/WHV   V/dm
Boa: play by William Gibson/book The Story Of My Life by Helen Keller.

MIRACLES: THE CANTON GODFATHER ***   PG
Jackie Chan HONG KONG   1989
Jackie Chan, Anita Mui, Richard Ng
A GODFATHER-inspired martial arts thriller that is set in Hong Kong in the 1930s, where a penniless country boy rescues a mortally wounded gangster boss and inadvertently becomes his successor. As his power increases, he heads towards a final

showdown with the rival gangster who murdered his predecessor. A well-made Jackie Chan vehicle: fast, furious and enlivened with occasional flashes of humour.
Aka: CANTON GODFATHER, THE
MAR   101 min   VIDrel: ONE/IMPENT   V

MIRAGE ***   18
Eric Edwards USA   1991
Ashlyn Gere, Peter North, Don Hart (Mike Horner), Nina Hartley, Taylor Wayne, Alice Springs, Eric Edwards, T.T. Boy, Greg Rome, Marc Wallice, Trinity Loren
Gere stars as the editor of a publishing house whose latest discovery is reclusive writer Edwards who, much to her annoyance, never shows up to help promote his books. She takes a trip to see him (he lives out in the desert in a trailer) and after a meeting, begins to experience strange sexual daydreams, doubtlessly inspired by reading too many of his erotic novels. An intriguing effort with quite a clever twist at the end.
A   80 min (ort 88 min) mVid   VIDrel: MIA/DISC/IMPENT V

MIRAGE **   18
Paul Williams USA   1995
Edward James Olmos, Sean Young, James Andronica, Bobalu, Mark Silverman, John Lizzi, Elsayed Badrya, Susan Helen Emerson, William Grillo, Apple Via, Tony King, Cauhtemoc Rivas, Walter Guarino, Patricia Sill, Chris De Rose, Joey Cook
A former policeman gets a job that involves protecting an attractive woman who suffers from a multiple personality disorder. Naturally, he soon falls for one of her more seductive personalities, which does not make it any easier for him to do his job.
DRA   87 min (ort 92 min)   VIDrel: 20VIS/SONOP   V

MIRANDA **   18
Tinto Brass ITALY   1985
Serena Grandi, Andrea Occhipinti
Erotic comedy with the attractive owner of a taverna getting to choose a lover from her many patrons. Lightweight nonsense, quite attractively put together.
COM   90 min dubbed   VIDrel: ARTPRO/RTM   V

MIRROR ****   U
Andrei Tarkovsky USSR   1974
Margarita Terekhova, L. Tarkovskaya, Philip Yankovsky, Ignat Danilisev, Oleg Yankovsky, Innokenti Smoktunovsky (narration only)
A very personal meditation on family and childhood, in which an artist (who is heard but not seen) ponders on his relationships with his parents (both as a child and an adult) and his wife and son. Set largely during WW2, this dreamlike, moody and hypnotic film works less as a straightforward narrative than as a homage to the past and the innocence of youth.
Aka: WHITE WHITE BOY, A; ZERKALO
DRA   102 min (ort 106 min) Col/B/W
VIDrel: ARTIF/20TH   V

MIRROR, THE **   15
Jimmy Lifton USA   1994
Sally Kellerman, Roddy McDowall, Veronica Cartwright, Tracy Wells, William Sanderson, Mark Ruffalo, Carlton Beener, Sarah Douglas, Lois Nettleton
Having recently lost her parents, a young dancer and her brother are placed in an orphanage, but do not find it to their liking but have greater worries when they are forced to combat evil forces. A sequel to an earlier film called MIRROR, MIRROR, it was soon followed by a further sequel.
Aka: MIRROR, MIRROR 2: RAVEN DANCE
DRA   97 min   VIDrel: NEWAGE/COLUM L/A   V

MIRROR CRACK'D, THE *   PG
Guy Hamilton UK   1980
Angela Lansbury, Geraldine Chaplin, Elizabeth Taylor, Rock Hudson, Tony Curtis, Edward Fox, Kim Novak, Marella Oppenheim, Charles Gray, Maureen Bennett, Carolyn Pickles, Eric Dodson, Charles Lloyd Pack, Richard Pearson, Thick Wilson
Agatha Christie's female detective, Miss Marple, investigates a series of murders being committed in an English village, where a film is being made with a star-studded cast. All the stereotyped English characters are here, as in so much of the British cinema. Flat, lifeless and extremely boring.
DRA   109 min (ort 105 min)   VIDrel: WHV   V
Boa: novel The Mirror Crack'd From Side To Side by Agatha Christie.

MIRROR HAS TWO FACES, THE *   15
Barbra Streisand USA   1996
Barbra Streisand, Jeff Bridges, Pierce Brosnan, George Segal, Mimi Rogers, Brenda Vaccaro, Lauren Bacall, Aiustin Pendleton, Elle Macpherson, Ali Marsh, Leslie Stefanson, Taina Elg, Lucy Avery Brooks, Amber Smith, David Kinzie
A remake of "Le Miroir A Deux Faces" that casts Streisand (who co-produced) as a college professor who is convinced she is too plain to find love, and winds up marrying a maths teacher, who having suffered numerous failed romances wants a sexless relationship. Various daft interludes and complications intrude, until the inevitable happy resolution. An irritating and self-indulgent vanity-vehicle for Streisand, who is out-acted at every turn by Bacall, cast as her mother.
COM   122 min (ort 126 min) cC   VIDrel: COLUM/SONOP V/sur

MIRROR IMAGES *   18
Alexander Gregory Hippolyte USA   1991
Delia Sheppard, Julie Strain, Nels Van Patten, Jeff Conaway, John O'Harley, Richard Arbolino, Korey Mall, Michael Meyer, Buck Flower, Deider Morrow, Lee Ann Beaman, Andre Rosey Brown, Juliet James, Janie Wilson, Sergio Salerno
Sheppard plays identical twins Kaitlin, a nymphomaniac, and Shauna, the bored wife of a ruthless political adviser. When Kaitlin tells her sister that she must "vanish" for a while, Shauna is understandably intrigued, and having decided to adopt her sister's identity, embarks on a double life of excitement and danger. A boring erotic-thriller that blends bad acting with an idiotic script. A sequel followed. See also SECRET GAMES and CARNAL CRIMES.
A   92 min (ort 94 min)   VIDrel: POLY/POLYREC/MED   V

MIRROR IMAGES 2 **   18
Alexander Gregory Hippolyte USA   1993
Shannon Whirry, Luca Bercovici, Tom Reilly, Sara Suzanne Brown, Eva Larue, Kristine Kelly, Frank Pesce, Jeremy McCollum, Ricahrd Eden, Kevin West, Crystal Jade Mckay, Lauren Hays, Ken Zavayn, J.P. Hubbell, Chip Campbell
When a sexually liberated woman dies under mysterious circumstances, her identical but more reserved twin sister takes on her identity in a bid to learn the truth. An adequate erotic thriller with enough twists and turns to maintain interest.
THR   88 min (ort 92 min)   VIDrel: MIA/DISC/IMPENT   V

MIRROR, MIRROR **   18
Marina Sargenti USA   1990
Karen Black, Rainbow Harvest, Kristin Dattilo, Yvonne De Carlo, William Sanderson, Ricky Paull Goldin, Charlie, Dorit Sauer, Stephen Tobolowsky, Ann Hearn, Tom Breznahan, Pamela Parfilli, Scott Campbell, Traci Lee Gold
Another tale of supernatural revenge, with a new, shy girl at school falling victim to a nasty bullying gang of cheerleaders, and discovering a strange mirror that proves to harbour a magical and deadly force. A confused horror saga that runs out of steam all too soon. A few gory touches do nothing to help the plot along. Followed by THE MIRROR.
HOR   104 min (ort 105 min)   VIDrel: L/A   V

MISCHIEF **   15
Mel Damski USA   1984
Doug McKeon, Catherine Mary Stewart, Kelly Preston, Chris Nash, D.W. Brown, Jami Gertz, Graham Jarvis, Maggie Blye, Terry O'Quinn, Dennis L. O'Connell, Darren Ewing, Jordan Baker, John Miranda, Cristen Kauffman, Bob McGuire
Predictable teenage sex comedy about a boy anxious to lose his virginity and gain a little sexual experience, just like his more suave and self-assured buddy. Set in the 1950s and strong on period flavour; if only the script was halfway decent and gave the appealing cast a little more to work with.
Aka: HEART AND SOUL
COM   92 min (ort 97 min)   VIDrel: 20TH/TECH   V

MISERY ***   18
Rob Reiner USA   1990
James Caan, Kathy Bates, Frances Sternhagen, Richard Farnsworth, Lauren Bacall, Graham Jarvis, Jerry Potter, Tom Brunelle, June Christopher, Julie Payne, J.T. Walsh, Archie Hahn III, Gregory Snegoff, Wendy Bowers
A highly successful romantic novelist has a car crash and is rescued and nursed by an obsessive fan of his novels, who is furious to learn that he has killed off her favourite character. A nasty battle of wits ensues, with the writer being held prisoner

by his murderous captor, who wants him to completely rewrite his latest manuscript. Wonderfully well acted, the film is hampered by weak direction and scripting. AA: Actress (Bates).
THR   103 min (ort 107 min)   VIDrel: VCC/DISC/FIRST V/sur
Boa: novel by Stephen King.

MISFITS, THE ** **   PG
John Huston USA   1961
Clark Gable, Marilyn Monroe, Montgomery Clift, Eli Wallach, Thelma Ritter, Estelle Winwood, James Barbon, Kevin McCarthy, Dennis Shaw, Walter Ramage, Peggy Barton, Philip Mitchell, J. Lewis Smith, Marietta Tree, Bobby La Salle
The story of a group of drifters in Arizona who are brought together by chance and rope some wild mustangs. An ill-fated melodrama in which the stars do their best but with little success. Scripted by Arthur Miller. This was the last film for both Gable (who did his own stunts) and Monroe.
DRA   120 min (ort 124 min) B/W   VIDrel: MGM/WHV   V

MISFITS OF SCIENCE *   PG
James D. Parriott USA   1985
Dean-Paul Martin, Kevin Peter Hall, Mark Thomas Miller, Courtney Cox, Mickey Jones, Jennifer Holmes, Eric Christmas, Edward Winter, Larry Linville, Kenneth Mars
An assorted band of strange characters join forces to prevent a research project from destroying the world. A feeble pilot for a failed TV series.
COM   90 min mTV   VIDrel: CIC/SONOP   V/h

MISHIMA: A LIFE IN FOUR CHAPTERS *   15**
Paul Schrader JAPAN/USA   1985
Ken Ogata, Kenji Sawada, Yasosuke Bando, Mashayuki Shionya, Junkichi Orimoto, Go Riju, Naoko Otani, Masato Aizawa, Yuki Nagahara, Hisako Manda, Kyuzo Kobayashi, Yasosuke Bando, Sachiko Hidari, Toshiyuki Nagashima
A stylised four-part account of the life and ritual suicide of a famous Japanese writer. In Japanese with sub-titles, and an English narration read by Roy Scheider. An exotic, difficult, powerful and generally self-indulgent work, that will be hard to follow for any except those who are familiar with the writings and/or the life of the character. The script is by Paul and Leonard Schrader.
DRA   116 min (ort 120 min) B/W/Col (some Japanese dialogue)   VIDrel: WHV   V/sur
Boa: novels Runaway Horses/Temple Of The Golden Pavilion by Yukio Mishima.

MISS ALL-AMERICAN BEAUTY ** **   (PG)
Gus Trikonis USA   1982
Diane Lane, Cloris Leachman, David Dukes, Jayne Meadows, Alice Hirson, Brian Kerwin, Norman Bennett, Jim Mills, Bobby Fite, Jane Roberts, Debra Lynn Rogers, Stephanie Dunnam, Michele Rusheene, Peggy Woody, Ralph Baker
Lane plays a naive, classical pianist who wins first prize in a beauty contest, only to learn that the organisers now intend to take over her life and remould her image. A solid and enjoyable TV drama, good in all departments.
DRA   96 min mTV   SATrel: UK LIVING

MISS ANNIE ROONEY *   U
Edwin L. Marin USA   1942
Shirley Temple, William Gargan, Peggy Ryan, Guy Kibbee, Dickie Moore, Gloria Holden, Jonathan Hale, Selmer Jackson, Mary Field, June Lockhart, Virginia Sale
An attempt to give Miss Temple a more adult role, this is the trite tale of a girl from a poor family who falls for a boy from a rich one. Not quite the success its makers hoped for, this vehicle for the former child star is as routine as it is forgettable.
DRA   87 min   VIDrel: RCA L/A   V

MISS FIRECRACKER *   PG**
Thomas Schlamme USA   1988
Holly Hunter, Tim Robbins, Mary Steenburgen, Alfre Woodard, Scott Glenn, Ann Wedgeworth, Trey Wilson, Amy Wright, Bert Remsen, Veanne Cox, Kathleen Chalfont, Robert Fieldsteel, Greg Germann, Avril Gentles, Angela Turner
A young Mississippi woman's desire for self-respect propels her into her hometown beauty queen competition – the Yagoo City Miss Firecracker Contest. Hunter is wonderful as the foolish and vulnerable girl, recreating the role she had in Henley's play, which is well transferred to the screen. A warm, zany and colourful comedy; of the excellent cast Steenburgen is

particularly memorable as a former beauty queen. Schlamme's feature debut.
COM   90 min (ort 102 min)   VIDrel: MGM L/A   V
Boa: play The Miss Firecracker Contest by Beth Henley.

MISS MARPLE: A CARIBBEAN MYSTERY ** **   PG
Robert Lewis USA   1983
Helen Hayes, Barnard Hughes, Brock Peters, Swoosie Kurtz, Season Hubley, Jameson Parker
Miss Marple murder mystery with our amateur sleuth off to the West Indies to take a break in the warmth, having just got over a bout of pneumonia. However, her holiday is cut short when a retired major is found dead in suspicious circumstances. A dullish effort indeed, not especially helped by the lead actress, whose personality is not well suited to playing this role.
Aka: CARIBBEAN MYSTERY, A
DRA   92 min mTV   VIDrel: WHV   V/h
Boa: novel by Agatha Christie.

MISS MARPLE: A MURDER IS ANNOUNCED **   PG**
David Gilies UK   1985
Joan Hickson, Ursula Howells, Renee Asherson, John Castle, Kevin Whately, Sylvia Syms, Joan Sims, Simon Shepherd, Samantha Bond, Mary Kerridge, Ralph Michael, Paola Dionosotti
Excellent adaptation to the small screen of the Christie mystery. Fine acting and a carefully recreated period atmosphere make this well worth watching
Aka: MURDER IS ANNOUNCED, A
DRA   154 min mTV   VIDrel: BBC L/A   V
Boa: novel by Agatha Christie.

MISS MARPLE: A POCKETFUL OF RYE *   PG**
Guy Slater UK   1985
Joan Hickson, Peter Davison, Clive Merrison, Annette Badland, Selina Cadell, Fabia Drake, Timothy West, Tom Wilkinson, Stacey Dorning, Susan Gilmore, Louis Mahoney, Jon Glover, Artyn Stanbridge, Rachel Bell
The murder of a London financier by poison sets in motion a macabre train of events in this excellent TV adaptation, one in a series. The script is by T.R. Bowen.
Aka: A POCKETFUL OF RYE
DRA   100 min mTV   VIDrel: BBC L/A   V/h
Boa: story by Agatha Christie.

MISS MARPLE: MURDER AHOY *   U**
George Pollock UK   1964
Margaret Rutherford, Lionel Jeffries, Charles Tingwell, William Mervyn, Joan Benham, Stringer Davis, Nicholas Parsons, Miles Malleson, Henry Oscar, Derek Nimmo, Francis Matthews, Gerald Cross
Miss Rutherford once again plays our lady detective in her own inimitable way, this time out to solve the deaths of trustees of a cadet training ship, set up to give wayward youngsters a dose of naval discipline. Despite Rutherford's considerable screen presence, this weak yarn (the last in the series) rapidly gets bogged down with excessive dialogue.
Aka: MURDER AHOY
DRA   89 min (ort 93 min) B/W   VIDrel: MGM/WHV   V

MISS MARPLE: MURDER AT THE GALLOP *   U**
George Pollock UK   1963
Margaret Rutherford, Robert Morley, Flora Robson, Charles Tingwell, Duncan Lamont, Stringer Davis, James Villiers, Robert Urquhart, Katya Douglas, Finlay Currie, Gordon Harris, Noel Howlett, Kevin Stoney
The second film in a series of four, this one has our elderly detective investigating the death of a wealthy cat-hating recluse and his sister, the man having been apparently frightened to death by a cat. Spirited performances and a decent script make this the best of the Rutherford-Miss Marple whodunits.
Aka: MURDER AT THE GALLOP
DRA   87 min B/W   VIDrel: MGM/WHV   V
Boa: novel After The Funeral (Funerals Are Fatal) by Agatha Christie.

MISS MARPLE: MURDER MOST FOUL ** **   U
George Pollock UK   1964
Margaret Rutherford, Ron Moody, Charles Tingwell, James Bolam, Francesca Anniss, Terry Scott, Andrew Cruickshank, Stringer Davis, Dennis Price, Megs Jenkins, Ralph Michael
Despite what appears to be conclusive evidence concerning the murder of an actress, Rutherford refuses to return a guilty verdict whilst serving as a juror and mounts her own

investigation. This takes her to a second-rate repertory company which she joins in order to unmask the real culprit. A fairly good murder mystery held back by a lack of vigour.
Aka: MURDER MOST FOUL
DRA   91 min (ort 100 min) B/W   VIDrel: MGM/WHV   V
Boa: novel Mrs McGinty's Dead by Agatha Christie.

MISS MARPLE: MURDER SHE SAID *   PG
George Pollock UK   1961
Margaret Rutherford, Charles Tingwell, Muriel Pavlow, Arthur Kennedy, James Robertson Justice, Thorley Walters, Gerald Cross, Conrad Phillips, Ronald Hoard, Joan Hickson, Stringer Davis, Ronnie Raymond, Peter Butterworth, Richard Briers
An elderly and inquisitive spinster spots a girl apparently being strangled when she happens to glance across to a passing train. Her attempt to solve this mystery takes her to a strange country mansion where she poses as the new housekeeper. The first of four Miss Marple murder mysteries, inspired casting gave Rutherford the chance to demonstrate her considerable screen presence, apart from which this tedious affair has little to offer.
Aka: MURDER SHE SAID
DRA   87 min B/W   VIDrel: MGM/WHV   V
Boa: novel 4.50 From Paddington (What Mrs McGillicuddy Saw!) by Agatha Christie.

MISS MARPLE: MURDER WITH MIRRORS **   PG
Dick Lowry USA   1985
Helen Hayes, Bette Davis, John Mills, Leo McKern, Liane Langland, Dorothy Tutin, Frances De La Tour Tim Roth
Miss Marple comes to the rescue of an old friend who is being menaced by a ruthless villain who is determined to steal her family home. A competent if distinctly unremarkable adaptation.
Aka: MURDER WITH MIRRORS
DRA   94 min (ort 100 min)   VIDrel: WHV   V/h
Boa: novel by Agatha Christie.

MISS MARPLE: THE BODY IN THE LIBRARY ***   PG
Silvio Narizzano UK   1985
Joan Hickson, Gwen Watford, Trudie Styler, Andrew Cruickshank, Anthony Smee, Moray Watson, Valentine Dyall, Frederick Jaeger, David Horovitch, Ian Brimble, Raymond Francis, John Moffat, Ciaran Madden, Hugh Walters, Jess Conrad
The discovery of the body of a young woman in the library of a retired colonel marks the start of a complex tale of murder. One in a series of fine adaptations of Christie mysteries, with Hickson playing Marple as if made for the part.
Aka: BODY IN THE LIBRARY, THE
DRA   154 min mTV   VIDrel: BBC L/A   V/h
Boa: novel by Agatha Christie.

MISS MARY **   15
Maria Luisa Bemberg ARGENTINA   1986
Julie Christie, Donald McIntire, Eduardo Pavlovsky, Nacha Guevara, Tato Pavlovsky, Sofia Viruboff, Barbara Bunge, Luisina Brando, Iris Marga, Gerard Romano
A British governess in Argentina in 1938 becomes deeply involved with the son of the family she is employed by. A steamy tale of repressed passions and desires, all done with great style but no more memorable on that account.
DRA   95 min (ort 100 min)   VIDrel: NWV L/A   V

MISSILES OF OCTOBER, THE ***   U
Anthony Page USA   1974
William Devane, Martin Sheen, Ralph Bellamy, Howard da Silva, John Dehner, Andrew Duggan
A tense and detailed recreation of the Cuban missile crisis of the 1960s when the USA and the USSR nearly came into conflict over the Soviet missile bases in Cuba. Devane particularly well chosen to play John F. Kennedy.
DRA   149 min (ort 155 min) mTV   VIDrel: RNK L/A   V

MISSING ***   15
Constantine Costa-Gavras USA   1982
Jack Lemmon, Sissy Spacek, Melanie Mayron, John Shea, Charles Cioffi, David Clennon, Janice Rule, Richard Venture, Richard Bradford, Ward Costello, Jerry Hardin, John Doolittle, Felix Gonzalez, Robert Hitt, Martin Lasalle
A father goes to a troubled Latin American country in search of his son who has disappeared, and finds that the American authorities are not telling him the truth. Based on the real-life experiences of Ed Horman, this gripping and well-made account never falters, and Lemmon's wonderful performance as the

dogged father is a gem. AA: Screen/adapt (Costas-Gavras/Donald Stewart).
DRA   116 min (ort 122 min)   VIDrel: CIC/SONOP L/A   V
Boa: novel The Execution Of Charles Horman by Thomas Hauser.

MISSING IN ACTION **   15
Joseph Zito USA   1984
Chuck Norris, M. Emmet Walsh, James Hong, David Tress, Lenore Kasdorf, Ernie Ortega, Pierrino Mascarino, E. Erich Anderson, Joseph Carberry, Avi Kleinberger, Willy Williams, Ric Segreto, Bella Flores, Gil Arceo, Roger Dantes, Nam Moore
A routine rescue-our-brave-boys film, with Norris as a walking arsenal rescuing American POWs held in Vietnam after the war there. A standard RAMBO-style outing, unalloyed with anything as superficial as a plot. And of course the obligatory sequels followed. Written (if that is the right word) by James Bruner.
A/AD   97 min (ort 101 min)   VIDrel: L/A   V

MISSING IN ACTION 2: THE BEGINNING *   18
Lance Hool USA   1985
Chuck Norris, Soon-Teck Oh, Cosie Costa, Steven Williams, Bennett Ohta, Joe Michael Terry, John Wesley, David Chung, Toru Tanaka, Christopher Cary, Dean Ferrandini, Pierre Issot, John Otrin, Joseph Hieu, Mischa Hausserman
A prequel to MISSING IN ACTION detailing the treatment of American POWs at the hands of the Vietnamese with the expected one-sidedness of portrayal. Flat and predictable, and followed by one that's (by way of a variation) predictable and flat.
Aka: BATTLE RAGE
WAR   91 min (ort 95 min)   VIDrel: VCC L/A   V

MISSING IN ACTION 3 *   18
Aaron Norris USA   1988
Chuck Norris, Aki Aleong, Yehuda Efroni, Roland Harrah III, Miki King, Ron Barker, Jack Rader, Floyd Levine, Melinda Betron
This time round our hero returns to Vietnam to rescue the wife he thought was dead, as well as his child. Norris co-scripted and his brother directed. A flat and one-dimensional affair.
Aka: BRADDOCK: MISSING IN ACTION 3
A/AD   98 min (ort 104 min)   VIDrel: WHV L/A   V/sh

MISSING PARENTS **   PG
Martin Nicholson USA   1993
Martin Mull, Blair Brown, Matt Frewer, Bobby Jacoby, Brigidm Conley Walsh, Seth Green, Chance Quinn, Kevin Meaney, Benjamin J. Stein, Peter Michael Goetz, Elena Wohl, Nick Toth, Charlotte Booker, Eric Poppick, Dana Kaminski
When a youngster's fugitive parents run off leaving him alone, he decides to celebrate by throwing a huge party. Average.
Aka: MATT MILLER: PARTY DUDE
COM   91 min mTV   VIDrel: MARQ/REFLEC L/A   V

MISSING PIECES **   PG
Leonard Stern USA   1991
Eric Idle, Robert Wuhl, Lauren Hutton, Bob Gunton, Richard Belzer, Bernie Kopell, Kim Lankford, Don Gibb, Leslie Jordan, Louis Zorich, Don Hewitt, John De Lancie, James Hong, Janice Lynde, Mary Fogarty, Bruce Kronenberg
Having just lost his job at a greetings car company, a man nearly knocks down an unemployed cellist, and the two strike up an unlikely friendship. Soon they are embroiled in a mad conspiracy, that appears to revolve around a legacy the former has been bequeathed by one of his numerous sets of foster parents. Various narrow escapes and complications now ensue, and all the while the two friends try to solve a vital riddle. A disjointed and awkward comedy-thriller.
COM   88 min (ort 92 min)   VIDrel: VCC/DISC   V/sur

MISSION, THE ***   PG
Roland Joffe USA   1986
Robert De Niro, Jeremy Irons, Ray McAnally, Aidann Quinn, Ronald Pickup, Liam Neeson, Cherie Lunghi, Monirak Sisowath, Chuck Low, Daniel Berrigan, Moya, Bercelio, Sigifredo Ismare, Asuncion Ontiveros, Alejandrino, Rolf Gray
Colourful historical drama set in South America in the 18th century, when the jurisdiction of a region is to be transferred from Spain to Portugal. A Jesuit missionary and a reformed slave trader join forces to stop the Portuguese enslaving the Guarani Indians. A sincere film, but rather patronising in its attitudes to the natives. Scripted by Bolt, it won the Best Picture

Award at the Cannes Film Festival. Music is by Ennio Morricone. AA: Cin (Chris Menges).
DRA 120 min (ort 126 min) wScrn VIDrel: WHV V/sur
Boa: novel by Robert Bolt.

MISSION FOR THE DRAGON * 18
Godfrey Ho HONG KONG
Dragon Lee, Martin Chiu, Sheila Kim, Roger Wong, Burt Lim
The fight sequences are the only thing of interest in this highly confusing martial arts film that has an evil man causing much grief and bloodshed by his desire to possess a priceless antique.
MAR 83 min Cut (2 min 27 sec) VIDrel: IMPENT V

MISSION GALACTICA: THE CYCLON ATTACK * U
Vince Edwards/Christian I. Nyby II USA 1980
Richard Hatch, Lorne Greene, Lloyd Bridges, Dirk Benedict, Herbert Jefferson Jr
A sequel to BATTLESTAR GALACTICA with Lorne Greene leading a few struggling survivors in a desperate battle against the human-hating Cyclons. This feeble effort, cobbled together from episodes of the "Battlestar Galactica" TV series, has our heroic band of humans embarking on a plan to destroy the Cyclons. Their efforts very nearly prove fatal, just as this film is.
FAN 102 min (ort 107 min) mTV VIDrel: CIC/SONOP L/A
V

MISSION IMPOSSIBLE ** PG
Brian De Palma USA 1996
Tom Cruise, Jon Voight, Emmanuelle Beart, Henry Czerny, Jean Reno, Ving Rhames, Kristin Scott-Thomas, Vanessa Redgrave, Dale Dye, Marcel Iures, Ion Caramitru, Ingeborga Dapkunaite, Valentina Yakunina, Marek Vasut
A travesty of the clever TV series created by Bruce Geller, in which a special team of secret agents took on assorted villains week after week, and so arranged things that no one ever knew they had been at work. Here, Cruise is operating solo (all the other agents having been killed off) and his mission is to clear his name and catch those responsible for killing his colleagues. Costing more than $60,000,000, this is flashy, over-complex and quite tiring.
A/AD 106 min (ort 110 min) VIDrel: CIC/SONOP; PION (LV only) V LV

MISSION IMPOSSIBLE: THE GOLDEN SERPENT * PG
Don Chaffey AUSTRALIA 1989
Peter Graves, Jane Badler, Phil Morris, Patrick Bishop, Rod Mullinar, Thaao Penghlis, Tony Hamilton, Greg Morris, Adrian Brown, Nadja Kostich, Michael Long, Max Fairchild, Tim Bell, John Lee, David Letch, Leong Lim
Pilot for the revived TV series that has Graves as the only familiar face and was produced on the cheap in Australia. The "Serpent" of the title refers to a heroin producer who Jim Phelps and the "Mission Impossible" team are ordered to neutralise. The technology is more advanced than in the original series, but the stunts and plotting are as improbable as ever, and do nothing to enliven this dull undercover adventure.
A/AD 90 min Cut (42 sec – some cuts subst)
VIDrel: CIC/SONOP V

MISSION KILL * 18
David Winters USA 1984
Robert Ginty, Merete Van Kemp, Cameron Mitchell, Olivia D'Abo
A man seeks revenge for the death of his friend during a gun-running mission in the jungle. A low-grade actioner.
A/AD 92 min VIDrel: SPEAR/SONOP/CALECO V

MISSION MANILLA * 18
Peter M. MacKenzie/Les Parrott USA 1987
Larry Wilcox, James Wainwright, Robin Eisenman, Tetchie Agabayani, Sam Hennings, Al Mancini, Neil French, Maria Isabel Lopez
A man goes to Manila to locate his missing brother who has vanished with $1,000,000 worth of heroin belonging to a drugs syndicate, who have sent their hit-men after him to recover the drugs and kill him as a warning to others. Eventually they meet up and get to take on the villains together in the final reel of this poorly-made action thriller, that resembles nothing so much as a formula TV movie.
Aka: WEB
A/AD 94 min Cut (29 sec – ort 98 min)
VIDrel: 20TH/TECH V

MISSION OF JUSTICE ** 18
Steve Barnett USA 1992
Jeff Wincott, Brigitte Nielsen, Luca Bercovici, Matthias Hues, Cyndi Pass, Karen Shepherd, Billy Sly Williams, Christopher Kriesa, Tom Wood, James Lew, Adrian Ricard, Tony Burton, Jeff Pruitt, Lenny Roebuck, Betsy Soo
A ruthless female politician attempts to use a urban vigilante group in her bid for power but is opposed by a cop who puts his badge on the line. An unremarkable actioner of severely limited entertainment value.
A/AD 95 min cC VIDrel: MIA/DISC V

MISSION OF THE SHARK *** PG
Robert Iscove USA 1990
Stacy Keach, Richard Thomas, Carrie Snodgress, Steve Landesberg, Don Harvey, David Caruso, Robert Cicchini, Bob Gunton, Cary-Hiroyuki Tagawa, Tim Giunee, Jeff Nordling, Neil Guintoli, Gordon Clapp, Joe Carberry, Dale Dye
The true story of the cruiser USS Indianapolis, which in July 1945 was sailing home after having delivered the two atomic bombs that were to end the war in the Pacific. Hit by a torpedo, it quickly sank, leaving its 1,200 survivors clinging to the wreckage, where they endured five days of horror watching as their comrades were devoured by sharks. Based on newly-released war records, this is a depressing story of military bungling and dishonesty.
DRA 93 min (ort 114 min) mTV VIDrel: VCC/DISC L/A
V/sh

MISSIONARY, THE * 15
Richard Loncraine UK 1983
Michael Palin, Maggie Smith, Trevor Howard, Phoebe Nichols, Denholm Elliott, Michael Hordern, Graham Crowden, Roland Culver, David Suchet, Tricia George, Valerie Whittington, Roland Culver, Rosamund Greenwood, Timothy Spall
Dreadfully stilted and unfunny story of a missionary who, newly returned from Africa, is given the job of running a refuge for fallen women ("What, women who have fallen over?" – illustrates the level of humour). A pathetic example of British comedy at its worst, with all the usual class-system cliches. Screenplay is by Palin.
COM 83 min (ort 86 min) VIDrel: CIC/SONOP V

MISSISSIPPI BURNING *** 18
Alan Parker USA 1988
Gene Hackman, Willem Dafoe, Brad Dourif, Frances McDormand, R. Lee Ermey, Gailand Sartain, Stephen Tobolowsky, Michael Rooker, Pruitt Taylor Vince, Park Overall, Badji Djola, Kevin Dunn, Frankie Faison, Geoffrey Nauffts, Tom Mason
A potent, vivid drama telling of two FBI agents, who are sent to investigate the mysterious disappearance of three civil rights workers, in Mississippi in the 1960s. The script (inspired by real-life events) sometimes falters, but there's nothing wrong with the period-detail or performances, especially Hackman's, as a small-town sheriff who finally solves the case. See also CAROLINA SKELETONS and MURDER IN MISSISSIPPI. AA: Cin (Peter Biziou).
DRA 121 min (ort 127 min) VIDrel: 4-FRONT/POLYREC
V/sur

MISSISSIPPI MASALA *** 18
Mira Nair USA 1991
Denzel Washington, Roshan Seth, Sarita Choudhury, Sharmila Tagore, Charles S. Dutton, Joe Seneca, Ranjit Chowdhry, Mohan Gokhale, Mohan Agashe, Tico Wells, Yvette Hawkins, Anjan Srivastava, Dipti Suthar, Varsha Thaker, Ashok Lath
Inspired (if that is the right word) by the expulsion of Uganda's Asians by Idi Amin in 1972, the film is set in the Deep South, to which many Asian entrepreneurs gravitated. At an Asian-owned motel, Choudhury (in her debut) enjoys a furtive affair with a young black man. Unfortunately, love is never enough, and the complications thrown up by this inter-racial Romeo and Juliet tale, do much to illustrate the narrowness of cultural roots.
DRA 112 min (ort 113 min) VIDrel: PAL/GUILD L/A
V/h

MISSOURI BREAKS, THE * 15
Arthur Penn USA 1976
Marlon Brando, Jack Nicholson, Kathleen Lloyd, Randy Quaid, Frederic Forrest, Harry Dean Stanton, John McLiam, John Ryan, Sam Gilman, Steve Franken, Richard Bradford, James Greene, Luana Anders, R.L. Armstrong, Dan Ades
Ranchers clash with rustlers and hire a gunman to help them

out. Supposedly a spoof of sorts, but it's hard to see where the jokes are in this disorganised and interminable mess.
WES   121 min (ort 126 min)   VIDrel: MGM/WHV   V

MR AND MRS BRIDGE ***
James Ivory USA
PG
1990
Paul Newman, Joanne Woodward, Blythe Danner, Simon Callow, Kyra Sedgwick, Robert Sean Leonard, Margaret Welsh, Saundra McClain, Diane Kagan, Austin Pendleton, Gale Garnett, Remak Ramsay
A look at the changing lifestyle of a reserved Kansas City couple, who have to cope with the demands of their children and various events that take place through the 1930s and 1940s. Not really a strongly narrated film, it's more a set of carefully produced vignettes illuminating the most important aspects of the characters' lives. A warm and moving work, scripted by Ruth Prawer Jhabvala; what it lacks in development it more than makes up for in acting.
DRA   125 min (ort 127 min)   VIDrel: 4-FRONT/POLYREC L/A   V/sh
Boa: novels: Mr Bridge and Mrs Bridge by Evan S. Connell.

MR BASEBALL **
Fred Schepisi USA
15
1992
Tom Selleck, Dennis Haysbert, Ken Takakura, Aya Takahashi, Toshi Shioya, Kohsuke Toyhara, Toshizo Fujiwara, Mak Takako, Kenji Morinaga, Haoki Fujii, Jon Nishimura, Korihide Goto, Kensuke Toiya, Tayakobu Kozumi, Leon Lee, Scott Plank
An ailing baseball player who has reached the end of his useful career is sent to Japan where he experiences the usual problems caused by cultural clashes and similar problems. Selleck's projects his customary amiable persona but his undeniable charm is not enough to give this film any real bite and the neat ending is both dull and uninspired.
COM   104 min (ort 108 min)   VIDrel: CIC/SONOP L/A   V

MISTER BIRD TO THE RESCUE ***
Paul Grimault FRANCE
U
1959
Voices of: Claire Bloom, Denholm Elliott, Peter Ustinov
Loosely inspired by Andersen's tale and scripted by Jacques Prevert, the famous poet and screenwriter, this complex and imaginative story follows the changing fortunes of the King of Tachacardia, who one day loses his power. An elegant and skilfully executed animation that is certain to have wide appeal.
Aka: CURIOUS ADVENTURES OF MISTER WONDERBIRD, THE; KING AND MISTER BIRD, THE; KING AND THE BIRD, THE; KING AND THE MOCKINGBIRD, THE; LE ROI ET L'OISEAU; MY BIRD TO THE RESCUE
ANIM   80 min (ort 87 min)   VIDrel: EIV/SONOP   V
Boa: short story The Shepherdess and the Chimneysweep by Hans Christian Andersen.

MISTER BLANDINGS BUILDS HIS DREAM HOUSE ***
H.C. Potter USA
U
1948
Cary Grant, Myrna Loy, Melvyn Douglas, Reginald Denny, Nestor Paiva, Jason Robards, Louise Beavers, Ian Wolfe, Harry Shannon, Sharyn Moffett, Connie Marshall, Lurene Tuttle, Lex Barker, Emory Parnell
A couple from the city attempt to build a house in the countryside, but find this no easy task. Though the film is nowhere near as subtle as the novel, this bright and cheerful comedy has the two stars in top form. The script is by Norman Panama and Melvin Frank. See also THE MONEY PIT.
COM   90 min B/W   VIDrel: VCC L/A   V
Boa: novel by Eric Hodgkin.

MISTER DEEDS GOES TO TOWN ****
Frank Capra USA
U
1936
Gary Cooper, Jean Arthur, Raymond Walburn, Lionel Stander, Walter Catlett, George Bancroft, Douglass Dumbrille, H.B. Warner, Ruth Donnelly, Margaret Seddon, Margaret McWade
A man must defend his sanity in a courtroom battle when he inherits a vast fortune, and proclaims his intention of using the money to help ordinary people suffering the effects of the Depression to make a fresh start. Vintage Capra, that's neither dated nor stilted, but simply a joy to watch. The script is by Robert Riskin. Later a brief TV series. AA: Dir.
COM   111 min (ort 118 min) B/W   VIDrel: RCA L/A   V
Boa: story Opera Hat by Clarence Budington Kelland.

MR DESTINY **
James Orr USA
PG
1990
James Belushi, Michael Caine, Linda Hamilton, Jon Lovitz, Hart Bochner, Bill McCutcheon, Rene Russo, Jay O. Sanders, Maury Chaykin, Pat Corley, Douglas Seale, Courtney Cox, Doug Barron, Jeff Weiss, Tony Longo, Kathy Ireland
When a mysterious stranger comes into the life of a junior executive, some dramatic changes are brought about and he sees what his life would have been like had he not struck out in a basketball game. A very poor variant on IT'S A WONDERFUL LIFE that is burdened with a transparent plot and a total lack of believable characters.
COM   106 min (ort 110 min)   VIDrel: TOUCH/SONOP L/A   V

MR FROST *
Philippe Setbon FRANCE/UK
15
1990
Jeff Goldblum, Alan Bates, Kathy Baker, Roland Giraud, Jean-Pierre Cassel, Daniel Gelin, Francois Negret, Maxime Leroux, Vincent Schiavelli, Catherine Allegret, Charley Boorman
A mass-murderer who's incarcerated in an insane asylum tells his female psychiatrist that's he's really an incarnation of Satan. As the doctor struggles to get at the truth, she learns that there's a police officer out there who already believes him. A pointless film, that appears to have been made without any clear idea of what direction it should take.
Aka: DEADLY MISTER FROST, THE
HOR   99 min   VIDrel: 20TH/TECH   V/sur

MR HOLLAND'S OPUS ***
Stephen Herek USA
PG
1995
Richard Dreyfuss, Glenne Headly, Jay Thomas, Olympia Dukakis, William H. Macy, Alicia Witt, Terrence Howard, Damon Whitaker, Jean Louisa Kelly, Alexandra Boyd, Nicholas John Renner, Joseph Anderson, Anthony Natale, Joanna Gleason
A composer becomes a music teacher at a high school and spends the next thirty years imbuing successive generations with his own love of music. And all this while he continues work on his symphony, which proves to be his salvation when he is sacked from his job. Dreyfuss and simply tremendous in this heartwarming and intriguing drama, and the finale is well worth waiting for.
DRA   136 min (ort 144 min) cC   VIDrel: POLY/POLYREC V

MISTER HORN **
Jack Starett USA
PG
1979
David Carradine, Richard Widmark, Karen Black, Richard Masur, Clay Tanner, Pat McCormick, Jack Starrett, John Durren, Jeremy Slate, Enrique Lucero, Stafford Morgan, Don Collier, James Oliver, George Reynolds, Ian McLean
Long, episodic Western about a legendary frontier figure, who captured the Apache chief Geronimo. Average. See also TOM HORN.
WES   134 min (ort 200 min) mTV   VIDrel: POLY L/A   V

MISTER JOHNSON ***
Bruce Beresford USA
PG
1990
Maynard Eziashi, Pierce Brosnan, Edward Woodward, Beatie Edney, Denis Quilley, Nick Reading, Bella Antonio, Femi Fatoba, Kwabena Manso, Hubert Ogunde, Sola Adeymi, Jerry Linus, George Menta, Steve James, Tunde Kelani, Albert Egbe
First-time actor Eziashi plays an educated Nigerian who works for a local magistrate and keeps getting into trouble, mostly on account of being a little too clever for his own good. Meanwhile, the country is being opened up, with the construction of a road about to link north and south. A broad and sweeping period-drama, set in the 1920s, that offers some insight into the history of Africa, but considerably less into the main character.
DRA   97 min (ort 105 min)   VIDrel: WHV   V/sh
Boa: novel by Joyce Carey.

MR JONES *
Mike Figgis USA
18
1993
Richard Gere, Lena Olin, Anne Bancroft, Tom Irwin, Delroy Lindo, Lauren Tom, Bruce Altman, Lisa Malkiewicz, Thomas Kopache, Peter Jurasik, Leon Singer, Anna Maria Horsford, Edward Padilla, Baha Jackson, Anne Lange, Kelli Williams
A woman psychotherapist takes on the case of a manic-depressive who enjoys flirting with death, and she is so successful that he is cured of his moods of deep despair, but also loses his bouts of euphoria and becomes quite normal. Unfortunately, she has fallen in love with him and comes to realise that she was far more attracted to him before she cured him. A sad and rather dreary love story that strains credibility and patience.
DRA   109 min (ort 114 min) cC   VIDrel: COLUM/SONOP V/sur

MISTER KLEIN **   12
Joseph Losey FRANCE/ITALY   1976
Alain Delon, Jeanne Moreau, Suzanne Flon, Michel Lonsdale, Juliet Berto, Louise Seigner, Francine Berge, Massimo Girotti
In 1942 a Parisian antiques dealer profits from the plight of Jews who are forced to sell him their priceless items for a trifle. However, by a series of strange coincidences, he is mistaken for a shadowy Jewish figure called Klein and begins to assume his identity, ultimately sharing the fate of Jews being sent to their deaths. A ponderous and unsatisfying drama, whose strongly surreal undertones are never fully developed.
Aka: MR KLEIN
DRA 118 min (ort 122 min) VIDrel: ARROW/RTM V

MISTER LOVE **   15
Roy Battersby UK   1986
Barry Jackson, Marcia Warren, Julia Deakin, Maurice Denham, Margaret Tyzack, Linda Marlowe, Christina Collier, Helen Cotterill, Donal McCann, Janine Roberts, Tony Melody, Patsy Byrne, Robert Bridges, Jacki Piper, Alan Starkey
A gentle comedy about a man in charge of Southport's municipal gardens who in his own quiet way becomes something of a Don Juan. The treatment is strangely old-fashioned and the theme never amounts to much. Made as one in a series for TV called "First Love".
Aka: FIRST LOVE... MISTER LOVE
COM 88 min (ort 91 min) VIDrel: MGM/WHV L/A V/h

MISTER MAJESTYK **   18
Richard Fleischer USA   1974
Charles Bronson, Al Lettieri, Linda Cristal, Lee Purcell, Paul Koslo, Alejandro Rey
The great stoneface plays a Colorado melon farmer, who runs into trouble with gangsters when he refuses to pay protection money. A well-paced, violent film in which the melons act beautifully. (Bronson comes a close second). The script is by Elmore Leonard.
DRA 100 min (ort 103 min) VIDrel: MGM/WHV L/A V

MR MUM **   PG
Stan Dragoti USA   1983
Michael Keaton, Teri Garr, Martin Mull, Ann Jillian, Christopher Lloyd, Frederick Koehler, Graham Jarvis, Taliesin Jaffe, Courtney White, Brittany White, Tom Leopold, Carolyn Seymour, Michael Alaimo, Jeffrey Tambor
A couple switch domestic roles when the husband gets fired and the wife finds a job. As is usual, the fun revolves around Hollywood's obsession with men being useless at housework.
Aka: MR MOM
COM 87 min (ort 91 min) VIDrel: LUMI/SPEAR L/A V

MR NANNY **   PG
Michael Gottlieb USA   1992
Terry "Hulk" Hogan, Sherman Hemsley, Austin Pendleton, Robert Gorman, David Johansen, Madeline Zima, Raymond O'Connor, Mother Love, Peter Kent, Arlie Marlesci, Afna Anoal Alfa, Brutus Beefcake, Butch Brickell, James Coffey
A wrestler who hates kids gets hired as a sort of glorified babysitter to a pair of insufferable rich brats who proceed to torment him at every turn. Things, change, however, when their father's business rival attempts to kidnap them in order to get his hands on a new and very secret microchip. A modestly amusing but totally insubstantial formula comedy.
Aka: ROUGH STUFF
COM 80 min (ort 84 min) VIDrel: EIV/SONOP V

MISTER NORTH **   PG
Danny Huston USA   1988
Anthony Edwards, Robert Mitchum, Anjelica Huston, Lauren Bacall, Harry Dean Stanton, Virginia Madsen, Mary Stuart Masterson, David Warner, Hunter Carson, Christopher Durang, Mark Metcalf, Katherine Houghton
In the 1920s, a young man takes Newport society by storm thanks to his charm and charisma. A likeable if somewhat patchily realised fantasy, with fine locations and a good cast. The feature debut for the director, who wrote the script with his father, John Huston.
COM 88 min (ort 92 min) VIDrel: RCA L/A V
Boa: novel Teophilus North by Thornton Wilder.

MR RELIABLE (A TRUE STORY) ***   15
Nadia Tass AUSTRALIA   1996
Colin Friels, Jacqueline McKenzie, Paul Sonkkila, Frank Gallacher,
Lisa Hensley, Aaron Blabey, Geoff Morrell, Neil Fitzpatrick, Gillian Stratham, Gilliam Graham, Rebecca Di Corpo, Ken Radley, Graham Rouse, Elaine Cusick
With a good dose of humour, this film attempts to examine a true incident from 1968, when an attempt to arrest a small-time crook turned into a major siege and a media circus (complete with hot dog stands and fairy lights). The person being held inside was at first thought to be a hostage, but was in fact the man's girlfriend, and the two got married whilst under siege. Quite endearing in its way, even if this oddball of a movie does go on for too long.
DRA 108 min (ort 113 min) VIDrel: POLFIL V/sh

MISTER ROBERTS ****   U
John Ford/Mervyn Le Roy USA   1955
Henry Fonda, James Cagney, William Powell, Jack Lemmon, Betsy Palmer, Ward Bond, Nick Adams, Phil Carey, Harry Carey Jr, Ken Curtis, Frank Aletter, Fritz Ford, Buck Kartalian, William Henry, William Hudson, Stubby Kruger
Story of a cargo ship in WW2, and an officer who is itching for transfer to where the fighting is. The film hovers between comedy and pathos, and has some enjoyable, if not really believable moments. Cagney as the martinet captain, provides a fine counterweight to Fonda, as the officer determined to obtain a transfer. This was Powell's last film. The vastly inferior sequel "Ensign Pulver" followed. AA: S. Actor (Lemmon).
COM 115 min B/W VIDrel: WHV V
Boa: play by Thomas Heggen and J. Logan.

MR SARDONICUS **   12
William Castle USA   1961
Ronald Lewis, Audrey Dalton, Guy Rolfe, Oscar Homolka, Vladimar Sokoloff, Erika Peters, Tina Woodward, Constance Cavendish, Mavis Neal, David Janti Charles H. Radilac, Franz Roehn, Annalena Lund, Ilse Burkert, Albert D'Arno
An offbeat horror tale about a wealthy baron whose face is disfigured by a permanent grin, which forces him to wear a mask. Obsessed by beauty, he attempts to force a surgeon to operate but the results fail to turn out as he had expected. This being a William Castle film, it had a gimmick when shown in the cinemas, this one allowing the audience to decide on the film's ending.
Aka: MISTER SARDONICUS; SARDONICUS
HOR 87 min (ort 89 min) B/W
VIDrel: ENCORE/SPEAR/COLUM V

MR SATURDAY NIGHT ****   15
Billy Crystal USA   1992
Billy Crystal, David Paymer, Julie Warner, Helen Hunt, Jerry Orbach, Ron Silver, Mary Mara, Sage Allen, Jackie Gayle, Carl Ballantine, Slappy White, Conrad Janis, Jerry Lewis, Lowell Ganz, Babaloo Mandel, Jason Marsden
Brilliant account of forty years in the life of a once successful but far from lovable comedian whose selfish nature prove hard to take for those closest to him, most especially his manager-brother. Crystal gives his finest performance to date in a film whose sparkling script is memorable for some of the sharpest one-liners ever. A superb directorial debut for Crystal who co-wrote the slightly schmaltzy script with Lowell Ganz and Babaloo Mandel.
COM 114 min (ort 119 min) VIDrel: FIRST/SONOP; PION (LV only) V LV

MISTER SISTER **   PG
Jimmy Zeilinger USA   1992
Jonathan Silverman, Alyssa Milano, George Newbern, Michele Matheson, Leilani Sarelle, Christine Healy, Jenny Gideon, Jesse Dabson, Moon Zappa, William Frankfather, Patrick Richwood, Michael Berryman, Damon Standifer, James Garfield
A young college student dresses up as a girl as part of a fraternity prank to infiltrate a sorority house but is forced by circumstances to maintain this masquerade. But in this guise he meets and falls for a sorority sister. A juvenile but harmless gender-swap comedy.
Aka: LITTLE SISTER
COM 94 min VIDrel: MARQ/QUANT V

MISTER SMITH GOES TO WASHINGTON ****   U
Frank Capra USA   1939
James Stewart, Jean Arthur, Edward Arnold, Claude Rains, Guy Kibbee, Thomas Mitchell, Eugene Palleyye, Beulah Bondi, Harry Carey, H.B. Warner, Charles Lane, Porter Hall, Jack Carson, Astrid Allwyn, Ruth Donnelly, Grant Mitchell

An idealistic young man becomes a U.S. senator and discovers that the seat of power is also a hotbed of corruption which he proceeds to reveal, even at the risk of ruining his own political career. A highly enjoyable and vintage Capra film that still feels fresh and original. AA: Story/orig (Lewis R. Foster).
DRA   125 min (ort 130 min) B/W   VIDrel: COLUM/SONOP
V/dm

MR WONDERFUL ***   15
Anthony Minghella USA   1992
Matt Dillon, Annabella Sciorra, Mary-Louise Parker, William Hurt, Vincent D'Onofrio, Paul Bates, Arabella Field, Renee Lippin, Brooke Smith, Peter Appel, Bruce Altman, James Gnadolfini, David Barry Gray, Geoffrey Grider, Luis Guzman
A married couple who were childhood sweethearts get divorced but find they are still very much attracted to each other. However, the husband has dreams of opening a bowling alley with some friends and needs to find her a new husband so he can stop paying alimony. Minghella's second film (after TRULY, MADLY, DEEPLY) has charming performances that sustain the thin plot. It's a pity the film is let down by a predictable happy ending that is pure Hollywood at its most trite.
COM   95 min (ort 99 min)   VIDrel: TOUCH/TECH L/A   V

MISTRAL'S DAUGHTER: PARTS 1, 2 AND 3 **   PG
David Hickox FRANCE/LUXEMBOURG/USA   1984
Stefanie Powers, Lee Remick, Stacy Keach, Timothy Dalton, Robert Urich, Stephane Audran, Ian Richardson, Stephanie Dunnam, Cotter Smith, Alan Adair, Pierre Malet, Philippine Leroy Beaulieu, Alexandra Stewart, Joanna Lumley
Overlong, dull as ditchwater saga, of a painter over a period from the 1920s to the 1950s, and mainly examining his relationship with his first model and the daughter she bears him. Not one scene carries dramatic conviction as the one-dimensional characters progress through their parts in a variety of unconvincing ways. Originally shown in three parts.
DRA   540 min (3 cassettes) mTV   VIDrel: BRAVE/SONOP
L/A   V
Boa: novel by Judith Krantz.

MISTRESS **   15
Barry Primus USA   1992
Danny Aiello, Robert De Niro, Martin Landau, Eli Wallach, Robert Wuhl, Jean Smart, Sheryl Lee Ralph, Tuesday Knight, Jace Alexander, Laurie Metcalfe, Ernest Borgnine, Roberta Wallach, Christopher Walken, Vasek C. Simek, Tomas R. Voth
A failed scriptwriter meets a producer down on his luck who claims that he has found a backers for a film based on one of his old scripts. The only stumbling block is that each of these three men has a mistress whom they want to play the lead female role. A good behind-the-scenes look at Hollywood that offers a great cast but is badly let down by the lack of a solid script.
COM   105 min (ort 112 min)   VIDrel: COLUM/SONOP
V/sh

MIXED BLESSINGS **   12
Bethany Rooney USA   199-
Bess Armstrong, Scott Biao, Gabrille Carteris, James Naughton
A set of three stories all of which take a look at the trials of parenthood. In the first a newly-married couple attempt to adopt when they find themselves unable to have children. In the second story an older couple have to decide whether the risks involved in having a child are worth taking. Finally, the last story charts the conflict between a couple where the husband wants kids but the wife does not. Average.
Aka: DANIELLE STEEL'S MIXED BLESSINGS
DRA   87 min   VIDrel: 20VIS/SONOP   V/h
Boa: stories by Danielle Steel.

MIXED NUTS *   12
Nora Ephron USA   1994
Steve Martin, Madeline Kahn, Robert Klein, Anthony LaPaglia, Juliette Lewis, Rob Reiner, Adam Sandler, Rita Wilson, Liev Schreiber, Parker Posey, Jon Stewart, Joely Fisher, Steven Randazzo, Christine Cavanaugh, Henry Brown
An unfunny comedy set on Christmas Eve in Venice, California, with the volunteers at a helpline for would-be suicides struggling with their complicated love lives and other personal problems. Feeble scripting fails to generate laughs from the serious and potentially promising subject matter. Written by

Ephron and her sister, this weak farce was inspired by the French film "Le Pere Est Une Ordure" (1982).
Aka: LIFESAVERS
COM   93 min (ort 97 min)   VIDrel: COLUM/SONOP
V/sur

MO' BETTER BLUES **   15
Spike Lee USA   1990
Denzel Washington, Spike Lee, Wesley Snipes, Giancarlo Esposito, Joie Lee, Robin Harris, Bill Nunn, John Turturro, Dick Anthony Williams, Cynda Williams, Nicholas Turturro, Ruben Blades, Abbey Lincoln, Mamie Louise Anderson
Written, produced and directed by Lee (with music by his dad Bill) this messy and unfocused tale has Washington playing a jazz trumpeter who only gives himself to his music, keeping everyone else at arm's length, including his two girlfriends. A colourful and often highly enjoyable effort, but neither consistent in tone nor clear in direction, and about thirty minutes too long.
DRA   124 min (ort 129 min)   VIDrel: CIC/SONOP   V/sur

MO' MONEY **   15
Peter MacDonald USA   1992
Damon Wayans, Marlon Wayans, Joe Santos, John Diehl, Stacey Dish, Harry J. Lennix, John Diehl, Mark Beltzman, Quincy Wong, Kevin Casey, Larry Brandenburg, Garfield, Almayvonne, Dick Butler, Matt Doherty, Evan Lionel Smith, James Deuter
A small-time confidence trickster decides to go straight in order to win the affection of a girl he has fallen for. He gets a job in a mail-room but finds that the wages this pays are not enough to allow him to achieve his goal. Screenplay is by Damon Wayans.
Aka: MO' MONEY: WHY SETTLE FOR LESS?
COM   86 min (ort 91 min) cC   VIDrel: VCC/DISC/COLUM
V/sh

MOB WAR **   18
J. Christian Ingvordsen   1989
Johnny Stumper, David Henry Keller, Jake La Motta, John Christian
A top New York gangster and a leading media executive devise a plan of action that involves them working together.
A/AD   92 min Cut (10 sec – ort 96 min)
VIDrel: COLUM/SONOP   V

MOBSTERS: THE EVIL EMPIRE *   18
Michael Karbelnikoff USA   1991
Christian Slater, Patrick Dempsey, Richard Grieco, Costas Mandylor, F. Murray Abraham, Lara Flynn Boyle, Michael Gambon, Christopher Penn, Anthony Quinn, Rodney Eastman, Jeremy Schoenberg, Miles Perlich, Alan Charof, Anto Nolan
Set in New York in 1917, this violent, expensive but unutterably dreary gangster movie charts the rise of Lucky Luciano, and takes us up to the 1930s, when he succeeded in putting organised crime on a more businesslike footing. The obligatory shoot-outs are much in evidence, as are a series of showgirl sequences, but there's really very little here to sustain interest, and in the main, the cast are just too young to convince as gangsters.
Aka: MOBSTERS
A/AD   115 min (ort 121 min)
VIDrel: 4-FRONT/POLYREC/CIC   V/sur

MOBY DICK ***   PG
John Huston USA   1956
Gregory Peck, Richard Basehart, Leo Genn, Friedrich Ledebur, Orson Welles, James Robertson Justice, Harry Andrews, Bernard Miles, Noel Purcell, Edric Connor, Joseph Tomelty, Mervyn Johns, Seamus Kelly, Philip Stainton, Royal Dano
A whaling skipper swears vengeance on the great white whale that cost him a leg in an earlier encounter. This slow-moving remake of the 1930 film, has some remarkable photography and scenes, but Peck makes a rather disappointing Ahab and the intrusive music is a severe distraction. Nevertheless, an enjoyable film, with screenplay by Huston and Ray Bradbury. A highlight is the church sermon given by Father Mapple, which Welles did in a single take.
A/AD   110 min (ort 116 min)   VIDrel: LUMI/SPEAR   V
Boa: novel by Herman Melville.

MODEL BY DAY **   (18)
Christian Duguay USA   1993
Famke Janssen, Shannon Tweed, Stephen Shellen, Sean Young, Gene Simmons, Traci Lind, Kim Coates, Nigel Bennett, Von Flores, David

Hemblen, Lovis Di Bianco, Henrittte Ivanans, Jefferson Marpin, Santino Buda, Tony De Santis
When a number of models fall victim to a vicious attacker, two of their colleagues use their karate skills to become night-time vigilantes and dress up in stylish, revealing but impractical costumes. A ludicrous assembly-line actioner like many of its kind but set for a change in the supposedly glamorous world of modelling.
A/AD   89 min   SATrel: MOVIE CHANNEL
Boa: comic strip by Kevin Taylor.

MODERN AFFAIR, A **   15
Vern Oakley USA   1995
Lisa Eichhorn, Caroline Aaron, Stanley Tucci, Vincent Young, Robert PuLong, Mary Jo Salerno, Robert Joy, Wesley Addy, Tammy Grimes
A very modern miss decides that she will not let the absence of a man prevent her having a child, and goes to a sperm bank to achieve this end. But when she learns that her donor has attributes that attract her, she sets out to locate him. See also MADE IN AMERICA and CARBON COPY, for another two comedies dealing with sperm-bank intrigues.
COM   91 min   VIDrel: COLUM/SONOP   V

MODERN GIRLS *   15
Jerry Kramer USA   1986
Daphne Zuniga, Virginia Madsen, Cynthia Gibb, Clayton Rohner, Chris Nash, Steve Shellen, Rich Overton, Pamela Springstein
Three beautiful girls have a wild night out in this tedious, low-brow exercise in murky unpleasantness and coarse humour.
COM   81 min (ort 84 min)   VIDrel: EIV/SONOP   V

MODERN LOVE *   15
Robby Benson USA   1990
Robby Benson, Karla DeVito, Rue McClanahan, Burt Reynolds, Frankie Valli, Kaye Ballard, Louise Lasser, Lyric Benson, Lori Tates, Beth Meadows Calvert, Cliff Bemis, Stan Brown, Lou Kaplan, Sharon Greene, Libby Campbell, Debra Port
Real-life marrieds Benson and DeVito play a couple of young lovers who experience the early trials of an eventful marriage, problems with in-laws and parents and a tragic bereavement. Written and produced by Benson, this clumsy blend of comedy and drama is both unfunny and manipulative.
DRA   104 min (ort 110 min)   VIDrel: MED/COLUM L/A
V/sh

MODERN ROMANCE ***   15
Albert Brooks USA   1981
Albert Brooks, Kathryn Harrold, Bruno Kirby, Jane Hallaren, James L. Brooks, George Kennedy, Bob Eisenstein, Tyann Means, Karen Chandler, Dennis Kort, Thelma Leeds, Virginia Feingold, Candy Castillo, Ed Weinberger, Cliff Einstein
A neurotic man is obsessed with his girlfriend, but is unable to maintain a normal relationship with her. A perceptive and generally witty exercise with some nice cameos, and a few in-jokes aimed at movie buffs. Written by Brooks and Monica Johnson.
COM   90 min (ort 93 min)   VIDrel: RCA L/A   V

MODERNS, THE ***   15
Alan Rudolph USA   1988
Keith Carradine, Linda Fiorentino, Genevieve Bujold, Geraldine Chaplin, Wallace Shawn, Kevin J. O'Connor, John Lone, Elsa Raven, Ali Giron, Gailard Sartain, Michael Wilson, Flora Bolzano, Veronique Bellegarde, Charlie Biddle
The story of a group of Americans living in Paris in 1926 provides the basis for this lavish and complex tale. Carradine (badly miscast) plays an art forger who meets a former lover, but finds she is now married to a ruthless businessman. They rekindle their romance, which takes place against the background of the cafes and salons that were the haunt of the intellectuals of the period. Ponderous and uneven, but visually engrossing.
DRA   121 min (ort 126 min)   VIDrel: FIRST/SONOP L/A
V

MODESTY BLAISE **   PG
Joseph Losey UK   1966
Monica Vitti, Terence Stamp, Dirk Bogarde, Harry Andrews, Michael Craig, Clive Revill, Scilla Gabel, Tina Marquand, Rosella Falk, Joe Melia, Lex Schoorel, Alexander Knox
A film version of a comic strip female super-heroine hired to protect a diamond shipment, with Bogarde extremely well cast as the villain out to get it. A high-camp spoof in the 1960s tradi-

tion, that loses its direction rather too often. The comic strip was by Peter O'Donnell.
COM   120 min   VIDrel: 20TH/TECH L/A   V

MOLL FLANDERS *   12
Pen Densham USA   1995
Robin Wright, Morgan Freeman, Stockard Channing, Brenda Fricker, John Lynch, Aisling Corcoran, Geraldine James, Jeremy Brett, Jim Sheridan
On the way to America, Freeman recounts the story of Moll Flanders to his daughter. This takes the form of a diary recording her exploits, from orphan through to prostitute and then finally wealth and comfort, but with a spell of prison in between. Defoe's bawdy 18th century romp is given a reverential, Hollywood treatment in this laborious retread, and the result is a work that is stilted, lifeless and remarkably lacking in humour.
DRA   120 min   CINrel
Boa: novel by Daniel Defoe.

MOLLY AND THE GHOST **   15
Don Jones USA
Ena Henderson, Lee Darling, Daniel Martine, P.K. Flamingo
An over-sexed and predatory seventeen-year-old has designs on her sister's husband, but when she concocts a plan to get rid of her sister, it backfires and she is killed instead. She returns as a succubus.
FAN   87 min   VIDrel: 20VIS/SONOP   V/sh

MOLLY MAGUIRES, THE ***   PG
Martin Ritt USA   1970
Sean Connery, Richard Harris, Samantha Eggar, Frank Finlay, Art Lund, Bethel Leslie, Anthony Zerbe, Anthony Costello, Philip Bourneuf, Brendan Dillon, John Alderson, Frances Heflin, Malachy McCourt, Susan Goodman, Peter Rogan
This tells the true story of a secret Irish terrorist group which conspired amongst Irish miners in Pennsylvania in 1876, and of the informer who infiltrated the group in order to betray them to the authorities. A sombre and detailed film, that lacks depth to the characters and an effective climax. Arthur Conan Doyle also made use of these events in his tale "The Valley Of Fear".
DRA   119 min (ort 123 min)   VIDrel: CIC/SONOP L/A   V
Boa: book Lament For Molly Maguires by A.H. Lewis.

MOM **   18
Patrick Rand USA   1989
Mark Thomas Miller, Jeanne Bates, Brion James, Mary McDonough, Stella Stevens
A young reporter is assigned his first big story, which involves investigating a series of gruesome murders of women, in which the remains of each victim was found to have been partially eaten. Finding a lead that taker him to his mother's blind lodger, he discovers that the man is really a demon from Hell who has turned his mother into a similar creature.
HOR   91 min   VIDrel: EIV/SONOP   V

MOM AND DAD SAVE THE WORLD *   PG
Greg Beeman USA   1991
Jon Lovitz, Teri Garr, Jeffrey Jones, Kathy Ireland, Thalmus Rasulala, Wallace Shawn, Kathy Ireland, Suzanne Ventulett, Michael Stoyanov, Danny Cooksey, Charlie Dell, Laurie Main, Jim Maniaci, Dennis Macalone, Eddie Paul
The ruthless and paranoid ruler of a distant planet plans to destroy Earth but postpones his plans when he spots a human woman he takes a strong fancy to. He abducts her and her husband in the family car, with the intention of killing him and making her his bride, but our intrepid pair finally get the upper hand. A dreadfully unfunny spoof that apart from striking visuals, has absolutely nothing else to offer.
COM   86 min (ort 88 min)   VIDrel: FIRST/SONOP   V/sur

MOM FOR CHRISTMAS, A **   U
George Miller USA   1990
Olivia Newton-John, Juliet Sorcey, Doug Sheehan, Doris Roberts, Carmen Argenziano, Aubrey Morris, James Piddock, Elliot Moss Greenbaum, Erica Mitchell, Jesse Vincent, Brett Harrelson, Steve Russell, Gregory Procaccino, Ron Lautore
A motherless young girl whose father has little time for her makes a secret wish that she should get a mother. Her dream is magically fulfilled when a department store mannequin comes to life and sets about bringing some joy into her life. A rather

typical Disney family-fable, sweet and superficial, but surprisingly entertaining. See also MANNEQUIN.
JUV 88 min (ort 100 min) mTV VIDrel: WDV/BUENA L/A V

Boa: novel A Mom By Magic by Barbara Dillon.

MOMENT OF ROMANCE, A *** 18
Benny Chan HONG KONG 1990
Andy Lau, Ng Sin Lin, Tommy Wong
In Hong Kong, a gang member has to break away from his old cronies and go on the run, in order to save an innocent girl from being killed after she is taken hostage by the gang following a robbery, and has now served her purpose. The girl plainly reciprocates the feelings her champion has expressed, but the two lovers must also contend with the police as well as the man's former comrades. Quite a good action tale, fast and well handled.
A/AD 88 min wScrn VIDrel: MADE/RTM V

MOMENT OF TRUTH ** 18
Larry Gross USA 1984
Adam Baldwin, Deborah Foreman, Ed Lauter, Rene Auberjonois, Danny De La Paz, Scott McGinnis, John Scott Clough, Mario Van Peebles
A gang member becomes a reformed character, falling out with the gang leader and refusing to help him beat a drugs charge. One of those standard teen-gang movies in which the story unfolds in an entirely predictable manner.
Aka: 3.15; 3:15: MOMENT OF TRUTH
DRA 81 min Cut (54 sec – ort 95 min)
VIDrel: VISVID/POLYREC L/A V

MOMENT OF TRUTH: CRADLE OF CONSPIRACY * 15
Gianni Amelio USA 1994
Dee Wallace Stone, Danica McKellar, Kurt Deutsch, Carmen Argenziano, Merle Kennedy, Geoffrey Thorne, Ellen Crawford, Shannon Fill, Jamie Lunar, Christine Healy, Burke Byrnes, Matthew Faison, Charlie Holliday, Owen Bush, Jana Arnold
This was one of the "Moment Of Truth" series of made-for-TV family dramas, all of which purport to offer some insights into families that experience some unusual stress or crisis. In this story, a budding athlete suffers undue pressure from her parents, which results in her running off with a smarmy conman, who deliberately gets her pregnant, intending to sell her child to a sinister adoption agency. Mawkish and implausibly scripted nonsense.
DRA 90 min VIDrel: TART V

MOMMIE DEAREST ** 15
Frank Perry USA 1981
Faye Dunaway, Diana Scarwid, Steve Forrest, Howard Da Silva, Mara Hobel, Rutanya Alda, Harry Goz, Michael Edwards, Jocelyn Brando, Priscilla Pointer, Joe Abdullah, Garry Allen, Selma Archered, Adrian Aron, Xander Berkeley, Russ Marin
Film version of a book by Crawford's adopted daughter describing her loveless upbringing at the hands of this star. The book caused a stir when published, as it upset many of Crawford's fans, this film may do the same. Despite being undeniably well put together and acted, it's hard to see what audience the film is aimed at as its appeal is so very limited.
Aka: MOMMY DEAREST
DRA 124 min (ort 129 min) VIDrel: CIC/SONOP L/A V
Boa: book by Christine Crawford.

MOMMY MARKET, THE ** PG
Tia Brelis USA 1993
Sissy Spacek, Anna Chlumsky, Aaron Michael Metchik, Asher Metchik, Maureen Stapleton, Merritt Yohnka, Andre The Giant, Sean MacLaughlin, Jemar Jeferson, Nancy Chlumsky, Eleni Schirmer, Fran Joseph, Catherine Paolono, Maria Fagan
A friendly neighbourhood witch gives three young kids the chance to use a spell that rids them of their mother and allows them to choose a new one from the "Mommy Market". Predictably, the replacement does not turn out to be quite so satisfactory, and after a series of disasters they are soon hankering after their real mom. Fair.
JUV 79 min VIDrel: 20TH/TECH V/sur

MON HOMME *** 18
Bertrand Blier FRANCE 1996
Anouk Grindberg, Gerard Lanvin, Valeria Bruni-Tedeschi, Olivier Martinez, Dominique Valadie, Jacques Francois, Michel Galabru,

Robert Hirsch, Bernard Fresson, Jacques Gamblin, Jean-Pierre Darroussin
A young and very successful prostitute comes across a homeless man and offers him shelter, eventually becoming so enamoured with him that she suggests he become her pimp. He takes to this new trade with great enthusiasm, and starts recruiting other women, but this leads to his arrest and imprisonment. A witty and stylish film whose inventive script is initially great fun, even if the plot twists get increasingly silly and the film ultimately lacks any direction.
COM 98 min CINrel

MONA LISA *** 18
Neil Jordan UK 1986
Bob Hoskins, Cathy Tyson, Michael Caine, Robbie Coltrane, Clarke Peters, Kate Hardie, Zoe Nathenson, Sammy Davis, Rod Bedall, Joe Brown, Pauline Melville, David Halliwell, Hossein Karimbeik, John Darling, Donna Cannon
A small-time crook fresh out of prison, is given the job of acting as chauffeur to a high-class prostitute, and becomes deeply involved with her, while not being aware of just how sordid her life is. Excellent performances are combined with a sound if squalid story. Written by Jordan and David Leland.
DRA 100 min (ort 104 min) VIDrel: CIC/SONOP V/h

MONDO TOPLESS * 18
Russ Meyer USA 1966
Lead dancers: Babette Bardot, Pat Barringer, Sin Lenee
A plotless examination of the big-breasted ladies of San Francisco, that tells us more about the director's absurd obsession with the upper-half of the female form than it does about San Francisco, women or indeed anything else.
A 60 min VIDrel: ALLIED/RTM/TROMA V

MONEY FOR NOTHING ** (15)
Ramon Menendez USA 1993
John Cusack, Debi Mazar, Michael Madsen, Benicio Del Toro, Michael Rapaport, Fionnula Flanagan, Maury Chaykin, James Gandolfini, Elizabeth Bracco, Ashleigh Dejon, Lenny Venito, Philip S. Hoffman, Currie Graham. Frankie Faison
A longshoreman finds $1,200,000 in unmarked bills and soon learns that his life is about to change out of all recognition. After some unwise and highly conspicuous spending, he has to cope with a nosy detective and responds by asking for Mob assistance in laundering his stash. A limp black comedy that never gets going. Based on a true story in which the individual involved escaped punishment by entering a plea of temporary insanity.
COM 97 min SATrel: MOVIE CHANNEL

MONEY PIT, THE * 15
Richard Benjamin USA 1986
Tom Hanks, Shelley Long, Alexander Godunov, Maureen Stapleton, Joe Mantegna, Philip Bosco, Josh Mostel, Yakov Smirnoff, Carmine Caridi, Brian Backer, Mia Dillon, Billy Lombardo, John Van Dreelan, Douglass Watson, Lucille Dobrin
Essentially an updating of MISTER BLANDINGS BUILD HIS DREAM HOUSE but without the latter's wit and charm. A young couple buy a fine house in the country at an unbelievably low price, and find themselves stuck with a worthless property that just swallows money. This strains their patience (and mine).
COM 88 min (ort 91 min) VIDrel: CIC/SONOP V/sur

MONEY, POWER, MURDER ** 15
Lee Philips USA 1989
Kevin Dobson, Blythe Danner, Josef Sommer, John Cullum, Paul McCrane, Wayne Tippit, Dion Anderson, Julianne Moore, Peter Bergman, Peter Maloney, Casey Sander, Julie Philips, Russ Anderson, Rocky Carroll, Alice Drummond
A reporter for a cable news station investigates the disappearance of a network anchor-woman and inadvertently uncovers a nasty conspiracy involving a network chief and a TV preacher. Familiar territory it may be, but on the whole it's more effective than one might have expected.
THR 91 min mTV VIDrel: 20TH V
Boa: novel Dead Air by Mike Lupica.

MONEY TALKS ** PG
James Scott UK 1989
Robert Lindsay, Molly Ringwald, John Gielgud, Frances De La Tour, Max Wall, Simon De La Brosse, Margi Clarke, Vladek Sheybal, Michel Blanc, Marianne Price, Jeffrey Robert, Willy Ross, Harriet Reynolds, Tim Seely, Marius Goring

A newly-married couple go to Monte Carlo for their honeymoon, and once there the husband is bitten by the gambling bug, wins a fortune and decides to use his winnings to buy shares in the company he works for and get rid of his boss.
Aka: LOSER TAKES ALL
DRA   81 min   VIDrel: 20VIS/SONOP   V
Boa: novel Loser Takes All by Graham Greene.

MONEY TRAIN **
Joseph Ruben USA
18
1995
Wesley Snipes, Woody Harrelson, Robert Blake, Jennifer Lopez, Chris Cooper, Joe Grifasi, Scott Sowers, Skipp Suddeth, Vincent Laresca, Nelson Vasquez, Aida Turturro, Vincent Patrick, Alvaleta Guess, Vinny Pastore, David Tawil
Two foster brothers who are constantly at loggerheads work as transit cops for the New York subway system and are assigned to protect the train that is used to transport the cash collected from all the ticket offices. However, while performing their duties, they also dream of robbing the train themselves. A solidly made action-comedy with no shortage of thrills; if only the plotting and characterisations had been a bit fresher.
A/AD   110 min cC   VIDrel: 20VIS/SONOP;
ENCORE/COLUM (LV only)   V/sur LV

MONKEY BUSINESS **
Howard Hawks USA
U
1952
Cary Grant, Ginger Rogers, Charles Coburn, Marilyn Monroe, Hugh Marlow, Henri Letondal, Robert Cornthwaite, Larry Keating, Douglas Spencer, Esther Dale, George Winslow, Emmett Lynn, Jerry Sheldon, Joseph Mell, Olan Soule
A research chemist invents a rejuvenation drug, which a laboratory chimpanzee accidentally pours into the water cooler, so that he, his wife and his boss all suffer from its effects. A laboured comedy, with plenty of star talent, but disappointingly few laughs. Written by Ben Hecht, Charles Lederer and I.A.L. Diamond.
Aka: DARLING I AM GROWING YOUNGER
COM   93 min (ort 97 min) B/W   VIDrel: 20TH/TECH   V

MONKEY SHINES: AN EXPERIMENT IN FEAR **   18
George A. Romero USA
1988
Jason Beghe, John Pankow, Melanie Parker, Kate McNeil, Joyce Van Patten, Christine Forrest, Stephen Root, Stanley Tucci, Janine Turner, William Newman, Tudi Wiggins, Tom Quinn, Chuck Baker, Patricia Tallman, David Early
Written by Romero, this horror-thriller concerns a young man who is confined to a wheelchair as a quadriplegic, following a serious accident. A super-intelligent monkey (it has been injected with human brain cells) is brought in as a companion, and begins to anticipate his every wish, but this includes a desire for revenge against those he blames for his predicament. A strange blend of horror and black humour. Music is by David Shire.
Aka: EXPERIMENT IN FEAR, AN
HOR   108 min (ort 115 min)   VIDrel: VISVID/POLYREC
L/A V
Boa: novel by Michael Stewart.

MONKEY TROUBLE **
Franco Amurri USA
U
1994
Harvey Keitel, Mimi Rogers, Thora Birch, Christopher McDonald, Adrian Johnson, Julian Johnson, Kevin Scannell, Alison Elliott, Robert Miranda, Victor Argo, Remy Ryan, Adam Lavorgna, Jo Champa, John Lafayette plus "Finster"
A young girl who is forbidden by her parents to have any pets, makes friends with a monkey belonging to a gypsy organ-grinder, and eventually provides a refuge for it when it runs away. This leads to all sorts of adventures but a happy outcome is never in doubt in this bright and entirely predictable confection. See also DUNSTON CHECKS IN.
COM   92 min (ort 96 min)   VIDrel: EIV/SONOP   V/sur

MONOLITH **
John Eyres USA
18
1993
Louis Gossett Jr, John Hurt, Lindsay Frost, Bill Paxton, Paul Ganus, Musetta Vander, Andrew Lamond, Mark Phelan, Alex Gaona, Angela Gordon, Boris Kurtonog, Jennifer Naud, Steve Barbo, Todd Jeffires, Bill Woodbridge, David St. James
Two L.A. cops have to cope with the havoc caused by a powerful alien being that is able to possess any human at will and also bring about the destruction of all mankind. The intriguing premise that starts off this movie promises much, but the movie soon gets bogged down in the usual "serial killer on the loose"

plotting, plus the obligatory conspiracy in high places sub-plot.
FAN   91 min (ort 96 min)   VIDrel: FIRST/SONOP   V/sur

MONSIEUR HIRE ***
Patrice Leconte FRANCE
15
1989
Michel Blanc, Sandrine Bonnaire, Andre Wilms, Luc Thuiller, Eric Berenger
A restrained and wisely underplayed adaptation of the original novel, whose central figure is a voyeuristic middle-aged tailor who becomes the chief suspect following the murder of a young girl. He spends much of his time spying through his window at a young girl who lives opposite, and a strange and touching relationship develops; the two disparate strands of this drama being eventually brought together in an absorbing and unusual tale.
DRA   80 min   VIDrel: BRAVE/SONOP L/A   V
Boa: novel Les Fiancailles De M. Hire by Georges Simenon.

MONSIEUR HULOT: MON ONCLE ***
Jacques Tati FRANCE
U
1956
Jacques Tati, Jean-Pierre Zola, Adrienne Servantie, Yvonne Arnaud, Alain Becourt, Lucien Fregis, Betty Schneider, Dominique Marie, J.F. Maritial, Andre Dino, Claude Badolle, Nicolas Bataille, Regis Fontenay, Denise Peronne
An accident-prone uncle visits his nephew at his ultra-modern gadget-ridden home, and causes havoc. This second Tati excursion suffers badly from its lack of dialogue and depends rather too heavily on a nearly continuous succession of sight gags. For all that, it is still well worth a look. Followed by MONSIEUR HULOT: PLAYTIME. AA: Foreign.
Aka: MON ONCLE; MY UNCLE; MY UNCLE MISTER HULOT
COM   104 min (ort 116 min)   VIDrel: CONNO/RTM   V

MONSIEUR HULOT'S HOLIDAY ***
Jacques Tati FRANCE
U
1953
Jacques Tati, Nathalie Pascaud, Michelle Rolla, Valentine Camax, Louis Perrault, Andre Dubois, Valentine Camax, Lucien Fregis, Marguerite Gerard, Rene Lacourt, Raymond Carl, Jean-Pierre Zola, Michele Brabo, Georges Adlin
An accident-prone bachelor reduces a seaside resort to chaos in the first Tati comedy (rather charmingly done in the style of a silent) to introduce this character. Beautifully timed routines make this a memorable work. Followed by MONSIEUR HULOT: MON ONCLE.
Aka: LES VACANCES DE MONSIEUR HULOT; MISTER HULOT'S HOLIDAY
COM   83 min (ort 114 min) B/W   VIDrel: CONNO/RTM
V

MONSIEUR HULOT: PLAYTIME ****
Jacques Tati FRANCE
U
1968
Jacques Tati, Barbara Dennek, Henri Piccoli, Jacqueline Lecomte, Valerie Camille, France Romilly, Leon Doyen, Jack Gautier, France Delahalle, Laure Paillete, Colette Proust, Erika Dentler, Yvette Ducreux, Rita Maiden
Tati's famous bumbling, fumbling character, Monsieur Hulot, wanders through a Paris he does not recognise, as he attempts to keep an appointment in this very Gallic assault on the confusing complexities of modern civilisation. Tati's understanding of the possibilities inherent in a 70 mm screen (largely lost on TV) is remarkable. Followed by MONSIEUR HULOT: TRAFFIC.
Aka: PLAYTIME
COM   115 min (ort 155 min) dubbed
VIDrel: CONNO/RTM L/A   V

MONSIEUR HULOT: TRAFFIC **
Jacques Tati/Bert Haanstra FRANCE/HOLLAND/
ITALY
15
1972
Jacques Tati, Maria Kimberly, Marcel Fravel, Honore Bostel, Tony Kneppers, Francois Maisongrosse, Franco Ressel, Mario Zanuelli
Our hero transports his new and innovative car model to a motor show and encounters various problems. A mildly amusing outing that tends to meander too much for its own good. This was the fourth and concluding film in the "Monsieur Hulot" series and the last film for Tati (excluding the dreary TV film "Parade"). Written by Tati and Jacques Legrange.
Aka: TRAFIC; TRAFFIC
COM   89 min (ort 96 min)   VIDrel: L/A   V

MONSIGNOR *
Frank Perry USA
15
1982
Christopher Reeve, Genevieve Bujold, Fernando Rey, Jason Miller, Joe Cortese, Adolfo Celi, Leonardo Cimino, Tomas Milian, Robert J. Prosky

Long, boring tale of an American Army chaplain, who rises to a position of great power within the Vatican by apparently breaking every moral rule in the book, including having an affair with a nun. An affront to both Catholics and film-lovers in equal measure.
DRA   116 min (ort 121 min)   VIDrel: 20TH/TECH L/A   V
Boa: novel by Jack Alain Leger

MONSTER CITY **   18
JAPAN
The man responsible for getting Tokyo rebuilt after it was devastated by earthquakes now has to deal with demons who are out to get him as they make ready for a "Day of Resurrection". Another standard Japanese Manga offering, full of the usual mayhem, destruction and supernatural battles.
ANIM   77 min dubbed   VIDrel: MANGA/SONOP   V/sh

MONSTER FROM GREEN HELL *   PG
Kenenth Crane USA   1958
Jim Davis, Barbara Turner, Robert E. Griffin, Eduardo Cianelli, Joel Fluellen, Vladimir Sokoloff, Tim Huntley, Frederic Potler, LaVerne Jones
This time, the world is threatened by giant wasps, the result of laboratory experiments in orbit where normal-sized insects were contaminated by radiation. Don't get stung by this one.
HOR   64 min (ort 71 min) B/W   VIDrel: FIRC/RTM   V

MONSTER FROM THE OCEAN FLOOR, THE *   (12)
Wyatt Ordung USA   1958
Anne Kimball, Stuart Wade, Wyott Ordung, Dick Pinner, Jack Hayes, Inez Palang, David Garcia
Abysmal first-time effort by legendary cheapie producer Roger Corman has a squid-like creature threatening a deep-sea diver. A film of curiosity value only.
Aka: IT STALKED THE OCEAN FLOOR; MONSTER MAKER
HOR   66 min B/W   SATrel: BRAVO MOVIES

MONSTER IN THE CLOSET, THE *   15
Bob Dahlin USA   1983 (released 1986)
Donald Grant, Denise DuBarry, Claude Akins, Howard Duff, Donald Moffat, Paul Dooley, John Carradine, Stella Stevens, Henry Gibson, Frank Ashmore, Paul Walker, Jesse White, Kevin Peter Hall, Stacy Ferguson, Arthur Berggen
A fine cast can do little with this silly 1950s horror spoof, in which an indestructible monster lurking in a closet, causes a national emergency as it goes on the rampage. Written by Dahlin. See also CAMERON'S CLOSET.
HOR   86 min (ort 100 min)   VIDrel: TROMA/RTM   V

MONSTER SQUAD, THE **   15
Fred Dekker USA   1987
Andre Gower, Duncan Regehr, Stan Shaw, Stephen Macht, Tom Noonan, Robby Kiger, Brent Chalem, Ryan Lambert, Ashley Bank
Some kids get the thrill of their lives when their favourite monsters make an unexpected visit to their home town, and all to find a magic amulet, desperately required by Dracula. A likeable homage to this genre, that remains a mite too insipid to really get going, although the ending is suitably exciting (courtesy of Richard Edlund's special effects).
COM   78 min (ort 82 min)
VIDrel: 4-FRONT/POLYREC/BRAVE   V/sur

MONTALVO AND THE CHILD **   (PG)
Claude Mourieras FRANCE   1988
Mathilde Altaraz, Christophe Delachaux, Robert Seyfried, Jean Claude Gallotta, Michel Ducret, Marceline Bertolot
Based on the dance "Pandora" choreographed by Jean Claude Gallotta, this is an unusual coming-of-age tale, that explores the lives of both characters by way of several linked short stories. Blending both realism and highly stylised elements, the simple story concerns a child's growing loss of illusions, chiefly revolving around his hero, Montalvo. With little dialogue, the film uses dance to illustrate the themes explored. An odd hybrid movie, exuberant but contrived.
Aka: MONTALVO ET L'ENFANT
DRA   76 min B/W   CINrel

MONTANA **   PG
Ray Enright USA   1950
Errol Flynn, Alexis Smith, S.Z. "Cuddles" Sakall, Douglas Kennedy, James Brown, Ian MacDonald, Charles Irwin, Paul E. Burns, Tudor Owens, Nacho Galindo, Lane Chandler, Lester Matthews, Monte Blue, Billy Vincent, Warren Jackson

An Australian sheep farmer comes to Montana in search of grazing for his hers but incurs the resentment of local cattle ranchers, headed by a wealthy woman. A range war almost erupts after a sheep farmer is murdered but love eventually saves the day, in this cliched and unimpressive oater, the main attraction of which is Karl Freund's splendid Technicolor photography and a cast of reliable stalwarts.
WES   76 min   VIDrel: WHV   V
Boa: story by Ernest Haycox.

MONTANA ***   PG
William A. Graham USA   1990
Richard Crenna, Gena Rowlands, Lea Thompson, Justin Deas, Scott Coffrey, Elizabeth Berridge, Darren Dalton, Michael Madsen, Jim Bishop, Dana Anderson, Kurt Bushnell, Dean Folkvord, A.J. Kallan, Tammy Lewis, Dean Norris
A modern-day Western in which Crenna and Rowlands play a squabbling couple who cannot agree on what to do with the family ranch, which they've always managed to keep out of the hands of creditors. When Crenna finds that he could sell it to an oil drilling company, he finds that his wife would prefer to keep it running. An appealing character study that's all about old-fashioned family values, its lack of a strong narrative is no handicap.
WES   87 min (ort 100 min) mCab   VIDrel: MGM/WHV L/A   V

MONTE CARLO OR BUST! *   PG
Ken Annakin FRANCE/ITALY/UK/USA   1969
Tony Curtis, Terry-Thomas, Peter Cook, Dudley Moore, Eric Sykes, Susan Hampshire, Gert Frobe, Jack Hawkins, Bourvil (Andre Raimbourg), Lando Buzzanca, Walter Chiari, Mireille Darc, Marie Dubois, Nicoletta Machiavelli
Limply directed and wildly unfunny account of a car journey through Europe to Monte Carlo. Set in the 1920s and misfiring on all cylinders.
Aka: QUEI TEMERARI SULLE LORO PAZZE, SCATENATE, SCALCINATE CARRIOLE; THOSE DARING YOUNG MEN IN THEIR JAUNTY JALOPIES
COM   132 min   VIDrel: 4-FRONT/POLYREC/CIC   V/h

MONTE CARLO: PARTS 1 AND 2 **   PG
Anthony Page USA   1986
Joan Collins, George Hamilton, Lauren Hutton, Malcolm McDowell, Robert Carradine, Leslie Phillips, Lisa Eilbacher, Philip Madoc, Peter Vaughn, Henri Garcin, Clement Harari, Rainer Harold, Steve Kalfa, David Quilter
Set in Monaco in 1940, this tells of a glamorous band of refugees, for whom the principality is a haven from the harsh realities of WW2. A beautiful international cabaret singer arrives, who leads a secret life as a double agent working for the Allies. Soon involved in a love affair with a refugee, she is arrested by the Gestapo when they learn of her exploits. A kind of big, glossy and superficial follow-on to SINS.
DRA   180 min (2 cassettes – ort 240 min) mTV
VIDrel: 4-FRONT/POLYREC/NWV   V
Boa: novel by Stephen Sheppard.

MONTENEGRO **   18
Dusan Makavejev SWEDEN/UK   1981
Susan Anspach, Erland Josephson, Per Oscarsson, John Zacharias, Svetozar Cvetkovic, Patricia Gelin, Bora Todorovic, Jamie Marsh, Marianne Jacobi, Lisbeth Zachrisson, Marina Zindahl, Nikola Janic, Lasse Aberg, Dragan Ilic, Jan Nygren
A bored American housewife living in Sweden, takes up with Yugoslavian workers at a noisy bar which they frequent, eventually finding herself a lover, but ultimately murdering him. A reasonably entertaining drama, that tries to be a good deal more meaningful than its script allows.
Aka: MONTENEGRO; OR PIGS AND PEARLS
DRA   92 min (ort 96 min)   VIDrel: ARROW/RTM   V

MONTH BY THE LAKE, A ***   PG
John Irvin UK/USA   1994
Vanessa Redgrave, Edward Fox, Uma Thurman, Alida Valli, Carlo Cartier, Alessandro Gassman, Natalia Bizzi, Frances Nacman, Paolo Lombardi, Riccardo Rossi, Sonia Martinelli, Veronica Wells, Carlotta Bresciani, Bianca Tognocchi
A spirited English spinster and stuffy army officer meet while on holiday on Lake Cosmo and, despite an initial mutual dislike, a love affair develops. Quite a good romantic comedy-drama, that though hardly the stuff of great drama, has a gentleness and endearing warmth about it.
DRA   88 min (ort 91 min) cC   VIDrel: TOUCH/TECH   V/sur
Boa: novella by H.E. Bates.

MONTH IN THE COUNTRY, A ***

Pat O'Connor UK
Colin Firth, Kenneth Branagh, Natasha Richardson, Patrick Malahide, Tony Haygarth, Jim Carter, Richard Vernon, Tom Baker, Vicki Arundale, Martin O'Neil, Ellen O'Brien, Elizabeth Anson, Barbara Marten, Kenneth Kitson

PG
1987

Two young men, both worn out veterans of the WW1 trenches, meet when they come to a church in a little Yorkshire village. One slowly uncovers a medieval fresco inside, whilst the other proceeds to excavate Saxon relics in the churchyard. An engrossing study, charting the slow recovery of two men from the rigours of war, and their developing friendships with the local people.
DRA   92 min (ort 96 min)   VIDrel: MGM/WHV   V/sh
Boa: novel by J.R. Carr.

MONTH OF SUNDAYS, A ***

Allan Kroeker USA
Hume Cronyn, Vincent Gardenia, Michelle Scarabelli, Tandy Cronyn, Esther Rolle, Barry Flatman, Edward McGibbin, Phil Jarrett, Murray Westgate, Anna Ferguson, Aaron Schwartz, David Harvey, Frummie Blatt

PG
1989

A poignant look at the problems of old age as seen in the lives of two friends who are spending their final days in a retirement home. The story is simple and lacking in action, but the refreshing lack of sentimentality makes for an absorbing drama whose tone is lightened by a few comic touches. Excellent performances abound, but Scarabelli as a warm and sympathetic nurse is especially worthy of note.
Aka: AGE OLD FRIENDS
DRA   98 min mCab   VIDrel: FUTUR L/A   V
Boa: play by Bob Larbey.

MONTY PYTHON AND THE HOLY GRAIL ***

Terry Gilliam/Terry Jones UK
Graham Chapman, John Cleese, Terry Gilliam, Eric Idle, Terry Jones, Michael Palin, Carole Cleveland, Connie Booth, Neil Innes, John Young, Bee Duffell, Rita Davies, Sally Kinghorn, Avril Stewart

15
1975

The first feature from the team who produced the TV series "Monty Python's Flying Circus". This is a surreal look at the legend of the Knights of the Round Table and their quest for the Holy Grail. No more than a linked series of sketches of uneven quality, but held together by a good feel for the period. Followed by MONTY PYTHON'S LIFE OF BRIAN.
COM   89 min   VIDrel: 20TH/FOXVID; ENCORE (LV only)
V/h LV

MONTY PYTHON'S LIFE OF BRIAN ***

Terry Jones UK
John Cleese, Graham Chapman, Eric Idle, Terry Gilliam, Terry Jones, Michael Palin, Carol Cleveland, Neil Innes, Spike Milligan, Ken Colley, Gwen Taylor, Terrence Bayler, Charles McKeown, Sue Jones, Bernard McKenna, Chris Langham

15
1979

Intermittently amusing satire on a failure whose life runs parallel to that of Christ, to the extent of even having a following (of sorts) of his own. When released, it caused considerable offence among some religious groups, and was even banned in Norway. Certainly not one for the devout. Written by John Cleese and the other members of the "Monty Python" team, and their most sustained effort to date. Followed by MONTY PYTHON'S MEANING OF LIFE.
Aka: LIFE OF BRIAN, THE
COM   89 min (ort 93 min)   VIDrel: CIC/SONOP   V/sh

MONTY PYTHON'S MEANING OF LIFE ***

Terry Jones UK
Graham Chapman, John Cleese, Terry Gilliam, Eric Idle, Terry Jones, Michael Palin, Carol Cleveland, Simon Jones, Patricia Quinn, Judy Loe, Mark Holmes, Andrew MacLachlan, Valerie Whittington, Jennifer Franks, Angela Mann

15
1983

The third film spin-off from the surrealistic TV comedy series "Monty Python's Flying Circus", that began with MONTY PYTHON AND THE HOLY GRAIL (if we exclude AND NOW FOR SOMETHING COMPLETELY DIFFERENT). This cheerfully chaotic look at life and death takes shots at a whole range of subjects, and has some remarkable (if nauseating) sequences. Best enjoyed on an empty stomach.
Aka: MEANING OF LIFE, THE
COM   86 min (ort 107 min)   VIDrel: CIC/SONOP   V/sur

MOOD INDIGO: MIND OF A KILLER **

John Patterson USA
Tim Matheson, Alberta Watson, Giancarlo Esposito, Claudia

15
1992

Christian, Robert Wisden, Tom McBeath, William S. Taylor, Forbes Angus, Roger Barnes, Matt Bennett, John Milton Branton, Sarah Chalke, Glynis Davis, Lovie Eli
A psychologist studies the case of a woman accused of murdering her husband, and finds that matters are not quite as clear-cut as they first appeared. Average.
DRA   86 min (ort 90 min)   mTV   VIDrel: CIC/SONOP   V

MOON 44 **

Roland Emmerich WEST GERMANY
Michael Pare, Malcolm McDowell, Lisa Eichhorn, Brian Thompson, Leon Rippy, Stephen Geoffreys, Roscoe Lee Browne, Dean Devlin, Mechmed Yilmaz, Jochen Nickel, John March, Drew Lucas, David Williamson, Calvin Burke, Andy Howarth

15
1989

High-budget, low-brow SF thriller set in a future that's controlled by interplanetary mining corporations who provide the chemicals that are now the only source of fuel on Earth. An undercover cop posing as a helicopter pilot is sent to stop ore freighters being hijacked, but the impressive special effects of Volker Engel fail to maintain interest in this elaborate Europroduction that's filmed in Germany.
FAN   95 min   VIDrel: MED/POLYREC L/A   V/h

MOON IN SCORPIO **

Gary Graver USA
Britt Ekland, John Philip Law, William Smith, Lewis Van Bergen, April Wayne, Jillian Kesner, Bruno Marcotulli, Ken Smolka, Thomas Bloom, James Booth, Donna Kei Benz, Don Scribner

18
1987

A woman goes on a honeymoon cruise with her Vietnam veteran husband but discovers that two of the passengers are former comrades, and that all three are haunted by memories of wartime atrocities they committed. After several gruesome murders are committed, the passengers become gripped with fear, in this competent thriller.
THR   83 min (ort 90 min)   VIDrel: MGM L/A   V

MOON IN THE GUTTER, THE **

Jean-Jacques Beineix FRANCE
Nastassia Kinski, Gerard Depardieu, Victoria Abril, Vittorio Mezzogiorno, Dominique Pinon, Bertice Reading, Gabriel Monnet, Milena Vukotic, Bernard Farcy, Anne-Marie Coffient, Katia Berger, Jacques Herlin, Rudo Alberti, Rosa Furneto

18
1983

A man seeks out the rapist responsible for his sister's suicide, and becomes involved with a mysterious femme fatale. A highly stylised film, whose flashy camerawork and visual effects totally swamp the plot. Made by the director of DIVA, a film of similar qualities.
Aka: LA LUNE DANS LE CANIVEAU
DRA   132 min wScrn   VIDrel: ARTIF/20TH   V/h

MOON OVER MIAMI ***

Walter Lang USA
Don Ameche, Betty Grable, Carole Landis, Jack Haley, Charlotte Greenwood, Robert Cummings, Cobina Wright Jr, Robert Greig

U
1941

Two sisters leave Texas and head for Florida in the hope of landing rich husbands, in this enjoyable musical remake of "Three Blind Mice". Not a great musical, but tuneful and nicely put together. Songs include "You Started Something", probably the most memorable number. Remade as "Three Little Girls In Blue" and HOW TO MARRY A MILLIONAIRE.
MUS   88 min (ort 91 min)   VIDrel: 20TH/TECH   V
Boa: play Three Blind Mice by Stephen Powys.

MOON OVER PARADOR **

Paul Mazursky USA
Richard Dreyfuss, Raul Julia, Sonia Braga, Jonathan Winters, Michael Greene, Polly Holliday, Charo, Marianne Sagebrecht, Sammy Davis Jr, Dick Cavett, Edward Asner, Ike Pappas

15
1988

An unemployed American actor is unwillingly persuaded to impersonate a recently-deceased Latin American dictator, and soon finds himself enjoying his role (and the late dictator's sexy mistress). Extremely likeable but hardly the stuff of classic comedy, with a few too many gags that misfire. Co-written by Mazursky (who has a cameo in drag).
COM   99 min (ort 105 min)   VIDrel: CIC/SONOP   V/sur

MOON SPINNERS, THE **

James Neilson UK/USA
Hayley Mills, Eli Wallach, Joan Greenwood, Pola Negri, Peter McEnery, John Le Mesurier, Irene Papas, Sheila Hancock, Michael Davis, Paul Stassino, George Pastell, Andre Morell, Tutte Lemkow, Steve Plytas, Harry Tardios, Pamela Barrie

U
1964

A young girl on holiday in Crete with her aunt, gets involved

in helping a young man track down a gang of jewel thieves. A colourful but muddled children's thriller, that goes on rather too long.
JUV 114 min (ort 118 min) VIDrel: WDV/TECH L/A V
Boa: novel by Mary Stewart.

MOON WARRIORS *** 18
Samo Hung Kam-Bo HONG KONG 1992
Andy Lau Tak-Wah, Anita Mui Yim-Fong, Maggie Cheung Man-Yuk, Kenny Bee (Chung Chun-To)
A fisherman skilled in the martial arts gets involved with a prince who has been driven off his throne by his evil brother. This marks the start of some incredible adventures as well as a romance with a spoiled princess, who is also loved by the prince's bodyguard.
FAN 82 min (ort 90 min) wScrn VIDrel: MADE/RTM V

MOONDIAL ** PG
Colin Cant UK 1990
Helen Cresswell, Siri Neal
A young girl goes to stay with her aunt who lives in an old gate-house, and a series of strange events begins with her arrival. First shown on the BBC.
JUV 112 min mTV VIDrel: PARADOX/TOTAL V/h
Boa: novel by Helen Cresswell.

MOONLIGHT AND VALENTINO *** 15
David Anspaugh USA 1995
Elizabeth Perkins, Gwyneth Paltrow, Jon Bon Jovi, Kathleen Turner, Whoopi Goldberg, Jeremy Sisto, Josef Sommer, Shadia Simmons, Erica Luttrell, Matthew Koller, Scott Wickware, Kelli Fox, Harrison Liu, Wayne Lam, Ken Wong
Unashamed weepie in which a woman is widowed when her husband is killed in a car crash, and can get little comfort from her family and friends, who all seem to have insecurities of their own. However, the arrival of a handsome young decorator moves things on somewhat. Scripted by Simon, this is a pleasing drama that manages to give even the male stars a few good lines, something of a rarity in films of this kind.
DRA 104 min VIDrel: POLFIL V
Boa: play by Ellen Simon.

MOONLIGHTING *** PG
Jerzy Skolimowski UK 1982
Jeremy Irons, Eugene Lipinski, Jiri Stanislaw, Eugeniusz Haczkiewicz, Dorothy Zienciowska, Edward Arthur, Denis Holmes, Renu Setna, David Calder, Judy Gridley, Claire Toeman, Catherine Harding, David Squire, Mike Sarne
Four Polish workers come to England to renovate a house for a rich ex-compatriot. Only their foreman speaks English, and he keeps them from learning about the declaration of martial law during the military takeover in their homeland. A film that is perceptive, wry and utterly cynical. The script is by Jerzy Haczkiewicz. Followed by "Success Is The Best Revenge".
DRA 93 min (ort 97 min) mTV
VIDrel: POLY/POLYREC/BRAVE L/A V

MOONLIGHTING: THE ORIGINAL TV MOVIE ** PG
Robert Butler USA 1985
Cybill Shepherd, Bruce Willis, Allyce Beasley, Robert Ellenstein, Jim Mackrell, James Karen, Rebecca Stanley, Dennis Lipscomb, John Medici, Dennis Stewart
A disjointed, overlong pilot for a popular American TV detective series. A former model is swindled by her business manager, and is reluctantly forced into a partnership with the head of the tax-loss agency she finds her manager has foisted on her. Their first case involves a confused hunt for Nazi loot, in the form of a cache of diamonds. A continual emphasis on the interaction between the two characters takes the place of a decent plot.
Aka: MOONLIGHTING
A/AD 93 min (ort 97 min) mTV VIDrel: FUTUR L/A V

MOONRAKER ** PG
Lewis Gilbert UK 1979
Roger Moore, Lois Chiles, Richard Kiel, Michael Lonsdale, Corinne Clery, Bernard Lee, Desmond Llewelyn, Lois Maxwell, Geoffrey Keen, Emily Bolton, Toshiro Suga, Blanche Ravalec, Jean-Pierre Castaldi, Leila Shenna, Walter Gotell
Another James Bond adventure, with 007 investigating cases of disappearing space shuttles, and discovering yet another baddie out to control the world and replace its population with his own genetically engineered race. Despite a much enlarged budget

and the usual gadgetry, this feels like a tired and mechanical rehash of all the previous films. Number eleven in a series that was really beginning to show its age. Followed by FOR YOUR EYES ONLY.
A/AD 121 min (ort 126 min) wScrn VIDrel: MGM/WHV V/dm V/s
Boa: novel by Ian Fleming.

MOONSTRUCK **** PG
Norman Jewison USA 1987
Cher, Nicholas Cage, Olympia Dukakis, Vincent Gardenia, Danny Aiello, Julie Bovasso, John Mahoney, Louis Guss, Feodor Chaliapin, Anita Gillette, Nadia Despotovich, Joe Grifasi, Gina DeAngelis, Robin Bartlett, Helen Hanft
Set in Brooklyn, this tells of an Italian-American woman who falls madly in love with the brother of the man she is engaged to marry. A refreshing and sharply observed Fellini-style look at life, transported to the New World. AA: Actress (Cher), S. Actress (Dukakis), Screen/orig (John Patrick Shanley).
COM 98 min (ort 103 min) VIDrel: MGM/WHV L/A V

MOONWALKER ** PG
Jerry Kramer/Colin Chilvers USA 1988
Michael Jackson, Sean Lennon, Joe Pesci, Brandon Adams, Kellie Parker
A musical fantasy vehicle for Jackson, that rather clumsily inter-cuts some of his best numbers with a series of bizarre images, that purport to show a struggle between good and evil. Really nothing more than a crudely exploitative pop video/fairytale that's worth hearing but not worth seeing.
Aka: MICHAEL JACKSON'S MOONWALKER
MUS 89 min VIDrel: POLY/POLYREC L/A V/sh

MORE THE MERRIER, THE *** U
George Stevens USA 1943
Jean Arthur, Joel McCrea, Charles Coburn, Richard Gaines, Bruce Bennett, Frank Sully, Clyde Fillimore, Stanley Clements, Don Douglas, Ann Savage, Don Barclay, Grady Sutton, Sugar Geise, Shirley Patterson, Ann Doran, Mary Treen
During WW2, a Washington girl does her bit to relieve the housing shortage by sub-letting to an older man, who takes pity on her loneliness and decides to fix her up with a boyfriend by sub-letting his half of the apartment. This very slight story receives a marvellous treatment, with fine acting and delicate and sensitive direction. Remade in 1966 as "Walk, Don't Run". AA. S. Actor (Coburn).
COM 100 min (ort 104 min) B/W
VIDrel: COLUM/SONOP V/dm

MORECOMBE AND WISE: NIGHT TRAIN TO MURDER * PG
Joe McGrath UK 1984
Eric Morecombe, Ernie Wise, Lysette Anthony, Fulton MacKay, Kenneth Haigh
Comic-duo Morecombe and Wise star in this parody of some of the writings of great thriller writers such as Edgar Wallace, Raymond Chandler and Agatha Christie, and get involved in a 1940s murder mystery when they decide to take care of Eric's niece.
Aka: NIGHT TRAIN TO MURDER
COM 70 min mTV VIDrel: THAMES/DISC V

MORECOMBE AND WISE: THAT RIVIERA TOUCH * PG
Cliff Owen UK 1966
Eric Morecombe, Ernie Wise, Suzanne Lloyd, Paul Stassino, Armand Mestral, George Eugeniou, George Pastell, Peter Jeffrey, Gerald Lawson, Michael Forest, Paul Danquah, Francis Matthews, George Moon
Dreadful, utterly dreadful. Our comedy duo of TV fame are now tourists on the Riviera, and become involved with jewel thieves. British comedy at its absolute rock bottom worst.
Aka: THAT RIVIERA TOUCH
COM 94 min (ort 98 min) VIDrel: VCC/DISC V

MORECOMBE AND WISE: THE INTELLIGENCE MEN * U
Robert Asher UK 1965
Eric Morecombe, Ernie Wise, William Franklyn, April Olrich, Richard Vernon, Gloria Paul, David Lodge, Jacqueline Jones, Terence Alexander, Francis Matthews, Warren Mitchell, Brian Oulton, Peter Bull, Joe Melia, Tutte Lemkow
Comedy duo Morecombe and Wise are here cast as two incom-

petent spies who go from mishap to mishap, eventually saving a Russian ballerina from assassination, with the help of a waitress. One of those ghastly British attempts at comedy that almost makes one squirm with embarrassment.
Aka: INTELLIGENCE MEN, THE; SPYLARKS
COM   99 min (ort 104 min)   VIDrel: VCC/DISC   V

MORECOMBE AND WISE: THE MAGNIFICENT TWO *
Cliff Owen UK   PG
                                                 1967
Eric Morecombe, Ernie Wise, Margit Saad, Cecil Parker, Isobel Black, Martin Benson, Virgilio Teixeira, Michael Godfrey, Sue Sylvaine, Andreas Malandrinos, Victor Maddern
Third, and thankfully last, film in the 1960s series employing the talents of the popular TV comedy duo. This time round, one of a pair of fairly useless travelling salesmen is mistaken for the leader of a coup in a Latin American country. Grade zero on the laugh scale.
Aka: MAGNIFICENT TWO, THE
COM   91 min (ort 100 min)   VIDrel: VCC/DISC   V

MORGAN: A SUITABLE CASE FOR TREATMENT **
Karel Reisz UK   PG
                                                 1966
Vanessa Redgrave, David Warner, Irene Handl, Robert Stephens, Newton Blick, Nan Munro, Bernard Bresslaw, Arthur Mullard, Graham Crowden, John Rae, Peter Collingwood, Peter Cellier, Angus McKay, John Garrie, Marvis Edwards
An artist refuses to accept that his wife is divorcing him, and does everything he can to win her back. Some very funny ideas swim about aimlessly in what is neither a comedy nor a satire. Screenplay is by David Mercer from his own play.
Aka: MORGAN!; SUITABLE CASE FOR TREATMENT, A
COM   93 min (ort 97 min) B/W   VIDrel: LUMI/SPEAR   V
Boa: TV play A Suitable Case for Treatment by David Mercer.

MORNING AFTER, THE *
Sidney Lumet USA   15
                                                 1986
Jane Fonda, Jeff Bridges, Raul Julia, Diane Salinger, Richard Foronjy, James Haake, Geoffrey Scott, Kathleen Wilhoite, Don Hood, Fran Bennett, Michael Flanagan, Bruce Vilanch, Michael Prince, Frances Bergen, Bob Minor
A fading actress with a drink problem wakes up to find a corpse on her bed, and is unable to understand how it got there. A star cast are saddled with a heavy and opaque script, that never develops beyond a few meaningless exchanges.
THR   102 min (ort 104 min)   VIDrel: GUILD/SONOP L/A   V/sh

MORTAL FEAR **
Larry Shaw USA   15
                                                 1994
Joanna Kerns, Gregory Harrison, Max Gail, Tobin Bell, Robert Englund, Judith Chapman, Katherina LaNasa, Rebecca Schull, Leslie Ackerman, Suzanne Barnes, Bus Riley, Jerome Butler, Amanda Bauer, Michael Robert Berger, Donre Sampson
The head of the medical staff at a clinic becomes increasingly concerned at the high mortality rate among the clinic's patients, and in her attempts to get at the truth, finds herself led to seedy strip club and a web of greed and corruption. As ever, there is a sinister fellow physician who has a sinister role in these proceedings. Adequately plotted but over-derivative, it inevitably puts one in mind of the earlier (and better) medical thriller COMA.
Aka: ROBIN COOK'S MORTAL FEAR
THR   86 min   VIDrel: ODY/SONOP   V/sh
Boa: novel by Robin Cook.

MORTAL KOMBAT **
Paul Andersson USA   15
                                                 1995
Linden Ashby, Cary-Hiryuki-Tagaawa, Robin Shou, Christopher Lambert, Talisa Soto, Bridgette Wilson, Trevor Goddard, Chris Casamassa, Francois Petit, Keith H. Cooke, Hakim Alston, Kenneth Edwards, John Fujioka, Daniel Haggard
Live-action version of a popular video game that has three powerful warriors fighting an evil wizard whose plans (as ever) involve an attempt to take over the world, something he will be able to do should he win just one more bout. However, he is restricted to fighting under the rules of the title form of combat. The special effects are good and there is plenty of action, but this is a film for the kids (despite the 15 certification) if ever there was one.
Aka: MORTAL KOMBAT: THE MOVIE
FAN   97 min (ort 101 min) wScrn   VIDrel: FIRST/SONOP; ENCORE (LV only)   V LV

MORTAL PASSION **
Michael Switzer USA   15
Michael Dudikoff, Tim Matherson, Susan Lucci
An erotic thriller that tells of a beautiful woman whose dinner date with a fashion photographer leads to her becoming embroiled in a murder.
THR   90 min   VIDrel: GENESIS   V

MORTAL PASSIONS ***
Andrew Lane USA   18
                                                 1989
Zach Galligan, Krista Errickson, Michael Bowen, Sheila Kelly, Luca Bercovici, David Warner
A cuckolded husband teams up with the girlfriend of a missing man to solve the mystery of his disappearance, and they eventually learn that both the man and his brother were his wife's lovers. Despite the flaws of the weakly plotted and developed script, this well acted thriller develops a good deal of tension and has several highly charged moments.
THR   92 min (ort 96 min)   VIDrel: VIR/RCA L/A   V

MORTAL SINS **
Bradford May USA   PG 13
                                                 1992
Christopher Reeve, Roxann Biggs, Francis Guinan, Weston McMillan, Phillip R. Allen, Lisa Vullaggio, George Touliatos, Mavor Moore, Karen Kondazian, Thomas Peacocke, Blu Mankuma, Julia Satterfield, Michael Jacobucci
A Catholic priest is placed in a terrible moral dilemma when he hears the confession of a serial killer who has been preying on the women of his parish. Since he feels unable to break the sanctity of the confessional, he goes after the killer himself. A by-the-numbers thriller that proceeds along familiar lines.
Aka: LEAVE HER TO HEAVEN
THR   89 min (ort 93 min) mCab
VIDrel: POLY/POLYREC/WHV   V/h

MORTAL THOUGHTS **
Alan Rudolph USA   18
                                                 1991
Demi Moore, Bruce Willis, Glenne Headly, Harvey Keitel, John Pankow, Billie Neal, Frank Vincent, Karen Shallo, Crystal Field, Maryane Leone, Christopher Scotellaro, Doris McCarthy, Marc Tantillo, Ron J. Amodea, Kelly Ginnante
Moore co-produced this rambling mystery, which is told entirely in flashback sequences, and deals with two women. Both unhappy with their marriages, they are linked by friendship and murder, with one of the wives killing her ill-mannered and thuggish spouse. Most of the film revolves around the interrogation carried out by homicide detectives, and though the acting is uniformly excellent, the film lacks both warmth and a well-defined structure.
THR   98 min (ort 104 min)   VIDrel: VCC/DISC/COLUM   V/sur

MOSCOW ON THE HUDSON ***
Paul Mazursky USA   15
                                                 1984
Robin Williams, Maria Conchita Alonso, Cleavant Derricks, Alejandro Rey, Savely Kramarov, Elya Baskin, Yakov Smirnoff, Oleg Rudnik, Alexander Beniaminov, Ludmilla Kramarevsky, Ivo Vrzal, Natalie Iwanov, Tiger Haynes, Edye Byrde
A Russian defector in New York tries to adjust to the problems of living in a new and strange country. A nicely observed tale, that's not without the inevitable dose of sentimentality, but is redeemed by a stand-out performance from Williams (who clearly worked hard on his pronunciation), ably assisted by a fine supporting cast. Written by Mazursky and Leon Capetanos.
COM   112 min (ort 117 min)   VIDrel: VCC/RCA L/A   V/sh

MOSES THE LAWGIVER *
Gianfranco DeBosio ITALY/UK   15
                                                 1975
Burt Lancaster, Ingrid Thulin, Anthony Quayle, Irene Papas, William Lancaster, Mariangela Melato, Laurent Terzieff, Simonetta Stefanelli, Aharon Ipale, Melba Englander, Mario Ferrari, Antonio Piovanelli, Marina Berti
A lethargic and ponderous account of the life of this biblical hero, made for TV, and filmed in Israel. Screenplay is by Anthony Burgess and Vittorio Bonicelli.
Aka: LAWGIVER, THE; MOSES
DRA   136 min Cut (1 sec – ort 300 min) mTV
VIDrel: 4-FRONT/POLYREC/ITC   V

MOSQUITO COAST, THE **
Peter Weir USA   PG
                                                 1986
Harrison Ford, River Phoenix, Helen Mirren, Jadrien Steele, Andre Gregory, Conrad Roberts, Martha Plimpton, Dick O'Neill, Hilary

Gordon, Rebecca Gorden, Butterfly McQueen, Jason Alexandre, Dick O'Neill, Alice Sneed, Tiger Haynes
An idealistic inventor turns his back on modern America with all its problems, and takes his family to live in a remote part of Central America, where he attempts to create an idyllic lifestyle. Ford is quite wonderful as a narrow-minded idealist who eventually destroys himself, but the subtleties of the novel do not translate well to the screen.
DRA   113 min (ort 119 min)   VIDrel: 20TH/TECH L/A   V
Boa: novel by Paul Theroux.

MOSQUITO SQUADRON ** U
Boris Sagal UK   1968
David McCallum, Charles Gray, David Buck, Suzanne Neve, David Dundas, Dinsdale Landen, Nicky Henson, Bryan Marshall, Michael Anthony, Peggy Thorpe-Bates, Peter Copley, Vladek Sheybal, Robert Urquhart
In 1944, a crack Mosquito squadron is given the job of eliminating a secret WW2 German rocket factory that is hidden deep beneath of French chateau. But there Germans are using this building to house Allied POWs, in the belief that this will prevent any attack. The first film to explore the use of Barnes Wallace's "bouncing bomb" (see also THE DAM BUSTERS), this is a very minor piece, adequate but hardly memorable.
WAR   87 min (ort 90 min)   VIDrel: MGM/WHV   V/h

MOST BEAUTIFUL BREASTS IN THE WORLD, THE *** 15
Rainer Kaufman GERMANY   1990
Eva Kryll, Dominic Raacke
Oddball short in which a man admires a woman's pretty bust, and then finds it magically transposed to his own chest. A comic exploration of gender typing and the expectations that go with it, that makes its observations and ends. A longer work would have been nothing like as effective.
Aka: DER SCHONSTE BUSEN DER WELT
COM   15 min; 68 min (2-film cassette)
VIDrel: ELPIC/POLYREC   V/s
Boa: short story by Roland Topor. Osca: MAKING UP

MOST DANGEROUS GAME, THE *** (18)
Ernest B. Schoedsack/Irving Pichel USA   1932
Joel McCrea, Fay Wray, Leslie Banks, Robert Armstrong, Steve Clemento, Noble Johnson, Hale Hamilton
A deranged Russian count living on an isolated island is a fanatical hunter, even to the point of stalking human prey. Two unfortunates stranded there provide him with a perfect opportunity to indulge in his favourite pastime. A much-imitated, fast-moving melodrama, it was remade as "Game Of Death" and "Run For The Sun". See also DEATH RING, THE NAKED PREY, FINAL ROUND, SURVIVING THE GAME, NO EXIT, THE WOMAN HUNT and ESCAPE 2000.
Aka: HOUNDS OF ZAROFF, THE
THR   78 min B/W   TVrel
Boa: short story by Richard Connell.

MOST DESIRED MAN, THE *** 18
Sonke Wortmann GERMANY   1994
Til Schweiger, Katja Riemann, Joachim Krol, Rufus Beck, Armin Rohde, Nico van der Knapp, Antonia Lang, Martina Gedeck, Judith Reinartz, Kai Wiesinger, Horst D. Scheel, Christof Wackernagel, Martin Armknecht, Heinrich Schafmeister
A woman kicks out her two-timing boyfriend and he goes off to live with a gay friend. But when she finds out she is pregnant she tries to get him back, only to learn that he is quite happy staying where he is. Yet further misunderstandings conspire to keep them apart, despite that fact that in truth they really do love each other. A goofy engaging comedy that is totally implausible and contrived, but has masses of charm and is hard not to like.
Aka: DER BEWEGTE MANN; MAYBE... MAYBE NOT
COM   93 min (ort 98 min)   VIDrel: TART/20TH   V/sh
Boa: comics Der Bewegte Mann and Pretty Baby by Ralf Konig.

MOST TERRIBLE TIME IN MY LIFE, THE *** (18)
Hayashi Kaizo JAPAN   1993
Masdatoshi Nagase, Shiro Sano, Kiyotaka Nanbara, Yang Haitin, Hou De Jian, Akaji Maro, Shinya Tsukamoto, Joe Shishido, Haruko Wanibuchi, Kaho Minami, Mika Ohmine, Housei Kondo, Masako Miyachi, Kenji Anan, Zen Rajiwara
The story of an accident prone private eye's adventures in a spoof on Spillane's Mike Hammer (the character's name is Maiku Hama) that has echoes of 1940s film noir, and blends

black comedy and violent action quite incongruously. The plot has the detective coming up against a criminal gang composed of Chinese and Korean immigrants, and some caustic observations in this direction are perhaps an attempt to widen the film's appeal to a more intellectual audience.
Aka: WAGA JINSEI SAIAKU NO TOKI
THR   92 min B/W   CINrel

MOTH, THE ** PG
Roy Battersby UK   1996
Juliet Aubrey, Jack Davenport, David Bradley, Justine Waddell, Janet Dale, Jeremy Clyde, Judy Loe, David Howey, Sally Grey, Delena Kidd, Margareta Scott, Rupert Penry-Jones, Ian Kelly, Aedin Maloney, Michael Gunn, Mike Rossi
Three-part period drama set in 1913, that tells of the friendship that develops between a young carpenter and a strange girl, who is kept very much under the thumb of her odious and conniving family. The blossoming relationship they enjoy has unexpected consequences. Another well mounted adaptation; if only the story were not so sombre. Adapted from the novel by Gordon Hann.
DRA   150 min mTV   VIDrel: FOCUS/DISC   V
Boa: novel by Catherine Cookson.

MOTHER **** PG
Vsevolod I. Pudovkin USSR   1926
Vera Baranovskaya, Nikolai Batalov, A.P. Khristialov, Ivan Koval-Samborski, Anna Zemtsova, Vsevolod Pudovkin
A mother inadvertently turns her son in to the authorities, after he leads an illegal strike, but eventually she comes to see the error of her ways and embraces Communism. A potent tale of a family caught up in the abortive 1905 revolution, and at the same time a brilliant piece of social-realist propaganda.
Aka: MAT
DRA   81 min (ort 84 min) B/W silent   VIDrel: HEND L/A   V
Boa: novel by Maxim Gorky.

MOTHER, JUGS AND SPEED *** PG
Peter Yates USA   1976
Bill Cosby, Raquel Welch, Harvey Keitel, Allen Garfield, Larry Hagman, Bruce Davidson, Dick Butkus, L.Q. Jones, Toni Basil, Milt Kamen, Barra Grant, Allan Warnick, Valerie Curtin, Rick Carrott, Severn Darden, Bill Henderson
An account of the incidents in the lives of private ambulance drivers at an L.A. based service that has seen better days, and is now forced to cut corners all round, in order to stay in business. A chaotic but generally highly amusing black comedy.
COM   98 min   VIDrel: 20TH/TECH   V

MOTHER NIGHT *** 15
Keith Gordon USA   1996
Nick Nolte, Sheryl Lee, Alan Arkin, John Goodman, Kirsten Dunst, Ayre Gross, Frankie Faison, David Strathairn, Bernard Behrens, Norman Rodway, Anna Berger, Henry Gibson
Nolte plays an American citizen who awaits trial in Israel as a Nazi war criminal and is unable to prove to them that far from being a supporter of the Nazis, was in fact secretly at work as a spy for the Allies, his pro-Nazi activities merely being a convenient cover. A strange black comedy of contrived scripting that is often engrossing, occasionally funny, but never believable, even if Nolte is superb in the lead role.
COM   114 min   CINrel
Boa: novel by Kurt Vonnegut Jr.

MOTHER OF THE BRIDE ** (15)
Charles Correll USA   1992
Rue McClanahan, Kristy McNichol, Paul Dooley, Anne Bobby
A middle-aged but feisty female estate agent is presented with a fait accompli by her daughter, who suddenly announces her intention to wed. As the mother sets about planning for the big day, matters take an unexpected and not entirely desirable turn when her ex-husband appears on the scene, and she realises she still (despite being happily remarried) retains some feelings for him. A formulaic FATHER OF THE BRIDE variant.
COM   91 min   SATrel: MOVIE CHANNEL

MOTHERHOOD ** 15
Jonathan Wacks USA   1992
Steve Buscemi, Ned Beatty, John Glover, Miriam Margoyles, Sam Jenkins
A man brings his dead mother back to life, but finds that a surprise awaits him in that she has now become a violent

monster, with a penchant for chainsaws and an appetite for cooked dog. A grotesque black comedy, so overblown it quickly loses its comic elements.
COM   86 min   VIDrel: ITC/POLYREC   V/h

MOTHER'S BOYS ** 15
Yves Simoneau USA 1993
Jamie Lee Curtis, Peter Gallagher, Joanne Whalley-Kilmer, Vanessa Redgrave, Luke Edwards, Joss Ackland, Paul Guilfoyle, J.E. Freeman, Colin Ward, John C. McGinley, Joey Zimmerman, Jill Freedman, Lorraine Toussaint, Ken Lerner
After walking out for her family three years before, a woman returns to them and mounts a campaign of lies and deception in a bid to regain her former position. A contrived but absorbing psychological thriller that is spoiled by a predictable ending which is meant to generate suspense but instead only manages incredulity.
THR   91 min (ort 95 min)
VIDrel: 4-FRONT/POLYREC/GUILD   V/sur
Boa: novel by Bernard Taylor.

MOTHER'S INSTINCT, A ** PG
Sam Pillsbury USA 1995
Lindsay Wagner, Debra Farentino, John Terry, Lynn Thigpen, Alana Austin
A divorced woman is suspicious about the past of her second husband, a man with two boys of his own who claims to be a widower. But when she attempts to tackle him about this he disappears, taking his children with him. Her search for him eventually uncovers his first wife, who has been trying to find her sons for six years. Fair.
DRA   84 min   VIDrel: 20TH/FOXVID   V/sh

MOTHER'S PRAYER, A *** 15
Larry Elikann USA 1995
Linda Hamilton, Noah Fliess, Bruce Dern, Kate Nelligan, RuPaul, S. Epatha, Merkeson, Corey Parker, Jenny O'Hara, Gail Strickland, McNally Sagal, Aaron Lustig, Jane Whitney, Nancy Cassaro, Alex Kapp, James Avone, Julie Garfield
A widow with an eight-year-old son is shattered to learn that she is HIV-positive and has to face an agonising choice, in her search to find someone who will take care of her son after she is gone. An unabashed tearjerker that is sustained by sensitive direction and fine acting, especially from Hamilton. Based on a true story, it provides a good examination of this problem. Drag star RuPaul is effective in a small supporting role. See also THE OTHER WOMAN.
DRA   89 min (ort 94 min) mTV   VIDrel: CIC/SONOP L/A   V/sh

MOTORAMA ** 15
Barry Shils USA 1991
Jordan Christopher Michael, Martha Quinn, Flea, Michael J. Pollard, Drew Barrymore, Meatloaf, Garrett Morris, Robin Duke, Sandy Baron, Mary Woronov, Susan Tyrell, John Laughlin, John Diehl, Robert Picardo, Jack Nance
A ten-year-old delinquent boy who becomes obsessed with collecting game cards in a gas station contest, leaves home and hits the road for a number of wild adventures, in this bizarre comedy that seems certain to acquire a cult following.
COM   86 min (ort 89 min)   VIDrel: 20VIS/SONOP L/A   V

MOTORCYCLE GANG ** (18)
John Milius USA 1993
Gerald McRaney, Jake Busey, John Cassini, Richard Edson, Carla Gugino, Elan Oberon, Marshall Teague, Robert Miranda, Gina Mastrogiacomo, Don McManus, Julia Mueller, Pete Antico, Peter Sherayko, Dawn Cody, Juan Devoto
A nasty encounter ensues when a married couple and their teenage daughter run into a gang of bikers along an isolated highway. Set in the 1950s, the film paid good attention to detail, but the plot fails to engage one's interest. Adequate.
A/AD   90 min mTV   SATrel: SKY MOVIES

MOTORPSYCHO * 18
Russ Meyer USA 1965
Haji, Holle K. Winters, Sharon Lee, Arshalouis Aivazian, Alex Rocco, Stephen Oliver, Joseph Cellini, Thomas Scott, Coleman Francis, Steve Masters, F. Rufus Owens, E.E. Meyer, George Costello, Richard Brummer
Early saga of a Vietnam vet's revenge on the gang of nasty Hell's Angels who, led by another Vietnam vet, were responsible for the rape of his wife. The veteran teams up with the widow of

one of their other victims (in a purely platonic relationship) and the pair finally mete out the expected violent justice. A ludicrous and distasteful Meyer offering.
DRA   74 min B/W   VIDrel: ALLIED/RTM   V

MOULIN ROUGE **** PG
John Huston USA 1952
Jose Ferrer, Suzanne Flon, Zsa Zsa Gabor, Eric Pohlmann, Colette Marchand, Christopher Lee, Peter Cushing, Katherine Kath, Claude Nollier, Mary Clare, Muriel Smith, Georges Lannes, Walter Crisham, Harold Kasket, Lee Montague
A lovingly made portrayal of the life and work of the French Impressionist painter Toulouse Lautrec, that has a remarkable performance by Ferrer (he also plays the painter's father). An accomplished and moving work, with all the excitement and glamour of the Moulin Rouge splendidly captured. The theme song is by Georges Auric. AA: Art/Set (Paul Sheriff/Marcel Vertes), Cost (Marcel Vertes).
DRA   114 min (ort 123 min)   VIDrel: FABFIL/SPEAR   V
Boa: novel by Pierre La Mure.

MOUNTAIN MEN, THE * 15
Richard Lang USA 1980
Charlton Heston, Brian Keith, Victoria Racimo, Stephen Macht, John Glover, Victor Jory, Seymour Cassell, David Ackroyd, John Glover, Carl Bellini, Bill Lucking, Ken Ruta, Danny Zapien, Tim Haldeman, Bob Terhume, Chuck Roberson
Two fur trappers have to fight off attacks by Indians, in a film that is dull and boring, but rather bloody in parts. Poor direction and dialogue are helped slightly by some good photography. Screenplay is by Chuck's son, Fraser Clarke Heston.
WES   96 min (Cut at film release – ort 102 min)
VIDrel: RCA L/A   V

MOUNTAINS OF THE MOON **** 15
Bob Rafelson USA 1989
Patrick Bergin, Iain Glen, Richard E. Grant, Fiona Shaw, John Savident, James Villiers, Adrian Rawlins, Delroy Lindo, Paul Onsongo, Bernard Hill, Roshan Seth, Anna Massey, Leslie Phillips
A colourful and enthralling account of Richard Burton's search for the source of the Nile in 1857. Beautifully photographed by Roger Deakins, and directed in a way that highlights both the Victorian sense of adventure and its foibles. Bergin as Burton and Glen as his companion John Hanning Speke both give outstanding performances. Scripted by Rafelson and Harrison, the film is based on the latter's biographical novel and the journals of the period.
A/AD   129 min (ort 135 min)   VIDrel: GHV/POLY L/A;
PION (LV only)   V/sh LV
Boa: novel Burton and Speke by William Harrison.

MOUSE THAT ROARED, THE *** U
Jack Arnold UK 1959
Peter Sellers, Jean Seberg, David Kossoff, William Hartnell, Leo McKern, Harold Kasket, Monty Landis, Colin Gordon, George Margo, Macdonald Parke, Robin Gatehouse, Jacques Cey, Stuart Saunders, Ken Stanley, Bill Nagy, Mavis Villiers
The near-bankrupt Duchy of Grand Fenwick plans to declare war on the USA, get defeated and thus become eligible for Marshall Aid to rebuild its economy. Delightfully dotty, with Sellers engagingly playing three parts and effectively carrying the entire film. Written by Roger Macdougall and Stanley Mann. The inferior "The Mouse On The Moon" followed.
COM   80 min (ort 85 min)   VIDrel: VCC L/A   V
Boa: novel The Wrath Of The Grapes by Leonard Wibberley.

MOVE OVER, DARLING ** PG
Michael Gordon USA 1963
Doris Day, Polly Bergen, James Garner, Chuck Connors, Thelma Ritter, Fred Clark, Don Knotts, Elliott Reid, John Astin, Pat Harrington Jr
Following a plane crash and five years spent on a desert island, Day returns home to find that her husband is about to remarry. A sporadically entertaining remake of MY FAVOURITE WIFE that has little going for it except the sheer professionalism of the cast. It originally started production as "Something's Got To Give" and was to have starred Marilyn Monroe.
COM   99 min (ort 103 min)   VIDrel: 20TH/TECH L/A   V/h

MOVERS AND SHAKERS * 15
William Asher USA 1985
Walter Matthau, Charles Grodin, Vincent Gardenia, Tyne Daly, Bill Macy, Earl Boen, Gilder Radner, Michael Lerner, Joe Mantell, William

Prince, Nita Talbot, Sandy Ward, Steve Martin, Penny Marshall, Luana Anders
A wild but unfunny spoof on Hollywood in the 1980s, as a studio head fighting to avoid bankruptcy, decides to make a film version of a best-selling sex manual. A plotless mess of gags and anecdotes, thrown together without care or thought. The script is by Grodin.
COM 76 min (ort 100 min) VIDrel: MGM/WHV L/A V

MOVIE, MOVIE **
Stanley Donen USA
George C. Scott, Trish Van Devere, Eli Wallach, Red Buttons, Barbara Harris, Harry Hamlin, Barry Bostwick, Art Carney, Ann Reinking, Kathleen Beller, Rebecca York, Michael Kidel, George Burns
A fair parody of a 1930s double bill, complete with a "coming attractions" trailer, introduced by George Burns. The two parodies are "Dynamite Hands", a boxing tale, and "Baxter's Beauties of 1933", a spoof on all those Busby Berkeley musicals. Generally quite agreeable to watch, but don't expect to be any more than mildly amused.
COM 102 min (ort 106 min) B/W/Col VIDrel: L/A V

PG
1978

MOVING IN **
Michael Apted USA
Teri Garr, Peter Weller, Christopher Collett, Corey Haim, Sarah Jessica Parker, Robert Downey, Christopher Gartin, James Harper, Richard Brandon, Gayle Harbor, Ellen Barber, Richard E. Szlasa, Beverly W. May, Brian Lima
A divorced woman and her two children find themselves having to cope with the fact that her lover is a drug-dealing psychotic, after he moves into their apartment and takes over. Convincing performances are drowned in a welter of unpleasantness, with an utterly over-the-top ending for good measure.
Aka: FIRSTBORN
DRA 96 min (ort 103 min) VIDrel: CIC/SONOP L/A V

15
1984

MOVING TARGET **
Chris Thompson USA
Jason Bateman, John Glover, Jack Wagner, Chynna Phillips, Richard Dysart, Tom Skerritt, Donna Mitchell, Claude Brooks, Bernie Coulson, William Lanteau, Robert Downey Sr, Tom Fridley, Arnold Turner, Javier Grajeda, Aimee Brooks
A youngster returns home from summer camp to find his family gone, his house cleaned out, and some killers after him. A good start is thrown away by poor acting and little development. Pity – it could have been really gripping.
THR 96 min (ort 100 min) mTV VIDrel: MGM L/A V

15
1988

MRS DOUBTFIRE ***
Chris Columbus USA
Robin Williams, Sally Field, Pierce Brosnan, Harvey Fierstein, Mara Wilson, Robert Prosky, Polly Holliday, Lisa Jakub, Matthew Lawrence, Ralph Peduto, Scott Beach, Juliette Marshall, Drew Letchworth, Jessica Myerson, Sharon Lockwood
A divorced man is separated from his kids when his spiteful wife is awarded custody, but is unable to accept this situation. When the wife decides she needs a housekeeper he disguises himself as an elderly woman and applies for the post. A surefire box-office hit, with a hugely enjoyable performance by Williams in the title role, to which he brings both humour and pathos. AA: Make (Greg Cannom, Ve Neill, Yolanda Toussieng).
COM 119 min Cut (ort 125 min) cC VIDrel: 20TH/TECH; ENCORE/FOXVID (LV only) V/sur LV
Boa: novel Alias Madame Doubtfire by Anne Fine.

PG
1993

MRS LAMBERT REMEMBERS LOVE ***
Charles Matthau USA
Ellen Burstyn, Walter Matthau, Ryan Todd, William Schallert, Kathleen Garrett, Sherry Hursey, Richard Zobel, Jeanne Reynolds, Tom McGraw, Victoria Richardson, Dustin Weaver, Zachary Brown, Mary Lins, Denny Delk, James Neila
An elderly woman puts up a fierce struggle when she finds that she may lose custody of her nine-year-old grandson. A touching and believable story. Charles Matthau is the son of Walter.
DRA 89 min (ort 96 min) mTV VIDrel: COLUM/SONOP V

PG
1991

MRS MINIVER ***
William Wyler USA
Greer Garson, Walter Pidgeon, Teresa Wright, Richard Ney, May Whitty, Henry Travers, Reginald Owen, Henry Wilcoxon, Helmut Dantine, Rhys Williams, Aubrey Mather
A classic WW2 morale-booster about an English village family

U
1942

learning to deal with their wartime experiences, but set in a laughable Hollywood version of rural England. However, despite its failings, it succeeded in arousing American sympathies for Britain. "The Miniver Story" followed. AA: Pic, Dir, Actress (Garson), S. Actress (Wright), Cin (Joseph Ruttenberg), Screen (Arthur Wimperis/George Froeschel/James Hilton/Claudine West).
DRA 128 min (ort 134 min) B/W VIDrel: MGM/WHV V
Boa: novel by Jan Struther.

MRS PARKER AND THE VICIOUS CIRCLE ***
Alan Rudolph USA
Jennifer Jason Leigh, Matthew Broderick, Campbell Scott, Andrew McCarthy, Tim McGowan, Nick Cassavetes, Gary Basaraba, Jake Johannsen, Chip Zien, Matt Malloy, Sam Robards, Martha Plimpton, Jane Adams, David Thornton, Leni Parker
In the 1920s, an acid-tongued woman writer called Dorothy Parker presided over a kind of literary salon that brought together a number of the era's most celebrated writers. First-rate performances and an excellent period atmosphere enhance this languid tale of the doing of a collection of disparate and disagreeable characters.
DRA 119 min (ort 124 min) wScrn VIDrel: ARTIF/20TH V/sh

15
1994

MRS SOFFEL **
Gillian Armstrong USA
Mel Gibson, Diane Keaton, Matthew Modine, Edward Hermann, Trini Alvarado, Jennie Dundas, Danny Corkill, Harley Cross, Terry O'Quinn, William Youmans, Maury Chaykin, John W. Carroll, Wayne Robson
A couple of brothers go to prison, where the warden's wife falls for one of them, eventually joining them on the run. Set in Pittsburgh of 1901, this is a detailed but sombre version of a real-life incident. Remarkably unappealing for all the care that went into production. The earlier "Molly And Lawless John" used much the same idea.
DRA 107 min (ort 113 min) VIDrel: MGM/WHV V/sur

PG
1984

MRS WINTERBOURNE **
Richard Benjamin USA
Ricki Lake, Shirley MacLaine, Brendan Fraser, Loren Dean, Susan Haskell
Lake plays a heavily pregnant working class waitress who gets mistaken for the widowed daughter-in-law of a wealthy society lady after she is pulled out of the wreckage of a train crash. Taken home by her new family, she learns that the woman's dead son also had a brother, and he is still single. A formulaic comedy of errors, very short on humour, but worth seeing if only for MacLaine's delightfully feisty performance as the eponymous Mrs Winterbourne.
COM 102 min VIDrel: COLUM/SONOP V/sur

12
1996

MUCH ADO ABOUT NOTHING ***
Kenneth Branagh UK/USA
Emma Thompson, Denzel Washington, Keanu Reeves, Michael Keaton, Kenneth Branagh, Robert Sean Leonard, Kate Beckinsale, Brian Blessed, Phyllida Law, Imelda Staunton, Gerald Horan, Jimmy Yuill, Richard Clifford, Ben Elton
Returning home from the wars in triumph, a man soon gets involved in various romantic and political complications, in this excellent and highly original version of Shakespeare's play that offers some bright performances and excellent direction.
DRA 106 min (ort 111 min) wScrn VIDrel: EIV/SONOP V
Boa: play by William Shakespeare.

PG
1992

MUD HONEY *
Russ Meyer USA
Lorna Maitland, Rena Horton, Lee Ballard, Hal Hopper, John Furlong, Stu Lancaster, Antoinette Cristiani, Frank Bolger, Rena Horten, Nick Wolcuff, Sam Hanna, Princess Livingston, Lee Ballard, Mickey Foxx, F. Rufus Owens
Another Meyers extravaganza set during the Depression and telling of an aged farmer who gives a job to a convict, recently released after serving five for manslaughter. The farmer's son is an evil brute, given to raping his wife and beating up the convict, who refuses to fight. When the farmer dies and leaves his property to the convict, the son is outraged and much violence ensues before virtue finally triumphs.
Aka: MUDHONEY; ROPE; ROPE OF FLESH
A 92 min B/W VIDrel: ALLIED/RTM V
Boa: novel Streets Paved With Gold by Raymond Friday Locke.

18
1965

MULHOLLAND FALLS ***   18
Lee Tamahori USA   1996
Nick Nolte, Melanie Griffith, Chazz Plaminteri, John Malkovich,
Michael Madsen, Chris Penn, Treat Williams, Jennifer Connelly,
Andrew McCarthy
After a woman with plenty of influential friends is murdered,
a group of L.A. cops notorious for their unconventional
methods, investigate the case. However, group loyalties are
strained to the limit when it appears that one of their number
may be implicated in this crime. Absorbing crime thriller set in
the 1950s.
THR   106 min   VIDrel: POLFIL   V

MULTIPLE MANIACS **   18
John Waters USA   1970
Divine (Glenn Milstead), Mary Vivian Pearce, Edith Massey, Mink
Stole, Paul Swift, David Lochary, Cookie Mueller, Susan Lowe, Rick
Morrow, Howard Gruber, Vince Peranio, Jim Thompson, Dee Vitolo,
Ed Peranio, Bob Skidmore
The notorious overweight female impersonator (Divine playing
him/herself) and her boyfriend, run a travelling sideshow at
which the audience is robbed and sometimes murdered.
Somewhat less disjointed than Waters's films usually are, this
one has some genuinely funny moments. Unfortunately the
sleaze of it all dilutes the comedy.
COM   90 min Cut (4 min 53 sec) B/W
VIDrel: CASPIC/TERRY L/A   V

MULTIPLICITY ***   12
Harold Ramis USA   1996
Michael Keaton, Andie McDowell, Harris Yulin, Richard Masur,
Eugene Levy, Ann Cusack, John deLancie, Brian Doyle-Murray, Julie
Bowen, Katie Schlossberg, Zack Duhame, Judith Kahan, Obba
Babatunde, Dawn Maxey, Kari Coleman, Robin Duke
A man hits on a novel solution to the stresses of modern life and
the lack of time to do everything he would like. This involves
the creation of a clone, thanks to a geneticist who is pleased to
have a human subject to work on. But pretty soon, matters get
out of hand and when the duplicate finds himself with too much
to do, he has few clones of his own created as well. Keaton is
really terrific in this anarchic, inventive and hugely enjoyable
comedy.
COM   113 min (ort 117 min) cC   VIDrel: COLUM/SONOP
V/sur

MUMMY, THE **   15
Karl Freund USA   1932
Boris Karloff, Zita Johann, David Manners, Arthur Byron, Edward
Van Sloan, Bramwell Fletcher, Noble Johnson, Leonard Mudie, Eddie
Kane, Henry Victor, Kathryn Byron, James Crane, Arnold Grey, Tony
Marlow
A mummy is accidentally revived, and goes off in search of its
ancient soul-mate. Some atmospheric sequences are present, but
on the whole the film is let down by stilted acting and ludicrous
dialogue. Followed by a large number of films that made use of
this theme.
HOR   72 min B/W   VIDrel: CIC/SONOP L/A   V/h

MUMMY, THE **   PG
Terence Fisher UK   1959
Peter Cushing, Christopher Lee, Yvonne Furneaux, Eddie Byrne, Felix
Aylmer, Raymond Huntley, John Stuart, Michael Ripper, George
Pastell, Dennis Shaw, Willoughby Gray, Stanley Meadows, Frank
Singuineau, Frank Sieman, Gerald Lawson
Imaginatively shot, but otherwise a typical Hammer horror
movie rooted in the same Egyptian background as in the 1932
classic and all the imitations it spawned. A priest executed
because of his love for a queen comes back to life when her tomb
is desecrated by a group of Egyptologists.
Aka: TERROR OF THE MUMMY
HOR   88 min   VIDrel: WHV   V

MUMMY LIVES, THE *   (PG)
Gerry O'Hara USA   1993
Tony Curtis, Greg Wrangler, Jack Cohen, Muhammed Bakri, Mosko
Alkelai, Leslie Hardy, Yose Chiloach, Uri Gavreil Yigal Naor, Eli
Danker, Yossi Graber, Charlie Buzgalo, Rafi Weinstock, Amos Lavie,
Uri Mauda, Rivka Bachar
When an Egyptian tomb is opened by archaeologists, the
preserved body of a former priest comes to life and sets about
chastising those responsible for this act of desecration. In many
ways this poorly acted and directed horror yarn recalls many a

similar movie from the 1960s and 1970s, to which it is noticeably
inferior in almost every way.
HOR   93 min   VIDrel: WHV L/A   V

MUMMY'S GHOST, THE *   PG
Reginald LeBorg USA   1944
Lon Chaney Jr, John Carradine, Ramsay Ames, Robert Lowery, Barton
MacLane, George Zucco
A sequel to THE MUMMY'S TOMB that has our bandaged one
on the trail of a women he believes is a reincarnation of his
beloved princess. A flat and hollow affair, directed without
energy or conviction. Followed by "The Mummy's Curse"
(1944), which ended this "Kharis" series.
HOR   60 min B/W   VIDrel: CIC/SONOP L/A   V/h

MUMMY'S HAND, THE ***   15
Christy Cabanne USA   1940
Dick Foran, Wallace Ford, Peggy Moran, Cecil Kellaway, George
Zucco, Tom Tyler, Eduardo Ciannelli, Charles Trowbridge
A mummy is revived thanks to an infusion of tanna leaves, and
seeks out the reincarnation of his former princess, wreaking
havoc on those who desecrated her tomb. A kind of partial
sequel to THE MUMMY that makes use of some footage from
that earlier film, and starts off weakly but builds up to an excellent and chilling climax. This was the first of the "Kharis"
mummy films. Several sequels followed, starting with THE
MUMMY'S TOMB (1942).
HOR   67 min B/W   VIDrel: CIC/SONOP L/A   V/h

MUMMY'S SHROUD, THE **   PG
John Gilling UK   1967
Andre Morell, David Buck, John Philips, Elizabeth Sellars, Michael
Ripper, Maggie Kimberley, Tim Barrett, Richard Warner, Roger
Delgado, Catherine Lacey, Dickie Owen, Bruno Barnabe, Eddie Powell
(the Mummy)
A group of archaeologists search for an ancient tomb, and eventually an exhumed mummy is brought to a city museum, but it
comes back to life and embarks on a rampage in revenge for the
desecration of its tomb. This was an attempt by Hammer Films
to revive the sub-genre, but for all the good production values
and gory realism, there were no new ideas on offer here. BLOOD
FROM THE MUMMY'S TOMB was only slightly more successful.
HOR   84 min   VIDrel: LUMI/SPEAR L/A   V

MUMMY'S TOMB, THE **   PG
Harold Young USA   1942
Lon Chaney Jr, Elyse Knox, John Hubbard, Turhan Bey, Dick Foran,
Wallace Ford, George Zucco, Mary Gordon
A dreary re-run of THE MUMMY'S HAND, not improved by
its borrowed footage from that earlier film, THE MUMMY
(1932) and FRANKENSTEIN (1931). This was the first sequel in
a series of "Kharis" films, in which our revived mummy
dispensed rough justice to those who violated his tomb and
spent the rest of the time searching from the reincarnation of his
ancient beloved. Followed by THE MUMMY'S GHOST.
HOR   61 min B/W   VIDrel: CIC/SONOP L/A   V/h

MUNCHIE **   PG
Jim Wynorski USA   1992
Loni Anderson, Andrew Stevens, Andrew Stevens, Arte Johnson,
Jaime McEnnan, Love Hewitt, Toni Napier, Ace Mask, Monique
Gabrielle, Scott Ferguson, Mike Simmrin, Angus Scrimm, Lenny
Juliang, Fred Olen Ray, Dom De Luis (voice only)
A young boy who has just moved to a new town with his
mother, finds a box in which a strange creature with large eyes
has been imprisoned. He releases him and gains a good friend
who protects him from bullies and grants his wishes. In return,
our young lad helps to save him from a professor who is trying
to capture him. A watchable if very juvenile comedy.
JUV   78 min (ort 80 min)   VIDrel: CIC/SONOP L/A   V

MUNCHIE STRIKES BACK **   PG
Jim Wynorksi USA   1993
Lesley-Anne Down, Andrew Stevens, Trenton Knight, Angus
Scrimm, John Byner, Steve Franken, Natanya Ross, Ace Mask, Cory
Mendelsohn, Howard Hesseman (voice only)
Sequel to the earlier "Munchie" in which our mischievous alien
finds himself on trial on his home planet and is forced to return
to Earth to help a family there, as part of his punishment.
JUV   90 min   VIDrel: MARQ/QUANT   V

MUPPET CHRISTMAS CAROL, THE ** U
Brian Henson USA 1992
Michael Caine, The Muppets (Kermit the Frog, Miss Piggy, The Great Gonzo, Rizzo the Rat, Fozzie Bear), Steven MacKintosh, Meredith Braun, Robin Weaver, Donald Austen, Edward Sanders, Theo Sanders, Kristopher Milnes, Russell Martin
The classic Dickens work gets the Muppet treatment with them playing all the major roles with the exception of Scrooge (Caine). A standard effort but somewhat overlong as a Christmas special.
JUV 85 min (ort 95 min) cC VIDrel: BUENA/TECH
V/sur

MUPPET MOVIE, THE ** U
James Frawley UK/USA 1979
The Muppets, Frank Oz, Jerry Nelson, Richard Hunt, Dave Goelz, Bob Hope, Charles Durning, Austin Pendleton, Scott Walker, Edgar Bergen, Milton Berle, Mel Brooks, Madeleine Kahn, Steve Martin, Dom DeLuise, Elliott Gould
A tale using puppets from a TV series, with this being the story of the rise to fame of Kermit from a swamp to Hollywood. Though the characters are cleverly transferred to the screen, the material is very weak, and the film just drags on and on. Followed by several more "Muppet" movies.
JUV 94 min (ort 98 min) cC VIDrel: WDV/TECH V/sur

MUPPET TREASURE ISLAND ** U
Brian Henson USA 1996
Tim Curry, Jennifer Saunders, Kevin Bishop, Billy Connolly
A loose (very loose) adaptation of the Robert Louis Stevenson classic tale of high-seas adventure and treasure, with many of the important parts filled by The Muppets. As ever, young Hawkins has to thwart the plans of the wicked Long John Silver and keep his map giving the location of the buried treasure out of that brigand's clutches. Good fun if you like that sort of thing.
JUV 95 min (ort 100 min) cC VIDrel: WDV/TECH V/sh

MUPPETS TAKE MANHATTAN, THE *** U
Franz Oz USA 1984
The Muppets, Joan Rivers, Liza Minelli, Linda Lavin, Louis Zorich, Lonny Price, Juliana Donald, Art Carney, James Coco, Dabney Coleman, John Landis, Gregory Hines, Elliott Gould, Brooke Shields, Edward I. Koch
Third in the series of "Muppet" adventures with these puppets from a popular TV series trying to make the big time in New York. Packed with a host of guest stars, and for the most part a lively and entertaining affair.
JUV 94 min VIDrel: 20TH/TECH L/A V

MUPPETS, THE: THE GREAT MUPPET CAPER ** Uc
Jim Henson UK 1981
Diana Rigg, Charles Grodin, John Cleese, Robert Morley, Peter Usinov, Jack Warden, Erica Creer, Kate Howard, Della Finch, Michael Robbins, Joan Sanderson, Tommy Godfrey, Katia Borg, Valli Kemp, Michele Ivan-Zadeh
Feature film written around the puppet characters from the TV series; "The Muppets". Here Kermit and Fozzie are newspaper reporters assigned to cover a jewel robbery. A large collection of guest stars make appearances and help buoy up this somewhat thin story.
Aka: GREAT MUPPET CAPER, THE
JUV 98 min cC VIDrel: WDV/TECH V/sur

MURDER *** PG
Alfred Hitchcock UK 1930
Herbert Marshall, Norah Baring, Phyllis Konstam, Edward Chapman, Esme Percy, Miles Mander, Donald Calthrop, Amy Brandon Thomas, Marie Wright, Hannah Jones, Una O'Connor, R.E. Jeffrey, Violet Farebrother, Kenneth Kove, Gus McNaughton
One of the jurors who convicted a woman of murder at her trial, turns amateur sleuth to prove her innocence in this interesting example of an early Hitchcock work.
DRA 100 min (ort 108 min) B/W VIDrel: LUMI/SPEAR
V
Boa: play Enter Sir John by Clemence Dane and Helen Simpson.

MURDER BETWEEN FRIENDS *** 15
Waris Hussein USA 1993
Timothy Busfield, Stephen Lang, Martin Kemp, Lisa Blount, O'Neal Compton, Alex Courtney, Karen Moncreiff, Sab Shimono, Stanley Anderson, Nicholas Pryor, Jay Robnson, Sharon Schlarth, Macon McCalman, Gregg Almquist, Lenny Wolpe
Fact-based murder thriller set in New Orleans, where a man and his best friend plot an elaborate hoax in order to do away

with the wife of the former. But one can never be quite certain who is really to blame, in a film of complex plotting that has them both blaming each other for the crime. Intriguing and absorbing, this is a film that demands careful attention.
THR 91 min (ort 100 min) mTV VIDrel: ODY/SONOP
V/sh

MURDER BY DECREE *** 15
Bob Clark CANADA/UK 1978
Christopher Plummer, James Mason, Donald Sutherland, Genevieve Bujold, Susan Clark, David Hemmings, Anthony Quayle, John Gielgud, Frank Finlay, Chris Wiggins, Teddi Moore, Catherine Kessler, Terry Duggan, Peter Jonfield
Sherlock Holmes and Watson investigate the Jack the Ripper killings, and come close to discovering the truth behind these crimes. An ambitious idea, with good casting of Plummer and Mason as Holmes and Watson respectively. Only the fumbling development of the narrative and a few grisly touches spoil it.
DRA 121 min VIDrel: L/A V

MURDER BY NIGHT ** 15
Paul Lynch USA 1989
Robert Urich, Kay Lenz, Michael Ironside, Jim Metzler, Michael Williams, Richard Monette, Geoffrey Bowes, Barbara Von Radicki, Christine Midges, Steve Mousseau, Dale Wilson, Cynthia Belliveau, Paula Barrett, Howard Jerome
A man found unconscious at the scene of a murder and suffering from amnesia, seems to hold the key to the murderer's identity, but as his treatment at the hands of a police therapist progresses, a suspicious cop becomes increasingly convinced that he is the killer. However, the real killer soon learns of the danger the man poses, and sets out to silence him. A thin, assembly-line thriller lacking a much-needed injection of tension.
THR 89 min (ort 100 min) mCab VIDrel: CIC/SONOP L/A
V

MURDER IN A COLLEGE TOWN ** 15
Bradley Wigor USA 1996
Kate Jackson, Drew Ebersole, Matthew Settle, Kristian Alfonso, Gary Basaraba, Neal Huff, Sean McCann
Fact-based story telling of a talented young man who gets embroiled in his friend's shady schemes for making a fast buck. Average.
DRA 95 min VIDrel: ODY/SONOP V/sh

MURDER IN MISSISSIPPI **** 15
Robert Young USA 1989
Tom Hulce, Blair Underwood, Josh Charles, Jennifer Grey, C.C.H. Pounder, Andre Braugher, John Dennis Johnson, Lou Walker, Scott Lawrence, Jill Jane Clements, Eugene Byrd, J. Don Ferguson, Walt Goggins, Tom Gossom, Bill Coates
The story of the murder of Mickey Schwerner, James Chaney and Andrew Goodman, three civil rights campaigners whose killing in 1964 did much to mobilise support for the movement. Unlike the largely fictional MISSISSIPPI BURNING, this excellent drama (written by Stanley Weiser) is both factual and incisive, and attempts to analyse the reasons for the white bigotry which led to this crime. Originally shown in two parts.
DRA 93 min (ort 200 min) mTV VIDrel: MGM/WHV L/A
V

MURDER IN NEW HAMPSHIRE: THE PAMELA SMART STORY ** 15
Joyce Chopra USA 1991
Helen Hunt, Larry Drake, Chad Allen, Howard Hesseman, Ken Howard, Michael Learned, Hank Stratton, Riff Rean, Michael Learned, Richard K. Olsen, V. Michael Hunter, Sean Bridges, Marilyn Carter, Bill Spencer, Jennifer Hayes
This fact-based drama follows the career of a married woman who embarked on an affair with a fifteen-year-old boy and used him as a tool in planning the murder of her husband. Both well made and acted, this is a straightforward account that works well without any unnecessary frills or surprises, and makes good use of ample flashbacks.
DRA 89 min (ort 93 min) mTV VIDrel: CAPIT/GUILD
V/sh

MURDER IN THE FIRST *** 15
Mark Rocco USA 1994
Christian Slater, Kevin Bacon, Gary Oldman, Embeth Davidtz, Brad Dourif, Bill Macy, Stephen Tobolowsky, R. Lee Ermey, Mia Kirshner, Ben Slack, Stefan Gierasch, Kyra Sedgwick, Alexander Bookston, Richie Allan, Herb Ritts

A young lawyer who has just started a career working as a public defender gets a tough case on his very first job, namely the defence of a prisoner at Alcatraz who is charged with murder. Investigating the background to this case, he learns that the man's crime was in part conditioned by a long history of severe brutality at the hands of guards that literally drove him insane. A well intentioned piece with first-rate acting and a good eye for detail.
DRA   117 min (ort 123 min)   VIDrel: GUILD/20TH   V/sur

MURDER IN THE HEARTLAND *   18
Robert Markowitz USA   1993
Brian Dennehy, Tim Roth, Fairuza Balk, Randy Quaid, Kate Reid, Ryan Cutrona, Angie Bolling, Roberts Blossom, Jake Carpenter, Heather Kafka, Dan Bloomfield, John Hussey, James Prince, John Davies, Mark Walters, Jennifer Griffin
A young couple go on the run, having already killed eleven people, in this unpleasant and unedifying story.
DRA   150 min mTV   VIDrel: NWV/HIFLI   V/h

MURDER IN THE MUSIC HALL **   (PG)
John English USA   1946
Vera Ralston, William Marshall, Helen Walker, Nanceu Kelly, William Gargan, Ann Rutherford, Julie Bishop, Jerome Cowan, Edward Norris, Frank Orth, Jack LaRue, James Craven, Fay McKenzie, Tom London, Joe Yule, Mary Field, Anne Nagel
When a blackmailer is murdered, suspicion falls on a number of stars at the music-hall that were the dead man's neighbours. The orchestra leader there decides to turn detective and eventually tracks down the real culprit. Chiefly designed to some extent to display the talents of ice-skating star Ralston, it remains surprisingly competent in all departments, and maintains a strong sense of tension right up to the end.
Aka: MIDNIGHT MELODY
DRA   95 min B/W   TVrel: BBC

MURDER IS EASY **   PG
Claude Whatham USA   1982
Bill Bixby, Lesley-Anne Down, Olivia De Havilland, Helen Hayes, Patrick Allen, Shane Briant, Freddie Jones, Leigh Lawson, Jonathan Pryce, Ivor Roberts, Timothy West, Anthony Valentine, Patrick M. Wright
An American computer expert takes a journey across the English countryside by train, in the course of which he meets a distressed elderly lazy who tells him that there is a killer at large in her village. Her death on arrival at her station convinces him of the truth of her claim, and he sets out to mount his own investigation, but as one might expect, this amateur sleuthing is not without its risks. Good-looking, absorbing but not especially memorable.
Aka: AGATHA CHRISTIE: MURDER IS EASY
DRA   95 min (ort 100 min) mTV   VIDrel: WHV   V/h
Boa: novel Easy To Kill by Agatha Christie.

MURDER, MY SWEET **   PG
Edward Dmytryk USA   1944
Dick Powell, Claire Trevor, Ann Shirley, Otto Kruger, Mike Mazurki, Miles Mander, Douglas Watson, Ralf Harolde, Don Douglas, Esther Howard, Jack Carr, John Indrisano, Shimen Ruskin, Ernie Adams, Dewey Robinson, Larry Wheat
Powell as the hard-boiled detective Philip Marlowe, has a new image in this version of Chandler's novel, with our 'tec getting mixed up in murder and blackmail. Well-acted and with a good supporting cast, this is a splendid example of its genre – moody, imaginative and totally absorbing. Remade in 1975 as FAREWELL, MY LOVELY.
Aka: FAREWELL, MY LOVELY
THR   95 min B/W   VIDrel: VCC L/A   V
Boa: novel Farewell, My Lovely by Raymond Chandler.

MURDER OF INNOCENCE *   18
Tom McLoughlin USA   1993
Valerie Bertinelli, Stephen Caffrey, Graham Beckel, Jerry hardin, Millie Perkins, Justin Whalin, Anne Ramsay, Steve Banks, Megan Cavanagh, John Scott Clough, Juanita Jennings, Nancy McLoughlin, Trishalee Hardy, Frank Novak
Based on a real-life case, this disturbing drama tells of an apparently happily married woman whose increasingly irrational behaviour leads her husband to tackling his in-laws about his wife's sanity, but they are unwilling to reveal details of her past mental history. Eventually her sanity snaps completely, and having obtained a gun she kills one child, wounds seven others

and takes a hostage. Very well made, but not exactly light entertainment.
DRA   94 min mTV   VIDrel: ODY/SONOP   V/sh
Boa: book by Joel Kaplan, George Papajohn and Eric Zorn.

MURDER OF MARY PHAGAN, THE **   15
Billy Hale USA   1988
Jack Lemmon, Richard Jordan, Robert Prosky, Peter Gallagher, Paul Dooley, Rebecca Miller, Kathryn Walker, Charles Dutton, Kevin Spacey, Cynthia Nixon, Wendy J. Cooke, Kenneth Walsh, Penny Allen, Thomas Anderson, Dylan Baker
Based on a real incident in Atlanta, Georgia in 1913, this tells of John Slaton, the governor of Georgia, who had to make a decision whether or not to allow the execution of Leo Frank, who was convicted of Phagan's murder. An excellent reconstruction of events, with fine performances and a literate script by George Stevens Jr and Jeffrey Lane. The acting debut of Miller, the daughter of playwright Arthur Miller. First shown in two parts.
DRA   112 min (ort 250 min) mTV
VIDrel: VISVID/POLYREC L/A   V
Boa: story by Larry McMurtry.

MURDER ON DEMAND *   15
Alan Metzger USA   1990
Patrick Duffy, William Devane, Chelsea Field, Mariette Hartley, Allan Miller, Alex Hyde-White, Harris Laskaway, Janet Margolin, Charles Robinson, Bert Kramer, David Cascadden, Mike Garvey, Walt Hoppert, Herb Hyde, Rick Jones
The new chief of a police department appears to be a man of the utmost ability, but hides secrets known to another officer, who is both an expert in surveillance techniques and a ruthless killer.
Aka: MURDER BY DEMAND; MURDER C.O.D.
THR   91 min mTV   VIDrel: GENESIS   V
Boa: novel Kill Fee by Barbara Paul.

MURDER ON THE ORIENT EXPRESS *   PG
Sidney Lumet UK   1974
Albert Finney, Lauren Bacall, Martin Balsam, Ingrid Bergman, Jacqueline Bisset, Sean Connery, Richard Widmark, Vanessa Redgrave, John Gielgud, Wendy Hiller, Jean-Pierre Cassel, Anthony Perkins, Rachel Roberts, Michael York
A pretty awful adaptation of an Agatha Christie whodunit, involving a murder that takes place on the Orient Express, eventually solved by her famous detective Monsieur Hercule Poirot (miscasting of Finney here). Slow moving and uninspired. As ever, a cast full of stars can do nothing to save it. AA: S. Actress (Bergman).
DRA   122 min (ort 131 min)   VIDrel: WHV   V
Boa: novel by Agatha Christie.

MURDER ON THE RIO GRANDE *   (18)
Robert Iscove USA   1993
Victoria Principal, Peter Onorati, Sean Murray, David Beecroft, Ari Meyers, Garry Grubbs, Lee Garlington, Dirk Blocker, Raymond Baker, Sal Lopez, Stan Ivar, Valerie Armstrong, Larisa Oleynik, Kristina Betts, Alexis Alexander, Tom Noga
A divorced mother takes a rafting holiday on the Rio Grande, but has to do battle with a psychopath who murders the rest of her party. DELIVERANCE it ain't.
Aka: RIVER OF RAGE: THE TAKING OF MAGGIE KEENE
THR   90 min mTV   SATrel

MURDER 101 *   15
Bill Condon USA   1990
Pierce Brosnan, Dey Young, Antoni Cerone, Raphael Sbarge, Kim Thomson, J. Kenneth Campbell, Mark L. Taylor, Todd Merrill, yorgo Constantine, Yavone Evans, Dianne Hull, Walter Klenhard, Terry Markwell, Kathe Mazur, Mary Lou Piccard
Writer-director Condon tells a ponderous tale of a college professor who gives his class an assignment that involves concocting a perfect crime, and then finds that he has been framed for a murder. Quite an intriguing premise, but the film never really gets going.
THR   88 min (ort 100 min) mCab   VIDrel: CIC/SONOP   V/h

MURDER OR MEMORY: THE TRUE STORY OF A TEENAGER'S DEADLY CONFESSION *   15
Christopher Leitch USA   1994
Leigh Taylor-Young, Michael Brandon, Karen Austin, Conor O'Farrell, Melinda Culea, Louis Giambalvo, Rebecca Budig, Melinda Culea,
True story of a mother's fight for justice, after her son is duped

by the police into confessing to a murder he was entirely innocent of. Fair.
Aka: MOMENT OF TRUTH: MURDER OR MEMORY?
DRA 87 min VIDrel: NWV/HIFLI V/h

MURDER SO SWEET ** PG
Larry Peerce USA 1994
Harry Hamlin, Helen Shaver, K.T. Oslin, Terence King, Eileen Brennan, Ed Lauter, Frances Lee McCain, Daphen Ashbrook, William Lucking, Faith Ford, Juliette Marshall, Michael Bowen, Steven Anthony Jones, Peter Anthony Jacobs
A woman finds a new boyfriend who seems to be the man of her dreams, but is unaware that he is a serial seducer of women, whom she marries and then murders. Based on a true story this is quite a good thriller, and for once casts Hamlin against type as the villain of the piece.
Aka: POISONED BY LOVE: THE KERN COUNTY MURDERS
THR 90 min (ort 94 min) mTV
VIDrel: NEWAGE/20VIS L/A V

MURDER WITHOUT MOTIVE *** PG
Kevin Hooks USA 1992
Cuba Gooding Jr, Curtis McClaren, Anna Maria Horsford, Carla Gugino, Tauren Blacque, Christopher Daniel Barnes, GuyKillum, Dakin Matthews, Hari Rhodes, Jason Christopher, Georg Stanford Brown, Jay Underwood, Vonelle Curtis-Hall
A straight-A black student with a bright future and a clean police record is gunned down and killed by an officer who claimed that he was forced to act in self-defence. A powerful drama based on a real-life story.
Aka: BEST INTENTION: THE EDUCATION AND KILLING OF EDMUND PARRY
DRA 91 min (ort 93 min) mTV VIDrel: ODY/SONOP V

MURDERED INNOCENCE ** 18
Frank Coraci USA 1994
Jason Miller, Jacqueline Macario, Fred Carpenter, Ellen Greene, Gary Aumiller
A cop has to live with the fact that he accidentally killed a handcuffed murder suspect who later proved to be innocent of any crime. When the real killer is eventually let out of prison, he faces the full wrath of the latter and has to go on the run.
THR 78 min (ort 88 min) VIDrel: 20VIS/SONOP V/sur

MURDERERS AMONG US: THE SIMON WIESENTHAL STORY *** PG
Henry Levin USA 1966
Dean Martin, Ann-Margret, Karl Malden, Camilla Sparv, James Gregory, Beverly Adams, Robert Eastman, Marcel Hillaire, Duke Howard, Tom Reese, Ted Hartley, Robert Terry, Corinne Cole, Mary Jane Mangler, Dale Brown, Mary Hughes
The second "Matt Helm" secret agent adventure in which he has to tackle a villain who threatens to melt Washington with his secret weapon. One of those pseudo-Bond espionage capers that were typical of the 1960s, that now looks unbelievably dated and dull. The script is by Herbert Baker.
COM 101 min (ort 108 min) VIDrel: RCA L/A V
Boa: novel by Donald Hamilton.

MURDEROUS AFFAIR, A: THE CAROLYN WARMUS STORY ** 15
Martin Davidson USA 1993
Virginia Madsen, Chris Sarandon, Ned Eisenberg, Tom Mason, Robert Picardo, William H. Macy, Olivia Burnette, Jay Acovone, Johnny Williams, Tracy Kolis, David Spielberg, Herb Mitchell, Lenore Kasdorf, Steven Marcus, Charley Lang
The brutal slaying of a woman leaves her husband the chief suspect, until the man's mistress appears on the scene and the police are obliged to start investigating her as well. A fact-based account of one of America's most recent crimes of passion, that achieved a notoriety of its own when all the nasty details came out. Fair.
DRA 89 min mTV VIDrel: ODY/SONOP V/sh

MURDEROUS INNOCENCE ** (18)
Frank Coraci USA 1993
Jason Miller, Fred Carpenter, Jacqueline Macario, Gary Aumiller, Bryant Holt, Bob Schlesinger, Donna Brittany, Craig Morris Weintraub, Aloysius Wilson, Victor Campos, Doug Hurst, Donald Graham, John Petti, Eddi Freeman, Matt Farrago
Denied parole, a former professor convicted of murdering his mistress twenty years ago, escapes before being transferred to another prison. At the same time the murder victim's son gets

out of jail in Alabama, and both men make their way to Long Island and the scene of the crime. Along the way, a young woman gets drawn into this forthcoming confrontation. A pretentious tale of guilt and innocence, terribly flashy and slick, but at the same time devoid of substance.
THR 77 min B/W/Col SATrel: SKY MOVIES

MURDEROUS INTENT ** 12
Gregory Goodell USA 1995
Tammy Arnold, Tushka Bergen, Corbin Bernsen, Sean Bridges, David Cutting, Lisa Darr, Aubrey Dollar, John Finn, Michael Goodwin, Stan Kang, Janell McLeod, Dash Mihok, Mark Jeffrey Miller, Alan Sader, Lesley Ann Warren, Alfred Wiggins
Real-life story of Gayle Bernish and her lover Brice Talbot. When the former wife of Talbot demands excessive child-support payments and yet denies him the opportunity to visit his children, the two lovers plot the murder of this woman.
DRA 88 min VIDrel: 20TH/FOXVID V/sh

MURDEROUS VISION * 15
Gary A. Sherman USA 1991
Bruce Boxleitner, Laura Johnson, Joseph D'Angerio, Glenn Plummer, Beau Starr, Dean Norris, Robert Culp, Elizabeth Kemp
A gruesome horror-thriller which deals with a maniacal serial killer who surgically cuts away the faces of his victims. Despairing of catching the culprit, a police detective turns to a psychic for assistance. A poorly put together effort, largely devoid of tension if not of nastiness.
HOR 89 min (ort 93 min) mCab VIDrel: CIC/SONOP V/h

MURIEL'S WEDDING ** 15
P.J. Hogan AUSTRALIA 1994
Toni Collette, Bill Hunter, Rachel Griffiths, Jeanie Drynan, Sophie Lee, Gennie Nevinson, Rosalind Hammond, Belinda Jarrett, Pippa Grandison, Daniel Wyllie, Gabby Millgate, Matt Day, Chris Haywood, David Lapaine, Susan Prior
A young girl, obsessed with white weddings and the music of Abba, grabs a chance to run away to Sydney and become a new person, after her mother rather unwisely gives her some blank cheques which she sets about cashing. A dark and downbeat black comedy that makes a blistering attack on Aussie materialism, replete with unattractive characters and depressing situations. Surprisingly successful, the film shows all the subtlety of a sledgehammer.
COM 101 min (ort 105 min) cC VIDrel: TOUCH/TECH V

MURPHY'S LAW ** 18
J. Lee Thompson USA 1986
Charles Bronson, Kathleen Wilhoite, Carrie Snodgress, Richard Romanus, Angel Thompkins, Robert F. Lyons, Bill Henderson, James Luisi, Janet MacLachlan, Lawrence Tierney
Formula Bronson vehicle, in which he plays a tough cop (surprise, surprise), framed for the murder of his ex-wife by a psychotic woman. Plenty of action and violence, and of course the usual Bronson stoneface acting.
A/AD 96 min (ort 101 min) VIDrel: MGM/WHV V

MUSIC BOX *** 15
Constantin Costa-Gavras USA 1989
Jessica Lange, Armin Mueller-Stahl, Frederic Forrest, Lukas Haas, Donald Moffat, Cheryl Lynn Bruce, Mari Torocsik, Michael Rooker, Elzbieta Czyzewska, Sol Frieder, Albert Hall, Felix Shuman, Tibor Kenderesi, Mitchell Litrofsky
A lawyer defends her Hungarian immigrant father, on trial for war crimes committed over fifty years ago. Partly based on the John Demjanuk case (the Ukrainian-American tried in Israel) this sluggish drama focuses on the interplay between the two leads, and barely looks at the issues at stake. The guessing game as to the man's guilt soon becomes quite tiresome. However, Lange is splendid. Winner of the 1990 Berlin Golden Bear Award.
DRA 120 min (ort 126 min) VIDrel: GUILD/POLYREC L/A; PION (LV only) V/sh LV

MUSIC LOVERS, THE * 18
Ken Russell UK 1970
Glenda Jackson, Christopher Gable, Richard Chamberlain, Max Adrian, Kenneth Colley, Maureen Pryor, Isabella Telezynska, Sabina Maydelle, Andrew Faulds, James Russell, Victoria Russell, Alexander Russell, Georgina Parkinson
A wildly overblown account of Tchaikovsky's life and career, with all the self-indulgent excesses one expects from the direc-

tor. This looks mostly at the composer's homosexuality and his wife's alleged nymphomania. The few visually memorable moments sink in the morass.
DRA 118 min (ort 123 min) VIDrel: WHV V
Boa: book Beloved Friend by Catherine Drinker Bowen and Barbara Von Meck.

MUSIC MAN, THE ****　　U
Morton Da Costa USA　　1962
Robert Preston, Shirley Jones, Buddy Hackett, Hermione Gingold, Paul Ford, Pert Kelton, Ron Howard, Ewart Dunlop, Oliver Hix, Jacey Squires, Timmy Everett, Olin Britt, Susan Luckey, Harry Hickox, Charles Lane, Mary Wickes
A con-man comes to a small town, and persuades its inhabitants to start a marching band with himself as its leader. An exuberant and highly successful transfer to the screen makes this one of the few screen musicals that really works. Numbers include: "76 Trombones", "Till There Was You" and "Trouble". AA: Score/adapt (Ray Heindorf).
MUS 146 min (ort 151 min) VIDrel: MGM/WHV L/A V/sh
Boa: musical by Meredith Wilson.

MUSIC OF CHANCE, THE **　　15
Philip Haas USA　　1993
James Spacer, Mandy Pantinkin, M. Emmet Walsh, Charles Durning, Joel Grey, Samanatha Mathis, Christopher Penn, Pearl Jones, Jordan D. Spainhour, Paul Auster
A couple of small-time gamblers get out of their depth when they are suckered into a high-stakes game with a couple of rich eccentrics who have been trained by a notorious card shark. Having incurred sizeable debts, these two unfortunates agree to work them off by becoming the servants of the former. A low-key mystery of intrigue and menace.
DRA 94 min (ort 97 min) VIDrel: POLY/POLYREC L/A V
Boa: novel by Paul Auster.

MUSIC TEACHER, THE **　　U
Gerard Corbiau BELGIUM/FRANCE　　1989
Jose Van Dam, Patrick Bauchau, Sylvie Fennec, Johan Leysen, Anne Roussel, Phillipe Volter, Dinah Bryant, Joachim Dallyroc
A great opera singer retires and devotes himself to teaching two promising students, unaware that a bitter professional rival is planning to use them in his warped revenge scheme. Pleasing drama set on the eve of WW1, but chiefly memorable for the good use it makes of the music of Verdi, Mahler, Mozart and Schubert.
DRA 93 min (ort 100 min) VIDrel: MAINPIC/RTM V

MUSSOLINI AND I **　　PG
Alberto Negrin FRANCE/ITALY/SPAIN/USA　　1985
Bob Hoskins, Anthony Hopkins, Susan Sarandon, Annie Girardot, Barbara De Rossi, Fabio Testi, Dietlinde Turban, Vittorio Mezzogiorno, Pier Paolo Capponi, Francesca Rinaldi, Kurt Raab, Oliver Dominick, Hans Dieter Asner
This production concentrates on the struggle for power between Mussolini and his son-in-law, with Hoskins playing the former in a lavish production that recreates the events which accompanied the dictator's downfall after ruling Italy for more than twenty years. Unfortunately, the film fails to organise its material in a coherent way. First shown in two parts.
Aka: IO E IL DUCE; MUSSOLINI: THE DECLINE AND FALL OF IL DUCE
DRA 130 min (ort 240 min) mCab VIDrel: L/A V
Boa: story by Nicola Badalucco.

MUTANT 2 *　　18
Deran Sarafian SPAIN　　1984
Martin Hewitt, Dennis Christopher, Lynn Holly-Johnson, Luis Prendes, J.O. Bosso, Yousaf Bokhari, Yolanda Palomo, Christina Augustin, Christina San Juan, Pablo Garcia, Carlos Ramirez
Rip-off of "The Mutations" with three friends discovering that alien microbes are taking over human beings, turning their unwitting hosts into hideous lizard-like creatures. They join forces with a scientist to combat this menace. A lacklustre effort with dismal effects.
Aka: ALIEN PREDATOR; ALIEN PREDATORS; FALLING, THE
HOR 85 min (ort 92 min) VIDrel: EIV/SONOP V

MUTANT HUNT **　　18
Tim Kincaid USA　　1987
Rick Gianasi, Mary Fahey, Ron Reynaldi, Taunie Vernon, Bill Peterson, Marc Umile, Stormy Spell, Doug De Vos, Warren Ulaner,

Mark Legan, Asle Kid, Leanne Baker, Nancy Arons, Adriane Lee, Ed Malia, Eliza Little, Owen Flynn
Originally designed for hazardous occupations, humanoid cyborgs have been turned into violent mutants, with a drug secretly administered by the head of a powerful corporation. He plans to sell them to the highest bidder, but meanwhile they are terrorising the city, and the inventor of the cyborgs calls in a private operative to destroy them. A BLADE RUNNER clone set in New York of the 21st century.
Aka: MATT RYKER: MUTANT HUNT
FAN 76 min (ort 90 min) mVid VIDrel: EIV/SONOP V

MUTANT ON THE BOUNTY **　　(18)
Robert Torrance USA　　1989
John Roarke, Deborah Benson, John Furey, Victoria Catlin, John Fleck, Kylie T. Heffner, Scott Williamson, John Durbin, Pepper Martin, Rob Paulson, Fox Harris, Tim Torrance, Rick Janov, Debbie Clemmer, Michael Lubin, Rick Milne
Life on a spaceship is rather dull, until the arrival of a crazy musician, who has been beamed aboard in a somewhat altered state. When two more strangers follow, chaos ensues. A silly blend of SF and comedy.
COM 90 min (ort 93 min) CABrel: HVC

MUTE WITNESS **　　18
Anthony Waller GERMANY/RUSSIA/UK　　1995
Marina Sudina, Fay Ripley, Evan Richards, Igor Volkow, Oleg Jankowskij, Sergej Karlenkov, Alexander Buriev, Alec Guinness, Alexander Piatov, Nikolai Pastuhov, Stephen Bouser, Valeri Barahtin, Nikolai Chindjaikin, Larisa Husnolina
After being left alone all night in a film studio, a mute make-up girl, who is working on a low-budget movie that is being shot in Moscow, witnesses a murder. In no time at all, both she and those whom she works with are being terrorised by the killer. The Moscow locations really add little of interest to what is in truth, just one more unremarkable thriller. A few moments of black comedy inject the occasional unsettling note.
THR 98 min (ort 96 min) VIDrel: COLUM/SONOP V/sur

MUTILATOR, THE *　　18
Buddy Cooper USA　　1983
Matt Mitler, Ben Moore, Frances Raines, Bill Hitchcock, Jack Chatham, Trace Cooper, Ruth Martinez, Morey Lampley, Connie Rogers, Pamela Wendle Cooper, Ben Moore
A male college student and his friends closing up a beach condo at the end of the season are stalked by a deranged slasher, who proves to be the former's father in this unbelievably dull, run-of-the-mill entry.
Aka: FALL BREAK
HOR 85 min VIDrel: VIPCO/SGSVID V/h

MUTINY ON THE BOUNTY ***　　15
Lewis Milestone USA　　1962
Marlon Brando, Trevor Howard, Richard Harris, Hugh Griffith, Richard Haydn, Gordon Jackson, Tim Seely, Percy Herbert, Tarita, Duncan Lamont, Chips Rafferty, Noel Purcell
Remake of the 1935 film which tells of the famous mutiny against Captain Bligh during a voyage to the South Seas. This one is a little fairer to Bligh (a good performance by Trevor Howard), but otherwise weaker and less interesting as a piece of drama. Brando is miscast as Fletcher Christian, but, if nothing else, the film is a visual treat. Another remake followed in 1984 with THE BOUNTY.
DRA 177 min (ort 185 min) VIDrel: MGM/WHV L/A V/sh
Boa: book by Charles Nordhoff and James Hall.

MUTINY ON THE BUSES *　　PG
Harry Booth UK　　1973
Reg Varney, Stephen Lewis, Bob Grant, Anna Karen, Michael Robbins, Doris Hare, Pat Ashton, Janet Mahoney, Caroline Dowdeswell, Kevin Brennan, Bob Todd, David Lodge, Tex Fuller, Jan Rennison, Damaris Hayman, Juliet Duncan
Another dreary spin-off for your collection, based on a popular British TV series "On The Buses" all about the lives of a bus company crew and their families. In this one Varney teaches his brother-in-law to drive and they enjoy a day out at Windsor Safari Park. Preceded by ON THE BUSES and followed by HOLIDAY ON THE BUSES.
COM 84 min (ort 88 min) VIDrel: WHV V/sh

MY AMERICAN COUSIN *** PG
Sandy Wilson 1986
Margaret Langrick, John Wildman, Richard Donat, Jane Mortifee,
J.T. Scott, Camille Henderson, Darcy Bailey, Allison Hail, Samantha
Jocelyn, Babs Chula, Terry Moore, Brent Severson, Carter Dunham,
Julie Nevlon, Alexis Peat
A 1950s drama about a seventeen-year-old girl, who finds the
isolated life at "Paradise Ranch" less than idyllic. Come the
summer of '59 and she is looking forward another season of
cherry picking, until the arrival of her cousin from California,
who draws up one night in a red Cadillac. A charming and
nostalgic coming-of-age tale, that won six "Genies" or Canadian
Film Academy Awards. The sequel, AMERICAN BOYFRIENDS
followed.
DRA 85 min (ort 94 min) VIDrel: VES L/A V

MY ANTONIA *** PG
Joseph Sargent USA 1994
Jason Robards, Eva Marie Saint, Neil Patrick Harris, Elina
Lowensohn, Anne Tremko, Travis Fine, Jan Triska, Norbert Weisser,
Mira Furlan, Boris Krutonog
An evocative portrayal of farm life in 1890s Nebraska as a couple
of teenagers from different backgrounds, come of age and
develop their own special friendship. Based on Cather's novel
which is itself very autobiographical in nature.
DRA 93 min mCab VIDrel: CIC/SONOP V
Boa: novel by Willa Cather.

MY APPRENTICESHIP *** PG
Mark Donskoi USSR 1939
Alexi Lyarsky, Varvara O. Massalitinova, V. Novikov, M.
Troyanovski, E. Lilina, I. Kudriavtsev, N. Berezovskaya, F. Sekleznev,
I. Zarubina, I. Chuvelev, N. Plotnikov, Chuganov, V. Terentiev, M.
Gorlov
Part 2 of Donskoi's Maxim Gorky trilogy that began with THE
CHILDHOOD OF MAXIM GORKY and continues with MY
UNIVERSITIES. At the age of 8 young Maxim becomes appren-
ticed to a dyer and although promised an education, is forced
to learn to read by his own efforts. Later on, he leaves his
employer and undertakes a series of journeys that bring home
to him the extent of suffering of the common people. A moving
and absorbing experience.
Aka: AMONG PEOPLE; ON HIS OWN; OUT IN THE WORLD; V LYUDAKH
DRA 95 min (ort 98 min) B/W VIDrel: HEND L/A V
Boa: autobiography of Maxim Gorky.

MY BEAUTIFUL LAUNDRETTE ** 15
Stephen Frears UK 1985
Saeed Jaffrey, Daniel Day Lewis, Shirley Anne Field, Roshan Seth,
Gordon Warnecke, Derreck Branche, Rita Wolf, Souad Faress, Richard
Graham, Winston Graham, Dudley Thomas, Gary Cooper, Charu Bala
Choksi, Neil Cunningham
A contemporary and offbeat look at the Asian community in
Britain, with a young Asian boy running a luxurious launderette
with the help of a former classmate, who has become his homo-
sexual lover. Amusing in parts, violent in others, with a message
buried in there somewhere.
DRA 93 min (ort 97 min) mTV VIDrel: VISVID/POLYREC
L/A V

MY BLUE HEAVEN ** PG
Herbert Ross USA 1990
Steve Martin, Rick Moranis, Joan Cusack, Melanie Mayron, Bill
Irwin, Carol Kane, William Hickey, Deborah Rush, Daniel Stern, Ed
Lauter, Julie Bovasso, Colleen Camp
Martin gets a change of hair colour in this lightweight romp
that deals with an Italian gangster, who moves from New York
to a suburb in California as part of an FBI witness protection
scheme. Moranis plays the government agent who is employed
to keep an eye on him, not an easy task as his charge proves to
be something of a handful. Excellent performances keep the film
going, but the script is a damp squib with a clear shortage of
good gags.
COM 92 min (ort 96 min) VIDrel: MGM/WHV V/sh

MY BOYFRIEND'S BACK ** (15)
Bob Balaban USA 1993
Andrew Lowery, Traci Lind, Edward Herrmann, Mary Beth Hurt,
Danny Zorn, Bob Dishy, Austin Pendleton, Jay O'Sanders, Paul
Dooley, Matthew Fox, Libby Villari, Cloris Leachman, Paxton
Whitehead, David Womack Galewsky, Zack Steeg
A teenage girl gets taken to the prom by her boyfriend but this
seemingly normal situation is complicated by the fact that he is

a zombie whose body is in an advanced state of decay (he was
gunned down by robbers just after he asked her for a date). A
silly comedy that fails to raise more than a few chuckles.
Aka: JOHNNY ZOMBIE
COM 90 min SATrel: SKY MOVIES

MY BREAST *** PG
Betty Thomas USA 1994
Meredith Baxter, Jamey Sheridan, James Sutorious, Barbara Barrie,
Sara Botsford, Graham Harley, Nicky Guadagni, Jank Azman, Chris
Bondy, Tyrone Benskin, Dale Wilson, Graham Harley
Based on a true story, this centres on a successful New York
journalist, whose forthcoming treatment for breast cancer is the
catalyst that impels her to an examination of her life, her friends
and her values. Quite a harrowing film, it recalls CLEO FROM
5 TO 7, and though not quite as skilfully made, still retains a
good deal of insight. As ever, Baxter is impressive in her ability
to bring her character to life.
DRA 92 min mTV VIDrel: ODY/SONOP V/sh

MY BROTHER'S KEEPER *** (PG)
USA 1994
John Lithgow, Annette O'Toole, Veronica Cartwright, Ellen Burstyn,
Michael Alldredge, Don Amendola, Adilah barnes, Raye Birk, Alanna
Brown, Wren T. Brown, Wanda Lee Evans, Julie Fulton, James Greene
Lithgow gives an outstanding double performance playing twin
brothers, one of whom is HIV positive. An engrossing and
unabashed tearjerker, and at the same time partly a courtroom
drama as well.
DRA 90 min mTV SATrel: MOVIE CHANNEL

MY COUSIN VINNY * 15
Jonathan Lynn USA 1992
Joe Pesci, Ralph Macchio, Marisa Tomei, Mitchell Whitfield, Fred
Gwynne, Lane Smith, Austin Pendleton, Bruce McGill, Maury
Chaykin, Pauline Meyers, James Rebhorn, Raynor Scheine, Chris Ellis,
Michael Simpson, Lou Walker
A couple of college students are wrongfully arrested for a
murder in Alabama and unable to afford decent representation,
settle for Macchio's idiotic lawyer cousin, who needed six goes
to pass his exams. This promising opening is sadly wasted by
the hackneyed script and dumb dialogue. AA: S. Actress
(Tomei).
COM 114 min (ort 119 min) VIDrel: 20TH/TECH L/A
V/sh

MY CRAZY LIFE ** 15
Allison Anders USA 1994
Seidy Lopez, Angel Aviles, Jacob Vargas, Marlo Marron, Jessie
Borrego, Magali Alvarado, Julian Reyes, Bertilla Damas, Art Esquer,
Christina Solis, Rick Salinas, Gabriel Gonzalez, Danny Trejo, Rosa
Segura, Salma Hayek, Noah Verduzeo
A rambling look at the lives and aspirations of the female
members of the Latino gangs of L.A. that attempts to show just
how normal their hopes and dreams are, despite the all-perva-
sive culture of violence that surrounds them.
Aka: MI VIDA LOCA
COM 92 min (ort 95 min) wScrn VIDrel: TART/20TH;
ENCORE (LV only) V LV

MY DARLING CLEMENTINE *** U
John Ford USA 1946
Henry Fonda, Linda Darnell, Victor Mature, Walter Brennan, Cathy
Downs, Tim Holt, Ward Bond, Alan Mowbray, John Ireland, Jane
Darwell, Grant Withers, Roy Roberts, Russell Simpson, Francis Ford,
J. Farrell MacDonald, Don Garner
An inaccurate but highly entertaining retelling of the events
leading up to that legendary gunfight at the O.K. Corral. A
classic Ford Western full of superb performances and excellent
photography. See also GUNFIGHT AT THE O.K. CORRAL,
TOMBSTONE and WYATT EARP.
WES 92 min (ort 97 min) VIDrel: 20TH/TECH V/h
Boa: book by Stuart N. Lake.

MY FAIR LADY *** U
George Cukor USA 1964
Rex Harrison, Audrey Hepburn, Stanley Holloway, Wilfrid Hyde
White, Gladys Cooper, Jeremy Brett, Theodore Bikel, Isobel Elsom,
Mona Washbourne, Walter Burke, John Alderson, John McLiam, Ben
Wrigley, Clive Halliday, Richard Peel
A sumptuous but clinical version of PYGMALION, with some
fine Lerner and Loewe songs (and some poor ones too). A
linguist teaches a cockney girl to talk posh and passes her off as

a lady. Good as light entertainment, but a trifle stilted. AA: Pic, Dir, Actor (Harrison), Cost (Cecil Beaton), Cin (Harry Stradling), Art/Set (Gene Allen and Cecil Beaton/George James Hopkins), Score/adapt (Andre Previn), Sound (George R. Groves).
MUS   163 min (ort 175 min) wScrn
VIDrel: 20TH/TECH L/A   V/dm
Boa: play Pygmalion by George Bernard Shaw/musical by Frederick Loewe and Allan J. Lerner.

MY FAMILY ***
Gregory Nava USA
15
1995
Edward James Olmos, Jimmy Smits, Esai Morales, Eduardo Lopez Rojas, Elpidia Carrillo, Jenny Gago, Enrique Castillo, Rafael Cortes, Ivette Reina, Amelia Zapata, Jacob Vargas, Emilio de Haro, Abel Woolrich, Leon Singer, Maria Canals
An immigrant chronicle covering seventy years in the lives of the various members of a Mexican-American family, as told by an author who settles in L.A. in the 1920s. Strong performances, a good measure of humour and a solid script are the main assets in this fine drama.
Aka: MI FAMILIA
DRA   121 min (ort 126 min)   CINrel

MY FAMILY AND OTHER ANIMALS ***
Peter Barber-Fleming AUSTRALIA/UK/USA
PG
1987
Hannah Gordon, Brian Blessed, Darren Redmayne, Anthony Calf, Sarah-Jone Holm, Guy Scantlebury, John Normington, Christopher Godwin, Yiorgas Dialegmanos, Edward Parsons, Angela Barrow
In 1935 the widow Durrell moves to Corfu with her four children, one of whom, young Gerald, is a budding naturalist at the age of ten. A leisurely and amusing adaptation of Durrell's popular book.
DRA   228 min (2 cassettes) mTV   VIDrel: BBC/TECH   V/h
Boa: book by Gerald Durrell.

MY FAMILY TREASURE **
Rolfe Kanefsky/Edward Staroelsky USA
(PG)
1993
Dee Wallace Stone, Theodore Bikel, Alex Vincent, Bitty Schram, Bill Weber, Alex Vincent, Melissa Perez, Boris Krutongo, V. Zamhartcenko, E. Morgunov, V. Krachkovskaya, Howard Wiseman, N. Merzlikin
A young boy learns that his grandfather was given a Faberge egg by the Czar and that his mother once travelled to the Soviet Union in an attempt to recover it.
JUV   94 min   SATrel: MOVIE CHANNEL

MY FATHER IS COMING ***
Monika Treut GERMANY/USA
18
1991
Alfred Edel, Shelly Kastner, David Bronstein, Annie Sprinkle, Mary Lou Graulau, Michael Massee, Fakir Musafar, Mario de Colombia, Dominique Gaspar, Flora Gaspar, Israel Marti, Bruce Benderson, Rebecca Lewin, Stephen Feld
A German actress in New York receives a visit from her father that gets her involved in some strange relationships with both transsexuals and lesbians. Meanwhile, her dad busies himself with former porno star Sprinkle. In a sense this is a film mainly about sexual identity and acceptance, that has both characters gaining an insight into themselves, and on that count at least, it offers a few interesting observations if not much of a plot.
DRA   81 min (English dialogue)   CINrel

MY FATHER, MY HERO ***
Gerard Lauzier FRANCE
PG
1991
Gerard Depardieu, Marie Gillain, Patrick Mille, Catherine Jacob, Charlotte de Turckheim, Jean-Francois Rangasamy, Koomaren Chetty, Evelyne Lagesse, Benoit Allemane, Nicolas Sobrido, Yan Brian, Franck-Olivier Bonnet, Harriet Batchelor
In this delightful film, Depardieu plays the divorced father of a precocious teenage girl, who picks her up from her mother's house with the intention of taking her on holiday. But at the resort, her budding interest in the opposite sex leads her into lying about her father's background, describing him as both lover and an adventurer in order to impress a local lad. Depardieu plays along with this ruse for as long as he can. Remade as MY FATHER, THE HERO.
Aka: MON PERE, CE HEROS; MY FATHER, THE HERO
COM   99 min (ort 104 min) wScrn   VIDrel: TART/20TH   V

MY FATHER, MY SON ***
Jeff Bleckner USA
15
1988
Karl Malden, Keith Carradine, Margaret Klenck, Michael Horton, Dirk Blocker, Jenny Lewis, Billy Sullivan, Grace Zabriskie, Mack Dryden, Tim Choate, Larry Larson, Marilyn Martin, Sandra Carradine, Libby Whitmore, Johnny Ziomek
Admiral Zumwalt was Secretary of the Navy at the time of the Vietnam War, and authorised the use of the defoliant Agent Orange, which caused his son (who was fighting in Vietnam at the time), and others to develop cancer. This poignant but remarkably restrained account focuses on their relationship and the guilt Zumwalt Sr felt. Zumwalt Jr died a short while after the film was released.
DRA   96 min (ort 100 min) mTV   VIDrel: MGM L/A   V
Boa: book by Admiral Elmo Zumwalt and Elmo III Zumwalt.

MY FATHER, THE HERO **
Steve Miner USA
PG
1994
Gerard Depardieu, Katherine Heigl, Dalton James, Lauren Hutton, Faith Prince, Stephen Tobolowsky, Ann Hearn, Robyn Peterson, Frank Renzulli, Jeffrey Chea, Manny Jacobs, Stephen Burrows, Michael Robinson, Robert Miner, Betty Miner
A precocious fourteen-year-old girl goes on vacation with her father but tells everyone whom she meets that he is really her lover, which leads to problems all round, in this Americanised version of the hit French comedy MY FATHER, MY HERO.
COM   86 min (ort 90 min)   VIDrel: TOUCH/TECH L/A   V

MY FATHER'S GLORY ****
Yves Robert FRANCE
U
1990
Philippe Caubere, Nathalie Roussel, Didier Pain, Therese Liotard, Julien Ciamaca, Victorien Delamare, Joris Molinas, Paul Crauchet, Pierre Maguelon, Michel Modo, Victor Garrivier, Jean Rougerie, Raoul Curet, Rene Loyon
Inspired by the first volume of Pagnol's autobiography, this utterly charming film takes as its starting point the birth of Marcel, whose happy childhood is marked by both his keen observation of the unaccountable ways of adults and his penchant for spinning tales. Though often rather artificial in structure, this chronicle of childhood is both intelligent and rewarding. Followed by MY MOTHER'S CASTLE.
Aka: LA GLOIRE DE MON PERE
DRA   106 min (ort 111 min) wScrn   VIDrel: ARTIF/20TH   V/sur
Boa: autobiography of Marcel Pagnol.

MY FAVOURITE BRUNETTE ***
Elliott Nugent USA
PG
1947
Bob Hope, Dorothy Lamour, Peter Lorre, Lon Chaney Jr, John Hoyt, Reginald Denny, Charles Dingle, Frank Puglia, Ann Doran, Willard Robertson, Jack LaRue, Charles Atndt, Garry Owen, Richard Keane, Anthony Caruso, Matt McHugh, Ray Teal
A photographer gets involved with gangsters when he chivalrously offers to help a lady in trouble. A good vehicle for Hope, with Lorre and Chaney adding some substance to an otherwise rather insubstantial piece of nonsense. The script is by Edmund Beloin.
COM   87 min (ort 90 min) B/W   VIDrel: ORBIT/DISC   V

MY FAVOURITE WIFE ***
Garson Kanin USA
U
1940
Cary Grant, Irene Dunne, Gail Patrick, Randolph Scott, Anne Shoemaker, Scotty Beckett, Donald MacBride, Mary Lou Harrington, Hugh O'Connell, Pedro De Cordoba, Granville Bates, Brandon Tynan, Leon Belasco, Harold Gerald
Witty comedy about a man whose first wife was presumed dead but returns home to find him remarried. Our hapless hubby now has to choose between them. Written by Sam and Bella Spewack and Leo McCarey. Remade as MOVE OVER, DARLING.
COM   87 min B/W   VIDrel: VCC/DISC/COLUM L/A   V

MY FAVOURITE YEAR ***
Richard Benjamin USA
PG
1982
Peter O'Toole, Mark Linn-Baker, Jessica Harper, Joseph Bologna, Bill Macy, Lainie Kazan, Anne De Salvo, Lou Jacobi, Adolph Green, George Wyner, Selma Diamond, Cameron Mitchell, Gloria Stuart, Basil Hoffman, Tony DiBenedetto
An aspiring young TV writer has to take care of an ageing but carousing screen swashbuckler, who is appearing on a 1950s TV show as that week's guest star. A cheerfully lighthearted romp, with O'Toole perfectly cast, and a mass of well researched period detail adding substance to the thin plot.
COM   88 min (ort 92 min)   VIDrel: MGM L/A   V

MY FIRST FORTY YEARS **
Carlo Vanzina ITALY
15
1989
Carol Alt, Elliott Gould, Jean Rochefort, Pierre Cosso, Massimo

Venturiello, Capucine, Isabel Russinova, Paolo Quattrini, Riccardo Garrone, Ted Teocoli, Sebatiano Somma, Elena Pompei, Giuseppe Pambieri, Martine Brochard, Cyrus Elias
A woman out to achieve a comfortable lifestyle marries an aristocrat and subsequently acquires a wealthy lover. However, her carefully laid plans go awry when she finds herself falling in love.
Aka: FIRST FORTY YEARS, THE; I MIEI PRIMI QUARANT'ANNI; MY WONDERFUL LIFE
DRA 104 min (ort 107 min) VIDrel: 20VIS/SONOP L/A V
Boa: novel by Marina Ripa Di Meana.

MY FOOLISH HEART ***   U
Mark Robson USA   1949
Dana Andrews, Susan Hayward, Kent Smith, Lois Wheeler, Jessie Royce Landis, Robert Keith, Gigi Perreau, Karin Booth, Todd Karns, Philip Pine, Martha Mears, Edna Holland, Jerry Paris, Marietta Canty, Barbara Woodell, Jerry Paris
The story of a WW2 romance between a soldier and a young girl that is adroitly handled and has winning performances that easily overcome the contrived nature of the plot. The memorable theme song is by Victor Young.
DRA 94 min (ort 98 min) B/W VIDrel: VCC/DISC V
Boa: story Uncle Wiggly in Connecticut by J.D. Salinger

MY FRIEND WALTER **   (U)
Gavin Millar UK   1992
Prunella Scales, Ronald Pickup, Polly Grant, Prunella Scales, Don Henderson, Louise Jameson, Constance Chapman, Richard Strange, Madge hindle, George Innes, Arthur Whybrow, Patrick Godfrey, Les Bolan, Brian Payser, Will Kenton
A young girl meets the ghost of Walter Raleigh and enjoys his help when it look like her family are about to lose their farm to their creditors. A lively and competent fantasy tale.
JUV 90 min SATrel: MOVIE CHANNEL

MY GIRL *   PG
Howard Zieff USA   1991
Macaulay Culkin, Anna Chlumsky, Dan Aykroyd, Jamie Lee Curtis, Richard Masur, Griffin Dunne, Ann Nelson, Peter Michael Goetz, Jane Hallaren, Anthony Jones, Tom Villard, Lara Steinick, Kristian Truelsen, Dave Caprita, Jody Wilson
Set in 1972, this tale opens in a fictitious Pennsylvanian town, where shy eleven-year-old Thomas (the son of the town's mortician) enjoys a friendship with his young sweetheart Vada. A little rivalry breaks out when Thomas' father gets a girlfriend of his own, and there are some funny moments. The film unwisely attempts to be both a grotesque comedy and a mawkish drama, ending on a distinctly sour note, with Thomas stung to death by killer bees.
COM 98 min (ort 102 min) cC
VIDrel: VCC/DISC/COLUM V/sur

MY GIRL 2 *   PG
Howard Zieff USA   1993
Dan Aykroyd, Jamie Lee Curtis, Anna Chlumsky, Austin O'Brien, Richard Masur, Christine Ebersole, John David Souther, Angeline Ball, Aubrey Morris, Gerrit Graham, Anthony R. Jones, Ben Stein, Keone Young, Richard Beymer, Jodie Markell
Maudlin and unsuccessful sequel to MY GIRL without charm or grace in which young Vada (Chlumsky – who played Culkin's girlfriend in the first film) is now a teenager. Encouraged by her dad and step-mother, she sets off for L.A. intent on finding out all she can about her dead mother, who died giving birth to her. This task brings her into contact with various colourful characters along the way. Contrived and over-sentimental nonsense.
DRA 95 min (ort 99 min) cC VIDrel: COLUM/SONOP V/sur

MY GIRLFRIEND'S BOYFRIEND ***   PG
Eric Rohmer FRANCE   1987
Emmanuelle Chaulet, Sophie Renoir, Anne-Laure Meury, Eric Viellard, Francois-Eric Gendron
Set in a new town just outside Paris, this film presents an ironic account of the relationship between two rather different young women, and follows the complex musical chairs they play with a couple of young men. Sixth in the series of Rohmer's "Comedies and Proverbs" this is a witty and often perceptive comedy of manners, not exactly profound, but quite charming for all that.
Aka: BOYFRIENDS AND GIRLFRIENDS; L'AMI DE MON AMI
DRA 99 min (ort 102 min) VIDrel: ARTIF/20TH V

MY GRANDAD'S A VAMPIRE **   PG
David Blyth NEW ZEALAND   1992
Al Lewis, Justin Gocke, Milan Borich, Noel Appleby, Pat Evison, Sean Duffy, S. Rands
A boy goes to New Zealand and discovers the title secret but that fact does not stop him and his grandpa from sharing some exciting adventures, in this feeble comedy.
Aka: MOONRISE; MY GRANDPA IS A VAMPIRE
COM 89 min (ort 90 min) VIDrel: MED/POLYREC L/A V/sh

MY LEFT FOOT ***   15
Jim Sheridan UK   1989
Daniel Day-Lewis, Ray McAnally, Brenda Fricker, Fiona Shaw, Cyril Cusack, Hugh O'Conor, Adrian Dunbar, Ruth McCabe, Alison Whelan, Declan Croghan, Eanna MacLiam, Marie Conmee, Phelim Drew, Patrick Laffan, Derry Power
A film based on the true story of Christy Brown, who was born with cerebral palsy into a large, working class Irish family, and endured years of humiliating helplessness. Eventually however, he learns to use his foot to both paint and write, and secures recognition and independence. A moving and inspiring tale. See also GABY: A TRUE STORY. AA: Actor (Day-Lewis), S. Actress (Fricker).
DRA 98 min (ort 103 min) VIDrel: ELPIC/POLYREC V/h

MY LIFE **   15
Bruce Joel Rubin USA   1993
Michael Keaton, Nicole Kidman, Haing S. Ngor, Brady Whitford, Queen Latifah, Michael Constantine, Toni Sawyer, Rebecca Schull, Lee Garlington, Romy Rosemont, Mark Lowenthal, Danny Rimmer, Ruth DeSosa, Richard Schiff, Stephen Taylor Knot
A man with terminal cancer makes a video of his life for his first child whom he will not live to see and in so doing comes to terms with the pain of his untimely demise. A sentimental and shallow effort, with Kidman adding little in her role as his saint-like wife. Overlong and disappointing, this was both written and directed by Rubin (in his directing debut) and is more than just a little reminiscent of GHOST – a film he also scripted.
DRA 112 min (ort 117 min) VIDrel: GUILD/20TH V/sur

MY LIFE AND TIMES WITH ANTONIN ARTAUD ***   (PG)
Gerard Mordillat FRANCE   1993
Sami Frey, Marc Barbe, Julie Jezequel, Valerie Jeannet, Clothilde de Bayser, Charlotte Valandrey, Anne Barbe, Philippe Baille, Philippe Beaufort, Alain Boissard, Jean-Marcel Bouguereau, Renee Cariven, Catherine Carree
Biopic on title character who in Paris in the 1940s had great influence on the theatre – founding the "Theatre of Cruelty", besides exploring his interest in the occult and experimenting with drugs. Not an easy subject for a film, the last two years of his life are explored through the eyes of a devotee and fellow poet Jacques Prevel. A compelling portrait of an unappealing albeit larger-than-life character, superbly performed and scripted.
Aka: EN COMPAGNIE D'ANTONIN ARTAUD
DRA 90 min B/W CINrel
Boa: book by Jacques Prevel.

MY LIFE AS A DOG ***   PG
Lasse Hallstrom SWEDEN   1985
Anton Glanzelius, Anki Liden, Tomas Von Bromssen, Melinda Kinnaman, Kicki Rundgren, Ing-Marie Carlsson, Manfred Serner, Lennart Hjulstrom, Christina Carlwind, Leif Ericsson, Ralph Carlsson
A twelve-year-old boy is sent to stay with relatives in the countryside while his sick mother recuperates. A lighthearted and touching look at childhood in Sweden of the 1950s, with the boy (who compares himself to the Russian dog shot into space) finding unexpected friendship and adventure among the town's warmhearted eccentrics. Jonsson's novel is largely autobiographical.
Aka: MITT LIV SOM HUND
DRA 97 min (ort 101 min) VIDrel: ARTIF/20TH V/h
Boa: novel by Reidar Jonsson.

MY LUCKY STARS ***   15/18
Samo Hung Kam-Bo HONG KONG   1985
Jackie Chan, Samo Hung Kam-Bo, Yuen Biao, Richard Ng Yil-Hon, Eric Tsang Chi-Wai, Andy Lau Tak-Wah, Michiko Nishikawa, Charlie Ching, Stanley Fung Shui-Fan, Lam Ching-Ying
Another Jackie Chan blend of comedy and martial arts action. This one is a police thriller in which he plays an undercover cop

assigned to capture a notorious criminal and destroy a powerful underworld organisation. Followed by TWINKLE TWINKLE LUCKY STARS.
A/AD   84 min Cut (2 sec – 15 ver); 90 min (18 ver) (ort 99 min)   VIDrel: IMPENT L/A   V

MY MOTHER'S CASTLE ****
Yves Robert FRANCE
Philippe Caubere, Nathalie Roussell, Didier Pain, Therese Liotard, Julien Ciamaca, Victorien Delamere, Joris Molinas, Julie Timmerman, Paul Crauchet, Philippe Uchan, Patrick Prejean, Pierre Maguelon, Michel Modo, Jean Carmet
This sequel to MY FATHER'S GLORY (it's based on the second volume of Pagnol's three-volume work) begins with ten-year-old Marcel now back in the city after his summer holidays. Though preparing for his entrance examinations for the Lycee, his mind inevitably drifts back to the idyllic summer he spent in Provence. Fortunately, his schoolteacher father decides to make a return trip to their holiday home each weekend. A warm and lyrical film.
Aka: LE CHATEAU DE MA MERE
DRA   95 min (ort 98 min) wScrn   VIDrel: ARTIF/20TH V/sur
Boa: autobiography of Marcel Pagnol.
U
1991

MY NAME IS KATE **
Rod Hardy USA
Donna Mills, Daniel J. Travanti, Nia Peeples, Eileen Brennan, Ryan Reynolds, Babs Chula, Linda Darlow, Deanna Milligan, Lovie Elie, Tara Frederick, Maxine Miller, Merrilyn Gann, Javeson Boyd, Stephen E. Miller, Cinnamon Bond
A successful woman battles both booze and drug addiction in this occasionally disturbing but never terribly original work. Films such as CLEAN AND SOBER, "My Name Is Bill W" and DAYS OF WINE AND ROSES covered the subject of addiction with a lot more conviction and impact.
DRA   90 min mTV   VIDrel: ODY/SONOP   V
15
1993

MY NEIGHBOUR TOTORO ***
Hayao Miyazaki JAPAN
Voices of: Lisa Michaelson, Cheryl Chase, Greg Snegoff, Natalie Core, Kenneth Hartman
In the countryside two young sisters encounter a giant furry forest spirit with which they become friends, and all three embark on some magical adventures. The very high production values of this leisurely tale ensure it stands out from the rest, especially with regard to the English language dubbing, which for once is really effective. The script is by Miyazaki.
ANIM   83 min   SATrel: MOVIE CHANNEL
(U)
1988

MY NEW GUN **
Stacy Cochran USA
Diane Lane, James LeGros, Tess Harper, Bruce Altman, Stephen Collins, Bill Raymond, Maddie Corman, Bill Raymond, Suzzy Roche, Phillip Seymour Hoffman, Patti Chambers, Stephen Pearlman, Leslie Brett Daniels, Paul J.Q. Lee, Kent Gash
After his friends purchase a gun, a young yuppie does likewise much to the chagrin of his wife who becomes highly nervous at the prospect of having a firearm in their home. Worse is to come, however, when their unbalanced neighbour borrows the weapon for reasons of his own. An undisciplined and bleak black comedy that neither frightens nor amuses despite the fine efforts of a very capable cast.
DRA   95 min (ort 99 min)   VIDrel: POLY/POLYREC L/A V
15
1992

MY NIGHT WITH MAUD ***
Eric Rohmer FRANCE
Jean-Louis Trintignant, Francoise Fabian, Marie-Christine Barrault, Antoine Vitez, Leonide Kogan, Anne Dubot, Guy Leger, Marie Becker, Marie-Claude Rauzier
In a small provincial town, a strait-laced Catholic engineer and an elegant divorcee share a mutual attraction. At her invitation he spends the night at her apartment, but is too bashful to court her and marries the girl he has already decided on. The fourth of Rohmer's "Six Moral Tales" is a comedy-of-manners, subdued, witty and elegant but too verbose and intellectual for its own good. Written by Rohmer, this was Barrault's last film. Followed by CLAIRE'S KNEE.
Aka: MA NUIT CHEZ MAUD; MY NIGHT AT MAUD'S
DRA   105 min (ort 111 min) B/W   VIDrel: CONNO/RTM V
U
1969

MY OWN PRIVATE IDAHO ***
Gus Van Sant Jr USA
River Phoenix, Keanu Reeves, James Russo, William Richert, Rodney Harvey, Chiara Caselli, Michael Parker, Jessie Thomas, Flea, Grace Zabriskie, Tom Troupe, Udo Kier, Sally Curtice, Robert Lee Pitchlynn, Mickey Cottrell
This idiosyncratic tale casts Phoenix and Reeves as a couple of drug-taking layabouts, one of whom, Scott, has chosen this life simply to spite his politician dad. His buddy Mike comes from a less privileged background, and is happy to sell his body to all and sundry. These two, part of a "family" of similar misfits, set out on a hopeful journey in search of Mike's mother, who abandoned him as a child. A highly original film, warm and sensitive.
DRA   99 min (ort 105 min)   VIDrel: 20TH/TECH   V/sh
18
1991

MY SAMURAI **
Fred Dresch USA
Julian Lee, Mako, Bubba Smith, Terry O'Quinn, Jim Turner, John Kallo, Tupper Callom, Carlos Palomino, Lynne Hart, Raymond Pearl, Christophe, Dale Girard, Barbara Champion, Keisuke Omaru, Jeff Austin, Steve Dalton, C. Edward McNeil
A young boy witnesses a murder and is pursued by the gang responsible. He asks a friend, a martial-arts expert for help, and after running from both his pursuers and the police, is able to conquer his fears and learn self-defence.
A/AD   85 min (ort 87 min)   VIDrel: COLUM/SONOP   V
15
1992

MY SISTER'S KEEPER **
David Saperstein USA
Peter Weller, Kathy Baker, John Glover, Bill Smitrovich, Rhetta Hughes
A woman finds her husband has been brutally murdered on their isolated island home and their boat is missing. There follows a terrifying life and death struggle with the murderer. Interestingly, the film COCOON was based on a novel by Saperstein.
THR   96 min   VIDrel: SPEAR/SONOP   V
15
1985

MY SON JOHN **
Leo McCarey USA
Robert Walker, Helen Hayes, Dean Jagger, Van Heflin, Minor Watson, Frank McHugh, Richard Jaeckel
The eldest son of a Catholic family is revealed to be a Communist, a revelation that leads to him becoming disowned by them. Not at all effective in terms of entertainment, but as a piece of history it shows just how very rabid anti-Communist feeling was at the time, a situation that the powers of Hollywood were not slow to exploit. Walker died before the film was completed.
DRA   122 min B/W   SATrel: MOVIE CHANNEL
(PG)
1952

MY SON JOHNNY **
Peter Levin USA
Michele Lee, Ricky Schroder, Corin Nemec, Rip Torn
Based on real events, this court-room drama examines the plight of a mother obliged to give testimony regarding her son, who was forced to kill his thuggish brother in self defence and now faces a possible life sentence. Fine performances all round make the most of this unedifying tale.
DRA   89 min (ort 96 min) mTV   VIDrel: ODY/SONOP V/h
15
1991

MY STEPMOTHER IS AN ALIEN *
Richard Benjamin USA
Dan Aykroyd, Kim Basinger, Jon Lovitz, Alyson Hannigan, Joseph Maher, Seth Green, Wesley Mann, Tony Jay, Peter Bromilow, Nina Henderson, Adrian Sparks, Juliette Lewis, Tanya Fenmore. Voieces of: Ann Prentiss, Harry Shearer
A widowed scientist inadvertently affects the gravitational field of a planet in another galaxy, and is sought out by a beautiful alien, whom he falls in love with and marries. However, her chief concern is to find a way to save her homeworld from imminent destruction. A witless blend of feeble special effects, comedy and SF, that does absolutely nothing with its single idea and ends on a sentimental and imbecilic note for good measure.
COM   103 min (ort 108 min)   VIDrel: VCC/COLUM L/A V/sh
15
1988

MY SUMMER STORY **
Bob Clark USA
Charles Grodin, Kieran Culkin, Mary Steenburgen, Christian Culkin, Troy Evans, Al Mancini, Roy Brocksmith, Glenn Shadix, Dick
PG
1994

O'Neill, Wayne Grace, Tedde Moore, Whitby Hertford, Wigdor Geoffrey, David Zahorsky, Darwyn Swalve
Sequel of sorts to A CHRISTMAS STORY that deals with the life of a country family in the 1940s and their various problems and conflicts. Once again Clark and Jean Shepherd collaborate but this offering is far weaker and less well than their first effort.
Aka: IT RUNS IN THE FAMILY
COM   81 min (ort 85 min) cC   VIDrel: MGM/WHV   V/sur

MY THERAPIST ***
Al Rossi (Gary Legon) USA
18
1983
Marilyn Chambers, David Winn, Judith Jordan, Kate Ward, Buck Flower, Robbie Lee
Film version of a play first performed in L.A. in 1971, with Chambers a psychologist who helps men with sexual problem, often acting as a fantasy model for them. However, she often becomes emotionally involved in their problems, and has a boyfriend who has fallen in love with her. But he finds he cannot accept her doing this kind of work, and the film ends with a sad parting on the beach. No great insights are offered, but Chambers does give a competent performance.
Aka: SEX SURROGATE
A   77 min Cut (4 min 34 sec plus cut at film release – ort 81 min)   VIDrel: MOPIC/QUANT   V

MY TUTOR *
George Bowers USA
18
1982
Martin Lattanzi, Caren Kaye, Kevin McCarthy, Arlene Golonka, Clark Brandon, Bruce Bauer, Crispin Glover, Amber Denyse Austin, John Vargas, Rex Ryon, Maria Melendez, Graem McGavin, Kathleen Shea, Brioni Farrel, Kitten Navidad, Mora Gray
A boy's parents get him a beautiful tutor to help him pass his French exams, but she gives him lessons in sex as well. A dire comedy that might well have been better as a straight drama. See also PRIVATE LESSONS and HOMEWORK for two films on much the same theme.
COM   97 min   VIDrel: STABL L/A   V

MY UNIVERSITIES ***
Mark Donskoi USSR
PG
1940
Nikolai Valbert, Stepan Kaloubov, Nicolas Dorokhine, N. Plotnikov, D. Segal, I. Fyedotova, L. Sverdin, V. Maruta, Pavlik Dojdev
The final part of Donskoi's 3-part work on the life of Gorky that began with THE CHILDHOOD OF MAXIM GORKY and then MY APPRENTICESHIP.
Aka: MOI UNIVERSITETI; MY UNIVERSITY; UNIVERSITY OF LIFE
DRA   95 min (ort 104 min) B/W   VIDrel: HEND L/A   V

MY YOUNG AUNTIE **
Liu Chia-Liang (Lau Ka-Leung) HONG KONG
15
1981
Liu Chia-Liang (Lau Kar-Leung), Kara Hui Ying-Hung, Hsiao Bo, Wang Lung-Wei
A Westernised Chinese man takes as a bride a very traditional Chinese woman, and they are obliged to fight against a family villain who is out to steal the land they have inherited. Typical martial arts mayhem from the Shaw Brothers studio.
MAR   115 min wScrn dubbed   VIDrel: MADE/RTM   V

MYRA BRECKINRIDGE **
Michael Sarne USA
18
1970
Raquel Welch, John Huston, Mae West, Rex Reed, Farrah Fawcett, Roger C. Carmel, George Furth, Calvin Lockhart, Jim Backus, John Carradine, Andy Devine, Grady Sutton, Tom Selleck
A film critic has a sex-change operation, and goes to Hollywood to get her own back on both men and mankind. A highly disjointed attempt to bring Vidal's caustic novel to the screen, making use both of old movie clips and stars such as West. The film's greatest failing is not its vulgarity, but simply its sheer lack of coherence and it hardly deserves the degree of critical abuse that has been heaped upon it.
COM   93 min (ort 100 min)   VIDrel: 20TH/TECH   V/h
Boa: novel by Gore Vidal.

MYSTERIOUS ISLAND ***
Cy (Cyril) Endfield UK
U
1961
Michael Craig, Herbert Lom, Joan Greenwood, Michael Callan, Gary Merrill, Beth Rogan, Percy Herbert, Dan Jackson, Nigel Green
Two Confederate officers escape from their Union captors by balloon and land on a strange island inhabited by prehistoric monsters (special effects are by Ray Harryhausen). An uneven but highly enjoyable adventure with excellent monsters. The

score is by Bernard Herrmann. Filmed once before in 1929 and as a cinema serial in 1951.
JUV   97 min (ort 101 min)   VIDrel: COLUM/SONOP L/A   V
Boa: novel by Jules Verne.

MYSTERY DATE **
Jonathan Wacks USA
15
1991
Ethan Hawke, Teri Polo, Brian McNamara, Fisher Stevens, B.D. Wong, Tony Rosato, Don Davis, James Hong, Victor Wong, Ping Wu, Duncan Fraser, Jerry Wasserman, Merrilyn Gaun, Stephen Chang, Russell Jung, Michelle Little
A young college kid is to shy to ask an attractive neighbour for a date, so he gets his elder brother to arrange things. She agrees and he gets all set for their big date but himself being chased by both gangsters and the police, in this so-so black comedy.
COM   93 min (ort 98 min)   VIDrel: 20VIS/SONOP   V/sur

MYSTERY MANSION, THE **
David E. Jackson USA
(PG)
1983
Dallas McKennon, Greg Wynne, Jane Ferguson, Randi Brown, Lindsay Bishop, David Wagner, Barry Hostetler, Joseph D. Savery, Robert Goldsberry, David Turner, John Knotts, Rolland Grubbe, Brian Burton, Lee Estes, Richard Poe
The descendants of a pioneer family who were killed by bank robbers after being taken hostage, go on a dangerous hunt for the money from the original robbery. A fair effort with a minor supernatural element.
A/AD   91 min (ort 95 min)   SATrel: SKY MOVIES

MYSTERY OF EDWIN DROOD, THE **
Timothy Forder USA
15
1993
Robert Powell, Nanette Newman, Finty Williams, Jonathan Phillips, Gemma Craven, Rupert Rainsford, Michelle Evans, Rosemary Leach, Ronald Fraser, Glyn Houston, Peter Pacey, Andrew Sachs, Freddie Jones, Kate Williams, Marc Sinden
A choir-master is so consumed with jealousy that he is driven to murder his nephew, hiding his body in a lime-pit. To conceal his guilt, he attempts to frame an innocent man, but his machinations are eventually revealed. An uninspired recent adaptation of an incomplete novel, reasonably watchable but a mite disappointing. The 1935 Claude Rains version was much better.
DRA   97 min (ort 112 min)   VIDrel: CURZON/20TH   V/sur
Boa: novel (unfinished) by Charles Dickens.

MYSTERY OF THE WAX MUSEUM, THE **
Michael Curtiz USA
PG
1933
Lionel Atwill, Fay Wray, Glenda Farrell, Allen Vincent, Frank McHugh, Arthur Edmund Carewe, Gavin Gordon, Edwin Maxwell, Holmes Herbert, DeWitt Jennings, Monica Bannister, Matthew Betz, Thomas E. Jackson, Bull Anderson, Pat O'Malley
This early horror film in two-colour Technicolor was lost for a good many years. Despite certain dated and contrived effects, its treatment of the theme of a disfigured sculptor covering his victims in wax is convincing and well handled, while good art direction is another plus. Remade some years later as "Wax Mask".
HOR   76 min   VIDrel: VISCOM/RTM   V

MYSTERY SCIENCE THEATER 3000: THE MOVIE *
Jim Mallon USA
PG
1996
Michael J. Nelson, Trace Beaulieu, Kevin Murphy
A feature-length spin-off from an intermittently amusing US TV show, that worked best in its shorter TV format. Here, we are offered the story of a mad scientist with plans for world domination, which he hopes to achieve by exposing everyone to the worst movies ever made. Since most of the story seems to consist of an astronaut watching THIS ISLAND EARTH, the whole purpose of the movie is revealed as a pointless sham in the face of this superior 1950s film.
COM   71 min cC   VIDrel: CIC   V/dm

MYSTERY TRAIN ***
Jim Jarmusch USA
15
1989
Screamin' Jay Hawkins, Masatoshi Nagase, Joe Strummer, Youki Kudoh, Cinque Lee, Nicoletta Braschi, Elizabeth Bracco, Rick Aviles, Steve Buscemi, Tom Noonan, Rockets Redglare, Rufus Thomas, Tom Waits (voice only)
Three interwoven tales from the director of DOWN BY LAW and STRANGER THAN PARADISE, and likewise marked by a

quirky set of observations and feel for the bizarre. The stories are all set at a seedy Memphis hotel, and revolve around the foreign guests who check in. A weird, minimalist comedy that's often surprisingly droll. Music is by John Lurie.
COM   110 min (ort 113 min)   VIDrel: ARTIF/20TH   V/h

MYSTIC PIZZA *** 15
Donald Petrie USA 1988
Annabeth Gish, Julia Roberts, Lili Taylor, Vincent Phillip D'Onofrio, Adam Storke, William R. Moses, Conchata Ferrell, Joanna Merlin, Arthur Walsh, Porscha Radcliffe, John Fiore, Gene Amorso, Sheila Ferrini, Janet Zarish, Ann Flood
Three pizza-house waitresses in the Connecticut resort of Mystic, discover that life offers more than they could have hoped for in this superficial but often quite touching coming-of-age comedy. The fine New England locations and an excellent cast help mask the basic shallowness of the script.
COM   103 min   VIDrel: 4-FRONT/POLYREC/VISVID   V

MYTH OF THE MALE ORGASM * (PG)
John Hamilton CANADA 1993
Bruce Dinsmore, Miranda De Pencier, Mark Camacho, Burke Lawrence, Ruth Marshall, Macha Grenon, Micheline Dahlander, Felicia Schulman, Gianpolo Bini, Genevieve Angers, Deena Aziz, Claudia Besson, Roy McLean, Jean-Claude Page
A thirty-year-old psychology lecturer volunteers to take part in a feminist experiment that takes the form of an interrogation in a darkened room, but despite all the rules a kind of rapport develops between him and his female interrogator. A lame attempt to redress sex stereotyping that never questions how it is possible to generalise about half of mankind. A confused and irritating effort, of contrived scripting and flat characterisation.
DRA   90 min   TVrel: C5

N

NADA *** 18
Claude Chabrol FRANCE/ITALY 1973
Fabio Testi, Lou Castel, Michel Aumont, Mariangela Melato, Maurice Garrel, Viviane Romance, Didier Kaminka, Michel Duchaussoy, Katia Romanoff
A group of French terrorists kidnap an American diplomat and hold him to ransom, their task being made fairly easy as he is in the habit of frequenting a brothel. However, this action leads to conflicts within the group, and there is the inevitable violent confrontation with the police. A very violent work, with occasional flashes of mordant humour, but no clear message. This unsettling film is made with enormous care, the title being the Spanish for "nothing".
THR   106 min wScrn   VIDrel: ARTPRO/RTM   V

NADJA *** 15
Michael Almereyda USA 1995
Suzy Amis, Galaxy Craze, Martin Donnovan, Peter Fonda, Jared Harris, Elina Lowensohn, Nic Ratner, Karl Geary, Jack Lotz, Isabel Gillies, David Lynch, Jose Zunga, Bernadette Jurkowski, Jeff Winner, Sean, Bob Gosse, Rome Neal, Anna Roma
A contemporary story of vampire twins, one a beautiful woman who revels in her nature and haunts the streets of Manhattan in search of victims, but the other a man who wishes to escape his destiny. Fonda plays the vampire-hunter who is out to put an end to them. The wise use of B/W photography helps recreate the unworldly atmosphere of Murnau's NOSFERATU, but the film's modern day setting soon breaks the spell. Flawed, but definitely worth a look. See also THE HUNGER.
HOR   100 min B/W   VIDrel: ICAPRO/MANGA   V/sh

NAILS ** 15
John Flynn USA 1992
Dennis Hopper, Anne Archer, Tomas Milian, Cliff De Young, Keith David, Jay Acovone, Carlos Carrasco, Charles Hallahan, Earl Billings, Luis Ramos, Raymond Cruz, Juan Fernandez, Teresa Crespo, Brian Markinson, John Hawkes, Danny Trejo
A tough action thriller with Hopper giving his usual overblown performance, this time as a mean and moody police detective nicknamed "Nails". When his partner is killed in a street ambush, he sets about tracking down the killers, a long process that eventually takes him to a confrontation with a drug-dealing master criminal. A production-line cops-and-robbers film with all the standard trimmings.
THR   96 min mCab   VIDrel: MED/POLYREC L/A   V

NAKED *** 18
Mike Leigh UK 1993
David Thewlis, Lesley Sharp, Katrin Cartlidge, Greg Cruttwell, Peter Wight, Claire Skinner, Ewen Bremner, Susan Vidler, Deborah Maclaren, Gina McKee, Darren Tunstall, Elizabeth Berrington, Carolina Giammetta, Robret Putt, Lynda Rooke
A bittersweet study of life in present-day London as seen through the eyes of an unemployed man who comes to London and stays with a former girlfriend. Having seduced her roommate, he proceeds to wander the streets, haranguing everyone he comes into contact with. A powerful, haunting and extraordinarily bleak view of life in contemporary Britain. One of the director's most mature works, it boasts an outstanding performance from Thewlis in the central role.
DRA   120 min; 125 min wScrn (FIRST/SONOP – ort 131 min) mTV   VIDrel: FIRST/SONOP L/A; IMAG/DISC   V

NAKED AND LUSTFUL * 18
Alfredo Rizzo ITALY 197-
Jacque Stanyslave, Pupo De Luca, Fiorella Galgano, Carlo Rizzo, Michelina Cavaliere, Consalvo Dell'Arti, Sonia Viviani, Mario Pisu, Mario De Rosa, Carlo Micolano, Clara Spatafora, Assunta Oglio, Anna Perego
The aged Count Orsini is expelled from his castle by the machinations of a philandering professor named Luciani. In revenge the Count sends a couple of over-sexed beauties to the castle with intention of putting the professor into an early grave. A mediocre softcore romp of little merit.
A   92 min   VIDrel: IMPENT   V

NAKED AND THE DEAD, THE ** PG
Raoul Walsh USA 1958
Cliff Robertson, Aldo Ray, Raymond Massey, William Campbell, Richard Jaeckel, James Best, Joey Bishop, L.Q. Jones, Robert Gist, Lili St Cyr, Barbara Nichols, Jerry Paris, Casey Adams, John Berardino, Edward McNally
A platoon is assigned to undertake a dangerous mission behind Japanese lines on a Pacific island during WW2. A diluted adaptation of the original novel with its intensity lost in the cliches and conventions of a standard war film.
WAR   130 min   VIDrel: ODY/SONOP   V
Boa: novel by Norman Mailer.

NAKED CIVIL SERVANT, THE *** 15
Jack Gold UK 1975
John Hurt, Patricia Hodge, Katherine Schofield, Stanley Lebor
A fine TV film based on the life and times of Quentin Crisp, an effeminate homosexual who recounts his experiences with both style and wit. Touching and surprisingly amusing, this is a work of considerable charm.
DRA   78 min (ort 80 min) mTV   VIDrel: THAMES/DISC   V
Boa: book by Quentin Crisp.

NAKED GUN, THE: FROM THE FILES OF POLICE SQUAD ** 15
David Zucker USA 1988
Leslie Nielsen, Priscilla Presley, O.J. Simpson, George Kennedy, Ricardo Montalban, Nancy Marchand, John Houseman, Reggie Jackson, Jeanette Charles, Curt Gowdy, Al Yankovic, Joyce Brothers, Ken Minyard, Bob Arthur, Susan Beaubian
An over-rated slapstick comedy based on a short-lived TV show, that fires off gags in all directions, few of which hit the viewer's funny-bone. A combined creative effort from the writing team responsible for AIRPLANE! that amply demonstrates how hard it is to take a thirty-minute TV format and turn it into a sustained feature-length comedy. The script is by Jim and Jerry Zucker, Jim Abrahams and Pat Profit. See also HOT SHOTS!
Aka: NAKED GUN, THE
COM   81 min (ort 85 min)   VIDrel: CIC/SONOP   V/dm V/sh

NAKED GUN 2½, THE: THE SMELL OF FEAR ** 15
David Zucker USA 1991
Leslie Nielsen, Priscilla Presley, George Kennedy, O.J. Simpson, Robert Goulet, Richard Griffiths, Jacqueline Brookes, Lloyd Bochner, Tim O'Connor, Fred Ward, Peter Mark Richman, Anthony James, Ed Williams, John Roarke, Margery Ross
In this sequel to the first film, Nielsen has to save the world from the machinations of a secret energy cartel, which is planning to prevent the government adopting a new energy policy. The usual rapid-fire collection of gags is on offer, but the film lacks vigour and runs out of steam quite soon.
COM   82 min (ort 85 min)   VIDrel: CIC/SONOP   V/sur V/dm

NAKED GUN 33¹/₃ – THE FINAL INSULT **   12
Peter Segal USA   1993
Leslie Nielsen, Priscilla Presley, Anna Nicole Smith, George Kennedy, O.J. Simpson, Fred Ward, Kathleen Freeman, Ellen Greene, Ed Williams, Raye Birk, Matt Roe, Wylie Small, Sharon Cornell, Earl Boen, Jeff Wright, Karen Segal
The second sequel to the first film in this series offers very much of the same brand of quick-fire gags but a very thin plot that soon becomes quite irritating. A number of big-name stars (e.g. Raquel Welch, James Earl Jones) put in an appearance in tiny cameos.
COM   79 min (ort 83 min) cC   VIDrel: CIC/SONOP; PION (LV only)   V/sur LV

NAKED IN NEW YORK **   15
Daniel Algrant USA   1993
Eric Stoltz, Mary-Louise Parker, Tony Curtis, Timothy Dalton, Ralph Maccio, Jill Clayburgh, Kathleen Turner, Whoopi Goldberg, Lynne Thigpen, Paul Guilfoyle, Roscoe Lee Browne, Chris Noth, Arabella Field, Eric Bogosian, Quentin Crisp
A young would-be playwright finds it difficult to decide whether or not to follow his inclinations and attempt a writing career or remain in the dull but secure academic world. Eventually he moves to New York while his girlfriend stays behind in Boston. An amiable and fairly entertaining comedy that would have been a whole lot better with sharper performances from the leads.
COM   87 min (ort 89 min)   VIDrel: IMAG/RTM   V/sur

NAKED KILLER **   18
Clarence Ford (Clarence Fok Yu) HONG KONG   1992
Chingamy Yau, My Yau Ching, Simon Yam, Carri Ng, Cynthia Simamgura
To avenge her father's death, a woman takes lessons in the martial arts from a female assassin. When her mentor is also murdered, she and her policeman boyfriend proceed to settle both these scores in the usual violent way. Just one more in a long line of Hong Kong revenge-driven movies of bad dubbing and worse scripting. Fortunately, the action sequences are handled with a good deal of flair.
A/AD   96 min wScrn dubbed   VIDrel: MIA/DISC; ENCORE (LV only)   V LV

NAKED LUNCH **   18
David Cronenberg CANADA/UK   1991
Peter Weller, Judy Davis, Ian Holm, Julian Sands, Roy Scheider, Monique Mercure, Nicholas Campbell, Michael Zelniker, Robert A. Silverman, Joseph Scorsiani, Peter Boretski, Yuval Daniel, John Friesen, Sean McCann, Jim Yip
Not so much an adaptation of the Burroughs novel as an "examination" of the process of writing it, this film, set in 1953, has an ex-junkie who works as a pest-controller coming home to find his wife high on bug powder. Matters degenerate further: giant talking insects appear, his wife is revealed to be an alien, and he is sent on a dangerous mission to a bizarre region known as the "Interzone". A gruesome, surreal shocker, it was scripted by Cronenberg.
HOR   110 min (ort 115 min) wScrn   VIDrel: FIRST/SONOP; PION (LV only)   V/sur LV
Boa: novel by William S. Burroughs.

NAKED PREY, THE ***   PG
Cornel Wilde SOUTH AFRICA/USA   1966
Cornel Wilde, Ken Gampu, Gert Van Den Bergh, Patrick Mynhardt, Bella Randels, Jose Sithole, Richard Mashiya, Eric Sabela, Joe Diamini, Morrison Gampu, Frank Mdhluli, Sandy Nkomo, Fusi Zazayokwe, John Marcus, Horace Gilman
A group of whites on a hunting safari are captured by an African tribe, and all of them are brutally murdered except one, who is given a sporting chance by being allowed to escape prior to natives being sent after him. A film made with enormous skill, that blends evocative African music with superb locations, to produce a work of considerable atmosphere and suspense. See also SURVIVING THE GAME, DEATH RING and FINAL ROUND.
A/AD   92 min (ort 94 min)   VIDrel: CIC/SONOP L/A V/h

NAKED ROBOT 4¹/₂ **   15
Philip J. Cook USA   199-
Hans Bachmann
A reporter investigates the murder of six Air Force officers in Virginia, in the course of which he comes up against eccentric military personnel, various FBI agents and a peculiar robot. An attempt at an anarchic SF comedy that fails to come off.
COM   91 min   VIDrel: 20VIS/SONOP   V

NAKED SCENTS **   18
Elissa Christine USA   1985
R. Bolla, Tish Ambrose, Sharon Kane, Taija Rae, Crystal Cox, Tasha Voux, Jerry Butler, Stephen Lockwood, Marita Exberg, Sarah Bernard, George Payne, Joey Silvera, Scott Baker
Bolla plays the head of a large pharmaceuticals company. He's about to marry for the second time, his fiancee being the head of a perfume company that has saved him from bankruptcy. Neither of the parties are virgin lovers, and prove as much on their wedding day, enjoying various couplings with all and sundry. Adequate.
A   51 min Cut (56 sec plus some cuts subst)
VIDrel: HAR/GOLD L/A   V

NAKED SOULS *   18
Lyndon Chubbuck USA   1995
Pamela Anderson (Lee), Brian Krause, Claython Rohner, Justina Vail, Dean Stockwell
A Pamela Anderson vehicle that attempts to exploit her enhanced physical charms (see also BARB WIRE) within the context of an SF thriller. It has her cast an artist, with a scientist boyfriend whose experiments in digitising human memory reach the point where he requires a suitable guinea pig. For this purpose he chooses the mind of a serial killer (well he would, wouldn't he?) and the expected nastiness and personality swaps take place. Dull dross.
FAN   98 min   VIDrel: POLFIL   V

NAKED TANGO **   18
Leonard Schrader USA   1991
Vincent D'Onofrio, Esai Morales, Mathilda May, Fernando Rey, Cipe Lincovski, Josh Mostel, Constance McCashin, Patricio Bisso, Ruben Szuchmacher, Marcos Woinski, Sergio Lerer, Javier Portalers, Jean-Pierre Reguerraz, Tony Payne
Steamy erotic tale set in 1924 in Buenos Aires, where the bored young wife of a middle-aged judge assumes the identity of a young Jewish "mail-order" bride, and eventually winds up working in her new husband's brothel. A wild farrago of implausible plot devices, that comes over as a clumsy pastiche of 1920s films, but is badly marred by its disagreeable anti-Semitic overtones. The overheated script (inspired by a work by Manuel Puig) is the work of Schrader.
THR   88 min (ort 92 min)   VIDrel: WHV   V/dm V/sur

NAME OF THE ROSE, THE ***   18
Jean-Jacques Annaud FRANCE/ITALY/
WEST GERMANY   1986
Sean Connery, F. Murray Abraham, Christian Slater, Elya Baskin, Feodor Chaliapin Jr, William Hickey, Michael Lonsdale, Ron Perlman, Helmut Qualtinger, Volker Prechtel, Michael Habeck, Urs Althaus, Valentina Vargas
A senior monk and his novice investigate a spate of deaths at a monastery, and reveal the seamier side of abbey life in the 14th century. Lovingly recreated exteriors and a wonderful set of grotesque figures provide a great feeling of authenticity, but their sheer visual impact detracts from the more appealing and intellectual ramifications of the novel.
DRA   123 min (ort 130 min)   VIDrel: 4-FRONT/POLYREC V/sur
Boa: novel by Umberto Eco.

NANA **   PG
Dorothy Arzner USA   1934
Anna Sten (Anjuschka Stenski), Lionel Atwill, Phillip Holmes, Richard Bennett, Mae Clarke, Muriel Kirkland, Reginald Owen, Jessie Ralph, Lawrence Grant, Helen Freeman, Ferdinand Gottschalk,
A dancer eventually commits suicide, in her desire to bring about a reconciliation between two estranged brothers, who have both fallen in love with her. A massively expensive flop, which presented a heavily censured version of Zola's novel, this movie represented Samuel Goldwyn's efforts to launch Russian-born Sten as the new Garbo. Unfortunately, her poor diction and inability to act in English, are massive drawbacks in this ill-conceived work.
Aka: LADY OF THE BOULEVARDS
DRA   87 min B/W   VIDrel: VCC/DISC   V
Boa: novel by Emile Zola.

NANA **
PG
Dorothy Arzner USA 1934
Anna Sten, Lionel Atwill, Phillip Holmes, Richard Bennett, Mae Clarke, Muriel Kirkland, Reginald Owen, Jessie Ralph
Story set in Paris in the 1890s, that follows the career of a dancer and courtesan, whose steadily falls from favour, eventually committing suicide in her desire to bring about a reconciliation between two brothers. Zola's tragic tale of a Parisian demi-mondaine was intended to make a star out of Sten, but she did little to shine in her role and the film was not a success.
Aka: LADY OF THE BOULEVARDS
DRA 87 min B/W VIDrel: VCC/DISC V
Boa: novel by Emile Zola.

NANNY, THE ***
15
Seth Holt UK 1965
Bette Davis, Wendy Craig, Jill Bennett, James Villiers, William Dix, Pamela Franklin, Maurice Denham, Jack Watling, Alfred Burke, Nora Gordon, Harry Fowler
A young boy discovers his nanny's guilty secret, and has to devise a means of thwarting her efforts to silence him. A taut psychodrama made with a degree of insight and subtlety not usually associated with Hammer productions. The script is by Jimmy Sangster.
DRA 89 min (ort 95 min) B/W VIDrel: WHV V
Boa: novel by Evelyn Piper.

NAOMI AND WYNONA: LOVE CAN BUILD A BRIDGE **
15
Bobby Roth USA 1994
Kathleen York, Viveka Davis, Bruce Greenwood, Melinda Dillon, Cari Shayne, Megan Ward, Nick Searcy, Travis Fine, Joshua Schaefer, Laura Morgan, Lisa Waltz, A.J. Langer, Elizabeth Moss, Dolly Parton, Mae Whitman, Rob Nillson, Billy Vera
Biopic on the life and career of Naomi and Wynona Judd, who have achieved fame as two of America's most successful Country-and-Western female singers. The soundtrack is by the Judds. Fair.
Aka: LOVE CAN BUILD A BRIDGE: THE JUDDS
DRA 117 min (ort 120 min) VIDrel: ODY/SONOP V/sh
Boa: book Love Can Build a Bridge by Naomi Judd and Budd Schaetzle.

NAPOLEON AND JOSEPHINE: A LOVE STORY – PARTS 1, 2 AND 3 ***
PG
Richard T. Heffron USA 1987
Armand Assante, Jacqueline Bisset, Jean-Pierre Stewart, Anthony Perkins, Anthony Higgins, Stephanie Beacham, Jane Lapotaire, William Lucking, Nickolas Grace, Patrick Cassidy, Leigh Taylor-Young
A lavish and glossy mini-saga, that follows the career of Napoleon from his rise to power to his defeat and exile on St Helena. Much of the tale was filmed at the original locations of the events portrayed, and this adds to the period atmosphere. However, the slow unfolding of the plot is a drawback and the film is marred by the gory execution sequence at the beginning.
DRA 262 min (2 cassettes) mTV VIDrel: WHV V/sh

NARROW MARGIN **
15
Peter Hyams USA 1990
Gene Hackman, Anne Archer, James B. Sikking, J.T. Walsh, M. Emmet Walsh, Susan Hogan, Nigel Bennett, J.A. Preston, B.A. "Smithy" Smith, Harris Yulin, Codie Lucas Wilbee, Barbara E. Russell, Antony Holland, Doreen Ramos
Written and directed by Hyams, this remake of the 1952 RKO thriller casts Hackman (giving a terrific performance) as a cop who is assigned to protect a female witness to a gangland slaying. He accompanies her on a train ride back to L.A. and a mobster trial, but has to contend with some hit-men out to silence her. Despite an excellent climax set on the roof of the train, the earlier film had an impact entirely lacking here.
DRA 93 min (ort 99 min)
VIDrel: 4-FRONT/POLYREC/GUILD; PION (LV only)
V/sur LV

NASTY BOYS **
15
Rick Rosenthal USA 1989
Benjamin Bratt, Don Franklin, Craig Hurley, Jeff Kaake, James Pax, William Russ, Melissa Leo, Sandy McPeak, Nia Peeples, Thomas Mikal Ford, Soon-Teck Oh, Whip Hubley, Nick Ramus, John Petlock, Mike Moroff, Nick Angotti
A crack squad of narcotics officers based in Las Vegas, don masks and the other trappings of Ninja warriors, going under-cover to combat local drug barons. The title and plot basis is

inspired by the "Nasty Boys" of North Las Vegas, a real-life elite drugs unit; but for all that, this is yet one more production-line cops adventure. The pilot for a series.
A/AD 88 min (ort 100 min) mTV VIDrel: CIC/SONOP
L/A V

NASTY BOYS 2: KILL OR BE KILLED **
15
Leo Penn USA 1990
James Pax, Jeff Kaake, Craig Hurley, Don Franklin, Benjamin Bratt
A formula sequel to the first film, in which an undercover cop sets out to nail a vigilante who has set himself up as judge and executioner, and now threatens the life of a local drug baron.
A/AD 86 min VIDrel: CIC/SONOP L/A V

NASTY BOYS 3: CRACK HOUSE **
15
Rob Cohen USA 1990
Benjamin Bratt, Don Franklin, Craig Hurley, Jeff Kaake, James Pax, Dennis Franz
The title group of cops are assigned to a brash new command-ing officer, and are given the job of assisting the FBI in smashing a drugs cartel by bringing a key witness to L.A. However, one of the officers gets drawn into a far more personal involvement when his daughter is abducted and offered in exchange for the release of the witness. The third outing in the NASTY BOYS series.
A/AD 94 min VIDrel: CIC/SONOP V

NASTY GIRL, THE ****
PG
Michael Verhoeven WEST GERMANY 1989
Lena Stolze, Hans-Reinhard Muller, Monika Baumgartner, Elisabeth Bertam, Michael Gahr, Roggert Giggenbach, Fred Stillkrauth, Udo Thomer, Barbara Gallauner, Kurt Weinzierl, Ottfried Fischer, Sandra White, Ludwig Wohr
A very sharp semi-comic work, that tells the true story of Anja Rosmus. As a young Bavarian woman, her decision to enter a national writing contest on the subject of "My Hometown During the Third Reich" leads to a long and difficult search for the truth, a quest that becomes both obsessive and unpleasant. A striking film in many ways, it was scripted by the director, and won nine international awards. See also HITLER, A FILM FROM GERMANY.
Aka: DAS SCHRECKLICHE MADCHEN
DRA 92 min (ort 95 min) Col/B/W VIDrel: VCC/DISC V

NATIONAL LAMPOON'S ANIMAL HOUSE ***
15
John Landis USA 1978
John Belushi, Tim Matheson, John Vernon, Stephen Furst, Verna Bloom, Thomas Hulce, Peter Riegert, Donald Sutherland, Mary Louise Weller, Karen Allen, Cesare Danova, James Daughton, Bruce McGill, Mark Metcalf, DeWayne Jessie, Martha Smith
Energetic spoof on American college life during the 1960s, with a mass of crazy pranks and coarse jokes to cover up a flimsy plot about the conflict between rival fraternities. Reminiscent of AIRPLANE! in terms of its wealth of visual gags, the film achieves a pleasing pre-Vietnam/pre-student protest era inno-cence. Gave rise to several inferior spin-offs and the TV series "Delta House". Written by Harold Ramis.
Aka: ANIMAL HOUSE
COM 104 min (ort 109 min) VIDrel: CIC/SONOP V/h

NATIONAL LAMPOON'S CHRISTMAS VACATION **
PG
Jeremiah S. Chechik USA 1989
Chevy Chase, Beverly D'Angelo, Randy Quaid, Diane Ladd, E.G. Marshall, Doris Roberts, John Randolph, Julia Louis-Dreyfuss, Mae Questel, William Hickey, Brian Doyle-Murray, Juliette Lewis, Johnny Galecki, Nicholas Guest
When the Griswold family decide to stay at home for the holi-days, they little suspect that their peace and tranquillity will be destroyed by an invasion of bickering relatives and a catalogue of disasters. A strained and often quite tasteless farce, with a few comic moments, a predictable outcome and a memorable perfor-mance from Quaid as an obnoxious cousin.
Aka: CHRISTMAS VACATION
COM 93 min (Cut at film release by 2 sec – ort 97 min)
VIDrel: MGM/WHV L/A V/sh

NATIONAL LAMPOON'S CLASS REUNION *
18
Michael Miller USA 1982
Gerrit Graham, Michael Lerner, Fred McCarren, Miriam Flynn, Stephen Furst, Marya Small, Shelley Smith, Zane Buzby, Anne Ramsey, Jacklyn Zeman, Blackie Dammett, Barry Diamond, Art Evans, Maria Pennington, Randolph Powers

Unfunny and unpleasant spoof on all those high school reunion films where the ex-classmates are terrorised by a maniacal killer. The level of humour can be gauged from the fact that the school is the Lizzie Borden High. The feeble script is by John Hughes.
Aka: CLASS REUNION
COM   81 min (ort 84 min)   VIDrel: ODY/SONOP   V/sur

NATIONAL LAMPOON'S EUROPEAN VACATION *   15
Amy Heckerling USA   1985
Chevy Chase, Beverly D'Angelo, Eric Idle, Dana Hill, Victor Lanoux, John Astin, Jason Lively, Sheila Kennedy, Paul Bartel, Cynthia Szigetti, Malcolm Danare, Tricia Lange, William Zabka, Wendy Goldman, Angus Mackay, Mel Smith
Feeble sequel to NATIONAL LAMPOON'S VACATION in which a strange and wimpish family of Americans decide to take a motoring holiday in Europe. Needless to say, everything goes disastrously wrong in a work reminiscent of the worst of the British "Carry On" films. Written by John Hughes and Robert Klane.
Aka: EUROPEAN VACATION
COM   90 min (ort 94 min)   VIDrel: WHV   V/h

NATIONAL LAMPOON'S JOY OF SEX *   15
Martha Coolidge USA   1984
Michelle Meyrink, Cameron Dye, Lisa Langlois, Charles Van Eman, Christopher Lloyd, Colleen Camp, Ernie Hudson, Joanne Baron, Darren Dalton, Heide Holicker, Cristen Kauffman, David H. MacDonald, Paul Tulley, Joe Unger
A girl believes that she has only weeks to live, and decides to lose her virginity before it's too late, since dying a virgin is almost as bad as death itself. So tasteless and unfunny that it insults Dr Alex Comfort's famous sex manual, whose title (and only that) it shares.
Aka: JOY OF SEX, THE
COM   89 min (ort 93 min)   VIDrel: CIC/SONOP L/A   V/h

NATIONAL LAMPOON'S LAST RESORT *   12
Rafal Zielinski USA   1993
Corey Feldman, Corey Haim, Geoffrey Lewis, Robert Mandan, Demetra Hampton, Milton Selzer, Eda Reiss Merin, John William Young, Marji Martin, Chris Barnes, Tony Longo, Michael Ralph, Zelda Rubenstein, Patrick Labyorteaux, Roger Clinton
Two men help the uncle of one of them protect his private resort from a former fellow actor who has taken a violent dislike to his colleague. Another wild hit-or-miss spoof with the usual unfunny gags.
Aka: LAST RESORT; NATIONAL LAMPOON'S SCUBA SCHOOL
COM   88 min (ort 91 min)   VIDrel: MARQ/QUANT   V/sh

NATIONAL LAMPOON'S LOADED WEAPON 1 **   PG
Gene Quintano USA   1992
Emilio Estevez, Jon Lovitz, William Shatner, Samuel L. Jackson, Tim Curry, Whoopi Goldberg, Bruce Willis, Charlie Sheen, Kathy Ireland, Frank McRae, James Doohan, Dhiru Shah, William Shatner, Bill Nunn, Gokul, F. Murray Abraham
Two cops are ordered to recover a microfilm that contains a formula for turning cocaine into cookies but this thin plot merely provides an excuse for a variety of tepid spoofs on such films as LETHAL WEAPON and BASIC INSTINCT.
Aka: LOADED WEAPON 1
COM   79 min (ort 83 min)   VIDrel: 4-FRONT/POLYREC   V

NATIONAL LAMPOON'S MOVIE MADNESS *   15
Henry Jaglom USA   1981
Bob Giraldi, Peter Riegert, Diane Lane, Candy Clark, Teresa Ganzel, Robert Culp, Ann Dusenberry, Bobby DiCenzo, Fred Willard, Joe Spinell, Mary Woronov, Dick Miller, Robby Benson, Richard Widmark, Christopher Lloyd
A collection of three parodies or spoofs on various movie genres, that deal with personal growth, soap operas and cop movies. Thankfully, a fourth parody on disaster movies was removed, leaving only these three unbelievably lame efforts.
Aka: NATIONAL LAMPOON GOES TO THE MOVIES
COM   89 min   VIDrel: MGM/WHV L/A   V

NATIONAL LAMPOON'S SENIOR TRIP *   15
Kelly Makin USA   1995
Matt Frewer, Tommy Chong, Valerie Mahaffey, Kevin McDonald, Rob Moore, Jeremy Renner
A group of high-school kids write to the President with their concerns over public education and become pawns in the machi-nations of a corrupt senator. A bottom-of-the-barrel effort of zero laughs.
Aka: SENIOR TRIP
COM   87 min (ort 91 min)   VIDrel: FIRST/SONOP   V

NATIONAL LAMPOON'S VACATION *   15
Harold Ramis USA   1983
Chevy Chase, Beverly D'Angelo, Anthony Michael Hall, Imogene Coca, Randy Quaid, Christie Brinkley, Dana Barron, John Candy, Eddie Bracken, Miriam Flynn, Brian Doyle-Murray, James Keach, Eugene Levy, Frank McRae, Jane Kralowski
A typical American family make a cross-country trip by car and, needless to say, everything goes wrong in true Disney style. But on the other hand, even Disney could probably have made a funnier and more interesting film than this episodic collection of predictable situations. Written by John Hughes.
Aka: VACATION
COM   95 min (ort 98 min)   VIDrel: WHV   V

NATIONAL VELVET ***   U
Clarence Brown USA   1944
Mickey Rooney, Elizabeth Taylor, Anne Revere, Donald Crisp, Angela Lansbury, Jackie Jenkins, Reginald Owen, Terry Kilburn, Norma Varden, Alec Craig, Arthur Shields, Dennis Hoey, Juanita Quigley, Aubrey Maher, Frederic Worlock
A young girl wins a horse in a raffle and trains it to run in the famous Grand National. A simple story, well told and acted, but far too sentimental to have much appeal today. Followed by INTERNATIONAL VELVET a good many years later. AA: S. Actress (Revere).
DRA   118 min (ort 125 min)   VIDrel: MGM/WHV   V/h
Boa: play by Enid Bagnold.

NATURAL, THE **   PG
Barry Levinson USA   1984
Robert Redford, Robert Duvall, Glenn Close, Kim Basinger, Wilford Brimley, Richard Farnsworth, Barbara Hershey, Joe Don Baker, Robert Prosky, Darren McGavin, John Finnegan, Alan Fudge, Paul Sullivan, Rachel Hall, Joe Van Ness
A man with a strange, almost supernatural talent for baseball, reaches the top, from where only one direction is left for him to go in this curiously oblique and bare adaptation of Malamud's strange updating of the King Arthur legend. Written by Roger Towne and Phil Dusenberry. Photography is by Caleb Deschanel.
DRA   118 min (ort 137 min)   VIDrel: 20TH/TECH L/A   V
Boa: novel by Bernard Malamud.

NATURAL BORN KILLERS **   18
Oliver Stone USA   1994
Woody Harrelson, Juliette Lewis, Robert Downey Jr, Tommy Lee Jones, Rodney Dangerfield, Tom Sizemore, O-Lan Jones, Edie McClurg, Sean Stone, Dale Dye, Eddy "Doogue" Conna, Evan Handler, Kirk Baltz, Terrylene, Maria Pitillo, Josh Richman
An over-the-top supposed satire on the power of the media that is built around the story of a couple of psychotics who celebrate their relationship by killing a large number of people, thereby achieving instant fame. Unfortunately, everything in this film is so exaggerated and overblown that the entire effort soon becomes exceedingly tiresome. The video tape was withdrawn by WHV for a period, in the wake of the Dunblane massacre. See also LOVE AND A .45.
DRA   114 min (ort 180 min)   VIDrel: WHV   V/h

NATURAL BORN THRILLERS *   18
Stuart Canterbury USA   1995
Melissa Hill, Rebecca Wilde, Sindee Coxx, Tess Newheart, Barbera Doll, Mike Horner, Tom Byron, Kyle Stone, Don Fernando
A couple of sex maniacs cause a sensation across the country, in this poorly made spoof.
A   80 min   VIDrel: MIA/DISC   V

NATURAL CAUSES **   12
James Becket USA   1993
Linda Purl, Ali McGraw, Cary-Hiroyuki Tagawa, Will Patton, Tim Thomerson, Janis Paige
A female doctor pays a visit to her mother who is living over-seas in Bangkok but gets herself entangled in a conspiracy to assassinate Henry Kissinger. The usual violent confrontations and plot twists ensue. An adequate spy thriller.
THR   85 min (ort 90 min)   VIDrel: FIRST/DISC   V

NATURAL COLD KILLER ** 18
Fred Olen Ray USA 1994
Michael Nouri, Margaux Hemingway, David Warner, Sandahl Bergman, Tracy Brooks Swope, Jennifer Ciesar, John Coleman, Suzanne Ager, Joe Estevez, James Booth, Fred Olen Ray, Peter Spellos, Eric Amiel, Gabriel Ortiz, Sigal Diamant
A coldhearted woman murders her husband and sets off for a relaxing holiday with the money she got from his life insurance policy, but does not realise that her activities have engaged the attention of others out to relieve her of her ill-gotten gains. A sequel of sorts to INNER SANCTUM.
Aka: INNER SANCTUM 2
THR 90 min VIDrel: NEWAGE/TECH L/A V

NATURAL ENEMY ** 18
Douglas Jackson USA 1996
Donald Sutherland, William McNamara, Lesley Ann Warren, Joe Pantoliano, Tia Carrere
A highly successful stockbroker gets invited to visit his boss at the latter's home, but once there he reveals his seriously disturbed nature and penchant for violence. An unpleasant psychological chiller, that does little to stretch a fine cast.
DRA 87 min VIDrel: GUILD/FOXVID V

NATURAL LIES *** (PG)
Ben Bolt UK 1992
Bob Peck, Sharon Duce, Rob Spendlova, Denis Lawson, Judi Maynard, Deborah Findlay, Brian Protheroe, Lynsey Baxter, Arkie Whiteley, Brian Croucher
Distraught at the news of his ex-girlfriend's suicide, a young advertising executive grows suspicious of the circumstances surrounding her death (she had been covering the first human case of mad cow disease) but his amateur sleuthing puts his life in danger. A concise and gripping thriller, whose conspiracy plot does not really hold water, but is nonetheless highly entertaining. The script is by David Pirie.
THR 120 min (ort 165 min) mTV VIDrel: SPEAR/SONOP V

NATURAL SELECTION *** 15
Jack Sholder USA 1993
C. Thomas Howell, Lisa Zane, Ethan Phillips, Richard Hamilton, Miko Hughes, Cameron Dye, Paul Eiding, Joanna Miles, Brenda Varda, John Mahon, J.P. Bumstead Nicolas Coster
A man makes the unhappy discovery that he was the result of a scientific experiment in cloning that went wrong, when he finds himself being followed one day by a duplicate. Eventually, he traces this affair to an aged genetic scientist, who thirty years ago cloned seven replicas of himself, of which our man is just one. But the other clone has dedicated himself to killing all the others. Tense and intriguing, its greatest asset is the unusual plot premise.
Aka: DARK REFLECTION
THR 87 min (ort 90 min) mCab VIDrel: MARQ/QUANT V

NAUGHTY BLUE KNICKERS ** 18
Andre Genoves FRANCE 1981
Marcha Grant, Andre Genoves, Caroline Aguilar, Charlotte Walior, Yves Massard, Marthe Mercadier, Bruno Du Louvat, Denyse Roland, Gerard Croce, Daniel Beretta, Sylvane Sannoy
Sex romp about the adventures of the various wearers of the title garment. Interestingly, in the original language version, it's the knickers that actually relate the story.
Aka: FOLIES OF ELODIE, THE; LES FOLIES D'ELODIE; SECRETS OF THE SATIN BLUES
COM 85 min Cut (1 min 6 sec – ort 90 min) dubbed VIDrel: ARTPRO/RTM L/A V

NAVIGATOR, THE *** PG
Vincent Ward AUSTRALIA/NEW ZEALAND 1988
Bruce Lyons, Chris Haywood, Hamish McFarlane, Marshall Napier, Noel Appleby, Paul Livingston, Sarah Peirse, Mark Wheatley, Tony Herbert, Roy Wesney, Jay Saausey, Jessica Cardiff-Smith, Kathleen-Elizabeth Kelly, Charles Walker
When a Cornish tin-mining village in 14th century England is struck by the plague, a young boy attempts to lead the survivors to safety by getting them to burrow into the ground. They do so and eventually emerge on the outskirts of a modern city. A strange and evocative tale, highly atmospheric and imaginative, with fine acting and direction. Screenplay is by Ward.
Aka: NAVIGATOR, THE: A MEDIEVAL ODYSSEY
FAN 87 min (ort 92 min) B/W/Col VIDrel: ARTPRO/RTM L/A V

NAVY SEALS ** 15
Lewis Teague USA 1990
Charlie Sheen, Michael Biehn, Joanne Whalley-Kilmer, Rick Rossovich, Bill Paxton, Cyril O'Reilly, Dennis Haysbert, Paul Sanchez, Ron Joseph, Nicholas Kadi, S. Epatha Merkerson, Greg McKinney, Rob Moran, Richard Venture, Ron Faber
Inspired by an actual commando unit formed during John F. Kennedy's presidency, this has an elite unit (Sea, Air and Land) setting out to do battle with a bunch of Middle Eastern terrorists, who never really stand a chance in the face of some overwhelming hi-tech weaponry. There's little characterisation, less plot, but a few exciting set-piece battles for those who like their entertainment simple and noisy.
A/AD 108 min (ort 114 min)
VIDrel: 4-FRONT/POLYREC/VISVID V/sur

NAVY VERSUS THE NIGHT MONSTERS, THE * PG
Michael Hoey USA 1966
Mamie Van Doren, Anthony Eisley, Pamela Mason, Bill Gray, Bobby Van, Walter Sande, Edward Faulkner, Phillip Terry, Edward Faulkner, Russ Bender
An Antarctic naval base is attacked by acid-secreting omnivorous walking vines in this dire and dull monster movie.
Aka: NIGHT CRAWLERS, THE
FAN 64 min (ort 87 min) VIDrel: FIRC/RTM V

NEAR DARK *** 18
Kathryn Bigelow USA 1987
Adrian Pasdar, Jenny Wright, Lance Hendriksen, Bill Paxton, Jenette Goldstein, Tim Thomerson, Marcie Leeds, Joshua Miller, Kenny Call, Ed Corbett, Troy Evans, Bill Cross, Roger Aaron Brown, Thomas Wagner
A young farmhand gets passionately involved with a young girl who proves to be a member of a band of small-town vampires. In no time, he finds himself a member of a strange band of wandering vampires, whose leader gives him one week in which to make his first kill and thus become truly one of them. A well-crafted and unusual horror film that skilfully blends elements of various genres.
HOR 90 min Cut (14 sec – ort 95 min) VIDrel: EIV/SONOP V

NEAR MRS * 15
Baz Taylor FRANCE/USA 1991
Judge Reinhold, Casey Siemaszko, Cecile Paoli, Rebecca Pauly, Muriel Combeau, Kashia Figura, Andrzej Jagora, Patrick Floersheim, Bruce McGuire, Dominic Gould, Stuart Seide, Stephan Meldegg, William Doherty, Olivier Pierre
A busy executive not only maintains two separate wives, but also a mistress for good measure. In an attempt to arrange a break from all this, he persuades his assistant to take his place at a two-week Territorial Army camp. Unfortunately, the KGB have been planning his abduction, and kidnap the assistant instead. A rather foolish one-joke comedy.
Aka: NEAR MISFITS; NEAR MISSES
COM 89 min (ort 95 min) VIDrel: BUENA/SONOP L/A V

NEAR ROOM, THE *** 18
David Hayman UK 1995
Adrian Dunbar, David O'Hara, David Hayman, Julie Graham, Tom Watson, James Ellis, Robert Pugh, Emma Faulkner, Andy Serkis, Peter McDoughall, Bill Gardiner, Gary Sweeney, Sean Scanlan
A foray into film noir territory from Scotland, with murder, blackmail and child prostitution very much on the menu, as a determined journalist attempts to unravel a mystery involving a missing girl, having been hired to do so by an ex-lover. Before very long, the trail has revealed corruption in high places, and the usual heated conflicts ensue. A deeply cynical film, its relentless air of sordidness takes quite some getting used to.
THR 89 min CINrel

'NEATH ARIZONA SKIES ** U
Harry L. Fraser USA 1934
John Wayne, Sheila Terry, Shirley Ricketts (Shirley Jane Rickert), Jack Rockwell, Yakima Canutt, Weston Edwards, Gabby Hayes, Buffalo Bill Jr, Phil Keefer, Frank Hall Crane, Earl Dwire, Artie Ortego, Tex Phelps, Eddie Parker
A cowboy with an Indian ward, who is heiress to oil lands, has to rescue her when she is kidnapped by cut-throat bandits. A rather low-budget entry in Wayne's "Lone Star" series.
Aka: BENEATH ARIZONA SKIES
WES 52 min (ort 54 min) B/W coVer VIDrel: CASPIC V

NECESSARY ROUGHNESS **   15
Stan Dragoti USA   1991
Scott Bakula, Robert Loggia, Harley Jane Kozak, Hector Elizondo,
Larry Miller, Sinbad, Fred Dalton Thompson, Rob Schneider, Jason
Bateman, Andrew Bryniarski, Duane Davis, Michael Dolan, Marcus
Giamatti, Kathy Ireland
An over-the-hill football player is hired to boost the fortunes of
an inept college team who are dogged by a relentless series of
drugs and bribery scandals. With a team consisting of misfits of
various nationalities (there is a blanket ban on colleges recruiting national champions) Bakula, as the central character, has a
major task on his hands. All ends happily, and this simple-
minded comedy provides a few laughs in the process.
COM   104 min (ort 108 min)   VIDrel: CIC/SONOP   V/sur

NECROMANCER *   18
Dusty Nelson USA   1989
Elizabeth Kaitan, Russ Tamblyn, John Tyler, Rhonda Durton, Stan
Hurwitz
A supernatural revenge tale in which a woman who is raped by
a couple of thugs resorts to witchcraft in order to have her
revenge. A tiresome affair that has a contrived and flimsy script
and a definite lack of fresh ideas.
Aka: NECROMANCER: SATAN'S SERVANT
HOR   84 min Cut (35 sec – ort 90 min)   VIDrel: 20TH   V

NECRONOMICON **   18
Brian Yuzna/Christophe Gans/Shusuke Kaneko USA   1994
Jeffrey Combs, Tony Azito, Juan Fernandez, Bruce Payne, Belina
Bauer, Maria Ford, Richard Lynch, David Warner, Bess Meyer, Millie
Perkins, Gary Graham, Curt Lowenves, Dennis Christopher, Signy
Coleman, Obba Babatunde, Don Calfa
The writer H.P. Lovecraft visits a library run by a strange cult
and takes an unauthorised book, which contains the secrets of
the universe. This leads on to three stories: "The Drowning",
"The Cold" and "Whispers" all of which are supposedly based
on his writings. The first deals with a demon from Hell in a
decaying ancestral home, the second offers a novel approach to
immortality and the third has a modern flavour, but the repulsive effects rather ruin it. Fair.
Aka: H.P. LOVECRAFT'S NECRONOMICON: BOOK OF THE DEAD
HOR   94 min (ort 97 min)   VIDrel: ITC/HIFLI L/A   V/h

NEEDFUL THINGS ***   15
Fraser C. Heston USA   1993
Max Von Sydow, Ed Harris, Bonnie Bedelia, J.T. Walsh, Amanda
Plummer, Ray McKinnon, Duncan Fraser, Valri bromfield, Shane
Meier, W. Morgan Sheppard, Don S. Davis. Campbell Lane, Eric
Schneider, Frank C. Turner, Gillian Barber
A small town is visited by an elderly and very helpful old
gentleman who seems too kind and good to be true, but his ulte-
rior purpose proves to be quite otherwise. He sets up an
antiques shop there and begins to sell small objects to the inhab-
itants that seem to bring bad luck and death to their owners,
forcing them to behave in a variety of evil ways. Totally silly
nonsense this may be, but the production values are high and
good pacing keeps boredom at bay.
Aka: STEPHEN KING'S NEEDFUL THINGS
FAN   120 min   VIDrel: POLY/POLYREC   V
Boa: novel by Stephen King.

NEIGHBOR, THE **   18
Rodney Gibbons USA   1992
Rod Steiger, Linda Kozlowski, Ron Lea, Benjamin Shirinian, Bruce
Boa, Harry Standjofski, Jane Wheeler, Sean McCann, Frances Bay,
Paulino Little, Claire Riley, Mark Comacho, Linda Singer, Philip
Spensley, Gordon Master, Sylvie Potvin
A couple buy a charming old house in a quaint Vermont town
and are ill-prepared for their encounter with their retired obste-
trician neighbour. The wife becomes pregnant and when her
husband is arrested for murder, she is left to his tender neigh-
bourly care. Another variation on the "psycho from hell" theme.
HOR   89 min (ort 93 min)   VIDrel: FIRST/SONOP L/A   V

NEIGHBORS *   15
John G. Avildsen USA   1981
John Belushi, Dan Aykroyd, Kathryn Walker, Cathy Moiarty, Lauren-
Marie Taylor, Tim Kazurinsky, Igors Gavon, Dru-Ann Chukron, Tino
Insana, P.L. Brown, Henry Judd Baker, Dale Two Eagle, Sherman
Lloyd, Bert Kittel, J.B. Friend
A stuffy man is irritated when new, swinging neighbours move
in but worse is to follow when their strange antics drive him to
the edge of insanity. Watch this one and it will do the same to

you. Scripted by Larry Gelbart. This poor effort was Belushi's
last film.
COM   94 min   VIDrel: MIA/DISC   V
Boa: novel by Thomas Berger.

NELL **   12
Michael Apted USA   1994
Jodie Foster, Liam Neeson, Natasha Richardson, Richard Libertini,
Nick Searcy, Robin Mullins, Jeremy Davies, O'Neal Compton,
Heather M. Bomba, Marianne E. Bomba, Sean Bridgers, Joe Inscoe,
Stephanie Dawn Wood, Mary Lynn Riner
After a woman hermit dies, a smalltown Southern doctor
discovers that she had a daughter whom she raised in total isola-
tion from the outside world. Since the mother's speech was
distorted because of paralysis, the girl has learnt to speak a
totally unintelligible language. When news of this "discovery"
reaches an ambitious university researcher, he sees her as an
object worthy of scientific study. A bizarre, rambling film,
sustained by Foster's fine acting.
DRA   108 min (ort 113 min) cC   VIDrel: POLY/POLYREC
V/sur
Boa: play Idioglossia by Mark Handley.

NELLY & MONSIEUR ARNAUD ***   PG
Claude Sautet FRANCE   1995
Emmanuelle Beart, Michel-Serrault, Jean-Hughes Anglade, Claire
Nadeau
An elderly judge who is writing his memoirs hires an attractive
young woman to act as his secretary. As their working relationship develops, the woman's employer finds himself growing
increasingly drawn to his young assistant. But with the arrival
of the man's publisher, the strains inherent in the unequal nature
of their relationship become apparent. A clever comedy of
manners that attempts to put the foibles of the French upper
class under the microscope.
COM   102 min (ort 106 min)   VIDrel: GUILD/20TH   V/sur

NEMESIS **   18
Albert Pyun USA   1992
Olivier Gruner, Deborah Shelton, Brion James, Tim Thomerson, Merle
Kennedy, Cary Hiroyuki-Tagawa, Yujo Okumoto, Marjorie
Monaghan, Nicholas Guest, Thom Mathews, Vince Klyn, Jennifer
Gatti, Borovnisa Blervaque, Toma Janes
Imaginative but terribly confused thriller set in Los Angeles in
2027 where cyborgs are both commonplace and accepted.
However, a police officer discovers a cyborg plan to replace
world leaders with artificial duplicates as a prelude to world
domination. Followed by the inevitable sequels. See also
CYBORG (this was also directed by Pyun) and its sequels, a
clutch of films also built around this idea.
FAN   91 min (ort 100 min)   VIDrel: POLY/POLYREC   V/sur

NEMESIS 2: NEBULA **   15
Albert Pyun USA   1994
Sue Price, Tina Cote, Earl White, Chad Stahelski
In the year 2077 A.D. the Cyborgs have finally conquered
humanity. However, a scientist working for the Resistance
develops a synthetic strain of super-DNA, and a rebel woman
is chosen to give birth to the first humans who may be able to
beat the cyborgs. An adequate first sequel to the original film,
it was soon followed by NEMESIS 3: TIME LAPSE.
FAN   81 min (ort 83 min)   VIDrel: MIA/DISC   V/sur

NEMESIS 3: TIME LAPSE *   18
Albert Pyun USA   1995
Sue Price, Norbert Weisser, Xavier Declie, Sharon Bruneau, Debbie
Muggli, Tim Thomerson
The Cyborgs are back in town, this time intent on destroying a
super-female whose synthetic DNA endows her with the power
to wipe out the Cyborg threat forever. The action involves time-
travel from 2077 to 1998 for our female heroine, but naturally
she is chased back in time by her adversaries. Former body-
building champion Bruneau adds a welcome touch of glamour
to this assembly-line entry in the series.
FAN   91 min   VIDrel: COLUM/SONOP   V/sh

NEON BIBLE, THE **   15
Terence Davies UK   1995
Gena Rowlands, Diana Scarwid, Denis Leary, Jacob Tierney, Leo
Burmester, Frances Conroy, Peter McRobbie, Joan Glover, Bob
Hannah, Tom Turbiville, Drake Bell, Dana Dick, Virgil Graham
Hopkins, Jill Jane Clements, Aaron Frisch
A young boy comes of age in the 1940s in America's Deep South

"Bible Belt" in this offbeat but not entirely satisfactory exploration of the trauma of a childhood spent in this claustrophobic environment. A series of slow depressing images are presented in an attempt to say something meaningful, but the message gets lost in the tedium of it all.
DRA   89 min (ort 91 min)   VIDrel: ARTIF/20TH   V
Boa: novel by John Kennedy Toole.

NEON CITY *** 15
Monte Markham USA 1991
Michael Ironside, Lyle Alzado, Vanity, Valerie Wildman, Nick Klar, Juliet Landau, Monte Markham, Sonny Trinidad, Richard Sanders, Curley Green, Jesse Bennett, Jeffrey Olson, Anthony Leger, John Perryman, Donna Todd, Russ McGinn
Former movie actor Markham turns writer-director for this unusual actioner, that's best described as a futuristic remake of STAGECOACH. Set in the year 2053, the world is a bleak and dangerous place that lacks an ozone layer, and is inhabited by various mutants, most particularly murderous "skins" who prey on travellers. Ironside plays a tough character who agrees to ride shotgun on a transporter that's making for the title town. Violent but fun.
A/AD   103 min (ort 107 min)   VIDrel: FIRST/SONOP L/A V

NERVOUS TICKS ** 15
Rocky Lang USA 1992
Bill Pullman, Julie Brown, Peter Boyle, Brent Jennings, James Le Gros, Josh Mostel, Cathy Ladman, Gerald Papasian, Claire Stansfield, Zoe Trilling, Jim Gray, Lauren Lane, Prince Hughes, Lenore Kasdorf, Kery Michales, Timothy Stack
A young man trying to achieve a successful career is dogged by a serious of mishaps, in this watchable comedy. The action is set during a span of ninety minutes in which our airline hero has finish his shift, pick up his married girlfriend and take a plane with her to Rio.
COM   91 min (ort 95 min) cC   VIDrel: 20VIS/SONOP   V

NET, THE ** 12
Irwin Winkler USA 1995
Sandra Bullock, Jeremy Northam, Dennis Miller, Diane Baker, Wendy Gazelle, Ken Howard, Ray McKinnon, Daniel Schorr, L. Scott Caldwell, Robert Gossett, Juan Garcia, Kristina Krofft, Julia Pearlstein, Tony Petez, Margo Winkler
A reclusive female computer expert with no friends and little contact with the outside world, is sent a disk containing a program for her to debug. A little later, she learns that her client has been murdered and all her official records deleted and replaced with a false criminal record, forcing her to fight alone against a faceless enemy. Made to cash in on the Internet hysteria, this disappointing thriller fails to capitalise on some very promising ideas.
THR   110 min (ort 114 min) wScrn cC
VIDrel: COLUM/SONOP; ENCORE (LV only)   V/sur LV

NETHERWORLD ** 18
David Schmoeller USA 1991
Michael C. Bendetti, Denise Gentile, Anjenette Comer, Holly Floria, Robert Sampson, Holly Butler, Alex Datcher, Robert Burr, George Kelly, Mark Kemble, Barret O'Briean, Michael Lowry, David Schmoeller, Candice Williams, Linda Ljoka
A man returns to his family ancestral estate in the bayou that he has inherited and discovers to his amazement that he is expected to bring back his dead father. However, the latter's involvement with a local coven poses a deadly threat, in this standard horror offering.
HOR   81 min (ort 87 min)   VIDrel: COLUM/SONOP   V

NETWORK **** 15
Sidney Lumet USA 1976
Peter Finch, Faye Dunaway, William Holden, Robert Duvall, Ned Beatty, Wesley Addy, Beatrice Straight, William Prince, Marlene Warfield, Arthur Burghardt, Kathy Cronkite, John Carpenter, Bill Burrows, Jordan Charney
With a biting script by Paddy Chayevsky and a superb performance from Finch in one of his last roles, this is a sharp and powerful satire on the world of broadcasting. A news commentator for a TV network becomes an overnight media prophet when he has a breakdown on the air. This was Finch's last film and it earned him a well deserved posthumous Oscar. AA: Actor (Finch), Actress (Dunaway), S. Actress (Straight), Screen/orig (Paddy Chayefsky).
DRA   117 min (ort 121 min)   VIDrel: WHV   V

NEVADA SMITH ** 15
Henry Hathaway USA 1966
Steve McQueen, Karl Malden, Brian Keith, Arthur Kennedy, Raf Vallone, Janet Margolin, Howard Da Silva, Pat Hingle, Martin Landau, Suzanne Pleshette, Paul Fix, Gene Evans, Josephine Hutchinson, John Doucette, Val Avery, Sheldon Allman
Western loosely based on the "early life" of a character in a Robbins novel, and dealing with his violent quest for revenge on the three outlaws who murdered his parents, a plot that also served as the basis for the 1975 TV movie of the same name. Scripted by John Michael Hayes.
WES   125 min (ort 135 min)
VIDrel: 4-FRONT/POLYREC/CIC L/A   V/h
Boa: novel The Carpetbaggers by Harold Robbins.

NEVERENDING STORY, THE *** U
Wolfgang Petersen UK/WEST GERMANY 1984
Noah Hathaway, Barret Oliver, Patricia Hayes, Tami Stronach, Moses Gunn, Sydney Bromley, Gerald McRaney, Drum Garrett, Darryl Cooksey, Nicholas Gilbert, Thomas Hill, Deep Roy, Tilo Pruckner, Alan Oppenheimer (voice only)
A young boy who doesn't much care for reading, is given a magical book that takes him of a journey to a fantasy kingdom which he alone can save from extinction. Good effects and visuals make this an enjoyable fantasy although the lack of any humour and an over-simplistic message are major irritants. This was Petersen's first English language film. See THE PAGEMASTER for another film that uses this idea.
JUV   90 min (ort 94 min)   VIDrel: WHV   V/sh
Boa: novel by Michael Ende.

NEVERENDING STORY 2, THE *** U
George Miller AUSTRALIA/USA/WEST GERMANY 1990
Jonathan Brandis, Kenny Morrison, Clarissa Burt, John Wesley Shipp, Martin Umbach, Alexandra Johnes, Thomas Hill, Helena Michell, Chris Burton, Patricia Fugger, Birge Schade, Claudio Maniscalco, Andreas Borcherding, Ralf Weikinger
With his mom dead and his dad too busy to pay him any attention, our young hero from the first film re-enters the storybook fantasy world of his imagination, where he finds a young empress in danger, and the forces of Good in dire need of a champion. Excellent special effects and some highly imaginative creatures cannot mask the shallowness of the plot, but the film will certainly appeal to young kids, and quite possibly a few adults too.
Aka: NEVERENDING STORY 2, THE: THE NEXT CHAPTER
FAN   86 min (ort 89 min)   VIDrel: WHV   V/dm V/sh

NEVERENDING STORY 3, THE * U
Peter Macdonald GERMANY/USA 1994
Jason James Richter, Melody Kay, Jack Black, Ryan Bollman, Freddie Jones, Julie Cox, Moya Brady, Tony Robinson, Thomas Petruo, Tracey Ellis, Kevin McNulty, Nicole Parker, P. Adrien Dorval, Keefan Shaw, Gorden Robertson
Our youngster returns once more to the mythical lands of Fantasia for another colourful adventure, but unfortunately his exploits fail to generate any excitement. What plot there is mostly revolves around some of the school bullies stealing the magical book that can transport our youngster into Fantasia, an act that imperils this imaginary refuge. A weak, dull and uninvolving affair, it will doubtless be followed by a sequel.
Aka: DIE UNENDLISCHE GESCHICHTE 3: RETTUNG AUS FANTASIA; NEVERENDING STORY 3, THE: ESCAPE FROM FANTASIA; NEVERENDING STORY 3, THE: RETURN TO FANTASIA
FAN   91 min (ort 95 min)   VIDrel: WHV   V/sh

NEVER GIVE AN INCH ** 15
Paul Newman USA 1971
Paul Newman, Henry Fonda, Lee Remick, Michael Sarrazin, Richard Jaeckel, Linda Lawson, Cliff Potts, Sam Gilman, Lee De Broux, Jim Burk, Roy Jenson, Joe Maross, Roy Poole, Charles Tyner, Bennie Dobbins, Alan Gibbs, Fred Lerner
Massive drama about a family of modern-day Oregon loggers whose members all have egos every bit as big as the trees they fell while the title (their family motto) sums up their attitude to life. Excellent performances and some good sequences enliven a dull film.
Aka: SOMETIMES A GREAT NOTION
DRA   110 min (ort 114 min)   VIDrel: L/A   V
Boa: novel Sometimes A Great Notion by Ken Kesey.

NEVER GIVE UP: THE JIMMY V STORY ** (PG)
Marcus Cole USA 1996
Anthony LaPuglia, Ashley Crow, Ronny Cox, Wendell Pierce, Wendy Hooper, Lou Criscuolo, Nikki Deloach, Amanda Minkus, Bob Hannah, Joe Inscoe, Paul Sincott, Jaime Eylsse, Syliva Harman, Dwane Adway, Sonny Shroyer
An abrasive but highly effective college basketball coach is so devoted to his job that he begins to risk his marriage through overwork and a tendency to exploit his personal life in his work. Tragedy strikes when his father suddenly dies in a heart and the arrest of a star player for theft causes an official investigation into his affairs. A curiously structured biopic on a real-life character, it will appeal mainly to fans of the game, but few others.
DRA   90 min mTV   SATrel: MOVIE CHANNEL

NEVER ON TUESDAY * 15
Adam Rifkin USA 1988
Charlie Sheen, Judd Nelson, Claudia Christian, Andrew Lauer, Pete Berg, Dave Anderson, Mark Garbarino, Melvyn Pears, Brett Seals, Nicolas Cage, Cary Elwes, Emilio Estevez, Gilbert Gottfried
Two naive but brash youngsters get a lesson in maturity and manners from a self-possessed young woman, when they wreck her car while driving across the desert on the way to California. A puerile coming-of-age comedy.
COM   87 min Cut (26 sec – ort 90 min)
VIDrel: 20TH/TECH   V

NEVER SAY DIE ** 18
Yossi Wein USA 1994
Frank Zagarino, Bill Drago, Jennifer Miller, Todd Jensen, Robin Smith, Hal Orlandini, Frank Notaro, Michael Brunner, Ted Le Plat, Simon Jones, Vanessa Pike, Skye Svorinic, Michelle Bowes, Graham Clarke, Victor Melleny, Jeff Fannell
A former member of a crack military unit sets out to locate the commander who betrayed him and left him for dead. However, he has to potpone his revenge mission when he is obliged to mount a rescue of the general's daughter, who has been abducted. Average.
A/AD   95 min (ort 99 min)   VIDrel: 20TH/FOXVID   V/sh

NEVER SAY NEVER AGAIN *** PG
Irvin Kershner UK 1983
Sean Connery, Edward Fox, Max Von Sydow, Klaus Maria Brandauer, Barbara Carrera, Alex McCowen, Kim Basinger, Bernie Casey, Rowan Atkinson, Milow Kirek, Valerie Leon, Anthony Sharp, Gavan O'Herlihy, Pat Roach, Prunella Gee
A routine Bond adventure which is a virtual remake of THUNDERBALL (well, they had to run out of Ian Fleming novels eventually). Happily, it also marked a much-needed return to an earlier style, where sophisticated gadgets played a less important role. All in all an enjoyable action film, and the return of Connery (Roger Moore was 007 in the preceding OCTOPUSSY) is more than welcome. The music is by Michel Legrand. Followed by A VIEW TO A KILL.
A/AD   128 min (Cut at film release – ort 134 min)
VIDrel: WHV   V/sur
Boa: novel Thunderball by Ian Fleming.

NEVER SAY NEVER: THE DEIDRE HALL STORY * 12
John Patterson USA 1995
Deidre Hall, Daniel Hugh Kelly, Raymond Baker, Eve Gordon, Connie Ray, Mark Lonow, Wayne Northrop, Lynn Herring, Elaine Bromka, Robert Curtis-Bronw, James Greene, Suzanne Rogers, Nolan Miller, Joel Polis, Kate Williamson, Emily Kuroda
A soap-opera queen has fame, money and success but still feels unsatisfied because she cannot have a child. When she is reunited with an old boyfriend, she tries to tackle this problem by means of a surrogate mother, but this leads to great personal loss. A flat and dull exploration of Hall's personal life (on which she also worked as executive producer) that will be of interest only to devoted fans. A very poor offering even for a mTV production.
Aka: DEIDRE HALL STORY, THE
DRA   90 min mTV   VIDrel: ODY/SONOP   V/sh

NEVER SLEEP ALONE ** 18
Kermal Horulu USA 1983
Honey Wilder, Joey Silvera, Joanna Storm, Victoria Jackson, Sharon Mitchell, Anna Victoria, Eleanor Liqoure, Chelsea Manchester, John Leslie, Ron Jeremy, George Payne, Clint Longley, Michael Knight, Kenneth M. Yontz, Alan Adrian
A woman goes on trial for murdering her husband and we see

in flashback how their open marriage gradually went sour as they spent a lot of their time at a club for swingers.
A   61 min   VIDrel: FIFTH/DISC   V

NEVER TALK TO STRANGERS * 18
Peter Hall CANADA/USA 1995
Rebecca De Mornay, Antonio Banderas, Harry Dean Stanton, Dennis Miller, Len Cariou, Beau Starr, Eugene Lipinski, Martha Burns, Phillip Jarrett, Tim Kelleher, Emma Corosky
A woman police psychologist comes under the influence of an attractive but mysterious young man who seduces her and begins to undergo a personality change. A series of threatening incidents now follow, all of which appear designed to get her to drop her involvement in a case that centres around a serial killer. However, the truth eventually turns out to be far from straightforward. A crude and convoluted psychological thriller, with a highly distasteful resolution.
THR   86 min   CINrel

NEW ADVENTURES OF PIPPI LONGSTOCKING, THE ** U
Ken Annakin USA 1988
Tami Erin, David Seaman Jr, Cory Crow, Eileen Brennan, Dick Van Patten, Dennis Dugan, Dianne Hull, George Di Cenzo, John Schuck, J.D. Dickinson, Chub Bailey, Branscombe Richmond, Evan Adam, Fay Masterson, Romy Mehlman,
The most colourful creation of the Swedish children's author Astrid Lindgren gets the full Hollywood treatment in this dull dud. Much spectacular action is on offer but not an ounce of the playful charm that endeared the earlier Pippi Longstocking films to adults and children alike, far beyond the author's native country.
AkA: PIPPI LONGSTOCKING: THE NEW ADVENTURES
JUV   97 min (ort 100 min) subH   VIDrel: COLUM/SONOP
L/A   V/sh
Boa: novels by Astrid Lindgren.

NEW AGE, THE *** 18
Michael Tolkin USA 1994
Judy Davis, Peter Weller, Adam West, Patrick Bauchau, John Diehl, Samuel L. Jackson, Corbin Bernsen, Jonatahn Hadary, Patricia Heaton, Audrea Lindley, Paula Marshall, Maureen Mueller, Tanya Pohlkotte, Bruce Ramsay, Rachel Rosenthal
A young couple with both business and marital problems get deeply involved with the New Age movement and find it a drain on their already strained finances as well as their shaky relationship. A well-acted portrayal of some incredibly unpleasant and totally self-obsessed yuppies. Watchable, but the lack of any sympathetic characters is a distraction.
COM   107 min cC   VIDrel: WHV   V/sur

NEW BARBARIANS, THE * R18/18
Henri Pachard USA 1990
Victoria Paris, Sabrina Dawn, Tianna, Natasha Skyler, Sheila Kelly, Nina Hartley, Sharon Kane, Randy West, Joey Silvera, Randy Spears, John Dough, Lauren Hall, Mia Powers, Michelle Monroe
Sex film with a sword-and-sandals flavour, telling of how a group of scantily clad female warriors spend their time when they are not engaged in fighting to survive. Very silly. A sequel followed.
Aka: CAVE SLAVES B.C.; ELECTRIC HOLLYWOOD: THE NEW BARBARIANS
A   VIDrel: ELV   V

NEW CENTURIONS, THE *** 15
Richard Fleischer USA 1972
George C. Scott, Stacy Keach, Jane Alexander, Rosalind Cash, Scott Wilson, Erik Estrada, Clifton James, Isabel Sanford, James B. Sikking, Ed Lauter, Roger E. Mosley, William Atherton, Richard Kalk, Beverly Hope, Burke Byrnes
A look at L.A. rookies being shown the ropes by an old cop who knows how to survive in a violent and dangerous city. An excellent adaptation of the author's novel (largely based on his own experiences) which has inspired many a TV police series, from "Police Woman" and "Police Story" to "Hill Street Blues". Inevitably episodic in structure but still worth a look. The script is by Stirling Silliphant.
Aka: PRECINCT 45: LOS ANGELES POLICE
DRA   99 min (ort 103 min)   VIDrel: RCA L/A   V
Boa: novel by Joseph Wambaugh.

NEW EDEN * PG
Alan Metzger USA 1993
Stephen Baldwin, Lisa Bonet, Tobin Bell, Michael Bowen, M.C.

Gainey, Abraham Verduzco, Kate McGregor-Stewart, Janet Hubert-Whitten, Fred Maio, Conrad Goode, Clark Sandrea, J. Michael Oliva, Bud Witte, Jeremy King, Duke Moosekian
Set on a post-WW3 prison planet in the year 2237, where prisoners are routinely expelled into a barren desert wasteland where they are easy prey for gangs of marauding barbarians. But one such prisoner manages to survive and becomes involved with a family and their struggle for existence. Another over-familiar foray into MAD MAX territory that is badly in need of a better script and some fresh ideas.
FAN 85 min (ort 89 min) VIDrel: CIC/SONOP V

NEW GALL FORCE: EPISODES 1 TO 5 ** **   PG
 JAPAN
In the post-apocalyptic wastelands of Australia, a woman and her trusty team battles killer robots, the bulk of humanity having fled them for a new life (and hopefully a new script) on Mars. Another standard Japanese Manga.
ANIM 250 min (2 cassettes – separately available) dubbed VIDrel: MANGA/SONOP V/sh

NEW JACK CITY ** **   18
Mario Van Peebles USA   1991
Wesley Snipes, Ice-T, Allen Payne, Chris Rock, Mario Van Peebles, Michael Michele, Bill Nunn, Russell Wong, Bill Cobbs, Christopher Williams, Judd Nelson, Vanessa Williams, Tracy Camilla Johns, Anthony DeSando, Nick Ashford
The story of a gangster's rise and fall provides the setting for this noisy morality tale, in which the bad guys are jive-talking black youngsters who, in no time at all, have taken control of the city's drugs trade. Snipes as the central character brings a tremendous amount of energy to his role, but this is an over-familiar tale, and the film needs a better plot to sustain it. However, as a directorial debut for Peebles, this is a work that shows great promise.
A/AD 96 min (ort 101 min) VIDrel: WHV V/sur

NEW JERSEY DRIVE ** **   18
Nick Gomez USA   1995
Sharron Corley, Koran C. Thomas, Saul Stein, Gwen McGee, Christine Baranski, Gabriel Casseus, Donald Adeosun Faison, Devin Eggleston, Samantha Brown, Robert Jason Jackson, Gary DeWitt Marchall, Conrad Meertins Jr
A rap soundtrack enlivens this tale of a bunch of kids from Newark, New Jersey, who use their free time (and they have plenty of it) to steal cars, mainly for the excitement it offers. However, they come up against a local cop who is ready to bend the rules in order to catch them. Harsh, bleak and uncompromising, this depressing foray makes little attempt to get the audience involved with the characters.
DRA 94 min (ort 98 min) cC VIDrel: CIC/SONOP V

NEW WAVE HOOKERS 2 ** **   18
Gregory Dark (Gregory Hippolyte) USA   1992
Madison, Danielle Rogers, Jamie Leigh, Savannah, Sandra Scream, Kelly Nicole, Patricia Kennedy, April Rayne, Amanda Stone, Peter North, Randy Spears
Sex film set in L.A. in 1995 where a private eye uncovers a strange plot by an alien being to take over the world through the activities of some sex-crazy women, having got them into his clutches thanks to his mind-controlling powers. As ever, our detective has to thwart these plans, an activity that brings him into contact with a succession of beautiful women. Quite cleverly scripted, even if the ending (it's actually alien women in charge) is a complete cop-out.
A 74 min (ort 80 min) VIDrel: GROHOM/MAXSCAN V

NEW YEAR'S DAY ** **   (12)
Henry Jaglom USA   1988
Maggie Jacobson, Gwen Welles, David Duchovny, Melanie Winter, Harvey Miller, Irene Moore, Michael Emil, Milos Forman, Maggie Jakobson, Tracy Reiner
A writer who has unwisely let his apartment to three women, tries to get it back and gets involved in much incessant talking. Another typical Jaglom look at relationships that offers little of real interest and is more than a little pretentious.
COM 92 min CINrel

NEW YORK COP ** **   18
Toru Murakawa ITALY/USA   1993
Chad McQueen, Mira Sorvino, Toru Nakamura, Conan Lee, Andreas Katsulas, Tony Sirico, Manny Perez, Manny Siverio
A Japanese martial artist comes New York to work undercover

with the local police who are trying to capture a a gang with links to both the Japanese yakuza and Colombian drug lords.
A/AD 84 min (ort 88 min) VIDrel: HIFLI/SONOP V/h

NEW YORK, NEW YORK *   PG**
Martin Scorsese USA   1977
Robert De Niro, Liza Minnelli, Lionel Stander, George Auld, Barry Primus, Dick Miller, Mary Kay Place, George Memmoli, Diahnne Abbott, Murray Moston, Lenny Gaines, Clarence Clemons, Kathi McGinnis, Norman Palmer, Frank Sivera
Excellent musical drama about a saxophonist and a vocalist and their troubled relationship, that was unfairly panned by the critics (always keen to echo each other's opinions), despite its fine music and excellent acting. Very loosely based on "The Man I Love", and scripted by Earl Mac Rauch and Mardik Martin. Musical direction was by Ralph Burns.
MUS 156 min (ort 163 min) VIDrel: MGM/WHV V

NEW YORK STORIES *   15**
Martin Scorsese/Francis Coppola/Woody Allen USA   1989
Nick Nolte, Rosanna Arquette, Steve Buscemi, Peter Gabriel, Deborah Harry, Heather McComb, Talia Shire, Giancarlo Giannini, Don Novello, Chris Elliott, Carole Bouquet, Woody Allen, Mia Farrow, Mae Questel, Julie Kavner
A trio of three separate tales. Scorsese's ponderous "Life Lessons" explores the obsessive love a famous painter develops for his protegee/lover. Coppola's "Life Without Zoe" is a charming but superficial fantasy that tells of a rich girl and the adventures she has in New York City whilst her wealthy parents globetrot. Finally Allen's hilarious "Oedipus Wrecks" gives us the tale of a lawyer who cannot escape the influence of his mother.
COM 119 min (ort 123 min) VIDrel: TOUCH L/A V

NEWS BOYS, THE ** **   PG
Kenny Ortega USA   1992
Christian Bale, Bill Pullman, Robert Duvall, David Moscow, Ann-Margret, Max Casella, Gabriel Damon, Luke Edwards, Michael Lerner, Marty Belafsky, Aaron Lohr, Trey Parker, Arvie Lowe Jr, Kevin Tighe, Charles Cioffi, Ele Keats
In 1899 a group of news boys went on strike in New York in a bid to raise the pittance they were paid by the newspaper owners. This event forms the basis for a very disappointing musical that never comes to life, despite the movie's generous budget. The film suffers badly from very poor songs and indifferent direction, although some of the dancing is lively enough.
Aka: NEWSIES
MUS 117 min (ort 121 min) VIDrel: WDV/TECH L/A V

NEXT DOOR ** **   15
Tony Bill USA   1994
James Woods, Randy Quaid, Kate Capshaw, Lucinda Jenney, Miles Feulner, Billy L. Sullivan, Joe Minares, Ivory Ocean, Dan Hildebrand, Temple Hammett, Chantal Coffey, Erik Taylor, Stella Choe
The competition between two suburban families erupts into open feuding in which they proceed to subject each other to a succession of nasty tricks. An unfocused black comedy-thriller that is too spiteful and uneven to amuse and has a trite and unclear message.
COM 90 min (ort 95 min) mTV VIDrel: COLUM/SONOP V/sur

NEXT KARATE KID, THE *   PG**
Christopher Cain USA   1994
Noriyui "Pat" Morita, Hilary Swank, Michael Ironside, Constance Towers, Chris Conrad, Arsenio Trinidand, Wilt COggins, Michael Carvilleri, Jim Ishida, Rodney Kagayama, Eugene Boles, Seth Sakat, Tom O'Brien, Tom Downey, Wayne Chou
Fourth film in the KARATE KID series, but without Macchio (who wisely went on to better things). This time our plucky pugilist is a teenage girl, whose parents have been killed in an auto crash. She has a hard time at school until a Japanese martial arts instructor (Morita) takes her on as his latest student, helping her develop both as a fighter and a person. A tedious rehash of all the earlier films, with the change of gender the only innovation.
MAR 103 min (ort 107 min) cC VIDrel: COLUM/SONOP V/sur

NEXT OF KIN *   PG**
Thorold Dickinson UK   1941
Mervyn Johns, Nova Pilbeam, Stephen Murray, Jack Hawkins, Reginald Tate, Basil Radford, Naunton Wayne, Geoffrey Hibbert, Mary Clare, Philip Friend

Originally made as a training film at Ealing Studios and then later expanded into a full-length feature, this WW2 thriller follows the exploits of a German spy, who arrives on the coast and sets about learning enough to thwart a commando raid being planned on France. Very effective both as a war-time propaganda film ("careless talk costs lives") and as a piece of entertainment, despite the downbeat ending being changed at the insistence of Churchill.
WAR   95 min B/W   VIDrel: DDVID   V

NEXT OF KIN *   15
John Irvin USA   1989
Patrick Swayze, Adam Baldwin, Helen Hunt, Liam Neeson, Bill Paxton, Andreas Katsulas, Ben Stiller, Michael J. Pollard, Ted Levine, Del Close, Valentino Cimo, Paul Greco, Vincent Gustaferro, Paul Herman, Brett Hadley, Rodney Hatfield
When his younger brother is killed by a local mobster, a Chicago cop from the Appalachian Hills dishes out some backwoods vengeance. Very much a standard revenger of little originality, spoilt (inasmuch as such a film can be) by a foolish ending. Music is by Jack Nitzsche.
A/AD   103 min (ort 109 min)
VIDrel: 4-FRONT/POLYREC/GUILD   V/sur

NIAGARA **   PG
Henry Hathaway USA   1952
Joseph Cotten, Jean Peters, Marilyn Monroe, Don Wilson, Casey Adams, Richard Allan, Dennis O'Dea, Lurene Tuttle, Russell Collins, Will Wright, Lester Matthews, Carleton Young, Sean Clory, Minerva Urecal, Nina Varela, Tom Reynolds
A couple spend their honeymoon at Niagara Falls, but the young wife is already devising ways in which to murder her husband. Monroe's first major role in a competently-made and suspenseful thriller. Written by Charles Brackett, Walter Reisch and Richard Breen.
THR   84 min (ort 89 min)   VIDrel: 20TH/TECH   V/h

NICE GIRLS DON'T EXPLODE *   PG
Chuck Martinez USA   1987
Barbara Harris, Michelle Meyrink, Wallace Shawn, William O'Leary, James Nardini, Margot Gray, Jonas Baughan, William Kuhlke, Belinda Wells, Irwin Keyes, Johnnie "Mac" McHaynes, Holmes Osborne, Peggy Friesen, Richard Kawecki
A girl suffers from a rare handicap in her dealings with the opposite sex: whenever she becomes aroused, fires break out spontaneously. Unfortunately, this film smoulders on dismally to its feeble ending, but will probably eventually become a cult favourite on account of its daft title.
COM   79 min (ort 92 min)   VIDrel: NWV L/A   V

NICHOLAS AND ALEXANDRA **   PG
Franklin J. Schaffner UK   1971
Michael Jayston, Janet Suzman, Tom Baker, Harry Andrews, Jack Hawkins, Fiona Fullerton, Laurence Olivier, Roderic Noble, Ania Marson, Ian Holm, Michael Bryant, Lynne Frederick, Maurice Denham, Curt Jurgens, Eric Porter, John McEnery
Flat and wooden account of the last fourteen years of Czarist Russia. The sets and photography are very good, but never has such an exciting and colourful period of history been treated in such a lifeless and uninspired way. By some curious oversight, Omar Sharif does not have a part in this film. The script is by James Goldman. AA: Art/Set (John Box, Ernest Archer, Jack Maxted and Gil Parrondo/Vernon Dixon), Cost (Yvonne Blake/Antonio Castillo).
DRA   165 min (ort 189 min)   VIDrel: VCC L/A   V
Boa: book by Robert K. Massie.

NICHOLAS NICKLEBY **   U
Alberto Cavalcanti UK   1947
Derek Bond, Cedric Hardwicke, Alfred Drayton, Bernard Miles, Sally Ann Howes, Mary Merrall, Sybil Thorndike, Cathleen Nesbitt, Aubrey Woods, Jill Balcon, Stanley Holloway, Cyril Fletcher, Vera Pearce, Athene Seyler
In 1831 a young schoolmaster struggles to gain his rightful fortune, and protect his family from a scheming uncle in this solid, workmanlike adaptation of the Dickens novel. Written by John Dighton.
DRA   107 min B/W   VIDrel: BRAVE/SONOP L/A   V
Boa: novel by Charles Dickens.

NICK AND JANE *   15
Rich Mauro USA   1996
James McCaffrey, Dana Wheeler-Nicholson, George Coe, David Johansen

A high-powered female executive with a well ordered lifestyle meets a unhappy artist who has lost his motivation, when they both go to the same club. He agrees to pose as her boyfriend to make her boyfriend jealous, and in due course becomes her new one. An unmemorable romantic comedy that remains completely uninvolving, mainly due to the characters being so very poorly delineated.
COM   100 min   VIDrel: MOSAIC/COLUM   V/sh

NICK OF TIME **   15
John Badham USA   1995
Johnny Depp, Courtney Chase, Charles S. Dutton, Christopher Walken, Roma Maffia, Marsha Mason, Peter Strauss, Gloria Reuben, Bill Smitrovich, G.D. Spradlin, Yul Vazquez, Edith Diaz, Armando Ortega, C.J. Bau, Cynthena Sanders
An accountant is forced to agree to become the unwilling assassin of the Governor of California when his young daughter is kidnapped soon after they arrive in L.A. The message he receives from her abductors makes it quite clear that she will be killed if he fails to follow their instructions to the letter. A breakneck suspense yarn that unfolds in "real time", though the plot is badly flawed and the opening premise seems absurdly contrived.
THR   89 min (ort 98 min) cC   VIDrel: CIC/SONOP   V/sur

NICKEL AND DIME **   PG
Ben Moses USA   1991
C. Thomas Howell, Wallace Shawn, Lise Cutter, Lynn Danielson, Roy Brocksmith
A con artist gets a nasty shock when the IRS impound all his worldly goods for non-payment of taxes and appoint an auditor. Anxious to have them returned, he tries to save the situation by earning the reward for finding a missing heiress.
COM   92 min (ort 96 min)   VIDrel: 20VIS/SONOP   V

NICKELODEON **   U
Peter Bogdanovich USA   1976
Ryan O'Neal, Burt Reynolds, Tatum O'Neal, Brian Keith, Stella Stevens, John Ritter, Jane Hitchcock, Harry Carey Jr
An attempt to recreate the early days of Hollywood silent film-making that is entertaining in parts but ultimately fails to get across any true feeling of the period. Scripted by Bogdanovich and W.D. Richter.
DRA   116 min (ort 122 min)   VIDrel: WHV   V/h

NICKY AND GINO **   15
Robert M. Young USA   1987
Tom Hulce, Ray Liotta, Jamie Lee Curtis, Robert Levine, Todd Graff, Bill Cobbs, Mimi Cecchini, David Strahairn
A sentimental but quite touching account of the relationship between a bright medical student and his more simple-minded twin brother, whose security is threatened when his brother (who looks after him) falls in love. Hulce is memorable as the retarded brother.
Aka: DOMINICK AND EUGENE
DRA   104 min (ort 111 min)   VIDrel: VISVID/POLYREC V/sur

NICO **   18
Andrew Davis USA   1988
Steven Seagal, Pam Grier, Sharon Stone, Daniel Faraldo, Henry Silva, Ron Dean, Nicholas Kusenko, Joe V. Greco, Chelcie Ross, Thalmus Rasulala, Jack Wallace, Joseph Kosala, John Drummond, Ronnie Barron
Real-life aikido master Seagal, uncovers a CIA plot in which former agents have become drug dealers. A non-stop action film that provides a fine showcase for his martial arts skills, and was a big hit at the USA box office.
Aka: ABOVE THE LAW; NICO: ABOVE THE LAW
A/AD   95 min Cut (15 sec)   VIDrel: WHV   V/sh

NIGHT AND DAY **   U
Michael Curtiz USA   1946
Cary Grant, Alexis Smith, Monty Woolley, Jane Wyman, Mary Martin, Ginny Simms, Eve Arden, Victor Francen, Alan Hale, Dorothy Malone, Tom D'Andrea, Slena Royle, Donald Woods, Henry Stephenson, Paul Cavanagh, Sig Rumann, Carlos Ramirez
Poor biography of Cole Porter with Grant well and truly miscast in the role of the great composer, rarely have we seen him give such an uncomfortable and stiff performance, although the musical numbers offer some compensation. Musical direction is by Max Steiner and Ray Heindorf.
MUS   122 min (ort 132 min)   VIDrel: WHV   V

**NIGHT AND DAY ** ** 15
Chantal Akerman BELGIUM/FRANCE/
SWITZERLAND 1991
*Guillaine Londez, Thomas Langman, Francois Negret, Nicole Colchat,
Pierre Laroche, Christian Crahay, Luc Fonteyn, Sandrine Laroche,
Yves Comeliau, Olindo Bolzan, Nicole Duret, Violette Leonard, Cecilia
Kankonda*
A young Parisian conducts her own version of a menage-a-trois
by having a relationship with two cab drivers who work oppo-
site shifts. This arrangement works well for a time, but
eventually the woman's desire to flout every accepted conven-
tion threatens to destroy her relationship with both men. This
starts out as a playful romantic comedy, but as it develops, one
becomes only too aware of the hollowness of it all; the film is
very much a triumph of style over content.
Aka: NUIT ET JOUR
COM 90 min CINrel

NIGHT AND DAY * 18
Wesley Emerson USA 1994
*Deidre Holland, Alicia Rio, Flame, Heather Hart, Lacy Rose, Jon
Dough, Cal Jammer, Joe Verducci*
A frustrated woman ventures a visit to a sex therapist, her hope
being that the treatment she receives there will enable her to live
out her fantasies. The predictable script then calls for a succes-
sion of sexual encounters, and neither the plot nor the settings
do anything to maintain attention.
A 52 min (ort 80 min) VIDrel: ONE V

NIGHT AND DAY 2 * 18
Wesley Emerson USA 1994
Debi Diamond, Tara Monroe, Jon Dough
Sex film built around the exploits of a couple of budding star-
lets, who have to do more than a simple screentest to satisfy the
demands of a sleazy Hollywood producer.
A 42 min VIDrel: ONE V

NIGHT AND THE CITY ** (15)
Jules Dassin UK 1950
*Richard Widmark, Gene Tierney, Herbert Lom, Francis L. Sullivan,
Hugh Marlowe, Googie Withers, Stanislaus (Stanley) Zbyszko, Mike
Mazurki, Charles Farrell, Ada Reeve, Ken Richmond, Elliott
Makeham, Betty Shale, Russell Westwood*
An American hustler working for the owner of a shady club,
hatches a scheme to make money by promoting wrestling
matches and attempts to persuade a famous former wrestler to
back it. This brings him into conflict with both the man's
mobster son and his own boss. A dark, brooding and depressing slice of film noir, set in a London full of unattractive
characters with no redeeming features. However, Widmark
gives an outstanding performance. Remade in 1992.
DRA 104 min B/W TVrel
 Boa: novel by Gerald Kersh.

NIGHT AND THE CITY ** 15
Irwin Winkler USA 1992
*Robert De Niro, Jessica Lange, Alan King, Eli Wallach, Jack Warden,
Cliff Gorman, Barry Primus, Gene Kirkwood, Pedro Sanchez, Gerry
Murphy, Anthony Canarozzi, Clem Caserta, David W. Butler, Byron
Utley, Margo Winkler*
Remake of the 1950 British film but set in New York City instead
of London, with De Niro playing the lead role of a self-obsessed
and fast-talking schemer. He hatches a plan to bring an old
boxing star out of retirement, but his hustling makes him quite
a few powerful enemies. Excellent acting, especially from De
Niro, and the New York locations compensate for the rather
thin plot, but on the whole, the film has nothing like the impact
of the original work.
DRA 103 min (ort 105 min) VIDrel: FIRST/SONOP; PION
(LV only) V/sur LV
 Boa: novel by Gerald Kersh.

NIGHT ANGEL * 18
Dominique Othenin-Girard USA 1990
*Isa Andersen, Linden Ashby, Karen Black, Debra Feuer, Helen Martin,
Doug Jones, Sam Hennings, Gary Hudson, Tedra Gabriel, Ben Ryan,
B.J. Turner,Ttwink Caplan, Phil Fondacaro, Susie Sparks, Jill Sparks,
Debra Cosey, Ira Levine*
A female demon in human form poses as a fashion model, which
gives her plenty of opportunity to attract suitable sacrificial
victims among the men who find themselves irresistibly drawn
to her. When the killings start, a male employee of the fashion
house where she works, eventually discovers the truth and

resolves to deal with her. A standard gory horror tale that
despite its efforts at eroticism is full of familiar cliches.
Aka: HELL BORN
HOR 90 min VIDrel: NEWAGE/20VIS L/A V

NIGHT BEAST * 18
Don Dohler USA 1983
*Tom Griffith, Richard Dyszel, Jaimie Zemarel, Don Leifert, Karin
Kardian, George Stover, Anne Frith, Eleanor Herman, Glenn Barnes,
Rose Wolfe, Richard Ruxton, Jerry Schuerholz, Hank Stuhmer, Fred
Gibmeyer, Bump Roberts*
An alien craft crash-lands near a small town and its pilot decides
to make a contribution towards population control by devour-
ing the townsfolk and zapping them with laser blasts. A
low-grade film from a company that specialises in such offer-
ings.
HOR 85 min VIDrel: TROMA/RTM V

NIGHT BEFORE, THE * 15
Thom Eberhardt USA 1988
*Keanu Reeves, Lori Loughlin, Theresa Saldana, Trinidad Silva,
Suzanne Snyder, Morgan Lofting, Gwil Greene, Chris Herbert,
Michael Greene, Pamela Gordon, David Sherril, Isabel Jurabe, Charles
Gruber, Michael Strasser*
A daft high school kid tries to recall what he got up to the night
before when he attended a concert. A feeble little comedy with
a few funny moments.
COM 86 min Cut (2 sec – ort 90 min) VIDrel: WHV V/sh

NIGHT CALLER * 18
Philip Chan HONG KONG 1987
Pauline Wong (Wang Xiao-Feng), Melvin Wong, Philip Chan, Pat Ha
When her lesbian lover puts an end to their relationship, a
woman is unable to come to cope with this and becomes a
crazed knife-wielding maniac. A well-produced and quite
graphic exploiter.
THR 100 min VIDrel: EAST/DISC V

NIGHT CALLER * 18
Fred Williamson USA 1992
*Fred Williamson, Gary Busey, Vanity, Peter Fonda, Robert Forster,
Sheree Devereaux, Henry Silva, Stella Stevens, Sam Jones, Mark Ross*
Two former football players enjoy the good life in Miami in
between assignments as part-time private eyes, but a mysteri-
ous woman hires them for a case involving a psycho that takes
up all their time. Unimpressive but watchable.
Aka: SOUTH BEACH
THR 88 min (ort 93 min) VIDrel: OPTIK/HIFLI V/h

NIGHT EYES * 18
Jag Mundhra USA 1990
*Tanya Roberts, Andrew Stevens, Warwick Sims, Karen Elise Baldwin,
Cooper Huckabee, Chick Vennera*
A security expert/private eye finds himself getting heavily
involved with the wayward wife he has been hired to spy on.
A highly derivative film that inevitably recalls BODY HEAT,
but with little to offer apart from the hot sex scenes.
DRA 94 min (ort 95 min) VIDrel: VIDMK/HIFLI V

NIGHT EYES 2 * 18
Rodney McDonald USA 1991
*Shannon Tweed, Andrew Stevens, Tim Russ, Richard Chaves, Geno
Silva*
An assassination attempt on the lives of a diplomat and his wife
is foiled by their bodyguard, and the would-be killer is shot
dead. With the safety of his wife uppermost in his mind, the
diplomat hires a security expert, but unknown to him another
assassination attempt is being planned.
Aka: HOUR OF DARKNESS
THR 94 min (ort 98 min) VIDrel: TRIM/HIFLI V

NIGHT EYES 3 * 18
Andrew Stevens USA 1992
*Shannon Tweed, Andrew Stevens, Tracy Tweed, Tristan Rogers, Todd
Curtis, Dan McVicar, Richard Portnow, Allison Mack, Leslie Sachs,
Mariane Muellerleile, Monique Parent, Phil Redrow, Alina Cenal,
Keith Hickes, J.D. Hinton, Jon Greene*
When a woman TV star finds herself being stalked, she calls in
a security expert who soon has her every activity taped, but he
has only done this in order to indulge in his penchant for
voyeurism.
THR 96 min (ort 101 min) VIDrel: EIV/SONOP V

NIGHT FIRE **   18
Mike Sedan USA   1994
Shannon Tweed, John Laughlin, Martin Hewitt
A couple whose marriage is on the rocks try to improve matters
by spying on another couple's amorous activities, but this
proves to be far more dangerous than they had anticipated.
THR   91 min (ort 95 min)   VIDrel: MARQ/QUANT   V

NIGHT GAME *   18
Peter Masterson USA   1988
*Roy Scheider, Karen Young, Richard Bradford, Paul Gleason, Carlin
Glynn, Lane Smith, Rex Linn, Alex Garcia, Michelle Cochran, Bob
Allen, Lisa Hart Carroll, Sarah Chattin, John Martin, Tony Frank,
Teresa Dell*
A homicide detective assigned to a case involving a series of
mutilation killings of young women, discovers that a murder
takes place every time a star player wins a game for his favourite
baseball team. When he finds that his girlfriend is likely to be
the next victim, matters are brought to a head in this slow and
unexciting thriller.
THR   91 min (ort 95 min)   VIDrel: EIV/SONOP L/A   V

NIGHT HUNTER **   18
Rick Jacobson USA   1995
*Don "The Dragon" Wilson, Melanie Smith, Nicholas Guest, Sid
Sham, Cash Casey, Maria Ford, Ron Winston Yuan, Michael
Cavanaugh, Vince Murdocco, James Lew, Marcus Aurelius, Vincent
Klyn, David "Shark" Fralik, Sofia Crawford*
A martial arts fighter attempts to track down a murderous
vampire but finds himself being sought by the police who
suspect him of these crimes. Meanwhile, the last of the world's
vampires are gathering in L.A., where they await an eclipse that
will allow them to increase in number. An inept blending of
two genres that does not really come off, possibly mainly due
to the film's poor production values.
A/AD   90 min   VIDrel: EIV/SONOP   V

NIGHT IN HEAVEN, A **   18
John G. Avildsen USA   1983
*Lesley Anne Warren, Christopher Atkins, Robert Logan, Carrie
Snodgress, Deborah Rush, Sandra Beall, Denny Terrio, Alix Elias,
Amy Levine, Fred Buch, Karen Margaret Cole, Don Cox, Veronica
Gamba, Joey Gian, Bill Hindman, Linda Lee*
A married woman teacher becomes obsessed with a boy pupil
who is an exotic dancer in a nightclub. A disappointing study
of an unusual romance.
DRA   80 min (ort 85 min)   VIDrel: 20TH   V

NIGHT IS YOUNG, THE **   15
Leos Carax FRANCE   1986
*Denis Lavant, Juliette Binoche, Michel Piccoli, Hans Meyer, Carroll
Brooks, Julie Delphy*
A petty criminal joins forces with a couple of other crooks, and
they plan the burglary of a laboratory, their intention being to
obtain supplies of an experimental drug used to stop the spread
of a disease that only attacks insincere lovers. With a plot akin
to something from pulp fiction, this peculiar film (a sort of loose
sequel to "Boy Meets Girl") recalls some of the earlier work of
Godard, but unlike that director breaks no new ground.
Aka: MAUVAIS SANG
DRA   114 min (ort 119 min)   VIDrel: ARTIF/20TH   V/h

NIGHT MOVES ***   18
Arthur Penn USA   1975
*Gene Hackman, Jennifer Warren, Susan Clerk, Edward Binns, Harris
Yulin, Melanie Griffith, James Woods, Janet Ward, Anthony Costello,
John Crawford, Ben Archibek, Maxwell Gail Jr, Victor Paul, Louis
Elias, Carey Loftin*
A Los Angeles private eye tracks a missing teenage girl to
Florida and returns her to her actress mother, only to learn of
her death during a movie stunt. After watching footage of the
accident, he comes to the conclusion that she was murdered and
decides to investigate. An intricate, downbeat thriller, scripted
by Alan Sharp.
THR   95 min (ort 99 min)   VIDrel: WHV   V

NIGHT OF SAN LORENZO, THE ***   12
Paolo Taviani/Vittorio Taviani ITALY   1981
*Omero Antonutti, Margarita Lozano, Claudio Bigagli, Massimo
Bonetti, Norma Martelli, Enrica Maria Modugno*
A women ponders the past events of her life, most especially the
time when, as a child in Tuscany during WW2, she was forced
to live in the forest to elude invading German soldiers, who

embarked on an orgy of destruction prior to the American
advance of 1944. The Taviani brothers were young teenagers at
the time, and use their experiences to bring to life one of the
turning points in modern Italian history. Unfortunately, the
film's rambling structure is not an asset.
Aka: LA NOTTE DI SAN LORENZO; NIGHT OF THE SHOOTING STARS
DRA   102 min (ort 105 min) wScrn   VIDrel: ARTPRO/RTM
V

NIGHT OF THE ALIEN *   18
Ronald W. Moore USA   1985
*Edwin Neal, Marilyn Burns, Gabriel Folse, Wade Reese, Baron Faulks,
Rob Rowley, Craig Kannet, Jeffrey Scott, Alice Villarreal, Dorothy
Grim, Karin Kay, Doug Davis, Elizabeth Henshaw, Cathy Durkin,
Kate Cadehead, Joe Abner*
Fraternity house members get caught on the wrong side of time
where they encounter anti-nuclear punks, one of whom has been
exposed to radiation. A futuristic SF thriller that is wildly over-
derivative and fails to deliver anything of real interest.
Aka: FUTURE KILL; SPLATTER
FAN   78 min (ort 90 min)   VIDrel: EIV/SONOP   V

NIGHT OF THE BLOODY APES *   18
Rene Cardona Jr MEXICO   1970
*Jose Elias Moreno, Armando Silvestre, Carlos Lopez Moctezuma,
Norma Lazareno, August Martin (Agustin Martinez Solares), Gina
Moret, Noelia Noel, Gerard Zapeda*
A boy is given transplanted organs taken from an ape, but this
experimental operation goes wrong and a mutant monster is
created. Crudely made and quite shoddy, it is replete with lash-
ings of gore plus some footage from a real-life operation to pad
out the proceedings. Unlikely to be available in an uncut form
in the UK.
Aka: GOMAR, THE HUMAN GORILLA; HORROR Y SEXO; LA HORRIPI-
LANTE BESTIA HUMANA
HOR   76 min (ort 82 min) partly dubbed
VIDrel: VIPCO/SGSVID   V

NIGHT OF THE CREEPS **   18
Fred Dekker USA   1986
*Jason Lively, Steve Marshall, Jill Whitlow, Tom Atkins, Wally Taylor,
Bruce Salomon, Vic Polizos, Allan J. Kayser, Ken Heron, Alice
Cadogan, June Harris, David Paymer, David Oliver, Evelyne Smith,
Ivan E. Roth, Dick Miller*
Alien parasites turn their human victims into zombies at a
college dance, but the care lavished on special effects would
have been put to better use if it had been applied to the script
of this lacklustre horror-comedy.
HOR   85 min (ort 89 min)   VIDrel: 20TH   V/sur

NIGHT OF THE DEMON ***   12
Jacques Tourneur UK   1958
*Dana Andrews, Peggy Cummins, Niall MacGinnis, Maurice
Denham, Athene Seyler, Liam Redmond, Reginald Beckwith, Ewan
Roberts, Peter Elliott, Brian Wilde, Rosamond Greenwood, Richard
Leech, Lloyd Lamble, Peter Hobbes, John Salew*
An American psychologist goes to Britain to assist a professor
in his struggle with a satanic cult but finds that the man has been
killed by the sect leader and is asked by his daughter to inves-
tigate his death. Despite some heavy-handed interference by
the producer, Tourneur's brilliant visual style and a literate
script make this a classic film of the supernatural.
Aka: CURSE OF THE DEMON; HAUNTED, THE NIGHT OF THE DEMON
HOR   95 min B/W wScrn
VIDrel: ENCORE/SPEAR/COLUM   V
Boa: short story Casting the Runes by Montagu R. James.

NIGHT OF THE DEMON *   18
James C. Wasson USA   1981
*Michael J. Cutt, Joy Allen, Bob Collins, Jodi Lazarus, Richard Fields,
Michael Lang, Shannon Cooper, Ray Jarris, Paul Kelleher, William F.
Nugent, Melanie Graham, Lynn Eastman, Eugene Doer, Don Hurst,
Terry Wilson*
The legend of "Big Foot" is found to be true and is linked to
devil worship as a survivor of a backwoods expedition tells his
story to the disbelieving police. A poor effort. Unlikely to be
available in an uncut form in the UK.
HOR   90 min (ort 97 min)   VIDrel: VIPCO/SGSVID   V

NIGHT OF THE DEMONS **   18
Kevin G. Tenney USA   1987
Linnea Quigley, Alvin Alexis, Lance Fenton, William Gallo, Hal

Havins, Mimi Kinkade, Cathy Podewell, Phillip Tanzani, Jill Terashita, Allison Barron, Donnie Jeffcoat
A routine shocker in which a bunch of dumb teens decide to hold a Halloween party in an abandoned funeral parlour and are picked off one at a time by a bunch of unpleasant and vicious demons. The special effects cannot be faulted, but the tired old storyline is no more than a vehicle for them.
Aka: HALLOWEEN PARTY
HOR   87 min Cut (4 sec – ort 90 min)   VIDrel: POPRO/RTM
V/h

NIGHT OF THE DEMONS 2 **   18
Brian Trenchard-Smith USA   1993
Bobby Jacoby, Amelia Kinkade, Zoe Trilling, Jennifer Rhodes, Merle Kennedy, Rod McCary, Christi Harris, Johnny Moran, Rick Peters, Christine Taylor, Ladd York, Darin Heames, Mark Neely, Rachel Longaker, Jim Quinn (voice only)
Routine sequel in which another bunch of teens on Halloween find themselves menaced by our friendly demons when they rather unwisely pay a visit to a haunted mansion. Plenty of overblown special effects and some rather clumsy (albeit inventive) comic sequences cannot mask the pitiful lack of fresh ideas.
Aka: NIGHT OF THE DEMONS: ANGELA'S REVENGE
HOR   95 min (ort 96 min)   VIDrel: MARQ/QUANT   V

NIGHT OF THE FOLLOWING DAY, THE **   18
Hubert Cornfield USA   1968
Marlon Brando, Richard Boone, Rita Moreno, Pamela Franklin, Jess Hahn
Having arrived in Paris to stay with her father, a young girl is kidnapped by a gang and held to ransom. A simple suspense yarn whose energy is rapidly dissipated by verbosity and whose cop-out ending (the whole episode may have been a dream) is a serious flaw. A few touches of that old Brando magnetism are scant compensation. Filmed in France.
THR   89 min (ort 100 min)   VIDrel: CIC/SONOP   V
Boa: novel The Snatchers by Lionel White.

NIGHT OF THE FOX **   15
Charles Jarrott FRANCE/UK/USA   1990
George Peppard, Deborah Raffin, Michael York, David Birney, John Mills, Gottfried John, Dieter Steinbrink, Andrea Ferreol, George Standing, Andrea Ferreol, Amadeus August, Niall O'Brien, George Mikel, Juliet Mills, Patachou
A film that recalls many a WW2 drama of earlier years. An Allied officer who possesses a detailed knowledge of the D-Day invasion plans falls into German hands. A team of agents is sent to rescue him, with specific instructions to kill him should their mission fail. The film "Breaking Point" dealt with a similar issue.
WAR   96 min (ort 192 min) mCab   VIDrel: ITC/POLYREC
V/h
Boa: novel by Jack Higgins.

NIGHT OF THE GENERALS *   15
Anatole Litvak FRANCE/UK   1967
Omar Sharif, Tom Courtenay, Peter O'Toole, Donald Pleasence, Joanna Pettet, Christopher Plummer, Coral Browne, Harry Andrews, John Gregson, Nigel Stock, Juliette Greco, Charles Gray, Yves Brainville, Gerald Buhr, Gordon Jackson
A Nazi intelligence officer tries to discover the identity of a disturbed general who has made a career out of murdering prostitutes. It's hard to believe that the Nazis cared tuppence about his behaviour, and by the end of this interminably boring film such considerations matter little. A clear waste of a good cast and large budget.
DRA   138 min (ort 148 min)   VIDrel: VCC/DISC/COLUM
V
Boa: novel by Hans Helmut Kirst.

NIGHT OF THE GHOULS *   PG
Edward D. Wood Jr USA   1959
Kenne Duncan, Duke Moore, Tor Johnson, Valda Hanesen, John Carpenter, Paul Marco, Criswell, Jeannie Stevens
This long-lost film from this cult director is merely dull without being amusing. A phoney psychic stages seances to bilk wealthy patrons of their money but accidentally revives a number of corpses who take revenge for being disturbed. Just about watchable although its slow pace and atrocious production values make it heavy going at times.
Aka: REVENGE OF THE DEAD
HOR   68 min B/W   VIDrel: CARL/TECH   V

NIGHT OF THE HUNTER, THE ****   12
Charles Laughton USA   1955
Robert Mitchum, Shelley Winters, Lillian Gish, Evelyn Varden, Peter Graves, James Gleason, Billy Chapin, Sally James Bruce, Don Beddoe, Gloria Castillo, Mary Elen Clemons, Cheryl Callaway, Corey Allen, Paul Bryar
Powerfully atmospheric thriller with an undercurrent of brooding menace, as a psychopathic fanatical preacher goes after two children whose father stole money from him. A wonderfully dark and nightmarish tale that impresses at every turn. Scripted by James Agee. Laughton's only venture into the field of directing is a marvellous one-off achievement.
THR   89 min (ort 94 min) B/W   VIDrel: MGM/WHV   V
Boa: novel by David Grubb.

NIGHT OF THE HUNTER **   15
David Greene USA   1991
Richard Chamberlain, Diana Scarwid, Amy Bebout, Reid Binion, Ray McKinnon, Mary Nell Santacroce, Burgess Meredith
Remake of the 1955 classic that casts Chamberlain as the fake preacher who's out to murder a couple of children for the money their dad stole. Scripted by Edmond Stevens, this adequate effort is spoilt by a new ending, but more importantly lacks a vital missing ingredient only to be found in the first film: Robert Mitchum.
THR   91 min (ort 100 min) mTV   VIDrel: CAPIT/GUILD   V
Boa: novel by David Grubb.

NIGHT OF THE IGUANA, THE ***   12
John Huston USA   1964
Richard Burton, Deborah Kerr, Ava Gardner, Sue Lyon, Skip Ward, Grayson Hall, Cyril Delevanti, Gladys Hill, Billie Matticks, Emilio Fernandes, Liz Rubey, Eloise Hardt, Thelda Victor, Betty Proctor, Dorothy Vance, C.G. Kim
Dramatic, eminently watchable filmisation of the Williams play in which the hounded figure of a defrocked clergyman reduced to working as a tour guide finds temporary solace in a broken-down Mexican hotel. A fine and surprisingly moving story that is given conviction by the cast, and though occasionally the performances are overblown and cruel, the story uplifts and has a satisfying resolution. AA: Cost (Dorothy Jeakins).
DRA   113 min (ort 125 min) B/W   VIDrel: MGM/WHV   V
Boa: play by Tennessee Williams.

NIGHT OF THE JUGGLER **   18
Robert Butler USA   1980
James Brolin, Richard Castellano, Cliff Gorman, Abby Bluestone, Linda G. Miller, Robert Butler, Julie Carmen, Barton Heyman, Mandy Patinkin, Dan Hedaya, Marco St John, Franj Adu, Nancy Andrews, Rick Anthony, Tony Anzito, Tito Goya
A mentally disturbed man mistakenly kidnaps an ex-cop's daughter, and the father begins to comb the city in a hunt for her. An average action thriller with no shortage of car chases.
THR   96 min (ort 101 min)   VIDrel: POLY/POLYREC   V

NIGHT OF THE LIVING DEAD ***   18
George A. Romero USA   1968
Judith O'Dea, Duane Jones, Karl Hardman, Russell Streiner, Marilyn Eastman, Keith Wayne, Judith Ridley, Kyra Schon, Bill Hinzman, Charles Craig, Frank Doak, George Kosana, Bill Cardille, Vince Survinski, John A. Russo
Flesh-eating zombies are activated by rays from a space rocket and ravage the countryside, trapping a small group of survivors in a deserted farmhouse. Followed by two sequels, ZOMBIES: DAWN OF THE DEAD (1979) and DAY OF THE DEAD (1985). A harsh and uncompromising chiller, done in a low-key, almost documentary style. Remade by Savini in 1990 it also inspired (if that is the right word) the spoof sequel RETURN OF THE LIVING DEAD.
Aka: NIGHT OF ANUBIS; NIGHT OF THE FLESH EATERS
HOR   96 min (ort 98 min) B/W coVer (4-FRONT/POLYREC)
VIDrel: TART/20TH; 4-FRONT/POLYREC   V/dm V/h

NIGHT OF THE LIVING DEAD **   18
Tom Savini USA   1990
Tony Todd, Patricia Tallman, Tom Towles, McKee Anderson, William Butler, Katie Finnerman, Bill Mosley, Heather Mazur, David Butler, Zachary Mott, Pat Reese, William Cameron, Pat Logan, Berte Ellis, Bill "Chully Billy" Cadille
A high-tech remake of the 1968 original that is virtually identical except for the ending and relies quite heavily on special effects, though Savini claims that much of the gory effects were excised in post-production. But the freshness and verve of the

earlier film are entirely lacking, despite the fact that Romero wrote the script. See also RETURN OF THE LIVING DEAD for a semi-humorous sequel to the original cult classic.

Aka: ALL NEW GEORGE A. ROMERO'S NIGHT OF THE LIVING DEAD: THE REMAKE; NIGHT OF THE LIVING DEAD: THE REMAKE

HOR   84 min (ort 96 min) wScrn   VIDrel: TART/20TH; ENCORE (LV only)   V/dm LV

NIGHT OF THE RUNNING MAN **
18
Mark L. Lester USA
1994
Scott Glenn, Andrew McCarthy, Alex Zonn, Matthew Laurance, Carl Ciarfalio, Antony Ponzini, Jeanna Michaels, Kathrin Lautner, Peter Iacangelo, Mayf Nutter, Don Stark, Damon Carr, Kim Lankford, Frank Novak, Terri Hawkes, Todd Susman
A Las Vegas cabbie is unlucky enough to take a man to the airport who has just stolen a fortune from a Mafia controlled casino. When his passenger is shot and run over, the cabbie flees for his life but takes the loot with him. Soon a Mafia hired contract killer is on his trail in a watchable if unoriginal actioner, with the usual quota of brutality etc.
A/AD   89 min (ort 100 min)   VIDrel: TRIM/HIFLI   V/sur
Boa: novel by Lee Wells.

NIGHT OF THE SCARECROW **
18
Jeff Burr USA
1995
Elizabeth Barondes, John Mese, Stephen Root, Bruce Glover, Dirk Blocker, Howard Swain, William Joseph Barker, Martine Beswick, Christi Harris, Bob Harvey, John Hawkes, Harri James, John Lazar, Gary Lockwood, Cynthia Merrill
A scarecrow becomes transformed into a raging monster and goes on the rampage in this predictable horror flick.
HOR   80 min (ort 90 min)   VIDrel: POLFIL   V

NIGHT OF THE SEAGULLS *
(18)
Amando de Ossorio SPAIN
1975
Victor Petit, Maria Kosti, Sandra Mazarosky, Julie James, Julia Saly, Jose Antonio Calvo
A doctor and his wife move to a seaside village with a sinister and deadly secret, and eventually learn of nightly sacrifices of virgins. One of a series of Spanish horror films, featuring the eyeless zombie figures of the Knights Templar. The powerful and moody opening is let down by a lack of plot development and the film tends to drag badly until the climax. See also THE RETURN OF THE EVIL DEAD and TOMBS OF THE BLIND DEAD.
Aka: BLOOD FEAST OF THE BLIND DEAD; LA NOCHE DE LOS GAVIOTAS; NIGHT OF THE DEATH CULT
HOR   85 min (ort 90 min)   VIDrel: L/A   V

NIGHT OF THE TWISTERS **
PG
Timothy Bond USA
1996
John Schneider, Devon Sawa, Amos Crawley, Lori Hallier, Laura Bertram, David Ferry, Helen Hughes, Irene Erwin, Alex Lastewka, Ronnie Vae, Graham McPherson, Ted Simonett, Don Allison, Markus Porilo, Rex Hagan, Deborah DeMille, Dave Berni
When a rural community is struck by a series of freak storms and tornados, a young boy and his friend are left alone with the former's younger sibling, while his stepfather goes to check on a relative. In the chaos that ensues after the house is struck by a tornado, the boys prove themselves, in this study of human beings coping with the forces of nature. A well researched film that was rushed out to capitalise on the success of TWISTER. Average.
A/AD   91 min mTV   VIDrel: CARL/TECH   V
Boa: novel by Ivy Ruckman.

NIGHT OF THE WARRIOR **
18
Rafal Zielinski USA
1990
Lorenzo Lamas, Anthony Geary, Kathleen Kinmont, Danny Kamekona, Arlene Dahl, Jeff Imada, Wilhelm Von Hamburg, Ray Sua, Felicity Waterman, Michael Catlin, Bill Erwin, Mary Ann Oedy, Richard Redlin, Mike Ballew, Mario Roberts
Action film revolving around the murky world of illegal street-fighting contests where anything is allowed, with a young man who wants to escape from such a life getting embroiled in a murder. An average blending of brutal shenanigans plus the obligatory romantic interest.
Aka: NIGHT WARRIOR
A/AD   95 min Cut (17 sec – ort 96 min)
VIDrel: MIA/DISC/FIRST   V

NIGHT ON EARTH ***
15
Jim Jarmusch USA
1991
Winona Ryder, Gena Rowlands, Giancarlo Esposito, Armin Mueller-

Stahl, Rosie Perez, Isaach De Bankole, Beatrice Dalle, Roberto Benigni, Paolo Bonacelli, Matti Pellonpaa, Kari Vaananen, Sakari Kuosmanen, Tomi Salmela, Lisanne Falk
Five cities – five stories, the starting point for each being a taxi-cab that serves as a confessional booth for those with the need to talk. From L.A. to New York, then to Paris, Rome and Helsinki, each tale works as a separate entity, deftly blending comedy and pathos, and all take place over the same span of time. Though none of the fragments are in the slightest way related to each other, this is intentional and is hardly a criticism.
COM   123 min (ort 129 min) partly subtitled
VIDrel: ELPIC/POLYREC   V

NIGHT PLEASURES **
18
Max Pecas FRANCE
1970
Sandra Julien, Janine Raynaud, Yves Vincent, Patrick Verde, Michel Lemoine, Alain Hitier, Bob Ingarao, Michel Charrel, Michel Vocoret, Helene Tossy, Saint Bris, France Noelle
The story of a nymphomaniac and her quest for pleasure, the central character being a strictly brought up French girl who is both bored with life and sexually unfulfilled, but has an accident that modifies her outlook.
Aka: CAROLE ET SES DEMONS; FORBIDDEN PASSION; I AM FRIGID, WHY?; LET ME LOVE YOU; LIBIDO; SENSUOUS TEENAGER, THE
A   90 min   VIDrel: ELV   V

NIGHT PORTER, THE *
18
Liliana Cavani ITALY/USA
1973
Dirk Bogarde, Charlotte Rampling, Philippe Leroy, Gabriele Ferzetti, Isa Miranda, Giuseppe Addobbati, Isa Miranda, Nino Bignamini, Marino Mase, Amedeo Amodio, Piero Vida, Geoffrey Copleston, Manfred Freiberger, Nora Ricci
A woman resumes her sado-masochistic affair with a former SS guard at a concentration camp, when she finds that he is now employed as the night porter at a hotel. A dismal and exploitative tale, set in the 1950s.
Aka: IL PORTIERE DI NOTTE
DRA   112 min (ort 117 min)   VIDrel: 4-FRONT/POLYREC   V

NIGHT RHYTHMS **
18
Alexander Gregory Hippolyte USA
1992
Martin Hewitt, Sam Jones, Deborah Driggs, Delia Sheppard, Tracy Tweed, Jamie Stafford, Patrice Leal, Julie Strain, Vincent Curto, David Carradine, Juliet James, Timothy G. Burns, Stephen Fiachi, Theresa Ring, Kristine Rose, Erika Nann
A talk-show host who is a hit with woman callers, who reveal their sexual problems and fantasies on the air, finds himself involved in a murder. Cast as the prime suspect, he has a hard time proving his innocence. A tepid attempt at an erotic thriller.
THR   98 min (ort 105 min)   VIDrel: COLUM/SONOP   V/sh

NIGHT SUN ***
15
Paolo Taviani/Vittorio Taviani FRANCE/ITALY/
GERMANY
1990
Julian Sands, Nastassja Kinski, Charlotte Gainsbourg, Massimo Bonetti, Margarita Lozano, Patricia Millardet, Rudiger Vogler, Pamela Villoresi, Geppy Gleijeses, Sonia Gessner, Gaetano Sperandeo, Matilde Piana, Vittorio Capotorto
In 18th century Italy, a nobleman learns that his fiancee was mistress to the King, and because of this he leaves the court and enters a monastery, but another romantic entanglement forces him to leave once more, and he ends up as a vagrant. A vigorous and full-blooded version of the Tolstoy story, that requires patience, but amply rewards it. Sands is dubbed in Italian, but this does nothing to detract from his fine performance.
Aka: IL SOLE ANCHE DI NOTTE
DRA   107 min (ort 113 min)   VIDrel: ARTIF/20TH   V/s
Boa: short story Father Sergius by Leo Tolstoy.

NIGHT TERRORS *
18
Tobe Hooper USA
1993
Robert Englund, Zoe Trilling, William Finley, Alona Kimhi, Juliano Merr
Presumably this confused epic was meant to be about the Marquis de Sade (it was originally entitled "De Sade") as it opens with England as the infamous Marquis getting his come-uppance. Then we are in modern-day Alexandria, where a teenage girl has arrived to visit her archaeologist father, only to learn that there is a killer on the loose (who it transpires is a

descendant of De Sade). A dire film, inspired by Jess Franco's "Eugenie – Her Voyage Into Perversion".
Aka: TOBE HOOPER'S NIGHT TERRORS
HOR   93 min   VIDrel: MGM/WHV   V/sh

NIGHT TO REMEMBER, A *** PG
Richard Wallace USA   1942
Loretta Young, Brian Aherne, Jeff Donnell, William Wright, Sidney Toler, Gale Sondergaard, Donald MacBride, Lee Patrick, Blanche Yurka, Don Costello, Richard Gaines, James Burke, Billy Benedict, Cyril Ring, Eddie Dunn
A woman married to a mystery writer tries to get him to write a romantic novel, and even finds them a quiet apartment where he can work. However, these good intentions go out the window when a body turns up and they turn to amateur sleuthing instead. A lively comedy-mystery with appealing madcap touches. Written by Richard Flournoy and Jack Henley.
COM   91 min B/W   VIDrel: RCA L/A   V

NIGHT TO REMEMBER, A *** PG
Roy Ward Baker UK   1958
Kenneth More, Honor Blackman, Michael Goodliffe, David McCallum, George Rose, Anthony Bushell, Ronald Allen, Robert Ayres, John Cairney, Jane Downs, Kenneth Griffith, Frank Lawton, Michael Bryant, Jill Dixon, James Dyrenforth
A straight and unadorned account of the sinking of the Titanic in 1912, with many star cameos in a film that makes highly effective use of its plain semi-documentary style. Puts the later films on this subject in the shade. Adapted from the book by Eric Ambler. See also S.O.S. TITANIC and TITANIC.
DRA   118 min (ort 123 min) B/W   VIDrel: CARL/TECH   V
Boa: book by Walter Lord.

NIGHT TRAIN TO VENICE ** 18
Carlo U. Quinterio USA   1993
Hugh Grant, Malcolm McDowell, Tahnee Welch, Kristina Soderbaum, Rachel Rice, Evelyn Opela, Samy Langs, Murphy Mclaren, Robinson Reichel, Ralph Herforth, Burkhard Kosminski, Matthias Kosminski, Renee Kuenzel, David Kehoe, Pascal Hess
A dull tale about the author of a treatise on neo-Nazism who finds himself being stalked by skinheads after coming to Venice to meet with a prospective publisher. Grant gives a flat and disappointing performance which contributes greatly to the film's lack of credibility.
THR   94 min (ort 98 min)   VIDrel: FABFIL/SPEAR   V

NIGHT TRAP ** 18
David A. Prior USA   1993
Robert Davi, Michael Ironside, Lesley-Anne Down, Lydie Denier, Margaret Avery, John Amos, Mike Starr, Keri-Anne Bilotta, David Dahlgren, Robert Engstrom, Jay Aikin, Michael Anderson, Steve Barker, Bryan Bolin, Guy D'Alema
A detective tries to crack a case involving a serial killer, with the latter specialising on attacks on women at the time of Mardi Gras and finds that these are the work of a man who sold his soul to the devil and has become a virtually unstoppable. When the latter challenges our detective to catch him by the end of Mardi Gras or forfeit his soul, the detective enlists the help of a voodoo priestess, but in so doing puts both his ex-wife and a new girlfriend at risk.
THR   90 min (ort 93 min)   VIDrel: GUILD/SONOP   V

NIGHT TRIPS ** 18
Andrew Blake USA   1989
Tori Welles, Porsche Lynn, Randy Spears, Victoria Paris, Peter North, Ray Victory, Jamie Summers, Tanja De Vries, Marc De Bruin
A woman undergoing sex therapy has dreams and fantasies which can be watched on a special screen by the doctors handling her case. Despite being acclaimed as an erotic classic, and for all its above-average photography, there really isn't enough in this film to sustain the thin plot, which to some extent is little more than a reworking of NIGHTDREAMS. A somewhat better sequel followed.
A   45 min (ort 75 min)   VIDrel: PURG/DANTE   V

NIGHT TRIPS 2 *** 18
Andrew Blake USA   1990
Paula Price, Randy Spears, Cheri Taylor, Erica Boyer, Bridgette Monroe, Lauren Hall, Randy West, Jon Dough, Nina Alexander, Tami Monroe, Eric Price, Cameo, Brianna Rai, Racquel Darrian, Derrick Lane
In this sequel, Price takes over the Tori Welles role, and seeks the professional help of Dr Randy Spears. This involves her

getting hooked up to a device that can monitor her dreams, thus leading to a display of her various sexual fantasies. An unusually well crafted film, that's made with a good deal of flair, and is something of an improvement on the earlier film.
A   52 min (ort 80 min)   VIDrel: HAR/QUANT   V

NIGHT VISITORS * 15
David Falk USA   1988
Daniel Hirsch, David Schroeder, Rochelle Savitt, Joe Whyte, Billy Wallace, Jeralyn Farbe, Richard Gaba, Richard Rifkin, Michele Winding, Gregory Carlton Battle
As a family enjoy their Christmas, four carol singers turn up and gain entrance, taking over the celebrations and forcing the family to take part in a series of bizarre and humiliating games. As each member of the family is degraded, they begin to see themselves in a new light. A sick little number, both written and directed by Falk.
DRA   90 min   VIDrel: IMPENT   V

NIGHT WATCH ** 15
David Jackson USA   1996
Pierce Brosnan, Ron Berglas, Harold Bone, William Devane, Kate Harper, Tom Jansen, Mark King, Hidde Maas, Irene Ng, Alexandra Paul, Tomaslav Raus, Natalie Roles, Rolf Saxon, Michael J. Shannon, Lim Kay Siu, Suncana Zelenika
When Rembrandt's famous painting "The Night Watch" is stolen, a couple of UN agents are sent by their boss to investigate this theft and locate it. The trail they follow, takes them from Amsterdam to Hong Kong, where that find themselves having to combat some high-tech crooks.
Aka: ALISTAIR MACLEAN'S NIGHT WATCH; DETONATOR 2: NIGHTWATCH; NIGHTWATCH
A/AD   95 min   VIDrel: TOUCH/TECH   V
Boa: novel by Alistair Maclean.

NIGHT WE NEVER MET, THE ** 15
Warren Leight USA   1992
Matthew Broderick, Annabella Sciorra, Kevin Anderson, Jeanne Trippelhorn, Justine Bateman, Christine Baranski, Doris Roberts, Dominic Chianese, Tim Guinee, Michael Mantell, Greg Germann, Dana Wheeler-Nicholson, Bill Campbell
A man rents out his apartment in New York to a couple of other people both of whom have problems with their love lives and some predictable complications ensue when his different tenants get involved with each other. With the various individuals all sharing the same apartment (albeit on different days of the week) one might have expected greater humour from this romantic comedy, but the yuppie characters are far from attractive and the direction is uninspired.
COM   94 min (ort 99 min)
VIDrel: POLY/POLYREC/GUILD   V/sur

NIGHTBREAKER *** 15
Peter Markle USA   1989
Martin Sheen, Emilio Estevez, Lea Thompson, Melinda Dillon, Joe Pantoliano, Nicholas Pryor, Geoffrey Blake, Paul Eidling, James Marshall, Leonard Post, Lance Slaughter, Tom Bower, Michael Laskin, Gerry Black, Jack Jozefson
Flashback style story of a neurologist whose successful career conceals his involvement with nuclear testing in which soldiers were used as unwitting guinea pigs. Somewhat overwrought but generally effective. The script is by T.S. Cook.
Aka: ADVANCE TO GROUND ZERO
DRA   95 min (ort 100 min) mCab   VIDrel: 20TH   V
Boa: novel Atomic Soldiers by Howard Rosenberg.

NIGHTBREED *** 18
Clive Barker USA   1990
Craig Sheffer, Anne Bobby, David Cronenberg, Charles Haid, Doug Bradley, Hugh Ross, Hugh Quarshie, Catherine Chevalier, Malcolm Smith, Bob Sessions, Oliver Parkert, Debora Weston, Nicholas Vince, Simon Bamford, Kim Robertson
Screenplay is by Barker in this futuristic fantasy that follows the adventures of a young man and a psychiatrist who become obsessed in locating a mythical and nearly-extinct race of shape-changing mutants said to live in a remote part of the country. The special make-up and visual effects (by Image Animation) are superb, all the more sad that the story is so weak and that the film suffers from clumsy post-production editing.
FAN   89 min (ort 102 min)   VIDrel: BRAVE/SONOP L/A   V
Boa: novel Cabal by Clive Barker.

NIGHTCOMERS, THE *
Michael Winner UK
Marlon Brando, Stephanie Beacham, Harry Andrews, Thora Hird, Verna Harvey, Christopher Ellis, Anna Palk

18
1971

An attempt to describe what happened to the children featured in the Henry James novel "The Turn Of The Screw" and how they became evil prior to the period at which the original story starts. An unappealing story is combined with bad direction to produce a work that is more wearisome than chilling. See also THE TURN OF THE SCREW.
HOR   92 min (ort 96 min)   VIDrel: L/A   V
Boa: novel by Michael Hastings.

NIGHTFLYERS *
T.C. Blake (Robert Collector) USA
Michael Praed, John Standing, Michael Des Barres, James Avery, Catherine Mary Stewart, Lisa Blount, Glenn Withrow, Helene Udy, Annabel Brooks

18
1987

The members of an expedition into deep space are confronted by an evil presence aboard an ancient space freighter, and a series of mysterious accidents begin to claim the lives of the crew. A pointless and derivative dud that the director virtually disowned by having his name removed from the credits.
FAN   86 min (ort 89 min)   VIDrel: 20TH/TECH   V
Boa: novel by George R. Martin

NIGHTHAWKS **
Ron Peck UK
Ken Robertson, Tony Westrope

18
1978

In London in the 1970s, a geography teacher troubled by his homosexuality finds himself forced to live a double life whose demands grow ever more difficult to reconcile. A low-budget effort, marred by weak performances, that though very sincere in its intentions, fails to get to grips with its promising subject matter. "What Can I Do With A Male Nude" is a twenty-four minute short in which a photographer struggles with his sexual inclinations during a shoot.
DRA   113 min; 137 min (2-film cassette)
VIDrel: CONNO/RTM   V
Osca: WHAT CAN I DO WITH A MALE NUDE

NIGHTHAWKS ***
Bruce Malmuth USA
Sylvester Stallone, Billy Dee Williams, Rutger Hauer, Nigel Davenport, Persis Khambatta, Lindsay Wagner, Hilarie Thompson, Joe Spinell, E. Brian Dean, Walter Mathews, Cesar Cordova, Charles Duval, Tony Munafo, Howard Stein

18
1981

Two New York cops are assigned to a special unit on the trail of an international terrorist, in this tense and exciting yarn.
THR   95 min (ort 99 min)
VIDrel: 4-FRONT/POLYREC/CIC   V/h

NIGHTMAN, THE **
Charles Haid USA
Joanna Kerns, Jenny Robertson, Ted Marcoux, Latanya Richardson, Lou Walker, Benji Wilhoite, Pamela Garmon, Penelope Windust, Hugh Jarrett, Janet McLeod, David Milford, Ed Grady, Kenny Leon, Jenny McGahee, Ron Leggett, Ben Tootle

(18)
1991

A mother and daughter both have a love affair with a night porter, in a tangled relationship that ends in tragedy when the lover kills the mother. When the killer is released on parole some eighteen years later, the daughter begins to feel that her life too is in danger. A powerfully acted psychological thriller with a few good twists although the true state of things becomes clear fairly early on in the proceedings.
THR   90 min mTV   TVrel

NIGHTMARE BEFORE CHRISTMAS, THE ***
Henry Selick USA
Voices of: Danny Elfman, Chris Sarandon, Catherine O'Hara, William Hickey, Glenn Shadix, Paul Reubens, Ken Page, Ed Ivory, Susan McBride, Debi Durst, Kerry Katz, Gregory Proops, Randy Crenshaw, Sherwood Ball, Carmen Twillie

PG
1994

A dark and disturbing animation that took two years to make and employs stop-motion animation. It tells of how the Pumpkin King of Halloweentown learns of the joys of Christmastown, but this prompts him to kidnap Santa. Visually most impressive (albeit unsuitable for very young children) this memorable fantasy has some excellent songs by way of a bonus.
Aka: TIM BURTON'S THE NIGHTMARE BEFORE CHRISTMAS
ANIM   73 min (ort 76 min) cC   VIDrel: TOUCH/TECH   V

NIGHTMARE COUNTY *
Sean McGregor USA
Beau Gibson, Sean McGregor, Gayle Hemingway

15
1977

A band of hippies face a serious confrontation with the locals, when they start buying up land to settle there. When the girlfriend of the local sheriff is raped, violence erupts. A low-grade effort.
Aka: STREET FORCE
A/AD   90 min   VIDrel: MOPIC/SGSVID   V

NIGHTMARE IN DAYLIGHT, A **
Lou Antonio USA
Jaclyn Smith, Christopher Reeve, Tom Mason, Glynnis O'Connor, Christina Pickles, Eric Bell, Wren T. Brown, John Ingle, Ricardo Guttierrez, Beth Anne Spanier, Shuko Akune, Patrick Culliton, Jack Kehler, Philip Moon

15
1992

A seemingly happily married schoolteacher has a six-year-old son and a contented life, but this is brought to an end by the activities of an obsessive individual who claims her to be the wife he thought had perished in a Mexico City earthquake seven years ago. It would appear that she faked her death to escape from him, as he was (and still is) a dangerous psychopath. Reeve is cast against type as the bad guy, and offers a chilling display of venom.
Aka: NIGHTNARE IN THE DAYLIGHT, A
THR   91 min mTV   VIDrel: COLUM/SONOP   V

NIGHTMARE ON ELM STREET, A ***
Wes Craven USA
Robert Englund, John Saxon, Heather Langenkamp, Ronee Blakely, Amanda Wyss, Nick Corri, Johnny Depp, Charles Fleischer, Joseph Whipp, Joe Unger, Lin Shaye, Mimi Meyer-Craven, Jack Shea, Ed Call, Sandy Lipton, David Andrews, Brian Reise

15
1984

A dead child-killer returns to murder the children of those responsible for burning him alive, and first appears to them in terrifying and identical nightmares. The horrifying special effects and the deliberate blurring of the line between reality and illusion made this a tremendous hit for the makers, and for Englund, who plays the delightful Freddy Krueger. Several sequels and a spin-off "Freddy" TV series followed. The script is by Craven.
HOR   88 min (ort 91 min)   VIDrel: 20TH/TECH L/A   V

NIGHTMARE ON ELM STREET, A: PART 2 – FREDDY'S REVENGE **
Jack Sholder USA
Mark Patton, Kim Myers, Robert Rusler, Clu Gulager, Hope Lange, Robert Englund, Marshall Bell, Melinda O. Fee, Thom McFadden, Sydney Walsh, Edward Blackoff, Christie Clark, Lyman Ward, Donna Bruce, Hart Sprager, Allison Barron

18
1985

Sequel to A NIGHTMARE ON ELM STREET that continues the story five years on. A teenage boy is tormented by nightmares after his family move into their new house, dreaming of a rotting corpse-like creature who has returned from the dead. The latter does indeed return, taking over the young man's body and forcing him to commit grisly crimes. Another slasher tale, not quite as effective but a lot bloodier than the earlier one. Written by David Chaskin.
Aka: FREDDY'S REVENGE
HOR   82 min (ort 87 min)   VIDrel: MGM/WHV L/A   V

NIGHTMARE ON ELM STREET, A: PART 3 – DREAM WARRIORS **
Chuck Russell USA
Robert Englund, John Saxon, Heather Langenkamp, Craig Wasson, Patricia Arquette, Larry Fishburne, Priscilla Pointer, Brooke Bundy, Rodney Eastman, Dick Cavett, Zsa Zsa Gabor, Bradley Gregg, Ira Heiden, Ken Sagoes, Michael Dick

18
1987

Second sequel in what promised to be a long-running study in terror akin to FRIDAY THE 13TH, with the same excellent if repulsive effects but little else. A small town is plagued by a series of suicides among teenagers, and a young psychiatrist finds a group of teenagers who are all suffering nightmares featuring the same horrific figure.
HOR   93 min (ort 96 min)   VIDrel: WHV L/A   V/sh

NIGHTMARE ON ELM STREET, A: PART 4 – THE DREAM MASTER ***
Renny Harlin USA
Robert Englund, Rodney Eastman, Ken Sagoes, Tuesday Knight, Hope Marie Carlton, Brooke Theiss, Danny Hassel, Andras Jones, Toy Newkirk, Lisa Wilcox, Brooke Bundy, Jeffrey Levine, Nicolas Mele, Richard Garrison, Joie Magidow

18
1988

Third sequel marks the arrival of a more playful and humorous attitude to the central figure of Freddy who, having finished off all the surviving children of his killers, now turns attention to their young friends. Good special effects and camerawork abound, but don't look for any logic in the plot. A TV series followed. See FREDDY'S NIGHTMARES: A NIGHTMARE ON ELM STREET.

HOR   88 min Cut (1 min 7 sec – ort 99 min)
VIDrel: 20TH/TECH L/A V

NIGHTMARE ON ELM STREET, A: PART 5 – THE DREAM CHILD **
18
Stephen Hopkins USA
1989
Robert Englund, Lisa Wilcox, Danny Hassel, Kelly Jo Minter, Erika Anderson, Beatrice Boepple, Whitby Hertford, Nick Mele, Joe Seely, Valorie Armstrong, Burr DuBenning, Clarence Felder, Matt Borlenghi, Stephen Grives
The novel idea of an unborn child communicating its dreams to the mother, is the one fresh element in this by now tired and repetitive series. Despite his defeat in previous films, Freddy returns to terrorise once again, preying on the heroine (Wilcox) and her lover. As ever, the special effects are truly memorable, but that's all there is. Followed by FREDDY'S DEAD: THE FINAL NIGHTMARE.

HOR   86 min (ort 91 min)   VIDrel: 20TH/TECH L/A V

NIGHTMARE ON ELM STREET, A: PART 6 – FREDDY'S DEAD, THE FINAL NIGHTMARE **
18
Rachel Talahay USA
1991
Robert Englund, Lisa Zane, Shon Greenblatt, Lezlie Deane, Ricky Dean Logan, Breckin Meyer, Yaphet Kotto, Mr Tom Arnold, Mrs Tom Arnold, Elinor Donahue, Oprah Noodlemantra, Cassandra Rachel Friel, David Dunard, Virginia Peters
This NIGHTMARE ON ELM STREET sequel (the fifth and hopefully the last) pits nasty old Freddy Krueger against his daughter, who now works with abused teenagers. As might be expected, the daughter's charges put up with a good deal more abuse than they normally meet, as the malign influence of Freddy makes itself felt. A gory rollercoaster of nastiness, with the final reel shot in 3-D, and now available on cassette (special glasses are advised).
Aka: FREDDY'S DEAD, THE FINAL NIGHTMARE
HOR   85 min (ort 96 min) cC   VIDrel: GUILD/SONOP L/A V/sur

NIGHTMARE VOYAGE *
18
Frank Mitchell USA
1976
Jonathan Lippe, Peter Kellett, Midori, Mora Modair, John Hart, Laurie Rose, Gene Tyburn
Eight people set out on a voyage, but are picked off one by one in this wholly predictable and unappealing yarn.
Aka: BLOOD VOYAGE
HOR   90 min Cut (4 min 30 sec)   VIDrel: MOPIC/SGSVID V

NIGHTS OF CABIRIA ****
15
Federico Fellini FRANCE/ITALY
1957
Francois Perier, Amedeo Nazzari, Giulietta Masina, Franca Marzi, Dorian Gray, Aldo Silvani, Mario Passante, Pina Gualandri, Polidor, Ennio Girolami, Christian Tassou, Jean Molier, Ricardo Fellini
A Rome prostitute fantasises about love and a normal life, but is betrayed when she puts her trust in others. A powerfully moving film that formed the basis for the Broadway musical and then later the film SWEET CHARITY. AA: Foreign.
Aka: CABIRIA; LE NOTTI DI CABIRIA
DRA   106 min (ort 110 min) B/W   VIDrel: L/A V

NIGHTSHIFT NURSES *
18
John Leslie USA
1988
Lois Ayres, Keisha, Sheena Horne, Bionca, Lynn Francis, D.J. Cone, Dana Dillon, Peter North, Steve Henessy, John Leslie, Josey Duval, Siobhan Hunter, Joey Silvera, Louis T.Beagle, Steve Hennessy
Leslie's directorial debut is the tale of a hospital doctor (Leslie) doing a nightshift. He would much rather play his harmonica and, strange to relate, his nurses would much rather be doing other things too. A very silly film, often quite bizarre, but not really able to sustain one's interest.
A   79 min Cut (20 sec – ort 88 min) mVid   VIDrel: IMPENT V

NIGHTWALK *
15
Jerrold Freedman USA
1989
Robert Urich, Lesley-Anne Down, Mark Joy, Ryan Urich, Bert Remsen, Michael Alldredge, Lawrence P. Casey, Richie Devaney,

Richard Butler, Freddie Koehler, Meredith Strange-Boston, Beverly Skinner, James Martin Jr, Jeanette Park Avery
A woman sees a murder take place on a lonely beach but fails to convince the police of this when the body disappears. However, the murderer decides to silence her for good by bringing in a contract killer. A weak thriller with an overly familiar plot.
THR   89 min (ort 100 min) mTV   VIDrel: 20TH/TECH V/sh

NIGHTWATCH **
18
Ole Bornedal DENMARK
1994
Nikolaj Coster Waldau, Sofie Grabol, Kim Bodnia, Lotte Andersen, Ulf Pilgaard, Rikke Louise Andersson, Stig Hoffmeyer, Gyrd Lofquist, Niels Anders Thorn, Jytte Rosholm, Leif Adolfsson, Henrik Fiig, Jesper Hyldegaard
Having become the night-watchman at the city morgue, a young law student finds himself getting involved in a series of brutal murders and the activities of a couple of daredevil youngsters. A hybrid movie that blends suspense with a strong dose of horror, it never quite finds an identity of its own, but entertains, albeit in a slightly unsettling way.
Aka: NATTEVAGTEN
THR   104 min (ort 107 min) wScrn   VIDrel: TART/20TH V

NIKITA ***
18
Luc Besson FRANCE/ITALY
1990
Anne Parillaud, Jean-Hugues Anglade, Tcheky Karyo, Roland Blanche, Jacques Boudet, Marc Duret, Jean Reno, Jeanne Moreau, Patrick Fontana, Laura Cheron, Alain Lathiere, Jacques Disses, Patrick Perez, Jean-Luc Caron, Stephane Fey
An anti-social punk who killed a policeman in the course of a robbery is sentenced to life imprisonment, but is given a chance to regain a measure of freedom by agreeing to work as a government assassin. Having acquired a change of identity, she embarks on her new career. Written by Besson, this cynical and downbeat film is sustained by some touching love scenes and Parillaud's strong presence. Moreau makes a guest appearance. Remade as THE ASSASSIN and BLACK CAT.
Ala: LA FEMME NIKITA
THR   112 min (ort 117 min) wScrn   VIDrel: ARTIF/20TH L/A; ENCORE (LV only)   V/sh LV

NINA TAKES A LOVER *
18
Alan Jacobs USA
1993
Laura San Giacomo, Paul Rhys, Michael O'Keefe, Fisher Stevens, Cristi Conaway
After her husband leaves on a trip that takes him away from home for a number of weeks, his wife gets herself involved with a Welsh artist and soon becomes his mistress. An empty-headed and pointless tale of little interest. Screenplay is by Jacobs.
DRA   94 min (ort 100 min)   VIDrel: COLUM/SONOP V/sh

9½ NINJAS **
15
Aaron Worth HONG KONG
1990
Michael Phenicie, Andree Gray, Robert Fieldsteel, Magda Harout, Tiny Lister, Don Stark, Barbara Lary, Sharon Lee Jones, Monty Hoffman, Rance Howard, Kitsan, Gerald Okamura, Keaton Simons, Ralph Manza, James Walker
A martial arts master becomes amorously interested in his latest pupil but finds that both he and she are being pursued by a band of deadly assassins. A sort of kung-fu spoof of 9½ WEEKS, so it would seem.
A/AD   82 min (ort 91 min)   VIDrel: VISVID/POLYREC V

9½ WEEKS *
18
Adrian Lyne USA
1986
Kim Basinger, Mickey Rourke, Margaret Whitton, David Margulies, Christine Baranski, Dwight Weist, Roderick Cook, Karen Young, William De Acutis, Victor Truro, Justine Johnson, Cintia Cruz, Kim Chan, Lee Lai Sing
The title refers to the duration of a steamy sex-obsessed relationship between a man and a woman, that seems to imply sado-masochistic aspects in a softcore film. Despite the great media mega-hype surrounding it on release, the film has nothing interesting to say in its threadbare plot.
DRA   112 min wScrn   VIDrel: 20TH/TECH V/dm V/sur
Boa: novel 9½ Weeks by Elizabeth McNeill.

NINE HOURS TO RAMA *
(PG)
Mark Robson UK
1962
Jose Ferrer, Diane Baker, Robert Morley, Horst Buchholz, Harry Andrews, J.S. Casshyap

Assassination drama built around the murder of Gandhi, that suffers badly from its heavily fictionalised and superficial approach, plus performances that at best are no more than competent. Buchholz is the hired killer, but his early romantic dalliances add nothing to the story. What a pity that the death of this remarkable man is dealt with in such a manner. Filmed in India.
DRA   120 min (ort 125 min)   SATrel: SKY MOVIES
Boa: novel by Stanley Wolpert.

NINE LIVES OF FRITZ THE CAT *   18
Ralph Bakshi USA   1974
Voices of: Skip Hinnant, Reva Rose, Bob Holt, Fred Smoot, Pat Harrington Jr
Animated sequel to FRITZ THE CAT with more adventures of our cool feline hero. Where the earlier film had some relevant and witty things to say about the 1960s, this one just sets out to shock with its crudity. A tiresome and banal effort. See also HEAVY TRAFFIC, WIZARDS, FIRE AND ICE and THE LORD OF THE RINGS – four more films by this director.
ANIM   73 min   VIDrel: POLY/POLYREC   V

NINE MONTHS *   12
Chris Columbus USA   1994
Hugh Grant, Julianne Moore, Robin Williams, Tom Arnold, Joan Cusack, Jeff Goldblum, Mia Cottet, Joey Simmrin, Ashley Johnson, Alexa Vega, Aislin Roche, Priscilla Alden, Edward Ivory, James M. Brady, Charles Martinet, Zelda Williams
A British child psychologist living in San Francisco is terrified by the thought of fatherhood and has his life completely disrupted when his live-in girlfriend tells him that she is pregnant. A hysterical, tiresome and unfunny parade of cliches about male incompetence that feels utterly stale and contrived. Based on the French movie "Neuf Mois" from 1994 that was equally bad.
COM   99 min (ort 103 min) cC   VIDrel: 20TH/TECH
V/sur

NINE TO FIVE ***   15
Colin Higgins USA   1980
Jane Fonda, Lily Tomlin, Dolly Parton, Dabney Coleman, Sterling Hayden, Elizabeth Wilson, Henry Jones, Lawrence Pressman, Marian Mercer, Ren Woods, Norma Donaldson, Roxanna Bonilla-Giannini, Peggy Pope, Richard Stahl
Three girls in an office work for a sexist boss who constantly makes lewd advances. They eventually find a way of teaching him a lesson. A surprisingly fresh and inventive comedy that never sags, despite its length. Tomlin gives a great performance and Coleman is in good form too as their odious boss. Later gave rise to a so-so TV series.
Aka: 9 TO 5
COM   105 min (ort 110 min)   VIDrel: 20TH   V

1900: PARTS 1 AND 2 ***   18
Bernardo Bertolucci FRANCE/ITALY/USA/WEST
GERMANY   1976
Burt Lancaster, Robert De Niro, Gerard Depardieu, Sterling Hayden, Dominique Sanda, Donald Sutherland, Laura Betti, Francesca Bertini, Werner Bruhns, Anna Henkel, Stefania Casini, Alida Valli, Stefania Sandrelli
Star-packed, ponderous and overlong account of Italian history from 1901 to 1945, as reflected in the lives of two men, one a peasant and the other the son of the local landowner. Firm friends as boys, they are inevitably separated on growing up by differences of class and wealth. Sadly, the film's slow pacing and the one-dimensionality of the characters deadens its political message. The fine photography is by Vittorio Storaro and the haunting score is by Ennio Morricone.
Aka: NOVECENTO
DRA   170 min (Part 1); 160 min (Part 2); (ort 360 min)
VIDrel: FOXVID L/A   V

1941 **   PG
Steven Spielberg USA   1979
John Belushi, Tim Matheson, Nancy Allen, Dan Aykroyd, Ned Beatty, Toshiro Mifune, Christopher Lee, Robert Stack, Treat Williams, Warren Oates, Murray Hamilton, Dianne Kay, Slim Pickens, Bobby DiCicco, Lorraine Gary, John Candy
Los Angeles is thrown into panic during a power cut, when a Japanese submarine is supposed to have been sighted off the coast following the Pearl Harbor attack. A vastly expensive and overblown slapstick comedy, where the time and effort yield a disappointing dividend of laughs.
COM   112 min (ort 118 min)   VIDrel: CIC/SONOP   V/h

1969 **   15
Ernest Thompson USA   1988
Robert Downey Jr, Kiefer Sutherland, Bruce Dern, Winona Ryder, Joanna Cassidy, Mariette Hartley, Christopher Wynne, Keller Kuhn, Steve Foster
A failed attempt to focus on the generation gap and social conflict that shook the USA in that year, when opposition to the Vietnam War began to grow, as seen in the lives of a group of teenagers. Features music by Cream, The Pretenders, Jimi Hendrix, The Animals, The Moody Blues etc. The directorial debut for writer Thompson, better known for having scripted ON GOLDEN POND, but here the strong script is weakened by haphazard direction.
DRA   91 min (ort 120 min)   VIDrel: POLY/POLYREC L/A
V/sh

1984 ****   15
Michael Bradford UK   1984
John Hurt, Suzanna Hamilton, Richard Burton Cyril Cusack, Gregor Fisher, James Walker, Andrew Wilde, David Trevena, David Cann, Anthony Benson, Peter Frye, Phyllis Logan, Pam Gems, Joscik Barbarossa, John Boswell, Bob Flag
A later version of Orwell's brilliantly prophetic satire on totalitarianism, examining the life of Winston Smith, a man living in the dictatorship of Oceania. Unlike the 1955 film, this work stays remarkably close to the book. Originally written in 1948, this nightmarish vision is accurately captured by sets that portray the future from the standpoint of the 1940s. Burton's last film in which he gives a superb portrayal of chief inquisitor O'Brien.
DRA   108 min (ort 110 min)   VIDrel: 4-FRONT/POLYREC
V
Boa: novel by George Orwell.

NINJA 2: THE REVENGE OF THE NINJA **   18
Sam Firstenberg USA   1983
Sho Kosugi, Keith Vitali, Virgil Frye, Arthur Roberts, Mario Gallo, Ashley Ferrare, Kane Kosugi, Grace Oshita, John La Motta, Melvin C. Hampton, Oscar Rowland, Toru Tanaka, Dan Shanks, Joe Pagliuso, Ladd Anderson, Steve Ketcher
Kung fu high kicks once more, as evil ninjas and drug smugglers get what they deserve from our black-clad hero (Kosugi). After his family is massacred in a ninja attack, he goes to the States with an American partner to open a string of Japanese art galleries, which the latter secretly plans to use as a front for his drugs business. A blend of well choreographed fights and bad acting, this one followed ENTER THE NINJA and gave rise in turn to NINJA 3: THE DOMINATION.
Aka: NINJA 2; REVENGE OF THE NINJA, THE
MAR   80 min Cut (57 sec – ort 88 min)   VIDrel: VCC L/A   V

NINJA 3: THE DOMINATION **   18
Sam Firstenberg USA   1984
Lucinda Dickey, Jordan Bennett, Sho Kosugi, David Chung, T.J. Castronova, Dale Ishimodo, James Hong, Bob Craig, Pamela Ness, Roy Padilla, Moe Mosley, John LaMotta, Ron Foster, Alan Amiel, Steve Lambert, Earl Smith, Karen Petty
In this sequel to NINJA 2: THE REVENGE OF THE NINJA, the spirit of a Ninja killed by the police takes over the body of a young girl, and proceeds to wreak havoc. Lots of action compensates for the mindlessness of it all.
MAR   84 min (ort 95 min)   VIDrel: L/A   V

NINJA AND THE WARRIORS OF FIRE *   18
Bruce Lambert   1973
Peter Davis, Jeff Houston, Glen Carson, Christine Warren, Julie Luk
The leader of a Ninja band rapes and then murders the fiancee of a man who refused to give them secret information. The woman's sister takes revenge.
A/AD   90 min Cut (1 min 24 sec)   VIDrel: MIA/DISC   V

NINJA CHAMPION *   18
Joseph Lai HONG KONG   1986
Bruce Baron, Pierre Tremblay, Nancy Chan, Richard Harrison, Jack Lam, Philip Ching, Dragon Lee
An Interpol agent's girlfriend is raped by a criminal gang, which gives our agent an excuse for a no-holds-barred high-kicking showdown with them.
MAR   87 min   VIDrel: IMPENT   V

NINJA COMMANDMENTS *   15
Joseph Lai HONG KONG   1987
Richard Harrison, Dave Wheeler, Peter Kjaer, Adam Frank, Eagle Lee, Louis Ruth

In this tale, some Ninja are expelled from the society for breaking some of its secret laws.
MAR 88 min Cut (6 sec) VIDrel: IMPENT V

NINJA DESTROYER ** 18
Godfrey Ho HONG KONG 1986
Stuart Smith, Bruce Baron, Sorapong Chatri, Na Yen Ha, Richard Berman, Timothy Nugent, Hai Lin Li, Hai Lo
Assorted Ninja fighters battle for control of an emerald mine, in this tale set on the Thai-Cambodian border.
MAR 84 min Cut (41 sec – ort 92 min) VIDrel: IMPENT V

NINJA FIST OF FIRE * 18
HONG KONG
Chor Yim Yung, Cheung Ching Ching, Kong Ming
Kung fu costume drama with a prince doing battle to secure his throne. Average.
MAR 90 min VIDrel: MOPIC/SGSVID V

NINJA IN THE KILLING FIELDS * 18
York Lam HONG KONG 198-
Stuart Steen, Louis Roth, Patricia Greenfield, Jane Kingsley
An American secret agent heads to Thailand to destroy a drugs trafficking operation and comes up against a ruthless Ninja band, which embarks on a series of terrorist attacks.
Aka: NINJA CONNECTION
A/AD 85 min Cut (37 sec) VIDrel: MIA/DISC V

NINJA KIDS: KISS OF DEATH ** 18
Joseph Kuo HONG KONG
Lo Yiu, Luk Yee Fung, Luk Fung Kwong Sang, Kong Ming, Loong Koon Mo
More high-kicking action but with a supernatural slant.
Aka: NINJA KISS OF DEATH KIDS
MAR 86 min Cut (ort 87 min) VIDrel: MOPIC/SGSVID V

NINJA KILL * 15
Joseph Lai HONG KONG
Mark White, Clayton Rice, Richard Harrison, Stuart Smith
A ruthless Ninja gang plots to assassinate a senator and replace him with a double who is under their control. Only two men can stop these devilish plans, but they must first take on an evil woman and her female gang. Another strictly routine blend of implausibility and fisticuffs.
MAR 86 min Cut (1 min 25 sec) VIDrel: IMPENT V

NINJA KILLER * 18
Lawrence Chan HONG KONG
Kau Ka Wen, Joey Louis, Carter Wong
A smuggler absconds to Istanbul with an associate's loot, but is followed b police agent who finds himself embroiled in a battle between rival gangs.
MAR 87 min Cut (3 sec) VIDrel: IMPENT V

NINJA MISSION, THE * 18
Mats Helge (Mats Helge Olsson) SWEDEN/UK 1984
Christopher Kohlberg, Hans Rosteen (Rastam), Hanna Pola, Curt Brober, John Qvants Von Ills, Sirka Sander, Wolf Linder, Leo Adolfson, Brett Barber, Nigel Bennett, Bob Gillberg, Lucy Frost, Derek Keith, Victor Lester, Terry Jason
A scientist who has devised a solution to the world's energy problems is kidnapped by the KGB and forced to co-operate when his daughter is also abducted. A specially trained ninja taskforce is assembled in order to mount a rescue. Bad dubbing, lousy direction, indifferent acting and lashings of gore all combine to make this a film of memorable ineptitude. However, this didn't stop it becoming a worldwide commercial success.
A/AD 95 min Cut (1 min 1 sec – ort 98 min)
VIDrel: MIA/DISC V

NINJA OPERATION: KNIGHT AND WARRIOR * 15
Joseph Lai HONG KONG 1987
Richard Harrison, Alphonse Beni, Stuart Smith, Grant Temple, Scott Smith, Geoffrey Brown, Chris Larris, Eric Brinner, Timothy Nugent, Rod Laurin, Peter King, Ian Harling, Mark Collins, Mandiere Nathalie
An Interpol agent is sent on a drugs busting operation and comes up against four assassins who murder his wife. He seeks help from an old friend. Formula mayhem followed by a whole series of sequels.
Aka: KNIGHT AND WARRIOR
A/AD 86 min Cut (1 min 7 sec) VIDrel: IMPENT V

NINJA OPERATION 2: WAY OF CHALLENGE * 15
Joseph Lai HONG KONG 1988
Richard Harrison, Geoffrey Ziebart, Gary Carter, John Ruthworth, Simon Heagan, Gaby Carter, Rick Bresmo, John Ruthworth, Simon Heuger
Another formula offering from the Ninja factory, this time telling of the efforts made to retrieve a sword of remarkable properties.
Aka: WAY OF CHALLENGE
A/AD 87 min Cut (20 sec) VIDrel: IMPENT V

NINJA OPERATION 3: LICENSED TO TERMINATE * 18
Joseph Lai HONG KONG 1987
Richard Harrison, Grant Temple, Paul Marshall, Jack McPeat, Louis Ruth, Bob Corridge, Alvin Blacksmith, Jeff Alberto, Tattooer Ma, Jane Lai, Chan Kwa Po, Sandra Lee, William Wai, Kent Taso
Good Ninja warriors take on bad ones as the Golden Ninja and the Prince of Justice confront the Black Ninja Empire. More mindless mayhem, made slightly ludicrous by the pretensions of the script.
Aka: LICENSED TO TERMINATE; NINJA OPERATION 3
MAR 87 min Cut (29 sec – ort 89 min) VIDrel: IMPENT L/A V

NINJA OPERATION 4: THUNDERBOLT ANGELS * 18
Joseph Lai HONG KONG 1988
Richard Harrison, George Ajex, Alan Cunningham, Bartel Markus, Chaplin George, Kenneth Smy, Jonathan Bould, Russell D. Bradshaw, Ricky Shaw, Sheila Lau, Nancy Yeh, Sidney Mo, Tony Yau, Chuan Yuan, Larry Hong, Adam Ko, Jack Tien
Two childhood friends become involved in a gang war and are jailed. On their release a new conflict develops and they soon find themselves on opposite sides when one is recruited by a cop to destroy a criminal gang. Adequate martial arts mayhem, barely sustained by the usual thin plot. Several of the actor's names were spelt incorrectly in the opening credits, which is a nice comment on the care lavished on the film.
Aka: THUNDERBOLT ANGELS
MAR 87 min Cut (51 sec) VIDrel: IMPENT V

NINJA OPERATION 5: GODFATHER THE MASTER * 18
Joseph Lai HONG KONG 1988
Richard Harrison, Grant Temple
The head of a criminal clan is preparing to hand over the family business to his son when he is murdered by rival gang leaders. The son now embarks on a course of training to become a Ninja fighter, and relentlessly pursues the culprits. A tired formula offering.
Aka: GODFATHER THE MASTER
MAR 85 min Cut (1 min 45 sec) VIDrel: IMPENT V

NINJA OPERATION 7: ROYAL WARRIORS ** 18
HONG KONG 1989
Richard Harrison, Mike Abbott
Martial arts adventure set near the Burmese border where the head of an evil Ninja gang seeks a fortune in gold, having murdered two explorers for their treasure map. When he brings in an old partner, who runs a sex-slave operation, he fails to realise that this may cause his downfall. Average entry in this long-running series.
Aka: ROYAL WARRIORS
MAR 73 min VIDrel: IMPENT V

NINJA OPERATION 8: CHAMPION ON FIRE ** 18
Joseph Lai HONG KONG 1988
Richard Harrison, Stuart Smith
Yet another story in this interminable series, this time telling of a peaceful priest who, after using a cross to great effect when caught in a train robbery, makes an enemy of an evil Ninja criminal. Later our priest befriends a martial arts expert, little realising that this friendship will save his life. Average production-line violence and mayhem tale.
Aka: CHAMPION ON FIRE
A/AD 79 min VIDrel: IMPENT V

NINJA SCROLL ** 18
JAPAN 199-
An animated tale set in Feudal Japan, with elements of the supernatural, that has a wandering Ninja fighter saving a female

Ninja from attack and capture, and by his actions getting drawn into a dangerous conflict.
ANIM 90 min (ort 94 min) dubbed
VIDrel: MANGA/SONOP V/sh

NINJA SQUAD **   15/18
Wo Kuo-Jen HONG KONG   1986
Alexander Lou, Xau Jin Thomas, Yang Sang, Lou Mei
When a New York cop arrives in Hong Kong seeking information as to the identity of the killer of his girlfriend's father, he soon finds that he has become a target for assassination.
Aka: KILLERS INVINCIBLE
MAR   VIDrel: IMPENT V

NINJA THUNDERBOLT *   18
Godfrey Ho HONG KONG/PHILIPPINES/TAIWAN   1985
Richard Harrison, Anna Lewis, Wang Tao, Jackie Chan, Randy To, Kulada Yusuaki, Barbara Yuen
A corrupt businessman plans an insurance fraud but is foiled by our nimble Ninja warriors in this standard kung fu yarn.
MAR 88 min Cut (34 sec – ort 90 min) VIDrel: IMPENT V

NINJA VAMPIRE BUSTERS **   15
Norman Law HONG KONG   1989
Jacky Cheung, Nick Chang, Kent Cheng
Title group are employed in saving the world from the deadly plague of Ninja Vampires who have escaped from imprisonment in a tomb after five-hundred years. A curious melding of martial arts and supernatural elements.
A/AD 89 min dubbed VIDrel: POPRO/RTM V

NINJA WARRIORS *   18
John Lloyd USA   1985
Ron Marchini, Paul Vance, Ken Watanabe, Romano Kristoff, Mike Cohen, Charlotte Cain
Routine story of revenge and in-fighting among a Ninja band, when a secret document is stolen from a research establishment.
MAR 83 min Cut (33 sec – ort 90 min)
VIDrel: MOPIC/SGSVID V

NINOTCHKA ***   U
Ernst Lubitsch USA   1939
Greta Garbo, Melvyn Douglas, Ina Claire, Bela Lugosi, Sig Rumann, Felix Bressart, Alexander Granach, Richard Carle, Edwin Maxwell, Rolfe Sedan, George Tobias, Dorothy Adams, Lawrence Grant, Charles Judels, Frank Reicher
In her first "laughing" role, Garbo plays a severe Russian emissary, sent to Paris to sell off some Tsarist jewels. Douglas is the American playboy who woos her, teaching her something about life and love in the process. A fairly engaging comedy of manners, albeit a trifle dated and mawkish. This work later formed the basis for the Broadway musical and film SILK STOCKINGS.
COM 108 min (ort 110 min) B/W VIDrel: MGM L/A V
Boa: short story by Melchior Lengyel.

NIXON ***   15
Oliver Stone USA   1995
Anthony Hopkins, James Woods, Mary Steenburgen, Joan Allen, Powers Boothe, Ed Harris, Bob Hoskins, E.G. Marshall, David Paymer, David Hyde Pierce, Paul Sorvino, J.T. Walsh, Madeline Kahn, Brian Bedford, Kevin Dunn, Annabeth Gish
Overlong, melodramatic and none too faithful film biography of America's most controversial president who was forced to leave his office in disgrace after the Watergate scandal. However, the real strength of this offering lies in a tour-de-force performance from Hopkins in title role as a man tortured by his ambitions and driven by forces he does not fully understand.
DRA 190 min (ort 192 min) wScrn VIDrel: EIV/SONOP V

NO CHILD OF MINE ***   (PG)
Michael Katleman USA   1994
Patty Duke, Tracy Nelson, G.W. Bailey, BurtA nderson, Megan Leitch, Marshall Teague, Leslie Carlson, Jan D'Arcy, Lachlan Murdoch, Lloyd Berry, Morgan Brayton, Dave Cameron, Lillian Carlson, Flavia Carrozi, John Destrey, Anne Hagan
The true-life story of a mother who gives birth to twins, one of whom is a Down's Syndrome child that she eventually decides to put up for adoption. However, her own mother has formed a close bond with both children and resolves to fight the adoption through the courts. A strongly acted weepie.
DRA 90 min mTV SATrel: SKY MOVIES

NO CONTEST **   18
Paul Lynch USA   1995
Andrew Dice Clay, Shannon Tweed, Robert Davi, Nichaolas Campbell, Roddy Piper, John Colicos, James Purcell, Judith Scott, Louis Wrightman, Kyrin Hall, Keram Malicki-Sanchez, Jack Nicholsen, Polly Shannon, Bridget Grigss, Hugo Dann
A group of ruthless terrorists take over a beauty pageant being held in a skyscraper and hold the contestants to ransom for the sum of ten million dollars and a helicopter (with which to escape). However, they reckon with the martial skills of one of the contestants (Tweed) who eventually foils their plans. A totally unoriginal and unimpressive actioner that bears more than a passing resemblance to DIE HARD.
A/AD 93 min (ort 98 min) VIDrel: COLUM/SONOP
V/sur

NO ESCAPE **   15
Martin Campbell USA   1993
Ray Liotta, Lance Henriksen, Kevin Dillon, Ernie Hudson, Stuart Wilson, Ian McNeice, Kevin J. O'Connor, Michael Lerner, Jack Shepherd, Don Henderson, Brian M. Logan, Michael Lerner, Russell Kiefel, Cheuk Fai-Chen, Machs Colombiani
In 2022 a Marine captain is sent to a penal island where the prisoners are divided into two factions, with one group of marauders preying on those who try to live a more settled and peaceful life. Another bleak futuristic fantasy with the emphasis firmly on the spectacular action sequences and little else. An overlong and hollow effort marred by an abrupt and unsatisfying resolution.
FAN 118 min VIDrel: ETL/POLYREC/GUILD V/sur
Boa: novel The Penal Colony by Richard Herley.

NO ESCAPE NO RETURN **   18
Charles T. Kanganis USA   1993
Maxwell Caulfield, Dustin Nguyen, Denise Loveday, John Saxon, Louis Anthony, Kevin Benton, Pamela Dixon, Joey Travolta, Michael Nouri, Anibal Lleras, Renee Bessone, Dallas Cole, Vinnie Curto, Jake Barker, Diana Barton
Three tough and bitter cops are forced to take part in a operation led by an FBI agent and a police captain, which is intended to put paid to the activities of a drugs cartel.
Aka: NO ESCAPE, NO RETURN
A/AD 91 min (ort 93 min) VIDrel: IMPENT V

NO EXIT **   18
Damiem Lee USA   1995
Jeff Wincott, Phillip Jarett, Richard Fitzpatrick, Sven Ole Thorsen, Joseph Di Mambro, Guylaine St Onge
A secretive millionaire sets up a combat camp inside the Arctic Circle, from where he transmits his own private show detailing martial arts contests to the death (the programme being entitled "No Exit"). However, he finds himself requiring the services of a final contestant: a top exponent of the martial arts. Yet another variant on THE MOST DANGEROUS GAME, although this one is for the most part effective. See also DEATH RING and FINAL ROUND.
A/AD 98 min VIDrel: EIV/SONOP V

NO GREATER LOVE **   PG
Richard T. Heffron USA   1994
Kelly Rutherford, Chris Sarandon, Simon MacCorkindale
Frothy Danielle Steel exercise in implausibility with a brave (if as ever, well dressed) woman surviving the sinking of the Titanic, while all her relatives and friends are lost. As one might expect, this being a Danielle Steel story, our heroine does not let such a catastrophe inhibit her rise to greatness one iota, and is soon well on the way to creating a media empire out of her family's newspaper business.
Aka: DANIELLE STEEL'S NO GREATER LOVE
DRA 87 min (ort 90 min) mTV VIDrel: MIA/DISC V
Boa: novel by Danielle Steel.

NO HOLDS BARRED **   15
Thomas J. Wright USA   1989
Hulk Hogan, Tom "Tiny" Lister Jr, Joan Severance, Kurt Fuller, Jesse (The Body) Ventura, Jesse Ventura, Bill Henderson, David De Vries, Bill Fleet, David Paymer, Armelia McQueen, Kathy Payne, Mike Scott, Bruce Taylor, Al Hamacher
A vehicle for WWF superstar Hogan, in which he appears as a TV wrestling star who takes on a corrupt TV executive whose thugs beat up his brother and girlfriend when he refused to take part in a ratings-boosting scheme, dreamt up by a greedy promoter who wanted him to switch networks. A decent if

rather crude and uninventive ROCKY 3-style film, that attempts to cash in on the growing popularity of TV wrestling.
A/AD 88 min (ort 93 min) VIDrel: 4-FRONT/POLY/MIA L/A V

NO LIMIT *** U
Montague Banks UK 1935
George Formby, Florence Desmond, Jack Hobbs, Edward Rigby, Peter Gawthorne, Alf Goddard, Betrix Fielden-Kaye, Howard Douglas, Evelyn Roberts, Florence Gregson, Ernest Sefton
Formby plays a chimney sweep's assistant who dreams of competing in the Isle of Man TT Race, eventually winning the race after a series of comic episodes. This was Formby's first film for ATP, and surprisingly it still remains quite fresh and entertaining, chiefly thanks to the star's lively performance. The later films were generally nowhere near as good.
COM 81 min (ort 79 min) B/W VIDrel: LUMI/SPEAR L/A V

NO MAN'S LAND ** 15
Peter Werner USA 1987
Charlie Sheen, D.B. Sweeney, Randy Quaid, Lara Harris, Bill Duke, Arlen Dean Snyder, R.D. Call, M. Emmett Walsh, Al Shannon, Bernard Pock, Linda Carol, James F. Kelly, Gary Riley, Claire Wren, Lori Butler, Ken Endosa, Philip Benichous
A rookie cop working underground to solve a murder gets drawn into a glamorous lifestyle and eventually faces an agonising choice between love and professional honour. Sluggish and uninspired treatment of a promising idea.
A/AD 101 min (Cut at film release by 1 min 5 sec – ort 107 min) VIDrel: VISVID/POLYREC V/sur

NO-ONE COULD PROTECT HER ** 18
Larry Shaw USA 1995
Joanna Kerns, Peter MacNeill, Anthony John Denison, Lori Loughlin, Dan Lett, Christina Cox, Dan Luria, Gerry Mendicino, Anna Louise Richardson, Dan Pawlick, Harold Burke, David Husband, Reg Dreger, Wayne Ward, Bruce Beaton, Adam Fleck
Fact-based tale of a woman who is raped and assaulted in her home, and finds that her assailant has decided to return and murder her. Average.
THR 93 min mTV VIDrel: ODY/SONOP V/sh

NO ORDINARY SUMMER *** (12)
Matty Rich USA 1994
Larenz Tate, Joe Morton, Suzzane Douglas, Glynn Turman, Vanessa Bell Calloway, Adrienne-Joi Johnson, Morris Chestnut, Jada Pinkett, Duane Martin, Mary Alice. Phyllis Yvonne Stickney, Markus Redmond, Perry Moore, Akia Victor
A troubled sixteen-year-old is obliged to accompany his parents on a two-week vacation to visit his relatives, who live in a beach-house at Martha's Vineyard. After accidentally burning down his home, the boy has no friends at all except a doll, and he has a lot of trouble adjusting to his new surroundings. However, he eventually learns a few useful lessons in this sensitive, coming-of-age tale, made with an all-black cast.
DRA 108 min SATmed: SKY MOVIES

NO PLACE LIKE HOME **** PG
Lee Grant USA 1989
Christine Lahti, Jeff Daniels, Lantz Landry, Kyndra Joy Casper, Scott Marlowe, Kathy Bates, C.C.H. Pounder, Steve Rankin, Rick Aviles, Ernest Eyth, Helena Routi, Michael Hume, Marcia Haufrecht, Brandon Greene, James Chapman
Events force a working-class Pittsburgh family onto the streets, where they experience at first hand the rigours of homelessness. Written by Ara Watson and Sam Blackwell, this excellent and moving tale incorporates some of the incidents found in Grant's earlier documentary, "Homeless". Watch Lahti, who gives one of the best performances of her career.
DRA 90 min (ort 100 min) mTV VIDrel: ODY/SONOP V/sh

NO PLACE TO HIDE ** (15)
Richard Danus USA 1992
Kris Kristofferson, Drew Barrymore, Martin Landau, O.J. Simpson, Dey Young, Zeev Revach, Bruce Weitz, Illana Shoshan, Seth Marten, Lydie Denier, Lilyan Chauvin, John Poposello, Michael Franco, Debra Ellis, Greg Sapel, Mike Stone
A detective who has hit rock bottom following the death of his entire family in a car crash is given the task of investigating the murder of a dancer. As his investigation proceeds, he rapidly finds himself becoming embroiled in corruption and intrigue.

However, having won the confidence of the victim's sister, he learns of a video-tape that may provide a key clue. A production-line drama – no more, no less.
DRA 93 min VIDrel: MGM/WHV L/A V

NO RETREAT, NO SURRENDER * 15
Corey Yuen USA 1985
Kurt McKinney, Jean-Claude Van Damme, Kathie Sileno, J.W. Fails, Kim Tai Chong, Kent Lipham, Ron Pohnel, Dale Jacoby, Pete Cunningham, Tim Baker, Gloria Marziano, Joe Vance, Fahid Pahani, Tom Harris, John Andes, Ty Martinez
A student learns karate in order to beat a bully, and ends up taking on a tough Russian boxer in this derivative nonsense that blends ROCKY with THE KARATE KID. The inevitable sequel followed.
MAR 79 min Cut (35 sec – ort 90 min)
VIDrel: 4-FRONT/POLYREC/EIV L/A V/h

NO RETREAT, NO SURRENDER 2: RAGING THUNDER ** 18
Corey Yuen USA 1989
Cynthia Rothrock, Loren Avedon, Max Thayer, Patra Wanthivanond, Matthias Hues, Nirut Sirijunya, Hwang Jang Lee, Perm Hongsakul, Chesda Smithuth, Ray Horan, Grisapong Hanviriyakitichai, Bunchaim Arunrak, Opisok Praechaya
When a man's fiancee is kidnapped and disappears in the dense jungles of South-East Asia, he enlists the aid of two martial arts experts in his efforts to locate and rescue her.
MAR 86 min (ort 92 min)
VIDrel: 4-FRONT/POLYREC/EIV L/A V/h

NO RETREAT, NO SURRENDER 3: BLOOD BROTHERS ** 18
Lucas Lo HONG KONG 1990
Loren Avedon, Keith Vitali, Joseph Campanella, Wanda Acuna, Luke Askew, Ron Hunter, David Michael Sterling, Mark Russo, Philip Benson, Sherrie Rose, Mike Genova, D. Christian Gottshall, Brett Ancell, B.J. McQueen, Tracy Bero
When a former CIA agent is murdered in South-east Asia, his two feuding sons (who are both martial arts experts) go their separate ways in an attempt to catch the international terrorist responsible. Eventually they learn that they must put their differences to one side in order to succeed. The usual collection of flying fists and bullets is on display.
A/AD 91 min (ort 97 min)
VIDrel: 4-FRONT/POLYREC/MIA L/A V

NO SECRETS ** 15
Dezso Magyar USA 1990
Adam Coleman Howard, Heather Fairfield, Traci Lind, Amy Locane, Jeff Yeagher, Daniel Beer, Bert Williams, Drew Snyder, Greg Wrangler, James Edgcomb, James Willett
A young psychotic drifter goes on the run and comes to a ranch where three well-off woman are taking their vacation. Needless to say, they all find our anti-hero absolutely irresistible, in this violent and undistinguished thriller.
THR 86 min (ort 92 min) VIDrel: COLUM/SONOP V

NO SEX PLEASE, WE'RE BRITISH * PG
Cliff Owen UK 1973
Ronnie Corbett, Beryl Reid, Arthur Lowe, Ian Ogilvy, Susan Penhaligon, David Swift, Cheryl Hall, Michael Bates, Valerie Leon, Margaret Nolan, Gerald Sim, Robin Askwith, John Bindon, Stephen Greif, Michael Ripper, Michael Robbins
Film adaptation of a strained and unfunny typically British farce that enjoyed a long stage run. It deals with the complications that arise when a parcel of pornographic material is delivered in error to a bank employee. Dated isn't the word – geriatric comes closer. Written by Marriott, Johnnie Mortimer and Brian Cooke.
COM 88 min (ort 96 min) VIDrel: VCC/DISC/COLUM V
Boa: play by Anthony Marriott and Alistair Foot.

NO SKIN OFF MY ASS ** 18
Bruce LaBruce CANADA 1991
G.B. Jones, Bruce La Bruce, Klaus Von Bruckner
A homosexual hairdresser meets a young skinhead in the park and finds him to be the man of his dreams, and he promptly takes him home and seduces him, hoping they can live happily ever after. An amusing underground tale that is often shown at fringe film festivals.
A 55 min (ort 108 min) VIDrel: DTK/RTM V

NO SURRENDER **   15
Jerry P. Jacobs USA   1994
*Ted Jan Roberts, Mako, Corey Feldman, Erin Gray, Marshall Teague,
Dean Cochran, William James Jones, Mark Riffon, Jason Majik, Dick
Van Patten*
A young man investigates the mysterious death of his brother,
a karate champ who was found hanging in the gym. However,
he finds that the only way to unmask the killers is to infiltrate
a nasty rival karate team, known as "The Scorpions", even if by
doing this he will be thought to have betrayed both family and
friends. A fairly decent foray into KARATE KID territory, with
Feldman making a surprisingly convincing baddie.
MAR   91 min   VIDrel: MIA/DISC V/sur

NO TIME FOR SERGEANTS **   U
Mervyn Le Roy USA   1958
*Andy Griffith, William Fawcett, Murray Hamilton, Nick Adams, Don
Knotts, Jamie Farr, Murray McCormick, Bartlett Robinson, Howard
Smith, Will Hutchins, Sydney Smith, James Milhollan, Jean Wiles,
Henry McCann, Dub Taylor*
The army adventures of a dumb Georgia farm boy conscript,
who gives his sergeant a hard time by his inability to carry out
the simplest tasks without making a mess of them. An agreeable
if contrived slapstick affair, that was adapted from a successful
stage show and still retains much of its theatrical stiffness. Gave
rise to a TV series some years later. The script is by John Lee
Mahin.
COM   114 min (ort 119 min) B/W   VIDrel: MGM/WHV
L/A V
Boa: novel by Mac Hyman/play by Ira Levin.

NO WAY BACK *   18
Frank Cappello JAPAN/USA   1996
*Russell Crowe, Helen Slater, Etushi Toyokawa, Michael Lerner, Ian
Ziering*
With an enemy posing a threat to both of them, the boss of the
Yakuza in Japan and an FBI agent are obliged to forget that they
operate on different sides of the law, and join forces to defeat
their common adversary. One of those "East meets West" action
tales, offering absolutely nothing new in plotting, but the usual
plethora of car chases, gunplay and fisticuffs, none of which are
handled with any notable verve.
A/AD   88 min   VIDrel: COLUM/SONOP V

NO WAY HOME ***   18
Buddy Giovanazzo USA   1996
Tim Roth, James Russo, Deborah Kara Unger
Having just got out of prison after serving six years for murder,
Roth returns to his family home, which has since his imprisonment been taken over by his brother and wife, following the
death of their mother. Hostility between the two brothers gives
way to a simmering resentment and a set of complications that
make this far-from-cosy set-up unlikely to last. A tightly
directed and well drawn drama, that eschews stereotypes in
favour of realistic characterisations.
DRA   93 min   CINrel

NO WAY OUT ***   15
Roger Donaldson USA   1987
*Kevin Costner, Gene Hackman, Sean Young, Will Patton, Howard
Duff, Iman, George Dzundza, Jason Bernard, Fred Dalton Thompson,
Leon Russom, Dennis Burkley, Marshall Bell, Chris D., Michael
Shillo, Nicholas Worth, Leo Geter*
A heavily disguised remake of "The Big Clock", which substi-
tutes the setting of a crime magazine for a tale of high-level
conspiracy and corruption in the Pentagon, where a naval officer
investigates the murder of his chief's mistress, but finds all the
evidence apparently pointing to his boss. A fairly decent thriller
let down by a foolish ending. Scripted by Robert Garland and
with music by Maurice Jarre.
THR   110 min (ort 114 min) wScrn (COLUM/SONOP)
VIDrel: ENCORE/SPEAR/COLUM; COLUM/SONOP
V/sur
Boa: novel The Big Clock by Kenneth Fearing.

NOBLE HOUSE: PARTS 1 AND 2 **   15
Gary Nelson USA   1987
*Pierce Brosnan, Deborah Raffin, Ben Masters, John Rhys-Davies,
Khigh Dhiegh, Julie Nickson, Gordon Jackson, Burt Kwouk, Nancy
Kwan, John Van Dreelan, Ping Wu, Kay Tong Lim, Damien Thomas,
Dudley Sutton, Ric Young, Harris Laskawy*
A sequel-of-sorts to TAI-PAN, about the tough head of a

modern-day Hong Kong corporation and his struggles with a
ruthless business rival. Average.
DRA   360 min (2 cassettes) mTV   VIDrel: BRAVE/SONOP
L/A V
Boa: novel by James Clavell.

NOBODY'S CHILDREN **   12
David Wheatley USA   1994
*Ann-Margret, Dominique Sanda, Reiner Schone, Clive Owen, Jay O.
Sanders, Allan Corduner, Latrin Cartlidge, Leon Lissek*
A couple who are desperate to have a child, decide to try for
adoption and travel to Romania. But once there, they find them-
selves having to cope with a mountain of obstacles, in the shape
of red-tape and general corruption. Based on a true story.
DRA   91 min (ort 95 min) mCab   VIDrel: ODY/SONOP
V/sur

NOBODY'S FOOL **   15
Evelyn Purcell USA   1986
*Rosanna Arquette, Eric Roberts, Mare Winningham, Louise Fletcher,
Jim Youngs, Gwen Welles, Stephen Tobolowsky, Charlie Barnett,
Lewis Arquette, Ann Hearn, Belita Moreno, Ronnie Claire Edwards,
Scott Rosenzweig, Sheila Paige*
A downtrodden waitress, shunned by the locals because she is
an unwed mother, finds acting and love when a visiting troupe
comes to town. A modern Cinderella tale that is neither very
credible nor entertaining, despite a nice performance from
Arquette. Written by Beth Henley.
COM   103 min (ort 107 min)   VIDrel: SONY L/A V

NOBODY'S FOOL ***   15
Robert Benton USA   1995
*Paul Newman, Jessica Tandy, Melanie Griffith, Bruce Willis, Dylan
Walsh, Pruitt Taylor Vince, Gene Saks, Josef Sommer, Philip Bosco,
Philip Seymour Hoffman, Catherine Dent, Alexander Goodwin, Carl
J. Matusovic, Jay Patterson*
An ageing construction worker returns to his small town and the
family that he abandoned many years before and starts to spend
time with his son as well a number of eccentric acquaintances.
An intriguing portrait of a loner that is much enhanced from
both Newman in the main role and the supporting cast. Written
by Benton and Russo, from the latter's novel. Newman's perfor-
mance got him a well-deserved Oscar nomination.
DRA   110 min   VIDrel: 20TH/FOXVID V/dm
Boa: novel by Richard Russo.

NOCE BLANCHE *   15
Jean-Claude Brisseau FRANCE   1989
*Vanessa Paradis, Bruno Cremer, Ludmila Mikael, Francois Negret,
Jean Daste, Veronique Silver, Philippe Tuin, Benoit Muracciole,
Arnaud Goujon, Pierre Gabaston*
A philosophy teacher has an affair with his most rebellious
student, and this leads to various anaemic musings on life, love
and destiny. Despite the undoubted charms of Paradis (who
initially found fame as a pop star) this sterile and pompous
effort says very little of interest.
DRA   88 min (ort 92 min) wScrn   VIDrel: TART/20TH V

NOI TRE ***   PG
Pupi Avati FRANCE   1984
*Christopher Davidson, Lino Capolicchio, Gianni Cavina, Ida De
Benedetto*
A whimsical and not especially factual look at the life of young
Mozart, who as a fourteen-year-old preparing for an important
music exam, stays at a villa in the woods near Bologna. There,
he enjoys rough-and-tumble with the local boys, develops a
crush on a pretty girl, and still finds time for his music. The
slight story may not be entirely true, but this is an exuberant
little work of enormous charm.
Aka: THREE OF US, THE
DRA   85 min (ort 90 min)   VIDrel: CONNO/RTM V

NOIR ET BLANC ***   18
Claire Devers FRANCE   1986
*Francis Frappat, Jacques Martial, Josephine Fresson, Marc Berman,
Claire Rigollier*
An accountant takes a temporary job at a health club and devel-
ops an intense relationship with the black masseur. But
sado-masochistic elements soon become apparent, and these are
explored in ever more extreme ways, leading to the inevitably
violent and chilling resolution. The feature film debut for
Devers, it is made with enormous skill, and if not graphic, a

general air of menace is never far below the surface. Winner of the Camera D'Or at Cannes.
DRA   80 min (ort 82 min) B/W   VIDrel: ELPIC/POLYREC
V

NOISES OFF! ** 15
Peter Bogdanovich USA 1992
Carol Burnett, Michael Caine, Denholm Elliott, Julie Hagerty, Marilu Henner, Mark Linn-Baker, Christopher Reeve, John Ritter, Nicollette Sheridan, Kate Rich, Zoe R. Cassavetes, Kim Sebastian, L.B. Straten, Kimberly Neville, Roger Michelon
Michael Frayn's successful stage-play painted a picture of a touring company slowly falling apart in the most chaotic manner imaginable, with the action portrayed from both sides of the stage. Much of anarchic script of the original has been lost in this film production, and with it has gone some of the humour. Yet despite this (and the inevitable "opening out" that filmed plays suffer) the end result works rather well. A partial success. See IN THE BLEAK MIDWINTER.
COM   99 min (ort 104 min)   VIDrel: BUENA/TECH L/A
V/sh
Boa: play by Michael Frayn.

NOMADS ** 18
John McTiernan USA 1985
Pierce Brosnan, Lesley-Anne Down, Adam Ant, Anna Maria Monticelli, Hector Mercado, Mary Woronov, Josie Cotton, Frank Doubleday, Frances Bay, Tim Walker, Alan Autry, Jeannie Elias, Nina Foch, J.J. Saunders, Paul Anselmo
A doctor gives emergency treatment to an apparently insane derelict and later learns that he is a French anthropologist, who has come to L.A. to study a strange gang of street people, who seem to be more than human. A supernatural thriller that fails to develop its initial premise to any degree.
HOR   89 min (ort 100 min)   VIDrel: 20TH   V

NONNI: PARTS 1, 2 AND 3 *** PG
Agust Gudmundsson 1988
Lisa Harrow, Luc Merenda, Stuart Wilson
An unusual tale set in Iceland, and telling of the career of two brothers who find themselves in danger when they attempt to clear an innocent friend of a a charge of murder.
A/AD   600 min (3 cassettes) mTV   VIDrel: ODY/SONOP
V
Boa: novel by Jon Svensson.

NORMA RAE *** PG
Martin Ritt USA 1979
Sally Field, Ron Leibman, Beau Bridges, Pat Hingle, Barbara Baxley, Gail Strickland, Lonny Chapman, Morgan Paull, Robert Broyles, John Calvin, Booth Colman, Lee DeBroux, James Luisi, Gilbert Green, Bob Minor, Mary Munday
The story of how a female textile worker became militant and helped the union gain a foothold in the factory where she worked. Very well made and enjoyable, but the triumph of unionisation is made to appear far too easy and the victory too complete. AA: Actress (Field), Song ("It Goes Like It Goes" – David Shire (m)/Norman Gimbel (l)).
DRA   110 min (ort 114 min)   VIDrel: 20TH/TECH L/A   V

NORMAL LIFE * 18
John McNaughton USA 1995
Luke Perry, Ashley Judd, Bruce Young, Jim True, Dawn Maxey, Scott Cummins, Kate Walsh, Penelope Milford, Tom Towles, Edmund Wyson, Michael Skewes, Kevin Mukherji, Brian McCann
Perry plays a cop who gets involved with a dizzy blonde (Judd) whose handling of the family finances after they get married leads to a mounting spiral of debt. When he is sacked after criticising his superiors, he embarks on a career change as a bank robber, a mode of employment that does their relationship a power of good. Presumably this was meant to be a crazy crime caper, it is really too bad that the script and acting is so very poor.
COM   102 min   CINrel

NORMAN WISDOM: A STITCH IN TIME * U
Robert Asher UK 1963
Norman Wisdom, Edward Chapman, Jerry Desmonde, Jeannette Sterke, Jill Melford, Glyn Houston, Hazel Hughes, Patsy Rowlands, Peter Jones, Ernest Clark, Lucy Appleby, Vera Day, Frank Williams, Penny Morell, Patrick Cargill
Butcher's assistant Norman is banned from visiting a little orphan girl who is in hospital, but hatches a plan to get into the

hospital disguised as an ambulanceman. Not one of the star's best comedies, it's thinly plotted and really quite tedious.
Aka: STITCH IN TIME, A
COM   90 min (ort 94 min) B/W   VIDrel: VCC/DISC   V

NORMAN WISDOM: FOLLOW A STAR ** U
Robert Asher UK 1959
Norman Wisdom, June Laverick, Jerry Desmonde, Hattie Jacques, Richard Wattis, Eddie Leslie, John Le Mesurier, Sydney Tafler, Fenella Fielding, Charles Heslop, Joe Melia, Ron Moody
A tailor has ambitions to be a singer and has a good voice, but is so nervous that he can only sing in the presence of his crippled girlfriend, a fact a fading singing star is not slow in exploiting but his unprincipled scheme is eventually exposed. A poorly made and rather unamusing comedy, an unworthy vehicle for Wisdom's talent.
Aka: FOLLOW A STAR
COM   99 min (ort 104 min) B/W   VIDrel: VCC/DISC   V

NORMAN WISDOM: GIRL TROUBLE * PG
Menahem Golan UK 1969
Norman Wisdom, Sally Geeson, Terence Alexander, Sarah Atkinson, Sally Bazeley, Derek Francis, David Lodge, Paul Whitsun-Jones, George Meaton, The Pretty Things
A banker on his way to a conference, falls for a a teenage hitchhiker he gives a lift to in this feeble sex comedy. Written by Wisdom.
Aka: GIRL TROUBLE; WHAT'S GOOD FOR THE GOOSE
COM   104 min   VIDrel: FABFIL/SPEAR   V

NORMAN WISDOM: JUST MY LUCK ** U
John Paddy Carstairs UK 1957
Norman Wisdom, Margaret Rutherford, Jill Dixon, Leslie Phillips, Edward Chapman, Delphi Lawrence, Joan Sims, Peter Copley, Michael Ward, Bill Fraser, Campbell Cotts, Robin Bailey, Sabrina, Marjorie Rhodes, Felix Felton, Sam Kydd
A man eager to get married to his girlfriend places a small bet that six horses will be ridden to victory by the same jockey in the course of a single day. After five wins, it seems that there is little chance that he will also win the final race too, but fate decrees otherwise. An okay little comedy, mildly amusing but a terrible waste of Wisdom's great comic talents.
Aka: JUST MY LUCK
COM   82 min (ort 86 min) B/W   VIDrel: VCC/DISC   V

NORMAN WISDOM: MAN OF THE MOMENT ** U
John Paddy Carstairs UK 1955
Norman Wisdom, Lana Morris, Belinda Lee, Jerry Desmonde, Karel Stepanek, Garry Marsh, Inia te Wiata, Evelyn Roberts, Violet Farebrother, Martin Miller, Lisa Gastoni, Charles Hawtrey, Man Mountain Dean, The Beverly Sisters
Norman plays a lowly United Nations clerk, who is catapulted into prominence when he gains the trust of the ruler of a Pacific island. Wisdom's third comedy film is helped along by an unusually clever plot, in which the intrigues of powerful politicians play no small part. Dated but quite good fun.
Aka: MAN OF THE MOMENT
COM   85 min (ort 88 min) B/W   VIDrel: VCC/DISC   V

NORMAN WISDOM: ON THE BEAT *** U
Robert Asher UK 1962
Norman Wisdom, Jennifer Jayne, Raymond Huntley, David Lodge, Esma Cannon, Eric Baker, Eleanor Summerfield, Ronnie Stevens, Terence Alexander, Maurice Kaufman, Dilys Laye
A car-park attendant at Scotland Yard dreams of becoming a policeman just like his dear old dad, but unfortunately is too short to be accepted. He is however, elevated to the ranks when it's discovered that he bears an uncanny resemblance to an Italian jewel thief, whom the police are anxious to catch. A lively Wisdom comedy with good routines making up for the deficiencies of the plot and the rather weak dialogue.
Aka: ON THE BEAT
COM   101 min (ort 105 min) B/W   VIDrel: VCC/DISC   V

NORMAN WISDOM: ONE GOOD TURN ** U
John Paddy Carstairs UK 1954
Norman Wisdom, Joan Rice, Shirley Abicair, Thora Hird, William Russell, Joan Ingram, Richard Caldicott, David Hurst, Harold Kasket, Keith Gilman
Norman plays an odd-job man who works at a children's orphanage. In his second film he divides his time between trying to get the money for a model car he has promised to one of the children and fighting against the closure of the orphanage. There

are some amusing mishaps on offer, plus a funny battle royal in which the kids see off the officials who have arrived to close the orphanage. The disjointed script and mawkish sentimentality are far less welcome features.
Aka: ONE GOOD TURN
COM   91 min B/W   VIDrel: VCC/DISC   V

NORMAN WISDOM: PRESS FOR TIME **   U
Robert Asher UK   1966
Norman Wisdom, Derek Bond, Angela Browne, Tracey Crisp, Allan Cuthbertson, Noel Dyson, Derek Francis, Peter Jones, David Lodge, Stanley Unwin, Frances White, Michael Balfour, Tony Selby, Totti TRuman, George Roderick, Hazel Coppin
The story of an accident-prone newspaper seller, who is sent by his prime minister grandfather to work as a cub reporter on a provincial newspaper, in an attempt to keep him out of trouble. A feeble comedy that makes poor use of Wisdom's comic gifts. However, there are some good things here, not least Wisdom playing three roles, and a nice score by Mike Vickers. This was the last of Wisdom's old-style comedy films.
Aka: PRESS FOR TIME
COM   98 min (ort 102 min) B/W/Col   VIDrel: VCC/DISC   V
Boa: novel Yea, Yea, Yea by Angus McGill.

NORMAN WISDOM: THE BULLDOG BREED **   PG
Robert Asher UK   1960
Norman Wisdom, Ian Hunter, David Lodge, Robert Urquhart, Edward Chapman, Eddie Byrne, Peter Jones, Liz Fraser, John Le Mesurier, Terence Alexander, Sydney Tafler, Brian Oulton, Johnny Briggs, Penny Morell, Claire Gordon
Slapstick comedy with a grocer bungling his way through life from the Navy to Outer Space. An adequate vehicle for a star whose talents always outshone his material.
Aka: BULLDOG BREED, THE
COM   93 min (ort 97 min) B/W   VIDrel: VCC/DISC   V

NORMAN WISDOM: THE EARLY BIRD *   U
Robert Asher UK   1965
Norman Wisdom, Edward Chapman, Jerry Desmonde, Paddie O'Neil, Bryan Pringle, Richard Vernon, John Le Mesurier, Penny Morell, Frank Thornton, Harry Locke, Dandy Nichols, Imogen Hassall, Eddie Leslie, Peter Jeffrey, Marjie Lawrence
A milkman gets involved in a feud between a large dairy company and a tiny one-horse outfit. A stiff and unfunny waste of Wisdom's considerable talents.
Aka: EARLY BIRD, THE
COM   93 min (ort 98 min)   VIDrel: VCC/DISC   V

NORMAN WISDOM: THE GIRL ON THE BOAT **   U
Henry Kaplan UK   1962
Norman Wisdom, Sheila Hancock, Bernard Cribbins, Millicent Martin, Richard Briers, Athene Seyler, Philip Locke
Two Englishmen travelling on a transatlantic liner in the 1920s have various mishaps and develop into rivals when they both fall for the same attractive female. A broad, good-natured comedy, that gave the star a chance to display his talents in an unusual setting. Unfortunately, a good deal of his thunder is stolen by the strong supporting cast, most of whom look far more at home than him.
Aka: GIRL ON THE BOAT, THE
COM   88 min (ort 91 min) B/W wScrn
VIDrel: FABFIL/SPEAR   V
Boa: novel by P.G. Wodehouse.

NORMAN WISDOM: THE SQUARE PEG ***   U
John Paddy Carstairs UK   1958
Norman Wisdom, Edward Chapman, Campbell Singer, Hattie Jacques, Brian Worth, Terence Alexander, Honor Blackman, John Warwick, Arnold Bell, Eddie Leslie, Andre Maranne, Frank Williams, Oliver Reed
Period comedy with a star whose immense talent was rarely done justice by his material. An army recruit turns out to be the exact double of a German general, so of course Norman gets to play both parts with his usual hilarious panache. A dated but enjoyable comedy. Look out for the mirror sequence that even rivals the Groucho Marx one in DUCK SOUP. Written by Jack Davies.
Aka: SQUARE PEG, THE
COM   85 min (ort 89 min) B/W   VIDrel: VCC/DISC   V

NORMAN WISDOM: TROUBLE IN STORE ***   U
John Paddy Carstairs UK   1953
Norman Wisdom, Margaret Rutherford, Moira Lister, Megs Jenkins,

Lana Morris, Jerry Desmonde, Derek Bond, Joan Sims, Michael Brennan, Michael Ward, John Warwick, Perlita Nielson, Eddie Leslie, Cyril Chamberlain, Ronan O'Casey
Wisdom's film debut has him as an accident-prone department store assistant who finally makes good. Probably one of the best of his films. Though dated, Wisdom's comic talent ensures that the film retains a charm of its own. Written by Carstairs with Maurice Cowan and Ted Willis. The music is by Mischa Spoliansky.
Aka: TROUBLE IN STORE
COM   82 min (ort 85 min) B/W   VIDrel: VCC/DISC   V

NORMAN WISDOM: UP IN THE WORLD **   U
John Paddy Carstairs UK   1956
Normam Wisdom, Maureen Swanson, Jerry Desmonde, Michael Caridia, Colin Gordon, Ambrosine Philpotts, Michael Ward, Jill Dixon, Edwin Styles, Hy Hazell, William Lucas, Lionel Jeffries, Cyril Chamberlain, Eddie Leslie
Norman starts a new job as a window cleaner and makes friends with a young millionaire, and by chance is on hand to save him from kidnappers. There is plenty of knockabout comedy here, plus no less than twenty-one songs from the star, but it's all just a bit too much, and cannot mask the sheer lack of thought given to the script plus a definite lack of pace.
Aka: UP IN THE WORLD
COM   86 min (ort 91 min) B/W   VIDrel: VCC/DISC   V

NORTH **   PG
Rob Reiner USA   1994
Elijah Wood, Julia Louis-Dreyfus, Jason Alexander, Jon Lovit, Alan Arkin, Dan Akroyd, Kathy Bates, Faith Ford, Graham Greene, Abe Vigoda, Reba McEntire, John Ritter, Kelly McGillis, Alexander Godunov, Pat "Noriyuki" Morita
A young boy is dissatisfied with his parents lack of attention and affection and decides to "divorce" them and get himself adopted by a more suitable couple. He thus goes in search of them, aided on his travels by a protective spirit (Bruce Willis). A ponderous and unsuccessful adaptation of Zweibel's novel, scripted by him and Reiner. A large number of cameo appearances add absolutely nothing of interest to this failed offering. Narration is also by Willis.
COM   83 min (ort 87 min)   VIDrel: VCC/DISC/COLUM
V/sur
Boa: novel by Alan Zweibel.

NORTH AND SOUTH: BOOK 1 – PARTS 1 TO 6 ***   15
Richard T. Heffron USA   1985
Patrick Swayze, James Read, Lesley-Anne Down, David Carradine, Kirstie Alley, George Stanford Brown, Philip Casnoff, Genie Francis, Terri Garber, Wendy Kilbourne, Jim Metzler, Lewis Smith, John Stockwell, Johnny Cash
Long, colourful, sprawling epic account of the events leading up to the American Civil War as seen through the eyes of several families, the Mains of South Carolina and the Hazards of Pennsylvania, whom fate has brought together as friends but whose relationship is compromised by the War. Look out for Robert Mitchum, Elizabeth Taylor and Jean Simmons, all of whom have cameos in this tale, as do many others.
DRA   524 min (3 cassettes) mTV   VIDrel: WHV   V
Boa: novel Love and War by John Jakes.

NORTH AND SOUTH: BOOK 2 – PARTS 1 TO 6 ***   15
Kevin Connor USA   1986
Patrick Swayze, James Read, Lesley-Anne Down, David Carradine, Kirstie Alley, Mary Crosby, Philip Casnoff, Terri Garber, Wendy Kilbourne, Jonathan Frakes, Jim Metzler, Lewis Smith, Parker Stevenson, Genie Francis
The continuing saga of the Mains and the Hazards, two families who find themselves on opposite sides during the American Civil War. A lavish and highly engaging production that follows where the first film left off, and recounts the changing fortunes of the characters in the aftermath of the War. Look out for Linda Evans, Olivia De Havilland, Lloyd Bridges, Jean Simmons, James Stewart, Anthony Zerbe and a host of other stars in cameos.
DRA   543 min (3 cassettes) mTV   VIDrel: WHV   V
Boa: novel Love And War by John Jakes.

NORTH AND SOUTH: BOOK 3 – PARTS 1 TO 6 **   PG
USA   1994
Peter O'Toole, Lesley-Anne Down, Philip Casnoff, Robert Wagner, Cathy Lee Crosby
Last part of the Civil War saga that follows the variable fortunes

of the Main and Hazard families, as they attempt to rebuild their lives. Quite enjoyable, even if it doesn't quite come up to the calibre of the earlier parts of this story.
DRA 261 min (3 cassettes) VIDrel: WHV V

NORTH BY NORTHWEST ****
PG
Alfred Hitchcock USA 1959
Cary Grant, Eva Marie Saint, James Mason, Leo G. Carroll, Martin Landau, Jessie Royce Landis, Philip Ober, Adam Williams, Josephine Hutchinson, Edward Platt, Robert Ellenstein, Les Tremayne, Philip Coolidge, Ken Lynch, Edward Binns
An innocent man becomes inextricably entangled in a complex espionage plot, in an exciting thriller full of non-stop action that drives the story forward at a furious pace, culminating in a memorable and justly famous fight sequence atop the Mount Rushmore Memorial. Scripted by Ernest Lehman and with a fine score by Bernard Herrmann.
THR 130 min (ort 136 min) VIDrel: MGM/WHV V/h

NORTH TO ALASKA ***
U
Henry Hathaway USA 1960
John Wayne, Stewart Granger, Ernie Kovacs, Fabian, Capucine, Mickey Shaughnessy, Karl Swenson, Joe Sawyer, John Qualen
Two gold prospectors strike it rich and one sends his partner south to bring back his fiancee. When the partner finds that the girl has since married, he sets out to find a substitute. Set in the 1900s, this vigorous and lighthearted adventure yarn has a lot going for it, excessive length is its only flaw.
WES 117 min (ort 122 min) VIDrel: 20TH/TECH L/A V
Boa: play Birthday Gift by Ladislas Fodor.

NORTHANGER ABBEY **
U
Giles Foster UK 1987
Peter Firth, Googie Withers, Robert Hardy
Careful adaptation of Austen's novel of mystery and suspense, with a woman arriving at the Abbey and discovering much intrigue hidden beneath the bland exterior.
DRA 89 min mTV VIDrel: BBC/TECH V/h
Boa: novel by Jane Austen.

NORTHERNERS, THE ***
15
Alex Van Warmerdam HOLLAND 1992
Leonard Lucieer, Jack Wouterse, Rudolf Lucieer, Alex Van Warmerdam, Annet Malherbe, Loes Wouterson, Veerle Dobbelaere, Dary Some, Jacques Commandeur, Janny Goslinga, Theo Van Gogh, Loes Luca, Rein Bloem, Wil Spoor, Anna Visser
Black comedy about the residents of a tightly-knit and terribly snobbish little Catholic community in Holland in the 1960s, who all live in a street that by virtue of its location, is more or less cut off from the rest of the world. A bizarre farce, detailing a set of comic episodes that generates a distinctive momentum of its own, even if the central story goes nowhere in particular.
Aka: DE NOORDERLINGEN
COM 102 min (ort 108 min) VIDrel: CURZON/20TH V/sh

NORTHWEST FRONTIER ***
PG
J. Lee Thompson UK 1959
Kenneth More, Lauren Bacall, Herbert Lom, Ursula Jeans, Wilfrid Hyde-White, I.S. Johar, Eugene Deckers, Ian Hunter, Govind Raja Ross, Jack Gwillim, Basil Hoskins, Moultrie Kelsall, Lionel Murton, S.M. Asgarelli, Homi Bode
A British officer has to get a young Indian prince to safety during a tribal rebellion and embarks on a hazardous train journey. Set in 1905, this enjoyable "Boy's Own"-style film offers just the right blend of action and suspense. The script is by Robin Estridge.
Aka: FLAME OVER INDIA
A/AD 87 min (ort 129 min) VIDrel: VCC/DISC/COLUM L/A V

NORTHWEST PASSAGE ***
PG
King Vidor USA 1940
Spencer Tracy, Robert Young, Walter Brennan, Ruth Hussey, Nat Pendleton, Louis Hector, Eugene Deckers, Robert Barrat, Lumsden Hare, Donald MacBride, Isabel Jewell, Douglas Walton, Addison Richards, Hugh Sothern, Regis Toomey, Montagu Love
A colourful account of the adventures of Major Rogers and his Rangers during the French and Indian War of 1759, as they sought to stem the Indian advance at St Francis in Canada. Part Two was never made, so the alternative title is a bit misleading, but this does not detract from the film's entertainment value.

This historical material also served as the basis for a 1958 TV series of the same name.
WES 121 min (ort 127 min) VIDrel: MGM/WHV L/A V
Boa: novel by Kenneth Rogers.

NOSFERATU ***
PG
F.W. Murnau GERMANY 1921
Max Schreck, Gustav Von Wangenheim, Greta Schroeder, Alexander Granach, G.H. Schnell, Ruth Landshoff, John Gottow, Max Nemetz
The original classic film on the Dracula theme. Written by Henrik Galeen and remade in 1979 as NOSFERATU THE VAMPYRE. Interestingly, it was later films that moved the vampire teeth to the sides of the mouth, a true vampire bat has them in the centre. An atmospheric and eerie film that has a number of imaginative touches absent in later works.
Aka: DIE ZWOLFTE STUNDE; DRACULA; EINE NACHT DES GRAUENS; EINE SYMPHONIE DES GRAUENS; NOSFERATU, A SYMPHONY OF TERROR; NOSFERATUR, EINE SYMPHONIE DES GRAUENS; NOSFERATU, THE VAMPIRE; TERROR OF DRACULA; TWELFTH NIGHT
HOR 80 min B/W silent VIDrel: EUREKA/GOLD V

NOSFERATU THE VAMPYRE ***
15
Werner Herzog WEST GERMANY 1979
Klaus Kinski, Isabelle Adjani, Bruno Ganz, Roland Topor, Walter Ladengast, Jacques Dufilho, Dan Van Husen, Jan Groth, Carsten Bodinus, Martje Grohmann, Ryk De Gooyer, Tim Beekman, Clemens Scheitz, Lo Van Hartingsveld
This is virtually a scene-for-scene remake of the silent 1921 classic that's a beautifully crafted work, but unfolds at such a slow pace that it seems more like a succession of static scenes that a living story. Written and directed by Herzog.
Aka: NOSFERATU THE VAMPIRE; NOSFERATU: PHANTOM DER NACHT
HOR 96 min (ort 107 min) VIDrel: 20TH/TECH V/h
Boa: novel Dracula by Bram Stoker.

NOSTALGIA ***
15
Andrei Tarkovsky ITALY/USSR 1983
Oleg Yankovsky, Domiziana Giordano, Erland Josephson, Patrizia Terreno, Delia Boccardo
At a spa in the hills of Tuscany, a Russian poet and musician is researching the life of an 18th century composer. He is challenged by a mysterious man, who believes the end of the world is nigh, to perform a difficult task as an act of faith. Tarkovsky's first film outside the USSR is an oppressive and doom-laden meditation, sustained by a few images of great beauty and opaque symbolism. Unfortunately, it's a bit too stylised and sluggish to be really effective.
Aka: NOSTALGHIA
DRA 120 min (ort 126 min) Col/B/W VIDrel: ARTIF/20TH V/h

NOSTRADAMUS **
15
Roger Christian GERMANY/UK 1994
Tcheky Karyo, F. Murray Abraham, Amanda Plummer, Rutger Hauer, Assumpta Serna, Julia Ormond, Anthony Higgins, Diana Quick, Michael Gough, Bruce Myers, Maia Morgenstern, Magdalena Ritter, Leon lissek, Micahel Byrne, Istefan Paroli
A less-than-impressive directorial debut for Christian (formerly an art director on such movies as STAR WARS and ALIEN) that tells the story of this French scholar who lived in the 16th century and wrote down a set of extensive but highly ambiguous prophecies that are claimed to foretell events several centuries after his death. Fortunately, the film is saved by good production values and Karyo's competent performance in the title role.
DRA 115 min (ort 120 min) VIDrel: FIRST/SONOP; ENCORE (LV only) V/sh LV

NOSTRIL PICKER, THE *
18
Patrick J. Matthews USA
Carl Zschering, Edward Tanner, Laura Cummings
An ugly nerd who is unable to attract women is taught a magical chant that enables him to turn into a girl at will, and he uses this power to satisfy his craving to murder women as a way of getting revenge for all the rejection he has suffered. A crude and clumsy horror-comedy, that was given a new title (based on the central character's favourite hobby) in an attempt to rescue it from a well deserved oblivion.
Aka: CHANGER, THE
HOR 76 min VIDrel: VIPCO/SGSVID V/h

NOSTROMO ***
12
Alastair Reid UK
1996
Albert Finney, Colin Firth, Serena Scott Thomas, Claudio Amendola, Claudia Cardinale, Brian Dennehy, Joaquim De Almeda, Paul Brooke
Passion and ambition vie for supremacy in this 19th century tale of a silver-mining company that operates in South America. Firth plays an Englishman who returns to his South American birthplace, his intention being to re-open a silver mine, the revolution having led the murder of his father and the destruction of the mine. First shown on the BBC in four parts, this is a handsome adaptation of one of Conrad's most ambitious and demanding works.
Aka: JOSEPH CONRAD'S NOSTROMO
DRA 321 min (2 cassettes) mTV VIDrel: BBC/TECH
V/sh
Boa: novel by Joseph Conrad.

NOT LIKE US *
(18)
Dave Payne USA
1995
Joanna Pacula, Peter Onorati, Rainer Grant, Morgan Englund, Billy Burnette, Annabelle Gurwitch, Paul bartel, Clint Howard, Doug Tract, Andrew Brut, Keith Contreras, Alexandra Picato, Kevin Austin, Bob Narland, Raf Elliott, Eb Lottimer
A woman's new neighbour appears to be a great hit with the local men who however, always seem to die or disappear. She proves to be a nasty alien. A black comedy SF offering, replete with low-budget, poor effects etc.
FAN 89 min SATrel: MOVIE CHANNEL

NOT OF THIS EARTH *
(18)
Terence H. Winkless USA
1995
Michael York, Parker Stevenson, Richard Belzer, Elizabeth Barondes, Ted Davis, Mason Adams, Julia Mueller, Bob McFarland, Wendy Buckner, Eddie Driscoll, Joshua D. Comen, Jennifer Coolidge, Mary Scher, Arthur Roberts, Diana Miranda
An alien comes to Earth in search of a cure for a blood disease that is destroying his world, and starts to prey on humans, killing them and draining their blood. The second remake of a classic 1950s B-movie, this one throws in ample special effects and a few extraneous sub-plots, but still remains an uninventive and uninspired dud. The ending is especially lame and totally predictable.
FAN 89 min SATrel: MOVIE CHANNEL

NOT OF THIS WORLD **
15
Jon Daniel Hess USA
1991
Lisa Hartman, A. Martinez, Pat Hingle, Luke Edwards, Michael Greene, Tracey Walter, Richard Grove, Stephen Prutting, Richard Epcar, Greg Natalie, J.B. Quon, Elizabeth Gill, Burr Middleton, Michele Palermo, Nicholas D. Bussey
A kind of SF-thriller that has a shapeless alien landing in a small Northwestern town in a meteor shower, and heading straight for a local power plant. Very much a 1990s celebration of the kind of cheap-looking films Corman and others churned out in the 1950s, this is a verbose but quite watchable effort, whose budget did not extend to much in the way of special effects. Not a spoof, but it might easily have been one.
FAN 89 min (ort 100 min) mTV VIDrel: CIC/SONOP
V/h

NOT OUR SON ***
15
Michael Ray Rhodes USA
1995
Neil Patrick Harris, Gerald McRaney, Tom Verica, Ari Meyers, Scott Allan Campbell, Tom McBeath, Stephen E. Miller, Cindy Pickett, Jason Gaffney, David Lovgren, Gary Chalk, Ken Camroux, Brenda Crichlow, Denalda Williams, Kim Restell
A father recognises an identikit picture of a wanted arsonist as his own son and has to grapple with his conscience before deciding to turn him in to the authorities. True-life drama built around the crimes of a serial-arsonist in Seattle.
DRA 84 min (ort 92 min) mTV VIDrel: ODY/SONOP
V/sh

NOT QUITE HUMAN *
U
Steve Hilliard Stern USA
1987
Alan Thicke, Robyn Lively, Robert Harper, Joseph Bologna, Jay Underwood, Brian Cole, Brandon Douglas, Lili Hadyn, Sasha Mitchell, Greg Monaghan, Judy Starr, Lonny Price, Carey Scott, Kristy Swanson, Bob Anthony, Gene Blakeley
A human android is given great intelligence and strength by its creator as well as the power of total recall, and is programmed to tell the truth under all circumstances, which leads to some awkward situations for all concerned. A juvenile comedy reminiscent of many a lame Disney effort. it was followed by two sequels.
JUV 87 min (ort 97 min) mCab VIDrel: WDV/TECH L/A
V

NOT QUITE HUMAN 2 **
U
Eric Luke USA
1989
Robyn Lively, Jay Underwood, Alan Thicke, Greg Mullavey, Katie Barberi, Mark Arnott, Dey Young, Scott Nell, Mike Russell, Ty Miller, Erik Burskotter, Bob Sorenson, Doug Cotner, Holly Robertson, Christine Baur, Doug Coleman
The further adventures of our lovable male android as he attends college, provides the story for this predictable sequel to the 1987 film. Agreeable as family entertainment but no more than that. Followed by "Still Not Quite Human".
COM 92 min (ort 105 min) mCab VIDrel: WDV/TECH
L/A V

NOT WITHOUT MY DAUGHTER **
15
Brian Gilbert USA
1990
Sally Field, Alfred Molina, Sheila Rosenthal, Roshan Seth, Sarah Badel, Mony Rey, Georges Corraface, Mary Nell Santacroce, Ed Grady, Marc Gowan, Bruce Evers, Jonathan Cherchi, Soudabeh Farrokhnia, Michael Morim, Gill Ben-Ozilio
An American woman takes a trip with her Iranian husband to his country, but once she is there she finds that she's treated as nothing more than a chattel. Although he's ready to allow her to leave, he won't release their daughter and she has to fight for the child's freedom. This might have been a potent tale, but relies heavily on crude caricatures, and is less an examination of foreign ways than a simple escape film. Music is by Jerry Goldsmith.
DRA 111 min (ort 116 min) VIDrel: MGM/WHV V/sur
Boa: book by Betty Mahmoody with William Hoffer.

NOTHING BUT THE BEST **
PG
Clive Donner UK
1963
Alan Bates, Denholm Elliott, Millicent Martin, Harry Andrews
One of those rather dated "swinging 60s" stories, this is a social satire about an amoral estate agent and his steady climb to the top. A fairly deft blend of comedy and drama, it is slightly reminiscent of ALFIE (a far better film) and offers a succession of amusing visual gags and some sharp dialogue. The script is by Frederic Raphael.
DRA 110 min VIDrel: LUMI/SPEAR V
Boa: short story The Best of Everything by Stanley Ellin.

NOTHING BUT TROUBLE *
15
Dan Aykroyd USA
1990
Chevy Chase, Dan Aykroyd, John Candy, Demi Moore, John Daveikis, Taylor Negron, Bertila Damas, Valri Bromfield, Brian Doyle-Murray, John Wesley, Earl Dixon, Peter Aykroyd, Daniel Baldwin, James Staskel, Deborah Lee Johnson
Chase and Moore take off for a weekend and get arrested for speeding whilst passing through a dump of a town, that's in the grip of a vicious old judge who delights in tormenting motorists. Most of the film revolves around their incessant attempts to escape and in short, this film is not funny, just plain nasty. Scripted by Aykroyd, this far-from-impressive effort was his directing debut.
Aka: VALKENVANIA
COM 89 min (ort 94 min) VIDrel: WHV V

NOTHING LASTS FOREVER *
15
Jack Bender USA
199-
Gail O'Grady, Brooke Shields, Vanessa Williams, Gregory Harrison, Stephen Caffrey, Lloyd Bridges
The tangled love lives of doctors are the subject matter in this typical soap opera, the story revolving around three female medics who work at the same hospital and share a house. One of them sleeps with a hospital administrator, another helps a dying man commit suicide and the third one gets pregnant, so we get the usual parade of anguish, heartbreak and intrigue. Glossy, contrived and very, very empty.
DRA 155 min VIDrel: 20TH/FOXVID V/sh
Boa: novel by Sidney Sheldon.

NOTHING PERSONAL ***
15
Thaddeus O'Sullivan EIRE/UK
1995
Ian Hart, John Lynch, James Frain, Michael Gambon, Gary Lydon, Ruaidhri Conroy, Maria Doyle Kennedy, Jeni Courtney, Gerard McSorley, Gareth O'Hare, Ciaran Fitzgerald, Antony Brophy, B.J. Hogg, Jim Duran, Cathy White, Lynne James

In the wake of a 1975 pub bombing in Belfast, both Loyalist and Republican bosses try to arrange a ceasefire, but their respective followers are not ready to lay down their weapons. The story mostly follows the activities of a couple of Protestant gunmen, their various bloody activities and an eventual confrontation with a Catholic childhood friend of one of them. Scripted by Mornin, this is an intense, focused and rather chilling work.
A/AD 81 min (ort 85 min) cC VIDrel: FILM4/RTM V/sh
Boa: novel All Our Fault by Daniel Mornin.

NOTHING SACRED ***   PG
William Wellman USA   1937
Carole Lombard, Frederic March, Walter Connolly, Charles Winninger, Sig Ruman, Frank Fay, Maxie Rosenbloom, Margaret Hamilton, Hedda Hopper, Monty Woolley, Hattie McDaniel, Olin Howland, John Qualen, Art Lasky, Alex Novinsky
A reporter exploits the case of a girl allegedly dying from a rare disease to increase his paper's circulation, but her imminent demise proves to be a stunt. Witty and bitter attack on the seamier side of journalism that has lost none of its bite over the years.
COM 73 min (ort 75 min) VIDrel: SECOND/RTM V
Boa: short story Letter to the Editor by James Street.

NOTHING TO LOSE **   18
Izidore K. Musallam USA   1993
Alexander Paul, Michael V. Gazzo, Paul Gleason, Yousef Abed-Alnour, Juliano Mer, David Campbell, William Corno, Chris Leavins, Greg Locke, Maureen Burgoyne, Mariliese Rizzardo, Vida Asante, Mary Feely, Roger Jaggernauth, Linda Barnett
When crime bosses murder his father, a young man learns how to fight and goes after them in a bid for revenge. A run-of-the-mill actioner that offers nothing new.
A/AD 92 min VIDrel: ENTREE/HIFLI V

NOTHING UNDERNEATH ***   18
Carlo Vanzina ITALY   1985
Donald Pleasence, Tom Schanley, Renne Simonsen, Nicola Perring, Ronald Lacey, Paolo Tomei, Maria McDonald, Catherine Noyes, Sonia Raule, Cyrus Elias, Anna Galiena, Big Laura, Bruce McGuire, Mimmo Sepe
A thriller with a telepathy theme in which a number of models are murdered and a young man experiences the death of his model sister, a twin with whom he shared a unique closeness. The police are unable to catch the killer and he resolves to seek him out himself. A film that is slow to start but generates considerable tension and has a neat twist.
Aka: INTO THE DARKNESS; SOTTO IL VESTITO NIENTE
THR 90 min VIDrel: SPEAR/SONOP V
Boa: novel by Marco Parma

NOTORIOUS ****   U
Alfred Hitchcock USA   1946
Cary Grant, Ingrid Bergman, Claude Rains, Louis Calhern, Reinhold Schunzel, Leopoldine Konstantin, Moroni Olsen, Ivan Triesault, Alex Minotis, Wally Brown, Gavin Gordon, Charles Mendl, Ricardo Costa, Fay Baker, Antonio Moreno
A classic period thriller about the daughter of a convicted Nazi spy who is persuaded to marry a suspected Nazi sympathiser to help US intelligence discover his plans. She agrees, but falls in love with the agent sent to contact her, which proves a dangerous complication in this highly exciting and enjoyable film. The superb script is by Ben Hecht.
THR 101 min B/W VIDrel: BRAVE/SONOP L/A; PION (LV only) V LV

NOTORIOUS **   15
Colin Bucksey USA   1991
John Shea, Jenny Robertson, Jean-Pierre Cassel, Marisa Berenson, Paul Guilfoyle, Ronald Guttman, Jean-Pierre Stewart, Stephane Meldegg, Marc Samuel, Bill Bailey, Laurence Bouvencourt, Dominique Figaro
Based on the Hitchcock classic, this updated remake dispenses with the WW2 background and Nazi villains, instead offering us the more up-to-date tale of an independent-minded heroine who's recruited by the CIA for a spot of espionage. This ultimately involves her in marrying a crooked arms dealer as part of a cunning plot to destroy him, but he grows suspicious and takes reprisals. An adequate yarn, it is best not compared to the earlier work.
THR 90 min mTV VIDrel: GUILD/POLYREC L/A V/sh

NOVEMBER MAN, THE ***   18
Paul Williams USA   1993
James Andronica, Leslie Bevis, Robert Davi, Beau Starr
Using footage from the 1992 Bush/Clinton presidential campaign, this story tells of an obsessive and self-centred filmmaker who recruits a bunch of malcontents to help stage a fake assassination attempt on Bush, his intention being to shoot this as a movie. But the stakes get raised and his project starts getting out of control. An interesting low-budget work, it has the look of an improvised effort, and though not completely successful is highly distinctive.
THR 98 min VIDrel: GUILD/SONOP L/A V

NOW AND THEN **   PG
Lesli Linka Glatter USA   1995
Christina Ricci, Thora Birch, Gaby Hoffman, Ashleigh Aston Moore, Demi Moore, Rosie O'Donnell, Rita Wilson, Melanie Griffith, Willa Glen, Bonnie Hunt, Devon Sawa, Travis Robertson, Justin Humphrey, Bradley Coryell, Ric Reitz
In the summer of 1970, four twelve-year-old girls swear undying loyalty to each other, promising to always be on hand to assist should one of their number ever be in need. Twenty years later they are reunited, and look back over their lives. Glatter's first theatrical feature is a pleasant enough film, with great acting from the child stars but less conviction on the part of the actresses who play the roles as adults. The memorable 1970s soundtrack is a highlight.
DRA 98 min (ort 102 min) VIDrel: FIRST/SONOP V

NOW, VOYAGER ***   PG
Irving Rapper USA   1942
Bette Davis, Claude Rains, Paul Henreid, Gladys Cooper, Bonita Granville, Janis Wilson, Ilka Chase, John Loder, Lee Patrick, Charles Drake, Franklin Pangborn
A lonely, frustrated spinster is transformed by her psychiatrist and becomes involved in an ill-fated love affair. Classy, romantic melodrama that wins through on the strength of the performances of its stars and the beauty of its music. Scripted by Casey Robinson. AA: Score (Max Steiner).
DRA 113 min (ort 117 min) B/W VIDrel: WHV V
Boa: novel by Olive Higgins Prouty.

NOWHERE TO HIDE **   PG
Bobby Roth USA   1994
Rosanna Arquette, Scott Bakula, Max Pomeranc, Clifton Powell, Robert Wisden, Jenny Gago, Richmond Arquette, Jerry Wasserman, Chris Mulkey, Laurie Paton, Nancy McClure, Peter Lacroiz, Jill Teed, Dee Jay Jackson
A woman risks her life in the pursuit of justice, despite being advised she should enter the Federal Witness Protection Program. Average.
THR 88 min cC VIDrel: CIC/SONOP V

NOWHERE TO RUN *   15
Robert Harmon USA   1992
Jean-Claude Van Damme, Kieran Culkin, Joss Ackland, Rosanna Arquette, Ted Levine, Tiffany Taubman, Edward Blatchford, Anthony Starke, Leonard Termo, Steve Chambers, Stephen Wesley Bridgewater, Christy Botkin, Luana Anders
An escaped convict who is a kickboxing expert comes to the rescue of a lonely widow and her two children who are facing eviction from their farm because of foreclosure by the bank. Naturally, the owner of the latter is the villain of the piece and is just as predictably in cahoots with a corrupt lawman. A tailor-made vehicle for Damme, who gets ample opportunity to display his fighting skills, the dire plot providing nothing else of note.
A/AD 91 min (ort 95 min) wScrn
VIDrel: COLUM/SONOP V/sur

NUDE VAMPIRE, THE **   18
Jean Rollin FRANCE   1969
Cathleen Raye, Felicia Denay, Olivier Martin, Maurice LeMaitre, Caroline Cartier, Ly Lestrong, Bernard Musson, Jean Aron, Christine Francois, Ursulle Pauly, Michel Delahaye, Nicole Isimat
A young man gets involved with a strange sect that performs bizarre suicide rituals. An involved and hard-to-follow, almost nightmarish tale with very striking camerawork and imagery that partially compensates for the slow plot and the rather poor acting.
Aka: LA VAMPIRE NUE; NAKED VAMPIRE, THE
HOR 81 min (ort 90 min) VIDrel: REDEM/RTM V

NUKIE ** **   U
Sias Odendaal   1992
Glynis Johns, Steve Railsback, Ronald France
The title creature is one of those ugly-yet-lovable aliens so
popular since E.T. THE EXTRA-TERRESTRIAL. Having crash-
landed with his brother Miko in the heart of Africa, Nukie finds
sanctuary with a couple of decent youngsters from a tribal
village. Unfortunately his brother was caught by the American
Space Research Foundation, and is now calling psychically for
help, which is not long in coming. Amiable kid's fantasy-adventure.
FAN   95 min   VIDrel: 20VIS/SONOP   V/sur

NUMBER SEVENTEEN *   U**
Alfred Hitchcock UK   1932
*Leon M. Lion, Anne Grey, John Stuart, Donald Calthrop, Barry Jones,
Garry Marsh, Ann Casson, Henry Caine, Herbert Langley*
A reformed female jewel thief helps the police catch the gang
she once belonged to. A remake of a 1928 British silent, that is
technically competent and moderately exciting, despite some
obviously crude effects. Written by Hitchcock, Alma Reville and
Rodney Ackland.
DRA   63 min B/W   VIDrel: LUMI/SPEAR   V
Boa: play by J. Jefferson Farjeon.

NUNS ON THE RUN *   15**
Jonathan Lynn UK   1990
*Eric Idle, Robbie Coltrane, Camille Coduri, Janet Suzman, Doris Hare,
Lila Kaye, Robert Patterson, Tom HIckey, Winston Dennis, Robert
Morgan, Colin Campbell, Richard Simpson, Nicholas Hewetson, Gary
Tang, David Foreman*
Written by Lynn, this wacky comedy tells of two career crooks
who, tired of working for their nasty boss, decide to pull off a
job and keep the loot for themselves. But the robbery is bungled
and they have to flee, and hide out in a convent disguised as
nuns. A bit patchy and slow at times, but Idle and Coltrane are
great together and the film is generally very funny.
COM   88 min (ort 94 min)   VIDrel: 20TH/TECH L/A   V

NUN'S STORY, THE *   PG**
Fred Zinnemann USA   1959
*Audrey Hepburn, Peter Finch, Edith Evans, Peggy Ashcroft, Dean
Jagger, Mildred Dunnock, Beatrice Straight, Colleen Dewhurst,
Patricia Collinge, Eva Kotthaus, Ruth White, Nial McGinnis, Patrica
Bosworth, Barbara O'Neil*
Well-made filmisation of the story of a Belgian girl who served
as a nun Congo, enduring hardship and discomfort, before
taking the decision to opt for a life outside the convent walls.
Highly acclaimed, but in many respects far from convincing,
although it does offer a fascinating insight into the routine of
convent life.
DRA   145 min (ort 151 min)   VIDrel: WHV   V
Boa: book by Kathryn C. Hulme.

NURSE, THE *   18**
Rob Malenfant USA   1996
*Lisa Zane, John Stockwell, Janet Gunn, William R. Moses, Michael
Fairman, Nancy Dussault, Sherrie Rose, Jay Underwood*
A woman sets out to get even with her father's former boss, as
she blames him for her family's troubles. When the latter
becomes incapacitated following a severe heart attack she inveigles her way into his home as his personal nurse, and hatches a
plan to murder his entire family in front of him. A ludicrous and
very spiteful thriller, treading the well-worn "psychotic woman
on the loose" pathway.
THR   90 min   VIDrel: FIRST/SONOP   V

NURSE ON WHEELS *   U**
Gerald Thomas UK   1963
*Juliet Mills, Ronald Lewis, Noel Purcell, Joan Sims, Raymond
Huntley, Athene Seyler, Jim Dale*
The new district nurse arrives at a rural community where she
falls for a local farmer and enjoys various other encounters. A
likeable if soppy comedy that attempts to give district nurses the
"Carry On" treatment.
COM   82 min (ort 86 min) B/W   VIDrel: WHV   V/h
Boa: novel Nurse is a Neighbour by Joanna Jones.

NURSES ON THE LINE *   PG**
Larry Shaw USA   1993
*Lindsay Wagner, Robert Loggia, David Clennon, Farrah Forke, Paula
Marshall, Hilary Edson, Jennifer Lopez, Tom Irwin, Joan McMurtrey,
Jay Patterson, Sergio Calderon, Pedro Armendariz, James Sutorious,*

Gary Frank, Bill Bolender
When a plane crashes in the Mexican jungle, the four doctors on
board are all seriously injured, and their only chance of surviving is to reach a remote medical clinic in the jungle. Fortunately,
the passengers include a group of student nurses, who now have
to attempt to save the lives of their colleagues.
Aka: RACE AGAINST THE DARK
DRA   89 min (ort 95 min) mTV
VIDrel: 4-FRONT/POLYREC/ODY   V/sh

NUTCRACKER, THE *   U**
Emile Ardolino USA   1993
*Darci Kistler, Damian Woetzel, Kyra Nichols, Bart Robinson Cook,
Heather Watts, Robert La Foss, Jessica Lynn Cohen, Macaulay Culkin,
Kevin Kline, Wendy Whelan, Margaret Tracy, Gen Horiuchi, Tom
Gold, William Otto, Peter Reznick*
A stagebound version of the famous story, that's set on
Christmas Eve, and tells of a little girl who is transported into
a fantasy world of animated toys. Adapted by Peter Martins
from a production put on by the New York City Ballet
Company, but not especially effective when brought to the
screen. Narration is by Kevin Kline.
Aka: GEORGE BALANCHINE'S THE NUTCRACKER
MUS   85 min (ort 92 min)   CINrel
Boa: ballet by Tschaikovsky.

NUTCRACKER PRINCE, THE *   U**
Paul Schibli CANADA/USA   1990
*Voices of: Kiefer Sutherland, Megan Follows, Peter O'Toole, Phyllis
Diller, Mike MacDonald*
A little girl shrinks in size and enters a magical land where all
her toys, nutcrackers and other articles become imbued with a
life of their own. A very weak and uninspired rendition of this
famous children's classic, that may have some limited appeal,
but only for the very young.
ANIM   72 min   VIDrel: EIV/SONOP   V/sur

NUTS *   18**
Martin Ritt USA   1987
*Barbra Streisand, Richard Dreyfuss, Karl Malden, Maureen Stapleton,
Eli Wallach, Robert Webber, James Whitmore, Leslie Nielsen, William
Prince, Dakin Matthews, Paul Benjamin, Warren Manzi, Elizabeth
Hoffman, Castulo Guerra*
A prostitute who killed one of her clients, goes to court to prove
herself sane and fit to stand trial, rather than be certified insane
and placed in an institution, which would suit her family, who
are anxious to hide their guilty shame. An absorbing piece that
is almost believable right up until the end, when a dreary mono-
logue from Streisand spoils it all. Scripted by Topor, Darryl
Ponicsan and Alvin Sargent.
DRA   111 min (ort 116 min)   VIDrel: WHV   V/sur
Boa: play by Tom Topor.

NUTS IN MAY *   PG**
Mike Leigh UK   1976
Roger Sloman, Alison Steadman, Anthony O'Donnell
An amusing BBC TV play about an ill-matched pair on a
camping holiday, that consists of a series of mishaps brought
about by the man's obsessive need to do everything "by the
book". A sharp character study, one of several tales, all of which
were written by Leigh.
COM   81 min mTV   VIDrel: PARADOX/TOTAL   V/h

NUTTY PROFESSOR, THE *   12**
Tom Shadyac USA   1996
*Eddie Murphy, Jada Pinkett, James Coburn, Larry Miller, Dave
Chappelle, John Ales, Patricia Wilson, Jamal Mixon, Nichole
McAuley, Hamilton Von Watts, Chao-Li Chi, Tony Carlin, Quinn
Duffy, Doug Williams, David Ramsey, Chaz Lamar Shepherd*
Incredible make-up turns slim Murphy into a 400-pound college
prof in this updated version of the famous Jerry Lewis comedy.
Having discovered a chemical that can transform him into a
more physically appealing character, our professor sets out to
have some fun. Sadly, the needs of the audience are forgotten
in the process and for all his charisma, Murphy has few chances
to shine with a script obsessed with bodily functions. See JERRY
LEWIS: THE NUTTY PROFESSOR.
COM   91 min (ort 95 min)   VIDrel: CIC; PION (LV only)   V
LV

NYMPHOID BARBARIAN IN DINOSAUR HELL *   18
Brett Piper USA   1993
Linda Corwin, Paul Guzzi, Alex Pirnie, Marc Dehsaies, K. Alan Hodder, Russ Greene, Rick Stewart, Scott Ferro, Ryan Piper, Dusty mcNeel, Quinn Piper, Melanie Pirnie, A. Strong, Jeneane De Prizio, Liz Prevett, Penny Townsend
In a dangerous world of monsters and mutants, a lone woman divides her time between battling for survival and fending off the attentions of amorous cavemen. A low-grade and unamusing spoof.
Aka: DARK FORTRESS
FAN   85 min (ort 90 min)   VIDrel: TROMA/RTM   V

O

O LUCKY MAN! **   15
Lindsay Anderson UK   1973
Malcolm McDowell, Arthur Lowe, Ralph Richardson, Rachel Roberts, Alan Price, Lindsay Anderson, Helen Mirren, Mona Washbourne, Dandy Nichols, Graham Crowden, Peter Jeffrey, Philip Stone, Mary MacLeod, Wallas Eaton, Anthony Nicholls
An Anderson feast of disjointed and strange images, as a trainee salesman journeys erratically through life on his way to the top before settling down to do various good deeds. Several actors have multiple roles and Anderson makes a brief appearance as himself. A film that's really got nothing to say once one digs below the surface. The music is by Alan Price. A "sequel" of sorts to the slightly more coherent IF ...
Aka: OH LUCKY MAN
COM   169 min (ort 174 min)   VIDrel: WHV   V

O PIONEERS! **   PG
Glenn Jordan USA   1991
Jessica Lange, David Strathairn, Reed Edward Diamond, Leigh Lawson, Heather Graham, Tom Aldredge, Anne Heche, Josh Hamilton, Graham Beckel, Adam Nelson, Darin Heames, Deborah May, Christopher Collet, John Durbin, Philip Heckman
In Nebraska in the late 1880s, the daughter of a Swedish immigrant father does everything she can to ensure the comfort of the other members of her family, but largely sacrifices her own happiness in the process. An enjoyable period drama with Lange making her TV movie debut. Interestingly, during her lifetime Carther banned any production based on her 1913 novel, so this film (and one other version) had to wait until the copyright on her work expired.
DRA   95 min (ort 100 min) mTV   VIDrel: MGM/WHV L/A   V
Boa: novel by Willa Carther

OBJECT OF BEAUTY ***   15
Michael Lindsay-Hogg UK/USA   1991
John Malkovich, Andie MacDowell, Rudi Davis, Lolita Davidovich, Joss Ackland, Bill Paterson, Ricci Harnett, Peter Riegert, Rosemary Martin, Roger Lloyd Pack, Andrew Hawkins, Pip Torrens, Stephen Churchett, Annie Hayes
A couple who enjoy life in the fast lane get stranded in London when they run out of cash, and reluctantly decide to raise some by selling their only possession of any value, a Henry Moore statuette. However, matters do not quite go as planned, and their desperate struggles to stay solvent teach them a thing or two about life and each other. A rambling comedy of manners, it is unusual, stylish and well acted, especially by Malkovich. First shown on BBC TV.
COM   98 min (ort 105 min) mTV   VIDrel: BUENA/SONOP L/A   V

OBJECT OF OBSESSION **   18
Alexander Gregory Hippolyte USA   1995
Scott Valentine, Erika Anderson, Liza Whitcraft, Robert Keith, Jane Rogers, Jane Higginson, Alisha Dub, Ashby Adams, Andrea Riare, DavidStrickland, Jennifer McDonald, J. Jay Cohen
A newly-divorced woman begins an affair with a handsome stranger, but a mistaken phone call leads to her getting drawn into a sexually fulfilling but obsessive relationship. Nothing new is on offer in this erotic thriller.
THR   96 min   VIDrel: ITC/HIFLI   V/sh

OBJECTIVE BURMA! ***   PG
Raoul Walsh USA   1944
Errol Flynn, James Brown, William Prince, George Tobias, Henry Hull, Warner Anderson, John Alvin, Stephen Richards, Tony Caruso,
Joel Allen, Dick Erdman, Richard John Whitney, George Tyne (Buddy Yarus), Rodric Redwing, William Hudson
Withdrawn for seven years in Britain following a public outcry over the fact it makes no mention of the British role in Burma, this one spawned the joke that Flynn won the war single-handed. A group of American paratroopers are dropped behind Japanese lines with orders to destroy a radio station. They eventually achieve their aim, but only at great cost in human lives, which effectively atones for any lack of historical verisimilitude.
WAR   135 min (ort 142 min) B/W   VIDrel: WHV   V

OBLONG BOX, THE **   15
Gordon Hessler UK/USA   1969
Vincent Price, Christopher Lee, Rupert Davies, Sally Geeson, Uta Levka, Peter Arne, Alastair Williamson, Hilary Dwyer, Maxwell Shaw, Carl Rigg, Michael Balfour, Harry Baird, Godfrey James, Ivor Dean, James Mellor
A man is horribly disfigured by an African witch doctor in revenge for his brother's killing of a native child. Upon their return to England, he is imprisoned in a tower room of the family mansion, from which he eventually escapes by faking his death. The inevitable revenge rampage then ensues. A gory horror offering that sails under false colours, by borrowing its title from an Edgar Allan Poe story to which it bears no resemblance.
HOR   95 min (Cut at film release)
VIDrel: VISVID/POLYREC L/A   V

OBSESSED **   15
Jonathan Sanger USA   1992
Shannen Doherty, William Devane, Clare Carey, James Handy, Lois Chiles, Lisa Ann Poggi, Albert Stratton, Julie Ariola, Jay Ingram, Tom Keena, Deanna Lund, Albie Selznick, Mark Kleid, Michael J. Martin, Ron Noble, Lynn Roth
An eligible unmarried man in his fifties gets involved with a beautiful younger woman, initially unaware that she has a dangerous personality disorder and an obsessively jealous streak. When he tries to end their affair the woman's love turns to hatred and she sets out to destroy him. An unedifying FATAL ATTRACTION clone.
THR   87 min mTV   VIDrel: MIA/DISC/IMPENT   V

OBSESSION ***   PG
Luchino Visconti ITALY   1942
Clara Calamai, Massimo Girotti, Juan De Landa, Elio Marcuzzo, Vittorio Duse, Dhia Christiani, Michele Riccardini, Michele Sakara
A handsome drifter falls for the unhappy wife of an elderly innkeeper and the two conspire to do away with him, after which their relationship proceeds to its inevitable conclusion. This unofficial adaptation of Cain's novel (later to be filmed in Hollywood in 1945 and 1981) marked the start of the Italian neo-realist period, but difficulties over the film rights to the novel kept the movie out of the States until 1975. Visconti's debut is clinical and effective.
Aka: OSSESSIONE
DRA   134 min (ort 140 min) B/W
VIDrel: CONNO/RTM; PION (LV only)   V LV
Boa: novel The Postman Always Rings Twice by James M. Cain.

OBSESSION ***   15
Edward Dmytryk UK   1948
Robert Newton, Sally Gray, Phil Brown, Naunton Wayne, Olga Lindo, Ronald Adam, Jaems Harcourt, Allan Jeayes, Russell Waters, Michael Balfour, Betty Cooper, Roddy Hughes, Lyonel Watts, Stanley Baker
A doctor plans to murder his wife's lover after imprisoning him in a cellar room and destroying all traces of the body with acid. However, his plan begins to go astray when the wife's dog follows him to the intended hiding place. A strong and effective thriller with ample suspense.
Aka: HIDDEN ROOM, THE
THR   93 min (ort 98 min) B/W   VIDrel: FABFIL/SPEAR   V
Boa: novel A Man About A Dog by Alec Coppel.

OBSESSION ***   15
Brian De Palma USA   1976
Cliff Robertson, Genevieve Bujold, John Lithgow, Sylvia Kuumba Williams, Wanda Blackman, Patrick McNamara, Stanley J. Reyes, Nick Kreiger, Stocker Fontelieu, Don Hood, Andrea Esterhazy
A wealthy businessman's wife and child disappear after he follows police instructions and substitutes blank paper for the ransom money. Some years later he and his partner are on a visit to Florence when he meets a woman who is his wife's

double and he becomes obsessed with discovering her true identity. A derivative thriller with a little violence and much suspense, although its best feature is the superb score by Bernard Herrmann.
THR   94 min (ort 98 min)   VIDrel: MIA/DISC   V

OCCASIONAL HELL, AN *
18
Salome Breziner USA
1996
Kari Wuhrer, Robert Davi, Valeria Golino, Tom Berenger, Stephen Lang
A former homicide detective now has a second career, teaching mystery writing at a local college. When a colleague is found murdered he decides to put his forensic skills to use by mounting his own investigation. However, his delvings are strangely complicated when the ghost of the dead man's mistress appears, whom only he can see. A laughable erotic thriller, saddled with a stolid script that fails to exploit the supernatural element.
THR   89 min (ort 92 min)   VIDrel: BMGVID/BMGREC   V

OCTAGON, THE **
18
Eric Karson USA
1980
Chuck Norris, Karen Carlson, Lee Van Cleef, Art Hindle, Jack Carter, Carol Bagdasarian, Kim Lankford, Tadashi Yamashita, Kurt Grayson, Yuki Shimoda, Larry D. Mann, John Fujioka, Richard Norton, Gerald Okamura, Redmond Gleeson
A secret terrorist training camp sets the scene for a battle royal between our hero and sundry villains, all of them versed in the martial arts. There is plenty of violent action to satisfy Norris fans but little else on offer here.
MAR   98 min Cut (32 sec – ort 103 min)
VIDrel: 4-FRONT/POLYREC   V

OCTOBER ***
PG
Sergei Eisenstein USSR
1927
Grigori Alexandrov, Boris Livanov, Nikandrov, Vladimir Popov
One of the films commissioned by the Soviet government to celebrate the tenth anniversary of the Revolution, this is a lavish recreation of the actual events and deservedly ranks as a masterly piece of propaganda. The use of montage is quite remarkable. Interestingly, Trotsky had figured prominently in the film, but with his fall from grace all sequences that included him were removed.
Aka: OKTYABR; TEN DAYS THAT SHOOK THE WORLD
DRA   101 min B/W silent   VIDrel: HEND L/A   V

OCTOBER MAN, THE ***
PG
Roy (Ward) Baker UK
1947
John Mills, Joan Greenwood, Edward Chapman, Kay Walsh, Joyce Carey, Felix Aylmer, Catherine Lacey, Patrick Holt, Frederick Piper, Adrianne Allen, Jack Melford, George Benson, John Salew, John Boxer, Esme Beringer, Ann Wilton
After suffering brain damage in an industrial accident that killed the child of a friend, a man develops suicidal tendencies. After his release from hospital he recuperates at a suburban hotel, and befriends a young model. When she is found strangled suspicion falls on him, and he begins to doubt his own innocence. A competent thriller, reminiscent of Hitchcock (Baker worked as the latter's assistant) with a strong script adapted by Ambler from his novel.
THR   91 min (ort 110 min) B/W   VIDrel: VCC/RTM   V
Boa: novel by Eric Ambler.

OCTOBER 32ND **
PG
Paul Hurt UK
1992
Richard Lynch, Peter Phelps, James Hong, Nadia Cameron, Ted Markland, Rodney Wood, Desmond Llewellyn, Pamela Mandell, Robert Padilla, John Stone
A Californian girl learns that she is a descendant of Merlin and as such must do battle with an evil magician, as the latter intends to steal a magical sword and use it to destroy the world.
Aka: MERLIN
FAN   84 min (ort 112 min)   VIDrel: COLUM/SONOP   V/sur

OCTOPUSSY ***
PG
John Glen UK
1983
Roger Moore, Maud Adams, Louis Jourdan, Kristina Wayborn, Kabir Bedi, Steven Berkoff, Desmond Llewellyn, Vijay Amritraj, David Meyer, Anthony Meyer, Lois Maxwell, Robert Brown, Walter Gotell, Geoffrey Keen, Cherry Gillespie
An excellent and entertaining Bond adventure with a complex Cold War plot, involving efforts by a Soviet general to create chaos by exploding a nuclear device at a NATO base in West Germany. High production values and the breathtaking Indian locations make this entry a winner. The script is by George McDonald Fraser (of the "Flashman" books fame), with music by John Barry. Number thirteen in the series, followed by NEVER SAY NEVER AGAIN.
A/AD   125 min (ort 131 min) wScrn   VIDrel: MGM/WHV   V/sur V/dm
Boa: short stories Octopussy/The Property Of A Lady by Ian Fleming.

ODD ANGRY SHOT, THE ***
18
Tom Jeffrey AUSTRALIA
1979
John Hargreaves, Graham Kennedy, Bryan Brown, John Jarratt, Graeme Blundell, Richard Moir, Ian Gilmour, John Allen, Brandon Burke, Graham Rouse, Johnny Garfield, Tony Barry, Max Cullen, John Fitzgerald, Ray Meagher, Brian Elvis
A group of professional Australian soldiers on service in Vietnam find that the reality of jungle warfare is far different from what they expected, and begin to question the reasons for their involvement. An unusually mature work that concentrates on personal relationships rather than combat.
Aka: DEATH FROM DOWN UNDER
WAR   87 min (ort 92 min)   VIDrel: ARTPRO/RTM   V

ODD COUPLE, THE ***
PG
Gene Saks USA
1967
Jack Lemmon, Walter Matthau, John Fielder, Herb Edelman, David Sheiner, Larry Haines, Monica Evans, Carole Shelley, Iris Adrian, Heywood Hale Broun, John C. Becher, Roberto Clemente, Matty Alou, Maury Willis, Vernon Law
An impossible fusspot is finally kicked out by his long-suffering wife and moves in with an unrepentant slob. From then on, they proceed inevitably to tear each other's nerves to shreds. Hilarious version of the Simon play, brought vividly to life by the excellent performances of the two leads. Later a TV series.
COM   101 min (ort 106 min)   VIDrel: CIC/SONOP   V/dm
Boa: play by Neil Simon.

ODD JOB, THE **
15
Peter Medak UK
1978
David Jason, Graham Chapman, Michael Elphick, Simon Williams, Diana Quick, Edward Hardwicke, Bill Paterson, Stewart Harwood, Carolyn Seymour, Joe Melia, Georges Innes, James Bree, Zulema Dene, Richard O'Brien, Carl Andrews, John Judd
When an insurance executive loses his wife to a friend, he hires a hitman to murder him, but following a change of heart has to persuade the killer to call the job off. The slight "Monty Python" flavour to the script is the work of Chapman, who co-scripted this sporadically amusing but generally strained effort.
COM   84 min (ort 100 min)   VIDrel: MIA/DISC   V
Boa: TV play by Bernard McKenna.

ODD MAN OUT ****
PG
Carol Reed UK
1946
James Mason, Robert Newton, Kathleen Ryan, Elwyn Brook-Jones, Robert Beatty, F.J. McCormick, William Hartnell, Fay Compton, Beryl Measor, Cyril Cusack, Dan O'Herlihy, Roy Irving, Maureen Delany, Kitty Kirwan, Min Milligan, W.G. Fay
An IRA man escapes from prison in Belfast and tries to make his way to the docks but is shot in a police raid on a mill. Hampered by his injuries, he encounters various individuals before meeting his inevitable and tragic end. A truly memorable work that is enormously effective despite a lack of warmth. Scripted by R.C. Sheriff and F.L. Green from the latter's novel, it was remade in 1969 as "The Lost Man".
Aka: GANG WAR
DRA   113 min (ort 115 min) B/W   VIDrel: L/A   V
Boa: novel by F.L. Green.

ODDBALL HALL **
PG
Jackson Hunsicker USA
1990
Don Ameche, Burgess Meredith, Bill Maynard, Tullio Moneta, Tiny Skefile, Tina Jaxa, Patrick Mynhardt, Phillip Wolfaardt, Graham Armitage, Ramolao Makhene, Patrick Ndlovu, Charles Stodel, Jonathon Meredith, Thelma Banya
A gang of jewel thieves hide out in a small African town to wait for the heat to cool and pretend to be members of a fraternity called the Oddballs. One day, they misunderstand a native's request and find themselves cast as powerful wizards able to end a long-lasting drought. Filmed on location in Southern Africa, this effort is slightly reminiscent of THE GODS MUST BE CRAZY series, but not nearly as entertaining.
COM   83 min (ort 87 min)   VIDrel: EIV/SONOP   V

ODESSA FILE, THE **
PG
Ronald Neame UK
1974
Jon Voight, Maximilian Schell, Maria Schell, Mary Tamm, Derek Jacobi, Peter Jeffrey, Klaus Lowitsch, Kurt Meisel, Hans Messemer, Garfield Morgan, Shmuel Rodensky, Ernst Schroeder, Gunter Strack, Noel William, Martin Brandt
A reporter infiltrates a secret Nazi organisation in West Germany in his search for some former members of the SS, after reading the diary of a camp survivor. A confused and slow-paced thriller that does little with material that should have guaranteed a high level of suspense. Written by Kenneth Ross and George Markstein.
THR   123 min (ort 129 min)   VIDrel: VCC L/A   V
Boa: novel by Frederick Forsyth.

ODETTE ***
PG
Herbert Wilcox UK
1950
Anna Neagle, Trevor Howard, Marius Goring, Peter Ustinov, Bernard Lee, Alfred Shieske, Gilles Queant, Maurice Buckmaster, Marianne Waller, Marie Burke, Frederick Wendhasusen, Guyri Wagner, Wolf Frees, Catherine Paul
A Frenchwoman is accidentally recruited as a WW2 British agent and parachutes into Occupied France where she spies for the Resistance. Eventually captured, she is tortured by the Nazis but reveals nothing, and is finally sent to a concentration camp for execution. A competent, low-key but rather wooden biography of a true heroine, done in a flat and unconvincing style. The film CARVE HER NAME WITH PRIDE tells of another such heroine.
WAR   118 min B/W   VIDrel: LUMI/SPEAR   V
Boa: book by Jerrard Tickell.

OEDIPUS REX **
15
Pier Paolo Pasolini ITALY
1967
Franco Citti, Silvana Mangano, Julian Beck, Carmelo Bene, Alida Valli, Pier Paolo Pasolini, Ninetto Davoli
A visually satisfying but sluggish and flat adaptation of two Sophocles tragedies, telling of the unfortunate Oedipus who murdered his father and married his mother in fulfilment of a prophecy.
Aka: EDIPO RE
DRA   100 min (ort 110 min)   VIDrel: CONNO/RTM L/A   V
Boa: plays Oedipus Rex/Oedipus At Colonus by Sophocles.

OF LOVE AND SHADOWS **
15
Betty Kaplan ARGENTINA/SPAIN
1996
Antonio Banderas, Jennifer Connelly, Stefania Sandrelli, Diego Wallraff, Camilio Gallardo, Patricio Contreras, Jorge Riviera Lopez
A conventional love story set in Chile in 1973 after the overthrow of the Allende government, when a reporter on a fashion magazine has an affair with an enigmatic photographer. Awkwardly plotted, the story is not exactly helped by its clumsy dialogue and the stars look uncomfortable and out of place. The novel is a great deal better, but its strengths are not brought out.
THR   105 min (ort 110 min)   VIDrel: EIV/SONOP   V/sh
Boa: novel by Isabel Allende.

OF MICE AND MEN **
PG
Gary Sinise USA
1992
John Malkovich, Gary Sinise, Ray Walston, Casey Siemaszko, Sherilyn Fenn, John Terry, Richard Riehle, Alex Arquette, Joe Morton, Noble Willingham, Joe D'Angerio, Tuck Milligan, David Steen, Moira Harris, Mark Boone Junior
Another competent, handsome but essentially unexciting film version of Steinbeck's powerful tale of two ranch hands, one a retarded giant, the other small, clever and protective. Together they wander America in the Depression years, looking for work and dreaming of owning their own home.
DRA   106 min (ort 111 min)   VIDrel: MGM/WHV   V/sur
Boa: novella by John Steinbeck.

OF PURE BLOOD **
15
Joseph Sargent USA
1986
Lee Remick, Patrick McGoohan, Robert Bowman, Gottfried John, Richard Munch, Katharina Bohm, Edith Schneider, Carolyn Nelson, Catherine McGoohan, Jem Wall, Hans Jurgen Schatz, Pascal Breuer, Beate Finckh, Thomas Kylau
The Nazi Lebensborn selective breeding programme, intended to eventually result in a master race, provides the factual background to this mediocre thriller set some forty years after the demise of the Third Reich, when a mother travels to West Germany to investigate the circumstances surrounding her son's violent death. See also THE BOYS FROM BRAZIL.
DRA   89 min (ort 100 min) mTV   VIDrel: MGM/WHV L/A   V/h
Boa: story by Del Coleman and Michael Zagor/book by Marc Hillel and Clarissa Henry.

OFF AND RUNNING **
15
Edward Bianchi USA
1990
Cyndi Lauper, David Keith, Johnny Pinto, David Thornton, Richard Belzer, Jose Perez, Anita Morris, Hazen Gifford, Linda Hart, Tracy Roberts, Dana Mark, Heather Davis, Sy Bondy, Tony Jones, George Richards, Steven Reibel
An aspiring actress flees for her life when her boyfriend, who was the wealthy owner of a stud-farm, is murdered by a business associate. A variable comedy-drama with a nasty and ever-present undercurrent of violence.
COM   87 min (ort 90 min)   VIDrel: COLUM/SONOP   V/sur

OFFENCE, THE ***
18
Sidney Lumet UK
1972
Sean Connery, Ian Brennen, Vivien Merchant, Trevor Howard, Derek Newark, John Hallam, Peter Bowles, Ronald Radd, Anthony Sagar, Howard Goorney, Richard Moore, Maxine Gordon
A tired and disillusioned police inspector goes over the top when confronted with a suspected child murderer. Connery gives a fine portrayal of a man at the end of his tether, in a fine psychological drama that was too serious and grim to be a hit. Scripted by John Hopkins from his play.
Aka: SOMETHING LIKE THE TRUTH
DRA   108 min (ort 113 min)   VIDrel: WHV   V
Boa: play This Story of Yours by John Hopkins.

OFFENSIVE SHAOLIN LONGFIST *
18
Philip Chan HONG KONG
1989
Natassa Tang, Lewis Ko, Chan Fung, Bruce Cheung
Two martial artists, one an expert and the other a student, meet in a bloody and violent confrontation. The set-piece battles are well-handled, but the lack of a strong plot allows one's attention to wane.
MAR   85 min   VIDrel: IMPENT   V

OFFERINGS *
18
Christopher Reynolds USA
1988
Loretta Leigh Brown, Jerry Brewer, Elizabeth Greene, G. Michael Smith
A man is released after spending ten years in the mental asylum to which he was committed, after a prank by the local kids both disfigured him and caused him to lose his mind. Another re-run of the avenging maniac theme.
HOR   93 min (ort 96 min)   VIDrel: MOPIC/QUANT   V

OFFICER AND A GENTLEMAN, AN ***
15
Taylor Hackford USA
1981
Richard Gere, Debra Winger, David Keith, Louis Gossett Jr, Robert Loggia, Lisa Blount, Lisa Eilbacher, Tony Plana, Harold Sylvester, David Caruso, Victor French, Grace Zabriskie, Tommy Petersen, Mara Scott Wood, David Greenfield
The story of a misfit becoming a US Naval Officer candidate and meeting a girl doing a dead-end job in a factory. The contrived plot, though violent and disjointed in parts, has touching moments as he is seen gaining maturity at the hands of his strict drill instructor (a superb performance from Gossett) and learning to care for others. AA: S. Actor (Gossett Jr), Song ("Up Where We Belong" – Jack Nitzche/Buffy Sainte-Marie (m)/Will Jennings (l)).
DRA   119 min (ort 126 min)   VIDrel: CIC/SONOP   V/h
Boa: novel by Stephen P. Smith.

OFFICIAL DENIAL **
PG
Brian Trenchard-Smith USA
1993
Parker Stevenson, Dirk Benedict, Erin Gray, Chad Everett, Nichael Pate, Christopher Pate, Natalie McCurry, Robert Mammone, Peter Curtin, Gina Gaigalas, Michael Edward-Stevens, Don hubert, Christopher Mayer, Chuck Perry, Kim Krejus
A man abducted by aliens finds his marriage under great strain when his wife flatly refuses to believe him. Unknown to them, the military are secretly monitoring their home and when an alien craft is shot down and crashes, he gets involved in the subsequent investigation. A few original ideas are not allowed to develop and the usual cliches abound. Filmed on location in

Australia with an American cast. See also COMMUNION and FIRE IN THE SKY.

FAN   84 min (ort 86 min) mTV   VIDrel: CIC/SONOP   V

OFFICIAL VERSION, THE **   15
Luis Puenzo ARGENTINA   1985
Norma Aleandro, Guillermo Battaglia, Patrizio Contreras, Hector Alterio, Analia Castro, Chunchuna Villafane
A woman discovers to her horror that her adopted child was taken from political prisoners, killed by the military junta who ruled Argentina up to 1984. A terribly disappointing film that demonstrates how good intentions alone cannot compensate for poor acting and an incoherent plot. Moreover, it fails to do justice to those whose loved ones perished at the hands of this brutal and benighted dictatorship. AA: Foreign.
Aka: LA HISTORIA OFICIAL; OFFICIAL STORY, THE
DRA   109 min   VIDrel: L/A   V

OH GOD! ***   PG
Carl Reiner USA   1977
George Burns, John Denver, Teri Garr, Ralph Bellamy, William Daniels, Paul Sorvino, George Furth, Barnard Hughes, Barry Sullivan, Dinah Shore, Jeff Corey, Donald Pleasence, David Ogden Stiers, Titos Vandis, Moosie Drier
God appears to the manager of a supermarket in the guise of an old man, and persuades him to deliver a message of hope and love. After a while, the joke of seeing God wearing a baseball cap wears a bit thin. A better story and dialogue would have helped, though Burns carries off the role of the Almighty with both charm and panache. The script is by Larry Gelbart and two sequels followed. Pleasence has a tiny cameo as a priest (no surprises there).
COM   93 min (ort 104 min)   VIDrel: MGM/WHV L/A   V
Boa: novel by Avery Corman.

OH GOD! BOOK 2 *   PG
Gilbert Cates USA   1980
George Burns, Suzanne Pleshette, David Birney, Louanne, Howard Duff, Hans Conried, Wilfrid Hyde-White, John Louie, Conrad Janis, Anthony Holland, Joyce Brothers, Bebe Drake Massey, Hugh Downs, Marian Mercer, Mari Gorman
Second of the OH GOD! films with the Almighty using a child to tell the world he is still there. A poor sequel with few ideas or laughs to offer.
COM   94 min   VIDrel: MGM/WHV L/A   V

OH GOD! YOU DEVIL **   15
Paul Bogart USA   1984
George Burns, Ted Wass, Ron Silver, Roxanne Hart, Eugene Roche, Robert Desiderio, John Doolittle, Julie Lloyd, Ian Giatti, Janet Brandt, Betita Moreno, Danny Ponce, Jason Wingreen, Danny Mora, Jane Dulo, Susan Peretz, Cynthia Tarr
Third in the OH GOD! series uses the plot device of a young singer who sells his soul to the Devil in return for success, but changes his mind after the initial euphoria wears off. God and Satan (both played by Burns) eventually face each other over a Las Vegas poker table. The blandness of it all is a serious handicap which not even Burns can overcome. Scripted by Andrew Bergman.
COM   92 min (ort 96 min)   VIDrel: MGM/WHV L/A   V/h

OH HEAVENLY DOG! *   PG
Joe Camp USA   1980
Jane Seymour, Chevy Chase, Omar Sharif, Donnelly Rhodes, Robert Morley, Alan Sues, Stuart Germain, John Stride, Barbara Leigh-Hunt, Frank Williams, Margaret Courtney, Albin Pahernik, Susan Kellerman, Lorenzo Music, Marguerite Corriveau
A private eye is killed and returns to solve his own murder – in the form of a dog. Reverse of "You Never Can Tell" from 1951 where a dog returned as a man. An idea that was used to good effect in THE SHAGGY D.A. with our mutt going through his paces well enough, but there is too little for him to chew on in this limp comedy. Joe Camp went on to make BENJI THE HUNTED.
COM   99 min (ort 103 min)   VIDrel: 20TH/TECH L/A   V/s

OH! MR PORTER ***   U
Marcel Varnel UK   1937
Will Hay, Moore Marriott, Graham Moffatt, Dave O'Toole, Dennis Wyndham, Sebastian Smith, Agnes Laughlan, Percy Walsh, Frederick Piper
An inept Irish railwayman lands a position as stationmaster thanks to the influence of his powerful brother-in-law, and finds

himself in charge of an isolated halt on the border between Northern Ireland and the Irish Free State. He and his two assistants are troubled by supernatural happenings, but the ghosts eventually prove to be a group of gunrunners whom they unmask and arrest. See also THE GHOST TRAIN.
COM   82 min (ort 84 min)   VIDrel: L/A   V
Boa: story The Ghost Train by Frank Launder.

OH! WHAT A LOVELY WAR ***   PG
Richard Attenborough UK   1969
Laurence Olivier, John Gielgud, Kenneth More, Ralph Richardson, Jack Hawkins, Michael Redgrave, John Mills, Dirk Bogarde, Susannah York, Maggie Smith, Vanessa Redgrave, Vincent Ball, Pia Colombo, Paul Daneman
An excellent and imaginative adaptation of a stage show that brings home the futility of WW1, with songs of the period imaginatively woven into both semi-comical dance routines and scenes of carnage, that are equally moving. Attenborough shows a fine flair for the cinematic in his directorial debut, though the excessive length is a handicap. Scripted by Len Deighton. See also PRIVATES ON PARADE for a similar (if inferior) work adopting this approach.
MUS   137 min (ort 144 min)   VIDrel: CIC/SONOP L/A   V
Boa: play The Long, Long Trail by Charles Tilton/musical by Joan Littlewood.

OH, WHAT A NIGHT! ***   18
John Leslie USA   1990
Cheri Taylor, Sean Michaels, Selena Steele, Nina Hartley, Jacqueline, Wayne Summers, Marc Wallice, Eric Price, Jon Dough, Joey Silvera, Randy Spears, A.L. Martin Jr, Jose Duval
A sexual version of "The Petrified Forest" that stars Michaels, Taylor and Steele in the roles originally taken by Leslie Howard, Bette Davis and Humphrey Bogart. The story tells of an unhappy wife whose uncouth husband is the owner of a cafe and filling station situated in the middle of nowhere. A drifter appears on the scene and gets involved with the woman, but matters are complicated by the arrival of some escaped convicts. A well plotted effort.
A   81 min (ort 89 min) mVid   VIDrel: MIA/DISC/IMPENT V

OH, WHAT A NIGHT ***   15
Eric Till CANADA   1992
Corey Haim, Genevieve Bujold, Robbie Coltrane, Barbara Williams, Keid Dullea
After the death of his mother, a seventeen-year-old boy goes with his father and stepmother to live on a chicken farm in Ontario. There, he eventually falls for a married woman with two children, in this touching 1950s coming-of-age tale.
DRA   88 min (ort 93 min)   VIDrel: MARQ/QUANT   V

O.H.M.S. **   U
Raoul Walsh UK   1936
John Mills, Wallace Ford, Anna Lee, Frank Cellier, Grace Bradley, Frederick Leister
An American gangster flees to the UK to escape justice, where he joins the British Army, goes off to fight in China, and redeems himself by his brave death. This film starts off with pace and vigour, but does tend to wind down after our hood reaches England. Some irritating stereotypes (e.g. the British are all jolly fine chaps) give the film a very dated look, and even the action sequences in China are nothing like as good as Walsh's Hollywood films.
Aka: YOU'RE IN THE ARMY NOW
DRA   86 min B/W   VIDrel: CARL/TECH   V

O.J. SIMPSON STORY, THE *   12
Alan Smithee (Jerrold Freedman) USA   1994
Bobby Hosea, Jessica Tuck, David Roberson, James Handy, Kimberly Russell, Bruce Weitz, Harvey Jason, Kimberly Russell, Mariann Alda, Steve Akahoshi, John Cann, Eliana H. Alexander, Jacques Apollo Bolton, Darwyn Carson, Martin Cassidy
A quickie that was rushed out while the record-busting trial of Simpson was still in progress, this dire mTV offering combines a conventional biopic with an attempt to take an objective view of the evidence that led to him being charged with the murder of his wife and her lover. The Alan Smithee pseudonym is a good indication of the director's feelings regarding the completed movie.
DRA   86 min (ort 120 min) mTV   VIDrel: 20TH/TECH   V

OKLAHOMA! *** U
Fred Zinnemann USA 1955
Gordon MacRae, Shirley Jones, Rod Steiger, Charlotte Greenwood, Gloria Grahame, Eddie Albert, Gene Nelson, James Whitmore, Barbara Lawrence, J.C. Flippen, Roy Barcroft, James Mitchell, Bambi Linn, Jennie Workman, Kelly Brown
A much-loved albeit stilted version of the classic Broadway musical, written by Sonya Levien and William Ludwig, with songs by Rodgers and Hammerstein, from their 1943 Broadway hit. The slight story has a cowboy almost lose his girl to a hired hand before eventually winning her back. Songs include "The Surrey With The Fringe On Top" and "Oh, What A Beautiful Morning". AA: Score (Robert Russell Bennett/Jay Blackton/Adolph Deutsch), Sound (Fred Hynes).
MUS 134 min (ort 143 min) wScrn VIDrel: 20TH/TECH L/A V
Boa: musical by Richard Rodgers and Oscar Hammerstein.

OKLAHOMA KID, THE *** PG
Lloyd Bacon USA 1939
James Cagney, Humphrey Bogart, Donald Crisp, Rosemary Lane, Harvey Stephens, Ward Bond, Hugh Sothern, Charles Middleton, Edward Pawley, Lew Harvey, John Miljan, Trevor Bardette, Arthur Aylesworth, Irving Bacon, Joe Devlin
Action-packed Western comedy in which Cagney plays a Robin Hood figure who seeks revenge for his father's murder. Bogart plays the bad guy with his customary aplomb.
WES 77 min (ort 82 min) B/W VIDrel: L/A V

OLD CURIOSITY SHOP, THE ** U
Thomas Bentley UK 1934
Ben Webster, Elaine Benson, Hay Petrie, Beatrix Thompson, Gibb McLaughlin, Reginald Purdell, Polly Ward, James Harcourt, J. Fisher White, Lily Long, Roddy Hughes, Amy Veness, Peter Penrose, Dick Tubb, Wally Patch, Vic Filmer
Fourth adaptation of the Dickens novel (versions appeared in 1912, 1913 and 1921) that follows the changing fortunes of a gambler and his sick grand-daughter who run a curiosity shop, but are persecuted by a heartless dwarf moneylender. When they are unable to repay a debt, the latter has them evicted. A workmanlike and over-sentimental adaptation that is spoilt by slow pacing and sluggish direction. A musical version appeared in 1975.
DRA 92 min (ort 96 min) B/W
VIDrel: BRAVE/SONOP L/A V
Boa: novel by Charles Dickens.

OLD CURIOSITY SHOP, THE ** U
Warwick Gilbert AUSTRALIA 1989
Voices of: Ross Higgins, Wallas Eaton, John Benton, Doreen Harrop, Sophie Horton, Jason Blackwell, Brian Harrison, Jennifer Mellett, Penne Hackforth-Jones
An enjoyable cartoon rendition of the Dickens tale of a young girl and her grandfather who become victims of a heartless and grasping moneylender.
ANIM 75 min VIDrel: TRING V
Boa: novel by Charles Dickens.

OLD CURIOSITY SHOP, THE *** (PG)
Kevin Conner UK 1994
Peter Ustinov, James Fox, Sally Walsh, Tom Courtenay, Julia McKenzie, Adam Blackwood, Christopher Ettridge, William Mannering
The lives of a young girl and her gambler grandfather are ruined when the latter is unable to repay a debt and they are evicted from title establishment by an evil moneylender. A solid and workmanlike adaptation that was filmed on location in Dublin.
DRA 240 min mCab SATrel: DISNEY CHANNEL
Boa: novel by Charles Dickens.

OLD DARK HOUSE, THE **** PG
James Whale USA 1932
Melvyn Douglas, Charles Laughton, Raymond Massey, Boris Karloff, Ernest Thesiger, Eva Moore, Gloria Stuart, Lilian Bond, Brember Wills, John Dudgeon (Elspeth Dudgeon)
A group of travellers seek shelter in title building which contains a household composed of very strange individuals. A classic horror-comedy replete with witty lines and wonderful vignettes.
HOR 69 min (ort 71 min) B/W VIDrel: VISVID/POLYREC V
Boa: novel Benighted by J.B. Priestley.

OLD DARK HOUSE, THE * PG
William Castle UK 1963
Tom Poston, Robert Morley, Janette Scott, Joyce Grenfell, Mervyn Johns, Fenella Fielding, Peter Bull, Danny Green
An American is invited to stay with his English friend, but his arrival marks the start of a series of fatalities, including that of his host in this creepy, creaking horror-comedy. Made by Hammer Films, this is a mutilated remake of a much better 1932 film, that dispenses with most of the comic elements so effective in the earlier work, yet offers us nothing fresh or original to fill the gap.
HOR 83 min (ort 86 min)
VIDrel: ENCORE/SPEAR/COLUM V
Boa: novel Benighted by J.B. Priestley.

OLD GRINGO *** 15
Luis Puenzo USA 1989
Gregory Peck, Jimmy Smits, Jane Fonda, Patricio Contreras, Jenny Gago, Jim Metzler, Gabriela Roel, Anne Pitoniak, Pedro Armendariz Jr, Serge Calderon, Guillermo Rios, Samuel Valadez (De La Torre), Stanley Grover, Pedro Damian
A complex epic adventure that sees an American schoolteacher with a thirst for adventure and the ageing writer Ambrose Bierce, crossing paths with each other and a young Mexican general when they find themselves caught up in the 1913 revolution of Pancho Villa. This ambitious and often spectacular tale was plagued with production difficulties and has many flaws, but remains quite memorable. Scripted by Puenzo and Aida Bortnik.
A/AD 116 min (ort 119 min) subH VIDrel: VCC L/A V/sh
Boa: novel by Carlos Fuentes.

OLD LADY WHO WALKED IN THE SEA, THE *** 18
Laurent Heynemann FRANCE 1991
Jeanne Moreau, Michel Serrault, Luc Thuillier, Geraldine Danon, J. Bochaud
A veteran female confidence trickster taking a vacation by the sea, becomes attracted to a young but rather unrefined adolescent and takes him on as a kind of apprentice. However, her growing attraction to this callow youth threatens her relationship with her longterm partner as well as her future career.
Aka: LA VIEILLE QUI MARCHAIT DANS LA MER; OLD WOMAN WHO WALKED INTO THE SEA, THE
COM 90 min (ort 95 min) wScrn VIDrel: TART/20TH V
Boa: novel by San Antonio.

OLD MAID, THE *** PG
Edmund Goulding USA 1939
Bette Davis, Miriam Hopkins, George Brent, Jane Bryan, Donald Crisp, Louise Fazenda, Jerome Cowan, Henry Stephenson, William Lundigan, Rand Brooks, Rod Cameron, Cecilia Loftus, Rand Brooks, Janet Shaw, DeWolf (William) Hopper,
After her lover is killed in the Civil War, a woman allows her cousin to adopt her baby daughter, an action which sparks of a bitter struggle between the two women for the child's love. An expertly crafted melodrama that is undeniably an old-fashioned tear-jerker, but still eminently enjoyable.
DRA 95 min B/W VIDrel: MGM/WHV V/h
Boa: novel by Edith Wharton/play by Zoe Akins.

OLD YELLER *** U
Robert Stevenson USA 1957
Dorothy McGuire, Fess Parker, Tommy Kirk, Kevin Corcoran, Jeff York, Chuck Connors, Beverly Washburn, Spike (the dog)
A mongrel mutt taken in by a Texas farming family proves his worth in Disney's very first boy-and-his-dog tale. Set in 1869, this has a father leaving his fifteen-year-old son in charge, when he takes part in a three-month cattle drive. The film gave rise to a sequel, SAVAGE SAM, in 1963.
JUV 80 min Cut (16 sec – ort 82 min) VIDrel: WDV/TECH L/A V
Boa: novel by Fred Gipson.

OLIVER! **** U
Carol Reed UK 1968
Ron Moody, Shani Wallis, Oliver Reed, Harry Secombe, Hugh Griffith, Jack Wild, Clive Moss, Mark Lester, Peggy Mount, Leonard Rossiter, Hylda Baker, Joseph O'Conor, Sheila White, Kenneth Cranham, Megs Jenkins, James Hayter
A splendid musical version of this classic tale with fine settings and songs by Lionel Bart, many of which are now classics. Photography is by Oswald Morris and a special prize was

awarded for choreography. Terribly uneven but great fun. AA:
Pic, Dir, Art/Set (John Box and Terence Marsh/Vernon Dixon
and Ken Muggleston), Score/adapt (John Green), Sound
(Shepperton Studio Sound Dept.), Hon Award (Onna White for
choreography).
MUS   140 min (ort 146 min) wScrn
VIDrel: COLUM/SONOP   V/sur
Boa: novel Oliver Twist by Charles Dickens/musical by Lionel
Bart.

OLIVER TWIST ****   U
David Lean UK   1948
*Alec Guinness, Robert Newton, Francis L. Sullivan, John Howard
Davies, Kay Walsh, Anthony Newley, Henry Stephenson, Mary Clare,
Ralph Truman, Josephine Stuart, Kathleen Harrison, Gibb
McLaughlin, Amy Veness, Diana Dors, W.G. Fay*
Lean's version of this classic is the best one ever; a brilliant tale
of a workhouse boy running away and getting drawn into a life
of crime but making good in the end. Several remakes have
followed (including OLIVER! – a 1968 musical), but none
approach it in terms of power. Guinness and Newley make a
fine team as Fagin and the Artful Dodger respectively. Written
by Lean and Stanley Haynes.
DRA   111 min (ort 116 min) B/W   VIDrel: CARL/TECH   V
Boa: novel by Charles Dickens.

OLIVER TWIST **   PG
Clive Donner USA   1982
*George C. Scott, Tim Curry, Michael Hordern, Timothy West, Eileen
Atkins, Cherie Lunghi, Oliver Cotton, Martin Tempest, Matthew
Drew, Eleanor David, Philip Locke, Spencer Rheult, Ann Tirard, Ann
Beach, Brenda Cowling*
Yet another version of this famous story which offers little
except a fine performance from Scott as Fagin. The script is by
James Goldman.
DRA   98 min (ort 100 min) mTV   VIDrel: LUMI/SPEAR
L/A   V
Boa: novel by Charles Dickens.

OLIVER TWIST **   U
Richard Slapczynski AUSTRALIA   1982
*Voices of: Barbara Frawley, Robin Stewart, Wallas Eaton, Derani
Scarr, Ross Higgins, Bill Conn, Sean Hinton, Faye Anderson, Robin
Ramsay*
Animated version of this famous story of a workhouse boy who
falls in with a gang of thieves but eventually finds his rightful
inheritance. Fair.
ANIM   79 min   VIDrel: TRING   V
Boa: novel by Charles Dickens.

OLIVER TWIST **   U
Al Guest/Jean Mathieseon EIRE   1986
*Voices of: Aiden Grennell, Collette Proctor, Joseph Taylor, Daniel
Reardon, Jim Reid*
Another competent but rather uninspiring attempt at a
condensed animated version of this classic tale.
ANIM   48 min (ort 50 min)   VIDrel: 4-FRONT/POLYREC
V
Boa: novel by Charles Dickens.

OLIVIA **   18
Ulli Lommel USA   1983
*Robert Walker, Suzanna Love, Bibbe Hansen, Jeff Winchester,
Nicholas Love, Amy Robinson*
A woman traumatised in her childhood by the killing of her
mother becomes a prostitute, but is driven by the compulsion
to murder her customers, which she does on or near London
Bridge. Made in Arizona and London.
Aka: BEYOND THE BRIDGE; DOUBLE JEOPARDY; FACES OF FEAR;
PROZZIE; TASTE OF SIN, A
HOR   80 min (ort 84 min)   VIDrel: VIPCO/SGSVID   V

OLIVIER, OLIVIER ***   15
Agnieszka Holland FRANCE   1992
*Francois Cluzet, Brigitte Rouan, Gregoire Colin, Maurice Golovine,
Fredric Quiring, Jean-Francois Stevenin, Faye Gatteau, Emmanuel
Morozof, Florian Billion, Carole Lemerle, Jean-Bernard Kosko, Lucrece
La Chenardiere, Luc Etienne*
When a nine-year-old boy disappears from his small-town
home, his family gradually fall apart, with his father running off
to North Africa. Six years later, a detective brings them a
teenager who has lost his memory whom they are all too willing
to accept as their beloved Oliver. A complex and carefully

mounted drama with fine performances. Screenplay is by
Holland.
DRA   104 min (ort 110 min) wScrn   VIDrel: TART/20TH
V/dm

OMEGA DOOM *   15
Albert Pyun USA   1996
Rutger Hauer, Tina Cote, Anna Katerina
Futuristic fantasy set (as ever) on a devastated Earth, where the
majority of the survivors from a catastrophic war are androids.
Enter a mysterious stranger: a synthetic soldier known as
"Omega Doom". More post-apocalyptic nonsense, watchable,
but no more than that.
FAN   81 min   VIDrel: COLUM/SONOP   V/sur

OMEGA MAN, THE **   PG
Boris Sagal USA   1971
*Charlton Heston, Rosalind Cash, Anthony Zerbe, Paul Koslo, Eric
Laneeuville, Lincoln Kilpatrick*
In post-apocalyptic Los Angeles of 1977, a scientist works fever-
ishly to develop a serum to protect against a plague that has
decimated the world's population. Another version of the orig-
inal novel that was filmed as "The Last Man On Earth".
A/AD   94 min (ort 98 min) wScrn   VIDrel: WHV   V/h
Boa: novel I Am Legend by Richard Matheson.

OMEN, THE **   18
Richard Donner USA   1976
*Gregory Peck, Lee Remick, Billie Whitelaw, David Warner, Patrick
Troughton, Leo McKern, Harvey Stephens, Martin Benson, Robert
Rietty, Holly Palance, Anthony Nicholls, Sheila Raynor, Robert
MacLeod, John Stride, Tommy Duggan*
A ridiculously overblown and unbelievable story of the coming
of the anti-Christ that spawned two sequels: DAMIEN: OMEN
2 and OMEN 3: THE FINAL CONFLICT. Peck inherited the
main role from Charlton Heston, who wisely turned it down.
The script is by David Seltzer. AA: Score/orig (Jerry Goldsmith).
HOR   106 min (ort 111 min)
VIDrel: 20TH/TECH; ENCORE/FOXVID (LV only)
V/dm LV
Boa: novel by David Seltzer.

OMEN 3: THE FINAL CONFLICT *   18
Graham Baker USA   1981
*Sam Neill, Rossano Brazzi, Don Gordon, Lisa Harrow, Mason Adams,
Robert Arden, Barnaby Holm, Tommy Duggan, Marc Boyle, Richard
Oldfield, Arwen Holm, Leueen Willoughby, Louis Mahoney, Milos
Kirek, Tony Vogel, Hugh Moxey*
The last part of THE OMEN cycle with a final battle between
Christ and the Antichrist taking place. The latter is now thirty-
two and an American ambassador in London. A tired and
dispirited sequel.
Aka: FINAL CONFLICT, THE; FINAL CONFLICT, THE: OMEN 3
HOR   109 min   VIDrel: 20TH/TECH   V/dm V/sur

OMEN 4: THE AWAKENING *   15
Jorges Montesi/Dominique Othenin-Girard USA   1991
*Faye Grant, Michael Woods, Michael Lerner, Madison Mason, Ann
Hearn, Jim Byrnes, Don S. Davis, Asia Vieira, Megan Leitch, Joy
Coghill, David Cameron, Duncan Fraser, Susan Chapple, Dana Still,
Andrea Mann, Camille Mitchell*
A late addition to the OMEN series, where the anti-Christ who
was killed in the previous film is found to have sired a baby girl,
who's adopted by an American politician and his wife. An over-
familiar parade of violent and fatal accidents marks her early
teens, in a film whose utter lack of fresh ideas and hackneyed
plot are everywhere in evidence.
HOR   93 min (ort 97 min) mTV   VIDrel: 20TH/TECH
V/sh

ON A CLEAR DAY YOU CAN SEE FOREVER ***   U
Vincente Minnelli USA   1970
*Barbra Streisand, Yves Montand, Bob Newhart, Larry Blyden, Jack
Nicholson, Simon Oakland, Roy Kinnear, Irene Handl, Pamela Brown,
John Richardson*
The colourful story of a woman who discovers she has ESP
powers, goes to a psychiatrist, and learns that she had a previ-
ous existence in 19th-century England. Under hypnosis, she gets
the chance to re-live her past life, and the bulk of the film (easily
its best section) is devoted to this. A sumptuous and extremely
handsome film, somewhat flawed by a lack of warmth.
MUS   124 min (ort 129 min)   VIDrel: CIC/SONOP   V/h
Boa: stage musical by Alan Jay Lerner and Burton Lane.

ON DEADLY GROUND **

Steven Seagal USA   15
1993

Steven Seagal, Michael Caine, Joan Chen, R. Lee Ermey, John C. McGinley, Shari Shattuck, Billy Bob Thornton, Richaed Hamilton, Chief Irvin Brink, Elsie Pistohead, Apanguluk Charlie Kairiavak, John Trudell, Mike Starr, Nany Kaguk

A white man living peacefully among the local Inuit inhabitants is forced to resort to violence to defend the area from a rapacious oil company and its evil plans. A by-the-numbers tale, tailor-made for Segal after the success of UNDER SIEGE. Its serious and admirable pro-environmental message is eclipsed all too soon by the excessive violence and special effects.

A/AD   96 min (ort 100 min) cC   VIDrel: WHV   V/sur

ON GOLDEN POND ***

Mark Rydell USA   PG
1981

Henry Fonda, Katharine Hepburn, Jane Fonda, Doug McKeon, Dabney Coleman, William Lanteau, Chris Rydell

Fonda's last film is a fine tribute in his portrayal of an eighty-year-old professor who has to cope with a crisis in his life and marriage. Plenty of sentimentality, a good script and the presence of three stars made this a sure hit and a firm contender for the Oscars. Simplistic it may be, but touching it is too. AA: Actor (Henry Fonda), Actress (Hepburn), Screen/adapt (Ernest Thompson).

DRA   104 min (ort 109 min)
VIDrel: 4-FRONT/POLYREC/ITC   V
Boa: play by Ernest Thompson.

ON HER MAJESTY'S SECRET SERVICE **

Peter Hunt UK   PG
1969

George Lazenby, Diana Rigg, Telly Savalas, Ilse Steppat, Yuri Borienko, Bernard Lee, Lois Maxwell, Gabriele Ferzetti, Bernard Horsfall, George Baker, Desmond Llewelyn, Angela Scoular, Catherina Von Schell, John Gay, Dani Sheridan

Lazenby had a great chance when Connery took a break from these Bond films, but muffed it as the most wooden secret agent one could imagine. Savalas plays a master criminal out, as they usually are, to take over the world and as such, gives a much stronger performance. This film, the sixth in the series, was followed by DIAMONDS ARE FOREVER. In fairness to Lazenby, he might well have been a lot better if his lines had been as good as those of Savalas.

A/AD   136 min (ort 142 min) wScrn cC
VIDrel: MGM/WHV   V/h V/dm
Boa: novel by Ian Fleming.

ON MOONLIGHT BAY ***

Roy Del Ruth USA   U
1951

Doris Day, Gordon MacRae, Jack Smith, Leon Ames, Rosemary DeCamp, Mary Wickes, Ellen Corby, Billy Gray, Jeffrey Stevens, Esther Dale, Suzanne Whitney, Eddie Marr, Sig Arno, Jimmy Dobson, Rolland Morris, Lois Astin

A tomboyish teenage girl in a small Indiana town gets involved with a local college student. Her bank employee father disapproves but gets worse problems when her younger brother claims that he is a drunk in order to deal with trouble at school. A so-so musical comedy with some strong performances. Songs include "Till We Meet Again" and "I'm Forever Blowing Bubbles". The success of the film led to BY THE LIGHT OF THE SILVERY MOON (with the same cast) two years later.

MUS   91 min (ort 95 min)   VIDrel: WHV   V/dm
Boa: novel Alice Adams and "Penrod" stories by Booth Tarkington.

ON SEVENTH AVENUE **

Jeff Bleckner USA   12
1995

Wendy Makkena, Stephen Collins, Faye Grant

Crime drama set in the world of fashion, with a man's daughter taking over his ailing designer-label company when the latter is beaten up by the Mob for failing to pay protection money. All the usual intrigues, jealousies and double-crosses are on offer here, in a film that unsurprisingly went straight to video.

DRA   85 min   VIDrel: 20TH/FOXVID   V/sh

ON THE BLACK HILL ****

Andrew Grieve UK   15
1987

Mike Gwilym, Robert Gwilym, Bob Peck, Gemma Jones, Nesta Harris, Benjamin Whitrow, Jack Walters, Nesta Harris, Huw Toghill, Gareth Toghill

Chatwin's fine novel chronicling the lives of a Welsh farming community over four generations is brought to life with great skill and enormous care. The story opens with tough farmer

Amos Jones getting married and working himself into an early grave on his farm, to be replaced by his twin sons who toil away in his stead. A passionate and poetic film, languid and mournful, it brings depth to a story of ordinary folk living in difficult circumstances.

DRA   110 min (ort 117 min)   VIDrel: CONNO/RTM   V
Boa: novel by Bruce Chatwin.

ON THE BLOCK **

Steve Yeager USA   18
1990

Marilyn Jones, Jerry Whiddon, Michael Gabel, Howard Rollins, Blaze Starr

A Baltimore stripper has to deal with both a crooked realtor, out to tear down the club where she performs, and a murderous lieutenant in the local vice squad. Of little interest, except for the presence of one-time stripper Starr.

THR   99 min   VIDrel: CAPIT/GUILD   V

ON THE BUSES *

Harry Booth UK   PG
1971

Reg Varney, Doris Hare, Anna Karen, Michael Robbins, Stephen Lewis, Bob Grant, Andrew Lawrence, Pat Ashton, Brian Oulton, Pamela Cundell, Pat Coombs, Wendy Richards, Peter Madden, David Lodge, Maggie McGrath

First spin-off from a popular TV comedy series about the lives and loves of the crews of a Northern bus depot. In this story female bus drivers cause problems. Hardly a celebration of life, more like a dreary and depressing night out in the rain. MUTINY ON THE BUSES and HOLIDAY ON THE BUSES soon followed and were both equally forgettable.

COM   84 min (ort 88 min)   VIDrel: WHV   V/h

ON THE TOWN ****

Gene Kelly/Stanley Donen USA   U
1949

Gene Kelly, Frank Sinatra, Jules Munshin, Vera-Ellen, Betty Garrett, Ann Miller, Tom Dugan, Florence Bates, Alice Pearce, George Meader, Bea Benaderet, Lester Dorr, Bern Hoffman, Walter Baldwin, Don Brodie, Sid Melton, Murray Alper

A marvellously enjoyable and vibrant musical built around the adventures of three sailors on a day's leave in New York. Gene Kelly choreographed the dance routines and Leonard Bernstein wrote the ballet "Fancy Free". Shot on location in New York City with "New York, New York" and the other songs by Roger Edens and Lennie Hayton. This memorable offering was Kelly's directorial debut. AA: Score (Roger Edens/Lennie Hayton).

MUS   94 min (ort 98 min)   VIDrel: MGM/WHV   V/dm

ON THE WATERFRONT ***

Elia Kazan USA   PG
1954

Marlon Brando, Lee J. Cobb, Eve Marie Saint, Karl Malden, Pat Henning, Leif Erickson, Rod Steiger, James Westerfield, John Hamilton, Fred Gwynne, Tony Galento, Tami Mauriello, John Hamilton, John Heldabrand, Rudy Bond, Don Blackman

A dramatic account of dockland corruption that's not without its sentimental moments, but has powerful performances and excellent direction. The music by Leonard Bernstein and the script by Budd Schulberg are two of its memorable features. Interestingly, the novel's ending was altered to allow Brando to survive. AA: Pic, Dir, Actor (Brando), S. Actress (Saint), Art/Set (Richard Day), Cin (Boris Kaufman), Edit (Gene Milford), Story/Screen (Schulberg).

DRA   103 min (ort 108 min) B/W   VIDrel: COLUM/SONOP V/dm
Boa: novel by Budd Schulberg.

ON WINGS OF EAGLES: PARTS 1 AND 2 **

Andrew V. McLaglen USA   PG
1986

Burt Lancaster, Richard Crenna, Paul Le Mat, Louis Giambalvo, Jim Metzler, Lawrence Pressman, Robert Wightman, Cyril O'Reilly, Esai Morales, Martin Doyle, Bob Delegall, Bill Bumiller, James Sutorius, Richard Crenna Jr

A millionaire mounts a private military expedition, to rescue two of his executives from prison in Iran after the downfall of the Shah. Apart from its well-executed action scenes, a typically undistinguished TV product. Originally shown in two parts.

A/AD   235 min (2 cassettes – ort 250 min) mTV
VIDrel: BRAVE/SONOP L/A   V
Boa: novel by Ken Follet.

ONCE A JOLLY SWAGMAN **

Jack Lee UK   U
1948

Dirk Bogarde, Bonar Colleano, Renee Asherson, Bill Own, Cyril Cussack, Thora Hird, James Hayter, Pauline Jameson, Stuart Lindsell,

Moira Lister, Sandra Dorne, Sidney James, Anthony Oliver, Dudley Jones, Russell Waters
A factory worker leaves his job to try his look as a speedway rider and achieves quite a measure of success. This good fortune soon goes to his head, resulting in a severely inflated ego that leads to his wife leaving him after he refuses to give up racing. However, he eventually does as she wishes and they are reconciled. A dull drama with uninspired performances but some exciting racetrack sequences.
Aka: MANIACS ON WHEELS
DRA   100 min B/W   VIDrel: CARL/TECH   V
Boa: novel by Montague Slater.

ONCE A THIEF ***   18
John Woo HONG KONG   1991
Chow Yun Fat, Cherie Cheung Cho-Hung, Leslie Cheung Kwok Wing, Kenneth Tsang Kong, David Wu, Pierre Yves Burton, Declan, Michael Wong, Patrick Hon, Tang Yat Kwun, Tong Kai Fat, Lina Kong
Three art thieves enjoy a good life on the French Riviera but things go wrong when they plan one last heist before two of them "retire" and get married to each other. A couple of double crosses and much action ensue before the final explosive climax. A less violent and more romantic effort from Woo, who uses a variety of styles that do not always blend well, though the insertion of several comic interludes is welcome. Loosely remade as the TV pilot VIOLENT TRADITION.
Aka: ZONHENG SIHAI
A/AD   104 min   VIDrel: MADE/RTM   V

ONCE AROUND ***   15
Lasse Hallstrom USA   1990
Richard Dreyfuss, Holly Hunter, Danny Aiello, Laura San Giacomo, Gena Rowlands, Roxanne Hart, Danton Stone, Tim Guinee, Greg Germann, Griffin Dunne, Cullen O. Johnson, John Gay, Lou Criscuolo, Myra Taylor, Michael Steve-Jones
In Boston, the eldest daughter of a strait-laced Italian family gets married to a horribly pushy Lithuanian salesman, whose less genteel ways create a rift between the girl and her family. This leads to an emotional conflict of loyalty and though uneven, the film blends comedy, romance and drama in a way that's both touching and incisive. The American film debut for Hallstrom.
COM   110 min (ort 115 min)   VIDrel: CIC/SONOP   V/sur

ONCE IN A LIFETIME **   PG
Michael Miller USA   1993
Lindsay Wagner, Barry Bostwick, Amy Aquino, Duncan Reghehr, Rex Smith, Fran Bennett, Debra Sullivan, Michael Mitz, Michael Laskin, Jessica Sinejal, Dawn Jeffory-Nelson, Nicholas Petrie, Spencer Garbett, Mark Davenport, Orlee Kenner
Wagner plays a bestselling authoress who loses her husband and daughter in an accident, and is left to bring up her deaf son alone, but eventually finds happiness in the arms of a dedicated speech therapist. Much of the film takes the form of an extended flashback, when she is hit by a car and reviews her life. Fair.
Aka: DANIELLE STEEL'S ONCE IN A LIFETIME
DRA   87 min (ort 90 min) mTV
VIDrel: MIA/DISC/IMPENT   V
Boa: novel by Danielle Steel.

ONCE UPON A CRIME *   PG
Eugene Levy USA   1991
John Candy, James Belushi, Sean Young, Cybill Shepherd, Richard Lewis, Ornella Muti, Giancarlo Giannini, George Hamilton, Roberto Sbaratto, Joss Ackland, Ann Way, Geoffrey Andrews, Caterina Boratto, Elsa Martinelli
Levy's directorial debut is a slackly plotted and peculiarly old-fashioned farce that tells of five daft Americans who are all implicated in the murder of an old dowager in Monte Carlo. This fatally flawed attempt to create a modern-day version of those classic Hollywood screwball comedies of old is by turns manic, stilted and ponderous, with ill-conceived dialogue and little apparent commitment from the cast.
COM   91 min (ort 94 min)   VIDrel: EIV/SONOP   V/sur
Boa: short story and screenplay by Rodolfo Sonego.

ONCE UPON A FOREST *   U
Charles H. Grosvenor et al. UK   1992
Voices of: Michael Crawford, Ben Vereen, Ellen Blain, Ben Gregory, Paige Gosney, Elizabeth Moss, Paul Eiding, Janet Waldo, Susan Silo, Will Nipper, Ricky Collins, Charlie Adler, Angle Harper, Don Reed, Robert David Hall, Haven Hartman
A preachy, poorly animated anti-pollution tale, with three

forest creatures mounting a desperate search for a cure for a seriously ill badger whose condition is the resulting of the damping of toxic waste. Their efforts bring them into conflict with human society and its mechanised civilisation. A total of five directors worked on this dud. (For this interested, the other four are: Roy Patterson, Kelly Ward, Mark Young and Dave Michener.)
ANIM   67 min (ort 71 min) cC   VIDrel: 20TH/TECH   V/sur
Boa: short story by Rae Lambert.

ONCE UPON A TIME IN AMERICA ****   18
Sergio Leone USA   1984
Robert De Niro, James Woods, Elizabeth McGovern, Tuesday Weld, Larry Rapp, William Forsythe, Treat Williams, Burt Young, Danny Aiello, Dutch Miller, Robert Harper, Joe Pesci, James Hayden, Darlanne Fleugel, Richard Bright
A compelling and powerful film tracing the career of four Jewish gangsters, from their beginnings as New York street punks to a final reckoning when one of them is apparently being sought for betraying the others. Hard to follow as it jumps back and forth in time, but it rewards close attention. Music is by Ennio Morricone. Originally the BBFC cut the film by 10 sec and though now available uncut, the new tape is inexplicably missing an important flashback sequence.
DRA   218 min (ort 228 min) wScrn (re-mastered special edition)   VIDrel: WHV   V/dm V/sh
Boa: novel by L. Hays.

ONCE UPON A TIME IN THE WEST ****   15
Sergio Leone ITALY/USA   1969
Charles Bronson, Henry Fonda, Claudia Cardinale, Jason Robards, Jack Elam, Woody Strode, Frank Wolff, Gabriele Ferzetti, Keenan Wynn, Paolo Stoppa, Marco Zuanelli, Lionel Stander, John Frederick, Dino Mele
Long, violent and highly atmospheric Western with a woman landowner being the target for a hired gun. An inspired piece of casting gave Fonda the chance to show his worth as one of the meanest, most sinister villains ever to hit the screen. The music is by Ennio Morricone and the film has one of the longest credits roll ever, twelve minutes in all. Scripted by Leone, Bernardo Bertolucci and Dario Argento.
Aka: C'ERA UNA VOLTA IL WEST
WES   158 min (ort 165 min) wScrn (CIC/SONOP)
VIDrel: CIC/SONOP L/A; 4-FRONT/POLYREC/CIC   V/h

ONCE WERE WARRIORS ***   18
Lee Matahori NEW ZEALAND   1994
Rena Owen, Temuera Morrison, Mamaengaroa Kerr-Bell, Julian Arahanga, Rachael Morris Jr, Taungaroa Emile, Joseph Kairau, Clifford Curtis, Pete Smith, George Henare, Mere Boynton, Shannon Williams, Calvin Tuteao, Ray Bishop, Ian Mune
An Auckland Maori family is brutalised by the father who, despite his love for his family, is unable to control his violent temper and find a less destructive outlet for his frustrations. A searing portrait of urban despair with a first-rate performances and a script that is both bleak and unrelenting. An impressive directorial debut for Matahori with outstanding cinematography of Stuart Dryburgh (who also worked on THE PIANO) an additional asset.
DRA   98 min (ort 103 min)   VIDrel: EIV/SONOP   V
Boa: novel by Alan Duff.

ONE AGAINST THE WIND ***   PG
Larry Elikann USA   1991
Sam Neill, Judy Davis, Kate Beckinsale, Denholm Elliott, Anthony Higgins, Christian Anholt, Frank Middlemass, David Ryall, Peter Cellier, Mark Wing-Davey, Stevan Rimkus, Tom Hodgkins, Michael Crossman, John Savident, Terry Taplin
This fact-based WW2 drama effectively recounts the heroic exploits of Mary Lindell, who was responsible for devising the many escape routes by which Allied servicemen got out of occupied France. Captured by the Nazis, she was sent to a concentration camp but survived and was twice awarded the Croix De Guerre by France. A workmanlike and unadorned tribute to an exceptional woman who died in Paris in 1987 aged ninety-two.
DRA   95 min mTV   VIDrel: ODY/SONOP   V/sh

ONE-ARMED BOXER **   18
Jimmy Wang Yu HONG KONG   1972
Jimmy Wang Yu, Tang Shin, Tien Yeh, Lung Fei
A student loses an arm in a fight, and learns a style of one-armed

combat in order to exact his revenge. Followed by a sequel and a series of "One-Armed Boxer" films.
Aka: DOP BEY KUAN WAN
MAR 88 min (ort 90 min)   VIDrel: MIA/DISC V

ONE DEADLY SUMMER *** 18
Jean Becker FRANCE 1983
Isabelle Adjani, Alain Souchon, Francois Cluzet, Manuel Gelin, Jenny Cleve, Suzanne Flon, Michel Galabru, Maria Machado
A sexy young woman comes to a village in France to exact a secret revenge, on the men who raped and assaulted her mother. Written by Sebastian Jarrisot from his novel, with the story seen (as in RASHOMON) through the eyes of several characters. Overlong but totally absorbing, and with Adjani giving one of her best performances.
Aka: DEADLY SUMMER, THE; L'ETE MEURTRIER
THR 127 min Cut (5 sec – ort 133 min)
VIDrel: ARTPRO/RTM V
Boa: novel by Sebastian Jarrisot.

ONE DOWN, TWO TO GO ** 18
Fred Williamson USA 1982
Fred Williamson, Jim Kelly, Jim Brown, Richard Roundtree, Paula Sillsa, Tom Signorelli, Laura Loftus, Joe Spinell, Louis Neglia, Peter Dane, John Guitz, Richard Noyce, Victoria Hale, Warrington Winters, Arthur Haggerty, Irwin Litvack
A professional karate fight promoter takes on the Mob who are rigging karate bouts, but the story merely serves to provide an excuse for the action. A typical Williamson product, it was a belated sequel to the earlier (and equally poor) "Three The Hard Way" (1974).
MAR 84 min (ort 87 min)   VIDrel: SUPVID/RTM V/sur

ONE-EYED JACKS *** PG
Marlon Brando USA 1961
Marlon Brando, Karl Malden, Pina Pellicer, Katy Jurado, Ben Johnson, Slim Pickens, Larry Duncan, Sam Gilman, Timothy Carey, Miriam Colon, Elisha Cook Jr, Rodolfo Acosta, Ray Teal, John Dierkes, Hank Worden, Nina Martinez
A convict released from prison seeks revenge on his former partner who betrayed him and has now become a sheriff. A slow-paced Western, replete with interesting psychological detail, and a strong visual impact, but seriously hampered by self-indulgent and sombre direction; a lighter touch was sorely needed.
WES 135 min (ort 141 min)   VIDrel: CIC/SONOP V
Boa: novel The Authentic Death of Hendry Jones by C. Neider.

ONE FALSE MOVE *** 18
Carl Franklyn USA 1992
Bill Paxton, Cynda Williams, Bob Thornton, Michael Beach, Jim Metzler, Earl Billings, Natalie Canerday, Robert Ginnaven, Kevin Hunter, Robert Anthony Bell, Phyllis Kirklin, Meredith "Jeta" Donovan, James D. Bridges, June Jones
Three drug dealers stage a rip-off in which they commit a number of brutal murders and then flee to Arkansas where they hide out. However, the L.A. cops soon pick up their trail and enlist the help of the local sheriff in mounting a stakeout. A well-acted and highly watchable effort, but the graphic violence is both exploitative and unnecessary and the opening is especially so.
DRA 101 min (ort 106 min)   VIDrel: COLUM/SONOP
V/sur

ONE FINE DAY ** PG
Michael Hoffman USA 1996
Michelle Pfeiffer, George Clooney, Mae Whitman, Alex D. Linz, Charles Durning, Jon Robin Baitz, Ellen Greene, Joe Grifasi, Pete Hamill, Anna Maria Horsford, Gregory Jbara
In the course of an event-filled twelve-hour period, a single mother architect (Pfeiffer) and a divorced journalist (Clooney) meet, find they do not get on at all, but ultimately find the sharing of common tasks brings them together. Relying too heavily on slapstick for its laughs, this romantic comedy could have been a whole lot better, as they verbal sparring matches between Pfeiffer and Clooney clearly show.
COM 109 min CINrel

ONE FLEW OVER THE CUCKOO'S NEST **** 18
Milos Forman USA 1975
Jack Nicholson, Louise Fletcher, William Redfield, Brad Dourif, Michael Berryman, Danny DeVito, Scatman Crothers, Will Sampson, Peter Brocco, Dean R. Brooks, Alonzo Brown, Christopher Lloyd, Sydney Lassick, Delos V. Smith

A brilliant film version of a powerful book describing the claustrophobic world of an insane asylum, its inmates and staff. The first film since IT HAPPENED ONE NIGHT to win all five top Oscars and without doubt it deserved them. AA: Pic, Dir, Actor (Nicholson), S. Actress (Fletcher), Screen/adapt (Bo Goldman/Lawrence Hauben).
DRA 128 min (ort 134 min) wScrn
VIDrel: 4-FRONT/POLYREC/BRAVE L/A V
Boa: novel by Ken Kesey.

ONE FULL MOON *** (15)
Endaf Emlyn UK 1991
Dyfan Roberts, Tudor Roberts, Betsam Llwyd, Delyth Einir, Cian Ciaran, Robin Griffith, Dilwyn Vaughn Thomas, Stewart Jones, Michael Povey, Elliw Haf, Wyn Bowen Harries, Sian Wheldon, Grey Evans, Endaf Emlyn, Llyr Dyfan, Dafydd Clwyd
A striking film that looks back over the childhood of the central character, who grew up in a Welsh slate-mining town in the 1920s, and now as a grown man in the 1950s, has decided to pay his home-town a return visit. But his childhood was an unhappy one, as he grew up without a father and his mother found it hard to support the family. An impressive work, that does not sentimentalise the characters or events they suffer, but is not without humour too.
Aka: UN NOS OLA LEUAD
DRA 98 min (Welsh dialogue) CINrel
Boa: novel by Caradog Prichard.

ONE GOOD TURN ** 18
Tony Randel USA 1995
Lenny Von Dohlen, Suzy Amis, James Remar, John Savage, Richard Michenberg
Four years after a businessman had his life saved by a stranger he bumps into him once more, and sees that he is now living as a down-and-out. Resolved to do something to help this fellow he takes him home, only to find that his actions have made him vulnerable to the latter's cunning schemes. A serviceable psychological thriller, helped by effective if overly derivative plotting.
DRA 85 min (ort 92 min)   VIDrel: MED/COLUM V

ONE HEAVENLY NIGHT ** U
George Fitzmaurice USA 1930
Evelyn Laye, John Boles, Leon Errol, Lilyan Tashman, Hugh Cameron, Lionel Belmore, Marion Lord, George Bikel, Vincent Barnett, Henry Victor, Henry Kolker, Luis Alberni
Story of a flower girl at a Budapest music-hall, who tries to seduce a man above her station, and does this by pretending to be a singer in the show, the real star having absented herself at this time. An attempt at a musical comedy in the style of Ernst Lubitsch that doesn't quite come off, remaining a minor work of slight charm. Gregg Toland's fine photography is one of its best aspects.
MUS 80 min (ort 82 min) B/W
VIDrel: VCC/DISC/COLUM V

ONE HUNDRED AND ONE DALMATIANS **** U
Wolfgang Reitherman/Hamilton S. Luske/Clyde Geromini
USA 1960
Voices of: Rod Taylor, Lisa Davis, Cate Bauer, Ben Wright, Fred Warlock, Tom Conway, J. Pat O'Malley, Betty Lou Gerson, Martha Wentworth, Barbara Beaird, Micky Maga, Queenie Leonard, Marjorie Bennett, Tudor Owen, Mimi Gibson
A pair of Dalmatians produce a litter of no less than 101 puppies that the owners refuse to sell to an insistent woman, Cruella De Vil, who secretly plans to make a coat from their skins. She has them stolen and when human efforts to locate them fail, all the dogs of London work against the clock to mount a rescue operation. One of Disney's best animated features, this outstanding work took three years and the work of three-hundred artists to produce it.
ANIM 79 min VIDrel: ENCORE LV
Boa: novel The Hundred and One Dalmatians by Dodie Smith.

ONE HUNDRED AND ONE DALMATIANS *** U
Stephen Herek USA 1996
Glenn Close, Jeff Daniels, Joely Richardson, Joan Plowright, Hugh Laurie, Mark Williams, John Shrapnel, Tim McInnerny, Hugh Fraser, Zohren Weiss, Mark Haddigan, Michael Percival, Neville Phillips, John Evans, John Benfield
Cruella De Vil (richly played by Close) is the fur-loving, ruthless boss of a fashion empire, who upon hearing of the title pups, sets her heart on having a fur coat made out of their skins. This

live-action version of the Walt Disney classic has a lot going for it, though it must be said that without the animal characters being able to speak (as in BABE for instance) the film is forced to rely a little too heavily on its breakneck action sequences.
JUV   103 min   CINrel
  Boa: novel The Hundred and One Dalmatians by Dodie Smith.

ONE HUNDRED DAYS BEFORE THE COMMAND ****
(18)
Khusein Erkenov RUSSIA   1990
Vladimir Zamanski, Armen Dzhigarkhanian, Oleg Vasilkov, Roman Grekov, Valerii Troshin, Aleksandr Chislov, Mikhail Solomatin, Sergei Rozhentsev, Sergei Bystritskii, Sergei Semenov, Elena Kondulainen, Oleg Khusainov
Having got the co-operation of the authorities with a fake script, Erkenov's film is an outright attack on the brutalising and degrading conditions suffered by the ordinary Russian soldier. A harrowing work that makes no concessions to the sensibilities of the viewer, it has no conventional narrative, relying instead on a sequence of powerful images. The film was not seen in the West until 1994, when a print was submitted to the Berlin Film Festival.
Aka: STO DNEI DO PRIKAZA
DRA   68 min   CINrel
  Boa: short story by Iurii Poliakov.

ONE MAN'S WAR **
(18)
Sergio Toledo UK/USA   1990
Anthony Hopkins, Norma Aleandro, Fernanda Torres, Ruben Blades, Leonardo Garcia, Rene Pereyra, Miah Michlle, Ana Ofelia Murguia, Sergio Bustamante, Jose Antonio Estrada, Fernesio De Bernal, Claudio Brook, Guillermo Rios
A Paraguayan doctor continues to fight for human rights even after government-paid thugs murder his son, struggling to bring these killers to justice. An earnest but rather flat dramatisation of the real-life efforts of Joel Filartiga and his family. Filmed on location in Mexico.
DRA   87 min mTV   TVrel: C4

ONE MILLION YEARS B.C. **
PG
Don Chaffey UK/USA   1966
Raquel Welch, John Richardson, Percy Herbert, Robert Brown, Martine Beswick, Jean Waldon, Lisa Thomas, Malaya Nappi, William Lyon Brown, Malya Nappi, Richard James, Frank Hayden, Terence Maidment, Mickey De Rauch, Yvonne Horner
Remake of a 1940 film with prehistoric tribes clashing in a landscape filled with fantastic monsters. A little laughable in parts with Richardson fleeing from his tribe and meeting up with a bikini-clad Welch before returning. Short on dialogue – grunts suffice. Ray Harryhausen handles the monsters and the unusual percussion score is by Mario Nascimbene.
A/AD   96 min (ort 100 min)   VIDrel: WHV   V/h

ONE OF HER OWN **
15
Armand Mastroianni USA   1994
Lori Loughlin, Martin Sheen, Greg Evigan, Valerie Landsberg, Jeff Yagher, Jason Schombing
True story of a rookie female police officer who is raped by a colleague and has to fight a battle for justice when he colleagues close ranks to protect the miscreant. A competent TV drama.
DRA   89 min mTV   VIDrel: ODY/SONOP   V/sh

ONE POLICE PLAZA **
15
Jerry Jameson USA   1986
Robert Conrad, George Dzundza, Jamey Sheridan, Larry Riley, James Olsen, Anthony Zerbe, Lisa Banes, Joe Grifasi, Earl Hindman, Janet-Laine Green, Peter McNeil, Barton Heyman, Nicholas Hormann, David Cryer, Alar Aedma
A simple case of murder proves to be linked to a major scandal involving the top echelons of the police force. A police lieutenant is determined to investigate, even at the cost of defying the orders and authority of his superiors. Sound but hardly memorable, and filmed mainly in Montreal despite the New York location. Followed by THE RED SPIDER.
THR   90 min (ort 100 min) Cut (10 sec) mTV
VIDrel: 20TH/TECH   V
  Boa: novel by William J. Caunitz.

ONE STEP BEYOND: THE SORCERER **
15
John Newland UK   1961
Christopher Lee, Martin Benson, Gabrielli Licudi, Alfred Burke, George Pravda, Peter Swanick, Frederick Jaeger, Edwin Richfield, Richard Shaw

The supernatural abilities of a farmer are used to enable a German officer in Britain to visit the flat of his fiancee in Berlin, as he has grown suspicious about her faithfulness. One in a series of 1960s supernatural dramas that attempted to do for British TV what "The Twilight Zone" did for American TV, but which met with only a limited degree of success. Several other stories in the series have also become available.
FAN   30 min; 80 min (3-episode cassette) B/W mTV
VIDrel: RETRO/20TH   V

ONE THAT GOT AWAY, THE ***
15
Paul Greengrass UK   1994
Paul McGann, David Morrissey, Nick Brimble, Sam Halpenny, Sandy Morton, Steve John Shepherd, SImon Mark, Davit Clatworthy, Jack Ellis, Andre Jacobs, Steve Waddington, Frank Opperman, Ron Semrczak, Stefan Kalipha, Jody Abrahams
During the Gulf War, an eight-man SAS team was sent behind enemy lines to help eliminate the Scud missiles that were being fired at Israel, since the USA feared that the coalition against Saddam would break up if Israel took military action. Four of the team were killed and the others captured but one of them, Corporal Chris Ryan managed to escape. This absorbing film follows his story.
WAR   104 min   VIDrel: TELVID/BMGREC   V

ONE THOUSAND AND ONE NIGHTS, THE *
15
Philippe De Broca   1990
Catherine Zeta Jones, Vittorio Gassman, Thierry Lhermitte
A wildly overblown adaptation of this classic tale that bears barely a trace of the original work (magic lamps and motorbikes are one of several uneasy combinations). Though having one or two fairly entertaining moments, this is a little bit too extravagant for its own good.
Aka: 1001 NIGHTS, THE
A/AD   94 min   VIDrel: 20VIS/SONOP   V/sh

ONE TOUCH OF VENUS ***
(U)
William A. Seiter USA   1948
Robert Walker, Ava Gardner, Dick Haymes, Eve Arden, Olga San Juan, Tom Conway, James Flavin, Sara Allgood, Hugh Herbert, Arthur O'Connell, Kenneth Patterson, Anne Nagel, Mary Benit, Russ Conway, Joan Miller, Jerry Marlowe
A young man working as a store window-dresser falls in love with a statue of Venus which comes to life. Music is by Kurt Weill in an amusing romantic comedy that still retains a considerable charm. The films MANNEQUIN and "Goddess Of Love" used much the same idea.
COM   82 min B/W   VIDrel: L/A   V
  Boa: play by S.J. Perelman and Ogden Nash.

ONE TOUGH BASTARD **
18
Kurt Wimmer USA   1995
Brian Bosworth, Bruce Payne, Jeff Korber, Dejuan Guy, Hammer, Neal McDonough, Robert Kotecki, Deborah Worthing, Rachel Duncan, Angela Brooks, Cyrus Farmer, M.C. Gainey, Robert L. Sardo, Leo Lee, Christopher Brown, K.B. Bowens
A tough drill sergeant survives an attack on his family that killed his wife and daughter, and when he has recovered from his injuries he sets out to track down those responsible, but receives a setback when he learns that the culprit is on the Federal Witness Protection program. Undaunted he continues, aided by a ten-year-old street kid. Bosworth takes out the bad guys with efficiency, but characterisation is virtually non-existent and the plot lacks originality.
A/AD   100 min   VIDrel: GUILD/FOXVID   V/sur

ONE, TWO, THREE ****
U
Billy Wilder USA   1961
James Cagney, Arlene Francis, Horst Buchholz, Pamela Tiffin, Lilo Pulver, Howard St John, Hans Lothar, Leon Askin, Red Buttons, Lois Bolton, Peter Capell, Ralf Wolter, Karl Lieffen, Henning Schluter, John Allen
This deliciously wicked Cold War satire never stops for breath, as it tells the story of a Coca-Cola executive posted to West Berlin, who is more than a little outraged at his daughter's involvement with a Communist. The gags come thick and fast and Cagney's performance is a utter joy to watch. This was his last screen role until RAGTIME in 1981. Written by Wilder and I.A.L. Diamond, and very, very loosely based on Molnar's one-act play.
COM   115 min B/W   VIDrel: MGM/WHV L/A   V
  Boa: play by Ferenc Molnar.

ONE WEDDING AND LOTS OF FUNERALS ** 18
Rodman Flender USA   1994
Warwick Davis, Shevonne Durkin
Supernatural black comedy with something of a BEETLEJUICE
flavour, in which a disagreeable leprechaun from Hell, who after
a thousand years has decided it's time to wed, the only problem
being that his prospective bride is less than willing. The funerals are for those foolhardy enough to try and stop him.
COM   81 min   VIDrel: MED/DISC   V/sh

ONE WOMAN'S COURAGE ** 15
Charles Robert Carner USA   1993
*Patty Duke, James Farentino, Margot Kidder, Dennis Farina, Keith
Szarabajka, Geoffrey Blake, Debra Sharkey, Titus Welliver, Jo
Anderson, Hattie Winston, Mary Donnelly-Haskell, Sarah Freeman,
Courtland Mead, Casey Biggs, Marjean Holden*
Having witnessed a murder, a grandmother gives evidence in
court, but by doing this she puts her life in danger, adding to
the burden she is already under with her family problems.
Thinly plotted and formulaic, it is sustained by Duke's strong
and convincing acting in the central role.
DRA   86 min (ort 90 min) mTV   VIDrel: ODY/SONOP
V/sh

ONEDIN LINE, THE – SERIES 1 AND 2 *** PG
G. Blake/D. Cunliffe/C. Coke/P. Scott/B. Rae/
R. Jenkins UK   1971
Peter Gilmore, Anne Stallybrass, Jane Seymour
A ship's captain dreams of starting up in business with his own
shipping line, and after many difficulties achieves his ambition.
Set at the time of the American Civil War, this handsome and
well mounted series charted the changing fortunes of James
Onedin, from his small beginnings through to his success
running a thriving company. Written by Cyril Abraham and
Alun Richards, its memorable theme music was taken from
Khatachurian's Spartacus Suite.
DRA   1,212 min (8 cassettes) mTV   VIDrel: BBC/TECH
V/h

ONIBABA *** 15
Kaneto Shindo JAPAN   1964
*Nobuko Otowa, Jitsuko Yoshimura, Kei Sato, Jukichi Uno, Taiji
Tonomura*
In medieval Japan a woman and her daughter-in-law live by
killing samurai and selling their armour. The girl
eventually takes one as a lover and her mother-in-law becomes
jealous. This slow and brooding folk-tale has moments of eerie
horror and good locations, but is hampered by excessive senti-
mentality. The script is by Shindo.
Aka: DEMON, THE; DEVIL WOMAN, THE; HOLE, THE
DRA   98 min (ort 105 min) B/W wScrn
VIDrel: TART/20TH   V

ONION FIELD, THE *** 18
Harold Becker USA   1979
*James Woods, Franklyn Seales, John Savage, Ted Danson, Ronny Cox,
Dianne Hull, David Huffman, Christopher Lloyd, Priscilla Pointer,
Beege Barkett, Richard Herd, Le Tari, Richard Venture, Lee Weaver,
Phillip R. Allen*
Factually based account of two cops held hostage by two thugs.
One is murdered but his killers escape justice, through legal
technicalities and clever manipulation of the judicial system.
The surviving policeman has to face accusations of cowardice
that result in a breakdown and his resignation from the force.
A film with a fine script by Joseph Wambaugh that asks some
uncomfortable questions about the nature of justice.
DRA   121 min (ort 126 min)   VIDrel: BMGVID   V
Boa: novel by Joseph Wambaugh.

ONLY ANGELS HAVE WINGS ** U
Howard Hawks USA   1939
*Cary Grant, Rita Hayworth, Richard Barthelmess, Jean Arthur, Allyn
Joslyn, Thomas Mitchell, Sig Ruman, Victor Kilian, John Carroll,
Donald (Don "Red") Barry, Noah Beery Jr, Melissa Sierra, Lucio
Villegas, Forbes Murray*
The tough male world of a struggling air freight line in Latin
America is invaded by a stranded showgirl who makes a play
for the outfit's boss. A rather studio-bound film that is too high
on talk and low on action to justify its excessive length.
DRA   116 min (ort 121 min) B/W   VIDrel: COLUM/SONOP
V

ONLY THE BRAVE *** 15
Ana Kokkinos AUSTRALIA   1994
Elena Mandalis, Dora Kaskanis, Maude Davey, Bob Bright
Two rebellious teenage girls use alcohol and drugs to deal with
life, and indulge in a spot of arson as a way of relieving their
boredom. But eventually the strains of their claustrophobic relationship begin to tell, as their different desires and needs start
to conflict and one of the girls develops a crush on her English
teacher. An interesting foray into the realm of teen angst that is
memorable if not especially profound.
DRA   59 min (ort 62 min)   VIDrel: DTK/TOTAL   V

ONLY THE LO\NELY ** 15
Chris Columbus USA   1991
*John Candy, Maureen O'Hara, James Belushi, Ally Sheedy, Anthony
Quinn, Kevin Dunn, Milo O'Shea, Bert Remsen, Joe V. Greco,
Macaulay Culkin, Kieran Culkin, Marvin McIntyre, Allen Hamilton,
Teri McEvoy, Bernie Landis, Les Podewell*
A lonely thirty-eight-year-old Chicago cop, who still lives at
home with his mom, falls for a woman who works as a cosmeti-
cian in a funeral parlour, and has to cope with opposition from
his possessive mother and the guilt he feels over leaving her.
O'Hara in her first role since the 1973 film "The Red Pony" is
splendid as the mother, but the film is just a mawkish updating
of a theme that was handled with more skill in MARTY. Written
by Columbus.
COM   100 min (ort 105 min)   VIDrel: 20TH/TECH   V/sur

ONLY THE STRONG ** 15
Sheldon Lettich USA   1993
*Mark Dacascos, Stacey Travis, Paco Christian Prieto, Todd Susman,
Geoffrey Lewis, Jeffrey Anderson Gunter, Richard Coca, Christian
Klemash, Ryan Bolman, Roman Cardwell, John Fionte, John Gregory
Kasper, Phyllis Sukoff, Antoni Corone*
A special forces officer comes back to Miami school and finds it
in the grip of drug dealers, so he trains twelve disaffected
students in a Brazilian form of the martial arts (Capoiera) and
goes after the bad guys. Some well-staged fight sequences fail
to compensate for the silliness of the plot and the lack of acting
ability on the part of Dacascos.
MAR   92 min (ort 125 min)   VIDrel: POLY/POLYREC
V/sur

ONLY THE VALIANT ** PG
Gordon Douglas USA   1950
*Gregory Peck, Barbara Payton, Ward Bond, Gig Young, Lon Chaney,
Warner Anderson, Jeff Corey, Steve Brodie, Neville Brand, Terry
Kilburn, Herbert Heyes, Art Baker, Hugh Sanders, Michael Ansara,
Nana Bryant, Harvey Udell*
Following a disastrous skirmish with Indians, a disciplinarian
cavalry officer finds that he has lost the respect of his men. When
he leads a detachment through hostile territory, he has to work
hard to regain their respect, eventually proving his courage by
holding off rampaging Apaches. A competent action Western,
technically well made but not terribly interesting.
WES   105 min (ort 107 min) B/W
VIDrel: 4-FRONT/POLYREC   V

ONLY TWO CAN PLAY ** PG
Sidney Gilliat UK   1962
*Peter Sellers, Mai Zetterling, Virginia Maskell, Richard
Attenborough, Raymond Huntley, Kenneth Griffith, John Le Mesurier,
Maudie Edwards, John Arnatt, David Davies, Meredith Edwards,
Frederick Piper, E. Eynon Evans*
A married Welsh librarian attempts to have a passionate affair
with the Norwegian-born wife of a local councillor, but is
prevented at every turn from consummating their relationship.
A dated and bitter-sweet comedy that is only occasionally
funny. Music is by Richard Rodney Bennett and the script is by
Bryan Forbes.
COM   102 min (ort 106 min) B/W   VIDrel: WHV   V/h
Boa: novel That Uncertain Feeling by Kingsley Amis.

ONLY WAY OUT, THE ** (12)
Rob Hardy USA   1994
*John Ritter, Harry Winkler, Stephanie Faracy, Sam Mancusco,
Matthew Walker, Jewel Shate, Jerry Wasserman, Caitlin Turner, Erik
Akai, Louise Vallance, Deryl Hayes, Julianne Phillips, Stephen
Dimopoulos, Robert Metclafe, Freda Perry*
In this terribly formulaic thriller, Ritter plays an architect who
is getting divorced from his wife, Winkler is the latter's psycho-
pathic boyfriend, who puts the family through hell.

Unappealing and derivative, its plot runs along the expected lines, and the eventual climax is more than welcome.
THR   92 min mTV   SATrel: MOVIE CHANNEL

ONLY YOU *   15
Betty Thomas USA   1992
Kelly Preston, Helen Hunt, Andrew McCarthy, Daniel Roebuck, Thom McFadden, Pepe Serna, Marc Lynn, Joel Murray, Monty Ash, Marsha Dietlein, Danny Mora, Liz Sheridan, Elvis Ballastreros, Michael Giotti, Christie Mellor, Pete Koch
A young man who has just been dished by his attractive girl-friend meets another girl, with whom he goes on a vacation trip. An attractive but far too predictable teenage romantic comedy, made watchable thanks to the efforts of a capable cast.
COM   87 min   VIDrel: COLUM/SONOP   V/sh

ONLY YOU *   PG
Norman Jewison USA   1994
Marisa Tomei, Robert Downey Jr, Bonnie Hunt, Billy Zane, Fisher Stevens, Joaquim De Almeida, Adam LeFevre, John Benjamin Hickey, Siobhan Fallon, Antonia Rey, Phyllis Newman, Denise Du Maurier, Tammy Minoff, Harry Barandes, Bob Tracy
Years ago an impressionable young woman was given a message via a Ouija board that spelt out the name of her future husband. Despite the fact that she is soon to be married, a trip to Italy brings her into contact with the man of her dreams, but she finds that his name does not match up to her prediction. Tomei's charm and Sven Nykvist's fine camerawork sustain this lightweight tale, which is to some extent a reworking of ROMAN HOLIDAY.
Aka: HIM; JUST IN TIME
COM   104 min (ort 108 min) cC   VIDrel: COLUM/SONOP
V/sur

OP CENTER **   PG
Lewis Teague USA   1994
Harry Hamlin, Kim Catrall, Wilford Brimley, Sherman Howard, Kabir Bedi, Rod Steiger, Lindsay Frost, George Alvarez, Patrick Bauchau, William Bumiller, Tom Breznahan, David garrison, Kabir Bedi, Victor Love, Bo hopkins
Three nuclear warheads are seized by Ukrainian rebels, who plan to sell them to a hostile power. An agent from a special unit set up to combat nuclear threats (the "op center") is given the task of preventing this and thus averting a possible nuclear war. A very gripping tale, well adapted from Clancy's equally good novel, that benefits from the meticulous attention given to detail. The weak sub-plot involving our agent in domestic problems is best ignored.
Aka: TOM CLANCY'S OP CENTER
THR   113 min (ort 170 min) mTV   VIDrel: FIRST/SONOP
V/h
Boa: novel by Tom Clancy.

OPEN FIRE *   18
Kurt Anderson USA   1994
Jeff Wincott, Patrick Kilpatrick, Lee De Broux, Mimi Craven, Anthony Dean Fields, Michael Shaner, Brenda Swanson, Bert Remsen, Ron Howard George, Lee Mathis, Bennet Guillory, Wesley Thompson, Joe Gaynor, Arthur Taxler
A group of mercenaries seize control of a town but are resisted and finally overcome by a one-man fighting force: a former FBI agent whose father is one of the hostages. Standard martial arts actioner.
MAR   88 min (ort 93 min)   VIDrel: REFLEC/FIRST   V

OPENING NIGHT **   15
John Cassavetes USA   1978
Gena Rowlands, John Cassavetes, Ben Gazzara, Joan Blondell, Paul Stewart, Zohra Lampert, Laura Johnson, John Tuell, Ray Powers, John Finnegan, Louise Fitch, Fred Draper, Katherine Cassavetes, Lady Rowland, Sharon Van Ivan
Intense and somewhat drawn out story of an actress who is pushed to the edge of a nervous breakdown by the death of her number one fan on the opening night of her new play. Written and directed by Cassavetes, this is a good example of his very personal style of directing – people who like his work as direc-tor will like this one a lot.
DRA   138 min (ort 144 min)   VIDrel: ELPIC/POLYREC   V

OPERATION DUMBO DROP *   PG
Simon Wincer USA   1995
Danny Glover, Ray Liotta, Dennis Leary, Doug E. Doug, Corin Nemec, Dinh Thien Le, Tscheky Karyo, Hoang Ly, James Hong, Vo

Trung Anh, Marshall Bell, Long Nguyen, Tim Kelleher, Scott N. Stevens, Kevin La Rosa, Christopher Ward
During the Vietnam War, an experienced local officer is replaced by a green and rule-bound authoritarian who inadvertently reveals to the Viet Cong that a village has been helping the Americans. In retaliation, they kill the village's elephant and our officer decides to obtain a replacement in just five days, in time for a major festival. A lightweight and bland comedy (despite its wartime setting) that is supposedly based on a true incident.
JUV   90 min   VIDrel: WDV/TECH   V

OPERATION INTERCEPT *   15
Paul Levine USA   1994
Lance Henriksen, Bruce Payne, Natasha Andreichenko, John Stockwell, Michael Champion, Dennis Christopher, Corbin Bernsen, Curt Lowens, Corinne Bohrer
A couple of top pilots are sent on a mission to eliminate global terrorists, who are led by the daughter of a dead space weapons expert, and from a secret base have taken control of a global satellite navigational system, creating havoc with the world's airlines and shipping.
Aka: AURORA: OPERATION INTERCEPT; PROJECT INTERCEPT
A/AD   90 min (ort 94 min)   VIDrel: GUILD/FOXVID
V/sur

OPERATION PACIFIC *   U
George Waggner USA   1951
John Wayne, Patricia Neal, Ward Bond, Scott Forbes, Philip Carey, Paul Picerni, William Campbell, Kathryn Givney, Vincent Forte, Martin Milner, Lewis Martin, Louis Mosconi, Carleton Young, Gordon Gebert, Steve Flagg
A tough Navy man who divorced his wife after the death of their child, meets up with her again but their reconciliation proves to be a long and painful process. Meanwhile, our hero undertakes a dangerous mission to attack the Japanese fleet. Both a drama and action war film, it incorporates some footage from DESTINATION TOKYO.
WAR   105 min (ort 115 min)   VIDrel: WHV   V

OPERATION PETTICOAT **   U
Blake Edwards USA   1959
Cary Grant, Tony Curtis, Joan O'Brien, Dina Merrill, Gene Evans, Arthur O'Connell, Virginia Gregg, Gavin McLeod, Marion Ross, Richard Sargent, Robert F. Simon, Robert Gist, George Dunn, Dick Crockett, Madlyn Rhue, Marion Ross
A crippled submarine has a captain who is determined to make his craft seaworthy again by fair means or foul. Grant and Curtis work together well in an enjoyable, lighthearted comedy. Remade as a TV movie in 1977 that in its turn served as a pilot for a brief TV series.
COM   115 min (ort 124 min)   VIDrel: 4-FRONT/POLYREC
V

OPERATION THUNDERBOLT **   (PG)
Menahem Golan ISRAEL   1977
Klaus Kinski, Assaf Dayan, Yehoram Gaon, Shai K. Ophir, Ori Levy, Sybil Danning (Sybelle Danninger), Mark Heath, Arik Lavi, Oded Teomi, Hi Kelos, Henry Czerniak, Gila Almagor, Rachel Marcus, Reuben Bar Yotam, Shoshik Shani
An account of the Entebbe raid when a group of hostages were rescued from Uganda by Israeli commandos, after being abducted by terrorists operating with the connivance of Idi Amin. Generally acknowledged to be the most realistic and authoritative of the three films dealing with this operation, as both RAID ON ENTEBBE and VICTORY AT ENTEBBE were rushed out quickly by US TV.
Aka: ENTEBBE: OPERATION THUNDERBOLT
THR   119 min (ort 125 min)   VIDrel: L/A   V

OPPOSITE SEX, THE *   15
Matthew Meshekoff USA   1992
Courtney Cox, Arye Gross, Kevin Pollack, Julie Brown, Mitch Ryan, B.J. Ward, Mitzi McCall, Phil Bruns, Jack Carter, Tess Foltyn, David DeCastro, Aaron Lustig, Connie Sawyer, Steven Brill, Davis Guggenheim, Craig Alan Edwards
A young Jewish man attempts to court a WASP girl but finds the differences in their backgrounds a major obstacle, in this failed romantic-comedy that is both flat and unfunny. It's not helped either by the use of asides to the camera from all four leads.
Aka: OPPOSITE SEX... AND HOW TO LIVE WITH THEM, THE
COM   83 min (ort 87 min)   VIDrel: GUILD/POLYREC L/A
V/s

ORBIT * 15
Mario Van Cleef USA 1996
Casper Van Dien, Betley Mitchum, Kelly Ann Sweeney, Joe Estevez, Carrie Mitchum, Nick Wilder, Jan-Michael Vincent, Christopher Mitchum
NASA launches a top-secret space-shuttle, but it then transpires that a traitor at mission control in Houston has been working with terrorists and has altered some vital computer codes, which will maroon the vessel in space unless a solution can be found. A cynical attempt to capitalise on the success of APOLLO 13, made without care or attention, and padded out with endless NASA footage and a host of irritating flashback sequences.
FAN 90 min mVid VIDrel: GUILD/FOXVID V/sur

ORCHESTRA REHEARSAL ** PG
Federico Fellini ITALY 1978
Balduin Bass, Clara Colosimo, Elisabeth Labi, Ronaldo Bonacchi, Ferdinando Villella, Giovanni Javarone, David Mauhsell, Francesco Aluigi, Andy Miller, Sibyl Mostert, Franco Mazzieri, Daniele Pagani, Filippo Trincia, Claudio Ciocca
The members of an orchestra argue with each other and the conductor, whilst they wait to be filmed performing in what is to be a television documentary. But while menacing vibrations shake the building and the electricity fails, the tyrannical conductor tries to exert his authority, but initially with only limited success. Clearly meant to be an allegory on the evils of Fascism, this clumsy work is partially redeemed by Nino Rota's score, the last one he wrote.
Aka: PROVA D'ORCHESTRA
DRA 69 min (ort 72 min) VIDrel: ARROW/RTM V

ORCHESTRA WIVES *** PG
Archie Mayo USA 1942
George Montgomery, Glenn Miller, Ann Rutherford, Carole Landis, Cesar Romero, Virginia Gilmore, Mary Beth Hughes, Jackie Gleason, Nicholas Brothers, Harry Morgan, Tamara Geva, Frank Orth, Grant Mitchell, Alec Craig, Tex Beneke
A fine example of a behind-the-scenes movie that examines the lives of the title group, as reflected in latest newcomer to that group, who has married the trumpet player. The sultry band singer spells danger for the newlyweds, but all ends happily in this lively and spirited musical. This was Miller's last screen appearance. The fine songs include "Serenade In Blue" and "I've Got A Gal In Kalamazoo".
MUS 93 min (ort 97 min) B/W cC VIDrel: 20TH/TECH V

ORDEAL IN THE ARCTIC ** PG
Mark Sobel CANADA 1993
Richard Chamberlain, Catherine Mary Stewart, Melanie Mayron, Scott Hylands, Page Fletcher, Christopher Bolton, Richard MacMillan, Robert Clinton, Blair Haynes, Stephen Sparks, Brian Jensen, Steve Adams, David Cameron, Mark Gibson
True story of a Canadian Air Force Hercules that crashed inside the Arctic Circle in October 1991 and of the struggle its crew of military personnel and civilians faced in order to survive until their rescue. See also ANGEL FLIGHT DOWN.
DRA 89 min mTV VIDrel: 20VIS/SONOP V
Boa: book Death and Deliverance by Robert Mason Lee.

ORDEAL OF PATTY HEARST, THE ** PG
Paul Wendkos USA 1979
Dennis Weaver, Lisa Eilbacher, Stephen Elliott, David Haskell, Dolores Sutton, Felton Perry, Tisa Farrow, Jonathan Banks, Anne De Salvo, Kathryn Butterfield, Karen Landry, Nancy Wolfe, Brendan Burns, Roy Poole
Patty Hearst's kidnapping by the Symbionese Liberation Front is related here by one of the FBI agents assigned to the case. Not a terribly impressive dramatisation of material that was later to form the basis of the 1988 film PATTY HEARST: HER OWN STORY.
DRA 138 min (ort 156 min) mTV VIDrel: CASPIC L/A V

ORDER OF THE BLACK EAGLE * 18
Worth Keeter USA 1987
Ian Hunter, Charles K. Bibby, William T. Hicks, Jill Donnellan, Anna Rappagna, Flo Hyman, Stephan Krayn
When a brilliant scientist is kidnapped by a bunch of neo-Nazis who want him to perfect a proton beam weapon, a secret agent and his assistant are sent on a rescue mission to South America. There they discover a sinister plot to use the body of Hitler (being kept in suspended animation) to usher in a new Nazi era. A film as ludicrous as it is incoherent.
A/AD 89 min Cut (13 sec – ort 93 min) VIDrel: 20TH V

ORDINARY PEOPLE **** 15
Robert Redford USA 1980
Donald Sutherland, Mary Tyler Moore, Judd Hirsch, Timothy Hutton, M. Emmet Walsh, Elizabeth McGovern, Dinah Manoff, Frederic Lehne, James B. Sikking, Basil Hoffman, Quinn Redeker, Mariclare Costello, Meg Mundy, Adam Baldwin
Redford's directorial debut (for which he won an Oscar) is an impressive drama about the hidden tensions and failings behind the facade of an ordinary "happy" family. Music is by Marvin Hamlisch. AA: Pic, Dir, S. Actor (Hutton), Screen/adapt (Alvin Sargent).
DRA 119 min (ort 124 min)
VIDrel: 4-FRONT/POLYREC/CIC V/h
Boa: novel by Judith Guest.

ORGY OF THE DEAD 18
A.C. Stevens (Stephen C. Apostoloff) USA 1965
Criswell, Fawn Silver, Pat Barringer, William Bates, Louis Ojena, John Andrews, Rod Lindeman, John Bealy, Arlene Spooner, Colleen O'Brian, Nadejda Dobrev, Mickey Jones, Barbra Norton, Dene Starnes, Texas Starr, Bunny Glaser
Screenplay is by our old friend Edward D. Wood Jr and the story, if one can call it such, involves two writers turning up at a cemetery where they are hoping to find some material for a novel. The Master of the Dead (Criswell) and his fetching assistant the Princess of Darkness (Silver) capture them, have them tied to posts, and force them to watch sinners being judged. A wild and endearingly bad offering, as ludicrous as it is laughable.
Aka: ORGY OF THE VAMPIRES
HOR 92 min VIDrel: WARMUS L/A V

ORIENTAL JADE * R18/18
G.W. Hunter USA 1987
Kristara Barrington, Lorri Smith, Raven, Suzie Hart, Treonna, Gabriella Fabray, Gina Valentino, Jade Nichols, Harry Reems, Ron Jeremy, Sasha Gabor, Greg Rome, Blake Palmer, Miles Long
A film producer shooting on location in the Philippines is faced with ruin when everything seems to go wrong, but he saves the day by importing a bevy of Hollywood beauties to spice up his film. Poor sex film that borrows its plot idea from S.O.B.
A 71 min (R18 ver); 44 min Cut (2 min 55 sec – 18 ver)
VIDrel: SHEP L/A (R18 ver); HAR/GOLD (18 ver) V

ORIGINAL GANGSTAS *** 18
Larry Cohen USA 1996
Fred Williamson, Jim Brown, Pam Grier, Paul Winfield, Isabel Saford, Ron O'Neal, Richard Roundtree, Christopher B. Duncan, Tim Rhoze, Eddie "Bo" Smity Jr, Oscar Brown Jr
A tale of urban violence in which a football coach returns to his home in Indiana after fifteen years to find the streets awash with crime. When his dad is attacked by a gang, he sets about assembling his old high school gang to take suitable reprisals. In some ways this harks back to those "blaxploitation" films of the 1970s, but here the movie cleverly works that idea into a kind of social comment on the changing times. Sometimes preposterous, but oddly effective.
A/AD 96 min (ort 99 min) VIDrel: FIRST/SONOP V

ORIGINAL SINS ** 18
Howard Berger/Matthew M. Howe USA 1993
Cheryl Clifford, Anqelique De Rochambeau, Faustina
A mysterious presence takes over the minds of three religious and highly neurotic girls who believe it to the spirit of Jesus Christ, but subsequent events prove that they were badly mistaken. A low-budget, independent horror movie with competent effects.
HOR 90 min VIDrel: SCEDGE/RTM V

ORLANDO *** PG
Sally Potter FRANCE/HOLLAND/ITALY/RUSSIA/UK 1992
Tilda Swinton, Billy Zane, Lothaire Bluteau, John Wood, Charlotte Valandrey, Heathcote Williams, Quentin Crisp, Peter Eyre, Thom Hoffman, Kathryn Hunter, Ned Sherrin, Jimmy Somerville, John Bott, Elaine Banham, Anna Farnworth
Imaginative and well-photographed adaptation of Wolf's novel about a unique human being who lives for several centuries first as a man and then as a woman. Her experiences take her from the court of Queen Elizabeth I of England to the world of the 20th century. Swinton is outstanding in the title role, in an

offbeat, leisurely movie that is certainly worth seeing. Screenplay is by Potter.
DRA 89 min (ort 93 min) wScrn VIDrel: ELPIC/POLYREC; ENCORE (LV only) V/sur LV
Boa: novel by Virginia Wolf.

ORPHEE **** PG
Jean Cocteau FRANCE 1949
Jean Marais, Maria Casares, Francois Perier, Marie Dea, Edouard Dermithe, Juliette Greco, Roger Blin, Henri Cremieux, Jacques Varennes, Rene Worms, Renee Cosima, Jean-Pierre Melville, Jean Cocteau
An updating of the famous legend with Orpheus a poet who meets and becomes obsessed with the Princess of Death. Encouraged by this, she carries off his wife, but Orpheus retrieves her and the unhappy princess sacrifices herself so as not to blight their mortal happiness. A display of cinematic tricks makes the film's evocation of the underworld as memorable as it is unique. The haunting score is by Georges Auric. Remade by Jacques Demy as "Parking".
Aka: ORPHEUS
FAN 90 min (ort 112 min) B/W VIDrel: CONNO/RTM V
Boa: play by Jean Cocteau.

OSCAR ** PG
John Landis USA 1991
Sylvester Stallone, Ornella Muti, Don Ameche, Peter Riegert, Vincent Spano, Tim Curry, Marisa Tomei, Eddie Bracken, Linda Gray, Chazz Palminteri, Yvonne DeCarlo, Kurtwood Smith, Ken Howard, William Atherton, Martin Ferrero
Stallone essays a comic role as a 1930s gangster who is determined to go straight, having been forced to promise to do so by his late father. But for all his efforts, he is soon facing trouble from rival gangsters, greedy financiers and a rebellious daughter. Based on a French stage farce (it was filmed before in 1957) this is a peculiarly ineffective film, yet it has some funny vignettes and a few good cameo performances (notably Kirk Douglas as Stallone's dad).
COM 105 min (ort 109 min) VIDrel: TOUCH/SONOP L/A V
Boa: play by Claude Magnier.

OSTERMAN WEEKEND, THE ** 18
Sam Peckinpah USA 1983
Rutger Hauer, John Hurt, Burt Lancaster, Craig T. Nelson, Dennis Hopper, Chris Sarandon, Meg Foster, Helen Shaver, Cassie Yates, Sandy McPeak, Cheryl Carter, Christopher Starr, John Bryson, Anne Haney, Kristen Peckinpah
The host of a TV show is persuaded by the CIA to help expose some of his friends who somehow just happen to be Soviet agents. The director's last film is one of those confused espionage thrillers where nobody seems to have a clue as to what is happening most of the time. Scripted by Alan Sharp and Ian Masters.
THR 102 min VIDrel: WHV L/A; ARENA V/h
Boa: novel by Robert Ludlum.

OTHELLO *** U
Orson Welles FRANCE/ITALY/USA 1949 (completed 1952)
Orson Welles, Michael MacLiammoir, Suzanne Cloutier, Robert Coote, Fay Compton, Michael Laurence, Doris Dowling, Nicholas Bruce, Jean Davis, Joseph Cotton, Joan Fontaine
Visually striking, immensely absorbing and highly underrated version of this famous play, with the title role taken by Welles. Hamstrung by a lack of funds, this low-budget production employs a fluid visual style with great use made of its varied locations. The bathhouse scene is especially memorable; this movie clearly ranks among the director's best work.
DRA 89 min (ort 93 min) B/W
VIDrel: 4-FRONT/POLYREC/CIC V/sur
Boa: play by William Shakespeare.

OTHELLO *** PG
Sergei Yutkevitch USSR 1955
Sergei Bondarchuk, Irina Skobotseva, Andrei Popov, A. Maximova, Vladimir Soshalsky, Evgeny Vesnik
Iago succeeds in convincing Othello that his wife Desdemona has been unfaithful, and our unhappy Moor takes the expected revenge. Shakespeare's masterpiece of jealousy and despair is well handled in visual terms, with costumes, sets and locations all contributing to the film's impact. For all that, Bondarchuk's performance in the title role is often unconvincing, though

Skobotseva makes an appealing Desdemona. Screenplay is by Yutkevitch.
DRA 103 min (ort 109 min) VIDrel: HEND/BMGREC L/A V
Boa: play by William Shakespeare.

OTHELLO ** U
Stuart Burge UK 1964
Laurence Olivier, Maggie Smith, Robert Lang, Frank Finlay, Joyce Redman, Derek Jacobi, Edward Hardwicke, Anthony Nicholls
Film version of Olivier's National Theatre triumph that doesn't really work particularly well as a piece of cinema, yet fills the role of a historical record of an acclaimed production, and has enough power and majesty to retain a measure of interest. Photography is by Geoffrey Unsworth.
DRA 170 min VIDrel: BRITHOM V
Boa: play by William Shakespeare.

OTHELLO ** U
Trevor Nunn UK 1989
Willard White, Ian McKellen, Imogen Stubbs, Zoe Wanamaker, Sean Baker, Clive Swift, Michael Grandage, Marsha Hunt, Phillip Sully, John Burgess
The Royal Shakespeare Company bring their talents to bear in this unusual adaptation that's set at the time of the American Civil War.
DRA 204 min VIDrel: CARL/TECH V
Boa: play by William Shakespeare.

OTHELLO *** 12
Oliver Parker USA 1995
Laurence Fishburne, Kenneth Branagh, Irene Jacob, Nathaniel Parker, Michael Maloney, Anna Patrick, Nicholas Farrell, Indra Ove, Michael Sheen, Pierre Vaneck, Gabriele Ferzetti, Andre Oumansky, Philip Locke, John Savident
A fairly satisfying (albeit severely pruned) version of Shakespeare's play about a brave but jealous Moorish warrior who is expertly manipulated by his second-in-command into believing that his beloved wife has been unfaithful. Branagh gives a stand-out performance as the villainous Iago, but Fishburne is terribly disappointing, a pity as in terms of appearance alone he is perfectly cast in the title role.
DRA 119 min (ort 123 min) VIDrel: COLUM/SONOP V/sur
Boa: play by William Shakespeare.

OTHELLO: PARTS 1 AND 2 *** U
Jonathan Miller UK 1981
Anthony Hopkins, Bob Hoskins, Penelope Wilton, Rosemary Leach, Anthony Pedley, Geoffrey Chater, Alexander Davion, David Yelland, Joseph O'Conor, Peter Walmsley, John Barron, Seymour Green, Howard Goorney
An excellent television adaptation, marked by good casting and sets, and without doubt one of the better Shakespeare renditions in this somewhat uneven series.
DRA 403 min (2 cassettes) mTV VIDrel: BBC V/h
Boa: play by William Shakespeare.

OTHER HELL, THE * 18
Stefan Oblowski (Bruno Mattei) ITALY 1981
Carlo de Meyo, Frances (Franca) Stoppi, Susan Ferget, Andrew Ray, Frank Garfield, Francesca Carmeno, Paola Montenero, Sandy Samuel
Mysterious events take place at a secluded convent, in a muddled effort that sees the nuns being stalked by terror, when one of their number is found murdered.
Aka: GUARDIAN OF HELL; L'ALTRO INFERNO; PRESENCE, THE
HOR 90 min wScrn dubbed VIDrel: REDEM/RTM V

OTHER MOTHER, THE *** PG
Bethany Rooney USA 1995
Frances Fisher, Deborah May, Corrie Clark, Cameron Bancroft, Gwyneth Walsh, Joe Coghill, Graham Beckel, Myles Ferguson, Greg Smith, Colleen Winton, Kevin McNulty, Karen Smith, Michelle Goodger, Malcolm Stewart, Lossen Chamber
A disturbing drama that tells of a woman's battle with bureaucracy in an attempt to regain contact with the son she was forced to give up for adoption when he was born. Based on the real-life experiences of Carol Schaefer, this well made story does not paint an especially agreeable portrait of American officialdom.
DRA 87 min (ort 90 min) mTV VIDrel: NWV/HIFLI V/h

OTHER PEOPLE'S MONEY **

Norman Jewison USA

15
1991

Danny DeVito, Gregory Peck, Penelope Ann Miller, Piper Laurie, Dean Jones, R.D. Call, Mo Gaffney, Bette Henritze, Tom Aldredge, Leila Kenzle, Cullen O. Johnson, William De Acutis, David Wells, Stephanie White, Jeffrey Hayenga

Peck plays the upright owner of a sluggish engineering company, whose days are numbered. DeVito plays the amoral asset-stripper who is out to "restructure" the company. The scene is soon set for a battle royal between these two opponents, though DeVito does find time for a little romance with a sharp female lawyer sent by Peck to block any takeover. For all its fine dialogue and acting, the film never really comes together. Pity.
COM   96 min (ort 102 min)   VIDrel: WHV   V/dm V/sur
Boa: play by Jerry Sterner.

OTHER SIDE OF MIDNIGHT, THE *

Charles Jarrott USA

18
1977

Marie-France Pisier, John Beck, Susan Sarandon, Raf Vallone, Clu Gulager, Christian Marquand, Sorrell Brooke, Michael Lerner, Louis Zorich, Anthony Ponzini, Charles Cioffi, Dimitra Arliss, Jan Arvan, Josette Banzet

A young Frenchwoman makes her way in the world from 1939 to 1947, during which period she exploits her charms in order to become a film star, and dallies with an American flier until her husband takes a cruel revenge. A faithful adaptation of a dull original novel with a script by Herman Raucher and music by Michel Legrand.
Aka: SIDNEY SHELDON'S THE OTHER SIDE OF MIDNIGHT
DRA   159 min (ort 166 min)   VIDrel: 20TH   V
Boa: novel by Sidney Sheldon.

OTHER SIDE OF PARADISE, THE ***

Renny Rye AUSTRALIA/NEW ZEALAND/UK

PG
1992

Jason Connery, Vivian Tan, Josephine Byrnes, Richard Wilson, Hywel Bennett, Terence Bayler, Jay Lavea Laga'aia, Judy Morris, Gerry McDonald, Robin Kora, June Bishop, Wi Kuki Kaa, Api McKinley, Edward Campbell, Derek Payne

A four-part TV drama, that's set in the late 1930s. A gifted young doctor is forced to leave England when he accidentally kills a thuggish Blackshirt. He decides to settle in the South Sea island of Koraloona, where it's not long before he finds romance with the daughter of the island doctor. A lavish and quite engaging effort, filmed in the South Pacific.
DRA   206 min (2 cassettes) mTV   VIDrel: CENTV/VCC L/A V
Boa: novel by Noel Barber.

OTHER WOMAN, THE **

Jag Mundhra USA

18
1992

Adrian Zmed, Lee Anne Beaman, Daniel Moriarty, Jenna Persaud, Craig Stepp, Melissa Moore, Allison Barron, Timothy C. Burns, Bill Bradshaw, Stephen Fiachi, Sam Jones, Beth Richards, Martine Muszek, Victoria Deuschle, Regina Geister

A journalist finds photographs of her husband making love with an unknown woman and starts to make her own enquiries, eventually learning that the person in question is a model with reputation for fast living. This soon leads her into a web of lies and deception, in this standard erotic thriller.
THR   87 min (ort 95 min)   VIDrel: MIA/DISC/IMPENT   V

OTHER WOMAN, THE **

Gabrielle Beaumont USA

PG
1995

Jill Eikenberry, Laura Leighton, Rosemary Forsyth, Monica Parker, Sarah Martineck, Lindsay Parker, James Read, Lloyd Bridges, Michael Covert, Willy Parson, Gloria Camden, Frank Von Zerneck Jr, David Jean Thomas, Michele Harrel

A woman splits up with her husband but is left with the two children. Her feelings of hatred to the new woman in her ex-husband's life are soon tempered by the knowledge that she has terminal cancer and will soon be unable to care for her children anyway, and this impels her to work towards getting her kids to accept this woman as their mother. Fair. See also A MOTHER'S PRAYER and WHO WILL LOVE MY CHILDREN? for two films exploring much the same idea.
DRA   88 min (ort 90 min) mTV   VIDrel: ODY/SONOP   V/sh

OTHER WOMEN'S CHILDREN **

Anne Wheeler CANADA

PG
1993

Melanie Mayron, Geraint Wyn Davies, Ja'net Du Bois, Mykelti Williamson, Eric Pospisil, Gabrielle Rose, Jerry Wasserman, Frederick Collins Jr., Linda Darlow, Sandra P. Grant, Kevin McNulty, Venus Terzo, P. Lynn Johnson, Merrilyn Gawn

A female doctor enjoys a successful career and a happy married life, but all this is put in peril when her son falls desperately ill, and she has to choose between her career and caring for her son. Adequate.
DRA   90 min mTV   VIDrel: ODY/SONOP   V/sh
Boa: novel by Perri Klass.

OUR DINNER WITH ANDREA *

Ona Zee/Frank Wind USA

R18/18
1987

Ona Zee, Sharon Kane, Robert Bullock, Shanna McCullough, Mike Horner, Jeremiah Logan, Richard Pacheco

A woman student of Tantric sex practices thaws a dinner party at which food is not the only thing on the menu. A flat sex film that tries to spoof "My Dinner With Andre" without much success.
A   VIDrel: SHEP L/A (R18 ver); HAR/GOLD (18 ver)   V

OUT FOR BLOOD *

Paul Thomas USA

18
1990

Torri Welles, Cheri Taylor, Kelly Royce, Raquel Darian, Jennifer O'Ryan, Randy Spears, Eric Price, Brad Derek, Alex Geradi, Nick Random, Tantala

A young girl grows up and goes off to college, blissfully unaware that her dad is a vampire, despite the fact that he only appears at night and is partial to day-time naps in a coffin. Though her mortal mother fears the girl has inherited his tendencies, dad is more than hopeful she will carry on the family tradition. A weak re-working of the Dracula legend, poorly shot on videotape, with flat jokes and flatter direction.
Aka: CAT HOUSE
A   60 min   VIDrel: MOPIC/QUANT   V

OUT FOR BLOOD **

Richard W. Munchkin USA

18
1992

Don Wilson, Shari Shattuck, Michael DeLano, Ron Steelman, Aki Aleong, Ken McLeod, Todd Curtis, Beua Billinsglea, Roberta Vasquez, Robert Miano, Addison Randall, Pam Dixon, Art Camacho, Deron McBee, Dino Homsey, Denise Y. Dowse

Echoes of DEATH WISH in this violent martial arts tale of a successful lawyer who turns vigilante after his family is murdered by drug dealers. An entirely predictable offering.
MAR   85 min (ort 90 min)   VIDrel: IMPENT   V/sh

OUT FOR JUSTICE *

John Flynn USA

18
1991

Steven Seagal, William Forsythe, Jerry Orbach, Joe Champa, Shareen Mitchell, Sal Richards, Gina Gershon, Jay Acovone, Nicky Corello, Robert Lasardo, John Toles-Bey, Joe Spataro, Ron Brumelow, Jack Ciolla, Charles Daniel, John Senger

A Brooklyn cop (Seagal) goes after a vicious hoodlum who the Mob would also like to see put out of circulation. Along the way, he breaks limbs, settles scores and dispenses rough justice to a variety of crooks, layabouts and social misfits. Seagal gives value for money in this violent outing, and brings brutal conviction to his role, yet for all that, the film never rises above the level of a simplistic revenger.
A/AD   86 min Cut (54 sec plus some cuts subst – ort 92 min)   VIDrel: WHV   V/sur

OUT OF AFRICA ***

Sydney Pollack UK/USA

PG
1985

Meryl Streep, Robert Redford, Klaus Maria Brandauer, Michael Kitchen, Joseph Thiaka, Mallick Bowens, Stephen Kinyanjui, Michael Gough, Suzanna Hamilton, Rachel Kempson, Graham Crowden, Leslie Phillips, Shane Rimmer

A long, heavily romanticised account of the life of Karen Blixen who wrote under the name of Isak Dinesen. Redford as an Englishman was an error but worse still was the use of Africa and Africans as an exotic backdrop to a mundane love story.
AA: Pic, Dir, Cin (David Watkin), Screen/adapt (Kurt Luedtke), Art/Set (Stephen Grimes/Josie MacAvin), Sound (Chris Jenkins/Gary Alexander/Larry Stensvold/Peter Handford), Score/orig (John Barry).
DRA   154 min (ort 161 min)   VIDrel: CIC/SONOP   V/sur
Boa: novel Silence Will Speak by Errol Trzebinski/novel Isak Dinesen by Judith Thurman/letters of Karen Blixen.

OUT OF ANNIE'S PAST **

Stuart Cooper USA

15
1995

Catherine Mary Stewart, Scott Valentine, Dennis Farina, Carsten Norgaard, Carlos Gomez, Ray Oriel, Michael Flynn, Brett Palmer, Christy Summerhuys, Bill Osborn, Lillian Cabal, Declan Pizzino, Tony Larimer, Shawn Nottingham

A New York gallery owner's lover is killed by gangsters, and when she flees for her life the police conclude that she must be a suspect. In Utah, she starts a new life with a fake identity as a gallery owner, having acquired a loving husband and family. One day a former detective appears, and demands $100,000 to keep silent, but she kills him by accident only to then find a former lover from her past turning up. A tangled, murky and totally implausible thriller.
THR   87 min (ort 91 min)   VIDrel: CIC/SONOP   V

OUT OF DARKNESS ***
Larry Elikann USA
15
1993
Diana Ross, Anne Weldon, Rhonda Stubbins White, Beah Richards, Carl Lumbly, Chasiti Hampton, John Marshall James, Juanita Jennings, Maura Tierney, Lindsay Crouse, Rusty Schmidt, Patricia Idlette, Barbara Howard, Cathy Raymond
A woman tries to regain control of her life and fights against the paranoid schizophrenia that is destroying it, having lived as a virtual recluse for twenty years. Fortunately, she is assisted in this by the efforts of a psychiatric social worker and a new experimental drug. Ross makes an impressive TV movie debut in this engrossing and often disturbing drama.
DRA   87 min (ort 90 min) mTV
VIDrel: 4-FRONT/POLYREC/ODY   V/sh

OUT OF THE PAST ****
Jacques Tourneur USA
PG
1947
Kirk Douglas, Robert Mitchum, Jane Greer, Richard Webb, Rhonda Fleming, Dickie Moore, Steve Brodie, Virginia Huston, Paul Valentine, Ken Niles, Lee Elson, Frank Wilcox, Mary Fields, Jess Escobar, James Bush, Hubert Brill
A private eye, hired to find a hoodlum's girlfriend, ends up falling in love with her. Film noir with plenty of intrigue and double dealing. The script is by Geoffrey Homes (Daniel Mainwaring). Later remade as AGAINST ALL ODDS.
Aka: BUILD MY GALLOWS HIGH
THR   93 min (ort 97 min) B/W   VIDrel: VCC L/A   V
Boa: novel Build My Gallows High by Geoffrey Homes.

OUT ON A LIMB **
Robert Butler USA
15
1986
Shirley MacLaine, Charles Dance, John Heard, Charles Dance, Anne Jackson, Jerry Orbach
Tiresome dramatisation of MacLaine's bestseller detailing her relationship with a British politician and her gradual involvement with all manner of supernatural beliefs, including UFOs and astral travelling. Of interest mainly to fans.
DRA   235 min (2 cassettes)   VIDrel: ODY/SONOP   V/sh
Boa: book by Shirley MacLaine.

OUT ON A LIMB *
Francis Veber USA
PG
1992
Matthew Broderick, Jeffrey Jones, Heidi King, John C. Reilly, Marian Mercer, Larry Hankin, David Margulies, Courtney Peldon, Michael Monks, Andy Kossin, Mickey Jones, Nancy Lenenhan, Noah Craig Andrews, Ben Diskin, Adam Wylie
A young stockbroker with a talent for making money returns home to help his sister, who has become convinced that their stepfather is a criminal. On the way there he is robbed by a female hitchhiker and encounters more than the usual quota of screwballs. A crude and unfunny effort with too many chase scenes and a marked absence of humour.
COM   79 min (ort 83 min)   VIDrel: CIC/SONOP   V

OUT ON BAIL **
Gordon Hessler USA
18
1988
Robert Ginty, Kathy Shower, Tom Badal, Sydney Lassick, Leo Sparrowhawk, Tom Aigner, Dewaal Stemmit, Christopher Dingle, James Whyle, Liam Cundill, Teuns Sanderman, Adrienne Pierce, Stephen Jennings, Brad Morris, Chris Chevez
A loner arrives at a small Southern town and gets entangled in the usual morass of stock situations, including the ubiquitous corrupt sheriff, murder frame-up and attractive young widow etc. Badal is good as the sadistic sheriff and a set of well-executed stunts adds a little zip to this unoriginal low-budget actioner. Filmed in South Africa.
A/AD   98 min Cut (31 sec – ort 102 min)   VIDrel: EIV   V

OUTBACK BOUND **
John Llewelyn Moxey USA
PG
1988
Donna Mills, John Schneider, John Meillon, Andrew Clarke, Colette Mann, Nina Foch, Robert Harper
A Beverly Hills woman cheated out of all she owns inherits a disused opal mine in Australia, and decides to move there to sell it and start a new life. Fair.
A/AD   89 min Cut (8 sec – ort 96 min) mTV
VIDrel: 20TH/TECH   V

OUTBREAK ***
Wolfgang Petersen USA
15
1995
Dustin Hoffman, Rene Russo, Donald Sutherland, Kevin Spacey, Morgan Freeman, Cuba Gooding Jr, Patrick Dempsey, Zakes Mokae, Malick Bowens, Susan Lee Hoffman, Benito Martinez, Bruce Jarchow, Leland Hayward III, Daniel Chodos, Dale Dye
Rather chilling film about a deadly (and highly contagious) virus that originates in a species of African monkey, but infects a small American town, leading to desperate measures to contain it. Hoffman plays the brilliant army medic upon whose success or failure the future of the entire U.S. population hangs. Though starting with promise, some typical Hollywood conspiracy cliches and scientific gobbledegook rather spoil it all. See also FORMULA FOR DEATH.
THR   123 min (ort 128 min) cC   VIDrel: WHV   V/sur

OUTCAST OF THE ISLANDS, AN ***
Carol Reed UK
PG
1951
Trevor Howard, Robert Morley, Kerima, Ralph Richardson, Wendy Hiller, George Coulouris, Frederick Valk, Wilfrid Hyde White, Betty Ann Davies, James Kenney, James Illing, A.V. Bramble, Dharma Emmanuel, Annabel Morley, Marne Maitland
At a Far Eastern trading post a selfish and misguided reprobate reveals a secret sea route in order to curry favour with an island princess, but his plan misfires and he is left with nothing. A full-blooded adaptation of Conrad's complex character study; Howard is compelling in the title role.
A/AD   95 min (ort 102 min) B/W   VIDrel: WHV   V/h
Boa: novel by Joseph Conrad.

OUTER LIMITS, THE: CORPUS EARTHLINGS **
Gerd Oswald USA
PG
1963
Robert Culp, Salome Jens, Barry Atwater
Two rocks contain life-forms able to control a human being and a doctor hears them plotting to take over the world. But is this real or just a figment of his imagination?
FAN   103 min (2-episode cassette) mTV B/W
VIDrel: MGM/WHV   V/h

OUTER LIMITS, THE: DEMON WITH A GLASS HAND ***
Byron Haskin USA
PG
1964
Robert Culp, Arline Martel, Steve Harris, Abraham Sofaer, Rex Holman, Robert Fortier
Doubtless one of the sources of inspiration for THE TERMINATOR, this complex story has alien soldiers from the future arriving on Earth in an attempt to prevent a future war taking place between their race and humanity. But a strange being with a false hand holds the key to these future events.
FAN   103 min (2-episode cassette) mTV B/W   VIDrel: WHV   V/h

OUTER LIMITS, THE: IT CRAWLED OUT OF THE WOODWORK ***
Gerd Oswald USA
PG
1963
Scott Marlowe, Kent Smith, Barbara Luna, Edward Asner, Michael Forrest, Joan Camden
When a ball of dust is sucked into a vacuum cleaner at an electronics research institute, it takes on a life of its own, growing in size as it absorbs electrical energy.
FAN   102 min (2-episode cassette) mTV B/W
VIDrel: MGM/WHV   V/h

OUTER LIMITS, THE – NEW SERIES: BIRTHRIGHT **
15
William Fruet USA
1995
After a freak accident, the life of an American senator is altered, as he gains access to memories that make him realise he is an alien in disguise, sent to Earth on a mission to alter the planet's atmosphere.
FAN   86 min (2-episode cassette) mTV cC
VIDrel: MGM/WHV   V/sh

OUTER LIMITS, THE – NEW SERIES: BLOOD BROTHERS **
Tibor Takacs USA
15
1995
Two very different brothers compete for control of a giant

biotechnology company, a vaccine that could hold the key to eternal life being the focus of their interest.
FAN   85 min (2-episode cassette) mTV cC
VIDrel: MGM/WHV   V/sh

OUTER LIMITS, THE – NEW SERIES: CAUGHT IN THE ACT ***

15
Mark Sobel USA   1995
Alyssa Milano, Jason London, Saul Rubinek
A female student becomes the host of an alien creature, and develops a voracious sexual appetite, but this is only exploited by the alien to bring her into close proximity with other people, who are then absorbed by the alien. Quite a strongly plotted entry, with one truly horrific moment.
FAN   86 min (2-episode cassette) mTV cC
VIDrel: MGM/WHV   V/h

OUTER LIMITS, THE – NEW SERIES: CORNER OF THE EYE ***

12
Stuart Gillard USA   1995
Len Cariou, Chris Sarandon, Justin Louis, Bill Croft, Callum Keith Rennie
A priest becomes able to detect the real appearance of horrific aliens passing themselves off as human beings, and when they learn of this they approach him, claiming that despite their fearsome appearance, they are alien emissaries on a mission of goodwill. However, doubts begin to emerge as to their real intentions. Chilling and quite effective, if a little overdone at the end.
FAN   86 min (2-episode cassette) mTV cC
VIDrel: MGM/WHV   V/sh

OUTER LIMITS, THE – NEW SERIES: FALLING STAR *

(PG)
Ken Girotti USA   1996
Sheena Easton, John Pyper-Ferguson, Kristin Lehman, Sarah Strange, Victoria Morsell, Xander Berkeley, Peter Bryant, Richard Lautsch, Rick Burgess, Rod Wilson
A rock singer with a waning career finds her cheating husband in bed with another woman, and this sends her into a cycle of depression that makes her contemplate suicide. But an image of a girl from the future repeatedly begs her not to do this, as her music is to be the key to the salvation of a future society. However, others from the future are out to persuade the rock singer to go through with it. Apart from the effects, this is a weak and implausible tale.
FAN   45 min mTV   TVrel: BBC2

OUTER LIMITS, THE – NEW SERIES: I, ROBOT **

12
Adam Nimoy USA   1995
Leonard Nimoy, Jake McKinnon, Cyndy Preston, Eric Schneider
A lawyer defends a robot accused of murder, that unaccountably ran amok, destroying the laboratory and killing the scientist that created it. However, the dead man's daughter feels a close bond with this creature, and argues that it should be given the chance to defend its actions in court. An updating of an episode from the original series, quite well done but not offering anything new.
FAN   85 min (2-episode cassette) mTV cC
VIDrel: MGM/WHV   V/sh

OUTER LIMITS, THE – NEW SERIES: IF THESE WALLS COULD TALK **

15
Tibor Takacs USA   1995
A scientist ventures into a haunted house to study the ghosts he hopes to find there, but in reality the house is occupied by a strange substance that feeds on anyone who occupies it.
FAN   86 min (2-episode cassette) mTV cC
VIDrel: MGM/WHV   V/sh

OUTER LIMITS, THE – NEW SERIES: INCONSTANT MOON **

15
Joseph L. Scanlan USA   1996
A physics professor becomes convinced that world is to end and embarks on a career of self-indulgence, and sets out to fulfil all his fantasies, including a desire to approach a woman he has secretly yearned after for years.
FAN   85 min (2-episode cassette) mTV cC
VIDrel: MGM/WHV   V/sh

OUTER LIMITS, THE – NEW SERIES: THE CHOICE **

(PG)
USA   1995
Thora Birch, Megan Follows, Frances Sternhagen, Matthew Walker, Page Fletcher, Sandra Nelson
After a strange telekinetic attack on another pupil, a young girl is suspended from school, and her worried parents decide to hire a nanny to look after her. This woman soon establishes a rapport with the child and begins and educational programme. Meanwhile, government agents are out looking for such people, especially those that belong to a secret organisation.
FAN   45 min mTV   TVrel: BBC2

OUTER LIMITS, THE – NEW SERIES: THE CONVERSION **

(PG)
USA   1995
Frank Whaley, John Savage, Kerry Sandormirsky, Rebecca De Mornay
A man who was sent to jail for fraud eventually gets out and sets out to plan his revenge. But matters do not go as planned and he is shot and wounded. Hiding out at an isolated restaurant he encounters a mysterious stranger who appears to know too much about him, but at the same time may be offering him a chance of salvation. A strange entry in the series, with a supernatural overtone.
FAN   45 min mTV   TVrel: BBC2

OUTER LIMITS, THE – NEW SERIES: THE MESSAGE **

(PG)
USA   1996
Marlee Matlin, Larry Drake
A woman who has been deaf since birth is given a new type of implant, but it does not appear to work. However, she then starts to hear signals that turn out to be in binary code, and it falls to a former astro-physicist (now working as a janitor) to work out their significance. A well realised entry, even if the final outcome is a little too neatly packaged to convince.
FAN   45 min mTV   TVrel: BBC2

OUTER LIMITS, THE – NEW SERIES: THE NEW BREED ***

15
Mario Azzopardi USA   1995
Richard Thomas, Peter Outerbridge, Tammy Isbell, Veena Sood
Having invented "nano-robots" – microscopic computer-driven machines that are capable of repairing damaged living tissue, a scientist finds that his brother-in-law (who is dying of cancer) has purloined a sample of these to inject into himself. But after saving his life, these devices continue to "improve" him with modifications that make him ever more grotesque in appearance.
FAN   85 min (2-episode cassette) mTV cC
VIDrel: MGM/WHV   V/sh

OUTER LIMITS, THE – NEW SERIES: THE QUALITY OF MERCY ***

15
Brad Turner USA   1995
Robert Patrick, Nicole DeBoer
One of the most chilling of the new series entries, this tells of an intergalactic war between humanity and a race of fearsome, reptilian creatures that are extremely hard to kill. Held in a prison cell for questioning, one of Earth's top fighter pilots finds her is sharing the cell with a human woman being forced to undergo an unpleasant mutation. But can he trust her?
FAN   85 min (2-episode cassette) mTV cC
VIDrel: MGM/WHV   V/sh

OUTER LIMITS, THE – NEW SERIES: THE SANDKINGS ***

(18)
Stuart Gillard USA   1995
Beau Bridges, Helen Shaver, Dylan Bridges, Kim Coates, Lloyd Bridges, Patricia Harns, Nathaniel Deveauz, Deryl Hayes, Mark Saunders, J.B. Bivens, David Cameron
A scientist heads a project breeding eggs taken from Mars and when it is decided that safety requires that the project be terminated, he takes a soil sample home. The eggs contained in it develop into insects of a high intelligence and voracious appetite. This strong feature-length pilot episode marked a return of the well-regarded 1960s SF series.
FAN   88 min (ort 93 min) mTV   VIDrel: MGM/WHV   V
Boa: novella by George R.R. Martin.

OUTER LIMITS, THE – NEW SERIES: TRIAL BY FIRE **

15
Jonathan Glasser USA   1996
An unknown object approaches the Earth, and the newly elected

American President learns that it is a vast alien armada, whose intentions are unknown.
FAN 85 min (2-episode cassette) mTV cC
VIDrel: MGM/WHV V/sh

OUTER LIMITS, THE – NEW SERIES: UNDER THE BED **
12
Rene Bonniere USA
1995
Timothy Busfield, Barbara Williams, Joel Palmer, Colleen Rennison
A little boy goes missing and his young sister claims he was taken away by something under their bed. While the children's estranged father comes under suspicion, the investigations of a doctor reveal other unaccountable disappearances have taken place, and that these are all linked to the same time of the month.
FAN 85 min (2-episode cassette) mTV cC
VIDrel: MGM/WHV V/sh

OUTER LIMITS, THE – NEW SERIES: VALERIE 23 ***
(PG)
USA
1995
Bill Sadler, Sofia Shinas, Nancy Allen, Don Butler, Kevin Conway
Confined to a wheelchair, a researcher who works for a company developing androids is given one such creation to take home as a temporary companion, his job being to assess its suitability for disabled individuals such as himself. But the creature has been given human thoughts and feelings, and soon becomes jealously possessive, so much so that it is prepared to kill to ensure it has no rivals for its affections. A sad and exceptionally well conceived entry.
FAN 45 min mTV TVrel: BBC2

OUTER LIMITS, THE – NEW SERIES: VIRTUAL FUTURE **
15
Joseph L. Scanlan USA
1995
Josh Brolin, Kelly Rowan, David Warner, Bruce French
A virtual reality device has been created that allows the user to take short "hops" into the near future. On one foray, a young researcher discovers the sinister truth about a top industrialist, and of his plans to use the device for his own ruthless purposes. Not especially well done, but the resolution is neat if nothing else.
FAN 85 min (2-episode cassette) mTV cC
VIDrel: MGM/WHV V/sh

OUTER LIMITS, THE – NEW SERIES: WHITE LIGHT FEVER **
(PG)
USA
1995
Bruce Davidson, William Hickey, Sonja Smit, Michelle Beaudoin, Benjamin Ratner, Kevin Conway
Another entry with a supernatural flavour. In this story an aged tycoon is so traumatised by his early experiences that he lives in permanent terror at the prospect of dying. Having devoted his life to building up a fortune, he now employs it for purely selfish ends, funding research into an artificial heart device with which he hopes to live forever. But forces are at work that appear designed to thwart his efforts.
FAN 45 min mTV TVrel: BBC2

OUTER LIMITS, THE – NEW SERIES: WORLDS APART **
(PG)
Brad Turner USA
1996
Bonnie Bedelia, Robert Ito, Michael MacRae, Chad Willett, Donnelly Rhodes
A signal is picked up from a space pilot whose ship vanished twenty years ago, but from his point of view the mission was but yesterday, as he has fallen into a region of space that distorts time. As the region is close to the Earth, a rescue attempt is mounted, but it may not reach him in time. Meanwhile the pilot, who appears to be floating on an unknown sea, has to fend of the attentions of a ferocious creature. Fair, but somewhat padded out.
FAN 45 min mTV TVrel: BBC2

OUTER LIMITS, THE: NIGHTMARE ***
PG
John Erman USA
1963
Ed Nelson, James Shigeta, John Anderson, Martin Sheen, David Frankham, Bill Gunn
An Earth strike force is captured by an alien race and held on their planet, where it would appear they are all to be subjected to a brutal interrogation. But appearances can be deceptive.
FAN 104 min (2-episode cassette) mTV B/W
VIDrel: MGM/WHV V

OUTER LIMITS, THE: O.B.I.T. **
PG
Gerd Oswald USA
1963
Peter Breck, Jeff Corey, Harry Townes, Alan Baxter, Joanne Gilbert
A murder investigation uncovers a surveillance machine that creates fear and hostility, and reveals the existence of an alien.
FAN 103 min (2-episode cassette) mTV B/W
VIDrel: MGM/WHV V/h

OUTER LIMITS, THE: SOLDIER ***
PG
Gerd Oswald USA
1964
LLoyd Nolan, Michael Ansara, Alan Jaffe, Tim O'Connor, Catherine McLeod, Jill Hill, Ralph Hart, Ted Stanhope
Two enemy soldiers from a bleak and brutal future conflict are thrown back in time to the present, and seek each other out to resolve their conflict. One of the most effective stories in this early SF series. Written by Ellison and Seeleg Lester, from a short story by the former.
FAN 104 min (2-episode cassette) mTV B/W
VIDrel: MGM/WHV V
Boa: short story by Harlan Ellison.

OUTER LIMITS, THE: THE ARCHITECTS OF FEAR **
PG
Byron Haskin USA
1963
Robert Culp, Geraldine Brooks, Leonard Stone, Martin Wolfson
An episode from a popular and often very well made SF series. "The Architects Of Fear" sees a group of scientists creating a mock alien to shock mankind into less-aggressive behaviour, but they find that fear alone is not the solution.
FAN 102 min (2-episode cassette) mTV B/W
VIDrel: MGM/WHV V/h

OUTER LIMITS, THE: THE BELLERO SHIELD **
PG
John Brahm USA
1963
Martin Landau, Sally Kellerman, Chita Rivera, Neil Hamilton, John Hoyt
A scientist experimenting with a laser device inadvertently attracts a benign alien creature from another world, which demonstrates its technological, and ultimately its moral superiority. A badly directed episode sustained by its unusual ideas.
FAN 103 min (2-episode cassette) mTV B/W VIDrel: WHV V/h

OUTER LIMITS, THE: THE GALAXY BEING ***
PG
Leslie Stevens USA
1963
Cliff Robertson, Jacqueline Scott, Lee Philips, William O. Douglas, Mavis Neal, Allyson Ames
The very first episode from a popular and well-mounted SF series. "The Galaxy Being" tells of a radio-station engineer whose experiments bring a benign creature of fearful appearance to Earth from a distant galaxy, where its arrival sparks off a wave of panic.
FAN 104 min (2-episode cassette) mTV B/W
VIDrel: MGM/WHV V/h

OUTER LIMITS, THE: THE HUNDRED DAYS OF THE DRAGON ***
PG
Byron Haskin USA
1963
Sidney Blackmer, Phil Pine, Mark Roberts, Nancy Rennick, Joan Camden, Clarence Lung, Richard Loo
The second episode from a popular and very well-mounted SF series. In this story the new American president is replaced by an Oriental despot who has obtained an experimental serum that gives him the ability to alter his appearance and cellular structure.
FAN 104 min (2-episode cassette) mTV B/W
VIDrel: MGM/WHV V/h

OUTER LIMITS, THE: THE MAN WHO WAS NEVER BORN ***
PG
Leonard Horn USA
1963
Martin Landau, Shirley Knight, John Considine, Karl Held, Maxine Stuart
A spaceman passes through a time barrier and finds the Earth of the 21st century a barren, blighted wasteland, but he returns home with one of its inhabitants in an effort to change the future.
FAN 103 min (2-episode cassette) mTV B/W
VIDrel: MGM/WHV V/h

OUTER LIMITS, THE: THE MAN WITH THE POWER **
PG
Laslo Benedek USA
1963
Donald Pleasence, Priscilla Morrill, Edward C. Platt, Fred Beir

A timid professor finds himself with awesome powers after a brain operation. A rather weak and implausible entry in the series.
FAN 102 min (2-episode cassette) mTV B/W
VIDrel: MGM/WHV V/h

OUTER LIMITS, THE – NEW SERIES: THE SECOND SOUL ** 12
Paul Lynch USA 1995
Mykelti Williamson, D.W. Moffett, Rae Dawn Chong, Richard Grove, Kevin Conway
Contact with aliens reveals that they wish to use human corpses to save themselves from a dying planet and thus preserve their culture. But a man whose beloved wife is used for such a purpose becomes convinced he has stumbled on a far more sinister purpose. Fair.
FAN 86 min (2-episode cassette) mTV cC
VIDrel: MGM/WHV V/sh

OUTER LIMITS, THE – NEW SERIES: THE VOYAGE HOME *** 15
Tibor Takacs USA 1995
Jay O. Sanders, Matt Craven, Michael Dorn
Men returning from Mars have a stowaway onboard in the form of a shapeless alien creature of voracious appetite. As it progressively works its way through the crew members, the final pilot realises that his actions will determine the future fate of humanity.
FAN 86 min (2-episode cassette) mTV cC
VIDrel: MGM/WHV V/h

OUTER LIMITS, THE: THE PRODUCTION AND DECAY OF STRANGE PARTICLES ** PG
Leslie Stevens USA 1964
George Macready, Leonard Nimoy, Rudy Solari, Joseph Ruskin, Allyson Ames, John Duke, Signe Hasso
A collection of strange atomic particles arrives on Earth, where they start to multiply until a nuclear catastrophe seems imminent.
FAN 102 min (2-episode cassette) mTV B/W
VIDrel: MGM/WHV V/h

OUTER LIMITS, THE: THE SIXTH FINGER ** PG
James Goldstone USA 1963
David McCallum, Edward Mulhare, Jill Haworth, Constance Cavendish, Nora Marlowe, Robert Doyle
"The Sixth Finger" follows the experiments a scientist conducts in the acceleration of evolution, and the changes he brings about in an illiterate young coal-miner, eventually evolving him into a super-intelligent but coldly ruthless being.
FAN 103 min (2-episode cassette) mTV B/W
VIDrel: MGM/WHV V/h

OUTLAND ** 15
Peter Hyams USA 1981
Sean Connery, Peter Boyle, Frances Sternhagen, James B. Sikking, Kika Markham, Clarke Peters, Steven Berkoff, John Ratzenberger, Nicholas Barnes, Manning Redwood, Pat Starr, Hal Galili, Angus MacInnes, Stuart Milligan
This SF version of HIGH NOON (complete with digital clock) has a tough investigator being sent out to a mining colony on one of Jupiter's moons, to solve a rash of apparent suicides. He becomes the target of a murder squad and awaits their arrival. A flashy but coldly unsatisfying film with a script by the director.
FAN 104 min wScrn VIDrel: WHV V/sur

OUTLAW, THE *** U
Howard Hughes USA 1941
Jane Russell, Walter Huston, Jack Buetel, Thomas Mitchell, Mimi Aguglia, Joe Sawyer, Gene Rizzi, Frank Darien, Pat West, Carl Stockdale, Nena Quartaro, Martin Garralaga, Julian Rivero, Dickie Jones, Ethan Laidlaw, Ed Brady
A pretentious but unusual version of the Billy the Kid legend, with our outlaw enjoying a little romance when he meets up with Doc Holliday and Pat Garrett, at a half-way station where they quarrel over a half-breed girl. This famous "sex" Western marked Russell's screen debut, and excessive interest in her bosom is certainly evident. The music is by Victor Young. Filmed in 1941 and directed mostly by Howard Hawks.
WES 114 min (ort 126 min) B/W VIDrel: NTV/TERRY V

OUTLAW BROTHERS, THE ** 18
Frankie Chan HONG KONG 1990
Frankie Chan, Yukari Oshima, Mok Sui Chung, Michiko Nishiwaki, Jeffrey Falcon, Mark Houghton, Michiko
A Porsche is stolen, but it belongs to a drug baron and contains a cache of cocaine. This leads to various violent episodes, fisticuffs (said to be have been choreographed by Jacki Chan) and plot twists. Oshima plays a tough female police inspector, who uses her martial arts prowess to dispatch a variety of adversaries, an activity she performs with considerable panache. An exciting, fast-paced and mindless actioner, only the plot and dreadful dubbing hamper it.
Aka: OUTLAWS, THE
A/AD 96 min dubbed VIDrel: MIA/DISC V

OUTLAW JOSEY WALES, THE *** 18
Clint Eastwood USA 1976
Clint Eastwood, Chief Dan George, Sondra Locke, John Vernon, Bill McKinney, Sam Bottoms, Paula Trueman, Geraldine Keams, Woodrow Parfrey, Joyce Jameson, Sheb Wooley, Royal Dano, Matt Clark, John Verros, Will Sampson, John Quade
A quiet man becomes an outlaw when Union soldiers murder his family in a tale that is long and violent but extremely well made. Followed by the dreadful "The Return Of Josey Wales" without Eastwood. The screenplay was by Philip Kaufman and Sonia Chernus, with the former directing until Eastwood took over.
WES 130 min (ort 135 min) VIDrel: WHV V
Boa: novel Gone to Texas by Forrest Carter.

OUTLAW OF GOR * 18
John (Bud) Cardos ITALY 1987
Urbano Barberini, Jack Palance, Donna Denton, Rebecca Ferrati, Nigel Chipps, Russel Savadier, Alex Heyns, Tulio Monetta, Larry Taylor, Michelle Clarke, Michael Brunner, Christobel D'Ortez, Natasha Piotrowski, Nicole De Gruchy
A sequel to GOR with our professor making a return trip to this primitive world for some further adventures. A dreary excursion into the world of sword-and-sorcery, only in this one the sorcery is in short supply.
Aka: OUTLAW
FAN 85 min (ort 89 min) VIDrel: L/A V
Boa: novel by John Norman.

OUTRAGED ** 18
Robert Anthony USA 1996
Frank Zagarino, Ayu Azhari, Frans Tumbuan, Martin Kove
A former US Marines officer who was betrayed by his country and forced to live abroad now earns his living fighting in backstreet kickboxing bouts. When a chance encounter causes him to recall the ordeal he suffered in the Far East, he is drawn into a vengeance-seeking mission, and this involves him in fighting drug dealers and other undesirables. A low-budget picture, with only a few moments of interest.
A/AD 88 min VIDrel: COLUM/SONOP V

OUTRAGEOUS FORTUNE *** 15
Arthur Hiller USA 1987
Shelley Long, Bette Midler, Peter Coyote, George Carlin, John Schuck, Robert Prosky, Anthony Heald, Ji-Tu Cumbuka, Florence Stanley, Jerry Zaks, Joe DiSanti, Diana Bellamy, Gary Morgan, Christopher McDonald, J.W. Smith, Robert Pastorelli
Two struggling actresses who detest each other on sight become bosom buddies, especially when they find themselves involved with the same worthless man. A screen celebration of female friendship with a calculated feminist slant. The script is by Leslie Dixon.
COM 95 min (ort 100 min) VIDrel: RNK L/A V

OUTRAGEOUS PARTY * 18
Jacki McKimmie AUSTRALIA 1986
Graeme Blundel, John Jarratt, Noni Hazelhurst, Barry Rugless, Margaret Lord, Jenny Mansfield, Caine O'Connell, Ruth Barraclough, James Ricketson, Lil Kelman, John Kerr, Jenny Kubler, Meg Simpson, Marlon Holden, Jill Loof
Neighbours take turns organising monthly parties. One such man invites his boss to his bash to curry favour, but the party gets wildly out of hand. A raucous and occasionally amusing adult comedy.
Aka: AUSTRALIAN DREAM
COM 82 min (ort 84 min) VIDrel: MIA/DISC L/A V

OUTSIDERS, THE *** PG
Francis Ford Coppola USA 1983
C. Thomas Howell, Matt Dillon, Ralph Macchio, Diane Lane, Patrick Swayze, Rob Lowe, Tom Cruise, Leif Garrett, Emilio Estevez, Tom Waits, Glenn Withrow, Darren Dalton, Michelle Meyrink, Gailard Sartain, William Smith, Tom Hillman
A study of teenagers growing up in Oklahoma in the 1960s as seen through the eyes of one boy. An ambitious if not exactly successful attempt to transpose Hinton's likeable novel. The score is by Carmine Coppola. Followed by RUMBLE FISH, another film based on the work of Hinton.
DRA 87 min (ort 95 min) wScrn VIDrel: SECOND/RTM V/sh
Boa: novel by S.E. Hinton.

OVER EXPOSED ** 15
Robert Markowitz USA 1992
Marcy Walker, Dan Luria, Terence Knox, Taylor Miller, Rod Sell, Howard Plat, Mike Nussbaum, Ann CUsack, Logan Hutt, Rick Snyder, Amanda Laughlin, Tom Amandes, Michael E. Myers, Rachel Patterson, Peter Syvertsen, Jason Wells
A woman embarks on an affair as a means of getting even with her husband, whom she suspects of adultery. But she eventually learns that she has been set up for blackmail, when her lover threatens to send cassettes of their lovemaking to her husband and friends. Average erotic thriller.
THR 89 min mTV VIDrel: 4-FRONT/POLYREC/ODY V/sh

OVER HER DEAD BODY * 15
Maurice Phillips USA 1989
Elizabeth Perkins, Judge Reinhold, Jeffrey Jones, Maureen Mueller, Rhea Perlman, Charles Tyner, Brion James, Henry Jones, Michael J. Pollard, James Lashly, Nicholas Love, Alex Chapman, Sharon Schwartz-Hartley, Deenie Dakota
Having accidentally killed her sister when she surprised her in bed with her husband, a woman finds herself lumbered with the job of disposing of the corpse, and resorts to a range of desperate but increasingly unsuccessful methods. A frenetic black comedy of little invention, its scant humour is derived from a one-joke format and a collection of idiotic characters, whose presence becomes increasingly annoying.
Aka: ENID IS SLEEPING
COM 98 min (ort 105 min) VIDrel: FIRST/SONOP L/A V

OVER THE HILL *** PG
Nadia Tass USA 1992
Olympia Dukakis, Sigrid Thornton, Derek Fowlds, Pippa Grandison, Bill Kerr, Steve Bisley, Martin Jacobs, Gerry Connolly, Andrea Moore, Aden Young, Gabriela Canlida, Joy Irvine, Rebecca Caswell, Craig Elliott, Anne Looby
An ageing widow decides to leave the confined atmosphere of her son's home and travels to Australia to visit her estranged and snobbish daughter, who is unwelcoming and resentful. Having acquired a car, she takes off into the Australian interior, experiencing some odd incidents and romantic encounters on something of a journey of self-discovery. A truly outstanding performance from Dukakis as the mother sustains the thin script.
DRA 98 min (ort 102 min) VIDrel: VCC/DISC/COLUM L/A V/sh

OVER THE LINE * 18
Oliver Hellman (Ovidio Assonitis)/Robert Barrett USA 1993
Lesley-Anne Down, John Enos, Lady B. Pearl, Carolyn Seymour, Hugo Napier, Michael Parks, Tomas Arana, Maurizio Fiorinia, Clyde Barrett, David Ianelto, Robert Messina, Michael Cavalleri, Armando Pucci, Chris Hyde, Gunther Simon
An attractive female professor is asked to teach a group of prisoners and falls for one of them, but he proves to be of the violent and possessive type. An undistinguished erotic thriller offering the conventional victim/aggressor gender-based stereotypes.
THR 104 min (ort 108 min) VIDrel: WHV L/A V

OVER THE TOP * PG
Menaham Golan USA 1986
Sylvester Stallone, Susan Blakely, Robert Loggia, David Mendenhall, Rick Zumwalt, Chris McCarty, Terry Funk, Chris McCarty, Terry Funk, Bob Bearttie, Allan Graf, Bruce Way, Jimmy Keegan, John Braden, Tony Munafo, Randy Raney
The title is an apt description for this story of a strong but determined truck driver who fights to gain custody of his young son,

against his wealthy and influential father-in-law. Stallone tries a softer image but winds up being merely inaudible.
A/AD 89 min (ort 100 min) VIDrel: MGM/WHV V/sur

OVERBOARD ** PG
Gary Marshall USA 1987
Goldie Hawn, Kurt Russell, Edward Herrmann, Katherine Helmond, Michael Hagerty, Roddy McDowall, Jared Rushton, Jeffrey Wiseman, Brian Price, Jamie Wild, Frank Campanella, Harvey Alan Miller, Frank Buxton, Carol Willard
An insufferably overbearing and arrogant rich woman falls off her yacht, losing her memory in the process, and ends up living the life of an ordinary mortal as a housewife and mother to a carpenter's four children. A pleasant comedy vehicle of appealing performances and banal scripting.
COM 107 min (ort 112 min) VIDrel: MGM/WHV V/sur

OVERKILL ** 18
Dean Ferrandini USA 1996
Aaron Norris, Michael Nouri, Pamela Dickerson, David Rose, Kenny Moscow
A narcotics officer is on holiday in San Carlos, but gets involved with mobsters when he saves a man's life. Unconvincing and formulaic actioner, that offers enough thrills to make it an adequate time-filler.
A/AD 89 min VIDrel: COLUM/SONOP V

OVERKILL: THE AILEEN WUORNOS STORY ** 15
Peter Levin USA 1993
Jean Smart, Park Overall, Tim Grimm, Ernie Lievely, Geoffrey Rivas, Erich Anderson, Anthony Dean Fields, T.C. Warner, Dave Florek, Brion James, Marc Alaimo, Arel Blanton, Joe Bratcher, Robert Alan Browne, Lee De Broux, Beth Grant
Unappealing and exploitative drama based on the criminal career of America's first known female serial killer. Despite good performances all round, this is a film that is very hard to like. The documentary "Aileen Wuornos: The Selling Of A Serial Killer" also dealt with this individual.
Aka: AILEEN WUORNOS STORY, THE
DRA 93 min mTV VIDrel: ODY/SONOP V/sh

OVERRULED: IN MY DAUGHTER'S NAME ** 15
Jud Taylor USA 1992
Donna Mills, Lee Grant, John Getz, John Rubinstein, Ron Frazier, Adam Storke
Loosely based on a true story, this tells of a woman who decides to take the law into her own hands when the man who raped and murdered her daughter gets acquitted at his trial. An unpleasant saga, quite well handled but hard to enjoy.
DRA 92 min mTV VIDrel: 4-FRONT/POLYREC/ODY V/sh

OVERTHROW, THE ** 18
Larry Ludman (Fabrizio de Angelis) ITALY 1989
Lewis Van Bergen, Roger Wilson, John Philip Law
A portrait of a country torn apart by civil strife, where a journalist becomes the target of a group plotting a military coup, after he inadvertently takes a photo of their leader.
Aka: GOING STRAIGHT
A/AD 80 min Cut (19 sec – ort 90 min) VIDrel: IMPENT V

OWD BOB ** U
Robert Stevenson UK 1938
Will Fyffe, John Loder, Margaret Lockwood, Moore Marriott, Graham Moffatt, Wilfred Walter, Eliot Mason, A. Bromley Davenport, H.F. Maltby, Edmund Breon, Wally Patch, Alf Goddard
In Cumberland, a young woman falls in love with the owner of the latest champion at the sheepdog trials, but a nasty rivalry develops between this man and the woman's father, eventually leading to the old man's dog being accused of killing sheep. Likeable outdoors yarn, a bit over-sentimental but generally competent in all departments. Much of the plot was used in the later film "Thunder In The Valley".
Aka: TO THE VICTOR
DRA 78 min B/W VIDrel: CARL/TECH V
Boa: novel by Alfred Olivant.

OWL, THE ** (18)
Michael Green/Alan Smithee USA 1991
Adrian Paul, Patricia Charbonneau, Brian Thompson, Erika Flores, Jacques Apollo Bollon, David Anthony Marshall, Billy "Sly"

Willaims, David Selburg, Mark Lowenthal, Alan Scarfe, Thomas Rosales Jr, Alejandro Quezada
Having lost his wife and daughter to a killer, the distraught husband finds himself unable to sleep at night, and thus becomes the title figure – a night-time vigilante who offers his help to those in need. This story revolves around his exploits on behalf of a girl whose father has been kidnapped by a drug ring. Fair.
DRA   81 min (ort 90 min)   SATrel: SKY MOVIES

OWL AND THE PUSSYCAT, THE *** 15
Herbert Ross USA 1970
Barbra Streisand, George Segal, Robert Klein, Allen Garfield, Roz Kelly, Jacques Sandulescu, Jack Manning, Grace Carney, Barbara Anson, Kim Chan, Stan Gottlieb, Evelyn Lang (Marilyn Chambers), Dominic T. Barto, Buck Henry
Beautifully acted zany comedy about an unlikely pair; a semi-illiterate prostitute and a bookseller who are thrown together by circumstance and fall in love. The script by Buck Henry is adapted from a hit Broadway comedy.
COM   96 min   VIDrel: CASPIC/BMGREC L/A   V/sh
Boa: play by Bill Manhoff.

OX, THE *** 15
Sven Nykvist SWEDEN 1991
Stellan Skarsgard, Ewa Froling, Lennart Hjulstrom, Max Von Sydow, Liv Ullman, Bjorn Granath, Erland Josephson, Rikard Wolff, Helge Jordal, Agneta Prytz, Bjorn Gustafson, Jaqui Safra
During a famine in the 1860s a man slaughters an ox belonging to his employer in order to feed his family and is sentenced to life imprisonment for this crime. Pardoned after six years he returns home to learn of how his family have managed in his absence. A slow but moving historical drama that gave Nykvist (Bergman's favourite cameraman) his directing debut.
Aka: OXEN
DRA   88 min (ort 92 min) wScrn   VIDrel: ARTIF/20TH V/sh

OXFORD BLUES * 15
Robert Boris USA 1984
Rob Lowe, Ally Sheedy, Julian Sands, Amanda Pays, Michael Gough, Aubrey Morris, Gail Strickland, Alan Howard, Julian Forth, Cary Elwes, Bruce Payne, Anthony Calf, Pip Torrens, Richard Hunt, Peter Jason, Peter-Hugo Daly
A poor remake of the 1938 film "A Yank At Oxford" with a young rough diamond pursuing a blue-blooded beauty, and trying to win her heart by volunteering for the rowing team. A curiously lifeless film that tries hard to please with a few comic episodes, but Lowe and the snobs he comes into conflict with are so obnoxious that this stilted work is all too easy to dislike.
COM   93 min (ort 98 min)   VIDrel: GUILD/POLYREC L/A V

OZONE ** 18
J.R. Bookwalter USA 1992
James Black, Tom Hoover
A botched drugs raid results in a police officer being injected with a deadly drug, and this results in him slipping into a nightmare world inhabited by zombie drug dealers. An interesting idea gets the usual overblown treatment, which is a great pity, as the opening premise had enormous potential.
HOR   83 min   VIDrel: SCEDGE/RTM   V

P

PACIFIC HEIGHTS ** 15
John Schlesinger USA 1990
Melanie Griffith, Michael Keaton, Matthew Modine, Mako, Laurie Metcalf, Nobu McCarthy, Carl Lumbly, Dorian Harewood, Luca Bercovici, Tippi Hedren, Sheila McCarthy, Dan Hedaya, Miriam Margoyles, Nicholas Pryor, Beverly D'Angelo
A yuppie couple buy a house in San Francisco, clean it up and rent out two of its apartments. However, one of their tenants proves to be a nutcase, and he makes their lives a misery. An implausible and disorganised mess, with a few bright moments, it veers wildly between tension and tomfoolery without ever developing into anything substantial. Music is by Hans Zimmer.
THR   98 min (ort 103 min)   VIDrel: 20TH/TECH L/A   V

PACK OF LIES ** 15
Anthony Page USA 1987
Ellen Burstyn, Teri Garr, Alan Bates, Ronald Hines, Clive Swift, Sammi Davis, Daniel Benzali

In the 1960s a London couple allow their home to be used by British Intelligence to spy on their Canadian neighbours, who unknown to them, are KGB agents. The film (very loosely inspired by real events) focuses on the strains the couple endure in having to keep up the pretence of normality under these difficult circumstances. Whitemore's clumsy play is improved by Ralph Gallup's adaptation, and the fine performances help slightly. Average.
DRA   103 min mTV   VIDrel: VCC/DISC L/A   V
Boa: play by Hugh Whitemore.

PACKAGE, THE ** 15
Andrew Davis USA 1989
Gene Hackman, Joanna Cassidy, Tommy Lee Jones, John Heard, Dennis Franz, Pam Grier, Reni Santoni, Kevin Crowley, Ike Pappas, Thalmus Rasulala, Ron Dean, Nathan Davis, Chelcie Ross, Joe Greco, Marco St John, Juan Ramirez, Miguel Nino
Political thriller that casts Hackman as an army sergeant who is reprimanded after a security breach at a disarmament conference and given the menial job of escorting a court-martialled soldier back to the States. Following the escape of his prisoner, he begins to discover that he has been used as a dupe by Soviet and American military dissidents. A film that initially develops tension, only to lose it in the face of progressive implausibility.
THR   103 min (ort 108 min)
VIDrel: 4-FRONT/POLYREC/VISVID   V/sur

PADRE PADRONE *** 15
Paolo Taviani/Vittorio Taviani ITALY 1977
Fabrizio Forte, Omero Antonutti, Saverio Marioni, Marcella Michelangeli, Gavino Ledda
The moving story of a Sardinian boy and his protracted rebellion against his brutish and domineering father. Despite the crushing ignorance of his surroundings, he contrives to study classical languages and eventually graduates from college. An absorbing example of autobiographical film-making, quite depressing but undeniably memorable.
Aka: FATHER AND MASTER
DRA   108 min (ort 114 min)   VIDrel: ARTIF/20TH   V/h
Boa: autobiography of Gavino Ledda.

PAGEMASTER, THE ** U
Maurice Hunt USA 1994
Macaulay Culkin, Christopher Lloyd, Ed Begley Jr, Mel Harris plus voices of: Whoopi Goldberg, Patrick Stewart, Frank Welker, Leonard Nimoy, George Heam, Dorian Harewood, Jim Cummings, B.J. Ward, Ed Gilbert, Richard Erdman
Caught in a storm, a young boy who is prone to nervousness takes refuge in a library, where the books come to life and he is taken on a magical journey to a world of fantasy and adventure. A live-action/animated tale that strays into NEVERENDING STORY territory, and sadly fails to exploit its opening premise, driving home a message about the joys of reading with a great lack of subtlety. The LV disc includes a short documentary on the making of the film.
FAN   97 min   VIDrel: COLUM/SONOP; PION (LV only) V/sur LV

PAINT ME A MURDER * (PG)
Alan Cooke UK 1984
Michelle Phillips, James Laurensen, David Robb, Alan Lake, Tony Steedman, Morgan Sheppard, Mark Heath, Michael Watkins, McKevitt, Richard Parmentier, Gerald Sim, Indira Ioshi, Jeillo Edwards, David Millett, Neil Morrissey
At the instigation of his grasping wife, a talented painter fakes his own death by sinking a rowing boat and throwing overboard a suitable painting as a suicide note. However, he soon regrets his rashness when he is forced to hide in his studio to avoid detection and his wife begins an affair with the art dealer who has been called in to sell his paintings. Predictably, this triangle soon ends in tragedy, in this downbeat and poorly made offering.
Aka: HAMMER HOUSE OF MYSTERY AND SUSPENSE: PAINT ME A MURDER
HOR   72 min mTV   SATrel: UK GOLD

PAINT YOUR WAGON *** PG
Joshua Logan USA 1969
Lee Marvin, Clint Eastwood, Jean Seberg, Harve Presnell, Ray Walston, Tom Ligon, Alan Baxter, William O'Connell, Paula Trueman, Robert Eastman, Geoffrey Morgan, H.B. Haggerty, Terry Jenkins, Karl Bruck
A glossy and expensive musical about the California Gold Rush,

with some fine Lerner and Loewe songs and plenty of action. Marvin and Eastwood share Seberg – the wife they bought at an auction. Marred in parts by poor singing, but Presnell's "They Call the Wind Maria" and Marvin's rendition of "Wanderin' Star" are highlights. The witty script is by Paddy Chayevsky.
MUS   153 min (ort 167 min) wScrn
VIDrel: 4-FRONT/POLYREC/CIC   V/sh
Boa: musical by Alan Jay Lerner and Frederick Loewe.

PAINTED HEART **   15
Michael Taav USA   1993
Will Patton, Bebe Nuewirth, Robert Pastorelli, Casey Siemaszko, Mark Boone Jr, Jayne Haynes, Richard Hamilton, Jeff Weiss, Everett Smith, John Diehl, Dale Rehfeld, Cody Dobson, Robert Breuler, Okoro H. Johnson, Jim Kruse, Wendy Bastrup
A house painter gets involved in an affair with his boss' wife and finds himself caught up in a web of passion and murder. A fine performance by Patton contributes greatly to the appeal of this otherwise routinely plotted movie.
Aka: PAINT JOB, THE
DRA   86 min (ort 90 min)   VIDrel: TART/20TH L/A   V/s

PAISAN **   PG
Roberto Rossellini ITALY   1946
Carmela Sazio, Robert Van Loon, Dots M. Johnson, Alfonsino Pasca, William Tubbs, Maria Michi, Renzo Avanzo, Harriet White, Carla Pisacane, Gar Moore, Dale Edmonds, Mats Carlson, Gigi Gori, Cigolani, Lorena Berg, Albert Heinz
Six vignettes of life in occupied Italy between 1843 and 1945, written by Rossellini and Federico Fellini, with a rough documentary feel that is enhanced by the use of improvisation and a largely non-professional cast. Interesting as a historical document but not on a par with the director's earlier ROME, OPEN CITY.
Aka: PAISA
DRA   120 min B/W   VIDrel: CONNO/RTM   V

PAJAMA GAME, THE ****   U
Stanley Donen USA   1957
Doris Day, John Raitt, Eddie Foy Jr, Reta Shaw, Carol Haney, Barbara Nichols, Thelma Pelish, Jack Straw, Ralph Dunn, Owen Martin, Buzz Miller, Jackie Kelk, Ralph Chambers, Mary Stanton, Kenneth LeRoy, Jack Waldron
Workers at a pajama factory go on strike for higher wages, but complications ensue when their female negotiator falls in love. A fine version of a Broadway musical that, despite the unlikely subject, works both as an entertaining musical and a fluid and effective piece of cinema. Songs (by Richard Adler and Jerry Ross) include: "There Once Was A Man", "Hey, There" and "Hernando's Hideaway". Choreography is by Bob Fosse.
MUS   97 min (ort 101 min)   VIDrel: WHV   V
Boa: musical/book Seven and a Half Cents by Richard Bissel.

PAL JOEY ***   PG
George Sidney USA   1957
Rita Hayworth, Frank Sinatra, Kim Novak, Bobby Sherwood, Hank Henry, Barbara Nichols, Elizabeth Patterson, Robin Morse, Frank Wilcox, Pierre Watkin, Barry Bernard, Ellie Kent, Mara McAfee, Betty Utey, Bek Nelson
A brash cabaret entertainer lands in San Francisco determined to make it big, but scores his biggest hits with a wealthy socialite and a chorus line cutie. The classic Rodgers and Hart score includes: "The Lady Is A Tramp", "There's A Small Hotel", "I Could Write A Book" and "My Funny Valentine".
MUS   105 min (ort 111 min) B/W   VIDrel: COLUM/SONOP   V
Boa: play/short stories by John O'Hara/musical by Richard Rodgers and Lorenz Hart.

PALAIS ROYALE **   15
Martin Lavut CANADA   1988
Kim Catrall, Matt Craven, Kim Coates, Brian George, Michael Hogan, Dean Stockwell, Jan Rubes, Dee McCafferty, Victor Ertmanis, Mario Romano, Sean Hewitt, Robin McCulloch, David Fox, Helen Hughes, Sam Malkin, Henry Alessan
Period melodrama circa 1959, with a Toronto advertising executive getting drawn into fronting for local crooks, when he meets the girl of his dreams, a gangster's moll whose face has appeared on cigarette advertising hoardings. Once enrolled, his money-making talents blossom as he develops the legitimate front his colleagues so badly need. An interesting film that's flawed by a disappointing climax.
Aka: SMOKE SCREEN
DRA   87 min (ort 100 min)   VIDrel: EIV/SONOP   V

PALE BLOOD **   18
V.V. Dachin Hsu USA   1989
Pamela Ludwig, Wings Hauser, George Chakiris, Diana Frank, Darcy Demoss
Armed with a video camera, a former vampire hunter walks the streets of L.A. in search of his quarry, little realising that the city also contains a lone vampire searching for some more of his kind.
HOR   89 min (ort 95 min)   VIDrel: CAPIT/GUILD   V

PALE RIDER **   15
Clint Eastwood USA   1985
Clint Eastwood, Michael Moriarty, Christopher Penn, Carrie Snodgress, John Russell, Sydney Penny, Richard Dysart, Richard Kiel, Doug McGrath, Charles Hallahan, Marvin J. McIntyre, Fran Ryan, Richard Hamilton, Graham Paul
Eastwood's first real Western since THE OUTLAW JOSEY WALES is essentially a re-run of HIGH PLAINS DRIFTER. A violent and mysterious stranger comes to the aid of a mining community threatened by the local baddies who are in the pay of a large mining corporation. Well made it may be, but the clumsy attempts at symbolism and shameless cloning of SHANE give the whole work an air of heavy-handed pretentiousness.
WES   111 min (ort 116 min)   VIDrel: WHV   V/sur
Boa: novel by A.D. Foster.

PALERMO CONNECTION, THE **   15
Francesco Rosi FRANCE/ITALY   1990
James Belushi, Mimi Rogers, Joss Ackland, Carolina Rosi, Philippe Noiret, Vittorio Gassman
A candidate in the New York mayoral race who is fighting on an anti-Mafia platform goes to Palermo to study the enemy at first hand, in this complex and dramatic adaptation of Roux's novel.
Aka: DIMENTICARE PALERMO; TO FORGET PALERMO
DRA   85 min (ort 110 min)   VIDrel: MARQ/QUANT   V
Boa: novel Oublier Palermo by Edmond Charles Roux.

PALMY DAYS **   U
Edward Sutherland USA   1931
Eddie Cantor, Charlotte Greenwood, Charles Middleton, George Raft, Walter Catlett, Spencer Charters, Barbara Weeks, Paul Page, Harry Woods, Charles B. Middleton, Loretta Andrews, Edna Callahan, Nadine Dore, Ruth Edding
A nervous man gets himself deeply involved with his neighbour's crooked schemes but prevents him from robbing the wealthy owner of a successful bakery. A frenetic Cantor vehicle that provides ample scope for his boundless energy and some nice song and dance routines. Look out for Betty Grable as one of the Goldwyn Girls.
MUS   77 min B/W   VIDrel: VCC/DISC/COLUM   V

PALOMINO **   PG
Michael Miller USA   1991
Lindsay Frost, Lee Horsley, Eva Marie Saint, Rod Taylor, Beau Gravitte, Michael Greene, Daniel Davis, Ryan Todd, Peter Bergman, Rose Portillo, Kelly Miller, Blake Conway, Andi Chapman, David Drummond, Ashley Bank
A beautiful female photographer suffers the break-up of her marriage and retreats to a friend's horse ranch for a break from work. Once there, she falls in love with a ranch-hand, who feels unable to respond as he feels out of his depth. However, when she is paralysed after being thrown from a horse, he makes a bid to return into her life. Very much a standard Danielle Steel offering: slick, expensive and instantly forgettable.
Aka: DANIELLE STEEL'S PALOMINO
DRA   91 min   VIDrel: 4-FRONT/POLYREC   V
Boa: novel by Danielle Steel.

PAMELA PRINCIPLE, THE **   18
Toby Phillips USA   1991
J.K. Dumont, Veronica Cash, Shelby Lane, Troy Donahue, Frank Pesce, Eugene Stevenson, Linnea Duvall, Kristine Wilde, Bobby Johnston, Tracey Wolfe, Todd Lemish, Darlene Sellers, Aestonisha, Walter McBride, Marilyn Mayfield
A middle-aged husband envies the sexual prowess of his son and being rather bored with his marriage, embarks on a wild affair with a sexy young model who is twenty years his junior. However, he soon realises that they have absolutely nothing in common, and starts to hanker for a return to his wife. Adequate erotic drama with an entirely predictable outcome. A sequel followed.
Aka: INDECENT ADVANCES
DRA   93 min   VIDrel: MIA/DISC/IMPENT   V

PAMELA PRINCIPLE 2, THE ** 18
Edward Holzman USA 1993
India Allen, Alina Thomson, Elizabeth Sandifer, Nick Rafter, Frank Pesce, Daniel Anderson, Shannon McLeod, Cathleen Raymond, Tonya Poole, Bobby Johnston, Shauna O'Brien, Sarah Bellomo
A successful architect experiences a mid-life crisis and abandons his chosen profession in favour of working as a glamour photographer. Unfortunately, he falls for one of the many scantily clad models he shoots and thereby puts his marriage in jeopardy. A tepid tale, with much nudity but little else to arouse the interest.
Aka: SEDUCE ME: THE PAMELA PRINCIPLE 2
DRA 101 min VIDrel: 20VIS/SONOP V

PANCHO BARNES ** (PG)
Richard T. Heffron USA 1988
Valerie Bertinelli, Ted Wass, James Stephens, Cynthia Harris, Geoffrey Lewis, Sam Robards, Bill Shaw, Melissa Ragsdale, Robert Manning, Randolph Tallman, Dan Ammerman, Luis Lemus, Russell Juelg, Norman Bennett, Vince Davis
A long and detailed account of the career of a famous woman pilot, charting her beginnings as a bored socialite who finds in flying something that gives her life the meaning it had lacked until then. Competently acted, its flying sequences are its best feature.
DRA 149 min mTV SATrel: SKY MOVIES

PANDORA'S BOX ** PG
Georg W. Pabst GERMANY 1928
Louise Brooks, Fritz Kortner, Francis Lederer, Gustav Diesse, Carl Gotz, Alice Roberts
After a murdering her lover, a temperamental woman turns to prostitution and eventually falls victim to Jack the Ripper. A curious, experimental fantasy melodrama, generally considered as marking the end of German Expressionism.
Aka: DIE BUCHSE DER PANDORA; LULU
DRA 104 min (ort 110 min) B/W silent
VIDrel: TART/20TH V/dm
Boa: plays Erdgeist/Pandora's Box by Franz Wedekind.

PANIC IN NEEDLE PARK, THE *** 18
Jerry Schatzberg USA 1971
Al Pacino, Kitty Winn, Richard Bright, Kiel Martin, Michael McClanathan, Warren Finnerty, Alan Vint, Marcia Jean Kurtz, Raul Julia, Larry Marshall, Gil Rogers, Paul Sorvino, Angine Ortega, Paul Mace, Nancy MacKay, Joe Santos
A young girl meets a small-time crook, who becomes hooked on heroin and drags her down with him. A sobering examination of the nether world of the junkie, scripted by Joan Didion and John Gregory Dunne. Pacino and Winn both give performances of remarkable conviction.
DRA 104 min Cut (57 sec – ort 110 min) VIDrel: 20TH V
Boa: novel by James Mills.

PANTHER ** 15
Mario Van Peebles USA 1995
Kadeem Hardison, Courtney B. Vance, Bokeem Woodbine, Marcus Chong, Richard Dysart, Anthony Griffith, Joe Don Baker, Bobby Brown, Nefertiti, James Russo, Jenifer Lewis, Chris Rock, Roger Guenveur Smith, Michael Wincott, Dick Gregory
A long but hardly complex and very two-dimensional look at the rise of the Black Panthers and their escalating confrontations with the authorities during the 1970s. Despite the efforts of a first-rate cast, this one fails to do justice to its subject and suffers greatly from the lack of a more intelligent approach. Screenplay is by Melvin Van Peebles.
A/AD 125 min cC VIDrel: POLY/POLYREC V/s
Boa: novel by Melvin Van Peebles.

PAPER, THE *** 15
Ron Howard USA 1993
Michael Keaton, Glenn Close, Robert Duvall, Randy Quaid, Marisa Tomei, Jason Robards, Jason Alexander, Spalding Gray, Catherine O'Hara, Lynn Thigpen, Jack Kehoe, Roma Maffia, Clint Howard, Geffrey Owens, Amelia Campbell, Jill Hennessy
Fly-on-the-wall account of twenty-four hours in the life of a New York tabloid and its harassed journalists as they juggle with the demands of both their jobs and private lives. A clutch of very watchable performances make this quite an enjoyable experience, but the absence of any credible discussion on newspaper ethics (or the lack of them) is a clear disappointment. See also FRONT PAGE STORY.
DRA 106 min (ort 112 min) cC VIDrel: CIC/SONOP; PION (LV only) V/sur LV

PAPER CHASE, THE *** PG
James Bridges USA 1973
Timothy Bottoms, Lindsay Wagner, John Houseman, Graham Beckel, Edward Herrmann, Craig Richard Nelson, James Naughton, Bob Lydiard, Regina Baff, David Clennon, Lenny Baker
The story of a Harvard law school as seen through the eyes of its first-year students. Bridges wrote the screenplay as well as directing. This film gave Houseman his first taste of stardom in the role of Professor Kingsfield, that he continued in a mediocre TV series. A blend of comedy and drama, with student Bottoms discovering that his girlfriend is Kingsfield's daughter. Photography is by Gordon Willis. AA: S. Actor (Houseman).
DRA 107 min (ort 111 min) VIDrel: 20TH/TECH L/A V
Boa: novel by John Jay Osborn Jr.

PAPER MARRIAGE *** 15
Krzysztof Lang POLAND/UK 1992
Gary Kemp, Joanna Trepchinska, Rita Tushingham, Richard Hawley, David Horovitch, William Ilkey, Martin McKellan, Ann Mitchell, Fred Pearson, Gary Whelan, Renata Breger, Charles Chadwick, Hans Chrudzynska, Art Davies
A polish girl comes to Britain in pursuit of an English doctor she met in Warsaw, but her former lover spurns her. Anxious to remain in the UK, she enters a marriage of convenience with a petty crook and gets caught up in his differences with some thuggish associates. A couple of strong character studies provide the bulk of interest in a well made but terribly depressing drama.
DRA 84 min (ort 90 min) VIDrel: CURZON/20TH V/sh

PAPER MASK *** 15
Christopher Morahan UK 1990
Paul McGann, Amanda Donohoe, Frederick Treves, Jimmy Yuill, Tom Wilkinson, Barbara Leigh-Hunt, Mark Lewis Jones, John Warnaby, Alexandra Mathie, Oliver Ford Davies, Frank Baker, Clive Rowe, Robert Oates, Karen Ascoe
An ambitious and amoral hospital porter decides to improve his standing in the community by adopting the identity of a recently-deceased doctor, and succeeds in bluffing his way into a job at a Bristol hospital. Near to giving up the hoax when his lack of skill leads to the death of a casualty patient, he is driven to carry on by his lust for an attractive nurse. A creepy and quite taut thriller, well scripted by former doctor John Collee.
THR 105 min VIDrel: L/A V
Boa: novel by John Collee.

PAPERBOY, THE ** 18
Douglas Jackson USA 1993
Alexandra Paul, Marc Marut, William Katt, Brigid Tierney, Krista Errikson, Frances Bay
An attractive divorcee returns home to deal with the details of her late mother's estate, and befriends a lonely, young paperboy who lives next door. The latter is soon looking upon her as a surrogate mother, and his interest soon becomes less than healthy. Meanwhile, the town is being terrorised by a wave of killings.
THR 90 min (ort 93 min) VIDrel: FIRST/SONOP V

PAPERHOUSE *** 15
Bernard Rose UK 1988
Charlotte Burke, Glenne Headly, Gemma Jones, Ben Cross, Elliot Spiers, Sarah Newbold, Jane Bertish, Samantha Cahill, Gary Bleasdale, Steve O'Donnell, Karen Gledhill, Barbara Keogh
Bored with her lessons, a young schoolgirl idly sketches an imaginary house. Later she begins to experience a series of fainting fits, and during each of these finds herself ever closer to an image of the house she has drawn. With time her visions of this dream-house begin to dominate her life, and she meets a crippled boy who inhabits the building. A highly atmospheric and rather chilling psychological fantasy, the feature debut of music-video director Rose.
FAN 88 min (ort 94 min) VIDrel: FIRST/SONOP L/A V
Boa: novel Marianne Dreams by Catherine Storr.

PAPILLON **** 18
Franklin J. Schaffner USA 1973
Steve McQueen, Dustin Hoffman, Victor Jory, Don Gordon, Anthony Zerbe, George Coulouris, Woodrow Parfrey, Bill Mumy, Gregory Sierra, Robert Deman, Ratna Assan, William Smithers, Barbara Morrison, Ellen Moss, Dalton Trumbo
An expensive and lovingly produced account of Henri Charriere, known as "Papillon", who made repeated attempts to escape from his incarceration on Devil's Island, the French

Caribbean penal colony off the coast of French Guyana, where he was sent after being convicted of murder. The colourful locations and attention to detail make this a most rewarding film. Written by Dalton Trumbo and Lorenzo Semple Jr, with music by Jerry Goldsmith.
DRA   144 min (ort 150 min)   VIDrel: VCC/DISC/COLUM V
Boa: novel by Henri Charriere.

PARADINE CASE, THE **
U
Alfred Hitchcock USA   1947
Gregory Peck, Charles Laughton, Louis Jourdan, Ann Todd, Charles Coburn, Ethel Barrymore, Alida Valli, Leo G. Carroll, Joan Tetzel, Isobel Elsom, Pat Aherne, Alfred Hitchcock, John Goldsworthy, Lewter Matthews, Colin Hunter
A barrister falls in love with the woman he is defending on a charge of murder. Despite the stilted and excessively verbose nature of the script, Hitchcock's treatment largely redeems it, though this is far from one of his best works. Produced and written by David O. Selznick.
DRA   109 min (ort 132 min) B/W   VIDrel: VCC/DISC; PION (LV only)   V LV
Boa: novel by Robert Hichens.

PARADISE **
PG
Mary Agnes-Donoghue USA   1991
Melanie Griffith, Don Johnson, Elijah Wood, Thora Birch, Sheila McCarthy, Eve Gordon, Louise Latham, Greg Davis, Sarah Trigger, Richard K. Olsen, Rick Andosca, Anthony Romano, Timothy Erskine, Chestley Price, Dave Hager
Jean-Loup Hubert's 1987 film "Le Grand Chemin" gets the Hollywood remake treatment, with a rather good screenplay from the director. The story has a ten-year-old boy being sent by his pregnant mother to spend a summer break with her former schoolfriend Lily. However, Lily and her husband are estranged since the death of their own child, but the boy's arrival forces them to confront their innermost feelings. Quite a good yarn, but far too sentimental.
DRA   107 min (ort 111 min)   VIDrel: TOUCH/TECH L/A V

PARADISE CANYON ***
U
Carl L. Pierson USA   1935
John Wayne, Marion Burns, Earle Hodgins, Yakima Canutt, Reed Howes, Perry Murdock, Gordon Clifford, Henry Hall, Gino Corrado, Tex Palmer, Earl Dwire, John Goodrich, Herman Hack
An undercover agent works along the Mexican border, his mission being to capture a gang of counterfeiters. This was Wayne's last film in the "Lone Star" series from Monogram, and is one of the better offerings of that period.
WES   55 min B/W coVer   VIDrel: ENTUK L/A   V

PARALLAX VIEW, THE ****
15
Alan J. Pakula USA   1974
Warren Beatty, Paula Prentiss, William Daniels, Walter McGinn, Hume Cronyn, Kenneth Mars, Kelly Thordsen, Earl Hindman, Chuck Waters, Bill Joyce, Bettie Johnson, Bill McKinney, Joanne Harris, Ted Gehring, Lee Purlford, Doria Cook
Taut thriller about a conspiracy to eliminate all the witnesses to a political killing. Script is by David Giler and Lorenzo Semple Jr. The same director-photographer-design team went on to make ALL THE PRESIDENT'S MEN, but that film is not nearly as exciting as this entertaining work.
THR   101 min   VIDrel: L/A   V
Boa: novel by Loren Singer.

PARALLEL LIVES **
15
Linda Yellen USA   1994
James Belushi, Liza Minnelli, Michael O'Rourke, James Brolin, Helen Slater, LeVar Burton, Jack Klugman, Grant Varjas, Patricia Wettig, Ben Gazzara, Lindsay Crouse, Mira Sorvino, Paul Sorvino, Matthew Perry, David Lansbury, Ally Sheedy
A fraternity/sorority reunion provides the background to this improvised tale, much on the lines of the director's previous CHANTILLY LACE, where various conflicts are played out as old friends and foes meet. The acting cannot be faulted but here the over-reliance on improvisation tends to result in a very sketchy plot (it revolves around a murder mystery) and dialogue that does not really hang together. See also THE BIG CHILL.
DRA   100 min (ort 105 min) mCab   VIDrel: ODY/SONOP V/sh

PARASITE *
18
Charles Band USA   1982
Robert Glaudini, Luca Bercovici, James Davidson, Demi Moore, Al Fann, Cherie Currie, Vivian Blaine, Tom Villard, James Cavan, Joanelle Romero, Freddie Moore, Natalie May, Cheryl Smith, Joel Miller
The work of a government scientist in creating a breed of killer parasites goes out of control after a nuclear explosion, and the inhabitants of a small town are soon falling victim to them, and these creatures not thrive on their hosts but leap out to attack others. An amateurish ALIEN clone that was filmed in 3-D just to make this experience a little more sickening, though this will be lost on the TV screen (a small blessing perhaps). See also SHIVERS.
HOR   84 min   VIDrel: MIA/DISC   V

PARENT TRAP, THE **
U
David Swift USA   1961
Hayley Mills, Maureen O'Hara, Brian Keith, Charlie Ruggles, Una Merkel, Leo G. Carroll, Joanna Barnes, Cathleen Nesbitt, Ruth McDevitt, Nancy Kulp, Crahan Denton, Linda Watkins, Frank DeVol
Twin sisters meet for the first time at a summer camp, and eventually hatch a plot to re-unite their divorced parents. A cute Disney comedy of the dated and sentimental kind, with Mills playing both parts. Written by Swift and followed by two TV sequels in 1986 and 1989. Filmed once before as "Twice Upon a Time".
COM   122 min (ort 124 min)   VIDrel: WDV/TECH L/A   V
Boa: novel Das Doppelte Lottchen (Lottie And Lisa) by Erich Kastner.

PARENT TRAP 2 **
U
Ronald F. Maxwell USA   1986
Hayley Mills, Tom Skerritt, Bridgette Anderson, Carrie Kei Heim, Alex Harvey, Gloria Cromwell, Judith Tannen, Janice Tesh, Duchess Tomasello, Daniel Brun, Antonio Fabrizio, Ted Science, Margaret Woodall
Sequel to THE PARENT TRAP, with Mills playing the dual role of twin sisters, now both grown up, divorced and moving from Florida to New York. She finds that her daughter and her best friend are both just as conniving as she and her sister were as children. Glossy, romantic nonsense that was followed by a further sequel in 1989. Filmed on location in Tampa, Florida.
DRA   81 min (ort 95 min) mCab   VIDrel: WDV/TECH L/A V

PARENT TRAP 3 **
U
Mollie Miller USA   1989
Hayley Mills, Barry Bostwick, Ray Baker, Patricia Richardson, Joy Creel, Leanna Creel, Monica Creel, Christopher Gartin, Jon Maynard Pennell, Loretta Devine, Richard Coca, Nancy Fish, Christopher Gartin, Joy Maynard Pennell
Third time round, this second sequel to the 1961 film has moved quite far from the original idea, with our grown-up twins falling out when they both decide to marry the same man, a widowed father of triplets. A light and frothy comedy, followed by PARENT TRAP: HAWAIIAN HONEYMOON.
COM   85 min (ort 100 min) mTV   VIDrel: WDV/TECH L/A V

PARENT TRAP: HAWAIIAN HONEYMOON **
U
Mollie Miller USA   1989
Hayley Mills, Barry Bostwick, Joy Creel, Leanna Creel, Monica Creel, John M. Jackson, Sasha Mitchell, Jayne Meadows, Lightfield Lewis, Glenn Shadix, Nancy Lenehan, Joe Mays, Wayne Federman, Al Acain, Mike Ebner, Kamika Nakanula
Third sequel to the original PARENT TRAP that has Mills playing twin sisters once more (they're now grown up) who enjoy some mild little adventures of an undemanding sort in Hawaii. An aimless and amiable Disney confection.
COM   86 min (ort 100 min) mTV
VIDrel: WDV/BUENA L/A   V

PARENTHOOD ****
15
Ron Howard USA   1989
Steve Martin, Mary Steenburgen, Dianne Wiest, Jason Robards, Rick Moranis, Tom Hulce, Martha Plimpton, Keanu Reeves, Harley Jane Kozak, Dennis Dugan, Leaf Phoenix, Paul Linke, Jasen Fisher, Eileen Ryan, Helen Shaw, Alisan Porter
A warm and insightful character study that focuses on the problems of parenthood, as seen through the eyes of several parents who belong to the same large family. While Martin and Steenburgen cope with careers and kids, and Moranis's three-

year-old copes with Kafka, Wiest finds herself the mother of a pair of problem teenagers and Robards learns that parenthood doesn't end when the kids grow up. The music is by Randy Newman. A TV series followed.
COM 118 min (ort 124 min) VIDrel: CIC/SONOP V/sur
Boa: story by Lowell Ganz, Babaloo Mandel and Ron Howard.

PARENTS ** 18
Bob Balaban USA 1989
Randy Quaid, Mary Beth Hurt, Sandy Dennis, Bryan Madorsky, Kathryn Grody, Juno Miles-Cockell, Deborah Rush, Graham Jarvis, Helen Carscallen, Warren Van Evera, Wayne Robson, Mariah Balaban, Larry Palef
A comedy-horror yarn set in the 1950s, and telling of an inquisitive ten-year-old boy who begins to wonder where the tasty leftovers come from that father brings home every night, and just what goes on in their cellar. Fairly well handled and there are one or two laughs, but the lack of a good story sinks this one. The directorial debut of actor Balaban.
HOR 78 min (ort 94 min) VIDrel: FIRST/SONOP L/A V

PARIS BLUES *** 12
Martin Ritt USA 1961
Paul Newman, Sidney Poitier, Joanne Woodward, Diahann Carroll, Andre Luguet, Louis Armstrong, Barbara Lange, Serge Reggiani, Marie Versini, Moustache, Niko, Aaron Bridgers, Guy Pederson, Marie Velasco, Roger Blin, Helene Dieudonne
Story of two jazz musicians in Paris, their efforts to achieve success and love, and the price they have to pay. A well acted affair with some nice vistas of the city and plenty of great music, which partially compensates for the weak script.
DRA 94 min (ort 100 min) B/W wScrn
VIDrel: MGM/WHV V

PARIS, FRANCE ** 18
Gerard Ciccoritti CANADA 1993
Leslie Hope, Peter Outerbridge, Victor Ertmanis, Raoul Taurilo, Dan Lett, Patricia Ciccoritti (voice only)
A female author who has been fighting writer's block catches a handsome stranger (the latest discovery of her publisher husband) going through her drawers. Seeing in him a chance to stimulate her creative abilities, she decides to get to know him better, but she finds she has a rival for his affections in the shape of her husband's business partner. Screenplay is by Walmsley, and quite what point this verbose effort is trying to make never becomes clear.
DRA 112 min VIDrel: FIRC/RTM V
Boa: novel by Tom Walmsley.

PARIS, TEXAS *** 15
Wim Wenders FRANCE/WEST GERMANY 1984
Harry Dean Stanton, Dean Stockwell, Nastassia Kinski, Aurore Clement, Hunter Carson, Bernard Wicki, Viva Auder, Socorro Valdez, Tom Farrell, John Lurie, Jeni Vici, Sally Norvell, Sam Berry, Claresie Mobley, Justin Hogg, Edward Fayton
The strange story of a man returning after a long absence, and looking for his wife in an effort to rebuild his life. He eventually finds her working in a brothel (of sorts). Richly atmospheric and beautifully photographed (by Robby Muller) and acted, but excruciatingly slow and murky. A big hit with the critics, who probably read their own meanings into the enigmatic Sam Shepard script. The excellent score is by Ry Cooder. Filmed in English.
DRA 138 min (ort 150 min) VIDrel: CONNO/RTM V

PARIS TROUT *** 18
Stephen Gyllenhaal USA 1991
Dennis Hopper, Barbara Hershey, Ed Harris, Ray McKinnon, Tina Lifford, Eric Ware, Darnita Henry, Ronreaco Lee, Gary Bullock, Sharlene Ross, Jim Peck, Dan Biggers, Ernest Dixon, Wallace Wilkinson, Ronn Leggett, Ed Grady
In Georgia of the 1950s, a totally ruthless businessman, hateful in both his professional and private life, shoots a black woman and daughter when they are unable to settle an outstanding debt. Having done this, he gets a lawyer to defend him in court, even though the man knows full well that his client should pay for his crime. A well crafted and searing indictment of greed and corruption, with a powerful and shocking ending.
DRA 95 min (ort 99 min) mCab VIDrel: CURZON/20TH V/sur
Boa: novel by Peter Dexter.

PARIS WHEN IT SIZZLES ** (PG)
Richard Quine USA 1964
William Holden, Audrey Hepburn, Gregoire Aslan, Noel Coward, Raymond Bussieres, Christian Duvallex, Tony Curtis, Marlene Dietrich, Mel Ferrer, Thomas Michel, Dominique Boschero, Evi Marandi, Frank Sinatra (voice only)
A screenwriter desperate to complete a manuscript within forty-eight hours hires a secretary to help him speed up this process and they try out a variety of ideas, imagining themselves as characters in different films. A dull and lifeless remake of the 1953 French film "Henriette's Holiday", the attractive Paris locations are the only item of real interest.
COM 106 min (ort 110 min) SATrel: MOVIE CHANNEL

PARK IS MINE, THE ** 15
Steven Hilliard Stern CANADA 1985
Yaphet Kotto, Tommy Lee Jones, Helen Shaver, Peter Dvorsky, Eric Peterson, Lawrence Dane, Dennis Simpson, Reg Dreger, Louis DiBianco, Gale Garnett, Carl Marotte, Dennis O'Connor, Tom Harvey, R.D. Reid, George Bloomfield
A frustrated Vietnam veteran takes over Central Park in New York by force, in order to give vent to his emotions, resulting in a violent conflict taking place. Despite its powerful emotional content, whatever point the film is trying to make never becomes clear. Filmed in Toronto, despite the setting of the tale.
A/AD 102 min (ort 105 min) mCab VIDrel: 20TH V
Boa: novel by Stephen Peters.

PART OF THE FAMILY, A ** 15
David Madden USA 1993
Robert Carradine, Elizabeth Arlen, Shirley Knight, George Newbearn, Kristen Cloke, Ronny Cox, James Calalhan, Lou Cutel, Allan Rich, Beth hogan, Whitney Lounsburg, Biggles Knight
Incest is the focus for this tale of intra-family rivalry, with a woman's new husband visiting his new in-laws, and finding that his wife has a father who is both jealous and highly manipulative. Quite intense in place, even though the plot is contrived and not especially well thought out.
DRA 85 min VIDrel: POLY/POLYREC L/A V

PARTING GLANCES ** 15
Bill Sherwood USA 1986
Richard Ganoung, John Bolger, Steve Buscemi, Adam Nathan, Kathy Kinney, Patrick Tull, Yolande Bavan, Andre Morgan, richard Wall, Jim Selfe, Kristin Moneagle, John Siemens, Bob Kokrherr
A none-too-enthralling look at the world of well-off Manhattan gays on the upper West Side, with the focus on the last twenty-four hours in the relationship between two lovers, before one of them goes abroad to Africa to work.
A 86 min (ort 90 min) VIDrel: PRIDE/PARADOX V

PARTY, THE ** PG
Blake Edwards USA 1968
Peter Sellers, Claudine Longet, Marge Champion, Denny Miller, Gavin MacLeod, Fay McKenzie, Steve Franken, Buddy Lester, Sharron Kimberly, Corinne Cole, Kathe Green, J. Edward McKinley, Carol Wayne, Tom Quine, Timothy Scott, Al Checco
A clumsy Indian actor is invited to a party at the home of a Hollywood studio boss in error, and proceeds to wreck his house in a series of accidents. The script is by Edwards, Tom and Frank Waldham and the film is every bit as loose and episodic as one might expect. There are however, several very funny sight gags.
COM 95 min (ort 99 min) VIDrel: MGM/WHV V

PARTY ANIMALS * 15
Hart Bochner USA
Jeremy Piven, Chris Young, David Spade, Megan Ward, Sarah Trigger, Jessica Walter
A model university is also home to a degenerate bar where loutishness and drunkenness is de rigeur. But all this is to change when a repair bill for over $7,000 arrives.
COM 77 min cC VIDrel: 20TH/TECH V

PARTY DOLL * 18
Alex De Renzy USA
Kelly Royce, Victoria Paris, Rebecca Steele, Joey Silvera, Rappolo, Tom Byron, Don Fernando, Brad Phillips, Mack Reynolds, T.T. Boy
The story of one woman's quest for sexual satisfaction provides an excuse for the usual parade of vignettes.
A 50 min (ort 90 min) VIDrel: IMPENT V

PARTY FAVORS *
Ed Hansen USA
18
1987
Gail Thackray, Jeanine Winters, Marjorie Miller, Jill Johnson, John F. Gott, Buck Flower, Kent Stoddart, Marjorie Miller, Don Edwards, Albert Lord, Alva Megowan, Tom Moses, Adam Hadum, Travis McKenna, P.J. brooks, Norman Sheridan
A group of strippers are made redundant when a local preacher's campaign leads to the closure of their club. They decide to solve their employment problems by forming a pizza delivery service with a difference. As might be expected from the opening, this blend of sex and comedy is little more than a loosely plotted set of striptease sequences. See also THE BIKINI CARWASH COMPANY and TAKIN' IT ALL OFF for two more films on the theme of stripping.
COM   83 min   VIDrel: FIRST/SONOP L/A   V

PARTY GIRLS **
Chuck Vincent USA
18
1989
Marilyn Chambers, Kurt Woodruff, Christina Veronica, Kimberly Taylor, Kurt Schwoebel
A woman becomes a professional party organiser when she is left penniless and in debt to the IRS, following the death of her husband.
Aka: PARTY INCORPORATED
A   76 min (ort 80 min)   VIDrel: MIA/DISC   V/sur

PASCALI'S ISLAND **
James Dearden UK
15
1988
Ben Kingsley, Charles Dance, Helen Mirren, George Murcell, Sheila Allen, Nadim Sawalha, Kevork Maikyan, Vernon Dobtcheff, T.P. McKenna, Danielle Allan, Nick Burnell, George Ekonomou, Alistair Campbell, Ali Abatsis, Brook Williams
A love and espionage triangle set on an Aegean island, in the last dying days of the Turkish Empire just prior to WW1, with a bogus English archaeologist plotting with a disaffected Turkish spy to smuggle out an archaeological treasure. A well-acted drama (written by the director) whose modest budget severely hampers it.
DRA   99 min (ort 106 min)   VIDrel: VISVID/POLYREC   V

PASSAGE TO INDIA, A ***
David Lean UK
PG
1984
Judy Davis, James Fox, Peggy Ashcroft, Victor Banerjee, Alec Guinness, Nigel Havers, Art Malik, Saeed Jaffrey, Richard Wilson, Michael Culver, Antonia Pemberton, Clive Swift, Ann Firbank, Roshah Seth, Sandra Holz, Ishaq Bux
An examination of the clash of cultures between the British and Indians in the 1920s, when a young and foolish woman goes to India accompanied by the mother of her fiance. Finding her honour compromised by a trip to the hills in the company of a native Indian, she accuses him of attempted rape. This overlong and stilted film is partially redeemed by its images of great beauty. AA: S. Actress (Ashcroft), Score/orig (Maurice Jarre).
DRA   157 min (ort 163 min)   VIDrel: MGM/WHV   V/sur
Boa: novel by E.M. Forster.

PASSAGE TO MARSEILLES **
Michael Curtiz USA
PG
1944
Humphrey Bogart, Michele Morgan, Claude Rains, Sidney Greenstreet, Philip Dorn, Peter Lorre, Helmut Dantine, George Tobias, John Loder, Victor Francen, Eduardo Ciannelli, Vladimir Sokoloff, Konstantin Shayne, Monte Bleu
Five convicts escape from Devil's Island in order to join up with the Free French forces. The film's complex structure (and use of flashbacks within flashbacks) belies its rather simplistic story, and despite the presence of many of the same actors and production crew, totally fails to recapture the feeling of CASABLANCA.
A/AD   105 min (ort 110 min) B/W   VIDrel: MGM/WHV   V

PASSAGES **
Paul Thomas USA
18
1991
Christy Canyon, Jennifer Stewart, Heather Hart, Marc Wallice, Blake Palmer, Peter North, Candice Hart
A sexual coming-of-age tale, first in a series of four, that is set on a college campus where an inexperienced virgin learns a thing or two from her more accomplished roommate.
A   63 min   VIDrel: VIVID/SCRN   V

PASSED AWAY *
Charlie Peters USA
15
1992
Bob Hoskins, Blair Brown, Tim Curry, Frances McDormand, William Petersen, Pamela Reed, Maureen Stapleton, Nancy Travis, Peter Reigert, Jack Warden, Don Brockett, Helen Lloyd Breed, Patricia O'Conell, Sally Gracie, Alice Eisner
After the head of a large Irish family dies, the funeral and subsequent wake brings together a host of colourful and somewhat eccentric characters, in this parade of uninspired cliches. A talented cast struggles in vain to inject some life and humour into this overlong attempt at a black comedy.
COM   97 min   VIDrel: HOLPIC/TECH L/A   V/sh

PASSENGER, THE **
Michelangelo Antonioni FRANCE/ITALY/SPAIN
PG
1975
Jack Nicholson, Maria Schneider, Jenny Runacre, Ian Hendry, Steven Berkoff, Ambrose Bia, Jose Maria Cafarel, James Campbell, Manfred Spies, Jean Baptiste Tiemele, Angel Del Pozo, Chuck Mulvehill
A disenchanted TV reporter in Africa, decides to take over the identity of an Englishman who died suddenly in a hotel room, and discovers that he is now a gunrunner being slowly drawn into a web of intrigue, that can only result in his death. An enigmatic and ponderous thriller whose opaque plotting makes it difficult to follow, but as an exercise in sinister mood creation it works perfectly.
Aka: PROFESSION: REPORTER; PROFESSIONE: REPORTER
THR   113 min (ort 119 min)   VIDrel: L/A   V

PASSENGER 57 **
Kevin Hooks USA
18
1992
Wesley Snipes, Bruce Payne, Tom Sizemore, Alex Datcher, Michael Horse, Bruce Greenwood, Robert Hooks, Elizabeth Hurley, Ernie Lively, Marc Macaulay, Cameron Roberts, Joel Fogel, Buchess Tomasello, James Short, Jane McPherson
After his wife is murdered, an anti-terrorist expert plans his retirement but finds himself on the same plane as a terrorist on his way to L.A. for trial. The latter's thugs attempt to hijack the plane and free him but reckon without our redoubtable hero. Essentially a by-the-numbers high-altitude actioner but greatly enhanced by strong performances from Snipes and Payne.
A/AD   84 min cC   VIDrel: WHV   V/sur

PASSION ***
Jean-Luc Godard FRANCE/SWITZERLAND
15
1982
Jerzy Radziwilowicz, Hanna Schygulla, Michel Piccoli, Isabelle Huppert, Laszlo Szabo
A Polish director who is engaged in making a film called "Passion" has an affair with the attractive owner of the motel where he and his crew are staying. Meanwhile, the woman's businessman husband has troubles enough of his own. A very difficult film to assess, this is not so much a straight narrative as an examination of some of Godard's own obsessions, principally film-making, religion and sexuality.
Aka: GODARD'S PASSION
DRA   83 min (ort 88 min)   VIDrel: ARTIF/20TH   V

PASSION FISH ***
John Sayles USA
15
1992
Mary McDonnell, Alfre Woodard, Nora Dunn, David Strathairn, Sheila Kelley, Angela Bassett, Vondie Curtis-Hall, Leo Burmester, Mary Portser, Maggie Renzi, Lenore Banks, William Mahoney, Michael Mantell, Marianne Muellerleile
A former soap-opera actress, confined to a wheelchair after an accident, wears out a succession of nurses with her impossible ways until she finally meets her match. However, this latest recruit has some problems of her own. A neatly directed and acted tale that gives both McDonnell and Woodard a chance to really shine. Screenplay is by Sayles.
DRA   129 min (ort 135 min)   VIDrel: CURZON/20TH   V/sur

PASSION FOR MURDER, A **
Neill Fearnley CANADA/USA
18
1992
Michael Nouri, Michael Ironside, Joanna Pacula, Lee. J. Campbell, Victor Cowie, Mickey Jones, Brent Heale, Hary Nelken, Arne Olsen, Gene Pyrz, Thomas Thomas Schioler, Rick Skene
A woman's married lover, a prominent politician, dies in suspicious circumstances and a cynical Detroit cabbie, who gave the woman involved in his death a ride to Seattle, finds himself involved. An overblown and implausible tale, her involvement with the politician turns out to have been part of her work for a government intelligence agency, leading to the inevitable "run for her life scenario" and final confrontation. Adequate, but far from original.
DRA   88 min (ort 90 min) mTV   VIDrel: COLUM/SONOP   V/sh

PASSION OF DARKLY NOON, THE ***   18
Philip Ridley BELGIUM/GERMANY/UK   1995
Brendan Fraser, Ashley Judd, Viggo Mortensen, Loren Dean, Grace Zabriskie, Lou Myers, Kate Harper, Mel Cobb, Josse de Pauw, Gabi Binder, Maximillian Paul, Knut Samel
A strange, dreamlike erotic drama, this was released directly to video and takes the form of a fable, with the title character an escapee from a fanatical and inward-looking religious cult, who is given refuge by a forest dwelling woman after he is discovered wandering aimlessly along the roadside. But she has a boyfriend, whose return sets in motion a series of violent events as the stranger reveals his obsessive traits. Screenplay is by Ridley.
Aka: DIE PASSION DES DARKLY NOON
DRA   96 min (ort 146 min)   VIDrel: EIV/SONOP   V

PASSION OF LOVE **   15
Ettore Scola FRANCE/ITALY   1981
Laura Antonelli, Bernard Giroudeau, Valerie D'Obici, Jean-Louis Trintignant, Bernard Blier, Massimo Girotti, Gerard Arnato, Sandro Ghiani, Alberto Inaocci, Francesco Piastra, Rosaria Schonumari, Saverio Vallone
In 1862 an army captain arrives at a small frontier post, where a woman of hideous appearance conceives a hopeless passion for him. When he leaves to visit his mistress, she falls dangerously ill, and a feeling of pity and moral obligation draws him into a relationship with her. A flawed and generally ineffective variant on the story of Beauty and the Beast.
Aka: PASSIONE D'AMORE
DRA   118 min   VIDrel: L/A   V

PASSION POTION, THE *   18
Jim Enright USA   1995
Marilyn Star, Nikki Arizona, Jessica James, Krista Maze, Dallas, Tony Tedeschi, Tom Byron, Vince Vouyer
A woman discovers a secret potion that can release sexuality and this proves to be the most powerful aphrodisiac in the world.
A   77 min   VIDrel: ONE   V

PASSION SEKA **   R18/18
Richard Pacheco USA
Seka (Dorothy Hundley Patton), Mike Horner, Kay Parker, Shanna McCullough, Misha Garr, Toni Brooks
A happily married housewife lives a double life, changing by night into an X-rated superstar called Molly Flame.
A   60 min Cut (5 min 16 sec – 18 ver); 69 min (R18 ver)
VIDrel: ELV/DISC   V

PASSIONATE FRIENDS, THE ***   PG
David Lean UK   1949
Claude Rains, Ann Todd, Trevor Howard, Wilfrid Hyde-White, Isabel Dean, Betty Ann Davies, Arthur Howard, Guido Lorraine, Marcel Poncin, Natasha Sokolova, Helen Burls, Jean Serrett, Frances Waring, Wanda Rogerson, Max Earl
A married woman has a chance meet with a former lover and puts her marriage to a wealthy banker in jeopardy, even though this time nothing improper transpired between them. A sensitively handled adaptation of the Wells novel.
Aka: ONE WOMAN'S STORY
DRA   87 min B/W   VIDrel: VCC/RTM   V
Boa: novel by H.G. Wells.

PASSION'S FLOWER **   (18)
Joe D'Amato ITALY   1990
Kristine Rose, Robert Labrosse, Kristine Frischhertz, Jack Ciolino
A man is no sooner released from jail then he is picked up by a beautiful woman, but she turns out to be his brother's new wife. However, they embark on an affair just the same, leading to the inevitable showdown with the justifiably annoyed husband. An adequate erotic effort.
A   89 min   SATrel: SKY MOVIES

PASSPORT TO MURDER **   (15)
David Hemmings USA   1993
Connie Sellecca, Ed Marinaro, Pavel Douglas, Peter Bowles, Lynda Baron, Marella Oppenheim, Arthur Cox, Mark Burns, Jilli Foot, Lazlo Borbely, Tamas Puskas, Gabi Fon, John Nadler, Antal Leisen, Peter Katz, Ricco Ross
In an attempt to get over a failed marriage, a woman takes off for Paris, but once there gets drawn into an exciting but dangerous adventure.
DRA   90 min mTV   SATrel: MOVIE CHANNEL

PASSPORT TO PIMLICO ***   U
Henry Cornelius UK   1949
Stanley Holloway, Basil Radford, Hermione Baddeley, Paul Dupuis, John Slater, Barbara Murray, Margaret Rutherford, Naunton Wayne, Raymond Huntley, Sidney Tafler, Betty Warren, Jane Hylton, Charles Hawtrey, James Hayter
An unexploded bomb from WW2 is detonated in the London district of Pimlico, exposing a hidden cache of treasure and an ancient document that reveals the area to be part of the old French kingdom of Burgundy. The inhabitants exploit this loophole in order to secede from the UK and various complications arise. A charmingly whimsical comedy, dated but very well executed. Screenplay was by Cornelius and T.E.B. Clarke.
COM   80 min (ort 84 min) B/W   VIDrel: WHV   V

PASSPORT TO TERROR **   PG
Lou Antonio USA   1989
Lee Remick, Norma Alexandro, Tony Goldwyn, John Standing, Suzanne Wouk, Shanit Keter, Jim Antonio, Christine Burke, Kim Lonsdale, Ian Abercrombie, Tuck Milligan, Vachik Mangassarian, Richard Balin, Pamela Kosh, Sirri Murad
Whilst on holiday in Turkey, an American woman is jailed on a charge of attempting to smuggle historical relics. Released on bail but fully expecting a long prison sentence she plans an escape, helped by the head of the American Consul. A trite and rather anaemic drama, based on a true story but inevitably too sanitised to really grip. See also MIDNIGHT EXPRESS for a more harrowing examination of the Turkish penal system.
Aka: DARK HOLIDAY
THR   90 min (ort 100 min) mTV   VIDrel: VISVID/POLYREC   L/A   V

PAST MIDNIGHT **   15
Jan Eliasberg USA   1991
Rutger Hauer, Natasha Richardson, Clancy Brown, Guy Boyd, Ernie Lively, Tom Wright, Kibbi Monie, Dana Eskelson, Ted D'Arms, Paul Giamatti, Charles Boswell, Brian T. Finney, Krisha Fairchild, Sarah Magnuson, Paul Mitre
A terribly derivative thriller, this opens with a social worker taking on the case of a man who has just been paroled after fifteen years in prison, having been jailed for the brutal murder of his wife. With her growing interest in her charge comes her belief that he could be innocent, but her life is endangered when she obtains evidence to this effect. Ambiguous in tone, this slow and verbose film takes far too long to reach its predictable outcome.
THR   96 min (ort 100 min)   VIDrel: GUILD/POLY L/A   V

PAST TENSE ***   18
Graeme Clifford USA   1993
Scott Glenn, Anthony LaPaglia, Lara Flynn Boyle, David Ogden Stiers, Marita Geraghty, Sheree J. Wilson, Duane Davis, Ned Van Zandt, Corey Gunnestad, Mark Phelan, Marianne Muellerleile, Joe Lala, Kevin Richardson, Babe Valera
A former cop now works as a thriller writer, and his interest in his mysterious neighbour seems to always lead to him putting her into his plots, until the day fiction and reality merge and she is found murdered. As he gets drawn into the hunt for her killer, he finds himself beginning to doubt his own sanity. A tense little work, this erotic thriller has ample nudity and the like, which really adds little to what is a good film in its own right.
THR   87 min (ort 91 min)   VIDrel: POLFIL L/A   V/s

PAT AND MARGARET **   12
Gavin Millar UK   1994
Victoria Wood, Julie Walters, Celia Imrie, Don Henderson, Deborah grant
A woman comes to the UK from the States to promote her latest blockbuster, but in none too pleased to be re-united with the sister she has not seen in twenty-seven years. Interesting and slightly poignant comedy-drama.
DRA   83 min (ort 90 min) mTV   VIDrel: BBC/TECH   V/h

PAT AND MIKE **   U
George Cukor USA   1952
Katherine Hepburn, Spencer Tracy, Jim Backus, Charles Buchinski (Bronson), William Ching, Sammy White, Phyllis Povah, Chuck Connors, Carl Switzer, Aldo Ray, William Self, Frank Richards, Owen McGiveney, Lou Lubin
A grouchy sports couch takes a talented female champion under his wing but his intimidating methods have an inhibiting effect on her performance. An "odd couple" comedy that fails to

exploit to the full the screen chemistry of its stars and seems
strangely tepid and contrived.
COM   91 min (ort 96 min) B/W   VIDrel: MGM/WHV   V

PAT GARRETT AND BILLY THE KID ***   18
Sam Peckinpah USA   1973
*James Coburn, Kris Kristofferson, Richard Jaeckel, Katy Jurado, Chill
Wills, Jason Robards, Bob Dylan, R.G. Armstrong, Luke Askew, John
Beck, Richard Bright, Matt Clark, Rita Coolidge, Jack Dodson, Jack
Elam, Emilio Fernandez*
A sombre retelling of this Western myth that has been filmed a
number of times before. This version has too many character
actors walking on for minor parts, and too little in the story
to make it a success, though the full-length version (which
only turned up in 1989) is slightly better, if a good deal more
bloody. A mutilated and despondent epic. Music is by Bob
Dylan.
WES   116 min (Cut at film release by 12 sec – ort 122 min)
VIDrel: MGM/WHV   V

PATERNITY *   15
David Steinberg USA   1981
*Burt Reynolds, Beverly D'Angelo, Norman Fell, Paul Dooley,
Elizabeth Ashley, Lauren Hutton, Peter Billingsley, Jacqueline Brooks,
Linda Gillin, Mike Kellin, Victoria Young, Elsa Raven, Carol Locatell,
Kay Armen, Juanita Moore*
A man who yearns to be a father, hires a waitress to act as a
surrogate mother, in this predictable comedy of few laughs and
much tedium. This was Steinberg's directing debut.
COM   89 min (ort 94 min)   VIDrel: CIC/SONOP   V

PATHER PANCHALI ****   U
Satyajit Ray INDIA   1955
*Subir Bannerjee, Runki Bannerjee, Uma Das Gupta, Chunibal Devi,
Reva Devi, Rama Gangopadhaya, Kanu Bannerjee, Karuna Bannerjee,
Tulshi Chakraborty, Harimoran Nag*
First in the director's justly celebrated "Apu" trilogy, this tells
of the life of a young boy, the son of an impoverished priest, and
follows his childhood in a Bengal village where he begins to
learn about life and death. The plot is confined to everyday
events, but these are told in such a lyrical and touching manner
that the result is never less than enthralling. Followed by
APARAJITO and THE WORLD OF APU. The music is by Ravi
Shankar.
Aka: LAMENT OF THE PATH, THE; SAGA OF THE ROAD, THE; SONG OF
THE ROAD, THE
DRA   119 min B/W   VIDrel: CONNO/RTM   V
Boa: novel by Bhibuti Bashan Bannerjee.

PATHFINDER ***   15
Nils Gaup NORWAY   1987
*Mikkel Gaup, Helgi Skulason, Nils Utsi, Sara Marit Gaup, Svein
Scharfenberg, Knut Walle, John Sigurd Kristensen, Ann-Marja Blind,
Sverre Porsanger, Ailu Gaup, Sven Birger Olsen, Nils-Aslek
Valkeapaa, Marius Muller*
This first film in the Lapp language, is the brutal tale of how a
young boy achieves revenge on the invading nomads who
slaughtered his parents and younger sister, when he offers to
become their pathfinder. A memorable tale, made more so by
the stark landscape in which it takes place. Only the dubbing is
an annoyance, sub-titles would have been far better.
A/AD   83 min dubbed   VIDrel: GUILD/SONOP L/A   V

PATHFINDER, THE **   PG
Donald Shebib CANADA   1994
*Kevin Dillon, Laurie Holden, Graham Greene, Dan macDonald,
Stephen Russell, Michelle St John, Lawrene Bayne, Bernard Behrens,
Russell Means, Stacy Keech*
In a remote part of the West, a young orphan boy is cared for
by a Mohican chief and returns years later as an adult, only to
find that a war between the French and British is about to erupt.
A/AD   91 min (ort 105 min)   VIDrel: 20TH/FOXVID   V/sh
Boa: novel by James Fenimore Cooper.

PATHS OF GLORY ****   (15)
Stanley Kubrick USA   1957
*Kirk Douglas, Adolphe Menjou, George Macready, Wayne Morris,
Ralph Meeker, Richard Anderson, Timothy Carey, Suzanne Christian,
Bert Freed, Joseph Turkel, Peter Capell, Emile Meyer, Kem Dibbs,
Jerry Hausner, Frederic Bell*
Three French soldiers in 1916 face a court-martial on the charge
of cowardice, which carries the death penalty, but are in fact
innocent of these charges, and have been selected as scapegoats

to save the reputation of a general. A harrowing, anti-war film,
intelligently written by Calder Willingham, Jim Thompson and
Kubrick, and based on Cobb's book which in turn was based on
a true incident.
DRA   80 min (ort 86 min) B/W   VIDrel: L/A   V
Boa: novel by Humphrey Cobb.

PATRICK ***   18
Richard Franklin AUSTRALIA   1978
*Susan Penhaligon, Robert Helpman, Rod Mullinar, Robert Thompson,
Bruce Barry, Julia Blake, Maria Mercedes, Helen Hemmingway,
Everett de Roche, Walter Pym, Frank Wilson, Peter Culpan, Peggy
Nichols, Carole-Ann Aylett*
A psychotic killer with telekinetic powers lies in a coma after
killing his mother, but strange things begin to happen around
him. A fair chiller with rather good effects but little develop-
ment. See also THE SENDER.
HOR   107 min (ort 115 min)   VIDrel: VIPCO/SGSVID   V

PATRIOT GAMES ***   15
Phillip Noyce USA   1992
*Harrison Ford, Anne Archer, Patrick Bergin, Sean Bean, Thora Birch,
James Fox, Samuel L. Jackson, Polly Walker, J.E. Freeman, James Earl
Jones, Richard Harris, Alex Norton, Hugh Fraser, David Threlfall,
Alun Armstrong, Hugh Ross*
A former CIA analyst is on holiday with his family in the UK
when he witnesses at close hand a terrorist bomb attack on the
Secretary of State for Northern Ireland. Having killed one of the
terrorists (and been awarded a knighthood for his efforts) he
finds that both he and his family are now the target for reprisals.
A fast and quite effective thriller, that's a little hampered by
extraneous supporting characters and far too many red herrings.
THR   117 min (ort 120 min) wScrn cC   VIDrel: CIC/SONOP;
PION (LV only)   V/sur LV
Boa: novel by Tom Clancy.

PATTON ****   PG
Franklin Schaffner USA   1970
*George C. Scott, Karl Malden, Stephen Young, Michael Strong, Frank
Latimore, James Edwards, Lawrence Dobkin, Michael Bates, Tim
Considine, Karl Michael Vogler, Cary Loftin, Albert Dumortier,
Morgan Paull, Bill Hickman, Paul Stevens*
A brilliant performance by Scott, as one of America's most
famous generals, plus a fine supporting cast, make this a most
enjoyable film and a classic in screen biographies. Scott won an
Oscar but didn't accept it. The score is by Jerry Goldsmith. AA:
Pic, Dir, Actor (Scott), Story/Screen (Francis Ford
Coppola/Edmund H. North), Art/Set (U. McCleary and G.
Parrondo/P. Thevenet and A. Mateos), Edit (H. Fowler), Sound
(Williams/Bassman).
Aka: PATTON: LUST FOR GLORY
DRA   164 min (ort 171 min) cC   VIDrel: 20TH/TECH   V/sh
Boa: book by L. Farago.

PATTY HEARST: HER OWN STORY *   18
Paul Schrader USA   1988
*Natasha Richardson, William Forsythe, Ving Rhamer, Dana Delany,
Frances Fisher, Jodi Long, Olivia Barash, Scott Kraft, Ermal
Washington, Gerald Gordon, Marek Johnson, Kitty Swink, Peter
Kowanko, Tom O'Rourke, Elaine Revard*
Dramatisation of the kidnap and subsequent brainwashing of
Patty Hearst, by the Symbionese Liberation Army in the 1970s.
An ill-advised attempt early on to portray her brainwashing
from the victim's point of view, kills any dramatic impact
the narrative (such as it is) might have achieved. Scripted
by Nicholas Kazan. See also THE ORDEAL OF PATTY
HEARST.
Aka: PATTY HEARST
DRA   105 min (ort 108 min)   VIDrel: EIV/SONOP L/A   V
Boa: book Every Secret Thing by Patty Hearst.

PAULINE AT THE BEACH ***   15
Eric Rohmer FRANCE   1982
*Amanda Langlet, Arielle Dombasie, Rosette, Pascal Gregory, Feodor
Atkine, Simon De La Brosse*
A wry account of the romantic entanglements of six people vaca-
tioning on the Normandy coast. The wafer-thin plot is studded
with fine performances and some keen observations, in this
excellent comedy of manners. The third film in director's
"Comedies and Proverbs" series. See also FULL MOON IN
PARIS and THE GREEN RAY.
Aka: PAULINE A LA PLAGE
COM   90 min (ort 95 min)   VIDrel: CONNO/RTM L/A   V

PAWNBROKER, THE *** (18)
Sidney Lumet USA 1965
Rod Steiger, Geraldine Fitzgerald, Brock Peters, Jaime Sanchez, Thelma Oliver, Marketa Kimbrell, Baruch Lumet, Juano Hernandez, Linda Geiser, Nancy R. Pollock, Raymond St Jacques, John McCurry, Charles Dierkop, Eusebia Cosme
A survivor of the Auschwitz death-camp, scratches a living from his pawnbroker's shop situated in Harlem, but is still haunted by his memories. A profoundly depressing but absorbing film for which Steiger received an Oscar nomination. The script is by David Friedkin and Morton Fine and the score is by Quincy Jones. Note the excellent inter-cutting of harrowing deathcamp sequences; the editing was done by Ralph Rosenblum.
DRA 109 min (ort 114 min) B/W VIDrel: L/A V
Boa: novel by Edward Lewis Wallant.

PAWNBROKER, THE ** 18
Ron Jeremy USA 1990
Renee Foxx, Ashley Dunn, Cameo, Chessy Moore, Tunisia, Kristarah Knight, Alexandria Quinn, Honey Rose, Randy West, Daryl Guard, T.T. Boy, Wayne Summers, Biff Malibu
A broke musician tries to sell his guitar, and discovers that the pawnbroker has a machine he uses to record and sell sexual fantasies. This promising opening is wasted as a framing device for five short vignettes. These are: a story set in a Wild West brothel, a farm story, a futuristic fantasy (set in the year 2090), a simple threesome and finally, an encounter that takes place at a massage parlour.
A 59 min VIDrel: FIFTH/DISC V

PAYBACK ** 18
Addison Randall USA 1988
Roger Rodd, Denise Dougherty, James Sweeney, Jean Carrol, Deron McBee, George Bahner
A senator's daughter and a Vietnam vet fight a vicious band of Neo-Nazis terrorists who are out to steal a new assault weapon, as part of their plans to take over the government of the USA. A standard action yarn that offers nothing fresh or original.
Aka: REVENGE
A/AD 90 min mVid VIDrel: MOPIC/SGSVID V

PAYBACK ** 18
Anthony Hickox USA 1994
Joan Severance, C. Thomas Howell, Wendy Abbott, Alex Doduk, Marshall Bell, Jason Gray-Stanford, Dean HAglund, Richard Burgi, John Toles-Bey, Steve Wilcox, Topaze Hasfal-Schou, R.G. Armstrong, Mark High, Lisa Robin Kelly, Byron Lucas
A newly-released prisoner agrees to the last wishes of a dying cell-mate, who has promised him a fortune in stolen money if he agrees to murder a sadistic former prison guard who caused his fatal injuries. Upon his release, he puts his plan into action, but is sidetracked when he learns that his intended victim is now crippled and he finds himself becoming irresistibly drawn to the man's beautiful wife.
DRA 89 min (ort 92 min) VIDrel: TRIM/HIFLI V/h

PAYDIRT ** 15
Bill Phillips USA 1991
Jeff Daniels, Catherine O'Hara, Hector Elizondo, Rhea Pearlman, Judith Ivey, Harris Yulin, Jonathan Banks, Chazz Palminteir, Dabney Coleman, Heidi Zeigler, Richard Portnow, Jeremy Piven
L.A. becomes a battleground when some very determined individuals attempt to locate $8,500,000 in hidden loot.
COM 85 min (ort 105 min) VIDrel: VCC/DISC/COLUM L/A V/sh

PAYOFF *** (18)
Stuart Cooper USA 1991
Keith Carradine, Harry Dean Stanton, Kim Greist, John Saxon, Robert Harper, Alan Blumenfield, William S. Taylor, Jeff Corey, Lawrence Monoson, Suki Kaiser, Stepehn E. Miller, Tom Heaton, Peter Radon, Don Crowe, John Cassine
Carradine plays a man shattered by past events, for as a child he was tricked by an underworld assassin into carrying into his home the bomb that killed his parents. However, when fate gives him the chance to settle the score with the mobster responsible for this outrage, he embarks on a complex scheme to expose, rob and then completely destroy his enemy. A well-mounted thriller with some lively action sequences.
THR 105 min (ort 111 min) mCab TVrel: BBC1
Boa: novel The Payoff by Ronald T. Owen.

PAYROLL *** PG
Sidney Hayers UK 1961
Michael Craig, Francoise Prevost, Billie Whitelaw, William Lucas, Kenneth Griffith, Tom Bell, Edward Cast, Andrew Faulds, Barry Keegan, William Peacock, Joan Rice, Vanda Godsell
In Newcastle, the wife of a murdered armoured-van driver sets out to catch those responsible for her husband's death in the course of a £100,000 robbery. A tough, gripping and very tense film, it moves along at a fair pace and benefits from George Baxt's strong script. One of those solid crime thrillers that Britain used to produce in large numbers in the 1950s and 1960s.
THR 102 min (ort 105 min) B/W VIDrel: WHV V/h
Boa: novel by Derek Bickerton.

PCU * (12)
Hart Bochner USA 1993
Jeremy Piven, Chris Young, David Spade, Megan Ward, Sarah Trigger, Jessica Walter, Jake Busey, Gale Mayron, Kevin Thigpen, George Clinton, Jon Favreau, Matthew Brandon Ross, Stivi Paskosi, Jody Racicot, Thomas Mitchell, Ted Kozma
A coarse campus comedy set in a university that revels in being "politically correct", it tells of a group of students who rebel against these half-baked ideas. Unfortunately, this is a tepid effort of juvenile antics that neither works as a comedy nor discredits a form of intellectual fascism that has done so much harm to the American educational system.
COM 78 min (ort 90 min) SATrel: SKY MOVIES

PEBBLE AND THE PENGUIN, THE *** U
Don Bluth USA 1995
Voices of: Shani Wallis, Scott Bullock, Martin Short, James Belushi, Louise Vallance, Pat Musick, Angeline Ball, Kendall Cunnigham, Alissa King, Michael Nunes, Tim Curry, Neil Ross, Philip Clarke, B.J. Ward, Annie Golden, Will Ryan
A shy penguin named Hubey finds a perfect pebble to give to his sweetheart, but she has another admirer who casts Hubey out to sea. A pleasing animation for children, with songs and music by Barry Manilow.
ANIM 71 min (ort 74 min) cC VIDrel: WHV V/sh

PEDICAB DRIVER ** 18
Samo Hung HONG KONG 1989
Samo Hung, Lam Ching Ying, Johnny Cheung, Bllly Chow, Lia Chia Liang
Action-comedy set in early 20th century Macao where title hero falls in love, which brings the usual complications. As ever, revenge plays no small part in the plot, with our central character finding himself obliged to avenge the murder of his blood-brother.
MAR 91 min (ort 98 min) wScrn VIDrel: MADE/RTM V

PEE-WEE'S BIG ADVENTURE ** U
Tim Burton USA 1985
Pee-Wee Herman (Paul Reubens), Elizabeth Daily, Mark Holton, Diana Salinger, Judd Owen, James Brolin, Morgan Fairchild, Irving Hellman, Monte Landis, Damon Martin, David Glasser, Gregory Brown, Mark Everett, Daryl Roach, Bill Cable
Feature-length vehicle for Herman, a famous American comic, whose humour involves a grown-up man acting and thinking like a child. Pee-Wee searches for his stolen bicycle and has other adventures, but despite a wonderfully overblown performance from Herman, the film has little of substance. The score is by Danny Elfman. See also BIG TOP PEE-WEE.
COM 88 min (ort 92 min) VIDrel: MGM/WHV L/A V/sh

PEEPING TOM *** 18
Michael Powell UK 1960
Carl Boehm, Moira Shearer, Anna Massey, Maxine Audley, Brenda Bruce, Martin Miller, Esmond Knight, Bartlett Mullins, Michael Goodliffe, Jack Watson, Shirley Ann Field, Pamela Green, Michael Powell, Nigel Davenport, Brian Wallace
A demented photographer is fascinated by the idea of murdering women and taking pictures of the fear on their faces, so he sets out with a camera and tripod (complete with concealed spike) to do this. Additionally he documents the police investigation that follows and finally his own suicide. A remarkable film in many ways, this powerful study of insanity created a storm of controversy and destroyed Powell's career. It's still strong stuff. See also FATAL EXPOSURE.
Aka: FACE OF FEAR
HOR 96 min Cut (1 min 6 sec – ort 109 min) VIDrel: WHV V

PERFECT FAMILY ** 15
E.W. Swackhamer USA   1992
*Bruce Boxleitner, Jennifer O'Neil, Juliana Hansen, Shiri Appleby,
Joanna Cassidy, Bill Birch, Colby Chester, William Earl Ray, Mary
Marsh, Robert S. Biheller, Martin Gross, Marvin L. Saunders, Chris
Mastrandrea, Sherilyn Lawson*
A recently widowed woman with two small daughters engages
a brother and sister as a nanny and handyman, but is completely
unaware of the man's psychotic tendencies. A standard psycho-
logical thriller with a predictable outcome.
DRA   88 min (ort 92 min) mTV   VIDrel: CIC   V

PEGGY SUE GOT MARRIED *** 15
Francis Ford Coppola USA   1986
*Kathleen Turner, Nicholas Cage, Barry Miller, Catherine Hicks,
Barbara Harris, Joan Allen, Kevin J. O'Connor, Don Murray,
Maureen O'Sullivan, Leon Ames, Helen Hunt, John Carradine, Wil
Shriner, Sofia Coppola, Randy Bourne*
A woman about to divorce her husband, gets taken back in time
to her last year at high school, and finds that she has a second
chance to make some important decisions. A pleasant and
nostalgic little fantasy that leaves too much unresolved, yet is
put together with considerable flair.
COM   98 min (ort 104 min)   VIDrel: 20TH/TECH L/A   V

PELICAN BRIEF, THE ** 12
Alan J. Pacula USA   1993
*Julia Roberts, Denzel Washington, Sam Shepard, John Heard, Tony
Goldwyn, James B. Sikking, Robert Culp, Hume Cronyn, John
Lithgow, William Atherton, Jake Weber, Stanley Tucci, Anthony
Heald, Cynthia Nixon, Stanley Anderson, John Finn*
A law student risks her life when she speculates about the
murders of two Supreme Court judges and puts forward the
theory that this was the result of a conspiracy. Fortunately, a
reporter investigating this story teams up with her and the
couple start to delve. A very well-made thriller that benefits
from the inspired pairing of Roberts and Washington but is
perhaps a trifle too long to maintain its suspense right up to the
end.
THR   135 min (ort 141 min) cC   VIDrel: WHV   V/sur
Boa: novel by John Grisham.

PELLE THE CONQUEROR **** 15
Billie August DENMARK/SWEDEN   1987
*Pelle Hvenegaard, Max Von Sydow, Erik Paaske, Kristina Tornqvist,
Morten Jorgensen, Axel Strobye, Astrid Villaume, Bjorn Granath,
Lena Pia Bernhardsson, Troels Asmussen, John Wittig, Nis Bank-
Mikkelsen*
A poor Swedish widower takes his young son Pelle to Denmark,
where he hopes they can find a better life. Once there they are
forced to face even greater hardships, but this serves to
strengthen the bond between father and son. Set in the 19th
century, this poignant period drama is as moving as it is memo-
rable. It deservedly won the Palme d'Or at the 1988 Cannes Film
Festival plus an Oscar in the States. AA: Foreign.
Aka: PELLE EROVRARE
DRA   150 min   VIDrel: ELPIC/POLYREC   V/sur
Boa: novel by Martin Anderson Nexo.

PENITENT, THE ** 15
Cliff Osmond USA   1986 (released 1987)
*Raul Julia, Armand Assante, Rona Freed, Julie Carmen, Lucy Reina,
Eduardo Lopez Rojas, Jose Gonzales Rodriguez, Paco Mauri, Justo
Martinez, Erique Novi, Valentina Hernandez, Martin Lasalle, Jose
Antonio Estrada, Tina Romero*
This unusual little fable is set in a remote village in the
Southwest, where each year an extreme religious sect re-enact
the Crucifixion, leaving a cult member nailed to a cross for a
whole day in the desert. An examination of this gives way to a
sub-plot, in which Julia has his sexually frustrated young wife
seduced by an old friend. Osmond's directing debut combines
interesting symbolism with weak plotting, ultimately saying
nothing of any great import.
DRA   90 min (ort 94 min)   VIDrel: 20TH   V

PENITENTIARY *** 18
Jamaa Fanaka USA   1979
*Leon Isaac Kennedy, Thommy Pollard, Hazel Speers, Badja Djola,
Gloria Delaney, Chuck Mitchell, Wilbur "Hi-Fi" White*
A young black man who has been wrongly imprisoned, uses his
skill as a boxer to improve his lot. Fanaka produced, directed
and wrote this one which, though predictable in development,
is carried along by its acute observation and unflinching

portrayal of prison brutality. Several less commendable sequels
followed.
DRA   96 min (ort 99 min)   VIDrel: L/A   V

PENITENTIARY 2 * 18
Jamaa Fanaka USA   1984
*Leon Isaac Kennedy, Ernie Hudson, Mr T, Glynn Turman, Peggy
Blow, Malik Carter, Cephaus Jaxon, Marvin Jones, Donovan Womack,
Ebony Wright, Eugenia Wright, rene Woods, Marci Thomas, Dennis
Lipscomb, Gerald Berns, Stan Kamber*
A sequel to the first film, with very much the same theme of a
young black prisoner using his boxing skills to make his life in
prison a little more bearable. Unfortunately, this one lacks the
care that was lavished on the earlier work, and represents little
more than a cynical exercise in exploitation.
DRA   103 min Cut (29 sec – ort 108 min)
VIDrel: STABL L/A   V

PENITENTIARY 3 ** 18
Jamaa Fanaka USA   1987
*Leon Isaac Kennedy, Anthony Geary, Steve Antin, Kessler Raymond,
Ric Mancini, Jim Bailey*
A further outing for this boxer-in-prison theme, with Kennedy
back in jail where he finds that both the warder, and the local
Mr Big who runs life inside, want him to fight on their boxing
teams. Slightly better than the preceding film but still puerile.
DRA   87 min Cut (25 sec – ort 91 min)   VIDrel: MGM/WHV
L/A   V

PENMARIC: PARTS 1, 2, 3 AND 4 ** PG
Tina Wakerell/Derek Martinus UK   1979
*Ralph Bates, Paul Darrow, June Ellis, Annabel Leventon, Angela
Scoular, Peter Blake, Shaughan Seymour, Shirley Steedman, Martin
C. Thurley*
The lives and loves of a family in Cornwall, from 1867 up to
1940, that opens with a young man coming into a mining fortune
by inheritance, but then causing no end of gossip by marrying
a local working class lass. However, she proves to be a considerable asset, taking over the mining business when he dies. As
the years flow by, varied episodes and intrigues take place, and
though the whole affair is always good to watch, it lacks an
over-riding sense of purpose.
DRA   619 min (4 cassettes) mTV   VIDrel: BBC/TECH   V/h
Boa: novel by Susan Howatch.

PENNIES FROM HEAVEN *** 15
Herbert Ross USA   1982
*Steve Martin, Bernadette Peters, Christopher Walken, Jessica Harper,
Vernel Bagneris, John McMartin, Jay Garner, Tommy Rall, Eliska
Krupka, Toni Kaye, Frank McCarthy, Raleigh Bond, Gloria LeRoy,
Nancy Parsons, Shirley Kirkes*
The story of a sheet-music salesman and his bleak life, is
compared to the carefree world of the songs of the 1930s. Superb
photography (by Gordon Willis) and sets (by Ken Adam) are
intelligently combined with the music of the period. Only the
transposition of the story (originally a BBC TV production with
Bob Hoskins) to the USA lets it down. Musical direction was by
Marvin Hamlisch.
MUS   103 min (ort 108 min)   VIDrel: MGM/WHV   V/h
Boa: TV play by Dennis Potter.

PENNY SERENADE *** U
George Stevens USA   1941
*Cary Grant, Irene Dunne, Beulah Bondi, Edgar Buchanan, Ann
Doran, Eva Lee Kuney, Leonard Willey, Wallis Clark, Walter
Soderling, Baby Biffle, Edmund Elton, Billy Bevan, Nee Wong Jr,
Michael Adrian Morris, Grady Sutton*
Weepy drama about a couple who adopt a baby when their own
child dies. A film that pulls out all the emotional stops and yet
still retains the power to retain interest and stimulate belief. The
script is by Morrie Ryskind.
DRA   117 min (ort 125 min) B/W
VIDrel: 4-FRONT/POLYREC   V

PENTATHALON ** 18
Bruce Malmuth USA   1994
*Dolph Lundgren, David Soul, Renee Coleman, Roger E. Mosely,
David Drummond, Erik Holland, Angelina Estrada, Michele Harrell,
Buddy Joe Hooker, Evan James, Gerald Hopkins, Barry Lynch, Bruce
Malmuth, Kristopher Logan, Brigette Nielsen*
During the run-up to the 1988 Seoul Olympics, a gifted East
German pentathlete is been pushed to the limit by his fanatical
coach, and eventually decides to defect rather than return home.

However, his obsessive coach is unwilling to allow this, and as a member of the Secret Police is quite prepared to do anything required to achieve his ends. An odd sports thriller, highly implausible but fairly absorbing.
THR   101 min   VIDrel: FIRST/SONOP   V

PEOPLE NEXT DOOR, THE **   15
Tim Hunter USA   1996
Faye Dunaway, Nicollette Sheridan, Michael O'Keefe, Ernie Lively, Rachel Duncan
A single mother with three kids learns to her cost that her friendly new neighbours are not quite as nice as they appear when two of her kids are abducted. A watchable thriller, but no more than that.
THR   90 min   VIDrel: ODY/SONOP   V/sh

PEOPLE UNDER THE STAIRS, THE **   18
Wes Craven USA   1991
Brandon Adams, Everett McGill, Wendy Robie, A.J. Langer, Ving Rhames, Sean Whalen, Bill Cobbs, Kelly Jo Minter, Jeremy Roberts, Conni Marie Brazelton, Joshua Cox, John Hostetter, John Mahon, Teresa Velarde, George R. Parker
This shocker is set in one of those classically creepy houses, that is owned by as nasty a set of weirdos as one could ever wish to avoid. However, the central character is not one of them, but a naive black boy who soon finds himself locked in a battle of wits with the murderous owners, in a house full of deadly traps and unpleasant secrets. Perhaps Craven was trying his hand at American Gothic, if so he failed.
HOR   97 min (ort 102 min)   VIDrel: CIC/SONOP   V/sur

PEOPLE VS. LARRY FLYNT, THE ***   18
Milos Forman USA   1997
Woody Harrelson, Courtney Love, Edward Norton, James Cromwell, Crispin Glover, James Carville, Brett Harrelson, Donna Hanover, Norm MacDonald, Vincent Schiavelli, Miles Chapin
Biopic loosely based on the career of Flynt, the controversial publisher of "Hustler Magazine", who as one of America's pornographer kings has seen the inside of more court-houses than many judges, not least thanks to his penchant for vigorously defending every action brought against him, generally using the First Amendment to support his right to freedom of expression. An acidic and amusing study of an unappealing character and America's free speech laws.
COM   130 min   CINrel

PEOPLE'S HERO **   18
Derek Yee (Yee Tung-shing) HONG KONG   1987
Ti Lung, Tony Leung Kar Fei, Tony Leung Chiu Wai, Chun Pui
Two criminals mess up a heist and find that a master criminal is attempting to exploit this situation for his own ends.
THR   90 min   VIDrel: EAST/DISC   V

PEPE LE MOKO ***   PG
Julien Duvivier FRANCE   1937
Jean Gabin, Mireille Ballin, Gabriel Gabrio, Lucas Gridoux, Line Noro, Roger Legris, Fernand Charpin, Saturnin Fabre, Gilbert Gil, Gaston Modot, Marcel Daliot, Frehel, Olga Lord, Renee Carl, Rene Bergeron, Charles Granval
A wonderfully evocative tale of a French gangster hiding out from the police in the Algiers Casbah, where he is safe, and how love brings about his ultimate downfall. Later remade in Hollywood in 1938 as "Algiers".
Aka: CASBAH
DRA   90 min B/W   VIDrel: ELPIC/POLYREC   V
Boa: novel by Roder d'Ashelbe (Henri La Barthe).

PEPI, LUCI, BOM *   18
Pedro Almodovar SPAIN   1980
Carmen Maura, Felix Rotaeta, Olvido "Alaska" Gara, Eva Siva, Diego Alvarez, Concha Gregori, Kiti Manver, Cecilia Roth, Julieta Serrano, Cristina S. Pascual, Jose Luis Aguirre, Carlos Tristancho, Eusebio Lazaro, Fabio de Miguel
A woman caught growing marihuana by a cop allows him sexual favours in order to avoid arrest, but is raped instead. She decides that a fitting retribution would be to seduce his wife, and recruits a girlfriend to assist her. Almodovar's directing debut is a film of much frenetic action (the director works on the principle that if it's fast-paced it must be funny) but little else besides.
Aka: PEPI, LUCI, BOM AND OTHER GIRLS ON THE HEAP; PEPI, LUCI, BOM Y OTRAS CHICAS DEL MONTON
COM   77 min (ort 86 min)   VIDrel: TART/20TH   V

PEPPERMINT SODA ***   15
Diane Kurys FRANCE   1977
Eleonore Klarwein, Odile Michel, Coralie Clement, Marie Veronique Maurin, Valerie Stano, Anne Guillard, Corinne Dacla, Veronique Vernon, Francoise Berlin, Arlette Nonnard, Jacquelien Boyen, Dora Doll, Tsila Chelton
A simple but highly effective coming-of-age tale of two teenage sisters who live with their divorced mother. These two discuss their hopes and fears but the younger girl of thirteen finds men something of a puzzle. Semi-autobiographical in nature, this was the director's debut and was followed by two similar works: "Cocktail Molotov" in 1980 and "Entre Nous" in 1983.
Aka: DIABOLO MENTHE
DRA   96 min (ort 101 min)   VIDrel: ARROW/RTM   V

PERCEVAL ***   PG
Eric Rohmer FRANCE   1978
Fabrice Luchini, Andre Dussollier, Arielle Dombasle, Marc Eyraud, Solange Boulanger, Marie-Christine Barrault, Catherine Schroeder
This drawn-out adventure, based on an unfinished 12th century poem, tells of a young knight who is granted a vision of the Holy Grail at a mysterious castle, only realises too late what he has seen, and embarks on a long quest to find the Grail once more. Dialogue takes the form of the original verses, and the whole affair is shot on stylised, pastel-coloured sets. A demanding and overlong work that requires patience and concentration.
A/AD   140 min   VIDrel: HEND/BMGREC   V
Boa: epic poem by Chretien De Troyes.

PERCY *   15
Ralph Thomas UK   1970
Hywel Bennett, Elke Sommer, Britt Ekland, Denholm Elliott, Cyd Hayman, Janet Key, Tracey Crisp, Antonia Ellis, Tracy Reed, Patrick Mower, Adrienne Posta, Julia Foster, Arthur English, Margaretta Scott
Tasteless story of the world's first penis transplant, with the recipient setting out to discover the identity of the owner. Written by Hugh Leonard and followed by PERCY'S PROGRESS. See also ME AND HIM.
COM   97 min (ort 103 min)   VIDrel: BRAVE/SONOP L/A   V
Boa: novel by Raymond Hitchcock.

PERCY'S PROGRESS *   15
Ralph Thomas UK   1974
Elke Sommer, Denholm Elliott, Leigh Lawson, Judy Geeson, Harry H. Corbett, Vincent Price, Adrienne Posta, Julie Ege, Barry Humphries, James Booth, Milo O'Shea, Ronald Fraser, Anthony Andrews, Bernard Lee, Madeline Smith
Sequel to PERCY, with all the men in the world becoming impotent except the owner of the world's first transplanted organ. Unalloyed dross of a very high order.
Aka: IT'S NOT SIZE THAT COUNTS; IT'S NOT THE SIZE THAT COUNTS
COM   96 min (ort 101 min)   VIDrel: LUMI/SPEAR   V

PEREZ FAMILY, THE **   15
Mira Nair USA   1995
Alfred Molina, Marisa Tomei, Chazz Palminteri, Angelica Huston, Celia Cruz, Trini Alvarado, Diego Wallraff, Angela Lanza, Ranjit Chowdhry, Ellen Cleghorne, Jose Felipe Padron, Lazaro Perez, Vincent Gallo, Billy Hopkins, Ruben Rabasa
During 1980, the Cuban government allow thousands of its citizens to escape by boat to Miami and even released large number of prisoners. Among them was a plantation owner held in custody over twenty years, whose wife and daughter had long since fled. A pleasant but relatively unfocused adaptation that fails to develop, although the competent cast give some strong performances that help to hold the interest.
COM   112 min (ort 135 min) cC   VIDrel: FILM4/RTM   V/sh
Boa: novel by Christine Belle.

PERFECT *   15
James Bridges USA   1985
John Travolta, Jamie Lee Curtis, Marilu Henner, Jann Wenner, Anne De Salvo, Stefan Gierasch, Laraine Newman, Murphy Dunne, Chelsea Field, Charlene Jones, David Paymer, Mathew Reed, Kenneth Welsh
A reporter for a music magazine, researching for an article on health clubs, finds his objectivity in danger when he falls for an aerobics instructress working in a club he was intending to write a bad piece on. A superficial, vain and pompous little film that

never found an audience – and no wonder. Scripted by Bridges and Aaron Latham.
DRA   115 min (ort 120 min)   VIDrel: VCC/DISC/COLUM
L/A   V/sh

PERFECT ALIBI *   15
Kevin Meyer USA   1994
Teri Garr, Hector Elizondo, Kathleen Quinlan, Anne Ramsey, Lydie Denier, Alex McArthur, Charles Martin Smith, Gedde Watanabe, Bruce McGill, Rex Linn, Estelle Harris, Robert Rockwell, Patrick Thomas, Jean-Paul Vignon, Max Ornstein
An au pair starts work for a doctor's wife and seems to be the perfectly suitable, but hides a devious nature and slowly starts to supplant the wife in the affections of the children. Garr plays a friend of the family, who with the help of a cop (Elizondo) sets out to investigate. A tired and totally derivative foray into territory covered amply elsewhere – THE HAND THAT ROCKS THE CRADLE being the first film that springs to mind.
THR   95 min   VIDrel: RYSHER/HIFLI   V/h

PERFECT BRIDE, THE **   15
Terence Meyer USA   1990
Sammi Davis, Kelly Preston, Linden Ashby, Marilyn Rockafellow, John McLaughlin, Jered Barclay, Ashley Tillman, John Agar, Cheryl Arutt, Patricia J. Wilson, Alison Mack, Tamara Clutterbuck, Peter Trencher, Lysa Hayland
A young woman about to get married has successfully concealed her psychopathic tendencies and murky past, the only person who might expose her being the bridegroom, everyone else having met untimely ends. A rather conventionally plotted thriller that runs along familiar lines, it's hardly helped by its tame and predictable ending and a far from inspired performance from Davis as the unbalanced central character.
THR   94 min   VIDrel: VISVID/POLYREC L/A   V

PERFECT SPY, A **   15
Peter Smith UK   1988
Alec Guinness, Peggy Ashcroft, Sarah Badel, Jane Booker, Alan Howard, Benedict Taylor, Rudiger Weigang, Peter Egan, Ray McAnally, Sarah Neville, Michael McStay, Tim Healy, Andy De La Tour, Jane Dooner, Jack Ellis
Espionage thriller that details the mysterious disappearance of a top secret agent and the efforts made to resolve this little difficulty. Average.
THR   376 min (2 cassettes) mTV   VIDrel: BBC/TECH L/A   V/h
Boa: novel by John Le Carre.

PERFECT STRANGER, A **   12
Michael Miller USA   1993
Robert Urich, Stacey Haiduk, Darren McGavin, Susan Sullivan, Holly Marie Combs, Marion Ross, Ron Gabriel, Tamaa Gorski, George Robertson, Patricia Brown, Denise McLeod, Margaret Ozols, Adrian Truss
The wife of a terminally-ill man eventually finds a new love in her life, with the encouragement of her husband, who does not want her to spend the rest of her life as a grieving widow. One of Steel's better stories, unusually for her, the characters achieve a measure of depth, even if the plot holds few surprises.
Aka: DANIELLE STEEL'S A PERFECT STRANGER
DRA   88 min (ort 90 min) mTV   VIDrel: MIA/DISC   V
Boa: novel by Danielle Steel.

PERFECT VICTIM **   18
Patrick Jamain USA
Jacques Penot, Teri Austin
A serial killer uses a video-dating agency to choose his victims, dispatching his victims by means of a lethal injection. A local reporter keen to make a name for herself adopts a fake identity and signs up with the agency, her hope being that she will expose the killer.
THR   84 min (ort 93 min)   VIDrel: 20VIS/SONOP   V

PERFECT WEAPON, THE **   18
Mark DiSalle USA   1991
Jeff Speakman, John Dye, Mako, James Hong, Mariska Hargitay, Dante Basco, Seth Sakai
A production-line adventure about drug dealers and a martial arts expert who sets out to destroy them after they foolishly kill his mentor. This forgettable martial arts caper gave Speakman his first starring role.
MAR   81 min (ort 112 min)   VIDrel: CIC/SONOP   V/sur

PERFECT WORLD, A **   15
Clint Eastwood USA   1993
Kevin Costner, Clint Eastwood, Laura Dern, T.J. Lowther, Keith Szarabajka, Paul Hewitt, Leo Burmester, Bradley Whitford, Ray McKinnon, Jennifer Griffin, Bruce McGill, Leslie Flowers, Brenda Flowers, Darryl Cox, Jay Whitaker
An escaped convict takes a small boy as his hostage but the two gradually begin to establish a mutually dependent relationship while they flee through the Texas countryside, with a Ranger (Eastwood) in dogged pursuit. A slow-paced character study that generates amazingly little in the way of suspense, but is partially redeemed by some fine acting, most surprisingly from Costner, who shows that he can act with grace and sensitivity when the occasion demands.
THR   132 min (ort 138 min) cC   VIDrel: WHV   V/sur

PERFECTIONIST, THE **   (PG)
Chris Thomson AUSTRALIA   1986
John Waters, Jacki Weaver, Adam Willits, Shane Tichner, Elliot Jord, Noel Ferrier, Jennifer Claire, Steven Vidler, Kate Fitzpatrick, Vic Hawkins, Linda Cropper, Maggie Dence
An obsessive university lecturer has been working on his thesis for nine years, all the while neglecting his wife, whose patience finally becomes exhausted. Deciding to go to college herself she hires a childminder, and he turns out to be young Danish man of charm and beauty. A variant on a theme first explored in "Sitting Pretty".
COM   90 min   SATrel: SKY MOVIES
Boa: play by David Williamson.

PERFECTLY NORMAL ***   15
Yves Simoneau CANADA   1990
Robbie Coltrane, Michael Riley, Deborah Duchene, Eugene Lipinski, Jack Nichols, Elizabeth Harpur, Patricia Gage, Kenneth Welsh, Kristina Nicoll, Peter Millard, Bryan Foster, Andrew Miller, Warren Van Evera, Graham Harley
A gentle and extremely likeable comedy about two lonely men, an introverted cab-driver and a entrepreneur/cook who's fallen on hard times. Their growing friendship is cemented by a mutual love of opera, and leads to some comical situations and a few sharp observations. Yet this quirky and unusual effort could have been dramatically improved with a stronger story, and fails to fully exploit the talents of its cast.
COM   105 min (ort 105 min)   VIDrel: PAL/GUILD L/A   V
Boa: story by Eugene Lipinski.

PERFORMANCE ***   18
Nicolas Roeg/Donald Cammell UK   1970
James Fox, Mick Jagger, Anita Pallenberg, Michele Breton, Johnny Shannon, Ann Sidney, John Bindon, Allan Cuthbertson, Stanley Meadows, Antony Morton, Anthony Valentine
Fox is remarkable as a vicious gangster, on the run from his boss who has decided to eliminate him. Arriving at the house of a former rock star, he is slowly drawn into a world of drugs and fantasy. A highly innovative film, full of powerful if somewhat pretentious images. Written by Donald Cammell and with musical direction by Randy Newman. Ry Cooder contributes several excellent musical numbers.
DRA   100 min (ort 105 min)   VIDrel: WHV L/A   V/h

PERMANENT RECORD ***   15
Marisa Silver USA   1988
Alan Boyce, Keanu Reeves, Richard Bradford, Jennifer Rubin, Michelle Meyrink, Pamela Gidley, Michael Elgart, Barry Corbin, Kathy Baker
A teenage boy's suicide, and the pressures that drove a seemingly happy high school kid and grade A student to end his life, form the basis for this powerful and unusual drama.
DRA   88 min (ort 92 min)   VIDrel: CIC/SONOP   V/sh

PERMISSIVE *   18
Lindsay Craig Shonteff UK   1970
Maggie Stride, Gay Singleton, Gilbert Wynne, Alan Gorrie, Robert Daubigny, Juliet Adams, John Allen, Nicola Austin, Samantha Bond, Debbie Bowen, Mary Collins, Madeleine Collinson, Stuart Cowell, Joyce Crossley, Linda Dean
Story of rock music groupies, with an examination of the life of a teenage girl who likes to live with pop groups. Wearisome drivel set in the 1970s, and now looking very dated.
DRA   86 min   VIDrel: JEZ/RTM   V

PERRY MASON RETURNS **   PG
Ron Satlof USA   1985
Raymond Burr, Barbara Hale, William Katt, Patrick O'Neal, Holland

Taylor, James Kidnie, Kerrie Keane, Roberta Weiss, Richard Anderson, Cassie Yates, David McIlwraith, Al Freeman Jr, Paul Hubbard, Lindsay Merrithew, Kathleen Laskey
Feature based on a popular 1960s series about a defence lawyer, who in this tale is now a judge. He makes a comeback appearance to defend his former female assistant on a charge of murdering her boss. A tedious affair that was followed by numerous further tales, starting with PERRY MASON: THE CASE OF THE NOTORIOUS NUN. Based on the character created by Erle Stanley Gardner and scripted by Dean Hargrove. A prime example of an idea that had passed its sell-by date.
Aka: DEFENSE NEVER RESTS, THE
DRA   89 min (ort 100 min) mTV   VIDrel: L/A   V

PERRY MASON: THE CASE OF THE ALL-STAR ASSASSIN *
Christian I. Nyby II USA   PG
                                                          1989
Raymond Burr, Barbara Hale, Alexandra Paul, William R. Moses, Deidre Hall, Shari Belafonte, Pernell Roberts, Bruce Greenwood, Jason Beghe, Julius J. Carry III, S.A. Griffin, James McEachin, Joe Horvath
A hockey star is accused of killing the team's owner, but is resolutely defended by Mason in another entry in this series, based on the earlier (and far more popular) television show. Like most of them, this overlong and painfully pedestrian drama has few bright moments to relieve the tedium.
DRA   90 min mTV   VIDrel: BRAVE/SONOP L/A   V

PERRY MASON: THE CASE OF THE AVENGING ANGEL **
Christian I. Nyby II USA   PG
                                                          1989
Raymond Burr, Barbara Hale, William Katt, Patty Duke, Erin Gray, Larry Wilcox, Charles Siebert, James Sutorius, James McEachin, Richard Sanders, David Ogden Stiers, James McIntire, Joel Colodner, Pam Ward, Tony Higgins
Mason gets a chance to help an innocent man he once sentenced to prison in his capacity as a judge. When a new witness steps forward, he gamely undertakes the man's defence at a second trial. Average.
Aka: PERRY MASON: THE CASE OF THE AVENGING ACE
DRA   91 min (ort 100 min) mTV   VIDrel: BRAVE/SONOP L/A   V

PERRY MASON: THE CASE OF THE FATAL FASHION **
Christian I. Nyby II USA   (PG)
                                                          1991
Raymond Burr, Valerie Harper, Scott Baio, Diana Muldaur, William R. Moses, Barbara Hale, Robert Clohessy, George Di Cenzo, Bruce Kirby, Gianni Russo, Robert Knepper, Ally Walker, Clair Yarlett
Mason undertakes the investigation of the murder of a fashion magazine editor, who had been threatening to expose a colleague's dark secret. Adequate.
DRA   96 min mTV   TVrel: BBC1

PERRY MASON: THE CASE OF THE GLASS COFFIN **
Christian I. Nyby II USA   (PG)
                                                          1991
Raymond Burr, Peter Scolari, Julie Sommars, Nancy Lee Grahn, Barbara Hale, William R. Moses, John Karlen, James McEachin, Kim Braden, David Ogden Stiers, Dennis Lipscomb, Betsy Jones-Moreland, Conor O'Farrell, Kate Vernon, Romy Walthall
This time round the task for Mason is out to prove that a flamboyant stage-illusionist did not intend to kill his assistant in the course of spectacular stunt. Another run-of-the-mill entry in a TV detective series not noted for its wit or plotting, this one presents an easy mystery plus the usual assortment of false leads and coincidences.
DRA   96 min mTV   TVrel: BBC1

PERRY MASON: THE CASE OF THE LADY IN THE LAKE *
Ron Satlof USA   PG
                                                          1988
Raymond Burr, Barbara Hale, William Katt, David Hasselhoff, John Beck, David Ogden Stiers, Doran Clark, John Ireland, Liane Langland, Audra Lindley, Darrell Larson, George Deloy, Terrance Evans, Ric Jury, Nadya Starr, Wendy MacDonald
An undistinguished mystery-style entry in the series, that revolves around a kidnap and murder plot, involving a young heiress to a vast fortune and her lakeside disappearance.
DRA   85 min mTV   VIDrel: BRAVE/SONOP L/A   V

PERRY MASON: THE CASE OF THE LETHAL LESSON **
Christian I. Nyby II USA   PG
                                                          1989
Raymond Burr, Barbara Hale, Alexandra Paul, William R. Moses, Brian Keith, Leslie Ackerman, Richard Allen, Karen Kopins, Brian Backer, John De Mita, John La Motta, Charley Lang, John Allen Nelson, Mark Rolston, Raye Birk
In this tale, Mason defends a young law student accused of murder, but his brief is complicated by the fact that the father of the victim is an old friend of many years' standing. Average.
DRA   100 min mTV   VIDrel: BRAVE/SONOP L/A   V

PERRY MASON: THE CASE OF THE LOST LOVE **
Ron Satlof USA   (PG)
                                                          1987
Raymond Burr, Barbara Hale, William Katt, Jean Simmons, Gene Barry, Robert Walden, Stephen Elliott, Robert Mandan, David Ogden Stiers, Jonathan Banks, Lucien Berrier, Stephanie Dunham, Stephen Elliott, Julian Gamble, Gordon Jump
Perry Mason is re-united with an old flame he once knew thirty years before, when she finds her husband accused of murder, and he gallantly comes to the rescue. A dull and overlong drama, which though watchable soon fades from the memory.
DRA   97 min (ort 100 min) mTV   VIDrel: BRAVE/SONOP L/A   V

PERRY MASON: THE CASE OF THE MALIGNED MOBSTER *
Ron Satlof USA   (PG)
                                                          1991
Raymond Burr, Paul Anka, Michael Nader, Mason Adams, Barbara Hale, William R. Moses, Anne Scheeden, Pamela Bowen, Betsy Jones-Moreland, Seth Kanen, Mitzi Kapture, Beverly Leech, Howard McGillin, Richard Portnow, Stephen Tobolowsky
Our celebrated defence attorney has his legal principles tested to the utmost when an old friend asks him to defend a gangster accused of murdering his wife. A routine entry in the series – very dull.
DRA   96 min mTV   TVrel: BBC1

PERRY MASON: THE CASE OF THE MURDERED MADAM *
Ron Satlof USA   (PG)
                                                          1987
Raymond Burr, Barbara Hale, William Katt, David Ogden Stiers, Ann Jillian, Anthony Geary, Daphne Ashbrook, John Rhys-Davies, Bill Macy, Vincent Baggetta, Kim Ulrich, Jamie Horton, Richard Portnow, Mike Moroff, Wendeline Harston
A scheming former brothel-keeper gets herself murdered and of course, Mason is on hand to defend the woman's husband (an old friend of his colleague Della Street). This involves a long investigation as he ferrets out the truth, but by the time this dreary story reaches its predictable outcome, one no longer cares.
DRA   97 min mTV   VIDrel: BRAVE/SONOP L/A   V

PERRY MASON: THE CASE OF THE MUSICAL MURDER **
Christian I. Nyby II USA   PG
                                                          1989
Raymond Burr, Barbara Hale, Alexandra Paul, William R. Moses, Jerry Orbach, Debbie Reynolds, Dwight Schultz, Mary Cadorette, Raymond Singer, Luis Avalos, Alexa Hamilton, Valerie Mahaffey, James McEachin, Jim Metzler, Lori Petty
A Broadway background adds a dash of much-needed colour to this story of a murdered director whose domineering and underhand methods earned him no shortage of enemies. For all that, this entry is no more than an average effort, with much talk but little else.
DRA   90 min (ort 100 min) mTV   VIDrel: BRAVE/SONOP L/A   V

PERRY MASON: THE CASE OF THE NOTORIOUS NUN **
Ron Satlof USA   PG
                                                          1986
Raymond Burr, Barbara Hale, William Katt, Timothy Bottoms, Jon Cypher, James McEachin, Michele Greene, Gerald S. O'Loughlin, William Prince, Edward Winter, Barbara Parkins, David Ogden Stiers, Tom Bosley, Arthur Hill, Donna Cox
An ageing lawyer helps a nun accused of killing a priest, who was alleged to have been her lover. The lead character is based on the popular American TV series of the 1960s – "Perry Mason". Second in the series that started with PERRY MASON RETURNS. Followed by "Perry Mason: The Case Of The Shooting Star". Average.
Aka: CASE OF THE NOTORIOUS NUN, THE
DRA   95 min (ort 100 min) mTV   VIDrel: IMPENT L/A   V

PERRY MASON: THE CASE OF THE SCANDALOUS SCOUNDREL **
PG
Christian I. Nyby II USA   1987
Raymond Burr, Barbara Hale, William Katt, David Ogden Stiers, Robert Guillaume, Morgan Brittany, Rene Enriquez, George Grizzard, Wings Hauser, Yaphet Kotto
The publisher of a muck-raking scandal sheet is silenced for ever, and a female reporter on his paper is charged with the murder. Another long and very tiresome murder mystery whose outcome is never in doubt.
DRA   95 min (ort 97 min) mTV
VIDrel: BRAVE/SONOP L/A   V

PERRY MASON: THE CASE OF THE SHOOTING STAR **
PG
Ron Satlof USA   1986
Raymond Burr, Barbara Hale, William Katt, Joe Penny, Ron Glass, Alan Thicke, Ivan Dixon, Wendy Crewson, David Ogden Stiers, Jennifer O'Neill, Ross Petty, Mary Kane, Lisa Howard, J. Kenneth Campbell, Lee Wilkof, Bryan Genesse
The host of a popular chat-show is murdered on prime-time TV, and a famous actor/director is the obvious culprit, having apparently shot the victim in front of millions of TV viewers. Mason and his associates inevitably demonstrate the weakness of the case, and triumph in yet another assembly-line production.
DRA   95 min (ort 97 min) mTV
VIDrel: BRAVE/SONOP L/A   V

PERRY MASON: THE CASE OF THE SINISTER SPIRIT *
PG
Richard Lang USA   1987
Raymond Burr, Barbara Hale, William Katt, Robert Stack, David Ogden Stiers, Dwight Schultz, Kim Delaney, Dennis Lipscomb, Jack Bannon, Leigh Taylor-Young, Matthew Faison, Percy Rodrigues, Ed O'Brien, Burt Douglas, Richard Jury
A novelist is thrown to his death from the top of a resort hotel and a publisher is accused of his murder. This yarn attempts to compensate for the lack of plot development by way of tired "haunted house" cliches, that add absolutely nothing to the story and are never resolved.
DRA   95 min (ort 104 min) mTV
VIDrel: BRAVE/SONOP L/A   V

PERSECUTION **
(15)
Don Chaffey UK   1974
Lana Turner, Trevor Howard, Ralph Bates, Olga Georges-Picot, Suzan Farmer, Ronald Howard, Patrick Allen, Mark Weavers, Shelagh Fraser
A rich, crippled American woman living in England dominates her son who both hates and fears her, but after years of misery he rebels and takes his revenge. A stilted and overblown melo-drama that's all atmosphere and no direction.
Aka: TERROR OF SHEBA, THE
HOR   91 min (ort 96 min)   VIDrel: ARTPRO/RTM   V

PERSONAL BEST ***
18
Robert Towne USA   1982
Mariel Hemingway, Scott Glenn, Patrice Donnelly, Kenny Moore, Jim Moody, Larry Pennell, Kari Gosswiller, Jodi Anderson, Maren Seidler, Martha Watson, Emily Dole, Pam Spencer, Deby LaPlante, Mitzi McMillan, Jan Glotzer
Two female athletes in training for the 1980 Olympics, fall in love in a rare and quite sensitive portrayal of a lesbian rela-tionship. Some clumsy camerawork is an annoyance, but the work as a whole remains engrossing and perceptive. Towne's directorial debut.
DRA   122 min (ort 124 min)   VIDrel: MGM/WHV L/A   V

PERSONAL SERVICES **
18
Terry Jones UK   1986
Julie Walters, Alec McGowan, Shirley Stelfox, Terry Jones, Danny Schiller, Victoria Hardcastle, Tim Woodward, Dave Atkins, Leon Lissek
Fictionalised account of the career of brothel keeper Cynthia Payne, who came to public notice by way of a prominent police prosecution. This film follows her life from humble beginnings, to the exalted position as purveyor of kinky sexual services to the rich and famous. An on-and-off comedy that handles the rather unfunny scenes of kinky sex in a remarkably stilted way. See also WISH YOU WERE HERE.
COM   100 min (ort 109 min)   VIDrel: 4-FRONT/POLYREC
V

PERSUASION ***
U
Howard Baker UK   1971
Ann Firbank, Bryan Marshall, Basil Dignam, Valerie Gearon, Marian Spencer, Georgine Anderson, Richard Vernon, Morag Hood, Rowland Davies, Mel Martin, Zhivilia Roche, Noel Dyson, William Kendall, Charlotte Mitchell, Helen Ryan
Early version of Austen's novel about a young woman who gets a second chance of happiness when she meets up with a former suitor she was once persuaded to reject because of his lowly social standing and lack of wealth. Now their positions are reversed as he has made his way in the world while her foolish family has squandered its wealth. A compilation of a BBC seri-alisation.
DRA   225 min (2 cassettes – ort 250 min) mTV
VIDrel: BBC/TECH   V
Boa: novel by Jane Austen.

PERSUASION ***
U
Roger Michell UK   1995
Corin Redgrave, John Woovine, Fiona Shaw, Phoebe Nicholls, Ciaran Hinds, Amanda Root, Sophie Thompson, Susan Fleetwood
Young Anne Elliott, who was the only person of any worth in the family of a baronet, had been courted by one Captain Wentworth, but family pressure forced her to give him up, as he was regarded to be a poor match. However, years later he returns, having made his fortune, and their romance is rekin-dled. Austen's gentle tale of love and conflict receives a rather conventional TV treatment, but one that fortunately lets the strengths of the novel come through.
DRA   102 min (ort 104 min) mTV   VIDrel: BBC/TECH
V/h
Boa: novel by Jane Austen.

PET SEMATARY ***
18
Mary Lambert USA   1989
Fred Gwynne, Dale Midkiff, Denise Crosby, Brad Greenquist, Michael Lombard, Blaze Berdahl, Miko Hughes, Suzanne Blommaert, Mara Clark, Kavi Raz, Andrew Hubatsek, Mary Louise Wilson, Liz Davies, Kara Dalke, Matthew August Ferrell
A family moves to a new home in the Maine woods that's sited near a cemetery, at the other end of which is an ancient Indian burial ground. With the death of the family cat, the father learns that the ancient site can be used to resurrect the dead, and when his son is killed in a tragic road accident, his distraught wish to have him back is given a horrible fulfilment. Scripted by King, this is one of his darkest and most effective fantasies.
HOR   98 min (ort 103 min)
VIDrel: 4-FRONT/POLYREC/CIC; PION (LV only)   V/sur
LV
Boa: novel by Stephen King.

PET SEMATARY 2 *
18
Mary Lambert USA   1992
Edward Furlong, Anthony Edwards, Clancy Brown, Jared Rushton, Darlanne Fluegel, Sarah Trigger, Lisa Waltz, Jason McGuire, Jim Peck, Len Hunt, Ken Fisher, Reid Binion, David Ratajczak, Wilbur Fitzgerald, Elizabeth Ziegler
Sequel to the previous film in which a widowed veterinarian and his teenage son move to Maine after his actress wife was acciden-tally electrocuted. When his friend's pet dog is killed, they bury it in the pet cemetery and so learn of its power to resurrect the dead. A gory freak show with sick humour and even sicker special effects, this lame effort has little to recommend it; it was produced minus the involvement of Stephen King.
HOR   96 min (ort 102 min)
VIDrel: 4-FRONT/POLYREC/CIC   V/sur

PETAIN **
15
Jean Marboeuf FRANCE   1992
Jacques Dufilho, Jean Yanne, Jean-Pierre Cassel, Jean-Claude Dreyfus, Julie Marboeuf, Antoinette Moya, Clovis Cornillac, Pierre Cognon, Roger Dumas, Eric Prat, Christian Charmetant, Andre Penvern, Denis Manuel, Jean-Francois Perrier
WW2 drama telling of the German occupation of France and the setting up of the puppet government under Marshal Petain, as seen from the perspective of the Vichy. A verbose history lesson, made with considerable care and attention to detail, yet hardly the stuff of great drama, not least due to the fact that the action is mostly confined to the Hotel du Parc in Vichy, which became Petain's headquarters.
DRA   129 min (ort 138 min) wScrn   VIDrel: TART/20TH
V/dm
Boa: book by Marc Ferro.

PETE KELLY'S BLUES *** PG
Jack Webb USA 1955
Jack Webb, Janet Leigh, Edmond O'Brien, Ella Fitzgerald, Peggy Lee
In Kansas City in the 1920s, a talented cornet player and his cronies get involved with gangsters. Produced as well as directed by Webb, this memorable film benefits (in the cinema at least) from the use of an early CinemaScope process. See also BIRD.
DRA 100 min wScrn VIDrel: TART/VISVID L/A V

PETER AND THE WOLF ** Uc
George Daugherty USA 1995
Lloyd Bridges, Kirstie Alley, Ross Malinger
A mixture of live-action and animation that tells this well-known fable, with a woman and her young son off to visit their grandfather in the country. Certainly worth hearing for the music of Prokofiev, but in no other way especially noteworthy.
ANIM 51 min (ort 60 min) VIDrel: BMGVID/BMGREC V/sh

PETER PAN *** U
Hamilton Luske/Clyde Geronimi/Wilfred Jackson USA 1953
Voices of: Bobby Driscoll, Kathryn Beaumont, Hans Conried, Bill Thompson, Heather Angel, Paul Collins
The story of a youngster who gets his wish to never have to grow up, becoming the title character, a boy with magical powers such as the ability to fly. When he returns to London, he takes three children back with him to Never Never Land, where they fight pirates and have other wonderful adventures. Barrie's charming fantasy on the magic of childhood receives a reverent if slightly prim treatment, but remains an invigorating and most enjoyable work.
ANIM 74 min (limited edition – ort 76 min)
VIDrel: WDV/TECH L/A V
Boa: play by James Matthew Barrie.

PETER THE GREAT * 15
Marvin K. Chomsky/Lawrence Schiller USA 1985
Maximilian Schell, Ursula Andress, Omar Sharif, Vanessa Redgrave, Trevor Howard, Laurence Olivier, Helmut Griem, Jan Niklas, Renee Soutendijk, Mel Ferrer, Hanna Schygulla, Mike Gwilym, Gunter-Maria Halmer, Jan Malmsjo
Enormously long and rather over-rated attempt to tell the story of the Czar who attempted to modernise Russia single-handedly, at vast expense in terms of human lives. Full of stars but still as dull as ditchwater, and what's worse, at the end of this overlong saga we emerge none the wiser as to Peter's personality or his achievements.
DRA 360 min (3 cassettes – ort 366 min)
VIDrel: BRAVE/SONOP L/A V
Boa: book by Robert K. Massie.

PETER'S FRIENDS ** 15
Kenneth Branagh UK 1993
Emma Thompson, Kenneth Branagh, Stephen Fry, Rita Rudner, Imelda Staunton, Alphonsia Emmanuel, Tony Slattery, Alex Lowe, Richard Briers, Phyllida Law, Hugh Laurie, Alex Scott, Edward Jewesbury, Hetta Charnley, Bill Parfitt, Ann Davies
Ten years after leaving university, an aristocrat who has just inherited the family mansion, arranges a reunion of the members of the musical troupe that he and his chums belonged to. However, the resulting weekend brings forth some very mixed emotions. A very bitter and over-rated comedy that features a group of highly unattractive, two-dimensional characters, especially Rudner as a typical self-obsessed Hollywood soap opera star.
COM 97 min (ort 102 min) VIDrel: EIV/SONOP V/sur

PETE'S DRAGON ** U
Don Chaffey USA 1977
Helen Reddy, Jim Dale, Shelley Winters, Mickey Rooney, Red Buttons, Jim Backus, Jeff Conaway, Sean Marshall, Jean Kean, Joe E. Ross, Ben Wrigley, Charles Tyner, Gary Morgan, Cal Bartlett, Charlie Callas (voice only)
Live action mixes with animation in this story of a lonely orphan boy befriended by a dragon named Elliott. A Disney production that's marred by poor animations and clumsy and plodding development. MARY POPPINS it ain't.
MUS 106 min (ort 134 min) cC VIDrel: WDV/TECH V/sur
Boa: story by Seton I. Miller and S.S. Field.

PETRIFIED FOREST, THE ** PG
Archie Mayo USA 1936
Bette Davis, Leslie Howard, Humphrey Bogart, Dick Foran, Charley Grapewin, Genvieve Tobin, Joe Sawyer, Porter Hall
A diner in the Arizona desert is captured by gangsters who hold the customers and employees hostage, while a drifter weary of his empty life gets a chance for glory by self-sacrifice. An empty, artificial and talky melodrama with Bogart's restrained performance as the gangster boss the only real attraction.
DRA 83 min B/W VIDrel: MGM/WHV V/h
Boa: play by Robert E. Sherwood.

PETTICOAT AFFAIR ** U
John Astin/Norman Abbott/William Asher USA 1977
Jamie Lee Curtis, John Astin, Richard Gilliland, Yvonne Wilder, Jackie Cooper, Richard Brestoff, Christopher Brown, Kraig Cassity, Wayne Long, Richard Marion, Michael Mazes, Jack Murdock, Peter Schuck, Raymond Singer
A pilot for a subsequent series that is essentially a remake of the 1959 film OPERATION PETTICOAT, which tells of a WW2 submarine and what happens to the crew when they rescue five nurses. Average.
Aka: LIFE IN THE PINK; OPERATION PETTICOAT
COM 89 min mTV VIDrel: L/A V
Boa: story by Paul King and Joe Stone.

PETULIA *** 15
Richard Lester USA 1968
Julie Christie, George C. Scott, Richard Chamberlain, Arthur Hill, Shirley Knight, Joseph Cotten, Pippa Scott, Kathleen Widdoes, Richard Dysart, Ruth Kobart, Ellen Geer, Lou gilbert, Nate Esformes, Maria Val, Vincent Arias
A woman married only six months to a playboy embarks on an affair with a recently divorced surgeon who is charmed by her kooky and easy-going personality. A tale of life in San Francisco in the self-styled swinging 1960s. A dated drama, admittedly quite well acted and put together, but not deserving the lavish critical acclaim it received.
DRA 100 min (ort 105 min) wScrn VIDrel: TART/20TH V/dm
Boa: novel Me And The Arch Kook Petulia by John Haase.

PEYTON PLACE *** 15
Mark Robson USA 1957
Lana Turner, Hope Lange, Russ Tamblyn, Arthur Kennedy, Diane Varsi, Lloyd Nolan, Terry Moore, Barry Coe, David Nelson, Betty Field, Mildred Dunnock, Leon Ames, Lorne Greene, Robert H. Harris, Tami Connor, Staats Cotsworth
A superlative if overlong soap opera, based on the bestseller, that deals with the lives of ordinary people in a New England town, whose prim facade conceals a morass of repressed sex, violence and tangled relationships. Much of the plot is far-fetched and contrived, but the excellent performances lend it credibility. Scripted by John Michael Hayes and with a fine score by Franz Waxman. A long-running TV series followed quite a few years later.
DRA 150 min (ort 167 min) VIDrel: 20TH/TECH V
Boa: novel by Grace Metalious.

PHANTASM * 18
Bud Lee USA 1996
Asia Carrera, Jenna Jameson, Bridgette Monroe, Anna Malle, Goldie, Paisley Hunter, Selena, T.T. Boy, Steve Drake, John Decker, Tom Byron
A nightclub offering perverted pleasures is the involuntary home of a couple who indulge their desires with total strangers. A dark and downbeat effort of no great distinction.
A 75 min VIDrel: ONE V

PHANTASM *** 18
Don Coscarelli USA 1977
Michael Baldwin, Bill Thornburg, Reggie Bannister, Kathy Lester, Angus Scrimm, Terrie Kalbus, Ken Jones, Susan Harper, Lynn Eastman, David Arntzen, Ralph Richmond, Bill Cone, Laura Mann, Mary Ellen, Myrtle Scrotton
A young boy is constantly drawn to a morgue that's patrolled by a strange sinister figure. Plucking up the courage to enter the building, he finds that it appears to be used for some horrific purpose. A rather flawed film that is hampered by a low budget but has some powerful and highly imaginative sequences. A couple of sequels have followed.
HOR 85 min (ort 90 min) VIDrel: GUILD/FOXVID V/sur

PHANTASM 2 ** 18
Don Coscarelli USA 1987
James Le Gros, Reggie Bannister, Angus Scrimm, Paula Irvine, Samantha Phillips, Kenneth Tigar, Ruth C. Engel, Mark Anthony Major, Rubin Kushner, Stacey Travis, J. Patrick McNamara, Michael Baldwin
This sequel to the 1977 film covers pretty much the same ground, but with the dubious benefits of bigger money and more explicit effects. The "Tall Man" and his deadly spheres return to raise another army of the dead for construction work on his home planet.
HOR 93 min (ort 97 min) VIDrel: GUILD/FOXVID V/sur

PHANTASM 3: LORD OF THE DEAD * 18
Don Coscarelli USA 1993
Angus Scrimm, Reggie Bannister, A. Michael Baldwin, Bill Thornbury, Gloria Lynne Henry, Kevin Connors, Cindy Ambuehl, John Chandler, Brooks Gardner, Irene Roseen, Sarah Davis, Duane Tucker, Claire Benedek, Wendy Way, Robert Beecher
Second sequel to the original film sees the by now thoroughly familiar figures of the "Tall Man" and his silver spheres hard at work on a scheme that bodes ill for humanity. After the sole survivor of a family is kidnapped, his friend tries to rescue him, and is obliged to travel through towns devastated by evil forces. This belated sequel has little plot, few ideas and a total absence of chills, its only asset is the competent special effects. Very poor.
HOR 86 min (ort 91 min) VIDrel: GUILD/FOXVID V/sh

PHANTOM, THE *** 12
Simon Wincer AUSTRALIA/USA 1996
Billy Zane, Kristy Swanson, Treat Williams, Catherine Zeta Jones, James REmar, Cary-Hiroyuki Tagawa, Bill Smitrovich, Casey Siemaszko, David Proval, Joseph Ragno, Samantha Eggar, Jon Tenney, Patrick McGoohan, Robert Coleby
Zane is the eponymous hero, a cave-dwelling crusader against crime who ventures out on regular crime-busting forays, much in the style of BATMAN (albeit considerably jollier in tone). Very much a homage to the 1930s comic-strip hero, it combines action, humour and over-ripe performances, recalling nothing so much as one of those Saturday morning kids' pictures. If it's pure escapism you want, with no intellectual demands, then this is perfect.
A/AD 100 min VIDrel: PION LV

PHANTOM CARRIAGE, THE *** 12
Victor Sjostrom SWEDEN 1920
Victor Sjostrom, Hilda Borgstrom, Astrid Holm, Tore Svennberg
A drunkard learns of the legend that the "Coachman of Death" is replaced each year by the last man to die before the New Year, and returns to his family a reformed character, An intense silent classic that effectively blends realism and fantasy, most especially thanks to the skilful use made of flashbacks and double-exposure sequences. The teetotal message is a bit unsubtle, but as an example of early fantasy film techniques it is well worth seeing.
Aka: KORKARLEN; PHANTOM CHARIOT, THE; STROKE OF MIDNIGHT; THY SOUL SHALL BEAR WITNESS
FAN 90 min B/W silent VIDrel: REDEM/RTM V
Boa: novel by Selma Lagerlof.

PHANTOM EMPIRE ** 15
Fred Olen Ray USA 1987
Sybil Danning (Sybelle Danninger), Ross Hagen, Jeffrey Combs, Robert Quarry, Susan Stokey, Michelle Bauer, Dawn Wildsmith, Russ Tamblyn
Whilst searching for a lost city, a group of scientists encounter a variety of dangers, including cannibals and an Amazon queen, in this slightly tongue-in-cheek adventure yarn.
A/AD 82 min Cut (7 sec – ort 85 min)
VIDrel: POPRO/RTM V

PHANTOM OF LIBERTY, THE *** 15
Luis Bunuel FRANCE 1974
Adreanna Asti, Jean-Claude Brialy, Michel Piccoli, Monica Vitti, Adolfo Celi, Milena Vukotic, Michael Lonsdale, Calude Pieplu, Julien Bertheau, Paul Frankeur, Paul Leperson, Helene Perdriere, Jean Rochefort
Another helping of assorted surrealistic happenings, where nothing is as it seems and logic and reason no longer prevail. It starts off with a firing squad in Napoleonic times and switches to a maid reading a story to a child. Many equally crazy scenes then follow.
Aka: LA FANTOME DE LA LIBERTE; SPECTER OF FREEDOM, THE
DRA 99 min (ort 104 min) wScrn
VIDrel: ELPIC/POLYREC V

PHANTOM OF THE CABARET, THE: PART 1 ** 18
Henri Pachard FRANCE/USA 1989
Vanessa, Dominique St Clair, Rick Savage, Bionca, Keisha, Sharon Kane, Randy Spears, Jamie Gillis, Caroline Laurie, Liza Robert, Annie Gerard, Michelle Darc, Jacques Croc, Christine Fields, Suzanne Folcq, Jean Servain,
An American writer goes to Paris to work on a book and gets involved with a night-club owner and his staff, while observing their tortured love-lives. He eventually discovers that the club is home to a strange, disfigured recluse whom he attempts to befriend. An ambitious but none-too-successful reworking of The Phantom Of The Opera within the context of a sex film.
A 55 min (ort 90 min) mVid VIDrel: HAR/GOLD L/A V

PHANTOM OF THE MALL: ERIC'S REVENGE ** 18
Richard Friedman USA 1989
Derek Rydall, Morgan Fairchild, Kari Whitman, Jonathan Goldsmith, Rob Estes, Pauly Shore, Kimber Sissons, Gregory Scott Cummins, Tom Fridley, Ken Foree, John Walter Davis, Dante D'Andre, Terrance Evans, Kelly Rutherford, Gary McGurk
A girl takes a job in a shopping centre built on the site of her late boyfriend's house, which was burnt down with him inside. A spate of strange murders soon occurs as the dead man unleashes his supernatural revenge for the act of arson which took his life. What might so easily have been an effective horror film degenerates into a confused welter of special effects.
HOR 90 min VIDrel: MOPIC/SGSVID V

PHANTOM OF THE OPERA, THE *** PG
Rupert Julian USA 1925
Lon Chaney, Norman Kerry, Mary Philbin, Gibson Gowland, Snitz Edwards, Arthur Edmond Carewe
An early version of this much-filmed story, of the disfigured composer who haunts the Paris Opera and lives in the catacombs. Chaney gives one of his most memorable performances, as the tormented creature who kidnaps a young girl to train as his protege. A flawed classic that still retains considerable power.
HOR 88 min (ort 101 min) B/W (tinted version available) silent VIDrel: CASPIC L/A V
Boa: novel by Gaston Leroux.

PHANTOM OF THE OPERA, THE *** PG
Arthur Lubin USA 1943
Claude Rains, Susanna Foster, Nelson Eddy, Edgar Barrier, Jane Farrar, Miles Mander, J. Edward Bromberg, Hume Cronyn, Fritz Leiber, Leo Carillo
An embittered and horribly scarred composer lives in the catacombs beneath the Paris Opera House, and abducts a promising young soprano to train as his protege. This was the first sound version of the famous tale, and though it spends too much time on the operatic sequences, remains a classic work. AA: Cin (Hal Mohr/W. Howard Greene), Art/Int (Alexander Golitzen and John B. Goodman/Russell A. Gausman and Ira S. Webb).
FAN 89 min (ort 92 min) VIDrel: CIC/SONOP V/dm
Boa: novel by Gaston Leroux.

PHANTOM OF THE OPERA, THE ** PG
Terence Fisher UK 1962
Herbert Lom, Heather Sears, Thorley Walters, Edward De Souza, Michael Gough, Ian Wilson, Martin Miller, John Harvey, Miriam Karlin, Miles Malleson
The third version of this classic tale, in which a hideously disfigured composer takes refuge in the sewers below the Paris Opera House, eventually abducting a girl to sing in his opera. A stolid remake that delivers a few shocks but conveys a general air of lifelessness.
HOR 81 min (ort 84 min) VIDrel: CIC/SONOP L/A V
Boa: novel by Gaston Leroux.

PHANTOM OF THE OPERA, THE *** 15
Robert Markowitz USA 1983
Jane Seymour, Maximilian Schell, Michael York, Jeremy Kemp, Diana Quick, Philip Stone, Paul Brooke, Andras Miko, Gellert Rakasanyi, Laszlo Nemeth, Jeno Kis, Laszlo Sos, Denes Ujlaky, Terez Bod, Agnes David, Sandor Halmagyi
This version of the story is set in Budapest but remains essen-

tially the same, telling of a disfigured composer who kidnaps a young girl to train as a singer. Schell is well cast and the excellent make-up of Stan Winston adds the appropriate chill. Adapted by Sherman Yellen.
HOR   92 min (ort 120 min) mTV   VIDrel: BRAVE/SONOP L/A   V
Boa: novel by Gaston Leroux.

PHANTOM OF THE OPERA, THE ** **   18
Dwight H. Little USA   1989
Robert Englund, Jill Schoelen, Alex Hyde-White, Bill Nighy, Terence Harvey, Stephanie Lawrence, Nathan Lewis, Peter Clapham, Molly Shannon, Emma Rawson, Mary Ryan, Yehuda Efroni, Robin Hunter, Virginia Fiol, Cathy Murphy
Despite some attempt to keep faithful to the original novel, this muddled remake is a gory, slow-moving film that is closer to the spirit of a slasher movie than anything else. Englund is alternately hammy and compelling, as a Phantom who has made a pact with the Devil and instead of a mask, uses the skin of those he murders to cover his ruined face. Schoelen makes a most appealing heroine. Filmed in Budapest but curiously sited in London.
HOR   88 min (ort 93 min)   VIDrel: 4-FRONT/POLYREC V/s
Boa: novel by Gaston Leroux.

PHANTOM OF THE OPERA, THE * **   15**
Tony Richardson UK/USA   1990
Charles Dance, Burt Lancaster, Teri Polo, Ian Richardson, Andrea Ferreol, Adam Storke, Jean-Pierre Cassel, Jean Rougerie, Andre Chauneau, Marie-Therese Orain, Marie-Christine Robert, Marie Lenoir, Anne Roumanoff, Catherine Erhady
A lush, sentimental and melodramatic re-telling of this much-filmed classic, that benefits from its French locations but teeters dangerously on the edge of parody. Dance makes an effective if somewhat romantic Phantom, though the poor dubbing of the opera sequences is an annoyance. Originally shown in two parts. Adapted by Kopit from his stage play.
FAN   210 min mTV   VIDrel: STYL L/A   V
Boa: novel by Gaston Leroux/play by Arthur Kopit.

PHANTOM SOLDIERS ** **   18
Irvin Johnson   1987
Max Thayer, Jack Yates, Corwyn Sperry, David Light, James Gaines, Jim Moss, Richard King, Mike Monty, David Anderson, John Fulch, Edward Bennett, Michael Welborne, Don Holtz, Jeff Griffith, Robert Clinton, Warren Smith, Howard Fineman
In Vietnam, a group of Green Berets come across a tiny fishing village, and are horrified at the carnage they find there. Having learnt that a unit of "phantom soldiers" are responsible, they follow them in an attempt to put a stop to their activities.
A/AD   91 min   VIDrel: IMPENT   V

PHANTOM TOLLBOOTH, THE ** **   U
Chuck Jones/Abe Levitow/David Monahan USA   1969
Butch Patrick plus voices of: Hans Conreid, Mel Blanc, Candy Candido, Shep Menken, Les Tremayne, Larry Thor, Daws Butler, June Foray, Patti Gilbert, Cliff Norton
Live-action changes to animation as a boy is catapulted into the strange world of the Kingdom of Wisdom, inhabited by numbers and letters that are now at war. A rather intellectual cartoon whose best aspects are derived from the novel. The slow start and poor songs are drawbacks. The animation sequences were directed by Levitow. Scripted by Jones and Sam Rosen.
Aka: ADVENTURES OF MILO IN THE PHANTOM TOLLBOOTH, THE
JUV   85 min (ort 90 min)   VIDrel: MGM L/A   V
Boa: novel by Norton Juster.

PHANTOMS ** **   18
Charles Band USA   1990
Sherilyn Fenn, Malcolm Jamieson, Hilary Mason, Alex Daniels, Vernon Dobtcheff, Phil Fondacaro
An American woman inherits an ancestral castle in Italy and goes there to take up residence. This sets the scene for a rather traditional horror tale (remarkably gore-free) that involves: ghosts, werewolves, a family curse, a haunted house and elements of Beauty and the Beast. And all of this has its place in a series of supernatural events that unfold when she invites a troupe of travelling players to dine at the castle. Fair.
Aka: MERIDIAN
HOR   82 min Cut (3 sec – ort 90 min)   VIDrel: EIV/SONOP V

PHASE 4 ** **   PG
Saul Bass USA   1973
Nigel Davenport, Michael Murphy, Lynne Frederick, Alan Gifford, Helen Horton, Robert Henderson
An unusual tale in which ants of normal size become super-intelligent, and attack a research station in the desert. Attempts to communicate with them fail and eventually a researcher enters the nest to kill the queen, but is "changed" by the insects instead. A film of striking images (documentary footage is cleverly used), opaque symbolism and stilted acting, in roughly that order. This was Bass's feature directing debut.
Aka: PHASE IV
FAN   80 min (ort 86 min)   VIDrel: CIC/SONOP L/A   V

PHENOMENON ** **   PG
Jon Turteltaub USA   1996
John Travolta, Kyra Sedgwick, Forest Whitaker, Robert Duvall, Richard Kiley, David Gallagher, Ashley Buccille, Tony Genaro, Sean O'Bryan, Bruce Young, Michael Milhoan, Vyto Ruginis, Elisabeth Nunziato, Jeffrey DeMunn, Brent Spiner
Apparently the victim of a bolt of energy from out of the blue, a car mechanic develops telekinetic powers and a photographic memory. But as he grows in both power and wisdom, a rift develops between him and his friends, only to be healed when the true cause of his abilities (a potentially fatal brain tumour) is revealed. A maudlin work that starts off as a delightful fantasy, only to get bogged down in an excess of sentimentality and contrivance.
DRA   118 min (ort 124 min) cC   VIDrel: TOUCH/TECH   V

PHILADELPHIA * **   12**
Jonathan Demme USA   1993
Tom Hanks, Denzel Washington, Antonio Banderas, Jason Robards, Ron Vawter, Joanne Woodward, Mary Steenburgen, Robert Ridgely, Charles Napier, Roger Corman, John Bedford Lloyd, Anna Deavere Smith, Tracey Walter, Kathryn Witt
A successful lawyer with an old and respectable firm learns that he is HIV-positive and loses his job when this comes to the notice of his employers. Being a lawyer, he decides to seek legal redress and brings a civil suit but has a hard time finding a fellow lawyer willing to represent him. An extremely well acted drama but one that's let down by a failure to create solid and believable AA: Actor (Hanks), Song (Bruce Springsteen).
DRA   120 min (ort 125 min) wScrn cC
VIDrel: COLUM/SONOP   V/sur

PHILADELPHIA EXPERIMENT, THE ** **   PG
Stewart Raffill USA   1984
Michael Pare, Nancy Allen, Eric Christmas, Bobby Di Cicco, Kene Holliday, Louise Latham, Michael Currie, James Edgcomb, Joe Dorsey, Gary Brockette, Debra Troyer, Stephen Tobolowsky, Miles McNamara, Ralph Manza, Ed Bakey
An experiment designed to make a battleship invisible goes wrong, and two sailors are drawn forward in time from 1943 to 1984 where they find themselves caught up in another experiment that has gone dreadfully wrong. Reasonably entertaining, but the SF elements are never explored and the obligatory romantic sub-plot pads the thin story out to a full-length feature. Scripted by William Gray and Michael Janover.
FAN   97 min (ort 102 min)
VIDrel: VCC/DISC/COLUM L/A   V/h
Boa: novel by William J. Moore and Charles Berlitz.

PHILADELPHIA EXPERIMENT 2, THE ** **   15
Stephen Cornwall USA   1993
Brad Johnson, Marjean Holden, Gerrit Graham, James Greene, Geoffrey Blake, John Christian Grass, Cyril O'Reilly, Lisa Robbins, David Wells, Larry Cedar, Al Pugliese, James Greene, Andrew Steel, Allen Perada, Mark Stone, Robert Gould
An experiment in making troops invisible goes wrong and sends a man into the future. He learns that a scientist used this technology to send a stealth bomber back to 1943, thus enabling the Nazis to win WW2 and enslave the USA. Naturally, it falls to him to set things to rights. An interesting premise is spoiled by lack of plot development and the film rapidly degenerates into just another futuristic actioner.
FAN   94 min (ort 98 min)   VIDrel: SPEAR/SONOP   V/sur

PHILADELPHIA STORY, THE ** **   U**
George Cukor USA   1940
Katherine Hepburn, James Stewart, Cary Grant, Ruth Hussey, Roland Young, John Halliday, Mary Nash, Virginia Weidler, John Howard,

Henry Daniell, Lionel Pape, Rex Evans, Russ Clark, Hilda Plowright, Lita Chevret, Hillary Brooke
A heiress about to be married for the second time gets cold feet when her ex-husband turns up, but his true purpose is to stop a scandal magazine from publishing damaging revelations about his former father-in-law. A marvellously witty and brilliantly directed film in which everything comes together in just the right way. Remade as HIGH SOCIETY. AA: Actor (Stewart), Screen (Donald Ogden Stewart).
COM   107 min (ort 112 min) B/W   VIDrel: MGM/WHV   V
Boa: play by Philip Barry.

PHOENIX ★★   15
Troy Cook USA   1995
Stephen Nichols, Brad Dourif, Billy Drago, Denice Duff, William Sanderson, Robert Gossett, Betsy Soo, Jeremy Roberts, Robert Clotworthy, Dan Kern
When a mining colony on a distant planetoid is taking over by androids, a special team of mercenaries are sent out there. However, when they arrive they soon learn that matters are not quite as they were represented, and it looks as though the androids are simply rebelling against their exploitation. Predictable SF fare, but fast-moving enough to keep one from bothering too much about the gaping holes in the plot.
FAN   87 min (ort 92 min)   VIDrel: GUILD/FOXVID   V

PHOENIX 2 ★   18
Fred Olen Ray USA   1996
Marc Singer, Matthias Hues
An android tracker is given the job of catching a space smuggler, who has abducted four dangerous female androids. Strictly by-the-numbers fantasy adventure.
FAN   90 min   VIDrel: GUILD/FOXVID   V

PHOENIX THE NINJA ★   15
Fong Ho HONG KONG   1986
Pearl Cheung, Chung Wah, Rose Kuei, Wang Shan, James Tyan
Kung fu revenger about an orphan girl and an evil monk, with our heroine doing battle with the murderers of her family.
MAR   87 min   VIDrel: IMPENT   V

PHOTO SCANDAL ★★   18
Jean-Claude Roy FRANCE   1979
Brigitte Lahaie, Marcel Charvey, Muriel Montosse, Sandra Flower, Maryse Huadebert, Thizou Durand, Tcheng Pham, Patrcia Suffert, Brigitte Bosquet, Daniel Berton, Gilles Mouroux
A woman enlists the help of her lover and her prostitute sister to seduce and then blackmail young heirs to fortunes, so arranging things that they are photographed in compromising positions.
Aka: PHOTOS SCANDALE
A   69 min   VIDrel: JEZ/RTM   V

PHYSICAL ATTRACTION ★★★   18
Richard Mailer USA   1984
Shanna McCullough, Pamela Mann, Bunny Bleu, Lisa Lake, David Cannon, Paul Thomas
A high-class hooker (McCullough) who wants to get away from her vicious pimp meets a sports coach while out jogging. The two fall in love, and with her new boyfriend's help, she breaks free and starts a new life. A well plotted and enjoyable work, that were it not for the obligatory sexual interludes, might have passed muster as a straight drama.
A   51 min   VIDrel: HAR/GOLD   V

PHYSICAL EVIDENCE ★   18
Michael Crichton USA   1989
Burt Reynolds, Theresa Russell, Ned Beatty, Kay Lenz, Kenneth Welsh, Tom O'Brien, Ted McGinley, Ray Baker, Ken James, Ken Baker, Michael P. Moran, Angelo Rizacos, Lamar Jackson, Paul Hubbard, Larry Reynolds, Peter MacNeill
Having already been suspended from the force, a violent and unorthodox cop is accused of murder, following the death of a notorious gangster. An ambitious female lawyer sets out to defend him. A flabby and unconvincing drama with precious little suspense and a poor performance from Russell, hampering a somewhat better one from Reynolds.
DRA   95 min (ort 99 min)   VIDrel: POLY/POLYREC L/A   V/sur

PIANO, THE ★★★   15
Jane Campion AUSTRALIA/FRANCE   1993
Holly Hunter, Harvey Keitel, Sam Neill, Anna Paquin, Kerry Walker, Genvieve Lemon, Tungia Baker, Ian Mune, Peter Dennett, Te
Whatanui Skipwith, Pete Smith, Bruce Alloress, Cliff Curtis, Carla Rupuha, Hahina Tunui, Hori Ahipene
At the turn of the century, a mute Scottish woman and her illegitimate daughter arrive in New Zealand where the mother is to become the mail-order bride of a local farmer. An accomplished pianist, she brings with her own piano, which plays a role in the relationship she comes to develop with a British immigrant (Keitel) who has been adopted by the Maori. A grotesque tale of adult passions. AA: Actress (Hunter), S. Actress (Paquin), Screen (Campion).
DRA   120 min wScrn   VIDrel: EIV/SONOP   V/sur

PICASSO TRIGGER ★★   15
Andy Sidaris USA   1987
Steve Bond, Hope Marie Carlton, Roberta Vasquez, John Aprea, Bruce Penhall, Harold Diamond, Guich Kook
A lethal hit-man is preparing to kill a number of federal employees, and is tracked by an agent in an attempt to thwart him – not an easy task as he is fond of using ingenious methods to kill his victims. A kind of follow-up to MALIBU EXPRESS with the same blend of action and pretty girls (who in this tale are employed as bait to trap the killer). Mildly entertaining, low-budget nonsense, followed by SAVAGE BEACH.
A/AD   96 min (ort 99 min)   VIDrel: RCA L/A   V

PICK-UP ARTIST, THE ★   15
James Toback USA   1987
Robert Downey Jr, Molly Ringwald, Dennis Hopper, Harvey Keitel, Danny Aiello, Mildred Dunnock, Brian Hamill, Vanessa Williams, Victoria Jackson, Tom Signorelu, Christine Baranski, Tamara Bruno, Clemenze Caserta, Tony Sirico
A young man whose romantic philosophy is of the love 'em and leave 'em kind, gets more than he bargained for when he takes up with the daughter of a gambler who owes money to the Mafia. A dull waste of time and talent.
COM   78 min (ort 81 min)   VIDrel: 20TH/TECH L/A   V/sh

PICKWICK PAPERS, THE ★★★   U
Noel Langley UK   1952
James Hayter, Nigel Patrick, James Donald, Kathleen Harrison, Hermione Baddeley, Joyce Grenfell, Hermione Gingold, Donald Wolfit, Harry Fowler, Sam Costa, George Robey, Mary Merrall, Athene Seyler, Alexander Gauge
A loosely-structured and episodic comedy built around the incidents that befall the members of the Pickwick Club in the 1830s. Humorous and well acted, but a little too insubstantial to work as anything more than a series of pleasing sketches.
COM   105 min (ort 115 min) B/W   VIDrel: BRAVE/SONOP L/A   V
Boa: book by Charles Dickens.

PICKWICK PAPERS, THE ★★   U
Warwick Gilbert/Jean Tych AUSTRALIA   1985
Voices of: Colin Borgonon, Margaret Christensen, Wallas Eaton, Brian Harrison, Phillip Hinton, Richard Meikle, Judy Nun, Henri Szeps
Animated version of this classic tale that deals with the various comic adventures of the members of the Pickwick Club in England in the 1830s.
ANIM   75 min   VIDrel: TRING   V
Boa: novel by Charles Dickens.

PICNIC ★★★★   U
Joshua Logan USA   1955
William Holden, Rosalind Russell, Kim Novak, Betty Field, Cliff Robertson, Arthur O'Connell, Verna Felton, Susan Strasberg, Nick Adams, Phyllis Newman, Reta Shaw
In a small Kansas town, a drifter looks up an old friend but winds up stealing the man's girl. Scripted by Daniel Taradash, this fine adaptation of Inge's play marked the start of a more realistic style of Hollywood melodrama. The tendency towards over-acting from the leads is more than made up for by the strong supporting cast. AA: Art/Set (William Flannery and Jo Mielziner/Robert Priestley), Edit (Charles Nelson/William A. Lyon).
DRA   108 min (ort 115 min)   VIDrel: COLUM/SONOP   V
Boa: play by William Inge.

PICNIC AT HANGING ROCK ★★★   PG
Peter Weir AUSTRALIA   1975
Rachel Roberts, Dominic Guard, Helen Morse, Jacki Weaver, Vivean Gray, Anne Lambert, Margaret Nelson, Kirsty Child, Karen Robson, Jane Vallis, Christine Schuler, John Jarratt, Ingrid Mason, Jack Fegan, Wyn Roberts

A beautifully-filmed but empty mystery about the disappearance of a party of Australian schoolgirls that took place in 1900. Scripted by Cliff Green and certainly not lacking in atmosphere. The haunting music is by Bruce Smeaton, and the pan-pipes are played by George Zamphir.
DRA   110 min (ort 115 min)   VIDrel: ELPIC/POLYREC   V
Boa: novel by Joan Lindsay.

PICTURE BRIDE *   12
Kayo Hatta USA   1995
Youki Kudoh, Akira Takayama, Tamlyn Tomita, Cary-Hiroyuki Tagawa, Toshiro Mifune, Yoko Sugi, Reverend Shoin Hoashi, Keiji Morita, Michael Hasegawa, Peter Clark, Lito Capina, Warren Fabro, Michael Ashby, Glenn Cannon, Kati Kuroda
After the death of both her parents, a Japanese woman embarks on a desperate course of action by agreeing to become the mail-order bride to a Hawaiian plantation owner with whom she exchanges portraits (hence the title). However, upon her arrival she learns that she has been fooled and that her bridegroom is considerably older than his picture. A solidly made human drama that won the Audience Award at the Sundance Festival. Screenplay is by Kayo and Mari Hatta.
DRA   95 min (ort 101 min)   CINrel   V

PIE IN THE SKY **   15
Bryan Gibson USA   1996
Josh Charles, Anne Heche, John Goodman, Christine Ebersole, Christine Lahti, Peter Riegert, Dey Young, Bob Balaban, Wil Wheaton, Jesse Stock, A.J. marton, Allison, Chalmers, Keven Scannell, Brian Davila, John Orofino, Nick Toth
A young man, conceived by his parents during a traffic jam, grows up to be totally obsessed by traffic and aspires to be a traffic reporter, who soars above the jams in a helicopter. He also has an equally great passion for the girl next door but they lose contact when he goes to California in search of his dream and they are separated for several years. A thin and not particularly amusing screwball comedy that would have worked better as a half-hour short.
COM   90 min   VIDrel: EIV/SONOP   V

PIERROT LE FOU **   15
Jean-Luc Godard FRANCE   1965
Jean-Pierre Belmondo, Anna Karina, Dirk Sanders, Raymond Devos, Graziella Galvani, Sam Fuller
The adventures of a young couple on the French Riviera form the basis for this offbeat and anarchic romantic drama, with Belmondo playing a bored Parisian who leaves wife and work far behind, getting involved with the criminal activities of a woman he meets while travelling across France. A very bloody and brutal film, clearly an allegory on the state of the modern world and a comment on the impermanence of love (Godard's own marriage was ending at the time).
DRA   106 min (ort 110 min) wScrn
VIDrel: ELPIC/POLYREC   V

PIG'S TALE, A *   U
Paul Tassie USA   1994
Graham Sack, Mike Damus, Sean Babb, Joe Flaherty, Lisa Jakub, Jonathan Hilario, Andrew Harrison Leeds, Olumiji Aina Olawumi, Chaz Lamar Shepherd, Jimmy Zepeda, Christopher Daniel Barnes, Jake Beecham, Jacki Hoffman, Annie McEnroe
A boy goes to summer camp and finds himself being bullied by a rich kid and his cronies. However, he and his friends form a rival group and use their superior brainpower to come out on top. An unfunny spoof, which resembles a sort of juvenile version of all those NATIONAL LAMPOON movies, with predictable sight gags and much broad humour revolving around bodily functions.
Aka: SUMMER CAMP
JUV   90 min (ort 94 min)   VIDrel: COLUM/SONOP   V

PILLOW BOOK, THE **   18
Peter Greenaway FRANCE/HOLLAND/UK   1995
Vivian Wu, Yoshi Oida, Ken Ogata, Hideko Yoshida, Ewan McGregor, Judy Ongg, Ken Mitsuishi, Yutaka Honda, Barbara Lott, Miwako Kawai, Chizuru Ohnishi, Shiho Takamatsu, Aki Ishimaru, Wichert Dromkert, Martin Tukker, Wu Wei, Tom Kane
A young Japanese girl fondly remembers her childhood, when for her birthday celebrations her father would paint her face with Chinese characters and read her extracts from the diary of a ninth century courtesan. Later, as an unhappily married woman living in exile in Hong Kong, these memories have a profound impact on her life. Scripted by Greenaway, this is

another visual tour-de-force in which the narrative lies deeply hidden within a set of intellectual games.
DRA   126 min   CINrel

PILLOW TALK **   (U)
Michael Gordon USA   1959
Rock Hudson, Doris Day, Tony Randall, Thelma Ritter, Nick Adams, Julia Meade, Allen Jenkins, Marcel Dalio, Lee Patrick, Mary McCarty, Alex Gerry, Hayden Rorke, Valerie Allen, Jacqueline Beer, Arlen Stuart, Don Beddoe
Innocent romantic comedy about two people who share a party-line and eventually fall in love despite their mutual dislike, with most of the film being given over to the twists and turns of their range strange courtship. Quite silly in some ways but very entertaining, with the two leads completely at ease in their roles. AA: Story/Screen (Russell Rouse and Clarence Greene/Stanley Shapiro and Maurice Richlin).
COM   102 min (ort 110 min)   VIDrel: CIC/SONOP   V

PIMPERNEL SMITH **   U
Leslie Howard UK   1941
Leslie Howard, Mary Morris, Francis L. Sullivan, Hugh McDermott, Raymond Huntley, David Tomlinson, Manning Whiley, Peter Gawthorne, Allan Jeayes, Dennis Arundell, Joan Kemp-Welch, Phillip Friend, Lawrence Kitchen, David Tomlinson
An updating of the Scarlet Pimpernel legend, set just before WW2, with an archaeology professor smuggling political prisoners out of Germany under the noses of the Gestapo. After a slow start, some good scenes (Howard disguised as a scarecrow) are worked into what is essentially a wartime propaganda film which shows the Nazis as being remarkably stupid. The script is by Anatole De Grunwald, Roland Pertwee and Ian Dalrymple.
Aka: FIGHTING PIMPERNEL, THE; MISTER V
DRA   116 min (ort 121 min) B/W   VIDrel: VCC L/A   V

PIN **   18
Sandor Stern CANADA   1988
David Hewlett, Cyndy Preston, John Ferguson, Terry O'Quinn, Bronwen Mantel, Jacob Tirney, Michelle Anderson, Steven Bernarski, Katie Shengler, Helene Udy, Patricia Collins, Joan Austen, David Gow, Terrence La Brosse
A psychological thriller telling of a young man's obsession with an anatomical dummy, that was the only means his cold and withdrawn father used to communicate with him and his sister. After the untimely death of their parents, the boy brings the dummy home and begins to plan a life of solitude for him and his sister, in which the dummy is to be their only companion.
Aka: PIN: A PLASTIC NIGHTMARE
HOR   99 min (ort 103 min)   VIDrel: VCC L/A   V/sh
Boa: novel by Andrew Neiderman.

PIN-UP GIRL **   U
Bruce Humberstone USA   1944
Betty Grable, John Harvey, Martha Raye, Joe E. Brown, Eugene Pallette, Dave Willcock, Mantan Moreland, Charlie Spivak Orchestra
One of Grable's less memorable vehicles in which she plays a Washington secretary who meets a navy hero and is catapulted to national fame. A poor attempt to exploit the star's musical and comic talents, this flimsy affair has moments of verve and some good sets, but for the most part remains curiously lifeless.
MUS   79 min (ort 82 min)   VIDrel: 20TH/TECH   V

PINK CADILLAC **   15
Buddy Van Horn USA   1989
Clint Eastwood, Bernadette Peters, Timothy Carhart, Tiffany Gail Robinson, Angela Louise Robinson, John Dennis Johnston, Geoffrey Lewis, William Hickey, Michael Des Barres
A man who works as a bail-bond bounty-hunter finds his latest assignment more complex than normal, when he has to rescue a mother and her child from neo-Nazi thugs, themselves the associates of the woman's useless husband. Don't expect any surprises in this knockabout comedy, and though Eastwood shows that he can do broad comedy when the need arises, neither he nor the rest of the cast are exactly challenged by the script.
COM   116 min Cut (27 sec – ort 122 min)   VIDrel: WHV   V/sur

PINK CHIQUITAS, THE **   (15)
Anthony Currie CANADA   1986
Frank Stallone, Claudia Udy, Bruce Pirrie, Elizabeth Edwards, Cindy Valentine, John Hemphill, Don Lake, McKinlay Robinson, Gearlad

Isaac, Diana Platts, Robert McBurnie, T.J. Scott, Kevin Frankoff plus voice of Eartha Kitt
A meteorite crashes near a small town, and turns the local women into nymphomaniacs in this cheerfully vulgar SF spoof. Not bad, but ten minutes is all that's required to get the gist of it. Written by Currie.
COM   80 min (ort 86 min)   CABrel: HVC

PINK FLAMINGOS *
John Waters USA
18
1973
Divine (Glenn Milstead), Mink Stole, Edith Massey, David Lochary, Mary Vivian Pearce, Danny Mills, Cookie Mueller, Channing Wilroy, Paul Swift, Susan Walsh
Two groups vie for the honour of being "the filthiest people in the world" in this Waters offering of low-budget nausea. There are one or two funny moments, but on the whole the show put on by Divine and friends is a trivial affair, celebrating high-camp dialogue and low-grade activities (coprophagia being but one of them). See also FEMALE TROUBLE for another Divine/Waters offering.
COM   86 min Cut (3 min 4 sec plus some cuts subst – ort 95 min)   VIDrel: CASPIC/BMGREC L/A   V

PINK FLOYD: THE WALL ***
Alan Parker UK
15
1982
Bob Geldof, Christine Hargreaves, James Laurenson, Eleanor David, Bob Hoskins, Kevin McKeon, David Bingham, Jenny Wright, Alex McAvoy, Ellis Dale, James Hazeldine, Marjorie Mason, Marie Passarelli, Winston Rose, Eddie Tagoe
An account of a pop star who has come to the end of the road, with some brilliant animated sequences. This combination of live-action, animation and concert footage was inspired by a Pink Floyd record album of the same name. Despite the potency of the images, the unrelenting tone of self-pity is a serious flaw. Animations were designed by political cartoonist Gerald Scarfe.
Aka: WALL, THE
MUS   91 min (ort 95 min)   VIDrel: POLY/POLYREC L/A   V/sh

PINK LIGHTNING **
Carol Monpere USA
15
1991
Sarah Buxton, Martha Byrne, Jennifer Guthrie, Jennifer Blanc, Rainbow Harvest, Anthony Palermo
An unpretentious coming-of-age drama that focuses on the lives of five young women growing up in a small California town in the 1960s.
DRA   88 min   VIDrel: 20TH/TECH   V

PINK PANTHER, THE *
Blake Edwards USA
PG
1963
Peter Sellers, David Niven, Capucine, Claudia Cardinale, Robert Wagner, John Le Mesurier, Fran Jeffries, Colin Gordon, Brenda de Banzie, Colin Gordon, James Lanphier, Guy Thomajan, Michael Trubshawe, Riccardo Bill, Meri Wells
An unfunny comedy about a bumbling French police inspector who creates chaos all around him. The opening credits of the film introduced the animated "Pink Panther" character, and this in turn inspired a long-running series of mediocre TV cartoons. Scripted by Maurice Richlen and Blake Edwards with the opening sequence animation by De Patie-Freleng. The music is by Henry Mancini. Followed by A SHOT IN THE DARK.
COM   110 min (ort 113 min)   VIDrel: MGM/WHV   V/h

PINK PANTHER STRIKES AGAIN, THE *
Blake Edwards UK
PG
1976
Peter Sellers, Herbert Lom, Colin Blakely, Leonard Rossiter, Lesley-Anne Down, Burt Kwouk, Andre Maranne, Richard Vernon, Michael Robbins, Briony McRoberts, Dick Crockett, Byron Kane, Paul Maxwell, Jerry Stovin, Phil Brown
I wish the actors would strike, then we would be rid of these endless idiotic sequels. Our bumbling Inspector Clouseau comes up against an organisation created by Chief Inspector Dreyfus that is designed to finally rid the world of Clouseau. This was the fourth of five sequels. Written by Frank Waldman and Blake Edwards with music by Henry Mancini. Followed by THE REVENGE OF THE PINK PANTHER.
COM   98 min (ort 103 min)   VIDrel: MGM/WHV   V

PINK PUSSYCAT, THE *
Dick Dumont USA
18
1992
Ashlyn Gere, Trixie Taylor, Carolyn Monroe, Flame, Melanie Moore,

Peter North, Bobby Sox, T.T. Boy, Tom Byron, Darryl Griffin, Tom Chapman
Title object plays a key role in this silly detective spoof which is nothing more than an excuse for yet another uninspired sex film. Gere plays the private eye hired to track it down, a task that takes her to the set of a porno movie, where she gets to watch some of the action.
A   58 min (Cut before video submission by 20 min 20 sec)
VIDrel: GROHOM/MAXSCAN   V

PINOCCHIO ****
Ben Sharpstein/Hamilton Luske USA
U
1940
Voices of: Dickie Jones, Christian Rub, Cliff Edwards, Evelyn Venable, Walter Catlett, Frankie Darro, Charles Judels, Don Brodie
The original classic Walt Disney animation of the tale of a carpenter whose dream of a son comes true, in the form of a wooden puppet. A splendid film that combines technical brilliance with an imaginative and atmospheric script. AA: Score/orig (Paul J. Smith/Ned Washington), Song ("When You Wish Upon A Star" – Leigh Harline (m)/Ned Washington (l)).
ANIM   84 min (ort 88 min)   VIDrel: WDV/TECH L/A   V
Boa: novel by Carlo Collodi (Carlo Lorenzini).

PINOCCHIO AND THE EMPEROR OF THE NIGHT *
Hal Sutherland USA
U
1987
Voices of: Edward Asner, Tom Bosley, Lana Beeson, Linda Gary, James Earl Jones, Jonathan Harris, Ricky Lee Jones, Don Knotts, William Windom, Scott Grimes, Frank Welker
This full-length cartoon is a flawed attempt to produce a sequel to the 1940 Disney classic, with Pinocchio celebrating his first birthday as a real boy. However, he is led into temptation by a mysterious carnival owner and experiences many adventures and perils. A flat and anaemic effort, of poor animation and banal dialogue.
ANIM   83 min (ort 95 min)   VIDrel: 4-FRONT/POLYREC   V/sur

PINOCCHIO'S CHRISTMAS **
Arthur Rankin Jr/Jules Bass USA
U
1988
Voices of: Todd Parker, George S. Irving, Alan King, Robert McFadden, Pat Bright, Allen Swift, Diane Leslie, Gerry Matthews, Ray Owens, Timothy Blake
Pinocchio and his creator put on a puppet show to earn money for Christmas, but things go wrong when a beautiful girl puppet appears on the scene. A pleasant and undemanding puppet-animation.
ANIM   47 min (ort 60 min)   VIDrel: WHV   V

PIRANHA ***
Joe Dante USA
18
1978
Bradford Dillman, Heather Menzies, Kevin McCarthy, Keenan Wynn, Dick Miller, Belinda Balaski, Barbara Steele, Melody Thomas, Bruce Gordon, Barry Brown, Paul Bartel, Shannon Collins, Shawn Nelson, Richard Deacon, Janie Squire
A stock of man-eating fish are accidentally released into local rivers in this fast-paced spoof on JAWS that attempts to parody SF films of the 1950s. The script is by John Sayles. Followed by a sequel in name only and remade in 1996.
COM   90 min (ort 94 min)   VIDrel: MGM/WHV L/A   V
Boa: novel by John Sayles.

PIRANHA 2: FLYING KILLERS *
James Cameron ITALY/USA
18
1981
Tricia O'Neal, Steve Marachuk, Lance Henriksen, Ricky G. Paull, Ted Richert, Leslie Graves, Carole Davis, Arnie Ross, Connie Lynn Hadden, Tracy Berg, Anne Pollack, Albert Sanders, Hildy Magnasun, Phil Colby, Lee Krug
Bears no relation to the original title film and tells of nasty killer fish that go on a rampage in the Caribbean. This was Cameron's debut as a director (he went on to make THE TERMINATOR) and is little more than a gratuitously gory mess.
Aka: PIRANHA 2: THE SPAWNING; SPAWNING, THE
HOR   91 min (ort 95 min)   VIDrel: MIA/DISC/COLUM   V

PIRANHA WOMEN *
J.D. Athens USA
15
1988
Adrienne Barbeau, Shannon Tweed, Barry Primus, Bill Maher, Karen Mistal
A famous anthropologist heads an expedition to find the long-lost Dr Kurtz and they discover a tribe ruled over by a demented

feminist. A low-budget jungle movie spoof, written and directed by Athens, its suggestive title is its best feature.
Aka: CANNIBAL WOMEN IN THE AVOCADO JUNGLE OF DEATH; JUNGLE HEAT
COM   85 min Cut (16 sec – ort 87 min)
VIDrel: ALLIED/TROMA   V

PIRATE, THE ***   U
Vincente Minnelli USA   1948
Judy Garland, Gene Kelly, Walter Slezak, Gladys Cooper, Reginald Owen, George Zucco, The Nicholas Brothers, Lester Allen, Lola Deem, Ellen Ross, Mary Jo Ellis, Jean Dean, Marion Murray, Ben Lessy, Jerry Bergen, Val Setz
A girl living in a West Indian port takes a circus clown for a pirate, and he happily strings her along to win her attention. Though studio-bound in conception, the film is saved by the songs and Kelly's dancing, in which he shows his usual exhilaration and zest. Music and lyrics are by Cole Porter and Lennie Hayton, with the script by Albert Hackett and Frances Goodrich.
MUS   97 min (ort 102 min)   VIDrel: MGM/WHV   V/dm V/h
Boa: play by S.N. Behrman.

PIRATE MOVIE, THE *   PG
Ken Annakin AUSTRALIA   1982
Kristy McNichol, Christopher Atkins, Ted Hamilton, Bill Kerr, Maggie Kirkpatrick, Garry McDonald, Linda Nagle, Kate Ferguson, Rhonda Burchmore, Chuck McKinney, Catherine Lynch, Marc Colombani, John Allansu, Paul Graham
An updated teenage version of The Pirates of Penzance, with a girl dreaming that she is in the musical, but changing some of the details. Not really very successful as it attempts to be both a silly bubblegum comedy and a parody of other films. Sequences of slapstick and swordplay are carelessly thrown together in the hope that this unlikely combination is sufficiently funny to not require a script.
COM   94 min (ort 105 min)   VIDrel: 20TH/TECH   V/ur
Boa: opera The Pirates of Penzance by Gilbert and Sullivan.

PIRATES *   PG
Roman Polanski FRANCE/TUNISIA   1986
Walter Matthau, Damien Thomas, Richard Pearson, Cris Campion, Charlotte Lewis, Michael Elphick, Bill Fraser, Roy Kinnear, Ferdy Mayne, Cardew Robinson, Olu Jacobs, Anthony Peck, Anthony Dawson, Richard Dieux
Polanski's homage to pirate adventure films but taken to extremes (the replica galleon cost $8,000,000 to build) and lacking the care he usually gives to the plot. Despite excellent performances, this expensive dud is no more than a lavish, robust, empty mess. The score is by Philippe Sarde. Cut before video submission by 1 min 30 sec.
A/AD   106 min (ort 124 min)   VIDrel: 4-FRONT/POLYREC V/sur

PIRATES OF DARK WATER, THE: THE SAGA BEGINS **   U
Don Lusk/Ray Patterson USA   1990
Voices of: Jodi Benson, Roscoe Lee Brown, Hector Elizondo, Lindra Gary, Dorian Harewood, Roddy McDowell, George Newbern, Brock Peters, Stanley Ralph Ross, Andre Stojka, Les Tremayne, Frank Welker
On a distant planet, a brave band of adventurers mount a quest for a number of magic crystals that are the only means of saving their world from destruction. A poorly animated but quite imaginative yarn that formed the pilot for a series.
ANIM   90 min   VIDrel: FIRST/SONOP L/A   V

PISTOL, THE: THE BIRTH OF A LEGEND **   PG
Frank C. Schroeder USA   1990
Nick Benedict, Millie Perkins, Adam Guier
A talented youngster is inspired by his father to attempt to become a basketball-playing prodigy in this adequate drama aimed at fans of the sport.
DRA   100 min   VIDrel: HIFLI/SONOP   V

PIT AND THE PENDULUM, THE **   15
Roger Corman USA   1961
Vincent Price, John Kerr, Barbara Steele, Luana Anders, Antony Carbone, Patrick Westwood, Lynne Bernay, Mary Menzies, Charles Victor, Larry Turner
Corman mutilates a fine Poe story, in this tale of a man who locks his sister and her lover in a torture chamber, after going mad and thinking he is his late father (who used to work for the

Inquisition). A slow and lacklustre film that's partially redeemed by good sets, especially the pendulum sequence. This was Corman's second foray into Poe territory. The script is by Richard Matheson.
HOR   77 min (ort 85 min)   VIDrel: VISVID/POLYREC L/A V
Boa: short story by Edgar Allan Poe.

PIT AND THE PENDULUM, THE ***   18
Stuart Gordon USA   1991
Lance Henriksen, Rona De Ricci, Jonathan Fuller, Jeffrey Combs, Tom Towles, Steven Lee, William J. Norris, Frances Bay, Oliver Reed, Barbara Bocci, Carolyn Purdy-Gordon, Geoffrey Coppleston, Benito Stefanelli, Larry Dolgin
Lavish and straightforward adaptation of Poe's tale, set during the period of the Spanish Inquisition. Toquemada the Grand Inquisitor, imprisons a young baker's wife who protests at his excesses, while reserving an especially hideous form of mental torture for her husband, after he attempts to engineer her release. Though both sets and costumes are excellent, the acting is pretty dire. A brave but flawed attempt to do Poe justice.
HOR   93 min (ort 97 min)   VIDrel: 4-FRONT/POLYREC L/A   V
Boa: short story by Edgar Allan Poe.

PIXOTE ****   18
Hector Babenco BRAZIL   1981
Fernando Ramos da Silva, Marilia Pera, Jorge Juliao, Gilberto Moura, Jose Nilson dos Santos, Edilson Lino, Zenildo Oliveira Santos, Claudio Bernardo, Tony Tornado, Jardel Filho, Rubens de Falco
A ten-year-old boy is abandoned and slowly drifts into crime, eventually becoming a murderous criminal, in this shocking indictment of the social conditions in the slums of Brazil. Not for the squeamish.
Aka: PIXOTE: LA LEY DEL MAS DEBIL
DRA   119 min (Cut at film release – ort 125 min)
VIDrel: L/A   V

PLACE FOR ANNIE, A ***   PG
John Gray USA   1993
Sissy Spacek, Mary-Louise Parker, Joan Plowright, S. Epatha Merkeson, Jack Noseworthy, David Spielberg, Richard Gilbert Hill, Stephen Mills, Robin Pearson, Rose, Lauree Berger, Linda Carlson, Wendy Robie, J.P. Bumstead
An abandoned HIV-infected baby is adopted by her nurse but her natural mother then challenges this decision, in this competently made tug-of-love tearjerker. Solid performances abound, especially from the very talented Spacek, and they help sustain this drama and counteract a slight tendency towards over-sentimentality. Based on a true case.
DRA   95 min (ort 98 min) mTV
VIDrel: 4-FRONT/POLYREC/ODY   V/sh

PLACE OF ONE'S OWN **   U
Bernard Knowles UK   1945
James Mason, Margaret Lockwood, Barbara Mullen, Dennis Price, Helen Haye, Michael Shepley, Dulcie Gray, Moore Mariott, Gus McNaughton, O.B. Clarence, Ernest Thesinger, John Turnball, Clarence Wright, Helen Goss, Edie Martin
A retired couple purchase a mansion that has been vacant for forty years but are unaware that it is said to be haunted by the spirit of a woman who died there. After they engage a housekeeper cum companion, strange events occur that culminate in her falling seriously ill, prompting them to send for the very same doctor who attended the dead woman many year before. A convincingly acted and well crafted supernatural tale.
FAN   89 min (ort 92 min) B/W   VIDrel: CARL/TECH   V
Boa: novel by Osbert Sitwell.

PLACE OF WEEPING ***   PG
Darrell Roodt SOUTH AFRICA   1986
James Whyle, Charles Comyn, Geina Mhlophe, Norman Coombes, Michelle du Toit, Ramolao Makhene, Patrick Shai
A clumsy but sincere appeal for change in South Africa, following the story of a black woman who fights for justice after a brutal white farmer beats one of his workers to death. This was South Africa's first home-grown indictment of Apartheid.
DRA   85 min (ort 88 min)   VIDrel: NWV L/A   V/h

PLACES IN THE HEART ***   PG
Robert Benton USA   1984
Sally Field, Lindsay Crouse, Ed Harris, Amy Madigan, John Malkovich, Danny Glover, Yankon Hatten, Gennie James, Lane Smith,

Terry O'Quinn, Bert Remsen, Jay Patterson, Raymond Baker, Toni Hudson, De Voreaux White, Jerry Haynes
A widow struggles to keep her farm and family together during the Depression years, enduring many hardships in this over-sentimental but generally effective tale. AA: Actress (Field), Screen/orig (Benton).
Aka: WAITING FOR MORNING
DRA   111 min   VIDrel: 20TH   V

PLAGUE, THE **
15
Luis Puenzo ARGENTINA/FRANCE/UK   1992
William Hurt, Robert Duvall, Raul Julia, Barbara Carrera, Ian Yule, Sandrine Bonnaire, Jean-Marc Barr, Victoria Tennant, Lautaro Murua, China Zorilla, Luz, Atilio Veronelli, Francisco Cocuzza, Laura Palmucci, Norman Erlich
In a South American city, an idealistic doctor struggles against the odds to help contain an outbreak of bubonic plague and becomes involved in the process with a host of other figures, including a woman TV journalist and a cameraman. An updated and altered adaptation of Camus' epic novel that dissipates all its virtues and fails to benefit from the change of location and the contemporary political references. An unworthy version of a great novel.
DRA   117 min   VIDrel: FIRST/SONOP   V
Boa: novel by Albert Camus.

PLAGUE DOGS, THE ***
PG
Martin Rosen UK/USA   1982
Voices of: John Hurt, James Bolam, Christopher Benjamin, Judy Geeson, Warren Mitchell, Barbara Leigh-Hunt, Bernard Hepton, Brian Stirner, Penelope Lee, John Bennet, Geoffrey Mathews, John Franklyn-Robbins, Bill Maynard
Animation telling of two dogs that are infected with a deadly virus at a research laboratory and escape, threatening to infect the entire countryside with an epidemic. Somewhat heavy-going to start with, but held together by fine animation and the sincere script, which is as much a plea to respect the rights of animals as it is an adventure tale.
ANIM   99 min (ort 103 min)   VIDrel: MGM/WHV L/A   V
Boa: novel by Richard Adams.

PLAGUE OF THE ZOMBIES, THE ***
15
John Gilling UK   1965
Andre Morell, Diane Clare, Jacqueline Pearce, John Carson, Brook Williams, Michael Ripper, Alex Davion, Marcus Hammond, Roy Royston, Dennis Chinnery, Ben Aris, Tim Condron, Bernard Egan, Norman Mann, Francis Willey, Jerry Verno
An epidemic in a Cornish village is traced to the local squire, who uses voodoo to create zombies who are then made to work in his tin mine. Quite an effective little chiller, made by Hammer Films and well above their usual quality.
Aka: ZOMBIE, THE; ZOMBIES, THE
HOR   87 min (ort 91 min)   VIDrel: MGM/WHV L/A   V

PLAIN CLOTHES **
PG
Martha Coolidge USA   1988
Arliss Howard, George Wendt, Seymour Cassel, Diane Ladd, Suzy Amis, Robert Stack
A local teacher is murdered, and a cop goes undercover in his high school in a bid to find the killer, experiencing once more all the things he had to contend with when he was a student. A messy mystery-comedy sustained by the fine cast.
COM   94 min Cut (1 sec – ort 98 min)   VIDrel: CIC/SONOP   V/sh

PLAN 9 FROM OUTER SPACE *
PG
Edward D. Wood Jr USA   1959
Gregory Walcott, Tom Keene, Duke Moore, Mona McKinnon, Dudley Manlove, Lyle Talbot, Bela Lugosi, Tor Johnson, Vampira (Maila Nurmi), Paul Marco, Carl Anthony, David DeMering, Norma McCaarty, Llynn Lemon plus Criswell (narration)
Here it is. One of the very worst films ever made. So bad that it achieves a kind of greatness. Lugosi died after only four days of shooting, but don't let that put you off; it's OK, a double took over. The story (based so our narrator assures us on "secret testimony") involves aliens re-animating corpses to use in an invasion of Earth – this is the dreaded "Plan 9". You have been warned. This gem was followed by "Night Of The Ghouls".
Aka: GRAVE ROBBERS FROM OUTER SPACE
FAN   78 min B/W   VIDrel: CARL/TECH   V/dm

PLAN OF ATTACK **
(18)
Fred Walton USA   1992
Loni Anderson, Anthony John Denison, Stephen Meadows, Coleby Lombardo, Candy Clark, John Apicella, Laura Koelsch, Len Myles, J. Hunter, Carol Kierman, Alix Clementa, Geoff Can, Jan Hoag, Barry Grayson, Mike Greenlee, Mika Sciba
Fairly adequate thriller about a rapist who gets out of jail and sets about terrorising his former victim. A general sense of tension is maintained, but the film lacks both originality and anything especially remarkable in the performances. A time-filler – no more than that.
Aka: PRICE SHE PAID, THE
THR   90 min mCab   TVrel: BBC1

PLANES, TRAINS AND AUTOMOBILES ***
15
John Hughes USA   1987
Steve Martin, John Candy, Laila Robbins, Michael McKern, Kevin Bacon, Dylan Baker, Carol Bruce, Olivia Burnette, Diana Douglas, Martin Ferrero, Larry Hankin, Richard Herd, Susan Kellerman, Matthew Lawrence, Edie McClurg
A businessman hurrying home to celebrate Thanksgiving with his family, gets inextricably involved with his boorish travelling companion, and together they experience an incredible succession of mishaps on their cross-country odyssey. Good characterisation and a certain measure of warmth sustain the rather thin plot.
COM   88 min (ort 93 min)
VIDrel: 4-FRONT/POLYREC/CIC   V/sur

PLANET OF THE APES ***
PG
Franklin J. Schaffner USA   1968
Charlton Heston, Roddy McDowall, Kim Hunter, Maurice Evans, James Whitmore, James Daly, Linda Harrison, Robert Gunner, Lou Wagner, Woodrow Parfrey, Jeff Burton, Buck Kartalian, Norman Burton, Wright King, Paul Lambert, Diane Stanley
Astronauts crash-land on a planet where apes rule and man is dumb and enslaved. An interesting and enjoyable film with a surprise ending. Scripted by Rod Serling and Michael Wilson, it spawned several sequels, a TV series and a cartoon. Followed by BENEATH THE PLANET OF THE APES, ESCAPE FROM THE PLANET OF THE APES, CONQUEST OF THE PLANET OF THE APES and BATTLE FOR THE PLANET OF THE APES. AA: Hon Award (John Chambers for make-up).
FAN   107 min (ort 119 min) wScrn
VIDrel: 20TH/TECH; ENCORE (LV only)   V LV
Boa: novel La Planete des Singes (Monkey Planet) by Pierre Boulle.

PLANK, THE **
U
Eric Sykes UK   1967
Eric Sykes, Tommy Cooper, Graham Stark, Stratford Johns, Jim Dale, Hattie Jacques, Jimmy Tarbuck
One of those endearingly idiotic comedies that is done without a word of dialogue and inevitably features several workmen making a mess of the simplest of tasks. This tale follows the havoc caused by two such men and their plank of wood. Films such as SAN FERRY ANN provide more of the same.
COM   46 min   VIDrel: THAMES/DISC   V

PLATOON ***
15
Oliver Stone USA   1986
Tom Berenger, Willem Dafoe, Charlie Sheen, Forest Whitaker, Francesco Quinn, John C. McGinley, Richard Edson, Kevin Dillon, Reggie Johnson, Keith David, Johnny Depp, David Neidorf, Mark Moses, Chris Pedersen, Oliver Stone
A realistic look at the experiences of a front-line American soldier in Vietnam that says what has been said many times before: war is hell. But what we still await is a film that puts this conflict in its historical context as THE BATTLE OF ALGIERS did for the Algerian liberation struggle. Followed by BORN ON THE FOURTH OF JULY. AA: Pic, Dir, Sound (John K. Wilkinson/Richard Rogers/Charles "Bud" Grenzbach/Simon Kaye), Edit (Claire Simpson).
WAR   114 min (ort 120 min) wScrn
VIDrel: COLUM/SONOP   V/sh

PLATOON LEADER **
18
Aaron Norris USA   1988
Michael Dudikoff, Robert F. Lyons, William Smith, Rick Fitts, Brian Libby, Michael De Lorenzo, Jesse Dabson
An inexperienced commander whose military knowledge is purely academic, receives a baptism of fire in the trenches, when

his unit comes into conflict with Communist forces in South-East Asia. Mindless high-action nonsense.
A/AD   92 min (ort 97 min)   VIDrel: WHV   V/dm

PLAY CHRISTY FOR ME ** **   18
Tina Marie USA   1990
Christy Canyon, Kitty Luv, Stacy Lords, Sasha Strange, Randy West, Tom Byron
Canyon and Luv run a detective agency that specialises in divorce cases. They get a client who is seeking a divorce from her husband, but their investigations soon reveal not only that both parties are enjoying affairs, but that it's Canyon's boyfriend who's involved with their new client. A fair effort whose plot is weakened by rather excessive cutting.
A   42 min Cut (16 min 31 sec)   VIDrel: IMPENT   V

PLAY DEAD * **   18
Peter Wittman USA   1982
Yvonne De Carlo, Stephanie Dunham, David Cullinane, Glenn Kezer, Ron Jackson, David Elizey, Carolyn Greenwood
A rich woman decides to murder her relatives one by one, and presents each in turn with the gift of a dog, one that has been duly imbued with Satanic evil. An incoherent and unmemorable yarn.
Aka: SATAN'S DOG
HOR   90 min   VIDrel: TROMA/RTM   V

PLAY IT AGAIN, SAM * **   15**
Herbert Ross USA   1972
Woody Allen, Diane Keaton, Susan Anspach, Tony Roberts, Jerry Lacy, Jennifer Salt, Joy Bang, Viva, Mari Fletcher, Diana Davila, Suzanne Zenor, Michael Greene, Ted Markland
An entertaining and charming Woody Allen comedy, with a film critic being helped in his love life by the figure of Humphrey Bogart, and finding that his life begins to follow Bogart's in CASABLANCA. A good deal less disjointed than most of Allen's work, this one actually has a coherent script.
COM   82 min (ort 86 min)   VIDrel: CIC/SONOP   V
Boa: play by Woody Allen.

PLAY IT COOL ** **   U
Michael Winner UK   1962
Billy Fury, Helen Shapiro, Bobby Vee, Dennis Price, Michael Anderson Jr, Richard Wattis, Anna Palk, Keith Hamshere, Ray Brooks, Jeremy Bulloch, Maurice Kaufmann, Peter Barkworth, Max Bacon, Felicity Young, Monty Landis
A worried father packs his young daughter off to Europe to get her away from a rock 'n' singer with whom she is completely infatuated. Unfortunately, the airport is fogbound and so she meets up with a rock group waiting for a flight to Brussels for a concert. They take her out in search of her boyfriend and she realises that dad was right when she finds him with another girl. A dreary effort intended solely as a vehicle for pre-Beatles pop idol Fury.
MUS   81 min   VIDrel: LUMI/SPEAR L/A   V

PLAY MISTY FOR ME * **   18**
Clint Eastwood USA   1971
Clint Eastwood, Donna Mills, Jessica Walter, John Larch, Irene Hervey, Jack Ging, Johnny Otis, James McEachin, Clarice Taylor, Donald Siegel, Duke Everts, George Fargo, Mervin W. Frates, Tim Frawley, Otis Kadani, Brit Lind
Eastwood's directorial debut is a highly-skilled piece of work with a good sense of suspense. He plays a late-night radio D.J. who is pursued by a deranged woman who has developed an obsessive interest in him. Don Siegel puts in an appearance as a bartender. Scripted by Jo Heims and Dean Reisner.
THR   98 min (ort 102 min)   VIDrel: CIC/SONOP L/A   V

PLAYBACK * **   18
Oley Sassone USA   1995
George Hamilton, Tawny Kitaen, Shannon Whirry, Harry Dean Stanton, Charles Grant, Quin Duffy
A couple like making their own sexy home-videos, only to find themselves compromised when some of these fall into the wrong hands, and the creation of a love quadrangle is the result. A softcore, erotic drama of little merit, from the same producers of the "Playboy" films.
DRA   90 min   VIDrel: 20TH/FOXVID   V/sh

PLAYBOY OF THE WESTERN WORLD, THE * **   PG**
Brian Desmond Hurst EIRE   1962
Siobhan McKenna, Gary Raymond, Elspeth March, Michael O'Brian,

Liam Redmond, Niall MacGinnis, Brendan Cauldwell, John Welsh, Eithne Lydon, Finnula O'Shannon, Anne Brogan, Katie Fitzroy
Film version of a famous play, telling of a boastful young man who arrives at a sleepy County Mayo village with a tale of how he killed his father, and other stories that generally charm and intrigue the inhabitants. A colourful romp that clearly shows its stage origins but is no less enjoyable for that.
COM   96 min (ort 100 min)   VIDrel: MGM/WHV L/A   V
Boa: play by John Millington Synge.

PLAYBOYS, THE * **   15**
Gillies MacKinnon UK   1992
Robin Wright, Aidan Quinn, Albert Finney, Milo O'Shea, Alan Devlin, Niamh Cusack, Ian McElhinney, Niall Buggy, Adrian Dunbar, Stella McCusker, Anna Livia Ryan, Niall Buggy, Lorcan Cranitch, Aine Ni Mhuiri, Doreen Hepburn, Tony Rohr
Small-town drama set in Eire in 1957, when a beautiful young woman causes a public outrage by becoming pregnant and steadfastly refusing to name the father. Matters become even more complicated when she gets herself involved with a member of a troupe of travelling actors. Undoubtedly well acted but badly in need of a more original script.
DRA   104 min (ort 113 min)   VIDrel: 20TH/TECH   V/sur

PLAYER, THE * **   15**
Robert Altman USA   1991
Tim Robbins, Greta Scacchi, Fred Ward, Whoopi Goldberg, Peter Gallagher, Brion James, Cynthia Stevenson, Vincent D'Onofrio, Dean Stockwell, Richard E. Grant, Sydney Pollack, Lyle Lovett, Dina Merrill, Angela Hall, Leah Ayres
A young studio executive receives a series of strangely worded postcards from a writer and later becomes a murder suspect when his tormentor turns up dead. A sharply observed satire on Hollywood life and mores with the usual over-complex and interwoven Altman plot as well as cameo appearances by over sixty stars. The LV disc includes some interviews with the director and the main cast members.
DRA   119 min (ort 124 min)   VIDrel: ETL/POLYREC; PION (LV only)   V/s LV
Boa: novel by Michael Tolkin.

PLAYING DANGEROUSLY ** **   15
Lawrence Lanoff USA   1995
David Keith Miller, George Shannon, Linda LoPorto, Ali Patrick, Jack Skillman, Andre Barran, Brett Davidson, John Norman Thomas, Adrian Vitoria, Mikey LeBeau
Five terrorists hold a boy's family hostage, and he realises that his only two choices are to watch them being murdered one-by-one or to take on the villains single-handedly.
A/AD   86 min   VIDrel: MARQ/FOXVID   V

PLAYIN' DIRTY * **   18**
John Leslie USA   1980
Selena Steele, Rachel Ashley, Sasha, Rene Morgan, Tom Byron, Heather Torrance, Jamie Gillis, Marc Wallice, Ron Jeremy, Joey Silvera
Sex-film with a comedy angle, that has Jeremy playing an embezzler and Gillis his tough boss, who nurses a secret passion. A goofy film that doesn't take itself seriously
A   64 min (ort 85 min) mVid   VIDrel: IMPENT   V

PLAYMAKER ** **   (18)
Yuri Zeltser USA   1993
Colin Firth, Jennifer Rubin, John Getz, Jeff Perry, Arthur Taxier, Dean Norris, Belinda Waymouth, Stephen Polk, Diane Robin, Clare Kirkconnell, William James Shaw, Alice Kushida
An aspiring actress gets involved with a drama coach and soon learns that she is dealing with a seriously disturbed individual who has made her a part of his schemes by involving her in a murder and so arranging things that she appears to be the culprit. An erotic thriller with little freshness, but a good sense of tension.
THR   88 min (ort 91 min)   SATrel: SKY MOVIES

PLAZA SUITE * **   PG**
Arthur Hiller USA   1971
Walter Matthau, Maureen Stapleton, Barbara Harris, Lee Grant, Louise Sorel, Jose Ocasio, Dan Ferrone, Jennie Sullivan, Tom Carey
Three stories set in the same hotel suite with Matthau playing all three parts: a successful businessman who has taken his wife out to celebrate an anniversary (poignant rather than funny), a sex-starved film director (acid-sharp but only slightly funny), and a pompous father whose daughter is about to get married

but has locked herself in the bathroom (extremely funny). Written by Neil Simon.
COM 109 min (ort 114 min) VIDrel: CIC/SONOP L/A V
Boa: play by Neil Simon.

PLEASE SIR! *   U
Mark Stuart UK   1971
John Alderton, Deryck Guyler, Joan Sanderson, Noel Howlett, Eric Chitty, Richard Davies, Patsy Rowlands, Peter Cleall, Carol Hawkins, Liz Gebhardt, David Barry, Peter Denyer, Malcolm McFee, Aziz Resham, Brinsley Forde
Spin-off from a British TV series about a teacher and his unruly class at a run-down school. Here, he takes the kids off to a summer camp; laughs are few and far between. The script is by John Esmonde.
COM 97 min (ort 101 min) VIDrel: VCC/DISC V

PLEASURE, THE **   18
Joe D'Amato (Aristide Massaccesi) ITALY   1985
Isabelle Andrea Guzon, Steve Wyler, Marco Mattioli, Lilli Carati, Laura Gemser, Dagmar Lassander
A widower is eventually seduced by his step-daughter, in a story set against the background of the carnival in pre-WW2 Venice.
Aka: IL PIACERE; PASSIONATE PLEASURES
A 90 min Cut (3 sec plus film cuts) VIDrel: JEZ/RTM V

PLEASURE IN PARADISE **   18
Boots Rakely USA   1992
John Paul Lorello, Linda Brown, Tom Case, Diane Colton, Eileen Anthony, S. Olin, Honey Smax, Jacqueline Jade, Daniel Bradford, Ed Blake, Gina Jourard, Pati Oppelt, Tony Alessandrini, Cyndi Cozzolino, James F, Danskin
Two sisters inherit a lakeside bar from their uncle, who drowned in mysterious circumstances and have to contend with a local businessman who has plans for their property and will not hesitate to use seduction to get his way. A well-photographed softcore erotic drama, with some pretty women, a thin plot and precious little else.
Aka: MURDER IN PARADISE LAKE
DRA 79 min (ort 85 min) VIDrel: NORVID/DISC V

PLEASURE PRINCIPLE, THE *   18
David Cohen UK   1991
Peter Firth, Hadyn Gwynne, Lysette Anthony, Lynsey Baxter, Sara Mair-Thomas, Ian Hogg, Francesca Folan, Liam McDermott, Stephen Finlay, Cliff Parisi, Gordon Warnecke, Patrick Tidmarsh, Chris Knowles, Lauren Tauben, Mark Carroll
Writer-director-producer Cohen makes his first feature the sporadically amusing tale of a divorced Lothario at large among the upwardly-mobile professional women of the 1990s. As he juggles three demanding (and very different) relationships, his increasingly implausible fibs lead to ever more complicated situations. With neither the strength of dialogue nor the conviction needed from the cast to sustain it, this is very poor fare indeed.
COM 96 min (ort 100 min) VIDrel: IMAG/RTM V/h

PLENTY **   15
Fred Schepisi UK   1985
Meryl Streep, Sam Neill, Charles Dance, John Gielgud, Tracy Ullman, Sting (Gordon Sumner), Ian McKellen, Andre Maranne, Tristram Jellinek, James Taylor, Peter Forbes-Robertson, Hugo De Vernier, Ian Wallace, Burt Kwouk
A former female intelligence agent is unable to adapt to civilian life, in post-WW2 Britain which she finds depressingly dull and meaningless. Remarkably convincing performances and a few sharp observations are hampered by the tedious nature of the story. The film's use of a wide-screen frame will be lost on TV. Scripted by Hare from his play.
DRA 119 min (ort 124 min) VIDrel: MGM/WHV V/sur
Boa: play by David Hare.

PLOT TO KILL HITLER, THE **   PG
Lawrence Schiller USA   1990
Brad Davis, Madolyn Smith, Ian Richardson, Mike Gwilyn, Helmut Griehm, Jonathan Hyde, Kenneth Colley, Michael Byrne, Helmuth Lohner, Rupert Graves, Jack Headly, John McEmery, Heather Chasen
Drama based on the events of the summer of 1944, when Germany was facing certain defeat and Hitler's underlings were convinced that the country would fare better with a conditional surrender to the Allies rather than wait until they were comprehensively beaten. An interesting but patchy affair that attempts to show some humanity within the German camp, but never

develops any tension and miscasts Davis (who died of AIDS in 1993) as the ringleader of the plotters.
DRA 86 min (ort 100 min) mTV VIDrel: MGM/WHV L/A V/sh

PLOUGHMAN'S LUNCH, THE ***   15
Richard Eyre UK   1983
Jonathan Pryce, Tim Curry, Rosemary Harris, Frank Finlay, Charlie Dore, David De Keyser, Nat Jackley, Bill Paterson, David Lyon, Pearl Hackney, Simon Stokes, Orlando Wells, Witold Schejbal, Libba Davies, Sandra Voe, Andrew Norton
Sharply observed study of contemporary Britain during the Falklands War, with Pryce excellent as an obnoxious and self-centred radio reporter. A minor piece that is confined within the narrow boundaries of its outlook, but is nonetheless pungent and thought-provoking.
DRA 102 min (ort 107 min) wScrn VIDrel: TART/20TH L/A V

PLUCKING THE DAISY *   15
Marc Allegret FRANCE   1956
Brigitte Bardot, Daniel Gelin, Robert Hirsch
An early Bardot vehicle, with her cast as the authoress of a scandalous bestseller who is forced to flee from her disapproving father to Paris, where she finds herself destitute. To earn some money she enters a striptease competition, and spots her supposedly strait-laced father among the judges. A dated and tame comedy of limited interest.
Aka: EN EFFLUEILLANT LA MARGUERITE; MADEMOISELLE STRIPTEASE; PLEASE! MISTER BALZAC; PLUCKING THE DAISIES
COM 97 min (ort 101 min) dubbed VIDrel: ARROW/RTM V

POCAHONTAS **   U
Mike Gabriel/Eric Goldberg USA   1994
Voices of: Irene Bedard, Judy Kuhn, Mel Gibson, David Ogden Stiers, John Kassir, Russell Means, Christian Bale, Linda Hunt, Danny Mann, Billy Connolly, Joe Baker, Frank Welker, Michelle St John, James Apaumut Fall, Gordon Tootoosis
An Indian princess falls in love with an English adventurer and goes against the wishes of her father, in this highly fictionalised account of true events. Despite several assets such as good music and a number of amusing talking animals, Pocahontas and her lover John Smith (who in reality was many years older than her) are drawn in a stilted and unrealistic way, and the lack of a strong script is all too apparent. AA: Score/orig (Menken/Schwartz), Song/orig.
ANIM 78 min (ort 81 min) VIDrel: WDV/TECH V

POETIC JUSTICE **   15
John Singleton USA   1992
Janet Jackson, Tupac Shakur, Tyra Ferrell, Regina King, Joe Torry, Maya Angelou, Roger Guenveur Smith, Q-Tip, Tone Loc, Billy Zane, Lori Petty, Khandi Alexander, Che J. Avery, Lloyd Avery II, Kimberley Brooks, Rico Bueno
After her boyfriend is murdered, a black woman gives up her college plans and becomes a beautician, retreating into a world of her own where she consoles herself by writing poetry. A postal worker tries to make contact with her but things do not turn out quite as he expects. Second film from the director of BOYZ N THE HOOD is more of the same albeit a bit lighter in tone. This dud was meant to launch Jackson's acting career – she deserved better.
DRA 105 min (ort 109 min) cC VIDrel: COLUM/SONOP V/sur

POINT, THE ***   U
Fred Wolf USA   1970
Ringo Starr (narration only)
A charming, animated parable about a young boy who is unique in having a rounded head, not a pointed one like everyone else in the Land of Point. This "crime" earns him banishment from the kingdom and he has a series of entertaining adventures on his travels. Inspired by a record album produced by rock musician Harry Nilsson, and featuring some of his bright tunes. The occasional dollops of moralising and the expected puns are minor failings.
ANIM 74 min mTV VIDrel: VES L/A V

POINT BLANK ***   18
John Boorman USA   1967
Lee Marvin, Angie Dickinson, Keenan Wynn, Carroll O'Connor, Lloyd Bochner, John Vernon, Michael Strong, Sharon Acker, James

Sikking, Sandra Warner, Roberta Haynes, Kathleen Freeman, Victor Creatore, Lawrence Hauben
A man is shot and left for dead by his wife and her lover, but survives the shooting to get his revenge two years later. This harsh, brutal and opaque thriller was a failure at the time, but is now regarded as a near classic. Written by Alexander Jacobs, David Newhouse and Rafe Newhouse.
THR   88 min (ort 92 min)   VIDrel: MGM/WHV   V
Boa: novel The Hunter by Richard Stark (Donald E. Westlake).

POINT BREAK *   18
Kathryn Bigelow USA   1991
Patrick Swayze, Keanu Reeves, Gary Busey, Lori Petty, John McGinley, James Le Gros, John Philbin, Bojesse Christopher, Julian Reyes, Daniel Beer, Chris Pederson, Vincent Klyn, Anthony Kiedis, Dave Olson, Lee Tergesen, Sydney Walsh
Reeves as a wayward FBI agent goes undercover in Southern California among the surfing set, where he hopes to discover the identity of a crafty bank robber. A set of eye-catching action sequences (mostly involving surfboards) and tight direction give the film give the appearance of moving along, but the limp storyline presents a handicap that proves to be impossible to overcome.
THR   122 min wScrn cC   VIDrel: 20TH/TECH; ENCORE (LV only)   V LV

POINTMAN *   PG
Robert Ellis Miller USA   1994
Jack Scalia, Roxann Biggs-Dawson, Bruce A. Young, Robert Knepper, Fritz Weaver, Brent Jennings, Annie Corley, Brandon Smith, Sean McGraw, Michael DeLorenzo, Derek De Lint, Johnny Dark, Tiffany Lawrence, Willy Rosario
Having been wrongfully imprisoned and then pardoned, a former Wall Street executive takes a new career path, working as bodyguard.
A/AD   87 min   VIDrel: WHV   V

POIROT: DEATH IN THE CLOUDS *   PG
Stephen Whittaker UK   1992
David Suchet, Philip Jackson, Sarah Woodward, Shaun Scott, Cathryn Harrison, David Firth, Amanda Royle, Richard Ireson, Jeremy Downham, Eve Pearce, Roger Heathcott, Guy Manning, Gabrielle Lloyd, John Bleasdale, Harry Audley
Poirot investigates a murder that took place whilst he was on board an aeroplane, the murder weapon possibly being a poisoned dart. As he carries out his investigation, it becomes apparent that any of his fellow passengers could have been the culprit, especially when he discovers that the poisoned dart was deliberately planted to throw any investigator off the track.
DRA   102 min (ort 115 min) mTV
VIDrel: CASPIC/BMGREC   V

POIROT: FOUR AND TWENTY BLACKBIRDS *   PG
Renny Rye UK   1989
David Suchet, Hugh Fraser, Philip Jackson, Pauline Moran, Richard Howard, Tony Aitken, Charles Pemberton, Geoffrey Larder, Denys Hawthorne, Holly De Jong, Clifford Rose, Philip Locke, Hilary Mason, Marjie Lawrence, Su Elliott
Taken to dinner by his dentist, Poirot observes an elderly eccentric artist at another table and learns a little about him from a helpful waitress. This being Agatha Christie, the artist is found dead a short while later, and Poirot's curiosity is such that he is able to show that the man was murdered. Another strong entry in this good quality series.
DRA   54 min; 100 min (2-film cassette) mTV
VIDrel: CASPIC/BMGREC   V
Boa: story by Agatha Christie.

POIROT: HOW DOES YOUR GARDEN GROW? *   (PG)
Brian Farnham UK   1991
David Suchet, Hugh Fraser, Philip Jackson, Pauline Moran, Anne Stallybrass, Tom Wylton, Margery Mason, Catherine Russell, Peter Birch, Ralph Nossek, John Burgess, Dorcas Morgan, Trevor Danby, John Rogan, Stephen Petcher
Dramatised by Andrew Marshall, this is another story from the high-quality series of BBC adaptations following the adventures of Christie's famous Belgian sleuth. In this one, Poirot attends the Chelsea Flower Show where a rose is named in his honour. But here, a packet of seeds provides the clue to the murder of a wheelchair-bound old lady, who was attending the flower show at the same time.
DRA   60 min mTV   TVrel: BBC1
Boa: story by Agatha Christie.

POIROT: MURDER IN THE MEWS *   PG
Edward Bennett UK   1988
David Suchet, Hugh Fraser, Philip Jackson, Pauline Moran, Juliette Mole, David Yelland, James Faulkner, Gabrielle Blunt, John Cording, Barrie Cookson
In this adventure, Poirot learns that an apparent suicide is not quite all it appears to be.
DRA   54 min; 100 min (2-film cassette) mTV
VIDrel: CASPIC/BMGREC   V
Boa: story by Agatha Christie.

POIROT: PERIL AT END HOUSE *   PG
Renny Rye UK   1990
David Suchet, Philip Jackson, Hugh Fraser, Polly Walker, Jeremy Young, John Harding, Mary Cunningham, Paul Geoffrey, Alison Sterling, Christopher Baines, Carol MacReady, Elizabeth Downes, Godfrey James, John Crocker
Suchet is excellent as Hercule Poirot, one of Christie's most popular sleuths, in this tale of mystery and murder set on the Cornish Riviera, to which our detective and his friend Captain Hastings have travelled for a holiday.
DRA   100 min (ort 115 min) mTV   VIDrel: VCC L/A   V
Boa: story by Agatha Christie.

POIROT: PROBLEM AT SEA *   (PG)
Renny Rye UK   1989
David Suchet, Hugh Fraser, Melissa Greenwood, Victoria Hastead, Roger Hume, Philip Jackson, Ben Aris, John Normington, Sheila Allen, Sheri Shepstone, Louisa Janes, Ann Firbank, James Ottaway, Geoffrey Beavers
Our plucky Belgian sleuth and his friend Captain Hastings are enjoying a cruise to the Egyptian port of Alexandria when a murder takes place on the vessel. Despite the fact that all of the suspects have alibis, it's not long before Poirot uncovers the culprit.
DRA   54 min mTV   VIDrel: CASPIC L/A   V
Boa: story by Agatha Christie.

POIROT: THE ABC MURDERS *   PG
Andrew Grieve UK
David Suchet, Hugh Fraser, Philip Jackson, Donald Sumpter, Nicholas Farrell, Donald Douglas, Pippa Guard, Cathryn Bradshaw, Nina Marc, David McAlister, Vivienne Burgess, Ann Windsor, Michael Mellinger, Miranda Forbes, Lucinda Curtis
Three bodies are found, and both the surname and location of each victim matches a letter of the alphabet. Beside each body is left a copy of the ABC Railway Guide, it becoming apparent that the killer intends to work his way through the alphabet. As ever, the police are baffled. Fortunately, Poirot is on hand to solve this mystery.
DRA   101 min mTV   VIDrel: CASPIC/BMGREC   V
Boa: story by Agatha Christie.

POIROT: THE ADVENTURE OF JOHNNIE WAVERLY *   PG
Renny Rye UK   1989
David Suchet, Hugh Fraser, Philip Jackson, Pauline Moran, Geoffrey Bateman, Julia Chambers, Dominic Rougier, Patrick Jordan, Carol Frazer, Sandra Freeman, Robert Putt, Patrick Connor, Phillip Marikum, Jona Jones, Jonathan Magnanti
An episode from the popular and well conceived TV series featuring Christie's famous Belgian detective. In this tale, Captain Hastings is loath to believe there is a plot to kidnap a country squire's son, but Poirot is on hand to investigate. Highly enjoyable and a pleasure to watch, but in this one the mystery is so transparent that most viewers will guess the truth well before the climax.
DRA   54 min; 100 min (2-film cassette) mTV
VIDrel: CASPIC/BMGREC   V
Boa: story by Agatha Christie.

POIROT: THE ADVENTURE OF THE CLAPHAM COOK *   PG
Edward Bennett UK   1988
David Suchet, Hugh Fraser, Philip Jackson, Pauline Moran, Brigit Forsyth, Dermot Crowley, Freda Dowie, Antony Carrick, Katy Murphy, Daniel Webb, Brian Poyser, Richard Bebb, Frank Vincent, Phillip Manikum, Jona Jones
An episode from the TV series that followed the exploits of Christie's famous sleuth. In this one, the strange disappearance of a cook leads Poirot to the discovery of a complex intrigue. Fair.
DRA   54 min; 100 min (2-film cassette) mTV
VIDrel: CASPIC/BMGREC   V
Boa: story by Agatha Christie.

POIROT: THE DREAM ***
PG
Edward Bennett UK
1989
David Suchet, Hugh Fraser, Philip Jackson, Pauline Moran, Joely Richardson, Alan Howard, Mary Tamm, Martin Wenner, Christopher Saul, Paul Lacoux, Neville Phillips, Tommy Wright, Fred Bryant, Donald Bissett, Arthur Howell
A story from a well-made and popular series of adaptations of Agatha Christie tales, in which Poirot is consulted by an industrialist who has been experiencing disturbing nightmares.
DRA   54 min; 100 min (2-film cassette) mTV
VIDrel: CASPIC L/A   V
Boa: story by Agatha Christie.

POIROT: THE INCREDIBLE THEFT ***
PG
Edward Bennett UK
1989
David Suchet, Hugh Fraser, Philip Jackson, Pauline Moran, John Stride, John Carson, Carmen Du Sautoy, Ciaran Madden, Phyllida Law, Guy Scantlebury, Albert Welling, Dan Hildebrand, Phillip Manikum
When the plans for a new fighter plane are stolen, Poirot undertakes an investigation, and learns that the case is a good deal more complex than it at first appeared.
DRA   54 min; 100 min (2-film cassette) mTV
VIDrel: CASPIC/BMGREC   V
Boa: story by Agatha Christie.

POIROT: THE KING OF CLUBS ***
PG
Renny Rye UK
1989
David Suchet, Hugh Fraser, Philip Jackson, Pauline Moran, Niamh Cusack, Dawn Grainger, David Swift, Jack Klaff, Jonathan Coy, Rosie Timpson, Mark Culwick, Avril Egar, Vass Anderson, Abigail Cruttenden, Sean Pertwee, Cathy Murphy
An episode from a popular series of adaptations based on Christie's stories of her Belgian detective, who in this story investigates the murder of a disreputable impresario, the chief suspect being an attractive and soon-to-be-married young starlet.
DRA   54 min; 100 min (2-film cassette) mTV
VIDrel: CASPIC L/A   V
Boa: story by Agatha Christie.

POIROT: THE MYSTERIOUS AFFAIR AT STYLES ***
PG
Ross Devenish UK
1988
David Suchet, Hugh Fraser, Philip Jackson, Beatie Edney, David Rintoul, Joanna McCallum, Anthony Calf, Morris Perry, Gilliam Barge, Michael Cronin, Lala Lord, Michael Goodley, Penelope Beaumont, David Saville, Tim Munro
The very first in this excellent series of TV adaptations, this tale is set in 1917, with Lt Hastings recovering from his war injuries. Having received an invitation to visit his friends the Cavendishs at Styles St Mary, he also finds himself meeting an old acquaintance – Hercule Poirot. When a rather nasty murder occurs nearby, the two friends work together to solve the crime, a successful partnership which they decide to continue.
DRA   100 min mTV   VIDrel: VCC L/A   V
Boa: novel by Agatha Christie.

POIROT: THE THIRD FLOOR FLAT ***
PG
Edward Bennett UK
1989
David Suchet, Hugh Fraser, Philip Jackson, Pauline Moran, Suzanne Burden, Nicholas Pritchard, Robert Hines, Amanda Elwes, Josie Lawrence, Susan Porrett, Alan Partington, James Aidan, Gillian Bush Bailey, George Little
Sitting at home at his apartment, Poirot frets because he has no murder case in hand. However, fate soon intervenes when a fatal shooting takes place two floors below.
DRA   50 min; 100 min (2-film cassette) mTV
VIDrel: CASPIC/BMGREC   V

POIROT: TRIANGLE AT RHODES ***
(PG)
Renny Rye UK
1989
David Suchet, Frances Low, Jon Cartwright, Annie Lambert, Peter Settelen, Angela Down, Timothy Kightley, Al Fiorentini, Anthony Benson, Patrick Monckton
On holiday on the island of Rhodes (then under Italian control) our Belgian sleuth finds his powers of deduction put to good use when a murder is committed. Fair.
DRA   54 min mTV   VIDrel: CASPIC L/A   V
Boa: story by Agatha Christie.

POISON **
18
Todd Haynes USA
1990
Edith Meeks, Millie White, Buck Smith, Anne Giotta, Lydia Lafleur,
Ian Nemser; Scott Renderer, James Lyons, John R. Lombardi, Tony Pemberton, Andrew Harpending; Larry Maxwell, Susan Norman, Al Quagliata, Michelle Sullivan
Allegedly inspired by the writings of Jean Genet, this underground-style study of alienation consists of three different stories. The first, entitled "Hero", is a fake documentary about a boy who shoots his father. "Horror" is a 1950s-style SF spoof and finally "Homo" tells of a gay thief who is sent to jail.
DRA   85 min B/W/Col (French dialogue) subs
VIDrel: VCC/DISC/COLUM L/A   V

POISON IVY **
18
Andy Ruben/Katt Shea Ruben USA
1992
Drew Barrymore, Cheryl Ladd, Tom Skerritt, Sara Gilbert, Alan Stock, Jeanne Sakata, E.J. Moore, J.B. Quon, Leonardo Di Caprio, Michael Goldner, Charley Hayward, Time Winters, Billy Kane, Tony Ervolina, Mary Gordon, Julie Jay
A disturbed teenage girl worms her way into her best friend's family, where she seduces the father and ingratiates herself with the dying mother, who almost comes to regard her as a daughter. Soon, the real daughter finds herself facing an almost life-and-death struggle, in this overblown and over-the-top melodrama. A sequel followed.
THR   89 min (ort 95 min)
VIDrel: 4-FRONT/POLYREC/GUILD   V/s

POISON IVY 2: LILY **
18
Anne Goursaud USA
1995
Alyssa Milano, Xander Berkeley, Belinda Bauer, Jonathon Schaech
Newly arrived in L.A., a student finds herself sharing rooms with four weird room-mates, and stumbles across the secret erotic diaries of the now deceased Ivy (from the first film). Aroused by the events it describes, she embarks on a relationship with her teacher and events proceed to their inevitable tragic end. Given much the same level of plotting as a host of similar works, but with none of the tension of the first film, it deservedly went straight to video.
DRA   90 min (ort 110 min)   VIDrel: GUILD/FOXVID   V/sur

POLAR BEAR KING, THE **
(U)
Ola Solum GERMANY/NORWAY/SWEDEN
1993
Maria Bonnevie, Jack Fjeldstad, Tobias Hoesel, Anna-Lotta Larsson, Jon Laxda, Helge Jordal, Marika Enstad, Kristen Mack, Rudiyer Kuhlbrodt, Ulrich Faulhaber, Bengt Ellis, Karen Randers-Pehrson, Karen hoje, Siw A. Andersen
A prince is changed into a polar bear by an evil witch's spell but is befriended by a princess who falls in love with him despite his ferocious exterior. A competently made children's fairy tale that benefits greatly from its Scandinavian locations.
FAN   84 min (ort 87 min)   SATrel: MOVIE CHANNEL

POLDARK: SERIES 1 AND 2 ***
PG
Paul Annett/Christopher Barry/Kenenth Ives/Philip Dudley UK
1975
Robin Ellis, Anghard Rees, Jill Townsend, Judy Geeson, Ralph Bates, Richard Morant, Clive Frances, John Baskcomb, Paul Curran, Tilly Tremayne, Mary Wimbush
After fighting on the British side in the Revolutionary War, a Cornishman returns home to find his family's fortunes at a low ebb, with their tin mine about to be sold to pay their debts. He starts a hard, long and bitter fight to turn this situation around, in this extremely well made and acted adaptation of Graham's historical novels.
DRA   1,376 min (8 cassettes) mTV   VIDrel: BBC/TECH   V/h
Boa: novels by Winston Graham.

POLDARK **
PG
Richard Laxton UK
1996
Mel Martin, Michael Attwell, Joan Gruffudd, Kelly Reilly, Sarah Carpenter, Amanda Ryan
Story set in 1810 at the time of Britain's war with France, with the title character away in London helping the government while his wife and children struggle on as best they can in Cornwall. A fair TV drama, slightly sluggish in places but for the most part, enjoyable and well mounted.
DRA   102 min mTV
VIDrel: FIRST/SONOP; ENCORE (LV only)   V LV
Boa: novel by Winston Graham.

POLICE ***
15
Maurice Pialat FRANCE
1985
Gerard Depardieu, Sophie Marceau, Richard Anconina, Pascale Rocard, Sandrine Bonnaire, Franck Karoui

Depardieu plays a cynical Paris police inspector who is out to smash a drugs ring operating from Marseilles, but gets involved with a girl at the centre of his enquiries. Much energy is expended on closely observing the demanding daily grind of police work, and this results in an uncompromising and compelling film that's sometimes drawn-out but is always realistic. Depardieu picked up the Best Actor prize at the Venice Film Festival.
DRA   109 min (ort 114 min)   VIDrel: ARTIF/20TH   V

POLICE ACADEMY ** 15
Hugh Wilson USA   1984
Steve Guttenberg, Kim Cattrall, G.W. Bailey, George Gaynes, Bubba Smith, Michael Winslow, David Graf, Donovan Scott, Andrew Rubin, Bruce Mahler, Leslie Easterbrook, Debralee Scott, Ted Ross, Don Lake, Marion Ramsey, Bill Lynn
A police commissioner lowers the recruiting standards of the police academy with the result that assorted misfits and weirdos enrol. Several sequels have followed, each one spinning out to gossamer thinness, a comic idea that was never more than moderately amusing to begin with.
COM   92 min (ort 96 min)   VIDrel: WHV   V/h

POLICE ACADEMY 2 * 15
Jerry Paris USA   1985
Steve Guttenberg, Bubba Smith, George Gaynes, David Graf, Michael Winslow, Bruce Mahler, Marion Ramsey, Colleen Camp, Howard Hesseman, Art Metrano, Ed Merlihy, Bobcat Goldthwait, Julie Brown, Peter Van Norden, Tim Kazurinsky
The continuing story of the world's most unlikely police graduates; funny if you like that sort of thing. In this film their main trouble comes from a gang of spray-paint "terrorists".
Aka: POLICE ACADEMY 2: THEIR FIRST ASSIGNMENT
COM   84 min (ort 87 min)   VIDrel: WHV   V/h

POLICE ACADEMY 3 ** PG
Jerry Paris USA   1986
Steve Guttenberg, Bubba Smith, David Graf, Michael Winslow, Marion Ramsey, Leslie Easterbrook, Art Metrano, Tim Kazurinsky, George Gaynes, Bobcat Goldthwait, Shawn Weatherly, Brian Tochi, Lance Kinsey, Brant Von Hoffman
Yet another attempt to squeeze a few more laughs (and some cash) out of a tired plot, revolving round a zany police training academy. In this one, the academy is faced with budget cuts that may close it; our rookie cops find a way to save it. This unmemorable affair was the last film of director Paris.
Aka: POLICE ACADEMY 3: BACK IN TRAINING
COM   80 min (ort 82 min)   VIDrel: WHV   V/h

POLICE ACADEMY 4 * PG
Jim Drake USA   1987
Steve Guttenberg, Bubba Smith, Michael Winslow, David Graf, Tim Kazurinsky, Sharon Stone, Leslie Easterbrook, Marion Ramsey, Lance Kinsey, G.W. Bailey, Bobcat Goldthwait, George Gaynes, Billie Bird, Brian Tochi, Tab Thacker
Another tired sequel, this time featuring police attempts to involve the local citizens in crime prevention.
Aka: POLICE ACADEMY 4: CITIZEN'S ON PATROL
COM   83 min Cut (8 sec – ort 87 min)   VIDrel: WHV   V/h

POLICE ACADEMY 5 * PG
Alan Myerson USA   1987
Bubba Smith, Michael Winslow, David Graf, Janet Jones, Matt McCoy, Leslie Easterbrook, Marion Ramsey, G.W. Bailey, Rene Auberjonois, George Gaynes, Lance Kinsey, Tab Thacker, Archie Hahn, James Hampton, Jerry Lazarus, Dana Mark
Bottom-of-the-barrel time, with this fourth sequel in which an accidental switch of luggage at an airport deprives some crooks of their stolen diamonds, which thus come into the possession of Chief Lassed, who's in Miami to address a police convention.
Aka: POLICE ACADEMY 5: ASSIGNMENT MIAMI BEACH
COM   87 min (ort 90 min)   VIDrel: WHV   V

POLICE ACADEMY 6 * PG
Peter Bonerz USA   1989
Kenneth Mars, G.W. Bailey, George Gaynes, Bubba Smith, Michael Winslow, David Graf, Leslie Easterbrook, Lance Kinsey, Marion Ramsey, Bruce Mahler, Matt McCoy, Gerrit Graham, George R. Robertson, Brian Seeman, Darwyn Swalve
Yet another tale in this increasingly dismal and strained series,

this time following the exploits of our cops as they try to solve a crime wave.
Aka: POLICE ACADEMY 6: CITY UNDER SIEGE
COM   80 min Cut (32 sec at UK cinema release – ort 84 min)
VIDrel: WHV   V

POLICE ACADEMY 7 – MISSION TO MOSCOW ** PG
Alan Metter RUSSIA/USA   1994
George Gaynes, Michael Winslow, David Graf, Leslie Easterbrook, Claire Forlani, Ron Pearlman, Christopher Lee, Charlie Schlatter, G.W. Bailey, Richard Israel, Gregg Berger, Vladimir Dolinsky, Valery Yavamenko, Vadim Dolgachov
This time a group of our inept cops are sent to Moscow to foil a Mafia plot to steal information through the use of a software game infected with a secret program. Apart from the scenes shot on location in Russia and the presence of Russian actors, no new ideas are in evidence here and it is a pity to see the Russians wasting their efforts on such a pitiful co-production.
COM   79 min (ort 83 min) cC   VIDrel: WHV   V/sur

POLICE ASSASSINS ** 18
David Chung HONG KONG   1986
Michelle Khan, Henry Sanada, Harry Khan, Michael Wong
A team of anti-terrorists foil an attempt to hijack a jet, killing two of the would-be hijackers in the process. However, they now find themselves targeted by an international group of terrorists, who have sworn revenge. A fast-paced and violent action tale, followed by a sequel.
A/AD   92 min   VIDrel: SCRN/DISC   V

POLICE ASSASSINS 2 * 18
HONG KONG   1987
Michelle Khan, Cynthia Rothrock
A tough Scotland Yard woman detective is sent to Hong Kong to take part in a hazardous investigation, and teams up with an equally tough female Hong Kong detective. Another formula blend of crime-busting and martial arts mayhem.
A/AD   83 min Cut (3 sec – ort 90 min) dubbed
VIDrel: SCRN/DISC L/A   V

POLICE GIRLS ACADEMY * 18
Richard Hieronymous USA   1987
Kirsten Baker, Perry Lang
A low-budget effort that attempts to cash in on the success of the POLICE ACADEMY series, by virtue of its title if not its plot, which is bland to the point of tedium. The script has two college girls helping the local police during their vacations and causing the expected mishaps.
COM   84 min Cut (1 min 59 sec)   VIDrel: IMPENT   V

POLICE STORY *** 15
Jackie Chan HONG KONG   1986
Jackie Chan, Briget Lin, Maggie Cheung, Cho Yuen, Bill Tung, Kenneth Tong, Lam Kok Hung, Lau Chi Wing, Charles Chao, Kam Hing Yin, Mars, Paul Wong, Wan Fat, Fung Hark On, Danny Chow, Tai Po, Wu Fang, Lau Ai Lai, Clarence Ford
A tough Hong Kong cop is given the assignment of protecting a key witness in a drugs trial, in this simple-minded but enjoyable mixture of martial arts, comedy and action genres. Chan does his own stunts, most of which are incredible. Switch the brain off and enjoy this one.
Aka: JACKIE CHAN'S POLICE FORCE; JACKIE CHAN'S POLICE STORY; JINGCHA GUSHI; POLICE FORCE
A/AD   84 min (ort 101 min)   VIDrel: 4-FRONT/POLYREC   V

POLICE STORY 2 *** 15
Jackie Chan HONG KONG   1987
Jackie Chan, Maggie Cheung, Bill Tung, Lam Kwok Hung, Charles Chao, Edward Tang
Following his one-man battle with drug dealers, Chan has been demoted from detective to traffic cop. When Hong Kong is hit by a wave of bomb threats, he finds himself caught up in a private investigation that results in the kidnapping of his girlfriend by those responsible. An entertaining and often spectacular follow-up to his earlier film. A further sequel followed.
A/AD   91 min (ort 92 min)   VIDrel: 4-FRONT/POLYREC   V

POLICE STORY 3: SUPERCOP ** 15
Stanley Tong HONG KONG   1992
Jackie Chan, Michelle Khan, Maggie Cheung, Ken Tsang, Yuen Wah, Bill Tung, Josephine Koo

Second sequel in this entertaining series sees Chan going to work for a foreign country where he gets involved with attempts to reform a notorious gangster.
A/AD   91 min   VIDrel: MIA/DISC   V

POLITICIAN'S WIFE, THE ** 18
Graham Theakston UK 1995
Juliet Stevenson, Trevor Eve, Ian Bannen, Anton Lesser, Frederick Treves, Minnie Driver
A faithful wife plans a complex revenge on her philandering politician husband, when his affair is exposed by the press. Initially supportive, she appears to have helped him save his career as a government minister, but this is only in order to make her planned revenge taste all the sweeter.
DRA   185 min mTV   VIDrel: CH4/RTM   V

POLLY *** U
Debbie Allen USA 1989
Phylicia Rashad, Keshia Knight Pulliam, Celeste Holm, Brock Peters, Dorian Harewood, Butterfly McQueen, Larry Riley, Ken Page, Barbara Montgomery, Vanessa Bell Calloway, T.K. Carter, George Anthony Bell, Tom McGreevey, Michael Peters
This bright and breezy Disney offering recasts their 1960 film POLLYANNA as a musical, with the action transferred to a small town in Alabama where virtually all the inhabitants are black. Choreographed by the director, and with a score by Joel McNeely. Followed by POLLY – COMIN' HOME.
MUS   90 min (ort 100 min) mTV   VIDrel: WDV/TECH L/A V

POLLY – COMIN' HOME! *** (U)
Debbie Allen USA 1990
Phylicia Rashad, Keshia Knight Pullman, Dorian Harewood, Barbara Montgomery, Branden Adams, Vanessa Bell Calloway, T.K. Carter, Anthony Newley, Celeste Holm, Ken Page, George Anthony Bell, Larry Riley, Vickilyn Reynolds, Geraldine Decker
Musical set in Alabama at a time of racial tension and Martin Luther King's Civil Rights movement, it follows the adventures of a feisty young black girl and her maiden aunt. A vigorous and joyful sequel to POLLY, it boasts strong direction, fine performances and excellent choreography (the latter is by Debbie Allen, who did such fine work on the TV series "Fame").
MUS   90 min mCab   SATrel: DISNEY CHANNEL

POLLYANNA ** U
David Swift USA 1960
Hayley Mills, Jane Wyman, Richard Egan, Karl Malden, Adolphe Menjou, Nancy Olson, Donald Crisp, Agnes Moorehead, Kevin Corcoran, Reta Shaw, James Drury, Leora Dana, Anne Seymour
A young girl goes to stay with her aunt and brightens up the lives of the inhabitants of a small New England town. Swift also wrote the script as well as directing this pleasant but rather humourless adaptation of Porter's tale. AA: Hon Award (Mills – for the most outstanding juvenile performance).
JUV   129 min (ort 134 min)   VIDrel: L/A   V
Boa: novel by Eleanor Porter.

POLLYANNA ** U
June Wyndham-Davies UK 1973
Elizabeth Archard, Elaine Stritch, Ray McAnally
An orphan girl goes to live with her strait-laced aunt and brings some joy into her life. Pleasant and undemanding kid's drama, very slightly reminiscent of ANNE OF GREEN GABLES.
JUV   154 min (2 cassettes) mTV   VIDrel: BBC/TECH   V/h
Boa: novel by Eleanor H. Porter.

POLTERGEIST ** 15
Tobe Hooper USA 1982
JoBeth Williams, Beatrice Straight, Dominique Dunne, Craig T. Nelson, Zelda Rubinstein, Heather O'Rourke, Oliver Robins, James Karen, Martin Casella, Richard Lawson, Michael McManus, Virginia Kiser, James Karen, Lou Perry
A family's home is invaded by hostile supernatural entities who kidnap their family's five-year-old daughter. It starts well but soon goes downhill as the special effects take over from the plot. Written by Steven Spielberg, Michael Grais and Mark Victor. As is so often the case, dazzling special effects cannot hide the lack of substance. The inevitable sequels followed. See also GRAVE SECRETS: THE LEGACY OF HILLTOP DRIVE and THE AMITYVILLE HORROR.
HOR   110 min (ort 114 min) wScrn   VIDrel: MGM/WHV V/sur

POLTERGEIST 2 ** 15
Brian Gibson USA 1986
JoBeth Williams, Craig T. Nelson, Heather O'Rourke, Oliver Robins, Zelda Rubinstein, Will Sampson, Julian Beck, Geraldine Fitzgerald, Noble Craig, John P. Whitecloud, Susan Peretz, Helen Boll, Kelly Jean Peters, Jaclyn Bernstein
This sequel is set four years on and finds the family facing a return encounter with these strange creatures from another world. As with the first film the special effects cannot be faulted, only the story lacks substance.
Aka: POLTERGEIST 2: THE OTHER SIDE
HOR   90 min wScrn (special edition)   VIDrel: MGM/WHV V/sur

POLTERGEIST 3 * 15
Gary Sherman USA 1988
Tom Skerritt, Nancy Allen, Heather O'Rourke, Zelda Rubinstein, Kip Wentz, Lara Flynn Boyle, Richard Fire, Nathan Davis, Rober May, Paul Graham, Stacy Gilchrist, Meg Weldon, Joey Garfield, Chris Murphy, Roy Hytower, Meg Thalken
Third time round, there's little that's new or fresh left to extract from the story of a family pursued by hideous ghosts. Here, they are summoned up by the daughter, when her therapist refuses to acknowledge the reality of the supernatural.
HOR   93 min (ort 99 min)   VIDrel: MGM/WHV   V/sur

POLYESTER ** 18
John Waters USA 1981
Divine (Glenn Milstead), Tab Hunter, Edith Massey, Mary Garlington, Ken King, David Samson, Mink Stole, Stiv Bators, Joni Ruth White, Hans Kramm, George Stover, Steve Yeager, Rick Breitenfeld, Michael Watson, Derek Neal, Jean Hill
Middle-class satire on a housewife's inability to cope with her life. For fans of Waters/Divine only. Originally released in "Odorama" with audience members being given smelly cards to scratch and sniff. This was Waters's first mainstream effort and the bad taste of his earlier films is but partially replaced by wit. There are however, some funny moments.
COM   79 min (ort 86 min)   VIDrel: CASPIC/TERRY L/A   V

PONTIAC MOON ** 12
Peter Medak USA 1994
Ted Danson, Mary Steenburgen, Cathy Moriarty, Eric Schweig, Ryan Todd, Max Gail, Lisa Jane Persky, J.C. Quinn, John Schuck, Don Swayze
An oddball dad decides to celebrate the 25th anniversary of the 1969 Apollo II Moon landing by taking his family on a 1,776 mile trip, his intention being to get the mileage on his old Pontiac Chief to match the 238,857 miles to the Moon. But this is a foolhardy venture, as his wife is an agoraphobic. A decidedly original comedy-drama that never gets going and lacks substance, even though the cast struggle valiantly to breathe life into their roles.
DRA   102 min (ort 108 min) cC   VIDrel: CIC/SONOP   V

POOR COW ** 15
Ken Loach UK 1967
Carol White, Terence Stamp, John Bindon, Kate Williams, Queenie Watts, Geraldine Sherman, Ellis Dale, Gerald Young, Gladys Dawson
While a thief is away in prison, his wife moves in with his best friend. A typical 1960s kitchen-sink drama full of squalor and despair but little else.
DRA   97 min (ort 101 min)   VIDrel: WHV   V/h
Boa: novel by Nell Dunn.

POOR LITTLE RICH GIRL: PARTS 1 AND 2 ** 15
Charles Jarrott USA 1987
Farrah Fawcett, Kevin McCarthy, Nicholas Clay, Bruce Davison, Sascha Hehn, Stephane Audran, Burl Ives, James Read, Anne Francis, David Ackroyd, Tony Peck, Zoe Wanamaker, Linden Ashby, Amadeus August, Fairuza Balk, Bruce Davison
A two-part soap telling the story of Woolworth heiress Barbara Hutton and her turbulent life. She inherited a vast fortune but frittered it away and died in 1979 almost penniless. Rather superficial but quite entertaining.
Aka: POOR LITTLE RICH GIRL: THE BARBARA HUTTON STORY
DRA   180 min (2 cassettes – ort 280 min) mTV
VIDrel: BRAVE/SONOP L/A   V
Boa: book by C. David Heymann.

POPE MUST DIE, THE ** 15
Peter Richardson USA 1991
Robbie Coltrane, Beverly D'Angelo, Herbert Lom, Alex Rocco, Paul

Bartel, Balthazar Getty, William Hootkins, Robert Stephens, Annette Crosbie, Steve O'Donnell, John Sessions, Salvatore Cascio, Peter Richardson, Khedija Sassi
This oddball satire was originally written for Steve Martin, but has been reworked to take in the rather more ample frame of Coltrane, who as a rock 'n' roll loving priest, finds himself elevated to the Papacy by a careless clerical error. Meanwhile, a criminally-minded priest is rather vexed, for he had planned to install his own man in the Vatican. A silly and superficial spoof that bears a faint resemblance to the rather better written KING RALPH.
Aka: POPE, THE; POPE MUST DIET, THE
COM   95 min (ort 99 min)   VIDrel: COLUM/SONOP L/A V/sh

POPE OF GREENWICH VILLAGE, THE *** 15
Stuart Rosenberg USA   1984
Mickey Rourke, Eric Roberts, Daryl Hannah, Geraldine Page, Kenneth McMillan, Tony Musante, Burt Young, M. Emmet Walsh, Jack Kehoe, Philip Bosco, Val Avery, Joe Grifasi, Tony DiBenedetto, Ronald Maccone, Betty Miller, Thomas A. Carlin
Two small-time crooks plan a robbery and find that they've stolen from the Mafia, but the story takes second place to a funny series of observations and acute character studies. The script is by Patrick from his novel.
COM   116 min (ort 120 min)   VIDrel: MGM/WHV   V/h
Boa: novel by Vincent Patrick.

POPE OF UTAH, THE *** 18
Chaim Bianco/Steve Saylor USA   1993
Lee Golden, Tom McCarthy, Ginny Brown Graham
A striking satire on TV evangelism set in Utah in the twenty-first century, where a TV evangelist is soon to go nationwide. However, a programme-censor has to find a way of stopping this, as his wife is so obsessed with the evangelist's show that she has started to put the family in debt. An audacious and often very funny attack on this peculiarly American phenomenon, is chiefly memorable for a wonderfully creepy character – the TV evangelist Melvis Pressin.
COM   82 min   VIDrel: SCEDGE/RTM   V

POPEYE * U
Robert Altman USA   1980
Robin Williams, Shelley Duvall, Ray Walston, Paul Smith, Paul Dooley, Richard Libertini, Wesley Ivan Hurt, Linda Hunt, Roberta Maxwell, Donald Moffat, MacIntyre Dixon, Donovan Scott, Allan Nicholes, Bill Irwin
An ambitious attempt to bring this famous comic-strip character (based on the work of E.C. Segar) to the screen. The script is by Jules Feiffer, and has the character returning to Sweethaven in search of the father who abandoned him. Unfortunately the film is nothing like as inventive as those old Max Fleischer shorts and very little action takes place until near the end.
COM   92 min (ort 114 min)   VIDrel: WDV/TECH L/A   V

PORKY'S *** 18
Bob Clark CANADA   1981
Dan Monahan, Mark Herrier, Wyatt Knight, Roger Wilson, Kim Cattrall, Chuck Mitchell, Scott Colomby, Kaki Hunter, Nancy Parsons, Alex Karras, Susan Clarke, Tony Ganios, Cyril O'Reilly, Boyd Gaines, Dough McGrath, Art Hindle
Set in South Florida circa 1954, where some high school kids try to patronise the local brothel as an attempt to discover sex. A number of sequels followed this film, that's essentially a recycling of the "American Graffiti" theme of adolescent pimply youth. A happy low-brow film that made a fortune and spawned several sequels. Written by Clark.
COM   94 min (ort 98 min)   VIDrel: 20TH/TECH   V

PORKY'S 2 * 18
Bob Clark CANADA   1983
Dan Monahan, Wyatt Knight, Mark Herrier, Roger Wilson, Kaki Hunter, Scott Colomby, Nancy Parsons, Edward Winter, Tony Ganios, Cyril O'Reilly, Joseph Running Fox, Eric Christmas, Cisse Cameron, Else Earl
The first sequel to the original bears little resemblance to it, and is set in a Florida high school where the inevitable high-jinks and low-brow comedy abound.
Aka: PORKY'S 2: THE NEXT DAY
COM   94 min   VIDrel: 20TH/TECH   V

PORKY'S REVENGE * 18
James Komack USA   1985
Dan Monahan, Wyatt Knight, Tony Ganios, Kaki Hunter, Mark Herrier, Scott Colomby, Nancy Parsons, Chuck Mitchell, Kimberly Everson, Rose McVeigh, Fred Buch, Wendy Feign, Eric Christmas, Ilse Earl, Bill Hindman, Donna Rosae
Third in this rather tedious PORKY'S series of teenage comedies. This time the school's basketball coach has gambling debts to the brothel owner. A witless parade of sex-obsessed antics with no shortage of nudity if nothing more.
Aka: PORKY'S 3
COM   88 min (ort 95 min)   VIDrel: 20TH/TECH   V/h

PORRIDGE ** PG
Dick Clement UK   1979
Ronnie Barker, Richard Beckinsale, Fulton Mackay, Brian Wilde, Peter Vaughan, Geoffrey Bayldon, Julian Holloway, Christopher Godwin, Barrie Rutter, Daniel Peacock, Sam Kelly, Ken Jones, Philip Locke, Gordon Kaye
A spin-off from a witty and observant British TV series, about the life of the prisoners at H.M. Slade Prison. Not as funny as the TV series, with a plot revolving around the attempt of old lags to arrange an escape for a first offender. The TV show came to the States as "On The Rocks". Scripted by Dick Clement and Ian La Frenais.
Aka: DOING TIME
COM   90 min (ort 96 min)   VIDrel: 4-FRONT/POLYREC V/h

PORT OF SHADOWS **** PG
Marcel Carne FRANCE   1938
Jean Gabin, Pierre Brasseur, Michele Morgan, Michel Simon, Aimos, Delmont
An army deserter flees to Le Havre after he has committed a murder, where he falls in love with a young girl, whom he attempts to rescue from the clutches of some crooks. But their attempt to run away together has a tragic conclusion. A brooding and melancholy work, inevitably quite depressing given the plot, but potent and often quite poetic. Much of the film's strength derives from the work of Alexandre Trauner, who designed the atmospheric port sets.
Aka: QUAI DES BRUMES
DRA   86 min (ort 89 min) B/W   VIDrel: ELPIC/POLYREC V

PORTERHOUSE BLUE: PARTS 1 AND 2 *** 15
Robert Knights UK   1987
David Jason, Ian Richardson, Charles Gray, John Sessions, Griff Rhys-Jones, Paul Rogers, John Woodnutt, Barbara Jefford, Harold Innocent, Paula Jacobs, Ian Wallace, Willoughby Goddard, Lockwood West, Bob Goody, Earl Rhodes
When the Master of Porterhouse, one of the colleges at Cambridge University, dies unexpectedly without naming a successor, a feverish power struggle is the result. An amusing blend of the comic and the bizarre, that preserves the flavour of Sharpe's novel of chaotic college life.
COM   220 min (2 cassettes) mTV   VIDrel: POLY/POLYREC V
Boa: novel by Tom Sharpe.

PORTNOY'S COMPLAINT * 18
Ernest Lehman USA   1972
Karen Black, Lee Grant, Richard Benjamin, Jack Somack, Jill Clayburgh, Jeannie Berlin, D.P. Barnes, Francesca De Sapio, Kevin Conway, Renee Lippin, Lewis J. Stadlen, Jessica Rains, Eleanor Zee, William Pabst, Tony Brande
The story of a neurotic Jewish boy's innumerable hangups and his downright peculiar relationship with his mother. The directorial debut for Lehman is a flawed film, with barely a trace of the humour that made the novel so successful. The music is by Michel Le Grand. Lehman also scripted this.
COM   101 min   VIDrel: MGM/WHV L/A   V
Boa: novel by Philip Roth.

PORTRAIT, THE *** (PG)
Arthur Penn USA   1992
Lauren Bacall, Gregory Peck, Cecilia Peck, Paul McCrane, Mitchell Laurance, Donna Mitchell, Joyce O'Connor, William Prince, Augusta Dabney, John Murphy, Marty McGaw, John Bennes, Ed Lillard, David Chandler, Lucille Patton, Jan Leslie
A woman artist preparing for an exhibition asks her aged parents to sit for a portrait and finds that this request opens up all manner of emotional confrontations. Bacall and Peck give

excellent performances as the still loving couple and appear together for the first time in thirty-seven years.
DRA 86 min (ort 89 min) mCab SATrel: SKY MOVIES
Boa: play Painting Churches by Tina Howe.

PORTRAIT OF A LADY, THE * 12
Jane Campion UK/USA 1996
Nicole Kidman, John Malkovich, Barbara Hershey, Mary Louise Parker, Martin Donovan, Shelley Winters, Richard E. Grant, Shelley Duvall, Christian Bale, Viggo Mortensen, Valentina Cervi
In Italy, two manipulative American dilettantes get their claws into a young heiress, as they see this as a means of permanently funding their lifestyle and providing a surrogate mother for the child of one of them. This adaptation of the Henry James novel always looks good, but is more akin to DANGEROUS LIAISONS than the original work, and though well acted does not offer any of the cast a chance to shine. A handsome film, but also an unusually boring one.
DRA 144 min CINrel
 Boa: novel by Henry James.

PORTRAIT OF JENNIE **** U
William Dieterle USA 1948
Jennifer Jones, Joseph Cotten, Ethel Barrymore, Lilian Gish, David Wayne, Henry Hull, Cecil Kellaway, Albert Sharpe, Clem Bevans, Felix Bressart, Florence Bates, Maude Simmons, Esther Somers, John Farrell, Robert Dudley
A struggling artist meets a strange girl in a park and falls in love with her, despite the fact that she is a ghost. A film rich in visual imagery, with remarkable photography and a fine score by Dmitri Tiomkin (after Debussy). Written by Peter Berneis, Paul Osborn and Leonard Bernovici, with David Selznick producing. AA: Effects (vis – Paul Eagler/J. McMillan/Russell Sherman/Clarence Slifer/aud – Charles Freeman/James G. Stewart).
Aka: JEANNIE; TIDAL WAVE
DRA 86 min B/W VIDrel: ENTUK/GOLD L/A; PION (LV only) V L V
Boa: novella by Robert Nathan.

POSEIDON ADVENTURE, THE *** PG
Ronald Neame USA 1972
Gene Hackman, Ernest Borgnine, Shelley Winters, Roddy McDowall, Red Buttons, Carol Lynley, Jack Albertson, Stella Stevens, Leslie Nielsen, Arthur O'Connell, Pamela Sue Martin, Eric Shea, Fred Sadoff, Sheila Matthews, Jan Arvan
An exciting drama of a band of survivors, trying to escape from a liner that has turned over completely, after a tidal wave has overwhelmed it. Harrowing in parts but for the most part quite gripping. Written by Stirling Silliphant and Wendell Mayes. The excellent score is by John Williams. AA: Song ("The Morning After" – Al Kasha/Joel Hirschhorn), Spec Award (L.B. Abbott/A.D. Flowers for visual effects).
DRA 112 min (ort 117 min) VIDrel: 20TH/TECH V
Boa: novel by Paul Gallico.

POSSE ** 15
Mario Van Peebles UK/USA 1992
Mario Van Peebles, Stephen Baldwin, Billy Zane, Tiny Lister Jr, Big Daddy Kane, Charles Lane, Blair Underwood, Melvin Van Peebles, Sali Richardson, Tone Loc, Pam Grier, Vesta Williams, Isaac Hayes, Robert Hooks, Richard Jordan
A group of black cavalrymen fighting during the Spanish-American War desert and head for a free frontier town, but find themselves caught up in the machinations of some white villains. Flaunted as an attempt to tell the story of the role of black people in the West, this is for all its good intentions, nothing more than another conventional big-budget Western.
A/AD 106 min (ort 113 min) VIDrel: VCC/DISC/COLUM; ENCORE (LV only) V/sur LV

POSSESSED ** (PG)
Clarence Brown USA 1931
Joan Crawford, Clark Gable, Wallace Ford, Skeets Gallager, John Miljan, Frank Conroy, Marjorie White, Clara Blandick
Inferior soap opera tale of an ambitious factory girl who becomes a kept woman but learns too late that she has traded both morals and happiness for wealth and security. Not to be confused with the film Crawford made seventeen years later.
DRA 77 min B/W VIDrel: MGM/WHV L/A V
Boa: play The Mirage by Edgar Selwyn.

POSSESSED *** PG
Curtis Bernhardt USA 1947
Joan Crawford, Van Heflin, Raymond Massey, Geraldine Brooks, Stanley Ridges, Moroni Olsen, Erskine Sanford, Gerald Perreau, Isabel Withers, Lisa Golm, Douglas Kennedy, Monte Blue, Don McGuire, Rory Mallinson, Clifton Young
After a woman is found in a dazed state on the streets of Los Angeles, she is hospitalised and given drug therapy. She eventually responds and begins to recall the events in her life that contributed to her breakdown. A strong and powerful drama, told mainly in flashbacks, with excellent acting and tight, effective direction. Often confused with another film of the same title from 1931, in which Crawford also starred.
DRA 108 min (ort 110 min) B/W VIDrel: MGM/WHV V/h
Boa: story One Man's Secret by Rita Weiman.

POSSESSED, THE * 18
Lucio Fulci ITALY 1982
Christopher Connelly, Martha Taylor, Brigitta Boccoli, Giovanni Frezza, Cinzia De Ponti, Andrea Bosic, Carlo DeMejo, Antonio Pulci, Lucio Fulci, Vincenzo Bellanich, Mario Moretti, Laura Lenzi
An American Egyptologist is struck by blindness whilst exploring a strange tomb, and after much hardship, returns to New York with a strange medallion in the shape of an eye. This proves to be a mistake, for when his daughter takes to wearing it she becomes an instrument of evil. The usual violent bloodshed results in a film chiefly remembered for the presence of De Ponti, who was Miss Italy of 1979.
Aka: EYE OF THE EVIL DEAD, THE; L'OCCHIO DEL MALE; MANHATTAN BABY
HOR 87 min (ort 91 min) VIDrel: 20TH/TECH L/A V

POSTCARDS FROM AMERICA *** 18
Steve McLean UK/USA 1994
James Lyons, Michael Tighe, Olmo Tighe, Michael Imperiolo, Michael Ringer, Maggie Low, Les "Linda" Simpson, Dean "Sissy Fit" Novotny, Tom Gilroy, Peter Byrne, Bob Romano, John Ventimiglia, Paul Germaine-Brown, Danny and Tony Urbino
An account of the life and times of homosexual writer, artist and AIDS campaigner David Wojnarowicz, with the film being split into three distinct sections, each of which focuses on various influential events in his life. Not very effective as a coherent whole, but some of the episodes (such as his miserable childhood) are portrayed in a way that is both realistic and touching. An episodic and sad film, its fluctuating style is its greatest flaw.
DRA 93 min (ort 95 min) VIDrel: DTK V

POSTCARDS FROM THE EDGE *** 15
Mike Nichols USA 1990
Meryl Streep, Shirley MacLaine, Dennis Quaid, Gene Hackman, Rob Reiner, Mary Wickes, Richard Dreyfuss, Conrad Bain, Annette Bening, Simon Callow, Gary Morton, C.C.H. Pounder, Sidney Armus, Robin Bartlett, Dana Ivey
Entertaining comedy-drama about the love-hate relationship between a superstar and her somewhat less successful actress daughter, who falls prey to drugs when the pressure get too great. An intense and bitter character study, which earned Streep an Oscar nomination for her excellent performance. Scripted by Fisher from her novel, the film moves along at a fair pace, and the music (by Carly Simon) is an added bonus.
DRA 97 min (ort 101 min) VIDrel: VCC/DISC/COLUM V/sur
Boa: novel by Carrie Fisher.

POSTMAN, THE *** U
Michael Radford ITALY 1994
Massimo Troisi, Philippe Noiret, Maria Grazia Cucinotta, Linda Moretti, Anna Bonaiuto, Renato Scarpa, Mariano Rigillo, Bruno Alessandro, Sergio Solli, Carlo di Maio, Nando Neri, Vincenzo di Sauro, Orazio Stracuzzi, Alfredo Cozzolino
Set on a tiny island in the Mediterranean, this gentle comedy has Noiret playing the exiled poet Pablo Neruda and the late comedian Troisi his protege, Mario. This latter is the awkward son of a local fisherman, hired by the postmaster to help deal with the writer's copious fan-mail. Mario (the postman of the title) is far too shy to win the hand of the woman he loves, so the poet agrees to instruct him in the use of language. A charming comedy of much warmth.
Aka: IL POSTINO
COM 105 min (ort 108 min) VIDrel: TOUCH/TECH V/sh
Boa: novel Il Postino di Neruda by Antonio Skarmeta.

POSTMAN ALWAYS RINGS TWICE, THE *** PG
Tay Garnett USA 1946
Lana Turner, John Garfield, Cecil Kallaway, Hume Cronyn, Leon Ames, Audrey Trotter, Alan Reed, Jeff York, Charles Williams, Cameron Grant, William Halligan, Wally Cassell, Morris Ankrum, Garry Owen, Dorothy Phillips
A drifter comes to a roadside cafe, gets a meal and a job, and because of his sexual passion for the wife is gradually entrapped in a scheme to murder her husband, which he eventually does. From then on, it is downhill all the way. A tense, dark and very disturbing film. Filmed before in France and Italy, and remade in 1981.
DRA 108 min (ort 113 min) B/W VIDrel: MGM/WHV V
Boa: novel by James M. Cain.

POSTMAN ALWAYS RINGS TWICE, THE ** 18
Bob Rafelson USA 1981
Jack Nicholson, Jessica Lange, John Colicos, Michael Lerner, John P. Ryan, Anjelica Huston, William Taylor, Tom Hill, Jon Van Ness, Brian Farrell, Raleigh Bond, William Newman, Albert Henderson, Ken Magee, Eugene Peterson, Don Calfa
A remake of the 1946 film, with a drifter taking up with the young wife of a middle-aged cafe owner and plotting with her to kill her husband. Far more faithful to the novel than the earlier film, but far less stylish too. Some vicious touches mark its 1980s origins. Written by David Mamet and with photography by Sven Nykvist.
DRA 116 min (ort 122 min) VIDrel: VCC L/A V
Boa: novel by James M. Cain.

POUND PUPPIES: THE LEGEND OF BIG PAW * U
Pierre DeCelles USA 1987
Voices of: George Rose, B.J. Ward, Ruth Buzzi, Brennan Howard, Cathy Cadavini, Nancy Cartwright
A rather mundane cartoon feature based on toys of the same name, with the title characters attempting to re-unite two fragments of a magical bone, that has been stolen by a villain out to take over the world. Done in the form of a 1950s rock 'n' roll adventure it may be, but this dull effort is unlikely to have much appeal for kids.
Aka: POUND PUPPIES AND THE LEGEND OF BIG PAW; POUND PUPPIES: THE MOVIE
ANIM 74 min (ort 76 min) VIDrel: 4-FRONT/POLYREC V/sh

POWDER * 12
Victor Salva USA 1995
Mary Steenburgen, Sean Patrick Flanery, Lance Henriksen, Jeff Goldblum, Brandon Smith, Bradford Tatum, Susan Tyrrell, Missy Crider, Ray Wise, Esteban Louis Powell, Reed Fredrichs, Chad Cox, Joe Marchman, Philip Maurice Hayes
A super-intelligent albino boy learns that his abilities (including some telekinetic powers) and appearance are largely the result of the freakish circumstances in which he was born. He struggles to come to terms with the factors that mark him out from all those around him, and the message in this dull film appears to be that we should be nice to people who are different. An over-sentimental dud, neither interesting nor terribly convincing.
DRA 112 min VIDrel: HOLPIC/TECH V/s

POWER ** 15
Sidney Lumet USA 1985
Richard Gere, Julie Christie, Gene Hackman, Kate Capshaw, Denzel Washington, E.G. Marshall, Beatrice Straight, Fritz Weaver, Michael Learned, J.T. Walsh, E. Katherine Kerr, Polly Rowles, Matt Salinger, Tom Mardirosian
The story of a political media man who markets his politicians like just another commodity, and prides himself on his shrewdness but discovers that others are a lot smarter. A flashy and insincere tale of implausibility and pompousness that makes a few sharp points, but even these are no longer fresh or original.
DRA 107 min (ort 111 min) VIDrel: 20TH V

POWER 98 ** 18
Jaime Hellman USA 1995
Eric Roberts, Jason Gedrick, Stephen Tobolowsky, Jennie Garth, Larry Drake
A Los Angeles D.J. who has an X-rated show listened to by millions attracts the attention of a caller who confesses to a series of vicious murders. When an L.A.P.D. detective learns of this, he attempts to discover the identity of the caller, but to do so, must first enter the bizarre and unconventional world of the

D.J. Engrossing for about the first half of the movie, it soon becomes just one more murder thriller, dispensing neither originality nor wit.
THR 86 min VIDrel: CURB/HIFLI V/h

POWER OF ONE, THE ** 15
John G. Avildsen USA 1991
Stephen Dorff, Guy Witcher, Simon Fenton, Armin Mueller-Stahl, John Gielgud, Morgan Freeman, Marius Weyers, Ian Roberts, Fay Masterson, Winston Ntshona, Dominic Walker, Robbie Bulloch, Daniel Craig, Clive Russell, Faith Edwards
A liberal-minded South African youth of British descent is sent to a Boer-run boarding school and is made to suffer for his beliefs, but is helped by a black boxing coach and a German pianist. An incredibly insensitive and cliche-filled attempt to exploit the injustices of the Apartheid system that is both overlong and unconvincing.
DRA 122 min (ort 127 min) cC VIDrel: WHV V/dm V/sur
Boa: novel by Bryce Courtenay.

POWER RANGERS: THE MOVIE ** PG
Bryan Spicer USA 1995
Karen Ashley, Johnny Yong Bosch, Steve Cardenas, Jason David Frank, David Yost, Paul Freeman
Workers on a building site uncover a strange object that releases an oozing substance of sinister powers. The title characters now have a battle on their hands to save the planet from being overcome by evil. A daft spin-off movie from a popular kid's TV series that revolved around the exploits of a bunch of youngsters with super-powers. Unfortunately, what works well as a weekly TV series needs more depth to succeed as a film. Simple fun for youngsters only.
JUV 100 min cC VIDrel: 20TH/TECH V/sur

POWER WITHIN, THE ** (PG)
Art Camacho USA 1995
Ted Jan Roberts, Karen Valentine, Keith Coogan, John O'Hurley, Gerald Okamura, Tracy Lindsey, P.J. Soles, Irwin Keyes, Jean Speegle Howard, Karen Kim, Ed O'Ross, William Zabka, Jacob Parker, Marc Riffon, Francis Fallen, Joe Hart
A young karate adept is given a ring by a mystic and finds that it possesses magic powers. In next to no time, he becomes the target of a skilled but utterly evil fighter who will stop at nothing to obtain this object. Apart from the unusual plot angle, there is nothing else of any originality in this standard martial-arts actioner.
MAR 92 min (ort 97 min) SATrel: SKY MOVIES

POWWOW HIGHWAY *** 15
Jonathan Wacks USA 1988
Gary Farmer, A. Martinez, Amanda Wyss, Joanelle Romero, Sam Vlahos, Wayne Waterman, Margo Kane, Geoff Rivas, Rosco Born, John Trudell, Chrissie McDonald, Wes Studi, Tony Frank, Sky Seals, Maria Antoinette Rogers
This humorous look at Indian mistreatment and the clash of cultures, follows a Cheyenne and an Indian activist friend as they take a trip to New Mexico. Despite the stereotyping of the racists they encounter along the way, this immensely likeable film is both mature and insightful.
COM 87 min (ort 90 min) VIDrel: PATHE L/A V

PRAGUE *** 12
Ian Sellar FRANCE/UK 1991
Alan Cumming, Sandrine Bonnaire, Bruno Ganz, Raphael Meiss, Henri Meiss, Hana Gregorova, Ladislav Lahoda, Nelly Gaierova, Luba Skorepova, Zdena Keclikova, Lubos Kafka, Jaroslav Jodl, Olga Michalkova, Ladislav Brothanek
A young Scotsman comes to Prague in the hope of finding some 1940s newsreel footage that contains shots of his grandparents, and while he is there he gets caught up in a romantic triangle with a Czech film archivist and her employer. A strong flavour of Kafka intrudes here, with the various references to the stifling bureaucracy of the country, but it is mainly the charming performances that keep one watching rather than the weird plot. Scripted by Sellar.
COM 89 min CINrel

PRANCER *** U
John Hancock USA 1989
Sam Elliott, Cloris Leachman, Rebecca Harrell, Abe Vigoda, Rutanya Alda, Michael Constantine, Ariana Richards, John Joseph Duda, Mark Rolston, Walter Charles, Michael Luciano, Johnny Galecki, Victor Truro, Marcia Porter

A family film revolving around a troubled young girl whose widowed father is in danger of losing their farm. When the girl finds an injured reindeer just before Christmas, she becomes convinced that it belongs to Santa Claus and sets out to nurse it back to health. Slow-moving, but a film that repays patience.
A/AD　98 min (ort 103 min)　VIDrel: POLY/POLYREC V/sh

PRANKS * 18
Jeffrey Obrow/Stephen Carpenter USA　　1982
David Snow, Laurie Lapinski, Stephen Sachs, Pamela Holland, Dennis Ely, Woody Roll, Jake Jones, Daphne Zuniga, Robert Frederick, Chris Morill, Chandre, Billy Criswell, Richard Cowgill, Kay Beth, Jimmy Betz
A near-derelict college building has been condemned and now awaits the bulldozers. When a group of students volunteer to clear it of its remaining furniture before demolition, they quickly succumb one by one to a psychotic killer, in this all too predictable stalk-and-slash movie.
Aka: DEATH DORM; DORM THAT DRIPPED BLOOD, THE
HOR　81 min (ort 83 min)　VIDrel: IMPENT　V

PRAYER OF THE ROLLERBOYS ** 15
Rick King USA　　1990
Corey Haim, Patricia Arquette, Christopher Collet, J.C. Quinn, Julius Harris, Devin Clark, Mark Pellegrino, Morgan Weisser, G. Smokey Campbell, Jake Dengel, John P. Connolly, Stanley Yale, Loren Lester, Tim Eyster, James Patrick
In an unfriendly and unwelcoming future, a gang of teenage roller-skating thugs make their violent way across an America that's now heavily in debt to the Japanese. Various brutal encounters take place, as good battles evil. A fast-moving, stylish but pointless effort, that borrows elements from SURF NAZIS MUST DIE and ROLLERBALL.
FAN　90 min (ort 94 min)　VIDrel: FIRST/SONOP　V

PRAYING MANTIS ** 15
James Keach USA　　1993
Jane Seymour, Barry Bostwick, Chad Allen, Frances Fisher, Colby Chester, Michael MaRae, Anne Schedden, John Martin, John Knots, Sherri Jensen, Roger Welch, Kathleen Randall, Barbara Kite, Mariah Milner, Victor Morris, Terry Ward
A beautiful woman imitates the behaviour of the title insect and gives cyanide to a succession of husbands on their wedding nights, in a far from erotic tale, that is both trite and predictable. What little tension there is derives from the investigation the latest victim's sister-in-law decides to make into this woman's past.
THR　85 min (ort 90 min) mTV　VIDrel: CIC/SONOP L/A V

PRAYING WITH ANGER ** PG
M. Night Shayamalan USA　　1992
M. Night Shyamalan, Mike Muthu, Christabel Howie, Richa Ahuja, Arun Balachandran
An American-Indian is persuaded by his mother to take part in a student exchange program with Madras University, in order for him to learn about his parents' homeland. He agrees, but finds the resulting cultural shock totally overwhelming as he struggles to cope with a totally unfamiliar environment. Well crafted but their is little here that has not been seen countless times before, while our hero's sudden change makes for an unconvincing ending.
DRA　103 min　VIDrel: IMAG/GUILD L/A　V

PREACHER'S WIFE, THE ** U
Penny Marshall USA　　1996
Denzel Washington, Whitney Houston, Courtney B. Vance, Gregory Hines, Jenifer Lewis, Loretta Devine, Lionel Ritchie, Paul Bates, Justin Pierre Edmund, Lex Monson, Darvel Davis Jr
A strident and underplotted remake of the Cary Grant hit THE BISHOP'S WIFE that has much the same idea of angelic intervention leading to good things for the title character and his family. The film is little more than a vanity vehicle for Houston who sings very prettily at every opportunity, unfortunately her acting is a good deal less accomplished, and this coupled with the dullness of the script does not work to the film's advantage.
COM　124 min　CINrel

PREACHING TO THE PERVERTED ** 18
Stuart Urban UK　　1996
Guinevere Turner, Christian Anholt, Tom Bell
Writer-director Urban's debut film takes a long, hard look at the world of fetish fashion and sexual perversion, by way of London's various underground fetish clubs. A civil servant (Emery) is given the job of infiltrating one such venue, run by a dominatrix (Turner). But this unlikely pair immediately form a strong rapport, which takes the form of a master/slave relationship. Weakly plotted, this is an oddly effective blending of the funny and the bizarre.
DRA　100 min　CINrel

PRECIOUS FIND ** 15
Philippe Mora USA　　1996
Rutger Hauer, Joan Chen, Brion James, Harold Pruett, Morgan Hunter, Philippe Mora, Don Hunter, Tim De Zarn, Anthony Guidera, Shay Duffin, Jonathan Ball, David Michael O'Neill, Douchan Gersi, J. Bryan McMillan, Buckley Morris
SF adventure set in the 21st century, with three humanoids who live and work on a mining colony setting out on a hunt for treasure that will take them on a long journey to the far reaches of the galaxy. Quite a likeable low-budget work, it plays like a space-age Western, ambling along to its satisfactory conclusion.
FAN　86 min　VIDrel: MED/20VIS　V/sh

PRECIOUS VICTIMS *** 15
Peter Levin USA　　1993
Park Overall, Robby Benson, Richard Thomas, Frederic Forrest, Eileen Brennan, Brion James, Tim Grimm, Cliff DeYoung, Robyn Lively, Nancy Cartwright, Molly McClure, Charles Noland, Don Amendolia, Kack Kenny, Louisa Abernathy
The true story of the kidnapping and murder of a young child in 1986, when the parents claimed their child was abducted but fell under suspicion due to inconsistencies in their story. However, there was insufficient evidence to charge them, until they carried out the same ploy with their second child three years later. A sickening chronicle of human depravity, the film is undeniably well made but given its subject matter is far from an enjoyable experience.
DRA　93 min mTV　VIDrel: ODY/SONOP　V/sh
Boa: book by Charles Bosworth Jr and Don W. Weber.

PREDATOR *** 18
John McTiernan USA　　1987
Arnold Schwarzenegger, Carl Weathers, Elpidia Carrillo, Bill Duke, Jesse Ventura, R.G. Armstrong, Kevin Peter Hall, Sonny Landham, Richard Chaves, Shane Black, Gregory Barnett, Bobby Bass, Gene Baxley, Steve Boyum
A crack team of agents are sent by the US Government to the jungles of South America on a rescue mission, but find themselves up against an elusive alien that is able to camouflage its appearance and pick them off one at a time. An implausible feast of gore and special effects (most notably the "invisibility" device the alien uses) but certainly not without considerable tension. A rather feeble sequel followed.
FAN　102 min (ort 107 min)　VIDrel: 20TH/TECH L/A; ENCORE (LV only)　V/sh LV

PREDATOR 2 * 18
Stephen Hopkins USA　　1990
Danny Glover, Gary Busey, Ruben Blades, Maria Conchita Alonso, Bill Paxton, Kevin Peter Hall, Robert Davi, Kent McCord, Morton Downey Jr, Calvin Lockhart, Adam Baldwin, Henry Kwigi, Elpidia Carrillo, Corey Rand, Teri Weigel
A daft sequel to the first film, it is set in 1997 and our invisible and bloodthirsty creature is now venturing onto the streets of Los Angeles. Pretty much a waste of time for all concerned, it has Glover and his fellow cops tackling the menace. Despite generally good production values, the lack of anything remotely resembling a plot is a major handicap (as is the sight of an apparently indestructible Glover continually falling from great heights).
A/AD　103 min (ort 108 min)　VIDrel: 20TH/TECH; ENCORE (LV only)　V LV

PREHYSTERIA ** PG
Albert Band/Charles Band USA　　1992
Brett Cullen, Austin O'Brien, Samantha Mills, Colleen Morris, Tony Longo, Stuart Fratkin, Stephen Lee , Stuart Fratkin, Tom Williams, Gill Gayle, Peter Mark Vasquez, Ellis Levinson, James Shanta, Jane Caldwell
A family find that some strange eggs in their basement and are astounded when they hatch into pygmy dinosaurs. An unscrupulous archaeologist gets wind of this and decides to steal them. Made to cash in on the short-lived dinosaur craze among kids in the wake of JURASSIC PARK, this is an assem-

bly-line effort, cliched and predictable in every department. Two sequels followed.
JUV   80 min (ort 90 min)   VIDrel: CIC/SONOP   V/sh

PREHYSTERIA 2 **
Albert Band USA
Kevin R. Connors, Dean Scofield, Jennifer Harte, Bettye Ackerman, Greg Lewis, Michael Hagiwara, Larry Hankin, Alan Palo, Owne Bush, Joey Andrews, Jason Dohring, Brian Wagner, Larry Pennell, Frank Welker (vocal effects)
Sequel to the first film and not a bit better, with our miniature dinosaurs playing a role in helping a young boy whose governess wants to send him away to a boarding school. The same sentimental offering as before combined with some excellent special effects in the form of the miniature dinosaur models.
FAN   77 min (ort 84 min)   VIDrel: CIC/SONOP   V

U
1993

PREHYSTERIA 3 **
Julian Breen USA
Fred Willard, Whitney Anderson, Pam Matteson, Dave Buzzotta, Matt Lescher, JohN Fujioka, Owen Bush, Bruce Weitz
Further adventures of our mini-dinosaurs, who help out the harassed couple a mini-putt course where business is not exactly flourishing.
FAN   82 min (ort 85 min) cC   VIDrel: CIC/SONOP   V/dm

U
1995

PRELUDE TO A KISS **
Norman Rene USA
Alec Baldwin, Meg Ryan, Kathy Bates, Ned Beatty, Patty Duke, Sydney Walker, Richard Riehle, Stanley Tucci, Rocky Carroll, Debra Moore, Kay Gill, Jobe Cerny, Ward Ohrman, Annie Golden, Frank Cariloo, Sally Murphy, Sali Richardson
After a whirlwind romance, a rather ill-matched couple decide to get married. At their wedding reception the wife kisses one of the guests and from that point on, her husband becomes obsessed by the notion that she is not the woman he originally fell in love with. A disappointing attempt to bring this delicate fantasy to life that does not survive its transition to the screen. The final resolution when all is made plain is both laughably inept and repetitive.
COM   98 min (ort 106 min)   VIDrel: 20TH/TECH L/A   V
Boa: play by Craig Lucas.

PG
1992

PREMONITION, THE ***
Rumle Hammerich SWEDEN
Tova Magnusson, Figge Norling, Bjorn Kjellman, Niklas Hjulstrom, Malin Berghagen, Liv Alsterlund, Lars Green, Agneta Ekmanner, Marie Goranzon, Reine Brynolfsson, Bjorn Granath, Catherine Hansson, Gunnel Fred, Josefin Ankarberg
A young girl develops an obsessive interest in her teacher and soon witnesses sado-masochistic acts and eventually murder. Fantasy and reality blend in the mind of this disturbed girl, who is the central figure in a film best described as a powerful examination of obsession and violence. Not very enjoyable but certainly memorable.
Aka: SVART LUCIA
HOR   109 min (ort 114 min)   VIDrel: MAINPIC/RTM   V

18

PRESCRIPTION FOR LUST **
USA
Blake Mitchell, Rebecca Bardoux, Dallas D'Amour, Nikki Arizona
At a hospital that has got into financial difficulties, one of the doctors comes up with a surefire money making scheme, when he hits on the idea of turning the hospital into a clinic/brothel, with the sex services on offer being paid for by the medical insurance of the patients. But a hospital inspector arrives to find out why there has been such an increase in longterm patients. However, before long he checks in for a stay there as well. Fair.
A   70 min   VIDrel: ONE   V

18
1995

PRESCRIPTION FOR MURDER **
Catherine Cyran USA
Adam Baldwin, Nina Siemasko, Don Harvey, Barbara Carrera
A failed medical student has an obsessive desire to work as a surgeon, and to this end starts kidnapping his intended victims, and operating on them without the benefit of anaesthetic. As the mutilated corpses begin piling up, the investigating detective realises that the killer shows some medical knowledge, and he teams up with a girl working at the medical school to help uncover the culprit. A gruesome thriller with a strong black comedy element.
THR   89 min   VIDrel: MED/COLUM   V/sh

18
1995

PRESIDENT'S CHILD, THE **
Sam Pilsbury USA
William Devane, Donna Mills, Trevor Eve, Thom Dillon, Michael Flynn, James Read, Nick Tate, David Valenza, Jesse Bennett, J. Scott Bronson
Political intrigue is afoot in this thriller telling of a foreign correspondent who several years ago had a child by a rising politician. When the latter decides to run for President, she finds herself the unwilling target of a CIA investigation and covert spying operation. Another tired and cliched thriller involving high-level conspiracies, corruption and intrigue.
THR   92 min mTV   VIDrel: 20TH/TECH   V
Boa: novel by Fay Weldon.

PG
1992

PRESIDIO, THE **
Peter Hyams USA
Sean Connery, Mark Harmon, Meg Ryan, Jack Warden, Dana Gladstone, Mark Blum, Jenette Goldstein, Don Calfa, Marvin J. McIntyre, John DiSanti, Robert Lesser, Rick Zumwalt, Kim Robillard, James Hooks Reynolds, Curtis W. Sims, Chuckie Davis
A cop clashes with the provost marshal at a military base in San Francisco, when he has to investigate a murder there. The latter used to be his commanding officer, and the fact that he is now dating the marshal's daughter adds yet more fuel to their old enmity. A formula plot that ill serves an actor of Connery's stature.
A/AD   94 min (ort 97 min)
VIDrel: 4-FRONT/POLYREC/CIC   V/dm V/sh

15
1988

PRESUMED INNOCENT **
Alan J. Pakula USA
Harrison Ford, Raul Julia, Greta Scacchi, Bonnie Bedelia, Brian Dennehy, Paul Winfield, John Spencer, Joe Grifasi, Tom Mardirosian, Anna Maria Horsford, Sab Shimono, David Wohl, Bradley Whitford, Christine Estabrook, Michael Polan
When his attractive female colleague is brutally murdered, the senior deputy D.A., unwillingly placed in charge of the murder investigation by his boss, finds himself the prime suspect, and becomes caught up in a web of deceit and lies. A slow-paced, moody thriller, replete with weak characterisations and topped off with an implausible ending. A disappointing film whose main focus of interest is Ford's restrained performance. Music is by John Williams.
DRA   121 min (ort 127 min)   VIDrel: WHV   V/sur
Boa: novel by Scott Turow.

18
1990

PRET-A-PORTER *
Robert Altman USA
Kim Basinger, Sophia Loren, Marcello Mastroianni, Tim Robbins, Anouk Aimee, Julia Roberts, Linda Hunt, Sally Kellerman, Lauren Bacall, Lyle Lovett, Tracey Ullman, Forest Whitaker, Lili Taylor, Stephen Rea, Danny Aiello, Teri Garr
Ostensibly a murder mystery set in the world of high fashion, but in reality a long, unstructured, rambling and highly episodic Altman extravaganza that is devoted to stating at great length that the world of high fashion is both ridiculous and superficial. Unfortunately, this simple and obvious truism cannot sustain a film over-loaded with cameo appearances that add nothing of interest to the story.
Aka: READY TO WEAR
DRA   133 min   VIDrel: TOUCH/TECH   V/sur

15
1994

PRETTY BABY *
Louis Malle USA
Brooke Shields, Keith Carradine, Susan Sarandon, Frances Faye, Antonio Fargas, Matthew Anton, Gerrit Graham, Mae Mercer, Diana Scarwid, Barbara Steele, Seret Scott, Cheryl Markowitz, Susan Manskey, Laura Zimmerman, Miz Mary
A photographer in New Orleans becomes obsessed by a twelve-year-old prostitute and eventually marries her. Carradine is badly cast in this forgettable WW1 tale, but Shields is rather appealing. This was Malle's first US work and was written by him and Polly Platt. The photography was by Sven Nykvist, but even that can do little for this film.
DRA   106 min (ort 109 min)   VIDrel: L/A   V

15
1978

PRETTY IN PINK **
Howard Deutch USA
Molly Ringwald, Andrew McCarthy, Harry Dean Stanton, Jon Cryer, Annie Potts, James Spader, Jim Haynie, Alexa Kenin, Andrew "Dice" Clay, Kate Vernon, Emily Longstreth, Margaret Colin, Jamie Anders, Gina Gershon
A poor but pretty girl falls in love with a handsome boy from a

15
1986

wealthy family, but their budding romance is not helped along by their differing social backgrounds. A pleasant tale of growing pains and teenage insecurities, but nothing memorable. Written by John Hughes.
COM   93 min (ort 99 min)   VIDrel: CIC/SONOP   V/sur

PRETTY POISON ***   15
Noel Black USA   1968
Anthony Perkins, Tuesday Weld, Beverly Garland, John Randolph, Dick O'Neill, Clarice Blackburn, Joseph Bova, Ken Kercheval, Don Fellows, Parker Fennelly, Paul Larson, Tim Callahan, George Fisher, William Sorrells, Dan Morgan
A psychotic arsonist teams up with a high school student to carry out sabotage at a chemicals factory, but soon discovers that she's a lot weirder than he is. A bizarre mixture of drama and black comedy, but quite effectively done. Written by Lorenzo Semple Jr.
DRA   85 min (ort 89 min)   VIDrel: FOXVID   V
Boa: novel She Let Him Continue by Stephen Geller.

PRETTY WOMAN ***   15
Garry Marshall USA   1990
Richard Gere, Julia Roberts, Ralph Bellamy, Jason Alexander, Laura San Giacomo, Hector Elizondo, Alex Hyde-White, Elinor Donahue, Larry Miller, Judith Baldwin, Jason Randal, Bill Applebaum, Tracy Bjork, Gary Greene, William Gallo
A runaway now working as a hooker is hired by a cold-blooded businessman, and the latter spruces her up so that she can act as his escort for a week while he wheels and deals. However, against all the odds an unlikely love affair blossoms. Roberts is delightful in a frothy and derivative variant on the PYGMALION theme. Despite managing to include just about every romantic cliche to be had, this appealing comedy retains great charm.
COM   115 min (ort 117 min)   VIDrel: TOUCH/TECH
V/sur

PREY *   18
Norman J. Warren UK   1977
Barry Stokes, Sally Faulkner, Glory Annen, Sandy Chimney, Eddie Stacey, Jerry Crampton
A strange young man has dinner with two lesbians, but unfortunately for them, he is in fact a replica of a man murdered by aliens who are searching for supplies of protein. Although the film does develop some atmosphere, its gratuitous dwelling on cannibalism and dismemberment makes it very hard to relate to. A kind of gruesome low-budget black comedy without the laughs.
Aka: ALIEN PREY
HOR   74 min Cut (10 sec – ort 85 min)
VIDrel: ARTPRO/RTM L/A   V

PREY, THE *   18
Edwin Scott Brown USA   1980
Steve Bond, Debbie Thureson, Lori Lethin, Robert Wald, Gayle Gannes, Philip Wenckus, Jackson Bostwick, Jackie Coogan, Connie Hunter, Ted Hayden, Garry Goodrow, Carel Struycken
Six teenagers hiking in the forests of the Colorado Rockies, are stalked by something connected with a terrifying tragedy that happened 30 years before. Routine kids-in-peril movie of the cheaper kind.
HOR   92 min   VIDrel: IMPENT   V

PREY OF THE CHAMELEON **   18
Fleming B. (Tex) Fuller USA   1991
Daphne Zuniga, James Wilder, Alexandra Paul, Don Harvey, Red West, Alexander Folk, Mark Carlton, Don Harvey, Michele McBride, Patricia Place, Julie St Clair, Kelly Gwinn, David Powledge, Bob Larkin, Paul Whitthorne, David Wells
This time, our escaped psychotic killer is a woman who takes on the personality traits of each of her victims after she murders them. Wilder is the unfortunate who gives her a lift, and is locked in the boot of a car, while she (dressed in his clothes) proceeds to rob a bank. Some FBI agents are not far behind, but she easily outwits them, and it falls to a female deputy sheriff to put an end to this dangerous criminal. Adequate, but no more than that.
THR   90 min mCab   VIDrel: IMPENT   V

PREY OF THE JAGUAR *   18
David De Coteau USA   1996
Stacy Keach, Maxwell Caulfield, Linda Bair, Paul Bartel
A man joins a witness relocation scheme after he gives evidence

that convicts a drugs baron, but the latter escapes from jail and comes after him, murdering his wife and child. This causes him to adopt a new guise in a battle against evil, that of a super-hero known as the "Jaguar". This ludicrous film would have played much better as a spoof, as matters stand its lack of drive is only matched by its foolish dialogue.
A/AD   89 min   VIDrel: FIRST/SONOP   V

PRICE OF LOVE, THE **   15
David Burton Morris USA   1995
Laurel Holloman, Harvey Silver, Steve Martini, Alexis Cruz, Ben Gould, Peter Facinelli
An innocent, young runaway winds up on the streets of Hollywood, where he turns to hustling to make a living. A foray into an area that has been explored many times before. Fortunately, the characters are interesting enough to keep us involved.
DRA   86 min   VIDrel: POLFIL   V/sh

PRICELESS BEAUTY *   15
Charles Finch USA   1987
Christopher Lambert, Diane Lane, J.C. Quinn, Francesco Quinn, Claudio Ohana, Monica Scattini, Joaquin D'Almeida
Following the accidental death of his brother, a musician turns his back on his fans and retreats to a house by the beach to live as a recluse. However, he finds a strange jar that contains a female genie which grants him three wishes, thus enabling him to rediscover his zest for living in this over-sentimental and rather mushy affair.
DRA   87 min (ort 88 min)   VIDrel: MIA/DISC/IMPENT   V

PRICK UP YOUR EARS ***   18
Stephen Frears UK   1987
Gary Oldman, Alfred Molina, Vanessa Redgrave, Julie Walters, Wallace Shawn, James Grant, Frances Barber, Lindsay Duncan, Janet Dale, Margaret Tyzack, Dave Atkins, John Bailey, Joan Sanderson, Roger Lloyd Pack, Garry Cooper
The story of Joe Orton and his stormy relationship with his lover Kenneth Halliwell, who eventually murdered him when he found himself being left in the shadows as Orton's fame grew. A convincing and absorbing piece, with Oldman and Molina giving remarkable performances as Orton and Halliwell respectively. The script is by Alan Bennett.
DRA   105 min (ort 111 min)   VIDrel: VISVID/POLYREC   V
Boa: book by John Lahr.

PRIDE AND PREJUDICE ****   U
Robert Z. Leonard USA   1940
Laurence Olivier, Greer Garson, Edmund Gwenn, Mary Boland, Melville Cooper, Edna May Oliver, Maureen O'Sullivan, Ann Rutherford, Frieda Inescort, Bruce Lester, Karen Morley, E.E. Clive, Heather Angel, Marsha Hunt, Edward Ashley
Story of five sisters hunting for husbands in Victorian England. Fine period flavour is retained in this elegant comedy of manners, and an excellent cast adds to the enjoyment. Written by Aldous Huxley and Jane Murfin. AA: Art (Cedric Gibbons/Paul Groesse).
DRA   113 min (ort 116 min) B/W   VIDrel: MGM/WHV   V
Boa: novel by Jane Austin/play by Helen Jerome.

PRIDE AND PREJUDICE ***   U
Cyril Coke AUSTRALIA/UK   1980
Elizabeth Garvie, David Rintoul
A faithful adaptation of Austen's classic, with good period detail and an excellent cast.
DRA   301 min (2 cassettes) mTV cC   VIDrel: BBC/TECH
V/dm V/sur
Boa: novel by Jane Austen.

PRIDE AND PREJUDICE ***   U
Simon Langton UK   1995
Colin Firth, Jennifer Ehle, Allison Steadman, David Bamber, Crispin Bonham Carter, Anna Chancellor, Susannah Harker, Adrian Lukis, Julie Sawalha, Benjamin Whitrow
Originally shown on BBC TV in six parts, this is an excellent adaptation of Austen's novel of early 19th century England, and the various fates that await five unmarried sisters. Firth is really outstanding as the arrogant and self-centred Darcy, whose wooing of the initially reluctant Elizabeth forms the major part of the story.
DRA   301 min (2 cassettes) mTV   VIDrel: BBC/TECH   V/h
Boa: novel by Jane Austen.

PRIDE OF THE YANKEES, THE *** U
Sam Wood USA 1942
Gary Cooper, Teresa Wright, Babe Ruth, Walter Brennan, Dan Duryea, Elsa Janssen, Ludwig Stossel, Virginia Gilmore, Bill Dickey, Ernie Adams, Pierre Watkin, Mark Koenig, Bob Meusel, Harry harvey, Bill Stern, Addison Richards
A classic baseball filmography with Cooper portraying Lou Gehrig, the legendary first baseman who was struck down and finally killed at the peak of his career by an incurable disease. A sensitively handled film that boasts superb performances and fine direction. AA: Edit (Daniel Mandell).
DRA 128 min B/W VIDrel: VCC/DISC V

PRIEST *** 15
Antonia Bird UK 1995
Linus Roache, Tom Wilkinson, Cathy Tyson, Robert Carlyle, James Ellis, Lesley Sharp, Robert Pugh, Christine Tremarco, Paul Barber, Rio Fanning, Jimmy Coleman, Bill Dean, Gilly Coman, Fred Pearson, Jimmy Gallagher, Tony Booth
A young idealistic priest, working in a poor Liverpool parish, attempts to reconcile his personal faith with the human weaknesses he finds in himself and in others, most especially in the area of sexual desire. As he struggles to overcome his homosexual leanings he comes to realise that he is far from alone in this respect. A forceful and fascinating look at Catholic religious life, that paints a flawed albeit realistic picture of both the Church and its flock.
DRA 104 min (ort 109 min) VIDrel: ELPIC/POLYREC V

PRIEST OF LOVE, THE ** 15
Christopher Miles UK 1981
Janet Suzmann, Ian McKellen, Ava Gardner, Penelope Keith, John Gielgud, Helen Mirren, Sarah Miles, James Faulkner, Jorge Rivero, Maruizio Merli, Mike Gwilym, Massimo Ranieri, Marjorie Yates, Jane Booker, Wendy Alnutt, Shane Rimmer
Dull biopic about the life of the writer D.H. Lawrence, written by Alan Plater and largely following his last years and the publication of "Lady Chatterly's Lover". A ponderous affair that fails to provide us with any insight into the character. The director also worked on Lawrence's "The Virgin And The Gypsy" some years before.
DRA 95 min (ort 125 min) VIDrel: CURZON/20TH V
Boa: book by Harry T. Moore.

PRIMAL FEAR ** 18
Gregory Hoblit USA 1996
Richard Gere, Laura Linney, John Mahoney, Edward Norton, Frances McDormand, Robert Jordan, Joanie Lum, Randy Salerno, Alfre Woodard, Terry O'Quinn, Andre Braugher, Steven Bauer, Randall Slavin, Mike Bacarella, Turk Muller, Joe Kosala
A Chicago-based lawyer with a real hunger for fame finds that when he takes on the case of an altar boy, charged with murdering a much-loved Catholic Archbishop, the case against the lad is far less convincing than it at first appeared. Quite an efficient albeit implausible thriller, but Gere does not convince as lawyer, though there are good moments of tension and a pleasing set of plot twists. This was the debut feature for both the director and Norton.
THR 125 min (ort 129 min) cC VIDrel: CIC/SONOP; PION (LV only) V/sur LV
Boa: novel by William Diehl.

PRIME OF MISS JEAN BRODIE, THE *** PG
Ronald Neame UK 1969
Maggie Smith, Gordon Jackson, Celia Johnson, Robert Stephens, Jane Carr, Pamela Franklin, Diane Grayson, Shirley Steedman, Margo Cunningham, Isla Cameron, Rona Anderson, Molly Weir, Helena Gloag, John Dunbar, Heather Seymour
A character study of a romantic Edinburgh lady schoolteacher who has a hypnotic effect on her pupils. Written by Jay Presson Allen, with music and lyrics by Rod McKuen. Filmed on location. The story later formed the basis for a TV mini-series. AA: Actress (Smith).
DRA 111 min (ort 116 min) VIDrel: 20TH/TECH V
Boa: novel by Muriel Spark.

PRIME SUSPECT ** 18
Christopher Menaul UK 1988
Helen Mirren, Tom Bell, Zoe Wanamaker, John Bowe, Tom Wilkinson, Ralph Fiennes, John Benfield, Bryan Pringle, John Forgeham, Craig Fairbrass, Jack Ellis, Mossie Smith, Jan Fitzgibbon, Andrew Tiernen, Philip Wright, Gary Whelan
A female Detective Chief Inspector takes over a murder case that's a good deal more complex than it at first appears, and her relentless pursuit of the truth unmasks a serial killer. An absorbing drama that is a little weakened by its emphasis on her continual conflicts with her male colleagues as well as a somewhat abrupt and satisfactory ending. Written by the ubiquitous Lynda La Plante, this TV series was followed by some sequels.
DRA 200 min (ort 230 min) mTV
VIDrel: CASPIC/BMGREC V

PRIME SUSPECT 2 *** 18
John Strickland UK 1993
Helen Mirren, Colin Salmon, John Benfield, Jack Ellis, Craig Fairbrass, Ian Fitzgibbon, Richard Hawley, Philip Wright, Andrew Tiernen, Lloyd Maguire, Dev Sagoo, Stephen Boxer, Shireen Shah, Claire Benedict, George Harris
The murder of a young girl proves a complex puzzle for a female Detective Chief Inspector who as ever, faces less than full co-operation from her male colleagues. Eminently watchable but the constant emphasis on petty bickering and unbridled nastiness makes for a very downbeat offering.
DRA 204 min (2 cassettes – ort 230 min) mTV
VIDrel: CASPIC/BMGREC V

PRIME SUSPECT 3 *** 18
David Drury UK 1993
Helen Mirren, Tom bell, John Benfield, Peter Capaldi, Ciaran Hinds, Struan Rodger, David Thewlis, Michael j, Shannon, Greg Saunders, Danny Dyer, Richard Rees, Terrence Hardiman, Mark Strong, Karen Tomlin, Struan Rodger, Lewis James
Detective Chief Inspector Tennison is called in when the partially charred body of a Soho rent-boy is discovered in a burnt-out flat of a transsexual. As ever, with a script by Lynda La Plante, we can be assured that there will be ample nastiness to be uncovered as well as incessant squabbles with her unpleasant male colleagues.
DRA 206 min mTV VIDrel: VCC/DISC V/sur

PRIME SUSPECT 4: INNER CIRCLES *** (18)
Sara Pia Anderson UK 1995
Helen Mirren, Jill Baker, James Laurenson, Helen Kvale, Anthony Bates, Kate Reilly, Nick Patrick, Richard Hawley, Phillada Sewell, Jonathan Copestake, Julie Rice, Roger Milner, Tony Spooner, Hamish McColl, Sam Rumbelow, Ralph Arliss
When an employee at a country club is found murdered not far from his place of work, ill-feeling is soon aroused among the locals when the chief suspects prove to be a couple living on a nearby housing estate. Superintendent Tennison investigates, and learns that there are plenty of guilty secrets behind the smug faces and chintz curtains of the well-to-do suburbanites.
DRA 102 min mTV TVrel

PRIME SUSPECT: THE LOST CHILD *** (18)
John Madden UK 1995
Helen Mirren, John Benfield, Beatie Edney, Jack Ellis, Robert Glenister, Richard Hawley, Adrian Lukis, Lesley Sharp, Mossie Smith, Stuart Wilson, Tracy Keating, David Phelan, Graham Seed, Tony Rohr, Mark Bazeley, Chris Brailsford
A woman is found badly injured in her home with her young daughter missing and Jane Tennison, just recently promoted to superintendent, takes on the case but complicates her private life by getting involved with a psychologist, whose patient proves to be one of the main suspects. While her rivals see this relationship as a useful chink in her armour, she and her colleagues try to solve this mystery and discover whether or not the child is still alive.
DRA 102 min mTV TVrel

PRIME SUSPECT: THE SCENT OF DARKNESS ***
(18)
Paul Marcus UK 1995
Helen Mirren, Tim Woodward, Stephen Boxer, Richard Haley, John Benfield, Christopher Fulford, Joyce Redman, Stuart Wilson
Once again there is a series of murders to solve, and Jane Tennison is soon on a collision course with her superiors (it's all getting a bit repetitive now) and is eventually suspended. Meanwhile, her scheming rivals are not slow to exploit her relationship with a psychologist for their own ends.
DRA 102 min mTV TVrel

PRIME TARGET * 15
David Heavener USA 1991
David Heavener, Isaac Hayes, Robert Reed, Andrew Robinson, Jenilee Harrison, Don Stroud, Tony Curtis, Michael Gregory

The hero in this action movie is a maverick cop whose over-zealous chastisement of criminals has led to his suspension from the force. He accepts an offer of work from the FBI which involves transporting a Mafia big-shot across the country to stand trial. However, there's a hidden agenda at work, and he soon realises that both he and the crook are to be eliminated. There's really very little here that's either fresh or stimulating.
A/AD   86 min (ort 95 min)   VIDrel: VGM/VCC L/A   V

PRINCE AND THE PAUPER, THE **
Richard Fleischer USA
PG
1978
Mark Lester, Oliver Reed, Raquel Welch, Ernest Borgnine, George C. Scott, Rex Harrison, David Hemmings, Charlton Heston, Harry Andrews, Murray Melvin, Julian Orchard, Graham Stark, Sybil Danning, Lalla Ward, Felicity Dean
In Tudor England, the young Prince Edward VI swaps places with a beggar boy who is his exact double, and this latter is instrumental in exposing a traitor. An adequate rendering of the Mark Twain tale (the third one) that doesn't improve on the earlier films, and is not helped by having a plethora of stars popping up in little a cameos that add nothing to the story. Directed in a jokey style, it might have worked better as a straight comedy.
Aka: CROSSED SWORDS
A/AD   121 min   VIDrel: ARROW/VCC   V
Boa: novel by Mark Twain.

PRINCE AND THE SHOWGIRL, THE **
Laurence Olivier UK
PG
1957
Marilyn Monroe, Laurence Olivier, Richard Wattis, Sybil Thorndike, Jeremy Spencer, Jean Kent, Esmond Knight, David Horne, Charles Victor, Daphne Anderson, Vera Day, Gillian Owen, Maxine Audley, Gladys Henson, Paul Hardwick
Set in London, this tells of an American showgirl who becomes the object of a foreign prince's attentions during the coronation of George V in 1911. A ponderous comedy with much lavishness but little mirth. The script is by Terence Rattigan, with music by Richard Addinsell and photography by Jack Cardiff.
COM   112 min (ort 117 min)   VIDrel: MGM/WHV   V
Boa: play The Sleeping Prince by Terence Rattigan.

PRINCE OF DARKNESS **
John Carpenter USA
18
1987
Donald Pleasence, Lisa Blount, Jameson Parker, Victor Wong, Dennis Dun, Anne Howard, Susan Blanchard, Alice Cooper, Ann Yen, Peter Jason, Dirk Blocker, Jesse Lawrence Ferguson, Ken Wright, Robert Grashmere, Thom Bray
All hell literally breaks loose, when a priest ventures into an abandoned church to investigate a glass cylinder found to contain a violently swirling green mist. Returning with some theology students to investigate further, he finds his young charges falling prey to demonic possession and horrific attacks. Written by the director (under the pseudonym of Martin Quatermass) this clumsy horror yarn has a several chilling moments but a remarkably weak resolution.
HOR   101 min (ort 110 min)   VIDrel: 4-FRONT/POLYREC/GUILD; PION (LV only)   V LV

PRINCE OF FOXES ***
Henry King USA
(PG)
1949
Tyrone Power, Orson Welles, Wanda Hendrix, Everett Sloane, Marina Berti, Felix Aylmer, Leslie Bradley, Joop Van Hulzen, James Carney, Rena Lennart, Eduardo Ciannelli, Giuseppe Faeti, Eugene Deckers, Eva Brauer, Frank Salvi
A marvellous evocation of 16th century Italy that centres on a subtle plot by Cesare Borgia (ably played by Welles) to seize control of a fortified mountain city state. He does so by ordering a young underling to seduce the wife of its elderly ruler but reckons without his integrity and sense of justice. Unable to wait any longer, Borgia attacks but final victory eludes him. Filmed on location, this excellent film should not be missed.
A/AD   105 min (ort 107 min) B/W   SATrel: MOVIE CHANNEL
Boa: novel by Samuel Shellabarger.

PRINCE OF JUTLAND *
Gabriel Axel DENMARK/FRANCE/GERMANY/UK
15
1993
Gabriel Byrne, Helen Mirren, Christian Bale, Brian Cox, Steve Waddington, Kate Keckinsake, Tony Haygarth, Freddie Jones, Tom Wilkinson, Brian Glover, Mark Williams, Saskia Wickham, Andy Sears, Philip Rham, Ewen Bremner, Ian Burns
The 6th century story that was Shakespeare's inspiration for Hamlet forms the basis for this tedious and cheap looking work,

that revolves around the passions and intrigues at the Danish royal court. An uneven and rambling film, it wastes the talents of a first-rate cast.
DRA   106 min (ort 115 min)   VIDrel: ARROW/RTM   V/sh

PRINCE OF PENNSYLVANIA, THE **
Ron Nyswaner USA
15
1988
Fred Ward, Keanu Reeves, Bonnie Bedelia, Amy Madigan, Jeff Hayenga, Tracey Ellis, Jay O. Sanders, Joseph De Lisi, Jessica Streiner, Lauren Camp, Pam Call, Demetria Mellot, Kari Keegan, Jeff Forman, Paul Palmer, Bob Tracey
A bizarre character study with a weird young man slowly finding himself, not really being helped in this by his equally unconventional dad. Despite many comic moments the film is too disorganised to work as anything more than a set of funny vignettes. Nyswaner's directorial debut.
COM   89 min (ort 93 min)   VIDrel: MIA/DISC/COLUM   V

PRINCE OF SHADOWS **
Pilar Miro SPAIN
18
1991
Terence Stamp, Patsy Kensit, Jose Luis Gomez, Geraldine James, Simon Andreu, Alexander Bardini, John McEnery, Jorge de Juan, Pedro Diez de Corral, Felipe Velez, Carlos Hipolito, Paco Casares, Queta Claver, William Job, Magda Wojcik
In 1960s Madrid, an exiled communist sets off for Lisbon, his mission being to eliminate a traitor who has turned police informer. On the way there, he thinks back to 1946, when he executed a similar traitor, despite his protestations of innocence. Later, he learns just how he has been manipulated all along by the real traitor. A dense and murky story, atmospheric but confusing, the poor English dialogue is its weakest feature.
Aka: BELTENEBROS
A/AD   86 min (English dialogue – ort 114 min)
VIDrel: TART/20TH   V
Boa: novel Beltenebros by Antonio Munoz Molina.

PRINCE OF THE CITY, THE ***
Sidney Lumet USA
15
1981
Treat Williams, Jerry Orbach, Richard Foronjy, Don Billett, Carmine Caridi, Kenny Marino, Paul Roebling, Lindsay Crouse, Michael Beckett, Norman Parker, Bob Balaban, James Tolkan, Steve Inwood, Matthew Laurance, Tony Turco
An undercover cop reveals police corruption in the department, but is frozen out when his colleagues close ranks against him. A powerful and provocative tale that is far longer than it need have been, but makes a number of sharp points all the more disturbing for being based on a real-life incident. Written by Jay Presson Allen and Lumet. See also SERPICO (which is just as effective but a lot shorter) and THE GLASS SHIELD.
DRA   160 min (ort 167 min)   VIDrel: MGM/WHV L/A   V
Boa: novel by Robert Daley

PRINCE OF THE SUN **
Weelson Chin HONG KONG
15
Cynthia Rothrock, Lam Ching Yin, Conan Lee, Jeffrey Falcon, Sheila Chan, Cheng Park Lam
Buddha has reincarnated as a young child but is in danger from evil priests. Only Rothrock stands in the way of those who wish to destroy him. Adequate enough, but by no means one of this star's best efforts.
MAR   90 min dubbed   VIDrel: MIA/DISC   V

PRINCE OF THIEVES **
Paul Wendkos USA
15
1990
Ray Sharkey, Edward Asner, James Keach, Robert F. Lyons
Based on real events, this drama details the 1980 Memorial Day robbery of the Depositor's Trust Bank of Boston, when $25,000,000 was stolen in one of the largest raids in US history, the culprits being found to be a group of Boston cops.
DRA   94 min   VIDrel: GENESIS   V

PRINCE OF TIDES, THE ***
Barbra Streisand USA
15
1991
Barbra Streisand, Nick Nolte, Blythe Danner, Kate Nelligan, Jeroen Krabbe, Melinda Dillon, Jason Gould, George Carlin, Brad Sullivan, Maggie Collier, Lindsay Wray, Brandlyn Whitaker, Justen Woods, Bobby Fain, Trey Yearwood
Nolte is an unemployed football coach who goes to New York to meet a psychiatrist (Streisand) whom he hopes can help him with his suicidal sister. As the doctor begins exploring the family background, she realises that Nolte has serious hang-ups of his own and, unhappily married herself, begins to fall in love with

him. A nicely handled drama that really could have done with a shorter running time and a more even script.
DRA   126 min (ort 132 min) wScrn cC
VIDrel: VCC/DISC/COLUM   V/sur
Boa: novel by Pat Conroy.

PRINCESS AND THE GOBLIN, THE **   U
Joszef Gemes/Les Orton/Imre Andras Nyerges
HUNGARY/UK   1992
Voices of: Joss Ackland, Peggy Mount, Peter Murray, Rik Mayall, Mollie Sugden, Ray Kinnear, Claire Bloom, Victor Spinetti, Sally Ann Marsh, Frank Rozzekaar Green, William Hootkins, Maxine Howe, Steven Lyons, Robin Lyons
An unremarkable animated version of this 1872 fairy tale about a princes who joins forces with a miner's son in an attempt to thwart a plot by a group of goblins who live underground. An overlong effort that does not hold together too well.
ANIM   79 min (ort 85 min)   VIDrel: EIV/SONOP   V
Boa: novel by George MacDonald.

PRINCESS AND THE PIRATE, THE ***   U
David Butler USA   1944
Bob Hope, Virginia Mayo, Walter Brennan, Walter Slezak, Victor McLaglen, Marc Lawrence, Maude Eburne, Hugo Haas, Adia Kuznetzoff, Brandon Hurst, Tom Kennedy, Stanley Andrews, Robert Warwick, Tom Tyler, Rondo Hatton, Ernie Adams
Our hero is on the run from a mean pirate, and gets into sundry scrapes in this lighthearted and crazy Hope vehicle. Brennan is well cast as a pirate, but many of the jokes will, unfortunately, seem a little dated now. The music is by David Rose.
COM   94 min   VIDrel: VCC/DISC/COLUM   V

PRINCESS BRIDE, THE ***   PG
Rob Reiner USA   1987
Cary Elwes, Peter Falk, Mandy Patinkin, Carol Kane, Billy Crystal, Chris Sarandon, Robin Wright, Fred Savage, Andre the Giant, Christopher Guest, Wallace Shawn, Mel Smith, Peter Cook, Willoughby Gray, Malcolm Story
A tongue-in-cheek fairytale aimed at adults and children alike, and telling of a man who must rescue his own true love from an evil villain. This uneven mixture of sparkling action and lack-lustre comedy, is loosely bound together by the muddled William Goldman script. There are however, some memorable moments.
A/AD   94 min (ort 98 min)   VIDrel: VCC/DISC   V/sur
Boa: novel by William Goldman.

PRINCESS CARABOO ***   PG
Michael Austin USA   1994
Phoebe Cates, Jim Broadbent, Wendy Hughes, Kevin Kline, John Lithgow, John Wells, Stephen Rea, John Sessions, Peter Eyre, Jacqueline Pearce, Roger Lloyd Pack, John Lynch, Arkie Whiteley, Kate Ashfield, Ewan Bailey, David Glover
Comic tale based on real events and set in 1817, when a young girl arrived on the doorstep of a house, dressed in a strange costume and speaking a language no-one could recognise. Narrowly avoiding imprisonment for vagrancy, she claimed to be a princess from Java who had escaped her kidnappers, and her charm and curiosity ensured that she was swiftly accepted by the local aristocrats, before ultimately being revealed as a fraud. A most enjoyable film.
COM   96 min (ort 97 min)   VIDrel: EIV/SONOP   V/sur

PRINCESS DAISY *   15
Warris Hussein USA   1983
Merete Van Kemp, Lindsay Wagner, Paul Michael Glaser, Robert Urich, Claudia Cardinale, Rupert Everett, Ringo Starr, Barbara Bach, Stacy Keach, Alexa Kenin, Nicolas Coster, Sal Viscuso, Lysette Anthony, David Haskell
Glossy, trashy story of a poor little rich girl who is the daughter of a Russian prince and one of a pair of twins. This overblown melodrama has our heroine overcoming many obstacles as she rises to a position of great wealth, but discovering that it counts for little without true love. As contrived as it sounds and a good deal more boring. Originally shown in two parts.
DRA   186 min (2 cassettes – ort 200 min) mTV
VIDrel: 4-FRONT/POLYREC/ODY   V/sh
Boa: novel by Judith Krantz.

PRINCESS IN LOVE *   12
David Greene UK   1996
Julie Cox, Christopher Villiers, Christopher Bowen, Julia St John

Another one in a seemingly endless series of films recounting the troubled marital affairs of members of the Royal Family. Here, the relationship between Diana the Princess of Wales and a Guards officer is put under the microscope, the film drawing much of its material from Pasternak's gushing and much-reviled masterpiece. See also DIANA: HER TRUE STORY and THE WOMEN OF WINDSOR.
DRA   95 min   VIDrel: ARENA/RTM   V
Boa: book by Anna Pasternak.

PRINCESS MADAM **   18
Godfrey Ho HONG KONG   1989
Moon Lee, Michiko Nishiwaki, Herbert Liu, Sharon Young
Two Hong Kong policewomen are given the job of protecting the mistress of a gangster who has agreed to become a police informant. When the gangster threatens to kill the stepfather of one of the cops as a ploy to get the police protection lifted, this cop chucks in her job to embark on a personal vendetta. A competent thrill-a-minute type action story.
A/AD   87 min wScrn   VIDrel: MIA/DISC; ENCORE (LV only)   V LV

PRINCESS WARRIOR *   18
Lindsay Norgard USA   1989
Sharon Lee Jones, Dana Fredsti, Tony Riccardi, Mark Pacific, Isibella Peralta, Laurie Warren, Sydney Coale Phillips, Augie Blunt, Cheryl Janecky, Lee N. Gerovitz, Stephen J. Cassarino, Selga Sanders, Diana Karankios
Space opera set in a future off-world society, where women are the aggressors and men serve as concubines. Two sisters enter a contest to determine the ruler of the planet Venus.
FAN   90 min   VIDrel: POPRO/RTM   V

PRISON **   18
Renny Harlin USA   1987
Lane Smith, Viggo Mortensen, Chelsea Field, Lincoln Kilpatrick, Andre De Shields, Ivan Kane, Steven Little, Tom Lister Jr, Michael Yablans, Larry Flash Jenkins, Arlen Dean Snyder, Hal Landon Jr, Matt Kanen, Rod Lockman
A prisoner executed in 1964, returns from the dead to take his revenge on the warden. Good special effects and atmosphere (the film was shot in a Wyoming prison) help bolster the thin plot. See also THE CHAIR for a dose of something similar.
HOR   99 min (ort 102 min)   VIDrel: EIV/SONOP   V

PRISON HEAT **   18
Joel Silberg USA   1993
Rebecca Chambers, Lori Jo Hendrix, Kena Land, Toni Naples, Gilya Stern, Uri Gavriel, Michael Yani, Diana Olearchik, Ahouva Kerem, Katia Zinbris, Ilana Reff, Inball Itskovich, Shamil Ben-Ari, Avi Cohen, Ofer Shiratsi, Rony Blits
A group of girls go to Turkey for a holiday, are framed on drug smuggling charges and are sent to prison, where they endure brutal treatment from both their lesbian fellow-inmates and the sadistic prison guards. Eventually, they have no option left but to plan an escape. Filmed in Israel (the Turkish government are very proud of their prison system) this is one of those standard "women in prison movies". Adequate, but MIDNIGHT EXPRESS is a lot better.
THR   83 min (ort 90 min)   VIDrel: WHV L/A   V/h

PRISONER, THE **   18
Chu Yen Ping HONG KONG   1991
Jackie Chan, Andy Lau, Samo Hung, Tony Leung, Jimmy Wang Yu, Ko Chuen Hiang, Tao Chung Hwa, Yip Chuen Chin, Kei Kei, Cheung Shui Chuk, Cheung Kwok, Tsui Sing Yee, Fong Ching, Chan Yin Yu, Ko Chit, Yu Bong, Chun Ho, Cho Kin
An investigation into an explosion that killed a top policeman reveals the fingerprints of a criminal who was supposedly executed a couple of years before. An undercover cop arranges to be jailed at the prison that housed (and was thought to have executed) the criminal in question. Stealing ideas used in COOL HAND LUKE (plus PAPILLON and NIKITA to name two others) the film still manages to entertain with its well staged action sequences.
Aka: ISLAND ON FIRE
A/AD   92 min   VIDrel: MIA/DISC   V

PRISONER, THE: EPISODES 1 TO 17 ***   PG
Pat Jackson/Don Chaffey/Peter Graham Scott and others UK
                                                      1967
Patrick McGoohan, Leo McKern, Angelo Muscat, Colin Gordon, Peter Wyngarde, Mary Morris, Guy Doleman, Alexis Kanner. Guest stars:

Finlay Currie, Paul Eddington, Eric Portman, Anton Rodgers, John Castle, Donald Sinden
This cult TV series tells of a British Intelligence agent, who after resigning, is kidnapped and taken to a strange village from which he cannot escape. There he is subjected to attempts to brainwash him and discover the reasons behind his resignation. This imaginative series began with great promise but eventually became dreary and repetitive. Some episodes were better than others, but the final one was inconclusive. Filmed at Portmeirion Hotel.
FAN   827 min (availability of individual episodes varies)
mTV   VIDrel: POLY/POLYREC L/A   V

PRISONER OF HONOUR **
Ken Russell UK
Richard Dreyfuss, Oliver Reed, Peter Firth, Jeremy Kemp, Brian Blessed, Peter Vaughan, Kenneth Colley, Lindsay Anderson
An overblown and ponderous tale, inspired by Dreyfus Affair of 1894, when a Jewish army officer was falsely accused of selling military secret and sent to the penal colony of Devil's Island. The French colonel Georges Picquart comes to believe in the man's innocence, but he is court-martialled for his pains. An unevenly scripted and episodic work, that was sorely in need of more disciplined direction. Dieterle's film "The Life Of Emile Zola" also examined this incident.
DRA   84 min (ort 115 min) mCab   VIDrel: 20TH/TECH
V/sur
PG
1991

PRISONER OF SECOND AVENUE, THE ***
Melvin Frank USA
Jack Lemmon, Gene Saks, Anne Bancroft, Elizabeth Wilson, Florence Stanley, M. Emmet Walsh, Maxine Stuart, Ed Peck, Gene Blakely, Ivor Francis, Stack Pierce
An executive faces a nervous breakdown when he suddenly becomes unemployed, in this unusual comedy that treads a thin line between comedy and pathos. Bancroft is cast as his understanding wife. Look out for an appearance by Sylvester Stallone as an alleged pickpocket. Written by Neil Simon and with music by Marvin Hamlisch.
COM   93 min (ort 105 min)   VIDrel: MGM/WHV L/A   V
Boa: play by Neil Simon.
PG
1975

PRISONER OF ZENDA, THE **
Richard Thorpe USA
Stewart Granger, Louis Calhern, Deborah Kerr, Robert Coote, Robert Douglas, James Mason, Jane Greer, Lewis Stone, Peter Brocco, Francis Pierlot, Tom Browne Henry, Eric Alden, Stephen Roberts, Bud Wolfe, Peter Mamakos, Joe Mell
A remake of the 1937 original, that lacks any feeling for this story of intrigue at the court of Ruritania, when a double has to stand in at the coronation of the country's new king. Granger plays both king and his English double – in both roles he is without presence or feeling.
A/AD   97 min (ort 100 min)   VIDrel: MGM/WHV   V
Boa: novel by Anthony Hope.
U
1952

PRISONER OF ZENDA, THE *
Richard Quine USA
Peter Sellers, Lynne Frederick, Lionel Jeffries, Elke Sommer, Gregory Sierra, Stuart Wilson, Jeremy Kemp, Catherine Schell, Simon Williams, Norman Rossington, John Laurie, Graham Stark, Michael Balfour, Arthur Howard
A dreadful remake of the famous Anthony Hope adventure that was played strictly for laughs but provided precious little amusement. Among his roles Sellers plays a London cabbie who is asked to impersonate the heir to the throne of Ruritania when the ruling monarch dies in a hot-air balloon. This painfully vacuous film came filled with its own supply of hot air, and did nothing to advance the careers of any of its passengers.
COM   104 min (ort 108 min)   VIDrel: CIC/SONOP   V
Boa: novel by Anthony Hope.
PG
1979

PRISONER OF ZENDA, THE **
Geoff Collins/Warwick Gilbert/George Stephenson
AUSTRALIA
Voices of: Christine Amor, Rober Loleby, Claire Crowther, John Fitzgerald, Phillip Hinton, Walter Sullivan, Frank Violi, David Whitney
Adequate animated version of this romantic historical novel, set in the kingdom of Ruritania, where a British visitor turns out the new king's double.
ANIM   48 min (ort 51 min)   VIDrel: CARL/TECH   V
Boa: novel by Anthony Hope.
U
1988

PRISONER OF ZENDA INC., THE *
Stefan Scaini USA
Jonathon Jackson, Jay Brazeau, Richard Lee Jackson, Don S. Davis, Katherine Isobel, John Tench, Jed Rees, Mark Acheson, William Shatner, Howard Dell, Sean Amsing, Demitir Goritsas, David Kaye, Alan Robertson, Stephen E. Miller
When the founder of a powerful computer company dies suddenly, his fourteen-year-old son, a brilliant programmer, is left the company but has to contend with the machinations of his evil uncle (Shatner) who has him kidnapped. By chance, a look-alike is found just in time to prevent our villain from seizing control of the company, while our kidnap victims frees himself in time to foil these plans. A dull, assembly-line offering with absolutely no sparkle.
Aka: DOUBLE PLAY
JUV   90 min mCab   SATrel: DISNEY CHANNEL
(U)
1996

PRIVATE BENJAMIN **
Howard Zieff USA
Goldie Hawn, Armand Assante, Eileen Brennan, Robert Webber, Sam Wanamaker, Barbara Barrie, Mary Kay Place, Hal Williams, Harry Dean Stanton, Albert Brooks, Gretchen Wyler, P.J. Soles, Sally Kirkland, Alan Oppenheimer, Maxine Stuart
Goldie Hawn vehicle, with Hawn playing a spoilt rich girl who enlists in the US Army and discovers to her surprise that it's not a bed of roses. After an initial shock she finds that life in the army does have certain advantages. Hawn gurgles and squeals her way through this one very pleasantly, but somehow it just isn't enough. A TV series followed.
COM   105 min (ort 110 min)   VIDrel: WHV   V
15
1980

PRIVATE FUNCTION, A **
Malcolm Mowbray UK
Maggie Smith, Michael Palin, Denholm Elliott, Richard Griffiths, Tony Haygarth, John Normington, Bill Paterson, Liz Smith, Alison Steadman, Jim Carter, Peter Postlethwaite, Eileen O'Brien, Rachel Davies, Reece Dinsdale
The story of the fate of a black-market pig, being secretly raised for the Royal Wedding celebrations in post-WW2 Britain. The humour, such as it is, focuses on the fact that rationing is still in force, making our grunter the centre of attention. Some nice character studies add a modicum of mirth. The acutely observant script is the work of Alan Bennett.
COM   92 min (ort 94 min)   VIDrel: CIC/SONOP   V
15
1984

PRIVATE LESSONS **
Alan Myerson USA
Sylvia Kristel, Howard Hesseman, Eric Brown, Pamela Bryant, Ed Begley Jr, Patrick Piccininni, Meredith Baer, Ron Foster, Peter Elbling, Don Barrows, Marian Gibson, Dan Greenburg
A young boy in a rich family is given a hand with his sexual initiation by the maid, unaware that she has more ambitious plans in this inconsequential little piece. The script is by Dan Greenburg. Cut before video submission by 3 min 5 sec. See also THE DARK SIDE OF LOVE.
Aka: PHILLY
COM   79 min Cut (15 sec – ort 87 min)
VIDrel: POLY/POLYREC/BRAVE   V/sur
Boa: novel Philly by Dan Greenberg.
18
1981

PRIVATE LIFE OF HENRY VIII, THE ****
Alexander Korda UK
Charles Laughton, Elsa Lanchester, Robert Donat, Merle Oberon, Wendy Barrie, Binnie Barnes, Everley Gregg, Lady Tree, Franklin Dyall, John Loder, Miles Mander, Claud Allister, William Austin, Gibb McLaughlin, Sam Livesey
A wonderfully lively and sparkling portrayal of the much-married British monarch, that is not always strictly true to historical fact but more than makes up for this, through its excellent cast and fine production values. Laughton is truly magnificent in the title role. AA: Actor (Laughton).
DRA   90 min (ort 97 min) B/W   VIDrel: CARL/TECH   V
U
1933

PRIVATE LIFE OF SHERLOCK HOLMES, THE ***
Billy Wilder UK/USA
Robert Stephens, Colin Blakely, Genevieve Page, Irene Handl, Eric Francis, Stanley Holloway, Christopher Lee, Clive Revill, Tamara Toumanova, George Benson, Catherine Lacey, Mollie Maureen, Peter Madden, Robert Cawdron
The discovery of a secret Watson manuscript, reveals that the famous detective did have relationships with women, in this uneven but affectionate and occasionally humorous film, constructed around two stories. The abrupt ending is the only
PG
1970

sour note in a work of great charm. Written by Wilder and I.A.L. Diamond, with music by Miklos Rozsa.

DRA 120 min (ort 125 min) VIDrel: MGM/WHV L/A V

PRIVATE LIVES OF ELIZABETH AND ESSEX, THE ***
U
Michael Curtiz USA 1939
Errol Flynn, Bette Davis, Donald Crisp, Vincent Price, Olivia De Havilland, Alan Hale, Nanette Fabray, Henry Stephenson, Henry Daniell, Leo G. Carroll, Robert Warwick, John Sutton
Solid account of the stormy relationship between Elisabeth I and the Earl of Essex, which led to his rebellion and subsequent execution as a traitor. British history gets the grand Hollywood treatment, but the inspired pairing of Davis and Flynn makes for excellent drama. Written by Norman Reilly Raine and with music by Erich Wolfgang Korngold.
Aka: ELIZABETH THE QUEEN
DRA 102 min (ort 106 min) VIDrel: MGM/WHV L/A V
Boa: play Elizabeth The Queen by Maxwell Anderson.

PRIVATE WARS **
18
John Weidner USA 1992
Steve Railsback, Michael Champion, Dan Tullis Jr, Holly Floria, Michael Delano, James Lew, Albert Garcia, Stuart Whitman, Brian Patrick Clarke, Cynthia Frost, Tim Colceri, John Robson, Vincent Murdocco, John Salvitti, Andy Lovett
An L.A. community where gang warfare has made life unbearable decides to hire a tough fighter to impose a little law and order, and eventually chooses a washed-up, alcoholic private eye for the job. Adequate action outing with one or two humorous touches.
A/AD 88 min (ort 90 min) VIDrel: IMPENT V/sh

PRIVATES ON PARADE **
15
Michael Blakemore UK 1982
John Cleese, Denis Quilley, Michael Elphick, Nicola Pagett, Joe Melia, Bruce Payne, Patrick Pearson, David Bamber, Simon Jones, Phil Tan, Vincent Wong, Neil Pearson, John Standing, John Quayle, Brigitte Kahn, Tim Barlow, Ishaq Bux
The story of an army entertainment troupe in Singapore in 1948, that is a clumsy mixture of drama, satire and farce. Quilley is excellent as the homosexual director of the troupe, but the film as a whole lacks substance, despite efforts to incorporate some realistic scenes into the camp concert setting. Written by Nichols, this isn't a patch on OH! WHAT A LOVELY WAR.
COM 107 min (ort 113 min) VIDrel: L/A V
Boa: play by Peter Nichols.

PRIVATE'S PROGRESS ***
U
John Boulting UK 1956
Ian Carmichael, Terry-Thomas, Richard Attenborough, Dennis Price, John Le Mesurier, Peter Jones, Thorley Walters, William Hartnell, Ian Bannen, Jill Adams, Victor Maddern, Kenneth Griffith, George Coulouris, Miles Malleson
An undergraduate with no visible talents joins the army in WW2, and soon learns the best ways of avoiding work, before getting involved in a scheme to steal German art treasures for his uncle. A lighthearted and splendid farce on army life, though it's not quite the satire it was intended to be. Look out for Christopher Lee in an early part – he plays a German.
COM 96 min (ort 102 min) B/W VIDrel: WHV V/h
Boa: novel by Alan Hackney.

PRIZE OF PERIL, THE **
18
Yves Boisset FRANCE/YUGOSLAVIA 1984
Gerrard Lanvin, Marie France Pisier, Michel Picolli
In an ultimate TV quiz of the near future, contestants stand to win a prize of $1,000,000 if they can elude capture by gangs of paid murderers. See THE RUNNING MAN and DEATHROW GAMESHOW for more films on this game-show theme.
Aka: LE PRIX DU DANGER
DRA 85 min (ort 95 min) VIDrel: L/A V
Boa: short story by Robert Sheckley.

PRIZZI'S HONOR ***
15
John Huston USA 1985
Jack Nicholson, Kathleen Turner, Anjelica Huston, Robert Loggia, William Hickey, Lawrence Tierney, John Randolph, Lee Richardson, Michael Lombard, Joseph Ruskin, George Santopietro, C.C.H. Pounder, Ann Selepegno
Skilfully directed account of the complex life of the Mafia. A hitman falls in love with the widow of his latest victim and

discovers that she's in the same business. Many bizarre twists transpire in a most enjoyable film in which John Huston made his last screen appearance. Written by Condon and Janet Roach and with music by Alex North. The photography is by Andrzej Bartkowiak. AA: S. Actress (Huston).
COM 124 min (ort 129 min) VIDrel: VCC/DISC; PION (LV only) V/sur LV
Boa: novel by Richard Condon.

PROBLEM CHILD *
PG
Dennis Dugan USA 1990
John Ritter, Michael Oliver, Jack Warden, Amy Yasbeck, Gilbert Gottfried, Michael Richards, Peter Jurasik, Charlotte Akin, Anna Marie Allred, Robert A. Anderson, Adam Anderly, Cody Beard, Jordan Burton, Eli Cummins, John Davies
A ghastly brat is adopted by a middle-class couple, and he does his best to make their lives hell. A spiteful black-comedy that was followed by a couple of sequels of similar attributes. There are one or two funny sequences, but nothing like enough to make the film worth watching. Music is by Miles Goodman.
COM 77 min (ort 81 min)
VIDrel: 4-FRONT/POLYREC/CIC L/A V/h

PROBLEM CHILD 2 **
PG
Brian Levant USA 1991
John Ritter, Michael Oliver, Jack Warden, Laraine Newman, Amy Yasbeck, Ivyann Schwan, Gilbert Gottfried, Paul Wilson, Alan Blumenfeld, Charlene Tilton, James Tolkan, Martha Quinn, Zach Grenier, Eric Edwards, Krystal Mataras
As if Ritter had not suffered enough in the first movie, here he is again adopting a second obnoxious brat, this time a little girl who looks set to outdo the antics of her stepbrother. Predictable from start to finish, this formula sequel boasts a few well choreographed sequences if nothing else.
COM 86 min (ort 91 min)
VIDrel: 4-FRONT/POLYREC/CIC L/A V

PROBLEM CHILD 3 *
PG
Greg Beeman USA 1995
William Katt, Justin Chapman
Junior has now enrolled at a dancing school, this being thought to be a good way of getting him calmed down, but instead he spreads his own unique brand of chaos, and makes a few enemies as he competes for the affections of a pretty classmate. An irritating film that is as painfully contrived as it is lacking in wit. At the end of it one feels like wringing the necks of the scriptwriters, but young children (of five and under) may be more appreciative.
COM 84 min VIDrel: CIC V

PRODUCERS, THE ***
PG
Mel Brooks USA 1967
Zero Mostel, Gene Wilder, Kenneth Mars, Dick Shawn, Lee Meredith, Estelle Winwood, Renee Taylor, Christopher Hewett, Andreas Voustinas, Bill Hickey, David Patch, Barney Martin, Frank Campanella, Madlyn Cates, John Zoller, Arthur Rubin
A funny black comedy that is marred by a terribly lame ending. A Broadway producer comes up with a surefire scheme for making money, on a musical production that he is absolutely certain must be a flop, as it's a hymn of praise to the Nazis. For sheer bad taste this one is hard to beat, a highlight being the "Springtime for Hitler" number, but from this point on it's downhill all the way. AA: Story/Screen (Brooks).
COM 84 min (ort 88 min) VIDrel: 4-FRONT/POLYREC L/A V

PROFESSIONAL, THE: GOLGO 13 *
18
JAPAN 199-
After his son is brutally murdered, a father seeks revenge but finds himself ranged against title hitman, but has to wait until the latter has returned from an assignment in Sicily before he can have his revenge. A brutal, tedious and low-quality animation, based on a comic-book tale, its level of sophistication is similar to those rubbishy 1960s espionage films from the Continent, with the hero shown to be virtually indestructible and almost super-human.
ANIM 91 min (ort 94 min) dubbed
VIDrel: MANGA/SONOP V/sur

PROFESSIONALS, THE ***
PG
Richard Brooks USA 1966
Burt Lancaster, Lee Marvin, Jack Palance, Robert Ryan, Claudia Cardinale, Ralph Bellamy, Woody Strode, Joe De Santis, Rafael

Bertrand, Jorge Martinez De Hoyos, Maria Gomez, Jose Chavez, Carlos Romero, Robert Conteras, Don Carlos
A wealthy rancher hires four gunslingers to rescue his wife from kidnappers. A well-paced film, with good action sequences, that may be a little implausible, but is still highly enjoyable. The fine photography is by Conrad Hall. Written by Brooks.
WES   112 min (ort 123 min)   VIDrel: RCA L/A   V
Boa: novel A Mule for the Marquesa by Frank O'Rourke.

PROFESSOR, THE ***   18
Giuseppe Tornatore ITALY   1986
Ben Gazzara, Laura Del Sol, Franco Interlenghi, Luciano Bartoli, Nicola Di Pinto, Leo Gullotta
A Mafia leader is sentenced to prison for murder, but continues to control all the criminal rackets of Naples from inside jail. An overlong and somewhat convoluted but absorbing saga.
Aka: CAMORRA MAN, THE; CAMORRA MEMBER, THE; IL CAMORRISTA
THR   145 min Cut (26 sec – ort 147 min)
VIDrel: SPEAR/SONOP   V
Boa: novel by Giuseppe Marrazzo.

PROFILE FOR MURDER **   18
David Winning USA   1996
Lance Henriksen, Joan Severance, Jeff Wincott
The former girlfriends of a millionaire are all being brutally killed, and a murder investigation is unable to make progress, despite a mounting collection of evidence implicating the man. As ever, there are the requisite red herrings and plot twists, in this adequate suspense tale.
THR   91 min   VIDrel: MED/SONOP   V/sh

PROGRAM, THE **   15
David S. Ward USA   1993
James Caan, Kristy Swanson, Craig Sheffer, Omar Epps, Halle Berry, Abraham Benrubi, Duane Davis, Jon Maynard Pennell, Andrew Bryniarski, J.C. Quinn, Joey Lauren Adams, J. Leon Pridgen II, Mike Flippo, Jeff Portell, Ernest Dixon
A college football coach faces a variety of problems as he tries to keep his young and unruly players in hand, in this well-intended but not quite successful attempt to examine the operation of college sports. A sequence in which one of the characters lies down in a busy road as a macho stunt, has been excised after a couple of kids were killed while imitating this piece of reckless stupidity.
Aka: PROGRAMME, THE
DRA   107 min (ort 115 min) cC   VIDrel: TOUCH/TECH
V/sur

PROJECT "A" ***   PG
Jackie Chan HONG KONG   1985
Jackie Chan, Samo Hung, Yuen Biao, Mars, Dick Wei, Isabella Wong, Lau Hak Suen, Winnie Wong
A high-action tale set at the turn of the century, and telling of a unit of marines who are sent to eradicate a gang of cutthroats who infest the China Seas. A sequel followed.
MAR   100 min   VIDrel: POPRO/RTM/IMPENT L/A   V

PROJECT A-KO: PARTS 1 TO 6 **   12/15/PG
Katsuhiko Nishijima JAPAN   1989/1994
Space opera type animation, in six episodes, that starts off with a city being rebuilt after its destruction by a meteor. Following this, the Earth has to face an imminent onslaught by a vast fleet of starships, their mission being one of total destruction. Further adventures follow, chiefly involving the exploits of the title A-Ko, a female space pilot, and her comrades. Animation is by Yuri Moriyama with art direction by Shinji Kimura. Adequate.
ANIM   300 min (six cassettes – separately available) dubbed
VIDrel: MANGA/SONOP   V/sh

PROJECT A: PART 2 ***   15
Jackie Chan HONG KONG   1987
Jackie Chan, Maggie Cheung, David Lam, Rosamund Kwan, Carina Lau, Bill Tung, Sam Lui, Regina Kent, Charlie Chan, Lau Siu Ming, Kenny Ho, Mars, Chris Li, Tai Bo, Father K, Wong Lung Wei, Ben Lam, Lo Wai Kwong
This sequel to the first film is another actioner set in Hong Kong at the turn of the century, when an intrepid police inspector fights crooks and a corrupt policeman. A lively and mindless blend of martial arts mayhem and comedy, with Chan as ever, choreographing and performing his own stunts.
MAR   94 min dubbed   VIDrel: IMPENT L/A   V

PROJECT S **   15
Gideon Amir USA
Edward Albert Jr, Leigh Taylor-Young, Jon Cypher
Thriller that focuses on a man working on a top secret project who learns that information about it has been tampered. All the signs point to this being an inside job, and as his colleagues begin to meet untimely deaths a race against time begins in order to find the culprit.
THR   86 min   VIDrel: EIV/SONOP   V

PROJECT SHADOWCHASER **   15
John Eyres USA   1992
Martin Kove, Meg Foster, Frank Zagarino, Paul Koslo, Joss Ackland, Raymond Evans, John Chancer, Robert Freeman, John Geurrasio, Robert Jezek, David Oliver, Andrew Lamond, John Pasternack, Angie Hill Richmond, Liza Ross, Ricco Ross
A powerful, top-secret robot escapes from the lab. and teams up with a small terrorist group who take a number of hostages in a hospital and demand a ransom of $50,000,00. A low-budget adventure yarn with elements borrowed from both DIE HARD and THE TERMINATOR. The inevitable sequels followed.
Aka: PROJECT: SHADOWCHASER; SHADOWCHASER
FAN   91 min (ort 97 min)   VIDrel: FIRST/SONOP L/A   V

PROJECT SHADOWCHASER: BEYOND THE
EDGE OF DARKNESS ***   18
John Eyres USA   1995
Sam Bottoms, Christopher Atkins, Christopher Neame, Ricco Ross, Aubrey Morris, Frank Zagarino
Sequel to earlier SHADOWCHASER films, that has a long-lost mining-ship crashing into a satellite space-station, where the crew are soon to face the wrath of a nasty, shape-changing android (this being Zagarino, who had much the same role in the earlier PROJECT SHADOWCHASER: NIGHT SIEGE). As this creature proceeded to slaughter the crew, those remaining have to contend with a possible nuclear meltdown as well. Competent, gore-laden and quite imaginative.
FAN   94 min   VIDrel: MED/COLUM L/A   V/sh

PROJECT SHADOWCHASER: NIGHT SIEGE **   18
John Eyres USA   1993
Frank Zagarino, Bryan Genesse, Beth Toussaint
In-name only sequel to the first film, that has a superhuman android leading terrorists, who are out to steal a nuclear weapon in order to blackmail a whole continent. But this story can be regarded as a completely separate work, as it has no point of reference to the earlier one, the slight plot calling for much mayhem and widespread destruction. Zagarino plays the blonde android reasonably well, and does what he can in a film whose script is mostly a parody of itself.
FAN   92 min   VIDrel: MED/POLYREC L/A   V/sh

PROM NIGHT **   18
Paul Lynch CANADA   1980
Leslie Nielsen, Jamie Lee Curtis, Casey Stevens, Robert Silverman, Eddie Benton, Antoinette Bower, Michael Tough, David Mucci, Pita Oliver, Marybeth Rubens, Joy Thompson, Jeff Wincott, George Touliatos, Melanie Morse MacQuarrie
A deranged killer threatens college students, who were responsible for the death of a little girl six years before. A standard slasher tale that's quite well handled, despite adding nothing original to this genre.
HOR   88 min (ort 91 min)
VIDrel: 4-FRONT/POLYREC/BRAVE   V

PROM NIGHT 2: HELLO MARY LOU **   18
Bruce Pitman CANADA   1987
Michael Ironside, Wendy Lyon, Justin Louis, Lisa Schrage, Richard Monette, Terri Hawkes, Beth Gondek, Brock Simpson, Beverly Hendry
Not really a sequel to the earlier PROM NIGHT, this tale has students being menaced by a murdered prom queen, who has returned from the grave after thirty years to have her revenge.
Aka: HAUNTING OF HAMILTON HIGH, THE; HELLO MARY LOU: PROM NIGHT 2
HOR   92 min (ort 97 min)   VIDrel: ENTUK L/A   V

PROM NIGHT 3: THE LAST KISS *   18
Ron Oliver/Peter Simpson CANADA   1990
Tim Conlon, Cyndy Preston, Courtney Taylor, David Stratton
Mediocre horror films tend to beget even worse series, and this one is no exception, with our murdered prom queen returning

from the grave to seduce a high school boy and embark on the expected bout of supernatural murder and mayhem.
HOR 95 min VIDrel: 20TH/TECH V/sh

PROMISE, THE *** (15)
Margarethe von Trotta FRANCE/GERMANY 1994
Meret Becker, Corinna Harfouch, Anian Zollner, August Zirner, Susann Uge, Eva Mattes, Jean-Yves Gaultier, Philippe Morier-Genoud, Tina Engel, Hans Kremer, Otto Sander, Hark Bohm, Dieter Mann, Ulrike Krumbiegel, Pierre Besson
A tragic love story that opens in the 1960s in East Berlin just after the Wall goes up, and follows the plight of a pair of young lovers, who make plans to get to the West. But they are unsuccessful and only the girl gets away, and as the years pass the political situation conspires to keep them apart. Finally, in 1989 the Wall comes down, and they are at last reunited. A sincere and deeply felt work, it takes a slow albeit melodramatic look at recent German history.
Aka: DAS VERSPRECHEN; LA PROMESSE
DRA 116 min SATrel: SKY MOVIES

PROMISE TO KEEP, A ** PG
Rod Holcomb USA 1990
Dana Delaney, William Russ, Mimi Kennedy, Frances Fisher, Adam Arkin, Richard Poe, David Seaman, Greg Young, David Cromwell, Devon O'Brian, Kyndra Joe Casper, Tiffany Ann Casper, J.R. Nutt, Jesse R. Tendler, Larry MacMullen
A couple who are just about coping with their own three kids get saddled with four more when a close relative dies. A fact-based drama that's little more than a contrived tearjerker, with a script by Susan Cooper and Carlton Cuse. See also A SON'S PROMISE.
DRA 92 min (ort 100 min) mTV VIDrel: MGM/WHV L/A V/sh

PROMISED A MIRACLE *** 15
Steven Gyllenhaal USA 1988
Rosanna Arquette, Judge Reinhold, Tom Bower, Vonni Ribisi, Gary Bayer, Maria O'Brien, Robin Pearson Rose, Shawn Elliott, John Vickery, Jennifer Puscus, America Marcin, Michael Cavanaugh, Wyatt Knights, Amy Michelson, Terry Wills
The true story of a deeply religious couple, who believed that they could cure their diabetic son through the power of prayer, and refused to allow him to be treated by doctors. When he died they were charged with manslaughter. Arquette and Reinhold give performances of great sincerity in this moving tale. The excellent script is by David Hill.
DRA 95 min (ort 100 min) mTV VIDrel: L/A V
Boa: book We Let Our Son Die by Larry Parker.

PROOF *** 15
Jocelyn Moorhouse AUSTRALIA 1991
Hugo Weaving, Genevieve Picot, Russell Crowe, Heather Mitchell, Jeffrey Walker, Daniel Pollock, Frankie J. Holden, Frank Gallacher, Saskia Post, Belinda Davey, Cliff Ellen, Tania Uren, Robert James O'Neill, Anthony Rawling
Strange tale of the triangular relationship between a blind man, his devoted housekeeper (whose love for him is totally unrequited) and a young kitchen-hand at his local restaurant, who blithely strays into this emotional minefield. A complex drama that conveys a nice sense of underlying tension, as well as some darkly satirical moments. The winner of no less than six awards in its homeland. Screenplay is by Moorehouse.
DRA 86 min (ort 90 min) VIDrel: ARTIF/20TH V/h

PROPHECY, THE ** (18)
Gregory Widen USA 1994
Elias Koteas, Virginia Madsen, Christopher Walken, Randy Adakai-Nez, Thomas "Doc" Bugoski, Jeff Cadiente, Emily Conforto, Adam Goldberg, Nicholas Gomez, Willian "Buck" Hart, Bobby Lee Hayes, Christina Holmes, Steve Hytner, J.C. Quinn
A simple murder investigation leads the detective in charge to the leader of a demonic cult and a battle of gigantic proportions between the forces of Good and Evil, as the cult leader is in reality a fallen angel now at war with the Almighty. An interesting if not entirely successful attempt at a Gothic fantasy, it relies on more atmosphere and acting than on special effects, and though Walken is chilling as the cult leader, his ludicrous dialogue is a handicap.
Aka: GOD'S ARMY
HOR 94 min (ort 97 min) SATrel: MOVIE CHANNEL

PROPHET OF EVIL *** 15
Jud Taylor USA 1993
Brian Dennehy, William Devane, Tracy Needham, Dee Wallace Stone, Danny Cooksey, Brian Reddy, Michael Watson, Jackie Earle Haley, Philip Abbott, Robert Pine, Michael Cavanaugh, Kale Brown, David Reichert, Bradford English
Dennehy is outstanding as the central character in this fact-based drama telling of a vicious Texan cult-leader whose philosophy was based on violence and murder. Devane plays investigator Dan Fields, who was instrumental in bringing him to justice.
Aka: PROPHET OF EVIL: THE ERVIL LeBARON STORY
DRA 89 min mTV VIDrel: ODY/SONOP V/sh

PROPRIETOR, THE ** 12
Ismail Merchant FRANCE/TURKEY/UK/USA 1996
Jeanne Moreau, Sean Young, Sam Waterston, Christopher Cazenove, Neil Carter, Jean-Pierre Aumont, Austin Pendleton, Charlotte de Turckheim, Pierre Vaneck, Marc Tissot, Josh Hamilton, Joanna Adler, James Naughton, J. Smith-Cameron
Having lived for years in self-imposed exile in New York, a successful French writer returns to her old family home in Paris, which was stolen by the Nazis in WW2. Both a tribute by the director to his longtime collaborator, the scriptwriter Ruth Prawer Jhabvala (who was forced to flee her native Cologne prior to WW2) and a complex and multi-layered tale, it attempts to explore the influence of the past on the present, but is saddled with a convoluted script.
Aka: LA PROPIETAIRE
DRA 113 min CINrel

PROSPERO'S BOOKS *** 15
Peter Greenaway FRANCE/HOLLAND/ITALY/UK 1991
John Gielgud, Michael Clark, Michel Blanc, Erland Josephson, Isabelle Pasco, Tom Bell, Kenneth Cranham, Mark Rylance, Gerard Thoolen, Pierre Bokma, Jim Van Der Woude, Michiel Romeyn, Orpheo, Paul Russell, James Thierree, Emil Wolk
Set in the early 17th century, this incredibly stylised and self-indulgent tapestry of a movie has Prospero, the deposed Duke of Milan, in his palace where he begins to improvise the text for Shakespeare's The Tempest. Soon the screen is swamped with a succession of complex computer-generated images, and one cannot but marvel at the stamina of eighty-seven-year-old Gielgud, who brings more depth and feeling to this beautiful but opaque film than it deserves.
FAN 120 min (ort 126 min) VIDrel: ELPIC/POLYREC V/sur

PROTEUS *** 18
Bob Keen UK 1995
Craig Fairbrass, Doug Bradley, Ricco Ross, Toni Barry, Jennifer Calvert, Robert Firth, William Marsh, Jordan Page, Margot Steinberg
Survivors of a shipwreck drift aimlessly on the high seas, but finally find shelter on a deserted oil-rig, quite unaware that it has been used as a site for top secret genetic experiments, and is now home to a shape-changing creature of murderous intent. Echoes of THE THING abound in this chilling story, the debut feature of special effects expert Keen, with the creature (which is rarely seen) able to both absorb and mimic the characteristics of its victims.
FAN 98 min VIDrel: POLFIL V
Boa: novel Slimmer by John Brosnan (Harry Adam Knight).

PROTOCOL * PG
Herbert Ross USA 1984
Goldie Hawn, Chris Sarandon, Richard Romanus, Andre Gregory, Ed Begley Jr, Gail Strickland, Chris Sarandon, Cliff De Young, Kenneth Mars, Kenneth McMillan, James Staley, Jean Smart, Maria O'Brien, Joel Brooks, Grainger Hines
A cocktail waitress accidentally saves the life of an Arab sheikh by being shot in the bottom. From such an unpromising beginning she is given an important-sounding diplomatic appointment, that is in reality a sinecure, and finds herself being used as a pawn in political shenanigans. The film even tries to make a point, but this gets lost in the silliness of it all.
COM 91 min (ort 96 min) VIDrel: MGM/WHV L/A V/sh

PROTOTYPE * 18
Phillip Roth USA 1992
Lane Lenhart, Robert Tossberg, Brenda Swanson, Paul Coulj, Mitchell Cox, Sebastien Scandiuzzi, Harold Cannon, Zack Nesis, Woon Park, Michael White, Bill Barshdorf, Eric Fedorin, Marcus Aurelius, Mark Holman, Max Holman, Rob Lee

After a devastating war in 2057, the great cities have all been destroyed and mechanical cyborgs called "Prototypes" hunt down the last few "Omegas" or genetically altered humans. A crippled ex-soldier is persuaded by an ambitious female scientist to become a prototype, but finds he has been programmed to kill his former lover. A derivative blend of MAD MAX, ROBOCOP and other such films, its few intriguing ideas are spoilt by poor acting and weak plot development.
Aka: PROTOTYPE X29A
FAN 94 min (ort 98 min) VIDrel: POLY/POLYREC L/A V/sh

PROUD MEN *
William A. Graham USA
15
1987
Charlton Heston, Peter Strauss, Belinda Belaski, Alan Autry, Nan Martin, Maria Mayenzet
A rancher and his son have a hard time getting along, and are hardly helped by the inadequacies of the script in this dreary offering. Heston as the dying, tough old rancher grinds his jaw a lot, but even this doesn't suffice.
DRA 90 min (ort 99 min) mTV VIDrel: GUILD/SONOP L/A V

PROUD REBEL ***
Michael Curtiz USA
U
1958
Alan Ladd, David Ladd, Olivia De Havilland, Dean Jagger, Cecil Kellaway, John Carradine, Henry Hull, James Westerfield
At the end of the Civil War, a Southern veteran wanders the country in search of a doctor who can cure his mute son (played by Ladd's real-life son David) and falls for the farm woman he goes to work for. A predictable piece of family entertainment, but wholesome and quite endearing in a simple-minded way.
WES 99 min (ort 103 min) VIDrel: CASPIC L/A V

PROUDHEART ***
Jack Cole USA
(PG)
1993
Lorrie Morgan, Nancy Moore Atchison, Mary Jo Deschanel Dominick LaRae, Darrell Larson, Collin Wilcox-Paxton, Norman Woodel, Grace Sanders, Tommy Jones, Albert S. Harris, Kevin Riley, Jeffrey Ford, Connye Florance, Larry Black
After her father dies, a single mother who works at an auto plant, returns with her daughters to the small town where she was born. There she undertakes the task of keeping her dad's garage business going, but finds it hard to cope when a friend leaves the company just as she needs him most. A sensitive little drama, with Country music star Morgan giving a strong and nicely judged performance.
DRA 45 min SATrel: SKY MOVIES

PROVIDENCE **
Alain Resnais FRANCE/SWITZERLAND
15
1976
John Gielgud, Dirk Bogarde, Ellen Burstyn, David Warner, Elain Stritch, Cyril Luckham, Denis Lawson, Kathryn Leigh-Scott
A novelist on his deathbed composes a spiteful last novel in his head, using as his material the daily events in the life of his family. An impenetrable and pretentious effort that is further burdened by David Mercer's wordy script.
DRA 102 min (ort 104 min) wScrn VIDrel: TART/20TH V/dm

P.S. I LOVE YOU **
Peter Hunt USA
15
1990
Connie Sellecca, Greg Evigan, Earl Holliman, Patrick Macnee, Rob Narita, Ken Howard, Tony Hamilton, George Morfogen, Dee Wallace Stone, Gregory Sierra, Jayne Frazer, Rebecca Staab, Sonny Bono, Dwight Yoaham, Steve Garrey
Accompanied by a police officer, a con-man is sent on a witness protection scheme after the arrest of a leading gangster. But when a lottery ticket worth $26,000,000 enters the picture, these carefully made arrangements fall apart.
A/AD 89 min mTV VIDrel: 20TH/TECH V

PSYCH-OUT **
Richard Rush USA
18
1968
Susan Strasberg, Dean Stockwell, Jack Nicholson, Adam Roarke, Max Julien, Bruce Dern, Henry Jaglom, Linda Gaye, I.J. Jefferson, Tommy Flanders, Garry Marshall, Ken Scott, Geoffrey Stevens, Susan Bushman, John Cardos, Gary Kent
An attractive deaf runaway, meets up with rock musicians and assorted weirdos in San Francisco, when she sets off in search of her missing brother. A nostalgia trip to the 1960s that serves up an extravaganza of psychedelic effects and some over-heated

acting and dialogue. Not so much a film as a gloriously dated parody of how it all was.
DRA 85 min (ort 95 min) VIDrel: MIA/DISC L/A V

PSYCHIC **
George Mihalka USA
18
1991
Zach Galligan, Catherine Mary Stewart, Michael Nouri, Albert Schultz, Ken James, Clark Johnson, Andrea Roth, Susan Horton, Lisa LaCroix, Geza Kovacs, Don Ritchie, Catherine Disher, Myra Fried, Sandi Stahlbrand, Bob Zidel
A college student suddenly finds that he has gained the ability to follow the activities of a serial killer and determine the identity of his victims. When the police fail to take him seriously he is forced to take the law into his own hands.
THR 88 min (ort 92 min) VIDrel: TRIM/HIFLI V

PSYCHIC KILLER *
Ray Danton/Mardi Rustam USA
18
1979
Jim Hutton, Paul Burke, Julie Adams, Aldo Ray, Neville Brand, Whit Bissell, Rod Cameron, Nehemiah Persoff, Della Reese
A patient at an insane asylum uses his newly-acquired psychic powers to have revenge on those he feels have wronged him and cause general mayhem. An unpleasant shocker that makes little of its interesting premise and wastes a host of cameo appearances for good measure.
Aka: DEATH DEALER, THE
HOR 85 min (ort 90 min) VIDrel: VIPCO/SGSVID V/h

PSYCHO ***
Alfred Hitchcock USA
15
1960
Anthony Perkins, Vera Miles, Janet Leigh, John Gavin, Martin Balsam, John McIntire, Patricia Hitchcock, Simon Oakland, Frank Albertson, Lurene Tuttle, Vaughn Taylor, John Anderson, Mort Mills, Francis De Sales, George Eldridge
A legendary shocker that may seem tame by today's standards, but reflects a skill in direction that cannot be denied. Though slow to develop, it retains moments of enormous power. Perkins is superb as the deranged owner of a motel to which a girl flees after she has defrauded her employer. Written by Joseph Stefano with music by Bernard Herrmann. Several sequels followed, but none ever equalled this one.
HOR 109 min B/W VIDrel: CIC/SONOP; PION (LV only) V/h LV
Boa: novel by Robert Bloch.

PSYCHO 2 ***
Richard Franklin USA
18
1983
Anthony Perkins, Vera Miles, Meg Tilly, Robert Loggia, Dennis Franz, Hugh Gillin, Claudia Bryar, Robert Alan Browne, Ben Hartigan, Lee Garlington, Tim Maier, Jill Carroll, Chris Hendrie, Tom Holland, Michael Lomazow
An competent sequel to the 1960 film, with Norman Bates now released from his mental hospital after twenty-two years, and returning to his motel where he tries to carry on running his business. Unfortunately, the murders soon start afresh and poor Norman begins to doubt his sanity. Suspense is nicely maintained right up to the surprise ending. Followed by PSYCHO 3.
HOR 108 min (ort 113 min) VIDrel: CIC/SONOP V/sur

PSYCHO 3 *
Anthony Perkins USA
18
1986
Diana Scarwid, Jeff Fahey, Anthony Perkins, Roberta Maxwell, Maureen Coyle, Hugh Gillin, Lee Garlington, Robert Alan Browne, Gary Bayer, Juliette Cummins, Patience Cleveland, Steve Guevara, Kay Heberle, Donovan Scott
Perkins's directing debut marks yet another return to the Bates Motel, where poor Norman Bates finds that one of his guests is intent on creating her own personal bloodbath. He begins to fall in love with her, but the disapproval of his "mother", and several other events soon drive him over the edge into his bloodiest massacre to date. A feast of gore and unpleasantness that is devoid of tension or style.
HOR 89 min (ort 96 min) VIDrel: CIC/SONOP V/sur

PSYCHO 4: THE BEGINNING **
Mick Garris USA
18
1991
Anthony Perkins, Olivia Hussey, Henry Thomas, C.C.H. Pounder, Warren Frost, Donna Mitchell, Thomas Schuster, Sharen Camille, Bobbi Evors, John Landis, Kurt Paul, Louis Crume, Cynthia Garris, Doreen Chalmers, Alice Hirson
A radio talk show host interviews Norman Bates and gets him to open up about his childhood and the influence of his highly disturbed mother and the role she played in turning him into a

deranged killer. A highly pedestrian effort, despite a script by Joseph Stefano who provided the screenplay for the first (and best) movie.
HOR   92 min (ort 96 min) mCab   VIDrel: CIC/SONOP V/h

PSYCHOCOP *   18
Wallace Potts USA   1989
Bobby Ray Shafer, Jeff Qualle, Palmer Lee Todd
Confused attempt by the same producers to exploit the appeal of MANIAC COP, that combines the concept of a murderous police enforcer with elements of horror and slasher films. The title figure pursues a group of college students and their girlfriends, who are on their way to a weekend in the country and the expected series of sinister events unfolds. A sequel followed.
Aka: PSYCHO COP
HOR   83 min (ort 100 min)   VIDrel: COLUM/SONOP   V

PSYCHOCOP RETURNS *   18
Rif Coogan USA   1993
Bobby Ray Shafer, Barbara Lee Alexander, Julie Strain, Roderick Darin, Dave Bean, Alexandria Lakewood, John Paxton
A psychopathic cop gatecrashes an office party and begins to slaughter the employees in a disagreeable shocker that adds nothing to the earlier film. The fact that this film (and the earlier one) was a spoof on the MANIAC COP series does little to make it any more palatable, and the jokey approach to the murderous proceedings soon grows wearisome.
Aka: PSYCHO COP 2
HOR   85 min (ort 90 min)   VIDrel: COLUM/SONOP   V/h

PT 109 **   U
Leslie Martinson USA   1963
Cliff Robertson, Ty Hardin, Robert Blake, Robert Culp, James Gregory, Leslie Martinson, Grant Williams, Lew Gallo, Errol John, Michael Pate, William Douglas, Norman Fell, Bill Elliott, Sam Gilman, Clyde Howdy, Buzz Martin, James McCallion
Recreation of the wartime career of John F. Kennedy in the Pacific that is strangely lacking in interest and excitement, with little action given its great length. Robertson as JFK brings little sparkle to this role. Part of a Night at the Movies series, this cassette includes coming attractions, a cartoon and a newsreel on JFK's assassination.
WAR   134 min (ort 140 min) wScrn   VIDrel: WHV   V/h
Boa: book The Wartime Adventures Of President John Kennedy by R.J. Donovan.

P'TANG, YANG, KIPPERBANG *   PG
Michael Apted UK   1982
John Albasiny, Alison Steadman, Garry Cooper, Abigail Cruttenden, Maurice Dee, Mark Brailsford, Chris Karallis, Frances Ruffelle, Robert Urquhart, Maurice O'Connell, Tim Seeley, Richenda Carey, Peter Dean
The story of young teenage love in Britain of 1948 with a tongue-tied teenager yearning to kiss his pretty classmate, an ambition he looks like achieving when they get cast in the school play together. Scripted by Jack Rosenthal, it offers occasional flashes of insight, but largely remains just one more trip down nostalgia lane (a frequent destination for British film-makers). See GREGORY'S GIRL for a slightly better study of teenage first love.
Aka: KIPPERBANG
COM   76 min (ort 80 min) mTV   VIDrel: LUMI/SPEAR   V

PUBLIC ACCESS **   18
Bryan Singer USA   1993
Ron Marquette, Dina Brooks, Burt Williams, Larry Maxwell, Charles Kavanaugh, Brandon Boyce, Jessie, Leigh Hunt, Jennifer McManus, Margaret Kerry, Liz Dilts, Randall Slavin, Heidie Van Lier, John Ellis, Shawn Ellis, Mark Norling
A new arrival in a small town makes his contribution to the community in the form of a call-in program on local cable TV that allows the citizenry to voice their discontentment with what they perceive to be wrong. Soon, this innovation has a disquieting effect on the populace, in this interesting but not entirely successful tale. This was Singer's directorial debut and the influence of David Lynch's view of smalltown America is clearly seen.
DRA   86 min (ort 90 min)   VIDrel: IMAG/RTM   V

PUBLIC ENEMY, THE ***   PG
William Wellman USA   1931
James Cagney, Edward Woods, Jean Harlow, Joan Blondell, Donald Cook, Mae Clarke, Beryl Mercer, Leslie Fenton, Mia Marvin, Robert Emmett O'Connor, Ben Hendricks Jr, Murray Kinnell, Rita Flynn, Clark Burroughs, Snitz Edwards
This powerful and entertaining film put Cagney on the road to stardom. It tells of two slum kids who grow up into gangsters and eventually get their just deserts. Contains the fondly remembered scene in which Cagney pushes a grapefruit into Mae Clarke's face. Written by Harvey Thew, Kubec Glasmon and John Bright.
Aka: ENEMIES OF THE PUBLIC
DRA   80 min (ort 96 min) B/W   VIDrel: MGM/WHV   V

PUBLIC EYE, THE ***   15
Howard Franklin USA   1992
Joe Pesci, Barbara Hershey, Stnaley Tucci, Jerry Alder, Jared Harris, Max Brooks, Bryan Travis Smith, Richard Riehle, Laura Ceron, Chuck Gillespie, Jack Denbo, Christian Solti, Ellen McElduff, Marge Kotlisky, Henry Bolzan
A hard-boiled press photographer in New York City in 1942, always looking for a good shoot, meets a young widow who is trying to escape her husband's Mob connections, and she decides that he would be the ideal man to assist her. A moody and stylish homage to film noir (and a tribute to New York photographer Weegee) that is weakened by the lack of a strong plot and the contrived relationship between the two leads. Franklin directed and did the screenplay.
DRA   94 min (ort 99 min)   VIDrel: CIC   V/sur

PUERTO ESCONDIDO ***   15
Gabriele Salvatores ITALY   1993
Diego Abatantuono, Valeria Golino, Claudio Bisio, Renato Carpentieri, Elena Callegari, Antonio Catania, Leonardo Gajo, Corinna Agustoni, Niki Mondellini, Son y la Rumba, Jorge Fegan, Francisco Mauri, Los Guajiros, Yolanda Orizaga
An Italian banker flees from his native soil after a narrowly escaping an assassination attempt and arrives at the title locale – a quiet Mexican village. Once there he is obliged to start pawning his possessions when his credit card is confiscated, and eventually faces poverty, but gamely struggles on from one crisis to the next. A contrived morality tale, uneven and disjointed, but one remains interested, chiefly thanks to Abatantuono's winning performance.
COM   106 min (ort 110 min)   VIDrel: CURZON/20TH V/sur

PULP FICTION **   18
Quentin Tarantino USA   1994
Tim Roth, John Travolta, Amanda Plummer, Uma Thurman, Samuel L. Jackson, Bruce Willis, Harvey Keitel, Rosanna Arquette, Ving Rhames, Eric Stoltz, Maria de Medeiros, Christopher Walken, Laura Lovelace, Robert Ruth, Burr Steers
Two L.A. hitmen philosophise about life as they go about their bloody work, in the course of which they cross paths with other equally nasty characters. An amoral, anarchic and very brutal tale that relaunched Travolta's career and gave Tarantino cult status thanks to the uncritical and lavish praise he received. The LV version includes some extra scenes. Won the Golden Globe Award at Cannes for Best Screenplay. AA: Screen/orig (Roger Avery/Quentin Tarantino).
DRA   154 min wScrn cC   VIDrel: BUENA/TECH; ENCORE (LV only)   V/sur LV

PULP FRICTION *   18
USA   1995
Shelby Stevens, Crystal Wilder, J.R. Carrington, Tina Tyler, Vince Vouyer, Jonathan Morgan, Alex Sanders, Blade Baran, Frank Towers, Dallas D'Amour
Stevens plays the girlfriend of a gangster who is refusing to pay a debt, an act that leads to her being kidnapped by Wilder and Carrington, a couple who are out to make sure the payment is made. These two take their hostage to a remote cabin, and the trio while away the time indulge in strenuous sex sessions. A dullish film with a misleading title, it wrongly suggests the movie is a kind of PULP FICTION sex spoof. It isn't.
A   68 min   VIDrel: ONE   V

PULSE *   18
Paul Golding USA   1988
Cliff De Young, Roxanne Hart, Joey Lawrence, Matthew Lawrence, Charles Tyner, Dennis Redfield, Robert Romanus, Myron D. Healey, Michael Rider, Jean Sincere, Tracy Beaver, Greg Norberg, Tim Russ
A family is terrorised by domestic appliances that become deadly, when a massive electrical surge becomes imbued with

a malevolent intelligence. This unusual idea is neither developed nor explained, and some tedious set-piece confrontations lead to the inevitable destruction of the house, by which time one hardly cares. In the main, the special effects are merely adequate, and the film has a strong mTV appearance.
HOR   87 min (ort 90 min)   VIDrel: RCA L/A   V

PUMMARO: LONG JOURNEY TO THE NORTH **   (15)
Michele Placido ITALY   1990
Thywill A.K. Amenya, Pamela Villoresi, Gerardo Scala, Jacqueline Williams, Nicola Di Pinto, Franco Interlenghi, Tom Felleghy, Ottaviano Dell'Acqua, Stephen Asenso Dunkor, Salvatore Billa, Mukuna Kabongo
Placido's directing debut follows the adventures of a young African, who having just got out of college, goes to Italy in search of his brother, who has been working there illegally in order to fund the education of the former. But his brother has moved on, necessitating an extended search. An interesting look at the problems immigrants face, it offers a few telling comments but not enough by way of a strong plot.
DRA   97 min   SATrel: MOVIE CHANNEL

PUMP UP THE VOLUME ***   15
Alan Moyle USA   1990
Christian Slater, Ellen Greene, Samantha Mathis, Scott Paulin, Annie Ross, Anthony Lucero, Andy Romano, Keith Stuart Thayer, Cheryl Pollack, Lala Sloatman, Jeff Chamberlain, Billy Morrissette, Holly Sampson, Annie Rusoff, Seth Green
A high school student starts up his own pirate radio station, and this proves to be a massive hit with his friends, as he offers a nightly diet of music and advice for angst-stricken teens. Nicely observed study of the trials of teenage life, that's a mite patchy, but is held together by a terrific performance from Slater in the lead role.
DRA   97 min (ort 105 min)   VIDrel: VCC/DISC/COLUM
V/sur

PUMPING IRON ***   PG
George Butler/Robert Fiore USA   1976
Arnold Schwarzenegger, Lou Ferrigno, Matty Ferrigno, Victoria Ferrigno, Mike Katz
This excellent documentary on male bodybuilding follows Schwarzenegger and Ferrigno as they compete for the Mr Universe title. An enjoyable and fascinating behind-the-scenes look at a very specialised sport, it benefits greatly from a strong sense of respect for the competitors and Schwarzenegger's engaging personality. Followed by PUMPING IRON 2: THE WOMEN, an altogether inferior film.
DOC   81 min (ort 85 min)   VIDrel: LUMI/SPEAR L/A   V
Boa: book by C. Gaines/George Butler.

PUMPING IRON 2: THE WOMEN **   PG
George Butler USA   1984
Rachel McLish, Bev Francis, George Plimpton
A sequel to PUMPING IRON in which a bevy of muscular ladies compete for the $50,000 prize in the 1983 Women's Bodybuilding Contest held at Caesar's Palace, Las Vegas in 1983. An oddity, it gives a glimpse of women's bodybuilding of the time, when quaint notions of femininity held the competitors back from competing on the basis of sheer muscular development. Sadly, unlike the first film, it fails to treat the subject matter in a serious and respectful way.
DOC   102 min   VIDrel: ACAD/RTM L/A   V/sur
Boa: book by C. Gaines/G. Butler.

PUNCHLINE ***   15
David Seltzer USA   1988
Tom Hanks, Sally Fields, John Goodman, Mark Rydell, Kim Griest, Pam Matteson, Taylor Negron, Barry Neikrug, Mae Robbins, Max Alexander, Paul Kozlowski, Barry Sobel, George Michael McGrath, Angel Salazar, Damon Wayans
An odd couple help each other make their way in the world of New York stand-up comedians. Fields is a housewife and aspiring comedienne and Hanks her mentor and would-be lover; both are excellent in this surprisingly effective if overlong look at two obsessed individuals. The perceptive and witty script is by Seltzer.
COM   117 min (ort 128 min)   VIDrel: COLUM/SONOP
V/sur

PUNISHER, THE *   18
Mark Goldblatt USA   1990
Dolph Lundgren, Louis Gossett Jr, Jeroen Krabbe, Nancy Everhard,

Kim Miyori, Barry Otto, Brian Rooney, Zoshka Mizak, Kenji Yamaki, Todd Boyce, Hirofumi Kanayama, Larry McCormick, Larney Tupu, John Negroponte
Inspired by the Marvel Comics character, this sees a cop going underground to take his revenge when his family is wiped out by gangsters. He appears to be all set for retirement until a Yakuza kidnaps the children of a Mafia leader in a bid for dominance, and he embarks on a rescue mission. A dark and humourless film of relentless violence and action.
A/AD
84 min Cut (1 min 21 sec at UK cinema release – ort 86 min)
VIDrel: COLUM/SONOP   V/sur

PUNK AND THE PRINCESS, THE *   18
Mike Sarne UK   1993
Charlie Creed-Miles, Vanessa Hadaway, David Shawyer, Jess Conrad, Jacqueline Skarvellis, Yolanda Mason, Alex Mollo, Peter Miles, R.J. Bell, Martin Harvey, Julian Wooley, David Doyle, Daniel Ilsley, Louie Simpson, Anthony Okunbowa
The timeless tale of Romeo and Juliet is given a new slant, being transplanted to London's seedy Notting Hill locality, where it is played out against a background of sex, drugs and heavy rock music. Made about fifteen years after the novel was written by young Sams (she was only fourteen at the time) the film betrays all the clumsy contrivances so very apparent in the book (which enjoyed brief cult following) and adds a few more of its own.
Aka: PUNK, THE
DRA   92 min (ort 96 min)   VIDrel: POLY/POLYREC   V
Boa: book The Punk by Gideon Sams.

PUPPET MASTER 5: THE FINAL CHAPTER **   (12)
Jeff Burr USA   1992
Gordon Currie, Chandra West, Jason Adams, Teresa Hill, Guy Rolfe, Ian Ogilvy, Nicholas Guest, Willard Pugh, Diane McBain, Duane Whitaker, Kaz Gavas, Clu Gulager, Harri James, Ron O'Neal, Chuck Williams, Christopher Hayes
A figure from another dimension arrives on Earth with a band of murderous totems and comes into conflict with our living puppets, whose young maker has to resist his evil schemes. Confused and annoying, this further sequel in the series has a few chilling moments but a general feeling of having taken the "living puppet" idea as far as it can go. Adequate.
HOR   79 min   SATrel: MOVIE CHANNEL

PUPPET MASTER ***   18
David Schmoeller USA   1989
Paul LeMat, Jimmie F. Scaggs, Irene Miracle, Robin Frates, Matt Roe, William Hickey, Barbara Crampton, Kathryn O'Reilly, David Boyd, Peter Frankland, Andrew Kimbrough, Marya Small
A master puppet-maker has used an ancient Egyptian power to imbue his creations with life, and following his suicide a group of psychics gather at a creepy hotel to locate them. However, the demonic creations have other ideas in this implausible but rather scary film. The special effects and animation work of David Allen are excellent. The inevitable sequels followed.
HOR   86 min Cut (3 sec – ort 90 min)
VIDrel: 4-FRONT/POLYREC/EIV L/A   V

PUPPET MASTER 2 **   18
David Allen USA   1989
Elizabeth MacLellan, Collin Bernsen, Greg Webb, Nita Talbot, Charlie Spradling, Steve Welles, Jeff Weston, Sage Allen, George "Buck" Flower, Sean B. Ryan, Ivan J. Rado, Michael Todd, Julianne Mazziotti, Taryn Band, Alex Band
A sequel to the first film that sees our murderous puppets back for another bout of murder, this time the victims are a team of scientific researchers. Average.
HOR   84 min (ort 90 min)
VIDrel: 4-FRONT/POLYREC L/A   V

PUPPET MASTER 3: TOULON'S REVENGE *   18
David Decoteau USA   1990
Guy Rolfe, Ian Abercrombie, Sarah Douglas, Walter Gotell, Matthew Faison, Richard Lynch, Aron Eisenberg, Kristopher Logan
A dull sequel that's set during WW2 and sees our malevolent Puppetmaster (who has discovered the secret of life) out to wreak havoc on all and sundry.
HOR   78 min (ort 86 min)   VIDrel: CIC/SONOP L/A   V

PUPPET MASTER 4 **   15
Jiff Burr USA   1993
Gordon Currie, Chandra West, Jason Adams, Teresa Hill, Guy Rolfe
In this sequel our evil puppets find themselves threatened by a

pair of equally malevolent twins who are out to steal their energy.
HOR 77 min (ort 80 min) VIDrel: CIC/SONOP L/A V

PUPPET MASTERS, THE ** 18
Stuart Orme USA 1994
Donald Sutherland, Eric Thal, Julie Warner, Yaphet Kotto, KeitH David, Will Patton, Richard Belzer, Tom Mason, Gerry Bamman, Sam Anderson, Marshall Bell, J. Patrick McCormack, Nicholas Cascone, Bruce Jarchow, Benjamin Mouton, Benj Thal
Aliens invade Earth in the form of parasitic creatures that take over their human victims by attaching themselves to their backs and inserting tendrils into their brains. As their number multiplies, the head of a government agency struggles to mobilise forces to defeat this menace. Given the strength of the basic premise, this is a remarkably ineffective adaptation with very few moments of horror or tension. THE INVASION OF THE BODY SNATCHERS did it so much better.
Aka: ROBERT A. HEINLEIN'S THE PUPPET MASTERS
FAN 105 min (ort 109 min) VIDrel: BUENA/TECH L/A V

Boa: novel by Robert A. Heinlein.

PUPPETMASTER, THE *** 15
Hou Hsio-Hsien (Hou Xiaoxian) TAIWAN 1993
Li Tianlu, Lim Giong (Lin Qiang), Cheng Kuei-Chung, Zuo Juwei, Hong Liu, Bai Minghua, Cai Zhennan, Gao Dongxiu, Yang Liyin, Chen Qianru, Wu Layun, Chen Bocan, Li Wenbin, Cai Yihua, Lu Fulu, Liu Nanyang, Lin Shuiqing, Chen Shufang
The life of one of Taiwan's most renowned puppeteers serves as a vehicle with which to explore the history of Taiwan itself, from 1909 up to 1945. In a troubled life Li Tianlu had to contend with pressures from family, friends and government, most especially the latter as Taiwan was occupied by Japan during the period in question. Part folk-tale, part history lesson and part drama, this unusual film conveys a feel for the culture and people like few other works.
Aka: HSIMENG RENSHENG
DRA 137 min (ort 142 min) VIDrel: ELPIC/POLYREC V/sur
Boa: biographical book Ximeng Rensheng by Li Tianlu, Hou Xiaoxian and Zeng Yuwen.

PURE COUNTRY ** PG
Christopher Cain USA 1992
George Strait, Isabel Glasser, Lesley Ann Warren, Rory Calhoun, John Doe, Kyle Chandler, Rory Calhoun, Molly McGuire, James Terry McIlvain, Toby Metcalf, Tom Christopher, Jeffrey K. Fontana, Jeff Prettyman, David Anthony, Gene Elders
A burnt-out country music star decides to return to his Texas home town and falls for a young woman who has her mind set on a rodeo career. Some good performances help to overcome the rather tedious plot.
MUS 107 min (ort 113 min) cC VIDrel: WHV V/sur

PURE HELL OF ST TRINIANS, THE * U
Frank Launder UK 1957
Cecil Parker, Joyce Grenfell, George Cole, Eric Baker, Thorley Walters, Irene Handl, Dennis Price, Sidney James, Julie Alexander, Lloyd Lamble, Raymond Huntley, Nicholas Phipps, Liz Fraser, John Le Mesurier, Lisa Lee
Yes, and it's pure hell watching this garbage, with our wayward schoolgirls back in their second sequel. In this one a rich Arab sheikh visits the school, looking for fresh recruits for his harem. Like so much British comedy, this film is tame, tedious and trite. Followed by THE GREAT ST TRINIANS TRAIN ROBBERY.
COM 90 min (Cut at film release – ort 94 min) B/W VIDrel: WHV V/h

PURE LUCK ** PG
Nadia Tass USA 1991
Martin Short, Danny Glover, Sheila Kelley, Scott Wilson, Sam Wanamaker, Harry Shearer, Jorge Russek, Rodrigo Puebla, John H. Brennan, Jorge Luke, Abel Woolrich, Patricia Gage, Ariane Pellicer, Alexandra Vicencio
A search for a missing mining heiress leads to a catalogue of disaster as an inept detective and his accident-prone companion stumble from one mishap to another. A simple slapstick film, with a basic plot that serves merely to tie a succession of visual gags together. A remake of the 1981 French film "La Chevre".
COM 91 min (ort 100 min) VIDrel: CIC/SONOP V/sur

PURPLE RAIN ** 15
Albert Magnoli USA 1984
Prince, Appollonia Kotero, Morris Day, Olga Karlatos, Jerome Benton, Billy Sparks, Clarence Williams III, Jill Jones, Charles Huntsberry, Susan, Dez Dickerson, Brenda Bennett, Sandra Claire Gershman, Kim Upsher, Alan Leeds
Really just a vehicle for singer Prince, this is the story of a performer struggling to gain acceptance in his fight to reach the top. Without the score, this dull effort would have little merit in its own right. Prince's film debut, and no more than a vehicle for this performer. See also GRAFFITI BRIDGE. AA: Score (Prince/John L. Nelson/The Revolution).
MUS 107 min (ort 113 min) VIDrel: WHV L/A V/sh

PURPLE ROSE OF CAIRO, THE *** PG
Woody Allen USA 1984
Mia Farrow, Danny Aiello, Jeff Daniels, Dianne Wiest, Van Johnson, Milo O'Shea, Zoe Caldwell, Edward Herrmann, Stephanie Farrow, Karen Akers, John Wood, Deborah Rush, Michael Tucker
The downtrodden housewife in a small American town in the 1930s, finds escape from reality by repeatedly visiting the cinema to see the title film. Things take an unexpected turn when the film's leading man falls in love with her and steps down from the screen. A clinically effective comedy, that would be cold indeed without the warmth Farrow and Aiello bring it.
COM 78 min (ort 82 min) VIDrel: VISVID/POLYREC V

PUSSY CALLED WANDA, A * 18
Eric Edwards USA 1992
Deidre Holland, Jon Dough, Melanie Moore, Cassidy, Mike Horner, Alicia Rio, Heidi Kat, Scott Irish, Eric Edwards
Politicians and police chiefs use the Pussycat Escort Agency and live to tell the tale. Daft sex film that was followed by a sequel of similar merits.
A 75 min (ort 80 min) VIDrel: MIA/DISC V

PUSSY CALLED WANDA 2, A ** 18
Eric Edwards USA 1993
Deidre Holland, Jon Dough, Chrissy Ann, Rocco Siffredi, Kiss, Cody O'Connor, Kriz Newz, Tom Byron
The owner of a call-girl service uses the services of a private eye, who used to be chief of police, to try to find some way to stop a new hotshot mayor who is threatening to close her down.
A 58 min (ort 80 min) VIDrel: PASSION/IMC V

PUSSYCAT SYNDROME ** 18
Irvin Miles WEST GERMANY 1983
Ajita Wilson, Jacqueline Marcan, Tina Eklund, Herbert Hofer, Cristus Nicoeul, Tony Woolf, Don Caverman, Nastassja Nataly, George Filipou, Carmella Bassi, Nina Georgiou, Marcella Banezi, Bobies Provatas
Two models in Greece enjoy a variety of adventures and end up staying on a millionaire's yacht.
A 73 min (Abridged at film release – ort 85 min) VIDrel: ELV V

PUSSYMAN'S HOUSE PARTY * 18
David Christopher USA 1996
Laura Palmer, Lovette, Sahara Sands, Nyrobi Knight, Nena Cherry, Taboo, Tia, Ashley Rene, Yasmine, Ruby Richards, Steve Hatcher, Dave Hardman, Nick East, Julian St. Jox, Mr. Marcus
A man who some have made more friends over but things soon get out of hand in this unremarkable offering.
A 83 min (ort) VIDrel: ONE V

PUTTIN' HER ASSETS ON THE LINE * 18
Jerry Ross USA 1991
Brandy Alexander, Casey Williams, Kim Wylde, Candice Hart, Angela Summers, Trixie Tyler, Steve Drake, Joey Silvera, Randy West
Alexander plays an unhappy housewife who has been married for eight years, a fact that has led to her husband growing sexually jaded and unfaithful. With divorce very much on her mind, she tries out a little infidelity herself with a repairmen, who has arrived to fix her dishwasher. However, all ends happily enough when our couple decide that their marriage is worth saving after all. A trite affair, the depressing dialogue gives it a decidedly anti-male bias.
Aka: PUTTING HER ASS ON THE LINE
A 57 min VIDrel: ONE V

PYGMALION **** U
Anthony Asquith/Leslie Howard UK 1938
Wendy Hiller, Leslie Howard, Wilfrid Lawson, Marie Lohr, Scott

Sunderland, Jean Cadell, David Tree, Everley Gregg, Leueen McGrath, Esme Percy, Violet Vanbrugh, Iris Hoey, Viola Tree, Irene Browne, Wally Patch, H.F. Maltby
An excellent straight version of Shaw's play, with sparkling, witty performances, following the exploits of a professor who, having boasted to a friend that he could teach a flower-girl to talk like a duchess, is presented with the opportunity to do so. Remade as the musical MY FAIR LADY. Shaw's screenplay and its adaptation both won Oscars. AA: Screen (George Bernard Shaw and Ian Dalrymple / Cecil Lewis / W.P. Lipscomb for adaptation).
COM   95 min B/W   VIDrel: L/A   V
Boa: play by George Bernard Shaw.

PYRATES *   18
Noah Stern USA   1991
Kevin Bacon, Kyra Sedgewick, Bruce Martin Payne, Kristin Datillo, Buckley Norris, Deborah Falconer, Raymond O'Conner, Byrne Piven, Ernie Banks, Mickey Jones, Bo Sharin, Daniel Ben Wilson, Jeff Silverman, William J. Murphy
A couple are literally so hot for each other that they cause fires every time they make love. A strictly one-joke film that soon grown unbearably tedious.
COM   91 min (ort 98 min)   VIDrel: VCC/DISC   V/sur

PYROMANIAC'S LOVE STORY, A **   PG
Joshua Brand CANADA/USA   1995
William Baldwin, John Leguizamo, Sadie Frost, Erika Eleniak, Michael Lerner, Joan Plowright, Armin Mueller-Stahl, Mike Starr, Julio Oscar Mechoso, Floyd Vivino, Babs Chula, Tony Perri, Randy Butcher, Lesley Kelly, Jennifer Rollin
The arson of a bakery has an unexpected effect of various people who become so passionately involved in romantic complications that they are all willing to confess to this crime. Sadly, this original idea for a romantic comedy does not get off the ground. This was Brand's directorial debut. Screenplay is by Morgan Upton Ward.
COM   91 min (ort 99 min)   VIDrel: HOLPIC/TECH   V

Q

Q & A **   18
Sidney Lumet USA   1990
Nick Nolte, Timothy Hutton, Armand Assante, Patrick O'Neal, Lee Richardson, Luis Guzman, Charles Dutton, Jenny Lumet, Paul Calderon, Leonard Cimino, Fyvush Finkel, International Chrysis, Dominick Chianese, Maurice Schell, Tommy A. Ford
Nolte plays a corrupt cop who shoots dead a street punk and finds himself the victim of an investigation conducted by Assistant D.A. Hutton. Lumet's attempt to detail the various pressures and tensions that make police work what it is has moments of power, but lacks a strong narrative and rapidly loses its way in a morass of sub-plots.
DRA   127 min (ort 134 min)   VIDrel: VISVID/POLYREC V/sur
Boa: book by Edwin Torres.

QUADROPHENIA **   18
Franc Roddam UK   1979
Phil Daniels, Mark Wingett, Philip Davis, Sting (Gordon Sumner), The Who, Kate Williams, Michael Elphick, John Bindon, Garry Cooper, Leslie Ash, Toyah Wilcox, Trevor Laird, Raymond Winstone, Gary Shail, Kim Neve, John Phillips
Highly enjoyable rock-musical combining the music of The Who with a nice (if a trifle inaccurate) feel for the mood of the 1960s, when Mods and Rockers battled it out on the beaches of Britain's seaside resorts. Written by Dave Humphries, Martin Stellman and Franc Roddam, this film saw Sting in his acting debut. Inspired by a record album of the same name.
DRA   114 min (ort 120 min)   VIDrel: POLY/POLYREC V/sur

QUARE FELLOW, THE **   15
Arthur Dreifuss EIRE   1962
Patrick McGoohan, Sylvia Syms, Walter Macken, Dermot Kelly, Jack Cunningham, Hilton Edwards
At a Dublin prison a new warder learns the reasons behind a condemned man's murder of his brother and comes to doubt the usefulness of capital punishment. A competent adaptation of the successful stage play, this engaging tragi-comedy retains only some of the humour but most of the melancholy of the

original work. McGoohan is well cast as the uncertain young warder and the articulate script is the work of Dreifuss.
DRA   86 min (ort 90 min) B/W   VIDrel: CONNO/RTM   V
Boa: play by Brendan Behan.

QUARTET **   PG
Ralph Smart/Harold French/Arthur Crabtree/Ken Annakin
UK   1948
Basil Radford, Naunton Wayne, Mai Zetterling, Angela Baddeley, Dirk Bogarde, Francoise Rosay, Honor Blackman, Irene Browne, Hermione Baddeley, Mervyn Johns, Susan Shaw, George Cole, Cecil Parker, Linden Travers, Nora Swinburne
Four short stories each with a different cast and director, each tale being introduced by Maugham. The episodes are of variable quality but the whole effort works extremely well. Stories are entitled: "The Facts Of Life", "The Alien Corn", "The Kite", and "The Colonel's Lady". The success of this film prompted the sequel "Trio".
Aka: SOMERSET MAUGHAM'S QUARTET
DRA   115 min (ort 120 min) B/W   VIDrel: VCC/RTM   V
Boa: short stories by William Somerset Maugham.

QUATERMASS AND THE PIT **   PG
Rudolph Cartier UK   1957 (shown 1958/59)
Andre Morell, Cec Linder, Christine Finn, Anthony Bushell, John Stratton, Harold Goodwin, Clifford Cox, Brian Worth, Michael Ripper, John Walker, Brian Gilmar, Richard Shaw, Richard Dare, Ian Ainsley, Allan McClelland
Compilation of the six 30-minute episodes, of the third series of a popular set of SF stories, that ran intermittently from 1953 until 1979. This tale deals with the discovery of an alien spacecraft at a new excavation for a London underground station. Closer examination reveals that the vessel is imbued by its makers with a hostile form of energy that has the potential to enslave mankind. Written by Nigel Kneale and remade in 1967. Fair.
FAN   178 min B/W mTV   VIDrel: PARADOX/TOTAL
V/h

QUATERMASS AND THE PIT **   15
Roy Ward Baker UK   1967
Andrew Keir, James Donald, Barbara Shelley, Julian Glover, Duncan Lamont, Bryan Marshall, Edwin Richfield, Peter Copley, Edwin Richfield, Grant Taylor, Maurice Good, Sheila Staefel, Thomas Heathcote
Hammer film version of a popular TV series in which the intrepid Professor Quatermass enjoyed a variety of encounters with alien forces. Based on the six TV episodes than ran from 1958 to 1959, it tells of the discovery of an alien vessel at a excavation for a new underground station, and the great danger it poses for mankind. Quite gripping in parts, but spoilt by plodding direction, stilted dialogue and some remarkably cheap looking effects.
Aka: FIVE MILLION YEARS TO EARTH
FAN   93 min (ort 97 min)   VIDrel: LUMI/SPEAR L/A   V
Boa: TV play by Nigel Kneale.

QUATERMASS CONCLUSION, THE **   15
Piers Haggard UK   1980
John Mills, Barbara Kellerman, Simon MacCorkindale, Margaret Tyzack, Ralph Arliss, Brewster Mason, Paul Rosebury, Jane Bertish, Rebecca Satire, Bruce Purchase, David Yip, Brenda Fricker, Tony Sibbald, Neil Stacy
A late addition to those "Quatermass" films of the 1950s, in which our intrepid professor had to battle it out with alien menaces. Here Professor Quatermass finds that aliens are trying to take over the Earth by affecting the minds of young people, and causing society to break down. Fairly imaginative but severely hampered by its low budget.
Aka: QUATERMASS
FAN   102 min (ort 200 min) mTV   VIDrel: VCC/DISC L/A
V

QUE LA BETE MEURE **   15
Claude Chabrol FRANCE   1969
Michel Duchaussoy, Caroline Cellier, Jean Yanne, Anouk Ferjac, Marc Di Napoli, Maurice Pialat
Whilst he is still trying to find the driver who knocked down and killed his son, a man deliberately starts an affair with a woman whose brother-in-law Paul appears to be the prime suspect. But he finds himself truly falling in love with the woman, and events take a far more serious turn when a murder occurs, it appearing to be the case that Paul has now killed his

own father. A convoluted and very oppressive film, compelling and extremely well acted.

Aka: KILLER!; THIS MAN MUST DIE

THR 107 min (ort 110 min) wScrn VIDrel: ARTPRO/RTM V

Boa: novel by Nicholas Blake (Cecil Day Lewis).

QUE VIVA MEXICO! *** PG
Sergei Eisenstein USSR 1930
Eduard Tisse
The director's unfinished opus on the history of Mexico was reconstructed by Grigora Alexandrov, and gives us some idea of the masterpiece it might have been. An ambitious, colourful and tragically incomplete paean of praise to Mexico and its people, that puts one in mind of "I, Claudius" and other such fragmented epics.
DOC 84 min (ort 85 min) B/W VIDrel: HEND L/A V

QUEEN BEE ** PG
Ranald MacDougall USA 1955
Joan Crawford, Barry Sullivan, Betsy Palmer, John Ireland, Lucy Marlow, Fay Wray, William Leslie, Katherine Anderson, Tim Hovey, Linda Bennett, Willa Pearl Curtis, Bill Walker, Olan Soule, Bob McCord, Juanita Moore
A mill owner is married to a woman whose congenial exterior conceals a manipulative and domineering personality. When a New York cousin comes to stay in their Southern mansion, a train of events is set in motion that ends in tragedy. A competent rather than inspired piece that provides a solid vehicle for Crawford's polished performance.
DRA 91 min (ort 94 min) B/W VIDrel: COLUM/SONOP V

Boa: novel by Edna Lee.

QUEEN CHRISTINA **** U
Rouben Mamoulian USA 1933
Greta Garbo, John Gilbert, Lewis Stone, Ian Keith, Elizabeth Young, C. Aubrey Smith, Gustav von Seyffertitz, Reginald Owen, Georges Tenavent, David Torrence, Ferdinand Munier, Akim Tamiroff, Cora Sue Collins, Edward Norris
Unable to face the prospect of an arranged marriage, a seventeenth-century Swedish queen dons man's clothes and goes off in search of adventure, only to fall in love with the newly-arrived Spanish ambassador. Stilted, over-sentimental and historically inaccurate it may be, but this charming romance shows the star to her best advantage; the tragic ending is quite moving. Garbo never looked more radiant.
DRA 100 min B/W VIDrel: MGM L/A V

QUEEN KELLY *** PG
Erich Von Stroheim USA 1928
Gloria Swanson, Walter Byron, Seena Owen, Wilhelm Von Brincken, Madge Hunt, Wilson Benge, Sidney Bracey, Lucille Van Lent, Ann Morgan, Tully Marshal, Florence Gibson, Sul Te Wan, Ray Gaggett, Sylvia Ashton
Extravagant chronicle of the ills that befall a convent girl who is bowled over by a prince and later becomes the madam of an African brothel. This was Stroheim's last directing effort, but one he never completed since he was fired by Swanson, who was annoyed by his spendthrift ways. Released to theatres in Europe, it was never shown commercially in the USA, but exists in a number of versions.
DRA 115 min B/W silent VIDrel: VISION/DISC V

QUEEN OF HEARTS *** PG
Jon Amiel UK 1989
Vittorio Duse, Joseph Long, Anita Zagaria, Eileen Way, Vittorio Amandola, Ian Hawkes, Tat Whalley, Jimmy Lambert, Anna Pernicci, Roberto Scateni, Alec Bregonzi, Stefano Spagnoli, Ronan Vibert, Matilda Thorpe, Anthony Manzoni
An account of the life of an Italian immigrant couple, from the days when they fled an arranged marriage in Italy to their arrival in England and eventual prosperity, running a cafe in London. Years later, a threat to their happiness appears in the shape of a figure from the past. An appealing blend of pathos, drama and comedy, with no real stars, but a lively story, mostly told as seen through the eyes of their ten-year-old son (Hawkes).
DRA 112 min VIDrel: MGM L/A V

QUEENS LOGIC ** 15
Steve Rash USA 1991
Kevin Bacon, Joe Mantegna, John Malkovich, Ken Olin, Linda Fiorentino, Jamie Lee Curtis, Tom Waits, Tony Spiridakis, Chloe

Webb, Kelly Bishop, Terry Kinney, Michael Zelniker, Ed Marinaro, Wendy Gazelle, Jodie Markell, Jenny Wright
A bunch of friends from Queens, New York, find that despite a separation over time and place, their roots are hard to forget. When a wedding brings them together again for a reunion, they experience the sadness and joy of former friendships. Scripted by Spiridakis, this is an uneven and bitter-sweet offering, sometimes poignant, but more often pointless.
DRA 108 min (ort 113 min) VIDrel: 20TH/TECH V/sur

QUERELLE ** 18
Rainer Werner Fassbinder FRANCE/WEST GERMANY 1982
Brad Davis, Jeanne Moreau, Franco Nero, Gunther Kaufmann, Hanno Poschl, Laurent Malet, Nadja Bunkhorst, Burkhard Driest, Dieter Schidor, Neil Bell, Roger Fritz, Michael McLernon, Harry Baer, Volker Sprengler, Isolde Barth
Fassbinder's final film tells of a French sailor who, after murdering a fellow seaman, goes to a notorious brothel in Brest and discovers his homosexuality. Here he comes to terms with his true nature, and the other habitues of the place fall victim to his charms. Shot in a garish series of studio sets, replete with macho icons, this film attempts to say something of interest but is ultimately overwhelmed by layers of tedium.

Aka: QUERELLE DE BREST

DRA 104 min (ort 120 min) wScrn VIDrel: ARTIF/20TH V/sur

Boa: novel Querelle De Brest by Jean Genet.

QUEST, THE ** 18
Jean-Claude Van Damme USA 1996
Jean-Claude Van Damme, Roger Moore, James Remar, Janet Gunn, Jack McGee, Aki Aleong, Louis Mandylor, Abdel Qissi, Brauno Belfiore, Tara Nichelle Biberstein, Anderson C. Bradshaw, Brick Bronsky, Michael Caloz, Cesar Carnerio, Ryan Cutrona
In the 1920s, a young New York pickpocket escapes from the police and takes ship for the Orient, but this vessel is captured by pirates and he is sold as a slave, but is purchased by the owner of a martial arts academy. Having become adept at the fighting arts, he makes ready to enter a tournament in a remote Tibetan city. In addition to Van Damme, some fifteen other martial artists provide plenty of action in this adventure.
A/AD 110 min CINrel

QUEST FOR FIRE *** 15
Jean-Jacques Annaud CANADA/FRANCE 1981
Everett McGill, Rae Dawn Chong, Ron Perlman, Nameer El Kadi, Gary Schwartz, Frank Olivier Bonnet, Kurt Schiegl, Brian Gill, Terry Fitt, Bibi Caspari, Peter Elliott, Michelle Leduc, Robert Lavoie, Matt Birman, Christian Benard
Saga of a Stone Age tribe which loses its fire and has to discover how it is made, so three members of the tribe go off in search of some. Despite some unintentionally funny moments, the film works well enough. The special sign language used was devised by Anthony Burgess, and Desmond Morris advised with regard to body language. The film was shot in Kenya, Canada, Scotland and Iceland. AA: Make (Sarah Monzani/Michele Burke).

Aka: LA GUERRE DU FEU

FAN 96 min (Cut at film release by 8 sec – ort 100 min) VIDrel: 20TH/TECH L/A V

Boa: novel Le Felin Geant (Quest Of The Dawn Man) by J.H. Rosny Sr.

QUEST FOR JUSTICE *** 15
James Keach USA 1993
Jane Seymour, Richard Kiley, D.W. Moffett, Michelle Joyner, Lou Walker, Starletta DuPois, Brett Rice, Jeanette Lane Bradburg, David Devries, Ralph Wilcox, Rebecca Wackler, Mary Nell Santacroce, Dwionne Dickerson, Derek Pruitt
In 1950s Mississippi a local newspaper owner supports school integration and the Civil Rights movement, and gets into trouble with the Ku Klux Klan and her more racist neighbours when she starts up a newspaper for Blacks. A strongly written and rather thoughtful piece, nothing like as shrill as one might have feared.
DRA 90 min mTV VIDrel: ODY/SONOP V/sh

QUEST FOR KING SOLOMON'S MINES, THE ** U
Kurt Neumann USA 1959
George Montgomery, Taina Elg, David Farrar, Rex Ingram, Dan Seymour, Robert Goodwin, Anthony M. Davis, Paul Thomas, Harold Dyrenforth
A dull sequel to KING SOLOMON'S MINES, made on the cheap

by using footage from the 1950 version of that film. Not quite as bad as one might have expected, but not all that good either.
Aka: WATUSI
A/AD   80 min (ort 85 min)   VIDrel: L/A   V

QUEST FOR THE MIGHTY SWORD **
David Hills (Aristide Massaccesi) ITALY
Eric Allen Kramer, Margaret Lenzey, Donal O'Brien, Dina Marrone, Chris Murphy, Laura Gemser, Marisa Mell
A sequel of sorts to ATOR, THE FIGHTING EAGLE and ATOR THE INVINCIBLE 2, with Ator dying and passing on his sword to his son, with instructions to seek out the dwarf who can restore its magical powers. This sets the scene for another series of fantastic adventures, very much in keeping with the style of the two earlier films.
Aka: HOBGOBLIN
A/AD   90 min (ort 94 min)   VIDrel: COLUM/SONOP   V

PG
1990

QUESTION OF ATTRIBUTION, A ***
John Schlesinger UK
James Fox, David Calder, Geffrey Palmer, Gregory Floy, Edward De Souza, John Cater, Jason Flemyng, Richard Bebb, Prunella Scales, Ann Beach, Julia St John, Mark Payton, Anne Jameson, Barbara Hicks
After the defections of Burgess, Maclean and Philby, suspicion continued that there was a possible "fourth man". Investigations cast suspicion on Sir Anthony Bunt, a respectable art historian with responsibility for the Queen's art collection. An excellent and fascinating dramatisation which details how the British Establishment succeeded in covering up Blunts's treachery and ensuring that he was never tried. Screenplay is by Bennett.
DRA   70 min (ort 75 min) mTV   TVrel: BBC
Boa: play by Alan Bennett.

(12)
1991

QUICK AND THE DEAD, THE **
Sam Raimi USA
Sharon Stone, Gene Hackman, Leonardo DiCaprio, Russell Crowe, Tobin Bell, Lance Henriksen, Gary Sinise, Roberts Blossom, Kevin Conway, Pat Hingle, Keith David, Mark Boone Junior, Olivia Burnette, Fay Masterson, Woody Strode
The corrupt mayor has made the town of Redemption a haven for gunmen and low-lifes, and every year asserts his rule by holding a shooting contest he always wins. However, a tough woman rides into town on a vengeance-seeking mission, as he is the man who killed her father many years before. Though Stone is hardly convincing as a gunslinger, this quirky effort is best enjoyed as a pastiche rather than as a straight movie. This was Strode's last film.
WES   103 min (ort 108 min) wScrn cC
VIDrel: COLUM/SONOP   V/sur

15
1994

QUICK CHANGE **
Howard Franklin/Bill Murray USA
Bill Murray, Geena Davis, Randy Quaid, Jason Robards, Bob Elliott, Philip Bosco, Phil Hartman, Kurtwood Smith, Jamey Sheridan, Kathryn Grady, Jack Gilpin, Richard Joseph Paul, Tony Shalhoub, Gary Klar, Susannah Blanchi, Ian Wheeler
An ill-assorted gang of bank robbers find that robbing their bank is nothing like as difficult as dealing with the day-to-day pettiness of New York cops, and have an impossibly tough time getting out of the city. A low-key and terribly rambling comedy of many gags and little plotting, whose appeal may be largely lost on people other than New Yorkers. Conley's novel was filmed once before in France as "Hold-Up" (1985).
COM   85 min (ort 89 min)   VIDrel: WHV   V/sur V/dm
Boa: novel by Jay Cronley.

15
1990

QUICKER THAN THE EYE **
Nicolas Gessner AUSTRIA/SWITZERLAND/WEST GERMANY
Mary Crosby, Ben Gazzara, Catherine Jarrett, Jean Vanne, Wolfram Berger, Dinah Hinz, Ivan Desny, Sophie Carle, Robert Liensol, Eb Lottimer, Laszlo J. Kish, Hans Leutenegger, Stefan Gubser, Christoph Walz, Norbert Schwientek
An American magician is invited to perform at a political summit being held in the Swiss Alps, but is unaware that the venue has been infiltrated by a set of dangerous terrorists on a mission of assassination. When the shooting starts, he finds himself compelled to use his professional skills in order to survive. An inconsequential but neat little film with an unusual premise.
THR   88 min (ort 94 min)   VIDrel: GENESIS   V
Boa: novel by Claude Cueni.

PG
1989

QUICKSAND: NO ESCAPE ***
Michael Pressman USA
Donald Sutherland, Tim Matheson, Felicity Huffman, Jay Acovone, Timothy Carhart, John Joseph Finn, Marc Alaimo, Al Pugliese, Margaret Reed, Colin Nelson, Amy Benedict, Bryan Clark, Jack Shearer, Steven Culp, Kaley Cuocco
Slightly offbeat thriller that stars Sutherland and Matheson as a corrupt private eye and a talented architect respectively. When the former learns of the latter's involvement in an accidental killing, he embarks on a spot of blackmail. However, his victim is unable to meet his demands, and instead is forced to involve himself in illegal schemes that draw both men ever closer together.
THR   88 min (ort 93 min)   VIDrel: CIC/SONOP   V

PG
1991

QUIET DAYS IN CLICHY ***
Claude Chabrol FRANCE/ITALY/GERMANY
Andrew McCarthy, Nigel Havers, Barbara De Rossi, Stephanie Cotta, Isolde Barth, Eva Grimaldi, Anna Galiena, Giuditta Del Vecchio, Stephane Audran, Mario Adorf, Eldie Mellis, Henri Attal, Jean-Marie Arnoux, Helene Benayon
This romantic period-drama has McCarthy as the writer Henry Miller, who in old age remembers Paris of the 1920s; the scene of his youth, his friendships and his passions. Having moved in with a photographer friend, he falls madly in love with a beautiful young girl who is also being given shelter. However, as both men desire her she agrees to marry the pair of them in a dummy "wedding", but disillusionment is not long in coming.
Aka: JOURS TRANQUILLES A CLICHY
DRA   100 min (ort 120 min)   VIDrel: ARTPRO/RTM   V
Boa: novel by Henry Miller.

18
1989

QUIET EARTH, THE ***
Geoff Murphy NEW ZEALAND
Bruno Lawrence, Alison Routledge, Peter Smith, Norman Fletcher, Tom Hyde
A haunting film in which a scientist wakes up to find that the Earth has been virtually depopulated as a result of an experiment that went wrong.
FAN   87 min (ort 100 min)   VIDrel: ARTPRO/RTM   V/sur
Boa: novel by Craig Harrison.

15
1985

QUIET MAN, THE ****
John Ford USA
John Wayne, Maureen O'Hara, Barry Fitzgerald, Victor McLaglen, Mildred Natwick, Francis Ford, Arthur Shields, Ward Bond, Eileen Crowe, Sean McClory, Jack McGowran, Ken Curtis, Mae Marsh, Joseph O'Dea, Eric Gorman
An Irish-American boxer goes to live in Ireland where he falls in love with a local girl. Though he has sworn never to fight again (a superb flashback sequence tells why), he is forced to in order to win her hand. A kind of Irish "Taming of the Shrew" of considerable charm and wit. Written by Frank Nugent and with a score by Victor Young. AA: Dir, Cin (Winton C. Hoch/Archie Stout).
DRA   124 min (ort 129 min)   VIDrel: 4-FRONT/POLYREC   V
Boa: novel by Maurice Walsh.

U
1952

QUIET VICTORY ***
Roy Campanella II USA
Michael Nouri, Pam Dawber, Bess Meyer, Peter Berg, James Handy, Dan Lauria, Gracie Harrison
The story of a former football star who developed Lou Gehrig's disease in the 1970s but refused to succumb to despair, going on to become a much-admired high school football coach. By the 1980s he was still teaching, despite being confined to a wheelchair. An inspiring and remarkably restrained account, with a literate script by Barry Morrow that (like the character portrayed) has no time for self-pity.
Aka: QUIET VICTORY: THE CHARLIE WEDEMEYER STORY
DRA   93 min (ort 100 min) mTV   VIDrel: ODY/SONOP   V/sh

PG
1987

QUIGLEY DOWN UNDER ***
Simon Wincer AUSTRALIA/USA
Tom Selleck, Laura San Giacomo, Alan Rickman, Chris Haywood, Ron Haddrick, Tony Bonner, Jerome Ehlers, Conor McDermottroe, Roger Ward, Ben Mendelsohn, Steve Dodd, Karen Dante, Kylie Foster, William Zappa, Jonathan Sweet, Jon Ewing
Highly enjoyable adventure set in the Australian Outback, where Selleck as a determined American of few words, is hired by a ruthless landowner, but turns against him and sets out to

15
1990

defend a community of Aborigines from the depredations of his former boss. A gripping action-thriller, of spectacular locations, excellent photography and strong performances.
A/AD   115 min (ort 120 min)   VIDrel: MGM/WHV   V/sur

QUILLER MEMORANDUM, THE ***   15
Michael Anderson UK/USA   1966
George Segal, Max Von Sydow, Alec Guinness, Senta Berger, George Sanders, Robert Helpmann, Robert Flemyng, Peter Carsten, Edith Schneider, Gunter Meisner, Robert Stass, Ernst Walder, Philip Madoc, John Rees
A British agent is given the task of investigating neo-Nazis in modern-day Berlin. Nothing sensational, but adroit and well handled. The script is by Harold Pinter with music by John Barry.
DRA   103 min (ort 105 min)   VIDrel: VCC L/A   V
Boa: novel The Berlin Memorandum by Adam Hall (Elleston Trevor).

QUINCE TREE SUN, THE ***   U
Victor Erice SPAIN   1991
Antonio Lopez, Maria Moreno, Enriqie Gran, Jose Carrtero, Maria Lopez, Carmen Lopez, Elisa Ruiz, Amalia Avia, Lucio Munoz, Esperanza Parada, Julio Lopez Fernandez, Janusz Pietrziak, Marek Domagala, Grzegorz Ponikwia
An artist spends the whole of the autumn working on paintings of the quince trees in his garden, endeavouring to capture a perfect image of the tree before the onset of winter. A slow-paced film of intensity and beauty, almost plotless, yet once one becomes accustomed to the lack of a straightforward narrative there is much here to enjoy. Not so much a film as a meditation on art, life and the passage of time.
Aka: EL SOL DE MEMBRILLO
DRA   132 min (ort 139 min)   VIDrel: ARTIF/20TH   V/h

QUIZ SHOW ***   15
Robert Redford USA   1994
Ralph Fiennes, John Turturro, Rob Morrow, David Paymer, Paul Scofield, Hank Azaria, Christopher McDonald, Johann Carlo, Elizabeth Wilson, Allan Rich, Mira Sorvino, George Martin, Paul Guilfoyle, Griffin Dunne, Michael Mantell
A true story set in 1956, with American TV audiences avidly following those big-dollar prize quiz shows such as "The $64,000 Question" and the like. Worried that their ratings will fall, the sponsors of one such show ease out a Jewish contestant by priming a rival on a warm-up eliminator. Fine performances compensate for stilted direction and over-length, making this a fascinating tale of how corruption became standard practice in the early days of television
DRA   127 min (ort 133 min) cC   VIDrel: HOLPIC/TECH   V/sur
Boa: book Remembering America: A Voice from the Sixties by Richard N. Goodwin.

QUO VADIS? ***   PG
Mervyn Le Roy USA   1951
Robert Taylor, Deborah Kerr, Peter Ustinov, Leo Genn, Patricia Laffin, Finlay Currie, Abraham Sofaer, Marina Berti, Buddy Baer, Felix Aylmer, Nora Swinburne, Ralph Truman, Norman Wooland, Peter Miles, Geoffrey Dunn
Describes Rome at the time of Nero, focusing on a Roman commander who falls in love with a Christian girl and becomes a convert himself. A spectacular film with many fine sequences but marred by a tendency to be tedious in some places, brutal in others. The film is at its most effective where it examines the lives of the early Christians. The pompous Miklos Rozsa score was based on music of the time. Remade for Italian TV in 1985.
DRA   162 min (ort 171 min)   VIDrel: MGM/WHV   V/s
Boa: novel by Henryk Sienkiewicz.

R

RA: THE PATH OF THE SUN GOD **   U
Lesley Keen UK   1990
Narrated by: Tamara Kennedy, Michael Mackenzie
The story of Egypt from its earliest mythical beginnings, that charts the mighty battle fought between Osiris and Isis (the forces of good and creation) and their wicked brother Set (who symbolises the forces of evil and destruction. First shown on Channel Four TV.
ANIM   71 min mTV   VIDrel: CONNO/RTM   V

RABID ***   18
David Cronenberg CANADA   1977
Marilyn Chambers, Frank Moore, Joe Silver, Patricia Gage, Susan Roman, Howard Ryspan, J. Roger Periard, Lynne Deragon, Victor Deay, Gary McKeehan, Terry Schonblum, Julie Anna, Terrence G. Ross, Robert O'Ree, Greg Von Riel
Following an accident, a woman motorcyclist receives an experimental skin graft of "morphologically neutral" skin. However, this develops into a repulsive growth that drains the blood from her lovers and gives rise to a rabies-like epidemic. Quite a forceful if terribly under-budgeted little chiller.
HOR   90 min (ort 94 min)   VIDrel: ARROW/RTM   V

RABID GRANNIES *   18
Emmanuel Kervyn BELGIUM/FRANCE/HOLLAND   1988
Catherine Aymerie, Danielle Daven, Anne-Marie Fox, Jack Mayar, Elliott Lison
Immensely bad movie, courtesy of Troma Productions (of TOXIC AVENGER fame), in which a birthday party for the elderly relatives of a large family turns a little sour, when they are sent a present by a devil-worshipping relative that causes them to mutate into murderous demons. Needless to say, the bodies soon start to pile up, in a badly dubbed and edited offering that verges on the unwatchable.
HOR   88 min dubbed   VIDrel: TROMA/RTM   V

RACE AGAINST TIME *   12
Fred Gerber USA   1996
Patty Duke, Richard Crenna, Katy Boyer, Missy Crider, Jon Gries, Ele Keats, Liza Snyder, Sydney Walsh, Brian Finney, Steve Eastin, Charles Grant, Caitlin Wachs
When a woman's car is found abandoned, the girl's parents mount a desperate bid to find their daughter. A muddled thriller, very loosely put together.
THR   95 min   VIDrel: ODY/SONOP   V/sh

RACE FOR GLORY **   15
Rocky Lang USA   1989
Alex McArthur, Peter Berg, Pamela Ludwig, Ray Wise, Burt Kwouk, Lane Smith, Jerome Dempsey, Teco Celio, Takashi Kawahara, Vincent Grass, Daniel Lombart, Frederic Darie, Chen Ching-Yi, Barbara Blossom, Sanford Gorodetsky
Buddies from a small town drift apart when one of them decides to pursue his dream of becoming an international motorcycle champion, and wins a race on a bike that was built by his two best friends. But though his growing obsession with the sport and determination to achieve worldwide fame is the cause of the rift, there's the expected happy ending. Production-line affair, slow to get going and with a no more than adequate plot.
Aka: AMERICAN BUILT
A/AD   102 min   VIDrel: 20VIS/SONOP   V/sh

RACE THE SUN **   PG
Charles T. Kanganis USA   1996
Halle Berry, James Belushi, Bill Hunter, Casey Afflect
A bored bunch of teenagers in Hawaii are inspired by their teacher, who gets them involved in a demanding project. Adequate feel-good movie.
DRA   96 min   VIDrel: COLUM/SONOP   V/sur

RACE TO FREEDOM: THE STORY OF THE UNDERGROUND RAILROAD ***   (PG)
Don McBrearty USA   1994
Tim Reid, Glynn Turman, Dawnn Lewis, Courtney B. Vance, Janet Bailey, Ron White, Michael Riley, Roy Lewis, Ron Small, Tom Butler, Susan Hogan, James B. Douglas, Ron Hartmann, Falconer Abraham, Kurt Reis, Gene Mack, Joseph Ziegler
A young slave couple face a perilous journey along title railroad, a escape route that led from slavery to freedom in Canada. Well meaning and quite competently made, but perhaps a trifle over-long and too slowly paced to make an effective drama of this fascinating story. In actual fact, almost 40,000 slaves undertook this perilous odyssey after the passing of the Fugitive Slave Act meant that an escapee was not safe anywhere in the United States.
DRA   90 min (ort 130 min)   SATrel: MOVIE CHANNEL

RACHEL PAPERS, THE ***   18
Damien Harris USA   1989
Dexter Fletcher, Ione Skye, Jonathan Pryce, James Spader, Bill Paterson, Michael Gambon, Shirley Anne Field, Lesley Sharp, Jared Harris, Pat Keen, Aubrey Morris, Claire Skinner, Nicola Kimber, Amanda De Cadenet, Gina McKee

A computer buff has programmed his machine with every chat-up line under the sun, and has become a master of seduction. However, when his best efforts are stymied by an aloof girl he seeks advice from his coarse brother-in-law (a great performance from Pryce). Eventually he wins her over and she moves in with him, but once the chase is ended he finds himself bored. A coy updating of the Amis novel, sustained by a few amusing cameos.
COM   90 min (ort 91 min)   VIDrel: VISVID/POLYREC
V/sur
Boa: novel by Martin Amis.

RACQUEL ON FIRE *   18
Gordon Vandermeer USA   1990
Racquel Darrian, Kimberley Kane, Debi Diamond, Derrick Lane, T.T. Boy, Bridgett Monroe, Jeff Golden
Together with some other girls, the mayor of a tiny hamlet (Darrian) hatches a scheme to attract business to the area. Her hope is that this will solve the town's unemployment problem, as even her secretary's husband has been reduced to selling encyclopaedias to make a living. However, this being a sex film, this activity seems to require him to make love to a succession of attractive women. A weakly plotted effort, cut before video submission by 17 minutes 14 seconds.
A   86 min   VIDrel: IMPENT   V

RADIANT CITY ***   (PG)
Robert Allan Ackerman USA   1996
Kirstie Alley, Clancy Brown, Larraine Newman, Adam Lamberg, Tori Petrie, Fab Filippo, Monica Parker, Gil Bellows, Maria Vacratis, Myra Fried, Jordan Hughes, Tony Shalhoub, Dov Tiefenbach, Julian E. Campbell, John E. Campbell
A young boy recalls his early childhood in 1950s Brooklyn at the title locality – a low-income housing project, and the safe but at times very stifling atmosphere of the neighbourhood. After eight years there, his mother feels terribly constrained, and starts dreaming of moving out and getting a job. Alley gives a fine performance, as do the rest of the cast in this touching and quite nostalgic film.
DRA   85 min   SATrel: MOVIE CHANNEL

RADIO DAYS ***   PG
Woody Allen USA   1986
Dianne Wiest, Julie Kavner, Michael Tucker, Mia Farrow, Danny Aiello, Tony Roberts, Diane Keaton, Josh Mostel, Seth Green, Jeff Daniels, Wallace Shawn, Tito Puente, Gina DeAngelis, Kitty Carlisle Hart, Julie Kurnitz, Leah Carrey
A nostalgic homage to the joys of a 1940s childhood and the mass medium of that era. Not much in the way of a plot, more a succession of episodes that largely focus on a young boy's upbringing, with a mass of period detail and host of affectionate vignettes. The photography is by Carlo Di Palma and the excellent sets are by Santo Loquasto.
COM   85 min   VIDrel: VCC L/A   V

RADIO FLYER ***   PG
Richard Donner USA   1992
Lorraine Bracco, John Heard, Ben Johnson, Elijah Wood, Joseph Mazzello, Adam Baldwin, Sean Baca, Robert Munic, Garette Ratliff, Thomas Ian Nichols, Noah Verzuo, Isaac Ocampo, Kaylan Romero, Abraham Verduzo, T.J. Evans, Daniel Bieber
After their divorced mother moves to a small town and eventually marries an alcoholic, her two young sons suffer severe abuse at his hands and retreat into a fantasy world of their own. There they conceive a plan to escape that leads to a tragedy. An appealing and well-acted view of childhood but one that seems completely unrealistic.
DRA   110 min (ort 114 min) cC
VIDrel: VCC/DISC/COLUM   V/sur

RADIO ON **   18
Christopher Pettit UK   1979
David Beames, Lisa Kreuzer, Sandy Ratcliff, Andrew Byatt, Sting, Sue Jones-Davies, Sabina Michael
A man travels from London to Bristol in the hope of learning something about his brother's death. A rambling but strangely absorbing little film, with a music soundtrack by David Bowie, Kraftwerk, Sting and Eddie Cochran, that is less a straightforward story than a depressing examination of loneliness and isolation in the modern world. The film's associate producer was Wim Wenders, and his influence can be clearly discerned here.
DRA   100 min B/W   VIDrel: IMAG/RTM   V

RADIOLAND MURDERS *   12
Mel Smith USA   1994
Mary Stuart Masterson, Brian Benben, Ned Beatty, George Burns, Scott Michael Campbell, Brion James, Michael Lerner, Michae McKean, Jeffrey Tambor, Corbin Bernsen, Stephen Tobolowsky, Christopher Lloyd, Rosemary Clooney, Candy Clark
Comedy-thriller set at a Chicago radio-station in the 1940s, where the staff put on a brave front, despite the fact that they are being stalked by an unseen assassin. Scripted by STAR WARS producer George Lucas, but in spite of that it is still a dreary and repetitive work.
COM   108 min   VIDrel: CIC/SONOP L/A   V/sh

RAFFLES ***   U
Sam Wood USA   1939
David Niven, Olivia De Havilland, Dudley Digges, Dame May Whitty, Douglas Walton, Lionel Pape, E.E. Clive, Peter Godfrey, Margaret Seddon, Gilbert Emery, Hilda Plowright, Vesey O'Davoren, George Cathrey, Keith Hitchcock, Eric Wilton
A remake of the earlier film, telling of a gentleman cricket player in 1930s England who lives a double life as a wily and incredibly skilful cat burglar, always one step ahead of the law. But when he falls for a respectable lady, he thinks of giving up crime after a final escapade. Will appeal to those who enjoy Hollywood films of this period, but Niven has little of Colman's magnetism. Written by John Van Druten and Sydney Howard and with music by Victor Young.
DRA   69 min (ort 72 min) B/W   VIDrel: VCC/DISC   V
Boa: novel Raffles the Amateur Cracksman by E.W. Hornung.

RAGE **   15
George C. Scott USA   1972
George C. Scott, Richard Basehart, Martin Sheen, Barnard Hughes, Nicolas Beauvy, Paul Stevens, Ed Lauter, Dabbs Greer, Ken Tobey, Lou Frizzell, John Dierkes, Stephen Young, William Jordan, Bette Heritze, Terry Wilson, Anna Aries
The US Army's tests of a new chemical agent, result in the accidental death of a young boy. The boy's father finds his grief turning to anger as he discovers that a vast cover-up has taken place. This was Scott's debut as a director, in a film that starts off well but goes gradually downhill in a welter of violent action, that eclipses whatever message it might have been intended to convey.
DRA   95 min (ort 104 min)   VIDrel: MGM/WHV L/A   V

RAGE **   18
Joseph Merhi USA   1995
Kenneth Tigar, Gary Daniels, Jillian McWhirter, Fiona Hutchinson, Dave Aranda-Richards, Judith-Marie Bergen. Luis Beckford, Gary Bullock, Chuck Butto, Tim Colceri, Ricky Cornell, Dorein Fein, Gill Ferrales, Raymond Fitzpatrick
A man is injected with a lethal combination of drugs that causes him to succumb to murderous rages, and with the cops on his trails, races against time to find an antidote.
A/AD   95 min   VIDrel: COLUM/SONOP   V/sh

RAGE AND HONOR **   18
Terence H. Winkless USA   1992
Cynthia Rothrock, Richard Norton, Brian Thompson, Catherine Bach, Teri Treas, Alex Datcher, Stephen Davies, Patrick Malone, Toshihiro Ogata, Tim Dezarn, Jon Van Ness, Matt O'Toole, Peter Cunningham, Roger Yuan, Faith Minton
Rothrock martial arts offering, that teams her up with Norton as a tough inner-city teacher and an undercover cop respectively, both of whom are equally determined to put a stop to the activities of the city's drug barons. A simple-minded actioner with a set of vigorous combat sequences but not a great deal else. A sequel followed.
MAR   89 min (ort 93 min)   VIDrel: POLY/POLYREC/MED
V/sh

RAGE AND HONOR 2: HOSTILE TAKEOVER **   18
Guy Norris USA   1992
Cynthia Rothrock, Richard Norton, Patrick Muldoon, Frans Tumbuan, Tanaka, Ron Vreeken, Alex Tumundo, John T. Soucy, Yenny Farida, David Cobb, Donald Paul Pemrick, Don Balfour, Graham Stumpf, Glenn Kuehland, John R. Melfi
Very much a rerun of the first film with our martial-arts expert hitting the revenge trail after the death of a colleague that was engineered by a group of drug dealers who had corrupted his younger brother. Full of the usual set-piece confrontations

that will satisfy action fans but badly in need of a stronger plot.
MAR   93 min (ort 98 min)   VIDrel: MED/POLYREC L/A V/sh

RAGE AT DAWN ***
Tim Whelan USA
U
1955
Randolph Scott, Forrest Tucker, Mala Powers, J. Carrol Naish, Edgar Buchanan, Kenneth Tobey, Howard Petrie, Myron Healey, Ralph Moody, Guy Prescott, Mike Ragan, Phil Chambers
In a daring attempt to capture a gang of outlaws, an undercover agent pretends to be a train-robber, a ruse that allows him to execute a clever plan. An exciting Scott vehicle that scores well in all departments.
Aka: SEVEN BAD MEN
WES   85 min (ort 87 min)   VIDrel: SCRN/DISC   V

RAGE IN HARLEM, A ***
Bill Duke UK/USA
18
1991
Forest Whitaker, Gregory Hines, Robin Givens, Danny Glover, Zakes Mokae, Badja Djola, John Toles-Bey, Ron Taylor, Stack Pierce, George Wallace, Willard E. Pugh, Samm-Art Williams, Wendell Pierce, T.K. Carter, Leonard Jackson
Set in 1956, this crime-comedy caper has Givens playing a gangster's moll on the run with a stash of gold, who hides out in Harlem. Once there, she gets involved with some wacky characters, not least a transvestite madam and a momma's boy who sees himself as her protector. The implausible plot is never meant to be taken seriously and though patchy, offers droll moments, good performances and a few nods in the direction of "Cotton Comes To Harlem".
A/AD   103 min (ort 115 min)   VIDrel: POLY/POLYREC V/sur
Boa: novel For Love Of Imabelle by Chester A. Himes.

RAGE OF ANGELS **
Buzz Kulik USA
15
1983
Jaclyn Smith, Ken Howard, Armand Assante, Ron Hunter, Kevin Conway, George Coe, Joseph Wiseman, Deborah May, Joseph Warren, Wesley Addy, Bill Cobbs, James Greene, Pauline Flanagan, Lois Smith, Art Vasil, Edward J. Lynch
A woman lawyer fights her way to the top, enjoying various love affairs as a diversion from the pressures of her profession. A glossy and unmemorable celebration of pulp fiction, followed by a sequel of similar merits. First shown in two parts.
DRA   176 min (ort 200 min) mTV
VIDrel: 4-FRONT/POLYREC/ODY   V
Boa: novel by Sidney Sheldon.

RAGE OF ANGELS 2: PARTS 1 AND 2 **
Paul Wendkos USA
15
1983
Jaclyn Smith, Ken Howard, Michael Nouri, Susan Sullivan, Mason Adams, Linda Dano, Brad Dourif, Paul Roebling, Michael Woods, Ronald Hunter, Paul Shenar, Angela Lansbury, Philip Bosco, Danny Gerard, Michael O'Hare, Jay O. Sanders
The continuing story of Jennifer Parker, a high-powered Manhattan lawyer who has fought her way to the top, enjoying various love affairs and a rather complex past. A former lover has just been elected US Vice President and is anxious to rekindle their relationship, much against his wife's wishes. The brother of yet another lover, now dead, turns up to blackmail her. Even her mother, who abandoned her as a child, re-appears. A busy day.
Aka: RAGE OF ANGELS: THE STORY CONTINUES
DRA   190 min (ort 240 min) mTV
VIDrel: 4-FRONT/POLYREC/ODY   V/sh

RAGE OF INNOCENCE **
Paul Wendkos USA
PG
1991
Martin Sheen
A clerical error leads to a man being arrested as a drugs dealer, convicted and unjustly imprisoned for fifteen years. Naturally, upon release the man sets about planning his revenge.
DRA   89 min (ort 94 min)   VIDrel: GENESIS   V

RAGE: RING OF FIRE 2 **
Richard W. Munchkin USA
18
1992
Don "The Dragon" Wilson, Maria Ford, Sy Richardson, Dale Jacoby, Eric Lee, Vince Murdocco, Ian Jacklin, Evan Lurie, Charlie Ganis, Ron Yuan, Elena Sahagun, Michael Delano, William Bassett, Diana Phipps, Victoria Hawley, Art Camacho
A sequel of sorts to the first RING OF FIRE movie, with our martial artist witnessing a robbery in which one of the criminals

is killed. This angers the gang-leader, who reacts by kidnapping our hero's girlfriend, forcing him to take matters into his own hands.
Aka: RING OF FIRE 2: BLOOD AND STEEL
MAR   91 min (ort 94 min)   VIDrel: MIA/DISC/IMPENT V

RAGEWAR *
Rose-Marie Turko et al. USA
15
1983
Jeffrey Byron, Leslie Wing, Richard Moll, Blackie Lawless, Danny Dick, Gina Calebrese, Daniel Dion, Bill Bestolarides, Scott Campbell, Ed Dorini, R.J. Miller, Don Moss, Alanna Roth, Kim Connell, Janet Welsh, Carol Soloman
A girl is held hostage by an alien, and a computer expert is given seven challenges to overcome in order to free her. Seven directors worked on this dismal effort. (For the curious, the other six were: John Buechler, Charles Band, David Allen, Steve Ford, Peter Manoogian and Ted Nicolau.)
Aka: DIGITAL KNIGHTS; DUNGEONMASTER, THE
FAN   76 min (ort 80 min)   VIDrel: NTV/TOTAL   V

RAGING BULL ****
Martin Scorsese USA
18
1980
Robert De Niro, Cathy Moriarty, Joe Pesci, Frank Vincent, Nicholas Colasanto, Theresa Saldana, Frank Adonis, Mario Gallo, Frank Topham, Joseph Bono, Lori Anne Flax, James V. Christy, Bernie Allen, Bill Mazer, Mike Miles
Biopic on the prizefighter Jake La Motta, who went from obscurity to fame and then back again. A powerful film with violent fight scenes that may disturb. The central character is remarkably unappealing but De Niro's superb performance and the care lavished on period detail, combine to produce a most engrossing work. Written by Paul Schrader and Mardik Martin. AA: Actor (De Niro), Edit (Thelma Schoonmaker).
DRA   124 min (ort 129 min) B/W/Col wScrn
VIDrel: MGM/WHV   V/sur
Boa: book by Jake La Motta.

RAGTIME ***
Milos Forman USA
15
1981
Elizabeth McGovern, Howard E. Rollins Jr, James Olson, Mary Steenburgen, James Cagney, Pat O'Brien, Norman Mailer, Brad Dourif, Moses Gunn, Donald O'Connor, Ken McMillan, Mandy Patinkin, Jeffrey DeMunn, Robert Joy, Bruce Boa
Film version of a novel that attempted to present a picture of American life in 1906. Fine performances are let down by early concentration on a single strand of the novel to the exclusion of all else – the obsessive search for justice undertaken by a black man, after humiliation at the hands of some racists. This was Cagney's last cinema role, after a 20-year absence. The score is by Randy Newman.
DRA   148 min (ort 155 min)   VIDrel: CASPIC L/A   V
Boa: novel by E.L. Doctorow.

RAID ON ENTEBBE **
Irvin Kershner USA
PG
1977
Peter Finch, Charles Bronson, Horst Buchholz, Martin Balsam, John Saxon, Jack Warden, Sylvia Sidney, Yaphet Kotto, Tige Andrews, Eddie Constantine, Warren Kemmerling, Robert Loggia, David Opatoshu, Allan Arbus, James Woods
Another version of the Entebbe hijack drama and subsequent rescue of the hostages, by a crack unit of Israeli commandos on July 4th 1976 (no thanks to Uganda's dictator Idi Amin). Finch's penultimate film, for which he received only an Emmy nomination (though Bill Butler did get an Emmy award for his cinematography). See also VICTORY AT ENTEBBE and OPERATION THUNDERBOLT.
THR   120 min mTV   VIDrel: VCC L/A   V

RAID ON ROMMEL *
Henry Hathaway USA
15
1971
Richard Burton, John Colicos, Clinton Greyn, Wolfgang Preiss, Danielle Demetz, Karl Otto Alberty, Christopher Cary, John Orchard, Brook Williams, Greg Mullavey, Ben Wright, Michael Sevareid, Chris Anders
A British officer posing as a Nazi, gathers together a group of POWs for an attack on Rommel's fuel dump, intended to be a prelude to a raid on the guns at Tobruk before the Allied invasion of North Africa. A dull dud, filmed in Mexico and with its battle footage borrowed from TOBRUK.
WAR   93 min (ort 99 min) wScrn
VIDrel: 4-FRONT/POLYREC L/A   V

RAIDERS OF THE LOST ARK ****   PG
Steven Spielberg USA   1981
Harrison Ford, Karen Allen, Wolf Kahler, Paul Freeman, Ronald Lacey, John Rhys-Davies, Denholm Elliott, Anthony Higgins, Alfred Molina, Vic Tablian, William Hootkins, Don Fellows, Fred Sorenson, Bill Reimbold, Patrick Durkin
A lavish blockbuster, with Ford as a freelance hunter of lost treasures, sent to find the lost Ark of the Hebrews before the Nazis do. Conceived by George Lucas, Philip Kaufman and Spielberg, with music John Williams and the script by Lawrence Kasdan. Followed by INDIANA JONES AND THE TEMPLE OF DOOM. AA: Art/Set (Reynolds and Dilley/Ford), Edit (Kahn), Sound (Varney et al.), Effects/vis (Edlund et al.), Spec Award (Burtt et al. for sound effects).
A/AD   112 min (ort 115 min)   VIDrel: CIC/SONOP; PION (LV only)   V LV

RAILWAY CHILDREN, THE ***   U
Lionel Jeffries UK   1970
Dinah Sheridan, William Mervyn, Jenny Agutter, Bernard Cribbens, Iain Cuthbertson, Gary Warren, Sally Thomsett, Peter Bromilow, Ann Lancaster, Gordon Whiting, David Lodge, Beatrix Mackey, Deddie Davies, Paddy Ward
Three children who live by a railway line, have various adventures whilst waiting for their father to be found innocent of the charge of spying. An aimless work redeemed by nice locations and a charming feel for the period. The script is by Jeffries.
JUV   104 min (ort 108 min)   VIDrel: WHV   V/h
Boa: novel by E. Nesbit.

RAILWAY STATION MAN, THE ***   (15)
Michael Whyte UK/USA   1992
Julie Christie, Donald Sutherland, Frank MacCusker, Mark Tandy, Ingrid Craighie, John Lynch, Niall Cusack, Maire Hastings, Peadar Lamb, Ann Callanan, Gary Walker, John Craig
After her husband is killed in an IRA bombing, a woman artist tries to rebuild her life and meets a mysterious American who is restoring the local railway station. She finds herself drawn to him and they becomes lovers, but some nasty surprises are in store when she discovers his true purpose. A strong and capable adaptation of Johnston's novel.
DRA   90 min mCab   TVrel: BBC
Boa: novel by Jennifer Johnston.

RAIN KILLER, THE ***   18
Ken Stein USA   1990
Ray Sharkey, David Beecroft, Tanio Coleridge, Maria Ford, Woody Brown
An L.A. detective investigates a series of baffling murders whose only common link is the fact that in each case the victim was a wealthy woman who met her end in the rain. A well-conceived thriller that offers a few rather intriguing plot twists.
THR   90 min (ort 94 min)   VIDrel: COLUM/SONOP   V

RAIN MAN ***   15
Barry Levinson USA   1988
Dustin Hoffman, Tom Cruise, Valeria Golino, Jerry Molden, Jack Murdock, Michael D. Roberts, Ralph Seymour, Lucinda Jenney, Bonnie Hunt, Beth Grant, Kim Robillard, Dolan Dougherty, Marshall Dougherty, John-Michael Dougherty
A brash young car-dealer goes home for his father's funeral and discovers that he has an autistic older brother, who quite unexpectedly has been made sole beneficiary of his father's estate. He kidnaps his brother, intending to have him declared incompetent and get control of the inheritance, but in time a bond of affection grows between them. A contrived but likeable film. AA: Pic, Dir, Actor (Hoffman), Screen/orig (Ronald Bass/Barry Morrow).
DRA   127 min (ort 140 min)   VIDrel: WHV   V/sh

RAINBOW **   U
Jackie Cooper USA   1978
Andrea McArdle, Don Murray, Piper Laurie, Martin Balsam, Michael Parks, Jack Carter, Rue McClanahan, Nicholas Pryor, Donna Pescow, Erin Donovan, Moosie Drier, Johnny Doran, Philip Sterling, Ben Frank, Peggy Walton, Carol Leigh
This follows Judy Garland's career, from her early days as one of three singing sisters up to her role in The Wizard of Oz. Fairly accurate, but McArdle is woefully miscast and has not one tinge of the Garland charisma. An Emmy went to the cinematographer, Howard R. Schwartz.
DRA   73 min (ort 100 min) mTV   VIDrel: L/A   V
Boa: book by Christopher Finch.

RAINBOW, THE ***   15
Ken Russell UK   1989
Sammi Davis, Paul McGann, Amanda Donohoe, Christopher Gable, David Hemmings, Glenda Jackson, Ken Colley, Dudley Sutton, Jim Carter, Judith Paris, Glenda McKay, Mark Owen, Ralph Nossek, Nicola Stephenson
A remarkably restrained film from Russell that serves as a prequel to WOMEN IN LOVE, and follows the career of a naive and sheltered young girl who learns of love and life at the hands of both her teacher and a soldier. This mature study of adolescence and growth is visually pleasing if a little disjointed. Jackson has a part as the mother of the character she played in the earlier work. Screenplay is by Ken and Vivian Russell.
DRA   106 min   VIDrel: FIRST/SONOP L/A   V
Boa: novel by D.H. Lawrence.

RAINBOW ***   PG
Bob Hoskins CANADA/UK   1995
Bob Hoskins, Dan Aykroyd, Saul Rubinek, Jacob Tierney, Willy Lavendal, Jonathan Schuman, Eleanor Misrahi, Jack Fisher, Norris Dominique, Babs Gadbois, Suzan Glover, Larry Day, Jane Gilchrist, Griffith Brewer, Gordon Masten
A youngster stumbles across the spot where the rainbow meets the ground and even gets to have a look inside, but when he investigates further with his friends he finds that someone has stolen the gold from it, and this poses a danger for the planet. With the help of his magician grandfather, he sets out to remedy this. A charming fantasy, with a strong ecological slant, it will be mainly of appeal to very young kids, being a bit too lightweight for adults.
JUV   94 min (ort 101 min)   VIDrel: FIRST/SONOP   V

RAINBOW BRITE AND THE STAR STEALER *   U
Bernard Degries/Kimio Yabuki USA   1985
Voices of: Bettina, Patrick Fraley, Peter Cullen, Robbie Lee, Andre Stojka, David Mendenhall, Rhonda Aldrich, Les Tremayne, Mona Marshall, Jonathan Harris, Marissa Mendenhall, Scott Menville, Charles Adler, David Workman
Animated feature that looks (and sounds) as if it was made on an assembly-line, all about a little girl who must stop the forces of evil from extinguishing the last star.
ANIM   81 min (ort 85 min)   VIDrel: WHV   V/sur

RAINBOW DRIVE **   18
Bobby Roth USA   1990
Peter Weller, Sela Ward, Bruce Weitz
A homicide detective stumbles across the scene of a multiple murder, but when he returns after chasing a suspect, one of the bodies has vanished. As he investigates this mystery, he begins to unearth a cover-up of major proportions.
A/AD   91 min (ort 100 min) mCab   VIDrel: 20VIS/SONOP   V
Boa: novel by Roderick Thorp.

RAINBOW VALLEY **   U
Robert North Bradbury USA   1935
John Wayne, Lucille Browne, LeRoy Mason, George Dillard, Lloyd Ingraham, Lafe McKee, Frank Ellis, Art Dillard, Frank Ball, Fern Emmett, Henry Rocquemore, Eddie Parker, Herman Hack
An undercover agent goes to prison and then escapes in order to penetrate an outlaw gang who have been sabotaging efforts to build a road into a mining area. A capable Wayne oater with a complex plot.
WES   50 min (ort 56 min) B/W   VIDrel: SCRN/DISC   V

RAINING STONES **   15
Ken Loach UK   1993
Bruce Jones, Julie Brown, Gemma Phoeniz, Ricky Tomlinson, Tom Hickey, Mike Fallon, Ronnie Ravey, Lee Brennan, Karen Henthorn, Christine Abbott, Geraldine Ward, William Ash, Matthew Clucas, Anna Jaskolka, Jonathan James, Anthony Bodell
A long-term unemployed man supplements his dole money with odd jobs, for which he depends on his van. But when this is stolen and he has to find money for the clothes his daughter needs for her communion, he resorts to a little cattle rustling and the services of a loan shark. A grim story of working class drudgery, from a director who specialises in such films, yet the tone is slightly lightened by a few dashes of humour. The script is by Jim Allen.
DRA   86 min (ort 91 min) mTV   VIDrel: FIRST/SONOP   V/sur

RAISE RAVENS ***   12
Carlos Saura SPAIN   1975
Geraldine Chaplin, Ana Torrent, Conchita Perez, Maite Sanchez, Hector Alerio
An exploration of a childhood world of fantasy and imagination, that casts Torrent as a nine-year-old girl who tries to make sense of everything around her, having become deeply traumatised by the tragic death of her mother from cancer a few years before. Three major episodes in the child's life are examined, but the movie as a whole cannot sustain its power, despite Torrent's wonderful performance. Winner of the Special Jury Prize at Cannes 1976.
Aka: CRIA CUERVOS; CRIA!
DRA 105 min (ort 115 min) wScrn   VIDrel: ARTPRO/RTM
V

RAISE THE RED LANTERN ***   PG
Zhang Yimou CHINA   1991
Gong Li, Ma Jingwu, Jin Shuyuan, Cao Cuifeng, He Caifei, Kong Lin, Ding Weimin, Cui Zhigang, Chu Xiao, Cao Zhengyin, Zhao Qi
In 1920s China, a young educated woman of nineteen is forced to become the fourth wife of a nobleman and goes to live in his palace, where she discovers a strange enclosed world of unfulfilled expectations and scheming servants. A deliberately slow-paced but highly fascinating study of social relationships that belong to a vanished era.
Aka: DAHONG DENGLONG GAOGAO GUA
DRA 119 min (ort 125 min) wScrn
VIDrel: ELPIC/POLYREC   V
Boa: short story by Su Tong.

RAISE THE TITANIC! *   PG
Jerry Jameson USA/UK   1980
Jason Robards, Richard Jordan, David Selby, Anne Archer, Alec Guinness, J.D. Cannon, Paul Carr, Michael C. Gwynne, Dirk Blocker, Norman Barfold, Bo Brundin, Charles Macauley, Elya Baskin, Harvey Lewis, M. Emmet Walsh
Lew Grade who produced this film, later said that it would have been cheaper to lower the Atlantic. One must agree, this dismal flop tells of the largest salvage operation ever; the ridiculous plot involving the presence on board the title vessel of a rare ingredient used in nuclear weapons. Written by Adam Kennedy and Eric Hughes and severely edited just prior to release but still a good hundred minutes too long.
A/AD 109 min (ort 121 min)   VIDrel: POLY/POLYREC
V/sur
Boa: novel by Clive Cussler.

RAISING ARIZONA ***   15
Joel Coen USA   1987
Nicolas Cage, Holly Hunter, Trey Wilson, John Goodman, William Forsythe, Sam McMurray, Frances McDormand, Randall "Tex" Cobb, T.J. Kuhn, Lynne Dumin Kitei, Peter Benedek, Charles "Lew" Smith, Warren Keith, Henry Kendrick
A rather odd couple decide to kidnap one child from a set of quintuplets as they are unable to have a child of their own. A crazy blend of unfocused energy and wild slapstick sequences, but highly enjoyable in its way. The script is by Ethan and Joel Cohen with the score by Carter Burwell and cinematography by Barry Sonnenfeld.
COM 90 min (ort 94 min)   VIDrel: 20TH/TECH   V/sur

RAISING CAIN ***   15
Brian De Palma USA   1992
John Lithgow, Lolita Davidovich, Steven Bauer, Frances Sternhagen, Gregg Henry, Tom Bower, Mel Harris, Teri Austin, Gabrielle Carteris, Amanda Pombo, Barton Heyman, Kathleen Callan, Ed Hooks, Jim Johnson, Riq Boogie Espinoza
Stylish, violent and extremely nasty tale of twin brothers and their supposedly dead father, a mad Norwegian psychiatrist. Unfortunately, he proves to be very much alive and wants their help to kidnap young children for some insane experiments he has in mind. Lithgow (in no less than five roles) provides the bulk of the interest, in this well-crafted thriller. Screenplay is by the director.
THR 87 min (ort 92 min)   VIDrel: CIC/SONOP   V/dm
V/sur

RAKE'S PROGRESS, THE ***   U
Sidney Gillat UK   1945
Rex Harrison, Lilli Palmer, Godfrey Tearle, Griffith Jones, Margaret Johnston, Guy Middleton, Jean Kent, Marie Lohr, Garry Marsh, David Horne, Alan Wheatley, Brefni O'Rorke, Charles Victor, Joan Maude, Patricia Laffan
In the 1930s a disgraced student and ne'er-do-well merrily continues his irresponsible ways, moving from an affair with a friend's wife to a life as a professional philanderer, but eventually attains some worth when he redeems himself in war. A pleasing and witty light comedy whose virtuous ending is clearly at odds with the general thrust of the narrative.
Aka: NOTORIOUS GENTLEMAN, THE
COM 116 min (ort 123 min) B/W   VIDrel: VCC L/A   V

RAMBLING ROSE **   15
Martha Coolidge USA   1991
Laura Dern, Robert Duvall, Diane Ladd, Lukas Haas, John Heard, Kevin Conway, Robert Burke, Lisa Jakub, Evan Lockwood, Matt Sutherland, D. Anthony Pender, David E. Scarborough, Robin Dale Robertson, General Fermon Judd Jr, Michael Mott
A man returns to his childhood home to visit his father and recalls the days of his youth in the Depression years and Rose, the young woman who came to work in their home. Though sexually experienced, she proves to be rather naive in her dealings with men, a failing which is destined to get her into a good bit of trouble. A restrained and rather over-sentimentalised chronicle of a young woman's coming-of-age. The LV disc includes an interview with the director.
COM 106 min (ort 115 min)   VIDrel: GUILD/POLYREC
L/A; PION (LV only)   V/s LV
Boa: novel by Calder Willingham.

RAMBO: FIRST BLOOD, PART 2 *   15
George Pan Cosmatos USA   1985
Sylvester Stallone, Richard Crenna, Steven Berkoff, Charles Napier, Julia Nickson, Martin Kove, Andy Wood, George Kee Cheung, William Ghent, Steve Williams, Don Collins, Chris Grant, John Sterlini, Alain Hocquenghem
Unbelievable and incredibly violent fantasy about one-man arsenal Stallone, a Vietnam veteran out to rescue POWs held by the Viet Cong. One of many gung-ho adventures produced since the Vietnam War – Chuck Norris and Arnold Schwarzenegger being other valued contributors to this genre. A callous film that totally fails to address the very real problem of those GIs still being held. See also FIRST BLOOD, the film that started the RAMBO series.
A/AD 92 min   VIDrel: POLY L/A   V/sh

RAMBO 3 *   18
Peter MacDonald USA   1988
Sylvester Stallone, Richard Crenna, Marc De Jonge, Kurtwood Smith, Spiros Focas, Sasson Gabai, Doudi Shoua, Randy Raney, Marcus Gilbert, Alon Abutbul, Mahmoud Assadollahi, Joseph Shiloach, Harold Diamond, Seri Mati, Shaby Ben-Aroya
This time the cartoon action is set in Russian-occupied Afghanistan, where our walking one-man arsenal has to rescue a former superior from captivity in an impregnable fortress. An almost non-stop series of loud bangs is a poor substitute for a plot.
A/AD 94 min Cut (3 min 3 sec – ort 101 min)
VIDrel: 4-FRONT/POLYREC   V/sh

RAN ****   15
Akira Kurosawa FRANCE/JAPAN   1985
Tatsuya Nakadai, Akira Terao, Jinpachi Nezu, Ryu Daisuke, Mieko Harada, Hisashi Igawa, Peter (Shinnosuke Ikehata), Hitoshi Veki, Jun Tazaki, Norio Matsui, Hisachi Ikawa, Kenji Kodama, Toshiya Ito, Takeshi Kato, Masayuki Yui
Brutal, bloody and brilliant. Kurosawa's epic version of King Lear is perfectly adapted to Japanese customs and history, and tells of a warlord who turns his kingdom over to his eldest son, a rash action that sows dissension and splits his family. The mutilated LUMI/SPEAR tape is best avoided, despite the inclusion of Chris Marker's short: "A.K. – The Making Of Ran". AA: Cost (Emi Wada).
DRA 71 min (ort 162 min)   VIDrel: LUMI/SPEAR   V
Boa: play King Lear by William Shakespeare.

RANCHO NOTORIOUS ***   PG
Fritz Lang USA   1952
Marlene Dietrich, Mel Ferrer, Arthur Kennedy, Lloyd Gough, Gloria Henry, William Frawley, Jack Elam, George Reeves, Lisa Ferraday, John Raven, Frank Ferguson, Francis McDonald, Dan Seymour, John Kellogg, Redd Redwing
A cowboy in search of the murderer of his girlfriend, arrives at an outlaw's hideout run by Dietrich, and proceeds to upset her

life. An unusual if rather slow Western, with a good script by Daniel Taradash.
WES   89 min   VIDrel: L/A   V

RANDY RIDES ALONE **   U
Harry Fraser USA   1934
John Wayne, Alberta Vaughn, George "Gabby" Hayes, Yakima Canutt, Earl Dwire, Tex Phelps, Artie Ortego, Herman Hack, Mack V. Wright
A drifter finds himself being blamed for robbery and a series of murders, and helped by a young girl, he sets out to clear his name and catch those responsible. A fairly standard offering from Monogram that is considerably enlivened by some nice work from Hayes as the killer.
WES   53 min B/W coVer   VIDrel: ENTUK/GOLD L/A   V

RANGE FEUD **   U
D. Ross Lederman USA   1931
Buck Jones, Susan Fleming, John Wayne, Harry Woods, Glenn Strange, Frank Austin, Lew Meehan, Jim Corey, Frank Ellis, Bob Reeves
A young man is falsely accused of killing a rancher and is arrested by his foster brother, the local sheriff, whose prospective father-in-law was the dead man. An OK Buck Jones oater with Wayne in a supporting role.
WES   58 min B/W   VIDrel: VCC/DISC/COLUM   V

RANSOM *   PG
Caspar Wrede UK   1974
Sean Connery, Ian McShane, Norman Bristow, John Cording, Isabel Dean, William Fox, Robert Harris, Richard Harrison, Harry Landis, Preston Lockwood, James Maxwell, John Quentin, Jeffrey Wickham, Knut Hansson
The British ambassador to a Scandinavian country is kidnapped by terrorists and has to be rescued. An unenthralling adventure thriller offering little excitement, though the splendid photography (by Sven Nykvist, a long-time associate of Ingmar Bergman) is a slight compensation. Filmed in Norway.
Aka: TERRORISTS, THE
THR   90 min (ort 98 min)   VIDrel: WHV L/A   V/h

RANSOM ***   15
Ron Howard USA   1996
Mel Gibson, Rene Russo, Brawley Nolte, Gary Sinise, Delroy Lindo, Lili Taylor, Liev Schreiber, Donnie Wahlberg, Evan Handler, Nancy Ticotin, Michael Gaston, Kevin Neil McCready, Paul Guilfoyle, Allen Bernstein, Jose Zuniga
A self-made millionaire who runs an airline has all the comforts his success can bring, has his young son kidnapped one day, whilst the family are out in the park. Ransom demands are received, but the FBI mess up the payment and the father goes nationwide on TV, offering the ransom payment to anyone who brings in the kidnappers. The are very few surprises in this super-slick but (unusually for Howard) vicious offering. Fortunately, a strong cast sustains interest.
THR   121 min   CINrel

RAPA-NUI ***   12
Kevin Reynolds USA   1993
Jason Scott Lee, Esai Morales, Sandrine Holt, Zac Wallace, George Henare, Eru Potaka-Dewes, Nathaniel Lees, Pete Smith, Rawiri Paratene, Emilio Tuki Hito, Gordon Hatfield, Faenza Reuben, Hori Ahipene, Chiefy Elkington, Huihana Rewa
Ambitious if not entirely successful attempt to create a historical drama set in the 18th century on Easter Island, that revolves around the events that led to the erection of its famous giant statues, and the rivalry between two men who compete for the hand of the same woman, the loser facing execution.
DRA   102 min (ort 107 min)   VIDrel: EIV/SONOP   V/sur

RAPE OF DR WILLIS, THE **   15
Lou Antonio USA   1991
Jaclyn Smith, Holland Taylor, Robin Thomas, Gregg Henry, Lisa Jakub, Sarah Hynley, Tina Lifford, Richard Balin, Bobby Brett, John Hawkins, Eddie De Harp Larry Dobkin, Harry Melching, Bennett Liss, Andrew Maler, Paul Hewitt
When her husband dies, a female doctor decides to start a new life and takes a post as resident surgeon at a hospital. Not only does she suffer a rape in the hospital car park on her first day at work, but has to save the life of her assailant when he is shot in the course of a police car chase a few days later. Perhaps this might have worked as a play, but here the contrived nature of the script seriously hampers this glossy soap opera-style effort.
DRA   88 min   VIDrel: ODY/SONOP   V/sh

RAPE OF EDEN **   (PG)
Sam Auster USA   1992
Phil Nordell, Francine Lapense, Jeff Conaway, Vernon Wells, Vallie Ullam, Allan Eu, Polly Firestone, Lance Buckman, Michael O'Connor, Mike Ciccolini, Mike Fitzpatrick, Jer E. Lee, Chuck Brewster, Sam Selvaggio, Marlan Glenn
After a deadly virus results in the death of almost the entire population of the world, a single uninfected female remains alive, and naturally, she has become a highly prized person. One of those standard post-apocalyptic action fantasies, adequate enough in its way.
Aka: BOUNTY HUNTER 2002
FAN   98 min   CABrel: HVC

RAPE OF THE VAMPIRE, THE **   18
Jean Rollin FRANCE   1967
Solange Pradel, Bernard Letrou, Ursule Pauly, Ariane Sapriel, Marquis Polho, Barbara Girard
Incomprehensible and unscary, despite good camerawork, this started life as a short that was made to accompany an American vampire movie. Expanded and then released as a feature, it purports to be about two young girls who may possibly be vampires, but the slight plot takes very much second place to a succession of dreamlike and bizarre images, in-jokes and meaningless references. Weird enough to now have a cult following, it is in truth a memorable, failed experiment.
Aka: LA REINE DES VAMPIRES; LE VIOL DU VAMPIRE; QUEEN OF THE VAMPIRES; VAMPIRE WOMEN, THE
HOR   90 min B/W wScrn   VIDrel: REDEM/RTM   V

RAPID FIRE **   18
Dwight H. Little USA   1992
Brandon Lee, Powers Boothe, Nick Mancuso, Raymond J. Barry, Kate Hodge, Tzi Ma, Tony Longo, Michael Pual Chan, Dustin Nguyen, Brigitta Stenberg, Basil Wallace, Al Leong, Francois Chau, Quentin O'Brian, D.J. Howard, Maurice Chasse
An art expert and martial artist in Chicago finds his work cut out when he is called in by the police to help stop the violent struggle between Italian and Asian gangs over the drugs trade. There is plenty of the stock-in-trade of every martial arts movie by way of great action sequences, and this plus the lively performances (and a welcome touch of humour) make this an enjoyable if mindless outing.
MAR   91 min (ort 95 min)   VIDrel: 20TH/TECH L/A   V

RAPTURE, THE ***   18
Michael Tolkin USA   1991
Mimi Rogers, David Duchovny, Patrick Bauchau, Kimberly Cullum, Terri Hanauer, Dick Anthony Williams, James Le Gros, Carole Davis, Will Patton, Darwyn Carson, Marvin Elkins, Stephanie Menuez, Sam Vlahos, Rustam Branaman
Writer-director Tolkin makes his debut with an exploration of West Coast American fundamentalism, that takes as its focus the story of the wanton lifestyle of a bored and emotionally unfulfilled telephonist. Promised salvation by a fundamentalist group, her faith grows. But when her husband is murdered some years later, she becomes convinced that she has a special mission, and her belief has a tragic result. A flawed, yet hypnotic work.
DRA   96 min (ort 110 min) cC
VIDrel: ENCORE/SPEAR/COLUM   V/sur

RASHOMON ***   12
Akira Kurosawa JAPAN   1950
Toshiro Mifune, Machiko Kyo, Masayuki Mori, Takashi Shimura, Scinobu Hascimoto, Minoru Chiaki, Kichijiro Ueda, Fumiko Homma, Daisuke Kato
In medieval Japan, a merchant and his wife travelling through a forest encounter a bandit, who rapes the wife and murders the husband. This incident is now retold in four different versions, initially by each of the protagonists. A vivid and fascinating film that established its director's international reputation. Remade poorly as a Western – "The Outrage". AA: Hon Award (most outstanding foreign language film released in the USA during 1951).
DRA   86 min (ort 90 min) B/W   VIDrel: CONNO/RTM   V
Boa: story by Ryunosuke Akutagawa.

RASPUTIN: THE MAD MONK *   15
Don Sharp UK   1966
Christopher Lee, Barbara Shelley, Richard Pasco, Francis Matthews, Suzan Farmer, Dinsdale Landen, Renee Asherson, Joss Ackland, Robert Duncan, John Bailey

A highly fictionalised account of the career of Rasputin, the notorious former monk, who for a time attained great influence at the Russian court by claiming, among other things, the power to heal. Lee throws himself into his role with great gusto, and this is some slight compensation for a muddled script and inept direction.
DRA   87 min (ort 91 min)   VIDrel: LUMI/SPEAR L/A   V

RAT PFINK A BOO-BOO *   18
Ray Dennis Steckler USA   1966
Vin Saxon, Carolyn Brandt, Titus Moede, George Caldwell, James Bowie, Mike Cannon, Keith Webster, Romeo Barrymore, Bob Burns, Dean Danger (narration only)
An abysmally inept film, that starts out as a thriller but turns into pure corny slapstick, as the two title superheroes attempt to rescue the kidnapped girlfriend of a singing idol. Despite the sometime cult status of the director, most will find this offering tedious in the extreme.
Aka: ADVENTURES OF RAT PFINK A BOO-BOO; RAT FINK AND BOBO; RAT FINK AND BOO-BOO
COM   69 min (ort 90 min) B/W   VIDrel: RTM/VCC L/A V

RATBOY **   PG
Sondra Locke USA   1986
Sondra Locke, Sharon Baird, Robert Townsend, Sydney Lassick, Christopher Hewitt, Gerrit Graham, Larry Hankin, Louie Anderson, S.L. Baird, Billie Bird, Gordon Anderson
A gentle, rodent-like alien is discovered by a department store window-dresser, who attempts to turn him into a media celebrity and exploit his notoriety by becoming his manager. Locke's directorial debut is a curious blend of comedy and pathos, that has no clear message but a powerful charge of earnestness. Written by Rob Thompson.
DRA   100 min Cut (3 sec – ort 104 min)
VIDrel: MGM/WHV L/A   V

RATS: NIGHT OF TERROR *   18
Vincent Dawn (Bruno Mattei)/Clyde Anderson
FRANCE/ITALY   1984
Richard Raymond, Richard Cross, Cindy Leadbetter, Janna Ryan, Alex McBride, Ann Gisel Glass, Christoph Brenner, Tony Lombardo, Chris Fremont, Moune Duvivier
Post-nuclear holocaust time again folks – now the survivors find that an uncontaminated city has been over-run by rats. A schlock-horror film from a famed director of many.
Aka: RATS; RATS: NOTTE DI TERROR
HOR   92 min (ort 98 min)   VIDrel: ARTPRO/RTM   V

RATTLED *   PG
Tony Randel USA   1996
William Katt, Shanna Reed, Ed Lauter, Bibi Besch
When quarrying operations lead to the destruction of a massive nest of snakes, hundred of these creatures invade a nearby community. An interminable B-movie that tries to does for reptiles what SLUGS did for molluscs, it never amount to much and only very occasionally delivers the promised chills.
HOR   85 min (ort 90 min) mTV cC   VIDrel: CIC   V/dm

RAVEN, THE ***   15
Roger Corman USA   1963
Boris Karloff, Vincent Price, Peter Lorre, Hazel Court, Jack Nicholson, Olive Sturgess, Connie Wallace, William Baskin, Aaron Saxon
Horror spoof about a contest that takes place between two powerful wizards. Set in the 15th century, this is one of the few Corman films that actually succeeds in telling a credible tale; one that's inspired by an Edgar Allan Poe poem. Richard Matheson wrote the screenplay.
HOR   86 min   VIDrel: L/A   V

RAVEN **   18
Russell Solberg USA   1996
David Ackroyd, Burt Reynolds, Matt Battaglia, Krista Allen
Unusually, Reynolds is cast against type playing a baddie, who is busy eliminating the folk who hired him to steal a top-secret decoder in Bosnia. But his former sidekick (Battaglia) now wants out. A competent time-filler, with a neat twist towards the end.
A/AD   90 min   VIDrel: MARQ   V/s

RAW DEAL *   18
John Irvin USA   1986
Arnold Schwarzenegger, Kathryn Harrold, Darren McGavin, Sam Wanamaker, Paul Shenar, Steven Hill, Joe Regalbuto, Robert Davi, Blanche Baker, Ed Lauter, Mordecai Lawner, Robey, Victor Argo, George Wilbur, Denver Mattson, John Malloy
Schwarzenegger plays a retired agent, blackmailed by the FBI into going after a Chicago Mob family. Plenty of gore as Arnie wipes 'em out, but the film suffers badly from its comic book approach. A few flashes of humour are a poor compensation.
A/AD   101 min (ort 106 min)   VIDrel: BMGVID/BMGREC V/sh

RAW NERVE **   18
David A. Prior USA   1990
Glenn Ford, Sandahl Bergman, Randall "Tex" Cobb, Ted Prior, Traci Lords, Jan-Michael Vincent, Red West, Graham Timbes, Jerry Douglas Simms, Yvonne Stancil, Doris Hearn, Trevor Hale, Brian J. Scott, Jim Aycock, Ken Kennedy
A dangerous psychopath who has already murdered seven women is still at large, but a racing driver who has gained psychic powers believes he can catch him. With the police unwilling to take him seriously, he gains a female journalist as an ally.
THR   91 min   VIDrel: 20VIS/SONOP   V/s

RAWHEAD REX **   18
George Pavlou EIRE/UK   1986
David Dukes, Kelly Piper, Niall Toibin, Niall O'Brian, Ronan Wilmot, Hugh O'Connor, Heinrich Von Schellendorf, Cora Lunny, Donal McCann, Gladys Sheehan, Madely Erskine, Gerry Walsh, Noel O'Donovan, John Olohan, Peter Donovan
An old-fashioned horror film, in which a farmer ploughing his field accidentally releases a huge demon who was trapped below ground for centuries. Having gained his freedom, our monster embarks on a rampage until a very nice American arrives to deal with him. The screenplay is by Clive Barker, in this moderately enjoyable effort that would have benefited from a better climax and a more convincing monster.
Aka: RAWHEAD
HOR   85 min (ort 103 min)   VIDrel: FIRST/SONOP L/A   V
Boa: novel by Clive Barker.

RAZORBACK **   18
Russell Mulcahy AUSTRALIA   1983
Gregory Harrison, Bill Kerr, Judy Morris, Arkie Whiteley, Chris Haywood, David Argue, John Howard, John Ewart, Mervyn Drake, Alan Beecher, Redmond Phillips, Peter Schwartz, Beth Child, Rick Kennedy, Chris Hession
Music video director Russell Mulcahy's directing debut is a story of a female animal rights campaigner who vanishes in the outback. Her husband sets out to find her, coming up against a monstrous wild boar that is terrorising the countryside. A hammy version of JAWS.
A/AD   91 min (ort 95 min)   VIDrel: WHV L/A   V/sur
Boa: novel by Paul Brennan.

RAZOR'S EDGE, THE **   PG
Edmund Goulding USA   1946
Tyrone Power, Gene Tierney, Clifton Webb, Herbert Marshall, John Payne, Anne Baxter, Lucile Watson, Frank Latimore, Elsa Lanchester, Fritz Kortner, John Wengraf, Cecil Humphreys, Harry Pilcer, Cobina Wright Jr, Albert Petit
Sickened by the slaughter he witnessed in WW1, a well-to-do young man leaves Chicago and travels the world in a spiritual quest for the meaning of life. Scripted by Lamar Trotti and remade in 1984, this ambitious adaptation of Maugham's philosophical novel is a rambling affair that captivates in parts, but is ultimately defeated by the emptiness at the heart of the original novel. AA: S. Actress (Baxter).
DRA   146 min B/W   VIDrel: 20TH/TECH   V
Boa: novel by Somerset Maugham.

RE-ANIMATOR **   18
Stuart Gordon USA   1985
Jeffrey Combs, Bruce Abbott, Barbara Crampton, Robert Sampson, David Gale, Gerry Black, Carolyn Purdy-Gordon, Peter Kent, Barbara Pieters, Ian Patrick Williams, Bunny Summers, Al Berry, Derek Pendleton, Gene Scherer, James Ellis
A grisly black comedy not without its moments of humour, as a young man experiments with a fluid that is able to re-animate the dead. Undeniably gross and overblown in terms of its effects, this film could never be accused of showing lightness of

touch. Gordon's directing debut. Written by Gordon, Dennis Paoli and William J. Norris. A sequel followed.
COM   81 min Cut (1 min 42 sec – ort 96 min)
VIDrel: POLY/POLYREC L/A   V
Boa: short stories in the series Herbert West – The Reanimator by H.P. Lovecraft.

RE-ANIMATOR 2 **   18
Brian Yuzna USA   1990
Jeffrey Combs, Bruce Abbott, David Gale, Kathleen Kinmont, Claude Earl Jones, Fabiana Udenio, Mel Stewart, Michael Strasser, Irene Forrest, Mary Sheldon, "Friday", Margie Turner, Johnny Legend, David Bynum, Noble Craig
Sequel to RE-ANIMATOR with our mad inventor moving on to greater things and creating living beings from dismembered limbs in the expected haphazard fashion. Inevitably his creations turn on him, but the flow of ideas has by now dried up and a few desperate attempts at black humour are all that is left to sustain the film.
Aka: BRIDE OF RE-ANIMATOR; BRIDE OF THE RE-ANIMATOR
HOR   93 min (ort 99 min)   VIDrel: POLY/POLYREC   V/sh

REACH FOR THE SKY **   U
Lewis Gilbert UK   1956
Kenneth More, Muriel Pavlow, Lyndon Brook, Alexander Knox, Sydney Tafler, Lee Patterson, Dorothy Alison, Howard Marion Crawford, Jack Watling, Michael Warre, Nigel Green, Anne Leon, Walter Hudd, Eddie Byrne, Charles Carson
The story of WW2 flying ace Douglas Bader, who put up a heroic struggle to prove himself capable of returning to active service, after he lost both his legs in a crash in 1931. He eventually succeeded but was shot down and imprisoned by the Germans, remaining in captivity for the duration of the war, despite making numerous attempts to escape. A sincere but stilted affair that is more boring than inspiring.
WAR   130 min (ort 136 min) B/W   VIDrel: CARL/TECH   V
Boa: book by Paul Brickhill.

REAL GLORY, THE ***   PG
Henry Hathaway USA   1939
Gary Cooper, Reginald Owen, David Niven, Andrea Leeds, Kay Johnson, Vladimir Sokoloff, Broderick Crawford, Henry Kolker, Charles Waldron, Russell Hicks, Roy Gordon, Benny Inocencio, Rudy Robles, Henry Kolker, Tetsu Komai, Elvira Rios
Shortly after the Spanish-American war of 1898, a group of Moro Moslem tribesmen in the Philippines rose up against the US-led forces, and fought so fiercely in their attacks that they seemed almost invulnerable to US firepower. An American forces doctor soon finds himself in the thick of the fighting, in a well-staged war film whose combat scenes are highly realistic.
WAR   92 min (ort 96 min) B/W
VIDrel: VCC/DISC/COLUM   V

REAL KUNG FU OF SHAOLIN: PART 1 *   18
C.Y. Yang HONG KONG   1986
Sing Lung, Kong Lai Lai, Ngan Lung, Yung Wai Lam
An undistinguished and mediocre kung fu actioner with the standard plot elements, as a young boy haunted by his father's murder joins the monks of the Shaolin Temple.
MAR   89 min   VIDrel: IMPENT   V

REAL McCOY, THE **   15
Russell Mulcahy USA   1993
Kim Basinger, Val Kilmer, Terence Stamp, Gailard Sartain, Zach English, Raynor Scheine, Deborah Hobart, Pamela Stubbart, Andy Stahl, Dean Rader-Duval, Nomran Max Maxwell, Marc Macauley, Peter Turner, David Dwyer, Frank Roberts
Unimpressive actioner with Bassinger cast as a top-rate cat burglar who is anxious to retire. However, the mob force her into pulling one last job by kidnapping her son, and, to make matters worse, lumber her with a totally inept partner. An unwholesome blend of comedy and violence, clearly made to a standard formula and offering very little in the way of plot or acting.
A/AD   100 min (ort 106 min)
VIDrel: 4-FRONT/POLYREC/GUILD   V/sh

REAL MEN *   15
Dennis Feldman USA   1987
James Belushi, John Ritter, Barbara Barrie, Bill Morey, Isa Anderson, Mark Herrier, Gail Barle, Marian Brooks, Marian Dobson, Steven Corvin, Charles Walker, Dyanne Thorne, Don Dolan, Mary E. Thompson, Suzee Slater, James LeGros

An unfunny spy spoof, with a secret agent on a mission to save the world from a deadly toxin, eventually obtaining the antidote from some friendly aliens.
COM   82 min (ort 86 min)   VIDrel: MGM/WHV L/A   V

REAL MEN EAT KEISHA **   18
Adele Robbins USA   1990
Keisha, Tess Ferre, Kristara Barrington, Tamara Longly, Randy West, Jerry Butler, Sharon Mitchell, Buddy Love, Joey Silvera
A man relates his fantasies to a female sex therapist, telling her of an imaginary Oriental maid, who is the only woman able to arouse him. However, at the end of the session she has of course succeeded in seducing him – and it's all part of his therapy. A dullish sex-fantasy.
Aka: MEGA JUGS
A   60 min (ort 75 min)   VIDrel: QUANT/TOTAL   V

REALITY BITES **   12
Ben Stiller USA   1993
Winona Ryder, Ethan Hawke, Ben Stiller, Janeane Garofalo, Steve Zahn, Joe Don Baker, Swoosie Kurtz, John Mahoney, Harry O'Reilly, Susan Norfleet, Keith David, David Pirner, Kevin Pollak, Karen Duffy, Helen Childress, Evan Dando
The story of some twenty-year-olds, who enter the real world after they leave college, much of the plot revolving around the love life of a girl who thinks she has found Mr Right. But her friend is not so sure, mainly because he is in love with her himself. Meanwhile, the other characters cope as best they can with the problems in their lives, and their unfulfilled hopes and dreams. A very 1990s romantic comedy, intermittently absorbing but also a little sour.
COM   94 min (ort 99 min) cC   VIDrel: CIC/SONOP   V/sur

REAP THE WILD WIND ***   PG
Cecil B. De Mille USA   1942
John Wayne, Susan Hayward, Ray Milland, Paulette Goddard, Raymond Massey, Lynne Overman, Charles Bickford, Walter Hampden, Louise Beavers, Elisabeth Risdon, Martha O'Driscoll, Hedda Hopper, Victor Kilian, Oscar Polk, Ben Carter
Fine, overblown 19th century adventure tale, set among salvagers in Georgia who fight over the same Southern belle. Solid entertainment, with excellent underwater scenes just one of its many attractive features. AA: Effects (vis – Gordon Jennings/Farciot Edouart/L. Pereira/aud – Louis Mesenkop).
A/AD   123 min B/W   VIDrel: 4-FRONT/POLYREC/CIC   V

REAR WINDOW ***   PG
Alfred Hitchcock USA   1954
James Stewart, Grace Kelly, Wendell Corey, Thelma Ritter, Raymond Burr, Judith Evelyn, Ross Bagdasarian, Georgine Darcy, Sara Berner, Frank Cady, Irene Winston, Jesslyn Fax, Harris Davenport, Marla English, Alan Lee
A news photographer confined to bed with a broken leg, spends his time looking through his window over to the windows of the rooms on the other side of the building, but this harmless pursuit changes when he witnesses a murder. A fascinating idea that's severely hampered by the restriction of action and location, though it is to Hitchcock's credit that he is able to maintain the tension so well. Written by John Michael Hayes, with music by Franz Waxman.
THR   107 min (ort 112 min)   VIDrel: CIC/SONOP L/A   V/h
Boa: novel by Cornell Woolrich.

REBECCA ****   PG
Alfred Hitchcock USA   1940
Laurence Olivier, Joan Fontaine, George Sanders, Judith Anderson, Nigel Bruce, Reginald Denny, C. Aubrey Smith, Gladys Cooper, Philip Winter, Edward Fielding, Florence Bates, Melville Cooper, Leo G. Carroll, Forrester Harvey
A country house holds some strange secrets that torment the life of the young second wife of a Cornish landowner. Hitchcock's first American film has a fine cast and photography. Remade after a fashion as "Under Capricorn" by the same director and then again in 1983. Screenplay is by Robert E. Sherwood and Joan Harrison, with music by Franz Waxman. AA: Pic, Cin (George Barnes).
DRA   130 min B/W   VIDrel: VCC/DISC; PION (LV only) V LV
Boa: novel by Daphne du Maurier.

REBECCA ***
UK
Charles Dance, Diana Rigg, Faye Dunaway, Emilia Fox, Timothy West, Geraldine James

PG
1995

Passion and jealousy are the order of the day when a widower brings his new wife back home, upsetting the housekeeper who was devoted to the first wife. A lavish remake of the famous story that manages to stay quite faithful to the text of the novel but was unaccountably not shot where the story is set – in Cornwall. A pity, as the rugged Cornish landscapes would have added immeasurably to the atmosphere of this haunting drama.
DRA   206 min   VIDrel: CARL/TECH   V
Boa: novel by Daphne Du Maurier.

REBECCA OF SUNNYBROOK FARM **
Allan Dwan USA
Shirley Temple, Randolph Scott, Gloria Stuart, Jack Haley, Phyllis Brooks, Helen Westley, Slim Summerville, Bill Robinson, J. Edward Bromberg, Alan Dinehart, Dixie Dunbar, Paul Hurst, William Demarest, Ruth Gillete, Paul Harvey

U
1938

A radio presenter finds a child star only to lose her again, but they are re-united once more. Not to be compared to the 1921 silent (with Mary Pickford) or the 1932 remake, this version does not follow Wiggin's novel at all. Screenplay is by Karl Tunberg and Don Ettinger.
COM   77 min (ort 80 min) B/W   VIDrel: 20TH/TECH L/A V
Boa: novel by Kate Douglas Wiggin.

REBECCA'S DAUGHTERS **
Karl Francis UK/WEST GERMANY
Paul Rhys, Peter O'Toole, Joely Richardson, Keith Allen, Simon Dormandy, Dafydd Hywel, Sue Roderick, Huw Ceredig, Desmond Barritt, Eiry Palfrey, William Lawford, Gwenllian Davies, Clive Merrison, Ray Gravell, Jack Walters, Neil Caple

15
1991

In 1843, a group of irate Welsh villagers dress as women as a way of avoiding a local landowner's toll-gate charges. Based on a screenplay by Dylan Thomas, this is a broad and lively farce, that is so very silly it is hard not to like it, even if it plays more like a stage pantomime than as a film. O'Toole's appearance in drag as an inebriated Queen Elizabeth I is a highlight.
COM   93 min (ort 97 min)   VIDrel: CURZON/20TH   V/sur
Boa: screenplay by Dylan Thomas.

REBEL *
Robert Allen Schnitzer USA
Sylvester Stallone, Antony Page, Rebecca Grimes, Roy White, Henry G. Sanders, Vickey Lancaster, Barbara Lee Govan

15
1973

A former terrorist has a change of heart and sets out on a lone crusade against those who manufacture weapons, but his good intentions are marred by his destructive methods. In 1998 the film was re-dubbed with a new soundtrack a released as "A Man Called Rainbo" (at ninety-eight minutes), the whole project being a failed attempt to make it into something of a joke, a bit like Woody Allen's WHAT'S UP, TIGER LILY?
Aka: MAN CALLED RAINBO, A; NO PLACE TO HIDE
A/AD   90 min (ort 86 min)
VIDrel: 4-FRONT/POLY/POLYREC   V/sur

REBEL ROUSERS, THE **
Martin B. Cohen USA
Cameron Mitchell, Jack Nicholson, Bruce Dern, Diane Ladd, Harry Dean Stanton, Neil Burstyn, Lou Procopio, Earl Finn, Phil Carey, Robert Dix, Jim Logan, Sid Lawrence, Johnny Cardes

18
1967

A road-movie that captures the spirit of the motorcycle gangs of the 1960s, with Dern holding a drag race with Mitchell to see who gets to have the girlfriend of the former. A mildly engaging and slightly amusing offering.
Aka: LIMBO; REBEL WARRIORS
DRA   73 min (ort 80 min)
VIDrel: 4-FRONT/POLYREC/MIA L/A   V

REBEL WITHOUT A CAUSE ****
Nicholas Ray USA
James Dean, Natalie Wood, Sal Mineo, Jim Backus, Ann Doran, Corey Allen, Edward Platt, Dennis Hopper, William Hopper, Rochelle Hudson, Nick Adams, Jack Simmons, Marietta Canty, Jack Grinnage, Beverly Long, Steffi Sidney, Tom Bernard

PG
1955

The film that propelled Dean to stardom. He plays an unhappy, insecure youth alienated from his parents. He hangs around with the local youths and causes the death of one in a car stunt. A classic story of youthful alienation that was a well-deserved

hit, Dean's wonderful portrayal is supported by a fine cast. Written by Stewart Stern.
DRA   106 min (ort 111 min) wScrn   VIDrel: WHV   V/s
Boa: novel Children of the Dark by I. Schulman.

RECKLESS KELLY **
Yahoo Serious AUSTRALIA
Yahoo Serious, Hugo Weaving, Melora Hardin, Alexei Sayle, John Pinette, Kathleen Freeman, John Pinette, Bob Maza, Martin Ferrero, Athony Ackroyd, Tracy Mann, Max Walker, Adam Bowen, Warren Coleman, Tyler Coppin, J. Andrew Bilgore

PG
1993

When his island paradise is threatened by developers, a Australian comes to the USA intending to rob a number of banks for the cash needed to save it. A weak albeit high-spirited spoof, very loosely inspired by the figure of Ned Kelly.
COM   76 min (ort 81 min) cC   VIDrel: WHV   V/sur

RECTOR'S WIFE, THE ***
Giles Foster UK
Lindsay Duncan, Ronald Pickup, Miles Anderson, Jonathan Coy, Joyce Redman, Lucy Dawson, Mary Macleod, Prunella Scales, Orla Brady, Rynagh O'Grady, Marie Louise McKenzie, Simon Fenton, Gabrielle Lloyd, Carol MacReady

15
1993

Story of an unhappy ill-fated marriage between a rector and his wife, who finds herself obliged to take a menial job to earn extra money when her husband fails to win the post he had expected. A sad little tale, of endearing performances and a downbeat ending.
DRA   198 min (ort 211 min) mTV   VIDrel: POLY/POLYREC V/sh
Boa: novel by Joanna Trollope.

RED BADGE OF COURAGE, THE ****
John Huston USA
Audie Murphy, Bill Mauldin, Douglas Dick, Royal Dano, John Dierkes, Andy Devine, Arthur Hunnicutt, Robert Easton Burke, Smith Ballew, Glenn Strange, Dan White, Frank McGraw, Tim Durant, Emmett Lynn, Stanford Jolley, House Peters

U
1951

An excellent drama dealing with the American Civil War, as seen through the eyes of a young and uncertain hero. Good performances and realistic battle scenes make this a powerful yet uncliched attack on the suffering caused by war. The troubled production of this film is told in Lillian Ross's book, "Picture". Vastly under-rated at the time, distributors were unwilling to handle it (the film showed American soldiers running away). Remade for TV in 1974.
WES   67 min (ort 69 min) B/W   VIDrel: MGM L/A   V
Boa: novel by Stephen Crane.

RED BALLOON, THE *****
Albert Lamorisse FRANCE
Pascal Lamorisse, Georges Sellier, Wladimir Popof

U
1955

On the way to school, a lonely little boy finds a balloon which has a life of its own. It becomes his faithful companion and when its life is cruelly cut short, the child receives unexpected comfort. First shown in Cannes (where it drew a standing ovation) this exquisite "silent" fantasy is one of the great film shorts. Written by Lamorisse and photographed by Edmond Sechan. See also STOWAWAY IN THE SKY. AA: Screen/orig (Albert Lamorisse).
Aka: LE BALLON ROUGE
JUV   34 min (ort 36 min)   VIDrel: BRAVE/SONOP L/A   V

RED BLOODED ***
David Blyth USA
Kari Salin, Kristoffer Ryan, Ryan Winters, Burt Young, Nicholas Pasco, David Keith

18
1996

A naive student comes to the rescue when he sees a hooker apparently about to be raped by a group of men, one of whom is the father who abused her as a child. This pair take off in his car, but with dad in hot pursuit. And to make matters worse, he is unaware than his companion (Salin) is so traumatised by her childhood that she has become a callous and manipulative psychopath. It is their relationship that makes this violent and exploitive road-movie so interesting.
THR   83 min   VIDrel: MED/20VIS   V/sh

RED BLOODED AMERICAN GIRL **
David Blyth USA
Andrew Stevens, Heather Thomas, Christopher Plummer, Kim Coates, Lydie Denier

18
1990

Disgraced for having produced "designer drugs" a brilliant genetic scientist takes a well paid job at a prestigious research

institute, only to find that its suave director is keeping a top secret project under wraps. When he inadvertently develops a drug that causes vampirism, he is obliged to find an antidote, in a frantic race against time. A tepid and aimless SF thriller that generates little suspense from its interesting premise.
THR 89 min VIDrel: 20VIS/SONOP L/A V

RED DAWN *   15
John Milius USA   1984
Charlie Sheen, Patrick Swayze, C. Thomas Howell, Lea Thompson, Powers Boothe, Ben Johnson, Harry Dean Stanton, William Smith, Vladek Sheybal, Ron O'Neal, Darren Dalton, Jennifer Grey, Brad Savage, Doug Toby, Roy Jensen
Ridiculous film with a real bubble-gum plot – Soviet forces invade the USA only to be eventually defeated by teenage guerillas. A violent and poorly thought out effort.
A/AD 109 min (ort 114 min) VIDrel: MGM/WHV V/sur

RED DESERT, THE *   15
Michelangelo Antonioni FRANCE/ITALY   1964
Monica Vitti, Richard Harris, Carlo Chionetti, Xenia Valderi, Rita Renoir, Aldo Grotti, Valerio Beroleschi, Giuliano Missirini, Lili Rheims, Emanuela Pala Carboni, Bruno Borghi, Beppe Conti, Giuli Cotignoli, Hiram Mino Madonia
An unhappy and suicidal young woman has a brief fling with a friend of her husband's, but this fails to relieve her depression. Despite the imaginative use of colour (this was the director's first such film) the sluggish and banal script does not make for a rewarding or convincing film.
Aka: IL DESERTO ROSSO; LE DESERT ROUGE
DRA 111 min (ort 116 min) VIDrel: CONNO/RTM V

RED DUST **   PG
Victor Fleming USA   1932
Clark Gable, Jean Harlow, Mary Astor, Gene Raymond, Doanld Crisp, Tully Marshall, Forrester Harvey, Willie Fung
Romantic melodrama set on an Indo-China rubber plantation whose overseer is pursued by the wife of a young engineer but falls himself for a stranded hooker, who is on the run from the Saigon authorities. A robust and well made romance that gives its two lead plenty of scope to make the most of their rather limited script.
DRA 79 min (ort 83 min) B/W VIDrel: MGM/WHV V
Boa: play by Wilson Collison.

RED FIRECRACKER, GREEN FIRECRACKER **   15
He Ping CHINA/HONG KONG   1993
Ning Jing, Wu Gang, Zhao Xiao Rui, Gao Yang, Xu Zhengyun, Zhao Liang, Ju Xingmao, Li Yushen, Lu Hui, Wang Liyuan, Zhang Bolin
At the turn-of-the-century, a young woman inherits her family's fireworks factory, but only because of the lack of a male heir. Despite her wishes, she is treated as a man and forbidden ever to marry, which places her in a difficult position when she falls in love with one of her employees. Leisurely paced but quite fascinating, with first-rate acting and camerawork.
Aka: PAODA SHUANG DENG
DRA 112 min (ort 117 min) VIDrel: ELPIC/POLYREC V
Boa: novel by Feng Jicai.

RED HEAT *   18
Walter Hill USA   1988
Arnold Schwarzenegger, James Belushi, Peter Boyle, Ed O'Ross, Gina Gershon, Larry Fishburne, Richard Bright, Oleg Vidov, Brent Jennings, Pruitt Taylor Vance, J.W. Smith, Gretchen Palmer, Michael Hagerty, Brion James, Peter Jason
A Chicago cop is teamed up with a Russian detective, in order to capture a Russian drugs dealer. Standard guns-and-oaths action film, whose only claim to posterity is that it represents the first time the Soviet authorities allowed an American camera-team to film in Red Square.
A/AD 99 min (ort 106 min)
VIDrel: VCC/DISC/COLUM V/sur

RED KING, WHITE KNIGHT *   15
Geoff Murphy CANADA/UK/USA   1988
Max Von Sydow, Helen Mirren, Tom Skerritt, Tom Bell, Neil Dudgeon, Gavan O'Herlihy, Barry Corbin, Clarke Peters, Lou Hirsch, Kerry Shale, David De Keyser, Ken Nelson, Shane Rimmer, Boris Isarov, Garrick Hagon
Confused thriller set in the era of glasnost, with the CIA helping to foil a plot by the KGB to assassinate Gorbachev by sending a former agent to the USSR to work with an old Soviet girlfriend, who (as one might expect) is played by Mirren. A film whose

single interesting idea rapidly gets bogged down in a morass of sub-plots and distractions. Filmed on location in Hungary.
THR 100 min VIDrel: FUTUR L/A V

RED PONY, THE *   PG
Lewis Milestone USA   1949
Myrna Loy, Robert Mitchum, Louis Calhern, Peter Miles, Shepperd Strudwick, Margaret Hamilton, Beau Bridges, Patty King, Jackie Jackson, Nino Tempo, Tommy Sheridan, Little Brown Jug, Wee Willie Davis, George Tyne, Max Wagner
The story of a young boy and his life on a farm, his love of his pony and how he loses faith in his father when it dies. Attractive, slow-moving and atmospheric, but somewhat thinly plotted – the story being rather too short to stretch into a full-length feature. The fine score is by Aaron Copeland. Remade for TV in 1973.
DRA 89 min VIDrel: STABL L/A V
Boa: short story by John Steinbeck.

RED RIDING HOOD *   PG
Adam Brooks USA   1987
Isabella Rossellini, Craig T. Nelson, Amelia Shankley, Rocco Sisto, Helen Elazary, Linda Kaye, Amnon Meskin, Julian Joy-Chagrin, Haim Zehavy, Stuart Kingston, Danny Segev, Arie Moscuna, Igor Borisov, Barbara Allen
A colourful live-action version of this fairytale, telling of wicked Sir Godfrey, who wants to marry the wife of his brother, but is opposed by his young niece. He hatches a fiendish plan to get rid of her.
Aka: CANNON MOVIE TALES: RED RIDING HOOD; LITTLE RED RIDING HOOD
JUV 78 min (ort 84 min) VIDrel: MGM/WHV L/A V/sh
Boa: short story by Jakob Ludwig Karl Grimm and Wilhelm Karl Grimm.

RED RIVER **   U
Howard Hawks USA   1948
John Wayne, Montgomery Clift, Joanne Dru, Walter Brennan, Coleen Gray, John Ireland, Noah Beery Jr, Harry Carey Sr, Harry Carey Jr, Paul Fix, Mickey Kuhn, Chief Yowlachie, Ivan Perry, Hank Worden, Hal Taliaferro, Paul Fiero
A vast, open-spaced and sprawling epic Western, set mainly on a cattle drive along the Chisholm trail, undertaken by a rancher and his rebellious son. Wayne gives one of his best performances as the tyrannical guardian, and Clift is excellent in his first major film role. The score is by Dmitri Tiomkin and photography is by Russell Harlan. Written by Borden Chase and Charles Schnee and remade for TV in 1988.
WES 127 min (restored director's cut – ort 133 min)
VIDrel: WHV V/sh
Boa: novel by Borden Chase.

RED RIVER *   PG
Richard Michaels USA   1988
James Arness, Gregory Harrison, Bruce Boxleitner, Ray Walston, Zachary Ansley, Laura Johnson, Ty Hardin, Robert Horton, John Lupton, Guy Madison, L.Q. Jones, Burton Gilliam, Jerry Porter, Johnmark Bradley, Donnie Jeffcoat
Based on the 1948 version starring John Wayne and Montgomery Clift, the story follows a cattle ranch owner who sets out with his adopted son and a bunch of tough cowboys to drive a herd across 1,000 dangerous miles of Texas landscape. A textbook illustration of how unnecessary remakes generally are. Richard Fielder adapted the Borden Chase and Charles Schnee screenplay from the original.
WES 91 min (ort 96 min) mTV VIDrel: MGM L/A V
Boa: novel by Borden Chase.

RED ROCK WEST *   15
John Dahl USA   1993
Nicolas Cage, Dennis Hopper, Lara Flynn Boyle, Timothy Carhart, J.T. Walsh, Dan Shor, Dwight Yoakam, Bobby Joe McFadden, Craig Reay, Vance Johnson, Robert Apel, Dale Gisbon, Ted parks, Babs Bram, Rpbert Guajardo, Sarah Sullivan
An unemployed man finds himself innocently embroiled in danger and deception when he is mistaken for a contract killer and learns that his victim is a highly attractive woman. A complex but at times quite illogical thriller that makes up for its lack of a coherent plot with a fine atmosphere and some solid performances, including a suitably unpleasant one by Hopper as the real contract killer.
THR 94 min (ort 98 min) cC VIDrel: COLUM/SONOP V/sur

RED SCORPION *

Joseph Zito USA

15
1989

Dolph Lundgren, M. Emmet Walsh, Al White, T.P. McKenna, Carmen Argenziano, Alex Colon, Brion James, Regopstann, Ruben Nthodi, Vuzi Dibukwana, James Mthola, Dinky Notsemme, Ernest Nidlovu, Tahpeld Mofokeng, Mxolisi Hulana, Nicky Rebelo

Lundgren plays a Russian Spetznaz agent, sent to Africa to murder a rebel African leader in this overlong and flat action tale. A couple of good action sequences cannot save it.

A/AD   100 min (ort 102 min)   VIDrel: VCC/DISC/COLUM V/sur

RED SCORPION 2 **

Michael Kennedy CANADA/USA

18
1994

Matt Malcolm, Michael Ironside, John Savage, Paul Ben-Victor, Real Andrews, Michael Covert, Duncan Fraser, George Touliatos, Vladimir Kulich, Suki Kaiser, Jerry Wasserman, Tong Lung, Antony Stamboulieh, Samantha Schubert, Wren Robertz

A white supremacist group is planning to cleanse the USA of all non-Aryan peoples, and to do this have produced a nationwide computer database. The Red Scorpions are assigned the task of eliminating both these terrorists and their computer system. Moderately entertaining nonsense of the mindless variety, that understandably went straight to video.

A/AD   90 min   VIDrel: FIRST/SONOP   V/sur

RED SHOE DIARIES **

Zalman King USA

18
1992

David Duchovny, Brigitte Bako, Billy Wirth, Kai Wulff, Bridgit Ryan, Brenda Vaccaro, Evi Sullivan, Anna Karin, Christina Canon, Leana Hall, Harry Cohn, Jonathan Zeichner, Rhonda Aldrich, Charlotte Blunt, Kelsey, Sherry Bilsing

Erotic-drama that serves as a spin-off to WILD ORCHID films, and follows the plight of an attractive woman who finds herself caught in a love triangle with two possessive men, neither of whom is aware of the other. This started off the RED SHOE DIARIES series, all the films in the set being much the same, i.e. glossy and empty romantic dross, albeit very well made.

Aka: WILD ORCHID: THE RED SHOE DIARY

A   104 min (ort 107 min) mCab   VIDrel: EIV/SONOP   V

RED SHOE DIARIES 2: DOUBLE DARE **

Zalman King/Tibor Takacs USA

18
1992

David Duchovny, Steven Bauer, Joan Severance, John Toles-Bey, Patrick Banta, Denise Crosby, Scott Lawrence, Sarah Luck Pearson, Denise Crosby, Robert Kneipper, Laura Johnson, Arnold Vosloo, Michael Woods

Second sequel to the first film takes the form of three separate stories, all with the common theme of a woman's passion and the lengths to which they are prepared to go to indulge their desires. Titles are: "Safe Sex", "You Have The Right to Remain Silent" and "Double Dare".

Aka: DOUBLE DARE; WILD ORCHID: RED SHOE DIARIES 2

A   91 min (ort 94 min) mCab   VIDrel: EIV/SONOP   V

RED SHOE DIARIES 3: ANOTHER WOMAN'S LIPSTICK **

Zalman King/Rafael Eisenman USA

(18)

1992

Richard Tyson, Maryam D'Abo, Nina Siemaszko, Zalman King, Matthew LeBlanc, Tcheky Karyo, Richard Tyson, Lydie Denier, Christina Fulton, Kevin Haley, David Duchovny

A trio of steamy erotic tales, all of which are tastefully photographed, but have little plot to hold one's interest. In the title story, a young wife learns that her husband has been having an affair, and becomes obsessed with finding out all she can about this other woman.

Aka: ANOTHER WOMAN'S LIPSTICK

A / 87 min (ort 90 min) mCab   SATrel: MOVIE CHANNEL

RED SHOE DIARIES 4: AUTO EROTICA **

Zalman King/Alaln Smithee/Michael Karbelnikoff USA   1993

(18)

David Duchovny, Ally Sheedy, Caitlyn Dulany, Scott Plank, Sheryl Lee, Maria Giulia Cavalli, Nick Chinlund, Kenneth A. Johnson, William Burns, Frederick Washburn

A further collection of vignettes, very much on the same lines as before. Titles are "Auto Erotica", "Accidents Happen" and "Jake's Story".

Aka: AUTO EROTICA

A   79 min (ort 83 min) mCab   SATrel: MOVIE CHANNEL

RED SHOE DIARIES 5: WEEKEND PASS **

Domenique Othenin-Girard/Peter Care/
Ted Kotcheff USA

(18)

1994

Francesco Quinn, Paula Barbieri, Ron Marquette, Anthony Addabbo, Sue Kiel, David Duchovny, Claire Stansfield, Nicholas Love, Dee McCafferty, Ely Pouget, Shashawnee Hall, Francoise Koster, Thom Haste

Three more erotic stories in the series, entitled: "Weekend Pass", "Double Or Nothing" and "Bounty Hunter". Very much the same as all the other entries in the series, which all consist of separate half-hour stories made for cable TV. Title story follows the adventures of a lonely woman who joins some friends for a wild time at a local bar.

Aka: WEEKEND PASS

A   84 min (ort 90 min) mCab   SATrel: MOVIE CHANNEL

RED SHOE DIARIES 6: HOW I MET MY HUSBAND **

Bernard Auroux/Philippe Angers/Anne Goursand USA   1994

(18)

David Duchovny, Luigi Amodeo, Neuth Hunter, Sue Kiel, Jennifer Leigh Burton, Alex Ardenti, Andrea Rave, Rave Snow, John Enos, Rudy De Rooy, California Ralph, Enzo, Charlotte Lewis, Carsten Norgaard, Lucas Leestemaker

Another trio of erotic tales, their titles being: "How I Met My Husband", "Naked In The Moonlight" and "Midnight Bells". Tastefully made they may be, but they make for very tepid viewing, with love as ever being portrayed as the exclusive property of the young and beautiful. In the title story, a young woman joins a dominatrix society as a way of livening up her sex life, and there she gets to meet a handsome gigolo.

Aka: HOW I MET MY HUSBAND

A   81 min (ort 85 min) mCab   SATrel: MOVIE CHANNEL

RED SHOE DIARIES 7: BURNING UP **

Rafael Eisenman USA

(18)

1995

Daniel Duchovny, Amber Smith, Daniel Blasco, Udo Kier, Jennifer Burton, Ron Marquette, Hartley Silver, Jennifer Ciesar, Matt Le Blanc, Robert Mailhouse, Mark Zuelke, Alexandra Tydings, Anthony guidera, Ivan Allen, Rhonda Aldrich

"Runaway", "Kidnap" and "Burning Up" are the titles of the three stories this time, and as ever they feature short tales where couples meet, fall in love and part. Beautifully photographed, but as superficial as they are glossy, forming yet another entry in this assembly-line series.

Aka: BURNING UP

A   85 min (ort 90 min) mCab   SATrel: MOVIE CHANNEL

RED SHOE DIARIES 8: NIGHT OF ABANDON **

Rene Manzor USA

(18)

1995

Daniel Leza, Ann Cockburn

A beautiful but lonely woman offers her most valued possession to a goddess named Lemenja, in return for the pleasures of the flesh. Lexa plays the attractive hunk she gets involved with. Erotic fantasy nonsense set during the Rio De Janeiro carnival. The other two stories that accompany the title one depict similar soft-core encounters, albeit without the supernatural element.

Aka: NIGHT OF ABANDON

A   82 min mCab   SATrel: SKY MOVIES

RED SHOE DIARIES 9: HOTLINE GINA **

Rafael Eisenman USA

(18)

1995

Lynette Walden, Stephane Bonnet

Another set of erotic stories, the title one being all about a woman who movies from L.A. to Paris, where she has an affair with an attractive stranger.

Aka: HOTLINE GINA

A   86 min mCab   SATrel: SKY MOVIES

RED SHOE DIARIES 10: SOME THINGS NEVER CHANGE **

Rafael Eisenmann USA

(18)

1997

David Duchovny, Sylvie Guelton, Guy Amum, Hero Kanasia

Another three stories of erotic love and romance among the young and beautiful. In "You Make Me Want To Wear A Dress" a tough-as-nails woman ranch owner finds a tough macho rival brings out her repressed feminine side. In "Some Things Never Change" a fashion designer finds her work with an ex-boyfriend a strain, despite the resurgence of past feelings. Finally, in "Alphabet Girl", a young woman gets a chance of fame, fortune and a little romance.

A   85 min mCab   SATrel: SKY MOVIES

RED SHOE DIARIES 11: THE GAME ** (18)
Philippe Angers / Brian Grant / Rafael Eisenman USA   1995
Caron Bernstein, Frederick Washburn, Michael T. Weiss, Allan Graf, Eugene Choy, Jennifer MacDonald, Christian Le Blanc, Saxon Trainor, Peter Sands, John Bergantine, Arielle Dombasle, Elodie Frenck, Will Stewart, Jean-Yves Gauthier
Another trio of erotic stories about the love lives of the young and beautiful. In "The Game" a woman who finds a lack of passion in her present relationship tells her partner how she and a former lover became obsessed with sexual games. In "The Cake" a woman who is collecting her stuffy husband's 35th birthday cake suffers temptation in the shape of the baker. Finally, "Like Father, Like Son" has young man falling for his father's mistress.
Aka: GAME, THE
A   74 min mCab   SATrel: MOVIE CHANNEL

RED SHOE DIARIES 12: GIRL ON A BIKE ** (18)
Stephen Halbert / Tibor Takacs / Lydie Callier USA   1996
David Duchovny, Alan Boyce, Sofia Shinas, Peter Gregory, Rhonda Aldrich, Brent Fraser, Geraldine Cotte, Jesse Harris, Peter Quartaroli, Robbie Chang, Adewale, Corrine Chuffart, Sally Le Boeuf
Another three stories in this long-running series about erotic encounters. In "The Written Word" a woman law professor gains a secret admirer. "Borders of Salt" tells of a woman who loves to travel by train, and of her meeting with a handsome stranger. Finally, "Girl On A Bike" describes how an American in Paris becomes obsessed with meeting a girl who rides a red bicycle. Similar in conception and tone to the earlier entries, these are pleasant if unmemorable.
Aka: GIRL ON A BIKE
A   76 min mCab   SATrel: SKY MOVIES

RED SHOE DIARIES 13: FOUR ON THE FLOOR ** (18)
David Womak / Rafael Eisenman / Zalman King USA   1996
David Duchovny, Denise Crosby, Georges Corrafece, Demetra Hamilton, Natasha Cashman, Paco Reconti, Manoellee Gaillard, Christopher Atkins, Nick Corri, Rachel Palmieri, Dominique Abel, Freedom Williams, Mary Morrow, Kent Master-King
A further trio of assembly-line, soft-focus erotic stories of soft-core sexual encounters. In "The Psychiatrist" a female shrink is so aroused by the actions of a patient that she follows her example. "Floor On The Floor" tells of two couples who shelter from a storm and experience a night of collective passion. Finally, in "Emily's Dance" a girl who comes to L.A. in the hope of achieving success a dancer has her life complicated by her love for a colleague.
A   82 min mCab   SATrel: MOVIE CHANNEL

RED SHOES, THE **** U
Michael Powell / Emeric Pressburger UK   1948
Anton Walbrook, Moira Shearer, Marius Goring, Robert Helpmann, Leonide Massine, Ludmilla Tcherina, Frederick Ashton, Albert Basserman, Esmond Knight, Irene Browne, Austin Trevor, Marcel Poncon, Jerry Verno, Jean Short
A great ballet dancer is torn between love for her art and her love of two men, finally being driven to a tragic end. A beautiful film that gives an intimate backstage view of the world of ballet. This was Shearer's screen debut. The photography is by Jack Cardiff and the script by Powell and Pressburger. AA: Score (Brian Easdale), Art/Set (Hein Heckroth/Arthur Lawson).
DRA   128 min (ort 136 min)   VIDrel: CARL/TECH   V

RED SONJA * 15
Richard Fleischer USA   1985
Arnold Schwarzenegger, Brigitte Nielsen, Sandahl Bergman, Paul Smith, Ernie Reyes Jr, Ronald Lacey, Pat Roach, Terry Richards, Janet Agren, Donna Ostabuhr, Lara Naszinsky, Hans Meyer, Tutte Lemkow, Tad Horino
Sword-and-sorcery adventure with all the standard ingredients, telling how Sonja avenges her sister's death and deposes a wicked queen. A pulp adventure, rather loosely based on the writings of Robert E. Howard of "Conan" fame. In this one the actors are as unconvincing as the monsters.
A/AD   85 min (Cut at film release – ort 89 min)
VIDrel: BMGVID/BMGREC   V/sh

RED SORGHUM *** 15
Zhang Yimou CHINA   1987
Gong Li, Jian Weng, Liu Ji, Ji Cun Hua, Teng Rujun, Cui Cun Hua, Qian Ming, Zhai Cun Hua
A pretty eighteen-year-old is betrothed to the unappealing fifty-year-old owner of a wine distillery, but falls in love with the porter who saved her from a bandit attack. With the death of her husband, her life improves and the distillery prospers, but the Japanese invasion brings suffering to the whole community. A lavish and intricately plotted film, often too arty for its own good, but equally both exotic and highly original.
Aka: HONG GAOLIANG
DRA   91 min   VIDrel: PAL/TERRY L/A   V
Boa: short stories by Mo Yan.

RED SPIDER, THE ** 18
Jerry Jameson USA   1988
James Farentino, Amy Steel, Jennifer O'Neill, Philip Casnoff, Soon-Teck Oh, Stephen Joyce, Earl Hindman, Kario Salem, Blu Mankuma, Brad Sullivan, Kevin O'Connor, Olivia Virgil Harper, Jim Byrnes, Jay Brazeau, Lelani Morrell
A sequel to ONE POLICE PLAZA with a different cast stepping into the same roles. A dogged New York police lieutenant investigates a series of bizarre Chinatown murders, in which the victims are all marked with a spider emblem, and a Vietnam link is discovered. A production-line thriller.
THR   89 min (ort 96 min) mTV   VIDrel: 20TH   V

RED SQUIRREL, THE *** 18
Julio Medem SPAIN   1993
Emma Suarez, Nancho Novo, Carmelo Gomez, Maria Barranco, Karra Elejalde, Christina Marcos, Monica Molina, Ana Gracia, Elena Irureta, Susana Garcia, Ane Sanchez, Eneko Irizar, Sarai Noceda, Maite Yerro, Txema Blasco, Chete Lera
Romantic thriller in which an ex-pop star thinking of suicide is snapped out of this despondency by his rescue of a woman in a motorbike crash. When he finds that she is suffering from amnesia, he takes her to a camp-site (the title establishment) where he passes her off as his girlfriend.
Aka: LA ARDILLA ROJA
THR   109 min (ort 114 min) wScrn
VIDrel: TART/20TH; ENCORE (LV only)   V LV

RED SUN RISING ** 18
Frances Megahy USA   1993
Don "The Dragon" Wilson, Terry Farrell, Mako, Michael Ironside, Soon-Teck Oh, James Lew, Edward Albert, Yuji Okumoto, Stoney Jackson, James Hatch, Forry Smith, Peter Mark Vasquez, Jacqueline Obradors, Leonard D. Turner, Diana Ipari
A Japanese cop is sent to the USA to track down one of the gangsters involved in the killing of his partner, and once there teams up with an L.A. female police officer when they realise the gangster is an enemy to both of them. A competent martial arts cop outing, well staged but not especially original.
MAR   99 min   VIDrel: GUILD/SONOP   V/sh

RED WIND ** 18
Alan Metzger USA   1991
Lisa Hartman, Philip Casnoff, Christopher McDonald, Deanna Lund, Antoni Corone, Camilla Moore, Rhonda Chambers, E. Thomas B. Nowicki, Tony Bolano, Tom Kouchlakos, Lori Creevay, Darlene Deicon, Rob Fuller, Michael Gioia
The treatment being administered by a female psychiatrist to disturbed woman backfires when the woman in question murders her husband and then comes after the doctor, whom she blames for her actions.
THR   89 min (ort 93 min) mCab   VIDrel: CIC/SONOP
V/h

REDEMPTION, THE ** 18
Kristine Peterson USA   199-
Mark Damascos, James Ryan, Tony Caprari, Greg Latter, Rulan Booth
A former kickboxing champ plans his revenge against those people who killed some former kickboxing buddies because they would not work for the local Mr Big. A simple and straightforward martial arts actioner, no better or worse than countless others.
MAR   90 min   VIDrel: 20TH/FOXVID   V/sur

REDHEADS ** 15
Danny Vendramini AUSTRALIA   1992
Claudia Karvan, Catherine McClements, Alexander Petersons, Mark Hembrow, Sally McKenzie, Anthony Phelan, Iain Gardiner, Jennifer Flowers, Malcolm Cork, Charlie Barry, Peter Grose, Craig Cronin, Alex Sweetman, Anthony Heffeman
A prostitute sees a murder and contrives to get herself arrested as a means of gaining protection, but her court appointed lawyer soon realises that the case involves police officers. Inspired by

the play "Say Thank You To The Lady" by Rosie Scott, this starts out with promise, as if it going to explore the iniquities of a legal system in which innocence often counts for very little. Unfortunately, this intention loses out to a rather clumsy crime thriller.
THR 102 min (ort 107 min) VIDrel: HIFLI/SONOP V/sur

REDS ** 15
Warren Beatty USA 1981
Warren Beatty, Diane Keaton, Jack Nicholson, Edward Herrmann, Paul Sorvino, Jerzy Kosinski, Maureen Stapleton, Gene Hackman, Nicolas Coster, M. Emmet Walsh, Ian Wolfe, Bessie Love, MacIntyre Dixon, Pat Starr, George Plimpton
Vast sprawling biopic on the life and loves of idealistic journalist John Reed; an eyewitness to the Russian Revolution. Heavily weighed down with its political content and dealing with Reed's affair with Louise Bryant in a superficial way, this represents a film ruined by over-ambition. Despite this it won three Oscars. AA: Dir, S. Actress (Stapleton), Cin (Vittorio Storaro).
DRA 187 min (Cut at film release by 3 sec – ort 200 min) VIDrel: 4-FRONT/POLYREC/CIC V/h

REDTOPS *** (U)
Svend Johansen DENMARK 1988
Sune Carlsson, Sara Danielle Arentsen, Michel Belli, Line Kruse, Trine Vildbaek, Karen-Lise Mynster, Kirsten Olseen, Peter Hesse Overgaard, Kirsten Lehfeldt, Peter Schroder, Lisbeth Dahl, Nils Vest, Henrik Kofoed, Nina Rosenmeir
A group of kids who used to bully each other unite in a common cause when they learn that a developer plans to bulldoze virgin forest to make way for a housing project. This involves them in spying on the baddies and keeping tabs on the helicopter that is being used to clear the site. A modern-day fairytale on the classical theme of good versus evil, well acted and made with a degree of care not usually lavished on films specifically aimed at kids.
Aka: RODTOTTERNE OG TYRANNOS
JUV 73 min SATrel: MOVIE CHANNEL

REEFER AND THE MODEL *** 15
Joe Comerford EIRE 1987
Ian McElhinney, Carol Scanlan, Eve Watkinson, Sean Lawlor, Ray McBride, Birdey Sweeney, Maire Chinselach
An unusual comedy-thriller in which a former member of the IRA and a petty crook get involved with a prostitute and decide to commit one last big crime, before going straight for good. Both scripted and directed by Comerford.
THR 89 min (ort 90 min) VIDrel: VISCOM/RTM V

REEFER MADNESS 15
Louis Gasnier USA 1936
Dave O'Brien, Dorothy Short, Warren McCollum, Lillian Miles, Carleton Young, Thelma White, Kenneth Craig, Pat Royale
Unbelievable, campy anti-marihuana warning, that is as ill-informed as it is exploitative and counter-productive, claiming as it does, that smoking dope turns kids into drug-crazed murderers and perverts. As might be expected, this laughable and moronic tale now enjoys cult status. ASSASSIN OF YOUTH and MARIHUANA: THE DEVIL'S WEED cover much the same ground.
Aka: BURNING QUESTION, THE; DOPE ADDICT; DOPED YOUTH; LOVE MADNESS; TELL YOUR CHILDREN
DRA 70 min B/W VIDrel: VISVID/POLYREC L/A V

REEL WORLD, THE: PART 2 – AFTER HOURS * 18
Frank Marino USA 1995
Christina Angel, Lacy Rose, Mickey Ray, Nick East, Dyanna Lauren, Debi Diamond, Alex Sanders, Gerry Pike, Jonathan Morgan
Seven strangers enjoy getting to know each other when they share a hayloft in Vienna. A sex spoof on "reality" movies.
A 82 min (ort) VIDrel: ONE V

REFLECTING SKIN, THE *** 15
Philip Ridley UK 1990
Viggo Mortensen, Lindsay Duncan, Jeremy Cooper, Sheila Moore, Duncan Fraser, David Longworth, Robert Koons, Sherry Bie, Evan Hall, Codie Lucas Wilbee, David Bloom, Jason Wolfe, Dean Haas, Guy Buller, Jason Brownlow, Jeff Walker
Written by Ridley, and slightly reminiscent of CELIA, this atmospheric psychological thriller takes as its focus a seven-year-old boy who lives in a remote community in 1950s Idaho. With the murder of his young buddy, the boy mistakenly comes

to believe that a vampire is responsible and further events conspire to convince him of this. Beautifully photographed, this is a tense if rather gruesome examination of the power of imagination and nightmares.
DRA 91 min (ort 93 min) VIDrel: VISVID/POLYREC V/sur

REFLECTIONS IN A GOLDEN EYE *** 15
John Huston USA 1967
Elizabeth Taylor, Marlon Brando, Brian Keith, Julie Harris, Robert Forster, Zorro David
A study of repression and intrigue in a peace-time army camp situated in Georgia, that features a private who likes to go horse-riding in the nude, and a homosexual major who nurses a secret passion for him. Meanwhile, the major's wife is carrying on with a neighbour whose own wife has a liking for self-mutilation. A ludicrous and overheated melodrama, rather unpleasant, but often quite fascinating in a morbid way.
DRA 104 min (ort 108 min) wScrn VIDrel: TART/20TH V/dm
Boa: novel by Carson McCullers.

REFLECTIONS ON A CRIME ** (12)
Jon Purdy USA 1994
Mimi Rogers, Billy Zane, John Terry
A death-row story with Rogers waiting the time of her execution, having been convicted of the murder of her husband. But as her time draws nearer, she realises that one of the guards (Zane) is beginning to fall in love with her, and must decide whether or not she is to exploit this situation. A little contrived in plotting, often quite touching, but ultimately spoilt by sluggish pacing and leaden direction.
THR 91 min SATrel: SKY MOVIES

REFORM SCHOOL GIRL * PG
Edward Bernds USA 1957
Gloria Castillo, Ross Ford, Edward Byrnes, Ralph Reed, Jack Kruschen, Sally Kellerman, Yvette Vickers
Having just about fought off the advances of her obnoxious uncle, a young girl finds herself sent to reformatory school for riding in a hit-and-run vehicle. Once there she has an even rougher time of it, in a film that's cheap, contrived and exploitative.
DRA 71 min B/W VIDrel: HEND/BMGREC L/A V

REFORM SCHOOL GIRL ** (18)
Jonathan Kaplan USA 1994
Aimee Graham, Matt LeBlanc, Teresa Dispina, Carolyn Seymour, Marissa Ribisi, Eleanor O'Brien, Samaria Graham, Erin Lesham Wiley, Catherine Paolone, Nick Chinlund, Ivan Eastman, Dino Anello, Harry Northrup, Ashley Laster, Bob Minor
Remake of a dreary 1950s movie all about a rebellious high school girl who gets in with the wrong crowd and is sent to reformatory. Here she enters a slow downward spiral of deepening depression, until she meets a boy from outside who takes an interest in her. The strong cast try to inject some life into this story, but the whole effort is depressingly derivative.
DRA 90 min SATrel: MOVIE CHANNEL

REFRIGERATOR, THE ** 18
Nicholas E. Jacobs USA 1993
Julia McNeal, David Simonds, Angel Caban, Phyllis Sunz, Nena Segal, Jaime Rojo, Alex Trisano, Peter Justinius, Karen Wesler, Michael Beltran, Jack Mason, Larry Tate, Darrell Smith, Phil Butard, Weston Blakesley, Jan Groff
A young couple move into an apartment in a none too salubrious neighbourhood and find themselves experiencing a series of strange and terrifying events, with the title appliance going on a nasty, gore-filled rampage. A low-budget and very tongue-in-cheek spoof that offers much enthusiasm and a pop music soundtrack to make up for a marked lack of wit. Good performances are a partial compensation.
HOR 96 min VIDrel: MIA/DISC V

REGARDING HENRY ** 15
Mike Nichols USA 1991
Harrison Ford, Annette Bening, Bill Nunn, Mikki Allen, Donald Moffat, Aida Linares, Elizabeth Wilson, Robin Bartlett, Bruce Altman, Rebecca Miller, R.M. Haley, Stanley H. Swerdlow, Julie Follansbee, John MacKay, Mary Gilbert
When a hot-shot lawyer suffers severe brain damage following an accident, he has to learn to walk and talk all over again, and it's a while before he can enjoy anything like a normal life. A

glossy and synthetic piece, it not only makes little of the struggles brain damaged people really have to face, but by virtue of the insipid script, fails to engage one's emotions. A shame to see actors of the calibre of Ford and Bening being wasted on this nonsense.
DRA   102 min (ort 108 min)   VIDrel: CIC/SONOP   V/sur

REILLY – ACE OF SPIES: AN AFFAIR WITH A MARRIED WOMAN **   15
Jim Goddard/Martin Campbell UK   1983
Sam Neill, Leo McKern, Norman Rodway, Peter Egan, Sebastian Shaw, Tom Bell, Jeananne Crowley, John Rhys Davies
First episode of a multi-part serial about a famous spy in the 1930s. Though detailed and carefully made, the stories proceeded so slowly that little tension was generated. Here, Reilly is inside Russia on the way to Baku, sent to gather information on the country's budding oil industry. But when the train he is on is stopped, he is detained by the authorities, who have grown suspicious. Some of the other stories were released on video at about the same time as this one.
Aka: REILLY – THE ACE OF SPIES
DRA   80 min mTV   VIDrel: VCC/DISC L/A   V
Boa: book by Robin Bruce-Lockhart.

RELATIVE FEAR **   15
George Mihalka CANADA   1993
Darlanne Fluegel, James Brolin, Martin Neufeld, Denise Crosby, M. Emmett Walsh, Bruce Dinsmore, Linda Sorensen, Jason Blicker, Vlasta Vrana, Matthew Dupuis, Liz McRae, Michael Caloz, Gisele Rousseau, Jenny Campbell, Alan West
A couple have an autistic son whose outwardly placid exterior hides a more sinister nature, and who appears to cause the deaths of those that upset him.
THR   90 min (ort 94 min) mTV   VIDrel: COLUM/SONOP   V/sur

RELENTLESS **   18
William Lustig USA   1989
Judd Nelson, Robert Loggia, Leo Rossi, Meg Foster, Patrick O'Bryan, Ken Lerner, Mindy Seeger, Angel Tompkins, Beau Starr, Harriet Hall, Ron Taylor, Roy Brocksmith, G. Smokey Campbell, Frank Pesce, Matt Bolduc, Lou Bonacki
A neat little police thriller that casts Rossi as a New York cop now working in L.A., where he sets out to solve a series of killings with his new partner. Nelson is memorable as the psychotic killer who, having been abused as a child, has taken to random killings by selecting his victims from phone books. Though it has its share of cliches, this well-acted tale benefits from a strong script, the work of Phil Alden Robinson. Several sequels followed.
THR   88 min (ort 93 min)   VIDrel: MGM/WHV L/A   V/sh

RELENTLESS 2: DEAD ON **   18
Michael Schroeder USA   1991
Ray Sharkey, Leo Rossi, Miles O'Keefe, Meg Foster, Marc Poppel, Dale Dye, Allan Rich, Steve Kahan, Art Kimbro, Sven-Ole Thorsen, Dawn Magnum, Leilani Jones, Frank Rossi, Mindy Seeger, Paul Hertzberg, Anthony Donato, Perry Lang
A police detective following up leads regarding a series of vicious murders learns that an FBI agent has been sent to shadow him, and that all the murder victims officially "died" twenty years ago. This eventually puts him on the track of a nasty international political conspiracy, in this sequel to the earlier film. RELENTLESS 3 and 4 followed.
Aka: DEAD ON; DEAD ON: RELENTLESS 2
THR   90 min (ort 95 min)   VIDrel: MGM/WHV L/A   V/sh

RELENTLESS 3 **   (18)
James Lemmo USA   1993
Leo Rossi, William Forysthe, Signy Coleman, Tom Bower, Robert Constanzo, Patricia Allison, Savannah Smith Boucher, Stacy Edwards, Jay Arlen Jones, Jack Knight, Catherine Paolone, Felton Perry, Mindy Seeger, George Tovar
A psychopath murders and mutilates young women and taunts the police by mailing them pieces of the bodies. When it seems as if he has turned his attention to the girlfriend of the detective in charge of the case, the latter is forced into a race against the clock in order to save her.
THR   82 min (ort 84 min)   VIDrel: WHV L/A   V

RELENTLESS: THE REDEEMER **   18
Oley Sassone USA   1994
Leo Rossi, Famke Janssen, Colleen Coffey, Ken Lerner, John Scott

Clugh, Christopher Pettiet, Lisa Kelly, John Kelly, Rainer Grant, Charlene Henryson, John Myers
A cop hunting yet another crazed serial murderer finds that these crimes are linked to an attractive woman psychiatrist, who for reasons of professional confidentiality refuses him the information he needs. A dark and very downbeat tale.
Aka: RELENTLESS 4: ASHES TO ASHES
THR   91 min   VIDrel: WHV L/A   V

RELIC, THE **   15
Peter Hyams USA   1996
Penelope Ann Miller, Tom Sizemore, Linda Hunt, James Whitmore, Clayton Rohner, Chi Muoi Lo, Thomas Ryan, Robert Lesser, Diane Robin, Lewis Van Bergen, Constance Towers, Francis X. McCarthy, Audra Lindley, John Kapelos, Tico Wells
This may be a big-budget horror pic, but none of the money went on the stale plotting, even if the opening premise is, to say the least, unusual. In the Brazilian rain-forests, explorers unwisely eat some berries that turn them into monsters. Fortunately, the owner of a museum specialising in South American folklore and myths has found a way to put paid to this peril.
HOR   110 min   CINrel

REMAINS OF THE DAY, THE ****   U
James Ivory UK/USA   1993
Anthony Hopkins, Emma Thompson, Christopher Reeve, James Fox, Peter Vaughn, Hugh Grant, Michael Lonsdale, Tim Pigott-Smith, Patrick Godfrey, Peter Cellier, Ben Chaplin Paul Copley, Peter Eyre, Lena Headey, Brigitte Kahn, Ian Redford
During the 1930s a butler serving at a stately home sacrifices himself and his personal feelings in the pursuit of his job, rejecting the affections of the housekeeper with whom he works. His loyalty to his aristocratic employer never wavers, even when the latter begins to associate with British supporters of the Nazis. Despite its literary origins, a very well acted effort, as ever Ivory's scriptwriter is the reliable Ruth Prawer Jhabvala.
DRA   128 min (ort 134 min) wScrn cC
VIDrel: COLUM/SONOP   V/sur
Boa: novel by Kazuo Ishiguro.

REMBRANDT ****   U
Alexander Korda UK   1936
Charles Laughton, Gertrude Lawrence, Elsa Lanchester, Edward Chapman, Walter Hudd, Roger Livesey, John Bryning, Sam Livesey, Herbert Lomas, Allan Jeayes, John Clements, Raymond Huntley, Abraham Sofaer, Lawrence Hanray, Austin Trevor
A lavish examination of the life of this 17th century Dutch painter, made as a set of vignettes exploring episodes in his life. A handsome, austere and gently tragi-comic work, loosely held together by a memorable performance from Laughton. Photography is by Georges Perinal and Richard Angst.
DRA   80 min (ort 84 min) B/W   VIDrel: CARL/TECH   V

REMEMBER **   15
John Herzfeld USA   1993
Donna Mills, Stephen Collins, Derek De Lint, Gail Strickland, Christoph M. Orht, Ian Richardson, Cathy Tyson, Lynsey baxter, Daphne Deckers, Claire Bloom, Rosaleen Linehan, Colin Stinton, Sabri Saas El Hamus, Eric Handler, Bob Van Tol
A news reporter learns that her fiance, who was believed to have committed suicide just prior to their wedding, is living in Barcelona.
Aka: BARBARA TAYLOR BRADFORD'S REMEMBER
DRA   175 min (ort 200 min) mTV
VIDrel: 4-FRONT/POLYREC/ODY   V/sh
Boa: novel by Barbara Taylor Bradford.

REMO: UNARMED AND DANGEROUS *   15
Guy Hamilton USA   1985
Fred Ward, Joel Grey, Wilford Brimley, J.A. Preston, George Coe, Charles Cioffi, Kate Mulgrew, Michael Pataki, William Hickey, Patrick Kilpatrick, Davenia McFadden, Cosie Costa, J.P. Romano, Joel J. Kramer, Frank Ferrara
A tough New York cop is drafted into a top-secret intelligence agency and assigned to the care of a Korean martial arts instructor for training. What promises to be an enjoyable adaptation, true to the spirit of the pulp "Destroyer" novels of Richard Sapir and Warren Murphy, rapidly goes downhill and is hardly helped by some remarkably cheap effects. Only Grey's splendid

performance as the Korean martial arts master makes this dud worth watching.
Aka: REMO WILLIAMS: THE ADVENTURE BEGINS; REMO WILLIAMS: THE ADVENTURE CONTINUES
A/AD   111 min (Cut at film release – ort ort 121 min)
VIDrel: VISVID/POLYREC L/A   V

REMOTE **   *U*
Ted Nicolaou USA   1992
Chris Carrara, Jessica Bowman, John Diehl, Derya Ruggles, Tony Longo, Stuart Fratkin, Jordan Belfi, Kenneth A. Brown, Lorna Scott, Ann Randolph, Michael Keys Hall, Robin Westphal, Robbie Chandler, Reno Goodale
A thirteen-year-old genius delights in designing and building remotely controlled toys capable of many feats, but gets himself into trouble with his parents and nearly everybody else, in this broad comedy. However, his undeniable resourcefulness stands him in good stead when he has a run-in with three inept crooks, in a finale that is highly reminiscent of a similar scene from HOME ALONE.
COM   77 min (ort 80 min)   VIDrel: CIC/SONOP L/A   V

REMOTE CONTROL **   *15*
Jeff Lieberman USA   1987
Kevin Dillon, Jennifer Tilly, Deborah Goodrich, Christopher Wynne, Frank Beddor, Bert Remsen, Jaimie McEnnan, Jerold Pearson, Jennifer Buchman, Will Nye, Deborah Downey, Marilyn Adams, Richard Warlock, Ann Walker, Ty Kelley
A spoof horror yarn, the title of which refers to a videotape that is supposedly a 1950s film looking forward to the 1980s, but is in reality a tape created by aliens in order to destroy mankind. Those who watch the tape are driven to violence, and with more and more people clamouring for it, a young man working in a video store realises that all copies must be destroyed.
HOR   85 min (ort 88 min)   VIDrel: 20TH/TECH L/A   V

RENAISSANCE MAN ***   *12*
Penny Marshall USA   1994
Danny DeVito, Gregory Hines, James Remar, Cliff Robertson, Ed Begley Jr, Stacy Dash, Mark Wahlberg, Lillo Brancato, Kadeem Harrison, Richard T. Jones, Khalil Kain, Peter Simmons, Jennifer Lewis, Greg Sporleder, Ann Cusack
An advertising executive who loses his job takes on the seemingly hopeless task of teaching a class of army recruits. His unorthodox methods meet with military disapproval but help his pupils to cope with learning their lessons. Meanwhile, they manage to teach him a thing or two in return. A breezy little comedy, quite uneven but sustained by DeVito's sheer charm and acting ability.
COM   123 min (ort 128 min)   VIDrel: GUILD/20TH   V/sur

RENDEZVOUS IN PARIS ***   *15*
Eric Rohmer FRANCE   1995
Clara Bellar, Antoine Basler, Mathias Megard, Judith Chancel, Malcolm Conrath, Cecile Pares, Olivier Poujol, Aurore Rauscher, Serge Renko, Michael Kraft, Benedicte Loyen, Veronika Johansson
Three touching and delightful love stories set in Paris. In the first one, a young woman discovers to her dismay that her boyfriend is unfaithful to her. The next story follows the despair of a young woman who wants to leave her husband for her lover, but in the end just deserts both. In the final story an arrogant artist meets a woman at a gallery and tries to seduce her when he takes her home, but gets nowhere. See also SIX IN PARIS.
Aka: LES RENDEZVOUS DE PARIS
DRA   96 min (ort 98 min) wScrn   VIDrel: ARTIF/20TH
V/h

RENEGADE MURDERER'S ROW **   *15*
Ralph Hemecker USA   1993
Lorenzo Lamas, Branscombe Richmond, Grand Bush, Don Gibb, Art la Fleur, Kathleen Kinmont, Deprise Brescia, Stephen J. Cannell
Pilot episode for a prospective series, with an ex-cop who is wrongly blamed for a murder setting out to have his revenge as a undercover vigilante.
Aka: RENEGADE
A/AD   93 min (ort 96 min) mTV   VIDrel: WHV   V

RENEGADES **   *18*
Jack Sholder USA   1989
Kiefer Sutherland, Lou Diamond Phillips, Rob Knepper, Bill Smitrovich, Jami Gertz, Floyd Westerman, Clark Hohnson, Peter MacNeill, John Di Benedetto, Joe Griffin, Kyra Harper, Dee McCafferty, Heide Van Palieske, Tom Butler

A pilot for a short-lived series that has a Philadelphia cop and a Lakota Indian teaming up to bust a gun-running operation and settle a few old scores at the same time. Written by David Rich, it combines a paper-thin plot with a series of well-handled action scenes.
A/AD   102 min (ort 106 min)
VIDrel: COLUM/SONOP L/A   V/sh

RENT-A-KID **   *U*
Fred Gerber USA   1995
Leslie Nielsen, Christopher Lloyd, Matt McCoy, Amos Crawley, Cody Jones, Tony Rosato, Kevin Hicks, Ed Healey, Judah Katz, Michael Anderson Jr, Walter Alza, Sherry Miller, Tabitha Lupien, Lisa LaCroix, Victoria Mitchell, Paul Brown
The supervisor of an orphanage goes on holiday, leaving his eccentric salesman father in charge. The latter applies all his professional skills to the problem of getting kids out of the orphanage and into families, and comes up with the idea of renting kids out "on approval", allowing a couple to take home three such kids as a way of sampling parenthood before making a commitment.
COM   85 min (ort 90 min)   VIDrel: MED/DISC   V/sh

REPLIKATOR **   *18*
G. Philip Jackson USA   1994
Michael St Gerard, Brigitte Bako, Ned Beatty, Lisa Howard, Ron Lea, Peter Outerbridge, David Hemblen, La Cicciolina, Mackenzie Gray, Lynne Deragon, Janet Bailey, Doug Bagot, Tim Lee, Kyra Harper, Andrew Lewarne, Erica Ehmn, R.D. Reid
A streetwise cop is forced to team up with two of his cyberpunk friends when cloning technology is stolen by a ruthless crook who is using it to replicate human beings. A bleak vision set in the year 2014 in a world where population growth and ozone layer depletion are major problems.
Aka: REPLIKATOR, THE: CLONED TO KILL
FAN   96 min   VIDrel: HIFLI/SONOP   V/h

REPO MAN **   *18*
Alex Cox USA   1984
Emilio Esterez, Harry Dean Stanton, Vonetta McGee, Olivia Barash, Sy Richardson, Susan Barnes, Tracey Walter, Fox Harris, The Circle Jerks, Tom Finnegan, Del Zamora, Eddie Velez, Zander Schloss, Jennifer Balgobin
A young punk gets a job repossessing cars but finds himself involved in a strange series of events. A curious blend of film noir and macabre fantasy that is intermittently engaging but is best enjoyed in small doses. Written by Cox (it has a minor cult following) this is probably his best work to date.
A/AD   88 min (ort 92 min)
VIDrel: 4-FRONT/POLYREC/CIC   V/h

REPOSSESSED **   *15*
Bob Logan USA   1989
Linda Blair, Leslie Nielsen, Ned Beatty, Anthony Starke, Thom J. Sharp, Thom Schwab, Benj Thall, Jacquelyn Masche, Melissa Moore, Willie Garson, Kathy Topia, Richard Helpern, Carol Shermer, Gary Howe Scott, Linda Brennan, Murray Langston
A spoof on THE EXORCIST that is written by Logan, and features Blair, who is now married with two kids, falling under the spell of Satan once more, but this time succumbing whilst watching TV. An attempt to do for demonic possession what AIRPLANE! did for passenger flights, it features an interminable series of juvenile gags, plenty of nauseous special effects and a manic performance from Nielsen as a priest that is easily the best thing in it.
COM   80 min (ort 89 min)
VIDrel: 4-FRONT/POLYREC; PION (LV only)   V/sh LV

REPTILE, THE **   *(15)*
John Gilling UK   1966
Noel Willaims, Jennifer Daniel, Ray Barrett, Jacqueline Pearce, Michael Pipper, John Laurie, Marne Maitland, David Baron, Charles Lloyd Pack, Harold Goldblatt, George Woodbridge
A Cornish village is suddenly afflicted by a outbreak of deaths in which the victims turn black. A man investigating his brother's demise meets a wall of silence but is finally able to determine that the cause is a woman who turns into a giant reptile as a result of a Malayan curse. A reasonably enthralling horror tale with some intelligent acting. See also THE LAIR OF THE WHITE WORM.
HOR   90 min   VIDrel: LUMI/SPEAR L/A   V

REPULSION **** 18
Roman Polanski UK 1965
Catherine Deneuve, Ian Hendry, John Fraser, Patrick Wymark, Yvonne Furneaux, Renee Houston, James Villiers, Valerie Taylor, Hugh Futcher, Helen Fraser, Mike Pratt, Monica Merlin, Imogen Graham, Wally Bosco, Roman Polanski
Polanski's first English language film is a brilliant but harrowing tour de force, charting the slow descent of a withdrawn young girl into madness, when left alone in a flat for a few days while her sister goes on holiday. An abundance of powerful images serve to underline her growing insanity, culminating in a nightmarish climax. Written by Polanski and Gerald Brach.
HOR 100 min (ort 105 min) B/W VIDrel: ODY/SONOP V

REQUIEM FOR A VAMPIRE ** 18
Jean Rollin FRANCE 1970
Marie-Pierre Castel, Mireille D'Argent, Philippe Gaste, Dominique Toussaint, Dominique, Louise Dhour, Michel Delsalle, Oliver Francois, Antoine Mausin, Paul Bisciglia
Two female fugitives seeks shelter at a castle and soon find themselves at the mercy of a vampire, in this strange melding of horror and sex genres, one of several such films from this director.
Aka: CAGED VIRGINS; CRAZED VAMPIRE, THE; DUNGEON OF TERROR; REQUIEM POUR UN VAMPIRE; SEX VAMPIRES; VIERGES ET VAMPIRES; VIRGINS AND VAMPIRES
HOR 76 min (ort 85 min) dubbed VIDrel: REDEM/RTM V

RESCUE OF JESSICA McCLURE, THE ** PG
Mel Damski USA 1989
Beau Bridges, Patty Duke, Pat Hingle, Roxana Zal, Will Oldham, Whip Hubley, Robin Gammell, Walter Olkewicz, Rudy Ramos, Jack Rader, Guy Stockwell, Daryl Anderson, Mills Watson, Bo Foxworth, Robin Frates, Don Hood
Dramatisation of the true story of an eighteen-month-old girl who fell into a well-shaft and was trapped there for fifty-six hours until her rescue. This seemingly simple task was revealed to be fraught with unforeseen perils and the rescue attempt attracted worldwide attention. An adequate TV film that generally maintains a good sense of momentum, despite the fact that the outcome is never in doubt.
Aka: EVERYBODY'S BABY: THE RESCUE OF JESSICA McCLURE
DRA 90 min mTV VIDrel: ODY/SONOP V/s

RESCUERS, THE *** U
Wolfgang Reitherman. USA 1977
Voices of: Bob Newhart, Geraldine Page, Eva Gabor, Joe Flynn, Jeanette Nolan, Pat Buttram, Jim Jordan, John McIntire, Michelle Stacy, Bernard Fox, Larry Celmmons, James MacDonald, George Lindsey, Bill McMillan, Dub Taylor
When a youngster named Penny gets lost in the swamp, she falls into the clutches of the evil Madame Medusa and her alligator bodyguards. Two mice are sent by the "Mouse Rescue Aid Society" to attempt a rescue. Though this feature cannot be compared with the work of the Disney studio at its peak, it marked a welcome return to the strong scripting that was so lacking in animations from the early 1960s onwards. THE RESCUERS DOWN UNDER followed.
ANIM 74 min (ort 77 min) VIDrel: WDV/BUENA L/A V
Boa: short stories of Margery Sharp.

RESCUERS DOWN UNDER, THE *** U
Hendel Butoy/Mike Gabrbiel USA 1991
Voices of: Bob Newhart, Eva Gabor, John Candy, George C. Scott, Tristan Rogers, Adam Ryen, Wayne Robson, Douglas Seale, Frank Welker, Bernard Fox, Peter Firth, Billy Barty, Ed Gilbert, Carla Meyer, Russi Taylor, Charlie Adler
Based on the heroic mice created by writer Margery Sharp, this has them being called to Australia where they have to save a little boy and various other creatures from a nasty villain. Memorable for its excellent animation (much of it computer-generated) and strong characterisations, if a good deal less so for the plot. See also THE RESCUERS, the earlier film to which this was a sequel.
ANIM 77 min VIDrel: WDV/SONOP L/A V

RESERVOIR DOGS *** 18
Quentin Tarantino USA 1992
Tim Roth, Harvey Keitel, Chris Penn, Steve Buscemi, Lawrence Tierney, Eddie Bunker, Michael Madsen, Quentin Tarantino, Randy Brooks, Kirk Baltz, Tony Cosmo, Rich Turner, David Steen, Steve Poliy, Robert Ruth, Steven Wright (voice only)

A group of thieves known only to each other by their code names assemble to pull off a heist but this goes terribly wrong and two of them are killed. The survivors meet back at their hide-out for a post-mortem and a bloody and brutal settling of accounts takes place. A sickeningly violent debut from Tarantino (who also scripted) that is also irritatingly over-stylised. The fine performances and sharp dialogue are its only redeeming features.
DRA 95 min (ort 99 min) wScrn cC
VIDrel: POLY/POLYREC; PION (LV only) V/sh LV

RESORT TO KILL * 18
Dan Neria USA 1993
Roddy Piper, Meg Foster, Tiny Lister
On a remote island scientists working on immortality use an ancient Mayan secret technique, and re-animate a number of corpses, creating an invincible and seemingly indestructible army of murderous zombies. Two special agents are dispatched to put an end to this nonsense.
A/AD 100 min VIDrel: MED/20VIS V/sur

RESTLESS NATIVES * PG
Michael Hoffman UK 1985
Vincent Friell, Joe Mullaney, Teri Lally, Bernard Hill, Mel Smith, Robert Urquhart, Ned Beatty, Bryan Forbes, Nanette Newman, Anne Scott-James, Rachel Boyd, Iain McColl, Dave Anderson, Eiji Kusuhara, Ed Bishop, Laura Smith
Two youths become bored with their jobs and make themselves unemployed, but decide to improve their dull and impecunious lifestyle by dubbing themselves the "Clown" and the "Wolfman" and robbing tourist coaches in Scotland. Soon these latter-day highwaymen find themselves becoming media celebrities. A flat and unsatisfactory comedy that fails to make the most of its one idea.
COM 89 min VIDrel: MGM/WHV L/A V

RESTORATION *** 15
Michael Hoffman USA 1996
Robert Downey Jr, Sam Neill, David Thewlis, Polly Walker, Meg Ryan, Ian McKellan, Hugh Grant, Ian McDiarmid, Mary Macleod, Mark Letheren, Sandy McDade, Rosalind Bennett, Willie Ross, David Gant, Benjamin Whitrow, Neville Watchurst
A blend of comedy and drama set in 17th century London, where a philandering medical student is coerced by the newly-restored King Charles II in a marriage of convenience to the King's mistress. However, things turn a good deal nastier when the student physician falls in love with his wife, and he is punished with banishment for his impudence. Visually this is a striking film, so it is all the more of a pity it is so unmoving. AA: Cost (J. Acheson), Art (E. Zanetti).
DRA 113 min (ort 118 min) cC VIDrel: TOUCH/TECH V/s
Boa: novel by Rose Tremain.

RETALIATOR * 18
Allan Holzman USA 1987
Robert Ginty, Sandahl Bergman, Louise Caire Clark, James Booth, Alex Courtney, Paul Walker, Peter Bromilow, George Fisher, Jim Turner, Bill Allard, Dan Coffey, Chaim Banai, Liliana Cameroni, Jason Cornado, John Gillespie
When two American kids are snatched during a terrorist outrage in Greece, a former CIA agent is brought out of retirement to mount a rescue attempt. The mission is successful and the terrorists are killed, but the body of a female terrorist is used in a gruesome experiment in which she is resurrected as a vengeance-seeking cyborg. Gory, derivative and mind-numbing nonsense.
Aka: PROGRAMMED TO KILL
A/AD 87 min (ort 91 min) VIDrel: IMPENT L/A V

RETURN ENGAGEMENT ** 18
Cheung Tung Joe HONG KONG 1989
Andy Lau, Tung Joe
A battle breaks out between two rival Triad gangs in this very ordinary martial arts adventure.
MAR 90 min dubbed VIDrel: POPRO/RTM V

RETURN FROM THE RIVER KWAI *** 15
Andrew V. McLaglen USA 1988
Christopher Penn, Edward Fox, Denholm Elliott, Tatsuya Nakadai, George Takei, Nick Tate, Timothy Bottoms, Michael Dante, Richard Graham, Etsushi Takahashi, Blaise Alexandre, Masato Nagamori
US Air Force planes destroy a set of bridges over the River Kwai in Japanese occupied Thailand during WW2, and a pilot para-

chutes to safety after his plane is hit by anti-aircraft fire. After joining a unit of Thai guerrillas, led by an English major, he takes part in an attempt to free POWs being held at a local camp, is himself captured but eventually escapes. A vigorous and well handled if stereotyped war film.
WAR 98 min VIDrel: 4-FRONT/POLYREC/BRAVE V/sur
Boa: book by Joan and Clay Blair Jr.

RETURN FROM WITCH MOUNTAIN ** *U*
John Hough USA 1978
Bette Davis, Christopher Lee, Ike Eisenmann, Kim Richards, Denver Pyle, Jack Soo, Anthony James, Ward Costello, Dick Bakalyan, Brad Savage, Christian Juttner, Jeffrey Jacquet
A sequel to ESCAPE TO WITCH MOUNTAIN, in which a group of criminals try to exploit the supernatural powers of the two alien children. Quite inventive in terms of its special effects, but coy and lacking in strong characterisation with regard to the two kids. An undemanding Disney outing.
FAN 89 min (ort 95 min) VIDrel: RNK L/A V

RETURN OF A MAN CALLED HORSE, THE *** *15*
Irvin Kershner USA 1976
Richard Harris, Gale Sondergaard, Bill Lucking, Geoffrey Lewis, Jorge Luke, Enrique Lucero, Ana DeSade, Claudio Brook, Jorge Russek, Pedro Damien, Humberto Lopez, Patricia Reyes, Regino Herrerra, Rigobert Rico, Alberto Mariscal
Sequel to A MAN CALLED HORSE, with Harris returning to the Yellow Hand Sioux to help them resist the onslaught of the white man. A harsh and brutal adventure, written by Jack De Witt. The memorable score is by Laurence Rosenthal. Followed by TRIUMPHS OF A MAN CALLED HORSE.
WES 119 min Cut (1 min 46 sec – ort 125 min)
VIDrel: MGM/WHV L/A V
Boa: novel by Dorothy M. Johnson.

RETURN OF ELIOT NESS, THE ** *15*
James A. Contner USA 1990
Robert Stack, Jack Coleman, Philip Bosco, Anthony De Sando, Lisa Hartman, J. Winston Carroll, Frank Adamson, Shaun Austin-Olsen, George Chuvalo, Michael Kirby, Dwight Bacque, Rummy Bishop, Walker Boone, Frank Canino, David Clement
This reasonably entertaining attempt to cash in on the popularity of the TV series "The Untouchables" follows the exploits of the FBI's most successful agent, who having brought Al Capone down, now enters the murky world of the gangster's old empire, to avenge the death of an old friend. A grim-faced (albeit ageing) Stack is the best thing in this effort.
DRA 90 min (ort 94 min) mTV VIDrel: BRAVE/SONOP L/A V

RETURN OF IRONSIDE, THE * *PG*
Gary Nelson USA 1993
Raymond Burr, Don galloway, Barabra Anderson, Elizabeth Baur, Don Mitchell, Dana Wynter, Perrey Revves, Eddie Jones, Jeff Kaake, Derek Webster, Cliff Gorman, Robin Sachs, Scott Patterson, Ed Lauter, Chick Booms, Darlene Vogel
With the murder of a police chief, suspicion falls on the police department, as it is riddled with corruption. Recently retired Commissioner Ironside investigates. A dull spin-off feature from the TV detective series "Ironside".
DRA 88 min mTV VIDrel: CIC/SONOP V

RETURN OF JAFAR, THE *** *U*
USA 1993
Voices of: Scott Weinger, Linda Larkin, Gilbert Gottfried, Val Bettin, Dan Castellaneta
Made for video (a first for Disney) this animation serves as a sequel to the hit ALADDIN, with the evil sorcerer Jafar being accidentally released from captivity within the magic lamp and seeking his revenge. Fortunately, all Aladdin's friends rally round to defeat this evil plan. Adequate, but not a patch on the earlier film. Followed by ALADDIN AND THE KING OF THIEVES, the last film in this cycle.
ANIM 66 min (ort 88 min) mVid cC VIDrel: WDV/TECH V

RETURN OF MARTIN GUERRE, THE *** *15*
Daniel Vigne FRANCE 1982
Gerard Depardieu, Nathalie Baye, Roger Planchon, Maurice Jacquemont, Bernard Pierre Donnadieu, Sylvie Meda, Maurice Barrier, Stephane Peau, Rose Thiery, Isabelle Sadoyan, Chantal Deruaz, Tcheky Karyo, Dominique Pinon, Andre Chaumeau
A 16th century peasant boy disappears from his village and returns seven years later as a man, and though there are some suspicions that he may be an imposter, most villagers (including his family) are initially content to take his word at face value. An absorbing study of self-deception that's a bit too meandering for its own good. Good direction and Michel Portal's atmospheric score are assets. Written by Vigne and Jean-Claude Carriere. See also SOMMERSBY.
Aka: LE RETOUR DE MARTIN GUERRE
DRA 106 min (ort 123 min) VIDrel: ARROW/RTM V

RETURN OF SHERLOCK HOLMES, THE ** *PG*
Kevin O'Connor USA 1987
Margaret Colin, Connie Booth, Michael Pennington, Lila Kaye, Barry Morse, Nicholas Guest
An American descendant of Dr Watson inherits a property in Britain and finds the frozen body of the famous sleuth in the cellar. She revives him and together they go back to the USA, where our detective has more than a little difficulty coming to terms with the 20th century. A good premise is spoilt by flippant treatment and poor development.
COM 90 min (ort 104 min) mTV VIDrel: 20TH/TECH V
Boa: short story by Arthur Conan Doyle.

RETURN OF SHERLOCK HOLMES, THE: THE MAN WITH THE TWISTED LIP *** *PG*
Patrick Lau UK 1986
Jeremy Brett, Edward Hardwicke, Clive Francis, Eleanor David, Terence Longdon, Denis Lill, Patricia Garwood, Rosalie Williams, Albert Moss, Dudley James
In this story Holmes is called in by the distraught wife of a missing businessman and finds that the clues lead him to a mysterious, unwashed beggar, who has been arrested as a possible suspect. Written by Alan Plater, this is one of the best entries in the entire series, with the air of mystery sustained right up to the final resolution.
Aka: ADVENTURES OF SHERLOCK HOLMES, THE: THE MAN WITH THE TWISTED LIP; MAN WITH THE TWISTED LIP, THE
DRA 104 min (2-film cassette) mTV
VIDrel: HEND/BMGREC V
Boa: short story by Arthur Conan Doyle.

RETURN OF SHERLOCK HOLMES, THE: SILVER BLAZE *** *PG*
Brian Mills UK 1986
Jeremy Brett, Edward Hardwicke, Peter Barkworth, Jonathan Coy, Rosalie Williams, Malcolm Storry, Manda-Jayne Beard, David John, Sally Faulkner, Russell Hunter, Nicholas Teare, Marcus Kimber, Geoffrey Banks
In this tale Holmes embarks on an attempt to discover exactly why a valuable race-horse vanished on the eve of a major race, and why its trainer was found dead, apparently murdered. Another highly enjoyable entry in this generally excellent series.
Aka: ADVENTURES OF SHERLOCK HOLMES, THE: SILVER BLAZE; SILVER BLAZE
DRA 102 min (2-film cassette) mTV
VIDrel: HEND/BMGREC V
Boa: short story by Arthur Conan Doyle.

RETURN OF SHERLOCK HOLMES, THE: THE ABBEY GRANGE *** *PG*
Peter Hammond UK 1986
Jeremy Brett, Edward Hardwicke, Paul Williamson, Conrad Phillips, Anne Louise Lambert, Oliver Tobias, Zulema Dene, Nicolas Chagrin
Holmes is called in to investigate the murder of a certain Sir Eustace Brackenstall and soon finds that what appears to be a simple case has deeper implications. Another episode in this detailed and entertaining series of adaptations. The script is by Trevor Bowen.
Aka: ABBEY GRANGE, THE; ADVENTURES OF SHEROLCK HOLMES, THE: THE ABBEY GRANGE
DRA 105 min (2-film cassette) mTV
VIDrel: HEND/BMGREC V
Boa: short story by Arthur Conan Doyle.

RETURN OF SHERLOCK HOLMES, THE: THE BRUCE PARTINGTON PLANS *** *15*
John Gorrie UK 1986
Jeremy Brett, Edward Hardwicke, Charles Gray, Denis Lill, Rosalie Williams, Jonathan Newth, Geoffrey Bayldon, Amanda Waring, Sebastian Stride, Robert Fyfe, John Rapley, Simon Carter, Derek Ware, Stephen Crane, John Laing
In this intriguing mystery, Holmes attempts to solve a baffling

case that revolves around the discovery of the body of a young man on a railway line. His pockets were found to contain secret plans for construction of a submarine and Holmes sets out to determine whether or not he was a traitor.

Aka: ADVENTURES OF SHERLOCK HOLMES, THE: THE BRUCE PARTINGTON PLANS; BRUCE PARTINGTON PLANS, THE
DRA   55 min; 104 min (2-film cassette) mTV
VIDrel: HEND/BMGREC   V
Boa: short story by Arthur Conan Doyle.

RETURN OF SHERLOCK HOLMES, THE: THE DEVIL'S FOOT **
Ken Hannam UK   PG
                                                                1986
Jeremy Brett, Edward Hardwicke, Peter Barkworth, Norman Bowler, Dennis Quilley, Damien Thomas, Michael Aitkins, Fred Dowie, Peter Shaw, Christine Collins, John Saunders, Frank Moorey
Holmes sets out for Cornwall and a much-needed holiday, but spends his time solving a mysterious murder in which one of the victims, a young woman, appears to have died of sheer fright, while her brothers, who have survived, have gone completely insane. Another excellent adaptation that's highly enjoyable right up to the final resolution.
Aka: ADVENTURES OF SHERLOCK HOLMES, THE: THE DEVIL'S FOOT; DEVIL'S FOOT, THE
DRA   55 min; 102 min (2-film cassette) mTV
VIDrel: HEND/BMGREC   V
Boa: short story by Arthur Conan Doyle.

RETURN OF SHERLOCK HOLMES, THE: THE EMPTY HOUSE ****
Howard Baker UK   PG
                                                                1986
Jeremy Brett, Edward Hardwicke, Patrick Allen, Colin Jeavons
After the success of the first set of Holmes tales this new series began, starting where the last episode (see THE ADVENTURES OF SHERLOCK HOLMES: THE FINAL PROBLEM) concluded. Three years have passed since the apparent death of his colleague at the Reichenbach Falls, and Watson attempts to solve a puzzling murder case. A mysterious stranger appears to offer him some help. Written by John Hawkesworth.
Aka: ADVENTURES OF SHERLOCK HOLMES, THE: THE EMPTY HOUSE; EMPTY HOUSE, THE
DRA   105 min (2-film cassette) mTV
VIDrel: HEND/BMGREC   V
Boa: short story by Arthur Conan Doyle.

RETURN OF SHERLOCK HOLMES, THE: THE HOUND OF THE BASKERVILLES ****
Brian Mills UK   PG
                                                                1988
Jeremy Brett, Edward Hardwicke, Raymond Adamson, Neil Duncan, Ronald Pickup, Rosemary McHale, Kristoffer Tabori, Edward Romfourt, James Faulkner, Philip Dettmer, Stephen Tomlin, Fiona Gillies, Bernard Horsfall, Donald McKillop
An excellent adaptation of the famous tale, one in a series of television productions that gave Brett an opportunity to demonstrate his outstanding interpretation of the part of Holmes, ably supported by Hardwicke as Watson. Was also available in a three-cassette set. This adventure was the final story in the RETURN OF SHERLOCK HOLMES series. The next set of tales began with THE CASEBOOK OF SHERLOCK HOLMES: THE DISAPPEARANCE OF LADY CARFAX.
Aka: HOUND OF THE BASKERVILLES, THE
DRA   105 min mTV   VIDrel: HEND/BMGREC   V
Boa: novel by Arthur Conan Doyle.

RETURN OF SHERLOCK HOLMES, THE: THE MUSGRAVE RITUAL ****
David Carson UK   PG
                                                                1986
Jeremy Brett, Edward Hardwicke, Michael Culver, James Hazeldine, Johanna Kirby, Teresa Banham, Ian Marter, Patrick Blackwell, Wayne Michaels
Holmes and Watson spend the weekend at the country home of one of the latter's college chums, and get involved in the disappearance of the butler and one of the maids. An ancient document dating from the Civil War appears to hold the key to this mystery, but Holmes learns of an even deeper one involving the lost crown of Charles I. Written by Jeremy Paul, this is one of the best episodes in the series.
Aka: ADVENTURES OF SHERLOCK HOLMES, THE: THE MUSGRAVE RITUAL; MUSGRAVE RITUAL, THE
DRA   55 min; 105 min (2-film cassette) mTV
VIDrel: HEND/BMGREC   V
Boa: short story by Arthur Conan Doyle.

RETURN OF SHERLOCK HOLMES, THE: THE PRIORY SCHOOL ***
John Madden UK   PG
                                                                1986
Jeremy Brett, Edward Hardwicke, Christopher Benjamin, Alan Howard, Nicholas Gecks, Michael Bertenshaw, Jack Carr, Brenda Elder, Rosalie Williams, Mark Turin, William Abne
When the son of a duke is kidnapped Holmes finds himself working on one of his most puzzling cases ever. Another excellent episode in this series of adaptations. The script is by T.R. Bowen.
Aka: ADVENTURES OF SHERLOCK HOLMES, THE: THE PRIORY SCHOOL; PRIORY SCHOOL, THE
DRA   55 min; 104 min (2-film cassette) mTV
VIDrel: HEND/BMGREC   V
Boa: short story by Arthur Conan Doyle.

RETURN OF SHERLOCK HOLMES, THE: THE SECOND STAIN ***
John Bruce UK   PG
                                                                1986
Jeremy Brett, Edward Hardwicke, Patricia Hodge, Harry Andrews, Colin Jeavons, Stuart Wilson, Sean Scanlon, Yves Beneyton, Yvonne Orengo, Rosalie Williams, Alan Benion
A call from the Prime Minister involves our detective in the hunt for a stolen letter whose contents could, were it made public, lead to a state of war. Written by John Hawkesworth.
Aka: ADVENTURES OF SHERLOCK HOLMES, THE: THE SECOND STAIN; SECOND STAIN, THE
DRA   105 min (2-film cassette) mTV
VIDrel: HEND/BMGREC   V
Boa: short story by Arthur Conan Doyle.

RETURN OF SHERLOCK HOLMES, THE: THE SIX NAPOLEONS ***
David Carson UK   PG
                                                                1986
Jeremy Brett, Edward Hardwicke, Colin Jeavons, Eric Sykes, Gerald Campion, Vincenzo Nicoli, Steve Plytas, Vernon Dobtcheff, Marina Sirtis, Emil Wolk, Nadio Fortune, Michael Logan, Jeffrey Gardiner
A burglar steals six busts of Napoleon from a statuette shop and smashes them to bits. Further similar incidents occur at the homes of people who had bought such busts. Eventually Holmes discovers that the culprit is searching for something he secreted in the plaster used to make the busts. The script is by John Kane, and though there is little action to be had here, the first-rate cast and strong atmosphere make this a most enjoyable episode.
Aka: ADVENTURES OF SHERLOCK HOLMES, THE: THE SIX NAPOLEONS; SIX NAPOLEONS, THE
DRA   52 min (ort 55 min) mTV   VIDrel: HEND/BMGREC   V
Boa: short story by Arthur Conan Doyle.

RETURN OF SHERLOCK HOLMES, THE: WISTERIA LODGE ***
Peter Hammond UK   PG
                                                                1986
Jeremy Brett, Edward Hardwicke, Freddie Jones, Charles Gray, Kika Markham, Donald Churchill, Basil Hoskins, Trader Faulkner, Arturo Venegas, Guido Adorni, Sonny Caldinez, Abigail Melia, Lorna Rossi
In "Wisteria Lodge" Holmes sets out to solve a set of mysterious events that seem to have culminated in a murder, and is soon deeply embroiled in a plot involving a deposed Central American tyrant. He has to act swiftly in order to save a woman's life and finds to his delight that for once the detective investigating the case is a remarkably astute fellow. Another fine adaptation in a series of consistently high quality.
Aka: ADVENTURES OF SHERLOCK HOLMES, THE: WISTERIA LODGE; WISTERIA LODGE
DRA   55 min; 104 min (2-film cassette) mTV
VIDrel: HEND/BMGREC   V
Boa: short story by Arthur Conan Doyle.

RETURN OF SUPERFLY, THE **
Sig Shore USA   18
                                                                1990
Nathan Purdee, Margaret Avery, Sam Jackson, Leonard Thomas, Christopher Curry, David Groh, Carlos Carrasco, John Gabriel, Luis Ramos, Kirk Taylor, O.L. Duke, Eric Payne, Tico Wells, Patrice Abalck, Arnold Mazer, Rutanya Graves
This belated sequel to SUPERFLY and SUPERFLY T.N.T. has our hero returning to New York from Paris, and finding himself caught up in the middle of a bloody war between drug runners and the DEA. The latter body has been spreading rumours that have resulted in his friends being killed and he is now ready to put an end to the carnage.
A/AD   95 min (ort 97 min)   VIDrel: MIA/DISC   V

RETURN OF THE EVIL DEAD, THE ** 18
Armando de Ossorio PORTUGAL/SPAIN   1973
Tony Kendall (Luciano Stella), Esther Ray (Esperanza Roy), Fernando Sancho, Lone Fleming, Frank Blake (Frank Brana), Loretta Tovar, Jose Canalejas, Ramon Lillo, Joseph Thelman, Maria Nuria Rodriguez, Juan Cazallila
An evil medieval sect is resurrected and wreaks havoc on a small Portuguese town. Second in this series of films about our dear old eyeless Knights Templar, who were executed in the Middle Ages but now have returned as eyeless, vengeance-seeking zombies. A competent, low-budget sequel to TOMBS OF THE BLIND DEAD. See also NIGHT OF THE SEAGULLS.
Aka: ATTACK OF THE BLIND DEAD; EL ATAQUE DE LOS MUERTOS SIN OJOS; EL RETORNO DEL TERROR CIEGO; RETURN OF THE BLIND DEAD
HOR   83 min (ort 91 min) wScrn dubbed
VIDrel: REDEM/RTM   V

RETURN OF THE FLY * 15
Edward L. Bernds USA   1959
Vincent Price, Brett Halsey, David Frankham, John Sutton, Danielle De Metz, Dan Seymour, Michael Mark, Janine Grandel, Pat O'Hara, Jack Daly, Barry Bernard, Richard Flato, Joan Cotton, Florence Strom, Gregg Martell, Ed Wolff
A rubbishy sequel to THE FLY, with the son of the ill-fated scientist reconstructing his dad's matter transporter and suffering a similar fate. Followed six years later by "The Curse Of The Fly".
HOR   75 min (ort 78 min) B/W   VIDrel: 20TH/TECH   V/h

RETURN OF THE JEDI *** U
Richard Marquand USA   1983
Mark Hamil, Harrison Ford, Billy Dee Williams, Carrie Fisher, Anthony Daniels, Dave Prowse, Alec Guinness, Peter Mayhew, Sebastian Shaw, Frank Oz, Ian McDiarmid, Kenny Baker plus James Earl Jones (voice of Darth Vader)
Third in the STAR WARS trilogy, with the plot a continuation of the battle between the heroic rebels and the decadent Galactic Empire, and a new Death Star poses a threat to freedom. An agreeable, juvenile fantasy that soon wears out its welcome and uses a few too many Muppet-like characters. Written by Lawrence Kasdan and George Lucas with music by John Williams. AA: Spec Award (Richard Edlund/Dennis Muren/Ken Ralston/Phil Tippett for visual effects).
FAN   126 min; 132 min (special edition) wScrn
VIDrel: 20TH/TECH   V/drm

RETURN OF THE KILLER TOMATOES! * 15
John DeBello USA   1988
Anthony Starke, George Clooney, Karen Mistal, Steve Lundquist, John Astin, Charlie Jones, Mike Villani, Harvey Weber, Rock Peave, John DeBello, Gordon Howard, C.J. "Clark" Dillon, Mark Wenzel, Spike Sorrentino, Devlin John
Sequel to ATTACK OF THE KILLER TOMATOES that promises more of the same, with one joke being over-extended ad nauseam. A mad scientist turns tomatoes into people and people into tomatoes, and our hero falls in love with a former tomato. Unfortunately, unless you're heavily into tomatoes, this just isn't very funny.
Aka: RETURN OF THE KILLER TOMATOES: THE SEQUEL; REVENGE OF THE KILLER TOMATOES
COM   94 min (rt 99 min)   VIDrel: VCC/DISC/COLUM L/A   V

RETURN OF THE LIVING DEAD ** 18
Dan O'Bannon USA   1984
Clu Gulager, James Karen, Don Calfa, Thom Mathews, Beverly Randolph, John Philbin, Linnea Quigley, Miguel A. Nunez Jr, Brian Peck, Jewel Shepard, Mark Venturini, Jonathan Terry, Cathleen Cordell, Drew Deigham
A semi-comic sequel to the original NIGHT OF THE LIVING DEAD in which the zombie plague is once more unleashed on humanity and the cinema-going public. Much comic mayhem mixes with gory effects, but what starts out as a spoof turns nasty about halfway through. This was the directorial debut for SF screenwriter O'Bannon. Followed by a couple more sequels.
HOR   86 min (ort 96 min) wScrn   VIDrel: TART/20TH   V/s

RETURN OF THE LIVING DEAD: PART 2 * 18
Ken Wiederhorn USA   1987
James Karen, Thom Mathews, Michael Kenworthy, Dana Ashbrook, Philip Bruns, Marsha Dietlein, Suzanne Snyder, Thor Van Lingen, Jason Hogan, Sally Smythe, Suzan Stander, Jonathan Terry, Allan Trautman, Don Maxwell, Reynold Cindrich

A sequel to the 1984 spoof that has a group of irresponsible kids causing the dead to rise up again, when they open strange barrels that have fallen off an army truck. Good effects are combined with a ponderous and uninventive script.
HOR   85 min (ort 89 min)
VIDrel: 4-FRONT/POLYREC/GUILD   V/sh

RETURN OF THE LIVING DEAD 3 ** 18
Brian Yuzna USA   1993
J. Trevor Edmond, Mindy Clarke, Sarah Douglas, Kent McCord, Sarah Douglas, James T. Callahan, Mike Moroff, Sal Lopez, Basil Wallace
Long-awaited second sequel to the first cult hit adopts a radically different and less gory slant as it attempts to tell the story of a youth, who having lost his girlfriend, proceeds to bring her back to life. Needless to say, the results are nothing less than catastrophic.
Aka: RETURN OF THE LIVING DEAD: PART 3
HOR   93 min (ort 97 min)   VIDrel: TRIM/HIFLI   V/sur

RETURN OF THE MUSKETEERS, THE ** PG
Richard Lester UK   1988
Michael York, Oliver Reed, Frank Finlay, C. Thomas Howell, Kim Cattrall, Roy Kinnear, Geraldine Chaplin, Christopher Lee, Richard Chamberlain, Philippe Noiret, Eusebio Lazaro, Alan Howard, David Birkin, Jean-Pierre Cassel
A lavish but dispirited follow-up to Lester's THE THREE MUSKETEERS/THE FOUR MUSKETEERS with many of the same stars in the same roles as the earlier film. The complex story is almost impossible to follow, largely consisting of kidnapping and murder plots plus a good dose of swashbuckling. It moves along rapidly, but has too many shortcomings to be all that effective. Kinnear died during shooting after falling from a horse – this film is dedicated to him.
A/AD   97 min   VIDrel: EIV/SONOP L/A   V
Boa: novel Twenty Years After by Alexandre Dumas.

RETURN OF THE PINK PANTHER, THE * PG
Blake Edwards UK   1974
Peter Sellers, Christopher Plummer, Herbert Lom, Catherine Schell, Peter Arne, Peter Jeffrey, Burt Kwouk, Gregoire Aslan, Andre Maranne, Victor Spinetti, David Lodge, Graham Stark, Eric Pohlmann, John Bluthal, Mike Grady, Jeremy Hawk
Another unfunny PINK PANTHER film, with Sellers once more playing our bumbling Inspector Clouseau, this time investigating the theft of the Pink Panther diamond. Music is by Henry Mancini. The opening titles, animated by Richard Williams and Ken Harris, are the best part of the film. (Benny Green wrote in Punch that this was the first movie to be upstaged by its own credits.) Followed by THE PINK PANTHER STRIKES AGAIN.
COM   108 min (ort 113 min)   VIDrel: POLY L/A   V

RETURN OF THE SHAGGY DOG, THE * U
Stuart Gillard USA   1987
Cindy Morgan, Todd Waring, Gary Kroeger, Michelle Little, Jane Carr, Gavin Reed, K. Callan
A feeble attempt to exploit the success of the 1959 Disney film, "The Shaggy Dog", with a young boy finding that he is able to change into a dog thanks to a magic ring. An uninspired and wearisome effort. See also THE SHAGGY D.A.
COM   85 min (ort 100 min) mTV   VIDrel: WDV/TECH L/A   V

RETURN OF THE SOLDIER, THE ** PG
Alan Bridges UK   1982
Alan Bates, Julie Christie, Glenda Jackson, Ann-Margret, Ian Holm, Frank Finlay, Jeremy Kemp, Edward De Souza, Jack May, Emily Irvin, William Booker, Elizabeth Edmonds, Hilary Mason, John Sharp, Valerie Aitken, Amanda Grinling
A soldier returns home in a state of shell-shock, with no memory of the past twenty years and his marriage to a haughty woman. Nor does he realise that two other women are in love with him. A delicate and touching love story by West, becomes a bleak and ponderous tale that, despite good performances, never develops into anything of substance. Scripted by Hugh Whitemore.
DRA   98 min (ort 102 min)   VIDrel: L/A   V
Boa: novel by Rebecca West.

RETURN OF THE SWAMP THING, THE ** 15
Jim Wynorski USA   1988
Heather Locklear, Sarah Douglas, Louis Jordan, Dick Durock, Joey Segal
Based on the D.C. Comics series, this sequel to THE SWAMP THING is a spoof horror tale, in which a girl searching for the

reasons behind her mother's death gains an unlikely ally in the shape of a man-vegetable creature, who helps her escape the clutches of her mad scientist stepfather and thwarts his fiendish plans. A thoroughly silly effort that tries too hard to parody early B-movie horror films and ends up merely parodying itself.
HOR 88 min wScrn VIDrel: IMAG/RTM V

RETURN OF THE TIGER *   18
Jimmy Shaw (James Fung Shaw) HONG KONG/TAIWAN 1973
Bruce Li (Ho Tsung-Tao), Paul Smith, Angela Mao Ying, Chang I, Lung Fei, Hsieh Hsing
An undercover agent combats narcotics smugglers in Bangkok, being assisted n his efforts by a female cop. Another cynical attempt to exploit the popularity of the late star by using a Bruce Lee clone.
Aka: SILENT KILLER FROM ETERNITY
MAR 92 min Cut (3 sec – ort 95 min) VIDrel: IMPENT V

RETURN OF TOMMY TRICKER, THE ***   (PG)
Michael Rubbo CANADA   1994
Michael Stevens, Adele Gray, Joshawa Mathers, Heather Goodsell, John Dapery, Paul Nicholls, Danette McKay, Daniel Richter, Charles V. Doucet, Michael Perran, Sam Stone
Jean-Raymond Charles, Danette McKay, Daniel Richter, Charles V. Doucet, Michael Perran, Sam Stone
A sequel to TOMMY TRICKER AND THE STAMP TRAVELLER that sees our young hero once again using his power to travel all over the world via the medium of postage stamps. In doing so he frees a man who was held captive on a stamp for sixty-five years, but faces an even greater challenge when his sister begins to age an accelerated rate.
JUV 97 min SATrel: SKY MOVIES

RETURN TO HORROR HIGH **   18
Bill Froehlich USA   1987
Alex Rocco, Vince Edwards, Brendan Hughes, Philip McKeon, Scott Jacoby, Lori Lethin, Andy Romano, Richard Brestoff, Al Fann, Pepper Martin, Maureen McCormick, Panchito Gomez, Michael Eric Kramer
A film crew go to a deserted high school to make a low-budget movie about a series of gruesome murders that took place there five years earlier. Soon, the murders start once more, in this fairly effective slasher tale that is hampered by poor direction and a muddled script.
HOR 90 min (ort 95 min) VIDrel: NWV L/A V/h

RETURN TO JUSTICE **   15
Vincent G. Cox USA
Richard Lynch, Griffin O'Neal, Cameron Mitchell
A few exciting moments are to be found in this predictable tale of a youth held in a Third World prison with little hope of release.
A/AD 88 min VIDrel: GENESIS V

RETURN TO LONESOME DOVE: PARTS 1, 2 AND 3 **   PG
Mike Robe USA   1992
Jon Voight, Barbara Hershey, Rick Schroder, Louis Gossett Jr, Oliver Reed, William Petersen, Dennis Haysbert, Reese Witherspoon, Tim Scott, C.C.H. Pounder, Chris Cooper, Nia Peeples, Barry Tubb, William Sanderson, Leon Singer
Sequel to the earlier LONESOME DOVE carries on the story where it left off, with most of the now familiar characters. After burying his friend, an ex-Texas Ranger sets out to capture some wild horses as he has decided to go in for mustang breeding. Along the way he meets with the usual stock situations and colourful characters.
WES 315 min (3 cassettes) mTV VIDrel: MARQ/QUANT

RETURN TO MACON COUNTY *   15
Richard Compton USA   1975
Don Johnson, Nick Nolte, Robin Mattson, Robert Viharo, Eugene Daniels, Matt Greene, Devon Ericson, Ron Prather, Philip Crews, Laura Sayer, Walt Guthrie, Mary Ann Hearn, Sam Kilman, Bill Moses, Pat O'Connor, Maurice Hunt, Kim Graham
A follow-up to MACON COUNTY LINE, in which our 1950s teenagers get involved in drag racing and murder and run into trouble from a brutal cop. A forgettable effort that has a few unintentionally funny moments, but nothing else of interest. Nolte's first film.
Aka: RETURN TO MACON COUNTY LINE
DRA 90 min VIDrel: L/A V

RETURN TO OZ **   PG
Walter Murch USA   1985
Fairuza Balk, Nicol Williamson, Jean Marsh, Piper Laurie, Matt Clark, Tim Rose, Michael Sundin, Stewart Larange, Justin Case, John Alexander, Deep Roy plus voices of: Sean Barrett, Denise Bryer, Brian Henson, Lyle Conway
This unattractive directing debut for sound technician Murch, is a partial sequel to THE WIZARD OF OZ. Oz now lies in ruins and Dorothy ventures there to save her old friends from petrification. Though loosely based on Baum's classic tales, this film is dark in tone and no more than coldly efficient. A highlight is the sequence in which rock faces come to life – courtesy of Will Vinton's Claymation technique. Written by Murch and Gill Dennis.
JUV 108 min (ort 110 min) VIDrel: WDV/TECH L/A V
Boa: novels The Land of Oz/Ozma of Oz by Lyman Frank Baum.

RETURN TO SALEM'S LOT, A ***   18
Larry Cohen USA   1987
Michael Moriarty, Samuel Fuller, Andrew Duggan, June Havoc, Evelyn Keyes, Richard Addison Reed, Ronee Blakely, Jill Gatsby, James Dixon, Tara Reid, David Holbrook, Natja Crosby, Brad Rijn, Robert Burr, Jacqueline Britton
A sequel to SALEM'S LOT with an anthropologist and his son going to the New England town where he has inherited a cottage from his aunt, and discovering that the area is the home of vampires who have kept their existence there secret for over 300 years. He eventually teams up with an ex-Nazi hunter in an effort to lift the curse from the town. A highly effective film that balances the chills with some nice touches of humour.
HOR 96 min (ort 101 min) VIDrel: WHV V

RETURN TO THE BLUE LAGOON *   15
William A. Graham USA   1991
Milla Jovovich, Brian Krause, Lisa Pelikan, Courtney Phillips, Garette Patrick Ratliff, Emma James, Jackson Barton, Nana Coburn, James Blain, Peter Hehir, Alexander Petersons, John Mann, Wayne Pygram, John Dicks, Todd Rippon
Supposedly a sequel to the 1980 film THE BLUE LAGOON, with the child born in the first film now a grown-up orphan, and enjoying the charms of an attractive widow. Together with her young daughter, all three find themselves on the island, where the mother meets an untimely end. The film now marks time until the girl is old enough to excite the interest of her companion. A boring film, unashamedly exploitative and prurient.
DRA 97 min (ort 102 min) cC VIDrel: COLUM/SONOP V/sur
Boa: novel The Garden of God by Henry de Vere Stacpoole.

RETURN TO THE LOST WORLD *   PG
Timothy Bond CANADA   1991
John Rhys-Davies, Tamara Gorski, Nathania Stanford, David Warner, Darren Peter Mercer, Eric McCormack, Sala Cane, Fideliz Chea, John Chinosiyani, Geza Kovacs, Innocent Choga, Brian Cooper, Charles David, Kate Egan, Mike Grey
Sequel to THE LOST WORLD that sees our two rival explorers returning to the plateau in Latin America and helping to save it from a group of oil prospectors who are prepared to stop at nothing in their search for quick riches. Another cardboard effort with low production values and indifferent acting.
Aka: JURASSIC ADVENTURES
A/AD 94 min (ort 180 min) mTV VIDrel: AUDIO/VCC V

RETURN TO THE 36TH CHAMBER **   PG
Liu Chia-Liang HONG KONG   1980
Lo Lieh, Liu Chia-Hui, Hui Ya-Hung, Wang Lung-Wei
A young mill worker learns the martial arts to avenge the death of a workmate killed by a gang. A sequel to THE THIRTY-SIXTH CHAMBER OF SHAOLIN.
MAR 100 min VIDrel: L/A V

RETURN TO TWO MOON JUNCTION **   18
Farhad Mann USA   1993
Melinda Clarke, John Clayton Schafer, Louise Fletcher, Wendy Davis, Yorgo Constantine, Molly Shannon, Montrose Hagins, Bill Hollis, Richard Keats, James Callahan, David Dunard, Brian Sanders, Suzanne Ircha, Pamela West
A sequel of sorts to TWO MOON JUNCTION, in which the sister of the main character in that film returns to her home town in Georgia to re-assess her life after her modelling career takes off. However, matters are complicated when she falls madly in

love and her former two-timing boyfriend turns up from New York. Adequate steamy goings on.
DRA   92 min (ort 96 min)   VIDrel: TRIM/HIFLI   V/h

REUBEN, REUBEN ***   15
Robert Ellis Miller USA   1983
Tom Conti, Kelly McGillis, Roberts Blossom, Cynthia Harris, E. Katherine Kerr, Joel Fabiani, Kara Wilson, Lois Smith, Ed Grady, Damon Douglas, Rex Robbins, Jack Davidson, Robert Nichols, Tom McGowan, Dan Doby, Barry Bell
A drunken, burnt-out Scottish poet lives off his female admirers in a small university town in New England, but then falls in love with a beautiful young student (McGillis in her screen debut). A curious and offbeat comedy in which mirth often gives way to a feeling of irritation. Written by Julius J. Epstein.
COM   95 min (ort 101 min)   VIDrel: POLY L/A   V
Boa: novel by Peter De Vries/play Spofford by Herman Shumlin.

REUNION ***   15
Jerry Schatzberg FRANCE/UK/USA/
WEST GERMANY   1989
Jason Robards, Christian Anholt, Samuel West, Alexander Trauner, Francoise Fabian, Maureen Kerwin, Barbara Jafford, Dorothea Alexander, Frank Baker, Tim Barker, Imke Burnstedt, Gideon Boulting, Alan Brunet, Jacques Brunet
Screenplay is by Harold Pinter in a story that casts Robards as an elderly businessman who is returning to his native Germany, a trip that revives memories of an aristocratic schoolmate he knew in 1933 and now wishes to find. His reminiscences are tied to a set of engrossing flashbacks in a languid tale that serves to focus on the anti-Semitism of Hitler's Germany and all it led to. A sharp and poignant film, yet also one that is surprisingly positive in tone.
DRA   105 min (ort 120 min)   VIDrel: COLUM/SONOP   V/sur
Boa: novel by Fred Uhlman.

REUNION **   (15)
Lee Grant USA   1993
Marlo Thomas, Peter Strauss, Frances Sternhagen, Courtney Chase, Matthew Kelly, Leelee Sobieski, Victor Slezak, Susan Kellerman, Stanley Anderson, Norris Domingue, Charly Simonyl
A supernatural thriller casting Thomas as a distraught mother who has lost her four-year-old son, but receives a measure of comfort when he returns to her as a ghost. A superficial and rather flashy yarn, it is not quite the touching ghost story its makers intended, chiefly thanks to heavy-handed direction and a lack of conviction.
FAN   90 min   SATrel: SKY MOVIES
Boa: novel Points of Light by Linda Guy Sexton.

REUNION IN FRANCE **   U
Jules Dassin USA   1942
John Wayne, Joan Crawford, Philip Dorn, Reginald Owen, Albert Basserman, John Carradine, Ann Ayars, J. Edward Bromberg, Henry Daniell, Moroni Olsen, Howard Da Silva, Charles Arnt, Morris Ankrum, Edith Evanson, Ernst Dorian
A pilot downed over France is helped to escape by a former Parisian dress designer who is now working for the Resistance. Inevitably, the two fall in love as they desperately try to stay one jump ahead of the Nazis. A pleasing WW2 morale-booster that is still quite watchable.
Aka: MADEMOISELLE FRANCE; REUNION
DRA   100 min B/W   VIDrel: MGM/WHV   V/h

REVELATIONS ***   18
Candida Royalle USA   1992
Amy Rapp, Colin Matthews, Ava Grace
A woman inhabits a future world where sex has been outlawed except for procreative purposes. Her discovery of a video-tape is about to change her attitude to this. This one benefits from both its intriguing opening premise and Royalle's penchant for taking a decidedly romantic view of sex, which tends to ensure the storyline is stronger than is usually the case.
A   73 min (ort 80 min)   VIDrel: MIA/DISC   V

REVENGE **   18
Tony Scott USA   1989
Kevin Costner, Madeleine Stowe, Anthony Quinn, Sally Kirkland, Tomas Milian, Joaquin Martinez, James Gammon, Miguel Ferrer, Joe Santos, Jesse Conti, Luis De Icaza, Gerardo Zapeda, John Leguizamo, Christopher De Oni, Daniel Rojo
Navy pilot Costner pays a visit to a Mexican gangster whose life

he once saved, but unwisely embarks on an affair with the man's beautiful wife. For this act of betrayal, a brutal revenge follows (hence the title) and this sordid tale rapidly descends into nastiness. A competently made film that blends good camerawork with unpleasant violence, and hides nothing beneath its flashy exterior.
DRA   118 min (ort 124 min)   VIDrel: VCC/COLUM   V/sur

REVENGE OF BILLY THE KID *   18
Jim Groom USA   1991
Michael Balfour, Samantha Perkins, Jackie D. Broad
The title character is the offspring of an unnatural coupling between a degenerate old farmer and his goat, and as he grows to adulthood, he becomes progressively more uncontrollable, until the inevitable violent, gory and sickening resolution. A difficult film to either describe or sit through, this repulsive and unashamedly offensive effort brings to mind BAD TASTE, a movie of remarkably similar attributes.
COM   87 min   VIDrel: MED/POLYREC L/A   V

REVENGE OF FRANKENSTEIN, THE **   15
Terence Fisher UK   1958
Peter Cushing, Francis Matthews, Eunice Grayson, Michael Gwynn, Lionel Jeffries, Oscar Quitak, John Welsh, Richard Wordsworth, Charles Lloyd Pack, John Stuart, Arnold Diamond, Margery Cresley, Anna Walmsley, George Woodbridge
After the death of his monster, the good doctor resumes his experiments in another town, under an assumed name after escaping the guillotine. A slow-moving horror yarn enlivened by a few touches of ghoulish humour. This was the second "Frankenstein" film from Hammer, being preceded by THE CURSE OF FRANKENSTEIN. "The Evil Of Frankenstein" followed.
HOR   87 min (ort 91 min)   VIDrel: COLUM/SONOP   V

REVENGE OF PUMPKINHEAD, THE **   18
Jeff Burr USA   1993
Ami Dolenz, Andrew Robinson, R.A. Mihailoff, Linnea Quigley, Steve Kanaly, J. Trevor Edmond, Caren Kaye, Kane Holder, Roger Clinton, Soleil Moon Frye
This sequel to VENGEANCE, THE DEMON has our monstrous demon on the loose once more, helped by an old witch who guards the soul of a murdered boy. A most unremarkable effort, notable solely because of an appearance by President Clinton's brother.
Aka: PUMPKINHEAD 2: BLOODWINGS
HOR   84 min   VIDrel: HIFLI/SONOP   V

REVENGE OF THE DRUNKEN MASTER **   18
Godfrey Ho HONG KONG   198-
Johnny Chan, Eagle Han, Wang Sao, Si-Fu
A young man masters the technique of Drunk Fist and use his newly acquired skills to fight an evil Ninja gang, who are forced to unite in an effort to destroy him.
MAR   81 min   VIDrel: IMPENT   V

REVENGE OF THE LIVING DEAD GIRLS, THE *   18
Pierre Reinhard FRANCE   1965
Kathryn Charly, Anthea Wyler, Veronik Catanzaro, Sylvie Novak, Laurence Mercier
Three young women die in a road accident, but the fourth passenger survives, and helps the others return by raising them up as zombies. They them embark on a sexual rampage, working their way through the unfortunate local inhabitants of a small town. A ridiculous film of inane plotting and clumsy gore-laden effects, helped along with a twist at the end so daft it defies description.
Aka: LA REVANCHE DES MORTES VIVANTES; LIVING DEAD, THE
HOR   75 min (ort 78 min) wScrn dubbed
VIDrel: REDEM/RTM   V

REVENGE OF THE NERDS **   18
Jeff Kanew USA   1984
Robert Carradine, Anthony Edwards, Julie Montgomery, Curtis Armstrong, Ted McGinley, Bernie Casey, Tim Busfield, Andrew Cassese, Larry B. Scott, Brian Tochi, Michelle Meyrink, Matt Salinger, Donald Gibb, James Cromwell
Brainy college students get tired of being humiliated by sporty types and form their own fraternity, leading to a humorous conflict on the campus. A simple-minded teen comedy of appealing performances and obvious gags. The script is by Tim Metcalfe, Miguel Tejada-Flores, Steve Zacharias and Jeff Buhai. And of course, some sequels followed.
COM   87 min (ort 90 min)   VIDrel: 20TH/TECH   V/h

REVENGE OF THE NERDS 2 *
Joe Roth USA
15
1987
Robert Carradine, Curtis Armstrong, Larry B. Scott, Timothy Busfield, Andrew Cassese, Barry Sobel, Courtney Thorne-Smith, Anthony Edwards, Ed Lauter, Donald Gibb, Bradley Whitford, Tom Hodges, Jack Gilpin, James Cromwell, James Hong
Once again the nerds prove themselves worthy of our respect, as they triumph over their traditional enemies at a fraternity convention held at Fort Lauderdale. A production-line comedy that is even less inspired than the original.
Aka: REVENGE OF THE NERDS 2: NERDS IN PARADISE
COM   85 min (ort 95 min)   VIDrel: 20TH/TECH   V/sur

REVENGE OF THE NERDS 3: THE NEXT GENERATION *
Roland Mesa USA
PG
1992
Gregg Binkley, Richard Israel, Morton Downey Jr, John Pinette, Henry Cho, Grant Heslov, Mark Clayman, Chi, K.T. Vogt, Tim Conlon, Laurel Moglen, Brian Tochi, Curtis Armstrong, Robert Carradine, Bernie Casey, Jamie Cromwell
Bottom-of-the-barrel offering that attempts to squeeze a few last laughs from the same stock situations as in the first two films. When the nephew of the founder of the nerd fraternity attends Adams College, he finds himself turning to his illustrious uncle for help after he starts to suffer the usual indignities.
COM   87 min (ort 93 min) mTV   VIDrel: 20TH/TECH   V

REVENGE OF THE NERDS 4: NERDS IN LOVE **
Steve Zacharias USA
PG
1994
Robert Carradine, Julia Montgomery, Curtis Armstrong, Corinne Bohrer, Robert Picardo, Jessica Tuck, Christina Pickles, Stephen Davies, Larry B. Scott, Brian Tochi, Jamie Cromwell, Don Gibb, Bernie Casey, John Pinette, Greg Binkley
Ladies man Booger Dawson is about to get married, but things do not go according to plan, especially as the girl's family are not too be pleased at the prospect. Meanwhile, one of his friends throws a bachelor's party, and this being a juvenile comedy, we are regaled with the sight of things getting wildly out of control. Another entry in the series, offering no shortage of dumb gags for fans of the series.
COM   88 min mTV   VIDrel: 20TH/TECH   V

REVENGE OF THE PINK PANTHER *
Blake Edwards USA
PG
1978
Peter Sellers, Herbert Lom, Robert Webber, Dyan Cannon, Burt Kwouk, Robert Loggia, Paul Stewart, Graham Stark, Andre Maranne, Sue Lloyd, Tony Beckley, Valerie Leon, Alfie Bass, Danny Schiller, Douglas Wilmer, Ferdy Mayne
Sellers' final portrayal (whilst alive that is) of the bumbling and totally unfunny Inspector Clouseau. In this painfully tedious yawn-inducer, our detective is supposedly murdered and attempts to find his "killer". A nice performance from Cannon is thrown away on this dud. As ever, Henry Mancini provides the music. THE TRAIL OF THE PINK PANTHER followed, Sellers's death notwithstanding.
COM   94 min (ort 99 min)   VIDrel: MGM/WHV   V

REVENGE OF THE RADIOACTIVE REPORTER *
Craig Pryce CANADA
18
David Scammell, Kathryn Boese, Richard Sali, Randy Pearlstein, Derrik Strange, Angelo Celesta, Michael Lebovic, Erich Arensen, Grant Vistorino, Stuart Dunsworth, Linda Findlay, Paul Lamothe, Jane Edmundson, Larry Baker
An investigative reporter examining alleged irregularities at a nuclear plant is deliberately plunged into a barrel of radioactive waste as a way of getting rid of him. However, just as in TOXIC AVENGER, far from this causing his demise, he emerges all the better equipped to fight evil. A low-budget spoof of flat jokes and questionable taste, that ends on a decidedly off-key note as our hero expires in the arms of his girlfriend.
FAN   80 min (ort 90 min) mTV   VIDrel: CIC/SONOP L/A   V

REVERSAL OF FORTUNE ***
Barbet Schroeder USA
15
1990
Jeremy Irons, Glenn Close, Ron Silver, Fisher Stevens, Uta Hagen, Anabella Sciorra, Jack Gilpin, Christine Baranski, Stephen Mailer, Julie Hagerty, Johann Carlo, Christine Dunford, Mano Singh, Felicity Huffman, Keith Reddin, Tom Wright
The Claus von Bulow affair serves as the basis for a complex account of the events that led up to his trial for attempted murder of his wife, plus a few sharp observations on the moral-ity of the wealthy. Built around a set of flashbacks, the film is

effective both as a semi-documentary and a mystery, and its examination of the antics of Bulow's defence counsel (Dershowitz and others) injects a bizarre note of black humour.
AA: Act (Irons).
DRA   106 min (ort 120 min)   VIDrel: VCC/DISC/COLUM   V/sur
Boa: book by Alan Dershowitz.

REVOLUTION *
Hugh Hudson NORWAY/UK
PG
1985
Al Pacino, Nastassja Kinski, Donald Sutherland, Joan Plowright, Dave King, Steven Berkoff, John Wells, Annie Lennox, Dexter Fletcher, Sid Owen, Richard O'Brien, Paul Brooke, Eric Milota, Felicity Dean, Jo Anna Lee, Cheryl Miller
A lavish but badly cast and clumsy account of the American War of Independence. An unmitigated disaster from start to finish, with a constant reliance on hand-held cameras giving the film a nice "shaky" feel that makes it painful to watch. Severely panned by the critics, and deservedly so, this film was one of the biggest flops in recent cinema history, and at £18,000,000 an expensive dud for British cinema.
DRA   121 min (ort 125 min)   VIDrel: WHV   V/sur

RG VEDA: PARTS 1 AND 2 **
Hiroyuki Ebata/Yoshimasa Ikegami JAPAN
PG
199-
When the rightful ruler of the Universe is slain, a legend says that six warriors will arise to put matters to right. However, they must first find each other and then defeat the evil Taishakuten.
ANIM   78 min (ort 86 min)   VIDrel: MANGA/SONOP   V

RHAPSODY IN AUGUST ***
Akira Kurosawa JAPAN
U
1990
Sachiko Murase, Hisashi Igawa, Narumi Kayashima, Tomoko Ohtakara, Mitsunori Isaki, Richard Gere, Toshie Negishi, Choichiro Kawarasaki, Hidetaka Yoshioka, Mie Suzuki
The director's follow-up to DREAMS has four Japanese children spending the summer with their grandmother near Nagasaki, while their self-seeking parents are off to Hawaii to visit a wealthy but dying uncle. Prompted by the children, the grand-mother slowly begins to remember life in both pre and post-WW2 Japan, and this contrasts sharply with the selfish cultural "amnesia" of her children. A rather contrived but fasci-nating study.
Aka: HAKIGATSU NO KYOSHIKYOKU
DRA   97 min   VIDrel: L/A   V
Boa: novel Nabe-no-Naka by Kiyoko Murata.

RHODES **
David Drury CANADA/SOUTH AFRICA/UK/USA
12
1996
Martin Shaw, Neil Pearson, Frances Barber, Ken Stott, Joe Shaw, Alex Ferns, Frantz Dombrowsky, Jeremy Crutchley, Andre Odendaal, Patrick Shai, Robin Smith, David Butler, Philip Godawa, Ernest Ndlovu, Alistair Prodgers, Nick Lorentz
A long, ambitious but not terribly convincing attempt to portray the life and career of this remarkable empire-builder who seri-ously believed that the English-speaking "race" were destined to rule the world. After coming to Africa at an early age and taking over his brother's diamond workings, he proceeded to ruthlessly carve own a country of his own. Filmed entire on location in South Africa, this lavish mini-series never comes to life.
DRA   466 min (3 cassettes) mTV   VIDrel: BBC/TECH   V/sur

RHYTHM THIEF ***
Matthew Harrison USA
18
1994
Jason Andrews, Eddie Daniels, Kevin Corrigan, Kimberley Flynn, Sean Hagerty, Mark Alfred, Chris Cooke, Alan Davidson, Bob McGrath, Paul Rodriguez, Cynthia Sley
A young hustler makes a living selling pirated music tapes on New York's Lower East Side and apart from this, has a pecu-liarly empty life. But all this changes when he comes into contact with a young girl. Shot in just eleven days on a budget of $11,000, Harrison demonstrates his ability as director, drawing out of the cast performances of depth and vigour. An impres-sive film, it was well received by the critics at the World Film Festival at Montreal.
DRA   84 min B/W   VIDrel: SCEDGE/RTM   V

RICE PEOPLE ***
Rithy Panh CAMBODIA/FRANCE
PG
1994
Peng Phan, Mom Soth, Chhim Naline, Va Simorn, Sophy Sodany, Muong Danyda, Pen Sopheary, Proum Mary, Sam Kourou, Noy

Samnang, Phang Chamroeun, Chou Saokhun, Meas Daniel, Sou Bottra, Say, Seang Sarin, Ros Yarann, Sam Maly
Story of the sufferings of the Cambodians, whose lives are totally dependent on their cultivation of rice – hence the title. Panh's debut feature is an impressive work, drawing fine performances from a mainly non-professional cast, but at over two hours, it is just a little too long and far too depressing to be an easy film to watch. Yet the film's sincerity could not be plainer. Panh escaped from a Khmer Rouge camp and almost certain death aged only sixteen.
Aka: LES GENS DE LA RIZIERE
DRA   130 min   CINrel
  Boa: novel Ranju Sepanjang Jalan (Le Riz) by Shannon Ahmad.

RICH AND FAMOUS *   18
George Cukor USA   1981
Jacqueline Bisset, Candice Bergen, David Selby, Hart Bochner, Steven Hill, Meg Ryan, Matt Lattanzi, Michael Brandon, Daniel Faraldo, Nicole Eggert, Joe Maross, Kres Mersky, Cloyce Morrow, Cheryl Robinson, Allan Warnick, Ann Risley
Cukor's last film (he was eighty-two when it was shot) is a remake of "Old Acquaintance" and deals with two women who maintain their friendship over 20 years, despite being rivals in both love and career. A painfully contrived and muddled effort, made worse by excessive dialogue and an ill-advised attempt to spice up a straightforward story with some 1980s sex.
DRA   112 min (ort 117 min)   VIDrel: L/A   V
Boa: play Old Acquaintance by John Van Druten.

RICH AND FAMOUS **   18
Wong Tai Loi HONG KONG   1987
Chow Yun Fat, Danny Lee, Any Lau, Pauline Wong
Tale of violence, cowardice, lost innocence and betrayal, all wrapped up in a gangster format.
A/AD   99 min wScrn   VIDrel: MADE/RTM   V

RICH AND STRANGE **   (15)
Alfred Hitchcock UK   1932
Henry Kendall, Joan Barry, Betty Amann, Percy Marmont, Elsie Randolph, Hananh Jones, Aubrey Dexter
A young couple who have inherited money learn a few important truths about themselves and their lives while on a luxury cruise, when they take turns in getting themselves involved with other partners. Rather the odd-man-out in this director's repertoire and no more than moderately interesting.
Aka: EAST OF SHANGHAI
DRA   83 min (ort 92 min) B/W   VIDrel: LUMI/SPEAR   V
Boa: novel by Dale Collins.

RICH IN LOVE **   PG
Bruce Beresford USA   1992
Albert Finney, Jill Clayburgh, Piper Laurie, Kyle MacLachlan, Kathryn Erbe, Ethan Hawke, Suzy Amis, Alfre Woodard, J. Leon Pridgren II, David Hager, Ramona Ward, D.L. Anderson, Wayne Dehart, Janell McLeod, Jennifer Banio, Anthony Burke
Rambling tale of a family of Southern eccentrics after the mother walks out, leaving her indolent husband and teenage daughter to fend for themselves. He contemplates rekindling an old romance while she tries hard to patch things up. A meandering and unfocused adaptation where the acting, especially Finney's, partly compensates for these major shortcomings.
DRA   100 min (ort 105 min)   VIDrel: MGM/WHV   V/sur
Boa: novel by Josephine Humphreys.

RICH MAN, POOR MAN ***   PG
David Greene/Boris Sagal USA   1976
Nick Nolte, Peter Strauss, Susan Blakely, Steve Allen, Edward Asner, Bill Bixby, Dick Butkus, Kim Darby, Andrew Duggan, Mike Evans, Norman Fell, Fionnuala Flanagan, Lynda Day George, Gloria Grahame, Murray Hamilton, Van Johnson
Originally shown in six-parts, this ambitious mini-series followed the fortunes of two brothers over a period of twenty years, from the end of WW2 up to the 1960s. As much an examination of changing social values as the story of the characters and their intertwined lives, it even led to a weekly series "Rich Man, Poor Man: Book 2". A moderately entertaining affair, but its lack of depth tended to put it on a par with far less well made soap operas.
DRA   551 min (3 cassettes – ort 720 min) mTV
VIDrel: CIC/SONOP   V
Boa: novel by Irwin Shaw.

RICH MAN'S WIFE, THE **   18
Amy Holden Jones USA   1996
Halle Berry, Peter Greene, Clive Owen, Christopher McDonald, Charles Hallahan, Frankie Faison, Clea Lewis
Unhappily married to a successful TV executive, Berry makes the mistake of telling a stranger that she wishes her husband were dead. Totally implausible this little thriller may be, but there are a clutch of strong performances, especially one from Greene who has a great time playing the nutter Berry confides in. The twist at the end is completely unexpected.
THR   91 min cC   VIDrel: HOLPIC/TECH   V/sur

RICHARD III ***   U
Laurence Olivier UK   1955
Laurence Olivier, John Gielgud, Ralph Richardson, Claire Bloom, Alex Clunes, Cedric Hardwicke, Stanley Baker, Pamela Brown, Michael Gough, John Laurie, Norman Wooland, Laurence Naismith, Mary Kerridge, Helen Heye, Clive Morton
This famous play is brought to the screen by a host of Shakespearean actors. Now considered to be a classic, this is a remarkably theatrical production, that refuses to rely on the marvellous language of the play, and resorts instead to an over-reliance on visual effects. Nor are the battle scenes all that impressive. A rather disappointing and faded epic.
DRA   150 min (ort 161 min)   VIDrel: VCC/DISC/COLUM L/A   V
Boa: play by William Shakespeare.

RICHARD III ***   15
Richard Loncraine UK   1995
Ian McKellan, Annette Bening, Kristin Scott-Thomas, Jim Broadbent, Robert Downey Jr, Maggie Smith, Nigel Hawthorne, Jim Carter, Dominic West, John Wood, Bill Paterson, Adrian Dunbar, Donald Sumpter, Tom McInnerny, Matthew Groom
A spirited and ambitious updated retelling of Shakespeare story, with the protagonist now portrayed as a fascist on the rise in England of the 1930s. Unfortunately, the wonderful dialogue simply serves to highlight the obvious awkwardness of it all, despite severe editing of the text. For all that, there are some memorable moments and the whole work is striking in the extreme.
DRA   100 min (ort 104 min)   VIDrel: GUILD/FOXVID V/sur
Boa: Richard Eyre's adaptation of the play by William Shakespeare.

RICHIE RICH **   PG
Donald Petrie USA   1994
Macaulay Culkin, John Larroquette, Edward Herrmann, Christine Ebersole, Jonathan Hyde, Michael McShane, Chelcie Ross, Mariangelo Pino, Stephi Lineburg, Michael Maccarone, Joel Robinson, Jonathan Hilario, Reggie Jackson, Stacy Logan
A young millionaire has everything he could want, except some friends of his own age. However, an attempted theft of the family wealth brings him help from an unexpected quarter. An enjoyable family film based on the popular comic-strip character, with Hyde memorable as Rich's loyal English butler, one of the few highlights in a film that is rather too obviously influenced by the HOME ALONE movies.
A/AD   90 min (ort 95 min) cC   VIDrel: WHV   V/sur

RICKY 1 ***   15
Bill Naud USA   1986
Michael Michaud, Maggie Hughes, Peter Zellars, Lane Montano, Jon Chaney, Steve Welles, James Herbert, Lisa Traficante
When his girlfriend walks out on him, leaving a note saying that she does not wish to stand in the way of his career, a boxer decides to hang up his gloves and gets a job as an executive trainee at a fish market. Slipping into debt, he is eventually forced to fight in a rigged fight and surprises everyone by winning, but this is only the start of his problems. A fresh and light-hearted comedy.
Aka: HEART TO WIN
COM   87 min   VIDrel: EIV/SONOP   V

RICOCHET **   18
Russell Mulcahy USA   1991
Denzel Washington, John Lithgow, Ice T, Kevin Pollak, Lindsay Wagner, Mary Ellen Trainor, Josh Evans, Victoria Dillard, John Amos, John Cothran Jr, Linda Don, Matt Landers, Lydell M. Cheshier, Starletta Dupois, Viveka Davis
This fast-moving blend of action and comedy starts off with ambitious cop Washington shooting an unbalanced criminal.

Said criminal (a splendidly demented performance from Lithgow) does not take this at all well, and spends the rest of the movie trying to get even. An over-violent and often very gory film, it cannot quite decide what it wants to be.
A/AD   98 min (ort 110 min) wScrn   VIDrel: FIRST/SONOP; ENCORE (LV only)   V/sur LV

RIDE 'EM HARD * 18
USA 1992
Patricia Kennedy, Brigitte Aime, Stacey Nichols, Jon Dough, Tom Byron, Christina Applelay
A husband does everything in his power to cure his wife's frigidity, enlisting the services of a mystic pill-pusher who convinces her that vitamin tablets she is giving her are actually an aphrodisiac. Predictably, the desired change takes place.
A   60 min   VIDrel: MOPIC/SGSVID   V

RIDE HIM COWBOY ** U
Fred Allen USA 1932
John Wayne, Ruth Hall, Henry B. Walthall, Harry Gribbon, Oyis Harlam, Charles Sellon, Frank Hagney
A cowboy saves a horse from being put down since it was though to have caused the death of a rancher and sets out to find the culprit. He proves to be a notorious outlaw whose capture proves more difficult than he anticipated.
Aka: HAWK, THE
WES   53 min (ort 55 min) B/W   VIDrel: MGM/WHV   V/h

RIDE IN THE WHIRLWIND ** PG
Monte Hellman USA 1966
Jack Nicholson, Cameron Mitchell, Millie Perkins, Katherine Squire, Harry Dean Stanton, Rupert Crosse, George Mitchell, Tom Filer, Brandon Carroll, Peter Cannon, John Hackett, B.J. Herzholz
Three cowboys on their way home from a trail drive are mistaken for outlaws by a posse which promptly give chase. An enjoyable if not memorable little film.
WES   77 min (ort 82 min)   VIDrel: CREMED/LABYY   V

RIDE THE HIGH COUNTRY *** 15
Sam Peckinpah USA 1962
Randolph Scott, Joel McCrea, Mariette Hartley, Ronald Starr, Edgar Buchanan, R.G. Armstrong, John Anderson, L.Q. Jones,, Warren Oates, James Drury, John Davis Chandler, Jenie Jackson, Carmen Phillips, Percy Helton
Two former lawmen are hired to protect a gold shipment but one of them may be hatching his own plans for a comfy retirement. Scott's last film is almost a classic Western and remains a highly watchable example of this genre.
Aka: GUNS IN THE AFTERNOON
WES   89 min (ort 93 min)   VIDrel: MGM/WHV   V/sh

RIDE WITH THE WIND ** (PG)
Bobby Roth USA 1993
Craig T. Nelson, Helen Shaver, Bradley Pierce, Max Gail, Tracey Walter, Henry G. Sanders, Travis McKenna, Steev Parlevecchio, Linda Donna, David Schickele, Doria Cook-Nelson, Gina Gallego, Stacey Nelkin, MacDonald Cook
After years spent sustaining serious injuries, a middle-aged motor cycle racer is strongly advised to give up the sport that has become his obsession. Determined to have just one more crack at a title race before retiring, he starts making his plans, but when he becomes romantically involved with a single mother he starts getting a new perspective on his life. Fair.
DRA   88 min (ort 90 min) mTV
SATrel: MOVIE CHANNEL

RIDERS ** 15
Gabrielle Beuamont UK 1993
Marcus Gilbert, Arabella Tjye, Michael Praed, Caroline Harker, Stephanie Beacham, Anthony Calf, John Standing, Anthony Valentine, Ian Hog, Brenda Bruce, Alexander Torriglia, Cecile Paoli, Serenea Gordon, Belinda Mayne, Andrew Hall
Drama set in the hot-house world of international show-jumping. Being based on a Jilly Cooper novel, one can expect the usual parade of torrid encounters, double dealing and complex intrigues. Mildly diverting, and more so if one has an interest in horses.
DRA   200 min (ort 210 min) mTV   VIDrel: FOCUS/DISC   V
Boa: novel by Jilly Cooper.

RIDERS OF DESTINY ** U
Robert North Bradbury USA 1933
John Wayne, Cecilia Parker, George "Gabby" Hayes, Forrest Taylor,

Al St John, Heinie Conklin, Earl Dwire, Yakima Canutt, Lafe McKee, Fern Emmett, Hal Price, Si Jenks, Horace B. Carpenter
A government agent poses as an outlaw, to aid settlers whose water rights have been usurped by a corrupt businessman. The only "Singin' Sandy" film made, with competent action sequences and dubbed singing on the part of the star.
WES   52 min (ort 59 min) B/W   VIDrel: CREMED/LABY   V

RIDICULE *** 15
Patrice Leconte FRANCE 1996
Fanny Ardant, Charles Berling, Bernard Giraudeau, Judith Godreche, Jean Rochefort, Carlo Brandt, Bernard Dheran, Albert Delpy, Jacques Mathou, Urbain Cancelier, Bruno Zanardi, Marie Pillet, Jacques Roman, Philippe Magnan
A period piece set in France in the 18th century, that on while level is a lavish costume drama, but on another, is a cynical exploration of deception and superficiality. The slight plot has a poor by honest aristocrat (Berling) attempting to gain access to the court of King Louis XVI, as he wishes to build a sewage system in the country, but is constantly sidetracked by the prevailing court intrigues. Clever, cruel and richly detailed.
COM   102 min   CINrel

RIDING THE EDGE ** 15
James Fargo USA 1990
Raphael Sbarge, Catherine Mary Sewart, Peter Haskell, Lyman Ward, Michael Sarne
When his scientist father is captured and held hostage be terrorists in the Middle East, his teenage biker son comes riding to the rescue. A limited action effort, memorable if only for its reversal of the usual father/son rescue premise.
A/AD   99 min (ort 110 min)   VIDrel: EIV/SONOP   V/sur

RIFF-RAFF *** 15
Ken Loach UK 1991
Robert Carlyle, Emer McCourt, Jimmy Coleman, George Moss, Ricky Tomlinson, David Finch, Richard Belgrave, Ade Sapara, Derek Young, Bill Moores, Luke Kelly, Garrie J. Lammin, Willie Ross, Dean Perry, Dylan O'Mahony, Brian Coyle
A biting and harsh look at working-class life, as seen through the eyes of a motley gang of labourers working on a London building site, who are in the process of transforming a derelict hospital into luxury apartments. Some weird and occasionally amusing episodes take place, accompanied a little awkwardly by the tale of a romance between the latest recruit to this group, and an unhappy would-be singer. The ending is contrived, depressing and very Ken Loach.
COM   95 min (ort 96 min)   VIDrel: FIRST/SONOP   V

RIFT, THE ** 15
Juan-Piquer Simon USA 1989
Jack Scalia, R. Lee Ermey, Ray Wise, Deboard Adair, John Toles Bey, Ely Pouget, Emilio Linder, Tony Isbert, Alvaro Labra, Luis Lorenzo, Frank Brana, J. Martinez Brodiu, Edmund Purdom, Garick Hogan, James Aubrey, Jed Downey
A rescue team faces unknown dangers as they search the depths for a missing nuclear sub. As time runs out, pressures and tensions mount, in this unoriginal amalgam of such recent films as THE ABYSS, THE DIVE and DEEPSTAR SIX, to name but three. Story is by Simon and Mark Klein.
Aka: ENDLESS DESCENT
FAN   79 min   VIDrel: RCA L/A   V

RIGHT STUFF, THE *** 15
Philip Kaufman USA 1983
Sam Shepard, Scott Glenn, Ed Harris, Dennis Quaid, Fred Ward, Charles Frank, Barbara Hershey, Kim Stanley, Veronica Cartwright, Scott Paulin, Pamela Reed, Donald Moffat, Levon Helm, Scott Wilson, Jeff Goldblum, Harry Shearer
A behind-the-scenes look at the development of the US space programme. Seen as a certain boost to Senator John Glenn's presidential ambitions, it failed to get off the ground; perhaps due to writer-director Kaufman making all the characters one-dimensional except for the pilots. A disappointment. AA: Sound (Berger/Scott/Thom/MacMillan), Edit (Farr/Rolf/Fruchtman/Rotter/Stewart), Score/orig (Bill Conti), Effects/aud (Jay Boekelheide).
DRA   185 min (ort 193 min)   VIDrel: WHV   V/sur
Boa: book by Tom Wolfe.

RIKKI AND PETE ** (12)
Nadia Tass AUSTRALIA 1988
Stephen Kearney, Nina Landis, Tetchie Agbayani, Bill Hunter, Bruce

Spence, Bruno Lawrence, Dorothy Alison, Don Reid, Lewis Fitz-Gerald, Peter Cummins, Peter Hehir, Ralph Cotterill, Roderick Williams, Denis Lees, Rob Baxter
A girl geologist and her zany inventor brother move to a small Outback town where they find other kindred souls. An average story of outsiders and misfits that relies more on the acting skills of the cast than on any innate virtues of the script.
DRA   99 min (ort 103 min)   SATrel: MOVIE CHANNEL

RING OF BRIGHT WATER ***
Jack Couffer UK   U
                                                                 1969
Bill Travers, Virginia McKenna, Peter Jeffrey, Jameson Clark, Helena Gloag, Roddy McMillan, Willie Joss, Jean Taylor-Smith, Archie Duncan, Kevin Collins, John Young, James Gibson, Michael O'Halloran, Philip McCall, June Ellis
A man buys an otter and takes it to live with him in the Scottish Highlands, where the pair enjoy various escapades. A wholesome and likeable tale, carefully made and well photographed. The script is by Jack Couffer and Bill Travers. See also TARKA THE OTTER.
JUV   102 min (ort 109 min)   VIDrel: ODY/SONOP   V/dm V/sh
Boa: book by Gavin Maxwell.

RING OF FIRE *
Richard W. Munckin USA   18
                                                                 1991
Don "The Dragon" Wilson, Maria Ford, Vince Murdocco, Dale Edmund Jacoby, Michael Delano, Eric Lee, Steven Vincent Leigh, Marta Merrifield, Rod Kei, Shirley Spiegler Jacobs, Jane Chung
In L.A. the rivalry between two of Chinatown's kickboxing clubs reaches such a pitch that a major showdown is called for, and a martial arts contest within a "ring of fire" is arranged. As ever, honour must be avenged, etc. Followed by RAGE: RING OF FIRE 2.
MAR   92 min Cut (19 sec – ort 96 min)
VIDrel: MIA/DISC/IMPENT   V

RING OF STEEL *
David Frost USA   15
                                                                 1993
Robert Chapin, Joe Don Baker, Carol Alt, Gary Kasper, Darlene Vogel
After accidentally killing an opponent in the ring, a fencer who hoped to take part in the Olympics becomes involved with a club whose respectable facade conceals a sinister secret, this being that the owner of the club runs a secret matches where the wealthy bet on contests to the death. This might have been a reasonably exciting effort, but a wholly unnecessary romantic sub-plot dissipates any tension the story tries to develop.
A/AD   94 min   VIDrel: MIA/DISC   V

RING OF THE MUSKETEERS **
John Paragon USA   15
                                                                1993
David Hasselhoff, Cheech Marin, Corbin Berensen, Alison Doody, Thomas Gottschalk, John Rhys-Davis
The descendants of the title heroes emulate their famous ancestors by coming together to deal with a nasty Mafia boss and save a small child. Updated comedy-action film (our heroes use motorbikes instead of horses etc.) that is essentially one long over-extended joke. Nice try – shame about the script.
COM   83 min (ort 120 min)   VIDrel: 20VIS/SONOP L/A   V

RIO BRAVO ***
Howard Hawks USA   PG
                                                               1959
John Wayne, Dean Martin, Ricky Nelson, Angie Dickinson, Walter Brennan, Ward Bond, Claude Akins, John Russell, Pedro Gonzalez, Estelita Rodriguez, Harry Carey Jr, Malcolm Atterbury, Bob Steele, Bing Russell, Myron Healey
A classic Western, with a brave sheriff trying to prevent a killer being sprung from jail. More or less remade in 1966 as "El Dorado", and in 1970 as RIO LOBO. Despite a fine cast who give credible performances, little suspense is generated, and the film's inordinate length makes this failing even more apparent. Written by Jules Furthman and Leigh Brackett.
WES   136 min (ort 141 min)   VIDrel: WHV   V

RIO CONCHOS ***
Gordon Douglas USA   (12)
                                                               1964
Richard Boone, Stuart Whitman, Tony Franciosa, Wende Wagner, Jim Brown, Vito Scotti, Edmond O'Brien, Warner Anderson, Rodolfo Acosta, Barry Kelly, House Peters Jr, Kevin Hagen, Timothy Carey
In Texas just after the end of the Civil War, two men are released from jail in return for agreeing to take part in an Army expedition to prevent a large consignment of stolen rifles from falling into the hands of the Indians. Their search eventually takes them

to the camp of a former Confederate general who plans to restore the vanished glory of the South. A well made and exciting film, with a strong cast and a good, solid script.
WES   107 min   TVrel
Boa: novel by Clair Huffaker.

RIO GRANDE ****
John Ford USA   U
                                                                1950
John Wayne, Maureen O'Hara, Ben Johnson, Harry Carey Jr, Victor McLaglen, Claude Jarman Jr, Chill Wills, J. Carol Naish, Grant Withers, Peter Ortiz, Steve Pendleton, Karolyn Grimes, Alberto Morin, Stan Jones, Jack Pennick
An excellent account of the US Cavalry and their struggle with the Apache in the post-Civil War period. The last of a Ford trilogy that dealt with the same subject matter, the other two being FORT APACHE and SHE WORE A YELLOW RIBBON. Written by James Kevin McGuinness, with music by Victor Young and songs by the Sons of the Pioneers. The splendid photography is by Bert Glennon and Archie Stout.
WES   102 min (ort 105 min) B/W
VIDrel: 4-FRONT/POLYREC   V
Boa: story by James Warner Bellah.

RIO LOBO ***
Howard Hawks USA   PG
                                                                1970
John Wayne, Jorge Rivero, Jennifer O'Neill, Jack Elam, Victor French, Chris Mitchum, Jim Davis, Susana Dosamantes, Mike Henry, David Huddleston, Edward Faulkner, Bill Williams, Sherry Lansing, Dean Smith, Robert Donner, Bob Steele
At the end of the Civil War, a Union colonel sets out to expose a traitor and recover a gold shipment. A long and rambling tale, but highly enjoyable for its good dialogue and action sequences. Written by Leigh Brackett and Burton Wohl. This was Hawks's final film.
WES   108 min (ort 114 min)   VIDrel: 20TH/TECH   V/h

RIO SHANNON **
Mimi Leder USA   (PG)
                                                                1992
Blair Brown, Patrick Van Horn, Michael Delvise, Shay Aitar, David Dunard, Michael Lowry, Gary Werntz, Bernard White, Derek Cravin
In this family drama Brown is cast as a wife and mother who suddenly loses her husband and after a period of grieving, decides to take herself and her three kids to New Mexico in the hope of making a new life there. Once there she buys up a dilapidated ranch with the intention of making it into a tourist hotel, but encounters some unexpected hostility to this plan from the locals. An enjoyable little drama.
DRA   89 min (ort 96 min)   SATrel: SKY MOVIES

RIOT IN CELL BLOCK 11 ***
Don Siegel USA   15
                                                                1954
Neville Brand, Emile Meyer, Leo Gordon, Frank Faylen, Robert Osterloh, Paul Frees, Don Keefer, Alvy Moore, Dabbs Greer, Whit Bissell, James Anderson, Carleton Young, Harold J. Kennedy, William Schallert, Jonathan Hale, Joe Kerr
A powerful and realistic prison movie, telling of three convicts who seize their guards and barricade themselves in their cellblock, intending to use the resulting press coverage to expose their brutal treatment. The script is by Richard Collins and the photography by Russell Harlan.
DRA   77 min (ort 80 min) B/W   VIDrel: VCC L/A   V

RIPPER MAN **
Phil Sears USA   18
                                                                1994
Mike Norris, Timothy Bottoms, Sofia Shinas, Robert F. Lyons, Charles Napier
A former cop wrongfully dismissed from the police force now works as a nightclub performer, using a hypnosis routine. He meets a stranger who believes in reincarnation, and this latter hires him to help in a regression experiment, so that he can explore his previous lives. The result is not pleasant.
THR   89 min   VIDrel: COLUM/SONOP   V/sur

RISE AND FALL OF LEGS DIAMOND, THE ***
Budd Boetticher USA   PG
                                                                1960
Ray Danton, Karen Steele, Elaine Stewart, Jessie White, Simon Oakland, Robert Lowery, Warren Oates, Judson Pratt, Frank De Kova, Gordon Jones, Buzz Henry, Joseph Ruskin, Dyan Cannon, Richard Gardner, Sid Melton, Nesdon Booth
An excellent account of the rise of a notorious gangster of the 1920s, with Danton well cast as an aspiring criminal, who takes shooting lessons from his old sergeant in a bid for the top. Part of a not altogether successful effort on the part of Warner, to

resurrect the gangster genre. Written by Joseph Landon and with photography by Lucien Ballard.
DRA   101 min B/W   VIDrel: L/A   V

RISE AND WALK: THE DENNIS BYRD STORY **   PG
Michael Dinner USA   1994
Peter Berg, Kathryn Morris, Wolfgang Bodison, Johnny Carlo, Carrie Snodgress, Steve Fitchpatrick, Patrick Warburton, Zakes Mokae, Steve Anderson, Allan, Royal, William Forward, Dennis Howard, Richie Allan, Bert Remsen
Fact-based drama telling of a New York football player who suffers appalling injuries in a freak accident, and is left a quadriplegic. Facing months if not years of intensive physiotherapy, he learns that his wife is pregnant. See also TO WALK AGAIN.
Aka: DENNIS BYRD STORY, THE
DRA   85 min (ort 90 min) mTV   VIDrel: 20TH/TECH   V
Boa: book by Dennis Byrd and Michael D'Orso.

RISE OF CATHERINE THE GREAT, THE **   U
Paul Czinner UK   1934
Douglas Fairbanks Jr, Elisabeth Bergner, Flora Robson, Joan Gardner, Gerald Du Maurier, Irene Vanbrugh, Griffith Jones, Lawrence Hanray, Dorothy Hale, Gibb McLaughlin, Clifford Heatherley, Allan Jeayes, Diana Napier
Creaky, melodramatic account of how an unsophisticated German princess became absolute ruler of Russia, through both marriage and military power, after forcing her husband to abdicate in her favour. Written by Biro, Arthur Wimperis and Marjorie Deans.
Aka: CATHERINE THE GREAT
DRA   84 min (ort 96 min) B/W   VIDrel: L/A   V
Boa: play The Czarina by Lajos Biro and Melchior Lengyel.

RISING DAMP *   PG
Joe McGrath UK   1980
Leonard Rossiter, Frances De La Tour, Don Warrington, Denholm Elliott, Christopher Strauli, Carrie Jones, Glyn Edwards, John Cater, Derek Griffiths, Ronnie Brody, Alan Clare, Jonathan Cecil
A feature-length spin-off from a mediocre British TV comedy, telling of the lodgers who live in a run-down boarding house owned by a seedy and disreputable landlord. Made without Richard Beckinsale, who appeared in the TV series but died prematurely. The thin and dreary little plot tells of a female tenant who develops an interest in one of the new arrivals, much to the chagrin of the landlord. Written by Eric Chappell.
COM   94 min (ort 98 min)   VIDrel: 4-FRONT   V

RISING SUN **   18
Philip Kaufman USA   1992
Sean Connery, Wesley Snipes, Harvey Keitel, Mako, Cary-Hiroyuki Tagawa, Ray Wise, Kevin Anderson, Stan Egi, Tia Carrere, Stan Shaw, Steve Buscemi, Clyde Kusatsu, Tatjana Patitz, Peter Crombie, Sam Lloyd, Alexandra Powers, Amy Hill
An experienced cop is teamed with an expert on things Japanese to deal with a tricky case of murder in which a dead hooker was found in the boardroom of a Japanese corporation. As he and his associate investigate, they find themselves caught up in a mystery of considerable complexity. A strained and simplified adaptation of the Crichton's novel that tones down its more controversial aspects to refute allegations of "Japan bashing".
THR   129 min   VIDrel: 20TH/TECH L/A; ENCORE (LV only)   V LV
Boa: novel by Michael Crichton.

RISKY BUSINESS *   18
Paul Brickman USA   1983
Tom Cruise, Rebecca De Mornay, Curtis Armstrong, Bronson Pinchot, Raphael Sbarge, Nicholas Pryor, Joe Pantoliano, Richard Masur, Janet Carroll, Shera Danese, Bruce A. Young, Kevin C. Anderson, Sarah Partridge, Nathan Davis
A boy's parents go out of town on a visit and leave him to look after the house. He begins to enjoy himself, but decides things have gone a little too far when a couple of hookers start using the house as a brothel. An utterly banal comedy, combining a stilted and witless script with wooden acting; the whole set off by a loud and monotonous dirge (courtesy of Tangerine Dream) that often drowns the dialogue out entirely.
COM   95 min (ort 99 min)   VIDrel: WHV   V/sur

RITES OF PASSION ***   18
Annie Sprinkle/Veronica Vera USA   1988
Nina Hartley, Jeanna Fine, Robert Bullock, Scott Baker, Roger T. Dodger, David Sandler

Two short sexual stories, each of which has a different director. In "Shady Madonna" Baker plays a fundamentalist preacher who has arrived to tape his weekly sermon, but can't keep his mind off sex. The second tale is "In Search Of The Ultimate Sexual Experience", which deals with the kinky but unsatisfying sex-life of a woman who has decided to seek guidance from an Indian guru. A couple of well produced albeit edited offerings.
A   80 min Cut (3 min 11 sec)   VIDrel: MIA/DISC   V

RITZ, THE **   15
Richard Lester USA   1976
Jack Weston, Rita Moreno, Jerry Stiller, Kaye Ballard, F. Murray Abraham, Treat Williams, Paul B. Price, John Everson, Christopher J. Brown, Dave King, Bessie Love, Tony De Santis, Ben Aris, George Coulouris, Hugh Fraser
A man on the run from his murderous brother-in-law, hides out in one of New York's gay Turkish baths. Moreno is memorable in a reprise of her Tony-winning role as the untalented entertainer Googie Gomez, but the whole affair has a very stagebound look to it. Filmed in the UK.
COM   86 min (ort 90 min) wScrn   VIDrel: TART/20TH   V
Boa: play by Terrence McNally.

RIVER OF DEATH *   15
Steve Carver USA   1988
Michael Dudikoff, Donald Pleasence, Robert Vaughn, Herbert Lom, L.Q. Jones, Cynthia Erland, Sarah Maur Thorpe, Foziah Davidson, Victor Melleney, Ian Yule, Rufus Swart, Gordon Mulholland, Alain Woolf, Lindsey Reardon, Crispin De Nuys
At the end of WW2, a Nazi doctor escapes to South America and hides out in the Amazon jungle, where he continues his work trying to perfect a virus that will only kill non-Aryans. Twenty years later, a plague is found to be killing off the Amazonian Indians living near a lost city, and a party of explorers become embroiled in a conflict with him. A loose adaptation of MacLean's novel, that is as flat as a cartoon but not as entertaining.
A/AD   98 min   VIDrel: WHV L/A   V/dm
Boa: novel by Alistair MacLean.

RIVER OF NO RETURN **   PG
Otto Preminger USA   1954
Robert Mitchum, Marilyn Monroe, Tommy Rettig, Rory Calhoun, Murvyn Vye, Wil Wright, Douglas Spencer, Ed Binns, Don Beddoe, Claire Andre, Jack Mather, Edmund Cobb, Jarma Lewis, Hal Baylor, Barbara Nichols, Fay Morley
Set at the time of the California Gold Rush, this tells of a widower and his young son, who meet a saloon singer whose husband has deserted her. She hires the man to take her downriver in pursuit of her husband, and all three experience various hazards along the way. The plot is insubstantial but the film moves along briskly and the lovely locations are an asset. An early Cinemascope film, whose impact will be diminished on TV.
WES   88 min (ort 91 min)   VIDrel: 20TH/TECH   V

RIVER RUNS THROUGH IT, A ***   PG
Robert Redford USA   1992
Brad Pitt, Craig Sheffer, Tom Skerritt, Emily Lloyd, Robert Redford, Brenda Blyth, Edie McClurg, Stephen Shellen, Vann Gravage, Nicole Burdette, Susan Taylor, Michael Cudlitz, Joseph Gordon-Levitt, Rob Cox, Buck Simmonds
A preacher teaches his sons about life and respect for Nature when they go fly-fishing in the unspoilt countryside of Montana, hoping the lessons they learn will stand them in good stead in later life. They pursue very different paths but still feel that this is the common bond that unites them. A well-realised adaptation of Maclean's autobiography but very slow paced and unfocused at times. AA: Cin (Philippe Rousselot).
DRA   118 min (ort 124 min) wScrn
VIDrel: POLY/POLYREC/GUILD; PION (LV only)   V/sur LV
Boa: autobiography of Norman Maclean.

RIVER WILD, THE ***   12
Curtis Hanson USA   1994
Meryl Streep, David Strathairn, Joseph Mazzello, Kevin Bacon, Buffy, John C. Reilly, Stephanie Sawyer, Buffy, Elizabeth Hoffman, Victor H. Galloway, Diane Delano, Thomas F. Duffy, William Lucking, Benjamin Bratt, Paul Cantelon
A husband and wife decide to go on a white-water rafting trip together with their son in a bid to patch their ailing marriage. Unfortunately, they meet up with a pair of psychotic criminals who kidnap the wife and force her to take them down-river.

Excellent performances from Streep and Bacon and many spectacular stunts add greatly to the enjoyment of this fast-paced thriller. See also DEAD AHEAD.
THR 106 min (ort 112 min) cC VIDrel: CIC/SONOP; PION (LV only) V/dm V/sh LV

ROAD GIRLS *   18
J.T. Monroe USA   1990
Kelly Royce, Krisstarah Knight, Eric Price, Victoria Paris, Danielle Rogers, Randy Spears, Rayne, Peter North, Randy West
A major league baseball player has a wild time with his mistress but eventually returns to his wife, who in the meantime has had a little fling of her own.
A 56 min (ort 65 min) VIDrel: PASSION/SGSVID V

ROAD HOME, THE *   15
Hugh Hudson USA   1989
Donald Sutherland, Adam Horovitz, Don Bloomfield, Amy Locane, Celia Weston, Graham Beckel, Kevin Tighe, John C. McGinley, Park Overall, Gary Riley, Ron Frazier, Patricia Richardson, Joseph D'Angerio, Kevin Corrigan, Jack Gold
Set at one of those private mental hospitals where well-to-do parents send their "difficult" offspring, this strange hybrid of a movie (it attempts both social comment and teen-oriented melodrama) deals with one such young man, and the relationship he develops with his fellow inmates and the psychiatrist in charge. A pointless film, that says nothing of great profundity and only memorable as the debut for Horovitz (a member of pop group The Beastie Boys).
Aka: LOST ANGELS
DRA 111 min (ort 116 min) VIDrel: 20VIS/SONOP V

ROAD TO BALI **   U
Hal Walker USA   1952
Bob Hope, Dorothy Lamour, Bing Crosby, Murvyn Vye, Ralph Moody, Jerry Lewis, Jane Russell, Dean Martin, Carolyn Jones, Bernie Gozier, Harry Cording, Herman Cantor, Michael Ansara, Jack Claus, Allan Nixon
A routine comedy in the "Road To" series (the only colour one), with Hope and Crosby rescuing Lamour from various perils in the jungle. The film debut for Jones. Number six in a series of seven films.
COM 91 min VIDrel: ORBIT/DISC V

ROAD TO GALVESTON, THE **   PG
Michael Uno Toshiyuki USA   1995
Cicely Tyson, Piper Laurie, Tess Harper, James McDaniel, Starlett DuPois, Penny Johnson, Brnadon Hammond, Stephen Root, Clarence Williams III
A woman who has the job of looking after three patients with Alzheimer's Disease, takes them on a trip of a lifetime as a way of relieving the pressure. Fair.
DRA 89 min (ort 93 min) cC VIDrel: CIC/SONOP V

ROAD TO MOROCCO ***   U
David Butler USA   1943
Bing Crosby, Bob Hope, Dorothy Lamour, Anthony Quinn, Vladimir Sokoloff, Monte Blue, Yvonne De Carlo, Andrew Tombes, Leon Belasco, Dan Seymour, Mikhail Rasumny, George Givot, Dona Drake, Jamiel Hasson, Dona Drake
Another entry in the series, with both heroes becoming involved with a beautiful Arab princess. A frothy and well paced excursion, with attractive locations and some good gags. This was the third "Road To" film.
COM 78 min (ort 83 min) B/W VIDrel: CIC/SONOP L/A V

ROAD TO RUIN **   15
Charlotte Brandstrom FRANCE   1991
Peter Weller, Carey Lowell, Michel Duchaussoy, Nathalie Auffret, Antoine Blanquefort, Jean-Michel Dagory, Philippe Deherdin, William Doherty, Frederique Feder, Steve Calder, Michael Goldman, Gilles Guarderas, Jean Guichard
A businessman finds the woman of his dreams and resolves to test her love by losing his money in a phoney bankruptcy. Things go wrong, however, when his partner decides that he would rather have the business all to himself. A well acted but unremarkable romantic comedy set in Paris.
COM 94 min VIDrel: MED/COLUM L/A V/sh

ROAD TO SINGAPORE **   U
Victor Schertzinger USA   1940
Bob Hope, Bing Crosby, Dorothy Lamour, Charles Coburn, Anthony

Quinn, Judith Barrett, Jerry Colonna, Johnny Arthur, Pierre Watkin, Gaylord (Steve) Pendleton, Miles Mander, Pedro Regas, Greta Granstedt, Edward Gargan
Two men hide out in Singapore, and forswear women until they meet one who makes them change their minds. The first in what proved to be a long series of light and frothy comedies starring Hope, Crosby and Lamour, all of which were invariably entitled "Road To...".
COM 81 min (ort 84 min) B/W VIDrel: CIC/SONOP L/A V/h
Boa: story by Harry Hervey.

ROAD TO UTOPIA **   PG
Hal Walker USA   1945
Bob Hope, Bing Crosby, Dorothy Lamour, Hillary Brooke, Douglass Dumbrille, Jack La Rue, Robert Barrat, Nestor Paiva, Will Wright, Billy Benedict, Alan Bridge, Robert Benchley, Stanley Andrews, Edgar Dearing, Arthur Loft
The fourth "Road To" movie set in the Klondike with talking animals, songs and the usual gags, plus Robert Benchley's witty commentary. The plot is flimsy but a few good gags are to be found.
COM 86 min (ort 90 min) B/W VIDrel: CIC/SONOP L/A V

ROAD TO WELLVILLE, THE **   18
Alan Parker USA   1994
Matthew Broderick, Bridget Fonda, John Cusack, Anthony Hopkins, Dana Carvey, Michael Lerner, Colm Meaney, John Neville, Lara Flynn Boyle, Traci Lind, Camryn Manheim, Roy Brocksmith, Norbert Weisser, Monica Parker, Jacob Reynolds
The story of cornflakes inventor Dr Kellogg and the sanatorium he founded in Battle Creek, Michigan in 1907, where he explored to the full his bizarre ideas regarding health, with basic principles such as the harmfulness of sex and the health benefits of cold showers doing nothing to discourage an army of wealthy patients who flocked to his clinic and became his willing guinea-pigs. A messy and unstructured film with little of the humour of the original novel.
DRA 115 min (ort 120 min) VIDrel: EIV/SONOP L/A V/s
Boa: novel by T. Coraghessan Boyle.

ROAD TO ZANZIBAR **   PG
Victor Schertzinger USA   1941
Bob Hope, Bing Crosby, Dorothy Lamour, Una Merkel, Eric Blore, Luis Alberni, Douglass Dumbrille, Charles Coburn, Anthony Quinn, Iris Adrian, Lionel Royce, Buck Woods, Leigh Whipper, Ernest Whitman, Noble Johnson, Leo Gorcey
This entry is set in Africa, where two circus artists go on a jungle safari in search of a diamond mine. The second film in this long-running series.
COM 88 min (ort 92 min) B/W VIDrel: CIC/SONOP L/A V/h

ROADFLOWER *   18
Deran Sarafian USA   1993
Christopher Lambert, Craig Sheffer, David Arquette, Josh Brolin, Michelle Forbes, Joseph Gordon-Levitt, Michael Grene, Alexondra Lee, John Pyper-Ferguson, Christopher McDonald, Richard Sarafian, Adrienne Shelley, Patrick Thompson
A family's vacation journey turns into a nightmare when they encounter a group of manic joyriders who take pleasure in terrorising other road users. Very slightly reminiscent of DUEL, this brutal and cliched revenger was released directly to video. As devoid of merit as it is amply endowed with gratuitous violence.
Aka: ROAD KILLERS
A/AD 85 min (ort 90 min) VIDrel: EIV/SONOP V

ROADHOUSE *   18
Rowdy Herrington USA   1989
Patrick Swayze, Kelly Lynch, Sam Elliott, Ben Gazzara, Marshall R. Teague, Julie Michaels, Kevin Tighe, John Doe, Kurt James Stepka, Travis McKenna, Jeff Healey, Roger Hewlett, Gary Hudson
A philosophy student is taken on as a bouncer at an ultra-tough Midwest bar and proceeds to break heads at speed, which ultimately leads to an encounter with the local crime boss. A feast of mindless violence that tires the eye and exhausts the mind.
Aka: ROAD HOUSE
A/AD 109 min Cut (10 sec – ort 114 min)
VIDrel: MGM/WHV V/sur

ROADHOUSE 66 *   18
John Mark Robinson USA   1984
Willem Dafoe, Judge Reinhold, Kaaren Lee, Kate Vernon, Stephen
Elliott, Alan Autry, Kevyn Major Howard, Peter Van Norden, Erica
Yohn, James Intveld, William T. Lane, Katie Graves, Dave Cass,
Roydon Clark, John Moio, Mike Adam, Joe Dunne
Two men on the road discover that the locals are none too
friendly, as they travel across the US in a 1955 Thunderbird on
route 66. Sounds reminiscent of the TV series "Route 66". A film
with little going for it, not least being the dismal plotting.
A/AD   90 min (ort 96 min)   VIDrel: EIV/SONOP   V

ROADRACERS *   (12)
Robert Rodriguez USA   1994
David Arquette, John Hawkes, Salma Hayek, Jason Wiles, William
Sadler, Kevin McCarthy, O'Neal Compton, Christian Kelmash, Aaron
Vaughn, Tammy Brady Conrad, Mark Lowenthal, Karen Landry,
Lance Legault, Tommy Nix, Gina Mari, Boti Bliss
Period drama set in the 1950s, with a rebellious, guitar-playing
youngster coming into conflict with an increasingly obsessive
police officer, who takes out his anger on the boy, having
become estranged from his own wayward son. Meanwhile, our
hero dreams of leaving town for good, but is kept there by the
need to support his alcoholic mother.
DRA   89 min mCab   SATrel: SKY MOVIES

ROADSIDE PROPHETS *   15
Abbe Wool USA   1992
John Doe, David Anthony Marshall, Adam Horovitz, Barton Heyman,
John Cusack, Jennifer Balgobin, Ellie Raab, Judith Thurman, Biff
Yeager, Sonna Chavez, J.D. Cullum, David Swinson, David
Carradine, Arlo Guthrie, Timothy Leary, Pam Lambert
An L.A biker freak undertakes to carry the ashes of an acquain-
tance to Nevada for final disposal. Along the way, he meets the
usual assortment of weirdos and eccentrics, most of whom have
been left over from the hippy era of the 1960s. A wry and somewhat self-conscious look at the power of nostalgia, with good
performances that compensate for the slight script and rambling
structure.
COM   93 min (ort 96 min)   VIDrel: 20VIS/SONOP   V/sur

ROARING TWENTIES, THE ***   PG
Raoul Walsh USA   1939
James Cagney, Priscilla Lane, Humphrey Bogart, Gladys George,
Jeffrey Lynn, Frank McHugh, Paul Kelly, Elizabeth Risdon, Joe
Sawyer, Abner Biberman, George Humbert, Clay Clement, Don
Thaddeus Kerr, Ray Cooke, Vera Lewis
A classic gangster movie, that shows how a WW1 veteran
returns to New York and builds a bootlegging empire. Fast-
paced, with fine acting and excellent direction. The script is as
predictable as they come, but the whole effort is done with
considerable panache. This was one of the last in a series of
gangster movies from Warner, and is generally regarded as one
of the best.
A/AD   102 min (ort 106 min) B/W   VIDrel: MGM/WHV
V
Boa: story by Mark Hellinger.

ROB ROY *   U
Geoff Collins/Warwick Gilbert/George Stephenson
AUSTRALIA   1987
Voices of: Tim Elliot, Ron Haddrick, Jane Harders, Phillip Hinton,
Simon Hinton, Andrew Inglis, Bill Kerr, Andrew Lewis, Bruce
Spence, Nick Tate
Animated version of Stevenson's children's classic telling of an
18th century Highland warrior and his battle against the forces
of the King of England.
ANIM   48 min (ort 50 min)   VIDrel: CARL/TECH   V
Boa: novel by Robert Louis Stevenson.

ROB ROY *   15
Michael Caton-Jones USA   1995
Liam Neeson, Jessica Lange, John Hurt, Tim Roth, Eric Stoltz, Andrew
Keir, Brian Cox, Brian McCardie, Gilbert Martin, Vicki Masson, Gilly
Gilchrist, Jason Flemyng, Ewan Stewart, David Hayman, Brian
McArthur, David Palmer, John Murtagh
This story of the 18th century Scottish hero was all but forgot-
ten due to its release at the same time as that of BRAVEHEART,
and casts Neeson as the title character, who seeks a justice of his
own after losing his money to a swindler. Neeson is not especially convincing, but Roth as a sinister fop is a good deal better
(and received an Oscar nomination). Quite well done, but a very

brutal rape sequence makes one wonder how the film ever got
a 15 certification.
A/AD   133 min (ort 139 min) cC   VIDrel: MGM/WHV
V/sur
Boa: novel by Robert Louis Stevenson.

ROBE, THE *   U
Henry Koster USA   1953
Richard Burton, Jean Simmons, Michael Rennie, Victor Mature, Jay
Robinson, Torin Thatcher, Richard Boone, Dean Jagger, Betta St John,
Richard Morrow, Dawn Addams, Ernest Thesiger, Leon Askin, Frank
Pulaski, David Leonard
First film in Cinemascope, but simultaneously shot flat (that's
what's seen on TV), with the robe worn by Jesus at his crucifix-
ion, attracting attention from both Romans and Christians. First
cast with Tyrone Power in Burton's role and Burt Lancaster in
Mature's. Written by Philip Dunne with music by Alfred
Newman. DEMETRIUS AND THE GLADIATORS followed.
AA: Art/Set (Lyle Wheeler and G.W. Davis/W. Scott and P.
Fox), Cost (C. LeMaire/E. Santiago).
DRA   129 min (ort 135 min) wScrn   VIDrel: 20TH/TECH
V/sh
Boa: novel by Lloyd C. Douglas.

ROBIN AND MARION *   PG
Richard Lester USA   1976
Sean Connery, Audrey Hepburn, Robert Shaw, Richard Harris, Nicol
Williamson, Denholm Elliott, Kenneth Haigh, Ian Holm, Ronnie
Barker, Bill Maynard, Esmond Knight, Peter Butterworth, Veronica
Quilligan, John Barrett
A kind of downbeat version of the legend of Robin of Sherwood,
with our hero returning after an absence of many years, and
finding the conditions in England somewhat depressing.
However, he rekindles his romance with Maid Marion and even
kills his old enemy the Sheriff of Nottingham, but dies in the
attempt. A coldly sterile and unappealing work, with not one
trace of magic or sparkle. Written by James Goldman.
DRA   102 min (ort 107 min)   VIDrel: VCC/DISC/COLUM
L/A   V

ROBIN AND THE SEVEN HOODS *   U
Gordon Douglas USA   1964
Frank Sinatra, Dean Martin, Sammy Davis Jr, Peter Falk, Barbara
Rush, Bing Crosby, Victor Buono, Sig Rumann, Allen Jenkins, Hans
Conried, Jack La Rue, Edward G. Robinson, Barry Kelley, Hank
Henry, Robert Carricart, Sonny King
The Robin Hood legend is updated to 1920s Chicago, where a
gangster and his cronies become local heroes by stealing from
the rich and giving to the poor. A set of sharp routines and some
good jokes keep this rather shallow gangster spoof on the move.
Music is by Nelson Riddle for which he received an AAN, as did
songwriter-lyricist team James Van Heusen and Sammy Cahn
for "My Kind Of Town". The unnecessary use of Panavision
will be lost on TV.
COM   118 min (ort 124 min)   VIDrel: MGM/WHV   V

ROBIN HOOD *   PG
Allan Dwan USA   1922
Douglas Fairbanks, Wallace Beery, Alan Hale Sr, Enid Bennett, Sam
De Grasse, William Lowery
The story of this legendary hero and his battles with the villain-
ous Sheriff of Nottingham. An excellent (albeit silent) version of
this tale, with ambitious sets and some exciting stunts and
swordplay.
A/AD   117 min (ort 127 min) B/W silent
VIDrel: VISVID/POLYREC   V

ROBIN HOOD *   U
Wolfgang Reitherman USA   1973
Voices of: Brian Bedford, Peter Ustinov, Terry-Thomas, Phil Harris,
Andy Devine, Pat Buttram, Roger Miller, George Lindsey, Carole
Shelley
An unusual animated version of this famous legend, with a fresh
approach in that all the parts are played by animals. However,
despite good dialogue, the songs and animation leave a lot to
be desired.
ANIM   80 min (ort 83 min) cC   VIDrel: WDV/TECH   V

ROBIN HOOD *   PG
John Irvin USA   1991
Patrick Bergin, Uma Thurman, Jurgen Prochnow, Edward Fox, Jeroen
Krabbe, Owen Teale, David Morrissey, Alex Norton, Gabrielle Reidy,

Cecily Hobbs, Conrad Asquith, Anthony O'Donnell, Barry Stanton, Jeff Nuttall, Daniel Webb
A well handled if overlong rendition of the Robin Hood legend that draws a competent if not exactly sparkling performance from Bergin as Robin and a surprisingly robust one from Thurman as Marion. Nicely photographed, it offers neither great insights nor interesting plotting, and suffered on release by being compared unfavourably with Costner's more successful ROBIN HOOD: PRINCE OF THIEVES, to which it stands up quite well.
A/AD 99 min (ort 150 min) mTV VIDrel: 20TH/TECH V

ROBIN HOOD: MEN IN TIGHTS *

| | | PG |
Mel Brooks USA 1993
Cary Elwes, Tracey Ullman, Richard Lewis, Dom DeLuise, Isaac Hayes, Mel Brooks, Amy Yasbeck, Dave Chappelle, Mark Blankfield, Eric Allan Kramer, Megan Cavanagh, Dom DeLuise, Dick Van Patten, Matthew Poretta, Patrick Stewart
Now the Robin Hood legend gets the Brooks treatment in this bottom-of-the-barrel parody that offers weak jokes and much silliness but nothing that is remotely amusing. Brooks plays one Rabbi Tuchman (a Jewish version of Friar Tuck) with his usual lack of grace or conviction.
Aka: MEN IN TIGHTS
COM 99 min (ort 105 min) VIDrel: VCC/DISC/COLUM V/sur

ROBIN HOOD: PRINCE OF THIEVES ***

PG
Kevin Reynolds USA 1990
Kevin Costner, Morgan Freeman, Christian Slater, Alan Rickman, Mary Elizabeth Mastrantonio, Geraldine McEwan, Michael McShane, Brian Blessed, Nick Brimble, Michael Wincott, Jack Wild, Soo Drouwt, Daniel Newman, Walter Sparrow
Costner, American accent and all, is well and truly miscast in the title role, and gives one of his most stolid performances, in a film that is additionally hampered by some Politically Correct revisions of history. That said, the film has ample adventures and spectacular feats of derring-do, and though slow to get going, is surprisingly entertaining. Rickman's distinctive presence as the Sheriff of Nottingham is as offbeat as it is welcome.
A/AD 137 min Cut (18 sec plus some cuts subst – ort 144 min) cC VIDrel: WHV V/dm

ROBOCOP ***

18
Paul Verhoeven USA 1987
Peter Weller, Nancy Allen, Daniel O'Herlihy, Ronny Cox, Kurtwood Smith, Ray Wise, Miguel Ferrer, Robert DoQui, Paul McCrane, Jesse Goins, Del Zamora, Calvin Jung, Rick Lieberman, Lee DeBroux, Mark Carlton
In the not-too-distant future, policing has become the responsibility of a private corporation who, in their search for greater efficiency, resurrect a murdered police officer and turn him into a bionic law enforcement officer. However, he still seeks revenge for his murder, in this brutal and effective fantasy. See also R.O.T.O.R. AA: Spec Award (Stephen Flick/John Pospisil for sound effects editing).
FAN 98 min Cut (36 sec – ort 103 min)
VIDrel: 4-FRONT/POLYREC V/sur

ROBOCOP 2 **

18
Irvin Kershner USA 1990
Peter Weller, Nancy Allen, Daniel O'Herlihy, Tom Noonan, Belinda Bauer, Gabriel Damon, Felton Perry, Robert Do'Qui, Willard E. Pugh, John Doolittle, Patricia Charbonneau, Galyn Gorg, Stephen Lee, Jeff McCarthy, Wanda De Jesus
Our half-human crime-fighter makes his expected return, but suffers some less welcome modifications, and for good measure has to do battle with the lethal cyborg that's meant to replace him, a massive creature known as "Robocop 2". An overwhelmingly violent and offensive sequel, its spectacular effects (courtesy of Phil Tippett) are its best feature, but the film has none of the ironic humour to be found in its predecessor.
FAN 111 min Cut (4 sec plus film cuts – ort 117 min)
VIDrel: 4-FRONT/POLYREC V/sh

ROBOCOP 3 *

15
Fred Dekker USA 1991
Robert Burke, Nancy Allen, Rip Torn, Jill Hennessy, Remy Ryan, John Castle, C.C.H. Pounder, Mako, Robert Do'Qui, Bruce Locke, Stanley Anderson, Daniel Von Bargen, Stephen Root, Felton Perry, Bradley Whitford, Mario Machado, John Nesci
The third film in the series has neither the humour of the first nor the violence of the second. Instead, the plot hinges on the

efforts of an evil and all-powerful Japanese-owned corporation to evict the poor residents of Detroit from their homes. Robocop responds by leaving the force to take up cudgels on behalf of this oppressed group. Some good ideas remain undeveloped while the strongly anti-Japanese stance is both contrived and unproductive.
A/AD 100 min (ort 104 min) wScrn (LASER/SPEAR)
VIDrel: VCC/DISC/COLUM; LASER/SPEAR V/sur

ROBOCOP – THE SERIES: THE FUTURE OF LAW ENFORCEMENT **

12
Paul Lynch CANADA/USA 1993
Richard Eden, Yvette Nipar, Blu Mankuma, Sarah Campbell, Andrea Roth, David Gardner, Cliff De Young, John Rubinstein, Peter Costigan, Dan Duran, Erica Ehm, Jennifer Dale, Chris Kennedy, Patrick McKenna, Chris Bondy, Neil Crone
Pilot episode for a series built around the adventures of our half-human, half-robot crimefighter. Here he has to defeat the plans of a ruthless villain who is scheming to take over the city. Low production values and a total lack of originality in the plot department as well as uninspired acting makes this one hardly worth seeing. Followed by an innumerable number of similar episodes.
A/AD 86 min (ort 90 min) mTV VIDrel: ONE V/sh

ROBOT CALLED GOLDDIGGER, A **

PG
Mark Richardson/Jack Schaoul USA 1993
Joe Pantoliano, John Rhys-Davies, Danny Gerard, Amy Wright
An eccentric inventor creates a robot that he believes will bring him recognition at last, its purpose being to locate deposits of gold and extract them. Accompanied by a youngster, all three set off on an adventure, their hope being to locate a priceless gold artefact in this undemanding kids' comedy.
Aka: GOLDDIGGER
JUV 100 min VIDrel: NEWAGE/20VIS L/A V

ROBOT JOX ***

15
Stuart Gordon USA 1989
Gary Graham, Anne-Marie Johnson, Paul Koslo, Robert Sampson, Danny Kamekona, Hilary Mason, Michael Alldredge, Jeffrey Combs, Jason Marsden, Ian Patrick Williams, Carolyn Purdy-Gordon, Michael Saad, Thyme Lewis, Cary Houston
In the post-WW3 future, war has been outlawed and all armed conflicts are decided by single combat between trained fighters in vast robot machines. The American champ, upset by the death of 300 spectators caused when his machine fell on them, refuses to fight the Soviet champ over the sovereignty of Alaska, but a femme fatale is sent to persuade him otherwise. A competent effort, inspired by "The Gladiators" – a Swedish film by Peter Watkins.
FAN 81 min VIDrel: EIV/SONOP V

ROBOT WARS **

(PG)
Albert Band USA 1993
Don Michael Paul, Barbara Crampton, James Staley, Lisa Rinna, Peter Haskell, Danny Kamekona, Yuji Okumoto, J. Downing, Sam Scarber, Steve Eastin, Peter Mark Vasquez, Juan Garcia, Burke Byrnes, Keith S. Payson
A major threat is posed when the last mega-robot falls into the hands of a hostile group and a small team led by a renegade pilot goes in search of another such machine that is rumoured to be buried in a secret location. Just one more fantasy actioner where most of the interest revolves around the special effects.
FAN 106 min VIDrel: CIC/SONOP L/A V

ROBOTECH: THE MOVIE ***

PG
Carl Macek/Ishiguro Noburo USA 1985 (released 1986)
Voices of: Ryan O'Flannigan, Brittany Harlowe, Muriel Fargo, Greg Sbow, Jeffrey Platt, Guy Garrett, Abe Lasser, Merle Pearson, Penny Sweet, Wendee Swan, Wayne Anthony, Spike Niblick, Bruce Nielson, Ike Medlick, Tom Warner
A full-length animation that's set in the year 2027, and tells of an alien space-fleet assembled to invade Earth, but requiring information held on a master computer to be certain of success. When the man in charge of the planet's defences is captured, he's replaced by a duplicate, but a young mechanic who has learnt of this tries to foil the duplicate's plans. A lively adventure helped along by a well thought-out script.
ANIM 82 min VIDrel: RNK L/A V

ROCCO AND HIS BROTHERS ****

15
Luchino Visconti ITALY 1960
Alain Delon, Renato Salvatori, Annie Girardot, Katina Paxinou,

Roger Hanin, Paolo Stoppa, Suzy Delair, Claudia Cardinale, Spiros Focas, Rocco Vidolazzi, Max Cartier, Corrado Pani, Alessandra Panaro, Claudia Mori, Adriana Asti
A moody, melodramatic saga of a family who leave their poverty-stricken existence in the South in search of a better life in the harsh urban jungle of Milan. Related in four parts, with each section following the fate of one of the brothers. A most powerful film that's a little too long to digest in one sitting.
Aka: ROCCO E I SUOI FRATELLI
DRA   170 min (ort 180 min) B/W   VIDrel: CONNO/RTM
V

ROCK, THE ***
Michael Bay USA   15   1996
Sean Connery, Nicolas Cage, Ed Harris, Michael Biehn, William Forsythe, David Morse, John Spencer, John C. McGinley, Tony Todd, Bokeem Woodbine, Danny Nucci, Claire Forlani, Vanessa Marcil, Gregory Sporleder, Anthony Clark
A dissatisfied general seizes Alcatraz Island as a protest against cuts in military spending, and threatens to fire poison-gas rockets at San Francisco unless his demands are met. A chemical weapons expert and former prisoner (the only man known to have successfully escaped from Alcatraz) are sent to deal with this danger. One of those "what if" scenarios, but done with enormous dash and flair, with the ever dependable Connery giving one of his best performances.
A/AD   120 min   VIDrel: HOLPIC/TECH; ENCORE (LV only)   V/sur LV

ROCK-A-DOODLE **
Don Bluth UK   U   1990
Voices of: Phil Harris, Glen Campbell, Eddie Deezen, Kathryn Holcomb, Toby Scott Granger, Stan Ivar, Christian Hoff, Jason Marin, Christopher Plummer, Will Ryan, Sandy Duncan, Charles Nelson-Reilly, Ellen Greene, Sorrell Booke
A farmyard cock known as Chanticleer has learnt to sing like Elvis and this, coupled with his unhappy realisation that the sun is able to rise without his assistance, prompts him to make a journey to the big city in search of fame and fortune. However, with his departure comes persistent bad weather, and a band of animals sets out bring him home. An uneasy blend of excellent voice-overs, adequate animation and ill-advised Elvis adulation.
ANIM   71 min (ort 74 min)   VIDrel: VCC/DISC/COLUM
V/sur

ROCKETEER **
Joe Johnston USA   PG   1991
William Campbell, Jennifer Connelly, Timothy Dalton, Alan Arkin, Paul Sorvino, Terry O'Quinn, Ed Lauter, James Handy, Tiny Ron, Robert Guy Miranda, John Lavachielli, Jon Polito, Eddie Jones, William Sanderson, Don Pugsley
The inspiration for this film came from a comic-strip "novel" ten years before, and the film (set in 1930s L.A.) centres on the fantastic exploits of a stunt-pilot who finds a lost experimental rocket-pack that enables him to fly. Assisted by his trusty mechanic, he sets out to use it to revive their flagging financial fortunes. Unfortunately, there are other parties out to obtain this invention. A fast-moving, stylish but strangely uninvolving adventure.
A/AD   104 min (ort 108 min)   VIDrel: TOUCH/SONOP
L/A   V

ROCKETSHIP X-M **
Kurt Neumann USA   PG   1950
Lloyd Bridges, Osa Massen, John Emery, Hugh O'Brien, Noah Beery Jr, Morris Ankrum, Patrick Ahern, John Dutra, Katherine Marlowe, Sherry Moreland, Judd Holdren
An expedition to the Moon misses its destination due to a malfunction and lands on Mars instead, where they discover that atomic war has devastated its once-advanced civilisation. This version contains added tinted sequences and special effects shot in 1976. A superior effort considering its low budget and scripted with an ending that makes no concessions to sentimentality.
FAN   77 min B/W   VIDrel: SCREAM/SPEAR   V

ROCKING HORSE WINNER, THE ***
Anthony Pelissier UK   PG   1949
Valerie Hobson, John Mills, John Howard Davies, Ronald Squire, Cyril Smith, Hugh Sinclair, Charles Goldner, Susan Richards, Antony Holles, Melanie McKenzie, Caroline Steer, Michael Ripper
A young boy develops a strange ability to pick winning horses after being inspired by the thrilling stories of a groom. At first everything goes well, but events begin to take an unexpected

turn. A competent little drama, put together with wit and care, but hampered by the thin plot. Remade as a thirty-three-minute short in 1983
Aka: D.H. LAWRENCE: THE ROCKING HORSE WINNER
DRA   88 min (ort 90 min) B/W   VIDrel: VCC L/A   V
Boa: short story by D.H. Lawrence.

ROCKY ****
John G. Avildsen USA   PG   1976
Sylvester Stallone, Talia Shire, Burt Young, Carl Weathers, Thayer David, Burgess Meredith, Joe Spinell, Jimmy Gambina, Bill Baldwin, George Memmoli, Al Silvani, Jodi Letizia, Diana Lewis, George O'Hanlon, Larry Carr
Stallone finally broke into the big time with this excellent story (written by him) of a fading small-time boxer who gets a chance to regain his self respect by taking a crack at the heavyweight championship. Music by Bill Conti. Several sequels followed, each more hollow than the one before. AA: Pic, Dir, Edit (Richard Halsey/Scott Conrad).
DRA   114 min (ort 119 min)   VIDrel: MGM/WHV   V/h

ROCKY 2 **
Sylvester Stallone USA   PG   1979
Sylvester Stallone, Talia Shire, Burt Young, Carl Weathers, Burgess Meredith, Tony Burton, Joe Spinell, Leonard Gaines, Sylvia Meals, Frank McRae, Al Silvani, John Pleashette, Stu Nahan, Bill Baldwin, Jerry Ziesmer
Sequel to ROCKY with our boxer marrying his girl and working towards another bout. There's nothing new in this one, although the film remains watchable. Written once more by Stallone and with music by Bill Conti.
DRA   114 min (ort 119 min)   VIDrel: MGM/WHV   V/sur

ROCKY 3 **
Sylvester Stallone USA   PG   1982
Sylvester Stallone, Talia Shire, Burt Young, Burgess Meredith, Carl Weathers, Tony Burton, Mr. T (Lawrence Tureaud), Hulk Hogan, Ian Friend, Al Silvani, Jim Hill, Tony Burton, Wally Taylor, Leslie Morris, Dennis James
The second sequel to ROCKY, with the champ beaten at his first match against a tough opponent. He trains under Weathers to get his revenge. More of the same with a script by Stallone.
DRA   95 min (ort 99 min)   VIDrel: MGM/WHV   V/sur

ROCKY 4 **
Sylvester Stallone USA   PG   1985
Sylvester Stallone, Dolph Lundgren, Carl Weathers, Talia Shire, Burt Young, Brigitte Nielsen, Michael Pataki, James Brown, R.J. Adams, Al Dandiero, Dominic Barto, Daniel Brown, Rose Mary Campos, Jack Carpenter, Marty Denkin
The fourth in the ROCKY series pits the champ against a mighty Soviet fighting-machine (admirably played by Lundgren – who's Swedish). When his buddy dies at the hands of this monster, our perennial pugilist swears revenge. Did I hear you ask about a new plot?
DRA   88 min (ort 91 min)   VIDrel: WHV   V/sur

ROCKY 5 **
John G. Avildsen USA   PG   1990
Sylvester Stallone, Talia Shire, Burt Young, Sage Stallone, Tommy Morrison, Burgess Meredith, Richard Gant
Having been left penniless by a crooked accountant, Rocky Balboa returns to his roots, opens up a gym in Southern Philadelphia and discovers a promising new boxer. At the same time, he learns about the responsibilities of fatherhood and marriage. A production-line sequel (scripted by Stallone) that has little tension but a more gentle tone, and is surprisingly agreeable. Music is by Bill Conti. Sage is Stallone's real-life son.
DRA   99 min (ort 105 min)   VIDrel: MGM/WHV   V/sur

ROCKY HORROR PICTURE SHOW, THE ***
Jim Sharman UK   15   1975
Tim Curry, Susan Sarandon, Barry Bostwick, Richard O'Brien, Jonathan Adams, Meatloaf, Charles Gray, Little Nell (Neil Campbell), Patricia Quinn, Peter Hinwood, Hilary Labow, Jeremy Newson, Koo Stark, Christopher Biggins
A camp spoof on horror movies, with a straight couple taking refuge in a house full of weirdos. An outrageous blend of sex, transvestism and rock, with music and lyrics by O'Brien. Now something of a cult film, it has become available as a digitally re-mastered video. The sequel SHOCK TREATMENT followed.
COM   99 min (ort 101 min) wScrn   VIDrel: 20TH/TECH
V/dm

ROGER & ME **   15
Michael Moore USA   1989
Michael Moore
Former investigative journalist Moore made this stimulating
study of how General Motors killed off Flint in Michigan, when
its boss Roger Smith closed a major car plant there. Despite the
depressing subject matter, this study of the death of a community and the American Dream is rich in irony, much of it derived
from the ineffectual reactions of public figures who promised to
help. One-sided it may be, but this is potent stuff.
DOC   87 min   VIDrel: MGM/WHV L/A   V

ROGOPAG **   PG
R. Rossellini/Jean-Luc Godard/Pier Passolini/Ugo Gregoretti
ITALY   1962
*Rosanna Schiaffino, Bruce Balaban, Alexandra Stewart, Jean-Marc
Bory, Orson Welles, Renato Salvatori, Lisa Gastoni, Ugo Tognazzi,
Maria Cipriani, Laura Betti, Edmonda Aldini, Ettore Garofolo, Ricky
Tognazzi*
A quartet of sketches of variable quality loosely linked by the
theme of comment on modern life and all its negative aspects.
An airline hostess has to cope with harassment from a passen-
ger, a couple fall in love after a nuclear war, an actor in a
religious play suffers death by crucifixion and in the final story,
a married couple have to cope with the horrors of modern
marketing.
Aka: LAVIAMOCI IL CERVELLO; LET'S HAVE A BRAINWASH
DRA   118 min (ort 122 min) B/W/Col wScrn
VIDrel: CONNO/RTM L/A   V

ROLLER BLADE *   18
Donald G. Jackson USA   1985
*Katina Garner, Suzanne Solari, Jeff Hutchinson, Robby Taylor, Sam
Mann, Shaunn Michelle, Chris Douglas-Olen Ray, Michelle Bauer,
Lisa Marie, Barbara Peckinpaugh*
A post-WW3 survival epic, with mutant baddies and the other
stock characters that crop up in this genre. This one is set in a
devastated L.A. but has nothing new on offer except the appear-
ance of roller-skating heroines – namely two psychic sisters from
a band of female crusaders whose mission is to save humanity.
Some sequels followed. See also PRAYER OF THE ROLLER-
BOYS.
Aka: ROLLER BLADE WARRIORS
FAN   80 min (ort 88 min)   VIDrel: MED/POLYREC L/A
V/h

ROLLER BLADE SEVEN, THE *   18
Donald G. Jackson USA   1992
*Scott Shaw, Frank Stallone, Karen Black, Joe Estevez, William Smith,
Rhonda Shear, Don Stroud*
Unless the "Queen of Light" can be rescued from the "Pharaoh"
who is holding her captive, the world will remain in perpetual
darkness. So her companion the "Master of Light" sends the
title warriors off to mount a rescue. Nonsensical sword and
sorcery outing, whose chief point of interest is that it all takes
place on roller-skates.
A/AD   90 min   VIDrel: MIA/DISC   V

ROLLERBALL *   15
Norman Jewison USA   1975
*James Caan, John Houseman, Ralph Richardson, Maud Adams, John
Beck, Moses Gunn, Pamela Hensley. Shane Rimmer, Bert Kwouk,
Barbara Trentham, Alfred Thomas, Burnell Tucker, Angus MacInnes,
Nancy Blair, Rick Le Permentier*
A bleak, pointless and hollow look at a futuristic society, where
aggression is channelled into the following of a violent specta-
tor sport – a kind of brutal hockey on roller-skates and
motorcycles. A hero of the sport has achieved such a devoted
following that the state sees him as a threat, and he is ordered
in no uncertain terms to retire. One of the most empty and sterile
films ever made. Written by William Harrison.
FAN   118 min (ort 129 min) wScrn   VIDrel: MGM/WHV
V/s
Boa: short story Rollerball Murder by William Harrison.

ROLLOVER **   15
Alan J. Pakula USA   1981
*Jane Fonda, Kris Kristofferson, Hume Cronyn, Josef Sommer, Bob
Gunton, Macon McCalman, Ron Frazier, Jodi Long, Crocker Nevin,
Martin Chatinover, Ira B. Wheeler, Paul Hecht, Norman Snow, Nelly
Hoyos, Lansdale Chatfield*
The murder of a financial tycoon leads to the discovery of a
major conspiracy, with far-reaching implications for the

economy of the West. The title refers to the use made of funds
by large institutions, when they are used to finance further
loans. A complex and barely intelligible tale that tries to be more
important than it is. Written by David Shaber.
DRA   112 min (ort 115 min)   VIDrel: MGM/WHV L/A   V

ROMA **   15
Federico Fellini ITALY   1972
*Peter Gonzales, Stefano Majore, Britta Barnes, Pia De Doses, Fiona
Florence, Renato Giovannoli, Federico Fellini, Gore Vidal, Marne
Maitland, Galliano Sbarra, Alvaro Vitali, Britta Barnes, Angela De
Leo, Elisa Mainardi*
Fellini's homage to the Eternal City is a confusing blend of
drama and documentary that also draws heavily on details on
details from the director's life. Unfortunately, the sheer number
of ideas and the lack of a coherent narrative structure make this
movie extremely hard to follow at times, although as a work of
poetry and atmosphere it is certainly memorable.
Aka: FELLINI'S ROMA
DRA   114 min (ort 128 min)   VIDrel: MGM/WHV   V/h

ROMAN HOLIDAY **   U
William Wyler USA   1953
*Audrey Hepburn, Gregory Peck, Eddie Albert, Hartley Power, Laura
Solari, Harcourt Williams, Margaret Rawlings, Tullio Caminati,
Paolo Carlini, Claudio Ermelli, Paolo Borboni, Heinz Nindrich,
Gorella Gori, Alfred Rizzo*
A teenage princess on a visit to Rome gives her entourage the
slip in a bid to experience ordinary life, and meets a newspa-
perman who takes her on a tour of Rome. He pretends to be
ignorant of her true identity and the two eventually fall in love.
A captivating, old-fashioned romantic comedy, it gave the star
a career boost and a well-deserved Oscar. Filmed in Rome and
remade as a TV movie in 1987. AA: Actress (Hepburn), Story
(Ian McLellan Hunter), Cost (Edith Head).
COM   113 min (ort 119 min) B/W   VIDrel: CIC/SONOP
V/h

ROMAN HOLIDAY *   U
Noel Nosseck USA   1987
*Tom Conti, Catherine Oxenberg, Ed Begley Jr, Eileen Atkins, Paul
Daneman, Patrick Allen, Francis Matthews, Shane Rimmer,
Christopher Munke, Tessa Hood, Andrew Bicknell, David Rolfe, Felipe
Ferrer, Peter Peterson, Michael Roubaix*
A dull remake of the William Wyler classic with neither charm
nor skill, and sorely missing the presence of Audrey Hepburn and
Gregory Peck, who made the thin little tale of a runaway princess
and her affair with a reporter so enjoyable in the original.
DRA   96 min (ort 100 min) mTV   VIDrel: CIC/SONOP L/A
V

ROMAN SCANDALS **   PG
Frank Tuttle USA   1933
*Eddie Cantor, Ruth Etting, Gloria Stuart, David Manners, Edward
Arnold, Verree Teasdale, Alan Mowbray, Jack Rutherford, Grace
Poggi, Willard Robertson, Harry Holman, Lee Kohlmar, Stanley
Fields, Charles C. Wilson, Clarence Wilson*
Hilarious musical, with Cantor dreaming that he's back in
ancient Rome, where he gets caught up in some crazy intrigues.
Great songs and Busby Berkeley numbers ("Keep Young and
Beautiful" features Lucille Ball) combine with brilliant comic
routines to produce a memorable film. The story is by George
S. Kaufman and Robert E. Sherwood, with photography by
Gregg Toland.
MUS   92 min B/W   VIDrel: VCC/DISC   V

ROMANCE ON THE HIGH SEAS **   (12)
Michael Curtiz USA   1948
*Jack Carson, Janice Paige, Don DeFore, Doris Day, Oscar Levant,
S.Z. Sakall, Fortunio Bonanova, Eric Blore, Franklin Pangborn, Leslie
Brooks, William Bakewell, Johnny Berkes, Kenneth Britton, Frank
Dae, John Holland*
A woman and her husband suspect each other of infidelity and
so set in motion a complex web of mistaken identities, with a
detective hired by the man shadowing an impersonator hired by
her. The action takes place aboard an ocean liner, where the two
meet and fall in love. Day's film debut is a light and frothy affair
with some pleasant songs. These include "It's Magic" and "It's
You Or No One".
Aka: IT'S MAGIC
MUS   99 min   TVrel: C4
Boa: story Romance in High C by S. Pondal Rios and Carlos A.
Olivari.

ROMANCING THE STONE ** PG
Robert Zemeckis USA 1984
Michael Douglas, Kathleen Turner, Danny DeVito, Zack Norman, Alfonso Arau, Mary Ellen Trainor, Manuel Ojeda, Holland Taylor, Eve Smith, Joe Nesnow, Jose Chavez, Camillo Garcia, Rodrigo Puebla, Paco Morayta, Jorge Zamora
A writer of romantic fiction gets into trouble in South America and is helped out by an American fortune-hunter. The action proceeds at a breakneck pace and both Turner and Douglas turn in credible performances, but this is just enough to gloss over the lack of a solid plot. The unnecessarily brutal climax also strikes a sour note. Written by Diane Thomas and produced by Michael Douglas. THE JEWEL OF THE NILE followed.
A/AD 101 min Cut (23 sec – ort 106 min)
VIDrel: 20TH/TECH; ENCORE/FOXVID (LV only) V/sh LV
Boa: novel by J. Wilder.

ROMANTIC COMEDY * 15
Arthur Hiller USA 1983
Dudley Moore, Mary Steenburgen, Frances Sternhagen, Janet Eilber, Robyn Douglass, Ron Leibman, Roziska Halmos, Alexander Lockwood, Erica Hiller, Sean Patrick Guerin, Dick Wieand, Brass Adams, Stephen Roberts, Tom Kubjak
A playwright tries to suppress his feelings for his female partner, in this unsuccessful attempt to transfer a hit Broadway play to the screen. Written by Slade.
COM 98 min (ort 103 min) VIDrel: MGM/WHV L/A V
Boa: play by Bernard Slade.

ROMANTIC ENGLISHWOMAN, THE ** 15
Joseph Losey FRANCE/UK 1975
Michael Caine, Glenda Jackson, Helmut Berger, Marcus Richardson, Kate Nelligan, Rene Kolldehof, Michel Lonsdale, Beatrice Romand, Anna Steele, Nathalie Delon, Bill Wallis, Julie Peasgood, David De Keyser, Phil Brown
The wife of a novelist goes on holiday, falling in love with another man and so initiating a strange, three-cornered relationship. An uncertain blend of drama and fantasy, that attempts to make a few portentous statements on the meaning of life, but ultimately becomes merely tiresome. Written by Tom Stoppard and Thomas Wiseman.
DRA 112 min (ort 117 min)
VIDrel: 4-FRONT/POLYREC/ODY V/h
Boa: novel by Thomas Wiseman.

ROMANTIC UNDERTAKING ** (PG)
Peter McCubbin CANADA 1995
Valerie Buhagiar, William Katt, Ishwar Mooljee, Paul Berry, Greg Blanchard, Simon Richards, Peter C. Ferri, Leetha Carroll, Sherrill Lion, Laura James, Susan Cooke, Barry Stone, Christopher Moore, Ed Fielding, Petty Kazmer
After her father dies, a daughter inherits his rundown funeral business and decides to carry it on. Short of cash, she is obliged to let out rooms and gets involved in an unexpected complication when a new and attractive tenant moves in, who appears to have a mysterious past. A gentle romantic comedy, of flimsy plotting, good performances and a pleasing (if totally expected) resolution.
COM 95 min SATrel: MOVIE CHANNEL

ROME EXPRESS *** U
Walter Forde UK 1932
Conrad Veidt, Esther Ralston, Joan Barry, Gordon Harker, Cedric Hardwicke, Harold Huth, Donald Calthrop, Hugh Williams, Finlay Currie, Frank Vosper, Muriel Aked, Eliot Makeham
A prototype for those train thrillers such as THE LADY VANISHES and MURDER ON THE ORIENT EXPRESS, with thieves and blackmail victims among the passengers on the Paris-Rome express. Much of the action revolving around the theft of a painting, with a French detective eventually unmasking the culprit. Sturdily directed, this entertaining but verbose thriller was Gaumont Studio's first film. Written by Clifford Grey, Sidney Gilliat, Frank Vosper and Ralph Stock.
THR 86 min (ort 94 min) B/W VIDrel: VCC/RTM V

ROME, OPEN CITY *** 15
Roberto Rossellini ITALY 1945
Aldo Fabrizi, Marcello Pagliero, Anna Magnani, Maria Michi, Harry Feist, Giovanna Galletti, Nando Bruno, Francesco Grandjacquet, Vito Annichiarico, Carla Revere, Carlo Sindic, Joop Van Hulzen, Akos Tolnay, Eduardo Passarelli
Classic neo-Realist account of a resistance leader in occupied Rome on the run from the Nazis, into whose hands he is even-tually betrayed. Grim and almost documentary in tone, this remains a gripping and powerful work. See also PAISAN.
Aka: OPEN CITY; ROMA, CITTA APERTA
DRA 97 min (ort 105 min) B/W VIDrel: CONNO/RTM V

ROMEO AND JULIET *** PG
Franco Zeffirelli ITALY/UK 1968
Leonard Whiting, Olivia Hussey, Milo O'Shea, Michael York, John McEnery, Pat Heywood, Robert Stephens, Bruce Robinson, Paul Hardwick, Natasha Perry, Antonio Piefederici, Esmeralda Ruspoli, Roberto Bisacco, Keith Skinner, Aldo Miranda
Whiting and Hussey were only seventeen and fifteen respectively when this was made, and are ideal as lovers in a glossy screen version of this famous play, telling of the doomed love affair between the children of feuding families. A brave attempt, but Shakespeare's verse suffers badly from the overly brisk pace. The beautiful score is by Nino Rota and the prologue is read by Laurence Olivier. AA: Cin (Pasqualino De Santis), Cost (Danilo Donati).
DRA 132 min (ort 152 min) VIDrel: CIC/SONOP V
Boa: play by William Shakespeare.

ROMEO AND JULIET ** PG
Alvin Rakoff UK 1979
Patrick Ryecart, Rebecca Saire, Celia Johnson, Michael Hordern, John Gielgud, Joseph O'Conor, Laurence Naismith, Anthony Andrews, Alan Rickman, Jacqueline Hill, Christopher Strauli, Christopher Northey, Paul Henry
A standard BBC adaptation, competent but unmemorable.
DRA 168 min mTV VIDrel: BBC L/A V/h
Boa: play by William Shakespeare.

ROMEO AND JULIET *** 12
Baz Luhrmann USA 1996
Leonardo DiCaprio, Claire Danes, John Leguizamo, Brian Dennehy, Pete Postlethwaite, Paul Sorvino, Diane Venora, Harold Perrineau, Paul Rudd, Jesse Bradford, Dash Mihok, Miriam Margoyles, Vondie Curtis-Hall
Set in modern-day L.A., this is (like WEST SIDE STORY) an updating of the famous play, but instead of the Montagus and Capulets, we have two rival street gangs, and guns are used in place of swords to settle differences. However, in most other respects the spirit of the play (and its text) is retained, and apart from one or two faintly ludicrous moments, the film works surprisingly well. Danes and DiCaprio are especially good as the ill-fated lovers.
Aka: WILLIAM SHAKESPEARE'S ROMEO AND JULIET
DRA 120 min CINrel
Boa: play by William Shakespeare.

ROMEO IS BLEEDING ** 18
Peter Medak USA 1993
Gary Oldman, Lena Olin, Annabella Sciorra, Juliette Lewis, Roy Scheider, David Proval, Will Patton, Larry Joshua, James Cromwell, Julia Migenes, Dennis Farina, Ron Pearlman, Gene Canfield, Paul Butler, Michael Wincott, Wallace Wood
A burnt-out cop looking for a quick and easy way to make big money allows himself to become corrupted. He makes contact with a gangster who charges him with the liquidation of a sadistic assassin whose physical charms play a certain role in her line of work. An over-the-top exercise in nastiness that strives too hard for both laughs and a film noir atmosphere. The voiceover narration is most irritating and in no way improves this unpleasant dud.
DRA 105 min (ort 109 min) VIDrel: VCC/DISC/COLUM V/sur

ROMERO *** 15
John Duigan USA 1989
Raul Julia, Richard Jordan, Ana Alicia, Eddie Velez, Tony Plana, Alejandro Bracho, Lucy Reina, Harold Gould, Al Ruscio, Robert Viharo, Tony Perez, Paco Mauri, Harold Cannon-Lopez, Claudio Brook, Eduardo Lopez-Rojas, Juan Pelaez
An absorbing movie with a message that attempts to chart the career of Oscar Romero, the Archbishop of El Salvador, who changed from obedient cleric to opponent of the state, and was assassinated by right-wing thugs in 1980. Ponderous and vague with regard to US support for the repressive government, but Julia's fine performance compensates. Scripted by John Sacret Young, this was the first feature financed by the US Roman Catholic Church.
DRA 101 min (ort 105 min) VIDrel: MGM/WHV L/A V/sh

ROMPER STOMPER ** *18*
Geoffrey Wright AUSTRALIA 1993
Russell Crowe, Jacqueline McKenzie, Daniel Pollock, Alex Scott, Leigh Russell, Daniel Wylie, James McKenna, Samantha Bladon, Josephine Keen, John Brumpton, Eric Mueck, Frank Magree, Christopher McLean, James Bridges
A group of racist skinheads in Melbourne attack local Asians but are surprised when they begin to encounter strong resistance and are eventually defeated. A potent concoction of sex and violence with a confused message. This was former film critic Wright's debut feature, and he also wrote the screenplay.
A/AD 89 min (ort 91 min) VIDrel: POLY/POLYREC/MED V/sur

ROMUALD AND JULIETTE *** *PG*
Coline Serreau FRANCE 1989
Daniel Auteuil, Firmine Richard, Pierre Vernier, Maxime Leroux, Gilles Privat, Muriel Combeau, Catherine Salviat, Sambout Tati, Alexandre Basse, Aissatou Bah, Mamdou Bah, Marina M'Boa Ngong, Nicolas Serreau, Alain Tretout
A top executive with a yoghurt company is the victim of both his wife's infidelity and a sinister plot to frame him for food poisoning. Fortunately, he is rescued from disaster by Juliette, a black cleaning lady at his company, who he has never even noticed before. While he hides out at her tiny apartment and plans his next move, a touching romance develops between this unlikely couple. A rather contrived but warm and engaging comedy.
Aka: MAMA, THERE'S A MAN IN YOUR BED; ROMUALD ET JULIETTE
COM 107 min (ort 112 min) wScrn VIDrel: TART/20TH V

ROOFTOPS ** *15*
Robert Wise USA 1989
Jason Gedrick, Troy Beyer, Eddie Velez, Tisha Campbell, Alexis Cruz, Allen Payne, Steve Love, Rafael Baez, Jaime Tirelli, Luis Guzman, Millie Tirelli, Jay M. Boryea, Robert LaSardo, Rockets Redglare, Edouard DeSoto, John Canada Terrell
The director of WEST SIDE STORY returns with another urban musical, dealing with the tough New York kids who dance, fight and survive on the streets. This time round, the starcrossed lovers are a white boy and his Hispanic girlfriend, and the unusual but rather awkwardly inserted "combat" dancing sequences blend kung fu movements with DIRTY DANCING sexuality. If only the dreary plot were not there to dissipate the energy of it all.
MUS 91 min (ort 98 min) VIDrel: 20TH/TECH V/sur

ROOKIE, THE * *18*
Clint Eastwood USA 1990
Clint Eastwood, Charlie Sheen, Raul Julia, Sonia Braga, Lara Flynn Boyle, Tom Skerritt, Pepe Serna, Mara Corday, Marco Rodriguez, Pete Randall, Donna Mitchell, Hal Williams, Xander Berkley, Tony Plana, David Sherril
Depressingly violent cop-actioner with little by way of a coherent plot; the best performances come from Julia and Braga, but they are really too Latin in looks to be convincing as German terrorists. A most dismal affair, with Eastwood well below his best and Sheen (as his new rich-kid partner) little better.
A/AD 116 min (ort 121 min) VIDrel: WHV V/sur

ROOKIE OF THE YEAR ** *PG*
Daniel Stern USA 1993
Thomas Ian Nicholas, Gary Busey, Dan Hedaya, Daniel Stern, Albert Hall, Amy Morton, Bruce Altman, Eddie Bracken, Robert Gorman, Patrick LaBreque, Daniel Stern, Colombe Hacobsen-Derstine, Kristie Davis, Tyler Ann Carroll, Ross Lehman
After breaking his arm, a twelve-year-old boy appears to acquire incredible baseball skills and becomes a pitcher for the Chicago Cubs, but in reality his prowess is the result of assistance from the ghost of a dead player. A likeable but rather thinly plotted fantasy clearly aimed at kids, its main idea being taken from a minor 1954 film entitled "Roogie's Bump".
FAN 99 min (ort 103 min) cC VIDrel: 20TH/TECH V/sh

ROOKIES, THE * *18*
Jake Craig USA 1991
Raven, Kandi, Alexandria Quinn, Judy Zee, Marc Wallice, T.T. Boy, Wayne Summers
A group of unemployed women decide to make their own amateur adult movie in the hope of winning a cash prize of $10,000. However, they become aware that this competition is a merely a means for a porno producer to get free footage.

A 80 min VIDrel: IMPENT V

ROOM AT THE TOP **** *15*
Jack Clayton UK 1958
Laurence Harvey, Simone Signoret, Heather Sears, Hermione Baddeley, Donald Wolfit, Donald Houston, Allan Cuthbertson, John Westbrook, Raymond Huntley, Ambrosine Phillpots, Richard Pasco, Beatrice Varley, Delena Kidd, Mary Peach
An ambitious young man decides to take a short cut in the struggle to reach the top by marrying his boss's daughter, ultimately ditching an older woman he had a loving relationship with. A memorable drama, excellently acted, and perfectly capturing the atmosphere of the grimy Northern town where the story is set. Followed by the inferior LIFE AT THE TOP, a TV series and then MAN AT THE TOP. AA: Actress (Signoret), Screen/adapt (Neil Paterson).
DRA 113 min (ort 117 min) B/W VIDrel: FABFIL/SPEAR V
Boa: novel by John Braine.

ROOM OF WORDS, THE * *18*
Joe D'Amato USA 1989
Martine Brochard, Linda Carol, David Brandon
A highly inferior imitation of HENRY & JUNE, that covers very much the same ground in its story of the relationship between the writer Henry Miller, his wife June, and the French diarist Anais Nin. However, virtually the entire film is given over to softcore couplings (accompanied by a relentless and intrusive jazz soundtrack) and soon becomes unbearably repetitive. A witless film of zero acting that's both dull and pretentious in equal measure.
DRA 97 min VIDrel: POLY/POLYREC/BRAVE V
Boa: novel by F. Mole.

ROOM WITH A VIEW, A *** *PG*
James Ivory UK 1985
Maggie Smith, Helena Bonham Carter, Judi Dench, Julian Sands, Daniel Day Lewis, Denholm Elliott, Simon Callow, Rosemary Leach, Joan Henley, Maria Britneva, Amanda Walker, Rupert Graves, Fabia Drake, Peter Cellier
Upper-class mores are put under the microscope, in one of those typically British class-conscious studies. The story has a refined young lady planning to marry a dull prig, but finding there is more to life when she meets a passionate young man while on holiday in Florence. A detailed period tale. AA: Art/Set (Gianni Quaranta and Brian Ackland-Snow/Brian Savegar and Elio Altramura), Screen/adapt (Ruth Prawer Jhabvala), Cost (Jenny Beavan/John Bright).
DRA 113 min (ort 115 min) VIDrel: 4-FRONT/POLYREC V/sur
Boa: novel by E.M. Forster.

ROOM-MATES * *18*
Gail Force/Jim Powers USA
Sierra, Trinity Lane, Dominique Bouche, Dave Hardman, Dave Dodge, Anthony Crane, Ron Jeremy
A group of guys advertise for room-mates and get more than they bargained for when they meet some randy country girls.
A 69 min VIDrel: ELV V/sur

ROOMMATES *** *(18)*
Alan Metzger USA 1993
Randy Quaid, Eric Stoltz, Elizabeth Pena, Charles Durning, Frank Buxton, Joe Maffei, Babz Chula, Jill Teed, Kean Hooker, David Michael Mulling, Michael Roberds, Angela Gann, Pat La Plant, Celia-Jose Martin, Richard Sali, John Scott
Quaid has contracted the AIDS virus via a contaminated blood transfusion while Stoltz has been infected by a homosexual relationship. Both men grow tired of the prejudice they encounter in the outside world, and when they move into a home for AIDS sufferers are put together as roommates. Soon their initial hostility has given way to mutual respect and they each set out to re-establish contact with their families. A nicely focused and touching drama.
DRA 90 min mTV SATrel: SKY MOVIES

ROOMMATES ** *PG*
Peter Yates USA 1995
Peter Falk, Julianne Moore, D.B. Sweeney, ellen Burstyn, Jan Rubens, Joyce Reehling, Ernie Sabella, John Cunningham
One of those terribly over-familiar tearjerkers, casting Sweeney as a medical student who has to contend with his irascible old grandfather, who has taken it upon himself to move in with the

lad. The plot runs along predictable lines, but one has to keep watching, if only to appreciate Falk's tremendous performance as the old grouch.
COM 109 min VIDrel: HOLPIC V

ROOSTER COGBURN **
Stuart Miller USA 1975
U

John Wayne, Katharine Hepburn, Richard Jordan, Anthony Zerbe, John McIntire, Paul Koslo, Tommy Lee, Strother Martin, Jack Colvin, Jon Lormer, Lane Smith, Richard Romancito, Warren Vanders, Jerry Gatlin, Mickey Gilbert, Chuck Hayward
A woman missionary teams up with a hard-fighting, hard-drinking marshal, to bring in the outlaws who killed her father. A kind of misguided sequel to TRUE GRIT by way of THE AFRICAN QUEEN, with neither the freshness nor verve of either.
Aka: ROOSTER COGBURN AND THE LADY
WES 103 min (ort 108 min)
VIDrel: 4-FRONT/POLYREC/CIC V/h

ROOT OF EVIL *
John Patterson USA 1990
15

Elizabeth Montgomery, Dale Midkiff, Heather Fairfield, Talia Balsam, Jerry Bossard, Mimi Kennedy
An apparently normal man who has a good job and many friends is in reality a twisted victim of his mother's dominance, who finds release as a multiple rapist. Unpleasant melodrama with a few Freudian overtones.
DRA 90 min VIDrel: GENESIS V

ROOTS: EPISODES 1 TO 6 **
Marvin J. Chomsky/John Erman/David Greene/Gilbert Moses USA 1977
PG

Edward Asner, Chuck Connors, Cicely Tyson, Lloyd Bridges, Doug McClure, Vic Morrow, George Hamilton, Lorne Greene, Louis Gossett Jr, Richard Roundtree, Maya Angelou, Moses Gunn, Thalmus Rasulala, Harry Rhodes, William Watson
A long, loose and rambling film version of a bestseller, that looks at black history from the capture of an African slave and his transportation to the USA up to the time of the Civil War. Well made and acted, but terribly idealistic in terms of the view of life portrayed in Africa. The impact is severely limited by the sluggish script and directing. Followed by ROOTS: THE NEXT GENERATIONS – VOLS. 1 AND 2 and ROOTS: KUNTA KINTE'S GIFT.
DRA 538 min (6 cassettes) Cut (16 sec – ort 570 min) mTV
VIDrel: WHV V
Boa: novel by Alex Hailey.

ROOTS: THE NEXT GENERATIONS – VOLS 1 AND 2 **
John Erman/Charles S. Dubin/Georg Stanford Brown USA 1978
PG

Georg Stanford Brown, Henry Fonda, Paul Koslo, Avon Long, Lynne Moody, Greg Morris, Marc Singer, Richard Thomas, Fay Hauser, Brian Mitchell, Ja'net Du Bois, Debbi Morgan, Kathleen Doyle, Slim Gaillard, Harry Morgan, Stan Shaw
Inspired by the success of ROOTS, this massive sequel continues the story of Haley's family from the post-Civil War Reconstruction era until 1967, when the author undertook his celebrated journey to West Africa. A lengthy pageant of people and events covering some of the most important milestones in Afro-American history.
DRA 658 min (4 cassettes – ort 685 min) mTV
VIDrel: WHV V/h
Boa: novel by Alex Haley.

ROOTS: KUNTA KINTE'S GIFT **
Kevin Hooks USA 1988
PG

LeVar Burton, Louis Gossett Jr, Avery Brooks, Shaun Cassidy, John McMartin, Jerry Hardin, Kate Mulgrew, Michael Learned, Fran Bennett, Annabella Price, Tim Russ
Another dose of glamorised history, that began with ROOTS and the tale of a young black boy who is taken into slavery. Three years have passed since an unsuccessful escape attempt, and Kunta Kinte now becomes involved in an ambitious scheme to lead his fellow slaves to freedom. Average.
Aka: KUNTA KINTE'S GIFT; ROOTS: THE GIFT
DRA 89 min (ort 100 min) mTV VIDrel: WHV V

ROOTS: ALEX HALEY'S QUEEN – THE ROOTS SAGA CONTINUES **
John Erman USA 1992
15

Ann-Margret, Hale Berry, Patricia Clarkson, Timothy Daly, Ossie Davis, Danny Glover, Jasmine Guy, Dennis Haysbert, Martin Sheen, Madge Sinclair, Paul Winfield
Continues the further story of Kunte Kinte and his family. Adequate.
Aka: QUEEN
DRA 270 min (2 cassettes) mTV VIDrel: WHV V/sh

ROPE **
Alfred Hitchcock USA 1948
PG

James Stewart, John Dall, Farley Granger, Cedric Hardwicke, Constance Collier, Joan Chandler, Edith Evanson, Douglas Dick, Dick Hogan
Two homosexuals murder a friend for kicks and hide his body in a trunk. They then proceed to throw a party, using the trunk as a cocktail bar. The entire action takes place in one room, and was inspired by the 1920s Leopold and Loeb murder. A really stagebound work, with irritatingly long takes (note that it runs in real-time) and performances that just don't convince. Written by Arthur Laurents, this over-rated experiment was Hitchcock's first colour film.
DRA 77 min (ort 81 min) VIDrel: CIC/SONOP V/h
Boa: play by Patrick Hamilton.

ROSALIE GOES SHOPPING ***
Percy Adlon WEST GERMANY 1989
15

Marianne Sagebrecht, Brad Davis, Judge Reinhold, William Harlander, Alex Winter, Erika Blumberger, Patricia Zehentmayr, John Hawkes, Courtney Kraus, Lisa Fitzhugh, Lori Fitzhugh, Dina Chandel, David Denney, Ed Geldart
Bavarian-born Sagebrecht and her weird family settle in Little Rock, Arkansas, where she becomes obsessed with the Great American Dream. Having taken this to mean an endless spending spree and a house full of possessions, she devises a crazy way to beat her creditors. A wonderfully screwy satire on American consumerism, that fires off gags in almost every direction and is as uneven as it is hilarious. The script is by Percy Adlon.
COM 90 min (ort 96 min) VIDrel: PAL/TERRY L/A V

ROSE, THE **
Mark Rydell USA 1979
15

Bette Midler, Alan Bates, Frederic Forrest, Harry Dean Stanton, Barry Primus, David Keith, Sandra McCabe, Will Hare, Rudy Bond, Don Calfa, James Keane, Doris Roberts, Sandy Ward, Michael Greer, Claude Sacha, Pearl Heart
A spin-off from the rather sad life of singer Janis Joplin, this has Midler starring in a powerful but overlong story, of a self-destructive singer who engineers her own downfall and death. The photography is by Vilmos Zsigmond, and is one of the best features of this depressing work.
DRA 129 min (ort 134 min) VIDrel: 20TH/TECH V/sur

ROSE GARDEN, THE ***
Fons Rademakers USA/WEST GERMANY 1989
15

Liv Ullmann, Maximilian Schell, Peter Fonda, Jan Niklas, Kurt Hubner, Hans Zischler, Georg Marischka, Gila Almagor, Katarina Lena Muller, Nicolaus Sombart, Ozay Fecht, Achim Ruppel, Friedhelm Lehmann, Mareike Carriere
A man accused of assault claims in his defence that his elderly victim was the commander of the death camp in which his family died. Fine performances and a dollop of sincerity fail to round out this shallow and rather oblique attempt to deal with the issues raised by the Holocaust. Loosely based on the career of SS commander Arnold Strippel, who by ill health escaped punishment for the murder of 20 children at his trial following WW2.
Aka: DER ROSENGARTEN
DRA 109 min (ort 111 min) VIDrel: PATHE L/A V
Boa: book Swastika Over Paris by Jeremy Josephs.

ROSE MARIE ***
Mervyn Le Roy USA 1954
U

Howard Keel, Ann Blyth, Fernando Lamas, Bert Lahr, Marjorie Main, Joan Taylor, Ray Collins, Chief Yowlachie, James Logan, Turl Ravenscroft, Abel Fernandez, Billy Dix, Al Ferguson, Frank Magney, Marshall Reed, Sheb Wooley
A remake of the 1936 musical in which a Mountie falls for a tomboy girl of the woods, eventually "taming" her and winning her love. Not as good as the earlier film but closer to the original operetta. A highlight is Lahr singing "I'm The Mountie Who Never Got His Man". The choreography is by Busby Berkeley.
MUS 97 min (ort 115 min) VIDrel: MGM/WHV V/sh
Boa: play by Rudolf Friml, Otto Harbach and Oscar Hammerstein II.

ROSEANNA McCOY **
PG
Irving Reis USA   1949
Joan Evans, Farley Granger, Charles Bickford, Raymond Massey, Richard Basehart, Aline MacMahon
Story of two feuding hillbilly families in Virginia and the love affair that develops between sweethearts Granger and Evans, who coming from either camp, do not find the path of true love an easy one. A stolid and rather humourless variant on the Romeo and Juliet story, with little to recommend it except some nice outdoors locations.
DRA   89 min B/W   VIDrel: VCC/DISC   V

ROSEANNE: AN UNAUTHORISED BIOGRAPHY **   12
Paul Schneider USA   1994
Denny Dillon, David Graf, John Walcutt, Judith Scarpone, Joycelyn O'Brien, Sara Melson, Dawn Zeek, John Karlen, Matt Landers, Ian Patrick Williams, Allison McMillan, Danielle Harris, Lee Magnuson, Jandi Swanson, J. Patrick McCromack
Exactly what it says, an account of the life and times of this TV star, her acrimonious divorce from her first husband and her allegations of childhood sexual abuse. Yet somehow, despite metal illness and other troubles she managed to achieve stardom if not happiness. Reasonably well made but of real interest only to ardent fans – others will be put off by her strident and aggressive personality. See also ROSEANNE AND TOM: BEHIND THE SCENES.
DRA   86 min (ort 90 min) mTV   VIDrel: POLFIL   V

ROSEANNE AND TOM: BEHIND THE SCENES ***   (12)
Richard A. Colla USA   1994
Patrika Darbo, Stephen Lee, Jan Hogg, Nancy Youngblood, Robert Nechez, Ria Davies, Andrew Hill Newman, Steven Andersson, Stephen Mendel, Scott Harlan, David Powledge, Heather Paige Kent, Harvey J. Alperin, Lily Mariye, Luise Heath
Biopic on the tumultuous wedding, marriage and various battles of the title characters, who made bad taste and noisy feuding something of a way of life. Some very clever touches are used in an attempt to provide an insight or two into these characters, and the two leads are perfectly cast as this larger than life couple. Unfortunately, the central characters are so very unlovable that the movie eventually becomes as painful to watch as their marriage was to live.
Aka: ROSEANNE AND TOM: A HOLLYWOOD MARRIAGE
DRA   86 min (ort 90 min) mTV   SATrel: SKY MOVIES

ROSELYNE AND THE LIONS *
PG
Jean-Jacques Beineix FRANCE   1989
Isabelle Pasco, Gerard Sandoz, Philippe Clevenot, Gunter Meisner, Gabriel Monnet, Wolf Harnisch, Jacques Le Carpanetier, Dimitri Furdui, Carlos Pavlidas, Jaroxlas Vizner, Jacques Mathou, Hans Myer, Francois Boum
Another slice of flashy, over-stylised emptiness from the director of DIVA. A young man falls in love with a beautiful girl and together they form a lion-taming act with a travelling circus. A sub-plot involving a schoolmaster keen to turn their exploits into fiction adds nothing of substance to this lightweight effort.
DRA   113 min (ort 131 min) wScrn   VIDrel: ARTIF/20TH
V/sur

ROSEMARY'S BABY ***
18
Roman Polanski USA   1968
Mia Farrow, John Cassavetes, Ruth Gordon, Sidney Blackmer, Maurice Evans, Ralph Bellamy, Patsy Kelly, Angela Dorian, Elisha Cook Jr, Charles Grodin, Emmaline Henry, Marianne Gordon, Phil Leeds, Hope Summers, Wendy Wagner
The frightening tale of a woman, the wife of an actor, who is befriended by Satanists and unwittingly becomes pregnant by the Devil. Written by Polanski. It may sound foolish but is actually very well done. Followed some years later by the truly ludicrous TV sequel, "Look What's Happened To Rosemary's Baby". AA: S. Actress (Gordon).
HOR   131 min (ort 137 min)
VIDrel: 4-FRONT/POLYREC/CIC   V
Boa: novel by Ira Levin.

ROSENCRANTZ AND GUILDENSTERN ARE DEAD **
PG
Tom Stoppard USA   1990
Gary Oldman, Tim Roth, Richard Dreyfuss, Iain Glen, Ian Richardson, Joanna Roth, Donald Sumpter, Joanna Miles, Brnako Zavrsan, Ljubo Zecevic, Tomislav Meretic, Mare Mlacnik, Srdjan Soric, Maladen Vasary, Zeljko Vukmirica
Written by Stoppard, this film version of his highly successful stage play deals with two minor characters found in Hamlet, who wander aimlessly around a castle, and indulge in a feast of verbal interplay. A lifeless work that has some inventive visual ideas, it's certainly worth listening to, if not especially rewarding in other ways. Despite winning the Best Film award at Venice, it should have been left on the stage.
COM   113 min (ort 118 min)   VIDrel: BUENA/TECH L/A
V
Boa: play by Tom Stoppard.

ROSES ARE FOR THE RICH: PARTS 1 AND 2 **   *PG*
Michael Miller USA   1987
Lisa Hartman, Bruce Dern, Joe Penny, Richard Masur, Howard Duff, Morgan Stevens, Jim Youngs, Betty Buckley, Kate Mulgrew, Rhonda Dotson, Dee Dee Rescher, Deidre Allan Taylor, Mary Conley, Helen Duffy, Jamie Greenspan
A two-part soap opera, telling of a determined Appalachian woman who sets out to have her revenge on the coal magnate whom she blames for the death of her husband. Hartman is unconvincing in her role, but this enjoyable piece of escapist nonsense has enough in it to sustain interest if not credibility.
DRA   177 min (ort 200 min) mTV   VIDrel: GUILD/SONOP
L/A   V
Boa: novel by Jonell Lawson.

ROSIE DIXON: NIGHT NURSE *
18
Justin Cartwright UK   1977
Debbie Ash, Caroline Argyle, Beryl Reid, John Le Mesurier, Arthur Askey, Liz Fraser, Lance Percival, John Junkin, Bob Todd, David Timson, Leslie Ash, Jeremy Sinden, Peter Mantle, Ian Sharp, Christopher Ellison, Patricia Hodge
A young nurse encounters various over-sexed individuals in her work, in this ghastly attempt at a sex comedy.
COM   84 min (ort 88 min)   VIDrel: VCC/DISC/COLUM   V
Boa: novel Confessions of a Night Nurse by Rosie Dixon.

ROSIE: THE ROSEMARY CLOONEY STORY ***   *PG*
Jackie Cooper USA   1982
Sondra Locke, Tony Orlando, Penelope Milford, Katherine Helmond, Kevin McCarthy, John Karlen, Cheryl Anderson, Robert Ridgely, Joey Travolta, Richard Berlin, Christopher Thomas, Eli Rill, Jeb Adams, Edson Stroll
A simple account of the career of singer Clooney, that charts her rise from an act on Cincinnati radio, through to stardom, a mental breakdown and a courageous struggle back into the limelight. This predictable rise-and-fall story benefits from Locke's fine performance (she lip-synchs to Clooney's singing voice) and an above-average script that's largely based on Clooney's autobiography. Poor casting of Orlando as Jose Ferrer is a minor failing.
DRA   93 min (ort 100 min) mTV   VIDrel: CASPIC L/A   V
Boa: autobiography This For Remembrance by Rosemary Clooney.

ROSWELL **
12
Jeremy Kagan USA   1994
Kyle MacLachlan, Martin Sheen, Dwight Yoakam, Xander Berkeley, Bob Gunton, Kim Griest, Peter MacNicol, John M. Jackson, Charles Martin, Nick Searcy, J.D. Daniels, Eugene Roche, Lisa Wlatz, Charles Hallan, Ray McKinnon, Doug Wert
When a UFO crashes in New Mexico in 1947, the military are not slow in imposing a news blackout and mounting a cover-up to keep secret what they found at the crash site. A dramatisation of a true-life incident which remains a mystery to this day, with many people firmly convinced that the USAF did indeed find a crashed UFO and its crew. Well acted if insubstantial, both MacLachlan's dreadful makeup as an old man and the interminable flashbacks are flaws.
Aka: ROSWELL: THE UFO COVER-UP
DRA   87 min (ort 91 min) mCab   VIDrel: POLY/POLYREC
V/sur
Boa: book UFO Crash At Roswell by Kevin D. Randle and Donald K. Schmitt.

ROUGE ***
15
Stanley Kwan HONG KONG   1987
Anita Mui, Leslie Cheung, Alex Man, Emily Chu, Man Tsz-leung
In the Hong Kong of the 1930s, a courtesan takes her own life for love but fails to find her lover in the next world. Fifty years later, her spirit is still earthbound and haunting the denizens of this modern and now highly overcrowded city. A tragic

romance with a complex flashback structure, it has considerable visual impact and a most moving climax.
Aka: INJI KAU
FAN   92 min (ort 99 min)   VIDrel: ICAPRO/MANGA   V/sh

ROUGH CUT **
Donald Siegel USA
PG
1980
Burt Reynolds, Lesley-Anne Down, David Niven, Timothy West, Patrick Magee, Joss Ackland, Al Matthews, Susan Littler, Isobel Dean, Wolf Kahler, Andrew Ray, Julian Holloway, Douglas Wilmer, Geoffrey Russell, Ronald Hines
A gentleman jewel thief, living in London and involved in a multi-million dollar diamond heist, becomes the prey of a Scotland Yard detective about to retire, who decides to trap his quarry with a pretty woman. An adequate romantic comedy that was plagued by production difficulties, and had four directors and a clumsily re-filmed ending. Written by Larry Gelbart (who used the pseudonym Francis Burns).
COM   107 min (ort 112 min)   VIDrel: CIC/SONOP   V/h
Boa: novel Touch the Lion's Paw by Derek Lambert.

ROUGH DIAMONDS **
Donald Crombie AUSTRALIA
PG
1994
Jason Donovan, Angie Milliken, Peter Phelps, Max Cullen, Hayley Toomey, Jocelyn Gabriel, Kit Taylor, Lee James, Roger Ward, Tim Gaffney, Kaye Stevenson, Jeff Truman, Kim Lynch, Steven Tandy, Gerry Skilton, Charles Barry, Jamie Rowe
Romantic comedy in which a cattleman and a former nightclub singer both find a means of achieving their ambitions in the great Australian Outback. Filmed on location in Queensland.
COM   85 min (ort 88 min)   VIDrel: ITC/HIFLI   V/sh

ROUGH MAGIC **
Clare Peploe FRANCE/UK
12
1995
Bridget Fonda, Russell Crowe, Jim Broadbent, D.W. Moffett, Paul Rodriguez, Euva Anderson, Gregory Avellone, Michael Ensign, Gabriel Pingarron Gonzalez, Santos Morales, Rene Pereyra, Mark del Castillo Morante, Chris Otto
Fonda plays a gifted young magician, who flees to Mexico to escape the attentions of a crooked senator, who intends to marry her. Once there, she gets involved in a scheme to steal an ancient magic potion from a secret sect of Aztec sorcerers. Set in the 1950s, this is a curiously dated looking effort, hardly helped by weak production values and a definite lack of screen presence from Fonda. Some gruesome black comedy moments offer a little tepid relief.
Aka: MISS SHUMWAY JETTE UN SORT
THR   100 min (ort 105 min)   VIDrel: 20TH/FOXVID   V/sur
Boa: novel Miss Shumway Waves a Wand by James Hadley Chase.

ROUJIN Z ***
Katsuhiro Otomo/Hiroyuki Kitakubo JAPAN
15
1991
Voices of: Allan Wenger, Toni Barry, Barbara Barnes, Adam Henderson, Jana Carpenter, Ian Thompson, John Jay Fitzgerald, Graydon Gould, Nicolette McKenzie, Sean Barrett, Blain Fairman, Nigel Anthony
In this futuristic animation the bulk of the world's population are elderly, and the Japanese government sets up a project to introduce a bed that can feed, clothe, exercise and entertain its user. However, it transpires that the bed has more sinister uses. A blend of sharp social comment and exciting action sequels, this unusual "anime" benefits from a lively and imaginative script, not often the case for works of this genre.
Aka: ROJIN Z
ANIM   80 min (ort 84 min) dubbed
VIDrel: MANGA/SONOP   V

'ROUND MIDNIGHT ***
Betrand Tavernier FRANCE/USA
15
1986
Dexter Gordon, Francois Cluzet, Lonette McKee, Gabrielle Haker, Martin Scorsese, Herbie Hancock, Sandra Reaves-Phillips, Christine Pascal, Bobby Hutcherson, Wayne Shorter, John Berry, Philippe Noiret, Liliana Rovere
The enjoyable story of an expatriate black jazz musician, and his struggle to create the be-bop sound in 1959 Paris. A kind of homage to the world of jazz and its musicians, and to some extent inspired by the lives of Bud Powell and Lester Young. Real-life tenor sax Gordon gives a great performance as an ageing sax player in his film debut, for which he received a nomination but unfortunately no Oscar. AA: Score/orig (Herbie Hancock).
MUS   126 min (ort 133 min)   VIDrel: MGM/WHV   V/sur

ROUND TRIP TO HEAVEN **
Alan Roberts USA
15
1992
Corey Feldman, Zach Galligan, Ray Sharkey, Julie McCullough, Rowanne Brewer, Steven Vinovich, Ivory Ocean, Pat Harrington, Miguel A. Nunez Jr, Billy Hufsey, Lilyan Chauvin, Debbie James, Lloyd Batista, Brioni Farrell, Brent Corman
A wild-living teenager and his less experienced cousin borrow a Rolls Royce to get to a supermodel contest where the former hopes to meet one of the contestants he idolises. Unfortunately, the trunk of their vehicle is the hiding place for some hidden cash, and the crook who stole will stop at nothing to get it back. A harmless little teen comedy that feels like countless others of its kind.
COM   92 min (ort 97 min)   VIDrel: 20VIS/SONOP   V/sh

ROVER DANGERFIELD: THE DOG WHO GETS NO RESPECT **
Jim George/Bob Seely USA
U
1991
Voices of: Rodney Dangerfield, Susan Boyd, Ronnie Schell, Ned Luke, Shawn Southwick, Dana Hill, Sal Landi, Tom Williams, Chris Collins, Robert Bergen, Danton Whitehead, Ron Taylor, Bert Kramer, Eddie Barth, Ralph Monaco
Amusing tale of the adventures of a freewheeling canine hero who lives by his wits in Las Vegas but feels that his lack of breeding is a definite social handicap.
ANIM   70 min (ort 74 min)   VIDrel: WHV   V/dm V/sur

ROW OF CROWS, A **
J.S. Cardone USA
15
1991
Katharine Ross, John Beck, Steven Bauer, Mia Sara, Phil Brock, Dedee Pfeiffer
When the remains of a murdered woman are found in the desert outside a small town, the pressure mounts on the inhabitants as a web of corruption begins to unravel.
Aka: CLIMATE FOR KILLING, A
DRA   98 min (ort 104 min)   VIDrel: CIC/SONOP   V

ROXANNE ***
Fred Schepisi USA
PG
1987
Steve Martin, Daryl Hannah, Rick Rossovich, Shelley Duvall, John Kapelos, Fred Willard, Max Alexander, Michael J. Pollard, Shandra Beri, Blanche Rubin, Jane Campbell, Jean Sincere, Claire Caplan, Thom Curley, Brian George
An updating of Rostand's famous play "Cyrano de Bergerac", with our romantic Captain of the Guard in 17th century France transformed into a modern-day fire-chief, played with considerable panache by Martin. Hannah is the object of his affections in this amiable comedy.
COM   102 min (ort 107 min)   VIDrel: VCC/DISC/COLUM   V/sur

ROXANNE: THE PRIZE PULITZER **
Richard A. Colla USA
(12)
1989
Perry King, Chynna Phillips, Courteney Cox, Betsy Russell, Sondra Blake, Amy Adams, Caitlin Brown, Michael Champlin, Richardson Morse, Jay Glick, Christian Cousins, Mary Fanaro, Thom Scoggins, Joseph Cousins, Jancie Benson, Don Sheldon
During the 1980s the heir to the Pulitzer publishing fortune fought a very nasty divorce battle against his much younger wife. A dull dramatisation that glosses over some of the more scandalous aspects of these events.
Aka: PRIZE PULITZER, THE
DRA   90 min (ort 95 min) mTV   TVrel: BBC1
Boa: book by Roxanne Pulitzer.

ROYAL WEDDING **
Stanley Donen USA
U
1951
Fred Astaire, Jan Powell, Sarah Churchill, Peter Lawford, Keenan Wynn, Viola Roche, Albert Sharpe, James Finlayson, Henri Letondal, Alex Frazer, William Cabanne, Jack Reilly, John Hedloe, Francis Bethancourt, Dee Turnell
A musical revolving around a group of journalists and performers in London for the Royal Wedding. Some good scenes are interspersed throughout a film that's rather tedious. Highlights are Astaire dancing on the ceiling and Powell singing "Too Late". Music is by Alan Jay Lerner (who also wrote the script) and Burton Lane.
Aka: WEDDING BELLS
MUS   89 min (ort 93 min)   VIDrel: MGM/WHV   V

ROYCE **
Rod Holcomb USA
15
1993
James Belushi, Chelsea Field, Miguel Ferrer, Peter Boyle, Anthony

Head, Michael J. Shannon, Marie Theodore, Peter Boyle, Paris Jefferson, Nevyn McKenna, Susan Denaker, Christopher Fairbank, William Marsh, Ralph Ineson, Daniel Kash

A secret agent is sent on a perilous mission to stop nuclear weapons from falling into the hands of terrorists who have abducted a US senator's daughter. A parade of the usual espionage film cliches abound in this action-comedy, and Belushi is bland and unconvincing as our tough hero. Filmed on location in Hungary.

A/AD   94 min   VIDrel: ITC/HIFLI   V/h

RUBDOWN **   15
Stuart Cooper USA   1993
Michelle Phillips, Jack Coleman, Kent Williams, Alan Thicke, Catherine Oxenberg, William Devane, Jack Angles, Michael Ray Miller, Ron Johnson, Kane Hodder, Patricia Clipper, Chris Cardona, Paul Nolan, David Partington

A masseur who works in Beverly Hills becomes the prime suspect when one of his wealthy clients is found murdered and encounters the usual difficulties in trying to establishing his innocence. An assembly-line thriller with the obligatory so-called erotic element.

THR   87 min mTV   VIDrel: CIC/SONOP   V

RUBY **   15
John MacKenzie USA   1992
Danny Aiello, Sherilyn Fenn, Frank Orsatti, Jeffrey Nordling, Jane Hamilton, Maurice Bernard, Joe Viterelli, Robert S. Telford, John Roselius, Lou Eppolito, J. Marvin Campbell, David Duchovny, Richard Sarafian, Joe Cortese

This film examines the role the killer of Kennedy's assassin played in the whole affair – there are many who believe he was hired by the Mob to silence Oswald. Long regarded as a "bit-player" with criminal connections (he died in jail in 1967) scriptwriter Stephen Davis offers a highly sympathetic if not especially convincing portrait. See also RUBY AND OSWALD, JFK and THE TRIAL OF LEE HARVEY OSWALD, three more accounts of that period.

DRA   106 min (ort 110 min)   VIDrel: VCC/DISC/COLUM   V/sur

RUBY AND OSWALD **   U
Mel Stuart USA   1978
Michael Lerner, Frederic Forrest, Doris Roberts, Lou Frizzell, Lanna Saunders, Brian Dennehy, Bruce French, Sandy McPeak, Sandy Ward, James E. Brodhead, Gwynne Gilford Jump, Erick Kilpatrick, Walter Mathews

A reconstruction of the four days prior to and following the assassination of President John Kennedy, in Dallas in November 1963, drawn from authenticated events and eyewitness accounts. A sincere effort that never rises above the level of a competent but unmemorable drama. See also RUBY, JFK and THE TRIAL OF LEE HARVEY OSWALD.

Aka: FOUR DAYS IN DALLAS

DRA   102 min (ort 156 min) mTV   VIDrel: ODY/SONOP   V/h

RUBY CAIRO *   15
Graeme Clifford USA   1992
Liam Neeson, Andie MacDowell, Viggo Mortnesen, Jack Thompson, Paul Spencer, Chad Power, Monica Mikala, Kaelynn Craddick, Sara Craddick, Luis Cortes, Amy Van Norstrand, Pedro Gonzalez-Gonzalez, Lucy Rodriguez, Jef Corey, Miriam Reed

A recently widowed woman takes a trip to Mexico to bury her husband, who was killed in a plane crash. Once there she begins to find evidence that suggests he was leading a double life. A dull and derivative thriller, of weak dialogue and plotting.

Aka: DECEPTION

THR   106 min (ort 111 min)   VIDrel: EIV/SONOP   V/sur

RUBY GENTRY **   PG
King Vidor USA   1952
Jennifer Jones, Charlton Heston, Karl Malden, Josephine Hutchinson, Bernard Phillips, Tom Tully, James Anderson, Phyllis Avery, Herbert Heyes, Myra Marsh, Charles Cane, Sam Flint, Frank Wilcox

Brought up as a tomboy on the wrong side of the North Carolina swamps, a young girl associates with a local aristocrat, but eventually settles for marriage to a wealthy businessman in order to spite those who looked down on her. A ponderous, overblown and remarkably unrealistic melodrama.

DRA   82 min B/W   VIDrel: VCC/DISC   V

RUBY IN PARADISE ***   15
Victor Nunez USA   1993
Ashley Judd, Todd Field, Bentley Mitchum, Allison Dean, Dorothy Lyman, Betsy Douds, Felicia Hernandez, Divya Satla, Bobby Barnes, Sharon Lewis, Brik Berkes, Paul E. Mills, Abigail McKelvey, Christina Daman, Mark Limmer, J.D. Roberts

A young Tennessee woman on her way to a new life in Florida finds herself being deeply affected by the experiences that this brings her. A finely acted independently made drama with a welcome honesty rarely seen in main line efforts, if slightly pretentious at times.

DRA   114 min   VIDrel: MAINPIC/RTM L/A   V

RUDE AWAKENING *   15
Aaron Russo/David Greenwalt USA   1989
Cheech Marin, Eric Roberts, Julie Haggerty, Robert Carradine, Buck Henry, Louise Lasser, Cindy Williams, Andrea Martin, Cliff De Young, Dion Anderson, Peter Boyden, Andy Glass, Beck Glass, Ed Fry, Nicholas Wyman, William C. Paulsen

After twenty years spent in a commune in Latin America, two ageing hippies return to New York and are, obviously enough, shocked by the changes they find around them. However, the transformation of a couple of old friends into yuppies proves to be their greatest disappointment. But for us, the entire movie is a big letdown.

COM   96 min (ort 100 min)   VIDrel: MGM/WHV   V/sur

RUDY ***   PG
David Anspaugh USA   1993
Sean Astin, Jon Favreau, Ned Beatty, Greta Lind, Scott Benjaminson, Mary Ann Thebus, Charles S. Dutton, Lili Taylor, Christopher Reed, Deborah Wittenberg, Christopher Erwin, Kevin Duda, Robert Benirschke, Luke Massery, Jake Armstrong

A young boy who is neither particularly brilliant nor a first-rate athlete cherishes a dream of become a football star and eventually overcomes a number of obstacles to achieve this. A familiar story but rendered eminently watchable thanks to its first-rate acting and solid direction. Based on the true story of Rudy Ruettiger.

DRA   109 min (ort 116 min)   VIDrel: 20VIS/SONOP   V/sur

RUE CASES NEGRES ****   PG
Euzhan Palcy FRANCE   1983
Garry Cadenat, Darling Legitimus, Douta Seck, Joby Bernabe, Francisco Charles, Marie-Jo Descas

An orphan lives with his grandmother in a squalid collection of shacks adjoining a cane plantation in Martinique, and has little to look forward to except a life of drudgery on the plantation. However, his grandmother plans to see him escape this fate by way of an education, but her need to earn enough money causes her death from overwork. Set in the 1930s, this moving account of poverty and hope was documentary-maker Palcy's feature debut.

Aka: BLACK SHACK ALLEY

DRA   102 min (ort 106 min)   VIDrel: ARTIF/20TH   V

RUGGED GOLD ***   (PG)
Michael Anderson CANADA/NEW ZEALAND   1994
Jill Eikenberry, Art Hindle, Ari Magder, Graham Greene, Davina Whitehouse, Tony Groser, June Bishop, Helen Moulder, Sam Tyson Hogg, Christopher Douglas

A young widow with a small son goes against her family's wishes when she remarries as she choose an Alaskan gold prospector with whom she ends up living in a log cabin in the wilderness and sharing a variety of unexpected hardships, although it takes an unexpected disaster for her to realise just what physical and mental strength she possesses. Set in 1954, this is a well acted true story, with interesting period detail and some nice scenic locations.

DRA   91 min mTV   SATrel: SKY MOVIES

Boa: novel O Rugged Land of Gold by Marthe Martin.

RUMBLE FISH ***   18
Francis Ford Coppola USA   1983
Matt Dillon, Mickey Rourke, Diane Lane, Dennis Hopper, Vincent Spano, Diana Scarwid, William Smith, Nicolas Cage, Christopher Penn, Tom Waits, Larry Fishburne, Michael Higgins, Glenn Withrow, Herb Rice, Maybelle Wallace

An account of two adolescent brothers who get involved in gang warfare. The title refers to one of the brothers, who works in a petshop and identifies with the bad-tempered and ferocious rumble fish. A powerful but unfocused effort, this was Dillon's third Hinton film and Coppola's second (following THE

OUTSIDERS). Written by Coppola and with music by Stewart Copeland.
DRA   90 min (ort 94 min) B/W/Col
VIDrel: 4-FRONT/POLYREC/CIC   V/sur
Boa: novel by S.E. Hinton.

RUMBLE IN HONG KONG **   15
Hdeng Tsu HONG KONG
Jackie Chan, John Chang, Charlie Chin, Phoenix Chen
A villain will do anything to protect his crime boss but comes up against a tough and determined policewoman who has set her mind on destroying his boss's crime syndicate. Fair.
MAR   75 min   VIDrel: MIA/DISC   V

RUMBLE IN THE BRONX ***   15
Stanley Tong USA   1996
Jackie Chan, Anita Mui, Francoise Yip, Bill Tung, Marc Akerstream, Garvin Cross, Morgan Lam
A Hong Kong man comes on a visit to his uncle in the Bronx on the occasion of the latter's wedding. But the wedding plans take very much second place when the family store comes under attack from assorted thugs and Mafia hoodlums. Fortunately, Chan is more then a match for them all, and sets off to dish out the chastisement they clearly deserve. Judge not this film by normal standards, the plot is rubbish but the martial arts action sequences are truly impressive.
MAR   91 min   CINrel

RUMPELSTILTSKIN: THIS AIN'T NO FAIRY TALE *   18
Mark Jones USA   1995
Kim Johnston Ulrich, Max Grodenchik, Tommy Blaze, Allyce Beasley
Banished from our world by a witch's curse five-hundred years ago, the title creature is brought back by a mother's wish for assistance, and now seeks the soul of a first-born baby boy, leading to a chase for a mother and her baby across the length of the USA. A modern updating of this well known fairytale that is slackly directed as a kind of inept, slapstick horror-comedy, and fails to entertain on both scores.
HOR   87 min (ort 91 min)   VIDrel: FIRST/SONOP   V

RUN OF THE COUNTRY, THE ***   15
Pete Yates EIRE   1995
Albert Finney, Matt Keeslar, Victoria Smurfit, Anthony Brophy, David Kelly, Dearbhla Molloy, Carole Nimmons, Vinnie McCabe, Trevor Clarke, Kevin Murphy, Michael O'Reilly, P.J. Brady, Miche Doherty, Declan Mulholland, Dawn Bradfield
Tragic love story set in County Cavan, Ireland where an eighteen-year-old boy's mother dies of a heart attack and he is unable to face life alone with only his tyrannical father for company, and moves out of their home. Scripted by Connaughton this honest and vigorous story gets better as it develops, and the film's high production values are very much in evidence. The later reconciliation between father and son is touching without being over-sentimental.
DRA   104 min (ort 110 min)   VIDrel: 20VIS/SONOP   V/sur
Boa: novel by Shane Connaughton.

RUN SILENT, RUN DEEP ***   U
Robert Wise USA   1958
Clark Gable, Burt Lancaster, Jack Warden, Brad Dexter, Don Rickles, Nick Cravat, Mary LaRoche, Eddie Foy III, Joe Maross, H.M. Wynant, John Bryant, Ken Lynch, Joel Fluellen, Jimmie Bates, John Gibson
A submarine captain pursues a single-minded campaign of revenge against the Japanese destroyer that sank the previous sub he commanded. A tense portrayal of men facing the stresses of combat. The script is by John Gay.
WAR   89 min (ort 93 min) B/W   VIDrel: WHV   V
Boa: novel by Commander Edward L. Beach.

RUN WILD, RUN FREE **   U
Richard C. Sarafian UK   1969
John Mills, Sylvia Syms, Mark Lester, Bernard Miles, Gordon Jackson, Fiona Fullerton
A ten-year-old mute and withdrawn boy, gains self confidence through his love of animals, when he encounters a wild colt on the moors. A pleasant enough family film with nothing of great consequence. Nicely photographed by Wilkie Cooper and written by Rook from his novel.
JUV   95 min (ort 97 min)   VIDrel: VCC/DISC/COLUM   V
Boa: novel The White Colt by David Rook.

RUNAWAY ***   15
Michael Crichton USA   1984
Tom Selleck, Cynthia Rhodes, Gene L. Simmons, Kirstie Alley, Stan Shaw, Joey Cramer, G.W. Bailey, Chris Mulkey, Anne-Marie Martin, Michael Paul Chan, Carol Teesdale, Elizabeth Norment, Paul Batten, Babs Chulla, Marilyn Schreffler
The head of a police squad that deals with runaway robots, discovers a sinister plot masterminded by an electronics genius, but his superiors refuse to believe him, forcing him to fight on alone. A well-paced and imaginative fantasy, scripted by Crichton and with an effective score by Jerry Goldsmith.
FAN   96 min (ort 100 min) wScrn   VIDrel:
ENCORE/SPEAR/COLUM; ENCORE (LV only)   V LV

RUNAWAY *   18
Judy Blue USA   1992
Savannah, Randy Spears, Tianna, T.T. Boy
A couple about take a trip across the USA get an unexpected visit from the man's old college friend and they indulge in a threesome. This is later expanded when the two men rope in a willing woman.
A   50 min   VIDrel: VIVID/SCRN   V

RUNAWAY DAUGHTERS **   (PG)
Joe Dante USA   1993
Julie Bowen, Paul Rudd, Holly Fields, Jenny Lewis, Chris Young, Dick Miller, Dee Wallace Stone, Christopher Stone, Robert Picardo, Wendy Schaal, Belinda Balaski, Joe Flanerty, Roger Corman, Julie Corman, Courtney Gains, Leo Rossi
An annoying spoof on the rubbishy teen movies of the 1950s, full of clever references and in-jokes that are meant to provide amusement, but instead just get on one's nerves. The slight plot takes the form of a road-movie that is set in the 1950s, and has three high school girls chasing after the boy who made one of them pregnant, to tell him he is a dad before he joins the navy. Some stars of the 1950s appear in cameo roles, so perhaps it is worth seeing on that count.
COM   75 min (ort 90 min) mTV   SATrel: SKY MOVIES

RUNAWAY DREAMS *   18
Michele Noble USA   1989
Jennifer Corey, Brian Tarantino, Cordis Heard, Kaitlin Hopkins
A fifteen-year-old girl runs away from home and the sexual abuse of her dad, and heads for the beaches of Florida. Once there she gets drawn into the sordid world of prostitution, and despite the interest of a caring cop, carries on with her lifestyle until the arrival of her younger sister. Simple and superficial, this wholly artificial exercise has little of merit.
DRA   90 min Cut (1 min 9 sec)   VIDrel: CAPIT/GUILD   V/sur

RUNAWAY EXPRESS: THE ENGINE **   (U)
Gerd Haag GERMANY   1992
Rolf Hoppe, Isabel Dotzauer, Marcus Fleischer, Christa Kitsch, Sebastian Kroenhert, Katharina Schuttler, Klaus Koleit, William Mang, Veronika Bayer, Ilse Page, Haro Neivelstein, Jochen Kalandra, Iva-Maria Kurz, Armin Rohde, Udo Helger
A group of kids are planning to travel by rail to Siberia using a hand-cart, but are forced to reconsider this plan when they are chased away from the engine shed by a mysterious stranger. After some initial skirmishes they finally make friends with him, and together they help restore an ancient steam locomotive. A mildly diverting children's film, with most of the grown-ups shown as nasty and stupid, while the kids are all wise and capable beyond their years.
Aka: DIE LOK
JUV   90 min dubbed   SATrel: SKY MOVIES

RUNAWAY FATHER **   PG
John Nicolella USA   1991
Jack Scalia, Donna Mills, Chris Mulkey, Jenny Lewis, J.C. Brandy, Amy Moore Davis, Jane Daly, Starletta DuPois, Priscilla Pointer, Nancy Lenehan, Lisanne Falk, Patricia Gaul, Russ Marin, Janet MacLachlan, Michael Talbot, Ned Bellamy
A sluggish tearjerker based on a true story, that tells of a woman who falls for a seemingly ideal man, only to discover that he cannot give up his old life as a pilot. Desperate to escape from married life he fakes his own death, and his wife makes a new life for herself. Seventeen years later she learns that he is still alive, and sets about locating him. This languid and over-sentimental tale does very little to hold one's attention.
DRA   90 min (ort 94 min) mTV   VIDrel: ODY/SONOP   V/h

RUNAWAY TRAIN ***
Andrei Konchalovsky USA   18   1985
Jon Voight, Eric Roberts, Rebecca DeMornay, Kyle T. Heffner, John P. Ryan, T.K. Carter, Kenneth McMillan, Stacey Pickren, Walter Wyatt, Edward Bunker, Reid Cruikshanks, Michael Lee Gogin, John Bloom, Norton E. Warden
Two escaped convicts stow away on a speeding freight train, out of control after the driver suffers a fatal heart attack, but find the chief warden hot on their trail. A solid exciting thriller, based on a script by master director Akira Kurosawa and adapted by Djordje Milicevic, Paul Zindel and Edward Bunker.
THR   109 min (ort 111 min)   VIDrel: L/A   V

RUNESTONE, THE **
Willard Carroll USA   15   1991
Peter Riegert, William Hickey, Joan Severance, Alexander Godunov, Tim Ryan, Mitchell Laurance, Lawrence Tierney, Dawn Scott, Chris Young, John Hobson, Erika Schikel, Donald Hotton, Bill Kalmenson, Arthur Malet, Anthony Cistaro
A supernatural creature imprisoned in a magical stone by the ancient Vikings, gains access to the real world when the stone is discovered in a Pennsylvanian coal-mine. Standard horror offering of little distinction but with fairly competent special effects.
HOR   99 min (ort 105 min)   VIDrel: EIV/SONOP   V

RUNNER, THE ****
Amir Naderi IRAN   PG   1984
Majid Nirumand, Musa Torkizadeh, A. Gholamzadeh, Reza Ramezani, Shirzan Bechkal
On the streets of a Gulf port lives a young vagabond, who barely ekes out a living scavenging amongst the garbage, but dreams of getting an education in order to escape this life. A rather beautiful and moving film with very little dialogue, a strong sense of realism and the touching hope and determination of the central character. The story is semi-autobiographical, being in part based on the director's own unhappy and deprived childhood.
Aka: DAWANDEH
DRA   94 min   VIDrel: ELPIC/POLYREC   V

RUNNING **
Steven Hilliard Stern CANADA   PG   1979
Michael Douglas, Susan Anspach, Lawrence Dane, Eugene Levy, Chuck Shamata, Philip Akin, Trudy Young, Murray Westgate, Jennifer McKinney, Lesleh Donaldson, Jim McKay, Lutz Brode, Deborah Burgess, Gordon Clapp
A man who has been a lifelong failure, decides to compete in the Olympic marathon, in a last attempt to regain his self respect. A forgettable little melodrama.
DRA   97 min (ort 103 min)   VIDrel: VISION/DISC   V

RUNNING AGAINST TIME ***
Bruce Seth Green USA   U   1990
Robert Hays, Catherine Hicks, Sam Wanamaker, James DiStefano, Brian Smiar, Wayne Tippit, Tracey Frain, Juanita Jennings, Mark Phelan, Russ Marin, Milt Tarver, Paul Scherrer, Julie Ariola, Duncan Gamble, Damion Stevens, Dean Hill
This clever time-travel story has Hays stumbling on the secret invention of a time machine, and using it to travel back to November 22nd, 1963, where he hopes to prevent J.F. Kennedy's assassination. He arrives in Dallas just as planned, but events do not develop quite as he had hoped. An entertaining story, well acted and scripted, that gives a few new insights into a familiar idea.
FAN   89 min (ort 100 min) mCab   VIDrel: CIC/SONOP   V
Boa: short story A Time To Remember by Stanley Shapiro.

RUNNING BRAVE **
D.S. Everett (Donald Shebib) CANADA   (PG)   1983
Robby Benson, Pat Hingle, Claudia Cron, Jeff McCracken, August Schellenberg, Denis Lacroix, Graham Greene, Michael J. Reynolds, Kendall Smith, Margo Kane, George Clutesi, Maurice Wolfe, Carmen Wolfe, Fred Keating
A Sioux Indian leaves his reservation to pursue a career as a runner, and wins a gold medal in the 1964 Tokyo Olympics. Based on the story of Billy Mills, a real-life champion. A corny exercise in good intentions and melodrama.
DRA   102 min (ort 105 min)   SATrel: SKY MOVIES

RUNNING COOL **
Beverly C. Sebastian/Ferd Sebastian USA   15   1992
Andrew Divoff, Tracy Sebastian, Dedee Pfeiffer, Paul Gleason, Bubba

Baker, Arlen Dean Snyder, James Gammon, B.J. Davis, Arnie Cox, Carolyn Gendron, Marlene Cameron, Virginia Light, Wayne Nardella, Jan Duncan, Mark Salem, Frank Perrotti
Two bikers come to the aid of an old buddy who is trying to defend his property from a ruthless developer. A standard actioner with ample violence and stunts to make up for the cliched and hackneyed plot.
A/AD   102 min (ort 106 min)   VIDrel: CIC/SONOP   V/sur

RUNNING DELILAH *
Richard Franklin USA   18   1993
Kim Cattrall, Billy Zane, Francois Guetary, Yorge Voyagis, Diana Rigg, Michael Francis Clarke, Dawn Comer, Rod LaBelle, Marilyn McIntyre, Philip Moon, Quentin O'Brien, Philip Sokoloff, Eric Stone, Richard Topol, Victor Touzie
A female secret agent is killed on active duty but is brought back to life as a cyborg killing-machine and assigned to track down the terrorists who are building a nuclear device with stolen plutonium. A poor effort with silly dialogue and a totally unimaginative ROBOCOP-style plot.
Aka: CYBORG AGENT
FAN   85 min (ort 90 min) mTV   VIDrel: COLUM/SONOP   V

RUNNING FREE **
Steve Kroschel USA   (PG)   1994
Jesse Montgomery Smythe, James Lee Misfeldt, Michael Pena, Michael Hood, Mike Meekim, Joe Corneille, Ristine Casagranda, Laura C. Tryon, Ed Rush, Tony Jones, Billy Jo Hopsack, Adline Marie Coffman, James McGill, William Forster
A young boy in Alaska befriends a wolverine cub and finds this to be the start of a wonderful adventure. An enjoyable outdoors tale.
JUV   88 min (ort 90 min)   SATrel: SKY MOVIES

RUNNING LATE **
Udayan Prasad UK   (PG)   1992
Peter Bowles, Carole Nimmons, Michael Byrne, Adrian Rawlins, Suzette Llewellyn, Jim McManus, Jack Chissick, Brian Bovell, Lucy Bayler, Susan Lynch, Gino Melvazzi, Vincenzo Ricotta, James Fleet, Anne Lambton, Marcia Layton
A self-centred TV presenter has to take stock of his life when he is forced to take a frantic journey across London to see his wife on an important matter. An over-extended comedy of errors, that grows ever more bizarre and implausible as it develops, eventually revealing a supposedly surprise ending that makes everything clear. However, there are good things in it, not least a very funny performance from Bowles as our hapless presenter.
COM   75 min mTV   TVrel: BBC

RUNNING MAN, THE **
Paul Michael Glaser USA   18   1987
Arnold Schwarzenegger, Richard Dawson, Maria Conchita Alonso, Yaphet Kotto, Jim Brown, Jesse Ventura, Mike Fleetwood, Dweezil Zappa, Erland Van Lidth, Gus Rethwisch, Marvin J. McIntyre, Toru Tanaka, Karen Leigh Hopkins
In a totalitarian USA of 2019, a framed murderer gets a chance at freedom when he competes on a TV show, in which convicted felons race for their lives through a ruined L.A. A violent and predictable tale, but done on a big budget. See also THE PRIZE OF PERIL and DEATHROW GAMESHOW.
A/AD   97 min (ort 101 min)
VIDrel: 4-FRONT/POLYREC/BRAVE   V/sur
Boa: novella by Richard Bachman (Stephen King).

RUNNING ON EMPTY ***
Sidney Lumet USA   15   1987
Christine Lahti, River Phoenix, Judd Hirsch, Martha Plimpton, Jonas Arby, Ed Crowley, L.M. Kit Carson, Steven Hill, Augusta Dabney, David Margulies, Lynne Thigpen, Marcia Jean Klutz, Sloane Shelton, Justine Johnston, Herb Lovele
Two former 1960s student radicals have been on the run from the FBI for seventeen years, but now have two children in tow. Unable to stay anywhere for very long they live out of suitcases, but are faced with a dilemma when their talented teenage son is accepted at a music college. A well-made and convincing story that unfortunately takes a wrong turning and concentrates on the plight of the son. Scripted by Naomi Foner, the film's co-executive producer.
DRA   111 min (ort 116 min)   VIDrel: GUILD/POLYREC
L/A   V

RUNNING SCARED ** **   15
Peter Hyams USA   1986
Gregory Hines, Billy Crystal, Steven Bauer, Darlanne Fluegel, Dan
Hedaya, Joe Pantoliano, Jimmy Smits, Jonathan Gries, Tracy Reed,
John DiSanti, Larry Hankin, Don Calfa, Robert Lesser, Betty Carvalho
Two unorthodox Chicago cops decide that it's time to retire to
Key West, and decide to nab a notorious drug runner as their
last case. A routine mixture of comedy and thrills, with a good
car chase and a nice performance from Crystal, but nothing else
of note.
COM   102 min (ort 106 min)   VIDrel: MGM/WHV   V/sh

RUNNING SCARED ** **   15
Lev Spiro USA   1995
Emile Levisetti, Kevin Contreras, Nikki Fritz, Bob McFarland, Kevin
Walker, Kevin Alber
A group of hostages gain their freedom when the plane they
were being abducted in crashes. Their two captors now attempt
to make their escape with the ransom money, but find their
former captives are now after them. Adequate.
A/AD   90 min   VIDrel: GUILD/DFOXVID   V

RUSH ** **   18
Lili Fini Zanuck USA   1991
Jason Patric, Jennifer Jason Leigh, Max Perlich, Sam Elliott, Gregg
Allman, Tony Frank, William Sadler, Special K. McCray, Dennis
Letts, Dennis Burkley, Glenn Wilson, Jimmy Pickens, Barbara Lasater,
Toni Pilgreen, Merrill Connally
A female undercover cop who is assigned to a drugs investiga-
tion gets too involved in her role-playing and becomes a victim
of the pushers she was supposed to help bust. Adequate urban
adventure that treads a wellworn path.
THR   115 min (ort 120 min) cC   VIDrel: MGM/WHV
V/sur
Boa: book by Kim Wozencraft.

RUSSIA HOUSE, THE ** **   15
Fred Schepisi USA   1990
Sean Connery, Michelle Pfeiffer, Roy Scheider, James Fox, John
Mahoney, J.T. Walsh, Klaus Maria Brandauer, Michael Kitchen, Ken
Russell, Ian McNeice, David Threlfall, Christopher Lawford, Mac
McDonald, Martin Clunes, Nicholas Woodeson
Opaque spy-thriller, scripted by Tom Stoppard, that has a saxo-
phone-playing English publisher being given a manuscript
containing Soviet state secrets. Location shots of Moscow and
Leningrad (it makes a nice change from Helsinki) help move the
film along, but the story is as tedious as it is irritating. Fox as a
British Intelligence officer provides one of the few bright
moments. Music is by Jerry Goldsmith.
DRA   118 min (ort 124 min)   VIDrel: MGM/WHV   V/sur
Boa: novel by John le Carre.

RUTHLESS PEOPLE ** **   18
Jim Abrahams/David Zucker/Jerry Zucker USA   1986
Danny DeVito, Bette Midler, Judge Reinhold, Helen Slater, Anita
Morris, Bill Pullman, William G. Schilling, Art Evans, Clarence
Felder, J.E. Freeman, Gary Riley, Phyllis Applegate, Jeannine
Bisignano, J.P. Bumstead, Jon Cutler
A clothing manufacturer refuses to pay his wife's ransom
because he's very happy since she's been kidnapped. In fact, he
was planning to have her bumped off anyway. An enjoyable
farce with a neat twist but a somewhat over-extended gag.
Written by Dane Launer.
COM   91 min (ort 93 min)   VIDrel: TOUCH L/A   V/s

RYAN WHITE STORY, THE ** **   15
John Herzfeld USA   1988
Judith Light, Lukas Haas, George C. Scott, George Dzundza, Valerie
Landsburg, Sarah Jessica Parker, Mitchell Ryan, Nikki Cox, Peter
Scolari, Grace Zabriskie
Based on a true story, this tells of a teenage haemophiliac who
is diagnosed as having AIDS, and whose mother is forced to
campaign for him to be allowed to attend school.
DRA   93 min (ort 100 min) mTV   VIDrel: ODY/SONOP
V/h

RYAN'S DAUGHTER ** ** **   15
David Lean UK   1970
Robert Mitchum, Sarah Miles, Christopher Jones, John Mills, Trevor
Howard, Leo McKern, Barry Foster, Archie O'Sullivan, Marie Kean,
Barry Jackson, Evin Crowley, Douglas Sheldon, Philip O'Flynn, Ed
O'Callaghan, Gerald Sim, Des Keogh
A flawed attempt to produce a romantic masterpiece, set in 1916

against the background of the Irish troubles, when a teacher's
wife has an affair with a British soldier. Well-acted and visually
impressive, but overlong and remarkably unmoving. Mills'
performance as the village idiot is the film's chief highlight. The
music is by Maurice Jarre and the script by Robert Bolt. AA: S.
Actor (Mills), Cin (Freddie Young).
DRA   186 min (ort 206 min) wScrn   VIDrel: MGM/WHV
V/s

S

SABOTAGE ** ** **   PG
Alfred Hitchcock UK   1936
Sylvia Sidney, Oscar Homolka, John Loder, Desmond Tester, Joyce
Barbour, Matthew Boulton, S.J. Warmington, William Dewhurst,
Austin Trevor, Torin Thatcher, Aubrey Maher, Peter Bull, Charles
Hawtrey, Pamela Bevan
A woman begins to suspect that her husband is a spy working
for a foreign power, and it soon transpires that he is a danger-
ous terrorist whose contempt for humanity extends to using his
wife's little brother as the unwitting messenger boy who deliv-
ers his bombs. A tense and well-realised adaptation of Conrad's
novel. Musical direction is by Louis Levy. Written by Charles
Bennett, Ian Hay and Helen Simpson.
Aka: WOMAN ALONE, A
THR   74 min (ort 76 min) B/W
VIDrel: VCC/DISC/COLUM L/A   V
Boa: novel The Secret Agent by Joseph Conrad.

SABOTAGE * **   18
Tibor Takacs CANADA   1996
Mark Dacascos, Carrie Anne Moss, Tony Todd, Graham Greene, John
Neville
The survivor of a bungled CIA attempt to rescue hostages from
Arab terrorists now works as a bodyguard to an arms dealer as
a way of exorcising the trauma of his past. But when his boss is
assassinated, he sets out to have his revenge, and as ever, there
is a tough, female FBI agent who is ready and willing to help
him do this. A cliched parade of explosions and heroics, weakly
drawn characters and the usual conspiracies in high places.
A/AD   99 min   VIDrel: BMGVID   V

SABOTEUR ** ** **   PG
Alfred Hitchcock USA   1942
Priscilla Lane, Robert Cummings, Otto Kruger, Alan Baxter, Clem
Bevans, Norman Lloyd, Alma Kruger, Vaughan Glaser, Dorothy
Peterson, Ian Wolfe, Frances Carson, Murray Alpher, Kathryn
Adams, Pedro de Cordoba, Billy Curtis
A worker in an aircraft factory is unjustly accused of arson, and
having lost his best friend in the fire, flies for his life. He embarks
on a cross-country odyssey, searching for the fifth-columnist
who are masterminding a campaign of sabotage. An extremely
tense, well-paced thriller that builds to a nerve-tingling climax
atop the Statue of Liberty.
THR   104 min (ort 108 min) B/W   VIDrel: CIC/SONOP   V/h

SABRINA ** ** **   U
Billy Wilder USA   1954
Humphrey Bogart, Audrey Hepburn, William Holden, Martha Hyer,
John Williams, Walter Hampden, Joan Vohs, Macrel Dalio, Francis
X. Bushman, Nancy Kulp, Nella Walker, Ellen Corby, Marhorie
Bennett, Emory Parnell, Kay Riehl
A chauffeur's daughter is sent to finishing school in France and
comes back home transformed into a sophisticated and beauti-
ful young woman. Unfortunately, her arrival spoils the plans of
her father's employer who is hoping to make a wealthy match
for his youngest playboy son. Despite the flimsy plot, the accom-
plished acting and directing ensure a highly amusing comedy
of enormous charm. Remade in 1995. AA: Cost (Edith Head).
Aka: SABRINA FAIR
COM   109 min (ort 113 min) B/W   VIDrel: CIC/SONOP   V
Boa: play by Samuel Taylor.

SABRINA ** **   12/PG
Sydney Pollack USA   1995
Harrison Ford, Julia Ormond, Greg Kinnear, Nancy Marchand, John
Wood, Angie Dickinson, Richard Crenna, Lauren Holly, Dana Levy,
Miriam Colon, Elizabeth Franz, Fanny Ardant, Valeria Lemercier,
Patrick Bruel, Becky Ann Baker
An updated version of the original which stars Ford as a tycoon
with no time for romance but who finds his interest growing in
his chauffeur's daughter. Ford is quite likeable in the part orig-

inally taken by Bogart, but Ormond is no Hepburn and generates not one trace of the lively charm that made the original Billy Wilder film such a success. This is a workmanlike remake, directed with clinical efficiency, but devoid of vigour and freshness. See also FORGET PARIS.
COM   127 min wScrn cC   VIDrel: CIC/SONOP; PION (LV only – PG cert)   V/sur LV

SACRED GROUND **   (PG)
Charles B. Pierce USA   1983
Tim McIntire, Jack Elam, Mindi Miller, Serene Hedin, Eloy Phil Casados, L.Q. Jones, Lefty Wild Eagle, Larry Kenoras, Vernon Foster, Franklin Fritz, Danny Wilson, Ben Mitchell, Jerald Jackson Jr,. Ronnie Wilson, Aaron Wright
An intrepid frontiersman and his wife settle on land that is the sacred burial ground of the Paiute Indians, an action they soon deeply regret, as trouble is not slow in coming. A competently made Western tale.
WES   98 min (ort 100 min)   SATrel: SKY MOVIES

SACRIFICE, THE ***   PG
Andrei Tarkovsky FRANCE/SWEDEN   1986
Erland Josephson, Susan Fleetwood, Valerie Mairesse, Allan Edwall, Gundun Gisladottir, Sven Wollter, Filippa Franzen, Tommy Kjellqvist, Per Kallman, Tommy Nordahl
In the face of an impending nuclear Armageddon, a man strikes a bargain with God, and promises to sacrifice everything he holds dear if the world is spared destruction. A ponderous epic, set on an isolated Swedish island, it's mostly an intense examination of various angst-ridden obsessions, but is sustained by excellent performances. Sven Nykvist's outstanding camerawork and some truly memorable sequences are additional merits. This was the director's final film.
Aka: OFFRET
DRA   142 min (ort 145 min)   VIDrel: ARTIF/20TH   V

SAD INHERITANCE **   15
Rod Hardy USA   1995
Susan Dey, Piper Laurie, Lorraine Toussaint, D.W. Moffett, Kathleen York, Guy Bond
A female drug addict has to prove herself able to be a mother, and has to battle officialdom and red-tape.
DRA   89 min   VIDrel: ODY/SONOP   V/h

SADDLETRAMP **   R18/18
USA   1988
Hyapatia Lee, Randy West, Nina Hartley, Shanna McCullough, Peter North, Eva Allen, Scott Irish
A beautiful Indian woman is sold as a sex slave but escapes this predicament and runs off with a masked bandit, in this sex Western enlivened by some broad humour. Lee (who is of Indian ancestry) looks convincing in the role and gives quite a decent performance.
A   VIDrel: SHEP L/A (R18 ver); HAR/GOLD (18 ver)   V

SADISTEROTICA **   18
Jess Franco WEST GERMANY   1967
Janine Reynaud, Rosanna Yanni, Adrian Hove
A sexy female private eye duo who use the imprint of rouged lips as their trademark (hence the alternative title) take on the case of a demented artist, whose penchant is to abduct young women and sketch them as they are being killed (see PEEPING TOM). An odd blending of eroticism and horror, very typical of this director's work. Followed by KISS ME MONSTER.
Aka: ROTE LIPPEN
HOR   75 min (ort 80 min) wScrn dubbed
VIDrel: REDEM/RTM   V

SAFE ***   15
Todd Haynes USA   1995
Julianne Moore, Xander Berkeley, Dean Norris, Julie Burgess, Ronnie Farer, Jodie Markell, Susan Norman, Martha Velez-Johnson, Chauncy Leopardi, Saachiko, Tim Gardner, Wendy Haynes, Alan Wasserman, Jean Pflieger, Steven Gilborn
A look at the modern-day syndrome of "multiple chemical sensitivity" whose victims develop an allergic reaction to all the artificial chemicals that permeate the products of a modern society. This story follows the plight of an affluent housewife, who having developed this complaint, is forced to seek sanctuary at a refuge for such sufferers in the Texan desert. A disturbing but ponderous work, its message is too ambiguous to have much impact.
DRA   118 min   VIDrel: TART/20TH   V/sh

SAFE PASSAGE **   15
Robert Allan Ackerman USA   1994
Susan Sarandon, Sam Shepard, Nick Stahl, Robert Sean Leonard, Marcia Gay Harden, Sean Astin, Priscilla Reeves, Joe Lisi, Matt Keeslar, Marvin Scott, Bill Boggs, Kathryn Kinley, Cindy Hom, Christopher Wynkoop, Philip Arthur Ross
The story of how a family face up to the tragic news that their son, a US Marine, may have been killed in the bomb attack on the US Marine base in Beirut. This should have been an intense and compelling drama but instead becomes quite chore to watch, thanks to both a lack of pace and a general sense of gloom.
DRA   94 min (ort 98 min)   VIDrel: EIV/SONOP   V
Boa: novel by Ellyn Bache.

SAGEBRUSH TRAIL ***   U
Armand L. Schaefer USA   1933
John Wayne, Nancy Shubert, Lane Chandler, Yakima Canutt, Henry Hall, Wally Wales, Art Mix, Bob Burns, Bill Dwyer, Earl Dwire, Hank Bell, Slim Whitaker, Hal Price
A man is falsely accused of murder and thrown into jail, but escapes and joins an outlaw gang in a bid to find the real culprit. A brisk and enjoyable film, the second offering in producer Paul Malvern's "Lone Star" series.
WES   53 min (ort 55 min) B/W coVer
VIDrel: ENTUK/GOLD L/A   V

SAHARA ***   PG
Zoltan Korda USA   1943
Humphrey Bogart, Bruce Bennett (Herman Brix), Dan Duryea, Lloyd Bridges, Rex Ingram, J. Carrol Naish, Richard Nugent, Kurt Krueger, Louis T. Mercier, Patrick O'Moore, Carl Harbord, Guy Kingsford, Kurt Krueger, John Wengraf
An American tank sergeant is cut off from his unit and forms his own from various Allied soldiers. With a total strength of thirteen men and one tank, they withstand the onslaught of a German battalion, cause considerable damage and help ensure victory at El Alamein. Based on the Soviet film "The Thirteen", the simple plot takes second place to strong characterisations and excellent action sequences. Later remade as the Western "Last of the Commanches" in 1953.
WAR   93 min (ort 97 min) B/W   VIDrel: COLUM/SONOP
V

SAHARA **   15
Brian Trenchard-Smith USA   1995
James Belushi, Mark Lee, Paul Empson, Jerome Elmers, Michael Massee, Robert Wisdom, Jerome Ehlers, Abgelo D'Angelo, Paul Empson
Trapped by the German army, an American tank crew prepare a last ditch attempt to defend themselves in the desert during WW2. A remake of the earlier Bogart film, that is competently directed in terms of the action sequences, even if the characters are never more than superficial.
WAR   106 min   VIDrel: COLUM/SONOP   V

SAIGON: THE YEAR OF THE CAT **   PG
Stephen Frears UK   1983
Judi Dench, Frederic Forrest, E.G. Marshall, Chic Murray, Yim Moontrakul, Pichit Bulkul, Roger Rees, Wallace Shawn, Rong Wongsawan, Manning Redwood, Somsak Seangwilai, Josef Sommer, Deborah Eisenberg, Thomasine Heiner, Malinee
Story of a romance between a British bank employee and a CIA agent, and set in Saigon in 1974 with the impending fall of South Vietnam an ever-present threat in the background.
DRA   105 min mTV   VIDrel: VCC/DISC L/A   V

SAILOR WHO FELL FROM GRACE WITH THE SEA, THE **   18
Lewis John Carlino UK   1976
Sarah Miles, Kris Kristofferson, Jonathan Kahn, Margo Cunningham, Earl Rhodes, Paul Tropea, Gary Lock, Stephen Black, Peter Clapham, Jennifer Tolman
In an English coastal town, a sailor courts a sexually repressed widow while her unhappy son looks on. The torrid love scenes climax in a repulsive ending when the woman's son castrates his mother's lover. A bizarre mixture of sexuality, torture and symbolism. Photography is by Douglas Slocombe and the locations are on the Dartmouth coast. Written and directed by Carlino.
DRA   100 min (ort 104 min)   VIDrel: L/A   V
Boa: novel Gogo No Eiko by Yukio Mishima.

SAINT, THE ** 12
Phillip Noyce USA 1997
Val Kilmer, Elisabeth Shue, Rade Serbedzija, Valery Nikolaev, Henry Goodman, Alun Armstrong, Michael Byrne, Evgeny Lazarev, Irina Apeximova, Lev Prigunov, Charlotte Cornwell, Emily Mortimer, Lucija Serbedzija
Kilmer chose not to star in a further BATMAN sequel, but instead appears in this $70,000,000 updating of the Leslie Charteris action hero: a professional burglar, con-man and thief. In this lavish adventure, he is hired by a sinister oil tycoon to steal a female scientist's cheap energy formula, but instead finds himself falling in love with her. A fast but slightly tepid spy caper, light on plotting, decent dialogue and excitement.
A/AD 116 min CINrel

SAINT ELMO'S FIRE ** 15
Joel Schumacher USA 1985
Rob Lowe, Emilio Estevez, Andrew McCarthy, Demi Moore, Judd Nelson, Ally Sheedy, Mare Winningham, Martin Balsam, Andie MacDowell, Joyce Van Patten, Jenny Wright, Blake Clark, Jon Cutler, Matthew Laurance, Gina Hecht, Andy Scott
Seven college graduates stick together as a group once they leave, out of fear of losing the cosiness of college life. An uninspiring and unimpressive attempt to dramatise the conflict between immature aspirations and the harsh realities of life. Written by Schumacher and Carl Kurlander.
DRA 103 min (ort 108 min) VIDrel: VCC/DISC/COLUM V/sh

SAINT IVES * 15
J. Lee Thompson USA 1976
Charles Bronson, John Houseman, Jacqueline Bisset, Harry Guardino, Elisha Cook Jr, Maximilian Schell, Harry Yulin, Dana Elcar, Michael Lerner, Dick O'Neill, Val Bisoglio, Burr DeBenning, Daniel J. Travanti, Joe Roman, Tom Pedi
A crime reporter and aspiring novelist is asked to recover stolen records that could trigger a violent gang war, and after reluctantly accepting this assignment finds himself faced with a labyrinthine conspiracy that baffles both him and the audience. Less violent, and somewhat better acted than is usual for Bronson's films, but the unnecessarily convoluted plot is a recipe for tedium. Written by Barry Beckerman.
DRA 90 min (ort 98 min) VIDrel: MGM/WHV L/A V
Boa: novel The Procane Chronicle by Oliver Bleeck.

SAINT OF FORT WASHINGTON, THE *** 15
Tim Hunter USA 1993
Danny Glover, Matt Dillon, Rick Aviles, Nina Siemaszko, Ving Rhames, Joe Seneca, Harry Ellington, Ralph Hughes, Bahni Turpin, Robert Beatty Jr, Reuben Schaefer, Louis Williams, Adam Trese, Kevin Corrigan, Brian Tarantina
Touching drama set among the homeless of New York, that follows the exploits of a mentally disturbed youth who is pushed out onto the streets when his hostel is pulled down. There he is befriended by an older man, a Vietnam veteran who teaches him how to survive both on the streets and in a night shelter, where they have to contend with the aggression of a brutal thug.
DRA 99 min (ort 103 min) wScrn VIDrel: 20TH/TECH L/A V

SAINT TAMMANY MIRACLE, THE ** (U)
Joy N. Houck Jr USA 1994
Mark-Paul Gosselaar, Jamie Luner, Soleil Moom Frye, Steve Allen, Jeffrey Meek, Julie McCullough, Harry Fleer, Sasha Spalding, Richard Folmer, Kenneth Spalding, Brian Townes, Ed Hearron, Dominic Cordaro, Daniel Johnson, Garry Allen
A troubled and very unsuccessful woman's basketball team at a Catholic school gets a new coach who, with the help of her assistant, gradually turns its fortunes right around. However, this is not without a fair share of difficulties and hardship, including the machinations of a former boyfriend. A competently made tale, with few surprises.
DRA 91 min SATrel: MOVIE CHANNEL

SAINTS AND SINNERS *** 18
Paul Mones USA 1994
Damian Chapa, Jennifer Rubin, Scott Plank, Damon Whitaker
A couple of petty crooks set out to form a team of drug-dealers and take over a Mafia operation, and they have to contend with the expected conflicts and double-crosses, plus some romantic rivalry by way of a sub-plot. An entertaining slice of urban

underworld life, directed with considerable verve. Written and produced by Mones, this film went straight to video.
THR 95 min mTV VIDrel: HIFLI/SONOP V/h

SAKURA KILLERS ** 18
Richard Ward USA 1986
Chuck Connors, George Nichols, Mike Kelly, John Ladalski, Manji Otsuki, Brian Wong, Thomas Lung
Standard Ninja capers involving the theft of a secret video tape, and the efforts of two top agents to recover it.
MAR 86 min Cut (40 sec) VIDrel: EIV/SONOP V

SALAAM BOMBAY! **** 15
Mira Nair INDIA/UK 1988
Shariq Syed, Anjaan, Amrit Patel, Murari Sharma, Ram Moorti, Sarfuddin Qurrassi, Raju Barnad, Raghubir Yadav, Aneeta Kanwar, Nana Patekar, Irshad Hashmi, Hansa Vithal, Mohanraj Babu, Chandrashekhar Naidu, Kishan Thapa
A ten-year-old boy goes to Bombay in the hope of earning the five-hundred rupees he needs in order to be allowed home, and once there becomes involved with various figures in the teeming street life of the city. A sad, touching and realistic study, filmed on location, and with the lead and some of the other roles played by homeless children. Nair's debut feature was partly funded by Channel Four TV and won the Camera d'Or at Cannes 1988 for Best First Feature.
DRA 109 min (ort 114 min) VIDrel: CONNO/RTM V

SALEM'S LOT *** 18
Tobe Hooper USA 1979
David Soul, James Mason, Lance Kerwin, Bonnie Bedelia, Lew Ayres, Reggie Nalder, Ed Flanders, Elisha Cook, Marie Windsor, Fred Willard, Clarissa Kaye, James Gallery, Kenneth McMillan, George Dzundza
Vampires run riot in a small New England town, where an old hilltop house forms the focus for the all-pervasive deadly evil. Soul plays a writer returning home to these strange events. Well-made and quite chilling, but Soul is somewhat uninspiring, though Mason is a good deal better as a malevolent antique dealer. Followed by A RETURN TO SALEM'S LOT.
Aka: SALEM'S LOT: THE MOVIE
HOR 184 min (ort 200 min) mTV VIDrel: WHV V/h
Boa: novel by Stephen King.

SALMONBERRIES *** 15
Percy Adlon GERMANY 1991
k.d. lang, Rosel Zech, Chuck Connors, Jane Lind, Oscar Kawagley, Wolfgang Steinberg, Christel Merian, Eugene Omiak, Wayne Waterman, Alvira H. Downey, George Barrill, Gary Albers
In Alaska, an orphaned Eskimo living an aimless existence latches onto a middle-aged German emigre working as a librarian, who has little in her life except her memories of past happiness. The two women become friends, and together embark on a voyage of self-discovery, that takes them to Berlin, where the sources of the older woman's past are examined. This quirky, evocative and slow-moving film offers much of value to the patient.
DRA 91 min (English dialogue – ort 95 min) wScrn VIDrel: ELPIC/POLYREC V

SALOME * 18
Claude D'Anna ITALY 1986
Tomas Millan, Pamela Salem, Joe Ciampa, Tim Woodward, Fabrizio Bentivoglio
This mangled and updated version of the Wilde play, tells of the erotic young dancer who obtains the promise from her stepfather, the King of Judea, that she will be rewarded with whatever she desires in return for performing the Dance Of The Seven Veils. A baffling and incoherent hotchpotch made worse by low-key lighting and lousy acting. Definitely an experience to miss.
A 91 min (ort 100 min) VIDrel: RNK L/A V
Boa: play by Oscar Wilde.

SALOME'S LAST DANCE ** 18
Ken Russell UK 1988
Imogen Millais-Scott, Glenda Jackson, Stratford Johns, Nickolas Grace, Linzi Drew, Douglas Hodge, Dennis Lill, Russell Lee Nash, Alfred Russell, David Doyle, Warren Saire, Kenny Ireland, Michael Van Wijk, Paul Clayton, Imogen Claire
Set in a Victorian London brothel, this adaptation of Wilde's play is a stiff studio-bound production, which finds the author (a miscast Grace) lounging about, as the brothel proprietor stages a production of the title work. A fairly typical flight of

Russellian self-indulgence, though slightly less oppressive than some of his other later works.
A   86 min (ort 89 min)   VIDrel: VES L/A   V
Boa: play Salome by Oscar Wilde.

SALON KITTY *   18
Giovanni Tinto Brass FRANCE/ITALY/
WEST GERMANY   1977
Helmut Berger, Ingrid Thulin, Therese Ann Savoy, Bekim Fehmiu, John Steiner, John Ireland, Rosemarie Lins, Sara Sperati, Maria Michaels, Stefa Sataflores Sataflores
Nazi decadence time again in this story of a Berlin brothel where the madam spies on the SS officers whose strange tastes she helps to satisfy. Pure camp from the director of CALIGULA.
A   112 min (ort 120 min) wScrn dubbed
VIDrel: REDEM/RTM   V

SALT ON OUR SKIN ***   (PG)
Andrew Birkin CANADA/FRANCE/GERMANY   1992
Greta Scacchi, Vincent D'Onofrio, Anais Jeanneret, Hanns Zischler, Claudine Auger, Rolf Illig, Petra Brandt, Shirley Henderson, Laszlo I. Kish, Sandra Voe, Barbara Jones, Charles Berling
Whilst on holiday in Scotland, a French intellectual falls in love with a local fisherman, and despite her belief that such a relationship cannot work due to their very different backgrounds, they enjoy a love affair that spans thirty years. A charming drama based on a bestselling novel.
DRA   105 min   VIDrel: WHV L/A   V/h
Boa: novel Les Vaisseaux du Couer by Benoite Groult.

SALUTE OF THE JUGGER, THE ***   18
David Webb Peoples AUSTRALIA/USA   1988
Rutger Hauer, Joan Chen, Vincent Phillip D'Onofrio, Anna Katerina, Delroy Linda
Echoes of MAD MAX and ROLLERBALL abound in this futuristic post-WW3 world where gladiatorial games are played by teams using a dog's skull instead of a ball and a few meagre scraps from the previous era serve as prizes. A flashy and heavily derivative film that soon exhausts its stock of ideas but retains considerable visual impact.
Aka: BLOOD OF HEROES
A/AD   99 min   VIDrel: 4-FRONT/POLYREC/VISVID
V/sur

SALVADOR ***   18
Oliver Stone USA   1986
James Woods, James Belushi, John Savage, Michael Murphy, Elpedia Carrillo, Tony Plana, Colby Chester, Cindy (Cynthia) Gibb, John Doe, Jose Carlos Ruiz, Jorge Luke, Juan Fernandez, Salvador Sanchez, Rosario Zuniga
Seduced by the lure of cheap drink, drugs and sex, an experienced photo-journalist goes with his friend to El Salvador in the early 1980s. Once there, they see the turmoil and brutality of the country as it is being destroyed by the conflict, and learn of American involvement. A patchy but effective piece of propaganda, largely based on the experiences of Richard Boyle, a journalist there in the 1980s.
WAR   117 min (ort 123 min)   VIDrel: VCC L/A   V

SALZBURG CONNECTION, THE *   PG
Lee H. Katzin USA   1972
Barry Newman, Anna Karina, Klaus-Maria Brandauer, Karen Jensen, Joe Maross, Wolfgang Preiss, Helmut Schmid, Udo Kier, Michael Hausserman, Whit Bissell, Raoul Retzer, Elisabeth Felchner, Bert Fortell, Alf Beinell, Patrick Jordan
An American lawyer on holiday in Salzburg gets mixed up with spies, when a chest of incriminating Nazi war documents is discovered in an Austrian lake. Soon a host of foreign agents are out to obtain it. A gimmicky and dated yarn, made worse by its annoying reliance on slow motion and freeze-frame photography.
THR   89 min (ort 93 min)   VIDrel: 20TH   V
Boa: novel by Helen MacInnes.

SAMANTHA **   15
Stephen La Rocque USA   1992
Martha Plimpton, Dermot Mulroney, Hector Elizondo, Mary Kay Place, Ione Skye, Marvin Silbersher, I.M. Hobson, Maryedith Burrell, Robert Picardo, Shay Astar, Cameron Williams, Dody Goodman, Randy Delish, Sandra Kinder
Informed by her parents on her 21st birthday that is really a foundling left on their doorstep, a spoiled and utterly self-centred young woman leaves home to find her roots, after a

failed suicide attempt. Plimpton is excellent as the highly unsympathetic title character, in a performance that recalls "Silence Like Glass", but her abrasive personality and the sheer silliness of the feeble attempts at comedy conspire to create a most unfunny movie.
COM   97 min (ort 101 min)   VIDrel: FIRST/SONOP L/A   V

SAME TIME, NEXT YEAR **   15
Robert Mulligan USA   1978
Ellen Burstyn, Alan Alda, Ivan Bonar, Bernie Kuby, Cosmo Sardo, David Northcutt, William Cantrell
A couple have an affair lasting twenty-six years, but only meet once a year for a weekend together. Through their meetings we see a cavalcade of recent American social history and changing ideas and attitudes. An entertaining trifle enlivened by an excellent performance from Burstyn. Scripted by Slade from his successful Broadway play.
COM   113 min (ort 119 min)   VIDrel: CIC/SONOP L/A
V/h
Boa: play by Bernard Slade.

SAMMY AND ROSIE GET LAID **   18
Stephen Frears UK   1987
Shashi Kapoor, Claire Bloom, Frances Barber, Ayub Khan Din, Roland Gift, Wendy Gazelle, Badi Uzzmann, Suzetta Llewellyn, Meera Syal, Tessa Wojtczak, Emer Gillespie, Lesley Manville, Mark Sproston, Cynthia Powell, Dennis Colon
The story of a couple whose open relationship is upset by the arrival of the man's old-fashioned father. A pointless attempt at social commentary built around the lives of Indians living in Britain. Written by Hanif Kureishi.
DRA   97 min (ort 100 min)   VIDrel: POLY/POLYREC L/A
V/h

SAM'S SON **   PG
Michael Landon USA   1984
Michael Landon, Eli Wallach, Timothy Patrick Murphy, Anne Jackson, Hallie Todd, Alan Hayes, Jonna Lee, Howard Witt, William Boyett, John Walcutt, David Lloyd Nelson, William H. Bassett, Harvey Gold, James Karen
An underprivileged boy finds that his athletic prowess eventually lands him an acting career in Hollywood. A typically sentimental Landon effort that is clearly autobiographical in nature (he also wrote the script) and if undistinguished is nonetheless mildly diverting, though its best feature is Wallach in the role of the boy's father.
DRA   107 min   VIDrel: GENESIS   V

SAMSON AND DELILAH ***   U
Cecil B. De Mille USA   1949
Victor Mature, Hedy Lamarr, Angela Lansbury, George Sanders, Henry Wilcoxon, Olive Deering, Fay Holden, Russ Tamblyn, William Farnum, Lane Chandler, Moroni Olsen, Francis J. McDonald, William "Wee Willie" Davis, John Miljan
An entertaining film version of the biblical story, with good costumes and sets. Mature as Samson is good, even if the lion fight is every bit as phoney as Lansbury's Philistine accent. However, the final scene in which the temple is destroyed is splendid. Written by Jesse L. Lasky Jr with music by Victor Young. AA: Art/Set (Hans Dreier and Walter Tyler/Sam Comer and Ray Moyer), Cost (Edith Head/Dorothy Jeakins/Elois Jenssen/Gile Steele/Gwen Wakeling).
DRA   122 min (ort 128 min) wScrn   VIDrel: CIC/SONOP
V/dm

SAMURAI COWBOY ***   PG
Michael Keusch CANADA   1993
Hiromi Go, Robert Conrad, Matt McCoy, Catherine Mary Stewart, Max Kirishima, Conchata Ferrell, Ian Tyson, Byron Chief Moon, Bradley M. Rapier, Tom Glass, Mark Acheson, Rick Harvey, Owen Smith, Ehud Ellman, Tom Bonny, Harald Ludwig
A Japanese man leaves the cramped conditions of his homeland and the rat race for the West where he hopes to realise his dream of becoming a cowboy. Once there he buys up a ranch and sets about making a go of it, only to find that life here has problems too, not least an avaricious landowner who wants to buy up the ranch. A gentle tale with a nice balance of comedy and drama and some charming performances.
WES   96 min (ort 101 min)   VIDrel: 20VIS/SONOP   V/sh

SAN ANTONIO **   PG
David Butler USA   1945
Errol Flynn, Alexis Smith, S.Z. Sakall, John Litel, Victor Francen,

Paul Kelly, Florence Bates, Robert Shayne, Monte Blue, John Alvin, Robert Barrat, Pedro De Cordoba, Tom Tyler, Chris-Pin Martin, Charles Stevens, Dan White

Having returned to San Antonio from Mexico, a reformed rustler sets out to prove that the owner of a saloon is the ringleader of a gang of cattle rustlers. He incurs the enmity of the latter when he starts taking an interest in a hostess who works at the saloon, but all his problems are resolved in the usual way in a conclusive gun battle. A thinly plotted but quite entertaining affair.

WES   104 min (ort 111 min)   VIDrel: WHV   V

SAN DEMETRIO, LONDON **   PG
Charles Frend USA   1943
Walter Fitzgerald, Mervyn Johns, Ralph Micheal, Robert Beatty, Charles Victor, Frederick Piper, Gordon Jackson, Arthur Young, Barry Letts, James McKechnie, Nigel Clarke, Lawrence O'Madden, David Horne, Neville Mapp

In 1940 a merchant marine tanker is badly damaged while at sea and its crew stage a heroic fight to take their vessel to safety. A dated and none too convincing propaganda tale, that is chiefly remembered for some arresting sequences, most especially the one where the crew are obliged to reboard their torpedoed vessel, having found that escape by lifeboat is impossible.

DRA   105 min B/W   VIDrel: LUMI/SPEAR L/A   V

SAN FERRY ANN *   PG
Jeremy Summers UK   1965
Rodney Bewes, Wilfrid Brambell, Lynne Carol, Fred Emney, Catherine Feller, David Lodge, Ron Moody, Andrea Malandrinos, Warren Mitchell, Hugh Paddick, Joan Sims, Joan Sterndale Bennett, Ronnie Stevens, Graham Stark, Barbara Windsor

One of those oh so terribly unfunny British "silent" comedies that follows the exploits of a bunch of British holidaymakers in France. See also THE PLANK.

COM   55 min B/W silent   VIDrel: LUMI/SPEAR   V

SAN FRANCISCO ***   U
W.S. Van Dyke II USA   1936
Clark Gable, Jeanette MacDonald, Spencer Tracy, Jack Holt, Ted Healy, Al Shean, Margaret Irving, Jessie Ralph, Harold Huber, William Ricciardi, Tom Dugan, Kenneth Harlan, Roger Imhof, Frank Mayo, Charles Judels, Bert Roach

A tale of romance as two men fight over the same girl in old San Francisco, which is soon to be devastated by the 1906 earthquake. A star-studded old-style melodrama that climaxes with some superb special effects that give a seemingly realistic portrayal of the havoc that this natural disaster brought. AA: Sound (Douglas Shearer).

DRA   110 min (ort 117 min) B/W   VIDrel: MGM/WHV L/A V

SAND PEBBLES, THE **   15
Robert Wise USA   1966
Steve McQueen, Richard Crenna, Richard Attenborough, Candice Bergen, Mako, Simon Oakland, Larry Gates, Marayat Andriane, Gavin MacLeod, Ford Rainey, Charles Robinson, Joe Turkel, Joseph Di Reda, Richard Loo, Barney Phillips

Follows the crew of a US gunboat on patrol in the Yangtze River in China in 1926, and their involvement with Chinese warlords. McQueen gives a fine performance in this film as a cynical sailor, but the mixture of action and romance is a recipe for confusion. Written by Robert Anderson, and well photographed by Joseph MacDonald.

WAR   174 min (ort 195 min)   VIDrel: 20TH/TECH   V/sh
Boa: novel by Richard McKenna.

SANDERS OF THE RIVER **   U
Zoltan Korda UK   1935
Leslie Banks, Paul Robeson, Nina Mae McKinney, Robert Cochran, Martin Walker, Richard Grey, Marquis De Portago, Eric Maturin, Allan Jeayes, Charles Carson, Orlando Martins

An adventure tale set in Nigeria in colonial times, and following the exploits of a river patrol officer and his black servant. A curious and dated tale, with Robeson giving a dignified performance that lends weight to an otherwise unmemorable effort.

Aka: BOSAMBO
A/AD   84 min (Abridged by distributor – ort 98 min) B/W
VIDrel: L/A   V
Boa: novel by Edgar Wallace.

SANDLOT KIDS, THE **   PG
David Mickey Evans USA   1992
Karen Allen, Denis Leary, James Earl Jones, Arliss Howard, Tom Guiry, Mike Vitar, Patrick Renna, Chauncey Leopardi, Marty York, Brandon Adams, Maury Wills, Art La Fleur, Grant Gelt, Shane Obedzinski, Victor DiMattia, Marlee Shelton

In 1962, a young boy moves to a new area in California and tries to make friends with the neighbourhood kids but is hampered by his lack of knowledge of baseball. A slow-paced and nostalgic look at kids and the problems of growing up that is okay to watch but rather lacking in substance.

Aka: SANDLOT, THE
COM   97 min (ort 101 min) cC   VIDrel: 20TH/TECH   V

SANDOKAN: THE AMAZING ADVENTURES OF
SANDOKAN **   U
Terry Wilson SPAIN   1995
Voices of: Stuart Organ, Steve Edwin, Gavin Muir, Richard Garnett, Steve Bent, Regina Reagan, Candida Gubbins

A cartoon version of Salgari's novel that uses animal figures to tell the story of the title hero, a 19th century adventurer who fights colonialist oppression in South-east Asia. Though quite watchable, the disappointingly poor quality of the animation is a drawback.

Aka: SANDOKAN: THE MOVIE
ANIM   96 min mTV   VIDrel: COLUM/SONOP L/A   V
Boa: novel by Emilio Salgari.

SANDPIPER, THE **   15
Vincente Minnelli USA   1965
Elizabeth Taylor, Richard Burton, Eva Marie Saint, Charles Bronson, Robert Webber, Torin Thatcher, Morgan Mason, Tom Drake

Story of a love triangle involving a female beatnik and a respected pillar of the community, whose passionate love affair comes close to breaking up his marriage and ruining his career. Written by Dalton Trumbo and Michael Wilson. Glossy but shallow. AA: Song ("The Shadow Of Your Smile" – Johnny Mandel (m)/Paul Francis Webster (l)).

DRA   112 min (ort 116 min)   VIDrel: MGM L/A   V

SANDS OF IWO JIMA ***   PG
Allan Dwan USA   1949
John Wayne, John Agar, Adele Mara, Forrest Tucker, Arthur Franz, Julie Bishop, Richard Jaeckel, Wally Cassell, James Brown, Martin Milner, Richard Webb, James Holden, Peter Coe, Bill Murphy, George Tyne, Hal Baylor

A young Marine gradually accepts the need for military discipline, in a film with excellent and realistic battle scenes. Wayne's role as a tough sergeant brought him his first nomination for an Oscar, although he was not to win one until TRUE GRIT.

WAR   108 min (ort 109 min) B/W
VIDrel: 4-FRONT/POLYREC   V

SANDWICH MAN, THE *   PG
Robert Hartford-Davies UK   1966
Michael Bentine, Suzy Kendall, Norman Wisdom, Dora Bryan, Bernard Cribbins, Harry H. Corbett, Stanley Holloway, Alfie Bass, Ian Hendry, Warren Mitchell, John Le Mesurier, John Junkin, Ron Moody, Diana Dors, Wilfrid Hyde-White

One of those daft British comedy-mimes, in which there's no dialogue, just a series of oh-so-funny encounters. This one is about the daily life of a sandwich-board man, and looks at the eccentrics he encounters.

COM   91 min (ort 95 min)   VIDrel: VCC/RTM   V

SANJURO ****   12
Akira Kurosawa JAPAN   1962
Toshiro Mifune, Tatsuya Nakadai, Takashi Shimura, Yuzo Kayama, Reiko Dan, Masao Shimizu, Yunosuke Ito, Takako Irie, Kamatari Fujuwara, Akihiko Hirata, Keiju Kobayashi, Kunie Tanaka, Hiroshi Tachikawa, Tatsuhiko Hari

Spoof samurai film, in which a shabby warrior helps young clan members root out corruption among the elders of their clan. A kind of sequel to YOJIMBO, with excellent fighting sequences, effective moments of humour and a mesmerising performance by Mifune in the title role. The clever script (our hero wins the day by guile rather than his fighting prowess) is just one more asset in a vivid and captivating film.

Aka: TSUBAKI SANJURO
COM   91 min (ort 95 min) wScrn   VIDrel: CONNO/RTM
V

SANKOFA *** 15
Haile Gerima BURKINA FASO/GERMANY/GHANA/
USA 1993
Kofi Ghanaba, Oyafunmike Ogunlano, Alexandra Duah, Nick Medley,
Mutabaruka, Afemo Omilami, Reginald Carter, Mzuri, Jimmy Lee
Savage, Hasinatu Camara, Jim Faircloth, Stanley Michelson, Louise
Reid, Alditz McKenzie, Christian Rigby
A black fashion model arrives in Ghana for a photo-shoot and
poses for her white photographer, an action that so enrages an
old drummer, Sankofa, that he uses a magical spell to transport
her back in time, where she experiences the evils of slavery at
firsthand on a sugar plantation in the States. The title is an Akan
word that means "returning to one's roots", and this film does
exactly that. Despite the plot contrivance, it is both absorbing
and impressive.
DRA 124 min CINrel

SANTA CLAUS ** U
Jeannot Szwarc UK 1984
David Huddleston, Dudley Moore, John Lithgow, Burgess Meredith,
Jeffrey Kramer, Judy Cornwell, Christian Fitzpatrick, Carrie Kei Heim,
John Barrard, Anthony O'Donnell, Peter O'Farrell, Tim Stern,
Christopher Ryan, Don Estelle
The Santa Claus legend is here milked for all it's worth, as Santa
sets off for New York from an unhappy elf from the grip of
a toymaker. A poorly executed film with an abysmally low level
of humour. Only the eye-catching opening, telling of the origin
of Santa, has any memorable sequences.
Aka: SANTA CLAUS: THE MOVIE
COM 103 min (ort 112 min) VIDrel: MGM/WHV V/sur

SANTA CLAUSE, THE *** U
John Pasquin USA 1994
Tim Allen, Judge Reinhold, Wendy Creson, Eric Lloyd, David
Krumholtz, Larry Brandenburg, Mary Gross, Paige Tamada, Peter
Boyle, Judith Scott, Melissa King, Jayne Eastwood, Bradley
Wentworth, Azura Bates, Joshua Satok, Zach McLemore
A toy salesman inadvertently causes a calamity for the kids at
Christmas when Santa falls from his rooftop. Luckily, he finds
an "emergency card" in Santa's pocket, instructing him to take
over. But when he puts on the Santa Claus outfit, he finds he has
taken on more than he realised. See also IN THE NICK OF
TIME.
COM 94 min (ort 98 min) cC VIDrel: WDV/TECH V/sh

SANTA FE TRAIL ** U
Michael Curtiz USA 1940
Errol Flynn, Olivia De Havilland, Raymond Massey, Ronald Reagan,
Alan Hale, Guinn "Big Boy" Williams, Van Heflin, William
Lundigan, Gene Reynolds, Henry O'Neill, Alan Baxter, John Litel,
Moroni Olsen, David Bruce, Joseph Sawyer
Long rambling Western. Massey as John Brown is tracked down
by J.E.B. Stuart (Flynn) and George Armstrong Custer (Reagan),
both of whom were classmates at West Point and are rivals for
De Havilland. The music is by Max Steiner.
WES 107 min (ort 110 min) B/W VIDrel: ORBIT/DISC V

SANTA SANGRE **** 18
Alejandro Jodorowsky ITALY/MEXICO 1989
Axel Jodorowsky, Bianca Guerre, Guy Stockwell, Thelma Tixou,
Sabrina Dennison, Adam Jododrowsky, Faviola Elenka Tapia, Teo
Jododrowsky, Ma De Jesus Aranzabal, Jesus Juaez, Sergio Bustamente,
Gloria Contreras, S. Rodriguez
After seventeen years, the director "El Topo" returns with this
profoundly disturbing and surrealistic tale of madness and
eventual liberation. A series of dreamlike images recount the
career of a child circus-magician who was driven insane by his
father's violent attack on his mother, but escapes his asylum to
rejoin his mother in a bizarre stage act. A film of violent
extremes and limited appeal, it is as memorable as it is shocking.
HOR 118 min (ort 123 min) VIDrel: FABFIL/SPEAR
V/sur LV

SARABAND FOR DEAD LOVERS *** PG
Basil Dearden UK 1949
Stewart Granger, Joan Greenwood, Francoise Rosay, Flora Robson,
Frderick Valk, Peter Bull, Anthony Quayle, Megs Jenkins, Michael
Gouch, Jill Balcon, Cecil Trouncer, David Horne, Mercia Swinburne,
Miles Malleson, Allan Jeayes
A woman marries the ruler of the German state of Hanover,
who later became King George I, but experiences nothing
but unhappiness until she meets a dashing Swedish count,

They become lovers and experience a brief period of bliss
until fate takes a hand. A well-realised, period romance with
lavish sets and costumes, in this first colour film from Ealing
Studios.
Aka: SARABAND
DRA 93 min (ort 96 min) VIDrel: LUMI/SONOP L/A V
Boa: novel by Helen Simpson.

SARAFINA! *** 15
Darrell James Roodt USA 1992
Miriam Makeba, Leleti Khumalo, Whoopi Goldberg, John Khani,
Mbongeni Ngema, Dumisani Dlamini, Sipho Kunene, Tertius
Meintjes, Robert Whithead, Somizi "Whacko" Mhlongo, Nhlanhla
Ngema, Faca Kulu, Wendy Mseleku, Mary Twala
Sweeping, uncertain but highly watchable adaptation of the
stage musical that attempts to tell of the lives of a group of
Soweto schoolchildren and their strong-willed teacher (played
with great charm and a sure hand by Goldberg) as they cope
with life under Apartheid and the all-pervasive violence.
Screenplay is by William Nicholson and Mbongeni Ngema (who
wrote the screenplay for the original work).
MUS 111 min (ort 116 min) VIDrel: WHV V/dm

SARAH, PLAIN AND TALL *** PG
Glenn Jordan USA 1990
Glenn Close, Christopher Walken, Margaret Sophie Stein, Lexi
Randall, John De Vries, Christopher Bell, James Rebhorn, Woody
Watson, Betty Laird, Marc Denney, Kara Beth Taylor
In 1910, a Maine spinster schoolteacher answers an ad from a
widowed Kansas farmer with two children, and travels to his
farm to take charge of his family. As they face day-to-day hard-
ships together, their growing friendship eventually blossoms
into love. A warm yet unsentimental adaptation of a children's
story, much enhanced by fine acting and a leisurely pace.
Screenplay is by MacLachlan in collaboration with Carl Sobieski.
Followed by SKYLARK.
DRA 90 min (ort 98 min) mTV VIDrel: PAL/GUILD L/A
V/sh
Boa: book by Patricia MacLachlan.

SATAN RETURNS * 18
Ah Lun HONG KONG 1996
Donnie Yen, Chingamy Yau, Yuen King Dan
In Hong Kong, a female police officer is stalked by a nutter who
believes her to be the daughter of Satan. Daft.
A/AD 95 min wScrn VIDrel: MIA/DISC V

SATANIC RITES OF DRACULA, THE * 18
Alan Gibson UK 1973
Peter Cushing, Christopher Lee, Michael Coles, William Franklyn,
Freddie Jones, Joanna Lumley, Richard Vernon, Patrick Barr, Barbara
Yu Ling, Richard Mathews, Lockwood East, Maurice O'Connell,
Valerie Van Ost, Peter Adair
An outbreak of vampirism is traced to the presence in London
of Dracula, who is masquerading as a property developer and
plans to unleash a deadly plague on the world. The modern
setting for this ancient myth adds absolutely nothing to a tired
and dispirited production.
Aka: COUNT DRACULA AND HIS VAMPIRE BRIDE; DRACULA IS DEAD
AND WELL AND LIVING IN LONDON; RITES OF DRACULA
HOR 84 min Cut (1 sec plus film cuts – ort 88 min)
VIDrel: WHV V

SATAN'S CHILD ** 15
Robert Lieberman USA 1990
Marita Geraghty, Shirley Knight, Anthony Zerbe, Pete Kowanko
A couple of newlyweds settle in a remote community, where
they learn to their horror that their newborn child has been
claimed by the Devil.
HOR 83 min (ort 90 min) VIDrel: GUILD/SONOP V/sh

SATISFACTION * 15
Joan Freeman USA 1988
Justine Bateman, Liam Neeson, Trini Alvarado, Britta Phillips, Scott
Coffey, Julia Roberts, Debbie Harry, Chris Nash, Michael De Lorenzo,
Tom O'Brien, Kevin Haley, Peter Craig, Steve Chopper, Sheryl Ann
Martin
Story of a rock band consisting of four girls and just one guy,
who get a chance to break into the big time when they land an
engagement at a summer resort. A cliched little tale that was
designed for the teen market but aimed too low.
DRA 89 min (ort 93 min) VIDrel: 20TH V

**SATISFACTION JACKSON ** **   R18/18
Henri Pachard USA   1988
*F.M. Bradley, Ona Zee, Shanna McCullough, Alicia Monet, Joey
Silvera*
A male sex therapist cum private eye (and Princeton graduate)
takes every job very seriously indeed, and always does his
utmost to satisfy his clients. But his latest assignment promises
to test his skills fully, as this involves finding a woman who has
run away from her boyfriend because he is such a lousy lover.
A lighthearted and very silly sex romp that offers a few modest
laughs if nothing else. The heavily cut versions are almost
impossible to follow.
A   57 min Cut (3 sec – ort 80 min); 40 min Cut (21 min 36 sec
– 18 ver)   VIDrel: SHEP L/A (R18 ver); HAR/GOLD (18 ver)
V

SATURDAY NIGHT AND SUNDAY MORNING **   PG
Karel Reisz UK   1960
*Albert Finney, Shirley Ann Field, Hylda Baker, Rachel Roberts, Bryan
Pringle, Norman Rossington, Peter Cawdron, Edna Morris, Elsie
Wagstaff, Frank Pettitt, Avis Bunnage, Colin Blakely, Irene
Richmond, Louise Dunn, Peter Madden*
The life of a coarse, fun-loving factory hand in Nottingham, who
has an affair with the wife of a workmate, gets beaten up for his
trouble but eventually finds a girl of his own. One of the first of
a batch of fresh, raw working-class dramas, this faithful adap-
tation broke new ground both in its frank attitude towards
sexuality and its unashamed anti-authority stance. Finney's
zestful performance deservedly shot him to stardom.
DRA   85 min (ort 89 min) B/W   VIDrel: CASPIC L/A   V
Boa: novel by Alan Sillitoe.

SATURDAY NIGHT FEVER *   18
John Badham USA   1978
*John Travolta, Karen Lynn Gorney, Joseph Cali, Barry Miller, Paul
Page, Bruce Ornstein, Donna Pescow, Julie Bovasso, Martin Shakar,
Sam J. Coppola, Nina Hansen, Lisa Pelluso, Denny Dillon, Bert
Michaels, Robert Costanza*
A young man doing a dead-end job only comes to life at the local
disco on Saturday nights, when he becomes the local dancing
sensation. A flashy and vigorous musical that skilfully blends
exciting dancing with foul-mouthed language. The PG rated
version dropped some scenes and dialogue. Written by Norman
Wexler, with songs written and performed by the Bee Gees. This
was Travolta's first starring film. Followed by STAYING ALIVE.
MUS   114 min (ort 119 min – abridged at film release)
VIDrel: 4-FRONT//POLYREC/CIC   V/sur
Boa: story by Nik Cohn.

**SATURN 3 ** **   15
Stanley Donen UK   1980
*Kirk Douglas, Farrah Fawcett, Harvey Keitel, Douglas Lambert, Ed
Bishop, Christopher Muncke*
Two research scientists enjoy an idyllic existence on a deserted
space station until the arrival of an unbalanced scientist, whose
sinister robot takes on some of the less desirable aspects of his
personality. A glossy and menacing exercise in emptiness.
Written by Martin Amis and scored by Elmer Bernstein.
FAN   84 min (ort 88 min)
VIDrel: 4-FRONT/POLYREC/ITC   V/sur
Boa: short story by John Barry.

SATYRICON **   18
Federico Fellini FRANCE/ITALY   1969
*Martin Potter, Hiram Keller, Max Born, Salvo Randone, Mario
Romagnoli, Magali Noel, Capucine, Alain Cuny, Lucia Bose, Tanya
Lopet, Gordon Mitchell, Fanfulla, Mario Romagnoli, Donyale Luna,
Giuseppe Sanvitale, Hylette Adolphe*
Scorned by intellectual critics, this magnificently realised evoca-
tion of Ancient Rome portrays a decadent and diverse empire
buzzing with life and activity. The screenplay (by Fellini) mostly
follows the adventures of a young student and his mainly sexual
encounters with various people in Roman society. Strikingly
original if a little self-indulgent, its a joy to watch even on TV,
though its full impact is best savoured in the cinema.
Aka: FELLINI SATYRICON; FELLINI'S SATYRICON
DRA   124 min (ort 130 min)   VIDrel: MGM/WHV   V/h
Boa: book by Gaius Petronius.

**SAVAGE! ** **   18
Cirio H. Santiago USA   1973
*James Inglehart, Carol Speed, Lada Edmond Jr., Sally Jordan, Ken
Metcalf, Rossana Ortiz, Vic Diaz*
A professional killer takes on both cops and cons with his all-
woman army, in this Filipino-lensed, low-budget effort. Typical
of much of this director's work and barely adequate in terms of
entertainment.
Aka: BLACK VALOR
A/AD   81 min   VIDrel: SUPVID/RTM   V

**SAVAGE BEACH ** **   18
Andy Sidaris USA   1989
*Dona Spier, Hope Marie Carlton, Bruce Penhall, Rodrigo Obregon,
John Aprea, Teri Weigel, Michael Shane, Al Leong, Eric Chen, James
Lew, Dann Seki, Mike Mikasa, Roy Summorselt, Patty Duffek, Paul
Cody, Lisa London, Maxine Ware*
A couple of narcotics agents, stranded on a remote island after
their plane crashes, see their chances of rescue going up in
smoke when they are taken prisoner by an armed group search-
ing for buried treasure. A sequel of sorts to PICASSO TRIGGER
by writer-director Sidaris, who in this film throws in a measure
of tongue-in-cheek humour.
A/AD   92 min (ort 94 min)   VIDrel: L/A   V

**SAVAGE HEARTS ** **   18
Mark Ezra UK   1995
*Maryam D'Abo, Richard Harris, Jamie Harris, Myriam Cyr, Jerry
Hall, Stephen Marcus, Angus Deayton*
A beautiful woman has an unusual job, as the Mafia's most
accomplished killer. Having been told by her medic that she has
only six months to live, she decides to make a splash by stealing $2,000,000 from her employers. A hybrid film that is at times
a black comedy, at others a thriller; the ludicrous plot is not its
chief asset. Harris is appealing as the cheeky con-artist who gets
involved with these proceedings.
THR   107 min   VIDrel: BMRVID/BMGREC   V

**SAVAGE INTRUDER ** **   18
Donald M. Wolfe USA   1975
*Miriam Hopkins, John David Garfield, Gale Sondergaard, Florence
Lake, Lester Mathews, Riza Royce, Joe Besser, Minta Durfee, Virginia
Wing*
An ageing, alcoholic former star falls for a handsome young
man whose pleasant exterior and manners belie the fact that he
is a psychopathic murderer of women, in this down-market,
gory version of SUNSET BOULEVARD.
Aka: HOLLYWOOD HORROR HOUSE
HOR   85 min (ort 90 min)   VIDrel: VIPCO/SGSVID   V/h

**SAVAGE ISLANDS ** **   (PG)
Ferdinand Fairfax NEW ZEALAND   1983
*Tommy Lee Jones, Jenny Seagrove, Michael O'Keefe, Max Phipps,
Grant Tilly, Peter Rowley, Bill Johnson, Kate Harcourt, Reg Ruka,
Roy Billing, David Letch, Bruce Allpress, Tui Teka, Pudji Waseso,
Peter Vere Jones, Tom Vanderlaan*
This attempt to recreate one of those Errol Flynn pirate movies
transfers the action to the South Pacific, where in the 1880s a
young missionary attempts to rescue his kidnapped wife. Far
from brilliant, though it offers a fair amount of action in spite
of the almost non-existent plot.
Aka: NATE AND HAYES
A/AD   96 min (ort 100 min)   SATrel: SKY MOVIES

SAVAGE NIGHTS *   18
Cyril Collard FRANCE/ITALY   1992
*Cyril Collard, Romane Bohringer, Carlos Lopez, Corine Blue, Claude
Winter, Rene-Marc Bini, Maria Schneider, Clementine Celarie, Laura
Favali, Jean-Jacques Jauffret, Denis D'Archangelo, Alissa Jabri,
Francisco Gimenez, Marine Delterme*
A promiscuous bisexual man learns that he is HIV-positive but
refuses to modify his lifestyle, even intensifying his self-destruc-
tive behaviour. He has sex with a young woman and neglects
to tell her or his condition until she has fallen in love with him.
A well made tale of a highly unattractive and unsympathetic
character that makes for compelling viewing. Winner of both the
1993 Cesar Best Film and Best First Film awards (the first entry
to do so).
Aka: LES NUITS FAUVES
DRA   122 min (ort 126 min) wScrn   VIDrel: ARTIF/20TH
V/sh

**SAVAGE SAM ** **   PG
Norman Tokar USA   1962
*Brian Keith, Tommy Kirk, Kevin Corcoran, Dewey Martin, Jeff York,
Royal Dano, Marta Kristen, Rafael Campos, Slim Pickens, Rodolfo
Acosta, Pat Hogan, Dean Fredericks, Brad Weston*

A sequel to OLD YELLER, with another mongrel mutt proving his worth in order to gain acceptance, by tracking down Apaches who have kidnapped some children. A fairly routine outing, but pleasant enough.
DRA   103 min   VIDrel: WDV/TECH L/A   V
Boa: novel by F. Gipson.

SAVATE **
Isaac Florentine USA
Olivier Gruner, James Brolin, Michael Palance, Don Gibb, Marc Singer, Ian Ziering, Ashley Laurence, Charles Santore, Rance Howard, Takis, Gene Wolande, Zale Kessler, Scott Schwartz, Eric Betts, George O'Mara, Stan Ward, Hien Nguyen
Martial arts tale in a Western setting, that sees the best friend of a tough exponent of savate getting killed after an encounter with a crooked landowner. As expected, our fighter sets out to avenge his murdered buddy.
MAR   90 min (ort 120 min)   VIDrel: NEWAGE/TECH L/A V

18
1994

SAVE ME **
Alan Roberts USA
Harry Hamlin, Lysette Anthony, Olivia Hussey, Bill Nunn, Steve Railsback, Jospeh Campanella, Neil Ronco, Steve Diamant, Reilly Murphy, Bill Smillie, Kato Kaelin, Stephen Landis, Randy Mermell, Grant Cramer, Kristine Ros, Greg Lewis
Complex and not entirely successful attempt at yet another erotic-style thriller in which a man abandoned by his wife falls prey to a sexy seductress. Predictably, she is not all she seems and his life is soon in danger. An over-the-top exercise in nastiness, with Anthony rather unconvincing in her role as the irresistible femme fatale.
THR   89 min   VIDrel: EIV/SONOP   V

18
1993

SAVING GRACE **
Robert M. Young USA
Tom Conti, Giancarlo Giannini, Donald Hewlitt, Fernando Rey, Angelo Evans, Erland Josephson, Patricia Mauceri, Edward James Olmos, Marta Zoftoli, Tom Felleghy, Guido Alberti, Massimo Sarchielli, Massimo Serato, Agnes Nobescourt,
A newly-elected Pope is overcome by a feeling of distaste for the meaningless routines that forms an important part of his job, and sneaks out of the Vatican in disguise. He winds up in an Italian village and enjoys various encounters along the way. An amusing idea, but fatally hampered by slow development.
COM   107 min (ort 112 min)   VIDrel: L/A   V
Boa: novel by Celia Gittelson.

PG
1985

SAVIOR OF THE EARTH *
Roy Thomas HONG KONG
A further fantasy-adventure from the producer of FALCON 7 and CAPTAIN COSMOS, in which three heroes are given the task of stopping a mad scientist from wrecking Earth's computer network in his bid for world domination. A contrived futuristic tale that runs along predictable lines.
ANIM   65 min   VIDrel: IMPENT   V

PG
1989

SAVIOUR OF THE SOUL **
Yuen Kwai/Corey Yuan/Jeffrey Lau/David Lai HONG KONG
Andy Lau, Aaron Kwok, Anita Mui, Gloria Yip, Kenny Bee, Carina Lau
Action-fantasy set in the 21st century, where a bunch of bounty hunters have to do battle with an evil, soul-stealing wizard. Good special effects and some well mounted stunts are about all that is on offer here.
Aka: SILVER FOX
A/AD   90 min wScrn   VIDrel: MADE/RTM   V

15
1992

SAY ANYTHING ***
Cameron Crowe USA
John Cusack, Ione Skye, John Mahoney, Lili Taylor, Amy Brooks, Pamela Segall, Jason Gould, Joan Cusack, Eric Stoltz, Lois Chiles, Loren Dean, Glenn Walker Harris Jr, Charles Walker, Russell Lunday, Polly Platt
Comedy-drama that takes a look at first love with an unexceptional high school kid falling for a clever and seemingly unattainable girl, and finding that she is not quite so unapproachable after all. Despite the use of a well-worn theme, this refreshing film has both charm and wit. Lois Chiles has an unbilled cameo as the girl's mother. Scripted by Crowe in his directing debut.
DRA   96 min (ort 100 min)   VIDrel: 20TH/TECH   V/sh

15
1989

SAY ONE FOR ME **
Frank Tashlin USA
Bing Crosby, Debbie Reynolds, Robert Wagner, Ray Walston, Les Tremayne, Connie Gilchrist, Frank McHugh, Joe Besser, Alena Murray, Stella Stevens, Nina Shipman, Sebastian Cabot, Judy Harriet, Dick Whittinghill, Murray Alper
A priest who runs a church in the New York theatre district agrees to keep a paternal eye on a young girl who is trying to raise the cash need for her father's operation by becoming a show business performer. Eventually our clergyman and several colleagues end up staging a charity TV show but not before our young lady seems set to wed a heel. Songs include "You Can't Love 'Em All". A poor work with some quite tasteless sequences.
COM   115 min (ort 119 min)   SATrel: MOVIE CHANNEL

(U)
1959

SAYONARA ***
Joshua Logan USA
Marlon Brando, Miiko Taka, Ricardo Montalban, Miyoshi Umeki, Red Buttons, Martha Scott, James Garner, Patricia Owens, Kent Smith, Douglas Watson, Reiko Kuba, Soo Young, Harlan Warde, Shochiku Kagedikan, Girls Revue
American servicemen in Japan become deeply involved with Japanese women, and learn that neither side is prepared to accept such relationships, as pilot Brando discovers when he falls for Japanese entertainer Taka. Written by Paul Osborn, with an Irving Berlin theme song and music by Franz Waxman. AA: S. Actor (Buttons), S. Actress (Umeki), Art/Set (Ted Hawarth/Robert Priestley), Sound (George Groves).
DRA   142 min (ort 147 min)   VIDrel: 4-FRONT/POLYREC V
Boa: novel by James A. Michener.

PG
1957

SCALPHUNTERS, THE **
Sydney Pollack USA
Burt Lancaster, Shelley Winters, Telly Savalas, Ossie Davis, Dan Vadis, Armando Silvestre, Dabney Coleman, Paul Picerni, Nick Cravat, John Epper, Jack Williams, Chuck Roberson, Tony Epper, Agapito Roldan, Gregorio Acosta
A fur trapper and his highly-educated slave, track down a gang who specialise in killing Indians for their scalps. A strange subject for a comedy, that meanders aimlessly despite one or two good moments.
WES   99 min (ort 102 min)   VIDrel: WHV   V

PG
1968

SCAM **
John Flynn USA
Lorraine Bracco, Christopher Walken, Miguel Ferrer, Martin Donovan, James McDaniel, Daniel Von Bargen, Erick Avari, Skipp Suduth, Maxi Priest, Edgar Allen Poe IV, James Walker, Rob Fuller, Carmen Lopez, Clarence Thomas, Xavier Coronel
An attractive female con-artist goes looking for suitable victims at Miami Beach but picks the wrong man. He turns out to be a former FBI agent who is now in the same line of work. He persuades her to accompany him to Jamaica, having offered to pay her $10,000 to help her mount a complex scam, his intention being to relieve a Mafia accountant of a pile of money. However, it soon transpires that there far easier ways to earn $10,000.
THR   98 min (ort 100 min)   VIDrel: POLY/POLYREC L/A V/s
Boa: novel Ladystinger by Craig Smith.

15
1992

SCANDAL **
Michael Caton-Jones UK
John Hurt, Joanne Whalley-Kilmer, Ian McKellen, Bridget Fonda, Britt Ekland, Daniel Massey, Roland Gift, Jeroen Krabbe, Leslie Phillips, Jean Aexander, Alex Norton, Ronald Fraser, Paul Brooke, Keith Allen, Ralph Brown, Ken Campbell
A recreation of a sex scandal of the early 1960s, when government minister John Profumo became involved with call-girl Christine Keeler, whom he had been introduced to by Stephen Ward, an osteopath and curious social-climbing figure of the time. The fact that Keeler (and her colleague Mandy Rice-Davies) had been involved with a Soviet diplomat, was felt to compromise Profumo's integrity and he was forced to resign. A competent directing debut for Caton-Jones.
DRA   110 min (ort 115 min)   VIDrel: 4-FRONT/POLYREC V/sur

18
1988

SCANDAL IN A SMALL TOWN *
Anthony Page USA
Raquel Welch, Christa Denton, Frances Lee McCain, Peter Van

15
1988

Norden, Robin Gammell, Ronny Cox, Mickey Jones, Ray Girardin, Peter Palmer, Ernie Lively, Arell Blanton, Katherine McGrath, Laurence Haddon, Helen Page Camp
The story of one woman's lone fight with the establishment, when she learns that her daughter's teacher is promoting anti-Semitism. An unconvincing and exploitative effort, that covers much the same ground as EVIL IN CLEAR RIVER.
DRA 96 min (ort 100 min) mTV VIDrel: L/A V

SCANDALOUS PHOTOS ** 18
Jean-Claude Roy FRANCE 1979
Brigitte Lahaie, Marcel Charvey, Muriel Montosse
A woman hatches a plan with her lover to blackmail young heirs by taking photos of them in compromising positions, but to do this she has to find some willing young girls to assist her.
A 81 min wScrn VIDrel: JEZ/RTM V

SCANNER COP ** 18
Pierre David USA 1993
Daniel Quinn, Darlanne Fluegel, Richard Grove, Mark Rolston, Hilary Shepard, Gary Hudson, Cyndi Pass, Luca Bercovici, Brion James, Christopher Kriesa, James Horan, Savannah Smith Boucher, Ben Reed, Richard Lynch, Elan Rothschild
A cop who is himself a scanner uses his abilities to defend the city from a Scanner army that is attempting to take over. A spin-off entry in the SCANNERS series that feels tired and derivative. A sequel followed.
FAN 90 min (ort 94 min) VIDrel: REFLEC/FIRST V

SCANNER COP 2: VOLKIN'S REVENGE ** 18
Pierre David USA 1994
Daniel Quinn, Patrick Kilpatrick, Khrystybe Haje, Jewel Shepard, Stephen Mendel, Brenda Swanson
Newly released from prison, a criminal Scanner is intent on revenge. But in order to deal with the cop who put him behind bars (who is also a Scanner) he sets about acquiring more power at the expense of other Scanners, by draining them of their life-force. Our good Scanner cop is the one who put this criminal away, and he must now deal with him again. A spectacular sequel, of exploding heads and other equally gruesome effects, but not all that strongly plotted.
Aka: SCANNERS: THE SHOWDOWN
FAN 91 min (ort 95 min) VIDrel: FIRST/SONOP V

SCANNER FORCE * 18
Christian Duguay CANADA 1991
Liliana Komorowska, Steve Parrish, Daniel Pilon, Valerie Valois, Collin Fox, Peter Wright, Sith Sekae, Harry Hill, Claire Cellucci, Michael Copeman, Chip Chuipka, Jean Frenette, Sylvan Beauchamps, Gaston Perreault, Michel Perran
A further spin-off from SCANNERS that sees groups of rival telepaths locked in a deadly battle for supremacy. An unappealing and derivative dud, that adds nothing new to a tired idea.
Aka: SCANNER 3: THE TAKEOVER; SCANNERS 3: THE TAKEOVER
FAN 97 min (ort 101 min) VIDrel: FIRST/SONOP L/A V

SCANNERS *** 18
David Cronenberg CANADA 1981
Stephen Lack, Jennifer O'Neill, Patrick McGoohan, Lawrence Dane, Charles Shamata, Michael Ironside, Robert Silverman, Lee Broker, Mavor Moore, Adam Ludwig, Lee Murray, Fred Doederlein, Geza Kovacs, Sony Forbes
A tranquilliser test-marketed on pregnant women in the 1940s, led to the creation of "scanners" – powerful telepaths, some of whom have telekinetic powers. The corporation responsible for their creation tracks down one with the intention of sending him to infiltrate a dangerous group of subversive scanners. Gory effects are coupled to a sound if rather convoluted story. Written by Cronenberg. Several sequels and spin-offs followed.
HOR 99 min (ort 103 min) VIDrel: ARROW/RTM V

SCANNERS 2: THE NEW ORDER ** 18
Christian Duguay USA 1990
David Hewlett, Deborah Raffin, Yvan Ponton, Raoul Trujillo, Isabelle Meijas, Tom Butler, Vlastra Vrana, Murray Westgate, Doris Petrie, Dorothee Berryman, Michael Rudder, David Francis, Stephen Zarou, Tom Harvey, Jason Cavalier
This sequel to the 1981 film takes up the story some eight years on, when a ruthless police chief obsessed with creating a new crime-fighting order, is exploiting the Scanners for his own ends. A young Scanner learns of his plans and runs for his life, before returning in time for the final showdown. A clockwork plot and

unnecessary gory effects ruin any intelligent handling of the theme of telekinesis. Very disappointing. See SCANNER FORCE.
FAN 100 min (ort 104 min) VIDrel: BRAVE/SONOP L/A V

SCARECROW, THE ** 15
Sam Pillsbury NEW ZEALAND 1981
John Carradine, Tracy Mann, Jonathan Smith, Daniel McLaren, Stephen Taylor, Anne Flannery, Des Kelly, Paul Owen-Lowe, Bruce Allpress, Greer Robson, Tracy Mann, Denise O'Connell, Greg Naughton, Jonathan Hardy
A small town in the early 1950s suddenly becomes the focus for a series of strange events. An intriguing but obscure and unsatisfying blend of the sinister and the bizarre, that might have worked as an effective study of evil if the film been blessed with a sharper script.
DRA 84 min (ort 97 min) VIDrel: ARTPRO/RTM V
Boa: novel by Ronald Hugh Morrieson.

SCARECROWS *** 18
William Wesley USA 1988
Ted Vernon, Michael Sims, Richard Vidan
Five army deserters hijack a plane carrying a large amount of money, but one of them decides to keep it all for himself. After throwing the loot out he parachutes down after it. However, he lands in a field of murderous, living scarecrows, and these creatures soon kill him and prepare a fitting welcome for the others. A gripping and quite scary film, it attempts no rational explanation, merely delivering a set of gruesome shocks. See also TOURIST TRAP.
HOR 80 min VIDrel: POLY L/A V

SCARED STIFF ** 18
Richard Friedman USA 1986
Andrew Stevens, Mary Page Keller, Josh Segal, David Ramsey, Nicole Fortier, Jackie Davis, William M. Hindman
The tale of a female singer who moves into a new house with her boyfriend and young son. She finds a strange talisman in the attic that soon begins to exert a sinister influence on their lives.
THR 81 min (ort 85 min) VIDrel: 20TH V

SCARED TO DEATH ** 18
William Malone USA 1980
John Stinson, Diana Davidson, Jonathan David Moses, Pamela Brown, Toni Jannotta, Kermit Eller, Walker Edminston, Mike Muscat, Pam Bowman, Freddie Dawson, Tracy Weddle, Joleen Porcaro, Joseph Daniels, Stephen Fenning
A man-eater-monster (the result of genetic experimentation) escapes into the sewers of LA, but surfaces from time to time to attack humans and suck out their bone marrow. A low-budget film that is mildly scary, but suffers from an over-reliance on that good old horror standby: the man in the rubber suit. Followed by a sequel in 1984 from the same director.
Aka: TERROR FACTOR, THE
HOR 78 min (ort 94 min) VIDrel: SPEAR/SONOP V/sur

SCARFACE *** 15
Howard Hawks USA 1931 (released 1932)
Paul Muni, Ann Dvorak, George Raft, Boris Karloff, Karen Morley, Vince Barnett, Osgood Perkins, C. Henry Gordon, Henry Armettc, Edwin Maxwell, Inez Palange, Tully Marshall, Harry J. Vejar, Bert Starkey, Henry Armetta, Paul Fix
A detailed chronicle of the life and death of Chicago gangster Al Capone, with Dvorak as an incestuous sister thrown in for good measure. This brutal and graphic film was made in 1931, but its release was delayed by the censors. Written by Ben Hecht, Seton I. Miller, John Lee Mahin, Fred Pasley and W.R. Burnett. Remade in 1983 with Al Pacino in the title role.
Aka: SCARFACE: THE SHAME OF THE NATION; SHAME OF A NATION, THE
DRA 86 min (ort 99 min) B/W VIDrel: CIC/SONOP V/h
Boa: novel by Armitage Traill.

SCARFACE * 18
Brian De Palma USA 1983
Al Pacino, Steven Bauer, Michelle Pfeiffer, Mary Elizabeth Mastrantonio, Robert Loggia, Paul Shenar, Harris Yulin, F. Murray Abraham, Miriam Colon, Angel Salazar, Arnaldo Santana, Pepe Serna, Michael P. Moran, Al Israel
An updated version of the 1932 film, with the central figure now a Cuban refugee who becomes a powerful drugs dealer, and the setting changed from Chicago to Miami. An abundance of

graphic brutality, coarse language and drug-taking, gives it an authentic 1980s flavour but we are offered no new insights. Not so much a film, as a static and disagreeable wallow in the gutter. Written by Oliver Stone.
A/AD 162 min (Cut at film release by 25 sec – ort 170 min) wScrn VIDrel: CIC/SONOP; PION (LV only) V/dm LV

SCARFACE MOB, THE *** PG
Phil Karlson USA 1962
Robert Stack, Neville Brand, Barbara Nichols, Keenan Wynn, Patricia Crowley, Bill Williams, Joe Mantell, Bruce Gordon, Peter Leeds, Eddie Firestone, Paul Picerni, Robert Osterloh, Abel Fernandez, Walter Winchell (narration only)
The original pilot episode for the TV series "The Untouchables", the story of FBI agent Elliot Ness, who heads a team of hand-picked men to break Capone's grip during the Prohibition years in America. Originally shown in 1959, but still retaining considerable force.
A/AD 91 min B/W mTV VIDrel: CIC/SONOP L/A V/h
Boa: book The Untouchables by Elliot Ness and O. Fraley.

SCARLET AND THE BLACK, THE *** PG
Jerry London ITALY/USA 1981
Gregory Peck, Christopher Plummer, Raf Vallone, John Gielgud, Barbara Bouchet, Edmund Purdom, Kenneth Colley, Walter Gotell, Julian Holloway, Angelo Infanti, Olga Karlatos, Michael Byrne, T.P. McKenna, Vernon Dobtcheff
In 1943, when Italy has capitulated and been taken over by the Nazis, an Irish priest founds an organisation to help Allied prisoners escape, and fights a cat-and-mouse game with a Nazi officer who is out to catch him. Peck's first dramatic starring role for TV, with Pope Pius XII played by Gielgud. A competent thriller, adapted by David Butler.
THR 138 min (ort 155 min) mTV VIDrel: L/A V
Boa: book The Scarlet Pimpernel Of The Vatican by J.P. Gallagher.

SCARLET LETTER, THE *** 15
Roland Joffe USA 1995
Demi Moore, Gary Oldman, Robert Duvall, Lisa Jolliff-Andoh, Robert Prosky, Edward Hardwicke, Roy Dotrice, Joan Plowright, Malcolm Storry, Jim Bearden, Amy Wright, Larissa Lapchinski, George Aguilar, Tim Woodward, Joan Gregson
Trapped in a New World colony in America, a woman finds love with a priest, but their affair creates problems for the other settlers, she is jailed for her pains and upon release forced to wear a scarlet "A" to mark her out. A stylish adaptation, atmospheric and strongly romantic, but seriously weakened by sluggish pacing and over-ripe dialogue.
DRA 129 min (ort 135 min) wScrn VIDrel: EIV/SONOP V
Boa: novel by Nathaniel Hawthorne.

SCARLET PIMPERNEL, THE **** U
Harold Young UK 1934
Leslie Howard, Merle Oberon, Raymond Massey, Nigel Bruce, Anthony Bushell, Bramwell Fletcher, Joan Gardner, Walter Rilla, Mabel Terry-Lewis, O.B. Clarence, Ernest Milton, Edmund Breon, Melville Cooper, Gibb McLaughlin
Classic story of the English aristocrat who rescued French compatriots from the guillotine, with Howard first class as a dashing Englishman playing a dangerous game with consummate skill. Written by Robert E. Sherwood, Sam Berman, Arthur Wimperis and Lajos Biro. The music is by Arthur Benjamin. Remade as "The Elusive Pimpernel" in 1950 and also as a 1982 TV series.
A/AD 93 min (ort 98 min) B/W VIDrel: L/A V
Boa: novel by Baroness Orczy.

SCARLET STREET *** PG
Fritz Lang USA 1945
Edward G. Robinson, Joan Bennett, Dan Daryea, Margaret Lindsay, Rosalind Ivan, Jess Barker, Samuel S. Hinds, Arthur Loft, Vladimir Sokoloff, Charles Kemperer, Russell Hicks, Lou Lubin, Anita Bolster, Cyrus W. Kendall, Fred Essler
A weak man becomes tragically involved with a prostitute and her pimp, in this fairly effective but ponderous tale. Interestingly, this was the first Hollywood film to show a crime going unpunished, for when the prostitute is murdered her pimp is wrongfully executed for the crime. Written by Dudley Nichols. The play on which this is based was filmed once before by Jean Renoir in 1932.
DRA 101 min (ort 103 min) B/W VIDrel: SECOND/RTM V
Boa: play La Chienne by George De La Fouchardiere.

SCARLETT: PARTS 1 AND 2 ** 15
John Erman AUSTRIA/FRANCE/GERMANY/UK 1994
Joanne Whalley-Kilmer, Timothy Dalton, Sean Ban, Julie Harris, Ann-Margret, Stephen Collins, John Gielgud, Pippa Guard, Jean Smart, Melissa Leo, Colm Meaney, John Fraser, Charles Gray, Mark Lambert, Ruth McCabe, Donald Pickering
Sequel to the novel/movie GONE WITH THE WIND, based on Ripley's much-hyped novel. In Part 1 Scarlett tries to recapture the love of her estranged husband Rhett, but at the same time finds herself falling under the spell of a handsome aristocrat. Part 2 sees her giving birth to Rhett's child, but without his knowledge, as the latter has now remarried. Adequate.
DRA 360 min (ODY/SONOP); 350 min (2 cassettes – ort 480 min) mTV VIDrel: ODY/SONOP (V/sh version); CARL/TECH V/sh
Boa: novel by Alexandra Ripley.

SCARS OF DRACULA, THE ** 18
Roy Ward Baker UK 1970
Christopher Lee, Dennis Waterman, Wendy Hamilton, Jenny Hanley, Christopher Matthews, Anoushka Hempel, Patrick Troughton, Michael Gwynn, Delia Lindsay, Michael Ripper, Bob Todd, Toke Townley
A young couple looking for the husband's missing brother, stumble across Count Dracula. Yet another reworking of this old legend, with our lovable vampire being resurrected by bat's blood and commencing his anti-social activities all over again. Plenty of irrelevant sex and sadism are thrown in, but there is nothing of lasting value in a film that was one more nail in the coffin of a series approaching its end. DRACULA A.D. 1972 followed.
HOR 91 min (ort 96 min) VIDrel: WHV V/h

SCATTERED DREAMS *** 15
Neema Barnette USA 1993
Tyne Daly, Gerald McRaney, Alicia Silverstone, Sonny Shroyer, Andrew Prine, Macon McCalman, Ramsay Midwood, Sean Briges, Elizabeth Grillo Barbeau, Robert Hannah, Blair Struble, Lindsay Broockman, Frank Taylor
When their parents are wrongfully imprisoned the children in a family are sent to a juvenile camp in America's Deep South. Unhappy family drama set in the 1950s.
DRA 89 min VIDrel: ODY/SONOP V

SCENES FROM A MALL * 15
Paul Mazursky USA 1991
Bette Midler, Woody Allen, Bill Irwin, Daren Firestone, Rebecca Nickels, Paul Mazursky, Gregory Moore, Michael Brown, Jonathan Guss, David Frye, Marc Shaiman, Joan Delaney, Amanda Bruce, Betsy Mazursky, Jack Brodsky, Glen Alterman
A married couple spend their anniversary at a shopping mall where they begin to tell each other the truth about their marriage, eventually torpedoing their relationship in the process. A well acted but essentially static film whose overly bitter tone makes it more tragic than comic. The unoriginal and derivative title hints at a perverse homage to Bergman that is entirely inappropriate. A disappointing and largely humourless effort.
COM 83 min (ort 87 min) VIDrel: TOUCH/BUENA L/A V

SCENES FROM A MARRIAGE *** 15
Ingmar Bergman SWEDEN 1973
Liv Ullman, Erland Josephson, Bibi Andersson, Jan Malmsjo, Anita Wall, Gunnel Lindblom, Barbro Hiort Af Ornas
Gloomy and depressing look at an unhappy marriage and its disintegration. An intimate and perceptive examination of human misery, with superb performances from Ullman and Josephson as a couple, who divorce and find themselves new partners. The upbeat ending does, however, see them eventually come together as lovers once more, if not as husband and wife. Originally shown on TV in six episodes, but edited into a feature by the director.
Aka: SCENER UR ETT AKTENSKAP
DRA 162 min (ort 300 min) mTV VIDrel: TART/20TH V

SCENES FROM THE CLASS STRUGGLE IN BEVERLY HILLS *** 18
Paul Bartel USA 1988
Jacqueline Bisset, Ray Sharkey, Mary Woronov, Robert Beltran, Ed Begley Jr, Wallace Shawn, Arnetia Walker, Paul Bartel, Paul Mazursky, Rebecca Schaeffer, Barret Oliver, Edith Diaz, Jerry Tondo, Susan Sanger, Michael Feinstein

A bizarre sex farce set in Beverly Hills where two gardeners each bet that they can be first to bed the other's employer. This scattershot comedy of musical beds among the rich and not-so-rich, tries hard to be a satire but remains just one more wacky sex romp. A highlight is Bisset as an ex-sitcom star whose dead husband keeps re-appearing to pledge his undying love. The screenplay is by Bruce Wagner.
COM 99 min (ort 103 min) VIDrel: MGM/WHV V/sh
Boa: story by Paul Bartel and Bruce Wagner.

SCENT OF A WOMAN ***   15
Martin Brest USA   1992
Al Pacino, Chris O'Donnell, Gabrielle Anwar, James Rebhorn, Richard Venture, Bradley Whitford, Ron Eldard, Philip S. Hoffman, Rocehelel Olivier, Margaret Eginton, Tom Riis Farrell, Nicholas Sadler, Todd Louiso, Matt Smith, June Squibb
A retired army colonel who is both embittered and virtually blind lives with his niece who engages a local prep school pupil to take care of him while her family is away over the Thanksgiving weekend. What no-one knows is that our colonel is planning one last trip to New York to live it up. Pacino's strong (if rather overblown) performance holds together this long and well crafted remake of a 1974 Italian film. AA: Actor (Pacino).
DRA 149 min (ort 157 min) cC VIDrel: CIC/SONOP; PION (LV only) V/dm V/sur LV
Boa: novel Il Buio E Il Miele by Giovanni Arpino.

SCENT OF GREEN PAPAYA, THE ***   U
Tran Anh Hung FRANCE/VIETNAM   1993
Yen-Khe Tran Nu, Man San Lu, Thi Loc Truong, Anh Hoa Nguyen, Hoa Hoi Vuong, Ngoc Trung Tran, Vantha Talisman, Keo Souvannavong, Van Oanh Nguyen, Gerard Neth, Nhat Do, Thi Hai Vo, Thi Thanh Tra Nguyen, Lam Huy Bui, Xuan Thu Nguyen
A ten-year old girl comes to work as a servant to a Hanoi household in 1951 and stays there for some ten years, during which we see how she relates to the various people who live there, most especially the eldest son, to whom she becomes devoted.
Aka: L'ODEUR DE LA PAPAYE VERTE; MUI DU DU XANH
DRA 99 min (ort 130 min) wScrn VIDrel: ARTIF/20TH V/h

SCENT OF HEATHER, A **   18
Phillip Drexler USA   1981
Tracy Adams, Christine Ford, Veronica Hart, Vanessa Del Rio, Paul Thomas, Jessica Teal, Lisa Be, Richard Bell, R. Bolla, Neil Peters, Felix Krull
A silly Gothic romance set in an English castle, with a naive young woman getting married and then discovering that her husband is really her brother. The story now has both parties remaining married but making alternative arrangements to obtain sexual satisfaction. Finally, the husband decides that they should have one night together and then commit suicide, but just before this happens they discover they are not brother and sister after all.
A 70 min Cut (1 min 38 sec plus film cuts – ort 99 min) VIDrel: ELV/DISC V

SCENT OF PASSION *   18
Pasquale Fanetti ITALY   1990
Malu, Agnes Lopez Barea, Giancarlo Teoderi, Zvonco Zrncic, Suada Herak, Melita Turisic, Mircea Hurdubea
A dance director who uses coarse and brutal methods on his ballet troupe, is obsessed with the quest for both artistic perfection and the perfect woman. When he rescues a young woman from an attack, he thinks he has found what he was seeking and sets about moulding her into his personal vision of an ideal woman. However, he soon learns that life is more complex than art. A turgid and hard-to-watch softcore tale of very little merit.
Aka: SAPORE DI DONNA
A 84 min (ort 90 min) VIDrel: FABFIL/SPEAR V
Boa: novel Violette by Theophile Gautier.

SCHEMES **   (18)
Derek Westervelt USA   1995
Allison Mackie, James McCaffrey, Leslie Hope, John Glover, Polly Draper, John DeLance, Debra Mooney, George D. Wallace, Henry Marshall, Amy Hunter, Dan Cashman, Michael John Baker, Thom Barry, Will MacMillan, Jessica Craven
After his wife is killed in a car accident, an architect receives a cheque for $500,000 from the company who insured her life. Soon, an attractive woman turns up on his doorstep claiming to be an acquaintance of hers who was unaware that she died.

However, in reality she is working with a private detective who is playing a little game of his own. A solidly made thriller with a neat plot twist that helps sustain interest right up to the climax.
THR 92 min SATrel: MOVIE CHANNEL

SCHINDLER'S LIST ****   15
Steven Spielberg USA   1993
Liam Neeson, Ben Kingsley, Ralph Fiennes, Caroline Goodall, Jonathan Sagalle, Embeth Davitz, Malgoscha Gebel, Shmulik Levy, Beatrice Macola, Andrzej Seweryn, Friedrich Von Thun, Krzysztof Luft, Harry Nehring, Norbert Weisser
In WW2 a German businessman named Oskar Schindler exploits his contacts with the Nazis to get Polish Jews for slave labour. After a time, he grows aware of what Nazism entails and devotes himself to saving as many of them as he can. A stark and searing account of heroism in the face of evil, this is Spielberg's most mature work . See also KORCZAK. AA: Art, Actor (Neeson), Dir, Cin (Janusz Kaminsky), Edit, Score/orig (John Williams), Screen/adapt (Steven Zaillian).
DRA 187 min (ort 195 min) B/W/Col wScrn cC VIDrel: CIC/SONOP; PION (LV only) V LV
Boa: book by Thomas Keneally.

SCHOOL DAZE **   18
Spike Lee USA   1988
Larry Fishburne, Giancarlo Esposito, Tisha Campbell, Kyme, Joe Seneca, Art Evans, Julian Eaves, Ellen Holly, Ossie Davis, Spike Lee, Branford Marsalis, Bill Nunn, James Bond III, Anthony Thompkins, Darryl M. Bell, Joie Lee
A comedy set at a black college in the South, where a serious-minded student fights against both the college establishment and the juvenile antics of his fellow students. An unusual comedy, written by Lee, that unfortunately fails to make the most of its opening premise.
COM 120 min (ort 121 min) VIDrel: CASPIC/BMGREC V/sur

SCHOOL FOR SCOUNDRELS **   U
Robert Hamer UK   1960
Ian Carmichael, Alastair Sim, Terry-Thomas, Janette Scott, Dennis Price, Edward Chapman, Kynaston Reeves, Irene Handl, John Le Mesurier, Hugh Paddick, Peter Jones, Gerald Campion, Hattie Jacques, Anita Sharp Bolster
The story of a failure in life who enrols at the College of One-Upmanship, in the hope that he will learn the secret of coming out on top in any situation. Not really a film, but a series of amusing sketches that are drawn from the humorous books of Potter. Written by Patricia Mayes and Hal E. Chester, this is mild entertainment, but no more than that.
COM 90 min (ort 94 min) B/W VIDrel: WHV V/h
Boa: books Gamesmanship/Oneupmanship/Lifemanship by Stephen Potter.

SCHOOL FOR SEX **   18
Pete Walker UK   1968
Derek Aylward, Rose Alba, Hugh Latimer, Francoise Pascal, Bob Andrews, Vic Wise, Wilfred Babbage, Robert Dorning, Dennis Castle, Edgar K. Bruce, Nosher Powell, Julie May, Cathy Howard, Gilly Grant
A much-married embezzler inherits a house, and sets up a school to teach women the arts of seduction so they can fleece rich old men, in this fairly forgettable adult comedy.
A 80 min VIDrel: JEZ/RTM V

SCHOOL TIES **   PG
Robert Mandel USA   1992
Brendan Fraser, Chris O'Donnell, Andrew Lowery, Matt Damon, Amy Locane, Ben Affleck, Cole Hauser, Peter Donat, Kevin Tighe, Ed Lauter, Michael Higgins, Anthony Rapp, Zelko Ivanek, Peter McRobbie, John Cunningham, Elizabeth Franz
When a Jewish youth wins a football scholarship to a snobby prep school, his father and his coach both advise him to conceal his non-WASP. Unfortunately, after he achieves a certain measure of sporting success, a rival learns about his secret and is not slow in whipping up an atmosphere of bigotry and hatred.
DRA 102 min (ort 110 min) VIDrel: CIC/SONOP V/sur

SCHTONK! ***   15
Helmut Dietl WEST GERMANY   1992
Gotz George, Christiane Horbiger, Uwe Ochsenknecht, Rolf Hoppe, Dagmar Manzel, Veronica Feres, Rosemarie Fendel, Ulrich Muhe, Hermann Luase, Harald Juhnke, Karl Schonbock, Georg Marischka, Peter Roggisch, Martin Benrath

Dietl's funny debut feature deals with the infamous forgeries known as the "Hitler Diaries" that fooled both Stern magazine and The Sunday Times. The title refers to the nonsense word Chaplin used in his anti-Nazi spoof THE GREAT DICTATOR. Nominated for the 1992 Best Foreign Film Oscar.
COM   106 min wScrn   VIDrel: ARTIF/20TH   V/sur

SCI-FIGHTERS *
18
Peter Svatek USA
1996
Roddy Piper, Jayne Heitmeyer, Billy Drago
Fantasy adventure set in the year 2009 A.D. with the dead inmate of an off-world prison being shipped back to Earth after apparently being killed in a riot. However, it would appear that only an empty coffin was returned. A tedious SF outing that throws away its one interesting idea (involving a deadly virus that causes people to explode).
FAN   91 min   VIDrel: HIFLI/SONOP   V

SCISSORS **
18
Frank De Felitta USA
1990
Sharon Stone, Steve Railsback, Michelle Philips, Ronny Cox, Lany Moss, Vicki Frederick, Austin Kelly, Albert Powell, Jesse Garcia, Will Leskin, Ivy Jones, Laura Ann Caulfield, Ed Crick, Hal Riddle, Ivy Bethune, Jim Shankman
A young woman is attacked in her apartment by an intruder but manages to stab her assailant with a pair of scissors, forcing him to flee. He survives to mount a campaign of terror that almost destroys her sanity. Unfortunately, very little suspense is generated by this intriguing albeit distasteful idea, in what is essentially a slow and feebly plotted thriller that appears to be ever so slightly inspired by GASLIGHT.
THR   91 min Cut (11 sec – ort 105 min)
VIDrel: FIRST/SONOP   V

SCOOP **
PG
Gavin Millar UK
1986
Denholm Elliott, Michael Hordern, Michael Maloney, Herbert Lom, Jack Shepherd, Nicola Pagett, Donald Pleasence, Renee Soutendijk, Robert Eddison, Helen Lindsay, Rosamund Greenwood, Victoria Hasted, Nicholas Le Prevost
In 1938 a young writer is mistaken for a successful novelist of the same name and takes a post as special correspondent to cover the war that's breaking out in the African country of Ishmaelia. A series of events are set in motion by a meeting with an old friend and he eventually emerges with a valuable scoop.
DRA   120 min   VIDrel: VCC L/A   V

SCORCHERS **
18
David Beaird USA
1991
Faye Dunaway, Denholm Elliott, James Earl Jones, Emily Lloyd, Jennifer Tilly, Leland Crooke, James Wilder, Anthony Geary, Luke Perry, Kevin Michael Brown, Michael Covert, Saxon Trainor, Steve LeFleur, Jonno Frisberg
In a sleepy bayou town, the lives of three women become interlinked due to their sexual and marital problems. One is the town whore, another the wife of one of her clients, and the third a newlywed who is terrified of sex. A fairly uninteresting stab at an erotic drama, with screenplay by Beaird.
DRA   77 min (ort 85 min)   VIDrel: 20TH/TECH   V/sh
Boa: play by David Beaird.

SCOTT OF THE ANTARCTIC ***
U
Charles Frend UK
1948
John Mills, Derek Bond, Harold Warrender, James Robertson Justice, Kenneth More, Reginald Beckwith, James McKechnie, John Gregson, Norman Williams, Barry Letts, Clive Morton, Anne Firth, Diana Churchill, Christopher Lee
An account of the ill-prepared and ill-fated British expedition to the South Pole in 1912, which perished largely because of a lack of professionalism as compared to the Norwegians. Ignore the stiff-upper-lip syndrome so typical of British films of this genre, and this becomes quite a decent film. Written by Mary Hayley Bell (the wife of Mills), Ivor Montagu and Walter Meade, and with music by Ralph Vaughan Williams.
DRA   105 min (ort 111 min)   VIDrel: WHV L/A   V/h

SCOUNDREL, THE ***
15
Jean-Paul Rappeneau FRANCE
1971
Jean-Paul Belmondo, Marlene Jobert, Michel Auclair, Sami Frey, Laura Antonelli, Pierre Brasseur
A Frenchman flees to the States for a spell after killing an aristocrat, but returns a while later with the intention of divorcing his wife. However, he has second thoughts about this when he sees others showing an interest in her. Eventually they are reconciled, and achieve a measure of security as titled aristocrats in their own right. Set in the 18th century, this boisterous action-comedy benefits greatly from strong period detail and a lively cast.
Aka: LES MARIES DE L'AN 2; LES MARIES DE L'AN DEUX
A/AD   109 min dubbed   VIDrel: ARROW/RTM   V

SCOUT, THE **
12
Michael Ritchie USA
1994
Albert Brooks, Brendan Fraser, Dianne Wiest, Anne Twomey, Lane Smith, Michael Rapaport, Barry Shabaka Henley, John Capodice, Louis Giovannetti, Ralph Drischell, Stephen Demek, Brett Rickaby, Jack Rader, Marcia Rood, Steve Eastin
So-so comedy set in the world of baseball where a talent discovers a phenomenally talented player when he goes south of the border to Mexico. He so arranged things that his protege is signed up almost immediately, despite the fact that he is more than a little mentally unbalanced, and thus requires regular psychiatric counselling.
COM   97 min cC   VIDrel: 20TH/FOXVID   V/sh

SCREAM **
18
Wes Craven USA
1996
David Arquette, Neve Campbell, Courtney Cox, Matthew Lillard, Rose McGowan, Skeet Ulrich, Drew Barrymore, Roger Jackson, Kevin Patrick Walls, Carla Hatley, Lawrence Hecht, W. Earl Brown, Lois Saunders, Joseph Whipp
Replete with all sorts of references to films of this genre, the story opens with a murderous prankster playing a nasty cat-and-mouse game with a young woman, that requires her to identify movie trivia from horror films. Having eventually murdered her, he then moves on to other victims, until the final resolution and obligatory plot twist. Technically excellent, it is let down by its slow pacing and illogicality. Screenplay is by Kevin Williamson.
HOR   111 min   CINrel

SCREAM AND SCREAM AGAIN **
18
Gordon Hessler UK
1969
Vincent Price, Christopher Lee, Peter Cushing, Judy Huxtable, Alfred Marks, Peter Sallis, Anthony Newlands, David Lodge, Uta Levka, Christopher Matthews, Judy Bloom, Clifford Earl, Kenneth Benda, Yutte Stensgaard
Confused and disjointed film about a mad scientist's attempts to create a race of super-beings, by using the limbs and organs taken from kidnapped people. Lurid and quite unpleasant, but there are occasional flashes of power. Written by Christopher Wicking.
Aka: SCREAMER
HOR   91 min (ort 95 min)   VIDrel: COLUM/SONOP   V
Boa: novel The Disorientated Man by Peter Saxon.

SCREAMERS ***
18
Christian Duguay CANADA/JAPAN/USA
1995
Peter Weller, Roy Dupuis, Jennifer Rubin, Andy Lauer, Charles Powell, Ron White, Michael Caloz, Liliana Komorowska, Jason Cavalier, Leni Parker, Sylvain Masse, Bruce Boa, Tom Berry, Henry Ramer
The year is 2078 and two opposing forces battle for supremacy on a distant mining planet, where one side has perfected small killer robots or "screamers" that have unwisely, now been given the power to self-replicate. Dick's brilliant Cold War story is wisely transplanted to avoid looking dated and there are some truly frightening moments, even though the film as a whole does have a low-budget look about it.
FAN   104 min (ort 109 min) cC
VIDrel: COLUM/SONOP; ENCORE/COLUM (LV only)
V/sur LV
Boa: short story Second Variety by Philip K. Dick.

SCREWBALL HOTEL *
18
Rafal Zielinski USA
1988
Michael Bendetti, Andrew Zeller, Jeff Greenman, Kelly Monteith
The manager of a seedy Miami hotel has a month to find $2,500,000 before gangsters take it over. His one faint hope is to stage a "Miss Purity" pageant, but unfortunately he finds "pure" girls to be in short supply. When three military school rejects take jobs there, they set up an illegal casino and then use an inhibition-releasing drug to stage a wild party. A brash, vulgar comedy, the third film in the "Screwball" series.
COM   97 min (ort 101 min)
VIDrel: POLY/POLYREC/BRAVE L/A   V

SCREWBALLS *
Rafal Zielinski CANADA   18   1982
Peter Keleghan, Linda Speciale, Alan Daveau, Linda Shayne, Kent Deuters, Jim Coburn, Jason Warren, Terrea Foster, Donnie Bowes, Kimberly Brooks, Nicky Fylan, Paula Farmer, Joe Crozier, Heather Smith, Nola Whale, John Fox
Routine comedy about sex-mad adolescents who are determined to get a topless view of a particular young girl. Made with neither wit nor care and followed by two sequels.
COM   76 min (ort 80 min)
VIDrel: SPEAR/SONOP/CALECO   V

SCREWBALLS 2: LOOSE SCREWS *
Rafal Zielinski CANADA   18   1985
Bryan Genesse, Lance Van Der Kolk, Annie McAutey, Alan Deveau, Jason Warren, Cyd Belliveau, Mike MacDonald, Karen Wood, Annie McAuley, Liz Green, Beth Gondek, Deborah Lobdan, Stephanie Sulik, Terrea Öster, Wayne Fleming
A production-line sequel with our four heroes going to a summer school and getting up to the usual pranks. Followed by SCREWBALL HOTEL.
Aka: LOOSE SCREWS; SUMMER SCHOOL
COM   88 min   VIDrel: SPEAR/SONOP/CALECO   V

SCROOGE ***
Brian Desmond-Hurst UK   U   1951
Alastair Sim, Mervyn Johns, Kathleen Harrison, Jack Warner, Hermione Baddeley, Clifford Millison, Michael Hordern, George Cole, Rona Anderson, John Charlesworth, Glyn Dearman, Francis De Wolff, Carol Marsh, Brian Worth
A lively, vivid and highly enjoyable version of this classic tale, with fine performances and intelligent direction and casting. Dickens's tale has been adapted many times, but this work is without doubt one of the most memorable.
DRA   86 min (ort 93 min) B/W coVer   VIDrel: DDVID   V
Boa: novella A Christmas Carol by Charles Dickens.

SCROOGE ***
Ronald Neame UK   U   1970
Albert Finney, Alec Guinness, Edith Evans, Kenneth More, Michael Medwin, Laurence Naismith, David Collings, Richard Beaumont, Kay Walsh, Anton Rodgers, Suzanne Neve, Frances Cuka, Derek Francis, Roy Kinnear, Mary Peach
An enjoyable, exuberant musical version of this seasonal story, with Finney giving a lovely overblown performance as dear old Mr Scrooge, who learns the meaning of Christmas from a trio of ghosts. Guinness makes a wonderful Marley, only the poor songs of Leslie Bricusse let it down. The sets were designed by Terry Marsh, and Ronald Searle did the title design. Richard Harris and Rex Harrison were both sought for the role before Finney.
MUS   117 min (ort 118 min)   VIDrel: 20TH/TECH   V/sur
Boa: novella A Christmas Carol by Charles Dickens.

SCROOGED ***
Richard Donner USA   PG   1988
Bill Murray, Karen Allen, John Forsythe, David Johanson, Carol Kane, John Glover, Bobcat Goldthwait, Robert Mitchum, Michael J. Pollard, Nicholas Phillips, Alfre Woodard, Mabel King, John Murray, Jamie Farr, Robert Goulet
This amusing updating of the Dickens tale has Murray cast as a ruthless and small-minded TV network executive out to spread gloom and misery amongst his employees and the public at large. Following a visit from his dead former boss, he makes a half-hearted effort to mend his ways, but is only really convinced of the need to change with the arrival of the Ghosts of Past, Present and Future. See also MIRACLE AT CHRIST-MAS: EBBIE'S STORY.
COM   97 min (ort 101 min)
VIDrel: 4-FRONT/POLYREC/CIC L/A   V/dm
Boa: novella A Christmas Carol by Charles Dickens.

SCRUBBERS *
Mai Zetterling UK   15   1982
Chrissie Cotterill, Amanda York, Kate Ingram, Elizabeth Edmonds, Eva Motley, Kathy Burke, Amanda Symonds, Debbie Bishop, Imogen Bain, Honey Bane, Camille Davis, Rachael Weaver, Dawn Archibald, Faith Tingle, Anna McKeown
An unrelenting look at the enclosed world of a girls' borstal, with all its squalor and hopelessness. A film that provides neither entertainment nor education. Written by Mai Zetterling, Roy Minton and Jeremy Watt. See SCUM, which covers similar ground.
DRA   89 min (ort 93 min)   VIDrel: FABFIL/SPEAR   V

SCRUPLES **
Alan J. Levi/Hy Averback USA   PG   1980
Lindsay Wagner, Barry Bostwick, Marie-France Pisier, Efrem Zimbalist Jr, Connie Stevens, Nick Mancuso, Robert Reed, Gene Tierney, Louise Latham, Paul Carr, Genvieve, Michael Callan, Gary Graham, Sarah Marshall, Milton Selzer
After her wealthy and elderly tycoon husband dies from illness, a woman tries to rebuild her life by opening a smart Beverly Hills boutique. To help her in this venture, she hires a fashion designer and a fashion photographer. The and loves of this trio provide the basis for this glossy and vacuous mini-saga. In 1981 a made-for TV film continued this enthralling story.
DRA   275 min (ort 279 min) mTV   VIDrel: WHV   V
Boa: novel By Judith Krantz.

SCULPTRESS, THE ***
UK   18   1996
Pauline Quirke, Caroline Goodall, Christopher Fulford
Story of a convicted murderess, nicknamed the "sculptress" due to the fact that she allegedly dismembered her mother and sister after having killed them. However, a journalist visits her in prison to gather material for a book, and becomes convinced of her innocence.
DRA   180 min mTV   VIDrel: CHRYS/CARL   V
Boa: novel by Minette Walters.

SCUM **
Alan Clarke UK   18   1979
Ray Winstone, Mick Ford, John Blundell, Julian Firth, Phil Daniels, John Fowler, Ray Burdis, Patrick Murray, Herbert Norville, George Winter, Alrick Riley, Peter Francis, Philip Da Costa, Perry Benson, Alan Igbon, Andrew Paul
A graphic account of the horrors of a boys' borstal that was orig-inally made as a TV play and then banned by the BBC. Not for the squeamish. See also SCRUBBERS for a look at a female insti-tution.
DRA   92 min (ort 98 min)
VIDrel: 4-FRONT/POLYREC/ODY   V/sh
Boa: TV play by Roy Minton.

SCUMBUSTERS *
Gorman Bechard USA   18   1987
Debi Thibeault, Karen Nielson, Lisa Schmidt, Ruth Collins, Simone, Griffin O'Neal, Christina Whitaker, Elizabeth Kaitan, Nick Cassavetes, Jamie Bozian, Tammara Souza, Mike Muscat, Patti Astor, David Marsch, Clayton Lancey
A group of cheerleaders take the law into their own hands when one of their number is raped. They enjoy the experience so much they decide to form a female vigilante group to rid the streets of undesirables, using hatchets and chainsaws for this purpose. A clumsy mixture of bloodshed and slapstick, this noisome film was intended to be a horror spoof, but winds up being merely repellent.
Aka: ASSAULT OF THE KILLER BIMBOS; CEMETERY HIGH; HACK 'EM HIGH
HOR   80 min (ort 90 min)   VIDrel: ALLIED/RTM   V

SEA CHASE, THE **
John Farrow USA   U   1955
John Wayne, Lana Turner, Tab Hunter, James Arness, David Farrar, Richard (Dick) Davalos, Lyle Bettger, John Qualen, Wilton Graff, Paul Fix, Luis Van Rooten, Peter Whitney, Alan Hale Jr, Lowell Gilmore, John Doucette, Alan Lee
At the outbreak of WW2 in 1939, a German freight ship gets trapped in Sydney harbour. Eventually the captain decides to make a run for it and heads for South America and freedom, but the Royal Navy is not far behind. An offbeat adventure, but one that suffers badly from its pedestrian treatment and serious miscasting, not least being Wayne as the freighter's anti-Nazi captain.
A/AD   112 min (ort 117 min) wScrn   VIDrel: WHV   V/sh
Boa: novel by Andrew Geer.

SEA HAWK, THE ****
Michael Curtiz USA   U   1940
Errol Flynn, Flora Robson, Brenda Marshall, Henry Daniell, Claude Rains, Donald Crisp, Alan Hale, Una O'Connor, James Stephenson, William Landigan, Gilbert Roland, Julien Mitchell, Montague Love, J.M. Kerrigan, Fritz Leiber
A splendid swashbuckling classic, with Flynn in fine form in this tale of intrigue at the court of Elizabeth I, where the Spanish Ambassador is plotting with the queen's enemies to seize her throne for his master, King Phillip II. Meanwhile, Flynn as a

privateer is causing both the Spanish and his sovereign no end of trouble, by capturing Spanish treasure ships. Written by Seton I. Miller, and with a rousing score by Erich Wolfgang Korngold.
A/AD   122 min B/W   VIDrel: WHV   V
Boa: novel by Rafael Sabatini.

SEA OF LOVE ***   18
Harold Becker USA   1989
Al Pacino, Ellen Barkin, John Goodman, Michael Rooker, William Hickey, Richard Jenkins, Christine Estabrook, Barbara Baxley, Patricia Barry, Jacqueline Brookes, Paul Calderon, Gene Canfield, John Spencer, Mark Phelan, Luis Ramos
When a smart New York detective in the throes of a mid-life crisis, begins investigating a series of killings apparently linked to lonely hearts small ads, he finds himself falling for the chief suspect. A tough and sexy urban thriller, held together by a knockout performance from Pacino. The script is by Richard Price.
THR   108 min (ort 112 min)
VIDrel: 4-FRONT/POLYREC/CIC   V/sur

SEA WIFE **   PG
Bob McNaught UK   1957
Joan Collins, Richard Burton, Basil Sydney, Ronald Squire, Cy Grant, Harold Goodwin, Roddy Hughes, Gibb McLaughlin, Lloyd Lamble, Nicholas Hannen, Ronald Adam, Beatrice Varley, Joan Hickson, Otokichi Ikeda, Tenji Takagi
Three men and a girl survive in a dinghy when their ship is torpedoed off Singapore during WW2, but are unaware that the lady in question is a nun. Done in a highly awkward flashback style, this uneven and fairly ineffectual film is a poor adaptation of a rather minor novel.
Aka: SEA WYF AND BISCUIT
DRA   82 min   VIDrel: 20TH/TECH L/A   V/h
Boa: novel Seawyf And Biscuit by J.D. Scott.

SEA WOLF, THE ***   PG
Michael Curtiz USA   1941
Edward G. Robinson, John Garfield, Ida Lupino, Alexander Knox, Gene Lockhart, Barry Fitzgerald, Stanley Ridges, David Bruce, Howard Da Silva, Frank Lackteen, Ralf Harolde, Louis Mason, Dutch Hendrian, Cliff Clark
The tyrannical and unbalanced captain of a freighter picks up three survivors from a ferry crash but holds them captive. Robinson gives an intense and utterly believable performance in London's brutal tale, but the action takes second place to the dialogue rather too often. Remade several times since, most notably as "Wolf Larsen".
DRA   84 min (ort 100 min) B/W   VIDrel: WHV   V
Boa: novel by Jack London.

SEA WOLF, THE **   PG
Michael Anderson USA   1992
Charles Bronson, Catherine Mary Stewart, Christopher Reeve, Len Cariou, Marc Singer, Clive Revill, Shane Kelly, Gary Chalk, Tom McBeath, Sean Barrett, Dee Jay Jackson, Eli Gabay, Russell J. Roberts, Bill Croft, John Novak
Unimpressed dramatisation of London's story of a demented sea captain who is animated by a brutal philosophy of survival of the fittest and most ruthless. Having rescued two survivors from a sunken ferry, he keeps them virtual prisoners aboard his freighter and subjects them to ample indignities until fate takes a hand. Vastly inferior to the 1941 version from which some footage was lifted and added after colourisation.
DRA   89 min (ort 93 min) mCab   VIDrel: FIRST/SONOP   V
Boa: novel by Jack London.

SEA WOLVES, THE **   PG
Andrew V. McLaglen SWITZERLAND/UK/USA   1980
Gregory Peck, Roger Moore, David Niven, Trevor Howard, Barbara Kellerman, Patrick Macnee, Patrick Allen, Bernard Archard, Martin Benson, Faith Brook, Allan Cuthbertson, Kenneth Griffith, Donald Houston, Glyn Houston, Percy Herbert
A retired British cavalry unit in India undertakes a hazardous sabotage operation against a German ship, in the neutral Portuguese enclave of Goa in 1943. More of a stiff-jointed romp than a war film, with a curiously old-fashioned look to it, but mildly entertaining in its way. Written by Reginald Rose and based to some extent on a true story.
WAR   115 min (ort 122 min)   VIDrel: VCC L/A   V
Boa: novel Boarding Party by James Leasor.

SEALED TRAIN, THE **   PG
Damiano Damiani FRANCE/ITALY/SPAIN/
WEST GERMANY   1987
Ben Kingsley, Jason Connery, Timothy West, Dominique Sanda, Leslie Caron, Gunther Maria Halmer, Robin Lermitte, Peter Whitman, Paolo Bonacelli, Xavier Elorriaga, Dagmar Schwarz, Jeannine Mestre, Martin Sbragia, Hans Michael Rehberg
A sluggish and none too interesting docu-drama about Lenin's return to Russia in 1917, when the German government sent him back in a sealed train in the hope that he would successfully lead an uprising that would take Russia out of the war. Though this was indeed an event that shaped history, little effort has been made to imbue the subject matter with the dramatic tension it deserves. Another one of those very typical Euro-pudding productions.
Aka: LENIN... THE TRAIN; LENIN, THE SEALED TRAIN
DRA   124 min mTV   VIDrel: L/A   V
Boa: book by Michael Pearson.

SEARCH AND DESTROY **   18
David Salle USA   1994
Griffin Dunne, Rosanna Arquette, Dennis Hopper, Illeana Douglas, Ethan Hawke, John Turturro, Christopher Walken
Pursued both by the I.R.S. and his wife, who is suing for divorce, a man tries to get himself set up as a film-maker, having become obsessed with a TV therapist's philosophy of winning, which he thinks he can turn into a successful movie. Not at all the action-thriller one might have expected from the title and publicity, this is a cluttered and unfocused attack on the American way of life, occasionally deeply sardonic, but not very often. This was Salle's debut film.
COM   87 min (ort 91 min)   VIDrel: 20TH/FOXVID   V/sh
Boa: play by Howard Korder.

SEARCH FOR GRACE **   15
Sam Pillsbury USA   1994
Lisa Hartman Black, Ken Wahl, Richard Masur, Suzzzane Douglas, Don Michael Paul, Lori Lindberg, Evan Wood, Ann Donnell, Delbert C. Taylor, Mark Miller, Alex Van, Charles McLawlorn, Mark Miller, Stephanie Wood, D.L. Anderson
A career woman starts to get anxiety attacks when her boyfriend proposes, and undergoes a session on a psychiatrist's couch, where she is found to have lived a previous life. An adequate variant on a theme explored in ON A CLEAR DAY YOU CAN SEE FOREVER.
DRA   89 min mTV   VIDrel: 20TH/TECH   V

SEARCH FOR JUSTICE **   15
Noel Nosseck USA
Peggy Lipton, Danica McKellar, Bruce Weitz, Susan Ruttan
A divorced high school graduate takes lodgings with a married couple, but is found dead. Her mother suspects foul play and mounts an investigation of her own.
DRA   92 min   VIDrel: ODY/SNOOP   V/sh

SEARCHERS, THE ****   U
John Ford USA   1956
John Wayne, Jeffrey Hunter, Natalie Wood, Vera Miles, Ward Bond, John Qualen, Henry Brandon, Harry Carey Jr, Olive Carey, Ken Curtis, Antonio Worden, Lana Wood, Walter Coy, Dorothy Jordan, Pippa Scott, Patrick Wayne
A classic film describing the years a man spends searching for his niece who has been kidnapped by Indians. A dramatic and rather solemn film, written by Frank S. Nugent and with music by Max Steiner. The longer running time of the wScrn version is due to the inclusion of a behind-the-scenes documentary.
WES   114 min (ort 119 min); 140 min wScrn (special edition)
VIDrel: WHV   V/h
Boa: novel by Alan Le May.

SEASON OF CHANGE **   (PG)
Robin P. Murray USA   1994
Nicholle Tom, Michael Madsen, Ethan Randall, Hoyt Axton, Jo Anderson, Bette Ford, Jeannie Bisgnano, Kira Endsley, Robyn Mundell, Kevin McBride, Charlene Campbell, Veronica Brown, Muzzy Lambert, Gloria Malgavini, Scott Turner
Drama set in Montana just after WW2, where a thirteen-year-old girl hopes that life will get back to normal now that her father is home, but finding that her dad cannot adjust to civilian life and has an affair. Meanwhile, she has problems of her own, having developed a crush on a young fellow her mother disapproves of. A pleasant coming-of-age story, nicely paced and well acted.
DRA   91 min   SATrel: SKY MOVIES

SEASON OF THE WITCH *
18
George A. Romero USA
1972
Jan White, Ray Laine, Joedda McClain, Anne Muffly, Bill Thunhurst, Neil Fisher, Esther Lapidus, Dan Mallinger, Ken Peters, Virginia Greenwald
A middle-aged woman becomes interested in a witches' coven which she sees as a way of curing herself of feelings of insecurity. She begins dabbling in black magic and later on mistakes her husband for a prowler and shoots him, whereupon she is received into said coven. An uncertain and slack story of poor acting and little atmosphere.
Aka: HUNGRY WIVES; JACK'S WIFE
HOR   104 min (ort 130 min)   VIDrel: REDEM/RTM   V

SEASONS OF THE HEART **
12
Lee Grant USA
1994
Carol Burnett, George Segal, Eric Lloyd, Harvey Atkin, Jill Teed, Malcolm McDowell, Lisa Richards, Jim Jansen, Krista Hawley, Margaret Sophie Stein, Nicu Branza, Florence Paterson, Scott Marlowe, Harvey Atkin, Tom Melissis, Bill Dow
An elderly book editor has lived happily with her partner for many years, but this cosy set-up is spoilt when she finds herself obliged to take care of her grandson, the boy's drug-addict mother having booked herself into a rehabilitation clinic. Initial hostility gives way to affection, and when the boy's mother gets out of the clinic, conflict over the youngster's future care ensues. Fair.
DRA   90 min (ort 92 min) mTV   VIDrel: MARQ/20TH   V/s

SEAVIEW KNIGHTS **
15
UK
James Bolan, Clive Darby, Sarah Alexander, Hildegard Neil
Ealing-style British comedy about a man who is struck on the head by a painting whist staying at a guest house in Blackpool, and wakes up thinking he is King Arthur.
COM   100 min wScrn   VIDrel: GUER/PINN   V

SEBASTIANE *
18
Derek Jarman/Paul Humfress UK
1976
Leonardo Treviglio, Barney James, Neil Kennedy, Richard Warwick, Donald Dunham, Ken Hicks, Junusz Romanov, Steffano Massari, David Finbar, Gerald Incandela, Robert Medley, Graham Cracker, Lindsay Kemp and Troupe
An account of the martyrdom of St Sebastiane with the emphasis on the homosexual aspects of Roman life. If for nothing else, memorable for being the only film made in Latin (it carries English subtitles).
DRA   81 min (Latin dialogue – ort 86 min)
VIDrel: TART/20TH   V

SECOND BEST **
12
Chris Menges UK
1993
William Hurt, Chris Cleary Miles, Nathan Yupp, Keith Allen, Doris Irving, James Warrior, Jane Horrocks, Alfred Lynch, Rachel Freeman, Gus Troakes, Mossie Smith, Shaun Dingwell, Paul Wilson, Alan Cumming, Jake Owen, Prunella Scales
An emotionally repressed postmaster who longs to be a father but is unmarried, solves this problem by adopting a disturbed ten-year-old boy. As both have their own emotional problems, matters do not initially look promising. A flat and unimpressive drama with a contrived feelgood factor that does very little to generate interest, despite fine work from a capable cast. Screenplay is by Cook.
DRA   101 min (ort 105 min) cC   VIDrel: WHV   V/sur
Boa: novel by David Cook.

SECOND SIGHT *
PG
Joel Zwick USA
1989
John Larroquette, Bronson Pinchot, Bess Armstrong, Stuart Pankin, John Schuck, James Tolkan, William Prince, Christine Estabrook, Cornelia Guest, Michael Lombard, Marisol Massey, Adam LeFevre, Andrew Mutnick, Ron Taylor
Larroquette is the boss of a detective agency that solves its cases with the help of psychic Pinchot. A love interest sub-plot involving a nun is thrown in as an afterthought. A fairly mindless film of amusing bits and pieces that deservedly died at the box office. As a TV sitcom it might well have done better.
COM   81 min (ort 84 min)   VIDrel: MGM/WHV L/A   V/sh

SECONDS OUT **
(PG)
Bruce MacDonald UK
1992
Steven Waddington, Tom Bell, Colum Convey, Derek Newark, Clive Russell, Jack Watson, Nick Brimble, Otis Munyang'iri, Frank Mills,
Amelda Brown, Thomas Craig, Timothy Barlow, Alan Talbot, Jane Wymark, Roy Heather, Julia Chambers
After he is falsely accused of rape (having been set up) a promising young boxer finds that the only venue for his talents is to be found in the murky world of unlicensed boxing. There is not enough of a story here to allow the characters to develop, though Bell gives a nice performance as our young pugilist's trainer and mentor. A tendency to show the fight sequences in slow motion is both irritating and pretentious.
DRA   90 min mTV   TVrel: BBC

SECRET ADVENTURES OF TOM THUMB, THE ***
12
The Bolexbrothers UK
1993
Nick Upton, Deborah Collard, Frank Passingham, John Schofield, Mike Gifford, Robert Heath, George Brandt, Andy Davis, Dave Alex Riddett, Andy Joyce, Richard Goleszowski plus voices of: Brett Lane, Helen Veysey, Paul Veysey, Tim Hands
A mixture of animation and pixilation (in which live-action images are animated frame-by-frame) that provides an alternative version of this famous fairytale. Here, young Tom is abducted from his home by people whose intention is to use him in some rather nasty experiments.
ANIM   60 min   VIDrel: MANGA/SONOP   V/sh

SECRET AGENT, THE **
U
Alfred Hitchcock UK
1936
Peter Lorre, Robert Young, Madeleine Carroll, John Gielgud, Percy Marmont, Lilli Palmer, Florence Kahn, Charles Carson, Michel Saint-Dennis, Andrea Malandrinos, Tom Helmore, Michael Redgrave, Howard Marion Crawford
A pair of secret agents, posing as husband and wife, are sent to Switzerland on a mission of murder, in this blend of comedy and thriller genres. A curious film for Hitchcock, and one that doesn't work all that well, although there are some funny moments. Written by Charles Bennett and with musical direction by Louis Levy. The photography is by Bernard Knowles.
THR   82 min (ort 86 min) B/W   VIDrel: VCC L/A   V
Boa: play by Campbell Dixon/short stories The Hairless Mexican and Triton from "Ashenden" by William Somerset Maugham.

SECRET AGENT CLUB, THE **
PG
John Murlowski USA
1996
Hulk Hogan, Richard Moll, Matthew McCurley, Edward Albert, Lyman Ward, James Hong, Barry Bostwick, Lesley-Anne Down, Jack Nance, M.C. Gainey, Maukice Woods, Jimmy Pham, Ashley Power, Danny McCue, Brian Yandrisovitz, Vachik Mangassarian
A youngster is totally unaware that his scruffy slob of a toy-store owning father is in reality a secret agent. But when his dad is kidnapped, he has to get together with a gang of other kids to mount a rescue. An irritatingly cute foray into the kid's action genre, with little of substance.
JUV   87 min (ort 90 min)   VIDrel: GUILD/FOXVID   V

SECRET EXECUTIONERS **
18
HONG KONG
198-
Wong Chen-Li, Jim Norris, Peggy Min
A martial arts expert forms a little band of fighters in an effort to free the city from the grip of a criminal gang. The gang responds by hiring some American assassins to remove this threat to their operations.
MAR   86 min (ort 90 min)   VIDrel: IMPENT   V

SECRET FRIENDS **
15
Dennis Potter UK
1992
Alan Bates, Gina Bellman, Frances Barber, Tony Doyle, Joanna David, Colin Jeavons, Rowena Cooper, Ian McNeice, Davyd Harries, Niven Boyd, Martin Whiting, Roy Hamilton, Nicholas Russell-Pavier, Colin Ryan, David Swift
A mentally unstable artist is haunted by murderous visions of his wife and events from his strict and repressive upbringing. On a train, he forgets his identity and as he struggles to find out who he is, finds himself tracing a path that may end in either enlightenment or madness. A blend of fantasy and reality whose boundaries are deliberately blurred, the film seems ever on the verge of revealing something profound, but never does so. Inspired by Potter's novel.
THR   97 min mTV   VIDrel: MIA/DISC   V/sur
Boa: novel Ticket To Ride by Dennis Potter.

SECRET GAMES **
18
Alexander Gregory Hippolyte USA
1991
Martin Hewitt, Billy Drago, Delia Sheppard, Michelle Brin, Catya

Sassoon, Sabrina Mesko, Kimberly Williams, Monique Parent, Juliet James, Gary Kaspar, Christian Bochner, Alison Armitage, Craig Stepp, Lana Wilson, John Tripp
As with MIRROR IMAGES and CARNAL CRIMES (two other films by Hippolyte) this is little more than a glossy sexploitation film. Its slight plot deals with an unfulfilled married woman who drifts into prostitution, and embarks on an odyssey of sexual adventure. By way of a sub-plot, one of her less well balanced clients develops an obsessive interest in her, and threatens to expose her when she rejects him. Nice locations help sustain one's interest.
DRA 92 min (ort 105 min) VIDrel: POLY/POLYREC/MED V/sh

SECRET GAMES 2: THE ESCORT **　　　　18
Alexander Gregory Hippolyte USA　　　　1993
Martin Hewitt, Marin Leroux, Amy Rochelle, Sara Suzanne Brown, Thomas Milan, Holly Spencer, Jennifer Place, Mark T. Paladini, Sherry Patterson, Gregg Christine, Bill Williams
While his wife is away on a vacation trip, a performance artist gets himself deeply embroiled with an escort service thanks to which he has a series of erotic encounters. However, this leads to an unexpected complications when he falls in love with a young prostitute who spurns his advances.
DRA 91 min VIDrel: COLUM/SONOP V

SECRET GAMES 3 **　　　　18
Alexander Gregory Hippolyte USA　　　　1993
Woody Brown, Rochelle Swanson, Brenda Swanson, May Karasun, Bob Delegall, Dean Scofield, Mark Davenport, Melinda Grieger, Jasae
A sexually bored housewife attends a club where she is paid to live out her fantasies. However, both she and her husband are put in danger when one of its clients becomes obsessed with her. Another average erotic soft-core drama.
DRA 87 min (ort 96 min) VIDrel: MIA/DISC V/h

SECRET GARDEN, THE ***　　　　U
Fred McLeod Wilcox USA　　　　1949
Margaret O'Brien, Dean Stockwell, Herbert Marshall, Gladys Cooper, Elsa Lanchester, Brian Roper, Reginald Owen, Aubrey Mather, George Zucco, Lowell Gilmore, Billy Bevan, Dennis Hoey, Matthew Boulton, Isobel Elsom, Norma Varden
In Victorian times, a young orphan girl goes to live with her morose and secretive uncle. Her friendship with a strange boy and their adventures in a walled garden whose gate is always kept locked, has unexpected consequences but a happy outcome. A charming moral fable; slow, careful and atmospheric. Remade for TV in 1975 and 1987 and for the cinema in 1993.
JUV 88 min (ort 92 min) B/W/Col VIDrel: MGM/WHV L/A V
Boa: novel by Frances Hodgson Burnett.

SECRET GARDEN, THE ***　　　　U
Dorothea Brooking UK　　　　1975
Sarah Hollis Andrews, David Patterson, John Woodnutt
The story of a young girl who lives with her uncle in Yorkshire, and learns of a secret garden where she makes some new friends. A pleasant adaptation of Burnett's famous classic.
JUV 106 min (ort 210 min) mTV VIDrel: BBC/TECH V/h
Boa: novel by Frances Hodgeson Burnett.

SECRET GARDEN, THE ***　　　　U
Agnieszka Holland UK　　　　1993
Kate Maberly, Heydon Prowse, Andrew Knott, Maggie Smith, Laura Crossley, John Lynch, Walter Sparrow, Irene Jacob, Frank baker, Valerie Hill, Andrea Pickering, Peter Moreton, Arthur Speckley, Colin Bruce, Pasan Singh, Eileen Page
Capable and strongly adapted version of this much-loved (and filmed) tale of a young girl who has an unforeseen effect on the lives of those around her when she sets about restoring an abandoned garden.
DRA 97 min (ort 102 min) cC VIDrel: WHV V/dm
Boa: novel by Frances Hodgson Burnett.

SECRET INVASION, THE ***　　　　(PG)
Roger Corman USA　　　　1964
Stewart Granger, Raf Vallone, Mickey Rooney, Edd Byrnes, Henry Silva, Mia Massini, William Campbell, Peter Coe
A WW2 story, telling of five convicts who are given the chance to do their bit by working behind enemy lines in Yugoslavia. An exciting and well-handled tale, exploring a theme that was to be used a little later in THE DIRTY DOZEN.
WAR 94 min (ort 98 min) VIDrel: WHV L/A V

SECRET LIES **　　　　18
Sam Irvin USA　　　　1993
C. Thomas Howell, Linda Forentino, Nancy Allen
A man falls in love with an actress, but does not suspect that she is wanted for murder and that his association is a dangerous one. An adequate thriller of the standard kind.
THR 89 min VIDrel: PROMARK/HIFLI V/h

SECRET LIFE OF WALTER MITTY, THE ***　　　　U
Norman Z. McLeod USA　　　　1947
Danny Kaye, Virginia Mayo, Boris Karloff, Fay Bainter, Ann Rutherford, Florence Bates, Thurston Hall, Gordon Jones, Reginald Denny, Fritz Feld, Henry Corden, Doris Lloyd, Frank Reicher, Milton Parsons, Mary Brewer
Danny Kaye is splendid as an incorrigible daydreamer in this very, very loose adaptation of a James Thurber short story. He spends his days dreaming of heroic deeds but is a little taken aback when he finds himself involved in a real-life spy plot. The early fantasy sequences are the highlight of the film, and the later complexities are far less amusing.
COM 110 min VIDrel: VCC/DISC/COLUM V
Boa: short story by James Thurber.

SECRET MISSION **　　　　U
Harold French UK　　　　1944 (relased 1947)
James Mason, Hugh Williams, Carla Lehmann, Roland Culver, Michael Wilding, Nancy Price, Percy Walsh, Anita Gombault, David Page, Betty Warren, Nicholas Stuart, Brefni O'Rorke, Karel Stepanek, F.R. Wendhausen, John Salew, Herbert Lom
A group of British intelligence agents are sent on a vital mission to Nazi-occupied France to determine the size and strength of enemy forces being assembled for the invasion of Britain. They adopt the guise of champagne sellers and penetrate Nazi headquarters but are revealed after they depart with the information they need. Standard wartime heroics, slow-paced but supported by capable performances.
WAR 94 min B/W VIDrel: CARL/TECH V

SECRET NINJA, ROARING TIGER **　　　　18
Godfrey Ho HONG KONG　　　　198-
Dragon Lee, Wong Cheng Li, Jack Lam, Winnie Lui, Petty Suh, Kon Yit So
When the daughter of a millionaire is kidnapped by a Ninja sect, a martial arts master undertakes a perilous mission to rescue her.
MAR 80 min Cut (1 min 37 sec) VIDrel: IMPENT V

SECRET OF MY SUCCESS, THE **　　　　PG
Herbert Ross USA　　　　1987
Michael J. Fox, Helen Slater, Margaret Whitton, Richard Jordan, John Pankow, Christopher Murney, Gerry Bamman, Fred Gwynne, Carol-Ann Susi, Elizabeth Franz, Drew Snyder, Susan Kellermann, Barton Heyman, Mercedes Ruehl, Rex Robbins
An ambitious and clever country boy comes to the big city where he works his way up in the world, encountering various complications along the way. A bright and cheerful comedy that is sustained by an attractive performance from Fox, but eventually runs out of ideas.
COM 106 min (ort 110 min) VIDrel: CIC/SONOP V/sur

SECRET OF NIMH, THE ***　　　　U
Don Bluth USA　　　　1982
Voices of: Derek Jacobi, Dom DeLuise, John Carradine, Peter Strauss, Aldo Ray, Elizabeth Hartman, Hermione Baddeley, Edie McClurg, Shannen Doherty, Wil Wheaton, Jodi Hicks, Lucille Bliss
Quality animated feature produced by a group of defectors from the Disney studios, all about a widowed mouse who fights to save her family homestead from imps and seeks help from super-intelligent rats. Written by Bluth (in his directorial debut), John Pomeroy, Gary Foldman and Will Finn. The music is by Jerry Goldsmith.
ANIM 79 min (ort 82 min) VIDrel: MGM/WHV V/sur
Boa: novel Mrs Frisby and the Rats of Nimh by Robert C. O'Brien.

SECRET OF ROAN INISH, THE ***　　　　PG
John Sayles USA　　　　1993
Mick Lally, Eileen Colgan, John Lynch, Jeni Courtney, Richard Sheridan, Cillian Byrne, Pat Howey, Dave Duffy, Decian Hannigan, Mairead Ni Ghallchoir, Eugene McHugh, Tony Rubini, Michael MacCarthaigh, Fergal McElherron
In a remote Irish fishing village in the 1940s, a young girl comes to stay with her grandparents and grows increasingly

absorbed in the fantastical stories that have grown up around the death of her brother, who as legend has it, has been raised by the Selkies, mythical half-human, half-seal creatures. Set in the 1940s, this unusual and engrossing fantasy is packed with Irish whimsy, is strongly acted and has fine landscape photography.
DRA   103 min   VIDrel: TART/20TH   V/sur
Boa: novel Secret of the Ron Mor Skerry by Rosalie K. Fry.

SECRET PASSION OF ROBERT CLAYTON, THE **   15
E.W. Swackhamer USA   1993
John Mahoney, Scott Valentine, Eve Gordon, Kevin Conroy, Donna Biscoe, Boyce Hobart, Donna Biscoe, Boyce Holleman, Elizabeth Swackhamer, Dana Lee, Joe Dorsey, Davide De Vries, David Dwyer, Tim Ware, Kathryn Firago, Bill Coates
A man returns home to become the local D.A. and finds himself saddled with a sensational case that involves the husband of a former girlfriend, who stands accused of the murder of a stripper. To make matters worse, he finds that his own father is acting as the defence lawyer. Fair.
THR   87 min (ort 91 min) mCab   VIDrel: CIC   V

SECRET RAPTURE, THE **   15
Howard Davies UK   1993
Juliet Stevenson, Joanne Whalley-Kilmer, Penelope Wilton, Neil Pearson, Alan Howard, Robert Howard, Robert Stephens, Milton McRae, Robert Glenister, Richard Long, Finty Williams, Saira Todd, Julia Lane, Philip Voss, Janet Steel
When their father dies, two sisters are left alone with their estranged stepmother, who is planning their destruction, in this capable film version of Hare's play.
DRA   96 min   CINrel
Boa: play by David Hare.

SECRET SEDUCTIONS: PART 1 ***   (18)
USA   1995
Rachel Love, Crystal Gold, P.J. Sparxx, Jill Kelly, Christina West, Ginger Graham, Steve Drake, Tony Martino, Mike Horner
A high-class hooker (Love) has a businessman client who is unable to decide whether to marry his fiancee or start a new life with the hooker, whom he is in love with. But matters are complicates by the fact that his future father-in-law is a wealthy man who is ready to help him with his business interests, but only if the wedding takes place. A good quality erotic story, but one that is only resolved in Part 2.
A   75 min   VIDrel: ONE   V

SECRET SINS OF THE FATHER **   15
Beau Bridges USA   1993
Beau Bridges, Lloyd Bridges, Lee Purcell, Frederick Coffin, Victoria Rowell, Patrika Darbo, Michael McManus, Ed Lauter, Bert Remsen, Mark L. Taylor, Andrew Parks, Lucinda Cunningham, Gene Ross, Robin Pearson Rose, Dorothy Dean
A police chief grows suspicious after the death of his mother and begins his own investigation, learning to his shock that she was murdered.
DRA   88 min (ort 93 min) mTV   VIDrel: HIFLI/SONOP   V/h

SECRET WEAPON *   15
Ian Sharp AUSTRALIA/UK/USA   1989
Griffin Dunne, Karen Allen, Jeroen Krabbe, Stuart Wilson, Joe Petruzzi, John Rhys-Davies, Beian Cox, Ian Mitchell, Patrick Bailey, Ronnie Stevens, Chris Constantinou, Emma Hitching, Oliver Lewis, Michael O'Cruze, Jeff Turman
A dullish dramatisation of the Mordecai Vanunu case, in which an Israeli technician working at a nuclear research facility revealed the existence of a large number of nuclear weapons, before fleeing abroad. The Israeli secret service then devised a plot to abduct him, using a beautiful female agent as bait. All the tension of the original story is dissipated by the film's slow pace and wooden performances.
THR   90 min (ort 119 min) mCab   VIDrel: VCC/DISC L/A   V

SECRET WORLD OF EMMANUELLE, THE **   18
Francis Leroi FRANCE   1992
Sylvia Kristel, Marcella Walerstein
Injured in a car crash, Emmanuelle has to recall her past loves with the aid of a psycho-analyst. This simple plot device is used to introduce us to Walerstein, who plays the title character as a young woman (and does so in all the other films in this extensive, made-for-cable "Emmanuelle" series). As one might

expect, there is precious little plot here, but Walerstein performs well enough and the production values are quite good.
A   77 min (ort 90 min) mCab   VIDrel: CREA/DISC   V

SECRETARY, THE **   15
Andrew Lane USA   1994
Mel Harris, Barry Bostwick, Sheila Kelley, James Russo
A secretary becomes pathologically jealous of her woman boss who is both a mother and a successful career women. Another variation on the psycho from hell theme. See also THE TEMP.
DRA   90 min (ort 94 min)   VIDrel: FIRST/SONOP   V

SECRETS **   15
Peter Hunt USA   1992
Stephanie Beacham, Christopher Plummer, Linda Purl, Gary Collins, Ben Browder, Josie Bissett
The wealthy and successful producer of a top-rated TV soap enjoys a fling with the star of the show. Meanwhile, the rest of the cast are set to suffer various intrigues, petty jealousies and rivalries, culminating in the latest cast member being charged with the murder of his wife. Standard Danielle Steel melodramatics, wrapped up in a script that's contrived, superficial and forgettable.
Aka: DANIELLE STEEL'S SECRETS
DRA   89 min   VIDrel: MIA/DISC/IMPENT   V
Boa: novel by Danielle Steel

SECRETS ***   (12)
Jud Taylor USA   1994
Veronica Hamel, Julie Harris, Richard Kiley, Thomas Gibson, Shae D'Lyn, Reed Diamond, Jessica Bowman, Shannon Eubanks, Ann Bronston, Beatrice Bush, Erin Williby, Rob Brownstein, Rick Warner
A young teenage girl finds herself torn between the love she feels for the elderly housekeeper Carline who raised her, and her charming but ruthlessly manipulative mother Etta. When Carline's granddaughter gets pregnant, Etta tries to convince her to have the girl committed and raise the child. But Etta's own daughter opposes this, and eventually discovers the secret that has made her mother the person she is. A well plotted period drama.
DRA   89 min (ort 92 min)   SATrel: SKY MOVIES
Boa: novel The Other Anna by Barbara Eastman.

SECRETS & LIES ***   15
Mike Leigh UK   1995
Timothy Spall, Phyllis Logan, Brenda Blethyn, Claire Rushbrook, Marianne Jean-Baptiste, Elizabeth Berrington, Michele Austin, Lee Ross, Lesley Manville, Ron Cook, Emma Amos, Brian Bovell, Trevor Laird, Clare Perkins, June Mitchell
Bickerers round the barbecue in another Mike Leigh study of dysfunctional families and friends, one of whom is an adopted black girl, who sets out to track down her real mother after the funeral of her adoptive parents. A bitter-sweet comedy in which the lives of the various odd characters are both intertwined and examined with considerable skill. The film was a hit at Cannes 1996 where it won the Palme D'Or and the Best Actress Award (Brenda Blethyn).
COM   138 min (ort 141 min) cC   VIDrel: FILM4/RTM   V/s

SECRETS OF LADY TRUCKERS *   18
Stu Segall USA   1976
Valdesta, Jake Barnes, Elke Van, Janice Jordan, Catherine Barkley, Edward Roehm, John Alderman
Hookers use a trucking business as a front for their more horizontal activities but find themselves clashing with the local sheriff. A strictly second-rate effort that tried to cash in on the C.B. of the 1970s.
Aka: C.B. HUSTLERS
DRA   73 min (ort 90 min)   VIDrel: FUNNY/SGSVID   V

SECT, THE **   18
Michele Soavi ITALY   1991
Herbert Lom, Kelly Curtis, Maria Angela Giordano, Tomas Arana, Michel Hans Adatte, Carla Cassola, Angelika Maria Boeck, Tomas Arana
As the millennium draws to an end the world is threatened by an evil sect, the members of which have made a pact with Satan to flood the world with evil, and to do this start off with a series of ritual killings. Curtis is the naive schoolteacher who is made a part of their fiendish plans. A charismatic performance from Lom as the cult leader and excellent direction are the chief assets

in a story that in other ways is quite mundane, if rather gruesome.
Aka: DEMON'S 4: THE SECT; DEVIL'S DAUGHTER, THE; LA SETTTA
HOR   112 min   VIDrel: MARQ/QUANT   V

SEDUCED BY EVIL ** (PG)
Tony Wharmby USA   1994
Suzanne Somers, James B. Sikking, John Vargas, Mindy Spence,Nnancy Moonves, Julie Carmen, Doug Coleman, Arthur Baranowski, Roberto Guajardo, Miguel Ortega, Ric San Nicholas, Brooks Tomb
A woman journalist accepts an assignment that takes her to a small town in the Southwest. There, she meets a self-proclaimed healer who becomes obsessed with her and resorts to the black arts, including the ability to change himself in a variety of creatures, in his efforts to possess her.
HOR   85 min (ort 88 min)   mTV   SATrel: MOVIE CHANNEL
Boa: novel Brujo by Lynn Arrington Wolcott.

SEDUCTION ** 18
Michael Ray Rhodes USA   1992
Victoria Principal, John Terry, John O'Hurley, W. Morgan Sheppard, Andreas Katsulas, Joseph Lambie, Richard Heard
A murderous femme fatale enjoys using her charms to seduce and then destroy men, this being her way of paying back the male half of the species for the various liberties they have taken with her in the past. See also BLACK WIDOW for another stab at this theme.
THR   92 min   VIDrel: L/A   V

SEDUCTION OF JOE TYNAN, THE *** 15
Jerry Schatzberg USA   1979
Alan Alda, Meryl Steep, Rip Torn, Barbara Harris, Charles Kimbrough, Melvyn Douglas, Blanche Baker, Adam Ross, Carrie Nye, Chris Arnold, Blanche Baker, Adam Ross, Maureen Anderman, John Badila, Robert Christian, Maurice Copeland
An examination of the corroding effect of political power on integrity and family relationships, with the central figure a liberal senator who has to contend with a variety of pitfalls in his climb to the top. Not quite in the same league as ALL THE KING'S MEN is not bad either, though the climax is a complete letdown. Written by Alda.
DRA   102 min (ort 107 min)   VIDrel: CIC/SONOP L/A   V/h

SEDUCTION OF MARY, THE * 18
Michael Craig USA   1992
Victoria Paris, Mike Horner, Cassidy, Jon Dough, Sharise, Teri Diver, Nick E., Chrissy Ann, Ron Dye
When a married lawyer chances on a book all about sexual fantasies, she gets becomes completely obsessed with the writer, who just happens to be one of his clients. Seeing herself in the role of one of his characters, she dreams about re-enacting some of his fantasies, an activity that may or may not be actually happening. There is an interesting twist at the end of this tale, but the plot is so threadbare one has guessed it long before then.
A   83 min   VIDrel: GROHOM/MAXSCAN   V

SEDUCTION: THE CRUEL WOMAN ** 18
Elfi Mikesch/Monika Treut WEST GERMANY   1985
Mechthild Grossmann, Udo Kier, Sheila McLaughlin, Carola Regnier, Georgette Dee, Peter Weibel
Portrait of a dominatrix and the sado-masochistic games she plays with her circle of friends, lovers and clients, principally a new female slave she puts through her paces. Inspired by Leopold Sacher-Masoch's "Venus in Furs", this is a lesbian reworking of the novel, hampered by a low budget, yet quite intriguing if only as a bizarre examination of perversion.
Aka: VERFÜHRUNG: DIE GRAUSAME FRAU
DRA   84 min   VIDrel: DTK/TOTAL L/A   V

SEE JANE RUN *** (15)
John Patterson USA   1994
Joanna Kerns, John Shea, Katy Boyer, Lee Garlington, Blaire Baron, Kurt Fuller, Cliff Potts, Macon McCalman, Denise Dowse, Tom Henschel, Laura Innes, Robert Mailhouse, Lora Staley, Tiffany Taubman, Melissa Weber, Bonnie Bartlett
A woman finds herself in supermarket suffering from amnesia, covered in blood and with her pockets stuffed with $10,000. Unable to remember anything about her past, she finds herself being "claimed" by a stranger, who assures her that she is his wife. But this may be untrue. A little convoluted in places, but

this lost-memory thriller generates a good degree of tension, and benefits from strong direction and good characterisation.
THR   88 min (ort 90 min)   mTV   SATrel: SKY MOVIES
Boa: book by Joy Fielding.

SEE NO EVIL, HEAR NO EVIL ** 15
Arthur Hiller USA   1989
Richard Pryor, Gene Wilder, Joan Severance, Kevin Spacey, Kirsten Childs, Alan North, Anthony Zerbe
Wilder and Pryor are two handicapped buddies who are respectively deaf and blind. Having witnessed a murder and become suspects, each tries to hide his disability and they go on the run from both killer and cops. The two stars work well together and although the comic sequences are never cruel, the utterly contrived and disagreeably foul-mouthed script gives them little chance to shine.
COM   97 min (ort 103 min)   VIDrel: VCC/DISC/COLUM   V/sur

SEE YOU IN THE MORNING ** 15
Alan J. Pakula USA   1988
Jeff Bridges, Alice Krige, Farrah Fawcett, Linda Lavin, Drew Barrymore, Lukas Haas, David Dukes, Frances Sternhagen, Theodore Bikel, George Hearn, Heather Lilly, Macaulay Culkin, William LeMassena, Tom Aldredge, Dorothy Dean
A New York psychiatrist struggles to get over the trauma of his failed marriage, and balance the obligations he still has to his former wife and kids, against his role as stepfather to his new wife's children. A touching character study that blends comedy and drama fairly agreeably, but is spoilt by contrived scripting and a disappointing ending. The screenplay is by Pakula.
COM   114 min (ort 119 min)
VIDrel: 4-FRONT/POLYREC/GUILD L/A   V/sh

SEED PEOPLE ** 15
Peter Manoogian USA   1991
Sam Hennings, Andrea Roth, Dane Witherspoon, Dave Dunard, Holly Fields, Bernard Kates, John Mooney, Anne Betancourt, Charles Bouvier, Sonny Carol Davis, J. Marvin Campbell, Matt Demeritt, Debbie Carrington
Some bloodthirsty plants arrive at Comet Valley, where they embark on a fiendish plan to pollinate and colonise the world. When a geologist is sent by the State to investigate reports of a meteorite, he realises that the local people are being taken over, and teams up with the deputy sheriff and a doctor to save both the townsfolk and the planet. A gory INVASION OF THE BODY SNATCHERS clone, based on an original idea by Charles Band.
Aka: SEEDPEOPLE
HOR   78 min (ort 87 min)   VIDrel: CIC/SONOP   V

SEEDS OF DECEPTION ** 15
Arlene Sanford USA   1994
George Dzundza, Melissa Gilbert, Tom Verica, Shanna Reed, R.H. Thompson, Shannon Lawson, Geoffrey Bowes, James B. Douglas, Jesse Collins, Michael Charles Roman, Damir Andrei, Jank Azman, Liza Blakan, David Blacker, Marilyn Boyle
A woman who is desperate to conceive goes to a renowned expert in artificial insemination, but learns to her cost that he uses unethical methods. Fact-based tale built around the scandal of Dr Cecil Jacobson, who duped a number of women until he was exposed.
DRA   90 min   VIDrel: ODY/SONOP   V/sh

SEEDS OF TRAGEDY *** 18
Martin Donovan USA   1991
Norbet Weisser, Jeff Kaake, Michael Fernandes, Conor O'Farrell, Terry Finn, Sarah Buxton, Page Moseley, Juan Ibarra, Guillermo Rios, Lahmard Tate, Lisa Lord, Robert D'Amico, Juan Ignacio Avanda, Roberto Susa Martinez, Larenz Tate
Docu-drama on the South American drugs trade that follows the path of a consignment from its harvesting in the Andes to its sale on the streets of L.A. The fast-moving account is sustained by an intelligent script, a strong if sometimes intrusive music soundtrack (by Gerlad Gouriet) and a welcome lack of moralising. The cast is uniformly excellent, and the film takes the form of a set of absorbing vignettes.
DRA   87 min (ort 90 min)   mTV   VIDrel: 20TH   V

SEEMS LIKE OLD TIMES *** PG
Jay Sandrich USA   1980
Goldie Hawn, Chevy Chase, Charles Grodin, Robert Guillaume, Harold Gould, George Grizzard, Yvonne Wilder, T.K. Carter, Judd

Omen, Marc Alaimo, Joseph Running Fox, Ray Tracey, Ray Hauser, Carolyn Fromson, Sandy Lipton
Film version of a play about a woman living in the 1930s, whose life with her new husband is jeopardised by the crazy antics of her old one. A lightweight farce that ambles along most enjoyably but with the occasional lapse. Written by Simon. This was the feature debut of TV director Sandrich.
COM 97 min (ort 121 min) VIDrel: MIA/VCC/COLUM V
Boa: play by Neil Simon.

SEIZE THE DAY ***
Fielder Cook USA
Robin Williams, Joseph Wiseman, Jerry Stiller, Glenne Headly, Tony Roberts, Richard Shull, John Fielder, Jo Van Fleet, William Hickey, Eileen Heckart, Fran Brill, Tom Aldredge, Jayne Heller, Katherine Borowitz, William Duel, Carl Low
The story of a middle-aged failure who has lost his job, and the manner in which he attempts to come to terms with his life. An absorbing but overly depressing tale of naive trust and broken promises. Despite the unappealing nature of the material, a fine performance from Williams gives it substance. Written by Ronald Ribman.
DRA 93 min VIDrel: VES L/A V
Boa: novel by Saul Bellow.

PG
1986

SUKEBAN DEKA: PARTS 1 AND 2 **
JAPAN
A ruthless criminal syndicate uses a high school as a front for its activities, but the cops recruit a reformed bad girl who, armed with a secret weapon and total freedom of action, goes undercover to destroy the crooks.
ANIM 120 min (2 cassettes) dubbed (subs ver. available)
VIDrel: ADVID V/sur

15
1991

SEMI-PRECIOUS **
Frank Arnold USA
Joanna Kerns, Stepahnie Zimbalist, Michael Shulman, Lee Garlington, Lawrence Presssman, Lisa Wilhoit, Jesse Bennett, Gaby Hoffmann, Leo Geter, David McConnell, Joey Mitashima, Brittney Rice, Ruth Hale, Derryl Yeager, Trent Hanson
A widow and her two step-children find their normal life is soon reduced to chaos when the children's biological mother appears on the scene and demands their custody, despite the fact that she walked out on them many years before when they were babies. A tug-of-love family drama whose workmanlike performances do a good deal to make for a watchable movie, despite the hackneyed nature of the plot.
Aka: WHOSE DAUGHTER IS SHE?
DRA 89 min mTV SATrel: MOVIE CHANNEL

(PG)
1995

SEMI-TOUGH **
Michael Ritchie USA
Burt Reynolds, Kris Kristofferson, Jill Clayburgh, Robert Preston, Bert Convy, Lotte Lenya, Roger E. Mosley, Richard Masur, Carl Weathers, Brian Dennehy, Mary Jo Catlett, Joe Knapp, Ron Silver, Jim McKrell, Peter Bromilow
A comic look at two football stars and their girlfriends, that is meandering and unstructured, though not without one or two mildly amusing moments. Adapted from the Jenkins novel by Ring Lardner Jr (who took his name off the credits) and Walter Bernstein. Later a brief TV series.
COM 105 min (ort 107 min) VIDrel: MGM/WHV L/A V
Boa: novel by Dan Jenkins.

15
1977

SENDER, THE ***
Roger Christian UK
Zeljko Ivanek, Kathryn Harrold, Shirley Knight, Paul Freeman, Sean Hewitt, Harry Ditson, Olivier Pierre, Tracy Harper, Al Matthews, Marsha Hunt, Angus MacInnes, Jana Sheldon, Monica Buferd, Manning Redwood, John Stephen Hill
A telepathic individual is unable to control his nightmares, and causes havoc in a hospital to which he has been confined with amnesia, after making a suicide attempt. An intriguing and quite gripping tale. See also PATRICK.
THR 87 min (ort 91 min) VIDrel: L/A V

18
1982

SENSATION **
Brian Gant USA
Eric Roberts, Kari Wuhrer, Ron Perlman, Paul Le Mat, Clare Stansfield, Tracey Needham, Kieran Mulroney, Ed Begley Jr
A co-ed with psychic powers willingly agrees to take part in experiments being conducted by a brilliant university professor,

18
1994

who is researching into the paranormal. She finds herself physically attracted to the man but receives a shock when she learns that he seems to be implicated in the sex murder of a former student.
DRA 90 min (ort 102 min) VIDrel: NEWAGE/TECH L/A V

SENSE AND SENSIBILITY **
Rodney Bennett UK
Irene Richard, Tracey Childs, Diana Fairfax
A compilation of a TV serial adaptation of the famous novel, telling of the hapless Dashwoods who suffer indignity and the loss of their family home at the hands of their snobbish relatives. A lavish and detailed saga that draws on the strengths of Austen's work, but is somewhat sluggish in development.
DRA 173 min; 131 min cC (V/sur version) mTV
VIDrel: BBC/TECH V/sur V/h
Boa: novel by Jane Austen.

U
1980

SENSE AND SENSIBILITY ****
Ang Lee UK/USA
James Fleet, Tom Wilkinson, Harriet Walter, Kate Winslet, Emma Thompson, Gemma Jones, Hugh Grant, Emile Francois, Elizabeth Spriggs, Robert Hardy, Ian Brimble, Isabelle Amyes, Alan Rickman, Greg Wise, Imelda Staunton
Tale set in rural Georgian England, where two sisters who are struggling by in reduced circumstances, dream of finding true love rather than making do with the suitable matches that family and convention demands. Wonderfully well crafted, all the subtle nuances and delicacies of the Austen novel are brought to life. Thompson won a well deserved Oscar for her fine work in adapting the novel. AA: Screen/adapt (Emma Thompson)
DRA 131 min (ort 136 min) wScrn cC
VIDrel: COLUM/SONOP; ENCORE/COLUM (LV only) V/sur LV
Boa: novel by Jane Austen.

U
1995

SENSE OF FREEDOM, A **
John McKenzie UK
David Hayman, Alex Norton, Fulton MacKay, Jake D'Arcy, Sean Scanlon
Partially based on the true story of Jimmy Boyle, this tells of the imprisonment and eventual rehabilitation of a convicted murderer. An abundance of unpleasant detail cannot mask the superficial treatment his story is given. Average. Written by Peter MacDougall.
DRA 85 min (ort 104 min) mTV VIDrel: LUMI/SPEAR L/A V/h
Boa: book by Jimmy Boyle.

18
1979

SENSUAL ESCAPE ***
Candida Royalle/Gloria Leonard USA
Richard Pacheco, Nina Hartley; Siobhan Hunter, Steve Lockwood
Two separate adult tales: "Fortune Smiles" in which we are given a look into the minds of two people whose romantic involvement is soon to lead to the bedroom, and "The Tunnel", where the recurring erotic dreams of a young artist lead her to an encounter with a mysterious man. Not exactly powerfully plotted stuff, but gentle, quite pleasing and (unusual for sex films) unashamedly romantic.
A 66 min (ort 80 min) mVid VIDrel: MIA/DISC V

18
1988

SENSUALIST, THE ***
Paul Verhoeven HOLLAND
Monique Van de Ven, Rutger Hauer, Tonny Huurdeman, Wim Van Den Brink, Hans Bokamp
A hippie sculptor marries a girl from a well-to-do family, but they eventually split up and he tries to find solace by sleeping with as many women as possible. Some time later they meet, but under sad circumstances. A vigorous, uncompromising and quite disturbing tale, hampered slightly by ponderous direction and a few clumsy attempts at satire.
Aka: TURKISH DELIGHT; TURKS FRUIT
DRA 101 min Cut (39 sec – ort 106 min) VIDrel: MIA/DISC V
Boa: novel by Jan Wolkers.

18
1973

SENTINEL, THE **
Michael Winner USA
Cristina Raines, Ava Gardner, Chris Sarandon, Burgess Meredith, Sylvia Miles, Jose Ferrer, John Carradine, Arthur Kennedy, Deborah Raffin, Eli Wallach, Christopher Walken, Jerry Orbach, Beverly D'Angelo, Tom Berenger
A fashion model discovers that the apartment block she has

18
1976

rented in Brooklyn Heights is inhabited by demons, and that she is destined to be the next sentinel guarding the gateway to Hell, over which the block has (rather unwisely it would seem) been built. A flashy but rather shallow yarn. See also THE BEYOND for something on a similar theme.
HOR   88 min (ort 92 min)   VIDrel: CIC/SONOP L/A   V
Boa: novel by Jeffrey Konvitz.

SEPARATE BUT EQUAL ***
George Stevens Jr USA
Burt Lancaster, Sidney Poitier, Richard Kiley, Cleavon Little, John McMartin, Graham Beckel, Ed Hall, Lynne Thigpen, Macon McCallman, Albert Hall, Gloria Foster
A carefully made and detailed reconstruction of the legal moves that led to the dismantling of the system of racially segregated schools in the southern states of the USA, following a landmark ruling of the Supreme Court in 1954. A powerful and convincing drama, in which Poitier gives a splendid portrayal of civil rights lawyer Thurgood Marshall. A literate script and fine direction are additional strengths.
DRA   184 min (2 cassettes – ort 200 min) mTV
VIDrel: ODY/SONOP   V/sh

PG
1991

SEPARATE LIVES ***
David Madden Jr USA
James Belushi, Linda Hamilton, Vera Miles, Elizabeth Moss, Drew Snyder, Mark Lindsay Chapman, Marc Poppel, Elizabeth Arlen, Joshn Taylor, Ken Kerman, Michael Whaley, Jackie Debatin, Joshua Makina, Craig Stepp, Linda Chess, Pat Delany
An ex-homicide detective who now works freelance, is hired by a female psychologist to keep watch on her, as she is beginning to doubt her own sanity, it appearing to be the case that she unknowingly leads a double life. But this is only the start of her problems as her detective finds begins to find himself irresistibly drawn to her alter ego. A taut and cleverly plotted thriller.
THR   98 min (ort 102 min)   VIDrel: TRIM/HIFLI   V/h

18
1995

SEPARATE VACATIONS **
Michael Anderson USA
David Naughton, Jennifer Dale, Mark Keyloun, Lally Cadeau, Blanca Cuerra, Susan Almgren, Laurie Holdeu, Tony Rosato
A husband with a seven-year-itch, goes on holiday by himself to Mexico in search of a quick fling, but makes a complete mess of it while his wife enjoys success without even trying. A hackneyed old formula comedy.
COM   90 min (ort 94 min)   VIDrel: MED/POLYREC L/A   V
Boa: novel by Eric Webber.

18
1986

SEPARATED BY MURDER ***
Donald Wrye USA
Sharon Gless, Steve Railsback, Ed Bruce, Mike McLaren, Mark W. Johnson, Jim Ostrander, Bob Penny, Paul Willis, Mimmye Goode, Steve Watson, Patrick Manning, Lisa Foster, Mari Askew, Denise Hickes, Lyb Cobb, Mike McLaren, Greg Fletcher
A woman tells her twin sister that she is fed up with her marriage and wants her wealthy husband dead, but the latter is shocked and will have no part of this. However, when the murder takes place, both sisters are charged, the police being convinced that they conspired together. Inspired by true events, this excellent TV movie benefits from a terrific and demanding performance from Gless in the dual role; the unexpected twist is an additional bonus.
THR   92 min (ort 94 min) mTV   VIDrel: MARQ/GUILD   V

12
1994

SEPTEMBER **
Woody Allen USA
Mia Farrow, Denholm Elliott, Sam Waterston, Dianne Wiest, Jack Warden, Elaine Strich, Ira Wheeler, Jane Cecil, Rosemary Murphy
Back to the serious side of life as seen through Allen's distorting mirror, as six characters on summer vacation in Vermont indulge their assorted anxieties and obsessions. Despite good performances and an intelligent script, this one becomes very tiresome very quickly.
DRA   79 min (ort 82 min)   VIDrel: VISVID/POLYREC   V

PG
1987

SEPTEMBER **
Colin Bucksey GERMANY/UK
Michael York, Jacqueline Bisset, Mariel Hemingway, Edward Fox, Virginia McKenna, Jenny Agutter, Paul Guilfoyle, Judt Parfitt, Angela Pleasence, Anna Cropper, Emily Hamilton, Jesse Birdsall, Thomas Szekeres, Sarah Winman

PG
1994

In the isolated world of the Scottish Highlands, a tale of intrigue and jealousy unfolds.
DRA   171 min (ort 200 min) mTV   VIDrel: ODY/SONOP   V
Boa: novel by Rosamunde Pilcher.

SGT BILKO *
Jonathan Lynn USA
Steve Martin, Dan Aykroyd, Phil Headly, Glenne Hartman, Max Casella, Daryl "Chill" Mitchell, Eric Edwards, Dan Ferro, John Marshall Jones, Brian Leckner, John Ortiz, Pamela Segall, Mitchell Whitfield, Austin Pendleton, Chris Rock
For some unaccountable reason this spin-off from the popular 1950s American TV series is updated to the 1990s and this, plus the unwise use of Martin in the Phil Silvers role, does not make for a very satisfying film. The story, such as it is, has Bilko fighting to prevent closure of the Army base, but in between this there is plenty time for mayhem and mishaps. Not funny, just rather tiring.
COM   90 min (ort 95 min) cC   VIDrel: CIC/SONOP   V/sur

PG
1995

SGT KABUKIMAN N.Y.P.D. **
Lloyd Kaufman/Michael Herz USA
Rick Gianesi, Susan Byun, Bill Weeden, Thomas Crnkovich, Noble Lee Lester, Brick Bromsky, Larry Robinson, Pamela Alster, Shaler McClure, Fumio Furuya
A sergeant on the New York police force becomes possessed by the spirit of a Japanese Kabuki actor and becomes a crimebusting superhero, in this low-budget spoof fantasy from the Troma studios. An occasionally very funny outing, marked as all films from Troma are by the usual tasteless excesses. Enjoyable for about the first thirty minutes, after which it becomes a lot less so.
Aka: KABUKIMAN
FAN   100 min   VIDrel: TROMA/RTM   V

18
1991

SERGEANT YORK ****
Howard Hawks USA
Gary Cooper, Walter Brennan, Joan Leslie, George Tobias, Noah Beery Jr, June Lockhart, David Bruce, Stanley Ridges, Margaret Wycherly, Dickie Moore, Ward Bond, George Tobias, Clem Bevans, Howard Da Silva, Charles Trowbridge
The true story of a one-time hillbilly and pacifist Sergeant Alvin York, who became a hero in WW1 by singlehandedly capturing 132 German soldiers. An excellent tale that never becomes maudlin yet still manages to make its points with considerable force. Written by Abem Finkel, Harry Chandler, Howard Koch and John Huston. AA: Actor (Cooper).
WAR   129 min (ort 134 min) B/W   VIDrel: MGM/WHV   V
Boa: book Sergeant York And His People by S.K. Cowan.

U
1941

SERIAL KILLER **
Pierre David CANADA
Kim Delaney, Gary Hudson, Tobin Bell, Pam Grier, Marco Rodriguez, Lyman Ward, Joel Polis, Cyndi Pass, Andrew Prine, Leonard Termo, Kimberly Faith Jones, Anne Bellamy, Jean Pfleiger, Jodi Karger, Gaby Nimier, Michael Briggs
The FBI assigns two agents to track down a serial killer, whose penchant it is to cut out the eyes of his victims, and who has now escaped from custody. As Delaney is the profiling expert whose work originally led to his capture, she is not only one of the agents given this job, but also the very person our psychopath is out to eliminate. A competent thriller.
THR   91 min   VIDrel: FIRST/SONOP   V

18
1994

SERIAL MOM **
John Waters USA
Kathleen Turner, Sam Waterston, Ricki Lake, Suzanne Somers, Matthew Lilard, Scott Wesley Morgan, Walt MacPherson, Justin Whalin, Patricia Dunnock, Mary Jo Catlett, Mink Stole, Traci Lords, Patricia Hearst, John Badila, Kathy Fannon
A perfect suburban mom with the correct ecological concerns, turns out to be a ruthless murderer, quite prepared to kill anyone who fails to conform to her own standards. Naturally, when her exploits become known, she becomes an instant media celebrity. Despite a marvellous performance by Turner in the title role, this satire is a hit-and-miss affair that never really catches fire, while the violence of the whole thing tends to nullify the expected comic effect.
COM   89 min (ort 93 min)   VIDrel: POLY/POLYREC/GUILD   V/sur

18
1993

SERIOUS CHARGE **
Terence Young UK   PG
                                                        1959
*Anthony Quayle, Andrew Ray, Sarah Churchill, Irene Browne, Percy
Herbert, Cliff Richard, Liliane Brousse, Noel Howlett, Wensley Pithey,
Leigh Madison, Wilfred Pickles, Jean Cadell, Oliver Sloane*
After a local layabout is accused by a priest of being instrumental
in causing the death of a girl, the former has his revenge by accus-
ing the priest of making homosexual advances towards him. One
of those films that was daring in its day and is therefore absorb-
ing on that count, but now looks rather dated. An additional
point of interest is Cliff Richard (in his first film) giving us his
rendition of "Living Doll", his first Number One hit in the UK.
Aka: TOUCH OF HELL, A
DRA   95 min (ort 99 min) B/W
VIDrel: 4-FRONT/POLYREC/ODY L/A   V
Boa: play by Philip King.

SERPENT AND THE RAINBOW, THE ***
Wes Craven USA   18
                                                        1987
*Bill Pullman, Cathy Tyson, Zakes Mokae, Paul Winfield, Brent
Jennings, Theresa Merritt, Michael Gough, Conrad Roberts, Paul
Guilfoyle, Dey Young, Badja Dola, Aleta Mitchell, William Newman,
Jaime Pina Gautier*
Supposedly based on the experiences of Wade Davis, this tells
of an anthropologist who is sent to investigate the phenomenon
of zombies, who have been created by the use of drugs and
voodoo rituals. Despite support from a local psychiatrist, he
soon finds that the tribal leaders are determined to protect their
secrets, in this predictable but highly effective shocker. Filmed
on location in Haiti.
HOR   94 min Cut (5 sec – ort 98 min)   VIDrel: CIC/SONOP
V/sur
Boa: book by Wade Davis.

SERPENT'S EGG, THE **
Ingmar Bergman USA/WEST GERMANY   18
                                                        1977
*Liv Ullmann, David Carradine, Gert Frobe, Heinz Bennett, Glynn
Turman, James Whitmore, Toni Berger, Christian Berkel, Paula
Braend, Edna Bruenell, Paul Buerks, Gaby Dohm, Emil Feist, Kai
Fischer, Georg Hartman, Edith Heerdegen*
An American Jewish trapeze artist in Berlin at the time of
Hitler's rise to power is surrounded by depravity and corrup-
tion. A cluttered mess of splendid photography (by Sven
Nykvist), interesting art direction and general nastiness. A big
misfire for the director, who also wrote the script.
Aka: DAS SCHLANGENEI
DRA   120 min   VIDrel: L/A   V

SERPENT'S LAIR ***
Jeffrey Reiner USA   18
                                                        1995
*Jeff Fahey, Lisa B, Heather Medway, Anthony Palermo, Jack Kehler,
Taylor Nichols, Patrick Bauchau*
A married couple unknowingly move into an apartment with a
sinister past, in which a previous tenant died an agonising
death. Shortly after their arrival, a woman appears on the scene,
claiming to be the dead man's sister, but sets about seducing the
man and drawing him into great danger. The limp storyline
may be more than a little predictable, but this erotic-supernat-
ural yarn develops quite effectively, even if the abrupt
resolution does rather spoil it all.
HOR   89 min   VIDrel: HIFLI/SONOP   V/h

SERPICO ***
Sidney Lumet USA   18
                                                        1973
*Al Pacino, John Randolph, Jack Kehoe, Biff McGuire, Barbara Eda-
Young, Tony Roberts, Cornelia Sharpe, John Medici, Allan Rich,
Norman Ornellas, M. Emmet Walsh, Ed Grover, Al Henderson, Hank
Garrett, James Tolkan, Lewis J. Stadlen*
Based on a true story, this tells of a New York cop who is so
shocked at the widespread corruption inside the force that he
stages a one-man crusade. Harsh and uncompromising, with
powerful performances and a fine script. Written by Waldo Salt
and Norman Wexler, with music by Mikis Theodorakis, it was
followed 1976 by the pilot for a TV series – "Serpico: The Deadly
Game". The title character resigned from the force in 1972. See
also THE PRINCE OF THE CITY.
DRA   125 min (ort 130 min)
VIDrel: 4-FRONT/POLYREC/CIC L/A   V/sh
Boa: book by Peter Maas.

SERVANT, THE ***
Joseph Losey UK   15
                                                        1963
Dirk Bogarde, James Fox, Sarah Miles, Wendy Craig, Catherine Lacy,
*Patrick Magee, Richard Vernon, Brian Phelan, Dorothy Bromiley,
Hazel Terry, Alison Seebohm, Philippa Hare, Alun Owen, Harold
Pinter, Derek Tansley, Hazel Terry*
A crafty servant eventually comes to dominate his weak-willed
master, in this fine drama with Fox well cast as the ineffectual
employer, gradually debased by servant Bogarde and his sister.
Written by Harold Pinter, the script generates a sense of unease
that never lets up. As ever with films of this period, the music
is by Johnny Dankworth.
DRA   110 min (ort 116 min) B/W   VIDrel: WHV   V/h
Boa: novel by Robin Maugham.

SERVANTS OF TWILIGHT **
Jeffrey Obrow USA   15
                                                        1991
*Bruce Greenwood, Belinda Bauer, Grace Zabriskie, Richard Bradford,
Jarrett Lennon, Jack Kehoe, Dale Dye, James Harper, Bruce Locke,
Kelli Maroney, Al White, Dante D'Andre, Patrick Massett, Jillian
McWhirter, Russel Lunday*
A look at the world of religious fanatics that follows the fortunes
of a six-year-old boy, who having been singled out by a church
leader as an instrument of Satan, faces a threat to his life. With
the help of a private detective, the boy's mother takes desper-
ate measures to ensure his safety.
HOR   92 min   VIDrel: FIRST/SONOP L/A   V
Boa: novel by Dean R. Koontz.

SERVING IN SILENCE: THE MARGUERITE
CAMMERMEYER STORY ***
Jeffrey A. Bleckner USA   PG
                                                        1995
*Glenn Close, Judy Davis, Jan Rubes, Wendy Makkena, Susan Barnes,
William Converse-Roberts, Colleen Flynn, William Allen Young,
Kevin McNulty, Vic Polizos, Eric Dane, Molly Parker, Trevor St.
John, Ryan Reynolds, Lance Robinson*
In the US Army, a female nursing officer is discharged from the
military, having revealed during a confidential security inter-
view that she is a lesbian. She is obliged to mount a campaign
to have this decision reversed and save her career. A strongly
plotted, fact-based drama, it is hampered by sluggish direction
and pacing, but fortunately the strong cast keep one watching.
DRA   90 min (ort 100 min) mTV   VIDrel: ODY/SONOP
V/sh

SET UP, THE **
Strathford Hamilton USA   PG
                                                        199-
Billy Zane, Mia Sara, James Russo, James Coburn
A computer wizard gets paroled from jail after serving six
months on a cat burgling charge. He soon starts work for a man
who wants him to help design a robbery proof bank and gets a
girlfriend too. But when the body of his girlfriend's former
policeman husband is found dumped in his car, it becomes clear
that he is about to be framed. A competent thriller offering few
surprises and an entirely predictable outcome.
THR   90 min mTV   VIDrel: MGM/WHV   V

SETTLE THE SCORE **
Edwin Sherin USA   18
                                                        1990
*Jaclyn Smith, Jeffrey DeMunn, Howard Duff, Richard Masur, Louise
Latham, Amy Wright, Frederick Coffin, Robert King, Monica Miller,
Donald Hotton, Dave Adams, Joseph-Gordon Levitt, Bob Larkin,
Lenore Harris*
A woman police officer from Chicago returns home to the small
town in Arkansas where she was raped twenty years before.
She is horrified to learn that her attacker is still at large and has
attacked and murdered three women. A dark and exploitative
mystery tale that is handled in an unrealistic and unconvincing
manner, with little effort being made to delineate either characters or motives.
THR   93 min (ort 100 min) mTV   VIDrel: NWV/HIFLI   V

SE7EN ***
David Fincher USA   18
                                                        1995
*Morgan Freeman, Brad Pitt, Kevin Spacey, Gwyneth Paltrow, John
C. McGinley, Richard Roundtree, Reg E. Cathey, R. Lee Ermey,
Hawthorne James, Julie Araskog, Richard Portnow*
Two homicide detectives investigate nasty serial killer whose
victims are linked to the Seven Deadly Sins. Pitt is good as the
eager young detective, but Freeman as his cynical and jaded
partner is even better. A brutal, intense and extremely chilling
film, and though mercifully the killings are not shown onscreen,
there is no shortage of gore in this movie. Some of the more
striking visual techniques used will be better appreciated on TV
if the lights are off.
THR   122 min (ort 127 min) wScrn   VIDrel: EIV/SONOP   V

SEVEN BEAUTIES ****
Lina Wertmuller ITALY   *18*
                                                          1975
Giancarlo Giannini, Fernando Rey, Shirley Stoler, Elena Fiore, Enzo Vitale, Piero Di Orio, Mario Conti, Ermelinda De Felice, Francesca Marciano, Lucio Amelio, Roberto Herlitzka, Doriglia Palmi
A small-town Don Juan has to learn to survive in a concentration camp during WW2, with all the attendant horrors. Written and directed by Wertmuller, this is an unforgettable and quite harrowing film, and one that mixes satire and unpleasantness with considerable skill.
Aka: PASQUALINO SETTEBELLEZZE; PASQUALINO: SEVEN BEAUTIES
DRA   115 min   VIDrel: MGM/WHV L/A   V

SEVEN BRIDES FOR SEVEN BROTHERS ****
Stanley Donen USA   *U*
                                                          1954
Howard Keel, Jane Powell, Jeff Richards, Russ Tamblyn, Tommy Rall, Virginia Gibson, Julie Newmeyer (Newmar), Ruta Kilmonis (Lee), Matt Mattox, Nancy Kilgas, Betty Carr, Jacques d'Amboise, Norma Doggett, Ian Wolfe, Earl Burton
Classic musical about seven brothers who decide on a radical method of settling down and getting married, and kidnap seven local girls. Written by Frances Goodrich, Albert Hackett and Dorothy Kingsley, with the lively songs by Johnny Mercer and Gere DePaul. The brilliant choreography is by Michael Kidd.
AA: Score (Adolph Deutsch/Saul Chaplin).
MUS   98 min (ort 104 min) wScrn (special edition)
VIDrel: MGM/WHV   V/dm V/sh
Boa: short story Sobbin' Women by Stephen Vincent Benet.

SEVEN DAYS IN MAY ****
John Frankenheimer USA   *(PG)*
                                                          1964
Kirk Douglas, Burt Lancaster, Fredric March, Ava Gardner, Martin Balsam, Edmond O'Brien, George Macready, John Houseman, Whit Bissell, Hugh Marlowe, Malcolm Atterbury, Bart Burns, Richard Anderson, Jack Mullaney, Andrew Duggan
Utterly absorbing and high watchable suspenser, about a plot by a US general to seize control of the country to prevent the President from signing an arms control agreement with the Soviet Union. Fine performances by the entire cast, good direction and a brisk pace maintain the sense of tension right up to the end. Remade as THE ENEMY WITHIN. TWILIGHT'S LAST GLEAMING covered similar ground, but with a nuclear threat.
THR   120 min B/W   TVrel
Boa: novel by Fletcher Knebel and Charles W. Bailey II.

SEVEN DIALS MYSTERY, THE *
Tony Wharmby UK   *PG*
                                                          1980
John Gielgud, Harry Andrews, Cheryl Campbell, James Warwick, Rula Lenska, Terence Alexander, Lucy Gutteridge, Leslie Sands, Christopher Scoular, Brian Wilde, Joyce Redman, James Griffiths, Henrietta Baynes, John Vine, Sandor Eles
The theft of a secret formula by a foreign power, and the murder of two Foreign Office officials, leads an amateur female sleuth to a secret society in Soho. An overlong and tiresome adaptation of one of Christie's minor novels, and featuring one of her less memorable characters, a certain Lady Brent.
DRA   117 min (ort 140 min) mTV   VIDrel: VCC L/A   V
Boa: novel by Agatha Christie.

SEVEN GRANDMASTERS **
                                                          15
 HONG KONG   1978
Jack Long, Lee Yim Min
An ageing kung fu master seeks new opponents in order to achieve his lifelong ambition of becoming the greatest kung fu exponent in all of China. The usual heroics are on offer for our edification.
MAR   90 min   VIDrel: EAST/DISC   V

SEVEN HOURS TO JUDGEMENT **
Beau Bridges USA   *15*
                                                          1988
Beau Bridges, Ron Leibman, Julianne Phillips, Reggie Johnson, Tiny Ron Taylor, Al Freeman Jr, Glenn-Michael Jones, Tony Lee Troy, Shawn Miller, Nick Granado, Albert Ybarra, Chris Garcia, John Billingsley, Don Creery, David Wasman
In a desperate bid to ensure the conviction of the men who murdered his wife, a grief-stricken husband kidnaps the wife of the judge trying the case, giving him just seven hours in which to find evidence that will ensure a guilty verdict. A gripping but utterly implausible tale.
DRA   89 min   VIDrel: EIV/SONOP   V

SEVEN MINUTES ***
Klaus Maria Brandauer AUSTRIA/WEST GERMANY   *15*
                                                          1989
Klaus Maria Brandauer, Rebecca Miller, Brian Dennehy, Elisabeth Orth, Nigel Le Vaillant, Vadim Glowna, Peter Andorai, Maggie O'Neill, Hans Michael Rehberg, Roger Ashton-Griffiths, Hans Stetter, Robert Easton, Janos Acs
A tightly plotted suspense film, that's set in Germany during the Nazi era, when in 1939 a shy clockmaker devises a plan to assassinate Hitler at a local beer-hall. Naturally enough, we know the plan is doomed to failure, but this knowledge does nothing to dampen the strong sense of tension this detailed and well paced film generates. The script is by Sheppard.
Aka: GEORG ELSNER: EINER AUS DEUTSCHLAND
THR   91 min   VIDrel: POLY/POLYREC/BRAVE   V
Boa: novel The Artisan by Stephen Sheppard.

SEVEN MINUTES IN HEAVEN **
Linda Feferman USA   *15*
                                                          1986
Jennifer Connelly, Bryon Thames, Alan Boyce, Maddie Corman, Billy Wirth, Polly Draper, Marshall Bell
A light comedy-drama in which a runaway teenager is invited by a female friend to move in whilst her father is away, but the latter finds that their platonic relationship leads to complications with her friends. This contrived formula tale has production-line teen appeal but little to offer an older audience.
COM   85 min (ort 91 min)   VIDrel: MGM/WHV L/A   V/h

SEVEN NIGHTS IN JAPAN *
Lewis Gilbert FRANCE/UK   *PG*
                                                          1976
Michael York, Hidemi Aoki, James Villiers, Peter Jones, Charles Gray, Anne Lonnberg, Eleonore Hirt, Lionel Murton, Yolande Donlon
Limp story of seven nights of passion, when the Crown Prince to the throne in the UK falls for a geisha he meets on a visit to Japan. A flat and unappealing tale of insipid dialogue and wooden performances.
DRA   100 min (ort 104 min)   VIDrel: MGM/WHV L/A   V

SEVEN PERCENT SOLUTION, THE ***
Herbert Ross USA   *15*
                                                          1976
Nicol Williamson, Alan Arkin, Robert Duvall, Vanessa Redgrave, Laurence Olivier, Joel Grey, Samantha Eggar, Jeremy Kemp, Charles Gray, Georgia Brown, Regine, Anna Quayle, Jill Townsend, John Bird, Alison Leggatt
Watson lures Holmes to Vienna so that Freud can cure him of his addiction to morphine (the title refers to the concentration he uses). An excellent idea is just thrown away by the slapstick antics of the last third of the film. Written by Nicholas Meyer from his novel.
DRA   110 min (ort 114 min)   VIDrel: CIC/SONOP L/A   V
Boa: novel by Nicholas Meyer.

SEVEN SAMURAI, THE *****
Akira Kurosawa JAPAN   *PG*
                                                          1954
Toshiro Mifune, Takashi Shimura, Yoshio Inaba, Kuninori Kodo, Isao Kimura, Seiji Miyaguchi, Minoru Chiaki, Daisuke Kato, Keiko Tsushima, Ko (Isao) Kimura, Kunihori Kodo, Kamataru Fujiwara, Yoshio Tsuchiya, Bokuzen Hidari
In 16th century Japan, poor villagers set out to hire samurai as a means of protecting themselves from bandits who steal most of each year's harvest. Since the samurai can expect neither payment nor glory, their attempts to hire warriors prove difficult. A superb recreation of medieval Japan, the film is both an exciting epic and an inspiring examination of human hopes and aspirations. Hollywood based THE MAGNIFICENT SEVEN on this masterpiece.
Aka: MAGNIFICENT SEVEN, THE; SHICHININ NO SAMURAI
A/AD   190 min (ort 204 min) B/W   VIDrel: CONNO/RTM V

SEVEN-STAR GRAND MANTIS **
Sunny Chin HONG KONG   *15*
                                                          198-
Benny Eagle, Man Tsui, Gerry Wong, Ronny Chan, Alex Hung, Stella Chang
A village under threat from a vicious criminal gets a chance to free itself for good, with the arrival of a poor beggar. His far from impressive appearance belies superb martial arts skills, for he proves to be a master of the title secret fighting style. Another typical offering with some well handled fight sequences.
Aka: 7-STAR GRAND MANTIS
MAR   86 min   VIDrel: IMPENT   V

SEVEN-UPS, THE *** 15
Philip D'Antoni USA 1973
Roy Scheider, Tony LoBianco, Bill Hickman, Richard Lynch, Victor Arnold, Jerry Leon, Larry Haines, Ken Kercheval, Ed Jordan, David Wilson, Robert Burr, Rex Everhart, Matt Russo
A secret unit is formed within the New York police, to combat drug peddlers, with Scheider playing a tough cop. Shot in New York City, this thriller boasts one of the best car chase sequences on film. Directed by the producer of THE FRENCH CONNECTION, in what is something of an unofficial sequel to that film.
THR 99 min (ort 103 min) VIDrel: 20TH/TECH L/A V

SEVEN YEAR ITCH, THE *** PG
Billy Wilder USA 1955
Marilyn Monroe, Tom Ewell, Evelyn Keyes, Sonny Tufts, Oscar Homolka, Victor Moore, Robert Strauss, Marguerite Chapman, Carolyn Jones, Doro Merande, Butch Bernard, Roxanne, Donald MacBride, Dorothy Ford, Mary Young, Ralph Sanford
Enjoyable classic comedy about a man alone in his New York apartment while his wife and kids are on holiday in the country. Monroe gives a marvellous performance as the girl in another apartment in the block that he almost has a wild fling with, but beneath the witty lines there is little of substance. Written by Wilder and George Axelrod and with music by Alfred Newman.
COM 100 min (ort 105 min) VIDrel: 20TH/TECH V/h
Boa: play by George Axelrod.

SEVENTH DAWN, THE ** (PG)
Lewis Gilbert USA 1964
William Holden, Susannah York, Capucine, Tetsuro Tamba, Michael Goodlife, Allan Cuthbertson, Maurice Denham, Sidney Tafler
In Malaya in the 1950s a rubber planter learns to his shock that his best friend is a Communist terrorist. A sombre adventure with romantic overtones, offered in the form of York and Capucine, both of whom find themselves drawn to Holden. Predictable in outcome, the story is really too gloomy to be effective as a piece of entertainment.
Aka: 7TH DAWN, THE
A/AD 119 min (ort 123 min) SATrel: SKY MOVIES
Boa: novel The Durian Tree by Michael Keon.

SEVENTH FLOOR, THE * (15)
Ian Barry AUSTRALIA/JAPAN 1994
Brooke Shields, Masayo Kato, Craig Pearce, Linda Cropper, Russell Newman, Malcolm Kennard, Joy Smithers, Patrick Thompson, Barry Lancashire, John Andrews, David Baldwin, Norry Constantian, Kate Gerathy, Andrew Harris
After the death of her husband, his young widow continues to run his advertising agency but soon comes into conflict with a jealous and unbalanced female executive. The latter eventually kidnaps her and subjects her victim to the usual cat-and-mouse games before the final confrontation. A barely watchable piece, not helped by poor acting and a climax worthy of any production by Hammer Films, which this clumsy effort puts one in mind of.
THR 99 min SATrel: MOVIE CHANNEL

SEVENTH SEAL, THE ***** PG
Ingmar Bergman SWEDEN 1956
Gunnar Bjornstrand, Max Von Sydow, Bibi Andersson, Nils Poppe, Bengt Ekerot, Inga Gill, Maud Hanson, Inga Landgre, Bertil Anderberg, Gunnel Lindblom, Ake Fridell, Erik Strandmark, Gunnar Olsson, Benkt-Ake Benktsson, Gudron Brost
A knight returning to Sweden after an absence of ten years fighting in the Crusades challenges Death to a game of chess to win a brief respite. Around them, thousands are falling victim to the Black Death and the end of the world seems imminent. As he makes his way home to his wife, the knight and his squire encounter a group of travelling players who accompany them. A brilliant, uplifting and absorbing masterpiece. Bergman's best film ever.
Aka: DET SJUNDE INSEGLET
DRA 92 min (ort 110 min) B/W VIDrel: TART/20TH V/dm
Boa: play Tramalning (Sculpture In Wood) by Bergman.

SEVENTH VEIL, THE *** PG
Compton Bennett UK 1945
James Mason, Ann Todd, Herbert Lom, Albert Lieven, Hugh McDermott, Yvonne Owen, David Horne, Manning Whiley, Grace Allardyce, Ernest Davies, Beatrice Varley, John Slater, Margaret Withers, Arnold Goldsborough
A well-acted melodrama in which a psychiatrist uses hypnosis, to probe the mind of a young woman who tried to drown herself in the Thames, gradually unearthing the secrets of her unhappy life and enabling her to be restored to the man she loves. The music is by Benjamin Frankel. AA: Screen/orig (Muriel Box/Sydney Box).
DRA 90 min (ort 94 min) B/W
VIDrel: 4-FRONT/POLYREC/ODY V

SEVENTH VOYAGE OF SINBAD, THE *** U
Nathan Juran USA 1958
Kerwin Matthews, Kathryn Grant, Richard Eyer, Torin Thatcher, Alec Mango, Danny Green, Alfred Brown, Harold Kasket, Nana de Herrera, Virgilio Teixeira, Luis Guuedes, Nino Falanga
When his fiancee is reduced to miniature size by an evil magician, Sinbad is obliged to seek out a magical roc's egg in order to restore her. A lively fantasy-adventure, with excellent Ray Harryhausen effects and a well-paced narrative. Music is by Bernard Herrmann.
JUV 84 min (ort 89 min) VIDrel: VCC/DISC/COLUM V

SEVERED TIES ** 18
Damon Santostefano/Richard Roberts USA 1992
Oliver Reed, Elke Sommer, Garrett Morris, Billy Morrissette, Johnny Legend, Denise Wallace, Roger Perkovich, Bekki Vallin, Gerald Shidell, Marilyn Penn, Julian Weaver, Betty Perkovich, James Clermont, Dave Consoer, Vincent Madeira
A genetic scientist invents a means of regenerating severed limbs but his experiments lead to the creation of a number of hideous creatures. A confused and inarticulate horror offering whose only asset lies in some excellent but very repulsive special effects. Produced by Fangoria magazine, this movie shows a juvenile approach that seems to delight in gore for its own sake, while its poor acting, weak direction and thin script are major drawbacks.
A/AD 91 min (ort 95 min) VIDrel: COLUM/SONOP V

SEWAGE BABY * 18
Francis Teri USA 1990
Frank Reeves, Marie Michaels
A pregnancy is terminated with an unwanted baby is flushed down the drain, but doesn't die and returns as a monster. Some echoes of IT'S ALIVE! abound in this thoroughly revolting offering.
HOR 85 min VIDrel: VIPCO/SGSVID V/h

SEX *** 18
Paul Thomas USA 1993
Nikki Dial, Mike Horner, Tom Byron, P.J. Sparxx, Crystal Wilder, Christy Canyon, Rasha Romana, Mimi Miyagi, Rocco Siffredi, Terry Thomas
Sex film set in the world of business, with the thin plot tied to the attempts a man makes to have it off with one of the office secretaries, who all the while, has a few ploys of her own to get on in the company. Well written, it was shot on film and for the most part, has a very professional look about it, especially the dialogue, which for a change is believable. Some annoying male stereotyping is clumsily at odds with the general quality of the movie.
A 70 min (ort 80 min) VIDrel: VIVID/SCRN V

SEX-A-VISION *** 18
Ned Morehead USA 1985
Joey Silvera, Gina Carrerra, Herschel Savage, Tamara Longly, Sheri St Clair, Melissa Melendez, Colleen Brennan
Daryl is a weedy nerd who has no friends except his dog, and spends most of his time watching sex videos. Whilst watching one such film he expresses a desire to change places with Dick, the porno star. And this does actually happen, with Dick helping Daryl climb in through the TV screen whilst Dick climbs out and goes in search of a lady who deserted the film. A silly, amusing sexy spoof, almost certainly inspired by THE PURPLE ROSE OF CAIRO.
A 44 min VIDrel: RAVEN/QUANT V

SEX, LIES AND VIDEOTAPE *** 18
Steven Soderbergh USA 1989
James Spader, Andie MacDowell, Peter Gallagher, Laura San Giacomo, Ron Vawter, Steven Brill, Alexandra Root, earl T. Taylor, David Foil
This bizarre morality tale has a selfish lawyer married to a frigid woman, but having an affair with her sister. An old college chum visits their home, and his strange project of videotaping women as they discuss their sexuality serves to put the lives of

all concerned onto a new footing. Writer-director Soderburgh's first feature is a disturbing blend of honesty and pretension. Winner of the 1989 International Critic's Prize and Palme d'Or at Cannes.
DRA   95 min (ort 130 min)
VIDrel: 4-FRONT/POLYREC/VISVID   V/sur

SEX, LOVE AND COLD HARD CASH **   15
Harry S. Longstreet USA   1992
JoBeth Williams, Anthony John Denison, Robert Forster, Eric Pierpoint, Tobin Bell, Randle Mell, Robert Gossett, John P. Connolly, Richard Sarafian, Stephen Rowe, Joel Swetow, Henry Brown, Bradford English, Carolyn J. Silas, Don Fischer
After a long prison term for a racetrack heist, a man is released from prison and falls in love with a hooker, who like him, has also been double-crossed by her criminal associates. A reasonably well made if unoriginal comedy-thriller with all the usual ingredients.
THR   82 min (ort 86 min) mTV   VIDrel: CIC/SONOP   V

SEX SYMBOL, THE **   15
David Lowell Rich USA   1973 (released 1974)
Connie Stevens, Shelley Winters, Jack Carter, William Castle, Don Murray, Nehemiah Persoff, James Olson, Madlyn Rhue, Milton Selzer, Tony Young, William Smith, Rand Brooks, Malachi Throne, Frank Loverde, Bing Russell
The story of a movie queen of the 1940s and 1950s, following her climb to super-stardom and her fall. A fictitious account, but inspired by the story of Monroe. The film was threatened by a libel suit just prior to its March 1974 premiere and was severely edited for USA release, with Throne's voice being redubbed by Eduard Franz. Average.
DRA   106 min mTV   VIDrel: RCA L/A   V
Boa: novel The Symbol by Alvah Bessie.

SEX THIEF, THE *   18
Martin Campbell UK   1973
David Warbeck, Diane Keen, Terence Edmond, Deirdre Costello, Christopher Neil, Michael Armstrong, Harvey Hall, Jenny Westbrook, Gerald Taylor, Gloria Walker, Eric Deacon, Christopher Mitchell, Christopher Biggins, Val Penny
A jewel thief makes love to his women victims, who then give the police deliberately misleading information. Tired and dated sex comedy.
COM   89 min   VIDrel: JEZ/RTM   V

SEXORCIST, THE ***   18
Mario Gariazzo ITALY   1974
Chris Avram, Lucretia Love, Stella Carnacina, Luigi Pistilli, Gianrico Tondinelli, Umberto Raho, Ivan Rassimov, Gabriele Tinti, Giuseppe Addobbati, Piero Gerlini, Elisa Mantellini, Maria Teresa Piaggio, Edoardo Toniolo
Sex film combining "The Exorcist" effects with sexual activities. An art student becomes possessed by an evil spirit that dwells within a crucifix taken from a desecrated church that his mother is restoring. This interesting combination of genres helps lift this one a cut or two above the usual run of such films.
Aka: DEVIL OBSESSION, THE; EERIE MIDNIGHT HORROR SHOW, THE; ENTER THE DEVIL; L'OSSESSA; OBSESSED, THE; OBSESSION, THE; SEXORCISTS, THE; TORMENTED, THE; TORMENTORS, THE
A   83 min (ort 92 min) Cut (XTASY – 40 sec); 81 min Cut (PHE – 45 sec)   VIDrel: L/A   V

SEXUAL INTENT **   18
Kurt Mac Carley USA   1992
Gary Hudson, Michelle Brin, Sarah Hill, Mary baldwin, Ron Drewed, Maria Faraldo, Milo Floeter, Varness Ann Giorgio, Stan Herman, Erica howard, Daniel Ilic, Karen Lacava, Shanyel Lee, Bruce McIntosh, Tammy Newbold, Christine Shelle
A woman psychiatrist gets herself involved with one of her patients, a notorious womaniser, but things get complicated when his former victims band together to seek revenge. Based on the true story of John Walsome, who seduced and exploited forty women across the whole of the USA.
DRA   87 min   VIDrel: POLY/POLYREC/MED   V/h

SEXUAL LIFE OF THE BELGIANS 1950-1978, THE ***
18
Jan Bucquoy BELGIUM   1993
Jean-Henri Compere, Isabelle Legros, Noe Francq, Sophie Schneider, Jacques Druaux, Sacha Jacques, Boris Bucquoy, Pascale Binnert, Michel Angely, Michele Shor, Stefan Lernous, Georgette Stulens, Raymond Vandermissen, Tegan Pick

A witty film that looks at the director's sexual experiences, from his early encounters through his unhappy teenage years and on to his later relationships. Covering a twenty-eight-year period, this is a character study of unflattering honesty, its one truly insightful observation being that the central figure's sexual preferences were in fact determined from his babyhood. Quite episodic and disjointed, its is nonetheless engrossing and often unexpectedly touching.
Aka: LA VIE SEXUELLE DES BELGES 1950-1978
DRA   78 min (ort 80 min) wScrn   VIDrel: TART/20TH   V

SEXUAL MALICE **   18
Jag Mudhra USA   1993
John Laughlin, Diana Barton, Samantha Phillips, Kathy Shower, Chad McQueen, Don Swayze, Edward Albert
A woman leaves her confining marriage for a passionate and lustful affair with a relative stranger but soon learns that there are hidden dangers in their relationship. Another stab at an erotic thriller on the hackneyed theme of sexual obsession.
DRA   98 min   VIDrel: 20VIS/SONOP   V/sh

SEXUAL OUTLAWS **   18
Edwin Brown USA   1994
Mitch Gaylord, Erika West, Kim Dawson, Nicole Grey, Mike McCollow, Eric Kohner, Michael Stanton, Jonathan Baker, Jonathan Peters, George St Thomas, Bill Williams, Don Fisher, Jennifer Peace, Shauna Yager, Monique Parent
A couple who run a magazine for sexual fantasies answer a reader's ad in a bid to add some passion to their dull marriage. This leads to their involvement in fulfilling the fantasies of a married couple, but they fail to realise this may have lethal results.
DRA   97 min   VIDrel: MIA/DISC   V/sh

SEXUAL RESPONSE **   18
Yaky Yosha USA   1992
Shannon Tweed, Catherine Oxenburg, Emile Levisetti, Vernon Wells, David Kriegel, David Pawledge, Jack Hamblin, Lori Thomas, Illana Shoshan, Kimberly Stryker, Debra Gleich, Debbie Waters, Jimmy Boeven, Sigal Diamant, Tonia George
A woman talk show host is unhappily married and so gets herself a new lover who soon suggests that they get rid of her husband.
DRA   90 min   VIDrel: COLUM/SONOP   V

SEXY DOZEN, THE **   18
Max Sieber/Norbert Terry SWITZERLAND/WEST GERMANY   1969
Barbro Hedstrom, Vincent Gauthier, Renee St Cyr, Noel Roquevert, Lovis Navarre, Julie Jordan, Bruno Kaspar, Marianik Revillon, Ulla-Britt Dahlberg, Harriet Eires, Hanneke Harmsen, Elisabeth Hedlund, Maria Frost, Ann Rosander
After being caught making love to her boyfriend on the carpet, a young girl is sent to a finishing school in the Swiss Alps. However, she gets her boyfriend to dress up as a girl and successfully smuggles him into the school in this dated and tepid farce. The opening sequences are filmed in Stockholm and Uppsala.
Aka: CHARLEY'S TANTE NACKT; DAS BUMSFIDELE TOCHTERINTERNAT
A   94 min (ort 104 min)   VIDrel: IMPENT   V

S.F.W. **   18
Jefery Levy USA   1994
Stephen Dorff, Reese Witherspoon, Jake Busey, Joey Lauren Adams, Pamela Gidley, David Barry Gray, Jake Noseworthy, Richard Portnow, Edward Wiley, Lela Ivey, Natasha Gregson Wagner, Annie McEnroe, Virgil Frye, Francesca P. Roberts
A man and his friends become national heroes after being held hostage for thirty-six days at a store that was being robbed, their declining health being videotaped by the robbers and aired nightly on TV as if it were a soap opera. However, becoming a TV celebrity proves to be very much a mixed blessing. A strident and pointless satire on fame and the cynicism of the media. The title stands for the obscene catchphrase used by the leader of the robbers.
DRA   91 min (ort 100 min)   VIDrel: COLUM/SONOP
V/sur
Boa: novel by Andrew Wellman.

SHADES OF GRAY **   (15)
Kevin James Dobson USA   1993
Valerie Bertinelli, George Dundza, Peter Dobson, David Marshall Grant, Micole Mercurio, Andrew May, Randle Mell, Joe Viterelli,

Vondie Curtis Hall, Bruce Kirkpatrick, Tom Meyers, Larry John Meyers, Jerry Lyden, Mark Joy
An ambitious young law graduate starts work in the office of the New York D.A. to the great disappointment of her father, who is a police detective. Adequate.
DRA   90 min   SATrel: MOVIE CHANNEL

SHADES OF L.A. **   15
Jim Johnston USA
John Di Aqiono, Sam Jones, Michael Parks, Warren Berlinger
Following a near-death experience, a detective gains the ability to communicate with the dead, which he finds of great use when a former police buddy returns to warn him that his girlfriend is in danger.
DRA   79 min   VIDrel: CIC   V/sh

SHADOW, THE **   12
Russell Mulcahy USA   1994
Alec Baldwin, John Lone, Penelope Ann Miller, Tim Curry, Peter Boyle, Ian McKellan, Jonathan Winters, Sab Shimono, Andre Gregory, Brady Tsurutani, James Hong, Arsenio "Sonny" Trnidad, Joseph Maher, John Kapelos, Max Wright
A wealthy New York playboy has a secret identity as title figure, a dedicated and mysterious crime-fighter, in a movie whose character originated in a 1930s magazine comic-strip (and led to a 1930s radio series and then a 1940s film serial). This flashy and stylish adaptation is at its best when showing off the expensive visual effects that took the biggest bite out of the budget. Unfortunately, it never rises above its pulp-fiction roots.
THR   103 min (ort 112 min) cC   VIDrel: CIC/SONOP; PION (LV only)   V LV
Boa: radio serial by Maxwell Grant (Walter Gibson).

SHADOW FORCE **   18
Darrell Davenport USA   1992
Dirk Benedict, Lise Cutter, Lance LeGault, Bob Hastings, Dixie King-Wade, Glenn Corbett, Jack Elam
After a D.A. and a cop are brutally murdered, a tough detective starts to investigate, assisted by a woman journalist. Soon he uncovers a plot to murder a number of key government officials and in so doing makes himself a target.
THR   76 min (ort 90 min)   VIDrel: CURB/HIFLI   V/h

SHADOW MAKERS **   PG
Roland Joffe USA   1989
Paul Newman, Dwight Schultz, Bonnie Bedelia, John Cusack, Laura Dern, John C. McGinley, Ron Frazier, Natasha Richardson, Michael Brockman, Del Close, John Considine, Roger Cubicciotti, James Eckhouse, Ron Frazier, Ed Lauter
The story of the world's first two atomic bombs (nicknamed "Fat Man" and "Little Boy") that details their creation at the Manhattan Project and their subsequent use in the destruction of Hiroshima and Nagasaki. An episodic and verbose drama of sporadic impact, that is weakened by its lack of authenticity and a failure to examine the Project's political background as a logical response to Japanese aggression. Music is by Ennio Morricone.
Aka: FAT MAN AND LITTLE BOY
DRA   121 min (ort 126 min)   VIDrel: CIC/SONOP   V

SHADOW OF A DOUBT ***   PG
Alfred Hitchcok USA   1943
Joseph Cotton, Teresa Wright, Macdonald Carey, Wallace Ford, Hume Cronyn, Patricia Collinge, Henry Travers, Edna Mae Wonacott, Irving Bacon, Charles Bates, Clarence Muse, Janet Shaw, Estelle Jewell, Minerva Urecal, Eily Malyon
A young girl living in a small Californian town begins to suspect that her visiting uncle may be a murderer. A well-acted drama whose tension builds at a controlled and credible pace.
DRA   103 min (ort 108 min) B/W   VIDrel: CIC/SONOP   V

SHADOW OF A DOUBT **   PG
Karen Arthur USA   1991
Mark Harmon, Margaret Welsh, Norm Skaggs, William Lanteau, Diane Ladd, Rick Lenz, Shirley Knight, Tippi Hedren, Sydney Walker, Bianca Rose, Seth Smith, Fran Lish, Michael Wisley, Oliver Charles, John Gavigan, Ronny Rosemont
A fairly good remake of Hitchcock's 1943 classic that is updated to the 1950s and has Harmon playing Uncle Charlie, who preys on wealthy, older women. Only his formerly adoring namesake suspects that he's a ruthless killer, and she may soon be silenced. An absorbing story, it's nothing like as exciting as the original

(a few nods are cast in Hitchcock's direction) but always holds one's attention, despite some over-acting from Harmon.
THR   96 min (ort 100 min) mTV   VIDrel: CIC/SONOP   V

SHADOW OF A DOUBT, A ***   15
Aline Issermann FRANCE   1992
Alain Bashung, Mireille Perrier, Sandrine Blancke, Emmanuelle Riva, Michel Aumont, Luis Issermann, Roland Bertain, Dominique Lavanant, Thiery Lhermitte, Jean-Pierre Sentier, Feodor Atkine, Isabelle Petit-Jacques, Cynthia Garas
When a young girl claims that her father has sexually abused her, she is not believed and finds she has no option but to flee the family, taking her younger brother with her. Fortunately, a social worker decides that she is telling the truth and sets out to investigate. Based on a collection of real-life cases, this is a carefully made tale, happily free of sentimentality or melodrama, but also one in which the narrative often alternates with a display of polemics.
Aka: L'OMBRE DU DOUTE; SHADOW OF DOUBT, A
DRA   101 min (ort 106 min) wScrn
VIDrel: ELPIC/POLYREC   V/sh

SHADOW OF A DOUBT ***   (12)
Brian Dennehy USA   1995
Brian Dennehy, Bonnie Bedelia, Fairuza Balk, Mike Nussbaum, Joe Grifasi, Michael Macrae, Ken Pogue, Donnelly Rhodes, Brent Jennings, Bruce McGil, Kevin Dunn, Mavor Moore, Henry beckman, Don S. Davis, Fulvio Cecere, Robert Clothier
An enjoyable courtroom drama in which Dennehy plays an alcoholic lawyer whose career is in terminal decline, but who gets a chance to shine once more and show his former brilliance when an old girlfriend hires him to defend her daughter, who is accused of murder. Dennehy acts with his usual conviction, and shows he is a fine director too.
DRA   89 min mTV   SATrel: MOVIE CHANNEL
Boa: novel by William J. Coughlin.

SHADOW OF A STRANGER **   15
Richard Friedman CANADA   1992
Emma Samms, Parker Stevenson, Michael Easton, John Pyper-Ferguson, Joan Chen, Deryl Hayes, Colleen Winton, Stephen E. Miller, Anthony Harrison, June Pentyliuk, John Scott, Brenda McConnell, Louie Ely
In this DEAD CALM clone Samms plays a top fashion model who is married to a successful lawyer. In an attempt to patch up their shaky marriage they take a break an isolated coastal resort, where they rescue a couple who appear to have got into difficulties with their boat. As expected, nothing is quite as it seems and this action proves to be the biggest mistake of their lives. Tired and derivative, this overwrought thriller soon grows quite predictable.
THR   87 min (ort 90 min) mTV
VIDrel: MIA/VCC/IMPENT   V

SHADOW OF THE PAST **   (12)
Terry Benedict USA   1995
Dwight Yoakam, Michelle Joyner, Kiersten Warren, Cindy Pcikett, John Getz, Bo Hopkins, Walton Gogiins, Terry McIlvain, Peter Fonda, Brent Anderson, Tony Metcalf, Bill Thurmna, Boyd Polhamus, Justin McKee, Jonathan Joss, Brad Leland
Seven years ago a rodeo star was unjustly blamed for a death and driven out of his hometown, but now makes a return in a bid to pick up the threads of his life.
WES   90 min   SATrel: MOVIE CHANNEL

SHADOW OF THE TIGER **   15
Yeung Kuen HONG KONG   198-
Cliff Look, Ka Sa Fa, Lam Man Wei
A group of top fighters join forces, in this standard kung fu action tale.
Aka: DUEL OF THE SEVEN TIGERS
MAR   92 min Cut (8 sec)   VIDrel: IMPENT   V

SHADOW OF THE WOLF **   15
Jacques Dorfmann CANADA/FRANCE   1992
Lou Diamond Philipps, Toshiro Mifune, Jennifer Tilly, Donald Sutherland, Bernard-Pierre Donnadieu, Nicholas Campbell, Raoul Trujillo, Qalingo Tokkalak, Julie Arnatiuk, Tamussie Sivuarapik, Harry Hill, David Okpik
An Eskimo couple leave the confining life of their village and enter the world of the white man where they have to face both the rigours of the landscape and the power of prejudice. When the man is accused of murder, the couple must fight hard to prove his innocence. A well-intentioned drama that despite the

efforts of all concerned never comes to life, although the location shooting does provide a little compensation.
WES   107 min (ort 108 min)   VIDrel: EIV/SONOP   V/sur
Boa: novel Agaguk.

SHADOWHUNTER ** 18
J.S. Cardone USA 1992
Scott Glenn, Angela Alvarado, Robert Beltran, Benjamin Bratt, Tim Sampson, George Aguilar, Beth Broderick, Frederick Flynn, Phil Brock, Lee De Broux, Nancy Lineham Charles, Gloria Reuben, Geraldine Keams, Betty Canyon
A tired and cynical cop from Los Angeles has to go to a Navajo reservation to arrest a murder suspect and take him back. However, when his prisoner escapes, our cop is forced into a strange game of cat and mouse.
Aka: SHADOW HUNTER
A/AD   93 min (ort 98 min)   VIDrel: MED/POLYREC   V/h

SHADOWLANDS *** PG
Norman Stone HOLLAND/UK/USA 1985
Joss Ackland, Claire Bloom, Philip Stone, Tim Preece, David Waller, Max Harvey, Rupert Baderman, Rhys Hopkins, Alan MacNaughton, Norman Rutherford, John Ringham, Henry Moxon, Michael Cunningham, Jim Kirby, Dilys Price
Beautifully acted, sensitive and restrained account, of how love came to the writer and theologian C.S. Lewis late in life, when he was sought out by an unhappily married American lady who was soon to die of cancer. Though the subject of his literary work is only touched on, and the film has a tendency to have Lewis mouthing pompous platitudes, the story is never less than wholly absorbing. Scripted by William Nicholson, it was a BBC production.
Aka: C.S. LEWIS: THROUGH THE SHADOWLANDS; THROUGH THE SHADOWLANDS
DRA   89 min mTV   VIDrel: PARADOX/TOTAL   V/h
Boa: book by P. Straub.

SHADOWLANDS *** U
Richard Attenborough UK 1993
Anthony Hopkins, Debra Winger, Edward Hardwicke, Michael Denison, Joseph Mazzello, John Wood, Robert Flemyng, Peter Howell, Peter Firth, Julia Fellowes, Roddy Maude-Roxby, Andrew Seear, Yim McMullen, Andrew Hawkins, James Frain
Remake of the excellent 1985 TV drama about the meeting and subsequent love affair between reserved Oxford don and writer C.S. Lewis and a forthright and blunt American woman poet. Fine performances all round and a literate script ensure that this one is worthwhile, although the earlier film still remains the better of the two.
DRA   126 min (ort 133 min)   VIDrel: CIC/SONOP   V/sur
Boa: book by P. Straub/play by William Nicholson.

SHADOWS *** PG
John Cassavetes USA 1959
Hugh Hurd, Leila Goldoni, Ben Carruthers, Anthony Ray, Rupert Crosse, David Pokitillow, Tom Allen, Dennis Sallas, David Jones, Pir Marini, Victoria Vargas, John Ackerman, Jacqueline Walcott, Cliff Carnell, Jay Grecc, Bob Rech
Shot in 16mm this (the director's first film) is an improvised, formless and stark piece of work, telling of a light-skinned black girl who lives in Manhattan with her two brothers. She becomes involved with a white man who is unaware of her race, but who ultimately rejects her upon learning the truth. With its improvised narrative, the film lacks dramatic compression and direction, yet conveys a strong sense of realism.
DRA   77 min (ort 81 min) B/W   VIDrel: ELPIC/POLYREC   V

SHADOWS AND FOG ** 15
Woody Allen USA 1992
Woody Allen, Kathy Bates, John Cusack, Mia Farrow, Jodie Foster, Madonna, Fred Gwynne, Julie Kavner, John Malkovich, Kenneth Mars, Kate Nelligan, Lily Tomlin, Donald Pleasence, Philip Bosco, Robert Joy, Wallace Shawn, Josef Summer
Unfocused and only sporadically funny exercise in style, with Allen playing the role of a timid clerk who is rudely woken from his sleep one night by a band of vigilantes out to catch the killer who is terrorising the city. Offbeat and very unpredictable, this homage to Kafka has little to offer, and the presence of many familiar stars in cameo roles does not to make the movie any more interesting.
DRA   82 min (ort 86 min) B/W
VIDrel: VCC/DISC/COLUM   V/sur

SHADOWS OF THE PAST ** 15
Gabriel Pelletier CANADA/FRANCE 1991
Erika Anderson, Nicholas Campbell, Berry Richard, Heidi Von Palleske, Dennis O'Connor, Lorne Brass, Jacques Herlin, Pierre LeBlanc, Jean l'Italien, Pierre Augier, Edgar Fruitier, Linda Smith, Liz McCrane, Amanda Strawn, Ronald France
Saved from death following a serious car crash, a female photojournalist finds that she is unable to remember the events prior to her accident. With the assistance of a detective, she attempts to solve this mystery, but only succeeds in placing her life in danger.
THR   91 min   VIDrel: GUILD/SONOP   V/sh

SHADOWZONE *** 18
J.S. Cardone USA 1989
David Beecroft, James Hong, Shawn Weatherly, Miguel Nunez, Lu Leonard, Frederick Flynn, Louise Fletcher
Scientists researching into sleep patterns accidentally bring back a being from another dimension. As in THE THING, the creature is a shape-changer that rapidly depletes the members of the research team, until a way is found to repatriate it. An intelligent, low-budget horror-thriller of minimal gore.
HOR   82 min (ort 89 min)   VIDrel: EIV/SONOP   V

SHAFT *** 15
Gordon Parks USA 1971
Richard Roundtree, Moses Gunn, Charles Cioffi, Christopher St John, Drew Bundini Brown, Gwenn Mitchell, Lawrence Pressman, Antonio Fargas, Victor Arnold, Sherri Brewer, Rex Robbins, Camille Yarbrough, Margaret Warncke, Joseph Leon
A black detective is hired to find a Harlem gangster's kidnapped daughter, and gets caught up in the usual rackets. Violent, flashy and competent. Written by Ernest Tidyman and John D.F. Black, and followed by SHAFT IN AFRICA and SHAFT'S BIG SCORE! plus a rather unmemorable TV series. AA: Song ("Theme From Shaft" – Isaac Hayes).
A/AD   96 min (ort 100 min)   VIDrel: MGM/WHV   V
Boa: novel by Ernest Tidyman.

SHAFT IN AFRICA * 18
John Guillermin USA 1973
Richard Roundtree, Frank Finlay, Vonetta McGee, Neda Arneric, Cy Grant, Jacques Marin
A rubbishy further sequel to SHAFT, with an African nation calling on our black detective to put an end to the local slave trade. A violent and brutal piece of dross. Written by Sterling Silliphant.
A/AD   112 min   VIDrel: MGM L/A   V

SHAFT'S BIG SCORE! ** 15
Gordon Parks USA 1972
Richard Roundtree, Moses Gunn, Drew Bundini Brown, Joseph Mascolo, Kathy Imrie, Wally Taylor, Joe Santos, Julius W. Harris, Rosalind Miles, Angelo Nazzo, Don Blakely, Melvin Green, Thomas Anderson, Evelyn Davis, Robert Kya-Hill
First sequel to SHAFT with our private eye getting into trouble when he sets out to investigate the murder of a friend. A fast and flashy film, quite enjoyable in its way, and containing an excellent chase sequence. Written by Ernest Tidyman and with music by Gordon Parks. SHAFT IN AFRICA followed.
A/AD   101 min (ort 105 min)   VIDrel: MGM L/A   V

SHAGGY D.A., THE ** U
Robert Stevenson USA 1976
Dean Jones, Suzanne Pleshette, Tim Conway, Keenan Wynn, Jo Anne Worley, Dick Van Patten, Vic Tayback, Shane Sinutko, John Myhers, Dick Bakalyan, Warren Berlinger, Ronnie Schell, Jonathan Daly, John Fiedler, Hans Conried
Sequel to "The Shaggy Dog" with a contender for the job of D.A. having to cope with the fact that he changes into a dog from time to time, whilst at the same time trying to expose corruption in high places. A mildly entertaining slapstick romp, very much a Disney film in both conception and execution. See also OH HEAVENLY DOG!
COM   91 min   VIDrel: WDV/TECH L/A   V

SHAKA ZULU: PARTS 1, 2 AND 3 ** 15
William C. Faure SOUTH AFRICA/USA 1986
Henry Cele, Dudu Mkhize, Edward Fox, Robert Powell, Fiona Fullerton, Trevor Howard, Christopher Lee, Roy Dotrice, Kenneth Griffith, Gordon Jackson, John Carson, Eric Rogers, Conrad Magwaza, Patrick Ndlovu, Roland Mqwebu
A lavish mini-epic set in South Africa of the early 18th century,

and telling of the rise to power of Shaka, who welded disparate tribes into the powerful Zulu nation that was later to battle the British. An expensive, earnest tale that's somewhat hampered by a ponderous script. Filmed on location where it all happened, in the rolling hills of Zululand. Originally shown in ten 50-minute episodes.
Aka: SHAKA ZULU
A/AD   448 min (3 cassettes – ort 500 min) mTV
VIDrel: MGM L/A   V

SHAKE, RATTLE AND ROCK! ***
(PG)
Allan Arkush USA   1993
Howie Mandel, Renee Zellweger, Jenifer Lewis, Max Perlich, Patricia Childress, John Doe, Neia Bray, Mary Woronov, Stepehen Furst, Ruth Brown, Gerrit Graham, Dick Miller Dey Young, P. J. Soles, William Schallert, Nora Dunn
A very loose remake of a 1956 B-movie with Mandel playing a rebellious teenager who sets out to help the other teens dance to the hottest sounds by forming a series of dance clubs. But in doing so, he comes into conflict with a materialistic society that is out to ban all forms of rock 'n' roll. A film that will not tax the intellect but is still enjoyable toe-tapping fun.
Aka: YOUNG AND THE RESTLESS, THE
MUS   80 min (ort 90 min)   SATrel: MOVIE CHANNEL

SHAKEDOWN ON THE SUNSET STRIP **
15
Walter Grauman USA   1987
Perry King, Season Hubley, Joan Van Ark, Vincent Baggetta, Michael McGuire, David Graf, Charles Siebert
A patchy blend of comedy and thriller genres, with an ambitious L.A. vice squad detective out to nab a notorious madam and finding himself coming up against some unexpected obstacles when he discovers that she has powerful friends in high places.
Aka: SHAKEDOWN ON SUNSET STRIP
THR   89 min (ort 96 min) mTV   VIDrel: 20TH/TECH   V

SHAKEDOWN: RETURN OF THE SONTARANS *
PG
Kevin Davies USA   1994
Brian Croucher, Jan Chappell, Sophie Aldred, Carole Ann Ford, Michael Wisher
A solar yacht carries onboard a stowaway whose mortal enemies are the Sontarans. These latter are ready to destroy the yacht if necessary, but Captain Lisa Deranne must do what she can to avoid this happening. Space opera nonsense of the simpler sort.
FAN   90 min wScrn   VIDrel: REEL/DISC   V/sh

SHAKES THE CLOWN *
18
Bobcat Goldthwait USA   1992
Julie Brown, Bobcat Goldthwait, Paul Dooley, Florence Henderson, Blake Clark, Adam Sandler, Tom Kenny, Sydney Lassick, Paul Dooley, Tim Kazurinsky, Florence Henderson, LaWanda Page, Robin Williams
An alcoholic clown receives the finishing blow to his career when he is accused of his boss by a rival and runs away in an attempt to prove his innocence. A confused and unfunny black comedy (set in the all-clown town of Palukaville) that is never clear about what it is meant to be satirising and consequently fails to work on any level. Goldthwait also wrote the script.
DRA   82 min (ort 87 min)   VIDrel: 20VIS/SONOP   V/sur

SHAKING THE TREE **
15
Duane Clark USA   1990
Arye Gross, Gale Hansen, Doug Savant, Steven Wilde, Courtenay Cox, Christina Haag, Michael Ababian, Dennis Cockrum, Nathan Davis, Ron Dean, Brittany Hansen, Terry "Turk" Muller, Ned Schmidtke, Maurice Chasse, Dick Sasso, John Malloy
Story of a group of friends from different backgrounds who stick together through thick and thin. Something of an updated version of DINER, this drama is set in Chicago towards the close of the 1980s, and what story there is mostly centres on the reassessment of their lives that is taking place as the various characters leave young adulthood behind and enter their thirties. Pleasant enough, despite lacking both insight and originality.
DRA   102 min (ort 107 min)   VIDrel: CURZON/20TH
V/sh

SHALAKO **
PG
Edward Dmytryck UK   1968
Sean Connery, Brigitte Bardot, Stephen Boyd, Jack Hawkins, Peter Van Eyck, Honor Blackman, Woody Strode, Eric Sykes, Alexander Knox, Valerie French, Julian Mateos, Donald (Don "Red") Barry, Rodd Redwing, Chief Tug Smith
A party of European aristocrats, on a hunting expedition in New Mexico in the 1880s, come into conflict with the Apache, but despite this encounter the film remains a slow-moving and ponderous Western. Written by J.J. Griffith, Hal Hopper and Scot Finch, and with music by Robert Farnon.
WES   108 min (ort 118 min)   VIDrel: WHV   V/h
Boa: novel by Louis L'Amour.

SHALL WE DANCE? ***
U
Mark Sandrich USA   1937
Fred Astaire, Ginger Rogers, Edward Everett Horton, Eric Blore, Jerome Cowan, Ketti Gallian, Ann Shoemaker, William Brisbane, Harriet Hoctor, Ben Alexander, Emma Young, Sherwood Bailey, Pete Theodore, Marek Windheim
A dancing duo pretend to be married to simplify their lives but eventually decide to do so. A flimsy little plot that has no bearing on this highly enjoyable musical. Songs are by George and Ira Gershwin and include "Let's Call The Whole Thing Off", "They All Laughed" and "They Can't Take That Away From Me".
MUS   116 min B/W   VIDrel: 4-FRONT/POLYREC   V

SHALLOW GRAVE ***
18
Danny Boyle UK   1994
Kerry Fox, Christopher Eccleston, Ewan McGregor, Keith Allen, Ken Stott, Colin McCredie, Victoria Nairn, Gary Lewis, Jean Marie Coffey, Peter Mullan, Leonard O'Malley, David Scoular, Grant Glendinning, Robert David MacDonald
When their flatmate dies, the remaining two attempt to dispose of the body themselves, as they found a load of money at the scene of the death. However, others are out to get this loot. An incredibly taut black comedy-thriller, quite grisly in places but done with enormous style and zest. This was Boyle's debut feature and says much for his abilities as a director, for though much of the film takes place on just the one set, it is never dull or oppressive.
THR   88 min (ort 92 min) wScrn cC
VIDrel: POLY/POLYREC   V/sur

SHAME **
15
Steve Jodrell AUSTRALIA   1989
Deborra-Lee Furness, Tony Barry, Simone Buchanan, Gillian Jones, Margaret Ford, Peter Aanensen, David Franklin, Bill McClusky, Allison Taylor, Graeme "Stig" Wemyss, Phil Dean, Douglas Walker, Matthew Quartermaine, Matt Hayden
A female attorney on vacation is stranded in a small outback town when her bike breaks down. She discovers that local women are being subjected to a reign of sexual violence and uncovers a conspiracy of silence over the rape of a young girl. Despite the dangers, she persuades her victim to bring charges against her assailants. An uneven and not terribly effective film, whose earlier conventional ending was wisely ditched for a dose of realism.
DRA   90 min   VIDrel: CBS L/A   V/sh

SHAMPOO *
18
Hal Ashby USA   1975
Warren Beatty, Julie Christie, Goldie Hawn, Jack Warden, Lee Grant, Carrie Fisher, Tony Bill, Jay Robinson, George Furth, Ann Weldon, Randy Sheer, Mike Olton, Susanna Moore, Luana Anders, Brad Dexter, William Castle, Jack Bernardi
A Beverly Hills hairdresser tries to borrow money from an investment counsellor, with whom he has a few things in common as the former is having an affair with the man's wife, mistress and daughter. A dismal comedy-drama that is sluggish when it should be funny, and feeble when it should be dramatic. AA: S. Actress (Grant).
DRA   106 min (ort 110 min)   VIDrel: VCC/COLUM L/A   V

SHAMROCK CONSPIRACY, THE **
(15)
James Frawley USA   1994
Edward Woodward, Elizabeth Hurley Jeffrey Nordling, Nigel Bennett, Terri Hawkes, Chris Benson, Silvo Oliviero, James Luk, Gerry Salsberg, Dick Murphy, Maria Vacratsis, Markus Parilo, Bruce McFee, Tony Craig, Tony De Santis
A widowed Scotland Yard chief inspector who has now retired, stops over in New York to bid farewell to his estranged daughter before leaving for Australia where he has bought a sheep farm. However, he soon gets involved with her cop boyfriend and assists him with what appears to be the re-emergence of a serial murderer. However, our wily and capable hero is soon able to get at the truth behind these killings. First in a series of films built around this figure.
DRA   89 min (ort 91 min)   SATrel: SKY MOVIES

SHAMUS **

18

Buzz Kulik USA

1972

Burt Reynolds, Dyan Cannon, Giorgio Tozzi, John Ryan, Joe Santos, Kevin Conway, Ron Weyland, Barry Beckerman, Kay Frye, Larry Block, Beeson Carroll, John Glover, Merwin Goldsmith, Melody Santangelo, Irving Selbst, Alex Wilson

A private eye has to solve a strange case, and suffers the usual indignities at the hands of assorted villains. A 1940s film done in 1970s style, with the obligatory violence and a few jokes for film buffs. Written by Barry Beckerman.

THR 94 min (ort 106 min) VIDrel: RCA L/A V

SHANE ****

PG

George Stevens USA

1953

Alan Ladd, Jean Arthur, Van Heflin, Jack Palance, Brandon De Wilde, Ben Johnson, Edgar Buchanan, Emile Meyer, Elisha Cook Jr, Douglas Spencer, John Dierkes, Ellen Corby, Paul McVey, John Miller, Edith Evanson, Leonard Strong

A beautiful classic Western, with Ladd as the quiet lone gunfighter who only wants to hang up his guns and retire, but is finally forced to aid the homesteaders he is living with and whose son idolises him. Palance gives a remarkably sinister performance as a gunfighter who is far from retired. Written by A.B. Guthrie Jr. AA: Cin (Loyal Griggs).

WES 113 min (ort 117 min) VIDrel: CIC/SONOP V/h

Boa: novel by Jack W. Schaefer.

SHANGHAI EXPRESS **

PG

Samo Hung HONG KONG

1987

Samo Hung, Yuen Biao, Olivia Cheng, Eric Tsang, Cynthia Rothrock, Lam Ching Ying, Kenny Bee

Fast-paced martial arts tale set aboard a luxurious new high speed train travelling from Shanghai, that is used by bank robbers as a means of making their getaway. But also onboard are another gang of crooks, and they have made plans of their own involving the train.

MAR 88 min dubbed VIDrel: IMPENT V

SHANGHAI SURPRISE *

15

Jim Goddard UK

1986

Sean Penn, Madonna (Madonna Ciccone), Paul Freeman, Richard Griffiths, Clyde Kusatsu, Philip Sayer, Kay Tong Lim, Sonserai Lee, Victor Wong, Toru Tanaka, Michael Aldridge, Sarah Lam, George She, Won Gam Bor, To Ghee Kan, David Li

Romantic adventure, set in 1930s China, following the exploits of a fortune hunter who is looking for a chance to earn his fare out of the country, and who meets up with a woman missionary looking for a missing cache of opium, needed for her clinic. A most uncharismatic pairing of erstwhile husband-and-wife team Penn and Madonna, each totally miscast and showing not a trace of the panache and acting ability so vital to films of this kind.

A/AD 93 min (ort 97 min) VIDrel: MGM/WHV L/A V/sh

Boa: novel Faraday's Flowers by Tony Kenrick.

SHANGHAI TRIAD ***

15

Zhang Yimou CHINA/FRANCE

1995

Gong Li, Li Baotian, Wang Xiao Xiao, Li Xuejian, Sun Chun, Fu Biao, Chen Shu, Liu Jiang, Jiang Baoyang, Yang Qianquan, Gao Ying, Gao Weiming, Lian Shuliang, Wang Ya'nan, Zhang Yayun, Guo Hao, Zheng Jiasen, Ni Zengshao

In 1930s Shanghai, a young man starts work for a cabaret singer who is the mistress of a gangster, and has to flee with her when a rival gangster decides to have revenge for the murder of comrade. A stylish and visually arresting morality tale, which aims for a more leisurely and restrained approach to its subject matter than is usual for such films, making the inevitable violent conflicts all the more effective.

Aka: YAO A YAO YAO DAO WAIPO QIAO

A/AD 112 min VIDrel: ELPIC/POLYREC V

SHAOLIN AND WU TANG **

15

Gordon Liu HONG KONG

1981

Gordon Liu, Adam Cheng, Wong Lung Wei, Ching Li

Liu's directing debut is this tale of a Shaolin trained martial artist having to combat a swordsman.

MAR 87 min (ort 90 min) VIDrel: EAST/DISC V

SHAOLIN CHASTITY KUNG FU **

18

Robert Tai HONG KONG

Alexander Lou, Lou Chun, Tien Lung, Yang Hsiung, Chien Hsun, Wang Hau

A gang of criminals set out to intercept a shipment of gold, but

pass through a village on their way to steal it and indulge in a senseless orgy of violence and murder. Having got most of the women and children to safety, a monk sets out to avenge the dead villagers.

MAR 90 min Cut (6 sec) VIDrel: AUDIO/DISC V

SHAOLIN DEVIL AND SHAOLIN ANGEL **

15

HONG KONG

198-

Chen Sing

When a series of inexplicable events leads the Emperor to conclude that one of his Shaolin warriors is a traitor, he sends a young fighter to discover his identity.

Aka: SHAOLIN DEVIL, SHAOLIN ANGEL

MAR 83 min (ort 90 min) VIDrel: IMPENT V

SHAOLIN DRUNKEN FIGHTER **

15

Tao Man Po HONG KONG

1986

Wong Tien Lung, Ting Lan

A man who survived an attack on himself and his family, hides in a Shaolin monastery where he is instructed by the priests, and acquires the skill needed to have his revenge.

Aka: SHAOLIN DRUNK FIGHTER

MAR 85 min Cut (1 min 42 sec – ort 95 min)

VIDrel: IMPENT V

SHAOLIN EX-MONK **

18

Chang Shin HONG KONG

Lai Tzong Liang, Wang Xing Hstu, Kao Shui Liang

A former Shaolin monk who is well versed in both the art of combat and the art of disguise turns to crime, and rapidly becomes a serious problem. A martial artist of similar prowess is sent to track him down.

MAR 86 min (ort 100 min) VIDrel: IMPENT V

SHAOLIN IRON CLAWS **

18

Ko Shih Hao HONG KONG

Wang Tao, Lee I Min, Chang Yh, Cheng Shing, Hwa Ling, Chu Li

A Shaolin assassin is hired to threaten members of the Ching Dynasty into signing a letter, calling for the restoration of the monarchy. When a police chief investigates this plot and his friend is murdered as a warning, it becomes clear that the treacherous assassin must be captured in order to save the Dynasty.

MAR 86 min VIDrel: IMPENT V

SHAOLIN MARTIAL ARTS **

18

Chang Cheh HONG KONG

1974

Fu Sheng, Chi Kuan Chun, Liu Chia Hui

The Manchu clan sets about ridding China of all opposition to its dominance, but only the ancient Shaolin school can defy them, but their best fighters have to contend with two individuals, who have each mastered special techniques that make them almost impossible to overcome. Much mayhem ensues until two Shaolin fighters find a master who can instruct them in the skills they need to beat their Manchu adversaries. Fair.

Aka: FIVE FINGERS OF DEATH

MAR 111 min dubbed VIDrel: MADE/RTM V

SHAOLIN RED MASTER *

18

Sung Ting Mei HONG KONG

Chi Kwan Chun, Tommy Lee, Hu Chin

An orphan expelled from the mighty Hang Saih Shaolin Temple, goes in search of the man who murdered his family. The usual revenge-based developments occur.

MAR 90 min dubbed VIDrel: EAST/DISC V

SHAOLIN TEMPLE **

18

Cheung Hsin Yen HONG KONG

1976

Wang Chung, Shan Mao, Wang Lung-Wei, Jet Li, Ti Lung, Shee Fong, Lin Kuang-Tseng, Li Lin Jei, Yue Chen Wei, Yue Hai, Din Nam

A secret that is of vital importance to the Emperor, is guarded by kung fu masters prepared to protect it with their lives. When two pupils, who learnt kung fu in order to avenge their fathers' deaths, return to their Shaolin temple, they find it under siege and unite with the other pupils to fight off the attackers. An enjoyable martial arts adventure, based on a legend from the folklore of Shaolin.

MAR 90 min Cut (8 sec – ort 117 min) wScrn dubbed

VIDrel: EAST/DISC V

SHAOLIN TEMPLE 2 **

18

Joseph Kuo HONG KONG

Chen Chien Chang, Sun Kuo Ming

Kung fu adventure set in the last days of the Ming Dynasty, and

telling of how the last Ming princess seeks refuge in a Shaolin mountain temple, where the priests prepare to defend her from her attackers.
Aka: SHAOLIN TEMPLE STRIKES BACK
MAR   86 min   VIDrel: IMPENT L/A   V

SHAOLIN TEMPLE 2: KIDS FROM SHAOLIN **   15
Cheung Hsin Yen HONG KONG
Jet Li, Din Nan, Yu Cheng Wai, Yu Hai
Two rival families living on either side of a wide river distrust each other but are obliged to join forces when they are both threatened by bandits.
Aka: KIDS FROM SHAOLIN
MAR   90 min wScrn dubbed   VIDrel: EAST/DISC   V

SHAOLIN: THE BLOOD MISSION ***   18
Leung Wing Chan HONG KONG   1984
Sun Kok Ming, Poou Cheung, Huang Tang Ming (Huang Cheng-Li), Lo Wah Sing, Wong Yin Fong
An evil general is sent to discover the names of any people who oppose the rule of the Manchu Dynasty, and becomes convinced that the peaceful monks of a Shaolin temple are hiding rebels. He has the temple burned to the ground, but this is only the prelude to a fast-paced martial arts adventure.
MAR   85 min   VIDrel: IMPENT   V

SHAOLIN VERSUS LAMA **   18
Lee Tso-Nan HONG KONG   198-
Lo Jui, Chang Shan, Li Wei Yun, Robert Yang, William Yen, Ching Kuo-Chang, Sun Yung-Chi
Shaolin monks fight against their Tibetan brethren, folk not normally renowned for their martial arts prowess. Apart from this unusual premise, this is in other respects a standard fighting tale, albeit a rather violent one.
Aka: SHAOLIN AGAINST LAMA
MAR   90 min   VIDrel: AUDIO/DISC   V

SHAOLIN VERSUS NINJA *   18
Robert Tai HONG KONG
Alexander Lou, Ian Hsu, George Chang, John Wu, James Tin, Jacky Hwang
The valiant members of a Shaolin Temple battle corrupt Japanese officials in an attempt to retain control of their sanctuary.
Aka: SHAOLIN: THE STORY OF SHAOLIN
MAR   81 min   VIDrel: IMPENT   V

SHARKY'S MACHINE **   18
Burt Reynolds USA   1981
Burt Reynolds, Rachel Ward, Brian Keith, Vittorio Gassman, Charles Durning, Bernie Casey, Henry Silva, Earl Holliman, Richard Libertini, John Fiedler, Darryl Hickman, Joseph Mascolo, Carol Locatell, Hari Rhodes, James O'Connell
Extremely bloody and violent film about a vice cop and his crusade against the underworld, falling for a beautiful prostitute in the process. Written by Gerald Di Pego.
THR   118 min (Cut at film release by 9 sec – ort 120 min)
VIDrel: WHV   V/sur
Boa: novel by William Diehl.

SHARON *   18
Navred Reef USA   1971
Zebedy Colt, David Christopher, Jean Jennings, Sharon Saunders, Jamie Gillis, Susan McBain
An ever-loving Atlanta father and daughter like nothing better than a spot of incestuous love-making in between their other engagements in this dull and extremely disagreeable dud.
A   90 min   VIDrel: MOPIC/SGSVID   V

SHARON'S SECRET **   15
Michael Scott USA   1995
Mel Harris, Candace Cameron, Gregg Henry, Alex McArthur, Paul Regiina, James Pickens Jr, Elaine Kagan
A psychiatrist tries to help a disturbed teenager who is accused of the murder of her wealthy parents, and gradually begins to realise that she is not guilty but is in fact shielding someone else. Unimpressive murder mystery.
DRA   86 min (ort 91 min) mTV cC   VIDrel: CIC   V/sh

SHARPE'S BATTLE **   12
Tom Clegg UK   1995
Sean Bean, Daragh O'Malley, Allie Byrne, Oliver Cotton, Jason Durr, Hugh Fraser, Hugh Ross, Ian McNeice, Siri Neal, John Tams, Jason

Salkey, Lyndon Davies, Liam Carney, Phelim Drew, Diana Perez, Robert Hands, Maria Petrucci*
In May 1811 Sharpe is given the job of training the Royal Irish Company for battle and encounters the loathsome General Guy De Loup. As the Battle of Fuentes de Onoro approaches, and with it the outcome of the Peninsula War, Sharpe is called upon to play a major part in defeating the plans of the French. The seventh entry, followed by SHARPE'S SWORD.
A/AD   101 min mTV   VIDrel: CTE/CARL   V
Boa: novel by Bernard Cornwell.

SHARPE'S COMPANY **   PG
Tom Clegg UK   1992
Sean Bean, Assumpta Serna, Brian Cox, Hugh Fraser, Michael Byrne, Clive Francis, Nicholas Jones, Peter Postlethwaite, Daragh O'Malley, Michael Mears, John Tams, Jason Salkey, Lyndon Davies, Clive Francis, Nicholas Jones
Sharpe is ordered to lay a siege to the fortified city of Badajos, but experiences a conflict of interest when he learns that his lover, the guerilla fighter Teresa, is trapped within the city. However, his task is made all the more difficult by the machinations of his mortal enemy Obadiah Hakeswill. The third entry in the series, it was followed by SHARPE'S ENEMY.
WAR   101 min mTV   VIDrel: CTE/CARL   V
Boa: novel by Bernard Cornwell.

SHARPE'S EAGLES **   15
Tom Clegg UK   1992
Sean Bean, Brian Cox, Assumpta Serna, Katia Caballero, Martin Jacobs, Katia Caballero, Daragh O'Malley, MIchael Mears, John Tams, Jason Salkey, Paul Trussell, Lyndon Davies, Michael Cochrance, Gavin O'Herlihy, David Ashton
Second feature-length episode from this series about a British officer fighting the French during the Napoleonic War. When the regiment's colours are captured because of the incompetence and inexperience of his superiors, Sharpe has to risk his life and that of his men to avenge this insult. Set during the Talavera campaign of July 1809, and followed by SHARPE'S COMPANY.
WAR   101 min mTV   VIDrel: CENTV/CARL   V
Boa: novel by Bernard Cornwell.

SHARPE'S ENEMY **   12
Tom Clegg UK   1994
Sean Bean, Daragh O'Malley, Michael Byrne, Jeremy Child, Hugh Fraser, Francois Guetary, Tony Haygarth, Elizabeth Hurley, Pete Postlethwaite, Assumpta Serna, Daragh O'Malley, Michael Mears, John Tams, Jason Salkey, Lyndon Davies
Sharpe risks his life when he has to deal with a gang of deserters whose leader proves to be none other than his old enemy Hakeswill. The latter have taken a number of people hostage, including a former lover, but this is only the start of Sharpe's troubles as he must prevent the French from occupying Portugal and thus defeating the allied armies. Set at Christmas time in 1812, this fourth entry was followed by SHARPE'S HONOUR.
A/AD   101 min mTV   VIDrel: CTE/CARL   V
Boa: novel by Bernard Cornwell.

SHARPE'S GOLD **   12
Tom Clegg UK   1995
Sean Bean, Deragh O'Malley, Jayne Linehan, Hugh Ross, Hugh Fraser, Rosaleen Linehan, Jayne Ashbourne, Peter Eyre, Abel Fork, Philip McGough, Ian Shaw, John Tams, Michael Mears, Jason Salkey, Lyndon Davies, Julian Sims, Diana Perez
Major Sharpe is given the unenviable task of trading guns for British deserters, and must fight a battle of wits with El Catolico, a ruthless guerilla leader. With Wellington's army short of funds, Sharpe is ordered to investigate rumours of Aztec gold hidden in the Portuguese hills, but finds himself trapped in the besieged town of Almeida. Set in August 1810, this sixth entry was followed SHARPE'S BATTLE.
WAR   103 min mTV   VIDrel: CTE/CARL   V
Boa: novel by Bernard Cornwell.

SHARPE'S HONOUR **   15
Tom Clegg UK   1992
Sean Bean, Diana Perez, Feodor Atkine, Alice Krige, Lyndon Davies, Michael Byrne, Jeremy Child, Hugh Fraser, Francois Guetary, Tony Haygarth, Elizabeth Hurley, Peter Postlethwaite, Assumpta Serna, Daragh O'Malley, John Tams
Major Sharpe gets caught up in a French plot, is captured by the enemy and sentenced to death for the alleged killing of a Spanish aristocrat. This results in a desperate attempt to prove his inno-

cence, by venturing deep behind enemy lines to capture a female spy. Set during the Victoria Campaign from February to June of 1813, this fifth entry was followed by SHARPE'S GOLD.
WAR 101 min mTV VIDrel: CTÉ/CARL V
Boa: novel by Bernard Cornwell.

SHARPE'S MISSION ** 15
Tom Clegg UK 1996
Sean Bean, Daragh O'Malley, Abigail Cruttenden, Hugh Fraser, James Laurenson, Mark Strong, Andrew Schofield, Jason Salkey, Nigel Betts, Aysun Metiner, Diana Perez, Warren Saire
Sharpe is sent on a mission behind the French lines, his task being to destroy an enemy gunpowder store. By way of a sub-plot, he also has the task of revealing a traitor in the British forces, and preventing valuable information from falling into the hands of the French. The eleventh episode in the series.
WAR 101 min mTV VIDrel: CARL/TECH V
Boa: novel by Bernard Cornwell.

SHARPE'S REGIMENT ** 12
Tom Clegg UK 1996
Sean Bean, Daragh O'Malley, Abigail Cruttenden, Michael Cochrane, Nicholas Farrell, Caroline Langrishe, James Laurenson, Mark Lambert, Julian Fellowes, Norman Rossington
Further exploits of Sharpe, with our gallant soldier returning in the company of Sergeant Harper to England, their intention being to gain fresh recruits for the now depleted South Essex Battalion. Here, he soon smells the stench of corruption as he moves about in society, not only encountering his future wife but also the Prince Regent himself. Ninth episode in the series, followed by SHARPE'S SIEGE.
WAR 102 min mTV VIDrel: CTE/CARL V
Boa: novel by Bernard Cornwell.

SHARPE'S RIFLES ** 15
Tom Clegg UK 1992
Sean Bean, Assumpta Serna, Brian Cox, Simon Andreu, David Throughton, Daragh O'Malley, Michael Mears, John Tams, Jason Salkey, Paul Trussell, Lyndon Davis, Julian Fellowes, Tim Bentinck, Richard Ireson, Martin Jacobs, Malcolm Jamieson
Napoleonic War tale of a British soldier in Spain and his various adventures. Having risen to the rank of lieutenant our hero is popular with his men but arouses the envy of the more aristo-cratic officers. In fighting the French invasion of Galicia in 1809, he wins both the admiration of the Spanish resistance fighters and the love of a woman. First in a set of films based on Cornwell's novels. Several sequences not seen when first tele-vised are included.
WAR 90 min mTV VIDrel: CTE/CARL V
Boa: novel by Bernard Cornwell.

SHARPE'S SIEGE ** 12
Tom Clegg UK 1996
Sean Bean, Daragh O'Malley, Abigail Cruttenden, Hugh Fraser, James Laurenson, Nicholas Farrell, Feodor Atkine, Christian Brendel, Christopher Villiers, Amira Casar, Philip Whitchurch
Having married his sweetheart, Sharpe is given little opportu-nity to enjoy married life, when he is ordered to capture a French fort in the Pyrenees, in a mission placed under the command of a reckless and over-ambitious colonel. Whilst he is away fight-ing, his wife falls seriously ill, but there are many dangers he must overcome before he can return to England. Fortunately, help arrives from an unexpected quarter. Number ten, with SHARPE'S MISSION following.
WAR 102 min mTV VIDrel: CARL/TECH V
Boa: novel by Bernard Cornwell.

SHARPE'S SWORD ** 12
Tom Clegg UK 1995
Sean Bean, Daragh O'Malley, Emily Mortimer, Michael Cochrane, Patrick Flerry, John Kavanagh, Stephen Moore, Hugh Ross, James Purefoy, Stephen Moore, Vernon Dobtcheff, John Tams, Jason Salkey, Diana Perez, Pat Laffan, Bob White
Set against the background of the Salamanca Campaign of June to July 1812, with Sharpe being ordered to prevent the assassi-nation by the French of a master spy known as El Mirador. He achieves this goal by capturing Colonel Lerouz, the French officer charged with this mission, but when the latter escapes, Sharpe is soon plunged into a deadly struggle, in which he loses his sword and almost his life. Number eight in the series, followed by SHARPE'S REGIMENT.
WAR 102 min mTV VIDrel: CTE/CARL V
Boa: novel by Bernard Cornwell.

SHATTER DEAD ** 18
Scooter McCrae USA 1993
Stark Raven, Flora Fauna, Larry "Smalls" Johnson, Marina Del Ray
Made for only $4,000 (it was shot on video) this low-budget film is set in the future, in a world where no-one ever dies, instead becoming members of a zombie-like underclass, who are reduced to begging. On the way home one evening, a young woman has several frightening encounters with these creatures. Despite being seriously hampered by its low production values, this is a genuinely creepy and chilling work, it's a pity its message is so ambiguous and uncertain.
HOR 84 min VIDrel: SCEDGE/RTM V

SHATTERED ** 15
Lamont Johnson USA 1989
Shelley Long, Tom Conti, John Rubinstein, Alan Fudge, Jamie Rose, Christine Healy, Frank Converse
A woman suffering from multiple personalities undergoes therapy and learns that this condition results from abuse she suffered as a child, when both she and her sister were sexually abused by their brutal stepfather. Having learnt that he is still alive, she leaves – apparently on a mission of vengeance. Allegedly based on a true story, but Long's distinct lack of range and the implausibilities of the script do not work to the film's advantage.
DRA 120 min VIDrel: NWV L/A V

SHATTERED ** 15
Wolfgang Petersen USA 1991
Tom Berenger, Greta Scacchi, Bob Hoskins, Joanne Whalley-Kilmer, Corbin Bernsen, Debi A. Monahan, Bert Robario, Scott Getlin, Kellye Nakahara, Frank Caveston, Dona Hardy, Jasmin Gabler, Charlene Hall, Derek Torsek, Theodore Bikel
Petersen's debut tries to revive the film noir genre of yesterday, and it opens with a car crashing down a mountainside. The female passenger is thrown clear, whilst the man awakens in a hospital bed suffering from amnesia. Told of his identity, he attempts to pick up the strands of his life, but serious doubts soon begin to emerge, and it's not long before a complex (and totally unbelievable) plot is uncovered. A ponderous offering.
THR 93 min (ort 98 min) VIDrel: VCC/COLUM L/A V/sh
Boa: novel The Plastic Nightmare by Richard Neely.

SHATTERED DREAMS ** 15
Robert Iscove USA 1990
Lindsay Wagner, Michael Nouri, Georgann Johnson, James Karen, Irene Miracle, Tim Ahern, Russ Anderson, Bud Anthony, Marianne Guran, Jeff Ellison, Joe Faust, Lou Felder, Guy Garnier, Preston Hanson, Virginia Hawkins
Factually based study of the wife of a high-ranking Washington official whose life was one endless round of domestic violence at the hands of her sadistic husband. A worthy and touching subject, but unfortunately one that is handled with neither inspi-ration nor insight, and is seriously hampered by a feeble script and flat direction.
DRA 90 min (ort 100 min) mTV VIDrel: GUILD/SONOP V
Boa: book by Charlotte Fedders and Laura Elliott.

SHATTERED FAMILY *** PG
Sandy Smolan USA 1993
Richard Crenna, Rhea Pearlman, Linda Kelsey, Cotter Smith, Joycelyn O'Brien, Cyril O'Reilly, Sam Anderson, Amy Aquino, Adilah Barnes, Gary Bater, Tom Guiry, Lezlie Deane, Elizabeth Dennehy, Hayley Carr, Chris Carrara
A court battle ensues when eleven-year-old Gregory Kingsley opts to live permanently with his foster parents, and leads eventually to the first American action in which a child "divorced" or freed itself from its natural parents. Based on a true story, but not very absorbing for all that. See also SWITCHING PARENTS.
DRA 88 min (ort 90 min) mTV VIDrel: ODY/SONOP V

SHATTERED PROMISES ** 15
John Korty USA 1995
Brian Dennehy, Treat Wiliams, Susan Ruttan, Embeth Davidtz
A brilliant criminal lawyer who has links with the Mafia becomes the chief suspect when his wife is murdered. Little new is on offer here, but thankfully the casting of the ever-reli-able Dennehy in the central role makes this effort worth watching.
DRA 173 min (ort 175 min) VIDrel: ODY/SONOP V/h

SHATTERED SILENCE ** 15
Linda Otto USA   1992
Bonnie Bedelia, Terence Knox, Kenneth Welsh, Nick Searcy, Pam Grier, Caroline Dollar, Harold Bergman, Grayce Spence, Patricia Neal, Rip Torn, Jim Sharp, Al Wiggins, Rick Warner, Patricia Gray, Rhoda Griffis, Helen Casey
In 1987 Dr Elizabeth Morgan sent her daughter into hiding rather than allow her ex-husband his court-ordered visiting rights, as she accused him of sexual abuse of her two daughters (a claim her husband, Dr Eric Foretich has strenuously and consistently denied). For her pains, this defiance of the courts resulted in a spell of imprisonment, but to date she has been awarded sole custody of her daughter Hilary by the New Zealand family court. Fair.
DRA   91 min   VIDrel: MIA/DISC/IMPENT   V

SHATTERED TRUST ** 15
Bill Corcoran USA   1993
Melissa Gilbert, Kate Nelligan, Ellen Burstyn, Shirley Dougls, Dick latessa, Keneth Walsh, Nigel Bennett, Geoffrey Bowes, Jed Dixon, Ron Hartman, Henritte Ivanans, Kirsten Kieferle, Michael Rhoades, Anna-Louise Richardson, Danika Banka
A woman lawyer handling a nasty case concerning a man accused of sexual abuse cracks under the strain and gets psychiatric help. But this therapy reveals to her that she was in fact sexually abused as a child by her father, who committed incest with her. Based on a true story, which led to a change in the law, namely removal in the U.S. of the statute of limitations in the case of such crimes.
DRA   88 min mTV   VIDrel: ODY/SONOP   V/sh

SHATTERING THE SILENCE *** 15
Linda Otto USA   1993
Joanna Kerns, Michael Brandon, Tony Roberts, George Grizzard, Shelley Hack, Richard Gailliland, Valerie Landsburg, Laura Owens, Tony Roberts, Diana Merrill, Molly Orr, Rose Weaver, Brett Kelly, Charles McLawhorn, Peter Gregory
A happily-married mother learns that she has for years repressed her early childhood memories of sexual abuse and realises that unless she acts, her niece may suffer the same fate.
Aka: NOT IN MY FAMILY
DRA   87 min (ort 90 min)   VIDrel: 4-FRONT/POLYREC   V/sh

SHAWSHANK REDEMPTION, THE **** 15
Frank Darabunt USA   1994
Tim Robbins, Morgan Freeman, Bob Gunton, William Sadler, James Whitmore, Gil Bellows, Clancy Brown, Mark Rolston, Jeffrey DeMunn, Larry Brandenburg, Neil Giuntoli, Brian Libby, David Proval, Joseph Ragno, Jude Ciccolella, Paul McCrane
Unusual, riveting and moving dramatisation of King's novel of prison life, as seen through the eyes of a prisoner (a banker) given a double life sentence for killing his unfaithful wife and lover. While in prison he forms an unlikely friendship with the prison "fixer", who can secure virtually anything for the right price. A high quality drama telling of hope and redemption, it was nominated for no less than seven Oscars.
DRA   136 min (ort 143 min) wScrn; 140 min (LV version)
VIDrel: VCC/DISC; PION (LV only)   V/sh LV
Boa: short story Rita Hayworth and the Shawshank Redemption by Stephen King.

SHE * 18
Avi Nesher ITALY   1982 (released 1985)
Sandahl Bergman, Harrison Muller, David Goss, Quinn Kessler, Elena Wiedermann, Gordon Mitchell, Laurie Sherman, Andrew McLeay, Cyrus Elias, David Brandon, Susan Adler, Gregory Snegoff, Mary D'Antin, Mario Pedone
Very loose adaptation of the classic story of a female warrior who leads her tribe against its enemies. In this variant, our female goddess leads her band of Amazons against mutants and various other bloodthirsty tribes, in order to establish her supremacy. A feeble and vacuous yarn.
A/AD   100 min (ort 104 min)   VIDrel: SPEAR/SONOP   V
Boa: novel by H. Rider Haggard.

SHE-DEVIL ** 15
Susan Seidelman USA   1989
Meryl Streep, Roseanne Barr, Ed Begley Jr, Sylvia Miles, Linda Hunt, Bryan Larkin, Elisebeth Peters, A. Martinez, Deborah Rush, Mary Louise Wilson, Maria Pitillo, Susan Willis, Jack Gilpin, Robin Leach, Nitchie Barrett, June Gable
A mutilated version of Weldon's comic novel, that follows the revenge plotted by dowdy Barr, as she sets out to destroy the life of her unfaithful husband after he goes off to live with a romantic novelist. This crude and grotesque comedy is hampered by sluggish plotting and Barr's lack of dramatic range, but Streep is quite wonderful as the glamorous writer, and does all she can with her limited part. Disappointing. See THE FIRST WIVES CLUB.
COM   95 min (ort 99 min)   VIDrel: VISVID/POLYREC L/A   V
Boa: The Life and Loves of a She Devil by Fay Weldon.

SHE-DEVILS ON WHEELS * 18
Herschell Gordon Lewis USA   1968
Betty Connell, Nancy Lee Noble, Pat Poston, Christie Wagner, Ruby Tuesday, John Weymer, Rodney Bedell, Joani Kramer, David Harris, Donna Testa, Laura Platz, Steve White, Roy Collodi, Rick Williams, Donna Stelzer, John Coffin
The story of a deadly female bike gang, told, as only he can, by the gore master of B-movies. A weekly cycle race which the gang leader is used to winning, sets in motion a cycle of revenge and death, when a male gang appears on the scene. A nasty and sickening parade of hard-core cruelty that is as meaningless as it is revolting.
DRA   83 min (ort 90 min)   VIDrel RTM/DISC   V

SHE FOUGHT ALONE ** 15
Christopher Leitch USA   1995
Tiffani-Amber Thiessen, Brian Austin Green, Isabella Hofmann, David Lipper, Maureen Flannigan, Keith MacKechnie, Jessie Robertson, Babs George, Mariella Marich, Jill Parker-Jones, Richard A. Jones, Ashley Jones, Rachel Wolfe
A young girl becomes a social outcast both at her high school and in town when she accuses another student of rape, and is obliged to go to great lengths to clear her name.
DRA   90 min mTV   VIDrel: ODY/SONOP   V/sh

SHE KILLED IN ECSTASY * 18
Frank Hollmann (Jesus Franco) SPAIN/
WEST GERMANY   1970
Soledad Miranda, Fred Williams, Paul Muller, Ewa Stroemburg, Howard Vernon, Horst Tappert, Germano Robles, Jesus Franco
Miranda plays a demented woman, who sets out to avenge the wrongs she feels her doctor father endured, when he was condemned by his peers for experimenting on embryos, and consequently killed himself. Three male and one female doctor are her victims, the men being first seduced and then tortured and castrated. A sick blend of sex and depravity, directed in a flat style (probably due to Franco working on no less than eight films in that year).
Aka: MRS HYDE; SIE TOTETE IN EKSTASE
HOR   86 min (ort 77 min – German dialogue) wScrn
VIDrel: REDEM/RTM   V

SHE LED TWO LIVES *** PG
Bill Corcoran USA   1994
Connie Sellecca, A. Martinez, Patricia Clarkson, J. Smith-Cameron, David Wohl, George Martin, J.C. Cutler, Sally Wingert, Signe Albertson, John Patrick Martin, Barbara Kingsley, Richard Anson, George P. Farr, Peter Thoemke
True story of a woman who meets a former lover and takes up where she left off, but unfortunately does not tell him that she is now married. One bigamous marriage later, she is obliged to lead two quite separate lives, but eventually the past catches up with her.
DRA   88 min (ort 90 min) mTV   VIDrel: ODY/SONOP   V/sh

SHE QUEST * 18
Mitchell Spinelli USA   1993
Leena, Jonathan Morgan, Jake Williams, Sahara Sands, Steve Austin, Angel Bust, Lawrie Cameron
A group of woman take part in an underwater quest to locate buried treasure, but this is no more than a device to allow some very busty actresses to parade about in scuba gear.
A   48 min (ort 80 min)   VIDrel: ONE   V

SHE SAID NO *** 15
John Patterson USA   1990
Veronica Hamel, Judd Hirsch, Lee Grant, Mariclare Costello, Ray Baker, Deborah White, Becky Ann Backer, Isabel Estorick, Loren Hayes, Emily Kuroda, Rosalea Mayeux, Elmarie Wendel, Andrea Garfield, Douglas Stark, Chance Boyer
A woman raped by a lawyer sues him but finds herself on the

receiving end, when he brings a counter-action alleging slander. Although hardly a balanced examination of the various short-comings inherent in the legal system, this absorbing court-room drama is extremely well done. Both Hamel and Hirsch give fine performances, the latter is especially memorable as the loath-some lawyer, but this would have been a far better drama with a few shades of grey.
DRA   91 min (ort 100 min) mTV   VIDrel: ODY/SONOP V/sh

SHE STOOD ALONE: THE TAILHOOK SCANDAL *   15
Larry Shaw USA   1995
Gail O'Grady, Bess Armstrong, Hal Holbrook, Rip Torn, Robert Urich
The true story of US Navy Lieutenant Paula Coughlin, who was assaulted whilst attending a convention, but found her complaints ignored and no attempt made to bring her attackers to book. Left with no other alternative, she publicly exposed her superiors, an action that failed to endear her to the US Navy top brass.
DRA   92 min   VIDrel: ODY/SONOP   V/sh

SHE WAS MARKED FOR MURDER *   15
Charles Thompson USA   1989
Stefanie Powers, Lloyd Bridges, Hunt Block, Debrah Farentino, Polly Bergen, Lou Liberatore, Randy Brooks, Eddie Jones, Dana Stevens, Rowena Balos, Steve Forleo, T. Patrick Cavanaugh, Michael Fox, Tom Henschell, Ben Kronen
A hot-shot publisher who recently lost her husband meets a nice young man, enjoys a whirlwind courtship, and winds up marry-ing him. However, she has time to repent when she learns that he's not quite as lovable as she thought he was. An anaemic little thriller whose plot was culled from a thousand similar pieces.
THR   100 min mTV   VIDrel: MGM/WHV L/A   V

SHE WORE A YELLOW RIBBON **   U
John Ford USA   1949
John Wayne, Joanne Dru, John Agar, Victor McLaglen, Ben Johnson, Harry Carey Jr, Mildred Natwick, George O'Brien, Arthur Shields, Francis Ford, Harry Woods, Chief Big Tree, Cliff Lyons, Noble Johnson, Michael Dugan
The second in Ford's cavalry trilogy, with Wayne as the dutiful officer reluctant to retire in the face of an imminent Indian upris-ing. An excellent work, followed by RIO GRANDE. Filmed in Technicolor and written by Frank Nugent and Lawrence Stallings. AA: Cin (Winton C. Hoch).
WES   98 min (ort 103 min) B/W
VIDrel: VCC/DISC/COLUM L/A   V
Boa: short story by James Warner Bellah.

SHELL SEEKERS, THE *   U
Waris Hussein UK/USA   1989
Angela Lansbury, Sam Wanamaker, Christopher Bowen, Anna Carteret, Michael Gough, Patricia Hodge, Dennis Quilley, Sophie Ward, Irene Worth, John Rowe, Mark Lewis Jones, Andrew Keir, Serena Gordon, Tracey Childs, William Hope
A reserved widow has a heart attack and finds that her inde-pendence is compromised, not least by her grown-up kids who now no longer feel she can look after herself. A strong support-ing cast helps make this one work, quite a feat in the face of the contrived and mawkish script, written by John Pielmeier from Pilcher's novel.
DRA   98 min (ort 100 min) mTV   VIDrel: ODY/SONOP   V
Boa: novel by Rosamunde Pilcher.

SHE'LL TAKE ROMANCE *   PG
Piers Haggard USA   1990
Linda Evans, Larry Poindexter, Tom Skerritt, Alan Blumenfield, Kenneth Mars, John Procaccino, Joe Restivo, Grant Goodeve, Robert Zenk, Walt Beaver, Frank Roberts, Matthew Billings, Tom Hammond, Scott Black, Larry Paulsen
Lightweight comedy in which a woman who appears to have all the good things in life finds excitement and romance outside the stable and safe relationship she enjoys with her fiance.
COM   90 min   VIDrel: NWV/HIFLI L/A   V

SHELTERING SKY, THE *   18
Bernardo Bertolucci USA   1990
Debra Winger, John Malkovich, Campbell Scott, Jill Bennett, Timothy Spall, Eric Vu-an, Sotigui Koyate, Amina Annabi, Paul Bowles, Tom Novembre, Ben Smail, Philippe Morier-Genoud, Kamel Cherif, Afifi Mohammed, Brahim Oubana
A visually striking but otherwise obscure tale of a triangle rela-tionship among three American intellectuals who, on a visit to North Africa in the late 1940s, argue, wander aimlessly and make love (all activities shown in some detail). Camerawork and music (by Ryuichi Sakamoto) are often haunting, but this film, though highly atmospheric, conveys little of the original and virtually unfilmable novel. This was Bennett's last film.
DRA   132 min (ort 139 min) wScrn   VIDrel: 20TH/TECH V/sur
Boa: book by Paul Bowles.

SHENANDOAH *   PG
Andrew V. McLaglen USA   1965
James Stewart, Doug McClure, Glenn Corbett, Rosemary Forsyth, Katherine Ross, Patrick Wayne, Phillip Alford, Charles Robinson, Denver Pyle, George Kennedy, Paul Fix, Tim McIntire, James McMullan, James Best, Warren Oates
Saga of the American Civil War and how it affects a Virginia family. Stewart as a widower is indifferent to the struggle between the states, until his family is reluctantly drawn into it. A moving and intelligent tale that formed the basis for a later Broadway musical. This was Ross's film debut. Written by James Lee Barrett.
WES   100 min (ort 105 min)   VIDrel: L/A   V

SHEPHERD ON THE ROCK *   (PG)
Bob Keen UK   1995
Bernard Hill, Betsy Brantley, Doug Bradley, Mary McKenna, Oliver Parker, John Bowles, Ronald Simon, Bill Denniston, Jean Faulds, Marilyn Gray, Don Anderson, James Bryce, Ron Paterson, John Stewart, Dave Stirruo, Simon Longworth
In Scotland a sheep farmer finds that his land has oil and has to contend with the developers who want to exploit it.
DRA   95 min   SATrel: MOVIE CHANNEL

SHERLOCK HOLMES AND THE INCIDENT AT VICTORIA FALLS *   PG
Bill Corcoran UK   1990
Christopher Lee, Patrick Macnee, Jenny Seagrove, Claude Akins, Richard Todd, Joss Ackland, John Indi, Stephen Gurney, Sonitha Singh, Claudia Udy, Anthony Fridjhon, Neil McCarthy, Pat Pillay, Dale Cutts, Alan Coates, Margaret John
Holmes comes out of retirement to supervise the transport of a priceless diamond from Africa to London, a mission which strange to relate, involves him in various dangerous encoun-ters. Adequate Sherlock Holmes adventure that is based on a story by Gerry O'Hara rather than an original Conan Doyle work.
DRA   180 min   VIDrel: POLY/POLYREC/BRAVE   V

**SHERLOCK HOLMES AND THE LEADING LADY *   PG
Peter Sasdy UK   1991
Christopher Lee, Patrick Macnee, Morgan Fairchild, John Bennett, Englebert Humperdinck, Tom Lahm, Ronald Hines, Nicolas Gecks, Jenny Quayle, Michael Siberry, Frank Middlemass, Charlotte Attenborough, James Bree, John Gower
An investigation leads Holmes and his friend Dr Watson to Vienna, and a meeting with a woman he once loved. Added to this is a sub-plot involving a ruthless villain who could plunge the world into war. A rather ineffective offering, not based on a Conan Doyle work, but instead on an original script, written by Bob Shayne and H.R.F. Keating.
DRA   180 min   VIDrel: POLY/POLYREC/BRAVE L/A   V

SHERLOCK HOLMES AND THE SECRET WEAPON *   PG
Roy William Neill USA   1942
Basil Rathbone, Nigel Bruce, Lionel Atwill, Kaaren Verne, William Post Jr, Dennis Hoey, Mary Gordon, Harry Woods, George Burr MacAnnan, Paul Fix, Henry Victor, Holmes Herbert, Harold De Becker, Harry Cording, Paul Bryar
The fourth film in a long series, this is set rather unexpectedly in London of the 1940s, with our sleuth now involved in foiling a Nazi attempt to steal an experimental bomb-sight. This one contains a few elements from the Arthur Conan Doyle story "The Dancing Men". SHERLOCK HOLMES IN WASHINGTON followed. Average.
DRA   68 min B/W   VIDrel: ORBIT/DISC   V

SHERLOCK HOLMES AND THE SPIDERWOMAN *   U
Roy William Neill USA   1943
Basil Rathbone, Nigel Bruce, Gale Sondergaard, Dennis Hoey, Mary Gordon, Arthur Hohl, Alec Craig

Holmes comes up against Sondergaard, in the guise of a femme fatale who is responsible for a series of unexplained "suicides", the victims all being found to have large insurance policies. One of the best in the series, this adventure has an alluring villainess and a gripping climax set in a shooting gallery.
Aka: SPIDER WOMAN
DRA   59 min (ort 62 min) B/W   VIDrel: 20TH/TECH L/A V/h

SHERLOCK HOLMES AND THE VOICE OF TERROR *   U
John Rawlins USA   1942
Basil Rathbone, Nigel Bruce, Hillary Brooke, Evelyn Ankers, Reginald Denny, Montagu Love, Mary Gordon, Thomas Gomez, Henry Daniell, Olaf Hytten, Harry Stubbs, Edgar Barrier, Robert O. Davies, Lon Chaney Jr
An outbreak of sabotage leads Holmes to a traitor in the Cabinet Office of His Majesty's Government. Number 3 in a long-running series, and followed by SHERLOCK HOLMES AND THE SECRET WEAPON. Fair. The script is by Lynn Riggs.
THR   62 min (ort 65 min) B/W   VIDrel: 20TH/TECH L/A V/h
Boa: short story His Last Bow by Arthur Conan Doyle.

SHERLOCK HOLMES AND THE WOMAN IN GREEN *   U
Roy William Neill USA   1945
Basil Rathbone, Nigel Bruce, Hillary Brooke, Henry Daniell, Paul Cavanagh, Matthew Boulton, Eve Amber, Frederic Worlock, Tom Bryson, Sally Shepherd, Mary Gordon, Percival Vivian, Olaf Hytten, Harold de Becker, Tommy Hughes
A number of strange murders baffle the police (don't they always?) and our sleuth is called in, discovering the existence of a blackmail outfit that makes use of a lady hypnotist. A fairly enjoyable adventure (the eleventh) in a long-running series, it marked the final appearance of an arch enemy of Holmes – Professor Moriarty.
Aka: WOMAN IN GREEN, THE
DRA   64 min (ort 66 min) B/W   VIDrel: ORBIT/DISC   V

SHERLOCK HOLMES: DRESSED TO KILL *   U
Roy William Neil USA   1946
Basil Rathbone, Nigel Bruce, Tom Dillon, Edmond Breon, Patricia Morison, Frederic Worlock, Carl Harbord, Patricia Cameron, Tom P. Dillon, Topsy Glyn, Harry Cording, Mary Gordon, Ian Wolfe, Lillian Bronson, Cyril Delevanti
A Sherlock Holmes adventure in a long-running series made in the 1940s. In this tale Holmes and Watson are on the trail of three wooden boxes. The last in a series of 14 films, that made use of Rathbone and Bruce as Holmes and Watson respectively, and started with SHERLOCK HOLMES: THE HOUND OF THE BASKERVILLES and SHERLOCK HOLMES: THE ADVENTURES OF SHERLOCK HOLMES. Fair, but by no means one of their best efforts.
Aka: DRESSED TO KILL; SHERLOCK HOLMES AND THE SECRET CODE
DRA   72 min B/W   VIDrel: ORBIT/VCC   V

SHERLOCK HOLMES FACES DEATH *   U
Roy William Neill USA   1943
Basil Rathbone, Nigel Bruce, Hillary Brooke, Milburn Stone, Halliwell Hobbes, Arthur Margetson, Dennis Hoey, Gavin Muir, Frederic Worlock, Olaf Hytten, Gerald Hamer, Mary Gordon, Vernon Downing
Watson asks Holmes to solve a series of bizarre deaths at a convalescent home for retired officers. A good film in the series that sticks fairly closely to the basics of the story. A highlight is Holmes's demonstration of his deductive powers, when he gathers all the suspects together and moves them about like so many chess pieces.
DRA   65 min (ort 68 min) B/W   VIDrel: 20TH/TECH L/A V/h
Boa: short story The Musgrave Ritual by Arthur Conan Doyle.

SHERLOCK HOLMES IN WASHINGTON *   U
Roy William Neill USA   1942
Basil Rathbone, Nigel Bruce, Marjorie Lord, Henry Daniell, George Zucco, John Archer, Gavin Muir
Another updated but enjoyable entry in this variable series, with Holmes now in Washington where he breaks a Nazi spy ring and prevents an important microfilmed document from falling into the wrong hands. A strong beginning gives way to a sluggish middle portion, but the film is redeemed by the gripping climax.
DRA   68 min (ort 71 min) B/W   VIDrel: 20TH/TECH L/A V/h

SHERLOCK HOLMES RETURNS:
THE ADVENTURE OF THE TIGER MURDERS *   PG
Kenneth Johnson USA   1994
Anthony Higgins, Debrah Farentino, Mark Adair Rios, Joy Coghill, Julian Christopher, Ken Pogue
After a hundred years in hibernation, our famous sleuth awakens in a secret vault in a house in San Francisco, from where he sallies forth to begin a new life of crime-solving, teaming up with a female doctor to assist him (sadly, Watson didn't make the journey). In this story, he is obliged to battle the descendants of his old adversary Dr Moriarty, who are behind a series of drug-related murders.
A/AD   90 min   VIDrel: ARENA/SPEAR   V

SHERLOCK HOLMES: TERROR BY NIGHT *   U
Roy William Neil USA   1946
Basil Rathbone, Nigel Bruce, Alan Mowbray, Dennis Hoey, Renee Godfrey, Mary Forbes, Billy Bevan, Frederic Worlock, Leyland Hodgson, Geoffrey Steele, Boyd Davis, Janet Murdoch, Skelton Knaggs, Gerald Hamer, Harry Cording
Routine Sherlock Holmes adventure, in which a stolen jewel is recovered and murders are solved on a speeding train, complete with interposed shots of the supposed scenery to be seen between London and Edinburgh. Number 13 in a long-running series.
Aka: TERROR BY NIGHT
DRA   60 min B/W   VIDrel: ORBIT/DISC   V

SHERLOCK HOLMES: THE ADVENTURES OF SHERLOCK HOLMES **   PG
Alfred L. Werker USA   1939
Basil Rathbone, Nigel Bruce, George Zucco, Ida Lupino, Alan Marshal, Terry Kilburn, E.E. Clive, Mary Gordon, Henry Stephenson, Arthur Hohl, May Beatty, Peter Willes, Holmes Herbert, George Regas, Mary Forbes, Frank Dawson
Released together with SHERLOCK HOLMES: THE HOUND OF THE BASKERVILLES, both films marked the start of a series based on Conan Doyle's famous sleuth. In this yarn, Holmes is sent on a false trail by his old enemy Moriarty, as the latter is planning to steal the Crown Jewels. Look out for the sequence where Rathbone sings a charming little ditty in disguise. Written by Edwin Blum and followed by SHERLOCK HOLMES AND THE VOICE OF TERROR.
Aka: ADVENTURES OF SHERLOCK HOLMES, THE
DRA   81 min (ort 86 min) B/W   VIDrel: L/A   V

SHERLOCK HOLMES: THE BASKERVILLE CURSE *   U
Alex Nicholas AUSTRALIA   1983
Voices of: Peter O'Toole, Ron Haddrick, Earle Cross, Helen Morse, Robin Stewart, Moya O'Sullivan, Phillip Hinton
Intended for kids, this animated version of the Conan Doyle tale of a monstrous dog said to curse the Baskerville family sticks reasonably well to the original.
Aka: BASKERVILLE CURSE, THE; SHERLOCK HOLMES AND THE BASKERVILLE CURSE
ANIM   69 min   VIDrel: TRING   V
Boa: novella The Hound of the Baskervilles by Arthur Conan Doyle.

SHERLOCK HOLMES: THE HOUND OF THE BASKERVILLES **   PG
Sidney Lanfield USA   1939
Basil Rathbone, Nigel Bruce, Richard Greene, Wendy Barrie, Lionel Atwill, John Carradine, Beryl Mercer, Mary Gordon, E.E. Clive, Morton Lowry, Ralph Forbes, Barlowe Borland, Eily Malyon, Ivan Simpson
Having inherited a Dartmoor estate from his uncle, who met a tragic end, a young man calls in Holmes in order to avoid a similar fate. Rathbone's first appearance as the sleuth gives ample demonstration of his suitability, and this careful though studio-bound adaptation is atmospheric if a little languid. A highlight is Rathbone's appearance as a tramp (Holmes was after all a master of disguise).
Aka: HOUND OF THE BASKERVILLES, THE
DRA   80 min B/W   VIDrel: L/A   V
Boa: novella by Arthur Conan Doyle.

SHERLOCK HOLMES: THE HOUND OF THE BASKERVILLES *   15
Douglas Hickox UK   1983
Ian Richardson, Martin Shaw, Denholm Elliott, Brian Blessed, Connie Booth, Donald Churchill, Nicholas Clay, Ronald Lacy, Peter

Rutherford, Francesca Gonshaw, Eric Richard, David Langton, Michael Burrell, Cindu O'Callaghan
Film version of the Sherlock Holmes tale, in which the famous detective has to solve an ancient family curse. Competent rather than memorable.
Aka: HOUND OF THE BASKERVILLES, THE
DRA   96 min (ort 101 min)   VIDrel: CREMED/LABY   V
Boa: novella by Arthur Conan Doyle.

SHERLOCK HOLMES: THE MASKS OF DEATH **   PG
Roy Ward Baker UK   1984
Peter Cushing, John Mills, Anne Baxter, Ray Milland, Gordon Jackson, Anton Diffring, Susan Penhaligon, Marcus Gilbert, Jenny Laird, Russell Hunter, James Cossins, Eric Dodson, Georgina Coombs, James Mead, Dominic Murphy
Set in 1913, this tale has Holmes coming out of retirement to solve a case that threatens the security of Britain, and whose trail of murder leads him to an old and dangerous enemy.
Aka: MASKS OF DEATH, THE
DRA   80 min mTV   VIDrel: L/A   V

SHERLOCK HOLMES: THE SCARLET CLAW ***   PG
Roy William Neill USA   1944
Basil Rathbone, Nigel Bruce, Gerard Hamer, Arthur Hohl, Miles Mander, Ian Wolfe, Paul Cavanaugh, Kay Harding
A series of nasty murders takes place at a remote Canadian village, but fortunately Holmes is on hand to nab the culprit. One of the best in this updated Holmes series, the mood is unashamedly patriotic and the settings realistic. Only the plot (loosely based on "The Hound Of The Baskervilles") is decidedly weak.
Aka: SCARLET CLAW, THE
DRA   70 min (ort 74 min) B/W   VIDrel: L/A   V

SHERLOCK HOLMES: THE SIGN OF FOUR ***   15
Desmond Davies UK   1983
Ian Richardson, David Healy, Thorley Walters, Cherie Lunghi, Joe Melia, Terence Rigby, Donald Churchill, Michael O'Hogan, John Pedrick, Clive Merrison, Darren Michael, Kate Binchy, Moti Makan, John Benfield, Robert Russell
A remake of the 1932 Sherlock Holmes mystery in which our sleuth finds himself working on a mystery that involves a secret pact, hidden treasure and a murderous pygmy. Richardson gives a most enjoyable performance that recalls much of the attraction of those Basil Rathbone B-movies of the 1940s.
Aka: SHERLOCK HOLMES' THE SIGN OF THE FOUR; SIGN OF FOUR, THE
A/AD   96 min (ort 100 min)   VIDrel: CREMED/LABY V/sur
Boa: short story by Arthur Conan Doyle.

SHERLOCK HOLMES: THE VALLEY OF FEAR **   U
Warwick Gilbert/Di Rudder AUSTRALIA   1984
Voices of: Peter O'Toole, Earle Cross, Brian Adams, Colin Borgonon, Judy Nunn, Henry Szeps, Ron Haddrick, Robin Stewart, John Stone
Holmes and Watson investigate a murder at Burlston House.
Aka: SHERLOCK HOLMES AND THE VALLEY OF FEAR; VALLEY OF FEAR, THE
ANIM   48 min   VIDrel: TRING   V
Boa: short story The Valley of Fear by Arthur Conan Doyle.

SHERLOCK: UNDERCOVER DOG **   PG
Richard Harding Gardner USA   1994
Benjamin Eroen, Brynne Cameron, Anthony Simmons, Margy Moore
A couple of kids find themselves working with a talking police-dog in their efforts to bring to justice a gang of smugglers and rescue the dog's kidnapped master. A bright and breezy juvenile romp.
Aka: UNDERCOVER DOG
JUV   80 min   VIDrel: EIV/SONOP   V

SHERWOOD'S TRAVELS **   (PG)
Steve Miner UK   1994
Jamey Sheridan, Serena Scott Thomas, Edward Fox, Prunella Scales, Julian Fellowes, Ben Gazzara, Girolami Aldieri, Natale Nazzareno, Pasquale Finicelli, Lawrence A. Kapust, Noa Meldy, Cyrus Ellias, Gianna Paola Scaffidi
Mystery drama in which two writers go in search of a missing cross and have many adventures and narrow escapes.
DRA   89 min   SATrel: SKY MOVIES

SHE'S BACK *   18
Tim Kincaid USA   1988
Carrie Fisher, Robert Joy, Matthew Cowles, Joel Swetow, Sam Coppola, Donna Drake, Anthony Mannino, Bobby Di Cicco, Erick Avari, Gary Yudman, Robert Bottone, Sam Cugnina, Michael Speero, Robert La Moia, Billy Morrissette
A tasteless comedy in which a man's wife is murdered by a gang of thugs, just after they move into their new home. Her ghost returns to nag her husband until he satisfies her desire for revenge.
COM   88 min   VIDrel: VES L/A   V

SHE'S GOTTA HAVE IT ***   18
Spike Lee USA   1986
Tracy Camila Johns, Tommy Redmond Hicks, John Canada Terrell, Spike Lee, Raye Dowell, Bill Lee
The story of a sexy young woman, her complicated love life and the three men who compete for her affections. A vigorous if uneven little film, quite charming in its way. The director's father, Bill Lee, provides the catchy jazz score.
COM   85 min B/W/Col   VIDrel: PAL L/A   V

SHE'S HAVING A BABY **   15
John Hughes USA   1988
Kevin Bacon, Elizabeth McGovern, Alec Baldwin, Isabel Lorca, William Windom, Cathryn Damon, Holland Taylor, James Ray, Dennis Dugan, John Ashton, Edie McClurg, Paul Gleason, Reba McInney, Bill Erwin, Anthony Mockus Jr, Larry Hankin
A lighthearted look at the problems facing a newly-married couple, all told from the point of view of the husband who is beginning to feel trapped by it all. Not quite the fresh and witty comedy it was intended to be, but despite the rather tedious plot, the stars perform well enough to make it watchable.
COM   101 min (ort 106 min)   VIDrel: CIC/SONOP   V/sur

SHE'S IN THE ARMY NOW **   PG
Hy Averback USA   1981
Kathleen Quinlan, Jamie Lee Curtis, Susan Blanchard, Julie Carmen, Melanie Griffith, Janet MacLachlan, Dale Robinette, Robert Peirce, Lynn Barbara Block, Rocky Bauer, Douglas Dirkson, Damita Jo Freeman, Susan Barnes
A PRIVATE BENJAMIN clone that formed the pilot for a TV series, and tells of the comic mishaps and romantic entanglements of five young women who join the army. A film as unfunny as it is unoriginal.
Aka: G.I. JOANS
COM   96 min Cut (7 sec) mTV   VIDrel: L/A   V

SHE'S OUT OF CONTROL *   15
Stan Dragoti USA   1989
Tony Danza, Catherine Hicks, Wallace Shawn, Dick O'Neill, Ami Dolenz, Laura Mooney, Derek McGrath, Dana Ashbrook
Danza plays a widower whose precocious and rapidly maturing daughter gives him a few headaches when she is transformed into a raving beauty via some beauty tips, courtesy of his girlfriend. A one-idea comedy that takes an idea more suited to a 30-minute TV sketch and stretches it just about as far as it will go.
COM   90 min (ort 95 min)   VIDrel: RCA/VCC L/A   V

SHE'S THE ONE **   15
Edward Burns USA   1996
Jennifer Aniston, Maxine Bahns, Edward Burns, Cameron Diaz, John Mahoney, Mike McGlone, Anita Gillette, Leslie Mann, Amanda Peet, Frank Vincent, Malachy McCourt, Beatrice Winde, Eugene Osborne Smith, Robert Weil, Tom Tammi
The sexual antics among a group of characters in Brooklyn form the basis for this rather tedious comedy of manners, which was scripted by Burns. Occasionally mildly amusing it may be, but the plot is so insubstantial one has to struggle to stay involved.
COM   100 min   CINrel

SHINING, THE **   18
Stanley Kubrick UK   1980
Jack Nicholson, Shelley Duval, Danny Lloyd, Scatman Crothers, Barry Nelson, Joe Turkel, Philip Stone, Lia Beldam, Billie Gibson, Barry Dennan, David Baxt, Lisa Burns, Alison Coleridge, Kate Phelps, Anne Jackson, Tony Burton
A man takes on the job of caretaker to a deserted hotel, in its closed season. As he begins to fall victim to evil supernatural forces, he undergoes an unpleasant personality change that threatens his wife and son. A moody but cumbersome tale, that has Nicholson over-acting so wildly that it all becomes rather

ludicrous. Nevertheless, there are some genuinely powerful moments.
HOR 114 min (ort 146 min) VIDrel: WHV V/sur
Boa: novel by Stephen King.

SHINING THROUGH *
David Seltzer USA   15
                                                                  1992
Michael Douglas, Melanie Griffith, Liam Neeson, Joely Richardson, John Gielgud, Francis Guinan, Patrick Winczewski, Anthony Walters, Victoria Shalet, Sheila Allen, Stanley Beard, Sylvia Syms, Ronald Nitschke, Hansi Jochmann
A dreadful WW2 spy thriller, following the exploits of a half-Irish, half-Jewish New York secretary, who falls in love with her boss but learns that he is a colonel in the OSS. Implausible as it may sound, this leads her to volunteer for a secret mission to Berlin, where she hopes to uncover the secrets of Hitler's rocket programme, and possibly rescue a trio of Jewish relatives at the same time. Daft from start to finish, with Douglas never more wooden.
THR 127 min (ort 133 min) cC VIDrel: 20TH/TECH V/sh
Boa: novel by Susan Isaacs.

SHIP THAT DIED OF SHAME, THE ***
Basil Dearden UK   PG
                                                                  1955
Richard Attenborough, George Baker, Bill Owen, Virginia McKenna, Roland Culver, Bernard Lee, Ralph Truman, John Chandos, Harold Goodwin, John Longden, David Langston, Stratford Johns
At the end of WW2, the former crew of a motor gunboat team up again, and buy their old ship in order to use it for smuggling. Unfortunately, they find that their increasingly grimy exploits lead to a "rebellion" on the part of their dishonoured vessel. A thinly-plotted moral fable with a predictable outcome, chiefly sustained by an excellent cast.
Aka: P.T. RAIDERS
DRA 89 min (ort 95 min) B/W VIDrel: BRAVE/SONOP L/A V
Boa: novel by Nicholas Monsarrat.

SHIRLEY VALENTINE ***
Lewis Gilbert UK/USA   15
                                                                  1989
Pauline Collins, Tom Conti, Alison Steadman, Julia McKenzie, Joanna Lumley, Bernard Hill, Sylvia Syms, George Costigan, Anna Keaveney, Tracie Bennett, Ken Sharrock, Karen Craig, Gareth Jefferson, Gillian Kearney, Cardew Robinson
Collins repeats her stage role in this quirky little comedy about a bored middle-aged housewife, who leaves her dull life behind when she heads for the Greek isles and a spot of extra-marital romance. Written by Russell, who also gave us EDUCATING RITA, this wry little film is sustained by its witty observations on life, delivered by Collins in a splendid Tony Award-winning performance. Not exactly a masterpiece, but quite charming.
COM 104 min (ort 108 min) VIDrel: CIC/SONOP V/sur
Boa: play by Willy Russell.

SHIVERS **
David Cronenberg CANADA   18
                                                                  1975
Paul Hampton, Joe Silver, Lyn Lowry, Alan Migicovsky, Barbara Steele, Susan Petrie, Ronald Mlodzik, Barrie Baldaro, Camille Ducharme, Al Rochman, Hanna Poznanska, Wally Martin, Vlastra Vrana, Charles Perley, Julie Wildman
The tenants of an apartment block fall victim to nasty parasites that cause violent sexual excesses. The first big film from the director, with an abundance of gruesome effects that have now become his trademark. Not an especially well made film, it is handicapped by uncertain direction and uneven pacing, but does have moments of power. A TV version was also made, but this was cut to 77 minutes. See also PARASITE.
Aka: PARASITE MURDERS, THE; THEY CAME FROM WITHIN
HOR 84 min (ort 86 min) VIDrel: ARROW/RTM V

SHOAH *****
Claude Lanzmann FRANCE   Ex
                                                                  1985
Eschewing all documentary footage, this masterpiece attempts to analyse both the how and why of the extermination of the Jews by the Nazis during WW2. Bystanders, victims and perpetrators are persuaded, cajoled or even tricked into giving details of the part they played in these events, and under the director's probing and masterly questioning a true and deeply disturbing picture of this evil begins to emerge. The director took ten years to make this landmark film.
DOC 550 min (four cassettes) VIDrel: ACAD/RTM L/A
V

SHOCK ***
Mario Bava/Lamberto Bava ITALY   18
                                                                  1977
John Steiner, Daria Nicolodi, David Colin Jr, Ivan Rassimov, Nicola Salerno
Bava's last feature tells of a woman's son who is possessed by the spirit of his late father, and used as a tool to drive his mother insane, as a way for the dead man to avenge his brutal murder at her hands. A highly atmospheric film, but one in which much gratuitous gore tends to detract from the narrative. A loose remake called UNTIL DEATH, was made by Bava's son in 1987.
Aka: ALL 33 DI VIA OROLOGIO FA SEMPRE FREDDO; BEYOND THE DOOR 2; SHOCK (TRANSFER SUSPENSE HYPNOS); SUSPENSE
HOR 87 min VIDrel: STABL L/A V

SHOCK CORRIDOR ***
Samuel Fuller USA   15
                                                                  1963
Peter Breck, Constance Towers, Gene Evans, James Best, Hari Rhodes, Philip Ahn, Paul Dubov, Frank Gerstle, William Zuckert, John Matthews, John Craig, Larry Tucker, Chuck Roberson, Neyle Morrow, Linda Randolph, Rachel Roman
A reporter pretends to be mad in order to solve the mystery surrounding the death of an inmate of an asylum, but soon finds himself overwhelmed by the experience. A taut and gripping drama that still retains considerable power. Scripted by Fuller and photographed by Stanley Cortez.
DRA 101 min B/W/Col VIDrel: PAL L/A V

SHOCK TO THE SYSTEM, A **
Jan Egleson USA   15
                                                                  1989
Michael Caine, Elizabeth McGovern, Peter Riegert, Swoosie Kurtz, Will Patton, Jenny Wright, John McMartin, Barbara Baxley, Haviland Morris, Kent Broadhurst, Philip Moon, Zach Grenier, David Schramm, Sam Schacht, Mia Dillon
A man passed over for promotion discovers that murder is an easy solution to his problems, and starts to apply it as a tool of corporate management. A well acted but disappointingly superficial black comedy that takes some predictable swipes at a familiar target.
DRA 84 min (ort 88 min) VIDrel: MED/POLYREC L/A V/sh
Boa: novel by Simon Brett.

SHOCK WAVES **
Ken Weiderhorn USA   18
                                                                  1977
Peter Cushing, Brooke Adams, John Carradine, Fred Buch, Jack Davidson, Luke Halpin, D.J. Sidney, Don Stout, Tony Moskal, Gary Levinson, Bob Miller, Bob White, Jay Meader, Talmadge Scott
Tourists land on a Caribbean island, and discover that a former SS officer has been conducting strange experiments in the creation of Nazi androids. A low-budget effort that has been somewhat ignored, but is well worth seeing for the imaginative handling of its ideas.
Aka: ALMOST HUMAN; DEATH CORPS; DEATH WAVES
HOR 86 min VIDrel: VIPCO/SGSVID V

SHOCKER ***
Wes Craven USA   18
                                                                  1989
Michael Murphy, Peter Berg, Mitch Pileggi, Cami Cooper, John Tesh, Heather Langenkamp, Jessica Craven, Richard Brooks, Sam Scarber, Virginia Morris, Emily Samuel, Keith Anthony, Lubow Bellamy, Peter Tilden, Bingham Ray, Theodore Raimi
A demented TV repairman with a penchant for hacking families to death in the evenings, is finally caught and executed, but continues to live on within TV sets, and terrorises the family of his last victim. Craven's new creation is a larger-than-life monster of massive personality, well placed to take over from Freddy of the NIGHTMARE ON ELM STREET films. Written by Craven, this gruesome shocker employs some remarkable state-of-the-art visual effects.
HOR 105 min (ort 111 min)
VIDrel: 4-FRONT/POLYREC/GUILD V/sh

SHOGUN ***
Jerry London JAPAN/USA   15
                                                                  1981
Richard Chamberlain, Toshiro Mifune, Yoko Shimada, Frankie Sakai, Yuki Meguro, John Rhys-Davies, Michael Hordern, Alan Badel, Damien Thomas, Leon Lissek, Nobuo Kaneko, Vladek Sheybal, Hideo Takamatsu, Hiromi Senno
An ambitious TV adaptation of Clavell's novel, loosely based on the true story of the rise of Toranaga, first of the Japanese Shoguns, and his relationship with Captain Blackthorne, one of a group of English sailors shipwrecked off the coast of Japan. A

colourful and detailed account, but one with much unnecessary footage. This was Badel's last film. Narration is by Orson Welles.
DRA   VIDrel: CIC/SONOP   V
Boa: novel by James Clavell.

SHOGUN ASSASSIN ***
Kenji Misumi/Robert Houston JAPAN/USA   18
                                                   1981
Tomisaburo Wakayama, Akihiro Tomikawa, Kayo Matsuo, Minoru Ooki, Shoji Kobayashi, Mori Kishida, Reiko Kasahara, Yukari Wakayama, Yurio Mishima Voices of: Lamont Johnson, Marshall Efron
A strangely compelling film (the second in a series but with 12 minutes from the first film) that features a master swordsman whose wife is killed by the Emperor's assassins, and is forced to flee with his son. Thereafter, he travels the country, wreaking vengeance on the Emperor's servants. A superior samurai film, that's marred by poor dubbing added in 1980, when the film was bought by Roger Corman's New World Company.
Aka: BABY CART AT THE RIVER STYX; KOSURE OOKAMI; KOSURE OOKAMI N. 2; SANZU NO KAWA NO UBAGURAMA
MAR   81 min (ort 89 min) dubbed   VIDrel: VIPCO/SGSVID
V/sur

SHOOT THE MOON **
Allan Parker USA   15
                                                   1981
Albert Finney, Diane Keaton, Karen Allen, Peter Weller, Dana Hill, Viveka Davis, Tracey Gold, Tina Yothers, George Murdock, Leora Dana, Lou Cutell, Irving Metzman, Kenneth Kimmins, Michael Aldredge, Robert Costanzo, James Cranna
A study of marital breakdown set in Marin County, California, that largely explores the effect this has on the children, and the new relationships the partners embark on. A competent but depressing piece offering no memorable insights or fresh ideas. Written by Bo Goldman.
DRA   118 min (ort 123 min)   VIDrel: MGM/WHV L/A   V

SHOOT THE PIANIST ***
Francois Truffaut FRANCE   15
                                                   1960
Charles Aznavour, Marie Dubois, Nicole Berger, Michele Mercier, Albert Remy, Jacques Aslanian, Richard Kanayan, Claude Mansard, Daniel Boulanger, Serge Davri, Claude Heymann, Alex Joffe, Boby Lapointe, Catherine Lutz
A former concert pianist, reduced to working in a bar for a living, becomes involved with gangsters in this homage by Truffaut to the American B-movie. A stylish New Wave pastiche with a plot that fluctuates wildly in mood, but also demonstrates Truffaut's power and innovation as a director.
Aka: SHOOT THE PIANO PLAYER; TIREZ SUR LE PIANISTE
DRA   78 min (ort 84 min) B/W wScrn   VIDrel: ARTIF/20TH
V/h
Boa: novel Down There by David Goodis.

SHOOTDOWN **
Michael Pressman USA   (PG)
                                                   1988
Angela Lansbury, George Coe, Molly Hagan, Kyle Secor, Jennifer Savidge, John Cullum, Diana Bellamy, Alan Fudge, Booth Colman, Richard McKenzie, Robin Curtis, Terri Hanauer, Haunani Minn, Paul Linke, Richard Green, George Grant
The relatives of passengers on a civilian airliner that strayed over Soviet territory and was shot down, press the American government for a full investigation into the circumstances surrounding the flight. The first of two TV films (CODED HOSTILE being the other) that attempted to examine this tragic event, when a South Korean aeroplane was shot down by Soviet fighter planes in 1983, killing all 269 onboard. A good quality thriller, competent and absorbing.
THR   93 min (ort 100 min) mTV   TVrel

SHOOTER, THE *
Ted Kotcheff USA   18
                                                   1994
Dolph Lundgren, Maruschka Detmers, Assumpta Serna, Gavan O'Herlihy, John Ashton, Simon Andreu, Pablo Scola, Petr Drozda, Roslav Walter, Michael Rogers, Pavel Vokoun, Martin Hub, Jana Altmanova, Jiri Kraus, Giulio Kukurugya
When the Cuban Ambassador to the UN is assassinated, a US Federal Marshal is sent to investigate, and finds the trail leading to Prague and a female assassin, whom he is ordered to eliminate. But as matters develop, he begins to have doubts about her complicity in this business, and much confusion follows before the truth is revealed. Seemingly partially inspired by NIKITA, this inept mess lacks originality and tension, and does little to entertain.
Aka: HIDDEN ASSASSIN
A/AD   103 min   VIDrel: POLY/POLYREC   V/sh

SHOOTFIGHTER **
Pat Alan USA   18
                                                   1992
Bolo Yeung, Martin Kove, William Zabka, Michael Bernardo, Edward Albert, Maryam D'Abo, James Pak, Sigal Diamant, Lang Yung, Sagiv Diamant, Alexia Damon, Richard Eden, Jack Ong, George Kee Cheung, Hakim Alston, Roger Yuan, Joe Son
Two practitioners of a brutal martial art continue a rivalry that began in childhood until they face each other in a final deadly showdown. One is a murderous thug, the other a man of honour. A repulsively violent and gory movie that opens with a murder in the ring and goes downhill from there on.
Aka: SHOOTFIGHTER: FIGHT TO THE DEATH
MAR   90 min (ort 96 min)   VIDrel: VCC/DISC L/A   V

SHOOTING, THE ***
Monte Hellman USA   PG
                                                   1966
Millie Perkins, Jack Nicholson, Will Hutchins, Warren Oates, B.J. Merholz, Charles Eastman, Guy El Tsosie
A former bounty hunter has a hired gunman on his trail in this offbeat and unconventional revenge yarn. Nicholson is excellent as the hired killer but the film's best moments and unusual climax are hampered by the enigmatic script.
WES   71 min (ort 82 min)   VIDrel: CREMED/LABY   V

SHOOTING ELIZABETH **
Baz Taylor FRANCE   15
                                                   1992
Jeff Goldblum, Mimi Rogers, Juan Echanove, Simon Andreu, Fernando Guillen Cuero, Cristina Higueras, Burt Kwouk, Ernesto Alterio, Alberto Jiminez Arias Arias, Santiago Alvarez, Ferran Audi, Alicia Borrachero, Richard Borras
A disgruntled husband has some nasty plans for his wife, but when she disappears without trace, he finds himself the only suspect in a murder investigation. His only recourse is to try and convince the police that he is a model husband. A frantic and unfunny comedy, with Goldblum giving a weak and unusually flat performance.
COM   92 min   VIDrel: MED/20VIS L/A   V/sh

SHOOTING PARTY, THE ***
Alan Bridges UK   15
                                                   1984
James Mason, Edward Fox, Dorothy Tutin, John Gielgud, Gordon Jackson, Cheryl Campbell, Robert Hardy, Aharon Ipale, Rupert Frazer, Judi Bowker, Joris Stuyck, Rebecca Saire, Sarah Badel, John Carney, Ann Castle, Daniel Chatto
A weekend party in 1913 recreates the vanished world of the British aristocracy, that is mistakenly thought to have ended with WW1. An absorbing but ultimately rather meaningless character study, made enjoyable by some appealing performances. The screenplay is by Julian Bond.
DRA   93 min (ort 108 min)   VIDrel: BRAVE/SONOP L/A
V/sh
Boa: novel by Isabel Colegate.

SHOOTIST, THE ****
Don Siegel USA   PG
                                                   1976
John Wayne, Lauren Bacall, James Stewart, Ron Howard, Harry Morgan, Richard Boone, John Carradine, Hugh O'Brian, Sheree North, Richard Lenz, Scatman Crothers, Bill McKinney, Gregg Palmer, Alfred Dennis, Dick Winslow
Set in 1901, Wayne's last film follows the final days of a legendary gunman who has learned that he has cancer and wishes to finalise his affairs. Unfortunately, his reputation precludes a peaceful death. From the clever opening sequence (in which good use is made of clips from old John Wayne movies) to the satisfactory conclusion, this superior Western is both a solid piece of entertainment and a touching tribute to a long career.
WES   95 min (ort 100 min)
VIDrel: 4-FRONT/POLYREC/CIC   V/h
Boa: novel by Glendon Swarthout.

SHOP ON THE HIGH STREET, THE ****
Jan Kadar/Elmar Klos CZECHOSLOVAKIA   15
                                                   1965
Ida Kaminska, Josef Kroner, Hana Slivkova, Frantisek Zvarik
During WW2 in occupied Czechoslovakia, a elderly Jewish shop-keeper thinks that the Aryan overseer placed in charge of her button shop is in fact a new assistant. Being completely deaf, this lady is quite unaware that there is even a war in progress, and as her "assistant" grows increasingly fond of her, he makes a clumsy attempt to protect her, but this has unhappy conse-

quences. A sharp, sad, witty and most moving satire on the tragic events of the time.
Aka: OBCHOD OD NA KORZE; SHOP ON MAIN STREET, THE
COM   119 min (ort 128 min) B/W   VIDrel: CONNO/RTM
V

SHOPPING *   18
Paul Anderson UK   1994
Sadie Frost, Sean Pertwee, Jude Law, Sean Bean, Fraser James, Danny Newman, Jonathan Pryce, Marianne Faithfull, Lee Whitlock, Ralph Nelson, Eammon Walker, Jason Isaacs, Chris Constantinou, Tilly Yosburgh, Melanie Hill, Grant Russell
A slightly futuristic "Bonnie and Clyde" story of a young couple who steal expensive cars and deliberately wreck them in between bouts of ram-raid shoplifting. Set against a sprawl of urban decay and grime, this is a noisy, violent and nihilistic exercise in the pointlessness of existence, all topped off with a suitably brutal ending.
A/AD   103 min (ort 107 min)   VIDrel: POLY/POLYREC
V/s

SHORT CIRCUIT ***   PG
John Badham USA   1986
Steve Guttenberg, Ally Sheedy, Fisher Stevens, Austin Pendleton, Brian McNamara, G.W. Bailey, Martin McIntyre, John Garaber, Penny Santon, Vernon Weddle, Barbara Tarbuck, Tom Lawrence, Fred Slyter, Tim Blaney (voice only)
A brilliant but totally reclusive inventor has constructed robots which possess the capacity to destroy entire cities. At a military demonstration, the fifth one in the series is struck by lightning and becomes self-aware. Eager to escape from those who wish to terminate it, it takes refuge with animal-lover Sheedy, who protects it from the commandos sent to track it down. An amusingly quirky tale, followed by the inevitable (and far weaker) sequel.
FAN   95 min (ort 99 min)   VIDrel: 20TH/TECH L/A; ENCORE (LV only)   V LV

SHORT CIRCUIT 2 **   PG
Kenneth Johnson USA   1988
Fisher Stevens, Cynthia Gibb, Michael McKean, Jack Weston, David Hemblen, Dee McCaffrey, Don Lake, Damon D'Oliveira, Tito Nunez, Jason Kuriloff, Lilli Franks, Robert La Sardo, Wayne Best, Adam Ludwig, Tim Blaney (voice only)
Sequel to the first film with Stevens and Number 5 now having further adventures as they meet up with several strange characters. This slightly overlong sequel lacks the freshness of the first film and has a flawed plot, but remains generally agreeable and fast-paced.
FAN   106 min Cut (1 sec plus some cuts substituted – ort 110 min)   VIDrel: VCC/COLUM L/A   V/sh

SHORT CUTS ***   18
Robert Altman USA   1993
Andie MacDowell, Ann Archer, Matthew Modine, Jack Lemmon, Jennifer Jason Leigh, Bruce Davison, Lily Tomlin, Tom Waits, Fred Ward, Julianne Moore, Tim Robbins, Huey Lewis, Madeleine Stow, Buck Henry, Peter Gallagher, Lori Singer
Nine stories by Carver form the basis for this tale of life among a group of residents of Southern California that takes quite a cynical and jaundiced view of human nature. Well directed and acted it may be, but this ensemble piece suffers from many of the faults as the director's earlier film NASHVILLE, uneven pacing and a fragmented narrative being just two of them. See also BITS AND PIECES.
DRA   180 min (ort 188 min) wScrn   VIDrel: ARTIF/20TH
V/sur
Boa: short stories by Raymond Carver.

SHORT FILM ABOUT KILLING, A ****   18
Krzysztof Kieslowski POLAND   1988
Miroslaw Baka, Krzysztof Globisz, Jan Tesarz
Part of the director's self-imposed task of making a film built around each of the Ten Commandments this tells of an aimless, amoral young thug, who kills a boorish taxi driver, is put on trial and finally executed. The state execution is every bit as disturbing as the killer's crime, and nothing is spared in the director's determination to fully examine his subject matter. See also DEAD MAN WALKING. Winner Prix du Jury and International Critic's Prize (Cannes 1988).
Aka: KROTKI FILM O ZABIJANIU; THOU SHALT NOT KILL
DRA   85 min mTV wScrn   VIDrel: TART/20TH   V

SHORT FILM ABOUT LOVE, A ****   15
Krzysztof Kieslowski POLAND   1988
Grazyna Szapolowska, Olaf Lubaszenko, Stefania Iwinska, Piotr Machalika
Two people are deprived of love in this perceptive and mature examination of human relationships, which opens with a lonely teenager who lives with a friend of his mother's, spying on a girl who lives across the street. Winner of the Grand Prix at the Gdansk Film Festival 1988, it formed part of the director's cycle of ten works based on the Ten Commandments. A superb exploration of human hopes and frailties, its title is an ironic comment on the whole sad affair. See also the two DEKALOG films.
Aka: KROTKI FILM O MILOSCI
DRA   83 min (ort 90 min) mTV   VIDrel: TART/20TH   V

SHORT TIME **   15
Gregg Champion USA   1990
Dabney Coleman, Matt Frewer, Teri Garr, Barry Corbin, Joe Pantoliano, Xander Berkeley, Rob Roy, Jak-Erik Eriksen
A few days before he is due to retire, a veteran cop is mistakenly informed that he has a terminal illness. Worried about providing for his wife and child, he decides that the best remedy is for him to be killed on duty, thus giving his family a large insurance payout. Thanks to a convincing performance from Coleman and a few well handled action sequences, this contrived film is both entertaining and mildly amusing.
COM   93 min (ort 97 min)   VIDrel: 20VIS/SONOP   V/sur

SHOT IN THE DARK, A **   PG
Blake Edwards USA   1964
Peter Sellers, Elke Sommer, George Sanders, Herbert Lom, Tracy Reed, Graham Stark, Moira Redmond, Vanda Godsell, Maurice Kaufmann, Ann Lynn, David Lodge, Andre Maranne, Martin Benson, Reginald Beckwith, Bryan Forbes
Sellers plays bumbling detective Clouseau, who attempts to clear Sommer of a murder charge in this so-so sequel to THE PINK PANTHER, the first in a line of progressively feebler comedies. Followed by THE RETURN OF THE PINK PANTHER. The script is by Edwards and William Peter Blatty and as ever, the score is by Henry Mancini. See also INSPECTOR CLOUSEAU.
COM   98 min (ort 101 min)   VIDrel: WHV   V
Boa: plays by Harry Kurnitz and Marcel Archard.

SHOUT, THE ***   15
Jerzy Skolimowski UK   1979
Alan Bates, Tim Curry, John Hurt, Susannah York, Robert Stephens, Julian Hough, Carol Drinkwater, Nick Stringer, John Rees, Susan Woolridge
A man visiting an asylum is told a strange tale by one of the inmates, of how after learning secret powers from the Aborigines, he came to dominate and ultimately destroy the lives of a married couple he stayed with. Bates gives a hypnotic performance as the inmate with these strange powers, that include the ability to kill by shouting (the film's best sequence). A pointless but exceptionally powerful film. Written by Skolimowski and Michael Austin.
DRA   82 min (ort 86 min)   VIDrel: CARL/TECH   V/sur
Boa: short story by Robert Graves.

SHOUT ***   PG
Jeffrey Hornaday USA   1991
John Travolta, Linda Fiorentino, James Walters, Heather Graham, Richard Jordan, Scott Coffey, Glenn Quinn, Frank Von Zerneck, Michael Bacall, Johnny Depp, Sam Hennings, Gwyneth Paltrow, Kristina Simonds, Charles Taylor
The director of FLASHDANCE presents this engaging study of teenage angst in which a rebellious adolescent is sent to a harsh approved school that's run by a stern and unbending principal. Matters improve when he strikes up a friendship with the pretty daughter of the principal, and with the arrival of a rock 'n' roll loving teacher, things get even better. Don't look for realism, this film is pure escapism. This was the screen debut for Walters.
MUS   85 min (ort 90 min)   VIDrel: CIC/SONOP   V/sur

SHOUT AT THE DEVIL **   15
Peter R. Hunt UK   1976
Lee Marvin, Roger Moore, Barbara Parkins, Ian Holm, Rene Koldehoff, Gernot Endemann, Karl Michael Vogler, Horst Janson, Gerard Pasquis, Maurice Denham, Jean Kent, Heather Wright, George Coulouris, Murray Melvin, Bernard Horsfall
In Zanzibar, a poacher, his daughter and an English adventurer,

set out to blow up a German cruiser at the beginning of WW1. A ponderous and overlong tale, that has a few good action sequences but a clutch of characters too disagreeable for one to identify with. Written by Wilbur Smith, Stanley Price and Alastair Reid, and with music by Maurice Jarre.
WAR 108 min (ort 147 min) VIDrel: MED L/A V/h
Boa: novel by Wilbur Smith.

SHOW BOAT *** (U)
James Whale USA 1936
Paul Robeson, Irene Dunne, Allan Jones, Helen Morgan, Donald Cook, Charles Winniger, Helen Westley, Hattie McDaniel, Sammy White, Queenie Smith, J. Farrell MacDonald, Arthur Hohl, Charles Middleton, Francis X. Mahoney
Remake of the 1929 film about life on a Mississippi river-boat that is replete with a rich gallery of characters, including a caddish gambler who finally returns to the girl he abandoned. This version wisely avoids the faults of the earlier film and is made with great care and professionalism. Written by Oscar Hammerstein II from his book for the musical based on Ferber's novel. A further remake appeared in 1951.
MUS 110 min (ort 113 min) B/W TVrel
Boa: novel by Edna Ferber/play by Oscar Hammerstein II/Jerome Kern.

SHOW BOAT ** U
George Sidney USA 1951
Ava Gardner, Howard Keel, Kathryn Grayson, Joe E. Brown, Marge Champion, Gower Champion, Robert Sterling, William Warfield, Agnes Moorehead, Adele Jergens, Leif Erickson, Owen McGiveney, Frances Williams, Regis Toomey
A reasonable remake of the film version of a famous musical about riverboat life along the Mississippi which, though no more than a pale imitation of the 1936 classic, is enjoyable for its songs. Written by John Lee Mahin.
MUS 103 min (ort 108 min) VIDrel: MGM/WHV V/dm
Boa: novel by Edna Ferber/musical by Jerome Kern and Oscar Hammerstein II.

SHOW OF FORCE, A * 15
Bruno Barreto USA 1990
Amy Irving, Andy Garcia, Robert Duvall, Lou Diamond Phillips, Kevin Spacey, Joe Campanella, Erik Estrada, Priscilla Pointer, Hattie Winston, Jorge Luis Ramos, Juan Fernandez, Lupe Ontiveros, Leon Singer, Fernando Quinones
Brazilian director Barreto's first US movie is very loosely based on a true 1978 incident, when two Puerto Rican radicals were apparently shot dead as terrorists with the connivance of the FBI. Irving plays a TV news reporter who sets out to get at the truth and in so doing puts her own life in danger. What might well have worked as an examination of US covert action serves as little more than a dreary political thriller of no great force.
THR 89 min Cut (20 sec – ort 93 min) VIDrel: CIC/SONOP L/A V
Boa: book Murder Under Two Flags by Ann Nelson.

SHOWDOWN ** 18
Henri Pachard USA 1986
Herschel Savage, Gina Carrera, Sharon Mitchell, Shannah McCullough, Patti Petite, Nina Hartley, Joey Silvera, Jamie Gillis, Mike Horner, Henri Pachard, Nick Random
Sex Western set in a desert whorehouse that masquerades as a dude ranch, established by its owner after the local sheriff had temporarily closed down her business. The arrival of some young men whose car has broken down on the way to a legitimate dude ranch brings the girls some much-needed custom. A harmless and silly romp, whose flimsy plot is held together by an abundance of rather feeble jokes.
A 46 min Cut (1 min 21 sec – ort 75 min)
VIDrel: HAR/GOLD V

SHOWDOWN ** 18
Robert Radler USA 1993
Christine Taylor, Ken McLeod, Linda Dona, Brion James, Patrick Kilpatrick, Billy Blanks, Ken Scott
A boy attending a new school makes a lethal enemy on his first there day, an unsavoury individual who is heavily involved in promoting illegal fights. In order to survive he is obliged to learn self defence, but luckily finds a mentor in the form of the caretaker, a former cop who also happens to be a kickboxing expert.
A/AD 95 min VIDrel: MIA/DISC V

SHOWDOWN IN LITTLE TOKYO ** 18
Mark L. Lester USA 1991
Dolph Lundgren, Brandon Lee, Tia Carrere, Cary Hiroyuki-Tagawa, Toshishiro Obata, Philip Tan, Rodney Kagayama, Ernie Lively, Renee Griffin, Reid Asato, Takayo Fischer, Simon Rhee, Vernee Watson-Johnson, Lenny Immamura, Roger Yuan
With the police clearly unable to control the city's crime-rate, a determined martial artist and his friend set about cleaning up the streets of L.A. A standard blood-and-thunder actioner with all the expected trimmings.
A/AD 75 min Cut (9 sec – ort 78 min) VIDrel: WHV V/sur

SHOWGIRLS * 18
Paul Verhoeven USA 1995
Elizabeth Berkley, Gina Gershon, Kyle MacLachlan, Glenn Plummer, Robert Davi, Alan Rachins, Gina Ravera, Lin Tucci, Greg Travis, Al Ruscio, Patrick Bristow, William Schockley, Michelle Johnston, Dewey Weber, Rena Riffel
An ex-hooker working as a stripper and lap dancer in a seedy nightclub is discovered by the star of a top erotic show in Las Vegas. A crude and not terribly interesting film, featuring the inevitable displays of naked flesh and a blank-faced Berkley who appears to be sleepwalking through the central role. Amazingly, this exceptionally dull film became a cult hit. See also LAP DANCER.
DRA 131 min VIDrel: GUILD/20TH V/sur

SHRUNKEN HEADS ** 18
Richard Elfman USA 1993
Aeryk Egan, Bodhi Elfman, Becky Herbst, Meg Foster, Julius W. Harris, A.J. Damato, Bo Sharon, Darris Love, Leigh-Allyn Baker, Troy Fromin
Three kids are gunned down when they attempt to rid their neighbourhood of criminal gangs but their heads are rescued by a voodoo doctor, who shrinks them in a cauldron. This has the effect of resurrecting them as vengeance-seeking undead beings who continue their crime-busting crusade. A crudely made horror spoof.
HOR 83 min (ort 86 min) VIDrel: EIV/SONOP V

SHY PEOPLE ** 15
Andrei Konchalovsky USA 1987
Jill Clayburgh, Barbara Hershey, Martha Plimpton, Merritt Butrick, John Philbin, Pruitt Taylor Vince, Don Swayze, Michael Audley, Brad Leland, Paul Landry, Mare Winningham, Tony Epper, Warren Battiste, Edward Bunker
A New York woman journalist, with her protesting daughter in tow, travels to the Louisiana backwoods to write an article on a distant branch of the family and discovers some dark secrets. A muddled, unconvincing and uninteresting melodrama, though the photography of Chris Menges offers a slight compensation.
DRA 114 min (ort 118 min) VIDrel: L/A V/sh

SIBLING RIVALRY ** 15
Carl Reiner USA 1990
Kirstie Alley, Bill Pullman, Carrie Fisher, Jami Gertz, Scott Bakula, Sam Elliott, Ed O'Neill, Frances Sternhagen, John Randolph, Paul Benedict, Bill Macy, Ed O'Neill, Matthew Laurance, Ron Orbach, Edward Escobar, Greg Collins
A woman trapped in a dull and conventional marriage spends an afternoon of passion in a hotel with an acquaintance, but unfortunately her lover is overtaxed by his exertions and promptly expires, leaving her to face the music. A frantic comedy of inane complications that strains for both laughs and credibility, but is saved from disaster by strong direction and a winning performance from Alley.
COM 84 min (ort 88 min) VIDrel: FIRST/SONOP V/sur

SICILIAN, THE * 18
Michael Cimino USA 1987
Christopher Lambert, Joss Ackland, Terence Stamp, John Turturro, Richard Bauer, Barbara Sukowa, Barry Miller, Aldo Ray, Giulia Boschi, Ray McAnally, Andreas Katsulas, Michael Wincott, Derrick Branche, Richard Venture
The story of Salvatore Giuliano and his struggle to achieve Sicily's secession from Italy in the 1940s. This dull dud fails to make anything of its material and unimaginably bad casting of Lambert ensures its failure. A haphazard blend of drama and pathos, with neither humour nor pace, this film is also available in a special "director's cut" version of 146 minutes, which is slightly better. The 1961 film "Salvatore Giuliano" is far better.
A/AD 110 min (ort 146 min) VIDrel: 20TH/TECH V/sur
Boa: novel by Mario Puzo.

SICILIAN CROSS *   18
Maurizio Lucidi/William Garroni ITALY/USA   1976
Roger Moore, Stacy Keach, Ivo Garrani, Fausto Tozzi, Ettore Manni,
Ennio Balbo, Pietro Martellanz, Romano Puppo
A San Francisco Mafia boss imports a jewel-encrusted cross as
a present for his church, but a rival gangster uses it for smug-
gling heroin. He sends Moore (laughably cast as a Sicilian) and
Keach to discover the culprit, in a flabby effort in which tedium
alternates with car chases.
Aka: EXECUTORS, THE; GLI ESECUTORI; LA CROCE SICILIANA; STREET
PEOPLE
DRA   99 min   VIDrel: IMPENT   V

SID AND NANCY *   18
Alex Cox UK   1986
Gary Oldman, Chloe Webb, David Hayman, Drew Schofield, Debby
Bishop, Tony London, Perry Benson, Gloria LeRoy, Xavier Berkeley,
Sandy Baron, Edward Tudor-Tudor, Sy Richardson, Biff Yeager,
Courtney Love, Rusty Blitz
An account of the bizarre relationship between Sid Vicious (of
the British punk rock group The Sex Pistols) and his American
groupie girlfriend Nancy Spungen. A kind of overblown docu-
drama, in which the two utterly repellent characters battle it out
on the way to self-destruction, culminating in the murder of
Spungen and the suicide of Vicious. Two powerful perfor-
mances cannot redeem this. See also THE GREAT ROCK 'N'
ROLL SWINDLE.
DRA   109 min (ort 111 min)   VIDrel: BMGVID/BMGREC
V/sh

SIDEKICKS *   15
Aaron Norris USA   1992
Beau Bridges, Chuck Norris, Jonathan Brandis, Mako, Joe Piscopo,
Danica McKellar, Julia Nickson-Soul, John Buchanan, Dennis
Burkley, Gerrit Graham, Dennis Letts, Christy Martin, David Born,
Lawrence Joe, Keefe Millard, David Cox
A young man who is being bullied at school gets little support
from his parents and falls back on fantasies in which he teams
up with Chuck Norris to beat the bad guys. This inspires him
to take up karate, which soon improves his situation beyond all
recognition and wins him the girl of his dreams., A flat and
insipid effort, something of a vehicle for Norris who gets to play
himself.
A/AD   92 min (ort 100 min)   VIDrel: EIV/SONOP   V/sur

SIEGE AT RUBY CAIRO, THE **   15
Roger Young USA   1996
.Laura Dern, Randy Quaid, Diane Ladd, Joe Don Baker, Darren E.
Burrows, Kirsten Dunst
Fact-based tale of the exploits of neo-Nazi Randy Weaver, who
having sold a sawn-off shot-gun to a government undercover
agent, faces the prospect of serving time or agreeing to become
an informer. He attempts to make a stand at the family's isolated
Ruby Ridge home, an FBI operation ensues, but this is bungled
and leads to the death of two of his children. A shallow and
exploitative film, that is uncertain exactly where its sympathies
lie.
A/AD   167 min mTV   VIDrel: BMGVID/BMGREC   V

SIEGE OF FIREBASE GLORIA, THE **   18
Brian Trenchard-Smith AUSTRALIA   1988
Wings Hauser, R. Lee Ermey, Robert Avevalo, Gary Hershberger,
Mark Neely, Clyde R. Jones, Margi Gerard, Richard Kuhlman, John
Calvin, Albert Popwell, Eric Hauser, Michael Cruz, Gael Romero,
Donald Wilson, Nick Nicholson, Ray Cedena
An account of a true incident that took place during the Tet
offensive in the course of the Vietnam War. During a 36-hour
truce in 1968, some US Marines carry out a patrol and learn of
an imminent attack against a US Army base, but the sergeant is
unable to get the base commander to believe him. Standard war
heroics, with the usual plotting, characterisation and dialogue.
Filmed on location in the Philippines.
WAR   99 min   VIDrel: SONY L/A   V

SIEGFRIED ***   U
Fritz Lang GERMANY   1924
Paul Richter, Margareta Schoen, Hanna Ralph, Bernhard Goetzke,
Theodore Loose, Hans Adalbert Von Schlettow, Rudolf Klein-Rogge,
Getrude Arnold, Erwin Biswanger, Frieda Richard, George Jurowski,
Iris Roberts
This first part of Lang's recreation of old Teutonic legends is a
vivid and exciting film that tells of how Siegfried journeys from
Iceland to Burgundy to bring back a bride for his brother-in-law,

and enjoys many adventures along the way. A lavish and finely
crafted adaptation, with stylish sets and good special effects.
Aka: DIE NIEBELUNGEN 1; NIEBELUNGEN SAGA, THE: PART 1
A/AD   85 min (ort 131 min) B/W silent
VIDrel: TART/20TH   V

SIESTA *   18
Mary Lambert UK/USA   1987
Ellen Barkin, Gabriel Byrne, Julian Sands, Martin Sheen, Jodie Foster,
Grace Jones, Isabella Rossellini, Alexei Sayle
A professional stunt-woman awakens on a Spanish run-way,
semi-nude and covered in blood. A deliberately arty and preten-
tious effort, in which our heroine slowly discovers the reasons
for her predicament, by way of a series of flashbacks and weird
encounters. Not so much a film as a failed experiment. The jazz
score by Miles Davis, is one bright spot in this mess.
THR   93 min (ort 96 min)   VIDrel: 4-FRONT/POLYREC
V/h
Boa: novel by Patrice Chaplin.

SIGN OF LEO, THE ***   PG
Eric Rohmer FRANCE   1959
Jess Hahn, Van Doude, Michele Giradon, Stephane Audran, Jean Le
Poulain, Francoise Prevost
Rohmer's debut feature is set in Paris, where a poor American
composer holds a lavish party when he hears that he has inher-
ited a fortune. Later, he learns that all the money has gone to a
cousin, and by this catastrophe he is slowly obliged to become
a beggar. A perceptive and detailed study of human frailty, the
film is weakened by its contrived happy ending, but is equally
memorable for some fine photography, the work of Nicolas
Hayer.
Aka: LE SIGNE DU LION
DRA   100 min B/W   VIDrel: HEND/BMGREC L/A   V

SILAS MARNER ***   PG
Giles Foster UK   1985
Ben Kingsley, Jenny Agutter, Patrick Ryecart, Jonathan Coy, Patsy
Kensit, Freddie Jones, Rosemary Martin, Frederick Treves, Angela
Pleasence
An adaptation of Eliot's classic story of a reclusive, miserly linen
weaver, who is consumed with bitterness at being cast out of a
religious sect, having been unjustly blamed for a theft. But when
he finds an abandoned baby girl he takes her home and brings
her up as his daughter, lavishing upon her all the love he is
capable of. A BBC film, this is a bright and moving tale, wonderfully well acted by Kingsley in the title role. See also A SIMPLE
TWIST OF FATE.
DRA   91 min mTV   VIDrel: PARADOX/TOTAL   V/h
Boa: novel by George Eliot (Marian Evans).

SILENCE, THE ***   18
Ingmar Bergman SWEDEN   1963
Ingrid Thulin, Gunnel Lindblom, Jorgen Lindstrom, Eduardo
Futierrez, Haken Jahnberg, Birger Malmsten, The Eduardino, Eduardo
Gutierrez, Lissi Alandh, Leif Forstenberg, Nils Waldt, Birger
Lensander, Eskil Kalling, Olof Widgren
The last film in Bergman's trilogy (following on from WINTER
LIGHT), this has a woman, her ten-year-old son and her sister
travelling to an unnamed country with an unrecognisable
language, where they experience various strange encounters. A
bleak and complex meditation on human life, the purpose of
which remains shrouded in obscurity. However, taut direction,
powerful acting and effective camerawork unite to make this a
most compelling work.
Aka: TYSTNADEN
DRA   95 min B/W   VIDrel: TART/20TH   V/dm

SILENCE OF ADULTERY, THE **   (PG)
Steven Stern USA   1995
Kate Jackson, Art Hindle, Patricia Gage, Kristin Fairlie, Robert
Desiderio, Kevin Zeger, Jeremy Harris, Reg Dreger, Caroline Yaeger,
Howard Jerome, Larry Reynolds, Corey Sevier, Craig Elldriger, George
Robertson, Sean Mconachie
A female psychotherapist who specialises in working with autis-
tic children sets up an equine centre for such kids, but then
becomes so obsessed with this project that she neglects her
marriage and career. These two concerns are put under even
greater strain when she then embarks on an affair with the father
of one of the children brought to the centre. Fair.
DRA   87 min (ort 90 min) mTV   SATrel: MOVIE CHANNEL
Boa: book Adultery the Forgivable Sin by Dr Bonnie Eaker Weil
with Ruth Winter.

SILENCE OF THE HAMS *   15
Ezio Greggio ITALY/USA   1993
Ezio Greggio, Dom DeLuise, Billy Zane, Joanna Pacula, Charlene Tilton, Martin Balsam, Stuart Pankin, John Astin, Phyllis Diller, Bubba Smith, Larry Storch, Rip Taylor, Shelley Winters, Nedra Volz, Rosey Brown, Henry Silva
Silly and tasteless spoof on THE SILENCE OF THE LAMBS, with the usual juvenile jokes that are as flat as can be, especially as one is hard put to find much mirth in a spoof built around a series of brutal murders.
Aka: IL SILENZIO DEI PROSCIUTTI
COM   81 min (ort 85 min)   CINrel

SILENCE OF THE HEART ***   PG
Richard Michaels USA   1984
Mariette Hartley, Howard Hesseman, Dana Hill, Chad Lowe, Silvana Gallardo, Elizabeth Berridge, Alexandra Powers, Charlie Sheen, Lynnette Mettey, Rick Fitts, Ray Giradin, Jaleel White, Rad Daly, Melissa Hayden, Jim Boyle
A poignant study of how one family is affected by a teenage suicide, with the parents unable to come to terms with the death of their son, whilst his sister tries to discover the reasons behind it. The perceptive and moving script is by Phil Penningroth.
DRA   91 min (ort 100 min) mTV   VIDrel: 20TH/TECH L/A V

SILENCE OF THE LAMBS, THE ****   18
Jonathan Demme USA   1991
Jodie Foster, Anthony Hopkins, Scott Glenn, Ted Levine, Anthony Heald, Diane Baker, Brooke Smith, Kasi Lemmons, Charles Napier, Tracey Walter, Roger Corman, Frankie Faison, Chris Isaak, Laurence T. Wrentz, Stuart Rudin, Jeffrie Lane
Trainee FBI agent Foster is recruited to help catch a ghastly serial killer, and goes to a top-security prison to interview a brilliant psychiatrist, the murderous Dr Lector, who may be able to provide some valuable insights. A powerful, intense and brilliantly acted nightmare, often repellent but hard to ignore. Lector also popped up in an earlier film, MANHUNTER. AA: Pic, Dir, Actor (Hopkins), Actress (Foster), Screen/adapt (Ted Tally).
THR   113 min (ort 118 min)   VIDrel: COLUM/SONOP V
Boa: novel by Thomas Harris.

SILENCER, THE **   18
Amy Goldstein USA   1992
Lynette Walden, Chris Mulkey, Paul Ganus, Jaime Gomez, Morton Downey Jr, Brook Parker, George Shannon, Anders Hove, Scott Kraft, Kamar Reyes, Reid Cruikshanks, Ava Dupree, Stephen P. Hart, Cole McKay, Jacqueline Jacobs
A female undercover agent is charged with the task of killing five criminals and relaxes after each murder with a brief love affair. However, she fails to realise that one of her former lovers is on her trail. A fast-paced action film of average merits, marred by intrusive rock music and rapid cutting, but spiced up with ample sex scenes.
A/AD   81 min (ort 85 min)   VIDrel: HIFLI/SONOP V/h

SILENCERS, THE *   18
Richard Pepin USA   1995
Jack Scalia, Dennis Christopher, Clarence Williams III; Lucinda Weist, Carlos Lauchu
A sinister group of assassins are out to eliminate anyone who saw a UFO, as the US government has been tricked by the aliens into agreeing to allow this, in return for promises of advanced technological knowledge. SF hokum of a high order, that has a strong feel of the 1950s, when those "aliens are out to take us over" movies were a good deal more popular than they are now. Singularly unimpressive.
FAN   103 min   VIDrel: MARQ/QUANT V/sh

SILENCES OF THE PALACE, THE ***   PG
Moufida Tlatli FRANCE/TUNISIA   1994
Ahmel Hedhili, Hend Sabri, Najia Ouerghi, Ghalia Lacroix, Sami Bouajila, Kamel Fazaa, Hichem Rostom, Helene Catzaras, Sonia Meddeb, Mechket Krifa, Kamel Touati, Fatma Ben Saidane, Zahira Ben Ammar, Sabah Bouzouita, Bechir Feni
Set in Tunisia in the 1950s, where the beautiful daughter of one of the country's last princes is brought up to be independent in mind, unlike her mother, who was obliged to give way to the sexual whims of the aristocracy. However, she both wonders who her real father is and has to contend with her own pregnancy by her long-term lover. Slow-paced and rather

moody, this touching story was an impressive directing debut for Tlatli.
Aka: LES SILENCES DU PALAIS; SAIMT EL QUSUR
DRA   127 min   VIDrel: ICAPRO/MANGA V/sh

SILENT FALL **   15
Bruce Beresford USA   1994
Richard Dreyfuss, Linda Hamilton, John Lithgow, J.T. Walsh, Ben Faulkner, Liv Tyler, Zahn McClarnon, Brandon Stouffer, Treva Moniik King, John McGee Jr, Ron Tucker, Catherine Shaffner, Heather M. Bomba, Marianne M. Bomba, Jane Beard
A child-psychologist is brought into a murder investigation nine years after he stopped practising and has to deal with the autistic son of a murdered couple who may have seen the killers of his parents. Sadly, the promising opening premise is not given the treatment it deserves and the film fails to exploit the potential inherent in the subject matter.
DRA   96 min (ort 101 min) cC   VIDrel: WHV V/sur

SILENT HEROES **   15
Richard Driscoll UK   1987
Robert Wilford, Martin Arlott, Bob Flag, Kym Ryan
Set during the Falklands War, this tale follows the exploits of a group of SAS soldiers, who are given the task of infiltrating islands prior to the British landings. Average.
WAR   90 min (ort 92 min)   VIDrel: MOPIC/SGSVID V

SILENT HUNTER **   18
Fred Williamson GERMANY/ITALY/USA   1994
Miles O'Keefe, Fred Williamson, Peter Colvey, Lynne Adams, Jason Cavalier, Sabine Karsenti, Frank Fontaine, Louis Pharand, Michael O'Reilly, Cindy Ellis, Sam Stone, Erika Rafuls, Ben Lawson, Brandon Cheatham, Anthony Giamo, Ted Vernon
A man seeking revenge for the death of his family's teams up with a local sheriff to hunt down the killers when they crash-land in a mountainous area. A standard Williamson offering with a brutal and abrupt ending.
A/AD   93 min (ort 97 min)   VIDrel: REFLEC/FIRST V

SILENT MADNESS *   18
Simon Nuchtern USA   1984
Belinda Montgomery, Viveca Lindfors, Sydney Zassick, David Greenan, Solly Max, Roderick Cook, Ed Van Nuys, Stanja Lowe, Dennis Helfend, Philip Levy, Toni Hartman, Katherine Kamhi, Katie Bull, Rick Aiello, Jeffrey Bingham
A computer error leads to the release of a homicidal mental patient instead of one who is fully recovered, and the doctor in charge goes to the town where this man committed a series of murders, in an attempt to catch him. However, two hospital orderlies follow with instructions to kill the doctor and cover up the mistake. As laughable as it is ludicrous. Originally shot in 3-D.
Aka: NIGHT KILLER; OMEGA FACTOR
HOR   87 min Cut (1 min 34 sec – ort 93 min)
VIDrel: CALECO/GOLD L/A V

SILENT MOVIE **   PG
Mel Brooks USA   1976
Mel Brooks, Marty Feldman, Dom DeLuise, Bernadette Peters, Sid Caesar, Liza Minnelli, James Caan, Madeline Kahn, Harold Gould, Fritz Feld, Paul Newman, Burt Reynolds, Anne Bancroft, Marcel Marceau, Carol Arthur, Chuck McCann
An attempt to revive silent film comedy, revolving around the efforts of a producer to get back into the big time. An interesting experiment that never really succeeds. Look out for Marcel Marceau who utters the only word in the entire movie – "non". Written by Brooks, Ron Clark, Rudy De Luca and Barry Levinson.
COM   83 min (ort 88 min)   VIDrel: 20TH/TECH V/h

SILENT PARTNER, THE ***   18
Daryl Duke CANADA   1978
Elliot Gould, Christopher Plummer, Celine Lomez, Susannah York, Michael Kirby, Kenneth Pogue, John Candy, Gail Dahms, Michael Donaghue, Jack Duffy, Sean Sullivan, Nancy Simmonds, Nuala Fitzgerald, Guy Sanvido
Slick and rather violent tale of a bank teller who, discovering that there is to be a robbery, secretes a large amount of money in order to pretend that a lot more has been stolen than was actually the case. A battle of wits now takes place, with Plummer excellent as the vicious and sadistic robber, bent on revenge. Written by Curtis Hanson with music by Oscar Peterson.
Aka: DOUBLE DEADLY
THR   105 min   VIDrel: L/A V
Boa: novel Think Of A Number by Anders Bodelson.

SILENT RAGE ** **   18
Michael Miller USA   1982
Chuck Norris, Ron Silver, Steven Keats, Toni Kalem, William Finley,
Brian Libby, Stephanie Dunnam, Stephen Furst, Joyce Ingle, Jay
DePland, Lillette Zoe Raley, Mike Johnson, Linda Tatum, Kathy Lee,
Desmond Dhooge, Joe Farago
Norris stars as the tough street-fighting sheriff of a small town,
which is being terrorised by a psychotic killer. An average blend
of mayhem and gore, with little new on offer in the plot depart-
ment except Norris' opponent – a super-human zombie. Written
by Joseph Fraley.
THR   96 min Cut (41 sec – ort 105 min)
VIDrel: MIA/DISC/COLUM   V

SILENT RUNNING **   U**
Douglas Trumbull USA   1971
Bruce Dern, Cliff Potts, Ron Rifkin, Jesse Vint, Mark Persons, Steven
Brown, Larry Whisenhunt, Cheryl Sparks, Roy Engel
In the future all that remains of Earth's plant-life is preserved
on huge spaceship-domes. When it is decided that these can
be destroyed, a "keeper" resorts to murder in order to save
one. An excellent film, with a beautiful opening and memo-
rable effects. Scripted by Michael Cimino, Steve Bochco and
Deric Washburn. The directorial debut for special effects man
Trumbull. The unusual score is by Peter Schickele, with Joan
Baez singing.
FAN   85 min (ort 90 min)   VIDrel: CIC/SONOP L/A   V

SILENT SCREAM *   15**
David Hayman UK   1989
Iain Glenn, Paul Samson, Anne Kristen, Tom Watson, David McKail,
Andrew Barr, Kenneth Glenaan, Steve Hotchkiss, John Murtagh
Based on life of Larry Winters, this story is set in Scotland, where
a convicted murderer takes a drugs overdose and finds himself
remembering the formative events of his life. A harrowing
portrait of a seriously disturbed and guilt-ridden individual,
who appears to have received no help for his mental condition
nor any sympathy from his jailors. A difficult film, it is not easy
or pleasant to watch, but Glenn's superb performance ensures
that one does so.
DRA   81 min (ort 85 min)   VIDrel: IMAG/RTM   V

SILENT THUNDER *   15**
Craig R. Baxley USA   1992
Stacy Keach, Lisa Barnes, Tom Bower, Thomas Wilson Brown, Dwier
Brown, Lenore Kasdorf, Deborah Mansy, Tym Thurston, Dennis
Hayden, Sandahl Bergman, Damu King, Matthew Hotsinpiller,
Steven Schwartz-Hartley, Jake Walker, Sam Gauny
A man sets out to avenge the death of his son at the hands of a
hit-and-run truck driver. Competent action-thriller based on a
real-life case.
A/AD   87 min (ort 90 min) mTV   VIDrel: IMPENT   V

SILENT TONGUE *   12**
Sam Shepard USA   1993
Alan Bates, Richard Harris, River Phoenix, Dermot Mulroney, Sheila
Tousey, Jeri Arrnedono, Tantoo Cardinal, Bill Irwin, David Shiner,
The Red Clay Ramblers, Arturo Gil, Joseph Griffo, Billy Beck, Phillip
Attmore, Al Lujan
When his bride dies, a young man is so inconsolable in his grief
that his father resolves to undertake a perilous journey across
the plains in search of her sister, whom he believes is the only
one who can help his son. A capable cast can to little to breathe
life into this ill-conceived tale, which was scripted by Shepard.
WES   97 min (ort 101 min)   VIDrel: EIV/SONOP   V

SILENT TOUCH, THE *   15**
Krzysztof Zanussi DENMARK/POLAND/UK   1992
Max Von Sydow, Lothaire Bluteau, Sarah Miles, Sofie Grabol,
Aleksander Bardini, Peter Hesse Overgaard, Lars Lunoe, Slawomira
Lozinska, Trevor Cooper, Stanislaw Brejdygant, Beata Tyszkiewicz,
Maja Plasczynska, Peter Thurrell
After remaining silent for some forty years, a noted composer
is visited by a young music student who has been having a
recurring dream that the former needs the tune the student has
been hearing in his head. This visit proves to have a strange and
powerful effect on the composer, who starts to work again.
Sydow's powerhouse performance helps to bolster the sagging
script, in this offbeat meditation on the problems of artistic
creation.
DRA   93 min (ort 96 min)   VIDrel: CURZON/20TH L/A
V/sh

SILENT TRIGGER *   18**
Russell Mulcahy USA   1996
Dolph Lundgren, Gina Bellman, Conrad Dunn, Christopher
Heyerdahl
A professional assassin teams up with a woman to eliminate his
next target, but finds his emotions clouding his judgement when
he starts falling for his new colleague. There is an interesting
character play between the two leads, plus Mulcahy's love of
mobile cameras and odd angles, so it really is a pity there is not
enough of a story here to hold it all together. However, one
cannot fault it in terms of technical expertise. Lundgren is
slightly less wooden than usual.
A/AD   96 min cC   VIDrel: HOLPIC/TECH   V/sur

SILENT VICTIM *   18**
Bradley Battersby USA   1992
D.B. Sweeney, Craig Sheffer, Courteney Cox
A woman tires of the violence of inner-city life in New York, and
flees to a remote town in the Southwest, but once there comes
in for some attention from a couple of admirers who may not
be quite as they appear.
THR   93 min   VIDrel: HIFLI/SONOP   V

SILHOUETTE *   15**
Carl Schenkel USA   1990
Faye Dunaway, David Rasche, John Terry, Carlos Gomez, Ron
Campbell, Margaret Blye, Talisa Soto, Ritch Brinkley, Kiersten
Warren, Perla Walter, Nancy Parsons, Glenn Quinn, Viola Kates
Stimpson, Jonathan Hale, Louie Leonardo
A sophisticated woman from the big city witnesses a murder in
a small backwoods town and is forced to go to ground there in
order to avoid the same fate at the hands of the culprit. An
unimaginative and melodramatic thriller that offers little in the
way of suspense, and with Dunaway sorely miscast in the role
of the terrified heroine.
THR   84 min (ort 96 min) mCab   VIDrel: CIC   V

SILHOUETTE *   18**
Lloyd A. Simandl USA   1991
Tracy Scoggins, Brion James, Marc Singer, Jason Scott, Marc Bennett,
Frank Wilson, Suzy Joachim,
Scoggins plays a public prosecutor who decides to take the law
into her own hands when the police are unwilling to pursue
an investigation regarding the murder of a local call-girl. This
necessitates her taking over aspects of the dead girl's identity
in a bid to get at the truth, a quest which exposes her to
considerable danger. A glossy and marginally erotic thriller,
that is replete with corny situations and unconvincing
dialogue.
THR   91 min (ort 93 min)   VIDrel: VCC/DISC L/A   V

SILHOUETTE *   15**
Eric Till CANADA   1994
Jobeth Williams, Stephanie Zimbalist, Corbin Bernsen, Winston
Rekert, Ken Pogue, Charles Boswell, Sandra Nelson, Justin Lewis,
Peter White, Roman Podhora, Sheila Moore, Deryl Hayes, Brenda
Crichchlow, Judith MAxie, Sheila Patterson
A pianist goes to L.A. to identify her murdered sister and meets
a man who claims he was her husband. With the help of a police
shrink she investigates his claim, but in so doing puts her life in
danger. Fair.
THR   87 min   VIDrel: ODY/SONOP   V/sh

SILK STOCKINGS *   U**
Rouben Mamoulian USA   1957
Fred Astaire, Cyd Charisse, Janis Paige, Peter Lorre, George Tobias,
Jules Munshin, Joseph Buloff, Barrie Chase, Belita, Wim Sonneveld,
Ivan TRiesault, Betty Barrie, Da Utti, Tybee Afra
A musical remake of NINOTCHKA, with Charisse a female
commissar on a mission to Paris, where she is to persuade a
Russian composer to return home. A lively but overlong effort.
Songs include "All Of You" and "Stereophonic Sound". Lyrics
are by Andre Previn with a Cole Porter score. Written by
Leonard Spiegelgass and Leonard Gershe. This was the direc-
tor's last film.
MUS   114 min (ort 116 min)   VIDrel: MGM/WHV   V/dm
Boa: musical by George S. Kaufman, Leueen McGrath and Abe
Burrows/play by Melchior Lengyel.

SILKEN SKIN *   PG**
Francois Truffaut FRANCE   1964
Jean Desailly, Nelly Benedetti, Francoise Dorleac, Daniel Ceccaldi,
Laurence Badie, Jean Lanier, Paule Emanuele, Philippe Dumat, Pierre

Risch, Dominique Lacarriere, Sabine Haudepin, Maurice Garrel, Gerard Poirot, Georges De Givray
A heel of a literary critic cheats on a his wife by having an affair with an airline stewardess, whom he treats in a very shabby fashion. A well-made dissection of marital infidelity spoilt only by a somewhat overly melodramatic ending.
Aka: LA PEAU DOUCE; SOFT SKIN, THE
DRA   112 min (ort 118 min) B/W   VIDrel: ARTIF/20TH V/h

SILKWOOD ***   15
Mike Nichols USA   1983
Meryl Streep, Kurt Russell, Cher, Craig T. Nelson, Diana Scarwid, Fred Ward, Ron Silver, Charles Hallahan, Josef Summer, Sudi Bond, Henderson Forsythe, Bruce McGill, E. Katherine Kerr, David Strathairn, J.C. Quinn, Les Lannom
Recreates the story of Karen Silkwood, who died in mysterious circumstances when she set out to expose the total disregard for statutory safety procedures at a nuclear facility she worked at. Fine performances and care in production sustain this film further than the thin little script would normally carry it. Written by Nora Ephron and Alice Arlen. See also THE PLUTO-NIUM INCIDENT.
DRA   125 min (ort 131 min)   VIDrel: VCC/DISC; PION (LV only)   V LV

SILVER BEARS ***   PG
Ivan Passer UK   1977
Michael Caine, Cybill Shepherd, Louis Jourdan, Martin Balsam, David Warner, Stephane Audran, Charles Gray, Tom Smothers, Jay Leno, Tony Mascia, Jeremy Clyde, Joss Ackland, Moustache, Mike Falco, Philip Mascellino, Steve Plytas
Complex but highly entertaining comedy-thriller, telling of a Mafia plan to launder money by the simple expedient of buying a bank in Switzerland, and then using it in a scheme to manip-ulate the world's silver market. The clever script is the work of Peter Stone. Filmed in Switzerland and Morocco.
Aka: GOLD STRIKE
COM   108 min (ort 114 min)   VIDrel: WHV   V/h
Boa: novel by Paul Erdman.

SILVER BRUMBY, THE **   U
John Tatoulis AUSTRALIA   1994
Caroline Goodall, Russell Crowe
Based on an Australian children's story, this opens with a farmer's wife telling her daughter folktales about a legendary wild horse or brumby that roams the wilds of Australia. Crowe plays the cattle merchant who dreams of capturing the horse for himself. A nice outdoors tale, a bit thinly plotted but quite diverting.
DRA   91 min   VIDrel: 20TH/FOXVID   V/sur

SILVER BULLET **   18
Daniel Attias USA   1985
Corey Haim, Gary Busey, Megan Follows, Everett McGill, Kent Bradhurst, Terry O'Quinn, Robin Groves, Leon Russom, Bill Smitrovich, James Gammon, Heather Simmons, Joe Wright, James A. Baffico, Rebecca Fleming, Lawrence Tierney
A small town is terrorised by a series of brutal murders that seem to be the work of a werewolf. A crippled boy, his sister and their uncle become involved in the hunt for the killer. A limp horror movie that could have done with a stronger storyline.
Aka: STEPHEN KING'S SILVER BULLET
HOR   90 min (ort 95 min)   VIDrel: MGM/WHV L/A   V
Boa: novelette Cycle Of The Werewolf by Stephen King.

SILVER DRAGON NINJA **   18
Don Kong HONG KONG   198-
Harry Caine, Sam Yosida, Jim Gross, Guy Samson, Eddy Chan, Conrad Chow, Anne McDonald
A young female cop goes undercover, in an attempt to learn the secrets of a sinister Ninja sect, and foil its plans for domination of the free world.
MAR   85 min Cut (55 sec)   VIDrel: IMPENT   V

SILVER FLEET, THE ***   U
Vernon Sewell/Gordon Wellesley UK   1943
Ralph Richardson, Esmond Knight, Googie Withers, Beresford Egan, Frederick Burtwell, Kathleen Byron, Willem Akkerman, Dorothy Gordon, Charles Victor, John Longden, Joss Ambler, Margaret Emden, Ivor Barnard, Valentine Dyall
Wartime propaganda film set in occupied Holland with Richardson a shipyard owner who seems to be collaborating

with the Nazis, but in fact has made secret plans to destroy one of his two new U-boats, allowing the Resistance to take the other one back to Britain. A film that is slow to capture one's interest but eventually does so, culminating in a dramatic, explosive finale.
DRA   84 min (ort 87 min) B/W   VIDrel: CONNO L/A   V

SILVER SEDUCTION ***   18
F.J. Lincoln USA   1992
Silver Forrest, Aja, Masison, Malia, Peter North, Randy West, Sean Michaels
The first of a set of three "Silver" movies, all of which star Forrest. This one casts her as an aspiring actress, who is intro-duced to the ploys she must use to get to the top by North. As he is married, this does little to endear him to his wife, but she has to get back her pre-nuptial agreement before a divorce can be considered, and hires an errant burglar to help her do this. Slightly over-complex, but well shot on high-definition video. SILVER SENSATION followed.
A   69 min mVid   VIDrel: ONE   V

SILVER SENSATION **   18
F.J. Lincoln USA   1992
Silver Forrest, Cassidy, Monique Hall, Marc Wallice, Ashley Nicole, Steve Drake, Felipe, Derrick Lane
This continues the story of Forrest's Hollywood career, but there's not much of a plot in this one, with much of it taken up with Forrest's sexual exploits, which now and then are inter-rupted by her need to go off on photo-shoots. As with the earlier SILVER SEDUCTION, this film is very well put together in tech-nical terms at least, being shot on high-quality video. It's a pity the plot is so weak. Followed by "Silver Elegance".
Aka: SILVER SENSATIONS
A   51 min (ort 70 min) mVid   VIDrel: ONE   V

SILVER STRAND **   18
George Miller USA   1995
Gil Bellows, Tony Plana, Nicollette Sheridan
A US Navy wife meets a SEAL cadet and they embark on a passionate affair, but a series of personal attacks makes them realise they are in danger. Scripted by Douglas Day Stewart, who also wrote the script for AN OFFICER AND A GENTLE-MAN, this film offers us much the same blend of romance, jealousy and conflict, all the while interspersed with the usual tough training sequences. Directed without enthusiasm, the cliched script generates little tension.
DRA   99 min (ort 104 min)   VIDrel: MGM/WHV   V

SILVER STREAK ***   PG
Arthur Hiller USA   1976
Gene Wilder, Jill Clayburgh, Richard Pryor, Patrick McGoohan, Ned Beatty, Clifton James, Richard Kiel, Ray Walston, Valerie Curtin, Stefan Gierasch, Scatman Crothers, Fred Willard, Len Birman, Lucille Benson, Delos Smith
A publisher on a train becomes involved in a murder, and finds himself menaced by the crooks, in this high-spirited and madcap mixture of comedy and mayhem. A rather patchy effort, with the humour distinctly strained in a few places, but generally a highly effective film - something of a spoof on Hitchcock-style thrillers. The script is by Colin Higgins.
COM   109 min (ort 113 min)   VIDrel: 20TH/TECH L/A   V

SILVER TWILIGHT, THE **   PG
Roy Thomas HONG KONG   1989
Another Joseph Lai/Betty Chan production, made very much in the same style as FALCON 7 or THE COSMOS CONQUEROR, and as ever, featuring a set of nasty rebel aliens out to conquer Earth. When the ruler of the home planet sends his son to help us, he thoughtfully provides him with a magical staff, but when this is stolen things begin to look very bleak indeed. Average.
ANIM   59 min Cut (4 sec)   VIDrel: IMPENT   V

SILVERADO **   PG
Lawrence Kasdan USA   1985
Kevin Kline, Scott Glenn, Brian Dennehy, John Cleese, Linda Hunt, Kevin Costner, Rosanna Arquette, Danny Glover, Jeff Goldblum, Marvin J. MacIntyre, Brad Williams, Todd Allen, Kenny Call Hartline, Rusty Meyers
As the result of a strange meeting, four strangers are drawn together in a violent conflict at a crooked Western town, in this attempt to revive the old-style Western. A lumbering affair that

has nothing noteworthy on offer, yet moves along with sufficient speed to mask its shortcomings.
WES   127 min (ort 132 min)   VIDrel: VCC/DISC/COLUM
V/sur

SIMON OF THE DESERT ** 12
Luis Bunuel SPAIN 1966
Claudio Brook, Silva Pinal, Hortensia Santovena, Enrique Alvarez Felix
Allegory about Man's fall from grace in which a man emulates a famous saint by standing on a pillar in the desert for six years, six weeks and six days, resisting the cunning attempts of the devil to make him abandon his self-imposed task. A well-realised parable whose brief length is due to the fact that the project ran out of money.
Aka: SIMON DEL DESIERTO
DRA   43 min (ort 45 min) B/W   VIDrel: ELPIC/POLYREC
V

SIMPLE JUSTICE ** 18
Deborah Del Prete USA 1988
Cesar Romero, Doris Roberts, John Spencer, Priscilla Lopez, Kevin Geer, Candy McClain, Matthew Galle, David Auerbach, Tony Cucci, Michael Sergio, Paul Lemos, Michael Genet, Antonia Rey, Gary Majchrzak
A happy couple are looking forward to the birth of their first child when, in the course of a bank raid, the wife is murdered. After the trial collapses due to the disappearance of a key witness, the culprits are set free, but soon begin to fall victim to a mysterious assailant. Simple-minded revenge tale with a twist ending.
A/AD   96 min Cut (2 sec)   VIDrel: IMPENT   V

SIMPLE MEN ** 15
Hal Hartley GERMANY/ITALY/UK/USA 1992
Martin Donovan, Karen Sillas, Robert Burke, William Sage, Elina Lowensohn, Mark Chandler Bailey, Margaret A. Bowman, Chris Cooke, Martin Donovan, Ed Geldart, Jeffrey Howard, Vivian Ianko, John Alexander MacKay, Matt Mallory
Two brothers on the run from the law decide to go in search of the father who used to be a baseball player for the Brooklyn Dodgers but gave up his career for violent political activism. Screenplay is by Hartley, and this is very much his work, being an offbeat comedy full of oddball characters that recalls nothing so much as an episode of the TV series "Twin Peaks".
COM   101 min (ort 105 min) wScrn   VIDrel: TART/20TH
V/dm

SIMPLE TWIST OF FATE, A ** PG
Gillies MacKinnon USA 1994
Steve Martin, Gabriel Byrne, Catherine O'Hara, Stephen Baldwin, Alana Austin, Alyssa Austin, Laura Linney, Anne Heche, Michael Des Barres, Byron Jennings, Michael des Barres, Tim Ware, David Dwyer, Tom Even, Ed Grady
An updated version of Silas Marner, with Martin cast as the disillusioned bachelor who finds a baby girl abandoned on his doorstep and decides to take her in, a decision that has far-reaching consequences when the child's biological father decides to mount a custody battle. Written by Martin (who is really at his best when working with material others have written) this is an ambitious but flawed and mawkish effort, and Martin is badly miscast in the central role.
DRA   101 min (ort 106 min) cC   VIDrel: TOUCH/TECH
V/sh
Boa: novel Silas Marner by George Eliot (Marian Evans).

SIN COMPASION *** 15
Francisco J. Lombardi FRANCE/MEXICO/PERU 1994
Diego Bertie, Adriana Davila, Jorge Chiarella, Marcello Rivera, Ricardo Fernandez, Carlos Onetto, Hernan Romero, Mariella Trejos, Jose Maria Salcedo, Augusto Modenesi, Humberto Modenesi, Juan Jose Criados, Isabel Solari
A poor but arrogant law student rejects the conventional values being taught at his philosophy class, and sets out to live according to his own code of ethics, but the result is a tragedy when he decides to murder his landlady for her savings. Dostoevsky's story is transplanted to Peru in a sombre, slow and skilful film, yet the central character is so very disagreeable that one feels little sympathy with him or his intellectual musings.
Aka: NO MERCY; SANS PITIE
DRA   120 min   CINrel
Boa: novel Crime and Punishment by Fyodor Mikhailovich Dostoevsky.

SINATRA *** PG
Jim Sadwith USA 1992
Adam Lavorgna, Philip Casnoff, Olympia Dukakis, Joe Santos, Gina Gershon, Marcia Gay Harden, Bob Gunton, David Raynor, Nina Siemaszko, Rod Steiger, James Kelly, Ralph Seymour, Andrew Bloch, Jeff Corey, Vincent Gustaferro, Thomas Ryan
Long, quite accurate and rather candid account of the life and career of this legendary singer, tracing his ups and downs from his early days in Hoboeken, New Jersey to his comeback concert at Madison Square Garden in 1974. Despite all the detail with which this film is packed, the lip-synched musical numbers are its most enjoyable feature.
DRA   228 min (2 cassettes) mTV   VIDrel: WHV   V/dm

SINBAD AND THE EYE OF THE TIGER ** U
Sam Wanamaker UK 1977
Patrick Wayne, Taryn Power, Jane Seymour, Margaret Whiting, Patrick Troughton, Kurt Christian, Nadim Sawalha, Damien Thomas, Bruno Barnabe, Bernard Kay, Samali Coker, David Sterne
Some further adventures of Sinbad, in this rather limp sequel to THE GOLDEN VOYAGE OF SINBAD, with our hero freeing a city from the grip of an evil spell. An overlong film that really drags after a while – even Ray Harryhausen's monsters begin to look bored.
FAN   108 min (ort 117 min)   VIDrel: VCC/DISC/COLUM
L/A   V

SINBAD THE SAILOR *** U
Richard Wallace USA 1947
Douglas Fairbanks Jr, Maureen O'Hara, Anthony Quinn, Walter Slezak, Jane Greer, George Tobias, Mike Mazurki, Sheldon Leonard, Alan Napier, John Miljan, Barry Mitchell, Glenn Strange, George Chandler, Louise-Jean Heydt
A lavish swashbuckling version of a much told story. A classic Hollywood film with Fairbanks well cast as Sinbad, who goes off on his eighth voyage in search of the lost treasure of Alexander. Only the occasional lack of pace lets it down.
A/AD   112 min (ort 117 min)   VIDrel: VCC L/A   V

SINCE YOU WENT AWAY *** PG
John Cromwell USA 1944
Claudette Colbert, Jennifer Jones, Joseph Cotten, Shirley Temple, Monty Wooley, Agnes Moorehead, Lionel Barrymore, Guy Madison, Hattie McDaniel, Robert Walker, Craig Stevens, Keenan Wynn, Albert Basserman, Nazimova
Story of a family on the home front in the USA, and their sufferings during WW2. A memorable flagwaver that has inevitably dated, but still demonstrates the power Hollywood could bring to bear on a subject when need be. Written by David O. Selznick and well photographed by Lee Garmes and Stanley Cortez. This was the film debut for Guy Madison and John Derek, with the latter having just a tiny part as an extra. AA: Score (Max Steiner).
DRA   172 min B/W   VIDrel: VCC/DISC   V
Boa: book by Margaret Buell Wilder.

SINDERELLA: PART 1 ** 18
Paul Thomas USA 1992
Savannah, Raquel Darian, P.J. Sparxx, Brittany Morgan, Melanie Moore, Randy Spears, Joey Silvera, Randy West, Mike Horner, T.T. Boy, Derrick Lane, Jake Steed
Sex spoof on the Cinderella story, that casts Savannah in the title role, who is left at the mercy of her wicked stepmother and two stepsisters when her father dies suddenly. Spears plays the Prince, and the story is similar to the original in most important essentials, in that there is a grand ball, a fairy godmother (Moore) and an eventual happy conclusion (see Part 2 for this). However, as this is a sex film the plot is really of secondary interest.
A   64 min (Cut before video submission by 3 min 43 sec)
VIDrel: GROHOM/MAXSCAN   V

SINDERELLA: PART 2 ** 18
Paul Thomas USA 1992
Racquel Darian, Derrick Lane, P.J. Sparxx, Savannah, Randy Spears, Randy West, Joey Silvera, T.T. Boy, Melanie Moore, Brittany Morgan
This sequel to SINDERELLA: PART 1 attempts to continue the Cinderella story, with most of the action revolving around the grand ball thrown by the Prince (Spears). However, whereas the first part at least attempted to stay within the confines of the story, this sequel more of less dispenses with it, in preference for a parade of couplings, the only concession to the original story being the twelve o'clock disappearance of our heroine.
Aka: STEPSISTER; STEPSISTERS
A   56 min (ort 60 min)   VIDrel: MAXSCAN/TERRY   V

SING *   15
Richard Baskin USA   1989
Peter Dobson, Lorraine Bracco, Jessica Steen, Patti La Belle, Louise Lasser, George DiCenzo, Susan Peretz, Laurnea Wilkerson, Rachel Sweet, Jank Azman, Jason Blicker, Cuba Gooding Jr, Adam Kostisky, Sam Moses, Ingrid Veninger
A hot-tempered but talented street punk is chosen to take part in a high school song-and-dance contest, and has to overcome his self-destructive urges. Scripted by Dean Pritchford (who worked on FAME: THE MOVIE and FOOTLOOSE), this mediocre musical has a host of stereotyped characters ambling through a tired and predictable script. Not really a movie, merely an excuse for a seemingly endless succession of forgettable musical numbers.
DRA   95 min (ort 98 min)   VIDrel: RCA L/A   V

SINGING DETECTIVE, THE: VOLS. 1 AND 2 *   15
Jon Amiel UK   1986
Michael Gambon, Patrick Malahide, Joanne Whalley-Kilmer, Janet Suzman
A writer's life is blighted by the severe psoriasis that has hospitalised him, and from his hospital bed he creates a fantasy-world in which he enjoys various adventures as a 1940s detective. Written by Dennis Potter, this is a complex, surrealistic odyssey, with some musical interludes and an unusually demanding structure. A sad, funny, poignant and highly memorable affair. Originally shown on TV in six parts.
DRA   393 min (2 cassettes) mTV   VIDrel: BBC/TECH   V/h

SINGIN' IN THE RAIN **   U
Gene Kelly/Stanley Donen USA   1952
Gene Kelly, Donald O'Connor, Debbie Reynolds, Jean Hagen, Cyd Charisse, Rita Moreno, Millard Mitchell, Douglas Fowley, Madge Blake, King Donovan, Bobby Watson, Kathleen Freeman, Jimmie Thompson, Dan Foster, Margaret Bert
Two friends go to Hollywood in the transitional period, as silent film is giving way to talkies. Their rise to fame is shown to good effect in this classic Hollywood musical, which contains one of the most famous dance sequences ever filmed, notably Gene Kelly (who did the choreography) singing the title song. Written by Adolph Green and Betty Comden. Music is by Nacio Herb Brown with lyrics by Arthur Freed.
MUS   98 min (ort 102 min)   VIDrel: MGM/WHV   V/dm

SINGLE WHITE FEMALE *   18
Barbet Schroder USA   1992
Bridget Fonda, Jennifer Jason Leigh, Steven Weber, Peter Friedman, Frances Bay, Tara Karsian, Stephen Tobolowsky, Michele Farr, Christiana Capetillo, Ken Tobey, Rene Estevez, Jessica Lundy, Tiffany Mataras, Krystle Mataras
After breaking up with her fiance, a self-employed computer consultant advertises for a roommate and takes in a quiet and introverted bookstore clerk. The two become firm friends but as their friendship progresses, the room-mate begins to show signs of some decidedly odd behaviour. An interesting premise is never really developed and the film culminates in a nasty and gory mess that greatly diminishes its impact as a psycho-thriller.
THR   103 min (ort 108 min) cC
VIDrel: VCC/DISC/COLUM   V/sur
Boa: novel SWF Seeks Same by John Lutz.

SINGLE WOMEN, MARRIED MEN *   15
Nick Havinga USA   1989
Michele Lee, Lee Horsley, Margaret Avery, Carrie Hamilton, Julie Harris
A woman deserted by her husband trains as a psychotherapist as a way of coming to terms with her life, but finds herself becoming involved with a married man.
DRA   91 min   VIDrel: 20TH/TECH   V

SINGLES *   15
Cameron Crowe USA   1992
Bridget Fonda, Campbell Scott, Kyra Sedgwick, Matt Dillon, Sheila Kelley, Jim True, Bill Pullman, Devon Raymond, Ally Walker, Jeremy Piven, Peter Horton, Eric Stoltz, Chuck McQuary, Matt Magnano, Jaffar Smith, Mykol Hazen, Michael Su
In a Seattle apartment block, the lives of a diverse group of young, single people gradually become intertwined, in this well-acted and comic look at the lifestyles and attitudes of this generation. Buoyed up by some attractive performances but the lack of a solid script (the work of the director) is a major drawback. The soundtrack features music by various Seattle groups.
COM   95 min (ort 120 min) cC   VIDrel: WHV   V/sur V/dm

SINK THE BISMARCK! *   U
Lewis Gilbert UK   1960
Kenneth More, Dana Wynter, Carl Mohner, Laurence Naismith, Geoffrey Keen, Karel Stepanek, Maurice Denham, Michael Goodliffe, Michael Hordern, Esmond Knight, Jack Watling, Sydney Tafler, John Stuart, Ed Morrow, Jack Gwillim
A workmanlike account of British efforts to trap and sink Germany's largest battleship in 1941, that is quite exciting, despite an over-reliance on newsreel footage and poor modelwork. The relationship between the director of naval operations and his secretary injects a note of personal drama into this otherwise standard war yarn. The script is by Edmund H. North.
WAR   97 min (ort 98 min) B/W   VIDrel: 20TH/TECH   V
Boa: book by C.S. Forester.

SINS OF DESIRE *   18
Jim Wynorski USA   1992
Delia Sheppard, Jay Richardson, Jan-Michael Vincent, Tanya Roberts, Nick Cassavetes, Gail Harris, Carrie Stevens, Pamela Pond, Jeana Wilson, Lou Bonacki, Becky LeBeau, Monique Parent, Roberta Vasquez, Ace Mask
When her sister dies while undergoing therapy at a sex clinic, a woman goes undercover as a nurse to probe the circumstances surrounding her death. To her horror, she discovers a world of strange and brutal rituals that take place behind locked doors.
DRA   87 m in (ort 90 min)   VIDrel: REFLEC/FIRST   V

SINS OF SILENCE *   (PG)
Sam Pillsbury USA   1995
Lindsay Wagner, Holly marie Combs, Cynthia Sikes, Sean McCann, Victor Argo, Brian Kerwin, Jason Cadieux, Richard Fitzpatrick, Chris Wiggins, Colin Fox, Christina Cox, Norma Dell'Agnese, larry Reynolds, Jason Weinberg, Suzanne Coy
A free-spirited girl is raped in a car by a man she met and goes to a rape crisis centre for help and support. The latter, run by a former nun, helps her to press charges but the accused's wealthy parents hire a smart lawyer who pursues a very aggressive line. Anxious to paint the victim as unreliable and of loose morals, he has the centre's records subpoenaed but the nun goes to jail rather than comply. Well-acted but never engaging, this one fails to shine.
DRA   90 min mTV   SATrel: MOVIE CHANNEL

SINS OF THE FATHER *   15
Kevin Dobson USA
George Dzundza, Peter Dobson, David Marshall Grant, Micole Mercurio, Randle Mell
An attractive young D.A. discovers not only that a powerful gangster is being shielded by local pillars of the establishment, but that her own training at law school was paid for out of mobster money.
DRA   92 min   VIDrel: ODY/SONOP   V/sh

SINS OF THE MOTHER *   (18)
John Patterson USA   1991
Elizabeth Montgomery, Dale Midkiff, Heather Fairfield, Talia Balsam, Mimi Kenendy, Jerry Bossard
Montgomery plays the mother of a serial rapist, and gives a powerful if overblown performance as a domineering tyrant who is largely to blame for the creation of her monstrous son. Based on the true story of one such individual, who was responsible for no less than forty attacks, but one cannot help but think that the film's characterisations owe more to the fevered imaginations of the scriptwriters than to the real-life characters supposedly portrayed here.
Aka: SOUTH HILL RAPIST
THR   90 min mTV   TVrel: BBC1

SINS OF THE NIGHT *   18
Alexander Gregory Hippolyte USA   1992
Deborah Shelton, Nick Cassavetes, Miles O'Keefe, Richard Roundtree, Matt Roe, Courtney Taylor Michelle, Moffett, Lee Anne Beaman, Julie James, MIchele Brin
An ex-con takes a job as an insurance investigator and finds himself investigating a claim made by an exotic dancer, with whom he becomes more than a little involved, initially unaware that she is married to a gangster. Average erotic thriller.
DRA   86 min (ort 90 min)   VIDrel: 20VIS/SONOP   V

SINS: PARTS 1, 2 AND 3 *   15
Douglas Hickox USA   1986
Joan Collins, Marisa Berenson, Jean-Pierre Aumont, Joseph Bologna, Steven Berkoff, Elizabeth Bourgine, Judi Bowker, Capucine, Timothy

Dalton, Arielle Dombasle, James Farentino, Paul Freeman, Allen Garfield, Giancarlo Giannini
A soapy mini-series about the terrible secret of a beautiful female millionairess and head of a business empire. Can she leave the past behind her and find true happiness? Sit through this for five hours or so and you may find out.
DRA   320 min (3 cassettes – ort 420 min) mTV
VIDrel: 4-FRONT/POLYREC/NWV   V
Boa: novel by Judith Gould.

SIR HENRY AT RAWLINSON'S END **   15
Steve Roberts UK   1980
Trevor Howard, Patrick Magee, Denise Coffey, J.G. Devlin, Sheila Reed, Harry Fowler, Vivian Stanshall, Jeremy Child, Susan Porrett, Liz Smith, Ben Aris, David Geroll, Suzanne Danielle, Gary Waldhorn, Simon Jones, Michael Crane
An aristocrat tries to rid his estate of a family ghost. The idea for this film came from the pop group The Bonzo Dog Doodah Band, who once released an album with a track entitled "Rawlinson's End". (Which in turn led to a long-running radio series of brief episodes performed by Vivien Stanshall.) This film is best described as a pleasant comedy – very British, somewhat laboured but often quite poetic. Written by Stanshall and Steve Roberts.
COM   68 min (ort 75 min) wScrn   VIDrel: TART/20TH
V/h
Boa: radio play by Vivien Stanshall.

SIRENS **   15
John Duigan AUSTRALIA   1993
Hugh Grant, Sam Neill, Tara Fitzgerald, Elle MacPherson, Portia De Rossi, Kate Fischer, Pamela Rabe, Ben Mendelsohn, John Polson, Mark Gerber, Julia Stone, Ellie MacCarthy, Vincent Ball, John Duigan, Lexy Murphy, Scott Lowe
An English minister takes a position in Australia and one of the first problems he encounters is an artist whose paintings are a major source of controversy due to their sexually charged content. In a bid to persuade said artist to see the error of his ways, our straitlaced vicar pays him a visit, but this brings him into contact with the artist's three models, who like the sirens of mythology lure all men to their doom.
COM   90 min (ort 95 min) cC   VIDrel: TOUCH/TECH
V/sur

SIROCCO **   PG
Curtis Bernhardt USA   1951
Humphrey Bogart, Marta Toren, Lee J. Cobb, Everett Sloane, Gerald Mohr, Zero Mostel, Vincent Renno, Marta Toren, Onslow Stevens, Ludwig Donath, Lee J. Cobb, Martin Wilkins, Peter Ortiz, Edward Colmans, Al Eben, Peter Brocco
Set in 1920s Damascus, this sluggish yarn has an American gunrunner helping the rebels and finding time for a little romance on the side. A turgid and unimpressive work.
A/AD   94 min (ort 111 min) B/W
VIDrel: COLUM/SONOP   V
Boa: novel Coup De Grace by Joseph Kessel.

SISTER ACT ***   PG
Emile Ardolino USA   1992
Whoopi Goldberg, Maggie Smith, Harvey Keitel, Mary Wickes, Wendy Makkena, Kathy Najimy, Bill Nunn, Robert Miranda, Richard Porthow, Ellen Albertini Dow, Carmen Zapata, Pat Crawford Brown, Georgia Creighton, Prudence Wright Holmes
After she witnesses a Mob murder, a Reno lounge singer becomes a marked women in need of police protection. They decide to hide her in a convent where her easy-going ways and extrovert nature lead to a clash with the Mother Superior. When she forms a gospel choir that becomes an overnight success, her cover is blown and she has to run for her life. A great performance by Goldberg and the good music make for an enjoyable movie. A sequel followed.
COM   90 min (ort 100 min) cC   VIDrel: TOUCH/TECH
V/sur

SISTER ACT 2: BACK IN THE HABIT **   PG
Bill Duke USA   1993
Whoopi Goldberg, Maggie Smith, Kathy Najimny, Wendy Makkena, Mary Wickes, James Coburn, Michael Jeter, Sheryl Lee Ralph, Robert Pastorelli, Barnard Hughes, Thomas Gottschalk, Lauryn Hill, Brad Sullivan, Alanna Usach, Ryan Toby
Disappointing and unnecessary sequel to the first film, in which our feisty heroine returns to the convent to coach the choir for a competition, while a nasty villain plots the closure of their

school. Excellent musical numbers and Goldberg's sparkling presence help sustain the movie despite its sentimental and highly derivative plot.
COM   102 min (ort 107 min) cC   VIDrel: TOUCH/TECH
V/sur

SISTER-IN-LAW, THE **   15
Noel Nosseck USA   1995
Shanna Reed, Kate Vernon, Craig Wasson, Kevin McCarthy, Kent Williams, Tonea Stewart, Al Wiggins, Barbara Niven, Jeromie Wilson, Marty Terry, Ralph Wilcox
A woman plots to get even with the man whom she blames for the loss of her parents in a car crash, and finds a way to do this when she meets a man who is married to the man's daughter. Claiming to be a long-lost sister-in-law, she inveigles her way into the family and makes plans to eliminate them, and get her hands on their $12,500,000 fortune. Perfectly watchable, but terribly unoriginal.
THR   95 min mTV   VIDrel: CIC   V/sh

SISTER MY SISTER ***   15
Nancy Meckler UK   1994
Joely Richardson, Julie Walters, Johdi May, Sophie Thursfield, Amelda Brown, Lucita Pope, Kate Gartside, Aimee Schmidt, Gabriella Schmidt
In Le Mans in the 1930s a wealthy widow and her frumpish daughter hire two sisters who agree to work as maids for the price of one, but gradually the claustrophobic setting works on their minds, and a double murder is the result. Meckler's debut feature is a strongly plotted and disturbing film, that sets out to minutely examine all the events that led to this notorious real-life murder case. An unusually controlled film, scripted by Kesselman.
DRA   89 min   VIDrel: ARROW/RTM   V
Boa: play My Sister In This House by Wendy Kesselman.

SISTERHOOD, THE *   18
A.K. Allen USA   1984
Diana Scarwid, Karen Austin, Christine Belford, Bruce Davidson, Shera Danese, Beverly Todd, Marilyn Kagan, Kit McDonough, Arliss Howard, Randee Heller, Paul Carafotes, Nicholas Worth, Scott Lincoln
A nasty and exploitative piece telling of a group of rape victims, who band together in order to castrate rapists whom the law has not punished. A version of DEATHWISH by way of women's lib, which totally fails to offer any insights into the true nature of rape and rapists or explore the threat that vigilante action poses for any society based on the rule of law. Something like ACT OF VENGEANCE, only more vicious.
Aka: LADIES' CLUB, THE; VIOLATED
DRA   79 min (ort 90 min)   VIDrel: MED/POLYREC L/A   V
Boa: novel by Betty Black and Casey Bishop.

SISTERHOOD, THE *   18
Cirio H. Santiago ITALY   1988
Rebecca Holden, Lynn-Holly Johnson, Barbara Hooper, Chuck Wagner
In the year 2021 A.D. the majority of women are held as mere chattels in a male-dominated society, except for a violent gang of women known as "The Sisterhood". Another silly blend of fantasy and exploitative violence that takes as its setting a shattered post-WW3 world.
FAN   82 min   VIDrel: MOPIC/QUANT   V

SISTERS ***   18
Brian De Palma USA   1972
Margot Kidder, Jennifer Salt, Charles Durning, Barnard Hughes, William Finley, Mary Davenport, Dolph Sweet, Lisle Wilson
Siamese twins are separated at birth, one becomes a psychopath, the other a normal person. Various complications follow in the wake of a series of murders. Salt plays a reporter who thinks she has witnessed one of the killings. A tense blend of mystery and horror, with an excellent score by Bernard Herrmann, but equally a film that borrows shamelessly from others (such as Hitchcock) and gives one little scope to identify with the plight of the central characters.
Aka: BLOOD SISTERS
HOR   90 min (ort 93 min)   VIDrel: DTK L/A   V

SISTERS **   (15)
Michael Hoffman USA   1987
Patrick Dempsey, Jennifer Connelly, Lila Kedrova, Florinda Bolkan
A young teenage boy goes to stay with his girlfriend's family in Quebec but finds himself ill at ease with the family, which comprises two other predatory sisters, a devoutly Catholic

mother, an atheist father (who is writing a biography of Pascal in the nude) and a dying granny, who has escaped from her hospital bed. A sporadically amusing comedy, littered with occasional bouts of irritating symbolism.
Aka: SOME GIRLS
COM 90 min SATrel: SKY MOVIES

SISTERS OF SATAN ** 18
P. Dominici (Domenico Paolella) ITALY 1973
Anne Heywood, David Silva, Claudio Brook, Ornella Muti, Luc Merenda, Muriel Catala, Martine Brochard, Claudia Gravi, Pier Paolo Capponi
Story of the corruption of a convent of nuns who fall prey to all manner of perversions in this run-of-the-mill horror tale. Slavery, torture and murder are well represented, as are the obligatory displays of naked flesh.
Aka: INNOCENTS FROM HELL; LE MONACHE DI SANT ARCANGELO; MONASTERY OF SAINT MICHAEL, THE; NUN AND THE DEVIL, THE; NUNS OF SAINT ARCHANGEL, THE
HOR 99 min wScrn dubbed VIDrel: REDEM/RTM V

SITTER, THE ** 15
Rick Berger USA 1991
Kim Myers, Brett Cullen, Eugene Roche, Susan Barnes, Kimberly Cullum, James McDonnell, Susanne Reed, Patricia George, Maria Montoya, Shabba-Doo, Gregory White, Dennis Paladino, Tyler Bowe, Jeff Jeffcoat, Frank Isles
A shy young woman given the job of babysitting for a guest at a hotel, but retreats into a bizarre fantasy world and starts to believe that she is the child's mother, becoming ready to kill to preserve her unhealthy and dangerous illusion.
THR 88 min VIDrel: 20TH/TECH V
Boa: novel Mischief by Charlotte Armstrong.

SITTING PRETTY * 18
Jerry Ross USA 1992
Sunny McKay, Jon Dough, Ashley Nickole, Carrie Breeze, Joey Silvera, Eric Taylor, Summer Knight, Robert Manning, Heather Mills
A kind of FATAL ATTRACTION adult movie but with the roles now reversed, as a top model is relentlessly pursued by a blue-collar worker who will stop at nothing to be near her. However, as the makers were not working on a film with the same size budget, they largely dispense with the plot after about the first thirty minutes, concentrating on the sex scenes, which without a story to bolster them, soon grow quite tiresome.
Aka: SITTIN' PRETTY
A 90 min VIDrel: FALCON/TOTAL V

SITTING TARGET *** 18
Douglas Hickox UK 1972
Oliver Reed, Jill St John, Ian McShane, Edward Woodward, Frank Finlay, Jill Townsend, Freddie Jones, Robert Beatty, Tony Beckley, Mike Pratt, Robert Russell, Joe Cahill, Robert Ramsey, Susan Shaw
An extremely well made but very brutal film, with Reed playing a psychopathic killer to perfection. He escapes from prison with the intention of punishing his unfaithful wife. Some neat twists in the plot lead to a violent if not unexpected climax. Written by Alexander Jacobs.
THR 89 min (ort 92 min) VIDrel: MGM L/A V
Boa: novel by Lawrence Henderson.

SIX DEGREES OF SEPARATION *** 15
Fred Schepisi USA 1993
Stockard Channing, Donald Sutherland, Will Smith, Mary Beth Hurt, Bruce Davison, Ian McKellan, Anthony Michael Hall, Richard Masur, Heather Graham, Eric Thal, Anthony Rapp, Osgood Perkins, Catherine Keller, Jeffrey Abrams
An upper-class New York couple of liberal views allow a smooth-talking black con-man to take them for ride, after he claims to be both a friend of their college student son and the son of Sidney Poitier. A successful adaptation by Guare of his hit stage play that offers excellent acting and a good use of location shooting.
Aka: 6 DEGREES OF SEPARATION
DRA 107 min (ort 112 min) cC VIDrel: MGM/WHV V/sur
Boa: play by John Guare.

SIX IN PARIS *** 15
J. Douchet/J. Rouch/J. Pollet/E. Rohmer/J. Godard/C. Chabrol FRANCE 1965
Barbara Wilkin, Jean-Francois Chappy, Jean-Pierre Andreani, Nadine Ballot, Barbet Schroeder, Micheline Dax, Claude Melki, Jean-Michel Rouziere, Marcel Gallon, Joanna Shimkus, Philippe Hiquilly, Serge Davri, Stephane Audran
Six-part compilation of slight stories, all set in Paris, which is their only common denominator. They are entitled "Saint-Germaines-des-Pares" (Jean Douchet), "Gare du Nord" (Jean Rouch), "Rue Saint-Denis" (Jean-Daniel Pollet), "Place De L'Etoile" (Eric Rohmer), "Montparnasse-Levallois" (Jean-Luc Godard) and "La Muette" (Claude Chabrol). See also RENDEZVOUS IN PARIS.
Aka: PARIS VU PAR
DRA 91 min (ort 93 min) VIDrel: CONNO/RTM L/A V

SIX MILLION DOLLAR MAN, THE ** U
Richard Irving USA 1973
Lee Majors, Darren McGavin, Martin Balsam, Barbara Anderson, Charles Knox Robinson, Dorothy Green, Ivor Barry, Anne Whitfield, George Wallace, Robert Cornthwaite, Olan Soule, Norma Storch, Maurice Sherbanee, John Mark Robinson
A test-pilot is so badly injured in a crash that only advanced technology can save him, albeit in a rebuilt form with several artificial limbs, plus superhuman strength and senses. The US government then persuades him to undertake a secret mission. The pilot for a long-running TV series that in turn gave rise to several more feature spin-offs (such as "Return Of The Six Million Dollar Man And The Bionic Woman"). Adequate.
Aka: CYBORG: THE SIX MILLION DOLLAR MAN
A/AD 72 min (ort 90 min) mTV VIDrel: CIC/SONOP L/A V
Boa: novel Cyborg by Martin Caidin.

633 SQUADRON ** PG
Walter Grauman UK 1964
George Chakiris, Maria Perschy, Cliff Robertson, Harry Andrews, Michael Goodliffe, Donald Houston, Angus Lennie, John Meillon, Scot Finch, John Bonney, Suzan Farmer, Barbara Archer, John Church, Sean Kelly, Johnny Briggs
WW2 story in which Mosquito aircraft are sent to destroy a munitions factory in occupied Norway, by attempting to cause the cliff that overhangs it to collapse. A standard stiff-upper-lip war film with plenty of heroics and action, but not much else. The script is by James Clavell and Howard Koch.
Aka: SIX THREE THREE SQUADRON
WAR 92 min (ort 102 min) VIDrel: MGM/WHV V/h
Boa: novel by Frederick E. Smith.

SIX WIVES OF HENRY VIII, THE *** PG/U
John Glenister/Naomi Capon UK 1970
Keith Michell, Annette Crosbie, Dorothy Tutin, Patrick Troughton, Anne Stallybrass, Sheila Burrell, Elvi Hale, Angela Pleasence, Rosalie Crutchley, John Ronane
A stylish and painstakingly-produced set of TV plays, that explored the character of Henry VIII and his relationship with his wives, with one play being devoted to each wife. Originally shown in six parts and later used as a basis for the film HENRY VIII AND HIS SIX WIVES. Music is by David Munrow and The Early Music Consort of London.
DRA 541 min (6 cassettes – certifications vary) mTV
VIDrel: BBC/TECH V/h

SIXTEEN CANDLES ** U
John Hughes USA 1984
Molly Ringwald, Anthony Michael Hall, Michael Schoeffling, Justin Henry, Paul Dooley, Carlin Glynn, Blanche Baker, Gedde Watanabe, Edward Andrews, Billie Bird, Carole Cook, Max Showalter, John Cusack, Joan Cusack
A girl becomes sixteen and wants to find her "Mister Right", but unknown to her he is already making plans of his own. Quite a perceptive little comedy, but flawed by muddled direction and some sequences of dubious appeal. A lovely performance from Ringwald carries the film. Written by Hughes.
COM 93 min VIDrel: CIC/SONOP V/h

SKATEBOARD KID, THE ** (PG)
Larry Swerdlove USA 1992
Timothy Busfield, Bess Armstrong, Cliff De Young, Dom De Luise, Rick Dean, Trevor Lissauer, Lee McLaughlin, Shanelle Workman, Jonathon Pemar, Kai Lennox, Lee Velazquez, Gerry Lock, Sindy McKay, Derek Mark Lochran, David Wells
A kid wants to join a gang of skateboarders but they refuse to have anything to do with him. A little later a chance meeting with a strange humanoid results in him learning to build a magical skateboard that has the ability to talk. This shallow

romp will be enjoyed by young kids of six or under. A sequel followed.
JUV 82 min SATrel: SKY MOVIES

SKATEBOARD KID 2, THE ** U
Andrew Stevens USA 1994
Dee Wallace Stone, Bruce Davison, Andrew Stevens, Trenton Knight, Andrew Keegan, Turhan Bey
Sequel to the first film detailing more adventures of a kid with a magic skateboard, who having lost his test pilot father in a crash, must battle a rival skateboard gang in order to win a cash prize and thus secure the future of his family.
Aka: SKATEBOARD KID, THE: A MAGICAL MOMENT
JUV 88 min (ort 90 min) VIDrel: MED/DISC V/sh

SKEETER ** 15
Clark Brandon USA 1994
Tracy Griffith, Jim Youngs, Charles Napier, Michael J. Pollard
In a small desert town, years of dumping toxic waste in a pool of stagnant water causes mosquitos to mutate into a new and far more dangerous species. Average.
HOR 91 min VIDrel: REFLEC/FIRST V

SKETCH ARTIST ** (18)
Phedon Papamichael USA 1992
Sean Young, Jeff Fahey, Drew Barrymoore, Frank McRae, Tcheky Karyo, James Tolkan, Charlotte Lewis, Ric Young, Stacey Haiduk
After a fashion designer is murdered, a police artist produces a sketch of the prime suspect and is horrified to see that he has drawn his own wife. Keeping this information to himself, he mounts his own investigation but soon finds himself in danger. An interesting premise receives a poor and uninspired treatment. A sequel followed. See EVIL HAS A FACE for another story built around the work of a police artist.
THR 86 min (ort 89 min) mCab VIDrel: WHV L/A V

SKETCH ARTIST 2: HANDS THAT SEE *** (18)
Jack Sholder USA 1994
Jeff Fahey, Courtney Cox, Brion James, James Tolkan, Jonathan Silverman, Michael Beach, Leilani, Ferrer, Michael Nicolosi, Scott Burjhilder, John Prosky, Robbie T. Robinson, Glenn Morshower, Li Shaye, Stephen Rappaport, Cindy Katz
Sequel to the first film, in which our artist finds that the only person who can identify a wanted serial killer is his only surviving victim, a blind woman. The defence attorney is not slow in claiming that the thirty seconds with which she was in contact with his face and head, cannot possibly be sufficient for her to point him out. However, events prove otherwise in a climactic and gripping finale.
THR 90 min (ort 100 min) VIDrel: MGM/WHV V

SKI SCHOOL * 18
Damian Lee USA 1990
Dean Cameron, Patrick Laborteaux, Ava Fabian, Tom Breznahan, Mark Thomas Miller, Stuart Fratkin, Spencer Rochefort, Darlene Vogel, Charlie Spadling, Gaetana Korbin, Mark High, John Pyper-Ferguson, Johnny Askwith, Alison Dobie
Clearly inspired by the success of the "Animal House" films, this tepid comedy deals with the exploits of a gang of inept and unappealing ski-bums who set out to enjoy life both on and off the piste. A witless effort of little mirth but much foolishness.
COM 84 min (ort 89 min) VIDrel: MIA/DISC V

SKI SCHOOL 2 * 18
David Mitchell USA 1994
Dean Cameron, Wendy Hamilton, Heather Campbell, Brent Sheppard, Bill Dwyer, Noah Heney, William Sasso, Jane Soweerby, Carrie-Lee Alhassan, Cinammon Bond, Malt Brasier, Shannon Lee Brown, Tiffany Burns, Shawn Carle, Jennifer Jasey
Sequel offering more of the same coarse humour and unfunny antics as our ski instructor returns to his home slopes and discovers to his consternation that his former girlfriend is about to marry a loathsome individual, who is after her money. Naturally, both he and his friends eventually manage to avert this catastrophe.
COM 88 min VIDrel: FIRST/SONOP V

SKI SLUTS GO SNOWBALLING * 18
Roger Cardinal CANADA 1970
Daniel Pilon, Mariette Levesque, Celine Lomez, Robert Arcand, Robert Demontigny, Jacques Desrosiers, Pierre Labell, Rene Angelil, Janine Sutto, Angele Coutu, Raymond Levesque, Carmen Champagne, Roger Michael, Tamora

This dire offering might just as well have been entitled "Sex On The Ski Slopes" as that's what it's all about, with a randy ski instructor out to bed as many of his female pupils as he possibly can.
Aka: APRES-SKI; SEX IN THE SNOW; SEX ON SKIES; SKI SLUTS GO SNOW-BALLIN'; SNOWBALLIN'; WINTER GAMES
A 85 min Cut (1 min 10 sec – ort 104 min)
VIDrel; MIA/DISC V

SKIN DEEP ** 18
Blake Edwards USA 1989
John Ritter, Vincent Gardenia, Alyson Reed, Julianne Phillips, Joel Brooks, Chelsea Field, Nina Foch, Denise Crosby, Michael Kidd, Bryan Genesse, Sheryl Lee Ralph, Dee Dee Rescher, Bo Foxworth, Raye Holliff, Jean Marie McKee
A slob still loves his ex-wife but is unwilling to reform in order to get her back. A patchy comedy, written by Edwards, that is very much a hit-and-miss affair.
COM 96 min (ort 102 min)
VIDrel: POLY/POLYREC/BRAVE L/A V

SKIN GAME, THE ** (PG)
Alfred Hitchcock UK 1931
Edmund Gwenn, Helen Hayes, Frank Lawton, John Longden, C.V. France, Phyllis Konstam, Jill Esmond, Edward Chapman, Ronald Frankau, Herbert Ross, Dora Gregory, R.E. Jeffrey, Geore Blanchof
Remake of a 1921 silent, about the conflict between a wealthy family and a builder with ambitious plans for some valuable land. An undistinguished epic, very untypical of the work of its director.
DRA 85 min (ort 87 min) B/W VIDrel: LUMI/SPEAR V
Boa: play by John Galsworthy.

SKIN GAME, THE ** PG
Paul Bogart USA 1971
James Garner, Louis Gossett Jr, Ed Asner, Susan Clark, Brenda Sykes, Andrew Duggan, Henry Jones, Royal Dano, Neva Patterson, Parley Baer, George Tyne, Pat O'Malley, Joel Fluellen, Napoleon Whiting, Juanita Moore, Cort Clark
In pre-Civil War days, a pair of con-men, one black and the other white, travel from town to town, repeating the same scam in which they pretend to be master and slave. Mild comedy western that was later remade as the TV movie "Sidekicks".
WES 98 min (ort 102 min) VIDrel: WHV V

SKINHEADS * 18
Greydon Clark USA 1989
Chuck Connors, Barbara Bain, Jason Culp, Elizabeth Sagal, Brian Brophy
An appallingly unattractive piece of dross, that deals with the activities of a group of L.A. neo-Nazis who flee to North California when one of them shoots a black man. It's not long before they start a vicious feud with a bunch of college kids, two of whom survive to mete out the expected (and deserved) vengeance.
Aka: SKINHEADS: THE SECOND COMING OF HATE
A/AD 86 min (ort 93 min) VIDrel: MOPIC/SGSVID V

SKYLARK ** (PG)
Joseph Sargent USA 1992
Glenn Close, Christopher Walken, Lexi Randall, Christopher Bell, Margaret Sophie Stein, Jon DeVries, Tresa Hughes, Elizabeth Wilson, Lois Smith, Lee Richardson
This weak sequel to SARAH, PLAIN AND TALL attempts to continue where the earlier film left off, with our reserved schoolteacher from Maine now firmly ensconced in the household of widowed farmer Walken. Unfortunately, the magic of the first film just isn't there and the contrived calamities (drought and a fire which wipes out the farm) are a little too excessive to carry conviction.
DRA 95 min (ort 100 min) mTV TVrel

SKY'S THE LIMIT, THE ** U
Edward H. Griffith USA 1943
Fred Astaire, Robert Ryan, Joan Leslie, Robert Benchley, Marjorie Gateson, Elizabeth Patterson, Clarence Kolb, Richard Davis, Peter Lawford, Eric Blore, Henri DeSoto, Dorothy Kelly, Norma Drury, Jerry Mandy, Clarence Muse
A flier on leave meets a lady photographer and they make sweet music together. Worthwhile episodes are Benchley's dinner speech and Astaire's "One For My Baby" and "My Shining Hour". Written by Frank Fenton and Lynn Root.
MUS 87 min (ort 89 min) B/W VIDrel: VCC L/A V

SKYSCRAPER *
Raymond Martino USA   18   1996
Anna Nicole Smith, Richard Steinmetz, Branko Cikatic, Calvin Levels, Charles Huber, Jonathan Fuller, Lee De Broux, Deirdre Imerhein
With her ample bosom very much to the fore, Smith fights off a gang of terrorists in a skyscraper, who have obviously made the mistake of thinking she would be a pushover. Trashy exploitation fare, promoted on the strength of the lead, who cannot act, has no understanding of dialogue and whose presence is the sole reason for this vanity vehicle ever seeing the light of day.
A/AD   90 min   VIDrel: MED   V

SLACKER **
Richard Linklater USA   15   1991
Richard Linklater, Rudy Basquez, Jean Caffeine, Jan hockey, Stephan Hockey, Mark James, Samuel Dietert, Bob Boyd, Terrence Kirk, Keith McCormack, Jennifer Schaudies, Dan Kratochvil, Maris Strautmanis, Brecht Andersch, Tom Pallotta
Low-budget and highly improvised account of a group of 1990s dropouts and the various ways they have of coping (or not) with modern problems. Some interesting ideas are presented but never amount to anything much. Screenplay is by Linklater, from whom one might have expected better.
DRA   100 min   VIDrel: FEATFIL/RTM   V

SLAM DUNK ERNEST **
John Cherry USA   U   1994
Jim Varney, Kareem Abdul-Jabbar, Jay Brazeau
A handyman at the local shopping mall becomes the odd one out when six of his workmates form a strong basketball team that gets a chance to play against the NBA. A good-natured if slightly moronic comedy outing.
COM   92 min   VIDrel: FIRST/SONOP   V

SLAP SHOT ***
George Roy Hill USA   18   1977
Paul Newman, Strother Martin, Jennifer Warren, Michael Ontkean, Lindsay Crouse, Jerry Houser, Andrew Duncan, Jeff Carlson, Steve Carlson, David Hanson, Yvon Barrette, Alan Nicholls, Brad Sullivan, Stephen Medillo
An ice hockey team that has fallen on hard times resorts to violence in a bid to recapture its popularity. Very slightly reminiscent of "The Deadliest Season", this sprawling and confused satire on the world of professional ice hockey is both vulgar and brutal, but is also occasionally very funny indeed. The uneven script is by Nancy Dowd.
COM   119 min (ort 123 min)
VIDrel: 4-FRONT/POLYREC/CIC   V

SLAUGHTER HIGH *
George Dugdale/Mark Ezra/Peter Litten USA   18   1985
Caroline Munro, Simon Scuddamore, Sally Cross, Donna Yeager, Dick Randal, Kelly Baker, Carmine Iannaccone, Gary Hartman, Billy Martin, Michael Saffran, John Segal, Josephine Scandi, Marc Smith, Jon Clark
An April Fool's Day prank that went wrong, left a student deformed and insane, and he was confined to a local asylum. Years later, a bunch of former students are invited to a school reunion and our lunatic escapes to have his revenge. There's nothing new in this formula shocker.
Aka: APRIL FOOL'S DAY
HOR   85 min Cut (32 sec – ort 89 min)
VIDrel: FIRST/SONOP L/A   V

SLAUGHTER IN SAN FRANCISCO *
William Lowe HONG KONG   18   1973
Don Wong, Chuck Norris, Sylvia Channing, Robert Jones, Dan Ivan, Bob Talbot, Robert J. Herguth, James Economides, Chuck Boyde
A former cop seeks revenge for the murder of his partner. Not released in the USA until 1981, when it added in some measure to the growing popularity of Norris, who here plays a baddie. A very poor effort that is flawed in almost every department.
Aka: KARATE COP; YELLOW FACED TIGER, THE
MAR   84 min (ort 87 min)   VIDrel: MIA/DISC   V

SLAUGHTER OF THE INNOCENTS **
James Glickenhaus USA   18   1993
Scott Glenn, Jesse Cameron-Glickenhaus, Zitto Kazann, Darlanne Fluegel, Sheila Tousey, Zakes Mokae, Kevin Sorbo, Henry Brown, Armin Shimerman, Jan Gardner, Elizabeth Johnson, Terri Hawkes, Tim Colceri, Leo Geter, Thom Dillon
An FBI agent is assigned to assist a local police force in the hunt for a serial murderer of young girls. His son, who is a computer

freak, uses his skills to pinpoint the killer, which, as might have been expected, makes him a target too. An assembly-line thriller, slickly made but highly unoriginal.
DRA   99 min (ort 144 min)   VIDrel: EIV/SONOP   V/sur

SLAUGHTERHOUSE FIVE ***
George Roy Hill USA   15   1972
Michael Sacks, Ron Leibman, Eugene Roche, Sharon Gans, Valerie Perrine, Sorrel Booke, John Dehner, Perry King, Roberts Blossom, Friedrich Ledebur, Holly Near, Lucille Benson, Kevin Conway, Nick Belle, Stan Gottlieb
Entertaining and polished rendition of a novel about an American optometrist who has become displaced in time. He finds himself constantly flitting back and forth between various important passages in his life, such as his time as a POW in Germany or the period he spent living on another planet. The music of J.S Bach and Vivaldi is used to good effect. Written by Stephen Geller.
FAN   99 min (ort 104 min)   VIDrel: CIC/SONOP L/A   V/h
Boa: novel Slaughter House Five; Or, The Children's Crusade by Kurt Vonnegut.

SLAVE GIRLS FROM BEYOND INFINITY **
Ken Dixon USA   18   1987
Elizabeth Clayton, Cindy Beal, Brinke Stevens, Don Scribner, Carl Horner, Kirk Grave, Randolph Roebling, Bud Graves, Jeffrey Blanchard, Mike Cooper, Greg Cooper, Sheila White, Fred Tate, Jacques Scardo
A low-budget SF sex fantasy done in the style of a 1950s Roger Corman film, with some amusing tongue-in-cheek touches and a plot similar to that of "The Most Dangerous Game". The story tells of a couple of slave girls who escape their cruel master and are forced to fight for survival on a planet full of androids, monsters and various other hazards.
FAN   71 min (ort 87 min)   VIDrel: ALLIED/RTM   V

SLAVES OF NEW YORK **
James Ivory USA   15   1989
Bernadette Peters, Adam Coleman Howard, Chris Sarandon, Mary Beth Hurt, Nick Corri, Madeleine Potter, Mercedes Ruehl, Betty Comden, Steve Buscemi, Jonas Abry, Michael Schoeffling, Tammy Grimes, Stephen Bastone, Charles McCaughan
Janowitz's set of tales about the arty inhabitants of downtown New York are cobbled together into an episodic adaptation, with the central character an intelligent woman who is constantly humiliated by her inability to fit in with her boyfriend's cliquey friends. Written by Janowitz, this long and cluttered film is all gloss and no substance.
COM   120 min (ort 125 min)   VIDrel: RCA L/A   V
Boa: book by Tama Janowitz.

SLAYER, THE **
J.S. Cardone USA   18   1981
Sarah Kendall, Frederik Flynn, Carol Kottenbrook, Alan McRae, Michael Holmes, Carl Kraines
Two couples vacationing on Georgia's Tybee Island meet up with the title character, but this meeting brings both danger and death to the party of four holiday-makers. A lukewarm, extremely gory and claustrophobic shocker. This was one of those VIPCO "video nasties" that dropped out of circulation for a while, but has now resurfaced, albeit in a slightly edited form.
Aka: NIGHTMARE ISLAND
HOR   85 min (ort 89 min)   VIDrel: VIPCO/SGSVID   V/h

SLEEP, BABY, SLEEP **
Armand Mastroianni USA   PG   1995
Tracey Gold, Kyle Chandler, Thomas Calabro, Joanna Cassidy
A woman with a congenital mental disorder becomes the chief suspect when her husband and baby vanish.
DRA   92 min   VIDrel: ODY/SONOP   V/sh

SLEEP MY LOVE ***
Douglas Sirk USA   PG   1948
Claudette Colbert, Robert Cummings, Don Ameche, Rita Johnson, George Coulouris, Raymond Burr, Hazel Brooks, Keye Luke
A husband tries to drive his wife mad in order to be rid of her, but his schemes are foiled. A hackneyed plot is given new life by an excellent cast. Written by St Clair McKelway. The memorable photography is by Joseph Valentine.
THR   97 min B/W   VIDrel: STABL L/A   V
Boa: novel by Leo Rosten.

SLEEP WITH ME ✶✶   18
Rory Kelly USA   1994
Eric Stoltz, Meg Tilly, Craig Sheffer, Todd Field, Susan Traylor, Dean Cameron, Thomas Gibson, Tegan West, Amaryllis Borrego, Parker Posey, Joey Lauren Adams, Vanessa Angel, Adrienne Shelly, Quentin Tarantino, June Lockhart
Stylised, six-part account of the complexities of a love triangle involving a man, his wife and his best friend, with the last character giving vent to his feelings on the eve of the couple's wedding. An uneven and highly artificial affair of minimal interest. Look out for Quentin Tarantino in a tiny part as an obnoxious, know-it-all film critic.
DRA   82 min (ort 117 min)   VIDrel: FIRST/SONOP   V/sur

SLEEPER ✶✶✶   PG
Woody Allen USA   1973
Woody Allen, Diane Keaton, John Beck, Mary Gregory, Don Keefer, John McLiam, Peter Hobbs, Bartlett Robinson, Marya Small, Spencer Milligan, Chris Forbes, Susan Miller, Brian Avery, Lou Picetti, Spencer Ross, Jessica Rains
One of Allen's more consistent comedies, in which he plays a health store owner who was deep frozen after a minor operation went wrong and wakes up two-hundred years later in a totalitarian society. A succession of splendid sight gags and some memorable lines keep this one moving along, despite one or two pauses and a rather feeble ending. The jazz score is by the Preservation Hall Jazz Band. As ever, Allen writes and directs.
COM   83 min (ort 88 min)   VIDrel: WHV   V

SLEEPERS ✶✶   15
Barry Levinson USA   1996
Kevin Bacon, Robert DeNiro, Dustin Hoffman, Bruno Kirby, Jason Patric, Brad Pitt, Brad Renfro, Ron Eldard, Billy Crudup, Vittorio Gassman, Terry Kinney, Frank Medrano, Jonathan Tucker, Geoffrey Wigdor, Joe Perrino, Minnie Driver
Four youths are sentenced to detention in a juvenile prison, and endure torments from several prison guards, but principally one who is a dedicated and evil sadist. Thirteen years later, two of the boys have their revenge on the ring-leader, and eventually all is revealed in the witness box. An absurd homophobic fantasy, based on a bestseller that was purportedly inspired by real events, but no real events could follow such a neatly contrived pattern.
DRA   140 min (ort 147 min)   VIDrel: POLFIL   V/s
Boa: book by Lorenzo Carcaterra.

SLEEPING BEAUTY ✶✶✶   U
Clyde Geronimi USA   1958
Voices of: Mary Costa, Bill Shirley, Eleanor Audley, Verna Felton, Barbara Jo Allen (Vera Vague), Bill Thompson, Taylor Holmes, Candy Candido, Pinto Colvig, Bob Amsberry, Marvin Miller (narration only)
One of Disney's most expensive and successful animations, this tells simply and without unnecessary embellishment, the story of the beautiful princess who pricks her finger on a spinning wheel and is cast into a deep sleep. A splendid example of a classic Disney animation, with a fine music soundtrack adapted from Tchaikovsky. Filmed in wide-screen, but this will be lost on conventional TV screens.
ANIM   72 min (ort 85 min)   VIDrel: WDV/TECH   V/drm

SLEEPING BEAUTY ✶✶   U
David Irving USA   1986
Morgan Fairchild, Tahnee Welch, Nicholas Clay, Sylvia Miles, Kenny Baker, David Holliday, Jane Wiedlin, Shari R. Ophir, Julian Chagrin, Orna Porat, Yankele Ben Sira, Danny Segev, Joseph Bee, Yehuda Efroni, Michael Schneider
A competent but unmemorable live-action version of this classic fairytale.
Aka: CANNON MOVIE TALES: SLEEPING BEAUTY
JUV   89 min   VIDrel: MGM/WHV   V/sh

SLEEPING BEAUTY ✶   18
Paul Thomas USA   1989
Hyapatia Lee, Tori Welles, Ariel Knights, April West, Randy West, Ona Zee, Tom Byron, Robert Bullocks, Bud Lee
A dreary softcore version of a US hardcore film that uses this famous fairytale as no more than a vehicle for the usual antics. Laughable attempts at "Olde Worlde" English and settings to match are complemented by indifferent acting and a non-exis-

tent plot. Shot on videotape for good measure, this feeble effort is definitely one to avoid.
Aka: KISS ON THE LIPS; SLEEPING BEAUTY; SLEEPING BEAUTY AROUSED
A   60 min Cut (6 sec) mVid   VIDrel: MOPIC/SGSVID L/A
V

SLEEPING TIGER, THE ✶✶   U
Joseph Losey UK   1954
Dirk Bogarde, Alexander Knox, Alexis Smith, Hugh Griffith, Maxine Audley, Glynn Houston, Billie Whitelaw, Patricia McCarron, Harry Towb, Russell Waters, Fred Griffiths, Esma Cannon
A psychiatrist tries to help a criminal change his ways but the man repays him seducing his wife, although he later relents when his benefactor provides him with an alibi. An implausible and overblown melodrama that fails to convince or entertain.
DRA   85 min (ort 89 min) B/W   VIDrel: WHV   V/h
Boa: novel by Maurice Moiseiwitch.

SLEEPING WITH STRANGERS ✶✶   (15)
William T. Bolson CANADA   1992
Adrienne Shelly, Neil Duncan, Kymberley Huffman, Shawn Alex Thompson, Scott McNeill, Gary Jones, Anthony Vic, Claire Caplan, Betty Linde, Jeff Cohen, Susan Wilkey, Gabrielle Rose, Tamsin Jones, Sarah Deakins, Tim Battle, Tom Heaton
A struggling resort is visited by a famous married couple, who soon begin to quarrel and exhibit signs of sexual jealousy in this fast-paced romantic comedy. Fair.
COM   100 min (ort 103 min)   SATrel: SKY MOVIES

SLEEPING WITH THE ENEMY ✶✶   15/18
Joseph Ruben USA   1991
Julia Roberts, Patrick Bergin, Kevin Anderson, Elizabeth Lawrence, Kyle Secor, Claudette Nevins, Tony Abatemarco, Marita Geraghty, Harley Venton, Nancy Fish, Sandi Shackleford, Bonnie Cole, Graham Harrington
Intended as a vehicle for Roberts after her success in PRETTY WOMAN, this film sees her playing the part of a battered wife who has escaped from her unbalanced husband and made a new life for herself. As might be expected, he turns up again and begins to plan fresh sufferings for her. A failed thriller of little tension. See THE STEPFATHER for another film by Ruben on the psychotic husband/father theme. Music is by Jerry Goldsmith.
THR   94 min (15 ver); (99 min – uncut 18 ver)
VIDrel: 20TH/TECH L/A   V
Boa: novel by Nancy Price.

SLEEPLESS IN SEATTLE ✶✶✶   PG
Nora Ephron USA   1992
Meg Ryan, Tom Hanks, Bill Pullman, Rob Reiner, Rosie O'Donnell, Rita Wilson, Ross Mallinger, Gaby Hoffman, Victor Garber, Barbara Garrick, Carey Lowell, Dana Ivey, Calvin Trillin, Bill Pullman, Kevin O'Morrison, Valerie Wright, Tom Tammi
After his mother dies, a young boy calls a national talkshow to say how unhappy his father is, convinced that he will never find such love again. He is heard by a woman who, although just engaged, decides that this is the man of her dreams and she promptly spends the rest of the movie searching for him. A witty, beautifully acted and unashamedly romantic tale, with fine camerawork by Sven Nykvist.
COM   100 min (ort 105 min) wScrn cC
VIDrel: COLUM/SONOP   V/sur

SLEEPSTALKER: THE SANDMAN'S LAST RITES ✶✶✶   18
Turi Meyer USA   1994
Jay Underwood, Kathryn Morris, William Lucking, Michael D. Roberts, Michael Harris
The title character started life as a psychopath, and slaughtered a small boy's parents. Placed on death row for this and other crimes, he is executed, but a black magic pact allows him to return as a creature that can transform itself into sand. Quite a decent yarn, that has a lot in common with the NIGHTMARE ON ELM STREET series, although the budget is a good deal smaller, forcing the director to opt for inventiveness rather than overblown effects.
Aka: SLEEP STALKER
HOR   102 min   VIDrel: MIA/DISC   V/h

SLEEPWALKERS ✶✶   18
Mick Garris USA   1992
Brian Krause, Madchen Amick, Alice Krige, Jim Haynie, Cindy Pickett, Ron Perlman, Lyman Ward, Dan Martin, Glenn Shadix,

Cynthia Garris, Monty Bane, John Landis, Joe Dante, Stephen King, Clive Barker, Tobe Hooper, Frank Novak

Stephen King's first completely original screenplay tells of a mother and son whose arrival in town brings with it the requisite measure of horror. For although they appear outwardly normal, both mother and son are shape-changing alien creatures that live by feeding on the life-force of human virgins. Amick plays one such girl who gets to attract their attention. A stilted and wholly derivative affair.

Aka: STEPHEN KING'S SLEEPWALKERS

HOR 85 min (ort 89 min) VIDrel: VCC/DISC/COLUM V/sur

SLEUTH ***   15
Joseph L. Mankiewicz USA   1972

Laurence Olivier, Michael Caine, Alec Cawthorne, Margo Channing

Based on a successful play, this tells of how a writer of detective stories decides to give his wife's lover the fright of his life, by playing a series of nasty tricks on him after he has been inveigled down to his country home. Both actors are in top form in this game of wits, but the terribly contrived resolution is something of a let-down. Written by Shaffer with sets by Ken Adam.

DRA 132 min (ort 139 min) VIDrel: MIA/DISC V
Boa: play by Anthony Shaffer.

SLICK HONEY **   18
John Leslie USA   1989

Selena Steele, Rachel Ashley, Sasha Strange, Heather Torrance, Marc Wallice, Jamie Gillis, Don Hart, Joey Silvera, Charli, Mike Horner, Gene Carrera

Slick (Wallice) is a private eye for whom business is so bad that he's had to give up his apartment and sleep in the office. Matters improve when a wealthy woman mistakes his office for the sex clinic next door. This leads to him being hired by the woman to check on her unfaithful husband, from whom she wants a divorce. A muddled effort that starts off quite well but is hampered by a badly thought out script.

A 62 min VIDrel: MIA/DISC V

SLIME, THE **   18
Gregory Lamberson USA   1989

Robert C. Sabin, Mary Hunter

A man takes over an apartment that was once the home of the leader of an occult sect, and wakes up one morning to find himself covered in slime and no longer the master of his actions.

HOR 81 min VIDrel: VIPCO/SGSVID V

SLIME PEOPLE, THE *   (PG)
Robert Hutton USA   1962

Robert Hutton, Les Tremayne, Susan Hart, Robert Burton, Judee Morton, John Close, William Boyce

Los Angeles is attacked by strange creatures from deep beneath the surface who have been awakened from their dormant state by nuclear tests. In order to take over our world, they start to modify the environment so that they can breed, and soon the city is enveloped in a strange kid of fog. A cheerfully inept fantasy, worth seeing to check just how far special effects have developed since then.

FAN 76 min B/W SATrel: BRAVO MOVIES

SLINGSHOT, THE ***   12
Ake Sandgren SWEDEN   1992

Jesper Salen, Stellan Skarsgard, Basia Frydman, Niclas Olund, Ernst-Hugo Jaregard, Reine Brynolfsson, Jacob Leygraf, Frida Hallgren, Axel Duberg, Ernst Gunther, Ing-Marie Carlsson, Tomas Norstrom, Rolf Lassgard, Jurek Sawka

Similar in outlook to MY LIFE AS A DOG, this rites-of-passage film is a bittersweet look at growing up in Sweden. In 1920s Sweden, a young and lively twelve-year-old boy becomes a target for bullies because of his Jewish mother and the radical politics of his father. To protect himself he fashions the title implement out of two condoms, in this warmhearted and very well acted autobiographical tale.

Aka: KADISBELLAN

DRA 102 min VIDrel: CONNO/RTM V
Boa: novel by Roland Schutt.

SLIPPER AND THE ROSE, THE *   U
Bryan Forbes UK   1976

Richard Chamberlain, Kenneth More, Gemma Craven, Michael Hordern, Edith Evans, Margaret Lockwood, Annette Crosbie, Christopher Gable, Lally Bowers, Julian Orchard, John Turner, Sherrie Hewson, Rosalind Ayres, Keith Skinner

An overlong and quite dreadful musical version of Cinderella. The actors are wooden, the dialogue stilted and the songs terrible. A film that is almost painful to watch. The end when it does come is a blessed relief. Written and directed by Forbes with Richard Sherman. Songs (such as they are) appear courtesy of Robert and Richard Sherman.

MUS 136 min (ort 146 min) VIDrel: CASPIC/BMGREC L/A V
Boa: short story Cinderella by Perrault.

SLIPSTREAM **   PG
Steven M. Lisberger USA   1988

Mark Hamill, Bob Peck, Bill Paxton, Kitty Aldridge, Ben Kingsley, Eleanor David, Robbie Coltrane, F. Murray Abraham, Roshan Seth, Richard Huggett, Rita Wolf, Susan Leong, Deborah Leng, Gay Baynes, Bruce Boa, Paul Reynolds

Fantasy set in a world where the laws of nature are in disarray, and the main mode of transport is in craft carried along by the powerful winds. A man being taken to prison escapes from his captors, and is kidnapped by a young adventurer who intends claiming a reward. Despite some highly imaginative ideas, this story soon degenerates into a standard chase movie, but one that's long on talk and short on action. Music is by Elmer Bernstein.

FAN 98 min (ort 101 min)
VIDrel: 4-FRONT/POLYREC/EIV L/A V

SLIVER **   18
Philippe Noyce USA   1992

Sharon Stone, William Baldwin, Tom Berenger, Polly Walker, Colleen Camp, Martin Landau, Amanda Foreman, C.C.H. Pounder, Nina Foch, Keene Curtis, Tony Peck, Annie Betangcourt, Nicholas Pryor, Melvin Kinder, Franz Turner, Jose Ray

A young professional woman moves into a luxurious apartment in a Manhattan building plagued by a series of mysterious deaths. She soon becomes romantically involved with two men, one of whom is the building's owner. The latter turns out to be a computer freak with voyeuristic tendencies, who has installed video cameras in every apartment. A dark and poorly developed affair that was not improved by having its ending re-shot.

DRA 103 min (ort 108 min) cC VIDrel: CIC/SONOP; PION (LV only) V/dm V/sur LV
Boa: novel by Ira Levin.

SLOW BURN **   18
John Eyres USA   1990

William Smith, Anthony James, Ivan Rogers, Scott Anderson

An overworked cop is given the job of stemming the gang war between the Mafia and a Chinese gang that has muscled in on the narcotics trade. With Chinatown at boiling point because of several brutal slayings, the only hope for peace seems to lie with an undercover cop who has becomes a trusted member of the Don's organisation. A violent and low-grade action film that is no more than watchable.

A/AD 94 min VIDrel: FIRST/SONOP L/A V

SLOW MOTION ***   18
Jean-Luc Godard FRANCE   1980

Isabelle Huppert, Jacques Dutronc, Nathalie Baye, Marguerite Duras, Roland Amstutz, Anna Baldaccini

A parody of sexuality in the modern world with three very different people finding themselves unhappy in their surroundings and trying to escape them. Huppert is a naive country girl who comes to the city but becomes a prostitute, Baye leaves the city for a life in the country, while finally, Dutronc would like to escape life in town but is unable to. A plotless, multi-stranded movie, that would never have worked in the hands of a lesser director.

Aka: EVERY MAN FOR HIMSELF; SAUVE QUI PEUT (LA VIE)

DRA 84 min (ort 87 min) VIDrel: ARTIF/20TH V

SLUGS *   18
Jean Piquer Simon SPAIN/USA   1988

Michael Garfield, Kim Terry, Philip McHale, Kris Mann, Patty Shepard, Alicia Maro, Santiago Alvarez, Concha Cuetos, Kari Rose, Andy Alsup, Emilio Linder, Stan Schwartz, Manuel De Blas, Frank Brana, Juan Majan, Lucia Prado

This first film version of one of Hutson's books, opens with several people meeting nasty accidents arising from their contact with deadly, mutated slugs. Despite a lack of belief on the part of the police, who put the deaths down to a series of accidents, a health inspector who has learnt the truth sets out to destroy

them (or at least keep them away from his lettuces). A repulsive and far-fetched story, of poor dialogue and little imagination.
Aka: SLUGS, THE MOVIE
HOR 85 min Cut (42 sec – ort 90 min)
VIDrel: VCC/DISC/COLUM L/A V
Boa: novel by Shaun Hutson.

SMALL BACK ROOM, THE ***   PG
Mike Powell/Emeric Pressburger UK   1949
David Farrar, Jack Hawkins, Leslie Banks, Robert Morley, Kathleen Byron, Cyril Cusack, Emrys Jones, Renee Asherson, Walter Fitzgerald, Anthony Bushell, Milton Rosmer, Michael Hough, Michael Goodliffe, Henry Caine
A withdrawn young scientist with an ill-fitting artificial foot that causes him constant pain, finds some meaning to his life when he is called upon to disarm a new type of German bomb. A low-key, human interest drama, it builds to a most gripping climax. A highly expressionistic dream sequence (our hero is an alcoholic tormented by his condition) is an unexpected sequence very typical of the innovative flourishes these directors liked to add to their films.
Aka: HOUR OF GLORY
DRA 107 min (ort 108 min) B/W VIDrel: LUMI/SPEAR V
Boa: novel by Nigel Balchin.

SMALL FACES **   15
Gillies MacKinnon UK   1995
Ian Robertson, Joseph McFadden, J.S. Duffy, Laura Fraser, Garry Sweeney, Ian McElhinney, Clare Higgins, Kevin McKidd, Mark McConnochie, Steven Singleton, David Walker, Paul Doonan, Colin Semple, Colin McCredie, Debbie Welsh
Gangster drama set in the 1960s in the inner city area of Glasgow, where two young brothers get caught up in a gang war beyond their control. Despite the slightly flashy direction this coming-of-age saga has a lot going for it, not least the endearingly natural performances.
DRA 104 min (ort 108 min) VIDrel: GUILD/FOXVID V/sur

SMALL SACRIFICES ***   15
David Greene USA   1988
Farrah Fawcett, Ryan O'Neal, John Shea, Gordon Clapp, Emily Perkins, Gary Chalk, Ken James, Sean McCann, Garwin Sanford, Tom Butler, Elan Ross Gibson, Lynn Cormack, Jayne Eastwood, Christopher Carvalho, Ann Bradley Jaeger
Fawcett gives a convincing performance in her role as Diane Downs, an Oregon mother who in 1983 was accused of shooting her three children, killing one outright and paralysing another. However, the evidence linking her to this crime was purely circumstantial. An intense and powerful melodrama that is often quite disturbing to watch. Originally shown in two parts.
DRA 186 min (ort 200 min) mTV VIDrel: ODY/SONOP V/s
Boa: book by Ann Rule.

SMALL TOWN MASSACRE *   18
Michael Laughlin AUSTRALIA/NEW ZEALAND   1981
Michael Murphy, Louise Fletcher, Arthur Dignam, Dan Shor, Fiona Lewis, Marc McClure, Scotty Brady, Dey Young, Charles Lane, Jim Boelsen, Beryl Te Wiata, B. Courtenay Leigh, William Hayward, Elizabeth Cheshire, Billy Al Bengston
A small town is terrorised by a brutal and crazed killer who leaves the usual trail of corpses, and a determined police chief traces the culprit back to a scientific establishment where secret experiments are being carried out. A low budget shocker that is supposedly set in the Midwest but was actually filmed in New Zealand.
Aka: DEAD KIDS; HUMAN EXPERIMENTS; STRANGE BEHAVIOUR
HOR 93 min (ort 97 min) VIDrel: VIPCO/SGSVID V

SMALLEST SHOW ON EARTH, THE **   U
Basil Dearden UK   1957
Virginia McKenna, Bill Travers, Peter Sellers, Margaret Rutherford, Bernard Miles, Leslie Phillips, Francis De Wolff, Sidney James, June Cunningham, George Cross, Stringer Davis, George Cormack, Sam Kydd, Michael Corcoran
Very British, slightly unreal comedy with a young couple inheriting a flea-pit cinema, complete with drunken projectionist and other eccentric characters. Against all the odds they manage to make it pay and the humour derives from the various under-

hand ploys they use. An uncomfortable blend of sentiment and farce that carries little force in either area.
Aka: BIG TIME OPERATORS
COM 81 min B/W VIDrel: CASPIC L/A V

SMASH PALACE **   18
Roger Donaldson NEW ZEALAND   1981
Bruno Lawrence, Anna Jemison, Greer Robson, Keith Aberdein, Desmond Kelly, Margaret Umbers, Sean Duffy, Bryan Johnson, Terence Donovan, Dick Rollo, Ian Barber, Mike Beytagh, Brian Chase, Ross Davies, Colin Fredericken, Chris Pasco
A junkyard owner is so engrossed in smashing up cars that he fails to notice the crisis affecting his family. A murky and muddled melodrama slightly redeemed by good performances.
DRA 108 min VIDrel: ARTPRO/RTM V

SMILES OF A SUMMER NIGHT ***   PG
Ingmar Bergman SWEDEN   1955
Gunnar Bjornstrand, Harriet Anderson, Ulla Jacobsson, Eva Dahlbeck, Margit Carlquist, Jarl Kulle, Bjorn Belvenstam, Ake Fridell, Naima Wifstrand, Gull Natrop, Birgitta Valberg, Bibi Andersson, Anders Wulff, Svea Holst, Hans Straat
Bergman's wry look at love and marriage among the upper classes at the turn of the century. Behind the sophisticated manners and talk can be sensed the emptiness and meanness of their comfortable lives, while only the working-class figures seem natural and in touch with their feelings. Despite the lightness of the summer night, the comedy remains bleak and bitter, with pain and sorrow just below a seemingly carefree surface. See also A LITTLE NIGHT MUSIC.
Aka: SOMMARNATTENS LEENDE
COM 104 min (ort 108 min) B/W VIDrel: TART/20TH V

SMILEY'S PEOPLE ***   15
Simon Langton UK   1982
Alec Guinness, Mario Adorf, Eileen Atkins, Anthony Bate, Michael Byrne, Rosalie Crutchley, Michael Elphick, Barry Foster, Michael Gough, Bernard Hepton, Curt Jurgens, Maureen Lipman, Michael Lonsdale, Vladek Sheybal
Based on another Le Carre novel, this sequel of sorts to TINKER, TAILOR, SOLDIER, SPY sees former master-spy Smiley involved in a plot to ensure the defection of Karl, his old Soviet adversary and opposite number. A thin plot is sustained by complex detail, some excellent acting and a few well chosen locations; the merciful absence of the usual gunfight/car chase espionage cliches is an additional bonus.
DRA 337 min (2 cassettes) mTV VIDrel: BBC/TECH V/h
Boa: novel by John Le Carre.

SMITHEREENS **   15
Susan Seidelman USA   1982
Susan Berman, Brad Rinn, Richard Hill, Roger Jett, Nada Despotovitch, Kitty Summerall
A selfish nineteen-year-old punk girl comes to New York and hopes to fulfil her dreams of becoming a rock star, but soon finds herself in desperate straits. A realistic and absorbing character study, often quite abrasive and not all that well made, but full of energy and a welcome measure of humour.
DRA 89 min (ort 93 min) VIDrel: ARROW/RTM V

SMOKE ***   15
Wayne Wang USA   1995
Ginacarlo Esposito, Jose Zuniga, Stephen Gevedon, Harvey Keitel, Jared Harris, William Hurt, Daniel Auster, Harold Perrineau, Deidre O'Connell, Victor Argo, Michelle Hurst, Forest Whitaker, Stockard Channing, Vincenzo Amelia
A look at the role chance events play in life, when five individuals, all of whom have their own problems, get caught up in a robbery at a Brooklyn cigar store, the owner of which is much given to flights of homespun wisdom. Scripted by Auster, this is a clever conversation piece of witty anecdotes and tall tales. The title comes from Paul Benjamin's story of how Sir Walter Raleigh got the weight of smoke by weighing a cigar both before and after it was smoked.
DRA 108 min VIDrel: BUENA V
Boa: short story Auggie Wren's Christmas Story by Paul Auster.

SMOKEY AND THE BANDIT ***   PG
Hal Needham USA   1977
Burt Reynolds, Jackie Gleason, Sally Field, Jerry Reed, Mike Henry, Paul Williams, Pat McCormick, Alfie Wise, George Reynolds, Macon McCalman, Linda McClure, Susan McIver, Michael Mann, Lamar Jackson, Ronnie Gay, Hank Worden

A bootlegging truck driver and his sidekick, outwit a redneck sheriff in hot pursuit across several states. The cars act beautifully but the overall effect is one of motion without progress. Followed by the inevitable trail of sequels. See also BANDIT 1: BANDIT GOES COUNTRY – the first film in a similar set of trucker adventures.
COM   92 min (ort 97 min)
VIDrel: 4-FRONT/POLYREC/CIC   V/h

SMOKEY AND THE BANDIT 2 *
Hal Needham USA
Burt Reynolds, Jackie Gleason, Sally Field, Jerry Reed, Dom DeLuise, Paul Williams, Pat McCormick, Mike Henry, Brenda Lee, Mel Tillis, John Anderson, David Huddleston, Phil Balsley, Lew DeWitt, Don Reid, Harold Reid
This time around our heroes agree to take a pregnant elephant to Texas. Unlike the first film, this one is short on action, with few gags to enliven the boring bits.
Aka: SMOKEY AND THE BANDIT RIDE AGAIN
COM   97 min (ort 101 min)   VIDrel: CIC/SONOP L/A
V/h

PG
1980

SMOKEY AND THE BANDIT: PART 3 *
Dick Lowry USA
Jackie Gleason, Paul Williams, Jerry Reed, Pat McCormick, Mike Henry, Burt Reynolds, Colleen Camp, Faith Minton, Sharon Anderson, Silvia Arana, Alan Berger, Ray Bouchard, Connie Brighton, Earl Houston Bullock, Ava Cadell
Third film in the series, with the redneck sheriff accepting a bet that he can drive from Miami to Texas in 24 hours. A vapid yawn-inducer. Originally filmed with Gleason playing two roles and then re-shot with Reed.
Aka: SMOKEY IS THE BANDIT
COM   84 min (ort 88 min)   VIDrel: CIC/SONOP L/A   V

18
1983

SMOKING/NO SMOKING ***
Alain Resnais FRANCE
Sabine Azema, Pierre Arditi
Screenplay by Ayckbourn in this attempt to bring his set of eight plays to the screen and compress their ideas into a mere two and a half hours. Set in a Yorkshire village, the film follows the interactions of nine characters over five years, the central idea being that each story has the same starting point and its course depends on mere chance remarks. An intellectual exercise best enjoyed in a theatre setting, with two actors playing all of the parts.
DRA   147 min   CINrel
Boa: plays Intimate Exchanges by Alan Ayckbourn.

PG
1993

SNAKE DEADLY ACT **
Wilson Tong HONG KONG
Ng Kun Lung, Wilson Tong, Fong Hak An, Angela Mao, Chan Wai Man, Phillip Kao
A man comes to the aid of a prostitute who is being attacked, but is badly injured. Fortunately, he is saved by a powerful fighter who teaches him a deadly combat style, and at the same time he learns the truth about his father.
MAR   91 min wScrn   VIDrel: EAST/DISC   V

15

SNAKE EATER, THE **
George Erschbamer CANADA
Lorenzo Lamas, Josie Bell, Ronnie Hawkins, Robert Scott, Cheryl Jeans, Larry Csonka, Ron Palilo, Ben Di Gregorio, Mona Pryor
When his sister is kidnapped and the other members of his family murdered by a psychotic, an ex-Marine takes off in pursuit. He ultimately tracks his quarry to a swamp where his military skills give him the edge he needs to survive and achieve his ends. A violent and disagreeable actioner that covers familiar territory. Lamas may be known to TV viewers from the series "Falcon Crest". Followed by SNAKE EATER'S REVENGE.
A/AD   89 min Cut (1 min 48 sec)   VIDrel: RCA L/A   V

18
1989

SNAKE EATER'S REVENGE **
George Erschbamer USA
Lorenzo Lamas, Larry B. Scott, Michelle Scarabelli, Kathleen Kinmont
A sequel to THE SNAKE EATER in which a suspended cop wages a lone war against a ruthless drugs baron who is responsible for supplying contaminated drugs to children. A standard assemblage of smoke, noise and gunfire. SNAKE EATER 3: HIS LAW followed.
Aka: SNAKE EATER 2: THE DRUG BUSTER
A/AD   89 min (ort 93 min)   VIDrel: RCA L/A   V

18
1990

SNAKE EATER 3: HIS LAW **
George Erschbamer CANADA
Lorenzo Lamas, Minor Mustain, Tracy Cook, Holly Chester, Chip Chuika, Tracy Hway, Una Kay, Gordon Atkinson, Walker Boone, Chris Benson, Scott "Bam Bam" Bigelow, Cary Lawrence, Jason Cavalier, Alan Keiping-Legros, Michael Scherer
Second sequel to THE SNAKE EASTER in which he and his cowboy friend rescue a girl who has been kidnapped and abused by a biker gang, most of whose members are eliminated in the process. Another assembly-line tale, full of bloodshed, violence and fury.
A/AD   87 min (ort 109 min)   VIDrel: CIC/SONOP   V
Boa: novel Rafferty's Rules by W. Glenn Duncan.

18
1992

SNAKE IN THE EAGLE'S SHADOW ***
Yuen Woo Ping HONG KONG
Jackie Chan, Juan Jon Lee, Simon Yuen, Shi Tien, Chiu Chi-Ling, Chen Hsia, Wang Chang, Louis Feng, Cheng Lung, Cheng Li
The last remaining master of the snake style of kung fu, decides to take on a student to ensure the survival of this style of combat. A vigorous blend of kung fu and comedy, the first film of this type starring Jackie Chan.
Aka: EAGLE'S SHADOW, THE; SHE-HSING ZIAO SHOU; SNAKE IN EAGLE SHADOW; SNAKE IN EAGLE'S SHADOW
MAR   92 min Cut (34 sec – ort 97 min) wScrn
VIDrel: MADE/RTM   V

18
1978

SNAKE IN THE EAGLE'S SHADOW: PART 2 **
HONG KONG
Long Fei
Four different kung fu styles are on display here in this competent sequel to the first film, that as expected, uses the formula honour-must-be-avenged plot as a vehicle for some fancy footwork.
MAR   89 min   VIDrel: ONE/IMPENT   V

15
1989

SNAKE IN THE MONKEY'S SHADOW **
Chang Shen (Cheng Sum) HONG KONG
John Chong (Chang Wu Lang), Wilson Tang (Tang Wei-Cheng), Hou Chao-Sheng, Charlie Chan, Domson Shi
Two rival fighting styles are put to the test in a series of bloody clashes
Aka: HOU HSING K'OU SHOU
MAR   81 min Cut (2 min 51 sec)   VIDrel: ONE/IMPENT   V

15
1979

SNAKE STRIKES BACK, THE **
HONG KONG
Eagle Han, Elton Chong, Kim Miou
A ruthless fighter plots to achieve control of a kung fu school, and thus gain possession of a secret book of invincible fighting techniques. Only the favourite pupil of the current master is able to oppose his wicked plans.
MAR   82 min Cut (4 sec)   VIDrel: IMPENT   V

18
198-

SNAPDRAGON **
Worth Keeter USA
Steven Bauer, Chelsea Field, Pamela Anderson, Matt McCoy, Kenneth Tigar, Larry Manetti, Rance Howard, Gloria Le Roi, Diana Lee Hsu, Irene Tsu, John F.O'Donahue, Michael Monks, Drew Snyder, Phillip Troy, Michael Young, John Green
A police psychologist is assigned to treat a beautiful woman amnesiac and gets implicated in a strange and deadly game. The title refers to the small concealed knife Chinese concubines used to murder their masters' enemies, and these weapons figure large in the story. Quite predictable in plotting, but taut and compelling.
THR   94 min (ort 96 min)   VIDrel: GUILD/SONOP   V/sh

18
1992

SNAPPER, THE ***
Stephen Frears UK
Colm Meaney, Tina Kellegher, Ruth McCabe, Colm O'Byrne, Eanna MacLiam, Ciara Duffy, Pat Laffan, Joanne Gerrard, Peter Rowan, Fionnuala Murphy, Karen Woodley, Brendan Gleeson, Virginia Cole, Denis Menton, Ronan Wilmot, Stuart Dunne
A young teenage girl causes a scandal in her working-class Irish family when she announces that she is pregnant and chooses not to reveal the name of the father. Needless to say, both her family and her community eventually rally round, in this fine and warm-hearted adaptation of Doyle's novel.
COM   91 min (ort 95 min) mTV   VIDrel: ELPIC/POLYREC
V/sur
Boa: novel by Roddy Doyle.

15
1993

SNEAKERS ** 15
Phil Alden Robinson USA 1992
Robert Redford, Dan Aykroyd, River Phoenix, Mary McDonnell, Ben Kingsley, Sidney Poitier, David Stathairn, Timothy Busfield, Eddie Jones, Donal Logue, Lee Garlington, George Hearn, Stephen Tobolowsky, Eddie Jones, James Earl Jones
A security analyst with a past involvement in computer fraud is blackmailed by government agents into stealing a sophisticated code-breaking device from its inventor. Needless to say, nothing is as its seems and both he and his team find themselves in great danger. Unfortunately, a strong plot idea is frittered away in a long and rambling thriller that takes far too much time to reach its disappointing conclusion.
THR 120 min (ort 125 min) cC VIDrel: CIC/SONOP; PION (LV only) V/sur LV

SNIPER ** 15
Luis Llosa USA 1992
Tom Berenger, Billy Zane, J.T. Walsh, Aden Young, Ken Rudly, Gary Swanson, Reinaldo Arenas, Hank Garrett, Frederick Miragliotta, Vanessa Steele, Carlos Alvarez, Tyler Copin, Teo Gebert, Edward Wiley, William Curtin, Howard Bose
A season Marine sergeant experienced in jungle warfare teams up with a crack marksman to bring down a drugs baron and his politician partner. A violent and unappealing actioner, poorly acted and indifferently directed.
A/AD 93 min (ort 99 min) VIDrel: EIV/SONOP V/sur

SNOOPY, COME HOME! *** U
Bill Melendez USA 1972
Voices of: Bill Melendez, Stephen Shea, David Carey, Chad Webber, Robin Kahn, Johanna Baer, Hilary Momberger, Chris De Faria, Linda Ercoli, Linda Mendelson
Animation based on a famous comic strip cartoon by Schulz – "Peanuts" – with Snoopy, the beagle, getting so annoyed at the number of "No Dog" signs, that he decides to run away with Woodstock, his feathered friend. A distraught Charlie Brown searches everywhere for him. An enjoyable second outing for the "Peanuts" gang (the first film being "A Boy Named Charlie Brown"). See also CHARLIE BROWN: BON VOYAGE.
ANIM 77 min VIDrel: 20TH/TECH V

SNOW QUEEN, THE ** (U)
Paivi Hartzell FINLAND 1986
Satu Sivo, Outi Vainiqnkulmu, Sebastian Kaatrasalo, Tuula Nyman, Esko Hukkanen, Pirjo Bergstrom, Paavo Westerberg, Julia Ukkonen, Saara Pakkasvirka, Marja Pyynko, Elina Salo, Ismo Alanko, Markku Huhtamo, Antti Litja, Jari Leino
In this appealing variant on the Hans Christian Andersen story, the Snow Queen kidnaps a young boy, but her intention is only to get him to help her find a jewel that has fallen from her crown.
JUV 86 min SATrel: MOVIE CHANNEL

SNOW WHITE AND THE SEVEN DWARFS **** U
Walt Disney (Ben Sharpsteen) USA 1937
Voices of: Andrea Caselotti, Harry Stockwell, Lucille La Verne, Moroni Olsen, Pinto Colvig, Otis Harlan, Scotty Mattraw, Roy Atwell, Stuart Buchanan, Billy Gilbert, Marion Darlington, The Fraunfelder Family, Jim Macdonald
Disney's first animated feature is nothing less than a brilliant and incredibly innovative retelling of this classic fairy story of a young orphaned princess and her jealous stepmother and how she comes to find her way to a cottage deep in the forest that is inhabited by seven dwarfs. Technically, this film sets a standard that has rarely been equalled and never surpassed. AA: Hon Award.
ANIM 83 min VIDrel: WDV/TECH L/A V/dm
Boa: fairytale Schneewitchen (Snow-Drop) from Kinder und Haus-marchen collected by Jakob Karl Grimm and Wilhelm Karl Grimm.

SNOW WHITE AND THE SEVEN DWARFS *** Uc
Peter Medak USA 1983
Elizabeth McGovern, Vanessa Redgrave, Rex Smith, Vincent Price, Michael Preston, Lou Carry, Tony Cox, Bill Curtis, Phil Fondacaro, Daniel Frishman, Peter Risch, Kevin Thompson, Patrick De Santis, Shelley Duvall
An adaptation of the classic tale from the "Fairie Tale Theatre" series, with an evil queen plotting to kill the beautiful Snow White when she learns that the latter is

more beautiful than her. Price makes a fine talking mirror in this effective rendition.
JUV 53 min VIDrel: MGM L/A V
Boa: fairytale Schneewitchen (Snow-Drop) from Kinder und Haus-marchen collected by Jakob Karl Grimm and Wilhelm Karl Grimm.

SNOW WHITE AND THE SEVEN DWARFS ** U
Michael Berz USA 1987
Diana Rigg, Sara Patterson, Douglas Sheldon, Nicola Stapleton, Billy Barty, Mike Edmunds, Ricardo Gil, Malcolm Dixon, Gary Friedkin, Tony Cooper, Dorit Adi, Ian James, Amnon Meskin, Julien Joy Chagrin, Azaria Rappaport
A live-action version of this famous fairytale, telling of how the daughter of a king was forced by the jealousy of her stepmother to flee to the forest for safety, where she was adopted by a family of dwarfs. A competent work.
Aka: CANNON MOVIE TALES: SNOW WHITE
JUV 81 min (ort 85 min) VIDrel: WHV V/sh
Boa: fairytale Schneewitchen (Snow-Drop) from Kinder und Haus-marchen collected by Jakob Karl Grimm and Wilhelm Karl Grimm.

SNOWBOUND: THE JIM AND JENNIFER STOLPA STORY *** (18)
Christian Duguay USA 1993
Neil Patrick Harris, Kellie Williams, Richard Cox, Duncan Fraser, Michael Gross, Susan Clark, Andrew Airlie, Joy Coghill, Kevin McNulty, Roger Barnes, J.B. Bivens, Ken Camroux, John B. Destry, Beverly Elliott, Tina Gilbertson
The compelling tale of a young couple and their son who were stranded in a snow-bound wilderness when their truck breaks down, the couple having attempted to make the one-thousand mile journey to attend the funeral of the man's grandmother. Having taken refuge in a cave, the husband is forced to undertake a perilous journey in search of help. Based on true events (the ordeal of the Stolpas lasted eight days) this engrossing survival saga is made with skill and care.
Aka: JIM AND JENNIFER STOLPA STORY, THE
A/AD 90 min mTV SATrel: MOVIE CHANNEL

SNOWMAN, THE *** U
Diane Jackson UK 1982
Delightful animation based on the Briggs storybook, that tells of a little boy who builds a snowman one winter's day, finds that it has come to life, and goes off with it for an adventure at the North Pole. There's no dialogue at all in this short, but this does nothing to detract from the charm of the piece, as in truth, none is needed.
ANIM 26 min (ort 30 min) mTV VIDrel: POLY/POLYREC V/sh
Boa: book by Raymond Briggs.

SNOWS OF KILIMANJARO, THE ** PG
Henry King USA 1952
Gregory Peck, Susan Hayward, Ava Gardner, Hildegarde Neff, Leo G. Carroll, Torin Thatcher, Marcel Dalio, Ava Norring, Helene Stanley, Vincente Gomez, Richard Allan, Leonard Carey, Paul Thompson, Emmett Smith, Victor Wood
In Africa, a wounded hunter has ample time to review his past life as he awaits rescue. Despite the star-studded cast, Hemingway's tale of a man trying to find a meaning to life falls victim to that common Hollywood failing, in short, a lack of conciseness. A muddled mess of unconvincing performances, spectacular locations and rambling dialogue.
DRA 109 min (ort 117 min) VIDrel: 20TH/TECH V/h
Boa: short story by Ernest Hemingway.

SO DEAR TO MY HEART ** U
Harold Schuster USA 1948
Burl Ives, Beulah Bondi, Harry Carey, Bobby Driscoll, Luana Patten, Raymond Bond, Daniel Haight, Walter Soderle, Matt Willis, Spelman B. Collins. Voices of John Beal, Ken Carson, Bob Stanton, The Rhythmaires
A small boy is determined to tame his pet black sheep and enter it in a country fair. A view of life on a country farm in 1903 that may appeal to some. Contains several animated sequences. Music is by Eliot Daniels with lyrics by Larry Morey. Written by John Tucker Battle.
JUV 84 min VIDrel: WDV/TECH L/A V
Boa: novel Midnight and Jeremiah by Sterling North.

**SO FINE ** *   15
Andrew Bergman USA   1981
*Ryan O'Neal, Jack Warden, Richard Kiel, Mariangela Melato, Fred
Gwynne, Mike Kellin, David Rounds, Joel Stedman, Angela Pietro
Pinto, Michael Lombard, Jessica James, Bruce Millholland, Merwin
Goldsmith, Irving Metzman*
Comedy set in the New York garment industry, where a profes-
sor of literature tries to save his garment manufacturer dad from
gangsters, and gets caught up in some strange adventures. The
directing debut for Bergman, who wrote the script. Music is by
Ennio Morricone. Fair, but somewhat chaotic.
COM   87 min (ort 91 min)   VIDrel: MGM/WHV L/A   V

**SO I MARRIED AN AXE MURDERER ** *   15
Thomas Schlamme USA   1993
*Mike Myers, Nancy Travis, Brenda Fricker, Amanda Plumme,
Anthony LaPaglia, Debi Mazar, Matt Doherty, Charles Grodin,
Steven Wright, Phil Hartman, Michael Richards, Cintra Wilson, Al
Nalbandian, George Mauricio, Kiki Douveas*
A man with a trail of broken relationships behind him finally
thinks he has found the right woman. Smart, sexy and the owner
of a butcher's shop, he cannot believe his luck but gradually
begins to suspect that she may in fact be a notorious "black
widow" murderess who has married and disposed of a succession of husbands. A failed black comedy that never realises the
potential of its premise.
COM   89 min (ort 93 min) cC   VIDrel: VCC/DISC/COLUM
V/sur

**SO PROUDLY WE HAIL ** *   15
Lionel Chetwynd USA   1989
*David Soul, Edward Herrmann, Chad Lowe, David Lowe, Gloria
Carlin, Raphael Sbarge, Kevin Conroy, Peter Dobson, Harley Jane
Kozak*
Dull examination of resurgent neo-Nazism that focuses on a
gullible university professor, whose propagation of a theory to
explain cultural differences makes him a useful tool for a white
supremacist group. Chad Lowe plays the naive working class
youngster who gets drawn into their net. A poor drama that
trivialises and exploits a disturbing issue. The script is by
Chetwynd.
DRA   90 min (ort 100 min) mTV   VIDrel: 20TH/TECH   V

**SOAPDISH ** *   15
Michael Hoffman USA   1991
*Sally Field, Kevin Kline, Robert Downey Jr, Cathy Moriarty, Whoopi
Goldberg, Carrie Fisher, Teri Hatcher, Paul Johansson, Elisabeth Shue,
Arne Nannestad, Tim Choate, Kathy Najimy, Costas Mandylor,
Cornelia Kiss, Robert Camiletti*
A soap opera star who's at the top of her profession feels herself
slipping, and various other personal problems start mounting
up at the same time. A frantic and over-heated farce, it attempts
to spoof countless soap operas, yet does little with its talented
cast and confuses motion with progress. It's a shame to see
someone as talented as Goldberg being given so little to do here.
Fortunately, Field (as the star) and Kline fare somewhat better.
COM   92 min (ort 97 min)
VIDrel: 4-FRONT/POLYREC/CIC   V/sur

**SOCIETY ** *   18
Brian Yuzna USA   1989
*Billy Warlock, Devin Devasquez, Evan Richards, Ben Meyerson, Ben
Slack, Patrice Jennings, Heidi Kozak, Pamela Matheson, Connie
Danese, Tim Bartell, Charles Lucia, David Wiley, David Wells, Heidi
Kozak, Brian Bremer*
A surreal horror-comedy that's based on the bizarre premise
that some of the rich folk of Beverly Hills are a separate race who
literally feed on the poor. An alienated youngster learns the
frightful truth about his family when his sister's jilted boyfriend
gives him a videotape showing the things the family get up to
with other like-minded individuals. Weird, horrific and unforgettable; the nauseating special effects are by Screaming Mad
George.
HOR   95 min (ort 99 min)   VIDrel: POLY/POLYREC/MED
V/sh

**SODBUSTERS ** *   (15)
Eugene Levy CANADA/USA   1993
*Kris Kristofferson, John Vernon, Max Gail, Fred Willard, Don Lake,
Wendel Meldrum, Steve Landesberg, John Hemphill, Don Lake, Lou
Wagner, Cody Jones, James Pickens Jr, Lela Ivey, Maria Vacratsis,
George Buza, Earl Pastko*
A further attempt to revive the long-gone Western genre by

spoofing some of its more beloved cliches, including
Kristofferson as a gunslinger, who arrives in the town of Marble
Hat, Colorado just in time to help the townsfolk make a stand
against a ruthless land developer. Moderate amusing but little
more.
WES   94 min (ort 97 min)   SATrel: MOVIE CHANNEL

**SOFIE ** *   PG
Liv Ullmann DENMARK/NORWAY/SWEDEN   1992
*Karen-Lise Mynster, Erland Josephsson, Ghita Norby, Jesper
Christiansen, Torben Zeller, Stig Hoffmeyer, Henning Moritzen,
Kirsten Rolffes, Lotte Herman, Jonas Oddermose, David Naym, Jacob
Allon, Kasper Barfoed, Anne Werner Thomsen*
A young Jewish girl in late 19th century Copenhagen is
prevented by her parents from marrying the man she loves and
eventually submits to their choice of husband. An overlong but
quite absorbing look at a vanished epoch. This was Ullmann's
directorial debut.
DRA   145 min (ort 152 min)   VIDrel: ARROW/RTM   V/sur
Boa: novel Mendel Philipsen and Son by Henri Nathansen.

**SOFT DECEIT ** *   18
Jorge Montesi CANADA   1994
*Patrick Bergin, Kate Vernon, John Wesley Shipp, Nigel Bennett, Ted
Dykstra, Gwynyth Walsh, Damir Andrei, Von Flores, Krista Bridges,
Jefferson Mappin, Carlo Savard, Joe Bellissimo, Victor Ertmanis, Len
Doncheff, Regianld Doresa*
A man is jailed for a major robbery, but the police are unable to
discover the whereabouts of the $6,000,000 he stole. However,
a female special agent is used in a clever plan that hinges on
getting the crook to lead her to the money. Unfortunately, our
crafty robber has made a few plans of his own.
DRA   91 min (ort 95 min)   VIDrel: HIFLI/SONOP   V/h

**SOFT KILL, THE ** *   18
Eli Cohen USA   1994
*Brion James, Corbin Bernsen, Matt McCoy, Michael Harris, Carrie-
Anne Moss, Kim Morgan Greene, Jimmy Medina, Annie James, Ted
Hayden, Maxine James, Louise Fitch, Alain Joel Silver, Judith Ziehn,
Robert Hoover, Gloria Hayes*
An LA private eye embarks on a relationship with the wife of a
politician, but when she is found murdered he finds himself
being framed for this crime. However, she turns out to be only
the first of a number of such victims, and our 'tec sets out to
catch the killer and thus prove his innocence before the cops
close in and prevent him clearing his name.
Aka: KILLING ME SOFTLY
THR   89 min (ort 95 min)   VIDrel: HIFLI/SONOP   V/h

**SOFT TOP, HARD SHOULDER ** *   15
Stefan Schwartz UK   1992
*Peter Capaldi, Francis Barber, Phyllis Logan, Catherine Russell,
Jeremy Northam, Richard Wilson, Elaine Collins, Peter Ferdinando,
Scott Hall, Simon Callow, Sophie Hall, Robert James, Andrew
Downie, Ann Scott James, Bill Gavin*
In Glasgow an unemployed art student has to get to London to
attend his dad's sixtieth birthday, where he hopes to get in his
uncle's good books as the latter has promised him a share of his
estate in his will. On the way there he picks up a weird hitch-
hiker and this odd couple have some even odder adventures
together. Screenplay is by Capaldi in this likeable road-movie.
COM   90 min (ort 95 min)   VIDrel: COLUM/SONOP L/A
V

**SOLAR ADVENTURE ** *   PG
Roy Thomas HONG KONG   1989
Yet another Joseph Lai/Betty Chan production (see FALCON 7
or CAPTAIN COSMOS) that blends rather nondescript animation techniques with an ambitious fantasy plot. Once again Earth
is under threat from evil aliens, and this time our would-be
conqueror has captured the only weapon that can stop him – the
"Canon Robot". Fortunately, a friendly alien is on hand, and
finds a group of children able to control the robot with their
minds.
ANIM   62 min   VIDrel: IMPENT   V

**SOLAR CRISIS ** *   15
Alan Smithee (Richard C. Sarafian) JAPAN/USA   1990
*Tim Matheson, Jack Palace, Charlton Heston, Peter Boyle, Annabel
Schofield, Corin "Corky" Nemec, Tetsuya Bessho, Dorian Harewood,
Paul Koslo, Dan Shor, Sandy McPeak, Silvana Gallardo, Scott Allan
Campbell, Frantz Turner, Eric James*
In the year 2050, Earth is threatened by increased solar activity

that seems like to result in its destruction by a giant flare. A mission is mounted to avert this danger by exploding a device on the sun's surface but a ruthless tycoon has other plans. Excellent special effects fail to compensate for a preposterous and unimaginative plot and poor acting. As is so often the case, the director found the finished result not to his liking, hence the pseudonym.
Aka: CRISIS 2050; KURAISHI NIJU-GOJU NEN; STARFIRE
FAN   107 min (ort 111 min)   VIDrel: EIV/SONOP   V
Boa: novel by Takeshi Kawata.

SOLARIS **** PG
Andre Tarkovsky USSR   1972
Donatas Banionis, Natalya Bondarchuk, Yuri Jarvet, Anatoly Solonitsin, Vladislav Dvorjetsky, Sos Sarkissian, Nikolia Grinko
A dour psychologist is sent to a space station that orbits a mysterious planet to find out why research there has ground to a halt, and discovers that the planet has the power to materialise one's innermost thoughts. A remarkable film in many ways, this long, hypnotic, brooding and visually impressive work taxes one's patience, but ultimately rewards careful attention. Photography is by Vadim Yusov.
FAN   159 min (ort 167 min) Col/B/W wScrn
VIDrel: CONNO/RTM   V
Boa: novel by Stanislav Lem.

SOLDIER BLUE * 18
Ralph Nelson USA   1970
Candice Bergen, Peter Strauss, Donald Pleasence, John Anderson, Dana Elcar, Jorge Rivero, Martin West, Jorge Russek, Marco Antonio Arzate, Ron Fletcher, Barbara Turner, Aurora Clavell
The love affair between the two survivors of an Indian attack, one a naive cavalry officer, the other a determined resourceful girl, serves as background to an account of the inhuman treatment meted out to the Indians courtesy of the US Cavalry. A meandering and cliched affair that culminates in a rather sickening massacre of an Indian village. Written by John Gay.
WES   109 min Cut (36 sec – ort 114 min)
VIDrel: 4-FRONT/POLYREC L/A   V
Boa: novel Arrow in the Sun by Theodore V. Olsen.

SOLDIER OF FORTUNE ** 15
Beau Davis USA   1989
Brandon Lee, Ernest Borgnine, Debi Monahan, Werner Pochath
A special agent has to locate a missing laser-weapon and has to make good use of his martial arts skills. A competent if unoriginal action film, one of the last that Lee worked on before his untimely death.
A/AD   84 min   VIDrel: 4-FRONT/POLYREC   V/h

SOLDIER'S STORY, A *** 15
Norman Jewison USA   1984
Howard E. Rollins Jr, Adolph Caesar, Art Evans, Dennis Lipscombe, Denzel Washington, Larry Riley, Wings Hauser, Patti LaBelle, David Alan Grier, David Harris
A university-trained black US army investigator, is assigned to tackle the case of a young black sergeant murdered at a training camp in Louisiana. Based on Fuller's Pulitzer Prize-winning play, which was in turn inspired by Melville's "Billy Budd", the film features most of the black cast from the original stage production. A somewhat stilted but absorbing drama.
DRA   97 min (ort 101 min)   VIDrel: VCC/DISC/COLUM V/sur
Boa: play by Charles Fuller.

SOLID GOLD CADILLAC, THE ** U
Richard Quine USA   1956
Judy Holliday, Paul Douglas, Fred Clark, John Williams, Arthur O'Connell, Hiram Sherman, Neva Patterson, Ralph Dumke, Ray Collins plus George Burns (narration only)
A woman takes on the board of the company she holds a few shares in when she discovers mismanagement at the most senior level. Holliday adds a lot of sparkle to this uneven comedy, which begins with a lot of promise and then loses its way. However, it's bright and cheerful and the cast almost pulls it off.
AA: Cost: (Jean Louis).
COM   96 min (ort 99 min) B/W   VIDrel: COLUM/SONOP V
Boa: play by George S. Kaufman and Howard Teichman.

SOLITAIRE FOR TWO *** 15
Gary Sinyor UK   1994
Mark Frankel, Amanda Pays, Roshan Seth, Jason Isaacs, Maryam

D'Abo, Helen Lederer, Malcolm Cooper, Annette Crosbie, Neil Mullarkey, Liza Walker, Right Said Fred, Kelly Salmon, Ricky Jones, Diana Eskell, Robert Harley, Carli Harris*
A woman who has ESP meets a man who is an expert in reading body language but is also a confirmed bachelor who enjoys using his skills to play the field. An enjoyable, lightweight romp, sustained by witty dialogue and pleasing performances.
COM   100 min (ort 106 min)   VIDrel: EIV/SONOP   V

SOLOMON AND SHEBA ** PG
King Vidor USA   1959
Yul Brynner, Gina Lollobrigida, Marisa Pavan, George Sanders, Alejandro Ray, John Crawford, Harry Andrews, David Farrar, Laurence Naismith, Jose Nieto, Julio Pena, Maruchi Fresno, William Devlin, Felix De Pomes, Jean Anderson
A disappointingly overblown Biblical epic, in which the battle scenes are the highlight, with little dramatic interest being generated by the limp account of the tortured relationship between the two title subjects. Tyrone Power died before the film was completed, and his footage had to be re-shot using Brynner. Filmed on location in Spain, this Vidor's last film.
DRA   135 min (ort 139 min) wScrn   VIDrel: MGM/WHV V/s

SOLOMON'S CHOICE ** PG
Andrew Tennant USA   1992
Joanna Kerns, Bruce Davison, Reese Witherspoon, Joseph Mazzello, Steven Gilborn, Bruce McGill, Phillip Curry
A close-knit family have to face a tough decision when they learn that the organs of their dead son can be used as a donor in a life-saving operation. Fair. See also THE GIFT OF LOVE.
DRA   92 min   VIDrel: ODY/SONOP   V/sh

SOME CAME RUNNING *** PG
Vincente Minnelli USA   1958
Frank Sinatra, Dean Martin, Shirley Maclaine, Martha Heyer, Arthur Kennedy, Nancy Gates, Leora Dana, Betty Lou Keim, Larry Gates, Steven Peck, Connie Gilchrist, Ned Wever, Carmen Phillips, John Brennan, William Schallert
After completing his military service, a writer returns home and encounters disillusionment and hypocrisy among its worthy citizens, and falls in with a hooker and a gambler. Good performances and Elmer Bernstein's score help to sustain the interest in this extended soap opera.
DRA   130 min (ort 137 min) wScrn   VIDrel: MGM/WHV   V
Boa: novel by James Jones.

SOME KIND OF HERO * 15
Michael Pressman USA   1981
Richard Pryor, Margot Kidder, Ray Sharkey, Ronny Cox, Lynne Moody, Olivia Cole, Paul Benjamin, David Adams, Martin Azarow, Shelly Batt, Susan Berlin, Tim Thomerson, Mary Betten, Herb Braha, Peter Jason, Anthony R. Charnota
A Vietnam veteran returns home after six years as a POW, only to find that things are not as he expected to find them. A patchy and irritating blend of comedy and drama, that is unappealing in both departments.
DRA   93 min (ort 97 min)   VIDrel: CIC/SONOP   V/h
Boa: novel by James Kirkwood.

SOME KIND OF WONDERFUL ** 15
Howard Deutch USA   1987
Eric Stoltz, Mary Stuart Masterson, Craig Sheffer, John Ashton, Lea Thompson, Elias Koteas, Maddie Corman, Jane Elliot, Candace Cameron, Chynna Phillips, Scott Coffrey, Carmine Caridi, Lee Garlington, Laura Leigh Hughes
A young man struggles to find his own identity, despite the pressure to conform on the part of both family and friends. Eventually, he finds true love in the company of the tomboy girl he ignored in favour of a flashy but empty-headed girl. A film exploring a similar theme to that of PRETTY IN PINK, but with a gender change.
DRA   91 min (ort 93 min)   VIDrel: CIC/SONOP   V/sur

SOME LIKE IT HOT **** U
Billy Wilder USA   1959
Tony Curtis, Jack Lemmon, Marilyn Monroe, Joe E. Brown, George Raft, Pat O'Brien, Nehemiah Persoff, Joan Shawlee, Mike Mazurki, George E. Stone, Dave Barry, Billy Gray, Beverly Wills, Barbara Drew, Edward G. Robinson Jr
Having inadvertently witnessed the St Valentine's Day Massacre, two unemployed musicians evade capture by posing as women and joining an all-girl dance band that is on its way

to Miami. A romance between one of them (out of female disguise) and the pretty singer and the other and a playboy millionaire, leads to incredible, hilarious complications. Wilder's best comedy is a trifle uneven but has sharp performances and terrific dialogue. AA: Cost (Orry-Kelly).
COM   117 min (ort 120 min) B/W   VIDrel: MGM/WHV   V/h

SOME LIKE IT SEXY *
18
Donovan Winter UK
1969
Christopher Matthews, Erika Bergmann, Penny Riley, Yolanda Turner, Maddy Smith, Valerie St Helene, Annabel Leventon, Nicola Pagett, Mary Collinson, Madeleine Collinson
Sexual adventures of a butcher's assistant. A film that has achieved a well deserved oblivion.
Aka: COME BACK PETER; IMPORTANCE OF BEING SEXY, THE; SEDUCERS, THE
A   89 min   VIDrel: JEZ/RTM   V

SOME MOTHER'S SON ***
15
Terry George EIRE/USA
1996
Helen Mirren, Fionnula Flanagan, Aidan Gillen, David O'Hara, John Lynch, Tom Hollander, Tim Woodward, Ciaran Hinds, Geraldine O'Rawe, Gerard McSorley, Dan Gordon, Grainne Delany, Ciaran Fitzgerald, Robert Lang, Stephen Hogan
After the imprisoned IRA terrorist Bobby Sands dies from a hunger strike in 1981, twenty-one of his comrades decide to follow his example. However, the mothers of two of them join forces and wage a grimly fought battle of their own to save the lives of their sons. Mirren and Flanagan are outstanding, and their performances are the best thing in this one-sided and rather simplistic affair.
DRA   107 min (ort 112 min)   VIDrel: COLUM/SONOP   V/sur

SOME WILL, SOME WON'T *
PG
Duncan Wood UK
1969
Ronnie Corbett, Thora Hird, Michael Hordern, Barbara Murray, Leslie Phillips, James Robertson Justice, Dennis Price, Wilfrid Brambell, Eleanor Summerfield, Arthur Lowe, Stephen Lewis
The four heirs to a fortune find that the will stipulates they are each obliged to perform uncharacteristic tasks in order to inherit. A flaccid remake of LAUGHTER IN PARADISE that is as unfunny as it is unmemorable.
COM   86 min (ort 90 min)   VIDrel: WHV   V/h

SOMEBODY HAS TO SHOOT THE PICTURE ***
15
Frank R. Pierson USA
1990
Roy Scheider, Bonnie Bedelia, Arliss Howard, Robert Carradine, Andre Braugher, Marc Macaulay, Antoni Corone, Tom Nowicki, Tom Schuster, Ginger Burgett, Tom Kouchalakos, Mark McCracken, Bob Barnes, Michael O'Smith
In Florida, a small-time drugs dealer is wrongly sentenced to death after his conviction for the murder of a cop, and makes a bizarre last request for his execution to be photographed. This assignment is given to a once-famous photo-journalist who, though now gone to seed, in former days won the Pulitzer Prize. However, stimulated by his strange task, he uncovers fresh evidence. A gripping drama with two powerful performances from the leads.
DRA   99 min (ort 105 min) mCab   VIDrel: L/A   V/h
Boa: book Slow Coming Dark by Doug Magee.

SOMEBODY KILLED HER HUSBAND *
PG
Lamont Johnson USA
1978
Farrah Fawcett-Majors (Fawcett), Jeff Bridges, John Wood, Tammy Grimes, John Glover, Patricia Elliott
A woman's lover becomes the prime suspect when her husband is killed in mysterious circumstances. A feeble little comedy-mystery that has very little going for it except the title. Written by Reginald Rose and filmed in Manhattan. This was the first starring role for Fawcett after her spell in the TV series "Charlie's Angels".
COM   92 min (ort 96 min)   VIDrel: MIA/DISC   V

SOMEBODY TO LOVE **
18
Alexandre Rockwell USA
1994
Rosie Perez, Harvey Keitel, Anthony Quinn, Michael DeLorenzo, Steve Buscemi, Stanley Tucci, Gerardo, Steven Randazzo, Paul Herman, Sam Fuller, Helena, Angel Aviles, Elizabeth Bracco, Lorelei Leslie, Julie Shannon, Francesco Messina
A naive young Mexican lad does his best to impress a girl who works as a dollar-a-dance partner at a club in East L.A., but in

doing so gets drawn into a life of crime. This tragic love story starts off well enough, but the chaotic nature of the script prevents the undoubted merits of the opening to develop and the cast have little chance to shine.
DRA   98 min (ort 102 min)   VIDrel: EIV/SONOP   V

SOMEBODY UP THERE LIKES ME ***
PG
Robert Wise USA
1956
Paul Newman, Sal Mineo, Robert Loggia, Steve McQueen, Pier Angeli, Everett Sloane, Eileen Heckart, Joseh Buloff, Harold J. Stone, Sammy White, Robert Lieb, Arch Johnson, Theodore Newton, Robert Easton, Ray Walker, Billy Nelson
How Rocky Graziano came from humble surroundings in New York and a spell in reform schools during the Depression and went on eventually to become the middleweight boxing champion of the world. A sentimental affair, redeemed by some strong performances. The film debut for both Loggia and McQueen. AA: Cin (Joseph Ruttenberg), Art/Set (Cedric Gibbons and Malcolm F. Brown/Edwin B. Willis and F. Keogh Gleason).
DRA   109 min (special edition – ort 97 min) B/W   VIDrel: MGM/WHV   V

SOMEBODY'S DAUGHTER **
15
Joseph Sargent USA
1993
Nicollette Sheridan, Nick Mancuso, Boyd Kestner, Michael Cavanaugh, Max Gail, Richard Lineback, Micole Mercurio, Bill Marcus, Elliott Gould, Cameron Kaz, Elena Stiteler, James Dyabs, Janice Ehrlich, Lenore Kasdorf, Mina Badiyi
A cop investigating police corruption drops out of sight after his informant and ex-girlfriend is murdered and is sought by his friend, an undercover agent. The former has tape recordings that are may be vital evidence in getting a corrupt police chief convicted. Quite a strongly plotted mystery thriller.
THR   92 min mTV   VIDrel: HIFLI/SONOP   V/h

SOMEONE ELSE **
18
Michael Craig USA
1992
Ashlyn Gere, Danielle Rogers, Bianca Trump, Randy West, Mona Lisa, Randy Spears, Mike Horner, Jesse Eastern
West and Gere play live-in lovers whose relationship has grown stale. The slight story has them splitting up for a short while, her to the mountains to visit a girlfriend, him to the beach to stay with a friend. In the course of this trial separation they sleep around with various different partners, but eventually get back together again. Both well acted and directed, but terribly deficient in terms of plotting.
A   75 min   VIDrel: GROHOM/MAXSCAN   V

SOMEONE ELSE'S CHILD ***
PG
John Power USA
1995
Lisa Hartman Black, Bruce Davison, Whip Hubley, Ken Pogue, Glynn Turman, Don Davis, Scott McNeil, Michael David Simms, Joel Palmer, Brandon Obray, Stephen E. Miller, Gillian Barber, Tom Butler, Louise Fletcher, Jonathan Hogg
A woman learns that her seven-year-old is not a her son, her real child having been switched at birth with that of another woman. Whilst fighting to retain custody of the child she now loves as her own, she attempts to learn more about her real son, discovering that he appears to be in the care of abusive parents. A crusading lawyer help her fight for custody of her own child. An engrossing and well acted real-life drama. See also SWITCHED AT BIRTH.
DRA   92 min mTV   VIDrel: ODY/SONOP   V/s

SOMEONE SHE KNOWS ***
(15)
Eric Laneuville USA
1994
Markie Post, Gerald McRaney, Jeffrey Nordling, Spencer Garrett, Don Hood, Sharon Lawrence, Kathleen Lloyd, Shawn Modrelli, Harold Sylvester, Jamie Renee Smith, Philip Van Dyke, Sarah Freeman, Alma Beltran, Jeff Doucette, L.A. Sargent
Based on a true murder case, this TV movie follows the efforts made by a distraught mother to uncover the identity of her five-year-old's killer, a task that the police seem to be unable to make progress with. Helped in her investigations by an old family friend, she begins to suspect that the culprit may indeed be known to her. Taut and well plotted.
DRA   89 min mTV   SATrel: SKY MOVIES

SOMEONE TO DIE FOR **
18
Clay Borris USA
1995
Corbin Bernsen, Ally Walker, Shell Danielson
When the daughter of a private detective is accidentally killed

in a police stakeout, the distraught father suffers a breakdown and spends some time in a mental hospital. There he is seduced by a woman who claims to be a visitor, but when he ends their relationship he learns she has a nasty side. A sub-plot has someone killing the cops who took part in the stakeout, and the police suspect him. A competent thriller with strong echoes of FATAL ATTRACTION.
THR   92 min   VIDrel: MED/DISC   V/sh

SOMEONE TO WATCH OVER ME ***
15
Ridley Scott USA
1987
Tom Berenger, Jerry Orbach, Mimi Rogers, Lorraine Bracco, John Rubinstein, Andreas Katsulas, Tony DiBenedetto, James Moriarty, Mark Moses, Daniel Hugh Kelly, Harley Cross
A New York cop is assigned to protect a wealthy woman who has received death threats, and gradually becomes infatuated with her. A gripping romantic thriller made with a good deal of flair.
THR   103 min (ort 106 min)   VIDrel: VCC/COLUM L/A   V/sh

SOMEONE'S WATCHING ***
18
Scott McGinnis USA
1993
Tim Daly, Mia Sara, Paul Le Mat, Clayton Rohner, Zach Galligan, Xander Berkeley, Stacey Travis, Thomas F. Wilson, Virginia Madsen, Judd Nelson, Lewis Van Bergen, Caroline Barclay, Hawthorne James, Ben Mayerson, Jay Baker
A reporter uncovers a murder by crooked cop, gets vamped by a femme fatale and finds his own life in danger after an ex-girlfriend is killed. An action-thriller that is in many ways an appealing homage to film noir, not exactly the most plausibly plotted work, but brooding and highly atmospheric.
THR   88 min (ort 90 min)   VIDrel: COLUM/SONOP   V/sh

SOMETHING IN THE SHADOWS **
18
Michael Preece USA
1993
Chuck Norris, Clarence Gilyard, Noble Willingham, Floyd "Red Crow" Westerman, Sheree J. Wilson
A student joins a karate class, but his instructor learns that the boy is being blackmailed into delivering drugs. Another action foray for Norris, who is cast in the role of Texas Ranger Cordell Walker, that delivers the requisite number of thrills, but precious little originality in the plot department. As ever, Norris brings little depth or conviction to his role.
A/AD   85 min   VIDrel: WHV   V/sh

SOMETHING IS OUT THERE **
18
Richard A. Colla USA
1988
Joe Cortese, Maryam D'Abo, Gregory Sierra, John Putch, Kim Delaney, George Dzundza, Robert Webber, Joseph Cali, John O'Hurley, Melani Jones, Matthew Faison, Ray Reinhardt, Michael Cutt, Lori Michaels, Dean Scofield
A police officer is baffled by a series of brutal, motiveless murders, where there are neither clues nor witnesses. Eventually he learns that the murderer is the last member of a dangerous alien species. Fortunately, help arrives in the form of an attractive alien woman. Worth seeing for the special effects of John Dykstra, and Rick Baker's monster, but rather poorly plotted. A pilot for a brief TV series, and first shown in two parts.
HOR   166 min (ort 192 min) mTV   VIDrel: EIV/SONOP   V

SOMETHING TO SING ABOUT **
U
Victor Schertzinger USA
1937
James Cagney, Evelyn Daw, William Frawley, Mona Barrie, Gene Lockhart, James Newill, Harry Barris, Candy Candido, Cully Richards, William B. Davison, Richard Tucker, Marek Windheim, Dwight Frye, John Arthur, Philip Ahn
A New York bandleader starts a new career in Hollywood where he becomes involved with a scheming producer, in this lightweight musical made by Cagney as an independent, during his rift with Warner Studios. The unmemorable songs are by Schertzinger with musical direction by Constantin Bakaleinikoff.
Aka: BATTLING HOOFER
MUS   88 min (ort 93 min) B/W   VIDrel: L/A   V

SOMETHING TO TALK ABOUT ***
15
Lasse Hallstrom USA
1995
Julia Roberts, Dennis Quaid, Robert Duvall, Kyra Sedgewick, Gena Rowlands, Brett Cullen, Haley Aull, Muse Watson, Anne Shropshire, Ginnie Randall, Terence P. Currier, Rebecca Koon, Rhoda Griffis, Lisa Roberts, Deborah Hobart
A woman who appears to have everything she could wish for

learns that her husband is having an affair. Roberts and Quaid work well together as the couple whose lives are put under examination here, and though the script is not especially profound, it manages to keep things moving along nicely.
DRA   100 min (ort 105 min) cC   VIDrel: WHV   V/sur

SOMETHING WICKED THIS WAY COMES ***
PG
Jack Clayton USA
1983
Shawn Carson, Vidal Peterson, Jason Robards, Jonathan Pryce, Royal Dano, Pam Grier, Diane Ladd, Mary Grace Canfield, James Stacy, Jake Dengel, Bruce M. Fischer, Richard Davalos, Brendan Klinger, Arthur Hill (narration)
A mysterious carnival visits a small American town in Illinois and fulfils some of the inhabitants' dreams, but at a heavy price. The intelligent and often poetic script is by Bradbury, with Pryce giving a remarkably intense performance as the demonic carnival owner who tries to ensnare two small boys who have discovered his true identity. An unusually mature Disney treatment of a fantasy tale.
FAN   91 min (ort 94 min)   VIDrel: WDV/TECH L/A   V
Boa: novel Ray Bradbury.

SOMETHING WILD **
18
Jonathan Demme USA
1986
Jeff Daniels, Melanie Griffith, Ray Liotta, Margaret Colin, Tracey Walter, Dana Preu, Jack Gilpin, Su Tissue, Kristin Olsen, John Sayles, John Waters, George Schwartz, Charles Napier, Robert Ridgely, Sister Carol East
Comedy thriller about a yuppie tax consultant, who gets embroiled in a series of adventures with an unconventional girl, eventually the pair finding themselves on the run from the girl's psychopathic ex-husband. A careless blend of comedy and melodrama that starts with great promise but by the halfway mark changes tone and becomes a feast of gratuitous violence, as ex-husband Liotta (in his movie debut) sets out to get even.
COM   109 min (ort 116 min)   VIDrel: VCC L/A   V/sh

SOMETIMES THEY COME BACK ***
15
Tom McLoughlin USA
1991
Tim Matheson, Brooke Adams, Robert Rusler, Chris Demetral, Nicholas Sadler, Robert Hy Gorman, William Sanderson, Bentley Mitchum, Matt Nolan, Matt Nolan, Tasia Valenza, T. Max Graham, Duncan McLeod, Nancy McLoughlin
Despite the daft title, this is quite a good effort. Together with his wife and young son, Matheson returns to his hometown to take up a teaching post. They settle into the house where he grew up, but it is not long before certain dark events from the past (his brother was killed by high school thugs) lead to unpleasantness for his son at the local school. The gore is wisely kept to a minimum in this atmospheric and scary yarn. A sequel followed.
Aka: STEPHEN KING'S SOMETIMES THEY COME BACK
HOR   97 min   VIDrel: EIV/SONOP   V
Boa: novel by Stephen King.

SOMETIMES THEY COME BACK... AGAIN **
18
Adam Grossman USA
1996
Michael Gross, Hilary Swank, Alexis Arquette, Morgan Sheppard, Jennifer Elise Cox
A man goes back to his home-town to bury his mother, but is haunted by the murder of his eldest sister, and the evil teenager who was responsible for this murder comes back onto the scene with the intention of seducing his daughter. Some unpleasant special effects plus Arquette's strong performance as the villain are the chief assets here, in other departments the film is simply a cliched horror sequel to the earlier (and better) SOMETIMES THEY COME BACK.
HOR   94 min (ort 98 min)   VIDrel: TRIM/HIFLI   V

SOMEWHERE IN SONORA **
U
Mack V. Wright USA
1933
John Wayne, Henry B. Walthall, Shirley Palmer, J.P. McGowan, Ann Fay, Frank Rice, Billy Franey, Paul Fix, Ralph Lewis, Slim Whitaker, Blackie Whiteford, Jim Corey
After being accused of cheating in a stagecoach race, a cowboy goes to Mexico. There he pretends to be an outlaw to infiltrate a gang who have kidnapped the son of an old rodeo boss and stops them from stealing a silver mine. A competent remake of a 1927 silent, one of six oaters made by Wayne for Warner Bros. that helped establish him as a Western star.
WES   55 min (ort 57 min) B/W   VIDrel: MGM/WHV   V/h
Boa: novel Somewhere South in Sonora by Will Levington Comfort.

SOMEWHERE IN TIME **
PG
Jeanot Szwarc USA   1980
Christopher Reeve, Jane Seymour, Christopher Plummer, Teresa Wright, Bill Erwin, George Voskovec, Susan French, John Alvin, Eddra Gale, Sean Hayden, W.H. Macy, Audrey Bennett, Laurence Coren
A young playwright becomes fascinated by a locket containing a portrait of an actress from seventy years ago. By an effort of will, he travels back to Chicago of 1912 in order to meet her, but they are soon parted. A lightweight romantic fantasy, written by Matheson from his novel. The pretty scenery is of Mackinac Island. Music is by John Barry.
DRA   98 min (ort 104 min)   VIDrel: CIC/SONOP   V
Boa: novel Bid Time Return by Richard Matheson.

SOMMERSBY ***
15
Jon Amiel FRANCE/USA   1992
Richard Gere, Jodie Foster, James Earl Jones, Bill Pullman, William Windom, Brett Kelley, Richard Hamilton, Maury Chaykin, Lanny Flaherty, Frankie Faison, Wendell Wellman, Clarice Taylor, Ronald Lee Ermey
A man returns home at the end of the Civil War but his wife is unsure that he is her husband and other doubts as to his identity gradually begin to emerge. This eventually leads to a trial for murder and an agonising dilemma for our hero and his family. In many ways, quite a successful Americanisation of THE RETURN OF MARTIN GUERRE and undoubtedly beautifully filmed and staged, but despite excellent performances, this movie fails to come alive.
DRA   108 min (ort 114 min) cC   VIDrel: WHV   V/sur

SON-IN-LAW, THE **
15
Steve Rash USA   1993
Pauly Shore, Carla Gugino, Lane Smith, Cindy Pickett, Mason Adams, Patrick Renna, Dennis Burkley, Dan Gauthier, Tiffani-Amber Thiessen, Ria Pavia, Lisa Lawrence, Graham Jarvis, Nick Light, Ernie Kinney, Troy Shire, Adam Goldberg
A pretty co-ed takes a college chum home for the Thanksgiving week-end and passes him off as her intended so as to fend off a marriage proposal from a local lad whom she finds unappealing. Unfortunately for all concerned, his strange ways cause no end of problems. A sort of vanity vehicle for Shore that is trite and predictable.
COM   92 min (ort 96 min)   VIDrel: HOLPIC/TECH L/A   V

SON IN LAW **
12
Steve Rash USA   1993
Pauly Shore, Carla Gugino, Lane Smith, Cindy Pickett, Mason Adams, Patrick Renna, Dennis Burkley, Dan Gauthier, Tiffani-Amber Thiessen, Ria Pavia, Lisa Lawrence, Graham Jarvis, Nick Light, Ernie Kinney, Troy Shire, Adam Goldberg
A pretty co-ed takes a college chum home for the Thanksgiving week-end and passes him off as her intended so as to fend off a marriage proposal from a local lad whom she finds unappealing. Unfortunately for all concerned, his strange ways cause no end of problems. A sort of vanity vehicle for Shore that is trite and predictable.
Aka: SON-IN-LAW, THE
COM   95 min   CINrel

SON OF KONG, THE **
15
Ernest B. Schoedsack USA   1933
Robert Armstrong, Helen Mack, Victor Wong, John Marston, Frank Reicher, Lee Kohlmar, Ed Brady, Noble Johnson, Clarence Wilson, Katherine Ward, Gertrude Sutton, Steve Clemento, Gertrude Short, James L. Leong, Frank O'Connor
A sequel to the classic KING KONG, with the original expedition going back to Skull Island and finding the giant ape's little son. A feeble offering rushed out in a hurry, with little attention paid to the script. The film was so ludicrous that it was promoted as a comedy, but in fact Willis O'Brien's special effects are still potent. The music is by Max Steiner.
A/AD   66 min (ort 69 min) B/W   VIDrel: L/A   V

SON OF THE MORNING STAR ***
PG
Mike Robe USA   1991
Gary Cole, Rosanna Arquette, Terry O'Quinn, David Strathairn, Rodney A. Grant, Dean Stockwell, George American Horse, Stanley Anderson, Ed Blatchford, George Dickerson, Michael Medeiros, Tom O'Brien, Tim Ransom, Robert Schenkkan
A long but well-paced account of the career of George Armstrong Custer, who perished at the Battle of the Little Bighorn. Cole gives an impressive and convincing performance in the role of this unattractive figure, and a literate script and excellent camerawork (by Kees Van Oostrum) are additional attractions in this epic-style biography. See also THEY DIED WITH THEIR BOOTS ON and CUSTER OF THE WEST.
WES   174 min Cut (9 sec – ort 192 min) mTV
VIDrel: ODY/SONOP   V/sh
Boa: novel by Evan S. Connell.

SON OF THE PINK PANTHER *
PG
Blake Edwards USA   1993
Roberto Benigni, Herbert Lom, Robert Davi, Shabana Azmi, Claudia Cardinale, Burt Kwouk, Debrah Farentino, Graham Stark, Jennifer Edwards, Anton Rogers, Mark Schneider, Kenny Spalding, Oliver Cotton, Aharon Ipale, Natasha Pavlova
A tasteless and unfunny attempt to find a way to continue this series after the death of Peter Sellers by having Italian comic Benigni play his son. A parade of unamusing gags pads out the thin story of his efforts to rescue a kidnapped princess. Davi, as ever, is suitably nasty and menacing as the chief villain but is totally wasted in this mess, while his straight acting clashes completely with the uncontrolled buffoonery of Benigni's character.
COM   88 min (ort 115 min) cC   VIDrel: MGM/WHV   V/sur

SON OF THE SHEIK, THE ***
U
George Fitzmaurice USA   1926
Rudolph Valentino, Vilma Banky, Agnes Ayres, Karl Dane, Bull Montana, Montague Love
A sequel to "The Sheik" with the star playing both father and son, in the tale of a desert leader who abducts a dancing girl he thinks has betrayed him, and then falls in love with her. One of Valentino's best films (it was also his last one) and done in a slightly jocular style. A 1934 re-release had a music soundtrack by Jack Ward.
DRA   65 min (ort 74 min) B/W silent
VIDrel: VISION/DISC   V

SONATINE ***
18
Takeshi Kitano JAPAN   1993
Takeshi Kitano, Aya Kokumai, Tetsu Watanabe, Masanobu Katsumura, Susumu Terashima, Ren Ohsugu, Tonbo Zushi, Kenichi Yajimia, Eiki Minakata
After several of his cronies are killed in a shootout with a rival gang on a trip to Okinawa, a gangster boss takes refuge with the rest of his buddies at a remote beach location. Once there, they play games, wrestle, paddle in the sea and generally sit about while they plan their next move. Written by Kitano, this is a slow, poignant and hypnotic film, and the eventual violent confrontation is done with great style, but even this is topped by the sad and moving climax.
A/AD   89 min (ort 93 min)   VIDrel: ICAPRO/MANGA   V/sur

SONG FOR BEKO, A ***
(PG)
Nizamettin Aric ARMENIA/GERMANY   1992
Nizamettin Aric, Bezara Arsen, Lusika Hesen, Cemale Jora, Fila Tital, Nuriye Tital, Temure Jora, Sirine Sinco, Xasea Rizgo, Berivanna Feqi, Leylea Guhar, Rusteme Cemal, Arsen Poladow, Agite Cimo, Rizgoye Resit, Aschot Abrahamian
A Kurd is forced into the army but deserts and sets out to find his brother, who has become a guerrilla fighter. A very strong anti-war movie that takes no sides and offers no simple solutions, just excellent performances and an intelligent script. This was the first movie to be made in Kurdish.
Aka: EIN LIED FUR BEKO; KLAMEK JI BO BEKO
DRA   100 min   CINrel

SONG IS BORN, A **
PG
Howard Hawks USA   1948
Danny Kaye, Virginia Mayo, Hugh Herbert, Steve Cochran, Felix Bressart, J. Edward Bromberg, Mary Field, Ludwig Stossel, Louis Armstrong, Benny Goodman, Charlie Barnet, Lionel Hampton, Tommy Dorsey, Mel Powell, O.Z. Whitehead
A stuffy professor working on an encyclopaedia of music with seven others, discovers jazz and life while becoming romantically entangled with a gangster's girlfriend, in this remake of the 1941 film BALL OF FIRE by the same director. A mild and fairly anodyne musical comedy. Look out for guest stars: Louis Armstrong, Benny Goodman, Tommy Dorsey, Lionel Hampton and Charlie Barnet.
COM   120 min B/W   VIDrel: VCC/DISC/COLUM   V

SONG OF BERNADETTE, THE ****   U
Henry King USA   1943
Jennifer Jones, William Eythe, Charles Bickford, Vincent Price, Lee J.
Cobb, Anne Revere, Gladys Cooper, Roman Bohnen, Patricia Morison,
Aubrey Mather, Charles Dingle, Mary Anderson, Edith Barrett, Sig
Rumann, Blanche Yurka
In the 1800s a peasant girl has a vision of the Virgin Mary
(Linda Darnell in an unbilled role) at the spot that becomes the
shrine of Lourdes. Despite its length and lack of historical
accuracy, fine production values and a wonderfully ethereal
performance from Jones ensured the film was an enormous
success. AA: Actress (Jones), Cin (Arthur C. Miller), Art/Int
(James Basevi and William Darling/Thomas Little), Score
(Alfred Newman).
DRA   156 min B/W   VIDrel: 20TH/TECH   V/h
Boa: novel by Franz V. Werfel.

SONG OF NORWAY *   U
Andrew L. Stone UK/USA   1970
Florence Henderson, Toralv Maurstad, Christina Schollin, Frank
Poretta, Edward G. Robinson, Harry Secombe, Robert Morley, Oscar
Homolka, Elizabeth Larner, Bernard Archard, Richard Wordsworth,
Frederick Jaeger, Henry Gilbert
A musical fantasia on the life of Grieg, in a variety of styles.
Weak as a biographical work, though the Super Panavision
photography helps enhance the beauty of the landscapes, but
this will be lost on TV. Written by Stone.
MUS   140 min (ort 143 min)   VIDrel: VCC/DISC   V
Boa: musical by Milton Lazarus, Robert Wright and George
Forrest.

SONG OF THE SOUTH ***   U
Harve Foster USA   1956
Ruth Warrick, Bobby Driscoll, James Baskett, Luana Patten, Lucile
Watson, Hattie McDaniel, Glenn Leedy, George Nokes, Gene Holland,
Erik Rolf, Mary Field, Anita Brown. Voices of: Nicodemus Stewart,
Johnny Lee, James Baskett
Live-action and animated musical fantasy, set on a Southern
plantation where an old "Uncle Remus" recounts "Brer Rabbit"
stories. The three excellent animated sequences feature Brer
Rabbit, Brer Fox and Brer Bear. Animation is by Wilfred Jackson,
with music by Daniele Amfitheatrof and photography by Gregg
Toland. Uneven, but great fun. AA: Song ("Zip-A-Dee-Doo-
Dah" – Allie Wrubel (m)/Ray Gilbert (l)), Spec Award (James
Baskett).
MUS   90 min (ort 94 min)   VIDrel: WDV/TECH   V
Boa: short stories Tales of Uncle Remus by Joel Chandler Harris.

SONG TO REMEMBER, A **   U
Charles Vidor USA   1944
Cornel Wilde, Merle Oberon, Paul Muni, Stephen Bekassy, Nina Foch,
George Coulouris, Sig Arno, Howard Freeman, George Macready,
Claire Dubrey, Frank Pulia, Fern Emmett, Sybil Merrit, Ivan
Triesault, Fay Helm, Dawn Bender
A lavish but stilted biopic on the life of Chopin, with Wilde
woefully inadequate for the part, but Oberon appealing as
George Sand. The ludicrous script tends to sink this one early
on. Worth listening to but not seeing.
MUS   108 min (ort 113 min)   VIDrel: RCA L/A   V
Boa: novel Polonaise by D. Leslie.

SONNY BOY *   18
Robert Martin Carroll USA   1988
David Carradine, Paul L. Smith, Brad Dourif, Conrad Janis, Sydney
Lassick, Savinna Gersak, Alexandra Powers, Steve Carlisle, Michael
Griffin
A sick film telling of a baby boy brought up in a nightmare of
cruelty by a psychotic, who had his parents murdered in order
to have the opportunity to create a monster he could unleash on
his enemies.
HOR   98 min   VIDrel: EIV/SONOP   V/sur

SONS OF KATIE ELDER, THE ***   U
Henry Hathaway USA   1965
John Wayne, Dean Martin, Earl Holliman, Michael Anderson Jr,
Martha Hayer, Paul Fix, George Kennedy, Jeremy Slate, James
Gregory, Dennis Hopper, John Litel, Sheldon Allman, John Doucette,
James Westerfield, Rhys Williams
Four brothers return home for their mother's funeral and must
defend the family honour after they are wrongly accused of
murder, in this entertaining and well-made oater.
WES   116 min (ort 122 min)
VIDrel: 4-FRONT/POLYREC/CIC   V/h

SON'S PROMISE, A **   U
John Korty USA   1990
Ricky Schroder, David Andrews, Veronica Cartwright, Stephen Dorff,
Donald Moffat, Boyd Gaines, Andrew Lowry, Ryan Marshall, Pierce
Baehr, Trey Yearwood, Grayson Fricke, Linda Pierce, Edith Ivey, Ken
Strong
When a woman with seven sons is struck down with terminal
cancer, her eldest son, a boy just in his teens, promises to do all
he can to keep the family together after her death. However, he
soon faces a long battle with the welfare authorities who doubt
his ability to cope. Though intermittently appealing and well
acted, this is really far too sentimental and self-indulgent to be
truly effective. See A PROMISE TO KEEP for another film on
this theme.
DRA   90 min (ort 100 min) mTV   VIDrel: CAPIT/GUILD
V/sh

SOPHIE'S CHOICE ***   PG
Alan J. Pakula USA   1982
Meryl Streep, Kevin Kline, Peter MacNicol, Rita Karin, Stephen D.
Newman, Josh Mostel, Greta Turken, Marcell Rosenblatt, Moishe
Rosenfeld, Robin Bartlett, Eugene Lipinski, John Rothman, Joseph
Leon plus Josef Sommer (narration)
A Polish woman who survived the death camps lives in New
York in the 1940s, but is still tormented by her experiences, and
the memory of an agonising decision forced on her in the camps.
A ponderous and gloomy piece about guilt and despair, with
Streep giving one of her finest performances. The music is by
Marvin Hamlisch and the photography by Nestor Almendros.
Written by Pakula. AA: Actress (Streep).
DRA   144 min (ort 157 min)
VIDrel: 4-FRONT/POLYREC/ITC   V/dm
Boa: novel by William Styron.

SORRY, WRONG NUMBER **   15
Tony Wharmby USA   1989
Loni Anderson, Carl Weintraub, Patrick Macnee, Hal Holbrook, Diana
D'Aquila, Miguel Fernandes
A flat and unconvincing remake of the minor 1948 classic, which
starred Barbara Stanwyck as an invalid overhearing a murder
plan on the telephone and then realising that she is to be the
victim. Here she is replaced by a surprisingly healthy-looking
Anderson as the worried woman, who gradually becomes aware
of the plans others have in store for her. The insertion of a sub-
plot all about drug dealing is an unnecessary distraction.
THR   85 min (ort 100 min) mCab   VIDrel: CIC/SONOP L/A
V
Boa: story by Lucille Fletcher.

S.O.S. TITANIC **   PG
William Hale UK/USA   1979
David Janssen, Cloris Leachman, Susan St James, David Warner, Ian
Holm, Helen Mirren, Harry Andrews, Beverly Ross, David Battley,
Ed Bishop, Tony Caunter, Nicholas Davies, Matthew Guinness, Jerry
Houser, Victor Langley
Documentary-style drama about the famous sea disaster of 1912,
which mirrors fictional elements of the 1958 film A NIGHT TO
REMEMBER and the 1953 one TITANIC, but still remains dull
and unconvincing, even in its shortened version. Scripted by
James Costigan.
DRA   98 min (ort 140 min) mTV   VIDrel: L/A   V

SOUL MAN **   15
Steve Miner USA   1986
C. Thomas Howell, Rae Dawn Chong, Arye Gross, James Earl Jones,
Ann Walker, Melora Hardin, Leslie Nielsen, James B. Sikking, Max
Wright, Jeff Altman, Jonathan Leonard, Julia Louis-Dreyfus, Wally
Ward, Eric Schiff, Ron Reagan
When his father refuses to support him through college, a young
boy takes an overdose of tanning pills in order to win a black
scholarship to Harvard Law School. There he is subject to the
usual ethnic stereotyping and finds himself falling in love with
a black student who, he learns, failed to win the scholarship
because of his ruse. A clumsy attempt at a social satire, surprisingly ineffective despite its extensive cast and credits list.
COM   100 min   VIDrel: VCC L/A   V

SOULTAKER **   15
Michael Rissi USA   1990
Joe Estevez, Robert Z'dar, Vivian Schilling, Gregg Thomsen, David
Shark, Jean Reiner, Chuck Williams, David Fawcett, Gary Kohler,
Dave Scott, Peter Dach, Cinda Lou Freman, Meschelle Manley,
Charles Bosworth, Jeff Deen

Having been involved in a serious car crash, a young couple learn that they must use every means possible to re-unite body and soul, or face eternal separation from each other.
FAN 90 min VIDrel: 20VIS/SONOP V

SOUND BARRIER, THE *** U
David Lean UK 1952
Ralph Richardson, Ann Todd, Nigel Patrick, Dinah Sheridan, John Justin, Joseph Tomelty, Denholm Elliott, Jack Allen, Ralph Michael, Leslie Phillips, Jolyon Jackley
An aircraft engineer who is obsessed with proving that the sound barrier can be broken, and takes many risks to prove it, in this taut story of the early days of jet flight. AA: Sound (London Films).
Aka: BREAKING THE SOUND BARRIER
DRA 111 min (ort 118 min) B/W VIDrel: WHV V/h

SOUND OF MUSIC, THE *** U
Robert Wise USA 1965
Julie Andrews, Christopher Plummer, Eleanor Parker, Peggy Wood, Anna Lee, Richard Hadyn, Marni Nixon, Heather Menzies, Charmian Carr, Duane Chase, Angela Cartwright, Nicholas Hammond, Debbie Turner, Kym Karath
In 1938 a novice nun becomes governess to a musical family, and falls in love with their widower father, helping the family escape from the Nazis when Austria is occupied. Based on the life of the Von Trapp family, this has many fine songs set in a syrupy confection. Written by Ernest Lehman, with music and lyrics are by Richard Rodgers and Oscar Hammerstein II. AA: Pic, Dir, Score/adapt (Irwin Kostal), Sound (F. Hynes et al.), Edit (W. Reynolds).
MUS 180 min wScrn (special edition including a 15 min doc – ort 172 min) VIDrel: 20TH/TECH L/A V/dm V/sh
Boa: book The Trapp Family Singers by Maria Augusta Von Trapp, Russell Crouse and Howard Lindsay.

SOUR GRAPES * PG
John De Bello USA 1986
Richard Gilland, Jamie Farr, Tawny Kitaen, Rich Little, Ty Henderson, Debbie Gates, Eddie Deezen
A secret formula developed to end world hunger but proving to be irresistibly addictive, provides the excuse for this routine chase film, as all and sundry attempt to steal this latest boon to mankind. Another flabby "food" comedy from the director of ATTACK OF THE KILLER TOMATOES.
Aka: HAPPY HOUR
COM 85 min (ort 88 min) VIDrel: L/A V

SOURSWEET ** 15
Mike Newell UK 1988
Sylvia Chang, Danny Dun, Soon-Teck Oh, Jodi Lang
A look at the struggles of a family of emigrants from Hong Kong, who come to London and struggle for success in a seedy part of London, with the husband working as a waiter to save enough money for his own business, and then losing it all at cards. A bleak, cheerless but rather touching tale.
DRA 110 min VIDrel: PAL L/A V/h
Boa: novel by Timothy Mo.

SOUTH BEACH ** PG
David Carson USA 1992
Yancy Butler, John Glover, Patty D'Arbanville, Wendee Pratt, Rob knepper
When a man gets involved in a diamond smuggling ring that is being run by some former KGB agents, his sister becomes a blackmail victim to someone who claims to be a government agent.
A/AD 89 min VIDrel: CIC/SONOP V

SOUTH CENTRAL *** 15
Steve Anderson USA 1992
Glenn Plummer, Byron Keith Minns, LaRita Shelby, Carl Lumbly, Christian Coleman, Kevin Best, Starletta Dupois, Ivory Ocean, Lexie D. Bingham, Vincent Craig Dupree, Baldwin C. Sykes, Rana Mack, Diane Manzo, Sal Landi, Tim DeZarn
A member of a black L.A. gang is sent to prison where he meets a charismatic leader who gradually manages to effect a remarkable change in him, so much so that he becomes determined to save his son from a life of crime. After his release, he tries to put this plan into effect but encounter great difficulties of all kinds. Excellent acting sustains this powerful drama.
Aka: SOUTH CENTRAL, L.A.
DRA 94 min (ort 99 min) VIDrel: WHV V/dm V/sur
Boa: novel Crips by Donald Bakeer.

SOUTH PACIFIC *** U
Joshua Logan USA 1958
Rossano Brazzi, Mitzi Gaynor, John Kerr, Ray Walston, Juanita Hall, France Nuyen, Russ Brown, Jack Mullany, Ken Clark, Floyd Simmons, Candace Lee, Tom Laughlin, Warren Hsieh, Beverly Aadland, Giorgio Tozzi (voice only)
Enjoyable musical, about life for US troops and natives on an island in the South Pacific during 1943, when an American Army nurse falls in love with a French planter (Brazzi). The fine songs include such gems as "Bali H'ai", "There Is Nothing Like A Dame", "Happy Talk" and others. Written by Paul Osborn, Richard Rodgers, Oscar Hammerstein II and Joshua Logan, and based on the successful stage show. AA: Sound (Fred Hynes).
MUS 143 min (ort 171 min) wScrn VIDrel: 20TH/TECH L/A V/sh
Boa: short stories Tales Of The South Pacific by James A. Mitchener/musical by Richard Rodgers and Oscar Hammerstein II.

SOUTHERN COMFORT ** 18
Walter Hill USA 1981
Keith Carradine, Powers Boothe, Fred Ward, Franklyn Seales, T.K. Carter, Lewis Smith, Peter Coyote, Les Lannom, Carlos Brown, Brion James, Sonny Landham, Allan Graf, Ned Dowd, Rob Ryder, Greg Guirard, June Borel, Orel Borle
National Guardsmen on an exercise in the Louisiana Everglades come into conflict with the local French-speaking Cajun Indians, and this initial clash rapidly escalates into a full-scale bloody conflict. A brutal DELIVERANCE-style yarn that is quite well made despite having not a shred of originality.
THR 102 min (ort 106 min) VIDrel: WHV V/sh

SOYLENT GREEN ** 15
Richard Fleischer USA 1973
Edward G. Robinson, Charlton Heston, Leigh Taylor-Young, Chuck Connors, Joseph Cotten, Brock Peters, Paula Kelly, Mike Henry, Leonard Stone, Lincoln Kilpatrick, Whit Bissell, Celia Lousky, Dick Van Patten, Morgan Farley
A grim view of an Earth so overpopulated in the year 2022, that life has become an endless struggle, with the resources of the planet utterly spent. The film soon degenerates into a routine murder mystery, involving Heston as a cop and Robinson his "book" (giving a fine performance in his last film). A pretentious, ponderous and studio-bound affair, scripted by Stanley R. Greenberg.
FAN 93 min (ort 100 min) wScrn (special edition)
VIDrel: MGM/WHV V/h
Boa: novel Make Room! Make Room! by Harry Harrison.

SPACE 1999: ALIEN ATTACK ** PG
Lee H. Katzin/Charles Crichton/Bill Lenny UK 1977
Barbara Bain, Martin Landau, Barry Morse, Roy Dotrice, Anthony Valentine, Nick Tate, Catherine Schell, Tony Anholt, Isla Blair, Zienia Merton, Yasuko Magazumi
A spin-off from the TV series "Space 1999", which dealt with the adventures encountered by the personnel on Moonbase Alpha, who were sent off on a journey into deep space, when a freak nuclear accident wrenched the Moon from its orbit. In this tale they explore a rogue planet, suffer a mysterious illness and experience a nuclear explosion that affects the Moon's orbit.
Aka: ALIEN ATTACK
FAN 99 min (ort 122 min) mTV VIDrel: POLY/POLYREC V

SPACE 1999: DESTINATION MOONBASE ALPHA * PG
Tom Clegg UK 1976
Martin Landau, Barbara Bain, Barry Morse, Nick Tate, Tony Anholt, Catherine Schell, Zienia Merton, Yasuko Magazumi, Stuart Damon, Jeremy Young, Drewe Henley, Cher Cameron, Earl Robinson, Patrick Westwood, Nicholas Young
Film version of a two-part episode from "Space 1999", a rather cheap looking SF TV series. Originally entitled "The Bringers Of Wonder" this has a spaceship from Earth bringing a group of visitors to Moonbase Alpha, where they claim they've been sent to rescue the personnel and take them home. But Commander Koenig discovers that the visitors are not as benign (or indeed as human) as they outwardly appear.
Aka: BRINGERS OF WONDER, THE; DESTINATION MOONBASE ALPHA
FAN 96 min mTV VIDrel: POLY/POLYREC V

SPACE: ABOVE AND BEYOND ** PG
David Nutter USA 1995
Morgan Weisser, Kristen Cloke, Rodney Rowland, Lanei Chapman,

Joel De la Fuente, James Morrison, Bill Hunter, Colin Friels, Amanda Douge, Peter Lent, Theresa Wong, Anja Coleby, Darrin Klimek, Chris Kirby, Robert Coleby, Alan Dale
Man finally starts colonising distant worlds, only to face the unknown in the form of a race of aggressive alien creatures intent on invading Earth. A cliched pilot for a TV series, that has little to recommend it except the state-of-the-art visual effects, which go some way towards compensating for the lack of fresh ideas. This is BATTLESTAR GALACTICA territory, but updated to the 1990s.
FAN   87 min (ort 90 min) mTV   VIDrel: 20TH/FOXVID
V/sh

SPACE ADVENTURE COBRA *   PG
JAPAN   1985
Voices of: John Guerrosio, Tamsin Hollo, Lorelei King, David McMiller, Lesley Morris, Allan Wegner
The adventures of a space pirate form the basis for this rather forgettable space opera fantasy.
ANIM   95 min dubbed   VIDrel: MANGA/SONOP   V

SPACE FIREBIRD *   PG
Sugu Sugiyama/Osamu Tezuku JAPAN   1980
Animated story of a space pilot who falls out with his superiors and is sent to a prison camp on Iceland, from which he escapes with an older prisoner.
Aka: HINOTORI 2772; HINOTORI 2772 AI NO COSMOZONE; PHOENIX 2772; SPACE FIREBIRD 2772
ANIM   90 min (ort 116 min) dubbed
VIDrel: WESCON/RTM   V

SPACE MARINES *   15
John Wiedner USA   1996
Billy Wirth, Edward Albert Jr, James Shigeta, Meg Foster
In the 21st century the title unit are charged with keeping the peace, but have to deal with the depredations of a bunch of inter-planetary criminals, who have hijacked a nuclear cargo vessel. Silly space opera nonsense, one of those movies that are churned out in the wake of bigger, more expensive SF films.
FAN   93 min   VIDrel: THIRD   V

SPACE RANGERS CHRONICLES **   12
Mikael Saloman USA   1993
Linda Hunt, Jeff Kaake, Marjorie Monaghan, Jack McGee, Clint Howard, Danny Quinn, Cary-Hiroyuki Tagawa
A group of police officers volunteer to maintain law and order in the remote outposts of space, in this derivative and unimaginative fantasy series set in the year 2104 A.D. A pity to see acting talent of the calibre of Hunt involved in such third-rate material.
FAN   300 min (89 min pilot plus five further episodes) mTV
VIDrel: MARQ/FOXVID L/A   V

SPACE RIDERS *   PG
Joe Massot UK   1983
Barry Sheene, Gavin O'Herlihy, Toshiya Ito, Stephanie McLean, Sayo Inaba, Caroline Evans, Hiroshi Kato, Jeff Harding, Marina Sirtis, Yuriko Tagaki, Maureen Moody, Steve Parrish, Andrew Marriott
The story of a motorcycle team and their bid to win the world championships, with Sheene playing a top rider who is signed up to ride for a Japanese corporation. Very dull.
DRA   93 min (ort 99 min)   VIDrel: LUMI/SPEAR   V/sh

SPACE TRANSFORMER **   PG
Johnny T. Howard HONG KONG   1989
Another film in a seemingly endless stream of Hong Kong animated space adventures (see SOLAR ADVENTURE, CAPTAIN COSMOS or FALCON 7) that inevitably features an Earth under attack from aliens. This time round, our only saviour is a genius who has succumbed to a virulent bacteriological weapon. In an idea borrowed from FANTASTIC VOYAGE, a rescue crew is miniaturised and injected into his bloodstream to save her.
ANIM   59 min   VIDrel: IMPENT   V

SPACEBALLS **   PG
Mel Brooks USA   1987
Mel Brooks, John Candy, Rick Moranis, Bill Pullman, Daphne Zuniga, Dick Van Patten, George Wyner, Michael Winslow, Lorene Yarnell, Ronny Graham, Leslie Bevis, Sal Viscuso, John Hurt. Voices of: Joan Rivers, Dom DeLuise
STAR WARS gets the Mel Brooks treatment in this soggy spoof, that has a plot loosely revolving around the rescue of a princess, but is really little more than a succession of visual and verbal gags, with most of the latter being simple parodies of names used in the earlier film. Quite amusing, but only in 10-minute doses.
COM   92 min Cut (1 sec – ort 96 min)   VIDrel: MGM/WHV
V

SPACECAMP *   PG
Harry Winer USA   1986
Kate Capshaw, Lea Thompson, Tom Skerritt, Kelly Preston, Larry B. Scott, Leaf Phoenix, Tate Donovan, Barry Primus, Terry O'Quinn, Mitchell Anderson, T. Scott Coffey, Daryl Roach, Peter Scranton, Holly Rebecca Suggs, Terry White
Five teenagers are chosen to train at a NASA summer camp and by accident get launched, together with their instructor, into space. They now have to get safely back to Earth. A vacuous little effort that isn't even memorable in the special effects department.
Aka: SPACE CAMP
FAN   107 min (ort 115 min)   VIDrel: VCC/DISC/COLUM
V/sur

SPACED INVADERS **   U
Patrick Read Johnson USA   1989
Douglas Barr, Ariana Richards, Royal Dano, Kevin Thompson, Jimmy Briscoe, J.J. Anderson, Gregg Berger, Tony Cox, Debbie Lee Carrington, Tommy Madden
A Halloween anniversary broadcast of Orson Welles' famous adaptation of Wells' "War Of The Worlds" results in a group of dimwitted Martians assuming they have been given invasion orders. They arrive at a small town in the Midwest with the intention of "kicking some Earthling butt", but strange to relate, are simply taken for kids (they are quite small) dressed up in Halloween costume. A one-joke kid's fantasy of disappointingly few ideas.
FAN   95 min (ort 102 min)   VIDrel: MED/POLYREC L/A
V

SPACEHUNTER *   (15)
Lamont Johnson CANADA   1983
Peter Strauss, Molly Ringwald, Ernie Hudson, Andrea Marcovicci, Michael Ironside, Beeson Carroll, Grant Alianak, Deborah Pratt, Aleisa Shirley, Cali Timmins, Paul Boretski, Patrick Rowe, Reggie Bennett
A space salvage man rescues three space maidens on a dangerous planet in the "Forbidden Zone" of the 22nd century, who are being held captive by a nasty mutant who goes under the title of "Overdog". The Elmer Bernstein score is the best thing in this bleak yarn. Written by Edith Ray, David Preston, Dan Goldberg and Len Blum.
Aka: SPACEHUNTER: ADVENTURES IN THE FORBIDDEN ZONE
FAN   87 min (ort 90 min)   SATrel: SKY MOVIES GOLD

SPANISH GARDENER, THE ***   U
Philip Leacock UK   1956
Dirk Bogarde, Michael Hordern, Jon Whiteley, Cyril Cusack, Geoffrey Keen, Maureen Swanson, Lyndon Brook, Josephine Griffin, Bernard Lee, Rosalie Crutchley, Ina De La Haye, Harold Scott, Jack Stewart
The son of a diplomat develops a strong friendship with their gardener, which is resented by the father. A slow, careful and interesting character study, that is a little too languid but remains quite absorbing. Written by Lesley Storm and John Bryan.
DRA   88 min (ort 97 min)   VIDrel: VCC L/A   V
Boa: novel by A.J. Cronin

SPANKING THE MONKEY ***   18
David O. Russell USA   1994
Jeremy Davies, Elizabeth Newitt, Benjamin Hendrickson, Alberta Watson, Carla Gallo, Liberty Jean, Archer Martin, Matthew Puckett, Zak Orth, Josh Weinstein, Judah Domke, Nancy Fields, Judette Jones, Carmine Paolini, Neil Connie Wallace
A pre-med freshman returns home for a vacation and finds himself forced to care for his mother, who has broken her leg. Initially his only escape from this chore is masturbation (the meaning of the title slang expression) but he is ultimately driven into the arms of a local girl, arousing the jealous ire of his patient. The sharp dialogue and low-key performances of the leads add greatly to the enjoyment of this offbeat and slightly unsettling comedy.
COM   98 min wScrn   VIDrel: TART/20TH   V/s

SPARE ME *** 18
Matthew Harrison USA 1993
Lawton Paseka, Christie MacFadyen, Mark Alfred, Sunny Weil
A man is suspended from the professional bowling circuit for
injuring a rival player, and returns to his home-town in search
of his father, a former bowling legend whom he hopes will be
able to help him get reinstated. But dad has problems of his own
with a partner whose son has just escaped from a mental institution. A real oddity, this low-budget film has a bizarre plot,
pleasingly low-key performances and a bundle of energy.
DRA 83 min (ort 88 min) VIDrel: SCEDGE/RTM V/sh

SPARKLING CYANIDE *** PG
Robert Lewis UK 1989
Anthony Andrews, Deborah Raffin, Christine Belford, H. Morgan
An absorbing murder mystery in which a woman's death during
an anniversary party is initially taken to be suicide, but proves
to be otherwise. Detective Kemp investigates.
Aka: AGATHA CHRISTIE: SPARKLING CYANIDE
DRA 92 min VIDrel: WHV V/h

SPARKS: THE PRICE OF PASSION * 15
Richard A. Colla USA 1990
Victoria Principal, Ted Wass, Hector Elizondo, William Lucking,
Elaine Stritch, Ralph Waite, Gary Farmer, Thomas Callaway, Gracie
Harrison, Radha Delamarter, Steven Tyler, Lois Geary, Tory Polone,
Patricia Van Ingen
A ludicrously implausible melodrama, that has Principal
playing the outspoken, newly-elected mayor of Albuquerque,
New Mexico, who's up to her pretty eyebrows in problems, not
least the activities of both a serial killer and a blackmailer, both
of whom intend to put paid to her political career, albeit in
markedly different ways.
DRA 92 min (ort 100 min) mTV VIDrel: GENESIS V

SPARROW * 12
Franco Zefferelli ITALY 1993
Angela Bettis, Jonathan Schlaech, Sara-Jane Alexander, Andrea
Cassar, John Castle, Valentina Cortese, Sinead Cusack, Frank Finlay,
Mia Fothergill, Pat Heywood, Janet Maw, Denis Quilley, Vanessa
Redgrave, Annabel Ryan, Gareth Thomas
Story set in the Sicilian town of Catania in 1854, where a beau-
tiful girl is released from twelve years of enforced isolation in a
convent (she was placed there by her stepmother) when the
town is evacuated due to an outbreak of plague. Taken to her
father's splendid villa near Mount Etna, she falls for a hand-
some friend of the family. A cliche-ridden melodrama,
unconvincingly acted, terribly disjointed, and badly dubbed for
good measure.
Aka: STORIA DE UNA CAPINERA
DRA 102 min (ort 106 min) dubbed
VIDrel: ELPIC/POLYREC V
Boa: novel A Sparrow's Tale by Giovanni Verga.

SPARROWS CAN'T SING * PG
Joan Littlewood UK 1962
James Booth, Barbara Windsor, Roy Kinnear, George Sewell, Barbara
Ferris, Avis Bunnage, Murray Melvin, Arthur Mullard
After a couple of years at sea, a sailor returns home to find that
his wife and baby have gone off to live with a bus driver. He
searches for them relentlessly. A cheerful, chirpy, cockney cari-
cature; the characters are generally a little too disagreeable for
one to sympathise. Barbara Windsor sings the title song.
COM 88 min (ort 94 min) B/W VIDrel: BRAVE/SONOP
L/A V
Boa: play Sparrers Can't Sing by Stephen Lewis.

SPARTACUS **** PG
Stanley Kubrick USA 1960
Kirk Douglas, Laurence Olivier, Jean Simmons, Charles Laughton,
Tony Curtis, Peter Ustinov, Herbert Lom, John Gavin, Woody Strode,
Nina Foch, John Ireland, John Dall, Charles McGraw, Joanna Barnes,
Harold J. Stone, Peter Brocco
A splendid film about an actual slave rebellion that took place
in 71 B.C., led by a famous gladiator who defeated several
Roman armies sent against him. Produced by Kirk Douglas and
written by Dalton Trumbo, with the fine score by Alex North.
AA: S. Actor (Ustinov), Cin (Russell Metty), Art/Set (Alexander
Golitzen and Eric Orbom/Russell A. Gausman and Julia Heron),
Cost (Valles/Bill Thomas).
A/AD 186 min wScrn (restored edition – ort 196 min)
VIDrel: CIC/SONOP V/sur
Boa: novel by Howard Fast.

SPECIALIST, THE * 18
Luis Llosa USA 1994
Sharon Stone, Sylvester Stallone, James Woods, Rod Steiger, Eric
Roberts, Mario Ernesto Sanchez, Sergio Dore Jr, Chase Randolph,
Jeana Bell, Britanny Paige Bouck, Emilio Estefan Jr, LaGaylia Frazier,
Ramon Gonzalez-Cuevas
A former CIA explosives expert is hired by a seductive woman
planning revenge for her parents' murder. A formula actioner
with big, box-office stars, plenty of explosive action and steamy
sex scenes, but absolutely nothing remotely resembling a new
idea. Well below average.
A/AD 105 min (ort 112 min) cC VIDrel: WHV V/sur
Boa: the "Specialist" novels of John Shirley.

SPECIES ** 18
Roger Donaldson USA 1995
Natasha Henstridge, Ben Kingsley, Forest Whitaker, Michael Madsen,
Marg Helgenberger, Alfred Molina, Michelle Williams, Jordan Lund,
Don Fischer, Scott McKenna, Virginia Morris, Jayne Luke, David K.
Schroeder, David Jensen
Scientists receive a communication from space containing
instructions on how to create alien DNA, which they make and
merge with a human sample. The result is an apparently human
female, who rapidly grows to adulthood, but then escapes and
starts to mutate into a murderous monster with an overwhelm-
ing desire to reproduce. A race against time ensues to stop this
happening. An underplotted, hi-tech variant on the 1961 BBC
TV film "A For Andromeda". See also EMBRYO.
HOR 104 min (ort 120 min) wScrn cC VIDrel: MGM/WHV
V/sur

SPECIMEN ** 18
John Bradshaw CANADA 1995
Mark Paul Gosselaar, Doug O'Keefe, Ingrid Kalevaars, David
Herman, Andrew Jackson, Michelle Jackson
A twenty-four-year old man has been under a strange influence
all his life, his mother having been abducted by aliens. He now
retreats to a remote community in an attempt to learn more
about his past, but his abilities (which includes the power to
cause spontaneous combustion) lead to the inevitable violent
confrontation. Adequate. See also FIRESTARTER and WILDER
NAPALM.
HOR 90 min VIDrel: GUILD/FOXVID V

SPECTERS * 18
Marcello Avallone ITALY 1987
Donald Pleasence, John Pepper, Katrine Michelsen, Massimo De Rossi,
Riccardo De Torrebruna, Lavinia Grizi, Riccardo Parisio Perrotti,
Giovanni Bilancia, Erna Schurer
A group of men working on a new subway line, unearth a
hidden Roman tomb that has been undisturbed for centuries,
and when archaeologists investigate, they awaken a demonic
creature. A cumbersome and derivative film that starts off well
enough, but gets rather silly once the creature is revealed. In
between the sparse second-rate effects are acres of tedium, effectively dissipating all tension before the story can develop.
Aka: SPECTRES; SPETTRI
HOR 90 min (ort 96 min) VIDrel: SPEAR/SONOP V

SPEECHLESS ** 12
Ron Underwood USA 1994
Michael Keaton, Geena Davis, Bonnie Bedelia, Ernie Hudson, Charles
Martin Smith, Gailard Sartain, Christopher Reeve, Ray Baker,
Mitchell Ryan, Willie Garson, Paul Lazar, Richard Poe, Harry
Shearer, Steven Wright, Jodi Carlisle
A screwball romantic comedy in which a pair of speech-writers
for opposing politicians meet at an all-night store and not
knowing what the other does for a living, embark on a rela-
tionship. However, when they do learn more about each other,
their mutual attraction becomes tinged with not a little rivalry.
The strength of the cast are this film's greatest asset, and the two
leads work wonderfully well together.
COM 100 min VIDrel: MGM/WHV V/sh

SPEED *** 15
Jan De Bont USA 1994
Keanu Reeves, Dennis Hopper, Sandra Bullock, Joe Morton, Jeff
Daniels, Alan Ruck, Glenn Plummer, Richard Lineback, Beth Grant,
Carlos Carrasco, David Kriegel, Hawthorne James, Natsuko Ohama,
Daniel Villarreal, Margaret Medina
A psychotic criminal who is an expert bomber, conceives a nasty
revenge on a member of an elite SWAT team by planting a bomb
on a bus on which the latter is travelling. To make things really

devilish, the device is primed to explode should the vehicle's speed ever fall below 50 mph. A non-stop actioner, with great stunts and effects, but lacking enough character development to flesh out the simple plot. Exciting, but a little mindless. AA: Sound, Effects/aud.

A/AD   111 min (ort 92 min) wScrn cC
VIDrel: 20TH/FOXVID; ENCORE (LV only)   V/sur LV

SPELLBINDER *
Janet Greek USA
(18)
1988
Timothy Daly, Kelly Preston, Rick Rossovich, Audra Lindley, Diana Bellamy, Cary-Hiroyuki Tagawa, Anthony Crivello, Roderick Cook, Stefan Gierasch, Kyle Heffner, James Louis Watkins, Rick Rossovich, M.C. Gainey, Sally Kemp
A man rescues a woman from a beating, but is horrified to learn that she is on the run from a ruthless cult who are prepared to use any amount of violence to get her back, as they are intending to sacrifice her. An unpleasant and highly predictable supernatural thriller.
HOR   96 min (ort 99 min)   SATrel: MOVIE CHANNEL

SPELLBOUND ***
Alfred Hitchcock USA
PG
1945
Ingrid Bergman, Gregory Peck, Leo G. Carroll, John Emery, Michael Chekhov, Wallace Ford, Rhonda Fleming, Jean Acker, Donald Curtis, Norman Lloyd, Steven Geray, Paul Harvey, Erskine Sanford, Janet Scott, Victor Kilian
Bergman as a female psychiatrist tries to probe Peck's mind, when he arrives to take over the asylum and then turns out to be an impostor and an amnesiac. All the evidence points to him having killed the previous asylum chief, but her efforts reveal the truth and clear him. An engrossing psychological mystery, with a dash of romance, imaginative dream sequences and a good cast. Written by Ben Hecht and Angus MacPhail. AA: Score (Miklos Rozsa).
DRA   110 min B/W   VIDrel: VCC/DISC; PION (LV only)
V LV
Boa: novel The House of Dr Edwardes by Francis Beeding.

SPELLCASTER **
Rafael Zielinski USA
18
1988
Adam Ant, Richard Balde, Gail O'Grady, Bunty Bailey, Teaci Lin, Harold Pruett
A group of rock 'n' roll fans receive an invitation to participate in a treasure hunt at a private estate, but learn to their horror that their host is a wizard, and his home is a haunted castle, plagued by demons and restless spirits. A low-budget horror yarn, bereft of thrills or wit.
HOR   79 min (ort 83 min)   VIDrel: COLUM/SONOP   V

SPENCER'S MOUNTAIN **
Delmer Daves USA
(PG)
1963
Henry Fonda, Maureen O'Hara, James MacArthur, Donald Crisp, Wally Cox, Mimsy Farmer, Virginia Gregg, Lillian Bronson, Whit Bissell, Hayden Rorke, Kathy Bennett, Dub Taylor, Hope Summers, Ken Mayer, Bronwyn Fitzsimmons, Larry Mann
The head of a large rural family nurtures a long-standing dream to build a house large enough for everyone including his parents who also live in their crowded quarters. However, problems occur when his eldest son is about to graduate from high college and sets his sights on going to college. A strong cast do a good job with the rather weak script, based on Hamner's novel, which later gave rise to the TV series "The Waltons".
DRA   113 min (ort 121 min)   SATrel: MOVIE CHANNEL
Boa: novel by Earl Hamner Jr.

SPENSER: A SAVAGE PLACE **
Joseph L. Scanlon USA
(PG)
1994
Robert Urich, Avery Brooks, Wendy Crewson, Cynthia Dale, Tyroen Berskin, Neil Crone, Richard Fitzpatrick, Jerry Levitan, Douglas Miller, Daniel Parker, Ross Pelty, Michael Ricupero, David Spooner, Hayley Tyson
Our tough detective helps an old girlfriend, a TV reporter who is trying to expose the shady activities of a top film studio, their investigations not being made easier by the murder of their informant. Fair.
Aka: SAVAGE PLACE, A
THR   87 min   SATrel: SKY MOVIES

SPENSER: CEREMONY **
Andrew Wild USA
(15)
1994
Robert Urich, Avery Brooks, Barbara Williams, J. Winston Carroll, Dave Nichols, Tanya Allen, Jefferson Mappin, Lynne Cormack, Lili Francks, Janet Bailey, Alexa Gilmour, William Colgate, Falconer Abraham, Henry Gomez
Feature-length episode of Urich's Spenser series in which our 'tec is asked to locate a missing teenage girl and discovers that she has become caught up in the world of teenage prostitution. Adequate.
Aka: CEREMONY
THR   85 min (ort 90 min) mTV   SATrel: SKY MOVIES

SPENSER: PALE KINGS AND PRINCES **
Victor Sarin USA
(15)
1994
Robert Urich, Avery Brooks, Barbara Williams, J. Winston Carroll, Matthew Ferguson, Sonja Smits
In this mystery thriller private eye Spenser joins forces with an enigmatic friend and his psychologist girlfriend, his task being to discover who killed a reporter in an isolated hick town. Despite the fact that his investigations come up against a wall of silence, bit by bit he gets at the truth and as expected, uncovers the web of corruption, intrigue etc. that always seems to be the stock-in-trade of such tales. Adequate.
Aka: PALE KINGS AND PRINCES
THR   87 min (ort 90 min) mTV   SATrel: SKY MOVIES

SPENSER: THE JUDAS GOAT **
Joe Scanlon USA
(PG)
1994
Robert Uhrich, Avery Brooks, Wendy Crewson, Leon Pownall, Falconer Abraham, Barclay Hope, Adam Bass, Adrian Truss, Daniel Parker, Alexandra Amini, Jonathan Welsh, Todd Schroeder, Anne Ritchie, David Crean, Sally Cahill, Caroline Yeager
When his family are killed in an explosion ostensibly meant to eliminate the head of an African state, a wealthy industrialist hires Spenser to find those responsible for this outrage. This proves to be a perilous and complex undertaking, as he soon finds himself investigating a nasty conspiracy involving international terrorists. Hardly the most original of plots, but a watchable feature-length episode from a popular TV detective.
Aka: JUDAS GOAT, THE
THR   87 min   SATrel: SKY MOVIES

SPIDER, THE **
Bert I. Gordon USA
PG
1958
Ed Kemmer, June Kenny, Gene Persson, Gene Roth, Hal Torey, June Jocelyn, Jack Kosslyn, Sally Fraser, Mickey Finn, Hank Patterson, Troy Patterson, Skip Young, Bill Giorgio, Howard Wright, Bob Garnet, Shirley Falls
When a high school biology teacher finds a giant spider that is apparently dead, he puts it for safe-keeping the school's gym. Unfortunately, it's only snoozing, and when some kids arrive to practice in their rock 'n' roll band, it wakes up, and is most vexed at the disturbance. This leads to a rampage, a retreat to its secret cave, and peril for a couple of teenagers who have unwisely ventured there. An entertaining, low-budget horror quickie.
Aka: EARTH VERSUS THE SPIDER
HOR   73 min B/W   VIDrel: HEND/BMGREC L/A   V

SPIDER AND THE FLY, THE **
Michael Katleman USA
12
1994
Mel Harris, Ted Schackelford, Kim Coates, Colm Feore, Frankie R. Faison, Cynthia Belliveau, Kenneth Welsh, Peggy Lipton
Two mystery writers embark on a passionate affair, and then make a wager over who can come up with the best idea for committing the perfect crime. Unfortunately, they find life beginning to imitate art, when they are both framed for a murder.
THR   84 min (ort 87 min)   VIDrel: CIC/SONOP   V

SPIDER-MAN: THE MOVIE *
E.W. Swackhamer USA
U
1976
Nicholas Hammond, Lisa Eilbacher, Thayer David, Michael Pataki, David White, Ivor Francis, Jeff Donnell, Hilly Hicks, Dick Balduzzi, Bob Hastings, Barry Cutler, Norman Rice, Len Lesser, Ivan Bonar, Carmelita Pope, George Cooper
A spider's bite gives a weedy student superhuman powers, which he uses to fight crime. In this pilot for a brief series, our hero sets out to expose a multi-million dollar extortion racket. A not terribly successful or well-conceived adaptation, of one of the most popular comic book characters to grace the pages of Marvel Magazine's comics. Followed by a couple of sequels.
Aka: AMAZING SPIDERMAN, THE; SPIDER-MAN
FAN   90 min (ort 92 min) mTV   VIDrel: L/A   V

SPIDER-MAN: SPIDER-MAN STRIKES BACK *   U
Ron Satlof USA   1978
Nicholas Hammond, Joanna Cameron, Robert F. Simon, Michael Pataki, Chip Fields, Robert Alda, Randy Powell, Sid Clute
In this adventure, our masked crusader against crime sets out to foil an attempt to destroy L.A. Another mediocre effort, based on a character created by Marvel Comics, but lacking both flair and imagination.
FAN   89 min Cut (17 sec) mTV
VIDrel: MIA/DISC/COLUM   V

SPIDER-MAN: THE DRAGON'S CHALLENGE *   U
Don McDougall USA   1979
Nicholas Hammond, Robert F. Simon, Benson Fong, Eileen Bry, Chip Fields, Myron Healey, Rosalind Chao, Hagan Beggs, Richard Erdman, John Milford, Ted Danson, Anthony Charnotta, George Cheung, Tony Clark, Michael Mancini
A sequel to SPIDER-MAN: THE MOVIE, with our masked crime-fighter setting out to clear the name of a Chinese official who has been accused of being a WW2 traitor. Very poor.
Aka: CHINESE WEB, THE; SPIDERMAN AND THE DRAGON'S CHALLENGE
FAN   93 min (ort 95 min) mTV
VIDrel: MIA/DISC/COLUM   V

SPIDER'S STRATAGEM, THE ***   PG
Bernardo Bertolucci ITALY   1970
Guilio Brogi, Alida Valli, Tino Scotti, Pino Campanini, Franco Giovanelli, Allen Midgett
A young man returns to his native village, where the defacement of a memorial to his father, murdered by the Fascists in 1936, marks the beginning of a twisted search for the truth about a man he had always worshipped as a hero. An over-elaborate but fascinating adaptation of the Borges story; the superb colour photography of Vittorio Storaro is an additional attraction.
Aka: LA STRATEGIA DEL RAGNO
DRA   95 min (ort 110 min) mTV   VIDrel: CONNO/RTM   V
Boa: short story The Theme of the Traitor and the Hero by Jorge Luis Borges.

SPIES, LIES AND NAKED THIGHS **   PG
James Frawley USA   1989
Harry Anderson, Ed Begley Jr, Wendy Crewson, Linda Purl, Rachel Ticotin
A madcap comedy telling of a strange character, a house-guest thanks to his resemblance to a former college chum, who embroils a UN translator and his wife in a crazy mystery revolving around the activities of a rather bizarre killer. The script is by Ed Self.
COM   87 min (ort 100 min) mTV   VIDrel: MOPIC/SGSVID V

SPIES LIKE US *   PG
John Landis USA   1985
Chevy Chase, Dan Aykroyd, Steve Forrest, Donna Dixon, Bruce Davison, Tom Hatten, William Prince, Bernie Casey, Michael Apted, Frank Oz, Constantin Costa-Gavras, Terry Gilliam, Ray Harryhausen, Martin Brest, Bob Swain
A silly farce about a pair of incompetent bumbling agents who are used as decoys in a complex plot. An infantile exercise made in the style of one of those Bing Crosby/Bob Hope ROAD TO... movies, but far too chaotic and muddled to deliver much in the way of humour. Written by Dan Aykroyd, Lowell Ganz and Babaloo Mandel.
COM   98 min (ort 109 min)   VIDrel: WHV   V/h

SPILL **   12
Allan Goldstein USA   1996
Brian Bosworth, Leah Pinsent, Eric Peterson
On the eve of an international conference on the environment, a company is responsible for the discharge of deadly toxic waste near Yellowstone Park, and then attempts to mount a cover-up. A competent plot with a strong ecological flavour, but severely constrained by its low budget, which gives all the proceedings a depressingly cheap look. Bosworth's wooden performance does little to make amends for this failing.
A/AD   86 min   VIDrel: EIV/SONOP   V/sh

SPIRAL STAIRCASE, THE ****   PG
Robert Siodmak USA   1945
Dorothy McGuire, George Brent, Ethel Barrymore, Kent Smith, Rhonda Fleming, Elsa Lanchester, Gordon Oliver, James Bell, Charles Wagenheim, Ellen Corby, Rhys Williams, Richard Tyler, Erville Alderson, Sara Allgood, Myrna Dell

In 1906, a New England town is terrorised by a psychopath who murders three handicapped girls. A mute servant working in a strange, dark household, has her own suspicions as to the identity of the murderer. An atmospheric thriller that doesn't miss a single chance to convey a feeling of tension and fear. Written by Mel Dinelli and remade in the UK in 1975.
THR   80 min (ort 83 min) B/W   VIDrel: VCC/DISC; PION (LV only)   V LV
Boa: novel Some Must Watch by Ethel Lina White.

SPIRIT OF ST LOUIS, THE *   U
Billy Wilder USA   1957
James Stewart, Marc Connelly, Patricia Smith, Murray Hamilton, Bartlett Robinson, Robert Cornthwaite, Sheila Bond, Marc Connelly, Arthur Space, Paul Birch, Harlan Warde, Dabbs Greer, David Orrick, Robert Burton, Maurice Manson
Dramatisation of Charles A. Lindbergh's epic 3,600-mile solo flight across the Atlantic, when he took off from Roosevelt Field, New York, and landed 33 hours later in France. Overlong and full of numerous boring flight sequences. Written by Billy Wilder and Wendell Mayes. The music is by Franz Waxman.
A/AD   129 min (ort 138 min)   VIDrel: L/A   V
Boa: book by Charles A. Lindbergh.

SPIRIT OF '76 *   15
Lucas Reiner USA   1991
Olivia D'Abo, David Cassidy, Leif Garrett, Geoff Hoyle, Jeff McDonald, Liam O'Brian, Steve McDonald, Barbara Bain, Moon Zappe, Julie Brown, Devo, Thomas Chong, Iron Eyes Cody, Kipper Kids, Don Novello, Carl Reiner, Rob Reiner
After American culture is destroyed by a magnetic storm in the 22nd century, an expedition is sent in a time machine to return to 1776 but a computer malfunction lands them in 1976 instead. A feeble and highly unfunny spoof of minimal interest, though the 1970s music soundtrack is pleasing enough.
COM   78 min (ort 110 min)   VIDrel: 20VIS/SONOP   V

SPIRIT OF THE BEEHIVE, THE ***   PG
Victor Erice SPAIN   1973
Ana Torrent, Isabel Telleria, Fernando Fernan Gomez, Teresa Gimpera, Jose Villasante, Lally Soldavilla, Juan Margallo, Miguel Picazo
A young girl in a Castilian village in 1940 sees the film Frankenstein and becomes obsessed with the figure of the monster (portrayed by Boris Karloff). She later befriends a fugitive soldier and sees him in her imagination as this famous creature. A slow-moving account of childhood that is annoyingly plotless but very well staged. This was Erice's directing debut.
Aka: EL ESPIRITU DE LA COLMENA
DRA   93 min (ort 98 min)   VIDrel: ARTPRO/RTM   V

SPITFIRE **   15
Albert Pyun USA   1994
Kristie Phillips, Sarah Douglas, Lance Henriksen, Tim Thomerson
A female gymnast and martial artist is drawn into the world of espionage and becomes a hired killer when her mother is killed and her father abducted. A standard actioner that was shot on location in the Bahamas, Hong Kong, Athens and Rome.
A/AD   91 min (ort 95 min)   VIDrel: TRIM/HIFLI   V/h

SPLASH **   PG
Ron Howard USA   1984
Daryl Hannah, Tom Hanks, Eugene Levy, Dody Goodman, John Candy, Richard B. Shull, Shecky Greene, Bobby Di Cicco, Howard Morris, Tony Di Benedetto, Patrick Cronin, Charles Walker, David Knell, Jeff Doucette, Tony Longo
Twice in his life a man is saved from drowning, once as a boy and then as a young man. His saviour is a beautiful and mysterious girl who one day comes in search of him. She is however a mermaid except when on dry land. Some comic moments vie with some touching ones, but a weak plot lets it all down. Looks a bit like an update of MIRANDA. Written by Lowell Ganz, Babaloo Mandel, Bruce Jay Friedman and Brian Grazer. Followed by SPLASH, TOO.
COM   105 min (ort 110 min) cC   VIDrel: TOUCH/TECH V/sur

SPLASH, TOO *   U
Greg Antonacci USA   1988
Todd Waring, Amy Yasbeck, Donovan Scott, Rita Taggart, Noble Willingham, Dody Goodman, Mark Blankfield
In this Disney-style sequel to the first film, Waring and his

mermaid wife Yasbeck settle into happy domesticity, until our mermaid takes off to save a dolphin friend. Unfortunately, nothing can save this ill-conceived and poorly executed effort.
COM   100 min mTV   VIDrel: BUENA L/A   V

SPLENDOR **   U
Elliott Nugent USA   1935
Miriam Hopkins, Joel McCrea, Paul Cavanagh, Helen Westley, Billie Burke, Katharine Alexander, Ruth Weston, David Niven, Ivan Simpson, Torben Meyer, Arthur Treacher, Reginald Sheffield, Willim R. Arnold, Maidel Turner
The son of a wealthy family whose fortunes are falling fast marries a poor girl of whom his parents naturally disapprove. However, when a financier comes on to the scene who may be able to restore their financial standing, they push her into having an affair. A routine melodrama with the usual happy ending, albeit one that leaves our couple sadder and wiser but at least re-united.
DRA   77 min B/W   VIDrel: VCC/DISC   V

SPLENDOR IN THE GRASS ***   15
Elia Kazan USA   1961
Natalie Wood, Pat Hingle, Audrey Christie, Barbara Loden, Warren Beatty, Zohra Lampert, Sandy Dennis, Joanna Roos, Jan Norris, Gary Lockwood, Crystal Field, Marla Adams, Lynn Loring, John McGovern, Martine Bartlett, Sean Garrison
A young couple in a small Kansas town in 1925 fall in love but are unable to cope with the results of their repressed physical feelings and their relationship ends, resulting in tragedy all around. A searing and powerful melodrama that gives a convincing account of the effects of the prejudice and narrow-mindedness of their local community. This was Beatty's screen debut and Inge's first work for the cinema. AA: Story/Screen (William Inge).
DRA   119 min (ort 124 min)   VIDrel: WHV   V

SPLIT DECISIONS **   15
David Drury USA   1986
Craig Sheffer, Jeff Fahey, Gene Hackman, Jennifer Beals, John McLiam, Eddie Velez, Carmine Caridi, James Tolkan, DeVoreaux White, David Labiosa, Harry Van Dyke, Anthony Trujillo, Victor Campos, Tom Bower, Julius Harris
A tough boxing family constantly argue, until the eldest son is murdered for refusing to fix a fight, at which point they close ranks. A violent and generally unappealing tale that covers much the same territory as ROCKY.
DRA   91 min (ort 95 min)   VIDrel: GUILD/SONOP L/A V/sh

SPLIT IMAGES ***   18
Sheldon Larry CANADA   1992
Gregory Harrison, Nicholas Campbell, Rebecca Jenkins, Maury Chaykin, Steve Whistance-Smith, Nahanni Johnstone, David Hewlett, Melody Ryan, Tom Holis, Dennis O'Connor, Sandi Ross, Dante Smith-Beze, Kristina Nicoll, Eugene Clark
A millionaire makes a hobby of committing murders in this tense and quite gripping psychological thriller.
THR   95 min   VIDrel: 20VIS/SONOP   V
Boa: novel by Elmore Leonard.

SPLIT INFINITY **   (U)
Stan Ferguson USA   1992
Trevor Black, Melora Slover, H.E.D. Redfords, Sean Nevs, Marcia Dangerfield, Devin Healey, Heath Ezell, Dave Jensen, Isaac Shamy, Jodi Webb. Mary Bishop, Jonathan Wilde, Vaioletti Purcell, Talia Argyle, Jennifer McDonald, Thom Dillon
A money-obsessed American teenage girl is consumed with a desire for the material things of life, but receives a salutary lesson when her grandmother tells her about life during the Depression. When the girl is knocked unconscious in an accident, she awakes to find herself magically transported back in time to 1929, where she learns a thing or two about life. Nicely realised if somewhat preachy, with a simple message quite clearly aimed at kids and young adults.
JUV   86 min (ort 90 min)   SATrel: SKY MOVIES

SPLIT SECOND *   18
Tony Maylam UK   1991
Rutger Hauer, Kim Cattrall, Neil Duncan, Michael J. Pollard, Alun Armstrong, Peter Postlethwaite, Ian Dury, Roberta Eaton, Tony Steedman, Steven Hartley, Ken Bones, Sarah Stockbridge, Colin Skeaping, Dave Duffy, Paul Grayson, Tina Smith
This derivative fantasy is set in the year 2008, when a runaway

greenhouse effect has resulted in an Earth of endless torrential rain. Hauer plays an overworked cop who has been after a ferocious serial killer (the victims' hearts are ripped out) since his partner was murdered. He finds that he is chasing a humanoid monster which lives by absorbing the DNA of its victims. The expected and appropriate messy climax takes place. A low-budget ALIEN-style rip-off.
FAN   86 min (ort 90 min)   VIDrel: EIV/SONOP   V/sur

SPLITTING HEIRS **   15
Robert Young UK/USA   1992
Eric Idle, Rick Moranis, Barbara Hershey, John Cleese, Catherize Zeta Jones, Sadie Frost, Stratford Johns, Brenda Bruce, William Franklyn, Richard Huw, Eric Sykes, Chawbala Chokshi, Jeremy Clyde, Bridget McConnel, Bill Stewart
A man raised by impoverished Pakistanis learns that he is really the heir to an aristocratic family and sets about eliminating the present duke, who is in reality an American impostor. A madcap comedy that is only sporadically funny, and overburdened with numerous sight gags, few of which work.
COM   83 min (ort 87 min)   VIDrel: CIC   V/sur

SPOILS OF WAR **   PG
David Greene USA   1993
Kate Nelligan, John Heard, Andrea Roth, Rhea Pearlman, Matthew Walker, George Touliatos, Morris Panych, Eric Keenlyside, Robert Wisden, James Bell, David Lovgren, Alf Humphreys, Krin Konoval, Roger Cross, Cara Labine, Tony Lung
Since the split-up of his parents, a youngster has lived on the road with his mother. But when they return home to New York, he is placed in the position of being able to act as matchmaker between his divorced parents. Lightweight, but quite good fun.
COM   89 min mTV   VIDrel: MARQ/GUILD   V
Boa: play by Michael Weller.

SPONTANEOUS COMBUSTION *   18
Tobe Hooper USA   1989
Brad Dourif, Cynthia Bain, Jon Cypher, William Prince, Melinda Dillon, Dey Young, Dick Butkus, John Landis
Not a serious examination of an alleged "scientific" reality, but another Hooper extravaganza of special effects, in this tale of a man whose strange pyrotechnic powers seem to be due to the exposure to radiation his parents underwent in the 1950s. The use of flamethrower effects cannot hide the sheer absurdity of it all in this poorly conceived dud.
HOR   92 min (ort 108 min)   VIDrel: BRAVE/SONOP L/A V

SPOOKIES **   18
Eugenie Joseph/Thomas Doran/Brendan Faulkner USA   1986
Felix Ward, Dan Scott, Alec Nemser, Maria Pechukas, A.I. Lowenthal, Pat Wesley Bryan, Peter Din, Nick Giorta, Lisa Fried, Joan Ellen Delaney, Peter Lasillor Jr, Charlotte Seely, Kim Merrill, Anthony Valburg, Soo Paek
A horror spoof that tells of a mysterious count, who lives in a coffin in a graveyard and needs some human sacrifices, to revitalise his dead bride. When a bunch of bored kids venture into his cemetery for fun, he decides to seize his chance. An utterly silly and overblown effort, but done with a lot of verve.
Aka: TWISTED SOULS
HOR   85 min (ort 98 min)   VIDrel: VIPCO/SGSVID   V

SPOOKY ENCOUNTERS **   15
Lau Chun Wei HONG KONG   199-
Samo Hung, Lam Ching Ying, Meng Hoi
Hung plays a martial artist who gets involved with ghosts, wizards, kung fu and zombies in a wild, overblown but exuberant blending of magic and martial arts mayhem. The slight plot revolves around a hunt for vampires, but that need not get in the way of the comic effects and high-kicking stunts. Good fun, but a bit breathless and more than a little underplotted. A sequel to ENCOUNTERS OF THE SPOOKY KIND.
Aka: CLOSE ENCOUNTERS OF THE SPOOKY KIND 2
MAR   94 min (ort 100 min)   VIDrel: EAST/DISC   V

SPOTSWOOD **   (PG)
Mark Joffe AUSTRALIA   1991
Anthony Hopkins, Ben Mendelsohn, Alwyn Kurts, Bruno Lawrence, Angela Punch McGregor, John Walton, Rebecca Riggs, Toni Collette, Russell Crowe, Daniel Wylie, John Flaus, Gary Adams, Jeff Truman, Ton Lamnond, Jillian Murray
Title figure is brought in to save a family-run moccasin factory that is on the verge of bankruptcy but he eventually learns a

thing or two about what money cannot buy. An moderately enjoyable if hardly original tale with a very predictable plot and an irritatingly neat resolution.

Aka: EFFICIENCY EXPERT, THE

COM   90 min (ort 95 min)   TVrel

SPRING AND PORT WINE **   PG
Peter Hammond UK   1970
James Mason, Diana Coupland, Rodney Bewes, Hannah Gordon, Susan George, Len Jones, Keith Buckley, Adrienne Posta, Avril Edgar, Frank Windsor, Arthur Lowe, Marjorie Rhodes, Bernard Bresslaw
In Bolton, the wife of a dictatorial engineer pawns her husband's coat in order to send their daughter to London when she becomes pregnant. A mildly amusing generation-gap comedy that reveals all too clearly its stage origins and never the conviction it so badly needs.
COM   97 min (ort 101 min)   VIDrel: BRAVE/SONOP L/A V
Boa: play by Bill Naughton.

SPRING FEVER USA *   18
William Milling USA   1988
Darrel Gullbeau, Michelle Kemp, Jeff Greenman, Lara Belmonte, Janine Lindenmulden
Two over-sexed teenagers spend their holiday in search of women, in this thinly plotted sex-comedy. Written by Milling.
COM   91 min (ort 92 min)   VIDrel: MIA/DISC/IMPENT   V

SPRING FLING! **   (PG)
Chuck Bowman USA   1995
James Eckhouse, Joyce DeWitt, Joseph Foss, Todd Denham McCormick, Bryan Karl Moeller, Monica Creel, Justin Burnette, Rachael De Azecedo, Steve Posner, Bob Roitblat, Greg Woodhill, Jeffrey Brown
Romantic comedy in which a recently widowed man decides to sell up and move to New York, a move not viewed with favour by his kids, as it requires him to sell the family hotel. However, matters take an unexpected turn with the arrival at the hotel of a busload of teenagers.
COM   89 min mTV   SATrel: MOVIE CHANNEL

SPRINGFIELD RIFLE **   U
Andre De Toth USA   1952
Gary Cooper, Phyllis Thaxter, David Brian, Paul Kelly, Philip Carey, Lon Chaney Jr, James Millican, Martin Milner, Guinn Williams, James Brown, Jack Woody, Alan Hale, Vince Barnett, Fess Parker, Richard Lightner, George Ross
Gary Cooper stars as an officer in the Union Army who gets himself cashiered, joins the Confederates as a spy, and finds out who is behind the theft of government arms. A solid Western with excellent production values (it was filmed on location on the slopes of California's Mount Whitney) but a lack of zest. However, the memorable WarnerColor photography (by Edwin DuPar) is a considerable asset.
WES   89 min Cut (3 sec – ort 93 min)   VIDrel: MGM/WHV L/A   V

SPRINGTIME IN THE ROCKIES ***   U
Irving J. Cummings USA   1942
Betty Grable, John Payne, Carmen Miranda, Cesar Romero, Charlotte Greenwood, Edward Everett Horton, Jackie Gleason, Frank Orth, Harry Hayden, Iron Eyes Cody, Chick Chandler, Trudy Marshall, Harry James and his Music Makers
At a mountain retreat a Broadway couple repeatedly bicker and make up, while others enjoy more romantic encounters. Grable is at her prettiest in this thinly plotted but colourful and very typical wartime musical. The feeble storyline is forgotten in the face of the gorgeous Technicolor photography and catchy numbers, that include: "Chattanooga Choo Choo" and "I Had The Craziest Dream". Songs are by Mack Gordon and Harry Warren.
MUS   87 min (ort 91 min)   VIDrel: 20TH/TECH   V

SPY **   15
Philip F. Messina USA   1989
Bruce Greenwood, Catherine Hicks, Jameson Parker, Michael Tucker, Ned Beatty, Tim Choate
An intelligence yarn that faintly echoes THREE DAYS OF THE CONDOR in its premise of a "freebooter" group at work inside the CIA. When a former operative stumbles across their existence, he soon finds himself paying a high price for this

unwelcome knowledge. A reasonably entertaining espionage thriller.
THR   81 min (ort 100 min) mCab   VIDrel: CIC/SONOP L/A V
Boa: novel by Norman Garbo.

SPY HARD **   PG
Rick Friedberg USA   1996
Leslie Nielsen, Nicollette Sheridan, Charles Durning, Marcia Gay Harden, Barry Bostwick, John Ales, Andy Griffith, Elya Baskin, Mason Gamble, Carlos Lauchu, Stephanie Romanov, Curtis Armstrong, Tina Arning, Gayle Obodzinski
Another Nielsen spoof (see the NAKED GUN series) this time on the secret agent theme. Here, he plays agent WD-40 whose mission is to save the world from the ambitions of a madman, who trades under the pseudonym of General Rancor and for good measure, has kidnapped the daughter of Nielsen's former partner. Corny and terribly derivative (it spoofs PULP FICTION, HOME ALONE and SPEED) but fast-paced enough to keep one amused if not hysterical.
COM   78 min (ort 90 min) cC   VIDrel: HOLPIC/TECH V/sur

SPY IN BLACK, THE ***   U
Michael Powell UK   1938
Conrad Veidt, Valerie Hobson, Sebastian Shaw, Marius Goring, June Duprez, Athole Stewart, Agnes Lauchlan, Helen Haye, Cyril Raymond, Hay Petrie, Grant Sutherland, Robert Rendel, Mary Morris, George Summers, Margaret Moffatt
In WW1, a German agent is pursued by the British in the Orkneys, after escaping from a trap, in which the wife of a naval officer replaced the schoolmistress he was supposed to contact. An enthralling story full of twists and turns, with a little romance added in for good measure. The terrific script is by Emeric Pressburger and Roland Pertwee. The atmospheric music is the work of Miklos Rosza.
Aka: U-BOAT 29
THR   78 min (ort 82 min) B/W   VIDrel: CARL/TECH   V
Boa: novel by J. Storer Clouston.

SPY WHO CAME IN FROM THE COLD, THE ****   PG
Martin Ritt UK   1966
Richard Burton, Claire Bloom, Oskar Werner, Sama Wanamaker, Peter Van Eyck, George Voskovec, Rupert Davies, Cyril Cusack, Michael Hordern, Robert Hardy, Bernard Lee, Beatrix Lehmann, Esmond Knight, Niall McGinnis, Warren Mitchell
A British spy apparently defects to East Germany in what later proves to be an intricate game of double and triple cross. The convoluted plot may be hard to follow, but the acting and atmosphere are both first-class and contribute greatly to the making of a fine film. One of the few adaptations of Le Carre's novels that really transfers well to the screen, the ending is both sad and believable.
THR   107 min (ort 112 min) B/W
VIDrel: 4-FRONT/POLYREC/CIC   V/h
Boa: novel by John Le Carre.

SPY WHO LOVED ME, THE ***   PG
Lewis Gilbert UK   1977
Roger Moore, Barbara Bach, Richard Kiel, Curt Jurgens, Caroline Munro, Lois Maxwell, Bernard Lee, Sydney Tafler, Walter Gotell, Geoffrey Keen, Olga Bisera, Shane Rimmer, Bryan Marshall, Michael Billington, Desmond Llewellyn
Another "James Bond" film (the tenth) with East joining West, in an attempt to prevent a megalomaniac from achieving world domination, by instigating a nuclear war between the USA and the USSR. Not one of the best but still good, Moore as suave as ever and the undersea battle is a high-spot. Music is by Marvin Hamlisch with Carly Simon singing. Written by Christopher Wood and Richard Maibaum. Followed by MOONRAKER.
A/AD   120 min (ort 125 min) wScrn   VIDrel: MGM/WHV V/sh V/dm
Boa: novel by Ian Fleming.

SPYMAKER: THE SECRET LIFE OF IAN FLEMING **   15
Ferdinand Fairfax UK   1990
Jason Connery, Kristin Scott Thomas, Joss Ackland, David Warner, Patricia Hodge, Richard Johnson, Colin Welland, Fiona Fullerton, Julian Firth, Marsha Fitzalan, Arkie Whiteley, Tara McGowran, Ingrid Held, Geoffrey Chater
Packed to the brim with in-jokes for 007 fans, this is a highly fictionalised portrait of the colourful life writer Fleming enjoyed

before he came to the Bond novels. A thrown together effort, it's adequately watchable, but doesn't compare especially well to other films of this genre, nor indeed to any regular James Bond movies. See also GOLDENEYE.

Aka: SECRET LIFE OF IAN FLEMING, THE
A/AD   96 min (ort 100 min) mCab   VIDrel: FIRST/SONOP
L/A   V

S*P*Y*S *   PG
Irvin Kershner USA   1974
Donald Sutherland, Elliott Gould, Joss Ackland, Zouzou, Vladek Sheybal, Kenneth Griffith, Kenneth J. Warren, Yuri Borisenko, Michael Petrovitch, Pierre Oudry, Jacques Marin, Shane Rimmer, George Pravda, Xavier Gelin, Alf Joint
Spy spoof about a defecting Russian dancer and a list of KGB agents in China. A lame-brained and ludicrous attempt to do for espionage what M*A*S*H did for war.
COM   99 min   VIDrel: WHV   V

SQUEEZE, THE *   15
Roger Young USA   1987
Michael Keaton, Rae Dawn Chong, John Davidson, Ric Abernathy, Danny Aiello, Bobby Bass, Jophrey Brown, Leslie Bevis, Lou Criscoulo, Ray Gabriel, George Gerdes, Ronald Guttman, Paul Herman, Richard E. Huhn, John Dennis Johnston
A man is tricked by his former wife into collecting a package containing a magnet, that crooks intend using to score with on a TV game show lottery. A scatterbrained comedy with very few laughs but a good many chases, as Keaton and Chong uncover this fiendish plot.
COM   98 min Cut (6 sec – ort 101 min)
VIDrel: 20TH/TECH   V

SQUEEZE PLAY *   (18)
Samuel Weil (Lloyd Kaufman/Michael Herz) USA   1980
Jim Harris, Jenni Hetrick, Rick Gitlin, Helen Campitelli, Alfred Corley, Jim Metzler, Melissa Michaels, Michael P. Morgan, Sonya Jennings, Sharon Kyle Bramblett, Zachary, Diana Valentien, Rick Kahn, Lisa Beth Wolf, Tony Hoty
A vulgar comedy about a man-woman softball match, with humour reminiscent of those saucy English seaside postcards of Donald McGill. Jokes about human anatomy and various bodily functions abound in this smutty and unrewarding experience.
Aka: SOFT BALLS
COM   82 min (ort 92 min)   SATrel: BRAVO MOVIES

S.S. GIRLS *   18
Vincent Dawn ITALY   1976
Gabriele Carrara, Marina Davnia, Thomas Rudy, Vassili Karis, Macha Magal, Ivano Staccioli, Rachel Demian, Maria Lindstrom, Della Mattei
After the attempt on Hitler's life in 1944, a special group of prostitutes is recruited to test the loyalties of the Wehrmacht generals. Low-grade nonsense.
DRA   82 min   VIDrel: MIA/DISC   V

STACY KEEPS IT UP *   18
Peter Kay UK   1993
Stacy Owen, Charry Owen, Karen Britain, Jackie Hunter, Marta Capriano, Melony Morgan, Randy Andy Smart
Filmed at a 16th century mansion, where a bevy of large-breasted women enjoy a party, not least because the host has hired the services of a stud to liven things up.
A   65 min (ort 80 min)   VIDrel: ELV/DISC   V

STAGE DOOR **   U
Gregory La Cava USA   1937
Katharine Hepburn, Ginger Rogers, Adolphe Menjou, Andrea Leeds, Ann Miller, Lucille Ball, Gail Patrick, Eve Arden, Samuel S. Hinds, Franklin Pangborn, Jack Carson, Constance Collier, Fred Santley, William Corson, Frank Reicher
A rich girl tries to make it as an actress on her own merits, while staying at a theatrical boardinghouse. Excellent performances and a good script carry the thin storyline, though the latter half of the movie strikes a more sombre note, that seems strangely at odds with the contrived comic aspects. Written by Morrie Ryskind and Anthony Veiller.
DRA   92 min (ort 100 min) B/W   VIDrel: VCC L/A   V
Boa: play by Edna Ferber and George S. Kaufman.

STAGE FRIGHT **   18
Michele Soavi USA   1986
David Brandon, Barbara Cupisti, Robert Gilgorov, John Morghen,

Mary Sellers, Jo Anne Smith, Lori Parrel, Martin Philips, Ulrike Schwerk, Clain Parker, Piero Vida, James E.R. Sampson, Don Fiore, Richard Berkley
A gripping but quite unpleasant slasher tale, in which a performance group are rehearsing in an obsolete theatre, for a horror musical that is inspired by the exploits of a mass murderer. What they don't realise is that the demented psycho, who inspired their play, has just escaped from an asylum and is heading their way.
Aka: AQUARIUS; BLOODY BIRD; DELIRIA; STAGEFRIGHT
HOR   86 min (ort 95 min) wScrn dubbed
VIDrel: REDEM/RTM   V

STAGECOACH **   U
John Ford USA   1939
John Wayne, John Carradine, Claire Trevor, Thomas Mitchell, Andy Devine, Tim Holt, Louise Platt, George Bancroft, Donald Meek, Berton Churchill, Francis Ford, Tom Tyler, Chris-Pin Martin, Elvira Ross, Yakima Canutt, Bill Cody
Classic film about a group of travellers on a stagecoach, their fears and conflicts as they are attacked by Indians etc. Highly entertaining but don't expect realism. The stuntwork is by a famous stuntman of the time – Yakima Canutt. Written by Dudley Nichols and remade in 1966 and 1986. AA: S. Actor (Mitchell), Score (Richard Hageman/Frank Harling/John Lepold/Leo Shuken).
WES   91 min (ort 99 min) B/W
VIDrel: 4-FRONT/POLYREC   V
Boa: short story Stage to Lordsburg by Ernest Lee Haycox.

STAGGERED **   15
Martin Clunes UK   1993
Martin Clunes, Michael Praed, John Forgeham, Anna Chancellor, Sylvia Syms, Sarah Winman, Griff Rhys Jones, Michele Winstanley, Kate Byers, David Kossoff, Helena McCarthy, Sandy Poustie, Virginia McKenna, Steve Ritchie, Frank Helner
A man's stag night turns out badly when he is dumped – drugged, naked and penniless at a remote Scottish island. He has only three days to get back to London and his wedding. Moderately amusing, although this is in truth a one-joke film that would have worked better as a short.
COM   95 min (ort 97 min)   VIDrel: EIV/SONOP   V/sur

STAKEOUT **   15
John Badham USA   1987
Richard Dreyfuss, Emilio Estevez, Madeleine Stowe, Aidan Quinn, Dan Lauria, Forest Whitaker, Ian Tracey, Earl Billings, Jackson Davies, J.J. Makaro, Scott Anderson, Tony Pantages, Beatrice Boepple, Kytle Woida, Jan Speck
Two cops are sent on a routine assignment to keep watch on a beautiful woman, but one of them falls in love with her. This fairly improbable and rather shallow plot is transformed into a witty and lively comedy, by fine performances from Dreyfuss and Estevez and some memorable dialogue. The script is by Jim Kouf. Followed by ANOTHER STAKEOUT.
COM   112 min (ort 115 min)   VIDrel: TOUCH/TECH
V/sur

STALAG 17 **   PG
Billy Wilder USA   1953
William Holden, Don Taylor, Otto Preminger, Robert Strauss, Harvey Lembeck, Peter Graves, Sig Ruman, Neville Brand, Richard Erdman, Michael Moore, Peter Baldwin, Robinson Stone, Robert Shawley, William Pierson, Gil Stratton Jr
A classic wartime POW story mixing comedy and pathos, as a group of American POWs begin to suspect that one of their number (Holden) is a traitor, because their escape attempts are so easily foiled. The blending of different genres is at times uneasy but it's a fine film just the same. Written by Wilder and Edwin Blum with music by Franz Waxman. Humour is supplied by Strauss and Lembeck repeating their Broadway roles. AA: Actor (Holden).
WAR   115 min (ort 120 min) B/W   VIDrel: CIC/SONOP
V/h
Boa: play by Donald Bevan and Edmund Trzinski.

STALIN **   PG
Ivan Passer CANADA/HUNGARY/RUSSIA/UK/USA 1992
Robert Duvall, Julia Ormand, Jeroen Krabbe, Joan Plowright, Maximillian Schell, Frank Finlay, Roshan Seth, Daniel Massey, Miriam Margolyes, Joanna Roth, Jim Carter, Andras Balint
Highly watchable account (thanks to Duvall's performance) of the life and bloody career of title dictator, covering the period

from 1917 to his death in 1953. It deals with both his personal relations and his affect on the lives of his subjects, millions of whom perished because of his policies. Filmed on location in Russia, which adds greatly to its authenticity.
DRA   166 min (ort 173 min) mCab   VIDrel: WHV   V/h

STALINGRAD ***
15
Joseph Vilsmaier GERMANY
1992
Dominique Horowitz, Thomas Kretschmann, Jochen Nickel, Seabstian Rudolph, Sylvester Groth, Martin Benrath, Dana Vavova, Karl Hermanek, Mark Kuhn, Heinz Emigholz, Eckhardt A. Wuchholz, Ferdinand Schuster, Oliver Broumis, Dieter Okras
Story of the brutal WW2 battle for possession of the title city, which left the German army severely depleted in number, but at the same time was enormously more costly for the Russian people. An epic account of a major conflict, told from the point of view of the Germans, but also a film with little sympathy for them, in that the atrocities committed by their troops are shown in graphic and disturbing detail. The striking photography is by Vilsmaier.
WAR   132 min (ort 138 min) wScrn dubbed
VIDrel: EIV/SONOP   V/sur

STALKED *
18
Douglas Jackson USA
1994
Maryam D'Abo, Tod Fennell, Jay Underwood, Lisa Blount, Karen Robinson, Alex Karzis, Vivian Reis
A ruthless and psychotic man becomes fixated on a beautiful woman and starts to stalk her, leaving a trail of carnage behind him. A dark, depressing and wholly unoriginal psychological thriller.
THR   91 min (ort 95 min)   VIDrel: FIRST/SONOP   V

STALKER, THE **
18
Louis Morneau USA
Steve Railsback, Erika Anderson, Dick Miller, Eb Lottimer, Burton Gilliam
In San Francisco, a woman becomes the target of a stalker during an earthquake. A familiar treatment of a familiar theme.
THR   80 min   VIDrel: 20VIS/SONOP   V

STALKER ***
PG
Andre Tarkovsky USSR/WEST GERMANY
1979
Aleksander Kaidanovsky, Anatoly Solonitsin, Nikolai Grinko, Alisa Freindlich, Natasha Abramova, F. Yurma, E. Kostin, R. Rendi
In a vaguely menacing future, the title character guides a writer and a scientist to a dangerous, forbidden region known as "The Zone", where a room exists within which can be found "Truth". Tarkovsky's long, opaque and difficult fantasy shows many flashes of brilliance, but the use of the film as a vehicle for the examination of differing philosophical viewpoints tends to weaken its impact in terms of pure entertainment.
FAN   155 min (161 min) B/W/Col   VIDrel: CONNO/RTM V
Boa: novel Roadside Picnic by Boris and Arkady Strugatsky.

STALKER: SHADOW OF OBSESSION **
18
Kevin Connor USA
1994
Veronica Hamel, Jack Scalia, Jonathan Banks, Page Mosely, Sam Behrens
For the past six years a female professor has been obliged to move from town to town, as a demented ex-student has plagued her. Finally, she arrives in L.A. where she settles down with her boyfriend, but her unsought admirer is not far away. This is routine stuff indeed, directed and acted with competence, but unlikely to make it on anyone's hundred best movies list.
Aka: SHADOW OF OBSESSION
THR   85 min mTV   VIDrel: MED/20VIS   V/sh

STALKING BACK **
15
Corey Allen USA
1993
Tom Kurlander, LuAnne Ponce, Shanna Reed, Cassie Yates, Paul Rudd, Mike Genovese, Gregory Itzin, Rif Hutton, Prisilla Lopez, Jeremy Russell, Diana Lamb, Rachel Tuve, Katherine Willis, Heather McNair, Samuel Hernandez, Ty Bass
True story of a thirteen-year-old girl being stalked by obsessed admirer. When the police reveal that they have no power to do anything about this, the girl's mother decides to take the law into her own hands. Fair.
DRA   88 min mTV   VIDrel: NWV/HIFLI   V/h

STALKING LAURA *
(18)
Michael Switzer USA
1993
Brooke Shields, Richard Thomas, Viveka Davis, William Allen Young, Richard Yniguez, Scott Bryce, Linda Edmond, T. Max Graham, Kevin Brief, Tim Snay, Dick Mueller, Mark McCarthy, Merle Moores, Caroline Vincguerra, Dean Vivian
A nasty variant on FATAL ATTRACTION that stars Shields as a graduate who has just started her first job with a top firm, only to find that the company's computer expert has developed an obsessive interest in her. Having turned him down when he asked for a date, she soon finds his attentions growing ever more extreme, and eventually he reveals himself to be a psychopath. A totally predictable film, unbelievable, superficial and wholly derivative.
THR   87 min (ort 95 min)   SATrel: MOVIE CHANNEL

STAND, THE **
15
Mick Garris USA
1993
Gary Sinise, Molly Ringwald, Jamey Sheridan, Laura San Giacomo, Ruby Dee, Ossie Davis, Miguel Ferrer, Corin Nemec, Matt Frewer, Adam Storke, Ray Walston, Rob Lowe, Bill Fagerbakke, Peter Van Norden, Bridgit Ryan, Kellie Overbey
After a man-made virus wipes out virtually the entire world population, the remaining survivors find themselves being assailed by frightening dreams and visions and eventually divide into camps. Soon the final apocalyptic struggle between good and evil begins in preparation for the end of the world. A chilling and imaginative mini-series adapted by King from his novel. First shown on TV in four parts.
Aka: STEPHEN KING'S THE STAND
FAN   345 min (2 cassettes – ort 360 min) mTV
VIDrel: WHV   V/sur
Boa: novel by Stephen King.

STAND AND DELIVER ***
PG
Ramon Menendez USA
1988
Edward James Olmos, Lou Diamond Phillips, Rosana De Soto, Andy Garcia, Will Gotay, Ingrid Oliu, Virginia Paris, Mark Eliot, Vanessa Marquez, Karla Montana, Patrick Baca, Lydia Nicole, Daniel Villareal, Carmen Argenziano
Based on a real-life story, this tells of a determined and dedicated teacher who instils in his unruly students a sense of pride and self-worth, when his teaching methods enable them to pass a difficult examination. Olmos gives a remarkable performance in this warmhearted and enthralling tale. The script is by Menendez and Tom Musca. See also DANGEROUS MINDS.
DRA   99 min (ort 105 min)   VIDrel: MGM/WHV L/A   V/h

STAND BY ME ***
15
Rob Reiner USA
1986
River Phoenix, Corey Feldman, Wil Wheaton, Jerry O'Connell, John Cusack, Richard Dreyfuss, Kiefer Sutherland, Casey Siemaszko, Gary Riley, Marshall Bell, Bradley Bregg, Jason Oliver, Frances Lee, Bruce Kirby, William Bronder
A trip back to a 1950s American boyhood, that follows the friendships and adventures of four young misfits, who take a trip into the woods to have a look at a dead body. An overly sentimental and contrived tale, based on King's autobiographical story, but sustained by some appealing performances. The constant stream of profanities seems somewhat out of place in the 1950s.
DRA   85 min (ort 87 min)   VIDrel: VCC/DISC/COLUM   V
Boa: novella The Body by Stephen King.

STAND-IN **
U
Tay Garnett USA
1937
Leslie Howard, Humphrey Bogart, Joan Blondell, Alan Mowbray, Marla Shelton
Satire on the movie industry that has Howard as an efficiency expert trying to balance the books and save a studio from bankruptcy. He decides that in order to save the studio he will have to learn all about its operations from the bottom up, and gets stand-in Blondell to show him the ropes. An amusing little trifle, that could have been more tightly scripted. Bogart's presence as the cynical studio boss is most welcome.
COM   87 min B/W   VIDrel: 4-FRONT/POLYREC   V

STAND UP VIRGIN SOLDIERS *
15
Norman Cohen UK
1977
Robin Askwith, Nigel Davenport, George Layton, Robin Nedwell, Warren Mitchell, Edward Woodward, John Le Mesurier, Irene Handl, Fiesta Mei Lung, Pamela Stephenson, Lynda Bellingham, David Auker, Robert Booth

Sequel to THE VIRGIN SOLDIERS, depicting more sexual adventures in 1950 for British recruits in Singapore. A vulgar farce lacking any of the authenticity of the 1969 film. Written by Leslie Thomas.
COM   90 min   VIDrel: MGM/WHV L/A   V
Boa: novel by Leslie Thomas.

STANLEY & IRIS *** 15
Martin Ritt USA 1989
Robert De Niro, Jane Fonda, Swoosie Kurtz, Feodor Chaliapin, Martha Plimpton, Harley Cross, Jamey Sheridan, Zohra Lampert, Loretta Devine, Julie Garfield
An unhappy and recently widowed woman discovers that a colleague at work is unable to read and sets about teaching him, only to find that their relationship begins to take on deeper and more intimate overtones. The story may be flat and manipulative, but a couple of outstanding performances from the leads convert what might have been mediocre into a warm and sensitive human drama. Music is by John Williams.
DRA   100 min (ort 107 min)   VIDrel: MGM/WHV   V/sur
Boa: novel Union Street by Pat Barker.

STANLEY AND THE WOMEN ** PG
David Tucker UK 1991
John Thaw, Geraldine James, Sheila Gish, Penny Downie, Michael Elphick, Sian Thomas, Michael Aldridge, David Lyon, Alun Armstrong, Samuel West, Donald Churchill, Doreen Mantle, Dafydd Hywel, Joanna Mays, Ronan Vibert
Originally shown in four parts, this intensely disagreeable drama (it was advertised as a black-comedy) details the ever-increasing despair of a man whose life is thrown into turmoil when his son becomes mentally ill. This event drags him into an unwilling conflict with several unpleasant women, most especially the psychiatrist in charge of his son's treatment, a twisted creature best described as an arrogant monster. An unashamedly misogynist offering.
DRA   205 min (2 cassettes) mTV   VIDrel: CENTV/VCC L/A   V
Boa: novel by Kingsley Amis.

STANLEY'S MAGIC GARDEN ** U
USA 1994
Voices of: Dom De Luise, Cloris Leachman, Charles Nelson REilly, Hayley Mills
A magic troll teaches two kids that anything is possible if one is determined in this pleasant if rather anodyne children's outing.
ANIM   72 min   VIDrel: WHV   V/sh

STAR! *** (U)
Robert Wise USA 1968
Julie Andrews, Richard Crenna, Michael Craig, Daniel Massey, Robert Reed, John Collin, Bruce Forsyth, Beryl Reed, Jenny Agutter, Alan Oppenheimer, Richard Karlan, Lynley Laurence, Garrett Lewis, Elizabeth St Clair
The story of Gertrude Lawrence and her rise from poverty to international stardom. This incredibly overlong saga boasts some huge musical numbers that are worth seeing for their sheer novelty. Apart from that the film has very little to offer. It was later cut to 120 minutes and retitled after it flopped at the box office. The script is by William Fairchild. The title song (which received an AAN) is by James Van Heusen and Sammy Cahn.
Aka: THOSE WERE THE HAPPY DAYS; THOSE WERE THE HAPPY TIMES
MUS   175 min (ort 194 min)   VIDrel: L/A   V

STAR * 15
Michael Miller USA 1993
Jennie Garth, Gary Bierko, Terry Farrell, Penny Fuller, Mitchell Ryan, Jim Haynie, Roxanne Reese, Albert Hall, Bryan Smith, Patrick Massett, John McCann, Melendy Britt, Jane Daly, Ted Wass, Bibi Osterwald, James Gleason, Mark LaMurs
After a young girl's father dies matters degenerate to the point that the girl finds herself unable to continue living with her abusive mother, so she starts a new life as a night club singer, and meets up with an old sweetheart who has since married. Years pass, her career is blessed with success, and we are all on the edge of our seats wondering if the lovers will ever be reunited. Glossy, trite and tedious.
Aka: DANIELLE STEEL'S STAR
DRA   93 min   VIDrel: MIA/DISC/IMPENT   V
Boa: novel by Danielle Steel.

STAR CHAMBER, THE ** 15
Peter Hyams USA 1983
Michael Douglas, Hal Holbrook, Yaphet Kotto, Sharon Gless, James B. Sikking, Joe Regalbuto, Don Calfa, John Di Santi, DeWayne Jessie, Jack Kehoe, Larry Hankin, Dick Anthony Williams, Margie Impert, Dana Gladstone, Fred McCarren
A young judge is so annoyed at having to release criminals because of legal niceties, that he joins a secret society that dispenses its own justice. A simplistic and implausible affair, written by Hyams and Roderick Taylor. (The title comes from the chamber, complete with star motifs on the ceiling, used by Henry VIII to conduct business with his ministers, which later became a court with summary and arbitrary powers.) See also EXTREME JUSTICE.
DRA   104 min (ort 109 min)   VIDrel: 20TH/TECH   V/sur

STAR 80 *** 18
Bob Fosse USA 1983
Mariel Hemingway, Eric Roberts, Cliff Robertson, Carroll Baker, Josh Mostel, Roger Rees, David Clennon, Sidney Miller, Jordan Christopher, Keenan Ivory Wyans, Stuart Damon, Ernest Thompson, Robert Fields, Tina Wilson
Biopic about the tragic life and death of Playboy magazine centrefold Dorothy Stratton, who was murdered by her jealous small-time hustler husband, Paul Snider. A well-handled but sordid tale that draws fine performances from all concerned, but provides no perceptive insights or interesting observations. This was Fosse's last film. The TV film "Death Of A Centrefold" covered similar ground.
DRA   99 min (ort 104 min)   VIDrel: MGM/WHV L/A   V

STAR IS BORN, A **** U
William Wellman USA 1937
Janet Gaynor, Fredric March, Adolphe Menjou, May Robson, Andy Devine, Lionel Stander, Owen Moore, Franklin Pangbourn, Edgar Kennedy, William Wellman, Clara Blandick, Elizabeth Jenns, A.W. Seweatt, Peggy Wood, Arthur Hoyt
The story of an actress on the way up, who marries a leading man whose career is on the way down. A classic film with the two stars in peak form. The sharply observant script is by Dorothy Parker, Alan Campbell and Robert Carson and is partly based on the 1932 film "What Price Hollywood". Remade in 1954 and 1976. AA: Story/orig (Robert Carson/William A. Wellman), Spec Award (W. Howard Greene for colour cinematography).
DRA   110 min   VIDrel: SECOND/RTM   V
Boa: story by William A. Wellman.

STAR IS BORN, A *** U
George Cukor USA 1954
Judy Garland, James Mason, Charles Bickford, Jack Carson, Tommy Noonan, Amanda Blake, Lucy Marlow, Irving Bacon, Hazel Shermet, James Brown, Lotus Robb, Joan Shawlee, Dub Taylor, Louis Jean Heydt, Leonard Penn, Olin Howland
First remake of the 1937 film done as a musical, with Garland and Mason in fine form as the ill-fated Hollywood lovers. A lack of depth in the story is noticeable in the second half and the film suffered from excessive editing, that took place after its premiere. The songs are by Harold Arlen and Ira Gershwin, and with the exception of "The Man That Got Away" are not all that good. A restored version put together in 1983 is now the one generally seen.
MUS   169 min (ort 181 min)   VIDrel: WHV   V/h

STAR IS BORN, A ** 15
Frank Pierson USA 1976
Barbra Streisand, Kris Kristofferson, Paul Mazursky, Gary Busey, Oliver Clark, Marta Heflin, M.G. Kelly, Sally Kirkland, Vanetta Fields, Clydie King, Joanne Linville, Uncle Rudy, Tony Orlando, Robert Englund
A remake of the 1937 original, telling of a Hollywood love affair that ends in tragedy. This one is updated to the 1970s and is well supplied with screaming crowds and some high decibel numbers, plus an overblown and insensitive performance from Streisand. However, the movie does come to life when she sings. A flashy but irritating and rather empty work. AA: Song ("Evergreen" – Streisand (m)/Paul Williams (l)).
DRA   134 min (ort 140 min)   VIDrel: WHV L/A   V

STAR '90 * 18
Tina Marie USA 1990
Christy Canyon, Chantel, Suzie Bartlett, Paulina Downs, Frank James, Joey Silvera, Marc Wallice, Don Fernando, Tony Montana

Very little plot is to be had in this film, most of which has Canyon talking incessantly to the camera about how much she enjoys sex. There are one or two interludes with other characters, but these do nothing to enliven a meagre and dull offering.
A 52 min Cut (7 min 3 sec) VIDrel: MIA/DISC/IMPENT V

STAR PACKER, THE ***
Robert North Bradbury USA 1934
John Wayne, Verna Hillie, George "Gabby" Hayes, Yakima Canutt, Earl Dwire, Ed (Eddie) Palmer, George Cleveland, Tom Lingham, Artie Ortego, Lloyd Whitlock, David Aldrich, Tex Palmer, Billy Franey, Virginia Brown Faire
A young cowboy discovers the identity of the outlaw who murdered his parents and kidnapped his baby brother years ago. He becomes a sheriff but poses as an outlaw in order to infiltrate the man's gang and have his revenge. A well-made second feature, with Whitlock as the outlaw especially memorable.
Aka: WEST OF THE DIVIDE
WES 52 min (ort 58 min) B/W coVer VIDrel: ENTUK L/A V

STAR STRUCK **
Jim Drake USA 1994
Kirk Cameron, Chelsea Noble, J.T. Walsh, Ned Eisenberg, Desiree Marie, Anne Haney, Jim Fyfe, D.W. Moffett, HadleyEvre,Keith Flippen, Marion Guyot, Robert Raiford, Eric paisley, Catherine Schaffner, Bob Tyson, Jacquelyn Rasool
When a man meets a famous movie star he realises she is his sweetheart from years ago, when they met at a summer camp. However, persuading her of this proves to be far from easy. A bland romantic comedy.
COM 89 min SATrel: DISNEY CHANNEL

STAR TREK: THE MOTION PICTURE ***
Robert Wise USA 1979
Willian Shatner, Leonard Nimoy, DeForest Kelley, Stephen Collins, Persis Khambatta, James Doohan, George Takei, Walter Koenig, Nichelle Nichols, Mark Lenard, Majel Barrett, Grace Lee Whitney, David Gautreaux, Marcy Lafferty
Earth is threatened by a massive energy field and the Starship Enterprise is sent to investigate, in this spin-off of the popular TV series. Wooden acting and stilted dialogue let down an excellent plot enlivened by some superb special effects. Screenwriters were Alan Dean Foster and Harold Livingstone with music by Jerry Goldsmith. Followed by several sequels.
FAN 126 min; 138 min (extended version – ort 132 min) wScrn VIDrel: CIC/SONOP; PION (LV only) V/h LV

STAR TREK 2: THE WRATH OF KHAN **
Nicholas Meyer USA 1982
William Shatner, Leonard Nimoy, DeForest Kelley, Ricardo Montalban, Walter Koenig, James Doohan, George Takei, Paul Winfield, Nichelle Nichols, Bibi Besch, Merritt Butrick, Kirstie Alley, Ike Eisenmann, Judson Scott
A further adventure of the crew of the Enterprise but this time the plot is an amplified version of a TV episode from 1967 ("Space Seed"), in which our heroes are menaced by a selectively bred superman. A laughable and wholly inferior sequel. Followed by STAR TREK 3.
Aka: WRATH OF KHAN, THE
FAN 108 min (ort 114 min) wScrn VIDrel: CIC/SONOP; PION (LV only) V LV

STAR TREK 3: THE SEARCH FOR SPOCK **
Leonard Nimoy USA 1984
William Shatner, DeForest Kelley, James Doohan, George Takei, Walter Koenig, Christopher Lloyd, Nichelle Nichols, Jane Wyatt, Mark Lenard, Robin Curtis, James B. Sikking, Catherine Hicks, Leonard Nimoy, Robert Ellenstein
Third "Star Trek" adventure continuing from the previous one. Captain Kirk goes on a mission to find the body of his dead lieutenant but has to overcome various problems, not least being a hostile Klingon warship and the rapid self-destruction of a barren world, artificially seeded with life. Seriously overlong and painfully contrived. STAR TREK 4: THE VOYAGE HOME followed.
Aka: SEARCH FOR SPOCK, THE
FAN 101 min (ort 105 min) wScrn VIDrel: CIC/SONOP; PION (LV only) V/sh LV

STAR TREK 4: THE VOYAGE HOME ***
Leonard Nimoy USA 1986
William Shatner, Leonard Nimoy, DeForest Kelley, James Doohan, Jane Wyatt, Walter Koenig, George Takei, Nichelle Nichols, Catherine Hicks, Mark Lenard, Robin Curtis, Robert Ellenstein, John Schuck, Brock Peters, Michael Snyder
Fourth outing for the crew of the Enterprise, with a comedy flavour and a contemporary setting. The crew travel back in time on a mission that involves them with a marine biologist and the saving of one of Earth's most precious life forms. A lightweight but fairly entertaining tale, followed by STAR TREK 5.
Aka: VOYAGE HOME, THE
FAN 117 min (ort 119 min) wScrn VIDrel: CIC/SONOP; PION (LV only) V/sh LV

STAR TREK 5: THE FINAL FRONTIER **
William Shatner USA 1989
William Shatner, Leonard Nimoy, DeForest Kelley, James Doohan, Walter Koenig, Nichelle Nichols, George Takei, David Warner, Laurence Luckinbill, Charles Cooper, Cynthia Gouw, Todd Bryant, Spice Williams, Rex Holman
The crew of the Enterprise are recalled from a camping trip, and head for a distant planet that has been taken over by a renegade Vulcan, in search of spiritual enlightenment. Having captured their ship, he forces them to set off in search of the ultimate goal – God. Shatner's feature directing debut (which he also co-wrote) starts off slowly, gets slightly better, but ultimately disappoints. Music is by Jerry Goldsmith.
Aka: FINAL FRONTIER, THE
FAN 102 min (ort 106 min) wScrn VIDrel: CIC/SONOP; PION (LV only) V/sh LV

STAR TREK 6: THE UNDISCOVERED COUNTRY ***
Nicholas Meyer USA 1991
William Shatner, Leonard Nimoy, DeForest Kelley, James Doohan, Nichelle Nichols, Walter Koenig, George Takei, Christian Slater, Kim Cattrall, Grace Lee Whitney, Mark Lenard, Brock Peters, Leon Russom, Kurtwood Smith, David Warner
This latest (and probably last) STAR TREK sequel has the same cast as the earlier films, and a definite Shakespearean flavour. It commences with the opening of peace overtures between the Federation and their enemies the Klingons, and an unwilling Kirk is given the job of transporting a Klingon ambassador to the negotiating table. Matters degenerate when the Enterprise appears to fire on a Klingon vessel. A most enjoyable outing.
FAN 108 min (ort 110 min) wScrn VIDrel: CIC/SONOP; PION (LV only) V/sh LV

STAR TREK FIRST CONTACT **
Jonathan Frakes USA 1996
Patrick Stewart, Jonathan Frakes, Brent Spiner, LeVar Burton, Michael Dorn, Alice Krige, Gates McFadden, Marina Sirtis, Alfre Woodard, James Cromwell, Neal McDonough, Michael Horton, Marnie McPhail, Robert Picardo, Dwight Schultz
Picard's worst nightmare (the film's superb opening shot is part of one) becomes a reality when the Borg attack Earth. Having defied Starfleet Command to join the fray, his ship is caught up in the wake of a Borg vessel, which is racing back in time, as the Borg intend to alter the Earth's future by changing a pivotal event in mankind's past. Several powerful ideas and a rather repulsive sequence are wasted on a film of weak direction and clumsy, cluttered scripting.
FAN 111 min VIDrel: PION LV

STAR TREK: GENERATIONS ***
David Carson USA 1994
Patrick Stewart, Jonathan Frakes, Brent Spiner, LeVar Burton, Micahel Dorn, Gates McFadden, Marina Sirtis, Malcolm McDowell, James Doohan, Walter Koenig, William Shatner, Whoopi Goldberg, Alan Ruck, Jacqueline Kim, Jenette Goldstein
An astronomical phenomena bridges two times, bringing two Enterprise captains face to face in the 24th century, where they must join forces to destroy an evil scientist who has gained the power to wipe out civilisation. This was the first feature-film spin-off from the STAR TREK THE NEXT GENERATION series and though disjointed and uneven, has a lot to offer by way of special effects and guest appearances.
FAN 113 min (ort 118 min) wScrn cC VIDrel: CIC/SONOP; PION (LV only) V/sur LV

STAR TREK: ALL OUR YESTERDAYS *** PG
Herb Wallerstein USA   1969
*William Shatner, Leonard Nimoy, DeForest Kelley, James Doohan,
George Takei, Nichelle Nichols, Walter Koenig, Mariette Hartley, Ian
Wolfe, Kermit Murdock, Johnny Haymer, Stan Barrett, Ed Bakey, Al
Cavens, Anna Karen*
A landing party beams down to the planet Sarpeidon to rescue
the inhabitants before the sun explodes. Not only do they find
a vast library, but also a time portal, which Kirk, Spock and
McCoy inadvertently step through. The former is transported to
a period of Salem-like witch trials, whilst Spock and McCoy are
carried back even further, to an ancient and far more savage
time, where Spock learns of emotions and love.
FAN   147 min (3-episode cassette) mTV
VIDrel: CIC/SONOP   V

STAR TREK: THE ALTERNATIVE FACTOR ** PG
Gerd Oswald USA   1967
*William Shatner, Leonard Nimoy, DeForest Kelley, James Doohan,
George Takei, Nichelle Nichols, Robert Brown, Janet MacLachlen,
Richard Derr, Eddie Paskey*
Two versions of a man known as Lazarus are encountered by
the Enterprise; one a rational and humane individual and the
other a violent madman from an anti-matter universe. The latter
plans the destruction of both universes by opening a channel
between them, but the rational Lazarus sacrifices himself in
order to stop him. One of the most awkwardly plotted episodes
in the series, this one was as unusual as it was over complex.
FAN   144 min (3-episode cassette) mTV
VIDrel: CIC/SONOP   V/h

STAR TREK: AMOK TIME ** PG
Joseph Pevney USA   1967
*William Shatner, Leonard Nimoy, DeForest Kelley, James Doohan,
George Takei, Nichelle Nichols, Arlene Martel, Celia Lovsky, Lawrence
Montaigne, Byron Morrow*
The Enterprise takes Spock back to the planet Vulcan, in order
for him to participate in a strange marriage ritual that occurs
every seven years. On the surface of the planet a strange
madness takes hold of Spock and Kirk finds himself fighting
him in a battle to the death.
FAN   98 min (2-episode cassette) mTV
VIDrel: CIC/SONOP   V

STAR TREK: AND THE CHILDREN SHALL LEAD ** PG
Marvin J. Chomsky USA   1968
*William Shatner, Leonard Nimoy, DeForest Kelley, James Doohan,
George Takei, Nichelle Nichols, Walter Koenig, Melvin Belli, James
Wellman, Craig Hundley, Pamela Ferdin, Mark Robert Brown, Brian
Tochi, Caesar Belli*
The crew of the Enterprise is brought to the planet Triacus by a
distress call, where they set about rescuing a group of orphaned
children. However, it soon transpires that the children are being
manipulated by a rather sinister being.
FAN   98 min (2-episode cassette) mTV
VIDrel: CIC/SONOP   V

STAR TREK: THE APPLE ** U
Joseph Pevney USA   1967
*William Shatner, Leonard Nimoy, DeForest Kelley, James Doohan,
George Takei, Nichelle Nichols, Keith Andes, Celeste Yarnall, Jay
Jones, Shari Nims, David Soul*
The Enterprise arrives at a planet whose inhabitants are content
to have their lives run by an all-powerful computer which they
worship as a god. Much to the chagrin of the computer ruler,
Kirk and co. set out to show the inhabitants of the planet that
they are better off living in freedom.
FAN   98 min (2-episode cassette) mTV
VIDrel: CIC/SONOP   V

STAR TREK: ARENA ** PG
Joseph Pevney USA   1967
*William Shatner, Leonard Nimoy, DeForest Kelley, James Doohan,
George Takei, Nichelle Nichols, Carole Shelyne, Jerry Ayers, Tom
Troupe, Grant Woods, Sean Kenney, James Farley*
A Federation starbase is destroyed and the Enterprise sets off
in pursuit of its attackers, a lizard-like race known as the Gorn.
However, an alien race intervenes and both Kirk and the
commander of the Gorn vessel are conveyed to a planet where
they are to fight each other in a battle to the death.
FAN   144 min (3-episode cassette) mTV
VIDrel: CIC/SONOP   V/h

STAR TREK: ASSIGNMENT – EARTH *** U
Marc Daniels USA   1968
*William Shatner, Leonard Nimoy, DeForest Kelley, James Doohan,
George Takei, Walter Koenig, Nichelle Nichols, Robert Lansing, Terri
Garr, Jim Keefer, Morgan Jones, Lincoln Demyan*
The Enterprise is sent to investigate the old world, which
according to the records, narrowly escaped destruction in 1968.
There, they encounter Gary Seven, a man whose mission is to
prevent mankind from destroying itself. An interesting entry
that was to have been a pilot for a quite separate series.
FAN   98 min (2-episode cassette) mTV
VIDrel: CIC/SONOP   V/h

STAR TREK: BALANCE OF TERROR *** U
Vincent McEveety USA   1966
*William Shatner, Leonard Nimoy, DeForest Kelley, James Doohan,
George Takei, Nichelle Nichols, Mark Lenard, Paul Comi, Lawrence
Montaigne, John Warburton, Stephen Mines, Barbara Baldwin, Gary
Walberg*
The Starship Enterprise comes up against an unseen alien ship
when their enemy the Romulans devise a means of making their
vessel invisible. A strong episode that was the first one to intro-
duce an enemy race, this tale took the form of an interstellar
cat-and-mouse game.
FAN   144 min (3-episode cassette) mTV
VIDrel: CIC/SONOP   V/h

STAR TREK: BREAD AND CIRCUSES ** PG
Ralph Senensky USA   1967
*William Shatner, Leonard Nimoy, DeForest Kelley, James Doohan,
George Takei, Nichelle Nichols, Walter Koenig, William Smithers,
Logan Ramsey, Ian Wolfe, Rhodes Reason, Lois Jewell, Bart LaRue,
Jack Perkins*
The crew of the Enterprise encounter a society similar to that of
ancient Rome, except insofar as advanced technology is known
and made use of. Kirk is obliged to engage in combat to save the
lives of his crew. By way of a sub-plot, a pacifist movement is
discovered whose followers appear to worship the sun.
FAN   98 min (2-episode cassette) mTV
VIDrel: CIC/SONOP   V/h

STAR TREK: BY ANY OTHER NAME ** U
Marc Daniels USA   1968
*William Shatner, Leonard Nimoy, DeForest Kelley, James Doohan,
George Takei, Nichelle Nichols, Walter Koenig, Warren Stevens,
Barbara Bouchet, Stewart Moss, Robert Fortier, Carol Byrd, Leslie
Dalton, Julie Cobb*
Members of the sinister Kelvan race take human form and gain
control of the Enterprise, turning most of the starship crew into
tetrahedral blocks. Their intention is to report to their rulers that
the worlds of the Federation are ripe for colonisation, but fortunately
Kirk and his crewmates devise a clever plan that involves using the
human emotions newly acquired by their Kelvan adversaries.
FAN   98 min (2-episode cassette) mTV
VIDrel: CIC/SONOP   V

STAR TREK: THE CAGE *** U
Robert Butler USA   1964
*Leonard Nimoy, John Hoyt, Peter Duryea, Jeffrey Hunter, Susan
Oliver, Majel Barrett, Meg Wylie*
The original pilot for the long-running TV series, in which the
USS Starship Enterprise "boldly goes where no man has gone
before". Here Captain Kirk's predecessor runs into a good deal
of trouble on a strange planet. It was never televised in its orig-
inal form, but was later edited down to provide the flashback
sequences for the TV series' only two-part episode – STAR
TREK: THE MENAGERIE from 1966.
FAN   73 min; 160 min (3-episode cassette) B/W mTV
VIDrel: CIC/SONOP; PION (LV only)   V/dm LV

STAR TREK: CATSPAW ** U
Marc Daniels USA   1967
*William Shatner, Leonard Nimoy, DeForest Kelley, James Doohan,
George Takei, Nichelle Nichols, Walter Koenig, Antoinette Bowers,
Theo Marcus, Michael Barrier, Jimmy Jones*
This tale concerns a landing on a sinister planet, where Kirk and
his officers find themselves involuntary guests of a group of
aliens hellbent on conquering the universe. Written by Robert
Bloch and D.C. Fontana, this strange story was notable both for
its Halloween-like flavour and as the episode that introduced
actor Walter Koenig as Chekov.
FAN   144 min (3-episode cassette) mTV
VIDrel: CIC/SONOP   V/h

STAR TREK: THE CHANGELING *   U
Marc Daniels USA   1967
*William Shatner, Leonard Nimoy, DeForest Kelley, James Doohan,
George Takei, Nichelle Nichols, Blaisdell Makee, Arnold Lessing, Vic
Perrin (voice only)*
When more than four billion people are annihilated in a star
system, the perpetrator is discovered to be an ancient Earth
probe that was lost centuries ago but was then found and
"modified" by an alien civilisation. Having come aboard the
Enterprise, the probe mistakenly identifies Kirk as its creator
and the captain attempts to trick it into destroying itself. A
memorable episode that's similar in plot to STAR TREK: THE
MOTION PICTURE.
FAN   98 min (2-episode cassette) mTV
VIDrel: CIC/SONOP   V/h

STAR TREK: CHARLIE X *   U
Lawrence Dobkin USA   1966
*William Shatner, Leonard Nimoy, DeForest Kelley, James Doohan,
George Takei, Nichelle Nichols, Robert Walker Jr, Abraham Sofaer,
Patricia McNulty, Charles J. Stewart, Dallas Mitchell*
After an adolescent is transferred from the Federation starship
Antares to the Enterprise the former ship explodes. The culprit
is found to be none other than the youngster, in whom imma-
turity and incredible telekinetic powers combine to form a
dangerous combination, all the more so when he falls in love
with a female crew member.
FAN   144 min (3-episode cassette) mTV
VIDrel: CIC/SONOP   V/h

**STAR TREK: CITY ON THE EDGE OF
FOREVER ****   U
Joseph Pevney USA   1967
*William Shatner, Leonard Nimoy, DeForest Kelley, James Doohan,
George Takei, Nichelle Nichols, Joan Collins, John Harmon, Bartell
LaRue (voice only)*
The Enterprise is investigating time disturbances in a particu-
lar region of space when a series of mishaps bring McCoy and
then Spock and Kirk to the surface of a planet that provides
the base for an ancient time portal. When McCoy passes
through it he changes the past and the Enterprise vanishes.
Kirk and Spock are obliged to follow him and attempt to
reverse this alteration in time. An excellent and cleverly plotted
entry in this series.
FAN   98 min (2-episode cassette) mTV
VIDrel: CIC/SONOP   V

STAR TREK: THE CLOUD MINDERS *   PG
Judd Taylor USA   1969
*William Shatner, Leonard Nimoy, DeForest Kelley, James Doohan,
George Takei, Nichelle Nichols, Walter Koenig*
The Enterprise arrives at the planet Ardana, where it has been
sent to obtain zienite – a sovereign remedy for the plague now
raging on Merak II. Unfortunately, Kirk finds a race on the brink
of civil war, with those who mine the zienite treated as underlings by an elite who live a life of ease and comfort in a city in
the clouds. Our doughty captain sets about resolving this little
difficulty.
FAN   98 min (2-episode cassette) mTV
VIDrel: CIC/SONOP   V

STAR TREK: THE CONSCIENCE OF THE KING *   PG
Gerd Oswald USA   1966
*William Shatner, Leonard Nimoy, DeForest Kelley, James Doohan,
George Takei, Nichelle Nichols, Arnold Moss, Barbara Anderson,
Bruce Hyde, Eddie Paskey*
Suspicions are aroused when a troupe of actors comes aboard
the starship, for their leader resembles a former governor of the
planet Tarsus, a notorious criminal who ruthlessly planned the
deaths of many of his people when starvation appeared to
threaten the planet. Soon a series of deaths begin aboard the
Enterprise, and Kirk sets out to trap the culprit.
FAN   144 min (3-episode cassette) mTV
VIDrel: CIC/SONOP   V/h

STAR TREK: THE CORBOMITE MANEUVER *   U
Joseph Sargent USA   1966
*William Shatner, Leonard Nimoy, DeForest Kelley, James Doohan,
George Takei, Nichelle Nichols, Anthony Hall, Clint Howard*
A hostile alien probe is destroyed by the Enterprise whilst the
starship is exploring an unknown region of space, but the
result of this action is to bring the ship into conflict with

another vessel. Kirk has to use all his ingenuity to overcome
his enemy.
FAN   160 min (3-episode cassette) mTV
VIDrel: CIC/SONOP   V/dm

STAR TREK: COURT-MARTIAL *   PG
Marc Daniels USA   1967
*William Shatner, Leonard Nimoy, DeForest Kelley, James Doohan,
George Takei, Nichelle Nichols, Percy Rodriguez, Joan Marshall, Elisha
Cook Jr, Richard Webb, Alice Rawlings, Hagan Beggs, Winston DeLugo*
A computer malfunction results in the death of a fellow officer
and Kirk is charged with negligence and put on trial. With the
help of Kirk's lawyer, Spock sets out to discover the truth.
FAN   144 min (3-episode cassette) mTV
VIDrel: CIC/SONOP   V/h

STAR TREK: DAGGER OF THE MIND *   PG
Vincent McEveety USA   1966
*William Shatner, Leonard Nimoy, DeForest Kelley, James Doohan,
George Takei, Nichelle Nichols, James Gregory, Morgan Woodward,
Marianna Hill, Suzanne Wilson*
The crew of the Enterprise is placed in grave danger, when the
ship arrives at the planet Tantalus, home to a penal colony and
a ruthless doctor, who has devised a method of imposing total
obedience on his patients. An unusual entry in the series, that
makes good use of a very sinister concept.
FAN   144 min (3-episode cassette) mTV
VIDrel: CIC/SONOP   V/h

STAR TREK: DAY OF THE DOVE *   PG
Marvin J. Chomsky USA   1968
*William Shatner, Leonard Nimoy, DeForest Kelley, James Doohan,
George Takei, Nichelle Nichols, Walter Koenig, Michael Ansara,
Susan Howard*
A strange presence invades the Enterprise, and nearly causes a
catastrophe with its ability to "feed" on hostility, for it arms
both the Klingons and Kirk's crew and sets them against each
other. Kirk and his opponent Kang find themselves co-operat-
ing in order to avoid mutual slaughter.
FAN   98 min (2-episode cassette) mTV
VIDrel: CIC/SONOP   V

STAR TREK: THE DEADLY YEARS *   U
Joseph Pevney USA   1967
*William Shatner, Leonard Nimoy, DeForest Kelley, James Doohan,
George Takei, Nichelle Nichols, Walter Koenig, Charles Drake, Sarah
Marshall, Beverly Washburn, Felix Locker, Carolyn Nelson, Laura
Woods*
This tales deals with a bizarre infection caught on a visit to a new
planet, that causes Kirk and his three chief officers to age
rapidly. As McCoy battles to find a cure before it's too late, the
Enterprise proceeds to a starbase situated in the Neutral Zone,
thus provoking the Romulans into an attack.
FAN   96 min (2-episode cassette) mTV
VIDrel: CIC/SONOP   V

STAR TREK: THE DEVIL IN THE DARK *   PG
Joseph Pevney USA   1967
*William Shatner, Leonard Nimoy, DeForest Kelley, James Doohan,
George Takei, Nichelle Nichols, Ken Lynch, Barry Russo, Brad
Weston, Biff Elliot, Janos Prohaska*
On Janus VI a mining operation is under threat from attacks by
an alien creature that is able to penetrate solid rock. Kirk leads
a team to the planet, where they hunt for the creature through-
out the underground caverns.
FAN   144 min (3-episode cassette) mTV
VIDrel: CIC/SONOP   V

STAR TREK: THE DOOMSDAY MACHINE *   PG
Marc Daniels USA   1967
*William Shatner, Leonard Nimoy, DeForest Kelley, James Doohan,
George Takei, Nichelle Nichols, William Copage, John Copage,
Elizabeth Rogers, Richard Compton, John Winston, Tim Burns*
An awesomely powerful "doomsday machine" created by a
long-dead race sets about destroying entire planets and the
Enterprise attempts to halt its relentless progress. A strongly
plotted tale, this was SF writer Norman Spinrad's first TV script.
FAN   98 min (2-episode cassette) mTV
VIDrel: CIC/SONOP   V

STAR TREK: ELAAN OF TROYIUS *   PG
Judd Taylor USA   1968
William Shatner, Leonard Nimoy, DeForest Kelley, James Doohan,

George Takei, Walter Koenig, Nichelle Nichols, France Nuyen, Jay Robinson, Tony Young, Lee Duncan, Victor Brandt, K.L. Smith
The Enterprise is assigned the task of transporting a woman to her marriage ceremony on the planet Troyius, as a means of creating an alliance between two planets. However, matters are complicated by the fact that the woman is an unwilling bride, the Klingons wish to see the alliance fail and Kirk has fallen madly in love with her.
FAN   98 min (2-episode cassette) mTV
VIDrel: CIC/SONOP   V/h

STAR TREK: THE EMPATH **   PG
John Erman USA   1966
William Shatner, Leonard Nimoy, DeForest Kelley, James Doohan, George Takei, Walter Koenig, Nichelle Nichols, Kathryn Hays, Alan Bergman, William Sage, Jason Wingreen, Davis Roberts
"The Empath" sees Kirk, Spock and McCoy being held on a doomed planet, where powerful aliens have devised a series of tests involving personal sacrifice, in order to determine the viability of another species and whether or not it should be saved from a forthcoming supernova.
FAN   98 min (2-episode cassette) mTV
VIDrel: CIC/SONOP   V

STAR TREK: THE ENEMY WITHIN **   U/PG
Leo Penn USA   1966
William Shatner, Leonard Nimoy, DeForest Kelley, James Doohan, George Takei, Nichelle Nichols, Jim Goodwin, Edward Madden, Garland Thompson
A transporter malfunction splits Captain Kirk into two distinct individuals with completely opposite temperaments, and each battles for supremacy at the expense of the other. Spock and Scotty attempt to devise a means of combining the two Kirks into their original captain.
FAN   144 min (3-episode cassette – PG cert) mTV
VIDrel: CIC/SONOP   V/h

STAR TREK: THE ENTERPRISE INCIDENT *   PG
John Meredyth Lucas USA   1968
William Shatner, Leonard Nimoy, DeForest Kelley, James Doohan, George Takei, Walter Koenig, Nichelle Nichols, Joanne Linville, Jac Donner
The Enterprise is sent on a top secret mission to the Romulan Neutral Zone, their task being to obtain one of the cloaking devices used on the Romulan vessels. Along the way, there's a romantic interlude as Spock, posing as a Romulan, forms a relationship with a female Romulan commander. A weak entry, illogical in conception and clumsy in execution.
FAN   98 min (2-episode cassette) mTV
VIDrel: CIC/SONOP   V/h

STAR TREK: ERRAND OF MERCY **   U
John Newland USA   1967
William Shatner, Leonard Nimoy, DeForest Kelley, James Doohan, George Takei, Nichelle Nichols, John Abbott, John Colicos, Peter Brocco, Victor Lundin, David Hillary Huge
When Kirk warns the rulers of a planet about an imminent Klingon invasion, he finds them strangely unconcerned. Matters take a turn for the worse when Klingons arrive to claim the planet as a colony, but this leads to the inhabitants of the planet revealing their true nature and the reason for their complacency.
FAN   98 min (2-episode cassette) mTV
VIDrel: CIC/SONOP   V/h

STAR TREK: FOR THE WORLD IS HOLLOW AND I HAVE TOUCHED THE SKY **   PG
Tony Leader USA   1968
William Shatner, Leonard Nimoy, DeForest Kelley, James Doohan, George Takei, Walter Koenig, Nichelle Nichols, Kate Woodville, Byron Morrow, John Lormer
What at first glance appears to be an asteroid is in fact a disguised spaceship that's heading directly for a Federation planet. Whilst Kirk and Spock work against time to adjust the vessel's computers and alter its course, McCoy learns that he has an incurable disease and no more than a year to live. As the doctor has fallen in love with an inhabitant of the spaceship, he chooses to stay with her until the end.
FAN   98 min (2-episode cassette) mTV
VIDrel: CIC/SONOP   V

STAR TREK: FRIDAY'S CHILD **   U
Joseph Pevney USA   1967
William Shatner, Leonard Nimoy, DeForest Kelley, James Doohan,

Nichelle Nichols, George Takei, Walter Koenig, Julie Newmar, Tige Andrews, Michael Dante, Ben Gage, Cal Boulder, Kirk Raymone, Robert Bralver
Captain Kirk becomes involved in a conflict with his old enemies the Klingons, when both them and the Federation attempt to negotiate an alliance with the people of Capella IV. Much intrigue results, with the wife of a former ruler of the planet becoming a pawn in these negotiations.
FAN   51 min; 144 min (3-episode cassette) mTV
VIDrel: CIC/SONOP   V/h

STAR TREK: THE GALILEO SEVEN **   PG
Robert Gist USA   1967
William Shatner, Leonard Nimoy, DeForest Kelley, James Doohan, George Takei, Nichelle Nichols, Don Marshall, Peter Marko, Rees Vaughn, Grant Woods, Phyllis Douglas, John Crawford
Spock gains his first independent command, but finds himself under attack from the inhabitants of Taurus II whilst the Enterprise is en route to Makus III with much-needed medical supplies. Although he initially relies on logic alone to combat the planet's hostile environment, he is eventually forced to use human intuition to resolve the situation.
FAN   144 min (3-episode cassette) mTV
VIDrel: CIC/SONOP   V/h

STAR TREK: THE GAMESTERS OF TRISKELLON ***   PG
Gene Nelson USA   1968
William Shatner, Leonard Nimoy, DeForest Kelley, James Doohan, George Takei, Nichelle Nichols, Walter Koenig, John Ruskin, Angelique Pettyjohn, Steve Sandor, James Ross, Victoria George, Mickey Horton
Kirk, Uhura and Chekov are abducted to a strange world whose powerful rulers enjoy using other races to stage set-piece tournaments. Ultimately, the captain makes a dangerous wager in a bid to gain freedom for himself and his colleagues.
FAN   98 min (2-episode cassette) mTV
VIDrel: CIC/SONOP   V

STAR TREK: I, MUDD **   U
Marc Daniels USA   1967
William Shatner, Leonard Nimoy, DeForest Kelley, James Doohan, George Takei, Nichelle Nichols, Walter Koenig, Roger C. Carmel, Richard Tatro, Michael Zaslow, Mike Howden, Rhae Andrece, Alyce Andrece, Tom LeGarde, Ted LeGarde
Kirk and his colleagues are forced to go to an unknown planet by an android where they meet up with Harry Mudd once more. It would appear that he now lives a life of ease with an army of android servants, but matters are not quite as agreeable as they at first appear.
FAN   98 min (2-episode cassette) mTV
VIDrel: CIC/SONOP   V/h

STAR TREK: THE IMMUNITY SYNDROME **   U
Joseph Pevney USA   1968
William Shatner, Leonard Nimoy, DeForest Kelley, James Doohan, George Takei, Nichelle Nichols, Walter Koenig
A vast, amoeba-like creature is encountered which has not only destroyed an entire star-system and a starship, but is about to reproduce. The crew of the Enterprise work on a plan to halt its life cycle before it can divide.
FAN   49 min; 98 min (2-episode cassette) mTV
VIDrel: CIC/SONOP   V/h

STAR TREK: IS THERE NO TRUTH IN BEAUTY? ***   PG
Ralph Senensky USA   1968
William Shatner, Leonard Nimoy, DeForest Kelley, James Doohan, George Takei, Nichelle Nichols, Walter Koenig, Diana Muldaur, David Frankham
Special visors are the order of the day when an ambassador representing the benign but hideous Medusans visits the Enterprise. Amidst an outbreak of insanity and murder, it transpires that one must look beyond superficial appearances. An absorbing episode with a strong moral tone.
FAN   98 min (2-episode cassette) mTV
VIDrel: CIC/SONOP   V

STAR TREK: JOURNEY TO BABEL **   PG
Joseph Pevney USA   1967
William Shatner, Leonard Nimoy, DeForest Kelley, James Doohan, George Takei, Nichelle Nichols, Walter Koenig, Jane Wyatt, Mark Lenard, William O'Connell, Reggie Nalder, John Wheeler, James X. Mitchell

Spock meets his father, from whom he parted bitterly many years before, when the Enterprise is used to carry various alien races to the planet Babel, where an important treaty is to be signed.
FAN 98 min (2-episode cassette) mTV
VIDrel: CIC/SONOP V/h

STAR TREK: LET THAT BE YOUR LAST
BATTLEFIELD **   U
Judd Taylor USA   1969
William Shatner, Leonard Nimoy, DeForest Kelley, James Doohan, George Takei, Nichelle Nichols, Walter Koenig, Frank Gorshin, Lou Antonio
The Enterprise is caught up in the middle of a 50,000-year-old dispute between two warring aliens, each of whom has a face that is half white and half black. However, each alien's pattern is the reverse of the other, and it transpires that both are the sole survivors of their race.
FAN 98 min (2-episode cassette) mTV
VIDrel: CIC/SONOP V

STAR TREK: THE LIGHTS OF ZETAR **   PG
Herb Kenwith USA   1969
William Shatner, Leonard Nimoy, DeForest Kelley, James Doohan, George Takei, Nichelle Nichols, Walter Koenig, Jan Shutan, John Winston, Libby Erwin
Scotty has fallen in love with a crew member, but to his chagrin she is taken over by "The Lights of Zetar" – a cloud-like being representing the consciousness of a long-dead race. Having experienced corporate existence once more, the being becomes unwilling to relinquish its hold on the woman.
FAN 98 min (2-episode cassette) mTV
VIDrel: CIC/SONOP V

STAR TREK: THE MAN TRAP **   PG
Marc Daniels USA   1966
William Shatner, Leonard Nimoy, DeForest Kelley, James Doohan, George Takei, Nichelle Nichols, Jeanny Bal, Francine Pyne, Alfred Ryder, Michael Zaslow
The Enterprise is sent to a distant planet, where medical supplies are needed by the planet's only two inhabitants, Doctors Robert and Nancy Crater. Unfortunately, Crater has chosen to ignore the fact that his "wife" is actually an alien that kills its victims by draining the salt from their bodies. When it gets aboard the Enterprise, the lives of the entire crew are threatened. This was the very first episode in the long-running Star Trek series.
FAN 144 min (3-episode cassette) mTV
VIDrel: CIC/SONOP V/h

STAR TREK: THE MARK OF GIDEON **   PG
Judd Taylor USA   1969
William Shatner, Leonard Nimoy, DeForest Kelley, James Doohan, George Takei, Nichelle Nichols, Walter Koenig, Sharon Acker, David Hurst, Gene Kynarski, Richard Derr
Kirk attempts to beam down to the planet Gideon, but instead finds himself back onboard the Enterprise, the only difference being that the ship is now deserted. It later transpires that this is no more than a duplicate of his vessel, created with the bizarre intention of holding Kirk and using him to infect an over-populated and disease-free society.
FAN 98 min (2-episode cassette) mTV
VIDrel: CIC/SONOP V

STAR TREK: THE MENAGERIE –
PARTS 1 AND 2 ***   PG
Marc Daniels USA   1966
William Shatner, Leonard Nimoy, Jeffrey Hunter, Susan Oliver, Malachi Throne, Julie Parrish, Hagan Beggs, Peter Duryea, Meg Wylie, John Hoyt, Majel Barrett
The only two-part episode from the original TV series, which has Spock on trial for mutiny, after he took control of the Enterprise and "kidnapped" Captain Pike, the ship's former commander. This episode makes considerable use of footage from the original pilot episode – STAR TREK: THE CAGE, in which the first captain of the Enterprise was imprisoned on a strange planet by creatures that live off the emotions of others.
FAN 93 min (ort 98 min); 144 min (2-episode cassette) mTV
VIDrel: CIC/SONOP V/h

STAR TREK: METAMORPHOSIS ***   U
Ralph Senensky USA   1967
William Shatner, Leonard Nimoy, DeForest Kelley, James Doohan, Nichelle Nichols, George Takei, Walter Koenig, Glenn Corbett, Elinor Donahue

Kirk, Spock and McCoy are taken prisoner when they visit a strange planet and encounter a famous scientist who should be long since dead but has been kept alive and youthful by his companion, a strange energy cloud. A most unusual Star Trek episode that takes the form of a bizarre love story.
FAN 51 min; 144 min (3-episode cassette) mTV
VIDrel: CIC/SONOP V/h

STAR TREK: MIRI **   PG
Vincent McEveety USA   1966
William Shatner, Leonard Nimoy, DeForest Kelley, James Doohan, George Takei, Nichelle Nichols, Kim Darby, Michael J. Pollard, Jim Goodwin
An Earth-like planet is discovered and a landing team is sent down to investigate. Once there, they learn of a strange disease which, though it prolongs youth, eventually causes madness. When the landing party becomes infected, they realise they will be trapped on the planet unless McCoy can come up with a cure.
FAN 144 min (3-episode cassette) mTV
VIDrel: CIC/SONOP V/h

STAR TREK: MIRROR, MIRROR **   U
Marc Daniels USA   1967
William Shatner, Leonard Nimoy, DeForest Kelley, James Doohan, George Takei, Nichelle Nichols, Barbara Luna, Vic Perrin
The Enterprise is caught up in an ion storm during transportation, and the captain and his buddies find themselves in a parallel universe where the Federation is as feared as the Klingons. Meanwhile, a set of doubles from that universe materialises aboard the ship. Whilst they attempt to fit in and avoid discovery, Scotty works on a plan to recreate the conditions that prevailed during the ion storm and get them back to their own ship.
FAN 96 min (2-episode cassette) mTV
VIDrel: CIC/SONOP V/h

STAR TREK: MUDD'S WOMEN **   U/PG
Harvey Hart USA   1966
William Shatner, Leonard Nimoy, DeForest Kelley, James Doohan, George Takei, Nichelle Nichols, Roger C. Carmel, Karen Steele, Susan Denberg, Maggie Three, Gene Dynarski, Jim Goodwin, Jon Kowal, Seamon Glass
The Enterprise takes onboard an interstellar confidence trickster and his three beautiful female companions. Unfortunately, the ship is crippled in the course of doing this, and Kirk is obliged to set a course for a mining planet in order to replenish his ship's dilithium crystals. Meanwhile, our con-man sets about manipulating the miners to his own advantage.
FAN 144 min (3-episode cassette – PG cert) mTV
VIDrel: CIC/SONOP V/h

STAR TREK: THE NAKED TIME **   U
Marc Daniels USA   1966
William Shatner, Leonard Nimoy, DeForest Kelley, James Doohan, George Takei, Nichelle Nichols, Bruce Hyde, Stewart Moss, John Bellah
The Enterprise is sent to rescue a group of scientists from the doomed planet Psi 2000, but in the course of this work, a strange disease is brought aboard the starship which has the effect of removing the inhibitions of the crew members. Meanwhile, the engines of the Enterprise are tampered with and the ship is threatened with destruction.
FAN 144 min (3-episode cassette) mTV
VIDrel: CIC/SONOP V/h

STAR TREK: OBSESSION **   U
Ralph Senensky USA   1967
William Shatner, Leonard Nimoy, DeForest Kelley, James Doohan, George Takei, Nichelle Nichols, Walter Koenig, Stephen Brooks, Jerry Ayres
Kirk becomes obsessed with destroying a malevolent gaseous creature that killed half the crew of a starship he served on many years ago, a catastrophe for which he has blamed himself ever since. A kind of interstellar version of "Moby Dick".
FAN 49 min; 98 min (2-episode cassette) mTV
VIDrel: CIC/SONOP V/h

STAR TREK: THE OMEGA GLORY **   U
Vincent McEveety USA   1968
William Shatner, Leonard Nimoy, DeForest Kelley, James Doohan, George Takei, Nichelle Nichols, Walter Koenig, Morgan Woodward, Roy Jensen, Irene Kelley, David L. Ross, Eddie Paskey, Ed McReady, Lloyd Kino, Morgan Farley

This story sees the Enterprise picking up a strange warning from a deserted starship, which leads Kirk to beam down onto a planet where a renegade Federation captain has interfered with a frontier society and changed the course of its development, setting one group of people against another. Kirk does his best to set matters straight.
FAN   98 min (2-episode cassette) mTV
VIDrel: CIC/SONOP   V

STAR TREK: OPERATION-ANNIHILATE! **   U
Herschel Daugherty USA   1967
William Shatner, Leonard Nimoy, DeForest Kelley, James Doohan, George Takei, Nichelle Nichols, Dave Armstrong, Maurishka Taliferro, Craig Hundley, Joan Swift
The Enterprise encounters a spaceship hurtling towards the sun, whose inhabitants prefer death to the insanity caused by an alien parasite. This leads to the planet Deneva where these parasites have caused mass outbreaks of insanity. When McCoy devises a method of destroying them, he finds he needs a volunteer as a test subject.
FAN   98 min (2-episode cassette) mTV
VIDrel: CIC/SONOP   V/h

STAR TREK: THE PARADISE SYNDROME **   PG
Judd Taylor USA   1968
William Shatner, Leonard Nimoy, DeForest Kelley, James Doohan, George Takei, Nichelle Nichols, Walter Koenig, Sabrina Scharf, Rudy Lolari, Richard Hale
The USS Enterprise comes to the rescue of a planet which is threatened with destruction by an asteroid, and Kirk leads a landing party with the intention of evacuating the population. Unfortunately, our captain suffers an accident that results in amnesia, and is eventually found by a tribe of Indians who look upon him as a god. A rather contrived story, chiefly remembered for its tragic ending.
FAN   98 min (2-episode cassette) mTV
VIDrel: CIC/SONOP   V

STAR TREK: PATTERNS OF FORCE **   U
Vincent McEveety USA   1968
William Shatner, Leonard Nimoy, DeForest Kelley, James Doohan, George Takei, Nichelle Nichols, Walter Koenig, David Brian, Skip Homier, Richard Evans, Valora Norand, William Wintersole, Patrick Horgan, Bart LaRue, Ralph Mauer
Spock and Kirk land on a remote planet held in the grip of a Nazi-like regime, where they learn that a misguided Federation historian is responsible for the development of this movement.
FAN   98 min (2-episode cassette) mTV
VIDrel: CIC/SONOP   V

STAR TREK: A PIECE OF THE ACTION **   U
James Komack USA   1968
William Shatner, Leonard Nimoy, DeForest Kelley, James Doohan, George Takei, Nichelle Nichols, Walter Koenig, Anthony Caruso, Vic Tayback, Lee Delano, John Harmon, Steve Arnold, Sheldon Collins, Dyanne Thorne, Sharyn Hillyer
This adventure takes Captain Kirk to a remote planet that has evolved into a society resembling Chicago in the Prohibition era. Having failed to deal effectively with the inhabitants on an intellectual level, Kirk is forced to employ the same mobster methods as his adversaries. A weakly plotted entry, slightly amusing but very uneven.
FAN   98 min (2-episode cassette) mTV
VIDrel: CIC/SONOP   V

STAR TREK: PLATO'S STEPCHILDREN **   PG
David Alexander USA   1968
William Shatner, Leonard Nimoy, DeForest Kelley, James Doohan, George Takei, Walter Koenig, Nichelle Nichols, Michael Dunn, Liam Sullivan, Barbara Babcock, Ted Scott, Derek Partridge
The Enterprise responds to a distress call from Platonius, a planet whose society is loosely based on that of ancient Rome. However, the Platonians have powerful telekinetic powers, and when their request for McCoy to stay behind is refused, they set about humiliating Kirk and his officers.
FAN   98 min (2-episode cassette) mTV
VIDrel: CIC/SONOP   V

STAR TREK: A PRIVATE LITTLE WAR ***   PG
Marc Daniels USA   1967
William Shatner, Leonard Nimoy, DeForest Kelley, James Doohan, George Takei, Nichelle Nichols, Walter Koenig, Nancy Kovak, Michael Witney, Booker Marshak, Arthur Bernard, Joe Romero

The Klingons have armed some of the natives of a primitive planet and Kirk is forced to intervene, despite a Starfleet Command ruling to the contrary. Possibly intended as a SF analogy of the Vietnam War, this was a well conceived and generally quite effective entry.
FAN   98 min (2-episode cassette) mTV
VIDrel: CIC/SONOP   V/h

STAR TREK: REQUIEM FOR METHUSALEH ***   PG
David Alexander USA   1969
William Shatner, Leonard Nimoy, DeForest Kelley, James Doohan, George Takei, Nichelle Nichols, Walter Koenig, James Daly, Louise Sorel
Kirk, McCoy and Spock beam down to Holberg 917-G, their mission being to find a cure for a rare disease. There they encounter an apparently immortal man whose daughter Kirk falls in love with, only to learn to his dismay that all the inhabitants of the planet are androids.
FAN   98 min (2-episode cassette) mTV
VIDrel: CIC/SONOP   V

STAR TREK: THE RETURN OF THE ARCHONS **   PG
Joseph Pevney USA   1967
William Shatner, Leonard Nimoy, DeForest Kelley, James Doohan, George Takei, Nichelle Nichols, Harry Townes, Torin Thatcher, Charles McCauley, Brioni Farrell, Christopher Heid, Sid Haig, Jon Lormer, Morgan Farley, Ralph Maurer
A planet inhabited by strange zombie-like people is investigated by Kirk, and he learns that they are being controlled by a computer that regards them as nothing more than an extension of itself. Our valiant captain sets out to destroy the computer and free the planet's inhabitants.
FAN   144 min (3-episode cassette) mTV
VIDrel: CIC/SONOP   V

STAR TREK: RETURN TO TOMORROW **   U
Ralph Senensky USA   1968
William Shatner, Leonard Nimoy, DeForest Kelley, James Doohan, George TAkei, Walter Koenig, Nichelle Nichols, Diana Muldaur
A distress call brings the Starship Enterprise to an apparently lifeless planet where three discorporate aliens are encountered who wish to construct android bodies as receptacles for their minds. In order to do this, they need the loan of three bodies, a request to which Kirk and his colleagues rather reluctantly agree.
FAN   98 min (2-episode cassette) mTV
VIDrel: CIC/SONOP   V/h

STAR TREK: THE SAVAGE CURTAIN *   PG
Marvin J. Chomsky USA   1969
William Shatner, Leonard Nimoy, DeForest Kelley, James Doohan, George Takei, Nichelle Nichols, Walter Koenig, Lee Bergere, Barry Atwater, Phillip Pine, Nathan Jung, Carol Daniels Dement, Robert Herron
A powerful rock-like alien abducts Kirk and Spock and conveys them to the surface of its planet, where they encounter Abraham Lincoln and Surak, a famed Vulcan pioneer. Here they are to fight four notorious villains from history, for the alien is curious as to whether Good is more powerful than Evil. A clumsily contrived story, weak in both plotting and resolution.
FAN   147 min (3-episode cassette) mTV
VIDrel: CIC/SONOP   V

STAR TREK: SHORE LEAVE **   PG
Robert Sparr USA   1966
William Shatner, Leonard Nimoy, DeForest Kelley, James Doohan, George Takei, Nichelle Nichols, Emily Banks, Oliver McGowan, Bruce Mars, Perry Lopez, James Gruzaf, Shirley Bonne, Sebastian Tom
The crew of the Enterprise enjoy a spot of leave on an idyllic Earth-like planet, but have a few surprises awaiting them on a planet where one's dearest wishes can be made reality. A rather silly episode that attempted to inject a note of comedy into this series.
FAN   144 min (2-episode cassette) mTV
VIDrel: CIC/SONOP   V/h

STAR TREK: SPACE SEED **   PG
Marc Daniels USA   1967
William Shatner, Leonard Nimoy, DeForest Kelley, James Doohan, George Takei, Nichelle Nichols, Ricardo Montalban, Madlyn Rhue, Blasidell Makee, Mark Tobin
Captain Kirk has to outsmart a race of genetic supermen when the Enterprise encounters a derelict "sleeper ship" and revives

its crew, a group of ruthless individuals who attempted to take over the Earth in the 20th century. Montalban later had the chance to repeat his role of Khan in STAR TREK 2: THE WRATH OF KHAN.
FAN 144 min (3-episode cassette) mTV
VIDrel: CIC/SONOP V
STAR TREK: THE DEVIL IN THE DARK/STAR TREK: THIS SIDE OF PARADISE

STAR TREK: SPECTRE OF THE GUN **   U
Vincent McEveety USA   1968
William Shatner, Leonard Nimoy, DeForest Kelley, James Doohan, George Takei, Nichelle Nichols, Walter Koenig, Ron Soble, Rex Holman, Bonnie Beecher, Sam Gilman, Charles Maxwell, Bill Zuckert, Ed McReady, Abraham Sofaer
Kirk decides to explore a forbidden planet, and finds himself and his officers trapped in a reconstruction of The Gunfight at the O.K. Corral. A lightweight episode, slightly reminiscent of STAR TREK: PATTERNS OF FORCE, but a good deal less sombre in tone.
FAN 98 min (2-episode cassette) mTV
VIDrel: CIC/SONOP V

STAR TREK: SPOCK'S BRAIN *   PG
Marc Daniels USA   1968
William Shatner, Leonard Nimoy, DeForest Kelley, James Doohan, George Takei, Walter Koenig, Nichelle Nichols, Marj Dusay, Sheila Leighton, James Daris
A mysterious woman causes Spock to suffer a severe brain-drain that may lead to his death when it is found by the Eymorgs that his brain would serve as a suitable power source for their planet. A bizarre basic premise that deserved a better treatment to be effective.
FAN 98 min (2-episode cassette) mTV
VIDrel: CIC/SONOP V/h

STAR TREK: THE SQUIRE OF GOTHOS **   PG
Don McDougall USA   1967
William Shatner, Leonard Nimoy, DeForest Kelley, James Doohan, George Takei, Nichelle Nichols, William Campbell, Richard Carlyle, Michael Barrier, Venita Wolf
An omniscient alien captures Kirk and a group of companions and conveys them to the surface of his planet, his plan being to use them as toys for his own amusement. He is eventually revealed to be no more than a very powerful and very bored "child".
FAN 144 min (3-episode cassette) mTV
VIDrel: CIC/SONOP V/h

STAR TREK: A TASTE OF ARMAGEDDON **   PG
Joseph Pevney USA   1967
William Shatner, Leonard Nimoy, DeForest Kelley, James Doohan, George Takei, Nichelle Nichols, Gene Lyons, David Opatoshu, Robert Sampson, Barbara Babcock, Miko Mayama
Kirk visits a planet where battles are fought with a rival world by computer, the losing side in each individual "conflict" being obliged to exterminate an appropriate number of its citizens. The Enterprise inadvertently becomes a target in this war, and when it is "hit" Kirk is expected to exterminate his own crew.
FAN 144 min (3-episode cassette) mTV
VIDrel: CIC/SONOP V

STAR TREK: THAT WHICH SURVIVES ***   U
Herb Wallerstein USA   1968
William Shatner, Leonard Nimoy, DeForest Kelley, James Doohan, George Takei, Walter Koenig, Nichelle Nichols, Arthur Batanides, Naomi Pollack, Lee Meriwether
"That Which Survives" has Kirk and McCoy stranded on a hostile planet when the Enterprise is flung light-years away. There, they encounter the lethal holographic image of a woman, created as a warning device by an ancient and now extinct race.
FAN 98 min (2-episode cassette) mTV
VIDrel: CIC/SONOP V

STAR TREK: THIS SIDE OF PARADISE **   PG
Ralph Senensky USA   1967
William Shatner, Leonard Nimoy, DeForest Kelley, James Doohan, George Takei, Nichelle Nichols, Jill Ireland, Frank Overton, Grant Woods, Dick Scotter
The Enterprise is sent to the planet Omicron Ceti III, where a group of Federation colonists are presumed dead after exposure to deadly radiation. However, they appear to have been saved by strange spores found on the planet's surface which

create feelings of bliss and harmony in those they infect. When said spores get onto the Enterprise, a state of chaos ensues.
FAN 144 min (3-episode cassette) mTV
VIDrel: CIC/SONOP V

STAR TREK: THE THOLIAN WEB ***   PG
Ralph Senensky USA   1968
William Shatner, Leonard Nimoy, DeForest Kelley, James Doohan, George Takei, Nichelle Nichols, Walter Koenig
Having set about exploring the remnants of a derelict starship, Kirk finds himself trapped onboard the Enterprise, unable to reach the Enterprise as he is trapped in another dimension. Whilst Spock works out a plan of rescue, the Tholians, into whose territory they have entered, create an energy field that will eventually destroy the Enterprise.
FAN 98 min (2-episode cassette) mTV
VIDrel: CIC/SONOP V

STAR TREK: TOMORROW IS YESTERDAY **   PG
Michael O'Herlihy USA   1978
William Shatner, Leonard Nimoy, DeForest Kelley, James Doohan, George Takei, Nichelle Nichols, Roger Perry, Ed Peck, Hal Lynch, Richard Merrifield, John Winston
The Enterprise encounters a black hole and is thrown backwards in time to the 20th century where it is spotted by an air force jet. Having rescued the pilot just before the jet was accidentally destroyed, Kirk has to return the pilot without changing the course of history.
FAN 144 min (3-episode cassette) mTV
VIDrel: CIC/SONOP V/h

STAR TREK: THE TROUBLE WITH TRIBBLES ***   U
Joseph Pevney USA   1967
William Shatner, Leonard Nimoy, DeForest Kelley, James Doohan, George Takei, Nichelle Nichols, Walter Koenig, William Schallert, William Campbell, Whit Bissell, Stanley Adams, Michael Pataki, Charlie Brill, Ed Reiners
An emergency signal brings the Enterprise to a space station whose commander fears a Klingon attack. Meanwhile, Uhura has brought onboard the Enterprise a cuddly little creature known as a "tribble", little suspecting that these creatures reproduce rapidly. An enjoyable Star Trek adventure, significant in being the only episode to be produced as a complete comedy from start to finish. See also STAR TREK DEEP SPACE NINE: TRIALS AND TRIBBLE-ATIONS.
FAN 98 min (2-episode cassette) mTV
VIDrel: CIC/SONOP V/h

STAR TREK: TURNABOUT INTRUDER ***   PG
Herschel Daugherty USA   1969
William Shatner, Leonard Nimoy, DeForest Kelley, James Doohan, George Takei, Nichelle Nichols, Walter Koenig, Sandra Smith, Harry Landers
A woman who was once romantically involved with Kirk has been unable to obtain a ship's command of her own. Blaming Kirk for her failure, she plots her revenge, using an alien device that causes the two to swap minds. The very last episode of the original series and though not brilliant, a fair story to bow out on.
FAN 147 min (3-episode cassette) mTV
VIDrel: CIC/SONOP V

STAR TREK: THE ULTIMATE COMPUTER ***   U
John Meredyth Lucas USA   1968
William Shatner, Leonard Nimoy, DeForest Kelley, James Doohan, George Takei, Nichelle Nichols, Walter Koenig, William Marshall, Barney Russo, Sean Morgan
A super-intelligent computer is installed aboard the Enterprise, to replace its human crew, but it goes berserk and becomes a dangerous menace. Despite having initially become convinced that he no longer serves any purpose, Kirk rises to the occasion and causes the computer to destroy itself.
FAN 98 min (2-episode cassette) mTV
VIDrel: CIC/SONOP V/h

STAR TREK: THE WAY TO EDEN *   PG
Murray Golden USA   1969
William Shatner, Leonard Nimoy, DeForest Kelley, James Doohan, George Takei, Nichelle Nichols, Walter Koenig, Skip Homeier, Charles Napier, Mary Linda Rapelye, Victor Brandt, Deborah Downey, Phyllis Douglas
A bunch of space-age hippies commandeer the Enterprise, their intention being to use it to reach a mythical paradise-planet

known as Eden. A very silly episode – the less said about it the better.
FAN   98 min (2-episode cassette) mTV
VIDrel: CIC/SONOP   V

STAR TREK: WHAT ARE LITTLE GIRLS MADE OF? **
PG
James Goldstone USA 1966
William Shatner, Leonard Nimoy, DeForest Kelley, James Doohan, George Takei, Nichelle Nichols, Michael Strong, Ted Cassidy, Sherry Jackson, Harry Basch, Vince Deadrick, Budd Albright
In order to assist a woman who wishes to be reunited with her scientist fiance, Kirk takes the starship Enterprise to a distant planet. However, the scientist has learned from the ancient technology of the planet and creates an android duplicate of Captain Kirk, his intention being to take over the Enterprise and use it to spread his androids throughout the galaxy.
FAN   144 min (3-episode cassette) mTV
VIDrel: CIC/SONOP   V/h

STAR TREK: WHERE NO MAN HAS GONE BEFORE ****
U
James Goldstone USA 1966
William Shatner, Leonard Nimoy, DeForest Kelley, James Doohan, George Takei, Nichelle Nichols, Gary Lockwood, Paul Carr, Sally Kellerman, Paul Fix, Andrea Dromm, Lloyd Haynes
The Enterprise is ordered to penetrate beyond the region of explored space, but after passing through a strange force-field a crew member begins to mutate into a being with unlimited power. This second pilot was one of the very best of the early Star Trek episodes. Interestingly Lockwood, who is admirably cast as the transformed crewman, went on to play astronaut Frank Poole in Kubrick's 2001: A SPACE ODYSSEY.
FAN   51 min; 160 min (3-episode cassette) mTV
VIDrel: CIC/SONOP; PION (LV only)   V/dm LV

STAR TREK: WHO MOURNS FOR ADONAIS? **
PG
Marc Daniels USA 1967
William Shatner, Leonard Nimoy, DeForest Kelley, James Doohan, George Takei, Nichelle Nichols, Michael Forest, Leslie Parrish, John Winston
Captain Kirk has to confront the god Apollo, the last of the Greek gods, who has decided that he requires the worship of mankind once more. Kirk makes use of both the Enterprise and a female crew member who has fallen in love with the god, in a bid to overcome his adversary.
FAN   98 min (2-episode cassette) mTV
VIDrel: CIC/SONOP   V/h

STAR TREK: WHOM GODS DESTROY **
PG
Herb Wallerstein USA 1969
William Shatner, Leonard Nimoy, DeForest Kelley, James Doohan, George Takei, Nichelle Nichols, Walter Koenig, Yvonne Craig, Steve Ihnat, Key Luke
Not only does the Enterprise have to deal with an asylum on Elba II whose inmates have taken it over, but also with a former Starfleet captain who has gained the ability to change his shape, and has taken Kirk and Spock hostage.
FAN   97 min (2-episode cassette) mTV
VIDrel: CIC/SONOP   V

STAR TREK: WINK OF AN EYE **
PG
Judd Taylor USA 1968
William Shatner, Leonard Nimoy, DeForest Kelley, James Doohan, George Takei, Nichelle Nichols, Walter Koenig, Kathie Browne, Jason Evers, Eric Holland, Geoffrey Binney
Living at a rate incomparably faster than human beings, the Scalosians have nearly become extinct, but have formed a plan to make use of the crew of the Enterprise to repopulate their planet. Fortunately, Kirk has drunk some Scalosian water and now finds himself living at an equally rapid rate, in which condition he does what he can to foil their plan.
FAN   98 min (2-episode cassette) mTV
VIDrel: CIC/SONOP   V

STAR TREK: WOLF IN THE FOLD **
PG
Joseph Pevney USA 1967
William Shatner, Leonard Nimoy, DeForest Kelley, James Doohan, George Takei, Nichelle Nichols, Walter Koenig, John Fiedler, Charles Macauley, Peter Seurat, Joseph Bernard, Charles Dierkop, Judy McConnell, Virginia Ladridge
Kirk, McCoy and Scotty take a break on Argelius II after the chief engineer injures his head in an accident. Once there, a

series of murders occurs, and all the evidence points to Scotty as the culprit. Fortunately, the real cause of these events is ultimately revealed.
FAN   98 min (2-episode cassette) mTV
VIDrel: CIC/SONOP   V

STAR TREK THE NEXT GENERATION: CAPTAIN'S HOLIDAY **
U
Chip Chalmers USA 1990
Patrick Stewart, Jonathan Frakes, LeVar Burton, Michael Dorn, Marina Sirtis, Wil Wheaton, Brent Spiner, Gates McFadden, Jennifer Hetrick, Karen Landry, Max Grodenchik, Michael Champion, Deirdre Imershein
Having arranged to take a well earned break, Captain Picard becomes involved in a search for a valuable treasure that has been stolen from the 27th century. Average.
FAN   88 min (2-episode cassette) mTV
VIDrel: CIC/SONOP   V/dm V/sh

STAR TREK DEEP SPACE NINE: THE ABANDONED **
PG
Avery Brooks USA 1994
Avery Brooks, Rene Auberjonois, Siddig El Fadil, Terry Farrell, Armin Shimerman, Colm Meaney, Nana Visitor, Cirroc Lofton, Bumper Robinson, Jill Sayre, Leslie Bevis, Matthew Kimbrough, Hassan Nicholas
Quark is made a seemingly irresistible business offer when he buys as scrap the remains of a spaceship that crashed on a planet in the Gamma Quadrant. Unfortunately, life gets complicated when he finds an alien baby aboard. It grows up at an astonishing race and proves to be a Jem'Hadar who is taken under Odo's protective wing. Very reminiscent of "I, Borg" from the STAR TREK THE NEXT GENERATION series but hampered by an ending that is pure anti-climax.
FAN   45 min; 88 min (2-episode cassette) mTV
VIDrel: CIC/SONOP   V/sur

STAR TREK DEEP SPACE NINE: ACCESSION **
PG
Les Landau USA 1994
Avery Brooks, Rene Auberjonois, Alexander Siddig, Terry Farrell, Armin Shimerman, Colm Meaney, Nana Visitor, Michael Dorn, Cirroc Lofton, Rosalind Chao, Robert Symonds, Camille Saviola, Richard Libertini, Hana Hatae
Sisko faces a problem of some delicacy when a strange-looking vessel emerges from the wormhole and proves to be a solar-powered craft piloted by a long-dead and much revered poet. Soon, he is taken to be the emissary foretold in a Bajoran religious tradition, a position which Sisko is initially relieved to be able to give up. However, it soon becomes clear that the stranger's attempts to restore former practices and customs are a source of growing tension.
FAN   88 min (2-episode cassette) mTV cC
VIDrel: CIC/SONOP   V/sur

STAR TREK DEEP SPACE NINE: THE ADVERSARY **
(PG)
Alexander Singer USA 1994
Avery Brooks, Rene Auberjonois, Siddig El Fadil, Terry Farrell, Armin Shimerman, Colm Meaney, Nana Visitor, Cirroc Lofton, Lawrence Pressman, Kenneth Marshall, Jeff Austin
Sisko's long-awaited promotion to the rank of captain finally comes through but the celebrations are muted when he is forced to undertake an urgent mission involving a former enemy race. Once aboard the Defiant, the situation grows critical when control of the ship is lost and a hostile shape-shifter runs amuck. A moderately exciting episode that ends the third series on a note of menace.
FAN   45 min; 88 min (2-episode cassette) mTV
VIDrel: CIC/SONOP   V/sur

STAR TREK DEEP SPACE NINE: THE ALTERNATE **
PG
David Carson USA 1993
Avery Brooks, Rene Auberjonois, Siddig El Fadil, Terry Farrell, Armin Shimerman, Colm Meaney, Nana Visitor, James Sloyan, Matt McKenzie
The Bajoran scientist who investigated Odo and helped him to integrate into human society, comes to DS9 with news of a planet in the Gamma Quadrant that might be the home of his species. An expedition is mounted but the landing on the planet's surface proves almost a disaster. After returning to DS9

with some specimens, strange events begin to occur. An episode of moderate interest but without much suspense.
FAN   45 min; 88 min (2-episode cassette) mTV
VIDrel: CIC/SONOP   V/sur

STAR TREK DEEP SPACE NINE: ARMAGEDDON GAME **
U
Winrich Kolbe USA
1993
Avery Brooks, Rene Auberjonois, Siddig El Fadil, Terry Farrell, Colm Meaney, Armin Shimerman, Nana Visitor, Rosalind Chao, Darleen Carr, Peter White, Larry Cedar, Bill Mondy
When the Federation arranges a peace treaty putting end to a centuries-long war between two alien races, DS9's doctor and chief engineer are put in charge of the destruction of their biological weapons. Just as the last batch is about to undergo this process, they are attacked and barely escape with their lives. A most unexciting episode with a contrived and wholly unsatisfactory plot.
FAN   45 min; 88 min (2-episode cassette) mTV
VIDrel: CIC/SONOP   V/sur

STAR TREK DEEP SPACE NINE: THE ASSIGNMENT ***
PG
Allan Kroeker USA
1994
Avery Brooks, Rene Auberjonois, Alexander Siddig, Terry Farrell, Armin Shimerman, Colm Meaney, Nana Visitor, Michael Dorn, Cirroc Lofton, Rosalind Chao, Max Grodenchik, Hana Hatae
Keiko O'Brien comes back from a visit to Bajor and shocks her husband by stating that she is an alien being that has taken control of Keiko's body. If he refuses to comply with her orders, Keiko will be killed. To prove the serious intent behind these threats, she jumps off the promenade. O'Brien tries to stop her from harming DS9 but is forced to engage in a tense game of cat-and-mouse. A well-acted tale of alien possession thanks to Chao's outstanding performance.
FAN   45 min; 88 min (2-episode cassette) mTV cC
VIDrel: CIC/SONOP   V/sur

STAR TREK DEEP SPACE NINE: BABEL **
U
Paul Lynch USA
1993
Avery Brooks, Rene Auberjonois, Siddig El Fadil, Terry Farrell, Colm Meaney, Cirroc Lofton, Armin Shimerman, Nana Visitor, Jack Kehler, Matthew Farrell, Ann Gillespie, Geraldine Farrell, Bo Zenga, Kathleen Wirt, Lee Brooks
As O'Brien is carrying out routine maintenance on the station's replicators he accidentally activates a Bajoran device planted there some eighteen years before. This device creates an artificial aphasia virus that randomly scrambles speech patterns and makes communication impossible. An interesting idea but its slow development leads to a very unimaginative resolution, capped for good measure with a hurried and contrived happy ending.
FAN   45 min; 87 min (2-episode cassette) mTV
VIDrel: CIC/SONOP   V/sur

STAR TREK DEEP SPACE NINE: BAR ASSOCIATION **
12
LeVar Burton USA
1996
Avery Brooks, Rene Auberjonois, Alexander Siddig, Terry Farrell, Armin Shimerman, Colm Meaney, Nana Visitor, Michael Dorn, Cirroc Lofton, Jason Marsden, Max Grodenchik, Chase Masterson, Emilio Borelli, Jeffrey Combs
A Bajoran festival in which they abstain from worldly pleasures gives Quark a chance to exploit his downtrodden employees even more by cutting their wages. But this includes his own brother Rom, who inspired by a chance remark sees an opportunity to organise a workers' association. Unfortunately, this is considered to represent a dishonourable break with Ferengi tradition, and trouble is not long in coming.
FAN   45 min; 88 min (2-episode cassette) mTV cC
VIDrel: CIC/SONOP   V/sur

STAR TREK DEEP SPACE NINE: BATTLE LINES **
PG
Paul Lynch USA
1993
Avery Brooks, Rene Auberjonois, Siddig El Fadil, Terry Farrell, Colm Meaney, Armin Shimerman, Nana Visitor, Camille Saviola, Paul Collins, Jonathan Banks, Majel Barrett (voice only)
DS9 is visited by the Bajoran spiritual leader whom Sisko takes on a courtesy tour of the wormhole. Approaching a barren moon for a closer look, the runabout is attacked by a defence satellite and crashes. Marooned on this world, they find themselves in the midst of a raging war and make an astounding discovery. A watchable but hardly outstanding episode.
FAN   45 min; 87 min (2-episode cassette) mTV
VIDrel: CIC/SONOP   V/dm V/sur

STAR TREK DEEP SPACE NINE: BLOOD OATH **
PG
Winrich Kolbe USA
1993
Avery Brooks, Rene Auberjonois, Siddiq El Faddil, Terry Farrell, Armin Shimerman, Colm Meaney, Nana Visitor, John Colicos, Michael Ansara, William Campbell, Bill Bolender, Christopher Collins
When two ageing Klingons comes back from a visit to DS9, Jadzia Dux finds herself re-united with some former comrades of Cruzon Dux, the previous Trill host. She soon learns that they are on a mission of revenge and have come to redeem Dux's blood oath to help them track down and murder the man who slaughtered their children. A silly episode, with much talk of honour and little action, it seems to be patterned on Japanese samurai films.
FAN   45 min; 88 min (2-episode cassette) mTV
VIDrel: CIC/SONOP   V/dm V/sur

STAR TREK DEEP SPACE NINE: BODY PARTS **
12
Avery Brooks USA
1994
Avery Brooks, Rene Auberjonois, Alexander Siddig, Terry Farrell, Armin Shimerman, Colm Meaney, Nana Visitor, Michael Dorn, Cirroc Lofton, Rosalind Chao, Max Grodenchik, Hana Hatae, Jeffrey Combs, Andrew Robinson
After visiting his home-world, Quark returns to DS9, having learnt that due to a rare disease, he has only one week left to live. This obliges him to put his remains up for sale, in line with Ferengi custom, in order to repay his debts before his demise. When he learns that the diagnosis was wrong and that his condition is far from fatal, he has to make a most difficult decision.
FAN   45 min; 88 min (2-episode cassette) mTV cC
VIDrel: CIC/SONOP   V/sur

STAR TREK DEEP SPACE NINE: BROKEN LINK **
12
Les Landau USA
1994
Avery Brooks, Rene Auberjonois, Alexander Siddig, Terry Farrell, Armin Shimerman, Colm Meaney, Nana Visitor, Michael Dorn, Cirroc Lofton, Salome Jens, Robert O'Reilly, Jill Jacobson, Leslie Bevis, Andrew Robinson
Odo falls dangerously ill ands is rushed to the infirmary where Bashir is forced to conclude that for reasons unknown his patient is no longer able to maintain a solid form. Increasingly unable to observe the complete rest that Bashir has prescribed, Odo begins to deteriorate and it would appear that his only hope is in making a visit to his own race – The Founders. But once he arrives back on his home-world, he is arrested on murder charges.
FAN   45 min; 88 min (2-episode cassette) mTV cC
VIDrel: CIC/SONOP   V/sur

STAR TREK DEEP SPACE NINE: CAPTIVE PURSUIT **
U
Corey Allen USA
1993
Avery Brooks, Rene Auberjonois, Siddig El Fadil, Terry Farrell, Colm Meaney, Armin Shimerman, Nana Visitor, Gerrit Graham, Scott MacDonald, Kelly Curtis
An alien arrives at the station having come through the wormhole in a ship that's badly in need of repair and shows signs of having been damaged in combat. However, the creature steadfastly refuses to volunteer any information as to its origins or purpose, but all is made clear when the station comes under attack from a huge alien vessel that lands an assault team on the space-station. A competent if not over-exciting entry that fails to live up to its early promise.
FAN   45 min; 87 min (2-episode cassette) mTV
VIDrel: CIC/SONOP   V/sur

STAR TREK DEEP SPACE NINE: CARDASSIANS ***
U
Cliff Bole USA
1993
Avery Brooks, Rene Auberjonois, Siddig El Fadil, Terry Farrell, Colm Meaney, Armin Shimerman, Nana Visitor, Rosalind Chao, Terrence Evans, Andrew Robinson, Vidal Peterson, Robert Mandan, Dion Anderson, Marc Alaimo, Sharon Conley
The presence aboard DS9 of a Cardassian war orphan who has been adopted by a Bajoran threatens to lead to a major political crisis when the boy proves to be the son of an important Cardassian politician. However, the doctor and a Cardassian friend who works aboard the station as a tailor, eventually

unearth a nasty little conspiracy. A strongly plotted episode that works quite well.
FAN   45 min; 88 min (2-episode cassette) mTV
VIDrel: CIC/SONOP   V/dm V/sur

STAR TREK DEEP SPACE NINE: THE CIRCLE **   U
Corey Allen USA   1993
Avery Brooks, Rene Auberjonois, Siddig El Fadil, Terry Farrell, Cirroc Lofton, Colm Meaney, Armin Shimerman, Nana Visitor, Richard Beymer, Stephen Macht, Bruce Gray, Philip Anglim, Louise Fletcher, Mike Genovese, Eric Server
After leading her rescue mission (see "The Homecoming") Kira is recalled to Bajor and promised promotion, although she is loathe to leave DS9. Meanwhile back on that planet, a coup seems to be in the making, and an armed extremist group that is hostile to all non-Bajorans appears poised to take over.
FAN   45 min; 88 min (2-episode cassette) mTV
VIDrel: CIC/SONOP   V/dm V/sur

STAR TREK DEEP SPACE NINE: CIVIL DEFENSE ***   U
Reza Badiyi USA   1994
Avery Brooks, Rene Auberjonois, Siddig El Fadil, Terry Farrell, Armin Shimerman, Colm Meaney, Nana Visitor, Cirroc Lofton, Andrew Robinson, Marc Alaimo, Danny Goldring
An attempt to restart the ore processing plant on DS9 inadvertently triggers an old Cardassian computer program that interprets this action as a threat and takes appropriate countermeasures. Efforts to overcome this situation only make matters worse while an offer of Cardassian help comes with a very price-tag. For once, a quite exciting and suspenseful episode that keeps one guessing almost to the end.
FAN   45 min; 88 min (2-episode cassette) mTV
VIDrel: CIC/SONOP   V/sur

STAR TREK DEEP SPACE NINE: THE COLLABORATOR **   PG
Cliff Bole USA   1994
Avery Brooks, Rene Auberjonois, Siddiq El Faddil, Terry Farrell, Armin Shimerman, Colm Meaney, Nana Visitor, Phlip Anglim, Bert Remsen, Camille Saviola, Louise Fletcher, Charles Park, Tom Villard
The selection of a new Kai on Bajor places Kira in a conflict of loyalties when one of the candidates tricks her into investigating her lover, a rival for this high office. The latter is said by a returning collaborator to have been responsible for betraying a resistance group to the Cardassians who staged an ambush in which they were all killed. Fletcher gives a fine performance in an episode that ends on a clear note of anti-climax.
FAN   45 min; 87 min (2-episode cassette) mTV
VIDrel: CIC/SONOP   V/dm V/sur

STAR TREK DEEP SPACE NINE: CROSSFIRE **   PG
Les Landau USA   1994
Avery Brooks, Rene Auberjonois, Alexander Siddig, Terry Farrell, Armin Shimerman, Colm Meaney, Nana Visitor, Michael Dorn, Cirroc Lofton, Duncan Regehr, Bruce Wright, Charles Tentindo
When an important Bajoran official visits DS9, Odo is required to take elaborate measures to protect him from assassination by a shadowy Cardassian group. As the visit progresses, he comes to realise just how attached he has become to Major Kira and even begins to feel jealousy at her growing involvement with the visitor. A well acted but essentially static episode, lacking in suspense and plot development.
FAN   88 min (2-episode cassette) mTV cC
VIDrel: CIC/SONOP   V/sur

STAR TREK DEEP SPACE NINE: CROSSOVER **   PG
David Livingston USA   1994
Avery Brooks, Rene Auberjonois, Siddiq El Faddil, Terry Farrell, Armin Shimerman, Colm Meaney, Nana Visitor, Andrew Robinson, John Cotharn Jr, Steven Gevedon, Jack R. Orend, Dennis Malone
Returning from a Bajoran colony in the Gamma Quadrant, Major Kira and Dr Bashir find themselves the victims of a freak accident that transports them to a parallel universe. Here, DS9 is in orbit around Bajor as an ore processing plant under the command of Kira, who is fascinated to meet herself in another guise. herself in another guise. A promising idea is virtually ignored in an excessively melodramatic episode that at times verges on parody.
FAN   45 min; 87 min (2-episode cassette) mTV
VIDrel: CIC/SONOP   V/dm V/sur

STAR TREK DEEP SPACE NINE: DAX **   U
David Carson USA   1993
Avery Brooks, Rene Auberjonois, Siddig El Fadil, Terry Farrell, Colm Meaney, Cirroc Lofton, Armin Shimerman, Nana Visitor, Gregory Itzin, Anne Haney, Richard Lineback, Fionnula Flanagan
An attempt to kidnap Dux that almost succeeded turns out to have been motivated by a warrant for her extradition to a planet where Cruzon Dux, the former host of the symbiont, is alleged to have committed treason and murder. A legal hearing then follows but in the midst of all this Jadzia keeps strangely silent and refuses to defend herself. A static and unsatisfying episode whose final resolution seems trite and predictable.
FAN   45 min; 88 min (2-episode cassette) mTV
VIDrel: CIC/SONOP   V/sur

STAR TREK DEEP SPACE NINE: DEFIANT **   PG
Cliff Bole USA   1994
Avery Brooks, Rene Auberjonois, Siddig El Fadil, Terry Farrell, Armin Shimerman, Colm Meaney, Nana Visitor, Jonathan Frankes, Marc Alaimo, Shannon Cochran, Tricia O'Neill, Majel Barrett (voice only)
DS9 receives a surprise visit from Ryker who charms Kira into allowing him on board the Defiant, the Federation's latest and highly top-secret ship. Suddenly, he kidnaps both her and this spacecraft and vanishes. It soon transpires that this is Ryker's double, created some years earlier in a transporter accident, although the purpose of his action remains unclear. A moderately exciting episode but with a weak and undeveloped plot.
FAN   45 min mTV   VIDrel: CIC/SONOP   V

STAR TREK DEEP SPACE NINE: DESTINY ***   U
Les Landau USA   1994
Avery Brooks, Rene Auberjonois, Siddig El Fadil, Terry Farrell, Armin Shimerman, Colm Meaney, Nana Visitor, Tracy Scoggins, Jessica Hendra, Wendy Robie, Erick Avari
A joint scientific project with the Cardassians arouses the opposition of a Bajoran cleric who informs Sisko that an ancient prophecy forecast doom should he refuse to cancel it. As events proceed, the prophecy seems to be becoming true but a major disaster is narrowly averted at the last moment. A neatly plotted episode that maintains interest to the end, and with (for once at least) a relevant sub-plot.
FAN   45 min; 88 min (2-episode cassette) mTV
VIDrel: CIC/SONOP   V/sur

STAR TREK DEEP SPACE NINE: THE DIE IS CAST **   PG
David Livingston USA   1994
Avery Brooks, Rene Auberjonois, Siddig El Fadil, Terry Farrell, Armin Shimerman, Colm Meaney, Nana Visitor, Andrew Robinson, Leland Orser, Kenneth Marshall, Leon Russom, Paul Dolley, Wendy Schenker
Having pledged his loyalty to his former mentor and joined his crusade to crush the Dominion, Garak finds himself being forced to torture Odo for some key information he is thought to have. Meanwhile, Sisko defies Starfleet orders and takes the Defiant into the Gamma Quadrant on a rescue mission. A not terribly exciting conclusion to a two-part story that began with "Improbable Cause".
FAN   45 min; 88 min (2-episode cassette) mTV
VIDrel: CIC/SONOP   V/sur

STAR TREK DEEP SPACE NINE: DISTANT VOICES **   PG
Alexander Singer USA   1994
Avery Brooks, Rene Auberjonois, Siddig El Fadil, Terry Farrell, Armin Shimerman, Colm Meaney, Nana Visitor, Andrew Robinson, Victor Rivers, Ann Gillespie, Nicole Forester
When Doctor Bashir refuses to sell a restricted medical substance to an alien visitor, the latter attempts to steal it. In the process the doctor is injured and falls into a coma from which he wakes up to find the station virtually deserted. As he begins to age at an ever-accelerating pace, he faces a struggle to survive. Unfortunately, a single strong idea fails to get the treatment it deserves and the story generates little suspense.
FAN   88 min (2-episode cassette) mTV
VIDrel: CIC/SONOP   V/sh

STAR TREK DEEP SPACE NINE: DRAMATIS PERSONAE **   PG
Cliff Bole USA   1993
Avery Brooks, Rene Auberjonois, Siddig El Fadil, Terry Farrell, Colm Meaney, Armin Shimerman, Nana Visitor, Tom Towles, Stephen Paur, Randy Pflug, Jeff Pruit
A dying Klingon taken aboard the station after arriving from the

Gamma Quadrant brings with him a strange alien energy field that soon takes hold of the crew and forces them to play out a deadly power struggle. A strongly acted episode where solid performances help support the rather rudimentary storyline, but as usual it's all spoiled by the hurried conclusion.
FAN   45 min; 87 min (2-episode cassette) mTV
VIDrel: CIC/SONOP   V/dm V/sur

STAR TREK DEEP SPACE NINE: DUET ***
James L. Conway USA   PG
                                                       1993
Avery Brooks, Rene Auberjonois, Siddig El Fadil, Terry Farrell, Colm Meaney, Armin Shimerman, Nana Visitor, Marc Alaimo, Robin Christopher, Norman Large, Tony Rizzoli, Ted Sorel, Harris Yulin
A Cardassian in need of medical treatment arrives at the space station and is taken into custody upon suspicion of having served at a labour camp during the occupation of Bajor. A game of wits then ensues as Major Kira tries to establish whether he is possibly a notorious war criminal. An inventive storyline with a good twist and an excellent performance by Yulin enhance this above-average episode.
FAN   45 min; 87 min (2-episode cassette) mTV
VIDrel: CIC/SONOP   V/dm V/sur

STAR TREK DEEP SPACE NINE: THE EMISSARY – PARTS 1 AND 2 **
David Carson USA   PG
                                                       1993
Avery Brooks, Rene Auberjonois, Siddig El Fadil, Terry Farrell, Colm Meaney, Cirroc Lofton, Armin Shimerman, Nana Visitor, Patrick Stewart, Lily Mariye, Marc Alaimo, Felecia Bell, Camille Saviola, Aron Eisenberg, Stephen Davies
Pilot episode in the latest STAR TREK series, set aboard a space station above a planet recently liberated from occupation by the warlike Cardassians who have withdrawn their forces. The recently appointed Star Fleet officer in charge of the station (Brooks) goes on a dangerous mission that brings unforeseen consequences for all concerned. The special effects may be excellent, but mediocre plotting and unconvincing acting are major handicaps.
FAN   87 min (ort 120 min) mTV
VIDrel: CIC/SONOP; PION (LV only)   V/sur LV

STAR TREK DEEP SPACE NINE: EQUILIBRIUM **
Cliff Bole USA   PG
                                                       1994
Avery Brooks, Rene Auberjonois, Siddig El Fadil, Terry Farrell, Armin Shimerman, Nana Visitor, Cirroc Lofton, Lisa Barnes, Jeff Magnus McBride, Harvey Vernon, Nicholas Cascone
At an informal dinner with some members of the crew and Sisko, Dux starts to suddenly exhibit some unbalanced behaviour that seems linked in some strange way to a half-remembered tune and later starts experiencing abrupt hallucinations, Dr Bashir examines her and concludes that treatment on her home world is essential. Eventually, this mystery is resolved but not before it would appear that she is doomed to die.
FAN   45 min mTV   VIDrel: CIC/SONOP   V/sur

STAR TREK DEEP SPACE NINE: EXPLORERS **   PG
Winrich Kolbe USA   1994
Avery Brooks, Rene Auberjonois, Siddig El Fadil, Terry Farrell, Armin Shimerman, Colm Meaney, Nana Visitor, Cirroc Lofton, Marc Alaimo, Bari Hochwald, Chase Masterson
Sisko returns from Bajor fired with enthusiasm for a strange project to build an ancient spacecraft that was said to have been used by the ancient Bajorans to explore the universe some 800 years earlier. He eventually succeeds in doing so and together with Jake, sets out on a trial voyage. Plenty of father-son bonding adds little to an episode that is seriously deficient in suspense despite the spectre of Cardassian involvement.
FAN   45 min; 88 min (2-episode cassette) mTV
VIDrel: CIC/SONOP   V/sur

STAR TREK DEEP SPACE NINE: FACETS **   (PG)
Cliff Boles USA   1994
Avery Brooks, Rene Auberjonois, Siddig El Fadil, Terry Farrell, Armin Shimerman, Colm Meaney, Nana Visitor, Cirroc Lofton, Jefrey Alan Chandler, Max Grodenchik, Aron Eisenberg, Chase Materson, Majel Barrett (voice only)
Jadzia Dux astounds her closest friends by requesting their participation in a unique ceremony that will make their bodies temporary homes for the personalities of former hosts. Most of them willing agree although a number of problems do arise as well as some unforeseen consequences for Odo. A disjointed

and unsatisfying episode that is badly in need of a stronger story.
FAN   45 min; 88 min (2-episode cassette) mTV
VIDrel: CIC/SONOP   V/sur

STAR TREK DEEP SPACE NINE: FAMILY BUSINESS **
Rene Auberjonois USA   PG
                                                       1994
Avery Brooks, Siddig El Fadil, Terry Farrell, Armin Shimerman, Colm Meaney, Nana Visitor, Andrea Martin, Perry Johnson, Max Grodenchik, Jeffrey Combs
Quark and his brother are forced to return to the Ferengi home world when word reaches them that their mother has been violating both law and custom by indulging in trade and earning a profit. Once there, they find themselves having to confront their own fate and deal with some very ambivalent feelings. An interesting if not entirely successful attempt to fill in the gaps concerning Ferengi culture.
FAN   90 min (2-episode cassette) mTV
VIDrel: CIC/SONOP   V/sur

STAR TREK DEEP SPACE NINE: FASCINATION *
Avery Brooks USA   (PG)
                                                       1994
Avery Brooks, Rene Auberjonois, Siddig El Fadil, Terry Farrell, Armin Shimerman, Colm Meaney, Nana Visitor, Cirroc Lofton, Majel Barrett, Philip Anglim, Rosalind Chao
After Lwaxana Troi travels to DS9 to attend a Bajoran festival and pursue Odo, a rash of passionate sexual attractions breaks out among the station's inhabitants. However, the reverse seems to be happening to Chief O'Brien and his wife, who has arrived there on a brief visit. A dull episode saddled with a transparent and over-simple plot and the usual pat resolution. Very poor.
FAN   45 min mTV   TVrel

STAR TREK DEEP SPACE NINE: THE FORSAKEN **
Les Landau USA   PG
                                                       1993
Avery Brooks, Rene Auberjonois, Siddig El Fadil, Terry Farrell, Colm Meaney, Armin Shimerman, Nana Visitor, Majel Barrett, Constance Towers, Michael Ensign, Jack Shearer, Benita Andre
A group of Federation ambassadors including Lwaxana Troi pay a visit to the station and are witnesses to the emergence from the wormhole of an unknown probe. The latter is scanned and its database downloaded to the station computer, which soon begins to suffer a series of mysterious malfunctions. An interesting idea is heavy developed and a romantic sub-plot involving Troi and Odo is both unnecessary and contrived.
FAN   45 min; 87 min (2-episode cassette) mTV
VIDrel: CIC/SONOP   V/dm V/sur

STAR TREK DEEP SPACE NINE: HEART OF STONE **
Alexander Singer USA   PG
                                                       1994
Avery Brooks, Rene Auberjonois, Siddig El Fadil, Terry Farrell, Armin Shimerman, Colm Meaney, Nana Visitor, Cirroc Lofton, Max Grodenchik, Aron Eisenberg, Majel Barrett (voice only)
A routine mission by Kira and Odo takes an unexpected turn when they pursue a Maquis ship after it attacks a freighter and then lands on a small moon. This chase takes them to an underground cavern where Kira becomes trapped by a strange crystal growth and Odo struggles desperately to free her. Meanwhile back on DS9, Nog takes a decision that surprises everyone. An adequate entry in the series, watchable but not very exciting.
FAN   88 min (2-episode cassette) mTV
VIDrel: CIC/SONOP   V/sur

STAR TREK DEEP SPACE NINE: THE HOMECOMING **
Winrich Kolbe USA   U
                                                       1993
Avery Brooks, Rene Auberjonois, Siddig El Fadil, Terry Farrell, Cirroc Lofton, Colm Meaney, Armin Shimerman, Nana Visitor, Richard Beymer, Michael Bell, Max Grodenchik, Marc Alaimo, Leslie Bevis, Paul Nakauchi
Major Kira learns that a famous Bajoran resistance leader is being held in a prison camp on the planet Cardassia 4. As Bajor is in the throes of civil strife, his return would do much to unify the people so she persuades Sisko to allow her to mount a rescue mission. The latter is successful but the unexpected response of the Cardassians comes as a surprise and awakens suspicions. The story continues in the following episode "The Circle".
FAN   45 min; 88 min (2-episode cassette) mTV
VIDrel: CIC/SONOP   V/dm V/sur

STAR TREK DEEP SPACE NINE: THE HOUSE OF QUARK **

Les Landau USA   (PG)
1994

Avery Brooks, Rene Auberjonois, Siddig El Fadil, Terry Farrell, Armin Shimerman, Colm Meaney, Nana Visitor, Rosalind Chao, Mary Kay Adam, Carlos Carrasco, Max Grodenchik, Robert O'Reilly, Joseph Ruskin

When a Klingon customer dies accidentally in his bar, Quark is not slow in exploiting this to attract custom by claiming that he killed him in self-defence. Unfortunately, the man's brother and wife come after him and he finds himself involved in a power struggle between two Klingon families. Quite well acted, but the plot does seem a little thin. A sub-plot deals with Mrs O'Brien's problems after she is forced to close the school.

FAN   45 min mTV   TVrel

STAR TREK DEEP SPACE NINE: IF WISHES WERE HORSES ***

Robert Legato USA   PG
1993

Avery Brooks, Rene Auberjonois, Siddig El Fadil, Terry Farrell, Colm Meaney, Cirroc Lofton, Armin Shimerman, Nana Visitor, Rosalind Chao, Keone Young, Hana Hatae, Michael John Anderson

The station is suddenly caught up in a strange phenomenon in which fairy-tale figures, human fantasies and holodeck characters suddenly come alive. At first, this is merely a strange inconvenience but when its cause is traced to a major rift in the fabric of space, the very survival of DS9 is placed in doubt. Highly imaginative if a little contrived, the plot of this episode has enough twists to maintain interest right to the end.

FAN   45 min; 87 min (2-episode cassette) mTV
VIDrel: CIC/SONOP   V/dm V/sur

STAR TREK DEEP SPACE NINE: IMPROBABLE CAUSES **

Avery Brooks USA   PG
1994

Avery Brooks, Rene Auberjonois, Siddig El Fadil, Terry Farrell, Armin Shimerman, Colm Meaney, Nana Visitor, Andrew Robinson, Carlos LaCamara, Joseph Ruskin, Darwyn Carson, Julianna McCarthy, Paul Dooley

A mysterious explosion in Garak's shop is investigated by Odo with his usual thoroughness but the mystery deepens when the only suspect is killed after his spaceship explodes. The evidence points to the Romulans who it seems have made an alliance with Garak's former mentor and are now planning a surprise attack on the Dominion. First in a two-part story that concludes with "The Die Is Cast".

FAN   88 min (2-episode cassette) mTV
VIDrel: CIC/SONOP   V/sur

STAR TREK DEEP SPACE NINE: IN THE HANDS OF THE PROPHETS **

David Livingston USA   PG
1993

Avery Brooks, Rene Auberjonois, Siddig El Fadil, Terry Farrell, Colm Meaney, Armin Shimerman, Nana Visitor, Cirroc Lofton, Rosalind Chao, Robin Christopher, Louise Fletcher, Philip Anglim, Michel Eugene Fairman

The station school receives an unwelcome visit from an important Bajoran spiritual leader who accuses its teacher of heresy in not adhering to their religious ideas in her classroom. However, the resulting confrontation and tensions that she causes are found to have a deeper and far more sinister purpose. Fletcher gives a first-rate performance as the religious fanatic in an otherwise unmemorable episode.

FAN   45 min; 87 min (2-episode cassette) mTV
VIDrel: CIC/SONOP   V/dm V/sur

STAR TREK DEEP SPACE NINE: INVASIVE PROCEDURES **

Les Landau USA   PG
1993

Avery Brooks, Rene Auberjonois, Siddig El Fadil, Terry Farrell, Colm Meaney, Armin Shimerman, Nana Visitor, John Glover, Megan Gallagher, Tim Russ, Steve Rankin

DS9 is struck by a massive plasma disturbance and most of its occupants are evacuated, leaving behind a skeleton crew. However, the station is soon invaded by a small group, led by a Trill who is out to steal Dux, the symbiont that lives inside Jadzia. When he threatens to kill everyone, the doctor has no alternative but to remove the symbiont from her and transfer it to him, but the usual complications eventually frustrate his plans.

FAN   45 min; 88 min (2-episode cassette) mTV
VIDrel: CIC/SONOP   V/dm V/sur

STAR TREK DEEP SPACE NINE: THE JEM HADAR **

Kim Friedman USA   PG
1994

Avery Brooks, Rene Auberjonois, Siddig El Faddil, Terry Farrell, Colm Meaney, Armin Shimerman, Nana Visitor, Cirroc Lofton, Alan Oppenheimer, Aron Eisenberg, Molly Hagan, Cress Williams, Michael Jace, Sandra Grundon

Sisko plans a field trap for his son to a planet in the Gamma Quadrant but is reluctantly forced to allow Nog and Quark along too. However, their little excursion becomes deadly serious when the two adults are captured by fierce and hostile warriors and a menacing confrontation looms. An attempt to liven up the series with an implacable foe on the lines of the Borg. The story continues in the two-part episode "The Search" that opens Series Three.

FAN   45 min; 87 min (2-episode cassette) mTV
VIDrel: CIC/SONOP   V/dm V/sur

STAR TREK DEEP SPACE NINE: LIFE SUPPORT **

Reza Badiyi USA   PG
1994

Avery Brooks, Rene Auberjonois, Siddig El Fadil, Terry Farrell, Armin Shimerman, Colm Meaney, Nana Visitor, Cirroc Lofton, Philip Anglim, Louise, Fletcher, Aron Eisenberg, Lark Voorhies, Ann Gillespie, Andrew Prine, Eva Loseth

When a Bajoran spacecraft suffers massive damage and explodes, Vedek Barell is taken aboard DS9 in a critical state. His survival seems hopeless but he is eventually saved by some unorthodox and dangerous methods. This allows him to continue his secret peace negotiations with the Cardassians, which may result in much improved relations between these two races. A rather well realised story that's spoilt by an unnecessary and irrelevant sub-plot.

FAN   88 min (2-episode cassette) mTV
VIDrel: CIC/SONOP   V/sur

STAR TREK DEEP SPACE NINE: A MAN ALONE **

Paul Lynch USA   PG
1992

Avery Brooks, Rene Auberjonois, Siddig El Fadil, Terry Farrell, Colm Meaney, Cirroc Lofton, Armin Shimerman, Nana Visitor, Edward Laurence Albert, Rosalind Chao, Max Godenchik, Peter Voight, Aron Eisenberg, Tom Klunis

Shape-changer Odo, who is in charge of security aboard the space station, becomes the victim of an ingenious plot to frame him for the murder of a former enemy. Fortunately, the station doctor eventually reveals the truth and proves his innocence. A mediocre episode with little of interest and a totally superfluous sub-plot about Chief Engineer O'Brien's wife, and how she becomes inspired to open a school.

FAN   45 min; 87 min (2-episode cassette) mTV
VIDrel: CIC/SONOP   V/dm V/sur

STAR TREK DEEP SPACE NINE: THE MAQUIS – PARTS 1 AND 2 **

David Livingston/Corey Allen USA   U
1994

Avery Brooks, Rene Auberjonois, Siddiq El Faddil, Terry Farrell, Armin Shimerman, Colm Meaney, Nana Visitor, Tony Plana, Bertila Damas, Richard Poe, Michael A. Krawic, Amanda Carlin, Marc Alaimo, Bernie Casey

After the destruction of a Cardassian ship proves to have been caused by sabotage, Sisko faces with a moral dilemma when he learns that an old friend is involved with the resistance movement on former Federation colonies that are now under Cardassian rule. To uphold the treaty and prevent a possible war, some tough decisions are needed. A well acted double episode that is let down by poor plot development. Parts 1 and 2 were available together or separately.

FAN   90 min mTV   VIDrel: CIC/SONOP   V/dm V/sur

STAR TREK DEEP SPACE NINE: MELORA **

Winrich Kolbe USA   U
1993

Avery Brooks, Rene Auberjonois, Siddig El Fadil, Terry Farrell, Colm Meaney, Armin Shimerman, Daphne Ashbrook, Peter Crombie, Don Stark, Ron Taylor

A new Federation cartographer assigned to DS9 is a member of a race living on a low-gravity world and has a hard time adapting to her new environment, while her initial over-assertiveness makes things difficult for all those around her. Meanwhile, Quark finds his life under threat when a former associate is released from prison.

FAN   45 min; 88 min (2-episode cassette) mTV
VIDrel: CIC/SONOP   V/dm V/sur

STAR TREK DEEP SPACE NINE: MERIDIAN ***   U
Jonathan Frakes USA   1994
Avery Brooks, Rene Auberjonois, Siddig El Fadil, Terry Farrell, Armin Shimerman, Colm Meaney, Nana Visitor, Brett Cullen, Christine Healy, Jeffrey Combs, Mark Humphrey
While exploring part of the Gamma Quadrant on an exploratory mission, Sisko and his crew are astonished when a planet suddenly appears from nowhere. They contact its inhabitants, who number no more than thirty, and learn that both they and their planet alternate at regular intervals between two dimensions. When Jadzia falls deeply in love, this presents her with a difficult choice. Quite well down but spoiled by a silly sub-plot back on DS9.
FAN   45 min; 88 min (2-episode cassette) mTV
VIDrel: CIC/SONOP   V/sur

STAR TREK DEEP SPACE NINE: MOVE ALONG HOME **   PG
David Carson USA   1993
Avery Brooks, Rene Auberjonois, Siddig El Fadil, Terry Farrell, Colm Meaney, Cirroc Lofton, Armin Shimerman, Nana Visitor, James Lashly, Joel Brooks, Clara Bryant
An alien race from the Gamma Quadrant (who look exactly like humans but with tattooed foreheads) arrive on DS9. They ignore the "first contact" welcome prepared by the crew and are interested only in Quark's bar. They play for hours and when ruin seems imminent, Quark tries to cheat them but is caught. In retaliation he is forced to play one of their games with the lives of the entire crew put at stake. Moderately amusing but cheaply done and unconvincing.
FAN   45 min; 87 min (2-episode cassette) mTV
VIDrel: CIC/SONOP   V/dm V/sur

STAR TREK DEEP SPACE NINE: NECESSARY EVIL **   PG
James L. Conway USA   1993
Avery Brooks, Rene Auberjonois, Siddig El Fadil, Terry Farrell, Armin Shimerman, Katherin Moffat, Marc Alaimo, Max Grodenchik, Robert Mackenzie, Tiny Ron
Quark goes to Bajor to meet with a Bajoran woman who five years before lived on DS9 with her husband who was murdered there. He agrees to retrieve a box hidden behind the wall in what used to be their shop, but is shot and almost fatally injured. As Odo investigates, he re-lives the events of the past in an attempt to solve this mystery. A well acted but rather dull episode with an unsatisfactory climax.
FAN   45 min; 88 min (2-episode cassette) mTV
VIDrel: CIC/SONOP   V/dm V/sur

STAR TREK DEEP SPACE NINE: THE NEGUS **   U
David Livingston USA   1993
Avery Brooks, Rene Auberjonois, Siddig El Fadil, Terry Farrell, Colm Meaney, Cirroc Lofton, Armin Shimerman, Nana Visitor, Max Grodenchik, Lou Wagner, Barry Gordon, Lee Arenberg, Aron Eisenberg, Tiny Ron, Wallace Shawn
A visit to Quark's bar by the Grand Negus, the head of the Ferengi business empire, leaves Quark feeling both surprised and agitated. However, worse is to come when he finds himself appointed as the man's successor and faces a threat to his life. A silly and unamusing episode, too much taken up with comic Ferengi antics and a sub-plot involving Jake and his Ferengi friend Nog.
FAN   45 min; 87 min (2-episode cassette) mTV
VIDrel: CIC/SONOP   V/dm V/sur

STAR TREK DEEP SPACE NINE: PARADISE **   (PG)
Corey Allen USA   1993
Avery Brooks, Colm Meaney, Nana Visitor, Terry Farrell, Julia Nickson, Steve Vinovich, Michael Buchman Silver, Erick Weiss, Gail Strickland, Majel Barrett (voice only)
While scouting suitable planets for colonisation, Sisko and the chief engineer discover an unknown human colony on a planet where no technological devices appear to work. The colony is led by a woman with a fanatical hatred of modern technology, who has found the perfect place to test her ideas of self-reliance. However, the truth proves to be more complex. A disappointing episode that fails to say anything about fanatics and their behaviour.
FAN   45 min mTV   TVrel

STAR TREK DEEP SPACE NINE: THE PASSENGER ***   PG
Paul Lynch USA   1993
Avery Brooks, Rene Auberjonois, Siddig El Fadil, Terry Farrell, Armin Shimerman, Nana Visitor, Caitlin Brown, James Lashly, Christopher Collins, James Harper
A DS9 runabout rescues the occupants of a burning spaceship, two of whom are an alien policewoman and her prisoner, a master criminal she has been tracking for twenty years. The latter seems to have died in the fire but subsequent events call this into doubt, as the time of a crucial cargo delivery approaches. A competently directed episode that manages to maintain suspense almost until the end.
FAN   45 min; 87 min (2-episode cassette) mTV
VIDrel: CIC/SONOP   V/dm V/sur

STAR TREK DEEP SPACE NINE: PAST PROLOGUE **   PG
Winrich Kolbe USA   1992
Avery Brooks, Rene Auberjonois, Siddig El Fadil, Terry Farrell, Colm Meaney, Nana Visitor, Jeffrey Nordling, Andrew Robinson, Gwynyth Walsh, Barbara Marsh, Susan Bay, Vaughn Armstrong
A terrorist being pursued by the Cardassians is beamed aboard the space station and given temporary refuge. He claims to have renounced his former violent ways but gradually reveals his involvement in a secret plan whose true intent does not become clear until much later. A banal and trite story, not much helped by the flat characterisations. A most disappointing first episode in this much-vaunted series.
FAN   45 min; 87 min (2-episode cassette) mTV
VIDrel: CIC/SONOP   V/sur

STAR TREK DEEP SPACE NINE: PAST TENSE PARTS 1 AND 2 **   PG
Reza Badiyi/Jonathan Frakes USA   1994
Avery Brooks, Rene Auberjonois, Siddig El Fadil, Terry Farrell, Armin Shimerman, Colm Meaney, Nana Visitor, Jim Metzler, Frank Military, Dick Miller, Al Rodrigo, Tina Lifford, Bill Smitrovich, Clint Howard, Richard Lee Jackson
A freak transporter accident sends Dux, Bashir and Sisko back in time to an era in Earth's history when the unemployed and homeless were forced to live in special compounds, separated from the rest of society. Unfortunately, our trio arrive just a few days before a pivotal event is due to happen that will change the course of history. A disappointing story that soon runs out of steam and proceeds along very conventional lines.
FAN   88 min (2-episode cassette) mTV
VIDrel: CIC/SONOP   V/sur

STAR TREK DEEP SPACE NINE: PLAYING GOD **   PG
David Livingston USA   1993
Avery Brooks, Rene Auberjonois, Siddiq El Faddil, Terry Farrell, Armin Shimerman, Colm Meaney, Nana Visitor, Cirroc Lofton, Geoffrey Blake, Ron Taylor, Richard Poe, Chris Nelson Norris, Majel Barrett (voice only)
A Trill host candidate comes to DS9 for evaluation by Jadzia Dux and shows himself ill-prepared to deal with his situation and very much out of his depth. However, a trip through the wormhole brings back an unexpected threat and he gets a chance to prove his worth and grow in self-esteem. A rather uninteresting episode whose single good idea never gets the treatment it deserves.
FAN   45 min; 88 min (2-episode cassette) mTV
VIDrel: CIC/SONOP   V/dm V/sur

STAR TREK DEEP SPACE NINE: PROFIT AND LOSS **   PG
Robert Wiemer USA   1993
Avery Brooks, Rene Auberjonois, Siddiq El Faddil, Terry Farrell, Armin Shimerman, Colm Meaney, Nana Visitor, Mary Crosby, Andrew Robinson, Michael Reilly Burke, Heidi Swedberg, Edward Willey
Quark gets a reminder of past amorous adventures when three Cardassians are transported to DS9 from their heavily damaged ship. One of them proves to be a former lover of his who is now deeply involved in the underground resistance to Cardassia's military government. For once, he finds himself acting for reasons not dictated by commercial gain. A competently acted episode but somewhat lacking in suspense.
FAN   45 min; 88 min (2-episode cassette) mTV
VIDrel: CIC/SONOP   V/dm V/sur

STAR TREK DEEP SPACE NINE: PROGRESS **   PG
Les Landau USA   1993
Avery Brooks, Rene Auberjonois, Siddig El Fadil, Terry Farrell, Colm Meaney, Armin Shimerman, Nana Visitor, Brian Keith, Aron Eisenberg, Nicholas Worth, Michael Bofshever, Terrence Evans, Annie O'Donnell, Daniel Riordan
A scheme to heat Bajor by tapping the core of one of its moons needs Major Kira to ensure the evacuation of its inhabitants. However, she discovers that a stubborn old man and his two mute companions refuse to leave at any price and risks her career to help him. Meanwhile, Jake and Nog undertake their first business venture. An unfocused episode that never really amounts to much. Watchable but unimpressive.
FAN   45 min; 87 min (2-episode cassette) mTV
VIDrel: CIC/SONOP   V/dm V/sur

STAR TREK DEEP SPACE NINE: PROPHET MOTIVE **   U
Rene Auberjonois USA   1994
Avery Brooks, Rene Auberjonois, Siddig El Fadil, Terry Farrell, Armin Shimerman, Colm Meaney, Nana Visitor, Max Grodenchik, Julian Donald, Tiny Ron, Wallace Shawn, Bennet Guillory
A surprise visit to DS9 by the Ferengi Grand Negus leaves Quark stunned when he begins to realise that the leader of his world has become a reformed character for whom profit is no longer the supreme goal in life. Perplexed by this change, he probes for its cause and eventually finds a way to put matters right. A silly and unamusing episode, burdened by slow development and a totally irrelevant sub-plot involving Dr Bashir.
FAN   88 min (2-episode cassette) mTV
VIDrel: CIC/SONOP   V/sur

STAR TREK DEEP SPACE NINE: Q-LESS **   U
Paul Lynch USA   1993
Avery Brooks, Rene Auberjonois, Siddig El Fadil, Terry Farrell, Colm Meaney, Cirroc Lofton, Armin Shimerman, Nana Visitor, Jennifer Hetrick, John De Lancie, Van Epperson, Tom McLesier, Laura Cameron
A runabout returns to DS9 almost totally depleted of energy and carrying three passengers, one of whom is the female archaeologist Vash who was once involved with Captain Picard. She says that she has spent the last two years in the Gamma Quadrant but refuses to explain how she got there. When Q is sighted on the station, he gets the blame for a series of mysterious power malfunctions, but the real cause finally proves to be far stranger.
FAN   45 min; 88 min (2-episode cassette) mTV
VIDrel: CIC/SONOP   V/sur

STAR TREK DEEP SPACE NINE: RETURN TO GRACE ***   PG
Jonathan West USA   1994
Avery Brooks, Rene Auberjonois, Alexander Siddig, Terry Farrell, Armin Shimerman, Colm Meaney, Nana Visitor, Michael Dorn, Cirroc Lofton, Marc Alaimo, Cyia Batten, Casey Biggs, John K. Shull
Major Kira reluctantly agrees to attend a Cardassian/Bajoran meeting to discuss the threat from the Klingon empire and is thrown together with an old enemy – Gul Dukat, who is now in disgrace and demoted to the rank of freighter captain. Events force them to co-operate in order to survive when the Klingons mount an attack that kills all the conference delegates. A well written and original story, with the emphasis on both characterisation and action.
FAN   88 min (2-episode cassette) mTV cC
VIDrel: CIC/SONOP   V/sur

STAR TREK DEEP SPACE NINE: RIVALS **   PG
David Livingston USA   1993
Avery Brooks, Rene Auberjonois, Siddig El Fadil, Terry Farrell, Armin Shimerman, Colm Meaney, Nana Visitor, Rosalind Chao, Barbara Bosson, R. Callan, Albert Henderson, Chris Sarandon
An alien con-man comes to DS9 and sets up a rival casino to Quark's and shortly afterwards the station is struck by a series of events that seem to defy the normal laws of probability. A slow-paced and very transparent episode with a most predictable plot. Quite poor.
FAN   45 min; 88 min (2-episode cassette) mTV
VIDrel: CIC/SONOP   V/sur

STAR TREK DEEP SPACE NINE: RULES OF ACQUISITION **   PG
David Livingston USA   1993
Avery Brooks, Rene Auberjonois, Siddig El Fadil, Terry Farrell, Colm Meaney, Armin Shimerman, Helen Udy, Brian Thompson, Max Grodenchik, Emilia Crow, Tiny Ron
The Ferengi Grand Negus appoints Quark his chief negotiator in trade talks with a race from the Gamma Quadrant that are to be held aboard DS9. These prove a lot trickier than expected and Quark is pleased to have the assistance of a bright and ambitious young waiter from his bar, but is totally unaware that this individual harbours a deep secret. A poorly plotted but rather well acted entry with good characterisations.
FAN   45 min; 88 min (2-episode cassette) mTV
VIDrel: CIC/SONOP   V/dm V/sur

STAR TREK DEEP SPACE NINE: RULES OF ENGAGEMENT *   PG
LeVar Burton USA   1994
Avery Brooks, Rene Auberjonois, Alexander Siddig, Terry Farrell, Armin Shimerman, Colm Meaney, Nana Visitor, Michael Dorn, Cirroc Lofton, Ron Canada, Deborah Strang, Christopher Michael
The Klingons demand the extradition of Worf to face charges arising from an incident when he fired on a Klingon vessel in the heat of battle. Fortunately, Sisko springs to his defence and attempts to prove his colleague is innocent of any wrongdoing, but it falls to Odo to provide the vital piece of evidence required. A thoroughly unconvincing affair, totally contrived and devoid of any dramatic tension.
FAN   88 min (2-episode cassette) mTV cC
VIDrel: CIC/SONOP   V/sur

STAR TREK DEEP SPACE NINE: SANCTUARY **   PG
Les Landau USA   1993
Avery Brooks, Rene Auberjonois, Siddig El Fadil, Terry Farrell, Armin Shimerman, Cirroc Lofton, Colm Meaney, Nana Visitor, William Schallert, Andrew Koenig, Aron Einseberg, Deborah May, Michael Durrell, Kitty Swink
A stricken alien craft comes through the wormhole and its passengers are rescued and taken aboard DS9. They prove to be members of a race from the Gamma Quadrant who have been freed from centuries-long oppression and are now in search of their legendary home. A disappointing episode that starts off quite well but soon degenerates to the usual predictable level.
FAN   45 min; 88 min (2-episode cassette) mTV
VIDrel: CIC/SONOP   V/dm V/sur

STAR TREK DEEP SPACE NINE: THE SEARCH – PARTS 1 AND 2 **   PG
Kim Friedman/Jonathan Frakes USA   1994
Avery Brooks, Rene Auberjonois, Siddig El Fadil, Terry Farrell, Armin Shimerman, Colm Meaney, Nana Visitor, Cirroc Lofton, Salome Jens, Martha Hackett, John Fleck, Kenneth Marshall, Andrew Robinson, Natalja Nogulich
Sisko obtains a prototype battleship with which to visit the Gamma Quadrant and attempt to meet the Founders, the beings rumoured to be the force behind the Dominion and their elite soldiers the Jem'Hadar. After a furious battle against overwhelming odds, Odo and Kira find their way to what appears to be Odo's home world. A hastily constructed but unconvincing twist ending does little to imbue this story with any real interest. Disappointing and dull.
FAN   88 min (2-episode cassette) mTV
VIDrel: CIC/SONOP   V/sur

STAR TREK DEEP SPACE NINE: SECOND SIGHT **   PG
Alexander Singer USA   1993
Avery Brooks, Rene Auberjonois, Siddig El Fadil, Terry Farrell, Armin Shimerman, Cirroc Lofton, Colm Meaney, Sallie Else Richardson, Richard Kiley, Mark Erickson
Commander Sisko has a brief romantic encounter with a strange woman who disappears without leaving him a clue as to her identity. When an alien scientist in charge of a project to re-ignite a dead star visits DS9, Sisko is invited aboard his craft. There he meets the woman once more, but she denies all knowledge of their relationship. An interesting idea is wasted in a dull story that develops in an overly predictable way.
FAN   45 min; 88 min (2-episode cassette) mTV
VIDrel: CIC/SONOP   V/dm V/sur

STAR TREK DEEP SPACE NINE: SECOND SKIN **

PG

Les Landau USA 1994

Avery Brooks, Rene Auberjonois, Siddig El Fadil, Terry Farrell, Armin Shimerman, Nana Visitor, Andrew Robinson, Gregory Sierra, Tony Rupenfuss, Cindy Katz, Lawrence Pressman, Christopher Carroll, Freyda Thomas, Billy Burke

When a historian asks Kira about her detention in a Cardassian prison camp, an event she fails to recall, she travels to Bajor to investigate but never arrives. Captured and surgically altered to appear Cardassian, she finds herself kidnapped and taken to the home world of Cardassia Prime. There she is told that she is in reality a Cardassian spy who was to infiltrate the underground. A nicely constructed plot spoiled by a hasty resolution.
FAN 45 min; 88 min (2-episode cassette) mTV
VIDrel: CIC/SONOP V/sur

STAR TREK DEEP SPACE NINE: SHADOWPLAY ** U

Robert Scheerer USA 1993

Avery Brooks, Rene Auberjonois, Siddiq El Faddil, Terry Farrell, Armin Shimerman, Colm Meaney, Nana Visitor, Cirroc Lofton, Kenneth Mars, Kenneth Tobey, Noley Thornton, Philip Anglim, Trula M. Marcus, Martin Cassidy

Dux and Odo investigate a strange particle field beyond the wormhole and discover that their emanate from one valley on a distant planet. They beam down and find themselves involved in solving the mystery behind a number of disappearing villagers. Meanwhile, back on DS9 Kira falls in love and in so doing helps stop one of Quark's scams. An unfocused episode with too many sub-plots and too little suspense.
FAN 45 min; 88 min (2-episode cassette) mTV
VIDrel: CIC/SONOP V/sur

STAR TREK DEEP SPACE NINE: SHAKAAR ** PG

Jonathan West USA 1994

Avery Brooks, Rene Auberjonois, Siddig El Fadil, Terry Farrell, Armin Shimerman, Colm Meaney, Nana Visitor, Duncan Regehr, Diane Salinger, William Lucking, Sherman Howard, Louise Fletcher, John Kenton Shull, Harry Hutchinson

The death of the head of the Bajoran government soon escalates into a full-blown crisis when the ambitious spiritual leader Kai Win takes over. A conflict over the use of some soil reclamation machinery brings Kira into contact with some old friends and the beginnings of a dangerous armed rebellion. An average episode with the usual presence of an unnecessary sub-plot involving O'Brien back on DS9.
FAN 45 min; 90 min (2-episode cassette) mTV
VIDrel: CIC/SONOP V/sur

STAR TREK DEEP SPACE NINE: THE SIEGE ** PG

Winrich Kolbe USA 1993

Avery Brooks, Rene Auberjonois, Siddig El Fadil, Terry Farrell, Cirroc Lofton, Colm Meaney, Armin Shimerman, Nana Visitor, Richard Beymer, Stephen Macht, Philip Anglim, Louise Fletcher, Rosalind Chao, Aron Esineberg

A continuation of "The Circle" that sees DS9 being given an ultimatum to evacuate all non-Bajorans and hand over the station to Bajor. While Sisko and a skeleton crew remain behind in hiding, Kira travels to Bajor to bring the of Ministers evidence of Cardassian involvement in the activities of the Circle.
FAN 45 min; 88 min (2-episode cassette) mTV
VIDrel: CIC/SONOP V/dm V/sur

STAR TREK DEEP SPACE NINE: SONS OF MOGH ** 12

David Livingston USA 1994

Avery Brooks, Rene Auberjonois, Alexander Siddig, Terry Farrell, Armin Shimerman, Colm Meaney, Nana Visitor, Michael Dorn, Cirroc Lofton, Tony Todd, Robert Doqui, Dell Young, Elliot Woods

Worf's brother suddenly arrives at DS9 and begs him to perform a Klingon ritual that will restore the lost family honour, but this calls for him to kill his disgraced brother. Meanwhile, the Klingon fleet is involved in war exercises just outside Bajoran space and matters are becoming extremely tense. A most uninspired episode, with much of the action taken up with this potential conflict, the other sub-plot being neatly resolved in next to no time.
FAN 88 min (2-episode cassette) mTV cC
VIDrel: CIC/SONOP V/sur

STAR TREK DEEP SPACE NINE: THE STORYTELLER ** PG

David Livingston USA 1993

Avery Brooks, Rene Auberjonois, Siddig El Fadil, Terry Farrell, Nana

Visitor, Colm Meany, Cirroc Lofton, Lawrence Monoson, Kay E. Kuter, Gina Philips, Jim Jansen, Aron Eisenberg, Jordan Lund

The space station doctor and chief engineer travel to Bajor to assist in a medical emergency, but this turns out to be something far more complex and involved, with a village coming under attack from a strange creature. Meanwhile, back on DS9, Sisko faces a series of negotiations that are fraught with difficulties. Average episode in this series whose lack of in-depth characterisation was a major drawback.
FAN 45 min; 87 min (2-episode cassette) mTV
VIDrel: CIC/SONOP V/dm V/sur

STAR TREK DEEP SPACE NINE: THROUGH THE LOOKING GLASS *

PG

Winrich Kolbe USA 1994

Avery Brooks, Rene Auberjonois, Siddig El Fadil, Terry Farrell, Armin Shimerman, Colm Meaney, Nana Visitor, Andrew Robinson, Felecia M. Bell, Max Grodenchik, Tim Russ, John Patrick Hayden, Dennis Madalone

A sequel of sorts to "Crossover" with Sisko being kidnapped by O'Brien's double from that parallel world. Forced to agree to impersonate his dead counterpart, he has to prevent the latter's scientist wife from finishing a surveillance device that will defeat the Terrans' revolt against their masters. An unfocused and terribly silly episode that is so melodramatic and over-the-top that at times it verges on parody.
FAN 45 min; 88 min (2-episode cassette) mTV
VIDrel: CIC/SONOP V/sur

STAR TREK DEEP SPACE NINE: TRIALS AND TRIBBLE-ATIONS **

PG

Jonathan West USA 1994

Avery Brooks, Rene Auberjonois, Alexander Siddig, Terry Farrell, Armin Shimerman, Colm Meaney, Nana Visitor, Michael Dorn, Cirroc Lofton, Jack Blessing, James W. Janesen, Charlie Brill

A Klingon agent uses a Bajoran orb to send Sisko and a number of the crew back in time to the moment at which Captain Kirk and the USS Enterprise encounter a call for help from a space station under Klingon but soon find themselves having to cope with a plague of furry creatures called "tribbles". Video technology was used to edit the DS9 characters into the original STAR TREK: THE TROUBLE WITH TRIBBLES episode to celebrate its 30th anniversary.
FAN 45 min; 88 min (2-episode cassette) mTV cC
VIDrel: CIC/SONOP; PION (LV only) V/sur LV

STAR TREK DEEP SPACE NINE: TRIBUNAL ** PG

Avery Brooks USA 1994

Avery Brooks, Rene Auberjonois, Siddiq El Faddil, Terry Farrell, Armin Shimerman, Colm Meaney, Nana Visitor, Rosalind Chao, John Beck, Richard Poe, Julian Christopher, Caroline Lagerfelt, Fritz Weaver, Majel Barrett (voice only)

O'Brien and his wife leaves DS9 to take a much-deserved vacation but their runabout is intercepted by Cardassian forces who beam aboard and arrest him. Taken by force to their home planet, he is eventually informed that he is to stand trial on unknown charges. When it discovered that twenty photon warheads are missing from DS9's arsenal, his prospects seem none too favourable. An unremarkable offering with the usual pat ending.
FAN 45 min; 87 min (2-episode cassette) mTV
VIDrel: CIC/SNOP V/dm V/sur

STAR TREK DEEP SPACE NINE: VISIONARY ** PG

Reza Badiyi USA 1994

Avery Brooks, Rene Auberjonois, Siddig El Fadil, Terry Farrell, Armin Shimerman, Colm Meaney, Nana Visitor, Jack Sheaer, Annette Halde, Ray Young, Bob Minor, Dennis Madalone

Slightly injured in an accident, Chief O'Brien begins to experience strange temporal dislocations in which he moves briefly into the future. This unpleasant phenomenon gives him unwelcome glimpses of both his own death and the destruction of the space-station. Meanwhile, a visit from a Romulan de egation provides plenty of other headaches. A well-staged episode but it hardly explores the potential of its time-travel theme.
FAN 45 min; 88 min (2-episode cassette) mTV
VIDrel: CIC/SONOP V/sh

STAR TREK DEEP SPACE NINE: VORTEX ** U

Winrich Kolbe USA 1993

Avery Brooks, Rene Auberjonois, Siddig El Fadil, Terry Farrell, Armin Shimerman, Nana Visitor, Cliff De Young, Randy Oglesby, Max

Grodenchik, Gordon Clapp, Kathleen Garrett, Leslie Engelberg, Majel Barrett (voice only)
An alien commits a murder in the course of a robbery instigated by Quark and is extradited back to his home planet in the custody of Odo, head of security on DS9. During the voyage he teases him by hinting that shape-shifters like Odo once lived on his home world and that a colony of them still exists on another planet, on which they are forced to land. Average episode in a series that never did find a winning formula.
FAN   45 min; 87 min (2-episode cassette) mTV
VIDrel: CIC/SONOP   V/dm V/sur

STAR TREK DEEP SPACE NINE: THE WAY OF THE
WARRIOR *   12
James L. Conway USA   1994
Avery Brooks, Rene Auberjonois, Alexander Siddig, Terry Farrell, Armin Shimerman, Colm Meaney, Nana Visitor, Michael Dorn, Cirroc Lofton, Marc Alaimo, Penny Johnson, Robert O'Reilly, Obi Ndefo, Christopher Darga, Andrew Robinson
This double-episode opened the fourth season on a note of high tension, with a massive Klingon taskforce coming to DS9, where some of the troops make themselves more obnoxious than usual. However, the purpose of their presence remains unclear, until Sisko persuades Worf to ferret out the truth. A vague, overlong and unconvincing effort, full of rather tiresome Klingon histrionics, and only very occasionally interesting.
FAN   88 min mTV   VIDrel: CIC/SONOP; PION (LV only) V/sur LV

STAR TREK DEEP SPACE NINE: WHISPERS ***   U
Les Landau USA   1993
Avery Brooks, Rene Auberjonois, Siddig El Fadil, Terry Farrell, Armin Shimerman, Colm Meaney, Nana Visitor, Cirroc Lofton, Rosalind Chao, Susan Bey, Philip Le Stange plus voices of: Majel Barrett, Han Hatae
After his return from a mission in the Gamma Quadrant, where he was to assist in arrange peace negotiations between two warring races, the chief engineer at DS9 begins to notice small details that suggest there is something amiss. He investigates, only to be plunged into a chain of strange and mysterious events. A better than average episode that climaxes with a good plot twist.
FAN   45 min; 88 min (2-episode cassette) mTV
VIDrel: CIC/SONOP   V/sur

STAR TREK DEEP SPACE NINE: THE WIRE ***   U
Kim Friedman USA   1994
Avery Brooks, Rene Auberjonois, Siddiq El Faddil, Terry Farrell, Armin Shimerman, Colm Meaney, Nana Visitor, Andrew Robinson, Jimmie F. Skaggs, Ann Gillespie, Paul Dooley
When Garak, the exiled Cardassian tailor, begins to exhibit some erratic and rather aggressive behaviour, Dr Bashir has a hard time persuading his friend to agree to treatment. After he collapses, a scan reveals the presence of an implant in his skull whose purpose is unknown and further investigations reveal a link to a murky past. Strong performances flesh out a rather thin plot in this above average episode.
FAN   45 min; 88 min (2-episode cassette) mTV
VIDrel: CIC/SONOP   V/dm V/sur

STAR TREK THE NEXT GENERATION: 11001001 **   U
Paul Lynch USA   1987
Patrick Stewart, Jonathan Frakes, LeVar Burton, Denise Crosby, Brent Spiner, Gates McFadden, Marina Sirtis, Wil Wheaton, Carolyn McCormick, Katy Boyer, Gene Dynarski, Alexandra Johnson, Iva Lane, Kelly Ann McNally, Jack Sheldon
The Enterprise is to have its computers upgraded by an alien race known as the Bynars, who have a natural affinity for computers. Unfortunately, Captain Picard and Commander Riker find that the Bynars have made plans of their own and have lured them into a trap.
FAN   88 min (2-episode cassette) mTV
VIDrel: CIC/SONOP   V/dm

STAR TREK THE NEXT GENERATION:
ACQUIEL *   (PG)
Cliff Bole USA   1992
Patrick Stewart, Jonathan Frakes, Brent Spiner, LeVar Burton, Michael Dorn, Gates McFadden, Marina Sirtis, Reg E. Cathey, Renee Jones, Wayne Grace, Majel Barrett (voice only)
A message relay station near the Klingon border is found abandoned and its occupants missing and foul play is suspected when some human remains are found. Later, the woman lieu-

tenant in charge turns up alive in Klingon custody. A poorly conceived murder mystery with a novel twist that fails to generate more than a modicum of suspense.
FAN   45 min mTV   TVrel

STAR TREK THE NEXT GENERATION: ALL GOOD
THINGS: PARTS 1 AND 2 ***   U
Winrich Kolbe USA   1994
Patrick Stewart, Jonathan Frakes, LeVar Burton, Michael Dorn, Brent Spiner, Gates McFadden, Marina Sirtis, John De Lancie, Denise Crosby, Colm Meaney, Patti Yasutake, Andreas Katsulas, Clyde Kusatsu, Pamela Kosh, Tim Brooks
A two-episode adventure in which Picard becomes lost in three different sections of time and finds himself constantly shifting back and forth, yet is unable to convince his colleagues that this is really happening. Meanwhile, he receives alarming news that the Romulans are massing near the Neutral Zone. An imaginative and well conceived story, with the Q-being making a re-appearance, but in a less abrasive role. This marked the end of the last series.
FAN   88 min mTV   VIDrel: CIC/SONOP; PION (LV only) V/sur LV

STAR TREK THE NEXT GENERATION:
ALLEGIANCE **   U
Winrich Kolbe USA   1990
Patrick Stewart, Jonathan Frakes, LeVar Burton, Michael Dorn, Brent Spiner, Gates McFadden, Wil Wheaton, Marina Sirtis, Stephen Markle, Reiner Schone, Jeff Rector, Joycelyn O'Brien, Jerry Rector
The captain of the Enterprise is kidnapped and replaced by an impostor whose intentions are far from virtuous. Meanwhile, Picard awakes from his sleep to find himself sharing a cell with three other prisoners, one of whom is a fierce alien predator who views the others as a source of food. However, if they are to escape they must all work together, overcoming their fear and mistrust. Fair.
FAN   88 min (2-episode cassette) mTV
VIDrel: CIC/SONOP   V/dm V/sh

STAR TREK THE NEXT GENERATION:
ANGEL ONE **   PG
Michael Ray Rhodes USA   1987
Patrick Stewart, Jonathan Frakes, LeVar Burton, Gates McFadden, Wil Wheaton, Brent Spiner, Michael Dorn, Marina Sirtis, Karen Montgomery, Sam Hennings, Patricia McPherson, Leonard John Crofoot
The Enterprise arrives at Angel One, a planet run entirely by women. Their mission is to pick up the survivors of a crashed space freighter, who being all male, are apparently none too willing to be rescued.
FAN   89 min (2-episode cassette) mTV
VIDrel: CIC/SONOP   V/dm

STAR TREK THE NEXT GENERATION:
THE ARSENAL OF FREEDOM **   (PG)
Les Landau USA   1988
Patrick Stewart, Jonathan Frakes, LeVar Burton, Gates McFadden, Wil Wheaton, Brent Spiner, Michael Dorn, Marina Sirtis, Vincent Schiavelli, Marco Rodriguez, Vyto Ruginis, Julia Nickson, Georges de la Pena
The Enterprise investigates the strange disappearance of another starship that was orbiting the planet Minos. When Picard and his colleagues beam down to the surface of the planet they are attacked by a variety of defensive weapons. At the same time the Enterprise itself comes under attack from some orbital systems. This tape was originally available with the episode WE'LL ALWAYS HAVE PARIS.
FAN   47 min mTV   VIDrel: CIC/SONOP   V

STAR TREK THE NEXT GENERATION:
ATTACHED *   PG
Jonathan Frakes USA   1993
Patrick Stewart, Jonathan Frakes, Brent Spiner, LeVar Burton, Michael Dorn, Gates McFadden, Marina Sirtis, Robin Gammell, Lenore Kasdorf, J.C. Stevens
Dr Crusher and the Captain beam down to a planet to confer with the representatives of a nation that wishes to join the Federation. But they are captured by the enemies of these people, who prepare them for interrogation by attaching devices to their necks. When Crusher and Picard escape, they find that these devices have created a telepathic link between them. Of

minimal interest, this episode failed to explore any of its interesting ideas.
FAN 45 min; 88 min (2-episode cassette) mTV
VIDrel: CIC/SONOP V/dm V/sur

STAR TREK THE NEXT GENERATION: THE BATTLE *
PG
Rob Bowman USA 1987
Patrick Stewart, Jonathan Frakes, LeVar Burton, Denise Crosby, Brent Spiner, Gates McFadden, Michael Dorn, Wil Wheaton, Elaine Nalee, William A. Wallace, Majel Barrett, Rob Knepper, Nan Martin, Robert Ellenstein, Carel Struycken
Picard's past catches up with him when the Enterprise encounters a derelict vessel known as "The Stargazer". It was a former command vessel of his and an old enemy awaits him, having never forgiven him for having caused the death of a Ferengi crew. A rather routine entry in this long-running and somewhat variable series.
FAN 89 min (2-episode cassette) mTV
VIDrel: CIC/SONOP V/dm

STAR TREK THE NEXT GENERATION: THE BEST OF BOTH WORLDS – PART 1 ***
PG
Cliff Bole USA 1990
Patrick Stewart, Jonathan Frakes, LeVar Burton, Michael Dorn, Brent Spiner, Marina Sirtis, Wil Wheaton, Gates McFadden, Elizabeth Dennehy, George Murdock, Whoopi Goldberg
An over-ambitious officer and Borg expert has assumed that she will step into Riker's shoes when he is promoted to captain and given his own vessel. But soon she is confronted with a more serious problem, for the Borg kidnap Picard and turn him into a cybernetic creature like themselves. Shelby, Data and Worf lead a rescue team to retrieve their captain, and meanwhile Riker sets about planning an attack on the Borg vessel.
FAN 88 min (2-episode cassette) mTV
VIDrel: CIC/SONOP; PION (LV only) V/dm V/h LV

STAR TREK THE NEXT GENERATION: THE BEST OF BOTH WORLDS – PART 2 ***
PG
Cliff Bole USA 1990
Patrick Stewart, Jonathan Frakes, Marina Sirtis, LeVar Burton, Gates McFadden, Michael Dorn, Brent Spiner, Wil Wheaton, Elizabeth Dennehy, George Murdock, Whoopi Goldberg
Having failed to disable the Borg ship, neither the Enterprise nor a fleet of starships seems able to prevent the Borg from making for Earth. However, another engagement with the Borg leads to Picard's rescue and Data immediately sets to work, using the altered captain to analyse the collective consciousness of the Borg and discover their weaknesses. This two-part story has also been made available on a single tape/disc.
FAN 88 min (2-episode cassette) mTV
VIDrel: CIC/SONOP; PION (LV only) V/sur LV

STAR TREK THE NEXT GENERATION: THE BIG GOODBYE **
PG
Joseph L. Scanlon USA 1987
Patrick Stewart, Gates McFadden, Brent Spiner, LeVar Burton, Denise Crosby, Jonathan Frakes, Marina Sirtis, Wil Wheaton, Michael Dorn, Lawrence Tierney, Harvey Jason, William Boyett, David Selburg, Gary Armagnal, Mike Genovese
A malfunctioning holodeck traps Picard, Data and Dr Crusher in a recreation of San Francisco of 1941, where they become embroiled in an adventure not unlike a story from the pen of Raymond Chandler.
FAN 89 min (2-episode cassette) mTV
VIDrel: CIC/SONOP V/dm

STAR TREK THE NEXT GENERATION: BIRTHRIGHT – PARTS 1 AND 2 **
U
Winrich Kolbe/Dan Curry USA 1993
Patrick Stewart, Jonathan Frakes, Brent Spiner, LeVar Burton, Michael Dorn, Gates McFadden, Marina Sirtis, Siddig El Fadil, James Cromwell, Christine Rose, Jennifer Gatti, Richard Herd, Sterling Macer Jr, Alan Scarfe
The Enterprise is sent to the Deep Space Nine space-station to assist in its renovation and a strange alien artefact found there causes Data to have a strange vision. Meanwhile, Worf has learnt that his father may still be alive in Romulan captivity and leaves in search of the truth. A tedious double-episode with two unrelated stories that spends too much of its time on Worf's

startling discovery. This story has also been split up and put onto two separate tapes.
FAN 83 min mTV VIDrel: CIC/SONOP; PION (LV only)
V/sur LV

STAR TREK THE NEXT GENERATION: BLOODLINES *
PG
Les Landau USA 1994
Patrick Stewart, Jonathan Frakes, LeVar Burton, Michael Dorn, Marina Sirtis, Gates McFadden, Brent Spiner, Ken Olandt, Peter Slusker, Lee Arenberg, Amy Pietz, Michelan Sisti, Majel Barrett (voice only)
Picard is troubled by a holographic message from Bok, an old Ferengi enemy who still seeks revenge for the death of his son. He threatens to kill Picard's own son, much to the amazement of the Captain, who didn't know that he had any offspring. Eventually, he locates the individual in question, and attempts to secure his safety aboard the Enterprise. Both pat and predictable, this weak entry offers few surprises.
FAN 45 min; 87 min (2-episode cassette) mTV
VIDrel: CIC/SONOP V/dm V/sur

STAR TREK THE NEXT GENERATION: THE BONDING **
PG
Winrich Kolbe USA 1989
Michael Dorn, Patrick Stewart, Jonathan Frakes, LeVar Burton, Gates McFadden, Wil Wheaton, Brent Spiner, Marina Sirtis, Susan Powell, Gabriel Damon, ColmMeaney
Worf leads a team on a mission to investigate a barren planet that was once the home of the Koinonians, a race that destroyed itself in the course of a protracted war. When the ship's archaeologist is accidentally killed whilst exploring an underground passage, Lieutenant Worf feels he is to blame, and performs a strange Klingon ritual with the dead woman's young son.
FAN 87 min (2-episode cassette) mTV
VIDrel: CIC/SONOP V/dm

STAR TREK THE NEXT GENERATION: BOOBY TRAP **
PG
Gabrielle Beaumont USA 1989
Patrick Stewart, Jonathan Frakes, LeVar Burton, Gates McFadden, Wil Wheaton, Brent Spiner, Michael Dorn, Marina Sirtis, Susan Gibney, Colm Meaney, Albert Hall, Julie Warner, Whoopi Goldberg
When the Starship Enterprise comes across a derelict vessel that appears to be caught in an asteroid belt, the Enterprise becomes trapped there too, and its energy reserves are depleted by a deadly radiation field that prevents the ship from breaking free. Desperately trying to find a solution, Geordi resorts to the holodeck, where he recreates the Federation's greatest expert on the warp drive, in the hope that this will provide the answer he needs. Average.
FAN 87 min (2-episode cassette) mTV
VIDrel: CIC/SONOP V/dm

STAR TREK THE NEXT GENERATION: BROTHERS **
PG
Rob Bowman USA 1990
Patrick Stewart, Jonathan Frakes, LeVar Burton, Marina Sirtis, Wil Wheaton, Michael Dorn, Gates McFadden, Cory Danziger, Brent Spiner, Adam Ryen
Data beams down to a mysterious planet where his creator, Dr Soong awaits him, his intention being to implant a chip in Data that will give him human emotions. Unfortunately, the android's evil twin "brother" dupes the doctor and succeeds in stealing the circuitry that was meant for Data.
FAN 87 min (2-episode cassette) mTV
VIDrel: CIC/SONOP V/sur

STAR TREK THE NEXT GENERATION: CAUSE AND EFFECT *
PG
Jonathan Frakes USA 1992
Patrick Stewart, Jonathan Frakes, Brent Spiner, LeVar Burton, Michael Dorn, Gates McFadden, Marina Sirtis, Michelle Forbes, Patt Yasutake, Kelsey Grammer
Exploring an uncharted region of space, the Enterprise and its crew find themselves trapped within a temporal loop, repeating the same period of time over and over again, until they find a way out of their predicament. A poorly handled variation on this theme.
FAN 45 min; 88 min (2-episode cassette) mTV
VIDrel: CIC/SONOP V/dm V/sur

STAR TREK THE NEXT GENERATION: CHAIN OF COMMAND – PART 1 **
Robert Scheerer USA   U
1992
Patrick Stewart, Jonathan Frakes, Brent Spiner, LeVar Burton, Michael Dorn, Gates McFadden, Marina Sirtis, Ronny Cox, Nataija Nogulich, John Durbin, Lou Wagner, David Warner, Heather Lauren Olson, Majel Barrett (voice only)
The crew of the Enterprise are taken completely by surprise when Captain Picard is relieved of duty, along with Dr Crusher and Worf, and a new captain takes over with a very different command style. However, our trio are in fact being sent of a perilous mission to a planet inside Cardassian territory where a new biological weapon is rumoured to be under development. However, once on the planet, they find themselves held hostage by the Cardassians.
FAN   45 min; 87 min (2-episode cassette) mTV
VIDrel: CIC/SONOP; PION (LV only)   V/dm V/sur LV

STAR TREK THE NEXT GENERATION: CHAIN OF COMMAND – PART 2 **
Les Landau USA   PG
1992
Patrick Stewart, Jonathan Frakes, Brent Spiner, LeVar Burton, Michael Dorn, Gates McFadden, Marina Sirtis, Ronny Cox, Nataija Nogulich, John Durbin, Lou Wagner, David Warner, Heather Lauren Olson, Majel Barrett (voice only)
Imprisoned by the Cardassians, Picard is tortured in an attempt to learn the secrets of Federation strategy should a conflict ensue. The new captain of the Enterprise has the delicate task of averting a war and rescuing Picard and the other crew members. Some cassettes have become available with Parts 1 and 2 together rather than the two other episodes (as has the LV disc).
FAN   45 min; 86 min (2-episode cassette) mTV
VIDrel: CIC/SONOP; PION (LV only)   V/dm V/sur LV

STAR TREK THE NEXT GENERATION: THE CHASE **
Jonathan Frakes USA   U
1993
Patrick Stewart, Jonathan Frakes, Brent Spiner, LeVar Burton, Michael Dorn, Gates McFadden, Marina Sirtis, Salome Jens, John Cothran Jr, Maurice Roeves, Linda Thorson, Norman Lloyd
Picard's old archaeology professor pays the ship an unexpected visit hoping to lure the captain into taking part in a project he claims is of truly cosmic importance. This soon involves the Enterprise on a chase from planet to planet as they seek to solve a mystery that has to do with the origins of human life itself. A fascinating idea gets a rushed and unsatisfying treatment.
FAN   45 min; 86 min (2-episode cassette) mTV
VIDrel: CIC/SONOP   V/dm V/sur

STAR TREK THE NEXT GENERATION: THE CHILD **
Rob Bowman USA   PG
1988
Patrick Stewart, Marina Sirtis, Jonathan Frakes, Whoopi Goldberg, Diana Muldaur, LeVar Burton, Gates McFadden, Wil Wheaton, Brent Spiner, Michael Dorn, Seymour Cassel, R.J. William, Colm Meaney, Dawn Armenian
While the Enterprise is on the way to pick up samples of highly contagious viruses, needed to combat an outbreak of plasma plague, Troi's body is taken over by a strange glowing light that leaves her pregnant. As her pregnancy proceeds at many times the normal rate, she soon gives birth to a son who continues to grow rapidly. An imaginative plot idea that is handled quite well, in an intriguing episode that has a neat resolution.
FAN   91 min (2-episode cassette) mTV
VIDrel: CIC/SONOP   V/dm V/sh

STAR TREK THE NEXT GENERATION: CLUES ***
Les Landau USA   PG
1991
Patrick Stewart, Jonathan Frakes, Michael Dorn, LeVar Burton, Gates McFadden, Marina Sirtis, Brent Spiner
Some strange anomalies convince the crew of the Enterprise that they have lost twenty-four hours in their lives since coming across a mysterious region of space. All the signs point to a deliberate cover-up, and Picard eventually learns that the ship came too close for comfort to a xenophobic race which erased the memories of the entire crew in a bid to preserve its solitude.
FAN   87 min (2-episode cassette) mTV
VIDrel: CIC/SONOP   V/sur

STAR TREK THE NEXT GENERATION: CODE OF HONOR **
Russ Mayberry USA   PG
1987
Patrick Stewart, Jonathan Frakes, LeVar Burton, Gates McFadden, Wil Wheaton, Karole Selmon, James Louis Watkins, Jessie Lawrence, Michael Rider, Brooke Bundy, David Renan, Skip Stellrecht, Kenny Koch
The security chief of the Enterprise is abducted, together with a supply of vital vaccine, and Captain Picard is obliged to take part in a battle of wits with a cunning adversary.
FAN   88 min (2-episode cassette) mTV
VIDrel: CIC/SONOP   V/dm V/sh

STAR TREK THE NEXT GENERATION: COMING OF AGE **
Michael Vejar USA   U
1987
Patrick Stewart, Jonathan Frakes, LeVar Burton, Gates McFadden, Brent Spiner, Michael Dorn, Marina Sirtis, Wil Wheaton, Ward Costello, Robert Schenkkan, John Putch, Robert Ito, Stephen Gregory, Tasia Valenza
While Ensign Crusher is away being tested in a Starfleet Academy exam, Captain Picard finds that he is the subject of an inquiry into his abilities as a starship captain.
FAN   91 min (2-episode cassette) mTV
VIDrel: CIC/SONOP   V/dm

STAR TREK THE NEXT GENERATION: CONSPIRACY **
Cliff Bole USA   PG
1988
Patrick Stewart, Jonathan Frakes, LeVar Burton, Gates McFadden, Wil Wheaton, Brent Spiner, Michael Dorn, Marina Sirtis, Henry Darrow, Ward Costello, Ray Reinhardt, Robert Schenkan, Jonathan Farwell, Michael Berryman
The captain of the Enterprise learns of a dangerous conspiracy that could destroy the competence of Starfleet Command, and travels to Earth to resolve this problem.
FAN   90 min (2-episode cassette) mTV
VIDrel: CIC/SONOP   V/dm V/sh

STAR TREK THE NEXT GENERATION: CONTAGION **
Joseph L. Scanlan USA   U
1988
Patrick Stewart, Jonathan Frakes, LeVar Burton, Gates McFadden, Wil Wheaton, Brent Spiner, Michael Dorn, Marina Sirtis, Diana Muldaur, Thalmus Rasulala, Carolyn Seymour, Dana Sparks, Colm Meaney, Folkert Schmidt
After a bizarre electronic virus cripples the computer system of the Enterprise, the ship is left stranded in the Neutral Zone, and a hostile Romulan vessel is found to be close at hand.
FAN   92 min (2-episode cassette) mTV
VIDrel: CIC/SONOP   V/dm V/sh

STAR TREK THE NEXT GENERATION: CONUNDRUM **
Les Landau USA   U
1992
Patrick Stewart, Jonathan Frakes, Brent Spiner, LeVar Burton, Michael Dorn, Gates McFadden, Marina Sirtis, Erich Andersen, Liz Vassey, Erick Weiss, Majel Barrett (voice only)
An encounter with an alien vessel leads to the entire crew loosing the memory of their own identity, while the computer system is also affected. Once the latter's functions are restored, it appears that the Enterprise is on a secret mission to destroy the base of an alien race with whom the Federation is at war.
FAN   45 min; 88 min (2-episode cassette) mTV
VIDrel: CIC/SONOP   V/dm V/sur

STAR TREK THE NEXT GENERATION: COST OF LIVING *
Winrich Kolbe USA   U
1992
Patrick Stewart, Jonathan Frakes, Brent Spiner, LeVar Burton, Michael Dorn, Gates McFadden, Marina Sirtis, Brian Bonsall, Majel Barrett, Tony Jay, Carel Struycken, David Oliver, Albie Selznick, Patrick Gronin, Tracey D'Arcy
After destroying an asteroid that was about to crash on an inhabited planet, the Enterprise begins to suffer a series of malfunctions that soon turn into a major crisis. At the same time, Roxana Troi decides to get married and hold the wedding ceremony onboard, which gives her a chance to annoy everyone. A very loosely plotted episode whose energy is dispersed among far too many sub-plots.
FAN   45 min; 92 min (2-episode cassette) mTV
VIDrel: CIC/SONOP   V/dm V/sur

STAR TREK THE NEXT GENERATION: DARK RAGE *

PG

Les Landau USA 1993

Patrick Stewart, Jonathan Frakes, Brent Spiner, LeVar Burton, Michael Dorn, Gates McFadden, Marina Sirtis, Majel Barrett, Norman Large, Kirsten Dunst, Amick Byram, Adriana Weiner

A new race of telepathic beings apply for Federation membership, and Troi's mother is sent to teach their representatives how to use spoken language. But the strain of this proves to be too much for her, resulting in emotional outbursts and eventual collapse. As Troi tries to save her mother's life she learns of a past family tragedy. An unexciting episode with a weak plot and poor development.

FAN 45 min; 88 min (2-episode cassette) mTV
VIDrel: CIC/SONOP V/dm V/sur

STAR TREK THE NEXT GENERATION: DARMOK **

PG

Winrich Kolbe USA 1991

Patrick Stewart, Jonathan Frakes, Brent Spiner, LeVar Burton, Michael Dorn, Gates McFadden, Marina Sirtis, Richard Allen, Colm Meaney, Paul Winfield, Ashley Judd, Majel Barrett

A meeting with a strange alien race whose speech consists of nothing but incomprehensible allusions proves a test of mental and physical stamina for Picard. When the aliens beam him and their captain down to a planet, he finds that an unknown challenge awaits them both as they struggle to devise a means of communicating in the face of a common danger.

FAN 45 min; 88 min (2-episode cassette) mTV
VIDrel: CIC/SONOP V/sur

STAR TREK THE NEXT GENERATION: DATALORE **

PG

Rob Bowman USA 1987

Patrick Stewart, Jonathan Frakes, LeVar Burton, Denise Crosby, Brent Spiner, Gates McFadden, Marina Sirtis, Wil Wheaton, Michael Dorn, Bill Yeager

The Starship Enterprise is sent on a mission to the Omicron Theta star system, the home of Data. But on their arrival they find a planet apparently devoid of life, and a sinister mystery awaits them.

FAN 89 min (2-episode cassette) mTV
VIDrel: CIC/SONOP V/dm

STAR TREK THE NEXT GENERATION: DATA'S DAY **

U

Robert Wiemer USA 1990

Patrick Stewart, Jonathan Frakes, Michael Dorn, LeVar Burton, Gates McFadden, Marina Sirtis, Brent Spiner, Rosalind Chao, Sierra Pecheur, Alan Scarfe

A one-day-in-the-life episode that follows Data as he prepares for Chief O'Brien's wedding and among other things, gets a tap dancing lesson from Dr Crusher. Meanwhile, a Romulan spy is posing as a Vulcan ambassador, and makes arrangements to beam onboard.

FAN 87 min (2-episode cassette) mTV
VIDrel: CIC/SONOP V/sur

STAR TREK THE NEXT GENERATION: THE DAUPHIN *

PG

Rob Bowman USA 1988

Patrick Stewart, Jonathan Frakes, LeVar Burton, Gates McFadden, Wil Wheaton, Brent Spiner, Michael Dorn, Marina Sirtis, Paddi Edwards, Jamie Hubbard, Colm Meaney, Peter Neptune, Madchen Amick, Cindy Sorenson, Jennifer Barlow

Young Ensign Crusher finds himself in love with a pretty alien princess being escorted by the Enterprise, but falls foul of her guardian: a shape-changing alien woman. However, his ardour is dampened by his realisation that the object of his affections also possessing this ability, and that her attractive exterior is nothing more than a pleasing illusion. Quite a weak story, with little being made of the ramifications of this idea.

FAN 91 min (2-episode cassette) mTV
VIDrel: CIC/SONOP V/dm V/sh

STAR TREK THE NEXT GENERATION: THE DEFECTOR *

PG

Robert Scheerer USA 1990

Patrick Stewart, Jonathan Frakes, LeVar Burton, Gates McFadden, Wil Wheaton, Brent Spiner, Michael Dorn, Marina Sirtis, James Stoyan, Andreas Katsulas, John Hancock, S.A. Templeton

A Romulan in a small scoutship requests asylum aboard the Enterprise and offers news of a secret Romulan offensive. When

this defector turns out to be a high-ranking military officer, notorious for his brutality, Picard suspects that his defection may be part of an elaborate trap. However, to investigate the truth of his claims, he is forced to take the ship to a Romulan base at which their fleet is supposed to be gathering.

FAN 88 min (2-episode cassette) mTV
VIDrel: CIC/SONOP V/dm

STAR TREK THE NEXT GENERATION: DEJA Q *

PG

Les Landau USA 1990

Patrick Stewart, Jonathan Frakes, LeVar Burton, Michael Dorn, Gates McFadden, John De Lancie, Whoopi Goldberg, Brent Spiner, Wil Wheaton, Marina Sirtis, Richard Cansino, Betty Muramoto

A Q-being who formerly possessed limitless power materialises aboard the Enterprise, whilst it is in the process of attempting to save a planet from its own moon, which has broken free from its orbit. A very lightweight yarn that injects a note of comedy.

FAN 88 min (2-episode cassette) mTV
VIDrel: CIC/SONOP V/dm

STAR TREK THE NEXT GENERATION: DESCENT – PARTS 1 AND 2 **

PG

Alexander Singer USA 1993

Patrick Stewart, Jonathan Frakes, LeVar Burton, Michael Dorn, Brent Spiner, Gates McFadden, Marina Sirtis, John Neville, Jim Norton, Natalija Nogulich, Brian J. Cousins, Stephen Hawking, Richard Gilbert-Hall, Stephen James Carver

The Borg attack a Federation outpost and the Enterprise comes to its aid, but find that the Borg have grown more powerful and are now starting to behave as if they are separate beings, exhibiting signs of an individual consciousness. This marks a dangerous development, in which Data plays a key role. A rather disappointing two-part entry, a pilot to the last season, with little work done on its strong ideas. Both parts 1 and 2 have also appeared with other episodes.

FAN 82 min mTV VIDrel: CIC/SONOP; PION (LV only)
V/sur LV

STAR TREK THE NEXT GENERATION: DEVIL'S DUE **

PG

Tom Benko USA 1991

Patrick Stewart, Jonathan Frakes, Michael Dorn, LeVar Burton, Gates McFadden, Marina Sirtis, Brent Spiner, Marta DuBois, Marcelo Tubert, Paul Lambert

An aborted 1970s "Star Trek" episode formed the basis for this unusual if rather talky story, in which Picard must argue in court that a woman claiming to be the devil is in fact an impostor, and that the contract she holds on an entire planet is therefore invalid.

FAN 87 min (2-episode cassette) mTV
VIDrel: CIC/SONOP V/sur

STAR TREK THE NEXT GENERATION: DISASTER **

PG

Gabrielle Beaumont USA 1991

Patrick Stewart, Jonathan Frakes, Brent Spiner, LeVar Burton, Michael Dorn, Gates McFadden, Marina Sirtis, Rosalind Chao, Colm Meaney, Michelle Forbes, Erika Flores, John Chrisitan Graas, Max Supera, Cameron Arnett

A collision with a quantum filament almost cripples the Enterprise, putting members of its crew to a test of endurance and ingenuity. With Picard trapped inside a turbolift, Troi is the most senior officer on the bridge and has to make some tough decisions, while Data literally puts his head to good use.

FAN 45 min; 88 min (2-episode cassette) mTV
VIDrel: CIC/SONOP V/sur

STAR TREK THE NEXT GENERATION: THE DRUMHEAD **

PG

Jonathan Frakes USA 1991

Patrick Stewart, Jonathan Frakes, LeVar Burton, Denise Crosby, Brent Spiner, Gates McFadden, Marina Sirtis, Michael Dorn, Jean Simmons, Bruce French, Spencer Garrett, Earl Billings, Henry Woronicz, Ann Shea

A Klingon onboard the Enterprise is found to be working for the Romulans and a witch-hunt rapidly develops as a search is made for other collaborators. A rather talky episode doubtlessly inspired (if that's the right word) by the McCarthy trials of the 1950s.

FAN 87 min (2-episode cassette) mTV
VIDrel: CIC/SONOP V/sur

STAR TREK THE NEXT GENERATION: ELEMENTARY, DEAR DATA ★★

Rob Bowman USA   PG
                                                   1988
Brent Spiner, Patrick Stewart, Jonathan Frakes, LeVar Burton, Michael Dorn, Gates McFadden, Wil Wheaton, Marina Sirtis, Daniel Davis, Alan Shearman, Biff Manard, Diz White, Anne Elizabeth Ramsey, Richard Merson
Data plays at Sherlock Holmes, using the holodeck to recreate Victorian London. However, seemingly harmless pastimes have a way of becoming dangerous after the computer is instructed to create a character capable of defeating Data, as it comes up with no less an adversary than Professor Moriarty. In no time at all, this villain has seized control of the Enterprise and forces a battle of wits before the situation is resolved.
FAN   92 min (2-episode cassette) mTV
VIDrel: CIC/SONOP   V/dm V/sh

STAR TREK THE NEXT GENERATION: EMERGENCE ★★★

Cliff Bole USA   U
                                                   1994
Patrick Stewart, Jonathan Frakes, LeVar Burton, Michael Dorn, Marina Sirtis, Gates McFadden, Brent Spiner, David Huddleston, Vinny Argiro, Thomas Kopache, Arlee Reed
Whilst rehearsing "The Tempest" on the holodeck, Picard and Data are almost run over by the Orient Express, when two separate computer programs appear to merge. But this proves to be just a prelude to far stranger events that see the ship apparently evolving into an intelligent and sentient being. One of the most imaginative stories in the series, even if the plot does get a bit bogged down in the contrived symbolism at times.
FAN   45 min; 87 min (2-episode cassette) mTV
VIDrel: CIC/SONOP   V/dm V/sur

STAR TREK THE NEXT GENERATION: THE EMISSARY ★★

Cliff Bole USA   PG
                                                   1989
Patrick Stewart, Michael Dorn, Suzie Plakson, Jonathan Frakes, LeVar Burton, Gates McFadden, Wil Wheaton, Brent Spiner, Marina Sirtis, Suzie Plakson, Lance LeGault, Georgann Johnson, Colm Meaney, Anne Elizabeth Ramsey
Lieutenant Worf finds himself becoming personally involved when a former lover of his comes aboard the Enterprise, her mission being to mediate between the Federation and some rebel Klingons who have just been released from suspended animation.
FAN   90 min (2-episode cassette) mTV
VIDrel: CIC/SONOP   V/dm

STAR TREK THE NEXT GENERATION: ENCOUNTER AT FARPOINT ★★

Corey Allen USA   U
                                                   1987
Patrick Stewart, Jonathan Frakes, LeVar Burton, Denise Crosby, Brent Spiner, Gates McFadden, Marina Sirtis, Wil Wheaton, John De Lancie, Michael Bell, Colm Meaney, DeForest Kelley, Cary Hiroyuki, Timothy Dang, David Erskine
The first in a series of new "Star Trek" adventures, "Encounter At Farpoint" is set eighty-five years after the time of Captain Kirk, and tells of how the crew of the new Enterprise are sent on a mission, to determine whether a space station can be used as a new Starfleet base.
FAN   91 min (ort 96 min) mTV   VIDrel: CIC/SONOP; PION
(LV only)   V/dm V/sh LV

STAR TREK THE NEXT GENERATION: THE ENEMY ★★★

David Carson USA   PG
                                                   1989
Patrick Stewart, Jonathan Frakes, LeVar Burton, Gates McFadden, Wil Wheaton, Brent Spiner, Michael Dorn, Marina Sirtis, John Snyder, Andreas Katsulas, Colm Meaney
Lieutenant Geordi LaForge is accidentally stranded on an inhospitable and storm-lashed planet, when he is taken prisoner after encountering a wounded Romulan Centurion. As the Enterprise attempts to retrieve him and avoid a confrontation with the Romulans, Geordi has to convince his captor that they have to work together if they are to survive. A well plotted episode with ample dramatic tension that makes some interesting points.
FAN   88 min (2-episode cassette) mTV
VIDrel: CIC/SONOP   V/dm

STAR TREK THE NEXT GENERATION: ENSIGN RO ★★

Les Landau USA   U
                                                   1991
Patrick Stewart, Jonathan Frakes, Brent Spiner, LeVar Burton,
Michael Dorn, Gates McFadden, Marina Sirtis, Michelle Forbes, Scott Marlowe, Ken Thorley, Jeffrey Hayenga, Frank Collison, Harley Venton, Cliff Potts, Whoopi Goldberg
A disgraced and court-martialled Bajoran ensign is assigned to the Enterprise as it is about to begin a delicate mission following an attack on a Federation outpost. However, Picard soon learns that things and people are not always what they seem and eventually uncovers a cunning conspiracy.
FAN   45 min; 88 min (2-episode cassette) mTV
VIDrel: CIC/SONOP   V/dm V/sur

STAR TREK THE NEXT GENERATION: ENSIGNS OF COMMAND ★★

Cliff Bole USA   PG
                                                   1989
Patrick Stewart, Jonathan Frakes, LeVar Burton, Gates McFadden, Wil Wheaton, Brent Spiner, Michael Dorn, Marina Sirtis, Eileen Seeley, Mark L. Taylor, Richard Allen, Colm Meaney, Mart McChesney
A group of humans have settled illegally on Tau Cygna Five, and are now threatened by some hostile aliens, who are simply following to the letter of the law their rights under a Federation treaty.
FAN   87 min (2-episode cassette) mTV
VIDrel: CIC/SONOP   V/dm

STAR TREK THE NEXT GENERATION: ETHICS ★★   PG

Chip Chalmers USA   1992
Patrick Stewart, Jonathan Frakes, Brent Spiner, LeVar Burton, Michael Dorn, Gates McFadden, Marina Sirtis, Caroline Kova, Patti Yasutake, Brian Bonsall
Worf is crippled in an accident in a cargo bay and faces the prospect of not being able to walk again without artificial aids. True to his Klingon principles, he prefers to commit suicide, but opts instead to risk everything in a new operation, pioneered by a brilliant medical researcher who is ruthless and cold-blooded about experimenting with patients' lives.
FAN   45 min; 88 min (2-episode cassette) mTV
VIDrel: CIC/SONOP   V/dm V/sur

STAR TREK THE NEXT GENERATION: EVOLUTION ★★

Winrich Kolbe USA   PG
                                                   1989
Patrick Stewart, Jonathan Frakes, LeVar Burton, Gates McFadden, Wil Wheaton, Brent Spiner, Michael Dorn, Marina Sirtis, Ken Jenkins, Mary McCusker, Randal Patrick, Whoopi Goldberg, Scott Grimes, Amy O'Neill
After a visiting astrophysicist starts conducting experiments in a binary star system, the Enterprise is shaken by a number of inexplicable system failures across the entire ship, and disaster is only narrowly averted. Wesley Crusher begins to suspect that this may be related to the escape of two "nanites" – tiny organic robots he had bred as part of an experiment, but it soon becomes clear that the situation is far worse than he could have foreseen.
FAN   87 min (2-episode cassette) mTV
VIDrel: CIC/SONOP   V/dm

STAR TREK THE NEXT GENERATION: EYE OF THE BEHOLDER ★★★

Robert Wiemer USA   U
                                                   1994
Patrick Stewart, Jonathan Frakes, LeVar Burton, Michael Dorn, Marina Sirtis, Gates McFadden, Brent Spiner, Mark Rolston, Nancy Harewood, Tim Lounibos, Nora Leonhardt, Johanna McCloy, Dugan Savoye, Majel Barrett (voice only)
The suicide of an apparently stable warp-drive technician presents Troi and Worf with a mystery that seems to be insoluble. However, Troi puts her empathic abilities to good use, but not before she has put her own life at risk. Quite a tense episode, though it is slightly spoilt by a climax that is both contrived and a little hard to follow.
FAN   45 min; 88 min (2-episode cassette) mTV
VIDrel: CIC/SONOP   V/dm V/sur

STAR TREK THE NEXT GENERATION: FACE OF THE ENEMY ★★

Gabriel Beaumont USA   (PG)
                                                   1992
Patrick Stewart, Jonathan Frakes, Brent Spiner, LeVar Burton, Michael Dorn, Gates McFadden, Marina Sirtis, Scott McDonald, Carolyn Seymour, Robertson Dean, Barry Lynch, Dennis Cockrum, Pamela Winslow
Troi awakes and finds herself aboard a Romulan warship, with her appearance surgically altered to that of a Romulan. She learns that she has been kidnapped from a symposium she was attending in order to play a key role in smuggling three high-

ranking dissidents into Federation space. However, when things go wrong, she has to use all her skills to survive. A reasonably entertaining episode but far from outstanding.
FAN   45 min mTV   TVrel

STAR TREK THE NEXT GENERATION: FAMILY **   PG
Les Landau USA   1990
Patrick Stewart, Jonathan Frakes, Marina Sirtis, LeVar Burton, Wil Wheaton, Gates McFadden, Jeremy Kemp, Samantha Eggar, Theodore Bikel, Georgia Brown, Whoopi Goldberg, Dennis Creaghan, David Tristan Birken, Doug Wert
While the Enterprise undergoes a refit, Worf enjoys a visit from his human adoptive parents, and Picard sets out to visit his brother on Earth, using his trip as a way of purging the painful memories of his encounter with the Borg.
FAN   87 min (2-episode cassette) mTV
VIDrel: CIC/SONOP   V/sur

STAR TREK THE NEXT GENERATION: FINAL MISSION ***   PG
Corey Allen USA   1990
Patrick Stewart, Jonathan Frakes, LeVar Burton, Marina Sirtis, Wil Wheaton, Michael Dorn, Gates McFadden, Brent Spiner, Nick Tate, Kim Hamilton
A shuttle carrying Picard and Wesley to a conference crashes on a desert planet whose supplies of water are defended by automatic weapons. When Picard is injured and the shuttle captain killed trying to obtain water from one such source, Wesley finds that the life of the captain depends on his resourcefulness. This was Wheaton's last episode playing Ensign Wesley Crusher.
FAN   87 min (2-episode cassette) mTV
VIDrel: CIC/SONOP   V/sur

STAR TREK THE NEXT GENERATION: FIRST CONTACT **   PG
Cliff Bole USA   1991
Patrick Stewart, Jonathan Frakes, Michael Dorn, LeVar Burton, Gates McFadden, Marina Sirtis, Brent Spiner, George Coe, Carolyn Seymour, Michael Ensign, George Hearn, Steven Anderson, Sachi Parker, Bebe Neuwirth
The honour of making "first contact" goes to the Enterprise, their mission being to offer Federation membership to a race that stands on the brink of developing warp drive. However, the ruler of the planet is forced to reject their offer when he realises that his people are not yet ready to embrace other cultures.
FAN   87 min (2-episode cassette) mTV
VIDrel: CIC/SONOP   V/sur

STAR TREK THE NEXT GENERATION: THE FIRST DUTY **   U
Paul Lynch USA   1992
Patrick Stewart, Jonathan Frakes, Brent Spiner, LeVar Burton, Michael Dorn, Gates McFadden, Marina Sirtis, Wil Wheaton, Ray Walston, Robert Duncan McNeill, Ed Lauter, Richard Fancy, Jacqueline Brookes
After Wesley Crusher is injured during a training exercise in which a cadet was killed, his mother and Picard go to Starfleet Academy on Earth to be with him during the board of inquiry into this incident. What transpires there leads Picard to suspect that the full truth has not yet emerged and he has to persuade his young protege to make a difficult decision.
FAN   45 min; 92 min (2-episode cassette) mTV
VIDrel: CIC/SONOP   V/dm V/sur

STAR TREK THE NEXT GENERATION: FIRSTBORN ***   PG
Jonathan West USA   1994
Patrick Stewart, Jonathan Frakes, LeVar Burton, Michael Dorn, Brent Spiner, Gates McFadden, Marina Sirtis, James Sloyan, Brian Bonsall, Gwynyth Walsh, Joel Swetow, Barbara Marsh, Clint Mitchell, Armin Shimerman, Michael Danek
Troubled by the unwillingness of his son to become a Klingon warrior, Worf takes the boy on a visit to a Klingon colony for a religious festival. There, an attempt on his life is foiled by a mysterious stranger who soon comes to exert quite an impact on the lives of both father and son. But when his identity is revealed, it comes as an utter surprise. Well written and thought-provoking, this was one of the more intelligent stories in the series.
FAN   45 min; 88 min (2-episode cassette) mTV
VIDrel: CIC/SONOP   V/dm V/sur

STAR TREK THE NEXT GENERATION: A FISTFUL OF DATAS **   U
Patrick Stewart USA   1992
Patrick Stewart, Jonathan Frakes, Brent Spiner, LeVar Burton, Michael Dorn, Gates McFadden, Marina Sirtis, Brian Bonsall, John Pyper-Ferguson, Joy Garrett, Jorge Cervera Jr, Majel Barrett (voice only)
Worf and his son Alexander head for the holodeck for a little Western adventure but find themselves facing real dangers when an experiment in using Data as a backup computer system goes slightly wrong. An enjoyable if slightly awkwardly contrived tale that seems to have been inspired by the feature film WESTWORLD.
FAN   45 min; 88 min (2-episode cassette) mTV
VIDrel: CIC/SONOP   V/dm V/sur

STAR TREK THE NEXT GENERATION: FORCE OF NATURE **   U
Robert Lederman USA   1993
Patrick Stewart, Jonathan Frakes, LeVar Burton, Michael Dorn, Marina Sirtis, Brent Spiner, Gates McFadden, Michael Corbett, Margaret Reed, Lee Arenberg, Majel Barrett (voice only)
An alien brother and sister are driven to desperate measures and beam aboard the Enterprise, where they beg the crew to listen to their pleas for assistance, claiming that the use of the warp drive is responsible for the impending destruction of their planet, which is threatened by a rift in the space-time continuum. Average.
FAN   45 min; 88 min (2-episode cassette) mTV
VIDrel: CIC/SONOP   V/dm V/sur

STAR TREK THE NEXT GENERATION: FRAME OF MIND ****   PG
James L. Conway USA   1993
Patrick Stewart, Jonathan Frakes, Brent Spiner, LeVar Burton, Michael Dorn, Gates McFadden, Marina Sirtis, David Selburg, Andrew Prine, Gary Wentz, Susanna Thompson, Allan Dean Moore
While preparing for his role in Chekhov's play Ward 47, Riker begins to suffer from strange hallucinations in which he is a patient in a mental asylum on an alien planet, and his life aboard the Enterprise is a mere delusion. An extremely well-realised and multi-layered episode (one of the best ever) where nothing is as it seems and the distinction between illusion and reality becomes extremely blurred.
FAN   45 min; 86 min (2-episode cassette) mTV
VIDrel: CIC/SONOP   V/dm V/sur

STAR TREK THE NEXT GENERATION: FUTURE IMPERFECT ***   PG
Les Landau USA   1990
Patrick Stewart, Jonathan Frakes, LeVar Burton, Marina Sirtis, Wil Wheaton, Michael Dorn, Gates McFadden, Brent Spiner, Andreas Katsulas, Chris Demetral
Following his trip to a mysterious planet, Riker awakes in sickbay and learns that he has apparently been in a coma for 16 years and that during this time, peace negotiations have begun with the Romulans. However, he soon grows suspicious and begins to suspect that he is a victim of a clever Romulan trick. A tightly plotted and intriguing episode with a neat surprise ending.
FAN   87 min (2-episode cassette) mTV
VIDrel: CIC/SONOP   V/sur

STAR TREK THE NEXT GENERATION: GALAXY'S CHILD **   PG
Winrich Kolbe USA   1991
Patrick Stewart, Jonathan Frakes, Gates McFadden, Michael Dorn, Lanei Chapman, LeVar Burton, Marina Sirtis, Brent Spiner, Susan Gibney, Whoopi Goldberg
Geordi receives a visit from the female warp-drive specialist whom he had re-created on the holodeck in the BOOBY TRAP episode, but is hurt to find that in real life she is abrasive, hostile and quite introverted. However, after a pregnant space creature is accidentally killed and its offspring attaches itself to the ship's hull, they have to work together to find a way of preventing it from draining the ship's power supply. A moderately exciting episode.
FAN   87 min (2-episode cassette) mTV
VIDrel: CIC/SONOP   V/sur

STAR TREK THE NEXT GENERATION: GAMBIT – PARTS 1 AND 2 ***   PG
Peter Lauritson/Alexander Singer USA   1993
Patrick Stewart, Jonathan Frakes, LeVar Burton, Michael Dorn, Gates

McFadden, Marina Sirtis, Brent Spiner, Richard Lynch, Robin Curtis, Cauitlin Brown, Alan Altschuld, Stephen Lee, Bruce Gray, Cameron Thor, Sabrina Lebeuf
Another full-length story that sees the crew of the Enterprise tackling a gang of mercenaries who are out to steal Romulan artefacts from various sites of historical interest throughout the galaxy. Picard and Ryker find their nerve and guile tested to the limit. An enjoyable entry, it maintains tension right to the end, even if the climax is disappointing. The non-LV versions of Parts 1 and 2 have also been made available with "Interface" and "Phantasms" respectively.
FAN   82 min; 88 min (2-episode cassette) mTV
VIDrel: CIC/SONOP; PION (LV only)   V/sur LV

STAR TREK THE NEXT GENERATION: THE GAME **
Corey Allen USA   PG
                                                       1991
Patrick Stewart, Jonathan Frakes, Brent Spiner, LeVar Burton, Michael Dorn, Gates McFadden, Marina Sirtis, Ashley Judd, Katherine Moffat, Colm Meaney, Patti Yasutake, Wil Wheaton, Diane M. Hurley, Majel Barrett (voice only)
Riker returns from leave on the planet Raisa and brings back with him a strange alien game device that spreads like wildfire aboard the Enterprise and soon has most of the crew addicted. However, Wesley Crusher and a bright female ensign eventually discover a far more sinister side to this entire affair.
FAN   45 min; 88 min (2-episode cassette) mTV
VIDrel: CIC/SONOP   V/dm V/sur

STAR TREK THE NEXT GENERATION: GENESIS ***
Gates McFadden USA   PG
                                                       1994
Patrick Stewart, Jonathan Frakes, LeVar Burton, Michael Dorn, Marina Sirtis, Gates McFadden, Brent Spiner, Patti Yasutake, Dwight Schultz, Carlos Ferro, Majel Barrett (voice only)
When a missile goes astray during trials, Picard and Data leave the ship on a shuttle in a bid to retrieve it. Upon their return, they find the Enterprise drifting through space, with most of its systems inactive and the crew transformed into a variety of strange animals. A highly imaginative and well staged episode, as ever let down by the supreme ease with which a solution is found to these difficulties.
FAN   45 min; 88 min (2-episode cassette) mTV
VIDrel: CIC/SONOP   V/dm V/sur

STAR TREK THE NEXT GENERATION: HALF A LIFE *
Les Landau USA   PG
                                                       1991
Patrick Stewart, Jonathan Frakes, LeVar Burton, Denise Crosby, Brent Spiner, Gates McFadden, Marina Sirtis, Michael Dorn, Majel Barrett, David Ogden Stiers, Michelle Frobes, Terence McNally, Carel Struycken
In order to save their dying sun, an alien scientist comes aboard the Enterprise to conduct a series of important experiments. There he encounters and (impossible to believe!) falls in love with Lwaxana Troi. Sadly, it soon becomes known that his time is almost up, since everyone in his culture is obliged to commit suicide once they reach sixty. Mrs Troi is outraged by this, presenting Picard with a major diplomatic crisis. A silly and moralistic episode.
FAN   87 min (2-episode cassette) mTV
VIDrel: CIC/SONOP   V/sur

STAR TREK THE NEXT GENERATION: HAVEN **   PG
Richard Compton USA   1987
Marina Sirtis, Patrick Stewart, Jonathan Frakes, LeVar Burton, Gates McFadden, Wil Wheaton, Brent Spiner, Michael Dorn, Majel Barrett, Rob Knepper, Nan Martin, Robert Ellenstein, Danitza Kingsley, Anna Katarina
Deanna Troi sets out to take part in an arranged marriage, a time-honoured Betazoid custom. However, all is not as it seems and even a disappointed Commander Riker could hardly guess at what is to follow. But when the Enterprise meets another vessel whose crew are suffering from a deadly disease all is made clear.
FAN   89 min (2-episode cassette) mTV
VIDrel: CIC/SONOP   V/dm

STAR TREK THE NEXT GENERATION: HEART OF GLORY **   U
Rob Bowman USA   1987
Michael Dorn, Patrick Stewart, Jonathan Frakes, LeVar Burton, Denise Crosby, Brent Spiner, Gates McFadden, Marina Sirtis, Wil

Wheaton, Vaughn Armstrong, Charles J. Hyman, David FRoman, Robert Bauer, Dennis Madalone
The USS Enterprise is taken over by a couple of rogue Klingons, and Lieutenant Worf finds himself locked in a conflict between loyalty to his comrades and loyalty to his kin.
FAN   91 min (2-episode cassette) mTV
VIDrel: CIC/SONOP   V/dm

STAR TREK THE NEXT GENERATION: HERO WORSHIP **   PG
Patrick Stewart USA   1991
Patrick Stewart, Jonathan Frakes, Brent Spiner, LeVar Burton, Michael Dorn, Gates McFadden, Marina Sirtis, Joshua Harris, Jarley Venton, Sheila Franklin, Steven Einspahr
A young boy taken off a derelict ship by Data starts to hero worship him so much that he eventually dresses and talks like his android rescuer. It soon becomes clear that this obsession is no mere childish whim, but is linked to what happened to the ship on which he was travelling.
FAN   45 min; 88 min (2-episode cassette) mTV
VIDrel: CIC/SONOP   V/dm V/sur

STAR TREK THE NEXT GENERATION: HIDE AND Q **   PG
Cliff Bole USA   1987
Patrick Stewart, Jonathan Frakes, LeVar Burton, Gates McFadden, Wil Wheaton, Brent Spiner, Michael Dorn, Marina Sirtis, John De Lancie, Elaine Nalee, William A. Wallace
Captain Picard's old friend Q, the mischievous alien with unlimited power, returns once more to plague the crew of the Enterprise, this time making Riker the object of his attentions.
FAN   89 min (2-episode cassette) mTV
VIDrel: CIC/SONOP   V/dm

STAR TREK THE NEXT GENERATION: THE HIGH GROUND **   PG
Gabrielle Beaumont USA   1990
Patrick Stewart, Jonathan Frakes, LeVar Burton, Gates McFadden, Wil Wheaton, Brent Spiner, Michael Dorn, Marina Sirtis, Kerrie Keane, Richard Cox, Marc Buckland, Fred G. Smith, Christopher Pettiet
Dr Crusher is taken hostage by a group of terrorists who, having developed a means of instantaneous teleportation, now find themselves stricken with a lethal degenerative disease. When they come aboard the Enterprise with the intention of destroying it, Crusher attempts to win them over by finding a cure.
FAN   88 min (2-episode cassette) mTV
VIDrel: CIC/SONOP   V/dm

STAR TREK THE NEXT GENERATION: HOLLOW PURSUITS **   PG
Cliff Bole USA   1990
Patrick Stewart, Jonathan Frakes, LeVar Burton, Michael Dorn, Marina Sirtis, Wil Wheaton, Brent Spiner, Gates McFadden, Dwight Schultz, Whoopi Goldberg, Charley Lang
Having created a fantasy world into which he can retreat, a painfully shy member of the engineering team finds that his invention poses a threat to the entire crew of the Enterprise.
FAN   87 min (2-episode cassette) mTV
VIDrel: CIC/SONOP   V/dm V/sh

STAR TREK THE NEXT GENERATION: HOME SOIL **   PG
Corey Allen USA   1987
Patrick Stewart, Jonathan Frakes, LeVar Burton, Gates McFadden, Wil Wheaton, Brent Spiner, Michael Dorn, Marina Sirtis, Walter Gotell, Elizabeth Lindsay, Gerard Prendergast, Mario Roccuzzo, Carolyne Barry
The Enterprise arrives at Velara III, a planet undergoing a massive transformation at the hands of its scientists. Whilst there, Data is attacked by a crystalline lifeforce that poses a deadly threat to the crew of the Enterprise.
FAN   89 min (2-episode cassette) mTV
VIDrel: CIC/SONOP   V/dm

STAR TREK THE NEXT GENERATION: HOMEWARD **   PG
Alexander Singer USA   1994
Patrick Stewart, Jonathan Frakes, LeVar Burton, Michael Dorn, Marina Sirtis, Gates McFadden, Brent Spiner, Penny Johnson, Brian Markinson, Edward Penn, Paul Sorvino, Susan Christy, Majel Barrett (voice only)
The Enterprise receives a distress call from Worf's foster brother

who is a cultural observer on a planet that is in the process of losing its atmosphere. Worf beams down and learns that his brother is so anxious to save the planet's inhabitants that he has gone native, thus ignoring the Federation's "Prime Directive" on non-interference. Predictably enough, a solution is found that satisfies all parties. Average.
FAN   45 min; 88 min (2-episode cassette) mTV
VIDrel: CIC/SONOP   V/dm V/sur

STAR TREK THE NEXT GENERATION: THE HOST **
Marvin V. Rush USA   PG
                                                          1991
Patrick Stewart, Jonathan Frakes, LeVar Burton, Michael Dorn, Gates McFadden, Marina Sirtis, Brent Spiner, Franc Luz, Barbara Tarbuck, William Newman, Nicole Orth-Pallavicini
Dr Crusher falls for a handsome Federation ambassador who has come aboard to negotiate a planetary dispute. When he is injured and needs medical attention, she discovers that he is host to a parasitic creature that's the real ambassador. One of the more unusual ideas in this variable series that lacks the impact a stronger script would have given it.
FAN   87 min (2-episode cassette) mTV
VIDrel: CIC/SONOP   V/sur

STAR TREK THE NEXT GENERATION: THE HUNTED **
Cliff Bole USA   PG
                                                          1989
Patrick Stewart, Jonathan Frakes, LeVar Burton, Gates McFadden, Wil Wheaton, Brent Spiner, Michael Dorn, Marina Sirtis, Jeff McCarthy, James Cromwell, Colm Meaney, J. Michael Flynn, Andrew Bickell
When the planet Angosia applies for membership of the Federation, they receive a visit from the crew of the Enterprise, who get involved in hunting and capturing an escaped prisoner, who is said to be highly dangerous. However, it soon becomes apparent that they have not been told the entire truth about this matter and that there is a deeper mystery awaiting resolution.
FAN   88 min (2-episode cassette) mTV
VIDrel: CIC/SONOP   V/dm

STAR TREK THE NEXT GENERATION: I, BORG **   U
Robert Lederman USA   1992
Patrick Stewart, Jonathan Frakes, Brent Spiner, LeVar Burton, Michael Dorn, Gates McFadden, Marina Sirtis, Jonathan Del Arco, Whoopi Goldberg
The discovery of a crashed Borg with one survivor causes moral problems for the crew of the Enterprise, when Picard devises a plan that is intended to result in the total destruction of that race. However, as work on this scheme progresses, the lone Borg seems to exhibit some signs of acquiring an individual consciousness. A well-paced, morality tale.
FAN   45 min; 87 min (2-episode cassette) mTV
VIDrel: CIC/SONOP   V/dm V/sur

STAR TREK THE NEXT GENERATION: THE ICARUS FACTOR **
Robert Iscove USA   PG
                                                          1989
Patrick Stewart, Jonathan Frakes, LeVar Burton, Michael Dorn, Mitchell Ryan, Marina Sirtis, Brent Spiner, Wil Wheaton, Diana Muldaur, Mitchell Ryan, Colm Meaney, Lance Spellerberg
Already uncertain whether or not to accept command of his own starship, Riker faces yet another problem when he comes face to face with his father. Meanwhile, Worf begins to act rather strangely but his friends find a way to cope with this crisis. A run-of-the-mill episode that concentrates on the origins of the father-son conflict between the Rikers.
FAN   90 min (2-episode cassette) mTV
VIDrel: CIC/SONOP   V/dm V/sh

STAR TREK THE NEXT GENERATION: IDENTITY CRISIS **
Winrich Kolbe USA   PG
                                                          1991
Patrick Stewart, Jonathan Frakes, LeVar Burton, Denise Crosby, Brent Spiner, Gates McFadden, Marina Sirtis, Michael Dorn, Maryann Plunkett
Five years after Geordi took part in an Away Team mission, his former comrades are beginning to mutate into alien creatures and then vanish. Geordi sets about using the holodeck to recreate the mission in an attempt to learn whether he can avoid suffering an identical fate.
FAN   88 min (2-episode cassette) mTV
VIDrel: CIC/SONOP   V/sh

STAR TREK THE NEXT GENERATION: IMAGINARY FRIEND ***
Gabrielle Beaumont USA   PG
                                                          1992
Patrick Stewart, Jonathan Frakes, Brent Spiner, LeVar Burton, Michael Dorn, Gates McFadden, Marina Sirtis, Noley Thornton, Shay Ahstar, Jeff Allin, Brian Bonsall, Patti Yasutake, Sheila Franklin, Whoopi Goldberg
An investigation of a cloud nebula causes unexpected dangers when an alien being comes aboard and takes the form of a girl's imaginary friend. This leads to it viewing human behaviour from a child's perspective, which threatens to have devastating consequences. An intriguing idea that is not really explored and is overlaid with a sentimental conclusion.
FAN   45 min; 87 min (2-episode cassette) mTV
VIDrel: CIC/SONOP   V/dm V/sur

STAR TREK THE NEXT GENERATION: IN THEORY *
Patrick Stewart USA   PG
                                                          1991
Patrick Stewart, Jonathan Frakes, LeVar Burton, Michael Dorn, Gates McFadden, Marina Sirtis, Brent Spiner, Michele Scarabelli, Rosalind Chao, Pamela Winslow, Whoopi Goldberg
As the Enterprise explores a dark matter nebula of great density, Data has to work alongside a female cadet who has just broken up with her rather undemonstrative boyfriend. She gradually begins to seen the android as a perfect mate despite his lack of emotion, and Data is uncertain how to react, and seeks advice from a succession of colleagues. By way of a sub-plot, a series of weird events show that there are more serious problems afoot. Barely adequate.
FAN   88 min (2-episode cassette) mTV
VIDrel: CIC/SONOP   V/sur

STAR TREK THE NEXT GENERATION: INHERITANCE ***
Robert Scheerer USA   U
                                                          1993
Patrick Stewart, Jonathan Frakes, LeVar Burton, Michael Dorn, Marina Sirtis, Gates McFadden, Brent Spiner, Fionnula Flanagan, William Lithgow
The Enterprise comes to the aid of a planet whose core is solidifying, causing earthquakes and various other seismic disturbances. A party of scientists beams up to the ship, and this includes an Earth woman who tells Data that she was the wife of his creator and thus in a sense is his mother. A competent story, well put together, and given a nice twist at the end.
FAN   45 min; 88 min (2-episode cassette) mTV
VIDrel: CIC/SONOP   V/dm V/sur

STAR TREK THE NEXT GENERATION: THE INNER LIGHT ***
Peter Lauritson USA   PG
                                                          1992
Patrick Stewart, Jonathan Frakes, Brent Spiner, LeVar Burton, Michael Dorn, Gates McFadden, Marina Sirtis, Margot Rose, Richard Riehle, Scott Jaeck, Patti Yasutake, Jennifer Nash, Daniel Stewart
An alien probe appears in front of the Enterprise and takes over control of Picard who fall unconscious. He awakes to find himself a member of a small village community on an alien planet, where in the course of time he comes to accept his new situation, while never forgetting his former life. A well-crafted and moving story that provides scope for some fine performances in one of the best episodes in the entire series.
FAN   45 min; 87 min (2-episode cassette) mTV
VIDrel: CIC/SONOP   V/sur

STAR TREK THE NEXT GENERATION: INTERFACE **
Robert Wiemer USA   PG
                                                          1993
Patrick Stewart, Jonathan Frakes, LeVar Burton, Michael Dorn, Marina Sirtis, Gates McFadden, Brent Spiner, Madge Sinclair, Warren Munson, Ben Vereen
Geordi becomes a guinea-pig in a new technique that connects his brain and senses to a remote probe. This is later used to recover a derelict science vessel whose crew are dead, but worries about his mother, who has been declared missing along with the starship she commanded, play a strange role in this task. Well acted and maintaining a good sense of tension, though as is so often the case, the ending is both contrived and unsatisfying.
FAN   45 min; 88 min (2-episode cassette) mTV
VIDrel: CIC/SONOP   V/dm V/sur

STAR TREK THE NEXT GENERATION: JOURNEY'S END **
Corey Allen USA   PG   1994
Patrick Stewart, Jonathan Frakes, LeVar Burton, Michael Dorn, Marina Sirtis, Gates McFadden, Brent Spiner, Wil Wheaton, Tom Jackson, Ned Romero, George Aguilar, Richard Poe, Eric Menyuk, Doug West
After arduous negotiations, the Federation concludes a treaty with the Cardassians, but this requires the evacuation of a number of colonies on both sides, among them a group of American Indians. At the same time, a visit to the ship by Wesley Crusher reveals him to be a much changed and deeply troubled young man. An over-complex episode that resolves all its varied storylines in an unconvincing manner.
FAN   45 min; 88 min (2-episode cassette) mTV
VIDrel: CIC/SONOP   V/dm V/sur

STAR TREK THE NEXT GENERATION: JUSTICE **
James L. Conway USA   PG   1987
Patrick Stewart, Wil Wheaton, Jonathan Frakes, LeVar Burton, Gates McFadden, Brent Spiner, Michael Dorn, Marina Sirtis, Brenda Bakke, Jay Louden, Josh Clark, David Q. Combs, Richard Lavin, Judith Jones, Eric Matthews
On Rican 3 Ensign Crusher finds himself sentenced to death for having unwittingly violated one of the planet's major laws. Captain Picard must now rescue his comrade without creating a serious conflict with the rulers of the planet.
FAN   89 min (2-episode cassette) mTV
VIDrel: CIC/SONOP   V/dm V/sh

STAR TREK THE NEXT GENERATION: THE LAST OUTPOST **
Richard Colla USA   PG   1987
Jonathan Frakes, Patrick Stewart, Wil Wheaton, LeVar Burton, Brent Spiner, Gates McFadden, Marina Sirtis, Darryl Henriques, Armin Shimerman, Tracey Walter, Jake Dengel, Mike Domez, Stanley Kamel, Eric Menyuk, Herta Ware
In this adventure Riker attempts to save both the Enterprise and a starship belonging to a ferocious race known as the Ferengi, when both become frozen in space by a mysterious and powerful life-form.
FAN   89 min (2-episode cassette) mTV
VIDrel: CIC/SONOP   V/dm V/sh

STAR TREK THE NEXT GENERATION: LEGACY **   U
Robert Scheerer USA   1990
Patrick Stewart, Jonathan Frakes, LeVar Burton, Marina Sirtis, Wil Wheaton, Michael Dorn, Gates McFadden, Brent Spiner, Beth Toussaint, Don Mirrault
A Federation emergency shuttle crashes on a planet on which a war is raging, and a woman of that world offers to help an Away Team rescue the survivors. Unfortunately, it transpires that she has made plans of her own, that involve using the Enterprise to defeat her clan's enemies.
FAN   87 min (2-episode cassette) mTV
VIDrel: CIC/SONOP   V/sur

STAR TREK THE NEXT GENERATION: LESSONS **
Robert Wiemer USA   U   1993
Patrick Stewart, Jonathan Frakes, Brent Spiner, LeVar Burton, Michael Dorn, Gates McFadden, Marina Sirtis, Wendy Hughes, Majel Barrett (voice only)
Picard falls in love with a forthright and attractive woman officer who shares his passion for music, but their growing affection comes into conflict with duty when he is forced to allow her to take part in a high risk mission. A watchable episode that benefits from a good performance by Hughes.
FAN   45 min; 86 min (2-episode cassette) mTV
VIDrel: CIC/SONOP   V/dm V/sur

STAR TREK THE NEXT GENERATION: LIAISONS *
Cliff Bole USA   PG   1993
Patrick Stewart, Jonathan Frakes, LeVar Burton, Michael Dorn, Marina Sirtis, Gates McFadden, Brent Spiner, Barbara Williams, Eric Pierpoint, Paul Eiding, Michael Harris, Rickey D'Shon Collins
The first contact between the Federation and a hitherto unknown alien race results in a visit to the Enterprise by a couple of their ambassadors, one of whom sorely tests Worf's patience with his arrogant and curt manner. Meanwhile, Captain Picard faces problems of his own after crashing onto a hostile and barren world. Well below par, this assembly-line story has the usual neat and all too predictable resolution.
FAN   45 min; 88 min (2-episode cassette) mTV
VIDrel: CIC/SONOP   V/dm V/sur

STAR TREK THE NEXT GENERATION: LONELY AMONG US **
Cliff Bole USA   PG   1987
Patrick Stewart, Jonathan Frakes, LeVar Burton, Denise Crosby, Brent Spiner, Gates McFadden, Wil Wheaton, Marina Sirtis, Kavi Raz, Colm Meaney, John Durbin
The Enterprise is engaged in a vitally important mission that involves transporting two adversarial life-forms to a peace conference. When the ship is invaded by a third alien life-form, a difficult job is made just that little bit harder.
FAN   89 min (2-episode cassette) mTV
VIDrel: CIC/SONOP   V/dm V/sh

STAR TREK THE NEXT GENERATION: THE LOSS **
Chip Chalmers USA   PG   1990
Patrick Stewart, Jonathan Frakes, LeVar Burton, Marina Sirtis, Michael Dorn, Gates McFadden, Brent Spiner, Kim Braden, Whoopi Goldberg
A cloud of alien creatures are drawn into the field of the Enterprise and causes Troi to lose her empathic abilities. She slowly begins to succumb to self pity, only recovering when the creatures are cleared away from the vicinity of the ship. Fortunately, through her trials she retains her ability to think logically, and is thus able to save the Enterprise from being sucked into the path of a cosmic string fragment. Average.
FAN   87 min (2-episode cassette) mTV
VIDrel: CIC/SONOP   V/sur

STAR TREK THE NEXT GENERATION: LOUD AS A WHISPER **
Larry Shaw USA   U   1988
Patrick Stewart, Jonathan Frakes, LeVar Burton, Michael Dorn, Marina Sirtis, Brent Spiner, Wil Wheaton, Diana Muldaur, Howie Seago, Marnie Mosiman, Leo Damian, Thomas Oglesby, Colm Meaney, Richard Lavin, Chip Heller
A highly skilled but mute mediator is aboard the Enterprise on a mission to secure peace between two warring factions, but loses his vital means of communication when his interpreters are killed.
FAN   92 min (2-episode cassette) mTV
VIDrel: CIC/SONOP   V/dm V/sh

STAR TREK THE NEXT GENERATION: LOWER DECKS **
Gabrielle Beaumont USA   PG   1994
Patrick Stewart, Jonathan Frakes, LeVar Burton, Michael Dorn, Marina Sirtis, Gates McFadden, Brent Spiner, Dan Gauthier, Shannon Fill, Alexander Enberg, Don Reilly, Bruce Beaty, Patti Yasutake
A group of junior officers awaits the results of evaluations for possible promotion, and each experiences the usual anxieties, but the recovery of a Federation shuttle close to the Cardassian border provides a distraction, involving one of their number in a mission of some importance. A competent entry in the series, though lacking dramatic focus and a little spoilt by its downbeat ending.
FAN   88 min (2-episode cassette) mTV
VIDrel: CIC/SONOP   V/sur

STAR TREK THE NEXT GENERATION: MAN OF THE PEOPLE **
Winrich Kolbe USA   PG   1992
Patrick Stewart, Jonathan Frakes, Brent Spiner, LeVar Burton, Michael Dorn, Gates McFadden, Marina Sirtis, Colm Meaney, Chip Lucia, Patti Yasutake, Lucy Broyer, George D. Wallace, Susan French, Rich Scarry, Stephanie Erb
An ambassador who is to mediate between two warring factions is taken aboard the Enterprise for his own safety after the ship taking him to peace negotiations is attacked. He finds himself drawn to Troi and when his mother dies, becomes increasingly dependent on her. However, this soon produces a strange change in her that soon turns into a major crisis. A clever idea, rather indifferently handled, that fails to deliver the expected suspense.
FAN   45 min; 88 min (2-episode cassette) mTV
VIDrel: CIC/SONOP   V/dm V/sur

STAR TREK THE NEXT GENERATION: MANHUNT *

PG

Rob Bowman USA 1989

Patrick Stewart, Jonathan Frakes, LeVar Burton, Gates McFadden, Wil Wheaton, Brent Spiner, Michael Dorn, Marina Sirtis, Majel Barrett, Robert Constanzo, Mick Fleetwood, Carel Struycken, Rod Arrants, Colm Meaney, Rhonda Aldrich

On her way to attend a conference on Pacifica, Troi's mother comes aboard, and proceeds to alienate just about everyone. She is now experiencing the Betazoid equivalent of the menopause, her sex drive is at its peak, and she pursues Picard unmercifully. The Captain takes refuge in the holodeck, where he is soon joined by Riker. Meanwhile, two other conference delegates present far more serious problems. A silly and unrewarding story of great triviality.

FAN   90 min (2-episode cassette) mTV
VIDrel: CIC/SONOP   V/dm

STAR TREK THE NEXT GENERATION: MASKS ***

U

Robert Wiemer USA 1994

Patrick Stewart, Jonathan Frakes, LeVar Burton, Michael Dorn, Marina Sirtis, Gates McFadden, Brent Spiner, Rickey D'Shon Collins

After encountering an unknown comet which the Enterprise begins to scan, strange things start to happen. Alien artefacts and symbols appear on board the ship, and Data is taken over by some unknown personalities. The comet proves to be an alien archive that is progressively turning the ship into an alien city, until Picard and his crew devise a way to halt the process. An imaginative story that develops in a satisfactory way.

FAN   45 min; 88 min (2-episode cassette) mTV
VIDrel: CIC/SONOP   V/dm V/sur

STAR TREK THE NEXT GENERATION: THE MASTERPIECE SOCIETY **

U

Winrich Kolbe USA 1991

Patrick Stewart, Jonathan Frakes, Brent Spiner, LeVar Burton, Michael Dorn, Gates McFadden, Marina Sirtis, Dey Young, Ron Canada, Sheila Franklin

On a mission to monitor a stellar core fragment, the Enterprise discovers a small human colony on a planet that is directly in the path of the fragment. Anxious to save the colonists, Picard makes contact, only to find that he is dealing with a closed society of genetically engineered individuals that wants to maintain its isolation at all costs.

FAN   45 min; 88 min (2-episode cassette) mTV
VIDrel: CIC/SONOP   V/dm V/sur

STAR TREK THE NEXT GENERATION: A MATTER OF HONOR **

PG

Rob Bowman USA 1988

Jonathan Frakes, Patrick Stewart, Levar Burton, Michael Dorn, Marina Sirtis, Brent Spiner, Wil Wheaton, Diana Muldaur, John Putch, Christopher Collins, Brian Thompson, Colm Meaney, Peter Parros, Laura Drake

An officer exchange programme results in Commander Riker volunteering to serve aboard a Klingon vessel, where after some difficulties he establishes his authority, only to face a more difficult test when the Enterprise is blamed for an infection that has entered the Klingon ship.

FAN   92 min (2-episode cassette) mTV
VIDrel: CIC/SONOP   V/dm V/sh

STAR TREK THE NEXT GENERATION: A MATTER OF PERSPECTIVE **

PG

Cliff Bole USA 1990

Patrick Stewart, Jonathan Frakes, LeVar Burton, Michael Dorn, Gates McFadden, Brent Spiner, Wil Wheaton, Marina Sirtis, Mark Margolis, Gina Hecht, Craig Richard Nelson, Juli Donald, Colm Meaney

Having just returned from a routine mission, Commander Riker is suspected of murder when a scientist who had accused him of seducing his wife is found dead. The holodeck is put to good use in recreating the relevant events.

FAN   88 min (2-episode cassette) mTV
VIDrel: CIC/SONOP   V/sh

STAR TREK THE NEXT GENERATION: A MATTER OF TIME **

U

Robert Scheerer USA 1991

Patrick Stewart, Jonathan Frakes, Brent Spiner, LeVar Burton, Michael Dorn, Gates McFadden, Marina Sirtis, Stefan Goerasch, Matt Frewer, Sheila Franklin, Shay Garner

A time traveller comes aboard the Enterprise and claims to

be a historian from the 26th century, keen to observe the completion of an important mission. Soon several crew members began to have doubts as to his true identity and true purpose.

FAN   45 min; 88 min (2-episode cassette) mTV
VIDrel: CIC/SONOP   V/dm V/sur

STAR TREK THE NEXT GENERATION: THE MEASURE OF A MAN ***

PG

Robert Scheerer USA 1988

Patrick Stewart, Brent Spiner, Jonathan Frakes, LeVar Burton, Gates McFadden, Wil Wheaton, Michael Dorn, Marina Sirtis, Amanda McBroom, Brian Brophy, Clyde Kusatsu, Colm Meaney, Whoopi Goldberg

A cybernetics expert is so enthusiastic about the benefits to Starfleet of an android such as Data that he wishes to dismantle him in order to learn how to mass-produce such a perfect robot. Data resists this idea and attempts to resign his commission, but since his legal status is in doubt, a hearing must take place, and Riker is reluctantly obliged to play the role of devil's advocate. A highly dramatic story similar in theme to the episode entitled THE OFFSPRING.

FAN   91 min (2-episode cassette) mTV
VIDrel: CIC/SONOP   V/dm V/sh

STAR TREK THE NEXT GENERATION: MENAGE A TROI **

U

Robert Legato USA 1990

Patrick Stewart, Jonathan Frakes, Marina Sirtis, LeVar Burton, Michael Dorn, Wil Wheaton, Brent Spiner, Gates McFadden, Majel Barrett, Frank Corsentino, Rudolph Willrich, Ethan Phillips, Peter Slutsker, Carel Struycken

Counsellor Troi, her mother and Riker are abducted by a lovestruck Ferengi, who desires the elder Troi for both her body and her telepathic abilities. With the Enterprise in pursuit, he agrees to release his other hostages only in return for her hand in marriage.

FAN   87 min (2-episode cassette) mTV
VIDrel: CIC/SONOP   V/dm

STAR TREK THE NEXT GENERATION: THE MIND'S EYE **

PG

David Livingston USA 1991

Patrick Stewart, Jonathan Frakes, LeVar Burton, Michael Dorn, Gates McFadden, Marina Sirtis, Brent Spiner, Larry Dobkin, Edward Wiley, John Fleck

On his way to attend a conference on Risa, LaForge is captured by the Romulans and subjected to their advanced brainwashing techniques. His memory of this event is erased and he is implanted with subconscious instructions to assassinate a Klingon governor and thus damage the Klingon-Federation alliance. A nicely plotted and well acted episode that maintains a good sense of tension, with the storyline recalling ideas used in THE MANCHURIAN CANDIDATE.

FAN   87 min (2-episode cassette) mTV
VIDrel: CIC/SONOP   V/sur

STAR TREK THE NEXT GENERATION: THE MOST TOYS **

PG

Timothy Bond USA 1990

Patrick Stewart, Brent Spiner, Jonathan Frakes, LeVar Burton, Saul Rubinek, Michael Dorn, Marina Sirtis, Wil Wheaton, Gates McFadden, Saul Rubinek, Jane Daly, Nehemiah Persoff

A space trader negotiates the sale of rare chemicals urgently needed by a Federation colony to control an outbreak of water contamination, but matters takes a bizarre turn when Data's shuttle explodes whilst being used to carry these unstable chemicals and Data himself is abducted by the trader.

FAN   87 min (2-episode cassette) mTV
VIDrel: CIC/SONOP   V/dm V/sh

STAR TREK THE NEXT GENERATION: THE NAKED NOW **

PG

Paul Lynch USA 1987

Patrick Stewart, Jonathan Frakes, LeVar Burton, Gates McFadden, Wil Wheaton, Brent Spiner, Michael Dorn, Marina Sirtis, Brooke Bundy, Benjamin W.S. Lum, Michael Rider, David Renan, Skip Stellrecht, Kenny Koch

When the Enterprise is sent to check the whereabouts of a starship with which all contact was lost, the crew find a ship full of corpses, all of whom seem to have been stricken with insanity just before death. LaForge becomes infected with the virus

responsible, and carries it back to the Enterprise, where it spreads throughout the crew causing chaos and madness.
FAN   88 min (2-episode cassette) mTV
VIDrel: CIC/SONOP   V/dm V/sh

STAR TREK THE NEXT GENERATION: THE NEUTRAL ZONE ** PG
James L. Conway USA   1988
Patrick Stewart, Jonathan Frakes, LeVar Burton, Gates McFadden, Wil Wheaton, Brent Spiner, Michael Dorn, Marina Sirtis, Marc Alaimo, Anthony James, Leon Rippy, Gracie Harrison, Peter Mark Richman
On its way to a meeting with a group of hostile Romulans, the Enterprise comes across a spaceship containing three people from the 20th century, all of whom are in a state of suspended animation, and prove to be the subjects of an early cryogenic experiment. After being successfully revived, they attempt to adjust to life 370 years in the future. Meanwhile, an encounter with the Romulans passes off without any major incident.
FAN   90 min (2-episode cassette) mTV
VIDrel: CIC/SONOP   V/dm V/sh

STAR TREK THE NEXT GENERATION: NEW GROUND ** U
Paul Lynch USA   1991
Patrick Stewart, Jonathan Frakes, Brent Spiner, LeVar Burton, Michael Dorn, Gates McFadden, Marina Sirtis, Jennifer Edwards, Richard McGonagle, Georgia Brown, Brian Bonsall, Sheila Franklin, Majel Barrett (voice only)
The Enterprise is sent to monitor an experiment into the possibility of warp drive without warp engines that goes badly wrong and threatens to destroy both the ship and an entire planet. Meanwhile, Worf has problems of his own with a young son whose behaviour leaves him angry and puzzled. A rather routine outing whose two main stories only merge towards the end without having generated much interest along the way.
FAN   45 min; 88 min (2-episode cassette) mTV
VIDrel: CIC/SONOP   V/dm V/sur

STAR TREK THE NEXT GENERATION: THE NEXT PHASE *** U
David Carson USA   1992
Patrick Stewart, Jonathan Frakes, Brent Spiner, LeVar Burton, Michael Dorn, Gates McFadden, Marina Sirtis, Kenneth Mesroll, Michelle Forbes, Thomas Kopache, Susanna Thompson, Shelby Leverington, Brian Cousins
While assisting a Romulan vessel that has suffered a massive onboard explosion, Geordi and Ro Laren find themselves stranded in a kind of limbo where they are invisible and impervious to the rest of the crew, who believe them to be dead. Forced to act to save the Enterprise from sabotage, they find a way to return to normal reality. An enjoyable episode with a strong and suspenseful plot.
FAN   45 min; 87 min (2-episode cassette) mTV
VIDrel: CIC/SONOP   V/sur

STAR TREK THE NEXT GENERATION: NIGHT TERRORS ** PG
Les Landau USA   1991
Patrick Stewart, Jonathan Frakes, LeVar Burton, Denise Crosby, Brent Spiner, Gates McFadden, Marina Sirtis, Michael Dorn, Rosalind Chao, John Vickery, Duke Moosekian
Trapped in a weird region of space, the crew of the Enterprise are unable to sleep and begin to suffer from hallucinations. With the help of a Betazoid survivor rescued from another vessel, Troi contacts a similarly stranded alien ship, and the two races co-operate in a plan of escape.
FAN   88 min (2-episode cassette) mTV
VIDrel: CIC/SONOP   V/sh

STAR TREK THE NEXT GENERATION: THE NTH DEGREE ** U
Robert Legato USA   1991
Patrick Stewart, Jonathan Frakes, LeVar Burton, Denise Crosby, Brent Spiner, Gates McFadden, Marina Sirtis, Michael Dorn, Dwight Schultz, Jim Norton, David Coburn
Barclay (from HOLLOW PURSUITS) accompanies Geordi on a mission to study an alien probe that appears to have damaged the control computer of a Federation telescope array. A sudden radiation bursts enhances Barclay's intellect to such an extent that he is able to build a device on the ship's holodeck, and takes the Enterprise 30,000 light-years through space to an unexplored

region. An excellent idea is let down by the rushed and slightly silly resolution.
FAN   87 min (2-episode cassette) mTV
VIDrel: CIC/SONOP   V/sur

STAR TREK THE NEXT GENERATION: THE OFFSPRING ** PG
Jonathan Frakes USA   1990
Patrick Stewart, Brent Spiner, Jonathan Frakes, LeVar Burton, Michael Dorn, Gates McFadden, Wil Wheaton, Marina Sirtis, Hallie Todd, Nicolas Coster, Judyann Elder, Diane Moser, Hayne Bayle, Maria Leone, James Becker
After attending a cybernetics conference, Data undertakes an experiment that will enable him to study the human phenomenon of reproduction and parenthood. Using his neural circuitry, he constructs an android offspring – a daughter that he names Lal. He undertakes her upbringing, but his plans are upset when a Starfleet admiral decides to remove her for further study. The episode entitled THE MEASURE OF A MAN also deals with the issue of android rights.
FAN   88 min (2-episode cassette) mTV
VIDrel: CIC/SONOP   V/dm V/sh

STAR TREK THE NEXT GENERATION: THE OUTCAST * PG
Robert Scheeer USA   1992
Patrick Stewart, Jonathan Frakes, Brent Spiner, LeVar Burton, Michael Dorn, Gates McFadden, Marina Sirtis, Cullan White, Megan Cole
Riker finds himself working closely with a member of a strange genderless race on a mission to rescue one of their craft from a kind of black hole in space. He learns that this race once had two sexes and that certain rare individuals still possess a sexual identity. However, his growing feelings for this female lead to major problems. A rather weak episode not helped by an abrupt and unsatisfying conclusion.
FAN   45 min; 88 min (2-episode cassette) mTV
VIDrel: CIC/SONOP   V/dm V/sur

STAR TREK THE NEXT GENERATION: THE OUTRAGEOUS OKONA ** PG
Robert Becker USA   1988
Patrick Stewart, Jonathan Frakes, LeVar Burton, Gates McFadden, Wil Wheaton, Brent Spiner, Michael Dorn, Marina Sirtis, William O. Campbell, Douglas Rowe, Albert Stratton, Rosalind Ingledew, Kieran Mulroney, Whoopi Goldberg
The Enterprise comes to the aid of a damaged cargo ship, but Captain Picard soon discovers that its captain is a notorious swindler with a price on his head. Meanwhile, Data gets a lesson in humour and some advice on how to deal with women, but as for our rogue, he finds himself in a tricky predicament when two ships arrive and both demand that he be handed over to them. A lighthearted episode that is intermittently amusing.
FAN   92 min (2-episode cassette) mTV
VIDrel: CIC/SONOP   V/dm V/sh

STAR TREK THE NEXT GENERATION: PARALLELS *** PG
Robert Wiemer USA   1993
Patrick Stewart, Jonathan Frakes, Brent Spiner, LeVar Burton, Michael Dorn, Gates McFadden, Marina Sirtis, Wil Wheaton, Patti Yasutake, Mark Burmall, Majel Barrett (voice only)
After winning a tournament, Worf returns to the Enterprise by shuttle on his birthday, where he has to endure the embarrassment of a surprise party. This marks the start of a series of bizarre and frightening events that see him travelling between a number of parallel realities. One of the best stories in the series, this tense and compelling tale makes the most of its unusual plot premise.
FAN   45 min; 88 min (2-episode cassette) mTV
VIDrel: CIC/SONOP   V/sur

STAR TREK THE NEXT GENERATION: PEAK PERFORMANCE ** PG
Robert Scheerer USA   1989
Patrick Stewart, Jonathan Frakes, LeVar Burton, Gates McFadden, Wil Wheaton, Brent Spiner, Michael Dorn, Marina Sirtis, Roy Brocksmith, Armin Shimerman, David L. Landers, Leslie Neale, Glenn Morshower
In order to cope with the threat of a Borg invasion, simulated wargames are held, in which a haughty and unpleasant master strategist is to play a key role. Riker is given command of a rundown spaceship and a crew of forty, and has to improvise

like never before, aided by the ever resourceful Wesley Crusher. But during all this, the intervention of a Ferengi vessel adds a dangerous and unforeseen element.
FAN 90 min (2-episode cassette) mTV
VIDrel: CIC/SONOP V/dm

STAR TREK THE NEXT GENERATION: THE PEGASUS **
LeVar Burton USA
Patrick Stewart, Jonathan Frakes, Brent Spiner, LeVar Burton, Michael Dorn, Gates McFadden, Marina Sirtis, Nancy Vawter, Terry O'Quinn
PG
1993
Riker's past returns to haunt him when his first commanding officer (now an admiral) comes aboard and sends the Enterprise on a dangerous mission to locate a wrecked Starfleet vessel. However, the admiral has a secret agenda of his own, and this forces Riker to deal with a severe test of his loyalty. Quite well written and acted, but not terribly exciting. The usual last-minute escape from danger seems both over-familiar and unrealistic.
FAN 45 min; 88 min (2-episode cassette) mTV
VIDrel: CIC/SONOP V/sur

STAR TREK THE NEXT GENERATION: PEN PALS **
Winrich Kolbe USA
Brent Spiner, Patrick Stewart, Jonathan Frakes, LeVar Burton, Michael Dorn, Marina Sirtis, Wil Wheaton, Diana Muldaur, Mitchell Ryan, Nicholas Cascone, Nikki Cox, Ann H. Gillespie, Colm Meaney, Whitney Rydbeck
U
1989
The life of an alien girl is in great danger as she is trapped on a doomed planet, and Data races to the rescue, ignoring the requirements of the non-interference Prime Directive.
FAN 90 min (2-episode cassette) mTV
VIDrel: CIC/SONOP V/dm V/sh

STAR TREK THE NEXT GENERATION: THE PERFECT MATE **
Cliff Bole USA
Patrick Stewart, Jonathan Frakes, Brent Spiner, LeVar Burton, Michael Dorn, Gates McFadden, Marina Sirtis, Frank Janssen, Tim O'Connor, Max Grodenchik, Mickey Cottrelll, Michael Snyder, Majel Barrel, Charles Gunning, April Grace
PG
1992
When a peace conference between two warring planets is due to be held aboard the Enterprise, Picard learns that his ship is transporting a special gift from one side to the other. This proves to be a beautiful female empath with very special abilities, and her presence causes more than its fair share of complications for all concerned.
FAN 45 min; 87 min (2-episode cassette) mTV
VIDrel: CIC/SONOP V/dm V/sur

STAR TREK THE NEXT GENERATION: PHANTASMS **
Patrick Stewart USA
Patrick Stewart, Jonathan Frakes, LeVar Burton, Marina Sirtis, Brent Spiner, Gates McFadden, Michael Dorn, Gina Ravarra, Bernard Kates, Clyde Kusatsu, David L. Crowley
PG
1993
After a new warp coil is fitted to the Enterprise, Data start to experience terrifying nightmares that eventually pass into waking dreams, making him unfit for service and even a potential danger to others. At the same time the ship lies crippled, as its engines have stopped functioning. Some imaginative dream sequences and a clever plot keep one watching, despite the rushed and simplistic resolution.
FAN 45 min; 88 min (2-episode cassette) mTV
VIDrel: CIC/SONOP V/dm V/sur

STAR TREK THE NEXT GENERATION: POWER PLAY **
David Livingston USA
Patrick Stewart, Jonathan Frakes, Brent Spiner, LeVar Burton, Michael Dorn, Gates McFadden, Marina Sirtis, Colm Meaney, Michelle Forbes, Ryan Reid, Majel Barrett (voice only)
PG
1992
After returning from a mission to investigate distress signals from a Federation ship that vanished nearly 200 years earlier, Data, Troi and Chief O'Brien attempt to seize control of the Enterprise. It soon becomes evident that their minds have been taken over by alien intelligences that appear to have a purpose all their own. A highly imaginative and unusual episode that maintains a good sense of suspense right up to the end.
FAN 45 min; 88 min (2-episode cassette) mTV
VIDrel: CIC/SONOP V/dm V/sur

STAR TREK THE NEXT GENERATION: PRE-EMPTIVE STRIKE **
Patrick Stewart USA
Patrick Stewart, Jonathan Frakes, LeVar Burton, Michael Dorn, Marina Sirtis, Gates McFadden, Brent Spiner, Michelle Forbes, John Franklyn-Robbins, Natalija Nogulich, William Thomas Jr, Shannon Cochran
U
1994
Ro Laren returns to serve on the Enterprise and is sent on a dangerous mission to infiltrate the resistance movement among Federation colonists now living under Cardassian rule. As a Bajoran, she finds herself suffering a conflict of loyalties, but is finally obliged to take a stand for her principles. Solidly made, this effort has a strong central performance from Forbes as Ro, though at times the story loses its dramatic focus.
FAN 45 min; 87 min (2-episode cassette) mTV
VIDrel: CIC/SONOP V/dm V/sur

STAR TREK THE NEXT GENERATION: THE PRICE *
Robert Scheerer USA
Patrick Stewart, Jonathan Frakes, LeVar Burton, Gates McFadden, Wil Wheaton, Brent Spiner, Michael Dorn, Marina Sirtis, Matt McCoy, Elizabeth Hoffman, Scott Thomson, Castulo Guerra, Kevin Peter Hall, Dan Shor, Colm Meaney
PG
1990
The inhabitants of a world called Chrysalia discover what they believe is a stable wormhole and when decide to auction it off the Federation takes charge of the proceedings. While Geordi and Data investigate the wormhole to determine whether or not it is truly stable, Troi learns that the Chrysalian representative is part Betazoid and has been using his empathic powers to manipulate the negotiations.
FAN 88 min (2-episode cassette) mTV
VIDrel: CIC/SONOP V/dm

STAR TREK THE NEXT GENERATION: Q-PID **
Cliff Bole USA
Patrick Stewart, Jonathan Frakes, LeVar Burton, Denise Crosby, Brent Spiner, Gates McFadden, Marina Sirtis, Michael Dorn, John De Lancie, Jennifer Hetrick, Clive Revill
U
1991
An old flame of Picard's comes aboard to take part in an archae-ological conference, but so does Picard's old friend Q. The alien takes on the role of unofficial matchmaker, creating a "Robin Hood" fantasy and placing crew members within it, his intention being to teach Picard something of love and honour.
FAN 87 min (2-episode cassette) mTV
VIDrel: CIC/SONOP V/dm V/sur

STAR TREK THE NEXT GENERATION: Q WHO? ***
Rob Bowman USA
Patrick Stewart, Jonathan Frakes, LeVar Burton, Michael Dorn, Marina Sirtis, Brent Spiner, Wil Wheaton, Diana Muldaur, Mitchell Ryan, John De Lancie, Lycia Naff, Colm Meaney, Whoopi Goldberg
U
1989
Our malevolent Q-being amuses himself by bringing the Enterprise into contact with the Borg – alien beings who are not as nice as they sound. One of the most effective episodes in the series, this later gave rise to a further encounter: THE BEST OF BOTH WORLDS.
FAN 90 min (2-episode cassette) mTV
VIDrel: CIC/SONOP V/dm

STAR TREK THE NEXT GENERATION: THE QUALITY OF LIFE ***
Jonathan Frakes USA
Patrick Stewart, Jonathan Frakes, Brent Spiner, LeVar Burton, Michael Dorn, Gates McFadden, Marina Sirtis, Ellen Bry, J. Downing, Majel Barrett (voice only)
U
1992
A project to develop new mining technology on an alien machine brings Data and Geordi in contact with the woman scientist in charge, who has developed amazingly adept and intelligent machines. These can handle a variety of complex and dangerous tasks but start to behave as if they have a minds of their own. Data suspects that this may indeed the case and sets out to prove that these mini-robots are a form of life akin to himself.
FAN 45 min; 87 min (2-episode cassette) mTV
VIDrel: CIC/SONOP V/dm V/sur

STAR TREK THE NEXT GENERATION: RASCALS *
Adam Nimoy USA
Patrick Stewart, Jonathan Frakes, Brent Spiner, LeVar Burton, Michael Dorn, Gates McFadden, Marina Sirtis, Colm Meaney, Rosalind Chao, Michelle Forbes, Whoopi Goldberg, David Tristan Birkin, Megan Parlen, Caroline Junko King
U
1992

Returning to the Enterprise aboard a shuttle, Picard, Ro Laren, Guinan and Keiko pass through an energy field that causes their vessel to break up. Beamed aboard in the nick of time, their bodies are regressed to the age of twelve. Fortunately, when the ship is attacked and board by Ferengi pirates, their predicament allows them some unexpected advantages. A silly action-style episode with poor acting from the juvenile leads.
FAN   45 min; 88 min (2-episode cassette) mTV
VIDrel: CIC/SONOP   V/dm V/sur

STAR TREK THE NEXT GENERATION: REALM OF FEAR ***

Cliff Bole USA   PG
                                                                  1992
Patrick Stewart, Jonathan Frakes, Brent Spiner, LeVar Burton, Michael Dorn, Gates McFadden, Marina Sirtis, Colm Meaney, Patti Yasutake, Dwight Schultz, Renata Scott, Majel Barrett (voice only), Thomas Belgrey
After an explosion on board a scientific study vessel kills all its crew, an away team from the Enterprise attempts to investigate. One of its members has a deep fear of transporting and experiences what he takes to be a hallucination upon his return to the ship. A nicely sustained episode that maintains the suspense and mystery right up the climax.
FAN   45 min; 88 min (2-episode cassette) mTV
VIDrel: CIC/SONOP   V/dm V/sur

STAR TREK THE NEXT GENERATION: REDEMPTION – PART 1 **

Cliff Bole USA   PG
                                                                  1991
Patrick Stewart, Jonathan Frakes, LeVar Burton, Michael Dorn, Marina Sirtis, Gates McFadden, Brent Spiner, Robert O'Reilly, Tony Todd, Whoopi Goldberg, Ben Slack, Barbara March, Gwyneth Walsh, Nicholas Kepros, J.D. Cullum
The concluding part of a trilogy that began with SINS OF THE FATHER and continued with REUNION, this episode sees Worf attempting to clear his family name, whilst Picard takes the Enterprise to the Klingon homeworld, where he is to take part in a Klingon Rite of Succession.
FAN   88 min (2-episode cassette) mTV
VIDrel: CIC/SONOP;PION (LV only)   V/sur LV

STAR TREK THE NEXT GENERATION: REDEMPTION – PART 2 ***

David Carson USA   PG
                                                                  1991
Patrick Stewart, Jonathan Frakes, Brent Spiner, LeVar Burton, Michael Dorn, Gates McFadden, Marina Sirtis, Denise Crosby, Tony Todd, Barbara March, J.D. Cullum, Gwynyth Walsh, Robert O'Reilly, Michael G. Hagerty, Fran Bennett
His honour restored, Worf resigns from Starfleet to fight in the Klingon war of succession, while Picard implements a daring scheme to reveal Romulan involvement with the Klingon rebels. A complex and action-packed episode that also sees Data briefly assuming the burden of command. This two-part episode has also become available on a single tape/disc.
FAN   45 min; 88 min (2-episode cassette) mTV
VIDrel: CIC/SONOP; PION (LV only)   V/dm V/sur LV

STAR TREK THE NEXT GENERATION: RELICS **   PG

Alexander Singer USA   1992
Patrick Stewart, Jonathan Frakes, Brent Spiner, LeVar Burton, Michael Dorn, Gates McFadden, Marina Sirtis, Lanei Chapman, Erick Weiss, James Doohan, Stacie Foster, Ernie Mirich, Majel Barrett (voice only)
A distress signal from a Federation ship that vanished seventy-five years before leads the Enterprise to a vast artificial sphere in space, millions of miles in diameter, on whose surface the ship has crashed. The sole survivor proves to be Scotty, the chief engineer from the first Enterprise, who found an ingenious way to survive all that time. A fascinating idea is not really well explored in this episode, whose plotting seems over-familiar and dull.
FAN   45 min; 88 min (2-episode cassette) mTV
VIDrel: CIC/SONOP   V/dm V/sur

STAR TREK THE NEXT GENERATION: REMEMBER ME ***

Cliff Bole USA   U
                                                                  1990
Patrick Stewart, Jonathan Frakes, LeVar Burton, Marina Sirtis, Wil Wheaton, Michael Dorn, Gates McFadden, Brent Spiner, Eric Menyuk, Bill Erwin
One of Wesley's experiments results in the creation of a warp bubble, and Dr Crusher finds herself apparently trapped in a strange parallel universe, in which the crew of the Enterprise are

seen to be vanishing one by one. But the Doctor appears to be the only one to notice these changes, and her remaining colleagues begin to question her sanity. As the situation deteriorates, events move on to a powerful climax in this tense and eminently watchable episode.
FAN   87 min (2-episode cassette) mTV
VIDrel: CIC/SONOP   V/sur

STAR TREK THE NEXT GENERATION: REUNION **

Jonathan Frakes USA   PG
                                                                  1990
Patrick Stewart, Jonathan Frakes, LeVar Burton, Marina Sirtis, Wil Wheaton, Michael Dorn, Gates McFadden, Brent Spiner, Suzie Plakson, Charles Cooper, Patrick Massett, Robert O'Reilly, Jon Steuer
The head of the Klingon High Council confides in Picard that he has been poisoned and that a power struggle between two rival Klingons will place the Klingon-Federation treaty in jeopardy. Meanwhile, Worf learns that he has fathered an illegitimate son, when a Klingon woman comes aboard the Enterprise.
FAN   87 min (2-episode cassette) mTV
VIDrel: CIC/SONOP   V/sur

STAR TREK THE NEXT GENERATION: RIGHTFUL HEIR **

Winrich Kolbe USA   PG
                                                                  1993
Patrick Stewart, Jonathan Frakes, Brent Spiner, LeVar Burton, Michael Dorn, Gates McFadden, Marina Sirtis, Alan Oppenheimr, Robert O'Reilly, Kevin Conway, Charles Esten, Norman Snow, Majel Barrett (voice only)
Worf experiences a deep spiritual crisis and travels to a planet where Klingons await the coming of a famous leader who lived 1,500 years before. This figure does in fact reappear but his presence brings division not unity with some Klingons denouncing him as a fraud. However, Worf eventually devises a solution that averts the threat of civil war.
FAN   45 min; 88 min (2-episode cassette) mTV
VIDrel: CIC/SONOP   V/sur

STAR TREK THE NEXT GENERATION: THE ROYALE **

Cliff Bole USA   U
                                                                  1988
Patrick Stewart, Jonathan Frakes, LeVar Burton, Gates McFadden, Wil Wheaton, Brent Spiner, Michael Dorn, Marina Sirtis, Sam Anderson, Jill Jacobson, Leo Garcia, Noble Willingham
An Away Team is sent down to a planet where they find a duplicate of a Las Vegas casino, copied down to every last detail. The team members learn that they are trapped within an alien simulation taken from a cheap gangster novel, originally created to entertain the lone survivor of an ancient Earth expedition. Only by taking on a set of roles found within the novel and playing out the recreation are the team members able to escape.
FAN   92 min (2-episode cassette) mTV
VIDrel: CIC/SONOP   V/dm V/sh

STAR TREK THE NEXT GENERATION: SAMARITAN SNARE **

Les Landau USA   PG
                                                                  1989
Wil Wheaton, Patrick Stewart, Jonathan Frakes, LeVar Burton, Gates McFadden, Brent Spiner, Michael Dorn, Marina Sirtis, Christopher Collins, Leslie Morris, Daniel Benzali, Lycia Naff, Tzi Ma
While Captain Picard fights for his life in surgery, Lieutenant Commander LaForge is held prisoner by a group of technologically incompetent aliens, who need his help in repairing their malfunctioning drive unit. At the same time, Ensign Crusher is given a task somewhat more daunting than his recent Starfleet exams.
FAN   90 min (2-episode cassette) mTV
VIDrel: CIC/SONOP   V/dm

STAR TREK THE NEXT GENERATION: SAREK **   U

Les Landau USA   1990
Patrick Stewart, Jonathan Frakes, LeVar Burton, Michael Dorn, Marina Sirtis, Wil Wheaton, Brent Spiner, Gates McFadden, Mark Lenard, Joanna Miles, William Denis, Rocco Sisto
The Enterprise is called upon to transport a renowned Vulcan ambassador to a mission of the utmost importance, but with his arrival aboard the ship, the crew are stricken with an illness that results in cycles of uncontrollable emotion.
FAN   87 min (2-episode cassette) mTV
VIDrel: CIC/SONOP   V/dm

STAR TREK THE NEXT GENERATION:
SCHISMS ***
Robert Wiemer USA *PG* 1992
Patrick Stewart, Jonathan Frakes, Brent Spiner, LeVar Burton,
Michael Dorn, Gates McFadden, Marina Sirtis, Lanei Chapman, Ken
Thorley, John Nelson, Lina Fiordelissi, Scott T. Trost, Angela McCabe,
Majel Barrett (voice only)
On a mapping mission in deep space, members of the crew start
to experience strange sleep disturbances that prove to be linked
to a sub-space rupture that has appeared in one of the cargo
bays. Comparing notes, they eventually stumble across the
horrifying truth that they are being kidnapped by aliens from
sub-space who are carrying out medical experiments on them.
FAN   45 min; 88 min (2-episode cassette) mTV
VIDrel: CIC/SONOP   V/dm V/sur

STAR TREK THE NEXT GENERATION:
THE SCHIZOID MAN **
Les Landau USA *U* 1988
Patrick Stewart, Jonathan Frakes, Levar Burton, Michael Dorn,
Marina Sirtis, Brent Spiner, Wil Wheaton, Diana Muldaur, W.
Morgan Sheppard, Barbara Alyn Woods, Suzie Plakson
A brilliant but self-centred scientist finds that he is dying, and
in a bid to survive death plans to use Data's body as a recepta-
cle for his consciousness after an away-team beams down to his
base from the Enterprise. He and Data spend much time
together and after the man's death the android begins to behave
in a way that shows all is not well. A capably realised episode.
FAN   92 min (2-episode cassette) mTV
VIDrel: CIC/SONOP   V/dm V/sh

STAR TREK THE NEXT GENERATION:
SECOND CHANCES *
LeVar Burton USA *PG* 1993
Patrick Stewart, Jonathan Frakes, Brent Spiner, LeVar Burton,
Michael Dorn, Gates McFadden, Marina Sirtis, Mae Jemison
After eight years, the Enterprise returns to a planet to retrieve
some data from a research laboratory whose crew were rescued
by Riker in an act of bravery that won him rapid promotion.
Leading an away-team, he returns there and is shocked to
encounter a duplicate of himself, who later proves to be the
result of a freak transporter accident. This novel plot adds little
of interest to a rather dull and pedestrian episode.
FAN   45 min; 88 min (2-episode cassette) mTV
VIDrel: CIC/SONOP   V/sur

STAR TREK THE NEXT GENERATION:
SHADES OF GRAY *
Rob Bowman USA *PG* 1989
Patrick Stewart, Jonathan Frakes, Diana Muldaur, LeVar Burton,
Gates McFadden, Wil Wheaton, Brent Spiner, Michael Dorn, Marina
Sirtis, Colm Meaney
Commander Data is stricken by a deadly virus that sends him
into a coma, and Dr Pulaski is obliged to carry out an intricate
piece of surgery that may kill or cure. A rather cheap-looking
entry that utilised clips from several earlier episodes.
FAN   90 min (2-episode cassette) mTV
VIDrel: CIC/SONOP   V/dm

STAR TREK THE NEXT GENERATION:
SHIP IN A BOTTLE ***
Alexander Singer USA *PG* 1992
Patrick Stewart, Jonathan Frakes, Brent Spiner, LeVar Burton,
Michael Dorn, Gates McFadden, Marina Sirtis, Daniel Davis,
Stephanie Bachamn, Dwight Schultz, Clement Von Franckenstein,
Majel Barrett (voice only)
The holodeck figure of Professor Moriarty is inadvertently
recalled from storage in the computer's memory and soon starts
to make demands, resorting to an ingenious deception in order
to seize control of the Enterprise. A highly inventive if some-
what confusing episode where nothing is as it seems and the
dividing line between reality and illusion becomes very blurred
indeed.
FAN   45 min; 86 min (2-episode cassette) mTV
VIDrel: CIC/SONOP   V/dm V/sur

STAR TREK THE NEXT GENERATION: SILICON
AVATAR **
Cliff Bole USA *U* 1991
Patrick Stewart, Jonathan Frakes, Brent Spiner, LeVar Burton,
Michael Dorn, Gates McFadden, Marina Sirtis, Ellen Geer, Susan
Diol
A world that is just about to be colonised by the Federation is

attacked and devastated by the same crystalline entity that
ravaged the planet on which Data was created. The Enterprise
and its crew take off in pursuit of the creature, assisted by a
woman scientist whose interest in this phenomenon is less than
purely professional.
FAN   45 min; 88 min (2-episode cassette) mTV
VIDrel: CIC/SONOP   V/dm V/sur

STAR TREK THE NEXT GENERATION:
SINS OF THE FATHER **
Les Landau USA *U* 1990
Michael Dorn, Patrick Stewart, Jonathan Frakes, LeVar Burton, Gates
McFadden, Brent Spiner, Wil Wheaton, Marina Sirtis, Tony Todd,
Charles Cooper, Patrick Massett, Thelma Lee, Teddy Davis
Lieutenant Worf finds himself thrust into a battle that can only
end with death for one of the protagonists, when he returns to
his Klingon homeworld in order to absolve his late father of
charges of treason.
FAN   88 min (2-episode cassette) mTV
VIDrel: CIC/SONOP   V/dm V/sh

STAR TREK THE NEXT GENERATION:
SKIN OF EVIL ***
Joseph L. Scanlan USA *U* 1989
Patrick Stewart, Jonathan Frakes, LeVar Burton, Gates McFadden,
Wil Wheaton, Brent Spiner, Michael Dorn, Marina Sirtis, Denise
Crosby, Mart McChesney, Walker Boone, Brad Zerbst, Raymond
Forchion, Ron Gans (voice only)
An Enterprise shuttlecraft crash-lands on Vagra 2, where the crew
find their lives placed in great danger by an oil-like alien entity
which is both sentient and evil. After it kills Tasha Yar, the
remaining crew members are hard put to defend their lives and
find a way to escape the planet. They do so, and a dignified memorial service is held for the fallen comrade. Quite absorbing, though
the plot idea is possibly borrowed from FORBIDDEN PLANET.
FAN   90 min (2-episode cassette) mTV
VIDrel: CIC/SONOP   V/dm V/h

STAR TREK THE NEXT GENERATION:
STARSHIP MINE ***
Cliff Bole USA *PG* 1993
Patrick Stewart, Jonathan Frakes, Brent Spiner, LeVar Burton,
Michael Dorn, Gates McFadden, Marina Sirtis, David Spielberg,
Maire Marshall, Tim Russ, Glenn Morshower, Tom Nibley, Tim
DeZarn, Patricial Tallman, Arlee Reed
The Enterprise is docked at a special array for a routine operation
to clear it of harmful radiation, but this proves anything but when
Picard finds that this procedure is being used as a cover for some-
thing far more sinister. A well paced and quite suspenseful
episode, though one with a routine adventure-style plot.
FAN   45 min; 86 min (2-episode cassette) mTV
VIDrel: CIC/SONOP   V/dm V/sur

STAR TREK THE NEXT GENERATION:
SUB ROSA **
Jonathan Frakes USA *PG* 1993
Patrick Stewart, Jonathan Frakes, LeVar Burton, Michael Dorn,
Marina Sirtis, Gates McFadden, Brent Spiner, Michael Keenan, Shay
Duffin, Duuncan Regher, Ellen Albertini Dow
While attending the funeral of her grandmother, who lived on
a colonised planet made to resemble Scotland, Dr Crusher
becomes embroiled in a dark family secret in the shape her
grandmother's young lover, who claims to be an 800-year-old
ghost. He soon transfers his attentions to the doctor, who falls
madly in love with him. A thinly plotted and wildly romantic
episode that relies too heavily on atmosphere and special effects.
FAN   45 min; 88 min (2-episode cassette) mTV
VIDrel: CIC/SONOP   V/dm V/sur

STAR TREK THE NEXT GENERATION:
SUDDENLY HUMAN **
Gabrielle Beaumont USA *PG* 1990
Patrick Stewart, Jonathan Frakes, LeVar Burton, Marina Sirtis, Wil
Wheaton, Michael Dorn, Gates McFadden, Sherman Howard, Chad
Allen, Barbara Townsend
A human teenager who was brought up by the Tellerians is
rescued from a damaged Tellerian ship and brought aboard the
Enterprise, where Picard tries to acquaint him with human
society. When the Tellerians demand his return Picard refuses
their request, but ultimately the youngster chooses of his own
accord to return to the people and culture he knows.
FAN   87 min (2-episode cassette) mTV
VIDrel: CIC/SONOP   V/sur

STAR TREK THE NEXT GENERATION: THE SURVIVORS **
U
Les Landau USA   1989
Patrick Stewart, Jonathan Frakes, LeVar Burton, Gates McFadden, Wil Wheaton, Brent Spiner, Michael Dorn, Marina Sirtis, John Anderson, Anne Haney
Only a man and his wife appear to have survived on Rana 4, a planet that came under an all out alien attack. Suspicious of the explanation given, Picard learns that the man's "wife" is in reality no more than an image, generated by a powerful shape-changing alien who has chosen to live his life as a human.
FAN   87 min (2-episode cassette) mTV
VIDrel: CIC/SONOP   V/dm

STAR TREK THE NEXT GENERATION: SUSPICIONS **
PG
Cliff Bole USA   1993
Patrick Stewart, Jonathan Frakes, Brent Spiner, LeVar Burton, Michael Dorn, Gates McFadden, Marina Sirtis, Patti Yasutake, Tricia O'Neil, Peter Slusker, James Horan, John S. Ragia, John Stuart Moore, Whoopi Goldberg, Majel Barrett
Doctor Crusher arranges a scientific conference aboard the Enterprise to investigate a new discovery by a Ferengi scientist, but after he is found murdered, she takes a course of action that threatens her future career. A competent whodunit-style episode, enjoyable to watch but hardly imaginative or outstanding.
FAN   45 min; 86 min (2-episode cassette) mTV
VIDrel: CIC/SONOP   V/dm V/sur

STAR TREK THE NEXT GENERATION: SYMBIOSIS **
PG
Win Phelps USA   1988
Patrick Stewart, Jonathan Frakes, LeVar Burton, Gates McFadden, Wil Wheaton, Brent Spiner, Michael Dorn, Marina Sirtis, Judson Scott, Merritt Butrick, Richard Lineback, Kimberly Farr
While investigating a star in the Delos system that shows abnormal fluctuations in its magnetic field, the Enterprise rescues four humanoids and their cargo from a stricken freighter. This cargo is found to be a special drug produced on one planet for supply to another, where it was formerly used to cure a plague but is now merely an addictive narcotic. Picard faces a moral dilemma, caught between his conscience and the Prime Directive on non-interference.
FAN   92 min (2-episode cassette) mTV
VIDrel: CIC/SONOP   V/dm V/sh

STAR TREK THE NEXT GENERATION: TAPESTRY ***
PG
Les Landau USA   1993
Patrick Stewart, Jonathan Frakes, Brent Spiner, LeVar Burton, Michael Dorn, Gates McFadden, Marina Sirtis, Ned Vaughn, J.C. Brandy, Clint Carmichael, John De Lancie, Rae Norman, Clive Church, Marcus Nash, Majel Barrett (voice only)
When Picard is badly injured and is on the verge of dying, he experiences a strange vision in which he meets Q and is given a chance to go back in time and change his actions. Though this saves his life, it is at a cost that he only becomes aware of when it is too late.
FAN   45 min; 86 min (2-episode cassette) mTV
VIDrel: CIC/SONOP   V/dm V/sur

STAR TREK THE NEXT GENERATION: THINE OWN SELF ***
PG
Winrich Kolbe USA   1994
Patrick Stewart, Jonathan Frakes, LeVar Burton, Michael Dorn, Marina Sirtis, Gates McFadden, Brent Spiner, Ronnie Claire Edwards, Michael Rothaar, Kimberly Cullum, Andy Kossin, Michael G. Hagerty, Richard Ortega-Miro, Majel Barrett
When a Federation probe crashes on a planet with a pre-technological civilisation, Data is sent to recover it since it contains radioactive components that could prove harmful to organic life. But when a fateful loss of memory occurs, it seems that his mission is to end in tragedy. Generally well acted but lacking in plot development, and with a contrived and very telescoped ending.
FAN   45 min; 88 min (2-episode cassette) mTV
VIDrel: CIC/SONOP   V/sur

STAR TREK THE NEXT GENERATION: TIME SQUARED **
PG
Joseph L. Scanlan USA   1989
Patrick Stewart, Jonathan Frakes, LeVar Burton, Michael Dorn,

Marina Sirtis, Brent Spiner, Wil Wheaton, Diana Muldaur, Mitchell Ryan, Colm Meaney
When the Enterprise rescues a disabled shuttlecraft an incoherent double of Captain Picard is found onboard. Investigations reveal that the double is from a future in which a lethal region of space awaits the starship. As time passes, Picard becomes ever more fearful of making a decision that could lead to the very situation he wishes to avoid.
FAN   90 min (2-episode cassette) mTV
VIDrel: CIC/SONOP   V/dm V/sh

STAR TREK THE NEXT GENERATION: TIME'S ARROW – PART 1 **
PG
Les Landau USA   1992
Patrick Stewart, Jonathan Frakes, Brent Spiner, LeVar Burton, Michael Dorn, Gates McFadden, Marina Sirtis, Jerry Hardin, Michael Aron, Barry Kivel, Ken Thorley, Whoopi Goldberg, Marc Alaimo, Milt Tarver, Pamela Kosh, James Gleason
An excavation on Earth reveals the presence of Data's head which was buried together with some 19th century artefacts in the vicinity of San Francisco. Other evidence takes the Enterprise to a distant planet when they learn of the activities of time-travelling aliens with a macabre secret. An unusual double-episode which unhappily fails to make the most of the paradoxes inherent in the plot and instead wastes time on a buffoonish portrayal of Mark Twain.
FAN   87 min (2-episode cassette) mTV
VIDrel: CIC/SONOP; PION (LV only)   V/sur LV

STAR TREK THE NEXT GENERATION: TIME'S ARROW – PART 2 **
PG
Les Landau USA   1992
Patrick Stewart, Jonathan Frakes, Brent Spiner, LeVar Burton, Michael Dorn, Gates McFadden, Marina Sirtis, Jerry Hardin, Michael Aron, Barry Kivel, Ken Thorley, Whoopi Goldberg, Marc Alaimo, Milt Tarver, Pamela Kosh, James Gleason
Conclusion of this two-part adventure, the whole of which has become available on a single cassette (as is the case with the LV disc).
FAN   88 min (2-episode cassette) mTV
VIDrel: CIC/SONOP; PION (LV only)   V/dm V/sur LV

STAR TREK THE NEXT GENERATION: TIMESCAPE ***
PG
Adam Nimoy USA   1993
Patrick Stewart, Jonathan Frakes, Brent Spiner, LeVar Burton, Michael Dorn, Gates McFadden, Marina Sirtis, Michael Bofshever, John DeMita, Joel Fredericks
Having attended some conferences, Picard, Troi, LaForge and Data are returning by shuttle to the Enterprise, when they begin to experience some phenomena indicating a breakdown in the flow of time. Having located the Enterprise, they find it frozen in time and locked in apparent conflict with a Romulan vessel. A tense and absorbing episode, with a cleverly plotted storyline.
FAN   45 min; 88 min (2-episode cassette) mTV
VIDrel: CIC/SONOP   V/sur

STAR TREK THE NEXT GENERATION: TIN MAN **
U
Robert Scheerer USA   1990
Patrick Stewart, Jonathan Frakes, LeVar Burton, Michael Dorn, Marina Sirtis, Wil Wheaton, Brent Spiner, Gates McFadden, Harry Groener, Michael Cavanaugh, Peter Vogt
A telepathic Betazoid comes aboard, his mission being to communicate with a newly discovered creature that takes the form of a living "spaceship". However, the Romulans are out to prevent this contact ever being made.
FAN   88 min (2-episode cassette) mTV
VIDrel: CIC/SONOP   V/dm V/sh

STAR TREK THE NEXT GENERATION: TOO SHORT A SEASON **
PG
Rob Bowman USA   1987
Patrick Stewart, Jonathan Frakes, LeVar Burton, Gates McFadden, Wil Wheaton, Brent Spiner, Michael Dorn, Marina Sirtis, Clayton Rohner, Marsha Hunt, Michael Pataki
The Enterprise is sent to Mordan IV, where a group of terrorists have taken hostages and are demanding that a seventy-year-old Federation admiral handle the negotiations. However, the man in question has recently taken a youth elixir whose side effects could be fatal.
FAN   88 min (2-episode cassette) mTV
VIDrel: CIC/SONOP   V/dm

STAR TREK THE NEXT GENERATION: TRANSFIGURATIONS **
PG

Tom Benko USA 1990

Patrick Stewart, Jonathan Frakes, LeVar Burton, Michael Dorn, Brent Spiner, Marina Sirtis, Wil Wheaton, Mark La Mura, Charles Dennis, Julie Warner

While exploring an unknown star system, the Enterprise rescues an injured humanoid from his wrecked ship and Dr Crusher immediately sets about treating him, but is amazed to find her patient recovering at an impossibly fast rate. He soon regains consciousness but is suffering from amnesia, and appears to be undergoing a mysterious and powerful evolutionary change at the cellular level. A watchable episode, built around some intriguing ideas.

FAN 87 min (2-episode cassette) mTV
VIDrel: CIC/SONOP V/dm V/h

STAR TREK THE NEXT GENERATION: TRUE Q ***
PG

Robert Scheerer USA 1992

Patrick Stewart, Jonathan Frakes, Brent Spiner, LeVar Burton, Michael Dorn, Gates McFadden, Marina Sirtis, Lanei Chapman, Olivia D'Abo, John P. Connolly John De Lancie

A young girl who joins the crew of the Enterprise as a sort of trainee proves to have super-human powers and soon Q is on hand to claim her as one of his kind, her parents having been renegades who took refuge on Earth. She finds herself caught in a terrible dilemma between the human world she knows and the lure of godlike abilities. A nicely dramatic episode that holds together well.

FAN 45 min; 88 min (2-episode cassette) mTV
VIDrel: CIC/SONOP V/dm V/sur

STAR TREK THE NEXT GENERATION: UNIFICATION, PARTS 1 AND 2 ***
U

Cliff Bole/Les Landau USA 1991

Patrick Stewart, Jonathan Frakes, LeVar Burton, Michael Dorn, Marina Sirtis, Gates McFadden, Brent Spiner, Leonard Nimoy, Joanna Miles, Stephen Root, Graham Jarvis, Malachi Thorne, Norman Large, Daniel Roebuck, Erick Avari

Eighty years after the original Enterprise was performing its duty for the Federation, it's learnt that an unauthorised peace initiative has been instigated with the Romulans, who are possibly holding the legendary Mr Spock on the planet Romulas. Captain Picard is given a specially equipped Klingon vessel and sent on an undercover mission to the planet to find out the truth. An enjoyable full-length episode with some very good effects.

FAN 88 min mTV VIDrel: CIC/SONOP; PION (LV only)
V/sur LV

STAR TREK THE NEXT GENERATION: UNNATURAL SELECTION **
PG

Paul Lynch USA 1988

Patrick Stewart, Jonathan Frakes, Levar Burton, Michael Dorn, Marina Sirtis, Brent Spiner, Wil Wheaton, Diana Muldaur, Patricia Smith, Colm Meaney, Patrick McNamara, Scott Trost

An episode similar in theme to STAR TREK: THE DEADLY YEARS, this one has a genetic mutation on an experimental colony leading to an outbreak of premature ageing among the scientists there. Dr Pulaski investigates in the hope of finding a cure, but is herself infected and begins to age rapidly.

FAN 92 min (2-episode cassette) mTV
VIDrel: CIC/SONOP V/dm V/sh

STAR TREK THE NEXT GENERATION: UP THE LONG LADDER **
PG

Winrich Kolbe USA 1989

Patrick Stewart, Jonathan Frakes, LeVar Burton, Gates McFadden, Wil Wheaton, Brent Spiner, Michael Dorn, Marina Sirtis, Barrie Ingham, Jon de Vries, Rasalyn Landor, Colm Meaney

This adventure has the Enterprise being sent to rescue the people from a doomed planet, namely the members of a primitive farming community and an advanced race of clones. The latter face extinction for lack of genetic material, but devise a plan to avoid this by duplicating the DNA structure of Riker and Pulaski.

FAN 90 min (2-episode cassette) mTV
VIDrel: CIC/SONOP V/dm

STAR TREK THE NEXT GENERATION: THE VENGEANCE FACTOR **
PG

Timothy Bond USA 1989

Patrick Stewart, Jonathan Frakes, LeVar Burton, Gates McFadden,

Wil Wheaton, Brent Spiner, Michael Dorn, Marina Sirtis, Lisa Wilcox, Joey Aresco, Nancy Parsons, Stephen Lee, Marc Lawrence, Elkanah J. Burns

Picard tries to resolve a dispute between a band of pirates and the leader of their homeworld, but matters are complicated when one of the pirates is murdered, the culprit being a woman with whom Picard has become romantically involved.

FAN 88 min (2-episode cassette) mTV
VIDrel: CIC/SONOP V/dm

STAR TREK THE NEXT GENERATION: VIOLATIONS **
PG

Robert Wiemer USA 1991

Patrick Stewart, Jonathan Frakes, Brent Spiner, LeVar Burton, Michael Dorn, Gates McFadden, Marina Sirtis, Rosalind Chao, Ben Lemmon, Daniel Sage, Rick Pitts, Eve Brenner, Doug Wert, Craig Benton, Majel Barrett (voice only)

After a group of telepathic aliens come aboard the Enterprise, a number of its officer fall into deep comas without any apparent reason, and all apparent causes of this condition appear to be ruled out. However, diligent work by Geordi and Data eventually unravel this mystery.

FAN 45 min; 88 min (2-episode cassette) mTV
VIDrel: CIC/SONOP V/dm V/sur

STAR TREK THE NEXT GENERATION: WE'LL ALWAYS HAVE PARIS **
PG

Robert Becker USA 1988

Patrick Stewart, Jonathan Frakes, LeVar Burton, Gates McFadden, Wil Wheaton, Brent Spiner, Michael Dorn, Marina Sirtis, Michelle Phillips, Rod Loomis, Isabel Lorca, Dan Kern, Jean-Pail Vignon, Kelly Ashmore, Lance Spellerberg

Picard is reunited with a former love when the Enterprise investigates a research post where a scientist has been conducting a series of time experiments. Unfortunately, this has resulted in a distortion of time that could destroy the fabric of space itself.

FAN 90 min (2-episode cassette) mTV
VIDrel: CIC/SONOP V/dm V/h

STAR TREK THE NEXT GENERATION: WHEN THE BOUGH BREAKS **
PG

Kim Manners USA 1987

Patrick Stewart, Jonathan Frakes, LeVar Burton, Gates McFadden, Wil Wheaton, Brent Spiner, Michael Dorn, Marina Sirtis, Jerry Hardin, Brenda Strong, Paul Lambert, Jandi Swanson, Ivy Bethune, Derek Torsek, Michele Marsh, Dan Mason

The people of Aldea have become sterile, and to avoid extinction they kidnap a group of children from the Enterprise, intending to use them to replenish their race. As Picard attempts to negotiate for their release Dr Crusher investigates the reasons for the race's sterility.

FAN 89 min (2-episode cassette) mTV
VIDrel: CIC/SONOP V/dm

STAR TREK THE NEXT GENERATION: WHERE NO ONE HAS GONE BEFORE **
PG

Rob Bowman USA 1987

Patrick Stewart, Jonathan Frakes, LeVar Burton, Gates McFadden, Wil Wheaton, Brent Spiner, Michael Dorn, Marina Sirtis, Eric Menyuk, Stanley Kamel, Herta Ware, Biff Yeager, Charles Dayton, Victoria Dillard

A Federation propulsion expert and his alien assistant come aboard and modify the drive units of the Enterprise, but this results in the ship being accidentally flung over 350,000,000 light years from home, without any means of returning. To add to their problems, the crew begin hallucinating as reality starts to break down in that sector of space.

FAN 89 min (2-episode cassette) mTV
VIDrel: CIC/SONOP V/dm

STAR TREK THE NEXT GENERATION: WHERE SILENCE HAS LEASE **
PG

Winrich Kolbe USA 1988

Patrick Stewart, Jonathan Frakes, LeVar Burton, Gates McFadden, Wil Wheaton, Brent Spiner, Michael Dorn, Marina Sirtis, Earl Boen, Charles Douglas, Colm Meaney

An omniscient alien creates a void in space and traps the Enterprise there, its intention being to conduct a series of experiments that will inevitably cause deaths among the starship crew. Picard is forced to threaten to self-destruct the ship in order to force the alien to release them.

FAN 91 min (2-episode cassette) mTV
VIDrel: CIC/SONOP V/dm V/sh

STAR TREK THE NEXT GENERATION: WHO WATCHES THE WATCHERS? **

Robert Wiemer USA   U
                                                      1989

Patrick Stewart, Jonathan Frakes, LeVar Burton, Gates McFadden, Wil Wheaton, Brent Spiner, Michael Dorn, Marina Sirtis, Kathryn Leigh Scott, Ray Wise, James Greene, John McLlam, Pamela Seagall, James McIntyre, Lois Hall

When the Enterprise sends a team to rescue a group of scientists that were studying a primitive society, a glimpse of some advanced technology is accidentally revealed to the race. Having rescued a wounded Federation officer, matters take an unexpected turn with the capture of Troi and the adoption of Picard as a god.

FAN   87 min (2-episode cassette) mTV
VIDrel: CIC/SONOP   V/dm

STAR TREK THE NEXT GENERATION: THE WOUNDED **

Chip Chalmers USA   U
                                                      1991

Patrick Stewart, Jonathan Frakes, Michael Dorn, LeVar Burton, Gates McFadden, Marina Sirtis, Brent Spiner, Bob Gunton, Rosalind Chao, Marc Alaimo, Tim Winters, John Hancock, Marco Rodriguez

Though a peace treaty has recently been concluded with the untrustworthy Cardassians, a starship captain whose wife was killed by them launches a one-man war. He is only talked out of his plan of vengeance by Chief O'Brien, who once served as a member of his crew.

FAN   87 min (2-episode cassette) mTV
VIDrel: CIC/SONOP   V/sur

STAR TREK THE NEXT GENERATION: YESTERDAY'S ENTERPRISE **

David Carson USA   PG
                                                      1990

Patrick Stewart, Jonathan Frakes, LeVar Burton, Michael Dorn, Gates McFadden, Brent Spiner, Marina Sirtis, Denise Crosby, Christopher McDonald, Tricia O'Neill, Whoopi Goldberg

A conflict with the Romulans results in the Enterprise being propelled forward through time into a parallel universe in which the Federation is still at war with the Klingons. Only Guinan realises this has occurred, and tries to persuade Picard to get the ship back in time and thus preserve the original events that led to the Federation-Klingon treaty.

FAN   88 min (2-episode cassette) mTV
VIDrel: CIC/SONOP   V/dm V/sh

STAR TREK VOYAGER: THE 37s **

James L. Conway USA   PG
                                                      1995

Kate Mulgrew, Robert Beltran, Roxann Biggs-Dawson, Jennifer Lien, Robert Duncan, Ethan Phillips, Robert Picardo, Tim Russ, Garrett Wang, John Rubinstein, Sharon Lawrence, David Graf, James Saito, Mel Winkler

A trail of rust particles leads the Voyager to a 1936 Ford pickup truck that is floating in space. They bring it aboard for examination and find their way to a nearby planet where they discover an ancient Earth aircraft as well as some cryogenically preserved bodies, including that of Amelia Earhart. An attempt to build a story around the idea of alien abductions that starts off well but ends in a hurried resolution that leaves the way open for the second series.

FAN   45 min; 88 min (2-episode cassette) mTV
VIDrel: CIC/SONOP   V/sur

STAR TREK VOYAGER: ALLIANCES **

Les Landau USA   (PG)
                                                      1995

Kate Mulgrew, Robert Beltran, Roxann Biggs-Dawson, Jennifer Lien, Robert Duncan, Ethan Phillips, Robert Picardo, Tim Russ, Garrett Wang, Martha Hackett, Charles O. Lucia, Anthony De Longis, Raphael Sbarge, Larry Cedar, Simon Billig

Increasingly ferocious attacks on Voyager by Kazon vessels lead to a tension situation aboard and the captain comes under pressure to relax her strict adherence to Federation principles. After some discussion, it is decided to seek an alliance with one of the Kazon clans but events eventually prove her right about the wisdom of non-interference in other cultures. A poorly plotted episode with little tension.

FAN   45 min mTV   TVrel

STAR TREK VOYAGER: BASICS, PART 1 **

Winrich Kolbe USA   PG
                                                      1995

Kate Mulgrew, Robert Beltran, Roxann Biggs-Dawson, Jennifer Lien, Robert Duncan, Ethan Phillips, Robert Picardo, Tim Russ, Garrett Wang, Brad Dourif, Anthony De Longis, John Gegenhuber, Martha Hackett, Henry Darrow, Scot Haven

Chakotay receives a message from Seska that his son has been born and that they are both in danger, and soon he and the rest of the crew go storming to the rescue. Though they have anticipated the possibility of a trap, they are both outfought and outsmarted. After a decisive battle with the Kazon, the crew are forced to surrender the ship to the Kazon leader, who abandons them without technology on a desolate planet. A two-part story of varying quality.

FAN   45 min; 88 min (2-episode cassette) mTV cC
VIDrel: CIC/SONOP   V/sur

STAR TREK VOYAGER: CARETAKER **

Winrich Kolbe USA   PG
                                                      1995

Kate Mulgrew, Robert Beltran, Roxann Biggs-Dawson, Jennifer Lien, Robert Duncan, Ethan Phillips, Robert Picardo, Tim Russ, Garrett Wang, Basil Langton, Gavan O'Herlihy, Angela Paton, Armin Shimerman, Alicia Coppola, Bruce French

Pilot episode to this new series opens against the background of the continuing revolt of Federation colonists against the Cardassians. When contact is lost with her first officer, working under cover on a Maquis ship, the Voyager's captain sets out to find him. However, the ship is mysteriously taken 70,000 light years to an unknown part of the galaxy. Long and uninspired, this flat story suffers from unimpressive effects and weak acting.

FAN   88 min mTV   VIDrel: CIC/SONOP; PION (LV only)
V/sur LV

STAR TREK VOYAGER: CATHEXIS ***

Kim Friedman USA   PG
                                                      1995

Kate Mulgrew, Robert Beltran, Roxann Biggs-Dawson, Jennifer Lien, Robert Duncan, Ethan Phillips, Robert Picardo, Tim Russ, Garrett Wang, Brian Markinson, Michael Cumpsty, Carolyn Seymour, Majel Barrett (voice only).

A long-range scan reveals that Chakotay and Tuvok suffered severe injuries after completing a routine trading mission. After they are beamed aboard, the Voyager begins to experience strange malfunctions that are eventually traced to an invisible alien entity that can take over its human host. Further incidents follow that culminate in a crisis which threatens to destroy the ship. For once, this fairly well-written episode maintains suspense right to the end.

FAN   45 min; 88 min (2-episode cassette) mTV
VIDrel: CIC/SONOP   V/sur

STAR TREK VOYAGER: THE CLOUD **

David Livingston USA   PG
                                                      1995

Kate Mulgrew, Robert Beltran, Roxann Biggs-Dawson, Jennifer Lien, Robert Duncan, Ethan Phillips, Robert Picardo, Tim Russ, Garrett Wang, Angela Dohrmann, Judy Geeson, Luigi Amodeo

When a nebula is spotted that seems to be rich in energy, the Voyager heads for it at full speed, since this offers hope of replenishing dwindling energy supplies. However, what they find proves to be something completely unexpected, as they actually encounter what appears to be a vast living being. An interesting concept receives very summary treatment and after putting right an unintentional mistake, our crew continue their odyssey. Woefully unimaginative.

FAN   45 min; 88 min (2-episode cassette) mTV
VIDrel: CIC/SONOP   V/sur

STAR TREK VOYAGER: COLD FIRE **

Cliff Bole USA   (PG)
                                                      1995

Kate Mulgrew, Robert Beltran, Roxann Biggs-Dawson, Jennifer Lien, Robert Duncan, Ethan Phillips, Robert Picardo, Tim Russ, Garrett Wang, Gary Graham, Lindsay Ridgeway, Norman Large, Majel Barrett (narration)

The remains of the Caretaker (see episodes 1 and 2) begin unexpectedly to show signs of life, which raises hopes in the captain that it might be possible to find his mate and ask her to send Voyager back home. However, what they first encounter is a small array inhabited by some 2,000 Ocampa, who have evolved some remarkable mental powers. A standard ship-in-peril-episode that creates no sense of menace and is resolved in a very matter-of-fact and anti-climactic way.

FAN   45 min mTV   TVrel

STAR TREK VOYAGER: DEADLOCK **

David Livingston USA   12
                                                      1995

Kate Mulgrew, Robert Beltran, Roxann Biggs-Dawson, Jennifer Lien, Robert Duncan McNeill, Ethan Phillips, Robert Picardo, Tim Russ, Garrett Wang

After it inadvertently enters Vidiian territory, the Voyager is forced to hide in a plasma drift when it is pursued by Vidiian

vessels. But unexpected circumstances result in the ship being drained of its limited supplies of anti-matter, and it would appear to have become trapped in a parallel universe, one of the stranger effects of which is the duplication of all matter, including the crew. Intriguing, but indifferently developed.
FAN 88 min (2-episode cassette) mTV cC
VIDrel: CIC/SONOP V/sur

STAR TREK VOYAGER: DEATH WISH *** PG
James L. Conway USA 1996
Kate Mulgrew, Robert Beltran, Roxann Biggs-Dawson, Jennifer Lien, Robert Duncan McNeill, Ethan Phillips, Robert Picardo, Tim Russ, Garrett Wang, Gerrit Graham, John De Lancie, Jonathan Frakes
A member of the Q race appears onboard after a sensor scan releases him from imprisonment inside a comet, where he was placed by his fellow beings in order to prevent his suicide. When the more notorious Q also arrives on the ship, a hearing is held to decide on the fugitive alien's request for asylum. Well scripted and acted, this interesting episode takes the form of a courtroom drama that probes the nature of these beings and the drawbacks of immortality.
FAN 45 min; 88 min (2-episode cassette) mTV cC
VIDrel: CIC/SONOP V/sur

STAR TREK VOYAGER: DREADNOUGHT ** PG
LeVar Burton USA 1996
Kate Mulgrew, Robert Beltran, Roxann Biggs-Dawson, Jennifer Lien, Robert Duncan McNeill, Ethan Phillips, Robert Picardo, Tim Russ, Garrett Wang, Susan Diol, Raphael Sbarge, Martha Hackett, Michael Spound, Rick Gianisi
B'Elanna Torres' Maquis past catches up with her when the Voyager detects a missile of Cardassian origin that appears to be on a collision course with a densely populated planet. Though she had originally programmed the missile to attack the Cardassians, for some reason it has found its way into the Delta Quadrant and in order to prevent a catastrophe, Torres has to beam aboard and attempt to reprogram its navigational computer. A flawed but tense episode.
FAN 45 min; 88 min (2-episode cassette) mTV cC
VIDrel: CIC/SONOP V/sur

STAR TREK VOYAGER: ELOGIUM ** (PG)
Winrich Kolbe USA 1995
Kate Mulgrew, Robert Beltran, Roxann Biggs-Dawson, Jennifer Lien, Robert Duncan, Ethan Phillips, Robert Picardo, Tim Russ, Garrett Wang, Nancy Hower, Gary O'Brien, Terry Correll
After the Voyager encounters a swarm of strange beings that live in space, Kes finds herself becoming sexually mature far ahead of her time, which poses quite a few problems for her and Neelix. Meanwhile, the presence of these aliens also represents a danger to the ship's survival. A disappointing episode whose two sub-plots do not meld very well. Average.
FAN 45 min mTV TVrel

STAR TREK VOYAGER: EMANATIONS ** PG
David Livingston USA 1995
Kate Mulgrew, Robert Beltran, Roxann Biggs-Dawson, Jennifer Lien, Robert Duncan, Ethan Phillips, Robert Picardo, Tim Russ, Garrett Wang, Jerry Hardin, Jefrey Alan Chandler, John Ciglino, Martha Hackett, Robin Groves, Cecile Callan
An away team is sent to investigate some unusual phenomena on a planet when a freak transporter accident lands Harry Kim in another dimension, on a world whose inhabitants believe firmly in a physical existence after death. They thus view him as having returned from this afterlife, which places him in a very delicate position. Some interesting ideas get the usually cursory treatment in a well made but rather unsatisfying episode.
FAN 45 min; 88 min (2-episode cassette) mTV
VIDrel: CIC/SONOP V/sur

STAR TREK VOYAGER: EX POST FACTO ** PG
Levar Burton USA 1995
Kate Mulgrew, Robert Beltran, Roxann Biggs-Dawson, Jennifer Lien, Robert Duncan, Ethan Phillips, Robert Picardo, Tim Russ, Garrett Wang, Robin McKee, Francis Guinan, Aaron Lustig, Ray Reinhardt, Henry Brown
A mission to an alien planet to enlist their scientific help lands Tom Paris in trouble when he falls for a femme fatale and finds himself framed for the murder of her scientist husband. Unfortunately, the case against him seems watertight as it consist of memories taken from his victim's brain. Pronounced guilty and sentenced to a cruel punishment, the captain and her

first officer work hard to prove his innocence. A ludicrous episode that verges on parody.
FAN 45 min; 88 min (2-episode cassette) mTV
VIDrel: CIC/SONOP V/sur

STAR TREK VOYAGER: EYE OF THE NEEDLE *** PG
Winrich Kolbe USA 1995
Kate Mulgrew, Robert Beltran, Roxann Biggs-Dawson, Jennifer Lien, Robert Duncan, Ethan Phillips, Robert Picardo, Tim Russ, Garrett Wang, Vaughn Armstrong, Tom Virtue
The search for a wormhole is finally rewarded but hope soon turns to disappointment when it proves to be no more than some thirty centimetres in width. Unable to take the ship through it, the captain orders the launching of a micro-probe by means of which they eventually make contact with a vessel on the other side of the wormhole. An above-average episode that maintains a sense of suspense right to the end and offers quite a few plot twists.
FAN 45 min; 88 min (2-episode cassette) mTV
VIDrel: CIC/SONOP V/sur

STAR TREK VOYAGER: FACES *** PG
Winrich Kolbe USA 1995
Kate Mulgrew, Robert Beltran, Roxann Biggs-Dawson, Jennifer Lien, Robert Duncan, Ethan Phillips, Robert Picardo, Tim Russ, Garrett Wang, Brian Markinson, Bob LaBell, Barton Tinapp
A scientific study team from Voyager are captured by the Vidiians whose scientists used advanced techniques to split B'Elanna into two separate individuals, one human, the other a full Klingon. They hope to make use of the latter's DNA in their efforts to rid themselves of the bacteriophage that afflicts them. A rescue mission is soon mounted, in this episode of average excitement, whose initial ideas are not all that well explored.
FAN 45 min; 88 min (2-episode cassette) mTV
VIDrel: CIC/SONOP V/sur

STAR TREK VOYAGER: HEROES AND DEMONS ** U
Les Landau USA 1995
Kate Mulgrew, Robert Beltran, Roxann Biggs-Dawson, Jennifer Lien, Robert Duncan, Ethan Phillips, Robert Picardo, Tim Russ, Garrett Wang, Marjorie Monaghan, Christopher Neame, Michael Keenan, Majel Barrett (voice only).
After samples of a kind of energy are beamed aboard Voyager as part of a scientific experiment, Harry Kim appears to be missing and his disappearance is linked to a holodeck malfunction. Eventually, the ship's doctor is forced to take on a most unlike role after he is transferred to the holodeck to investigate. Some interesting ideas are barely developed in this rather silly tale that takes too long to get to the point.
FAN 45 min; 88 min (2-episode cassette) mTV
VIDrel: CIC/SONOP V/sur

STAR TREK VOYAGER: INITIATIONS ** (PG)
Winrich Kolbe USA 1995
Kate Mulgrew, Robert Beltran, Roxann Biggs-Dawson, Jennifer Lien, Robert Duncan, Ethan Phillips, Robert Picardo, Tim Russ, Garrett Wang, Aron Eisenberg, Patrick Kilpatrick, Tim De Dard, Majel Barrett (voice only)
Chakotay is returning to Voyager in a shuttle craft when he comes under fire from a Kazon ship piloted by a young boy who, in the warrior traditions of his world, must earn his name by making his first kill. Forced to defend himself, he soon gets involved in giving our young hothead a lesson in friendship. A quite strong story does not really get the treatment it deserves, in this fairly absorbing episode.
FAN 45 min mTV TVrel

STAR TREK VOYAGER: INNOCENCE ** 12
James L. Conway USA 1995
Kate Mulgrew, Robert Beltran, Roxann Biggs-Dawson, Jennifer Lien, Robert Duncan McNeill, Ethan Phillips, Robert Picardo, Tim Russ, Garrett Wang
Forced to crash-land their shuttle on a small moon, Tuvok and Bennett encounter three children who belong to the Dryans, a mysterious alien race about whom very little is known. They explain that this planetoid is a scarred site on which they have been left to die by a figure known as "Morrock". Both fascinated and repelled by a society that kills its own young, Tuvok investigates and learns a strange secret about these creatures. Fair.
FAN 88 min (2-episode cassette) mTV cC
VIDrel: CIC/SONOP V/sur

STAR TREK VOYAGER: INVESTIGATIONS **   PG
Les Landau USA   1996
Kate Mulgrew, Robert Beltran, Roxann Biggs-Dawson, Jennifer Lien, Robert Duncan, Ethan Phillips, Robert PIcardo, Tim Russ, Garrett Wang
Neelix develops an enthusiasm for investigative reporting and adopts the epithet of journalist after becoming convinced there is a traitor aboard Voyager. Unfortunately, his suspicions fall on Paris, whose growing disaffection causes him to resign from his post and take up work as a pilot for a Talaxian convoy. But when this convoy it attacked by Kazon vessels, Paris is captured and interrogated by Seska. An adequate entry, hampered by over-complexity.
FAN   45 min; 88 min (2-episode cassette) mTV cC
VIDrel: CIC/SONOP   V/sur

STAR TREK VOYAGER: JETREL ***   (PG)
Kim Friedman USA   1995
Kate Mulgrew, Robert Beltran, Roxann Biggs-Dawson, Jennifer Lien, Robert Duncan, Ethan Phillips, Robert Picardo, Tim Russ, Garrett Wang, James Sloyan, La Larry Hankin
Neelix receives an unexpected visitor whose presence revives painful memories since this is the scientist who devised a doomsday weapon that killed his family but brought a devastating war to close. He claims that Neelix may be suffering from a deadly disease and needs immediate examination but his real purpose only becomes apparent much later. A nicely dramatic and intriguing episode with some fine plot twists, that keeps one guessing till the end.
FAN   45 min mTV   TVrel

STAR TREK VOYAGER: LEARNING CURVE **   (PG)
David Livingston USA   1995
Kate Mulgrew, Robert Beltran, Roxann Biggs-Dawson, Jennifer Lien, Robert Duncan, Ethan Phillips, Robert Picardo, Tim Russ, Garrett Wang, Armand Shultz, Kenny Morrison, Catherine MacNeal, Thomas Aleander Dekker
Growing disciplinary problems among a small group of former Maquis crew members lead Tuvok to resort to a tough training program that is intended to give them a firm grounding in Starfleet procedures. At the same time, the ship faces a serious crisis due to an escalating series of component failures. A very routine episode that is watchable but provides no great surprises.
FAN   45 min mTV   TVrel

STAR TREK VOYAGER: LIFESIGNS **   PG
Cliff Bole USA   1996
Kate Mulgrew, Robert Beltran, Roxann Biggs-Dawson, Jennifer Lien, Robert Duncan, Ethan Phillips, Robert Picardo, Tim Russ, Garrett Wang, Susan Diol, Raphael Sbarge, Martha Hackett, Michael Spound, Rick Gianisi
After encountering a damaged spacecraft, a seriously ill Vidiian woman is beamed aboard directly to sickbay. Clearly suffering badly from the ravages of the phage, she is given a radically new treatment that involves transferring her consciousness to a holographic body. As time progresses, the holographic doctor becomes aware that he is falling in love with her, thanks to an adaptive program that allows him to feel emotions. Watchable but not overly interesting.
FAN   45 min; 88 min (2-episode cassette) mTV cC
VIDrel: CIC/SONOP   V/sur

STAR TREK VOYAGER: MANEUVERS **   (PG)
David Livingston USA   1995
Kate Mulgrew, Robert Beltran, Roxann Biggs-Dawson, Jennifer Lien, Robert Duncan, Ethan Phillips, Robert Picardo, Tim Russ, Garrett Wang, Martha Hackett, Anthony De Longis, Terry Lester, John Gegenhuber, Majel Barrett (narration)
A message from what appears to be a Federation beacon lures the Voyager into an ingenious trap set by the Kazon, who are now being assisted by the traitor Seska (see STATE OF FLUX). Humiliated by this defeat, Chakotay takes off on a one-man mission to recover the transporter technology that was stolen and gets himself captured, which forces the captain to make a most difficult choice. A lacklustre episode that does little with its strong storyline.
FAN   45 min mTV   TVrel

STAR TREK VOYAGER: MELD **   PG
Cliff Bole USA   1996
Kate Mulgrew, Robert Beltran, Roxann Biggs-Dawson, Jennifer Lien, Robert Duncan McNeill, Ethan Phillips, Robert Picardo, Tim Russ,

Garrett Wang, Brad Dourif, Angela Hohrmann, Simon Belling, Majel Barrett (voice only)
When an apparently motiveless murder is committed in engineering, Security Chief Tuvok locates the killer, but finds himself unable to deal effectively with a human being as unbalanced as the culprit. Determined to find a deeper and more logical reason for the crime, he resorts to a mind-meld but suffers unforeseen complications. Some interesting ideas are half-heartedly explored but never developed, while the two sub-plots are needless distractions.
FAN   45 min; 88 min (2-episode cassette) mTV cC
VIDrel: CIC/SONOP   V/sur

STAR TREK VOYAGER: NON SEQUITUR **   (PG)
Winrich Kolbe USA   1995
Kate Mulgrew, Robert Beltran, Roxann Biggs-Dawson, Jennifer Lien, Robert Duncan, Ethan Phillips, Robert Picardo, Tim Russ, Garrett Wang, Louis Giambalvo, Jennifer Gatti, Jack Shearer, Mark Kiely, Majel Barrett (voice only)
While piloting his shuttle craft, Ensign Kim encounters a temporal anomaly that sends him back to Earth. However, it soon transpires that he has landed in a changed reality in which he learns that he was not accepted for service on Voyager. Despite the attraction of being back with his fiancee, he realises that he must return to his reality and gets some unexpected help. A wholly conventional episode with few surprises and a notable lack of tension.
FAN   45 min mTV   TVrel

STAR TREK VOYAGER: PARALLAX **   PG
Kim Friedman USA   1995
Kate Mulgrew, Robert Beltran, Roxann Biggs-Dawson, Jennifer Lien, Robert Duncan, Ethan Phillips, Robert Picardo, Tim Russ, Garrett Wang, Martha Hackett, Josh Clark, Justin Williams
Integration aboard ship between the Starfleet crew and the Maquis survivors is causing problems when the Voyager stumbles across a strange phenomenon, a breach in space-time that appears to have trapped another vessel. Rescue attempts soon lead to a crisis, forcing the captain to disregard normal procedures and appoint officers solely on grounds of merit. An unexciting episode that focuses on Biggs-Dawson as a surly but brilliant engineer.
FAN   45 min; 88 min (2-episode cassette) mTV
VIDrel: CIC/SONOP   V/sur

STAR TREK VOYAGER: PARTURITION **   (PG)
Jonathan Frakes USA   1995
Kate Mulgrew, Robert Beltran, Roxann Biggs-Dawson, Jennifer Lien, Robert Duncan, Ethan Phillips, Robert Picardo, Tim Russ, Garrett Wang
With its food supplies seriously depleted, Voyager makes for an M-class planet that scans show to be rich in plant life. Neelix and Paris are ordered to land there in a shuttle and collect as much food as they can, but the strained relationship between the two men does not make this an easy mission for either of them. Forced to crash-land, they find themselves cast in the unlikely role of midwives. An undeveloped episode with minimal tension.
FAN   45 min mTV   TVrel

STAR TREK VOYAGER: PERSISTENCE OF VISION ***   (PG)
James L. Conway USA   1995
Kate Mulgrew, Robert Beltran, Roxann Biggs-Dawson, Jennifer Lien, Robert Duncan, Ethan Phillips, Robert Picardo, Tim Russ, Garrett Wang, Michael Cumpsty, Stan Ivar, Carolyn Seymour, Warren Munson, Lindsey Hart, Thomas Dekker
Voyager enters a region of space where, according to Neelix, a number of vessels have disappeared and the local race is hostile to outsiders. Challenged by an alien craft, the captain requests permission to enter their space. At the same time, she starts to experience parts of her holodeck fantasy in real life. After the ship is attacked, the crew begin to see visions from their past. A well handled episode with plenty of tension that lasts right until the end.
FAN   45 min mTV   TVrel

STAR TREK VOYAGER: PHAGE **   PG
Winrich Kolbe USA   1995
Kate Mulgrew, Robert Beltran, Roxann Biggs-Dawson, Jennifer Lien, Robert Duncan, Ethan Phillips, Robert Picardo, Tim Russ, Garrett Wang, Stephen B. Rappaport, Cully Fredricksen, Martha Hackett, Majel Barrett (voice only)
Neelix tells the captain of an asteroid that appears to be rich in

dilithium, which, if true, could spell an end to the power short-ages aboard Voyager. He convinces her to let him take part in the away team but suffers an attack in which his lungs are surgically removed. While the doctor racks his holographic brains to keep him alive, a desperate search is conducted for those responsible. A moderately exciting story.
FAN   45 min; 88 min (2-episode cassette) mTV
VIDrel: CIC/SONOP   V/sur

STAR TREK VOYAGER: PRIME FACTORS ** PG
Les Landau USA   1995
Kate Mulgrew, Robert Beltran, Roxann Biggs-Dawson, Jennifer Lien, Robert Duncan, Ethan Phillips, Robert Picardo, Tim Russ, Garrett Wang, Greg Elliot, Michael Perricone, Jeri Taylor, David R. George III, Eric A. Stillwell
The crew of the Voyager receive a visit from the representative of a highly evolved pleasure-loving race for whom hospitality is a sacred duty. He invites them to visit their planet for some much-need recreation but once there they learn that these aliens possess a technology that could send nearly 40,000 light years in an instant. However, their refusal to share this secret, places the captain in a delicate moral dilemma. An OK, moderately interesting episode.
FAN   45 min; 88 min (2-episode cassette) mTV
VIDrel: CIC/SONOP   V/sur

STAR TREK VOYAGER: PROJECTIONS ** (PG)
Jonathan Frakes USA   1995
Kate Mulgrew, Robert Beltran, Roxann Biggs-Dawson, Jennifer Lien, Robert Duncan, Ethan Phillips, Robert Picardo, Tim Russ, Garrett Wang, Dwight Shultz, Majel Barrett (voice only)
Working in sick-bay, Voyager's holographic doctor discovers that most of the crew are missing after an attack by the Kazons and that he appears to have become human. Worse is to come when it appears that the entire crew were nothing more than holographic projections. Unable to make any meaning of this situation, he cannot decide whether what is happening to him is real or imaginary. A confused and rather ineffective entry.
FAN   45 min mTV   TVrel

STAR TREK VOYAGER: PROTOTYPE ** (PG)
Jonathan Frakes USA   1995
Kate Mulgrew, Robert Beltran, Roxann Biggs-Dawson, Jennifer Lien, Robert Duncan, Ethan Phillips, Robert Picardo, Tim Russ, Garrett Wang, Rick Worthy, Hugh Hodgin
A damaged robot is found floating in space and taken aboard Voyager, where B'Eleanna tries desperately to save its artificial life by repairing its power source. She eventually succeeds but runs into great trouble when it asks for her help in building more of its kind. Forced reluctantly to refuse this request, our robot abducts her. Some good ideas are not really aired in any depth, in this fairly absorbing effort.
FAN   45 min mTV   TVrel

STAR TREK VOYAGER: RESISTANCE * (PG)
Winrich Kolbe USA   1995
Kate Mulgrew, Robert Beltran, Roxann Biggs-Dawson, Jennifer Lien, Robert Duncan, Ethan Phillips, Robert Picardo, Tim Russ, Garrett Wang, Alan Scarfe, Tom Todoroff, Glenn Morshower, Joel Grey
A clandestine mission to a hostile planet is mounted in order to obtain a supply of a substance vital to the maintenance of Voyager's energy system. Unfortunately, two of the away team fall into the hands of the local secret police, while the captain is sheltered by a deluded old man who seems to believe that she is his daughter. A very poor episode indeed, that appears to have been thrown together without much thought or care.
FAN   45 min mTV   TVrel

STAR TREK VOYAGER: RESOLUTIONS * PG
Alexander Singer USA   1995
Kate Mulgrew, Robert Beltran, Roxann Biggs-Dawson, Jennifer Lien, Robert Duncan McNeill, Ethan Phillips, Robert Picardo, Tim Russ, Garrett Wang, Susan Diol, Simon Billing, Bahni Turpin
Infected by an insect-borne viral disease when they landed on an unknown planet, Janeway and Chakotay are obliged to return there in the hope of finding out enough to stop the progress of the disease, when all efforts to do this onboard the ship have failed. Command passes to Tuvok, who is ordered not to request help from the Vidiians and the ship continues on its way, a decision the crew members begin to question. A predictable and terribly contrived entry.
FAN   88 min (2-episode cassette) mTV cC
VIDrel: CIC/SONOP   V/sur

STAR TREK VOYAGER: STATE OF FLUX * U
Robert Scheerer USA   1995
Kate Mulgrew, Robert Beltran, Roxann Biggs-Dawson, Jennifer Lien, Robert Duncan, Ethan Phillips, Robert Picardo, Tim Russ, Garrett Wang, Martha Hackett, Josh Clark, Athony De Longis, Majel Barrett (voice only).
An away mission is aborted after the team get into a firefight with some Kazon warriors and later a called for help is received from one of their ships. Voyager responds but gets there after an explosion on the bridge has killed all but one of these aliens. Analysis reveals that an item of Federation technology was involved in this disaster, and the possibility of a traitor among the crew poses problems for the senior staff. A predictable effort of little imagination.
FAN   45 min; 88 min (2-episode cassette) mTV
VIDrel: CIC/SONOP   V/sur

STAR TREK VOYAGER: TATTOO *** (PG)
Alexander Singer USA   1995
Kate Mulgrew, Robert Beltran, Roxann Biggs-Dawson, Jennifer Lien, Robert Duncan, Ethan Phillips, Robert Picardo, Tim Russ, Garrett Wang, Richard Fancy, Douglas Spain, Nancy Hower, Richard Chaves
An away mission to a moon in search of a much-needed mineral brings Chakotay face to face with an amazing discovery that is intimately linked to both his personal history and that of his Indian tribe. Further investigations take Voyager to another moon, where this intriguing mystery is eventually resolved. A well realised episode (apart from a silly sub-plot involving the doctor) that holds the interest and maintains a good sense of suspense.
FAN   45 min mTV   TVrel

STAR TREK VOYAGER: THE THAW *** PG
Marvin V. Rush USA   1996
Kate Mulgrew, Robert Beltran, Roxann Biggs-Dawson, Jennifer Lien, Robert Duncan, Ethan Phillips, Robert Picardo, Tim Russ, Garrett Wang, Thomas Kopache, Carel Struycken, Patty Maloney, Tony Carlin, Shannon O'Hurley, Michael McKean
The crew of the USS Voyager answers an automated distress call that brings them to a planet recovering from a major environmental breakdown, where below its surface they find three survivors held in suspended animation. They are taken aboard the vessel but cannot be revived, and it is found they are trapped in a virtual reality world created by a sophisticated computer. A nightmarish episode of little logic, but one that works extremely well.
FAN   88 min (2-episode cassette) mTV cC
VIDrel: CIC/SONOP   V/sur

STAR TREK VOYAGER: THRESHOLD *** PG
Alexander Singer USA   1996
Kate Mulgrew, Robert Beltran, Roxann Biggs-Dawson, Jennifer Lien, Robert Duncan McNeill, Ethan Phillips, Robert Picardo, Tim Russ, Garrett Wang, Raphael Sbarge, Mirron E. Willis, Majel Barrett (voice only)
After getting a little advice from Neelix, a research crew seems to achieve the impossible in breaking the transwarp barrier by achieving a stable flight at warp ten. Lieutenant Paris volunteers to be the first pilot to test this discovery, but soon after returning from the test flight he becomes ill and starts to undergo a frightening transformation. A most enjoyable and strongly written story, with enough plot twists to maintain tension right up to the end.
FAN   88 min (2-episode cassette) mTV cC
VIDrel: CIC/SONOP   V/sur

STAR TREK VOYAGER: TIME AND AGAIN ** PG
Les Landau USA   1995
Kate Mulgrew, Robert Beltran, Roxann Biggs-Dawson, Jennifer Lien, Robert Duncan, Ethan Phillips, Robert Picardo, Tim Russ, Garrett Wang, Nicholas Surovy, Joel Polis, Brady Bluhm, Ryan MacDoanld, Steve Valiant, Jerry Spicer
A massive shock-wave in space results in an away mission to a planet where a massive explosion has destroyed all organic life. The fabric of time has also been disrupted, causing the captain and first officer to travel back to the day before. This lands them at the centre of the events that result in this catastrophe. A poor and unimaginative handling of time paradoxes that offers far too little excitement.
FAN   45 min; 88 min (2-episode cassette) mTV
VIDrel: CIC/SONOP   V/sur

STAR TREK VOYAGER: TUVIX *** PG
Cliff Bole USA 1996
Kate Mulgrew, Robert Beltran, Roxann Biggs-Dawson, Jennifer Lien, Robert Duncan, Ethan Phillips, Robert Picardo, Tim Russ, Garrett Wang, Tom Wright, Simon Billing
An alien plant interferes with the operation of the transporter and Tuvok and Neelix emerge from a mission to study plant life on a nearby planet as a single, merged individual. Although this creature incorporates the DNA and memories of both crew members, it also appears to possess a consciousness and identity of its own. As the medical staff attempt to find a solution, this new creature struggles to adapt to life on the ship. Well handled and dramatic.
FAN 88 min (2-episode cassette) mTV cC
VIDrel: CIC/SONOP V/sur

STAR TREK VOYAGER: TWISTED ** PG
Kim Friedman USA 1995
Kate Mulgrew, Robert Beltran, Roxann Biggs-Dawson, Jennifer Lien, Robert Duncan, Ethan Phillips, Robert Picardo, Tim Russ, Garrett Wang, Judy Geeson, Larry A. Hankin, Tom Virtue
A spatial anomaly causes unexpected and very strange problems for the crew of Voyager when the commanding officers themselves trapped near the holodeck. A singularly unimpressive episode with the usual rapidly contrived resolution.
FAN 45 min; 88 min (2-episode cassette) mTV
VIDrel: CIC/SONOP V/sur

STAR WARS *** U
George Lucas USA 1977
Mark Hamill, Harrison Ford, Carrie Fisher, Alec Guinness, Dave Prowse, Peter Cushing, Anthony Daniels, Kenny Baker, Peter Mayhew, Phil Brown, Eddie Byrne, Shelagh Fraser plus James Earl Jones (voice of Darth Vader)
SF blockbuster with heroic rebels fighting an evil galactic empire. Shallow but fun, with aliens, robots, spaceships and much more. Followed by THE EMPIRE BACK and RETURN OF THE JEDI, plus the spin-off CARAVAN OF COURAGE: AN EWOK ADVENTURE. Scripted by Lucas. AA: Art/Set (Barry et al./Christian), Cost (Mollo), Edit (Hirsch/Lucas/Chew), Score/orig (Williams), Effects/vis (Stears et al.), Sound (MacDougall et al.), Spec Award (B. Burtt Jr for sound effects).
FAN 116 min (ort 121 min); 126 min (special edition) wScrn
VIDrel: 20TH/TECH V/dm

STARCHASER: THE LEGEND OF ORIN *** PG
Steven Hahn USA 1985
Voices of: Joe Colligan, Carmen Argenziano, Noelle North, Anthony Delongis, Les Tremayne, Tyke Carvelli, Ken Sanson, John Moschitta Jr, Mickey Morton, Herb Vigran, Dennis Alwood, Mona Marshall, Tina Romanus, Ryan MacDonald
Animated story of an evil intergalactic ruler, who tries to enslave Earth but is opposed by our hero. A solid and entertaining kid's adventure. The imaginative use of 3-D will not translate well to TV.
ANIM 100 min (ort 107 min) VIDrel: EIV/SONOP L/A V

STARDUST *** 15
Michael Apted UK 1974
David Essex, Adam Faith, Larry Hagman, Keith Moon, Dave Edmunds, Ines Des Longchamps, Rosalind Ayres, Marty Wilde, Edd Byrnes, Paul Nicholas, Rick Lee Parmentier, Karl Howman, Peter Duncan, John Normington, Dave Daker
An excellent sequel to THAT'LL BE THE DAY charting the career and inevitable decline of a rock group's lead singer. Well made, well acted and extremely enjoyable. The script is by Ray Connolly and musical direction is by Dave Edmunds and David Puttnam.
DRA 106 min (ort 111 min) VIDrel: WHV V/sur

STARDUST MEMORIES *** 15
Woody Allen USA 1980
Woody Allen, Charlotte Rampling, Jessica Harper, Marie-Christine Barrault, Tony Roberts, Daniel Stern, Laraine Newman, Amy Wright, Helen Hanft, Louise Lasser, John Rothman, Ann DeSalvo, Joan Neuman, Ken Chaplin, Leonard Cimino
A Woody Allen look at life – a film producer attends a weekend conference and is pursued by fans, producers, lovers, relatives etc. One of Allen's most uneven and self-indulgent films, but there are some sharp and wonderfully witty moments. As ever, Allen writes and directs.
DRA 85 min (ort 88 min) B/W VIDrel: MGM/WHV L/A V

STARGATE ** PG
Roland Emmerich USA 1994
Kurt Russell, James Spader, Jaye Davidson, Viveca Lindfors, Alexis Cruz, Mili Avital, Leon Rippy, John Diehl, Carlos Lauchu, Djimon, Erick Avari, French Stewart, Gianin Loffler, Christopher John Fields, Derek Webster, Jack Moore
While excavating in Egypt close to the pyramids, an archaeologist discovers a strange artefact that proves to be a gateway to a world light years away from Earth that is ruled by a ruthless tyrant. Stunning special effects provide much visual excitement but fail to compensate for the major failings of weak plotting and characterisation, in this imaginative but flawed fantasy-adventure.
FAN 116 min (ort 121 min) wScrn; 122 min (LV version)
VIDrel: POLY/POLYREC; PION (LV only) V/sur LV

STARLIGHT HOTEL *** PG
Sam Pillsbury NEW ZEALAND 1987
Peter Phelps, Greer Robson, Marshall Napier, The Wizard, Alice Fraser, Elric Hooper, Patrick Smyth, John Watson, Vanessa Young, Teressa Bonney, Duncan Anderson, Russell Gibson, Marshall Napier, Norman Forsey, Craig Halkett
Interesting drama set in New Zealand during the Depression, with an unhappy teenager leaving home in search of her father who is looking for work and has been forced to go on the run after killing a bailiff. Along the way she strikes up an unlikely friendship with a disturbed veteran from WW1 in this warm-hearted and unusual tale. The beautiful photography is by Warrick Atwell.
DRA 93 min VIDrel: L/A V
Boa: novel The Dream Monger by Grant Hinden Miller.

STARMAKER, THE *** 18
Giuseppe Tornatore ITALY 1994
Sergio Castellitto, Tiziana Lodato, Leopoldo Trieste, Nicola Di Pinto, Tony Sperandeo, Franco Scaldati, Clelia Rondinella, Jane Alexander, Tano Cimarosa, Costantino Carrozza, Luigi Burruano, Antonella Attili, Carmelo Di Mazzarelli
In Sicily in the years just after WW2, a man tours the countryside posing as a talent scout for a major Roman studio and makes a living by charging the locals for "screen tests". As they perform in front of his camera (which has no film) they reveal to him their innermost thoughts and feelings. A companion piece to the director's earlier CINEMA PARADISO, with which it shares many similarities, this is a thoughtful and warmhearted work.
Aka: L'UOMO DELLE STELLE
DRA 107 min CINrel

STARMAN *** PG
John Carpenter USA 1984
Jeff Bridges, Karen Allen, Richard Jaeckel, Charles Martin Smith, Robert Phalen, Tony Edwards, John Water Davis, Ted White, Dirk Blocker, Sean Faro, M.C. Gainey, George Buck Flower, Russ Benning, Ralph Cosham, Jim Deeth
Some splendid special effects are used in this story of an alien visitor sent to investigate the Earth as a result of the 1970's Voyager II space probe. He assumes the form of a woman's dead husband and she falls in love with him, and shelters him from inquisitive government agents. Bridges plays to perfection the part of a creature slowly learning to be human. The action is set in Wisconsin. Later a TV series.
FAN 110 min (ort 115 min) wScrn
VIDrel: CASPIC/BMGREC V/sur

STARS FELL ON HENRIETTA, THE *** PG
James Keach USA 1995
Robert Duvall, Jason Wagner, Aidan Quinn, Frances Fisher, Brian Dennehy
A poor farmer who can barely scratch a living on his farm in Texas pins his hopes on a travelling prospector who assures him that his land is rich in oil. Produced by Clint Eastwood, this amusing and offbeat film is set in 1935, and provides a fascinating glimpse of the period, when itinerant prospectors really did make their way up and down the country in search of a quick fortune.
DRA 105 min (ort 110 min) cC VIDrel: WHV V/sur

STARS LOOK DOWN, THE *** U
Carol Reed UK 1939
Margaret Lockwood, Michael Redgrave, Edward Rigby, Emlyn Williams, Nancy Price, Allan Jeayes, Cecil Parker, Linden Travers, Milton Rosmer, Desmond Tester, Ivor Barnard, Olga Lindo, George Carney, David Markham, Clive Baxt

A realistic coal mining drama, set in a town in Northern England. A coalminer's son returns from university, and is forced to take a teaching job to support his faithless wife, who soon takes up with an old boyfriend. When plans are put forward to re-open an unsafe mine, he leads the miners in a protest, as his father had previously done. A grim and tragic film, well crafted in all departments. The script is by Cronin and J.B. Williams.
DRA 100 min (ort 104 min) B/W
VIDrel: 4-FRONT/POLYREC/ODY V
Boa: novel by A.J. Cronin.

STARTING AGAIN ** (15)
Oz Scott USA 1995
Joan Rivers, Melissa Rivers, Dorothy Lyman, Mark Keiley, Jon Kean, Denis Arndt, Merrill Karpf, Sheila Moore, Ken Kramer, Jay Brazeau, Matthew Bennett, Don MacKay, Roger Allford, Rebecca Toolan, Jerry Wasserman, Tasha Simms
Story of comedienne Joan Rivers and her daughter, and how they coped after her estranged husband committed suicide in 1987 and the pressures of work began to create a rift between the two women. Surprisingly effective, the fact that Rivers and her daughter were prepared to play themselves in this brutally frank story adds considerably to its impact.
DRA 91 min mTV SATrel: MOVIE CHANNEL

STATE FAIR *** U
Walter Lang USA 1945
Dana Andrews, Jeanne Crain, Dick Haymes, Vivian Blaine, Fay Bainter, Charles Winninger, Frank McHugh, Percy Kilbride, Donald Meek, Henry (Harry) Morgan, Jane Nigh, William Marshall, Phil Brown, Paul E. Burns, Tom Fadden
A charming remake of the 1933 film telling of the adventures of a family out for the day at the Iowa State Fair. A cheerful and attractive film, its fine Rodgers and Hammerstein songs (their only film score), include "A Grand Night For Singing" and "That's For Me" as well as the Oscar-winner. The film was retitled for TV. AA: Song ("It Might As Well Be Spring" – Richard Rodgers (m)/Oscar Hammerstein II (l)).
Aka: IT HAPPENED ONE SUMMER
MUS 96 min (ort 100 min) VIDrel: 20TH/TECH V/sh
Boa: novel by Philip Stong.

STATE OF EMERGENCY ** (12)
Lesli Linka Glatter USA 1993
Joe Mantegna, Lynn Whitfield, Melinda Dillon, Paul Dooley, Richard Beymers, Robert Beltran, Jay O. Sanders, Christopher Birt, Dean Cameron, Deborah Unger, Paul Benvictor, F. William Proctor, Lucy Butler, John Considine, Gearld Castillo
A day in the life of the head of a hospital emergency room doctor as he struggles to save lives and battle the heartless financial bureaucrats. A worthy subject badly let down by its flat made-for-TV treatment.
DRA 83 min (ort 98 min) mCab SATrel: MOVIE CHANNEL

STATE OF GRACE *** 18
Phil Joanou USA 1990
Sean Penn, Ed Harris, Gary Oldman, Robin Wright, John Turturro, R.D. Call, Joe Vitorelli, Burgess Meredith, John C. Reilly, Dierdre O'Connell, Marco St John, Thomas G. Waites, Brian Burke, Jaime Tirelli, Sandra Beall, Mo Gaffney
A gangster movie that is set in the 1970s, among the tough close-knit Irish community of New York's Hell's Kitchen, where various local mobsters are constantly obliged to defend their turf from both developers and other encroaching gangs. A strongly scripted film, that centres on the return to his old turf of a young man, the rekindling of an earlier romance and his unhappy conflicts of loyalty. Music is by Ennio Morricone.
A/AD 128 min (ort 134 min) VIDrel: VISVID/POLYREC V/sur

STATE OF SIEGE *** 15
Costa-Gavras FRANCE 1973
Yves Montand, Renato Salvatori, Jacques Weber, O.E. Hasse, Jean-Luc Bideau, Evangeline Peterson
An American advisor in an unnamed Latin American country is kidnapped by left-wing guerillas, and it essentially transpires that has been sent to train the military in torture. A controversial attack on US government involvement with certain unsavoury regimes, the film was also criticised by the left

because Montand's torturer is portrayed as a civil and dignified individual. See also Z.
AKA: ETAT DE SIEGE
DRA 116 min (ort 120 min) VIDrel: ARROW/RTM V

STATE SECRET *** PG
Sidney Gilliat UK 1950
Douglas Fairbanks Jr, Glynis Johns, Jack Hawkins, Herbert Lom
An eminent American surgeon is lured to a small European dictatorship to receive a prize, but once there is tricked into performing a life-saving operation on the country's hated military ruler. However, when this man is assassinated his cronies replace him with a double, forcing the surgeon to flee as his knowledge of this has put his life in danger. Similar to the Cary Grant film "Crisis", this is a carefully plotted and fairly effective Cold Warn yarn.
Aka: GREAT MANHUNT, THE
THR 104 min B/W VIDrel: LUMI/SPEAR V
Boa: novel Appointment with Fear by Ross Huggins

STATIC ** 15
Mark Romanek USA 1986
Keith Gordon, Amanda Plumber, Bob Gunton, Barton Heyman, Lily Knight, Jane Hoffman, Reathel Bean, Kitty Mei-Mei, Eugene Lee, Joel K. Rehbeil, Jack Murakami, Mike Murakami, Uma Ridenhour, Janice Abbott, Tamma Allgood
Critically acclaimed off-beat film, involving an obsessive inventor who claims to have produced a TV that tunes into Heaven, but most people only see a blank screen and hear static. An incoherent comedy-drama that has some poignant and funny moments but for most of the time has a lot in common with that blank screen. Written by Gordon and Romanek.
Aka: HOTLINE TO HEAVEN
DRA 90 min (ort 93 min) VIDrel: L/A V

STAY HUNGRY ** 18
Bob Rafelson USA 1976
Jeff Bridges, Sally Field, Arnold Schwarzenegger, R.G. Armstrong, Roger E. Mosley, Helena Kallianiotes, Scatman Crothers, Ed Begley Jr, Gary Godorow, Joanna Cassidy, Robert Englund, Fannie Flagg, Richard Gilliland
The wealthy young heir to an Alabama estate is not interested in the family business, but becomes involved with body builders when he is sent on a crooked mission to buy up a health club. A curious and rather meaningless story, with good performances but no clear direction. Written by Rafelson and Charles Gaines.
DRA 103 min VIDrel: MGM/WHV L/A V
Boa: novel by Charles Gaines.

STAY THE NIGHT ** 15
Harry Winer USA 1991
Barbara Hershey, Jane Alexander, Morgan Weisser
Having begun an affair with a woman years older than himself, a naive high school student is cleverly manipulated into murdering the woman's husband. This leads to his conviction for murder and a sentence of life imprisonment, with the boy's mother the only person able to prove his innocence.
DRA 149 min (ort 180 min) mTV VIDrel: NWV/HIFLI V

STAY TUNED ** PG
Peter Hyams USA 1992
John Ritter, Pam Dawber, Don Calfa, Bob Dishy, Jeffrey Jones, David Tom, Heather McComb, Erik King, Susan Blommaert, Eugene Levy, John Blackwell Destrey, Maurice Yerkaar, Ken Douglas, Dale Wilson, Don Fargo, Lou Albano, George Gray
After buying a satellite TV system, a young couple find themselves caught inside their TV set and forced to take place in a variety of hideous shows. Unwittingly the husband has made a deal with the devil and if they survive this situation for twenty-four hours, they will be returned to the real world. An unfunny spoof that never does anything with an inventive premise.
COM 84 min (ort 89 min)
VIDrel: POLY/POLYREC/BRAVE V/sur

STAYING ALIVE ** PG
Sylvester Stallone USA 1983
John Travolta, Cynthia Rhodes, Finola Hughes, Stevie Inwood, Julie Bovasso, Frank Stallone, Charles Ward, Steve Bickford, Patrick Brady, Jesse Doran, Norma Donaldson, Joyce Hyser, Deborah Jensen, Robert Martini, Sarah Miles
An aspiring Broadway dancer has problems with both his girlfriend and his leading lady, in this sequel to the hugely

successful SATURDAY NIGHT FEVER that offers little of the vigour of the earlier film, except for the "Satan's Alley" finale. Written by Stallone and Norman Wexler.
MUS 92 min (ort 96 min) VIDrel: CIC/SONOP V/sur

STAYING TOGETHER ** 15
Lee Grant USA 1989
Sean Astin, Stockard Channing, Melinda Dillon, Levon Helm, Dermot Mulroney, Jim Haynie, Dinah Manoff, Tim Quill, Keith Szarabajka, Daphne Zuniga
Three brothers face an uncertain future after their father sells the family restaurant. A well acted comedy-drama, that's completely let down by flat and often laughable dialogue, a contrived storyline, and a set of unresolved sub-plots that betray much post-production tampering.
DRA 88 min (ort 91 min) VIDrel: 20VIS/SONOP V

STEAL, THE * (PG)
John Hay UK 1994
Alfred Molina, Helen Slater, Peter Bowles, Dinsdale Landen, Heathcote Williams, Stephen Fry, Brian Pringle, Patricia Hayes, Jack Dee, Ian Porter, Lindsay Holiday, Rob Freeman, R.J. Bell, Jason Salkey, Ann Bryson, Sara Crowe
Anodyne comedy revolving around the theft of a high-tech computer from a top City bank in London. Computer genius Slater is the key to this operation, while Molina does his best to assist. As is the case with most British comedies, a succession of crazy encounters, misunderstandings and chaotic mishaps starts to pile up, but very few of them are funny.
COM 91 min SATrel: MOVIE CHANNEL

STEAL BIG, STEAL LITTLE ** 12
Andrew Davis USA 1995
Andy Garcia, Alan Arkin, Rachel Ticotin, Joe Pantoliano, Holland Taylor, Ally Walker, David Ogden Stiers, Charles Rocket, Richard Bradgord, Kevin McCarthy, Nathan Davis, Dominik Garcia-Lorido, Mike Nussbaum, Rita Taggart
After the death of their adopted mother, two identical twins contend for the ownership of the 400,000 acre ranch which she left to just one of them. As ever, in the soap operas on which this film seems to be based, one of our characters is good and the other totally ruthless and evil, doing his utmost to steal his brother's inheritance. A few good performances provide welcome relief to a film replete with all the usual melodramatic cliches.
DRA 134 min CINrel

STEALING BEAUTY * 15
Bernardo Bertolucci FRANCE/ITALY/UK 1995
Sinead Cusack, Jeremy Irons, Jean Marais, Donal McCann, D.W. Moffett, Rachel Weisz, Stefania Sandrelli, Liv Tyler, Carlo Cecchi, Joseph Fiennes, Anna Maria Gherardi, Jason Flemyng, Ignazio Oliva, Francesca Siciliano, Leonardo Treviglio
After the death of her mother, a nineteen-year-old girl returns to the artist's villa in Italy she visited four years before. There she hopes to finally resolve the mystery surrounding her paternity and be reunited with the handsome young man she fell in love with. A fatuous example of a voyeuristic albeit well photographed skin flick, with the beautiful heroine mainly interested in losing her virginity, and finding no shortage of willing helpers.
Aka: BEAUTE VOLEE; LO BALLO DA SOLA
DRA 113 min (ort 118 min) cC VIDrel: 20TH/FOXVID V/sur

STEALING HEAVEN *** 15
Clive Donner USA 1989
Kim Thomson, Derek De Lint, Denholm Elliott, Rachel Kempson, Kenneth Cranham, Bernard Hepton, Patsy Byrne, Cassie Stuart, Philip Locke, Angela Pleasence, Slavica Maras, Niki Hewitt, Yvonne Bryceland, Vjencslav Kapurai
A lavishly-shot film version of the 12th century love story of Abelard and Heloise, with the former a young cleric who falls hopelessly in love with Heloise, who has been promised in marriage to the highest bidder by her rich uncle. When she becomes pregnant, a gruesome revenge is taken on her lover. Always a pleasure to watch (well, nearly always), this film is perhaps a trifle overlong and is not quite as well acted as it might have been.
DRA 111 min (ort 116 min) VIDrel: 20VIS/SONOP L/A V
Boa: novel by Marion Meade.

STEAMIE, THE ** PG
Haldane Duncan UK 1989
Eileen McCallum, Dorothy Paul, Katy Murphy
Comedy set in 1953 and built around a communal wash-house and the women using it on the eve of their annual Hogmanay celebration.
COM 84 min (ort 85 min) VIDrel: POLY/POLYREC V
Boa: play by Tony Roper.

STEAMING * 18
Joseph Losey UK 1985
Vanessa Redgrave, Sarah Miles, Diana Dors, Patti Love, Brenda Bruce, Sally Sagoe, Felicity Dean, Anna Tzelniker
A dreary adaptation of a play set in a women's Turkish baths, in which a motley collection of women unite across class (always a feature of British films) and age barriers to prevent the closure of their baths. The last film for both Losey and Dors. Scripted by Patricia Losey.
DRA 92 min (ort 95 min) VIDrel: RCA L/A V
Boa: play by Nell Dunn.

STEAMY WINDOWS * 18
Alex De Renzy USA 1989
Rachel Ryan, Randy Spears, Danielle Rogers, Rocco Siffredi, Joey Silvera, Sunny McKay, Randy West, Debi Diamond, Tianna, Peter North
An art critic finds willing bed partners among the female employees at a friend's gallery
A 57 min (ort 75 min) VIDrel: FIFTH/DISC V

STEEL DAWN ** 18
Lance Hool USA 1987
Patrick Swayze, Lisa Niemi Swayze, Anthony Zerbe, Christopher Neame, John Fujioka, Brion James, Brett Hool, Marcel Van Heerden, Arnold Vosloo, James Whyle, Rusell Savadier, Joe Ribeiro, David Sherwood, Brad Morris, Tullio Moneta
A re-run of SHANE in a futuristic setting, with Swayze protecting an attractive farmer and her son from a bunch of villainous thugs. A derivative and fairly unappealing effort, filmed in the deserts of southern Africa.
FAN 96 min Cut (21 sec – ort 102 min)
VIDrel: 4-FRONT/POLYREC L/A V/sh

STEEL FRONTIER ** 18
Paul G. Volk USA 1994
Bo Svenson, Joe Lara, Jim Cody Williams, Brion James, Stacie Foster, Robert O'Reilly
Post-WW3 tale in the form of a futuristic Western set in the 21st century, with a stranger coming to a small town that a colony of survivors have built on the radioactive wastelands. Once there, he finds himself fighting to liberate it from the rule of a gang of brutal thugs and their evil leader.
FAN 101 min VIDrel: COLUM/SONOP V/sur

STEEL JUSTICE ** PG
Christopher Crowe USA 1992
Robert Taylor, J.A. Preston, Roy Brocksmith, John Finn, Neil Giuntoli, John Toles-Bey, Geoffrey Rivas, Season Hubley, Joan Chen, Jacob Vargas, Garvin Funches, Ken Thorley, Augie Blunt, Vincent Chase, Maxwell Crowe, Henry Kingi
A cop in a crime-ridden city of the future gains an assistant in the form of a virtually indestructible forty-five foot robot, in this derivative but quite enjoyable ROBOCOP-style action-fantasy.
FAN 87 min (ort 90 min) mTV VIDrel: CIC/SONOP V

STEEL MAGNOLIAS *** PG
Herbert Ross USA 1989
Sally Field, Dolly Parton, Shirley MacLaine, Daryl Hannah, Olympia Dukakis, Julia Roberts, Tom Skerritt, Dylan McDermott, Kevin J. O'Connor, Bibi Besch, Sam Shepard, Bill McCutcheon, Ann Wedgeworth, Knowl Johnson, Jonathan Ward
A comedy-drama featuring an all-star cast, and following the lives of six friends who over several years, congregate at Parton's beauty-salon in a small Louisiana town. The film is light and cheerful to begin with, but becomes progressively more poignant and culminates with an emotional climax. Written by Harling (who has a cameo as a minister) from his one-set play, it lacks some of the wit of the original, but remains an entertaining work.
DRA 113 min (ort 118 min) VIDrel: VCC/DISC/COLUM V/sur
Boa: play by Robert Harling.

STEELYARD BLUES **
15
Alan Myerson USA
1972
Jane Fonda, Donald Sutherland, Peter Boyle, Garry Goodrow, Howard Hesseman, John Savage, Richard Schaal, Melvin Stewart, Morgan Upton, Roger Bowen, Howard Storm, Jessica Myerson, Dan Barrows, Nancy Fish, Lynn Bernay
An assorted group of misfits plan to rebuild an abandoned seaplane and fly away to a better life. Their efforts are hampered by the local D.A. who is the brother of one of them. Funny in parts but disappointing overall. The script is by David S. Ward who also write THE STING.
Aka: FINAL CRASH, THE
COM   89 min (ort 93 min)   VIDrel: MGM/WHV L/A   V

STELLA *
15
John Erman USA
1989
Bette Midler, John Goodman, Trini Alvarado, Stephen Collins, Marsha Mason, Eileen Brennan, Linda Hart, Ben Stiller, William McNamara, John Bell, Ashley Peldon, Alison Porter, Kenneth Kimmins, Bob Gerchen, Willie Rosari
Third version of Prouty's tear-jerker about a poor mother who gives up everything so that her spoilt daughter can enjoy the good things in life. A really unmoving and tedious remake, that doesn't even boast especially good performances. Not a patch on the 1937 film STELLA DALLAS, it's slightly updated, but for the most part remains strangely anachronistic. Music is by John Morris.
DRA   104 min (ort 109 min)   VIDrel: TOUCH/BUENA L/A
V
Boa: novel by Olive Higgins Prouty.

STELLA DALLAS ***
U
King Vidor USA
1937
Barbara Stanwyck, John Boles, Anne Shirley, Barbara O'Neil, Alan Hale, Tim Holt, Marjorie Main, George Walcott, Gertrude Short, Nella Walker, Jimmy Butler, Bruce Satterlee, Jack Egger, Dickie Jones, Ann Shoemaker, Jessie Arnold
A remake of a 1925 silent weepie, dealing with a woman who has to make great sacrifices for the sake of her daughter's happiness. Despite the mawkish nature of the script, an excellent performance from Stanwyck turns a rather mundane soap opera into a touching film of considerable power. See also STELLA.
DRA   106 min B/W   VIDrel: VCC/DISC   V
Boa: novel by Olive Higgins Prouty.

STENDAHL SYNDROME, THE **
18
Dario Argento ITALY
1996
Asia Argento, Thomas Kretschmann, Marco Leonardi
This was the first time computer-generated effects were used in an Italian horror/fantasy or "giallo" film, and follows a female police detective who is hunting a serial rapist and killer. Unfortunately, she suffers from a medical condition that causes her to hallucinate whenever she is deeply moved by a work of art. A vicious and very intense horror outing, saddled with the usual mad slasher plot, and a few flashes of visual brilliance.
HOR   110 min   VIDrel: GUILD/FOXVID   V

STEP LIVELY ***
U
Tim Whelan USA
1944
Frank Sinatra, George Murphy, Adolphe Menjou, Gloria De Haven, Eugene Pallette, Anne Jeffreys, Walter Slezak, Wally Brown, Alan Carney, Grant Mitchell, Frances King, Harry Noble, George Chandler, Rosemary La Planche
A musical remake of "Room Service", with a producer having to wheel and deal as never before to get his show off the ground. A fast and furious romp with witty dialogue and good performances. Songs are by Julie Styne and Sammy Cahn with musical direction by Constantin Bakaleinikoff.
MUS   85 min (ort 88 min) B/W   VIDrel: STABL L/A   V
Boa: play Room Service by John Murray, Allan Boretz and Sammy Cahn.

STEPFATHER, THE ***
18
Joseph Ruben USA
1987
Terry O'Quinn, Shelley Hack, Jill Schoelen, Stephen Shellen, Charles Lanyer, Stephen E. Miller, Robyn Stevan, Jeff Schultz, Lindsay Bourne, Anna Hagan, Gillian Barber, Blu Mankuma, Jackson Davies, Sandra Head, Gabrielle Rose
An apparently mild-mannered family man is in reality a ruthless and demented killer, who marries widows with children in a constant search for a "perfect" family, inevitably erupting into a murderous rage when they disappoint him. A frightening film

with O'Quinn giving a chilling performance in his lead debut. Written by Donald Westlake.
HOR   85 min (ort 89 min)   VIDrel: 4-FRONT/POLYREC   V

STEPFATHER 2 **
18
Jeff Burr USA
1989
Terry O'Quinn, Meg Foster, Jonathan Brandis, Caroline Williams, Mitchell Laurance, Miriam Byrd-Methery, Leon Martell, Renata Scott, John O'Leary, Glen Adams, Eric Brown
A highly inferior sequel to the earlier film, devoid of the tension and bite of the original, and saddled instead with some silly dialogue that attempts a few feeble touches of offbeat humour, but merely succeeds in being tedious and dreary.
Aka: STEPFATHER 2: MAKE ROOM FOR DADDY
THR   84 min (ort 93 min)   VIDrel: 4-FRONT/POLYREC
V/sh

STEPFATHER 3 **
18
Guy Magar USA
1992
Priscilla Barnes, Robert Wightman, Season Hubley, David Tom, John Ingle, Dennis Paladino, Stephen Menedl, Jay Acovone, Chrita Miller, Mario Rocuzzo, Joan Dareth, Jennifer Bassey, Adam Ryen, Mindy Ann Martin, Joel Carlson
Our third STEPFATHER offering opens with a new actor (Wightman) in the lead role, cosmetic surgery being advanced for the explanation of his changed appearance. In all other respects he's the same however, and it's not long before he has found a new conquest in a quiet Californian suburb, where he once again believes he has found his "perfect" family. Another stalk 'n' slash shocker, but rather more carefully plotted than the two earlier films.
Aka: STEPFATHER 3: FATHER'S DAY
HOR   105 min (ort 115 min)
VIDrel: 4-FRONT/POLYREC/ITC   V/h

STEPFORD HUSBANDS, THE **
15
Fred Walton USA
1996
Donna Mills, Michael Ontkean, Cindy Williams, Sarah Douglas, Louise Fletcher
This belated sequel to THE STEPFORD WIVES now has the menfolk getting the docility treatment. Mills and Ontkean are the new arrivals in town who uncover the secrets behind this state of affairs. An interesting feminist variant, but once one has accepted the initial plot premise, there are not enough ideas here to keep it going.
FAN   90 min   VIDrel: ODY/SONOP   V/sh

STEPFORD WIVES, THE ***
15
Bryan Forbes USA
1974
Katherine Ross, Paula Prentiss, Peter Masterson, Nanette Newman, Patrick O'Neal, Tina Louise, Carol Rossen, William Prince, Carole Mallory, Barbara Rucker, Tonie Reid, Judith Baldwin, George Coe, Michael Higgins
A couple newly arrived at a small Connecticut town find many of the young wives obsessively houseproud and totally obedient to their husbands, with new arrivals changing after a time too. A good idea with some genuinely horrific moments, but one that fails to work completely. Followed by two trashy sequels – "The Revenge Of The Stepford Wives" and "The Stepford Children". The script is by William Goldman.
FAN   110 min (ort 115 min)   VIDrel: VISVID/POLYREC
L/A   V
Boa: novel by Ira Levin.

STEPKIDS *
PG
Joan Micklin Silver USA
1992
Griffin Dunne, Dan Futterman, Patricia Kalember, Jenny Lewis, Ben Savage, Adrienne Shelly, David Strathairn, Trenton Teigen, Margaret Whitton, Hillary Wolf, Jessica Seely, Jim Haynie, Sean Blackman, Denis Heames, Cory Danziger
In these days of multiple divorces and marriages, thirteen-year-old Laura bravely attempts to give us an account of the complex relationships around her. With Mom onto her third marriage and Dad acting like an overgrown kid, the pressure is intense, and eventually Laura runs away to a lakeside cabin occupied by her older brother. But her family are not far behind. For all its sharp dialogue, this contrived effort makes little use of a most promising opening.
Aka: BIG GIRLS DON'T CRY... THEY GET EVEN
COM   101 min (ort 104 min)   VIDrel: COLUM/SONOP
V/sh

STEPMOTHER, THE *** 15
Jorge Montesi USA 1993
Diane Ladd, Wendel Meldrum, Geraint Wyn Davies
In this tense, psychological thriller, a woman finds that the
mother who rejected her as a child has decided to make a return
into her daughter's life.
THR 87 min VIDrel: MIA/DISC V

STEPPING OUT ** PG
Lewis Gilbert USA 1991
*Liza Minnelli, Shelley Winters, Robyn Stevan, Jane Krakjowski, Bill
Irwin, Ellen Greene, Sheila McCarthy, Andrea Martin, Julie Walters,
Carol Woods, Luke Reilly, Nora Dunn, Eugene Robert Glazer, Geza
Kovacs, Raymond Rickman*
A former professional tap dancer turned teacher coaches her
pupils for a forthcoming charity performance, inspiring them
to overcome their fears and generally bringing out their collec-
tive best in preparation for the "big day". An interesting
variant on the "let's put on a show" musical, that founders on
Gilbert's uncertain direction, poor dialogue and most annoyingly, an over-emphasis on the personal problems of the
characters.
COM 104 min (ort 110 min) VIDrel: CIC/SONOP V/sur
Boa: play by Richard Harris.

STEPTOE AND SON * PG
Cliff Owen UK 1972
*Wilfrid Brambell, Harry H. Corbett, Carolyn Seymour, Arthur
Howard, Victor Maddern, Fred Griffiths, Queenie Watts, Patsy
Smart, Alec Mango, Perri St Clare, Lon Satton, Vivien Lloyd, Mike
Reid, Barry Ingham, Joan Heath*
A popular and very witty TV series about a father and son junk-
yard business, is brought rather unsuccessfully to the screen.
When the son falls for a stripper it looks as though true love is
only a whisker away, but taking his father on the honeymoon
is not a very good idea. Pitifully weak and unfunny. Written by
Ray Galton and Alan Simpson and followed by STEPTOE AND
SON RIDE AGAIN.
Aka: STEPTOE AND SON: THE FEATURE
COM 93 min (Cut at film release – ort 98 min)
VIDrel: WHV V/h

STEPTOE AND SON RIDE AGAIN * PG
Peter Sykes UK 1973
*Harry H. Corbett, Wilfrid Brambell, Milo O'Shea, Diana Dors, Neil
McCarthy, Bill Maynard, George Tovey, Sam Kydd, Yootha Joyce,
Olga Lowe, Henry Woolf, Geoffrey Bayldon, Frank Thornton, Peter
Thornton, Grazina Frame*
A follow-on to STEPTOE AND SON with our rag-and-bone man
using his father's savings to buy a greyhound. A feeble attempt
to build a full-length feature out of material, that would just
about sustain an episode of the original and well-loved TV series
"Steptoe And Son" (1964-73).
COM 95 min (ort 99 min) VIDrel: WHV V/h

STICK * 18
Burt Reynolds USA 1985
*Burt Reynolds, Candice Bergen, George Segal, Charles Durning, Jose
Perez, Richard Lawson, Dar Robinson, Tricia Leigh Fisher, Castulo
Guerra, Alex Rocco*
An ex-con released from prison seeks revenge for his friend's
murder in a drugs deal that went wrong. A murky melodrama
of little action but much introspection. Written by Leonard.
THR 104 min Cut (13 sec – ort 109 min)
VIDrel: CIC/SONOP L/A V/sh
Boa: novel by Elmore Leonard.

STING, THE *** PG
George Ray Hill USA 1973
*Paul Newman, Robert Redford, Robert Shaw, Charles Durning, Eileen
Brennan, Harold Gould, Ray Walston, Dana Elcar, Jack Kehoe,
Dimitra Arliss, Charles Dierkop, Robert Earl Jones, John Heffernan,
James J. Sloyan, Sally Kirkland*
Clever film about two con-men who set up a completely fake
scenario in order to fleece a gangleader (played by a miscast
Robert Shaw), who had one of their friends killed. Scott Joplin's
music is used to good effect. THE STING 2 followed. AA: Pic,
Dir, Art/Set (Henry Bumstead/James Payne), Edit (William
Reynolds), Cost (Edith Head), Story/Screen (David S. Ward),
Score (Marvin Hamlisch).
COM 129 min VIDrel: 4-FRONT/POLYREC/CIC L/A
V/h

STING 2, THE ** PG
Jeremy Paul Kagan USA 1983
*Jackie Gleason, Mac Davis, Karl Malden, Teri Garr, Oliver Reed, Bert
Remsen, Kathalina Veniero, Jose Perez, Larry Bishop, Frank
McCarthy, Richard C. Adams, Ron Rifkin, Harry James, Frances
Bergen, Monica Lewis, Val Avery*
Sequel to THE STING with our two con-artists arranging a
complicated fraud, this time a rigged boxing match, in order
to put one over on a bigwig. A flat and unappealing tale in
which the complexities of the plot overshadow the perfor-
mances. Written by David S. Ward and with music by Lalo
Schifrin.
COM 98 min (ort 102 min) VIDrel: CIC/SONOP L/A V

STINGRAY: THE INCREDIBLE VOYAGE *** U
Alan Patillo/David Elliott/John Kelly UK 1965 (re-edited 1980)
Voices of: Don Mason, Robert Easton, Lois Maxwell, Ray Barrett
Four episodes from this children's puppet series are compiled
into a single feature-length adventure. The crew of underwater
vessel "Stingray" are menaced by the denizens of the deep as
they fight an evil underwater emperor who plots to take over
the world. Fair kid's adventure.
Aka: INCREDIBLE VOYAGE OF STINGRAY, THE
ANIM 95 min mTV VIDrel: 4-FRONT/POLYREC/ITC V

STIR CRAZY ** 15
Sidney Poitier USA 1980
*Gene Wilder, Richard Pryor, Georg Stanford Brown, JoBeth Williams,
Craig T. Nelson, Barry Corbin, Erland van Lidth de Jeude, Lee Purcell,
Miguelangel Suarez, Charles Weldon, Nicolas Coster, Joel Brooks,
Jonathan Banks*
Two unfortunate men are sent to jail by mistake, when the
woodpecker outfits they were hired to wear to promote the
opening of a new bank, are stolen and used to rob the bank.
Once there they suffer the usual indignities, more so when the
warden discovers that they can be persuaded to ride for him in
a rodeo contest. An over-extended farce whose highlights are
appearances by the massive (and melodious) de Jeude. Written
by Bruce Jay Friedman.
COM 106 min (ort 111 min) VIDrel: VCC/DISC/COLUM
V

STOCKADE *** 15
Martin Sheen USA 1990
*Charlie Sheen, Martin Sheen, Larry Fishburne, Michael Beach, Ramon
Estevez, F. Murray Abraham, Blu Mankuma, John Toles-Baey, Harry
Stewart, Matt Clark, Jay Brazeau, Tom McBeath, David Glyn Jones,
Samantha Langevin, Tony Pantages*
A U.S. Army stockade in West Germany in 1965 is the setting
for this tense story of a clash of wills between the tough sergeant
in charge and his latest prisoner, a young private serving ninety
days for assaulting a military policeman. A well acted drama
that boasts fine performances from both Sheens, but is impaired
by an ending that's far too pat to convince.
Aka: CADENCE
A/AD 93 min (ort 97 min)
VIDrel: 4-FRONT/POLYREC/EIV V
Boa: novel Count A Lonely Cadence by Gordon Werner.

STOLEN BABIES ** PG
Eric Laneuville USA 1993
*Lea Thompson, Kathleen Quinlan, Mary Tyler Moore, Mary Nell
Santacroce, Brett Rice, Tom Nowicki, Jenny krochmal, Judson
Vaughn, kenny Jones, Jillian Boyd, Ed Grady, Clarinda Ross, Edith
Ivey, Janekk McLeod, Terry Beaver*
A true story set in Tennessee in the 1940s, where a young social
worker uncovers an illegal adoption racket that is about to tear
apart thousands of loving families. Fair.
DRA 90 min mTV VIDrel: ODY/SONOP V/sh

STOLEN CHILDREN, THE *** 15
Gianni Amelio FRANCE/ITALY 1992
*Enrico Lo Verso, Valentina Scalici, Giuseppe Ieracitano, Renato
Carpentieri, Vitalba Andrea, Grignani, Vincenzo Peluso, Massimo de
Lorenzo, Celeste Brancato, Santo Santonocito, Agostino Zumbo,
Maria Pia Di Giovanni, Florence Darel*
Tragic tale (based on a true case) of a brother and sister who
grow up on a seedy Milan housing estate, mostly home to
various immigrant families. When their neglectful mother is
arrested, a young soldier is given the job of transporting them
across Italy to an orphanage, but grows increasingly fond of
them. This simple story is told with warmth if not originality,

and in many ways recalls the strengths of films from the Italian neo-realist period.
Aka: IL LADRO DI BAMBINI
DRA 110 min (ort 114 min) VIDrel: CURZON/20TH
V/sh

STOLEN HEARTS ** 15
Bill Bennett USA 1996
Denis Leary, Sandra Bullock, Stephen Dillane, Yaphet Kotto, Mike Starr, Jonathan Tucker, Wayne Robson, Michael Badalucco, Lenny Clarke, Jonny Fido, Don Gavin, Shaun R. Clark, Markus Parillo, John Friesen, Ian White, Jane Moffat
The bored wife of a petty thief agrees to go with him to a luxury resort where he intends to carry out one last robbery, but whilst there, she finds she enjoys the luxurious lifestyle on offer. Minor slapstick antics are interspersed with interminable sequences of ranting dialogue, as the couple argue and rant their way from one disaster to the next. It all becomes more tiresome than funny after a short while.
Aka: TWO IF BY SEA
COM 92 min (ort 96 min) cC VIDrel: WHV V/sur

STOLEN KISSES *** 15
Francois Truffaut FRANCE 1969
Jean-Pierre Leaud, Delphine Seyrig, Michel Lonsdale, Claude Jade, Harry-Max, Daniel Ceccaldi, Claire Duhamel, Catherine Lutz, Andrew Falcon, Paul Pavel, Serge Rousseau, Marie-France Pisier, Jean-Francois Adam, Jacques Robiolles
Third episode in the continuing story of Antoine Doinel, the central figure in THE FOUR-HUNDRED BLOWS, THE. Here, he is dishonourably discharged from the Army but finds solace in his final successes with women. An entertaining and well made comedy-drama, but too lightweight to be really memorable. Followed by BED AND BOARD. The 25-minute short "Antoine And Colette" has also been made available on the same tape.
Aka: BAISERS VOLES
DRA 91 min VIDrel: ARTIF/20TH V

STOMPIN' AT THE SAVOY ** PG
Debbie Allen USA 1992
Lynn Whitfield, Vanessa Williams, Jasmine Guy, Mario Van Peebles
Period drama set in 1939, and following the lives and loves of four young black women who share an apartment in Brooklyn, and add a little to their mundane lives by regular weekly trips to the title ballroom. Good ensemble acting sustains this pleasant movie, whose plot is not its strongest feature.
DRA 91 min VIDrel: CIC V

STONE COLD ** 18
Craig R. Baxley USA 1991
Brian Bosworth, Lance Henriksen, William Forsythe, Arabella Holzbog, Sam McMurray, Richard Gant, Paula Tocha, David Tress, Evan James, Tony Pierce, Billy Million, Robert Winley, Gregory Scott Cummins, Demetre Phillips, Magic Schwartz
Having been suspended, an undercover cop is blackmailed by the FBI into joining a gang of thuggish bikers who have aspirations of moving into the criminal big-time. Our cop's mission is to stop them. There is little of note in this lukewarm blend of noise and fury, except perhaps a riveting performance from Henriksen as the gang's brutal leader, and a few interesting attempts at painting a supposedly authentic picture of biker life.
A/AD 88 min (ort 92 min) VIDrel: VCC/DISC/COLUM
V/sur

STONE KILLER, THE ** 18
Michael Winner USA 1973
Charles Bronson, Martin Balsam, David Sheiner, Norman Fell, Ralph Waite, Christina Raines, Stuart Margolin, John Ritter, Frank Campanella, Kelly Miles, Paul Koslo, Jack Colvin, Eddie Firestone, David Moody, Walter Burke
A police officer unravels a complex plot involving Vietnam veterans and a gangland boss. One of the director's customary violent offerings. Written by Gerald Wilson.
THR 91 min (ort 96 min) VIDrel: RCA L/A V
Boa: novel A Complete State Of Death by John Gardner.

STONEWALL ** 15
Nigel Finch UK 1995
Guillermo Diaz, Frederick Weller, Brendan Corbalis, Duane Boutte, Bruce MacVittie, Peter Ratray, Dwight Ewell, Matthew Faber, Michael McElroy, Luis Guzman, Joey Dedio, Tim Artz, John Doman, Isaiah Washington, Candis Cayne
The lives of a handful of gay New Yorkers provide the focus for this story, that is set in the weeks just prior to the infamous Stonewall riots of 1969. The problem with this film is that it should have been a documentary; the director's decision to build a fictitious story around the events portrayed just does not convince, and though some of the characters are truly memorable (Diaz is wonderful as a Puerto Rican drag queen) the film feels contrived and awkward.
DRA 98 min VIDrel: TART/20TH V
Boa: book by Martin Duberman.

STONING, A ** PG
Larry Elikann USA 1988
Ken Olin, Jill Eikenberry, Ron Perlman, Olivia Burnette, Maureen Mueller, Gregg Henry, Nicholas Pryor, Noble Willingham, Peter Michael Goetz, Theodore Bikel, Brad Pitt, Bill Allen, Michael Criscuolo, Michael Johnston, Mert Hatfield
Story set in the Amish community with an Amish baby being accidentally killed when the parents' buggy is stoned by four youths from outside the community. Though the D.A. is keen to prosecute, the parents are unwilling to allow their young daughter (who witnessed the attack) to testify as they do not believe in secular courts. A rather clinical offering that remains aloof and unmoving, despite good work from the cast.
Aka: STONING IN FULHAM COUNTY, A
DRA 94 min (ort 97 min) mTV VIDrel: ODY/SONOP V/h

STOP AT NOTHING *** 15
Chris Thompson USA 1989
Veronica Hamel, Lindsay Frost, Annabella Price, Robert Desiderio, Caroline McWilliams, David Ackroyd, Joseph Hacker, Deborah Anne Gorman, Lou Beatty Jr, Francis X. McCarthy, Ingrid Oliu, Al Ruscio, Roger La Page, Alan Koss
A look at the inability of the law to protect children from sexual abuse, this has Frost playing a private eye working with professional child stealer Hamel and an underground network, and setting out to protect a little girl from the sexual advances of her dad. A strongly scripted work (written by Stephen W. Johnson) it moves along at a brisk pace and is both well acted and gripping.
DRA 94 min (ort 100 min) mCab VIDrel: FIRST/SONOP
L/A V

STOP! OR MY MOM WILL SHOOT * PG
Roger Spottiswoode USA 1992
Sylvester Stallone, Estelle Getty, JoBeth Williams, Roger Rees, Martin Ferrero, Gailard Sartain, John Wesley, Al Fann, Ella Joyce, J. Kenneth Campbell, Nicholas Sadler, Dennis Burkley, Ving Rhames, Jana Arnold, Christopher Collins
Stallone plays an L.A. cop whose meddlesome mom pays a visit from New Jersey, complete with pet pooch. Mother and son may not be Jewish, but all the standard Jewish-mother cliches are to be found here, and some more besides, as Mom begins to play an ever increasingly active part in her son's life. Not so much a movie as a crash-bang, shoot-'em-up, car chasing, guntoting celebration of stale scripting and dumb humour.
COM 83 min (ort 87 min) VIDrel: CIC/SONOP V/sur

STOPOVER TOKYO ** U
Richard L. Breen USA 1957
Joan Collins, Robert Wagner, Edmond O'Brien, Ken Scott, Reiko Oyama, Larry Keating, Sarah Selby, Solly Nakamura, H. Okhawa, K.J. Seijto, Demmei Susuki, Yuki Kaneko, Michei Miura
An American intelligence agent in Japan discovers a plot to assassinate the US ambassador there, but the intended victim refuses to take the threat seriously. The Japanese locations are very pretty, but the flabby script and unconvincing performances get in the way. Written by Breen and Walter Reisch.
THR 100 min B/W VIDrel: 20TH/TECH L/A V/h
Boa: novel by John P. Marquand.

STORM OVER ASIA *** PG
Vsevolod Pudovkin USSR 1928
I. Inkizhinov, Valeri Inkizhanov, A. Dedintsev, V. Tzoppi, Paulina Belinskaya
The British intervene in Mongolia in 1918 and attempt to create a puppet government, headed by a fur trapper who claims to be heir to Genghis Khan, but who eventually rejects his imperialist role and leads a rebellion. An unusual film of great beauty but made with the expected political overtones. This was Pudovkin's last major silent work, but a sound track was added under his direction in 1950.
Aka: HEIR TO GENGHIS KHAN, THE; POTOMOK CHINGIS-KHANA
DRA 87 min (ort 149 min) B/W silent VIDrel: HEND L/A
V

STORMRIDER **   18
Giancarlo Santo FRANCE/ITALY/WEST GERMANY   1972
Lee Van Cleef, Peter O'Brien, Marc Mazza, Elaus Grunberg, Horst Frank, Jess Hahn, Antony Vernon, Sandra Sandrini
A veteran gunfighter becomes the protector of an innocent young man, who was wrongfully convicted of the murder of a gang leader, and together the pair fight to free the town from the grip of this gang. Average.
Aka: BIG SHOWDOWN, THE; DREI VATERUNSER FUR VIER HALUNKEN; GRAND DUEL, THE; IL GRANDE DUELLO; LE GRAND DUEL
WES   89 min Cut (3 sec – ort 93 min)
VIDrel: MOPIC/SGSVID   V

STORMY MONDAY ***   15
Mike Figgis UK   1988
Melanie Griffith, Tommy Lee Jones, Sting (Gordon Sumner), Sean Bean, James Cosmo, Mark Long, Brian Lewis
An atmospheric thriller set in Newcastle where Sting is the owner of a nightclub, and a ruthless American businessman arrives with the intention of making a fortune from re-development of the area. A slowly-paced but stylish film of little substance, with all the leads performing well. Written by Figgis (who also did the score) and with photography by Roger Deakins.
THR   89 min (ort 93 min)   VIDrel: VISVID/POLYREC L/A V/sh

STORMY WEATHER ***   U
Andrew L. Stone USA   1943
Lena Horne, Bill Robinson, Fats Waller, Dooley Wilson, Cab Calloway and His Orchestra, The Nicholas Brothers, Katherine Dunham and her Dance Troupe, Ada Brown, The Tramp Band, Babe Wallace, Ernest Whitman, Zutty Singleton
An amazing cavalcade of black talent makes this movie a documentary record of virtually every major musical star. The almost non-existence plot has Robinson looking back at his long career and remembering his fellow artistes and their show-stopping numbers. Lena Horne sings the title song, and the other numbers of note are: "Ain't Misbehavin'", "I Can't Give You Anything But Love, Baby" and "That Ain't Right".
MUS   77 min B/W   VIDrel: 20TH/TECH   V

STORY OF A CLOISTERED NUN *   18
Domenico Paolella ITALY   1973
Catherine Spaak, Eleonora Giorgi, Suzy Kendall, Martine Brochard, Antonio Falsi
Story of a sexually obsessed Mother Superior and a nun who falls in love with the wrong man.
Aka: STORIA DI UNA MONACA DI CLAUSURA
A   93 min (ort 97 min) wScrn dubbed
VIDrel: REDEM/RTM   V

STORY OF QIU JU, THE ***   15
Zhang Yimou CHINA/HONG KONG   1992
Gong Li, Lei Lao Sheng, Liu Pei Qi, Ge Zhi Jun, Yang Liu Xia, Liu Chun, Cui Louwen, Yang Huiqin
When the village headman assaults and injures her husband, a pregnant woman stubbornly pursues her a crusade for justice, which forces her to fight her way upwards through the obstructive bureaucratic hierarchy. A strong but simple story that is filmed with great conviction and supported by first-rate acting and direction. Voted Best Foreign Language Film at the Venice Film Festival.
Aka: QIUJU DA GUANSI
DRA   96 min (ort 100 min)   VIDrel: ELPIC/POLYREC V/sur
Boa: novel The Wan Family's Lawsuit by Chen Yaun Bin.

STORY OF ROBIN HOOD AND HIS MERRIE MEN, THE **   U
Ken Annakin UK   1952
Richard Todd, Joan Rice, James Hayter, James Robertson Justice, Peter Finch, Hubert Gregg, Marita Hunt, Anthony Forwood, Elton Hayes, Patrick Barr, Bill Owen, Michael Hordern, Reginald Tate, Hal Osmond, Anthony Eustrel
Yet another version of a famous legend, in which an excellent cast do their best to breathe some life into this story of the outlawed Earl of Sherwood, who leads his band against the corrupt Sheriff of Nottingham in the 12th century. Quite a colourful romp, but rather weakly plotted, the legend having received the standard Disney treatment.
Aka: STORY OF ROBIN HOOD, THE
A/AD   83 min   VIDrel: WDV/TECH L/A   V

STORY OF VERNON AND IRENE CASTLE, THE ***   U
H.C. Potter USA   1939
Fred Astaire, Ginger Rogers, Edna May Oliver, Walter Brennan, Lew Fields, Etienne Girardot, Janet Beecher, Rolfe Sedan, Leonid Kiskey, Robert Strange, Douglas Walton, Clarence Derwent, Sonny Lamont, Frances Mercer
The last major film starring Astaire and Rogers, in which they appear as a husband and wife team, who achieve fame in Paris just before the outbreak of WW1. Somewhat cramped by slow pacing but still a pleasure to watch. Numbers include: "Little Brown Jug", "Too Much Mustard" and "Only When You're In My Arms".
MUS   89 min (ort 93 min) B/W   VIDrel: VCC   V
Boa: books by Irene Castle.

STORYBOOK *   (U)
Lorenzo Doumani USA   1995
William McNamara, Swoosie Kurtz, Robert Constanzo, James Doohan, Richard Moll, Brenda Epperson-Doumani, Gary Morgan, Jack Scalia, Sean Fitzgerald, Milton Berle, Vinny Agiro, Kathrin Lautner, Billy Stamp, Zachary Benjamin
After his Air Force father goes missing while on an overseas assignment, a young boy and his mother move to a house whose attic contains a gateway to another world. Soon, our young hero finds himself embarking on a quest to find a magic sword and defeat an evil queen who has usurped her nephew's throne. A laughably poor fantasy, with inappropriate adult humour, poor animated puppets, pedestrian direction and indifferent acting. Best avoided.
JUV   84 min (ort 88 min)   SATrel: SKY MOVIES

STORYVILLE **   15
Mark Frost USA   1992
James Spader, Joanne Whalley-Kilmer, Jason Robards, Charlotte Lewis, Michael Warren, Michael Parks, Chuck McCann, Charlie Haid, Chino Fats Williams, Jeff Perry, Woody Strode, Galyn Gorg, Justine Arlin, George Kee Cheung, Piper Laurie
A young and very wealthy New Orleans lawyer decides to run for Congress but finds a blackmailing mistress only one of many problems that threaten to ruin his plans. These include a murder investigation and a number of well-kept family secrets, in this well acted but rather confused and dark tale. This was Frost's directorial debut.
DRA   113 min   VIDrel: HIFLI/SONOP   V/h
Boa: novel Juryman by Frank Galbally and Robert Macklin.

STOWAWAY ***   U
William A. Seiter USA   1936
Shirley Temple, Alice Faye, Robert Young, Eugene Pallette, Helen Westley, Arthur Treacher, J. Edward Bromberg, Astrid Allwyn, Allan Lane, Robert Greig, Jayne Regan, Julius Tannen, Willie Fung, Philip Ahn, Paul McVey
The orphan daughter of an American missionary smuggles herself aboard a ship and has various adventures. Young and Faye provide the romantic interest. A good vehicle for Temple in which she performs some enjoyable numbers, including a song in Chinese. Songs are by Mack Gordon and Harry Revel, with musical direction by Louis Silvers.
MUS   86 min B/W   VIDrel: 20TH/TECH L/A   V

STOWAWAY IN THE SKY ***   U
Albert Lamorisse FRANCE   1962
Pascal Lamorisse, Andre Gille, Maurice Baquet, Jack Lemmon (narration only)
Having sneaked aboard his grandfather's latest invention, a sixty foot orange balloon, a youngster enjoys a scenic tour of France. A pleasantly diverting tale from the director of THE RED BALLOON.
JUV   82 min   VIDrel: POLY/POLYREC/BRAVE L/A   V

STRAIGHT OUT OF BROOKLYN **   15
Matty Rich USA   1991
George T. Odom, Ann D. Sanders, Lawrence Gilliard Jr, Barbara Sanon, Reana E. Drummond, Matty Rich, Mark Malone, Ali Shahid Abdul Wahha, Joseph A. Thomas, James McFadden, Dorise Black, Robert N. Nash, Fran Sperling, Booker T. Matthews
An award-winning story of life in the black slum ghettos of Brooklyn, mostly concentrating on the lot of a family and the tribulations they suffer. Shot on location in Brooklyn's Red Hook district. The noisy hip-hop and rap soundtrack is an added distraction.
DRA   82 min (ort 88 min)   VIDrel: ARTIF/20TH   V/h

STRAIGHT TALK ***
PG
Barnet Kellman USA 1992
Dolly Parton, James Woods, Griffin Dunne, Michael Madsen, Deidre O'Connell, John Sayles, Teri Hatcher, Spalding Gray, Jerry Orbach, Philip Bosco, Keith MacKechnie, Charles Fleischer, Jay Thomas, Amy Morton, Paula Newsome
Parton plays Shirlee Kenyon in this ludicrous but enjoyable rags-to-riches tale, which sees Shirlee leaving her dead-end job in Arkansas and heading for Chicago, where she lands a plum job as a radio psychologist through a case of mistaken identity (highly reminiscent of THE COUCH TRIP). Her growing popularity attracts the attentions of a nosy reporter, but even he is eventually won over by her Southern charm and honesty.
COM 91 min VIDrel: HOLPIC/TECH L/A V

STRAIGHT TIME **
18
Ulu Grosbard USA 1978
Dustin Hoffman, Theresa Russell, Gary Busey, Harry Dean Stanton, M. Emmet Walsh, Rita Taggart, Kathy Bates, Edward Bunker, Fran Ryan, Jacob Busey, Tina Menard, Stephanie Ericson-Baron, Dave Kelly, Don Sommese, Kit Lee Wong
A man released from prison goes steadily downhill despite his best efforts to go straight. An absorbing but rather unattractive tale that's sustained by its performers. Hoffman started directing this one but Grosbard took over. Scripted by Alvin Sargent, Edward Bunker and Jeffrey Boam.
DRA 114 min VIDrel: L/A V
Boa: novel No Beast So Fierce by E. Bunker.

STRANDED **
15
Tex Fuller USA 1987
Ione Skye, Joe Morton, Maureen O'Sullivan, Susan Barnes, Cameron Dye, Michael Green, Brendan Hughes, Michael Burke, Flea, Spice Williams, Dennis Vero, Florence Schaffer, Gary Swanson, Harry Caesar, Kevin Haley
Aliens escape from one planet and risk capture on Earth, but are given shelter by an independent-minded old lady and her grand-daughter. Dye is the young man who comes to their rescue, helping them escape the clutches of a sheriff's posse and an other-worldly assassin. Not a story that amounts to much, though it has better characterisation than one usually finds in an SF film.
FAN 78 min (ort 80 min) VIDrel: 20VIS/SONOP L/A V

STRANDED **
15
Paul Tucker USA 1991
Deborah Wakeham, Ryan Michael, Stephen E. Miller, Blu Mankuma, Gabrielle Rose, Ric Reid, Bill Reiter, Ian Forsyth, Rod Menzies, Sue Olafson, Holly Chester, Viktoria Langton
A couple take a trip to their own private island in an attempt to save their marriage, but danger threatens in the shape of psychopath. One of those overblown and underplotted psychological thrillers, entertaining enough if not exactly original. The film is dedicated to the director, who died shortly after completion.
Aka: LIGHTHOUSE
THR 87 min (ort 90 min) VIDrel: IMPENT V
Boa: story by Daniel D. Williams.

STRANGE AFFAIR, A **
15
Ted Kotcheff USA 1996
Judith Light, Jay Thomas, Linda Sorensen, Robin Dunne, Rachel Wilson, William Russ
Fact-based drama that follows the career of a love-triangle, within which a married couple and the woman's male lover attempt to live together, the wife refusing to leave her cheating husband, as he has just suffered a seriously disabling stroke.
DRA 89 min VIDrel: ODY/SONOP V/sh

STRANGE AFFAIR OF UNCLE HARRY, THE **
15
Robert Siodmark USA 1945
George Sanders, Geraldine Fitzgerald, Ella Raines, Sara Allgood, Moyna MacGill, Samuel S. Hinds, Harry Von Zell, Ethel Grimes
A man falls in love, but cannot escape from his overbearing sisters, and plans the murder of one of them. A stilted film that develops a modicum of suspense, but is then completely spoilt by a poor resolution, tacked on to comply with the constraints of 1940s censorship. Produced by Joan Harrison, one of Hitchcock's associates and written by Stephen Longstreet and Keith Winter.
Aka: UNCLE HARRY
THR 77 min (ort 82 min) B/W VIDrel: STABL L/A V
Boa: play by Thomas Job.

STRANGE DAYS ***
18
Kathryn Bigelow USA 1995
Ralph Fiennes, Angela Bassett, Juliette Lewis, Tom Sizemore, Michael Wincott, Vincent D'Onofrio, Glenn Plummer, Brigitte Bako, Richard Edson, William Fichtner, Josef Sommer, Joe Urla, Nicky Katt, Michael Jace, Louise LeCavalier
A visually impressive thriller with a strong technological slant, that casts Fiennes as a street hustler in the dangerous L.A. underworld of 1999. He makes a living making sensory "recordings" of fantasies to sell to voyeurs, but learns of a plot to record real murders (rather than fantasy ones). Despite a strong futuristic feel (almost as good as BLADE RUNNER in this respect) the film relies far too heavily on violence for its impact. Screenplay is by James Cameron.
THR 139 min Cut (13 sec) wScrn cC VIDrel: CIC/SONOP; PION (LV only) V/sur LV

STRANGE INVADERS **
PG
Michael Laughlin USA 1983
Paul Le Mat, Nancy Allen, Diana Scarwid, Michael Lerner, Louise Fletcher, Fiona Lewis, Kenneth Tobey, Wallace Shawn, Charles Lane, June Lockhart, Lulu Sylbert, Joel Cohen, Dan Shor, Dey Young, Jack Kehler, Mark Goddard
A spoof on SF films with a Midwestern town being taken over by aliens, that parodies INVASION OF THE BODY SNATCHERS and other such 1950s films. Patchy and disorganised but generally agreeable. The script is by William Condon and Laughlin, and the evocative score is supplied by John Addison.
FAN 90 min (ort 93 min) VIDrel: LUMI/SPEAR L/A V

STRANGE LOVE OF MARTHA IVERS, THE ***
PG
Lewis Milestone USA 1946
Kirk Douglas, Barbara Stanwyck, Lizabeth Scott, Judith Anderson, Van Heflin, Roman Bohnen, Darryl Hickman, Janis Wilson, Ann Doran, Frank Orth, Charles D. Brown, James Flavin, Max Wagner, Mickey Kuhn, Matt McHugh, Walter Baldwin
Brooding and atmospheric film noir, a melodramatic study of human evil in which a wealthy woman is desperate to conceal the truth about an incident in her past. This intriguing film gave Douglas his film debut. The music is by Miklos Rosza.
DRA 115 min B/W VIDrel: SECOND/RTM V
Boa: short story Love Lies Bleeding by John Patrick.

STRANGE PLACE TO MEET, A *
PG
Francois Dupeyron FRANCE 1988
Catherine Deneuve, Gerard Depardieu, Jean-Pierre Sentier, Andre Wilms, Nathalie Cardone, Alain Rimoux
Following a row, a woman is dumped by her husband at a motorway lay-by, and she meets up with another stranded motorist, a doctor who is attempting to fix his car. Despite his initial reluctance to get involved with her, this man grows increasingly attracted to her, but meanwhile she awaits the arrival of her husband at a motorway cafe, having heard that he has been in looking for her. Dupeyron's feature film debut is plotless, sterile and very disappointing.
Aka: DROLE D'ENDROIT POUR UNE RENCONTRE
DRA 93 min (ort 98 min) VIDrel: ARTIF/20TH V

STRANGE VOICES **
PG
Arthur Allan Seidelman USA 1988
Valerie Harper, Nancy McKeon, Stephen Macht, Tricia Leigh Fisher, Millie Perkins, Robert Krantz, Jack Blessing
McKeon plays a happy teenager who develops schizophrenia, in this dreary clone of the infinitely superior I NEVER PROMISED YOU A ROSE GARDEN.
DRA 92 min (ort 100 min) mTV VIDrel: ODY/SONOP V/h

STRANGER, THE ***
PG
Orson Welles USA 1946
Edward G. Robinson, Orson Welles, Loretta Young, Philip Merivale, Richard Long
A top Nazi lives quietly with a fake identity as a professor at a school in New England, but a determined Nazi-hunter is not far behind. A restrained film, it carefully builds its atmosphere, and though lacking the sheer style of other works by Welles, has no shortage of tension. The climax is memorable and totally unexpected.
THR 95 min B/W VIDrel: SECOND/RTM V

STRANGER, THE ***
(PG)
Satyajit Ray FRANCE/INDIA 1991
Dipankar De, Mamata Shankar, Bikram Bhattacharya, Utpal Dutt,

Dhritiman Chatterjee, Rabi Ghosh, Subrata Chatterjee, Promode Ganguly, Ajit Bannerjee
A middle-class couple living in Calcutta are surprised when a stranger comes to their home and announces that he is the wife's long-lost uncle, whose existence had remained unknown to her. He moves in with them and it is not long before they find that lives are being turned virtually upside down because of his presence. Ray's final film is a graceful if overly stylised satire, well worth watching, even if it never reaches the heights of his earlier works.
Aka: AGANTUK
DRA   120 min   CINrel

STRANGER, THE **   18
Fritz Kiersch USA   1994
Kathy Long, Robin Lynn Heath, Eric Pierpoint, Andrew Divoff, Jason Adams, Ginger Lynn Allen, David Antony Marshall, Nils Allen Stewart, Danny Trejo, Faith Minton, Jeff Cadiente, Randy Vasquez, Billy Maddox, Robert Winley, Paul Hampton
Former martial arts champ Long stars in a tale of a woman out to destroy a criminal gang, in a fairly enjoyable but unoriginal action outing.
A/AD   98 min   VIDrel: 20VIS/SONOP   V/sur

STRANGER AND THE GUNFIGHTER, THE **   18
Anthony M. Dawson (Antonio Margheriti) HONG
KONG/ITALY/SPAIN   1976
Lee Van Cleef, Lo Lieh, Patty Shepard, Julian Ugarte, Karen Yeh, Femi Benussi, Erika Blanc, George Rigaud, Richard Palacios, Goyo (Gregorio) Peralta, Al Tung, Alfred Boreman, Bart Barry, Paul Costello
A clumsy spoof attempt to marry two diverse genres that is only partially successful, and tells of a gunfighter who joins forces with a kung fu expert in order to regain a missing fortune, the whereabouts of which is tattooed on the backsides of several girls.
Aka: BLOOD MONEY
WES   90 min (ort 107 min)   VIDrel: MOPIC/SGSVID   V

STRANGER BY NIGHT **   18
Gregory H. Brown USA   1994
Steven Bauer, Jennifer Rubin, William Katt, Michael Parks, Michelle Greene, J.J. Johnston
A serial killer seems to have the police baffled, and one of the detectives who is trying to catch him is still haunted by the memory of his father murdering his stepmother. This has left him prone to have violent rages and blackouts, and he becomes worried by the fact that he experienced blackouts on the nights several of the victims met their gruesome deaths.
THR   91 min (ort 96 min)   VIDrel: HIFLI/SONOP   V/h

STRANGER IN THE MIRROR, A ***   15
Charles Jarrott USA   1992
Perry King, Lori Loughlin, Christopher Plummer, Paula Shaw, Janie Woods-Morris, George Touliatos, Robert Clothier, Terence Kelly, Phil Peters, Kimberly Sheppard, Suki Kaiser, Eli Gabay, Lochlyn Munro, Brent Stait
A woman dreams of enjoying the good things in life, and gets her chance to do so when she marries an up-and-coming comedian, but her affection for him creates jealousy on the part of the man's agent, who has devoted himself solely to the career of his protege, and this latter attempts to devise ways of destroying their happiness. An unusually well plotted drama, with a depth of characterisation not usually seen in films of this kind.
Aka: SIDNEY SHELDON'S A STRANGER IN THE MIRROR
DRA   94 min mTV   VIDrel: MIA/VCC/IMPENT L/A   V
Boa: novel by Sidney Sheldon.

STRANGER IS WATCHING, A **   (18)
Sean S. Cunningham USA   1981
Kate Mulgrew, Rip Torn, James Naughton, Shawn Von Schreiber, Barbara Baxley, Stephen Joyce, James Russo, Frank Hamilton, Maggie Task, Roy Poole, Stephen Strimpell, Jason Robards III
A deranged murderer kidnaps a young girl and a female TV reporter, holding them hostage in part of the New York subway system, below Grand Central Station. A tense but over-complex tale done with the obligatory 1980s-style touches of viciousness, from the director of FRIDAY THE 13TH. Written by Earl MacRauch and Victor Miller.
THR   92 min   SATrel: TNT MOVIES
Boa: novel by Mary Higgins Clark.

STRANGER THAN PARADISE ***   15
Jim Jarmusch USA/WEST GERMANY   1984
John Lurie, Eszter Balint, Richard Edson, Cecillia Stark, Danny Rosen, Tom Docillo, Richard Boes, Rammellzee, Rockets Redglare, Harvey Per, Brian J. Burchill, Sara Driver, Paul Sloane
A grey view of America seen through the eyes of a young girl from Budapest, who meets up with her cousin and his best friend. Together they visit New York, Cleveland and Florida. An offbeat tale, divided into three segments, that is alternately funny, perceptive and dull. Written by Jarmusch, this is essentially an expanded version of a thirty-minute short. The music is by John Lurie.
DRA   85 min (ort 90 min) B/W   VIDrel: ARTIF/20TH   V/h

STRANGER, THE: THE AIRZONE SOLUTION **   PG
Bill Baggs UK   1993
Colin Baker, Peter Davison, Sylvester McCoy, Jon Pertwee
An episode from the British SF series "The Stranger", this tells of a sinister corporation that claims to have the technology needed to cope with all forms of environmental pollution.
FAN   62 min (ort 65 min) mTV   VIDrel: CARL/TECH   V

STRANGER WITHIN, THE ***   18
Tom Holland USA   1990
Kate Jackson, Chris Sarandon, Ricky Schroder, Clark Sandford, Peter Breitmayer, Pamela Danser, Ross Swanson, Dale Dunham, Ollie Osterberg, Kelsey Rose, Zachary Hunke, Susanne Egli, James Harris
With the death of her husband in the Vietnam War being followed by the abduction of her three-year-old son, a woman's life is shattered. However, she slowly rebuilds it, and fifteen years later has buried her trauma and entered a new relationship. But the appearance of a stranger who claims to be her long-lost son threatens to unhinge her mind, especially as his behaviour becomes ever more erratic. A tense psychological thriller.
THR   89 min (ort 93 min) mTV   VIDrel: NWV/SONOP
V/h

STRANGERS ON A TRAIN ****   PG
Alfred Hitchcock USA   1951
Robert Walker, Farley Granger, Ruth Roman, Leo G. Carroll, Marion Lorne, Patricia Hitchcock, Laura Elliott, Jonathan Hale, Howard St John, John Brown, Norma Varden, Robert Gist, John Doucette, Dick Wessel
When an obsessive young man (brilliantly played by Walker) meets a stranger on a train, he discusses how they might each be rid of their respective problems by collaborating in a murder plan. They part and the former assumes that he has the latter's agreement. A marvellously taut and exciting film and one of Hitchcock's best. Written by Raymond Chandler and Czenzi Ormonde. Remade in 1970 as "Once You Kiss A Stranger". See also ACCIDENTAL MEETING.
THR   96 min (ort 101 min) B/W   VIDrel: WHV   V
Boa: novel by Patricia Highsmith.

STRANGERS: THE STORY OF A MOTHER AND
DAUGHTER **   (PG)
Milton Katselas USA   1979
Bette Davis, Gena Rowlands, Ford Rainey, Donald Moffat, Whit Bissell, Royal Dano, Kate Riehl, Krishan Timberlake, Renee McDonell, Sally Kemp, Don Fosse, Don Sale, John Zumino, Jay Coffman, Grail Dawson
After an absence of twenty years, a dying woman returns home to her widowed mother in a bid to heal the bitter rift that separated them for so long. A powerful drama with standout performances from both Davis (who won an Emmy) and Rowlands in the leads. Scripted by Michael DeGuzman.
Aka: STRANGERS
DRA   92 min mTV   SATrel: SKY MOVIES

STRANGERS WHEN WE MEET **   15
Richard Quine USA   1960
Kirk Douglas, Kim Novak, Walter Matthau, Barbara Rush, Ernia Kovacs, Virginia Bruce, Helen Gallagher, Kent Smith
An unhappily married architect falls in love with his neighbour, but faces a conflict between his emotional needs and his career, when he is suddenly presented with an attractive job opportunity in Hawaii. A glossy and quite superficial soap opera, with lots of misery but no real problems. The script is by Hunter.
DRA   113 min (ort 117 min)   VIDrel: RCA L/A   V
Boa: novel by Evan Hunter.

STRANGLERS OF BOMBAY, THE ** 15
Terence Fisher UK 1959
Guy Rolfe, Allan Cuthbertson, Jack Holden, Marne Maitland, Andrew Cruikshank, George Pastell, Paul Stassino, Tutti Lemkow, David Spenser, Marie Devereaux
Later remade as THE DECEIVERS, this is the story of how an army captain sets about exposing the officials behind the Kali-worshipping sect of killers that plagued the British in India in the 1820s. Made by Hammer Films it benefits from strong direction and plotting, but is a little marred by some sadistic touches, very typical of films from the Hammer studios.
A/AD 77 min (ort 80 min) B/W VIDrel: ENCORE/SPEAR V

STRAPLESS *** 15
David Hare UK 1988
Blair Brown, Bruno Ganz, Bridget Fonda, Alan Howard, Hugh Laurie, Alexandra Pigg, Billy Roche, Camille Coduri, Michael Gough, Gary O'Brien
Scripted by playwright Hare, this complex and rather slow-moving character study tells of an American doctor who lives and works in London and reaches a mid-life crisis on her 40th birthday, just at the time her younger sister is paying a visit. Brown and Fonda are excellent as respectively, the older and younger woman, and this unusual adult romantic-drama makes up in atmosphere what it lacks in momentum.
DRA 95 min VIDrel: VISVID/POLYREC V

STRAPPED *** (18)
Forest Whitaker USA 1993
Kia Joy Goodwin, Michael Biehn, Bokeem Woodbine, Paul McCrane, Craig Wasson, Fredro, Jermaine Hopkins, Samuel E. Wright, Chi Ali, Starletta Dupois, Nzingha Monie Love, Tangyanika, Don Blakely, Isaiah Washington, Dynamite
A black youth living in a housing project hears shots and finds that a young boy has shot and fatally killed a companion with whom he argued. Leaving the victim to die alone, he tells the killer to run for it and takes care of his gun. When his pregnant girlfriend is arrested on drugs charges, he tries to raise bail money by selling guns but is soon forced into a fateful deal with the cops. A bleak, intense and utterly depressing tale.
DRA 98 min (ort 102 min) mCab SATrel: MOVIE CHANNEL

STRAW DOGS ** (18)
Sam Peckinpah UK 1971
Dustin Hoffman, Susan George, Peter Vaughan, T.P. McKenna, Peter Falk, David Warner, Del Henney, Ken Hutchinson, Colin Welland, Jim Norton, Len Jones, Sally Thomsett, Donald Webster, Michael Mundell, Peter Arne, Robert Keegan
An American academic and his wife settle in a Cornish village but find that the atmosphere is far from welcoming. An extremely unpleasant film in which the locals are portrayed as a bunch of brutish thugs and rapists. The ending is violent and bloody. Written by Peckinpah and David Zelag. Filmed in England.
THR 113 min (ort 118 min) VIDrel: L/A V
Boa: novel The Siege Of Trencher's Farm by Gordon M. Williams.

STRAWBERRY AND CHOCOLATE *** 18
Thomas Gutierrez Alea/Juan Carlos Tabio CUBA 1993
Vladimir Cruz, Jorge Perrugoria, Mirta Ibarra, Francisco Galtorno, Joel Angelino, Marilyn Solaya, Andres Crotina, Antonio Carmona, Ricardo Avila, Maria Elena Del Toro, Zolnada Ona, Diana Iris Del Puerto
A poor student jilted by his girlfriend makes friends with an older man, to whom he feels a homosexual attraction. However, their relationship is threatened by differences of political outlook. Screenplay is by Paz.
Aka: FRESA Y CHOCOLATE
COM 105 min (ort 111 min) wScrn VIDrel: TART/20TH V
Boa: novella El Lobo, El Bosque y El Hombre Nuevo by Paz Senel.

STRAWBERRY BLONDE, THE *** U
Raoul Walsh USA 1941
James Cagney, Olivia De Havilland, Rita Hayworth, George Tobias, Alan Hale, Jack Carson, Una O'Connor, George Reeves
Set in New York in the 1890s, this tells of a dentist who becomes infatuated with one woman but marries another, and then has doubts as to whether he made the right choice. A lively and witty remake of "One Sunday Afternoon", with Cagney's bouncy performance sustaining the thin plot. Remade in 1948 by Walsh, but using the original title once more.
COM 94 min (ort 97 min) B/W VIDrel: MGM/WHV L/A V
Boa: play One Sunday Afternoon by James Hagan.

STRAYS ** 15
John McPherson USA 1991
Kathleen Quinlan, Timothy Busfield, Claudia Christian, William Boyett, Heather Lilley, Jessica Lilley, Gary McGurk, Eve Brenner, Michael McNab
A young couple leave the big city and head for the country, where they have bought a new home. However, they are soon to learn that the building is heavily occupied by a horde of dangerous wild cats. The expected dose of moggy mayhem soon follows.
HOR 79 min (ort 83 min) VIDrel: CIC/SONOP V/h

STREAMERS *** 18
Robert Altman USA 1983
Matthew Modine, Michael Wright, Mitchell Lichtenstein, David Alan Grier, Guy Boyd, George Dzundza, Albert Macklin, B.J. Cleveland, Bill Allen, Phil Ward, Paul Lazar, Terry McIlvain, Todd Savell, Mark Fickert, Dustye Winniford
Film version of a powerful play ostensibly about four soldiers in an army barracks, waiting to be sent to Vietnam and facing mounting tension at the training camp of the 83rd Airborne Division. Written by Rabe, this overlong and sombre tale is a little too claustrophobic for its own good, but has many striking moments.
DRA 113 min (ort 118 min) VIDrel: ARROW/RTM V/h
Boa: play by David Rabe.

STREET CRIMES * 18
Stephen Smoke USA 1991
Dennis Farina, Michael Worth, Max Gail, James T. Morris, Shawn Shimoda, Patricia Zehentmayr, Joe Banks, Ron Winston Yuan, Max Gail, Mayah McCoy, Mark Kaufman, Doug Franklin, Angus Duncan, Michelle Smith, Walter Cox, Fran Joseph
Another one of those all-action films revolving around a tough, streetwise cop (there must be a factory somewhere) who has to contend with the activities of a psychopathic gangster, the latter having begun a campaign of terror, gunning down cops and civilians alike. As ever, there's a brutal and bloody final reckoning in this strictly by-the-numbers effort.
A/AD 95 min VIDrel: MIA/DISC V/sh

STREET FIGHTER ** 15
Steven E. de Souza USA 1994
Jean-Claude Van Damme, Raul Julia, Ming-Na Wen, Damian Chapa, Kylie Minogue, Simon Callow, Roshan Seth, Wes Studi, Bryan Mann, Grand L. Bush, Jay Tavare, Peter Tuiasosopo, Andre Bryniarski, Gregg Rainwater, Miguel A. Nunez Jr
Movie version of a popular computer game in which our intrepid hero takes on an evil Latin American dictator who has taken civilian workers of the Allied Nations peacekeeping force hostage. As the politicians attempt to negotiate, the head of this organisation's army and a fellow agent make plans of their own to ensure justice is done. A cliched blend of action and spoof that is not helped by the severely limited acting skills of its leads. This was Julia's last film.
Aka: STREET FIGHTER: THE BATTLE FOR SHADALOO; STREET FIGHTER: THE MOVIE; STREET FIGHTER: THE ULTIMATE BATTLE
FAN 100 min (ort 102 min) VIDrel: COLUM/SONOP V/sur

STREET HUNTER ** 18
John A. Gallagher USA 1989
Steve James, Reb Brown, John Leguizamo, Valarie Pettiford, Frank Vincent
After his son is brutally killed by a Colombian drugs gang, a Mafioso gives a bounty hunter 24 hours in which to locate the killers before he initiates his own revenge crusade.
A/AD 90 min (ort 96 min) VIDrel: CAPIT/GUILD V

STREET JUSTICE ** 18
Richard Safarian CANADA 1987
Michael Ontkean, William Windom, Catherine Bach, Joanna Kerns
Following his escape from a Siberian prison camp after twelve years, a CIA agent returns home to a rather less than warm welcome. Deliberately set up years ago to be captured, he was long thought to be dead, and his return is a severe political

embarrassment. Sent to a high-security prison, he escapes and heads for his home town in this standard revenge tale. Watchable, but no more than that.
THR   89 min (ort 93 min)   VIDrel: GUILD/SONOP L/A V/sh

STREET KNIGHT **   18
Albert Magnoli USA   1992
Jeff Speakman, Christopher Neame, Lewis Van Bergen, Jennifer Gatti, Richard Allen, Ramon Franco, Richard Coca, Marco Rodriguez, Stephen Liska, Grainger Hines, Sal Landi, Hank Stone, Ketty Lester, Jeremiah Birkett, Bernie Casey
A cop left for dead after a shootout meets an attractive woman who asks him to help her, as her sixteen-year-old brother witnessed a gang killing and has now gone missing. As the cop sets out to track the boy down, he learns that the killing is part of a sinister plot to escalate the level of violent crime in L.A. Adequate.
A/AD   87 min (ort 90 min)   VIDrel: WHV   V/sur

STREET OF SHADOWS **   PG
Richard Vernon UK   1953
Cesar Romero, Kay Kendall, Edward Underdown, Victor Maddern, Simone Silva, Liam Gaffney,m Robert Cawdron, John Penrose, Bill Travers, Molly Hamley Clifford, Eileen Way, Paul Hardtmuth, Annaconda
Adequate thriller set in and around some of London's less salubrious locations, with a Soho arcade-owner finding himself being accused of the murder an ex-mistress, who was found stabbed in his apartment. He asks his assistant for help in clearing his name, but is unaware that the latter is the real culprit. This fairly routine underworld outing has a few good moments and an effective air of seediness.
Aka: SHADOW MAN
THR   80 min (ort 84 min) B/W   VIDrel: WHV   V
Boa: novel The Creaking Chair by Lawrence Meynall.

STREET SMART **   18
Jerry Schatzberg USA   1987
Christopher Reeve, Mimi Rogers, Kathy Baker, Jay Patterson, Andre Gregory, Morgan Freeman, Anna Maria Horsford, Frederick Rolf, Erik King, Michael J. Reynolds, Shari Hilton, Donna Bailey, Ed Van Nuys, Daniel Nalbach
A down-on-his-luck magazine writer tries to get back in favour by cobbling together a fake study of New York pimps, but finds that its success embroils him in a murder investigation. Written by David Freeman and based to some extent on his own experiences, but something of a hit-and-miss affair that tries to make a few sharp observations about journalistic corruption. A great performance from Freeman as a ruthless pimp is a highlight.
THR   93 min (ort 97 min)   VIDrel: MGM/WHV L/A   V

STREET SOLDIERS **   18
Lee Harry USA   1990
Jun Chong, Jeff Rector, David Homb, Jonathan Gorman, Katherine Armstrong, Joon Kim, Jude Gerard
Tired of the domination of their locality by a vicious street gang, a group of high school kids forms a rival gang, in a desperate attempt to reclaim the streets. An stale, assembly-line actioner, replete with violence if nothing else.
A/AD   92 min (ort 98 min)   VIDrel: REDEYE/DISC   V/sh

STREET TRASH **   18
Jim Muro Jr USA   1986
Karen Krawiec, Ellen O'Neil, Dave Weckerman, Vic Noto, Nicole Potter, Bill Chepil, Mike Lackey, Mark Sferrazza, Jane Arakawa, Clarenze Jarmon, Bernard Perlman, Miriam Zucker, M. D'Jango Krunch, James Corinz, Morty Storm
Two brothers and a cop help street tramps get their own back on a ruthless gang, but a strange rot-gut liquor presents new problems. An incoherent, low-grade mess of melting bodies and exploding heads, that is both difficult to follow and watch. An effective shocker with little plotting but a plethora of nauseating effects and violent encounters. This grade 10 shocker brings to mind films such as ERASERHEAD and TETSUO: THE IRON MAN. See also BODY MELT.
HOR   96 min Cut (6 sec – ort 102 min)
VIDrel: SPEAR/SONOP   V

STREETCAR NAMED DESIRE, A ****   15
Elia Kazan USA   1951
Marlon Brando, Vivien Leigh, Kim Hunter, Karl Malden, Rudy Bond,

Nick Dennis, Peg Hillias, Wright King, Richard Garrick, Anne Dere, Edna Thomas, Mickey Kuhn, Chester Jones, Marietta Canty, Charles Wagenheim, Maxie Thrower
Powerful screen version of a famous play about steamy Southern passions, with Brando memorable as Stanley Kowalski and Leigh equally good as Blanche Dubois, his sister-in-law. The fine jazz score is by Alex North. Written by Williams from his play and photographed by Harry Stradling. Remade for TV in 1984. AA: Actress (Leigh), S. Actor (Malden), S. Actress (Hunter), Art/Set (Richard Day/George James Hopkins).
DRA   119 min (ort 122 min) B/W   VIDrel: WHV   V
Boa: play by Tennessee Williams

STREETCAR NAMED DESIRE, A ***   15
John Erman USA   1984
Treat Williams, Ann-Margret, Beverly D'Angelo, Randy Quaid, Rafael Campos, Erica Yohn, Ric Mancini, Fred Sadoff, Elsa Raven, Tina Menard, Raphael Sbarge
A remake of the 1951 classic of repressed emotions and blighted lives that, although nowhere near the earlier film in terms of sheer power, has many good aspects, not least a fine performance from Ann-Margret as Blanche. The adaptation is by Oscar Saul.
DRA   115 min (ort 124 min) mTV   VIDrel: L/A   V
Boa: play by Tennessee Williams.

STREETCAR NAMED DESIRE, A ***   15
Glenn Jordan USA   1995
Jessica Lange, Alec Baldwin, John Goodman, Diane Lane
Another screen adaptation of the Williams play, with the same story of a spoilt Southern belle who is obliged to stay with her poor sister and husband. But once there, she is at the mercy of her bullying brother-in-law. Lange won the Golden Globe Best Actress Award for her portrayal, and though the film is very good in terms of its claustrophobic atmosphere, it offers no new insights into the play.
Aka: TENNESSEE WILLIAMS' A STREETCAR NAMED DESIRE
DRA   150 min mTV cC   VIDrel: 20TH   V/sh
Boa: play by Tennessee Williams.

STREETFIGHTER, THE **   15
Walter Hill USA   1975
Charles Bronson, James Coburn, Strother Martin, Maggie Blye, Michael McGuire, Jill Ireland, Robert Tessier, Nick Dimitri, Felice Orlandi, Bruce Glover, Edward Walsh, Frank McRae, Maurice Kowalewski, Noami Stevens
Bronson plays a barefist fighter in the Depression years in a story set in New Orleans. An interesting if rather unsatisfying film, although Bronson is certainly believable in his role. Written by Hill (his directing debut), Bryan Gindoff and Bruce Henstell.
Aka: HARD TIMES
A/AD   89 min (ort 97 min)
VIDrel: ENCORE/SPEAR/COLUM   V

STREETS OF FIRE **   15
Walter Hill USA   1984
Michael Pare, Diane Lane, Rick Moranis, Amy Madigan, Willem Dafoe, Elizabeth Daily, Deborah Van Valkenburgh, Richard Lawson, Rick Rossovich, Lee Ving, Marine Jahan, Bill Paxton, Ed Begley Jr, Stoney Jackson, Grand L. Bush, The Blasters
Damsel in distress theme of a rock singer kidnapped by a bike gang. Plenty of violence tastefully set to music with her boyfriend setting out to rescue her. A kind of superficial, comic strip style melodrama that's all sham gilt and pulsating music. The stylish photography is by Andrew Laszlo.
DRA   90 min (ort 94 min)   VIDrel: CIC/SONOP L/A   V/sh

STREETS OF GOLD **   15
Joe Roth USA   1986
Klaus Maria Brandauer, Adrian Pasdar, Angela Molina, Wesley Snipes, Elya Baskin, Rainbow Harvest
A ROCKY type action movie about a former Soviet boxing champion, who defects to the USA and becomes the coach of two promising young boxers. A routine outing that not even a great performance from Brandauer can enliven.
DRA   89 min (ort 95 min)   VIDrel: VCC L/A   V

STREETWALKER, THE **   18
Walerian Borowczyk FRANCE   1976
Sylvia Kristel, Joe Dallesandro, Mireille Audibert, Denis Manuel, Andre Falcon, Louise Chevalier
After his wife dies, a businessman becomes emotionally

involved with a prostitute in the Parisian red light district. A bleak examination of the life of such people.

Aka: LA MARGE

A   80 min (ort 86 min) (dubbed version available)
VIDrel: LUMI/SPEAR L/A   V

STREETWISE *
Rafal Zielinski USA
C. Thomas Howell, Renee Humphrey

18
199-

Urban actioner in which a police detective crosses swords with a criminal gang who specialise in kidnapping girls to supply to a prostitution ring. The cop is drawn into this affair when he tries to find a prostitute who has gone on the run from the gang, being aided in this quest by the girl's younger sister. Normally, action movies get better towards the end, but this one is an exception to that rule, and is only worth watching about halfway through.

A/AD   96 min   VIDrel: EIV/SONOP   V

STRICTLY BALLROOM ***
Baz Luhrmann AUSTRALIA

PG
1992

Paul Mercurio, Tara Morice, Bill Hunter, Pat Thomson, Gia Carides, Peter Whitford, Barry Otto, John Hannan, Sonia Kruger, Kris McQuade, Pip Mushin, Leonie Page, Antonio Vargas, Armonia Benedito, Jack Webster, Lauren Hewett

A deliciously light-hearted comedy about competitive ballroom dancing, that is set in Australia, where a talented youngster falls foul of both the controlling Dance Federation and his peers by inventing some novel steps of his own. Undaunted by the obstacles he encounters, he teams up with a new partner and makes a determined bid for the championship title. A delightful piece of pure escapism, spectacular and endearingly absurd at the same time.

COM   90 min (ort 94 min) cC   VIDrel: VCC/DISC/COLUM V/sur

STRICTLY BUSINESS **
Kevin Hooks USA

15
1991

Tommy Davidson, Joseph C. Phillips, Halle Berry, Anne Marie Johnson, David Marshall Grant, Jon Cypher, Sam Jackson, Kim Coles, James McDaniel, Paul Provenza, Annie Golden, Kevin Hooks, Paul Butler, Sam Rockwell, Ira Wheeler

A black yuppie has turned his back on his cultural heritage and devoted himself to his estate agency business, but his priorities alter when he falls in love with a black woman who wants to break into show business. As she finds him too square for her tastes, he is forced to take a few quick lessons in street credibility from his more hip friend. A none-too-funny comedy that does little to develop its Pygmalion-inspired story.

Aka: GO NATALIE

COM   80 min (ort 83 min)   VIDrel: FIRST/SONOP   V

STRIKE ****
Sergei Eisenstein USSR

15
1924

Maxim Shtraukh, Grigori Alexandrov, Mikhail Gomarov, I. Ivanov, I. Kluvkin

The feature debut for Eisenstein, this remarkable film tells of how a strike on the part of factory workers in 1912, is brutally suppressed by the authorities. Much use is made of visual montage and caricature, with disturbing (if a little crude) slaughterhouse sequences used to highlight the brutal tactics employed by the police. A brilliant, harrowing and difficult film.

Aka: STACHKA; TOWARDS THE DICTATORSHIP OF THE PROLETARIAT

DRA   82 min B/W silent   VIDrel: TART/20TH   V

STRIKE A POSE *
Dean Hamilton USA

(18)
1994

Rober Eastwick, Michelle Lamothe, Margie Peterson, Calrk Katz, Michelle Garrin, Veronica Saners, Eric Paul, Shelby ane, Jeanine Antoine, Sindy Tenens, Thomas Prisco, Christina Gancevitch, Gregory Vlahakis, Aaron "Rocky" Hamilton

A female glamour photographer, an ex-model, goes to the desert for a photo shoot with her models. She looks forward to a reunion with her cop boyfriend but unknown to them both, a demented killer with a grudge is stalking the entire group and his potential victims have to work hard to stay alive. An unremarkable effort, seasoned with a little tepid eroticism, courtesy of the real-live models who feature in this film.

THR   87 min (ort 90 min)   SATrel: SKY MOVIES

STRIKE FORCE *
Barry Shear USA

PG
1975

Cliff Gorman, Donald Blakely, Richard Gere, Edward Grover, Joe Spinell, Marilyn Chris, Mimi Cecchini, Allan Rich, Billy Longo, Arnold Soboloff, Carl Don, Randy Jurgensen

Set in New York, this crime story has a Federal agent, a New York cop and a state trooper joining forces to smash a drugs ring. A sluggish and boring effort that is just one more failed TV pilot.

A/AD   74 min (ort 90 min) mTV
VIDrel: MIA/DISC/IMPENT   V
Boa: story by Sonny Grosso.

STRIKE ME PINK **
Norman Taurog USA

U
1935

Eddie Cantor, Ethel Merman, Sally Eilers, Parkyakarkus (Harry Einstein), William Frawley, Helen Lowell, Gordon Jones, Brian Donlevy, Jack LaRue, Rita Rio (Dona Drake), Sunnie O'Dea, Edward Brophy, Sidney H. Fields, Don Brodie

A meek tailor changes his personality after reading a self-confidence manual and takes to running an amusement park, but is unaware that the Mob control all the slot machines. They try to muscle in on his action but are eventually dealt with, in this brisk formula comedy, the last and weakest of the five films this gifted comedian made for Goldwyn.

COM   100 min B/W   VIDrel: VCC/DISC   V
Boa: novel Dreamland by Clarence Buddington Kelland.

STRIKE OF THUNDERKICK TIGER **
Henry Wong HONG KONG

18

Casanova Wong, Charles Han, Billy Yuen

A girl's father is murdered, and $1,000,000 is stolen from him and his criminal accomplices, who set out to regain it. However, only the dead man's daughter has a clear idea where to search, and so she sets out with her friends to recover the money before they do.

MAR   80 min Cut (33 sec)   VIDrel: IMPENT   V

STRIKING DISTANCE **
Rowdy Herrington USA

18
1993

Bruce Willis, Sarah Jessica Parker, Dennis Farina, Robert Pastorelli, Brion James, Tom Sizemore, Timothy Busfield, John Mahoney, Andre Braugher, Jodi Long, Tom Atkins, Mike Hodge, Roscoe Orman, Robert Gould, Gareth Williams, Ed Hooks

A tough Pittsburgh cop is demoted to the river patrol for claiming that his father (also a cop) was murdered by a colleague and a serial killer. When a number of women linked to his female partner fall victim to such a killer, he finds himself fighting the department in order to solve this mystery. An assembly-line offering whose action sequences help to plug the many holes in the plot.

A/AD   97 min (ort 102 min) wScrn cC
VIDrel: COLUM/SONOP   V/sur

STRIP SHOW **
Richard Styles USA

18
1995

Kimberley Kelly, Jack Turturici, Tane McClure, Ross Hagen, Julie Smith

Erotic thriller in which a woman goes undercover in a bid to discover the identity of the killer of her sister, who worked as an exotic dancer. This quest takes her on an odyssey through the seedy world of strip clubs, but the closer she gets the truth the more murders take place.

THR   88 min   VIDrel: HIFLI/SONOP   V/h

STRIPES **
Ivan Reitman USA

15
1981

Bill Murray, Harold Ramis, Warren Oates, P.J. Soles, Sean Young, John Candy, John Larroquette, John Voldstad, Lance LeGault, Roberta Leighton, Nick Toth, Judge Reinhold, Glenn-Michael Jones, Bill Lucking, Fran Ryan, Dave Thomas

This box office success is no more than a routine army comedy, about a loser who enlists and ends up in a platoon of volunteer misfits that a sergeant is doing his best to train. Written by Len Blum, Dan Goldberg and Harold Ramis and with music by Elmer Bernstein.

COM   101 min (ort 105 min)   VIDrel: VCC/DISC/COLUM V

STRIPTEASE **
Andrew Bergman USA

15
1996

Demi Moore, Burt Reynolds, Armand Assante, Ving Rhames, Robert Patrick, Paul Guilfoyle, Jerry Grayson, Rumer Willis, Robert Stanton, William Hill, Stuart Pankin, Dina Spybey, Pasean Wilson, Pandora Peaks, Barbara Alyn Woods

A married woman loses both her job and then her child as a

judge hands over the latter to her feckless, criminal husband. Badly in need of funds and hoping to have this decision challenged, she takes a job at a strip-bar and becomes embroiled with both the staff and the patrons. Moore's much-hyped nude scenes really do nothing to enhance a film in which the earnestness of the central character is so at odds with the satirical thrust and comic capers of the story.
COM   112 min (ort 117 min) cC   VIDrel: 20VIS/SONOP V/sur
Boa: novel Strip Tease by Carl Hiaasen.

STRONG MAN, THE *** U
Frank Capra USA   1926
Harry Langdon, Priscilla Bonner, Robert McKim, Gertrude Astor, William V. Mong, Robert McKim, Arthur Thalasso
At the end of WW1, a Belgian soldier comes to the USA in search of the blind girl he corresponded with, as he has fallen in love with her. When the circus show he is appearing in comes to a small town, he finally finds her, but must first win over her father, who is no friend of show-business types. Langdon's last film with Capra and one of his best features.
COM   75 min (ort 78 min) B/W silent
VIDrel: VISION/DISC   V

STRONGEST MAN IN THE WORLD, THE ** U
Vincent McEveety USA   1976
Kurt Russell, Joe Flynn, Eve Arden, Cesar Romero, Phil Silvers, Dick Van Patten, Harold Gould, James Gregory, William Schallert, Roy Roberts, Fritz Feld, Raymond Bailey, Eddie Quillan, Burt Mustin
A student attracts the attention of crooks when he invents a formula that gives him super-strength, in this predictable Disney romp. Written by Joseph L. McEveety and Herman Groves.
JUV   88 min (ort 92 min)   VIDrel: WDV/TECH L/A   V

STROSZEK *** 15
Werner Herzog WEST GERMANY   1976
Bruno S, Eva Mattes, Clemens Scheitz, Tom Paxton, Chet Atkins, Sonny Terry
Three Berlin misfits find life in the USA is not all roses in their pursuit of The American Dream, in a film that is clearly an outsider's view of the States, largely reflecting the director's own feelings towards the country.
DRA   103 min (ort 107 min)   VIDrel: TART/20TH   V

STRYKER * 18
Cirio H. Santiago PHILIPPINES/USA   1983
Steve Sandor, Andria Savio, William Ostrander, Michael Lane, Julie Gray, Monique St Pierre, Jon Harris III, Ken Metcalfe, Joe Zucchero, Michael De Mesa, Catherine Schroeder, Tony Carreon, Pete Cooper, Corey Casey
Survivors of the nuclear war fight desperately for the remaining supplies of uncontaminated water, in yet another variation on an all too familiar theme. Cut before video submission by 1 min 12 sec.
FAN   81 min (ort 86 min)   VIDrel: LUMI/SPEAR L/A   V

STUART SAVES HIS FAMILY * 12
Harold Ramis USA   1995
Al Franken, Laura San Giacomo, Vincent D'onofrio, Shirley Knight, Harris Yulin
When a cranky promoter of self-help on cable TV loses his one-man show he finds himself back at home with his family, but they are a dysfunctional bunch of weirdos who are even stranger than he is. This should have been a very funny film debut for former "Saturday Night Live" comedian Franken, who also wrote the script. Unfortunately, there is a severe shortage of good gags and the film is remarkably uninvolving.
COM   93 min (ort 97 min)   VIDrel: CIC   V

STUCK ON YOU * 18
Michael Herz/Samuel Weil USA   1982
Irwin Corey, Virginia Penta, Mark Mikulski, Albert Pia, Norma Pratt, Daniel Harris, Denise Silbert, Eddie Brill, June Martin, John Bigham, Robin Burroughs, Julie Newdow, Pat Tallman, Mr Kent, Barbie Kielian, Louis Homyak, Ben Kellman
A review of lovers and their problems through the ages, from Adam and Eve onwards.
COM   81 min (ort 90 min)   VIDrel: TROMA/RTM   V

STUD, THE * 18
Quentin Masters UK   1978
Joan Collins, Oliver Tobias, Sue Lloyd, Mark Burns, Walter Gottell,

Emma Jacobs, Tony Allyn, Doug Fisher, Peter Lukas, Natalie Ogle, Constantin De Goguel, Guy Ward, Sarah Lawson, Jeremy Child, Franco de Rosa, Minah Bird
The sexual cavortings of the super-rich and highly unpleasant are looked at in this successful soap opera, telling of a waiter who works his way up in the world by sleeping with his boss's wife, who has an insatiable sexual appetite. Pure dross this one may be, but it resurrected Collins' career. Written by Joan's sister Jackie Collins and followed by THE BITCH.
DRA   92 min   VIDrel: ENTUK L/A   V
Boa: novel by Jackie Collins.

STUDY IN TERROR, A *** 15
James Hill UK/WEST GERMANY   1965
John Neville, Donald Houston, John Fraser, Anthony Quayle, Robert Morley, Cecil Parker, Barbara Windsor, Georgia Brown, Barry Jones, Adrienne Corri, Frank Finlay, Judi Dench, Charles Regnier, Dudley Foster, Peter Carsten
Sherlock Holmes tracks down Jack the Ripper but is always one step behind, in this clever and entertaining film. Written by Donald and Derek Ford. See also MURDER BY DECREE.
Aka: FOG; SHERLOCK HOLMES' GROSSTER FALL
DRA   95 min   VIDrel: ARTPRO/RTM L/A   V
Boa: novel by Ellery Queen.

STUFF, THE ** 15
Larry Cohen USA   1985
Michael Moriarty, Paul Sorvino, Andrea Marcovicci, Garrett Morris, Scott Boom, Danny Aiello, Alexander Scourby, Russell Nype, James Dixon, Gene O'Neill, James Dukas, Peter Hock, Colette Blonigan, Frank Telfer
A strange white substance lands on Earth and is exploited commercially as a breakfast food. It does prove, however, to have rather unpleasant side effects. A somewhat chaotic and messy blend of comedy and horror, done in the style of 1950s SF films.
HOR   83 min (ort 93 min)   VIDrel: VCC/DISC/COLUM L/A   V

STUFF STEPHANIE IN THE INCINERATOR * 15
Lloyd Kaufman (Michael Herz/Samuel Weill) USA   1989
M.R. Murphy, Catherine Dee, William Dame, Dennis Cunningham, Paul Nielsen, Andy Milk, Don Nardo, Peter Jones, Herb Fuller
A wealthy couple attempt to banish their boredom by playing a succession of games where they take on a variety of roles and act out different situations. A dull dud of zero laughs.
COM   97 min   VIDrel: TROMA/RTM   V

STUNT MAN, THE *** 15
Richard Rush USA   1978
Peter O'Toole, Steve Railsback, Barbara Hershey, Chuck Bail, Allen Goorwitz, (Garfield), Adam Roarke, Alex Rocco, Sharon Farrell, Philip Bruns, Jim Hess, John Garwood, John B. Pearce, Michael Railsback, George D. Wallace, Dee Carroll
A Vietnam veteran on the run accidentally kills a stuntman, and has to take his place in return for not being handed over to the police by the director. An offbeat and rather witty comedy-drama that almost looks as if it has some deeper meaning. Written by Lawrence B. Marcus and with music by Dominic Frontiere.
DRA   129 min   VIDrel: MIA/DISC L/A   V
Boa: novel by Paul Brodeur.

STUPIDS, THE * PG
John Landis USA   1995
Tom Arnold, Jessica Lundy, Bug Hall, Alex McKenna, Mark Metcalf, Matt Keeslar, Christopher Lee, Scott Kraft, Victor Ertmanis, Earl Williams, George Chiang, Arthur Eng, Max Landis, Carol Ng, Jennifer Dean, Garry Robbins
A suburban American family are seriously weird, being complete idiots totally unfamiliar with almost every aspect of modern technology. When the father finds that his garbage has vanished from the bin, he senses a conspiracy and goes off on a wild goose chase. A complete misfire of a film, it mistakes stupidity for wit and wastes the talents of a strong cast. Not even Arnold's comic skills can salvage this unfunny mess. Based on a series of kid's books.
COM   89 min (ort 94 min)   VIDrel: POLFIL   V
Boa: book by Harry Allard and James Marshall.

SUBSPECIES * 18
Ted Nicolaou USA   1991
Michael Watson, Laura Tate, Anders Hove, Michelle McBride

This dull story opens with a 16th century vampire being imprisoned in a cage by his father. Having escaped and chastised dad in a suitable manner the film proper starts. We are now in modern times, and three dimwitted women come to Transylvania to study the local legends. They are soon victims of our vampire, but by good fortune he has a younger brother who has fallen in love with one of them. A tired dud with a few good effects. BLOODSTONE: SUBSPECIES 2 followed.
HOR   80 min   VIDrel: EIV/SONOP   V

SUBSTITUTE, THE *   15
Glenn Jordan USA   1986
Bruce Dern, Lee Remick, Piper Laurie, Jason Patric, Eric Schiff, Dee Dee Pfeiffer
A young man's life could not be more promising, but he decides to experiment with drugs, with predictable results. A typically sanitised Hollywood contribution to the anti-drugs lobby.
DRA   92 min   VIDrel: 20TH   V

SUBSTITUTE, THE **   15
Martin Donovan USA   1993
Amanda Donohoe, "Marky" Mark Wahlberg, James Dalton, Natasha Gregson Wagner, Eugene Glazer, Brigitte, David Frank, Christian Svensson, Cusse Mankuma, Molly Parker, Lossen Chambers, D. Neil Mark, Martin Martinuzzi, Justine Priestley
A woman substitute teacher has an appealing exterior but is in reality a cold and ruthless psychotic murderess, who having murdered both her husband and his mistress, changes her identity and moves away, and then gets a job at a new school. But once here, she embarks on a shortlived affair with a student, then dumps him in favour of a string of others. Pretty soon the bodies are starting to pile up in this overwrought and preposterous psycho-thriller.
THR   83 min (ort 90 min) mTV   VIDrel: CIC   V

SUBSTITUTE, THE *   18
Robert Mandel USA   1996
Tom Berenger, Ernie Hudson, Diane Venora, Glenn Plummer, Marc Anthony, Cliff De Young, Sharron Corley, Richard Brooks, Raymond Cruz, Rodney A. Grant, Luis Guzman, Willis Sparks, Peggy Pope, Maria Celedonio, Vincent Laresca, Ana Azcuy
A tough mercenary who has just got back from a drugs raid in Havana goes all out to get the thugs who beat up his teacher girl-friend, and as a means of achieving this end takes over her classes at the rough Miami high-school where she taught. However, in the course of these proceedings he learns that both the Principal and a corrupt former cop are mixed up in a nasty drugs network. A preposterous film, violent, moralising and totally unbelievable.
A/AD   90 min Cut (ort 114 min)   VIDrel: EIV/SONOP   V

SUBSTITUTE WIFE, THE ***   15
Peter Weller USA   1994
Peter Weller, Farrah Fawcett, Lea Thompson, Karis Bryant, Cory Lloyd, Colton Conklin, Annie Suite, Babs George, Gena Sleete, Gail Gronauer, Lou Perryman, Jonathan Joss, Blue Deckert, Wally Wlech, Guich Koock, John S. Davies
After a frontierswoman falls ill and learns that she is dying, her thoughts constantly turn to her husband and children and their fate after her demise. In order to help them, she devises a novel if rather unconventional solution. Fine performances enhance this touching period drama which thankfully is handled in a low-key and unsentimental way.
DRA   88 min (ort 89 min) mTV   VIDrel: SPEAR/SONOP V/s

SUBURBAN COMMANDO **   PG
Burt Kennedy USA   1991
Hulk Hogan, Christopher Lloyd, Shelley Duvall, William Ball, Laura Mooney, Michael Faustino, Larry Miller, Tony Longo, JoAnn Dearing, Roy Dotrice, Dennis Burkley, Mike Ballew, Dave Efron, Mark Calaway, Jack Elam, Luis Contreras
An intergalactic bounty hunter takes a spot of well-deserved leave on Earth, renting a room with a suburban family. Unfortunately, his holiday is interrupted by a series of mishaps, culminating in the arrival of an old enemy. In spite of Hogan's very pleasing personality, this vehicle-film (made with an eye to his enormous popularity as a W.W.F. superstar) is a tacky and rather childish spoof, despite competent direction by veteran Western director Kennedy.
COM   86 min (ort 90 min)   VIDrel: EIV/SONOP   V

SUBURBAN SEX SECRETS *   18
Paulo Ricotini UK   1991
Kerry Kristensen, Lunt Grunberger, Spiv Eldridge, Sabrina Morrell, Zoe Peters, Davina Williams, Kristy Kerr, Paul Van Duren, Stacey Owens
Three-part account of British sex life in suburbia, allegedly based on a factual survey. The different episodes are introduced and linked by an onscreen commentator and feature: a lesbian encounter between three bored housewives, a wife's adventures with the handyman and the tale of two girls and a willing pizza delivery boy. Turgid, unimaginative and very, very British.
A   56 min Cut (3 min 45 sec plus some cuts subst)
VIDrel: TOTAL   V

SUBWAY **   15
Luc Besson FRANCE   1985
Christopher Lambert, Isabelle Adjani, Richard Bohringer, Jean-Hughes Anglade, Jean Bouise, Michel Galabru, Jean Reno, Jean-Pierre Bacri, Eric Serra, Pierre-Ange Le Pogam, Arthur Simms, Constantin Alexandrov
A stylish thriller concerning the strange drop-outs who live in the Paris subway system, and the young thief who takes refuge there after robbing a wealthy couple. A well-made triumph of style over content. Written by Besson and others.
THR   98 min (ort 104 min) wScrn   VIDrel: ARTIF/20TH; ENCORE (LV only)   V/sh LV

SUCCESS **   15
William Richert USA   1979
Jeff Bridges, Bianca Jagger, Ned Beatty, Belinda Bauer, Steven Keats, John Glover
A dissatisfied businessman decides to drop out from success and all its trappings, and acts out a fantasy by becoming a street-wise tough guy. Richert twice re-edited and re-issued this film – in 1981 as "An American Success" and in 1983 as SUCCESS, but for all that it remains decidedly uneven and loose, though there are funny moments.
Aka: AMERICAN SUCCESS; AMERICAN SUCCESS COMPANY, THE; RINGER, THE
COM   80 min (ort 88 min)   VIDrel: MIA L/A   V
Boa: story by Larry Cohen.

SUCCESS IS THE BEST REVENGE   15
Jerzy Skolimowski FRANCE/UK   1984
Michael York, Joanna Szczerbic, Michael Lyndon, Jerry Skol (George Skolimowski), Michel Piccoli, John Hurt, Anouk Aimee, Ric Young, Claude Le Sache, Malcolm Sinclair, Hilary Drake, Jane Asher, Adam French, Mark Sarne
Explores the conflicts between first and second generation Polish immigrants in the UK. By inference, an indirect comment on the state of Poland today. Written by Jerzy Skolimowski and Michael Lyndon and something of a surreal follow-up to MOONLIGHTING.
COM   90 min (ort 95 min)   VIDrel: ARROW/RTM   V

SUCCUBUS ***   18
Jess (Jesus) Franco WEST GERMANY   1967
Howard Vernon, Jack Taylor, Adrian Hoven, Janine Reynaud, Nathalie Nord, Michel Lemoine, Pier A. Caminnecci, Rosanna Yanni, Chris Howland, Amerigo Coimbra
A female nightclub performer whose act has a strong sado-masochistic content falls under the control of a mysterious stranger and is compelled to commit real murders. A well made and photographed horror yarn with a genuine sense of atmosphere that blurs the distinction between fantasy and reality.
Aka: GETRAUMTE SUNDEN; NECRONOMICON; NECRONOMICON: DREAMED SINS; NECRONOMICON: GETRAUMTE SUNDEN
HOR   86 min (ort 90 min) wScrn dubbed
VIDrel: RTM/PINN L/A   V

SUDDEN DEATH **   18
Peter Hyams USA   1995
Jean-Claude Van Damme, Powers Boothe, Raymond J. Barry, Whittni Wright, Ross Malinger, Dorian Harewood, Kate McNeil, Michael Gaston, Audra Lindley, Brian Delate, Steve Aronson, Michael Aubele, Karen Baldwin, Jennifer D. Bowser
As a bomb ticks away at a hockey stadium, one man struggles to free his daughter, one of a number of people threatened by a ruthless extortionist. Van Damme with the bad guys with his usual brutal efficiency in this formula action-thriller, that is not notable for any originality of ideas.
A/AD   104 min (ort 111 min) cC   VIDrel: CIC/SONOP; PION (LV only)   V LV
Boa: story by Karen Baldwin.

SUDDEN FURY **
Craig R. Baxley USA
15
1993
Neil Patrick Harris, Johnny Galecki, Linda Kelsey, John M. Jackson, Lisa Banes, James Handy, Tim Kelleher, Timothy Bass, Eric Lloyd, Gregory Harrison, Larry Black, Rhoda Griffis, Catherine Larson, Rosemary Newcott, Chris McKenzie
When the bodies of a couple are found brutally murdered in their backyard, all the evidence points to the culprit being one of their three adopted sons. A tense drama based on a true story.
DRA   96 min mTV   VIDrel: MARQ/QUANT   V

SUDDEN IMPACT **
Clint Eastwood USA
18
1983
Clint Eastwood, Sondra Locke, Pat Hingle, Bradford Dillman, Paul Drake, Jack Thibeau, Audrie J. Neenan, Michael Currie, Albert Popwell, Mark Keyloun, Kevyn Major Howard, Bette Ford, Nancy Parsons, Joe Bellan, Wendell Wellman, Nancy Fish
Our tough cop from DIRTY HARRY aids and abets a rape victim in her crusade of vengeance, and the couple leave a trail of corpses behind them. Music is by Lalo Schifrin. This was the third sequel, with THE DEAD POOL following.
A/AD   112 min (ort 117 min)   VIDrel: WHV   V/s

SUDDENLY ***
Lewis Allen USA
PG
1954
Frank Sinatra, Sterling Hayden, James Gleason, Nancy Gates, Kim Charney, Christopher Dark, Paul Frees, Willis Bouchey, Charles Smith
Effective suspenser dealing with a trio of hitmen (headed by Sinatra) who hold a family hostage as part of their preparations for an attempt on the life of the US President when he passes through a small town in the Midwest. The film has uncomfortable echoes of the Kennedy assassination, and was out of circulation for many years, but has emerged as a gripping if dated tale. The script is by Richard Sale, and Sinatra's performance is truly outstanding.
THR   72 min (ort 77 min) B/W   VIDrel: VISCOM/RTM   V

SUDDENLY, LAST SUMMER **
Joseph L. Mankiewicz UK
15
1959
Elizabeth Taylor, Katherine Hepburn, Montgomerey Cliff, Mercedes McCambridge, Albert Dekker, Gary Raymond, Mavis Villiers, Patricia Marmont, Joan Young, Maria Britneva, Sheila Robbins, David Cameron, Roberta Woolley
A psychiatrist probes the reasons for a young girl's breakdown and discovers that her aunt has reasons of her own for wanting her to be lobotomised. A shocking, powerful but very depressing adaptation of the original play that suffers from an excessively verbose script and indifferent performances from some of the cast.
DRA   110 min (ort 114 min) B/W   VIDrel: COLUM/SONOP V
Boa: play by Tennessee Williams.

SUDDENLY, LAST SUMMER ***
Joseph L. Mankiewicz UK
15
1960
Katharine Hepburn, Elizabeth Taylor, Montgomery Clift, Albert Dekker, Gary Raymond, Mercedes McCambridge, Mavis Villiers, Patricia Marmont, Joan Young
An adaptation of a one-act play that tells of a young woman, who went mad when she saw her homosexual cousin raped and murdered by beach boys, whose attentions he had courted. However, this is only revealed in the climax, with the bulk of the film consisting of a talky examination of her plight. A flawed but engrossing tale, scripted by Gore Vidal.
DRA   109 min (ort 114 min) B/W   VIDrel: RCA L/A   V
Boa: play by Tennessee Williams.

SUGAR HILL **
Leon Ichaso USA
18
1993
Wesley Snipes, Michael Wright, Clarence Williams III, Theresa Randle, Ernie Hudson, Joe Vigoda, Leslie Uggams, Khandi Alexander, Raymond Serra, Bryan Clark, Joe Dallesandro, Vondie Curtis-Hall, John Pittman, Steve J. Harris, Andre Lamal
Two brothers eventually become powerful drug dealers in title district but one of them decides that he wants out, after meeting an actress with whom he wants to become involved. A standard crime drama that is quite well made but hardly original and as such has nothing new to say.
DRA   118 min (ort 123 min)   VIDrel: EIV/SONOP   V/sur

SUGARLAND EXPRESS, THE ***
Steven Spielberg USA
PG
1974
Goldie Hawn, Ben Johnson, Michael Sacks, William Atherton, Gregory Walcott, Steve Kanaly, Louise Latham, Harrison Zanuck, A.L. Camp, Jessie Lee Fulton, Dean Smith, Ted Grossman, Bill Thurman, Kenneth Hudgins, Jim Harrell
Though he has only a short time left to serve in jail, a young man is persuaded by his girlfriend to escape, to help prevent their baby being forcibly adopted. On the way they abduct a cop as hostage and with the police in pursuit, try to rescue the child. Based on a true story, it was scripted by Hal Barwood and Matthew Robbins. A blend of drama and occasional comic elements, with a very downbeat ending. This was Spielberg's first theatrical venture.
DRA   109 min   VIDrel: CIC/SONOP L/A   V

SUGARTIME **
John N. Smith USA
18
1995
John Turturro, Mary-Louise Parker, Maury Chaykin, Elias Koteas
A fact-based drama that recounts the relationship between notorious Mafia boss Sam Giancoma and the female singer who caused his downfall, but is so superficial and sentimentalised that we learn next to nothing about the characters and any interest in their activities quickly evaporates. A cheaply made time-filler.
DRA   106 min (ort 110 min) mCab   VIDrel: MOSAIC/20VIS V/sur

SUICIDE CLUB, THE *
James Bruce USA
18
1988
Mariel Hemingway, Robert Joy, Lenny Henry, Madeleine Potter, Michael O'Donoghue, Anne Carlisle, Sullivan Brown, Leta McCarty
A bored and somewhat alienated young heiress becomes involved with a decadent group of wealthy dilettantes who delight in rituals and bizarre games. Hemingway looks very appealing, which is more than can be said for this incomprehensible mess.
Aka: WELCOME TO THE SUICIDE CLUB
DRA   89 min   VIDrel: VISION/DISC   V

SUITE 16 *
Dominique Deruddere BELGIUM/UK
18
1994
Peter Postlethwaite, Geraldine Pailhas, Antonie Kamerling, Tom Jansen, Bart Slegers, Suzanne Colin, Viviane de Muynck, Dirk Roofthooft, Corinne Rivierre, Henri Masini, Stephane Leveque, Valerie van Nitsen, Jean-Paul Ferrari
Trapped in his hotel, a wealthy but disabled man hatches a plan to indulge his sexual fantasies with women, when a comely youth stumbles by accident into his room. A shoddy attempt to make an erotic drama, this dreary film wastes its energy in various pointless conflicts and encounters, and without the strength of a clear narrative drive becomes very dull indeed.
DRA   106 min   VIDrel: POLFIL   V/sh

SUM OF US, THE ***
Kevin Dowling/Geoff Burton AUSTRALIA
15
1994
Jack Thompson, Russell Crowe, John Polson, Deborah Kennedy, Rebekah Elmaloglou, Joss Moroney
The story of the relationship between a beer-drinking, earthy widower and his gay son forms the basis for this surprisingly appealing work.
DRA   100 min   VIDrel: DTK/RTM   V
Boa: play by David Stevens.

SUMMER CITY **
Christopher Fraser AUSTRALIA
15
1977
John Jarrat, Phil Avalon, Steve Bisley, Mel Gibson, James Elliot, Debbie Forman, Abigail, Ward "Pally" Austin, Judith Woodroofe, Carl Rorke, Ross Bailey, Hank Tyck, Bruce Cole, Vicki Hekimian, Karen Williams, Peter McGovern
An Australian version of AMERICAN GRAFFITI with four young men setting out for a weekend at the beach and the usual innocent fun and frolics. Set in the 1950s, this mindless juvenile romp gave Gibson his first starring role.
Aka: COAST OF TERROR; REIGN OF FEAR
DRA   82 min (ort 90 min)   VIDrel: MIA/DISC   V

SUMMER DREAMS: THE STORY OF THE BEACH BOYS **
Michael Switzer USA
15
1989
Bruce Greenwood, Greg Kean, Bo Foxworth, Arlen Dean Snyder, Casey Sander, Andrew Myler
An unauthorised biopic on one of America's most successful bands, who promoted a lightweight and happy style of music that became inextricably bound up with sun, sand and surfing in California. Mostly the film concentrates on drummer Dennis

and the group's obsessive father/manager, all but ignoring both Carl and Brian (who was the creative force behind the band). Very disappointing. Features many of their most popular songs, but these are not performed by them.
MUS 90 min (ort 92 min) mTV VIDrel: POLY L/A V

SUMMER FEVER ** **   15
William Webb USA   1987
Leif Garrett, Martin Landau, Denver Pyle, Wendy Jo Sperber, Katherine Kelly Lange, Tom Eplin, Will Bledsoe
The son of the owner of a boat house gives little thought to his future and is content to enjoy life as a water ski instructor. With the arrival of his attractive cousin his attitudes change, and the death of his father gives him the opportunity to turn the boat house into a successful ski resort, with the help of a friend. However, to do this he has to win a water skiing contest. A cheerful, inane and pleasantly forgettable summer movie.
DRA 90 min VIDrel: NWV L/A V/h

SUMMER HEAT *   15**
Michie Gleason USA   1987
Lori Singer, Anthony Edwards, Bruce Abbott, Kathy Bates, Clu Gulager
A woman who is neglected by her farmer husband falls for the hired hand in this dreary effort.
DRA 77 min (ort 93 min) VIDrel: EIV/SONOP V

SUMMER INTERLUDE *   PG**
Ingmar Bergman SWEDEN   1950
Maj-Britt Nilsson, Birger Malmsten, Alf Kjellin, Stig Olin, Georg Funkquist, Mimi Pollack, Annalisa Ericsson, Gunnar Olsson, John Botvid, Douglas Hage, Julia Caesar, Carl Strom
A young ballerina at the height of her career recalls a summer when she met and fell in love with a young boy, and how she was completely shattered by his death in an accident. Her loss has left her numb and bitter but she eventually meets a young journalist who re-kindles her passion. A memorable and nicely balanced work, with some excellent scenes of the Swedish summer.
Aka: ILLICIT INTERLUDE; SOMMARLEK; SUMMERPLAY
DRA 91 min (ort 97 min) B/W VIDrel: TART/20TH V

SUMMER JOB *   15**
Paul Madden USA   1988
Sherrie Rose, James Summer, Amy Baxter, Can Mayor, Renee Shugart, Fred Bourdin, Chantal, Dave Clouse, Kirt Earhardt, George O.
Another predictable and carefree comedy about a group of indolent and lustful teenage students who take summer jobs at a holiday resort, but set out to do as little work as possible. A puerile piece of nonsense that offers minimal entertainment.
COM 88 min (ort 92 min) VIDrel: MED/POLYREC L/A V

SUMMER OF '42 *   15**
Robert Mulligan USA   1971
Jennifer O'Neill, Gary Grimes, Jerry Houser, Oliver Conant, Katherine Allentuck, Christopher Norris, Lou Frizell, Walter Scott, Maureen Stapleton (voice only), Robert Mulligan (narration only)
Touchingly nostalgic look at a young boy's relationship with the young wife of a soldier. Skilfully put together if rather soft-focus and followed by CLASS OF '44. Written by Raucher. AA: Score/orig (Michel Legrand).
DRA 99 min (ort 103 min) VIDrel: WHV V
Boa: novel by Herman Raucher.

SUMMER OF THE FALCON, THE *   (U)**
Arendt Agathe WEST GERMANY   1988
Andrea Losch, Janos Crecelius, Hermann Lause, Rolf Zacher, Volker Brandt, Heide Joshko, Babara Stanek, Johannes Thannheuser, Peter Zimmermann, Kurt Lanthaler, Hans-Werner Olm, Rudolf Hisel, Heinz Kammer, Heinz Ostermann
A couple of youngsters attempt to thwart the activities of a ruthless collector who is stealing the eggs from the nests of falcons. An enjoyable children's outdoors adventure.
JUV 90 min SATrel: MOVIE CHANNEL

SUMMER SCHOOL *   15**
Carl Reiner USA   1987
Mark Harmon, Kirstie Alley, Dean Cameron, Gary Riley, Shawnee Smith, Patrick Labyorteaux, Ken Olandt, Courtney Thorne Smith, Robin Thomas, Kelly Minter, Nels Van Patton, Lucy Lee Flippin, Frank McCarthy, Richard Steven Horvitz, Amy Stock

A high school teacher has his work cut out when he has to cancel his vacation, to take over a remedial class teaching assorted misfits and no-hopers. An amiable, mindless comedy, quite enjoyable if one isn't too demanding.
COM 93 min Cut (48 sec – ort 98 min)
VIDrel: CIC/SONOP L/A V/h

SUMMER VACATION *   18**
Bob Logan USA   1992
Corey Feldman, Jack Nance, Sarah Douglas
The holiday resort of Lakeside faces a bleak future owing to the machinations of the ruthless owner of Twin Oaks, a rival resort on the opposite side of the lake. In desperation, they hire a hot-shot water-ski instructor to lead them to victory over Twin Oaks in the annual water-ski competition. The usual blend of high school japes, romantic sub-plots and action, which in this case is some spectacular water-ski stunts.
COM 86 min VIDrel: FIRST/SONOP L/A V

SUMMER WITH MONIKA *   PG**
Ingmar Bergman SWEDEN   1952
Harriet Andersson, Lars Ekborg, John Harryson, Georg Skarstedt, Dagmar Ebbesen, Ake Gronberg, Ake Fridell, Naemi Briese
A young couple's brief and defiant affair results in pregnancy and marriage, and when the girl finds married life not to her liking she leaves her husband and child. A standard Nordic gloomy look at life, made with both care and sensitivity by a master director.
Aka: MONIKA; SOMMAREN MED MONIKA
DRA 91 min (ort 90 min) B/W VIDrel: TART/20TH V
Boa: novel by Per Anders.

SUMMER'S TALE, A *   U**
Eric Rohmer FRANCE   1996
Melvil Poupaud, Amanda Langlet, Gwenaelle Simon, Aurelia Nolin, Aime Lefevre, Alain Guellaff, Evelyne Lahana, Yves Guerin, Franck Cabot
As ever, screenplay is by Rohmer in this story of a young man who goes to a seaside restaurant to meet a girl with whom he believes himself to be in love. However, he spends most of his time discussing the pros and cons of his relationship with another girl. Despite the gossamer-thin plot and ponderous pacing, this is an insightful and carefully observed look at love, and was the third film in the director's series based on the four seasons.
Aka: CONTE D'ETE
DRA 113 min wScrn VIDrel: ARTIF/20TH V

SUN BUNNIES *   18**
Duck Dumont USA   1992
Alicyn Sterling, Angela Summers, Candice Hart, Jamie Leigh, Tom Byron, Peter North, T.T. Boy, Woody Long, Ed Powers
Two female volleyball teams are competing for the world championship, but we never get to see them play, as almost the entire film is taken up with an endless series of sexual frolics. A sequel followed.
A 38 min (ort 60 min) VIDrel: GROHOM/MAXSCAN V

SUN BUNNIES 2 *   18**
C.B. DeMille USA   1992
P.J. Sparxx, Heather Hart, Gail Force, Crystal Wilder, Jerry Butler, Ron Jeremy, Tery Rocks
This film continues where the first one left off, despite the fact that the cast are different. Much of the story seems to consist of the various ploys by which the opposing teams attempt to gain the upper hand, including attempts to ensure victory by seducing the referee. Against all expectations, the movie does indeed end with a volleyball match, not that this adds much to the plot.
A 60 min (ort 90 min) VIDrel: FALCON/TOTAL V

SUN DRAGON *   18**
Hwa I Hung HONG KONG   1981
Billy Chong, Louis Neglia, Carl R. Scott, Hau Chin Sing, Joseph Jennings, Ma Shung Tak, Liang Siao Sung, Kim Bill
Three hoodlums kill a coloured rancher when he refuses to sell his property to them, but the man's young son escapes and is befriended by a Chinese man newly arrived in America, who takes him to see his uncle, a martial arts expert. Naturally this leads to several further battles, with the youngster eventually gaining sufficient proficiency to avenge the death of his father. A very ordinary martial arts film that's given an unnecessary Western flavour.
Aka: HARD WAY TO DIE, A
MAR 86 min (ort 92 min) VIDrel: EAST/DISC V

SUN VALLEY SERENADE **
H. Bruce Humberstone USA
Sonja Henie, John Payne, Glenn Miller, Milton Berle, Lynn Bari, Joan Davis, Dorothy Dandridge, Nicholas Brothers, William Davidson, Dorothy Dandridge, Mel Ruick, Almira Sessions, Forbes Murray, Ralph Dunn, Chester Clute
A Norwegian refuge girl, sponsored by the pianist of a band appearing at the title ski resort in Idaho, proves to be a champion ice skater and eventually triumphs in an ice show. A lightweight but enjoyable vehicle for both Miller and Henie, with the latter showing her aptitude for comedy. The score by Harry Warren and Mack Gordon includes songs such as "I Know Why And So Do You" and "It Happened in Sun Valley".
MUS   83 min (ort 86 min) B/W cC   VIDrel: 20TH/TECH V/sh

U
1941

SUNCHASER, THE **
Michael Cimino USA
Woody Harrelson, Jon Seda, Anne Bancroft, Alexandra Tydings, Matt Mulhern, Talisa Soto, Richard Bauer, Victor Aaron, Lawrence Pressman, Michael O'Neill, Harry Carey Jr, Carmen Dell'Orefice, Brooke Ashley, Andrea Roth, Bob Minor
A cancer specialist on the fast track to career success is kidnapped by a wild sixteen-year-old Navajo delinquent. The latter is suffering from a rare tumour and forces the doctor to drive him out to meet a legendary medicine man who he believes will effect a cure by having him bathe in a sacred lake. Predictably, the two characters find their initial hostility turning to friendship as they elude potential captors. A trite and superficial story.
DRA   122 min   CINrel

15
1996

SUNDAY, BLOODY SUNDAY ***
John Schlesinger UK
Peter Finch, Glenda Jackson, Murray Head, Peggy Ashcroft, Maurice Denham, Vivian Pickles, Frank Windsor, Thomas Baptiste, Tony Britton, Harold Goldblatt, Hannah Norbert, Richard Pearson, Caroline Blakiston, Bessie Love
A bi-sexual artist conducts simultaneous affairs with a lover of each sex. An intelligent and sensitive exploration of human relations with Finch cast as a kindly Jewish doctor whose homosexuality is portrayed in a realistic way.
DRA   105 min (ort 110 min) wScrn   VIDrel: MGM/WHV   V

15
1971

SUNDAY IN THE COUNTRY ***
Betrand Tavernier FRANCE
Louis Ducreux, Sabine Azema, Michel Aumont, Genevieve Mnich, Monique Chaumette, Claude Winter, Thomas Duval
A touching account of an elderly painter waiting for his weekly Sunday visit from his family and musing on the events of his life. Though both well filmed and acted, it suffers from a lack of strong characterisation and a excessively leisurely pace. A stylish, mannered and superficial film which for all its failings, was winner of the Best Director award at Cannes 1984.
Aka: UN DIMANCHE A LA CAMPAGNE
DRA   90 min (ort 94 min)   VIDrel: ARTIF/20TH   V/h

PG
1984

SUNDOWN: THE VAMPIRE IN RETREAT ***
Anthony Hickox USA
David Carradine, Maxwell Caulfield, Bruce Campbell, Morgan Brittany, Jim Metzler, Deborah Foreman, M. Emmet Walsh, John Ireland, Dana Ashbrook, John Hancock, Marion Eaton, Dabbs Greer, Bert Remsen, Sunshine Parker
At long last a vampire spoof with an original idea, that sees the entire population of the isolated desert town of Purgatory a community of vampires, who have devised both a synthetic blood substitute and a means of surviving in direct sunlight. When a group of traditionalist vampires demand a return to the old ways the inevitable conflict results. A well-paced film that unfortunately doesn't make the most of its amusing premise.
Aka: SUNDOWN
HOR   99 min   VIDrel: FIRST/SONOP L/A   V

15
1988

SUNFLOWER **
Vittorio De Sica FRANCE/ITALY
Sophia Loren, Marcello Mastroianni, Ludmilla Savelyeva, Anna Carena, Galina Andreyeva, Germano Longo, Nadya Serednichenko, Glauco Onorato, Marisa Traversi, Silvano Tranquili, Gunar Zilinski, Carlo Ponti Jr, Giorgio Basso
A woman searches for her lost husband who failed to return from Stalingrad. After much effort she traces him to Moscow,

15
1970

but is dismayed to find that he has since remarried. A foolish and plodding melodrama.
Aka: LE FLEURS DU SOLEIL; I GIRASOLI
DRA   103 min   VIDrel: L/A   V

SUNNYBOY AND SUGARBABY *
Franz Josef Gottlieb WEST GERMANY
Sabine Wollin, Gina Janssen, Bernie Paul, Claus Obalski, Ekkehardt Belle, Orestes Ojeda, Ike Lozada, Frieda Von Giese, Dschinghis Khan
A girl living in a remote Alpine village with her two boyfriends inherits her uncle's taxi and fast-food businesses in the Far East. Our trio immediately fly off to take control of these enterprises, but find them to be worthless and are forced to find work to pay for their passage home. A trite romantic-comedy of pleasant locations and occasional humour.
Aka: SUNNYBOY UND SUGARBABY
COM   86 min   VIDrel: IMPENT L/A   V

18
1979

SUNRISE ****
F.W. Murnau USA
George O'Brien, Janet Gaynor, Bodil Rosing, Margaret Livingston, J. Farrell MacDonald, Ralph Sipperly, Jane Winton, Arthur Housman, Eddie Boland, Sally Ellers, Gino Corrado, Barry Norton, Robert Kortman, Sidney Bracey
A simple man from the country is seduced by a sophisticated but evil big-city siren who persuades him to murder his wife so that they can be together. He almost does so but relents just in time and gets a second chance to prove his devotion to his true love, in a extended and enchanting fantasy sequence. Murnau was a given a free hand to direct as he wanted, which resulted in this brilliant film. AA: Actress (Gaynor), Cin (Charles Rosher/Karl Struss).
Aka: SONG OF TWO HUMANS
DRA   91 min (ort 108 min) B/W silent   VIDrel: TART/20TH L/A   V
Boa: novel Die Reise nach Tilsit by Hermann Sudermann.

U
1927

SUNSET **
Blake Edwards USA
Bruce Willis, James Garner, Malcolm McDowall, Mariel Hemingway, Jennifer Edwards, Kathleen Quinlan, Patricia Hodge, Richard Bradford, M. Emmet Walsh, Dermot Mulroney, Joe Dallesandro
Lighthearted comedy-thriller with 1920s actor Tom Mix cast to play Wild West legend Wyatt Earp, this latter having come along as an advisor. Besides making a movie, they wind up solving a murder that shocks all of Hollywood. Often irritating, generally disjointed, this disorganised effort is largely saved by a good performance from Garner.
COM   102 min (ort 107 min)   VIDrel: COLUM/SONOP V/sur

15
1988

SUNSET BEAT **
Sam Weisman USA
George Clooney, Michael De Luise, Markus Flanagan, Erik King, James Tolkan, Anthony Geary, Ami Dolenz, Marshall Teague, Sydney Walsh, Gary Frank, David Pymer, Richard Roat, David Raynr, Clabe Hartley, Jack McGee, Clint Howard
A teen-oriented actioner that has a bunch of young crimefighters doing battle with a psychotic extortionist and still finding time to deal with various family problems. A simplistic and synthetic offering, it has much in common with TWENTY-ONE JUMP STREET and also the TV series that earlier movie led to.
A/AD   95 min (ort 96 min) mTV   VIDrel: 20VIS/SONOP L/A   V

15
1990

SUNSET BOULEVARD ****
Billy Wilder USA
William Holden, Gloria Swanson, Eric Von Stroheim, Fred Clark, Nancy Olson, Jack Webb, Hedda Hopper, Anna Q. Nilsson, H.B. Warner, Sidney Skolsky, Bernice Mosk, Ray Evans, Jay Livingston, Franklyn Farnum, Larry Blake, Charles Dayton
Brilliant classic film looking at the relationship between a struggling young writer and an ageing faded movie star. Full of powerful psychological insights helped along by a great script and fine acting; Swanson was never better. Hedda Hopper, Buster Keaton and Cecil B. De Mille appear briefly. AA: Art/Set (H. Drier and J. Meehan/S. Comer and R. Moyer), Score (Franz Waxman), Story/Screen (Wilder/Charles Brackett/D.M. Marshman Jr).
DRA   105 min   B/W   VIDrel: CIC/SONOP V/dm

PG
1950

SUNSET GRILL ** 18
Kevin Connor USA 1992
Peter Weller, Lori Singer, Stacy Keach, John Rhys-Davies, Alexandra Paul
After both his wife and brother are murdered, a cop discovers that similar killings have taken place and that they are all linked to a sinister organisation that is trading in human parts. As his investigation proceeds, he meets up with a singer and is led on a trail that takes him to Mexico. The plot in this one is about as hard to understand as Weller's mumbling delivery, and it never becomes clear whether this is a comedy or a straightforward thriller.
THR 99 min (ort 105 min) VIDrel: EIV/SONOP V/sur

SUNSHINE BOYS, THE *** PG
Herbert Ross USA 1975
Walter Matthau, George Burns, Richard Benjamin, Lee Meredith, Carol Arthur, Howard Hesseman, Ron Rifkin, Fritz Feld, Jack Bernardi, F. Murray Abraham, Rosetta LeNoire, Jim Cranna, Jennifer Lee, Garn Stephens, Santos Morales
A witty look at what happens when two old and cantankerous comedians, once a successful stage duo, are brought together for a TV special. Burns stepped in when Jack Benny fell ill and died, and this film was largely responsible for relaunching his career. Remade in 1996. AA: S. Actor (Burns).
COM 106 min (ort 111 min) VIDrel: MGM L/A V
Boa: play by Neil Simon.

SUNSHINE BOYS, THE ** U
John Erman USA 1996
Woody Allen, Peter Falk, Sarah Jessica Parker, Michael McKean, Tyler Noyes, Olga Merendiz, Andy Taylor, Kirk Acevedo, William Hill, Herbert Rubens, David Lipman, Roy Anthony Thomas, Peter Appel, Jennifer Esposito, Michael Badalucco
An ageing pair of comics, who once worked as a comedy duo, are re-united for one last outing. As unnecessary a remake as one could imagine, this adaptation of one of Simon's most successful plays tries hard to please, but Allen and Falk are no replacement for Burns and Matthau. However, difficult as it is to forget about the original, if one can do so the film is often very funny, chiefly thanks to Neil Simon's updating of his original script.
Aka: NEIL SIMON'S THE SUNSHINE BOYS
COM 85 min VIDrel: 20TH/FOXVID V/sh
Boa: play by Neil Simon.

SUNSTROKE * 15
James Keach USA 1992
Jane Seymour, Stephen Meadows, Don Ameche, Steve Railsback, Ray Wise, Mark Davenport, Kristina Betts, Bobby Joe McFadden, Suzanne Alyn, Ann Risley, George Salazar, Heather McNair, Matt Gavin, Sam Hernandez, Mick Young
A woman attempting to rescue her kidnapped child, gives a ride to a male hitcher who is found murdered a few days later. After she learns that she is suspected of this crime, she confides her secret to an architect she meets, but its seems that he too may have reasons of his own for not telling her the whole truth about himself. A rather dire TV thriller, just about watchable, but certainly not entertaining.
THR 86 min (ort 90 min) mTV VIDrel: CIC V

SUPER, THE ** 15
Rod Daniel USA 1991
Joe Pesci, Vincent Gardenia, Ruben Blades, Madolyn Smith Osborn, Stacey Travis, Carole Shelley, Kenny Blank, Paul Benjamin, Beatrice Winde, Carol Jean Lewis, Anthony Heald, Daniel Saltzman, Jack Hallett, Steven Rodriguez
The life of a slum landlord who has ignored every housing regulation, is turned upside down when a court order forces him to take up residence in one of his own apartment blocks and implement the necessary repairs or go to jail. A well acted but wholly superficial affair, over-sentimental to the point of mawkishness (our disreputable property magnate has a change of heart and carries out the required repairs at the last moment). Enjoyable if unremarkable.
COM 86 min VIDrel: 20TH/TECH V/sur

SUPER MARIO BROTHERS * PG
Rocky Morton/Annabel Jankel USA 1992
Bob Hoskins, John Leguizamo, Dennis Hopper, Samantha Mathis, Fiona Shaw, Fisher Stevens, Richard Edison, Dana Kaminski, Mojo Nixon, Lance Henriksen, Gianni Rossi, Francesca Roberts, Sylvia Harman, Desiree Marie Velez, Don Lake
A couple of Brooklyn plumbers find themselves in a world populated by evolved dinosaurs to which a kidnapped princess has been taken. There they have many adventures and narrow escapes before triumphing over a villainous master criminal. A spectacular, expensive, garish and highly noisy screen spin-off from the famous Nintendo computer game aimed squarely at teenagers. It will only have limited appeal for adults.
Aka: SUPER MARIO BROS
FAN 99 min (ort 104 min) VIDrel: EIV/SONOP V

SUPER POWER ** 18
HONG KONG
Billy Chong, Hau Chiu Sing, Chiang Tao, Lou An Li
The son of an Imperial minister plans his revenge against the kung fu fighters who defeated his Manchu boxers. Only one powerful young martial artist can stop him. Adequate adventure set at the time of the Ching Dynasty.
MAR 89 min VIDrel: EAST/DISC V

SUPERDRAGON ** 18
Lin Chan Wai HONG KONG
Billy Chong, Hau Chiu Sing, Liu An Li, Chiang Tao
Kang, the son of an Imperial minister, plots his revenge against the Chinese fighters who defeated his Manchu Boxers, and takes lessons from his rivals in order to perfect his fighting skills.
MAR 86 min Cut (3 sec) VIDrel: IMPENT V

SUPERFLY ** 18
Gordon Parks Jr USA 1972
Ron O'Neal, Carl Lee, Sheila Frazier, Julius W. Harris, Charles McGregor, Nate Adams, Polly Niles, Yvonne Delaine, Henry Shapiro, K.C., Sig Shore, Jim Richardson, The Curtis Mayfield Experience
Flashy black consciousness film all about a heroin dealer trying to retire after one last big deal. All the white cops are wicked racists and the blacks are invariably portrayed sympathetically, whether drug dealing or not. For all that, this is a very slick and well acted film with some good moments and a fine score by Curtis Mayfield. SUPERFLY T.N.T. followed.
A/AD 87 min (ort 98 min) VIDrel: WHV V/h
Boa: novel by Philip Penty.

SUPERFLY T.N.T. * 18
Ron O'Neal USA 1973
Ron O'Neal, Roscoe Lee Brown, Sheila Frazier, Jacques Sernas, William Berger, Roy Bosier, Silvio Nardo, Olga Bisera, Dominic Barto, Minister Dem, Jeannie McNeill, Dan Davis, Luigi Orso, Ennio Catalfamo, Francesco Rachini
A sequel to SUPERFLY with our ex-pusher getting involved in an African revolution. A messy and generally worthless yarn, followed (after some considerable break) by THE RETURN OF SUPERFLY.
A/AD 83 min (ort 87 min) VIDrel: WHV L/A V

SUPERGIRL * PG
Jeannot Szwarc USA 1984
Helen Slater, Faye Dunaway, Peter O'Toole, Mia Farrow, Brenda Vaccaro, Peter Cook, Simon Ward, Marc McClure, Hart Bochner, Maureen Teefy, David Healy, Sandra Dickinson, Robyn Mandell, Diana Ricardo, Nancy Lippold, Sonya Leite
Superman's female cousin comes to Earth to regain a lost power source, that has fallen into the hands of arch-villainess Dunaway. A lethargic and disappointing follow-on to SUPERMAN that just drags and drags.
FAN 111 min (ort 124 min) VIDrel: MGM/WHV L/A V/sh

SUPERGRASS, THE *** 15
Peter Richardson UK 1985
Adrian Edmondson, Jennifer Saunders, Peter Richardson, Dawn French, Nigel Planer, Keith Allen, Robbie Coltrane, Daniel Peacock, Alexei Sayle, Michael Elphick, Ronald Allen, Patrick Durkin, Marika Rivera, Rita Treisman
A careless boast in a pub leads to a teenage boy being forced to work as a police informer, a job not without its share of dangerous complications. A funny but fairly lightweight comedy, from the same team that produced all those COMIC STRIP PRESENTS... episodes for Channel Four TV.
COM 93 min (ort 107 min) VIDrel: FIRST/SONOP V/sur

SUPERMAN **** PG
Richard Donner UK/USA 1978
Christopher Reeve, Margot Kidder, Marlon Brando, Gene Hackman, Glenn Ford, Jackie Cooper, Ned Beatty, Susannah York, Phyllis

Thaxter, Trevor Howard, Marc McClure, Valerie Perrine, Jeff East, Terence Stamp, Maria Schell
Our DC Comics superhero is updated for the screen in a sparkling adventure, and has to thwart arch-criminal Lex Luthor's scheme to activate the San Andreas fault. Music is by John Williams and the witty script is by Mario Puzo, David Newman, Robert Benton and Leslie Newman. The superb special effects (principally the brilliant flying sequences) won an Oscar. AA: Spec Award (Les Bowie/Colin Chilvers/Denys Coop/Roy Field/Derek Meddings/Zoran Perisic for visual effects).
Aka: SUPERMAN: THE MOVIE
FAN   119 min (ort 144 min) wScrn   VIDrel: WHV L/A V/sh

SUPERMAN 2 *** PG
Richard Lester USA 1980
Christopher Reeve, Margot Kidder, Gene Hackman, Terence Stamp, Ned Beatty, Sarah Douglas, Jackie Cooper, Valerie Perrine, Susannah York, E.G. Marshall, Jack O'Halloran, Marc McClure, Clifton James, Shane Rimmer, Michael Shannon
Newly escaped from imprisonment in the Phantom Zone, three super-villains (a great performance from Stamp) arrive on Earth and battle Superman for world domination. The story may be weaker but the special effects are even better. Slightly more tongue-in-cheek than before but well worth seeing.
FAN   127 min   VIDrel: WHV V/sur

SUPERMAN 3 * PG
Richard Lester USA 1983
Christopher Reeve, Richard Pryor, Robert Vaughn, Annette O'Toole, Jackie Cooper, Marc McClure, Annie Ross, Pamela Stephenson, Margot Kidder, Gavan O'Herlihy, Nancy Roberts, Graham Stark, Henry Woolf, Gordon Rollings
A further "Superman" film that is such a spoof that the good ideas in it are really spoilt. The story is inane (an attempt to control the world's oil tankers) and the dialogue is flat. An unworthy successor to the two previous films that's possibly worth seeing for the special effects, but only just.
FAN   120 min (ort 125 min)   VIDrel: ARROW/RTM V/sur

SUPERMAN 4: THE QUEST FOR PEACE * PG
Sidney J. Furie USA 1987
Christopher Reeve, Gene Hackman, Jackie Cooper, Marc McClure, Jon Cryer, Sam Wanamaker, Mark Pillow, Mariel Hemingway, Margot Kidder, Damian McLawhorn, Jim Broadbent, William Hootkins, Stanley Lebor, Don Fellows, Robert Beatty
Superman is out to rid the world of nuclear weapons, but Lex Luthor, now a nuclear arms entrepreneur, creates a superhero of his own, a genetic clone whose powers are equal to those of Superman. The two do battle and though the final outcome is never in doubt, things look bad for our hero at times. The weakest entry in the series, with shoddy, second-rate effects and a plot with holes large enough to push Krypton through. Hard to enjoy on any level.
Aka: QUEST FOR PEACE, THE
FAN   87 min (ort 90 min)   VIDrel: WHV V/sur

SUPERVIXENS * 18
Russ Meyer USA 1975
Shari Eubank, Charles Pitts, Charles Napier, Henry Rowland, Jack Provan, Ushi Digard, Christy Hartsburg, Haji, Ann Marie, Sharan Kelly, John Steen, Henry Blitz, Christy Melon, Sharon Black, John Zee, Stuart Mudd, Glen Musso
First sequel to "The Vixens" has Clint working at a gas station and living with insanely jealous Angel. Clint beats her up so badly the cops arrive. Later, when Clint is out one calls to see her. She teases the cop because he is impotent and in a rage he kills her. Clint now flees but months later meets a "reincarnation" of Angel running a gas station! The cop arrives and tries to kill them both but kills himself instead. A ludicrous and violent tale.
A   106 min   VIDrel: ALLIED/RTM/TROMA V

SUPPORT YOUR LOCAL GUNFIGHTER ** U
Burt Kennedy USA 1971
James Garner, Suzanne Pleshette, Jack Elam, Harry Morgan, John Dehner, Joan Blondell, Chuck Connors, Marie Windsor, Henry Jones, Dub Taylor, Kathleen Freeman, Willis Bouchey, Walter Burke, Gene Evans, Dick Haymes, Ellen Corby
A con-man passes off bumbling Elam as a notorious gunfighter so as to benefit from a mining dispute. A sequel to SUPPORT YOUR LOCAL SHERIFF that disappoints despite a good cast.
Aka: LATIGO
COM   88 min (ort 92 min)   VIDrel: L/A V

SUPPORT YOUR LOCAL SHERIFF! *** PG
Burt Kennedy USA 1968
James Garner, Walter Brennan, Joan Hackett, Harry Morgan, Jack Elam, Bruce Dern, Henry Jones, Walter Burke, Dick Peabody, Gene Evans, Willis Bouchey, Kathleen Freeman, Gayle Rogers, Richard Hoyt, Marilyn Jones
Western spoof about a sheriff bringing law to a tough town by any means he can muster. Though well acted, the film is only slightly amusing rather than really funny, despite some clever parodies from countless Westerns. Brennan's role as Old Man Clanton from "My Darling Clementine" adds a nice touch. Written and produced by William Bowers and followed in 1971 by SUPPORT YOUR LOCAL GUNFIGHTER.
COM   89 min (ort 93 min)   VIDrel: L/A V

SURE FIRE ** 18
Yuen Woo Ping HONG KONG 1987
Donnie Yen, Simon Yam, Do Do Cheng, Jacky Cheung
Having succeeded in breaking up a notorious drugs gang in Hong Kong, a police officer and his girlfriend decide to quit the force, but at their farewell party the cop is shot dead by the gang's leader, who had escaped custody and infiltrated the party. When the dead cop's former colleagues embark on a hunt for the killer, their investigations reveal the complicity of some senior police officials. A formula revenger, well paced and watchable.
aka: TIGER CAGE
MAR   89 min dubbed   VIDrel: POPRO/RTM L/A V

SURE FIRE *** (PG)
Jon Jost USA 1990
Tom Blair, Kristi Hager, Robert Ernst, Kate Dezina, Phillip R. Brown, Dennis R. Brown, Rick Blackwell, Robert Nalwalker, Haley Westwood, Kaye Evans, Henry M. Blackwell, J.T. Reynolds, Thomas D.A. Smith, John Betenson
In the Utah desert, a man with get-rich-quick dreams believes that he has found the perfect scheme, which involves vacation homes. Naturally, his efforts impose quite a burden on his wife and family, but it is not long before he too begins to feel the strain. A fine example of Jost's non-Hollywood approach to film-making, with a cast of unknown but capable actors giving of their best in this unusual drama.
DRA   83 min (ort 88 min)   CINrel

SURE THING, THE ** 15
Rob Reiner USA 1985
John Cusack, Daphne Zuniga, Anthony Edwards, Viveca Lindfors, Lisa Jane Persky, Boyd Gaines, Tim Robbins, Nicollette Sheridan, Fran Ryan, George Memmoli, Sunshine Parker, Teresa Baxter, Joshua Cadman, Carmen Filpi
Story of an unlikely romance that develops between two college students, who are thrown together by chance on a cross-country journey. Something of an attempt to update IT HAPPENED ONE NIGHT, that's contrived and predictable, but reasonably enjoyable. However, Lindfors with her heavy accent is just not believable as a college English professor and her presence in the film is a waste of her talents. The film HAPPY TOGETHER essays something similar.
COM   91 min (ort 94 min)   VIDrel: POLY L/A V

SURF NAZIS MUST DIE * 18
Peter George USA 1987
Barry Brenner, Gail Neely, Michael Sonye, Dawn Wildsmith, Tom Shell, Dawne Ellison, Bobbie Bresee, Robert Harden, Joel Hile, Gene Mitchell, Terry Lee, Brian Krutoff, Ted Prior, Andrew Bick, Berta Dahl, Willa Reynolds
In the near future, Californian beaches are taken over by various gangs following a devastating earthquake. The most vicious gang consists of a bizarre bunch of ruthless punk neo-Nazis. When an innocent black youth is murdered by the gang his mother swears vengeance. A ludicrous and rather boring mess, that despite its much-hyped release was neither shocking nor especially interesting. See PRAYER OF THE ROLLERBOYS for something similar.
Aka: SURF NAZIS
A/AD   78 min Cut (31 sec – ort 90 min)
VIDrel: TROMA/RTM V

SURF NINJAS ** PG
Neal Israel USA 1992
Ernie Reyes Jr, Rob Schneider, Tone Loc, John Karlen, Ernie Reyes Sr, Keone Young, Nicolas Cowan, Kelly Hu, Tad Horino, Leslie Nielsen,

Olivier Mills, Jonathan Schmock, Vladimir Parra, Brandon Karrer, Phillip Bayless, Phillip Tan
The incompetent dictator of a small kingdom in the South Seas discovers that two distant heirs to its throne are living in California where they work as lifeguards. Naturally, he takes measures to ensure that they will never be able to return to claim their birthright. A silly and frantically paced spoof, with unfunny jokes and mock martial art fights but little to amuse an adult audience.
Aka: SURF WARRIORS
COM 83 min (ort 87 min)   VIDrel: EIV/SONOP   V/sur

SURRENDER *
PG
Jerry Belson USA
1987
Sally Field, Michael Caine, Steve Guttenberg, Peter Boyle, Jackie Cooper, Julie Kavner, Louise Lasser
A struggling artist and a successful author embark on an affair, in this banal comedy that not even good performances can do anything for. The script is by Belson.
COM 91 min (ort 96 min)   VIDrel: MGM/WHV L/A   V/sh

SURROGATE, THE ***
18
Don Carmody CANADA
1983
Art Hindle, Carole Laure, Shannon Tweed, Michael Ironside, Jackie Burroughs, Marilyn Lightstone, Barbara Law, Gary Reineke, Jean-Claude Poitras, Jonathan Welsh, Tony Scott, Dean Hagopian, Mark Burns, Jim Hanley
A couple whose marriage is on the rocks are advised to seek a sex therapist, who seems to be linked in some way with a series of murders. A contrived but intriguing tale that suffers slightly from over-complexity.
Aka: BLIND RAGE
THR 96 min (ort 99 min)   VIDrel: POLY/POLYREC/MED V/h

SURROGATE, THE **
12
Jan Egleson USA
1995
Alyssa Milano, Connie Selleca, David Dukes, Vincent Ventresca, Polly Bergen, Scott Hylands, Lorena Gale
An outwardly kindly couple rent a struggling art student a cottage at a nominal charge, but make it a condition that she must provide them with a child, as they are unable to do so themselves. At first she finds matters progressing well enough, but eventually realises that they have not been entirely honest with her, and harbour a sinister secret that could threaten her unborn child. Highly predictable, this is an adequate time-filler.
DRA 88 min mTV   VIDrel: NWV/HIFLI   V/h

SURVIVAL GAME *
15
Herb Freed USA
1987
Mike Norris, Deborah Goodrich, Seymour Cassel, Ed Bernard, Arlene Golonka, John Sharp, Rick Grassi, Lee Paul, Michael Halton, Ken Grantham, John Vick, Paul Samuelson, Daniel O'Conner, Michael Wittmers, Mike Vaughan, Sandy Bull
A combat expert employed by a war games resort, is forced to put his skills to the test when he finds himself up against a gang of ruthless criminals. Very poor.
A/AD 87 min (ort 91 min)   VIDrel: EIV/SONOP   V/sur

SURVIVAL RUN *
15
Larry Spiegel USA
1980
Peter Graves, Ray Milland, Vincent Van Patten, Pedro Armendariz Jr, Alan Conrad, Marianne Savage, Anthony Charnota, Gonzalo Vega, Cosie Costa, Randi Meryl, Robby Weaver, Danny Ades, Susan Pratt O'Hanlon
Teenagers on an excursion into the desert stumble across drug smugglers and have to escape on foot after their vehicle is destroyed. Unfortunately, they cannot escape from this banal film.
THR 85 min (ort 90 min)   VIDrel: MOPIC/SGSVID   V

SURVIVAL ZONE *
18
Percival Reubens USA
1984
Gary Lockwood, Camilla Sparv, Morgan Stevens, Zoli Markey, Ian Steadman, Arthur Hall, Karl Eric Kostlin, Elizabeth Meyer, Lillian Randall, Jeanne Combrink, Mimi Kheswa
Post-WW3 survivors are menaced by motorcycle maniacs in this painfully unoriginal yarn.
A/AD 90 min   VIDrel: MOPIC/SGSVID   V

SURVIVE THE SAVAGE SEA ***
U
Kevin James Dobson USA
1991
Ali MacGraw, Robert Urich, Danielle Von Zerneck, Mark Ballou,

Ryan Urich, David Franklin, Peter Sumner, David Daniels, Dylan Daniels, John Heywood, Gus Mercurio, Betty Bobbit, Michael Bishop, David Tipoki, Pahnie Jantzen
A couple sell off all their possessions in pursuit of their dream, which is to buy a boat and sail the Pacific. Forced to abandon their vessel when an encounter with a whale damages it, they face a long ordeal until their rescue by a Japanese fishing boat some 38 days later. An enjoyable survival saga, loosely based on a true story.
A/AD 88 min mTV   VIDrel: MGM/WHV L/A   V
Boa: book by Dougal Robertson.

SURVIVING DESIRE **
(15)
Hal Hartley USA
1991
Martin Donovan, Mary Ward, Matt Malloy, Rebecca Nelson, Julie Sukman, Thomas J. Edwards, George Feaster, Lisa Gorlitsky, Emily Kunstler, John MacKay, Jim McCauley, Vinny Rutherford, Gary Sauer, Steve Schub, Patricia Sullivan
A female student falls in love with her professor but both spend their time intellectually analysing their short-lived affair instead of indulging in any real passion. A stylised offering (it was shot on 16 mm as an American Playhouse production for TV) that is very well made but has far too narrow a focus to be effective, with sharp dialogue initially pleases but soon begins to irritate. The two Hartley shorts included on the tape are of little consequence.
DRA 57 min; 84 min (3-film cassette) mTV
VIDrel: TART/20TH L/A   V
Osca: AMBITION/THEORY OF ACHIEVEMENT

SURVIVING PICASSO **
15
James Ivory UK/USA
1996
Anthony Hopkins, Natascha McElhorne, Julianne Moore, Joss Ackland, Peter Eyre, Jane Lapotaire, Joseph Maher, Bob Peck, Diane Venora, Joan Plowright, Tom Fisher, Andreas Wisniewski, Allegra Di Capegna, Nigel Whitmey, Leon Lissek
An aptly named attempt to portray the effect of the fiery title artist on all those about him, notably his long-term mistress who stayed with him for some ten years, bearing him two children. Hopkins uses his chameleonlike talent to great effect, giving a superb performance, but the film's episodic structure and the unwise use of a voice-over are distractions, as is the stilted dialogue. Sadly, Picasso stays an enigma, ands his innermost drives remain unexplored.
DRA 125 min   VIDrel: WHV   V
Boa: book Picasso: Creator and Destroyer by Arianna Stassinopoulos Huffington.

SURVIVING THE GAME **
15
Ernest R. Dickerson USA
1993
Rutger Hauer, Ice-T, Charles S. Dutton, Gary Busey, F. Murray Abraham, John C. McGinley, William McNamara, Jeff Corey, Bob Minor, Lawrence C. McCoy, George Fisher, Jacqui Dickerson, Victor Morris, Frederic Collins Jr, Steven King
A homeless man is tricked by a group of hunters into coming to their remote base in the countryside and learns that his true role is that of the prey in a hunt to the death. A very confused and highly derivative variant on a much-used theme (e.g. THE NAKED PREY and THE MOST DANGEROUS GAME) but with the dubious benefit of some scenes of graphic violence but a total absence of suspense. See also DEATH RING, FINAL ROUND and THE WOMAN HUNT.
A/AD 92 min (ort 96 min)   VIDrel: EIV/SONOP   V/sur

SURVIVORS, THE **
15
Michael Ritchie USA
1983
Walter Matthau, Robin Williams, Jerry Reed, James Wainwright, Annie McEnroe, Kristen Vigard, Anne Pitoniak, Bernard Barrow, Marian Hailey, Joseph Carberry, Skip Lynch, Marilyn Cooper, Meg Mundy, Yudie Bank, Michael Moran
Two men, an executive and a gas station attendant, find their lives changed out of all recognition when they identify a robber, and are forced as a result of their civic courage to take refuge in the mountains of Vermont. A patchy black comedy, written by Michael Leeson.
COM 98 min Cut (20 sec – ort 102 min)
VIDrel: VCC/COLUM L/A   V

SUSPECT ***
15
Peter Yates USA
1987
Dennis Quaid, Cher, Liam Neeson, John Mahoney, Joe Mantegna, Philip Bosco, E. Katherine Kerr, Fred Melamed, Lisbeth Bartlett, Paul D'Amato, Bernie McInerney, Thomas Barbour, Katie O'Hare, Rosemary Knower

A defence lawyer takes on a seemingly hopeless case when she agrees to represent a deaf tramp accused of murder. Convincing performances lift the film above its contrived plot.
DRA 117 min (ort 121 min) VIDrel: VCC/COLUM L/A V/sh

SUSPECT DEVICE **
18
Rik Jacobson USA
1995
C. Thomas Howell, Stacey Travis
A man with a happy and settled life finds himself on the run after his workplace is raided by masked assassins, and finds that his wife no longer recognises him, and that his friends are out to kill him. A competent psychological thriller.
THR 90 min VIDrel: GUILD/FOXVID V

SUSPICION ***
PG
Alfred Hitchcock USA
1941
Cary Grant, Joan Fontaine, Cedric Hardwicke, Nigel Bruce, Dame Mary Whitty, Isabel Jeans, Heather Angel, Leo G. Carroll, Auriol Lee, Reginald Sheffield, Maureen Roden-Ryan, Carol Curtis-Brown, Constance Worth, Violet Shelton
A wife believes that her husband is out to murder her, but the sense of suspense in this film is muted by the need to observe the moral restraints that film-makers had to contend with at that time. Written by Samson Raphaelson, Alma Reville and Joan Harrison and with music by Franz Waxman. Remade for TV in 1988. AA: Actress (Fontaine).
THR 95 min (ort 99 min) B/W
VIDrel: 4-FRONT/POLYREC V
Boa: novel Before the Fact by Francis Iles (Anthony Berkeley Cox).

SUSPICIOUS AGENDA **
18
Clay Borris USA
1995
Richard Grieco, Nick Mancuso, Jim Byrnes, Frank Crudele, Zachary Throne
A cynical and hardbitten cop has still not recovered after the death of his partner and is assigned to a crack team hunting a killer who is eliminating criminals who have escaped justice. An unnecessarily gruesome flourish is that the victims are murdered by having molten lead poured down their throats, a touch that makes the film more akin to a horror yarn than the action-thriller it was meant to be.
A/AD 96 min VIDrel: MED/20VIS V/sh

SUSPIRIA ***
18
Dario Argento ITALY
1976
Jessica Harper, Stefania Casini, Udo Kier, Joan Bennett, Alida Valli, Flavio Bucci, Miguel Bose, Barbara Magnolfi, Susanna Javicoli, Margherita Horowitz, Jacopo Mariani, Fulvio Mingozzi, Renato Zamengo, Rudolf Schundler, Eva Axen
A new pupil at a German ballet school discovers a hotbed of Satanism and other diabolical practices. A kind of melding of PSYCHO and "The Exorcist", with several good moments and an effective score by Argento and the rock group Goblin. Written by Argento and Dario Nicolodi.
HOR 93 min (ort 99 min)
VIDrel: 4-FRONT/POLYREC/EIV L/A V

SUTURE ***
15
Scott McGhee/David Siegel USA
1995
Dennis Haysbert, Mel Harris, Sab Shimono, Michael Harris, David Graf, Dina Merrill, Fran Ryan, John Ingle, Sandy Gibbons, Mark Demichele, Sandra Lafferty, Capri Darling, Carol Kiernan, Laura Groppe, Lon Carli, Ann Van Wey, Jack Rubens
A white man plans to kill his black half-brother and take over his life in this highly unusual and interesting modern example of film noir. However, this murder attempt (his car was booby trapped) fails, and the victim is left with amnesia, eventually coming to believe that he is in fact the other brother. A truly distinctive debut feature for the directors, it never quite convinces, but is certainly memorable, and makes a few salient points about race and identity.
THR 92 min (ort 96 min) B/W VIDrel: ICAPRO/MANGA V/sh

SVENGALI **
(12)
Anthony Harvey USA
1983
Peter O'Toole, Jodie Foster, Elizabeth Ashley, Larry Joshua, Pamela Blair, Barbara Byrne, Ronald Wyand, Robin Thomas, Brian Carney, Madeline Potter, Holly Hunter, Vera Mayer, Paul O'Keefe, Stu Charno, Peter Boruchowitz
Modern updating of the story of the hypnotic and domineering

voice teacher and his female protegee, who rises to stardom under his guidance. An offbeat performance from O'Toole makes this film watchable, but Foster is so lacking in charisma that it's hard to see how anyone could turn her into a rock star with a devoted following. Written by Frank Cucci.
DRA 100 min mTV VIDrel: L/A V
Boa: novel Trilby by George du Maurier/story by Sue Grafton.

SWALLOWS AND AMAZONS **
U
Claude Whatham UK
1974
Virginia McKenna, Ronald Fraser, Simon West, Sophie Neville, Zanna Hamilton, Stephen Grendon, Kit Seymour, Lesley Bennett, John Franklyn-Robbins, Jack Woolgar, Mike Prat, David Blagden, Brenda Bruce
A boring film version of a famous children's book about six young children on holiday in the Lake District in 1929. Much of the action takes place at night when one can hardly see a thing, but just hear the voices of the children as they go about their adventures. Written by David Wood and nothing like as good as the novel.
JUV 88 min (ort 92 min) VIDrel: WHV V
Boa: novel by Arthur Ransome.

SWAMP THING ***
15
Wes Craven USA
1981
Louis Jourdan, Adrienne Barbeau, Ray Wise, Don Knight, David Hess, Nicholas Worth, Dick Durock, Al Ruban, Ben Bates, Tommy Madden, Nannette Brown, Reggie Bates, Mimi Meyer, Karen Price, Bill Erickson, Dov Gottesfeld
A scientist becomes a monster after being contaminated with an experimental growth inducing chemical. However, our scientist still remains benevolent, even though he has become a walking vegetable. A ludicrous but enjoyable offering. Followed by THE RETURN OF THE SWAMP THING.
HOR 88 min (ort 90 min) VIDrel: L/A V

SWAN PRINCESS, THE **
U
Richard Rich/Mike Hodgson USA
1994
Voices of: Jack Palance, Howard McGillin, Michelle Nicastro, John Cleese, Steven Wright, Steve Vinovich, Mark Harelik, James Arrington, Joel McKinnon Miller, Dakin Matthews, Sandy Duncan, Brian Nissen, Adam Wylie, Adrian Zahiri
Prince Derek and Princess Odette were plighted in troth from an early age, but the two fall out over a foolish chance remark just before their marriage. She is then abducted by a wizard with ambitions to become her kingdom's new ruler, and he changes her into a daytime swan when she refuses to marry him, an act that requires the prince to save her. Inspired by the ballet "Swan Lake", this is an interesting animation, but alas one of variable quality.
ANIM 86 min (ort 89 min) cC VIDrel: COLUM/SONOP V/sur

SWANN IN LOVE ***
18
Volker Schlondorff FRANCE
1983
Jeremy Irons, Ornella Muti, Alain Delon, Fanny Ardant, Marie-Christine Barrault, Anne Bennent
In Paris at the turn of the century, an aristocrat gains entry into upper class society, despite the fact that he is a Jew, and embarks on a passionate affair with a beautiful woman. An ambitious and fairly successful adaptation of a segment of Proust's great work, the film cannot possibly preserve the richness of the novel, and instead achieves a curious, elegant lifelessness. The sumptuous photography is the work of Sven Nykvist.
Aka: UN ARMOUR DE SWANN
DRA 108 min (ort 111 min) wScrn VIDrel: ARTIF/20TH V
Boa: novel Swann's Way (first volume of novel Remembrance Of Things Past) by Marcel Proust.

SWAP, THE *
15
Jordan Leondopoulos (John Shade/John C. Broderick) USA
1969
Robert De Niro, Jennifer Warren, Jered Mickey, Martin Kelley, Anthony Charnato, Viva, Terrayne, Crawford, Lisa Blount, Sybil Danning, John Medici, Jim Brown, Sam Anderson
A man just out of prison goes in search of his brother's killers. Of minor interest for an early appearance by De Niro, this film was re-shot and new characters played by Danning and Blount were added – it was then re-issued under the above title. It now largely concentrates on a weekend Long Island party, attended by a New York film editor and a highly disruptive

character. Despite these changes, it remains a boring and point-less dud.
Aka: LINE OF FIRE; SAM'S SONG
DRA   83 min (ort 104 min)
VIDrel: 4-FRONT/POLYREC/MIA   V

SWARM, THE *
PG
Irwin Allen USA
1978
Michael Caine, Katherine Ross, Richard Widmark, Richard Chamberlain, Henry Fonda, Olivia De Havilland, Lee Grant, Fred MacMurray, Patty Duke Astin, Ben Johnson, Jose Ferrer, Slim Pickens, Cameron Mitchell, Bradford Dillman
Cliche-ridden effort about an unstoppable swarm of killer bees, invading the USA from South America and stinging all and sundry (except the director unfortunately). Just as with "The Bees", a good cast is utterly wasted on a dreary script. Written by Stirling Silliphant (from whom we would have expected better).
HOR   111 min (ort 116 min)   VIDrel: MGM/WHV L/A   V
Boa: novel by Arthur Herzog.

SWEET BIRD OF YOUTH ***
15
Richard Brooks USA
1962
Paul Newman, Geraldine Page, Shirley Knight, Ed Begley Sr, Rip Torn, Mildred Dunnock, Madeleine Sherwood, Philip Abbott, Corey Allen, Barry Cahill, James Douglas, Dub Taylor, Barry Atwater, Charles Arnt, Dorothy Konrad, James Chandler
A drifter returns home in the company of an ageing movie actress whom he sees as his ticket to a film career, but he reckons without the machinations of the corrupt town "boss", who seeks vengeance for the seduction of his daughter. A well acted but emasculated version of a powerful stage play, with the ending changed to a happy one. Scripted by Brooks. AA: S. Actor (Begley).
DRA   115 min (ort 120 min)   VIDrel: MGM/WHV   V
Boa: play by Tennessee Williams.

SWEET BIRD OF YOUTH ***
(12)
Nichoas Roeg USA
1989
Elizabeth Taylor, Mark Harmon, Rip Torn, Valeire Perrine, Kevin Geer, Ruta Lee, Michael Wilding Jr
A drifter hopes to get into the movies by associating with a faded movie star and takes her back to his home town in Florida but learns too late that was not a wise career move. A well-stage version of the Williams play that has the added virtue of retaining the original ending that was dropped from the 1962 Paul Newman movie.
DRA   95 min mTV   TVrel: BBC1
Boa: play by Tennessee Williams.

SWEET CHARITY ***
PG
Bob Fosse USA
1968
Shirley MacLaine, Chita Rivera, John McMartin, Paula Kelly, Sammy Davis Jr., Ricardo Montalban, Stubby Kaye, Barbara Bouchet, Alan Hewitt, Dante D'Paulo, John Wheeler, John Craig, Dee Carroll, Tom Hatten, Sharon Harvey, Ceil Cabot
A New York dance hall hostess dreams of love but finds only heartbreak and loneliness. The musical numbers are stage with great panache, but the meandering and virtually plotless narration makes large stretches of the film difficult to sit through. Loosely based on Fellini's NIGHTS OF CABIRIA. The fine songs include the title number, "If My Friends Could See Me Now" and "Rhythm Of Life".
MUS   142 min (ort 148 min)   VIDrel: 4-FRONT/POLYREC
V
Boa: play by Neil Simon, Cy Coleman and Dorothy Fields.

SWEET DREAMS ***
15
Karel Reisz USA
1985
Ed Harris, Jessica Lange, Ann Wedgeworth, David Clennon, James Staley, P.J. Soles, Gary Basabara, John Goodman, Teri Gardner, Caitlin Kelch, Robert L. Dasch, Courtney Parker, Colton Edwards, Bruce Kirby, Jerry Haynes
A biopic on the life of Patsy Cline, an American singer of the 1950s, that tends to focus on her marital problems to the detriment of the more interesting performing sequences. Some gutsy performances and memorable songs redeem it. Lange mimes to Cline's original recordings.
DRA   110 min (ort 114 min)   VIDrel: WHV   V/sur

SWEET EMMA, DEAR BOBE ***
18
Istvan Szabo HUNGARY
1992
Johanna Ter Steege, Eniko Borcsok, Peter Andorai, Eva Kerekes, Erszi Gaal, Hedi Temessy, Iren Bodis, Zoltan Musci, Jolanta Mielech,

Jurgen Mai, Gerd Blahuschek, Irma Patkos, Erzsi Pasztor, Hedi Temessy, Tamas Jordan, Gabor Mate
Two female Russian teachers arrive in Budapest but are unable to get work in their field, and it would seem that prostitution is the only option they have. A very depressing film that largely explores the major changes that have taken place in modern-day Hungary after the death of Communism, many of which convey a sense of foreboding rather than optimism.
Aka: EDES EMMA, DRAGA BOBE
DRA   78 min (ort 90 min) wScrn   VIDrel: TART/20TH
V/dm

SWEET 15 ***
U
Victoria Hochberg USA
1989
Tony Plana, Karla Montana, Jenny Gago, Panchito Gomez, Susan Ruttan, Leonard Camarillo, Robert Covarrublas, Ernie Fuentes, Vanessa Marquez, Jan Merlin, William Marquez, Tuck Milligan, Humberto Ortiz, Giselle Rubino, Liz Torres
A low-budget, multi-stranded story, mostly of youthful growing-pains, family love and the simple things in life, that revolves around teenager Martha, who has learnt that her Mexican father is in the process of making a secret immigration application. This must not fail, or the entire family could be deported, so Martha begins collecting signatures for a petition. Despite an awkward start, this is quite a pleasant little drama.
Aka: WONDERWORKS: SWEET 15
DRA   106 min (ort 110 min)   VIDrel: MGM/WHV L/A   V

SWEET HEARTS DANCE ***
15
Robert Greenwald USA
1988
Don Johnson, Susan Sarandon, Jeff Daniels, Elizabeth Perkins, Kate Reid, Justin Henry, Holly Marie Combs, Heather Coleman, Matthew Wohl, Stephen Stabler, Laurie Corbin, Lanie Conklin, Jock MacDonald, Frits Momsen, Paul Schnabel
Set in a small town in Vermont, this bittersweet comedy tells of a couple whose marriage is just about ending, while their close friend is at the beginning of a serious relationship. A charming little tale, not very profound or funny, but quite endearing. Scripted by Ernest Thompson.
COM   96 min (ort 100 min)   VIDrel: RCA L/A   V

SWEET JUSTICE **
15
Allen Plone USA
1991
Marc Singer, Finn Carter, Kathleen Kinmont, Frank Gorshin, Mickey Rooney, Catherine Hickland
A woman organises a group of all-female fighters to get her revenge for the murder of a friend. Apart from this one original twist, this is nothing more that the usual standard violent actioner with much gunplay and martial arts mayhem.
A/AD   90 min (ort 92 min)   VIDrel: COLUM/SONOP
V/sh

SWEET KILLING **
15
Eddy Matalon FRANCE/USA
1992
F. Murray Abraham, Leslie Hope, Andrea Ferreol, Michael Ironside
Having created what he believes is a fake identity in order to murder someone, a man appears to have committed the crime with impunity. But his satisfaction at leaving the police baffled is shortlived when a person bering his invented name turns up at his home.
THR   86 min   VIDrel: TRIM/HIFLI   V/sur
Boa: novel Qualthrough by Agnes Hall.

SWEET LIBERTY **
15
Alan Alda USA
1986
Alan Alda, Saul Rubinek, Michael Caine, Lise Hilboldt, Bob Hoskins, Lillian Gish, Michelle Pfeiffer, Lois Chiles, Linda Thorson, Diane Agostini, Alvin Alexis, Christopher Bergman, Leo Burmester, Cynthia Burr, Timothy Carhart
A history teacher in a small town writes a book on the American Revolution that becomes a bestseller. When the film rights are sold to Hollywood a film crew are sent out to make a film on location. In charge of them is a bad-tempered and anarchic director who proceeds to turn the book into a gag-filled sex romp, clashing with the writer in the process. A few laughs cannot hide the emptiness of this one. The script is by Alda.
COM   102 min (ort 107 min)   VIDrel: CIC/SONOP L/A
V/h

SWEET LITTLE SISTER *
18
John Leslie USA
1991
Savannah, Taylor Wayne, Angela Summers, Brigitte Ami, Joey Silvera, Tom Byron, Joey Murphy, Peter North, T.T. Boy

Although her older sister sleeps in the adjoining room, a young girl is still terrified of the dark, as she is plagued by strange supernatural visitors. A failed attempt to adopt a fresh approach in just one more run-of-the-mill sex film.
Aka: LAYING THE GHOST
A   55 min (ort 80 min)   VIDrel: GROHOM/MAXSCAN   V

SWEET MURDER **
Percival Rubens USA
18
1990
Helene Udy, Embeth Davidtz, Russell Todd
Two girls from very different backgrounds become room-mates and friends, but when one inherits a fortune and gets herself a handsome boyfriend, the other has a brainstorm and decides to do away with her.
THR   101 min   VIDrel: NWV/SONOP   V/sur

SWEET POISON **
Brian Grant USA
18
1991
Edward Herrmann, Steven Bauer, Patricia Healy, Pruitt Taylor Vince, Lyman Ward
A woman who has married for money befriends an escaped convict who has kidnapped both her and her husband on the way to a funeral. As events proceed, she gradually begins to see in him a means of getting rid of her unwanted spouse.
THR   96 min (ort 101 min)   VIDrel: CIC/SONOP   V/h

SWEET REVENGE **
Charlotte Brandstrom FRANCE/USA
PG
1990
Rosanna Arquette, Carrie Fisher, John Sessions, Francois Eric Gendron, John Hargreaves, Myriam Moszko, Consuelo De Haviland, Carina Barone, Dominique MacAvoy, Yves Brainville, Andrea Schieffer, Susan Carlson, Claire Marsden
While in Paris, a successful corporate lawyer finds that her ex-husband has filed an alimony claim against her. She hires an eccentric, out-of-work actress to be trained as his "perfect woman", hoping that this will lead to a quick marriage and no more problems with alimony. A madcap romantic-comedy that has something of the flavour of a nutty 1940s caper, it could have achieved more with a stronger script.
COM   84 min (ort 100 min) mCab   VIDrel: BUENA L/A   V

SWEET SIXTEEN *
Jim Sotos USA
18
1982
Bo Hopkins, Susan Strasberg, Aleisa Shirley, Don Stroud, Patrick MacNee, Dana Kimmell, Steve Antin, Sharon Farrell, Don Shanks, Logan Clarke, Michael Pataki, Henry Wilcoxon, Larry Storch, Michael Cutt
A sensuous young girl has a string of boyfriends who, one by one are found murdered, in this predictable thriller in which the identity of the murderer comes as no surprise.
Aka: SWEET 16
THR   84 min (ort 90 min)   VIDrel: IMPENT   V

SWEET SMELL OF SUCCESS, THE ****
Alexander MacKendrick USA
PG
1957
Burt Lancaster, Tony Curtis, Martin Milner, Sam Levene, Susan Harrison, Barbara Nichols, Emile Meyer, Chico Hamilton, Jeff Donnell, Joseph Leon, Edith Atwater, Joe Frisco, David White, Lawrence Dobkin, Lurene Tuttle, Queenie Smith
An opportunistic press agent plays up to a powerful columnist and makes use of the latter's influence in order to destroy his sister's marriage. A close and stylised study of the life of some highly unpleasant characters. The incisive script is by playwright Clifford Odets and Ernest Lehman, and it's backed up by Elmer Bernstein's fine jazz score and the excellent photography of James Wong Howe.
DRA   92 min (ort 96 min) B/W   VIDrel: MGM/WHV   V
Boa: short story Tell Me About It Tomorrow by Ernest Lehman.

SWEET TALKER ***
Michael Jenkins AUSTRALIA
PG
1991
Bryan Brown, Karen Allen, Chris Haywood, Bill Kerr, Bruce Spence, Bruce Myles, Paul Chubb, Peter Hehir, Justin Rosniak, Don Barker, Bruno Lucia, Bob Steele, Benjamin Franklin, Andrew S. Gilbert, Gary Waddell, Brian McDermott
A con-man straight out of prison comes to a coast resort in the belief that its inhabitants will be easy victims but learns to his coast that they are by no means as gullible as he supposed. Very shortly, he finds himself deeply involved with a widow and her son. Amiable performances (especially from Brown in the lead role) give this pleasant comedy plenty of sparkle.
Aka: CONFIDENCE
COM   84 min (ort 91 min)   VIDrel: 20TH/TECH   V/sur

SWEETIE **
Jane Campion AUSTRALIA
15
1989
Genevieve Lemon, Karen Colston, Tom Lycos, Jon Darling, Dorothy Barry, Michael Lake
Campion's first feature is an offbeat tale of a demanding and romantically mixed-up young woman, who finds that a visit from her weak and neurotic sister does little to help matters. Written by Campion and Gerard Lee, this is a dark and surreal satire on the difficulties of family relationships that is inconsistent in tone, and veers rather too wildly between sharp and profound insights and muddled, opaque passages.
COM   95 min (ort 105 min)   VIDrel: ELPIC/POLYREC   V

SWIMMER, THE ****
Frank Perry USA
PG
1968
Burt Lancaster, Janice Rule, Janet Landgard, Diana Muldaur, Kim Hunter, Tony Bickley, Marge Champion, Bill Fiore, Joan Rivers, Nancy Cushman, John Garfield Jr, Rose Gregorio, Charles Drake, Bernie Hamilton, House Jameson, Jimmy Joyce
One hot afternoon in suburban Connecticut, a middle-aged man makes his way home via his neighbours' swimming pools. This self-imposed odyssey plunges us into his past life, as a rich member of a shallow and snobbish community. A sad and immensely moving work, this minor masterpiece is beautifully filmed and acted and with Lancaster giving a superlative performance. The ending is both harrowing and disturbing.
DRA   91 min (ort 94 min)   VIDrel: FABFIL/SPEAR   V
Boa: short story by John Cheever.

SWIMMING WITH SHARKS **
George Huang USA
15
1994
Kevin Spacey, Frank Whaley, Michelle Forbes, Benicio del Toro, T.E. Russell, Roy Dotrice, Matthew Flynt, Patrick Fischler, Jerry Levine
The lives of a ruthless Hollywood executive, his downtrodden assistant and a and a cynical producer are put under the microscope in this variant on THE PLAYER, that mostly revolves around the unedifying relationship between Spacey and Whaley, as spiteful boss and unlucky underling respectively. Meant to be a sardonic attack on Hollywood and its lack of values, this dark and acidic film lacks the impact and cleverness to make the whole thing work.
COM   95 min   VIDrel: IMAG/RTM   V

SWING KIDS **
Thomas Carter USA
15
1992
Robert Sean Leonard, Christian Bale, Barbara Hershey, Tushka Bergen, David Tom, Julia Stemberger, Jayce Bartok, Noah Wyle, Johan Leysen, Martin Clunes, Douglas Roberts, Jessic Stevenson, Carl Brincat, Mary Fogarty, Karl Belohradsky
In Hamburg in 1939, three friends who share a common passion for American swing music find themselves on opposing sides of the political divide as the Nazis ban this kind of music as degenerate and un-German. Some well staged musical set-pieces sit uncomfortably with a realistic re-recreation of the brutalities of this period. Filmed on location in Prague.
DRA   109 min (ort 114 min) cC   VIDrel: HOLPIC/TECH   L/A   V/sh

SWING SHIFT *
Jonathan Demme USA
PG
1984
Goldie Hawn, Kurt Russell, Christine Lahti, Ed Harris, Fred Ward, Sudie Bond, Roger Corman, Holly Hunter, Patty Maloney, Belinda Carlisle, Lisa Pelikan, Susan Peretz, Joey Aresco, Morris "Tex" Biggs, Reid Cruikshanks
During WW2, American housewives were recruited to work in the factories. One such woman finds her life complicated when she dons overalls for Uncle Sam. A feeble romantic comedy written by several top writers (Ron Nyswaner, Bo Goldman and Nancy Dowd) under the pseudonym Rob Morton.
Aka: SWINGSHIFT
COM   96 min Cut (2 sec – ort 112 min)   VIDrel: L/A   V

SWING TIME ****
George Stevens USA
U
1936
Fred Astaire, Ginger Rogers, Victor Moore, Helen Broderick, Eric Blore, Betty Furness, George Metaxa, Landers Stevens, John Harrington, Harry Bowen, Pierre Watkin, Abe Reynolds, Gerald Hamer, Edgar Dearing, Harry Bernard
The love affair between the two partners of a dance team is complicated by his engagement to another girl. Forget the plot and just enjoy the dance routines, in one of the best of the

Astaire-Rogers musicals. AA: Song ("The Way You Look Tonight" – Jerome Kern (m)/Dorothy Fields (l)).
MUS   103 min (ort 114 min) B/W
VIDrel: 4-FRONT/POLYREC L/A   V

SWINGER'S INK ***
18
Michael Craig USA
1990
Tracey Adams, Sharon Kane, Champagne, Jacqueline, Rick Daniels, Scott St James, Jon Dough, Renee Summers, Randy Spears
A rather cleverly plotted effort that takes a look at the title enterprise, a near-bankrupt magazine aimed at swingers, and its crooked owner and staff of wackos and eccentrics. Most of the film is taken up with various sexual episodes, but the plot (it deals with the magazine's dishonest efforts to find a buyer) is complex enough to sustain interest. Written, produced and directed by Craig.
A   72 min Cut (2 min 13 sec – ort 85 min) mVid
VIDrel: IMPENT   V

SWISS FAMILY ROBINSON ***
U
Ken Annakin USA
1960
John Mills, Dorothy McGuire, James MacArthur, Tommy Kirk, Kevin Corcoran, Janet Munro, Sessue Hayakawa, Cecil Parker, Andy Ho, Milton Reid, Larry Taylor
A family are shipwrecked on an island and have various adventures, including a clash with a horde of pirates that they fight off with supreme ease. Lightweight but enjoyable nonsense. First made in 1940 and remade for TV in 1975.
JUV   126 min cC   VIDrel: WDV/TECH   V
Boa: novel by Johann David Wyss.

SWITCH *
15
Blake Edwards USA
1991
Ellen Barkin, Jimmy Smits, JoBeth Williams, Lorraine Bracco, Tony Roberts, Perry King, Bruce Martyn Payne, Lysette Anthony, Victoria Mahoney, Catherine Keener, Basil Hoffman, Kevin Kilner, David Wohl, James Harper, Joe Flood
One of those gender-change comedies, that has a womaniser (King) finding that he has been reincarnated in the body of a sexy woman (Barkin) but has retained his male consciousness. Barkin gives a great performance in a very demanding role, but this is essentially a one-joke film, and clumsy, last-minute editing resulted in a weak and ineffectual ending. The film "Goodbye, Charlie" doubtlessly inspired this dud.
COM   99 min (ort 114 min)   VIDrel: COLUM/SONOP L/A V/sh

SWITCH, THE ***
PG
Bobby Roth USA
1992
Gary Cole, Craig T. Nelson, Kathleen Nolan, Chris Mulkey, L. Scott Caldwell, Max Gail, David Purdham, Henry Sanders, Hinton Battle
A man left quadriplegic after a bad motorcycle crash fights for the right to end his own life, eventually devising a "suicide switch" that he can use to turn off the respirator that keep him alive. He now faces a long court battle before he gains the right to have it installed. Based on a true case, this is an incisive and deeply moving examination of the euthanasia theme, also explored in films such as "Right To Die" and WHOSE LIFE IS IT, ANYWAY?
DRA   91 min   VIDrel: ODY/SONOP   V/sh

SWITCH IN TIME, A **
15
Paul Donovan CANADA
1987
Tom McCamus, Laurie Paton, Brian Downey, Jacques Lussier, Lee Broker, David Hemblen, Marcos Woinsky, Armando Capo, Gabriela Salas, Ennique La Torre, Jorge Luis Estrella, Jacques Arndt, Theodore McNabney, Carlos Weber, Bill Carr
The story of three people who are accidentally transported back in time to Switzerland of the 1st century. Whilst two of them are content to bask in adulation from a barbarian tribe, the third begins to introduce them to the secrets of technology, much to the dismay of the Romans. A lighthearted and fairly amusing romp.
Aka: NORMAN'S AWESOME EXPERIENCE
COM   86 min   VIDrel: 20TH/TECH   V

SWITCHBOARD OPERATOR, THE ***
18
Dusan Makavejev YUGOSLAVIA
1967
Eva Ras, Slobodan Aligrudic, Ruzica Sokic, Miodrag Andric
A liberated young woman from the Hungarian minority in Yugoslavia embarks on an affair with a boy from a straitlaced Moslem family. A stimulating drama that examines both the stresses and strains within the relationship and the tensions

within Yugoslavia that ultimately led to its fragmentation. The clumsily contrived attempt to examine the story by having a couple of psychologists add their commentary, are irritating and dated touches that are best ignored.
Aka: AFFAIR OF THE HEART, AN; LJUBAVNI SLUCAJ; LOVE AFFAIR: OR THE CASE OF THE MISSING SWITCHBOARD OPERATOR; TRAGEDIJA SLUZBENICE P.T.T.; TRAGEDY OF A SWITCHBOARD OPERATOR, THE
DRA   69 min B/W   VIDrel: CONNO/RTM   V

SWITCHED AT BIRTH **
PG
Waris Hussein USA
1991
Bonnie Bedelia, Brian Kerwin, Edward Asner, Ariana Richards, Caroline McWilliams, John M. Jackson, Lois Smith, Judith Hoag, Eve Gordon, Jacqueline Scott, Kelli Williams, Erika Flores, Beth Grant, Rance Howard, John Wesley
A fact-based tearjerker that tells of two Florida girls who were accidentally swapped in hospital soon after being born, and spend the next ten years being brought up by each other's parents. Written by Michael O'Hara, this is a reasonably engrossing tale, though a bit too drawn out. See also SOMEONE ELSE'S CHILD.
DRA   176 min (ort 200 min) mTV   VIDrel: ODY/SONOP V/sh

SWITCHING CHANNELS ***
PG
Ted Kotcheff USA
1987
Kathleen Turner, Christopher Reeve, Burt Reynolds, Ned Beatty, Henry Gibson, George Newbern, Al Waxman, Ken James, Barry Flatman, Ted Simonett, Anthony Sherwood, Joe Silver, Charles Kimbrough, Monica Parker, Allan Royal, Fiona Reid
Another re-run of "His Girl Friday" but updated to the present day and set in a TV news network, where a cunning editor will stop at nothing to keep his ace reporter from leaving to get married. Quite enjoyable, but somewhat bland, although the stars do their best to overcome this.
COM   100 min (ort 108 min)   VIDrel: POLY/POLYREC L/A V/sh

SWITCHING PARENTS ***
PG
Linda Otto USA
1992
Kathleen York, Joseph Gordon-Levitt, Bill Smitrovich, Robert Joy, Kristin Griffith, Joyce Reehling, Geoffrey Bowes, Janet Bailey, Elizabeth Berman, Brett Halsey, Kathleen Laskey, Jamie Rainey, Maria Ricossa, Don Allison
An abused twelve-year-old boy makes legal history by suing his own parents to have their rights of custody over him removed, thus enabling him to stay with a loving foster family who want to adopt him. A perceptive fact-based drama. See also SHATTERED FAMILY.
DRA   88 min (ort 90 min) mTV   VIDrel: ODY/SONOP V/h

SWOON ***
18
Tom Kalin USA
1991
Daniel Schlachet, Craig Chester, Ron Vawter, Michael Kirby, Michael Stumm, Valda Z. Drabla, Natalie Stanford, Isabela Araujo, Mona Foot, Trash, Nashom Wooden, Trasharama, Peter Bowen, Ryan Landry, Christopher Hoover, Paul Connor
A murder-for-kicks-tale that's based on the Leopold/Loeb trial of 1924, with the two main characters homosexual lovers who live with their affluent families in the better part of Chicago. Taking pleasure in committing crimes for thrills, they graduate to a carefully-planned child abduction. When this goes wrong and the child is murdered, they continue the charade and send a ransom note, but are caught and jailed. A dark and unedifying offering.
THR   93 min (ort 95 min) B/W   VIDrel: CONNO/RTM   V

SWORD AND THE ROSE, THE ***
U
Ken Annakin USA
1952
Richard Todd, Glynis Johns, James Robertson Justice, Michael Gough, Jane Barrett, Peter Copley, Rosalie Crutchley, D.A. Clarke-Smith, Ernest Jay, Phillip Lennard, John Vere, Bryan Coleman, Jean Mercure, Gerard Oury
Period drama depicting the ill-fated romance between the captain of Henry VII's palace guard and Princess Mary Tudor, played out against the backdrop of court intrigue. A minor historical yarn, thinly plotted but quite colourful. Written by Laurence E. Watkin and filmed in England.
Aka: WHEN KNIGHTHOOD WAS IN FLOWER
DRA   87 min (ort 93 min)   VIDrel: WDV/TECH L/A   V
Boa: novel When Knighthood was in Flower by Charles Major.

SWORD IN THE STONE, THE **
Wolfgang Reitherman USA   *U*
                                                     1963
Voices of: Ricky Sorenson, Sebastian Cabot, Karl Swenson, Junius
Matthews, Alan Napier, Norman Alden, Martha Wentworth, Ginny
Taylor, Barbara Jo Allen, Richard Reithermann, Robert Reithermann
An animated version of the Arthurian legend following the
adventures of a youngster called "Wart", who is destined to
become King Arthur. One of the poorer Disney animations, fast
paced but flat, with ponderous dialogue and unmemorable
Sherman Brothers songs.
ANIM   76 min (ort 80 min) cC   VIDrel: WDV/TECH   V
Boa: novel The Once and Future King by T.H. White.

SWORD OF BUSHIDO, THE **
Adrian Carr USA   *18*
                                                     1989
Richard Norton, Rochelle Asana, Toshishiro Obata, Judy Green, Kovik
Wattankoon
Martial arts mayhem with an American karate expert in
Thailand to find out what happened to his grandfather, who
was onboard a plane that crashed there during WW2. When he
learns that his grandpa survived the crash but was executed by
the Japanese for stealing a sacred sword, he resolves to locate
and return it to the original owners. Abundant action sequences
(and a few softcore ones) partially bolster the flabby and mindless script.
MAR   100 min   VIDrel: MIA/DISC   V

SWORD OF HONOR **
Robert Tiffe USA   *18*
                                                     1994
Steven Vincent Leigh, Sophia Crawford, Angelo Tiffe, Jeff Pruitt, Jerry
Tiffe, Debbie Scofield
When his partner is killed, a Las Vegas cop with martial arts skills
plots his revenge and prepares to infiltrate the gang responsible,
using an ancient sword to gain the attention of its leader.
MAR   95 min   VIDrel: IMPENT   V

SWORD OF JUSTICE: BLACK JACK **
Daniel Haller USA   *(15)*
                                                     1978
Dack Rambo, Bert Rosario, George Hamilton, Lara Parker, Alex
Courtney
A man framed for tax evasion uses his time in prison to master
the secrets of white-collar crime and upon release, employs his
talent to make money from the rich to give to deserving causes.
In this tale he takes on a Las Vegas casino that's part-owned by
a corrupt and ruthless union leader. A highly implausible
adventure-thriller, this was a pilot for a short-lived 13-episode
series from NBC.
A/AD   80 min mTV   VIDrel: CIC/SONOP L/A   V

SWORDKILL **
Jo Larry Carroll USA   *15*
                                                     1984
Janet Julian, Charles Lumpkin, Hiroshi Fujioka, John Calvin, Frank
Schuler, Bill Morey, Andy Wood, Robert Kino, Joan Foley, Peter
Liapis, Mieko Kobayashi
The deep-frozen body of a samurai is discovered and he is
brought back to life. He now finds himself having to survive on
the streets of L.A. and deal with the criminals he comes into
contact with. A predictable tale enlivened by some good special
effects.
Aka: GHOSTWARRIOR
A/AD   77 min Cut (21 sec – ort 81 min)
VIDrel: NTV/TOTAL/EIV   V

SWORDSMAN, THE **
Michael Kennedy CANADA   *15*
                                                     1992
Lorenzo Lamas, Claire Stansfield, Michael Champion, Nicholas Pasco,
Raoul Trujillo, Eugene Clark, Michael Copeman, George Touliatos,
Frank Crudele, Ian White, Scott Davis, Robert Seale, Zack Kotlyar,
Tamara Bernier, Simon Fon
A detective is hired to recover a famous sword said to have
belonged to Alexander the Great, that has been stolen from a
museum. He soon learns that it has been stolen by a ruthless
millionaire and that both of them have been sworn enemies who
have fought down the centuries. A standard actioner with some
fantasy elements. A sequel soon followed.
A/AD   97 min (ort 98 min)   VIDrel: MARQ/QUANT   V

SWORN TO VENGEANCE **
Peter Hunt USA   *18*
                                                     1993
Robert Conrad, William McNamara, Gary Bayer, Sharon Farrell,
Peter Breck, Tom Atkins, Michael Cavanaugh, Kurt McKinney, James
McEachin, Meg Wittner, Dori Brenner, Ramon Franco, Clifton
Gonzalez Gonzalez, Paul Scherrer, La Velda Fann

A determined cop investigates the murder of three teenagers,
whose bodies were found in the woods, and with the help of a
psychic woman, eventually finds that the culprit is a dangerous
psychopath given to bragging about his crimes. Adequate
thriller base on a true case.
DRA   90 min   VIDrel: ODY/SONOP   V/sh

SYLVIA SCARLETT ***
George Cukor USA   *U*
                                                     1935
Katharine Hepburn, Cary Grant, Brian Aherne, Edmund Gwenn,
Natalie Paley, Lennox Pawle, Dennie Moore
First time teaming of Hepburn and Grant in a screwball comedy,
where she has to leave town disguised as a boy when her no-
good father gets into trouble. A rambling and offbeat tale that
was a failure on release, mainly due its languid pace. Written
by Gladys Unger, John Collier and Mortimer Offner.
COM   86 min (ort 94 min) B/W   VIDrel: VCC L/A   V
Boa: novel by Compton Mackenzie.

T

T BONE 'N' WEASEL **
Lewis Teague USA   *PG*
                                                     1992
Gregory Hines, Christopher Lloyd, Ned Beatty, Rip Torn, Larry
Hankin, Graham Jarvis, Rusty Schwimmer, Sam Whipple, Candy
Aston, Denise S. Bass, J. Michael Hunter, Trina D. Olson, Lloyd
Wilson
A couple of ex-cons decide that they are going to take things
easy so they steal a car and go to a rural area where they think
the locals will be an easy target for the usual scams. However,
they soon learn that their intended victims are far from gullible.
A failed attempt at a star comedy that makes poor use of its two
talented leads.
COM   93 min (ort 95 min) mCab   VIDrel: ITC/HIFLI   V/h

T-FORCE *
Richard Pepin USA   *18*
                                                     1994
Jack Scalia, Evan Lurie, Erin Gray, Bobby Johnston, Deron McBee,
Jennifer McDonald, Martin E. Brooks, Vernon Wells, Nick De Munro,
Sean Moran, Clement Van Franckenstein, Duke Stroud, R. David
Smith, Robert Gallo, Chris Jackson, Thom Bo
In the 21st century, the state has created a powerful anti-terrorist squad whose members are all androids endowed with human
traits (such machines having reached a high enough level of
development to be used widely in industry). When it is decided
to disband this unit after they make a mess of a hostage rescue
mission, all but one of these machines set out to destroy their
creators. It falls to the remaining android to stop them doing
this. Noisy and quite poor.
FAN   89 min   VIDrel: MARQ/QUANT   V

TABLE FOR FIVE ***
Robert Lieberman USA   *PG*
                                                     1983
John Voight, Richard Crenna, Millie Perkins, Marie-Christine
Barrault, Robby Kiger, Roxana Zal, Son Hoang Bui, Maria O'Brien,
Nelson Welch, Bernie Hern, Cynthia Kania, Kevin Costner, Marion
Russell, Gustaf Unger
A divorced man takes his three children on a European cruise
as a way of making up for his absence and has to tell them that
their mother has been killed. A slow and quite competent tear-
jerker, written by David Seltzer.
DRA   116 min (ort 123 min)   VIDrel: 20TH/TECH L/A   V

TABU: A STORY OF THE SOUTH SEAS **
Robert Flaherty/F.W. Murnau USA   *U*
                                                     1929
Anna Chevalier, Matahi, Hitu, Jean, Jules, Kong Ah
Docu-drama set in Tahiti where it was filmed on location. A
pearl fisherman falls in love with a woman whom the local
priests have declared as "taboo" and as such, is forbidden to
have any relations with a "man".
DRA   76 min (ort 81 min) B/W silent   VIDrel: TART/20TH
V

TAGGART: DEAD RINGER ***
Laurence Moody UK   *PG*
                                                     1985
Mark McManus, Neil Duncan, James MacPherson, Harriet Buchan,
Jake D'Arcy, Colette O'Neil, Alexander Morton, John McGlynn,
Maureen Beattie
When the remains of a body are found beneath the floor of a
house in Glasgow, DCI Taggart begins his murder investiga-
tion. The body is soon identifies as being that of a woman who
went missing ten years ago, which led to the woman's husband

being convicted of murder and given life, a case which was also handled by Taggart. Whilst Taggart fends of newspaper accusations of incompetence, he soon realises that he is the victim of a complex plot.
DRA 131 min mTV   VIDrel: CLEAR/DISC V

TAGGART: DEATH CALL *** 12
Haldane Duncan UK 1986
Mark McManus, James MacPherson, Stuart Hepburn, John Cairney, Bridgette Kahn, Jill Meagher, Michael Carter, Juliet Cadzow, Alan Cumming, Anne Kristen, Charles Kearney, Irene Sunters, Julie Graham, Harriet Buchan
Taggart investigates the death of woman whose body is recovered from a reservoir. When she is identified as the German-born wife of a wealthy landowner who has gone missing after withdrawing £40,000 from his account, the latter soon becomes an obvious suspect, but the focus of the police investigation eventually switches to a young shop assistant. Complex plotting with ample twists and solid performances are the chief assets here.
DRA 129 min mTV   VIDrel: CLEAR/DISC V

TAGGART: DEATH COMES SOFTLY *** 15
Laurence Moody UK 1990
Mark McManus, James MacPherson, Iain Anders, Harriet Buchan, David Rintoul, Eve Pearce, Georgine Anderson, Phyllida Hewat, Russell Hunter, Liam Brennan, Alec Westwood, Mandy Matthews, Joanne Bett, Blythe Duff, Norman Eshley
When a pensioner is found murdered, the police suspect the man's daughter and son-in-law, who appear to have been on holiday abroad. Our patient Detective Chief Inspector sets out to uncover the real culprits.
DRA 144 min mTV   VIDrel: COLUM/SONOP V/sh

TAGGART: DOUBLE JEOPARDY *** 15
Jim McCann UK 1988
Mark MacManus, James Macpherson, James Laurenson, Sheila Ruskin, Valerie Gogan, Rose McBain, Herbert Trattnigg, Claus Ellsman, Harriet Buchan, Alec Heggie, Barbara Rafferty, John Shedelen, Iain Anders, Robert Robertson
A woman commits suicide in a wood but the absence of a note and other circumstances convinces Taggart that she was murdered by her common-law husband. Unable to find evidence to prove the man's guilt, he mounts a dogged campaign of surveillance that takes him to Germany and finally back to Glasgow. A well-plotted murder mystery that moves along at a fast pace, maintaining interest right up to the end.
DRA 130 min mTV   VIDrel: CLEAR/DISC V

TAGGART: FLESH AND BLOOD *** 15
Alan Macmillan UK 1989
Mark McManus, Iain Anders, James MacPherson
A cache of explosives falls into the wrong hands just prior to Christmas, and our Glaswegian sleuth has to face the possibility of terrorist outrages being planned. His investigations only become more complex when the body of a murdered social worker is discovered.
DRA 148 min mTV   VIDrel: COLUM/SONOP V/sh

TAGGART: FUNERAL RITES ** 15
Alan MacMillan UK 1987
Mark McManus, James MacPherson, Annette Crosbie, Paul Young, Vincent Friell, Colin Gourley, Isabella Jarrett, Gavin Brown, Iain Anders, Harriet Buchanan, Stuart Hepburn, Nic D'Avirro, Sandra Clark, Mara Bruce, Katy Hale, Mel Donald
When a badly burnt corpse is found in an old railway tunnel, Taggart sets out to catch those responsible, and this quest brings him into contact with various occult groups since the victim, a sleazy private eye, belonged to one of them. However, this lead proves to be a dead-end, and the case is only solved after several more deaths have taken place.
DRA 127 min (ort 180 min) mTV   VIDrel: CLEAR/DISC V

TAGGART: GINGERBREAD ** 15
Sarah Hellings UK 1993
Mark McManus, James Macpherson, Blythe Duff, Mary MacLeod, Gary Hollywood, Iain Anders, Harriet Buchanan, Robert Robertson, Anne-Marie Timoney, Christopher Robbie, Fiona Gillies, Hugh Fraser, Vivien Heilbron, Malcolm Rennie, Sheila Reid
A twelve-year-old boy thinks he has discovered the identity of the mysterious intruder who murdered his private detective father, but has great difficulty getting the police to pay any attention to his claim that a woodland cottage holds the key to this

crime. Soon, a number of disappearances point in the direction of a possible serial killer. A harsh and brutal story, well acted but (typically for this series) downbeat and rather unappealing.
DRA 145 min (ort 150 min) mTV
VIDrel: CAMERON/BMGREC V

TAGGART: IN COLD BLOOD *** 15
Haldane Duncan UK 1988
Mark McManus, Diane Keen, James Macpherson, Harriet Buchan, Iain Anders, Freddie Boardley, Leonard O'Malley, Mona Bruce, Patricia Ross, Robert Robertson
Jim Taggart finds his approaching 24th wedding anniversary having to take a back seat to a baffling case in which a woman shot a man in a parked car and then tried to take her own life. However, it soon transpires that she cannot be charged with murder as a forensic investigation reveals that her intended victim was already dead. A competent movie-length feature based on the popular TV detective series.
DRA 77 min mTV   VIDrel: CLEAR/DISC V

TAGGART: KILLER *** 15
Laurence Moody UK 1983
Mark McManus, Neil Duncan, Vincent Friell, Tom Watson, Harriet Buchan, Geraldine Alexander, Robert Robertson, Gerard Kelly, Linda Mucha, Frank Wylie, Bertie Scott, Roy Hanlon, Anne Kidd, Colette O'Neill, Jake D'Arcy, John McGlynn
The very first episode in this series, set in Glasgow, the beat of a dour and cynical detective. In this story, a serial killer is loose on the streets, having murdered three women. Taggart begins to suspect the owner of a video store, but is prevented from taking action by his boss as the man is a golfing chum. Our detective's suspicions are soon vindicated, but not before another murder takes place. A downbeat story that set the tone for the entire series.
DRA 125 min mTV   VIDrel: CLEAR/DISC V

TAGGART: MURDER IN SEASON *** 15
Peter Barber-Fleming UK 1984
Mark McManus, Neil Duncan, Iain Anders, Isla Blair, Andrew Keir, Martin Cochrane, Douglas Sananchan, Eileen Nicholas, Leigh Biagi, Danny Hignett, Ronnie Letham, Katherine Stark, Ken Stott, Dorothy Paul, Derek Anders, Brenda Haldane
A Scottish opera singer comes home from abroad hoping for a reconciliation with her husband, but catches him in the midst of an affair with his secretary. When this woman is found murdered the singer becomes the chief suspect, but further killings soon convince Taggart that the truth is far more complex.
DRA 131 min mTV   VIDrel: CLEAR/DISC V

TAGGART: NEST OF VIPERS *** 15
Graham Theakston UK 1991
Mark McManus, James MacPherson, Blythe Danner, Ian Anders, Ann Mitchell, Ken Drury, Lorna Heilbron, Michael Cochrane, Juliet Cadzow, Geoffrey Beevers, Leone Connery, Harriet Buchan, Robert Robertson, Jeremy Young, Grant Cathro
When some skulls are found on an excavation site foul play is strongly suspected, and Taggart calls on the aid of a brilliant scientist. When this man is bitten by a poisonous spider, it would appear that someone wishes to impede any examination and possible identification. Another well thought out and carefully made mystery in this detective series.
DRA 146 min mTV   VIDrel: COLUM/SONOP V/sh

TAGGART: THE KILLING PHILOSOPHY *** 15
Haldane Duncan UK 1987
Mark McManus, James MacPherson, Iain Anders, Harriet Buchan, Sheila Grier, Philip Dupuy, Richard Jamieson, Rod Culbertson, Kika Mirylees, Kenneth Bryans, Jenny Lee, Harriet Reid, Patricia Ross, Gerda Stevenson
A rapist's brutal attacks and later murders present Taggart with a tough case, and his efforts are not helped when a convicted rapist escapes from jail. In a desperate bid to gather evidence, Taggart swallows his pride and accepts help from a woman psychic, while his subordinates are sent off to check out a local gay bar. A strong entry in the series, harsh and realistic, with a convoluted (if implausible) plot that keeps one guessing right up to the end.
DRA 128 min mTV   VIDrel: CLEAR/DISC V

TAI-PAN * 18
Daryl Duke USA 1986
Bryan Brown, John Stanton, Joan Chen, Tom Guinee, Bill Leadbitter,

Russell Wong, Katy Behan, Kyra Sedgwick, Janine Turner, Norman Rodway, John Bennett, Derrick Branche, Chang Cheng, Patrick Ryecart, Nicholas Grace, Carol Gillies
An overlong and rambling film based on Clavell's book of similar merits, telling of the adventures of a 19th century trader who is based in Hong Kong. A film that looks as if it should be a major epic but isn't really about anything in particular. Written by John Briley and Stanley Mann and followed by NOBLE HOUSE.
DRA   122 min (ort 127 min)   VIDrel: GAME/SPEAR V/sur
Boa: novel by James Clavell.

TAILOR OF GLOUCESTER, THE ***
John Michael Philips UK
Ian Holm, Thora Hird, Benjamin Luxon, Barry Ingham
A charming live-action recreation of one of Potter's timeless stories. Here, a poor tailor is at his wits' end to finish a silk coat that was ordered by the mayor for his wedding. However, his mood soon brightens when some tiny friends lend a hand.
JUV   45 min   VIDrel: ABBEY/POLYREC   V
Boa: story by Beatrix Potter.

U
1990

TAILS YOU LIVE, HEADS YOU'RE DEAD **
Tim Matheson USA
Corbin Bernsen, Ted McGinley, Tim Matheson, Maria Del Mar, Jeff Pustil
A psychotic murderer has turned stalking and killer into a game, but his next intended victim is a streetwise private eye who is more than a match for him.
THR   88 min (ort 91 min) cC   VIDrel: CIC/SONOP   V/sh
Boa: short story Liar's Dice by Bill Pronzini.

15
1996

TAINTED BLOOD **
Matthew Patrick USA
Raquel Welch, Alley Mills, Keri Green, Joan Van Ark, Natasha Gregson Wagner, John C. Mooney, Justin Isfehel, Richard Gross, Lewis Arquette, Molly McLure, Jack Owen, Janet Hale, Mary Egan, Tom Ashworth, David St James, Nancy Warren
A famous writer is investigating families in which homicidal tendencies are passed on by parents to their offspring and finds herself trying to locate a family who may have adopted her twin brother, from whom she was separated at birth.
DRA   83 min mCab   VIDrel: CIC/SONOP L/A   V

15
1992

TAKE A GIRL LIKE YOU *
Jonathan Miller UK
Hayley Mills, Oliver Reed, Sheila Hancock, Noel Harrison, John Bird, Ronald Aimi MacDonald, Geraldine Sherman, Imogen Hassall, Penelope Keith, John Fortune, Pippa Steel, Nicholas Courtney, George Woodbridge, Nerys Hughes
A sexually inexperienced Northern girl moves to the South to work as a teacher, and becomes the target of small-town womanisers. A dated little tale of very minor appeal. Written by George Melly.
DRA   94 min (ort 101 min)   VIDrel: RCA L/A   V
Boa: novel by Kingsley Amis.

15
1970

TAKE CARE OF YOUR SCARF, TATJANA ***
Aki Kaurismaki FINLAND
Kati Outinen, Matti Pellonpaa, Kirsi Tykklainen, Mato Valtonen, Elina Salo, Irma Junnilainen, Veikko Lavi, Pertti Husu, Viktor Vassel, Carl-Erik Calamnius, Atte Blom, Mauri Sumen, The Regals, The Renegades, Anu Aalto, Matti Ahjoniemi
A couple of sullen Finns go on the road and meet up with two women, one Russian and the other Estonian. While the women are clearly attracted to them, the men's lack of communication skills and penchant for swilling vodka makes things difficult for all concerned. A bleak and dismal road-movie, very well made in the director's inimitable style, but hardly a celebration of life. Will certainly find favour with devotees of this cult director.
Aka: PIDA HUIVISTA KIINNI, TATJANA
DRA   65 min   CINrel

(12)
1994

TAKE ME OUT TO THE BALL GAME ***
Busby Berkely USA
Frank Sinatra, Esther Williams, Gene Kelly, Betty Garrett, Edward Arnold, Jules Munshin, Richard Lane, Tom Dugan, Murray Alper, Wilton Graf, Charles Regan, Mack Gray, Douglas Fowley, Eddie Parkes, James Burke
Bright and cheerful musical set in the 1890s, telling of a young woman who takes over a baseball team. Full of humour and

U
1949

enjoyable lines, this served as something of a trial run for ON THE TOWN. Numbers include "O'Brien To Ryan To Goldberg" and "The Hat My Father Wore On St Patrick's Day". Songs are by Betty Comden, Adolph Green and Roger Edens.
Aka: EVERYBODY'S CHEERING
MUS   89 min (ort 93 min)   VIDrel: MGM/WHV   V/dm V/h

TAKE-OVER, THE **
Troy Cook USA
Billy Drago, John Savage, Nick Mancuso, Eric Dare, David Amos, Cali Timmins, Gene Mitchell
In Los Angeles, a notorious gangster boss comes into conflict with another mobster, who is determined to supplant him. A by-the-numbers actioner.
A/AD   87 min   VIDrel: HIFLI/SONOP   V

18
1996

TAKE THE MONEY AND RUN **
Woody Allen USA
Woody Allen, Janet Margolin, Marcel Hillaire, Jacquelyn Hyde, Lonny Chapman, Louise Lasser, Jan Merlin, James Anderson, Howard Storm, Mark Gordon, Minnow Moskowitz, Micil Murphy, Nate Jacobson, Jackson Beck (narration only)
A look at the career of a compulsive thief, with plenty of visual gags of variable quality. Louise Lasser appears briefly. This was Allen's first film as writer, director and star, and shows a few flashes of brilliance, but is far too episodic to be effective.
COM   85 min   VIDrel: BRAVE/SONOP L/A   V

PG
1968

TAKING BACK MY LIFE: THE NANCY ZIEGENMEYER STORY ***
Harry Winer USA
Patricia Wettig, Stephen Lang, Ellen Burstyn, Shelley Hack
A fact-based tale that examines the actions of a rape victim who decides to reveal details of her ordeal to the media despite keeping silent for nine months, after she reads an article about rape victims that persuades her to come out into the open. But her decision initiates a wave of similar admissions from women across the country, who prior to this have been too humiliated and ashamed to reveal their trauma. A disturbing film, of sincerity and conviction.
DRA   92 min   VIDrel: MGM/WHV L/A   V/s

15
1992

TAKIN' IT ALL OFF **
Ed Hansen USA
Jean Poremba, Kitten Natividad, Gail Thackiray
A dimwitted and mildly titillating sequel to "Takin' It Off", in which a girl wants to be a stripper but is too shy. She gets some help via hypnosis, the treatment is highly successful, but she now feels an uncontrollable urge to do a striptease whenever she hears music. See also PARTY FAVORS for something rather similar.
A   90 min   VIDrel: FIRST/SONOP L/A   V

18
1987

TAKING OF FLIGHT 847, THE ***
Paul Wendkos USA
Lindsay Wagner, Eli Danker, Sandy McPeak, Ray Wise, Leslie Easterbrook, Laurie Walters, Joseph Nasser
A gripping account of the real-life drama that took place in 1985, when a plane was hijacked and the coolness of the flight attendant was generally reckoned to have been instrumental in averting a bloodbath. Written by Norman Morrill.
Aka: FLIGHT, THE; TAKING OF FLIGHT 847, THE: THE ULI DERICKSON STORY
DRA   96 min (ort 100 min) mTV   VIDrel: EIV/SONOP L/A V

15
1988

TAKING OF PELHAM 123, THE ***
Joseph Sargent USA
Walter Matthau, Robert Shaw, Martin Balsam, Hector Elizondo, Earl Hindman, Dick O'Neill, James Broderick, Doris Roberts, Tony Roberts, Jerry Stiller, Lee Wallace, Kenneth McMillan, Julius Harris, Sal Viscuso
Taut, well made thriller about the ruthless hijacking of a New York subway train. Shaw and three accomplices take over a train, and demand a ransom pay-off of one million dollars to be delivered in one hour. The script is by Peter Stone and the music by David Shire.
THR   100 min (ort 104 min)   VIDrel: MGM/WHV   V/h
Boa: novel by John Godey.

15
1974

TAKING THE HEAT **
Tom Mankiewicz USA
Tony Goldwyn, Lynn Whitfield, George Segal, Peter Boyle, Will Patton, Joe Grifasi, Allan Arkin, Greg Germann, Rachel York, Alex

18
1992

Carter, Eddie Mekka, Paul Lazar, Frankie Crocker, Santino Buda, Larry McLean, Marco Bianco, Ken Quinn
After witnessing a gangster commit murder, a New York yuppie has to run for his life from both the police and the Mob, but a female detective helps him find a way out of this dilemma.
A/AD   86 min (ort 110 min)   VIDrel: POLY/POLYREC
V/s

TALE OF A VAMPIRE *** 18
Shimako Sato JAPAN   1988
Julian Sands, Suzanna Hamilton, Kenneth Cranham, Marian Diamond, Michael Kenton, Catherine Blake, Mark Kempneer, Nik Myers, David King, Adrianne Alexander, Jake Omega, Mark Motileb, Keri Motileb, Lisa Motileb, Lois Beattie
A vampire in London falls in love with a woman who reminds him of his long-lost love but this involves in a deadly game with the latter's husband, also a vampire, who is plotting his revenge. Sands is well cast in the lead role, while the good camerawork is another plus. This low-budget film was Sato's directing debut, and together with Jane Corbett, he co-wrote the screenplay.
Aka: MANIKA; VAMPIRE
HOR   97 min (ort 99 min)   VIDrel: COLUM/SONOP   V/sur

TALE OF SPRINGTIME, A *** U
Eric Rohmer FRANCE   1990
Anne Teyssedre, Sophie Robin, Florence Darel, Hugues Quester, Eloise Bennett
A woman philosophy teacher meets a young student who introduces her to his father in the hope that the two of them will hit it off. A gentle and warm comedy-drama with fine performances and assured. The first in a new series of movies entitled "Tales of the Four Seasons". Screenplay is by Rohmer.
Aka: CONTE DE PRINTEMPS
DRA   102 min (ort 107 min)   VIDrel: ARTIF/20TH   V

TALE OF THE FOX, THE **** (U)
Wladyslaw Starewicz FRANCE   1930/31
Voices of: Claude Dauphin, Romain Bouquet, Sylvain Itkine, Leon Larive, Robert Seller, Edy Debray, Nicolas Amato, Suzy Dornac, Raine, Pons, Sylvia Bataille
An early example of puppet animation that brings to life a traditional folk tale about the fox's legendary cunning. After the Lion, king of all the beasts, receives numerous complaints about the misdeeds of the former, he decides to have him imprisoned by way of a punishment, but clever Renard has other ideas. A simple work, but an utterly charming one, with an appeal that is timeless.
Aka: LE ROMAN DE RENARD
ANIM   65 min   CINrel
Boa: story Doe Reineke Fuchs by Goethe (from the medieval story Le Roman de Renart).

TALE OF TWO CITIES, A **** U
Jack Conway USA   1935
Ronald Coleman, Elizabeth Allen, Edna May Oliver, Basil Rathbone, Donald Woods, Blanche Yurka, Henry B. Walthall, Reginald Owen, Walter Catlett, H.B. Warner, Fritz Leiber, Mitchell Lewis, Claude Gillingwater, Billy Beva
A classic film version of the famous novel, with its exciting story set in the turbulent years of the French Revolution, when a cynical British lawyer finally finds a purpose in life by sacrificing himself to save another man from the guillotine.
DRA   121 min (ort 128 min) B/W   VIDrel: MGM/WHV   V
Boa: novel by Charles Dickens.

TALE OF TWO CITIES, A ** U
Ralph Thomas UK   1958
Dirk Bogarde, Dorothy Tutin, Stephen Murray, Athene Seyler, Christopher Lee, Rosalie Crutchley, Ernest Clark, Paul Guers, Cecil Parker, Donald Pleasence, Ian Bannen, Marie Versini, Alfie Bass, Freda Jackson, Duncan Lamont
A faithful adaptation of the famous novel. Bogarde as Sydney Carton does his best as the lawyer who saves a man from the guillotine by taking his place, but the film cannot compare to the 1935 Ronald Colman classic. Remade for TV in 1980. Written by T.E.B. Clarke.
DRA   112 min (ort 117 min) B/W   VIDrel: CARL/TECH   V
Boa: novel by Charles Dickens.

TALE OF TWO CITIES, A *** 15
Philippe Monnier UK   1991
John Mills, James Wilby, Xavier Deluc, Serena Gordon, Jean-Pierre Aumont, Anna Massey

Detailed, well-acted and high-class filming of Dickens' classic tale of a young lawyer who only finds a meaning for his life when he sacrifices it for the sake of another. A Masterpiece Theatre production with very high values and a fine cast.
DRA   188 min (2 cassettes – ort 240 min) mTV
VIDrel: CASPIC/BMGREC   V
Boa: novel by Charles Dickens.

TALENT FOR THE GAME ** PG
Robert M. Young USA   1991
Edward James Olmos, Lorraine Bracco, Jeff Corbett, Jamey Sheridan, Terry Kinney
A baseball scout travels the country ever in search of new talent, but suddenly learns that the owner of club he works for has decided to make him redundant. A moderately enjoyable drama, somewhat limited in scope but generally satisfying.
DRA   87 min (ort 91 min)   VIDrel: CIC/SONOP   V/sur

TALES FROM THE CRYPT ** 18
Freddie Francis UK   1972
Ralph Richardson, Geoffrey Bayldon, Peter Cushing, Joan Collins, Ian Hendry, Richard Greene, Chloe Franks, Oliver MacGreevey, Susan Denny, Angie Grant, Robin Phillips, David Markham, Robert Hutton, Barbara Murray, Roy Dotrice
Five visitors exploring catacombs are confronted by a strange monk who predicts their futures. Episodes are entitled: "And All Through The House", "Reflection Of Death", "Poetic Justice", "Wish You Were Here" and "Blind Alley". A sequel of sorts: VAULT OF HORROR followed. Fair. Written by Milton Subotsky and based on the comic strips of William Gaines.
HOR   92 min (Cut at film release)   VIDrel: 20TH/TECH
L/A   V

TALES FROM THE CRYPT: BORDELLO OF BLOOD * 18
Gilbert Adler USA   1996
Dennis Miller, Angie Everhart, Erika Eleniak, Chris Sarandon, Corey Feldman, John Kasir
A limp and entirely stupid entry in this series, stealing plot ideas from several other movies (such as THE FRIGHTENERS) and following the exploits of Miller, a sleazy private eye who learns that a woman's brother was killed by vampires at the local brothel. Everhart makes an attractive vampire madame, but the film, which ends with the usual wild farrago of overblown effects, is hardly worth watching on that account.
HOR   96 min cC   VIDrel: CIC   V/sur

TALES FROM THE CRYPT: DEMON KNIGHT ** 18
Ernest R. Dickerson USA   1994
Billy Zane, Willaim Sadler, Jada Pinkett, C.C.H. Pounder, Dick Miller, John Schuck, Thomas Haden Church, Gary Farmer, Charles Fleischer, Tim deZarn, Sherrie Rose, Ryan Sean O'Donohue, Tony Salome, Ken Baldwin, John Kassir (voice only)
An evil presence stalks a seedy hotel and the guests have to face an army of ghouls. The lack of a strong plot is a handicap in this horror outing, which for the most part is only worth seeing for the flashy direction and effects. However, Zane is a pleasing presence as the Lord of Darkness, a part he plays with immense gusto.
Aka: TALES FROM THE CRYPT PRESENTS DEMON KNIGHT
HOR   88 min (ort 92 min) mCab cC   VIDrel: CIC/SONOP
V

TALES FROM THE CRYPT: VOL. 1 ** 18
Walter Hill/Robert Zemeckis/Richard Donner USA   1989
Bill Sadler, J.W. Smith, Roy Brocksmith, David Wohl, Mary Ellen Trainor, Larry Drake, Marshall Bell, Lindsey Whitney Barry, Joe Pantoliano, Robert Wuhl, Kathleen York, Gustav Vintas, Stephen Kahan, Michael Ray Bower
Three simplistic horror tales, with each having a rather childish introduction by a corpse. In "The Man Who Was Death" a state executioner takes matters into his own hands when capital punishment is abolished, in "And All Through The House" an adulterous wife is pursued by a maniac disguised as Santa Claus, and finally "Dig That Cat...He's Real Gone" tells of a drifter who is given nine lives by a mad scientist.
HOR   78 min (ort 81 min)   VIDrel: CIC/SONOP   V

TALES FROM THE CRYPT: VOL. 2 18
Tom Holland/Mary Lambert/Howard Deutch USA   1989
Amanda Plummer, Stephen Shellen, M. Emmet Walsh, Audra Lindley, Martin Garner, Lea Thompson, Britt Leach, Brett Cullen
Three enjoyable supernatural yarns, introduced by a cackling,

demonic host. In "Lover Come Back To Me" a plain rich girl and her handsome gold-digging husband stop the night at a haunted house, when their car breaks down. "Collection Completed" has a man taking revenge on his wife's beloved private menagerie; a most unwise action on his part. Finally, in "Only Skin Deep" a hooker trades beauty for wealth, an exchange she comes to regret.
Aka: ALL NEW TALES FROM THE CRYPT; TALES FROM THE CRYPT 2
HOR   84 min (ort 87 min)   VIDrel: WHV   V/sh

TALES FROM THE CRYPT: VOL. 3 **   18
Arnold Schwarzenegger/Walter Hill/Howard Deutch USA 1990
Demi Moore, William Hickey, Kelly Preson, Jeffrey Tambor, Rick Rossovich, Lance Henriksen, Kevin Tighe
Another instalment of three horror tales from the TV series. In "Dead Right" a woman visits a fortune-teller and takes her advice about whom to marry, in "Cutting Cards" two men play for very high stakes indeed and finally "The Switch" has an old man in love with a young girl, and now on the look-out for a new body.
Aka: TALES FROM THE CRYPT 3
HOR   75 min (ort 80 min) mCab   VIDrel: WHV   V/sh

TALES FROM THE DARKSIDE: THE MOVIE **   18
John Kent Harrison USA   1990
Deborah Harry, David Forrester, Matthew Lawrence; Christian Slater, Robert Sedgwick, Steve Buscemi, Donald Van Horn; David Johansen (Buster Poindexter), Paul Greene, William Hickey; James Remar, Ashton Wise, Philip Lenkowsky
Three competent stories, well acted and mounted, and liberally supplied with gore, the linking device being a youngster telling tales to a suburban cannibal to avoid being eaten. "Lot 249" follows the exploits of a deadly mummy, "Cat From Hell" is about a fiendish feline and "Lover's Vow" deals with a promise made to a gargoyle. An adequate horror outing, of which the last tale is easily the best.
HOR   89 min (ort 93 min)   VIDrel: VCC/DISC/COLUM V/sh
Boa: short stories Lot 249 by Arthur Conan Doyle/Cat From Hell by Stephen King.

TALES OF HOFFMAN, THE **   U
Michael Powell/Emeric Pressburger UK   1951
Moira Shearer, David Rounseville, Ludmilla Tcherina, Ann Ayars, Robert Helpmann, Pamela Brown, Leonide Massine, Frederick Ashton, Meinhart Maur, Edmond Audran, John Ford, Robert Goldring, Philip Leaver, Thomas Beacham, Mogens Wreth
A poet hopelessly in love with a ballerina, tells of past encounters with three women, all of which ended in sadness. A lavish opera and ballet film with stunning design and imaginative dancing but seriously flawed by its excessive length.
MUS   119 min (ort 138 min)   VIDrel: WHV L/A   V/h
Boa: opera by Jacques Offenbach.

TALES OF MYSTERY AND IMAGINATION ***   18
Roger Vadim/Federico Fellini/Louis Malle FRANCE/ITALY 1967
Jane Fonda, Terence Stamp, Peter Fonda, Brigitte Bardot, Alain Delon, James Robertson Justice, Carla Marliea; Francoise Prevost, Anny Duperey, Philippe Lemaire, Andoin De Bardot; Katia Christina, Salvo Randone, Fabrizio Angeli
Three short films involving the supernatural that are very loosely based on Poe stories and are entitled: "Don't Bet Heads", "Willliam Wilson's Sketch" and "Metzengerstein". A variable collection, of which the last one directed by Fellini is easily the best, though all three are reasonably enjoyable.
Aka: HISTOIRES EXTRAORDINAIRES; POWERS OF EVIL; SPIRITS OF THE DEAD; TALES OF MYSTERY; TRE PASSI NEL DELIRIO
FAN   117 min Cut (1 min 4 sec – ort 120 min) dubbed
VIDrel: ARROW/RTM   V
Boa: short stories by Edgar Allan Poe.

TALES OF ORDINARY MADNESS *   18
Marco Ferreri FRANCE/ITALY   1981
Ben Gazzara, Ornella Muti, Tanya Lopert, Susan Tyrrell, Roy Brocksmith, Katia Berger, Hope Cameron, Judith Drake, Patrick Hughes, Wendy Welles, Jay Julien, Leopold Stratton, Anthony Pitillo, Peter Jarvis
Describes the life and fantasies of a Los Angeles poet (Gazzara) who boozes his life away and meets a variety of women. A disorganised wallow in self-indulgence and platitudes, mostly

set on Venice Beach, California. The script (such as it is) was written by Ferreri and others.
DRA   97 min (ort 108 min)   VIDrel: ARTPPO/RTM   V
Boa: short stories from Erections, Ejaculations, Exhibitions And Tales Of Ordinary Madness by Charles Bukowski.

TALES OF TERROR **   18
Roger Corman USA   1961
Vincent Price, Peter Lorre, Basil Rathbone, Debra Paget, Maggie Pierce, Leona Gage, Joyce Jameson, Ed Cobb, Lenny Weinrib, John Hacketl, Wally Campo, Alan Dewit, David Frankham, Scotty Brown
Three Poe stories get the Corman treatment. In "Morella" a dying woman comes home to find her father brooding over the mummified body of his wife. In "The Black Cat" (which contains elements of "The Cask Of Amontillado") Lorre walls up his faithless wife and her lover in a vault. Finally "The Facts In The Case Of M. Valdemar" has a man being kept alive by hypnosis, following his apparent death. A variable display of atmospheric effects and sluggish plotting.
Aka: POE'S TALES OF TERROR
HOR   90 min   VIDrel: VISVID/POLYREC L/A   V
Boa: short stories Morella, The Black Cat and The Facts In The Case Of M. Valdemar by Edgar Allan Poe.

TALES OF THE CITY ***   18
Alastair Reed UK   1993
Olympia Dukakis, Donald Moffat, Chloe Webb, Laura Linney, Marcus D'Amico, William Campbell, Thomas Gibson, Paul Gross, Barbara Garrick, Nina Foch, Meagen Fay, Lou Libertore, Mary Kay Place, Parker Posey, Lou Cutell, Amy Ryder
A house in San Francisco in the 1970s provides the setting for this filmisation of Maupin's work, as its various inhabitants mingle and interact and their stories gradually unfold.
Aka: ARMISTEAD MAUPIN'S TALES OF THE CITY
DRA   305 min (2 cassettes – ort 360 min) mTV
VIDrel: POLY/POLYREC   V/sh
Boa: novels by Armistead Maupin.

TALK DIRTY TO ME PLEASE! *   18
Jerry Ross USA   1991
Ashlyn Gere, Randy West, K.C. Williams, Shaime, P.J. Sparxx, Peter North, Tom Chapman, Steve Drake, Brandy Alexandre, Angel Summers, Kym Wilde, Candice Heart, Trixy Taylor
Sexual adventures of the woman, the host of a radio show who dispenses advice to her many listeners. One of a number of entries in a set of similarly-titled films, all of which revolve around the same talk-show idea.
Aka: TALK DIRTY TO ME PART 8
A   57 min   VIDrel: ONE   V

TALK OF THE TOWN, THE ***   U
George Stevens USA   1942
Cary Grant, Ronald Coleman, Edgar Buchanan, Jean Arthur, Glenda Farrell, Rex Ingram, Charles Dingle, Emma Dunn, Tom Tyler, Lloyd Bridges, Leonid Kinskey, Don Beddoe, George Watts, Clyde Fillmore, Frank M. Thomas
A condemned (but innocent) murderer escapes from jail and goes on the run, hiding out at a boarding house where he engages in debate with an unsuspecting law professor, while making up to the attractive landlady. A delightful comedy-thriller with the three leads making the most of their thin material. The script is by Irwin Shaw and Sidney Buchman.
COM   113 min (ort 118 min) B/W
VIDrel: COLUM/SONOP   V/dm

TALK RADIO ***   18
Oliver Stone USA   1988
Eric Bogosian, Alec Baldwin, Ellen Greene, Leslie Hope, John C. McGinley, John Pankow, Michael Wincott, Linda Atkinson, Tony Frank, Bill Johnson, Kevin Howard, Harlan Jordan, Bruno Rubeo, Teresa Bell, Chris Moody, David Poynter
The story of a controversial and abrasive talk-show host, whose hectoring and controversial style leads to a notoriety that puts his life in danger. Powerfully acted and directed, this fascinating film is set mainly in the confines of a radio studio, but is never claustrophobic. A perfect vehicle for writer Bogosian, who co-wrote the screenplay with Stone, adding elements of the true story of Alan Berg – a DJ assassinated by white supremacists.
THR   104 min (ort 110 min)   VIDrel: 20TH/TECH   V/sur
Boa: play by Eric Bogosian and Tad Savinar/book Talked to Death: The Life and Murder of Alan Berg by Stephen Singular.

TALL BLOND MAN WITH ONE BLACK SHOE, THE *

PG

Yves Robert FRANCE 1973
Pierre Richard, Bernard Blier, Jena Rochefort, Mireille Darc, Jean Carmet, Colette Castrel, Paul Le Person, Jean Obe, Robert Castel, Roger Caccia, Robert Dalban, Jean Saudray, Arlette Balkis, Yves Robert, Maurice Barrier
An innocent man becomes a decoy and thus a target on account of a power struggle within the French Intelligence service. A dull and seemingly interminable spoof.
Aka: LE GRAND BLOND AVEC UNE CHAUSSURE NOIRE
COM   85 min (ort 89 min)   VIDrel: ARROW/RTM   V

TALL, DARK AND DEADLY **

15

Kenneth Fink USA 1995
Jack Scalia, Kim Delaney, Todd Allen, Gina Mastrogiacomo, Ely Pouget, James O'Sullivan, John S. Davies, Nik Hagler, Cynthia Dorn, Randy Means, Alissa Alban, Blue Deckert, Esteban Powell, Cliff Stephens, Eleese Lester, Brady Coleman
Just getting over a broken relationship, a female architect falls for a handsome stranger, but ultimately he proves to by a dangerous psychopath. Sounds familiar?
THR   85 min (ort 88 min)   VIDrel: CIC/SONOP   V

TALL GUY, THE **

15

Mel Smith USA 1988
Jeff Goldblum, Emma Thompson, Rowan Atkinson, Peter Kelly, Emil Wolk, Kim Thomson, Geraldine James, Anna Massey, Harold Innocent, John Inman, Jonathan Ross, Melvyn Bragg, Mel Smith, Joanna Kanska, Peter Kelly, Timothy Barlow
A talented but accident-prone American actor works as nothing more than a stooge to an obnoxious and self-centred comedian. When he falls for a nurse who is treating him for hay fever, he misses a performance and gets fired. Many fruitless interviews later, he gets his break playing in a musical version of "The Elephant Man". Smith's directorial debut is an original but tepid work, that fails to stretch the comic talents of all concerned.
COM   88 min   VIDrel: 4-FRONT/POLYREC/VISVID V/sh

TALL IN THE SADDLE ***

U

Edwin L. Marin USA 1944
John Wayne, Ella Raines, Gabby Hayes, Elisabeth Risdon, Raymond Hatton, Ward Bond, Audrey Long, Don Douglas, Russell Wade, Frank Puglia, Emory Parnell, Raymond Hatton, Paul Fix, Harry Woods, Cy Kendall, Bob McKenzie
In his first film for RKO, Wayne plays a woman-hating cowhand who goes to work at a ranch and suddenly finds himself saddled with a female boss when the owner dies. He soon finds himself knee-deep in a mystery revolving two villains with land-grabbing plans. A rip-roaring, well-acted and directed film, much enlivened by the excellence of the cast.
WES   87 min B/W   VIDrel: VCC/DISC/COLUM L/A   V

TALL TALE: THE UNBELIEVABLE ADVENTURES OF PECOS BILL **

PG

Jeremiah S. Chechik USA 1995
Patrick Swayze, Oliver Platt, Roger Aaron Brown, Scott Glenn, Nick Stahl, Catherine O'Hara
A ruthless property developer/industrialist attempts to take over an entire valley, but his efforts are opposed by a young lad, who dreams up three characters to help him: a wily cowboy (the title figure), a tough mountain man and a powerful black Southerner. This fantasy owes some of its inspiration to THE WIZARD OF OZ, but there the similarity ends and the muddle begins, with the film presenting an anti-progress message that is both contrived and clumsy.
FAN   93 min (ort 97 min)   VIDrel: WDV/TECH   V

TALONS OF THE EAGLE **

18

Michael Kennedy USA 1992
Billy Blanks, Jalal Merhi, James Hong, Priscilla Barnes, Matthias Hues, Pan Qing Fu
A martial arts expert teams up with a tough detective in order to infiltrate a drugs syndicate responsible for the deaths of three DEA agents. A standard, fast-paced action tale, with no shortage of violent action.
MAR   92 min (ort 96 min) dubbed
VIDrel: POLY/POLYREC/BRAVE L/A   V

TAMARIND SEED, THE ***

15

Blake Edwards USA 1974
Julie Andrews, Omar Sharif, Anthony Quayle, Daniel O'Herlihy, Sylvia Sims, Oscar Homolka, Bryan Marshall, David Baron, Celia Bannerman, Roger Dann, Sharon Duce, George Mikell, Kate O'Mara, Constantin De Goguel
An innocent romance between an attractive widow and a Russian military attache in Barbados, leads to complications for the intelligence community when he decides to defect. A blend of romance and espionage that's well handled and fairly modest in scope. Written by Edwards.
THR   119 min (ort 125 min)   VIDrel: L/A   V
Boa: novel by Evelyn Anthony.

TAMING OF THE SHREW, THE ***

U

Franco Zeffirelli ITALY/USA 1967
Richard Burton, Elizabeth Taylor, Michael Hordern, Vernon Dobtcheff, Natasha Pyne, Michael York, Cyril Cusack, Alan Webb, Victor Spinetti, Alfred Lynch, Roy Holder, Mark Dignam, Bice Valori, Giancarlo Cobelli, Ken Parry
Excellent film version of the famous comedy with Burton and Taylor in fine form, ably supported by a good cast and splendid photography. The musical score is by Nino Rota. More of a film in the true sense than is generally the case with adaptations of this playwright's works, which so often become mere filmed plays. A colourful and bawdy romp.
COM   122 min (ort 126 min)   VIDrel: VCC/DISC/COLUM V
Boa: play by William Shakespeare.

TAMING OF THE SHREW, THE ****

U

Jonathan Miller UK 1980
Simon Chandler, Anthony Pedley, John Thornton-Roberts, Frank Thornton, Sarah Badel, John Cleese, Jonathan Cecil, Susan Penhaligon, Harry Waters, Bev Willis, David Kincaid, Angus Lennie, Harry Webster, Gil Morris, Leslie Sarony
A full-blooded and diverting version of this play of love and conquest, with Cleese effective as Petruchio and Badel equally good as Katharine, the woman he pursues and ultimately wins. An inspiring interpretation, funny, vigorous and utterly absorbing.
DRA   126 min (ort 128 min) mTV   VIDrel: BBC   V/h
Boa: play by William Shakespeare.

TAMPOPO ***

18

Juzo Itami JAPAN 1987
Tsutomu Yamazaki, Nobuko Miyamoto, Koji Yakusho, Ken Watanabe, Rikiya Yasuoka, Kinzo Sakura, Manpei Ikeuchi, Yoshi Kato, Shuji Otaki, Fukumi Kuroda, Setsuko Shinoi, Yoriko Doguchi, Masahiko Tsugawa, Tadakazu Kitami
The story of a truck driver who helps a widow both make a success of her noodle shop and discover the secret of the perfect noodle, forms the basis for this bizarre comedy. Not so much a film as a series of comical vignettes, all of which revolve around food.
Aka: DANDELION
COM   109 min (ort 117 min)   VIDrel: ELPIC/POLYREC   V

TANAMERA: PARTS 1, 2 AND 3 ***

15

Kevin Dobson/John Power AUSTRALIA/UK 1989
Christopher Bowen, Lewis Fiander, Kay Tong Lim, Anne-Louise Lambert, Kyhm Lam, Ed Devereaux, Gary Sweet, Penny Hackforth Jones, Robert Coleby, Brian Marshall, Anthony Calf, Tushka Bergen
A lavish mini-saga that tells of the rich English expatriates who lived a life of luxury in Singapore up to its fall to the Japanese in 1942. The son of a wealthy aristocratic family falls for a Chinese girl but his parents do all they can to keep them apart, and when WW2 breaks out they are soon separated. A well-acted and detailed film that delivers all the usual soap opera-style cliches.
DRA   336 min (2 cassettes)   VIDrel: CENTV/VCC L/A   V
Boa: novel by Noel Barber.

TANDEM **

18

Toshiki Sato JAPAN 1994
Kino Mahito, Hazuki Hotaru, Ishiwara Yuri
Late one night, two men of different ages and backgrounds meet by chance at a coffee bar and begin talking about their respective sex lives, a discussion that continues (with various flashback sequences) as they take a motorbike journey together. An ambitious attempt to examine the conventions and themes of Japan's so-called "pink cinema", and place them in their cultural context. But for Westerners, the movie is demanding, disorganised and not a little confusing.
A   60 min   VIDrel: VISCOM/RTM   V

TANGO *** 15
Patrice Leconte FRANCE 1993
Richard Bohringer, Thierry Lhermitte, Philippe Noiret, Miou Miou,
Judith Godreche, Carole Bouquet, Jean Rochefort, Michele Laroque,
Maxine Leroux, Laurent Gamelon, Jacques Mathou, Isabelle Wolfe,
Caroline Clerc, Yamine Dib
When his nephew's wife walks out on him, Noiret (who is a top
judge and woman-hater to boot) hires an assassin (Bohringer)
to get rid of her, his hold on the man being the fact that years
ago he was tried for the murder of his cheating wife and her
lover, but got off thanks to the judge. All three embark on a long
journey to Valence in search of this woman, in a caustic road
movie that takes swipes in all directions, and is both assured and
idiosyncratic.
COM 86 min (ort 90 min) wScrn VIDrel: ARTIF/20TH
V/sh

TANGO & CASH ** 18
Andre Konchalovsky/Albert Magnoli USA 1989
Sylvester Stallone, Kurt Russell, Terri Hatcher, Brion James, Geoffrey
Lewis, Jack Palance, James Hong, Marc Alaimo, Michael J. Pollard,
Philip Tan, Robert Z'dar, Lewis Arquette, Eddie Bunker, Leslie
Morris, Roy Brocksmith
Stallone and Russell team up as rival L.A. detectives who are
framed by a drugs mobster and have to unite in order to clear
their names. Reputed to have had a $55,000,000 budget, this
juvenile buddy-movie starts off well with excellent action
sequences (a highlight is the prison break) but soon descends
into self-parody. Mostly directed by Konchalovsky, it was
completed by Magnoli after the former quit in a dispute over the
film's ending.
A/AD 99 min (ort 104 min) VIDrel: WHV V/sur

TANK GIRL * 15
Rachel Talay USA 1995
Lori Petty, Malcom McDowell, Ice-T, Iggy Pop, Naomi Watts, Ann
Magnuson, Jeff Korber
Based on a comic-strip created by Alan Martin and James
Hewlett this MAD MAX clone is set in the Australian desert of
in 2033, where a powerful dictator controls the entire water
supply. The title character revolts against his rule, and with the
aid of a tank and some mutant kangaroo men sets out to
dispense her own brand of justice. Much cartoon violence and
the pop soundtrack (which was available in two versions) does
not make this wild spoof any more watchable.
FAN 99 min (ort 120 min) cC VIDrel: MGM/WHV V/sur
Boa: comic book by Alan Martin and Jamie Hewlett.

TAP *** PG
Nick Castle USA 1989
Gregory Hines, Suzzanne Douglas, Sammy Davis Jr, Savion Glover,
Joe Morton, Dick Anthony Williams, Terrence McNally, Sandman
Sims, Bunny Briggs, Steve Condos, Jimmy Slyde, Pat Rico, Arthur
Duncan, Harold Nicholas, Etta James
A trite but highly enjoyable story of an ex-con who has shunned
the use of his dancing skills for a life of crime and easy pickings.
When the pull of tap dancing becomes too strong, he forsakes
his old ways and learns a new set of values from an old-time
hoofer. A loving tribute to tap, with little plot but some great
dance sequences and a charming performance from Davis as
"Little Mo", a tap dancing legend.
DRA 106 min Cut (7 sec at UK cinema release – ort 110 min)
VIDrel: RCA L/A V/h

TAPEHEADS ** 15
Bill Fishman USA 1987
Tim Robbins, John Cusack, Doug McClure, Connie Stevens, Clu
Gulager, Mary Crosby, Katy Boyer, Lyle Alzado, Jessica Walter, Susan
Tyrrell, Junior Walker, Sam Moore, King Cotton, Ebbe Roe Smith,
Keith Joe Dick, John Durbin
A brash, energetic but puerile comedy in which con-artist
Cusack teams up with video genius Robbins to make some
music videos. As they struggle to make it in the L.A. rock music
scene they encounter a succession of wacky characters. A disor-
ganised and cluttered film, with some enjoyable celebrity
cameos and a rather good soundtrack.
COM 88 min Cut (48 sec – ort 97 min) VIDrel: RCA L/A
V

TAPS ** PG
Harold Becker USA 1981
Timothy Hutton, George C. Scott, Ronny Cox, Sean Penn, Tom
Cruise, Brendan Ward, Evan Handler, John P. Navin Jr, Billy Van
Zandt, Giancarlo Esposito, Donald Kimmell, Tim Wahrer, Tim Riley,
Jeff Rochlin, Rusty Jacobs
Army cadets occupy their academy by force, in order to
prevent the building from being demolished, but the dispute
escalates into violence. A kind of modern moral fable that
climaxes too soon and then has nowhere to go. Written by
Darryl Ponicsan and Robert Mark Kamen. This was Penn's
screen debut.
Aka: T.A.P.S.
DRA 121 min (ort 126 min) VIDrel: 20TH/TECH V/sh
Boa: novel Father Sky by Devery Freeman.

TARGET * 15
Arthur Penn USA 1985
Gene Hackman, Matt Dillon, Josef Sommer, Guy Boyd, Gail
Strickland, Gayle Hunnicutt, Victoria Fyodorova, Ilona Grubel,
Herbert Berghof, Richard Munch, Ray Fry, Jean-Pol Dubois, Werner
Pochath, Ulrich Haupt, James Selby
A son discovers a different side to his ex-CIA father, when the
latter is galvanised into action following the kidnapping of his
wife in Europe, by a former rival bent on revenge for an inci-
dent that took place eighteen years ago. A stolid and
cliche-ridden tale, that is so overlong that the few good action
scenes are able to do nothing for it.
THR 112 min (ort 117 min) VIDrel: 20TH/TECH V/sh

TARGET: HARRY ** 18
Henry Neill (Roger Corman) USA 1969
Vic Morrow, Victor Buono, Stanley Holloway, Suzanne Pleshette,
Cesar Romero, Charlotte Rampling, Michael Ansara, Katy Fraysse,
Christian Barbier, Fikret Hakan, Milton Reid, Jack Leonard, Ellen
Gilbert, Tony Barnum, Joanna Clerc
The owner of a seaplane is given the job of taking some stolen
banknote plates to Istanbul, in this mediocre reworking of THE
MALTESE FALCON. Originally made for TV but released
theatrically.
Aka: HOW TO MAKE IT; WHAT'S IN IT FOR HARRY?
THR 85 min mTV VIDrel: VCC/DISC V

TARGET WITNESS ** 18
Armand Garabidian USA 1994
Marc Singer, Debeorah Shelton, Michael Des Barres, Adrienne
Barbeau, Charles Napier, Gilbert Gottfried, Mark Hamill
Young actress who witnesses a brutal murder finds her own life
in danger in this straightforward action-thriller.
A/AD 81 min VIDrel: COLUM/SONOP V

TARKA THE OTTER *** PG
David Cobham UK 1978
Peter Bennett, Edward Underdown, Brenda Cavendish, John Leeson,
Reg Lye, George Waring, Stanley Lebor, Max Faulkner plus Peter
Ustinov (narration)
A tale following the life of an otter, from birth to adulthood, and
telling of the man who attempted to keep it as a household pet.
A warm and appealing film for nature lovers, set in Devon of
the 1920s. The script is by Gerald Durrell. See also RING OF
BRIGHT WATER.
DRA 87 min (ort 91 min) VIDrel: VCC/DISC/COLUM V
Boa: novel by Henry Williamson.

TARZAN, THE APEMAN * 15
John Derek USA 1981
Bo Derek, Richard Harris, Miles O'Keeffe, John Phillip Law, Akushula
Selayah, Steven Strong, Maxime Philoe, Leonard Bailey, Wilfrid Hyde
White, Laurie Mains, Harold Ayer
This unbearably dreary remake is merely a vehicle for Bo
Derek's talents. A dull and wearisome production. See also
GREYSTOKE: THE LEGEND OF TARZAN, LORD OF THE
APES.
A/AD 107 min (ort 112 min) VIDrel: L/A V

TARZAN THE FEARLESS ** 15
Robert Hill USA 1933
Buster Crabbe, Jacqueline Wells (Julie Bishop), E. Alyn Warren,
Edward Woods, Philo McCullough, Mathew Betz, Frank Lackteen,
Mischa Auer, Carlotta Monti, Symonia Boniface, Darby Jones, Al
Kikume, George De Normand
Tarzan aids a young girl in the search for her missing father
who has been kidnapped by followers of Zar, God of the
Jewelled Fingers. A fast-paced and actionful effort, with Crabbe
well cast in the title role. This is the feature version of a twelve-
chapter serial from Principal, and unusually, is based on an

actual Tarzan story by Buroughs. Unfortunately, it covers only the first four parts and thus ends on an inconclusive note.
A/AD 85 min B/W VIDrel: EUREKA/GOLD V
Boa: story by Edgar Rice Burroughs.

TASTE FOR KILLING, A ** 15
Lou Antonio USA 1992
Michael Biehn, Jason Bateman, Henry Thomas, Helen Cates, Blue Deckert, Dru Moser, Woody Watson, Brandon Smith, Harry Melching, Fred Lerner, Richard Dillard, Renee Zellwegger, Dan Ammermann, Diane Perella, Dennis Letts
A couple of rich kids go to work on an oil rig but have to cope with a hostile foreman who resents their privileged background. However, they learn to cope and are befriended by an older man. But as events progress they begin to have doubts about their new friend, who eventually proves to be a ruthless and scheming psychopath.
DRA 83 min (ort 87 min) mTV VIDrel: CIC/SONOP V

TASTE OF AMBROSIA, A *** 18
Candida Royalle/Veronica Hart USA 1987
Jeanna Fine, Randy Paul, Alexis Firestone, Rugby Rhodes
The first of three shot-on-video efforts from Royalle, notable for the romantic way in which they deal with sexuality. In "Nine Lives Hath My Love" Fine plays an artist married to a man who's jealous of the loving attention she lavishes on her cats. She finds a novel way of resolving his resentment. Hart's story is "The Pick Up", and follows the preparations being made by a woman, apparently a hooker, as she gets ready to meet a new customer.
A 67 min mVid VIDrel: MIA/DISC V

TASTE OF HONEY, A *** 15
Tony Richardson UK 1961
Rita Tushingham, Robert Stephens, Dora Bryan, Murray Melvin, Paul Danquah, David Boliver, Moira Kaye, Herbert Smith, Valerie Scarden, Rosalie Scase, Jack Yarker, Veronica Howard, Margo Cunningham, John Harrison, A. Goodman
In Salford the wayward teenage daughter of a widow has an affair with a black sailor and is helped through her pregnancy by a homosexual friend. A poignant and moving performance from Tushingham in her film debut, is complemented by good location work and strong characterisation.
DRA 96 min (ort 100 min) B/W VIDrel: CASPIC L/A V
Boa: play by Shelagh Delaney.

TASTE THE BLOOD OF DRACULA * 18
Peter Sasdy UK 1969
Christopher Lee, Linda Hayden, Geoffrey Keen, Gwen Watford, Peter Sallis, Anthony Corlan, Isla Blair, John Carson, Michael Jarvis, Ralph Bates, Roy Kinnear, Michael Ripper
This was the fourth Hammer stab at the Dracula legend (preceded by DRACULA HAS RISEN FROM THE GRAVE) and has a bunch of bored seekers of excitement murdering a servant of the Count, when the decadence he promised them turns out to be more than they can stomach. Dracula sets out to have his revenge, using a variety of people for this purpose. A flabby yarn that gets bogged down very quickly. Followed by THE SCARS OF DRACULA.
HOR 87 min (ort 95 min) VIDrel: WHV V/h

TATIE DANIELLE *** 15
Etienne Chatiliez FRANCE 1990
Tsilla Chelton, Catherine Jacob, Isabella Nanty, Neige Dolsky, Eric Prat, Laurence Fevrier, Virgine Pradal, Mathieu Foulon, Garry Cedoux, Andre Wilms, Patrick Bouchitey, Christine Pignet, Evelyne Didi, Bradley Harryman
When her faithful and much-tormented maid finally dies, a cantankerous old woman moves in with her unsuspecting nephew and his family. In next to no time she has transformed their lives into a living hell, inflicting vast mental and physical damage. A grotesque black comedy that is far too bleak to really amuse, and presents no interesting insights either. The performances are however, quite splendid and compensate for the lame ending.
DRA 106 min (ort 115 min) VIDrel: ELPIC/POLYREC V

TATTLE TALE ** PG
Baz Taylor USA 1992
C. Thomas Howell, Ally Sheedy
An actor on the brink of a career breakthrough finds his chances ruined when his ex-wife publishes an autobiography that contains damaging allegations about their married life.

Desperate to get her to tell the truth, he adopts the guise of a passionate Italian and sets out seducing her as part of a clever plan he has devised.
COM 89 min (ort 93 min) VIDrel: MED/COLUM L/A V/sh

TATTOO * 18
Bob Brooks USA 1981
Bruce Dern, Maud Adams, Leonard Frey, Rikke Borge, John Getz, Alan Leach, John Iachangelo, Cynthia Nixon, Trish Doolin, Anthony Mannino, Patricia Roe, Lex Monson, Jane Hoffman, Robert Burr, John Snyder, B.J. Cirell, Kevin O'Rourke
A mentally ill artist becomes obsessed with a famous model, and kidnaps her in order to use her body as a canvas. A sordid and offbeat tale that ends with our girl covered in tattoos and our artist dead. Written by Joyce Bunuel, the daughter-in-law of director Luis Bunuel.
DRA 98 min (ort 103 min) VIDrel: CASPIC/BMGREC L/A V
Boa: novel by Earl Thompson.

TAXI DRIVER *** 18
Martin Scorsese USA 1976
Robert De Niro, Cybill Shepherd, Harvey Keitel, Jodie Foster, Peter Boyle, Leonard Harris, Albert Brooks, Joe Spinell, Martin Scorsese, Diahnne Abbott, Vic Argo, Frank Adu, Gino Ardito, Gareth Avery, Harry Cohn, Cooper Cunningham
A very bleak look at life in New York City viewed through the eyes of one man, a crazed and lonely Vietnam veteran now working as a taxi driver. Written by Paul Schrader, who says that the story was inspired (if that is the right word) by the diaries of would-be assassin Arthur Bremer. The music is by Bernard Herrmann. Scorsese appears briefly as a passenger in De Niro's cab.
DRA 109 min (Cut at film release by 1 sec – ort 116 min)
wScrn VIDrel: COLUM/SONOP V/sur

TAXI TO THE TOILET *** 18
Frank Ripploh WEST GERMANY 1981
Frank Ripploh, Bernd Broderup, Magdalena Montezuma, Tabea Blumenschein, Gitte Lederer, Orpha Termin, Peter Fahrni, Dieter Gidde
A sleazy look at the world of Berlin homosexuals, and most particularly, the adventures of a teacher who is driven by his lover's jealousy to go out in search of fresh encounters, an action which eventually costs him his job. This brave and slightly comic film was Ripploh's directing debut, and though it is set in the pre-AIDS era, many of the observations made about Berlin's gay scene remain valid. A highly personal, rather oppressive but memorable work.
Aka: TAXI TO THE JOHN; TAXI TO THE LOO; TAXI ZUM KLO
A 90 min (ort 92 min) dubbed VIDrel: PRIDE/PARADOX V

T.C. 2000 * 18
T.J. Scott USA 1992
Bolo Yeung, Jalal Merhi, Billy Blanks, Bobbie Phillips, Matthias Hues, Harry Mok, Kelly Gallant, Ramsay Smoth, Gregory Philpott, M.J. Kang, Alex Appel, Scott Hogarth, Harold Howard, Douglas J. Lennox, Andy Pandoff, Kevin Lund
Martial arts action tale set in a future where the remnants of civilisation are under constant attack by roving bands and are defended by specially trained police officers. After a female cop is killed, her body is used as the basis for a super-cyborg (hence the title) that is to be used against the gangs. A variant of sorts on ROBOCOP that is full of intriguing but undeveloped ideas, and becomes increasingly confused and hard to follow. Very disappointing.
MAR 90 min (ort 99 min) VIDrel: GUILD/SONOP V/s

TEA FOR TWO *** U
David Butler USA 1950
Doris Day, Gordon Macrae, Gene Nelson, Eve Arden, Billy De Wolfe, S.Z. Sakall, Bill Goodwin, Virginia Gibson, Crauford Kent, Mary Eleanor Donahue, Johnny McGovern, Harry Harvey, George Baxter, Herschel Dougherty, Buddy Shaw
A wealthy heiress plans to star in a Broadway show she is prepared to finance but has to enter into a strange bargain when she asks her uncle and guardian for the cash. He promises it to her but only on condition that she agrees to spend the next twenty-four hours answering "no" to every question. A light-hearted and charming tale, with rap dialogue and much

enjoyable music. Songs include "I Only Have Eyes For You" and "I Want To be Happy".
MUS   93 min (ort 98 min)   VIDrel: WHV   V/dm

TEAMSTER BOSS: THE JACKIE PRESSER STORY ***
15
Alastair Reid CANADA/USA
1992
Brian Dennehy, Jeff Daniels, Maria Conchita Alonso, Eli Wallach
A real-life drama providing a fascinating glimpse of the movers and shakers during Reagan's presidency, most especially the title figure. As Jimmy Hoffa's successor he led of one of America's most powerful unions and was able to exert immense pressure on the Washington administration, but was tempted to resort to bribery and murder when he found it to his advantage. Dennehy gives a thoroughly convincing performance in the demanding central role. See also HOFFA.
Aka: JACKIE PRESSER STORY, THE; POWER PLAY: THE JACKIE PRESSER STORY
DRA   106 min (ort 111 min)   VIDrel: WHV   V/sh
Boa: book Mobbed Up by James Neff.

TEASE, THE *
18
John Leslie USA
1990
Lauren Hall, Selena Steele, Ashlyn Gere, Rayne, Randy West, Randy Spears, Tom Byron, T.T. Boy, Peter North, Marc Wallice
A man is chosen to ghost-write a woman's sex fantasies. She promises herself to him as his reward when he completes this assignment, but he grows impatient and wants her sooner. A badly scripted dud that rapidly gets bogged down in irrelevant interludes and meaningless flashbacks.
A   57 min Cut (11 min 44 sec – ort 75 min) mVid
VIDrel: IMPENT   V

TECHNO SAPIENS **
15
Lamar Card USA
199-
Terry O'Quinn, Timothy Patrick Cavanaugh, Evan Lurie, Ashley Anne Graham
A company develops a method a making the ultimate body-guards, the only drawback being that this process requires human bodies. This leads to a good deal of conflict, as there are others out to stop the efforts of the crazed scientist behind it all, but not before he has created an almost indestructible cyborg. A succession of well mounted stunts, explosions and shoot-outs are offered here, but the threadbare plot is painfully unoriginal.
FAN   90 min   VIDrel: NEWAGE/TECH L/A   V

TED AND VENUS **
18
Bud Cort USA
1990
Bud Cort, Carol Kane, Martin Mull, Vincent Schiavelli, Rhea Pearlman, Woody Harrelson, Andrea Martin, Timothy Leary, Kim Adams, James Brolin, Cassandra Peterson, Gena Rowlands
A cult poet becomes obsessed with a beautiful girl who rejects his advances but this only serves to fuel his passion until he pursues her mercilessly. A barely interesting tale of some distinctly odd characters, set in Southern California. This poor effort is Cort's directorial debut.
COM   100 min   VIDrel: 20VIS/SONOP   V

TEDDY: THE TRUE COURAGE OF A KENNEDY **
PG
Delbert Mann USA
1986
Craig T. Nelson, Susan Blakely, Kimber Shoop
This fact-based biopic tells of the courage with which the son of Senator Edward Kennedy faced the amputation of his leg when cancer was discovered. Though a reasonably absorbing drama, the subject matter would probably never have received a film treatment on its own merits.
DRA   86 min   VIDrel: ODY/SONOP   V

TEEN AGENT *
PG
William Dear USA
1991
Richard Grieco, Linda Hunt, Roger Rees, Robin Bartlett, Gabrielle Anwar, Geraldine James, Michael Siberry, Carole Davis, Frederick Coffin, Tom Rack, Roger Daltrey, Oliver Dear, Cyndy Preston, Michael Sinelnikoff, Travis Swords
When a young man fails to graduate from high school, his parents send him to France to attend a summer course in French. However, upon arrival he's met by agents of British Intelligence, who whisk him off to a secret rendezvous, where he is briefed on the nature of his mission. A dreary teen parody of innumerable secret agent-type movies, whose "mistaken identity" plot

device is both lame and contrived, in a film of little wit but much tedium.
Aka: IF LOOKS COULD KILL
A/AD   84 min (ort 88 min)   VIDrel: MGM/WHV   V/dm

TEEN VAMP *
18
Samuel Bradford USA
1988
Clu Gulager, Karen Carlsen, Angie Brown, Beau Bishop, Evans Dietz, Edd Anderson, Jude Gerard, Joey Terracina, Art Ruggles, David Lewis, Rose Smith, Mike Lane, Sherry Farmer, Ginger Folmer, Michele Kimpler, Jennifer Kimpler
A young jerk, constantly bullied at school, is bitten by a vampire prostitute and is magically transformed into an attractive hunk who now stands an excellent chance with the girl of his dreams. A poorly plotted and under-developed cousin to TEEN WOLF that lacks Fox's engaging personality to give it the sparkle it so badly needs.
COM   86 min   VIDrel: 20VIS/SONOP L/A   V

TEEN WITCH **
PG
Dorian Walker USA
1989
Robyn Elaine Lively, Dan Gauthier, Joshua Miller, Dick Sargent, Lisa Fuller, Zelda Rubinstein, Shelley Berman, Mandy Ingber, Noah Blake, Tina Marie Caspary, Megan A. Gallivan, Alssari Al-Shehail, Shelly Burman, Marcia Wallace, Dan Carter
A virtuous little teen comedy in which a girl discovers just before her 16th birthday that she is descended from a long line of Salem witches. She promptly uses her powers to attract the school's football hero. Quite a pleasant little effort but nothing substantial.
COM   88 min Cut (1 min 37 sec – ort 105 min)
VIDrel: POLY/POLYREC L/A   V

TEEN WOLF ***
PG
Rod Daniel USA
1985
Michael J. Fox, James Hampton, Susan Ursitti, Jerry Levine, Jim MacKrell, Scott Paulin, Lorie Griffin, Mark Arnold, Matt Adler, Mark Holton, Elizabeth Gorcey, Jay Tarses, Melanie Manos, Doug Savant, Charles Zucker, Clare Peck
A teenager finds that he is a werewolf, just like his dad, who now has some explaining to do. But this change gives him a certain popularity with his classmates. Fox is very appealing in his role and this is what largely carries the aimless story along. Followed by a sequel and an animated TV series.
Aka: TEEN WOLF 1
COM   92 min (Cut at film release by 13 sec)
VIDrel: POLY/POLYREC L/A   V

TEEN WOLF TOO *
PG
Christopher Leitch USA
1987
Jason Bateman, Kim Darby, John Astin, Paul Sand, James Hampton, Mark Holton, Estee Chandler, Robert Neary, Stuart Fratkin, Beth Ann Miller, Rachel Sharp
In this moronic sequel, the adventures of Teen Wolf's cousin are explored when the family trait makes itself apparent. Not a rewarding experience for him or for us.
COM   90 min (ort 95 min)
VIDrel: 4-FRONT/POLYREC/EIV L/A   V

TEENAGE BRIDE **
18
Gary Troy USA
1970
Sharon Kelly (Coleen Brennan), Cyndee Summers, Don Summerfield, Jayne Louise
A college drop-out stops off to visit his stepbrother en route to Florida, and gets involved with both the latter's wife and his mistress, in this turgid softcore sex romp. A minor sub-plot involving a private eye and his all-too-willing secretary provided a little much needed levity.
A   47 min (ort 75 min)   VIDrel: RAVEN/LWV   V

TEENAGE CATGIRLS IN HEAT **
15
Scott Perry USA
1993
Gary Graves, Carrie Vanston, Dave Cox, Helen Griffiths, Dorothy Layne, Chad Strader, Pamela Strader, Carly Strader, Raymond Zaplatar, Marissa Mireur, Chris Wineinger, Steve Johansen, Robin Biggs, Michael Monroe, Tina Martorell
A typical Troma offering, taking the form of a spoof, with cats succumbing to a magic Sphinx statue, killing themselves and returning as sex-crazed but murderous females. This puzzles both a local cat-warden and a young drifter, neither of whom can understand where all the cats in town have gone. As this is a Troma movie, we also get lots of topless shots of said catgirls.

Cheerful, inept nonsense. See INVASION OF THE BEE GIRLS for something similar.
A/AD 91 min (ort 93 min) VIDrel: TROMA/RTM V

TEENAGE INNOCENCE *
18
Chris Warfield USA 1982
John Alderman, Sandra Dempsey, Judy Medford
Two young girls decide to hitch-hike their way to pleasure, and after being picked up by a middle-aged man they spend two days trying to love him to death. A repulsive mixture of boredom and humiliation that takes an awfully long time to say absolutely nothing.
Aka: INNOCENT SEX; LITTLE MISS INNOCENCE; LITTLE MISS INNOCENT; TEENAGE INNOCENTS
A 70 min (ort 79 min) VIDrel: MOPIC/SGSVID V

TEENAGE MONSTER *
PG
Jacques Marquette USA 1958
Anne Gwynne, Stuart Wade, Gloria Catillo, Charles Courtney, Gilbert Perkins, Norman Leavitt, Stephen Parker, Jim McCullough, Gaybe Moradian, Frank Davis, Arthur Berkeley
A meteor transforms a youngster into a hairy, murderous monster, but fortunately, his mother still loves him. When she does her best to hide him from the authorities, who are investigating his rampages, she falls victim to the blackmailing attempt of a waitress. Pure, bottom-of-the-barrel SF dross.
Aka: METEOR MONSTER
HOR 64 min (ort 73 min) B/W VIDrel: FIRC/RTM V

TEENAGE MUTANT NINJA TURTLES: THE MOVIE **
PG
Steve Barron USA 1990
Judith Hoag, Elias Koteas, James Saito, Michael Turney, Raymond Serra, Jay Patterson, Toshishiro Obata, Jock Pais, Michael Sisti, Leif Tilden, David Foreman plus voices of: Robbie Rist, Kevin Clash, Brian Tochi, David McCharen
Based on comic strip characters created by Kevin Eastman and Peter Laird, this ambitious adaptation features four mutant turtles who started life in the sewers of New York. Having learnt the martial arts whilst living there, they emerge as crime-fighting, pizza-loving, jive-talking superheroes. The turtle costumes (courtesy of Jim Henson) are fine, but the other effects are shoddy and the film suffers badly from its poor production values.
JUV 87 min (Cut at UK cinema release – ort 95 min) VIDrel: 4-FRONT/POLYREC/VISVID V/sur

TEENAGE MUTANT NINJA TURTLES 2: THE SECRET OF THE OOZE **
PG
Michael Pressman USA 1991
Francois Chau, Paige Turco, David Warner, Ernie Reyes Jr, Michelan Sisti, Leif Tilden, Kenn Troum, Mark Caso, Kevin Clash, Raymond Obata, Raymond Sierra, Mark Ginther, Kurt Bryant, Kevin Nash, Joseph Amodei, Nick DeMarinis
A hastily assembled sequel to the first film that magnifies all the faults of its predecessor. Once again the heroic title creatures face the evil "Shredder", who plans to use on them the radioactive ooze that originally made them mutate. A thin plot and lousy dialogue kill this attempt at a live-action spin-off from the popular animated series. Dull, depressing and overlong; Vanilla Ice and their ninja rap music just make matters worse.
JUV 87 min (Cut at UK cinema release – ort 88 min) VIDrel: 20TH/TECH L/A V

TEENAGE MUTANT NINJA TURTLES 3 **
PG
Stuart Gillard USA 1992
Corey Feldman, Brian Tochi, Robbie Rist, Time Kelelher, Paige Turco, Elias Koteas, Stuart Wilson, Sab Shimono, Vivian Wu, Mark Corso, Malt Hill, Jima Raposa, David Frazer, James Murray, Henry Havashi, John Aylwood, Mak Takano
Second sequel in this series sees the Turtles in 17th century Japan where they aid villagers in their struggle against some ruthless pirates and their leader. A few gestures in the direction of anti-violence do little to make this entry any more acceptable.
JUV 91 min (ort 96 min) VIDrel: 20TH/TECH L/A V

TEENAGE TRAMP *
18
Anton Holden USA 1975
Robin Lee, Alisha Fontaine, Anthony Massena
A young blonde girl on the run from the law finds that she has to use her body in order to survive, but soon learns that she has become trapped in a life of degradation. A softcore exploitation film of little worth.
A 90 min Cut (49 sec) VIDrel: MOPIC/SGSVID V

TEKWAR 1: TEKWAR **
12
William Shatner USA 1993
Greg Evigan, William Shatner, Sheena Easton, Eugene Clark, Torri Higginson, Ray Jewers, Von Flores, David Hemblen, Marc Marut, Barry Morse, Sonja Smits, Maurice Dean Wint, Catherine N. Blythe, Dee McCafferty, Lena Doig, Joan Henry
One in a series of quickly churned-out SF movies (based on Shatner's novels) that are set in a frightening high-tech future of 2044, where evil TekLords use a black market drug (Tek) and special headgear to gain ever more control of society. After four years in a cryogenic prison for drug dealing, framed ex-cop Jake Cardigan is revived and sets out to locate the professor who created a device capable of destroying Tek. A battle with ruthless drug dealers ensues.
FAN 87 min mTV VIDrel: CIC/SONOP V
Boa: novel by William Shatner.

TEKWAR 2: TEKLORDS **
12
George Bloomfield USA 1994
William Shatner, Greg Evigan, Eugene Clarke, Torri Higginson
In this follow-up tale, an imprisoned TekLord still exerts his malign influence whilst awaiting trial. Meanwhile, a deadly computer virus is spreading misery throughout the Cyber-Matrix system. For the most part this is no more than an average SF adventure, but some excellent computer-generated effects add a modicum of interest.
FAN 86 min mTV VIDrel: CIC/SONOP V
Boa: novel by William Shatner.

TEKWAR 3: TEKLAB *
12
Timothy Bond USA 1994
William Shatner, Greg Evigan, Eugene Clarke, Torri Higginson
Further adventure set in London in 2044, where it would appear that the TekLords have been interfering in the British elections, and may possibly also been responsible for the murder of Crown Prince Albert and the theft of King Arthur's sword Excalibur. Former cop Jake Cardigan investigates. Probably the weakest entry in the series, the ludicrous plot painting a downright peculiar picture of a Britain of the future.
FAN 85 min mTV VIDrel: CIC/SONOP V
Boa: novel by William Shatner.

TEKWAR 4: TEKJUSTICE **
PG
Gerard Ciccoritti USA 1994
William Shatner, Greg Evigan, Eugene Clarke, Torri Higginson
This final episode sees Jake Cardigan apparently confessing to the murder of his ex-wife's new husband. Meanwhile, our old friends the TekLords have come up with a fiendish scheme to take control of the worldwide Cyber-Matrix computer system. An adequate conclusion to a fairly unmemorable set of tales.
FAN 86 min mTV VIDrel: CIC/SONOP V
Boa: novel by William Shatner.

TELEFON **
PG
Don Siegel USA 1977
Charles Bronson, Lee Remick, Tyne Daly, Donald Pleasence, Patrick Magee, Alan Badel, Sheree North, Frank Marth, Helen Page Camp, Roy Jenson, Iggie Wolfington, Jacqueline Scott, Ed Bakey, John Mitchum, Kathleen O'Malley
A renegade Stalinist hardliner reactivates some sleeper agents in the USA, who have been hypnotically conditioned to perform acts of sabotage. To avoid a super-power conflict, the Kremlin send their top agent to deal with this menace. A tedious yarn devoting too much time to the relationship between the agent (Bronson) and his assistant (Remick). As ever, Finland provides the Soviet locations. Written by Peter Hyams and Stirling Silliphant.
THR 94 min (ort 103 min) VIDrel: MGM/WHV V
Boa: novel by Walter Wager.

TELEGRAPH TRAIL, THE **
U
Tenny Wright USA 1933
John Wayne, Marcelien Day, Frank McHugh, Otis Harlan, Albert J. Smith, Yakima Canutt, Lafe McKee, Clarence Geldert
After his best friend is killed by Indians, an Army scout tries to complete the erection of a telegraph line by calling on the local citizens to help out. He soon learns that a man has been fomenting this Indian unrest for reasons of his own and sets out to deal with him. A poor Wayne oater that incorporates footage from the 1926 silent "The Red Raiders".
WES 52 min (ort 60 min) B/W VIDrel: MGM/WHV V/h

TELEPHONE, THE *   15
Rip Torn USA   1987
Whoopi Goldberg, Severn Darden, Elliott Gould, John Heard, Amy Wright, John Hatton, Ronald J. Stallings, Linda Chu. Voices of: Don Blakely, James Victor, Robin Menken, Danae Torn, Herve Villechaize
The story of an out-of-work actress and her many problems, mostly psychological, and her penchant for conducting her life over the telephone. With a talent like Goldberg's, it's really a shame to see her working with such poor material; she sued to prevent this version being released, and no wonder. A dismal film that goes nowhere. Written by Terry Southern and Harry Nilsson. This was Torn's directorial debut.
COM   80 min (ort 96 min)   VIDrel: VCC/DISC/COLUM
L/A   V

TELL ME A RIDDLE ***   (U)
Lee Grant USA   1980
Lila Kedrova, Melvyn Douglas, Brooke Adams, Dolores Dorn, Bob Elross, Jon Harris, Zalman King, Lilli Valenty, Winnifred Mann, Peter Owens, Deborah Sussel, Nora Heflin, Peter Coyote, Nora Bendich
An elderly Russian-Jewish couple have virtually touched bottom in their many years of marriage, but bickering gives way to re-kindled emotions as the wife learns that she is dying from cancer, and sets off with her husband to make one last visit to her children. A somewhat slow-paced but well-acted and sincere movie. This was Grant's directorial debut.
DRA   94 min   SATrel: MOVIE CHANNEL
Boa: novel by Tillie Olsen.

TELL THEM WILLIE BOY IS HERE ***   15
Abraham Polonsky USA   1969
Robert Redford, Katherine Ross, Robert Blake, Susan Clark, Barry Sullivan, John Vernon, Charles McGraw, Charles Aidman, Shelly Novack, Ned Romero, John Day, Lee De Broux, George Tyne, Robert Lipton, Steve Shermayne, Lloyd Gough
A slow-paced re-enactment of a true incident in 1909, when an Indian was pursued by a massive posse after an accidental killing. Undoubtedly sincere in its desire to redress the balance in favour of the Indians, this film is not helped by casting Ross and Blake as Indians. Written and directed by Polonsky – his first film since making "Force Of Evil" in 1948 and his subsequent blacklisting.
DRA   93 min (ort 98 min)
VIDrel: 4-FRONT/POLYREC/CIC L/A   V/h
Boa: novel Willie Boy by Harry Lawton.

TELLING SECRETS ***   15
Marvin J. Chomsky USA   1993
Cybill Shepherd, Ken Olin, Christopher McDonald, G.D. Spradlin, Dylan Walsh, Nicolas Surovy, James McCaffrey, Andrew Robinson, Gray Grubbs, Laura Innes, Nada Despotovich, Melora Walters, Mary Kay Place, Anne Haney, Denise Gentile
Fact-based drama telling the story of Faith Kelsey, a ruthless and cold-hearted woman who is prepared to do anything to achieve her ends, including so arranging things that someone else is blamed for a murder she committed.
Aka: CONTRACT FOR MURDER
DRA   129 min mTV   VIDrel: ODY/SONOP   V/h

TEMP, THE **   15
Tom Holland USA   1992
Timothy Hutton, Laura Flynn Boyke, Dwight Schultz, Oliver Platt, Colleen Flyn, Faye Dunaway, Steven Weber, Scott Coffey, Dakin Matthews, Lin Shaye, Maura Tiereny, Michael Winters, Danny Swanson, Demene E. Hall, Jessuie Vint
An office temp turns out to be both a highly ambitious and skilled and a totally ruthless psycho who starts arranging a series of "accidents" for all those who get in her way. An over-the-top tale that focuses all its efforts on the title figure but neglects important aspects such as dialogue and plot. See also THE SECRETARY.
THR   93 min (ort 99 min)   VIDrel: CIC/SONOP   V/sur

TEMPEST, THE **   U
George Schaefer USA   1963
Richard Burton, Maurice Evans, Tim Poston, Lee Remick, Roddy McDowall, Liam Redmond, Ronald Radd, William H. Bassett, Geoffrey Lumb, Paul Ballantyne
Competent rather than exciting TV version of Shakespeare's play, first broadcast as a segment of "The Hallmark Hall of Fame".
DRA   76 min mTV   VIDrel: LEIS/DDVID   V
Boa: play by William Shakespeare.

TEMPEST, THE ***   U
John Gorrie UK   1980
Michael Hordern, Derek Godfrey, David Waller, Warren Clarke, David Dixon, Nigel Hawthorne, Andrew Sachs, John Nettleton, Alan Rowe, Pippa Guard, Kenneth Gilbert, Christopher Guard, Edwin Brown
A fairly well mounted and enjoyable television adaptation, one in a series produced by the BBC in the 1970s and 1980s.
DRA   124 min (ort 126 min) mTV   VIDrel: BBC   V/h
Boa: play by William Shakespeare.

TEMPEST, THE **   15
Paul Mazursky USA   1982
John Cassavetes, Gena Rowlands, Susan Sarandon, Vittorio Gassman, Raul Julia, Sam Robards, Molly Ringwald, Paul Stewart, Jackie Gayle, Lucianne Buchanan, Jerry Hardin, Tony Holland, Vassilis Glezakos, Sergio Nicolai
An architect runs away to a Greek island in an attempt to solve his personal problems, but events there seem to echo the famous play by Shakespeare. An ill-advised attempt to update Shakespeare, that sputters with a couple of jokes early on and then just settles down to smoulder. Written by Mazursky and Leon Capetanos.
COM   136 min (ort 142 min)   VIDrel: RCA L/A   V

TEMPTATION **   18
Strathford Hamilton USA   1994
Jeff Fahey, David Keith, Alison Doody, Philip Casnoff, Patricia Durham
A couple decide to take a tropical vacation but find their lives in danger when they become the prey of a ruthless killer. Average.
DRA   87 min (ort 91 min)   VIDrel: POLFIL   V

TEMPTRESS *   18
Lawrence Lanoff USA   1994
Kim Delaney, Chris Sarandon, Corbin Berensen, Dee Wallace Stone, Jessica Walter, Katrina McNeal, Gregory Wallace, Ben Cross
Supernatural thriller that sees a woman getting a tattoo on her breast while on holiday in India, leading to her being taken over by the Goddess Kali, who intends to use her to seduce and murder as many men as possible. Promoted on the strength of its erotic content, this is a tired effort, of uncertain direction and poor scripting. Its gory moments would have been more suited to a conventional horror yarn.
THR   89 min (ort 93 min)   VIDrel: 20TH/FOXVID   V/sh

TEN *   18
Blake Edwards USA   1979
Dudley Moore, Julie Andrews, Bo Derek, Robert Webber, Dee Wallace, Sam Jones, Brian Dennehy, Max Showalter, Don Calfa, Nedra Volz, James Noble, John Hawker, Virginia Kiser, Deborah Rush, Walter George Alton, John Hancock
A successful songwriter suddenly meets his perfect woman, and determines to woo and win her at any cost. The title refers to his habit of scoring his girlfriends out of ten. Another tiresomely awful Edwards offering. See THE WOMAN IN RED for a similar and slightly funnier treatment of this idea. The music is by Henry Mancini with Ravel's Bolero a recurrent theme.
Aka: "10"
COM   118 min (ort 123 min)   VIDrel: WHV   V/h

TEN COMMANDMENTS, THE ***   U
Cecil B. De Mille USA   1956
Charlton Heston, Yul Brynner, Anne Baxter, Edward G. Robinson, Yvonne De Carlo, Debra Paget, Cedric Hardwicke, John Derick, Nina Foch, Martha Scott, Judith Anderson, Vincent Price, John Carradine, Eduard Franz, Olive Deering
A stilted and flat retelling of the events described in Exodus, but well worth seeing for the good effects and some moments of genuine power. The story traces the life of Moses from his birth and abandonment through to manhood, and his leading of the Israelites out of Egypt. A dated but enjoyable and vivid epic, filmed once before by De Mille as a silent in 1923. The music is by Elmer Bernstein. AA: Effects (John Fulton).
A/AD   220 min wScrn   VIDrel: CIC/SONOP   V/dm V/sh

TEN DAYS' WONDER ***   15
Claude Chabrol FRANCE   1971
Orson Welles, Anthony Perkins, Michel Piccoli, Marlene Jobert, Guido Alberti, Ermano Casanova, Tsilla Chariton, Mathilde Ceccarelli, Eric Frisdal, Aline Montonvani, Corinne Koenigswarter, Giovanni Sciuto, Vittorio Sanipoli

Perkins is excellent as the troubled adopted son of a millionaire, who begins to grow increasingly worried about his bouts of amnesia, especially when he wakes up one day in a strange hotel with his hands and clothes bloodstained. When he goes to the estate of his adoptive father, he promptly falls in love with his young stepmother. A strange, perplexing and complex film of much atmosphere, it hovers precariously on the edge of pretentiousness.
Aka: LA DECADE PRODIGIEUSE
THR   105 min   VIDrel: ARTPRO/RTM   V
Boa: short story by Ellery Queen.

TEN MAGNIFICENT KILLERS **   18
Fong Yeh HONG KONG
Chang Li, Liu Yung, Chu Chih Ming, Yang Szu
Two gangsters abduct the infant son of a man they wish to destroy, their intention being to train the child and use him to kill his own father. Twenty years later, the hoped-for conflict does take place, but the father survives and it's not long before the son discovers his true parentage. And when he does so, he becomes rather annoyed with his former comrades.
MAR   86 min (ort 97 min)   VIDrel: IMPENT   V

TEN MILLION DOLLAR GETAWAY, THE **   15
James A. Contner USA   1991
John Mahoney, Karen Young, Terrence Mann, Tony Lo Bianco, Gary Bamman, Joseph Carberry, Terrence Mann, Kenneth John McGregor, Christopher Murray, Tom Noonan, Wendell Pierce, Tom Signorelli, Louis Spenser, Chuck Aber
Like an earlier German 1986 film, this one tells the true story of the planning and execution of the robbery of a Lufthansa airliner in 1978, at New York's Kennedy Airport; a coup that netted the thieves bank-notes to the value of $10,000,000. Initially a tense and extremely well acted heist film, the second half deals with the aftermath of the robbery and sadly reads like a tenth-rate Mafia novel, effectively dispelling any suspense.
A/AD   89 min (ort 93 min) mCab   VIDrel: CIC/SONOP   V
Boa: book by Doug Feiden.

TEN RILLINGTON PLACE ***   15
Richard Fleischer UK   1970
Richard Attenborough, Judy Geeson, John Hurt, Pat Heywood, Isobel Black, Phyllis MacMahon, Geoffrey Chater, Robert Hardy, Bernard Lee, Andre Morell, Robert Keegan, David Jackson, Edward Evans, Sam Kydd, Gabrielle Daye
A restrained and sombre account of the murderer Christie, whose tenant, Timothy Evans, was hanged for his deeds. Both Attenborough and Hurt give remarkably fine performances, in a film that is gruesome but of undeniable power. Scripted by Clive Exton and with photography by Denys Coop.
DRA   106 min (ort 110 min)   VIDrel: VCC/DISC/COLUM L/A   V
Boa: book by Ludovic Kennedy.

TEN SHAOLIN DISCIPLES **   18
HONG KONG
Elton Chong, Eagle Han, Sue Lee, Mike Wong, Susana Chan
An assassin is sent by the Imperial Court to kill the Abbott of Shaolin, and having done this, he steals a valuable Buddhist scripture. A group of female fighters now resolve to get the sacred text back.
Aka: 10 SHAOLIN DISCIPLES
MAR   85 min   VIDrel: IMPENT   V

TENANT, THE ****   18
Roman Polanski FRANCE/USA   1976
Roman Polanski, Isabelle Adjani, Melvyn Douglas, Jo Van Fleet, Shelley Winters, Bernard Fresson, Lila Kedrova, Claude Dauphin, Claude Pieplu, Jacques Monod, Patrice Alexandre, Josiane Balasko, Jean Pierre Bagot
Polanski plays a withdrawn bank clerk who moves into a furnished apartment whose previous tenant is in hospital after a suicide attempt. Little by little, he becomes drawn into a strange and nightmarish world in which he adopts the previous tenant's identity. Stylish, well acted and very, very intense. Written by Polanski and Gerard Brach.
Aka: LE LOCATAIRE
HOR   120 min Cut (6 sec – ort 125 min)
VIDrel: CIC/SONOP L/A   V
Boa: novel by Roland Topor.

TENANT OF WILDFELL HALL, THE ***   (PG)
Mike Barker UK   1996
Tara Fitzgerald, Rupert Graves, Toby Stephens, Beatie Ednie, Pam Ferris, Linda Marlowe, Miranda Pleasence, James Purefoy
A fine looking adaptation of Bronte's tale of Victorian England of 1848, where an unconventional woman becomes the latest tenant of the title property. When one of the local farmers falls in love with her, she allows him to read her private diary, which details the tragic events in the last few years of her married life. A very atmospheric adaptation, slightly spoilt by occasionally flashy camerawork.
DRA   159 min mTV   VIDrel: BBC/TECH   V/sh
Boa: novel by Anne Bronte.

TENDER, THE *   15
Robert Harmon USA   1990
John Travolta, Ellie Raab, Tito Larriva, Richard Edson, Vincent Gustaferro, Jeffrey DeMunn, LIsa Ziegler, R. Ruddell Weatherwax, Gene LeBell, Robert Stitzel, Robert Rigamonti, John Duda, Mick One, Sydney Chankin, Bob Zrna
A modern-day Lassie-type story in which the mutt makes his first appearance in a dog-fighting arena where he loses out to a ferocious Pit Bull Terrier. Dumped in the river by the owner, who believes his dog is dead, he's rescued by Travolta's daughter and nursed back to health. A few fairly unimportant complications follow, and this undemanding film wends its weary way to the obligatory happy ending.
Aka: EYES OF AN ANGEL
DRA   91 min   VIDrel: POLY/POLYREC L/A   V

TENDER AND PERVERSE EMMANUELLE ***   18
James P. Johnson (Jess Franco) FRANCE   1973
Lina Romay, Norma Castel, Alice Arno, Jack Taylor
Castel plays the title character in a film that has no connection at all with the well known EMMANUELLE films. Found at the bottom of a cliff, the police naturally suspect suicide, but the investigation reveals that she may have been murdered. As she had numerous lovers of both sex, most of the film is then taken up with a set of flashbacks that explore her background, plus that of her sister (played by Franco's wife Romay). An arty albeit rather sleazy effort.
A   76 min wScrn dubbed   VIDrel: REDEM/RTM   V

TENDER MERCIES **   PG
Bruce Beresford USA   1982
Robert Duvall, Tess Harper, Allan Hubbard, Betty Buckley, Ellen Barkin, Wilford Brimley, Lenny Von Dohlen, Paul Gleason, Allan Hubbard, Norman Bennett, Michael Crabtree, Andrew Scott Hollon, Rick Murray, Stephen Funchess
An attractive widow and her young son help a Country and Western singer rebuild his life, in a sweet but insipid concoction that's easy to watch but leaves barely a trace on the memory. Duvall wrote his own songs for this film. AA: Actor (Duvall), Screen/orig (Horton Foote).
DRA   88 min (ort 93 min) mTV   VIDrel: WHV   V/h

TENKO: PARTS 1 TO 10 **   (PG)
Pennant Roberts UK   1981
Stephanie Beacham, Louise Jameson, Stephanie Cole, Renee Asherson, Burt Kwouk, Claire Oberman, Joanna Hole, Wendy Williams
First shown in three series of ten parts each, this tedious story detailed the daily lives of European women held in a Japanese POW camp. Very much a standard BBC drama, it was quite watchable and a big hit with viewers, despite being marred by the occasional shrill histrionics and a woeful lack of imagination.
DRA   1,280 min (4 double-cassette packs) mTV
VIDrel: BBC/TECH   V/h

TENNESSEE NIGHTS **   15
Nicolas Gessner ITALY/SWITZERLAND/USA/WEST GERMANY   1989
Julian Sands, Stacey Dash, Ned Beatty, Ed Lauter, Denise Crosby, Mary Kane, Brian McNamara, Johnny Cash, Rod Steiger, Gary Grubbs, Vince Williams, Billy Ray Sharkey, Lew Hopson, Michael Ean Reid
A self-centred young British lawyer comes to Tennessee intending to take a relaxing fishing trip, but instead gets involved with a young girl and her father, and is later implicated in a murder. An overlong, convoluted thriller that fails to stir the imagination.
Aka: BLACK WATER; MINNIE
DRA   100 min (ort 105 min)   VIDrel: 20VIS/SONOP   V/sur
Boa: novel Minnie by Hans Werner Kettenbach.

TENNIS COURT, THE * (PG)
Cyril Frankel UK 1984
Peter Graves, Hannah Gordon, Ralph Arliss, Isla Blair, Jonathan Newth, Cyril Shaps, Marcus Gilbert, Peggy Sinclair, Annis Joslin, George Little, David Cheesman
A woman takes over her mother's home and finds that the indoor tennis court appears to be the centre of terrifying events, and that these are linked in some way with the family's past history. With the local clergyman strangely unhelpful, she attempts to deal with this matter herself. A Hammer film that was made for TV as one of several, and exhibits all the usual failings. Very poor.
HOR 72 min mTV SATrel: UK GOLD

TENTH MAN, THE * PG
Jack Gold UK 1988
Anthony Hopkins, Kristin Scott Thomas, Derek Jacobi, Cyril Cusack, Brenda Bruce, Paul Rogers, Timothy Watson, Peter Jonfield, Geoffrey Bayldon, Jim Carter, Michael Attwell, Robert Morgan, Patrice Valota, John Bennett
The story of a successful French lawyer who becomes a Nazi prisoner during WW2, and signs over all his property to another prisoner so that the latter will take his place in front of a firing squad. After the war he returns to his villa, which has now become the property of the dead man's family, and takes a job there. A languid and rather murky fable, attractively staged, but ultimately pointless.
DRA 94 min (ort 100 min) mTV VIDrel: MGM L/A V
Boa: novel by Graham Greene

TEQUILA SUNRISE * 15
Robert Towne USA 1988
Mel Gibson, Michelle Pfeiffer, Kurt Russell, Raul Julia, J.T. Walsh, Arliss Howard, Ann Magnuson, Ayre Gross, Gabriel Damon, Garret Pearson, Tom Nolan, Eric Thiele, Dawn Mantel, Lala, Budd Boetticher, Ann Magnuson, Bob Swain
Two lifelong buddies go their separate ways, with one becoming a drug dealer and the other a cop. They come into conflict when they both start taking an interest in the same woman. A film that starts off as if it's really going somewhere, and then just peters out. Written by Towne.
DRA 110 min (ort 116 min) VIDrel: MGM/WHV V/sur

TERMINAL FORCE * 15
William Mesa USA 199-
Brigitte Nielsen, Richard Moll, John H. Brennan, Craig Fairbrass
A warrior from a far-off world is sent to Earth to locate a magical gem that holds the key to creation and represents the only means of saving her planet from a ruthless invader. However, she has to contend with an enemy who has followed her. Nielsen struts and pouts her way through this SF actioner with reasonable aplomb, but it is really the special effects (the director was formerly a specialist in this area) that keep one entertained. Fair.
A/AD 86 min VIDrel: MED/DISC V/sh

TERMINAL JUSTICE * 18
Rick King USA 1995
Lorenzo Lamas, Chris Sarandon, Peter Coyote, Kari Salin
Virtual reality simulation has now replaced enjoyment of the real pleasures life has to offer, and is used to provide its users with simulated sex, drugs or games. A new criminal has developed to exploit this situation, and so has a new type of cop, one such officer being given the task of protecting a virtual reality sex goddess. An unusual and innovative effort, not quite given the development its opening premise requires.
FAN 91 min VIDrel: PROMARK/HIFLI V/h

TERMINAL MAN, THE * 15
Mike Hodges USA 1974
George Segal, Joan Hackett, Richard Dysart, Jill Clayburgh, Donald Moffat, Matt Clark, James B. Sikking, Michael C. Gwynne, Donald Moffatt, Ian Wolfe, Jason Wingreen, Normann Burton, William Hansen, Robert Ito
A computer scientist becomes a violent killer, when a computer program is implanted in his brain after he volunteers to undergo an experimental form of treatment, in an attempt to control his violent outbursts of rage. This adaptation of a best-selling novel starts with promise, but goes downhill in the second half, when the film concentrates heavily on the murders the subject commits after escaping from the hospital. See THE BLACK CAT for a similar idea.
FAN 100 min (ort 107 min) wScrn VIDrel: TART/20TH V
Boa: novel by J. Michael Crichton.

TERMINAL RUSH * 18
Damian Lee USA 1996
Don Wilson, Roddy Piper
A group of terrorists attack a dam in California, and only one man, a former soldier but now a prisoner, has the expertise to stop them. Standard action outing, with the usual heroics.
A/AD 90 min VIDrel: GUILD/FOXVID V

TERMINAL VELOCITY * 15
Deran Serafian USA 1994
Charlie Sheen, Natassja Kinski, James Gandolfini, Christopher McDonald, Gary Bullock, Hans R. Howes, Melvin Van Peebles, Suli McCullough, Cathryn de Prume, Richard Sarafian Jr, Lori Lynn Dickerson, Terry Finn, Martha Vazquez
After the death of a student leads to the closure of his parachute school, an instructor begins his own investigation. He gradually learns that he has been set up and that the woman's death was faked in order to protect her from the Russian Mafia. Very much a standard action tale but burdened by an over-complex espionage plot. However, in terms of sheer visual impact, this film can stand alongside any others of this genre.
A/AD 98 min (ort 132 min) cC VIDrel: HOLPIC/TECH V

TERMINATOR, THE ** 18
James Cameron USA 1984
Arnold Schwarzenegger, Linda Hamilton, Paul Winfield, Michael Biehn, Lance Henriksen, Rick Rossovich, Earl Boen, Dick Miller, Bess Motta, Shawn Shepps, Bruce M. Kerner, Franco Columbo, Bill Paxton, Brian Thompson, Tom Oberhaus
A humanoid killer cyborg is sent back in time from a future society ruled by machines. Its task is to track down and kill a woman whose as yet unborn son will overthrow the rule of the robots. Schwarzenegger as the unstoppable (and murderous) man-machine is frightening. A violent action fantasy, it was the director's debut, and within the narrow boundaries of its plot is an impressive work. Produced by Gale Anne Hurd, who co-wrote with Cameron. A sequel followed.
FAN 102 min (ort 108 min) VIDrel: GUILD/POLYREC L/A V

TERMINATOR 2: JUDGEMENT DAY ** 15
James Cameron USA 1991
Arnold Schwarzenegger, Linda Hamilton, Robert Patrick, Edward Furlong, Earl Boen, Joe Morton, S. Epatha Merkerson, Castulo Guerra, Danny Cocksey, Jenette Goldstein, Xander Berkeley, Leslie Hamilton Gearren, Ken Gibbel, Robert Winley
Said to be the world's first $100,000,000 film, with Arnie cast once more as the deadly killer-robot from the future (a duplicate of the original) but now programmed to back humanity against the machine that is to supersede him, an advanced "liquid-metal" creation. Incredible special effects keep one watching, but the film is well plotted too. (The 139 min "director's cut" includes a highly memorable sequence). The LV disc includes short documentary on the film.
FAN 130 min Cut (18 sec plus film cuts – ort 139 min)stickins VIDrel: POLY/POLYREC; PION (LV only) V/sur LV

TERMS OF ENDEARMENT * 15
James L. Brooks USA 1983
Shirley MacLaine, Debra Winger, Jack Nicholson, John Lithgow, Jeff Daniels, Danny DeVito, Lisa Hart Carroll, Betty King, Huckleberry Fox, Megan Morris, Troy Bishop, Shane Serwin, Jennifer Josey, Tara Yeakey, Norman Bennett, Tom Wees
The relationship between a mother and her daughter over the years has its ups and downs in this finely balanced mixture of comedy and drama, which tips over into melodrama when the daughter dies of cancer. Nicholson gives an offbeat performance as a former astronaut unable to cope with the lack of excitement in everyday life. Followed by THE EVENING STAR. AA: Pic, Dir, Actress (MacLaine), S. Actor (Nicholson), Screen/adapt (Brooks).
DRA 126 min (ort 132 min) VIDrel: CIC/SONOP L/A V/h
Boa: novel by Larry McMurtry.

TERROR AT THE OPERA * 18
Dario Argento ITALY 1987
Cristina Marsillach, Ian Charleson, Urbano Barberini, William McNamara, Daria Nicolodi, Cristina Giachino, Antonella Vitale, Coralina Cataldi Tassoni, Barbara Cupisti, Gyorgy Gyorgiwanyi, Francesca Cassola
La Scala opera house in Milan is the appropriate setting for this gory tale, in which a young director of horror films is about to produce an avant-garde version of Verdi's Macbeth, when disas-

ter strikes and his prima donna is injured in a car crash. A series of brutal killings soon follow. The camerawork is impressive but the sheer emphasis on violence at the expense of atmosphere makes for very depressing viewing indeed.
Aka: DARIO ARGENTO'S OPERA; OPERA
HOR   91 min (Cut at UK cinema release by 47 sec – ort 107 min)   VIDrel: VIR/RCA L/A   V

TERROR IN BEVERLY HILLS **
John Myhers USA
18
1988
Frank Stallone, Behrouz Vossoughi, Cameron Mitchell, William Smith, Lysa Hayland
An ex-Marine is sent to rescue the daughter of the US President, when she is taken hostage by Middle East terrorists who have also kidnapped his family to stop the mission. Standard formula heroics that serves as a good vehicle for Stallone.
A/AD   88 min   VIDrel: GUILD/SONOP   V

TERROR IN THE NIGHT ***
Colin Bucksey USA
18
1995
Joe Penny, Valerie Landsburg, Matt Mulhern, Justine Bateman, Al Wiggins, John Bennes, Barry Bell, Pat Miller, Josh Copelan, Lindsay Brookman, Greg Hohn, Marc Macaulay, James Martin Jr, Howard Kinglade, Lisa Colley
A young couple out camping are kidnapped at gunpoint by a crazed killer, a psychotic police officer who is on the run with his girlfriend after having committed a murder. A disturbing fact-based story.
THR   88 min (ort 90 min)   VIDrel: 20TH   V

TERROR IN THE SHADOWS **
William A. Graham USA
(15)
1995
Genie Francis, Leigh J. McCloskey, Marcy Walker, Victoria Wyndham, Mark Damon Espinoza, Jacob Lyst, Marcy Walker, Frankie Ingrassia, Dan Manning, Cally Fredricksen, Kevin Cooney, Annie Fitzgerald, Reed Means, Nicholas Pappone
A couple find their cosy domestic life with their young son placed in danger when a severely disturbed former girlfriend of the husband escapes from mental hospital and heads their way, intent on revenge for perceived past injustices.
THR   87 min   SATrel: MOVIE CHANNEL
Boa: novel Night of Reunion by Michael Allegretto.

TERROR ON HIGHWAY 91 **
Jerry Jameson USA
15
1988
Ricky Schroder, George Dzundza, Matt Clark, Brad Dourif, David Sherrell, Lara Flynn Boyle, Frederic Lehne
A young man joins the local police force in a small Southern town but is dismayed to find it riddled with greed and corruption. Supposedly based on a true incident, but no more interesting for that.
DRA   90 min (ort 100 min) mTV   VIDrel: 20TH/TECH   V

TERROR ON TRACK 9 **
Robert Iscove USA
(18)
1992
Richard Crenna, Joan Van Ark, Cliff Gorman, Joseph Campanella, Ving Rhames, Bertila Damas, Stephen McHattie, Bethel Leslie, Swoosie Kurtz, August Schellenberg, Susan Blommaert, Peter Dvorsky, Harvey Atkin, Lawrence Bayne
Crenna plays New York cop Frank Janek in this entry in the series, who in this story is after a serial killer who selects his victims at Grand Central Station. When the killer picks up the niece of Janek's boss it creates a stir in the media and the pressure to crack the case builds up. The strong cast work well together to keep up the sense of suspense, but this average thriller has nothing new to offer in the plot department.
THR   90 min mTV   TVrel

TERROR TRAIN **
Roger Spottiswoode CANADA/USA
18
1980
Jamie Lee Curtis, Ben Johnson, Hart Bochner, David Copperfield, Sandee Currie, Derek MacKinnon, Timothy Webber, Anthony Sherwood, Howard Busgang, Greg Swanson, D.D. Winters, Joy Boushel, Victor Knight, Donald Lamoureux
A masked fancy-dress party aboard a train chartered by a college fraternity, is the unusual setting for a run-of-the-mill entrant to the teen slasher stakes. The culprit turns out to be a former student, demented by a fraternity initiation prank, and out to get his revenge. One interesting twist is that the dons disguise of each successive victim. The photography of John Alcott slightly redeems this routine affair.
Aka: TRAIN OF TERROR
HOR   93 min (ort 98 min)   VIDrel: MGM/WHV   V/h

TERROR VISION *
Ted Nicolaou USA
15
1986
Diane Franklin, Gerrit Graham, Mary Woronov, Chad Allen, Jonathan Gries, Jennifer Richards, Alejandro Rey, Bert Remsen, Randi Brooks, Sonny Clark Davis, Ian Patrick Williams, John Leamer, William Paulson
A family of strange characters find their satellite TV antenna attracts the unwelcome attentions of an alien creature, which comes to Earth and takes up residence in their television set. A comedy-horror yarn that sinks under the weight of its overblown special effects and clumsy humour.
Aka: TERRORVISION
COM   82 min (ort 85 min)   VIDrel: EIV/SONOP   V

TERROREYES **
Eric Parkinson USA
18
1988 (released 1991)
Daniel Roebuck, Vivian Schilling, Lance August
Gruesome horror spoof about a demon sent from Hell to recruit new scriptwriters. Having murdered the husband of a woman writer, he takes the man's place in order to help her complete her script for the Devil's next horror film – leading to three horror tales of variable quality. An over-the-top effort that tries for a type of humour reminiscent of BEETLEJUICE, but without much success.
HOR   90 min Cut (1 sec)   VIDrel: VIPCO/SGSVID   V/h

TERRY ON THE FENCE **
Frank Godwin UK
(U)
1985
Jack McNicholl, Neville Watson, Tracey-Ann Morris, Jeff Ward, Matthew Baker, Brian Coyle, Susan Jameson, Martin Fisk, Margery Mason
A schoolboy runs away from home and falls in with a band of thieves who force him to join them in their criminal activities. Produced by the Children's Film Foundation, this is a most undistinguished effort, watchable, but no more than that.
JUV   60 min   SATrel: MOVIE CHANNEL

TESS **
Roman Polanski FRANCE/UK
PG
1979
Nastassia Kinski, Leigh Lawson, Peter Firth, John Collin, John Bett, Tony Church, Tom Chadbon, Rosemary Martin, Sylvia Coleridge, Richard Pearson, Fred Bryant, Carolyn Pickles, Arielle Dombasle, David Markham
Film version of a classic novel about a peasant girl who tries to prove her aristocratic origins but comes unstuck when confronted by her betters. A solid, overlong and glossy work, quite evocative but not especially entertaining. The script is by Polanski, Gerard Brach and John Brownjohn. AA: Cin (Geoffrey Unsworth/Ghislain Cloquet), Art/Set (Pierre Guffroy and Jack Stephens), Cost (Anthony Powell).
DRA   164 min (ort 180 min) wScrn
VIDrel: GUILD/FOXVID   V/sur
Boa: novel Tess Of The D'Urbervilles by Thomas Hardy.

TESTAMENT ***
Lynne Littman USA
PG
1983
Jane Alexander, William Devane, Roxana Zal, Ross Harris, Lukas Haas, Lilia Skala, Leon Ames, Philip Anglim, Lurene Tuttle, Rebecca DeMornay, Kevin Costner, Mako, Mico Olmos, Gerry Murillo
A restrained view of how a small town is affected by a nuclear attack, that focuses on the effect this catastrophe has on one family. Quite harrowing, and infinitely better than THE DAY AFTER, but not well known as it was not well publicised. Written by John Sacret Young. See also THE LAST WAR.
DRA   86 min (ort 89 min)   VIDrel: L/A   V
Boa: short story The Last Testament by Carol Amen.

TESTAMENT OF ORPHEUS, THE ***
Jean Cocteau FRANCE
PG
1959
Jean Marais, Edouard Dermithe, Maria Casares, Francois Perier, Henri Cremiuex, Yul Brynner, Jean-Pierre Leaud, Daniel Gelin, Jean Marais, Pablo Picasso, Charles Aznavour, Francoise Christophe, Nicole Courcel
A surreal journey with the director through time and space as a poet seeks the meaning of existence. This self-indulgent and slightly mocking exercise of Cocteau's fails as a statement of his values, but is of interest for the melancholy array of bizarre images it presents. Written by Cocteau.
Aka: LE TESTAMENT D'ORPHEE
FAN   83 min B/W; 79 min (HOMVIS)
VIDrel: CONNO/RTM L/A   V

TESTIMONY ** PG
Tony Palmer UK 1987
Ben Kingsley, Sherry Baines, Peter Woodthorpe, Robert Stephens, Magdalen Asquith, Mark Asquith, Terence Rigby, Ronald Pickup, John Shrapnel, Mark Thrippleton, Robert Reynolds
Biopic on the life of composer Dmitri Shostakovich, with music played by the London Philharmonic, which is really the best thing in a movie that is distinctly overstretched, the script having too little in it for the film to be effective as a narrative. However, for devotees, there is ample footage of concerts, a strong sense of period and a sensitive and restrained performance from Kingsley, who demonstrates just very versatile he is.
DRA 150 min (ort 157 min) B/W/Col
VIDrel: CONNO/RTM V/sur

TETSUO: THE IRON MAN *** 18
Shinya Tsukamoto JAPAN 1992
Tomoroh Taguchi, Kei Fujiwara, Nobu Kanaoka, Shinya Tsukamoto, Naomasa Musaka, Renji Ishibashi
A weird cult fantasy tale about a Japanese office worker who, after a road traffic accident, begins to undergo a strange transmutation, rapidly turning into a metallic monster. This virtually indescribable blend of surrealism and sexuality (our central character still desires relations with his girlfriend) makes fascinating use of stop-action animation and other such techniques. A sequel followed.
FAN 67 min (ort 92 min) B/W VIDrel: ICAPRO/MANGA V

TETSUO 2: BODY HAMMER * 18
Shinya Tsukamoto JAPAN 1991
Tomoroh Taguchi, Nobu Kanaoka, Shinya Tsukamoto, Keinosuke Tomioka, Sujin Kim, Min Tanaka, Hideaki Tezuka, Tomoo Asada, Toraemon Utazawa
This sequel to the earlier film offers us some more bizarre bodily mutations, as a man whose young son is kidnapped from a Tokyo department store starts to transform into a human killing-machine. In transpires that he was one of two brothers, both of whom have inherited this ability, but to very different degrees. This nightmarish film is nothing like as coherent as the first one. A noisy, fast and violent journey nowhere. Screenplay is by the director.
HOR 79 min (ort 83 min) VIDrel: ICAPRO/MANGA V

TEXAS *** 15
Richard Lang USA 1993
Patrick Duffy, Maria Conchita Alonso, Ricky Schroder, John Schneider, Grant Show, Chelsea Field, David Keith, Stacy Keach, Randy Travis, Benjamin Bratt, Anthony Michael Hall,
Mini-series filmisation of Michener's epic recreation of the major events in the history of the state of Texas and its struggle to break free of Mexico. Set in 1821, this is a well made and acted film, albeit one with few plot twists or surprises.
Aka: JAMES A. MICHENER'S TEXAS
WES 180 min (ort 200 min) mTV VIDrel: POLFIL V/sh
Boa: novel by James A. Michener.

TEXAS ACROSS THE RIVER ** (PG)
Michael Gordon USA 1966
Alain Delon, Dean Martin, Joey Bishop, Rosemary Forsyth, Peter Graves, Tina (Aumont) Marquand, Andrew Prine, Michael Ansara, Linden Chiles, Stuart Anderson, Roy Barcroft, George Wallace, Richard Farnsworth, John Harmon, Don Beddoe
Western spoof with an assortment of characters and a complex plot, that has a Texan, an Indian and a Spanish nobleman going on the run to escape their enemies and experiencing various encounters along the way. A rambling affair to which the gags seem to have been added as an afterthought.
COM 97 min (ort 101 min) SATrel: SKY MOVIES

TEXAS ADIOS ** 15
Ferdinando Baldi ITALY/SPAIN 1966
Franco Nero, Luigi Pistilli, Cole Kitosch, Elisa Montes, Alberto Dell'Acqua, Jose Suarez, Hugo Blanco, Livio Lorenzoni
A Texas sheriff and his brother pursue their father's killer across the border in Mexico and finally track him down, only to experience a bitter disappointment, when the man's identity is revealed. A brutal spaghetti Western, atmospheric, well photographed (courtesy of Enzo Barboni) and predictable in its eventual outcome.
Aka: ADIOS, TEXAS; AVENGER, THE; GOODBYE TEXAS; TEXAS, ADDIO
WES 88 min wScrn dubbed (AKTIV/RTM)
VIDrel: AKTIV/RTM L/A; 4-FRONT/POLYREC V

TEXAS LIGHTNING ** 18
Gary Graver USA 1981
Cameron Mitchell, Peter Jason, Channing Mitchell, Maureen McCormick, Danone Camden, J.L. Clark
A boy goes on a hunting trip with his father and soon learns to chase more than game when he meets a shapely waitress.
COM 93 min VIDrel: IMPENT V

TEXASVILLE *** 15
Peter Bognanovich USA 1990
Jeff Bridges, Cybill Shepherd, Annie Potts, Cloris Leachman, Randy Quaid, Eileen Brennan, Timothy Bottoms, William McNamara, Su Hyatt, Angie Bolling, Harvey Christiansen, Pearl Jones, Loyd Catlett, Jimmy Howell, Romi Synder
An interesting sequel to THE LAST PICTURE SHOW that takes up the story some twenty years on, with our troubled teens now middle-aged parents with kids of their own. Good performances help retain one's attention, but the episodic nature of the narrative is a major handicap. Hardly as good as the original, but certainly not deserving of the critical panning it received on release. Screenplay is by Bogdanovich.
DRA 120 min (ort 123 min) VIDrel: GUILD/POLYREC L/A; PION (LV only) V LV
Boa: novel by Larry McMurtry.

THANK GOD IT'S FRIDAY * 15
Robert Klane USA 1978
Valerie Landsburg, Jeff Goldblum, Donna Summer, Terri Nunn, Chick Vennera, Debra Winger, Andrea Howard, Paul Jabara, Ray Vitte, Mark Lonow, Robin Menker, Marya Small, Chuck Sacci, The Commodores (with Lionel Ritchie)
The story of a Hollywood disc-jockey and his problems, set against the background of the disco where he works. A pounding, uninspiring and monotonous experience. The script is by Barry Armyan Bernstein. AA: Song ("Last Dance" – Paul Jabara).
MUS 86 min (ort 90 min) VIDrel: SUPVID/RTM V

THANK YOUR LUCKY STARS **** U
David Butler USA 1943
Eddie Cantor, Dinah Shore, Joan Leslie, Errol Flynn, Bette Davis, Humphrey Bogart, Edward Everett Horton, John Garfield, Alan Hale, Ann Sheridan, Jack Carson, Dennis Morgan, Olivia De Havilland, Ida Lupino, S.Z. Sakall
Cantor and his double (also, amazingly enough, played by Cantor) get involved in mounting a patriotic show. Not so much a film as a parade of talented performers. A lively, funny and most satisfying musical. Numbers include: "They're Either Too Young Or Too Old", "Love Isn't Born, It's Made", "I'm Going North" and "That's What You Jolly Well Get". Songs are by Frank Loesser and Arthur Schwartz.
MUS 123 min (ort 127 min) B/W VIDrel: WHV V/h

THAT GIRL IS A TRAMP ** 18
Jack Guy USA
Bente Nielsen, Sam Maree, Laura Viala, Yves Collignan
After running away from home, a young girl meets up with another girl and the two of them take on a job as nude models for photographers. Soon they graduate to more ambitious enterprises, such as fleecing the customers of their wallets.
Aka: LADY IS A TRAMP, THE; THAT LADY IS A TRAMP; THIS GIRL IS A TRAMP
A 85 min Cut (10 min 53 sec) VIDrel: IMPENT V

THAT HAMILTON WOMAN! *** PG
Alexander Korda USA 1941
Laurence Olivier, Vivien Leigh, Gladys Cooper, Alan Mowbray, Sara Allgood, Henry Wilcoxon, Heather Angel, Halliwell Hobbes, Gilbert Emery, Miles Mander, Ronald Sinclair, Luis Alberni, Norma Drury, Juliette Compton
An excellent film about Nelson's lifelong relationship with Emma Hamilton, it was made in the early days of WW2 as something of a patriotic flag-waver that it was hoped would ameliorate the marked hostility to Britain prevalent at the time. Though somewhat glamorised and stiff, it is nonetheless extremely effective, and Leigh is ravishing in the title role. The stirring music is by Miklos Rozsa. AA: Sound (Jack Whitney).
Aka: LADY HAMILTON
DRA 124 min (ort 128 min) B/W VIDrel: CARL/TECH V

THAT MAGIC MOMENT * 15
Daryl Duke USA 1990
Jace Alexander, Lindsay Frost, Cynthia Gibb, Jane Krakowski, Eric La Salle, Charles Hunter Walsh, Steven Weber, Ronny Cox, Wayne

Tippit, John Lehne, Don Reilly, Rebecca Cross, Galyn Gorg, Spencer Garrett, Tony Palone, Cary Pitts
Soap opera-style story of a group of American high school teenagers about to graduate in a small town during the summer of 1959, whose tight-knit camaraderie is shattered by the tragic deaths of two of its members. Further events conspire to split the group still further, with some of them going off to college whilst others stay behind. A dreary coming-of-age drama that never comes up with a strong and coherent narrative.
DRA   96 min   VIDrel: NWV/SONOP   V

THAT NIGHT *** 12
Craig Bolotin USA   1992
Juliette Lewis, C. Thomas Howell, Eliza Dushku, Helen Shaver, John Dossett, J. Smith-Cameron, Katherine Heigl, Sarah Jay Stevenson, Ben Terzulli, Thomas Terzulli, Michael Costello, Kathryn Meisle, Adam Lefevre, Carolyn Swift
In a small-town in 1961, a ten-year-girl becomes closely involved with a teenager whom she hero-worships. The latter, however, has no qualms about using her as a go-between in her romance with a local tearaway. Bolotin's debut is a leisurely paced and nicely handled low-key drama that offers some excellent performances.
DRA   85 min (ort 89 min) cC   VIDrel: WHV   V/sur
Boa: novel by Alice McDermott.

THAT NIGHT IN VARENNES *** 15
Ettore Scola FRANCE   1982
Marcello Madtroianni, Jean-Pierre Barrault, Hanna Schygulla, Harvey Keitel, Jean-Claude Brialy, Daniel Gelin, Jean-Louis Trintignant, Michel Piccoli, Andrea Ferreol, Elenore Hirt, Michel Vitold, Laura Betti, Enzo Jannacci
Historical drama set in pre-Revolutionary France, and based on the flight of Louis XVI and Marie Antoinette from Paris to Varennes, where their bid to rally the Royalists failed. However, the actual events form the background to the story of various characters (such as Thomas Paine and Casanova) whose fates intermingle as they spend the journey discussing politics, life and history. An impressive intellectual exercise, but not all that convincing.
Aka: LA NUIT DE VARENNES; NIGHT OF VARENNES, THE
DRA   150 min   VIDrel: ARROW/RTM   V

THAT OBSCURE OBJECT OF DESIRE *** 15
Luis Bunuel FRANCE/SPAIN   1977
Fernando Rey, Carole Bouquet, Angela Molina, Julien Bertheau, Andre Weber, Pieral (Pierre Aleyrangues), Milena Vukotic, Maria Asquerino, Ellen Bahl, Valerie Blanco, Auguste Carriere, Jacques Debary, Antonio Duque, Andre Lacombe
A middle-aged man is continually rebuffed in his obsessive attempts to seduce his maid, portrayed here by two actresses, a device Bunuel resorted to, after Maria Schneider left the production, This was the director's last film, based on a novel filmed twice before as a showcase for both Bardot and Dietrich. In essence, another well-formulated attack on the morality of the privileged classes.
Aka: CET OBSCUR OBJECT DU DESIR
DRA   99 min (ort 105 min) wScrn
VIDrel: ELPIC/POLYREC   V
Boa: novel La Femme Et Le Pantin by Pierre Louys.

THAT SECRET SUNDAY ** 15
Richard A. Colla USA   1986
James Farentino, Parker Stevenson, William Lucking, Daphne Ashbrook, Michael Lerner, Dan Hedaya, Charles Frank, Robert Romanus, Joe Regalbuto, Patrick Dollaghan, George Grizzard, Sondra Blake, Wendy Van Riesen, Lesley Ewen
Two reporters investigate an attempt to cover up the deaths of two young women, at a party attended mainly by police officers. A standard drama that sags rather badly, despite a strong performance from Farentino.
DRA   90 min (ort 92 min) mTV   VIDrel: 20TH/TECH L/A   V

THAT SINKING FEELING ** PG
Bill Forsyth UK   1979
Tom Mannion, Eddie Burt, Richard Demarco, Alex McKenzie, Margaret Adams, Kim Masterson, Danny Benson, Robert Buchanan, Drew Burns, Billy Greenlees, John Hughes, Eric Joseph, Alan Love, Derek Millar, James Ramsey, Janette Rankin
Very unfunny account of a gang of unemployed Glaswegian teenagers who think they've broken into the big time, when their leader unveils a scheme to boost their morale by stealing

stainless steel sinks from a warehouse. Released in the US after the success of GREGORY'S GIRL and LOCAL HERO, both of which were also by Forsyth, though this film marked his directorial debut.
COM   90 min (ort 93 min)   VIDrel: ODY/SONOP   V

THAT SUMMER OF WHITE ROSES *** 15
Rajko Grlic USA/YUGOSLAVIA   1989
Tom Conti, Susan George, Rod Steiger, Nitzan Sharron, Alun Armstrong, John Gill, John Sharp, Geoffrey Whitehead, Miljenko Brlecic, Vanja Drach, Slobodan Sembera, Stanka Gjuric
An atmospheric drama set in Nazi-occupied Yugoslavia where a naive and kindhearted lifeguard puts his life in danger by taking in the young widow of a partisan and her son, both of whom are on the run from the Nazis. His later action in saving a drowning Nazi officer leads to him being branded as a traitor, and his innocent world is brought to a shattering end. A film of improbabilities that's redeemed by fine performances and direction.
DRA   99 min   VIDrel: L/A   V/sh

THAT THING YOU DO! *** PG
Tom Hanks USA   1996
Tom Everett Scott, Liv Taylor, Johnathon Schaech, Steve Zahn, Ethan Embry, Tom Hanks, Charlize Theron, Obba Babatunde, Giovanni Ribisi, Chris Ellis, Alex Rocco, Bill Cobbs, Peter Scolari, Rita Wilson, Chris Isaak, Kevin Pollak
The story of a 1960s pop group and the various ups and downs, that unwisely has the band breaking up far too early on in the movie, hampering any excitement the lively script and catchy music might generate. That said, there is much to be enjoyed here (even if the title number does tend to get played ad nauseam). The script is by Hanks, and this was his directing debut.
MUS   107 min   CINrel

THAT TOUCH OF MINK ** U
Delbert Mann USA   1962
Cary Grant, Doris Day, Gig Young, Audrey Meadows, John Astin, Dick Sargent, Alan Hewitt, Joey Faye, John Fiedler, Willard Sage, Jack Livesey, Laurie Mitchell, John McKee, June Ericson, Laiola Wendorff, Roger Maris
Typical period sex comedy. Grant as his usual wooden self plays an unmarried executive who goes after his chaste secretary, the lovely Doris. Mildly amusing to start with, but as the film progresses it rapidly becomes quite tiresome. Written by Stanley Shapiro and Nate Monaster.
COM   95 min (ort 99 min)   VIDrel: 4-FRONT/POLYREC   V

THAT'LL BE THE DAY *** 15
Claude Whatham UK   1973
David Essex, Ringo Starr, Rosemary Leach, James Booth, Billy Fury, Keith Moon, Deborah Watling, Robert Lindsay, Brenda Bruce, Verna Harvey, Rosalind Ayres, James Ottoway, Beth Morris, Daphne Oxenford, Kim Braden, Ron Hackett
This tells of the beginnings of a young man's involvement in the world of rock music, and how it affects his life. Essex and Starr are splendid in a piece of well crafted entertainment. Followed by STARDUST which continues his story. Written by Ray Connolly, with musical direction by Keith Moon and Neil Aspinall.
DRA   87 min (ort 91 min)   VIDrel: WHV   V/h

THAT'S DANCING *** U
Jack Haley Jr USA   1985
Narrated by: Gene Kelly, Sammy Davis Jr, Mikhail Baryshnikov, Liza Minnelli, Ray Bolger
A selection of dance sequences from a wide range of movie musicals, with a host of poor sequences but several memorable ones too. A fascinating compilation in which the earlier numbers such as the Astaire-Rogers routines show just how deficient 1980s musicals are in comparison. Of added interest is a Ray Bolger number cut from THE WIZARD OF OZ.
MUS   103 min (ort 105 min)   VIDrel: MGM/WHV   V/sur

THAT'S ENTERTAINMENT: PART 1 *** U
Jack Haley Jr USA   1974
Fred Astaire, Bing Crosby, Gene Kelly, Peter Lawford, Liza Minnelli, Donald O'Connor, Debbie Reynolds, Frank Sinatra, James Stewart, Mickey Rooney, Elizabeth Taylor, Judy Garland, Esther Williams, Eleanor Powell (and others)
A compilation movie consisting of a selection of highlights from MGM musicals released between 1929 and 1958. A celebration

of nostalgia and talent, slightly spoilt by being clumsily organised and edited and sentimentally narrated. Nevertheless, there are many fine sequences. Written by Haley Jr and followed by a sequel. The selections are introduced by many of the stars that appeared in them.
MUS   122 min (ort 137 min)   VIDrel: MGM/WHV   V

THAT'S ENTERTAINMENT: PART 2 ***
U
Gene Kelly USA   1976
Jeanette MacDonald, Nelson Eddy, the Marx Brothers, Laurel and Hardy, Jack Buchanan, Judy Garland, Ann Miller, Mickey Rooney, Oscar Levant, Louis Armstrong (and others)
Introduced by Fred Astaire and Gene Kelly, this second compilation film has highlights from another collection of MGM movies, but lacks the verve of the first film. However, much of the material is memorable, and this time it wisely includes sequences from comedies and dramas.
MUS   121 min (ort 133 min)   VIDrel: MGM/WHV   V

THAT'S ENTERTAINMENT: PART 3 ***
U
Bud Friedgen/Michael J. Sheridan USA   1994
June Allyson, Cyd Charisse, Lena Horne, Howard Keel, Gene Kelly, Ann Miller, Debbie Reynolds, Mickey Rooney, Esther Williams (and others) plus Granville Van Dusen (narration)
Highlights from film classics such as SINGIN' IN THE RAIN, AN AMERICAN PARIS, KISS ME KATE and SHOW BOAT are just a few of the treats on offer here, as sequences from more than one-hundred musicals make their way across the screen. In addition, many discarded out-takes are also used, and the documentary provides a fascinating glimpse of Hollywood of days gone by.
DOC   108 min (ort 113 min) cC   VIDrel: MGM/WHV V/sur

THAT'S LIFE! ***
15
Blake Edwards USA   1986
Jack Lemmon, Julie Andrews, Sally Kellerman, Robert Loggia, Chris Lemmon, Emman Walton, Rob Knepper, Jennifer Edwards, Matt Lattanzi, Cynthia Sikes, Dana Sparks, Felicia Farr, Theodore Wilson, Nicky Blair, Jordan Christopher
A sharp albeit sour look at the emotional problems afflicting the members of an affluent family, most especially the father, who has just turned sixty, and his wife who thinks she may have cancer. Often quite touching, but hampered by the contrived script and an unrelenting tone of bitterness. However, this remains one of the director's more substantial efforts.
DRA   101 min   VIDrel: VCC/DISC/COLUM L/A   V

THEATRE OF BLOOD ***
18
Douglas Hickox UK   1973
Vincent Price, Diana Rigg, Ian Hendry, Harry Andrews, Coral Browne, Jack Hawkins, Michael Hordern, Arthur Lowe, Robert Morley, Dennis Price, Milo O'Shea, Diana Dors, Robert Coote, Joan Hickson, Renee Asherson, Eric Sykes
A mad actor in classical theatre decides to take a gruesome revenge on all the critics who have panned him over the years, by engineering a series of elaborate murder schemes, each of which recalls a scene from one of Shakespeare's plays. A black comedy of inventive plotting and repulsive effects.
HOR   100 min (ort 102 min)   VIDrel: WHV   V/h

THELMA & LOUISE ***
15
Ridley Scott USA   1991
Susan Sarandon, Geena Davis, Harvey Keitel, Christopher McDonald, Michael Madsen, Brad Pitt, Stephen Tobolowsky, Timothy Carhart, Lucinda Jenney, Jason Beghe, Sonny Carl Davis, Shelly Desai, Carol Mansell, Stephen Polk, Jack Lindine
Labelled the first female "buddy movie", this follows the exploits of two women who go on a fishing trip, leaving family and responsibilities behind. When one shoots a would-be rapist, they find themselves on the run from the law. Well acted and stylish, this diverting variant on the road-movie theme is weakened by the strident feminist script, in which the male characters are either shallow or stereotyped. See also LEAVING NORMAL.
AA: Screen/orig (Callie Khouri).
A/AD   124 min (ort 129 min) wScrn   VIDrel: WHV   V/sur

THEM! *
PG
Gordon Douglas USA   1954
Edmund Gwenn, James Arness, Joan Weldon, James Whitmore, Onslow Stevens, Don Shelton, Sean McClory, Chris Drake, Sandy Descher, Mary Ann Hokanson, Olin Howlin, William Schallert, Dub Taylor, Fess Parker, Leonard Nimoy

Radiation from a nuclear test causes ants to grow to giant size; the plot and acting are shrunk in proportion, thus achieving a kind of balance. The climax is in the sewers of L.A. Written by Ted Sherdeman, this is an example of a perfectly good idea that is a hostage to those "this is a job for the army" cliches of the 1950s. Fess Parker has a small role, as does Leonard Nimoy. The music is by Bronislau Kaper.
FAN   94 min B/W   VIDrel: L/A   V
Boa: short story by George Worthing Yates.

THEMROC ***
15
Claude Faraldo FRANCE   1972
Michel Piccoli, Beatrice Romand, Marilu Tolo, Francesca R. Coluzzi, Madame Herviale, Cafe De La Gare Theatre Group
Anarchic New Wave comedy (all the lines are spoken in gibberish) about the title character, who suffers a complete breakdown due to his dull life, and rejecting all modern values, reverts to a Stone Age style existence, smashing down the walls of his apartment in order to be able to live in a "cave". As matters progress, ever greater absurdities flow from this rebellion, and this attack on conformity and normality, though dated still remains very potent.
COM   104 min (ort 110 min)   VIDrel: ARTPRO/RTM   V

THEODORE REX *
PG
Jonathan Betuel USA   1995
Whoopi Goldberg, Armin Mueller-Stahl, Juliet Landau, Richard Roundtree, Bud Cort plus voice of George Newbern
In New York of the year 2013 dinosaurs are a commonplace sight, in fact they live and work alongside humans. When a female cop is assigned the task of solving a series of murders of these saurians, she is given one such creature as a partner. It's not that the idea of talking lizards is so bad (it works quite well in the TEENAGE MUTANT NINJA TURTLES movies) but simply that this terribly dull film has neither a decent story nor any good dialogue. Avoid it.
COM   88 min (ort 92 min)   VIDrel: EIV/SONOP   V

THEOREM ***
15
Pier Paolo Pasolini ITALY   1968
Silvana Mangano, Terence Stamp, Massimo Girotti, Anne Wiazemsky, Laura Betti, Andras Jose Cruz, Ninetto Davoli
A handsome stranger worms his way into the home of a wealthy industrialist and proceeds to seduce every member of his family in turn, before continuing on his way. A well-acted but obscure political parable that was banned by the Italian government who indicted the director on charges of obscenity (he was eventually acquitted). However, today it is hard to see what all the fuss was about.
Aka: TEOREMA
DRA   94 min   VIDrel: CONNO/RTM   V

THERE ARE NO CHILDREN HERE ***
15
Anita Addison USA   1993
Oprah Winfrey, Keith David, Mark Lane, Norman Golden II, Maya Angelou, Vonte Sweet, Crystal Laws Green, Ellis Peal, Eric McNeal, Relious Webb, Ashley Magby, Tiffany Magby, Phillip Edward Van Lear, Earl Johnson, Cheryl Lynn Bruce
A black woman does her best to protect her two sons from a life of crime in the slums of inner-city Chicago. A frank and unsentimental view of one family and their efforts to stick together, refreshingly free of all cliches and stereotypes. An excellent cast do their best to flesh out the thin storyline.
DRA   91 min mTV   VIDrel: 4-FRONT/POLYREC/ODY V/sh
Boa: book by Alex Kotlowitz.

THERE GOES MY BABY **
(12)
Floyd Mutrux USA   1994
Dermot Mulroney, Rick Schroder, Kelli Williams, Noah Wyle, Jill Schoelen, Kristin Minter, Seymour Cassel, Lucy Deakins, Kenny Ransom, Paul Xavier Gleason, Frederick Coffin, Janet MacLachlan, Andrew Robinson, Shon Greenblatt, Ele Keats
In the summer of 1965, a group of high-school seniors have to face up to the question of what to do with the rest of their lives. An unashamed nostalgic look at young people in this era, with all the expected themes and stock situations. Not terribly entertaining, but the fine period soundtrack adds some much-needed interest.
COM   95 min   SATrel: MOVIE CHANNEL

THERE WAS A CROOKED MAN ***   15
Joseph L. Mankiewicz USA   1970
Kirk Douglas, Henry Fonda, Warren Oates, Hume Cronyn, Burgess Meredith, John Randolph, Arthur O'Connell, Martin Gabel, Michael Blodgett, Claudia McNeil, Alan Hale, Victor French, Lee Grant, C.K. Yang, Pamela Hensley, Bert Freed
A murderer tries to escape from jail and recover his loot, but finds this rather difficult. An entertaining comedy that derives its humour from a battle of wits between Douglas and Fonda, as inmate and warden respectively. Written by David Newman and Robert Benton.
WES   118 min (ort 125 min) wScrn (special edition)
VIDrel: WHV   V/sh

THERE'S A GIRL IN MY SOUP **   15
Roy Boulting UK   1970
Peter Sellers, Goldie Hawn, Tony Britton, Nicky Henson, John Comer, Diana Dors, Judy Campbell, Gabrielle Drake, Nicola Paget, Geraldine Sherman, Thorley Walters, Christopher Cazenove, Raf de la Torre, Avril Angers, Eric Barker
A TV personality who is something of an ageing Don Juan, finds his life is vastly complicated when he picks up a young girl. A slightly amusing comedy that ends with all concerned back at the point they started. Sellers is woefully miscast in this flimsy screen version of the long-running play. The script is by Frisby.
COM   92 min (ort 96 min)   VIDrel: VCC/COLUM L/A   V
Boa: play by Terence Frisby.

THERE'S NO BUSINESS LIKE SHOW BUSINESS ***   U
Walter Lang USA   1954
Ethel Merman, Donald O'Connor, Marilyn Monroe, Dan Dailey, Johnnie Ray, Lee Patrick, Mitzi Gaynor, Hugh O'Brian, Frank McHugh, Rhys Williams, Chick Chandler, Eve Miller, Robin Raymond, Lyle Talbot, George Melford, Alvy Moore
An account of a family stage-act and their ups and downs, built around the songs of Irving Berlin. An entertaining musical that's well worth a look, despite the lack of plotting and a tendency to show off the fact that it was made in Cinemascope (a series of lavish and crowded numbers serving to fill up the screen). The music is by Lionel and Alfred Newman.
MUS   115 min (ort 117 min)   VIDrel: 20TH/TECH   V

THESE FOOLISH THINGS ***   PG
Bertrand Tavernier FRANCE   1990
Dirk Bogarde, Jane Birkin, Odette Laure, Emmanuelle Bataille, Charlotte Kady, Michele Minns, Sophie Dalezio, Sylvie Segalas, Helene Lafumat, Andree Duranson, Raymond Defendente, Fabrice Roux, Gilbert Guerrero, Louis Du Creux
When her father is admitted to hospital dying of cancer, his estranged daughter travels to France to patch up her relationship with him, and the two eventually work through their differences, developing a comfortable relationship based on mutual respect. Simple, very well acted indeed and highly enjoyable.
Aka: DADDY NOSTALGIE
DRA   102 min (ort 107 min) wScrn   VIDrel: TART/20TH   V

THESE THREE ***   U
William Wyler USA   1936
Miriam Hopkins, Merle Oberon, Joel McCrea, Bonita Granville, Alma Kruger, Margaret Hamilton, Walter Brennan, Catherine Doucet, Marcia Mae Jones, Mary Ann Durkin, Carmencita Johnson, Margaret Hamilton, Mary Louise Cooper
A schoolgirl falsely accuses one of her two teachers of having an affair and her vicious lies have a powerful effect on their lives. The lesbianism charges of the original play proved too shocking a theme to be retained, but the film is quite effective nonetheless.
DRA   92 min B/W   VIDrel: VCC/DISC   V
Boa: play The Children's Hour by Lillian Hellman.

THEY CALL ME MACHO WOMAN **   18
Patrick G. Donahue USA   1990
Debra Sweaney, Sean P. Donahue, Brian Oldfield, Jerry Johnson, Paul Henri, Roger Arildson
Oddly titled film that follows the exploits of a determined woman who takes on the might of some powerful corporations.
A/AD   80 min   VIDrel: TROMA/RTM   V

THEY DIED WITH THEIR BOOTS ON ***   U
Raoul Walsh USA   1941
Errol Flynn, Olivia De Havilland, Gene Lockhart, Regis Toomey,
Stanley Ridges, Arthur Kennedy, John Litel, Sydney Greenstreet, Hattie McDaniel, Anthony Quinn, Charles Grapewin, Walter Hampden, G.P. Huntlet Jr
An attempt to retell the story of the events leading up to the Battle of the Little Big Horn, with Custer cast in a romantic and honourable light, and the blame for this disaster firmly pinned on gunrunners. An episodic account, but mounted with a good deal of style. Photography is by Bert Glennon. The films CUSTER OF THE WEST and SON OF THE MORNING STAR also covered much the same ground.
WES   135 min (ort 140 min) B/W   VIDrel: MGM/WHV   V

THEY GOT ME COVERED **   U
David Butler USA   1943
Bob Hope, Dorothy Lamour, Lenore Aubert, Otto Preminger, Eduardo Ciannelli, Donald Meek, Marion Martin, Donald MacBride, Walter Catlett, Philip Ahn, John Abbott, Florence Bates, Mary Treen, Bettye Avery, Margaret Hayes
Wartime comedy tale of a bungling newspaper reporter who gets mixed up in a foreign spy ring. A good mixture of laughs and thrills done in the expected manic style. Written by Harry Kurnitz.
COM   90 min (ort 95 min) B/W   VIDrel: VCC/DISC   V

THEY LIVE **   18
John Carpenter USA   1988
Roddy Piper, Keith David, Meg Foster, George "Buck" Flower, Peter Jason, Raymond St Jacques, Jason Robards III, Larry Franco, John Lawrence, Susan Barnes, Sy Richardson, Susan Blanchard, Wendy Rainard, Lucille Meredith
This film's unusual premise is that aliens have succeeded in dominating society and walk undetected among the inhabitants, who are constantly bombarded with subliminal messages of obedience and consumption. A drifter who arrives in L.A. discovers this when he acquires a special pair of glasses enabling him to see the true appearance of the aliens. A good idea is thrown away, with the film becoming just one more urban actioner.
FAN   90 min (ort 97 min)
VIDrel: 4-FRONT/POLYREC/GUILD; PION (LV only)
V/sur LV
Boa: short story Eight O'Clock In The Morning by Ray Faraday Nelson.

THEY MADE ME A CRIMINAL ***   PG
Busby Berkeley USA   1939
John Garfield, Claude Rains, Ann Sheridan, Gloria Dickson, May Robson, Billy Halop, Leo Gorcey, Huntz Hall, Bobby Jordan, Gabriel Dell, Bernard Punsey, John Ridgely, Robert Gleckler, Barbara Pepper, William B. Davidson
Following a drunken brawl in which a man was killed, a boxer hides at a strange ranch out West, where he meets Robson and the "Dead End Kids". An effective remake of the 1933 film "The Life Of Jimmy Dolan", that is sustained by Garfield's sensitive performance rather than by any merits to be found in the script.
DRA   88 min (ort 92 min) B/W   VIDrel: VISVID/POLYREC   V

THEY MET IN THE DARK ***   U
Karel Lamac UK   1944
James Mason, Joyce Howard, Tom Walls, Phyllis Stanley, Edward Rigby, Ronald Ward, David Farrar
A respected naval commander suffers the disgrace of being found guilty of insubordination, and uncovers a Nazi spy-ring when he attempts to prove he is innocent. The plot may be full of holes, but one's attention is held by the actors, especially Mason in fine form as the naval officer in question and Howard as the naive young girl he gets involved with.
THR   91 min B/W   VIDrel: CARL/TECH   V
Boa: short story by Anthony Gilbert/novel by Basil Bartlett.

THEY SHALL HAVE MUSIC **   PG
Archie Mayo USA   1939
Walter Brennan, Joel McCrea, Gene Reynolds, Jascha Heifetz, Marjorie Main, Andrea Leeds, Porter Hall, Dolly Loehr (Diana Lynn), Terry Kilburn, Tommy Kelly, Chuck Stubbs, Walter Tetley, Alfred Newman, Jacqueline Nash
A young tearaway finds a ticket to a Heifetz concert and is so entranced by the music he hears that he enrols in a music school for slum kids. Finding that it is about to close for lack of funds, he asks Heifetz to play at a benefit concert. Worth seeing only

for the music, as Heifetz is certainly no actor and the rest of the film is a complete production-line effort.
Aka: MELODY OF YOUTH; RAGGED ANGELS
MUS   102 min B/W   VIDrel: VCC/DISC   V
Boa: novel by Charles I. Clifford.

THEY SHOOT HORSES, DON'T THEY? ***   15
Sydney Pollack USA   1969
Jane Fonda, Michael Sarrazin, Susannah York, Gig Young, Red Buttons, Bruce Dern, Bonnie Bedelia, Allyn Ann McLerie, Michael Conrad, Al Lewis, Robert Fields, Severn Darden, Jacquelyn Hyde, Felice Orlandi, Art Metrano
A dance marathon in 1932 in California is used as a microcosm of life in the Depression years, intertwined with many sub-plots built around the lives of the central characters. Young as the unctuous promoter is especially memorable. Written by James Poe and Robert E. Thompson. AA: S. Actor (Young).
DRA   113 min (ort 129 min)   VIDrel: 4-FRONT/POLYREC   L/A   V
Boa: novel by Horace McCoy.

THEY WATCH **   (PG)
John Korty USA   1993
Valerie Mahaffey, Patrick Bergin, Vanessa Redgrave, Nancy Moore Atchinson, Brandlyn Whitaker, Ken Strong, Christina Keefe, Rutanya Alda, Benji Wilhoite, Bill Bender, Jean Louisa Bradford, Charles McLawhorn, Genevieve Baens
After his daughter is killed in an auto accident, a father is overcome by grief and guilt at having let his work come between them. In this desperate state, he turns to a blind medium in the hope of contacting her in the afterlife.
Aka: THEY
DRA   95 min (ort 100 min) mTV   SATrel: MOVIE CHANNEL
Boa: short story They by Rudyard Kipling.

THEY WERE EXPENDABLE ***   U
John Ford USA   1945
Robert Montgomery, John Wayne, Donna Reed, Jack Holt, Ward Bond, Marshall Thompson, Paul Langton, Leon Ames, Arthur Walsh, Donald Curtis, Cameron Mitchell, Jeff York, Murray Alper, Harry Tenbrook, Jack Pennick
During the closing days of the Japanese invasion of the Philippines, the commander of a motor torpedo boat squadron eventually manages to prove to the reluctant top brass the military value of these PT boats. As Bataan and Corregidor fall, the squadron sinks many enemy ships although it is now too late to avoid final defeat. A sombre and downbeat film, untouched by flag-waving heroics.
WAR   129 min (ort 135 min) B/W   VIDrel: MGM/WHV   V/h
Boa: book by William L. White.

THEY WERE SISTERS ***   PG
Arthur Crabtree UK   1945
James Mason, Phyllis Calvert, Hugh Sinclair, Anne Crawford, Peter Murray Hill, Dulcie Gray, Barrie Livesey, Pamela Kellino
The story of three married sisters and their various domestic troubles forms the basis for this enjoyable drama set in a sleepy hamlet in the Home Counties. With each sister married to fellows of entirely different dispositions (one is a bounder, one is decent but dull and the third is a kind fellow) the story has ample scope to move about among the various characters, providing both entertainment and an examination (albeit dated) of the English middle-class.
DRA   109 min B/W   VIDrel: CARL/TECH   VV
Boa: novel by Dorothy Whipple.

THEY WHO DARE **   U
Lewis Milestone UK   1953
Dirk Bogarde, Denholm Elliott, Akim Tamiroff, Eric Pohlmann, Gerd Oury, Alec Mango, Kay Callard, Russell Enoch, David Peel, Sam Kydd, Lisa Gastoni
In WW2 a British raiding party are sent on a mission to destroy German airfields in occupied Rhodes. A noisy and explosive war film, quite gloomy and depressing, it is directed with competence rather than especial flair, a little disappointing as Milestone was the director of ALL QUIET ON THE WESTERN FRONT.
A/AD   103 min (ort 107 min)   VIDrel: WHV   V/h

THICK AS THIEVES **   15
Steve Di Marco USA   1990
Gerry Quigley, Carolyn Dunn, Amber-Lee Weston, Karl Pruner, Sara Botsford

A criminal couple enjoy the fruits of their chosen profession but their idyllic existence is spoilt by the arrival of their teenage cousin, who makes his living as a pickpocket.
A/AD   87 min   VIDrel: CAPIT/GUILD   V/sh

THICKER THAN BLOOD **   PG
Michael Diner USA   1993
Peter Strauss, Rachel Ticotin, Lynn Whitfield, Bob Dishy, Brenda Bazinet, Damir Andrei, Jacob Zelik Penn, Susan Hogan, Carolyn Dunn, John Robinson, Patricia Gage, Mark Wilson, Booth Savage, Maia Filar, Tara Charendoft
A man seeks custody of his girlfriend's son, whom he raised as his own over a three-year period, despite the fact that his relationship with the woman had virtually come to an end. Matters are not helped by an attempted abduction by the mother, and the case inevitably goes to court, where a blood test proves that he is not the boy's biological father. Good production values and acting raise this TV drama above the level usually found in such productions.
DRA   89 min mTV   VIDrel: ODY/SONOP   V

THIEF OF BAGDAD, THE ***   U
Raoul Walsh USA   1924
Douglas Fairbanks Sr, Julanne Johnston, Anna May Wong, Snitz Edwards, Charles Belcher, Brandon Hurst, Sojin, Winter Blossom, Etta Lee, Tote Du Crow, K. Nambu, Sadakichi Hartman, Noble Johnson, Mathilde Comont, Charles Stevens
The first version of this famous adventure story, with the title character having many adventures before rescuing Bagdad from the Mongol hordes, and winning the hand of the Caliph's daughter. A dated but vigorous and faithful rendition of the tale, written by Fairbanks and Lotta Woods. The sets are by William Cameron Menzies. Remade in 1940, 1961 and 1978. The newly recorded music track is the work of Carl Davis and the Philharmonia Orchestra.
A/AD   210 min (tinted version) B/W silent
VIDrel: THAMES/DISC   V

THIEF OF BAGDAD, THE ****   U
L. Berger/M. Powell/T. Whelan/Z. Korda/
W.C. Menzies/A. Korda UK   1940
Sabu, Conrad Veidt, June Duprez, John Justin, Morton Selten, Rex Ingram, Miles Malleson, Mary Morris, Bruce Winston, Hay Petrie, Roy Emerton, Allan Jeayes, Miki Hood, David Sharpe, Adelaide Hall
An entertaining Arabian Nights-style adventure which, though bearing very little resemblance to the original film, is a colourful blend of action and magic, as Sabu (aided by Ingram's splendid genie) outwits a wicked magician and saves a princess. Produced by Korda with music by Miklos Rozsa. AA: Cin (George Perrival), Art (Vincent Korda), Effects (vis – Lawrence Butler/aud – Jack Whitney).
A/AD   102 min (ort 109 min)   VIDrel: CARL/TECH   V

THIEF OF BAGDAD, THE **   U
Clive Donner FRANCE/UK   1978
Roddy McDowall, Kabir Bedi, Frank Finlay, Marina Vlady, Peter Ustinov, Terence Stamp, Pavla Ustinov, Ian Holm, Daniel Emilfork, Ahmed El-Shenawi, Kenji Tanaki, Neil McCarthy, Vincent Wong, Leon Greene, Bruce Montague
A fourth attempt at this Arabian Nights fantasy, that has Ustinov as the Caliph of Bagdad trying to find a suitable match for his daughter. Quite pleasant to look at, but hardly an inspired version.
FAN   86 min Cut (5 sec – ort 104 min) wScrn mTV
VIDrel: BRAVE/SONOP L/A   V

THIN ICE **   12
Fiona Cunningham Reid UK   1994
Sabra Williams, Charlotte Avery, James Dreyfus, Clare Higgins, Ian McKellen, Guy Williams, Barbara New, Martha Freud, Suzanne Bertish, Cathryn Harrison, Eamon Geoghegan, Gwyneth Strong, Nimmy March, Jimmy Gardner, Patsy Chilton
A couple of aspiring journalists plan to skate together in the Gay Games in New York, but when the man dumps his female partner just weeks before the event, the woman swiftly acquires a female partner. An unusual comedy with a strong lesbian flavour.
COM   85 min (ort 92 min) wScrn   VIDrel: DTK/RTM   V

THIN LINE BETWEEN LOVE AND HATE, A **   18
Martin Lawrence USA   1996
Martin Lawrence, Lynn Whitfield, Regina King, Bobby Brown, Della Reese, Malinda Williams, Daryl Mitchell, Roger E. Mosley, Simbi

Khali, Tangie Ambrose, Wendy Robinson, Stacii Jae Johnson, Miguel A. Nunez Jr, Faizon Love
A FATAL ATTRACTION-style clone with a man who runs a nightclub learning the dangers of spurning a woman after a one-night stand with her, and falling victim to her desire for revenge. A black comedy-drama of daft dialogue and rather crude slapstick antics, the film stands in need of a decent script and a clear direction.
DRA 104 min (ort 108 min) VIDrel: EIV/SONOP V

THING, THE *** 18
John Carpenter USA 1982
Kurt Russell, A. Wilford Brimley, Richard Dysart, Richard Masur, Donald Moffat, T.K. Carter, David Clennon, Charles Hallahan, Peter Maloney, Keith David, Joel Polis, Thomas Waites, Norbert Weisser, Larry Franco, Nate Irwin
A remake of the 1951 film THE THING FROM ANOTHER WORLD, which tells of the discovery at a frozen polar station of an alien creature, able to invade and control the bodies of those it comes into contact with. More faithful than its predecessor to the original story, this is a dullish film, but contains some of the most frightening and repulsive special effects ever put on film. Written by Bill Lancaster and with music by Ennio Morricone. See also PROTEUS.
Aka: JOHN CARPENTER'S THE THING
FAN 103 min VIDrel: 4-FRONT/POLYREC/CIC; PION (LV only) V/sur LV
Boa: short story Who Goes There? by John W. Campbell.

THING CALLED LOVE, THE *** 15
Peter Bogdanovich USA 1993
River Phoenix, Samantha Mathis, Dermot Mulroney, Sandra Bullock, K.T. Oslin, Anthony Clark, Webb Wilder, Earle Poole Ball, Wayne Grace, Micole Mercurio, Starletta Dupois, Larry Black, O'Neal Compton, Zoe Cassavetes, Lenae King
Four young people come to Nashville in the hope of breaking into the Country music business and face the usual problems and disappointments in their struggle for fame. A strictly average romantic drama, enhanced by a number of fine performances.
DRA 110 min (ort 116 min) cC VIDrel: CIC/SONOP V/sur

THING FROM ANOTHER WORLD, THE * PG
Christian Nyby USA 1951
Kenneth Tobey, Margaret Sheridan, Robert Cornthwaite, Dewey Martin, Douglas Spencer, Paul Frees, Robert Nichols, Eduard Franz, John Dierkes, William Self, Everett Glass, Tom Steele, James Arness (the Thing)
Scientists at a polar base discover a frozen alien and unwisely proceed to defrost it. A disappointing and rather moralistic adaptation of Campbell's story, promising great excitement early on and then simply failing to deliver. Produced by Howard Hawks (who is often credited with direction). Music is by Dmitri Tiomkin and the script by Charles Lederer. Remade (and very effectively too) as THE THING in 1982.
Aka: THING, THE
FAN 86 min (ort 87 min) B/W
VIDrel: 4-FRONT/POLYREC V
Boa: short story Who Goes There? by John W. Campbell Jr.

THINGS CHANGE ** PG
David Mamet USA 1988
Don Ameche, Joe Mantegna, Robert Prosky, J.J. Johnson, Ricky Jay, Mike Nussbaum, Jack Wallace, Dan Conway, J.T. Walsh, Vari Hausman, Gail Silver, Len Hodera, Josh Conescu, Merrill Holtzman, Adam Bitterman, W.H. Macy
A small-time gangster is given the job of "minding" a simple Italian cobbler, who has agreed for a fee, to take the rap for a Mob killing. However, instead of merely keeping him out of trouble, the gangster takes him out for one last wild fling. A minor comedy tale, sustained by appealing performances. The script is by Mamet and Shel Silverstein.
COM 96 min (ort 105 min) VIDrel: CONNO/RTM V

THINGS CHANGE ** 18
Paul Thomas USA 1993
Nikki Dial, Deidre Holland, Paula Harlow, Flame, Micky Ray, Woody Long, Jon Dough, Paul Thomas
Lisa has no knowledge of men and lives with her girlfriend Denise, but this is about to change when she initiates an affair with a man to whom she has long been attracted. Having moved out in order to find a place of her own, she then has a variety of one-night stands with several men. The first part of a longer movie that was cut in two for video release.
Aka: FOR THE LOVE OF LISA; THINGS CHANGE 1: MY FIRST TIME
A 83 min VIDrel: VIVID/SCRN V

THINGS OF LIFE, THE *** 15
Claude Sautet FRANCE/ITALY 1969
Romy Schneider, Michel Piccoli, Lea Massari, Gerard Lartigau, Jean Bouise, Boby Lapointe, Herve Sand, Jacques Richard, Betty Beckers, Dominique Zardi, Gabrielle Boulget, Roger Crouzet, Hanri Nassiet, Claude Conforates, Jean Gras
A man is torn between his new girlfriend and his estranged wife and their former life together. A carefully crafted study of the ordinary things that go to make up a person's life and habits. Well acted (especially from Piccoli in the lead) but hard to take on account of its plotless and circular structure (the opening sequence showing the road accident in which the central character is to die being one such example). Remade in 1993 as INTERSECTION.
Aka: LES CHOSES DE LA VIE
DRA 82 min (ort 89 min) VIDrel: ARROW/RTM V
Boa: novel Les Choes de la Vie by Paul Guimard.

THINGS TO COME **** PG
William Cameron Menzies UK 1936
Raymond Massey, Ralph Richardson, Cedric Hardwicke, Edward Chapman, Ann Todd, Margaretta Scott, John Clements, Maurice Braddell, Sophie Stewart, Derrick De Marney, Pearl Argyle, Kenneth Villiers, Ivan Brandt, Anne McLaren
The script is by H.G. Wells, in this look at the evolution of a World State after a devastating war. Despite a studio-bound feel, this is one of the most ambitious films in British cinema, which had great impact when first shown. This dated but classic work takes the form of a sequence of verbose episodes. The vibrant music of Arthur Bliss and the striking Menzies sets are its greatest strengths. Produced by Alexander Korda.
FAN 89 min (ort 113 min) B/W VIDrel: CARL/TECH V
Boa: novel The Shape of Things to Come by Herbert George Wells.

THINGS TO DO IN DENVER WHEN YOU'RE DEAD *** 18
Gary Fleder USA 1995
Andy Garcia, Christopher Walken, Gabrielle Anwar, William Forsythe, Treat Williams, Christopher Lloyd, Bill Nunn, Jack Warden, Steve Buscemi, Michael Nicolosi, Fairuza Balk, Josh Charles, Cheree Jaeb, Michael Nicolosi
Gore-laden melodrama with a gang doing one last job that goes horribly wrong, thus drawing vengeance down on them, with their wrathful boss sending an assassin to sort things out. Slightly similar to RESERVOIR DOGS, in that this is a crime-gone-wrong saga, but here the characters are drawn with considerably more skill.
THR 111 min (ort 115 min) cC VIDrel: TOUCH/TECH V/sh

THINK BIG ** PG
Jon Turtletaub USA 1990
Peter Paul, David Paul, Martin Mull, Ari Meyers, Richard Moll, David Carradine, Richard Kiel, Claudia Christian, Michael Winslow, Peter Lupus, Thomas Gottschalk
Twin truckers on their way to fetch a cargo of toxic waste get involved in a cross-country chase when they discover a girl hidden away on their truck. A A lame-brained action comedy that is an innocuous and unfunny vehicle for the Paul brothers, former professional wrestlers.
COM 86 min (ort 90 min) VIDrel: EIV/SONOP V

THIRD MAN, THE **** PG
Carol Reed USA 1949
Joseph Cotten, Alida Valli, Orson Welles, Trevor Howard, Bernard Lee, Ernst Deutsch, Wilfrid Hyde White, Paul Hoerbiger, Erich Ponto, Siegfried Breuer, Geoffrey Keen, Hedwig Bleibtreu, Annie Rosar, Herbert Halbik, Alexis Chesnakov
An American writer of cheap Westerns arrives in Vienna shortly after WW2, and finds that his friend whom he was to have met, has been killed in mysterious circumstances. His attempts to solve this mystery are constantly thwarted by strange incidents. A brilliant atmospheric film with memorable zither music by Anton Karas. Written by Graham Greene. The US version has a different opening narration. AA: Cin (Robert Krasker).
DRA 99 min (ort 104 min) B/W VIDrel: WHV V/h
Boa: novel by Graham Greene.

THIRST ***   18
Rod Hardy AUSTRALIA   1979
David Hemmings, Henry Silva, Chantal Contouri, Max Phipps,
Shirley Cameron, Rod Mullinar, Robert Thompson, Walter Pym, Rosie
Sturgess, Amanda Muggleton, Lulu Pinkus
A vampire cult kidnaps a woman executive descended from a
Hungarian countess (who bathed in the blood of virgins in order
to preserve her beauty) and brainwashes her, in an attempt to
ensure that she has the right views in order to become their
leader. A well-handled and quite unusual chiller.
Aka: THIRST: THE TASTE FOR BLOOD
HOR   91 min (ort 98 min)   VIDrel: VIPCO/SGSVID   V

THIRTEEN AT DINNER *   U
Lou Antonio USA   1985
Peter Ustinov, Faye Dunaway, David Suchet, Jonathan Cecil, Bill
Nighy, John Stride, Diane Keen, Benedict Taylor, Lee Horsley, Allan
Cuthbertson, Glyn Baker, John Barron, Peter Clapham, Lesley Dunlop,
Avril Elgar, Roger Milner
Yet another adaptation of an Agatha Christie novel, with
Hercule Poirot (the familiar mellifluous tones as Ustinov
attempts another accent) investigating a murder among English
aristocrats. Set in the 1980s, it opens with a TV interview of
Poirot. Limp, contrived, tiresome and tedious. Adapted from
the novel by Rod Browning and filmed in England.
Aka: AGATHA CHRISTIE'S THIRTEEN AT DINNER
DRA   91 min (ort 100 min) mTV   VIDrel: WHV   V
Boa: novel by Agatha Christie.

THIRTEEN COLD BLOODED EAGLES **   15
Choy Fat HONG KONG
Cynthia Khan, Waise Lee, Choy Fat
Several top martial artists attempt to locate a book of secret
skills, their master (who brought them up as orphans) having
explained that the book is so deadly it must never fall into the
"wrong" hands. But this involves them in killing off rival martial
artists at his request, until one of the fighters learns of his
mentor's duplicity and wickedness. A period kung fu actioner,
entertaining if a little simpleminded.
Aka: 13 COLD BLOODED EAGLES
MAR   90 min   VIDrel: EAST/DISC   V

THIRTEEN GHOSTS **   PG
William Castle USA   1960
Charles Herbert, Jo Morrow, Martin Milner, Rosemary De Camp,
Margaret Hamilton, Donald Woods, John Van Dreelan
A family of four move into a house they have inherited and
find that it is haunted by a dozen ghosts. They receive a
warning that a death will occur soon to bring this number to
thirteen. Another gimmicky Castle production for which
specially tinted glasses were distributed to an audience who
could then choose whether or not to see the ghosts. This effect
is absent here as is the introduction describing this device. Dull
and not the least bit scary.
HOR   88 min B/W/Col   VIDrel: ENCORE/SPEAR   V

THIRTEEN RUE MADELAINE ***   U
Henry Hathaway USA   1946
James Cagney, Annabella, Richard Conte, Frank Latimore, Melville
Cooper, Walter Abel, Sam Jaffe, E.G. Marshall, Blanche Yurka, Karl
Malden, Red Buttons, Marcel Rousseau, Peter Von Zerneck, Alfred
Linder, Judith Lowry, Richard Gordon
During WW2, OSS agents attempt to locate a German missile
site in occupied France. Cagney takes over as leader of the
mission when one of his agents is killed by the Nazis. One of
several films inspired by "The March Of Time" and done in
appropriate semi-documentary style. A taut and gripping espionage tale.
A/AD   95 min B/W   VIDrel: 20TH/TECH L/A   V/h

THIRTY-NINE STEPS, THE ****   U
Alfred Hitchcock UK   1935
Robert Donat, Madeleine Carroll, Lucy Mannheim, Godfrey Tearle,
John Laurie, Peggy Ashcroft, Helen Haye, Wylie Watson, Frank
Cellier, Peggy Simpson, Gus McNaughton, Jerry Verno, Hilda
Trevelyan, John Turnbull, Miles Malleson
An innocent man is embroiled in murder and espionage, in this
sophisticated comedy-thriller, that deviates quite considerably
from the original novel. Donat and Carroll make a fine pair, and
in one memorable sequence find themselves being chased across
Scotland while chained together. The climax is suitably
dramatic. Written by Charles Bennett and Alma Reville, and

with musical direction by Louis Levy. Remade several times
since.
THR   78 min (ort 87 min) B/W   VIDrel: CARL/TECH   V
Boa: novel by John Buchan.

THIRTY-NINE STEPS, THE **   U
Ralph Thomas UK   1959
Kenneth More, Taina Elg, Brenda de Banzie, Barry Jones, Reginald
Beckwith, Faith Brook, Michael Goodliffe, James Hayter, Duncan
Lamont, Jameson Clark, Andrew Cruikshank, Leslie Dwyer, Betty
Henderson, Joan Hickson, Sidney James
A virtually scene-for-scene remake of the original film, with the
dubious added attraction of colour. The cast do their best in this
curiously flat and wooden adaptation, telling of a young man
who finds himself embroiled in murder and intrigue.
Guaranteed suspense free.
A/AD   91 min (ort 93 min)   VIDrel: VCC L/A   V
Boa: novel by John Buchan.

THIRTY-NINE STEPS, THE **   15
Don Sharp UK   1978
Robert Powell, David Warner, Eric Porter, Karen Dotrice, John Mills,
George Baker, Ronald Pickup, Donald Pickering, Timothy West, Miles
Anderson, Andrew Keir, Robert Flemyng, William Squire, Paul
McDowell, David Collings
A remake of the 1935 film that is more faithful to the original
novel, and tells of a man pursued by crooks who think he has
obtained information on their plot to instigate WW1. Not a terri-
bly suspenseful film, and spoilt by the climax on the face of Big
Ben, which no doubt was inspired by the Will Hay film MY
LEARNED FRIEND. The script is by Michael Robson.
THR   102 min   VIDrel: VCC L/A   V
Boa: novel by John Buchan.

THIRTY-SIX FIST, THE **   15
Jackie Chan HONG KONG   198-
Jackie Chan
Martial arts film on the all-too-familiar revenge theme. When
Wang's father is murdered, the son enrols in a martial arts
academy, and after the usual series of tribulations gains the skill
needed to mete out his revenge.
Aka: 36 CRAZY FIST, THE; 36 CRAZY FISTS; 36 FIST, THE; JACKIE CHAN
AND THE THE 36 CAZY FISTS; THIRTY-SIX CRAZY FIST, THE
MAR   88 min   VIDrel: PARADE/SCRN   V

THIRTY-SIXTH CHAMBER OF SHAOLIN, THE **   18
Liu Chia-Liang HONG KONG   1977
Liu Chia-Hui, Huang Yu (Young Wang Yu), Lo Lieh, Liu Chia-Yung,
Hsu Shao-Chiang, Yu Yong
A young man masters the secrets of kung fu, so as to avenge his
family's death at the hands of the Manchu invaders. A compe-
tent film that spends rather too much time following the training
the hero is obliged to undergo, although for devotees of the
martial arts this will be one of the film's assets. Followed in 1980
by RETURN TO THE 36TH CHAMBER.
Aka: 36TH CHAMBER OF SHAOLIN, THE; MASTER KILLER
MAR   89 min (ort 109 min)   VIDrel: LUMI/SPEAR L/A   V

THIS BOY'S LIFE ***   15
Michael Caton USA   1992
Robert De Niro, Ellen Barkin, Leonard DiCaprio, Jonah Blechman,
Eliza Dusku, Chris Cooper, Carla Gugino, Zachary Ashley, Tracey
Ellis, Kathy Kinny, Bobby Zameroski, Tobey Maguire, Tristan Tait,
Travis MacDonald, Gerrit Graham
In the 1950s a woman and her teenage son flee from her abusive
lover and settle in Washington State, just outside the town of
Concrete. There she meets up with a former military man who
proves to be every bit as nasty as his predecessor. A fine if overly
depressing character study, with a suitable soundtrack of
vintage hits.
DRA   110 min (ort 115 min) cC   VIDrel: WHV   V/sur
Boa: book by Tobias Wolff.

THIS CAN'T BE LOVE **   (U)
Anthony Harvey USA   1994
Katherine Hepburn, Anthony Quinn, Jason Bateman, Jami Gertz,
Maxine Miller, Michael Feinstein, Lynda Boyd, Morris Panych, Lori
Ann Triolo, David Lovgren, Bob Metcalfe, Maria Herrera, Gary Jones,
French Tickner, Andrew Johnston
Pleasant romantic drama that casts Hepburn as an elderly screen
legend who finds that she experiences emotions when a former
lover re-enters her life, as she has now become a recluse and has
not seen this man since they were in a film together, fifty years

ago. But her lover's return has a purpose, in that he has fallen on hard times and intends to sell his autobiography, in which he reveals that they were secretly married all those years ago.
DRA   89 min mCab   SATrel: MOVIE CHANNEL

THIS CHILD IS MINE **
David Greene USA
Lindsay Wagner, Chris Sarandon, Michael Lerner, John Philbin, Kathleen York, Frank Dent, Joan McMurtrey, Matthew Faison, Nancy McKeon, Carolyn Coates, Eve Roberts, Jennifer Parsons, Leigh French, Paul Teurpe, Susan Peretz
The story of a teenager who gives her child up for adoption, and then has to battle the childless couple who have gained custody, when she has second thoughts. Average TV movie melodramatics.
DRA   93 min (ort 100 min) mTV   VIDrel: VCC L/A   V

PG
1985

THIS GUN FOR HIRE **
Lou Antonio USA
Robert Wagner, Nancy Everhard, Fredric Lehne, John Harkins, John "Spud" McConnell, Joe Warfield, Kristina Loggia, Patrik Baldauff, Lenore Banks, Ron Flagge, James Borders, Dean Cochran, David Dahlgren, Margaret Dubuisson
A hired assassin is forced to go on the run after he's framed, and spends the bulk of the film searching for those responsible. A good looking and leisurely film, with Wagner giving a strong if sullen performance in the title role. This was the second remake of the 1942 Alan Ladd film, following the 1957 remake "Short Cut To Hell", and does quite a good job of recreating the film noir feeling of all those 1940s thrillers.
THR   85 min (ort 100 min) mCab   VIDrel: CIC/SONOP   V
Boa: novel A Gun for Sale by Graham Greene.

15
1991

THIS HAPPY BREED ***
David Lean UK
Robert Newton, Celia Johnson, Stanley Holloway, John Mills, Kay Walsh, Amy Veness, Alison Leggatt, Eileen Erskine, Guy Verney, Merle Tottenham, Betty Fleetwood, John Blythe plus Laurence Olivier (narration only)
The story of a South London working-class family, with interesting detail on the period it covers, this being from 1919 to 1939. A hit with the public but a good deal less so with the critics, it deftly blends sentiment, drama, humour and patriotism. Scripted by Lean, Ronald Neame and Anthony Havelock-Allan, and with photography by Neame.
DRA   112 min (ort 114 min)   VIDrel: VCC/RTM   V
Boa: play by Noel Coward.

U
1944

THIS IS ELVIS ***
Malcolm Lee/Andrew Solti USA
David Scott, Paul Boensh III, Johnny Harra, Lawrence Koller, Rhonda Lyn, Debbie Edge, Larry Raspberry, Furry Lewis
Both documentary footage and acted sequences are used, in this intriguing and carefully made examination of the famous rock star, that ultimately leaves us none the wiser as to his motivation and private thoughts. An exploitative and flawed biopic, but quite fascinating. See also ELVIS AND ME and ELVIS: THE MOVIE.
DOC   144 min Col/B/W   VIDrel: MGM/WHV L/A   V

PG
1981

THIS IS MY LIFE ***
Nora Ephron USA
Julie Kavner, Dan Aykroyd, Carrie Fisher, Samantha Mathis, Gaby Hoffmann, Danny Zorn, Bob Nelson, Marita Geraghty, Welker White, Caroline Aaron, Kathy Ann Najimy, Renee Lippin, Joy Behar, Estelle Harris, Sidney Armus, David Eisner
A female stand-up comic who dreams of the big time achieves a sudden and most unexpected breakthrough, but as her career blossoms, she finds that the price of fame includes neglecting her two young daughters. Fine acting by Kavner and the rest of the cast lift this movie beyond the confines of its sketchy plot. Screenplay is by Nora and Delia Ephron.
DRA   90 min (ort 94 min) cC   VIDrel: 20TH/TECH   V/sur
Boa: novel This is Your Life by Meg Wolitzer.

15
1992

THIS IS SPINAL TAP **
Rob Reiner UK
Rob Reiner, Michael McKean, Christopher Guest, Harry Shearer, Tony Hendra, June Chadwick, R.J. Parnell, David Kaff, Paul Benedict, Patrick Macnee, Billy Crystal, Fred Willard, Ed Begley Jr, Howard Hesseman, Paul Shaffer
An uneven spoof about a heavy-metal rock band, whose ageing members have embarked on an American tour. The use of

15
1984

documentary techniques successfully satirises many of the aspects of the music business, but for the most part the end result is just not all that funny.
MUS   79 min (ort 82 min)   VIDrel: POLY/POLYREC
V/sur

THIS ISLAND EARTH **
Joseph M. Newman USA
Jeff Morrow, Rex Reason, Faith Domergue, Russell Johnson, Douglas Spencer, Robert Nichols, Karl Lindt, Lance Fuller, Reg Parton, Eddie Parker, Olan Soule, Richard Deacon, Robert B. Williams, Mark Hamilton, Guy Edwards
In a desperate attempt to save their civilisation, benevolent aliens plan to kidnap some Earth scientists, but arrive at their embattled planet too late to save it. A film that starts off with great promise, showing how the scientists are tricked into agreeing to work at a remote base, prior to their abduction. Unfortunately, the film is spoilt by clumsy development and cliche-ridden dramatics, but at least some striking ideas are on offer.
FAN   83 min (ort 86 min)   VIDrel: PION   LV
Boa: novel by Raymond F. Jones.

PG
1955

THIS MAN STANDS ALONE **
Jerrold Freedman USA
Louis Gossett Jr, Clu Gulager, Mary Alice, Barry Brown, James McEachin, Barton Heyman, Lonny Chapman, Philip Michael Thomas, Helen Martin, John Crawford, Burton Gilliam, Clebert Ford, Nick Smith, Mary Kay Place
In Alabama, a black man attempts to resist the powerful white family that dominates the town. An interesting film let down by poor characterisations, and largely based on the true story of a black civil rights activist, who ran for sheriff in a Southern town.
DRA   78 min (ort 90 min) mTV   TVrel

(PG)
1979

THIS PROPERTY IS CONDEMNED **
Sydney Pollack USA
Natalie Wood, Robert Redford, Kate Reid, Charles Bronson, Robert Blake, Mary Badham, Jon Provost, John Harding, Alan Baxter, Dabney Coleman, Ray Hemphill, Brett Pearson, Quentin Sondergaard, Michael Steen, Bruce Watson
An out of town railway official falls in love with his landlady's daughter, and has to face both her anger and that of the railway workers he has laid off. An implausible story, with unconvincing dialogue and aimless plotting, but quite stylishly done. Written by Francis Ford Coppola, Fred Coe and Edita Sommer. Photography is by James Wong Howe.
DRA   105 min (ort 110 min)   VIDrel: L/A   V
Boa: play by Tennessee Williams.

15
1966

THIS SPORTING LIFE ****
Lindsay Anderson UK
Richard Harris, Rachel Roberts, Alan Badel, William Hartnell, Colin Blakely, Arthur Lowe, Vanda Godsell, Anne Cunningham, Jack Watson, Harry Markham, George Sewell, Leonard Rossiter, Frank Windsor, Peter Duguid, Wallas Eaton
Powerful adaptation of a fine novel, telling of the career of a successful rugby player and his ambiguous relationship with his landlady. Retains much of the feeling of Northern life that came through so well in the book, and is helped along by splendid performances all round. A sombre but excellent piece, with Harris perfectly cast as the sullen rugger player. Written by Storey.
DRA   128 min (ort 134 min) B/W   VIDrel: CARL/TECH   V
Boa: novel by David Storey.

15
1963

THOMAS CROWN AFFAIR, THE ***
Norman Jewison USA
Steve McQueen, Faye Dunaway, Paul Burke, Jack Weston, Biff McGuire, Yaphet Kotto, Todd Martin, Sam Melville, Addison Powell, Sidney Armus, Jon Shank, Allen Emerson, Harry Cooper, John Silver, Astrid Heeren, Carol Corbett
An interesting and rather beautiful if empty film, which makes good use of the split-screen technique (which will suffer on TV). A millionaire amuses himself by carrying out a bank robbery, and then playing a cat-and-mouse game with an attractive female insurance investigator. Written by Alan R. Trustman. AA: Song ("The Windmills Of Your Mind" – Michel Legrand (m)/Alan and Marilyn Bergman (l)).
Aka: CROWN CAPER, THE; THOMAS CROWN AND COMPANY
DRA   98 min (ort 102 min)   VIDrel: MGM/WHV   V

PG
1968

THORNBIRDS, THE: VOLS 1 TO 4 **   15
Daryl Duke/David L. Wolper/Stan Margulies USA   1983
Richard Chamberlain, Rachel Ward, Christopher Plummer, Bryan Brown, Barbara Stanwyck, Richard Kiley, Jean Simmons, John Friedrich, Mare Winningham, Earl Holliman, Piper Laurie, Philip Anglim, Allyn Ann McLerie, Richard Venture
Long and glossy adaptation of a novel dealing with a period of forty years in the life of an Australian family and the complexities of their relations with a charismatic and ambitious Catholic priest. Never rises about the soap-opera level of the average TV movie and suffers from not having been shot on location in Australia.
DRA   467 min (3 cassettes – ort 600 min) mTV
VIDrel: WHV   V/h
Boa: novel by Colleen McCullough.

THOSE BEDROOM EYES **   18
Leon Ichason USA   1992
Tim Matheson, Mimi Rogers, Wlliam Forsythe, Carlos Gomez, Carroll Baker, Roy Tatum. Eddie Billups, Tommy Cresswell, Nina Jones, Chris Byrd, Susie Spear, Johnny Popwell, Challen Gates, J. Don Ferguson, Deborah Hobart, Shilla Benning
A widowed psychology professor meets a disturbed young woman who has a secret life as a hooker, and may be a murderous serial killer as well. But despite his growing suspicions, he finds himself falling ever more deeply under her seductive spell. A murky thriller that's strong on atmosphere but weak on plotting.
THR   87 min (ort 90 min) mTV   VIDrel: VCC/DISC L/A   V

THOSE MAGNIFICENT MEN IN THEIR FLYING MACHINES *   U
Ken Annakin UK   1965
Terry-Thomas, James Fox, Stuart Whitman, Sarah Miles, Robert Morley, Alberto Sordi, Gert Frobe, Eric Sykes, Benny Hill, Sam Wanamaker, Flora Robson, Fred Emney, Gordon Jackson, William Rushton, Tony Hancock, John Le Mesurier
A totally unfunny comedy about a sponsored air race from London to Paris in the early days of flying. A work of enormous tedium, it ever gets going, though Red Skelton has a mildly amusing cameo in a prologue that traces the history of aviation. Written by Annakin and Jack Davies, the film offers ample proof that with a few exceptions, British film-makers have never developed any feel or understanding of comedy. Judging by this film they probably never will.
Aka: HOW I FLEW FROM LONDON TO PARIS IN 25 HOURS AND 11 MINUTES
COM   126 min (ort 138 min)   VIDrel: 20TH/TECH   V/sh

THOSE SHE LEFT BEHIND **   PG
Waris Hussein USA   1989
Gary Cole, Joanna Kerns, Mary Page Keller, Colleen Dewhurst, George Coe, Maryedith Burrell, Harry Stevens, Jim Edgecomb, Folkert Schmidt, Wyatt Knight, Stephane Kain, Vicky Dawson, Julie Ariola, Amy Buffington, Mike Doukas
Weepy drama about a young father whose life is shattered by the death of his wife in childbirth and his feelings of inadequacy in the face of the prospect of having to bring up his baby daughter alone. Good performances sustain a film that is hampered by excessive sentimentality and spoilt by the contrived ending.
DRA   90 min (ort 100 min) mTV   VIDrel: ODY/SONOP   V/h

THREADS ***   15
Mick Jackson AUSTRALIA/UK   1985
Karen Meagher, Reece Dinsdale, Rita May, David Brierly, Harry Beety, June Broughton, Sylvia Stoker, Nicholas Lane, June Hazelgrove, Ashley Barker, Victoria O'Keefe, Henry Moxon
A critically acclaimed but chilling story of a nuclear strike on Britain, that explores the aftermath of the conflict through the eyes of two Sheffield families. Extremely restrained in its examination of grief, hypothermia and starvation, and certainly not ending on a hopeful note. See also TESTAMENT and THE DAY AFTER.
DRA   114 min (ort 125 min) mTV   VIDrel: BBC L/A   V

THREE AMIGOS *   PG
John Landis USA   1986
Steve Martin, Chevy Chase, Martin Short, Alfonso Rau, Patrice Martinez, Jon Lovitz, Tony Plana, Joe Mantegna, Fred Asparagus, Ned Nederlander, William Kaplan, Gene Hartline, Sophia Lamour, Santos Morales, Philip E. Hartman
Three friends involved in a silent screen comedy-act in the 1920s,

find themselves broke and out of work. When they are offered a personal appearance at a Mexican village, they soon realise the locals expect them to be the heroes they portray, and fight a band of desperadoes terrorising the area. A dullish one-joke comedy whose stars look distinctly out of place in a Western setting.
COM   99 min (ort 105 min)   VIDrel: POLY/POLYREC L/A   V/sh

THREE CABALLEROS, THE ***   U
Norman Ferguson USA   1945
Aurora Miranda, Carmen Molina, Dora Luz, Nestor Amarale, Almirante, Trio Calcaveras, Ascencio Del Rio Trio, Padua Hill Players plus voices of: Sterling Holloway, Clarence Nash, Jose Oliveira, Joaquin Garay, Fred Shields
A mixture of live-action and animation, and largely consisting of a set of short features paying homage to Latin America, with Donald Duck acting as tourist and guide. Part of the 1940s Good Neighbour Policy that also gave rise to the earlier and shorter "Saludos Amigos" of 1943. Not really a film in the true sense, but worth seeing for its fascinating combination of superb cartoons and colourful performances.
JUV   85 min   VIDrel: WDV/TECH L/A   V

THREE COINS IN THE FOUNTAIN ***   U
Jean Negulesco USA   1954
Clifton Webb, Dorothy McGuire, Louis Jourdan, Jean Peters, Rossano Brazzi, Maggie McNamara, Howard St John, Katherine Givney, Cathleen Nesbitt, Mario Siletti, Vincente Padula, Alberto Morin, Dino Bolognese, Tony De Mario
Three American girls working in Rome throw coins into the Trevi fountain and have their wishes come true when all three meet handsome Italians. A glossy and romantic tale, sustained by the exotic settings of Rome and Venice, and a lilting title song. A big hit that was remade (albeit set in Madrid) by Negulesco in 1964 as "The Pleasure Seekers". AA: Cin (Milton Krasner), Song ("Three Coins In The Fountain" – Julie Styne (m)/Sammy Cahn (l)).
DRA   97 min (ort 102 min)   VIDrel: 20TH   V/h
Boa: novel by John B. Secondari.

THREE COLOURS: BLUE ***   15
Krzysztof Kieslowski FRANCE   1993
Juliette Binoche, Benoit Regent, Florence Pernel, Charlotte Very, Helene Vincent, Philippe Volter, Claude Duneton, Hughes Quester, Emmanuelle Riva, Vacek Ostaszewski, Florence Vignon, Yann Trecourt, Isbaelle Sadoyan, Philippe Manesse
First part of a thematic trilogy, inspired by the colours of the French flag, with blue representing freedom. Here, this idea is worked into the story of a woman who, after her son and husband are killed in an accident, must come to terms with her grief and rebuild her life. A sterling performance by Binoche won a Cesar for Best Actress as well as a similar award at the Venice Film Festival.
Aka: BLUE; FILM BLEU; THREE COLORS: BLUE; TROIS COULEURS: BLEU
DRA   94 min (ort 98 min) wScrn   VIDrel: ARTIF/20TH   V/h

THREE COLOURS: RED ***   15
Krzysztof Kieslowski FRANCE/POLAND/SWITZERLAND   1994
Irene Jacob, Jean-Louis Trintignant, Frederique Feder, Jean-Pierre Lorit, Samuel Lebihan, Marion Stalens, Teco Celio, Bernard Escalon, Jean Schlegel, Elzbieta Jasinska, Paul Vermeulen, Jean-Marie Daunas, Roland Carey
The culminating film in Kieslowski's masterful trilogy opens with model Jacob accidentally running over a dog that belongs to an elderly, retired judge, a event that draws her into the latter's orbit and enables her to learn some disturbing facts about her new acquaintance. Not an especially easy film to appreciate without having seen the previous works, but its power and the skilful way the loose ends from the earlier films are resolved can only be admired.
Aka: TROIS COULEURS: ROUGE
DRA   95 min (ort 99 min) wScrn   VIDrel: ARTIF/20TH   V/sh

THREE COLOURS: WHITE ***   15
Krzysztof Kieslowski FRANCE/POLAND   1993
Zbigniew Zamachowski, Julie Delpy, Janusz Gajos, Jerzy Stuhr, Juliette Binoche, Florene Pernel, Aleksander Bardini, Grzegorz Warzehol, Jerzy Nowak, Cezary Harasimowicz, Jerzy Trela, Cezary Palura, Michel Usowski, Piotr Machura

A Polish hairdresser returns to Warsaw after being divorced by his French wife and works hard to regain his wealth and find a suitable form of revenge in this offbeat black-comedy. The second part of the director's THREE COLOURS trilogy, based on the colours of the French flag and their meanings. Here, white stands for equality.
Aka: THREE COLORS: WHITE; TROIS COULEURS: BLANC; WHITE
DRA   87 min (ort 92 min)   VIDrel: ARTIF/20TH   V/sh

THREE DAUGHTERS ***
Candida Royalle USA
18
1986
Siobhan Hunter, Robert Bullock, Nina Preta, Clark Sharp, Annette Heinz, Johnny Nineteen, Gloria Leonard, Ashley Moore, Carol Cross, Candace De Carlo
A sensitive and romantic look at the sex lives of three girls: one an inexperienced teenager and the other two her more worldly wise older sisters. All three encounters are seen in the context of warm and caring relationships (this is generally Royalle's hallmark) but the lack of a strong storyline is a major drawback in this overlong and leisurely paced film.
A   108 min Cut (49 sec)   VIDrel: MIA/DISC   V

THREE DAYS OF THE CONDOR ***
Sydney Pollack USA
(PG)
1975
Robert Redford, Faye Dunaway, Cliff Robertson, Max Von Sydow, John Houseman, Addison Powell, Walter McGinn, Tina Chen, Michael Kane, Don McHenry, Michael Miller, Jess Osuna, Dino Narizzano, Helen Stenborg, Patrick Gorman
A CIA researcher goes on the run after all his fellow workers are brutally murdered and he narrowly escapes himself death after he contacts his employers. The pieces gradually fall into place to reveal an unclear picture of conspiracy at the highest level. An excellent thriller that maintains a sense of suspense all the way through, but marred slightly by the hero's romantic involvement and the ambiguous ending. See also SPY.
THR   117 min   TVrel
Boa: novel Six Days of the Condor by James Grady.

THREE FACES OF EVE, THE ***
Nunnally Johnson USA
PG
1957
Joanne Woodward, David Wayne, Nancy Culp, Lee J. Cobb, Edwin Jerome, Vince Edwards, Alena Murray, Douglas Spencer, Terry Ann Ross, Ken Scott, Mimi Gibson plus Alistair Cooke (narration only)
A female schizophrenic being treated by a psychiatrist is found to have three completely separate personalities in this intriguing but dated story. Cooke's introduction carries a conviction that's somewhat lacking in the script, but Woodward's powerful performance ensured the film enjoyed considerable success at the box office. Screenplay is by Johnson, who also produced.
AA: Actress (Woodward).
DRA   87 min (ort 95 min) B/W   VIDrel: 20TH/TECH   V
Boa: book by Corbett H. Thigpen and Hervey M. Cleckley.

THREE FOR THE ROAD **
Bill W.L. Norton USA
15
1987
Charlie Sheen, Kerri Green, Alan Ruck, Sally Kellerman, Raymond J. Barry, Blair Tefkin, Alexa Hamilton, James Avery, Bert Remsen, Eric Bruskotter, Jackie D. Stewart, J.C. Mullins, Robert Ginnaven, Scott Edmonds
The selfish young daughter of a senator is packed off to boarding school, with the senator's aide and his buddy escorting her, but instead all three set forth on a series of wild adventures. A silly teen outing.
COM   88 min (ort 90 min)   VIDrel: 20TH/TECH   V

THREE FUGITIVES **
Francis Veber USA
15
1989
Nick Nolte, Martin Short, Sarah Rowland Doroff, James Earl Jones, Alan Ruck, Kenneth McMillan, Bruce McGill, David Arnott, Lee Garlington, Sy Richardson, Rocky Giordani, Rick Hall, Bill Cross, John Procaccino, Kathy Kinney
A reformed bank robber is forced to go on the run with an amateur thief and his six-year-old daughter, when he inadvertently gets caught up in a bungled bank raid. Veber's first US film is largely a re-run of his earlier work "Les Fugitifs", and is little more than a fast-paced blend of knockabout slapstick and cloying sentimentality, that soon becomes quite tiresome.
COM   92 min (ort 96 min)   VIDrel: TOUCH/BUENA L/A   V

THREE GODFATHERS, THE **
John Ford USA
U
1948
John Wayne, Pedro Armendariz, Harry Carey Jr, Ward Bond, Mildred

Natwick, Charles Halton, Jane Darwell, Mae Marsh, Guk Kibbee, Dorothy Ford, Michael Dugan, Ben Johnson, Fred libby, Hank Worden, Jack Pennick, Francis Ford
Three outlaws fleeing from a posse encounter a woman in labour. She finally dies after her baby is safely delivered, but not before she has exacted a promise from them to take the child to safety in a nearby town aptly named New Jerusalem. Sentimental story with a contrived biblical analogy, filmed many times before and remade as recently as 1975 as a TV movie entitled "The Godchild".
WES   102 min (ort 106 min)   VIDrel: MGM/WHV   V/sh
Boa: short story by Peter B. Kyne.

THREE KINGS, THE **
Mel Damski USA
(U)
1987
Lou Diamond Phillips, Jack Warden, Stan Shaw, Jane Kaczmarek, Vic Tayback, Charles Nelson Reilly, Rick Lenz, Tirana Alexandra
A seasonal tale about the spirit and meaning of Christmas, in which three mental patients break out of their asylum dressed as the Three Wise Men, and go off in search of a star in central L.A. A pleasant enough time-filler of no great consequence. Screenplay is by Stirling Silliphant.
COM   89 min mTV   VIDrel: GUILD/SONOP   V

THREE LITTLE NINJAS AND THE LOST TREASURE *
Emmett Alston USA
PG
1992
Douglas Ivan, Steven Nelson, Jonathan Anzaldo, Alan Godshaw, Takeshi Garcia, Cynthia Cheston, Ted Tabura, Bob Hunt, Casey Tabura, Bernie Garcia, Baron Tabura, Sam Plunkett, Royce O'Donnel, Robert Matheney, Phil Williams, Tera Lane
A trio of martial arts trained kids battle the evil Sarek and his criminal gang, the latter being the ruthless prime minister of a South Sea island. In between, they find time for a little treasure hunting. A juvenile blend of low-brow comedy without thrills or humour.
Aka: LITTLE NINJAS
JUV   88 min   VIDrel: NEWAGE/20VIS L/A   V

THREE MEN AND A BABY ***
Leonard Nimoy USA
PG
1987
Tom Selleck, Ted Danson, Steve Guttenberg, Nancy Travis, Margaret Colin, Philip Bosco, Celeste Holm, Lisa Blair, Michelle Blair, Barbara Budd, Paul Guilfoyle, Earl Hindman, Gary Klar, Joe Lynn, Thomas Quinn, Colin Quinn
A lively remake of the French film from 1985, THREE MEN AND A CRADLE, with three bachelors suddenly finding themselves in possession of a baby and a parcel of heroin. They manage to get rid of the drugs but are less willing to give up the child. The film really has very little plot, but the three stars carry it off most enjoyably. This was the feature film debut for Travis. The sequel THREE MEN AND A LITTLE LADY followed shortly.
COM   98 min (ort 101 min)   VIDrel: TOUCH/TECH   V/sur

THREE MEN AND A CRADLE ***
Coline Serreau FRANCE
PG
1985
Roland Gireaud, Michel Boujenah, Andre Dussolier, Philippine Leroy Beaulieu, Dominique Lavanant, Marthe Villalonga, Annik Alane, Josine Comelas, Francois Domange, Gwendoline Mourlet, Christian Zanetti, Gilles Cohen, Bernard Stancy
The lives of three bachelors who share an apartment are turned upside down when a baby in a cradle is left at their front door. In learning to accept and care for it, they also learn more about themselves. Written by Serreau, this gentle and witty comedy that won a Cesar for Best Film. Remade as THREE MEN AND A BABY in 1987.
Aka: TROIS HOMMES ET UN COUFFIN
COM   106 min   VIDrel: ARROW/RTM   V

THREE MEN AND A LITTLE LADY **
Emile Ardolino USA
PG
1990
Tom Selleck, Steve Guttenberg, Ted Danson, Nancy Travis, Christopher Cazenove, Fiona Shaw, Sheila Hancock, Lisa Blair, Michelle Blaire, Barbara Budd, Paul Guilfoyle, Earl Hindman, Gary Klar, Joe Lyn, Thomas Quinn
A lame sequel to THREE MEN AND A BABY that is set a few years on. Our three surrogate fathers are still bachelors and find their cosy set-up threatened when the mother of their little girl falls in love with a British theatre director. To stop her and the child disappearing to Britain for good, they resort to drastic measures. Overlong and not very funny, though there are some amusing moments.
COM   99 min (ort 103 min)   VIDrel: TOUCH/TECH   V/sur

THREE MUSKETEERS, THE ** U
Rowland V. Lee USA 1935
Walter Abel, Paul Lukas, Moroni Olson, Ian Keith, Onslow Stevens, Ralph Forbes, Margot Grahame, Heather Angel, Rosamond Pinchot, John Qualen, Nigel De Brulier, Murray Kinnell, Lumsden Hare, Miles Mander
Film version of the famous novel with Abel miscast as D'Artagnan. The good supporting cast can do little in the lifeless rendition of this tale. The script is by Dudley Nichols and Rowland V. Lee.
A/AD 92 min (ort 97 min) B/W VIDrel: VCC L/A V
Boa: novel by Alexandre Dumas.

THREE MUSKETEERS, THE *** U
George Sidney USA 1948
Gene Kelly, Lana Turner, June Allyson, Van Heflin, Angela Lansbury, Robert Coote, Frank Morgan, Vincent Price, Keenan Wynn, Gig Young, John Sutton, Reginald Owen, Ian Keith, Patricia Medina
In France at the time of Louis XIII three friends become involved in intrigue, treachery and swordplay. A rumbustious version of the classic tale, done in burlesque style, and remarkable for its vigour if not its realism.
A/AD 120 min (ort 125 min) VIDrel: MGM/WHV V
Boa: novel by Alexandre Dumas.

THREE MUSKETEERS, THE ** PG
Richard Lester UK 1973
Oliver Reed, Michael York, Raquel Welch, Richard Chamberlain, Frank Finlay, Christopher Lee, Geraldine Chaplin, Jean-Pierre Cassel, Spike Milligan, Roy Kinnear, Gitty Djamel, George Wilson, Simon Ward, Sybil Danning
Meant to be funny, this version of the famous tale sometimes goes well over the top in a desperate search for laughs. A good performance from Welch is a highlight. The second half of the film was released as a sequel – THE FOUR MUSKETEERS. Scripted by George MacDonald Fraser. Look out for Faye Dunaway, who appears very briefly in a small part. See also THE RETURN OF THE MUSKETEERS.
Aka: THREE MUSKETEERS, THE: THE QUEEN'S DIAMONDS
COM 102 min (ort 107 min) VIDrel: ARROW/RTM V/sur
Boa: novel by Alexandre Dumas.

THREE MUSKETEERS, THE ** PG
Stephen Herek USA 1993
Charlie Sheen, Kiefer Sutherland, Chris O'Donnell, Oliver Platt, Rebecca De Mornay, Gabrielle Anwar, Tim Curry, Julie Delpy, Michael Wincott, Paul McGann, Hugh O'Conor, Christopher Adamson, Philip Tan, Erwin Leder, Axel Anselm
Lavishly staged version of this Dumas classic that fails to treat its subject material with the respect it deserves. Unfortunately, its young stars do not bring any weight to their characterisations while the villains are pure cardboard. A sort of cartoon version with only moderate amounts of suspense and excitement and far too much emphasis on clowning around.
A/AD 101 min (ort 105 min) cC VIDrel: WDV/TECH V/sur
Boa: novel by Alexandre Dumas.

THREE NINJA KIDS ** PG
Jon Turteltaub USA 1992
Chad Power, Max Elliott Slade, Michael Treanor, Rand Kingsley, Victor Wong, Margarita Franco, Alan McRae, Kate Sergeant, Joel Swetow, Toru Tanaka, Race Nelson, Patrice Labyorteaux, D.J. Harder, Baha Jackson, Scott Caudill
Trained in the martial arts by the grandfather, three young kids put these skills to good use when a criminal gang is foolish enough to try to kidnap them. A juvenile actioner aimed clearly at the kiddie market that was quite a box-office hit. A couple of sequels followed.
Aka: 3 NINJAS; THREE NINJAS
JUV 79 min (ort 85 min) cC VIDrel: TOUCH/TECH L/A V/sur

THREE NINJAS KICK BACK ** U
Charles T. Kanganis USA 1993
Sean Fox, Max Elliott Slade, Evan Bonifant, Victor Wong, Sab Shimono, Jason Schombing, Dustin Nguyen, Caroline Juno King, Angelo Tiffe, Margarita Franco, Don Stark, Kellye Nakahara-Wallett, Scott Campbell, Tommy Clark, Jeremy Linson
In this sequel, our three young heroes find that an old adversary of their grandfather is out to steal a ceremonial knife which he won in a contest fifty years before. To help him in this venture, our villain enlists the aid of his three American grandkids, who have their own garage band. A dire and unfunny tale, with low-brow humour and a feeble plot. A further sequel followed.
JUV 88 min (ort 93 min) VIDrel: COLUM/SONOP V/sur

THREE NINJAS KNUCKLE UP ** PG
Simon S. Sheen USA 1995
Jeff Cadiente, Wayne Collins, Gary Epper, Stuart Grant, Dennis Holakan, Janie Melissa Gunderson, Michael Hungerford, Patrick Kilpatrick, Donal Logue, Crystle Lightning, Eric Mansker, Charles Napier, Cathy Perry, Amanda Nicole
Our three intrepid fighters are back in action once again, this time getting caught up in a conflict between an American Indian tribe and a ruthless businessman, who has polluted their land with toxic waste. Adequate family fare with a strong martial arts flavour.
MAR 85 min VIDrel: COLUM/SONOP V/sur

THREE O'CLOCK HIGH ** 15
Phil Joanou USA 1987
Casey Siemaszko, Ann Ryan, Richard Tyson, Jeffrey Tambor, Philip Baker Hall, John P. Ryan, Stacey Glick, Jonathan Wise, Liza Morrow, John Rothman, Stirling E. Gardner, Shirley Stoler, Alice Nunn, Brooke Stevens
HIGH NOON comes to teenage Southern California, when the class hero has to face up to the recently arrived bully whose legendary reputation has preceded him. Despite some great performances, this slight little effort has nothing of substance, and the stars have little material to work with.
COM 86 min (ort 101 min) VIDrel: CIC/SONOP V/sur

THREE OF HEARTS *** 18
Yurek Bogayevicz USA 1992
William Baldwin, Kelly Lynch, Sherilyn Fenn, Joe Pantoliano, Cec Verrell, Gail Strickland, Claire Callaway, Marek Johnson, Monique Mannen, Timothy D. Strickney, Frank Ray Perilli, Tony Amendola, Keith MacKechine, Ann Ryerson
When a lesbian is dumped by her bisexual lover, she hires an attractive young man to first seduce and then abandon her, in the hope that she will return. Predictably enough, however, things do not go exactly according to plan. A thinly plotted drama with good performances but burdened by an unnecessary sub-plot and a pat happy ending, which in Europe was changed for a sadder but more realistic one.
DRA 105 min (ort 110 min) VIDrel: GUILD/SONOP V/sur

THREE SISTERS, THE **** U
Laurence Olivier/John Sichel UK 1970 (released in USA 1974)
Jeanne Watts, Joan Plowright, Alan Bates, Derek Jacobi, Louise Purnell, Laurence Olivier, Kenneth Mackintosh, Sheila Reid, Ronald Pickup, Frank Wylie, Daphne Heard, Harry Lomax, Judy Wilson, Mary Griffiths, Richard Kay
An excellent version of Chekhov's famous play, about the three daughters of a deceased Russian colonel who dream of exchanging their dull provincial existence for life in the big city, at the turn of the century. Translated by Moura Budberg, this lavish and evocative film is one of the best adaptations of this writer's work.
DRA 155 min (ort 165 min) VIDrel: WHV V/h
Boa: play by Anton Chekhov.

THREE SOVEREIGNS FOR SARAH: THE SALEM WITCH HUNT *** 15
Philip Leacock USA 1985
Vanessa Redgrave, Patrick McGoohan, Kim Hunter, Phyllis Thaxter, Ronald Hunter, Shay Duffin, Will Lyman
A mini-series on the 17th century Salem witch hunts and subsequent trials, that tells of the survivor of a witchcraft trial who tries to clear the names of her two sisters, who were hanged as witches. A slow, careful and absorbing drama, with Redgrave giving one of her best performances.
DRA 128 min (ort 152 min) mTV VIDrel: STABL L/A V

3:10 TO YUMA **** PG
Delmer Daves USA 1957
Glenn Ford, Van Heflin, Felicia Farr, Richard Jaeckel, Henry Jones, Robert Emhardt
A poor rancher decides to go after a dangerous outlaw in the hope of getting the bounty money. Having captured the man he attempts to take him by train to Yuma where he is to stand trial, but has to contend with the man's gang. A verbose but very

tense film, probably the best Western Daves ever made, its literate script and strong performances give the film a depth not usually found in this genre. Frankie Laine sings the catchy theme song.
WES   88 min B/W   VIDrel: VCC/DISC/COLUM   V
Boa: novel by Elmore Leonard.

THREE WISHES * **   PG**
Martha Coolidge USA   1995
Patrick Swayze, Mary Elizabeth Mastrantonio, Joseph Mazzello, Seth Mumy, David Marshall Grant, Jay O. Sanders, Michael O'Keefe, John Diehl, Diane Venora, David Zahorsky, Brian Flannery, Brock Pierce, David Jacob Carey, David Hart
A strange drifter who wanders from place to place with his dog is knocked down by a car and breaks his leg, and after treatment the driver takes him back with him to his home to give him the chance to recuperate. Once there, he begins to reveal his ability to change people's lives for the better. A warmhearted and slightly over-sentimental film, but an enjoyable one, despite the lack of a clear narrative thrust.
JUV   110 min (ort 115 min)   VIDrel: EIV/SONOP   V

THREE WORLDS OF GULLIVER, THE ** **   U
Jack Sher UK   1960
Kerwin Mathews, Jo Morrow, June Thorburn, Lee Patterson, Gregoire Aslan, Basil Sydney, Mary Ellis, Charles Lloyd Pack, Martin Benson, Peter Bull
Having been washed overboard, ship's doctor Gulliver first encounters the tiny folk of Lilliput and then the giants of Brobdingnag. A well-mounted juvenile adaptation of Swift's classic satire, lacking the wonderful sharpness and vigour of the novel. The trick photography and special effects of Ray Harryhausen are adequate and the score is by Bernard Herrmann.
A/AD   94 min (ort 98 min)   VIDrel: VCC/DISC/COLUM   V
Boa: novel Gulliver's Travels by Jonathan Swift.

THREE x THREE EYES: PARTS 1 TO 5 * **   15
JAPAN   1995
A five-part Japanese serial in which a young lad finds himself battling the forces of evil and helping a three-hundred-year-old girl, the last member of a race of immortals, find a supernatural object she needs in her desire to become a human being. Truly dire in that the animation is probably the very worst example from this genre, with hardly any movement of the characters at all, apart from occasional simple gestures and the like. The story is little better.
ANIM   260 min (5 cassettes – approx 50 min each) dubbed
VIDrel: MANGA/SONOP   V/sh

THREESOME ** **   18
Andrew Fleming USA   1993
Lara Flynn Boyle, Stephen Baldwin, Josh Charles, Alexis Arquette, Martha Gehman, Mark Arnold, Michelle Mathieson, Joanne Baron, Jennifer Lawler, Jack Breschard, Jillian Johns, Amy Ferioli, Jason Workman, Katherine Kousi
A woman student finds herself sharing living quarters with two men, one of whom is gay, as the result of a clerical error (due to the fact that her first name is Alex). She soon adjusts to this bizarre setup and the trio become firm friends, although sexual tensions constantly lurk beneath the surface. A simplistic and puerile tale, which is made less agreeable by the distinctly unappealing personalities of the three main characters. Scripted by Fleming.
DRA   90 min (ort 93 min) cC   VIDrel: COLUM/SONOP   V/sur

THRILL KILLERS, THE * **   18
Ray Dennis Steckler USA   1965
Cash Flagg (Ray Dennis Steckler), Liz Renay, Brick Bardo, Carolyn Brandt, Ron Burr, Garay Kent, Keith O'Brien, Herb Robins, Laura Benedict, Atlas King, George J. Morgan, Erina Enyo, Titus Moede
Three maniacs escape from their asylum and subject Los Angeles to a reign of terror. (Probably threatening the populace with mass showings of this film.) See also THE INCREDIBLY STRANGE CREATURES WHO STOPPED LIVING AND BECAME MIXED UP ZOMBIES, an earlier gem from this director.
Aka: MANIACS ARE LOOSE, THE; MONSTERS ARE LOOSE, THE
HOR   90 min B/W   VIDrel: RTM/DISC   V

THRILL OF THE VAMPIRES, THE ** **   18
Jean Rollin FRANCE   1970
Sandra Julien, Jean-Marie Durand, Michel Delahaye, Jacques

Robiolles, Nicole Nancel, Marie-Pierre Tricot, Kuelan, Dominique
A honeymooning couple come to a castle inhabited by two men and their mistress as well as a seductive female vampire who emerges from a grandfather clock as it strikes midnight. A visually striking film that is hampered by bad dialogue and the lack of any clear direction. Those who like Rollin's special blending of sadism and eroticism will like this one a lot.
Aka: LE FRISSON DES VAMPIRES; SEX AND THE VAMPIRE; TERROR OF THE VAMPIRES, THE; VAMPIRE THRILLS; VAMPIRE'S THRILL
HOR   90 min dubbed   VIDrel: REDEM/RTM   V

THRONE OF BLOOD ** **   PG**
Akira Kurosawa JAPAN   1957
Toshiro Mifune, Isuzu Yamada, Minoru Chiaki, Akira Kubo, Takamaru Sasaki, Yoichi Tachikawa, Takashi Shimura, Chieko Namira
A brilliant reworking of the Macbeth story, successfully transposed to Japan, where a samurai is impelled to murder his friend and lord through the urging of both his wife and a spirit apparition. Although the plot of the original is pared down to the minimum, the film's powerful visual elements predominate and succeed completely in creating a genuinely eerie and doomladen atmosphere. A superb adaptation.
Aka: CASTLE OF THE SPIDER'S WEB; COBWEB CASTLE; KUMONOSU-JO
A/AD   105 min B/W   VIDrel: CONNO/RTM   V
Boa: play Macbeth by William Shakespeare.

THROUGH A GLASS DARKLY * **   18**
Ingmar Bergman SWEDEN   1961
Harriet Anderson, Gunnar Bjornstrand, Max Von Sydow, Lars Passgard
On a remote island, four members of a family spend the summer demonstrating that life is without meaning and that God is absent from a harsh and unjust world. A bleak, emotionally demanding but ultimately depressing view of human existence – but as ever, Bergman displays his consummate skill as a director. The first film in a Bergman trilogy dealing with loss of faith in a cruel world: WINTER LIGHT and THE SILENCE were to follow. AA: Foreign.
Aka: SASOM I EN SPEGEL
DRA   85 min (ort 91 min)   VIDrel: TART/20TH   V

THROUGH THE EYES OF A KILLER * **   18**
Peter Markle USA   1992
Marg Helgenberger, Richard Dean Anderson, David Marshall Grant, Melinda Culea, Tippi Hedren, Joe Pantoliano, Monica Parker, Joyce Seeley, John McCann, Pat Bremel, Bobby Stewart, William Taylor, Nancy Banks, Frank Ferucci
Having finished with her boyfriend, a woman goes flat-hunting, and eventually takes a rundown and rat-infested apartment after finding herself strangely drawn to a triangular pane of glass with an eye painted on it. She starts a relationship with the builder hired to redecorate the apartment, but things are not quite as they seem, and the body count soon starts to rise. Quite a chilling little affair.
THR   94 min mTV   VIDrel: CIC/SONOP L/A   V
Boa: short story The Master Builder by Christopher Fowler.

THROUGH THE OLIVE TREES * **   U**
Abbas Kiarostami IRAN   1994
Mohammad Ali Keshavarz, Farhad Kheradmand, Zarifeh Shiva, Hossein Rezai, Tahereh Ladanian, Hocine Redai, Zahra Nourouzi, Barastou Abbassi, Nasret Betri, Azim Aziz Nia, Astadouli Babani, N. Boursadiki, Kheda Barech Defai
A film director goes to an area recently devastated by a major earthquake to start organising the shooting of his film. Despite the massive dislocation, his activities arouse much interest, but his use of local acting talent threatens the project owing to the personal conflicts among those concerned. Scripted by Kiarostami, this is a solid look at the trials and tribulations of film-making, and at the same time it offers some insights into Iranian society.
Aka: UNDER THE OLIVE TREES; ZIR-E DARAKHTAN-E ZEYTON
DRA   103 min   CINrel

THROW MOMMA FROM THE TRAIN * **   15**
Danny DeVito USA   1987
Danny DeVito, Billy Crystal, Kim Greist, Anne Ramsey, Kate Mulgrew, Branford Marsalis, Rob Reiner, Bruce Kirby, Joey De Pinto, Annie Ross, Oprah Winfrey, Stu Silver, Philip Perlman, J. Alan Thomas, Randall Miller, Tony Ciccone
A comedy hostage to Hitchcock's STRANGERS ON A TRAIN, with a creative writing professor finding that one of his students has misunderstood his teaching and is convinced the professor

wants them to swap murders. An unusual and entertaining film, of uneven scripting and wild, overblown performances. Written by Stu Silver and with photography by Barry Sonnenfeld. This was the last film for Ramsey (who had battled against throat cancer for several years).
COM   84 min (ort 88 min)   VIDrel: 4-FRONT/POLYREC V/sur

THUMBELINA **   U
Don Bluth/Gary Goldman EIRE/USA   1994
Voices of: Jodi Benson, Gary Imhoff, Gino Conforti, Barbara Cook, Gilbert Gottfried, Will Ryan, June Foray, Kenneth Mars, Gary Imhoff, Joe Lynch, Charo, Danny Mann, Loren Michaels, Kendall Cunningham, John Hurt, Carol Channing
A young girl reads about title character in her storybook and is magically transported to her world where she has all manner of exciting adventures. A rather loose adaptation of Anderson's tale, not as well made as might have expected.
Aka: DON BLUTH'S THUMBELINA
ANIM   83 min (ort 87 min) cC   VIDrel: WHV   V/sur
Boa: story by Hans Christian Andersen.

THUNDER AND LIGHTNING **   15
Corey Allen USA   1978
David Carradine, Kate Jackson, Roger C. Carmel, Sterling Holloway, Ed Barth, Ron Feinberg, Pat Cranshaw, Charles Napier, George Murdock, Hope Pomerance, Malcolm Jones, Charles Willeford, Christopher Raynolds, Claude Jones
Two rival moonshiners in the Florida Everglades, fight it out in a struggle that offers ample scope for the obligatory chase sequences.
A/AD   94 min   VIDrel: 20TH/TECH   V

THUNDER IN PARADISE **   15
Douglas Schwartz USA   1993
Hulk Hogan, Felicity Waterman, Carol Alt, Robin Weisman, Chris Lemmon, Sam Jones, Charlotte Rae, Patrick Macnee, Lisa Stahl, Sandra Thigpen, Jim "The Ant" Neidhart, Russ Blackwell, Giant Gonzalez, Michael Edwards, Fred Ottaviano
Pilot for this TV series starring Hogan as a devil-may-care adventurer with a truly amazing high-speed racing boat equipped with all kinds of weapons and gadgets. Here, he and his companions go to Cuba to rescue a woman and her child. A simple piece of mindless entertainment.
A/AD   104 min mTV   VIDrel: POLY/POLYREC   V

THUNDER IN PARADISE 2 **   PG
Douglas Schwartz USA   1994
Hulk Hogan, Chris Lemmon, Carol Alt, Patrick Macnee
Yet another helping of high-speed action with our heroes right more wrongs with the help of their high-tech boat.
A/AD   81 min (ort 88 min)   VIDrel: COLUM/SONOP V/sh

THUNDER IN PARADISE 3 *   PG
Douglas Schwartz USA   1994
Terry "Hulk" Hogan, Chris Lemmon, Carol Alt, Ashley Gorrell, Patrick Macnee, Kiki Shepard, Carlos Lauchu, Jorge Gonzalez, Gerald Martin, Fred Wayne Ottman, Brutus "The Barber" Beefcake, Jimmy hart, Michael Marzella, Russ Wheeler
After the ruler of a small country loses his throne to a brutal general and his army of pirate mercenaries, our heroes rush to the rescue. Average outing for ex-wrestler Hogan, which has little of the freshness or appeal of the first film.
Aka: THUNDER IN PARADISE 3: SEALED WITH A KISMET
A/AD   78 min (ort 90 min)   VIDrel: IMPENT   V

THUNDER PRINCE **   PG
Lenny Washington HONG KONG   1989
Produced by Joseph Lai (and making a pleasant change from his usual animated space operas such as CAPTAIN COSMOS and the like) this martial arts tale tells of the master of a secret technique, whose son vows to avenge his father when he is murdered by three thugs. Despite being captured by one of those responsible, he eventually escapes and has his revenge. Fair.
ANIM   65 min Cut (33 sec)   VIDrel: IMPENT   V

THUNDER WARRIOR **   18
Larry Ludman (Fabrizio De Angelis) ITALY   1984
Mark Gregory, Bo Svenson, Raymond Harmstorf
A young Indian takes revenge for the desecration of a sacred

burial site in Arizona, when it is turned into a building site. Followed by two sequels. Average.
Aka: THUNDER
A/AD   79 min (ort 83 min)   VIDrel: EIV/SONOP L/A   V

THUNDER WARRIOR 2 **   15
Larry Ludman (Fabrizio De Angelis) ITALY   1986
Mark Gregory, Karen Reel, Raymond Harmstorf, Bo Svenson
A sequel to the first film, with a young Navajo Indian returning to his homeland as a police officer, and investigating corruption in the local force. A crooked sheriff frames him and he is sent to a desert prison camp, but soon starts plotting his escape.
Aka: THUNDER 2
A/AD   89 min (ort 114 min)   VIDrel: MED L/A   V

THUNDER WARRIOR 3 *   18
Larry Ludman (Farbizio de Angelis) ITALY   1988
Mark Gregory, Horts Schon, Werner Pochath, Ingrid Lawrence, John Phillip Law
Third in this series about a much-wronged Indian who is pushed too far by his racist tormentors. Violent and exploitative nonsense.
Aka: THUNDER 3
A/AD   82 min Cut (22 sec – ort 90 min)   VIDrel: IMPENT   V

THUNDERBALL ***   PG
Terence Young UK   1965
Sean Connery, Adolfo Celi, Claudine Auger, Luciana Paluzzi, Rik Van Nutter, Martine Beswick, Bernard Lee, Lois Maxwell, Guy Dolman, Molly Peters, Roland Culver, Desmond Llewelyn, Earl Cameron, Paul Stassino, Rose Alba
The fourth Bond film, with our secret agent up against the criminal organisation SPECTRE. There are some memorable sequences, as Miami is threatened with imminent destruction, but the story does tend to lose out to the special effects. Written by Richard Maibum and John Hopkins, with music by John Barry. Followed by YOU ONLY LIVE TWICE. Remade with Connery eighteen years later, as NEVER SAY NEVER AGAIN.
AA: Effects/aud (Tregoweth Brown).
A/AD   125 min (ort 132 min) wScrn cC
VIDrel: MGM/WHV   V/dm
Boa: novel by Ian Fleming.

THUNDERBIRDS ARE GO: THE MOVIE **   U
David Lane UK   1966
Voices of: Ray Barrett, Sylvia Anderson, Peter Dyneley, Neil McCallum, Shane Rimmer, Charles Tingwell, Bob Monkhouse, David Graham, Jeremy Wilkin, Matt Zimmerman, Alexander Davion, Christine Finn
A lively but dated puppet animation inspired by a popular TV series, that detailed the exploits of International Rescue, a sophisticated and elite organisation. In this story, they are called on to save a Martian exploration ship that's in danger of crashing. The effects are by Derek Meddings, who worked on SUPERMAN.
ANIM   89 min (ort 92 min)   VIDrel: MGM/WHV   V

THUNDERBIRDS: COUNTDOWN TO DISASTER **   U
David Elliott/David Lane/Desmond Saunders UK   1968
Voices of: Peter Dyneley, Shane Rimmer, Matt Zimmerman, David Holliday, David Graham, Ray Barrett, Sylvia Anderson
Our hi-tech puppets battle the clock to save the Empire State Building from collapse, when an attempt to move it to another part of New York City goes seriously wrong. A TV reporter and his cameraman get trapped underground and have to be rescued, in a highly hazardous operation. At the same time, fire breaks out near an oil rig. A full-length feature compiled from two episodes: "Trapped In New York City" and "Atlantic Inferno".
ANIM   100 min mTV   VIDrel: 4-FRONT/POLYREC/ITC V

THUNDERBIRDS SIX: THE MOVIE **   U
David Lane UK   1968
Voices of: Peter Dyneley, Catherine Finn, Sylvia Anderson, Keith Alexander, Jeremy Wilkin, John Carson, David Graham, Shane Rimmer, Gary Files, Matt Zimmermann, Geoffrey Keen
Puppet animation based on a popular TV series, telling of the efforts of a sophisticated scientific organisation to combat crime and avert disasters. In this tale International Rescue fights an old enemy, the "Black Phantom". Though OK for its time it now looks distinctly dated. The effects are by Derek Meddings.
Aka: THUNDERBIRD SIX
ANIM   89 min   VIDrel: MGM/WHV   V

THUNDERBIRDS TO THE RESCUE ***
U
Alan Pattillo UK
1966
Voices of: Peter Dyneley, Shane Rimmer, David Holliday, Matt Zimmerman, David Graham, Ray Barrett, Sylvia Anderson
The pilot episode for a very popular TV puppet series (it's actually a feature-length version of "Trapped In The Sky") that sees the villainous "Hood" plant a bomb aboard a supersonic atom-powered airliner, his intention being to photograph a Thunderbird craft when International Rescue arrive. As Virgil and Co. battle to land the stricken plane, Lady P. pursues the Hood in her specially equipped Rolls Royce.
ANIM   91 min mTV   VIDrel: 4-FRONT/POLYREC/ITC   V

THUNDERBOLT AND LIGHTFOOT ***
18
Michael Cimino USA
1974
Clint Eastwood, Jeff Bridges, George Kennedy, Geoffrey Lewis, Catherine Bach, Gary Busey, Vic Tayback, Dub Taylor, Bill McKinney, Jack Dodson, Burton Gilliam, Gene Elma, Lila Teigh, Roy Jenson, Claudia Lennear, Gregory Walcott
Two drifters team up to regain the loot stashed away by one of them from a previous robbery. A well-constructed blend of comedy and action, written by Cimino in his debut as director, but slightly marred by the unexpectedly downbeat ending.
A/AD   110 min (ort 115 min)   VIDrel: WHV   V

THUNDERCATS-HO: THE MOVIE **
U
Masaki Iizuka USA
1986
Voices of: Larry Kenney, Robert McFadden, Lynne Lipton, Earle Hyman, Peter Newman, Earl Hammond, Gerriane Raphael
Full-length animated adventure, the pilot for the second US series, in which our valiant band of feline humanoids must engineer a cunning rescue mission when other survivors from their doomed planet are captured by their arch enemy. Another instalment in this Good versus Evil saga, and as ever, set on a Planet Earth of the distant future, where mankind have long been extinct. An adequate assembly-line product.
ANIM   91 min Cut (5 sec)   VIDrel: VCC L/A   V

THUNDERHEART ***
15
Michael Apted USA
1992
Val Kilmer, Sam Shepard, Graham Greene, Fred Ward, Fred Dalton Thompson, Sheila Trousey, Chief Ted Thin Elk, John Trudell, Julius Drum, Sarah Brave, Allan R.J. Joseph, Sylvan Pumpkin Seed, Patrick Massett, Rex Linn, Duane Brewer
By virtue of his American Indian ancestry, an FBI agent is given the job of investigating a murder on a Sioux reservation. A most absorbing film that conveys a nice sense of tension in terms of the revelations our agent uncovers, offers some intriguing mystical insights, and yet is directed with such a light touch that it never becomes overblown or implausible.
THR   114 min (ort 119 min) cC
VIDrel: VCC/DISC/COLUM   V/sur

THUNDERING MANTIS, THE ***
18
Tsia Wing Choy HONG KONG
1976
Liang Chia-Yen, Huang I Lung, Hsia Chun, Chao Tung Shan, Cheng Feng, Chien Yueh-Sheng
After his best friend is killed, a delivery boy is inexplicably transformed into a giant mantis; a metamorphosis he is not slow to exploit in his thirst for revenge. A bizarre martial arts fantasy quite unlike any other film of this genre.
Aka: MANTIS FIST FIGHTER
MAR   89 min dubbed   VIDrel: EAST/DISC/IMPENT   V

THX 1138 ****
15
George Lucas USA
1970
Robert Duvall, Donald Pleasence, Maggie McOmie, Don Pedro Colley, Ian Wolfe, Sid Haig, John Pearce, Marshall Efrom, Irene Forrest, Claudette Bessing, John Seaton, Eugene L. Sullivan, Gary Alan Marsh, Raymond J. Walsh
A starkly brilliant look at an underground city of the future, with life so dehumanised that the inhabitants take drugs in order to work efficiently. Cutting and voice-over are used to great effect. Following an illegal love affair with his female room-mate, THX 1138 is imprisoned but escapes. An expanded version of a prize-winning feature Lucas made as a student at U.S.C. The script is by Lucas and Walter Murch, with music by Lalo Schifrin.
FAN   82 min (ort 95 min) wScrn   VIDrel: WHV   V

TIARA TAHITI **
PG
William (Ted) Kotcheff UK
1962
James Mason, John Mills, Claude Dauphin, Herbert Lom, Rosenda Monteros, Jacques Marin, Libby Morris, Madge Ryan, Gary Cockrell, Peter Barkworth
Conflict erupts between two army officers of different backgrounds, who settle on the island of Tahiti. An awkward blend of comedy and drama, that has engaging performances hampered by the shortcomings of the script. Written by Geoffrey Cotterell and Ivan Foxwell.
DRA   98 min (ort 100 min)   VIDrel: L/A   V
Boa: novel by Geoffrey Cotterell.

TICKET TO HEAVEN ***
15
Ralph L. Thomas CANADA
1981
Nick Mancuso, Meg Foster, Kim Cattrail, Saul Rubinek, R.H. Thomsom, Jennifer Dale, Guy Boyd, Paul Soles, Harvey Atkin, Robert Joy, Stephen Markle, Marcia Diamond, Timothy Webber, Patrick Brymer, Michael Zelniker, Denise Naples
A youth joins a weird Californian cult and undergoes brainwashing, but is eventually snatched from its clutches by his friends, who mount a kidnapping attempt in order to rescue him. Absorbing and quite chilling, this is a thinly veiled attack on the Moonies. Written by Ralph L. Thomas and Anne Cameron. See also CULT RESCUE.
THR   107 min   VIDrel: VISION/DISC   V
Boa: novel Moonwebs by Josh Freed.

TICKS **
18
Tony Randel USA
1993
Rosalind Allen, Ami Dolenz, Peter Scolari, Seth Green, Virginya Keehne, Ray Oriel, Alfonso Ribeiro, Dina Dayrit, Michael Medeiros, Barry Lynch, Clint Howard, Rance Howard, Judy Dean Berns, Timothy Landfield, J.D. Stone
When steroids are dumped in the local water supply, they cause woodticks to mutate and growing to giant size. Unfortunately, these lethal creatures have a liking for human flesh and start to prey on the inhabitants of a local campground. A familiar horror tale with the choice of giant insect the only new thing about this assembly-line offering.
Aka: INFESTED
HOR   81 min (ort 85 min)   VIDrel: COLUM/SONOP   V

TIDE OF LIFE, THE **
12
David Wheatley UK
1995
Gillian Kearney, Ray Stevenson, Patricia Dunn, John Bowler, Berwick Kaler, Susie Burthon, Charlie Hardwick, Analice Kate Ramsay, Jordan Lorimer, Anne Owen, Michael Hodgson, Shirley Waters, Michael Gunn, Chris Connaughton, Garry Catlin
Period drama following the variable fortunes of a young housekeeper and the very different men who come into her life. As ever, happiness eludes to heroine until the very end, and though extremely satisfying to watch, there are few surprises here.
DRA   150 min (ort 180 min) mTV   VIDrel: FOCUS/DISC   V
Boa: novel by Catherine Cookson.

TIE ME UP! TIE ME DOWN! ***
18
Pedro Almodovar SPAIN
1990
Victoria Abril, Antonio Banderas, Loles Leon, Francisco Rabal, Julieta Serrano, Maria Barranco, Rossy De Palma, Lola Cardona, Montse G. Romeu, Emiliano Redondo
Written by cult director Almodovar, this strange black comedy concerns a young man, his recent release from mental hospital, and his invasion of the apartment of a former porno movie star. Having tied her up, he keeps her as prisoner, convinced that she will eventually fall in love with him. Despite the distasteful premise, Almodovar's subdued approach and the explicit script give this film considerable impact. Music is by Ennio Morricone.
Aka: ATAME!
COM   97 min (ort 105 min)   VIDrel: VCC/DISC/COLUM L/A   V

TIE THAT BINDS, THE *
18
Wesley Strick USA
1995
Daryl Hannah, Keith Carradine, Moira Kelly, Vincent Spano, Julia Devin, Ray Reinhardt, Barbara Tarbuck, Ned Vaughn, Kerrie Cullen, Bob Minor, George Marshall Ruge, Tommy Rosales Jr, Laura Lee Kelly, Marquis Nunley, Jenny Gago
Having adopted a six-year-old girl a middle-class couple are horrified to find that the girl's natural parents, both criminals, are out to get her back by any means, having lost custody of the little girl when they abandoned her after a bungled robbery. An overwrought thriller in which the adopting couple are subjected to a catalogue of nasty torments, it wastes the talents of a fine

cast, says nothing new and is far too violent to make any valid points.
THR   94 min (ort 98 min) cC   VIDrel: POLY/POLYREC V/sur

TIGER BAY ***   PG
J. Lee Thompson UK   1959
John Mills, Hayley Mills, Horst Buchholz, Yvonne Mitchell, Megs Jenkins, Anthony Dawson, George Selway, George Pastell, Marne Maitland, Meredith Edwards, Shari, Paul Stassino, Marianne Stone, Rachel Thomas, Brian Hammond
A child witnesses a murder by a Polish seaman of his girlfriend, and is abducted by the seaman but proves to be more than a match for him. A box office success for Hayley Mills in her first major role. The script is by John Hawkesworth and Shelley Smith.
DRA   102 min (ort 105 min) B/W   VIDrel: MIA/VCC L/A V
Boa: novel Rodolphe Et Le Revolveur by Noel Calef.

TIGER CAGE **   18
Yuen Woo Ping HONG KONG   1990
Donnie Yen, Rosamund Kwan, David Wu, Gary Chau, Do Do Cheng, Cynthia Khan, Michael Woods, Law Lit, Lee Ka Sing, Cha Chen Yee, Leung Lam Ling, Lee Yuen Wah, Willie Tang
A lawyer who launders money for a crooked acquaintance is robbed whilst carrying a large sum, but unknown to him, this was planned by the latter. Helped by one of the crook's former employees, he starts planning the downfall of his untrustworthy and now murderous former colleague. An over-complex blend of action and martial arts capers. When a film called SURE FIRE was confusingly retitled "Tiger Cage", this one became known by its alternative title.
Aka: TIGER CAGE 2
MAR   91 min   VIDrel: POPRO/RTM/IMPENT L/A; MADE/RTM V

TIGER HEART **   15
Georges Chamchoun USA   1995
T.J. Roberts, Carol Potter, Robert LaSardo, Carol Potter, Timothy Williams, Jennifer Lyons, Rance Howard
A man meets a beautiful woman and is very taken with her, and soon learns that both her and her uncle are being forced out of their store by unscrupulous property developers. Fortunately, he just happens to be a martial arts expert, and with a little help from his friends, decides to take on the baddies. Ninety minutes of action, so one can hardly complain about not getting value for money, even if the plot is about as cliched as they come.
MAR   90 min   VIDrel: GUILD/FOXVID V

TIGER ON THE BEAT **   18
Lau Kar Leung HONG KONG   1988
Chow Yun Fat, Conan Lee, Liu Chia-Hui
A cop who is too fond of the ladies and his rookie partner are assigned to investigate the murder of a drugs baron and encounter the usual narrow escapes, in this fast-paced and violent comedy actioner. A sequel followed.
COM   89 min (ort 96 min) wScrn   VIDrel: MADE/RTM V

TIGER ON THE BEAT 2 **   15
Liu Chia Liang HONG KONG   1990
Conan Lee, Danny Lee, Ellen Chan, Roy Cheung, Maria Cordero
A cop with a mid-life crisis finds himself being asked by his nephew for help in finding himself a girlfriend. A possible candidate turns up in the shape of an attractive, young police informer, who has to be protected from the gangster who is out to silence her.
A/AD   85 min   VIDrel: EAST/DISC V

TIGER WARSAW **   15
Amin Q. Chaudhri USA   1988
Patrick Swayze, Barbara Williams, Lee Richardson, Mary McDonnell, Bobby Di Cicco, Piper Laurie, Jenny Chrisinger, Kaye Ballard, James Patrick Gillis, Kevin Bayer, Michelle Glaven, Beeson Carroll, Sally-Jane Heit, Sloane Shelton
A former junkie returns home with the intention of getting his life straightened out and healing an old wound, this being the fact that he nearly killed his father many years ago. A murky and confused melodrama that neither convinces nor entertains.
Aka: TIGER, THE
DRA   89 min (ort 92 min)
VIDrel: 4-FRONT/POLYREC/BRAVE V

TIGERS, THE ***   18
Eric Tsang HONG KONG
Andy Lau, Tony Leung, Miu Ki Wei, Wong Yat Wah, Leung Kar Yan, Tong Chen Yip
Action tale dealing with police corruption in Hong Kong that is in fact an adaptation of one of Hong Kong's most popular TV series. When several of these officers accept a bribe from a drug dealer they ruin their careers, and at the same time find that they are now in the power of the gangster, who blocks every effort they make to break free of his malign influence. Not a conventional action film, it attempts a deeper analysis of the events portrayed.
A/AD   109 min wScrn   VIDrel: EAST/DISC V

TIGHT SPOT ***   PG
Phil Karlson USA   1955
Ginger Rogers, Edward G. Robinson, Brian Keith, Lucy Marlow, Lorne Greene, Katherine Anderson, Allen Nourse, Peter Leeds, Doyle O'Dell, Eve McVeagh, Helen Wallace, Frank Gerstle, Gloria Ann Simpson, Robert Shield, Norman Keats
A model is released into the custody of a US attorney in the hope that she will testify at the forthcoming trial of a major gangster, with whom she was associated. She steadfastly refuses to risk her life, since many other potential witnesses have met violent ends, but begins to feel attracted to the cop assigned to guard her. A solid, crime programmer that gave Rogers a decent chance to show the range of her acting ability.
DRA   92 min (ort 97 min) B/W   VIDrel: COLUM/SONOP V
Boa: play Dead Pigeon by Lenard Kantor.

TIGHTROPE **   18
Richard Tuggle USA   1984
Clint Eastwood, Genevieve Bujold, Dan Hedaya, Alison Eastwood, Jennifer Beck, Marco St John, Rebecca Perle, Regina Richardson, Randi Brooks, Janet MacLachlan, Margaret Howell, Rebecca Clemons, Graham Paul, Donald Barber
A New Orleans policeman discovers that he has the same inclinations as the sex murderer he is after. A sordid and generally unappealing effort, not helped by the fact that much of the story takes place in darkness. Written by Tuggle.
THR   110 min (ort 114 min)   VIDrel: MGM/WHV L/A V

TILAI ***   PG
Idrissa Ouedraogo BURKINA FASO/FRANCE/
SWITZERLAND   1990
Rasmane Ouedraogo, Ina Cisse, Roukietou Barry, Assane Ouedraogo, Mariam Ouedraogo, Sibido Sidibe, Moumouni Ouedraogo, Mariam Barry, Seydou Ouedraogo, Daouda Porgo, Kogre Warma, Mamadou Ouedraogo, Noufou Ouedraogo
A man returns to his village after a long absence and finds that his father has married his betrothed. According to tribal law this now places them in a position where sexual relations between them would be considered incestuous and his life would be forfeit. However, the young couple choose to defy custom and face the consequences. A fascinating look at African society and its dilemmas. Written and produced by Idrissa Ouedraogo, who directed "Yaaba".
Aka: LAW, THE
DRA   78 min   VIDrel: ARTIF/20TH V

TILL DEATH US DO PART *   PG
Norman Cohen UK   1968
Warren Mitchell, Dandy Nicholls, Anthony Booth, Una Stubbs, Liam Redmond, Bill Maynard, Sam Kydd, Brian Blessed
A wretched spin-off from a TV series, that looks at the life of an obnoxious bigot from the 1930s to the 1960s. A sequel, "The Alf Garnett Saga" followed in 1972. Written by Johnny Speight.
COM   96 min (ort 99 min)   VIDrel: WHV V/h

TILL DEATH US DO PART **   15
Yves Simoneau USA   1991
Treat Williams, Arliss Howard, Rebecca Jenkins, J.E. Freeman, Pruitt Taylor Vince, Embeth Davidtz, John Schuck, Valerie Mahaffy, Louis Giambalvo, Jennifer Runyon, Dean Norris, Leilani Sarelle, Barry "Shabaka" Henley, Ashley Judd
Based on the true story of D.A. Vincent Bugliosi, this drama details his prosecution of a multiple wife murderer, against whom the state mounted a case based on nothing more than circumstantial evidence.
DRA   94 min mTV   VIDrel: ODY/SONOP   V/sh
Boa: book by Vincent Bugliosi and Ken Hurvitz.

TILL I KISSED YA' *** 15
Michael Zinberg USA 1990
Corin Nemec, Deidre Hall, Rebecca Cross, Cheryl Pollak, Madchen Amick, Tim Griffith, Ken-Laurence John, Donovan Leitch, Gerardo Mejia, David Sheinkopf, Ernie Lively, Christine Rose, Joe Spano, David Hayward, Christian Hoff
A slight but entertaining look at the life of high school teens in the far off halcyon days of the 1960s, when things seemed so much less complicated than they are in today's hi-tech world. Nothing's here that hasn't been covered in countless other films, though the story is told with a vitality and charm that is quite infectious.
Aka: FOR THE VERY FIRST TIME
DRA   91 min   VIDrel: MGM/WHV L/A   V
Boa: short story by Michael Zinberg.

TILL MURDER DO US PART *** 15
Dick Lowry USA 1992
Meredith Baxter, Stephen Collins, Michelle Johnson, Kelli Williams, Stephen Root, Lori Hallier, Christine Jansen, Debra Jo Rupp, Tricia O'Neill, Clayton Landey, Ralph Bruneau, Jordan Christopher Michael, Jandi Swanson, Aaron Freeman
After her husband divorces her and remarries, his former wife becomes so possessed by hate and bitterness that she is driven to murder both him and his new wife. Baxter gives a riveting performance of the real-life figure of Betty Broderick, whose televised trial for murder and subsequent acquittal caused a sensation at the time. Part 2 continues the story.
Aka: WOMAN SCORNED, A: THE BETTY BRODERICK STORY
DRA   95 min mTV   VIDrel: POLY/POLYREC/BRAVE L/A V

TILL MURDER DO US PART: PART 2 *** 15
Dick Lowry USA 1992
Meredith Baxter, Judith Ivey, Ray Baxter
The true story of Betty Broderick, who shot dead her former husband and his new wife. When an earlier trial fails to convict her of premeditated murder, a judgement of mistrial allows the local D.A. to make another attempt at having her convicted, but it's another two years before a final verdict is reached. Quite engrossing, but one really needs to see the earlier film to take in the whole story.
Aka: PASSION FOR INNOCENCE, A
DRA   90 min (ort 95 min) mTV   VIDrel: ODY/SONOP   V/s

TILL THE END OF THE NIGHT ** 18
Larry Brand USA 1994
Scott Valentine, Katherine Kelly Lang, John Enos, David Keith, Mary Fanaro
After a woman's obsessive ex-husband gets out of jail, the woman and her second husband are terrorised by this figure in this average erotic thriller, derivative in conception and predictable in development.
THR   87 min (ort 90 min)   VIDrel: 20VIS/SONOP   V/sur

TILL THERE WAS YOU * PG
John Sealey AUSTRALIA 1990
Mark Harmon, Deborah Unger, Jeroen Krabbe, Shane Briant, Ivan Kesa, Chief, Lech Mackiewicz, Meriana Obed, Kistina Nehm, Ritchie Singer, Philip Dodd, Jeff Truman, Helen O'Connor, Ira Seidenstein, Terry Davis, Kate Ceberano, Willy Roy
A weakly plotted adventure in which a jazz musician who's something of a free spirit gets a letter from his brother, who lives on a South Pacific island, telling him to hurry over. Upon arrival, he finds that his brother has been murdered, and that the killing is linked to the discovery of some sunken treasure. Despite threats from a local villain, he recovers the gold, and all ends happily, but by then one's interest has entirely evaporated.
A/AD   91 min (ort 95 min)   VIDrel: COLUM/SONOP V/sur

TILL WE MEET AGAIN: PARTS 1, 2 AND 3 * 15
Charles Jarrott USA 1989
Bruce Boxleitner, Barry Bostwick, Mia Sara, Courteney Cox, Lucy Gutteridge, Michael York, Hugh Grant, Maxwell Caulfield, Juliet Mills, John Vickery, Charles Shaughnessy, Denis Arndt, Keith Anderson
Overlong TV adaptation of Krantz's novel dealing with three French women: a former Paris music-hall star and her two daughters, and the major events in their lives over a period of fifty years. As one might expect, WW2 serves as little more than a backdrop to this artificial and saccharine confection.
DRA   300 min (2 cassettes) mTV
VIDrel: 4-FRONT/POLYREC   V
Boa: novel by Judith Krantz.

TIM *** PG
Michael Pate AUSTRALIA 1979
Piper Laurie, Mel Gibson, Alwyn Kurts, Pat Evison, Peter Gwynne, Deborah Kennedy, David Foster, Margo Lee, James Condon, Michael Caulfield, Brenda Senders, Brian Barrie, Kevin Leslie, Louise Pago, Arthur Faybes, Geoff Usher
The story of an affair between an older woman and her handsome gardener, who unfortunately is slightly retarded. The various difficulties their growing romance creates are handled with sensitivity and the film, though undeniably sentimental and predictable in outcome, is both touching and insightful. The script is by Pate. See also BONDS OF LOVE for another film on this theme.
DRA   104 min (ort 108 min)   VIDrel: SGSVID/GOLD   V/h
Boa: novel by Colleen McCullough.

TIME AFTER TIME *** 15
Nicholas Meyer USA 1980
Malcolm McDowell, David Warner, Mary Steenburgen, Charles Cioffi, Kent Williams, Patti D'Arbanville, Laurie Main, Andonia Katsaros, Leo Lewis, Keith McConnell, Geraldine Baron, James Garrett, Bryon Webster, Joseph Maher
Using his time machine, H.G. Wells pursues Jack the Ripper from 19th century England to present-day New York. An amusing fantasy with many loopholes in the story. Steenburgen's great performance, as the writer's 20th century girlfriend, is a highlight. The script is by Karl Alexander and Steve Hayes, with a memorable score by Miklos Rozsa.
FAN   108 min (ort 112 min)   VIDrel: L/A   V
Boa: novel by Karl Alexander and Steve Hayes.

TIME BANDITS ** PG
Terry Gilliam UK 1981
John Cleese, Sean Connery, Shelley Duvall, David Warner, Ralph Richardson, Katherine Helmond, Ian Holm, Michael Palin, Peter Vaughan, Craig Warnock, David Rappaport, Jack Purvis, Malcolm Dixon, Kenny Baker, Mike Edmonds
Long and totally unfunny fantasy about some dwarfs who steal a cosmic plan from God to use for their own advantage. They take an eleven-year-old boy on an adventure through space and time, eventually culminating in a meeting with the Devil (played with great panache by Warner). Written by Michael Palin and Terry Gilliam of "Monty Python" TV fame.
COM   116 min wScrn   VIDrel: CIC/SONOP   V/sur

TIME BOMB ** 18
Avi Nesher USA 1990
Michael Biehn, Patsy Kensit, Robert Culp, Billy Blanks, Richard Jordan, Tray Scoggins, Raymond St Jacques, Jim Manniaci, Stephen Oliver, Ray Mancini, Carlos Palomino, David Arnott, Jeanine Riley, Harvey Fisher, David Belafonte
A watchmaker convinced that he is the object of a murder plot goes to the police, but they refuse to believe him. He is eventually referred to a woman psychiatrist who dismisses his claims as hallucinations until a serious attempt on his life provides vindication. A mediocre action tale, with the overworked theme of international terrorism serving as a motive for the ample violence.
A/AD   92 min (ort 96 min)   VIDrel: EIV/SONOP   V

TIME GUARDIAN, THE * PG
Brian Hannant AUSTRALIA 1987
Tom Burlinson, Nikki Coghill, Dean Stockwell, Carrie Fisher, Henry Salter, Jo Fleming, Tim Robertson, Jim Holt, Peter Merill, Tom Robertson, Wan Thye Liew, Damon Sanders, Tom Karpanny, Peter Healy, Adrian Shirley, Don Baker
In yet another post-nuclear holocaust world of the future (A.D. 4039 to be exact) scattered bands of survivors fight off man-made cyborgs who have rebelled against their masters. Burlinson is the tough guy leader in this one, who takes off with his followers back into the 20th century, only to find his enemies waiting there for his arrival. This is not SF for the discerning, but a lightweight adventure for the desperate.
FAN   84 min (ort 90 min)   VIDrel: GUILD/SONOP L/A   V

TIME IN THE SUN ***   *PG*
Sergei Eisenstein USSR   1930/40
A collection of fragments of QUE VIVA MEXICO, consisting of some striking sequences following peasant and Indian life. In the final section "Death Day" is celebrated with appropriate displays of fireworks, skulls etc. Not so much a film as an incomplete set of visually pleasing images, some portions of which turned up in other works.
DRA   60 min B/W   VIDrel: HEND L/A   V

TIME MACHINE, THE ***   *PG*
George Pal UK/USA   1960
Rod Taylor, Yvette Mimieux, Alan Young, Sebastian Cabot, Tom Helmore, Whit Bissell, Doris Lloyd, Bob Barran, Paul Frees
A Victorian scientist builds a time machine and travels forward to the year 802,701, where he finds human beings have evolved into two separate races. A lavish and entertaining version of the H.G. Wells story, but considerably simplified and lacking the social perspective of the original. The special effects are the film's highlight, and it was remade for TV in 1978. AA: Effects (Gene Warren/Tim Baar).
FAN   98 min (ort 103 min)   VIDrel: MGM/WHV   V/h
Boa: short story by Herbert George Wells.

TIME OF DESTINY, A **   15
Gregory Nava USA   1988
William Hurt, Timothy Hutton, Melissa Leo, Stockard Channing, Francisco Rabal, Conchata Hidalgo, Megan Follows, Frederick Coffin, Kelly Pacheco, John O'Leary, Justin Gocke, John Thatcher, Peter Palmer, Sam Vlahos
Set in 1943, this tells of the favourite daughter of Basque immigrants, who is forbidden by her strict father to marry her soldier boyfriend. The lovers elope, but their wedding night is interrupted by the wrath of the girl's father, who bundles her into his car but is killed in a tragic accident. A dreary tale of passion and revenge. Written by Anna Thomas and with music by Ennio Morricone.
DRA   112 min (ort 118 min)   VIDrel: VES L/A   V

TIME OF THE BEAST **   15
John R. Bowey USA   1989
Brion James, Carolyn Ann Clark, Milton Raphael
A former doctor who once took part in genetic experiments at a research station, returns there as a security guard and witnesses the horrific outcome of his work.
HOR   87 min   VIDrel: CIC/SONOP   V

TIME OF THE GYPSIES ***   15
Emir Kusturica YUGOSLAVIA   1990
Davor Dujmovic, Ljubica Adzovic, Elvira Sali, Hudsnija Hasmovic, Sinolicka Trpkova, Boroa Todorovic
A bitter-sweet look at gypsy life in Yugoslavia as seen through the eyes of a young boy growing up in a shanty-town. Taken to Italy by a petty gangster known as "The Sheik", he is gradually taught to play his part in all the traditional rackets. Filmed at breakneck pace in a variety of styles, this is a moving but overlong tale, and the amateur gypsy cast give some fine performances. The film won Best Director award at Cannes.
Aka: DOM ZA VESANJE
DRA   136 min (ort 142 min)   VIDrel: COLUM/SONOP   V

TIME RUNNER **   15
Michael Mazo USA   1992
Mark Hamill, Rae Dawn Chong, Brion James, Gordon Tipple, Marc Baur, Barry W. Levy, Allen Forget, John Thomas, John McLaren, Suzy Joachim, J. Gregory McLaren, Clif Kosterman, Dale Moore, Gerlad Paetz, Wilf Wilson, Rachel Hayward, Jill Watt
In the year 2022 Earth is being overrun by invading aliens and a man travels back in time, hoping to change the course of history and save the world. He arrives in 1992 just before he was born but is pursued by the aliens who attempt to liquidate him. A confused film that makes very little of its time-travel theme and in many ways seems like just another violent actioner. The special effects are adequate but add little to the movie as a whole.
Aka: IN EXILE
FAN   92 min   VIDrel: MED/20VIS   V/sh

TIME SLIP *   18
Kossei Saito JAPAN   1983
Sonny (Shinichi) Chiba, Isao Natsuki, Raita Ryu, Miyuki Ono, Jana Okada
A Japanese army unit on manoeuvres gets sent back four-hundred years to the era of the samurai, when they fall into a time-warp.
Aka: DAY OF THE APOCALYPSE; SENGOKU JIEITAI; TELE; TIME WARS
FAN   88 min Cut (23 sec – ort 100 min)
VIDrel: MOPIC/SGSVID   V

TIME TO DIE, A **   18
Charles T. Kanganis USA   1991
Traci Lords, Jeff Conaway, Robert Miano, Jesse Thomas, Nitchie Barrett, Richard Roundtree
A female photographer who is desperately trying to earn enough money to retain custody of her small son, takes a P.R. job with the police, but finds that one of the photographs she has taken for them has placed her life in danger.
A/AD   89 min   VIDrel: CAPIT/GUILD   V/h

TIME TO HEAL, A ***   *PG*
Michael Toshiyuki Uno USA   1994
Nicollette Sheridan, Gary Cole, Mara Wilson, Annie Corley, Ken Jenkins, Ben Masters, Tim Ransom. Lorraine Toussaint, Doris Roberts, Bryan Clark, Toni Sawyer, Scott Smith, Trisha Simmons, Wesley Hayashi, Sarah Freeman, Lauren Kopit
The birth of her second child paralyses a woman for life when she suffers a stroke. A painful and quite moving drama, based on a real case.
DRA   89 min mTV   VIDrel: ODY/SONOP   V/h

TIME TO KILL **   15
Guilliano Montaldo ITALY   1990
Nicolas Cage, Giancarlo Giannini, Ennio Morricone, Robert Liensol
During WW2 an Italian officer in Ethiopia goes AWOL in search of a dentist when his raging toothache becomes unbearable, and on his travels meets and has a passionate affair with an African woman. But a freak accident results in him shooting her and he spends the rest of the movie trying to come to terms with his guilt, and later begins to suspect that she may have infected him with leprosy. A morbid and brooding curiosity, its soundtrack is by Ennio Morricone.
Aka: TEMPO DI UCCIDERE
DRA   102 min (ort 103 min)   VIDrel: ARROW/RTM   V

TIME TO KILL, A ***   15
Joel Schumacher USA   1996
Matthew McConaughey, Sandra Bullock, Samuel L. Jackson, Kevin Spacey, Oliver Platt, Charles Dutton, Brenda Fricker, Donald Sutherland, Kiefer Sutherland, Patrick McGoohan, Ashley Judd, Tonea Stewart, Raeven Larrymore Kelly
In a southern town, the father of a ten-year-old black girl brutally raped and almost killed by two drunken whites takes the law into his own hands through his lack of faith in local justice. Armed with a gun, he bursts into the courtroom and shoots the defendants in cold blood, which opens up the possibility of racial conflict when the Ku Klux Klan threaten to retaliate. A strong courtroom drama, hampered by stereotyping and a lack of conviction.
DRA   143 min (ort 149 min) cC   VIDrel: WHV   V/sur
Boa: novel by John Grisham.

TIME TO LIVE, A ***   *(PG)*
Rick Wallace USA   1985
Liza Minnelli, Scott Schwartz, Swoosie Kurtz, Jeffrey DeMunn, Corey Haim, Janine Manatis, Karen Shallo, Francois Klanfer, Henry G. Sanders, David Connor, Chuck Shamata, Ken Pogue, Samatha Langevin, Alain Goulen, Kurt Reis
Hollywood version of the true story of a woman writer who has to come to terms with the fact that her two-year-old son is suffering from muscular dystrophy. A wonderful performance from Minnelli saves this from becoming just another tearjerker. Scripted by John McGreevey. This was Minnelli's TV movie debut.
DRA   91 min (ort 100 min) mTV   TVrel: C4
Boa: book Intensive Care by May-Lou Weisman

TIME TRAVELERS, THE ***   *PG*
Ib Melchior USA   1964
Preston Foster, Philip Carey, Merry Anders, John Hoyt, Steve Franken, Joan Woodbury, Dolores Wells, Dennis Patrick, Forrest J. Ackerman, Gloria Leslie, Margaret Seldeen, Peter Strudwick
Scientists travel into the future to find the Earth devastated by nuclear war and inhabited by mutants. However, a colony of survivors live beneath the surface and are building a spaceship to take them to another planet. Quite enjoyable, if rather bleak.

The photography of Vilmos Zsigmond is an asset. Remade as JOURNEY TO THE CENTRE OF TIME in 1967.
FAN 80 min (ort 84 min) VIDrel: L/A V

TIME TRAX *   PG
Lewis Teague USA   1993
Dale Midkiff, Peter Donat, Elizabeth Alexander, Mia Sara, Michael Warren
Having murdered the President of the United Nations in the year 2193, a time-travelling criminal escapes back to 1993, pursued by a police officer.
FAN 88 min (ort 90 min) mTV VIDrel: WHV V/sh

TIMECOP **   18
Peter Hyams USA   1994
Jean-Claude Van Damme, Ron Silver, Mia Sara, Bruce McGill, Gloria Reuben, Scott Bellis, Jason Schombing, Scott Lawrence, Kenneth Welsh, Brent Woolsey, Brad Loree, Shane Kelly, Richard Faraci, Steve Lambert, Kevin McNulty
By the year 2004, time travel has become a reality and a special police force is created to prevent illicit use of this discovery for personal gain. A politician of that period flouts this law and travels back to the 1990s but is followed by a police officer who is determined to apprehend him. A violent and predictable action effort that does little with the time travel theme and all too soon is overshadowed by its special effects.
FAN 94 min (ort 98 min) cC; 97 min (LV version)
VIDrel: CIC/SONOP; PION (LV only) V/sur LV

TIMEMASTER *   12
James Glickenhaus USA   1995
Jesse Cameron-Glickenhaus, Noriyuki "Pat" Morita, Joanna Pacula, Duncan Regehr, Michael Dorn, Michelle Williams, Scott Colomby, Zelda Rubinstein, Nils Adam Stewart, Lindsey Ginter, Veronica Cameron-Glickenhaus, Shannon Kenny
A youngster who lost his parents when they were abducted by aliens, gains a mentor in the form of a wise old man, after he absconds from his orphanage. Taught to travel trough time, he ultimately saves his parents and the fate of the world. A muddled film that moves about all over the place (its plot revolves around cosmic timelords altering history as part of a bet) it has a few good ideas, but nowhere near enough to make it worth watching.
FAN 95 min (ort 100 min) VIDrel: FIRST/SONOP V

TIMES SQUARE *   15
Alan Moyle USA   1980
Tim Curry, Trini Alvarado, Robin Johnson, Peter Coffield, Herbert Berghof, David Margulies, Miguel Pinero, Elizabeth Pena, Anna Maria Horsford, J.C. Quinn, Michael Margota, Ronald Stevens, Billy Mernit, Paul Sass, Tim Choate
Two runaway girls in New York eventually make good by forming their own rock band. Unpleasant and unbelievable, with Times Square appearing a friendly (and safe!) rendezvous. The music is by Blue Weaver for those who are hard of hearing. Written by Jacob Brackman.
MUS 106 min (ort 113 min) VIDrel: MGM/WHV L/A
V/sh

TIMESCAPE **   15
David N. Twohy USA   1991
Jeff Daniels, Ariana Richards, Emilia Crow, Jim Haynie, Marilyn Lightstone, George Murdock, Nicholas Guest, David Wells, Robert Colbert, Tim Winters, Anna Neil, Willie Rack, Mimi Craven, Jacque McClure, Steven Gilborn
Written by script-writer Twohy (who also wrote WARLOCK) this workmanlike and fairly entertaining effort casts Daniels as a single dad who is in the process of renovating a rundown guesthouse. Some peculiar newcomers arrive and rent the building. When he inadvertently discovers that they are bored "tourists" from the 21st century, he attempts to find out what the future holds. An intriguing idea remains largely unexplored; adequate but certainly not outstanding.
Aka: DISASTER IN TIME; GRAND TOUR, THE; GRAND TOUR: DISASTER IN TIME
FAN 90 min (ort 99 min) VIDrel: MED/POLYREC L/A
V/sh
Boa: novella Vintage Season by Laurence O'Donnell and C.L. Moore.

TIMESTALKERS ***   PG
Michael Schultz USA   1986
William Devane, Lauren Hutton, Klaus Kinski, John Ratzenberger,
Forrest Tucker, John Considine, James Avery, Gail Youngs, Danny Pintauro, Tracey Walter, R.D. Call, Patrik Baldauff, Ritch Brinkley, J. Michael Flynn
A professor comes across an authentic 1886 photograph that shows a modern gun in a man's belt, and is mystified by this until he meets a girl from the future. Together, they embark on a journey into the past to track down a demented scientist from her time, who is out to change the course of history. A bizarre and highly imaginative tale, written by Brian Clemens. This was the last film for Tucker.
FAN 87 min (ort 100 min) mTV VIDrel: L/A V
Boa: novel (unpublished) The Tintype by Ray Brown.

TIN CUP ***   15
Ron Shelton USA   1996
Kevin Costner, Rene Russo, Cheech Marin, Don Johnson, Linda Hart, Dennis Burkley, Rex Linn, Lou Myers, Richard Lineback, George Perez, Mickey Jones, Michael Milhoan, Gary McCord, Craig Stadler, Peter Jacobsen, Jim Nantz
A golfing pro at the end of his career is reduced to running a driving range in the middle of nowhere. But a perceptive therapist succeeds in unblocking his potential as a player, and he qualifies for the US Open, breaking the course record. An unbelievable, illogical and contrived movie, sustained by clever dialogue and winning characterisations. The script is by Shelton and John Norville.
DRA 129 min (ort 134 min) cC VIDrel: WHV V/sur

TIN DRUM, THE ***   15
Volker Schlondorff FRANCE/POLAND/WEST
GERMANY/YUGOSLAVIA   1979
David Bennent, Mario Adorf, Angela Winkler, Daniel Olbrychski, Katharina Talbach, Heinz Bennent, Charles Aznavour, Berta Drews, Tina Engel, Andrea Ferreol, Fritz Hakl, Mariella Oliveri, Roland Beubner, Ernst Jacobi
An attempt to render in film terms a complex novel/allegory of German history, which takes as its central figure a three-year-old boy in Danzig who stops growing and constantly bangs his toy drum. Powerful and disturbing, it has an outstanding performance from twelve-year-old Bennent, yet remains a film that ultimately fails to give substance to the multi-layered complexity of the brilliant original work. AA: Foreign.
Aka: DIE BLECHTROMMEL
DRA 142 min VIDrel: CONNO/RTM V
Boa: novel by Gunter Grass.

TIN MEN ***   15
Barry Levinson USA   1987
Richard Dreyfuss, Barbara Hershey, Danny DeVito, John Mahoney, Jackie Gayle, Stanley Brock, Seymour Cassel, Bruno Kirby, J.T. Walsh, Richard Portnow, Matt Craven, Alan Blumenfield, Brad Sullivan, Michael Tucker
Set in 1963, this tells of two "tin men" or aluminium cladding salesmen, who meet by way of a traffic accident and find their lives linked, when they both engage in an obsessive series of tit-for-tat incidents. A detailed work with sharp dialogue, good characterisations and totally implausible situations.
COM 108 min (ort 112 min) VIDrel: TOUCH L/A V

TIN SOLDIER, THE **   (PG)
Jon Voight USA   1995
Jon Voight, Ally Sheedy, Dom DeLuise, Trenton Knight, Bethany Richards, Brandon harper, Stephen Harper, Dion Basco, Anita Gregory, Wanya Green, Jason Strickland, Kelly Lewis, Dean Simone, Bonnie Paul, Eric Simon, Steven Paul
A boy who is in danger of taking the wrong path in his life is given advice and guidance when a tin soldier, brought to him by his mother, comes to life. An updating of the Hans Christian Andersen story that, given its basic idea, should have been quite charming but unfortunately isn't, mostly because of slack direction. However, at least DeLuise gets a chance to shine as the toy-shop owner.
FAN 99 min SATrel: SKY MOVIES

TINGLER, THE **   12
William Castle USA   1959
Vincent Price, Judith Evelyn, Darryl Hickman, Patricia Cutts, Pamela Lincoln, Philip Coolidge
A scientist who believes that screaming releases a potentially lethal build-up of tension discovers that fear causes a strange insect to attach itself to the spinal cord where it sucks out the bone marrow. Fortunately, this nasty creature can be killed by screaming but panic ensues when it gets loose in a movie

theatre. An over-hyped shocker that was promoted by the use of gimmicks (a favoured ploy of the director) such as vibrating seats in selected cinemas.
HOR   79 min B/W/Col
VIDrel: ENCORE/SPEAR/COLUM   V

TINKER, TAILOR, SOLDIER, SPY ***
PG
John Irvin UK   1979
Joss Ackland, Michael Aldridge, Ian Bannen, Anthony Bate, Hywel Bennett, Bernard Hepton, Michael Jayston, Alexander Knox, Sian Phillips, Beryl Reid, Alec Guinness, Ian Richardson, Terence Rigby, George Sewell, John Standing
An exceptionally well acted tale of the search for a mole within the ranks of MI5, with George Smiley (ably played by Guinness) being brought out of retirement to lead the hunt. Although the final unmasking of the culprit comes as no surprise, the deliberately slow pacing allows ample time for the complex plot to develop, and also gradually reveal the natures and motivations of the characters. SMILEY'S PEOPLE followed in 1982.
DRA   288 min (2 cassettes – ort 350 min) mTV
VIDrel: BBC/TECH   V/h
Boa: novel by John Le Carre.

TINTIN: THE LAKE OF SHARKS **
(U)
Raymond Leblanc BELGIUM/FRANCE   1972
A pleasant animated story based on the popular comic book character, with Tintin and his friends becoming involved in an adventure which leads them to an underwater village, and a chase through the streets aboard some very strange vehicles. One of a number of Tintin adventures.
Aka: ADVENTURES OF TINTIN, THE: THE LAKE OF SHARKS; LAKE OF SHARKS, THE; TINTIN ET LE LAC AUX REQUINS
ANIM   74 min   SATrel: MOVIE CHANNEL

'TIS A PITY SHE'S A WHORE **
15
Giuseppe Patroni Griffi ITALY   1971
Charlotte Rampling, Oliver Tobias, Fabio Testi, Antonio Falsi, Rick Battaglia, Angela luce, Rino Imperio
A so-so film version of this bawdy Restoration drama with its theme of an incestuous brother-sister relationship. An average costumer, although the fine score by Ennio Morricone and the excellent photography by Vittorio Storaro do provide some points of interest.
Aka: ADDIO, FRATELLO CRUDELE
DRA   100 min dubbed   VIDrel: REDEM/RTM   V
Boa: play by John Ford.

TIT AND THE MOON, THE **
18
Bigas Luna FRANCE/SPAIN   1994
Biel Duan, Mathilda May, Gerard Darmon, Miguel Poveda, Abel Folk, Genis Sanchez, Laura Mana, Xavier Masse, Xus Estruch, Victoria Lepori, Jane Harvey, Venessa Isbert, Jordi Busquets, Diego Fernandez, Salvador Anglada
A sex comedy about a youngster and his obsession with the breasts of a new girl in town. Both puerile and macho, though a few typically Spanish touches of surrealistic humour help sustain one's wandering attention. Something of a follow-on to the director's JAMON, JAMON, which was an altogether better film.
Aka: LA LUNE ET LE TETON; LA TETA I LA LLUNA; LA TETA Y LA LUNA
COM   85 min (ort 91 min) wScrn   VIDrel: TART/20TH   V/sh

TITANIC ***
PG
Jean Negulesco USA   1953
Clifton Webb, Barbara Stanwyck, Robert Wagner, Audrey Dalton, Thelma Ritter, Brian Aherne, Richard Basehart, Allyn Joslyn, James Todd, Frances Bergen, William Johnstone, Christopher Severn, Charles FitzSimmons, Barry Bernard
A Hollywood version of the sinking of this ill-fated liner in 1912 that focuses on the fictional story of a family conflict and depictions of some of the more famous passengers. A well-acted studio production that is hampered to some extent by a rather poor script. The films A NIGHT TO REMEMBER and S.O.S. TITANIC covered this disaster. AA: Story/Screen (Charles Brackett/Walter Reisch/Richard Breen).
DRA   93 min (ort 97 min) B/W cC   VIDrel: 20TH/TECH   V/h

TITFIELD THUNDERBOLT, THE ***
U
Charles Crichton UK   1952
Stanley Holloway, George Relph, John Gregson, Naunton Wayne, Godfrey Tearle, Hugh Griffith, Sidney James, Edie Martin, Gabrielle

Brune, Nancy O'Neil, Michael Trubshawe, Reginald Beckwith, Jack MacGowran, Ewan Roberts
Villagers take over their threatened branch line and run it themselves. An Ealing comedy which was seriously undervalued at the time but has some fine touches, not least being the colour photography of Douglas Slocombe. Written by T.E.B. Clarke, with music by Georges Auric.
COM   80 min (ort 84 min)   VIDrel: WHV   V

TITS *
18
J.B. USA   1995
Stacy Nichols, Sally Layd, Sofia Ferrari, Nikki Siin, Peter North, T.T. Boy, Kyle Stone
A couple go through all the usual agonies of divorce, but this being a sex film, they indulge themselves with various partners. Plotless and very dull.
A   45 min (ort 80 min)   VIDrel: PARADOX/TOTAL   V

TITTY BAR *
18
Stuart Canterbury USA   1995
Nastasia, Steen St. Croik, Kimberly Cupps, Peter North, Leanna Fox, Joey Silvera, Rebecca Wilde, T.T. Boy
Sex film built around a sleazy nightclub whose low-life owner is not above abusing and generally molesting the unfortunate women who perform there.
A   47 min   VIDrel: ONE   V

TITTY SLICKERS *
18
Scotty Fox USA   1992
Randy West, Tonisha Mills, Angela Summers, Teri Diver, Jamie Lee, Tom Chapman, T.T. Boy
Dire sex film with a Western flavour, that has three hookers meeting up with the staff of a dude ranch, who mistake them for some female clients they are expecting. The usual encounters take place. A sequel followed. Title is a spoof on the movie CITY SLICKERS, but don't expect any laughs here.
A   51 min (ort 80 min)   VIDrel: ONE   V

TITTY SLICKERS 2 *
18
Scotty Fox USA   1994
Kylie Ireland, Lyndon Johnson, Shelby Stevens, Alex Jordan, Mike Horner, Ian Daniels, Steve Drake
A sequel to the first film, detailing the further sexual adventures of some fun-loving folk in the Great Outdoors.
A   48 min (ort 80 min)   VIDrel: ONE   V

T.N.T. JACKSON **
18
Cirio H. Santiago USA   1974
Jeanne Bell, Stan Shaw, Pat Anderson, Ken Metcalf, Leo Martin, Chris Cruz, Percy Gordon, June Gamble
Low-budget martial arts adventure with a black heroine, played by shapely former Playboy Playmate Wells starring as a karate expert who is looking for her missing brother. Filmed in the Philippines. Fair.
MAR   73 min (ort 87 min)   VIDrel: SUPVID/RTM   V

TO BE OR NOT TO BE ****
U
Ernst Lubitsch USA   1942
Carole Lombard, Jack Benny, Robert Stack, Felix Bressart, Lionel Atwill, Sig Ruman, Stanley Ridges, Tom Dugan, Charles Halton, Peter Caldwell, Helmut Dantine, Otto Reichow, Miles Mander, George Lynn, Henry Victor, Maude Eburne
A Warsaw theatre troupe find their own way to resist the Nazis, culminating in a mass impersonation that allows great scope for their acting skills. A combination of witty dialogue, tip-top performances and finely balanced comedy, enabling tragic events to be depicted in a light-hearted but sensitive way. Written by Edwin Justus Mayer, it was remade by Mel Brooks in 1983. This was Lombard's last film and was released after her death.
COM   94 min (ort 100 min) B/W
VIDrel: 4-FRONT/POLYREC   V

TO BE OR NOT TO BE ***
PG
Alan Johnson USA   1983
Mel Brooks, Anne Bancroft, Charles Durning, Tim Matheson, Jose Ferrer, Christopher Lloyd, George Gaynes, James Haake, George Wyner, Jack Riley, Lewis J. Stradlen, Ronny Graham, Estelle Reiner, Zale Kessler, Earl Boen
Faithful remake of an earlier film of 1942, based on an Ernst Lubitsch story that follows the first film almost scene for scene, and tells of a Polish theatre troupe who use their talents to resist the Nazi invaders. Let down by an irritating degree of preten-

tiousness and rather flat direction, though both Brooks and Bancroft are in fine form. Written by Thomas Meehan and Ronnie Graham.
COM   103 min (ort 108 min)   VIDrel: 20TH/TECH   V/sh

TO BE THE BEST: PARTS 1 AND 2 **   15
Tony Wharmby UK   1991
Lindsay Wagner, Anthony Hopkins, Stephanie Beacham, Christopher Cazenove, Stuart Wilson, James Saito, Fiona Fullerton, Gary Cady, Claire Oberman, David Robb, Christopher Blake, Thomas Ewbank, Julian Fellowes, Rob Freeman, Kate Spiro
Third instalment in the continuing saga of the Harte dynasty, that started with A WOMAN OF SUBSTANCE and continued with HOLD THE DREAM. Here, the present head of the company faces a treacherous plot by her cousin to seize control of her business empire. A flat and lifeless extravaganza whose glamorous locations (in England, Hong Kong and the USA) contrast sharply with the dull plot, limp acting and sheer emptiness of it all.
DRA   200 min mTV   VIDrel: 4-FRONT/POLYREC/ODY V/h
Boa: novel by Barbara Taylor Bradford.

TO CATCH A KILLER ****   15
Eric Till USA   1991
Brian Dennehy, Michael Riley, Margot Kidder, Meg Foster, Martin Julien, Mark Humphrey, David Eisner, Tony DeSantis, Gary Reineke, Scott Hylands, Tim Progosh, Bruce Ramsay, Brenda Bazinet, Liliane Clune, Toby Proctor
At three hours long, this fact-based film is still totally absorbing. In Des Plaines, Illinois, building contractor John Gaines was a respected figure. He was also a serial killer who raped and murdered thirty-three young men. Riley plays Detective Joe Kozenczak who assigned the case of a missing boy, uncovered the truth despite many obstacles. Dennehy is superb as the killer. An awful tale that brings to mind a similar film: HENRY: PORTRAIT OF A SERIAL KILLER.
DRA   177 min (ort 180 min) mTV   VIDrel: ODY/SONOP V/sh

TO CATCH A THIEF **   PG
Alfred Hitchcock USA   1955
Cary Grant, Grace Kelly, Jessie Royce Landis, John Williams, Charles Vanel, Brigitte Auber, Jean Martinelli, Georgette Anys, Roland Lessaffre, Wee Willie Davis, Rene Blancard, Dominique Davray, Edward Manouk, Russell Gaige
A retired cat burglar has to catch someone who is imitating his style. A slow and boring film that fortunately takes place on the French Riviera, so at least the scenery is pretty. Written by John Michael Hayes. AA: Cin (Robert Burks).
DRA   102 min (ort 106 min)   VIDrel: CIC/SONOP   V/h
Boa: novel by David Dodge.

TO CATCH A YETI **   PG
Bob Keen CANADA   1993
Meat Loaf, Chantallese Kent, Richard Howland, Jim Gordon, Leigh Lewis, Jeff Moser, Mike Panton, Mona Matteo, Ria Franchuk, Reginald Doresa, Andreas M. Haradampides, David Walberg, Audrey Barraclouth, Rob Rutter, Neil Verburg
A big-game hunter is determined to capture a specimen of the Yeti and takes part in an unsuccessful expedition to the Himalayas. Later, when sightings of a similar beast are reported in Manhattan, he investigates but gradually undergoes a change of heart. Pleasant comedy outing, rather more suited to a juvenile than an adult audience.
COM   90 min   VIDrel: DDVID   V

TO DANCE WITH THE WHITE DOG ***   PG
Glenn Jordan USA   1993
Hume Cronyn, Jessica Tandy, Christine Baranski, Amy Wright, Esther Rolle, Albright, David Dwyer, Terry Beaver, Harley Gross, Frank Whaley, Dan Albright, Dan Biggers, Janette Lane Bradbury, Warde Q. Butler, Ed Grady, Bob Hannah
When his wife dies after fifty-seven years of marriage, a man finds solace by befriending a strange white dog that only he can see. Unfortunately, his two fusspot daughters incline to the opinion that grief has unhinged their dad's mind. A strong performance by Cronyn in the central role greatly enhances this Hallmark Hall Of Fame production, which in fact takes a light and uplifting view of dying.
DRA   100 min mTV   VIDrel: ODY/SONOP   V/sh
Boa: novel by Terry Kay.

TO DIE FOR **   18
Peter Mackenzie Litten UK   1994
Thomas Arklie, Ian Williams, Tony Slattery, Dilly Keane, Jean Boht, John Altman, Caroline Munro, Gordon Milne, Nicholas Harrison, Ian McKellan, Sinitta, Paul Cottingham, Lloyd Williams, Robert Sturz, Benjamin Sterz, Brian Carter
Distraught at the death of his drag queen lover from AIDS, a homosexual man derives some initial comfort when the ghost of his dead friend returns, but as time goes by, they start to squabble. A kind of gay version of TRULY, MADLY, DEEPLY, this should have been very funny indeed, but a cloying and mawkish script holds it back, and the dialogue just isn't as sharp as one would have wished.
COM   97 min (ort 101 min)   VIDrel: TART/20TH   V/sur

TO DIE FOR ***   15
Gus Van Sant USA   1995
Nicole Kidman, Matt Dillon, Joaquin Phoenix, Casey Affleck, Illeana Douglas, Alison Folland, Dan Hedaya, Wayne Knight, Kurtwood Smith, Holland Taylor, Susan Traylor, Maria Tucci, Tim Hopper, Michael Rispoli, Buck Henry, Gerry Quigley
An ambitious woman talks her way into a job on a cable TV station and then proceeds to fight her way to the top, but finds the journey is not easy. But when her husband tells her that he want children, she tries to have him killed off. A spiteful black comedy that makes clever use of the various activities common to TV (such as chat shows and interviews) to get its points across, and manages to do this with style, wit and a great deal of gusto.
COM   102 min (ort 107 min)   VIDrel: POLFIL   V/sh
Boa: novel by Joyce Maynard.

TO GILLIAN ON HER 37TH BIRTHDAY **   15
Michael Pressman USA   1996
Peter Gallagher, Claire Danes, Kathy Baker, Wendy Crewson, Bruce Altman, Michelle Pfeiffer
One of those painful tearjerkers in which a man is unable to let go of his dead wife, who died in an accident two years ago. Totally immersed in his memories, he is unable to see that his obsession is blighting his relationship with his daughter. Fair.
DRA   89 min   VIDrel: COLUM/SONOP   V

TO GRANDMOTHER'S HOUSE WE GO **   (U)
Jeff Franklin USA   1992
Ashley Olsen, Mary-Kate Olsen, Rhea Pearlman, Jerry Van Dyke, Cynthia Geary, J. Eddie Peck, Stuart Margolin, Rick Poltaruk, Leslie Carlson, Venus Terzo, Florence Patterson, Andrew Wheeler, Walter Marsh, Frank Lewis, Doreen Ramus
Twin five-year-old girls have so worn out their mother with their constant demands for attention that she tells them she needs a break from them. Taking her at her word, these kids make a three-hour journey to their grandmother's house and have various adventures along the way.
JUV   86 min (ort 90 min) mTV   SATrel: SKY MOVIES

TO HAVE AND HAVE NOT ****   PG
Howard Hawks USA   1945
Humphrey Bogart, Lauren Bacall, Walter Brennan, Hoagy Carmichael, Dolores Moran, Sheldon Leonard, Dan Seymour, Marcel Dalio, Walter Sande, Aldo Nadi, Paul Marion, Pat West, Sir Lancelot, Eugene Borden, Elzie Emanuel, Pedro Regas
An American charter boat operator living on Vichy-run Martinique, is reluctantly drawn into the French resistance movement while becoming entangled with a young femme fatale. The plot has more holes than a Swiss cheese, but the sparkling performance of the leads (especially Bacall's catlike sensuality) and the great dialogue, make this a memorable film. Remade as "The Breaking Point" and "The Gun Runners".
DRA   96 min (ort 100 min) B/W   VIDrel: MGM/WHV   V
Boa: novel by Ernest Hemingway.

TO HEAL A NATION ***   U
Michael Pressman USA   1987
Eric Roberts, Glynnis O'Connor, Scott Paulin, Marshall Colt, Brock Peters, Lee Purcell, Laurence Luckinbill, Linden Chiles
The story of Vietnam veteran Jan Scruggs, and his campaign to have the Vietnam Veterans' War Memorial built in Washington. A sincere and effective film that all too often is sidetracked by self-righteousness and histrionics. The generally effective script is the work of Lionel Chetwynd.
DRA   105 min mTV   VIDrel: L/A   V
Boa: book by Jan Scruggs and Joel L. Swerdlow.

TO HELL AND BACK **
Jesse Hibbs USA
Audie Murphy, Marshall Thompson, Richard Castle, Charles Drake, Felix Noriego, Jack Kelly, Susan Kohner, Gregg Palmer, David Janssen, Mary Field, Paul Picerni, Paul Langton, Bruce Cowling, Julian Upton, Denver Pyle
Murphy, America's most decorated soldier holding no less than twenty-eight decorations, recreates his WW2 experiences in this cliche-ridden film. Excellent battle scenes make up for a poor script.
WAR   102 min (ort 106 min) wScrn
VIDrel: 4-FRONT/POLYREC/CIC   V/sur
Boa: autobiography by Audie Murphy.

PG
1955

TO KILL A MOCKINGBIRD ****
Robert Mulligan USA
Gregory Peck, Mary Badham, Philip Alford, John Megna, Brock Peters, Robert Duvall, Frank Overton, Collin Wilcox, William Windom, Rosemary Murphy, Paul Fix, Ruth White, Alice Ghostley, Estelle Evans, James Anderson
A white liberal lawyer in a small Southern town, defends a black man accused of rape, and finds that the locals are hostile to him and his family. A slow but careful and extremely absorbing study of bigotry and ignorance. This was Duvall's first film. The script was by Horton Foote and won a well-deserved Oscar.
AA: Actor (Peck), Screen/adapt (Horton Foote), Art/Set (Alexander Golitzen and Henry Bumstead/Oliver Emert).
DRA   124 min (ort 129 min) B/W   VIDrel: CIC/SONOP   V
Boa: novel by Harper Lee.

PG
1962

TO KILL A PRIEST **
Agnieszka Holland FRANCE/USA
Christopher Lambert, Ed Harris, David Suchet, Joss Ackland, Tim Roth, Joanne Whalley, Peter Postlethwaite, Timothy Spall, Cherie Lunghi, Tom Radcliffe, Wojtek Pszoniak, Johnny Allen, George Ray, Jerome Flynn, Paul Crauchet
A drama inspired by the career and subsequent death of Polish priest Father Jerzy Popieluszko, who fought for trade unionists in his native country. An undoubtedly sincere but shallow and unconvincing effort, with Lambert severely miscast as the priest caught up in the political troubles afflicting his country.
Aka: LE COMPLOT; TO KILL A PRIEST: A TRUE STORY
DRA   112 min Cut (21 sec – ort 130 min)   VIDrel: RCA L/A   V

18
1988

TO LIVE ***
Zhang Yimou HONG KONG
Ge You, Gong Li, Niu Ben, Guo Tao, Jiang Wu, Ni Dahong, Liu Tianchi, Zhang Lu, Xiao Cong, Dong Fei, Huang Zongluo, Liu Yanjing, Li Lianyi, Zhao Yuxiu, Zhang Kang, Pan Jingle, Liu Guangming, Liu Hua, Qu Zhishao, Yang Kuanhou
Story of a couple and the turmoil they endure during the time of the Chinese Civil War, it begins in the late 1940s and spans several decades, dealing with the couple's separation, the death of their children and the grim oppression they suffer at the time of Mao's Cultural Revolution. Unashamedly melodramatic in tone, it was not appreciated by mainland China's Communist dictators, who doubtless would have preferred a more rosy portrait of these turbulent times.
Aka: HUOZHE
DRA   127 min (ort 132 min) wScrn
VIDrel: ELPIC/POLYREC   V
Boa: novel Lifetimes by Yu Hua.

12
1994

TO LIVE AND DIE IN L.A. **
William Friedkin USA
Willem Dafoe, William L. Peterson, John Pankow, Debra Feuer, John Turturro, Darlanne Fluegel, Dean Stockwell, Steve James, Robert Downey, Christopher Allport, Jack Hoar, Val De Vargas, Dwier Brown, Michael Chong, Michael Zand
Tough, harsh portrayal of a federal agent looking for counterfeiters who murdered his partner. A depressingly realistic work from the director of THE FRENCH CONNECTION, and slightly enlivened by the obligatory car chase. The script is by Friedkin and Petievich.
THR   114 min (ort 116 min)   VIDrel: VCC L/A   V
Boa: novel by Gerald Petievich.

18
1985

TO MY DAUGHTER **
Larry Shaw USA
Rue McClanahan, Michele Greene, Samantha Harris, Ty Miller, George Coe, Tom Irwin, James Avery, Ellen Blake, Jeff Corey, Sean Six, Jennifer McComb, A.J. Stephans, Chris Reynolds, Douglas Rowe, Noah Blake, Tanya Howard, Zak Schwartz

PG
1990

A mother is absolutely grief-stricken when her oldest daughter dies and faces a long struggle to make sense of this tragedy and find the will to go on living. A wallow in misery, this soap opera treatment of bereavement suffers badly from its trite made-for-TV cliches.
DRA   91 min (ort 96 min) mTV   VIDrel: MGM/WHV L/A V/sh

TO PLAY THE KING ***
Paul Seed UK
Ian Richardson, Michael Kitchen, Kitty Aldridge, Colin Jeavons, David Ryall, Rowena King
This sees the return of Urquhart as the cunning and disreputable Prime Minister, who is determined to cling to power as long as possible. See HOUSE OF CARDS for the start of this saga. Followed by "The Final Cut".
DRA   210 min (2 cassettes) mTV   VIDrel: BBC/TECH   V/h
Boa: novel by Michael Dobbs.

12
1993

TO PROTECT AND SERVE **
Eric Weston USA
C. Thomas Howell, Lezlie Deane, Richard Romanus, Zoe Trilling, Joe Cortese, Randy Pelish, Rustam Branaman, Jon Edward Ross, Rolando Molina, Kurt Lott, Tracy Ray, Nina Taylor, Curt Barrett, Dyna Winston, Alliso Sie, Joann Avers
When a number of corrupt cops turn up dead, two young members of the force mount an investigation and begin to suspect that the culprit may be a young rookie. An effectively realised police thriller with acceptable performances and dialogue.
THR   88 min (ort 93 min)   VIDrel: EIV/SONOP   V

18
1992

TO SAVE THE CHILDREN **
Steven Hilliard Stern USA
Richard Thomas, Wendy Crewson, Jessica Steen, Joanne Vannicola, James Purcell, Robert Bednarski, Cecilley Carroll, Michael Copeman, Robert Urich, Barclay Hope, Elizabeth Lennie, Jospeh Ziegler, Shannon Lawson, Jenny Parsons
A criminal kidnaps a bus of over 150 school-kids, and wires himself to a bomb to prevent capture, his intention being to hold them to ransom. A fact-based story.
THR   91 min mTV   VIDrel: ODY/SONOP   V/sh
Boa: book When Angels Intervene by Hartt and Judene Wixom.

15
1994

TO SIR, WITH LOVE **
James Clavell UK
Sidney Poitier, Judy Geeson, Suzy Kendall, Lulu, Christian Roberts, Faith Brook, Geoffrey Bayldon, Edward Burnham, Gareth Robinson, Graham Charles, Fiona Duncan, Patricia Routledge, Adrienne Posta, Ann Bell, Mona Bruce
The story of a West Indian teacher who comes to a tough London school, and gradually wins the respect and affection of his unruly pupils. Somewhat over-sentimental and unrealistic, but mildly enjoyable. The film led to the creation of the British TV series "Please Sir". Written by James Clavell. See also THE CLASS OF MISS MacMICHAEL. A sequel followed a good many years later.
DRA   101 min (ort 105 min)   VIDrel: L/A   V
Boa: novel by E.R. Braithwaite.

PG
1967

TO SIR WITH LOVE 2 **
Peter Bognanovich USA
Sidney Poitier, Fernando Lopez, Daniel J. Travanti, Christian Payton, Dana Eskelson, Casey Lluberes, Michael Gilio, L.Z. Granderson, Bernadette L. Clarke, Jamie Kolacki, Saundra Santiago, Cheryl Lynn Bruce, Lulu, Judy Geeson
An experienced teacher takes on a tough assignment teaching at a run-down, multi-racial high school in Chicago in this belated sequel to the first film. Fair to middling, but the idea of a teacher who retired after thirty years in England coming to the States to teach at a run-down school is, to put it mildly, implausible. Poitier does however, have an impressive screen presence which makes the bland script (this is a TV movie, remember) a little more bearable.
DRA   92 min mTV   VIDrel: ODY/SONOP   V/sh

15
1996

TO THE DEATH **
Darrell James Roodt SOUTH AFRICA/USA
John Barrett, Michel Qissi, Robert Whitehead, Michelle Bestbier, Norman Anstey, Greg Latter, Ted Platte, Claudia Udy-Harris, Nick Lorentz, Ernest Mranze, Dan Robbertse, Warwick Grier, Martine Le Maitre, Neville Thomas
A kickboxer tires of tournament fights but finds his retirement

R
1991

plans being stymied by the corrupt promoters who control the game. Another ultra-violent collection of cliches with no shortage of gore. A disturbing and unpleasant film.
MAR   86 min (ort 100 min)   VIDrel: WHV L/A   V

TO THE DEVIL A DAUGHTER **
18
Peter Sykes UK/WEST GERMANY   1976
Richard Widmark, Christopher Lee, Nastassja Kinski, Denholm Elliott, Honor Blackman, Michael Goodliffe, Eva Marie Meineke, Anthony Valentine, Derek Francis, Isabella Telezynska, Irene Prador, Brian Wilde, Frances De La Tour
A girl is promised to a group of Satanists who want to dedicate her to the Devil, but an American occultist novelist intervenes in the struggle for her soul, aided by the rituals he finds described in an ancient book. A muddled and cumbersome adaptation lacking both force and pace. Written by Chris Wicking.
Aka: CHILD OF SATAN; DIE BRAUT DES SATANS
HOR   89 min (ort 93 min)   VIDrel: WHV   V/h
Boa: novel by Dennis Wheatley.

TO THE LIMIT *
18
Raymond Martino USA   1995
Anna Nicole Smith, Joey Travolta, John Aprea, David Proval, Michael Nouri, Branscombe Richmond, Jack Bannon, Lydie Denier, Floyd Levine, Gino Dentie, Melissa Martino, Alexander Marshall, Rebecca Ferrati, Kathy Shower, Paul Martino
A sexy CIA agent tries to expose the double-dealings of her boss, whilst all the while contending with the crooked partners and trained assassins that surround her. Former "Playboy" pin-up Smith is no actress, and walks through her part as if in a trance, but as this is an erotic-actioner, she does get to take her clothes off whenever the pace slackens. Often funny, but never intentionally so.
A/AD   96 min   VIDrel: MIA/DISC/COLUM   V

TO WALK AGAIN **
PG
Randall Risk USA   1994
Blair Brown, Ken Howard, Cameron Bancroft, Carmen Argenziano, Gabrielle Miller, Joely Collins, Chris Martin, Roman Podhora, Ken Camroux, Sheila Larkin, Malcolm Stewart, Catherine Louch, Jon Cuthbert, Silvio Pollo, Peter Lucroix
Having walked out on his wife and kids, a rebellious young man is persuaded by his parents to join the U.S. Marines, and becomes a model soldier. His achievements give him a sense of value for the first time, but a stray bullet destroys his hopes when he is paralysed. Fair. See also RISE AND WALK: THE DENNIS BYRD STORY.
Aka: MOMENT OF TRUTH: TO WALK AGAIN; WALK AGAIN
DRA   87 min   VIDrel: NWV/HIFLI   V/h

TO WONG FOO, THANKS FOR EVERYTHING! JULIE NEWMAR **
PG
Beeban Kidron USA   1995
Wesley Snipes, Patrick Swayze, John Leguizamo, Stockard Channing, Blythe Danner, Arliss Howard, Jason London, Chris Penn, Melinda Dillon, Beth Grant, Alice Drummond, Marceline Hugot, Jennifer Milmore, Jamie Harrold, Mike Hodge
Three drag queens are on their way from New York to Hollywood to take part in a beauty pageant when their car breaks down in a small town in the Midwest. Finding themselves stranded there for the weekend whilst awaiting spare parts, they attempt to win over the wary locals. See also THE ADVENTURES OF PRISCILLA, QUEEN OF THE DESERT, to which this anodyne film bears just a few superficial similarities.
COM   104 min (ort 109 min) cC   VIDrel: CIC/SONOP
V/sur

TOBRUK **
PG
Arthur Hiller USA   1967
Rock Hudson, George Peppard, Nigel Green, Guy Stockwell, Jack Watson, Leo Gordon, Liam Redmond, Norman Rossington, Percy Herbert, Heidy Hunt, Robert Wolders, Anthony Ashdown
A wildly inaccurate WW2 North African action film, based on the British raid on the port of Tobruk, in a desperate bid to destroy Rommel's fuel dumps. This factual basis is however, soon left behind in favour of a fictional view of how Hollywood won the war. As German Jews, Peppard and Stockwell are never less than ludicrous, the other characters faring little better. Battle footage was later used in RAID ON ROMMEL. The screenplay is by Leo Gordon.
WAR   105 min (ort 110 min)
VIDrel: 4-FRONT/POLYREC/CIC   V/h

TODAY IT'S ME, TOMORROW IT'S YOU! **
15
Tonino Cervi ITALY   1968
Montgomery Ford (Brett Halsey), Bud Spencer, Jeff Cameron, Wayde Preston, Stanley Gordon, Tatsuya Nakada
Fresh out of jail after five years, a revenge-seeking man sets out to recruit a team of gunmen to help him find the gang who slaughtered his Indian wife. A full-blooded spaghetti Western, filmed across the lush countryside of Italy's fields and woods, with ample gunplay and the like, but strangely deficient in atmosphere.
Aka: OGGI A ME... DOMANI A TE
WES   92 min   VIDrel: AKTIV/RTM   V

TOKYO BABYLON: VOLS. 1 AND 2 **
15
JAPAN   199-
Two "anime" tales about individuals with psychic powers. In the first one, a yuppie climbs the corporate ladder by using his powers to eliminate his rivals. In the second, a psychic aids the police in solving a series of murders, when the body of a girl is found on the Tokyo subway. Both stories are set in a future city that is rife with crime and depravity – something of an obligatory feature for these Japanese futuristic animations.
ANIM   94 min (2 cassettes) dubbed
VIDrel: MANGA/SONOP   V

TOKYO DRIFTER ***
12
Seijun Suzuki JAPAN   1966
Tetsuya Watari, Chieko Matsubara, Hideaki Natani, Ryuji Kita, Hideaki Esumi, Tamio Kawachi, Isao Tamagawa, Michio Hino, Tomoko Hamakawa, Takeshi Izumi, Tsuyoshi Yoshida, Hiroshi Cho, Eiji Go, Yuzo Kiura, Shinzo Shibata
A top martial arts fighter is called back to Tokyo by his old employer, in order to help the latter fight a gang. This might so easily have been just one more gangster action movie, but instead the director opts for an emphasis on the weird, giving the story so many irrational complications and inappropriate settings that the film plays like a spoof, which to some extent it was clearly meant to be. A fast-paced action-thriller, bizarre and highly distinctive.
Aka: TOKYO NAGAREMONO
A/AD   90 min (ort 83 min)   VIDrel: ICAPRO/MANGA
V/sh
Boa: story by Yasunori "Kohan" Kawauchi.

TOKYO STORY ****
U
Yashujiro Ozu JAPAN   1953
Chishu Ryu, Chieko Higashiyama, Satoshi Yamamura, Haruko Sugimura, Setsuko Hara, Kyoko Kagawa, Kuniko Miyake, Nobuo Nakamura, Shiro Osaka, Eijiro Tono, Teruko Nagaoka, Zen Murase, Mitsuhiro Mori, Hisao Toake, Toyoko Takahashi
An elderly couple visiting their children in Tokyo begin to be aware of the gulf between them and the younger generation, who tend to view them merely as a burden. They then return home, after which the wife dies. A perceptive and sensitive film that offers flawless acting and fine direction. One of the director's best offerings.
Aka: TOKYO MONOGATARI
DRA   130 min (ort 136 min) B/W   VIDrel: ARTIF/20TH
V/h

TOKYO: THE LAST MEGALOPOLIS **
15
Akjo Jissoji JAPAN   1987
A man plans to disturb a slumbering demon, which if awakened would destroy the city of Tokyo. Fortunately, a female warrior spirit and descendant of this demon is prepared to thwart him. Amiable nonsense in the familiar style of Japanese animations, despite being live-action.
FAN ANIM   102 min dubbed   VIDrel: MANGA/SONOP   V

TOM AND HUCK **
PG
Peter Hewitt USA   1995
Jonathan Taylor Thomas, Brad Renfro, Rachel Leigh Cook
Tom Sawyer and his chum Huckleberry Finn hunt for a pirate's treasure map and try to save an innocent man from being wrongfully convicted. A bright, lively and good-natured Disney outing. See also THE ADVENTURES OF HUCK FINN.
A/AD   89 min (ort 91 min) cC   VIDrel: WDV/TECH   V/sh
Boa: novel The Adventures of Tom Sawyer by Mark Twain.

TOM AND JERRY: THE MOVIE **
U
Phil Roman USA   1992
Voices of: Richard Kind, Dana Hill, Andi McAfee, Tony Jay, Rip Taylor, Henry Gibson, Michael Bell, Ed Gilbert, David L. Lander,

Tino Insana, Howard Morris, S Sydney Lassick, Raymond McLeod, Mitchell D. Moore, Scott Wojahn, Tino Insana
Feature-length animation that lacks all the charm of the cartoon characters created by William Hanna and Joseph Barbera. Here our two protagonists are unwisely given the ability to speak, and help re-unite a girl with her father. A big let-down, it lacks all the energy and pace (and furious pump-ups) of the shorts, becoming just a bland outing which sees the characters unbelievably becoming singing and dancing buddies!
ANIM   81 min (ort 84 min)   VIDrel: FIRST/SONOP L/A   V

TOM & VIV ** 15
Brian Gilbert UK   1994
Willem Dafoe, Miranda Richardson, Rosemary Harris, Tim Dutton, Nicholas Grace, Geoffrey Baylden, Clare Holman, Philip Locke, Joanna McCullm, Joseph O'Connor, John Savident, Michael Attwell, Sharon Bower, Linda Spurrier
An interesting portrait of the poet T.S. Eilliot and his relationship with his first wife, whom he meet when they both were students at Oxford. Despite the fact that their temperaments were completely opposite, they fell in love and married. However, her increasingly strange behaviour struck his friends in the Bloomsbury group as an obstacle to his literary success and he had her confined to a mental home where she died eleven years later.
DRA   120 min (ort 125 min)   VIDrel: EIV/SONOP   V/sur
Boa: play by Michael Hastings.

TOM BROWN'S SCHOOL DAYS *** U
Gordon Parry UK   1951
John Howard Davies, Robert Newton, Diana Wynyard, Hermione Baddeley, James Hayter, Kathleen Byron, John Charlesworth, John Forrest, Michael Hordern, Max Bygraves, Francis De Wolff, Amy Veness, Brian Worth, Rachel Gurney
A later version of this novel of Victorian school life, with a youngster finding his first years at Rugby harsh and brutal, but eventually exerting a civilising influence on those around him. Quite competently made, but clinical and annoyingly stilted – the 1940 film is definitely the better of the two. Written by Noel Langley.
DRA   94 min (ort 96 min) B/W coVer   VIDrel: LEIS/DDVID   V
Boa: novel by Thomas Hughes.

TOM, DICK AND HARRY *** U
Garson Kanin USA   1941
Ginger Rogers, George Murphy, Alan Marshal, Burgess Meredith, Jane Seymour, Joe Cunningham, Phil Silvers, Lenore Lonergan, Vickie Lester, Sid Skolsky, Betty Breckenridge, Edna Holland, Gus Glassmire, Netta Packer
A sharp and lively comedy about a girl so popular that she has no less than three boyfriends, and they all want to marry her. A highlight of the film is Silvers, memorable as an offensive ice-cream seller. Written by Paul Jarrico and later remade in 1957 as THE GIRL MOST LIKELY.
COM   83 min (ort 86 min) B/W   VIDrel: VCC L/A   V

TOM HORN *** 15
William Wiard USA   1979
Steve McQueen, Linda Evans, Richard Farnsworth, Billy Green Bush, Slim Pickens, Peter Canon, Elisha Cook Jr, Roy Jenson, James Kline, Geoffrey Lewis, Harry Northup, Steve Oliver
An ex-cavalry scout gets a job working with cattle, and suffers an unjust end when he is hanged for murder after being framed. A kind of semi-Western with the film revolving around the last days of the title character. McQueen lacks much of his old charisma in this, his penultimate role, but the film is beautifully shot. Written by Thomas McGuane and Bid Shrake, and to some extent based on the alleged autobiography of Tom Horn. See also MISTER HORN.
WES   93 min (Cut at film release by 39 sec – ort 98 min)   VIDrel: WHV   V/h
Boa: book by Tom Horn.

TOM JONES **** PG
Tony Richardson UK   1963
Albert Finney, Susannah York, Hugh Griffith, Edith Evans, Joyce Redman, Joan Greenwood, Diane Cilento, David Tomlinson, Rosalind Atkinson, David Warner, George A. Cooper, Angela Baddeley, Rosalind Knight, Peter Bull
A bawdy romp through 17th century England, with foundling Finney marrying the daughter of the squire after many adventures. An excellent cast and superb settings made this the huge

box office success it deserved to become. Written by John Osborne. AA: Pic, Dir, Screen/adapt (John Osborne), Score/orig (John Addison).
COM   117 min (ort 129 min)   VIDrel: 4-FRONT/POLYREC   V/sur
Boa: novel by Henry Fielding.

TOM SAWYER ** U
Don Taylor USA   1973
Johnnie Whitaker, Jeff East, Jodie Foster, Warren Oates, Celeste Holm, Noah Keen, Lucille Benson, Henry Jones, Dub Taylor, Richard Eastham, Susan Joyce, Sandy Kenyon, Joshua Hill Lewis, Steve Hogg, Sean Summers, Kevin Jefferson
A slow, careful, detailed and unmoving version of this classic tale of boyhood adventures. This fourth adaptation of Twain's yarn is given the dubious benefit of Richard and Robert Sherman songs to help the action along.
JUV   95 min (ort 103 min)   VIDrel: MGM/WHV   V/h
Boa: novel The Adventures Of Tom Sawyer by Mark Twain.

TOMAHAWK ** 15
Christopher Cain USA   1993
Rodney A. Grant, Richard Tyson, Barbara Carrera, Gordon Tootoosis
Fair Western set in 1820, with two young Sioux Indians trying to find their friend's killer.
WES   89 min   VIDrel: MARQ/QUANT   V

TOMB, THE * 15
Fred Olen Ray USA   1985
Cameron Mitchell, John Carradine, Richard Alan Hench, Susan Stokey, Michelle Bauer, David Pearson, George Hoth, Sybil Danning (Sybelle Danninger), Stu Weltman, Victor Von Wright, Frank MacDonald, Jack Frankel, Peter Conway
Loosely based on Stoker's horror story, this tells of an Egyptian princess who reincarnates in order to obtain a magic amulet she needs to ensure her survival. A campy horror yarn of three parts tedium to one part chills. The films BLOOD FROM THE MUMMY'S TOMB and THE AWAKENING were two other versions of this story.
HOR   84 min (ort 89 min)   VIDrel: VCC L/A   V
Boa: novel Jewel Of The Seven Stars by Bram Stoker.

TOMBS OF THE BLIND DEAD *** 18
Amando de Ossario PORTUGAL/SPAIN   1971
Cesar Burner, Lone Fleming, Helen Harp (Maria Elena Arpon), Maria Silva, Joseph Telman, Juan Cortes, Rufino Ingles, Antonio Orengo, Veronica Limera, Simon Arriaga Garibaldi, Francisco Sanz, Carmen Gir, Andres Speizer
Devil worshipping Knights Templars were executed in the 13th century, and initially left unburied so that birds pecked out their eyes. When some youngsters stray into an abandoned cemetery, the Templars rise from their graves in search of blood, locating their victims by sound. An effective film of uneven scripting but chilling detail. Followed by THE RETURN OF THE EVIL DEAD, the first of three sequels. See also NIGHT OF THE SEAGULLS.
Aka: A NOITE DO TERROR CEGO; BLIND DEAD, THE: CRYPT OF THE BLIND DEATH, THE; LA NOCHE DE LA MUERTA CIEGA; LA NOCHE DEL TERROR CIEGO; NIGHT OF THE BLIND DEAD
HOR   80 min Cut (1 min 57 sec – ort 93 min) wScrn dubbed   VIDrel: REDEM/RTM   V

TOMBSTONE *** 15
George P. Cosmatos USA   1994
Kurt Russell, Val Kilmer, Sam Elliot, Bill Paxton, Michael Biehn, Charlton Heston, Powers Boothe, Dana Delany, Dana Wheeler-Nicholson, Billy Zane, Jason Priestly, Stephen Lang, Thomas Haden Church, Joanna Pacula, Michael Rooker
Lavish if overlong retelling of the tale of Wyatt Earp and his two brothers who move to Tombstone after imposing order on Dodge City. They hope to find some peace but are soon plunged into a head-on confrontation with a violent gang of cutthroats. Russell is excellent as Earp and Kilmer is memorable as Doc Holliday, but Delany's performance as Earp's love interest lacks sparkle. WYATT EARP – the Kevin Costner version, came out several months later.
WES   120 min (ort 130 min) wScrn   VIDrel: EIV/SONOP   V/sur

TOMMY *** 15
Ken Russell UK   1975
Oliver Reed, Ann-Margret, Roger Daltrey, Elton John, Eric Clapton, Keith Moon, Robert Powell, Tina Turner, Jack Nicholson, Paul

Nicholas, Victoria Russell, Barry Winch, Ben Aris, Mary Holland, Jennifer Blake, Susan Baker

The idea for this film was developed from a rock opera album by the rock group The Who. It tells of a deaf blind and dumb boy who has an amazing ability to play on pinball machines, and who rises to the position of a kind of modern day saviour to the young. The use of multi-channel sound will be lost on TV, but the customary assault on the senses on the part of this director is thankfully muted. Written by Russell.
MUS 106 min (ort 111 min) VIDrel: POLY/POLYREC V/sh

TOMMY BOY ***   12
Peter Segal USA   1995
Chris Farley, David Spade, Brian Dennehy, Bo Derek, Dan Akroyd, Julie Warner, Sean McCann, Zach Greiner, James Blendick, Clinton Turnbull, Ryder Britton, Paul Greenberg, Philip Williams, David "Skippy" Malloy, Roy Lewis

A crazy, anarchic comedy with some of the characters from the "Saturday Night Live" TV show, the paper-thin plot mostly revolving around the exploits of the title figure, a dimwit who takes to the road with his partner, their intention being to sell brake pads and save the family business from ruin. A moronic comedy that is occasionally very funny indeed, thanks to the cast, who play their parts with considerable gusto.
COM 98 min cC VIDrel: CIC/SONOP V

TOMMY THE TOREADOR **   U
John Paddy Carstairs UK   1959
Tommy Steele, Janet Munro, Sidney James, Noel Purcell, Kenneth Williams, Warren Mitchell, Virgilio Texera, Pepe Nieto, Ferdy Mayne, Harold Kasket, Eric Sykes, Manolo Blazquez, Francis De Wolff, Tutte Lemkow, Bernard Cribbins

Lightweight musical-comedy in which Steele plays a sailor who's stranded in Seville and takes the place of a bullfighter who has been framed on smuggling charges. Pleasant enough in its way, and sustained for the most part by Steele's chirpy personality.
MUS 83 min (ort 98 min) VIDrel: WHV V/h

**TOMMY TRICKER AND THE STAMP
TRAVELLER *****   (U)
Michael Rubbo CANADA   1988
Lucas Evans, Anthony Rogers, Jill Stanley, Andrew Whitehead, Paul Popowich, Han Yan, Chen Yuan Tao, Catherine Wright, Cree Rubbo, Rufus Wainwright, Ron Lea, Lynda Smith, John Dapery, William James, Etienne De Maisonneuve

A young boy who is a passionate stamp collector discovers a way to travel the world by making himself so tiny that he can fit on a postage stamp stuck to a letter. A lively and extremely well-realised fantasy that can be enjoyed by adults and children alike. Followed by THE RETURN OF TOMMY TRICKER.
JUV 101 min SATrel: MOVIE CHANNEL

TOMMYKNOCKERS, THE ***   15
John Power USA   1992
Jimmy Smits, Mar Helgenberger, Robert Carradine, Cliff De Young, E.G. Marshall, Pail McIver, John Ashton, Allyce Beasley, Traci Lords, Chuck Henry, Leon Woods, Yvonne Lawley, Bill Johnson, Jon Steemson, Rick Leckinger

A strange object in the woods near a small Maine town begins to have a dark and mysterious effect on the lives of the inhabitants, in this effective and quite chilling shocker. The strong performances and excellent special effects are major assets.
Aka: STEPHEN KING'S THE TOMMYKNOCKERS
HOR 169 min cC VIDrel: WHV V/dm
Boa: story by Stephen King.

TOMORROW MAN, THE **   12
Bill D'Elia USA   1995
Julian Sands, Giancarlo Esposito, Craig Wasson

Whilst government officials investigate an alien crash site, a computer expert has a visitation from an artificial being, sent from the future to prevent a catastrophe taking place. An interesting idea is not really exploited here, and apart from a couple of stimulating moments there is little on offer here.
FAN 85 min VIDrel: 20TH/FOXVID V/sh

TONYA AND NANCY: THE INSIDE STORY ***   (15)
Larry Shaw USA   1993
Alexandra Powers, James Wilder, Heather Langenkamp, Susan Clark, Claude Earl Jones, Dennis Boutsikaris, Bonnie Burroughs, Raymond O'Keefe, Steve Eastin, Michael Cavenaugh, Jeanna Michaels, Christopher-Michael Moore, Christopher Grey

Fact-based story of one of sport's nastiest scandals, when ice-skating champ Tonya Harding was accused of conspiring to break the legs of her chief rival, fellow ice-skater Nancy Kerrigan, who suffered severe injuries in an attack that was clearly intended to permanently disable her. Not an edifying film, but one that does at least provide a fascinating glimpse of human obsession and ruthlessness.
DRA 87 min (ort 90 min) mTV SATrel: MOVIE CHANNEL

TOO BEAUTIFUL TO DIE ***   18
Dario Piana ITALY   1989
Francois Eric Gendron, Florence Guerin, Randy Ingermann, Giovanni Tombieri, Mikena Jesus, Norhana Amifin, Francois Marthouret, Gioia Maria Scola

A video producer in Milan holds a party to celebrate the end of a project, but during the celebrations a young model is raped while everyone looks on and applauds. The victim flees in the producer's car, but the following day her remains are found in the abandoned and burnt-out vehicle. Soon all the party guests find their lives in danger in this stylish and clever thriller that never reveals its hand until the very end.
THR 93 min Cut (1 min 49 sec – ort 97 min)
VIDrel: POPRO/RTM V

TOO FAST TOO YOUNG **   18
Tim Everitt USA   1995
Michael Ironside, Kasia Figura, James Wellington, Patrick Tiller

A man who makes his living as a con-artist shares his home with his attractive girlfriend. However, their happy idyll is destroyed with the arrival of the man's violent cousin, who has broken out of jail and wants the help of his cousin in planning a robbery.
DRA 91 min VIDrel: HIFLI/SONOP V/h

TOO GOOD TO BE TRUE **   15
Christian I. Nyby II USA   1988
Loni Anderson, Patrick Duffy, Glynnis O'Connor, Larry Drake, Neil Patrick, Carmen Argenziano, Elizabeth Norment, James B. Sikking, Julie Harris, Lorinne Vozoff, Carl Franklin, Daniel Baldwin, Arnold Turner, Ted Gehring

Ignoring the warnings of his friends, a lonely widower becomes involved with a seductive blonde who soon begins to show possessive tendencies, and whose beauty masks a murderous nature. A competent but overstretched remake of the 1945 film LEAVE HER TO HEAVEN.
THR 95 min (ort 100 min) mTV VIDrel: 20TH V
Boa: novel Leave Her To Heaven by Ben Ames Williams.

TOO HOT TO HANDLE **   15
Jerry Rees USA   1991
Kim Basinger, Alec Baldwin, Elisabeth Shue, Armand Assante, Robert Loggia, Paul Reiser, Fisher Stevens, Peter Dobson, Steve Hyter, Jeremy Roberts, Big John Studd, Tony Longo, Tom Milanovich, Tim Hauser, Carey Eidel, Karen Medak

A screwball comedy set in the late 1940s, with a likeable but very flimsy plot. A millionaire playboy paying a visit to Las Vegas with his friends for his stag night, falls for a sultry nightclub singer, but finds this poses a risk to his health as the woman is the mistress of a ruthless gangster (modelled on the real-life Bugsy Siegel). An amusing, bawdy romp that is given little scope to develop – tighter direction would have helped. Screenplay is by Neil Simon.
Aka: MARRYING MAN, THE
COM 112 min (ort 116 min) VIDrel: L/A V/sh

TOO LATE THE HERO **   15
Robert Aldrich USA   1969
Michael Caine, Cliff Robertson, Henry Fonda, Ian Bannen, Harry Andrews, Denholm Elliott, Ronald Fraser, Percy Herbert, Lance Percival, Ken Takakura, Patrick Jordan, Sam Kydd, William Beckley, Martin Horsey, Harvey Jason

Two soldiers are sent on a suicide mission, to an island held partly by the Japanese and partly by the Allied troops during WW2. A cat-and-mouse game takes place between them and a Japanese officer, in this well-handled but unremarkable actioner. The script is by Aldrich and Lukas Heller.
Aka: SUICIDE RUN
WAR 127 min (ort 144 min) VIDrel: VCC/DISC/COLUM L/A V
Boa: novel by W. Hughes.

TOO MANY CROOKS ***   U
Mario Zampi UK   1958
George Cole, Sidney James, Bernard Bresslaw, Brenda De Banzie, Terry-Thomas, Vera Day, Delphi Lawrence, John Le Mesurier, Sydney Tafler, Rosalie Ashley, Nicholas Parsons, Terry Scott, John Stuart, Vilma Ann Leslie, Edie Martin
A gang of inept crooks plot a kidnapping in an effort to get a businessman to part up with some money. A lightweight black comedy with a good chase sequence and a nice performance from Terry-Thomas. Written by Michael Pertwee.
COM   82 min (ort 87 min) B/W   VIDrel: VCC L/A   V

TOO MUCH SUN **   18
Robert Downey Sr USA   1990
Eric Idle, Robert Downey, Ralph Macchio, Jim Haynie, Laura Ernst, Andrea Martin, Leo Rossi, Howard Duff, Jennifer Rubin, Jim Haynie, Allan Arbus, Francis R. Hall, Lara Harris, James Hong, Melissa Jenkins, Marvin Kanter, Jon Korkes
When their father dies, a brother and sister find themselves rivals for his inheritance, and unless one can produce a grandchild within a set time, the entire fortune will pass to a shady priest. Unfortunately, as brother and sister are homosexual and lesbian respectively, they find this a daunting task. A silly one-joke film, redeemed by some nice performances and pretty locations.
COM   94 min (ort 110 min)   VIDrel: FIRST/SONOP L/A   V

TOO YOUNG THE HERO **   15
Buzz Kulik USA   1988
Ricky Schroder, John De Vries, Debra Mooney, Mary Louise Parker, Rick Warner
A dramatised account of the exploits of Calvin Graham, who joined the US Navy at the age of twelve and saw a good deal of action during WW2. Despite recreating some of Calvin's more harrowing experiences, including a brutal sexual attack whilst in prison, the film conveys little more than a sense of blandness – Schroder simply fails to convince in the lead role.
WAR   93 min (ort 100 min) mTV   VIDrel: ODY/SONOP V/h

TOO YOUNG TO DIE: A TRUE STORY ***   18
Robert Markowitz USA   1989
Michael Tucker, Juliette Lewis, Michael O'Keefe, Brad Pitt, Alan Fudge, Emily Longstreth, Laurie O'Brien, Yvette Heyden, Tom Everett
Harrowing account (based on a true story) of a fifteen-year-old girl who was raped by her stepfather and abused by just about everyone else, and finally found herself in a condemned cell after her murder of her soldier boyfriend. Told by the girl's lawyer in a series of flashbacks, this is a solid and well-handled drama.
DRA   88 min (ort 92 min) mTV   VIDrel: ODY/SONOP   V/h

TOOTSIE ***   15
Sydney Pollack USA   1982
Dustin Hoffman, Teri Garr, Jessica Lange, Dabney Coleman, Charles Durning, Bill Murray, Sydney Pollack, Geena Davis, George Gaynes, Estelle Getty, Christine Ebersole, Doris Belack, Ellen Foley, Peter Gatto, Lynne Thigpen
A failed actor is desperate to raise money to finance his buddy's play. He dresses as a woman and lands a role in a hospital soap. He falls for one of the stars, but doesn't dare tell the girl he's not half the woman she thinks he is. Entertaining and lightweight, with Hoffman giving a fine performance, but Lange a somewhat less memorable one. The script is by Larry Gelbart and Murray Shisgal. See also HOLLYWOOD SUPERSTAR. AA: S. Actress (Lange).
COM   112 min (ort 115 min) wScrn
VIDrel: VCC/DISC/COLUM   V
Boa: story by Don McGuire.

TOP DOG **   12
Aaron Norris USA   1994
Chuck Norris, Michelle Lamar Richards, Clyde Kusatsu, Herta Buttoms, Herta Ware, Carmine Caridi, Peter Savond Moore, Francesco Quainn, Erik Von Detten, Kai Wulff
A sheriff gets a canine partner after its owner is murdered and this crime-fighting duo get involved in a terrorist plot. A fast-paced blend of action and comedy with the occasional forays into the martial arts, with our dog proving his worth by helping prevent a terrorist plot to bomb San Diego. Not exactly original in conception (see TURNER & HOOCH and K-9) but good fun just the same.
A/AD   93 min   VIDrel: MIA/DISC/IMPENT   V

TOP GUN ***   15
Tony Scott USA   1986
Tom Cruise, Kelly McGillis, Anthony Edwards, Val Kilmer, Tom Skerritt, John Stockwell, Michael Ironside, Barry Tubb, Rick Rossovich, James Tolkan, Meg Ryan, Anthony Edwards, Whip Hubley, Clarence Gilyard Jr, Tim Robbins
Action tale filmed at the Miramar Naval Base in San Diego, where a bunch of F-14 pilots compete for the honour of becoming "Top Gun". Cruise as one such pilot, falls for McGillis whilst training at the weapons school. A contrived blend of memorable action sequences and stilted romantic ones. Written by Jim Cash and Jack Epps Jr, with music by Harold Faltermeyer. AA: Song ("Take My Breath Away" – Giorgio Moroder (m)/Tom Whitlock (l)).
A/AD   105 min (ort 110 min) wScrn cC
VIDrel: CIC/SONOP; PION (LV only)   V/sur LV

TOP HAT ****   U
Mark Sandrich USA   1935
Fred Astaire, Ginger Rogers, Helen Broderick, Edward Everett Horton, Eric Blore, Erik Rhodes, Lucille Ball, Leonard Mudie, Donald Meek, Florence Roberts, Edgar Norton, Gino Corrado, Peter Hobbes, Frank Mills, Tom Ricketts
The charming story of a couple and their romance, which suffers various complications arising from mistaken identity. A classic musical with Astaire and Rogers at their liveliest. Memorable numbers include "Cheek To Cheek", "Top Hat, White Tie, And Tails" and "Isn't This A Lovely Day To Be Caught In The Rain". Music and lyrics are by Irving Berlin, with choreography by Hermes Pan. The script is by Dwight Taylor and Allan Scott.
MUS   93 min (ort 101 min) B/W
VIDrel: 4-FRONT/POLYREC   V
Boa: play by Alexander Farago and Alasdar Laszlo.

TOP SECRET! *   15
Jim Abrahams/David Zucker/Jerry Zucker USA   1984
Val Kilmer, Lucy Gutteridge, Christopher Villiers, Jeremy Kemp, Omar Sharif, Warren Clarke, Peter Cushing, Michael Gough, Harry Ditson, Jim Carter, Billy J. Michell, Major Wiley, Gertan Klauber, Tristram Jellinek, Sydney Arnold
A rock star on tour in East Germany becomes involved in espionage, in this uninventive and patchy spoof that has Kilmer embroiled in the schemes of spies from both sides of the Iron Curtain. From the team that produced the far superior AIRPLANE! and written by Jerry and David Zucker, Abrahams and Martin Burke.
COM   86 min Cut (8 sec – ort 90 min)   VIDrel: CIC/SONOP V/sur

TOP SQUAD ***   18
Chin Sing Wai HONG KONG   1988
Cynthia Rothrock, Sibelle Hu, Vanessa Chan, Regina Kent, Ann Bridgewater, Anthony Carpio, Ng Kwan Ya, Billy Lau
Produced by Jackie Chan (who always keeps a close eye on the stunt sequences) this POLICE ACADEMY-style blend of comedy and action has Rothrock assigned to a team of female officers being trained to protect an important politician. Sent to a harsh training camp, they join forces with some male S.W.A.T. trainees, and set out to round up a gang of thieves. Lively, simpleminded and fun.
Aka: INSPECTOR WEARS SKIRTS, THE
MAR   91 min   VIDrel: 4-FRONT/POLYREC/MIA L/A   V

TOPAZ **   PG
Alfred Hitchcock USA   1969
John Forsythe, Philippe Noiret, Karin Dor, Michel Piccoli, John Vernon, Dany Robin, Frederick Stafford, Roscoe Lee Browne, Claude Jade, Michel Subor, Edmon Ryan, Per-Axel Arosenius, Sonja Kolthoff, Tina Hedstrom, Don Randolph
Dull thriller set against the background of the Cuban missile crisis as the CIA enlist the aid of a French agent in order to destroy a Soviet spy ring. Made with three different endings, this is a most disappointing offering from the director.
THR   120 min (ort 126 min)   VIDrel: CIC/SONOP   V/h
Boa: novel by Leon Uris.

TOPKAPI **   U
Jules Dassin USA   1964
Melina Mercouri, Peter Ustinov, Maximilian Schell, Robert Morley, Akim Tamiroff, Gilles Segal, Jess Hahn, Titos Vandis, Ege Ernart, Senih Orkan, Ahmet Danyal Topatan, Joseph Dassin, Amy Dalby, Despo Diamantidou
An international gang of thieves try to steal a jewelled dagger

from an Istanbul museum. An overlong and only slightly amusing comedy caper, filmed in Istanbul, with initial excitement soon evaporating until a finale that is reminiscent of "Rififi". The attractive score is by Manos Hadjidakis. Written by Monja Danischewsky. AA: S. Actor (Ustinov).
COM   115 min (ort 122 min)   VIDrel: MGM/WHV V
Boa: novel The Light of Day by Eric Ambler.

TORA! TORA! TORA! ***
U
Richard Fleischer/Ray Kellogg/Toshio Masuda/Kinji
Fukasuki JAPAN/USA   1970
Martin Balsam, Joseph Cotten, James Whitmore, Jason Robards, E.G. Marshall, Soh Yamamura, Takahiro Tamura, Edward Andrews, Leon Ames, George Macready, Toshio Masuda, Kinji Fukasuki, Tatsuya Mihashi, Wesley Addy
Pearl Harbour as seen from both viewpoints. An overlong and over-elaborate recreation of the actual attack, that badly needed the light touch of Kurosawa. (He wanted to do the Japanese parts but was not given the chance.) Written by Larry Forrester, Hideo Oguni and Ryuzo Kikushima. AA: Effects/vis (A.D. Flowers/L.B. Abbott).
WAR   138 min (ort 144 min) wScrn   VIDrel: 20TH/TECH; ENCORE (LV only)   V/h LV

TORCH SONG ***
15
Michael Miller USA   1993
Raquel Welch, Jack Scalia, George Newbern, Alicia Silverstone, Laura Innes, Nada Despotovitch, Stan Ivar, Lee Garlington, Lyon Clark, Joshua Abramson, Vinny Argiro, Steve Artiaga, Tom Badal, Marie Narrientos, Ralph Chelli, Chris Connelly
A female alcoholic is forced by her daughter to attend a clinic in this effective drama, whose chief asset is Welch's strong central performance if not the soap opera-style script.
Aka: JUDITH KRANTZ'S TORCH SONG
DRA   92 min mTV   VIDrel: GUILD/SONOP V/sh
Boa: novel by Judith Krantz.

TORCH SONG TRILOGY ***
15
Paul Bogart USA   1988
Matthew Broderick, Anne Bancroft, Harvey Fierstein, Brian Kerwin, Karen Young, Charles Pierce
An examination of nine years in the life of gay New York female impersonator Arnold Beckoff and his search for understanding in the heterosexual world. A touching drama that follows his struggles with his domineering mother, his unfaithful lover and his adopted son. Adapted by Fierstein from his Broadway play.
DRA   114 min (ort 120 min)   VIDrel: VCC/DISC/COLUM V/sur
Boa: play by Harvey Fierstein.

TORMENT ***
15
Claude Chabrol FRANCE   1993
Emmanuelle Beart, Francois Cluzet, Nathalie Cardone, Andre Wilms, Christiane Minazzoli, Marc Lavoine, Dora Doll, Mario David, Jean-Pierre Cassel, Sophie Artur, Thomas Chabrol, Noel Simsolo, Yves Verhoeven, Amaya Antolin
A recently married man becomes increasingly possessive and starts suspecting his beautiful and vivacious wife of infidelity, his insecurity being so great that he cannot believe she would marry him for himself. But his suspicions are almost certainly just figments of his fevered imagination. A bitter film that examines just how a happy relationship can be blighted by a lack of trust. Based on Henri-Georges Clouzet's screenplay for an aborted 1964 project.
Aka: HELL; L'ENFER
THR   98 min (ort 103 min)   VIDrel: CURZON/20TH V/sur

TORN BETWEEN TWO LOVERS **
PG
Delbert Mann USA   1979
Lee Remick, Joseph Bologna, George Peppard, Giorgio Tozzi, Molly Cheek, Kay Hawtrey, Derrick Jones, Murphy Cross, Martin Shakar, Andrea Martin, Mary Long, Lois Markle, Tom Harvey, Jess Osuna, Sean McCann, David Hughes
A married woman meets a man at an airport, and an affair develops which threatens her marriage. A pleasant but undemanding romantic drama.
DRA   95 min (ort 100 min) mTV   VIDrel: ODY/SONOP V
Boa: story by Doris Silverton and Rita Lakin.

TORN CURTAIN **
15
Alfred Hitchcock USA   1966
Paul Newman, Julie Andrews, Lila Kedrova, Hans-Jorg Felmy,
Tamara Toumanova, Ludwig Donath, Wolfgang Kieling, Gunter Strack, David Opatoshu, Gisela Fischer, Mort Mills, Carolyn Cromwell, Arthur Gould-Porter, Gloria Gorvin
An American nuclear scientist pretends to defect to East Germany in order to obtain a secret formula, but his mission is threatened by the arrival of his girlfriend/assistant. A long and dull spy yarn only sporadically enlivened by touches of suspense; it is not on films such as this that the director's reputation was built.
THR   122 min (ort 125 min)   VIDrel: CIC/SONOP V

TORRENTS OF SPRING **
PG
Jerzy Skolimowski CZECHOSLOVAKIA/FRANCE/
ITALY   1989
Timothy Hutton, Nastassia Kinski, Valerie Golino, William Forsythe, Urbano Barberini, Francesca De Sapio, Jacques Herlin, Antonio Cantafora, Alexia Korda, Christopher Janczar, Christian Dottorini, Marinella Anacterio, Piero Bontempo
A Russian aristocrat is engaged to an innocent young girl, but finds himself coming under the spell of a seductive married woman who is negotiating the purchase of his estate. A beautifully filmed adaptation of the original novel, yet strangely lifeless and uninteresting, and with Hutton seriously miscast for good measure.
DRA   101 min (ort 102 min)   VIDrel: L/A V
Boa: novel by Ivan Turgenev.

TORRID WITHOUT A CAUSE **
R18/18
Paul Thomas USA   1989
Tori Welles, April West, Tom Byron, Randy West, Ariel Knight, Marc Wallice
A married woman has fantasies about a former pop star who is believed to be dead, but who now makes a comeback (in her mind at least) in a film whose scenes of passion take place to the pounding beat of a rock score. A thinly-plotted film with a dull parade of softcore vignettes, clearly intended to be a feeble spoof on the "Elvis lives" idea, but devoid of humour.
Aka: EROTIC ADVENTURE; TORY WITHOUT A CAUSE
A   60 min Cut (1 min 39 sec – ort 80 min) mVid
VIDrel: ELV L/A (R18 version); RAVEN/QUANT V

TORSO *
18
Sergio Martino ITALY   1973
Suzy Kendall, Tina Aumont, Luc Merenda, John Richardson, Roberto Bisacco, Angela Covello, Carla Brait, Cristina Airoldi, Patricia Adiutori
Six glamour models are brutally strangled and their bodies cut to pieces with a hacksaw. An English model and her friends are also threatened by this lunatic, who in a terrifying climax traps Kendall, who has a broken leg that's in plaster, in a villa. A repulsive, exploitative and sick little film.
Aka: BODIES BEAR TRACES OF CARNAL VIOLENCE, THE; CORPSES SHOW SIGNS OF CARNAL VIOLENCE, THE; I CORPI PRESENTANO TRACCE DI VIOLENZA CARNALE
HOR   84 min (ort 91 min)   VIDrel: VIPCO/SGSVID V

TORTURE GARDEN **
15
Freddie Francis UK   1967
Jack Palance, Brugess Meredith, Peter Cushing, Beverly Adams, Bernard Kay, Robert Hutton, John Phillips, Michael Ripper, Maurice Denham, Barbara Ewing, Michael Bryant, Nicole Shelby, John Standing, Catherine Finn, Ursula Howells
Visitors to a carnival sideshow have their futures foretold by a mysterious barker in this average anthology of four horror tales. In "Enoch", a young playboy gives a new meaning to the word cat-lover. In "Terror Over Hollywood" a girl pays a high price for fame. "Mr Steinway" tells of the romance between a reporter and a concert pianist, and "The Man Who Collected Poe" deals with the perils of collector's mania.
HOR   96 min   VIDrel: COLUM/SONOP V
Boa: short stories Enoch, Terror Over Hollywood, Mr Steinway and The Man Who Collected Poe by Robert Bloch.

TOTAL RECALL ***
18
Paul Verhoeven USA   1990
Arnold Schwarzenegger, Rachel Ticotin, Sharon Stone, Michael Ironside, Ronny Cox, Marshall Bell, Mel Johnson Jr
A spectacular expansion of Dick's story, that adds a dose of gory violence and some remarkable effects. Schwarzenegger plays a man whose urge to travel to Mars leads him to purchase a memory of one such trip, from which he learns that his true identity has been erased and that he must return there to discover why. Music is by Jerry Goldsmith, and Rob Bottin of

ROBOCOP did the special make-up effects. AA: Spec Award for visual effects (Dream Quest).
FAN   108 min (ort 113 min)   VIDrel: POLY/POLYREC; PION (LV only)   V/sur LV
Boa: short story We Can Remember It For You Wholesale by Philip K. Dick.

TOTALLY F***ED UP
18
Gregg Araki USA
1993
James Duval, Roko Belic, Susan Behshid, Jenee Gill, Gilbert Luna, Lance May, Alan Boyce, Craig Gilmore, Nicole Dillenberg, Johanna Went, Robert McHenry, Brad Minnich, Michael Costanza, Joyce Brouwers, Clay Walker, Aymee Valdes
Gay movie that explores the lives of six teenagers in L.A.
DRA   80 min (ort 88 min)   VIDrel: DTK/RTM   V

TOTO THE HERO **
15
Jaco Van Dormael BELGIUM/FRANCE/ WEST GERMANY
1991
Michel Bouquet, Mireille Perrier, Hugo Harold Harisson, Jo De Backer, Thomas Godet, Sandrine Blancke, Didier Ferney, Gisela Uhlen, Peter Bohlke, Didier Ferney plus voices of: Michel Robin, Patrick Waleffe, Francois Toumarkine
An old man maintains his childhood fantasies including the belief that he was switched at birth with his rich neighbour as well as a fantasy about the adventures of a secret agent called Toto. A complex and at times confusing tale.
Aka: TOTO LE HEROS
DRA   87 min (ort 91 min) wScrn   VIDrel: ELPIC/POLYREC V

TOUCH, THE **
15
Ingmar Bergman SWEDEN/USA
1970
Elliott Gould, Bibi Andersson, Max Von Sydow, Sheila Reid, Maria Nolgard, Steffan Hallerstam, Barbro Hiort Af Ornas, Ake Lindstrom, Mimi Wahlander, Elsa Ebbesen.
Story of a woman who falls for an American academic and leaves her husband. Gould is miscast and this film, Bergman's first English language one, was not a success. Unfortunately, the characters seem contrived and out of place, mouthing their lines in a foreign tongue and from a script that lacks the usual flair of this director's writing.
Aka: BERORINGEN
DRA   112 min   VIDrel: VCC/DISC   V

TOUCH AND GO **
15
Robert Mandel USA
1985
Michael Keaton, Maria Tucci, Maria Conchita Alonso, Ajay Naidu, Max Wright, Lara Jill Miller
The story of an ice-hockey player who is mugged by a gang of youths, but catches one of the boys and escorts him home to his divorced mother, to whom he finds himself so attracted that a romance soon blossoms. Sincere performances sustain a thin and unconvincing blend of humour and sentiment.
COM   97 min (ort 101 min)   VIDrel: FIRST/SONOP L/A   V

TOUCH OF ADULTERY, A **
15
Gene Saks ITALY
1991
Julie Andrews, Marcello Mastroianni, Jonathan Cecil, Ian Fitzgibbon, Maria Machado, Jean-Pirre Castaldi, Jean-Jacques Dulon, Jean-Michel Cannone, Denise Grey, Catherine Jarret, Francoise Michaud, Herve Holle, Yvette Petit
The boss of a construction company learns that his wife is running off with her doctor, and at the same time the wife of said medic learns of his infidelity too. When the two innocent partners meet to discuss this situation, it is only a matter of time before they embark on an affair with each other.
Aka: TCHIN TCHIN
COM   94 min (ort 95 min)   VIDrel: VCC/DISC/COLUM L/A   V/sh

TOUCH OF CLASS, A **
PG
Melvin Frank UK
1973
George Segal, Glenda Jackson, Paul Sorvino, Hildegard Neil, Cec Linder, Mary Barclay, K. Callan, Michael Elwyn, Samantha Weston, Michael McVey, Edward Kemp, Lisa Vanderpump, Ian Thompson, Donald Hewlett, David Healy
A married man has a casual affair with a dress designer, but what was initially no more than a brief fling becomes more complex when his passion deepens into real love. A highly over-rated and contrived romantic comedy, that leaves one curiously

detached from the proceedings. Written by Melvin Frank and Jack Rose. AA: Actress (Jackson).
COM   106 min   VIDrel: L/A   V

TOUCH OF EVIL ****
12
Orson Welles USA
1958
Charlton Heston, Orson Welles, Janet Leigh, Jospeh Calleia, Akim Tamiroff, Marlene Dietrich, Dennis Weaver, Mercedes McCambridge, Ray Collins, Zsa Zsa Gabor, Joanna Moore, Mort Mills, Victor Millan, Lalo Rios, Michael Sargent
Moody, atmospheric story of a Mexican narcotics investigator whose wife is kidnapped on their honeymoon when he starts digging a little too deeply into a murder he witnessed. He encounters a police chief who seems to have no scruples when it comes to framing those he believes to be guilty. A brilliant performance by Welles and good support from the rest of a fine cast are coupled with highly imaginative camerawork to produce a memorable, stylish and enjoyable film.
DRA   104 min (ort 108 min) B/W   VIDrel: PION   LV
Boa: novel Badge of Evil by Whit Masterson.

TOUCH OF FROST, A: CARE AND PROTECTION **
PG
Don Leaver UK
1992
David Jason, Bruce Alexander, Matt Bardock, Claire Hacket, Ralph Nossek, Lindy Whiteford, Helen Blatch. Neil Phillips, Stuart Barren, Paul Moriarty, Bill Stewart, Sion Tudor Owen, Tim Wylton, David Gooderson, Arbel Jones
In this episode, D.I. Frost comes across a case that involves a thirty-year-old crime and sets out to locate a missing girl. A tolerable police drama, one of a number based on the novels of Wingfield.
DRA   102 min (ort 120 min) mTV   VIDrel: POLY/POLYREC V
Boa: novel by R.D. Wingfield.

TOUCH OF GOLD, A *
18
Henri Pachard USA
1992
Aisha, Michelle Monroe, Shawn Michaels, Suzie Bartlett, Randy Spears, Randy West, John Dough, Sunny McKay
A couple of teenage teasers get more than they bargained for when they become involved with some older men. Very dull.
A   58 min (ort 90 min)   VIDrel: GROHOM   V

TOUCH OF THE SUN, A **
U
Gordon Parry UK
1956
Frankie Howerd, Ruby Murray, Reginald Beckwith, Gordon Harker, Richard Wattis, Dennis Price, Alfie Bass, Dorothy Bromiley, Katherine Kath, Pierre Dudan, Colin Gordon, Naomi Chance
An eccentric porter inherits a small fortune and after taking a holiday on the Riviera, returns to London where he buys up the hotel that once employed him, rescuing it from bankruptcy in the process. Howerd manages to bring a lot of charm to his role, helping make up for the deficiencies of the weak script.
COM   81 min B/W   VIDrel: FABFIL/SPEAR   V

TOUCH OF TRUTH ***
12
Michael Switzer USA
1994
Patty Duke, Melissa Gilbert, Bradley Pierce, Markus Flanagan, Lisa Banes, Roger Aaron Brown, Peter Spears, Joe Chrest, Troy Evans, Shelly Morrison, Raye Birk, Christopher Conrad, Terrie Snell
Based on a true story, in which a mother and psychologist lock horns over the future of the mother's autistic son. Matters take a more serious turn when the child uses a keyboard to reveal that he has been sexually abused by his male live-in attendant. A very earnest piece, it makes no boners about dealing with its unsettling subject matter, and fortunately never strays into melodrama.
DRA   90 min mTV   VIDrel: ODY/SONOP   V/sh

TOUGH AND DEADLY **
18
Steve Cohen USA
1994
Billy Blanks, Roddy Piper, James Karen, Sal Landi
After being attacked in Paris, a special agent suffers amnesia, but upon his return to the USA becomes involved in combating an international ring of smugglers.
A/AD   90 min (ort 92 min)   VIDrel: MARQ/QUANT   V

TOUGH ENOUGH *
15
Richard Fleischer USA
1981 (released 1983)
Dennis Quaid, Stan Shaw, Carlene Watkins, Pam Grier, Warren Oates, Wilford Brimley, Bruce McGill, Fran Ryan, Christopher

Norris, Terra Perry, Big John Hamilton, Steve Ward, Susan Benn, Mark Edson, Steve "Monk" Miller
An aspiring Country singer turns amateur boxer, as a way of getting enough exposure to boost his singing career. A bland and predictable musical clone of ROCKY, that takes a fine cast, and does nothing with it.
COM 102 min (ort 107 min) VIDrel: 20TH/TECH L/A V

TOUGH GUYS *   15
Jeff Kanew USA   1986
Kirk Douglas, Burt Lancaster, Charles Durning, Alexis Smith, Dana Carvey, Darlanne Fluegel, Eli Wallach, Monty Ash, Bill Barty
A blend of action and comedy that tells of two bank robbers who are released on parole after 30 years and have difficulties adjusting to the 1980s, so they set about planning their next heist. Douglas and Lancaster work well together, but the film slowly runs out of steam and becomes both sentimental and puerile. Written by James Orr and Jim Cruickshank.
COM 100 min (ort 104 min) VIDrel: RNK L/A V

TOUGH GUYS DON'T DANCE *   18
Norman Mailer USA   1987
Ryan O'Neal, Isabella Rossellini, Debra Sandlund, Wings Hauser, Lawrence Tierney, Frances Fisher, John Bedford Lloyd, Penn Jillette, John Snyder, R. Patrick Sullivan, Stephen Morrow, Clarence Williams III, Ira Lewis
A black comedy-thriller telling of a writer who is unable to clear himself of murder charges with regard to his wife and mistress, and sets out to discover those responsible. An overwrought and self-indulgent display of purple prose and weird characterisations, but held together by a manic logic all of its own.
THR 105 min (ort 110 min) VIDrel: MGM/WHV L/A V/sh
Boa: novel by Norman Mailer.

TOURIST TRAP, THE *   18
David Schmoeller USA   1978
Chuck Connors, Jon Van Ness, Jocelyn Jones, Robin Sherwood, Tanya Roberts, Keith McDermott, Dawn Jeffory, Shailar Coby
A group of people are menaced by sinister mannequins, at a seedy petrol station in a run-down desert holiday resort. A restrained and generally gore-free film, with no reason ever given for the ability of the dummies to move, but several quite chilling moments that successfully exploit this. See also SCARE-CROWS.
Aka: NIGHTMARE OF TERROR
HOR 86 min (ort 90 min) VIDrel: ALLIED/RTM V

TOUT VA BIEN *   15
Jean-Luc Godard/Jean-Pierre Gorin FRANCE   1972
Jane Fonda, Yves Montand, Vittorio Caprioli, Jean Pignol, Anne Wiazemsky
An American journalist and her film-director boyfriend become involved with a French factory worker who stages a strike that develops into a bitter confrontation. Something of a polemic, it skilfully explores the politics and labour relations of the time, albeit within the conventional context of a straightforward drama.
Aka: ALL IS WELL
DRA 95 min wScrn VIDrel: TART/20TH V

TOWER, THE *   15
Richard Kletter USA   1993
Paul Reiser, Susan Norman, Richard Grant, Annabelle Gurwitch, Roger Rees, Dee Dee Rescher, Charmaine Cruise, Dee Dee Wilkinson, Jennifer Richards, Carlos Allen, Chris Doyle
An out-of-work musician takes a job with a hi-tech company, and has a fine old time breaking all the rules in the building. But the central computer that controls everything grows aware of this, and when our man stays behind to work late one night, he finds himself trapped inside the building, and at the mercy of a murderous computer-controlled security system that is going to chastise him for his infractions. Quite ludicrous, but unusually tense.
HOR 86 min (ort 90 min) VIDrel: 20TH/TECH V

TOWER OF EVIL *   18
Jim O'Connolly UK   1972
Bryant Halliday, Anna Palk, Jill Howarth, Jack Watson, William Lucas, Derek Fowlds, Mark Edwards, Anthony Valentine, John Hamill, Gary Hamilton, Candace Glendenning, Dennis Price, George Coulouris, Robin Askwith, Serretta Wilson
Horror story set on an eerie island where a search for Phoenician

treasure is taking place whilst a mad lighthouse keeper plans a series of brutal murders. A banal and second-rate dud, with weak direction and a set of performances best described as overblown.
Aka: BEYOND THE FOG; HORROR OF SNAPE ISLAND
HOR 89 min VIDrel: VIPCO/SGSVID V

TOWERING INFERNO, THE *   15
Irwin Allen/John Guillermin USA   1974
Steve McQueen, Paul Newman, William Holden, Faye Dunaway, Fred Astaire, Richard Chamberlain, Susan Blakely, Jennifer Jones, O.J. Simpson, Robert Vaughn, Robert Wagner, Susan Flannery, Gregory Sierra, Dabney Coleman
The world's tallest building catches fire on its inaugural night. This good idea is not helped by a lousy script, and though the sets are excellent, the pyrotechnics overpower most other aspects of the film. Scripted by Stirling Silliphant and with music by John Williams. See also FIRE! TRAPPED ON THE 37TH FLOOR. AA: Cin (Joseph Biroc/Fred Koenekamp), Edit (Harold K. Kress/Carl Kress), Song ("We May Never Love Like This Again" – Al Kasha/Joel Hirschhorn).
A/AD 158 min (ort 165 min) VIDrel: MGM/WHV V/sh
Boa: novels The Tower by Richard Martin Stern/The Glass Inferno by Thomas M. Scortia and Frank M. Robinson.

TOWN LIKE ALICE, A *   PG
Jack Lee UK   1956
Peter Finch, Virginia McKenna, Marie Lohr, Maureen Swanson, Jean Anderson, Renee Houston, Vincent Ball, Nora Nicholson, Takagi, Tran Van Khe, Marie Lohr, Eileen Moore, John Fabian, Tim Turner, Vi Ngoc Tuan, Geoffrey Keen
A group of British women taken on a forced march through the Malayan jungle by their Japanese captors during WW2, face great privation and suffering, but after the war the woman who led them is eventually re-united with the Australian soldier who was tortured when he stole food for them. A well-made and harrowing film that enjoyed great success. Written by W.P. Lipscomb and Richard Mason and later remade as a TV mini-series in 1981.
Aka: RAPE OF MALAYA
DRA 111 min (ort 117 min) B/W VIDrel: CARL/TECH V
Boa: novel by Nevil Shute.

TOWN TORN APART, A *   PG
Daniel Petrie USA   1992
Michael Tucker, Jill Eikenberry, Carole Galloway, Linda Griffiths, Bernard Behrens, Lindsey Connell, Noam Zylberman, Nicholas Van Burek, Leah Saloma, Patrick Gillen, Bruce Boa, Nicholas Rice, Victoria Mitchell, Linda Goranson
Small-town drama (based on the life of Dennis Littky) that tells the true story of one determined teacher's lone fight to bring some discipline and respect for education to a rundown high school, and at the same time win over the parents with his radical approach to education.
DRA 86 min (ort 91 min) mTV VIDrel: ODY/SONOP V/h

TOWN WITHOUT PITY *   15
Gottfried Reinhardt SWITZERLAND/USA/WEST
GERMANY   1961
Kirk Douglas, E.G. Marshall, Robert Blake, Richard Jaeckel, Frank Sutton, Alan Gifford, Christine Kaufmann, Barbara Rutting
A German girl is raped and four American GIs are accused, but the skilful tactics of their defence counsel cause the girl to commit suicide. A grim melodrama offering no surprises and generating little interest. Written by Silvia Reinhardt and George Hurdalek.
DRA 99 min (ort 103 min) B/W VIDrel: WHV V
Boa: novel The Verdict by Manfred Gregor.

TOXIC AVENGER, THE *   18
Michael Herz/Sam Weil (Lloyd Kaufman) USA   1986
Andree Maranda, Mitchell Cohen, Jennifer Baptist, Cindy Manion, Mark Torgi, Robert Pritchard, Gary Scjneider, Pat Ryan Jr, Dick Martinsen, Chris Liano, Dan Snow, David Weiss, Doug Isbecque, Charles Lee Jr, Pat Kilpatrick
A deliciously funny SF spoof in which a weedy janitor falls into a barrel of toxic waste and is turned into a powerful, but incredibly good-natured monster. He sets about doing good deeds, and cleaning up the corrupt and lawless town. Followed by a couple of sequels.
Aka: HEALTH CLUB
COM 76 min Cut (4 min 13 sec – ort 100 min)
VIDrel: TROMA/RTM V

TOXIC AVENGER PART 2, THE * 18
Samuel Weil (Michael Herz/Lloyd Kaufman) USA   1989
Ron Fazio, Phoebe Legere, John Altamura, Rick Collins, Rikiya Yasuoka, Lisa Gaye, Tsutomu Sekine, Mayako Katsuragi, Shino Buryu, Jessica Dublin, Jack Cooper, Erika Schikel, Bonnie Garvin, Karen King, Dick Mancuso
In this sequel to the first film, toxic hero Melvin Junko goes after the Japanese conglomeration responsible for destroying a home for the blind, and a trip to Japan gives him the chance to locate the man who may be his father. A zany comedy that exploits an issue of real concern, but has some hilarious moments. Filmed at the same time as "The Toxic Avenger 3: The Last Temptation Of Toxie".
COM   90 min Cut (1 min 7 sec – ort 103 min)
VIDrel: TROMA/DISC   V/sh

TOY, THE * PG
Richard Donner USA   1982
Richard Pryor, Jackie Gleason, Ned Beatty, Scott Schwarz, Teresa Ganzel, Wilfrid Hyde-White, Annazette Chase, Tony King, Don Hood, Virginia Capers, Karen Leslie-Lyttle, B.J. Hopper, Linda McCann, Ray Spruell, Orwin Harvey
Remake of the French film "Le Joue" in which a millionaire buys a black man to be a toy for his spoilt son. A feeble and tasteless romp that yields neither laughs nor social comment. Written by Carol Sobleski.
COM   97 min (ort 102 min)   VIDrel: MIA/VCC/COLUM V

TOY SOLDIERS * 15
Daniel Petrie Jr USA   1991
Wil Wheaton, Sean Astin, Keith Coogan, Andrew Divoff, Louis Gossett Jr, Denholm Elliott, T.E. Russell, George Perez, Mason Adams, R. Lee Ermey, Tracy Brooks Swope, Jerry Orbach, Michael Champion, Shawn Phelan, Joe Inscoe
A prep school designed to take on the unruly rich kids no-one else wants, is taken over by terrorists who are seeking the release of a drugs baron. Too bad the gang were not warned what they were letting themselves in for. An unbelievable and almost unwatchable dud, badly acted, badly scripted and way over-the-top.
A/AD   107 min (ort 112 min)   VIDrel: VCC/DISC/COLUM V/sur
Boa: novel by William P. Kennedy.

TOY STORY * PG
John Lasseter USA   1995
Voices of: Tom Hanks, Tim Allen, Don Rickles, Jim Varney, Wallace Shawn, John Ratzenberger, Annie Potts, John Morris, Erik Von Detten, Laurie Metcalf, R. Lee Ermey, Sarah Freeman, Penn Jillette
A computer-generated nursery adventure (the first animation in history to be done entirely so) in which a bunch of toys come to life and have various adventures. Spaceman Buzz Lightyear thinks he is a real astronaut, but it falls to cowboy-doll Woody to show him his limitations, and together they overcome hardship and danger when they are stranded in the world outside the nursery. The LV disc includes a short documentary on the making of the film. AA: Spec Award.
ANIM   84 min wScrn cC   VIDrel: WDV/TECH; ENCORE (LV only)   V LV

TOYS * PG
Barry Levinson USA   1992
Robin Williams, Robin Wright, Michael Gambon, Joan Cusack, L.L. Cool J., Donald O'Connor, Arthur Malet, Jack Warden, Debi Mazar, Wendy Melvoin, Blake Clark, Art Metrano, Julio Oscar Mechoso, Jamie Foxx, Shelly Desai, Manny Portel
After his father dies, a young man must fight his uncle's plan to turn the family toy factory into an armaments plant. Some impressive visual effects do little to help this heavy-handed and preachy anti-war satire. A major disappointment, especially considering the talent available in this movie.
COM   116 min (ort 121 min) wScrn   VIDrel: 20TH/TECH L/A   V

TRACES OF RED * 15
Andy Wolk USA   1992
James Belushi, Lorraine Bracco, Tony Goldwyn, William Russ, Faye Grant, Joe Lisi, Michelle Joyner, Victoria Bass, Melanie Tomlin, Jim Piddock, Harriet Grinnell, Ed Amatrudo, Lindsey Jayde Sapp, Daniel Tucker Kamin, Joseph C. Hess
In the somewhat unreal atmosphere of Palm Beach, a dead cop narrates the circumstances that lead up to his murder. Meant to be an erotic thriller, it is so badly directed and acted that it plays more like an unintentional spoof, full of deliberate red herrings and in dire need of a cohesive and understandable storyline.
THR   100 min (ort 105 min)   VIDrel: EIV/SONOP   V/sur

TRACK 29 * 18
Nicolas Roeg UK   1987
Theresa Russell, Gary Oldman, Colleen Camp, Sandra Bernhard, Seymour Cassel, Christopher Lloyd, Leon Rippy, Vance Colvig, Kathryn Tomlinson, Jerry Rushing, Tommy Hull, J. Michael Hunter, Richard K. Olsen, Ted Barrow
An offbeat black comedy telling of a frustrated and unloved woman, her selfish husband and a stranger who turns up claiming to be her long lost son. Written by Roeg and Dennis Potter, this bizarre film has a few moments of mirth but nothing more. Filmed in Carolina.
COM   87 min (ort 90 min)   VIDrel: PATHE L/A   V/sh

TRACKS OF A KILLER * 18
Harvey Frost USA   1995
Kelly Le Brock, Wolf Larson, James Brolin, Courtney Taylor, George Touliatos, Akiko Morrison, Howard Storey, Ken Camroux
A retiring company executive devises a severe mental and physical test for his would-be successor, in which the candidates and their wives are forced to spend a weekend at an isolated chalet atop a mountain, which they must first scale. However, this exercise results in madness and death, in this totally overblown psychological shocker.
A/AD   86 min (ort 110 min)   VIDrel: MED/20VIS   V/sh

TRADING PLACES * 15
John Landis USA   1983
Dan Aykroyd, Eddie Murphy, Jamie Lee Curtis, Don Ameche, Denholm Elliott, Paul Gleason, Ralph Bellamy, Jim Belushi, Kristin Holby, Alfred Drake, Bo Diddley, Frank Oz, Al Franken, Tom Davis, Jim Gallagher, Bonnie Behrend
Two rich brothers conduct an experiment by making a street punk and a rich kid swap places. Murphy in his second film gives a great performance, but the film, though often very funny, suffers badly from self-indulgent direction. Written by Timothy Harris and Herschel Weingrod.
COM   111 min (ort 116 min)   VIDrel: CIC/SONOP   V

TRAGIC HERO * 18
Taylor Wang HONG KONG   1987
Chow Yun Fat, Alex Man, Andy Lau
Three former gangsters try to escape their violent past. Naturally, there are others who are not especially willing for this to take place.
A/AD   88 min wScrn   VIDrel: MADE/RTM   V

TRAIL BEYOND, THE * U
Robert N. Bradbury USA   1934
John Wayne, Verna Hillie, Noah Beery Sr, Irish Lancaster, Noah Beeery Jr, Robert Frazer, Earl Dwire, Eddie Parker, Artie Ortego, James Marcus, Reed Howes
Two gold-seekers discover a map giving the location of a cache of gold in the Northwest, but the expedition they are about to mount is interrupted when they come to the rescue of a kidnapped girl. A rousing adaptation of the original work, also filmed in 1926 and 1949.
WES   54 min B/W coVer   VIDrel: ENTUK/GOLD   V
Boa: novel The Wolf Hunters by James Oliver Curwood.

TRAIL OF TEARS * 15
Donald Wrye USA   1995
Pam Dawber, Katey Sagal, William Russ, Jeffrey Nordling, Miko Hughes, D. David Marin, Christopher Cass, Eileen Brennan, Ryan Hoeck, Ilana Weiss, Sean Dunn, John Coney, Jacqueline Neves, Michael Senecca, Ron Quinto, Tom Plunkett
True story of a couple of women from different backgrounds who find they have a common bond, as they have both suffered betrayal by their husbands and the loss of their children when their errant spouses snatched them. When on of the women learns through a missing child network that both men have been spotted at Oregon, the pair set off on a cross-country trip in the hope of getting their kids back.
DRA   88 min (ort 92 min) mTV   VIDrel: ODY/SONOP V/sh

TRAIL OF THE PINK PANTHER * PG
Blake Edwards USA   1982
David Niven, Richard Mulligan, Herbert Lom, Joanna Lumley,

Capucine, Burt Kwouk, Harvey Korman, Robert Loggia, Leonard Rossiter, Peter Sellers, Graham Stark, Peter Arne, Andre Maranne, Ronald Fraser, Marne Maitland, Liz Smith
A lousy attempt to cash in on the success of the PINK PANTHER series after the death of Sellers, using footage discarded from earlier films. Sellers's widow strongly objected to this film – when you see it you'll understand why. The story has journalists seeking out Clouseau's colleagues for a TV feature. The film uses previously unseen footage and linking material – THE CURSE OF THE PINK PANTHER being filmed at the same time.
COM   92 min Cut (21 sec – ort 97 min)   VIDrel: WHV   V

TRAIN, THE *** PG
Arthur Penn/John Frankenheimer FRANCE/ITALY/ USA   1964
Burt Lancaster, Jeanne Moreau, Paul Scofield, Michel Simon, Albert Remy, Wolfgang Preiss, Suzanne Flon, Richard Munch, Charles Millot, Jacques Marin, Paul Bonifas, Jean Bouchard, Donald O'Brien, Jean-Pierre Zola, Art Brauss
Taut, well acted and highly realistic account of the attempts by the French resistance, to prevent a train loaded with national art treasures from going to Germany during the last days of WW2. The music is by Maurice Jarre.
WAR   127 min (ort 140 min) B/W wScrn
VIDrel: MGM/WHV   V/sur
Boa: novel Le Front De L'Art by Rose Valland.

TRAIN ROBBERS, THE ** U
Burt Kennedy USA   1973
John Wayne, Ann-Margret, Rod Taylor, Ben Johnson, Christopher George, Bobby Vinton, Ricardo Montalban, Jerry Gatlin
An outlaw's widow asks two cowboys to help her recover some stolen gold so that she can clear the family name. A low-key Western remarkably lacking in action. Written by Kennedy and with music by Dominic Frontiere.
WES   88 min (ort 92 min)   VIDrel: WHV   V/h

TRAINSPOTTING *** 18
Danny Boyle UK   1995
Ewan McGregor, Ewen Bremner, Jonny Lee Miller, Kevin McKidd, Robert Carlyle, Kelly Macdonald, Peter Mullan, James Cosmo, Eileen Nicholas, Susan Vidler, Pauline Lynch, Shirley Henderson, Stuart McQuarrie, Irvine Welsh, Dale Winton
The intertwined lives of assorted losers, petty thieves, drug addicts and psychopaths are closely examined as they head towards self-destruction. Often shocking, sometimes hilarious and always compelling, this is a memorable film, with a clever and witty script (by Ewan McGregor) and a set of standout performances. However, be warned, given the subject matter, the movie is occasionally shocking and not a little revolting.
DRA   89 min Cut (14 sec – ort 93 min) wScrn cC
VIDrel: POLY/POLYREC   V/sh
Boa: novel by Irvine Welsh.

TRANCERS ** 15
Charles Band USA   1984
Tim Thomerson, Helen Hunt, Michael Stefani, Art La Fleur, Thelma Hopkins, Richard Herd, Biff Manard, Anne Seymour, Miguel Fernandez, Pete Schrum, Brad Logan, Barbara Perry, Minnie Lindsay, Richard Erdman, Valey Harker
This low-budget variant of BLADE RUNNER, has a cop from the 23rd century returning to L.A. of the 20th, and taking over the body of one of his ancestors in order to prevent a charismatic megalomaniac from creating a future totalitarian state. An adequate SF excursion, of intriguing but undeveloped ideas. A series of sequels followed.
Aka: FUTURE COP
FAN   73 min (ort 83 min)   VIDrel: EIV/SONOP   V

TRANCERS 2 ** 15
Charles Band USA   1991
Tim Thomerson, Helen Hunt, Megan Ward, Richard Lynch, Barbara Crampton, Jeffrey Combs, Telma Hopkins, Martine Beswick, Biff Manard
Sequel to the 1985 film that's set in contemporary L.A., with our cop from the 32rd century back once more to do battle with the title creatures, mindless automatons best described as killer zombies. Our cop's task is made terribly difficult when his wife arrives on a similar mission. A complex and highly illogical low-budget affair lacking the freshness and verve of the first film, but reasonably enjoyable, despite the mediocre acting.
Aka: TRANCERS 2: THE RETURN OF JACK DETH
FAN   84 min (ort 86 min)   VIDrel: EIV/SONOP   V

TRANCERS 3 ** 15
C. Courtney Joyner USA   1992
Tim Thomerson, Melanie Smith, Andrew Robinson, Tony Pierce, Dawn Ann Billings, Helen Hunt, Megan Ward, Stephen Macht, Telma Hopkins, Ed Beechner, R.A. Mihailoff, Hunter Von Leer, Don Dove
In this further instalment in the TRANCERS series, our time-travelling cop finds himself battling these killer zombies once more – and this time they actually have brains (which is more than can be said for the scriptwriters).
Aka: TRANCERS 3: DETH LIVES
FAN   72 min (ort 83 min)   VIDrel: CIC/SONOP   V

TRANCERS 4: JACK OF SWORDS ** 15
David Nutter USA   1993
Tim Thomerson, Stacie Randall, Ty Miller, Stephen Macht, Alan Oppenheimer, Terri Ivens, Mark Arnold, Clabe Hartley
In this further adventure, our time-travelling cop from the future travels to a strange dimension where these vampirelike creatures live in a sort of medieval world and use the local population as a food source. Shot on location in Romania, this was quickly followed by a sequel that continued a distinctly uninspired series.
FAN   71 min (ort 74 min)   VIDrel: CIC/SONOP   V

TRANCERS 5: SUDDEN DETH * 15
David Nutter USA   1993
Tim Thomerson, Stacie Randall, Ty Miller, Stephen Macht, Terri Ivens, Mark Arnold, Clabe Huntley, Alan Oppenheimer, Jeff Moldovan, Luana Stoica, Rona Hartner, Ion Haiduc, Mihal Dinvales, Mihai Coman, Berti Barnieru, Laura Ilica
In an attempt to return to his own dimension, Deth teams up with a group of freedom fighters in their struggle to overthrow a disagreeable despot – the evil Lord Caliban. Very much a continuation of the previous film with the same shopworn medieval flavour and acting to match.
FAN   70 min (ort 73 min)   VIDrel: CIC/SONOP   V

TRANSFORMATIONS * 18
Jay Kamen USA   1988
Rex Smith, Lisa Langlois, Patrick Macnee, Christopher Neame, Michael Hennessy, Cec Verrell
An interplanetary smuggler discovers a girl hidden on his ship by his friends as a birthday present, but after an enjoyable night with her he finds himself slowly mutating into a lizard-like creature.
HOR   77 min (ort 88 min)   VIDrel: 20TH/TECH   V

TRANSFORMERS: THE MOVIE ** (U)
Kozo Morishita/Nelson Shin USA   1986
Voices of: Eric Idle, Lionel Stander, Orson Welles, Leonard Nimoy, Robert Stack, Judd Nelson, John Moschitta, Norm Alden, Jack Angel, Michael Bell, Gregg Berger, Susan Blu, Arthur Burghardt, Corey Burton, Roger C. Carmel
A spin-off film inspired by robot toys that are able to change their shape. This one has the mighty Autobots slugging it out with the Decepticons. When a sinister new planet appears, all Transformers, both good and bad, are placed in peril. Mediocre.
ANIM   85 min   SATrel: MOVIE CHANNEL

TRANSGRESSION ** 18
Michael DiPaolo USA   1993
Molly Jackson, Marc St Camille, Julio Rodriguez
A female TV reporter attempts to understand the motives of a serial killer, but having done so finds it impossible to go back to her old way of life. A low-budget serial killer yarn, gruesome but totally implausible, in that the reporter starts to emulate the activities of her subject.
THR   77 min   VIDrel: SCEDGE/RTM   V

TRANSPARENT DESIRE * 18
Jean-Pierre Ferrand   1992
Erica Boyar, Brigitte Aime, Cal Jammer, Jenna Weels, Randy West, Courtney, Scott Irish, Austin Moore
A couple move into a house where the wife begins to have visits from ghosts indulging in sex, which soon comes to have a major impact on their love life.
A   59 min   VIDrel: WEST/SGSVID   V

TRANSYLVANIA 6-5000 * PG
Rudy De Luca USA   1985
Jeff Goldblum, Ed Begley Jr, Joseph Bologna, Carol Kane, Jeffrey Jones, John Byner, Geena Davis, Michael Richards, Norman Fell, Donald Gibb, Teresa Ganzel, Rudy DeLuca, Ino Apelt, Bozidar Smiljanovic

A feeble attempt to produce a vampire spoof, with two journalists investigating rumours of vampires in modern day Transylvania. Shot in Yugoslavia.
COM   93 min   VIDrel: NWV L/A   V

TRANSYLVANIA TWIST ** 15
Jim Wynorski USA 1989
Robert Vaughn, Steve Altman, Teri Copley, Ace Mask, Jay Robinson, Angus Scrimm, Steve Franken, Howard Morris
A young couple undertake a perilous journey to Transylvania in order to steal a sacred book from the castle of a notorious vampire count. A low-budget spoof that provides minimal mirth but much in the way of references to other horror films. A poor offering from Roger Corman's studios. See also TRANSYLVANIA 6-5000 for another attempt at a parody in a similar vein.
HOR   79 min (ort 82 min)   VIDrel: MGM/WHV L/A   V

TRAP, THE ** 15
Sidney Hayers CANADA/UK 1966
Oliver Reed, Rita Tushingham, Rex Sevenoaks, Barbara Chilcott, Linda Goranson, Blain Fairman, Walter Marsh, Jo Golland, Jon Granik, Merv Campone, Reginald McReynolds
A rather stilted production, with the undeveloped story of a trapper in British Columbia who takes a mute girl as a wife. When he catches his leg in one of his own traps, she is obliged to amputate it. She then runs away but later returns to him. What could have been fine and touching, remains largely unexplored. The photography is by Robert Krasker.
DRA   102 min (ort 106 min)   VIDrel: CARL/TECH   V

TRAPEZE *** U
Carol Reed USA 1956
Burt Lancaster, Tony Curtis, Gina Lollabrigida, Katy Jurado, Thomas Gomez, Johnny Puleo, Minor Watson, Gerard Landry, Sidney James, Gabrielle Fontan, Jean-Pierre Kerrien, Pierre Tabard, Gamil Ratab, Michel Thomas
A young aspiring acrobat becomes the protege of an established professional but their friendship turns to rivalry when they both fall for their new female partner. Filmed almost entirely at the Cirque d'Hiver in Paris, this movie really does capture the circus atmosphere, while the exciting camerawork provides some tense moments. These two factors more than make up for the generally thin and unoriginal plot.
DRA   102 min (ort 105 min)   VIDrel: MGM/WHV   V

TRAPPED ** 15
Frank De Felitta USA 1973
James Brolin, Earl Holliman, Susan Clark, Robert Hooks, Ivy Jones, Bob Hastings, Tammy Harrington, Marco Lopez, Erica Hagen, Mary Robinson, Elliott Lindsey
After being attacked in a department store washroom, a man recovers and finds that he has been locked in overnight, and is now at risk from vicious guard dogs. A sound idea never develops into more than a sluggish and unrewarding thriller.
Aka: DOBERMAN PATROL
THR   83 min (ort 88 min) mTV   VIDrel: CIC/SONOP L/A   V

TRAPPED *** PG
Fred Walton USA 1989
Kathleen Quinlan, Bruce Abbot, Ben Loggins, Tyress Allen, Miles Mutchler, Wirt Cain, Bill Whitehead, Julius Tennon, Katy Boyer
When the building manageress of a new industrial office block decides to work late with a friend, they find themselves trapped there for the night. Unfortunately, so is a revenge-seeking killer, who blames the company for the death of his son. After her friend is murdered, a battle of wits ensues, made more complex when the woman stumbles across an industrial spy. Despite the hackneyed plot, this is a tense and highly absorbing thriller.
THR   88 min (ort 96 min) mTV   VIDrel: L/A   V

TRAPPED AND DECEIVED ** 12
Robert Iscove USA 1994
Jennie Garth, Jill Eikenberry, Tom Irwin, Johnny Galecki, Eric Close, Gene Lythgow, Michael Marich, Doug McKeon, Julius Tennon, Helen Shaver, Paul Sorvino
The parents of a problem child find they have no recourse but to send her to a unit that specialises in dealing with such children, but find that this is the biggest mistake of their lives.
Aka: WITHOUT CONSENT
DRA   88 min (ort 91 min)   VIDrel: ODY/SONOP   V/sh

TRAPPED IN PARADISE ** PG
George Gallo USA 1994
Nicolas Cage, Richard B. Shull, Jon Lovitz, Dana Carvey, Madchen Amick, Jack Heller, Mike Steiner, Greg Ellwand, Kirk Dunn, Blanca Jansuzian, Cherie Ewing, Florence Stanley, Jeff Levine, Sandra Myers, Frank Berardino, Mable & Sarge
Three brothers who are all thieves and inept ones at that, find themselves caught in a town whose citizens they had intended to rob. However, the overwhelming kindness and good nature of the inhabitants conspires to thwart their plans. A suitably sentimental Christmas yarn that has slight echoes of more than a few Capra films. Sadly, this overlong work never fulfils its early promise and proves a major waste of some talented actors. Written by Gallo.
COM   107 min (ort 111 min) cC   VIDrel: 20TH/FOXVID V/sur

TRAPPED IN SPACE ** 15
Arthur Allan Seidelman USA 1994
Jack Wagner, Jack Coleman, Kay Lenz, Craig Wasson, Sigrid Thornton, Kevin Colson, Mark Lee Kevin Copeland, Ian Stenlke, Francine Beel, Tania Martin, Jamie Stewart, Michael Mills
A collision with a meteor reduces the oxygen aboard a spaceship en route for Venus, and the five crew members find that their captain has deserted the ship in a lifepod, leaving them only sufficient oxygen for only three to complete the journey. Tediously sexist, with the two female crew members showing all the discipline and courage, while the men are mostly portrayed virtually as Neanderthals. Strong direction does what it can with Clarke's flat prose.
FAN   83 min (ort 95 min) mTV   VIDrel: CIC/SONOP   V
Boa: short story Breaking Strain by Arthur C. Clarke

TRASH * 18
Paul Morrissey USA 1970
Joe Dallesandro, Holly Woodlawn, Jane Forth, Michael Sklar, Geri Miller, Andrea Feldman
The sleazy saga of a heroin addict and his efforts to raise the cash for his habit, that is neither edifying, entertaining nor instructive, but represents a depressing and exploitative look at New York low-lives. Rarely was a film so aptly titled.
DRA   93 min Cut (1 min 12 sec plus some cuts substituted – ort 110 min)   VIDrel: FIRST/SONOP   V

TRAUMA ** 18
Dario Argento ITALY 1993
Christopher Rydell, Brad Dourif, Piper Laurie, James Russo, Asia Argento, Frederic Forrest, Deborah Barrymore, E.A. Violet Boor, David Chase, Cory Garvin, John Harding, Nick Holder, Godfrey James, Michael Jacques, Laura Johnson
After witnessing the murder of her parents by a psychotic killer who decapitates his victims, a young girl suffers a massive trauma. She responds by running away and meets a man who helps her track down this killer.
Aka: AURA'S ENIGMA; MOVING GUILLOTINE
HOR   102 min (ort 106 min)   VIDrel: OVER/HIFLI   V/sur

TRAVELING MAN *** (12)
Irvin Kershner USA 1989
John Lithgow, Jonathan Silverman, John Glover, Chynna Phillips, Margaret Colin, John M. Jackson, Dawn Arnemann, Paul Ambruster, Suzi Bass, Jerry Campbell, Marc Clement, David Dwyer, Christopher Ekholm, J. Don Ferguson
Lithgow is wonderful in this story of a burned-out salesman who travels the road selling foam insulation. When his ruthless boss decides that he isn't selling enough, he allocates him a young hotshot to train, but doesn't tell him that the young man is really a rival out to grab all his accounts. A downbeat comedy that tries to inject realism into its portrait of life on the road, and generally succeeds right up to the obligatory happy ending.
Aka: TRAVELLING MAN
DRA   100 min (ort 105 min) mCab   VIDrel: MGM/WHV L/A   V

TRAXX ** 18
Jerome Gary USA 1988
Shadoe Stevens, Priscilla Barnes, Willard E. Pugh, Robert Davi, Hugh Gillin, John Hancock, Michael Kirk, Raymond O'Connor, Hershal Sparber, Jonathan Lutz, Lucius Houghton, Darrow Igus, Arlene Lorre, Wally Amos, Jerry Colker
An ex-mercenary and former policeman decides to try a more peaceful mode of life and moves to the country to set up his own cookie company. However, when the local sheriff pleads

for help in dealing with a crime wave, our man cannot resist this opportunity and offers his services. Enter various mobsters who now arrange for him to be dealt with by a hit-man. A rather bizarre spoof on all those tough-cop-cleans-up-town movies.
A/AD   81 min Cut (9 sec – ort 84 min)   VIDrel: 20TH   V

TREACHEROUS **   18
Kevin Brodie USA   1993
C. Thomas Howell, Tia Carrere, Adam Baldwin, Randi Ingerman, Kirk Fox, Kevin Bernhardt, James Philips, Anastascia Belmonte, Joanan Dierck Brodie, Jo Barnett, Steve Monarque, Leo Napolitano, Al Cutillo, James R. Walker, Jose Amate Perez
Two young men run a hotel at a Mexican paradise resort, but both their peaceful lives and their friendship are endangered when a man and woman turn up with a suitcase full of money. A standard erotic-thriller offering few surprises.
THR   90 min   VIDrel: MED/COLUM L/A   V/sh

TREACHEROUS CROSSING **   PG
Tony Wharmby USA   1992
Lindsay Wagner, Angie Dickinson, Grant Show, Joseph Bottoms, Karen Medak, Charles Napier, Erick Avari, Cameron Watson, Jeffrey De Munn, Scott McCray, Scott Roberts, Paul Meadmore, Sharon Cornell, Terri Apple, Laurel Adams
A couple of newlyweds take a honeymoon cruise on an ocean liner but their married life come to an abrupt halt when the husband suddenly disappears, leaving his bride to mount her own investigation. A standard TV suspenser that is nothing more than a pale remake of the 1953 film "Dangerous Crossing".
THR   84 min (ort 88 min) mCab   VIDrel: CIC/SONOP   V
Boa: radio play Cabin B-13 by John Dickson Carr.

TREAD SOFTLY, STRANGER **   PG
Gordon Parry USA   1958
Diana Dors, Terence Morgan, George Baker, Patrick Allen, Jane Griffiths, Maureen Delany, Betty Warren, Thomas Heathcote, Russell Napier, Wilfrid Lawson, Norman MacOwan, Joseph Tomlety, Chris Fay, Terry Baker, Timothy Bateson
An embezzler whose girlfriend has expensive tastes plans a robbery at a steel-mill together with his brother who has considerable gambling debts. Unfortunately, the former accidentally kills a nightwatchman whose son searches for a witness to this crime, who proves however to blind. A poor thriller, set in the English town of Rotherham.
THR   91 min B/W   VIDrel: EMPIRE/TOTAL   V
Boa: play Blind Alley by Jack Poplewell.

TREASURE HUNTERS ***   15
Lau Kar Wing HONG KONG   1982
Fu Sheng, Chang Chan Peng, Gordon Liu, Wang Lung Wei
A con-man and a rich kid team up in a search for a lost treasure, but have to contend with a Shaolin monk who regularly appears to berate them about their materialistic ways. They also find themselves having to deal with a powerful swordsman who has been taking an interest in their quest. A fast-paced and light-hearted kung fu caper, blending both broad comedy and martial arts action.
MAR   105 min wScrn dubbed   VIDrel: MADE/RTM   V

TREASURE ISLAND ***   U
Victor Fleming USA   1934
Wallace Beery, Jackie Cooper, Lewis Stone, Lionel Barrymore, Otto Kruger, Nigel Bruce, Douglass Dumbrille, Chic Sale
Stevenson's tale of 18th century pirates, buried treasure and adventure receives a slow but careful treatment in this nicely mounted production. Beery makes a splendid Long John Silver, whose rumbustious presence is somewhat let down by Cooper's colourless performance as young Jim Hawkins.
A/AD   99 min (ort 105 min) B/W   VIDrel: MGM L/A   V
Boa: novel by Robert Louis Stevenson.

TREASURE ISLAND ***   U
Byron Haskin USA   1950 (re-issued 1975)
Bobby Driscoll, Robert Newton, Basil Sydney, Walter Fitzgerald, Dennis O'Dea, Ralph Truman, Finlay Currie, John Laurie, Francis de Wolff, Geoffrey Wilkinson, David Davies, Andrew Blackett, Paddy Brannigan, Ken Buckle
A charming Disney version of the classic tale. Filmed in England, with Newton perfectly cast as Long John Silver and Driscoll good as Jim Hawkins. The script is by Lawrence

Edward Watkin. Some "objectionable" violence was removed from the film when it was re-issued in 1975.
Aka: LONG JOHN SILVER'S RETURN TO TREASURE ISLAND
A/AD   96 min cC   VIDrel: WDV/TECH   V
Boa: novel by Robert Louis Stevenson.

TREASURE ISLAND **   U
Zoran Janjic/Leif Gram AUSTRALIA   1971
Voices of: Ron Haddrick, John Kingley, John Llewellyn, Bruce Montague, Brenda Senders, Colin Tilley
A spirited adaptation of this rousing adventure, one in a series of illustration-style animations made for Australian TV.
ANIM   60 min   VIDrel: 4-FRONT/POLYREC   V
Boa: novel by Robert Louis Stevenson.

TREASURE ISLAND *   PG
John Hough FRANCE/SPAIN/UK/WEST GERMANY   1971
Orson Welles, Kim Burfield, Walter Slezak, Lionel Stander, Rik Battaglia, Angel Del Pozo, Maria Rohm, Paul Muller, Michael Garland, Aldo Sambrelli, Jean Lefevbre, Alibe, Chinchilla
A very poor remake of a much filmed classic that lacks both pace and vigour, despite Welles's best efforts at hamming up his role as Long John Silver. The poor dubbing of a talented international cast does not help matters either. The script is by Wolf Mankowitz and O.W. Jeeves (Orson Welles).
A/AD   85 min (ort 95 min)   VIDrel: VCC L/A   V
Boa: novel by Robert Louis Stevenson.

TREASURE ISLAND **   PG
Michael E. Briant UK   1977
Alfred Burke, Anthony Bate, Patrick Troughton, Jack Watson, David Collings, Thorley Walters, Paul Copley, Richard Beale
A youngster sets sail with a bunch of brigands in the hope of finding the treasure whose location was revealed on a map he caught a glimpse of. Quite a passable work, though the constraints of producing it in the format of a long-running TV series means it was unnecessarily padded out.
A/AD   199 min (2 cassettes) mTV   VIDrel: BBC   V/sur
Boa: novel by Robert Louis Stevenson.

TREASURE ISLAND ***   PG
Fraser C. Heston USA   1989
Charlton Heston, Christian Bale, Oliver Reed, Christopher Lee, Julian Glover, Richard Johnson, Clive Woods, Michael Thoma, John Abbot, Isla Blair, Nicholas Amer, John Cosmo, James Coyle, Michael Halsey, Peter Postlethwaite
A spirited and colourful adaptation of the classic story of buried treasure and a young boy's adventures. An enjoyable remake (bizarrely re-titled by Warner Brothers) that is notable for its strong characterisations. Heston is splendid as Long John Silver, and Reed and Lee as Captain Bones and Blind Pew are also memorable, but Bale as Jim Hawkins is a disappointment. Written and produced by Fraser Heston, Charlton's son.
Aka: DEVIL'S TREASURE
A/AD   126 min (ort 131 min) mTV   VIDrel: MGM/WHV   V/sh
Boa: novel by Robert Louis Stevenson.

TREASURE ISLAND **   (12)
Raul Ruiz FRANCE/USA   1991
Melvil Poupaud, Martin Landau, Vic Tayback, Lou Castel, Jeffrey Kime, Anna Karina, Sheila, Jean-Francois Stevenin, Charles Schmidt, Jean-Pierre Leaud, Yves Afonso, Pedro Armendariz Jr, Tony Jessen, Michael Ferber
This variant on the Stevenson tale starts off promisingly enough, with a youngster watching his favourite TV serial, but then having to imagine how the story develops when a power-cut interrupts the program. A bunch of old seafarers arrive at the hotel where he lives, and he is soon whisked off on an adventure, but one that is saddled with innumerable allusions, contrivances and metaphors. Highly ambitious it may be, but this film is too clever for its own good.
Aka: L'ILE AU TRESOR
A/AD   115 min (English version)   CINrel
Boa: novel by Robert Louis Stevenson.

TREASURE OF SWAMP CASTLE, THE **   U
Attila Dargay/Tim Reid HUNGARY/WEST GERMANY   1988
Voices of: Adrian Knight, Michele Trumel, Bronwen Mantel, Vlasta Vrana, A.J. Henderson, Arthur Grosser, Michael Rudder, Dean Hagopian, Mark Hellma, Dave Patrick, Susan Glover, Steven Bednarski, Tim Webber, Jane Woods, Rob Roy
Enjoyable cartoon adventure that tells of a land of long ago in

which a protracted war has virtually destroyed the country. From this ruined land venture two children of noble birth, who overcome many obstacles to reach and lay claim to their rightful inheritance, an ancient castle in which they hope to find a hidden treasure more precious than gold.
ANIM   80 min (ort 90 min)   VIDrel: L/A   V
Boa: novel The Gypsy Baron by Jokai Moor.

TREASURE OF THE SIERRA MADRE **** PG
John Huston USA   1948
Humphrey Bogart, Walter Huston, Tim Holt, Bruce Bennett, Barton MacLane, Alfonso Bedoya, Robert Blake, John Huston, Martin Garralaga, A. Soto Rangel, Manuel Donde, Jose Torvay, Margarito Luna, Jacqueline Dalya, Spencer Chan
Excellent film adaptation of a story of three men who join up to search for gold in the wilds of Mexico, and of the difficulties and hazards they have to face. Walter Huston is outstanding as the experienced old-timer who leads the expedition. Written by John Huston, who appears briefly to give Bogart a series of handouts. Music is by Max Steiner with musical direction by Leo F. Forbstein. AA: Dir, S. Actor (Huston), Screen (John Huston).
A/AD   121 min (ort 126 min) B/W   VIDrel: MGM/WHV
V
Boa: novel by B. Traven.

TREE GROWS IN BROOKLYN, A *** (PG)
Elia Kazan USA   1945
Peggy Ann Garner, James Dunn, Dorothy McGuire, Joan Blondell, Lloyd Nolan, Ted Donaldson, James Gleason, Ruth Nelson, John Alexander, Adeline de Walt Reynolds, Charles Halton, B.S. Pully, Ferike Boros, Patricia mcFadden
A sensitive and well-made drama of tenement life in Brooklyn at the turn-of-the-century that concentrates on the fortunes of one Irish family, where the husband is a likeable but rather unsuccessful singing waiter. Undeniably episodic and somewhat sentimental (but not excessively so), it paints a moving portrait of the bustle and poverty of big city life. This was Kazan's debut movie. AA: S. Actor (Dunn), Spec Award (Peggy Ann Garner).
DRA   123 min (ort 128 min) B/W   SATrel: MOVIE CHANNEL
Boa: novel by Betty Smith.

TREE OF HANDS ** 18
Giles Foster UK
Helen Shaver, Paul McGann, Peter Firth, Lauren Bacall, Malcolm Stoddard, Kate Hardie, Tony Haygarth, Phyllida Law, David Schofield
A successful authoress and single parent is joined by her mother in London. When her young son dies tragically, the mother abducts a young boy in an effort to comfort her, telling her daughter that he is the son of a friend. But by the time the daughter has realised this is untrue and that her mother is mentally unbalanced, she finds herself embroiled in a nasty blackmail attempt. Screenplay is by Gordon Williams.
Aka: INNOCENT VICTIM
THR   89 min (ort 100 min)   VIDrel: PATHE L/A   V/sh
Boa: novel The Tree Of Hands by Ruth Rendell.

TREES LOUNGE *** 15
Steve Buscemi USA   1996
Steve Buscemi, Chloe Sevigny, Anthony LaPaglia, Elizabeth Bracco, Mark Boone Jr, Seymour Cassel, Carol Kane, Bronson Dudley, Michael Buscemi, Mimi Rogers, Samuel L. Jackson, Steven Randazzo, Suzanne Shepherd, Rockets Redglare, Joe Lisi
A look at a set of losers who live on Long Island, principally Buscemi as an unemployed and alcoholic mechanic who spends most of his time drinking with his companions at the title establishment, having been dumped by his girlfriend. Scripted by Buscemi, who gives a terrific central performance (drawing on his personal experiences) this is a touching and mercifully completely unsentimental portrait of wasted lives and blighted hopes.
DRA   95 min   CINrel

TREMORS *** 15
Ron Underwood USA   1989
Kevin Bacon, Fred Ward, Finn Carter, Michael Gross, Reba McEntire, Rhonda Le Beck, Bobby Jacoby, Charlotte Stewart, Tony Genaros, Victor Wong, Ariana Richards, Richard Margus, Sunshine Parker, Michael Dan Wagner, Conrad Bachmann
This effective updating of the old-fashioned 1950s-style monster movies, combines a sharp and witty script with some genuinely suspenseful moments. Bacon and Ward are a couple of likeable

handymen who live in a small Nevada town. The discovery of a decapitated sheep farmer and his partially devoured flock leads to the discovery of giant wormlike creatures that live in the sands and our duo lead the town's inhabitants in a battle against them.
HOR   92 min (ort 96 min)   VIDrel: CIC/SONOP L/A   V

TREMORS 2: AFTERSHOCKS ** 12
S.S. Wilson USA   1995
Fred Ward, Helen Shaver, Christopher Cartin, Michael Gross
Horror-comedy sequel to the first film, that went straight to video without a cinema release, and has the same carnivorous wormlike creatures from the first film now chomping their way through the folks at an oilfield in Mexico. The same mixture of shocks and humour is used as before, but this time round it has no freshness; the fact that the now appear able to mutate into surface creatures also robs the film of much of its tension.
HOR   96 min cC   VIDrel: CIC/SONOP   V/sh

TRENCHCOAT ** PG
Michael Tuchner USA   1983
Margot Kidder, Robert Hays, Daniel Faraldo, Gila Von Weitershausen, David Suchet, Ronald Lacey, John Justin, Pauline Delany, Leopoldo Trieste, P.G. Stephens, Brizio Montinaro, Martin Sorrentino, Luciano Crovato
A female stenographer with literary ambitions, takes a holiday on Malta in the hope of finding material for her novel, and gets more adventure than she bargained for when she finds a map that leads her into involvement with a gang of international terrorists. A lacklustre Disney studio production with little to recommend it.
COM   90 min (ort 95 min)   VIDrel: WDV/TECH L/A   V

TRENCHCOAT IN PARADISE ** PG
Martha Coolidge USA   1989
Dirk Benedict, Sydney Walsh, Bruce Dern, Catherine Oxenberg, Michelle Phillips, Jeremy Slate, Kim Zimmer
A 1940s-style pilot for a prospective series, that follows the adventures of a small-time detective who has to flee New Jersey when he incurs the wrath of a local crime boss. Arriving in Hawaii, he finds life there no quieter and is soon involved in an investigation into a local land swindle. An amiable if unremarkable detective story with a few amusing touches.
DRA   89 min (ort 100 min) mTV   VIDrel: MGM L/A   V

TRENT'S LAST CASE ** U
Herbert Wilcox UK   1952
Michael Wilding, Margaret Lockwood, Orson Welles, Hugh McDermott, John McCallum, Miles Malleson
Despite appearing to have committed suicide, a business tycoon may in fact have been murdered by his secretary. An inquisitive journalist investigates. A watery adaptation of a famous novel, its absorbing moments are diluted by excessive dialogue. Filmed once before in 1920.
DRA   86 min (ort 90 min) B/W   VIDrel: WHV   V
Boa: novel by E.C. Bentley.

TRESPASS ** 18
Walter Hill USA   1992
Bill Paxton, Ice T, Ice Cube, William Sadler, Art Evans, De'Voreaux White, Bruce A. Young, Glenn Plummer, Soney Jackson, T.E. Russell, Tiny Lister, John Toles-Bey, Byron Minns, Tico Wells, Hal Landon Jr, James Pickens Jr
An unspeakably violent tale of two greedy fireman who go looking for some valuable gold items in an abandoned building and are unfortunate enough to stumble across a gang of ruthless drug dealers who set out to murder them. There is no shortage of action to satisfy the most ardent fans, but the overall impact of this movie is profoundly disturbing. Understandably, after the L.A. riots that summer, its release was postponed to the winter.
Aka: LOOTERS
A/AD   96 min (ort 101 min) cC   VIDrel: CIC/SONOP
V/sur

TRIAL, THE *** PG
Orson Welles FRANCE/ITALY/WEST GERMANY   1963
Anthony Perkins, Orson Welles, Jeanne Moreau, Romy Schneider, Elsa Martinelli, Akim Tamiroff, Arroldo Foa, William Kearns, Jess Hahn, Suzanne Flon, Madelaine Robinson, Wolfgang Reichman, Thomas Holtzman, Maydra Shore
A moody, atmospheric adaptation of Kafka's novel of a man arrested for a crime that is never specified, yet against which he

must mount a defence. Not entirely effective, yet full of touches of brilliance with an intense performance from Perkins set against some remarkable photography. Written and directed by Welles, and very much aimed at admirers of Kafka. The photography is by Edmond Richard. Remade about thirty years later.

Aka: DER PROZESS; IL PROCESSO; LE PROCES
DRA   120 min B/W   VIDrel: ARTPRO/RTM   V
Boa: novel by Franz Kafka.

TRIAL, THE ** 15
David Jones UK 1992
Kyle MacLachlan, Anthony Hopkins, Jason Robards, Juliet Stevenson, Polly Walker, Alfred Molina, Michael Kitchen, David Thewlis, Tony Haygarth, Douglas Hodge, Jiri Schwartz, David Schneider, Ondrej Vetchy, Valerie Kaplanova
An innocuous bank clerk one day finds himself under arrest for an unspecified crime and struggles to understand the reason for his predicament. A well realised and acted adaptation of Kafka's nightmarish masterpiece (it was published after his death) which though perfectly watchable, really fails to convey either the tension or oppression so very evident in the 1963 Orson Welles version. Screenplay is by Harold Pinter.
DRA   115 min (ort 120 min)   VIDrel: IMAG/RTM   V
Boa: novel by Franz Kafka.

TRIAL AND ERROR ** U
James Hill UK 1962
Peter Sellers, Richard Attenborough, Beryl Reid, David Lodge
An incompetent lawyer defends a man accused of murdering his wife and despite the man wishing to plead guilty, sets about trying to get him off as a way of boosting his flagging career. Sellers is in fine form as the lawyer and Attenborough is convincing as the murderer, but the film's clumsy flashback structure does not work to its advantage.
Aka: DOCK BRIEF, THE
COM   78 min B/W   VIDrel: ARROW/RTM   V
Boa: play by John Mortimer.

TRIAL AND ERROR ** 15
Mark Sobel CANADA 1992
Tim Matheson, Helen Shaver, Eugene C. Clark, Sean McCann, Page Fletcher, Ian D. Clark, Michael J. Reynolds, Gene Mack, Frank Crduele, David Gardner, Ron Small, Reginald Huc, Delores Etienne, Maurice Dean Wint, Conrad Coates
A prosecutor builds his career on the successful conviction of a petty criminal for murder. Five years later, new evidence comes to light that may prove his innocence and save him from execution. Unfortunately, such a revelation might damage our DA's chances of a political nomination; meanwhile the real killer proceeds to stalk his wife. An effective if somewhat contrived crime drama.
DRA   90 min   VIDrel: BRAVE/SONOP L/A   V

TRIAL BY FIRE ** 15
Alan Metzger USA 1995
Gail O'Grady, Keith Carradine, Ken Lerner, Michael Bowen, Mariangela Pino, Devon Odessa, Richard Riehle, Brandon O'Connell, Richard Grant, Nicholas Pryor, James Pickens Jr, Brandon Douglas, Chance Quinn, Jay Paulson, Andrew Kavovit
A local community hold a young teacher responsible for the death of a male student, for when she tried to give this troubled youth some extra tuition, his mistakes her dedication for interest, and tells his friends they are having an affair. The questions surrounding his later suicide are only finally resolved in the course of a court case. Average.
DRA   89 min mTV   VIDrel: ODY/SONOP   V

TRIAL BY JURY *** 15
Heywood Gould USA 1994
Joanne Whalley-Kilmer, William Hurt, Armand Assante, Gabriel Byrne, Kathleen Quinlan, Margaret Whitten, Ed Lauter, Richard Portnow, Lisa Arindell Anderson, Jack Gwaltney, Graham Jarvis, William R. Moses, Joe Santos, Beau Starr
A crime boss on trial in New York decides to ensure his acquittal by making threats to a woman juror who refuses to play along, thus putting her life and that of her family in danger. She eventually does as he demands, but even after he is acquitted, both she and her young son remain in danger. A violent and deeply cynical affair of violence and sudden death, but it is greatly helped by strong performances and good camerawork. See also THE JUROR.
DRA   102 min (ort 107 min) cC   VIDrel: WHV   V/sur

TRIAL OF LEE HARVEY OSWALD, THE ** 15
David Greene USA 1976
Ben Gazzara, Lorne Greene, Frances Lee McCain, Lawrence Pressman, Charlie Robinson, George Wyner, Mo Malone, John Pleshette, Annabelle Weenick, William Jordan, Charles Cyphers, Marisa Pavan, Jack Collins, Ed Abry
Little of the 1967 off-Broadway play remains in this TV adaptation, which tries to get to grips with the mentality of the man responsible for shooting John F. Kennedy. Pleshette (whose cousin is actress Suzanne Pleshette) bears a surprisingly strong resemblance to Oswald, and this is the most notable feature in an otherwise unmemorable effort. A B/W film with the same title appeared in 1964. See also RUBY, JFK and RUBY AND OSWALD.
DRA   178 min (ort 240 min) mTV
VIDrel: CASPIC/BMGREC L/A   V
Boa: play by Amram Ducovny and Leon Friedman.

TRIAL OF THE INCREDIBLE HULK, THE *** PG
Bill Bixby USA 1989
Bill Bixby, Lou Ferrigno, Rex Smith, John Rhys-Davies, Marta DuBois, Nancy Everhard, Nicholas Hormann, Joseph Mascolo, Richard Cummings Jr, Linda Darlow, Dwight Koss, Meredith Woodward, Mark Acheson, Don Mackay
The second full-length feature inspired by the popular TV series, in which a brilliant scientist finds that after conducting various experiments with himself as subject, he turns into an irascible green giant in moments of rage. In this story the Hulk teams up with Daredevil, another character from Marvel Comics, and together they take on a powerful gangster. Harmless, mindless, and fun.
A/AD   93 min Cut (15 sec – ort 100 min) mTV
VIDrel: NWV L/A   V

TRIALS OF TRACI, THE ** R18/18
Gerry Ross USA
Traci Lords, John Leslie, Amber Lynn, Ginger Lynn
A mermaid gets washed up on the beach of a nature reserve and her first encounter is with a couple making love. Thinking that this is the normal method of communication between humans, she sets off to find a man of her own. A porno reworking of SPLASH enlivened by its unusual setting.
Aka: TALK DIRTY TO ME: PART 3
A   64 min Cut (2 sec – R18 ver); 53 min Cut (41 sec – 18 ver)
VIDrel: ELV/DISC   V

TRIBES *** 15
Joseph Sargent USA 1970
Darren McGavin, Jan-Michael Vincent, Earl Holliman, John Gruber, Danny Goldman, Richard Yniguez
When a hippie is drafted into the US Marines, a tough drill instructor finds he is unable to break his spirit, and his problems mount when the hippie teaches his meditation techniques to the other guys in the barracks. An absorbing and convincingly portrayed film, whose thoughtful, Emmy Award-winning script (by Tracy Keenan Wynn and Marvin Schwartz) combines drama with sharp social comment.
Aka: SOLDIER WHO DECLARED PEACE, THE
DRA   90 min mTV   VIDrel: 20TH/TECH   V

TRICK OR TREAT ** 18
Charles Martin Smith USA 1986
Marc Price, Tony Fields, Doug Savant, Gene Simmons, Ozzy Osbourne, Elaine Joyce, Lisa Orgolini, Glen Morgan, Elise Richards, Richard Pachorek, Clare Nono, Alice Nunn, Claudia Templeton, Denny Price, Ray Shaffer, Brad Thomas
A dead heavy-rock singer is accidentally resurrected by an ardent fan who then makes use of him as a means of asserting himself against his bullying classmates. Another supernatural revenge fantasy, no better or worse than a hundred others.
HOR   93 min (ort 97 min)   VIDrel: POLY/POLYREC   V/sh

TRIGGER FAST ** 15
David Lister USA 1993
Martin Sheen, Corbin Bernsen, Christopher Atkins, Jurgen Prochnow
Unremarkable Western offering that describes how some gunfighters set about imposing their own brand of law and order on the old West. Despite its recent vintage, this average oater has nothing new to offer.
WES   96 min   VIDrel: AUDIO/DISC   V

TRIGGER HAPPY **   15
Larry Bishop USA   1996
Ellen Barkin, Gabriel Byrne, Richard Dreyfuss, Jeff Goldblum, Diane Lane, Larry Bishop, Gregory Hines, Kyle MacLachlan, Burt Reynolds, Christopher Jones, Henry Silva, Michael J. Pollard
Writer-director Bishop's debut feature is a gangster spoof built around the much-feared return of a criminal boss to his nightclub, where it is expected he will settle scores with one of his henchmen, who has been sleeping with his wife in his absence. But matters do not progress in anything like so straightforward a fashion, and though a succession of big-name stars put in an appearance, their efforts are hampered by the coy, self-mocking script and a shortage of ideas.
Aka: MAD DOG TIME
COM   93 min   CINrel

TRIP TO BOUNTIFUL, THE **   (U)
Peter Masterton USA   1985
Geraldine Page, John Heard, Rebecca De Mornay, Carlin Glynn, Kevin Cooney, Richard Bradford, Norman Bennett, Harvey Lewis, Kirk Sisco, Dave Tanner, Gil Glasgow, Mary Kay Mars, Wezz Tildon, Peggy Ann Byers, David Romo
A widow living with her son and his wife in Houston hankers after her home town, and decides to revisit it before it is too late. After some difficulties she manages to get there, but finds only ruins. However, her trip does at least have a beneficial effect on everyone. Originally a play shown on US TV in 1953, it was later transferred to Broadway. A well acted but very intense tearjerker, that gave Page a long overdue Oscar. AA: Actress (Page).
DRA   103 min (ort 106 min)   TVrel: C5
Boa: play by Horton Foote.

TRIPLE BOGEY ON A PAR FIVE HOLE ***   (15)
Amos Poe USA   1991
Eric Mitchell, Daisy Hall, Angela Goethals, Jesse McBride, Alba Clemente, Robbie Coltrane, Olga Bagnasco, Phil Hoffman, Tom Cohen, John Heys, Avital Dicker, May Au, John Schmerling, Lee Nagrin, Chic Streetman
A low-budget, independently made film with an endearing title, that is all about a group of orphans whose parents were murdered by bank robbers whilst in the course of playing a foursome at a golf course. After this event the kids have taken to living on a boat in New York, near Manhattan, but become instant media celebrities when interviewed by a journalist in a bid to avoid being sent to an orphanage. An absorbing and most unusual story, well worth a look.
DRA   88 min B/W   CINrel

TRIPLE CROSS **   18
Ackyl Anwary   1991
Cynthia Rothrock, Chris Barnes, Roy Marten, Peter O'Brian
A female martial arts expert who's in charge of security at a large industrial corporation, is sent from the States to supervise transport of a top-secret computer. She soon finds herself up against a ruthless terrorist, who having been tricked into hijacking two dummy computers, is now determined to get his hands on the real one. A rough, tough, brawling Rothrock offering.
A/AD   81 min Cut (1 min 6 sec)
VIDrel: 4-FRONT/POLYREC   V

TRIPLE CROSS, THE **   18
Kinji Fukasaki JAPAN   1992
Sonny Chiba, Kenichi Hagiwara, Kazuya Kimura
Three ageing gangsters come together for one last job, but it all goes wrong and the young crook who set it up tries to betray them, but he in turn falls foul of his cunning and self-seeking girlfriend. Cartoon-like in its portrayal of the inevitable bloodshed that ensues, this is a slickly made and occasionally self-parodying gangster thriller, totally unoriginal, but done with much style if not depth.
THR   110 min wScrn   VIDrel: MIA/DISC   V/dm

TRIPODS, THE: PARTS 1, 2 AND 3 *   PG
Graham Theakston/Christopher Barry
AUSTRALIA/UK/USA   1984
John Shackely, Jim Baker, Ceri Seel, Robin Hayter, Roderick Horn, Lucinda Curtis, Michael Gilmour, Peter Dolphin, John Scott Martin, Eddie Caswell, Peter Stockbridge, John Michael McCarthy, James Staddon, Harry Meacher
In the year 2089 A.D., Earth has reverted to a pre-industrial state, ruled over by three-legged alien machines called tripods that are revered and worshipped. Mind control is assured by means of strange silvery caps that all humans are fitted with on their 16th birthday. One boy revolts and goes in search of a mythical area where men are still free. A confusing and poorly made low-budget series, originally shown in nine dreary episodes.
FAN   225 min (3 cassettes approx 75 min each – ort 325 min)
mTV   VIDrel: BBC/TECH L/A   V/h
Boa: novels by John Christopher.

TRIPWIRE *   18
James Lemmo USA   1988
Terence Knox, David Warner, Isabella Hofmann, Charlotte Lewis, Yaphet Kotto, Andras Jones, Sy Richardson, Viggo Mortensen, Tommy Chong, Meg Foster, Bobby Cummings, Dean Tokuno, Jon Platten, Richard Stay, Craig Clyde
A federal agent foils a weapons robbery by the world's most lethal terrorist, a deadly figure known as "Tripwire". In revenge, the latter kills the agent's ex-wife and kidnaps their son, thereby sparking off the inevitable cycle of revenge and general destruction. One of those utterly predictable dime-a-dozen action-thrillers.
THR   86 min (ort 120 min)   VIDrel: MED/POLYREC L/A   V/sh

TRISTANA **   PG
Luis Bunuel FRANCE/ITALY/SPAIN   1970
Catherine Deneuve, Fernando Rey, Franco Nero, Jesus Fernandez, Lola Gaos, Vincente Soler, Antonio Casas, Jose Calvo, Fernando Cebrian, Candida Losada, Mary Paz Pondal, Juan Jose Menendez, Sergio Mendizabal, Antonio Ferrandis
A young woman placed in the charge of an aristocratic nobleman is eventually seduced by him, despite his avowed intention to resist such a temptation. She leaves his household at one point when she falls in love with an artist but returns after a tumour causes the amputation of her leg, and finally agrees to marry him. A complex and rather opaque drama with a few odd comic touches, but one that lacks the cutting edge of Bunuel's earlier works.
DRA   94 min (ort 98 min) wScrn   VIDrel: ELPIC/POLYREC V
Boa: novel by Benito Perez Galdos.

TRIUMPH OF THE HEART, A: THE RICKY BELL STORY **   U
Richard Michaels USA   1991
Mario Van Peebles, Susan Ruttan, Lane Davis, Lynn Whitfield, Polly Holliday
Coached by a professional football-playing legend, a determined youngster battles to overcome his physical disabilities, only to learn that his mentor has developed health problems of his own. One of those feel-good sporting prowess dramas, with likeable characters making the best of the mawkish script, albeit one based on a true-life football hero. See also BRIAN'S SONG for something rather similar.
DRA   93 min mTV   VIDrel: ODY/SONOP   V

TRIUMPH OF THE SPIRIT ***   15
Robert M. Young USA   1989
Willem Dafoe, Edward James Olmos, Robert Loggia, Wendy Gazelle, Kelly Wolf, Costas Mandylor, Kario Salem, Edward Zentara, Burkhard Heyl, Sofia Saretok, Grazyna Kruk-Schejbal, Karolina Twardowska, Juranda Krol, Wikto Mlynarczyk
A Jewish boxer and his family are transported from Thessalonika in Greece to Auschwitz by the Nazis and are separated. His captors discover that he is a boxer and give him a chance of survival if he will fight an endless series of bouts with other prisoners for their amusement. The harrowingly true story of Salamo Arouch receives a weak treatment that, for all its location filming at Auschwitz and Birkenau, lacks both conviction and force.
A/AD   115 min (ort 121 min)   VIDrel: GUILD/POLYREC L/A   V/sh

TRIUMPH OVER DISASTER: THE HURRICANE ANDREW STORY ***   (PG)
Marvin J. Chomsky USA   1993
Ted Wass, Veronica Cartwright
In August of 1992 a massive hurricane left large parts of the south-eastern tip of the USA devastated. This extremely competent docu-drama attempts to tell the story of this catastrophe, and provides a glimpse of the numerous tales of courage and heroism that took place then. By no means a depressing work, it is at times quite the opposite.
DRA   90 min mTV   TVrel

TRIUMPHS OF A MAN CALLED HORSE ** 15
John Hough MEXICO/USA 1982
Richard Harris, Michael Beck, Ana De Sade, Vaughn Armstrong, Anne Seymour, Buck Taylor, Simon Andreu, Lautaco Murua, Roger Cudney, Jerry Gatlin, John Chandler, Jacquelline Evans
Third in the series of "Horse" films, about a man who has been accepted by the Sioux as one of them. In this disappointing sequel, Harris dies early on in the film, leaving his half-breed son to defend the Yellow Hand Sioux from the brutal and greedy prospectors who threaten the tribe's very existence. See A MAN CALLED HORSE, the first film in the series.
WES   86 min Cut (7 sec – ort 89 min)   VIDrel: L/A   V

TROLL ** 15
John Carl Buechler USA 1985
Michael Moriarty, Shelley Hack, Noah Hathaway, Jenny Beck, Sonny Bono, June Lockhart, Anne Lockhart, Phil Fondacaro, Brad Hall, Gary Sandy, James Beck, Julia Louis-Dreyfus, Robert Hathaway, Dale Wyatt, Barbara Sciorilli
A family's home is invaded by trolls who possess their younger daughter, and use a magical ring to turn people into mythical creatures. A dullish horror fantasy that attempts to incorporate elements of GREMLINS. The inevitable sequel followed.
HOR   79 min (ort 82 min)   VIDrel: NTV/TOTAL   V/sur

TROLL 2 ** 18
Drago Floyd (Aristide Massacessi) ITALY 1989
Connie McFarland, Michael Stephenson, George Hardy, Margo Prey, Deborah Reed, Robert Ormsby, Jason Wright, Darren Ewing, Jason Steadman, Gary Carlson, David McConnell
Sequel to the first film that has a family taking a holiday in a small town that is not quite as it seems, for the woods are inhabited by Trolls who disguised as peasants, offer people food that turns them into vegetables the creatures can eat. A strange blend of horror and black humour that does justice to neither genre.
HOR   94 min (ort 95 min)   VIDrel: MOPIC/SGSVID   V

TROMA'S WAR * 18
Samuel Weil (Lloyd Kaufman/Michael Herz) USA 1988
Carolyn Beauchamp, Sean Bowen, Jessica Dublin, Patrick Weathers, Michael Ryder, Steven Crossley, Ara Romanoff, Loryn Lane De Luca, Charles Kay Hume, Lisa Patruno, Aleida Harris, Brenda Brock, Patrick Weathers, Rick Collins, Dan Snow
One of the earliest films from Troma Studios, who gave us THE TOXIC AVENGER and RABID GRANNIES. A plane crashes on a Caribbean island and most of the survivors are captured by soldiers. Those remaining form themselves into an efficient combat force and attack the army base to free their fellow-passengers. A silly, overlong spoof on action movies, with each actor taking at least two roles.
COM   90 min (ort 105 min)   VIDrel: TROMA/RTM   V

TROMEO & JULIET * 18
Lloyd Kaufman USA 1996
Jane Jensen, Will Keenan, Valentine Miele, Maximillian Shawn, Sean Gunn, Debbie Rocchon, Steven Blackehart, Flip Brown, Patrick Connor, Earl McKoy, Gene Tevinoni, Wendy Adas, Tamara Craig Thomas, Antonin Curie, Jacquelien Tavarez
An utterly dire attempt to create a spoof version of Shakespeare's immortal play, with a contemporary setting, moronic punk characters and buckets of blood in place of wit. Virtually unwatchable, this effort from Troma marks a new low in their productions, and is light years away from their earlier cult successes such as THE TOXIC AVENGER.
COM   107 min   CINrel
Boa: play Romeo and Juliet by William Shakespeare.

TRON ** PG
Steven Lisberger USA 1982
Jeff Bridges, Bruce Boxleitner, David Warner, Cindy Morgan, Barnard Hughes, Dan Shor, Peter Jurask, Tony Stephano, Craig Chudy, Vince Deadrick, Sam Schatz, Jackson Bostwick, Dave Cass, Gerald Burns, Bob Neill, Ted White
A computer whizz-kid who lives by writing games, attempts to obtain some information contained within a master computer. The machine resists and absorbs him into an inner world full of strange electronic marvels. Despite great effects, the film never progresses beyond the juvenile in its plotting and characterisation. Written and directed by Lisberger.
FAN   92 min (ort 96 min)   VIDrel: WDV/TECH L/A   V

TROOP BEVERLY HILL$ ** PG
Jeff Kanew USA 1989
Shelley Long, Craig T. Nelson, Betty Thomas, Mary Gross, Stephanie Beacham, Audra Lindley, Edd Byrnes, Jenny Lewis, Frankie Avalon, Pia Zadora, Cheech Marin, David Gautreaux, Karen Kopins, Annette Funicello, Joyce Brothers
Long plays a silly Beverly Hills mom who agrees to act as a troop leader for her daughter's Wilderness group and ends up teaching her charges all about self-respect and the things that really matter in Beverly Hills. An amiable comedy vehicle for Long that offers the occasional flash of wit.
COM   101 min (ort 106 min)   VIDrel: RCA L/A   V

TROP BELLE POUR TOI! *** 18
Bertrand Blier FRANCE 1989
Gerard Depardieu, Josiane Balasko, Carole Bouquet, Roland Blanche, Francoise Cluzet
A car dealer seems to have the perfect wife, a woman who is beautiful, witty and cultivated, but nonetheless embarks on an affair. Strange to relate, her rival is an office temp who is none of the things she is, being rather plain and dowdy. A well acted trifle that generates more laughs than the slight nature of the script would suggest, but soon runs out of steam, and little is made of the deliberate contrast between wife and mistress.
Aka: TOO BEAUTIFUL FOR YOU
COM   87 min (ort 91 min) wScrn   VIDrel: ARTIF/20TH V/sur

TROPICAL NIGHTS ** 18
Jag Mudhra USA 1992
Rick Rossovich, Lee Anne Beaman, Maryam D'Abo, Asha Siewkumar, Ashok Rao, Govind Rao, Prakash Ral, Brian Tracy
An Indian maharajah is apparently killed on a safari and the man's widow files a $5,000,000 insurance claim. But the insurance company has suspicions of foul play, and sends a an investigator to India in a bid to learn the truth surrounding the circumstances of the man's death. Predictably enough, he is soon drawn into a relationship with the man's attractive, young widow. A routine erotic drama.
Aka: TROPICAL HEAT
DRA   86 min (ort 92 min)   VIDrel: MARQ/QUANT   V

TROUBLE BOUND ** 18
Jeffrey Reiner USA 1992
Michael Madsen, Patricia Arquette, Seymour Cassel, Sal Jenco, Damen Epton, Gregory Sporleder, Paul Ben-Victor, Billy Bob Thornton, Rustam Branaman, Florence Stanley, Syenour Cassell, Patrick Cupo, Monty Hoffman, Billy Bastiani
A convict just released from jail wins a car in a competition but is unaware that a body has been concealed in its trunk. Meanwhile to add to his troubles, his gives a lift to a young waitress who is planning to murder a Mafia boss. A wacky adventure tale that does not really work as a thriller.
A/AD   86 min (ort 90 min)   VIDrel: ITC/POLYREC   V/h

TROUBLE IN MIND *** 15
Alan Randolph USA 1985
Kris Kristofferson, Divine (Glenn Milstead), Lori Singer, Keith Carradine, Genevieve Bujold, Joe Morton, George Kirby, John Considine, Albert Hall, Dirk Blocker, Gailard Sartain, Robert Gould, Antonia Dauphin, Billy Silva
An ex-cop serves a jail sentence for several murders, and on his release returns to the town where the woman he killed for still lives, only to find himself deeply involved with a group of youngsters. A strange, highly charged drama, with a style that recalls 1940s film noir.
A/AD   107 min (ort 111 min)   VIDrel: ARROW/RTM V/sur

TROUBLE IN PARADISE ** 15
Di Drew AUSTRALIA/USA 1988
Raquel Welch, Jack Thompson, Nicholas Hammond, John Gregg
The elegant widow of a diplomat sets out for the USA and the promise of a new life, but when her ship is wrecked, she finds herself alone on a tropical island with a happy-go-lucky Australian seaman. Together they face various hazards, such as the arrival of drug smugglers. Welch is quite appealing in this undemanding tale, that has much in common with both "The Admirable Crichton" and "Swept Away". A pleasant time-filler.
DRA   89 min (ort 100 min) mTV   VIDrel: BRAVE/SONOP L/A   V

TROUBLE WITH HARRY, THE ***
PG
Alfred Hitchcock USA   1955
Shirley MacLaine, John Forsythe, Edmund Gwenn, Mildred Natwick, Mildred Dunnock, Jerry Mathers, Royal Dano, Parker Fennelly, Dwight Marfield, Leslie Woolf, Philip Truex, Ernest Curt Bach
A corpse found in a copse, causes problems in a peaceful rural community in New England, in this unusual black comedy, chiefly because several people believe themselves to be responsible for its demise. Among them is a young woman (MacLaine in her acting debut) who recognises it as her husband. A twee and unreal black comedy, with a musical score (the first to be used by Hitchcock) by Bernard Herrmann. Written by John Michael Hayes.
COM   95 min (ort 99 min)   VIDrel: CIC/SONOP   V
Boa: novel by Jack Trevor Story.

TROUBLE WITH SPIES, THE *
PG
Burt Kennedy USA   1984 (released 1987)
Donald Sutherland, Ned Beatty, Lucy Gutteridge, Ruth Gordon, Michael Hordern, Robert Morley, Gregory Sierra
A dreadful secret agent spoof that casts Sutherland as a bumbling spy who runs into the expected treachery and intrigue on an espionage mission. An inane comedy that has so many plot twists it rapidly becomes completely incomprehensible, and even Sutherland's charm fails to make it sparkle. The unsubtle script is the work of Kennedy.
COM   86 min (ort 91 min)   VIDrel: VCC/DISC/COLUM
L/A   V
Boa: novel Apple Spy in the Sky by Marc Lovell.

TROUBLESHOOTERS: TRAPPED BENEATH THE EARTH **
PG
Bradford May USA   1993
Kris Kristofferson, David Newsom, Leigh J. McCloskey, Frank McRae, Caitlin Dulany, Gary Graham, Leslie Ackerman, Shawn Levy, Mitch Pileggi, Barbara Rhoades, Cynthia Allison, Camilla Belle, Mary Betten, Frank Birney, Noel Conlon
A family of troubleshooters are given the difficult assignment of rescuing the people trapped under the ruins of a collapsed apartment block. One in a series of dull TV movies that were released directly to video, it puts one in mind of the popular THUNDERBIRDS series, except that the acting is not as good.
A/AD   88 min (ort 90 min) mTV   VIDrel: CIC/SONOP   V

TRUE BETRAYAL **
15
Roger Young USA   1990
Mare Winningham, Peter Gallagher, M. Emmet Walsh, G.W. Bailey, Caroline Williams, Robert Harper, William Shockley, James Harper, Garth D. Williams, William Raulerson, Brendan Ryan, Sean B. Ryan, Dennis Bowen, Karl Rumberg
An inexperienced female private eye relies heavily on her acting skills in order to solve her first case, the double murder of an old couple. Her best chance to crack the case involves her in a dangerous game of deception with the prime suspect. This prospective pilot has a few interesting ideas, but they fail to develop in a film that's essentially just one more standard TV 'tec offering. Based on the real-life Houston investigator Kim Paris.
Aka: LOVE AND LIES
DRA   93 min (ort 103 min) mTV   VIDrel: ITC/POLYREC
V/h

TRUE BLUE *
15
Ferdinand Fairfax UK   1996
Johan Leysen, Dominic West, Dylan Baker, Geraldine Somerville, Josh Lucas, Brian McGovern, Ryan Bollman, Andrew Tees, Robert Bogue, Noah Huntley, Edward Atterton, Nicholas Rowe, Jonathan Cake, Alexis Denisof, Patrick Malone
Ostensibly about the annual boat-race between Oxford and Cambridge universities, this limply directed and flaccid movie attempts to take a look at the role played behind the scenes at each seat of learning by some American students. Portrayed as averse to mixing with the locals, they inject an element of "winning at all costs" intensity into this formerly gentlemanly event, and this leads to the expected conflicts. A dull and unexciting piece of work.
DRA   117 min   VIDrel: L/A   V
Boa: book by Daniel Topolski and Patrick Robinson.

TRUE COLORS **
15
Herbert Ross USA   1991
James Spader, John Cusack, Imogen Stubbs, Mandy Patinkin, Richard Widmark, Dina Merrill, Philip Bosco, Paul Guilfoyle, Brad Sullivan,
Russell Dennis Baker, Don McManus, Karen Jablons-Alexander, Wendee Pratt, Anthony Fuscio
A dreary tale of political intrigue and over-ambition, with Cusack playing a law school drop-out who gets a job working for a US Senator, and proving that he has no qualms about exploiting his friendship with a former law school room-mate, who now works in the Justice Department. A film of few surprises and little impact, that's largely a waste of time for all concerned.
DRA   105 min (ort 111 min)   VIDrel: CIC/SONOP   V/sur

TRUE CONFESSIONS ***
15
Ulu Grosbard USA   1981
Robert De Niro, Robert Duvall, Charles Durning, Ed Flanders, Cyril Cusack, Burgess Meredith, Rose Gregorio, Kenneth McMillan, Dan Hedaya, Gwen Van Dam, Tom Hill, Jeanette Nolan, Jorge Cervera Jr, Susan Meyers, Louisa Moritz
An account of the complex relationship between two brothers, one a Catholic priest and the other a cop. Set in the late 1940s in California, and supposedly based on real life murder case the "Black Dahlia", which formed the basis for the TV movie "Who Is The Black Dahlia?" Written by John Gregory Dunne and Joan Didion, who present a far from flattering view of the Catholic church and its servants. De Niro gives a remarkably restrained and convincing performance.
DRA   104 min (ort 108 min)   VIDrel: MGM/WHV   V
Boa: novel by John Gregory Dunne.

TRUE CRIME **
18
Pat Verducci USA   1995
Alicia Silverstone, Kevin Dillon, Michael Bowen, Jennifer Savidge, Joshua Schaeffer, Tara Subkoff, Marla Sokoloff, Bill Nunn, Ann Deraney, David Packer, Alissa Dowdy, Aime Brooks, Brian Wankum, Melissa DeLizia, Jack Rader, Sean Moran
A teenage girl who devours crime magazines turns to real detection when one of her classmates is brutally murdered, and joins forces with a police cadet. She discovers that this killing was one of many, but in so doing attracts the unwelcome attention of the murderer. Blending brutality (plus an unpleasant sex sequence) with kid's derring-do was not a good idea, and this hybrid movie tries hard to find a clear identity, often descending into farce in the process.
THR   90 min (ort 94 min)   VIDrel: TRIM/HIFLI   V/h

TRUE GAME OF DEATH *
18
Chen Tien Tai Steve HONG KONG   197-
Bruce Le (Huang Kin Lung), Hsao Lung, Ali Taylor
A martial arts master tries to form his own movie production company, but runs into opposition from the owners of the major studios, who hatch a plot to stop him, by having him drugged with a deadly aphrodisiac. However, when he dies the spirit of Bruce Lee takes over the body of the actor who was to play his double in the movie, and sets out to chastise those responsible. A shoddy and exploitative copy of GAME OF DEATH, with the requisite Bruce Lee clone.
MAR   90 min Cut (3 min 33 sec)   VIDrel: BLUE/QUANT
V

TRUE GRIT ****
PG
Henry Hathaway USA   1969
John Wayne, Kim Darby, Glen Campbell, Jeremy Slate, Robert Duvall, Strother Martin, Dennis Hopper, Alfred Ryder, Jeff Corey, Ron Soble, John Fielder, James Westerfield, John Doucette, Donald Woods, Edith Atwater, Carlos Rivas
A plucky fourteen-year-old girl hires an ageing US marshal with a reputation for doing things his own way, in order to track down her father's killer. Wayne really finds his form in this film, as the cynical hardbitten marshal whose growing fondness for the girl brings some joy into his life. Wayne's fine portrayal won him his only Oscar. Written by Marguerite Roberts with music by Elmer Bernstein. ROOSTER COGBURN followed in 1975.
AA: Actor (Wayne).
WES   128 min   VIDrel: 4-FRONT/POLYREC/CIC   V/h
Boa: novel by Charles Portis.

TRUE IDENTITY **
15
Charles Lane USA   1991
Lenny Henry, Frank Langella, Charles Lane, J.T. Walsh, Anne-Marie Johnson, Andreas Katsulas, Michael McKean, Peggy Lipton, Bill Raymond, James Earl Jones, Darnell Williams, Christopher Collins, Melvin Van Peebles, Ruth Brown, Jim Gavin
This film started life as a "Saturday Night Live" sketch (the work of Andy Breckman) for Eddie Murphy. Expanded (or

padded) into a feature, it has an aspiring black actor learning (via an impulsive "confession") that the man he is seated next to on a plane that's about to crash is a notorious gangster, now living a respectable life thanks to cosmetic surgery. The plane rights itself, and Henry has to run for his life. An enjoyable, rather silly romp.
COM   89 min (ort 93 min)   VIDrel: TOUCH / BUENA L / A   V

TRUE LIES **   15
James Cameron USA   1993
Arnold Schwarzenegger, Jamie Lee Curtis, Tom Arnold, Bill Paxton, Tia Carrere, Eliza Dushku, Charlton Heston, Grant Heslov, Mitchell Manesh, James Allen, Deiter Rauter, Jane Morris, Katsy Chappell, Crystina Wyler, Ofer Samra
Taking as its inspiration a French film "La Total", this tongue-in-cheek blockbuster spoof tells of an intrepid secret agent who finds his failure to tell what he does for a living, has unexpected complications. Plenty of special effects will satisfy hard-bitten action fans, but for others the adolescent humour soon grows wearisome.
A / AD   135 min (ort 141 min) wScrn cC
VIDrel: CIC / SONOP; PION (LV only)   V / sur LV

TRUE LOVE ***   15
Nancy Savoca USA   1989
Annabella Sciorra, Ron Eldard, Aida Turturro, Roger Rignack, Star Jasper, Michael J. Wolfe, Kelly Cinnante, Rick Shapiro, Suzanne Costallos, Vinny Pastore, Marianne Leone, Marie Michaels, Anna Vergani, John Nacco, Ann Tucker
A rather noisy and profane piece about an argumentative Italian-American couple who are making preparations for their wedding, amidst a relentless welter of conflicting advice from friends and family. Foul-mouthed it may be, but Savoca's directorial debut is also acutely observant, and often very funny indeed.
COM   100 min   CINrel

TRUE ROMANCE **   18
Tony Scott USA   1993
Christian Slater, Patricia Arquette, Dennis Hopper, Val Kilmer, Gary Oldman, Brad Pitt, Christopher Walken, Samuel L. Jackson, Christopher Penn, Bronson Pinchot, Michael Rappapot, Saul Rubinek, Conchata Ferrell, James Gandolfini
A young couple marry and find themselves inadvertently involved with drug dealers from whom they flee to L.A. Written by Quentin Tarantino, this black comedy is full of violence and unconventional characters but totally devoid of real humour. However, excellent performances provide a good deal of interest, and help sustain flagging interest.
A / AD   113 min (ort 119 min) wScrn cC
VIDrel: WHV   V / sur

TRUE SIN *   18
R.T. Longhampton USA   1990
Ashlyn Gere, Lauren Brice, Jeannie Pepper, Ashly Dunn, Mike Horner, Randy West
A TV evangelist cannot keep his hands off women, and hires an endless succession of hookers to satisfy his cravings, all the while rationalising this as a perfectly acceptable way to be carrying on. An immensely tedious and verbose effort, that was presumably meant to explore the follies of all those real-life TV evangelists who have been caught with their pants down. Unfortunately, both the plot and dialogue are painfully trite.
A   49 min Cut (1 min 21 sec plus some cuts subst)
VIDrel: ONE   V

TRUE STORIES ***   PG
David Byrne USA   1986
David Byrne, John Goodman, Annie McEnroe, Swoosie Kurtz, Spalding Gray, Roebuck "Pop" Staples, Humberto "Tito" Larriva, John Ingle, Matthew Posey, Jo Harvey Allen, Alix Elias, Amy Buffington, Richard Dowlearn, Capucine DeWulf
Musical exploration of a fictional Texas town and the lives of its more than averagely eccentric inhabitants, with a narrator introducing a succession of figures. An offbeat, aimless and curious blend of comedy, music and social comment, all bundled up in the form of a set of vignettes. The excellent score is by Byrne and New Wave group Talking Heads.
MUS   86 min (ort 89 min)   VIDrel: WHV   V / sur

TRULY, MADLY, DEEPLY ***   PG
Anthony Minghella UK   1990
Juliet Stevenson, Alan Rickman, Bill Paterson, Michael Maloney,

Jenny Howe, Carolyn Choa, Christopher Rozycki, Keith Bartlett, David Ryall, Stella Maris, Henry James, Deborah Findlay, Ian Hawkes, Vania Vilers, Arturo Venegas
With the death of her fiance, a young woman is so distraught she is barely able to carry on living, so her boyfriend returns in order to comfort her. However, he has decided to tarnish her idealised memories of him and thus enable her to start a new life. Written by Minghella in his directing debut, this touching and unusual fantasy shows flashes of brilliance, but lacks consistency. Stevenson however, gives a remarkable performance. See also TO DIE FOR.
FAN   102 min (ort 106 min)   VIDrel: BUENA / TECH   V

TRUMAN ***   15
Frank F. Pierson USA   1995
Gary Sinise, Diana Scarwid, Ricahrd Dysart, Colm Feore, James Gammon, Tony Goldwyn, Pat Hingle, Harris Yulin, Leo Burmeister, Melia Campbell, Virginia Capers, John Fin, Zeljko Ivanek, David Lansburg, Remak Ramsey, Marion Seldes
Slow-moving and richly detailed depiction of the troubled presidency of Harry S. Truman, perhaps once of the most under-estimated presidents that the USA ever had. It follows the main events of his career, including the impact of the Korean War, but although the ground may be familiar, Sinise gives a sterling performance, as do the other cast members. A highly enjoyable and above-average TV biopic.
DRA   132 min mCab   VIDrel: THIRD   V
Boa: biography by David McCullough.

TRUST ***   15
Hal Hartley USA   1990
Adrienne Shelley, Martin Donovan, Merritt Nelson, Edie Falco, John MacKay, Gary Sauer, Matt Malloy, Suzanne Costollos, Jeff Howard, Karen Sillas, Tom Thon, M.C. Bailey, Patricia Sullivan, Marko Hunt, John St James, Kathryn Mederos
A surreal black-comedy about the unlikely relationship between two unhappy people. One is a pregnant high school student (whose father dropped dead when she announced the good news) and the other is an electronics wizard who makes his living repairing TV sets. They come together and embark on a desperate endeavour – a normal life of domesticity for both of them. Touching, if not entirely convincing.
COM   101 min (ort 107 min) wScrn   VIDrel: TART / 20TH   V

TRUST IN ME **   (12)
Bill Corcoran CANADA   1994
Stacy Keach, Currie Graham, Sandra Nelson, Ian tracey, Duncan Fraser, Tom McBeath, Akiko Morrison, Jason Schombing, Jackson Davies, Jerry Wasserman, Tegan Moss, Beverly Elliott, Tamsin Kelsey, Robert Kerr, Raimund Stamm, Ed Hong Louie
In the seedy underworld of Vancouver, a tough cop sets out to smash an arms smuggling operation. Average.
THR   88 min (ort 90 min) mTV   SATrel: SKY MOVIES

TRUST ME **   15
Robert Houston USA   1988
Adam Ant, Talia Balsam, David Packer, Barbara Bain, Karen Black, Joyce Van Patten, Ken Olofson, Alma Beltran, Rance Howard, Bill Saito, Barbara Perry, Tony Payne, Virgil Frye, Brigitte Burdine, Wiliam DeAcutis, Marilyn Tokuda, Kenia
The owner of an art gallery is heavily in debt, and has hatched a scheme to solve his financial worries by buying up the works of a great artist, and then killing him so that the paintings will increase in value.
Aka: TRUST ME: MURDER IS A DYING ART
COM   85 min (ort 91 min)   VIDrel: FABFIL / SPEAR   V

TRUTH ABOUT CATS & DOGS, THE ***   15
Michael Lehmann USA   1996
Uma Thurman, Janeane Garofalo, Ben Chaplin, Jamie Foxx, James McCaffrey, Richard Coca, Stanley DeSantis, Antoinette Valente, Mitch Rouse, La Tanya M. Fisher, Faryn Einhorn, David Cross, Mary Lynn Rajskub, Bob Odenkirk
Romantic comedy built around the idea of mistaken identity. A young female vet has her own radio show but her love life is far less satisfying due to her lack of self-confidence. When a male caller to the show tries to make a date with her, she persuades her neighbour to impersonate her. A strongly scripted work, it slightly recalls the screwball comedies of yesteryear and benefits from terrific performances by Thurman and Garofalo.
COM   93 min (ort 97 min)   VIDrel: 20TH / FOXVID   V / sur

TRUTH ABOUT SPRING, THE ** **   (U)
Richard Thorpe UK   1964
Hayley Mills, John Mills, Jamers McArthur, Lionel Jeffries, Harry
Andrews, Niall MacGinnis, Lionel Murton, David Tomlinson
A widowed American and his tomboyish daughter live a care-
free life aboard their houseboat in the Caribbean. But a young
lawyer, who is finding his stay on the yacht of his millionaire
uncle extremely dull, makes their acquaintance and takes an
immediate fancy to the girl. However, romance has to take
second place to a hunt for sunken treasure, in this lightweight,
romantic comedy.
COM   99 min   SATrel: MOVIE CHANNEL
Boa: short story Satan: A Romance of the Bahamas by Henry De
Vere Stacpoole.

TUCKER *   PG**
Francis Ford Coppola USA   1988
Jeff Bridges, Martin Landau, Joan Allen, Frederic Forrest, Dean
Stockwell, Mako, Elias Koteas, Nina Siemaszko, Christopher Slater,
Corky Nemec, Marshall Bell, Don Novello, Peter Donat, Dean
Goodman, Patti Austin
The story of Preston Tucker, who attempted to build a car for
the future (he first conceived the seatbelt) in the 1940s, but was
crushed by opposition from the giant automobile corporations.
A big, wholesome and stylish film in which the basically absorb-
ing tale tends to get swamped by flashy direction. The
performances are however, outstanding. The score is by Joe
Jackson.
Aka: TUCKER: THE MAN AND HIS DREAM
DRA   106 min (ort 110 min)   VIDrel: MGM/WHV L/A   V

TULSA **   U
Stuart Heisler USA   1949
Susan Hayward, Robert Preston, Pedro Armendariz, Chill Wills, Ed
Begley, Lloyd Gough, Lola Albright, Roland Jack, Harry Shannon,
Jimmy Conlin, Paul E. Burns, Roland Jack, Chief Yowlachie, Pierre
Watkin, Lane Chandler, Tom Dugan
A rancher's daughter fights to avenge the death of her father in
this so-so melodrama set in the tough world of the oil industry.
WES   87 min   VIDrel: SCRN/DISC   V

TUNE, THE *   PG**
Bill Plympton USA   1992
Voices of: Daniel Neiden, Maureen McElheron, Marty Nelson, Emily
Bindiger, Chris Hoffman
A songwriter with a failing career loses his way in his car and
ends up in a strange town called Flooby Nooby when he gains
some fresh inspiration for his songs. A fresh and enjoyable first
feature from animator Plympton that makes use of some 30,000
ink and watercolour drawings that he produced.
ANIM   67 min (ort 80 min) wScrn   VIDrel: TART/20TH   V

TUNES OF GLORY **   PG**
Ronald Neame UK   1960
Alec Guinness, John Mills, Dennis Price, Kay Walsh, John Fraser,
Gordon Jackson, Susannah York, Duncan Macrae, Allan Cuthbertson,
John MacKenzie, Keith Faulkner, Peter McEnery, Paul Whitsun-
Jones, Gerald Harper, William Marlowe
A callous, hard-drinking and lazy colonel is replaced by a
younger, far more disciplined officer, and a battle of wills takes
place. A superb character study, with fine performances and
realistic confrontations. Set in Scotland and written by
Kennaway. Photography is by Arthur Ibbetson.
DRA   107 min   VIDrel: L/A   V
Boa: novel by James Kennaway.

TUNNEL VISION **   18
Chris Fleury AUSTRALIA   1995
Patsy Kensit, Robert Reynolds, Rebecca Rigg, Gary Day, Shane
Bryant, Justin Monjo, David Woodley, Vanessa Steele, Craig Ashley,
Dean Nottle, Anthony Phelan, Liz Burch, Jonathan Hardy, Nathan
McGregor, Paul Denny, Brad Buckley
While having to deal with the case of a serial killer, together with
his female partner, a cop finds himself tormented by suspicions
that his wife is having an affair with one of his colleagues. When
the latter is found murdered, our cop naturally becomes the
prime suspect.
THR   90 min (ort 100 min)   VIDrel: HIFLI/SONOP   V

TURK 182! **   15
Bob Clark USA   1985
Timothy Hutton, Robert Urich, Robert Culp, Kim Cattrall, Darren
McGavin, Peter Boyle, Steven Keats, Paul Sorvino, James S. Tolkan,

Dick O'Neill, Thomas Quinn, Norman Parker, Maury Chaykin,
Richard Zobel, David Wohl
A teenage boy mounts a graffiti campaign to protest against the
shabby treatment of his fireman brother, injured in a fire but
denied compensation and a pension. Given the vast expanse of
graffiti (much highly artistic) that once covered the New York
subway system, it's hard to accept the basic premise of this film,
that it's OK to deface property provided it's in a good cause. In
addition, neither Hutton nor the script provide much humour.
COM   92 min (ort 98 min)   VIDrel: 20TH/TECH   V/sur

TURN BACK THE CLOCK **   15
Larry Elikann USA   1989
Connie Sellecca, David Dukes, Wendy Kilbourne, Jere Burns, Gene
Barry, Joan Leslie, Dina Merrill, Franc Luz, Pat Cupo, Kim Terry
Costin, Frank Coppola, Christopher Judges, Thomas H. Middleton,
Dennis Paladino, Carmela Rioseco
After shooting her husband to death a woman's wish for a
second chance is miraculously granted, and she finds herself
reliving the events of the previous year that led up to his
murder. An undistinguished remake of the 1947 film "Repeat
Performance" that fails to sparkle. Leslie (the lead in the earlier
film) has a small role in this one.
DRA   91 min (ort 96 min) mTV   TVrel

TURN OF THE SCREW, THE **   (15)**
Jack Clayton UK/USA   1961
Deborah Kerr, Michael Redgrave, Peter Wyngarde, Megs Jenkins,
Pamela Franklin, Martin Stephens, Clytie Jessop, Isla Cameron
Highly atmospheric tale of two children apparently possessed by
evil spirits. Kerr is the newly-appointed governess, who suspects
that their precocious and lascivious behaviour has its origin in a
tragic love affair between two of the master's former employees.
Remade as a TV movie in 1974. Written by William Archibald
and Truman Capote with music by Georges Auric and photogra-
phy by Freddie Francis. See also THE NIGHTCOMERS.
Aka: INNOCENTS, THE
DRA   100 min B/W   VIDrel: L/A   V
Boa: story by Henry James.

TURN OF THE SCREW **   18
Rusty Lemorande FRANCE/UK   1992
Patsy Kensit, Julian Sands, Stephane Audran, Marianne Faithfull
A young governess takes up a new position at an isolated
mansion and finds that her two young charges are possessed by
the evil spirits of two deceased former servants. A slack and
lifeless adaptation of the James story, filmed before in 1974 and
1989 as well as under the title "The Innocents" in 1961. The story
is updated to the 1960s, but this really adds nothing of interest
to the plot.
THR   92 min (ort 95 min)   VIDrel: 20VIS/SONOP   V/sur
Boa: story by Henry James.

TURN OF THE SCREW, THE *   15**
Tom McLoughlin USA   1995
Valerie Bertinelli, Aled Roberts, Florence Heath, Diana Rigg, Michael
Gough, Paul Rhys, Christopher Guard, Elizabeth Morton, Tricia
Thorns, Aisling Flitton, Flip Webster, Mark Longhurst
A governess takes up a position caring for a young boy and a
girl and gradually realises that these two small children are
being possessed by the evil spirits of two adults. These prove to
be those of her predecessor and a groom who are trying to renew
their affair. An insipid adaptation that relies more on state-of-
the-art special effects than any psychological depth.
Aka: HAUNTING OF HELEN WALKER, THE
HOR   88 min   VIDrel: ODY/SONOP   V
Boa: story by Henry James.

TURN OF THE TIDE *   U**
Norman Walker UK   1935
John Garrick, Geraldine Fitzgerald, Niall McGinnis, J. Fisher White,
Joan Maude, Sam Livesey, Wilfrid Lawson, Moore Marriott
A small-scale drama set in a Yorkshire fishing village where the
established way of life is upset by the arrival of an outside family
with modern ideas, such as the use of a motorboat for deep-sea
fishing. Despite their frosty welcome, the resulting antagonism
is short-lived, being eventually healed via the founding of a new
firm as well as a Romeo/Juliet style romance. A dated but
watchable piece of entertainment.
DRA   80 min; 126 min (2-film cassette) B/W
VIDrel: CONNO/RTM   V
Boa: novel Three Fevers by Leo Walmsley. Osca: MAN AT THE
GATE, THE

TURNAROUND **   15
Ola Solum USA   1986
Tim Maier, Doug McKeon, Jonna Lee, Ed Bishop, Eddie Albert, Gayle
Hunnicutt, Edward McClarity, Ramon Sheen
A gang harasses a young boy out with his girlfriend in his dad's
car, and the following night they gatecrash a party being given
at his house and terrorise the guests. The girlfriend escapes and
calls on the boy's uncle, a world famous illusionist, and he lures
the gang to his mansion where he nearly frightens them to death
by way of retribution. A thin little thriller rendered enjoyable
by the attractive cast.
THR   87 min (ort 97 min)   VIDrel: GUILD/SONOP L/A
V/sh

TURNER & HOOCH **   PG
Roger Spottiswoode USA   1989
Tom Hanks, Mare Winningham, Craig T. Nelson, Reginald
VelJohnson, Scott Paulin, J.C. Quinn, John McIntire, David Knell,
Ebbe Roe Smith, Kevin Scannell, Joel Bailey, Mary McCusker, Ernie
Lively, Clyde Kusatsu, Beasley the dog
Hanks plays a fastidious detective whose one chance to catch a
murderer resides in the only witness: the dead man's dog.
Unfortunately, the beast is an ugly, slobbering mutt of vora-
cious appetite and disagreeable habits. A kind of distant cousin
to K-9, that is sustained by Hanks's likeable personality if not
by the shallow plot, though there are several laughs to be had
before the lame ending. A weak film it may be, but a sure hit
with the kids. See also TOP DOG.
COM   95 min (ort 99 min)   VIDrel: TOUCH/TECH   V/sur

TURNING, THE **   18
L.A. Puopolo USA   1993
Karen Allen, Raymond Barry, Michael Dolan, Tess Harper, Gillian
Anderson
Racial bigotry and hatred in a small American town, where a
young man comes home determined to break up his dad's new
relationship.
DRA   87 min   VIDrel: UNIQUE/RTM   V/h

TURNING POINT, THE **   PG
Herbert Ross USA   1977
Shirley MacLaine, Anne Bancroft, Mikhail Baryshnikov, Leslie
Browne, Tom Skerritt, Martha Scott, Marshall Thompson, Antoinette
Sibley, Alexandra Danilova, Starr Danias, James Mitchell, Anthony
Zerbe, Phillip Saunders
After many years have elapsed, two friends from the world of
ballet meet up again. One has achieved stardom as a ballerina,
while the other teaches dance and now has a daughter who has
set her heart on a career in ballet. A tedious soap opera-style
drama, partially redeemed by the excellent ballet sequences.
This was Baryshnikov's debut. Written by Arthur Laurents.
DRA   114 min (ort 119 min)   VIDrel: L/A   V

TURTLE BEACH **   15
Stephen Wallace AUSTRALIA   1992
Greta Scacchi, Joan Chen, Jack Thompson, Art Malik, Norman Kaye,
Victoria Longley, Martin Jacobs, William McInnes, George Whaley,
Andrew Ferguson, Kee Chan, Daniel de Leur, Celia Wong, Monroe
Reimers, Stuart Campbell, Sean Scully
A woman photojournalist comes to Malaysia to do a piece on
the Vietnamese boat people and finds herself joining forces with
a hooker to prevent a massacre. A very poor adaptation of the
original novel, flatly directed and unconvincingly acted.
Aka: KILLING BEACH, THE
DRA   84 min (ort 88 min)
VIDrel: ENCORE/SPEAR/COLUM   V/sur
Boa: novel by Blanche d'Alpuget.

TURTLE DIARY ***   PG
John Irvin UK   1985
Glenda Jackson, Ben Kingsley, Richard Johnson, Michael Gambon,
Eleanor Bron, Rosemary Leach, Harriet Walter, Jeroen Krabbe, Nigel
Hawthorne, Michael Aldridge, Ron Anderson, Tony Melody, Gary
Olsen, Peter Capaldi
An odd couple are drawn together by their concern for the
welfare of the giant turtles at the local zoo, and hatch a plan to
return them to the ocean. An uneven but highly unusual char-
acter study, with offbeat dialogue and good performances.
Scripted by Harold Pinter and produced by Richard Johnson.
Look out for Pinter who appears very briefly as a customer in
Kingsley's bookshop.
DRA   90 min (ort 97 min)   VIDrel: 20TH/TECH   V/sur
Boa: novel by Russell Hoban.

TUSKEGEE AIRMEN, THE ***   PG
Robert Markowitz USA   1995
Laurence Fishburne, Allen Payne, Malcolm-Jamal Warner, Courtney
B. Vance, Andre Braugher, Christopher McDonald, John Lithgow,
Cuba Gooding Jr, Christopher Bevins, Daniel Hugh-Kelly, Mekhi
Phifer, Eddie Braun, Max Daniels, Jacy Dwyer
During WW2, the first squadron of black US Army combat pilots
is formed (the "Fighting 99th") who are obliged to overcome
enmity both on the ground and in the air. A high quality TV
movie, it mostly concentrates on the story of Hannibal Lee, who
was ultimately decorated for valour. It took the producers ten
years to raise the money to make this film. Based on the true
story of the 332nd Fighter Squadron, the title referring to the
airfield where they trained.
DRA   101 min (ort 107 min) mTV
VIDrel: MOSAIC/COLUM   V/sur

TUXEDO WARRIOR **   15
Andrew Sinclair USA   1985
John Wyman, Carol Royle, James Coburn Jr, Holly Palance, John
Terry, Mike Samson, Beth Adams, Ken Gampu
Routine thriller set in an African country, with a hard drinking,
hard living anti-hero type, involved in a hunt for stolen
diamonds.
Aka: AFRICAN RUN, THE
A/AD   90 min (ort 93 min)   VIDrel: MOPIC/SGSVID   V

TWELFTH NIGHT ***   U
John Gorrie UK   1980
Alec McCowen, Felicity Kendal, Sinead Cusack, Annette Crosby,
Robert Hardy
Excellent adaptation of this Shakespearean comedy of passion,
intrigue, disguise and jealousy, with a group of aristocrats
indulging themselves in a series of romantic interludes. Set at
the time of the Cavaliers, this play has some of Shakespeare's
finest lyrics. One in a series made for BBC TV.
DRA   127 min mTV   VIDrel: BBC   V/h
Boa: play Twelfth-Night; Or, What You Will by William
Shakespeare.

TWELFTH NIGHT ***   U
John Sichel UK   1980
Ralph Richardson, Alec Guinness, Tommy Steele, Joan Plowright,
Gary Raymond, John Moffat
An enjoyable adaptation Shakespeare's famous comedy-drama.
DRA   103 min   VIDrel: POLY/POLYREC   V
Boa: play Twelfth-Night; Or, What You Will by William
Shakespeare.

TWELFTH NIGHT ***   U
Trevor Nunn UK/USA   1996
Helena Bonham Carter, Richard E. Grant, Nigel Hawthorne, Ben
Kingsley, Mel Smith, Imelda Staunton, Toby Stephens, Imogen
Stubbs, Steven MacKintosh, Sid Livingstone, Nicholas Farrell, Alan
Mitchell, Peter Gunn, James Walker
Scripted by Nunn, this is a full-blooded and star-studded
rendition of Shakespeare's famous play of romance, jealousy
and intrigue. However, for all the fine work from a talented
cast (Stubbs is delightful as Viola and Smith makes a terrific
Sir Toby Belch) there are long periods of tedium in between
the sparkling set-piece confrontations. Kingsley as the jester
Feste is another actor worthy of mention in this flawed but
ambitious work.
DRA   128 min (ort 133 min)   VIDrel: EIV/SONOP   V/sh
Boa: play by William Shakespeare.

TWELVE ANGRY MEN ****   U
Sidney Lumet USA   1957
Henry Fonda, Lee J. Cobb, Ed Begley, E.G. Marshall, Jack Warden,
Martin Balsam, Jack Klugman, John Fiedler, George Voskovec, Robert
Webber, Edward Binns, Joseph Sweeney, Rudy Bond, James A. Kelly,
Bill Nelson, John Savoca
After a jury retire to consider their verdict in an apparently
simple murder case, one man stands out against the others, not
because he believes the suspect to be innocent, but simply
because he is not totally certain as to his guilt. In the course of
several hours, he convinces them of the soundness of his position. Despite the contrived script, this is a brilliant and
rewarding rewarding film, with memorable dialogue and
performances. Written by Rose.
DRA   92 min (ort 95 min) B/W   VIDrel: WHV   V
Boa: TV play by Reginald Rose.

TWELVE CHAIRS, THE *
U
Mel Brooks USA
1970
Ron Moody, Dom DeLuise, Mel Brooks, Frank Langella, Bridget Brice, Robert Bernal, Diana Coupland, David Lander, Andreas Voutsinas, Vlada Petric, Nicholas Smith, Elaine Garreau, Will Stampe, Paul Wheeler Jr, Mavid Popovic
An unfunny comedy written and directed by Brooks, but drawing its inspiration from a Russian satirical novel of the 1920s. A poor Russian aristocrat sells a set of twelve chairs only to discover that one of them contains a fortune in jewels sewn into the seat, thereby producing the perfect recipe for a frantic chase. Filmed in Yugoslavia and most tiresome. The film "Twelve Plus One" was based on the same story.
COM 89 min (ort 93 min) VIDrel: 20TH/TECH L/A V
Boa: novel Diamonds To Sit On by Ilya Arnoldovich Ilf and Evgeni Petroff.

TWELVE MONKEYS ****
15
Terry Gilliam USA
1995
Bruce Willis, Madeleine Stowe, Brad Pitt, Christopher Plummer, Joseph Melito, Jon Seda, Michael Chance, Vernon Campbell, H. Michael Walls, Bob Adrian, Simon Jones, Carol Florence, Bill Raymond, Ernest Abuba, Irma St Paule
Gilliam's ambitious expansion of LA JETEE works incredibly well on every level. Willis plays a convict sent back in time from a diseased and depopulated planet to the present day, where he hopes to solve a riddle and thus find a means of tracing the source of a virus that is to wipe out the bulk of the world's population. Incarcerated in an insane asylum, he gains assistance from psychiatrist Stowe and fellow inmate Pitt.
FAN 123 min (ort 129 min) wScrn cC
VIDrel: POLY/POLYREC V/sur V/dm

12. 01 **
15
Jack Sholder USA
1993
Jonathan Silverman, Helen Slater, Martin Landau, Nicolas Surovy, Robin Bartlett, Jeremy Piven, Constance Marie, Glenn Morshower, Paxton Whitehead, Cheryl Anderson, Joey Andrews, Frank Collison, Ed Crick, Jonthan Emerson
A young man who is a low-level clerical worker at a research institute wakes up to find himself endlessly repeating the previous day. He learns that a secret project at his place of work has caused a "time bounce" effect and sets to stop it, falling in love with a female scientist along the way. An utterly ruined adaptation of Lupoff's story that was made into a brilliant short (also by Sholder) the year before. Most disappointing. See also GROUNDHOG DAY.
Aka: 12:01
FAN 90 min mTV VIDrel: GUILD/SONOP V/sh
Boa: short story by Richard Lupoff.

TWELVE O'CLOCK HIGH ***
U
Henry King USA
1949
Gregory Peck, Hugh Marlowe, Gary Merrill, Millard Mitchell, Dean Jagger, Paul Stewart, Robert Arthur, John Kellogg, Robert Patten, Lee MacGregor, Sam Edwards, Roger Anderson, John Zilly, William Short, Richard Anderson
The story of a US bomber group based in England during WW2, and of the inevitable tensions the alternating periods of combat and boredom cause among the men. Peck gives an especially good performance. Later recycled as a TV series. The script was by Sy Bartlett and Beirne Lay Jr. AA: S. Actor (Jagger), Sound (Fox Studios).
WAR 128 min (ort 132 min) B/W VIDrel: 20TH/TECH V/h
Boa: novel by Beirne Lay Jr and Sy Bartlett.

TWENTIETH CENTURY ***
U
Howard Hawks USA
1934
John Barrymore, Carole Lombard, Walter Connolly, Roscoe Karns, Edgar Kennedy, Etienne Girardot, Ralph Forbes, Dale Fuller, Herman Bing, Lee Kohlmar, James P. Burtis, Billie Seward, Charles Lane, Mary Jo Mathews
A Broadway producer spots a talented young girl, makes her a star but loses her love in process. She departs for a career in Hollywood leaving him to face his rapidly declining fortunes. However, as fate decrees, they meet up again on the title train and he sets about winning back her love, mustering all the skill at his command. A tour-de-force performance from Barrymore sustains this entirely madcap and whirlwind comedy right to the end.
COM 89 min (ort 91 min) B/W VIDrel: COLUM/SONOP V
Boa: play Napoleon On Broadway by Charles Bruce Milholland.

TWENTY BUCKS **
15
Keva Rosenfeld USA
1993
Christopher Lloyd, Linda Hunt, Brendan Fraser, Steve Buscemi, Elisabeth Shue, Spalding Gray, Gladys Knights, Diane Baker, David Rasche, George Morfogen, Kamal Holloway, Melora Walters, William H. Macy, Matt Frewer, Nina Seimaszko
The title currency bill passes from one person to another and as it does so we see how it affects the lives of its various owners. An interesting and original study, based on a screenplay that is almost sixty years old but was updated by the screenwriter's son. See also THE GUN.
DRA 88 min (ort 91 min) cC VIDrel: COLUM/SONOP V

TWENTY-FOUR HOURS TO MIDNIGHT **
18
Leo T. Fong USA
1991
Cynthia Rothrick, Stack Pierce, Leo Fong
A woman martial artist takes revenge for the death of her husband and goes after the criminal gang responsible, in this standard and appropriately brutal urban actioner.
Aka: 24 HOURS TO MIDNIGHT
MAR 83 min (ort 91 min) VIDrel: MARQ/QUANT V

TWENTY MILLION MILES TO EARTH **
PG
Nathan Juran USA
1957
William Hopper, Joan Taylor, Frank Puglia, Thomas B. Henry, Jan Arvan, George Khoury, Bart Bradley, Tito Vuolo, George Pelling, Arthur Space, Rollin Moriyama, Don Orlando, Dale Van Sickel
A small dinosaur-like creature is brought back to Earth by a returning Venus space probe and grows to monstrous dimensions, thereby threatening Italy with total devastation, but getting its comeuppance in a battle on top Rome's Coliseum. A dated and cheap looking effort, with competent modelwork by Ray Harryhausen that was good enough in its day, but now looks quite unconvincing.
FAN 82 min B/W VIDrel: COLUM/SONOP V

TWENTY-NINTH STREET ***
15
George Gallo USA
1991
Danny Aiello, Anthony LaPaglia, Lainie Kazan, Frank Pesce, Rick Aiello, Ron Karabatos, Robert Forster, Leonard Termo, Vic Manni, Paul Lazar, Darren Bates, Tony Sirico, Hope Alexander-Willis, Sam Shamshak, Donna Mangani, Pete Antico
A slice-of-life study of an Italian family in New York, built around the true story of Frank Pesce, an actor who won $6,000,000 dollars on New York's first state lottery in 1976. Aiello gives a charming performance as a man for whom winning the lottery is just one more event in a life so plagued by such unnatural good fortune that it has caused endless family squabbles.
Aka: 29TH STREET
COM 97 min (ort 101 min) VIDrel: FIRST/SONOP L/A V
Boa: book by Frank Pesce and James Franciscus.

TWENTY-ONE **
15
Don Boyd UK
1991
Patsy Kensit, Jack Shepherd, Patrick Ryecart, Maynard Eziashi, Rufus Sewell, Sophie Thompson, Susan Wooldridge, Julia Goodman, Julian Firth, Robert Bathurst, Guy Oliver-Watts, Ben Murphy, Shelley Borkum, Veronica Clifford, Donald Tandy
Kensit gets one of her first decent roles in this arty film that goes for realism and quirky camerawork. She plays a young girl experiencing her first taste of adulthood, in the shape of a Scottish junkie she is living with. But with another couple of relationships going, she finally tires of the whole scene and takes off for the States. A bold and brash experiment that should have had better plotting to sustain interest.
DRA 97 min (ort 101 min) VIDrel: EIV/SONOP V/sur

TWENTY THOUSAND LEAGUES UNDER THE SEA **
PG
Stuart Paton USA
1916
Matt Moore, Allen Hollubar, June Gail, William Welsh, Curtis Benton, Dan Hamlon, Edna Pendleton, Howard Crampton, Wallace Clark, Martin Murphy, Leviticus Jones, Louis Alexander
Captain Nemo captures a professor and his daughter and takes them onboard his underwater vessel, in this early silent screen version of Verne's classic adventure. Though much restricted by the technical limitations of the day, this loose adaptation boldly attempts to capture some of the story's atmosphere and excitement (plus a few elements from Verne's other novel). Now

terribly dated, but the special effects and undersea sequences are still worth a look.
Aka: 20,000 LEAGUES UNDER THE SEA
A/AD   84 min (ort 100 min) B/W silent
VIDrel: SCREAM/SPEAR   V
Boa: novels 20,000 Leagues Under The Sea/Mysterious Island by Jules Verne.

TWENTY THOUSAND LEAGUES UNDER THE SEA ***
Richard Fleischer USA
U
1954
Kirk Douglas, James Mason, Peter Lorre, Paul Lukas, Robert J. Wilke, Ted De Corsia, Carleton Young, Percy Helton, Fred Graham, J.M. Kerrigan, Edward Marr, Harry Harvey, Herb Vigran, Ted Cooper
An entertaining film version of the story of Captain Nemo, who uses his submarine to wage a personal battle against ships used for warfare. Douglas gives a vigorous performance as the harpooner in the group taken on board the vessel, when their own ship is sunk by the Nautilus. Mason is equally good as Nemo in an enjoyable Disney film with excellent effects. Written by Earl Felton. AA: Art/Set (John Meehan/Emil Kuri), Effects (Walt Disney Studios).
Aka: 20,000 LEAGUES UNDER THE SEA
A/AD   118 min (ort 127 min)   VIDrel: WDV/TECH   V
Boa: novel by Jules Verne.

TWENTY-THREE HOURS FIFTY-EIGHT MINUTES ***
Pierre William-Glenn FRANCE
(PG)
1993
Jean-Francois Stevenin, Jean-Pierre Malo, Gerald Garnier, Yan Epstein, Kader Boukanef, Amelie Glenn, Jean-Charles Frappin, Sophie Tellier, Isabelle Maltese, Pierre-Octave Arrighi, Emmanuel Pinda, Georges Zsiga, Jean-Jacques Birod
Having drawn their inspiration from Kubrick's "The Killing", two racers plan the robbery at a stadium during the Le Mans motorcycle race. But having done this, they find themselves trapped in the stadium, and are forced to hide the money before splitting up. A well focused tale, its upbeat plotting is strangely at odds with its subject matter, but this is clearly quite deliberate. The title refers to the concluding moments of the twenty-four hour Le Mans race.
Aka: 23h58
THR   85 min   CINrel

TWICE IN A LIFETIME ***
Bud Yorkin USA
15
1985
Gene Hackman, Ann-Margret, Ally Sheedy, Ellen Burstyn, Amy Madigan, Brian Dennehy, Stephen Lang, Darrell Larson, Chris Parker, Rachel Street, Kevin Bleyer, Nicole Mercurio, Doris Hugo Drewien, Lee Corrigan, Ralph Steadman
A comedy-drama telling of a middle-aged man with a dull marriage who falls in love with a younger woman, causing his family a great deal of pain. This powerful examination of the mid-life crisis generally avoids both cliche and over-sentimentality, and has memorable performances from all concerned. Written by Colin Welland.
DRA   106 min (ort 117 min)   VIDrel: FIRST/SONOP L/A   V

TWICE ROUND THE DAFFODILS **
Gerald Thomas UK
PG
1962
Juliet Mills, Donald Sinden, Donald Houston, Kenneth Williams, Ronald Lewis, Jill Ireland, Joan Sims, Andrew Ray, Lance Percival, Sheila Hancock, Nanette Newman
This uneven collection of observations and misadventures revolves around the lives of the patients and staff at a male TB sanatorium, where an attractive nurse has a difficult time fending off her amorous patients. A bawdy adaptation of the stage play, done in appropriate "Carry On" style.
COM   85 min (ort 89 min) B/W   VIDrel: WHV   V/h
Boa: play Ring for Catty by Patrick Cargill and Jack Beale.

TWICE SHY ***
Deirdre Friel CANADA/EIRE
PG
1989
Ian McShane, Patrick Macnee, Niall Toibin, Stephen Brennan, Kate McKenzie, Dearbhla Molloy, Geraldine Fitzgerald, Karl Hayden, Conor Mullen, David Herlihy, John Keegan, Oliver McGuire, Pat Leavy, Jim Reid, Deidre O'Kane
Another mystery to solve for David Cleveland, our investigator from the Jockey Club. In this story, the death of a computer

expert and a schoolteacher who has been living beyond his means are found to be linked. See also IN THE FRAME.
Aka: DICK FRANCIS MYSTERIES: TWICE SHY
DRA   95 min mTV   VIDrel: IMC/VCC   V
Boa: novel by Francis.

TWILIGHT ZONE: THE MOVIE ***
Steven Spielberg/Joe Dante/George Miller/John Landis USA
15
1983
Vic Morrow, John Lithgow, Scatman Crothers, Bill Quinn, Dan Aykroyd, Albert Brooks, Kevin McCarthy, Kathleen Quinlan, Selma Diamond, Jeremy Licht, Abbe Lane, Donna Dixon, Doug McGrath, Charles Hallahan, Martin Garner
Four tales (three of which were remakes from the TV series), of varying power on the theme of the unknown. In one old folk regain their youth, in another an odious bigot suffers racism, in the third a boy has horrifying powers and in the last a passenger on an aeroplane sees a nasty creature perched outside. Narrated by Burgess Meredith, Dan Aykroyd and Albert Brooks.
Aka: TWILIGHT ZONE, THE
FAN   98 min (ort 101 min)   VIDrel: MGM/WHV L/A   V/sh
Boa: short stories Kick The Can by George Clayton Johnson/Nightmare At 20,000 Feet by Richard Matheson.

TWILIGHT ZONE, THE: A HUNDRED YARDS OVER THE RIM **
Buzz Kulik USA
PG
1960
Cliff Robertson, Miranda Jones, John Crawford, Evan Evans
A 19th century settler obtains medicine for his sick son from an strange journey, coming back with a supply of life-saving penicillin.
FAN   92 min (4-episode cassette) B/W mTV
VIDrel: 20TH/TECH L/A   V/h

TWILIGHT ZONE, THE: DEATH'S HEAD REVISITED **
Don Medford USA
PG
1960
Oscar Beregi, Ben Wright, Joseph Schildkraut, Karen Verne, Chuck Fox, Robert Boone
"Death's Head Revisited" has a former Nazi returning to visit the death-camp where his word was law, but finding himself being put on trial by a jury of his dead victims.
FAN   92 min (4-episode cassette) B/W mTV
VIDrel: 20TH/TECH L/A   V/h

TWILIGHT ZONE, THE: FIVE CHARACTERS IN SEARCH OF AN EXIT **
Lamont Johnson USA
PG
1960
Bill Windom, Susan Harrison, Clark Allen, Murray Matheson, Kelton Garwood, Mona Houghton, Carol Hill
Five disparate people are apparently trapped in a pit – a ballet dancer, a soldier, a tramp, a clown and a musician. But there are aspects of their predicament that suggest a very different reason for their presence together.
FAN   92 min (4-episode cassette) B/W mTV
VIDrel: 20TH/TECH L/A   V/h
Boa: short story by Marvin Petal.

TWILIGHT ZONE, THE: IN PRAISE OF PIP **
Joseph M. Newman USA
PG
1960
Jack Klugman, Billy Mumy, Kreg Martin, Ross Elliott, Stuart Nisbet, Russell Horton, Connie Glichrist, Bob Diamond, John Launer, Gerald Gordon
In Praise Of Pip" has a father praying to be allowed to die instead of his son, who has been critically wounded in Vietnam.
FAN   92 min (4-episode cassette) B/W mTV
VIDrel: 20TH/TECH L/A   V/h

TWILIGHT ZONE, THE: JUDGEMENT NIGHT **
John Brahm USA
PG
1960
Nehemiah Persoff, Patrick Macnee, Leslie Bradley, Kendrick Huxhum, Ben Wright, Hugh Sanders, Deidre Owen, James Franciscus
"Judgement Night" tells of a passenger onboard a wartime freighter who has a premonition that the ship will be torpedoed by a German submarine. But he can get no-one to take his warnings seriously.
FAN   92 min (4-episode cassette) B/W mTV
VIDrel: 20TH/TECH L/A   V/h

TWILIGHT ZONE, THE: LIVING DOLL ** PG
Richard C. Sarafian USA 1960
Telly Savalas, Tracy Stratford, Mary LaRoche
A man plans to dispose of an expensive doll that appears to
have a mind of its own, and is planning a revenge of its own.
FAN 92 min (4-episode cassette) B/W mTV
VIDrel: 20TH/TECH L/A V/h

TWILIGHT ZONE, THE: NIGHTMARE AT 20,000 FEET *** 15
Richard Donner USA 1960
*William Shatner, Edward Kemmer, Christine White, Asa Maynor,
Nick Cravat*
"Nightmare At 20,000 Feet" has an airline passenger and former
mental patient spotting a grotesque and malevolent creature on
the wing of the aeroplane, where it delights in fiddling about
with the innards of the engine. This story was used to great
effect as one of the tales in TWILIGHT ZONE: THE MOVIE.
FAN 92 min (4-episode cassette) B/W mTV
VIDrel: 20TH/TECH L/A V/h
Boa: short story by Richard Matheson.

TWILIGHT ZONE, THE: ONE FOR THE ANGELS ** PG
Robert Parrish USA 1960
*Ed Wynn, Murray Hamilton, Dana Dillaway, Jay Overholts, Merritt
Bohn, Mickey Maga*
A kindly salesman learns that his time on Earth is up, but talks
Death into sparing his life for one last big deal, but when he
reneges on the agreement, a small child is selected to die in his
place. Written by Rod Serling (the creator of the series) and like
most of his contributions, rather spoilt by whimsy and senti-
mentality.
FAN 99 min (4-episode cassette) B/W mTV
VIDrel: 20TH/TECH L/A V/h

TWILIGHT ZONE, THE: PEOPLE ARE ALIKE ALL OVER *** PG
Mitchell Leisen USA 1960
*Roddy McDowall, Paul Comi, Vic Perrin, Susan Oliver, Byron
Morrow, Vernon Gray*
"People Are Alike All Over" has the first man on Mars becom-
ing an object of great interest to the Martians, but the great
attention they give to making him comfortable is soon revealed
to have a more sinister purpose.
FAN 92 min (4-episode cassette) B/W mTV
VIDrel: 20TH/TECH L/A V/h
Boa: short story by Paul Fairman.

TWILIGHT ZONE, THE: STEEL ** PG
Don Weis USA 1960
*Lee Marvin, Joe Mantell, Merritt Bohn, Frank London, Tipp McClure,
Chuck Hicks, Larry Barton*
"Steel" is the story of a small-time promoter of robot fights who
is obliged to take his robot's place when it is damaged, as he
badly needs the purse from the fight, and cannot afford have the
match cancelled.
FAN 99 min (4-episode cassette) B/W mTV
VIDrel: 20TH/TECH L/A V/h

TWILIGHT ZONE, THE: THE AFTER HOURS ** PG
Douglas Heyes USA 1960
*Anne Francis, Elizabeth Allen, James Millhollin, John Cornwell,
Nancy Rennick*
In "The After Hours" a woman finds that the floor of a depart-
ment store where she just made a purchase does not exist, and
that the girl who served here is a store mannequin.
FAN 99 min (4-episode cassette) B/W mTV
VIDrel: 20TH/TECH L/A V/h

TWILIGHT ZONE, THE: THE CHANGING OF THE GUARD * PG
Robert Ellis Miller USA 1960
*Donald Pleasence, Liam Sullivan, Philippa Bevans, Kevin O'Neal,
Jimmy Baird, Kevin Jones, Bob Biheller, Tom Lowell, Russell Horton,
Buddy Hart, Darryl Richard, James Browning, Pat Close, Dennis
Kerlee*
"The Changing Of The Guard" explores the turmoil of a retir-
ing teacher who feels that his usefulness to society is now over,
but has a visitation from the ghosts of some of his former
students. Pure whimsy this one, and as one might expect, it was
written by Rod Serling.
FAN 99 min (4-episode cassette) B/W mTV
VIDrel: 20TH/TECH L/A V/h

TWILIGHT ZONE, THE: THE EYE OF THE BEHOLDER ** PG
Douglas Heyes USA 1960
*Joanna Hayes, Jennifer Howard, William D. Gordon, Maxine Stuart,
Donna Douglas*
The people of a future society consider ugliness to represent
normality, and a beautiful girl longs for the cosmetic surgery she
hopes will make her just like everyone else.
Aka: PRIVATE WORLD OF DARKNESS, THE
FAN 99 min (4-episode cassette) B/W mTV
VIDrel: 20TH/TECH L/A V/h

TWILIGHT ZONE, THE: THE HITCH-HIKER *** PG
Alvin Ganzer USA 1960
*Inger Stevens, Leonard Strong, Adam Williams, Dwight Townsend,
Mitzi McCall, Eleanor Audley, Lew Gallo, Russs Bender, George
Mitchell*
A female driver on a lonely road keeps seeing the same man
grimly beckoning, and ultimately the sinister reason for his pres-
ence is revealed. An entry whose final resolution is easy to
guess, yet one with a genuine sense of the spooky.
FAN 99 min (4-episode cassette) B/W mTV
VIDrel: 20TH/TECH L/A V/h
Boa: radio play by Lucille Fletcher.

TWILIGHT ZONE, THE: THE INVADERS *** PG
Douglas Heyes USA 1960
Agnes Moorehead
A woman on a remote farmhouse fights off midget aliens, even-
tually finding and destroying their space vessel. But the truth
about the origin of these creatures is then revealed.
FAN 99 min (4-episode cassette) B/W mTV
VIDrel: 20TH/TECH L/A V/h

TWILIGHT ZONE, THE: THE LONELY *** PG
Jack Smight USA 1960
Jack Warden, John Dehner, Jim Turley, Jean Marsh, Ted Knight
"The Lonely" tells of a man convicted of murder and exiled to
a barren asteroid, but later he is given a "female" robot for
company. However, when his sentence is commuted a ship
arrives to take him off the planetoid, but there is no room for
his companion. A very effective entry, believable and quite
touching.
FAN 99 min (4-episode cassette) B/W mTV
VIDrel: 20TH/TECH L/A V/h

TWILIGHT ZONE, THE: THE MIDNIGHT SUN ** 15
Anton Leader USA 1960
*Lois Nettleton, Betty Garde, Ned Glass, Jason Wingreen, June Ellis,
John McLiam, William Keene, Robert J. Stevenson, Tom Reese*
"The Midnight Sun" tells of how the Earth slowly begins
to move out of it orbit and get ever closer to the Sun. Yet this
may be no more than a dream, but one that has a disturbing
cause.
FAN 92 min (4-episode cassette) B/W mTV
VIDrel: 20TH/TECH L/A V/h

TWILIGHT ZONE, THE: THE MONSTERS ARE DUE ON MAPLE STREET ** PG
Ron Winston USA 1960
*Claude Akins, Jack Weston, Barry Atwater, Jan Handzlik, Burt
Metcalfe, Mary Gregory, Anne Barton, Lea Waggner, Ben Erway,
Lyn Guild, Sheldon Allman, William Walsh*
On Maple Street a power failure sparks off paranoid fears
of aliens among the residents, who quickly descend into
mutual antagonism, hatred and violence. Their experiences
are ultimately revealed to be nothing more than the result
of experiments being conducted into human behaviour by
aliens.
FAN 99 min (4-episode cassette) B/W mTV
VIDrel: 20TH/TECH L/A V/h

TWILIGHT ZONE, THE: THE ODYSSEY OF FLIGHT 33 ** PG
Justus Addiss USA 1960
*John Anderson, Paul Comi, Sandy Kenyon, Harp McGuire, Wayne
Heffley, Nancy Rennick, Beverly Brown, Jay Overholt, Betty Garde*
In "The Odyssey Of Flight 33" a passenger plane on a routine
flight to New York suddenly materialises in the prehistoric past.
FAN 99 min (4-episode cassette) B/W mTV
VIDrel: 20TH/TECH L/A V/h

TWILIGHT ZONE, THE: THE PURPLE TESTAMENT *** 15

Richard Bare USA   1960
William Reynolds, Dick York, Barney Phillips, Warren Oates, Ron Masak, William Phipps, Marc Cavell, Paul Mazursky
This story explores the actions of a soldier suddenly given the ability to see death in the faces of those comrades whose lives are soon to be cut short.
FAN   92 min (4-episode cassette) B/W mTV
VIDrel: 20TH/TECH L/A   V/h

TWILIGHT ZONE, THE: TIME ENOUGH AT LAST **
PG
John Brahm USA   1960
Burgess Meredith, Vaughn Taylor, Jacqueline de Wit, Lela Bliss
"Time Enough At Last" sees a shortsighted bank clerk apparently the only survivor of a nuclear war, a position the allows him to indulge his passion for reading.
FAN   99 min (4-episode cassette) B/W mTV
VIDrel: 20TH/TECH L/A   V/h
Boa: short story by Lynn Venable.

TWILIGHT ZONE, THE: TO SERVE MAN **   *PG*
Richard L. Bare USA   1960
Richard Kiel, Hardie Albright, Robert Tafur, Lloyd Bochner, Lomax Study, Theodore Marcuse, Susan Cummings, Nelson Olmstead
In "To Serve Man", apparently friendly alien emissaries prove to have an appetite for more than mere friendship with the human race, as thet clearly demonstrate when one of their books is stolen and translated.
FAN   92 min (4-episode cassette) B/W mTV
VIDrel: 20TH/TECH L/A   V/h
Boa: short story by Damon Knight.

TWILIGHT ZONE, THE: TWO **   *PG*
Montgomery Pittman USA   1960
Charles Bronson, Elizabeth Montgomery
A man and a woman are the only survivors of a nuclear holocaust, and attempt to overcome their mutual fear and suspicion. Not the strongest of the entries, the basic idea has a development that is all too predictable.
FAN   99 min (4-episode cassette) B/W mTV
VIDrel: 20TH/TECH L/A   V/h

TWILIGHT ZONE, THE: WALKING DISTANCE ** 15
Robert Stevens USA   1960
Gig Young, Michael Montgomery, Byron Foulger, Joseph Corey, Frank Overton, Irene Tedrow, Buzz Martin
"Walking Distance" sees a man slipping back in time to escape the pressures of work, his chosen escape route being a return to his own childhood.
FAN   92 min (4-episode cassette) B/W mTV
VIDrel: 20TH/TECH L/A   V/h

TWILIGHT ZONE – NEW SERIES: A LITTLE PEACE AND QUIET **   *PG*
Wes Craven USA   1986
Melinda Dillon
A harried and overworked housewife dreams of a way to have an easier life, but her wish comes true in a most unpleasant way, when she finds a way of freezing time.
FAN   95 min (4-episode cassette) mTV
VIDrel: 20TH/TECH L/A   V

TWILIGHT ZONE – NEW SERIES: A MESSAGE FROM CHARITY ***   *PG*
Paul Lynch USA   1986
Robert Duncan McNeil, Kerry Noonan
A young boy living in present-day Massachusetts makes contact with a Puritan girl from the past, and each is able to learn about the other's world. However, the girl unwisely speaks of the knowledge she has gained about the future, an action that makes her vulnerable to a charge of witchcraft, from which she is only saved by the boy's knowledge of events in her future.
FAN   97 min (2-episode cassette) mTV
VIDrel: 20TH/TECH L/A   V
Boa: short story by William M. Lee.

TWILIGHT ZONE – NEW SERIES: ACT BREAK ** 15
Ted Flicker USA   1986
James Coco, Bob Dishy
An unsuccessful and untalented playwright is given a magic amulet by his dying lover, and uses it to make a wish for the

only thing he truly cares about – dramatic immortality.
FAN   95 min (4-episode cassette) mTV
VIDrel: 20TH/TECH L/A   V

TWILIGHT ZONE – NEW SERIES: CHILDREN'S ZOO **   *PG*
Robert Downey USA   1986
Steven Keats, Lorna Luft
A withdrawn little girl who lives unhappily with her constantly bickering parents receives an invitation to visit a strange zoo, where she can take her pick of new parents.
FAN   96 min (4-episode cassette) mTV
VIDrel: 20TH/TECH L/A   V

TWILIGHT ZONE – NEW SERIES: DEAD WOMAN'S SHOES **   *PG*
Peter Medak USA   1986
Helen Mirren, Theresa Saldana, Jeffrey Tambor
In "Dead Woman's Shoes" a timid clerk slips on a pair of shoes that were once owned by a murdered woman, and immediately takes on this woman's personality and her thirst for vengeance.
FAN   95 min (4-episode cassette) mTV
VIDrel: 20TH/TECH L/A   V
Boa: short story by Charles Beaumont.

TWILIGHT ZONE – NEW SERIES: DREAMS FOR SALE **   *PG*
Tommy Lee Wallace USA   1986
Meg Foster, David Hayward
A couple and their two go daughters enjoy a picnic in the country, where everything is so perfect it is too good to be true.
FAN   95 min (4-episode cassette) mTV
VIDrel: 20TH/TECH L/A   V

TWILIGHT ZONE – NEW SERIES: EXAMINATION DAY **   *PG*
Paul Lynch USA   1986
Christopher Allport, Elizabeth Normant, David Mendenhall
"Examination Day" is a futuristic tale revolving around a twelve-year-old boy and his birthday wish to do well in a special government exam, a prospect his parents do not view with relish.
FAN   97 min (4-episode cassette) mTV
VIDrel: 20TH/TECH L/A   V
Boa: short story by Henry Slesar.

TWILIGHT ZONE – NEW SERIES: HEALER **   *PG*
Sig Neufeld Jr USA   1986
Eric Bogosian, Vincent Gardenia
In "Healer" a young drifter gains magical powers when he steals an ancient Mayan stone, and becomes a much sought figure, but finds there is a price to pay for his new powers.
FAN   96 min (4-episode cassette) mTV
VIDrel: 20TH/TECH L/A   V

TWILIGHT ZONE – NEW SERIES: HER PILGRIM SOUL **** 15
Wes Craven USA   1986
Kristofer Tabori, Gary Cole, Anne Twomey
In "Her Pilgrim Soul" (probably the very best tale in this new series) a computer scientist has to contend with a phantom image, that starts off life as an infant floating in the beam of a holographic device, but rapidly grows to adulthood as a beautiful woman from the past. As the scientist learns more about her background she continues to age, until as an elderly woman she reveals the true purpose of her visit. A really outstanding and beautifully made story.
FAN   95 min (2-episode cassette) mTV
VIDrel: 20TH/TECH L/A   V

TWILIGHT ZONE – NEW SERIES: I, OF NEWTON * 15
Ken Gilbert USA   1986
Sherman Hemsley, Ron Glass
A mathematician inadvertently invokes a demon when he works on a complex equation, and this latter attempts to claim his soul.
FAN   95 min (4-episode cassette) mTV
VIDrel: 20TH/TECH L/A   V

TWILIGHT ZONE – NEW SERIES: IF SHE DIES **   *PG*
John Hancock USA   1986
Tony Lo Bianco
A supernatural tale in this series. "If She Dies" tells of a visit by

a ghostly apparition to a man whose daughter lies in a coma, and which draws the desperate father towards a long-abandoned orphanage.
FAN   97 min (4-episode cassette) mTV
VIDrel: 20TH/TECH L/A   V

TWILIGHT ZONE – NEW SERIES: KENTUCKY RYE **
John Hancock USA
Jeffrey DeMunn

PG
1986

"Kentucky Rye" tells of a salesman with a serious drink problem, who gets involved in a car crash and retreats to a bar, where he is offered a most unusual opportunity.
FAN   96 min (4-episode cassette) mTV
VIDrel: 20TH/TECH L/A   V

TWILIGHT ZONE – NEW SERIES: ONE LIFE, FURNISHED IN EARLY POVERTY ***
Don Dunway USA
Peter Reigert, Jack Kehoe, Chris Hebert

15
1986

A writer returns to his childhood home in an effort to remember something of his youth and the hopes and dreams he had then, but this trip results in him taking a journey back in time and becoming a boy once again.
FAN   95 min (4-episode cassette) mTV
VIDrel: 20TH/TECH L/A   V
Boa: short story by Harlan Ellison.

TWILIGHT ZONE – NEW SERIES: OPENING DAY **
John Milius USA
Jeffrey Jones, Martin Kove

PG
1986

A man murders the husband of his lover, but then finds that he has inexplicably got caught up in a time-loop in which their lives of the two men somehow become exchanged.
FAN   95 min (4-episode cassette) mTV
VIDrel: 20TH/TECH L/A   V

TWILIGHT ZONE – NEW SERIES: PALADIN OF THE LOST HOUR **
Alan Smithee (Gilbert Cates) USA
Danny Kaye, Glynn Turman

15
1986

In "Paladin Of The Lost Hour" an old man is found to hold the last hour of the world in a magical timepiece, but the time has now come for him to find a new keeper to take over. However, acceptance of this position does confer on the recipient one rather unusual gift. An interesting idea that really needed more time to be developed fully.
FAN   95 min (4-episode cassette) mTV
VIDrel: 20TH/TECH L/A   V

TWILIGHT ZONE – NEW SERIES: SHATTERDAY **
Wes Craven USA
Bruce Willis

PG
1986

A rather callous and selfish man telephones home and hears himself answering, and becomes so fearful of meeting his double that he refuses to venture home. Bit by bit this double (who represents the man he might have been) takes over his life. A strange entry with an odd and unsatisfying conclusion.
FAN   95 min (4-episode cassette) mTV
VIDrel: 20TH/TECH L/A   V
Boa: short story by Harlan Ellison.

TWILIGHT ZONE – NEW SERIES: TEACHER'S AIDE *
Bill Norton USA
Adrienne Barbeau

15
1986

In "Teacher's Aide" a meek English teacher is terrorised by some of her students, but then comes under the baleful influence of a stone gargoyle that sits atop the school building, and which confers on her great physical strength and a malevolent disposition. A very silly entry indeed, implausible and poorly developed.
FAN   95 min (4-episode cassette) mTV
VIDrel: 20TH/TECH L/A   V

TWILIGHT ZONE – NEW SERIES: THE BEACON **
Gerd Oswald USA
Martin Landau, Cheryl Anderson, Charles Martin Smith

15
1986

"The Beacon" is the story of an eerie lighthouse whose beam shines inland, where on shines on the houses in an isolated community, picking out the homes where a death is soon to occur. But when a doctor inadvertently stumbles on this community, he upsets things by using modern drugs to save the

life of a dying child, an action the community at large do not look on kindly.
FAN   95 min (4-episode cassette) mTV
VIDrel: 20TH/TECH L/A   V

TWILIGHT ZONE – NEW SERIES: THE BURNING MAN **
J.D. Feigelson USA
Piper Laurie, Roberts Blossom

15
1986

On an unusually hot day an aunt and her young nephew go for a drive in the country, and pick up an unkempt old man who frightens the life out of them, having claimed that it is possible for a human-like creature to hibernate for years, and hatch out at regular intervals when it needs to satisfy its voracious appetite. On the way back home they encounter a similar figure, but this one is a young child.
FAN   95 min (4-episode cassette) mTV
VIDrel: 20TH/TECH L/A   V
Boa: short story by Ray Bradbury.

TWILIGHT ZONE – NEW SERIES: THE SHADOW MAN *
Joe Dante USA
Jonathan Ward, Heather Haase, Jeff Calhoun

PG
1986

A timid thirteen-year-old discovers a shadowy creature that lives under his bed, from where it forays out each evening on regular prowls. Though he doesn't exactly make friends with it, he learns that it will never harm him, and realises he can use this knowledge to frighten the wits out of the classmate who has been bullying him. Too short to be effective, even if the idea is rather neat.
FAN   95 min (4-episode cassette) mTV
VIDrel: 20TH/TECH L/A   V

TWILIGHT ZONE – NEW SERIES: THE UNCLE DEVIL SHOW **
David Steinberg USA
Murphy Dunne

PG
1986

A young boy starts watches a video his parents have got out for him, but they cannot guess just what he will get up to once he comes under its baleful influence.
FAN   95 min (4-episode cassette) mTV
VIDrel: 20TH/TECH L/A   V

TWILIGHT ZONE – NEW SERIES: WISH BANK **
Rick Friedberg USA
Dee Wallace

PG
1986

"Wish Bank" follows a bargain hunter who finds a magic lamp at a jumble sale and discovers that it will grant her three wishes. As one might guess, there are some serious consequences attached to making use of it.
FAN   96 min (4-episode cassette) mTV
VIDrel: 20TH/TECH L/A   V

TWILIGHT ZONE – NEW SERIES: WONG'S LOST AND FOUND EMPORIUM **
Paul Lynch USA
Brian Tochi, Anna Maria Poon

PG
1986

In "Wong's Lost And Found Emporium" a lonely young man who is thoroughly discontented with his life stumbles on a strange warehouse that exists in another dimension, and where all the unfulfilled dreams and precious memories of humanity are to be found. It would appear that he has been selected to become the new manager of this establishment.
FAN   95 min (4-episode cassette) mTV
VIDrel: 20TH/TECH L/A   V
Boa: short story by William F. Wu.

TWILIGHT ZONE – NEW SERIES: WORD PLAY *
Wes Craven USA
Robert Klein, Annie Potts

PG
1986

A man wakes up one morning to find that everyone about him is talking gibberish, yet he is the only one who can no longer understand it. One of the weakest entries in this new series, with neither and explanation nor a resolution offered for this state of affairs.
FAN   95 min (4-episode cassette) mTV
VIDrel: 20TH/TECH L/A   V

TWILIGHT ZONE – NEW SERIES: YE GODS *
Peter Medak USA
David Dukes, Robert Morse

PG
1986

"Ye Gods" has an executive with an unsatisfactory love life

discovering that the ancient Greek Gods are alive, if not all that well, and are not really in a good position to offer him help with his own love life.

FAN   97 min (4-episode cassette) mTV
VIDrel: 20TH/TECH L/A   V

TWILIGHT ZONE: ROD SERLING'S LOST CLASSICS ** 12
Robert Markowitz USA 1993
James Earl Jones, Amy Irving, Gary Cole, Heide Swedburg, Priscilla Pointer, Scott Burkholder, Don Bloomfield, Alex Van, Patrick Bergin, Jeann Stern, Jack Palance, Peter McRobbie, Bill Bolender, Mlachy McCourt, Ricahrd K. Olsen
Two Rod Serling tales that never used on original show: "The Theatre" in which an actress sees her future on the screen, but also learns of her death, and "Where The Dead Are" which tells of a surgeon who has developed a serum that can prolong life. Fair.

FAN   89 min   VIDrel: NTV/TOTAL   V

TWILIGHT'S LAST GLEAMING ** 15
Robert Aldrich USA/WEST GERMANY 1977
Burt Lancaster, Richard Widmark, Joseph Cotten, Charles Durning, Melvyn Douglas, Paul Winfield, Roscoe Lee Brown, Burt Young, Vera Miles, Richard Jaeckel, William Marshall, Charles Aidman, Charles McGraw, Leif Erickson
An ex-general captures a missile base and blackmails the White House into revealing some political truths about its policies during the Vietnam War. A mildly gripping little yarn, but seriously diluted by excessive length. The music is by Jerry Goldsmith. See also SEVEN DAYS IN MAY and THE ENEMY WITHIN.

Aka: NUCLEAR COUNTDOWN; VIPER THREE
THR   116 min (ort 146 min)   VIDrel: L/A   V
Boa: novel Viper Three by Walter Wager.

TWIN ACTION * 18
Frank Marino USA 1993
Lynn Lemay, Rebecca Wild, Lyndon Johnson, Angel Bust, Randy Speas, John Dough, J.D.
A woman evolves a secret identity in order to enhance her sex life, in this unutterably silly offering.

A   43 min   VIDrel: ONE   V

TWIN DRAGONS *** PG
Tsui Hark/Ringo Lam HONG KONG 1992
Jackie Chan, Maggie Cheung, Teddy Robin Kwan, Tsui Hark, Nina Li Chi, John Woo, Chu Yuan
Owing the actions of a wounded gangster who stole a child from a maternity hospital, the lives of a pair of twins take very different paths. One achieves success as a famous conductor whilst the other mixes with thieves and other such characters. When the latter appears on the scene seeking refuge from gangsters out to kill him, the life of the respectable twin is also placed in peril. An action-filled adventure, quite well plotted for films of this genre.

A/AD   100 min (ort 115 min)   VIDrel: ONE/IMPENT   V

TWIN PEAKS ** 15
David Lynch USA 1989
Michael Ontkean, Joan Chen, Kyle MacLachlan, Piper Laurie, Dan Ashbrook, Jim Marshall, Madchen Amick, Richard Beymer, Laura Flynn Boyle, Sherilyn Fenn, Peggy Lipton, Warren Frost, Everett McGill, Jack Nance, Ray Wise
A bizarre story set in a small North American logging town, where the discovery of the body of a dead girl is a prelude to a series of sinister events. This tale was the pilot for a massively over-hyped TV series and despite many wonderfully bizarre (and very typical) Lynch touches, is a disappointing and slack affair.

THR   108 min (ort 112 min) mTV   VIDrel: WHV   V/sur

TWIN PEAKS: FIRE WALK WITH ME ** 18
David Lynch USA 1992
Sheryl Lee, Kyle MacLachlan, David Bowie, Chris Isaak, Harry Dean Stanton, Ray Wise, Madchen Amick, Dana Ashbrook, Phoebe Augustice, Eric Dare, Miguel Ferrer, Pamela Gidley, Heather Graham, Chris Isaak, Mira Kelly, Peggy Lipton
Feature-length prequel to this highly overrated TV series that purports to describes the events that took place during the week prior to the death of Laura Palmer, the character around whose death the TV series seemed to revolve. Full of the oddball characters and impressively imaginative sequences that are very

much the hallmark of Lynch's work, but deficient in all other respects.

DRA   129 min (ort 134 min)
VIDrel: 4-FRONT/POLYREC/GUILD; PION (LV only)
V/sur LV

TWIN PEEKS * 18
Gary Wells USA 1993
Keisha, Stacy Nichols, Jon Dough, Nikki Wilde, Trixy Tyler, Blake West, Woodie Blain, Brandy Alexandre, Sebastian, Eric Edwards
An estranged couple have been separated for two years, but are unable to forget each other and eventually get together again.

A   57 min (ort 80 min)   VIDrel: ONE   V

TWIN SISTERS ** 15
Tom Berry CANADA 1991
Frederic Forest, James Brolin, Stepfanie Kramer, Susan Almgren, Geza Kovacs, Richard Zeman, Robert Morelli, Vlasta Vrana, Larry Dane, Terry Haig, Donald Lamoureux, Michael D'Amico, Norris Domingue, Jim McNabb, Ralph Allison
A woman investigating the disappearance of her twin sister discovers that she was a high-price call-girl who had got herself involved with some very nasty criminals. Average.

DRA   89 min (ort 92 min)   VIDrel: FIRST/SONOP L/A   V

TWINKLE TWINKLE LUCKY STARS *** 18
Samo Hung HONG KONG 1985
Jackie Chan, Samo Hung, Yuen Biao, Richard Ng, Andy Lau
A sequel to MY LUCKY STARS with Jackie Chan and his team still out to smash organised crime in Hong Kong, but now finding themselves forced to cut short a holiday to deal with assassins from Thailand who are out to kill a drug boss. Fast, furious and entertaining nonsense.

MAR   89 min   VIDrel: EAST/DISC   V

TWINS *** PG
Ivan Reitman USA 1988
Arnold Schwarzenegger, Danny DeVito, Chloe Webb, Kelly Preston, Bonnie Bartlett, Marshall Bell, Trey Wilson, Hugh O'Brian
A crazy and totally implausible comedy telling of genetically designed twins who were separated at birth and only learn of each other thirty-five years later, whereupon they embark on a search for their mother. Schwarzenegger and DeVito work perfectly together, and it is their fine performances that largely sustain the film.

COM   102 min (ort 112 min)   VIDrel: CIC/SONOP   V/sur

TWINS OF EVIL * 18
John Hough UK 1971
Peter Cushing, Dennis Price, Isobel Black, Mary Collinson, Madeleine Collinson, Kathleen Byron, David Warbeck, Damien Thomas, Alex Scott, Luan Peters, Katya Keith, Harvey Hall, Roy Stewart, Maggie Wright, Inigo Jackson
Identical twins become victims of a vampire cult. Only a crucifix-wielding vampire hunter can save them before they are burnt alive by a Puritan sect. Hammer used twin Playmates from Playboy magazine, but this novel aspect adds little to a film whose only outstanding asset is its excellent sets.

Aka: GEMINI TWINS, THE; TWINS OF DRACULA; VIRGIN VAMPIRES
HOR   83 min (Cut at film release – ort 86 min)
VIDrel: MIA L/A   V

TWISTED NERVE * 18
Roy Boulting UK 1968
Hayley Mills, Hywel Bennett, Billie Whitelaw, Phyllis Calvert, Frank Finlay, Barry Foster, Salmaan Peer, Gretchen Franklin, Christian Roberts, Timothy West, Thorley Walters, Russell Napier, Clifford Cox, Robin Parkinson, Brian Peck
A psychopathic young man with a retarded brother disguises himself in order to kill his hated stepfather. A tasteless and exploitative film that almost says that mental retardation and psychopathic violence are linked. A waste of a good cast. Written by Leo Marks and Roy Boulting. The music is by Bernard Herrmann.

HOR   113 min (ort 118 min)   VIDrel: MGM/WHV L/A   V

TWISTER * 15
Michael Almereyda USA 1988
Suzy Amis, Crispin Glover, Harry Dean Stanton, Dylan McDermott, Jenny Wright, Charlaine Woodard, Lois Chiles, Tim Robbins, William Burroughs, Ralf D. Reber, Lindsay Christman, David Brown, Joyce Cavarozzi, Sarah Cochran
Patriarch Stanton is the head of a large family of spoilt brats

and disagreeable misfits, and this rather pointless film looks at a few days in their lives. Written by Almereyda, this offbeat character study has no central point of view to give it direction, nor enough wit to make it enjoyable.
COM 89 min (ort 93 min) VIDrel: FIRST/SONOP L/A V
Boa: novel Oh by Mary Robison.

TWISTER * PG
Jan de Bont USA 1996
Helen Hunt, Bill Paxton, Jami Gertz, Cary Elwes, Lois Smith, Philip Seymour Hoffman, Alan Ruck, Sean Whalen, Scott Thomson, Todd Field, Joey Slotnick, Zach Grenier, Wendle Josepher, Jeremy Davies, Gregory Sporleder, Patrick Fischler
Aptly named after its principal player, this disaster movie never pauses for breath as it tells of how a weatherman, about to serve divorce papers on his estranged wife, joins in her hunt for a massive cyclone. Along with him for the ride is his girlfriend, who would rather be somewhere else. Meanwhile, husband and wife work together to perfect a sensor device they had developed. Worth watching for the superb effects if nothing else. See also NIGHT OF THE TWISTERS.
A/AD 108 min (ort 113 min) VIDrel: CIC; PION (LV only)
V/sur LV/cav

TWO CRIPPLED HEROES * 18
HONG KONG 1980
Wang Hop, Ka Hai, Yu Heng
A woman incurs the wrath of a powerful warlord when she learns the secret source of his weapons, but is saved from death by the two title characters, one of whom has no legs and the other no arms. Apart from the unusual spectacle of disabled fighters proving their worth, this routine martial arts film has little else of originality.
MAR 92 min VIDrel: IMPENT V

TWO DAUGHTERS * (PG)
Satyajit Ray INDIA 1961
Anil Chatterjee, Chandana Bannerjee, Nriparti Chatterjee, Kagen Rathak, Gopal Roy, Soumitra Chatterjee, Aparna Das Gupta, Sita Mukherji, Gita Dey, Santosh Dutt, Mihir Chakravarty, Debi Neogy
Two separate stories in a film made to celebrate the centenary of the author's birth. The first story, "The Postmaster" sees a city boy taking up a job as the postmaster of a small village, and befriending a young orphan girl. "The Conclusion" tells of the ups and downs of a couple's arranged marriage and how they are resolved. The Indian title means Three Daughters but the third story – "Monihara", was cut for the film's release outside India.
Aka: TEEN KANYA
DRA 114 min B/W TVrel
Boa: short stories by Rabindranath Tagore.

TWO DAYS IN THE VALLEY * 18
John Herzfeld USA 1996
Danny Aiello, Greg Cruttwell, Jeff Daniels, Teri Hatcher, Glenne Headly, Peter Horton, Marsha Mason, Paul Mazursky, James Spader, Eric Stoltz, Charlize Theron, Keith Carradine, Louise Fletcher, Austin Pendleton, Kathleen Luuong
In the San Fernando Valleys, the lives off a motley collection of individuals intersect and interact when a couple of vice cops, who aspire to better things, find themselves investigating a contract killing. Reminiscent of PULP FICTION and several Altman films (notably SHORT CUTS) this is a complex piece that manages to work convincingly thanks to skilful direction and a good cast. The strong script is by Herzfeld.
DRA 104 min VIDrel: EIV V

TWO DEATHS * 18
Nicolas Roeg UK 1994
Sonia Braga, Patrick Malahide, Ion Caramitru, Seville Delofski, Nickolas Grace, Michael Gambon, Ravil Isyanov, Matthew Terdre, John Shrapnel, Karl Tessler, Lisa Orgolini, Niall Refoy, Andrew Tiernan, Rade Serbedzva
An oblique and highly idiosyncratic adaptation of the novel, set in Bucharest during the fighting that led to the overthrow of Romania's much-hated Communist dictatorship. A wealthy doctor organises a lavish dinner party at which the few guests who are able to attend gradually reveal their inner drives and obsessions. A verbose exercise in acting, extremely well made in its way, but doing little to engage our sympathies or encourage our interest.
DRA 102 min VIDrel: TART V
Boa: novel The Two Deaths of Senora Puccini by Stephen Dobyns.

TWO EVIL EYES * 18
George A. Romero/Dario Argento USA 1989
Adrienne Barbeau, Harvey Keitel, Ramy Zada, Sally Kirkland, Martin Balsam, E.G. Marshall, John Amos, Kim Hunter
Two perfectly good Poe stories are mangled in the service of two directors. In the first tale a wife plots the murder of her rich husband with the help of a doctor and in the second a jealous husband kills his musician wife and her beloved pet cat (or so he thinks).
HOR 115 min (ort 121 min) VIDrel: MED/POLYREC L/A V/sh
Boa: short stories: The Facts In The Case Of M. Valdemar and The Black Cat by Edgar Allan Poe.

TWO-FISTED LAW * U
D. Ross Lederman USA 1932
Tim McCoy, Alice Day, Tully Marshall, Wheeler Oakman, Wallace MacDonald, John Wayne, Richard Alexander, Walter Brennan
A man loses his ranch to a land-grabber who loaned him money and then stole his cattle so that he could not pay. Fortunately, he manages to make some money from gold prospecting and is able to put a stop to this villain's activities with the help of a sheriff's posse. Adequate early Western, of curiosity value mostly.
WES 59 min B/W VIDrel: VCC/DISC/COLUM V

TWO-FISTED TALES * (18)
Richard Donner/Tom Holland/Robert Zemeckis USA 1991
William Sadlers, Neil Gray Giuntoli, David Morse, Thomas E. Duffy, Roderick Cook, Raymond J. Barry, Brad Pitt, Jack Kehler, Michele Brown, Alva L. Petway, Kirk Douglas, Eric Douglas, Dan Ackroyd, Lance Henriksen, Dominick Morra
A trio of short action movies by three directors. "Showdown" is a Western dealing with outlaws and shootouts, "King Of The Road" follows the plight of a cop with a dark secret who has to relive the events of twenty-seven years ago. Finally, in "Yellow" tells of a man forced to enlist during WW1 by his father, but for whom the horror of the trenches proves to be too much. A mixed bag of films, watchable but not especially memorable.
A/AD 89 min SATrel: MOVIE CHANNEL

TWO FOR THE ROAD * PG
Stanley Donen UK 1967
Audrey Hepburn, Albert Finney, Eleanor Bron, William Daniels, Nadia Gray, Claude Dauphin, Jacqueline Bisset
On a motoring holiday through France, an architect and his wife reminisce over their 12 years of marriage, doing their best to come to terms with what each has come to dislike about the other. An absorbing comedy-drama whose rather trite plot serves as little more than a backdrop to splendid performances from the leads. The score is by Henry Mancini.
DRA 111 min (ort 113 min) VIDrel: 20TH/TECH V/h

TWO-HUNDRED MOTELS * 18
Frank Zappa/Tony Palmer USA 1971
Frank Zappa, Ringo Starr, Theodore Bikel, The Mothers of Invention, Keith Moon, Flo and Eddie, Janet Ferguson, Lucy Offerall, Don Preston, Dick Barber, Jimmy Carl Black, Pamela Miller, Martin Lickert, Don Preston
Rambling musical fantasy featuring Frank Zappa of rock group fame, and his associates, that does not really work as a movie, as no discipline is applied to the filming. If you like the music of Zappa, this one is worth hearing if not worth seeing.
Aka: 200 MOTELS; FRANK ZAPPA: 200 MOTELS
MUS 95 min (ort 98 min) VIDrel: MGM/WHV L/A V/h

TWO JAKES, THE * 15
Jack Nicholson USA 1990
Jack Nicholson, Harvey Keitel, Meg Tilly, Madeliene Stowe, Eli Wallach, Ruben Blades, Frederic Forrest, David Keith, Richard Farnsworth, Tracey Walter, James Hong, Joe Mantell, Perry Lopez, Rebecca Broussard, Luana Anders
This belated follow-up to CHINATOWN is set in 1948 L.A., with speculation in oil rather than water the central theme, and adultery instead of incest the subsidiary issue. A private eye sets out to investigate some shady land deals and crafty intrigues. An over-complex and ponderous saga, barely watchable, mainly thanks to the excellent photography of Vilmos Zsigmond. Faye Dunaway makes a fleeting (and welcome) appearance. Broussard was briefly Mrs Nicholson.
THR 132 min (ort 138 min) wScrn
VIDrel: 4-FRONT/POLYREC/CIC V/sur

TWO LEFT FEET *   15
Roy Ward Baker UK   1963
Michael Crawford, Nyree Dawn Porter, Julia Foster, David Hemmings, Bernard Lee, Dilys Watling, David Lodge
A callow and inexperienced teenager embarks on an affair with a waitress, has a fight with her fiance but eventually achieves some maturity. A rather pointless stab at a comedy with sexual overtones, that looked dated at the time.
COM 89 min (ort 93 min) B/W VIDrel: WHV V
Boa: novel In My Solitude by David Stuart Leslie.

TWO MOON JUNCTION *   18
Zalman King USA   1988
Sherilyn Fenn, Richard Tyson, Louise Fletcher, Burl Ives, Kristy McNichol, Millie Perkins, Don Galloway, Herve Villechaize, Dabbs Greer, Screamin' Jay Hawkins, Martin Hewitt, Milla, Nicole Rosselle, Kerry Remsen, Chris Pederson
Just two weeks before she is due to be married, a well-bred Southern girl falls for a handsome carnival worker and runs off with him. Her rich granny and the local sheriff conspire to stop them. A dumb romance with a lot of humour on offer, but most of it of the unintentional kind. Followed by RETURN TO TWO MOON JUNCTION.
DRA 100 min (ort 104 min) VIDrel: 20TH/TECH L/A V

TWO MRS CARROLLS, THE ***   PG
Peter Godfrey USA   1945 (released 1947)
Humphrey Bogart, Barbara Stanwyck, Alexis Smith, Nigel Bruce, Isobel Elsom, Ann Carter, Creighton Hale, Peter Godfrey, Pat O'Moore, Anita Bolster, Colin Campbell, Leyland Hodgson, Barry Bernard
Bogart stars as a psychopathic artist who paints his wives as the Angel of Death, then disposes of them with poisoned milk. This disjointed film climaxes with a fair degree of tension, but is hampered by implausible casting of the star. Written by Thomas Job.
THR 95 min (ort 99 min) B/W VIDrel: WHV V
Boa: play by Martin Vale.

TWO MRS GRENVILLES, THE: PARTS 1 AND 2 ***   15
John Erman USA   1987
Ann-Margret, Claudette Colbert, Stephen Collins, John Rubinstein, Elizabeth Ashley, Alan Oppenheimer, Margaret Courtney, Delena Kidd, Penny Fuller, Sian Philips, Peter Eyre, Sam Wanamaker, Kate Harper, Jana Shelden, Toria Fuller
A chorus girl with a dubious past, marries a member of a wealthy and influential family, that's headed by a redoubtable matriarch who cannot accept her for her humble origins. Scandal erupts however, when the wife shoots her husband in what she claims is a tragic accident. Based on a true case, this fictionalised account offers great scope for a moving performance by Ann-Margret, in a restrained and well-crafted drama.
DRA 190 min (2 cassettes) mTV VIDrel: WHV V/h
Boa: novel by Dominick Dunne.

TWO MUCH TROUBLE *   (PG)
Michael James MacDonald USA   1994
Beverly D'Angelo, Ed Begley Jr., Carol Kane, Phil Hartman, Nell Carter, Sean Whalen, Steve Landesberg, Mink Stole, Brady Bluhm, Rachel Duncan, Tim Bagley, Eric Allan Kramer, Sheila Traviss, Christopher Darga, William G. Schilling
A wealthy upper-class couple can spare no time for their badly behaved twin son and daughter, and have left their upbringing to a succession of nannies. When this post becomes vacant, the kids conspire to get it filled by a lazy and slovenly female who is just out of jail. At first, this woman sees her job as a way of making easy money, but begins to develop an emotional attachment to the kids. A fine cast do what they can in a moronic PROBLEM CHILD-style comedy.
Aka: CRAZYSITTER, THE
DRA 88 min (ort 100 min) SATrel: MOVIE CHANNEL

TWO MULES FOR SISTER SARA **   15
Don Siegel USA   1969
Clint Eastwood, Shirley MacLaine, Manolo Fabregas, Alberto Morin, Armando Silvestre, John Kelly, Enrique Lucero, Jose Chavez, David Estuardo, Jose Chavez, Ada Carrasco, Pancho Cordova, Pedro Galvan, Jose Angel Espinosa, Xavier Marc
A tough American mercenary in Mexico meets a prostitute posing as a nun, and gradually becomes involved with her and the struggle against the French forces, who occupied the country in the 1860s. Together they embark on a daring scheme to seize control of the enemy's garrison, in this sporadically

entertaining but seriously overlong story. The script is by Albert Maltz.
WES 109 min (ort 116 min)
VIDrel: 4-FRONT/POLYREC/CIC V
Boa: story by Budd Boetticher.

TWO OF A KIND *   PG
John Herzfield USA   1983
John Travolta, Olivia Newton-John, Charles Durning, Beatrice Straight, Scatman Crothers, Castulo Guerra, Oliver Reed, James Stephens, Richard Bright, Vincent Bufano, Toni Kalem, Jack Kehoe, Gene Hackman (voice only)
A fantasy about God's angels persuading him to give humanity another chance and sparing the Earth from a second flood, provided they can find two people prepared to make a great sacrifice for each other. A silly yarn that lacks the light touch of those 1930s and 1940s films of the same genre, such as "Here Comes Mr Jordan". Written by Herzfield.
COM 84 min (ort 87 min) VIDrel: 20TH/TECH V/sur

TWO OF US ***   15
Roger Tonge USA   1986
Jason Rush, Lee Whitlock, Jenny Jay, Zoe Nathenson, Kathy Burke
A look at the lives of a couple of male students, one of whom has already left school whilst family pressure has encouraged the other to stay on. Eventually both lads decide to run away together and take off for the coast, where they embark on a love affair. Written by Leslie Stewart, this gay film benefits from a strong script and an attempt to bring some depth to the main characters.
DRA 58 min VIDrel: DTK/RTM V/sh

TWO OR THREE THINGS I KNOW ABOUT HER ***   15
Jean-Luc Godard FRANCE   1966
Marina Vlady, Anne Duperey, Roger Montsoret, Raoul Levy, Jean Narboni
This sharp and potent satire on commercialism takes a look at the life of a bored housewife who, once a week, works the city centre as a prostitute to earn money which she then spends on various household items. The film also serves as a political statement, and sympathises with the plight of the middle-classes, who through the force of advertising, so often find themselves living beyond their means.
Aka: DEUX OU TROIS CHOSES QUE JE SAIS D'ELLE
DRA 83 min (ort 95 min) wScrn VIDrel: CONNO/RTM V

TWO RODE TOGETHER **   PG
John Ford USA   1961
James Stewart, Richard Widmark, Shirley Jones, Linda Crystal, Andy Devine, John McIntire, Paul Birch, Willis Bouchey, Henry Brandon, Harry Carey Jr, Ken Curtis, Olive Carey, Chet Douglas, Annelle Hayes, David Kent, Anna Lee
A lawman and a cavalry officer form an uneasy alliance when they set out to rescue a group of settlers kidnapped by Comanches. A competent Western sustained by convincing performances rather than any intrinsic merits within the plot.
WES 104 min (ort 109 min) VIDrel: VCC L/A V

2001: A SPACE ODYSSEY ****   U
Stanley Kubrick UK   1968
Keir Dullea, William Sylvester, Gary Lockwood, Leonard Rossiter, Robert Beatty, Daniel Richter, Margaret Tyzack, Frank Miller, Alan Gifford, Penny Brahms, Edwina Carroll, Sean Sullivan, Douglas Rain (voice of HAL 9000)
A brilliant, innovative but opaque film, in which Man's evolution is seen as being shaped by alien forces which then waited until Mankind achieved space travel. The Clarke/Kubrick script has some memorable ideas, notably "HAL", a super-computer. Richard Strauss' opening to "Also Spake Zarathustra" and music by Ligeti, Johann Strauss Jr and others is used effectively. The film was cut by 17 min after its premiere. See also 2010. AA: Effects (Kubrick).
FAN 133 min (ort 141 min) VIDrel: MGM/WHV V/sh
Boa: short story The Sentinel by Arthur C. Clarke.

2010 ***   PG
Peter Hyams USA   1984
Roy Scheider, John Lithgow, Helen Mirren, Bob Balaban, Keir Dullea, Dana Elcar, James McEachin, Madolyn Smith, Elya Baskin, Savely Kramarov, Natasha Schneider, Oleg Rudnik, S. Newton Anderson, Douglas Rain (voice of HAL 9000)
Interesting sequel to Kubrick's famous original which builds on some of the ideas explored in the earlier film. A combined

US/USSR mission is launched nine years after Discovery's voyage, to find out what happened to the first mission. By way of a subplot, both participating countries are drawing progressively closer to nuclear war. Good effects, and a sound if unimaginative story, lead to a contrived ending. Written by Hyams.
Aka: 20 THE YEAR WE MAKE CONTACT
FAN   111 min (ort 114 min)   VIDrel: MGM/WHV   V/sur
Boa: novel Odyssey Two by Arthur C. Clarke.

TWO THOUSAND MANIACS *   18
Herschell Gordon Lewis USA   1964
Connie Mason, Thomas Wood, Jeffrey Archer, Shelby Livingston, Jerome Eden, Ben Moore, Vincent Santo, Gary Bakeman, Mark Douglas, Michael Korb, Yvonne Gilbert, Linda Cochran, Andy Wilson, The Pleasant Valley Boys
A legendary low-grade horror tale, telling of a sinister Southern ghost town that comes to life 100 years after it was sacked in the Civil War, and takes a grisly revenge on six young swingers who find themselves trapped there. As poorly supplied with acting talent as it is well supplied with gore. Better than "Blood Feast", an earlier Lewis film, but that's not saying much.
Aka: 2,000 MANIACS
HOR   80 min Cut (4 min 27 sec – ort 88 min)
VIDrel: IMPENT   V

TWO THOUSAND WOMEN ***   PG
Frank Launder UK   1944
Phyllis Calvert, Flora Robson, Patricia Roc, Renee Houston, Anne Crawford, Jean Kent, Thora Hird, Dulcie Gray, Reginald Purcell, James McKechnie, Carl Jaffe, Robert Arden, Muriel Aked, Kathleen Boutall, Hilda Campbell-Russell
Two English women interned by the Nazis in a camp at a French hotel during WW2 form a resistance group, and get their chance to do their bit for old England when they find three British pilots hiding out in the camp grounds. A competent wartime flag-waver with a few lighthearted touches.
Aka: 2,000 WOMEN; HOUSE OF A THOUSAND WOMEN
WAR   93 min (ort 97 min) B/W   VIDrel: CONNO/RTM
L/A V

TWO WAY STRETCH ***   U
Robert Day UK   1960
Peter Sellers, Lionel Jeffries, Wilfrid Hyde-White, Liz Fraser, Beryl Reid, Maurice Denham, Bernard Cribbins, David Lodge, Irene Handl, Thorley Walters, Walter Hudd, George Woodbridge, Cyril Chamberlain, John Wood
Amusing prison comedy with a certain period charm, that is blessed by the happy notion of prisoners devising a means of absenting themselves temporarily, in order to commit a robbery for which they will have a perfect alibi. Some of the plot is borrowed from the 1938 film "Convict 99".
COM   83 min (ort 87 min) B/W   VIDrel: WHV   V/h

TWO WONDROUS TIGERS ***   15
Cheung Sum HONG KONG   1980
Philip Ko, John Chang, Tiger Young, Wilson Tong, Yang Pan Pan, Charlie Chan, Mung Kwun Ha, Kim Woo-Suk, Lee Suk-Koo, Jang Jung-Kuok, Ahn Jin-Soo
A young man tries to kidnap a girl he wants to marry, but is opposed and told that only if he can defeat the girl, her sister and her brother will he gain her hand. When these three beat him, he posts a large reward to find some champions to fight on his behalf. A fast-paced action tale set in Nationalist China.
MAR   88 min Cut (38 sec)   VIDrel: IMPENT   V

TWO WORLDS OF JENNY LOGAN, THE ***   PG
Frank De Felitta USA   1979
Lindsay Wagner, Marc Singer, Alan Feinstein, Linda Gray, Henry Wilcoxon, Joan Darling, Irene Tedrow, Peter Hobbs, Constance McCashin, Charles Thomas Murphy, Allen Williams, Pat Corley, John Hawker, Gloria Stuart
A woman and her husband move to their new home and she finds a 19th century dress in the attic. Putting it on transports her back to that period, where she finds she has a life in that time too, and where two bitter rivals vie for her affections. She returns to the present, but with research learns of an "imminent" tragedy in the past, and ultimately leaves her husband in an attempt to avert it. An unusual and entertaining fantasy.
FAN   92 min (ort 100 min) mTV   VIDrel: L/A   V
Boa: novel Second Sight by David Williams.

TWOGETHER **   18
Andrew Chiaramonte USA   1992
Nick Cassavetes, Brena Bakke, Jeremy Piven, Jim Beaver, Tom Dugan, Damian London, William Bumiller, Jennifer Bussey, Jerry Bossard, Deborah Driggs, Lauren Grey, Christian Bocher, Margaret Muse, Stanley Grover, Thomas Knickerbocker
A look at relationships in the 1990s, with a Los Angeles struggling artist enjoying a weekend of drunken abandon with a perfect stranger, following which both individuals learn that they have somehow become married.
DRA   120 min   VIDrel: COLUM/SONOP   V/h

TYCOON **   U
Richard Wallace USA   1947
John Wayne, Anthony Quinn, Cedric Hardwicke, Laraine Day, Judith Anderson, James Gleason, Grant Withers, Paul Fix, Fernando Alvarado, Harry Woods, Michael Harvey, Charles Trowbridge, Martin Garralaga, Sam Lufkin, Wayne McCoy
An engineer engaged to build a railway tunnel in the Andes, comes into conflict with his boss when he falls for the latter's daughter. An overlong but mildly enjoyable tale. Written by Borden Chase and John Twist.
DRA   123 min (ort 129 min)   VIDrel: VCC L/A   V
Boa: novel by C.E. Scoggins.

TYSON: THE TRUE STORY ***   15
Uli (Ulrich) Edel USA   1994
Michael Jai White, George C. Scott, Paul Winfield, Malcolm-Jamal Warner, Tony Lo Bianco, James B. Sikking, Georg Stanford Brown, Clark Gregg, Joe Santos, Charles Napier, Tico Wells, Holt McCallany, Kristen Wilson, Rogal Hanley
Competent biopic on the life of boxer Mike Tyson that describes his troubled youth, his rise to fame and the circumstances that led up to his dubious conviction for rape. Told with skill and care, this is a well acted biopic, and White in the title role is both impressive and convincing.
Aka: TYSON
DRA   105 min mCab   VIDrel: WHV   V/sur
Boa: book Fire and Fear by Jose Torres.

U

U.F.O. CAFE **   PG
Paul Schneider USA   1990
Richard Mulligan, Beau Bridges, Barbara Barrie, Paul Dooley, James McEachin, John Furey, Betsy Randle, Michael Patrick Carter, Dion Anderson, Gary Bayer, Ronnie Claire Edwards, Troy Evans, Charles Stransky, Janni Brenn, James Lashly
The kindly owner of a smalltown hardware store befriends a travelling salesman, who eventually reveals that he's really an alien. A sentimental, sugary, Disney-style comedy with a lame plot, made watchable if not especially rewarding, thanks largely to the excellent efforts of Mulligan and Bridges.
Aka: GUESS WHO'S COMING FOR CHRISTMAS?
COM   96 min   VIDrel: ITC/HIFLI L/A   V

U.F.O. – INVASION U.F.O. **   PG
Gerry Anderson/David Lane/David Tomblin UK   1972
Ed Bishop, George Sewell, Michael Billington, Wanda Ventham, Vladek Sheybal, Gabrielle Drake, Antonia Ellis, Dolores Mantez, Peter Gordeno, Harry Baird, Grant Taylor, Jeremy Wilkin
A top-secret organisation is formed to protect Earth from an alien invasion. A mediocre feature compiled from TV series of similar virtues.
Aka: INVASION U.F.O.
FAN   95 min mTV   VIDrel: ITC/POLYREC L/A   V

U.F.O. – THE MOVIE   18
Tony Dow UK   1993
Roy "Chubby" Brown, Sara Stockbridge, Roger Lloyd Pack, Sue Lloyd, Elizabeth Hickling, Jackie Downey, Shirley Anne Field, Kiran Shah, Kenny Baker, Rushy Goffe, Anthony Georghiou, James Culshaw, Paul Sarony, Ben Aris, Laura Jackson
Brown is kidnapped by a man-hating female lesbian – lavatory humour at its crudest revolving around bodily functions. Unbelievable. If there were minus stars this one would get five. Just about watchable, at a few minutes a time. This gem was the comedian's film debut.
Aka: U.FO.
COM   75 min (ort 79 min)   VIDrel: POLY/POLYREC   V

UFORIA *** PG
John Binder USA 1981
Cindy Williams, Harry Dean Stanton, Fred Ward, Harry Carey Jr, Beverly Hope Atkinson, Darrell Larson, Robert Gray, Peggy McCay, Ted Harris, Diane Diefendorf, Hank Worden, Alan Beckwith, Andrew Winner, Pamela Lamont
Gentle look at a collection of Southwest misfits, focusing on a checkout cashier who believes in UFOs and is eagerly awaiting contact with visiting aliens.
COM 90 min (ort 92 min) VIDrel: CIC/SONOP L/A V

UGLY AMERICAN, THE *** PG
George Englund USA 1962
Marlon Brando, Eiji Okada, Sandra Church, Pat Hingle, Arthur Hill, Jocelyn Brando, Kurkit Pramoj, Judson Pratt, Reiko Sato, George Shibata, Judson Laire, Philip Ober, Yee Tak Yip, Stefan Schnabel, Pock Rock Ann, John Day
A publisher is appointed ambassador to a small Asian nation in the throes of a Communist insurgency, and his arrival leads to an escalation in the fighting. A sincere but naive and verbose attempt to bring to the screen the complexities of the novel, and its thinly disguised references to the Vietnam War. Brando is excellent, in an otherwise unmemorable film which neither enlightens nor entertains. The script is by Stewart Stern.
DRA 115 min (ort 120 min) VIDrel: CIC/SONOP V
Boa: novel by William Lederer.

UGLY, DIRTY AND BAD *** 15
Ettore Scola ITALY 1976
Nino Manfredi, Maria Luisa Santella, Francesco Anniballi, Maria Bosco, Giselda Castrini, Alfred D'Appolito
An impoverished family lives in a shanty town on the edge of Rome. Where another film-maker might give us a portrait of their poverty and struggles, this film is instead a comic celebration of their individuality, the central figure being the miserly and irascible patriarch who spends most of his time in conflict with the rest of his family. A zestful and life affirming work, it won the Best Director award at Cannes.
Aka: BRUTTI, SPORCHI E CATTIVI; DOWN AND DIRTY; UGLY, DIRTY AND MEAN
COM 115 min VIDrel: ARROW/RTM V

U.H.F. ** PG
Jay Levey USA 1989
Victoria Jackson, Weird Al Yankovic, Kevin McCarthy, Michael Richards, David Bowe, Stanley Brock, Anthony Geary, Trinidad Silva, Gedde Watanabe, Billy Barty, John Paragon, Fran Drescher, Sue Ann Langdon, Emo Philips, Jay Levey
Taking a break from his parodies of music videos, Yankovic gets a starring feature film debut, when he plays the new manager of Channel 62, a UHF station with the lowest ratings in the country. The station soon climbs the ratings ladder with its wacky programming, and this unfocused film provides some very funny parodies and a good role for Richards as an ex-janitor turned kiddie-show host. The disjointed script is by Yankovic and Levey.
COM 93 min (ort 97 min) VIDrel: VISVID/POLYREC V

ULTERIOR MOTIVES ** 18
James Beckett USA 1992
Thomas Ian Griffith, Mary Page Keller, Joe Yamanaka, Ellen Crawford, Tyra Ferrell, M.C. Gainey, Ken Howard, Craig Shugart, Dove Efron, Cameron Michael Erwin, Hayward Nishioka, Bill Ryusaki, Masami Saito, Toshishio Obata, Jeff Imada
A private detective agrees to protect a woman reporter from the Japanese yakuza and finds himself plunged into the usual web of deception and violence. A tough, assembly-line actioner that offers some undemanding entertainment and some entertaining martial arts battles.
A/AD 91 min VIDrel: MIA/DISC/FIRST V

ULTIMATE BETRAYAL, THE *** 15
Donald Wyre USA 1993
Marlo Thomas, Mel Harris, Eileen Heckart, Kathryn Dowling, Henry Czerny, Ally Sheedy, Ally Goodhand, David B. Nichols, Joanne Vannicola, Justin Louis, Kim Schraner, Bret Person, Valerie Valerie Buhagdar, Nigel Bennett
The true story of an abused daughter and her long fight to bring her father to justice for the years of mental and physical torture he inflicted on her and her two sisters. Quite a disturbing story, it is finely acted, most especially by Czerny who is totally convincing as the abusive parent.
DRA 93 min mTV VIDrel: ODY/SONOP V/sh

ULTIMATE NINJA, THE *** 18
Godfrey Ho HONG KONG 1986
Stuart Smith, Bruce Baron, Timothy Nugent, Sorapong Chatri, Anne Aswatep, Pedro Ernyes
A twenty-year-old feud between an evil tyrant and the benevolent leader of a village, results in the death of the latter and the village falling under the control of ruthless thugs. The eldest child of the murdered leader returns one day to have his revenge. A kind of Ninja version of YOJIMBO with the hero playing off the two clans who now battle for supremacy.
MAR 83 min Cut (38 sec – ort 92 min) VIDrel: IMPENT V

ULTIMATE TEACHER ** 15
JAPAN
Comedy animation set at Earth's toughest school whose extremely unruly pupils finally meet their match in the form of a bio-engineered teacher. See the CLASS OF 1999 films for live-action (and a good deal more violent) films on this theme.
ANIM 55 min (ort 60 min) dubbed
VIDrel: MANGA/SONOP V/sh

ULTIMATE WARRIOR, THE ** 15
Robert Clouse USA 1975
Yul Brynner, Max Von Sydow, Joanna Miles, William Smith, Richard Kelton, Stephen McHattie, Lane Bradbury, Darrell Zwerling, Mel Novak, Nate Esformes, Mickey Caruso, Gray Johnson, Susan Keener, Stevie Meyers, Fred Siyter
In the 21st century, scattered bands eke out a meagre existence following a worldwide ecological disaster. One group has developed resistant plants that may one day replenish the Earth, but is constantly threatened by another. The arrival of an expert fighter, who works as a kind of roving mercenary, changes all that. Good ideas are badly let down by stilted dialogue and poor acting. Written by Clouse.
FAN 94 min VIDrel: MGM/WHV L/A V

ULTRA ** 18
Ricky Tognazzi ITALY 1990
Ricky Memphis, Gianmarco Tognazzi, Claudio Amendola, Giuppy Izzo
The story of loyalty and fanaticism set in the world of Italian football supporters, with the leader of a particularly nasty gang coming out of prison just in time to discover that his leadership is being challenged by his second-in-command, who has embarked on an affair with his girlfriend. An uneven and irritatingly moralistic film, it never really gets to grips with its subject matter; the climax is suitably brutal. See also I.D. and THE FIRM.
Aka: HOOLIGANS
DRA 90 min VIDrel: ARTPRO/RTM V

ULTRAMAN: THE ALIEN INVASION ** PG
Andrew Prowse AUSTRALIA/JAPAN 1992
Dore Krause, Gia Carides, Ralph Cotterill, Grace Parr, Lloyd Morris, Rick Adams, Steve Adams, Mike REad, Johnny Halliday
Earth is invaded by fearsome alien monsters and a special team is formed to combat them. Ultraman lends his help but has to rely on an astronaut recently returned from a space mission, who is hiding a crucial secret. Endearing space opera nonsense.
FAN 94 min VIDrel: MANGA/SONOP V/sur

ULTRAVIOLET ** 18
Mark Griffiths USA 1991
Esai Morales, Patricia Healy, Stephen Meadows, Sorells Pickard, Louise Baker
A couple motoring through Death Valley rescue a man they find half-dead on the road and he repays their kindness by subjecting them to a reign of terror. Average chase thriller.
THR 90 min VIDrel: CIC/SONOP L/A V

ULYSSES *** 18
Joseph Strick UK 1967
Maurice Roeves, Milo O'Shea, Barbara Jefford, T.P. McKenna, Anna Manahan, Maureen Potter, Martin Dempsey, Graham Lines, Joe Lynch, Fionnuala Flanagan, Maire Hastings, Geoffrey Golden, Dave Kelly, Maureen Toal, O.Z. Whithead
An attempt to capture the feeling if not the sweep of Joyce's mammoth story of Dublin life, hampered by the lack of any visual images able to do the fine prose justice. O'Shea is wonderful as Bloom, and skilfully brings to life this central character. The film's release generated controversy and it was banned in parts of the UK, on account of an expected lewdness that never

materialised, the ribald language and a naked bottom, which
was all there was to upset the prudes.
DRA   132 min B/W   VIDrel: LUMI/SPEAR L/A   V
Boa: novel by James Joyce.

ULYSSES' GAZE ****   18
Theo Angelopoulos FRANCE/GREECE/ITALY   1995
*Harvey Keitel, Maia Morgenstern, Erland Josephson, Thanassis
Vengos, Yorgos Michalakopoulos, Dora Volanaki, Mania
Papadimitriou, Angel Ivanof, Ljuba Tadic, Gert Llanaj, Agni Vlahou,
Giannis Zavradinos, Vaguelis Kazan, Dimitris Kaberidis*
During the 1990s a film-maker travels across the Balkans in
search of a piece of film he can incorporate into a documentary
about the early Balkan cinematographers. In the skilled hands
of the director, this remarkable film becomes both a personal
journey for the central character and a leisurely but enthralling
odyssey through the political and cultural history of the region,
visually striking, nostalgic and memorable.
Aka: REGARD D'ULYSSE; TO VLEMMA TOU ODYSSEA
DRA   170 min (ort 176 min) wScrn   VIDrel: ARTIF/20TH
V/sh

ULZANA'S RAID ***   18
Robert Aldrich USA   1972
*Burt Lancaster, Bruce Davison, Jorge Luke, Richard Jaeckel, Lloyd
Bochner, Joaquin Martinez, Karl Swenson, Douglas Walton, Dran
Hamilton, John Pearce, Gladys Holland, Margaret Fairchild, Aimee
Eccles, Richard Bull, Otto Reichow*
The story of a violent conflict between Indians and the US
Cavalry with Lancaster as an experienced Indian fighter being
sent out with a raw officer and an Indian scout, to destroy a
marauding Apache band that is terrorising the citizens. Written
by Alan Sharp this violent and bloody film benefits from a taut
script and has Lancaster giving one of his best performances.
WES   96 min Cut (45 sec – ort 103 min)
VIDrel: 4-FRONT/POLYREC/CIC L/A   V/h

UMBERTO D ***   PG
Vittorio De Sica ITALY   1952
*Carlo Battisti, Maria Pia Casilio, Lina Gennari, Alberto Albani
Barbieri, Elena Rea, Ileana Simova, Memo Carotenuto*
An unsentimental examination of the plight of a retired civil
servant who is unable to pay his rent and cannot bear to be
parted from his beloved dog. He comes up with a plan to kill
himself and his pet but fails and is forced to go on living, accepting whatever fate has in store. A bleak and downbeat movie
that took three years to get a showing in the USA, and was seen
by the Italian authorities as painting too black a picture of Italy.
DRA   84 min (ort 125 min) B/W   VIDrel: FABFIL/SPEAR
V

UMBRELLAS OF CHERBOURG, THE ***   PG
Jacques Demy FRANCE/WEST GERMANY   1963
*Catherine Deneuve, Anne Vernon, Nino Castelnuovo, Marc Michel,
Ellen Farner, Mireille Perrey, Jean Champion, Jean Caden, Harald
Wolff, Dorothee Blank, Jean-Pierre Dorat, Bernard Fradet, Michel
Benoist, Philippe Dumat, Jane Carat*
Two lovers are parted by circumstances and only reunited for
a brief moment some years later when they are both married to
other people, which is the way they remain as the film ends. A
charming little musical without a single word of spoken
dialogue (everything is sung instead) that despite this limitation
manages to tell a strong and moving story. Screenplay is by
Demy.
Aka: DIE REGENSCHIRME VON CHERBOURG; LES PARAPLUIES DE CHER-
BOURG
MUS   90 min   CINrel

UN LUGAR EN EL MUNDO **   (PG)
Adolfo Aristarain ARGENTINA   1992
*Joe Sacristan, Federico Luppi, Leonor Benedetto, Cecilia Roth, Rodolfo
Ranni, Hugo Arana, Gaston Batyi, Mario Alarcon, Lorena Del Rio*
A young man returns to his home-town in Argentina and remi-
nisces about his early life there.
DRA   118 min   SATrel: MOVIE CHANNEL

UNABOMBER: A TRUE STORY **   15
Jon Purdy USA   1996
Dean Stockwell, Robert Hays, Tobin Bell
One of those fact-based tales, this one is based on the exploits
of the self-styled "Unabomber", who took his grievance against
the Federal government to irrational lengths, indulging in a
bombing campaign that lasted for many years. His final capture

only resulted after years of nationwide co-operation between the
various law enforcement agencies. Fair.
DRA   90 min   VIDrel: ODY/SONOP   V/sh

UNBEARABLE LIGHTNESS OF BEING, THE ***   18
Philip Kaufman USA   1987
*Daniel Day-Lewis, Juliette Binoche, Lena Olin, Erland Josephson,
Daniel Olbrychski, Derek De Lint, Paul Landovsky, Donald Moffat,
Stellan Skarsgard, Tomek Bork, Bruce Meyers, Pavel Slaby, Pascale
Kalensky, Jacques Ciron*
This intelligent and absorbing adaptation of Kundera's novel
tells of a young Czech doctor of the 1960s who is more interested
in women than in politics, but finds himself caught up in
his country's turmoil. A joyful, exuberant and zestful film, beau-
tifully acted and employing the talents of top photographer
Sven Nykvist. The script is by Kaufman and Jean-Claude
Carriere.
DRA   165 min (ort 172 min)   VIDrel: 20TH/TECH   V/sur
Boa: novel by Milan Kundera.

UNBELIEVABLE TRUTH, THE ***   15
Hal Hartley USA   1989
*Adrienne Shelly, Robert Burke, Christopher Cooke, Julia McNeal,
Gary Sauer, Mark Bailey, Katherine Mayfield, Matt
Malloy, Edie Falco, Jeff Howard, Kelly Reichardt, Ross Turner, Paul
Schultze, Mike Brady, Bill Sage*
A witty satire about small-town life, written by Hartley, that
tells of a young man who returns home after a spell in prison,
where he plans to spend some time reflecting on life in
general and his own life in particular. However, his return is
greeted with mixed feelings among the locals. An offbeat and
rewarding black comedy, with funny dialogue, comic characters and sharp little insights. Amazingly, it was all shot in
eleven days.
COM   86 min (ort 96 min)   VIDrel: ELPIC/POLYREC   V

UNBORN, THE *   18
Rodman Flender USA   1991
*Brooke Adams, Jeff Hayenga, James Karen, Jane Cameron, K. Callan,
Kathy Cameron, Kathy Griffin*
A young wife with a history of miscarriages gets help from an
unorthodox doctor, but ultimately learns that his bizarre
methods tend to produce homicidal progeny. A variant on a
theme handled more convincingly in IT'S ALIVE! that makes up
in needless gore what it lacks in plotting. A disagreeable and
slackly-narrated dud, with a few scary sequences. A sequel
followed.
HOR   81 min (ort 85 min)   VIDrel: 20VIS/SONOP   V/sh

UNCANNY, THE **   18
Denis Heroux CANADA/UK   1977
*Peter Cushing, Joan Greenwood, Donald Pleasence, Ray Milland,
Roland Culver, Susan Penhaligon, Simon Williams, Alexandra
Stewart, Chloe Franks, Katrina Holden, Donald Pilon, Samantha
Eggar, John Vernon, Sean McCann, Jean Leclerc*
Three rather foolish tales, based around author Cushing explain-
ing to Milland his obsessive fear that cats plot against and
control humanity. Apart from one rather funny moment when
a cat really does "get someone's tongue", this is a superficial and
insipid effort. The script is by Michael Parry.
HOR   84 min   VIDrel: VCC L/A   V

UNCERTAIN GLORY *   U
Raoul Walsh USA   1944
*Errol Flynn, Paul Lukas, Jean Sullivan, Lucile Watson, Faye Emerson,
James Flavin, Douglass Dumbrille, Dennis Hoey, Sheldon Leonard*
In Occupied France a disreputable philanderer pretends to be a
saboteur, sacrificing himself for the benefit of his country.
Cliched WW2 heroics with a lack of action and an uncertain
script.
DRA   97 min (ort 102 min) B/W   VIDrel: WHV   V

UNCHAINED ***   15
Daniel Mann USA   1987
*Val Kilmer, Charles Durning, Sonia Braga, Kyra Sedgewick, William
Sanderson, James Keach, Clancy Brown, Elisha Cook, Taj Mahal, Paul
Benjamin, Bill Bolender, Esther Benson, Burt Conway, Charles
Carroll, Ransom Andrews*
A fictionalised account loosely based on the life of Robert Burns,
who was sentenced to a chain gang in Georgia, after becoming
an unwilling accomplice during a robbery. After escaping, he
started a new life but was recaptured. He escaped once more,
and eventually wrote a book that helped expose the brutality of

the chain gang system. A solid and absorbing tale, made once before as the classic "I Am A Fugitive From A Chain Gang".
Aka: MAN WHO BROKE 1,000 CHAINS, THE
DRA 91 min (ort 113 min) VIDrel: GUILD/SONOP V
Boa: Robert Elliott Burns.

UNCLE BUCK **   15
John Hughes USA   1989
John Candy, Amy Madigan, Jean Louisa Kelly, Macaulay Culkin, Gaby Hoffman, Elaine Bromka, Garrett M. Brown, Laurie Metcalf, Jay Underwood, Mike Starr, Brian Tarantina, Suzanne Shepherd, William Windom, Dennis Cockrum
A untidy slob is called up by his brother to look after his two kids for a few days so the latter can attend a funeral. His inability to cook and clean generates an obvious slapstick humour, but his unexpected abilities as a loving uncle inject a note of warmth. Writer-director Hughes provides an uneven and sentimental script that blends light and dark humour; the flirtatious behaviour of Candy's sullen niece strikes a rather sour note. A TV series followed.
COM 95 min (ort 100 min)
VIDrel: 4-FRONT/POLYREC/CIC V/sur

UNCLE SAM *   18
William Lustig USA   1995
Timothy Bottoms, Christopher Ogden, David "Shark" Frahlich, Isaac Hayes, William Smith, Robert Forster, Bo Hopkins
The body of a soldier killed in the Gulf War is brought back to the States, where it returns to life, mounting a campaign of vengeance against those it sees as being disrespectful to American flag. The witty script was the work of Larry Cohen, but it needed a director with a far lighter touch to bring out its comic potential, and this one is not quite the spoof it should have been.
HOR 86 min VIDrel: BMGVID/BMGREC V

UNCOMMON VALOR ***   15
Ted Kotcheff USA   1983
Gene Hackman, Robert Stack, Fred Ward, Randall "Tex" Cobb, Reb Brown, Tim Thomerson, Patrick Swayze, Harold Sylvester, Alice Lau, Kwan Hi Lim, Gail Strickland, Kelly Yunkerman, Todd Allen, Jeremy Kemp
A grief stricken retired Army officer gathers together a motley collection of Vietnam veterans, and mounts an expedition to Laos when he learns that his son, reported as missing in action, may in fact be alive in a POW camp. Good performances lift this formula Vietnam actioner above the usual run.
A/AD 100 min VIDrel: 4-FRONT/POLYREC/CIC V/sur

UNCONQUERED, THE **   PG
Dick Lowry USA   1988
Peter Coyote, Dermot Mulroney, Tess Harper, Jenny Robertson, Bud Gunton, Larry Riley
The family of a D.A. are caught up in the violence surrounding the struggle to achieve desegregation, in the Southern states of the USA in the 1960s. At the same time the film tells the true story of the man's son (Richmond Flowers Jr) who overcomes physical disabilities to become a sporting star. An attempt to provide two stories for the price of one, making for a messy and disjointed affair. The script is by Pat Conroy.
DRA 113 min (ort 130 min) mTV VIDrel: 20TH/TECH V

UNDAUNTED WU DANG, THE **   18
Sun Sha HONG KONG
Lin Chen, Chao Cheng, Lee Yue Ven
A woman learns special form of combat to avenge her father's death in a typical formula revenger.
MAR 103 min dubbed VIDrel: SCRN/DISC V

UNDEAD, THE **   15
Roger Corman USA   1957
Pamela Duncan, Richard Garland, Richard Devon, Allison Hayes, Billy Barty, Mel Welles, Bruno VeSota, Dorothy Newman, Val Dufour, Aaron Saxon, Dick Miller
After regressing a patient back to a former life in which she was burnt at the stake as a witch, a psychiatrist contrives to travel back in time with her in order to prevent this event from occurring, thus changing the future. Several intriguing ideas make their presence felt but are soon swamped by a lack of plot development and some pseudo-Shakespearean dialogue which is as ineffective as it is laughable.
HOR 71 min (ort 76 min) B/W VIDrel: HEND/BMGREC L/A V

UNDEFEATABLE **   18
Godfrey Hall USA   1993
Cynthia Rothrock, Don Niam, John Miller, Donna Jason, Emilie Davazac, Hang Yip Kim, Gerald Klein
A female martial artist takes part in illegal bouts to earn enough money to put her younger sister through college. But when the girl is attacked by a murderous rapist, big sister swears to have her revenge, teaming up with a tough cop to do exactly that. The expected violent set-pierce battles now take place, and though Rothrock shows little flair for acting, she works hard to give value for money in all other departments.
MAR 88 min VIDrel: MED L/A V

UNDEFEATED, THE **   PG
Andrew V. McLaglen USA   1969
John Wayne, Rock Hudson, Tony Aguilar, Roman Gabriel, Marian McCargo, Lee Meriwether, Merlin Olsen, Melissa Newman, Bruce Cabot, Michael Vincent, Ben Johnson, Edward Faulkner, Harry Carey Jr, Paul Fix, Royal Dano, Carlos Rivas
Routine post-Civil War Western. Former colonels Wayne and Hudson from the Union and Confederate Armies respectively, form an uneasy alliance as they head for Mexico in order to start new lives. Apart from the unusual pairing of the two leads, this one has little new to say. Introduces former football stars Gabriel and Olsen.
WES 114 min (ort 119 min) VIDrel: 20TH/TECH V/h

UNDER FIRE ***   15
Roger Spottiswoode USA   1983
Nick Nolte, Gene Hackman, Joanna Cassidy, Jean-Louis Trintignant, Ed Harris, Richard Masur, Rene Enriquez, Hamilton Camp, Jorge Santoyo, Lucina Rojas, Raul Garcia, Victor Alocer, Eric Valdez, Andaluz Russel, E. Villaviciencio
An account of the experiences of three American TV journalists in Nicaragua in 1979 before the fall of the Somoza dictatorship. The film has a few good moments but they are a long time coming. Trintignant and Harris are both excellent in support. Written by Ron Shelton and Clayton Frohman with music by Jerry Goldsmith.
DRA 123 min (ort 128 min) VIDrel: VISVID/POLYREC V/sur

UNDER INVESTIGATION **   (18)
Kevin Meyer USA   1992
Harry Hamlin, Joanna Pacula, Richard Beymer, Ed Lauter, John Mese, Lydie Denier, Patrick Thomas, Robert Knott, Murray Rubin, Carol Ann Susi, Michael Wiles, Jeffrey Reed, Megan Blake, Marian Collier, Marcia Del Mar, Julie Baltay
An L.A. cop suffering from a bad case of burn-out investigates the murder of an artist whose wife inherited his fortune. Although she is the prime suspect, he unwisely finds himself being drawn into a steamy relationship with her. An insubstantial effort that is badly in need of a much stronger (and more original) plot.
DRA 96 min SATrel: SKY MOVIES

UNDER LOCK AND KEY **   18
Henri Charr USA   1994
Wendi Westbrook, Barbaa Niven, Taylor Leigh, Stephanie Ann Smith
A woman FBI agent goes under cover in a prison in order to befriend the former mistress of a major drugs baron. However, when she is killed, our agent's cover is blown and she has to confront this criminal, who has kidnapped the daughter of his late mistress.
A/AD 87 min VIDrel: MIA/DISC V

UNDER MILK WOOD ***   15
Andrew Sinclair UK   1973
Richard Burton, Elizabeth Taylor, Peter O'Toole, Glynis Johns, Sian Philips, Vivien Merchant, Victor Spinetti, Ryan Davies, Angharad Rees, Ray Smith, Glyn Edwards, Bridget Turner, Talfryn Thomas, Wim Wylton, Bronwen Williams
A film version of a wonderfully evocative radio play, weaving a magical tapestry of words around the lives and dreams of the inhabitants of the tiny Welsh fishing village of Llareggub. The visual images add nothing to the marvellous writing of Thomas, and in the end serve only as a distraction and an annoyance. Just close your eyes and listen as Burton narrates.
DRA 87 min (ort 90 min) VIDrel: 20TH/TECH L/A V/h
Boa: radio play by Dylan Thomas.

UNDER MILK WOOD **   PG
UK   1992
Narrated by Richard Burton
The original BBC soundtrack is used in this competent anima-
tion that as one might expect, really adds very little to the
wonderful writing of Thomas. Music is by Trevor Herbert.
ANIM   50 min mTV   VIDrel: PARADOX/TOTAL   V
Boa: radio play by Dylan Thomas.

UNDER PRESSURE ***   15
Peter Levin USA   1992
*Teri Garr, Harry Hamlin, Keith Coulouris, Terry O'Quinn, Gary
Frank, Georgia Emelin, Joycelyn O'Brien, Britt Sady, Michael Flynn,
John Daryl, Jeff Olson, Jan Turner, George Sullivan, Margo Swena,
Tip Boxell, Craig Clyde, Richard Clark*
One of those hostage-in-peril films in which a religious man
takes the staff of a clinic hostage, in a bid to get revenge on the
doctor who sterilised his wife (who had given birth to eight chil-
dren) two years before. Based on the true story of Richard
Worthington, whose actions led to a tense stand-off situation
involving the FBI, police and specialist SWAT teams. Quite
compelling, though of the cast only Garr stands out as one of the
nurses held hostage.
Aka: DELIVER THEM FROM EVIL: THE TAKING OF ALTA VIEW
DRA   89 min (ort 96 min)   VIDrel: VCC/DISC   V

UNDER SIEGE ***   15
Roger Young USA   1986
*Peter Strauss, Hal Holbrook, E.G. Marshall, Lew Ayres, Fritz Weaver,
Mason Adams, George Grizzard, Stan Shaw, Paul Winfield, Beatrice
Straight, David Opatoshu, Frederick Coffin, Victoria Tennant, Ann
Sweeny, David Opatoshu*
The US President has to deal with a series of terrorist outrages
that culminate in an attack on the Capital Building in
Washington. Despite the use of clever special effects (especially
in blowing the dome off the State Building, the one in Arkansas
being used), this remains a stilted and talky film. Scripted by a
host of writers, among them being Bob Woodward of ALL THE
PRESIDENT'S MEN fame.
A/AD   150 min mTV   VIDrel: GUILD/SONOP L/A   V

UNDER SIEGE **   15
Andrew Davis USA   1992
*Steven Seagal, Tommy Lee Jones, Gary Busey, Patrick O'Neal, Erika
Eleniak, Glenn Morshower, Raymond Cruz, Duanve Davis, Michael
Des Barres, Dale Dye, Troy Evans, David McKnight, Sandy Ward,
Micahel Welden, Tom Wood, Brad Rea*
A group of terrorists hijack a US battleship, taking its crew
captive and capturing its arsenal of nuclear missiles. However,
they make one mistake by overlooking a humble cook with
exceptional martial arts abilities. An ultra-violent variation on
the DIE HARD theme with no shortage of action, it is sorely
hampered by Seagal's limited acting range. By contrast, Jones
as ever makes a most effective villain.
A/AD   98 min (ort 102 min) cC   VIDrel: WHV   V/dm

UNDER SIEGE 2 **   18
Geoff Murphy USA   1995
*Steven Seagal, Eric Bogosian, Katherine Heigl, Morris Chesnut,
Everett McGill, Peter Greene, Patrick Kilpatrick, Scott Sowers, Afifi,
Andy Romano, Brenda Bakke, Sandra Taylor, Jonathan Banks, David
Gianopoulos, Nick Mancuso*
When a team of mercenaries threaten to destroy Washington
City with a nuclear satellite system, special agent Seagal (an ex-
SEAL who now works as a chef) is given the job of locating and
destroying their secret command centre. He achieves this by
killing lots of the bad guys, all of whom are travelling on a train
that has been hijacked by Bogosian – the power-crazed scientist
at the centre of this plot. A mindless action film, noisy, brutal
and pointless.
Aka: UNDER SIEGE 2: DARK TERRITORY
A/AD   95 min (ort 99 min) cC   VIDrel: WHV   V/sur

UNDER SUSPICION **   18
Simon Moore UK   1991
*Liam Neeson, Laura San Giacomo, Kenneth Cranham, Maggie
O'Neill, Alan Talbot, Malcolm Storry, Martin Grace, Kevin Moore,
Stephen Oxley, Colin Dudley, Richard Graham, Alison Ruffell,
Victoria Alcock, Michael Almaz, Tony Hughes*
Writer-director Moore's feature debut is set in Brighton in the
late 1950s, where the owner of a seedy private detective agency
specialises in helping married female clients get the "evidence"
of adultery they want by setting up the other party with a

strange woman. Matters take an ugly turn when the detective's
own wife is murdered, and our sleuth becomes the victim of a
frame-up. An intriguing opening is sadly spoilt by implausible
development.
THR   97 min (ort 100 min)   VIDrel: COLUM/SONOP L/A
V

UNDER THE BOARDWALK **   15
Fritz Kiersch USA   1987
*Keith Coogan, Danielle Von Zerneck, Richard Joseph Paul, Roxana
Zal, Tracey Walter, Dick Miller, Sonny Bono, Hunter Von Leer,
Elizabeth Kaitain*
A modern Romeo and Juliet tale set among the surfing gangs of
Southern California, with a girl from Venice falling in love with
a boy from San Fernando Valley. An empty-headed comedy of
little humour, interspersed with acres of tedium. The only diver-
sions (apart from the pretty girls and scenery) are the fine
performances from character actors in bit parts. Very much a
reversion to the tried and largely discarded beach-movie
formula.
A/AD   99 min (ort 102 min)   VIDrel: 20VIS/SONOP L/A
V/sh

UNDER THE CHERRY MOON **   15
Prince USA   1986
*Prince, Jerome Benton, Kristin Scott-Thomas, Steven Berkoff,
Francesca Annis, Victor Spinetti, Alexandra Stewart, Emmanuelle
Sallet*
Empty-headed attempt to build a dramatic vehicle for this
talented singing star. He plays an American entertainer on the
French Riviera whom women find irresistible. An eminently
resistible vanity vehicle.
MUS   96 min (ort 98 min) B/W   VIDrel: WHV   V/sur

UNDER THE PIANO **   15
Stefan Scaini CANADA   1996
*Amanda Plummer, Megan Follows, Teresa Stratas, John Juliani,
Carroll, Deborah Grover, Richard McMillan, Louisa Martin, Simone
Rosenberg, Dan Lett, Jackie Richardson, Richard Blackburn, Julian
Orban, Andrew Tabet, Tara Macri*
A woman comes home to find that her fading opera star mother
can no longer cope with her other daughter (who is autistic)
and has attempted suicide. As the normal daughter is a caring
person, she gives up her own life to look after them both. Set in
the 1930s, this overwrought tearjerker is based on a true story.
Fair.
DRA   90 min (ort 92 min)   VIDrel: ODY   V

UNDER THE RAINBOW *   PG
Steve Rash USA   1981
*Chevy Chase, Carrie Fisher, Eve Arden, Mako, Pat McCormick, Joseph
Maher, Robert Donner, Billy Barty, Adam Arkin, Cork Hubbert,
Richard Stahl, Freeman King, Peter Isacksen, Jack Kruschen, Bennett
Ohta, Gary Friedkin, Pam Vance*
A spy comedy set in a hotel during the filming of the "Wizard
of Oz". Features midgets, undercover agents, assassination
attempts on a duke and the nefarious activities of Nazi and
Japanese spies, all parties concerned crossing paths at the hotel.
A tasteless and breakneck trip to boredom.
COM   98 min   VIDrel: MGM/WHV L/A   V

UNDER THE VOLCANO ***   15
John Huston USA   1984
*Albert Finney, Jacqueline Bisset, Anthony Andrews, Ignacio Lopez
Tarso, Katy Jurado, James Villiers, Dawson Bray, Carlos Riquelme,
Jim McCarthy, Rene Ruiz, Eliazar Garcia Jr, Salvador Sanchez, Sergio
Calderon, Hugo Stiglitz*
A love triangle story that revolves around the last days of an
alcoholic retired British diplomat, his estranged wife and one of
her former lovers (his half-brother). Set in Mexico in 1938 on the
eve of WW2, this sombre and disturbing story has a plum part
for Finney, but is hardly a full rendering of the complexities of
the novel. Written by Guy Gallo and with a score by Alex North.
DRA   107 min (ort 112 min)   VIDrel: 20TH/TECH L/A   V
Boa: novel by Malcolm Lowry.

UNDER THREAT **   15
Rod Holcomb USA   1993
*Charles Bronson, Dana Delany, Xander Berkley, Jenette Goldstein,
Tom Verica, Bonnie Bartlett, Louis Giambalvo, Marc Alaimo, Robert
Gossett, Richard Kuss, Michael Cavanaugh, Julianna McCarthy, Kim
Weeks, Patti Yasutake, Sam Vincent*
A homicide cop is assigned to the same case his daughter is

working on, that of a vicious serial killer. But his interest becomes more than purely professional when he realises all the signs point to his daughter becoming the killer's next victim.
Aka: DEAD TO RIGHTS; DONATO AND DAUGHTER
A/AD 89 min mTV VIDrel: MIA/DISC/IMPENT V
Boa: novel by Jack Early.

UNDERCOVER ** 18
Alexander Gregory Hippolyte USA 1994
Athena Massey, Tom Tayback, Anthony Guidera, Rena Riffel, Jeffrey Dean Morgan, Meg Foster
In order to catch a murderer, a woman police detective goes undercover at a luxury brothel, where she learns more about her own secret passions than she bargained for. Unashamedly exploitative, this sleazy erotic thriller has the usual set-piece sexual shenanigans, but little attempt is made to create a credible storyline, and the murder is solved all too easily in the last ten minutes.
THR 94 min VIDrel: HIFLI/SONOP V/h

UNDERCOVER ANGEL ** 18
George Axmith USA 1994
Darlene Vogel, Shane Fraser, Mark De Carlo, Kerrie Clark, Peter JUrasik, Patrick Kilpatrick, Roddy McDowell
A former prostitute now works as a forensic photographer, but finds herself obliged to take to the streets in search of clues when she becomes involved in the case of a murdered prostitute. Erotic thriller that sets much of the action at the various strip clubs our photographer ventures into, in her bid to get at the truth, and essays a slightly more serious stance than is usual for films of this genre.
THR 94 min VIDrel: HIFLI/SONOP V/h

UNDERCOVER BLUES * 15
Herbert Ross USA 1993
Kathleen Turner, Dennis Quaid, Fiona Shaw, Stanley Tucci, Larry Miller, Obba Babatunde, Tom Arnold, Park Overall, Ralph Brown, Jan Triska, Marshall Bell, Richard Jenkins, Dennis Lipscomb, Saul Rubinke, Dakin Mathews, Michael Greene
A married couple who are both intelligence agents, take a vacation in New Orleans with their baby daughter but find their parental leave brutally cut short by their boss. Pressed back into service, they are order to deal with an old enemy who is planning to sell some stolen weapons. Despite the best efforts of its cast, this unamusing effort remains firmly below the level of all those POLICE ACADEMY films. An utter waste of time.
COM 86 min (ort 100 min) cC VIDrel: MGM/WHV V/sur

UNDERCOVER CAPER ** PG
Ed Montague USA
Richard Kiel, Tim Conway, Chuck McCann, Reni Santoni, Dub Taylor
Two misfit cops enter a prison as inmates on an undercover mission, but the only prison official who is a party to this scheme is the governor, and when he dies the two cops have no alternative but to plan a breakout.
COM 90 min VIDrel: MOPIC/SGSVID V

UNDERCURRENT ** (15)
Brian O'Flaherty USA 1994
Tina Kellegher, Barry Barnes, Liam Cunningham
Complex multi-stranded story set in Dublin, that revolves around a couple of hardbitten police officers, two sisters, a government official and an adulterous husband.
DRA 80 min VIDrel: HIFLI/SONOP V

UNDERGRADS, THE ** U
Steven Hilliard Stern USA 1984
Art Carney, Chris Makepeace, Len Birman, Dawn Greenhall, Lesleh Donaldson, Jackie Burroughs, Alfie Scopp, Angela Fusco, Nerene Virgin, Adam Ludwig, Ron James, Peter Spence, Wendy Bushell, Gary Farmer, Pat Patterson
An old man whose family want to put him in a home, enrols at his grandson's college instead. This idea was later used in a TV series starring Mickey Rooney. Made by Disney for cable TV and a typically bland offering, though Carney's considerable presence goes some way towards redeeming it.
DRA 98 min mCab VIDrel: WDV/TECH L/A V
Boa: story by Paul W. Shapiro and Michael Wisman.

UNDERGROUND *** 15
Emir Kusturica FRANCE/GERMANY/HUNGARY 1995
Miki Manojlovic, Lazar Ristovski, Mirjana Jokovic, Slavko Stimac,
Ernst Stotzner, Srdan Todorovic, Mirjana Karanovic, Milena Pavlovic, Bata Stojkovic, Bora Todorovic, Davor Dujmovic, Dr Nele Karajlic, Branislav Lecic, Erol Kadic*
The story of a couple of criminals from their days in Belgrade in 1941 to war-torn Yugoslavia of today. Screenplay is by Kovacevic and Kusturica, and the film generated enormous controversy on completion, being seen by some (quite unfairly) as nothing more than a piece of pro-Serb propaganda. In fairness to the director, this heavily stylised film is far more than that, being as much a for the country as it is a simple story. It won the Palme d'Or at Cannes.
Aka: IL ETAIT UNE FOIS UN PAYS; ONCE UPON A TIME THERE WAS A COUNTRY
DRA 161 min (ort 170 min) wScrn VIDrel: ARTIF/20TH V/sh
Boa: play by Dusan Kovacevic.

UNDERNEATH, THE *** 15
Steven Soderbergh USA 1995
Peter Gallagher, Alison Elliott, William Fichtner, Adam Trese, Joe Don Baker, Paul Dooley, Elisabeth Shue, Anjanette Comer, Dennis Hill, Harry Goaz, Mark Fletch, Jules Sharp, Kenneth D. Harris, Vincent Gaskins, Cliff Haby
A remake of CRISS CROSS that tells of a drifter, who upon returning home makes a dangerous attempt to recapture the love of the woman he left behind years before. Done in flashback form it explains how, as an armoured-truck driver, he got involved in a heist with the husband of the woman he loved, and was inevitably set on a slippery downward path. A concise and sombre character study, often ambiguous in tone but always compelling.
THR 95 min (ort 100 min) cC VIDrel: CIC/SONOP V/sur
Boa: novel Criss Cross by Don Tracy.

UNE BELLE FILLE COMME MOI ** 15
Francois Truffaut FRANCE 1972
Bernadette Lafont, Claude Brasseur, Charles Denner, Guy Marchand, Andre Dussollier, Philippe Leotard
A naive sociologist becomes the confidante of a deceitful young woman who has got involved in murder, and she tells him all about her various relationships. An awkward black comedy that essays a little social satire but is quickly sunk by the absurd plot and Lafont's irritating penchant for over-acting.
Aka: GORGEOUS BIRD LIKE ME, A; SUCH A GORGEOUS KID LIKE ME
DRA 98 min VIDrel: ARTIF/20TH V
Boa: novel by Henry Farrell.

UNE FEMME EST UNE FEMME *** PG
Jean-Luc Godard FRANCE 1961
Anna Karina, Jean-Claude Brialy, Jean-Paul Belmondo, Marie Dubois, Jean Moreau
Anxious to have a baby by her boyfriend, a nightclub stripper finds her lover less than agreeable to her plans, so she turns instead to his best friend. Full of clever editing and various literary references, the film mostly takes place in a Parisian apartment, but the whole is so very lighthearted that it never becomes claustrophobic. Moreau has a tiny cameo playing herself. The director married Karina soon after this film was made.
Aka: WOMAN IS A WOMAN, A
DRA 81 min (ort 85 min) wScrn VIDrel: ELPIC/POLYREC L/A V

UNE FEMME FRANCAISE ** 18
Regis Wargnier FRANCE 1995
Emmanuelle Beart, Daniel Auteuil, Gabriel Barylli, Jean-Claude Brialy
An army lieutenant returns to his wife after five years spent fighting during WW2 and the couple, who now have twins, move to Berlin. However, once there the woman embarks on an affair with the son of the landlord. A very well mounted film of high production values, but equally one that fails to engage one's interest, with the characters seemingly dwarfed by the events taking place around them.
DRA 95 min wScrn VIDrel: GUILD/20TH/FOXVID V/sur

UNE PARISIENNE ** PG
Michel Boisrond FRANCE 1957
Brigitte Bardot, Henri Vidal, Andre Luguet, Charles Boyer, Nadia Gray, Madeleine Lebeau, Noel Roquevert
Gossamer-light, dated sex-comedy that's clearly intended as a vehicle for Bardot, who plays the daughter of the French Prime

Minister. When she and his principal secretary are discovered in bed together, a quick and involuntary marriage seems inevitable. Without the charm of Bardot, this would have been forgettable indeed.
Aka: LA PARISIENNE
COM   86 min (ort 90 min)   VIDrel: CASPIC L/A   V

UNE PARTIE DE PLAISIR *** 15
Claude Chabrol FRANCE/ITALY   1974
Paul Gegauff, Daniele Gegauff, Paula Moore, Pierre Santini, Michel Valette, Clemence Gegauff
A happily married man encourages his wife to take a lover and promptly takes a male lover for himself as well. However, this cosy set-up does not last very long, as the husband's begins to grow increasingly jealous of his wife. A sad tale that examines the breakdown of a relationship in all its painful detail, a task made less than edifying by the hateful personality of the husband (a superb performance from Gegauff, who was murdered by his second wife in 1983).
Aka: LOVE MATCH; PLEASURE PARTY
DRA   96 min (ort 100 min) wScrn   VIDrel: ARTPRO/RTM
V

UNFAITHFUL ** 15
Steven Schachter USA   1992
Tom Skerritt, Blythe Danner, Roma Downey, Gary Frank, Julianne Phillips, Dorian Harewood
A middle-aged man embarks on an affair, but finds he has made a serious misjudgement about his new lady-friend.
DRA   88 min   VIDrel: GUILD/SONOP   V

UNFAITHFUL ENTRY * 18
Eric Edwards USA   1993
Summer Knight, Alicia Rio, Kym Wilde, Stacy Nichols, Lacy Rose, Celeste, Sheila Stone, Nicholas Rage, Kris Newz, Jonathan Morgan, Jeff Scott
A young wife is tempted to have an affair with her husband's best friend when she is propositioned by him.
A   94 min   VIDrel: FALCON/TOTAL   V

UNFAITHFUL WIVES *** 15
Claude Chabrol FRANCE/ITALY   1969
Stephane Audran, Michel Bouquet, Maurice Ronet, Stephen Di Napolo, Michel Duchaussoy
An atmospheric and very taut thriller telling of how a wife comes to have more respect for her husband when he murders her lover. A near classic tale of adultery and murder: elegant, sensuous and dramatic. This was one of the works in the director's excellent "Helene cycle", with his real-life wife Audran playing a character named Helene, in films that mostly dealt with adultery and murder. This is the first and best of them.
Aka: LA FEMME INFIDELE; UNFAITHFUL WIFE
THR   94 min (ort 98 min) wScrn   VIDrel: ARTPRO/RTM
V

UNFAITHFULLY YOURS *** 15
Howard Zieff USA   1984
Dudley Moore, Nastassja Kinski, Armand Assante, Albert Brookes, Cassie Yates, Richard Libertini, Richard B. Shull, Jan Triska, Jane Hallaren, Bernard Behrens, Leonard Mann, Estelle Omens, Penny Peyser, Nicholas Mele
A romantic comedy involving an orchestra conductor who plans to murder his wife whom he suspects of having an affair with a violinist. A remake of the 1948 Preston Sturges film that is funny at times but begins to run down about halfway through.
COM   92 min (ort 96 min)   VIDrel: 20TH/TECH   V

UNFINISHED PIECE FOR MECHANICAL PIANO *** U
Nikita Mikhalkov USSR   1977
Aleksandr Kalaigin, Yelena Solovei, Eugenia Glushenko, Antonina Shuranova, Yuri Bogatyrev, Nikita Mikhalkov
This rambling and slow-moving film tells of a set of guests who are invited to the remote country estate of a widow. Once there, efforts are made to recapture the magic of past relationships, insults are traded and various old resentments and petty jealousies emerge. An atmospheric work, well directed and often though not always, absorbing. The handsome photography of Pavel Lebeshev is a major asset.
Aka: NEOKONCHENNAYA PYESSA DLYA MEKHANICHESKOGO PIANIN; UNFINISHED PIECE FOR PLAYER PIANO, AN
DRA   97 min (ort 100 min)   VIDrel: CONNO/RTM   V
Boa: play Platonov by Anton Chekhov.

UNFORGETTABLE * 18
Jim Enright USA   1993
Alexis DeVell, Jonathan Morgan, Peter North, Lois Ayers, Melanie Moore, Meekah
An attractive woman makes an indelible impression on her many partners, none of whom can forget her, unlike this film, which is all too easy to forget.
A   35 min (ort 78 min)   VIDrel: ONE   V

UNFORGIVABLE ** 15
Graeme Campbell USA   1989
Harley Jane Kozak, John Ritter
A young girl threatens to call the police after she sees her father beating her mother, but he takes flight and hides out in the home of his mistress. But pretty soon her starts abusing her too.
DRA   89 min   VIDrel: ODY/SONOP   V/sh

UNFORGIVEN, THE ** PG
John Huston USA   1959
Burt Lancaster, Audrey Hepburn, Lillian Gish, Audie Murphy, John Saxon, Joseph Wiseman, Albert Salmi, Charles Bickford, Doug McClure, June Walker, Kipp Hamilton, Arnold Merritt, Carlos Rivas
A disappointing effort to bring alive the subject matter of the original novel, in which the suspected Indian origins of a rancher's daughter cause trouble with the local inhabitants when the Kiowa go on the warpath. Not without some exciting moments, but still very muddled nonetheless.
WES   116 min (ort 123 min) wScrn   VIDrel: MGM/WHV
L/A   V
Boa: novel by Alan Le May.

UNFORGIVEN **** 15
Clint Eastwood USA   1992
Clint Eastwood, Gene Hackman, Morgan Freeman, Jaimz Woolvett, Saul Rubinek, Richard Harris, Frances Fisher, Anna Thomson, David Mucci, Rob Campbell, Anthony James, Tara Dawn Frederick, Beverley Elliott, Josie Smith, Liisa Repo-Martell
A remarkable film in every way, it casts Eastwood as an impoverished pig farmer and ex-gunslinger who is tempted out of retirement by poverty, and agrees to work as a bounty hunter, tracking down the thugs who slashed a prostitute's face. Rich in atmosphere and imagery (photography is by Jack Green) this film offers an honest appraisal of the Wild West, that is violent and memorable. The script is by David Webb Peoples. AA: Pic, Dir, S. Actor (Hackman).
WES   124 min (ort 131 min) wScrn (special edition) cC
VIDrel: WHV   V/dm V/sur

UNGENTLEMANLY ACT, AN *** (PG)
Stuart Urban UK   1992
Ian Richardson, Rosemary Leach, Ian McNeice, James Warrior, Marc Warren, Elizabeth Bradley, Kate Spiro, Holly Barker, Claire Slater, Hugh Ross, Bob Peck, Ian Embleton, Richard Graham, Matthew Ashforde, Richard Long, Aiden Gillen
Dramatised reconstruction of the Falklands invasion by the Argentinians, that covers the first thirty-six hours of their occupation, for the most part exploring the actions of the Governor and the local population. A BBC production that in some ways appears to be a flawed attempt to use the real events portrayed to fashion a satire on the stupidity of war. Ultimately, it is Richardson's outstanding performance that maintains interest in the proceedings.
DRA   120 min (ort 130 min) mTV   VIDrel: SPEAR/SONOP
V

UNHOLY, THE ** 18
Camilo Vila USA   1988
Ben Cross, Hal Holbrook, Jill Carroll, William Russ, Trevor Howard, Claudia Robinson, Ned Beatty, Nicole Fortier, Peter Frechette, Ruben Rabasa, Phil Becker, Susan Bearden, Xavier Barquet, Larl White, Jeff D'Onofrio
A Catholic priest appointed to a parish where his two predecessors have been brutally murdered, discovers a welter of Satanic practices as well as a real-life demon, that sustains its existence by killing sinners in the act of sinning. An ambitious but somewhat ludicrous tale with rather good special effects. Scripted by Philip Yordan and Fernando Fonseca, the film's designer.
HOR   97 min Cut (10 sec plus some cuts subst)
VIDrel: VCC/DISC/COLUM L/A   V

UNHOLY GARDEN, THE ** U
George Fitzmaurice USA   1931
Ronald Colman, Fay Wray, Estelle Taylor, Tully Marshall, Ulrich

Haupt, Henry Armetta, Lawrence Grant, Warren Hymer, Mishca Auer, Morgan Wallace, Kit Guard, Lucille LaVerne, Arnold Korff, Charles Hill Mailes, Nadja, Henry Kolker
A gentleman jewel thief hides out in a hotel in Algeria that is full of criminals of various kinds. There he becomes involved with the grand-daughter of an embezzler and helps to foil a plan by some of the inmates to steal his loot. To complicate matters further, the local police are out to arrest him and to do this they bait their trap with an attractive woman. A dreary and unconvincing tale, poorly scripted and indifferently acted.
DRA   75 min (ort 85 min) B/W   VIDrel: VCC/DISC   V

UNHOLY MATRIMONY ***   (18)
Jerrold Freedman USA   1988
Patrick Duffy, Michael O'Keefe, Charles Durning, Lisa Blount
A detective investigating the death of a young newly married woman in a hit-and-run accident learns that this was murder and soon begins to suspect that a so-called pastor may be responsible. Eventually is transpires that this is part of a cunning scheme to collect $300,000 from a life insurance policy that was taken out on the woman. A tense thriller based on a true story, its strong direction and excellent cast ensure one's attention never wanders.
THR   100 min mTV   TVrel

UNINHIBITED *   (18)
Charles S. Allen USA   1991
Charles Allen, Tracey Wolfe, Asha, Julio Gonzalez, Rocco Tano, Randall Oliver, Andre Allen, Robert Miano, Theo Coumbis, Victor J. Pancerev, Leigh Betchley, Danielle Grammer, Mimi Faillace, Chaundry Southern, Stanley Benders
Standard soft-core with all the usual elements.
A   92 min   SATrel: MOVIE CHANNEL

UNINVITED **   18
Michael Derek Bohusz USA   1992
Jack Elam, Christopher Boyer, Erin Noble, Bari Buckner, Jerry Rector, Zane Paolo, Dennis Gibbs, Ted Haler, Eno Brutto
A motley group of treasure-seekers, led by a strange old man, goes in search of a cache of gold, but violate the sanctity of an Indian burial ground, paying a high price for this crime. An odd supernatural Western that is only occasionally effective.
WES   90 min   VIdrel: HIFLI/SONOP   V/h

UNIVERSAL SOLDIER ***   18
Roland Emmerich USA   1992
Jean-Claude Van Damme, Dolph Lungren, Ally Walker, Ed O'Ross, Jerry Orbach, Leon Rippy, Tico Wells, Ralph Moeller, Robert Trebor, Gene Davis, Drew Snyder, "Tiny" Lister Jr, Simon Rhee, Eric Norris, Michael Winther, Joseph Malone
Van Damme and Lundgren are soldiers brought back from the dead, brainwashed and turned into invincible cyborg-warriors, to be used as part of a secret anti-terrorist strike force. Van Damme goes AWOL when past memories (mostly of his time in Vietnam) begin re-surfacing. Lundgren is sent after him. The expected bloody confrontation takes place, in this fast-moving, futuristic action film. Fans of loud bangs and ample action will like this one a lot.
A/AD   99 min (ort 103 min) cC wScrn (GUILD/POLYREC)
VIDrel: GUILD/POLYREC L/A; POLY/POLYREC; PION (LV only)   V/sur LV

UNKNOWN SUBJECT **   15
Bill Corcoran/Gus Trikonis USA
David Soul, M. Emmet Walsh, Jennifer Hetrick, Kent McCord, Andrea Mann, Joe Maruzzo
A thriller that follows the work of a specialised unit within the Federal Justice Department, where the clues derived from unsolved crimes are used to create psychological profiles of the perpetrators.
THR   89 min   VIDrel: COLUM/SONOP   V

UNLAWFUL ENTRY *   18
Jonathan Kaplan USA   1992
Kurt Russell, Madeleine Stowe, Ray Liotta, Roger E. Mosley, Ken Lerner, Andy Romano, Deborah Offner, Carmen Argenziano, Johnny Ray Mehee, Bing Agnello, Sonny Davis, Harry Northrup, Sherri Rose, Alicia Ramirez, Ruby Salazar, Eduardo Migre
A cop called to assist a young couple whose home was burgled, begins a nasty game of cat-and-mouse when he becomes irresistibly attracted to the wife, and conspires to have hubby imprisoned so that he can have her to himself. A faintly ludicrous variation on the psycho-from-Hell theme, with our cop

seeming to have superhuman strength as he survives shooting and falling down stairs in order to suffer a truly gory ending. An unpleasant and manipulative farrago of nonsense.
DRA   111 min   VIDrel: 20TH/TECH L/A   V

UNLAWFUL VENGEANCE **   15
Paul Krasny USA   1993
George Hamilton, Robert Conrad
A wealthy financier and a night watchman both lose sons to a killer, who with the help of clever lawyers, walks free from court. They unite in their desire for revenge.
DRA   90 min (ort 94 min)   VIDrel: MARQ/GUILD   V

UNLIKELY SUSPECTS **   15
Joseph L. Scanlan USA   1996
Shanna Reed, Sarah Clarke, Lochlyn Munro, Josh Taylor
When a high school quarterback takes advantage of his position to sexually abuse his cheerleaders, the scene is set for a lot of angst and recrimination. Average.
DRA   90 min   VIDrel: ODY/SONOP   V/sh

UNMARRIED WOMAN, AN ****   18
Paul Mazursky USA   1978
Jill Clayburgh, Alan Bates, Michael Murphy, Cliff Gorman, Pat Quinn, Kelly Bishop, Lisa Lucas, Michael Tucker, Jill Eikenberry, Linda Miller, Andrew Duncan, Daniel Seltzer, Matthew Arkin, Penelope Russianoff, Novella Nelson
A woman has to start life over again when her husband leaves her, and takes up with several men, but her daughter gives her trouble. Written by Mazursky who has a good eye for the problems that beset a single mother trying to cope in New York City. Clayburgh gives a memorable performance in the title role. See LOVE AND BETRAYAL for a bleaker treatment of this theme plus the movie THE LAST TO GO, which also deals with this subject.
DRA   119 min   VIDrel: 20TH/TECH L/A   V

UNNAMABLE, THE *   18
Jean-Paul Quellette USA   1988
Charles King, Alexandra Durrell, Mark Kinsey Stephenson, Laura Albert, Eben Ham, Blane Wheatley, Mark Parra, Delbert Spain, Colin Cox, Paul Farmer, Paul Pajor, Marcel Luissier, Lisa Wilson, Nancy Kreisel, Katrin Alexandre
Very loosely based on the Lovecraft tale, this tells of a haunted house, and a demonic creature that gained its freedom from a locked room there, when the former warlock owner foolishly released it 300 years ago. The excellent special effects are a poor compensation for the incoherent plot. A sequel followed.
HOR   84 min Cut (21 sec – ort 87 min)
VIDrel: BRAVE/SONOP L/A   V
Boa: short story The Shuttered Room by H.P. Lovecraft.

UNNAMABLE RETURNS, THE **   18
Jean-Paul Quellette USA   1992
David Warner, John Rhys-Davies, Mark Kinsey Stephenson, Maria Ford
A researcher into the occult and a professor team up in a bid to apprehend the killer of four students, whom they have reason to believe is something other than human. Their search takes them to a deserted graveyard and a secret network of tunnels found below ground. An adequate sequel to the earlier film.
Aka: UNNAMABLE 2, THE: THE STATEMENT OF RANDOLPH CARTER
HOR   99 min (ort 104 min)   VIDrel: VCC/DISC L/A   V

UNSINKABLE MOLLY BROWN, THE **   U
Charles Walters USA   1964
Debbie Reynolds, Harve Presnell, Ed Begley, Jack Kruschen, Martita Hunt, Hermione Baddeley, Vassili Lambrinos, Fred Essler, Harvey Lembeck, Kathryn Card, Lauren Gilbert, Hayden Rorke, Harry Holcombe, Amy Douglass, Anna Lee
In Denver during the late 19th century, a girl rises from rags to riches. Broadway musical adaptation with Reynolds as the determined backwoods girl who knows what she wants out of life and aims to become accepted by Denver society. This lively musical comedy-western is likeable but unmemorable. Based on a true story (the lady in question survived the Titanic) and written by Helen Deutsch. The score is by Meredith Willson.
MUS   123 min (ort 128 min)   VIDrel: MGM/WHV   V/dm V/sh
Boa: musical by Richard Morris.

UNSPEAKABLE ACTS ***

PG

Linda Otto USA   1989

Jill Clayburgh, Brad Davis, Season Hubley, Gary Frank, Gregory Sierra, Bebe Neuwirth, Valerie Landsburg, Terence Knox, Sam Behrens, James Handy, Maureen Mueller, Mark Harelik, Jeff Seymour, Dave Wilson, Laura Owens

Clayburgh and Davis play real-life child advocates Laurie and Joe Braga in this powerful, fact-based dramatisation of a famous 1984 child abuse trial in Miami. The film focuses on the efforts made by child psychologists to gather evidence from their young witnesses, but although the Alan Landsburg script is based on both court transcripts and Hollingsworth's book, it raises important issues that it fails to explore. See also INDICT-MENT: THE McMARTIN TRIAL.

DRA   94 min (ort 100 min) mTV

VIDrel: BRAVE/SONOP L/A   V

Boa: book by Jan Hollingsworth.

UNSPOKEN TRUTH, THE ***

15

Peter Werner USA   1995

Lea Thompson, James Marshall, Robert Englund, Dick O'Neill, Karis Paige Bryant, Patricia Kalember, Gail Cronauer, Freda Willaims, Mona Lee Fultz, Tony Frank, Marina G. Palmier, Gil Glasgo, Gary Carter, Derek Cecil, Barry Thompson

A woman goes against the wishes of her parents and marries a man who turns out to be a brutal thug. Having been ground down by her husband's brutality, she takes the blame when he shoots a man dead in a bar, but they are both imprisoned. When her father dies she faces the further torment of seeing her only child put up for adoption. Based on a true story.

DRA   92 min mTV   VIDrel: ODY/SONOP   V/sh

UNSTRUNG HEROES **·

PG

Diane Keaton USA   1995

Andie MacDowell, John Turturro, Michael Richards, Maury Chaykin, Nathan Watt, Kendra Krull, Joey Andrews, Celia Weston, Jack McGee, Candice Azzara, Anne DeSalvo, Lillian Adams, Lou Cutell, Sumer Stamper, Sean P. Donahue

Keaton's directorial debut (excluding earlier documentaries and TV movies) is a flawed but interesting work in which a young boy gains comfort and support from his two eccentric uncles during his mother's serious illness. An uneven film that starts out as a potent drama, it all too soon changes into a rather light-weight comedy when the boy goes to his relatives, inevitably dissipating much of the film's earlier momentum.

COM   90 min (ort 94 min) cC   VIDrel: HOLPIC/TECH V/sh

Boa: book by Franz Lidz.

UNSUITABLE JOB FOR A WOMAN, AN ***

15

Christopher Petit UK   1981

Billie Whitelaw, Paul Freeman, Pippa Guard, Dominic Guard, David Horovitch, Elizabeth Spriggs, Dawn Archibald, Bernadette Short, James Gilbey, Kelda Holmes, Alex Guard

A secretary takes over the running of a detective agency after her boss commits suicide and investigates an unusual murder case. An interesting example of British film noir that is not without its moments.

THR   90 min (ort 94 min)

VIDrel: POLY/POLYREC/BRAVE L/A   V

Boa: novel by P.D. James.

UNTAMED, THE ***

PG

Geoff Burrowes AUSTRALIA   1988

Tom Burlinson, Sigrid Thornton, Brian Dennehy, Nicholas Eadie, Bryan Marshall

This sequel to THE MAN FROM SNOWY RIVER has Burlinson once more having to prove himself, with the villainous Eadie engaged to his sweetheart. Not as good as the earlier film but full of good action shots, including more of those herds of galloping wild horses.

Aka: RETURN TO SNOWY RIVER; RETURN TO SNOWY RIVER: PART 2

A/AD   99 min Cut (7 sec)   VIDrel: TOUCH L/A   V

UNTAMED HEART ***

15

Tony Bill USA   1992

Christian Slater, Marisa Tomei, Rosie Perez, Kyle Secor, Willia Garson, Gary Groomes, James Cada, Claudia Wilkens, Pat Clemons, Lotis Key, Vanessa Hart, Charley Bartlett, Vincent Kartheiser, Wendy Feder, Nancy maury, Paul Douglas Law

A shy young man is hopelessly in love with a waitress at a local diner but is unable to attract her attention until she saves her from some assailants. Thrown together by circumstances, they

embark on a relationship and fall blissfully in love within a short time, unaware of what fate has in store. A touching and well-made comedy-drama, with Slater and Tomei most appealing in the lead roles.

COM   97 min (ort 104 min)   VIDrel: MGM/WHV   V/sur

UNTAMED LOVE **

15

Paul Aaron USA   1994

Cathy Lee Crosby, Mel Winkler, Richard Fancy, Betty K. Bynum, Haunani Minn, Lois Foraker, Raye Birk, Frderick Collins Jr, Buck Haddix, Jarrett Lennnon, Paige Tamada, Abraham Verduzco, Ashley Woolfolk, Cain DeVore, Ellen Crawford

Fact-based drama in which a teacher of children with special needs, attempts to create a rapport between herself and an introverted, aggressive and seriously disturbed six-year-old girl. Crosby gives a strong performance in the central role as the teacher, but the film is far too sombre, and though the script is terribly sincere, the relentless earnestness of it all makes for a rather gruelling experience.

DRA   91 min mTV   VIDrel: ODY/SONOP   V/sh

Boa: book One Child by Tovey Hayden.

UNTIL SEPTEMBER **

15

Richard Marquand USA   1984

Karen Allen, Thierry Lhermitte, Christopher Cazenove, Marie-Catherine Conti, Nitza Saul, Hutton Hobb, Michael Mellinger, Rochelle Robertson, Raphaelle Spencer, Johanna Pavlis, Helene Desbiez, Steve Gadler, Edith Perret

Stranded temporarily in Paris, Allen meets married banker Lhermitte and falls in love with him. A predictable romantic drama that even manages to make Paris look boring.

DRA   95 min   VIDrel: MGM/WHV L/A   V/h

UNTIL THE END OF THE WORLD ***

15

Wim Wenders GERMANY   1991

William Hurt, Solveig Dommartin, Sam Neill, Rudiger Vogler, Max Von Sydow, Ernie Dingo, Chuck Ortega, Jeanne Moreau, Elena Smirnova, Ryu Chisu, Allen Garfield, Lois Chiles, David Gulpilil, Justine Saunders, Charlie MacMahon

Set in the year 1999, this futuristic fantasy sees the world imperilled by a nuclear satellite that's about to explode and cover the planet in deadly fall-out. Hurt plays a scientist being chased by government agents who want to make use of his electronic-vision invention for military purposes. This noisy, bizarre SF thriller inevitably reminds one of BLADE RUNNER (it even has Sam Neill narrating) but even if derivative, is still a look.

Aka: BIS ANS ENDE DER WELT

FAN   151 min (ort 158 min) (English version)

VIDrel: EIV/SONOP   V

UNTOUCHABLES, THE ****

15

Brian De Palma USA   1987

Kevin Costner, Sean Connery, Charles Martin Smith, Andy Garcia, Robert De Niro, Richard Bradford, Jack Kehoe, Brad Sullivan, Billy Drago, Patricia Clarkson

Writer David Mamet has updated the popular TV series that followed the career of FBI agent Eliot Ness, in his battle against organised crime in prohibition-era Chicago. A powerful and exciting production, with Connery winning an Oscar for his role as an experienced cop and De Niro truly memorable as a gangster boss. The music is by Ennio Morricone. AA: S. Actor (Connery).

A/AD   115 min (ort 119 min) wScrn   VIDrel: CIC/SONOP, PION (LV only)   V/sh LV

UNVEILED **

18

William Cole USA   1993

Lisa Zane, Nick Chinlund, Whip Hubley, Martha Gehman, Amidou

A woman on vacation in Marrakesh goes in search of her friend's killer after the latter is found murdered, and her investigations bring her into contact with the CIA. Average thriller whose exotic locations add little of interest.

THR   102 min   VIDrel: 20VIS/SONOP   V

UP! *

18

Russ Meyer USA   1976

Robert McLaine, Janet Wood, Mary Gavin, Francesca "Kitten" Natividad, Raven de la Croix, Monte Bane, Foxy Lae, Margo Winchester

Done in Meyer's inimitable style, this is a Shakespearean-style murder mystery, with Nazis, Southern rednecks, lesbians and one very large-chested woman (Natividad), who gets to caper

about in the woods with no clothes on. An obscure Russ Meyer film, this is one that did not achieve any cult following.
A   80 min   VIDrel: ALLIED/RTM   V

UP 'N' COMING ***   R18/18
Godfrey Daniels USA   1982
Marilyn Chambers, Lisa De Leeuw, Herschel Savage, Loni Sanders, Cody Nicole, Richard Pacheco, John C. Holmes, John La Zar, Clay Tanning, Ferris Weal, Tiny May, Doug Rossi
A second-billed Country singer lands the job as the opening act for a female singer and proves to be the more popular of the two. Not unnaturally, this antagonises the latter who, hooked on booze and drugs, is coming to the end of her career. Chambers plays the aspiring singer who is out for success even if she has to use her body to get it. De Leeuw gives good support as the singer she challenges. A better than average sex film.
Aka: CASSIE
A   61 min Cut (14 min 38 sec – 18 ver); 76 min (R18 ver) (ort 83 min)   VIDrel: MIA/DISC/IMPENT   V

UP CLOSE & PERSONAL **   15
Jon Avnet USA   1996
Robert Redford, Michelle Pfeiffer, Stockard Channing, Joe Mantegna, Kate Nelligan, Glenn Plummer, James Rebhorn, Scott Bryce, Raymond Cruz, DeDee Pfeiffer, Miguel Sandoval, Noble Willingham, James Karen, Brian Markinson
Suggested by Alanna Nash's book "Golden Girl", this is a rather glossy romantic drama that casts Michelle Pfeiffer as a TV news reporter who finally, after much hard work, comes to the notice of her boss Redford. However, her climb upwards is matched by his slow A STAR IS BORN-style descent. Quite weepy in places, even if the superficiality of it all is apparent from the word go.
DRA   120 min (ort 124 min)   VIDrel: EIV/SONOP   V

UP IN ARMS ***   U
Elliott Nugent USA   1944
Danny Kaye, Dinah Shore, Constance Dowling, Dana Andrews, Louis Calhern, Lyle Talbot, Margaret Dumont, Elisha Cook Jr, George Mathews, Benny Baker, Walter Catlett, George Meeker, Margaret Dumont, Edward Earle, Harry Hayden
A reluctant hypochondriac recruit in the US Army eventually becomes a hero. Kaye's first film is an untidy mess that hasn't worn well; only his singing of some great patter songs such as "The Lobby Number" and "Melody In 4F" makes it well worth seeing. Shore is attractive as the woman he doesn't love, as is Dowling as the one that he does. Look out for Virginia Mayo as one of the chorus girls. Filmed before as "Whoopee" with Eddie Cantor in 1930.
COM   105 min B/W   VIDrel: VCC/DISC/COLUM   V
Boa: play The Nervous Wreck by Owen Davis.

UP POMPEII **   15
Bob Kellett UK   1971
Frankie Howerd, Patrick Cargill, Lance Percival, Michael Hordern, Barbara Murray, Bill Fraser, Adrienne Posta, Julie Ege, Bernard Bresslaw, Royce Mills, Madeleine Smith, Rita Webb, Ian Trigger, Aubrey Woods, Hugh Paddick
Spin-off of a popular British TV series about a lascivious and lazy slave living in ancient Pompeii. Nero is outwitted and our slave effects his escape from Pompeii just before the volcano goes up. Unfortunately, Howerd's talent for leering and smutty innuendo does not suffice in a film that shows precious little imagination in plotting. Written by Sid Colin with music by Carl Davis.
COM   86 min (ort 90 min)   VIDrel: WHV   V/h

UP THE ACADEMY *   15
Robert Downey USA   1983
Ron Leibman, Ralph Macchio, Wendell Brown, Tom Citera, Tom Poston, Antonio Fargas, Stacey Nelkin, Barbara Bach, Leonard Frey, J. Hutchison, Barry Teinowitz, Ian Wolfe, Luke Andreas, Candy Ann Brown, King Coleman
A truly awful teen comedy set in a military academy peopled by unbelievably gross slobs, whose antics are clearly modelled on films like those in the "National Lampoon" series. Mad Magazine financed this one, and although it was made for Pay-TV release, all references to the magazine and their mascot Alfred E. Neuman were excised. Note that Ron Leibman kept his name off the credits. This was Macchio's film debut.
Aka: MAD MAGAZINE PRESENTS UP THE ACADEMY
COM   85 min (ort 88 min) mPay   VIDrel: MGM/WHV L/A
V/sh

UP THE CHASTITY BELT *   PG
Bob Kellett UK   1971
Frankie Howerd, Graham Crowden, Bill Fraser, Roy Hudd, Hugh Paddick, Anna Quayle, Eartha Kitt, Dave King, Fred Emney, Lance Percival, Nora Swinburne, Godfrey Winn, David Batley, Royce Mills, Veronica Clifford, Rita Webb
A serf and his master go off to the Crusades, in this broad and unfunny farce that stumbles along with precious few gags but much vulgarity.
Aka: CHASTITY BELT, THE; EROTIC ADVENTURES OF DON QUIXOTE, THE; NAUGHTY KNIGHTS
COM   90 min (ort 94 min)   VIDrel: WHV   V/h

UP THE CREEK ***   15
Robert Butler USA   1984
Tim Matheson, Stephen Furst, Dan Monatian, Jennifer Runyon, Sandy Helberg, Jeff East, Blaine Novak, James B. Sikking, John Hillerman, Mark Andrews, Will Bledsoe, Grant Wilson, Julie Montgomery, Jena Tomasina, Romy Windsor
A college team takes part in a desperate bid to win a whitewater rafting race against teams of preppies and military cadets. Matheson and Furst cut their teeth on NATIONAL LAMPOON'S ANIMAL HOUSE, and are reunited in this silly but generally entertaining college-kids comedy. The same level of brash humour is very much in evidence.
COM   92 min (ort 99 min)   VIDrel: RNK L/A   V

UP THE FRONT **   PG
Bob Kellett UK   1972
Frankie Howerd, Zsa Zsa Gabor, Stanley Holloway, Hermione Baddeley, Robert Coote, Bill Fraser, Lance Percival, Peter Bull, Jonathan Cecil, Percy Herbert, William Mervyn, Linda Gray, Madeleine Smith, Nicholas Bennett
In 1914 the footman of a lord is hypnotised into enlisting in the army, and helps a spy steal German plans by getting them tattooed on his buttocks. A tired and dated romp. The script is by Sid Colin and Eddie Braben.
COM   84 min (Cut at film release – ort 89 min)
VIDrel: WHV   V/h

UP THE SANDBOX ***   15
Irvin Kershner USA   1972
Barbra Streisand, David Selby, Jane Hoffman, Ariane Heller, John C. Becher, Jacobo Morales, Paul Benedict, George Irving, Jane House, Pitt Herbert, Janet Brandt, Pearl Shear, Carl Gottlieb, Joseph Bova, Mary Louise Wilson
A young mother living in New York is largely ignored by her piggish husband and finds that she needs her daydream world in order to survive. An uneasy blend of comedy and fantasy that's touching in parts and worth seeing for its surreal sequences. Not a great film, but Streisand's performance redeems it. Written by Paul Zindel.
COM   97 min   VIDrel: STABL L/A   V
Boa: novel by Anne Richardson Roiphe.

UP WORLD **   PG
Stan Winston USA   1990
Anthony Michael Hall, Jerry Orbach, Claudia Christian, Eli Danker, Robert Z'Dar, Mark Harelik, Joseph R. Sicari, Greg Kean, Reuben Grady, Guy Garner, Elizabeth Bliss, Michael Faustino, Tisha Putman, Deanna Oliver, Don Woodard
One of the most bizarre of recent cop-and-his-partner pairings, that teams a police detective with an endearing gnome, when both find themselves with a common enemy. A curious blend of comic, action and fantasy elements, that fails to forge its disparate parts into an effective whole.
Aka: ADVENTURES OF A GNOME CALLED GNORM, THE; UPWORLD
COM   88 min   VIDrel: FIRST/SONOP L/A   V

UPSTAIRS NEIGHBOUR, THE **   18
James Meredino USA   1994
Sebastian Gutierrez, Rustam Branaman, Christina Fulton, Kane Piccoy
A novelist suffering from writer's block finds himself irritated by noises being made by his upstairs neighbour, but on investigating this finds himself being victimised. It transpires that the neighbour is a Satanist who has placed a curse on him, and this will soon lead to his demise unless he can thwart it. A terribly derivative thriller, whose dark supernatural overtones promise many chills, few of which are ever delivered.
THR   89 min   VIDrel: SCEDGE/RTM   V

UPTOWN SATURDAY NIGHT ***

PG

Sidney Poitier USA   1974
Sidney Poitier, Bill Cosby, Harry Belafonte, Calvin Lockhart, Flip Wilson, Richard Pryor, Rosalind Cash, Roscoe Lee Browne, Lee Chamberlain, Paula Kelly, Johnny Sekka, Calvin Lockhart, Lincoln Kilpatrick, Ketty Lester
Two men trying to get a stolen lottery ticket back, run up against the local godfather. A broad and happy comedy with Belafonte hilarious as a gangster boss. Followed by "Let's Do It Again" and A PIECE OF THE ACTION.
COM   100 min (ort 104 min)   VIDrel: MGM/WHV L/A   V

URANUS ***

15

Claude Berri FRANCE   1990
Philippe Noiret, Gerard Depardieu, Jean-Pierre Marielle, Michel Blanc, Michel Galabru, Gerard Desarthe, Fabrice Luchini, Daniel Prevost, Florence Darel, Daniele Lebrun, Myriam Boyer, Josiane Leveque, Dominique Bluzet
Shortly after the end of WW2, a badly damaged French village comes under the control of the Communists, who use their power to ferret out former supporters of Petain and the Vichy government. When one such person seeks sanctuary with an old friend, a witch-hunt is mounted to which the village innkeeper falls victim. A stark account of one of the darker chapters in French history, and a look at how indifference and the instinct to survive can lead to injustice.
DRA   99 min   CINrel
Boa: novel by Marcel Ayme.

URBAN COWBOY **

15

James Bridges USA   1980
John Travolta, Debra Winger, Scott Glenn, Madolyn Smith, Barry Corbin, Brooke Alderson, Mickey Gilley, Charles Daniel Band, Bonnie Raitt, Cooper Huckabee, James Gammon, Betty Murphy, Ed Geldart, Leah Geldart
The story of a young Texan "cowboy" and his macho life style in Pasadena, with Travolta frequenting a honky-tonk bar and working hard to perfect his image. OK if you like Country music, others may find the mechanical bull that is taken on by would-be rodeo riders, one of the film's few highlights. Written by Aaron Latham and Bridges.
DRA   125 min (ort 135 min)   VIDrel: CIC/SONOP L/A   V
Boa: magazine story by Aaron Latham.

URBAN HEAT **

18

Candida Royalle USA
Sharon Kane, Cassandra Leigh, David Sandler, Taija Rae, Scott Baker, Chelsea Blake, Carol Cross, Bernard Daniels, Tish, David Ambrose
A couple of torrid vignettes set in the city, over a sultry weekend. First we see two dancers at the disco carrying their passion off of the dance-floor, and we then follow the developments that arise from a chance meeting between a handsome man and an attractive older woman. Very typical of Royalle, there erotic mini-dramas score well in romantic terms, but fall down heavily in terms of plotting.
A   77 min (ort 92 min)   VIDrel: MIA/DISC   V

URGA ***

PG

Nikita Mikhalkov FRANCE/RUSSIA   1990
Bayaertu Badema, Vladimir Gostukhin, Babushka, Larissa Kuznetsova, Jon Bochinski, Bao Yongyan, Wurinile, Wang Zhiyong, Baoyinhexige, Nikolai Vachtchiline
Drama set in Chinese-controlled Mongolia, where the local inhabitants attempt to adapt to the changes a more modern way of life brings. Fable and fact are skilfully blended here, and a series of arresting images are presented in a film that dispenses with a straightforward narrative structure in preference for a strong sense of time and place.
DRA   114 min (ort 120 min)   VIDrel: CURZON/20TH V/sur

UROTSUKIDOJI 1: LEGEND OF THE OVERFIEND **

18

Hideki Takayama JAPAN   1993
Highly imaginative and grotesque animated tale of a super-demon that seeks to unite humans, man-beasts and monster-demons in a single land every three-thousand years but is opposed by a man who is determined to prevent this from happening. An ultra-violent and sexually explicit offering, suitable for adults only. A couple of sequels followed.
Aka: WANDERING KID, THE
ANIM   103 min (ort 108 min) dubbed
VIDrel: MANGA/SONOP   V

URUSEI YATSURA: REMEMBER MY LOVE **

PG

Kazuo Yamazaki JAPAN   1984
Feature-length animation based on an incredibly popular TV series devoted to the doings of an alien princess and the hapless Earth youth she has fallen madly in love with. The original dialogue is full of multi-layered puns and many cultural references hardly accessible to non-Japanese. In this adventure, Lum tries to find the magician who turned her boyfriend into a pink hippopotamus. A number of the other episodes have also become available.
ANIM   93 min   VIDrel: ANIME/SGSVID L/A   V

USED CARS ***

15

Robert Zemeckis USA   1980
Kurt Russell, Jack Warden, Gerrit Graham, Frank McRae, Deborah Harmon, David L. Lander, Joseph P. Flaherty, Michael McKean, Michael Talbott, Andrew Duncan, Marvey Northup, Alfonso Arau, Al Lewis, Woodrow Parfrey, Dub Taylor
Two car dealers engage in great rivalry with no holds barred when it comes to pulling the customers in. The coarse but hilarious script is by Zemeckis and Bob Gale.
COM   107 min (ort 113 min)   VIDrel: RCA L/A   V

USED PEOPLE **

15

Beeban Kidron USA   1992
Shirley MacLaine, Kathy Bates, Marcello Mastroianni, Marcia Gay Hayden, Bob Dishy, Jessica Tandy, Sylvia Sidney, Bob Dishy, Joe Pantoliano, Matthew Branton, Louis Guff, Charles Cioffi, Doris Roberts, Helen Hanft, Lee Wallace, Asia Vieira
When her husband dies, a Jewish family is scandalised by the presence of an old admirer of the wife, who now sees his chance after waiting some twenty years. Set in Queens in 1969, this is a messy and clumsily directed effort of slight appeal. Despite doing their best, Bates, MacLaine and Tandy are seriously miscast and unconvincing in a work that calls for believable Jewish characters, and instead provides us with the usual parade of weirdos and stereotypes.
DRA   110 min (ort 116 min)   VIDrel: 20TH/TECH L/A   V
Boa: play The Grandma Plays by Todd Graff.

USUAL SUSPECTS, THE ***

18

Bryan Singer USA   1995
Gabriel Byrne, Suzy Amis, Kevin Spacey, Pete Poslethwaite, Steven Baldwin, Chazz Palminteri, Kevin Pollack, Benico Del Toro, Giancarlo Esposito, Dan Hedaya, Paul Bartel, Carl Bressler, Phillip Simon, Jack Shearer, Clark Gregg
There are only two survivors when twenty-seven bodies are recovered from a gutted ship after an attempt to steal $9,000,000 worth of cocaine goes badly wrong. Writer-director Singer has produced a clever but very violent film that opens with small-time crook Spacey squealing on his buddies. An extended flashback sequence then shows how the crooks were brought together by a police line-up in the first place. AA: S. Actor (Spacey), Screen/orig (C. McQuarrie).
THR   101 min (ort 105 min) wScrn cC
VIDrel: POLY/POLYREC   V/sh

UTZ **

12

George Sluizer GERMANY/ITALY/UK   1992
Armin Mueller-Stahl, Brenda Fricker, Peter Riegert, Paul Scofield, Gaye Brown, Miriam karlin, Pauline Melville, Vera Soukupova, Adrian Brine, Peter MacKriel, Caroline Guthrie, Clark Dunbar, Christian Mueller-Stahl, Jakub Zdenek
An art dealer travels to Prague to visit an old friend who his an avid collector of valuable figures and finds that his collection has disappeared, together with his housekeeper. He sets out to solve this mystery and learns that the Communist government is involved.
DRA   94 min (ort 101 min) mTV   VIDrel: WESCON/RTM V
Boa: novel by Bruce Chatwin.

V

V – THE ORIGINAL SERIES: PARTS 1 AND 2 ***

15

Kenneth Johnson USA   1983
Michael Durrell, Faye Grant, Peter Nelson, David Packer, Michael Wright, Jane Badler, Neva Patterson, Tommy Peterson, Blair Tefkin, Bonnie Bartlett, Leonard Cimino, Richard Herd, Evan Kim, Richard Lawson, George Morfogen
Earth is invaded by hordes of aliens whose saucer-like craft appear in the skies above most of its major cities. These "visi-

tors" as they like to be known, claim to be on a mission of peace but soon reveal a more sinister side as they engineer a takeover. An intriguing idea whose power is soon diminished by the self-conscious Nazi references. However, the special effects are generally quite impressive. A much weaker sequel followed.
Aka: V: THE ORIGINAL SERIES: PARTS 1 AND 2
FAN 193 min (ort 208 min) mTV VIDrel: WHV V/dm V/h

V – THE FINAL BATTLE: PARTS 1, 2 AND 3 **   15
Richard T. Heffron/Stuart Heisler USA   1984
Jane Badler, Michael Durrell, Robert Englund, Faye Grant, Richard Herd, Thomas Hill, Michael Ironside, Peter Nelson, David Packer, Neva Patterson, Sandy Simpson, Andrew Prine, Marc Singer, Blair Tefkin, Michael Wright
This mini-series follow-up to V: PARTS 1 AND 2 continues the story of how the Earth resists takeover by an alien race of ruthless lizards, who have concealed both their real plans for humanity and their reptilian nature. The special effects are adequate but hardly exhilarating, while the plot seems to get bogged down in a repetitive series of set-piece confrontations. Overlong and disappointing. A TV series followed, the episodes of which are now availaible.
FAN 258 min (3 cassettes – separately available) mTV VIDrel: WHV V/dm

VACILLATIONS OF POPPY CAREW, THE **   15
James Cellan Jones UK   1994
Sian Phillips, Tara Fitzgerald, Charlotte Coleman, Edward Atterton, Joseph Fiennes, Daniel Massey
A young and flirtatious girl inherits a fortune, and proceeds to enjoy the attentions of four possible suitors, but is unable to discover which of them truly loves her for herself.
COM 100 min mTV VIDrel: ODY/SONOP V

VAGRANT, THE **   18
Chris Walas USA   1992
Michael Ironside, Marshall Bell, Bill Paxton, Mitzi Kapture, Colleen Camp, Patrika Darbo, Stuart Pankin, Marc McClure, Teddy Wilson, Deek Mark Lochran, Mildred Brion, Brett Marston, Ken Love, Katherine Gosney, Steve Gates
A up-and-coming young executive finds an unpleasant surprise waiting for him when he moves into his new home, eventually learning that a repulsive and murderous tramp is living there and will not go, and to make matters worse, appears to have committed murders for which our executive is likely to be blamed. A chilling battle of wits now ensues in this ludicrously plotted but occasionally quite tense movie. Average.
THR 88 min (ort 90 min) mTV VIDrel: 20TH/TECH V/sh

VALACHI PAPERS, THE ***   18
Terence Young ITALY   1972
Charles Bronson, Lino Ventura, Joseph Wiseman, Fred Valleca, Walter Chiari, Jill Ireland
A tense retelling of the life of a Mafia godfather who told all to a crime commission. Despite its many merits, it was overshadowed by THE GODFATHER. Bronson as famed informer Joseph Valachi, spills the beans from his prison whilst we see his reminiscences as a series of flashbacks. Absorbing but untidy. Based on actual events and written by Stephen Geller with music by Riz Ortolani.
Aka: CARTEGGIO VALACHI; COSA NOSTRA; JOE VALACHI: I SEGRETI DI COSA NOSTRA
DRA 120 min (ort 129 min)
VIDrel: 4-FRONT/POLYREC/BRAVE V
Boa: book by Peter Maas.

VALDEZ: THE HALF-BREED ***   15
John Sturges USA   1973 (released in USA in 1976)
Charles Bronson, Vincent Van Patten, Marcel Bozzoffi, Jill Ireland, Melissa Chimenti, Fausto Tozzi, Ettore Manni, Adolfo Thous, Florencia Amarilla, Corrado Gaida, Diana Lorys
Set in 1880, this tells of a half-breed who befriends a teenage boy who helps him on his ranch. However, the man falls in love with the sister of a neighbour who vows to destroy him. An entertaining film that was shot in Spain.
Aka: CHINO; VALDEZ HORSES, THE
WES 93 min (ort 98 min)
VIDrel: 4-FRONT/POLYREC/BRAVE V

VALENTINO ***   18
Ken Russell USA   1977
Rudolf Nureyev, Leslie Caron, Michelle Phillips, Felicity Kendal,

Carol Kane, Seymour Cassel, Peter Vaughan, Anton Diffring, Alfred Marks, Jennie Linden, John Justin
Nureyev stars as Rudolph Valentino in this story of the life of the famous silent screen actor who had an immense female following. This lavish and painstakingly made work has all the makings of a fine film, except the merits which more disciplined, less self-indulgent direction would have brought it. Also filmed in 1951 as "Valentino" and in 1975 as THE LEGEND OF VALENTINO.
DRA 123 min (ort 127 min) VIDrel: WHV V/sur
Boa: book by Brad Steiger and Chaw Mane.

VALERIE AND HER WEEK OF WONDERS **   15
Jaromil Jires CZECHOSLOVAKIA   1970
Jaroslava Schallerova, Jan Klusak, Helena Anyzkova, Petr Kopriva, Jiri Prymek
In Czechoslovakia a young girl grows to adulthood, and has various Freudian fantasies along the way, mostly revolving around a series of imagined adventures that start with her grandmother (with whom she lives) turning into a vampire and being put on trial for witchcraft. Further magical episodes follow, and though the film has great visual charm, its rambling and incoherent structure leaves one feeling unsatisfied and very disappointed.
Aka: VALERIE A TYDEN DIVU
FAN 72 min (ort 77 min) VIDrel: REDEM/RTM V

VALLEY OF GWANGI, THE ***   12
James O'Connolly USA   1969
James Franciscus, Gila Golan, Richard Carlson, Laurence Naismith, Dennis Kilbane, Freda Jackson, Gustavo Rojo, Mario De Barros, Curtis Arden, Jose Burgos
A circus promoter takes a team in pursuit of a tiny prehistoric horse and they find their way to a strange valley that is inhabited by much larger creatures of various kinds. They eventually capture a dinosaur and take it back to their circus but needless to say it escapes and goes on a rampage. First-rate effects by animator Ray Harryhausen compensate for shortcomings such as wooden acting and an unimaginative plot.
A/AD 91 min (ort 95 min) VIDrel: WHV V/h

VALLEY OF THE DOLLS *   15
Mark Robson USA   1967
Barbara Perkins, Patty Duke, Sharon Tate, Susan Hayward, Paul Burke, Martin Milner, Tony Scotti, Lee Grant, Joey Bishop, George Jessel, Charles Drake, Alexander Davion, Jacqueline Susann, Robert Viharo, Mikel Angel, Barry Cahill
A story of corruption and drug-taking with a young actress innocent of the dangers falling victim to the system, intertwined with the stories of two other women trying to cope. Remade as a TV film in 1981, this is a glossy and exploitative adaptation of a similar novel. Written by Helen Deutsch and Dorothy Kingsley with music by Andre Previn and musical direction by John Williams.
DRA 118 min (ort 124 min) VIDrel: 20TH/TECH L/A V
Boa: novel by Jacqueline Susann.

VALLEYS OF THE MOON *   18
Charles Grey USA   1991
Debbie Diamond, Jeanna Fine, Shaunna Stevens, T.T. Boy, Marc Rider, Caesar Waylons, Debbie Vale, Wayne Summers, Tracey Adams, Monique Hall, Bianca Trump, Sikki Nix, Buck Adams, Anastasia, Barbara Madison
A sex therapist uses a bizarre form of therapy, that includes allowing his patients to make love in his waiting room. An over-stylised and very arty adult movie, full of contrived effects, that is hard to make sense of and even harder to watch.
A 62 min VIDrel: GROHOM V

VALMONT ***   15
Milos Forman FRANCE/UK   1989
Colin Firth, Annette Bening, Meg Tilly, Henry Thomas, Fairuza Balk, Sian Phillips, Jeffrey Jones, Fabia Drake, T.P. McKenna, Ian McNeice, Isla Blair, Aleta Mitchell, Ronald Lacey, Vincent Schiavelli, Sandrine Dumas, Antony Carrick
Third retelling of the classic epistolary novel from 1782 that describes the sexual games played by a blase aristocrat and his wife who encourage each other to seduce and manipulate their unwitting victims. A looser adaptation than DANGEROUS LIAISONS (1988) and a good deal lighter in tone. See also LES LIAISONS DANGEREUSES and WHEN A WOMAN IS IN LOVE.
DRA 137 min CINrel
Boa: novel Les Liaisons Dangereuses by Chodelos De Laclos.

**VAMP ** **   18
Richard Jones USA   1986
Grace Jones, Chris Makepeace, Robert Rusler, Gedde Watanabe, Sandy Baron, Edee Pfeiffer, Billy Drago, Brad Logan, Lisa Lyon, Jim Boyle, Larry Spinak, Eric Welch, Stuart Rogers, Gary Swailes, Ray Ballard, Paunita Nichols
Three college boys visit a strange club looking for a stripper for their fraternity party, and fall victim to some modern-day female vampires. Starts out as a very tongue-in-cheek comedy-horror, but this early promise soon gives way to a kinky and ghoulish nightmare that's not made any better by mumbled dialogue and muddled plotting. Very disappointing, despite the best efforts of Jones as outrageous head vampire/stripper Katrina.
HOR 93 min VIDrel: VCC L/A V

**VAMPING ** **   15
Frederick King Keller USA   1985
Patrick Duffy, Christine Hyland, Fred A. Keller, David Booze, Rod Arrants
An out-of-work saxophonist takes part in the robbery of a widow, but his only interest in the crime is to raise enough money to get his instrument out of the pawnshop. However, he finds himself falling in love with his victim. A strange and messy tale of passion, deceit and revenge.
DRA 105 min VIDrel: EIV/SONOP V

**VAMPIRE AT MIDNIGHT ** **   18
Gregory McClatchy USA   1987
Jason Williams, Gustav Vintas, Lesley Milne, Jeanie Moore, Ester Alise, Robert Random, Tom Friedman
Los Angeles witnesses the work of a grisly serial murderer, whose victims are found drained of blood and with their throats slashed. An excitable and unconventional cop is given the task of bringing the killer to justice, little realising what this will entail in this macabre tale.
HOR 89 min (ort 93 min) VIDrel: POPRO/RTM V/h

VAMPIRE BAT, THE * **   PG
Frank Strayer USA   1933
Melvyn Douglas, Fay Wray, Lionel Atwill, Dwight Frye, Maude Eburne, George E. Stone, Carl Stockdale, Robert Frazer, Harrison Greene, Lionel Belmore, Rita Carlisle, Stella Adams, William Humphrey, William V. Mong, Fern Emmett
A remote Bavarian village is the setting for a series of baffling murders of villagers. They have their origin in the efforts of mad doctor Atwill to find a blood substitute. A low budget horror tale that has a few eerie moments and benefits greatly from a strong cast. Beware of the many shorter prints.
HOR 60 min (ort 71 min) B/W VIDrel: REDEM/RTM V

**VAMPIRE CIRCUS ** **   18
Robert Young UK   1971
John Moulder Brown, Adrienne Corri, Laurence Payne, Thorley Walters, Lynne Frederick, Elizabeth Seal, Anthony Corlan, Richard Owens, Domini Blythe, Robin Hunter, Robert Tayman, Mary Wimbush, Lalla Ward, Robin Sachs
Set in 1825, this tells of a travelling circus entirely composed of vampires, including the animals. A slightly unusual Hammer film, but unfortunately the plot, which told of a vampire's curse and a subsequent plague, was spoilt by last minute cuts. The music is by Philip Martel.
HOR 83 min (Cut at film release – ort 87 min)
VIDrel: VCC/DISC/COLUM L/A V

**VAMPIRE COP ** *   18
Joel Bender USA   1993
Michelle Owens, Gregory A. Greer
A female cop is given a last chance to prove her worth when she is assigned the case of a serial killer, but it transpires that she is chasing a vampire, and when she is bitten by him, she has to kill him before he too is claimed by the Undead. A truly dire effort, crudely plotted and badly acted, its tacky sub-plot about sexism in the police force is as contrived as it is insincere.
HOR 81 min VIDrel: OVER/HIFLI V/h

**VAMPIRE HUNTER D ** **   15
Toyoo Ashida JAPAN   1985
Futuristic animation set in the year 12090 AD, with the Earth now barely recognisable, having undergone a cataclysmic changes that have left the planet infested with vampires. The title character is hired to challenge the might of the vampires.
ANIM 80 min (ort 90 min) dubbed
VIDrel: MANGA/SONOP V

**VAMPIRE IN BROOKLYN ** **   15
Wes Craven USA   1995
Eddie Murphy, Angela Bassett, Allen Payne, Kadeem Hardison, Zakes Mokae, John Witherspoon, Joanna Cassidy, Simbi Khali, Messiri Freeman, Kelly Cinnante, Nick Corri, W. Earl Brown, Ayo Adeyemi, Troy Curvey Jr, Vickilyn Reynolds
Vampire spoof in which a handsome Brooklyn-based vampire starts making plans to get his teeth into an attractive cop (Bassett) as he believes she is his soul-mate. There are thrills of the ghoulish sort to be had here, but not much humour and Murphy's great talent for comedy is never fully exploited, the film mostly relying on clumsy slapstick to generate its humour.
HOR 98 min (ort 102 min) VIDrel: CIC/SONOP; PION (LV only) V LV

**VAMPIRE IN VENICE ** *   18
Augusto Caminito ITALY   1987
Klaus Kinski, Donald Pleasence, Christopher Plummer, Anne Knecht, Barbara De Rossi, Elvire Audray, Giuseppe Mannajuolo, Yorgo Voyagis, Clara Colosimo, Micaela Flores Amaya ("La Chunga")
A slow-moving, unofficial sequel to NOSFERATU THE VAMPYRE, with Kinski now resurrected and haunting the canals of Venice during carnival time. As he works his way through various revellers, Plummer as our intrepid vampire hunter is not far behind. An arty exercise in glossiness that is well and truly overdone, with an irritating operatic score and a ludicrous climax for good measure. This self-indulgent mess was written by Caminito.
Aka: NOSFERATU A VENZIA; NOSFERATU IN VENICE
HOR 89 min (ort 106 min) VIDrel: FIRST/SONOP L/A V

**VAMPIRE JOURNALS ** *   18
Ted Nicolaou USA   1996
Jonathan Morris, Starr Andreef, Kirsten Cerre, David Gunn, Star Andreeff, Ilinka Goya, Dan Condurache
Strong on atmosphere, this is another entry in the vampire sub-genre, but kid's TV host Morris is so very bland that one cannot take it seriously. The plot is almost a re-run of INTERVIEW WITH THE VAMPIRE, the dialogue is stuffy and the acting leaves much to be desired. In short, a dud.
HOR 92 min VIDrel: EIV/SONOP V/sh

VAMPIRE LOVERS, THE * **   15
Roy Ward Baker UK   1970
Ingrid Pitt, Peter Cushing, Madeleine Smith, George Cole, Kate O'Mara, Dawn Addams, Pippa Steel, Douglas Wilmer, Jon Finch, Ferdy Mayne, Harvey Hall, Janey Key, Charles Farrell, Kirsten Betts, John Forbes-Robertson, Shelagh Wilcox
A fairly faithful version of "Carmilla", with the voluptuous Pitt a lesbian vampire who seduces many a pretty female victim. The touch of eroticism that Hammer studios added to a dated formula helped prolong the life of this genre, but by this time the paucity of possible plot variations had become apparent. Followed by LUST FOR A VAMPIRE and TWINS OF EVIL.
HOR 91 min (Cut at film release) VIDrel: RNK L/A V
Boa: short story Carmilla by Sheridan Le Fanu.

**VAMPIRE PRINCESS MIYU: CHAPTERS
1 TO 4 ** **   PG/12
JAPAN   1988
Japanese style Gothic horror tale in which a spiritualist seeks the truth about the title figure in a series of chilling confrontations.
ANIM 112 min (two cassettes – approx 55 min each) dubbed
VIDrel: MANGA/SONOP V/sh

**VAMPIRES AND OTHER STEREOTYPES ** **   18
Kevin Lindenmuth USA   1992
Bill White, Wendy Bednarz, Ed Hubbard, Rick Poli, Anna Dipace, Suzanne Scott, Mick McCleery
Two cops investigate a gruesome murder and find that a group of women have inadvertently opened an entrance to Hell which they find populated by giant mutant rats, demons and other terrifying monsters.
HOR 83 min (ort 90 min) VIDrel: SCEDGE/RTM V

**VAMPIRE'S KISS ** **   18
Robert Bierman USA   1989
Nicolas Cage, Maria Conchita Alonso, Jennifer Beals, Elizabeth Ashley, Kasi Lemmons, Bob Lujan, Jennifer Lundy
After a passionate encounter in a singles bar, a young and rather unpleasant literary agent comes to believe he was bitten by a vampire. He soon becomes obsessed with the belief that he is turning into one himself, as his behaviour becomes ever more

erratic. A hysterical, confused black comedy whose satanic barbs are not aimed at any clear targets. However, Cage is outstanding as the demented yuppie victim.
COM 99 min VIDrel: COLUM/SONOP L/A V/sh

VAMPYR *** PG
Carl Theodore-Dreyer FRANCE/GERMANY 1932
Julien West (Nicholas de Gunzburg), Sybille Schmitz, Henrietta Gerard, Jan Hieronimko, Maurice Schutz, Rena Mandel, Jane Mora, Albert Bras, N. Babanini
Dreyer's first sound film, shot on location near Paris in French, German and English, is an eerie and atmospheric tale of a young man who arrives at a village and goes to the local in where he learns that a room has already been made ready for him. This proves the prelude to a host of strange and terrifying events. A rich and evocative film that relies entirely on the power of the imagination to create a mood of doom and horror.
Aka: CASTLE OF DOOM; NOT AGAINST THE FLESH; STRANGE ADVENTURE OF DAVID GREY, THE; VAMPIRE, THE; VAMPYR, DER TRAUM DER DAVID GREY; VAMPYR OU L'ETRANGE AVENTURE DE DAVID GREY
HOR 62 min (ort 83 min) B/W dubbed
VIDrel: REDEM/RTM V
Boa: short stories from In A Glass Darkly by Joseph Sheridan Le Fanu.

VAMPYROS LESBOS ** 18
Jess (Jesus) Franco SPAIN/WEST GERMANY 1970
Soledad Miranda, Dennis Price, Paul Muller
The Countess Nadine lures young girls to her island to satisfy her perverted lusts in this sleazy and occasionally erotic variant on the Dracula legend. Eventually, an attractive female lawyer is sent to the island, her task being to obtain some information about an inheritance, but the Countess gets her fangs into her too. Weakly plotted, low-budget nonsense, it now and then displays a little power, but on the whole is far too hampered by its low production values.
HOR 86 min wScrn VIDrel: REDEM/RTM V

VAN, THE ** 15
Stephen Frears EIRE/UK 1996
Colm Meaney, Donal O'Kelly, Ger Ryan, Caroline Rothwell, Neili Conroy, Ruaidhri Conroy, Brendan O'Carroll, Stuart Dunne, Jack Lynch, Laurie Morton, Marie Mullen, Jon Kenny, Moses Rowen, Linda McGovern, Eoin Chaney, Jill Doyle
Having just lost his job as a baker, O'Kelly spends his redundancy cheque on a mobile fish-and-chip outlet in need of a good deal of repair work. Having got this done, he goes into business with a friend, parking it outside the pub on the nights that see Ireland heading towards the 1990 World Cup finals. Scripted by Doyle from the last of his three "Barrytown" novels of Dublin life, this is an appealing work – if only the dialogue and jokes had been a bit sharper.
COM 100 min CINrel
Boa: novel by Roddy Doyle.

VAN GOGH *** 15
Maurice Pialat FRANCE 1991
Jacques Dutronc, Alexandra London, Bernard Le Coq, Gerard Sety, Corinne Bourdon, Elsa Zylberstein, Leslie Azoulai, Jacques Vidal, Chantal Barbarit, Claudine Ducret, Frederic Bonpart, Maurice Coussoneau, Didier Barbier
Set in a tiny village twenty miles outside Paris, this restrained and thoughtful film (minus the standard "tortured 'genius'" cliches) examines the last few months of the painter's life, up to his tragic death at thirty-seven. Images of great beauty abound, as do fine performances, but it's a pity that the film is both clumsily edited and unfair, with the painter's brother being blamed for our artist's dementia. See also VINCENT & THEO and LUST FOR LIFE.
DRA 152 min (ort 159 min) wScrn VIDrel: ARTIF/20TH V/h

VANESSA *** 18
Hubert Frank WEST GERMANY 1976
Olivier Pascal, Anton Diffring, Uschi Zech, Gunter Clemens, Eva Eden, Henry Heller, Eva Leuze, Astrid Bohner, Gisela Krauss, Peter M. Kruger, Tom Garven
A convent-educated girl has no knowledge of sex except that gained from magazines. When a rich uncle in Hong Kong dies, she discovers on visiting his estate that she is now the owner of a string of brothels. Taken by a new acquaintance to the estate of a pervert, she is sexually abused but is soon rescued and returns to Paris to enjoy her new found wealth. A well-made

and quite sophisticated sex fantasy in the EMMANUELLE mould.
A 83 min Cut (4 min 53 sec plus some cuts subst)
VIDrel: IMPENT V

VANISHED ** PG
George Kaczender USA 1995
George Hamilton, Lisa Rinna, Robert Hayes
Soon after her son is killed in an accident, a divorced woman goes to the States in the hope of starting a new life. She remarries and has another son, but he is kidnapped by her first husband. Standard Danielle Steel shenanigans, quite watchable, and all wrapped up in a predictably happy outcome.
Aka: DANIELLE STEEL'S VANISHED
DRA 120 min mTV VIDrel: MIA/DISC V/sur
Boa: novel by Danielle Steel.

VANISHED WITHOUT A TRACE *** PG
Vern Gillum USA 1992
Karl Malden, Travis Fine, Julie Harris, Tim Ransom, Tom Hodges, Jonathan Hernandez, Bobby Zamorski, Debra Bluford, Kevin Brief, Brendan McCundy, Peggy Friesen, Ken Buehr, Cara Coffman, John Dunbar, Dick Solowicz, Barbara Houston
Drama based on true events and set in a small town in California, where twenty-six schoolkids and their bus driver are held at gunpoint by three young thugs, and the driver is forced to drive the bus into an underground hide-out, where they are imprisoned. While a nationwide hunt is in progress for them, the hostages struggle to find a way of escaping from their predicament before they starve. Strongly acted and absorbing, this is a well above average TV movie.
DRA 90 min mTV VIDrel: BRAVE/SONOP L/A V
Boa: book Why Have They Taken Our Children? by Jackie Bough and Jefferson Morgan

VANISHING, THE **** 15
George Sluizer FRANCE/HOLLAND 1988
Bernard-Pierre Donnadieu, Gene Bervoets, Johanna Ter Steege, Gwen Eckhaus, Bernadette Le Sache, Tania Latarjet, Lucille Glenn, Roger Souza, Caroline Appere, Pierre Forget, Didier Rousset, Raphaeline, Robert Lucibello, Doumee
A young Dutch couple on a motoring holiday through France stop for a few moments at a service station, from which the woman vanishes without trace, having been abducted by a maniac. Three years later the distraught boyfriend begins to receive tantalising postcards that promise to lead to his lost love ultimate confrontation with her sinister abductor. Despite the contrived script, this is a chilling and intelligent examination of madness and evil. Excellent.
THR 102 min (ort 106 min) VIDrel: CONNO/RTM L/A V
Boa: novel The Golden Egg by Tim Krabbe.

VANISHING, THE ** 15
George Sluizer USA 1992
Kiefer Sutherland, Jeff Bridges, Sandra Bullock, Nancy Travis, Park Overall, Maggie Linderman, Lisa Eichhorn, George Hearn, Lynn Hamilton, George Bennett, Frank Giraudeau, George Catalano, Stephen Wesley Bridgewater, Susan Barnes
Three years after the mysterious disappearance of his girlfriend, a man starts to hear from someone who may or may not be her abductor. A well-acted remake of Sluizer's 1988 film but hampered by all the constraints of Hollywood, as typified by an unnecessary happy ending (unlike the original) that severely weakens its dramatic impact.
DRA 105 min (ort 110 min) VIDrel: 20TH/TECH L/A V
Boa: novel The Golden Egg by Tim Krabbe.

VANISHING ACT ** PG
David Greene USA 1986
Margot Kidder, Mike Farrell, Elliott Gould, Fred Gwynne, Graham Jarvis, John Bluethner, Heather Ward Siegel, Wally McSween, Paul Jolicoeur, Grant Lowe, Linda Mackay, Larry Musser, Tony Totino, Reg Glass
A couple married one week have a row and the wife walks out. She returns in the company of the local priest after her husband has reported her as missing. However, he insists that the woman is not his wife. This third TV movie to be made from the play has a few moments of suspense but little else, while Kidder's mannerisms soon begin to irritate. The final resolution is disappointingly contrived. Scripted by Richard Levinson and William Link.
THR 93 min Cut (30 sec – ort 134 min) mTV
VIDrel: 20TH/TECH L/A V
Boa: play Trap For A Lonely Man by Robert Thomas.

VANISHING POINT **   18
Richard C. Sarafian USA   1971
*Barry Newman, Cleavon Little, Dean Jagger, Victoria Medlin, Paul
Koslo, Timothy Scott, Bob Donner, Gilda Texter, Anthony James, Karl
Swenson, Severn Darden, Lee Weaver, Tom Reese, Arthur Malet, Lee
Weaver, Owen Bush*
Given the job of driving a car from Denver to San Francisco,
Newman decides to do it in 15 hours. He is soon being chased
across several states by an army of law enforcement agencies
who finally mount a co-ordinated operation to stop him. A
strange, moody and ultimately pointless film that remains a cult
movie. Little as blind D.J. Super Soul and the rock score (with
musical direction by Jimmy Brown) are high-spots in a fast
journey to nowhere.
DRA   98 min (ort 107 min)   VIDrel: 20TH/TECH L/A   V

VANISHING SON **   15
John Nicolella USA   1993
*Russell Wong, Chi Muoi Lo, Vivian Wu, Rebecca Gayheart, Marcus
S. Chung, Paul Butler, Haing S. Ngor, Eric Frederick, Gary Wade
Morton, Brian Fong, Terry Laughlin, Gianni Lapis, Jim Grollman,
Stele McGoengal, Karen K. Kirschenbauer*
Two brothers are members of China's pro-democracy move-
ment, and are obliged to flee the country when the authorities
set out to arrest them. Given asylum in the USA, they settle in
California, but find it difficult to escape their past. The first of
four films detailing the adventures and exploits of these two
characters. Fair.
Aka: VANISHING SON: THE KLANSMAN
MAR   88 min   VIDrel: CIC/SONOP   V

VANISHING SON 2 **   15
John Nicolella USA   1994
*Russell Wong, Chi Muoi Lo, Tamlyn Tomita, Vivian Wu, James
Walters, Marcus S. Chong, Paul Butler, Ming-Na Wen, Haing S.
Ngor, Dean Stockwell, Harry Lennix, Dustin Nguyen, Eric Frederick,
Mitchell Thomas, Kim Chan, Bev L. Appleton*
Our two brothers have now arrived in the States, where one
chooses a life of crime and membership of a gang whilst the
other is content to live an honest life as a fisherman. But the
activities of both the Triads and the Ku Klux Klan are soon to
disrupt the lives of our brothers.
MAR   86 min (ort 90 min)   VIDrel: CIC/SONOP   V

VANISHING SON 3 **   15
John Nicolella USA   1994
*Russell Wong, Chi Muo Lo, Vivian Wu, Paul Butler, Rebecca
Gayheart, Haing S. Ngor, Jack Gwattney, David Bryce, Gianni Lapis,
Luoyong Wang, Al Leong, Jeff Imada, John David Cadiente, Merritt
Yohnka, Ron Yuan, Alvie A. Gulanding*
A further instalment of this saga, with our two brothers now
finding themselves forced to pit their wits and martial arts skills
against Triad gang members, having been pressured to infil-
trate the gang by the FBI.
MAR   87 min (ort 90 min) mTV   VIDrel: CIC/SONOP   V

VANISHING SON 4 **   15
John Nicolella USA   1995
*Russell Wong, Chi Muo Lo, Mark Valley, Matthew Lillard, Dee
Wallace Stone, Kat Tong Lim, Paul Butler, Gary Wade Morton, Terry
Loughlin, Thomas H. Thompson, Jeff Gibson, Vivian Wu, Marcus S.
Ching, Shewling M. Wong. Rebeca Gayheart*
The final episode of this story, that sees one of the brothers killed
in a shoot-out and the remaining brother now having to face life
alone. But when he is injured by a street gang, he finds romance
in the shape of the woman who takes care of him.
MAR   87 min (ort 90 min) mTV   VIDrel: CIC/SONOP   V

VANYA ON 42ND STREET **   U
Louis Malle USA   1994
*Wallace Shawn, Julianne Moore, Andre Gregory, George Gaynes,
Brooke Smith, Phoebe Bran, Lynn Cohen, Jerry Mayer, Larry Pine,
Madhur Jaffrey*
A group of actors put on a rehearsal of Gregory's play at a 42nd
street theatre (hence the title) and bring it to life despite the
absence of sets and costumes. An interesting minimalist exper-
iment.
DRA   119 min   VIDrel: ARTIF/20TH   V/sur
Boa: play Vanya by Andre Gregory

VAULT OF HORROR **   15
Roy Ward Baker UK   1973
Terry-Thomas, Glynis Johns, Curt Jurgens, Daniel Massey, Anna

*Massey, Dawn Addams, Denholm Elliott, Michael Pratt, John Forbes-
Robertson, Jasmina Hilton, Michael Craig, Edward Judd, Arthur
Mullard, Tom Baker*
Another set of horror stories in this quasi-sequel to TALES
FROM THE CRYPT, borrowed from the stories in E.C. Comics
of the early 1950s. Screenplay is by Milton Subotsky with the five
stories of murder, torture, vampirism and voodoo the work of
Al Feldstein and William Gaines. A good cast works hard, but
this poor effort is definitely short of atmosphere.
Aka: FURTHER TALES FROM THE CRYPT; TALES FROM THE CRYPT 2
HOR   83 min (ort 87 min)   VIDrel: VIPCO/SGSVID   V

VAULT OF HORROR 1 **   18
Tobe Hooper/Russell Mulcahy/Joel Silver USA   199-
James Remar, Whoopi Goldberg, Bill Paxton, Brad Dourif, Joe Pesci
Three short horror tales of varying quality. In "Dead Wait", a
con-artist steal a priceless black pearl from a voodoo witch, but
comes to regret having done so. "People Who Live In Brass
Hearses" tells of two brothers who make a serious mistake when
they rob an ice-cream seller. Finally, "Split Personality" has a
con-artist marrying two identical sisters, who take a fitting
revenge when they learn of his two-timing ways.
Aka: TALES FROM THE CRYPT 1
HOR   80 min (3-story cassette)   VIDrel: FIRST/SONOP   V

VAULT OF HORROR 2 **   18
William Friedkin/Russell Mulcahy/Michael J. Fox USA   199-
*Tia Carrere, Greg Allman, Billy Wirth, Brion James, Teri Garr, Bruno
Kirby, MIchael J. Fox*
Another three short horror tales: "On A Dead Man's Chest",
"Split Second" and "The Trap". In the first one, a woman gets
under the skin of a tattooed rock musician. The second tells of
some discontented lumberjacks, who take a grisly revenge on
their employer and his wife. Finally, the last story is all about a
man who makes his own death in order to get the insurance
money, but finds out too late that this was not a wise move. Fair.
Aka: TALES FROM THE CRYPT 2
HOR   80 min (3-story cassette)   VIDrel: FIRST/SONOP   V

VAULT OF HORROR 3 ***   18
Tom Hanks/Steven De Souza/Stephen Hopkins USA   199-
Beau Bridges, Tony Goldwyn, Kyle MacLachlan
These three tales are entitled: "None But The Lonely", "Abra
Cadaver" and "Carrion Death". In the first one, a wife killer
finds that he has an appointment at the cemetery. The next tale
essays a little black comedy, but with little success. Finally, the
third story tells of a man who is handcuffed to a corpse, this last
offering being the best one in the set if not the entire series.
Aka: TALES FROM THE CRYPT 3
HOR   80 min (3-story cassette)   VIDrel: FIRST/SONOP   V

VAULT OF HORROR 4 **   18
Paul Abascal/Steve Perry/Gilbert Adler USA   1994
*Timothy Dalton, Beverly D'Angelo, Priscilla Presley, Lou Diamond
Phillips, Christopher Reeve, Meat Loaf,*
"Werewolf Concerto" is the first story in this set, and is a weak
attempt to follow the affairs of a werewolf hunter, who arrives
at a remote inn. In "Oil's Well That Ends Well", a couple of con-
artists devise a scheme to convince everyone they have struck
oil at a cemetery, but then fall out when the man plays around
with another girl. Finally, "What's Cookin'" has a down-on-his-
luck restaurateur finding a novel way to deal with his
troublesome landlord.
Aka: TALES FROM THE CRYPT 4
HOR   72 min (3-story cassette)   VIDrel: NWV/HIFLI   V

VEGAS IN SPACE **   15
Phillip R. Ford USA   1993
Doris Fish, Ginger Quest, Miss X, Jennifer Blowdryer
Four male astronauts are ordered by the Empress of the Earth
to infiltrate the title resort, which is reserved solely for women.
In order to accomplish this, they are obliged to take gender
reversal pills. Adult comedy featuring transvestites, a lot of
banter but little in the way of a real plot. However, the work is
so good-humoured it is hard not to dislike it.
COM   86 min   VIDrel: PRIDE/PARADOX   V

VELVET DREAMS **   18
Vincenzo Salviani ITALY/SPAIN
*Kathy Shower, Brett Halsey, Alicia Moro, Ezio Prosperi, Alicia Moro,
Raquel Evans, Susana Egea, Gisele Echevaria, Lucia De Sousa, Daniel
Kennelly, Maria Luque, Gunther Scholoh, Stiwe Lontghans, Jordi
Batalla, Roger Merchant*

An erotic thriller on the lines of FATAL ATTRACTION starring ex-Playboy playmate Shower as a writer of romantic fiction, whose marriage is less than idyllic, as her obsessive husband has difficulty distinguishing between dreams and reality. Fair.
A 78 min Cut (2 min 10 sec – ort 93 min)
VIDrel: MIA/DISC V

VENDETTA **   15
Stuart Margolin USA   1990
Eric Roberts, Carol Alt, Eli Wallach, Burt Yuong, Nick Mancuso, Billy Barty, Thomas Calabro, Gianni Nazzaro, Mario Tadisco, Jason Allen, Federico Capara, Max Margolin, Susan Spafford, Stelio Candelli, Josie Bell, Enrico Lo Verso
After the Mob has her father killed, a young woman swears revenge. But before she can put her plans into operation, circumstances force both her and her mother to seek help from the crime boss responsible. This twist of fate leads to her eventually falling in love with her father's murderer. A long, complex and highly melodramatic tale, full of blood and violence. Watchable, but essentially covering much the same ground as many other such films..
Aka: BRIDE OF VIOLENCE; DONNA D'ONORE; FAMILY MATTER, A; VENDETTA: SECRETS OF THE MAFIA
A/AD 120 min (ort 173 min) mTV VIDrel: MIA/DISC V

VENGEANCE **   18
Anthony M. Dawson (Antonio Margheriti) ITALY/WEST GERMANY   1968 (released in USA in 1971)
Richard Harrison, Claudio Camaso, Allan Collins, Sheyla Rozin, Freddy Unger, Werner Pochath, Paolo Gozlino, Alberto Dell'Acqua, Guido Lollobrigida, Pedro Sanchez
Routine Euro-Western. A man seeks revenge on the outlaw gang who murdered his friend, whose body was town into five pieces by being drawn and quartered. Stylishly realised, but this is an extremely violent and brutal tale.
Aka: JOKO, INVOCO DIO... E MUORI
WES 96 min (ort 100 min) wScrn
VIDrel: 4-FRONT/POLYREC V

VENGEANCE OF FU MANCHU **   PG
Jeremy Summers UK   1967
Christopher Lee, Douglas Wilmer, Tsai Chin, Horst Frank, Maria Rohm, Noel Trevarthen, Howard Marion-Crawford, Tony Ferrer, Peter Carsten, Wolfgang Kieling, Susanne Roquette, Mona Chong
A sequel to THE BRIDES OF FU MANCHU that is the third in a generally poor series of films devoted to the exploits of Sax Rohmer's nefarious villain, who is ever plotting for world domination. Here, our Chinese master criminal has devised a means of cloning key police officials, including his arch enemy Nayland Smith of Scotland Yard. A good cast are wasted in a film that is both lifeless and deficient in action. THE BLOOD OF FU MANCHU followed.
THR 88 min (ort 89 min) VIDrel: BRAVE/SONOP L/A V

VENGEANCE, THE DEMON ***   18
Stan Winston USA   1987
Lance Henriksen, John Di Aquino, Kerry Remsen, Devon Odessa, Jeff East, Joel Hoffman, Cynthia Bain, Kimberly Ross, Florence Schauffler, George "Buck" Flower, Madeleine Taylor Holmes, Tom Woodruff Jr, Mayim Bialik, Briam Bemer
When a man's child is killed in an accident caused by a biker gang, the distraught father goes to a backwoods witch for help, and a monstrous demon is called up to execute revenge. However, the father begins to have second thoughts about what he has done but is unable to control the creature. The directorial debut of make-up expert Winston, this is a film rich in effects that is held back by predictability. Followed by THE REVENGE OF PUMPKINHEAD.
Aka: PUMPKINHEAD; VENGEANCE OF THE DEMON
HOR 82 min (ort 87 min) VIDrel: 20TH/TECH V

VENOM *   15
Piers Haggard UK   1982
Klaus Kinski, Oliver Reed, Susan George, Nicol Williamson, Sterling Hayden, Sarah Miles, Cornelia Sharpe, Lance Holcomb, Mike Gwilym, Rita Webb, John Williamson, Michael Gough, Peter Porteous, Maurice Colbourne, Moti Makan
The ten-year-old son of a wealthy family has been given a deadly black mamba snake as a present. The story has an international terrorist group holding a house full of hostages while the snake slithers through the ventilation ducts, stalking its victims. This clumsy combination of a police thriller and a horror yarn was a waste of the talents of a good cast. Watch it and you can waste your time too. Written by Robert Carrington.
HOR 89 min (ort 98 min) VIDrel: VCC L/A V
Boa: novel by Alan Scholefield.

VENUS IN FURS *   18
Massimo Dallamano ITALY/WEST GERMANY   1968
Laura Antonelli, Ewing Loren, Renate Kasche, Peter Heeg, Mady Rahl, Werner Pochath
The memories of a man who was punished as a child for his voyeuristic leanings forms the basis for this work, which has the central character craving domination, and attempting to realise his dream of being the slave to a cruel and sexually unattainable woman. A tedious, sleazy and often clumsy attempt to bring a virtually unfilmable work to the screen, that does little to get under the skin of its characters. The dreadful dubbing is a further annoyance.
Aka: VENUS IM PELZ
A 81 min (ort 84 min) wScrn dubbed
VIDrel: REDEM/RTM V
Boa: novel by Leopold Von Sacher-Masoch.

VENUS WARS **   15
Yoshikazu Yasuhiko/Shichiro Kabayashi/Kohichi Chiba JAPAN   1989
Voices of: Ben Fairman, Anna Alba, Denia Fairman, Jocelyn Cunningham, Bob Sessions, Bradley Cole, Michael Morris, William Dufries, Peter Marinker
Animated space-age adventure that's set in the year 2089, when the planet is divided between the states of Ishtal and Aphrodia, and a fierce battle rages between them. Adequate.
ANIM 103 min VIDrel: MANGA/SONOP V/sur

VERBOTEN! ***   PG
Samuel Fuller USA   1959
James Best, Susan Cummings, Tom Pittman, Paul Dubov, Harold Daye, Stuart Randall, Dick Kallman, Steven Geray, Anna Hope, Robert Boon, Sasha Harden, Paul Busch, Neyle Morrow, Joseph Turkel, Charles Horvath
An American soldier in Germany just after WW2 leaves his post to marry the woman who saved him from the Gestapo. Unfortunately, their relationship is overshadowed by her fifteen-year-old brother, a member of a neo-Nazi group known as the Werewolves. Sadly, this film is as relevant as ever and its footage of the death-camps is just as shocking. Another excellent work from a neglected and almost forgotten director.
DRA 86 min (ort 93 min) B/W VIDrel: ODY/SONOP V/h

VERDICT, THE ****   PG
Sidney Lumet USA   1982
Paul Newman, Charlotte Rampling, Jack Warden, James Mason, Milo O'Shea, Lindsay Crouse, Edward Binns, Julie Bovasso, Roxanne Hart, James Handy, Wesley Addy, Joe Seneca, Lewis Stadlen, Kent Broadhurst, Colin Stinton
A lawyer who's gone to seed, gets a chance to recover his self respect by fighting a medical incompetence case. Newman gives a powerful performance as the drunken Boston lawyer who has the chance to fight for the values he still believes in; Mason is his wily opponent. Written by David Mamet and one of Lumet's best films. Look out for Bruce Willis in one of his earliest appearances.
DRA 122 min VIDrel: 20TH/TECH V/h
Boa: novel by Barry Reed.

VERNON JOHNS STORY, THE ***   (15)
Kenneth Fink USA   1993
James Earl Jones, Mary Alice, Nicole Leach, Joe Senva, Cissy Houston, Tommy Hollis, Ashanti Blaze, Moses Gibson
Biopic on the life of civil rights leader Vernon Johns, a black pastor who in the 1940s grew increasingly outraged at the brutality the police showed towards Blacks, eventually denouncing this and all other forms of racial discrimination from his pulpit. But his actions lead to increasing tension, and when a deacon dies as a result of this, he is blamed and sacked from his post. Jones is charismatic in the central role of an absorbing and inspiring work.
DRA 91 min SATrel: SKY MOVIES

VERTIGO ****   PG
Alfred Hitchcock USA   1958
James Stewart, Kim Novak, Barbara Bel Geddes, Tom Helmore, Henry Jones, Ellen Corby, Raymond Bailey, Konstantin Shayne, Lee Patrick,

Paul Bryar, William Remick, Julian Petruzzi, Sara Taft, Fred Graham, Mollie Dodd
A retired policeman who has a fear of heights is drawn into a complex web of deceit and betrayal. Stewart plays the cop, hired to keep an eye on Novak, the wife of an old school friend. One of Hitchcock's best thrillers, this eerie and atmospheric film demands repeated viewing to unravel its complexities. Written by Alex Coppel and Samuel Taylor with a fine score by Bernard Herrmann.
THR   122 min (ort 128 min)   VIDrel: CIC/SONOP   V
Boa: novel D'Entre Les Morts by Pierre Boileau and Thomas Narcejac.

VERY BRITISH COUP, A *** 15
Mick Jackson UK   1988
Ray McAnally, Marjorie Yates, Geoffrey Beevers, Hugh Martin, Keith Allen, Oliver Ford Davies, Bernard Kay, Alan McNaughton, Tim McInnery, Christine Kavanaugh, David McKail, Oscar Quitak, Michael Godley, Shane Rimmer
The election of a radical left-wing Labour government dedicated to taking Britain out of NATO sends shock waves through the Establishment who soon find common cause with the Americans. Before long, plans are afoot to bring the government down. First-rate performances and a literate script are the hallmarks of this excellent drama.
DRA   155 min (ort 180 min) mTV
VIDrel: CASPIC/BMGREC L/A   V
Boa: novel by Chris Mullen.

VERY POLISH PRACTICE, A ** (PG)
David Tucker UK   1992
Peter Davison, Joanna Kanska, David TRoughton, Alfred Molina, Trevor Peacock, Adam Przedrzymirski, Dariusz Odija, Polly Hemingway, Nina Marc, Maria Quoss, Agnieszka Robotka, Jelena Budimir, Eileen Maciejewska, Gertan Klauber
This was a feature-length sequel to the earlier "A Very Peculiar Practice" (show on TV in 1986 and 1988) that followed the changing fortunes of an idealistic and rather naive doctor (Davison), who arrives in Poland to take up a job at a run-down hospital in Warsaw. Meanwhile, his Polish wife is on the brink of rekindling an affair with an old flame, who makes his living as a black marketeer. A quirky little satire, atmospheric but strangely ineffective.
DRA   95 min mTV   TVrel: BBC

VESTIGE OF HONOR *** PG
Jerry London USA   1991
Gerald McRaney, Michael Gross, Season Hubley, Kenny Lao, Cliff Gorman, Cary-Hiroyuki Tagawa, Kaiulani Lee, Ernest Abuba, R. Pickett Bugg, Angie Davis, J. Don Ferguson, David Hall, George L. Hasenstab, J. Michael Hunter
Seventeen years after the US pulled out of Vietnam and left the Montagnard tribesmen who fought loyally on their side to their fate, a former US serviceman returns to the region to see what he can do to ease the plight of those that survived. A moving and well made drama that offers an interesting insight into a lesser known (but equally tragic) consequence of the Vietnam War. Based on the real-life exploits of Don Scott.
A/AD   89 min (ort 100 min) mTV   VIDrel: CIC/SONOP V/h

V.I. WARSHAWSKI * 15
Jeff Kanew USA   1991
Kathleen Turner, Jay O. Sanders, Charles Durning, Angela Goethals, Frederick Coffin, Nancy Paul, Charles McCaughan, Stephen Meadows, Lynnie Godfrey, Wayne Knight, Anne Pitoniak, Stephen Root, Robert Clotworthy, Tom Allard, Lee Arenberg
Cobbled together from the detective novels of Sara Paretsky, this has a very glamorous looking Turner playing a hardbitten, but sexy, private investigator. When she takes on a case involving the murder of a hockey player, she gets more involved than she initially wanted to be. Turner deftly blends toughness and sensuality, more's the pity that her efforts are wasted in a film of weak scripting and poor production values.
A/AD   85 min (ort 89 min)   VIDrel: HOLPIC/TECH   V

VIBES * PG
Ken Kwapis USA   1987
Cyndi Lauper, Jeff Goldblum, Julian Sands, Googy Gress, Peter Falk, Michael Lerner, Ramon Bieri, Elizabeth Pena, Bill McCutcheon, Karen Akers, Hercules Vilchez, Jerry Vichi, Leo V. Finnie III, Ray Stoddard, Harvey J. Goldenberg
A silly adventure comedy, in which two psychics team up to

hunt for a lost treasure in the mountains of Ecuador. Written by Lowell Ganz and Babaloo Mandel, this unfortunate excursion probably sounded funnier on paper.
COM   96 min (ort 99 min)   VIDrel: RCA L/A   V

VICE ACADEMY * 18
Rick Sloane USA   1989
Linnea Quigley, Ginger Lynn Allen, Karen Russell, Jayne Hamil, Ken Abraham, Scott Layne
Two women police cadets go undercover and expose a variety of lawbreakers in a bid to secure their graduation. A flat and dreary adult comedy that relies heavily on sex and nudity. Several sequels followed.
COM   88 min (ort 90 min)   VIDrel: MED/POLYREC L/A V/h

VICE VERSA *** PG
Brian Gilbert USA   1988
Judge Reinhold, Fred Savage, Swoozie Kurtz, David Proval, Corinne Bohrer, Jane Kaczmerek, William Prince, Gloria Gifford, Beverly Archer, Harry Murphy, Kevin O'Rourke, Richard Kind, Chip Luca, Ajay Naidu, Elya Baskin, James Hong
A semi-remake of the 1947 British film with a magical Thai skull enabling a workaholic department store boss to swap bodies with his eleven-year-old son. This idea has been done to death, but the Dick Clement and Ian La Frenais script plus the comical Reinhold-Savage duo help keep this one on its toes.
COM   94 min (ort 97 min)   VIDrel: VCC/DISC/COLUM V/sur

VICTIM *** 15
Basil Dearden UK   1961
Dirk Bogarde, Sylvia Sims, Dennis Price, John Barrie, Derren Nesbitt, Peter McEnery, Donald Churchill, Nigel Stock, Anthony Nicholls, Hilton Edwards, Norman Bird, Alan McNaughton, Noel Howlett, Charles Lloyd Pack
Tense and well-handled drama in which a group of homosexuals find themselves the victims of a nasty blackmailing couple. A successful barrister breaks off his relationship with a young boy, who later commits suicide. Soon a blackmail victim himself, he decides to co-operate with the police. Written by Janet Green and John McCormack this was a courageous film (for the time) that will now seem much less daring. Photography is by Otto Heller.
DRA   96 min (ort 100 min) B/W   VIDrel: VCC/RTM   V

VICTIM, THE ** 15
Samo Hung HONG KONG   1981
Samo Hung, Chang Yi, Leung Ka Yan, Fanny
A martial arts student attaches himself to an unwilling master whose main concern is to get away from his new wife.
Aka: PRAY DEATH COMES SIFTLY
MAR   89 min (ort 92 min) wScrn   VIDrel: EAST/DISC   V

VICTIM OF BEAUTY ** 15
Paul Lynch CANADA/USA   1991
Jennifer Rubin, Peter Outerbridge, Sally Kellerman, Stephen Shellen, Michael Ironside, Lindsay Merrithew, Dwight Bucquie, Robert Bockstrel, Andrew Gillies, Tracey Cook, Ross Manson, Chapelle Jaffe, Ho Chow, Carolyn Scott, Alissa Berg
A woman working for a top modelling agency fends off a series of unwelcome suitors, only to discover that each rejected man meets an untimely end. It soon transpires that she has a secret admirer who is so obsessed with her that he has taken to murdering all potential rivals.
THR   86 min (ort 90 min) mCab   VIDrel: CAPIT/GUILD   V

VICTIM OF BEAUTY: THE DAWN SMITH STORY ** 15
Roger Young USA   1991
William Devane, Jeri Lynn Ryan, Michele Abrams, Nick Searcy, Linda Pierce, J. Don Ferguson, Butch Slade, Joyce Bowden, Stephen Preusse, Joe Maggard, Mike Flippo, Chuck Kinlaw, Mark Miller, Alan Sader, Don Tilley, John Bennes
This fact-based drama sees a woman whose younger sister has been kidnapped hearing from the abductor, who takes a perverse delight in taunting her. She sets out to trap him, beginning a dangerous battle of wits in a bid to secure his apprehension.
DRA   94 min   VIDrel: ODY/SONOP   V/sh

VICTIM OF LOVE ** 15
Jerry London USA   1990
Pierce Brosnan, JoBeth Williams, Virginia Madsen, Georgia Brown

A female psycho-analyst who finds herself falling for a charming British professor has her illusions shattered when one of her patients reveals that she is the man's current lover, and that the professor's suave exterior hides a ruthless and deadly killer. An implausible love-triangle tale, so far-fetched that one finds it impossible to take the film seriously. In addition, the extended erotic sequences early on add nothing of value to the story.

DRA   88 min (ort 95 min) mTV   VIDrel: CAPIT/GUILD   V

VICTIM OF LOVE *   18
Paul Thomas USA   1992
Christy Canyon, Jamie Summers, Kim Wylde, Danielle Rogers, Mickey Ray, Randy Spears, Tim Lake, T.T. Boy
When the friend of a married man tells him about his marital infidelities, the latter is so inspired by these tales that he decides to do likewise. Very much a thrown together effort, made without regard for the cast or viewers.

A   63 min (ort 80 min) mCab
VIDrel: GROHOM/MAXSCAN   V

VICTIM OF LOVE: THE SHANNON MOHR
STORY ***   PG
John Cosgrove USA   1993
Dwight Schultz, Bonnie Bartlett, Andy Romano, Sally Murphy, Michael O'Neil, Dennis Boutsikaris, Eddie Jones, Wendy Gazelle, Cara Buono, Dion Anderson, Robert Schenkkan, Matthew Posey, Gregg Henry, Darren Dalton, Robert Apisa, Ann
A woman marries a handsome charmer who turns out be devious and evil. When she is found dead her parents investigate, and ultimately learn that their daughter's husband is a ruthless villain who preys on lonely women, and that prior to their daughter's death, he had taken out no less than six life insurance policies on her. A long court battle now ensues in a bid to see justice is done, in this compelling, fact-based drama.

DRA   88 min (ort 90 min) mTV   VIDrel: ODY/SONOP
V/sh

VICTIM OF RAGE ***   18
Armand Mastroianni USA   1993
Jaclyn Smith, Brad Johnson, Hilary Swank, David Lascher, Carolyn McCormick, Andrew Laurence, Ramsey Midwood, Jason Kristofer, David Gianopoulos, Tom Nibley, David Bryan, Lisa Robin Kelly, Gary Hudson, Rosanna Huffman, David Cromwell
A woman who is married to a cop who uses steroids as a bodybuilder, and learns to her cost that the abuse of these drugs often results in the users experiencing bouts of irrational aggression. Eventually, she grows so terrified of her husband that she plots his murder. Based on the true story of Donna Yaklich, this is a disturbing tale indeed, given conviction and intensity by Smith's outstanding performance as the long-suffering wife.

DRA   91 min (ort 94 min) mTV   VIDrel: ODY/SONOP
V/sh

VICTIMS ***   15
Jerrold Freedman USA   1981
Kate Nelligan, Ken Howard, Madge Sinclair, Jonelle Allen, Pamela Dunlap, Amy Madigan, Rose Portillo, Bert Remsen, Michael C. Gwynne, Sherry Hursey, Karmin Murcelo, Howard Hesseman, Alex Henteloff, Charles Sweigart
When a rapist is freed by the courts on a technicality, his latest victim bands together with three former victims in order to stalk him. Nelligan is good as the woman intent on having her revenge, as is Hesseman as the villain. However, the poor story and plot development dissipate the tension as it develops.

THR   92 min (ort 100 min) mTV   VIDrel: MGM/WHV L/A
V

VICTOR/VICTORIA ***   15
Blake Edwards USA   1982
Julie Andrews, James Garner, Robert Preston, Lesley Ann Warren, Alex Karras, John Rhys-Davies, Graham Stark, Peter Arne, Sherloque Tanney, Norman Chacer, Michael Robbins, David Gant, Maria Charles, Malcolm Jamieson
A struggling singer dresses up as a man and masquerades as a drag artist, becoming a great success in Paris. A stylish and often hilarious version of the 1933 German film "Viktor Und Viktoria", first remade in 1936 as FIRST A GIRL. Preston gives a gem of a performance as the ageing drag queen who put her up to it. See also JUST ONE OF THE GUYS and HER LIFE AS A MAN. AA: Score (Henry Mancini/Leslie Bricusse)

COM   129 min (ort 133 min)   VIDrel: MGM/WHV   V/sur

VICTORIA THE GREAT ***   U
Herbert Wilcox UK   1937
Anna Neagle, Anton Walbrook, H.B. Warner, Walter Rilla, Mary Morris, James Dale, Felix Aylmer, Charles Carson, C.V. France, Gordon McLeod, Arthur Young, Grete Wegener, Paul Leyssac, Percy Parsons, Derrick De Marney
Biopic on the life of Britain's longest reigning monarch that examines some episodes in her life including her romance with Prince Albert. The Jubilee celebration in the final reel is in Technicolor. A solid film, well made and acted, with all the virtues (and faults) of those traditional stolid British biopics. Followed by "Sixty Glorious Years". Written by Robert Vansittart and Miles Malleson.

DRA   106 min (ort 112 min) B/W/Col
VIDrel: BRAVE/SONOP L/A   V
Boa: play Victoria Regina by Laurence Houseman.

VICTORY AT ENTEBBE **   PG
Marvin J. Chomsky USA   1976
Helmut Berger, Theodore Bikel, Linda Blair, Kirk Douglas, Richard Dreyfuss, Stefan Gierasch, David Groh, Julius Harris, Helen Hayes, Anthony Hopkins, Burt Lancaster, Christian Marquand, Elizabeth Taylor, Jessica Walter
A poor treatment of the actual events surrounding the daring June 4 1976 rescue of hostages held, with the connivance of Idi Amin, by terrorists at Entebbe airport. A star cast cannot save a dull movie rushed out after the actual events, originally on tape and then transferred to film. Harris as Amin replaced Godfrey Cambridge who died during filming. See also the far better films OPERATION THUNDERBOLT and RAID ON ENTEBBE.

A/AD   150 min mTV   VIDrel: MGM/WHV L/A   V

VIDEODROME **   18
David Cronenberg CANADA   1982
James Woods, Sonya Smits, Debbie Harry, Peter Dvorsky, Les Carlson, Jack Creley, Lynne Gorman, Julie Khaner, Reiner Schwartz, David Bolt, Lally Cadeau, Henry Gomez, Harvey Chao, David Tsubovchi, Kay Hawtrey, Sam Malkin
The boss of a cable TV station producing broadcasts for an underground market, picks up a series of strange, untraceable transmissions from his own TV set. These transmissions have the power to create hallucinations that initially fascinate, but ultimately destroy him. Despite its intelligent start, this dull film gets slower, wordier and sicker as it develops. The repulsive and strangely pointless special effects are by Rick Baker. Cut before video submission.

HOR   83 min Cut (1 min 12 sec – ort 90 min)
VIDrel: CIC/SONOP; PION (director's cut – LV only)   V/h LV

VIETNAM: PARTS 1 AND 2 **   18
John Duigan/Chris Noonan AUSTRALIA   1987
Barry Otto, Veronica Noonan, Nicholas Eadie, Nicole Kidman, Noel Ferrier, Imogen Annersley, Mark Lee, John Polson, Graeme Blundell, Alan Cassell, Alyssa Cook, Francesca Raft
An account of the Vietnam War as seen through the eyes of a young conscript and telling of the four years he spends fighting. A harsh and uncompromising tale, with a satisfactory if somewhat predictable script.

WAR   360 min (2 cassettes)   VIDrel: SCRN/DISC   V

VIETNAM, TEXAS *   18
Robert Ginty USA   1990
Robert Ginty, Tim Thomerson, Haing S. Ngor, Tamlyn Tomita, Kieu Chinh, Bert Remsen, Kieu Chinh, John Pleashette, David Chow, Chi-Muo Lo, Michelle Chan, Sachi Parker, Peter Pan, Steven Leigh, Steven L. Buckingham, Sydney Lassick
Ginty plays a priest who sets out to find the woman he abandoned (while pregnant) when he was in Vietnam. He finds her living in a Vietnamese community in Texas, where she has become the wife of the local drugs baron. One of those films that starts off with a social conscience and then rapidly descends into the standard cliches and violent action sequences.

A/AD   90 min (ort 92 min)   VIDrel: EIV/SONOP   V

VIETNAM WAR STORY ***   15
Luis Soto/Sandy Smolan/David Burton Morris USA   1989
Steve Antin, Haing S. Ngor, Chris Mulkey, Chau Mau Doan, Will Gotay, Tom Wright, California Xuan Tran, Charles A. Tamburro, Michael E. Tamburro, Jade Hoang, Phong Atwood Vo, Chi-Muoi Lo, Joseph Hieu, Sydney Walsh
Three absorbing tales set in Vietnam in April 1975 when US soldiers and CIA agents were preparing to withdraw just before the fall of Saigon to the Communists. "The Last Outpost" is set

at a South Vietnamese army post, that must survive without its US advisors. In "The Last Soldier" the same story is told from the Vietcong point of view. Finally, "Dirty Work" takes place inside the American Embassy compound twenty-four hours before the final airlift.
Aka: VIETNAM WAR STORY: THE LAST DAYS
WAR   92 min   VIDrel: MGM/WHV L/A   V

VIEW TO A KILL, A **
PG
John Glen UK/USA
1985
Roger Moore, Christopher Walken, Grace Jones, Tanya Roberts, Patrick Macnee, Lois Maxwell, Fiona Fullerton, Patrick Bauchau, David Yip, Manning Redwood, Alison Doody, Willoughby Gray, Desmond Llewelyn, Robert Brown, Walter Gotell
A James Bond adventure with remarkably few gadgets but some memorable stunts as 007 battles it out with a madman out to dominate the world. Walken is a remarkably bland and unconvincing villain, out to destroy California's lucrative "Silicon Valley". This one really does drag, only the spectacular stunts keep it (and Bond) alive. A weak story and a poor note to bow out on for Roger Moore as our secret agent. Followed by THE LIVING DAYLIGHTS.
A/AD   126 min (ort 131 min) wScrn   VIDrel: MGM/WHV
V/sur

VIGIL ***
15
Vincent Ward NEW ZEALAND
1984
Fiona Kay, Bill Kerr, Gordon Shields, Penelope Stewart, Frank Whitten, Bob Morrison, Arthur Sutton, Snow Turner, Lloyd Grundy, Bill Liddy, Joseph Ritaj, Maurice Trewern, Josie Herlihy, Éric Griffin, Sadie Marriner
An eleven-year-old girl has a hard time accepting her father's death, which happened at about the same time as a stranger arrived at their farm. She gradually withdraws into an inner world of fantasy. A sparse and moody drama that often demands one's attention.
Aka: FIRST BLOOD, LAST RITES
DRA   86 min (ort 90 min)   VIDrel: ARTPRO/RTM   V

VIKING SAGAS, THE **
15
Michael Chapman ICELAND
1995
Ralf Moeller, Sven-Ole Thorsen, Ingibjorg Stefandottir, Thorir Waagfjord, Hinrik Olafson, Raimund Harmstorf, Magnus Jonsson, Magnus Olafsson, David Kristjansson, Rurik Haraldsson, Bryndis Detursdottir, Valgedur Runarsdottir
Kjartan, the Prince of Iceland seeks to avenge the murder of his father and sets off on this quest, journeying in search of a man who can show him the true nature of a Viking warrior. An interesting blend of swordplay and mysticism, and former Mr Universe Moeller does his best in this fantasy-adventure, which attempts to rival genuine Icelandic movies on this theme. Unfortunately, a plodding script and an over-reliance on gore are serious handicaps.
A/AD   80 min   VIDrel: EIV/SONOP   V

VIKINGS, THE ***
PG
Richard Fleischer USA
1958
Tony Curtis, Janet Leigh, Kirk Douglas, Ernest Borgnine, Alexander Knox, Frank Thring, James Donald, Maxine Audley, Eileen Way, Edric Connor, Dandy Nichols, Per Buckhoj, Almut Berg
Big-budget historical tale with an-all star cast set in the Viking era and filmed on location in Brittany and Norway, with two men (who are half-brothers but do not know it) contend for the hand of a beautiful princess, and possibly the throne of Northumbria. The plot is hardly imaginative, but the well staged battle scenes and other high production values make this a very fine epic indeed.
A/AD   111 min (ort 116 min) wScrn (special edition)
VIDrel: MGM/WHV   V
Boa: novel The Viking by Edison Marshall.

VILLA RIDES! **
15
Buzz Kulik USA
1968
Yul Brynner, Robert Mitchum, Charles Bronson, Grazia Buccella, Frank Wolff, Robert Viharo, Herbert Lom, Alexander Knox, Diana Lorys, Robert Carricart, Fernando Rey, Regina De Julian, Andres Monreal, Antonio Ruiz, John Ireland
Superficial account of the Mexican Revolution and Pancho Villa's part in it. Set in 1912 this has Mitchum as a captured gunrunner deciding to throw in his lot with the rebels. An untidy and overblown action tale which offers us a rare chance

to see a hisute Brynner. Written by Sam Peckinpah and Robert Towne.
WES   117 min (ort 125 min)   VIDrel: CIC/SONOP L/A   V
Boa: book by William Douglas Lansford.

VILLAGE OF THE DAMNED ***
(PG)
Wolf Rilla UK
1960
George Sanders, Barbara Shelley, Michael Gwynne, Martin Stephens, Laurence Naismith, John Phillips, Richard Vernon, Jenny Laird, Richard Warner, Thomas Heathcote, Charlotte Mitchell, Alexander Archdale, Jenny Laird, Sarah Long
An English village is briefly and mysteriously isolated from the outside world. After this incident, twelve of the women are found to be pregnant and give birth to children who prove to have strange hypnotic powers. An intelligent, low-key approach helps to highlight the feeling of alien menace. Followed by an inferior sequel – CHILDREN OF THE DAMNED. Remade in 1995.
FAN   78 min B/W   TVrel
Boa: novel The Midwich Cuckoos by John Wyndham.

VILLAGE OF THE DAMNED **
15
John Carpenter USA
1995
Christopher Reeve, Kirstie Alley, Linda Kozlowski, Micahel Pare, Mark Hamill, Meredith Salenger, Pippa Pearthree, Peter Jason, Constance Forslund, Karen Kahn, Thomas Dekker, Lindsey Haun, Cody Dorkin, Trishalee Hardy
A wholly unnecessary remake of the 1960 British film, that tells of the arrival in a small village of a clutch of blonde babies, all of whom have deadly telekinetic powers. Transplanted to small-town America, it adds superior special effects and is very good for the first hour, when a strong sense of tension is maintained just prior to the birth of these mutant children. Unfortunately, from this point on the special effects take over to the great detriment of the film.
FAN   94 min (ort 99 min)   VIDrel: CIC/SONOP; PION (LV only)   V/sur LV
Boa: novel The Midwich Cuckoos by John Wyndham.

VILLAIN **
18
Michael Tuchner UK
1971
Richard Burton, Ian McShane, Nigel Davenport, Donald Sinden, Fiona Lewis, T.P. McKenna, Joss Ackland, Cathleen Nesbitt, Elizabeth Knight, Colin Welland, Tony Selby, John Hallam, Del Henney, Ben Howard, James Cossins
Burton plays a vicious homosexual cockney gangster, so feared that the police can never get anyone to testify against him. Heavy doses of gratuitous violence cannot hide the fact that Burton is woefully miscast; Reed does it so much better in SITTING TARGET. Written by Dick Clement and Ian La Frenais.
DRA   93 min (ort 98 min)   VIDrel: WHV   V/h
Boa: novel The Burden Of Proof by James Barlow.

VINCENT & THEO ****
15
Robert Altman FRANCE/HOLLAND/UK/WEST GERMANY 1990
Tim Roth, Paul Rhys, Johanna Ter Steege, Wladimir Yordanoff, Anne Canovas, Jip Wijngaarden, Hans Kesting, Jean-Pierre Cassel, Adrian Brine, Bernadette Giraud, Jean-Francois Perrier, Vincent Vallier
This fine film (originally a 4-hour mini-series) on Van Gogh mostly tells of his later years, when as a poverty-stricken recluse, he was largely dependent on the support of his brother, an art dealer, who tried for years to bring the work of the painter to the attention of the public. A moving portrait that avoids over-sentimentality, with the director giving an object lesson in how to make a biopic. Photography is by Jean Lepine. See VAN GOGH and LUST FOR LIFE.
DRA   134 min (ort 138 min) mTV
VIDrel: MANGA/SONOP   V

VINEYARD, THE **
18
James Hong/Bill Rice/Michael Wong USA
1989
James Hong, Karen Winter (Karren Witter), Michael Wong, Lars Wanberg, Cheryl Lawson, Rue Douglas, Sean P. Donahue, Karl Heinz Teuber, Ruth Lin, Michael Quion, Lissa Zappardino, Mark De Alessandro, Vivian Lee, Quincy Loo
A scientist who requires blood and bodies to remain immortal, lures unsuspecting victims to his island, drinking the blood of young women, and preparing a special wine from vines fertilised by the bodies of young men. Quite stylishly done, but this vampire variant is woefully lacking in fresh ideas.
HOR   89 min (ort 103 min)   VIDrel: VCC/DISC/COLUM L/A   V

**VIOLATION OF TRUST ** ** 15
Charles Correll USA 1991
Katey Sagal, Jameson Parker, Robert Picardo, Alan Rachins, David Lascher, Charlotte Ross, Linda Pierce, Kimberley Hooper
Having recently been divorced, a woman faces an uphill struggle re-establishing a relationship with her daughter. Unfortunately, just when mother and daughter develop a degree of mutual trust, the latter is arrested for the murder of a classmate.
DRA 90 min VIDrel: GUILD/SONOP V/sh

VIOLENT COP * ** 18
"Beat" Takeshi (Takeshi Kitano) JAPAN 1989
"Beat" Takeshi (Takeshi Kitano), Maiko Kawakami, Makoto Ashigawa, Shiro Sano, Shigeru Hiraizumi, Mikiko Otonashi, Kaku Ryu, Ittoku Kishibe, Ken Yoshizawa, Hiroyuki Katsube, Noboru Hamada, Yuuki Kawai, Ritsuko Amano
A maverick police detective dispenses his own form of justice, completely oblivious to the questionable ethics of his methods. He ultimately takes on both a Yakuza gang and his own superiors, who might baulk at his methods, but in truth are a good deal less honest. In the hands of the director, this simple story becomes a fast-paced and engrossing offering, quite brutal and sadistic, but also not without a measure of wry humour.
Aka: SONO OTOKO, KYOBO NI TSUKI
A/AD 98 min (ort 103 min) VIDrel: ICAPRO/MANGA V/sh

**VIOLENT MEN, THE ** * PG
Rudolph Maté USA 1955
Barbara Stanwyck, Glenn Ford, Edward G. Robinson, Dianne Foster, Brian Keith, Richard Jaeckel, Brian Keith, May Wynn, Warner Anderson, Basil Ruysdael, Lita Milan, James Westerfield, Jack Kelly, Willis Bouchey, Harry Shannon
A small rancher is forced to abandon his pacifist beliefs when he clashes with a ruthless land baron who is using terror tactics to expand his empire. The murder of one of his hands pushes him over the edge and he starts setting brush fires in a bid for revenge. Good photography does little to compensate for a distinct lack of pace and indifferent direction.
Aka: ROUGH COMPANY
WES 92 min (ort 95 min) VIDrel: VCC/DISC/COLUM V
Boa: novel Rough Company by Donald Hamilton.

**VIOLENT STREETS ** ** 18
Michael Mann USA 1981
James Caan, Tuesday Weld, Willie Nelson, James Belushi, Robert Prosky, Tom Sinorelli, Dennis Farina, Nick Nickeas, W.R. Bill Brown, Norm Tobin, John Santucci, Gavin MacFayden, Chuck Adamson, Sam Cirone, Spero Anast
Mann's directorial debut is a sick and violent story of a professional thief who gives up his dream of leading a normal life, in order to have his revenge on his mobster boss. The score is by Tangerine Dream.
Aka: THIEF
DRA 118 min VIDrel: MGM/WHV L/A V/s
Boa: novel The Home Invaders by Frank Hohimer.

**VIOLENT TRADITION ** ** 15
John Woo CANADA/HONG KONG 1996
Ivan Sergei, Michael Wong, Nicholas Lea, Sandrine Holt, Jennifer Dale
In Hong Kong, the adopted son of a gangster goes against his family, and together with his girlfriend, joins a secret law enforcement agency. Intended to be a pilot for a US TV series, this clumsy action-comedy is mostly a remake of ONCE A THIEF, but is weak in plotting, dialogue and acting. Flashes of Woo's imaginative touches are present, especially in the gunfights, but there is little to show that this is the same director who made films like THE KILLER.
Aka: JOHN WOO'S VIOLENT TRADITION
A/AD 97 min mTV VIDrel: BMGVID/BMGREC V

**VIOLENT YEARS, THE ** * (15)
Franz Eichorn USA 1956
Barbara Weeks, Glenn Corbett, Gloria Farr, Jean Moorehead, Arthur Milan, Theresa Hancock, JoAnne Cangi, Lee Constant, I. Stanford Jolley
Camp exploiter written by Edward D.Wood Jr. is a rather graphic account of leather-clad female gangs, with plenty of gratuitous toughness and violence.
Aka: FEMALE; TEENAGE THEATER: THE VIOLENT YEARS
DRA 70 min (ort 80 min) VIDrel: WARMUS L/A V/sh

**VIPER ** ** 18
Tibor Takas USA 1994
Lorenzo Lamas, Frankie Thorn, Hank Cheyne, Joe Son, Kimberley Kates, Beau Starr, John P. Ryan, S. Diamant
A man's black sheep brother is released after five years in jail, and promptly gets into trouble again when he steals money from South American drug barons.
A/AD 90 min VIDrel: NEWAGE/TECH L/A V

**VIRGIN AMONG THE LIVING DEAD ** ** 18
Jess (Jesus) Franco SPAIN 1971
Christina Von Blanc, Britt Nichols, Paul Muller, Anne Libert (Josiane Gibert), Rose Kiekens, Howard Vernon
A woman goes to British Honduras for the reading of her father's will and discovers plenty of nasty things in the eerie family house.
Aka: AMONG THE LIVING DEAD; EROTIC RITES OF FRANKENSTEIN, THE; ZOMBIE 4:
HOR 85 min (ort 90 min) wScrn VIDrel: REDEM/RTM V

**VIRGIN HUNTERS ** ** 18
Ellen Cabot USA 1993
Morgan Fairchild, Brian Bremer, Christopher Wolf, Ian Abercrombie, Michelle Matheson, Don Dowe, Sarah Suzanne Brown, Tamara Tohill
In a world of the future, sexual intercourse has been outlawed and two students attempt to seek out a part of the world free from the restraints of the "Moral Police". Adult comedy of the nudge-nudge, wink-wink variety.
COM 76 min VIDrel: MED/DISC V/sh

**VIRGIN MACHINE ** ** 18
Monika Treut WEST GERMANY 1988
Ina Blum, Dominque Gaspar, Susie Sexpert (Susie Bright), Marcelo Uriona, Gad Klein, Peter Kern, Hans-Christoph Blumenberg, Shelly Mars, Fritz Mikesch, Wolfgang Raach, George Lannan, Mona Mur, Erica Marcus, Rhonda Jarvis
A woman journalist comes to the USA and undertakes an investigation of the world of lesbians in San Francisco. Some amusing sequences (including Bright as a sex therapist) are offered, but otherwise little that seems fresh or original. Screenplay is by Treut and this was her first solo-directed feature.
Aka: DIE JUNGFRAUENMASCHINE
DRA 82 min mTV VIDrel: DTK L/A V

VIRGIN ON THE RUN * ** 18
Jack Remy USA 1992
Savannah, Jeanna Fine, Taylor Wayne, Rachel Ryan, Buck Adams, Marc Wallice, Mike Horner
After she is ditched by her former lover who stole all her money, a young girl gets involved with a famous boxer, who cannot understand why women need a romantic approach. Eventually he mends his ways and they fall in love, but she still returns to her old boyfriend, and our pugilist, now a wiser man, does likewise in respect of an old flame. A convincingly acted adult film that gave Savannah her debut.
Aka: ROXIE; ROXY
A 77 min (ort 80 min) VIDrel: JOKER/FALCON L/A V

VIRGIN QUEEN, THE * ** U
Henry Koster USA 1955
Bette Davis, Richard Todd, Joan Collins, Herbert Marshall, Jay Robinson, Dan O'Herlihy, Rod Taylor, Robert Douglas, Romney Brent, Marjorie Hellen, Lisa Daniels, Lisa Davis, Barry Bernard, Robert Adler, Noel Drayton, Ian Murray
Davis gives a pleasing performance as Queen Elizabeth I, in this costume drama that examines her relationship with Sir Walter Raleigh. Of little value as a history lesson, but an engaging piece of cinema. Written by Harry Brown and Mindret Lord. The music is by Franz Waxman.
DRA 87 min (ort 92 min) VIDrel: 20TH/TECH V/h

VIRGIN SOLDIERS, THE * ** 15
John Dexter UK 1969
Hywel Bennett, Lynn Redgrave, Nigel Davenport, Nigel Patrick, Tsai Chin, Rachel Kempson, Michael Gwynn, Jack Shepherd, Christopher Timothy, Geoffrey Hughes, Don Hawkins, Roy Holder, Jolyon Jackley, Riggs O'Hara
The story of National Service recruits and their experiences in Singapore, set during the time of the Communist insurgency in Malaya. A frothy comedy-drama, whose title refers to the recruits' lack of experience in bed as well as in battle. Written

by John Hopkins and followed by the inferior sequel STAND UP VIRGIN SOLDIERS in 1977.
DRA   90 min (ort 96 min)   VIDrel: VCC/DISC/COLUM   V
Boa: novel by Leslie Thomas.

VIRGIN SPRING, THE **** 15
Ingmar Bergman SWEDEN   1959
Max Von Sydow, Brigitta Pettersson, Brigitta Valberg, Gunnel Lindblom, Axel Duberg, Tor Isedal, Ove Porath, Allan Edwall, Gudrun Brost, Oscar Ljung, Axel Slangus, Tor Borong, Leif Forstenberg
A medieval drama telling of the rape and murder of a farmer's daughter by three brigands, and of her father's revenge. The title refers to the spring that gushes forth at the spot of her murder. Written by Ulla Isaakson, this is a stark and severely beautiful rendition of a medieval legend. The photography of Sven Nykvist is no small asset. Remade (after a fashion) as THE LAST HOUSE ON THE LEFT. AA: Foreign.
Aka: JUNGFRUKALLAN
DRA   85 min (ort 88 min) B/W   VIDrel: TART/20TH   V

VIRGIN WIFE, THE ** 18
Franco Martinelli ITALY   1976
Edwige Fenech, Carroll Baker, Ray Lovelock, Renzo Montagnani, Michele Gammino, Florence Barnes, Antonio Guidi, Maria Rosaria Rivizzi, Gianfranco De Angelis, Dino Matielli, Rosaura Marchi, Gastone Rescucci
A newly married couple have sexual problems when the husband finds that he is impotent in this slight Italian sex-comedy. Desperate to cure himself, he enlists the aid of a variety of call girls, waitresses, sexy girls etc., even turning for help to his twin cousins.
Aka: LA MOGLIE VERGINE; YOU'VE GOT TO HAVE HEART
A   94 min (ort 98 min)   VIDrel: IMPENT   V

VIRGIN WITCH ** 18
Ray Austin UK   1970
Ann Michelle, Vicki Michelle, Neil Hallett, James Chase, Patricia Haines, Keith Buckley, Paula Wright, Christopher Strain, Esme Smythe, Garth Watkins, Helen Downing
Two girls go to London to become models but get involved in Satanism when one of them, a psychic lesbian, tries to make her girlfriend a witch. Dull nonsense.
Aka: LESBIAN TWINS
HOR   85 min (ort 93 min)   VIDrel: REDEM/RTM   V
Boa: novel by Klaus Vogel.

VIRIDIANA *** 15
Luis Bunuel MEXICO/SPAIN   1961
Silvia Pinal, Francisco Rabal, Fernando Rey, Margarita Lozano, Victoira Zinny, Teresa Rabal, Jose Calvo, Joaquin Roa, Luis Heredita, Jose Manuel Martin, Lola Gaos, Juan Garcia Tienda, Maruda Isbert, Joaquin Mayol
Bunuel's first film to be made in his native country after twenty-nine years is a savage but none too clear parable that attacks Christianity. A novice about to take her vows inherits her uncle's home, and opens it to the poor and the homeless, but this motley collection of beggars and cripples repays her kindness with nothing but sullen ingratitude. A complex and controversial work that provides demanding if rewarding viewing. Won the Golden Palm at Cannes.
DRA   87 min (ort 90 min) B/W   VIDrel: ELPIC/POLYREC   V

VIRTUAL DESIRE ** (18)
Noble Henri USA   1993
Julie Strain, Michael Meyer, Gail Harris, Hoke Howell, Marcia Gray, Tammy Parks, Catherine Weber, Larissa McComas, Annette Bruger, Kimberly Knight, Peggy Trentini, Ross Hagen
Erotic drama built around a deadly game of computer dating, that tells of a bored husband who thinks the Internet is a safe way to add some spice to his life, by indulging in virtual reality without the drawbacks of the real thing. When his wife is found drowned in his swimming pool the police suspect him of murder, but our man becomes convinced one of his computer conquests is responsible, and he attempts to track her down.
DRA   95 min   SATrel: SKY MOVIES

VIRTUAL REALITY 69 * 18
Jim Enright USA   1995
Jenna Jameson, Rebecca Lord, Kristy Waay, China Lee, Christine Tyler, Steven St Croix, Peter North, Alex Sanders

Cyberspace sex in this story of a female hacker who discovers that she can add some reality to this virtual experience.
A   84 min   VIDrel: ONE   V

VIRTUOSITY *** 18
Brett Leonard USA   1995
Denzel Washington, Kelly Lynch, Russell Crowe, Stephen Spinella, William Forsythe, Louise Fletcher
It is now 1999 A.D. and a former cop languishes in jail for the murder of the man who wiped out his family. But when a virtual-reality being that has been programmed with the minds of two-hundred killers and demagogues escapes into the real world, our ex-cop is released from jail as (just like in DEMOLITION MAN) only he has the skill to deal with this menace. A very simple and derivative story, but Crowe makes a memorable villain and there is ample action on offer.
FAN   102 min (ort 105 min) cC   VIDrel: CIC/SONOP; PION (LV only)   V/sur LV

VISION, THE * 18
Scotty Fox USA   1991
Savannah, Candice Walker, Domonique, K.C. Wiliams, Randy West, Mike Horner, Tom Byron
An author becomes obsessed with his fictional girl, but his wife does not mind at all as she has a real lover of her own and in any case is thinking about getting a divorce. Meanwhile, the author's agent has become obsessed with his client's creation too. Nonsensical and badly developed, it is not helped much by its abrupt ending.
A   60 min   VIDrel: MOOM/RIO L/A   V

VISION OF CHRISTY, A * 18
Paul Thomas USA   1990
Christy Canyon, Lois Ayres, Rick Savage, Peter North, Patricia Kennedy, T.T. Boy, Alice Springs, Joey Silvera, Martin Danielles
A standard adult movie vehicle for Canyon, in which she has a number of encounters with assorted lovers. Undistinguished.
A   64 min Cut (2 min 19 sec – ort 80 min)   VIDrel: FIFTH/DISC   V

VISITING HOURS * 18
Jean-Claude Lord CANADA   1980
Michael Ironside, Lee Grant, William Shatner, Linda Purl, Harvey Atkin, Helen Hughes, Lenore Zann, Michael J. Reynolds, Kirsten Bishopric, Dustin Waln, Debra Kirschenbaum, Elizabeth Leigh Milne, Maureen McRae, Neil Afflec
Ironside as a psychopath stalks woman journalist Grant inside a hospital. An utterly distasteful film that's little more than an exploitative journey of humiliation, degradation and mutilation.
DRA   99 min (Cut at film release – ort 105 min)
VIDrel: 20TH/TECH L/A   V

VISITORS OF THE NIGHT * 15
Jorge Montesi USA   1996
Stephen McHattie, Candace Cameron, Dale Midkiff, Markie Post, Todd Allen, Charles Dutton, Faith Ford, Thomas Gibson, Christopher Gray
A scientist is killed shortly after he managed to break into a top-secret lab where an alien body was being stored. He leaves both his young son and some highly incriminating evidence with his sister, but the aliens abduct the boy and try to capture her, in a bid to eliminate the last witness to their presence on Earth. A feeble and very predictable affair, derivative in plotting and flatly directed.
FAN   90 min mTV   VIDrel: MARQ/QUANT   V

VISTA VALLEY P.T.A. ** 18
Anthony Spinelli USA   1980
Jamie Gillis, John Leslie, Jesie St James, Dorothy Le May, Desiree West, Aaron Stuart, Dewey Alexander, Juliet Anderson, Shirley Woods, Kay Parker
A new female English teacher arrives at a high school where two former teachers have been raped. The film now moves on to the sexual activities of the townsfolk and their wayward kids. The new teacher tries to make the kids study but doesn't have much success and is eventually raped herself. A kind of spiced up version of "Blackboard Jungle". Cut before video submission by 24 min 6 sec.
A   53 min Cut (9 sec – ort 77 min)   VIDrel: MOPIC/SGSVID   V

VITAL SIGNS * 15
Marisa Silver USA 1990
Adrian Pasdar, Diane Lane, Jimmy Smits, Norma Aleandro, Jack Gwaltney, Laura San Giacomo, Jane Adams, Tim Ransom, Bradley Whitford, Lisa Jane Persky, William Devane, James Kanre, Telma Hopkins
A medical melodrama that focuses on the lives and traumas of a set of third-year students as they struggle with both private and professional pressures and compete for an internship post. Hardly incisive, barely entertaining and painfully cliched.
DRA 98 min (ort 103 min) VIDrel: 20TH V/sur

VIVA KNIEVEL! * PG
Gordon Douglas USA 1977
Evel Knievel, Gene Kelly, Red Buttons, Lauren Hutton, Leslie Nielsen, Marjoe Gortner, Cameron Mitchell, Eric Shea, Frank Gifford, Albert Salmi, Dabney Coleman, Sheila Allen, Eric Olson, Ernie Orsatti, Sidney Clute
Unlike "Evel Knievel", this is nothing more than a shameless promo-film for our famous stunt-driver; the feeble plot has drug dealers planning to kill him in order to use his car to ship drugs from Mexico to the USA. Ample car stunts are on offer in a mindless but unintentionally funny piece of junk. The utterly ludicrous opening has Knievel paying a midnight visit to an orphanage to distribute models of his car. One boy throws down his crutches and walks!
Aka: SECONDS TO LIVE
DRA 100 min VIDrel: MGM/WHV L/A V

VIVA ZAPATA! **** PG
Elia Kazan USA 1952
Marlon Brando, Jean Peters, Anthony Quinn, Joseph Wiseman, Arnold Moss, Alan Reed, Margo, Frank Silvera, Mildred Dunnock, Harold Gordon, Lou Gilbert, Nina Varela, Florenz Ames, Bernard Gozier, Frank De Kova
The career of Mexican revolutionary Emiliano Zapata, who rose from peasant to leader of his country, but met his death when betrayed by a friend. Scripted by John Steinbeck, this romanticised look at history serves as a fine star vehicle for Brando, who is on top form in the title role. The AAN score is by Alex North. AA: S. Actor (Quinn).
A/AD 108 min (ort 113 min) B/W VIDrel: 20TH/TECH V/h

VIVACIOUS LADY *** U
George Stevens USA 1938
Ginger Rogers, Fred Astaire, James Stewart, Charles Coburn, Beulah Bonoi, Frances Mercer, James Ellison, Phyllis Kennedy, Franklin Pangborn, Grady Sutton, Jack Carson, Alec Craig, Willie Best
This entertaining comedy has Rogers as a nightclub singer marrying Stewart, who plays a professor. He tries to break the news to his conservative family and his fiancee back home. Overlong but quite good fun; the brawl between bride and ex-sweetheart is a little gem.
COM 87 min (ort 90 min) B/W VIDrel: VCC L/A V

VIXEN * 18
Russ Meyer USA 1968
Erica Gavin, Harrison Page, Garth Pillsbury, Jon Evans, Michael Donovan O'Donnell, Vincent Wallace, Robert Aiken, Jackie Ilman
Another in a long series of unfunny Meyer spoofs in which the sex is as phoney as the humour, this time concerning a voluptuous young woman with an unlimited appetite for sex. The talents of Meyer were employed as writer, producer, director and cameraman.
A 85 min VIDrel: ALLIED/RTM V

VOICE FROM THE GRAVE ** 15
David S. Jackson USA 1996
Kevin Dobson, Megan Ward, John Terlesky, Michael Riley
After a nasty murder, the only evidence that could convict the culprit comes from the grave. A supernatural tale of ghostly possession and retribution. Fair.
FAN 89 min VIDrel: ODY/SONOP V/sh

VOICE OF THE HEART: PARTS 1 AND 2 ** 15
Tony Wharmby USA 1989
Lindsay Wagner, Victoria Tennant, Honor Blackman, Richard Johnson, James Brolin, Stuart Wilson, Leigh Lawson, Neil Dickson, Kathryn Leigh Scott, David Baxter, Pip Torrens, Timothy Carlton, Jerry Harte, Trudy Weiss, Danny Schiller
A long, lavish soap opera, telling of an actress who is destined

for stardom and a novelist who becomes her friend, but falls in love with one of her leading men. A glossy piece of escapism.
DRA 196 min (2 cassettes)
VIDrel: 4-FRONT/POLYREC/ODY V/h

VOICES FROM BEYOND ** 18
Lucio Fulci ITALY 1991
Karina Hoff
A typical low-budget horror film from a notorious director of many such epics. This one involves the strange death of a rich financier, the visions and nightmares of his daughter, and the thirst the spirit of the dead man feels for revenge. The usual gore is on offer here, but little else.
Aka: VOCI DEL PROFONDO; VOICES FROM THE DEEP
HOR 85 min (ort 90 min) VIDrel: VCC/DISC L/A V

VOICES IN THE GARDEN ** (PG)
Pierre Boutron FRANCE/UK 1992
Anouk Aimee, Joss Ackland, Samuel West, Kashia Figura, Gayle Hunnicutt, Andre Oumanski, Nael Kervoas, Lisa Roy, Isabelle Mamann, Maureen Zufferey, Michael Davies, Jorgen Falk, Lou Ann Graham, Andreas Geiss, Afra Rosbach
A British academic and his wife find their staid and boring lifestyle thrown into turmoil by the arrival of a back-packing couple at their home in the South of France. A sluggish character study that slowly relinquishes its secrets, but by the time one has learnt all there is to know about the two couples one has long since ceased to care. However, the acting cannot be faulted.
DRA 86 min (ort 90 min) mTV TVrel: BBC
Boa: novel by Dirk Bogarde.

VOLCANO: FIRE ON THE MOUNTAIN ** (18)
Graeme Campbell USA 1996
Dan Cortese, Cynthia Gibb, Don Davis, Lynda Boyd, Colin Cunningham, Micah Gardner, John Novak, Kendall Cross, Brian Kerwin, Jano Frandsen, April Telek, Wiil DeVry, Kerry McPherson, Tasha Simms, Johnatahn Walker, Lorena Gale
A busy ski resort, located on the slopes of a dormant volcano in California, is expecting an excellent season, when a seismologist becomes convinced that an eruption is imminent. His superior scoff at the idea and nobody else seems prepared to believe him, despite increasing seismic activity and the disappearance of a newlywed couple. Some dramatic sub-plots hardly enhance this disaster tale. See also DANTE'S PEAK for a big-screen version of the same idea.
A/AD 90 min mTV SATrel: SKY MOVIES

VOLERE, VOLARE *** 18
Maurizio Nichetti/Guido Manuli ITALY 1991
Angela Finocchiaro, Maurizio Nichetti, Mariella Valentini, Patrizio Roversi, Remo Remotti, Luigi Gravier, Mario Gravier, Renato Scarpa, Massimo Sarchielli, Osvaldo Salvi, Lidia Biondi, Enrico Grazioli, Mario Pardi, Sergio Cosentino
Something of a grown-up variant on WHO FRAMED ROGER RABBIT, this has a sound-effects man who works on animations finding himself so smitten by a pretty girl that he starts to change into a cartoon character. A weird fantasy that blends live-action and animation with considerable skill, but is somewhat lacking in the plot department. See also COOL WORLD for another interesting foray into this field.
COM 91 min (ort 96 min) wScrn VIDrel: TART/20TH V/sur

VOLUNTEERS ** 15
Nicholas Meyer USA 1986
Tom Hanks, John Candy, Rita Wilson, Tim Thomerson, Gedde Watanabe, George Plimpton, Ernest Harada, Allan Arbus, Ji-Tu Cumbaka, Jacqueline Evans, Chick Hearn, Pamela Gula, Philip Guilmann, Virginia Kiser, Clyde Kusatsu
A playboy joins the Peace Corps to escape his Mafia creditors, and is sent to Thailand to educate the locals. Set in 1962, this film has a few good moments for Candy but lacks anything of real substance to hold it together. A dull and dismal comedy that replaces humour with profanity.
COM 102 min (ort 107 min) VIDrel: WHV V/sur

VON RYAN'S EXPRESS *** PG
Mark Robson USA 1965
Frank Sinatra, Trevor Howard, Raffaella Carra, Brad Dexter, Sergio Fantoni, Edward Mulhare, John Leyton, James Brolin, Wolfgang Preiss, Adolfo Celi, Vito Scotti, John Van Dreelen, Michael Goodliffe, Ivan Triesault
An unpopular American captain leads a daring escape of POWs

from a WW2 Italian prison camp by taking over a freight train. An exciting war thriller marred by some flat spots and an unexpectedly downbeat ending. Written by Wendell Meyers and Joseph Landon.
WAR   112 min (ort 117 min)   VIDrel: 20TH/TECH   V/h
Boa: novel by David Westheimer.

VOODOO **   18
Rene Eram USA   1995
Corey Feldman, Jack Nance, Sara Douglas, Joel J. Edwards, Diane Nadeau, Ron Melendez, Amy Raasch, Bryan Michael McGuire, Christopher Kriesa, Clark Tufis, Maury Ginsburg, Darren Eichhorn, Brendan Hogan, John David Ward, Aaron Kuhr
A young man finds himself missing his girlfriend who is away at an L.A. college, so he decides to enrol there too. In need of accommodation, he joins a fraternity only to find that it is a front for more sinister activities, with its members pursuing immortality by means of voodoo. The expected bloodbath is not long in coming.
HOR   89 min   VIDrel: FIRST/SONOP   V

VOODOO DAWN *   15
Steven Fierburg USA   1989
Raymond St Jacques, Theresa Merrit, Gina Gershon, Kirk Baily, J. Grant Albrecht, Tony Todd
Two college friends travel to the Deep South to pay a visit to a third, but find that he has vanished and that his girlfriend believes him to be a victim of voodoo.
HOR   90 min   VIDrel: 20VIS/SONOP L/A   V

VOW TO KILL, A **   PG
Harry S. Longstreet USA   1994
Julianne Phillips, Richard Grieco, Gordon Pinsent, Peter MacNeill, Nicole Oliver, Tom Cavanagh, Larissa Lapchinski, Patrick Paterson, Eva Mai Hoover, Lili Francks, Daniel Warry-Smith, Damir Andrei, Ron Van Hart, Jed Dixon, C.J. Silas
After her fiance dies, a woman lawyer falls for a photographer who convinces her to have a quick wedding. However, their honeymoon in a remote location soon shows that his intentions towards her are hardly loving and worst is to come when she learns that he is already married. A trite and predictable thriller of the conventional kind, with adequate acting.
THR   87 min (ort 91 min) mTV   VIDrel: CIC/SONOP L/A   V

VOWS OF DECEPTION **   15
Bill L. Norton USA   1996
Cheryl Ladd, Nick Mancuso, Mike Farrell, Michael Woolson, Nancy Cartwright
A woman who marries a rich businessman sets out to seduce the man's son. Another one of those psychological thrillers that slowly unravels as the inconsistencies in the plot start to make themselves apparent. Nonetheless, it remains a perfectly good time-filler.
THR   90 min mTV   VIDrel: ODY/SONOP   V/sh

VOYAGE, THE **   15
Fernando E. Solanas ARGENTINA/FRANCE   1991
Walter Quiroz, Soledad Alfaro, Ricardo Bartis, Cristina Vecerra, Chiquinho Brandao, Marc Berman, Franklin Caicedo, Carlos Carella, Angela Correa, Lilliana Flores, Juana Hidalgo, Justo Martinez, Kike Mendive, Francisco Napoli, Fito Paez
A young boy sets out on a long journey in the hope of finding his father, an anthropologist who is working in Brazil. This odyssey serves as a vehicle with which to take a long, sweeping and fascinating examination of the culture and peoples of South America, with the central character travelling overland on bicycle from the southern tip of Argentina right up to Mexico. Ultimately however, its sheer length and lack of structure are its undoing.
Aka: EL VIAJE
DRA   133 min (ort 150 min) wScrn   VIDrel: TART/20TH   V

VOYAGE **   15
John MacKenzie USA   1992
Rutger Hauer, Eric Roberts, Karen Allen, Connie Nielsen, Hazel Ellenby, Larry Powell, Peter Baldacchio, Martin Corrado, Joe Zarb Cousin, Phyllis Carlysle, Sue Ellen Denisen, Betty Mitchell
When their marriage hits the rocks, a couple decide to take a sailing vacation but unwisely take along another couple whom they have only just met. Unfortunately, their newly acquired friends prove to have psychopathic tendencies. Highly remi-

niscent of DEAD CALM but without any real suspense, while the performances are adequate but far from outstanding.
DRA   86 min (ort 90 min) mCab   VIDrel: EIV/SONOP   V/sur

VOYAGE OF TERROR: THE ACHILLE LAURO
AFFAIR ****   15
Alberto Negrin FRANCE/ITALY/USA/
WEST GERMANY   1990
Burt Lancaster, Eva Marie Saint, Robert Culp, Dominique Sanda, Joseph Nasser, Bernard Fresson, Brian Bloom, Gabriele Ferzetti, Renzo Montagnani, Foued Nassah, Rebecca Schaeffer, Adriana Innocenti, Joschen Horst, Yoshi Ashdot
Lancaster and Saint play the couple Leon and Marilyn Klinghoffer, whose monstrous treatment at the hands of Palestinian terrorists when the title vessel was hijacked in 1985 (wheelchair-bound Leon was murdered and thrown overboard) caused an international wave of revulsion. An excellent account, superior to THE HIJACKING OF THE ACHILLE LAURO, with Nasser quite chilling as the terrorist leader. The script is by Negrin and Sergio Donati.
Aka: IL SEQUESTRO DELL'ACHILLE LAURO
THR   174 min (2 cassettes – ort 200 min) mTV
VIDrel: ODY/SONOP   V

VOYAGE OF THE DAMNED **   PG
Stuart Rosenberg SPAIN/UK   1976
Max Von Sydow, Faye Dunaway, James Mason, Oskar Werner, Orson Welles, Luther Adler, Katherine Ross, Malcolm McDowell, Lee Grant, Ben Gazzara, Denholm Elliott, Jose Ferrer, Julie Harris, Wendy Hiller, Fernando Rey, Janet Suzman
In 1939 a liner carrying 937 Jewish refugees sailed from Germany for Havana, but once there they were denied asylum by the Cuban authorities. The liner then returned to Europe and was allowed to dock in Belgium, where some 600 of the passengers ultimately met their deaths during the Holocaust. Despite Von Sydow's fine performance this shameful story is poorly and insensitively told. Written by Steve Shagan and David Butler, with music by Lalo Schifrin.
DRA   175 min (ort 178 min)   VIDrel: L/A   V
Boa: book by Gordon Thomas and Max Morgan-Witts.

VOYAGE TO ITALY ***   PG
Roberto Rossellini ITALY   1953
George Sanders, Ingrid Bergman, Natalia Ray, Leslie Daniels, Maria Mauban, Paul Muller, Anna Proclemer, Jackie Frost
A couple go by car to Naples to sell a house left to them by a relative and as the journey progresses the strains in their marriage become increasingly evident. A well-made account of a troubled relationship that for once ends on an optimistic note.
Aka: JOURNEY TO ITALY; LONELY WOMAN, THE; STRANGERS; TRIP TO ITALY, A; VIAGGIO IN ITALIA
DRA   81 min (ort 83 min) B/W   VIDrel: CONNO/RTM   V

VOYAGE TO THE BOTTOM OF THE SEA, A ***   U
Irwin Allen USA   1961
Walter Pidgeon, Joan Fontaine, Barbara Eden, Peter Lorre, Robert Sterling, Henry Daniell, Michael Ansara, Frankie Avalon, Regis Toomey, John Litel, Howard McNear, Henry Daniell, Mark Slade, Charles Tannen, Delbert Monroe, David McLean
A giant experimental submarine is sent to prevent the Earth being destroyed by firing a missile at the Van Allen radiation belts in space, these having caught fire. Enjoyable and colourful nonsense, with Pidgeon at his bullying best as the admiral out to save the world from being fried. Written by Allen and Charles Bennett, this was later made into a rather dreary TV series.
FAN   100 min (ort 105 min) wScrn   VIDrel: 20TH/TECH L/A; FOXVID; ENCORE/FOXVID (LV only)   V/sh LV

VOYAGER ***   15
Volker Schlondorff FRANCE/GERMANY   1992
Sam Shepard, Julie Delpy, Barbara Sukowa, Dieter Kirchlechner, Traci Lind, Deborah Lee-Furness, August Zirner, Thomas Heinze, Bill Dunn, Peter Berling, Lorna Farrar, Kathleen Matiezen, Lou Cuttell, Charles Hayward, Irwin Wynn
A globe-trotting engineer whose life revolves around facts, has sacrificed all emotional commitment to his work. However, a series of inexplicable and powerful coincidences impress upon his mind the power of fate, and he ultimately meets a young woman who is destined to lead him back to a woman he left

behind before WW2. This pleasing meditation on life, love and fate is bleakly honest in tone and visually quite spectacular.
DRA   108 min (ort 115 min) (English dialogue)
VIDrel: CURZON/20TH   V/sh
Boa: novel Homo Faber by Max Frisch.

VOYAGER FROM THE UNKNOWN **   U
James D. Parriott/Rick Colby (Winrich Kolbe) USA 1982/1983
Jon-Erik Hexum, Meeno Peluce, Faye Grant, Donald Petrie, Ed Begley, Sondra Currie, Peter Frechette, Jeanie Bradley, Terence O'Hara, Suzanne Barnes, Sam Chew Jr, Lee De Broux, Will Kuluva, John McLiam, Hugh Reilly
SF adventure featuring two episodes from "Voyagers!" – a TV series that ran from 1982 to 1983. A young orphan becomes lost in time together with a time traveller from the future. He accompanies him on his travels, meeting a number of historical figures such as Mary Pickford, the Wright Brothers and Louis Pasteur.
FAN   87 min mTV   VIDrel: CIC/SONOP L/A   V

VROOM **   15
Beeban Kidron UK   1988
Clive Owen, David Thewlis, Diana Quick, Bill Rodgers, Tim Potter, Rosalind Bennett, Melanie Kilburn, James Duggan, Tricia Penrose, Dicken Ashworth, Louis Mellis, Jackie D. Broad, Christine Cox, Martin Oldfield
Two friends and the older woman who lives next door take off from their small town in search of happiness and adventure.
DRA   84 min (ort 89 min) mTV   VIDrel: FIRST/SONOP   V

W

WACKIEST SHIP IN THE ARMY, THE **   PG
Richard Murphy USA   1960
Jack Lemmon, Ricky Nelson, Chips Rafferty, John Lund, Patricia Driscoll, Tom Tully, Joby Baker, Warren Berlinger, Mike Kellin, Richard Anderson, Alvy Moore, Teru Shimada, George Shibata, Richard Torrence, Naaman Brown
A battered WW2 ship and its inexperienced crew still manage to pull off some heroic exploits, in this standard military comedy whose material is completely run-of-the-mill and predictable. Lemmon, however, brings such an air of frenzied energy to his role that his outstanding performance carries the entire film. Later a brief TV series.
COM   95 min (ort 99 min)   VIDrel: COLUM/SONOP   V

WAGES OF FEAR, THE ****   PG
Henri-Georges Clouzot FRANCE   1955
Yves Montand, Charles Vanel, Peter Van Eyck, Folco Lulli, Vera Clouzot, William Tubbs, Dario Moreno, Jo Dest, Centa, Luis De Lima, Jeronimo Mitchell
In a broken-down Central American town, four impoverished individuals agree to undertake a suicide mission for a fee of $2,000 each. Their task is to transport a cargo of nitro-glycerine to the site of an oil-well fire. This classic suspense epic is a magnificent study of human nature in the face of great danger and was a much deserved hit at the 1953 Cannes Film Festival. Remade in 1977 as "Sorcerer" by William Friedkin.
Aka: LE SALAIRE DE LA PEUR
DRA   147 min (ort 155 min) B/W   VIDrel: AROW/RTM   V

WAGNER **   15
Tony Palmer HUNGARY/UK/WEST GERMANY   1983
Richard Burton, Vanessa Redgrave, Laurence Olivier, Ralph Richardson, John Gielgud, Franco Nero, Ronald Pickup, Gemma Craven, Laszlo Galffi, Ekkehard Schall, Richard Pasco, Marthe Keller, Bill Fraser, Arthur Lowe, Cyril Cusack
A vast, sprawling biopic of this famous ego-maniacal composer whose music does not appeal to all tastes. An overlong and superficial study that covers most of the main events in the composer's long life, although it never really seems to get to grips with its subject. Music is performed by the orchestras of the London, Vienna and Budapest Philharmonics.
Aka: WAGNER: THE COMPLETE EPIC
DRA   488 min (3 cassettes – 540 min) mTV
VIDrel: CONNO/RTM   V

WAGONS EAST! **   PG
Peter Markle USA   1994
John Candy, Richard Lewis, John C. McGinley, Robert Picardo, Ellen Greene, William Sanderson, Ed Lauter, Melinda Culea, Joe Bays, Rodney A. Grant, Michael Horse, Abe Benrubi, Jill Boyd, Ryan Cutrona, Stuart "Proud Eagle" Grant
A group of people decide that they no longer wish to suffer the rigours of life in a frontier town and engage a wagon-master to take them back East to the joys of civilisation. This so-so western spoof was Candy's last film appearance and he died in the middle of shooting the film, forcing the director to make some unplanned modifications to the script, reducing Candy to little more than a peripheral character.
COM   103 min (ort 107 min)   VIDrel: GUILD/FOXVID   V/sur

WAITING **   15
Jackie McKimmie AUSTRALIA   1990
Noni Hazlehurst, Deborra-Lee Furness, Frank Whitten, Helen Jones, Denis Moore, Fiona Press, Ray Barrett, Nogo Bernstein, Peter Tu Tran, Brian Simpson, Matthew Fargher, Alan Glover, Kaye Stevenson, Justin King, Jeanette Cronin
A pregnant woman gets help with her imminent birth when her friends decide to use this event as an excuse for a reunion and return home from abroad in order to be with her.
DRA   94 min   CINrel

WAITING FOR THE LIGHT ***   PG
Christopher Monger USA   1990
Teri Garr, Shirley MacLaine, Vincent Schiavelli, Jeff McCracken, Jack McKee, John Bedford Lloyd, Clancy Brown, Colin Baumgartner, Hilary Wolf, William Dore, Robert Ginsburg, Mark Drusch, Eric Helland, Arthur H. Cahn, Lyn Terrell
A woman inherits a rundown diner in a small town and in a bid to get over a broken romance, moves there in the hope of making a fresh start. She packs up the family, complete with eccentric aunt (MacLaine) and sets to work, but various complications ensue. A slimly plotted comedy, with good performances from Garr and MacLaine and an undeniable measure of charm.
COM   90 min (ort 95 min)   VIDrel: EIV/SONOP   V

WAITING TO EXHALE **   15
Forest Whitaker USA   1995
Whitney Houston, Angela Bassett, Loretta Devine, Lela Rochon, Gregory Hines, Dennis Haysbert, Mykelti Williamson, Michael Beach, Leon, Wendell Pierce, Donald Adeosun Faison, Jeffrey D. Sams, Jazz Raycole, Brandon Hammond, Kenya Moore
Four professional black women all search for meaningful relationships, but none of their men can ever come up to their expectations. A shallow comedy-drama of superficial scripting and overly plush sets (strongly emphasising the women's middle-class status). Pleasant enough in its way, but no profound insights are offered, the male figures are depressingly stereotyped, and for good measure the film shows romance as just another aspect of the American dream of success.
DRA   124 min cC   VIDrel: 20TH/TECH   V/sur
Boa: novel by Terry McMillan.

WAITRESS! *   18
Samuel Weil (Lloyd Kaufman)/Michael Herz USA   1982
Jim Harris, Carol Drake, Carol Bevar, June Martin, David Hunt, Calvert DeForest, Renata Majer, Carl Sturmer, Bonnie Horan, Anthony Sarrero, Augie Grompone, Ed Fenton, Fred Salador, Wendy Stuart, Bill Kirksey, Katya Colman
A comedy about the strange staff of a high-class restaurant. Unpleasant, tasteless and moronic in roughly equal proportions.
Aka: SOUP TO NUTS
COM   82 min (ort 93 min)   VIDrel: EIV/SONOP   V

WAKE OF THE RED WITCH ***   PG
Edward Ludwig USA   1948
John Wayne, Gail Russell, Luther Adler, Gig Young, Adele Mara, Eduard Franz, Grant Withers, Henry Daniell, Paul Fix, Dennis Hoey, Jeff Corey, John Wengraf, Erskine Sanford, Duke Kahanamoku, Henry Brandon, Myron Healey, John Pickard
An adventure story set in the East Indies and involving a hunt for treasure and the rivalry between an East Indies shipping boss and an adventurous ship's captain. The good action sequences and fine photography by Reggie Lanning are compensations for the flashback style of the narrative, which is confusing at times. Written by Harry Brown and Kenneth Gamet.
A/AD   102 min (ort 106 min)   VIDrel: 4-FRONT/POLYREC   V
Boa: novel by Garland Roark.

WALK IN THE CLOUDS, A **   PG
Alfonso Arau USA   1995
Keanu Reeves, Aitana Sanchez-Gijon, Giancarlo Giannini, Anthony

Quinn, Angelica Aragon, Evangelina Elizondo, Freddy Rodriguez, Debra Messing, Febronio Covarrubias, Roberto Huerta, Juan Jimenez, Ismael Gallegos, Alejandra Flores
At the end of WW2 a young soldier comes home to his unloving wife and gets a job as a confectionery salesman. On a bus he meets a pregnant single woman whom he befriends, agreeing to pose as her husband so that she can avoid in incurring the wrath of her dad, and to carry this off he accompanies her to Southern California where her family has a vineyard. A pleasant romantic comedy-drama that would have been considerably better had Reeves been considerably better.
DRA 96 min (ort 102 min) cC
VIDrel: 20TH/TECH; ENCORE/FOXVID (LV only)
V/sur LV

WALK THE PROUD LAND **
Jesse Hibbs USA
Audie Murphy, Anne Bancroft, Pat Crowley, Chalres Drake, Tommy Rall, Robert Warwick, Jay Silverheels, Eugene Mazzola, Anthony Caruso, Victor Milian, Ainslie Pryor, Eugene Iglesias, Morris Ankrum, Addison Richards, Maurica Jarr
An ex-soldier takes on the difficult task of securing some measure of self-government for the Apache as their Indian agent and is eventually able to persuade Geronimo to surrender. Competent if less than enthralling account of the true-life figure John P. Clum.
WES 84 min (ort 88 min)
VIDrel: 4-FRONT/POLYREC/CIC V
Boa: novel Apache Agent by Woodworth Clum.

U
1953

WALKER *
Alex Cox USA
Ed Harris, Richard Masur, Rene Auberjonois, Marlee Martin, Peter Boyle, Migual Sandoval, Gerrit Graham, Keith Szarabajka, Sy Richardson, Xander Berkeley, John Diehl, Peter Boyle, Alfonso Arau, Pedro Armendariz, Alan Bolt
An account of one of the more bizarre cases of US intervention in Central America, when a soldier of fortune in 1855 briefly became the virtual ruler of Nicaragua. This messy account of the exploits of William Walker fails to make the most of its material. The script is by Rudy Wurlitzer.
DRA 91 min (Cut at film release by 6 sec – ort 95 min)
VIDrel: CIC/SONOP L/A V

18
1987

WALKER, TEXAS RANGER: DEADLY REUNION ** PG
Michael Preece USA   1995
Chuck Norris, Clarence Gilyard, Noble Willingham, Sheree J. Wilson, Floyd Red Crow Westermann
In this feature-length episode of the TV series, our brave ranger swings into action to prevent the assassination of an ambitious US senator with presidential ambitions. To do so, he has to stop a hired killer who is posing as cop from Kansas.
Aka: DEADLY REUNION
A/AD 87 min (ort 92 min) mTV VIDrel: WHV V/sh

WALKER, TEXAS RANGER: FLASHBACK *
USA
Chuck Norris, Sheree Wilson, Clarence Gilyard
Texas Ranger Cord Walker mounts an investigation when he is contacted by the spirit of a 19th century Ranger. This involves him venturing into the desert in search of some missing gold, the locating of which would help clear up a 130-year-old mystery. A truly dire entry in a series of adventures involving this boring character. As ever, Norris demonstrates that acting is not his strong point.
Aka: FLASHBACK
A/AD 92 min mTV VIDrel: WHV V/h

12
1995

WALKER, TEXAS RANGER: ONE RIOT, ONE RANGER **
Virgil W. Vogel USA   1993
Chuck Norris, Gailard Sartain, Clarence Gilyard, James Drury, Sheree J. Wilson, Floyd Red Crow Westerman, Marshall Teague, Elya Baskin, James Drury, Rhoda Gemignani, Marco Perella, Steven Ruge, Michael Crabtree, Woody Watson
A Texas Ranger with some martial-arts skills teams up with an athlete and a female DA in order to take on some very nasty criminals. A by-the-numbers actioner that serves as pilot for the TV series and offers no surprises.
Aka: ONE RIOT, ONE RANGER
A/AD 90 min (ort 96 min) mTV VIDrel: WHV V/sh

15

WALKER, TEXAS RANGER: ROAD TO VENGEANCE *
Michael Preece USA
Chuck Norris, Sheree Wilson, Clarence Gilyard
Texas Ranger Norris is on holiday with his partner in the Louisiana Bayou, where they are enjoying a fishing trip, that is until they get caught up in a conflict.
Aka: ROAD TO VENGEANCE
A/AD 90 min mTV VIDrel: WHV V

15
199-

WALKING AND TALKING ***
Nicole Holofcener UK/USA
Catherine Keener, Anne Heche, Todd Field, Liev Schreiber, Kevin Corrigan, Randall Batinkoff, Joseph Siravo, Vincent Pastore, Lynn Cohen, Amy Braverman, Miranda Stuart Rhyne, Brenda Thomas Denmark, Rafael Alvarez, Ritamarie Kelly
Two young women in their late twenties have been friends since childhood, but one of them feels herself to be very much in the shadow of the other as far as relationships with the opposite sex are concerned. A bittersweet tale of love and friendship that is quite appealing despite the tendency (so typical for films of this kind) to stereotype most of the male characters as both selfish and insensitive. The script is by Holofcener.
COM 85 min CINrel

15
1996

WALKING THUNDER **
Craig Clyde USA
James Read, David Tom, Klara Irene Miracle, Chritopher Neame, Chief Ted Thin Elk, Kevin Conners, Billy Oscar, Don Shanks, Robert Doqui, John Denver, Kasey Clyde, David Kirk Chambers, Carolyn Hurlburt, Wayne Brennan, Brian Keith
A youngster comes across the old diaries kept by his grandfather in his pioneer days, and learns just how he survived winter in the Rockies and also about his encounter with an Indian, whose mystical beliefs included the certainty that the soul of his dead son now inhabited the title creature – a grizzly bear. A most enjoyable outdoors adventure.
A/AD 95 min SATrel: SKY MOVIES

(PG)
1993

WALL OF SILENCE **
Philip Saville UK
Warren Mitchell, Bill Paterson, John Bowe, Brian Bovell, Suzanna Bertish, Juliette Caton, Tusse Silberg, Helen Kluger, Harry Landis, Clive Panto, Ken Bones, Jonathan Deverall, Saul Reichlin, Jack Klaff, Brigitte Kahn, Iris Russell
A murder among the orthodox Jewish Chasidic community in the North London district of Stamford Hill strikes fear into the other members of this close-knit sect. A downbeat murder tale, whose unusual setting does little to help generate a sense of tension. See CLOSE TO EDEN for a film that makes a slightly better use of this theme and WITNESS for a film that has a similar idea – but is set among the Amish.
DRA 90 min mTV TVrel: BBC

(15)
1993

WALL STREET ***
Oliver Stone USA
Michael Douglas, Charlie Sheen, Martin Sheen, Daryl Hannah, James Spader, Terence Stamp, Sylvia Miles, Hal Holbrook, Sean Young, Richard Dysart, Saul Rubinek, Annie McEnroe, Tamara Tunie, Franklin Cover, Josh Mostel
A young broker whose career is going nowhere manages to gain the confidence of a powerful Wall Street broker, and thus gain admittance to the exclusive world of financial dealing. However, in the process he is forced to "sell his soul" and compromise his principles. A shallow and simplistic morality tale updated to a modern setting; the noisy opening sequences and a general weakness of characterisation are major drawbacks. AA: Actor (Douglas).
DRA 120 min VIDrel: 20TH/TECH L/A V/sh

15
1987

WALTON WEDDING, A **
Robert Ellis Miller USA
Richard Thomas, Ralph Waite, Michael Learned, Jon Walmsley, Judy Norton, Mary Elizabeth McDonough, Eric Scott, David Harper, Kami Cotler, Kate McNeil Joey Conley, Ronnie Claire Edwards, Holland Taylor, Ellen Corby, Diane Baker
A feature-length TV movie that brings together the cast from long-running TV series "The Waltons". In this story the family gathers to celebrate a forthcoming wedding, but trouble threatens when the bride's rich family start trying to take over the proceedings. An amiable and well mounted if rather sentimental outing, that plays very much like an extended episode from the series (which came to an end in 1981).
DRA 88 min (ort 90 min) mTV SATrel: SKY MOVIES

(U)
1994

WALTONS THANSKGIVING REUNION, A ** (PG)
Harry Harris USA 1993
Richard Thomas, Ralph Waite, Michael Learned, Ellen Corby, Jon Walmsley, Joe Conley, Judy Norton, Mary McDonough, Eric Scott, David Harper, Kami Cotler, Ronnie Claire Edwards, Tony Becker, Steven Culp, Lisa Harrison, Mary Jackson
Another feature-length movie spin-of from the TV series. In this story the family are brought together for Thanksgiving of 1963, but their celebrations are blighted by the news of President Kennedy's assassination, an event that forces each family member to take stock of their lives. Fair.
DRA 90 min mTV SATrel: SKY MOVIES

WANDA *** 15
Barbara Loden USA 1970
Barbara Loden, Michael Higgins, Jerome Thier, Dorothy Shupenes, Charles Dosinen, Frank Jourdano, Valerie Manches
A feckless young woman gets divorced, and drifts into a life of petty crime, becoming the accomplice of a small-time crook. The only film ever made by Loden, this is a low-key, realistic work (done in semi-documentary style) that is set in a small town in Pennsylvania. The film won the International Critics' prize for best film at the 1970 Venice Film Festival.
DRA 98 min VIDrel: CONNO/RTM V

WANDERERS, THE **** 18
Philip Kaufman USA 1979
Ken Wahl, John Friedrich, Karen Allen, Toni Kalem, Alan Rosenberg, Erland Van Lidth de Jeude, Linda Manz, Olympia Dukakis, Jim Youngs, Tony Ganios, Val Avery, Dolph Sweet, Michael Wright, Burtt Harris, Samm-Art Williams
A wonderfully fresh and witty look at kids and their street gangs in the early 1960s. Full of fine performances with the music of the period used to great effect, often in highly apt ways. The story is really one of the relationship between several friends, set against attempts to stave off a conflict by holding a football match between rival gangs. A little gem. The script is by Rose and Phillip Kaufman.
DRA 114 min VIDrel: ARROW/RTM V
Boa: novel by Richard Price.

WANTED DEAD OR ALIVE ** 18
Gary A. Sherman USA 1986
Rutger Hauer, Gene Simmons, Robert Guillaume, Mel Harris, William Russ, Hugh Gillin, Susan McDonald, Jerry Hardin, Robert Harper, Eli Danker, Joe Nasser, Suzanne Wouk, Gerald Papasian, Nick Falatas, Hamman Shafie, Tyler Tyhurst
When an international terrorist begins a ruthless bombing attack on Los Angeles, the CIA call on a former operative. The terrorist discovers that he is being followed and attempts to kill the operative, but mistakenly kills one of his old friends from the CIA days. Our CIA operative now goes after said terrorist as a personal vendetta. A routine action-tale that offers little of interest except the explosive finale.
Aka: WANTED: DEAD OR ALIVE
A/AD 101 min (ort 104 min) VIDrel: VCC/DISC V

WAR, THE ** 12
Jon Avnet USA 1994
Elijah Wood, Kevin Costner, Mare Winningham, Lexi Randall, LaToya Chisholm, Christopher Fennell, Donald Sellers, Leon Sills, Will West, Brennan Gallagher, Adam Henderson, Charlette Julius, Jennifer Tyler, Lucas Black, Justin Lucas
A coming-of-age tale set in Mississippi in the summer of 1970, when two brothers help their Vietnam veteran father build a tree-house. The boys do it out of a sense of adventure, but for the father it becomes a refuge from the torments he still suffers from his time in Vietnam and provides a chance for him to regain a sense of family closeness. Quite well mounted, but often moralising, it miscasts Costner as the father and lacks an authentic feel for the period.
DRA 120 min (ort 126 min) cC VIDrel: CIC/SONOP V/sur

WAR AND PEACE *** U
King Vidor ITALY/USA 1956
Henry Fonda, Herbert Lom, Audrey Hepburn, Mel Ferrer, Vittorio Cassman, John Mills, Oscar Homolka, Anita Ekberg, Helmut Dantine, Mai Britt, Milly Vitale, Barry Jones, Anna Maria Ferrero, Lea Seidl, Wilfred Lawson, Sean Barrett
Competent version of this epic which attempts to condense both action and plot into manageable proportions, unlike the over-rated 1968 Soviet version. Not completely successful, the awkward script (six writers were credited, including Vidor) is a serious drawback as is some of the casting. However, the spectacular battle scenes are a compensating feature. Filmed once more in 1968.
DRA 208 min (ort 211 min)
VIDrel: 4-FRONT/POLYREC/CIC V
Boa: novel by Leo Nikolaevich Tolstoy.

WAR AND PEACE *** PG
Sergei Bondarchuck USSR 1967
Ludmila Savelyeva, Sergei Bondarchuck, Vasily Lanovoi, Vyacheslav Tikhonov, Andrei Bolkonsky, Hira-Ivanov Golovko, Irina Gubanova, Antonia Shuranova, Anastasia Vertinskaya, Natasha Rostova, Pierre Bezukov, Irina Skobotseva
An overlong, over-ambitious attempt to film this epic masterpiece; so faithful to the original novel that one can almost see the pages turning. In terms of a realistic recreation of the period this monumental film cannot be faulted. But to sit through the entire uncut version is an exhausting experience. This took five years to make and cost between 50 and 70 million dollars. Bondarchuk also wrote the screenplay. AA: Foreign.
Aka: VOINA I MIR
DRA 373 min Cut (24 sec – 3 cassettes – ort 434 min)
VIDrel: HEND/BMGREC L/A V
Boa: novel by Leo Nikolaevich Tolstoy.

WAR AND PEACE: VOLS 1 TO 6 *** PG
John Howard Davies UK 1972
Anthony Hopkins, Rupert Davies, Alan Dobie, Faith Brook, Morag Hood, Sylvester Morand, Angela Down, David Swift, Joanna David, Michael Billington
Tolstoy's magnificent tale of two families: the Rostovs and the Bolkonskys, and the manner in which their lives are affected at the time of the Napoleonic Wars in the early 1800s. A sombre, leisurely and over-reverent adaptation, hard to fault technically but somewhat uninvolving. Originally shown in twenty 45-minute episodes.
DRA 751 min (6 cassettes) mTV VIDrel: BBC/TECH V/h
Boa: novel by Leo Tolstoy.

WAR AND REMEMBRANCE: PARTS 1 AND 2 *** 15
Dan Curtis USA 1986
Robert Mitchum, Victoria Tennant, Polly Bergen, John Gielgud, David Dukes, Michael Woods, Robert Hardy, Hart Bochner, Barry Bostwick, Ralph Bellamy, Ian McShane, Jane Seymour, Sharon Stone, Robert Morley, Steven Berkoff, Jeremy Kemp
A blockbuster sequel to THE WINDS OF WAR that was reputed to be the most expensive mini-series ever. Watching this mammoth effort feels almost like living through the whole of WW2, and the overall impression is of a very mixed effort, ranging from harrowing death-camp scenes to the mundane and tangled domestic affairs of the main characters. However, the episodes dealing with the war in the Pacific are undeniably exciting. First shown in seven episodes.
WAR 1,302 min (7 cassettes – ort 1,410 min) mTV
VIDrel: ODY/SONOP V/s
Boa: novels by Herman Wouk.

WAR FOR BABY JESSICA, THE *** PG
John Kent Harrison USA 1993
Susan Dey, Michael Ontkean, David Keith, Amanda Plummer, Anan Ferguson, Joy Coghill, Linda Darlow, Barbara Tyson, Quinci Camazzola, Tom Heaton, Tom McBeath, Alf Humphreys, Iris Bernard, Marilyn Norry, Howard Storey, Morris Panych
A woman who has agreed to give her baby up for adoption changes her mind after the child is born, but the adoptive parents start a custody battle. The film "The Fight For Baby Jessica" also covers this true story.
Aka: WHOSE CHILD IS THIS? – THE WAR FOR BABY JESSICA
DRA 91 min VIDrel: ODY/SONOP V/sh

WAR IS OVER, THE ** 15
Alain Resnais FRANCE 1966
Yves Montand, Genevieve Bujold, Ingrid Thulin, Michel Piccoli, Jean Daste, Paul Crauchet, Jean Bouise
Montand plays an ageing Spanish revolutionary who lives in exile in Paris, where he struggles to come to terms with both his personal and political disappointments, chiefly deriving from the fact that after twenty-five years of opposition, Franco is still in power and nothing has changed. A study of disillusionment and regret, which though made in a most

assured manner does not have a strong enough story to hold our interest.
Aka: LA GUERRE EST FINIE
DRA 116 min (ort 122 min) VIDrel: ARTPRO/RTM V

WAR LOVER, THE ***   PG
Philip Leacock UK/USA   1962
Steve McQueen, Robert Wagner, Shirley Ann Field, Gary Cockrell, Michael Crawford, Bill Edwards, Chuck Julian, Jerry Stovin, Edward Bishop, Richard Leech, Bernard Braden, Sean Kelly, Neil McCallum, Al Waxman, Robert Easton
War story set in 1943, with a USAAF Flying Fortress captain stationed in Britain finding himself so in love with war that he cannot form normal relationships with women, viewing them as merely another kind of conquest. Trouble looms when he sets his sights on his co-pilot's girl on the eve of an especially hazardous mission. An unusual and ambitious study of the mind of a psychotic, it never really gets to grips with the issues raised.
DRA 83 min (ort 105 min) B/W VIDrel: VCC L/A V
Boa: novel by John Hershey.

WAR OF THE BUTTONS ***   PG
John Roberts FRANCE/UK   1993
Gregg Fitzgerald, Gerard Kearney, Darragh Naughton, Brendan McNamara, Kevin O'Malley, John Cleere, Anthony Cunnigham, Thomas Kavanagh, Eveanna Ryan, John Crowley, Stuart Dannell Foran, Danielle Tuite, Helen O'Leary, Yvonne McNamara
The directing debut for Roberts is a remake of a 1962 French film that is transplanted to Eire, where two rivals gangs of boys from different villages have a set-to, when the leader of one of the gangs has all his buttons removed by way of a humiliating prank. A splendidly daft family film, offering good fun for all ages.
JUV 90 min (ort 94 min) cC VIDrel: WHV V/sur
Boa: novel La Guerre des Boutons by Louis Pergaud.

WAR OF THE COLOSSAL BEAST *   PG
Bert I. Gordon USA   1958
Sally Fraser, Dean Parkin, Roger Pace, Russ Bender, George Beewar, Robert Hernandez, Roy Gordon, Charles Stewart, Rico Alaniz, Jack Kosslyn, Howard Wright, George Navarro, George Millan, Bill Giorgio, Warren Frost, June Jocelyn
In this sequel to THE AMAZING COLOSSAL MAN, giant sixty-foot Colonel Manning is still growing, having survived a confrontation with the army at Boulder Dam, which left him with facial injuries. After being confined to an aircraft hangar, he escapes and gets a chance to enjoy one last rampage before the inevitable end. A low-budget dud with special effects that even for the time, are far from impressive.
Aka: TERROR STRIKES, THE
FAN 68 min (ort 69 min) B/W VIDrel: HEND/BMGREC L/A V

WAR OF THE ROSES, THE ***   15
Danny DeVito USA   1989
Michael Douglas, Kathleen Turner, Danny DeVito, Marianne Sagebrecht, Sean Astin, Heather Fairfield, G.D. Spradlin, Peter Donat, Dan Castellaneta, Harlan Arnold, Gloria Cromwell, Mary Fogerty, Rika Hofmann, Patricia Allison, Sue Palka
A spiteful black comedy about an affluent couple whose marriage goes sour, leading to a vicious and escalating binge of destruction, when they are unable to reach an amicable property settlement. Acting and direction are top notch, and the bitterness of it all is softened by making the story a cautionary fable told by divorce lawyer DeVito to a prospective client. Unappealing after the first hour, but the film's viewpoint is certainly unusual. Scripted by Adler.
COM 111 min (ort 116 min) subH VIDrel: 20TH/TECH L/A V
Boa: novel by Warren Adler.

WAR OF THE WILDCATS ***   U
Albert S. Rogell USA   1943
John Wayne, Martha Scott, Albert Dekker, Gabby Hayes, Marjorie Rambeau, Dale Evans, Sidney Blackmer, Grant Withers, Paul Fix, Cecil Cunningham, Irving Bacon, Gloria Bacon, Anne O'Neal, Richard Graham, Robert Warwick
A wildcatter takes on a powerful oil baron who wants to cheat a tribe of Indians in Oklahoma out of their fair share of profits from oil wells on their land. Rip-roaring, fast-paced and entertaining Wayne vehicle.
Aka: IN OLD OKLAHOMA
WES 102 min B/W VIDrel: 4-FRONT/POLYREC V/h

WAR OF THE WORLDS, THE **   PG
Byron Haskin USA   1953
Gene Barry, Ann Robinson, Les Tremayne, Robert Cornthwaite, Henry Brandon, Jack Kruschen, Sandro Giglio, William Phipps, Paul Birch, Vernon Rich, House Stevenson, Paul Frees, Carolyn Jones, Pierre Cressoy, Nancy Hale
A thoroughly disappointing attempt to film a famous novel that tells of an alien invasion from Mars. The action is transplanted from the UK to the USA and brought forward to the 1950s – little of Wells' fine novel remains. Fortunately, at least the strong special effects and well handled crowd scenes make it worth a look. Produced by George Pal, it inspired a later (and rather dire) Canadian TV series. AA: Effects (George Pal).
FAN 82 min (ort 85 min) VIDrel: CIC/SONOP; PION (LV only) V LV
Boa: novel by H.G. Wells.

WAR OF THE WORLDS: THE RESURRECTION *   15
Colin Chilvers CANADA   1988
Jared Martin, Lynda Mason Green, Philip Akin, Richard Chaves, Richard Comar, Gwyneth Walsh, Ilse Van Glatz, Eugene Clark, Michael Rudder, Corinne Conley, Larry Reynolds, Rachel Blanchard, Frank Pellegrino, Martin Neufeld, Ric Sarabia
Pilot for the dull TV series that's remotely based on the H.G. Wells novel. Thirty-five years after the Martian invaders were destroyed by germs, a radiation leak revives six of the original creatures, that were unwisely being stored in canisters (an idea no doubt borrowed from RETURN OF THE LIVING DEAD: PART 2). Naturally, the man who has discovered this disagreeable piece of news cannot get anyone to believe him. An innumerable number of episodes followed.
FAN 91 min mTV VIDrel: CIC/SONOP L/A V

WAR PARTY **   18
Franc Roddam USA   1989
Kevin Dillon, Billy Wirth, Tim Sampson, Jimmie Ray Weeks, M. Emmet Walsh, Dennis Banks, Kevyn Major Howard, Bill McKinney, Jerry Hardin, Guy Boyd, Tantoo Cardinal, R.D. Call, William Frankfather, Saginaw Grant, Peggy Lipton
A town stages a re-enactment of a battle fought 100 years before between the US Cavalry and the Blackfeet that ended in a massacre of the Indians. This event takes a tragic turn, however, when four Indian youths happen to kill a white boy who was toting a loaded gun. They head for the hills pursued by a posse of angry townsfolk. An exploitative action yarn that claims to be a serious examination of Indian problems but in truth is nothing of the kind.
A/AD 93 min (ort 99 min) VIDrel: COLUM/SONOP V/sur

WAR STORY **   15
M. Uno/R. King/J. Sholder/T. Holland/D. Morris/
L. Glatter USA   1988
Tim Guinee, Wesley Snipes, William Frankfather, Joseph Hieu, Andrew Huy Nguyen; Tom Hodges, Francois Chau, Kieron Mulroney, Lee Debroux, Patricia Smith; Ronald William Lawrence, Andrew Roperto, Courtney Gains, Hutton Cobb
Three stories of the Vietnam War. In "An Old Ghost Walks The Earth" a young GI is shocked by his comrades' attitude to the locals and befriends a young girl. "Dusk To Dawn" sees a youngster who has enlisted enjoy one last evening with his family. In "The Fragging" a platoon decides to kill their incompetent commander. The 2-tape package has six separate films plus some extra footage of the actual conflict. Fair. See also VIETNAM WAR STORY.
WAR 87 min VIDrel: ODY/SONOP V/h

WAR STORY 2 ***   15
Todd Holland/David Burton Morris/
Leslie Linka Glatter USA   1989
Todd Graff, Raymond Cruz, Gregory Cooke, Kent Williams, Bryant Bradshaw; Cynthia Bain, Tate Donovan, Keith MacKechie, Jeris Lee Poindexter (Chris Pederson); Stacy Edwards, Laura Harrington, Pippa Pearthree, Glen Plummer
A set of three stories written to illustrate the brutality of the Vietnam War. In "Separated", a soldier finds himself cut off from his unit, in "R & R" an officer on leave is unable to tell his wife of his experiences, and in "The Promise", a nurse sent to a front-line hospital encounters hostility from the colleague she is replacing. Three short, sharp and literate stories.
Aka: VIETNAM WAR STORY 2
WAR 90 min VIDrel: ODY/SONOP V/h

WAR WAGON, THE * U
Burt Kennedy USA   1967
John Wayne, Kirk Douglas, Howard Keel, Robert Walker Jr, Keenan Wynn, Bruce Dern, Bruce Cabot, Valora Noland, Gene Evans, Joanna Barnes, Terry Wilson, Don Collier, Sheb Wooley, Ann McCrea, Emilio Fernandez, Frank McGrath
In revenge for past wrongs, a cowboy plans to steal an armoured gold-laden stage coach belonging to a crooked mine owner. A light-hearted and fast-paced action Western where everything combines to provide a highly entertaining time. The script is by Huffaker.
WES   96 min (ort 101 min)
VIDrel: 4-FRONT/POLYREC/CIC   V/h
Boa: novel Badman by Clair Huffaker.

WARGAMES * PG
John Badham USA   1983
Matthew Broderick, Ally Sheedy, Dabney Coleman, John Wood, Barry Corbin, Juanin Clay, Kent Williams, Dennis Lipscomb, Joe Dorsey, Irving Metzman, Michael Ensign, Susan Daws, James Tolkan, David Clover, Drew Snyder
A youngster who spends most of his time playing with his home computer, accidentally hacks into a military computer that is both used to simulate nuclear war and to control the conduct of a real one. When he begins playing with the computer he unwittingly sets in motion a chain of events that may lead to a nuclear exchange. Corbin as a wilfully retired scientist, adds some charm to an otherwise forgettable juvenile effort.
FAN   108 min (ort 113 min)   VIDrel: WHV   V/sur

WARLOCK * 15
Steve Miner USA   1989
Julian Sands, Lori Singer, Richard E. Grant, Mary Woronov, Kevin O'Brien, Richard Kuss, Allan Miller, Anna Levine, David Carpenter, Kay E. Kuter, Ian Abercrombie, Kenneth Danziger, Art Smith, Robert Breeze, Frank Renzulli
A 17th century warlock is about to be executed by the town elders, when a thunderbolt strikes, and he is catapulted into the 20th century. Once there, he embarks on a frantic search for the missing parts of the Devil's Bible, which would give him the power to undo all creation. Fortunately, he is followed into the 20th century by a witchfinder, who thwarts his plans. A lively and entertaining yarn. The music is by Jerry Goldsmith.
HOR   98 min   VIDrel: 4-FRONT/POLYREC/MED   V/sur

WARLOCK: THE ARMAGEDDON * 18
Tony Hickox USA   1992
Julian Sands, Paula Marhsall, Chris Young, Joanna Pacula, Steve Kahan, Charles Halalhan, R.G. Armstrong, Craig Hurley, Bruce Glover, Davis Gaines, Rebecca Street, Dawn Ann Billings, Zach Galligan, Wren Brown, Gary Cervantes
An inferior sequel to WARLOCK in which this sinister figure revisits Earth in search of six runestones that will enable him to surrender our world to Satan. His quest eventually takes him to a small California town where he must battle two young champions, who have been chosen by the last surviving members of a Druidic cult. Some good special effects are to be had, but too much gore and the weak storyline are major drawbacks. This film went straight to video.
HOR   94 min (ort 98 min)   VIDrel: FIRST/SONOP   V

WARLORDS * 18
Fred Olen Ray USA   1989
David Carradine, Ross Hagen, Sid Haig, Dawn Wildsmith, Fox Harris, Robert Quarry, Victoria Sellers, Michelle Bauer
In the usual devastated post-nuclear holocaust world so beloved by low-budget SF film-makers, Carradine plays a warrior out to rescue his wife from the clutches of a local despot. Wildsmith adds a little colour as the feisty woman he is accompanied by, but this dismal film is one more low-budget exercise in tedium.
FAN   85 min Cut (45 sec)   VIDrel: POPRO/RTM   V

WARLORDS OF ATLANTIS * PG
Kevin Connor UK   1978
Doug McClure, Peter Gilmore, Shane Rimmer, Lea Brodie, Cyd Charisse, Daniel Massey, Michael Gothard, Hal Galili, John Ratzenberger, Derry Power, Donald Bisset, Ashley Knight, Robert Brown
An expedition to find an underwater city encounters a mixture of mystery and monsters, when McClure joins British scientists in the search for Atlantis. Set in Victorian times, this is a whole-some but unremarkable period adventure, quite badly acted, but redeemed by the enjoyable special effects.
A/AD   92 min (ort 96 min)   VIDrel: WHV   V/h

WARM NIGHTS ON A SLOW MOVING TRAIN * 18
Bob Ellis AUSTRALIA   1987
Wendy Hughes, Colin Friels, Norman Kaye, John Clayton, Lewis Fitz-Gerald, Rod Zuanic, Peter Whitford, Grant Tilly, Steven J. Spears, Peter Sullivan, Chris Haywood, Eileen Price, Neil Thompson, John Flaus, Peter Carmody
A schoolteacher makes some extra money as a part-time prostitute aboard a train, in order to support her sick brother, but her cosy set-up is disturbed when she meets a man who plans to make use of her to kill a prominent politician. An engrossing if rather contrived drama, whose resolution is both flawed and disappointing. Hughes obtained the Best Actress Award at the 1987 Rio International Film Festival.
DRA   90 min   VIDrel: VES L/A   V

WARM SUMMER RAIN * 18
Joe Gayton USA   1989
Kelly Lynch, Barry Tubb, Ron Sloan, Larry Poindexter, Lupe Amador, Vanessa Conti, Peter MacPherson, Dianne Turley Travis, Stanley Grover, Gene Knight, Tony Markes, Cindy Guyer, Queenie, Susan Sherriffe, Jean Pflieger, Kimberly Hall
A small-town secretary with a botched suicide attempt behind her escapes from hospital and meets up with a romantic stranger. From a seedy bar they make their way to bed, and in between the steamy sex scenes the characters discuss their problems via a sequence of emotional monologues. An interesting idea for a drama that's hurt by self-indulgence and a lack of flow.
DRA   78 min (ort 82 min)   VIDrel: EIV/SONOP   V

WARNING, THE * 15
Greydon Clark USA   1980
Jack Palance, Martin Landau, Cameron Mitchell, Tarah Nutter, Neville Brand, Christopher S. Nelson, Ralph Meeker, Larry Storch, Sue Ann Langdon, Lynn Theel, Darby Hinton, David Caruso
Alien creatures attack four young people on a trip to a lake after they disregard warnings about strange happenings there. An utterly silly horror yarn which couples foolish ideas (the aliens make use of carnivorous disks) with a poor script and acting. The few shudders generated are soon dissipated.
Aka: IT CAME WITHOUT WARNING; WITHOUT WARNING
HOR   89 min (ort 96 min)   VIDrel: VIPCO/SGSVID   V/h

WARNING SIGN * 15
Hal Barwood USA   1985
Sam Waterson, Yaphet Kotto, Kathleen Quinlan, Jeffrey De Munn, Jerry Hardin, Richard Dysart, G.W. Bailey, Rick Rossovich, Cynthia Carle, Kavi Raz, Scott Paulin, Keith Szarabajka, Jack Thibeau, J. Patrick McNamara
An accident in a research laboratory leads to the escape of a deadly virus and the creation of a breed of dangerous mutants of hideous appearance. Slightly reminiscent of Romero's THE CRAZIES, but the serious first half gives way to an unfunny parody that it would have been wiser to avoid. This was the directorial debut of screenwriter Barwood.
Aka: BIOHAZARD
FAN   95 min   VIDrel: 20TH/TECH   V/sur

WARRIOR SPIRIT * PG
Rene Manzor USA   1994
Lukas Haas, Jimmy Herman, Allan Musy
A man and his Native American companion go in search of a gold treasure in the Yukon and must face a variety of problems, including a hostile native tribe.
A/AD   94 min   VIDrel: TRIM/HIFLI   V

WARRIORS, THE * 18
Walter Hill USA   1979
Michael Beck, James Remar, Thomas Waites, Dorsey Wright, Brian Tyler, David Harris, Deborah Van Valkenburgh
A New York youth gang is forced to cross enemy turf on its way home, in this violent and action-packed film. In the hands of the director, Yurick's incisive and powerful book becomes a simplified fast-paced actioner. However, the photography of Andrew Laszlo is an asset. Written by Hill and David Shaber.
A/AD   89 min (ort 94 min)
VIDrel: 4-FRONT/POLYREC/CIC   V/h
Boa: novel by Sol Yurick.

WARRIORS **
15
Shimon Dotan CANADA/ISRAEL
1994
Gary Busey, Michael Pare, Wendii Fulford, Catherine MacKenzie, Liz Macrae, Griffith Brewer, Peter Colvey, Richard Zeman, Desmond Campbell, Ricardo Juarez, Aaron Tager, Minor Mustain, Andy Bradshaw, Vlasta Vrana, Isaac Anbar
An anti-terrorist operative is considered to be so dangerous that he is held in a military prison. But he escapes and goes on the run with a young hooker as his hostage. The only person able to tackle him is his former protege, and he is sent to eliminate him. Naturally, our hero faces the usual conflict of loyalties before finally accomplishing his task. An assembly-line actioner of little distinction.
A/AD 96 min (ort 100 min) VIDrel: MED/DISC V/sh

W.A.S.P - WHITE ANGLO-SAXON PROSTITUTE ***
18
Jim Enright ITALY
1992
Taylor Wayne, Sierra, Tom Chapman, Leanna Foxx, Steve Drake, Melanie Moore, Nick E.
Wayne starts out in this odd film as the leader of a Moral Majority movement, who is struck dead by accident, and goes to the hereafter. In an attempt to straighten out this mess, an angel gives her the body of a hooker who has just died (hence the title) and she puts this to good use, experiencing all the pleasures she never knew in her previous life. A clever and quite witty sex variant on a theme explored in movies such as HEAVEN CAN WAIT.
Aka: WASP
A 47 min (ort 65 min) VIDrel: MIA/DISC V

WATCH IT ***
15
Tom Flynn USA
1992
Peter Gallagher, Suzy Amis, John C. McGinley, Tom Sizemore, John Tenney, Cynthia Stevenson, Teri Hawkes, Jordana Capra
A man returns to Chicago and goes to live with his cousin and his friends who are all in their earlier twenties and much addicted to staging cruel and immature practical jokes. His strained relationship with his cousin is soon made even worse when they became rivals for the same girl.
COM 98 min (ort 102 min) VIDrel: EIV/SONOP V/sur

WATCH ON THE RHINE ****
U
Herman Shumlin USA
1943
Bette Davis, Paul Lukas, Geraldine Fitzgerald, Lucile Watson, Beulah Bond, George Coulouris, Donald Woods, Henry Daniell, Donald Buka, Eric Roberts, Janis Wilson, Helmut Dantine, Mary Young, Kurt Katch, Erwin Kalser
A German man and his wife live in Washington but are harassed by Nazi spies. For once Davis is over-shadowed by another actor as Lukas gives what was probably the finest performance of his career. This excellent adaptation of the Hellman play was written by Dashiell Hammett. The music was by Max Steiner.
AA: Actor (Lukas).
DRA 107 min (ort 114 min) B/W VIDrel: WHV V
Boa: play by Lillian Hellman.

WATCH YOUR STERN *
PG
Gerald Thomas UK
1960
Kenneth Connor, Eric Barker, Leslie Phillips, Joan Sims, Noel Purcell, Hattie Jacques, Spike Milligan, Eric Sykes, Sid James
A ship's steward fools an admiral when he pretends to be the inventor of a homing torpedo. A predictable navy lark with all the expected complications and farcical situations. A "Carry On" film in all but name.
COM 84 min (ort 88 min) B/W VIDrel: WHV V/h
Boa: play Something about a Sailor by Earle Couttie.

WATCHERS *
18
Jon Hess USA
1988
Corey Haim, Barbara Williams, Michael Ironside, Lala
An accident at a biological research station that was engaged in experiments to breed an efficient killing machine, causes the release of a highly intelligent canine and its bloodthirsty and telepathic "twin", which has been programmed to find and destroy the former. A low-grade and incoherent mess.
HOR 87 min (ort 99 min)
VIDrel: 4-FRONT/POLYREC/GUILD L/A V
Boa: novel by Dean Koontz.

WATCHMAKER OF SAINT-PAUL, THE ***
PG
Bernard Tavernier FRANCE
1973
Philippe Noiret, Jean Rochefort, Jacques Denis, Julien Bertheau,
William Sabatier, Andree Tainsy, Sylvain Rougerie, Christine Pascal, Cecile Vassort, Yves Fonso, Jacques Hilling, Clotilde Joano, Julien Bertheau, Johnny Wesseler
A clockmaker has his life turned upside down when his son commits the violent murder of a factory foreman and comes to realise how very little he knows about his own flesh and blood. A grim and remorseless examination of tangled relationships forms the basis of this excellent debut by Tavernier.
Aka: CLOCKMAKER, THE; L'HORLOGER DE SAINT-PAUL; WATCHMAKER, THE; WATCHMAKER OF LYON, THE
DRA 100 min (ort 105 min) VIDrel: ARTPRO/RTM V
Boa: novel The Clockmaker of Everton by George Simenon.

WATER **
15
Dick Clement UK
1985
Michael Caine, Valerie Perrine, Brenda Vaccaro, Leonard Rossiter, Jimmie Walker, Billy Connolly, Fred Gwynne, Maureen Lipman, Dennis Dugan, Dick Shawn, George Harrison, Eric Clapton, Ringo Starr, Fulton McKay
The discovery of a mineral water spring on a British Caribbean island on the eve of independence, results in international complications with Caine as the embattled governor trying to rein in American oil interests, a rebellion and various malcontents. Slightly funny, more often not, this uneven film has Lipman giving a nasty caricature of Margaret Thatcher, and Vaccaro a pointless one of Carmen Miranda.
COM 93 min (ort 97 min) VIDrel: LUMI/SPEAR L/A
V/sur

WATER BABIES, THE **
U
Lionel Jeffries POLAND/UK
1978
James Mason, Billie Whitelaw, Bernard Cribbins, Joan Greenwood, David Tomlinson, Tommy Pender, Samantha Gates, Paul Luty plus voices of: Jon Pertwee, Olive Gregg, Lance Percival, David Jason, Una Stubbs
The story of a twelve-year-old chimney-sweep who discovers a marvellous underwater world, when he falls into the river having fled after being falsely accused of a theft. Written by Michael Robson, this bland adaptation of the Kingsley classic mixes animation with live-action, but makes the error of adding a virtually unrelated tale of Victorian London to its underwater adventure section.
Aka: SLIP SLIDE ADVENTURES
JUV 95 min VIDrel: FABFIL/SPEAR V
Boa: novel by Charles Kingsley.

WATER MARGIN, THE **
12/15
Nobuo Nakagawa JAPAN
1978
Kei Sato, Atsuo Nakamura, Hajime Hana, Toshiyo Matsuo, Takashi Obayashi, Isamu Nagato, Ko Sato, Teruhiko Aoi, Tsutomu Yamagata, Yoshiro Kitahara, Mieko Aoyyagi, Takahiro Tamura, Daijiro Harada, Shino Kawai, Toshio Kurosawa
Incrediby long-winded adventure set at the time of the Tsung Dynasty of over a thousand years ago, when a corrupt and despotic government oppressed the people of China. The story (based on the Chinese folk novel "Shui Hu Chuan") follows the exploits of a brave band of fighters who oppose the government, are branded traitors and seek refuge in a remote part of the country. Stilted and in truth, quite dull. Made for Japanese television and shown in twenty-six parts.
A/AD 1,120 min (six 180 min cassettes) mTV dubbed
VIDrel: FABFIL/SPEAR; ENCORE (LV only – Part 1 and 2)
V LV

WATERBABIES *
18
Jim Enright USA
1992
Ashlyn Gere, P.J. Sparks. Melanie Moore, Tianna, Tom Byron, T.T. Boy, Mike Horner
When a health club re-opens as a women-only establishment, all the male staff are kept on, and soon discover that their jobs carry some very attractive fringe benefits.
A 40 min (ort 80 min) VIDrel: ONE V

WATERDANCE, THE ***
15
Neal Jiminez/Michael Steinberg USA
1992
Eric Stoltz, Helen Hunt, Wesley Snipes, William Forsythe, Elizabeth Pena, Grace Zabriskie, William Allen Young, Starletta Du Pois, Tony Genaro, Erick Vigil, Susan Gibney, Henry Harris, Badja Djola, Elizabeth Dennehy, Eva Rodrigues
A young author is terribly injured in a hiking accident and becomes a paraplegic, a condition that he finds extremely hard to adjust to. But when he attends a rehabilitation centre, he meets others in the same predicament and sees how they cope.

An insightful and moving film (Jiminez himself suffered such injuries in 1984) that is unsentimental in tone and blessed with some fine performances, plus a few moments of sharp humour. A most endearing film.
DRA   102 min (ort 107 min)   VIDrel: COLUM/SONOP
V/sur

WATERLAND ***   15
Stephen Gyllenhaal UK   1992
Jeremy Irons, Ethan Hawke, Sinead Cusack, Grant Warnock, Lena Headey, Ross McCall, Sean McGuire, Callum Dixon, Camilla Hebditch, David Morrissey, John Heard, Maggie Gyllenhaal, Cara Buono, Peter Postlethwaite, Stewart Richman
A British teacher comes to the USA to teach history at a high school but starts to suffer from a nervous breakdown and starts to reveal to his class the dark secrets of his past. Unable to communicate with his wife, who is herself suffering from depression, his students become an increasingly important part of his life. Excellent performances by Irons and Cusack sustain this gloomy and meandering tale.
DRA   90 min (ort 95 min)   VIDrel: POLFIL   V
Boa: novel by Graham Swift.

WATERLOO **   U
Sergei Bondarchuk ITALY/USSR   1971
Rod Steiger, Christopher Plummer, Orson Welles, Jack Hawkins, Virginia McKenna, Ian Ogilvy, Dan O'Herlihy, Rupert Davies, Michael Wilding, Terence Alexander, Ivor Garrani, Gianni Garko, Eughenj Samoilov Zakhariadze
Clumsy historical drama that concentrates on Napoleon's defeat at Waterloo. The spectacular recreation of the battle scenes cannot compensate for the lack of drama and the confused plot. The original Soviet version was nearly twice as long as this one. Written by Bondarchuk and H.A.L. Craig, with music by Nino Rota.
DRA   128 min Cut (20 sec – ort 240 min)
VIDrel: VCC/COLUM L/A   V/sh
Boa: novel War And Peace by Leo Nikolaevich Tolstoy.

WATERLOO BRIDGE ***   (18)
Mervyn Le Roy USA   1940
Vivien Leigh, Robert Taylor, Lucile Watson, Virginia Field, C. Aubrey Smith, Maria Ouspenskaya, Steffi Duna, Leo G. Carroll, Clara Reid, Leonard Mudie, Herbert Evans, Halliwell Hobbes, Ethel Griffies, Gilbert Emery
During a WW2 blackout an officer on his way to Waterloo Station recalls the story of his lost love, a ballerina with whom he fell instantly in love. However, he is called to the front before they can wed, and having her job, she is forced into a life of prostitution. A fine remake of the 1931 film, with Leigh pulling out all the stops in a powerful performance.
DRA   104 min (ort 109 min) B/W   SATrel: SKY MOVIES
Boa: play by Robert E. Sherwood

WATERSHIP DOWN ***   U
Martin Rosen UK   1978
Voices of: John Hurt, Richard Briers, Roy Kinnear, Denholm Elliott, Michael Hordern, Ralph Richardson, Simon Cadell, Terence Rigby, Richard O'Callaghan, Lyn Farleigh, Mary Maddox, Zero Mostel, Harry Andrews, Joss Ackland
Animated version of the highly successful novel about a community of rabbits and the dangers they face as they embark on a perilous journey in search of a new home. This excellent animation has a fine score by Angela Morley. Not really lighthearted enough to be regarded as wholly a children's film, the only comic element is provided by Zero Mostel's bird character.
ANIM   88 min (ort 92 min)   VIDrel: POLY/POLYREC
V/sur
Boa: novel by Richard Adams.

WATERWORLD *   15/12
Kevin Reynolds USA   1995
Kevin Costner, Chaim Jeraffi, Ric Aviles, Henry Kapono Ka'aihue, Tracy Anderson, R.D. Call, Zitto Kazann, Leonardo Cimino, Sab Shimono, Zakes Mokae, Jeanne Tripplehorn, Jack Kehler, Lanny Flaherty, Dennis Hopper, Robert Silverman
This dreary adventure is set at a time in the future the Earth is largely covered with water, the ice-caps having melted, and where solid land is unknown. The scattered remnants of humanity survive as best they can on floating islands built up from anything buoyant. Costner plays "The Mariner", a mutated human with the ability to breathe underwater, who sets out to

defeat a gang of marauding pirates and if possible find dry land. A lavish, flashy, tedious flop.
A/AD   130 min (ort 136 min) wScrn cC
VIDrel: CIC/SONOP; PION (LV only – 12 cert)   V/sur LV

WAXWORK *   18
Anthony Hickox USA   1986
Zach Galligan, Deborah Foreman, Michelle Johnson, Miles O'Keeffe, David Warner, Dana Ashbrook, Patrick Macnee, Charles McCaughan, J. Kenneth Campbell, John Rhys-Davies, Joe Baker, Jennifer Bassey, Eric Brown
A group of teenagers inside a wax museum, fall victim to a killer in this patchy and generally unsatisfactory blend of comedy and horror. A highlight should have been the point at which the teenagers had to battle some of the famous fictitious monsters of history, but even this lacked impact. A sequel of sorts – LOST IN TIME, soon followed.
HOR   92 min Cut (34 sec at UK cinema release – ort 100 min)
Col/B/W   VIDrel: FIRST/SONOP L/A   V

WAY AHEAD, THE ***   U
Carol Reed UK   1944
David Niven, Stanley Holloway, William Hartnell, Jimmie Hanley, John Laurie, Hugh Burden, Peter Ustinov, Raymond Huntley, Trevor Howard, Tessie O'Shea, Penelope Dudley Ward, James Donald, Leo Genn, Mary Jerrold, Renee Asherson
Originally intended as a training film, this project grew into the exciting story of how civilians from varied backgrounds came together as a unit of raw recruits, and were then shaped into a fighting force. Written by Peter Ustinov and Eric Ambler, this film retains a strong documentary feel. The music is by William Alwyn with photography by Guy Green. Trevor Howard's screen debut.
Aka: IMMORTAL BATTALION, THE
DRA   110 min (ort 115 min) B/W   VIDrel: VCC/RTM   V

WAY OF THE DRAGON, THE ***   18
Raymond Chow HONG KONG   1973
Bruce Lee, Nora Miao, Chuck Norris, Huang Chung Hsun, Chin Ti, Jon T. Benn, Robert Wall, Liu Yun, Chu'eng Li, Little Unicorn, Ch'eng Pin Chih, Ho Pieh, Wel P'ing Au, Huang Jen Chih, Mali Sha
Bruce Lee arrives in Rome to help relatives, whose Chinese restaurant has attracted the unwelcome attentions of gangsters who run a protection racket. They soon learn the futility of trying to use normal methods to deal with him, and decide instead to engage a number of martial artists of their own to fight him. Some splendidly choreographed fights are embedded in a flabby script. A few humorous touches demonstrate Lee's unexplored comic gifts.
Aka: RETURN OF THE DRAGON
MAR   86 min Cut (1 min 11 sec – ort 90 min) dubbed
VIDrel: 4-FRONT/POLYREC   V

WAY TO THE STARS, THE ***   U
Anthony Asquith UK   1945
Michael Redgrave, John Mills, Rosamund John, Douglass Montgomery, Renee Asherson, Stanley Holloway, Felix Aylmer, Basil Radford, Bonar Colleano Jr, Trevor Howard, Joyce Carey, Jean Simmons, Bill Rowbotham, Charles Victor
A look at the relationships between the staff at a British Bomber Command base in 1942 at the height of WW2, that concentrates on the ground rather than the air (there are very few aerial sequences) but is no less enjoyable on that account. Despite the serious subject matter, there are some welcome comic moments, and a general air of brisk efficiency about the whole enterprise. One of the few WW2 films to really capture the homefront atmosphere.
Aka: JOHNNY IN THE CLOUDS
WAR   104 min (ort 109 min)   VIDrel: VCC/DISC/COLUM
L/A   V

WAY WE WERE, THE ***   PG
Sydney Pollack USA   1973
Barbra Streisand, Robert Redford, Bradford Dillman, Viveca Lindfors, Herb Edelman, Murray Hamilton, Patrick O'Neal, Lois Chiles, Allyn Ann McLerie, Sally Kirkland, Diana Ewing, Marcia Mae Jones, Don Keefer, George Gaynes
The story of a romance from the 1930s to the 1950s between two people of opposing political views, with Redford playing a good-natured square against Streisand's political activist. The intelligent script by Laurents survives despite some clumsy pre-release cutting that left Hamilton and Lindfors with only small

parts. AA: Score/orig (Marvin Hamlisch), Song ("The Way We Were" – Hamlisch (m)/Alan and Marilyn Bergman (l)).
DRA   113 min (ort 118 min)   VIDrel: VCC/DISC/COLUM L/A   V
Boa: novel by Authur Laurents.

WAYNE'S WORLD ***
15/PG
Penelope Spheeris USA   1992
Mike Myers, Dana Carvey, Rob Lowe, Tia Carrere, Brian Doyle-Murray, Lara Flynn Boyle, Michael DeLuise, Dan Bell, Lee Tergesen, Kurt Fuller, Sean Gregory Sullivan, Colleen Camp, Donna Dixon, Frederick Coffin, Chris Farley, Meat Loaf
A "Saturday Night Live" TV sketch is stretched into a full-length feature about a couple of nerd-like, ageing teenagers who run their own cable show from Wayne's basement in Aurora, Illinois. The show features their various disjointed musings on life, music and girls (mostly girls) and proves to be so popular that it attracts the attentions of a TV executive. Rambling, exasperating, but often very funny, it gave Carrere her screen debut. A sequel followed.
COM   90 min (ort 95 min)   VIDrel: CIC/SONOP; PION (LV only – PG cert)   V/sur LV

WAYNE'S WORLD 2 **
15/PG
Stephen Surjik USA   1993
Mike Myers, Dana Carvey, Lee Tergesen, Dana Bell, Tia Carrere, Heather Locklear, Gavin Glazer, Michael Mickles, Olivia D'Abo, Kevin Pollack, Jay Leno, Ed O'Neil, Duke Valente, Benny Graham, Goggy Gress, Joe Liss, Richard Epper
A rather tired sequel to the first film in which our two over-the-hill teens play a concert of their own (Waynestock) after Jim Morrison appears t them in a dream. However, other problems soon loom large when a nasty record promoter goes after Wayne's girlfriend. A few bright sparks enliven an otherwise pretty dreary affair.
COM   90 min (ort 95 min)   cC   VIDrel: CIC/SONOP; PION (LV only – PG cert)   V/sur LV

W.B. BLUE AND THE BEAN *
15
Max J. Kleven USA   1989
David Hasselhoff, Tony Brubaker, Thomas Rosales, Linda Blair, John Vernon, Gregory Scott Cummins, Wayne Montanio, Jack Jozefson, Sheree Bodoff, Valerie Swift, Joe Tornatore, Bob Minor, Caroline Barday, Clay M. Lilley, Bob Hoy
Adventure tale about a man with an unusual profession: a tennis pro who goes after bail-jumpers, ably assisted by his sidekicks "Blue" and "The Bean". When a suspect in a drugs case (an unusually demanding role for Blair) is kidnapped after bail is posted, our trio swing into action and mount a rescue. Mindless entertainment that looks and sounds like a TV pilot.
Aka: BAIL OUT; WINGS OF FREEDOM
A/AD   84 min (ort 88 min)   VIDrel: EIV/SONOP   V

WE ARE FROM KRONDSTADT ***
PG
Yefim Dzigan USSR   1936
Vasili Zaichikov, Grigori Bushuyev
Drama that revolves around a famous episode of the Russian Civil War, when the sailors of Krondstadt left their ships to defend the town from an attack by the White Russians. A dated but absorbing work.
DRA   88 min B/W   VIDrel: HEND L/A   V

WE ARE SEVEN ***
PG
Ken Horn UK   1991
Helen Roberts, Dafydd Hywell, Christopher Mitchum, Andrew Powell, Elen C. Jones, Julianne Barron, James Bird, Terence Bennett, Beth Robert, Jurgen Morche, Gudrun Gabriel, Christopher Villiers, Beth Morris, Howell Evans
The exploits of the inhabitants of a tiny Welsh village, most especially the large Morgan family, the various members of whom all boast different fathers. A very carefully made and detailed period drama, consisting for the most part of a set of interlinked stories. The script is by Robert Pugh.
DRA   312 min (2 cassettes) mTV   VIDrel: VCC L/A   V
Boa: novel by Una Troy.

WE DIVE AT DAWN ***
U
Anthony Asquith UK   1943
Eric Portman, John Mills, Reginald Purdell, Joan Hopkins, Josephine Wilson, Niall MacGinnis, Louis Bradfield, Ronald Millar, Jack Watling, Caven Wilson, Leslie Weston, Norman Williams, Lionel Grose, Beatrice Varley, Marie Ault
A low-key submarine drama telling of a crew assigned to attack

a German battleship, the Brandenberg, that concentrates on the tensions and pressures the men suffer after the attack leaves their vessel disabled, short of fuel and far from home. Written by J.P. Williams, Val Valentine and Frank Launder.
WAR   92 min (ort 97 min) B/W   VIDrel: CARL/TECH   V

WE DON'T WANT TO TALK ABOUT IT ***
PG
Maria Luisa Bemberg ARGENTINA/ITALY   1993
Marcello Mastroianni, Luisina Brando, Alejandra Podesta, Betiana Blum, Roberto Caranghi, Monica Lacoste, Jorge Luz, Monica Villa, Juan Manuel Tenuta, Tina Serrano, Veronica Llinas, Susan Cortinez, Martin Kalmill, Jorge Ochoa
In a small town in South America, the unhappy mother of a dwarf child refuses to acknowledge her daughter's disability, instead embarking on an ambitious educational program that turns her into a child prodigy. In fact, her daughter wins the heart of a wealthy, elderly bachelor. Done in the style of a modern fairytale, this is a unusual allegory, that takes a slow, careful but clinical look at the central characters. The music is by Nicola Piovani.
Aka: DE ESO NO SE HABLA
DRA   94 min (ort 105 min) wScrn   VIDrel: ARTIF/20TH V/h
Boa: short story by Julio Llinas.

WE LIVE AGAIN ***
PG
Rouben Mamoulian USA   1934
Fredric March, Anna Stern, Jane Baxter, C. Aubrey Smith, Sam Jaffe, Ethel Griffies. Gwendolyn Logan, Mary Forbes, Jessie Ralph, Leonid Kinskey, Davison Clark, Dale Fuller, Morgan Wallace, Crauford Kent, Barron Hesse, Jessie Arnold
A prince is raised with a peasant girl whom he treats as his equal and the two deeply in love but he seduces and then abandons her after getting her pregnant. He forgets all about her until seven years later when he becomes a juror in the trial of a prostitute accused of murder who proves to be none other than his former love. Overwhelmed by guilt, he sets about finding a way to make amends. A capable adaptation of Tolstoy's much-filmed novel.
DRA   82 min (ort 85 min) B/W   VIDrel: VCC/DISC   V
Boa: novel Resurrection by Leo Tolstoy.

WEB OF DECEIT ***
(12)
Bill Corcoran USA   1993
Corbin Bernsen, Amanda Pays, Don Enright, Les Alexander, Mimi Kozyk, Neve Campbell, TomMcCamus, Dawn Greenhalgh, Kim Coates, Albert Schultz, Carlton Watson, Nigel Bennett, Peter Krantz, Jeremy Harris, Helen Beavis, Sveta Kohli
A couple are torn apart when their baby son goes missing and is presumed murdered, and the police focus their attention on the mother, whom they suspect of this crime. A gripping work that keeps one's interest right the way through, thanks to tight direction and strong performances from real-life husband-and-wife team Bernsen and Pays.
THR   88 min (ort 90 min) mTV   SATrel: SKY MOVIES

WEB OF DECEPTION **
(18)
Richard A. Colla USA   1994
Powers Boothe, Pam Dawber, Lisa Collis, Rosalind Chao, Bradley Whitford, Paul Ben-Victor, Abigail Van Alyn, Robin Karfo, Victor Talmadge, L. Peter Callender, Rebecca Wink, Juliette Marshall, Elena Praskin, Jennifer Founds
An unbalanced woman commits suicide but cleverly makes it look like murder to blame the eminent but philandering psychiatrist who rejected her when he decided to end their affair. Meanwhile, at least his wife is prepared to stand by him in the ensuing scandal and inevitable court case. A moody thriller that plays like an odd variant of PLAY MISTY FOR ME, and though never very convincing remains worth watching.
THR   89 min mTV   TVrel: C4

WEDDING BANQUET, THE ***
15
Ang Lee CHINA/USA   1993
Ah-Leh Gua, Sihung Lung, May Chin, Winston Chao, Mitchell Lichtenstein, Dion Birney, Jeanne Kuo Chang, Paul Chen, May Chin, Chung-Wei Chou. Yun Chung, Jean Hu, Ho-Mean Pu, Michael Gaston, Jeffrey Howard, Theresa Hon, Yung-Teh Hsu
A gay American-Chinese conceals his homosexuality from his parents and agrees to an arranged marriage for their sake and to give his bride the chance to acquire a green card. His actions, however, lead to the usual complications, in this charming study

of cultural clashes that was made on a very limited budget but is greatly enjoyable nonetheless.

Aka: XIYAN

DRA   103 min (ort 110 min)   VIDrel: MAINPIC/RTM L/A V

WEDDING DAY BLUES **   PG
Paul Lynch USA   1988
Barbara Billingsley, Eileen Brennan, Cloris Leachman, Dick Van Patten, Scott Valentine, Joel Brooks, Michele Green, Max Wright, John Ratzenberger, Mark-Linn Baker
A comic variation on the Romeo and Juliet theme, with two rival families being forced to suspend hostilities when a young man and a girl from each camp fall in love and plan to marry. The cast of TV stars takes this one through its paces with customary professionalism, though there is little they can do to enliven the unimaginative script.

Aka: GOING TO THE CHAPEL

COM   91 min (ort 96 min) mTV   VIDrel: GENESIS V

WEDDING NIGHT, THE **   PG
King Vidor USA   1935
Gary Cooper, Anna Sten, Ralph Bellamy, Helen Vinson, Siegfried "Sig" Rumann, Esther Dale, Leonid Snegoff, Eleanor Wesselhoeft, Milla Davenport, Agnes Anderson, Hilda Vayghn, Walter Brennan, Douglas Wood, George Meeker, Hedi Shope
A New York novelist retreats to his country home after the failure of his latest book and gets to know a simple Polish country girl who impresses him with her home-spun wisdom. Inspired by their friendship, he uses her as the main character in his next book, but their relationship causes major problems with both her father and her fiance. A well crafted movie with a downbeat ending that failed to appeal to cinema audiences of the time.

DRA   83 min B/W   VIDrel: VCC/DISC V

WEDDING REHEARSAL **   U
Alexander Korda UK   1932
Roland Young, George Grossmith, John Loder, Lady Tree, Wendy Barrie, Maurice Evans, Joan Gardner, Merle Oberon, Kate Cutler, Edmund Breon, Morton Selten, Diana Napier, Lawrence Hanray, Rodolfo Mele
An aristocratic Guards officer survives his grandmother's matchmaking attempts by introducing the young ladies in question to his friends, but he is finally hooked by his mother's secretary. A faltering attempt at comedy that is too trite to be more than mildly amusing. Written by Biro and Arthur Wimperis.

COM   75 min (ort 84 min) B/W   VIDrel: L/A V

Boa: story by Lajos Biro and George Grossmith.

WEDLOCK **   18
Lewis Teague USA   1990
Rutger Hauer, Mimi Rogers, Joan Chen, James Remar, Stephen Tobolowsky, Basil Wallace, Grand L. Bush, Denis Forest, Glenn Plummer, Belle Avery, Ismael (East) Carlo, Ed Crick, Zaid Farid, Colin Hamilton, Harry Johnson, Rob Moran
In the near future, the brains behind a jewel theft that netted $25,000,000 is sent to a prison without walls. Pairs of prisoners are "linked" by collars that explode should one of the pair try to escape, and the identity of each "wedlock" mate is kept secret from the other. However, a woman claiming to be the thief's mate involves him in a daring escape plan. An intriguing idea is lost all too soon in the usual welter of action antics.

Aka: DEADLOCK

FAN   98 min (ort 103 min)   VIDrel: EIV/SONOP V

WEEDS **   18
John Hancock USA   1987
Nick Nolte, William Forsythe, Lane Smith, Joe Mantegna, Mark Rolston, John Tols-Bey, Ernie Hudson, Rita Taggart, Anne Ramsey, Charlie Rich, Orville Stoeber, J.J. Johnson, Essex Smith, Ray Reinhardt, Amanda Gronich, Felton Perry
A lifer in San Quentin who is also an aspiring playwright, is released from prison following pressure from a newspaper critic on his behalf. He then goes on to form his own theatre troupe, employing ex-convicts. This strange and uneasy blend of comedy and drama came from the real-life experiences of Rick Cluchey and his San Quentin Drama Group. A brave but muddled portrayal.

DRA   114 min   VIDrel: 20TH V/sur

Boa: play The Cage by Rick Cluchey.

WEEKEND ***   18
Jean-Luc Godard FRANCE/ITALY   1967
Mireille Darc, Jean Yanne, Jean-Pierre Leaud, Juliette Berto, Jean-Pierre Kalfon, Anne Wiazemsky, Yves Beneyton, Paul Gegauff, Daniel Pommereulle, Yves Alfonso, Blandine Jeanson, Virginie Vignon, Jean Eustach
A couple set out on a weekend to visit the woman's mother and embark on an odyssey of death and destruction, in a bleak, savage but ultimately ineffectual attack on the horrors of contemporary society. Godard vents his spleen on a variety of soft targets, attacking consumerism and the motor car with equal venom, but behind all this furious energy lies nothing but despair and resignation.

Aka: LE WEEK-END

DRA   99 min (ort 105 min)   VIDrel: CONNO/RTM V/sur

WEEKEND AT BERNIE'S **   15
Ted Kotcheff USA   1989
Andrew McCarthy, Jonathan Silverman, Catherine Mary Stewart, Terry Kiser, Don Calfa, Catherine Parks, Eloise Broady, Gregory Salata, Louis Giambalvo, Ted Kotcheff, Margaret Hall, Timothy Perez, Mark Kenneth Smaltz, Dolly Segal
Silverman and McCarthy play two ambitious young executives who get an invitation to spend the weekend at their boss' luxurious beach house. When they find he has been murdered, they realise they could be the next victims, and spend the rest of the movie trying to convince the unknown assassin that their boss is still alive. Written by Robert Klane, this raucous slapstick comedy lacks bite, but the two leads make the most of the simple-minded script.

Aka: HOT AND COLD

COM   94 min (ort 101 min)   VIDrel: MGM/WHV V/sur

WEEKEND AT BERNIE'S 2 **   PG
Robert Klane USA   1992
Andrew McCarthy, Jonathan Silverman, Terry Kiser, Tom Wright, Steve James, Troy Beyer, Barry Bostwick, Novella Nilson, Phil Coccioletti, Gary Dourdan, James Lally, Michael Rogers, Stack Pierce, Constance Shulman, Jennie Moreau
Untalented sequel in which our two young executives now go on a frantic hunt for Bernie's money that takes them to the Caribbean. Naturally, their well-preserved dead companion comes along too, providing an excuse to repeat all his tiresome antics. Silly and unfocused, this one offer nothing fresh and very few real laughs.

COM   85 min (ort 97 min)   VIDrel: EIV/SONOP V/sur

WEEKEND IN VEGAS **   18
Gary Graver USA   1997
Tane McClure, Tim Abell, Gabriella Hall, Richard Gabal
When a married couple go to Las Vegas in the hope of putting their finances on a better footing, their activities lead to much strain in the marriage.

DRA   90 min   VIDrel: THIRD V

WEEKEND WAR **   PG
Steven Hilliard Stern USA   1988
Stephen Collins, Daniel Stern, Evan Mirand, Michael Beach, Scott Paulin, James Tolkan, Charles Haid
A unit of part-time soldiers find themselves engaged in an all too real war with guerilla fighters, whilst doing two weeks National Guard duty in Honduras. Competent action-film that makes a number of salient points about Central America.

A/AD   91 min mTV   VIDrel: EIV/SONOP V

WEIRD SCIENCE **   15
John Hughes USA   1985
Kelly LeBrock, Anthony Michael Hall, Ian Mitchell-Smith, Bill Paxton, Judie Aronson, Suzanne Snyder, Robert Downey, Robert Rusler, Vernon Wells, Michael Berryman, Ivor Barry, Anne Coyle, Barbara Lang, Britt Leach
Silly teen-comedy with two shy kids using their computer to create a woman from a Barbie Doll to fill the gap in their lives. Engaging dialogue and good performances get lost in the totally moronic plot. Written by Hughes.

FAN   90 min (ort 94 min)

VIDrel: 4-FRONT/POLYREC/CIC V/sh

WELCOME HOME, ROXY CARMICHAEL *   15
Jim Abrahams USA   1990
Winona Ryder, Jeff Daniels, Laila Robins, Dinah Manoff, Thomas Wilson Brown, Joan McMurtrey, Frances Fisher, Graham Beckel,

Robin Thomas, Robby Kiger, Sachi Parker, Stephen Tobolowski, Micole Mercurio, John Short, Valerie Landsberg
Word gets out that the title character, a sexy woman who's been working in Hollywood, is to make a return trip to her home-town. This results in some frantic preparations as the whole town makes ready for her arrival. Meanwhile, young Ryder, who's rebellious, alienated and adopted, starts to believe that Roxy is her real mom. A quirky satire that goes nowhere, it is barely worth watching, and then only for a touching perfor-mance from Ryder.
COM   92 min (ort 98 min)   VIDrel: 20VIS/SONOP   V/sur

WELCOME II THE TERRORDOME **   18
Ngozi Onwurah UK   1994
Suzette Llewellyn, Saffron Burrows, Felix Joseph, Valentine Nonyela, Ben Wynter, Sian Martin, Jason Traynor, Brian Bovell, Cynthia Powell, Natasha Romulus, Marica Myrie, Olu Taiwo, Charlotte Moore, Tom Geoghean, John Adewole
In a Britain fifteen years hence, pollution and gang warfare are the most prevalent problems, from which a black family attempts to escape their claustrophobic ghetto life. An angry, crude and overblown affair, that is meant to be a provocative political allegory within a black context, but is far too weakened by its shrillness, its low budget and its lack of clarity. A pity, as it scores well in terms of imagination.
FAN   90 min (ort 94 min)   VIDrel: TART/20TH   V/sh

WELCOME TO THE DOLLHOUSE ***   15
Todd Solondz USA   1995
Heather Matarazzo, Victoria Davis, Christina Brucato, Christina Vidal, Siri Howard, Brendan Sexton Jr, Telly Pontidis, Herbie Duarte, Scott Coogan, Daria Kalinina, Matthew Faber, Josiah Trager, Ken Leung, Dimitri Iervolino, Bill Buell
Scripted by Solondz in his directing debut, this is a low-budget movie all about a plain and fairly nondescript eleven-year-old, at a school in New Jersey. Her life is examined over a single school year, when her lack of social skills leads to much taunt-ing by her more savvy classmates, most of the humour in the film being generated by her naive and pathetic attempts to fit in with her more style-conscious peers. Quite a cruel black comedy, but also a very honest one.
COM   88 min   CINrel

WE'LL MEET AGAIN *   PG
Tony Wharmby UK   1982
Susannah York, Michael J. Shannon, Ronald Hines, LIse-Ann McLaughlin, Ray Smith, Christopher Malcolm, Joris Stuyck
A WW2 drama set in a Suffolk town whose quiet life is turned upside down by the arrival of a group of US airmen. All the hoary old cliches of wartime Anglo-American relationships are put to use here, together with a plethora of stock characters and some exceptionally flat dialogue. Very poor, and not helped in the least by a lack of attention to detail and an over-reliance of modelwork and documentary footage for the aviation sequences.
DRA   343 min (3 cassettes – ort 700 min) mTV
VIDrel: SCRN/DISC   V

WENDY CRACKED A WALNUT ***   PG
Michael Pattinson AUSTRALIA   1987
Rosanna Arquette, Bruce Spence, Hugo Weaving, Kerry Walker, Desiree Smith, Dorren Warburton, Susan Lyons, Betty Lucas, Dennis Hoy, Douglas Hedge, Barry Jenkins
A neglected wife comforts herself through her avid consump-tion of romantic fiction, but gets an unexpected chance to experience love at first hand when she finally meets her own Prince Charming. An offbeat and very enjoyable comedy-drama, with Arquette giving a splendid performance in the title role.
Aka: ALMOST
COM   82 min (ort 87 min)   VIDrel: 20VIS/SONOP   V/sur

WENT THE DAY WELL? **   PG
Alberto Cavalcanti UK   1942
Leslie Banks, Elizabeth Allen, Frank Lawton, Basil Sydney, Valerie Taylor, Mervyn Jones, Edward Rigby, Marie Lohr, C.V. France, David Farrar, Muriel George, Thora Hird, Harry Fowler, John Slater, Johnnie Schofield
A British village is taken over by German paratroopers disguised as Royal Engineers but the villagers eventually learn the truth (the visitors cross their sevens in the Continental manner) and are able to turn the tables on their enemy. An intriguing idea that is handled in a rather unimaginative way.

See also THE EAGLE HAS LANDED for a similar exploration of this theme.
Aka: FORTY-EIGHT HOURS
DRA   90 min (ort 96 min) B/W   VIDrel: LUMI/SPEAR L/A V
Boa: short story The Lieutenant Died Last by Graham Greene.

WE'RE BACK! – A DINOSAUR'S STORY **   U
Phil Nibbelink/Simon Wells/Dick Zondag/Ralph Zondag
USA   1993
Voices of: Walter Cronkite, Julida Child, Martin Short, John Leno, John Goodman, Rhea Pearlman, Felicity Kendal, Kenneth Mars, Rene LeVant, Blaze Bergdahl, Charles Fleischer, Yeardley Smith
A space traveller comes to Earth in prehistoric times and enhances the intelligence of four dinosaurs whom he then trans-ports forward in time to present-day New York where they befriend some young humans.
ANIM   67 min (ort 71 min) cC   VIDrel: CIC/SONOP
V/sur
Boa: novel by Hudson Talbott.

WE'RE NO ANGELS **   U
Michael Curtiz USA   1954
Humphrey Bogart, Aldo Ray, Joan Bennett, Peter Ustinov, Basil Rathbone, Leo G. Carroll, John Baer, Gloria Talbott, Lea Penman, John Smith, George Dee, Louis Mercier, Torben Meyer, Paul Newlan, Ross Gould, Victor Romito
Three escaped Devil's Island convicts become involved with the family of a storekeeper, and help him outwit his greedy rela-tives. An amusing piece of period whimsy, that's both well made and acted, though the film's stage origins are very plain to see. Remade in 1989 with Robert De Niro and Sean Penn.
COM   106 min   VIDrel: CIC/SONOP   V
Boa: play La Cuisine Des Anges by Albert Husson.

WE'RE NO ANGELS **   15
Neil Jordan USA   1989
Robert De Niro, Sean Penn, Demi Moore, Hoyt Axton, Bruno Kirby, James Russo, Ray McAnally, Wallace Shawn, John C. Reilly
Two dumb cons inadvertently break out of jail and find them-selves being taken for priests on their way to a local shrine. This parody of the earlier Bogart movie, spoofs gangster films of the period, and Penn and De Niro give nice restrained performances, but the clumsy blending of comedy and drama and a violent sequence near the end are serious flaws. Russo's performance as a Cagney-style gangster is a highlight. Scripted by David Mamet.
COM   102 min (ort 110 min)   VIDrel: CIC/SONOP L/A   V

WE'RE TALKIN' SERIOUS MONEY **   15
James Lemmo USA   1992
Dennis Farina, Leo Rossi, Fran Drescher, Cynthia Frost, John La Motta, Peter Iacangelo, Anthony Powers, Lou Bonacki, John Cade, Catherine Paolone, Robert Constanzo, John Josef Spencer, Maria Cariani, Len Pera, Dona Hardy, John Kapelos
Two small-time con artists whose get-rich quick schemes have proved dismal failures in the past, borrow some cash from the Mob and get themselves into deep water. They take off for California with both the FBI and these gangsters in hot pursuit.
COM   88 min (ort 92 min)   VIDrel: 20VIS/SONOP   V/sh

WEREWOLF OF WASHINGTON, THE *   15
Milton Moses Ginsberg USA   1973
Dean Stockwell, Biff McGuire, Clifton James, Beeson Carroll, Thayer David, Jane House, Michael Dunn, Barbara Siegel, Stephen Cheng, Nancy Andrews, Ben Yaffe, Jacqueline Brooks, Thurman Scott, Tom Scott, Dennis McMullen
Horror film set against the Watergate scandal of 1972, that attempts to mix the elements in that political scandal with more traditional features of such films. An oddity that is neither useful nor decorative.
Aka: WEREWOLF AT MIDNIGHT
HOR   85 min (ort 90 min)   VIDrel: VIPCO/SGSVID   V

WES CRAVEN'S NEW NIGHTMARE **   15/18
Wes Craven USA   1994
Heather Langenkamp, Robert Englund, Miko Hughes, David Newsom, John Saxon, Wes Craven, Marianne Maddalena, Sam Rubin, Sara Risher, Robert Shaye, Nick Corri, Tuesday Knight, Tracy Middendorf, Fran Bennett, Matt Winston, Rob LaBelle
Film-within-a-film horror effort with Englund reprising his role as "Freddy" from all those NIGHTMARE ON ELM STREET films. This time the latter figure leaves the world of celluloid and enters the real world, where he forces an actress who was in one

of those earlier movies to go to work on a new horror film. And it would seem that it is only Craven who can put "Freddy" back where he belongs.
HOR   108 min (ort 112 min)   VIDrel: VCC/DISC; PION (LV only – 18 cert)   V/sh LV

WEST OF THE DIVIDE ***   U
Robert North Bradbury USA   1934
John Wayne, Virginia Browne Faire, Yakima Canutt, Gabby Hayes, Lafe McKee, Lloyd Whitlock, Billy O'Brien, John Whiteford, Earl Dwire, Tex Palmer, Dick Dickinson, Artie Ortego, Horace B. Carpenter, Hal Price, Archie Ricks
A young man discovers the identity of the outlaw who murdered his parents and kidnapped his baby brother years ago. He poses as an outlaw in order to infiltrate the man's gang and have his revenge. A well-made second feature, with Whitlock as the outlaw especially memorable.
WES   53 min B/W coVer   VIDrel: ENTUK L/A   V

WEST SIDE STORY ***   PG
Robert Wise/Jerome Robbins USA   1961
Natalie Wood, Russ Tamblyn, Rita Moreno, George Chakaris, Richard Beymer, Tucker Smith, David Winters, Tony Mordenete, Simon Oakland, John Astin, Burt Michaels, Eliot Feld, Robert Banas, Scooter Teague, Tommy Abbott, David Bean
This updated Romeo and Juliet tale, has a Puerto Rican girl and white boy falling in love despite peer pressures. The dancing and Leonard Bernstein score are dynamic but characterisation is weak. Scripted by Ernest Lehman, with Stephen Sondheim lyrics and choreography by Robbins. AA: Pic, Dir, Cin (Fapp), S. Actor (Chakiris), S. Actress (Moreno), Art/Set (Levin/ Gangelin), Score (Chaplin et al.), Cost (Sharaff), Sound (Hynes), Hon Award (Robbins).
MUS   145 min (ort 151 min) wScrn (special edition)
VIDrel: MGM/WHV   V/dm V/sh
Boa: play by Arthur Laurents.

WESTERNER, THE ***   U
William Wyler USA   1940
Gary Cooper, Walter Brennan, Forrest Tucker, Doris Davenport, Fred Stone, Paul Hurst, Chill Wills, Dana Andrews, Tom Tyler, Lillian Bond, Charles Halton, Arthur Aylesworth, Lupita Tovar, Julian Rivero, Roger Gray, Jack Pennick
A classic Western about the struggle over land claims and the problems which this causes. Brennan won an Oscar for his portrayal of Judge Roy Bean, and deservedly so. The music is by Dmitri Tiomkin with photography by Gregg Toland and the script by Jo Swerling and Niven Busch. Tucker's film debut. AA: S. Actor (Brennan).
WES   95 min (ort 100 min) B/W
VIDrel: VCC/DISC/COLUM   V
Boa: short story by Stuart N. Lake.

WESTWARD HO! **   Uc
George Stephenson AUSTRALIA   1988
Voices of: Bob Baines, Peter Beck, Claire Crowther, Phillip Minton, Larissa Lambert, Robert Menzies, Lloyd Morris, Noel Trevarthen
Adequate animated version of Kingsley's adventure classic that's set at the time of the Spanish Armada, and follows the fortunes of an adventurer who sets out to rescue a girl kidnapped from Ireland after a Spanish landing there.
ANIM   48 min (ort 50 min)   VIDrel: CARL/TECH   V
Boa: novel by Charles Kingsley.

WESTWORLD ***   15
Michael Crichton USA   1973
Richard Benjamin, Yul Brynner, James Brolin, Norman Bartold, Victoria Shaw, Alan Oppenheimer, Dick Van Patten, Majel Barrett, Steve Franken, Linda Scott, Nora Marlowe, Terry Wilson, Norman Bartold, Michael Mikler
A futuristic holiday camp offers trips to three recreated past periods, such as a Wild West town. Each setting employs robots programmed to satisfy the whims of the visitors, even to the point of staging mock duels. All goes well until the robots start to malfunction and the fights become real. An excellent idea spoilt by lack of development. Written and directed by Crichton. FUTUREWORLD followed in 1976.
FAN   85 min (ort 90 min) wScrn (special edition)
VIDrel: MGM/WHV   V/sh

WET NURSES *   18
Stuart Canterbury USA   1995
Tara Monroe, Keisha, Nicole London, Rebecca Wilde, Angel Bust,

Bunny Bleu, Mike Horner, Tony Tedeschi, Steve Drake, David Hardman
Dreary sex film in which a group of nurses employ some unorthodox treatments on their patients.
A   49 min   VIDrel: ONE   V

WHALE FOR A KILLING, A: PARTS 1 AND 2 ***   (PG)
Richard T. Heffron USA   1981
Peter Strauss, Richard Widmark, Dee Wallace, Kathryn Walker, Bruce McGill, Ken James, David Ferry, David Hollander, Bill Calvert, Larry Reynolds, Kent Barrett, George Morner, Reuven Bar-Yotam, Arthur Rosenberg, Colin Fox
A stranded humpback whale is the object of one man's care and attention, in the face of hostility from those who want to use it for commercial gain. Strauss is well cast as the campaigner, in this verbose but absorbing study of the slaughter of whales off the Newfoundland coast. Written by Lionel Chetwynd.
DRA   142 min (ort 150 min) mTV   SATrel: SKY MOVIES
Boa: book by Farley Mowat.

WHALES OF AUGUST, THE ***   U
Lindsay Anderson UK   1987
Bette Davis, Lillian Gish, Vincent Price, Ann Sothern, Harry Carey Jr, Frank Grimes, Frank Pitkin, Margaret Ladd, Tisha Sterling, Mary Steenburgen, Mike Bush
Two elderly sisters live out their last days on an island off the coast of Maine, in this gentle meditation on the inevitability of old age and death. Screenplay is by Berry from his touching play, with music by Alan Price.
DRA   86 min (ort 91 min)   VIDrel: VCC/DISC/COLUM L/A   V
Boa: play by David Berry.

WHAT A CARVE UP! *   U
Pat Jackson UK   1962
Sidney James, Kenneth Connor, Shirley Eaton, Dennis Price, Donald Pleasence, Michael Gough, Valerie Taylor, Emma Cannon, Michael Gwynn, Philip O'Flyn, George Woodbridge, Adam Faith, Timothy Bateson
A series of murders occur in conjunction with the reading of a man's will but the killer eventually proves to be the deceased who faked his death in order to get back at his relatives. An insipid and badly executed spoof of few laughs and uninspired acting, it is largely a spoof remake of THE GHOUL, a film that was little better.
Aka: NO PLACE LIKE HOMICIDE
COM   84 min (ort 88 min) B/W   VIDrel: FABFIL/SPEAR V
Boa: novel The Ghoul by Frank King.

WHAT ABOUT BOB? **   PG
Frank Oz USA   1991
Bill Murray, Richard Dreyfuss, Julie Hagerty, Charlie Korsmo, Kathryn Erbe, Tom Aldredge, Susan Willis, Roger Bowen, Fran Brill, Doris Belack, Melinda Mullins, Marcella Lowery, Margot Welch, Barbara Andres, Aida Turturro
Variant on a theme explored in "The Man Who Came To Dinner", with Murray a troublesome oddball who follows his psychiatrist on his vacation, makes the life of the latter a misery, but wins over the man's family. An under-plotted comedy that starts off very well, but gets bogged down and ever more irritating as it progresses, despite the best efforts of a really top-notch cast.
COM   95 min (ort 99 min)   VIDrel: TOUCH/SONOP L/A V

WHAT HAVE I DONE TO DESERVE THIS? **   18
Pedro Almodovar SPAIN   1985
Carmen Maura, Chus Lampreave, Julian Martinez, Luis Hostalof
A much put-upon housewife accidentally kills her chauvinist husband and begins to rebel against other oppressors, in this over-the-top black comedy that is full of the usual Almodovar touches. Unfortunately, these serve only to obscure the lack of any real plot or genuine sentiment.
Aka: QUE HE HECHO YO PARA MERECER ESTO?
COM   97 min (ort 100 min) wScrn   VIDrel: TART/20TH   V

WHAT HAVE YOU DONE TO SOLANGE? **   18
Massimo Dallamano ITALY   1971
Fabio Testi, Karin Baal, Christine Galbo, Joachim Fuchsberger, Camille Keaton, Gunther Stoll
In London, a married lecturer enjoys dating the students he teaches, and is with one such girl when they see a classmate

being murdered on the riverbank. Afraid of a scandal he keeps silent, but becomes a suspect when more murders take place. However, all is revealed when the mysterious Solange finally makes her appearance. A murky tale of vice rings, brutality and the inevitable erotic elements, it was banned in the UK on its initial release.
HOR   90 min (ort 100 min)   VIDrel: REDEM/RTM   V
Boa: novel by Edgar Wallace.

WHAT LOVES SEES: A TRUE STORY **   PG
Michael Switzer USA   1996
Annabeth Gish, Richard Thomas
A young blind girl from a rich family falls in love with a blind rancher, but he is from a much poorer family. Naturally, her parents are not in favour of this match.
DRA   92 min   VIDrel: ODY/SONOP   V/sh

WHAT PRICE GLORY **   U
John Ford USA   1952
James Cagney, Dan Dailey, Corinne Calvet, William Demarest, James Gleason, Robert Wagner, Casey Adams (Max Showalter), Craig Hill, Marisa Pavan, Wally Vernon, Henri Letondal, Fred Libby, Ray Hyke, James Lisburn, Henry Morgan
In WW1 France, two American Marines fight side by side in the trenches, but are rivals for the affections of a beautiful French girl. A remake of the 1926 silent, that was intended to be a musical (hence the strange colour scheme of the sets). However, Ford ignored the studio's orders apart from a few musical numbers at the beginning, and instead tried to inject a dose of broad comedy, resulting in a weird hybrid film that never comes to life.
A/AD   109 min   VIDrel: 20TH/TECH   V
Boa: play by Maxwell Anderson and Laurence Stallings.

WHAT SCHOOLGIRLS DON'T TELL *   18
Clause Rott WEST GERMANY   1974
Susie Atkins
Sex film done in that pseudo-documentary style used in the 1970s, by German film-makers with such feeble offerings.
Aka: SECRETS OF SWEET SIXTEEN: WHAT SCHOOLGIRLS DON'T TELL
A   71 min Cut (3 min 20 sec – ort 77 min)
VIDrel: RAVEN/LWV   V

WHATEVER HAPPENED TO BABY JANE? ****   18
Robert Aldrich USA   1962
Bette Davis, Joan Crawford, Victor Buono, Marjorie Bennett, Anna Lee, Dave Willock, Maidie Norman, Wesley Add, Bert Freed, Gina Gillespie, Ann Barton, Julie Allred, Barbara Merrill
Crawford as a crippled film star is at the mercy of her sadistic sister Davis. A macabre first venture into the black-comedy genre for these two stars. Buono supports as a pianist accompanist for Davis's hoped for return to films. Far-fetched, grisly and utterly absorbing. Written by Lukas Heller. AA: Cost (Norma Koch).
DRA   128 min (ort 132 min) B/W   VIDrel: WHV   V
Boa: novel What Ever Happened To Baby Jane Hudson by Henry Farrell.

WHAT'S EATING GILBERT GRAPE **   12
Lasse Hallstrom USA   1993
Johnny Depp, Juliette Lewis, Mary Steenburgen, Leonardo DiCaprio, Darlene Cates, John C. Reilly, Laura Harrington, Mary Kate Schellhardt, Kevin Tighe, Crispin Glover, Penelope Branning, Tim Green, Susan Loughran, Robert B. Hedges
A young boy faces an almost super-human task in taking care of his mentally retarded brother and his grotesquely overweight mother whose 500 lbs have virtually made her a prisoner in her own home. Unbelievably enough, he meets a female free spirit who gives him her love. First-rate performance are the major attraction, in this rambling and overlong work, which was scripted by Hedges from his own novel. A decidedly dark, oddball effort.
Aka: GILBERT GRAPE
DRA   112 min (ort 118 min)   VIDrel: EIV/SONOP   V/sur
Boa: novel by Peter Hedges.

WHAT'S LOVE GOT TO DO WITH IT ***   18
Brian Gibson USA   1993
Angela Bassett, Laurence "Larry" Fishburne, Vanessa Bell Calloway, Jenifer Lewis, Phyliis Yvonne Stickney, Khandi Alexander, Pamela Tyson, Penny Johnson, Rae'ven Kelly, Robert Miranda, Chi, Virginia Capers, Rob La Belle, Cora Lee Day
A highly energetic film version of Turner's biography, buoyed up by Bassett's sparkling performance and the many musical numbers. Her early life is covered briefly with the bulk of the film concentrating on her tortured relationship with Ike (well played by Fishburne). This movie suffers from many of the drawbacks of biopics but is hugely enjoyable to watch nonetheless.
Aka: TINA: WHAT'S LOVE GOT TO DO WITH IT
DRA   112 min (ort 118 min) cC   VIDrel: TOUCH/TECH   V
Boa: book I, Tina by Tina Turner and Kurt Loder.

WHAT'S NEW PUSSYCAT? **   15
Clive Donner USA   1965
Peter Sellers, Peter O'Toole, Woody Allen, Romy Schneider, Capucine, Paula Prentiss, Ursula Andress, Louise Lasser, Eddra Gale, Katrin Schaake, Jacques Balutin, Eleonore Hirt, Jean Paredes, Jess Hahn, Howard Vernon, Michel Subor
A silly, confused romp involving a disturbed fashion editor who pays a visit to a psychiatrist but finds that he is crazier than he is. Woody Allen's first film script is heavily flawed but has the occasional flash of wit. The catchy title song is by Burt Bacharach and Hal David.
Aka: QUOI DE NEUF, PUSSYCAT?
COM   104 min (ort 108 min)   VIDrel: MGM/WHV   V/h

WHAT'S UP DOC? ***   U
Peter Bogdanovich USA   1972
Barbra Streisand, Ryan O'Neal, Kenneth Mars, Austin Pendleton, Madeline Kahn, Sorrell Booke, Mabel Albertson, Randy Quaid, John Hillerman, Stefan Geirasch, Michael Murphy, Graham Jarvis, Liam Dunn, Phil Roth, Eleanor Zee
A crazy female student latches onto a musicologist and makes life unbearable for him and his dowdy fiancee. A plot involving spies and a jewel theft adds to the general confusion, culminating in an extremely funny chase and a splendid courtroom scene. Written by Buck Henry, David Newman and Robert Benton.
COM   90 min (ort 94 min)   VIDrel: WHV   V

WHAT'S UP NURSE? *   18
Derek Ford UK   1977
Nicholas Field, Felicity Devonshire, John Le Mesurier, Graham Stark, Kate Williams, Cardew Robinson, Barbara Mitchell, Angela Grant, Julia Bond, Peter Butterworth, Ronnie Brody, Sheila Bernette, Keith Smith, Chic Murrary
A new doctor takes up his first hospital post. The fact that the daughter of one of the consultants can only have sex whilst on the move (e.g. in a car), helps compound his problems. One more to add to the dated and tedious collection of British doctor-nurse sex comedies.
COM   77 min (ort 95 min)   VIDrel: EIV/SONOP   V

WHAT'S UP SUPERDOC? *   18
Derek Ford UK   1978
Christopher Mitchell, Julia Goodman, Harry H. Corbett, Bill Pertwee, Chic Murray, Angela Grant, Hughie Green, Julie Kirk, Beth Porter, Sheila Staefel, Marianne Stone, Melvyn Hayes, Nova Llewellyn, Ronnie Brody, Milton Reid
A doctor who has fathered 837 children by taking part in a project using artificial insemination, finds himself besieged by women claiming that he is the father of their children. A standard British medi-comedy – short on laughs and long on yawns.
Aka: WHAT'S UP NO. 2
COM   89 min (ort 96 min)   VIDrel: EIV/SONOP   V

WHAT'S UP, TIGER LILY? ***   PG
Senkichi Taniguchi/Compiled by Woody Allen USA   1966
Tatsuya Mihashi, Mie Hana, Akikio Wakabayashi, Tadao Nakamaru, Susumu Kurobe, Woody Allen, China Lee, The Lovin' Spoonful, Kumi Mizuno; Voices of: Louise Lasser, Frank Buxton, Len Maxwell, Mickey Rose, Julie Bennett
A Japanese agent film from 1964 was redubbed by Allen into an international hunt for a recipe for egg salad. An interesting project that provides a number of laughs whilst poking fun at all those James Bond movies. The music is by The Lovin' Spoonful who also appear in the film. The original film was entitled "Kagi No Kagi" (or Key Of Keys) and was released in 1964.
COM   79 min Cut (4 sec)   VIDrel: L/A   V

WHEELS OF TERROR **   15
Gordon Hessler USA   1986
Bruce Davidson, David Patrick Kelly, D.W. Moffett, Jay O. Sanders, Oliver Reed, David Carradine, Keith Szarabajka, Slavko Stimac, Andrija Maricic, Boris Komnenic, Bane Vidakovic, Irena Prosen, Svetlana Popovic, Branko Vidak

A WW2 story that follows the exploits of some soldiers from the German Penal Regiment, who are sent on a suicidal mission behind the Russian lines. A harsh and brutal account, scripted by Nelson Gidding from a book of similar attributes.
Aka: MISFIT BRIGADE, THE
WAR   100 min   VIDrel: POLY/POLYREC/MED   V/sur
Boa: novel by Sven Hassel.

WHEN A MAN LOVES A WOMAN **
Luis Mandoki USA
15
1994
Meg Ryan, Andy Garcia, Ellen Burstyn, Tina Majorino, Lauren Tom, Philip S. Hoffman, Eugene Roche, Latanya Richardson, Mae Whitman, Gail Strickland, Steven Brill, Susanna Thompson, Erinn Canovan, Bari K. Willerford, James Jude Courtney
A seemingly perfect family conceals a dark secret in the shape of the mother's alcoholism which she has tried too long to hide. Eventually, she agrees to treatment and her husband soon learns that he too has a major role to play. Quite weak in the plot department, this film is saved from mediocrity by its strong performances (especially Ryan's).
Aka: SIGNIFICANT OTHER; TO HAVE AND TO HOLD
DRA   121 min (ort 126 min) cC   VIDrel: TOUCH/TECH
V/sur

WHEN A STRANGER CALLS **
Fred Walton USA
15
1979
Carol Kane, Charles Durning, Colleen Dewhurst, Tony Beckley, Rachel Roberts, Ron O'Neal, Carmen Argenziano, Rutanya Alda, Kirsten Larkin, Bill Boyett, Michael Champion, Heetu, Joe Beale, Ed Wright, Louise Wright, Carol O'Neal
A crazed killer who was incarcerated seven years ago when he murdered two children and terrorised their babysitter, escapes from his asylum with the intention of repeating his crimes. A disagreeable and not really convincing shocker, that rapidly runs out of steam after the first ten minutes. Followed a good many years later by WHEN A STRANGER CALLS BACK.
HOR   93 min (ort 97 min)   VIDrel: MIA/DISC   V
Boa: short film The Sitter.

WHEN A STRANGER CALLS BACK **
Fred Walton USA
15
1993
Carol Kane, Charles Durning, Jill Schoelen, Gene Lythgow, Karen Austin, Babs Chula, John Blackwell Destry, Duncan Fraser, Jennifer Griffin, Gary Jones, Kevin McNulty, Terence Kelly, Michael Lonsdale-Smith, Sheelagh Megill, Rebecca Mullen
A sequel of sorts to the 1979 film WHEN A STRANGER CALLS in which our babysitter (now at college) finds herself being terrorised and her apartment is The police prove less than sympathetic and dismiss her complaints as hysteria but fortunately her student advisor comes to her rescue.
THR   90 min (ort 94 min) mTV   VIDrel: CIC/SONOP L/A
V

WHEN A WOMAN IS IN LOVE **
Roger Vadim FRANCE
18
1976
Sylvia Kristel, Nathalie Delon, Jon Finch, Gisele Casadesus, Marie Lebee, Jean Mermet
A mediocre film version of this classic 18th century tale of seduction and sexual conquests among the aristocracy. See also DANGEROUS LIAISONS and LES LIAISONS DANGEREUSES.
Aka: UNE FEMME FIDELE
DRA   82 min (ort 85 min)   VIDrel: L/A   V
Boa: novel Les Liaisons Dangereuses by Choderlos de Laclos.

WHEN DINOSAURS RULED THE EARTH ***
Val Guest UK
PG
1969
Victoria Vetri, Patrick Allen, Robin Hawdon, Drewe Henley, Imogen Hassall, Sean Caffrey, Magda Konopka, Patrick Holt, Jan Rossini, Carol-Anne Hawkins, Maria O'Brien, Connie Tilton, Maggie Lynton, Jimmy Lodge, Billy Cornelius
The story of a Stone Age romance between two people of different backgrounds who are themselves rejected by their respective tribes. The beautiful locations and good special effects (by Jim Danforth) help the J.G. Ballard story along nicely. See also ONE MILLION YEARS B.C.
FAN   96 min (ort 100 min)   VIDrel: MGM/WHV L/A   V

WHEN EIGHT BELLS TOLL **
Etienne Perier UK
15
1971
Anthony Hopkins, Robert Morley, Jack Hawkins, Corin Redgrave, Ferdy Main, Derek Bond, Nathalie Delon
A naval secret agent, sent to investigate the piracy of gold bullion from ships in the Irish Sea, poses as a biologist. A clumsy

adaptation of the novel, scripted by MacLean and devoid of wit or inspiration, but made a little more bearable by some beautiful locations.
A/AD   90 min (ort 94 min)   VIDrel: VCC L/A   V
Boa: novel by Alistair MacLean.

WHEN FATHER WAS AWAY ON BUSINESS ***
Emir Kusturica YUGOSLAVIA
15
1985
Moreno De Bartolli, Miki Manojlovic, Mirjana Karanovic, Mustafa Nadarevic, Mira Furlan, Pedrag Lakovic, Pavle Vujisic, Slobodan Aligrudic
The effect of the rift between Stalin and Tito as reflected in the fate of one family where the father is accused of pro-Stalinism and treated accordingly. These events are seen through the eyes of the young son, the film's central character. It won the Palme d'Or at Cannes and is a work of great charm, but criticism of the clear abuses that occur when the state is all-powerful are muted indeed. Set in early 1950s Sarajevo.
Aka: OTAC NA SLUZBENOM PUTU
COM   130 min (ort 136 min)   VIDrel: L/A   V

WHEN HARRY MET SALLY ****
Rob Reiner USA
15
1989
Billy Crystal, Meg Ryan, Carrie Fisher, Bruno Kirby, Steven Ford, Lisa Jane Persky, Michelle Nicastro, Harley Kozak, Gretchen Palmer, Robert Alan Beuth, David Burdick, Joe Vivianu, Joseph Hunt, Frank Luz, Kevin Rooney, Peter Day
A romantic comedy somewhat reminiscent of ANNIE HALL, that has Crystal and Ryan playing long-time acquaintances who drift from mild dislike through platonic friendship to love. Nora Ephron's semi-autobiographical script takes a funny and serious look at romantic relationships in a way that is both poignant and incisive. A topnotch cast and some great dialogue make it all the better.
COM   91 min (ort 96 min)   VIDrel: 4-FRONT/POLYREC
V/sur

WHEN HE'S NOT A STRANGER ***
John Gray USA
15
1989
Annabeth Gish, Kevin Dillon, John Terlesky, Kim Meyers, Stephen Elliott, Paul Dooley, John Jackson, Annabella Price, Jordan Charney, Janet Carroll, Allan Arbus, Micole Mercurio, Paul Dooley, Michael Oscar, John Lacy, Eugene Lee
A college girl suffers the trauma of being raped by her best friend's boyfriend – the top sportsman on the campus. Scripted by Gray and Beth Sullivan, and finely acted by Gish, this compelling drama attempts an honest look at a difficult and disturbing subject.
DRA   95 min (ort 100 min) mTV   VIDrel: MOPIC/SGSVID
V

WHEN NIGHT IS FALLING **
Patricia Rozema CANADA
18
1995
Pascale Bussieres, Rachael Crawford, Henry Czerny, David Fox, Don McKellar, Tracy Wright, Clare Coulter, Karyne Steben, Sarah Steben, Jonathan Potts, Tom Melissis, Stuart Clow, Richard Farrell, Fides Krucker, Thom Sokoloski
A straitlaced female professor at a Christian college is engaged to be married and has a well organised life until the day she meets a sexy circus performer and finds herself growing increasingly drawn to this charismatic woman. Potent and exotic, this lesbian love story is often melodramatic (lacking the light touch of Rozema's earlier I'VE HEARD THE MERMAIDS SINGING) and displays an irritating penchant for clumsy symbolism to get its points across.
DRA   93 min (ort 94 min)   VIDrel: TART/20TH   V/sh

WHEN NO-ONE WOULD LISTEN **
Armand Mastroianni USA
15
1993
Michele Lee, James Farentino, John Spencer, Lee Garlington, Andi McAfee, Chris Nash, Ron Perkins, Damion Stevens, Cameron Thor, Jill Tracy, Cicely Tyson, Dean Hill, David Farentino, Richard Penn, Ellen Crawford, Vincent Dale
A woman married to a brutal man tries to get some help from the authorities, but predictably her pleas fall on deaf ears. Another one of those overwrought dramas dealing with the unedifying subject of domestic violence. Well acted it may be, but the plot has few surprises.
DRA   90 min mTV   VIDrel: WHV L/A   V

WHEN SATURDAY COMES **
Maria Giese UK
15
1995
Sean Bean, Emily Lloyd, Pete Postlethwaite, Craig Kelly, John

McEnery, Ann Bell, Melanie Hill, Chris Walker, John Higgins, Tim Gallagher, Peter Gunn, Nick Waring, James McKenna, Tony Currie, Mel Sterland, Steve Huison, David Leland
In a bleak Northern town, a brewery worker attempts to escape a lifetime of drudgery by becoming a professional footballer, but has to do without the support of his family. A terribly stereotyped view (by American writer-director Giese) of the life of working class folk up North, for anyone who has ever lived there the cliched script is as patronising as it is unconvincing. Fortunately, the lively and often humorous performances are virtues that sustain the film.
DRA   94 min (ort 98 min)   VIDrel: GUILD/20TH   V/sur

WHEN THE BOUGH BREAKS **   18
Michael Cohn USA   1993
Martin Sheen, Ally Walker, Ron Perlman, Tara Subkoff, Robert Knepper, Scott Lawrence, John P. Connolly, Dick Welsbacher, Jimmy Medina, Ron Recasner, Juan Antonio Devoto, Christopher Doyle, Mark Daneri, Michael Raysses, Taylor Brock
Several severed children's hands, each with a tattooed number, are brought to light by a sudden flood and prove to be the handiwork of a serial killer. Having little else to go on, a police forensic expert battles to solve this case and learns of a young mute psychiatric patient who has some kind of mental link with the killer.
Aka: VICTIMS: IN THE MIND OF A SERIAL KILLER
THR   102 min (ort 103 min)   VIDrel: MED/DISC   V/sh

WHEN THE BULLET HITS THE BONE **   18
Damian Lee USA   1995
Jeff Wincott, Michelle Johnson, Doug O'Keeffe, Richard Fitzpatrick
A doctor becomes obsessed with stopping a drugs epidemic that is being fostered by a power-crazed maniac who has bribed a number of senators to help him achieve his ends. This leads to a series of set-piece battles and confrontations before our hero finally achieves his wish, but apart from the slightly unusual device of making the action hero a doctor, there is really nothing else of note here.
A/AD   88 min   VIDrel: EIV/SONOP   V

WHEN THE CAT'S AWAY ***   15
Cedric Klapisch FRANCE   1996
Garance Clavel, Zinedine Soualem, Renee Lecalm, Olivier Py, Simon Abkarian, Frederic Aufray, Olivier Bamy, Jane Bradbury, Joel Brisse, Olympe Brugeille, Franck Bussi, Marilyne Canto, Aline Chantal Pascal Chardin, Arapimou (Gris-Gris)
A charming if overly simplistic French tale of the misadventures of Chloe, a young girl, whose cat Gris-Gris vanishes from the flat of the old lady who was supposed to be looking after it while she went on holiday. Her determined attempt to run her moggy to ground leads her into an odyssey of the weird and unexpected, at the end of which she even finds a little romance. Scripted by Klapisch, this won the International Critics' Prize at the Berlin Film Festival.
Aka: CHACUN CHERCHE SON CHAT
DRA   95 min   VIDrel: ARTIF/20TH   V

WHEN THE LEGENDS DIE **   (PG)
Stuart Miller USA   1972
Richard Whitmark, Frederic Forrest, Luana Anders, Vito Scotti, Hebert Nelson, John War Eagle, John Gruber, Garry Walberg, Jack Mullaney, Malcolm Curley, Roy Engel, Rex Holman
After the death of his parents, a young Ute Indian is brought up by whites who exploit his exceptional riding ability on the rodeo circuit. A fair Western melodrama in which Widmark gives one of his best performances.
WES   101 min (ort 105 min)   SATrel: SKY MOVIES

WHEN THE PARTY'S OVER **   15
Matthew Irmas USA   1992
Elizabeth Berridge, Sandra Bullock, Rae Dawn Chong, Kris Kamm, Brian McNamara, Fisher Stevens, Michael Landes, S. Meadows
A group of people in their twenties share a house in L.A. in this of career-hungry and self-obsessed young things whose personal interaction is far from smooth. Solid characterisations help to sustain the interest.
DRA   110 min (ort 114 min)   VIDrel: FIRST/SONOP   V

WHEN THE TIME COMES ***   15
John Erman USA   1987
Bonnie Bedelia, Brad Davis, Terry O'Quinn, Karen Austin, Donald Moffat, Wendy Schaal, Corey Carrier, Annabelle Weenick, Mike Shanks, Judith Doty

A thirty-four-year-old woman discovers that she has terminal cancer and decides to kill herself, but is opposed by her husband and best friend. An engrossing drama with Bedelia giving a fine performance. See also THE LAST BEST YEAR.
DRA   104 min mTV   VIDrel: VISVID/POLYREC L/A   V

WHEN THE WHALES CAME ***   (12)
Clive Rees UK   1988
Paul Scofield, Helen Pearce, Max Rennie, Helen Mirren, John Hallam, David Threlfall
The story of an old recluse who lives on an island and spends his time whittling bird sculptures. A couple of inquisitive children get to know him and learn of a secret that has haunted him for years. An engaging tale set on the Scilly Isles just before WW1.
Aka: WHY THE WHALES CAME
DRA   96 min (ort 100 min)   VIDrel: 20TH/TECH   V/sur
Boa: story by Michael Morpurgo.

WHEN THE WIND BLOWS **   PG
Jimmy T. Murakami UK   1986
Voices of: Peggy Ashcroft, John Mills
An adequate adaptation of Briggs's morbid semi-comic tale of a rather naive middle-aged couple, and their pathetic attempts to prepare for an imminent nuclear conflict. Not quite the satire it was intended to be, but not bad either.
ANIM   85 min   VIDrel: 20TH/TECH L/A   V/sh
Boa: book by Raymond Briggs.

WHEN TIME RAN OUT *   PG
James Goldstone USA   1980
Paul Newman, Jacqueline Bisset, William Holden, James Franciscus, Edward Albert, Red Buttons, Ernest Borgnine, Burgess Meredith, Valentina Cortese, Alex Karras, Barbara Carrera, Veronica Hamel, John Considine, Sheila Allen
Routine disaster movie about a volcanic eruption on a tropical island peopled by the rich and famous. A disaster movie in every sense of the word with a rehashed and cliched plot courtesy of Carl Foreman and Stirling Silliphant. The music is by Lalo Schifrin. The original running-time is for versions that contain extra footage not shown theatrically.
THR   121 min (ort 144 min)   VIDrel: MGM/WHV L/A   V
Boa: novel The Day the World Ended by Max Morgan Witts and Gordon Thomas.

WHEN WORLDS COLLIDE ***   U
Rudolph Mate USA   1951
Barbara Rush, Richard Derr, Peter Hanson, Larry Keating, John Hoyt, Stephen Chase, Hayden Rorke, Sandro Giglio, Laura Elliot, Frank Cady, Judith Ames, Jim Congdon, Frances Sanford, Freeman Lusk, Joseph Nell
Scientists prepare a spaceship to save a selected band of people to start life on another world, before the Earth is destroyed by a giant planetoid hurtling towards it. Poor characterisation and pedestrian pacing are flaws, but the special effects (including the submerging of Manhattan) won a well-deserved Oscar. Written by Sidney Boehm. AA: Effects (George Pal).
FAN   79 min   VIDrel: CIC/SONOP L/A; PION (LV only) V/h LV
Boa: novel by Philip Wylie and Edwin Balmer.

WHEN YOU REMEMBER ME ***   15
Harry Winer USA   1990
Fred Savage, Ellen Burstyn, Kevin Sapcey, Lee Garlington
True-life story of a fourteen-year-old disabled boy who is callously treated at a badly-run nursing home and initiates a long court battle to win better treatment for himself and others like him.
DRA   91 min   VIDrel: ODY/SONOP   V/sh

WHERE ANGELS FEAR TO TREAD ***   PG
Charles Sturridge UK   1991
Helena Bonham Carter, Judy Davis, Rupert Graves, Giovanni Guidelli, Barbara Jefford, Helen Mirren, Thomas Wheatley, Sophie Kullman, Vass Anderson, Sylvia Barter, Eileen Davies, Siria Betti, Anna Lelio, Luca Lazzareschi, Gaetano Piro
Forster's first novel examines the feelings of the British to the greater world outside, with a widow taking a trip to Italy where, despite the disapproval of her former in-laws, she marries a handsome Italian. Yet she dies in childbirth, and her sudden death leads to a struggle on the part of various parties to adopt the child and bring it back to England; events which

prove to have a tragic outcome. A most absorbing and moving work.
DRA   108 min (ort 112 min)   VIDrel: COLUM/SONOP L/A V/sh
Boa: novel by E.M. Forster.

WHERE ARE MY CHILDREN? ***   (PG)
George Kaczender USA   1994
Marg Helgenberger, Corbin Bernsen, Christopher Noth, H. Richard Green, Jerry Hardin, Cynthia Martells, Angela Paton, Lois De Banze, John Cothran Jr, David Spielberg, Paul Linke, Ellen Crawford, Paxton Whitehead, Bonnie Bartlett
The true story of a single mother who is arrested by the FBI and has her three young children taken into care, and is denied access to them. It would appear that she is the victim of a high-level conspiracy, a fact that she campaigns to have exposed over the next twenty-five years. A deeply disturbing and profoundly depressing film.
DRA   90 min mTV   SATrel: MOVIE CHANNEL

WHERE EAGLES DARE ***   PG
Brian G. Hutton USA   1969
Richard Burton, Clint Eastwood, Mary Ure, Michael Hordern, Patrick Wymark, Robert Beatty, Ferdy Mayne, Anton Diffring, Donald Houston, Ingrid Pitt, Derren Nesbitt, Peter Barkworth, William Squire, Brook Williams, Neil McCarthy
Commandos have to mount a rescue operation for an American general who is being held in an impregnable Nazi mountain fortress. The tension in this one is maintained throughout whilst the plot is serpentine to say the least. Written by MacLean and with music by Ron Goodwin.
WAR   148 min (ort 158 min) wScrn (special edition) VIDrel: MGM/WHV   V/sh
Boa: novel by Alistair MacLean.

WHERE IS MY FRIEND'S HOUSE? ***   (12)
Abbas Kiarostami IRAN   1987
Babek Ahmed Poor, Ahmed Ahmed Poor, Kheda Barech Defai, Iran Outari, Ait Ansari, Sadika Taouhidi, Biman Mouafi, Ali Djamali, Aziz Babai, Nader Ghoulami, Akbar Mouradi, Teba Slimani, Mohamed Reda Berouana, Farahanka Brothers
A young schoolboy attempts to return a textbook to a friend who lives in another village close by, and along the way there enjoys a variety of interesting encounters. Scripted by the director, this is a fine look at life in modern-day Iran, shot on location in the north of the country and made with the participation of many non-professional actors. See also AND LIFE GOES ON, which attempts to provide a follow-up to the lives of some of the actors.
Aka: KHANEH-JE DOOST KOJAST?; WHERE IS THE FRIEND'S HOUSE
DRA   85 min   CINrel

WHERE PIGEONS GO TO DIE **   (U)
Michael Landon USA   1990
Art Carney, Michael Landon, Robert Hy Gorman, Cliff De Young, Ronne Troup, Bruce French
Told mostly in flashbacks, this is a sentimental look at the relationship a ten-year-old boy enjoys with his grandfather, and their shared interest in raising racing pigeons. A pleasant enough outing, though the script calls for major doses of sentimentality that do not work to the film's advantage.
DRA   85 min (ort 100 min) mTV   TVrel: BBC1
Boa: novel by R. Wright Campbell.

WHERE SLEEPING DOGS LIE ***   15
Charles Finch USA   1992
Dylan McDermott, Tom Sizemore, Sharon Stone, Charles Finch, Ren Karabatsos, Mary Woronov, Vanna Bonta, Liza Whitcraft, Phoebe Stone, Shawne Rowe, Richard Zavaglia, Julian McWhirter, Brett Cullen
A writer investigating the disappearance of a family moves into the house where this crime took place five years before. Hoping to write a novel about this event, he is disturbed by the arrival of a stranger who offers evidence of how the family were all murdered and the bodies buried beneath the swimming pool. A dark and very chilling tale, perhaps not so original in its plotting, but very well acted, especially by all three leads.
THR   87 min (ort 92 min)   VIDrel: COLUM/SONOP   V/sh

WHERE THE BOYS ARE ***   PG
Henry Levin USA   1960
Dolores Hart, George Hamilton, Yvette Mimieux, Jim Hutton, Barbara Nichols, Paula Prentiss, Connie Francis, Frank Gorshin, Chill

Wills, Frank Gorshin, Rory Harrity, John Brennan, Ted Berger, Vito Scotti, Sean Flynn
Story of four girls on holiday in Florida during Easter looking for sexual conquests. Fairly enjoyable light-hearted nonsense, with Connie Francis in her first film (she also gets to sing the title song). Look out for Nichols in a funny role as a flashy blonde. Written by George Wells and remade in 1984.
MUS   95 min (ort 99 min)   VIDrel: L/A   V
Boa: novel by Glendon Swarthout.

WHERE THE DAY TAKES YOU **   18
Marc Rocco USA   1992
Sean Astin, Lara Flynn Boyle, Peter Dobson, Balthazar Getty, Ricki Lake, James Le Gros, Dermot Mulroney, Will Smith, Adam Baldwin, Nancy McKeon, Alyssa Milano, Leo Rossi, Rachel Ticotin, Laura San Giacomo, Christian Slater
A parolee relates details of his life to a psychiatrist who records these interview sessions, which alternate with footage of the streets where he spends his time. A depressing look at the problems of the homeless in L.A. that pulls no punches.
DRA   99 min (ort 107 min) cC   VIDrel: 20VIS/SONOP V/sur

WHERE THE HEART IS **   15
John Boorman USA   1989
Dabney Coleman, Uma Thurman, Joanna Cassidy, Crispin Glover, Suzy Amis, Christopher Plummer, Maury Chaykin, David Hewlett, Dylan Walsh, Ken Pogue, Sheila Kelley, Michael Kirby, Dennis Strong, Timothy D. Stickney, Emma Woollard
A self-made millionaire who has grown rich by working as a demolition contractor takes some radical measures to deal with his sponging family, in the hope of teaching them the value of self-reliance. He simply throws them out and sends them off to live in a slum. An idea that would have worked well enough in the 1940s fails to sparkle today. The weak script was the work of John Boorman (who also produced the film) and his daughter.
COM   103 min (ort 107 min)
VIDrel: GUILD/POLYREC L/A   V/sh

WHERE THE RED FERN GROWS: PART 2 **   (U)
Jim McCullough USA   1991
Doug McKeon, Wilford Brimley, Chad McQueen, Lisa Whelchel, Adam Faraizl, Karen Carlson, Devin Payne, Tom Bertino, Jessie Turner, Patricia Meeks, Maggie McCullough, Daniel Glover, Philip Dale, Sherry Spurrier, Cindy Garrett
A belated sequel to the 1974 film, set this time in the Louisiana forests after WW2 where a young man grows to manhood thanks to the help and advice of his grandfather.
DRA   89 min (ort 105 min)   SATrel: MOVIE CHANNEL

WHERE THE RIVERS FLOW NORTH ***   15
Jay Craven USA   1993
Rip Torn, Tantoo Cardinal, Michael J. Fox, Bill Raymond, Treat Williams, Amy Wright, Mark Margolis, John Griesemer, Yusef Bulos, Dennis Mientka, Rusty De Wees, John Rothman, Sam Lloyd Sr., George Woodard, Jeri Lynn Cohen
In the 1920s, a stubborn Vermont man refuses to sell his land so that a much-needed dam can be constructed. However, it eventually becomes clear that his actions are motivated by reasons that are far from obvious. A balanced and careful adaptation of the original novel.
DRA   106 min   VIDrel: EIV/SONOP   V
Boa: novel by Howard Frank Mosher.

WHEREABOUTS OF JENNY, THE **   PG
Gene Reynolds USA   1990
Ed O'Neill, Debrah Farentino, Eve Gordon, Mike Farrell, Cassy Friel, Dakin Matthews, Lee Garlington, David Graf, Tony Danza, Michael Crabtree, Abraham Alvarez, Vinny Argiro, Harold Ager, Catherine Rusoff, Kathleen Coyne, Al Mancini
A divorced man tries to gain access to his daughter but finds that she has gone underground as part of the witness relocation programme, her mother having provided testimony that gave her the right to relocation and a new identity. The father's efforts soon bring him into contact with the officials who administer the scheme, but his request for help is treated with callous indifference. See also HIDE IN PLAIN SIGHT.
DRA   94 min (ort 100 min) mTV   VIDrel: ODY/SONOP V/sh

WHERE'S THE MONEY, NOREEN? **   12
Artie Mandelberg USA   1995
Julianne Phillips, A. Martinez, Nigel Bennett, Nancy Warren, Simon Reynolds, Jeremy Ratchford, Colm Feore, Stuart Clow, Gerry Quigley, Paul Lee, Anna Louise Richardson, Billy Otis, Sarah Goodwill, Steven Hill, Maggie Huculuk, Don Allison
After helping her brother rob an armoured car of $3,000,000, a woman leaves jail on parole after twelve years, but finds herself being stalked by a sinister group who are convinced she knows the location of the hidden loot. For her part, she believes there was an inside man on the job who was never caught and who may lead her to the money. A dull work that fails to generate much tension. See also GIRLS IN PRISON.
THR   88 min (ort 93 min) cC   VIDrel:CIC/SONOP   V/sh

WHEREVER YOU ARE **   18
Krzysztof Zanussi POLAND/UK/WEST GERMANY   1988
Julian Sands, Renee Soutendijk, Joachim Krol, Marcus Vogelbacher, Vadim Glowna, Maciej Robakiewicz, Tadeusz Bradescki, Maja Komorowksa, Milva
A newly-married couple arrive in Poland in 1938, full of hope and looking forward to spending their lives together. Unfortunately, they fail to realise that Europe is hovering on the brink of war and that the Germans are about to invade the country.
DRA   103 min   VIDrel: POLY/POLYREC/BRAVE   V

WHILE JUSTICE SLEEPS **   15
Alan Smithee (Ivan Passer) USA   1994
Cybill Shepherd, Tim Matheson, Karis Paige Bryant, Dion Anderson, Henry Beckman, Anna Ferguson, Kurtwood Smith, Robyn Stevan, Elan Ross Gibson, Crystal Verge, Chaelsey G. Marshall, Gabrielle Rose, Paul Couier, Valerie Pearson
A mother learns to her horror that a close family friend has been sexually abusing her eight-year-old daughter. The film opens with her shooting him in the course of a courtroom examination, and then moves into flashback mode. Rather overwrought, this film was not a happy experience for the director, who had his name removed from the credits (hence the notorious pseudonym Alan Smithee) after a dispute with the film's producers.
DRA   88 min mTV   VIDrel: ODY/SONOP   V/sh

WHILE THE CITY SLEEPS ***   PG
Fritz Lang USA   1956
Dana Andrews, Ida Lupino, Rhonda Fleming, George Sanders, Vincent Price, John Drew Barrymore, Thomas Mitchell, Sally Forrest, Howard Duff, Robert Warwick, Mae Marsh, Ralph Peters, Sandy White, Larry Blake, Celia Lovsky
A mad killer is hunted by both police and three newspaper reporters who are rivals for the post of editor-in-chief of their paper. This plum job has been promised as a reward for solving the case by the new owner who inherited the paper from his father. A corrosive and very cynical look at the dark side of human nature, with these journalists completely ruthless in their competitive efforts. An excellent and gripping film that never disappoints.
DRA   95 min (ort 100 min) B/W   VIDrel: ODY/SONOP   V
Boa: novel The Bloody Spur by Charles Einstein.

WHILE YOU WERE SLEEPING ***   PG
Jon Turteltaub USA   1995
Sandra Bullock, Bill Pullman, Peter Gallagher, Peter Boyle, Jack Warden, Glynis Johns, Micole Mercurio, Jason Bernard, Michael Rispoli, Ally Walker, Monica Keena, Ruth Rudnick, Marcia Wright, Dick Cusack, Thomas Q. Morris
Having saved a man from being killed by a train, a lonely female subway worker accompanies him to hospital, where in the resulting chaos his family mistakenly take her to be his fiancee. With her "fiance" in a deep coma, she finds herself in a difficult predicament and cannot bring herself to tell them the truth. A touching and tender romantic comedy, it gave Bullock a chance to demonstrate her gifts as an actress, putting her on the road to stardom.
COM   100 min (ort 103 min) cC   VIDrel: HOLPIC/TECH   V

WHISKY GALORE ***   PG
Alexander Mackendrick UK   1948
Basil Radford, Joan Greenwood, James Robertson Justice, Jean Cadell, Gordon Jackson, John Gregson, A.E. Matthews, Catherine Lacey, Bruce Seton, Wylie Watson, Gabrielle Blunt, Morland Graham, Henry Mollison
Scottish islanders plot to get hold of a load of whisky from a ship, that has foundered on a small island in the Hebrides during WW2. A fast-paced and witty comedy that helped establish the reputation of Ealing studios. Written by Compton Mackenzie and Angus MacPhail, and followed in 1957 by "Mad Little Island".
Aka: TIGHT LITTLE ISLAND
COM   79 min (ort 81 min) B/W   VIDrel: WHV   V/h
Boa: novel by Compton Mackenzie.

WHISPERS **   18
Doug Jackson CANADA   1989
Victoria Tennant, Chris Sarandon, Jean Leclerc, Pete McNeill, Keith Knight, Linda Sorenson, Eric Christmas, Jackie Burroughs, Tom Rack, Vlasta Vrana, Chris Britton, Mark Comancho, Walter Massey, Linda Singer, Stephan Perron
A woman tenant is attacked by a compulsive murderer whom she is forced to kill in self-defence. However, it would seem he is indestructible for he reappears once more to continue his reign of terror, his apparent immortality perplexing both her and a detective assigned to the case. An effective horror thriller, with moments of genuine terror, but a trifle too conventional to hold all that much in the way of surprises.
HOR   90 min (ort 96 min)   VIDrel: 20VIS/SONOP   V
Boa: novel by Dean R. Koontz.

WHISPERS IN THE DARK **   18
Christopher Crowe USA   1992
Annabella Sciorra, Jamey Sheridan, Jill Clayburgh, John Leguizamo, Deborah Unger, Anthony LaPaglia, Anthony Heald, Alan Alda, Jacqueline Brookes, Gene Canfield, Joseph Balducco Jr, Albert Pisarenkov, Malik, Bo Dietl
A psychiatrist becomes romantically involved with a man who later proves to have had a relationship with one of her current patients. When this woman is found murdered, she gets caught up in a very dangerous game. Good performances can do little to straighten out the confused and ponderous narrative.
THR   98 min (ort 103 min)   VIDrel: ETL/POLYREC/CIC   V/sur

WHISTLE BLOWER, THE ***   PG
Simon Langton UK   1986
Michael Caine, Nigel Havers, John Gielgud, James Fox, Gordon Jackson, Barry Foster, Felicity Dean, Kenneth Colley, David Langton, Dinah Stabb, James Simmons, Andrew Hawkins, Trevor Cooper, Katherine Reeve, Bill Wallis
Caine is well-cast as a former Korean War pilot, whose son was a highly talented linguist working for British Intelligence until he met an untimely end. He attempts to get at the truth behind his son's death and uncovers a sinister secret in the process. A fairly gripping spy thriller which takes a cynical point of view of the work of the intelligence services, but is let down by a contrived ending. The literate script is by Julian Bond.
THR   100 min (ort 104 min)   VIDrel: L/A   V
Boa: novel by John Hale.

WHISTLE DOWN THE WIND ***   PG
Bryan Forbes UK   1961
Hayley Mills, Alan Bates, Bernard Lee, Norman Bird, Alan Barnes, Hamilton Dyce, Elsie Wagstaff, Diane Holgate, Diane Clare, Patricia Heneghan, Roy Holder, John Arnatt, Howard Douglas, Gerald Sim, Ronald Hines
An escaped murderer hides out in a barn and is discovered by a group of kids who, because of his beard, take him to be Jesus. A sad, poignant and sometimes funny look at childhood innocence and fantasy. Written by Willis Hall and Keith Waterhouse, with music by Malcolm Arnold. This was the directorial debut of Forbes. See also ELENYA.
DRA   94 min (ort 99 min) B/W   VIDrel: VCC/DISC   V
Boa: novel by Mary Hayley Bell.

WHITE ANGEL **   18
Chris Jones UK   1993
Peter Firth, Harriet Robinson, Don Henderson, Anne Catherine Arton, Harry Miller, Joe Collins, Caroline Staunton, Mark Stevens, Inez Thorn, Jade Hansbury, Suzanne Sinclair, Chris Sullivan, Ken Sharrock, Samantha Norman, Caron Darwood
A serial killer is stalking the streets of London, attacking women dressed in white, and proves to be a transvestite dentist. He takes a room in the home of an American crime writer, but it transpires that she is also a killer, having disposed of her missing husband. A strange rapport slowly develops between them, but all the while the police are getting closer. An oddball thriller,

quite tense and claustrophobic, but so implausible it all becomes rather ludicrous.

THR   95 min (ort 96 min)   VIDrel: FEATFIL/RTM   V/sur

WHITE BALLOON, THE ****
Jafar Panahi IRAN
Aida Mohammadkhani, Mohsen Kafili, Fereshteh Sadr Orfani, Anna Bourkowska, Mohammad Shahani, Mohammad Bakhtiari, Aliasghar Samadi, Hamidreza Tahery, Asghar Barzegar, Hasan Neamatolahi, Bosnali Bahary, Mohammadreza Baryar, Shaker Hayely
The story of a seven-year-old girl and her attempts to find the money she needs to buy a goldfish she has set her heart on, after seeing it in a Tehran street-market. Unfortunately, this is not to be, as she is diverted from her quest and loses her money. An enthralling tale of childhood and its wonder and curiosity, the film has a charm that is universal and a simplicity that is timeless. Winner of the Camera d'Or at Cannes 1995.
Aka: BADKONAK-E SEFID
DRA   81 min (ort 84 min)   VIDrel: ELPIC/POLYREC   V

U
1995

WHITE CHRISTMAS **
Michael Curtiz USA
Bing Crosby, Danny Kaye, Rosemary Clooney, Dean Jagger, Vera-Ellen, Mary Wikes, Sig Ruman, Grady Sutton, John Brascia, Ann Whitfield, George Chakiris, Richard Shannon, Robert Crosson, Herb Vigran, Dick Keene, Johnny Grant
Two ex-army buddies team up and go into show business, helping to boost the popularity of a winter resort run by their ex-officer. More or less a vehicle for some fine Irving Berlin numbers (and some forgettable ones too), this partial reworking of the 1942 film HOLIDAY INN could have done with a stronger story. Written by Norman Krasna, Norman Panama and Melvin Frank.
MUS   115 min (ort 120 min) wScrn   VIDrel: CIC/SONOP   V

U
1954

WHITE FANG ***
Randal Kleiser USA
Ethan Hawke, Klaus Maria Brandauer, Seymour Cassel, James Remar, Susan Hogan, Bill Moseley, Clint B. Youngreen, Pius Savage, Aaron Hotch, Charles Jimmie Sr, Clifford Fossman, Irvin Sogge, Tom Fallon, Dick Mackey, Suzanne Kent
A young man travels to Alaska during the turn of the century goldrush, and encounters many hardships and dangers from both the elements and his fellow man, but once there becomes involves with a tame wolf and his owner. An exciting carefully wrought film, with fine performances and magnificent locations combine to produce an appealing film for all ages.
A/AD   104 min (ort 109 min) cC   VIDrel: WDV/TECH   V/sur
Boa: novel by Jack London.

PG
1991

WHITE FANG 2: MYTH OF THE WHITE WOLF **
Ken Olin USA
Scott Bairstow, Charmaine Craig, Alfred Molina, Geoffrey Lewis, Victoria Racimo, Paul Coeur, Anthony Michael Ruivivar, Al Harrington, Ethan Hawke, Matthew Cowles, Woodrow W. Morrison, Reynold Russ
A disappointing sequel in which our hero and his faithful wolf dog come to the aid of a tribe of Indian who are being attacked by some nasty prospectors, eventually helping them to avert the threat of starvation. A very black-and-white Disney affair that is almost unbearably wholesome and political correct. The beautiful scenery of British Columbia where this movie was shot on location, offers some compensation.
A/AD   102 min (ort 107 min)   VIDrel: WDV/TECH   V

U
1993

WHITE HEAT ****
Raoul Walsh USA
James Cagney, Virginia Mayo, Edmond O'Brien, Margaret Wycherly, Fred Clark, Steve Cochran, John Archer, Wally Cassell, Ford Rainey, Fred Coby, G. Pat Collins, Mickey Knox, Paul Guilfoyle, Robert Osterloh, Ian MacDonald
Immensely powerful story of a callous psychopath with a mother-fixation. Cagney is brilliant as the vicious hoodlum; this film marked his explosive return in a gangster role. O'Brien plays the undercover cop who infiltrates his gang leading to a dramatic final reckoning. A remarkable film that improves with repeated viewing. Written by Ivan Goff and Ben Roberts and with music by Max Steiner. Computer-coloured versions have been available.
DRA   109 min (ort 114 min) B/W   VIDrel: MGM/WHV   V
Boa: story by Virginia Kellogg.

15
1949

WHITE HUNTER, BLACK HEART ***
Clint Eastwood USA
Clint Eastwood, Jeff Fahey, George Dzundza, Alun Armstrong, Marisa Berenson, Richard Vanstone, Timothy Spall, Mel Martin, Charlotte Cornwell, Norman Lumsden, Edward Tudor Pole, Roddy Maude-Roxby, Richard Warwick, John Rapley
A film director working on location in Africa becomes almost completely obsessed with the need to hunt and kill an elephant, merely as a means of affirming his macho personality. Loosely based on the events surrounding John Huston's making of THE AFRICAN QUEEN, this is a quirky and absorbing film. Unfortunately, for all the real acting talent present, Eastwood fails to really explain his central character's motivations.
DRA   107 min (ort 112 min)   VIDrel: WHV   V/sur
Boa: book by Peter Viertel.

PG
1990

WHITE LIE **
Bill Condon USA
Gregory Hines, Annette O'Toole, Bill Nunn, Gregg Henry, Marc Macaulay, John Lawhorn, Carol Mitchell Leon, Tom Even, Ed GRady, C. Harrison Avery, Rick Andosca, Nat Adler, Dan Biggers, Tim Black, Jerry Campbell, Bill Coates
The arrival of an old photograph in the mail marks the beginning of a man's quest for the truth, since it shows the lynching of his father in the South, allegedly for the rape of a white woman. In an effort to learn the truth behind this appalling crime, he returns to his hometown but finds the locals both secretive and hostile. A strongly plotted and absorbing drama.
DRA   88 min (ort 93 min)   VIDrel: CIC/SONOP   V
Boa: novel Louisiana Black by Samuel Charters.

15
1991

WHITE LIGHT **
Al Waxman USA
Martin Kove, Martha Henry, Heidi Von Palleske, James Purcell, Bruce Boa, Allison Hossack, George Sperdakos, Aaron Schwartz, Raoul Trujillo, Harry Booker, Heath Lamberts, Larry Reynolds, Michelyn Emelle, Eve Crawford
Some echoes of GHOST are to be found in this agreeable romantic-thriller in which an undercover cop gains psychic powers after he nearly dies in a shooting incident. Complications abound, most especially when he finds himself falling in love with the spirit of a dead woman.
FAN   93 min (ort 96 min)   VIDrel: COLUM/SONOP   V

15
1989

WHITE LIGHTNING **
Joseph Sargent USA
Burt Reynolds, Jennifer Billingsley, Ned Beatty, Bo Hopkins, Matt Clark, Louise Latham, Diane Ladd, R.G. Armstrong, Conlan Carter, Dabbs Greer, Lincoln Demyan, John Steadman, Iris Korn, Stephanie Burchfield, Barbara Muller
A moonshiner goes after the crooked sheriff who drowned his brother. This lively actioner is helped along by a good cast, but is as predictable in outcome as it is contrived in development. Followed by GATOR in 1976.
A/AD   97 min (ort 101 min)   VIDrel: MGM/WHV   L/A   V

15
1973

WHITE MAN'S BURDEN **
Desmond Nakano FRANCE/USA
John Travolta, Harry Belafonte, Kelly Lynch, Margaret Avery, Tom Bower, Andrew Lawrence, Bumper Robinson, Tom Wright, Sheryl Lee Ralph, Judith Drake, Robert Gossett, Wesley Thompson, Tom Nolan, Willie Carpenter, Wanda Lee Evans
A kind of "reversal" fantasy set in a world where the Blacks are rich, educated and form the upper class of society, leaving the poor and ill-educated Whites to do all the menial jobs. When Travolta delivers a package to his boss he accidentally sees the man's wife in the nude, loses his job and later kidnaps his boss in reprisal. Scripted by Nakano, this is a flawed and clumsy stab at a social satire, with a violent ending that is contrived and pointless.
DRA   89 min   CINrel

15
1995

WHITE MEN CAN'T JUMP ***
Ron Shelton USA
Woody Harrelson, Wesley Snipes, Rosie Perez, Cylk Cozart, Tyra Ferrell, Eloy Casados, Kadeem Hadison, Marques Johnson, Nigel Miguel, Bill Henderson, Sonny Carver, Jon Hendricks, John Marshall Jones, David Roberson, Kevin Benton
A white minor-league con artist accepts a challenge to play a black guy at basketball and discovers to his surprise that he has some aptitude for the game. The two men become friends and decide to make some money by hustling other challengers, in this fast-pace urban comedy which offers good performances by

15
1992

the leads, especially Perez as Harrelson's hot-blooded girlfriend. Screenplay is by Shelton, and the lively dialogue is one of the film's best assets.
COM   111 min   VIDrel: 20TH/TECH L/A   V

WHITE MILE ** 15
Robert Butler USA 1993
Alan Alda, Peter Gallagher, Robert Loggia, Bruce Altman, Fionnula Flanagan, Jack Gilpin, Ken Jenkins, Dakin matthews, Don McManus, Robert Picardo, Alice Barden, Max Wright, Tim Choate, Kevin Cooney, Cab Covay, Denny Delk, Nigel Gibbs
A white-water rafting trip organised by an overbearing and unpleasant executive ends in tragedy and a court case where he stands accused of negligence. This confronts his young associate with a severe conflict of loyalties as he is forced to choose between his career and the need to tell the truth. An average TV drama, based on a true incident.
DRA   96 min mTV   VIDrel: GUILD/SONOP L/A   V

WHITE MISCHIEF *** 18
Michael Radford UK 1987
Sarah Miles, Joss Ackland, John Hurt, Greta Scacchi, Alan Dobie, Hugh Grant, Gregor Fisher, Ray McAnally, Charles Dance, Susan Fleetwood, Trevor Howard, Murray Head, Geraldine Chaplin, Catherine Neilson, Jacqueline Pearce
The tale of a bunch of British ex-pats living gloriously hedonistic lives out in Kenya during WW2. Fact-based (in 1941 Lord Erroll was murdered in Kenya following his very public affair with a married woman) this charts the response of a husband to his wife's infidelity and the scandal that arose at the time. An elegantly bizarre story made with great care and patience but in many ways rather clinical and unmoving.
DRA   104 min   VIDrel: VCC L/A   V/sh
Boa: book by James Fox.

WHITE NIGHTS *** 15
Taylor Hackford USA 1985
Mikhail Baryshnikov, Gregory Hines, Isabella Rossellini, Jerzy Skolimowski, Helen Mirren, Geraldine Page, John Glover, Stefan Gryff, Shane Rimmer, David Savile, Florence Faure, Ian Liston, Benny Young, Hilary Drake, Megumi Shimanuku
A Russian ballet dancer who defected finds himself home again after a flight he was on makes an emergency landing. He now has to find a way of escaping in an unbelievable film, that's only redeemed by some good dance sequences and the Lionel Richie score. Film director Skolimowski appears as the KGB agent who hounds Baryshnikov, and (as ever) there's a part for Mirren as a Russian. AA: Song ("Say You, Say Me" – Lionel Richie).
DRA   131 min (ort 135 min)   VIDrel: CASPIC/BMGREC L/A   V/sh

WHITE PALACE ** 18
Luis Mandoki USA 1990
Susan Sarandon, James Spader, Kathy Bates, Eileen Brennan, Rachel Levin, Renee Taylor, Jason Alexander, Spiros Focas, Gina Gershon, Steven Hill, Corey Parker, Kim Myers, Barbara Howard, Jonathan Penner, Hildy Brooks
The younger man meets older woman theme receives a few new twists in this story of a repressed young man with a good job who falls for a down-to-earth waitress. After an initial infatuation, he comes to realise that he will have to work very hard indeed to overcome the many differences between them. Although many aspects are undeniably superficial, this enjoyable drama shows a welcome absence of familiar cliches. See also IN LOVE WITH AN OLDER WOMAN.
DRA   99 min (ort 103 min)   VIDrel: CIC/SONOP   V/sur
Boa: novel by Glenn Savan.

WHITE SANDS * 15
Roger Donaldson USA 1992
Willem Dafoe, Mary Elizabeth Mastrantonio, Mickey Rourke, Samuel L. Jackson, M. Emmet Walsh, Mimi Rogers, James Rebhorn, Maura Tierney, Beth Grant, Fredrick Lopez, Alexander Nicksay, Miguel Sandoval, John Lafayette, Ken Thorley
This almost incomprehensible mystery thriller is set in a small town in New Mexico, where the local sheriff investigates an apparent suicide, and finds that the victim had a briefcase containing $500,000. Our plucky detective takes on the identity of the man in a bid to solve the mystery, and uncovers a complex plot that involves the purchase of military hardware on the black market and corruption within the FBI. Not a film for the impatient.
THR   97 min (ort 101 min)   VIDrel: WHV   V/sur

WHITE SHEIK, THE *** U
Federico Fellini ITALY 1951
Alberto Sordi, Brunella Bovo, Leopoldo Trieste, Giuletta Masina, Lilia Landi, Ernesto Almirante, Fanny Marchio, Gina Mascetti, Enzo Maggio, Ettore M. Margadonna, Jole Silvani, Anna Primula, Nino Billi, Armando Libianchi
A mismatched couple honeymoon in Rome where the wife spends time with the actor who plays the title figure in a popular romantic photo strip magazine. She soon discovers that her romantic hero is totally unlike his screen persona, while her straitlaced husband has to exercise his imagination for once, by devising excuses to explain her absence to friends and relations. An interesting early effort from this master director.
Aka: LO SCEICCO BIANCO
DRA   86 min B/W (ort 105 min)   VIDrel: CONNO/RTM L/A   V

WHITE SQUALL *** 12
Ridley Scott USA 1996
Jeff Bridges, Caroline Goodall, John Savage, Scott Wolf, Jeremy Sisto, Ryan Phillippe, David Lascher, Eric Michael Cole, Jason Marsden, David Selby, Julio Mechoso, Zeljko Ivanek, Balthazar Getty, Ethan Embry, Jordan Clarke, Jill Larson
A high-seas adventure that tells the largely true story of how a bunch of teenage landlubbers are shaped by experienced skipper Bridges into a capable crew whilst aboard a training vessel. All goes well until a tornado strikes, leading to loss of life and a court case. An absorbing and often rousing adventure than would have benefited from a tighter script; CAPTAIN'S COURAGEOUS it ain't. See also WIND for another sailing adventure.
A/AD   123 min (ort 129 min)   VIDrel: FIRST/SONOP; ENCORE (LV only)   V LV

WHITE TERROR ** 15
Paul Krasny USA
Robert Conrad, Dee Wallace, Chad McQueen, Wil Shriner
Two skiers are trapped on a mountain and face certain death unless a search and rescue reaches them in time. A simply plotted outdoors adventure.
A/AD   93 min   VIDrel: MARQ/REFLEC   V

WHITE TIGER ** 18
Richard Martin USA 1995
Gary Daniels, Matt Craven, Cary-Hiroyuki Tagawa, Julia Nickson-Soul, John Cassini, George Cheung, Philip Granger, Max Kirishima
Two special agents are sent to combat a drugs baron whose has obtained supplies of a powerful new designer drug that is intended take over the heroin and cocaine trade. But when one of the partners is killed, his partner swears to get even. A plodding thriller that has nothing original to offer and for good measure is not exactly helped by a weak and lifeless performance from Daniels as the surviving agent.
THR   90 min (ort 93 min)   VIDrel: GUILD/FOXVID   V/sur

WHITE ZOMBIE *** PG
Victor Halperin USA 1932
Bela Lugosi, Madge Bellamy, John Harron, Joseph Cawthorn, Robert Frazer, Clarence Muse, Brandon Hurst, Frederick Peters, George Burr McAnnan, Dan Crimmins, John Printz, John Fergusson, Claude Morgan, Annette Stone
Classic horror film set in Haiti where zombies toil at a sugar mill under the watchful eye of an evil overseer, who turns a young woman, about to be married, into the creature of the title. An undeniably powerful and atmospheric film, largely thanks to its good use of music and some striking photography, but let down by appalling acting.
HOR   64 min (ort 73 min) B/W   VIDrel: REDEM/RTM   V
Boa: novel The Magic Island by William Seabrook.

WHO DONE IT? * U
Basil Dearden UK 1956
Benny Hill, Belinda Lee, David Kossoff, Garry Marsh, George Margo, Ernest Thesinger, Denis Shaw, Frederick Schiller, Thorley Walters, Nicholas Phipps, Gibb McLaughlin, Ernest Jay, Harold Scott, Irene Handl, Charles Hawtrey
An ice-rink sweeper wins some money and a bloodhound in a competition and goes into business as a private detective, but gets more than he bargained for when he has to deal with spies who are planning to assassinate a number of key British scientists. A contrived and messy effort of minimal laughs that gave Benny Hill his film debut.
COM   85 min B/W   VIDrel: LUMI/SONOP L/A   V

WHO FRAMED ROGER RABBIT **** PG
Robert Zemeckis USA 1988
Bob Hoskins, Christopher Lloyd, Joanna Cassidy, Stubby Kaye, Alan Tilvern, Richard Le Parmentier, Joel Silver. Voices of: Charles Fleischer, Lou Hirsch, Mel Blanc, Mac Questel, Tony Anselmo, June Foray, Wayne Allwine,
This amazing blend of live-action and animation has seedy detective Hoskins investigating a murder, with chief suspect Roger Rabbit along to help. The dullish story takes second place to the incredible animation, especially the bit where Hoskins ventures into "Toontown". See also COOL WORLD and VOLERE, VOLARE. AA: Edit (Arthur Schmidt), Effects/aud (Charles L. Campbell/Louis L. Edemann), Effects/vis Ken Ralston/Richard Williams/Edward Jones/George Gibbs).
FAN 99 min (ort 105 min) cC VIDrel: TOUCH/TECH V/sur
Boa: novel Who Censored Roger Rabbit by Gary K. Wolf.

WHO GETS THE FRIENDS? *** PG
Lila Garrett USA 1988
Jill Clayburgh, James Farentino, Lucie Arnaz, Leigh Taylor-Young, Robin Thomas, James Sloyan, Greg Mullavey, Norman Parker, Melody Rogers, Thomas Galloway, Jenny Lewis, Laura Waterbury, Tim Russ, Anna Maria Horsford, Stan Ivar
A bitter-sweet comedy, that follows the changes in a couple's relationship with their circle of friends after they get divorced. Touches of despair are wrapped up neatly in a happy, candyfloss ending. Scripted by Garrett and Sandy Krinski.
COM 90 min (ort 96 min) mTV VIDrel: 20TH/TECH V

WHO IS JULIA? *** PG
Walter Grauman USA 1986
Mare Winningham, Jameson Parker, Jeffrey DeMunn, Jonathan Banks, Bert Remsen, Mason Adams, James Handy, Philip Baker Hall, Tracy Brooks Swope, Judy Ledford, Ford Rainey, Joel Colodner, Bruce French, Clare Nono
Following a car crash, the brain of a beautiful model is transplanted into the body of a plain housewife. Not as silly as it sounds, this literate tale of double identity is given credibility by the acting of Winningham in a dual role.
DRA 90 min (ort 120 min) mTV VIDrel: 20TH/TECH V
Boa: novel by Barbara S. Harris.

WHO WILL LOVE MY CHILDREN? *** PG
John Erman USA 1983
Ann-Margret, Frederic Forrest, Cathryn Damon, Donald Moffat, Lonny Chapman, Patricia Smith, Jess Osuna, Christopher Allport, Patrick Brennan, Soleil Moon Frye, Tracey Gold, Joel Graves, Rachel Jacobs, Robbie Kiger
A woman with only a year to live has to find loving homes for her ten kids. Based on the true story of farm wife Lucille Fray, who finding that she has terminal cancer, tried to find homes for her children rather than leave them in the care of her alcoholic husband. Written by Michael Bortman, this moving film won ten Emmy awards and gave Ann-Margret her TV film debut; she gives the central role both depth and conviction. See also THE OTHER WOMAN.
DRA 91 min (ort 120 min) mTV VIDrel: ODY/SONOP V/s

WHOOPS APOCALYPSE ** 15
Tom Bussman USA 1986
Loretta Swit, Peter Cook, Michael Richards, Rik Mayall, Ian Richardson, Alexei Sayle, Herbert Lom, Joanne Pearce, Christopher Malcolm, Shane Rimmer, Ian McNeice, Daniel Benzali, Richard Wilson, Richard Pearson, Marc Smith
Based on a popular TV series, this satire follows the events leading up to a nuclear conflict and though fun is very occasionally made of the Russians, most of the barbs are reserved for British and American institutions. Quite amusing in places, but undeniably partisan.
COM 88 min (Cut at film release – ort 93 min)
VIDrel: 4-FRONT/POLYREC/CIC V

WHORE, THE *** 18
Alex de Renzy/Henri Pachard USA 1990
Lili Carati, Jamie Gillis, Joey Silvera, Steven Vegas, Jeannie Pepper, Mike Horner, Tracey Adams, Raven Richards, Blake Palmer, Debi Diamond, Susan Vegas, Marc Wallice, Tom Byron
Carati plays an Italian woman who takes an American tourist for a lover, an act that leads to her Mafia family to murder him and then have her shipped of to the States, where she is sent to live with her uncle. But he has problems of his own, which she

inevitably gets caught up in. An unusually well plotted sex film, whose good production values (it was shot on 35 mm) contrast sharply with most of the material available in the genre.
A 54 min Cut (14 sec plus some cuts subst)
VIDrel: FIFTH/DISC V

WHORE ** 18
Ken Russell USA 1990
Theresa Russell, Antonio Fargas, Benjamin Mouton, Sanjay, Elizabeth Morehead, Michael Crabtree, John Diehl, Jack Nance, Tom Villard, Ginger Lynn Allen, Robert O'Reilly, Charles Macauley, Jason Kristoffer, Frank Smith
Although the very title of this film seems to have caused offence in parts of the USA (where it was changed to the one below) this grim look at the world of prostitutes and their customers in L.A. is neither more shocking nor more illuminating than similar works on this theme. We learn a good deal about the mechanics of the trade but precious little about the motivations of the players, and Russell's performance is far from convincing.
Aka: IF YOU CAN'T SAY IT, JUST SEE IT
DRA 81 min (ort 85 min) VIDrel: 4-FRONT/POLYREC V/sur
Boa: novel/play Bondage by David Hines.

WHORES FROM CHINA ** 18
Barry Lee HONG KONG 1997
Pauline Chan, Chow Hung
A naive Chinese woman arrives in Hong Kong and falls prey to men who exploit her innocence, their intention being to draw her into the world of organised prostitution.
DRA 90 min wScrn VIDrel: EAST/DISC V

WHO'S AFRAID OF VIRGINIA WOOLF? **** 15
Mike Nichols USA 1966
Richard Burton, Elizabeth Taylor, George Segal, Sandy Dennis
Two couples engage in a complex session of all-night conversation that leads to much bitterness and recrimination. Burton and Taylor were never better together than in this utterly absorbing but ultimately depressing film. Written and produced by Ernest Lehman this was Nichols's first film as director. AA: Actress (Taylor), S. Actress (Dennis), Cin (Haskell Wexler), Art/Set (Richard Sylbert/George James Hopkins), Cost (Irene Sharaff).
DRA 124 min (ort 129 min) B/W VIDrel: WHV V/h
Boa: play by Edward Albee.

WHO'S HARRY CRUMB? ** PG
Paul Flaherty USA 1989
John Candy, Jeffrey Jones, Annie Potts, Tim Thomerson, Barry Corbin, Shawnee Smith, Valri Bromfield, Renee Coleman, Joe Flaherty, Lyle Alzado, James Belushi, Stephen Young, Doug Steckler, Wesley Mann, Tamsin Kelsey, Fiona Roeske
The story of a pompous and accident-prone private detective, who tries to solve a kidnapping case, in the face of opposition from his boss. With but a few funny moments, this creaky and empty-headed comedy wastes the considerable talents of Candy on a plot that is more foolish than it is amusing.
COM 86 min Cut (2 sec at UK cinema release – ort 98 min)
VIDrel: VCC/DISC/COLUM V/sur

WHOSE LIFE IS IT, ANYWAY? *** 15
John Badham USA 1981
Richard Dreyfuss, John Cassavetes, Christine Lahti, Bob Balaban, Kenneth McMillan, Kaki Hunter, Thomas Carter, George Wyner, Alba Oms, Janet Eilber, Kathryn Grody, George Wyner, Mel Stewart, Ward Costello, Alston Ahern
A disabled man argues for his right to decide to end his own life, after he has become totally paralysed following a car crash. Based on Clark's hit play, this has Dreyfuss turning in an excellent performance, ably supported by a fine cast. Written by Brian Clark and Reginald Rose. The films THE SWITCH and "Right To Die" also explore this theme.
DRA 114 min VIDrel: MGM L/A V
Boa: play by Brian Clark.

WHO'S THAT GIRL * PG
James Foley USA 1987
Madonna (Madonna Ciccione), Griffin Dunne, John Miles, Haviland Morris, John McMartin, Bibi Besch, Robert Swan, Drew Pillsbury, Coat Mundi, Jim Dietz, Dennis Birkley, Cecile Callan, Karen Baldwin, Kimberlin Brown
This trite and tiresome attempt at a screwball comedy has Dunne being assigned to escort Madonna (who has just been

sprung from jail) out of town, and finding his life turned upside down in the process. A film that tries so hard one might award it points for effort, were it not so unremittingly unfunny.
Aka: SLAMMER
COM   88 min (Cut at film release by 1 min 43 sec)
VIDrel: MGM/WHV   V/sur

WHO'S THE MAN? ** 15
Ted Demme USA 1992
Doctor Dre, Ed Lover, Denis Leary, Colin Quinn, Heavy D., Kriss Kross, Run DmC, Flavor Flav, Cheryl "Salt" James, Ice-T, Jim Moody, Kim Chan, Rozwill Young, Badja Djola, Richard Bright, Andre B. Blake, Bill Bellamy, Queen Latifah
Two incompetent friends who work in a Harlem barbershop finds them becoming members of the police force, in which capacity they investigate a murder that proves to be linked to the plans of a ruthless property developer. A brash and over-loud piece of slapstick, with numerous cameos by hip-hop artists but of little interest to anyone else.
COM   85 min (ort 90 min)   VIDrel: TART/20TH   V/sur

WHY DID BODHI-DHARMA LEAVE FOR THE EAST? *** (12)
Bae Yong-Kyun SOUTH KOREA 1989
Yi Pan-Yong, Sin Won-Sop, Huang Hae-Jin, Ko Su-Myoung, Kim Hae-Yong
A leisurely and atmospheric story of a wise old Zen master who brings his two proteges, a young boy and an older monk, to a monastery in the countryside where he tries to instil in them some of his understanding and serenity.
Aka: DHARMAGA TONGJOGURO KAN KKADALGUN?; WHY HAS BODHI-DHARMA LEFT FOR THE EAST?
DRA   135 min   CINrel

WHY MY DAUGHTER? ** 15
Chuck Bowman USA 1993
Linda Gray, Jamie Luner, Joseph Burker, Louis A. Loturto, Lisa Sigell, Jan Burrell, Michael Lucas, kevin Quigley, Andrea White, John (J.R.) Knotts, Vana Vana O'Brien, Jodi Taylor, Paul Nada, Bob Roitblat, Mark Allen, Gretchen Corbett
A mother seeks to understand why her daughter became a hooker who was then murdered and tries to find the people responsible for her death.
DRA   90 min   VIDrel: NWV/HIFLI   V/h

WICKED AT HEART * 18
Paul Norman USA 1995
Chasey Lain, Tera Heart, Seth Damien, Sindee Coxx, P.J. Sparxx, Steve Drake, Tom Byron, Nick East, Jay Ashley, Micky Lynn, Jill Kelly
A largely plotless sex film dealing with the adventures of a woman whose behaviour is summed up by the title. Average for its genre.
A   80 min   VIDrel: ONE   V

WICKED CITY ** 18
Yoshiaki Kawajir/Michael Bakewell JAPAN
Voices of: Stuart Miller, Tammy Holloway, George Littlewood, Bill Richards, Ronald Baker, Lisa Robinson, Philip Gough
An uneasy peace exists between the world of mankind and that of demonic creatures, but this is about to change as our world is invaded by radical elements from the world of the supernatural.
Aka: MONSTER CITY; SUPERNATURAL BEAST CITY; YOJU TOSHI; YOUIJI TOSHI
ANIM   80 min (ort 83 min) dubbed
VIDrel: MANGA/SONOP   V

WICKED CITY, THE ** 18
Mai Tai Kit HONG KONG 1992
Leon Lai, Tatsuya Nakadai, Jacky Cheung, Michelle Li
Reptilian creatures arrive on Earth and assume human form, with one of them achieving power as a Hong Kong tycoon and seeking peaceful co-existence with humans. Unfortunately, his son is bent on world domination. A couple of special agents investigate a spate of killings which prove to be linked to this menace. Average.
A/AD   91 min dubbed   VIDrel: E2WEST/CREMED   V

WICKED LADY, THE ** 18
Michael Winner UK 1983
Faye Dunaway, Alan Bates, John Gielgud, Denholm Elliott, Prunella Scales, Oliver Tobias, Jane Purcell, Glynis Barber, Joan Hockson,

Helena McCarthy, Millie Maureen, Derek Francis, Nicholas Geeks, Hugh Millais, John Savident
Remake of the 1945 film about a lady highway robber updated with nudity and violence. Intended to be a costumed romp but lacking the style or wit to carry it off, this has Dunaway camping up her role as the wicked Lady Skelton, who robs the rich for the sheer fun of it, in the over-acting experience of the decade. The photography is by Jack Cardiff.
A/AD   94 min Cut (13 sec – ort 99 min)   VIDrel: VCC L/A V
Boa: novel The Life And Death Of The Wicked Lady Skeleton by Magdalen King-Hall.

WICKED STEPMOTHER * PG
Larry Cohen USA 1989
Barbara Carrera, Colleen Camp, Lionel Stander, David Rasche, Bette Davis, Tom Bosley, Richard Moll, Shawn Donahue, Evelyn Keyes, James Dixon, Seymour Cassel, Susie Garrett, Laurene London, Bob Goen, Robert Frank Telfer
An old woman with evil powers moves in with a family and causes havoc in this lame horror spoof. Reputedly intended to be a straight horror yarn, but a disagreement between Davis and writer-producer Cohen led to her replacement with Carrera and an extensively re-written script. But this film is a mess, and almost certainly would have been anyway. Davis's final film.
COM   91 min   VIDrel: MGM L/A   V

WICKER MAN, THE **** 18
Robin Hardy UK 1973
Edward Woodward, Britt Ekland, Christopher Lee, Ingrid Pitt, Diane Cilento, Lindsay Kemp, Ian Campbell, Aubrey Morris, Russell Waters, Walter Carr, Irene Carr, Irene Sunters, Geraldine Cowper, Jennifer Martin, Roy Boyd
A detective investigating the disappearance of a child on a remote Scottish island is slowly drawn into a web of intrigue by the inhabitants, who are worshippers of a strange and sinister cult. The original 103 minute version was believed lost, but still survives together with a 95 minute reconstruction made in 1980 by Hardy. This bizarre and erotic horror film has a strong script by Anthony Shaffer.
DRA   84 min (ort 103 min)   VIDrel: WHV   V/h

WIDE EYED AND LEGLESS *** (PG)
Richard Loncraine UK 1993
Julie Walters, Jim Broadbent, Thora Hird, Sian Thomas, Dinah Handley, Peter Whitfield, Candida Rundle, Andrew Nicholson, Martin Wenner, Graham Turner, Andrew Lancel, Ann Rye, Moya Brady, Carry Clubb, Saima Chaudry, Angela Walsh
A woman's life is completely dominated by a debilitating illness her doctors seem powerless to understand or treat. However, her loyal and loving husband is a source of strength and comfort, and helps sustain her spirits. One day he meets a friendly lady novelist who lends a sympathetic ear, but their growing friendship creates a difficult situation. A sensitive drama, both believable and well acted.
DRA   90 min mTV   TVrel: BBC
Boa: books: Diana's Story and Lost For Words by Deric Longden.

WIDE SARGASSO SEA *** 18
John Duigan AUSTRALIA 1992
Rachel Ward, Karina Lombard, Michael York, Nathaniel Parker, Huw Christie Williams, Claudia Robinson, Martine Beswike, Casey Berna, Ben Thomas, Paul Campbell, Audbrey Pilatus, Ancile Gloudin, Dominic Needham, Kevin Thomas
In the 1840s a young and correct British gentleman travels to Jamaica for an arranged marriage to a beautiful mulatto girl and is plunged into an unfamiliar world of eroticism and voodoo. Soon, dark doubts begin to assail his mind as to his wife's mental state. A lush adaptation of a fine novel, which is best described as a sort of prequel to "Jane Eyre" that tries to create a past for Rochester's mad wife, whom he kept locked in the attic.
DRA   95 min (ort 100 min)   VIDrel: POLY/POLYREC V/sur
Boa: novel by Jean Rhys.

WIDOWS ** 15
Ian Tonyton UK 1983
Fiona Hendley, Maureen O'Farrell, Eva Mottley, Ann Mitchell, David Calder, Paul Jesson, Maurice O'Connell, Stanley Meadows
Three widows of criminals killed while pulling off a job, decide to carry out what was to have been the gang's next robbery themselves, using the notes left behind by their leader. More

clever than appealing, this unusual heist tale was sustained by good performances if not by a believable script. A sequel followed in 1985. Originally shown in seven 52-minute episodes.
DRA   288 min (ort 364 min – 2 cassettes) mTV
VIDrel: THAMES L/A   V

**WIDOW'S KISS **   18
Peter Foldy USA   1996
Beverly D'Angelo, Mackenzie Astin, Dennis Haysbert, Bruce Davison
Not long after he gets married again, the father of a teenager dies suddenly, but not before changing his will and leaving everything to his wife, who as it turns out, has played no small part in his demise.
THR   98 min   VIDrel: HIFLI/SONOP   V/h

WIDOWS PEAK *   PG
John Irvin UK   1993
Mia Farrow, Joan Plowright, Natasha Richardson, Adrian Dunbar, Anne Kent, Jim Broadbent, John Kavanagh, Gerard McSorley, Rynagh O'Grady, Michael James Ford, Garrett Keogh, Britta Smith, Sheila Flitton, Marie Conmee, Ingrid Craven
In 1920s Ireland, a woman (Farrow) lives in a small town that is dominated by a clique of wealthy widows. Her arrival coincides with the son of the head of this group falling for a young and attractive American widow who has just arrived there. Soon, a raft of complications and intrigue is in full swing. Strong performances help enhance this bittersweet tale but its climax comes as no great surprise. Well-filmed with a good period atmosphere.
DRA   97 min (ort 101 min)   VIDrel: GUILD/SONOP L/A   V

**WIFE, MOTHER, MURDERER **   15
Mel Damski USA   1991
Judith Light, David Ogden Stiers, Keir Dullea, Kellie Overby, David Dukes, Whip Hubley, Jessie Jones, Mary Nell Santacroce, Joe Inscoe, Robin Florence, Janette Lane Bradnury, Maury Covington, Dan Biggers, Terrence Gibney
Having been imprisoned for the attempted murder of her husband and daughter, a woman escapes from jail with the intention of starting a new life. A competently directed thriller.
Aka: MARIE HILLEY STORY, THE; WIFE, MOTHER, MURDERER: THE MARIE HILLEY STORY
THR   90 min mTV   VIDrel: CIC/SONOP   V

WILD ANGELS, THE *   18
Roger Corman USA   1966
Peter Fonda, Nancy Sinatra, Bruce Dern, Lou Procopio, Michael J. Pollard, Diane Ladd, Gayle Hunnicutt, Coby Denton, Marc Cavell, Buck Taylor, Norman Alden, Joan Shawlee, Art Baker, Frank Mawell, Frank Gerstle, Kim Hamilton
Story of a nasty motorbike gang and their escapades that predates the far more stylish EASY RIDER and has none of that later film's wit or charm. A film of considerable unpleasantness that was the subject of much work on the part of Peter Bogdanovich, who was responsible for a good deal of the writing and editing, and even turns up briefly in one of the fights.
DRA   83 min (ort 124 min) Cut   VIDrel: VISVID/POLYREC L/A   V

WILD AT HEART *   18
David Lynch USA   1990
Nicolas Cage, Laura Dern, Willem Dafoe, Crispin Glover, Diane Ladd, Isabella Rossellini, Harry Dean Stanton, Grace Zabriskie, J.E. Freeman, Calvin Lockhart, Marvin Kaplan, W. Morgan Sheppard, David Patrick Kelly, Jack Nance
Writer-director Lynch brings his own inimitable style to this adaptation of Gifford's offbeat novel. Newly released from prison after killing a man in a brawl, Cage takes off with a free-spirit Dern, hotly pursued by the mother's lover and a vicious hit-man she has hired. As they travel to New Orleans, they meet a succession of weirdos and misfits. A bizarre, violent and patchy film, but certainly unforgettable. It won the Palme d'Or at Cannes in 1990.
A/AD   119 min (ort 127 min) wScrn
VIDrel: ELPIC/POLYREC   V/sur
Boa: novel by Barry Gifford.

WILD BILL: THE LEGENDARY TRUE STORY *   15
Walter Hill USA   1995
Jeff Bridges, Ellen Barkin, John Hurt, Diane Lane, Keith Carradine, Bruce Dern, Christina Applegate, James Gammon, David Arquette, Marjoe Gortner

An attempt to demystify the myths that have grown up around this character, the film opens with Wild Bill's funeral. An intricate flashback follows, that tells of Bill's return to Deadwood after three years, where he tries to pick up the threads of his life, and renews his friendship with Calamity Jane. Various deeds of derring-do follow, and the bustle and feel for the period is well captured, even if some jarring touches mark this as clearly a film of the 1990s.
WES   93 min (ort 98 min) cC   VIDrel: MGM/WHV   V/sur

WILD BUNCH, THE *   18
Sam Peckinpah USA   1969 (re-released in 1981 at 142 min)
William Holden, Ernest Borgnine, Robert Ryan, Edmond O'Brien, Warren Oates, Ben Johnson, Jaime Sanchez, Strother Martin, L.Q. Jones, Emilio Fernandez, Bo Hopkins, Albert Dekker, Dub Taylor, Paul Harper, Jorge Russek, Alfonso Arau
A group of outlaws who realise that they are becoming an anachronism as the world is changing, decide to pull off one last robbery in 1913. An extremely violent film whose ending puts one in mind of an abattoir, all concerned getting shot to pieces and blood spraying out everywhere in true Peckinpah style. Made with infinite care and written by Peckinpah and Walon Green, this brutal actioner is a classic of sorts.
WES   138 min wScrn   VIDrel: WHV   V

**WILD CACTUS **   18
Jag Mundhra USA   1992
India Allen, David Naughton, Gary Hudson, Michelle Moffett, Kathy Shower, Anna Karin, Robert Z'dar, Paul Gleason, David Wells
A couple attempt to re-ignite the passion in their relationship by taking a desert vacation but fall in with a dangerous criminal and his female companion, thereby putting their lives in danger. A standard thriller with no shortage of erotic sequences and the usual quota of violence.
THR   90 min   VIDrel: 20VIS/SONOP   V

**WILD CARD **   PG
Mel Damski USA   1992
Powers Booth, Cindy Pickett, Terry O'Quinn, Rene Auberjonois, M. Emmet Walsh, John Lacy, Don Hood, Kamala Lopez, Bert Remsen, Michael Shaner, Jake Jacobs, Royce D. Applegate, Diane Delano
A man with a colourful past, known as the "Preacher" (because he used to be one) is called to a small town in New Mexico because the local priest suspects that a landowner's death was no an accident. Fortunately, our hero is both an accomplished cardplayer and a former Vietnam veteran, so he is more than able to take care of himself. A weakly plotted action film with a Western flavour, but fairly entertaining thanks to the presence of Booth in the central role.
A/AD   81 min (ort 95 min) mCab   VIDrel: CIC   V
Boa: novel Preacher by Ted Thackrey Jr.

WILD CHILD: BEVERLY HILLS 38-24-36 *   18
Dominique Moreau USA   1991
Jennifer Irwin, Trystan Moore
A Hollywood starlet invites some friends home for a party, but naturally enough, this being Hollywood, said party soon degenerates into an orgy.
Aka: WILD CHILD
A   89 min   VIDrel: MIA/DISC   V

WILD GEESE, THE *   15
Andrew V. McLaglen UK   1978
Richard Burton, Roger Moore, Richard Harris, Hardy Kruger, Stewart Granger, Jack Watson, Winston Ntshona, Ronald Fraser, Frank Finlay, Kenneth Griffith, Jeff Corey, Barry Foster, Patrick Allen, John Kani, Ian Yule, Brook Williams
Action adventure yarn with a bunch of mercenaries being assembled, and trained for a mission to rescue a kidnapped African leader. The pace of the first half, in which our intrepid team are being trained, soon gives way to more plodding direction, as the implausible story quickly runs down. Not bad but definitely not great. Written by Reginald Rose with the first sequel following in 1985.
A/AD   129 min (ort 134 min)   VIDrel: L/A   V
Boa: novel by Daniel Carney.

WILD GEESE 2 *   18
Peter Hunt USA   1985
Scott Glenn, Barbara Carrera, Edward Fox, Laurence Olivier, Robert Webber, Robert Freitag, Kenneth Haigh, Stratford Johns, Derek Thompson, Paul Antrim, John Terry, Ingrid Pitt, Patrick Stewart, Michael Harbour, David Lumsden

In this sequel to the original our mercenaries set out on another mission, this time to spring Hess from Spandau Prison. A foolish and unbelievable film, burdened by a messy and disjointed script. A definite dud.
A/AD   120 min (ort 125 min)   VIDrel: L/A   V
Boa: novel The Square Circle by Daniel Carney.

**WILD GOOSE CHASE **   18
John Stagliano USA   1992
Julianne James, Joey Silvera, Angela Summers, Patricia Kennedy,
Tamara Lee, Champagne, Randy West, Candic, Jeanna Fine, Sean
Michaels, K.C. Williams, T.T. Boy, Christine Keith, Delores Delux,
Mal O'Ree, Delta Force, Robert J. Boisvert
A detective accepts an assignment to find a missing girl and is soon caught up in one bizarre encounter after another. Quite an average sex film, but with high production values for works of this genre, including the fact that it was shot on 35 mm.
A   80 min   VIDrel: HAR/GOLD   V

WILD HEART, THE *   PG
Michael Powell/Emeric Pressburger UK   1950
Jennifer Jones, David Farrar, Cyril Cusack, Sybil Thorndike, Edward
Chapman, Esmond Knight, George Cole, Hugh Griffith, Beatrice
Varley, Frances Clare, Raymond Rollett, Gerald C. Lawson, Valentine
Dunn
A 19th Century country romance, with a superstitious girl marrying a local minister whilst all the while being desired by the squire. Set in Victorian times with Jones well-cast as the wild and strange Shropshire lass. The beautiful photography is by Christopher Challis. Re-edited from an original and superior 110 minute version, which had extra scenes directed by Robert Mamoulian, and was first released in the UK under its alternative title.
Aka: GONE TO EARTH
DRA   82 min (ort 110 min)   VIDrel: VCC/DISC   V
Boa: novel by Mary Webb.

**WILD INNOCENCE **   18
Paul Norman USA   1994
Shayla, Kiss, Bionca, Alexis Deville, Celeste, Patricia Kennedy, Woody
Long, E.Z. Ryder, Tom Byron, Cal Jammer, Nick East
A ruthless woman worms her way into a couple's apartment, where she sets about seducing the husband and blackmailing the wife, although her ultimate intention remains unclear until the end. A strong and unusual plot distinguishes what is in other respects a standard adult film.
A   80 min   VIDrel: ONE   V

**WILD JUSTICE **   18
Paul Turner UK   1995
Nia Medi, Dafydd Emyr, Nick McGaughey, Christine Pritchard,
Trefor Selway, Patricia Millardet, Roy Scheider, Sam Wanamaker,
Huw Charles, Manon Pruson, Athen Constantine, Geraint Wyn,
Tracy Spottiswode, Ceri Tudno, Yoland Williams
In this psychological thriller, the rape and murder of the youngest daughter all but destroys a family, as they become ever more obsessive in their desire for revenge. Not the most appealing of films, but the strong cast ensure the film does not lack conviction.
Aka: COVERT ASSASSIN; DIAL
THR   88 min (ort 97 min)   VIDrel: FIRST/SONOP   V

WILD ONE, THE *   PG
Laslo Benedek USA   1954
Marlon Brando, Lee Marvin, Mary Murphy, Robert Keith, Jay C.
Flippen, Jerry Paris, Alvy Moore, Gil Stratton, Peggy Maley, Hugh
Sanders, Ray Teal, John Brown, Will Wright, Robert Osterloh, Robert
Bice, William Yedder
A powerful performance from Brando as a motorbike gang-leader terrorising a small town, helps keep this grand-daddy of all biker films from looking too dated. Produced by Stanley Kramer with a script by John Paxton.
DRA   76 min (ort 79 min) B/W   VIDrel: COLUM/SONOP
V/dm
Boa: short story The Cyclists' Raid by Frank Rooney.

WILD ORCHID *   18
Zalman King USA   1990
Mickey Rourke, Carre Otis, Jacqueline Bisset, Bruce Greenwood,
Assumpta Serna, Oleg Vidov, Milton Goncalves, Jens Peter, Antonio
Mario Da Silva, Paul Land, Michael Villela, Bernardo Jablonsky, Luiz
Lobo, Lester Berman
A beautiful law graduate is hired by a major brokerage firm and

goes to Rio de Janeiro to assist experienced lawyer Bisset. When the latter is called away she takes her place for a meeting with a wealthy businessman. She is soon embroiled in a strange, one-sided relationship with Rourke, an aloof weirdo who speaks in dull cliches and combines attention with indifference. A vastly over-hyped film, short on plotting and eroticism. A sequel followed.
DRA   106 min (ort 117 min)
VIDrel: 4-FRONT/POLYREC/VISVID L/A   V/h

WILD ORCHID 2: TWO SHADES OF BLUE *   18
Zalman King USA   1991
Wendy Hughes, Tom Skerritt, Robert Davi, Brent Fraser, Nina
Siemaszko, Joe Dallesandro, Christopher McDonald, Casey Sander,
Stafford Morgan, Bridgit Ryan, Don Bloomfield, Liane Curtis, Lyle
Denier, Gloria Reuben, Kathy Hartsell
Hardly a sequel, this sterile movie details the double life of a young teenager who goes to work in a brothel. There one of her customers is the same high school athlete whom she adores. Disguised in a black wig, he fails to recognise her, which leads to a number of complications. A lot of steamy sex scenes take the place of a coherent plot and strong characterisations, but add little of interest. Followed by the RED SHOE DIARIES TV series.
DRA   103 min (ort 111 min) cC
VIDrel: VCC/DISC/COLUM   V/sh

WILD PAIR, THE *   18
Beau Bridges USA   1987
Beau Bridges, Bubba Smith, LLoyd Bridges, Gary Lockwood, Danny
De La Paz, Raymond St Jacques, Lela Rochon, Ellen Geer, Angelique
De Windt, Creed Bratton, Randy Boone, Greg Finley, Andrew Parks,
Jack Eiseman, Brian Davis
Bridges's first theatrical film as director, is a highly forgettable little tale of a cop and FBI agent teaming up to bring some nasty drug dealers to justice. Disappointing in just about every area.
Aka: DEVIL'S ODDS; HOLLOW POINT
A/AD   92 min   VIDrel: EIV/SONOP   V

WILD PALMS: THE DREAM BEGINS/THE DREAM
CONCLUDES *   15
Kathryn Bigelow/Keith Gordon/Peter Hewitt/Phil Joanou
USA   1992
James Belushi, Kim Cattrall, Dana Delany, Robert Loggia, Angie
Dickinson, Ernie Hudson, Bebe Neuwirth, Nick Mancuso, Charles
Hallahan, Robert Morse, David Warner, Ben Savage, Bob Gunton,
Brad Dourif
In Los Angeles in the year 2007 a ruthless sect tries to gain control of a newly started virtual-reality TV station and exploit it for its own ends. A lawyer joins this channel as one of its executives and gets caught up in a new and frightening world. A stylish mini-series whose irritatingly oblique approach recalls "Twin Peaks", and though this offering is long on visual impact and acting it's rather disappointing in terms of actual content.
FAN   134 min (Part 1); 137 min (Part 2); (ort 300 min) mTV
VIDrel: BBC/TECH L/A   V/sh

WILD SEARCH *   18
Ringo Lam HONG KONG   1989
Chow Yun Fat, Cherrie Chung, Lau Kong, Tommy Wong, Roy
Cheung
A cynical cop given the job of protecting a little girl witness from the killer who wants her dead.
A/AD   92 min (ort 95 min) wScrn   VIDrel: MADE/RTM   V

WILD STRAWBERRIES *   15
Ingmar Bergman SWEDEN   1957
Victor Sjostrom, Ingrid Thulin, Bibi Andersson, Gunnar Bjornstrand,
Folke Sundquist, Bjorn Bjelvenstam, Naima Wifstrand, Gertrud Fridh,
Ake Fridell, Gunnel Brostrom, Gunnar Sjoberg, Per Sjostrand, Sif
Ruud, Yngve Nordwall
An ageing professor reviews the past tragedies of his rather empty life on a car journey to receive an honorary degree at his old university. This sharp and poignant look at life is really rather bleak, though many critics seem to regard it as one of the director's more joyful films. The script is by Bergman.
Aka: SMULTRONSTALLET
DRA   87 min (ort 93 min) B/W   VIDrel: TART/20TH
V/dm

WILD TARGET *   15
Pierre Salvadori FRANCE   1993
Jean Rochefort, Marie Tritignant, Guillaume Depardieu, Patachou,

Charlie Nelson, Wladimir Yordanoff, Serge Riaboukine, Philippe Girard, Daniel Laloux, Christophe Odent, Olga Poliakoff, Francois Toumarkine, Julien Cafaro
Salvadori's debut film tells of the mishaps of a professional assassin who apart from his mother (from whom he gets tips) has no other human contacts. When a messenger boy catches him at work, he has to take the youngster on as an accomplice. But the failure of their first joint mission to kill a beautiful con-artist is not successful, and he soon finds this elusive figure to be more than a match for him. A distinctive black comedy, written by the director.
Aka: CIBLE EMOUVANTE
COM   84 min (ort 88 min) wScrn
VIDrel: TART/20TH; ENCORE (LV only)   V LV

WILD THING ** **   15
Max Reid USA   1987
Rob Knepper, Kathleen Quinlan, Robert Davi, Maury Chaykin, Betty Buckley, Gillaume Lemay-Thivierge, Robert Bedarski, Clark Johnson, Sean Hewitt, Teddy Abner, Cree Summer Francks, Shawn Levy, Rod Torchia, Christine Jones
A kind of urban variation on the Tarzan theme, with a boy whose parents were murdered growing up wild on the streets, and swinging down from tower blocks as he protects the public from the depredations of criminals. Badly scripted and directed, this untidy effort has "Wild Thing" by The Troggs as its theme tune – one of the better things in it.
A/AD   88 min   VIDrel: EIV/SONOP   V

WILD WEST ** * **   15
David Attwood UK   1992
Naveen Andrews, Sarita Choudhury, Ronny Jhutti, Ravi Kapoor, Ameet Chana, Bhasker, Lalita Ahmed, Shaun Scott, Neran Persaud, Nrinder Dhudwar, Parv Bancil, Paul Bhattacharjee, Dinesh Shukla, Lou Hirsch, Rolf Saxon, Gurdial Sira
In London a trio of Pakistani brothers get together to form a band, start playing a distinctive form of Country and Western and set about trying to get into the big time. The amusing and inventive script is the work of Harwant Bains, and though the film is often uneven and uncertain in direction, the sheer exuberance of it all is most appealing.
COM   80 min (ort 85 min)   VIDrel: FIRST/SONOP   V/sur

WILD WOMEN OF WONGO ** *   (U)
James L. Wolcott USA   1958
Johnny Walsh, Zuni Dyer, Jean Hawkshaw, Mary Ann Webb, Cande Gerrard, Mary Ann Webb, Adrienne Borbeau, Ed Fury, Red Richards, Burt Williams, Olga Suarez, Marie Goodhart, Michelle Lamanck, Joyce Nizzari, Val Phillips
Stone Age tale of two tribes. In one all the men are beautiful and the woman ugly while exactly the opposite applies to the second. Sure enough, a mass mate-swap is soon initiated. A ludicrous film made with an amateur cast (except for Fury) and director which is so bad that its cult status is beyond question.
FAN   73 min   SATrel: BRAVO MOVIES

WILD ZONE ** **   18
Percival Rubens USA   1990
Philip Brown, Carla Herd, Edward Albert
A former Green Beret takes up the gun again when his father is captured and held hostage by a group of mercenaries plotting to free their imprisoned leader. A standard rescue actioner of average merit.
A/AD   88 min Cut (11 sec – ort 92 min)
VIDrel: EIV/SONOP   V

WILDCATS ** **   15
Michael Ritchie USA   1986
Goldie Hawn, James Keach, Swoosie Kurtz, Robyn Lively, Brandy Gold, Jan Hooks, Bruce McGill, M. Emmet Walsh, Nipsey Russell, Tab Thacker, Mykel T. Williamson, Wesley Snipes, Nick Corri, Woody Harrelson, Willie J. Walton
A woman P.E. teacher becomes coach to a football team at a rough inner-city high school, thereby fulfilling a long cherished dream of following in her father's footsteps. Routine formula comedy with Hawn gurgling and cooing her way through a film that's OK for easy laughs but instantly forgettable.
COM   102 min (ort 106 min)   VIDrel: MGM/WHV   V/sur

WILDCATS OF ST TRINIANS, THE **   (U)
Frank Launder UK   1980
Sheila Hancock, Michael Hordern, Rodney Bewes, Maureen Lipman, Joe Melia, Thorley Walters, Julia McKenzie, Veronica Quilligan,

Deborah Norton, Rose Hill, Luan Peters, Ambrosine Philpotts, Bernadette O'Farrell, Barbara Hicks
Another of these painfully unfunny comedies about an unruly girl's school, that started with THE BELLES OF ST TRINIANS. Each film in the series was less funny than the last, and all epit-omised everything that's wrong with British cinema, namely: NO NEW IDEAS. In this tale, the girls form a union and abduct the daughter of an Arab prince. Written by Launder.
COM   92 min   VIDrel: L/A   V

WILDER NAPALM ** **   15
Glenn Gordon Caron USA   1993
Debra Winger, Dennis Quaid, Arliss Howard, M. Emmet Walsh, Jim Varney, Mimi Lieber, Marvin J. McIntyre, Justin LeBlanc, Lance Lee Bailey, Peter Willie, Eric Whitmore, Daniel Hagen, Robert Peters, Buck Nolan, Rae L. Lawerence
Two feuding brothers share a strange family trait, a psychic ability to cause spontaneous combustion in objects on which they concentrate. One of the reasons for their disagreement is that one of them is in love with the wife of the other. A confused and hard-to-follow tale that does little with its capable cast. See also FIRESTARTER and SPECIMEN.
COM   104 min (ort 109 min) cC   VIDrel: COLUM/SONOP   V/sur

WILL ANY GENTLEMAN...? ** **   U
Michael Anderson UK   1953
George Cole, Veronica Hurst, Alan Badel, Joan Sims, Jon Pertwee, James Hayter, Heather Thatcher, William Hartnell, Diana Decker, Sidney James, Brian Oulton, Alexander Gauge, Peter Butterwirth, Wally Patch, Lionel Jeffries
A mild-mannered bank clerk attends a stage show and reluc-tantly agrees to take part in a stage hypnotist's act. However, he fails to come out of the trance and his personality has now become that of a devil-may-care womaniser who takes a rather casual attitude to his employer's money. An adequate adapta-tion of the successful stage play. The script is by Sylvaine.
COM   79 min (ort 84 min) B/W   VIDrel: WHV   V/h
Boa: play by Vernon Sylvaine.

WILLING TO KILL: THE TEXAS CHEERLEADER
STORY ** *   PG
David Greene USA   1992
Lesley Ann Warren, Tess Harper, Dennis Christopher, Olivia Burnette, William Forsythe, Lauren Woodland, Joanna Miles, Casey Sander, Arlen Dean Snyder, Dale Swann, Ann Walker, Lee Chamberlin, Alan Stepp, Will McMillan, Jim Calvert
Based on true events, this story covers the events that surrounded the murder of one woman by another, and all because of a rivalry between their respective daughters over who was going to get chosen as cheerleader. See also DEATH OF A CHEERLEADER and CHEERLEADER-MURDERING MOM.
DRA   89 min mTV   VIDrel: WHV L/A   V

WILLOW ** * **   PG
Ron Howard USA   1988
Val Kilmer, Joanne Whalley, Warwick Davis, Jean Marsh, Patricia Hayes, Billy Barty, Pat Roach, Gavan O'Herlihy, David Steinberg, Phil Fondacaro, Tony Cox, Robert Gillibrand, Mark Northover, Kevin Pollack, Rick Overton
Fantasy-adventure based on a George Lucas story, with an elf-like creature undertaking the task of taking an abandoned baby to its place of destiny, where it is to destroy the powers of evil. A high-spirited and exciting romp with numerous special effects and a fine performance from Marsh as the evil Queen Bavmorda.
FAN   120 min (ort 125 min) wScrn
VIDrel: ENCORE/SPEAR/COLUM   V/sur

WILLY WONKA AND THE CHOCOLATE
FACTORY ** **   U
Mel Stuart USA   1971
Gene Wilder, Jack Albertson, Peter Ostrum, Aubrey Woods, Michael Bollner, Roy Kinnear, Ursula Reit, Leonard Stone, Julie Dawn Cole, Denise Nickerson, Paris Themmen, Dodo Denny, Diana Sowle, David Battley
Clever but only slightly amusing tale of the owner of a fabulous sweet factory who runs a competition the winners of which get to visit his premises. A nasty edge to this fantasy and the vari-able quality songs (by Leslie Bricusse and Anthony Newley) spoil what might have been a most pleasing story. The script is by Dahl from his children's story.
JUV   95 min (ort 100 min)   VIDrel: WHV   V/sh
Boa: novel Charlie and the Chocolate Factory by Roald Dahl.

WILT * 15
Michael Tuchner UK 1988
Griff Rhys Jones, Mel Smith, Alison Steadman, Diana Quick, Jeremy Clyde, Roger Allam, David Ryall, Roger Lloyd Pack, Dermot Crowley, John Normington, Tony Mathews, Charles Lawson, Gabrielle Blunt, Adam Bareham, Edward Clayton
Sharpe's satirical novel of black humour and zest is reduced to yet another failed British farce, that revolves around the pathetic figure of a liberal arts teacher who becomes suspected of having murdered his over-bearing wife when she mysteriously disappears. Given the comic possibilities of the novel, to see it made into such a dreary and woodenly directed failure is a major disappointment.
Aka: MISADVENTURES OF MISTER WILT, THE
COM 89 min (ort 93 min) VIDrel: VCC/DISC/COLUM
V/sur
Boa: novel Henry Wilt by Tom Sharpe.

WINCHESTER '73 *** U
Anthony Mann USA 1950
James Stewart, Shelley Winters, Stephen McNally, Dan Duryea, Charles Drake, Millard Mitchell, John McIntire, Will Geer, J.C. Flippen, Rock Hudson, John Alexander, Steve Brodie, James Millican, Abner Biberman, Tony Curtis
A cowboy wins a prize Winchester rifle in a competition in Dodge City in 1873, only to have it stolen by the man who murdered his father. He embarks on a chase after the culprit that culminates in a shoot-out among the hills. A most memorable Western that is beautifully photographed by William Daniels and is scripted by Robert L. Richards and Borden Chase.
WES 82 min (ort 92 min)
VIDrel: 4-FRONT/POLYREC/CIC V/h
Boa: story by Stuart N. Lake.

WIND * PG
Carroll Ballard USA 1992
Matthew Modine, Jennifer Grey, Cliff Robertson, Stellan Skarsgard. Rebecca Miller, Ned Vaughn, Peter Montgomery, Elmer Ahlwardt, Saylor Creswell, James Rebhorn, Michael Higgins, Ron Colbin, Ken Kensei, Bill Buell, Tom Fervoy
A group of young sailors work together to build a yacht so that they can win back the America's Cup from Australia in 1983. An overlong and wholly unoriginal tale of little dramatic interest, although its breath-taking sailing sequences do much to enhance its appeal. See also WHITE SQUALL.
A/AD 120 min (ort 140 min) VIDrel: EIV/SONOP V/sur

WIND AND THE LION, THE * PG
John Milius USA 1975
Sean Connery, Candice Bergen, Brian Keith, John Huston, Geoffrey Lewis, Steve Kanaly, Vladek Sheybal, Nadim Sawalha, Roy Jenson, Simon Harrison, Polly Gottesman, Deborah Baxter, Shirley Rothman, Jack Cooley, Chris Adler
Turn-of-the-century adventure about an Arab sheik in Tangiers who abducts an American governess and her children, an action that leads to an international incident. A badly scripted and leaden-footed story with a confused and inconsistent plot that fails to do justice to the true events that inspired it.
A/AD 113 min Cut (ort 120 min)
VIDrel: ENCORE/SPEAR/COLUM; ENCORE (LV only) V LV

WIND CANNOT READ, THE * PG
Ralph Thomas UK 1958
Dirk Bogarde, Yoko Tani, Ronald Lewis, Anthony Bushell, Michael Medwin, John Fraser, Donald Pleasence
Romantic drama that is set in India and Burma during WW2, with a pilot who has escaped from a POW camp falling in love with a Japanese language teacher, but then learning that she has an inoperable brain tumour. Shot on location in India, this enjoyable tale is helped by strong casting – Bogarde is good as the dashing pilot and Tani is ravishing as his love interest. It's all the more of a pity that the direction is no more than workmanlike.
DRA 108 min (ort 115 min) VIDrel: CARL/TECH V
Boa: novel by Richard Mason.

WIND DANCER * (PG)
Craig Clyde USA 1993
Mel Harris, Matt McCoy, Brian Keith, Raenin Simpson, Nicholas Guest, Don Shanks, Craig Clyde, Pamela Guest, Hans Peterson, Jana Filimore, Byron Kennedy, Kate E. Brady, KeriAnn Brady, Sonja Torring, Kelee Ventura, Michael Ventura
The true story of a young girl who is so traumatised by the injuries she received in a riding accident that she loses the power of speech. However, as part of a new form of treatment she is sent to a horse ranch to recuperate, and develops a rapport with the title horse, a high-spirited stallion. A simple and quite touching little drama.
DRA 91 min SATrel: MOVIE CHANNEL

WIND IN THE WILLOWS, THE * U
Brian Cosgrave/Mark Hall UK 1983
Voices of: Ian Carmichael, Michael Hordern, David Jason, Richard Pearson, Beryl Reid, Una Stubbs, Jonathan Cecil
A lively puppet animation of this classic story, told with considerable zest.
ANIM 78 min VIDrel: THAMES/DISC V
Boa: novel by Kenneth Grahame.

WIND IN THE WILLOWS, THE * Uc
UK 1995
Voices of: Rik Mayall, Michael Palin, Alan Bennett, Michael Gambon, Vanessa Redgrave
Animation that stays quite faithful to Grahame's classic work, telling of the various adventures of Mole, Rat and Badger, and of how the rescue Toad from his own foolishness.
ANIM 72 min cC VIDrel: CARL/TECH V
Boa: novel by Kenneth Grahame.

WIND IN THE WILLOWS, THE * U
Terry Jones UK 1996
Steve Coogan, Eric Idle, Terry Jones, Anton Sher, Nicol Williamson, John Cleese, Stephen Fry, Bernard Hill, Michael Palin, Nigel Planer, Julia Sawalha, Victoria Wood, Robert Bathurst, Don Henderson, Richard James, Keith-Lee Castle
A pleasing live-action version of this much-loved classic, telling of the lives of a number of riverside creatures, including the pompous, self-centred but lovable Toad, whose foolish exploits cause his friends no little trouble. Solid characterisations and an intelligent script make this film a delight to watch, even if it doesn't quite succeed in capturing the magic of the book.
JUV 84 min (ort 87 min) VIDrel: GUILD/FOXVID V/sur
Boa: novel by Kenneth Grahame.

WIND OF AMNESIA, THE * 15
JAPAN 1993
In a world devastated by a nuclear war, a strange wind blows across the land, erasing human memory and with it all the knowledge that sustains civilisation. However, a young man manages to re-learn what has been forgotten and attempts to discover the cause of this catastrophe.
Aka: WIND NAMED AMNESIA, A
ANIM 78 min dubbed VIDrel: MANGA/SONOP V/sh

WINDPRINTS * 15
David Wicht UK 1989
John Hurt, Sean Bean, Marius Weyers, Lesley Fong, Eric Nobbs, Kurt Egelhof, Dania Niehaus, Trudie Taljaard, Goliath Davids, Johan Kruger, Sarel Bocks, Freda Kaptein, Jonathan Goliath, Lida Botha, Julia Burzain, Rod Alexander
The fact-based story of the hunting down of Nhadiep, a Namibian outlaw of legendary cunning. The film revolves around the investigations mounted by a Johannesburg cameraman who in 1982, arrives in pre-independence Namibia to gather information for a feature on Nhadiep, and finds himself working with a cynical English journalist. Though masquerading as a political thriller, this mature film attempts an analysis of South Africa's many problems.
THR 95 min (ort 100 min) VIDrel: VIR/RCA L/A V

WINDS OF JARRAH, THE * PG
Mark Egerton AUSTRALIA 1983
Susan Lyons, Terence Donovan, Emil Minty, Harold Hopkins, Steve Bisley, Martin Vaughan, Dorothy Alison, Isabelle Anderson, Steven Grives, Emily Minty, Nikki Gemmell, Mark Kounnas, Les Foxcroft, Michael Long, Ray Marshall
A love story set in post-WW2 Australia with a woman arriving to be the governess to a planter's three unruly children and finding that their father's frosty exterior hides his growing affection for her. A warm and meticulously detailed romantic drama.
DRA 78 min (ort 84 min) VIDrel: 20TH/TECH L/A V
Boa: novel The House in the Timberlands by Joyce Dingwall.

WINDS OF THE WASTELAND * U
Mack V. Wright USA 1936
John Wayne, Phyllis Fraser, Yakima Canutt, Lane Chandler, Sam Flint, Lew Kelly, Bob Kortman, Douglas Cosgrove, Robert Kortman,

Ed Cassidy, Merrill McCormack, Bud McClure, Jack Ingram, Charles Lochner (Jon Hall), Art Mix
Two men purchase a stagecoach and decide to go into business and take on a rival line in the bidding for a government mail contract. An excellent early Wayne Western.
WES   51 min (ort 54 min) B/W   VIDrel: SCRN/DISC   V

WINDS OF WAR, THE: VOLS 1 TO 5 ***   PG
Dan Curtis USA   1983
Robert Mitchum, Ali McGraw, Jan-Michael Vincent, John Houseman, Polly Bergen, Lisa Eilbacher, David Dukes, Topol, Ben Murphy, Peter Graves, Jeremy Kemp, Ralph Bellamy, Victoria Tennant
An expensive, handsome and very detailed drama following the fortunes of a family at the time of Hitler's rise to power and the Japanese Pearl Harbour attack. First shown in eight 96-minute episodes, the main story deals with a US naval officer (Mitchum) who finds himself taking part in high-level discussions on the eve of WW2. Despite its lack of depth and some tedious sub-plots, this is a most effective work. Followed by WAR AND REMEMBRANCE: PARTS 1 AND 2.
DRA   458 min (5 cassettes) mTV   VIDrel: CIC/SONOP L/A V/h
Boa: novel by Herman Wouk.

WINDWALKER ***   (PG)
Kieth Merrill USA   1980
Trevor Howard, Nick Ramus, James Remar, Serene Hedin, Dusty Iron Wing McCrea, Silvana Gallardo, Billy Drago, Rudy Diaz, Harold Goss-Cayote, Roy J. Cahoe, Jason Stevens, Emerson John, Roberta Deherrera, Curtis Powers
An old Cheyenne chief is unable to die until he has helped to rescue his grandson's family from the Crow. A captivating film set in the early part of the 19th century and performed entirely in Cheyenne and Crow by native American Indians (with the exception of Howard). For once, whites are almost entirely and the portrayal of Indian life and customs seems totally convincing and authentic.
WES   107 min subs   SATrel: SKY MOVIES

WINGLESS BIRD, THE *   PG
UK   1996
Claire Skinner, Julian Wadham, Edward Atterton
Depressing drama set at the start of WW1, with the strong-willed daughter of a sweets manufacturer falling in love with the son of a snobbish, upper-class couple. When the death of her father leads to her inheriting the business, her social position changes, but apparently not enough to satisfy her prospective in-laws. A terribly contrived story now unfolds, replete with convenient deaths and various other sad happenings. One of the weakest of Cookson's novels.
DRA   146 min mTV   VIDrel: FOCUS/DISC   V
Boa: novel by Catherine Cookson.

WINGS OF DESIRE ****   PG
Wim Wenders FRANCE/WEST GERMANY   1987
Bruno Ganz, Solveig Dommartin, Otto Sander, Curt Bois, Peter Falk, Elmer Wilms, Hans Martin Stier, Sigurd Rachman, Beatrice Manowski, Lajos Kovacs, Bruno Rosaz, Laurent Petigand, Chico Roja Ortega, Otto Kuhnle, Peter Werner
The director's follow-up to PARIS, TEXAS tells of two angels who wander West Berlin, observing and comforting humanity. When one of them tires of his role, he re-enters the world of mortals and his monochrome view of life (angels do not see in colour) takes on human attributes. Scripted by Wenders and Peter Handke, this opaque yet haunting fantasy was inspired by the poetry of Rainer Maria Rilke. Photography is by Henri Alekan. Followed by FARAWAY, SO CLOSE.
Aka: DER HIMMEL UBER BERLIN
FAN   122 min (ort 130 min) Col/B/W
VIDrel: CONNO/RTM   V

WINGS OF EAGLES, THE **   U
John Ford USA   1957
John Wayne, Dan Dailey, Maureen O'Hara, Ward Bond, Ken Curtis, Edmund Lowe, Kenneth Tobey, Sig Rumann, James Todd, Barry Kelley, Henry O'Neill, Dan Borzage, Willis Bouchey, Dorothy Jordan, Peter Ortiz, Louis Jean Heydt, Tige Andrews
Screen profile of the varied career of Frank "Spig" Wead, a WW1 flying ace who later become a successful screenwriter and eventually the commander of an aircraft carrier in WW2. A routine biopic that is enjoyable to watch but one that hardly stretched the talents or imagination of either Ford or Wayne.
A/AD   105 min   VIDrel: MGM/WHV   V/h

WINGS OF FAME ***   15
Otakar Votocek HOLLAND   1990
Peter O'Toole, Colin Firth, Marie Trintignant, Andrea Ferreol, Ellen Umlauf, Ken Campbell, Mark Tandu, Dagmar Schwarz, Phil Warnett, Herman Lanse, Terry Raven. Jonathan Hackett, Eva Kryll, Emilio Linder, Pat Roach, Jerry Di Giacomo
An actor and the writer who murdered him both arrive in the after-life, which takes the form of a luxury hotel. There they learn that the standard of accommodation on offer depends on how well they are remembered by the living. As this memory fades, their level of comfort declines accordingly. Quite unique in its way, this is a strange and not entirely successful black comedy, whose fantasy elements seem somewhat out of place. Won the Critics' Prize at Cannes.
FAN   109 min   CINrel

WINGS OF HONNEAMISE, THE ***   15/PG
Hiroyuki Yamaga JAPAN   1987/1994
Voices of: Robert Matthews, Melody Lee, Lee Stone, Alfred Thev, Stevie Beeline, Steve Blum, Warren Daniels, Rudy Luzion, Arnie Hinks, Sunley Gurd Jr, Simon Issacson, Anthony Mordy, Christophe de Groot, Jonathan Charles
On an Earth-like world, a young cadet looks forward to his journey into space, it being hoped that this will raise the standing of the space agency, which after twenty years has still not succeeded in putting an astronaut into space. However, the launch is intended to be a cynical ploy to bring about a conflict with a rival country. Technically a most impressive work, there remains an oddness about it all that has its roots in its country of origin.
Aka: ONEAMISU NO TSUBASA
ANIM   119 min subs; 125 min dubbed
VIDrel: MANGA/SONOP   V/sh

WINGS OF THE APACHE ***   15
David Green USA   1989
Nicolas Cage, Tommy Lee Jones, Sean Young, Bryan Kestner
British director Green makes his Hollywood debut in this helicopter-based adventure that inevitably reminds one think of TOP GUN. Having been nearly shot down while on a anti-drugs reconnaissance mission in Latin America, Cage gets to train in an outfit of crack pilots put together by the Pentagon to ensure the defeat of the drug barons. A fairly mindless blend of action and occasional romantic interludes, yet put together with considerable zest.
Aka: FIREBIRDS
A/AD   82 min (ort 89 min)
VIDrel: 4-FRONT/POLYREC/GUILD   V/sh

WINSLOW BOY, THE ***   U
Anthony Asquith UK   1948
Robert Donat, Cedric Hardwicke, Margaret Leighton, Frank Lawton, Neil North, Jack Watling, Francis L. Sullivan, Basil Radford, Walter Fitzgerald, Marie Lohr, Wilfrid Hyde-White, Kynaston Reeves, Ernest Thesiger, Lewis Casson
A courtroom drama based on a real incident, where a young naval cadet was expelled for the theft of a postal order from one of his fellow cadets. His family battled for years to clear his name, the case becoming something of a cause celebre of the time. Written by Terence Rattigan and Anatole De Grunwald with music by William Alwyn.
DRA   112 min (ort 117 min) B/W   VIDrel: BRAVE/SONOP L/A   V
Boa: play by Terence Rattigan.

WINTER LIGHT ****   18
Ingmar Bergman SWEDEN   1962
Ingrid Thulin, Gunnar Bjornstrand, Gunnel Lindblom, Max Von Sydow, Allan Edwall, Kolbjorn Knudsen, Olof Thunberg, Ella Ebbesen
Following on from THROUGH A GLASS DARKLY, this second film in Bergman's trilogy of despair tells of a widowed priest whose loss of faith leads to his inability to offer either comfort or guidance to his flock, one of whom later commits suicide. A beautifully filmed and acted study of human frailty, readily understandable and devoid of any hidden meanings or opaque symbolism. This moving gem was followed by THE SILENCE. See also DEVIL, PROBABLY, THE.
Aka: NATTVARDSGASTERNA
DRA   77 min (ort 80 min) B/W   VIDrel: TART/20TH   V/dm

WINTER OF OUR DREAMS **   15
John Duigan AUSTRALIA   1981
Judy Davis, Bryan Brown, Cathy Downes, Baz Luhrmann, Peter

Mochrie, Mervyn Drake, Margie McCrae, Mercie Deane-Johns, Joy Hruby, Kim Deacon, Caz Lederman, Jenny Ludlam, Virginia Duigan, Rosemary Lenzo, Alex Pinder
A lonely female prostitute with a drug habit meets a bookshop-owner who is unhappily married, and the two become romantically involved. An interesting drama that moves along somewhat fitfully, with an unsatisfying and confused ending.
DRA 90 min VIDrel: ARTPRO/RTM V

WINTER PEOPLE ** 15
Ted Kotcheff USA 1989
Kurt Russell, Kelly McGillis, Lloyd Bridges, Mitchell Ryan, Amelia Burnette, Eileen Ryan, Jeffrey Meek, Lanny Flaherty, Don Michael Paul, David Dwyer, Bill Gribble, Wallace Meek, Walker Averitt, Dashiel Coleman, Gary Bullock
During the Depression, clockmaker Russell leaves his home-town in search of work and arrives at an Appalachian town, where he falls for single mother McGillis and inadvertently sparks off a feud between rival families. McGillis is excellent in this overheated and underplotted romantic drama, and contrives to sustain a film that is given no support by its weak script.
DRA 105 min (ort 110 min) VIDrel: FIRST/SONOP L/A V
Boa: novel by John Ehle.

WINTER'S TALE, THE **** PG
Jane Howell UK 1981
Jeremy Kemp, Anna Calder-Marshall, John Welsh, David Burke, Robert Stephens, Jeremy Dimmick, Merelina Kendall, Susan Brodrick, Leonard Kavanagh, John Bailey, Cyril Luckham, William Relton, Margaret Tyzack, John Benfield
An excellent television production of this play, easily one of the best adaptations produced by the BBC.
DRA 172 min mTV VIDrel: BBC V/h
Boa: play by William Shakespeare.

WINTER'S TALE, A *** 15
Eric Rohmer FRANCE 1992
Charlotte Very, Frderic Van Den Driessche, Herve Furic, Ava Loraschi, Christiane Desbois, Rosette, Jean-Luc Revol, Haydee Caillot, Jean-Claude Biette, Marie Riviere, Roger Dumas, Daniele Lebrun, Diane Lepvrier, Edwig Navarro
Screenplay is by Rohmer in this strange romantic drama follow-ing the efforts made by an unhappy hairdresser to recapture a lost love, hopefully by finding the man by whom she had a daughter. Ultimately, a staged performance of Shakespeare's famous play provides her with the clue she needs to solve a mystery to her own background.
Aka: CONTE D'HIVER; TALE OF WINTER, A
DRA 109 min (ort 114 min) wScrn VIDrel: ARTIF/20TH V

WIRED * 18
Larry Peerce USA 1989
Michael Chiklis, Ray Sharkey, J.T. Walsh, Patti D'Arbanville, Alex Rocco, Lucinda Jenney, Gary Groomes, Jere Burns, Billy Preston, Matthew Faison, Brooke McCanter, Jon Snyder, Finis Henderson, Amy Michelson, Blake Clark, Scott Plank
Loosely based on Woodward's bestseller, this ill-conceived film purports to examine the life of John Belushi, offering us a cautionary tale of his self-destructive urges and death from a drugs overdose at thirty-three. After his untimely death Chiklis (who does look a little like Belushi) is escorted through his past by an angel/taxi driver, but the film fabricates as much as it examines and this tasteless fantasy is neither funny nor instruc-tive.
DRA 105 min (ort 112 min) mTV
VIDrel: 4-FRONT/POLYREC L/A V
Boa: book by Bob Woodward.

WISDOM ** 18
Emilio Estevez USA 1986
Emilio Estevez, Demi Moore, Tom Skerritt, Veronica Cartwright, William Allen Young, Richard Minchenberg, Ernie Brown
A young man who has recently graduated is unable to obtain work because of a past felony. Frustration drives him and his girlfriend into committing a series of bank robberies across the country, destroying the records of loans and mortgages as a way of helping debt-ridden farmers. First-time director Estevez was assisted by Robert Wise, but even so, this remains a shallow and unappealing effort.
A/AD 105 min (ort 109 min) VIDrel: MGM/WHV L/A V/sh

WISE BLOOD ** 15
John Huston USA/WEST GERMANY 1979
Harry Dean Stanton, Brad Dourif, Ned Beatty, Daniel Shor, John Huston, Amy Wright, Mary Nell Santacroce
Uneven, offbeat and semi-comic tale of a moody young man, whose obsessive dislike of religion drives him to attempt to found a church without Christ. Various individuals exploit his beliefs for their own ends. The music is by Alex North.
COM 102 min (ort 108 min) VIDrel: CIC/SONOP L/A V
Boa: novel by Flannery O'Connor.

WISE GUYS *** 15
Brian De Palma USA 1986
Danny DeVito, Joe Piscopo, Harvey Keitel, Ray Sharkey, Dan Hedaya, Captain Lou Albamo, Julie Bovasso, Patti LuPone
A couple of bumbling no-hopers are unhappy working for a gangster, so when he gives them the money to place a bet they try to take him for a ride by backing a horse of their own. However, their plan backfires when their horse loses and they are forced to flee in a stolen car with the gang in hot pursuit. A fast and furious gangster spoof whose deficiencies of plot are remedied by a dynamic performance from DeVito and some very funny moments.
COM 89 min (ort 91 min) VIDrel: MGM L/A V

WISH YOU WERE HERE *** 15
David Leland UK 1987
Emily Lloyd, Tom Bell, Jesse Birdsall, Geoffrey Durham, Clare Clifford, Barbara Durkin, Geoffrey Hutchings, Charlotte Barker, Chloe Leland, Pat Heywood, Charlotte Ball
Inspired by the early life of celebrated brothel-keeper Cynthia Payne, this is a bittersweet tale of a young girl's sexual matur-ing set against the austere backdrop of post-war Britain. A good debut as director for Leland, who also wrote the screenplay for PERSONAL SERVICES.
COM 92 min VIDrel: POLY/POLYREC L/A V/h

WISHFUL THINKING * 15
Murray Langston USA 1990
Murray Langston, Michelle Johnson, Ruth Buzzi, Billy Barty, Ray "Boom Boom" Mancini, Johnny Dark, Melissa Shear
A recently divorced writer with agoraphobia finds his dreams and fantasies coming true, after he is given a magic notepad by a midget. He gets involved in a bank robbery, and meets the girl of his dreams, but faces a devious plot by his sister and his agent to gain control of his assets. A crudely devised, smutty comedy that's flawed by poor scripting and a wooden performance from Langston in the lead role.
COM 90 min (ort 94 min) VIDrel: MED/RCA L/A V

WISHFUL THINKING ** 18
Michael Craig USA 1992
Heather Hart, Diedre Holland, Porsche Lynn, Holly Ryder, Kym Wilde, Teri Diver, Jon Dough, Scott Irish, T.T. Boy, Mike Horner
When a man releases a genie from a bottle, he finds that it's a lusty female one, but she is a bit mean with the wishes, being only prepared to grant him two, which he must make before midnight and her return to the bottle. He uses up his first wish on a sexual encounter, but his second wish is to find a perfect wife, which the genie grants by becoming a mortal. A simple-minded and cheerful variant on the famous Arabian Nights fantasy tale.
A 80 min (Cut before video submission by 5 min 19 sec – ort 85 min) VIDrel: GROHOM/MAXSCAN V

WITCH ACADEMY ** 18
Fred Olen Ray USA 1992
Robert Vaughn, Michelle Bauer, Veronica Carothers, Suzanne Ager, Ruth Collins, Don Dowe, Jay Richardson, Priscilla Barnes
An unhappy woman of high intelligence but zero sexual attrac-tiveness would do anything to become more appealing to men, including selling her soul to the Devil. This is the story of her adventures. See also HUNK for an exploration of this idea, but from a male perspective.
Aka: LITTLE DEVILS
DRA 83 min VIDrel: POPRO/RTM V

WITCH HUNT ** 18
Paul Schrader USA 1994
Dennis Hopper, Penelope Ann Miller, Julian Sands, Eric Bogosian, Sheryl Lee Ralph
In a future Hollywood where the use of magic is almost univer-sal, a starlet engages a private eye by the name of H.P. Lovecraft

(!) to investigate the murder of a producer whose death was brought about by black magic. See also CAST A DEADLY SPELL for a movie with a very similar theme.
FAN 98 min (ort 100 min) VIDrel: AUDIO/DISC V

WITCHBOARD ***
15
Kevin S. Tenney USA
1986
Tawny Kitaen, Todd Allen, Stephen Nichols, Kathleen Wilhoite, Rose Marie, Burke Byrnes, James W. Quinn, Judy Tatum, Gloria Hayes, J.P. Luebsen, Susan Nickerson, Ryan Carroll, Kenny Rhodes, Clare Bristol
A trio of friends are having fun playing with a Ouija board when they succeed in contacting the spirit of a young boy. However, their troubles really begin when they contact the spirit of an evil turn-of-the-century mass murderer, who takes over one of their bodies with the intention of finding fresh victims. Supernatural stalk 'n' slash film that was followed by a couple of sequels.
HOR 94 min (ort 97 min) VIDrel: GUILD/SONOP L/A V

WITCHBOARD: THE RETURN **
15
Kevin S. Tenney USA
1992
Ami Dolenz, Christopher Michael Moore, Laraine Newman, Timothy Gibbs, John Gatins, Julie Michaels, Sarah Kaite Coughlan, Marvi kaplan, Jeff Feringa, Foster, Todd Allen, Kenny Rhodes
A would-be artist moves into a studio apartment and discovers a Ouija board in a closet that she is unwise enough to use. This brings her in contact with a spirit who claims to be that of a former tenant who was murdered. Soon, her fellow tenants are turning up dead and her sleep is being disturbed by terrifying nightmares. A sequel to the first film that proceeds along entirely familiar lines.
Aka: WITCHBOARD 2: THE DEVIL'S DOORWAY
HOR 94 min (ort 98 min) VIDrel: MED/20VIS L/A V

WITCHBOARD: THE POSESSION *
18
Peter Svatek CANADA
1995
David Nerman, Locky Lambert, Cedric Smith, Donna Sarrasin
An out-of-work commodity broker has a change of fortune when he plays about with a ouija board, but the influence he becomes open to is far from benign, and he is prevailed upon to sell his soul to the Devil in return for material benefits. A terribly dull, in-name-only sequel to the earlier films, it displays a lack of work on the script and has little to redeem it, even the special effects and inevitable violent sequences have a stale and dated look about them.
HOR 90 min (ort 93 min) VIDrel: HIFLI/SONOP V

WITCHCRAFT **
18
Robert Spera USA
1988
Gary Sloan, Lee Kisman, Anat Topol-Barzilai, Deborah Scott, Mary Shelley, Alexander Kirkwood, Edward Ross Newman, Charles Grant, Lilian Lane, Karen Michaels, Cynthi Bell, Victoria Ressurection, Ofelia Montano, Steve Decker
A pregnant woman and her husband are burnt at the stake by an enraged mob who are convinced they are Satanists. Some 300 years later their spirits return to have their revenge. The slow script allows time for tension to build up in this satisfactory chiller. A sequel followed.
HOR 85 min (ort 90 min) VIDrel: CASPIC L/A V

WITCHCRAFT **
18
Sean Barton USA
1991
Jenilee Harrison, Christopher Lee, Henry Cele
A young American nurse is married to farmer in Africa, and enrages a local witchdoctor when she releases a goat that was to have been used as a sacrifice. Having been cursed by the furious native, she finds herself being stalked by an unseen creature that appears to have risen out of the sea. A moderately absorbing tale of horror and black magic.
Aka: CURSE 3: BLOOD SACRIFICE; PANGA
HOR 87 min (ort 91 min) VIDrel: COLUM/SONOP V

WITCHES, THE ***
PG
Nicolas Roeg USA
1989
Anjelica Huston, Mai Zetterling, Jason Fisher, Charlie Potter, Rowan Atkinson, Bill Paterson, Brenda Blethyn, Jane Horrocks, Jenny Runacre, Annabel Brooks, Anne Lambton, Sukie Smith, Rose English, Rosamund Greenwood, Emma Relph
Thanks to his granny, a nine-year-old boy knows a witch when he sees one, and on a trip to London he stumbles on a convention of them, and overhears their dastardly plans to change all of the country's children into mice. This delightful collaboration between Roeg and Muppets-creator Jim Henson (who produced

it) is a spooky tale of excellent scripting and superb effects. Intended for children but perhaps a touch more scary than it need have been.
FAN 88 min (ort 92 min) VIDrel: WHV V/sur V/dm
Boa: novel by Roald Dahl.

WITCHES OF EASTWICK, THE ***
18
George Miller USA
1987
Jack Nicholson, Cher, Susan Sarandon, Michelle Pfeiffer, Richard Jenkins, Veronica Cartwright, Keith Joachim, Carel Struycken, Helen Lloyd Breed, Caroline Struzik, Michele Savage, Heather Coleman, Carolyn Ditmars
Three bored and sex-starved ladies living in a New England town get together with the intention of bringing some male company into their lives, but find that they have unwittingly conjured up the Devil in the shape of the town's newest arrival. This lively and inventive fantasy has a terrific part for Nicholson and is sustained throughout by fine performances, despite often veering wildly between comedy and horror. The score is by John Williams.
FAN 114 min (ort 118 min) VIDrel: WHV V/sur
Boa: novel by John Updike.

WITCHFINDER GENERAL ***
18
Michael Reeves UK
1968
Vincent Price, Ian Ogilvy, Rupert Davies, Patrick Wymark, Wilfrid Brambell, Robert Russell, Nicky Henson, Hilary Dwyer, Tony Selby, Michael Beint, Peter Haigh, Bernard Kay, Godfrey James, John Trenaman, Bill Maxwell, Morris Jar
A superior horror film about a vindictive witch-hunter living at the time of Cromwell, who is not averse to using torture in as a means of extracting confessions. Price as Matthew Hopkins is excellent in this atmospheric period film set in 1645. The script is by Reeves and Tim Baker. See also MARK OF THE DEVIL.
Aka: CONQUEROR WORM, THE; EDGAR ALLAN POE'S CONQUEROR WORM
HOR 81 min (Cut at film release – ort 87 min)
VIDrel: REDEM/RTM V
Boa: novel by Ronald Bassett.

WITCHING HOUR **
18
Henri Pachard USA
1992
Ashlyn Gere, Brittany Morgan, Francesca Lee, Tom Byron, Peter North, Randy West, Joey Silvera
A repulsive witch has to resort to calling pay-phones to get a date, and is overjoyed when she finally meets a man who seems to fancy her. However, she unwisely uses her powers to test his fidelity, but this leads to great disappointment. An interesting variant on the "Beauty and the Beast" theme that is not fully explored; a number of well mounted romantic scenes help enhance interest.
A 45 min (ort 60 min) VIDrel: GROHOM V

WITCHTRAP **
18
Kevin S. Tenney USA
1988
James W. Quinn, Kathleen Bailey, Judy Tatum, Rob Zapple, Linnea Quigley, Jack W. Thompson, Clyde Talley II, Hal Havins, Kevin S. Tenney, J.F. Luebsen, Richard Fraga, Lynn McRae, Greg Lewolt, Virginia Miller
A Satanist who practised his dark rituals in his attic dies in mysterious circumstances. However, his evil spirit continues to haunt the house, and when the new owner calls in a group of psychics to rid the building of its entity, their attempts to do so merely unleash a series of horrifying events. An adequate chiller with a few good moments but a general lack of conviction.
HOR 86 min (ort 92 min) VIDrel: MIA/DISC/IMPENT V

WITH A VENGEANCE **
15
Michael Switzer USA
1992
Melissa Gilbert Brinkman, Michael Gross, Jack Scalia, Matthew Lawrence, John Cullum, Roger Aaron Brown, Robert Donner, Russell Johnson, Victoria Otto, Beata Jachulski Baker, Frank Buxton, Gayle Bellows, Dixie Lemley, Robert Nadir
Amnesia blots out a young woman's tragic past, but nightmare memories continue to surface, and she eventually sets out on a journey of self-discovery.
THR 93 min VIDrel: BRAVE/SONOP L/A V

WITH DEADLY INTENT **
15
Ruben D. Preuss USA
1991
Scott Valentine, Chris Mulkey, Joan Severance, G.W. Bailey, Ray Wise, Howard George, Andreas Katsulas, Frances Nuyen, Ray

Stricklyn, Peter MacLean, Renata Scott, Michael Blue, Loren James, Peter Radon, Gil Newsome, George Simons
A young writer whose brother is murdered by a counterfeiting gang sets out have his revenge. A predictable and unoriginal revenge yarn, of violence and bloodshed.
Aka: WRITE TO KILL
A/AD   95 min (ort 120 min)   VIDrel: 20VIS/SONOP
V/sur

WITH HARMFUL INTENT **   15
Richard Friedman USA   1994
Joan Van Ark, Christopher Noth, Rick Springfield, Dey Young, Daniel J. Travanti, Earl Billings, Michael Patrick Carter, Patrick Dollaghan, Bert Remsen, Ashley Johnson, Eileen Seeley, Jordan Davis, Melanie MacQueen, Mary Portser
A series of attacks on children brings fear to a quiet community in this unpleasant psychological thriller.
THR   92 min mTV   VIDrel: COLUM/SONOP   V
Boa: novel Someone's Watching by Judith Kelman.

WITH HONORS **   PG
Alek Keshishian USA   1993
Joe Pesci, Brendan Fraser, Moira Kelly, Patrick Dempsey, Josh Hamilton, Gore Vidal, Deborah Lake Fortson, Marshall Hambro, Melinda Chilton, Harve Kolzow, M. Lynda Robinson, James Deuter, Richard Auguste, Patricia B. Butcher
A Harvard senior student loses his thesis, which falls into the hands of a homeless man, and this leads to complications all round. The latter proves to be a natural philosopher who teaches the over-privileged students a thing or two about the trials of real life. Pesci's performance goes a long way towards overcoming the plot deficiencies of this formula comedy.
COM   97 min (ort 101 min) cC   VIDrel: WHV   V/sur

WITH HOSTILE INTENT: SISTERS IN BLACK AND
BLUE ***   15
Paul Schneider USA   1993
Melissa Gilbert, Mel Harris, Peter Onorati, Holland Taylor, Saundra Von Bargen, Paul Hecht, Saundra Santiago, Cotter Smith, Bob Barnes, Timothy Bates, Rus Blackwell, Ariane Brandt, Larry Bucklan, Kevin Corrigan
True story of two female officers who sued the Long Beach Police Department for compensation, after their careers were blighted by sexual harassment. When one of them stops dating a superior, virtually all the other officers unite in a campaign designed to make their lives unbearable. Despite a lack of tension and a convoluted plot, this strongly acted story provides an all too believable account of police behaviour and group mentality.
DRA   88 min mTV   VIDrel: 20TH/TECH   V

WITH SAVAGE INTENT **   15
Michael Tuchner USA   1992
Elizabeth Montgomery, Robert Foxworth, Howard Rollins Jr, Lee Richardson, Maureen O'Sullivan, Paul McCrane, Danton Stone, Tom Mardirosian, Ronny Cox, Jude Ciccolella Mary Ann Hayan, Adam LeFevre, Kevin O'Rourke, Seret Scott
Fact-based drama that follows the tribulations of a woman who discovers to her horror, that an assailant who tried to kill her is in fact a police officer. Fair.
Aka: WITH MURDER IN MIND
DRA   93 min (ort 95 min) mTV
VIDrel: MIA/DISC/IMPENT   V

WITHIN THE ROCK **   18
Gary J. Tunnicliffe USA   1996
Xander Berkely, Caroline Barclay, Bradford Tatum
A top-level scientific team is assembled to intercept a giant asteroid that is on a collision course with the Earth, and a landing is successfully carried out on it, the intention being to alter its course. But their operations on the asteroid re-animate a fossilised and virtually indestructible creature that embarks on the obligatory ALIEN-style rapage.
HOR   84 min   VIDrel: HIFLI/SONOP   V/h

WITHNAIL & I **   15
Bruce Robinson UK   1986
Paul McGann, Richard E. Grant, Richard Griffiths, Ralph Brown, Michael Elphick, Daragh O'Malley, Michael Wardle, Una Brandon-Jones, Noel Johnson, Irene Sutcliffe
Set in the closing months of 1969, this semi-autobiographical tale (the script is by Robinson) has two unemployed actors who share a room, setting off for what is to be a disastrous holiday

in the country. Amusing in parts, but the lack of pace or direction eventually begins to tell.
COM   103 min (ort 108 min) wScrn   VIDrel: CIC/SONOP
V

WITHOUT A CLUE **   PG
Thom Eberhardt UK/USA   1988
Michael Caine, Ben Kingsley, Jeffrey Jones, Lysette Anthony, Paul Freeman, Nigel Davenport, Pat Keen, Peter Cook, Tim Killick, Matthew Savage, John Warner, Matthew Sim, Harold Innocent, George Sweeney, Murray Ewan
This unusual little farce starts with the premise that Sherlock Holmes was a fictitious character invented by Dr Watson. When the fame of the former begins to be such that people wish to meet him, Watson is forced to hire an actor to impersonate him. Mildly amusing, but by no means hilarious.
COM   102 min (ort 106 min)   VIDrel: VISVID/POLYREC
L/A   V

WITHOUT MERCY **   18
Robert Anthony USA   1995
Frank Zagarino, Frans Tumbuan, Martin Kove, Ayu Azari
A US Marine is goes on a secret mission, but this brings him back to his homeland, a place where Triad gangs are in control.
A/AD   90 min   VIDrel: THIRD   V

WITHOUT RESERVATIONS ***   U
Mervyn Le Roy USA   1946
John Wayne, Claudette Colbert, Don DeFore, Anne Triola, Frank Puglia, Phil Brown, Louella Parsons, Thurston Hall, Dona Drake, Fernando Alvarado, Jose Alvarado, Michael Economides, Miguel Tapia, Rosemary Lopez, Henry Mirelez
This wartime comedy has a successful authoress meeting a soldier, whom she decides is perfect to play the leading man in a forthcoming film of her book. A light and frothy comedy that has a number of guest appearances by assorted Hollywood celebrities.
COM   102 min B/W   VIDrel: SECOND/RTM   V
Boa: novel Thanks God, I'll Take It From Here by J. Allen and M. Livingston.

WITHOUT WARNING ***   (15)
Robert Iscove USA   1994
Sander Vanocur, Jane Kaczmarek, Bree Walker Lampley, Dwier Brown, Brian McNamara, James Morrison, Ashley Peldon, James Handy, Kario Salem, Spencer Garrett, Gina Hecht, John De Lancie, Patty Toy, Dennis Lipscomb, Ron Canada
Done in documentary style as if it is really taking place, this is an account of what happens after the Earth is struck by three giant asteroids, a catastrophe suggesting an alien visitation. Tension grows when the impact sites begin to emit radio waves that disrupt civil aviation, and the news of further approaching asteroids is less than welcome. Real-life journalist Vanocur plays the TV anchorman who covers these events in this compelling disaster movie.
FAN   89 min mCab   SATrel: MOVIE CHANNEL

WITHOUT WARNING: TERROR IN THE
TOWERS ***   15
Alan J. Levi USA   1994
James Avery, Andre Braugher, George Clooney, Fran Drescher, John Karlen, Scott Plank, Susan Ruttan, Michael Stoyanov, Robin Thomas, Charles Bernard, Carl Clissmeyer, Tim Delaney, Aaron Fitzgerald, Kenneth Forrester, David Gomes
Done in the style of a news broadcast, this shot-on-video docudrama chronicles the events leading up to the bombing of the Manhattan World Trade Center and the resultant nationwide hunt for the perpetrators of this outrage. Solid acting and a literate script are nicely complemented by the low-key approach, and combine to ensure a most absorbing account.
DRA   93 min mTV   VIDrel: CIC/SONOP L/A   V

WITHOUT WARNING: THE JAMES BRADY
STORY ***   (PG)
Michael Toshiyuki Uno USA   1991
Beau Bridges, Joan Allen, David Strathairn, Bryan Clark, Steven Flynn, Susan Brown, Christopher Bell, Gary Grubbs, Christine Healy, Timothy Langfield, Alan Ackles, Tyress Allen, Rosemary Baxter, Gerry becker, James Belcher
Dramatisation of the failed attempt on President Reagan's life in 1981, in which his press secretary James Brady was seriously wounded. Pronounced dead on arrival at the hospital, he suffered a damaging head wound but courageously back to

make a recovery. A tightly scripted and workmanlike account, mercifully free of needless sentiment or other distractions. Bridges in the central role gives one of his strongest performances in years.
DRA   95 min (ort 120 min) mTV   TVrel: C4

WITHOUT YOU I'M NOTHING ***   18
John Boskovich USA   1990
Sandra Bernhard, John Doe, Steve Antin, Lu Leonard, Ken Foree, Cynthia Bailey, Grace Broughton, Kimberli Williams, Axel Vera, Estuardo M. Volty, Kevin Dorsey, Arnold McCuller, Oren Waters, Paul Thorpe, Djimon Hounson, Roxanne Reese
Playing herself, Bernhard is first seen backstage of an L.A. dinner club just prior to going on stage with her one-woman show. Intercut with various other sequences, this then leads on to her routine, which consists mostly of her reflections on life, sexuality and her middle-class childhood, with a few funny songs thrown in for good measure. As much a filmed performance as comedy feature, it's irreverent, energetic and often very funny.
COM   85 min (ort 90 min)   VIDrel: ELPIC/POLYREC
V/sur

WITNESS ***   15
Peter Weir USA   1985
Harrison Ford, Kelly McGillis, Lukas Haas, Josef Sommer, Alexander Godunov, Jan Rubes, Danny Glover, Patty LuPone, Brent Jennings, Angus MacInnes, Frederick Rolf
A little Amish boy in the big city sees a murder and the cop assigned to protect him and his widowed mother has to hide in the Amish community when his own life is threatened. The cultural clash between Harrison and the Amish and his growing affection for the boy's mother are skilfully handled, but the violent climax is thoroughly contrived. A flawed but unusual film. AA: Screen/orig (W. Kelley/P. Wallace/E.W. Wallace), Edit (T. Noble).
THR   108 min (ort 112 min)   VIDrel: CIC/SONOP; PION (LV only)   V/sur LV

WITNESS FOR THE PROSECUTION ****   U
Billy Wilder USA   1957
Charles Laughton, Marlene Dietrich, Tyrone Power, Elsa Lanchester, John Williams, Henry Daniell, Una O'Connor, Norma Varden, Ian Wolfe, Molly Rose, Torin Thatcher, Francis Compton, Philip Tonge, Ruta Lee, Ottola Nesmith
A powerful courtroom drama based on a Christie play with Laughton quite superb as a defence barrister intent on getting an acquittal for his client, an alleged murderer. Dietrich is excellent as the faithful wife, prepared to do anything to save Power from the gallows. Scripted by Wilder and Harry Kurnitz and remade in 1982.
DRA   113 min B/W   VIDrel: WHV   V
Boa: play by Agatha Christie.

WITNESS FOR THE PROSECUTION ***   PG
Alan Gibson USA   1982
Ralph Richardson, Deborah Kerr, Diana Rigg, Beau Bridges, Wendy Hiller, Donald Pleasence, David Langton, Richard Vernon, Peter Sallis, Frank Mills, Patricia Leslie, Zulema Dean, Peter Copley, Barbara New, John Kidd
Passable remake of the 1957 film in which a crusty old barrister takes on the case of a man charged with the murder of a wealthy widow. Made for TV but not shown in the UK until 1985 on BBC. Adapted by John Gay from the Christie play and the original Billy Wilder/Harry Kurnitz screenplay. This was Kerr's TV debut.
DRA   94 min mTV   VIDrel: MGM L/A   V
Boa: play by Agatha Christie.

WITNESS TO THE EXECUTION **   15
Tommy Lee Wallace USA   1993
Sean Young, Tim Daly, Len Cariou, Dee Walalce, Stone, George Newbern, Alan Fudge, Brian Markinson, Marina Gonzalez Palmier, Constance Jones, Alex Allen Morris, Lourdes Regala, Brandon Smith, Blue Deckert, Kurt Rhoads, Tony Norris
In a crime-ridden near-future, a hotshot woman TV executive comes up with a surefire scheme to boost the network's flagging ratings by televising publicexecutions of condemned criminals. However, her initial enthusiasm becomesmuted when she begins to suspect that the first victim may not be guilty.
DRA   88 min (ort 100 min) mTV   VIDrel: SPEAR/SONOP
V

WITTGENSTEIN ***   15
Derek Jarman UK   1993
Karl Johnson, Michael Gough, Tilda Swinton, John Quentin, Clancy Chassay, Kevin Collins, Jill Balcon, Sally Dexter, Gina Marsh, Vanya del Borgo, Ben Scantlebury, Howard Sooley, David Radzinowicz, Jan Latham-Koenig, Tony Peake
Biopic on the life of the title philosopher, who leaves his native Vienna to settle in Cambridge. The homosexuality of Wittgenstein may have been the main aspect that prompted Jarman to explore his life, but this extremely effective film is no less accessible on that account, being both enjoyable and well mounted. Shot on a tiny budget of ú300,000 in a mere two weeks, but one would never have thought so to look at it.
DRA   69 min (ort 75 min)   VIDrel: CONNO/RTM   V/sur

WIZ, THE ***   U
Sydney Lumet USA   1978
Michael Jackson, Diana Ross, Richard Pryor, Nipsey Russell, Ted Ross, Lena Horne, Thelma Carpenter, Theresa Merritt, Stanley Greene, Mabel King, Clyde J. Barrett, Carlton Johnson, Henry Madsen, Vicki Baltimor, Glory Van Scott
Despite many flaws, this updated version of THE WIZARD OF OZ with an all-black cast has many good things, not least some good music and dance routines. Undeservedly panned by the critics on release, this film version of the successful Brown/Smalls musical is an entertaining film, despite a poor performance from Ross. Written by Joel Schumacher with photography by Oswald Morris and musical direction by Quincy Jones.
MUS   128 min (ort 133 min) wScrn   VIDrel: CIC/SONOP
V/sur
Boa: novel The Wonderful Wizard Of Oz by Lyman Frank Baum/musical by William F. Brown and Charlie Smalls.

WIZARD OF LONELINESS, THE **   15
Jenny Bowen USA   1988
Lukas Haas, Lea Thompson, John Randolph, Jeremiah Warner, Lance Guest, Anne Pitoniak, Dylan Baker, Steve Hendrickson, Jeffrey Dreisbach, Dorothy Yates, Andrea Matheson, Alan Wright, Jerome Dempsey, David Moscow, Jason Cook
During WW2 a young boy is sent to live with his grandparents in Vermont when his father is away fighting. He finds his new life difficult to adjust to at first, but soon warms to the small town, especially when it seems that his aunt and cousin are in danger from a mysterious stranger. Fine acting is weakened by the disorganised nature of the narrative.
DRA   106 min (ort 110 min)   VIDrel: VISVID/POLYREC
L/A   V
Boa: novel by John Nichols.

WIZARD OF OZ, THE ***   PG
Larry Semon USA   1925
Larry Semon, Bryant Washburn, Dorothy Dwan, Virginia Pearson, Charles Murray, Oliver Hardy, G. Howe Black
This endearing silent version of Baum's classic follows Dorothy's amusing adventures in Oz, to which she's carried by a whirlwind: but strangely our heroine is now eighteen years old rather than eight. That said, the film mostly follows the original story, and Semon, who was a highly inventive slapstick comedian, both directs and gives a pleasing performance as Dorothy's scarecrow friend.
JUV   70 min (ort 94 min) silent B/W
VIDrel: EUREKA/GOLD   V
Boa: novel by Lyman Frank Baum.

WIZARD OF OZ, THE ****   U
Victor Fleming USA   1939
Judy Garland, Ray Bolger, Bert Lehr, Jack Haley, Frank Morgan, Billie Burke, Margaret Hamilton, Charley Grapewin, Clara Blandick, The Singer Midgets, Pat Walsh, Mitchell Lewis
This classic American fantasy is as enjoyable today as when it was made. Garland is perfect as Dorothy who finds herself taken to Oz by a tornado. The sequels are "Journey Back To Oz" and RETURN TO OZ. Written by E.A. Wolfe, F. Ryerson and N. Langley. MGM originally wanted Shirley Temple for Dorothy but luckily 20th Century Fox refused to release her. AA: Score (H. Stothart), Song ("Over The Rainbow" – H. Arlen (m)/E.Y. Harburg (l)).
MUS   97 min (ort 118 min) cC   VIDrel: MGM/WHV   V/dm
Boa: novel by Lyman Frank Baum.

WIZARDS **   PG
Ralph Bakshi USA   1977
Voices of: Mark Hamill, Bob Holt, Jesse Wells, Richard Romanus,

David Proval, James Connel, Christopher Tayback, Barbara Sloane, Hyman Wien, Tina Bowman, Angelo Grisanti, Peter Hobbs
An animated view of Earth 3,000 years after a nuclear war from the creator of FRITZ THE CAT, and telling a turgid tale of warring factions that battle for control of the planet. See also FIRE AND ICE, HEY GOOD-LOOKIN' and HEAVY TRAFFIC, three other films by this director.
ANIM   80 min   VIDrel: 20TH/TECH L/A   V

WIZARDS OF THE DEMON SWORD ** 15
Fred Olen Ray USA   1991
Lyle Waggoner, Russ Tamblyn, Blake Bahner, Heidi Paine, Dan Speaker, Michael Berryman, Jay Richardson, Lawrence Tierney, Bill Edwards, Hoke Howell, Hank Levy, Dan Golden, Heather Wilk, Gladys Jennet, Caryle Waldman, Melinda Golden
A low-budget sword-and-sorcery tale, as is usually the case, the setting being a remote period in the Earth's past, with the usual Good-versus-Evil theme.
FAN   81 min   VIDrel: TROMA/RTM   V

WOLF *** 15
Mike Nichols USA   1994
Jack Nicholson, Michelle Pfeiffer, James Spader, Kate Nelligan, Christopher Plummer, Dick O'Neill, Dehi Berti, David Hyde Pierce, Om Puri, Ron Rifkin, Prunella Scales, Brian Markison, Peter Gerety, Bradord English, Tom Offenheim
A book editor in the throes of a mid-life crisis runs over a wolf and is bitten when he tries to help the injured animal. After being treated for the bite, he gradually becomes aware of strange changes that affect both his body and his personality. An original and imaginative reworking of the werewolf myth that combines elements of romance and black comedy but is hampered by excessive length.
HOR   120 min (ort 127 min) wScrn
VIDrel: COLUM/SONOP   V/sur

WOLF MAN, THE **** PG
George Waggner USA   1941
Lon Chaney Jr, Evelyn Ankers, Claude Rains, Maria Ouspenskaya, Ralph Bellamy, Patric Knowles, Warren William, Bela Lugosi, Fay Helm, Leyland Hodgson, Forrester Harvey, J.M. Kerrigan, Kurt Hatch, Doris Lloyd
One of the great original horror stories, this has Chaney being bitten by werewolf Lugosi during the full moon, and surviving to continue life as a werewolf himself. His fate is foretold by a gypsy woman and eventually he is released from the curse by death. Numerous sequels followed but this one, with superb make-up by Jack Pierce and a good score by Charles Previn and Hans J. Salter, stands above them all.
HOR   70 min B/W   VIDrel: CIC/SONOP L/A   V/h

WOLFEN *** 18
Michael Wadleigh USA   1981
Albert Finney, Diane Venora, Edward James Olmos, Gregory Hines, Tom Noonan, Dick O'Neill, Peter Michael Goetz, Ralph Bell, Sam Gray, Max M. Brown, Anne Marie Photamo, Sarah Felder, Reginald Vel Johnson, Chris Manor
A New York detective investigates strange attacks by mysterious animals and starts to track them down, eventually discovering that after they have mutilated their victims they retreat underground. Scripted by Wadleigh and David Eyre, this moody and atmospheric horror tale is memorable for the fine photography of Gerald Fisher.
HOR   109 min (ort 114 min) wScrn (special edition)
VIDrel: WHV   V/sh
Boa: novel by Whitley Strieber.

WOMAN CHASES MAN ** U
John G. Blystone USA   1937
Miriam Hopkins, Joel McCrea, Charles Winninger, Erik Rhodes, Ella Logan, Leona Maricle, Broderick Crawford, Charles Halton, William Jaffrey, George Chandler, Alan Bridge, Monte Vandrgrift, Jack Baxley, Walter Soederling
A highly ambitious female architect, convinced that prejudice has been blighting her career, mounts an elaborate deception in order to convince a young millionaire to bail out his near-bankrupt father. A thinly plotted and terribly limp comedy that is almost, but not quite redeemed by some bright performances.
COM   69 min (ort 71 min) B/W
VIDrel: VCC/DISC/COLUM   V
Boa: story The Princess And The Pauper by Lynn Root/Frank Fenton.

WOMAN HUNT, THE * 15
Eddie Romero PHILIPPINES/USA   1972
John Ashley, Sid Haig, Laurie Rose, Lisa Todd, Eddie Garcia, Pat Woodell, Ken Metcalfe, Charlene Jones
A rotten and exploitative version of THE MOST DANGEROUS GAME, with kidnapped women being used as prey and hunted down like animals in the jungle. Story focuses on one such victim, a young stewardess who along with three other women, is abducted for this purpose. Todd is particularly unendearing as a black-clad lesbian sadist in this dull and trashy effort. A Roger Corman production. See also FINAL ROUND, DEATH RING and SURVIVING THE GAME.
Aka: ESCAPE
A/AD   72 min (ort 81 min)   VIDrel: ALLIED/RTM   V

WOMAN IN RED, THE *** 15
Gene Wilder USA   1984
Gene Wilder, Kelly LeBrock, Judith Ivey, Gilda Radner, Joseph Bologna, Charles Grodin, Michael Huddleston, Kyle T. Heffner, Michael Zorek, Billy Beck, Kyra Stempel, Robin Ingnico, Viola Kates Stimpson, Danny Wells
A middle-aged happily married executive falls in love with a model and pursues her relentlessly. Nothing more than a transplanted remake of the 1977 French film "Pardon Mon Affaire". Gilda Radner was married to Wilder until her untimely death from cancer. See also TEN. AA: Song ("I Just Called To Say I Love You" – Stevie Wonder).
Aka: FESTIVE DESSERTS
COM   83 min (ort 86 min)
VIDrel: 4-FRONT/POLYREC/VISVID   V/sur

WOMAN IN THE WINDOW, THE *** PG
Fritz Lang USA   1944
Edward G. Robinson, Joan Bennett, Raymond Massey, Dan Duryea, Bobby (Robert) Blake, Edmund Breon, Thomas Jackson, Dorothy Peterson, Arthur Loft
While his family is on holiday, a staid professor becomes involved with a pretty girl whose portrait he stopped to admire in a shop window. However, his new friendship leads to murder and he witnesses an investigation that can only lead to his certain apprehension. An absorbing melodrama, its surprise ending was criticised at the time, but should not be seen as a cop out, being both logical and artistically satisfying.
THR   95 min (ort 99 min) B/W   VIDrel: WHV   V
Boa: novel Once Off Guard by J.H. Wallis.

WOMAN NEXT DOOR, THE ** 15
Francois Truffaut FRANCE   1981
Gerard Depardieu, Fanny Ardant, Henri Garcin, Michele Baumgartner, Veronique Silver, Roger Van Hool, Philippe Morier-Genoud, Noel Pacquin, Olivier Becquaert, Nicole Vauthier, Muriel Combe
When an old flame and her husband move in next door, a happily married man is unable to resist the temptation to renew their relationship, and they soon become lovers again. Each tries to end their the affair but their obsessive love spirals out of control. A dark and tragic tale, with a stand-out performance from Ardant.
Aka: LA FEMME D'A COTE
DRA   101 min (ort 106 min)   VIDrel: ARTIF/20TH   V/h

WOMAN OF DESIRE * 18
Robert Ginty USA   1993
Bo Derek, Robert Mitchum, Jeff Fahey, Steven Bauer, John Carson, Michael McCabe, Elia Thompson, Craig Urbani, Warrick Grier, Peter Holden, Thomas Hall, John Matshikiza, Todd Jensen, James Whyle, David Lee, Alain Benatar, Ted Leplat
A rich woman goes yachting with her millionaire boyfriend but their vessel sinks during a storm. She and the captain are washed ashore and after being rescued she claims that he killed her lover and then raped her. While the dead man's brother presses for justice, an experienced lawyer undertakes the captain's defence. A poor excuse of a film that offers ample nude shots of Derek but nothing else.
DRA   98 min   VIDrel: MED/POLYREC L/A   V/sur

WOMAN OF SUBSTANCE, A: PARTS 1 AND 2 *** 15
Don Sharp UK/USA   1984
Jenny Seagrove, Barry Bostwick, Deborah Kerr, John Mills, Peter Egan, Gayle Hunnicutt, Barry Morse, Nicola Pagett, Miranda Richardson, Diane Baker, John Duttine, George Baker, Peter Chelsom, Mick Ford, Dominic Guard, Del Henny
A better than average mini-series that examines the life of an ill-

treated kitchen maid who works her way up to become the owner of a chain of department stores. Seagrove is attractive as the young Emma Hart in a performance that rises above the banality of the script. Part 1 is entitled "The Challenge" and Part 2 "The Fortune". Followed by the inferior HOLD THE DREAM in 1986.
DRA   300 min (2 cassettes) mTV
VIDrel: 4-FRONT/POLYREC/ODY   V/sh
Boa: novel by Barbara Taylor Bradford.

WOMAN OF THE DUNES ***   15
Hiroshi Teshigahara JAPAN   1964
Eiji Okada, Kyoko Kishida, Koji Mitsui, Sen Yano, Hiroko Ito, Ginzo Sekigushi, Kiyochiko Ichiha, Tumutsu Tamura, Hiroyuki Nishimo
An amateur entomologist is offered hospitality by some villagers whilst on a field trip, but finds himself trapped at the bottom of a sandpit with a young and attractive widow. He soon learns that their task is to continually shovel sand to stop it engulfing the village and after a futile attempt to escape, eventually accepts and takes pride in his strange fate. A bizarre, obscure and strangely compelling meditation on human hope and destiny.
Aka: SUNA NO ONNA; WOMAN IN THE DUNES
DRA   119 min (ort 127 min) B/W   VIDrel: CONNO/RTM   V
Boa: novel by Kobe Abe.

WOMAN OF THE YEAR ***   U
George Stevens USA   1942
Spencer Tracy, Katharine Hepburn, Reginald Owen, Fay Bainter, William Bendix, Dan Tobin, Minor Watson, Roscoe Karns, Gladys Blake, Dan Tobin, William Tannen, Ludwig Stossel, Sara Haden, Edith Evanson, George kezas
A sports columnist marries a woman political journalist but their marriage hits the rocks almost immediately since they have little in common, and she has no understanding of such mundane matters as baseball. The first pairing of Tracy and Hepburn was a great success and helped win the picture its Academy Award.
AA: Screen/orig (Michael Kanin/Ring Lardner Jr).
COM   114 min B/W   VIDrel: MGM/WHV   V

WOMAN ON THE LEDGE **   (15)
Chris Thomson USA   1992
Deidre Hall, Leslie Charleson, Colleen Zenk Pinter, Josh Taylor, Kate Browne, Peter Bergman, Michael Zaslow, Garry Chalk, Emily Perkins, Linda Darlow, Roger Gross, Shane Meier, Sarah Chalke, Charlene Fernetx, Kelli Fox, David Perry
This odd film starts with the image of a woman perched precariously high up on a ledge, and about to leap to her death. Then in flashback form we are given a look at the events that led to this crisis, when the lives of three unhappy women are examine in detail, it never being made clear until the climax exactly which woman is about to jump. A soppy drama that plays like three films rolled into one, and stretches both credulity and patience.
DRA   90 min mTV   TVrel: BBC1

WOMAN ON THE RUN: THE TRUE STORY OF
LAWRENCIA BEMBENEK **   15
Sandor Stern CANADA   1992
Tatum O'Neal, Bruce Greenwood, Peggy McCay, Colin Fox, Kenneth Welsh, Victor Garber, Catherine Disher, Alan Jordan, Saul Rubinek, Alex McArthur, Ron White, Barbara Eve Harris, Ari Magder, Graham Losee, Gail Webster
Dramatisation of a true incident, when the escape of a female prisoner sparked off a massive, nationwide hunt to recapture her. Average.
DRA   181 min (2 cassettes) mTV   VIDrel: ODY/SONOP   V

WOMAN OR TWO, A *   12
Daniel Vigne FRANCE   1978
Gerard Depardieu, Sigourney Weaver, Jean-Pierre Bisson, Ruth Westheimer, Michel Aumont, Zabou, Yann Babilee, Robert Blumenfeld, Michael Goldman, Adrian Howard, Tanis Vallely, Jean-Quentin Chatelain, Axel Bobovsslavsky
Having found the remains of what he thinks is the world's first French woman, an archaeologist arranges a meeting with the head of an American research foundation. However, by mistake his meeting turns out to be with a scheming female executive from an advertising agency, who wants to exploit this discovery to sell a new perfume. A silly and pointless film that should be avoided, not least thanks to an appearance by the tiresome sexologist Dr Westheimer.
Aka: ONE WOMAN OR TWO; UNE FEMME OU DEUX
COM   91 min   VIDrel: ARROW/RTM   V/s

WOMAN SCORNED, A **   18
Andrew Stevens USA   1993
Shannon Tweed, Andrew Stevens, Dan McVicar, Kim Morgan Greene, Michael D. Arenz, Stephen Young, Perla Walters, Anna Siena-Schwartz, Paul W. Carr, Ron Roy Melendez, Will H. Shriner, Robyn Scott, Anya Longwell, Teresa A. Hawkes
A widow torments the man she holds responsible for her husband's suicide after the collapse of his business, and sets out to destroy everything he holds. She does this by entering their home as a tutor to their teenage son, and then embarks on the seduction of each family member, with all of them soon in her clutches. A derivative revenger, whose plot idea is not a million miles away from the one used in THE HAND THAT ROCKS THE CRADLE. A sequel followed.
DRA   102 min   VIDrel: MED/DISC   V/sh

WOMAN SCORNED 2, A **   18
Rodney McDonald USA   1996
Tane McClure, Myles O'Brien, Wendy Schumacher, Andrew Stevens
With her bloody past hidden beneath amnesia and a fake identity, a woman married to a philandering university professor loses control when she catches him canoodling with a young student. Her devious scheme involves him being framed for a murder, which she proceeds to set up by seducing a variety of suitable individuals. An overheated sequel that spends too much time trying to clear up all the loose ends that were left over from the first film.
THR   93 min   VIDrel: MED/COLUM   V/sh

WOMAN UNDER THE INFLUENCE, A ***   15
John Cassavetes USA   1974
Peter Falk, Gena Rowlands, Katherine Cassavetes, Lady Rowlands, Fred Draper
The marriage of a lower middle-class couple turns sour, as the wife teeters on the brink of insanity, and her husband is able to offer neither support nor comfort. An over-long and overwrought example of the Cassavetes style of improvised film-making (he also wrote the script), the film is a difficult and harrowing work, for the most part sustained by Rowlands in a performance of remarkable power.
DRA   140 min (ort 155 min)   VIDrel: ELPIC/POLYREC   V

WOMAN WHO LOVED ELVIS, THE **   (PG)
Bill Bixby USA   1993
Roseanne Arnold, Tom Arnold, Cynthia Gibb, Sally Kirkland, Danielle Harris, Joe Guzaldo, Kimberly Dal Santo, Monica McCarthy, Liz Muckley, Sam Derence, James Andelin, Dottie Arnold, W. Earle Brown, Patrick Clear, John Duda
Having become estranged with her husband, a devoted Presley fan turns her home into a shrine to this singer, but all the while dreams of her husband's return, even though he has been living with another woman for the last five years. But her life takes a turn for the worse when the welfare agency begins to suspect her of fraud.
DRA   90 min mTv   SATrel: SKY MOVIES
Boa: novel by Laura Kalpakian.

WOMAN'S GUIDE TO ADULTERY, A **   15
David Hayman UK   1993
Theresa Russell, Sean Bean, Amanda Donahoe, Fiona Gillies, Ingrid Lacey
A woman and her three female friends all get involved in adulterous affairs.
DRA   148 min (ort 160 min) mTV
VIDrel: VISCORP/BMGREC   V
Boa: novel by Carol Clewlow.

WOMEN AND MEN: STORIES OF SEDUCTION ***   15
Frederic Raphael/Ken Russell/Tony Richardson USA   1990
Elizabeth McGovern, Beau Bridges, Liza Ross, Louis Mahoney, Phillip O'Brien, Dominic Hawksley, Paul Raphael, Molly Ringwald, Peter Weller, Melanie Griffith, James Woods, Carmen Segarra, Felipe Garcia Velez, Luis Maluenda
Three stories whose common theme is male-female relationships are presented in a well-crafted film that revives the anthology format. Though the quality of the stories varies, in each the acting is uniformly good. The stories are: "The Man In The Brooks Brothers Shirt", "Dusk Before Fireworks" and "Hills Like White Elephants".
DRA   84 min (ort 90 min) mCab   VIDrel: MGM/WHV L/A V/sh
Boa: short stories by Mary McCarthy, Dorothy Parker and Ernest Hemingway.

WOMEN IN CAGES *   18
Gerry De Leon PHILIPPINES/USA   1972
Judy Brown, Pam Grier, Roberta Collins, Jennifer Gan
A lurid and exploitative prison melodrama with Grier as a sadistic lesbian guard whose pleasure it is to torture prisoners in a
chamber she has fitted up for the purpose. One such woman
becomes her latest victim, but has an overwhelming determination to survive and take her revenge. A Roger Corman
production.
Aka: BAMBOO DOLLS HOUSE; WOMEN'S PENITENTIARY 3
THR   78 min Cut (3 min 19 sec)   VIDrel: ALLIED/RTM   V

WOMEN IN LOVE ***   18
Ken Russell UK   1969
*Alan Bates, Glenda Jackson, Oliver Reed, Jennie Linden, Eleanor Bron,
Alan Webb, Michael Gough, Vladek Sheybal, Catherine Wilmer,
Norman Shebbeare, Sarah Nicholls, Sharon Gurney, Christopher
Gable, Nike Arrighi*
The story of an intense and passionate love affair in a small
mining town is told with style and feeling, in this full-blooded
adaptation of the Lawrence novel. A highlight is the nude
wrestling scene between Bates and Reed which broke new
ground in the cinema. Written by Larry Kramer and one of
Russell's more mature works. AA: Actress (Jackson).
DRA   125 min (ort 130 min)   VIDrel: MGM/WHV   V
Boa: novel by D.H. Lawrence.

WOMEN OF WINDSOR, THE *   PG
Steven Hilliard Stern USA   1992
*Sallyanne Law, Nicola Formby, James Piddock, Robert Meadmore,
Nigel Bennett, Carolyn Sadowska, Dixie Seatle, Eugene Robert Glazer,
David Fox, Barbara Gordon, Deborah Duchene, Neil Munro, Don
Carrier, Dan Lett, Adam Bramble, Allan Morley*
A dull film purporting to be a behind-the-scenes look at the
private lives of Sarah Ferguson and Diana Spencer. See also DIANA:
HER TRUE STORY, ANDREW AND FERGIE: BEHIND THE
PALACE DOORS and PRINCESS IN LOVE.
DRA   140 min (ort 180 min) mTV
VIDrel: MIA/DISC/IMPENT   V

**WOMEN ON THE VERGE OF A NERVOUS
BREAKDOWN ***   15
Pedro Almodovar SPAIN   1988
*Carmen Maura, Fernando Guillen, Julieta Serrano, Maria Barranco,
Antonio Banderes, Rossy De Palma, Kitty Manver, Juan Lombardero,
Jose Antonio Navarro, Ana Leza, Ambite, Mary Gonzalez, Lupe
Barrado, Joaquin Climent, Mary Gonzalez*
An actress is thrown off balance when her lover walks out, and
the efforts she makes to win him back only result in ludicrous
complications. A highly stylised and incredibly frantic comedy
that's all surface gloss without a jot of depth, with all the
female characters being shown as poor creatures unhinged by
their attraction to worthless men. A great disappointment
despite the rave reviews it received from many critics upon
release.
Aka: MUJERES AL BORDE DE UN ATAQUE DE NERVIOS
COM   85 min (ort 88 min)   VIDrel: VISVID/POLYREC
V/sh

WONDER MAN *   U
Bruce Humberstone USA   1945
*Danny Kaye, Virginia Mayo, Vera-Ellen, Steve Cochran, S.Z. Sakall,
Allen Jenkins, Ed Brophy, Donald Woods, Otto Kruger, Richard Lane,
Natalie Schaefer, Huntz Hall, Virginia Gilmore, Edward gargan,
Grant Mitchell*
After his death at the hands of gangsters, the spirit of a brash
nightclub entertainer compels his timid twin brother to track
down and reveal the killer, but not until complications too
numerous to mention have been overcome. A crazy comedy that
is a perfect vehicle for Kaye's comic gifts and a joy to watch. AA:
Effects (vis – John Fulton/aud – Arthur W. Johns).
COM   97 min   VIDrel: VCC/DISC/COLUM   V

WONDERFUL VISIT *   12
Marcel Carne FRANCE   1975
Gilles Kohler, Deborah Berger, Roland Lesaffre, Jean Pierre Castaldi
An unknown young man is found unclothed and unconscious
on a Brittany beach by the rector of a small village, and learning about his background proves to be far from easy, especially
as he insists he is an angel fallen from heaven. Initially, the
general opinion is that he is a worthy candidate for the insane
asylum, but as his stay continues, many strange and incredible
events start to occur. An uncertain allegory, opaque but quite
charming and poetic.
Aka: LA MERVEILLEUSE VISITE
DRA   101 min (ort 105 min)   VIDrel: WESCON/RTM   V
Boa: novel by H.G. Wells.

**WONDERFUL WORLD OF THE BROTHERS
GRIMM, THE ***   U
Henry Levin/George Pal USA   1962
*Laurence Harvey, Claire Bloom, Jim Backus, Yvette Mimieux, Buddy
Hackett, Karl Boehm, Walter Slezak, Barbara Eden, Oscar Homolka,
Arnold Stang, Betty Garde, Martita Hunt, Bryan Russell, Ian Wolfe,
Tammy Marihugh, Walter Rilla*
Originally filmed in Cinerama, this film brings to life three
Grimm tales linked together by a very fictional account of the
lives of the brothers and their struggle to get their stories
published. Some nice German locations give a strong flavour of
authenticity, and the acting is quite good, yet for all this the
overall impression made by the final result is a little disap-
pointing. AA: Cost (Mary Wells).
DRA   130 min   VIDrel: MGM/WHV L/A   V/sh
Boa: book Die Bruder Grimm by Hermna Gerstner.

WONDERWORKS: WORDS BY HEART *   (U)
Robert Thompson USA   1986
*Alfre Woodard, Charlotte Rae, Robert Hooks, Fran Robinson, Ed Call,
Rance Howard, Leo Geter, Bobby Jacoby, Bill Zuckert, Michael Byers,
Gino De Mauro, Elyse Donalson, Dennis Robertson, Cory Taylor,
Scott Cameron, Hannah Cutrona*
A shattering look at prejudice as seen in this tale of a small
Missouri town in the early 1900s where the members of the only
black family have to cope with their racist neighbours. A
convincing and moving effort that is evocative and quite faith-
ful to the original novel.
Aka: WORDS BY HEART
JUV   108 min (ort 116 min)   SATrel: SKY MOVIES
Boa: novel by Ouida Sebestyen.

WOODEN HORSE, THE *   U
Jack Lee UK   1950
*Leo Genn, David Tomlinson, Anthony Steel, David Greene, Michael
Goodliffe, Bryan Forbes, Peter Finch, Peter Burton, Patrick
Waddington, Anthony Dawson, Franz Schaftheitlin, Hans Meyer,
Jacques Brunius, Dan Cunningham, Ralph Ward*
Taut story of POWs escaping from their camp by digging a
tunnel beneath a wooden horse they use in their gym. Exciting
and dramatic with a good script by Eric Williams. The photog-
raphy is by C. Pennington-Richards.
WAR   98 min (ort 101 min) B/W   VIDrel: LUMI/SPEAR
L/A   V
Boa: book by Eric Williams.

WORKING GIRLS *   18
Lizzie Borden USA   1986
*Louise Smith, Amanda Goodwin, Ellen McElduff, Marusia Zach, Jane
Peters, Helen Nicholas, Boomer Tibbs, Richard Davidson, Ronald
Willoughby, Patience Pierce, Paul Sumak, Fred Neuman, Ellen
McElduff, Grant Wheaton, Martin Haber*
A look at a day in the life of Molly, a photographer, who supple-
ments her income working as a prostitute in a small New York
brothel. A very laid-back account this one, with the sex shown
more as a viable alternative to poorly paid jobs than as anything
to be ashamed of. Not without its moments of humour and
extremely well made.
DRA   89 min   VIDrel: CONNO/RTM   V

WORKING GIRL *   15
Mike Nichols USA   1988
*Harrison Ford, Melanie Griffith, Sigourney Weaver, Alec Baldwin,
Joan Cusack, Philip Bosco, Nora Dunn, Oliver Platt, James Lally,
Kevin Spacey, Robert Easton, Olympia Dukakis, Ricki Lake, Jeffrey
Nordling, Elizabeth Whitcraft*
A working girl makes her way in Wall Street and discovers that
this requires all her resourcefulness and cunning, when she
finds herself involved in outsmarting her boss and closing a
business deal. A likeable, lightweight comedy, with pleasing
performances all round. AA: Song ("Let The River Run" – Carly
Simon).
COM   109 min (ort 115 min)   VIDrel: 20TH/TECH L/A   V

WORKING IT OUT *   18
Philip Jem USA
Joanna Storm, Janey Robbins, Erica Boyer, Danielle, Faustie, Paula

Di S, Francois, Eric Edwards, Ron Jeremy, Mike Horner, Herschel Savage, Dave Bannon
A woman turns a decrepit gym into a thriving work-out centre by dint of sheer hard work and much dedication, although we don't get to see much of these qualities at work. A silly and harmless sex film, that devotes itself instead to a detailed account of her more energetic leisure pursuits. Cut before video submission by 14 min 19 sec.
A   47 min Cut (8 min 41 sec – ort 70 min)
VIDrel: HAR/GOLD   V

WORKING TRASH **
Alan Metter USA
George Carlin, Ben Stiller, Leslie Hope, Buddy Ebsen, Jack Blessing, Dan Castellanata, Michael J. Pollard, Ellen Ratner, Mindy Sterling, George Wallace, Lisa Montgomery, Michael Gregory, Julian Christopher, Phil Rizzuto
Two garbage collectors get hold of a valuable stock market tip and decide on a change of profession. They soon plunge into a strange new world, gambling on the markets in the hope of making a quick killing. Average.
COM   87 min (ort 96 min) mTV   VIDrel: 20TH/TECH   V

U
1990

WORLD ACCORDING TO GARP, THE ***
George Roy Hill USA
Robin Williams, Mary Beth Hurt, Glenn Close, John Lithgow, Hume Cronyn, Jessica Tandy, Swoosie Kurtz, Amanda Plummer, James McCall, Peter Michael Goetz, George Ede, Mark Soper, Nathan Babcock, Ian MacGregor, Susan Browning
The story of a young man making his way in the world, with his unusual views on life owing not a little to the influence of his unorthodox and unmarried mother. Absorbing and entertaining, this loose adaptation of Irving's novel, gave Close her acting debut. The script is by Steve Tesich. Lithgow appears as a transsexual.
DRA   131 min   VIDrel: WHV L/A   V
Boa: novel by John Irving.

15
1982

WORLD CUP, THE: A CAPTAIN'S TALE ***
Tom Clegg UK
Dennis Waterman, Nigel Hawthorne, Andrew Keir, Derek Francis, Marjorie Bland, Richard Griffiths
Loosely based on real events, this engaging drama follows the fortunes of a football team from the mining town of Durham who get their chance to represent Britain in the first World Cup competition of 1910. Quite appealing, it offers ample humour of the broad sort, most of which revolves around the varied exploits of the players once they get out of Britain.
DRA   82 min (ort 90 min) mTV   VIDrel: NTV/TOTAL   V

PG
1982

WORLD IS FULL OF MARRIED MEN, THE *
Robert Young UK
Anthony Franciosa, Carroll Baker, Anthony Steel, Sherrie Lee Cronin, Gareth Hunt, Georgina Hale, Paul Nicholas, John Nolan, Jean Gilpin, Moira Downie, Alison Elliott, Eva Louise, Joanne Ridley, Emma Ridley, Roy Scammell
A model on her way to the top meets a forty-year-old advertising man who is married but cannot stop himself having affairs. A shoddy and exploitative drama based on a poor Collins screenplay, which is in turn based on a worse book.
DRA   102 min (ort 107 min)   VIDrel: ODY/SONOP   V
Boa: novel by Jackie Collins.

18
1979

WORLD OF APU, THE ****
Satyajit Ray INDIA
Soumitra Chatterjee, Sharmila Tagore, Shapan Mukerjee, S. Alke Chakravarty
The third and final part of the "Apu" trilogy that began with PATHER PANCHALI, deals with the joys and sadness of adult life for Apu, after he enters an arranged marriage, comes to deeply love his wife, but becomes an aimless and grieving wanderer after she dies in childbirth. The eventual meeting and reconciliation with his son is one of the most moving sequences in modern cinema.
Aka: APU SANSAR
DRA   100 min (ort 117 min) B/W   VIDrel: CONNO/RTM
V
Boa: novel Aparajito by Bibhutibhusan Bandopadhaya.

U
1959

WORLD WITHOUT PITY, A ***
Eric Rochant FRANCE
Hippolyte Girardot, Mireille Perrier
The story of a young drifter who lives off his brother's drug-

15
1989

dealing activities and his own talent for poker. Interested in nothing except chasing girls, this shiftless young man is the despair of his parents, and when he meets an independent and attractive woman, it's not long before he has thought out a careful plan of seduction. A mocking and leisurely tragi-comedy, well scripted and performed.
Aka: UN MONDE SANS PITIE
DRA   84 min wScrn   VIDrel: ARTIF/20TH   V

WORLD'S GREATEST LOVER, THE **
Gene Wilder USA
Gene Wilder, Carol Kane, Dom DeLuise, Carl Ballantine, Michael Huddleston, Fritz Feld, Ronny Graham, Matt Collins, Cousin Buddy, Hannah Dean, Candice Azzara, Carl Ballantine, Lou Cutell, James Gleason, Florence Sundstrom
Inspired by Fellini's "The White Sheik", this period comedy is set in the 1920s and concerns the ambitions of a rival studio, which screen-tests a man they hope will replace Valentino. Unfortunately, his wife falls for the star it's his ambition to supplant. An uneven slapstick romp with a few touching moments and a general over-abundance of energy. Written and produced by Wilder.
COM   88 min   VIDrel: 20TH/TECH L/A   V

15
1977

WORLD'S OLDEST LIVING BRIDESMAID, THE **
Joseph L. Scanlon USA
Donna Mills, Brian Winner, Beverly Garland, Laura Press, Winston Rekert
A successful woman attorney is dismayed by her inability to find a suitable partner and grows weary of attending her friends' weddings (hence the title). However, things set to improve when she employs a young and attractive male secretary. A lightweight and none-too-realistic romantic comedy.
COM   95 min mTV   TVrel: C4

(U)
1990

WORTH WINNING *
Will Mackenzie USA
Mark Harmon, Madeleine Stowe, Lesley Ann Warren, Maria Holvoe, Mark Blum, Andrea Martin, David Brenner, Alan Blumenfield, Brad Hall, Jon Korkes, Joan Severance, Devin Ratray, Tony Longo, Shannon Lawrence, Todd Cameron Brown
A stale and tasteless farce that focuses on the exploits of an obnoxious and smugly successful TV weatherman, who is convinced he is God's gift to women. This leads to a wager with a friend that he can get a trio of attractive women to accept his proposal of marriage, and the expected set of humorous episodes unfolds. A vapid and unappealing comedy that wastes both time and talent.
COM   98 min (ort 103 min)   VIDrel: 20TH/TECH   V/sur

15
1989

WOYZECK **
Werner Herzog WEST GERMANY
Klaus Kinski, Eva Mattes, Wolfgang Reichmann, Willy Semmelrogge, Josef Bierbichler, Paul Burian, Volker Prechtl, Dieter Augustin, Wolfgang Bachler, Irm Hermann, Rosy-Rosy Heinikel, Herbert Fux, Thomas Mettke, Maria Mettke
The story of an army private who is obliged to undertake menial work to support his family, and endures abuse from all those around him, until finally he goes mad and becomes a murderer. Kinski gives his usual intense and staring performance in this minor Herzog feature.
DRA   77 min (ort 82 min)   VIDrel: TART/20TH   V
Boa: play by Georg Buchner.

15
1978

W.R. – MYSTERIES OF THE ORGANISM ***
Dusan Makaveyev WEST GERMANY/YUGOSLAVIA
Milena Dravic, Jagoda Kaloper, Betty Dodson, Ivica Vidovic, Zoran Radmilovic, Miodrag Andric, Tuli Kupferberg
The theories of sexologist Wilhelm Reich are illustrated in a witty and inventive blend of newsreel footage and a fictional story of a Yugoslav girl and attempts to bring sexual liberation to a Soviet skating star. A provocative and thought-provoking work.
Aka: W.R. – MISTERIJE ORGANIZMA
DRA   86 min B/W/Col   VIDrel: CONNO/RTM L/A   V

18
1971

WRESTLING ERNEST HEMINGWAY ***
Randa Haines USA
Robert Duvall, Richard Harris, Shirley MacLaine, Piper Laurie, Sandra Bullock, Micole Mercurio, Marty Belafsky, Harold Bergman, Ed Amatrudo, Rudolph X. Herrera, Jag Davies, Persephone Felder, Stephen G. Anthony, Greg Paul Myers
Two completely different seventy-five-year-old men, living the

12
1993

empty lives of retirees in Florida, become friends of a kind and share some small and simple pleasures. A work that is more bitter than sweet, quite episodic in form, but offering some very fine performances by Duvall and Harris. Bullock offers a fine cameo as a good-natured waitress.
DRA   117 min (ort 123 min) cC   VIDrel: WHV   V/sur

WRITER'S BLOCK **   18
Charles Correll USA   1991
Morgan Fairchild, Michael Praed, Cheryl Anderson, Douglas Rowe, David Grant Wright, Joe Regalbuto, Danae Torn, Anthony Herrer, Tokeli Le Claire, Marnie Andrews, Janet Haley, John Flaiz, David J. Partington, Debi Fares
The author of a series of successful murder mysteries decides to kill off her central character and create a completely new work. However, when a real-life murder occurs, it appears that events are beginning to mirror her writing.
THR   86 min (ort 88 min) mCab   VIDrel: CIC/SONOP   V

WRONG ARM OF THE LAW, THE ***   15
Cliff Owen UK   1962
Peter Sellers, Lionel Jeffries, Bernard Cribbins, Davy Kaye, Nanette Newman, John Le Mesurier, Dennis Price, Bill Kerr, Irene Browne, Martin Boddey, Ed Devereaux, Arthur Mullard, Reg Lye, Dermot Kelly, Graham Stark, Dick Emery
British crooks collaborate with Scotland Yard in tracking down three Aussie criminals, who disguise themselves as policemen and are in the habit of confiscating loot from apprehended criminals. A lively comedy yarn with some very funny moments. Written by John Warren and Len Heath with music by Richard Rodney Bennett.
COM   94 min B/W   VIDrel: FABFIL/SPEAR   V

WRONG MAN, THE ***   PG
Alfred Hitchcock USA   1956
Henry Fonda, Vera Miles, Anthony Quayle, Harold J. Stone, Nehemiah Persoff, Esther Minciotti, Charles Cooper, Laurinda Barr, Norma Connolly, Doreen Lang, Frances Reid, Lola D'Annunzio, Robert Essen, Kippy Campbell
A musician is accused of robbery but protests his innocence. Though the semi-documentary style is spoilt by jerky camerawork and inappropriate music, the film is not without its moments; Miles as the wife who cracks under the strain is especially good. Hitchcock based this one on a true story. Written by Angus MacPhail and Maxwell Anderson, and with music by Bernard Herrmann.
DRA   101 min (ort 105 min) B/W   VIDrel: WHV   V
Boa: book The True Story of Christopher Emmanuel Balestrero by Maxwell Anderson.

WRONG MAN, THE ***   18
Jim McBride USA   1993
John Lithgow, Rosanna Arquette, Kevin Anderson, Jorge Cervera Jr, Ernesto Laguardia, Robert Harper, Dolores Heredia, Jose Escandon, Ted Swanson, Paco Moraya, Gerado Zepeda "Chiquilin", Pedro Altamirando, Alejandro Bracho
A seaman goes on leave in Mexico and is mistaken for a smuggler wanted for murder. With the authorities close on his heels, he manages to escape but in so doing is forced to join up with an ill-matched couple who soon manage to embroil him in their own devious games. Excellent acting and the Mexican locations provide the bulk of the interest in this modern-style film noir.
THR   98 min   VIDrel: POLY/POLYREC L/A   V

WRONG MOVE, THE **   15
Wim Wenders WEST GERMANY   1975
Rudiger Vogler, Hann Schygulla, Natassja Nakszynski (Kinski), Ivan Desny, Marianne Hoppe, Hans-Christian Blech, Peter Kern
Another bleak road-movie from a master of the ponderous anti-statement. A discontented writer journeys through the wasteland of modern West Germany, in an attempt to resolve the problems of the country's past. This was the second film in the director's road-movie trilogy, that began with ALICE IN THE CITIES and concluded with KINGS OF THE ROAD.
Aka: FALSCHE BEWEGUNG; WRONG MOVEMENT
DRA   99 min (ort 103 min)   VIDrel: CONNO/RTM   V
Boa: story Wilhelm Meister by Goethe.

WRONG WOMAN, THE **   15
Douglas Jackson USA   1995
Nancy McKeon, Chelsea Field, Stephen Shellen, Gary Hudson, Lyman Ward, Michelle Scarabelli
A new temp is invited to dinner by the womanising company

president, and beats a hasty retreat from his apartment when he makes advances towards her. However, when he is found murdered she becomes the chief suspect and although the evidence against her is very strong, the man's widow is convinced she is innocent. Meanwhile, she has to go on the run from both the police and the real killers, who intend to silence her. Fair.
THR   90 min   VIDrel: FIRST/SONOP   V

WUTHERING HEIGHTS ****   U
William Wyler USA   1939
Merle Oberon, Laurence Olivier, David Niven, Flora Robson, Donald Crisp, Geraldine Fitzgerald, Leo G. Carroll, Cecil Kellaway, Miles Mander, Hugh Williams, Cecil Humphreys, Sarita Wooton, Rex Downing, Douglas Scott
A splendid version of this classic romance, set in Victorian England and telling of the passionate but doomed love affair between a girl and a gypsy. The story stops at chapter seventeen of the novel, but the fine direction and performances sustain it. Scripted by Charles MacArthur and Ben Hecht, the story has been filmed several times since. The atmospheric climax is a highlight, and the wonderful photography won a well-deserved Oscar. AA: Cin (Gregg Toland).
DRA   99 min (ort 104 min) B/W
VIDrel: VCC/DISC/COLUM   V
Boa: novel by Emily Bronte.

WUTHERING HEIGHTS ***   U
Robert Fuest UK   1970
Anna Calder-Marshall, Timothy Dalton, Ian Ogilvy, Harry Andrews, Pamela Brown, Judy Cornwell, Hilary Dwyer, Hugh Griffith, Julian Glover, Rosalie Crutchley, James Cossins, Peter Sallis, Aubrey Woods, Morag Hood, John Comer
Another version of this famous tale, this is a solid and competent work, making good use of authentic locations. Dalton and Calder-Marshall are good as doomed lovers Heathcliff and Cathy but the film's compressed pace is a drawback. Written by Patrick Tilley and with music by Michel Legrand.
DRA   105 min   VIDrel: L/A   V
Boa: novel by Emily Bronte.

WUTHERING HEIGHTS **   PG
Peter Kosminsky USA   1992
Juliette Binoche, Ralph Fiennes, Janet McTeer, Sophie Ward, Simon Sheperd, Jeremy Northam, Jason Riddington, Simon Ward, Robert Demeger, Paul Geoffrey, John Woodvine, Jennifer Daniel, Janine Wood, Jonathan Firth, Jon Howard
When the daughter of a middle-class family falls in love with the gypsy lad she has been brought up with, circumstances conspire to keep them apart, until a tragedy re-unites them in death. A stylish and extremely ambitious remake, which for all the strenuous efforts made at imparting atmosphere and realism, never becomes more than a cold and unmoving collection of handsome set-pieces. A pity, as the opening sequence promises great things.
Aka: EMILY BRONTE'S WUTHERING HEIGHTS
DRA   102 min (ort 106 min)   VIDrel: CIC/SONOP   V/sur
Boa: novel by Emily Bronte.

WYATT EARP ***   12
Lawrence Kasdan USA   1993
Kevin Costner, Dennis Quaid, Gene Hackman, Jeff Fahey, Mark Hamon, Michael Madsen, Joanna Going, Isabella Rossellini, Catherine O'Hara, Bill Pullman, Tom Sizemore, JoBeth Williams, Mare Winningham, Betty Buckley, Adam Baldwin
Long, rambling but exceptionally well-acted account of the life and times of this famous western figure, which portrays him as a man whose passion for justice deteriorated into a mean-spirited thirst for revenge. Costner not very convincing as Earp in his younger days, but Quaid is really magnificent as Doc Holliday. The film was originally designed as a mini-series and its excessive length is a major drawback. See also TOMBSTONE and MY DARLING CLEMENTINE.
WES   183 min (ort 190 min) wScrn cC   VIDrel: WHV   V/sur

WYNNE AND PENKOVSKY ***   15
Paul Seed UK/USA   1984
David Calder, Christopher Rozycki, Fiona Walker, Frederick Treves, Paul Geoffrey
A recreation of the career of Greville Wynne as a secret agent working in the USSR during the 1960s, and concentrating on his relationship with Oleg Penkovsky, a high-ranking KGB

officer who passed secrets to him until he was exposed. Originally shown in three 55-minute episodes, this is a good, solid and unadorned account.

DRA   135 min (ort 165 min) mTV   VIDrel: BBC L/A   V
Boa: book The Man From Moscow by Greville Wynne.

X

X DREAMS **
R18/18
USA
Victoria Paris, Busty Belle, Tracey Adams
A young woman is plagued by uncontrollable dreams that effect those around her in some very strange ways. Very much a standard sex film, despite a few half-hearted attempts to incorporate a supernatural element.

A   75 min (R18 ver); 49 min Cut (1 min 5 sec – 18 ver)
VIDrel: SHEP L/A (R18 ver); HAR/GOLD (18 ver)   V

X-FILES, THE: TUNGUSKA **
12
Kim Manners/Rob Bowman USA
1996
David Duchovny, Gillian Anderson, Nicholas Lea, Mitch Pileggi
This feature-length story has our two FBI agents investigating a massive explosion that took place over the Tunguska region of Siberia in 1980, it being though to be the result of a massive meteor. Pretty soon, the two agents are captured by the Russians, and subjected to inhumane medical experiments which involve them being injected with alien life-forms. A average SF outing from the cult TV series. Many of the other episodes have become available.

FAN   91 min mTV   VIDrel: 20TH/FOXVID; 20TH/TECH
V/sur

XALA ***
12
Ousmane Sembene SENEGAL
1974
Tierno Leye, Seuen Samb, Miriam Niang, Yournoyss Seye, Dieynaba Niang
A wealthy businessman takes a third wife but is unable to satisfy her after suddenly being struck by "xala" or impotence. As a consequence his former high social standing begins to nose-dive. A satirical look at the self-indulgent and wasteful lifestyle of the post-colonial elite, especially those Africans who turn their back on their own heritage by adopting a Westernised lifestyle.

Aka: CURSE, THE
DRA   118 min (ort 123 min)   VIDrel: CONNO/RTM   V

XANADU **
PG
Robert Greenwald USA
1980
Olivia Newton-John, Gene Kelly, Michael Beck, James Sloyan, Dimitra Arliss, Katie Hanley, Sandahl Bergman, Fred McCarren, Ren Woods, Melinda Phelps, Lynn Latham, Cherise Bate, Juliette Marshall, Marilyn Tokuda, Teri Beckerman
A muse comes down to Earth to inspire a young musician and the owner of a nightclub who are planning to launch a roller-disco. Very much a vehicle for Newton-John, this flashy and empty film only serves to highlight her total lack of screen presence. A saccharine sweet confection leaving no aftertaste, that is essentially a remake of the 1947 film DOWN TO EARTH. Music is by The Electric Light Orchestra.

MUS   92 min (ort 96 min)   VIDrel: CIC/SONOP   V/sur

X-TRA PRIVATE LESSONS **
18
Dominique Othenin-Girard USA
1993
Maria Morgan, Ray Garaza, Theresa Morris, Deidre Imershein, Graham Gathright, Nancy Strandberg, Richard Callinan, Florence McGee, Sixto Nolasco, Nicole, Manuel Cimadevilla, Carole Cortland, Frank Marty, Kathryn Klvana
A top fashion model on the lookout for new talent gets involved with a Cuban man, despite already being married, and at the same time develops an obsessive interest in a woman she sees at a nightclub. A steamy and convoluted thriller.

Aka: PRIVATE LESSONS: ANOTHER STORY
THR   83 min   VIDrel: MED/DISC   V/sh

XTRO ***
18
Harry Bromley Davenport UK
1982
Philip Sayer, Bernice Stegers, Simon Nash, Danny Brainin, Anna Wing, Maryam D'Abo, David Cardy, Peter Mandell, Robert Fyfe, Arthur Whybrow, Katherine Best, Robert Moreno, Anna Mottram, David Henry, Robert Austin, Vanya Seager
A man kidnapped by aliens is returned to Earth three years later but is not quite human. Revolving around a series of repulsive

effects, this has our Mister Average returning home where he infects his son, kills countless people and turns the au pair into an alien-breeding chamber. A grotesque and unpleasant film. A sequel followed some years later.

Aka: JUDAS GOAT; XTRO: A NEW DIMENSION IN FEAR
FAN   82 min (ort 86 min)   VIDrel: FIRST/SONOP L/A
V/s

XTRO 2 **
18
Harry Bromley Davenport USA
1990
Paul Koslo, Jan-Michael Vincent, Tara Buckman, Jano Frandsen, Nicholas Lea, W.F. Wadden, Rolf Reynolds, Nic Amoroso, Tracy Westerholm, Rachel Hayward, Gerry Nairn, Bob Wilde, Nicola Crosbie, Michael Metcalfe, Thom Schioler, Steve Wright
This sequel from the director of the rather weakly plotted if gruesome earlier film, has a tighter script but less originality, and is set at a remote scientific research station where experiments are afoot that involve exploring a parallel universe. When three human guinea pigs fail to make the return journey, Vincent (the only man to survive such a trip) is called in. A messy and highly derivative ALIENS-style clone.

Aka: XTRO 2: THE SECOND ENCOUNTER
HOR   90 min (ort 92 min)   VIDrel: FIRST/SONOP L/A   V

XTRO 3: WATCH THE SKIES *
18
Harry Bromley Davenport USA
1995
Sal Landi, Andrew Divoff, Jim Hanks, Karen Moncreiff, Robert Culp
A remote island is the scene for a confrontation between a group of marines and an alien, the survivor of an expedition to Earth that landed there but was concealed by the government to prevent panic. After arriving at the island, the soldiers learn that they have been sent on a suicide mission. A weak and totally unimaginative second sequel that is virtually unrelated to the first film.

FAN   92 min   VIDrel: MIA/DISC   V/sur

XXX'S AND OOO'S **
(PG)
Allan Arkush USA
1995
Debrah Farentino, Andrea Parker, Nia Peeples, Susan Walters, Brad Johnson, Paul Gross, John Allen Nelson, David Lee Smith, Lisa Zane, Jesse Vint, Eddie Mills, Karen Carlson, Paul Tillis, David Keith, Jaime Nicole Dudney, Ed Bruce
Four divorced women in Nashville have made a new life for themselves after their husbands betrayed them in one way or another but are unable to get over their remaining feelings. Since this is Nashville, most of the characters are linked to the music scene and the film at times feels like a Country-and-Western video. Trite and unconvincing, with a distinct lack of balance, as our wayward husbands are all weak and repentant, while the women are strong and capable.

DRA   90 min mTV   SATrel: MOVIE CHANNEL

Y

YAKUZA, THE *
15
Sydney Pollack USA
1975
Robert Mitchum, Richard Jordan, Takakura Ken, Brian Keith, Okada Eiji, Herb Edelman, Kishi Keiko, James Shigeta, Kyosuke Mshida, Christina Kpkubo, Go Eiji, Lee Chrillo, M. Hisaka, William Ross, Akiyama, Harada
East meets West in this cross-cultural thriller set in modern Japan where an ex-GI sets about rescuing a friend's daughter kidnapped by gangsters. A flat and almost unwatchable yarn that just drags and drags.

Aka: BROTHERHOOD OF THE YAKUZA
THR   107 min (ort 112 min)   VIDrel: WHV

YANKEE DOODLE DANDY ****
U
Michael Curtiz USA
1942
James Cagney, Joan Leslie, Walter Huston, Rosemary De Camp, Irene Manning, Richard Whorf, Jeanne Cagney, S.Z. Sakall, Walter Cartlett, Eddie Foy Jr, Frances Langford, George Tobias, Jack Young, Minor Watson
The life and times of one of America's most talented songwriters George M. Cohan, as portrayed by Cagney in one of his finest performances. A lavish, enjoyable and patriotic production. Photographed by James Wong Howe, written by Robert Buckner and Edmund Joseph and with the songs by Cohan, naturally. AA: Actor (Cagney), Score (Heinz Roemheld/Ray Heindorf), Sound (Nathan Levinson).

MUS   121 min (ort 126 min)   VIDrel: MGM/WHV   V/dm
V/h

**YANKEE ZULU ** **   PG
Gray Hofmeyr SOUTH AFRICA   1995
Jon Matshikiza, Leon Schuster, Wilson Dunster, Terri Treas, Michelle Bowes, Skye Svormic
A young Zulu boy and his white friend enjoy a childhood friendship on an isolated farm, far removed from the hatred and bitterness engendered by Apartheid. Sadly, events of the time eventually force them apart, but they meet up again twenty-five years later, and enjoy some crazy adventures.
Aka: THERE'S A ZULU ON MY STOEP
COM   87 min (ort 89 min)   VIDrel: COLUM/SONOP V/sur

YANKS *   15
John Schlesinger USA   1979
Richard Gere, Lisa Eichhorn, Vanessa Redgrave, William Devane, Wendy Morgan, Chick Vennera, Rachel Roberts, Joan Hickson, John Ratzenberger, Anthony Sher, Tony Melody, Martin Smith, Philip Whileman, Derek Thompson, Annie Ross
Romantic story of three American GIs in Britain and their love-lives whilst billeted in a Lancashire town. An overlong and lavish production that would have benefited from some judicious editing. Written by Colin Welland and Walter Bernstein.
DRA   133 min (ort 139 min)   VIDrel: MGM/WHV L/A V

**YARN PRINCESS, THE ** **   (PG)
Tom McLoughlin USA   1993
Jean Smart, Robert Pastorelli, Dennis Boutsikaris, Justine Burnette, Karl David Djerf, Luke Edwards, Lee Garlington, Shirley Knight, Jesse D. Goins, Peter Crook, Steven Banks, Jack Shearer, Pierre Epstein, Nancy McLoughlin, Valeri Ross
When her husband dies suddenly, a simple-minded woman finds herself living on welfare and facing a battle with officials, who want to take her six children into care as she seems unable to cope. Fair.
DRA   90 min mTV   SATrel: SKY MOVIES

YEAR IN PROVENCE, A *   (U)
David Tucker UK   1989
John Thaw, Lindsay Duncan, Bernard Spiegel, Jean-Peirre Delage, Maryse Kuster, Louis Lyonet
A retired British couple buy an old farmhouse in the south of France in the hope of making a new life in some tranquil surroundings but are overwhelmed by subsequent events. A well-acted adaptation of Mayle's autobiographical novels.
DRA   239 min (2 cassettes – ort 360 min) mTV
VIDrel: BBC/TECH L/A V/h
Boa: novels A Year in Provence and Tojours Provence by Peter Mayle.

YEAR MY VOICE BROKE, THE *   PG
John Duigan AUSTRALIA   1987
Noah Taylor, Loene Carmen, Ben Mendelsohn, Graeme Blundell, Lynette Curran, Malcolm Roberts, Judi Farr, Bruce Spence, Tim Robertson, Harold Hopkins, Anja Colby, Kyle Ostara, Kelly Dingwall, Dorothy St Heaps, Colleen Clifford
An affectionate coming-of-age film set in a small Australian town in the early 1960s and telling of a young boy's friendship and growing infatuation with a troubled girl. The intelligent script is by Duigan, with the fine performances and perceptive dialogue major assets. Followed by FLIRTING.
DRA   100 min (ort 103 min)   VIDrel: PAL/TERRY L/A V/h

YEAR OF LIVING DANGEROUSLY, THE *   PG
Peter Weir AUSTRALIA   1982
Mel Gibson, Sigourney Weaver, Linda Hunt, Michael Murphy, Bill Kerr, Noel Ferrier, Paul Sonkkila, Bembol Roco, Domingo Landicho, Hermino De Guzman, Ali Nur, Joel Agona, Mike Emperio, Bernardo Nacilla, Coco Marantha
An Australian journalist and a cameraman are assigned to cover exciting but dangerous events in Indonesia in 1965, just prior to the fall of Sukarno. More effective as a political drama than as a romance, with diminutive Hunt playing a man, a role for which she deservedly won an Oscar. AA: S. Actress (Hunt).
DRA   110 min (ort 115 min)   VIDrel: MGM/WHV V
Boa: novel by Christopher J. Koch.

**YEAR OF THE COMET ** **   15
Peter Yates USA   1992
Penelope Ann Miller, Timothy Daly, Louis Jordan, Art Malik, Ian Richardson, Ian McNeice, Timothy Bentinck, Julia McCarthy, Jacques

Mathou, Nick Brimble, Arturo Venegas, Chapman Roberts, Andrew Robertson, Shane Rimmer, Wilfred Bowman
A young woman working in her family's very traditional wine merchant business is sent to locate a bottle of the rarest vintage in the world. She succeeds in this task but finds herself plunged into a mass of intrigue as others try to cash in on her discovery. An attempt at a romantic comedy that for all the strong acting of its two leads, founders on a poor and undeveloped script.
COM   86 min (ort 135 min)   VIDrel: FIRST/SONOP V/sur

YEAR OF THE DRAGON *   18
Michael Cimino USA   1985
Mickey Rourke, John Lone, Ariane, Leonard Termo, Ray Barry, Caroline Kava, Eddie Jones, Joey Chin, Mr. Lee, Victor Wong, Pao Han Lin, Rosang, K. Dock Yip, Dennis Dun, Way Dong Woo, Jimmy Sun, Daniel Davin
A former Vietnam veteran now working as a New York cop mounts his own violent anti-drugs campaign in Chinatown. Overblown, unrealistic scenes of violence spoil what might have been a good film. Written by Oliver Stone. Interestingly, the New York settings were all recreated in North Carolina on location.
DRA   128 min (ort 136 min)   VIDrel: 4-FRONT/POLYREC V/sur
Boa: novel by Robert Daley.

**YEAR OF THE GUN ** *   15
John Frankenheimer USA   1991
Andrew McCarthy, Valeria Golino, Sharon Stone, John Pankow, George Murcell, Mattia Sbragia, Roberto Posse, Thomas Elliot, Carla Cassola, Darren Modder, Ron Williams, Carol Schneider, Antonio Degli Schiavi, Aldo Mengolini
The film opens in Rome in 1978, when the notorious Red Brigade were at war with the state, and seemed able to carry out a host of brutal assassinations and kidnappings with impunity. The film focuses on a young American journalist, whose complex relationships with various figures only serve to suck him into the confusion of the period. An encounter with an ambitious ex-Vietnam photographer exacerbates his problems. A dull and opaque dud.
THR   108 min (ort 111 min)   VIDrel: FIRST/SONOP L/A V
Boa: novel by Michael Mewshaw.

**YEAR OF THE KICKBOXER, THE ** **   18
Eric Tsui HONG KONG   1990
Steve Brettingham, Corrie Thompson, Eric Tsui, Roger Thomas, Richard Brown, Steven Lee, Robin Merrith
A look at the world of underground kickboxing, where an apparently unbeatable and murderous champion awaits new challengers. Some well handled combat sequences are about all that's on offer here.
Aka: YEAR OF THE KING BOXER
MAR   87 min   VIDrel: VCC/DISC L/A V

YEARLING, THE *   U
Clarence Brown USA   1946
Gregory Peck, Jane Wyman, Claude Jarman Jr, Chill Wills, Henry Travers, Forest Tucker, June Lockhart, Clem Bevans, Margaret Wycherly, Donn Gift, Daniel White, Matt Wills, George Mann, Arthur Hohl, Joan Wells, Jeff York
A young boy is captivated by a yearling fawn and attempts to raise this wild deer as a pet, with all the problems that this entails. A fine and moving family film, about childhood, growth and maturity. AA: Cin (Charles Rosher/Leonard Smith/Arthur Arling), Art/Int (Cedric Gibbons and Paul Groesse/Edwin B. Willis) Willis) plus an honorary Oscar that was awarded to Jarman Jr (Outstanding Child Actor of 1946).
DRA   123 min (ort 128 min)   VIDrel: MGM/WHV V/h
Boa: novel by Marjorie Kinnan Rawlings.

YEARLING, THE *   PG
Rod Hardy USA   1993
Peter Strauss, Jean Smart, Wil Horneff, Phillip Seymour Hoffman, Nancy Moore Atchinson, Jarred Blancard, Brad Greenquist, Mary Nell Santacroce, Bart Hansard, Richard Hamilton, Scott Sowers, Ed Grady, Susan F. Allen, Anderson Martin
Updated made-for-TV version of this 1946 classic about a young boy who tries to keep a young fawn and turn into a family pet only to find that this far from easy, especially since is family has to struggle hard to survive. Excellent camerawork and fine acting enhance this movie, set in the Everglades in the 1930s.
DRA   94 min (ort 98 min) mTV   VIDrel: MARQ/QUANT V/sh
Boa: novel by Marjorie Kinnan Rawlings.

YENTL ***
Barbra Streisand USA PG
1983
Barbra Streisand, Mandy Patinkin, Amy Irving, Nehemiah Persoff,
Steven Hill, Allan Corduner, Ruth Goring, David De Peyser, Bernard
Spear, Doreen Mantle, Jack Lynn, Anna Tzelniker, Miriam
Margolyes, Mary Henry, Robbie Barnett
In order to get an education, a young girl disguises herself as
a boy in this tale of Eastern European Jewry. An unusual offering
from Streisand, her first film as director/producer, and written
by her and Jack Rosenthal. This loose adaptation of Singer's sad
little tale is made with great care, but suffers from excessive
length. AA: Score (Michel Legrand/Marilyn Bergman/Alan
Bergman).
DRA   128 min (ort 134 min)   VIDrel: MGM/WHV   V/sh
Boa: short story Yentl, The Yeshiva Boy by Isaac Bashevis Singer.

YES, GIORGIO **
Franklin J. Schaffner USA PG
1982
Luciano Pavarotti, Kathryn Harrold, Eddie Albert, Paolo Borboni,
James Hong, Beulah Quo, Norman Steinberg, Rod Colbin, Kathryn
Fuller, Joseph Mascolo, Karen Kondazarian, Leona Mitchell, Kurt
Adler, Emerson Buckley
On a tour of America, an opera singer loses his voice and
requires the services of a throat specialist. Smitten by the attrac-
tive woman doctor who helps him, he decides to court her. A
dull romantic comedy that's only partially redeemed by the
singing of the star.
COM   106 min (ort 110 min)   VIDrel: MGM/WHV   V/sh
Boa: novel by Anne Piper.

YESTERDAY, TODAY AND TOMORROW **
Vittorio De Sica FRANCE/ITALY PG
1963
Sophia Loren, Marcello Mastroianni, Aldo Giuffre, Agostino Salvietta,
Tina Pica, Armando Trovaioli, Giovanni Ridolfi, Gennaro Di Gregorio
Three spicy and slightly comic tales all written around Loren,
who in each one plays a determined lady exploiting sex to her
advantage. In "Adelina of Naples" she's a black marketeer
using pregnancy to keep out of jail, in "Anna of Milan" she's
the flirtatious wife of an industrialist, and finally "Mara of
Rome" sees her as a call girl whom a seminary student
attempts to reform. A brash and vulgar exercise, as noisy as it
is dated. AA: Foreign.
Aka: IERI, OGGI, DOMANI
DRA   119 min dubbed   VIDrel: EUREKA/GOLD   V/h

YESTERDAY'S TARGET **
Barry Samson USA 15
1995
Daniel Baldwin, LeVar Burton, Malcolm McDowell, Stacy Haiduk,
T.K. Carter
Three special agents equipped with extra-sensory powers are
sent back in time in a bid to change history, but find themselves
stranded there after they lose their memories. To make matters
worse, they become the quarry of a government agent who is
being assisted in his search by a tracker endowed with similar
ESP abilities.
A/AD   83 min   VIDrel: MARQ   V

YIELD TO THE NIGHT **
J. Lee Thompson UK 15
1956
Diana Dors, Yvonne Mitchell, Michael Craig, Geoffrey Keen, Olga
Lindo, Mary Mackenzie, Joan Miller, Marie Ney, Liam Redmond,
Marjorie Rhodes, Athene Seyler, Molly Urquhart, Harry Locke,
Michael Ripper
A condemned woman ponders on the events that led to her
shooting the rich mistress of her pianist lover. Both sombre and
thoughtful, this sad little moral tale was loosely based on the
story of Ruth Ellis. See also DANCE WITH A STRANGER.
Aka: BLONDE SINNER
DRA   95 min (ort 99 min) B/W   VIDrel: MGM/WHV L/A
V
Boa: novel by Joan Henry.

YOJIMBO ****
Akira Kurosawa JAPAN PG
1961
Toshiro Mifune, Eijiro Tone, Seizaburo Kawazu, Izuzu Yamada,
Hiroshi Tachikawa, Kyu Sazanka, Daisuke Kato, Katsuya Nakadai,
Kamatari Fujiwara, Ikio Sawamura, Taksahi Shimura, Atsushi
Watanabe, Yoshio Tsuchiya
A fine film by Kurosawa, in which a wandering samurai exploits
two gangs who are fighting for control of a village. Very much
a tongue-in-cheek samurai film, with Mifune quite superb as
the warrior who teaches both warring factions a lesson they
won't soon forget. Followed by SANJURO and without a doubt

the inspiration for A FISTFUL OF DOLLARS; a memorable
albeit less inventive film. Followed by a sequel – SANJURO.
Aka: BODYGUARD
A/AD   105 min (ort 110 min) B/W wScrn
VIDrel: CONNO/RTM   V

YOL ****
Serif Goren SWITZERLAND/TURKEY 15
1982
Tarik Akan, Serif Sezer, Halil Ergun, Necmettin Cobanoglu, Meral
Orhonsoy, Hikmet Celik, Sevda Aktolga, Tuncay Akca, Hale Akinli,
Turgut Savas, Hikmet Tashdemir, Engin Celik, Osman Bardakci,
Envery Guney, Erdogan Seren
A harrowing account of five prisoners out on a week's parole,
and of the circumstances that lead to tragedy. The work of
Turkish film-maker and political prisoner Yilmaz Gurney (who
worked secretly on the film from prison and escaped to edit it),
this is both a study of the oppression of totalitarian govern-
ments and an examination of the brutality that women can suffer
at the hands of men. Winner of the Cannes Film Festival Grand
Prix.
Aka: WAY, THE
DRA   109 min (ort 111 min)   VIDrel: ARTIF/20TH   V/h

YOU CAN'T TAKE IT WITH YOU ***
Frank Capra USA U
1938
Jean Arthur, Lionel Barrymore, James Stewart, Edward Arnold,
Mischa Auer, Ann Miller, Spring Byington, Eddie "Rochester"
Anderson, Donald Meek, Dub Taylor, Halliwell Hobbes, Samuel S.
Hinds, Harry Davenport
Entertaining if somewhat corny story of a highly eccentric New
York household, and of how the daughter falls for the son of a
wealthy man. Lacking the wit or warmth of the original play,
this film tends to concentrate far too much on taking swipes at
the world of big business. AA: Pic, Dir.
COM   121 min (ort 126 min) B/W
VIDrel: COLUM/SONOP   V
Boa: play by George S. Kaufman and Moss Hart.

YOU CAN'T WIN 'EM ALL **
Peter Collinson UK PG
1970
Tony Curtis, Charles Bronson, Michele Mercier, Patrick Magee,
Gregoire Aslan, Fikret Hakan, Salih Guney, Tony Bonner, John
Acheson, John Alderson, Leo Gordon
Set in Turkey at the time of the 1922 Civil War, this has two
rival American mercenaries venturing there, and finding them-
selves thrown together when one is obliged to take refuge on a
boat owned by the other. A daft comedy-actioner, it was
directed on location by Collinson, who took over when Howard
Hawks wisely dropped out of directing.
A/AD   96 min (ort 99 min)
VIDrel: ENCORE/SPEAR/COLUM   V

YOU LIGHT UP MY LIFE **
Joseph Brooks USA PG
1977
Didi Conn, Joe Silver, Michael Zaslow, Stephen Nathan, Melanie
Mayron, Amy Letterman, Jerry Keller, Lisa Reeves, John Gowans,
Simmy Bow, Bernice Nicholson, Ed Morgan, Joe Brooks, Marty
Zagon, Martin Gish, Brian Byers
An aspiring woman singer finds her life and work difficult,
until she meets a young man who prompts her into putting
her life in order. A painfully thin story produced, written and
musically supervised by Brooks. Followed by IF EVER I SEE
YOU AGAIN. AA: Song ("You Light Up My Life" – Joseph
Brooks).
DRA   87 min (ort 90 min)   VIDrel: RCA L/A   V

YOU ONLY LIVE TWICE ***
Lewis Gilbert UK/USA PG
1967
Sean Connery, Donald Pleasence, Karin Dor, Tetsuro Tamba, Akiko
Wakabayashi, Bernard Lee, Lois Maxwell, Charles Gray, Mie Hama,
Teru Shimada, Desmond Llewellyn, Tsai Chin, Alexander Knox,
Robert Hutton, Burt Kwouk
The fifth Bond adventure, with Secret Service agent 007 fight-
ing SPECTRE, a nasty organisation bent on world domination.
As always, the novels of Ian Fleming form the basis (if not the
plot) for these films. Set in Japan and memorable for some exotic
and colourful locations. Written by Roald Dahl, it has music by
John Barry and photography by Freddie Young. The spectacu-
lar sets were designed by Ken Adam. Followed by ON HER
MAJESTY'S SECRET SERVICE.
A/AD   112 min (ort 116 min) wScrn cC
VIDrel: MGM/WHV   V/dm
Boa: novel by Ian Fleming.

YOU TALKIN' TO ME? ** (15)
Charles Winkler USA 1987
Jim Youngs, James Noble, Mykel T. Williamson, Faith Ford, Bess Motta, Rex Ryon, Brian Thompson, Alan King, James Noble, Mayo Winkler, Carl D. Parker, Niles Brewster, Boll Wood, Kris Beren, Doug Cox, Webster Williams, Janet Ditz
Two actors go to Hollywood to take their shot at fame and fortune. One is black and the other white, and the latter is obsessed with his idol Robert De Niro and his role in TAXI DRIVER (hence the title). A satire on movie fame, its trappings and the superficiality of it all, that has an uneven script and uncertain direction, yet is distinctive enough to have achieved a minor cult following, despite being not a terribly good film.
COM   92 min (ort 97 min)   SATrel: MOVIE CHANNEL

YOU WERE NEVER LOVELIER *** U
William A. Seiter USA 1942
Fred Astaire, Rita Hayworth, Adolphe Menjou, Leslie Brooks, Adele Mara, Xavier Cugat and his Orchestra, Gus Schilling, Isobel Elsom, Barbara Brown, Larry Parks, Douglas Leavitt, Catherine Craig, Kathleen Howard, Mary Field
Pleasant musical with a Buenos Aires setting as a worried father engages a gambler to woo his daughter and so encourage her to think seriously about the idea of getting married. The second and final screen pairing of Hayworth and Astaire serves as a follow-on to YOU'LL NEVER GET RICH. The dance sequences are a pure delight and the fine score is the work of Jerome Kern and Johnny Mercer. Remade from an earlier Argentinian film "The Gay Senorita".
MUS   93 min (ort 97 min) B/W   VIDrel: COLUM/SONOP V
Boa: story The Gay Senorita by Carlos Olivari and Sixto Pondal Rios.

YOU'LL NEVER GET RICH *** U
Sidney Lanfield USA 1941
Fred Astaire, Rita Hayworth, Robert Benchley, John Hubbard, Osa Massen, Frieda Inescort, Guinn Williams, Donald MacBride, Cliff Nazarro, Marjorie Gateson, Ann Shoemaker, Boyd Davis, Mary Currier, Robert E. Homans
A dance director gets drafted into the Army in the middle of producing a Broadway show but decides to try and complete his preparations, resulting in many breaches of military discipline. Astaire and Hayworth appear together for the first time and make a thrilling dance team, while the Cole Porter score is another asset. Songs include: "Since I Kissed My Baby Goodbye", "So Near And Yet So Far" and "Dream Dancing". Followed by YOU WERE NEVER LOVELIER.
MUS   88 min B/W   VIDrel: COLUM/SONOP V

YOUNG AGAIN ** U
Steven Hilliard Stern USA 1986
Lindsay Wagner, Robert Urich, Jack Gilford, Keanu C. Reeves, Jessica Steen, Jason Nicolof, Jeremy Ratchford, Peter Spence, Jonathan Welsh, Louise Vallance, Vincent Murray, John Friesen, Marshall Perlmutter, Barbara Kyle
Angelic intervention transforms a thirty-year-old bachelor into a teenager, effectively placing his high-school sweetheart beyond reach as she is now twice his age. A cute idea that attempts to cash in on films like BACK TO THE FUTURE and PEGGY SUE GOT MARRIED, but lacking the real talent a 1940s Hollywood studio would have brought to such a film.
COM   85 min mTV   VIDrel: WDV/TECH L/A V
Boa: story by Steven Hilliard Stern and David Simon.

YOUNG AMERICANS, THE *** 18
Danny Cannon UK 1993
Harvey Keitel, Iain Glen, John Wood, Terence Rigby, Keith Allen, Craig Kelly, Thandie Newton, Viggo Mortensen, Sheila Trezise, Miles Peti, Dave Duffy, Chris Adamson, David Doyle, Norman Roberts, Toni Palmer, Steve Aston
Violent tale of gangland warfare in London where title group cause havoc among the local underworld as an American drug dealer attempts to monopolise this trade. An L.A. cop assigned to assist Scotland Yard in this somewhat unequal struggle, has a personal interest as he has scores of his own to settle with this old adversary. Well acted and directed but the relentless violence of the film is a major drawback.
DRA   99 min (ort 108 min)
VIDrel: VCC/DISC/COLUM; ENCORE (LV only) V/sur LV

YOUNG AND INNOCENT *** U
Alfred Hitchcock UK 1937
Derrick De Marney, Percy Marmont, Mary Clare, Edward Rigby, Basil Radford, Nova Pilbeam, George Curzon, John Longden, George Merritt, J.H. Roberts, Jerry Verno, H.F. Maltby, Pamela Carme, Torin Thatcher, Gerry Fitzgerald
When a man finds himself implicated in a murder he attempts to find the real culprit, aided by a young girl with whom he goes on the run. Echoes of THE THIRTY-NINE STEPS abound with a revelation taking place in a memorable nightclub scene. Written by Charles Bennett and Alma Reville with music by Louis Levy.
Aka: GIRL WAS YOUNG, THE
THR   79 min (ort 82 min) B/W
VIDrel: VCC/DISC/COLUM L/A V
Boa: novel A Shilling for Candles by Josephine Tey.

YOUNG AT HEART *** U
Gordon Douglas USA 1954
Frank Sinatra, Doris Day, Gig Young, Ethel Barrymorre, Dorothy Malone, Alan Hale Jr, Robert Keith, Elizabeth Fraser, Lonny Chapman, Frank Ferguson, John Maxwell, Marjorie Bennett, William McLean, Barbara Pepper, Tito Vuolo
Romantic musical comedy all about a cynical musician who falls in love. Sinatra plays the musician, a bitter and self-pitying pianist who unaccountably wins the heart of Day, the daughter of a small town music teacher. Meanwhile, her other two sisters enjoy romances of their own. Less cheerful than Day's usual films, this is a musical remake of the 1938 film "Four Daughter". It was written by Julius J. Epstein and Lenore Coffee. The score is by Cole Porter.
MUS   112 min (ort 117 min)   VIDrel: 4-FRONT/POLYREC V
Boa: novel by Fannie Hurst.

YOUNG AT HEART *** (U)
Allan Arkush USA 1994
Olympia Dukakis, Joe Penny, Philip Bosco, Yannick Bisson, Louis Zorich, Tony Longo, Audrey Landers, Richard Cox, Chelsea Altman, Louis Vanaria, Frank Sinatra, Arlene Meadows, Chandra West, Joseph Griffin, Lynne Cormack, Tony Perri
The story of three generations of one family, all bound together by a strong sense of loyalty, the authority of their elderly matriarch Rose (Dukakis in fine form) and the Italian restaurant where they work. But Rose's husband dies, and the family struggle to hang onto their restaurant, which no longer proves to be so profitable. Despite all this, it re-opens, even getting a visit from Frank Sinatra on opening night. A simple, enjoyable but overly sentimental story.
DRA   89 min mTV   SATrel: SKY MOVIES

YOUNG AVENGER, THE ** 18
HONG KONG 1989
Wang Yu, Tsui Shao Keung, Huang Hsing Shou
A young man working in a funeral parlour is addicted to gambling, and often robs the dead to provide his stake. One night a ghost appears to him and relates such a tale of injustice that he vows to seek revenge on its behalf, and the spirit becomes his martial arts mentor. This unusual fantasy element does much to enliven the standard revenge-seeking plot.
MAR   86 min   VIDrel: IMPENT V

YOUNG BRUCE LEE, THE * 18
HONG KONG 1983
Bruce Lee, Bruce Li (Ho Tsung-Tao), Hon Kwok Choi, Chan Kwok Kue, Shek Kin
After defeat in a match, the young Bruce learns a style called "The Three Cobras" and puts it to good use when his friend is murdered. A feature cobbled together using footage from four obscure Bruce Lee films: "Kid Cheung", "Bad Boy", "Carnival" and "Orphan Ah Sam". The tape is also padded out with a press conference from 1972.
MAR   85 min (ort 90 min)   VIDrel: FABFIL/SPEAR V

YOUNG CONNECTICUT YANKEE IN KING ARTHUR'S COURT, A ** (PG)
R.L. Thomas CANADA/FRANCE/UK 1995
Michael York, Theresa Russell, Nick Mancuso, Philippe Ross, Polly Shannon, Jack Lengedijk, Paul Hopkins, Ian Falconer, David Schaeffer, Michael Nelson, Romauld Weber, Robert Russell, Lisa Flores, Chase Stewart, Christophe Clarke
Ross is an American youngster who is transported back in time and space to England of the 6th century, where he enjoys many

varied adventures with King Arthur, Merlin and some of the Knights of the Round Table. A pleasant variant on the Mark Twain classic, quite engaging and colourful.
JUV   90 min mTV   SATrel: MOVIE CHANNEL
Boa: novel by Mark Twain.

YOUNG EINSTEIN *   PG
Yahoo Serious (Greg Pead) AUSTRALIA   1989
Yahoo Serious, Su Cruickshank, Peewee Wilson, Odile Le Clezio, John Howard, Lulu Finkus, Kaarin Fairfax, Michael Lake, Jonathan Coleman, Ray Forgo, Johnny McCall, Michael Blaxland, Terry Pead, Alice Pead, Tom Harvey
Silly, self-indulgent and childish, but this crazy screwball comedy has a lot going for it. Written by David Roach and Serious (who also produced), it details the career of young Albert, who when not discovering physics theorems, finds time to invent rock 'n' roll and surfing. The flimsy plot concerns the retrieval of his nuclear brewery before it blows the world up. Hilarious after a few beers and understandably a great success Down Under.
COM   87 min (ort 91 min)   VIDrel: MGM/WHV V/sur

YOUNG FRANKENSTEIN *   15
Mel Brooks USA   1974
Gene Wilder, Marty Feldman, Teri Garr, Peter Boyle, Madeline Kahn, Cloris Leachman, Gene Hackman, Richard Hadyn, Kenneth Mars, Leon Askin, Liam Dunn, Oscar Beregi Jr, Lou Cutell, Danny Goldman, Richard Roth, Rusty Blitz
An uneven spoof on those monster films of the 1930s, with Wilder as young brain surgeon Frederick Frankenstein returning to his roots in Transylvania. Not terribly funny though Hackman's spoof on the blind-man sequence from THE BRIDE OF FRANKENSTEIN is a joy, and the use of sets and lab equipment from the 1930s was a masterstroke. Scripted by Brooks and Wilder with photography by Gerald Hirschfield.
COM   106 min B/W   VIDrel: 20TH/TECH L/A V

YOUNG GUNS *   18
Christopher Cain USA   1988
Emilio Estevez, Kiefer Sutherland, Lou Diamond Phillips, Dermot Mulroney, Charlie Sheen, Casey Siemaszko, Jack Palance, Terry O'Quinn, Terence Stamp, Sharon Thomas, Brian Keith, Patrick Wayne
Set in New Mexico in 1878, this story tells of how a British ranch-owner befriends six young rebels and hires them to look after his property. When he is murdered and the law fails to act, the boys set out to avenge him, but their vendetta soon turns into a bloody rampage. Branded as outlaws they now find themselves the objects of a huge manhunt. All too soon this enjoyable and successful action film was followed by the inevitable, exploitative sequel.
WES   102 min (ort 107 min) wScrn (ENCORE/SPEAR)
VIDrel: VCC/DISC; ENCORE/SPEAR V/sur

**YOUNG GUNS 2: BLAZE OF GLORY **   15
Geoff Murphy USA   1990
Emilio Estevez, Kiefer Sutherland, Lou Diamond Phillips, Christian Slater, William Petersen, Alan Ruck, Balthazar Getty, R.D. Call, James Coburn, Jack Kehoe, Robert Knepper, Tom Kurlander, Viggo Mortensen, Leon Rippy, Scott Wilson
This sequel to the first film finds Billy the Kid and his gang fleeing for their lives to Mexico, with Pat Garrett and his posse in hot pursuit. The fine music and photography are enjoyable in themselves, but hardly compensate for the weak script and deliberate angling of the film towards the teen market. The theme song "Blaze Of Glory" was composed and performed by Jon Bon Jovi.
WES   99 min (ort 105 min) subH
VIDrel: 20TH/TECH; ENCORE (LV only)   V/sur LV

**YOUNG HERO, THE **   18
Lo Chia Po HONG KONG   1982
Wang Chiang, Wang Shia, Huang Cheng Li, Kwon Young Moon, Yuan Wu, Yuan Chu
Two powerful criminal clans struggle for control of the same territory, with no holds (or kicks) barred. Fairly standard kung fu actioner.
MAR   88 min   VIDrel: IMPENT V

**YOUNG INDIANA JONES: TRAVELS WITH FATHER **   (PG)
Deepa Mehta/Michael Schultz USA   1996
Corey Carrier, Sean Patrick Flanery, Lloyd Owen, Ruth De Sosa,

Margaret Tyzack, Michael Gough, Robyn Lively, George Yadsoumi, Raad Rawi, Thanassis Sakellarious, George Jackos, Jiri Konicke, Milan Riehs, Jiri Knot, Milan Gul
Indiana visits his widowed dad, but finds him in a state of severe depression, but gets him to snap out of it by recalling some of their previous adventures together. Rather tiresome feature-length episode from the TV series, which itself was inspired by the central character from those RAIDERS OF THE LOST ARK and INDIANA JONES movies.
JUV   93 min mTV   SATrel: MOVIE CHANNEL

**YOUNG INDIANA JONES: TREASURE OF THE PEACOCK'S EYE **   (PG)
Carl Schultz USA   1996
Sean Patrick Flanery, Ronny Coutteure, Adrian Edmondson, Jayne Ashbourne, Tom Courtney, Pip Torrens, William Osborne, Anthony Chin, Alice Lau, Karl Seth, Matthew Solon, Colleen Passard, Frederic Treves, Riz Abassi, Nick Lucas
One of several feature-length entries from the TV series, built around the adventures of the title figure. In this one, he embarks on a quest to locate the whereabouts of a stolen diamond. An assembly-line product. Several of the other episodes have become available.
JUV   91 min mTV   SATrel: MOVIE CHANNEL

**YOUNG IVANHOE **   (U)
Ralph L. Thomas CANADA/FRANCE/UK   1995
Stacy Keach, Margot Kidder, Nick Mancuso, Kris Holdenried, Rachel Blanchard, Matthew Daniels, Tom Rack, James Bradford, Louis Vicnent, Ian Falconer, Heather Ondersma, Marek Vasut, Gavin Stewart, Lisa Flores, Gordon Lovitt, Steven Fisher
When Ivanhoe's father is taken prisoner by a Norman count, the youngster with his head full of dreams about chivalry and noble deeds, sets out to rescue his dad. This involves him in thwarting a plot to overthrow the King Richard the Lionheart, who is (as ever) out of the country at the time. Filmed in Slovakia, this is a wholly conventional rehash of various over-familiar themes.
FAN   89 min mTV   SATrel: SKY MOVIES

**YOUNG LADY CHATTERLEY **   18
Alan Roberts USA   1976
Harlee McBride, Peter Ratray, William Beckley, Joi Stanton, Mary Forbes, Patrick Wright, Henry Charles, Edgar Daniels, Lawrence Montaigne, Ray Myles, Anne Michelle, Lindsay Freeman, Kelly Ann Page, Ray Martin, Michael Hearne
Lady C's niece inherits her estate and coming across her diary, decides to follow in her footsteps. A sequel followed in 1984.
A   100 min Cut (5 sec plus cut at film release)
VIDrel: CREMED/LABY V

**YOUNG LADY CHATTERLEY 2 **   18
Alan Roberts USA   1984
Harlee McBride, Sybil Danning, Adam West, Brett Clerk, Monique Gabrielle, Ed Quinlan, John St Angelo, Stephen Kean Mathews, Alex Sheafe, Mike Reynolds, Joi Staton, Brandon Brady, Henry Charles, Winston Richard, Barbara Stewart
Sequel to the 1976 film from the same director and with a largely identical cast. Here, Lady C fights to save the family estate, but still finds time for more athletic pursuits.
A   82 min (ort 87 min)   VIDrel: 4-FRONT/POLYREC/MED V

YOUNG LIONS, THE **   PG
Edward Dmytryk USA   1958
Marlon Brando, Montgomery Clift, Dean Martin, Hope Lange, Barbara Rush, Mai Britt, Maximilian Schell, Lee Van Cleef, Arthur Franz, Hal Hal Baylor, Dora Doll, Liliane Montevecchi, Parley Baer, Richard Gardner, Herbert Rudley
A study of WW2 as seen from both sides with Martin and Clift as American soldiers, and Brando quite superb as a Nazi officer who has come to doubt his most cherished beliefs. Written by Edward Anhalt with a fine score by Hugo Friedhofer and photography by Joe MacDonald.
DRA   167 min B/W   VIDrel: 20TH/TECH V/h
Boa: novel by Irwin Shaw.

YOUNG MAN WITH A HORN **   PG
Michael Curtiz USA   1950
Kirk Douglas, Lauren Bacall, Doris Day, Juano Hernandez, Hoagy Carmichael, Jerome Cowan, Nestor Paiva, Mary Beth Hughes, Orley Lindgren, Walter Reed, Jack Kruschen, Alex Gerry, Jack Shea, James Griffith, Deanm Hearne
A trumpet player is torn between his love for two women, one

nice the other nasty. The story is based on the life of Bix Biederbecke, with Harry James dubbing for Douglas. An excellent and beautifully acted film. Written by Edmund H. North and Carl Foreman with music by Ray Heindorf. On release in the UK it was retitled, the American title not being acceptable at that time.
Aka: YOUNG MAN OF MUSIC
DRA   110 min (ort 112 min) B/W   VIDrel: MGM/WHV L/A V
Boa: novel by Dorothy Baker.

YOUNG MASTER, THE ***   15
Jackie Chan HONG KONG   1979
Jackie Chan, Yuan Biao, Lily Li (Li-Li Li), Wei Pei, Shek Kin, Whang In Sik
A man tries to prevent his best friend from leading a life of crime and in the process finds himself on the wrong side of the law. He now is obliged to clear his name and bring the real criminals to justice.
MAR   105 min Cut wScrn subs; 87 min dubbed
VIDrel: MIA/DISC V

YOUNG NURSES IN LOVE **   18
USA
Nicole Blanc, Beverly Bliss, Crystal Blue
Three young girls make a pact to find themselves rich husbands, but despite their intentions of finding themselves some rich young doctors, are unable to resist the charms of just about any handsome men they happen to encounter. A harmless sex-romp, that panders to a recurrent male fantasy, and serves as a vehicle for three very attractive girls to play-act in a hospital setting.
A   46 min   VIDrel: IMPENT V

YOUNG NURSES IN LUST *   (18)
Harold Lime/Jane Walters USA   1995
Misty Rain, Vanessa Chase, Olivia, Barbara Doll, P.J. Sparxx, Jon Dough, Marc Wallice, Peter North
A porno version of all those medical dramas, with the patients, nurses and doctors perfecting the sexual techniques instead of fulfilling their usual roles. One of a goodly number of such films, predictable in both development and outcome.
A   75 min   SATrel

YOUNG PHILADELPHIANS, THE **   PG
Vincent Sherman USA   1959
Paul Newman, Barbara Rush, Alexis Smith, Billie Burke, Brian Keith, Robert Vaughn, Adam West, Diane Brewster, John Williams, Otto Kruger, Paul Picerni, Robert Douglas, Frank Contory, Adam West, Fred Esiely, Richard Deacon
A young man, brought up to believe that his father was a member of a wealthy and prominent Philadelphian family, graduates from Princeton law school and embarks on a career in law. Utterly ruthless in his determination to reach the top, he is resorts to intrigue and seduction to further his plans. A tepid and melodramatic stew whose only point of interest is a superb performance from Newman. Sadly, the film's prosaic direction is its major drawback.
Aka: CITY JUNGLE, THE
DRA   136 min B/W   VIDrel: WHV V

YOUNG POISONER'S HANDBOOK, THE ***   15
Benjamin Ross FRANCE/GERMANY/UK   1994
Hugh O'Conor, Antony Sher, Ruth Sheen, Roger Lloyd Pack, Charlotte Coleman, Paul Stacey, Samantha Edmonds, Vilma Hollingberry, Tobias Arnold, Frank Mills, Charlie Creed-Miles, Arthur Cox, John Thomson, Jean Warren, Simon Kunz
A fourteen-year-old with an obsessive interest in poisons starts experimenting on family and friends. Initially this plays like a black comedy, but after daring us to laugh at the antics of this demented monster, it offers us unsettling moments of revulsion. A stylish film, it skilfully develops its story to its inevitable conclusion. Based to some extent on the real-life career of 1960s poisoner Graham Young. This was the feature debut for Ross.
COM   95 min   VIDrel: ELPIC/POLYREC V/sh

YOUNG SHERLOCK HOLMES AND THE PYRAMID OF FEAR ***   PG
Barry Levinson USA   1985
Nicholas Rowe, Alan Cox, Sophie Ward, Anthony Higgins, Nigel Stock, Susan Fleetwood, Freddie Jones, Michael Hordern, Earl Rhodes, Brian Oulton, Roger Ashton-Griffith, Patrick Newell, Donald Eccles, Matthew Ryan, Jonathan Lacy
An interesting idea; Holmes and Watson meet at public school

and together solve their first case – the deaths of the elderly members of a previous archaeological expedition to Egypt. Unfortunately the careful detail and intellectual appeal of the story is drowned in a welter of spectacular effects, reducing the film to the level of RAIDERS OF THE LOST ARK. Written by Chris Columbus, with Steven Spielberg as one of the executive producers.
Aka: YOUNG SHERLOCK HOLMES
A/AD   105 min   VIDrel: CIC/SONOP   V/sh

YOUNG SOUL REBELS **   18
Isaac Julien UK   1991
Valentine Nonyela, Mo Sesay, Dorian Healy, Frances Barber, Sophie Okonedo, Jason Durr, Gary McDonald, Debra Gillet, Eamon Walker, James Bowyers, Billy Braham, Wayne Norman, Danielle Scillitoe, Ray Shell, Nigel Harrison, John Wilson
The winner of the Critics' Prize at Cannes, this film is set in 1977, when a couple of black childhood friends find themselves running a pirate radio station. One of the friends discovers that he inadvertently recorded a murder on his portable cassette-player, but soon finds that he has been set up. Sadly, the film then promptly degenerates, and its flawed attempts to explore issues of racism and violence are badly hampered by stereotyping.
DRA   100 min (ort 105 min)   VIDrel: POLY/POLYREC L/A V

YOUNG STRANGER, THE ***   U
John Frankenheimer USA   1958
James McArthur, Kim Hunter, Marian Seldes, James Daly. James Gregory, Whit Bissell, Jeff Silver, Jack Mullaney, Eddie Ryder, Jean Corbett, Gary Vinson, Charles Davis, Terry Kelman, Edith Evanson, Tom Pittman, Howard Price
The son of a prominent film executive gets into a fight with a movie house manager and gets himself arrested for hitting the man. His father manages to get the charges dropped but refuses to listen to his son's claim that he was acting in self-defence. An absorbing study of a father-and-son relationship and its complexities, acted with conviction by a fine cast.
DRA   84 min B/W   VIDrel: ODY/SONOP   V
Boa: TV play Deal a Blow by Robert Dozier.

YOUNG WARRIORS, THE **   18
Lawrence D. Foldes USA   1983
Ernest Borgnine, Richard Roundtree, Lynda Day George, James Van Patten, Anne Lockhart, Mike Norris, Dick Shawn, Tom Reilly, Linnea Quigley, Ed DeStefane, John Alden, Britt Helfer, Don Hepner, April Dawn, Nels Van Patten
After his sister is raped and murdered, a young boy and his college chums decide to take their own revenge, in this violent exploitation movie. Written by Foldes and Russell W. Colgin, and a sequel of sorts to MALIBU HIGH.
DRA   98 min (ort 103 min)   VIDrel: GUILD/SONOP L/A V

YOUNG WINSTON ***   PG
Richard Attenborough UK   1972
Simon Ward, Robert Shaw, Anne Bancroft, John Mills, Jack Hawkins, Anthony Hopkins, Patrick Magee, Edward Woodward, Ian Holm, Peter Cellier, Ronald Hines, Raymond Huntley, Russell Lewis, Pat Heywood, Laurence Naismith
A careful and enjoyable if somewhat glamorised account of the early years of Churchill, told in the form of a series of flashbacks and taking us up to his election to Parliament. The semi-narrative style and use of contrived "interviews", are flaws in an otherwise excellent film. Ward is well-chosen (he also narrates) and the film was scripted and produced by Carl Foreman.
DRA   120 min (ort 157 min)   VIDrel: VCC/DISC/COLUM V
Boa: book My Early Life by Winston Spencer Churchill.

YOUNGBLOOD **   15
Peter Markle USA   1986
Rob Lowe, Patrick Swayze, Cynthia Gibb, Ed Lauter, Eric Nesterenko, George Finn, Fionnula Flanagan, Jim Youngs, Ken James, Keanu Reeves, Harry Spiegel, Peter Faussett, Walter Boone, Martin Donlevy, Rob Spaiensze, Bruce Edwards
Dull account of a country lad who joins a Canadian hockey team and receives a fierce baptism. Writer/director Markle adds a few contrived comedy bits and some love interest, but this remains very much a production-line effort, and one that's very clearly aimed at the teen market.
DRA   106 min (ort 110 min)   VIDrel: MGM/WHV   V/sur

YOUNGER AND YOUNGER ** 15
Percy Adlon GERMANY/USA 1993
Donald Sutherland, Julie Delpy, Lolita Davidovich, Brendon Fraser, Sally Kellerman, Linda Hunt, Pit Kruger, Jay Brooks, Sabrina Weber, Milton Clark Jr., Davida Williams, Nicholas Quinn, Ellen Blake, Chris Warfield, Erick Weiss
A chronic womaniser loses his wife in a heart attack after she witnesses him being unfaithful, forcing him to assume control of their family business. However, he proves totally incompetent and as things go to rack and ruin he is haunted by visions of his dead wife in which she grows increasingly young with every apparition. A strained and unfunny comedy-drama, whose fine performances offer the only interest.
DRA 93 min VIDrel: ARROW/RTM V

Z

Z *** 15
Costa-Gavras ALGERIA/FRANCE 1969
Yves Montand, Irene Papas, Jean-Louis Trintignant, Jacques Perrin, Francois Perier, Charles Denner, Bernard Fresson, Jean Bouise, Georges Geret, Magali Noel, Coltilde Joano, Maurice Baquet, Gerard Darrieu, Jose Artur, Van Doude
The real-life assassination of a Greek opposition deputy in 1963 forms the basis for this political thriller set in an unidentified Mediterranean country. After the killing, an examining magistrate begins to unravel the threads of evidence that point towards a right-wing conspiracy. A compelling film that concentrates on the more exciting aspects of the story, it was also a major box-office hit. See also STATE OF SIEGE. AA: Foreign, Edit (Francoise Bonnot).
THR 121 min (ort 128 min) VIDrel: ARROW/RTM V/s
Boa: novel by Vassili Vassilikos.

ZABRISKIE POINT * 15
Michelangelo Antonioni USA 1970
Mark Frechette, Daria Halpin, Rod Taylor, Paul Fix, Harrison Ford, G.D. Spradlin, Bill Garaway, Kathleen Cleaver, The Open Theater Of Joe Chaikin
A boring and utterly meaningless series of loosely connected images that passes for a film, following a couple and their meanderings in the desert. Endless slow-motion shots were raved over by certain critics who, as ever, saw deep hidden meanings, too obscure to be spotted by lesser mortals. The exasperating script is by Antonioni and Sam Shepard. Harrison Ford is credited but his scenes were deleted.
DRA 112 min VIDrel: MGM L/A V

ZACHARIAH *** 15
George Englund USA 1971
John Rubinstein, Pat Quinn, Don Johnson, Country Joe and The Fish, Elvin Jones, The James Gang, Dick Van Patten, Doug Kershaw, William Challee, Robert Ball, White Lightnin', The New York Rock Ensemble, Lawrence Kubik
A satirical and offbeat rock musical with a Western setting that takes place in the 1870s, with two young men falling in with a bunch of rock 'n' rolling bandits. An odd little work, containing elements of a morality play, with no real direction but a host of songs to suit every taste. Now a minor cult classic. Co-scripted by members of the Firesign Theater.
MUS 93 min VIDrel: VCC/DISC V

ZANDALEE ** 18
Sam Pillsbury USA 1991
Nicolas Cage, Judge Reinhold, Erika Anderson, Viveca Lindfors, Aaron Neville, Joe Pantoliano, Steve Buscemi, Ian Abercrombie, Marisa Tomei, Jo-El Sonnier, Newell Alexander, Blaise Delacroix, Elliot Keener, Richard Greenberg
A young poet finds himself deserted by the muse, and begins to withdraw into his shell. This places a strain on his fragile marriage, which soon begins to deteriorate when his wife meets and falls in love with one of his childhood friends. The two embark on a passionate affair that seems doomed to end in tragedy for all concerned. A well-acted but essentially shallow drama of very limited appeal and with a singularly contrived ending.
DRA 100 min VIDrel: 4-FRONT/POLYREC/ITC V/h

ZAPPED AGAIN ** 15
Douglas Campbell USA 1990
Todd Eric Andrews, Kelli Williams, Ross Harris, Ira Heiden, Linda Larkin, Linda Blair, Reed Rudy, Maria McCann, Karen Black, Lyle Alzado, Sue Ann Langdon, M.K. Harris, Michael K. Colvar, Brent Hinkley, David Donah, Karen Black
A belated sequel to the 1982 film that copies much of its plot. A young high school student gains telekinetic powers by swallowing a magical potion, and uses his new-found abilities in such activities as undressing girls and helping his pals fight off the bullies trying to take over their science club meeting room. A bright and innocent juvenile romp that offers little more than a few modest chuckles.
COM 88 min (ort 94 min) VIDrel: COLUM/SONOP V

ZARA'S REVENGE * 18
Patti Rhodes USA 1991
Zara Whites, Lois Avers, Brigitte Monroe, Sikki Nixx, James Lewis, Buck Adams, Joey Murphy, Tracey Wynn
Zara is on the look-out for a new Master to inflict on her the corporal punishment and discipline she adores. An unappealing examination of domination and bondage games, this episodic film largely follows the life of the title figure, her encounters with like-minded individuals and her flirtation with her role as a dominatrix.
A 42 min (ort 75 min) VIDrel: MIA/DISC V

ZARDOZ *** 15
John Boorman UK 1974
Sean Connery, Charlotte Rampling, Sara Kestelman, Sally Anne Newton, John Alderton, Niall Buggy, Bosco Hogan, Reginald Jarman, Bairbe Dowling, Christopher Casson, Jessica Swift
In 2293 most of the world has reverted to barbarism following a cataclysm, and civilisation is preserved in small enclaves inhabited by self-indulgent groups. An inhabitant of the outer wastelands gains entry to one of these sheltered enclaves, ultimately finding that he was brought there on purpose. A visually impressive film whose many ideas needed a stronger story. Written by Boorman. The haunting scoring of Beethoven's 7th is by David Munrow.
FAN 105 min VIDrel: 20TH/TECH L/A V

ZAZIE DANS LE METRO *** 15
Louis Malle FRANCE 1960
Catherine Demongeot, Philippe Noiret, Carla Marlier, Vittorio Caprioli, Hubert Deschamps, Annie Fratellini, Yvonne Clech, Nicholas Bataille, Jacques Dufilho, Odette Picquet, Marc Doelnitz, Jacques Gheusi, Hubert Deschamps
Eleven-year-old Zazie, who appears wise beyond her years (and rather vulgar) arrives in Paris where she's to spend a couple of days in the care of her uncle, who makes his living as a female impersonator. Her only wish is to ride on the subway, but this is never fulfilled as an increasingly insane set of events draw her and her uncle into some crazy adventures. An ingenious, surreal, outlandish but now extremely dated French New Wave movie.
Aka: ZAZIE; ZAZIE IN THE SUBWAY; ZAZIE IN THE UNDERGROUND
COM 88 min (ort 92 min) VIDrel: ELPIC/POLYREC V
Boa: novel by Raymond Queneau.

ZED AND TWO NOUGHTS, A *** 15
Peter Greenaway HOLLAND/UK 1986
Andrea Ferreol, Brian Deacon, Eric Deacon, Joss Ackland, Frances Barber, Jim Davidson, Geoffrey Palmer, Agnes Brulet, Guusje Van Tilborgh, Ken Campbell, Wolf Kahler, Gerard Thoolwn, David Attenborough (narration)
Twin brothers lose their wives in an accident, and become obsessed with the driver of the car, who lost her leg in the accident. A strange and obsessive film that almost defies description, revolving around the study of decay. Beautifully photographed by Sacha Vierny and well worth a look, though never rising above the level of an interminable intellectual game.
Aka: ZOO: A ZED AND TWO NOUGHTS
DRA 112 min (ort 115 min) VIDrel: PAL L/A V/h

ZEGUY: PARTS 1 AND 2 ** PG
JAPAN 199-
Two girls are passengers on a bus that is accidentally projected into another dimension and receive strange powers as a result, in this average anime offering.
ANIM 65 min (ort 74 min) dubbed
VIDrel: MANGA/SONOP V

ZELIG *** PG
Woody Allen USA 1983
Woody Allen, Mia Farrow, Garrett Brown, Stephanie Farrow, Will Holt, Sol Lomita, John Buckwalter, Marvin Chatinover, Stanley

Swerdlow, Paul Nevens, Ellen Garrison, Mary Louise Wilson, Howard Erskine
A fake documentary that uses photomontage techniques, this follows the life of a 1920s character who can change his appearance to merge with whatever surroundings he is placed in. More clever than funny, the photography of Gordon Willis and music of Dick Hyman are assets. As ever, Allen writes and directs.
COM   76 min B/W/Col   VIDrel: WHV   V

ZELLY AND ME ***   15
Tina Rathborne USA   1988
Isabella Rossellini, Glynis Johns, Alexandra Johnes, Kaiulani Lee, David Lynch, Joe Morton, Courtney Vickery, Lindsay Dickson, Jason McCall, Aaron Boone, Lee Lively, John Raynes, Lynn Hallowell, Michael Stanton Kennedy
A perceptive but highly introspective study of the strains suffered by a wealthy but over-protected orphan. Intense and compelling but ultimately unsatisfying.
DRA   88 min (ort 90 min)   VIDrel: RCA L/A   V

ZENOBIA **   U
Gordon Douglas USA   1939
Oliver Hardy, Harry Langdon, Billie Burke, Alice Brady, James Ellison, Jean Parker, June Lang, Chester Conklin, Stepin Fetchit, Hattie McDaniel
In a sleepy Southern town, the local doctor finds himself saddled with the care of a performing elephant, for after he agrees to treat it the grateful animal refuses to leave him. Hardy's only solo starring venture (made during a break-up of the Laurel and Hardy contract) is a fairly straightforward comedy, rather dated but mildly amusing.
Aka: ELEPHANTS NEVER FORGET
COM   69 min (ort 71 min) B/W   VIDrel: VISVID/POLYREC   V

ZERAM **   18
Keita Ameniya JAPAN   1991
Yuko Moriyama, Yukihiro Hotanu, Kunihiko Ida
A female bounty hunter goes after a dangerous alien that has left a trail of murdered victims behind it. However, she is forced to carry her pursuit into another dimension. An unremarkable space opera of the more juvenile kind.
FAN   97 min dubbed   VIDrel: MANGA/SONOP   V/sh

ZERO DE CONDUITE **   PG
Jean Vigo FRANCE   1933
Jean Daste, Louis Lefevre, Gilbert Pruchon, Robert Le Flon, Delphin, Coco Goldstein, Constantin Kelber, Gerard De Bedarieux
Four boys at an oppressive boarding school organise a rebellion, in this anarchic and surreal film, that is told from the viewpoint of the pupils, all of whom see themselves as inmates trapped in a meaningless system of conventions and rules. A sporadically amusing experimental trifle that ill deserves the lavish praise heaped upon it. After its first showing, it was banned until 1945, and possibly inspired IF. Vigo's only other available film is L'ATALANTE.
Aka: ZERO FOR CONDUCT
DRA   41 min (ort 49 min) B/W   VIDrel: ARTIF/20TH   V

ZERO PATIENCE **   18
John Greyson CANADA   1993
John Robinson, Norman Fauteux, Dianne Heatherington, Richardo Keens-Douglas, Bernard Behrens, Charlotte Boisjoli, Marla Lukofsky, Michael Callen, Brenda Kamino, Scott Hurst, Von Flores, Duncan McIntosh, Cassel Miles, Peggy Baker
A musical built around gay themes, the 19th century figure of explorer Richard Burton and "patient zero", the first individual alleged to have brought the AIDS virus North America. A complex and absorbing work.
MUS   96 min (ort 100 min)   VIDrel: DTK/RTM   V

ZERO POPULATION GROWTH **   (PG)
Michael Campus UK   1972
Oliver Reed, Geraldine Chaplin, Don Gordon, Diane Cilento, Aubrey Woods, David Markham, Sheila Reid, Bill Nagy, Lotte Tholander, Ditte Maria, Wayne John Rhodda, Lone Lindorff, Belinda Donkin, Birgitte Federspiel, Paul Sceon
In an over-populated world of the future, tough government laws are used to restrict childbirth and the penalty for flouting them is death. In spite of this a couple have an illegal baby, and threatened with exposure by their neighbours are forced to share it with them, but ultimately flee to a remote island. An

unconvincing affair, made faintly ludicrous by the use of android-dolls that serve as substitutes in a child-starved world. See also FORTRESS.
Aka: Z.P.G.
FAN   95 min   VIDrel: L/A   V

ZERO TOLERANCE **   18
Joseph Merhi USA   1993
Robert Patrick, Mick Fleetwood, Titus Welliver, Kristen Meadows, Barbara Patrick, Billy Hufsey, Miles O'Keefe
After he is assigned to the case of a Mexican drug dealer, the family of an FBI agent are brutally murdered. The agent promptly heads for Las Vegas, where he sets about having his revenge on those responsible. Quite a nasty revenge yarn, full of gunfire and senseless slaughter. However, Patrick as the agent does very well with the overheated script, and almost makes this piece of nonsense believable.
DRA   85 min (ort 90 min)   VIDrel: NEWAGE L/A   V

ZERTIGO DIAMOND CAPER, THE **   U
Paul Asselin USA   1980
Adam Rich, David Groh, Jane Elliott, Jeffrey Tambor
When a priceless diamond is stolen from a museum, a young blind boy, who was present at the time, finds that his museum administrator mother is suspected. He embarks on a scheme to clear her name.
A/AD   45 min   VIDrel: START/DISC   V

ZETA ONE *   18
Michael Cort UK   1969
Anna Gael, Valerie Leon, Robin Hawdon, Yutte Stensgaard, Charles Hawtrey, James Robertson Justice, Brigitte Skay, Dawn Adams, Rita Webb, Steve Kirby, Yolande Del Mar, Rose Howlett
Almost certainly inspired by BARBARELLA, this unutterably foolish British sex farce tells of an alien race of Amazons who are searching for available men to propagate their race. Luckily, British Intelligence is soon on to them, having learned of their dastardly plans when they got a Soho stripper to go undercover and spy on these women. Barely watchable softcore nonsense, enlivened by a parade of extremely attractive girls, it is of nostalgia value only.
FAN   84 min (ort 82 min)   VIDrel: JEZ/RTM   V

ZIEGFELD FOLLIES ***   U
Vincente Minnelli USA   1946
William Powell, Judy Garland, Lucille Ball, Fred Astaire, Fanny Brice, Lena Horne, Red Skelton, Victor Moore, Virginia O'Brien, Cyd Charisse, Edward Arnold, Gene Kelly, Esther Williams, James Melton, Marion Bell, Avon Long
A musical extravaganza in which Ziegfeld mounts a spectacular revue in Heaven (!). Powell plays Ziegfeld, with a good supporting cast who mount a number of entertaining numbers. Of variable quality but bursting at the seams with life. Some segments were directed by George Sidney, Norman Taurog, Roy Del Ruth and others.
MUS   105 min (ort 109 min)   VIDrel: MGM/WHV   V/dm

ZOLTAN, HOUND OF HELL *   18
Albert Band USA   1977
Jose Ferrer, Reggie Nalder, Michael Pataki, Jan Shutan, Libbie Chase, John Levin, Arleen Martell, Simmy Bow, Jojo D'Amore, Roger Schumacher, Katherine Fitzpatrick, Cleo Harrington
The Count's dead servant comes back to life and tries to find the last member of the Dracula family. This takes him and the Count's dog, which has acquired the tastes of its dead master, on a trip to Los Angeles. A silly, tedious and totally unfrightening piece of nonsense.
Aka: DRACULA'S DOG; ZOLTAN; ZOLTAN, HOUND OF DRACULA
HOR   83 min   VIDrel: WHV   V/h

ZOMBIE FLESHEATERS ***   18
Lucio Fulci ITALY   1979
Tisa Farrow, Ian McCulloch, Richard Johnson, Al Cliver (Pier Luigi Conti), Auretta Gay, Stefania D'Amerio, Olga Karlatos, Ugo Bologna, Monica Zanchi
A woman travels to a remote island from which her father's boat returned with only zombies onboard and discovers a zombie epidemic. Though unlikely to ever be available in an uncut form in the UK, this unusual shocker has now been

released in a censored form that matches the version originally made available in the cinema.
Aka: ISLAND OF THE DEAD, THE; ISLAND OF THE LIVING DEAD, THE; ZOMBI 2; ZOMBIE; ZOMBIE FLESH-EATERS; ZOMBIE: THE DEAD AMONG US; ZOMBIES 2
HOR   85 min (Cut at UK cinema release – ort 98 min) wScrn
VIDrel: VIPCO/SGSVID   V/h

ZOMBIE HIGH *   15
Ron Link USA   1987
Virginia Madsen, Richard Cox, Kay E. Kuter, James Wilder, Sherilyn Fenn, Paul Feig, T. Scott Coffey, Paul Williams, Henry Sutton, Clare Carey, John Sack, Christopher Crews, Susan Barnes, Abigail Hanness, Arvid Holmberg
Ridiculous shocker with the students at a sleepaway academy being lobotomised, by their teachers and turned into zombies. Sit through this one and you'll think it happened to you.
HOR   87 min   VIDrel: FIRST/SONOP L/A   V

ZOMBIE ISLAND MASSACRE *   18
John N. Carter USA   1983
David Broadnax, Rita Jenrette, Ian McMillan, Tom Cantrell, Debbie Ewing, Tom Fitzsimmons, Diane Clayre Holub, George Peters, Dennis Stephenson, Kristina Wetzel, Harriet Rawlings, Christopher Ferris, Ralph Monaco
Fourteen tourists take a trip to a Caribbean island but far fewer return, in this bottom-of-the barrel horror effort.
Aka: LAST PICNIC, THE; PICNIC, THE
HOR   85 min (ort 89 min)   VIDrel: TROMA/RTM   V

ZOMBIE NOSH **   18
Bill Hinzman USA
John Mowod, Leslie Ann Wick, Kevin Kindlin, Lisa Smith, Rick Billock
Despite the title this is no comedy, and tells of a group of kids out for a picnic who are attacked by zombies, becoming the "nosh" as far as these latter are concerned.
HOR   86 min (ort 88 min)   VIDrel: VIPCO/SGSVID   V/h

ZOMBIES: DAWN OF THE DEAD ***   18
George A. Romero USA   1979
David Emge, Ken Foree, Scott H. Reiniger, Gaylen Ross, Tom Savini, David Early, George A. Romero, Richard France, Howard Smith, Daniel Dietrich, Fred Baker, David Crawford, Jim Baffico, Rod Stouffer, Jese Del Gre, Clayton McKinton
The zombie community increases and threatens to engulf the entire US population. Four people take refuge in a barricaded shopping mall. The grisly effects and genuinely frightening moments should please zombie-lovers everywhere. A film of considerable power if not charm, and both a sequel and a semi-remake of the 1968 film NIGHT OF THE LIVING DEAD. Followed by DAY OF THE DEAD.
Aka: DAWN OF THE DEAD
HOR   120 min; 137 min (director's cut – ort 142 min)
VIDrel: BMGREC/BMGVID   V

ZONE, THE *   18
Barry Zetlin USA   1995
Robert Davi, Alexander Godunov, Ben Gazzara, Lara Harris, David Gautreaux, Geza Kaszas, Patricia Rive, Kathleen Gati, Robin Dalglish, Josef Szekhelyi, Karoly Korogani, Balazs Galko, Akos Istvan Sinko, Tibor Felszeghy, Marta Bako
A former Gulf War hero and special forces agent now lives a life of luxury as a drug smuggler. But this has made him vulnerable, a position the CIA can exploit by framing him for murder, in order to blackmail him into taking part in a hazardous mission to destroy a nuclear weapons plant in a country controlled by a dangerous tyrant. Once there, he gets involved

with the tyrant's mistress. A turgid film of much tedium, it was shot in Hungary.
A/AD   90 min   VIDrel: GUILD/FOXVID   V

ZONE TROOPERS **   15
Danny Bilson USA   1985
Tim Thomerson, Timothy Van Patten, Art La Fleur, Biff Manard, Peter Boom, William Paulsen, Max Turilli, Eugene Brell, John Leamer, Bruce McGuire, Mike Manderville, Alviero Martin, Archille Brunini, Ole Jorgensen, Peter Hinz
In Italy in 1943 an American Army unit is trapped behind German lines, until it receives help from aliens off a ship that has crash-landed in the woods. The soldiers are now faced with the task of preventing the aliens being captured by the Germans. Little more than a gimmicky WW2 story, in which the plain fact of aliens being around during the war rapidly becomes somewhat irrelevant.
FAN   82 min (ort 86 min)   VIDrel: EIV/SONOP   V

ZORBA THE GREEK ****   PG
Michael Cacoyannis USA   1964
Anthony Quinn, Alan Bates, Irene Papas, Lila Kedrova, Sofiris Moustakas, Anna Kyriakou, Eleni Anousaki, George Viyadjis, Takis Emmanuel, George Foundas
Excellent adaptation of a novel about a stuffy young Englishman, who becomes involved with the title character on his arrival in Crete to open a family mine. Quinn is excellent as Zorba, a man with a sad past but a wonderful zest for life, which he attempts to instil into Bates. A rich, lively and joyful film. Written by Cacoyannis with a memorable Mikis Theodorakis score. AA: S. Actress (Kedrova), Cin (Walter Lassally), Art/Set (V. Fotopoulos).
DRA   136 min (ort 146 min) B/W   VIDrel: 20TH/TECH   V/h
Boa: novel by Nikos Kazantzakis.

ZORRO, THE GAY BLADE ***   PG
Peter Medak USA   1981
George Hamilton, Lauren Hutton, Brenda Vaccaro, Ron Leibman, Donovan Scott, James Booth, Helen BUrns, Clive Revill, Carolyn Seymour, Eduardo Noriega, Jorge Russek, Eduardo Alcaraz, Carlos Bravo, Robert Dumont, Jorge Bolito
A spoof on the Zorro films with some hints of homosexuality. This odd idea for a comedy has Hamilton playing Don Diego Vega, the foppish son of the famous adventurer plus his gay brother Bunny Wigglesworth. Dedicated to the director of THE MARK OF ZORRO – Rouben Mamoulian, this loving send-up of the Zorro films delivers a few laughs but never really gets into its stride. A high-spot is the villainous Leibman who over-acts outrageously.
COM   90 min (ort 96 min)   VIDrel: 20TH/TECH   V

ZULU ****   PG
Cy Endfield UK   1963
Stanley Baker, Jack Hawkins, Ulla Jacobbson, James Booth, Michael Caine, Nigel Green, Ivor Emmanuel, Paul Daneman, Glynn Edwards, Neil McCarthy, David Kernan, Gary Bond, Patrick Magee, Peter Gill, Chief Buthelezi
The story of how a handful of British soldiers resisted a ferocious Zulu attack at Rorke's Drift in 1879. The bulk of the film is taken up with the massive battle, when wave after wave of warriors were driven back, and on that level it is hard to fault. Characterisation tends towards cliche, but for all that it remains a spectacular and exciting film. Written by Endfield and the historian John Prebble. "Zulu Dawn" followed in 1979.
A/AD   132 min (ort 135 min) wScrn
VIDrel: CIC/SONOP; PION (LV only)   V/h LV

INDEX OF ALTERNATIVE TITLES

"10"
See TEN
$1,000,000 DUCK
See MILLION DOLLAR
DUCK
10 SHAOLIN DISCIPLES
See TEN SHAOLIN
DISCIPLES
1001 NIGHTS, THE
See ONE THOUSAND AND
ONE NIGHTS, THE
**11 DAYS 11 NIGHTS:
PART 3 – THE FINAL
CHAPTER**
See ELEVEN DAYS
ELEVEN NIGHTS: PART 3
12:01
See 12. 01
**13 COLD BLOODED
EAGLES**
See THIRTEEN COLD
BLOODED EAGLES
14 GOING ON 30
See FOURTEEN GOING ON
THIRTY
18 AGAIN!
See EIGHTEEN AGAIN!
**1990: THE BRONX
WARRIORS**
See BRONX WARRIORS
2,000 MANIACS
See TWO THOUSAND
MANIACS
2,000 WOMEN
See TWO THOUSAND
WOMEN
**20,000 LEAGUES
UNDER THE SEA**
See TWENTY THOUSAND
LEAGUES UNDER THE
SEA
200 MOTELS
See TWO-HUNDRED
MOTELS
**2010: THE YEAR WE
MAKE CONTACT**
See 2010
23h58
See TWENTY-THREE
HOURS FIFTY-EIGHT
MINUTES
24 HOURS TO MIDNIGHT
See TWENTY-FOUR
HOURS TO MIDNIGHT
29TH STREET
See TWENTY-NINTH
STREET
3 NINJAS
See THREE NINJA KIDS
3.15
See MOMENT OF TRUTH
**3.15: MOMENT OF
TRUTH**
See MOMENT OF TRUTH
36 CRAZY FIST, THE
See THIRTY-SIX FIST, THE
36 CRAZY FISTS
See THIRTY-SIX FIST, THE
36 FIST, THE
See THIRTY-SIX FIST, THE
**36TH CHAMBER OF
SHAOLIN, THE**
See THIRTY-SIXTH
CHAMBER OF SHAOLIN,
THE
**37.2 DEGREES IN THE
MORNING**
See BETTY BLUE
37.2 DEGRES LE MATIN

See BETTY BLUE
37.2° LE MATIN
See BETTY BLUE
42ND STREET
See FORTY-SECOND
STREET
52 PICK-UP
See FIFTY-TWO PICK-UP
55 DAYS AT PEKING
See FIFTY-FIVE DAYS IN
PEKING
**6 DEGREES OF
SEPARATION**
See SIX DEGREES OF
SEPARATION
7-STAR GRAND MANTIS
See SEVEN-STAR GRAND
MANTIS
7TH DAWN, THE
See SEVENTH DAWN, THE
8½
See FELLINI'S 8½
**800 LEGUAS POR EL
AMAZONAS**
See EIGHT HUNDRED
LEAGUES DOWN THE
AMAZON
**84 CHARING CROSS
ROAD**
See EIGHTY-FOUR
CHARING CROSS ROAD
84 CHARLIE MOPIC
See EIGHTY-FOUR
CHARLIE MOPIC
9 TO 5
See NINE TO FIVE
**976: EVIL 2 – THE
ASTRAL FACTOR**
See NINE SEVEN SIX: EVIL
2
A BOUT DE SOUFFLE
See BREATHLESS
**A DAME DO CINE
SHANGHAI**
See LADY FROM THE
SHANGHAI CINEMA, THE
A KOLDUM KLAKA
See COLD FEVER
**A NOITE DO TERROR
CEGO**
See TOMBS OF THE BLIND
DEAD
**A-HAUNTING WE WILL
GO**
See LAUREL AND HARDY:
A-HAUNTING WE WILL
GO
**A-TEAM, THE: TRIAL BY
FIRE**
See A-TEAM, THE: THE
COURT MARTIAL
**A.W.O.L. ABSENT
WITHOUT LEAVE**
See A.W.O.L.
ABBEY GRANGE, THE
See RETURN OF
SHERLOCK HOLMES, THE:
THE ABBEY GRANGE
ABGESCHMINKT
See MAKING UP
ABOVE THE LAW
See GOOD DIE FIRST FOR
A HANDFUL OF SILVER,
THE
ABOVE THE LAW
See NICO
**ABSOLUTE
CONVICTION**
See INSPECTOR MORSE:

ABSOLUTE CONVICTION
ACE UP MY SLEEVE
See ACE HIGH
ACES: IRON EAGLE 3
See IRON EAGLE 3
ACHILLES HEEL
See INSPECTOR
WEXFORD: ACHILLES
HEEL
ACTS OF OBSESSION
See BLINDFOLD: ACTS OF
OBSESSION
**ADDIO, FRATELLO
CRUDELE**
See TIS A PITY SHE'S A
WHORE
ADIOS, TEXAS
See TEXAS ADIOS
**ADVANCE TO GROUND
ZERO**
See NIGHTBREAKER
**ADVENTURES IN
DINOSAUR CITY**
See DINOSAURS: THE
MOVIE
**ADVENTURES OF A
GNOME CALLED
GNORM, THE**
See UP WORLD
**ADVENTURES OF
BLACK FEATHER, THE**
See BLACK FEATHER
**ADVENTURES OF
BUCKAROO BANZAI
ACROSS THE EIGHTH
DIMENSION**
See BUCKAROO BANZAI
**ADVENTURES OF
BUCKAROO BANZAI,
THE**
See BUCKAROO BANZAI
**ADVENTURES OF MILO
IN THE PHANTOM
TOLLBOOTH, THE**
See PHANTOM
TOLLBOOTH, THE
**ADVENTURES OF RAT
PFINK A BOO-BOO**
See RAT PFINK A BOO-
BOO
**ADVENTURES OF
SHERLOCK HOLMES,
THE**
See SHERLOCK HOLMES:
THE ADVENTURES OF
SHERLOCK HOLMES
**ADVENTURES OF
SHERLOCK HOLMES,
THE: SILVER BLAZE**
See RETURN OF
SHERLOCK HOLMES, THE:
SILVER BLAZE
**ADVENTURES OF
SHERLOCK HOLMES,
THE: THE ABBEY
GRANGE**
See RETURN OF
SHERLOCK HOLMES, THE:
THE ABBEY GRANGE
**ADVENTURES OF
SHERLOCK HOLMES,
THE: THE BRUCE
PARTINGTON PLANS**
See RETURN OF
SHERLOCK HOLMES, THE:
THE BRUCE PARTINGTON
PLANS
**ADVENTURES OF
SHERLOCK HOLMES,**

THE: THE DEVIL'S FOOT
See RETURN OF
SHERLOCK HOLMES, THE:
THE DEVIL'S FOOT
**ADVENTURES OF
SHERLOCK HOLMES,
THE: THE EMPTY
HOUSE**
See RETURN OF
SHERLOCK HOLMES, THE:
THE EMPTY HOUSE
**ADVENTURES OF
SHERLOCK HOLMES,
THE: THE MAN WITH
THE TWISTED LIP**
See RETURN OF
SHERLOCK HOLMES, THE:
THE MAN WITH THE
TWISTED LIP
**ADVENTURES OF
SHERLOCK HOLMES,
THE: THE MUSGRAVE
RITUAL**
See RETURN OF
SHERLOCK HOLMES, THE:
THE MUSGRAVE RITUAL
**ADVENTURES OF
SHERLOCK HOLMES,
THE: THE PRIORY
SCHOOL**
See RETURN OF
SHERLOCK HOLMES, THE:
THE PRIORY SCHOOL
**ADVENTURES OF
SHERLOCK HOLMES,
THE: THE SECOND
STAIN**
See RETURN OF
SHERLOCK HOLMES, THE:
THE SECOND STAIN
**ADVENTURES OF
SHERLOCK HOLMES,
THE: THE SIX
NAPOLEONS**
See RETURN OF
SHERLOCK HOLMES, THE:
THE SIX NAPOLEONS
**ADVENTURES OF
SHERLOCK HOLMES,
THE: WISTERIA LODGE**
See RETURN OF
SHERLOCK HOLMES, THE:
WISTERIA LODGE
**ADVENTURES OF THE
GOLDEN BEAR, THE**
See GOLDY 3: THE MAGIC
OF THE GOLDEN BEAR
**ADVENTURES OF THE
GREAT MOUSE
DETECTIVE**
See BASIL, THE GREAT
MOUSE DETECTIVE
**ADVENTURES OF
TINTIN, THE: THE LAKE
OF SHARKS**
See TINTIN: THE LAKE OF
SHARKS
ADVENTURESS, THE
See I SEE A DARK
STRANGER
**ADVENTUROUS ANTICS
OF ASTERIX IN BRITAIN,
THE: ONE LITTLE GUY
AGAINST ONE**
See ASTERIX IN BRITAIN
**AFFAIR OF THE HEART,
AN**
See SWITCHBOARD
OPERATOR, THE

AFRICA SCREAMS
See ABBOTT AND
COSTELLO: AFRICA
SCREAMS
AFRICAN FURY
See CRY THE BELOVED
COUNTRY
AFRICAN RUN, THE
See TUXEDO WARRIOR
AFTER DARK
See AFTER DARKNESS
**AFTERMATH: A TEST
OF LOVE**
See AFTERMATH
AFTERNOON
See ELEVEN DAYS
ELEVEN NINGHTS: PART
3
**AGAINST HER WILL: AN
INCIDENT IN
BALTIMORE**
See AGAINST HER WILL
AGANTUK
See STRANGER, THE
**AGATHA CHRISTIE:
MURDER IS EASY**
See MURDER IS EASY
**AGATHA CHRISTIE:
SPARKLING CYANIDE**
See SPARKLING CYANIDE
**AGATHA CHRISTIE'S
DEAD MAN'S FOLLY**
See DEAD MAN'S FOLLY
**AGATHA CHRISTIE'S
ENDLESS NIGHT**
See ENDLESS NIGHT
**AGATHA CHRISTIE'S
THIRTEEN AT DINNER**
See THIRTEEN AT DINNER
AGE OF OLD FRIENDS
See MONTH OF SUNDAYS,
A
**AGUIRRE, DER ZORN
GOTTES**
See AGUIRRE, WRATH OF
GOD
**AGUIRRE, THE WRATH
OF GOD**
See AGUIRRE, WRATH OF
GOD
AHFEI ZHENJUANG
See DAYS OF BEING WILD
**AILEEN WUORNOS
STORY, THE**
See OVERKILL: THE
AILEEN WUORNOS STORY
AIR RAID WARDENS
See LAUREL AND HARDY:
AIR RAID WARDENS
**AKIRA KUROSAWA'S
DREAMS**
See DREAMS
AL DI LA DELLA LEGGE
See GOOD DIE FIRST FOR
A HANDFUL OF SILVER,
THE
**AL DI LA DELLE
NUVOLE**
See BEYOND THE CLOUDS
**ALEXA: A
PROSTITUTE'S OWN
STORY**
See ALEXA
ALI
See FEAR EATS THE SOUL
**ALI: FEAR EATS THE
SOUL**
See FEAR EATS THE SOUL
ALICE IN DEN STADTEN
See ALICE IN THE CITIES
ALIEN ATTACK
See SPACE 1999: ALIEN
ATTACK

ALIEN PREDATOR
See MUTANT 2
ALIEN PREDATORS
See MUTANT 2
ALIEN PREY
See PREY
**ALIENS: THIS TIME IT'S
WAR**
See ALIENS
**ALISTAIR MacLEAN'S
DEATH TRAIN**
See DEATH TRAIN
**ALISTAIR MACLEAN'S
NIGHT WATCH**
See NIGHT WATCH
**ALL 33 DI VIA
OROLOGIO FA SEMPRE
FREDDO**
See SHOCK
ALL IS WELL
See TOUT VA BIEN
**ALL NEW GEORGE A.
ROMERO'S NIGHT OF
THE LIVING DEAD: THE
REMAKE**
See NIGHT OF THE LIVING
DEAD
**ALL NEW TALES FROM
THE CRYPT**
See TALES FROM THE
CRYPT: VOL. 2
ALL SHOOK UP!
See HEXED
ALL WEEKEND LOVERS
See KILLING GAME, THE
ALL YOU NEED IS CASH
See COMPLEAT RUTLES,
THE
ALLEGHENY UPRISING
See FIRST REBEL, THE
ALMOST
See WENDY CRACKED A
WALNUT
ALMOST HUMAN
See SHOCK WAVES
ALONE TOGETHER
See CRISS CROSS
**ALPHAVILLE, A
STRANGE CASE OF
LEMMY CAUTION**
See ALPHAVILLE
**ALS JE BEGRIJPT WAT
IK BEDOEL**
See DRAGON THAT
WASN'T... OR WAS HE?,
THE
AMANTES
See LOVERS
AMATEUR HOUR
See I WAS A TEENAGE TV
TERRORIST
AMAZING QUEST
See AMAZING
ADVENTURE, THE
**AMAZING QUEST OF
ERNEST BLISS, THE**
See AMAZING
ADVENTURE, THE
**AMAZING SPIDERMAN,
THE**
See SPIDERMAN: THE
MOVIE
**AMBUSH AT WACO: IN
THE LINE OF DUTY**
See IN THE LINE OF DUTY:
AMBUSH AT WACO
AMERICAN BUILT
See RACE FOR GLORY
AMERICAN DRAGON
See GUNS OF DRAGON
AMERICAN JUSTICE
See JACKALS
AMERICAN MASSEUSE

See MASSEUSE
AMERICAN NINJA 2
See AMERICAN NINJA 2:
THE CONFRONTATION
AMERICAN NINJA 3
See AMERICAN NINJA 3:
BLOODHUNT
AMERICAN SHAOLIN
See AMERICAN SHAOLIN:
KING OF THE
KICKBOXERS 2
AMERICAN STORY, AN
See AFTER THE GLORY
AMERICAN SUCCESS
See SUCCESS
**AMERICAN SUCCESS
COMPANY, THE**
See SUCCESS
AMERICAN WARRIOR
See AMERICAN NINJA
**AMITYVILLE HORROR,
THE: THE EVIL
ESCAPES PART 4**
See AMITYVILLE 4: THE
EVIL ESCAPES
**AMITYVILLE: THE EVIL
ESCAPES**
See AMITYVILLE 4: THE
EVIL ESCAPES
AMONG PEOPLE
See MY APPRENTICESHIP
**AMONG THE LIVING
DEAD**
See VIRGIN AMONG THE
LIVING DEAD
AMY
See AMY JOHNSON
STORY, THE
**AMY FISHER STORY,
THE**
See BEYOND CONTROL:
THE AMY FISHER STORY
**ANASTASIA: THE
MYSTERY OF ANNA**
See ANASTASIA: PARTS 1
AND 2
ANATOLIAN SMILE, THE
See AMERICA, AMERICA
**AND ONCE UPON A
LOVE**
See FANTASIES
**AND THE LITTLE
PRINCE SAID**
See LE PETIT PRINCE A
DIT
**AND WOMAN WAS
CREATED**
See AND GOD CREATED
WOMAN
**AND YOU'LL LIVE IN
TERROR! THE BEYOND**
See BEYOND, THE
ANDY WARHOL'S HEAT
See HEAT
**ANDY WARHOL'S
YOUNG DRACULA**
See ANDY WARHOL'S
DRACULA
ANGEL
See ANGELS
ANGEL
See IRON ANGELS
ANGEL 2
See IRON ANGELS 2
ANGEL 3
See ANGEL 3: THE FINAL
CHAPTER
ANGEL STREET
See GASLIGHT
ANGEL WARRIORS 2
See HELL'S ANGELS ON
WHEELS
ANGELS

See IRON ANGELS
ANGELS 2
See IRON ANGELS 2
**ANGELS HARD AS THEY
COME**
See ANGEL WARRIORS
**ANGELS IN THE
OUTFIELD**
See ANGELS
ANGLAGARD
See HOUSE OF ANGELS
**ANGST ESSEN SEELE
AUF**
See FEAR EATS THE SOUL
**ANGST ISS DIE SEELE
AUF**
See FEAR EATS THE SOUL
ANIMAL HOUSE
See NATIONAL
LAMPOON'S ANIMAL
HOUSE
**ANNE OF A THOUSAND
DAYS**
See ANNE OF THE
THOUSAND DAYS
ANNE OF AVONLEA
See ANNE OF GREEN
GABLES: THE SEQUEL
**ANNE OF GREEN
GABLES: THE SERIES**
See ANNE OF GREEN
GABLES
**ANNETTE FUNICELLO
STORY, THE**
See DREAM IS A WISH
YOUR HEART MAKES, A:
THE ANNETTE
FUNICELLO STORY
**ANOTHER MIDNIGHT
RUN**
See MIDNIGHT RUN 2:
ANOTHER MIDNIGHT
RUN
**ANOTHER WOMAN'S
LIPSTICK**
See RED SHOE DIARIES 3:
ANOTHER WOMAN'S
LIPSTICK
ANSIKTET
See MAGICIAN, THE
ANTEFATTO
See LAST HOUSE ON THE
LEFT PART 2, THE
ANTONIA
See ANTONIA'S LINE
**ANXIETY OF THE
GOALIE AT THE
PENALTY, THE**
See GOALKEEPER'S FEAR
OF THE PENALTY, THE
ANYTHING FOR LOVE
See JUST ONE OF THE
GIRLS
**APARTMENT ON THE
13TH FLOOR**
See CANNIBAL MAN, THE
APEX
See A.P.E.X.
APPLEGATES, THE
See MEET THE
APPLEGATES
**APPOINTMENT FOR
KILLING**
See APPOINTMENT FOR A
KILLING
APRES-SKI
See SKI SLUTS GO
SNOWBALLING
APRIL FOOL'S DAY
See SLAUGHTER HIGH
APU SANSAR
See WORLD OF APU, THE
AQUARIUS

See STAGE FRIGHT
ARIA ON GAZES, AN
See BEDROOM, THE
ARMISTEAD MAUPIN'S TALES OF THE CITY
See TALES OF THE CITY
ARMY OF DARKNESS: THE MEDIEVAL DEAD
See ARMY OF DARKNESS
ARMY OF ONE
See JOSHUA TREE
ARRIVA! IL CROW
See FISTFUL OF DEATH, A
ARS ARMANDI
See ART OF LOVE, THE
ARTHUR 2: ON THE ROCKS
See ARTHUR 2
ASCENSEUR POUR L'ECHAFAUD
See LIFT TO THE SCAFFOLD
ASHANTI SANKET
See DISTANT THUNDER
ASSAULT OF THE KILLER BIMBOS
See SCUMBUSTERS
ASSIGNMENT IN ISTANBUL
See CASTLE OF FU MANCHU, THE
ASSO PIGLIA TUTTO
See ACE UP MY SLEEVE
ASTERIX CHEZ LES BRETONS
See ASTERIX IN BRITAIN
ASTERIX ET LA SURPISE DE CESAR
See ASTERIX VERSUS CAESAR
ASTERIX IN AMERICA
See ASTERIX CONQUERS AMERICA
ASTERIX IN AMERIKA
See ASTERIX CONQUERS AMERICA
ASTERIX IN BRITAIN: THE MOVIE
See ASTERIX IN BRITAIN
ASTERIX LE GAULOIS
See ASTERIX THE GAUL
ASYLUM EROTICA
See COLD BLOODED BEAST
AT WAR WITH THE ARMY
See JERRY LEWIS: AT WAR WITH THE ARMY
ATAME!
See TIE ME UP! TIE ME DOWN!
ATLANTIC CITY, USA
See ATLANTIC CITY
ATOLL K
See LAUREL AND HARDY: UTOPIA
ATOR
See ATOR, THE FIGHTING EAGLE
ATOR L'INVINCIBLE
See ATOR, THE FIGHTING EAGLE
ATOR L'INVINCIBLE 2
See ATOR THE INVINCIBLE 2
ATOR THE INVINCIBLE
See ATOR, THE FIGHTING EAGLE
ATOR, THE RETURN
See ATOR THE INVINCIBLE 2
ATTACK OF THE BLIND DEAD

See RETURN OF THE EVIL DEAD, THE
ATTACK OF THE MARCHING MONSTERS
See GODZILLA: DESTROY ALL MONSTERS
AURA'S ENIGMA
See TRAUMA
AURORA BY NIGHT
See AURORA
AURORA: OPERATION INTERCEPT
See OPERATION INTERCEPT
AUSTRALIAN DREAM
See OUTRAGEOUS PARTY
AUSTRIA 1700
See MARK OF THE DEVIL
AUTO EROTICA
See RED SHOE DIARIES 4: AUTO EROTICA
AUX YEUX DU MONDE
See AUTOBUS
AVENGER, THE
See TEXAS ADIOS
AWAKE TO MURDER
See AWAKE TO DANGER
AWFUL DOCTOR ORLOFF, THE
See AWFUL DR ORLOFF, THE
BA WANG BIE JI
See FAREWELL, MY CONCUBINE
BABE RUTH
See BABE, THE
BABE, THE GALLANT PIG
See BABE
BABES IN TOYLAND
See LAUREL AND HARDY: BABES IN TOYLAND
BABETTES GAESTEBUD
See BABETTE'S FEAST
BABY CART AT THE RIVER STYX
See SHOGUN ASSASSIN
BACHELOR JAMBOREE
See BACHELOR PARTY 2
BACHELOR'S BABY, THE
See HERE COMES THE SON: A TRUE STORY
BACKTRACK
See CATCHFIRE
BAD TIMING: A SENSUAL OBSESSION
See BAD TIMING
BADKONAK-E SEFID
See WHITE BALLOON, THE
BAIL OUT
See W.B. BLUE AND THE BEAN
BAISERS VOLES
See STOLEN KISSES
BALBOA: MILLIONAIRE'S PARADISE
See BALBOA
BALLAD OF KID DIVINE, THE: THE COCKNEY COWBOY
See KID DIVINE
BAMBOO DOLLS HOUSE
See WOMEN IN CAGES
BANDIT BANDIT
See BANDIT 2: BANDIT BANDIT
BANDIT GOES COUNTRY
See BANDIT 1: BANDIT

GOES COUNTRY
BANDIT'S SILVER ANGEL
See BANDIT 4: BANDIT'S SILVER ANGEL
BARBARA TAYLOR BRADFORD'S REMEMBER
See REMEMBER
BARBARELLA, QUEEN OF THE GALAXY
See BARBARELLA
BARBARIAN QUEEN 2: THE EMPRESS STRIKES BACK
See BARBARIAN QUEEN 2
BARE BREASTED COUNTESSS
See FEMALE VAMPIRE
BARNABO DELLE MONTAGNE
See BARNABO OF THE MOUNTAINS
BASKERVILLE CURSE, THE
See SHERLOCK HOLMES: THE BASKERVILLE CURSE
BATMAN
See BATMAN: THE MOVIE
BATMAN: THE ANIMATED MOVIE
See BATMAN: MASK OF THE PHANTASM
BATTLE FOR ANZIO, THE
See ANZIO
BATTLE OF THE ASTROS
See GODZILLA: INVASION OF THE ASTRO-MONSTERS
BATTLE RAGE
See MISSING IN ACTION 2: THE BEGINNING
BATTLE STRIPE
See MEN, THE
BATTLING BELLHOP, THE
See KID GALAHAD
BATTLING HOOFER
See SOMETHING TO SING ABOUT
BAY OF BLOOD
See LAST HOUSE ON THE LEFT PART 2, THE
BEANS OF EGYPT, MAINE, THE
See FORBIDDEN CHOICES
BEASTMASTER 2: THROUGH THE PORTAL OF TIME
See BEASTMASTER 2
BEAUMARCHAIS L'INSOLENT
See BEAUMARCHAIS
BEAUTE VOLEE
See STEALING BEAUTY
BEAUTY AND THE BANDIT
See BANDIT 3: BEAUTY AND THE BANDIT
BEAUTY AND THE BEAST: ONCE UPON A TIME IN NEW YORK
See BEAUTY AND THE BEAST: ONCE UPON A TIME IN THE CITY OF NEW YORK
BEAUTY'S REVENGE
See MIDWEST OBSESSION
BEFORE THE FACT
See LAST HOUSE ON THE LEFT PART 2, THE

BEHIND THE CONVENT WALLS
See BEHIND CONVENT WALLS
BEHIND THE IRON MASK
See FIFTH MUSKETEER, THE
BEIJO DA A MUHER ARANNHA
See KISS OF THE SPIDER WOMAN
BEIQING CHENGSI
See CITY OF SADNESS, A
BELARUS FILE, THE
See KOJAK: THE BELARUS FILE
BELLBOY, THE
See JERRY LEWIS: THE BELLBOY
BELTENEBROS
See PRINCE OF SHADOWS
BENEATH ARIZONA SKIES
See 'NEATH ARIZONA SKIES
BERORINGEN
See TOUCH, THE
BERRY GORDON'S THE LAST DRAGON
See LAST DRAGON, THE
BERSERKER: THE NORDIC CURSE
See BERSERKER
BEST INTENTION: THE EDUCATION AND KILLING OF EDMUND PARRY
See MURDER WITHOUT MOTIVE
BEST MAN TO DIE, THE
See INSPECTOR WEXFORD: THE BEST MAN TO DIE
BETHUNE
See BETHUNE: THE MAKING OF A HERO
BETTER TOMORROW, A: PART 2
See BETTER TOMORROW 2, A
BEYOND THE BRIDGE
See OLIVIA
BEYOND THE DOOR 2
See SHOCK
BEYOND THE FOG
See TOWER OF EVIL
BEYOND THE LAW
See FIXING THE SHADOW
BEYOND THE LAW
See GOOD DIE FIRST FOR A HANDFUL OF SILVER, THE
BEYOND THE LAW
See CITY COPS
BEZHIN LUG
See BEZHIN MEADOW
BIAN ZOU BIAN CHANG
See LIFE ON A STRING
BIBLE, THE: IN THE BEGINNING, THE
See BIBLE, THE
BICYCLE THIEF, THE
See BICYCLE THIEVES
BIG BOSS, THE
See FISTS OF FURY
BIG COUNTRY
See GREAT OUTDOORS, THE
BIG DEAL AT DODGE CITY, A
See BIG HAND FOR A LITTLE LADY, A

BIG FEAST, THE
See LA GRANDE BOUFFE
**BIG GIRLS DON'T CRY...
THEY GET EVEN**
See STEPKIDS
BIG MAN ON CAMPUS
See HUNCHBACK
HAIRBALL OF L.A., THE
**BIG MAN, THE:
CROSSING THE LINE**
See BIG MAN, THE
BIG MOUTH, THE
See JERRY LEWIS: THE BIG
MOUTH
**BIG ONE, THE: THE
GREAT LOS ANGELES
EARTHQUAKE**
See GREAT LOS ANGELES
EARTHQUAKE, THE
BIG SHOWDOWN, THE
See STORMRIDER
BIG STORE, THE
See MARX BROTHERS:
THE BIG STORE
BIG TIME OPERATORS
See SMALLEST SHOW ON
EARTH, THE
BIGGLES
See BIGGLES GETS OFF
THE GROUND
**BIGGLES:
ADVENTURES IN TIME**
See BIGGLES GETS OFF
THE GROUND
**BILLION DOLLAR BOYS'
CLUB**
See BILLIONAIRE BOYS'
CLUB
BILLY BRONCO
See BRONCO BILLY
**BILLY ROSE'S
DIAMOND HORSESHOE**
See DIAMOND
HORSESHOE
BIOHAZARD
See WARNING SIGN
**BIONIC SHOWDOWN:
THE SIX MILLION
DOLLAR MAN AND THE
BIONIC WOMAN**
See BIONIC SHOWDOWN,
THE
BIRDS OF A FEATHER
See LA CAGE AUX FOLLES
**BIS ANS ENDE DER
WELT**
See UNTIL THE END OF
THE WORLD
BLACK CAT, THE
See KURONEKO
BLACK EYES
See DARK EYES
**BLACK MASSES OF
EXORCISM, THE**
See DEMONIAC
BLACK SHACK ALLEY
See RUE CASES NEGRES
BLACK TOWER, THE
See DALGLIESH: THE
BLACK TOWER – PARTS 1
AND 2
BLACK VALOR
See SAVAGE!
BLACK WATER
See TENNESSEE NIGHTS
BLACK WEREWOLF
See BEAST MUST DIE, THE
BLACKBELT
See BLACK BELT
**BLACKBOARD
MASSACRE**
See MASSACRE AT
CENTRAL HIGH

BLADE OF STEEL
See FAR PAVILIONS, THE
BLADEMASTER, THE
See ATOR THE
INVINCIBLE 2
**BLAKE'S 7: THE
BEGINNING**
See BLAKE'S 7: THE WAY
BACK
BLANK CHECK
See BLANK CHEQUE
BLIND DEAD, THE
See TOMBS OF THE BLIND
DEAD
BLIND RAGE
See SURROGATE, THE
**BLINFOLDED: ACTS OF
OBSESSION**
See BLINDFOLD: ACTS OF
OBSESSION
BLINK OF AN EYE
See FIRST LIGHT
BLOCKHEADS
See LAUREL AND HARDY:
BLOCKHEADS
BLOND FURY, THE
See ABOVE THE LAW 2:
THE BLOND FURY
BLOND JUSTICE
See BLONDE JUSTICE
BLONDE FURY
See ABOVE THE LAW 2:
THE BLOND FURY
BLONDE SINNER
See YIELD TO THE NIGHT
BLOOD BARON, THE
See BARON BLOOD
**BLOOD CAMP
THATCHER**
See ESCAPE 2000
**BLOOD FEAST OF THE
BLIND DEAD**
See NIGHT OF THE
SEAGULLS
BLOOD FOR DRACULA
See ANDY WARHOL'S
DRACULA
BLOOD HORSE
See BLACK FOX: BLOOD
HORSE
**BLOOD IN, BLOOD OUT:
BOUND BY HONOR**
See BLOOD IN BLOOD
OUT
BLOOD MONEY
See KILLER'S EDGE, THE
BLOOD MONEY
See STRANGER AND THE
GUNFIGHTER, THE
BLOOD OF HEROES
See SALUTE OF THE
JUGGER, THE
**BLOOD OF THE
DRAGON PERIL**
See BLOOD OF DRAGON
PERIL
**BLOOD OF THE
INNOCENT**
See BEYOND
FORGIVENESS
BLOOD SISTERS
See SISTERS
BLOOD SUCKERS, THE
See DOCTOR TERROR'S
HOUSE OF HORRORS
BLOOD VOYAGE
See NIGHTMARE VOYAGE
BLOOD WEDDING
See LES NOCES ROUGES
**BLOODBATH BAY OF
BLOOD**
See LAST HOUSE ON THE
LEFT PART 2, THE

BLOODSILVER
See GOOD DIE FIRST FOR
A HANDFUL OF SILVER,
THE
BLOODTIDE
See BLOOD TIDE
BLOODY BIRD
See STAGE FRIGHT
BLOODY SUNDAY
See BLOODY BIRTHDAY
BLOODY, THE
See DRACULA: PRINCE OF
DARKNESS
BLOW-OUT
See LA GRANDE BOUFFE
BLUE
See THREE COLOURS:
BLUE
BLUE CARBUNCLE, THE
See ADVENTURES OF
SHERLOCK HOLMES, THE:
THE BLUE CARBUNCLE
BLUE HAWAII
See ELVIS PRESLEY: BLUE
HAWAII
BLUE THUNDER
See BLUE THUNDER: THE
MOVIE
BLUT AN DEN LIPPEN
See DAUGHTERS OF
DARKNESS
BODAS DE SANGRE
See BLOOD WEDDING
**BODIES BEAR TRACES
OF CARNAL VIOLENCE,
THE**
See TORSO
BODIES IN HEAT 2
See BODIES IN HEAT: THE
SEQUEL
BODY FIRE
See LUST IN THE WOODS
**BODY IN THE LIBRARY,
THE**
See MISS MARPLE: THE
BODY IN THE LIBRARY
**BODY SNATCHERS:
THE INVASION
CONTINUES**
See BODY SNATCHERS
BODYGUARD
See YOJIMBO
BOGEYMAN, THE
See BOGEY MAN, THE
BOGUS BANDITS
See LAUREL AND HARDY:
BOGUS BANDITS
BOHEMIAN GIRL, THE
See LAUREL AND HARDY:
THE BOHEMIAN GIRL
BOMBSIGHT STOLEN
See COTTAGE TO LET
**BON VOYAGE, CHARLIE
BROWN**
See CHARLIE BROWN:
BON VOYAGE
**BON VOYAGE, CHARLIE
BROWN (AND DON'T
COME BACK!)**
See CHARLIE BROWN:
BON VOYAGE
BONNIE SCOTLAND
See LAUREL AND HARDY:
HEROES OF THE
REGIMENT
BOOGEY MAN, THE
See BOGEY MAN, THE
**BOOK OF THE DEAD,
THE**
See EVIL DEAD, THE
BORN FOR GLORY
See FOREVER ENGLAND
BOSAMBO

See SANDERS OF THE
RIVER
**BOSCOMBE VALLEY
MYSTERY, THE**
See CASEBOOK OF
SHERLOCK HOLMES, THE:
THE BOSCOMBE VALLEY
MYSTERY
**BOUDU SAUVE DES
EAUX**
See BOUDU SAVED FROM
DROWNING
BOUND BY HONOR
See BLOOD IN BLOOD
OUT
BOUNTY HUNTER 2002
See RAPE OF EDEN
BOXER ADVENTURE
See BOXER'S ADVENTURE,
THE
BOY FRIEND, THE
See BOYFRIEND, THE
**BOYAR'S PLOT, THE:
IVAN THE TERRIBLE –
PART 2**
See IVAN THE TERRIBLE
**BOYFRIENDS AND
GIRLFRIENDS**
See MY GIRLFRIEND'S
BOYFRIEND
**BOYZ N THE HOOD:
INCREASE THE PEACE**
See BOYS N THE HOOD
**BRADDOCK: MISSING IN
ACTION 3**
See MISSING IN ACTION 3
BRAINSMASHER
See BRAIN SMASHER: A
LOVE STORY
**BRAM STOKER'S
DRACULA**
See DRACULA
**BRAM STOKER'S
DRACULA**
See DRACULA
BRAVESTARR
See BRAVESTARR: THE
LEGEND
**BRAVESTARR: THE
MOVIE**
See BRAVESTARR: THE
LEGEND
BREAKIN'
See BREAKDANCE: THE
MOVIE
**BREAKIN' 2 ELECTRIC
BOOGALOO**
See BREAKDANCE 2:
ELECTRIC BOOGALOO
BREAKING POINT
See DOUBLE SUSPICION
**BREAKING THE SOUND
BARRIER**
See SOUND BARRIER, THE
BREAKOUT
See DANGER WITHIN
BREAKTHROUGH, THE
See LIFEFORCE
EXPERIMENT, THE
BREAKUP, THE
See LA RUPTURE
BRENN HEXE BRENN
See MARK OF THE DEVIL
**BRIDE OF RE-
ANIMATOR**
See RE-ANIMATOR 2
BRIDE OF THE ATOM
See BRIDE OF THE
MONSTER
**BRIDE OF THE RE-
ANIMATOR**
See RE-ANIMATOR 2
BRIDE OF VIOLENCE

See VENDETTA
BRIGHT SUMMER DAY, A
See BRIGHTER SUMMER DAY, A
BRINGERS OF WONDER, THE
See SPACE 1999: DESTINATION MOONBASE ALPHA
BRONENOSETS POTEMKIN
See BATTLESHIP POTEMKIN, THE
BROTHERHOOD OF THE YAKUZA
See YAKUZA, THE
BROWN ON RESOLUTION
See FOREVER ENGLAND
BRUCE CURTIS STORY, THE: JOURNEY INTO DARKNESS
See JOURNEY INTO DARKNESS: THE BRUCE CURTIS STORY
BRUCE LE VERSUS NINJA
See BRUCE THE SUPERHERO
BRUCE LEE VERSUS THE BLACK DRAGON
See BRUCE THE SUPERHERO
BRUCE LEE: DEADLY STRIKE
See DEADLY STRIKE, THE
BRUCE LEE: GAME OF DEATH
See GAME OF DEATH
BRUCE LEE: SUPERHERO
See BRUCE THE SUPERHERO
BRUCE LEE: THE TRUE STORY
See BRUCE LEE: THE MAN – THE MYTH
BRUCE PARTINGTON PLANS, THE
See RETURN OF SHERLOCK HOLMES, THE: THE BRUCE PARTINGTON PLANS
BRUTTI, SPORCHI E CATTIVI
See UGLY, DIRTY AND BAD
BUCK ROGERS
See BUCK ROGERS IN THE 25TH CENTURY
BUFFALO BILL
See BUFFALO BILL AND THE INDIANS, OR SITTING BULL'S HISTORY LESSON
BUILD MY GALLOWS HIGH
See OUT OF THE PAST
BULLDOG BREED, THE
See NORMAN WISDOM: THE BULLDOG BREED
BULLETPROOF HEART
See KILLER
BULLFIGHTERS, THE
See LAUREL AND HARDY: THE BULLFIGHTERS
BURN WITCH BURN
See MARK OF THE DEVIL
BURNING QUESTION, THE
See REEFER MADNESS
BURNING UP

See RED SHOE DIARIES 7: BURNING UP
BUSH WACKERS
See HUSTLER, THE
C.B. HUSTLERS
See SECRETS OF LADY TRUCKERS
C.H.U.D. 2: BUD THE CHUD
See C.H.U.D. 2
C.I.A. – TARGET ALEXA
See C.I.A. – CODENAME ALEXA
C.S. LEWIS: THROUGH THE SHADOWLANDS
See SHADOWLANDS
CABIRIA
See NIGHTS OF CABIRIA
CADENCE
See STOCKADE
CAGE 2: THE ARENA OF DEATH
See CAGE 2, THE
CAGED FEMALES
See CAGED HEAT
CAGED SEDUCTION: THE SHOCKING TRUE STORY
See CAGED SEDUCTION
CAGED VIRGINS
See REQUIEM FOR A VAMPIRE
CAHILL
See CAHILL: U.S. MARSHAL
CAHILL, UNITED STATES MARSHAL
See CAHILL: U.S. MARSHAL
CALIFORNIA HOLIDAY
See ELVIS PRESLEY: SPINOUT
CALIGULA'S FUNNIEST HOME VIDEOS
See CARRY ON CLEO
CALL ME A CAB
See CARRY ON CABBY
CALL ME GENIUS
See HANCOCK: THE REBEL
CAMORRA MAN, THE
See PROFESSOR, THE
CAMORRA MEMBER, THE
See PROFESSOR, THE
CAN CAN
See CAN-CAN
CAN I DO IT TILL I NEED GLASSES?
See CAN I DO IT 'TIL I NEED GLASSES?
CANDYMAN: FAREWELL TO THE FLESH
See CANDYMAN 2: FAREWELL TO THE FLESH
CANNIBAL WOMEN IN THE AVOCADO JUNGLE OF DEATH
See PIRANHA WOMEN
CANNIBALS
See EATEN ALIVE
CANNON MOVIE TALES: HANSEL AND GRETEL
See HANSEL AND GRETEL
CANNON MOVIE TALES: RED RIDING HOOD
See RED RIDING HOOD
CANNON MOVIE TALES: SLEEPING BEAUTY
See SLEEPING BEAUTY
CANNON MOVIE TALES: SNOW WHITE

See SNOW WHITE AND THE SEVEN DWARFS
CANNONBALL 2
See CANNONBALL RUN 2
CANTON GODFATHER, THE
See MIRACLES: THE CANTON GODFATHER
CANTONEN IRON KUNG FU
See CANTON IRON KUNG FU
CANTONESE IRON KUNG FU
See CANTON IRON KUNG FU
CANVAS: THE FINE ART OF CRIME
See CANVAS
CAPTAIN AMERICA: THE MOVIE
See CAPTAIN AMERICA
CAPTAIN POWER AND THE SOLDIERS OF THE FUTURE: THE LEGEND BEGINS
See CAPTAIN POWER AND THE SOLDIERS OF THE FUTURE
CAPTAIN POWER: THE SOLDIERS OF THE FUTURE
See CAPTAIN POWER AND THE SOLDIERS OF THE FUTURE
CARAVAN OF COURAGE
See CARAVAN OF COURAGE: AN EWOK ADVENTURE
CARE BEARS IN WONDERLAND
See CARE BEARS' ADVENTURE IN WONDERLAND, THE
CARIBBEAN MYSTERY, A
See MISS MARPLE: A CARIBBEAN MYSTERY
CARNABY M.D.
See DOCTOR IN CLOVER
CARNAGE
See LAST HOUSE ON THE LEFT PART 2, THE
CARNE PER FRANKENSTEIN
See ANDY WARHOL'S FRANKENSTEIN
CARO DIARIO
See DEAR DIARY
CAROLE ET SES DEMONS
See NIGHT PLEASURES
CARRIED AWAY
See ACTS OF LOVE
CARRY ON ROUND THE BEND
See CARRY ON AT YOUR CONVENIENCE
CARRY ON SAILOR
See CARRY ON JACK
CARRY ON VAMPIRE
See CARRY ON SCREAMING
CARRY ON VENUS
See CARRY ON JACK
CARS THAT ATE PEOPLE, THE
See CARS THAT ATE PARIS, THE
CARTEGGIO VALACHI
See VALACHI PAPERS, THE

CASBAH
See PEPE LE MOKO
CASE OF THE NOTORIOUS NUN, THE
See PERRY MASON: THE CASE OF THE NOTORIOUS NUN
CASSIE
See UP 'N' COMING
CASTLE OF DOOM
See VAMPYR
CASTLE OF THE DOOMED
See KISS ME MONSTER
CASTLE OF THE SPIDER'S WEB
See THRONE OF BLOOD
CAT WITH JADE EYES
See CAT'S VICTIM, THE
CATCH THE HEAT
See FEEL THE HEAT
CATHERINE THE GREAT
See RISE OF CATHERINE THE GREAT, THE
CATHY TIPPEL
See KATIE'S PASSION: A GIRL CALLED KATIE TIPPEL
CAT'S PAW, THE
See HAROLD LLOYD: THE CAT'S PAW
CAVE DWELLERS
See ATOR THE INVINCIBLE 2
CAVE SLAVES B.C.
See NEW BARBARIANS, THE
CELIA CHILD OF TERROR
See CELIA
CELINE ET JULIE VONT EN BATEAU
See CELINE AND JULIE GO BOATING
CEMETERY HIGH
See SCUMBUSTERS
C'ERA UNA VOLTA IL WEST
See ONCE UPON A TIME IN THE WEST
CEREMONY
See SPENSER: CEREMONY
CERTAIN MR SCRATCH, A
See ALL THAT MONEY CAN BUY
C'EST ARRIVE PRES DE CHEZ VOUZ
See MAN BITES DOG
CET OBSCUR OBJECT DU DESIR
See THAT OBSCURE OBJECT OF DESIRE
CHACUN CHERCHE SON CHAT
See WHEN THE CAT'S AWAY
CHAINS, THE: WARRIORS OF CHICAGO
See CHAINS
CHAINSAW HOOKERS
See HOLLYWOOD CHAINSAW HOOKERS
CHALLENGE OF NINJA
See CHALLENGE THE NINJA
CHALLENGE OF THE NINJA
See CHALLENGE THE NINJA
CHAMBER OF TORTURES

See BARON BLOOD
CHAMELEON
See MAY THE BEST MAN
WIN
CHAMPION ON FIRE
See NINJA OPERATION 8:
CHAMPION ON FIRE
CHAMPIONS
See MIGHTY DUCKS, THE
CHANGE OF HABIT
See ELVIS PRESLEY:
CHANGE OF HABIT
CHANGER, THE
See NOSTRIL PICKER, THE
**CHARLEY'S TANTE
NACKT**
See SEXY DOZEN, THE
**CHARLIE'S GHOST
STORY**
See CHARLIE'S GHOST:
THE SECRET OF
CORONADO
CHASTITY BELT, THE
See UP THE CHASTITY
BELT
CHATORAN
See ADVENTURES OF
MILO AND OTIS, THE
CHAUTAQUA
See ELVIS PRESLEY: THE
TROUBLE WITH GIRLS
**CHERUBIM AND
SERAPHIM**
See INSPECTOR MORSE:
CHERUBIM AND
SERAPHIM
**CHERYL HANSON:
COVER GIRL**
See COVER GIRL
**CHIKIYU KOGERI
MEIREI**
See GODZILLA: WAR OF
THE MONSTERS
CHILD OF SATAN
See TO THE DEVIL A
DAUGHTER
**CHILDREN OF
PARADISE**
See LES ENFANTS DU
PARADIS
**CHILDREN OF THE
CORN 3: URBAN
HARVEST**
See CHILDREN OF THE
CORN 3
**CHILDREN OF THE
NIGHT**
See DAUGHTERS OF
DARKNESS
**CHILD'S PLAY 3: LOOK
WHO'S STALKING**
See CHILD'S PLAY 3
**CHINESE CONNECTION,
THE**
See FISTS OF FURY
CHINESE WEB, THE
See SPIDERMAN: THE
DRAGON'S CHALLENGE
CHING-WU MEW SU-TSI
See FISTS OF FURY 2
CHINO
See VALDEZ: THE HALF-
BREED
**CHLOE IN THE
AFTERNOON**
See LOVE IN THE
AFTERNOON
CHONGQING SENLIA
See CHUNGKING EXPRESS
CHRISTMAS VACATION
See NATIONAL
LAMPOON'S CHRISTMAS
VACATION

**CHRONICLES OF
BENJAMIN KNIGHT, THE**
See INVISIBLE: THE
CHRONICLES OF
BENJAMIN KNIGHT
CHUMP AT OXFORD, A
See LAUREL AND HARDY:
A CHUMP AT OXFORD
CHUNG ON TSOU
See CRIME STORY
CIAO MASCHIO
See BYE BYE MONKEY
CIBLE EMOUVANTE
See WILD TARGET
CINDERFELLA
See JERRY LEWIS:
CINDERFELLA
CIRCUS, THE
See CHARLIE CHAPLIN:
THE CIRCUS
CITY JUNGLE, THE
See YOUNG
PHILADELPHIANS, THE
CITY LIGHTS
See CHARLIE CHAPLIN:
CITY LIGHTS
**CITY SLICKERS 2: THE
LEGEND OF CURLY'S
GOLD**
See CITY SLICKERS 2
CLAMBAKE
See ELVIS PRESLEY:
CLAMBAKE
**CLASS OF 1999 PART 2:
THE SUBSTITUTE**
See CLASS OF 1999 2
CLASS REUNION
See NATIONAL
LAMPOON'S CLASS
REUNION
CLEO DE 5 A 7
See CLEO FROM 5 TO 7
CLERKS, THE
See CLERKS
**CLIMATE FOR KILLING,
A**
See ROW OF CROWS, A
**CLIMAX 2000 PART 1:
THE PHANTOM'S
CURSE**
See CLIMAX 2000
**CLIVE BARKER'S LORD
OF ILLUSIONS**
See LORD OF ILLUSIONS
CLOCKMAKER, THE
See WATCHMAKER OF
SAINT-PAUL, THE
**CLOSE ENCOUNTERS
OF THE SPOOKY KIND 2**
See SPOOKY
ENCOUNTERS
**CLOSELY OBSERVED
TRAINS**
See CLOSELY WATCHED
TRAINS
CLOUD WALTZING
See CLOUD WALTZER
COAST OF TERROR
See SUMMER CITY
COASTWATCHER
See LAST WARRIOR, THE
COBRA VERSUS NINJA
See COBRA AGAINST
NINJA
COBWEB CASTLE
See THRONE OF BLOOD
COCOON 2
See COCOON: THE
RETURN
CODE NAME: TRIXIE
See CRAZIES, THE
COLD CUTS
See BUFFET FROID

COLLEGE
See BUSTER KEATON:
COLLEGE
COLONEL BLIMP
See LIFE AND DEATH OF
COLONEL BLIMP, THE
**COLUMBO AND THE
MURDER OF A ROCK
STAR**
See COLUMBO: MURDER
OF A ROCK STAR
COME BACK PETER
See SOME LIKE IT SEXY
**COMO AGUA PARA
CHOCOLATE**
See LIKE WATER FOR
CHOCOLATE
**COMO SER MUJER Y
NO MORIR EN EL
INTENTO**
See HOW TO BE A
WOMAN AND NOT DIE IN
THE ATTEMPT
COMPANION, THE
See BETRAYAL
**COMPANY 2:
SACRIFICES**
See COMPANY 2, THE
COMPUTER KILLERS
See HORROR HOSPITAL
**COMRADES OF
SUMMER, THE**
See COMRADES OF
SUMMER
COMTES IMMORALS
See IMMORAL TALES
CONAGHER
See LOUIS L'AMOUR'S
CONAGHER
**CONAN, KING OF THE
THIEVES**
See CONAN THE
DESTROYER
**CONCORDE AFFAIRE
SEVENTY-NINE**
See CONCORDE AFFAIR,
THE
**CONCRETE JUNGLE,
THE**
See CRIMINAL, THE
CONCRETE WAR
See LAST HOUR, THE
**CONFESSIONS OF A
HITMAN**
See FALLEN ANGELS
CONFIDENCE
See SWEET TALKER
**CONFIDENTIALLY
YOURS**
See FINALLY SUNDAY!
CONFRONTATION, THE
See AMERICAN NINJA 2:
THE CONFRONTATION
**CONQUEROR WORM,
THE**
See WITCHFINDER
GENERAL
CONTAMINATION 7
See CREEPERS
CONTE DE PRINTEMPS
See TALE OF SPRINGTIME,
A
CONTE D'ETE
See SUMMER'S TALE, A
CONTE D'HIVER
See WINTER'S TALE, A
CONTES IMMORAUX
See IMMORAL TALES
**CONTRACT FOR
MURDER**
See TELLING SECRETS
**CONVICTION: THE
KITTY DODDS STORY**

See CONVICTION OF
KITTY DODDS, THE
COPPER BEACHES, THE
See ADVENTURES OF
SHERLOCK HOLMES, THE:
THE COPPER BEACHES
**CORPSES SHOW SIGNS
OF CARNAL VIOLENCE,
THE**
See TORSO
CORRIDORS OF EVIL
See CARNIVAL OF SOULS
**CORSICAN BROTHERS,
THE**
See CHEECH AND
CHONG'S THE CORSICAN
BROTHERS, THE
COSA NOSTRA
See VALACHI PAPERS,
THE
**COUNT DRACULA AND
HIS VAMPIRE BRIDE**
See SATANIC RITES OF
DRACULA, THE
COURT MARTIAL
See CARRINGTON V.C.
COVER HER FACE
See DALGLIESH: COVER
HER FACE – PARTS 1 AND
2
COVERT ASSASSIN
See WILD JUSTICE
CRACKING UP
See JERRY LEWIS:
SMORGASBORD
CRACKSMAN, THE
See CHARLIE DRAKE: THE
CRACKSMAN
CRASH OF SILENCE
See MANDY
CRAZE
See CREEPING FLESH, THE
CRAZED VAMPIRE, THE
See REQUIEM FOR A
VAMPIRE
CRAZY HONG KONG
See GODS MUST BE
CRAZY 4, THE
CRAZYSITTER, THE
See TWO MUCH TROUBLE
CREATED TO KILL
See EMBRYO
CREEPING MAN, THE
See CASEBOOK OF
SHERLOCK HOLMES, THE:
THE CREEPING MAN
CREEPS
See BLOODY BIRTHDAY
CRIA CUERVOS
See RAISE RAVENS
CRIA!
See RAISE RAVENS
CRIES IN THE NIGHT
See AWFUL DR ORLOFF,
THE
CRIMEBROKER
See CORRUPT JUSTICE
CRIMEWAVE
See L.A. CRIMEWAVE
CRISIS 2050
See SOLAR CRISIS
CRISSCROSS
See CRISS CROSS
**CRISTO SI E FERMATO
A EBOLI**
See CHRIST STOPPED AT
EBOLI
**CRITTERS 2: THE MAIN
COURSE**
See CRITTERS 2
**CRITTERS 4: CRITTERS
IN SPACE**
See CRITTERS 4

CRONACA DI UN AMORE
See CHRONICLE OF A LOVE
CRONOS: IMMORTAL CURSE
See CRONOS
CROOKED MAN, THE
See ADVENTURES OF SHERLOCK HOLMES, THE: THE CROOKED MAN
CROOKLYN: A SPIKE LEE JOINT!
See CROOKLYN
CROSSED SWORDS
See PRINCE AND THE PAUPER, THE
CROW, THE: CITY OF ANGELS
See CROW CITY OF ANGELS, THE
CROWN CAPER, THE
See THOMAS CROWN AFFAIR, THE
CRY IN THE WILD: THE TAKING OF PEGGY ANN
See CRY IN THE WILD
CRYPT OF THE BLIND DEATH, THE
See TOMBS OF THE BLIND DEAD
CRYSTAL EYE
See CURSE OF THE CRYSTAL EYE, THE
CURIOUS ADVENTURES OF MISTER WONDERBIRD, THE
See MISTER BIRD TO THE RESCUE
CURSE 2: THE BITE
See BITE, THE
CURSE 3: BLOOD SACRIFICE
See WITCHCRAFT
CURSE OF THE DEMON
See NIGHT OF THE DEMON
CURSE, THE
See XALA
CYBORG 2: THE GLASS SHADOW
See CYBORG 2
CYBORG 3: THE CREATION
See CYBORG 3: THE RECYCLER
CYBORG AGENT
See RUNNING DELILAH
CYBORG SOLDIER
See CYBORG COP 2
CYBORG: THE SIX MILLION DOLLAR MAN
See SIX MILLION DOLLAR MAN, THE
D.H. LAWRENCE: THE ROCKING HORSE WINNER
See ROCKING HORSE WINNER, THE
D.M.J.V. – THE INFERNO
See DEVIL IN MISS JONES: PART 5
DADDY NOSTALGIE
See THESE FOOLISH THINGS
DADDY WANTED
See LOOK WHO'S TALKING
DAD'S ARMY: THE MOVIE
See DAD'S ARMY
DAHONG DENGLONG GAOGAO GUA

See RAISE THE RED LANTERN
DALEKS: INVASION EARTH 2150 A.D.
See DOCTOR WHO: INVASION EARTH 2150 A.D.
DANCING MASTERS, THE
See LAUREL AND HARDY: THE DANCING MASTERS
DANCING MEN, THE
See ADVENTURES OF SHERLOCK HOLMES, THE: THE DANCING MEN
DANDELION
See TAMPOPO
DANGEROUS LIAISONS 1960
See LES LIAISONS DANGEREUSES
DANGEROUS LOVE AFFAIRS
See LES LIAISONS DANGEREUSES
DANGEROUS MAN, A: LAWRENCE AFTER ARABIA
See DANGEROUS MAN, A: LAWRENCE OF ARABIA
DANGEROUS RELATIONS
See FATHER AND SON
DANIEL AND THE DEVIL
See ALL THAT MONEY CAN BUY
DANIELLE STEEL'S A PERFECT STRANGER
See PERFECT STRANGER, A
DANIELLE STEEL'S CHANGES
See CHANGES
DANIELLE STEEL'S CROSSINGS
See CROSSINGS
DANIELLE STEEL'S DADDY
See DADDY
DANIELLE STEEL'S FAMILY ALBUM
See FAMILY ALBUM
DANIELLE STEEL'S FINE THINGS
See FINE THINGS
DANIELLE STEEL'S HEARTBEAT
See HEARTBEAT
DANIELLE STEEL'S KALEIDOSCOPE
See KALEIDOSCOPE
DANIELLE STEEL'S MESSAGE FROM NAM
See MESSAGE FROM NAM
DANIELLE STEEL'S MIXED BLESSINGS
See MIXED BLESSINGS
DANIELLE STEEL'S NO GREATER LOVE
See NO GREATER LOVE
DANIELLE STEEL'S ONCE IN A LIFETIME
See ONCE IN A LIFETIME
DANIELLE STEEL'S PALOMINO
See PALOMINO
DANIELLE STEEL'S SECRETS
See SECRETS
DANIELLE STEEL'S STAR
See STAR
DANIELLE STEEL'S

VANISHED
See VANISHED
DANNY BOY
See ANGEL
DARIO ARGENTO'S DEMONS
See DEMONS
DARIO ARGENTO'S OPERA
See TERROR AT THE OPERA
DARIO ARGENTO'S WORLD OF HORROR
See WORLD OF HORROR
DARK BACKWARD, THE
See MAN WITH THREE ARMS, THE
DARK EYES OF LONDON, THE
See HUMAN MONSTER, THE
DARK FORTRESS
See NYMPHOID BARBARIAN IN DINOSAUR HELL
DARK HERO
See GUYVER, THE: DARK HERO
DARK HOLIDAY
See PASSPORT TO TERROR
DARK OBSESSION
See DIAMOND SKULLS
DARK REFLECTION
See NATURAL SELECTION
DARLING I AM GROWING YOUNGER
See MONKEY BUSINESS
DARYL
See D.A.R.Y.L.
DAS AMULETT DES TODES
See COLD BLOOD
DAS BOOT
See BOAT, THE
DAS BUMSFIDELE TOCHTERINTERNAT
See SEXY DOZEN, THE
DAS CABINET DES DR CALIGARI
See CABINET OF DOCTOR CALIGARI, THE
DAS SCHLANGENEI
See SERPENT'S EGG, THE
DAS SCHRECKLICHE MADCHEN
See NASTY GIRL, THE
DAS VERSPRECHEN
See PROMISE, THE
DATE WITH DEATH, A
See HIGH BRIGHT SUN, THE
DAVY CROCKETT, KING OF THE WILD FRONTIER
See DAVY CROCKETT
DAWANDEH
See RUNNER, THE
DAWN OF THE DEAD
See ZOMBIES: DAWN OF THE DEAD
DAY AT THE RACES, A
See MARX BROTHERS: A DAY AT THE RACES
DAY OF ANGER
See GUN LAW
DAY OF THE APOCALYPSE
See TIME SLIP
DAY OF THE DEVIL, THE
See INSPECTOR MORSE: THE DAY OF THE DEVIL
DAY OF WRATH
See GUN LAW
DAYBREAK

See BLOODSTREAM
DAYBREAK
See LE JOUR SE LEVE
DAYS OF WRATH
See GUN LAW
DE ESO NO SE HABLA
See WE DON'T WANT TO TALK ABOUT IT
DE LA PART DES COPAINS
See COLD SWEAT
DE LIFT
See LIFT, THE
DE NOORDERLINGEN
See NORTHERNERS, THE
DE ONFATSOENLIJKE VROUW
See INDECENT WOMAN, THE
DEAD AHEAD: THE EXXON VALDEZ DISASTER
See DISASTER AT VALDEZ
DEAD BY DAWN
See EVIL DEAD 2
DEAD CONNECTION
See FINAL COMBINATION
DEAD END: CRADLE OF CRIME
See DEAD END
DEAD IMAGE
See DEAD RINGER
DEAD KIDS
See SMALL TOWN MASSACRE
DEAD MAN SEEKS HIS MURDERER, A
See BRAIN, THE
DEAD OF JERICHO, THE
See INSPECTOR MORSE: THE DEAD OF JERICHO
DEAD ON
See RELENTLESS 2: DEAD ON
DEAD ON TIME
See INSPECTOR MORSE: DEAD ON TIME
DEAD ON: RELENTLESS 2
See RELENTLESS 2: DEAD ON
DEAD PLANET, THE
See DOCTOR WHO: THE DALEKS
DEAD TIME
See FREAKY FAIRY TALES
DEAD TO RIGHTS
See UNDER THREAT
DEADBOLT
See DEAD BOLT
DEADLIEST SIN, THE
See CONFESSION
DEADLOCK
See WEDLOCK
DEADLY DECEPTION
See DEADLY DECEPTIONS
DEADLY INVASION
See DEADLY INVASION: THE KILLER BEE NIGHTMARE
DEADLY MISSION, THE
See DIRTY DOZEN, THE: THE DEADLY MISSION
DEADLY MISTER FROST, THE
See MR FROST
DEADLY REUNION
See WALKER, TEXAS RANGER: DEADLY REUNION
DEADLY SIN, THE
See CHINA WHITE
DEADLY SLUMBER

See INSPECTOR MORSE:
DEADLY SLUMBER
DEADLY SUMMER, THE
See ONE DEADLY
SUMMER
**DEADLY TREASURE OF
THE PIRANA**
See KILLERFISH
DEATH CAGE, THE
See BLOODFIGHT 2: THE
DEATHCAGE
DEATH CORPS
See SHOCK WAVES
DEATH DEALER, THE
See PSYCHIC KILLER
DEATH DORM
See PRANKS
**DEATH DUEL OF
MANTIS**
See STRIKE OF MANTIS
FIST
**DEATH FROM DOWN
UNDER**
See ODD ANGRY SHOT,
THE
DEATH LEGACY
See GIRL'S SCHOOL
SCREAMERS
**DEATH OF AN EXPERT
WITNESS**
See DALGLIESH: DEATH
OF AN EXPERT WITNESS –
PARTS 1 AND 2
**DEATH OF THE SELF,
THE**
See INSPECTOR MORSE:
THE DEATH OF THE SELF
DEATH STALKER
See DEATHSTALKER
DEATH WAVES
See SHOCK WAVES
**DEATHSTALKER 2:
DUEL OF THE TITANS**
See DEATHSTALKER 2
**DEATHSTALKER 3: THE
WARRIORS FROM HELL**
See DEATHSTALKER 3
**DEATHSTALKER 4:
CLASH OF THE TITANS**
See DEATHSTALKER 4:
MATCH OF TITANS
**DEATHSTALKER AND
THE WARRIORS FROM
HELL**
See DEATHSTALKER 3
DECEIVED BY FLIGHT
See INSPECTOR MORSE:
DECEIVED BY FLIGHT
DECEMBER
See INNOCENT WAR, AN
DECEPTION
See RUBY CAIRO
DEEP DARK SECRETS
See INTIMATE BETRAYAL
DEEP IN THE HEART
See HANDGUN
**DEEP RED: HATCHET
MURDERS**
See DEEP RED
**DEFENSE NEVER
RESTS, THE**
See PERRY MASON
RETURNS
**DEFY TO THE LAST
PARADISE**
See EATEN ALIVE
**DEIDRE HALL STORY,
THE**
See NEVER SAY NEVER:
THE DEIDRE HALL STORY
**DELINQUENT
SCHOOLGIRLS**
See DELINQUENTS

DELIRIA
See STAGE FRIGHT
**DELIVER THEM FROM
EVIL: THE TAKING OF
ALTA VIEW**
See UNDER PRESSURE
**DELLAMORTE,
DELLAMORE**
See CEMETERY MAN
**DELTA FORCE 2:
OPERATION
STRONGHOLD**
See DELTA FORCE 2
**DELTA FORCE 2: THE
COLOMBIAN
CONNECTION**
See DELTA FORCE 2
DELUSION
See HOUSE WHERE
DEATH LIVED, THE
DEMON, THE
See ONIBABA
DEMONI
See DEMONS
DEMONI 2
See DEMONS 2
**DEMONI 2: L'INCUBO
RITORNA**
See DEMONS 2
DEMONS 1
See DEMONS
**DEMONS 2: THE
NIGHTMARE BEGINS**
See DEMONS 2
**DEMONS 2: THE
NIGHTMARE
CONTINUES**
See DEMONS 2
**DEMONS 2: THE
NIGHTMARE RETURNS**
See DEMONS 2
DEMONS 3: THE OGRE
See DEMONS 3
DEMONS 4: THE SECT
See SECT, THE
DEMON'S MASK, THE
See BLACK SUNDAY
DEMONWORLD
See DEMONWORLD,
MESSENGER OF DEATH
DEN GODA VILJAN
See BEST INTENTIONS,
THE
**DENNIS BYRD STORY,
THE**
See RISE AND WALK: THE
DENNIS BYRD STORY
DENNIS THE MENACE
See DENNIS
DENTRO IL CIMITERO
See GRAVEYARD
DISTURBANCE
DER AANSLAG
See ASSAULT, THE
**DER AMERIKANISCHE
SOLDAT**
See AMERICAN SOLDIER,
THE
DER BEWEGTE MANN
See MOST DESIRED MAN,
THE
DER BLAUE ENGEL
See BLUE ANGEL, THE
**DER GOLEM, WIE ER IN
DIE WELT KAM**
See GOLEM, THE
**DER HIMMEL UBER
BERLIN**
See WINGS OF DESIRE
DER PROZESS
See TRIAL, THE
DER ROSENGARTEN
See ROSE GARDEN, THE

**DER SCHONSTE BUSEN
DER WELT**
See MOST BEAUTIFUL
BREASTS IN THE WORLD,
THE
DERSU USARA
See DERSU UZALA
DESERT ATTACK
See ICE-COLD IN ALEX
DESIGN FOR LOVE
See CONFESSIONS OF A
SEX MANIAC
DESIGN FOR LUST
See CONFESSIONS OF A
SEX MANIAC
DESIRE
See DIRTY LOVE 2: THE
LOVE GAMES
**DESTINATION
MOONBASE ALPHA**
See SPACE 1999:
DESTINATION
MOONBASE ALPHA
**DESTROY ALL
MONSTERS**
See GODZILLA: DESTROY
ALL MONSTERS
DET SJUNDE INSEGLET
See SEVENTH SEAL, THE
DETONATOR
See DEATH TRAIN
**DETONATOR 2:
NIGHTWATCH**
See NIGHT WATCH
DETSTVO GORKOVO
See CHILDHOOD OF
MAXIM GORKY, THE
**DEUX OU TROIS
CHOSES QUE JE SAIS
D'ELLE**
See TWO OR THREE
THINGS I KNOW ABOUT
HER
DEVICES AND DESIRES
See DALGLIESH: DEVICES
AND DESIRES
**DEVIL AND DANIEL
WEBSTER, THE**
See ALL THAT MONEY
CAN BUY
**DEVIL AND DR
FRANKENSTEIN, THE**
See ANDY WARHOL'S
FRANKENSTEIN
**DEVIL AND THE DEAD,
THE**
See LISA AND THE DEVIL
DEVIL IN A CONVENT
See DEVILS IN THE
CONVENT
**DEVIL IN THE HOUSE
OF EXORCISM**
See LISA AND THE DEVIL
DEVIL OBSESSION, THE
See SEXORCIST, THE
DEVIL WOMAN, THE
See ONIBABA
**DEVIL'S ADVOCATE,
THE**
See SORCERESS
DEVIL'S BRIDE, THE
See DEVIL RIDES OUT,
THE
DEVIL'S BROTHER, THE
See LAUREL AND HARDY:
BOGUS BANDITS
**DEVIL'S DAUGHTER,
THE**
See SECT, THE
DEVIL'S FOOT, THE
See RETURN OF
SHERLOCK HOLMES, THE:
THE DEVIL'S FOOT

DEVIL'S ODDS
See WILD PAIR, THE
DEVIL'S TREASURE
See TREASURE ISLAND
**DHARMAGA
TONGJOGURO KAN
KKADALGUN?**
See WHY DID BODHI-
DHARMA LEAVE FOR
THE EAST?
**DIABOLICAL DOCTOR
SATAN, THE**
See AWFUL DR ORLOFF,
THE
DIABOLIQUE
See LES DIABOLIQUES
DIABOLO MENTHE
See PEPPERMINT SODA
DIAL
See WILD JUSTICE
DIAMOND FLEECE, THE
See GREAT DIAMOND
ROBBERY, THE
DIAMOND MOUNTAIN
See SHADOW OF
CHIKARA, THE
DIAMOND'S EDGE
See JUST ASK FOR
DIAMOND
DIARY OF OHARU
See LIFE OF OHARU, THE
**DICK FRANCIS
MYSTERIES: BLOOD
SPORT**
See BLOOD SPORT
**DICK FRANCIS
MYSTERIES: IN THE
FRAME**
See IN THE FRAME
**DICK FRANCIS
MYSTERIES: TWICE SHY**
See TWICE SHY
DICK TRACY
See DICK TRACY: THE
SPIDER STRIKES
**DIDN'T YOU KILL MY
BROTHER?**
See COMIC STRIP
PRESENTS: DIDN'T YOU
KILL MY BROTHER?
**DIE ANGST DES
TORMANNS BEIM
ELFMETER**
See GOALKEEPER'S FEAR
OF THE PENALTY, THE
**DIE ANGST ESSEN
SEELE AUF**
See FEAR EATS THE SOUL
**DIE BITTEREN TRANEN
DER PETRA VON KANT**
See BITTER TEARS OF
PETRA VON KANT, THE
DIE BLECHTROMMEL
See TIN DRUM, THE
**DIE BRAUT DES
SATANS**
See TO THE DEVIL A
DAUGHTER
**DIE BUCHSE DER
PANDORA**
See PANDORA'S BOX
**DIE EHE DER MARIA
BRAUN**
See MARRIAGE OF MARIA
BRAUN, THE
**DIE FOLTERKAMMER
DES DR FU MANCHU**
See CASTLE OF FU
MANCHU, THE
**DIE HARD 2: DIE
HARDER**
See DIE HARD 2
DIE

JUNGFRAUENMASCH–
INE
See VIRGIN MACHINE
DIE KAMELIENDAME
See LADY OF THE
CARMELIAS, THE
DIE LOK
See RUNAWAY EXPRESS:
THE ENGINE
DIE MARQUISE VON O
See MARQUISE OF O, THE
DIE NEUNSCHWANZIGE
KATZE
See CAT O'NINE TAILS
DIE NIEBELUNGEN 1
See SIEGFRIED
DIE PASSION DES
DARKLY NOON
See PASSION OF DARKLY
NOON, THE
DIE REGENSCHIRME
VON CHERBOURG
See UMBRELLAS OF
CHERBOURG, THE
DIE RUCKKEHR DER
WILDGANSE
See COBRA MISSION
DIE SAGE DES TODES
See BLOODY MOON
DIE UNENDLISCHE
GESCHICHTE 3:
RETTUNG AUS
FANTASIA
See NEVERENDING STORY
3, THE
DIE VERLORENE EHRE
DER KATHARINA BLUM
See LOST HONOUR OF
KATHARINA BLUM, THE
DIE XUE SHUANG
XIONG
See KILLER, THE
DIE ZWOLFTE STUNDE
See NOSFERATU
DIGGSTOWN
See MIDNIGHT STING
DIGITAL KNIGHTS
See RAGEWAR
DIM SUM: A LITTLE BIT
OF HEART
See DIM SUM
DIMENTICLARE
PALERMO
See PALERMO
CONNECTION, THE
DINAH EAST
See HOLLYWOOD
SUPERSTAR
DIRTY DOZEN 3: THE
DEADLY MISSION
See DIRTY DOZEN, THE:
THE DEADLY MISSION
DIRTY LOVE 2
See DIRTY LOVE 2: THE
LOVE GAMES
DISAPPEARANCE OF
LADY CARFAX, THE
See CASEBOOK OF
SHERLOCK HOLMES, THE:
THE DISAPPEARANCE OF
LADY CARFAX
DISAPPEARANCE OF
NORA, THE
See DEADLY RECALL
DISASTER IN TIME
See TIMESCAPE
DISCIPLINE OF
DRACULA
See DRACULA: PRINCE OF
DARKNESS
DISTANT COUSINS
See DESPERATE MOTIVE
DIVERTIMENTO

See LA BELLE NOISEUSE
DIXIE RAY:
HOLLYWOOD STAR
See IT'S CALLED MURDER
BABY
DJANGO RIDES AGAIN
See KEOMA
DJANGO'S GREAT
RETURN
See KEOMA
DJAVULENS OGA
See DEVIL'S EYE, THE
DOBERMAN PATROL
See TRAPPED
DOCK BRIEF, THE
See TRIAL AND ERROR
DOCTEUR PETIOT
See DOCTOR PETIOT
DOCTOR BETHUNE
See BETHUNE: THE
MAKING OF A HERO
DOCTOR BLOODBATH
See HORROR HOSPITAL
DOCTOR FROM SEVEN
DIALS
See CORRIDORS OF
BLOOD
DOCTOR MABUSE: THE
FATAL PASSION
See DOCTOR MABUSE:
THE GAMBLER
DOCTOR PHIBES RISES
FROM THE GRAVE
See DOCTOR PHIBES RISES
AGAIN
DOCTOR
STRANGELOVE, OR
HOW I LEARNED TO
STOP WORRYING AND
LOVE THE BOMB
See DR STRANGELOVE
DOES, THE
See BAD GIRLS
DOING TIME
See PORRIDGE
DOKTOR MABUSE: DER
SPIELER
See DOCTOR MABUSE:
THE GAMBLER
DOM ZA VESANJE
See TIME OF THE GYPSIES
DOMICILE CONJUGAL
See BED AND BOARD
DOMINICK AND
EUGENE
See NICKY AND GINO
DON BLUTH'S
THUMBELINA
See THUMBELINA
DON JUAN
See IF DON JUAN WERE A
WOMAN
DON JUAN 73
See IF DON JUAN WERE A
WOMAN
DON JUAN 73 OR IF
DON JUAN WERE A
WOMAN
See IF DON JUAN WERE A
WOMAN
DON JUAN 73 OU SI
DON JUAN ETAIT UNE
FEMME
See IF DON JUAN WERE A
WOMAN
DON JUAN AND THE
CENTREFOLD
See DON JUAN DE MARCO
DONA FLOR E SEUS
DOS MARIDOS
See DONA FLOR AND HER
TWO HUSBANDS
DONA HERLINDA Y SU

HIJO
See DONA HERLINDA
AND HER SON
DONATO AND
DAUGHTER
See UNDER THREAT
DONG-CHUN DE RIZI
See DAYS, THE
DONGFANG SAN XIA
See HEROIC TRIO, THE
DONNA D'ONORE
See VENDETTA
DON'T LOSE YOUR
HEAD
See CARRY ON DON'T
LOSE YOUR HEAD
DON'T MOVE, DIE AND
RESUSCITATE!
See DON'T MOVE, DIE
AND RISE AGAIN!
DON'T TOUCH MY
DAUGHTER
See KIDNAPPED
DON'T TOUCH WHITE
WOMEN
See DON'T TOUCH THE
WHITE WOMAN!
DONZOKO
See LOWER DEPTHS, THE
DOOMED
See LIVING
DOOMED TO DIE
See EATEN ALIVE
DOP BEY KUAN WAN
See ONE-ARMED BOXER
DOPE ADDICT
See REEFER MADNESS
DOPED YOUTH
See REEFER MADNESS
DOPPELGANGER: THE
EVIL WITHIN
See DOPPELGANGER
DORM THAT DRIPPED
BLOOD, THE
See PRANKS
DOTTIE WEST STORY,
THE
See BIG DREAMS AND
BROKEN HEARTS: THE
DOTTIE WEST STORY
DOUBLE DARE
See RED SHOE DIARIES 2:
DOUBLE DARE
DOUBLE DEADLY
See SILENT PARTNER, THE
DOUBLE JEOPARDY
See OLIVIA
DOUBLE PLAY
See PRISONER OF ZENDA
INC., THE
DOUBLE TROUBLE
See ELVIS PRESLEY:
DOUBLE TROUBLE
DOWN AND DIRTY
See UGLY, DIRTY AND
BAD
DR CRIPPEN
See DOCTOR CRIPPEN
DRACULA
See HORROR OF
DRACULA
DRACULA
See NOSFERATU
DRACULA AND THE
SEVEN GOLDEN
VAMPIRES
See LEGEND OF THE
SEVEN GOLDEN
VAMPIRES, THE
DRACULA CERCA
SANGUE DI VERGINE E
MORI DI SETE
See ANDY WARHOL'S

DRACULA
DRACULA IS DEAD AND
WELL AND LIVING IN
LONDON
See SATANIC RITES OF
DRACULA, THE
DRACULA TODAY
See DRACULA A.D. 1972
DRACULA VUOLE
VIVERE: CERCA
SANGUE DI VERGINA
See ANDY WARHOL'S
DRACULA
DRACULA'S BRIDE
See DRACULA SUCKS
DRACULA'S DOG
See ZOLTAN, HOUND OF
HELL
DRAGON BALL: THE
MOVIE
See DRAGON BALL: THE
MAGIC BEGINS
DRAGON PEARL
See DRAGON BALL: THE
MAGIC BEGINS
DRAGON THAT WASN'T
(OR WAS HE?), THE
See DRAGON THAT
WASN'T... OR WAS HE?
DREAM MAKER, THE
See IT'S ALL HAPPENING
DREI VATERUNSER FUR
VIER HALUNKEN
See STORMRIDER
DRESSED TO KILL
See SHERLOCK HOMES:
DRESSED TO KILL
DRIPPING DEEP RED
See DEEP RED
DRIVEN TO
DISTRACTION
See INSPECTOR MORSE:
DRIVEN TO DISTRACTION
DROLE D'ENDROIT
POUR UNE RENCONTRE
See STRANGE PLACE TO
MEET, A
DRUMS
See DRUM, THE
DUCK SOUP
See MARX BROTHERS:
DUCK SOUP
DUEL OF THE DRAGON
See DUEL OF THE
DRAGONS
DUEL OF THE SEVEN
TIGERS
See SHADOW OF THE
TIGER
DUEL OF THE TITANS
See DEATHSTALKER 2
DUNGEON OF TERROR
See REQUIEM FOR A
VAMPIRE
DUNGEONMASTER, THE
See RAGEWAR
DUOLUO TIANSHI
See FALLEN ANGELS
DUSTED
See DEATH WARRANT
DUTCH
See DRIVING ME CRAZY
DYING TRUTH
See DISTANT SCREAM, A
DYNASTY OF BLOOD
See BLOOD BROTHERS
DYNASTY OF FEAR
See FEAR IN THE NIGHT
E LA NAVE VA
See AND THE SHIP SAILS
ON
E TU VIVRAI NEL
TERRORE! L'ALDILA

See BEYOND, THE
EAGLE'S CLAW
See EAGLE FIST, THE
EAGLE'S SHADOW, THE
See SNAKE IN THE
EAGLE'S SHADOW
EARLY BIRD, THE
See NORMAN WISDOM:
THE EARLY BIRD, THE
EARTH MOVED, THE
See LA TERRA TREMA
EARTH TREMBLES, THE
See LA TERRA TREMA
**EARTH VERSUS THE
SPIDER**
See SPIDER, THE
**EARTH WILL TREMBLE,
THE**
See LA TERRA TREMA
EAST OF SHANGHAI
See RICH AND STRANGE
EASY COME, EASY GO
See ELVIS PRESLEY: EASY
COME, EASY GO
**EATEN ALIVE BY THE
CANNIBALS**
See EATEN ALIVE
EATEN ALIVE!
See DEATH TRAP
EBBIE
See MIRACLE AT
CHRISTMAS: EBBIE'S
STORY
**EBIRAH – HORROR OF
THE DEEP**
See GODZILLA: EBIRAH –
HORROR OF THE DEEP
EBOLI
See CHRIST STOPPED AT
EBOLI
ECHOES OF WIZARDRY
See IRON WARRIOR
**ECOLOGIA DEL
DELITTO**
See LAST HOUSE ON THE
LEFT PART 2, THE
**ECOLOGY OF A CRIME,
THE**
See LAST HOUSE ON THE
LEFT PART 2, THE
**EDES EMMA, DRAGA
BOBE**
See SWEET EMMA, DEAR
BOBE
**EDGAR ALLAN POE'S
BURIED ALIVE**
See BURIED ALIVE
**EDGAR ALLAN POE'S
CONQUEROR WORM**
See WITCHFINDER
GENERAL
**EERIE MIDNIGHT
HORROR SHOW, THE**
See SEXORCIST, THE
**EFFICIENCY EXPERT,
THE**
See SPOTSWOOD
**EGGS FROM 70 MILLION
B.C.**
See JOSH KIRBY: TIME
WARRIOR! – EGGS FROM
70 MILLION B.C.
EGYMASRA NEZRE
See ANOTHER WAY
EIN LIED FUR BEKO
See SONG FOR BEKO, A
**EIN TOTER SUCHT
SEINEN MORDER**
See BRAIN, THE
**EINE FRANZOSISCHE
FRAU**
See FRENCH WOMAN, A
EINE NACHT DES

GRAUENS
See NOSFERATU
**EINE SYMPHONIE DES
GRAVENS**
See NOSFERATU
**EL ANGEL
EXTERMINADOR**
See EXTERMINATING
ANGEL, THE
**EL ATAQUE DE LOS
MUERTOS SIN OJOS**
See RETURN OF THE EVIL
DEAD, THE
**EL CASTILLO DE FU
MANCHU**
See CASTLE OF FU
MANCHU, THE
**EL CAZADOR DE LA
MUERTE**
See DEATHSTALKER
EL CUCHO QUIEN SABE
See BULLET FOR THE
GENERAL, A
EL DIA DE LA BESTIA
See DAY OF THE BEAST,
THE
**EL DIABOLO SE LLEVA
A LOS MUERTOS**
See LISA AND THE DEVIL
**EL ESPIRITU DE LA
COLMENA**
See SPIRIT OF THE
BEEHIVE, THE
**EL MAESTRO DE
ESGRIMA**
See FENCING MASTER,
THE
EL PATRULLERO
See HIGHWAY
PATROLMAN
**EL RETORNO DEL
TERROR CIEGO**
See RETURN OF THE EVIL
DEAD, THE
**EL SADICO DE NOTRE
DAME**
See DEMONIAC
EL SOL DE MEMBRILLO
See QUINCE TREE SUN,
THE
EL VIAJE
See VOYAGE, THE
ELECTRA
See ELECTRA GLIDE IN
BLUE
**ELECTRIC BOOGALOO,
BREAKIN' 2**
See BREAKDANCE 2:
ELECTRIC BOOGALOO
ELECTRIC DREAMS
See ELECTRIC DREAMS:
THE MOVIE
**ELECTRIC
HOLLYWOOD: THE NEW
BARBARIANS**
See NEW BARBARIANS,
THE
**ELENYA: IN
KRIEGSZEITEN**
See ELENYA
**ELEPHANTS NEVER
FORGET**
See ZENOBIA
**ELEVATOR TO THE
GALLOWS**
See LIFT TO THE
SCAFFOLD
**ELIGIBLE BACHELOR,
THE**
See ADVENTURES OF
SHERLOCK HOLMES, THE:
THE ELIGIBLE BACHELOR
ELIZABETH THE QUEEN

See PRIVATE LIVES OF
ELIZABETH AND ESSEX,
THE
ELLES N'OUBLIENT PAS
See LOVE IN THE
STRANGEST WAY
ELVIS
See ELVIS: THE MOVIE
**ELVIS AND THE
COLONEL: THE UNTOLD
STORY**
See ELVIS AND THE
COLONEL
**EMANUELLE AROUND
THE WORLD**
See EMMANUELLE IN
AMERICA
**EMANUELLE IN
AMERICA**
See EMMANUELLE IN
AMERICA
EMERALD JUNGLE, THE
See EATEN ALIVE
**EMILY BRONTE'S
WUTHERING HEIGHTS**
See WUTHERING
HEIGHTS
EMMANUELLE 7
See EMMANUELLE IN
SEVENTH HEAVEN
**EMMANUELLE GOES
EAST**
See BLACK EMMANUELLE
GOES EAST
**EMMANUELLE IN
AFRICA**
See BLACK EMMANUELLE
**EMMANUELLE IN
BANGKOK**
See BLACK EMMANUELLE
GOES EAST
**EMMANUELLE
L'ANTIVIERGE**
See EMMANUELLE 2
**EMMANUELLE NEGRA
EN AMERICA**
See EMMANUELLE IN
AMERICA
EMMANUELLE NERA
See BLACK EMMANUELLE
**EMMANUELLE NERA:
ORIENT REPONTAGE**
See BLACK EMMANUELLE
GOES EAST
EMMANUELLE SEVEN
See EMMANUELLE IN
SEVENTH HEAVEN
**EMMANUELLE, THE
JOYS OF A WOMAN**
See EMMANUELLE 2
**EMMANUELLE: PERCHE
VIOLENZA ALLE
DONNE?**
See CONFESSIONS OF
EMMANUELLE
**EMMANUELLE'S 7TH
HEAVEN**
See EMMANUELLE IN
SEVENTH HEAVEN
**EMPEROR OF THE
NORTH POLE**
See EMPEROR OF THE
NORTH
EMPIRE OF PASSION
See AI NO BOREI
EMPTY HOUSE, THE
See RETURN OF
SHERLOCK HOLMES, THE:
THE EMPTY HOUSE
**EN COMPAGNIE
D'ANTONIN ARTAUD**
See MY LIFE AND TIMES
WITH ANTONIN ARTAUD

**EN EFFLUEILLANT LA
MARGUERITE**
See PLUCKING THE DAISY
ENCINO MAN
See CALIFORNIA MAN
**ENDGAME, GIOCCIO
FINALE**
See ENDGAME
**ENDGAME: BRONX
LOTTA FINALE**
See ENDGAME
**ENDGAME: FINAL
BRONX STRUGGLE**
See ENDGAME
ENDGAMES
See ENDGAME
ENDLESS DESCENT
See RIFT, THE
**ENEMIES OF THE
PUBLIC**
See PUBLIC ENEMY, THE
**ENID BLYTON'S THE
SECRET OF SPIGGY
HOLES**
See SECRET OF SPIGGY
HOLES, THE
ENID IS SLEEPING
See OVER HER DEAD
BODY
**ENTEBBE: OPERATION
THUNDERBOLT**
See OPERATION
THUNDERBOLT
ENTER THE DEVIL
See SEXORCIST, THE
ENTRE TINIEBLAS
See DARK HABITS
EPISODA DEL MARE
See LA TERRA TREMA
**ERAN RIKLIS' CUP
FINAL**
See CUP FINAL
**ERNEST HEMINGWAY'S
THE KILLERS**
See KILLERS, THE
EROTIC ADVENTURE
See TORRID WITHOUT A
CAUSE
**EROTIC ADVENTURES
OF DON QUIXOTE, THE**
See UP THE CHASTITY
BELT
**EROTIC MENAGE A
TROIS**
See EMILIENNE
**EROTIC RITES OF
FRANKENSTEIN, THE**
See VIRGIN AMONG THE
LIVING DEAD
ERRAND BOY, THE
See JERRY LEWIS: THE
ERRAND BOY
ERZEBETH
See DAUGHTERS OF
DARKNESS
**ESCAPE FROM
JUPITER: THE MOVIE**
See ESCAPE FROM
JUPITER
**ESCAPE FROM
SAFEHAVEN**
See INFERNO IN
SAFEHAVEN
**ESCAPE OF
MEGAGODZILLA, THE**
See GODZILLA:
MONSTERS FROM AN
UNKNOWN PLANET
ESCAPE TO FREEDOM
See JUDGEMENT IN
BERLIN
**ET DIEU CREA LA
FEMME**

See AND GOD CREATED
WOMAN
ETAT DE SIEGE
See STATE OF SIEGE
ETERNAL RETURN, THE
See LOVE ETERNAL
**ETERNELLE
EMMANUELLE**
See EMMANUELLE
FOREVER
EUROPEAN VACATION
See NATIONAL
LAMPOON'S EUROPEAN
VACATION
**EVERY MAN FOR
HIMSELF**
See SLOW MOTION
**EVERY MAN FOR
HIMSELF AND GOD
AGAINST ALL**
See ENIGMA OF KASPAR
HAUSER, THE
**EVERYBODY'S BABY:
THE RESCUE OF
JESSICA McCLURE**
See RESCUE OF JESSICA
McCLURE, THE
**EVERYBODY'S
CHEERING**
See TAKE ME OUT TO THE
BALL GAME
EVIDENCE OF LOVE
See KILLING IN A SMALL
TOWN, A
**EVIL DEAD 2: DEAD BY
DAWN**
See EVIL DEAD 2
EVIL DEAD 3
See ARMY OF DARKNESS
EVIL OBSESSION
See ILLEGAL ENTRY
**EWOK ADVENTURE,
THE**
See CARAVAN OF
COURAGE: AN EWOK
ADVENTURE
**EXCESS AND
PUNISHMENT**
See EGON SCHIELE
**EXCESSIVE FORCE 2:
FORCE ON FORCE**
See EXCESSIVE FORCE 2
EXECUTORS, THE
See SICILIAN CROSS
EXILED IN AMERICA
See EXILED
EXORCISME
See DEMONIAC
**EXORCISME ET
MESSES NOIRES**
See DEMONIAC
EXORCIST 3: LEGION
See EXORCIST 3, THE
EXPEDITIONS, THE
See MARTIAN
CHRONICLES, THE
**EXPERIMENT IN FEAR,
AN**
See MONKEY SHINES: AN
EXPERIMENT IN FEAR
EXTASE
See ECSTASY
**EYE OF THE EVIL DEAD,
THE**
See POSSESSED, THE
EYES OF AN ANGEL
See TENDER, THE
EYEWITNESS
See JANITOR, THE
F/X
See FX: MURDER BY
ILLUSION
F/X2

See FX2: THE DEADLY ART
OF ILLUSION
F.B.I. MURDERS, THE
See IN THE LINE OF DUTY:
THE F.B.I. MURDERS
FACE OF FEAR
See PEEPING TOM
FACE, THE
See MAGICIAN, THE
FACES OF FEAR
See OLIVIA
**FALCON'S MALTESER,
THE**
See JUST ASK FOR
DIAMOND
FALL BREAK
See MUTILATOR, THE
FALLING, THE
See MUTANT 2
FALSCHE BEWEGUNG
See WRONG MOVE, THE
FAME
See FAME: THE MOVIE
FAMILY MATTER, A
See VENDETTA
FAMILY PRAYERS
See FAMILY DIVIDED, A
FAREWELL, MY LOVELY
See MURDER, MY SWEET
FARINELLI
See FARINELLI IL
CASTRATO
FARM, THE
See CURSE, THE
FAT CHANCE
See INSPECTOR MORSE:
FAT CHANCE
**FAT MAN AND LITTLE
BOY**
See SHADOW MAKERS
**FATAL PASSION OF
DOCTOR MABUSE, THE**
See DOCTOR MABUSE:
THE GAMBLER
FATAL WOMAN
See FEMME TALE
FATHER AND MASTER
See PADRE PADRONE
**FATHER AND SON:
DANGEROUS
RELATIONS**
See FATHER AND SON
**FAUSTRECT DER
FREIHEIT**
See FOX
**FEAR IN THE CITY OF
THE LIVING DEAD**
See CITY OF THE LIVING
DEAD
FEAR, THE
See CITY OF THE LIVING
DEAD
**FEARLESS VAMPIRE
KILLERS OR: PARDON
ME, BUT YOUR TEETH
ARE IN MY NECK, THE**
See DANCE OF THE
VAMPIRES
**FEARLESS VAMPIRE
KILLERS, THE**
See DANCE OF THE
VAMPIRES
FEET FIRST
See HAROLD LLOYD: FEET
FIRST
FELLINI SATYRICON
See SATYRICON
FELLINI'S CASANOVA
See CASANOVA
FELLINI'S ROMA
See ROMA
FELLINI'S SATYRICON
See SATYRICON

FEMALE
See VIOLENT YEARS, THE
**FEMALE PLASMA
SUCKERS**
See BLOOD ORGY OF THE
SHE-DEVILS
FERNGULLY
See FERNGULLY: THE
LAST RAINFOREST
FESTIVE DESSERTS
See WOMAN IN RED, THE
FIENDS, THE
See LES DIABOLIQUES
FIGHTER, THE
See KICK FIGHTER
**FIGHTING PIMPERNEL,
THE**
See PIMPERNEL SMITH
FILM BLEU
See THREE COLOURS:
BLUE
**FILTHY SLEAZY
SCOUNDRELS**
See FILTHY ROTTEN
SCOUNDRELS
FINAL CONFLICT, THE
See OMEN 3: THE FINAL
CONFLICT
**FINAL CONFLICT, THE:
OMEN 3**
See OMEN 3: THE FINAL
CONFLICT
FINAL CRASH, THE
See STEELYARD BLUES
FINAL EXECUTOR
See FINAL EXECUTIONER
FINAL FRONTIER, THE
See STAR TREK 5: THE
FINAL FRONTIER
FINAL PROBLEM, THE
See ADVENTURES OF
SHERLOCK HOLMES, THE:
THE FINAL PROBLEM
**FINAL ROLL OF THE
DICE, THE**
See MacSHAYNE: THE
FINAL ROLE OF THE DICE
FINDERS KEEPERS
See CLIFF RICHARD:
FINDERS KEEPERS
FIREBIRDS
See WINGS OF THE
APACHE
**FIRST BLOOD, LAST
RITES**
See VIGIL
FIRST BORN
See MOVING IN
**FIRST FORTY YEARS,
THE**
See MY FIRST FORTY
YEARS
**FIRST LOVE... MISTER
LOVE**
See MISTER LOVE
FIRSTBORN
See MOVING IN
FIST
See F.I.S.T.
FIST OF DRAGON
See FISTS OF DRAGONS
FIST OF FURY
See FISTS OF FURY
FIST OF FURY 2
See FISTS OF FURY 2
FIST OF FURY: PART 2
See FISTS OF FURY 2
FIST OF NORTH STAR
See FIST OF THE NORTH
STAR
FIST OF POWER
See KARATE WARRIOR
FIST OF SHAOLIN

See FISTS OF SHAOLIN
FIST OF VENGEANCE
See FISTS OF VENGEANCE,
THE
**FIST-RIGHT OF
FREEDOM**
See FOX
FIVE DEADLY VENOMS
See FIVE VENOMS
**FIVE FINGERS OF
DEATH**
See KING BOXER
**FIVE FINGERS OF
DEATH**
See SHAOLIN MARTIAL
ARTS
**FIVE MILLION YEARS
TO EARTH**
See QUATERMASS AND
THE PIT
FLAME OVER INDIA
See NORTHWEST
FRONTIER
FLAMING STAR, THE
See ELVIS PRESLEY: THE
FLAMING STAR
FLASHBACK
See WALKER, TEXAS
RANGER: FLASHBACK
FLATFOOT IN AFRICA
See FLATFOOT
FLAVIA
See FLAVIA THE HERETIC
**FLAVIA LA MONACA
MUSULMANA**
See FLAVIA THE HERETIC
FLEISCH
See SPARE PARTS
**FLESH FOR
FRANKENSTEIN**
See ANDY WARHOL'S
FRANKENSTEIN
**FLESH GORDON MEETS
THE COSMIC
CHEERLEADERS**
See FLESH GORDON 2
FLIGHT 007
See CODED HOSTILE
**FLIGHT OF BLACK
ANGEL**
See FLIGHT OF THE
BLACK ANGEL
FLIGHT, THE
See TAKING OF FLIGHT
847, THE
**FLIPPER AND THE
PIRATES**
See FLIPPER'S NEW
ADVENTURE
FLYING ACES
See LAUREL AND HARDY:
THE FLYING DEUCES
FLYING DEUCES, THE
See LAUREL AND HARDY:
THE FLYING DEAUCES
FOG
See STUDY IN TERROR, A
FOLIES OF ELODIE, THE
See NAUGHTY BLUE
KNICKERS
FOLLOW A STAR
See NORMAN WISDOM:
FOLLOW A STAR
FOLLOW THAT CAMEL
See CARRY ON FOLLOW
THAT CAMEL
FOLLOW THAT DREAM
See ELVIS PRESLEY:
FOLLOW THAT DREAM
FONTAN
See FOUNTAIN, THE
**FONTANE'S EFFI
BRIEST**

See EFFI BRIEST
FOOL'S NIGHT
See KILLER PARTY
FOR KEEPS
See MAYBE BABY
FOR LOVE OR MONEY
See CONCIERGE, THE
FOR THE LOVE OF LISA
See THINGS CHANGE
FOR THE VERY FIRST TIME
See TILL I KISSED YA'
FORBIDDEN DANCE, THE
See LAMBADA: THE FORBIDDEN DANCE
FORBIDDEN LOVE
See FREAKS
FORBIDDEN PASSION
See NIGHT PLEASURES
FORBIN PROJECT, THE
See COLOSSUS: THE FORBIN PROJECT
FORCED TO FIGHT
See BLOODFIST 3: FORCED TO FIGHT
FORD FAIRLANE
See ADVENTURES OF FORD FAIRLANE, THE
FORESKIN GUMP
See FORREST HUMP
FORTY-EIGHT HOURS
See WENT THE DAY WELL?
FOUR DAYS IN DALLAS
See RUBY AND OSWALD
FOUR MUSKETEERS, THE: THE REVENGE OF MILADY
See FOUR MUSKETEERS, THE
FOURTH STORY
See BASIC DECEPTION
FOX AND HIS FRIENDS
See FOX
FOXFIRE STORY, THE
See FOXFIRE
FRA DIAVOLO
See LAUREL AND HARDY: BOGUS BANDITS
FRANK ZAPPA: 200 MOTELS
See TWO-HUNDRED MOTELS
FRANKENSTEIN
See ANDY WARHOL'S FRANKENSTEIN
FRANKENSTEIN EXPERIMENT
See ANDY WARHOL'S FRANKENSTEIN
FRANKENSTEIN MADE WOMAN
See FRANKENSTEIN CREATED WOMAN
FRANKIE AND JOHNNY
See ELVIS PRESLEY: FRANKIE AND JOHNNY
FRANKIE'S WAR
See FRANKIE'S HOUSE
FRANTIC
See LIFT TO THE SCAFFOLD
FRATERNALLY YOURS
See LAUREL AND HARDY: SONS OF THE DESERT
FRAU WIRTINS TOLLE TOCHTERLEIN
See DEVILS IN THE CONVENT
FREDDIE THE FROG
See FREDDIE AS F.R.0.7.
FREDDY'S DEAD, THE

FINAL NIGHTMARE
See NIGHTMARE ON ELM STREET, A: PART 6 – FREDDY'S DEAD, THE FINAL NIGHTMARE
FREDDY'S REVENGE
See NIGHTMARE ON ELM STREET, A: PART 2 – FREDDY'S REVENGE
FREE SPIRIT
See BELSTONE FOX, THE
FREE SPIRIT
See MAXIE
FREE TO LIVE
See HOLIDAY
FREE WILLY 2: THE ADVENTURE HOME
See FREE WILLY 2
FREEZE, DIE, COME TO LIFE!
See DON'T MOVE, DIE AND RISE AGAIN!
FRESA Y CHOCOLATE
See STRAWBERRY AND CHOCOLATE
FRESHMAN, THE
See HAROLD LLOYD: THE FRESHMAN
FRIED GREEN TOMATOES
See FRIED GREEN TOMATOES AT THE WHISTLESTOP CAFE
FRIENDS AND LOVERS: THE SEQUEL
See FRIENDS AND LOVERS: PART 2
FRIGHTMARE 2
See FRIGHTMARE
FROG DREAMING
See GO-KIDS, THE
FROM DOON WITH DEATH
See INSPECTOR WEXFORD: FROM DOON WITH DEATH
FROZEN TERROR
See MACABRE
FU MANCHU AND THE KISS OF DEATH
See BLOOD OF FU MANCHU, THE
FU MANCHU'S CASTLE
See CASTLE OF FU MANCHU, THE
FUGITIVE, THE: THE MOVIE
See FUGITIVE, THE
FUN IN ACAPULCO
See ELVIS PRESLEY: FUN IN ACAPULCO
FUNNY MAN
See FUNNYMAN
FUNSEEKERS, THE
See COMIC STRIP PRESENTS: THE FUNSEEKERS
FURTHER TALES FROM THE CRYPT
See VAULT OF HORROR
FUTURE COP
See TRANCERS
FUTURE KILL
See NIGHT OF THE ALIEN
G.I. BLUES
See ELVIS PRESLEY: G.I. BLUES
G.I. JOANS
See SHE'S IN THE ARMY NOW
GADAD LENIN
See LEAVING LENIN
GADAEL LENIN

See LEAVING LENIN
GALACTICA 3: CONQUEST OF THE EARTH
See CONQUEST OF THE EARTH
GAME, THE
See RED SHOE DIARIES 11: THE GAME
GAMES THAT NURSES PLAY
See SECRETS OF YOUNG NURSES
GANASHATRU
See ENEMY OF THE PEOPLE, AN
GANG WAR
See ODD MAN OUT
GAS-S-S-S
See GAS
GAS, FOOD AND LODGING
See GAS FOOD LODGING
GAS... OR IT MAY BECOME NECESSARY TO DESTROY THE WORLD IN ORDER TO SAVE IT
See GAS
GASSSSSSS OR IT BECAME NECESSARY TO DESTROY THE WORLD IN ORDER TO SAVE IT
See GAS
GASU-NINGEN DAI ICHIGO
See HUMAN VAPOUR, THE
GATE 2: RETURN TO THE NIGHTMARE
See GATE 2
GATES OF HELL, THE
See CITY OF THE LIVING DEAD
GAVRE PRINCIP – HIMMEL UNTER STEINEN
See DEATH OF A SCHOOLBOY
GAY DIVORCE, THE
See GAY DIVORCEE, THE
GAZON MAUDIT
See FRENCH TWIST
GEMAR GAVIA
See CUP FINAL
GEMINI TWINS, THE
See TWINS OF EVIL
GENERAL, THE
See BUSTER KEATON: THE GENERAL
GENERALNAYA LINYA
See GENERAL LINE, THE
GEORG ELSNER: EINER AUS DEUTSCHLAND
See SEVEN MINUTES
GEORGE BALANCHINE'S THE NUTCRACKER
See NUTCRACKER, THE
GEORGE TAKES THE AIR
See IT'S IN THE AIR
GERMANIA, ANNO ZERO
See GERMANY, YEAR ZERO
GERONIMO
See GERONIMO: AN AMERICAN LEGEND
GET DOWN AND BOOGIE
See DARKTOWN

STRUTTERS
GETRAUMTE SUNDEN
See SUCCUBUS
GETTYSBURG
See GETTYSBURG: PARTS 1 AND 2
GHOST BRIGADE
See KILLING BOX, THE
GHOST BRIGADE: THE KILLING BOX
See KILLING BOX, THE
GHOST IN THE MACHINE, THE
See INSPECTOR MORSE: THE GHOST IN THE MACHINE
GHOST STORIES
See KWAIDAN
GHOSTWARRIOR
See SWORDKILL
GHOULIES GO TO COLLEGE
See GHOULIES 3: GHOULIES GO TO COLLEGE
GILBERT GRAPE
See WHAT'S EATING GILBERT GRAPE
GIRL HAPPY
See ELVIS PRESLEY: GIRL HAPPY
GIRL ON A BIKE
See RED SHOE DIARIES 12: GIRL ON A BIKE
GIRL ON THE BOAT, THE
See NORMAN WISDOM: THE GIRL ON THE BOAT
GIRL SCHOOL SCREAMERS
See GIRL'S SCHOOL SCREAMERS
GIRL SHY
See HAROLD LLOYD: GIRL SHY
GIRL TROUBLE
See NORMAN WISDOM: GIRL TROUBLE
GIRL WAS YOUNG, THE
See YOUNG AND INNOCENT
GIRLFRIEND, THE
See LA AMIGA
GIRLFRIENDS, THE
See BAD GIRLS
GIRLS! GIRLS! GIRLS!
See ELVIS PRESLEY: GIRLS! GIRLS! GIRLS!
GIULIA E GIULIA
See JULIA AND JULIA
GIULIETTA DEGLI SPIRITI
See JULIET OF THE SPIRITS
GLENORKY
See MAGIC IN THE WATER, THE
GLI ESECUTORI
See SICILIAN CROSS
GLI ORRORI DEL CASTELLO DI NORIMBERGA
See BARON BLOOD
G'MAR GIVIYA
See CUP FINAL
GO TO THE LIGHT
See GO TOWARD THE LIGHT
GO WEST
See MARX BROTHERS: GO WEST
GOALIE'S ANXIETY AT THE PENALTY KICK, THE

See GOALKEEPER'S FEAR
OF THE PENALTY, THE
GODARD'S PASSION
See PASSION
GODFATHER 3, THE
See GODFATHER PART 3,
THE
**GODFATHER THE
MASTER**
See NINJA OPERATION 5:
GODFATHER THE
MASTER
GOD'S ARMY
See PROPHECY, THE
GODZILLA
See GODZILLA: THE
LEGEND IS REBORN
GODZILLA 1985
See GODZILLA: THE
LEGEND IS REBORN
**GODZILLA FIGHTS THE
GIANT MOTH**
See GODZILLA: GODZILLA
VERSUS MOTHRA
**GODZILLA ON
MONSTER ISLAND**
See GODZILLA: WAR OF
THE MONSTERS
GODZILLA TAI GAIGAN
See GODZILLA: WAR OF
THE MONSTERS
**GODZILLA VERSUS
GIGAN**
See GODZILLA: WAR OF
THE MONSTERS
**GODZILLA VERSUS
MECHAGODZILLA**
See GODZILLA: GODZILLA
VERSUS THE COSMIC
MONSTER
**GODZILLA VERSUS
MEGALON**
See GODZILLA: GODZILLA
VERSUS MEGALON
**GODZILLA VERSUS
MONSTER ZERO**
See GODZILLA: INVASION
OF THE ASTRO-
MONSTERS
**GODZILLA VERSUS
MOTHRA**
See GODZILLA: GODZILLA
VERSUS MOTHRA
**GODZILLA VERSUS THE
BIONIC MONSTER**
See GODZILLA: GODZILLA
VERSUS THE COSMIC
MONSTER
**GODZILLA VERSUS THE
GIANT MOTH**
See GODZILLA: GODZILLA
VERSUS MOTHRA
**GODZILLA VERSUS THE
SEA MONSTER**
See GODZILLA: EBIRAH –
HORROR OF THE DEEP
**GODZILLA VERSUS THE
THING**
See GODZILLA: GODZILLA
VERSUS MOTHRA
**GODZILLA, WAR OF
THE MONSTERS**
See GODZILLA: WAR OF
THE MONSTERS
GOING PLACES
See LES VALSEUSES
GOING STRAIGHT
See OVERTHROW, THE
GOING TO THE CHAPEL
See WEDDING DAY BLUES
GOING UP
See LIFT, THE
GOJIRA

See GODZILLA: THE
LEGEND IS REBORN
GOJIRA NO MUSUKO
See GODZILLA: SON OF
GODZILLA
GOJIRA TAI GAIGAN
See GODZILLA: WAR OF
THE MONSTERS
GOJIRA TAI MEGARO
See GODZILLA: GODZILLA
VERSUS MEGALON
**GOJIRA TAI MEKA
GOJIRA**
See GODZILLA: GODZILLA
VERSUS THE COSMIC
MONSTER
GOJIRA TAI MOSURA
See GODZILLA: GODZILLA
VERSUS MOTHRA
GOLD RUSH, THE
See CHARLIE CHAPLIN:
THE GOLD RUSH
GOLD STRIKE
See SILVER BEARS
GOLDDIGGER
See ROBOT CALLED
GOLDDIGGER, A
**GOLDEN NINJA
WARRIORS**
See GOLDEN NINJA
WARRIOR
**GOLEM, THE: HOW HE
CAME INTO THE WORLD**
See GOLEM, THE
**GOMAR, THE HUMAN
GORILLA**
See NIGHT OF THE
BLOODY APES
GONE TO EARTH
See WILD HEART, THE
GOOD DIE FIRST, THE
See GOOD DIE FIRST FOR
A HANDFUL OF SILVER,
THE
GOOD LIFE, THE
See LA VIE DU CHATEAU
GOOD LITTLE GIRLS
See GOOD LITTLE GIRL
GOOD MARRIAGE, A
See LE BEAU MARIAGE
GOOD MEN AND BAD
See BLACK FOX: GOOD
MEN AND BAD
**GOOD MORNING,
BABYLONIA**
See GOOD MORNING,
BABYLON
GOOD OLD BOY
See RIVER PIRATES, THE
**GOOD OLD BOY: A
DELTA BOYHOOD**
See RIVER PIRATES, THE
GOODBYE TEXAS
See TEXAS ADIOS
GOODBYE, CHILDREN
See AU REVOIR, LES
ENFANTS
**GOON SHOW MOVIE,
THE**
See DOWN AMONG THE
Z-MEN
**GORGEOUS BIRD LIKE
ME, A**
See UNE BELLE FILLE
COMME MOI
**GOSPEL ACCORDING
TO VIC, THE**
See HEAVENLY PURSUITS
GOTTERDAMMERUNG
See DAMNED, THE
GRAIL, THE
See LANCELOT DU LAC
GRAND DUEL, THE

See STORMRIDER
GRAND ILLUSION
See LA GRANDE ILLUSION
**GRAND MANEUVER,
THE**
See LES GRANDES
MANOEUVRES
GRAND TOUR, THE
See TIMESCAPE
**GRAND TOUR:
DISASTER IN TIME**
See TIMESCAPE
**GRAVE ROBBERS
FROM OUTER SPACE**
See PLAN 9 FROM OUTER
SPACE
GRAVEYARD TRAMPS
See INVASION OF THE
BEE GIRLS
GREAT DICTATOR, THE
See CHARLIE CHAPLIN:
THE GREAT DICTATOR
**GREAT ESCAPE 2, THE:
THE FINAL CHAPTER**
See GREAT ESCAPE 2, THE
**GREAT ESCAPE 2, THE:
THE UNTOLD STORY –
PARTS 1 AND 2**
See GREAT ESCAPE 2, THE
GREAT FEED, THE
See LA GRANDE BOUFFE
GREAT GUNS
See LAUREL AND HARDY:
GREAT GUNS
GREAT MANHUNT, THE
See STATE SECRET
**GREAT MOUSE
DETECTIVE, THE**
See BASIL, THE GREAT
MOUSE DETECTIVE
**GREAT MUPPET
CAPER, THE**
See MUPPETS, THE: THE
GREAT MUPPET CAPER
**GREAT WALL IS A
GREAT WALL, THE**
See GREAT WALL, A
GREAT WALL, THE
See GREAT WALL, A
**GREEK INTERPRETER,
THE**
See ADVENTURES OF
SHERLOCK HOLMES, THE:
THE GREEK INTERPRETER
**GREEKS BEARING
GIFTS**
See INSPECTOR MORSE:
GREEKS BEARING GIFTS
GREEN HORNET, THE
See FURY OF THE
DRAGON
GREY NIGHT
See KILLING BOX, THE
GREYSTOKE
See GREYSTOKE: THE
LEGEND OF TARZAN,
LORD OF THE APES
GRIM REAPER 2
See ANTHROPOPHAGOUS
2
GRITOS EN LA NOCHE
See AWFUL DR ORLOFF,
THE
GROWING PAINS
See HOMEWORK
GUARDIAN OF HELL
See OTHER HELL, THE
**GUESS WHO'S COMING
FOR CHRISTMAS?**
See U.F.O. CAFE
**GUILTY THING
SURPRISED, A**
See INSPECTOR

WEXFORD: A GUILTY
THING SURPRISED
**GULING JIE SHAONIAN
SHA REN SHIJIAN**
See BRIGHTER SUMMER
DAY, A
GUNS FOR DOLLARS
See THEY CALL ME
HALLELUJAH
**GUNS IN THE
AFTERNOON**
See RIDE THE HIGH
COUNTRY
GUYVER 2: DARK HERO
See GUYVER, THE: DARK
HERO
**H.G. WELLS' THE FIRST
MEN IN THE MOON**
See FIRST MEN IN THE
MOON, THE
**H.P. LOVECRAFT'S
NECRONOMICON:
BOOK OF THE DEAD**
See NECRONOMICON
HACK 'EM HIGH
See SCUMBUSTERS
**HAKIGATSU NO
KYOSHIKYOKU**
See RHAPSODY IN
AUGUST
HALFAOUINE
See HALFAOUINE: CHILD
OF THE TERRACES
**HALFAOUINE:
L'ENFANT DES
TERRASSES**
See HALFAOUINE: CHILD
OF THE TERRACES
HALLOWEEN PARTY
See NIGHT OF THE
DEMONS
HAMBONE AND HILLIE
See ADVENTURES OF
HAMBONE AND HILLIE,
THE
HAMLET
See HAMLET, PRINCE OF
DENMARK
**HAMMER HOUSE OF
HORROR: CHARLIE BOY**
See CHARLIE BOY
**HAMMER HOUSE OF
HORROR: GUARDIAN
OF THE ABYSSS**
See GUARDIAN OF THE
ABYSS
**HAMMER HOUSE OF
MYSTERY AND
SUSPENSE: A DISTANT
SCREAM**
See DISTANT SCREAM, A
**HAMMER HOUSE OF
MYSTERY AND
SUSPENSE: AND THE
WALL CAME TUMBLING
DOWN**
See AND THE WALL
CAME TUMBLING DOWN
**HAMMER HOUSE OF
MYSTERY AND
SUSPENSE: PAINT ME A
MURDER**
See PAINT ME A MURDER
HAMMER OF GOD
See CHINESE BOXER, THE
**HANS CHRISTIAN
ANDERSEN'S MAGIC
ADVENTURE**
See MAGIC ADVENTURE,
THE
HAPPY FAMILIES
See INSPECTOR MORSE:
HAPPY FAMILIES

HAPPY HOUR
See SOUR GRAPES
HAPPY NEW YEAR CAPER, THE
See HAPPY NEW YEAR
HARD DAY'S NIGHT, A
See BEATLES: A HARD DAY'S NIGHT
HARD DRIVER
See LAST AMERICAN HERO, THE
HARD RAIN
See INNOCENT MAN, AN
HARD TIMES
See STREETFIGHTER, THE
HARD WAY TO DIE, A
See SUN DRAGON
HAREM HOLIDAY
See ELVIS PRESLEY: HAREM HOLIDAY
HARRY AND THE HENDERSONS
See BIGFOOT AND THE HENDERSONS
HARRY PALMER RETURNS
See FUNERAL IN BERLIN
HARUM SCARUM
See ELVIS PRESLEY: HAREM HOLIDAY
HATCHET MURDERS, THE
See DEEP RED
HATRED
See LA HAINE
HAUNTED, THE NIGHT OF THE DEMON
See NIGHT OF THE DEMON
HAUNTING OF HAMILTON HIGH, THE
See PROM NIGHT 2: HELLO MARY LOU
HAUNTING OF HELEN WALKER, THE
See TURN OF THE SCREW, THE
HAVE GUN, WILL TRAVEL
See ACE HIGH
HAWK, THE
See RIDE HIM COWBOY
HAWKS AND THE SPARROWS, THE
See HAWKS AND SPARROWS
HE OR SHE
See GLEN OR GLENDA?
HEALTH CLUB
See TOXIC AVENGER, THE
HEART AND SOUL
See MISCHIEF
HEART TO WIN
See RICKY 1
HEAVEN FELL THAT NIGHT
See LES BIJOUTIERS DU CLAIR DE LUNE
HEAVEN HELP US
See CATHOLIC BOYS
HEIDI FLEISS STORY, THE
See MAKING OF A HOLLYWOOD MADAM, THE: THE HEIDI FLEISS STORY
HEIR TO GENGHIS KHAN, THE
See STORM OVER ASIA
HELL
See TORMENT
HELL BORN
See NIGHT ANGEL

HELL, HEAVEN AND HOBOKEN
See I WAS MONTY'S DOUBLE
HELLO MARY LOU: PROM NIGHT 2
See PROM NIGHT 2: HELLO MARY LOU
HELP!
See BEATLES: HELP!
HER LAST BEST YEAR
See LAST BEST YEAR, THE
HER WICKED WAYS
See LETHAL CHARM
HERCULES AND THE MAZE OF THE MINOTAUR
See HERCULES IN THE MAZE OF THE MINOTAUR
HERCULES GOES BANANAS
See HERCULES IN NEW YORK
HERCULES THE MOVIE
See HERCULES IN NEW YORK
HERE IS A MAN
See ALL THAT MONEY CAN BUY
HERETIC, THE
See EXORCIST 2: THE HERETIC
HERO
See ACCIDENTAL HERO
HERO OF THE PEOPLE
See BETHUNE: THE MAKING OF A HERO
HEROES OF THE REGIMENT
See LAUREL AND HARDY: HEROES OF THE REGIMENT
HERZ AUS GLAS
See HEART OF GLASS
HETEROSEXUALS, THE
See BAD GIRLS
HEXEN BIS AUFS BLUT GEQUAELT
See MARK OF THE DEVIL
HIDDEN ASSASSIN
See SHOOTER, THE
HIDDEN RAGE
See HIDDEN OBSESSION
HIDDEN ROOM, THE
See OBSESSION
HIDE AND GO SHRIEK
See CLOSE YOUR EYES AND PRAY
HIDEOUS SUN DEMONS
See HIDEOUS SUN DEMON, THE
HIFAZAAT
See IN CUSTODY
HIGH ART
See EXPOSURE
HIGH ENCOUNTERS OF THE ULTIMATE KIND
See CHEECH AND CHONG'S NEXT MOVIE
HIIDEN ASSASSIN
See SHOOTER, THE
HIM
See ONLY YOU
HINOTORI 2772
See SPACE FIREBIRD
HINOTORI 2772 AI NO COSMOZONE
See SPACE FIREBIRD
HIROSHIMA: OUT OF THE ASHES
See HIROSHIMA
HISTOIRES EXTRAORDINAIRE

See TALES OF MYSTERY AND IMAGINATION
HITLER, EIN FILM AUS DEUTSCHLAND
See HITLER, A FILM FROM GERMANY
HITLERJUNGE SALOMON
See EUROPA, EUROPA
HITZ
See JUDGEMENT
HOBGOBLIN
See QUEST FOR THE MIGHTY SWORD
HOBGOBLIN, THE
See ATOR, THE FIGHTING EAGLE
HOLD ME, THRILL ME, KISS ME
See HOLD ME THRILL ME KISS ME
HOLD THAT DREAM
See HOLD THE DREAM
HOLE, THE
See ONIBABA
HOLIDAY
See JOUR DE FETE
HOLLOW POINT
See WILD PAIR, THE
HOLLYWOOD HOOKERS
See HOLLYWOOD CHAINSAW HOOKERS
HOLLYWOOD HORROR HOUSE
See SAVAGE INTRUDER
HOMICIDAL IMPULSES
See KILLER INSTINCT
HONEY BOY
See HONEYBOY
HONEYMOON OF FEAR
See FEAR IN THE NIGHT
HONG GAOLIANG
See RED SORGHUM
HOOLIGANS
See ULTRA
HOOSIERS
See BEST SHOT
HORI, MA PANENKO
See FIREMAN'S BALL, THE
HORROR CHAMBER OF DR FAUSTUS
See EYES WITHOUT A FACE
HORROR HOTEL
See DEATH TRAP
HORROR HOTEL MASSACRE
See DEATH TRAP
HORROR IN THE MIDNIGHT SUN
See INVASION OF THE ANIMA PEOPLE
HORROR OF DEATH, THE
See ASPHYX, THE
HORROR OF SNAPE ISLAND
See TOWER OF EVIL
HORROR PLANET
See INSEMINOID
HORROR STAR
See FRIGHTMARE
HORROR Y SEXO
See NIGHT OF THE BLOODY APES
HOSTILE ADVANCES: THE KERRY ELLISON STORY
See HOSTILE ADVANCES
HOT AND COLD
See WEEKEND AT BERNIE'S
HOT CURVES

See ELEVEN DAYS ELEVEN NIGHTS: PART 2
HOT SPOT
See I WAKE UP SCREAMING
HOT SWEAT
See KATIE'S PASSION: A GIRL CALLED KATIE TIPPEL
HOTEL OKLAHOMA
See INNOCENT YOUNG FEMALE
HOTLINE GINA
See RED SHOE DIARIES: HOTLINE GINA
HOTLINE TO HEAVEN
See STATIC
HOU HSING K'OU SHOU
See SNAKE IN THE MONKEY'S SHADOW
HOUND OF THE BASKERVILLES, THE
See SHERLOCK HOLMES: THE HOUND OF THE BASKERVILLES
HOUND OF THE BASKERVILLES, THE
See RETURN OF SHERLOCK HOLMES, THE: THE HOUND OF THE BASKERVILLES
HOUND OF THE BASKERVILLES, THE
See SHERLOCK HOLMES: THE HOUND OF THE BASKERVILLES
HOUNDS OF ZAROFF, THE
See MOST DANGEROUS GAME, THE
HOUR OF DARKNESS
See NIGHT EYES 2
HOUR OF GLORY
See SMALL BACK ROOM, THE
HOURS AND THE TIMES, THE
See HOURS AND TIMES, THE
HOUSE 4: HOME DEADLY HOME
See HOUSE 4
HOUSE 4: THE REPOSSESSION
See HOUSE 4
HOUSE CALL
See DE FLAT
HOUSE OF A THOUSAND WOMEN
See TWO THOUSAND WOMEN
HOUSE OF CRAZIES
See ASYLUM
HOUSE OF DOOM
See BLACK CAT, THE
HOUSE OF EXORCISM, THE
See LISA AND THE DEVIL
HOUSE OF FRIGHT
See BLACK SUNDAY
HOUSE OF THE SPIRITS, THE
See HOUSE OF SPIRITS, THE
HOUSE OF USHER
See FALL OF THE HOUSE OF USHER, THE
HOUSE OUTSIDE THE CEMETERY, THE
See HOUSE BY THE CEMETERY, THE
HOUSE PARTY 2: THE PAJAMA JAM

See HOUSE PARTY 2
HOUSEKEEPER, THE
See JUDGEMENT IN
STONE, A
**HOW I FLEW FROM
LONDON TO PARIS IN
25 HOURS AND 11
MINUTES**
See THOSE MAGNIFICENT
MEN IN THEIR FLYING
MACHINES
**HOW I MET MY
HUSBAND**
See RED SHOE DIARIES 6:
HOW I MET MY HUSBAND
**HOW MUCH LOVING
DOES A NORMAL
COUPLE NEED?**
See COMMON-LAW
CABIN
**HOW TO FILL A WILD
BIKINI**
See HOW TO STUFF A
WILD BIKINI
HOW TO MAKE IT
See TARGET: HARRY
**HOW TO STEAL A
DIAMOND IN FOUR
UNEASY LESSONS**
See HOT ROCK, THE
**HOWARD BEACH:
MAKING A CASE FOR
MURDER**
See MAKING A CASE FOR
MURDER: THE HOWARD
BEACH STORY
HOWARD THE DUCK
See HOWARD: A NEW
BREED OF HERO
**HOWLING 2, THE: YOUR
SISTER IS A
WEREWOLF**
See HOWLING 2, THE
**HOWLING 3: THE
MARSUPIALS**
See HOWLING 3
HSIMENG RENSHENG
See PUPPETMASTER, THE
HUEVOS DE ORO
See GOLDEN BALLS
**HULLO HONEY, I'M
DEAD**
See HI HONEY, I'M DEAD
HUMAN BEAST, THE
See LA BETE HUMAINE
HUMAN EXPERIMENTS
See SMALL TOWN
MASSACRE
**HUMAN HIGHWAY: A
FILM BY NEIL YOUNG**
See HUMAN HIGHWAY
HUMAN PETS, THE
See JOSH KIRBY: TIME
WARRIOR! – THE HUMAN
PETS
**HUMAN TARGET:
BLOODFIST 5**
See BLOODFIST 5: HUMAN
TARGET
HUNGRY WIVES
See SEASON OF THE
WITCH
HUNTED, THE
See BENJI THE HUNTED
HUOZHE
See TO LIVE
**HUSBANDS AND
LOVERS**
See IN EXCESS
HUSTLER
See MAPANTSULA
HUSTLER SQUAD
See DOLL SQUAD, THE

**HYPER SAPIEN:
PEOPLE FROM
ANOTHER STAR**
See HYPER SAPIEN:
PEOPLE FROM ANOTHER
PLANET
HYPERSPACE
See GREMLOIDS
I ACCUSE
See DARK AVENGER
I AM A FUGITIVE
See I AM A FUGITIVE
FROM A CHAIN GANG
**I AM CURIOUS
(YELLOW)**
See I AM CURIOUS –
YELLOW
I AM FRIGID, WHY?
See NIGHT PLEASURES
I CAN'T LOSE
See DREAM RIDER
**I CAVALIERI DEL
BRONX**
See BRONX WARRIORS
I CHANGED MY SEX
See GLEN OR GLENDA?
**I CORPI PRESENTANO
TRACCE DI VIOLENZA
CARNALE**
See TORSO
I DO 3
See I DO PART 3: THE
OTHER WOMAN
I GIORNI DELL'IRA
See GUN LAW
I GIRASOLI
See SUNFLOWER
I HATE YOUR GUTS!
See INTRUDER, THE
I LED TWO LIVES
See GLEN OR GLENDA?
I LOVE THE RUTLES
See COMPLEAT RUTLES,
THE
I LOVE YOU, I DON'T
See I LOVE YOU NO MORE
**I MIEI PRIMI
QUARANT'ANNI**
See MY FIRST FORTY
YEARS
I PASCOLI ROSSI
See MASSACRE AT
GRAND CANYON
**I POSED FOR PLAYBOY
BEHIND THE SCENES**
See I POSED FOR PLAYBOY
**I QUATTRO DELL'AVE
MARIA**
See ACE HIGH
**I WAS A TEENAGE SEX
MUTANT**
See DOCTOR ALIEN
I, MADMAN
See HARDCOVER
ICE
See ED McBAIN'S 87TH
PRECINCT: ICE
**IDENTIFICAZIONE DI
UNA DONNA**
See IDENTIFICATION OF A
WOMAN
IERI, OGGI, DOMANI
See YESTERDAY, TODAY
AND TOMORROW
IF LOOKS COULD KILL
See TEEN AGENT
**IF YOU CAN'T SAY IT,
JUST SEE IT**
See WHORE
IKIRU
See LIVING
**IL BUONO, IL BRUTO, IL
CATTIUO**

See GOOD, THE BAD, AND
THE UGLY, THE
IL CAMORRISTA
See PROFESSOR, THE
**IL CASTELLO DI FU
MANCHU**
See CASTLE OF FU
MANCHU, THE
IL CONFORMISTA
See CONFORMIST, THE
IL DECAMERONE
See DECAMERON, THE
IL DESERTO ROSSO
See RED DESERT
IL DIAVOLO E I MORTI
See LISA AND THE DEVIL
IL DIAVOLO E IL MORTO
See LISA AND THE DEVIL
**IL ETAIT UNE FOIS UN
PAYS**
See UNDERGROUND
**IL GATTO A NOVE
CODE**
See CAT O'NINE TAILS
**IL GATTO DAGLI OCCHI
DI GIADA**
See CAT'S VICTIM, THE
**IL GATTO DI PARK
LANE**
See BLACK CAT, THE
IL GATTO NERO
See BLACK CAT, THE
IL GRANDE DUELLO
See STORMRIDER
**IL GRANDE RITORNO DI
DJANGO**
See DJANGO STRIKES
AGAIN
IL LADRO DI BAMBINI
See STOLEN CHILDREN,
THE
IL MERCENARIO
See DJANGO
IL MESSAGERO
See MESSENGER OF
DEATH
**IL MOSTRO E IN
TAVOLA... BARONE
FRANKENSTEIN**
See ANDY WARHOL'S
FRANKENSTEIN
IL PIACERE
See PLEASURE, THE
IL PORTIERE DI NOTTE
See NIGHT PORTER, THE
IL POSTINO
See POSTMAN, THE
IL PROCESSO
See TRIAL, THE
IL PROIEZIONISTA
See INNER CIRCLE, THE
**IL RAGAZZO DAL
KIMONO D'ORO**
See KARATE WARRIOR
**IL RICHIAMO DELLA
FORESTA**
See CALL OF THE WILD,
THE
**IL SEQUESTRO
DELL'ACHILLE LAURO**
See VOYAGE OF TERROR:
THE ACHILLE LAURO
AFFAIR
**IL SILENZIO DEI
PROSCIUTTI**
See SILENCE OF THE
HAMS
**IL SOLE ANCHE DI
NOTTE**
See NIGHT SUN
**IL VANGELO SECONDO
MATTEO**
See GOSPEL ACCORDING

TO ST MATTHEW, THE
**I'LL MEET YOU IN
HEAVEN**
See MAXIE
**ILLEGAL ENTRY:
FORMULA FOR FEAR**
See ILLEGAL ENTRY
ILLICIT BEHAVIOR
See CRIMINAL INTENT
ILLICIT INTERLUDE
See SUMMER INTERLUDE
**ILLUSTRIOUS CLIENT,
THE**
See CASEBOOK OF
SHERLOCK HOLMES, THE:
THE ILLUSTRIOUS CLIENT
ILS VONT TOUS BIEN
See EVERYBODY'S FINE
I'M ALRIGHT JACK
See I'M ALL RIGHT JACK
IM LAUF DER ZEIT
See KINGS OF THE ROAD
**IMMORTAL BATTALION,
THE**
See WAY AHEAD, THE
**IMPORTANCE OF BEING
SEXY, THE**
See SOME LIKE IT SEXY
**IN A YEAR WITH 13
MOONS**
See IN A YEAR OF 13
MOONS
**IN EAGLE SHADOW
FIST**
See EAGLE SHADOW FIST
**IN EINEM JAHR MIT 13
MONDEN**
See IN A YEAR OF 13
MOONS
IN EXILE
See TIME RUNNER
IN LOVE
See FALLING IN LOVE
AGAIN
IN LOVE AND WAR
See LOVE AND WAR
IN OLD OKLAHOMA
See WAR OF THE
WILDCATS
**IN THE BEST FAMILES:
MARRIAGE, PRIDE AND
MADNESS**
See BITTER BLOOD
**IN THE COMFORT OF
STRANGERS**
See COMFORT OF
STRANGERS, THE
**IN THE COURSE OF
TIME**
See KINGS OF THE ROAD
**IN THE LINE OF DUTY:
THE TWILIGHT
MURDERS**
See IN THE LINE OF DUTY
3: TIME TO KILL
**IN THE LINE OF FIRE:
THE MORRIS DEES
STORY**
See BLIND HATE
**IN THE REALM OF THE
PASSIONS**
See AI NO BOREI
**IN WEITER FERNE, SO
NAH!**
See FARAWAY, SO CLOSE
**INCIDENT IN A SMALL
TOWN**
See AGAINST HER WILL
**INCREDIBLE VOYAGE
OF STINGRAY, THE**
See STINGRAY: THE
INCREDIBLE VOYAGE
INDAGINE SU UN

CITTADINO AL DI SOPRA DI OGNI SOSPETTO
See INVESTIGATION OF A CITIZEN ABOVE SUSPICION

INDECENT ADVANCES
See PAMELA PRINCIPLE, THE

INFERNAL SERPENT, THE
See INSPECTOR MORSE: THE INFERNAL SERPENT

INFERNO '80
See INFERNO

INFESTED
See TICKS

INHUMANOIDS, THE
See INHUMANOIDS: THE EVIL THAT LIES WITHIN

INJI KAU
See ROUGE

INNER SANCTUM 2
See NATURAL COLD KILLER

INNOCENT OBSESSION
See EDEN: VOLS. 1 TO 6

INNOCENT SEX
See TEENAGE INNOCENCE

INNOCENT VICTIM
See TREE OF HANDS

INNOCENTS FROM HELL
See SISTERS OF SATAN

INNOCENTS, THE
See TURN OF THE SCREW, THE

INSATIABLE
See CALENDAR GIRL MURDERS, THE

INSPECTOR WEARS SKIRTS, THE
See TOP SQUAD

INSTANT JUSTICE
See MARINE ISSUE

INTELLIGENCE MEN, THE
See MORECOMBE AND WISE: THE INTELLIGENCE MEN

INTERIOR OF A CONVENT
See BEHIND CONVENT WALLS

INTERMEZZO: A LOVE STORY
See INTERMEZZO

INTERVIEW WITH THE VAMPIRE: THE VAMPIRE CHRONICLES
See INTERVIEW WITH THE VAMPIRE

INTO THE DARKNESS
See NOTHING UNDERNEATH

INTO THE WEST: WHERE MYTH AND MAGIC WALK THE EARTH
See INTO THE WEST

INTRUDER, THE
See INNOCENT, THE

INVADERS, THE
See FORTY-NINTH PARALLEL, THE

INVASION EARTH 2150 A.D.
See DOCTOR WHO: INVASION EARTH 2150 A.D.

INVASION FORCE
See HANGAR 18

INVASION OF THE ASTRO-MONSTERS
See GODZILLA: INVASION OF THE ASTRO-MONSTERS

INVASION U.F.O.
See U.F.O. – INVASION U.F.O.

INVESTIGATION INTO A CITIZEN ABOVE SUSPICION
See INVESTIGATION OF A CITIZEN ABOVE SUSPICION

INVESTIGATION OF A PRIVATE CITIZEN
See INVESTIGATION OF A CITIZEN ABOVE SUSPICION

INVINCIBLE BOXER
See KING BOXER

INVINCIBLE FIGHTER
See INVINCIBLE OBSESSED FIGHTER

IO E IL DUCE
See MUSSOLINI AND I

IP5: L'ILE AUX PACHYDERMES
See IP5

IP5: THE ISLAND OF PACHYDERMS
See IP5

IRON EAGLE 2: BATTLE BEYOND THE FLAG
See IRON EAGLE 2

IRON FIST OF KWANGTUNG
See CANTON IRON KUNG FU

IRON ROOSTER VERSUS THE CENTIPEDE
See LAST HERO IN CHINA, THE

ISLAND OF THE ALIVE
See IT'S ALIVE 3: ISLAND OF THE ALIVE

ISLAND OF THE DEAD
See ZOMBIE FLESHEATERS

ISLAND OF THE LIVING DEAD
See ZOMBIE FLESHEATERS

ISLAND ON FIRE
See PRISONER, THE

ISTANBUL: KEEP YOUR EYES OPEN
See ISTANBUL

IT CAME WITHOUT WARNING
See WARNING, THE

IT HAPPENED AT THE WORLD'S FAIR
See ELVIS PRESLEY: IT HAPPENED AT THE WORLD'S FAIR

IT HAPPENED ONE SUMMER
See STATE FAIR

IT LIVES AGAIN
See IT'S ALIVE 2

IT STALKED THE OCEAN FLOOR
See MONSTER FROM THE OCEAN FLOOR, THE

IT'S A 2 FOOT 6 INCH ABOVE THE GROUND WORLD
See ANYONE FOR SEX

IT'S MAGIC
See ROMANCE ON THE HIGH SEAS

IT'S NOT SIZE THAT COUNTS
See PERCY'S PROGRESS

IT'S NOT THE SIZE THAT COUNTS
See PERCY'S PROGRESS

IVAN GROZNYI
See IVAN THE TERRIBLE

IVAN THE TERRIBLE: PART 1
See IVAN THE TERRIBLE

IVANA TRUMP'S FOR LOVE ALONE
See FOR LOVE ALONE

JACK AND THE BEANSTALK
See ABBOTT AND COSTELLO: JACK AND THE BEANSTALK

JACKIE CHAN AND THE 36 CRAZY FISTS
See THIRTY-SIX FIST, THE

JACKIE CHAN'S POLICE FORCE
See POLICE STORY

JACKIE CHAN'S POLICE STORY
See POLICE STORY

JACKIE PRESSER STORY, THE
See TEAMSTER BOSS: THE JACKIE PRESSER STORY

JACK'S WIFE
See SEASON OF THE WITCH

JADE CLAW, THE
See CRYSTAL FIST

JADE JUNGLE, THE
See ARMED RESPONSE

JAG AR NYFIKEN – GUL
See I AM CURIOUS – YELLOW

JAILBIRDS
See LAUREL AND HARDY: PARDON US

JAILHOUSE ROCK
See ELVIS PRESLEY: JAILHOUSE ROCK

JAMES A. MICHENER'S TEXAS
See TEXAS

JAMI SONTEE
See FURY IN SHAOLIN TEMPLE

JANUARY UPRISING IN KIEV IN 1918, THE
See ARSENAL

JASON GOES TO HELL
See FRIDAY THE 13TH: JASON GOES TO HELL

JASON GOES TO HELL: THE FINAL FRIDAY
See FRIDAY THE 13TH: JASON GOES TO HELL

JASON LIVES: FRIDAY THE 13TH PART 6
See FRIDAY THE 13TH, PART 6: JASON LIVES

JAWS 3-D
See JAWS 3

JAWS: THE REVENGE
See JAWS 4

JAYNE MANSFIELD: A SYMBOL OF THE 50s
See JAYNE MANSFIELD STORY, THE

JE T'AIME, MOI NON PLUS
See I LOVE YOU NO MORE

JEANNE
See KING'S WHORE, THE

JEANNE LA PUCELLE
See JOAN OF ARC

JEANNIE
See PORTRAIT OF JENNIE

JEDER FUR SICH UND GOTT GEGEN ALLE
See ENIGMA OF KASPAR HAUSER, THE

J'EMBRASSE PAS
See I DON'T KISS

JENSEITS DER WOLKEN
See BEYOND THE CLOUDS

JESUS DE MONTREAL
See JESUS OF MONTREAL

JEU DE MASSACRE
See KILLING GAME, THE

JIM AND JENNIFER STOLPA STORY, THE
See SNOWBOUND: THE JIM AND JENNIFER STOLPA STORY

JINGCHA GOSHI
See POLICE STORY

JOE VALACHI: I SEGRETI DI COSA NOSTRA
See VALACHI PAPERS, THE

JOHN CARPENTER'S ESCAPE FROM L.A.
See ESCAPE FROM L.A.

JOHN CARPENTER'S IN THE MOUTH OF MADNESS
See IN THE MOUTH OF MADNESS

JOHN CARPENTER'S THE THING
See THING, THE

JOHN STEINBECK'S EAST OF EDEN
See EAST OF EDEN

JOHN WOO'S VIOLENT TRADITION
See VIOLENT TRADITION

JOHNNY IN THE CLOUDS
See WAY TO THE STARS, THE

JOHNNY VAGABOND
See JOHNNY COME LATELY

JOHNNY ZOMBIE
See MY BOYFRIEND'S BACK

JOKO, INVOCO DIO... E MUORI
See VENGEANCE

JOSEPH CONRAD'S NOSTROMO
See NOSTROMO

JOURNEY TO ITALY
See VOYAGE TO ITALY

JOURNEY TO THE MAGIC CAVERN
See JOSH KIRBY: TIME WARRIOR! – JOURNEY TO THE MAGIC CAVERN

JOURS TRANQUILLES A CLICHY
See QUIET DAYS IN CLICHY

JOY OF SEX, THE
See NATIONAL LAMPOON'S JOY OF SEX

JUDAS GOAT
See XTRO

JUDAS GOAT, THE
See SPENSER: THE JUDAS GOAT

JUDAS WAS A WOMAN
See LA BETE HUMAINE

JUDITH KRANTZ'S DAZZLE
See DAZZLE

JUDITH KRANTZ'S TORCH SONG

See TORCH SONG
JUGIN YUKIOTOKO
See HALF HUMAN
JULES ET JIM
See JULES AND JIM
JULES VERNE'S EIGHT
HUNDRED LEAGUES
DOWN THE AMAZON
See EIGHT HUNDRED
LEAGUES DOWN THE
AMAZON
JULIA UND DIE
GEISTER
See JULIET OF THE SPIRITS
JULIETTE DES ESPRITS
See JULIET OF THE SPIRITS
JULY PORK BELLIES
See FOR PETE'S SAKE
JUNGFRUKALLAN
See VIRGIN SPRING, THE
JUNGLE FIGHTERS
See LONG AND THE
SHORT AND THE TALL,
THE
JUNGLE HEAT
See PIRANHA WOMEN
JURASSIC
ADVENTURES
See RETURN TO THE LOST
WORLD
JURY OF ONE, A
See FROM THE FILES OF
JOSEPH WAMBAUGH: A
JURY OF ONE
JUST ANOTHER
MIRACLE
See HEAVENLY PURSUITS
JUST IN TIME
See ONLY YOU
JUST MY LUCK
See NORMAN WISDOM:
JUST MY LUCK
JUSTICE WOMEN, THE
See MIDNIGHT ANGEL
JUSTIFIABLE HOMICIDE
See AGAINST HER WILL
JUSTINE
See CRUEL PASSION
KABUKIMAN
See SGT KABUKIMAN
N.Y.P.D.
KADAICHA: THE DEATH
STONE
See KADAICHA
KADISBELLAN
See SLINGSHOT, THE
KAGEMUSHA THE
SHADOW WARRIOR
See KAGEMUSHA
KAIJU DAI SENSO
See GODZILLA: INVASION
OF THE ASTRO-
MONSTERS
KAIJU SOSHINGEKI
See GODZILLA: DESTROY
ALL MONSTERS
KAKUSHI TORIDE NO
SAN-AKUNIN
See HIDDEN FORTRESS,
THE
KARATE COP
See SLAUGHTER IN SAN
FRANCISCO
KARATE KID PART 3,
THE
See KARATE KID 3, THE
KARATE OLYMPIA
See KILL OR BE KILLED
KASPAR HAUSER
See ENIGMA OF KASPAR
HAUSER, THE
KATIE
See BORN TO BE WILD

KEEPING SECRETS:
SUZANNE SOMERS IN
HER OWN TRUE STORY
See KEEPING SECRETS
KEETJE TIPPEL
See KATIE'S PASSION: A
GIRL CALLED KATIE
TIPPEL
KEOMA, THE VIOLENT
BREED
See KEOMA
KEYS TO FREEDOM
See DEATH DEALERS
KHANEH-JE DOOST
KOJAST?
See WHERE IS MY
FRIEND'S HOUSE?
KICK BOXER
See KICKBOXER
KICKBOXER 2: THE
ROAD BACK
See KICKBOXER 2
KICKBOXER 2: THE
ROAD HOME
See KICKBOXER 2
KICKBOXING KING
See KICKBOXER KING
KID BROTHER, THE
See HAROLD LLOYD: THE
KID BROTHER
KID GALAHAD
See ELVIS PRESLEY: KID
GALAHAD
KID, THE
See CHARLIE CHAPLIN:
THE KID
KIDS FROM SHAOLIN
See SHAOLIN TEMPLE 2:
KIDS FROM SHAOLIN
KILLBOTS
See CHOPPING MALL
KILLER FISH
See KILLERFISH
KILLER OF KILLERS
See MECHANIC, THE
KILLER!
See QUE LA BETE MEURE
KILLERS INVINCIBLE
See NINJA SQUAD
KILLING BEACH, THE
See TURTLE BEACH
KILLING MAN, THE
See KILLING MACHINE,
THE
KILLING ME SOFTLY
See SOFT KILL, THE
KILLING TIME
See MACON COUNTY
LINE
KING AND MISTER
BIRD, THE
See MISTER BIRD TO THE
RESCUE
KING AND THE BIRD,
THE
See MISTER BIRD TO THE
RESCUE
KING AND THE
MOCKINGBIRD, THE
See MISTER BIRD TO THE
RESCUE
KING CREOLE
See ELVIS PRESLEY: KING
CREOLE
KING KONG VERSUS
GODZILLA
See GODZILLA: KING
KONG VERSUS GODZILLA
KING KONG: THE
EIGHTH WONDER OF
THE WORLD
See KING KONG
KING OF THE

KICKBOXERS 2
See AMERICAN SHAOLIN:
KING OF THE
KICKBOXERS 2
KINGFISH
See KINGFISH: A STORY
OF HUEY P. LONG
KINGU KONGU TAI
GOJIRA
See GODZILLA: KING
KONG VERSUS GODZILLA
KINJITE: FORBIDDEN
SUBJECTS
See KINJITE
KIPPERBANG
See P'TANG, YANG,
KIPPERBANG
KISS AND KILL
See BLOOD OF FU
MANCHU, THE
KISS ON THE LIPS:
SLEEPING BEAUTY
See SLEEPING BEAUTY
KISSIN' COUSINS
See ELVIS PRESLEY:
KISSIN' COUSINS
KISSING THE GUNNER'S
DAUGHTER
See INSPECTOR
WEXFORD: KISSING THE
GUNNER'S DAUGHTER
KLAMEK JI BO BEKO
See SONG FOR BEKO, A
KNICKERS AHOY
See DEVILS IN THE
CONVENT
KNIFE FIGHTER, THE
See EXPOSURE
KNIFE, THE
See EXPOSURE
KNIGHT AND WARRIOR
See NINJA OPERATION:
KNIGHT AND WARRIOR
KNOCK OUT COP, THE
See FLATFOOT
KOKAKU KIDOTAI
See GHOST IN THE SHELL
KOMMANDO LEOPARD
See COMMANDO
LEOPARD
KONEKO MONOGATARI
See ADVENTURES OF
MILO AND OTIS, THE
KONYETS SANKT-
PETERBURGA
See END OF SAINT
PETERSBURG, THE
KORKARLEN
See PHANTOM
CARRIAGE, THE
KOROL LIR
See KING LEAR
KOSURE OOKAMI
See SHOGUN ASSASSIN
KOSURE OOKAMI N. 2
See SHOGUN ASSASSIN
KREITZEROVA SONATA
See KREUTZER SONATA,
THE
KROTKI FILM O
MILOSCI
See SHORT FILM ABOUT
LOVE, A
KROTKI FILM O
ZABIJANIU
See SHORT FILM ABOUT
KILLING, A
KUMONOSU-JO
See THRONE OF BLOOD
KUNG FU COMMANDOS
See INCREDIBLE KUNG FU
MISSION
KUNG FU MASTER

NAMED DRUNK CAT
See KUNG FU MASTER
KUNTA KINTE'S GIFT
See ROOTS: KUNTA
KINTE'S GIFT
KURAISHI NIJU-GOJU
NEN
See SOLAR CRISIS
KURT VONNEGUT'S
HARRISON BERGERON
See HARRISON
BERGERON
LA ARDILLA ROJA
See RED SQUIRREL, THE
LA BELLE ET LA BETE
See BEAUTY AND THE
BEAST, THE
LA BELLE NOISEUSE:
DIVERTIMENTO
See LA BELLE NOISEUSE
LA BESTIA UCCIDE A
SANGUE FREDDO
See COLD BLOODED
BEAST
LA BIBLIA
See BIBLE, THE
LA BONNE ANNEE
See HAPPY NEW YEAR
LA CADUTA DEGLI DEI
See DAMNED, THE
LA CAGE AUX FOLLES
3: ELLES SE MARIENT
See LA CAGE AUX FOLLES
3
LA CAGE AUX FOLLES
3: THE WEDDING
See LA CAGE AUX FOLLES
3
LA CARNE
See FLESH, THE
LA CASA
DELL'EXORCISMO
See LISA AND THE DEVIL
LA CEREMONIE
See JUDGEMENT IN
STONE, A
LA CHIESA
See CHURCH, THE
LA CITE DES ENFANTS
PERDUS
See CITY OF LOST
CHILDREN, THE
LA CITTA DELLE
DONNE
See CITY OF WOMEN
LA COLLECTIONNEUSE
See COLLECTOR, THE
LA COMMARE SECCA
See GRIM REAPER, THE
LA CORSA
DELL'INNOCENTE
See FLIGHT OF THE
INNOCENT
LA CROCE SICILIANA
See SICILIAN CROSS
LA DAMA DELLE
CAMELIE
See LADY OF THE
CARMELIAS, THE
LA DAME AUX
CAMELIAS
See LADY OF THE
CARMELIAS, THE
LA DECADE
PRODIGIEUSE
See TEN DAYS' WONDER
LA DENTELLIERE
See LACEMAKER, THE
LA DOUBLE VIE DE
VERONIQUE
See DOUBLE LIFE OF
VERONIQUE, THE
LA FANTOME DE LA

LIBERTE
See PHANTOM OF
LIBERTY, THE
LA FEMME D'A COTE
See WOMAN NEXT DOOR,
THE
**LA FEMME DE
L'AVIATEUR**
See AVIATOR'S WIFE, THE
LA FEMME INFIDELE
See UNFAITHFUL WIVES
LA FEMME NIKITA
See NIKITA
**LA FILLE DE
D'ARTAGNAN**
See D'ARTAGNAN'S
DAUGHTER
**LA FLEUR DE MON
SECRET**
See FLOWER OF MY
SECRET, THE
LA FLOR DE MI SECRET
See FLOWER OF MY
SECRET, THE
**LA FRACTURE DU
MYOCARDE**
See CROSS MY HEART
**LA GLOIRE DE MON
PERE**
See MY FATHER'S GLORY
LA GUERRE DU FEU
See QUEST FOR FIRE
LA GUERRE EST FINIE
See WAR IS OVER, THE
LA HISTORIA OFICIAL
See OFFICIAL VERSION,
THE
**LA HORRIPILANTE
BESTIA HUMANA**
See NIGHT OF THE
BLOODY APES
**LA LEGGENDA DEL
SANTO BEVITORE**
See LEGEND OF THE
HOLY DRINKER, THE
LA LEY DEL DESEO
See LAW OF DESIRE
**LA LUNE DANS LE
CANIVEAU**
See MOON IN THE
GUTTER, THE
LA LUNE ET LE TETON
See TIT AND THE MOON,
THE
LA MACHINE
See MACHINE, THE
LA MADRE MUERTA
See DEAD MOTHER
**LA MAGIE
D'EMMANUELLE**
See EMMANUELLE'S
MAGIC
**LA MARCA DEL
MUERTO**
See CREATURE OF THE
WALKING DEAD
LA MARGE
See STREETWALKER, THE
LA MARQUISE D'O
See MARQUISE OF O, THE
**LA MASCHERA DEL
DEMONIO**
See BLACK SUNDAY
**LA MERVEILLEUSE
VISITE**
See WONDERFUL VISIT
LA MOGLIE VERGINE
See VIRGIN WIFE, THE
LA MORTE VIVANTE
See LIVING DEAD GIRL,
THE
LA MOTORCYCLETTE
See GIRL ON A

MOTORCYCLE
**LA NOCHE DE LA
MUERTA CIEGA**
See TOMBS OF THE BLIND
DEAD
**LA NOCHE DE LOS
GAVIOTAS**
See NIGHT OF THE
SEAGULLS
**LA NOCHE DEL
TERROR CIEGO**
See TOMBS OF THE BLIND
DEAD
**LA NOTTE DI SAN
LORENZO**
See NIGHT OF SAN
LORENZO, THE
LA NUIT AMERICAINE
See DAY FOR NIGHT
LA NUIT DE VARENNES
See THAT NIGHT IN
VARENNES
LA PARISIENNE
See UNE PARISIENNE
LA PARTITA
See GAMBLE, THE
LA PEAU DOUCE
See SILKEN SKIN
LA PETITE BANDE
See LITTLE GANG, THE
LA PLANETE SAUVAGE
See FANTASTIC PLANET
LA PROMESSE
See PROMISE, THE
LA PROPIETAIRE
See PROPRIETOR, THE
**LA REINE DES
VAMPIRES**
See RAPE OF THE
VAMPIRE, THE
**LA REVANCHE DES
MORTES VIVANTES**
See REVENGE OF THE
LIVING DEAD GIRLS, THE
LA ROUGE AUX LEVRES
See DAUGHTERS OF
DARKNESS
**LA SADIQUE DE NOTRE
DAME**
See DEMONIAC
**LA SEMANA DEL
ASESINO**
See CANNIBAL MAN, THE
LA SETTTA
See SECT, THE
**LA STRATEGIA DEL
RAGNO**
See SPIDER'S STRATAGEM,
THE
LA TAREA
See HOMEWORK
LA TETA I LA LLUNA
See TIT AND THE MOON,
THE
LA TETA Y LA LUNA
See TIT AND THE MOON,
THE
LA VAMPIRE NUE
See NUDE VAMPIRE, THE
LA VIA LATTEA
See MILKY WAY, THE
**LA VIE EST UN LONG
FLEUVE TRANQUILLE**
See LIFE IS A LONG QUIET
RIVER
**LA VIE SEXUELLE DES
BELGES 1950-1978**
See SEXUAL LIFE OF THE
BELGIANS 1950-1978, THE
**LA VIEILLE QUI
MARCHAIT DANS LA
MER**
See OLD LADY WHO

**WALKED IN THE SEA,
THE**
LA VOIE LACTEE
See MILKY WAY, THE
**LABERINTO DE
PASIONES**
See LABYRINTH OF
PASSION
LABOUR OF LOVE
See LABOR OF LOVE: THE
ARLETTE SCHWEITZER
STORY
L'ACCOMPAGNATRICE
See ACCOMPANIST, THE
LADIES' CLUB, THE
See SISTERHOOD, THE
LADIES OF THE PARK
See LADIES OF THE BOIS
DE BOULOGNE
LADRI DI BICICLETTE
See BICYCLE THIEVES
LADRI DI SAPONETTE
See ICICLE THIEF, THE
LADY HAMILTON
See THAT HAMILTON
WOMAN!
LADY IS A TRAMP, THE
See THAT GIRL IS A
TRAMP
**LADY OF THE
BOULEVARDS**
See NANA
LADY REPORTER
See ABOVE THE LAW 2:
THE BLOND FURY
LAERERINDEN
See LOVE LESSONS
**LAKE CONSEQUENCE:
A MAN AND TWO
WOMEN**
See LAKE CONSEQUENCE
LAKE OF SHARKS, THE
See TINTIN: THE LAKE OF
SHARKS
L'ALCOVA
See ALCOVE, THE
L'ALDILA
See BEYOND, THE
L'ALTRO INFERNO
See OTHER HELL, THE
L'AMANT
See LOVER, THE
**L'AMANT DE LADY
CHATTERLEY**
See LADY CHATTERLEY'S
LOVER
**LAMENT OF THE PATH,
THE**
See PATHER PANCHALI
L'AMI DE MON AMI
See MY GIRLFRIEND'S
BOYFRIEND
**L'AMOUR
D'EMMANUELLE**
See EMMANUELLE'S LOVE
L'AMOUR EN FUITE
See LOVE ON THE RUN
L'AMOUR L'APRES-MIDI
See LOVE IN THE
AFTERNOON
LAN FENGZHENG
See BLUE KITE, THE
**LANCELOT OF THE
LAKE**
See LANCELOT DU LAC
**L'ANNEE DERNIERE A
MARIENBAD**
See LAST YEAR AT
MARIENBAD
**L'ANNO SCORSO A
MARIENBAD**
See LAST YEAR AT
MARIENBAD

L'APPAT
See BAIT, THE
L'ART D'AIMER
See ART OF LOVE, THE
**LAS GARRAS DE
LORELEI**
See LORELEI'S GRASP,
THE
LASHOU SHENTAN
See HARD-BOILED
**LASKY JEDNE
PLAVOVSKY**
See BLONDE IN LOVE, A
**LAST BATTLE FOR THE
UNIVERSE**
See JOSH KIRBY: TIME
WARRIOR! – LAST BATTLE
FOR THE UNIVERSE
LAST BLOOD, THE
See HARD-BOILED 2: THE
LAST BLOOD
**LAST BUS TO
WOODSTOCK**
See INSPECTOR MORSE:
LAST BUS TO
WOODSTOCK
**LAST COURT MARTIAL,
THE**
See A-TEAM, THE: THE
COURT MARTIAL
LAST DRAGONS, THE
See LAST DRAGON, THE
LAST ENEMY, THE
See INSPECTOR MORSE:
THE LAST ENEMY
**LAST MAFIA
MARRIAGE, THE**
See LOVE, HONOR AND
OBEY: THE LAST MAFIA
MARRIAGE
**LAST OF THE FINEST,
THE**
See BLUE HEAT
LAST PICNIC, THE
See ZOMBIE ISLAND
MASSACRE
LAST RESORT
See NATIONAL
LAMPOON'S LAST
RESORT
LAST RIDE, THE
See F.T.W.
LAST SEEN WEARING
See INSPECTOR MORSE:
LAST SEEN WEARING
LAST VAMPIRE, THE
See ADVENTURES OF
SHERLOCK HOLMES, THE:
THE VAMPIRE OF
LAMBERLEY
LAST WARRIOR, THE
See FINAL EXECUTIONER
LATIGO
See SUPPORT YOUR
LOCAL GUNFIGHTER
**LAUREL AND HARDY IN
TOYLAND**
See LAUREL AND HARDY:
BABES IN TOYLAND
**LAVIAMOCI IL
CERVELLO**
See ROGOPAG
LAW 627
See L.627
LAW, THE
See TILAI
LAWGIVER, THE
See MOSES THE
LAWGIVER
**LAWNMOWER MAN 2:
JOBE'S WAR**
See LAWNMOWER MAN 2:
BEYOND CYBERSPACE

LAYING THE GHOST
See SWEET LITTLE SISTER
LE BALLON ROUGE
See RED BALLOON, THE
LE BIG BANG
See BIG BANG, THE
LE BOUCHER
See BUTCHER, THE
LE CAPTIVE DU DESERT
See CAPTIVE OF THE DESERT
LE CHALAND QUI PASSE
See L'ATALANTE
LE CHANT DES SIRENES
See I'VE HEARD THE MERMAIDS SINGING
LE CHARME DISCRET DE LA BOURGEOISIE
See DISCREET CHARM OF THE BOURGEOISIE, THE
LE CHAT
See CAT, THE
LE CHATEAU DE MA MERE
See MY MOTHER'S CASTLE
LE COMPLOT
See TO KILL A PRIEST
LE CONFESSIONNAL
See CONFESSIONAL, THE
LE CRI DU HIBOU
See CRY OF THE OWL
LE DECLIN DE L'EMPIRE AMERICAIN
See DECLINE OF THE AMERICAN EMPIRE, THE
LE DERNIER METRO
See LAST METRO, THE
LE DESERT ROUGE
See RED DESERT, THE
LE DIABLE, PROBABLEMENT
See DEVIL, PROBABLY, THE
LE FEU FOLLET
See FIRE WITHIN, THE
LE FRISSON DES VAMPIRES
See THRILL OF THE VAMPIRES, THE
LE GENOU DE CLAIRE
See CLAIRE'S KNEE
LE GRAAL
See LANCELOT DU LAC
LE GRAND BLEU
See BIG BLUE, THE
LE GRAND BLOND AVEC UNE CHAUSSURE NOIRE
See TALL BLOND MAN WITH ONE BLACK SHOE, THE
LE GRAND DUEL
See STORMRIDER
LE HUITIEME JOUR
See EIGHTH DAY, THE
LE HUSSARD SUR LE TOIT
See HORSEMAN ON THE ROOF, THE
LE JOURNAL DE LADY M
See DIARY OF LADY M, THE
LE JOURNAL D'UNE FEMME DE CHAMBRE
See DIARY OF A CHAMBERMAID
LE JUGE ET L'ASSASSIN
See JUDGE AND THE ASSASSIN, THE
LE LOCATAIRE
See TENANT, THE
LE MARI DE LA COIFFEUSE
See HAIRDRESSER'S HUSBAND, THE
LE MONACHE DI SANT ARCANGELO
See SISTERS OF SATAN
LE NEUVEU DE BEETHOVEN
See BEETHOVEN'S NEPHEW
LE NOTTI DI CABIRIA
See NIGHTS OF CABIRIA
LE NUIT AMERIC
See DAY FOR NIGHT
LE PARFUM D'EMMANUELLE
See EMMANUELLE'S PERFUME
LE PRIX DU DANGER
See PRIZE OF PERIL, THE
LE PROCES
See TRIAL, THE
LE RAYON VERT
See GREEN RAY, THE
LE RETOUR DE MARTIN GUERRE
See RETURN OF MARTIN GUERRE, THE
LE ROI ET L'OISEAU
See MISTER BIRD TO THE RESCUE
LE ROMAN DE RENARD
See TALE OF THE FOX, THE
LE SALAIRE DE LA PEUR
See WAGES OF FEAR, THE
LE SANG D'UNE POETE
See BLOOD OF A POET, THE
LE SIGNE DU LION
See SIGN OF LEO, THE
LE SOUFFLE AU COEUR
See DEAREST LOVE
LE TESTAMENT D'ORPHEE
See TESTAMENT OF ORPHEUS, THE
LE VIOL DU VAMPIRE
See RAPE OF THE VAMPIRE, THE
LE WEEK-END
See WEEKEND
LEADER OF THE PACK, THE
See HELL'S ANGELS ON WHEELS
LEATHER GIRLS
See FASTER PUSSYCAT, KILL... KILL
LEAVE HER TO HEAVEN
See MORTAL SINS
LEAVING HOME
See MILES TO GO
L'ECLISSE
See ECLIPSE, THE
L'EFFRONTEE
See IMPUDENT GIRL, AN
LEGACY OF LIES: THREE GENERATIONS ON TWO SIDES OF THE LAW
See LEGACY OF LIES
LEGACY OF MAGGIE WALSH, THE
See LEGACY, THE
LEGACY OF SIN: THE WILLIAM COIT STORY
See LEGACY OF SIN
LEGEND OF KASPAR HAUSER, THE
See ENIGMA OF KASPAR HAUSER, THE
LEGEND OF SPIDER FOREST, THE
See VENOM
LEGEND OF THE BAYOU
See DEATH TRAP
LEGEND OF THE SURAM FORTRESS, THE
See LEGEND OF SURAM FORTRESS, THE
LEGENDA SURAMSKOI KREPOSTI
See LEGEND OF SURAM FORTRESS, THE
LEKCE FAUST
See FAUST
LEN DEIGHTON'S BULLET TO BEIJING
See BULLET TO BEIJING
LENA: MY 100 CHILDREN
See LENA: MY HUNDRED CHILDREN
L'ENFER
See TORMENT
LENIN, THE SEALED TRAIN
See SEALED TRAIN, THE
LENIN... THE TRAIN
See SEALED TRAIN, THE
LENSMAN
See LENSMAN: THE POWER OF THE LENS
LES 400 COUPS
See FOUR-HUNDRED BLOWS, THE
LES AMANTS
See LOVERS, THE
LES BICHES
See BAD GIRLS
LES CHOSES DE LA VIE
See THINGS OF LIFE, THE
LES DAMES DU BOIS DE BOULOGNE
See LADIES OF THE BOIS DE BOULOGNE
LES DEUX ANGLAISES EN LE CONTINENT
See ANNE AND MURIEL
LES DOUZE TRAVAUX D'ASTERIX
See ASTERIX AND THE TWELVE TASKS
LES FLEURS DU SOLEIL
See SUNFLOWER
LES FOLIES D'ELODIE
See NAUGHTY BLUE KNICKERS
LES GENS DE LA RIZIERE
See RICE PEOPLE
LES INNOCENTS AUX MAINS SALES
See INNOCENTS WITH DIRTY HANDS
LES LEVRES ROUGE
See DAUGHTERS OF DARKNESS
LES MARIES DE L'AN 2
See SCOUNDREL, THE
LES MARIES DE L'AN DEUX
See SCOUNDREL, THE
LES MISERABLES DU VINGTIEME SIECLE
See LES MISERABLES
LES NUITS DE LA PLEINE LUNE
See FULL MOON IN PARIS
LES NUITS FAUVES
See SAVAGE NIGHTS
LES PARAPLUIES DE CHERBOURG
See UMBRELLAS OF CHERBOURG, THE
LES PETITES FILLES MODELES
See GOOD LITTLE GIRL
LES QUATRE CENT COUPS
See FOUR-HUNDRED BLOWS, THE
LES RENDEZVOUS DE PARIS
See RENDEZVOUS IN PARIS
LES RIPOUX
See LE COP
LES SILENCES DU PALAIS
See SILENCES OF THE PALACE, THE
LES VACANCES DE MONSIEUR HULOT
See MONSIEUR HULOT'S HOLIDAY
LES VEUFS
See ENTANGLED
LES YEUX SANS VISAGE
See EYES WITHOUT A FACE
LESBIAN TWINS
See VIRGIN WITCH
LET ME LOVE YOU
See NIGHT PLEASURES
L'ETE MEURTRIER
See ONE DEADLY SUMMER
L'ETERNEL RETOUR
See LOVE ETERNAL
LETHAL ATTRACTION
See HEATHERS
LET'S HAVE A BRAINWASH
See ROGOPAG
LETZE AUSFAHRT BROOKLYN
See LAST EXIT TO BROOKLYN
L'EVANGILE SELON SAINT MATTHIEU
See GOSPEL ACCORDING TO ST MATTHEW, THE
L'EVENTREUR DE NOTRE DAME
See DEMONIAC
L'HOMME DE MA VIE
See MAN IN MY LIFE, THE
L'HORLOGER DE SAINT-PAUL
See WATCHMAKER OF SAINT-PAUL, THE
L'HORRIBLE DOCTEUR ORLOFF
See AWFUL DR ORLOFF, THE
LI HSIAO-LUNG CH'UAN-CH'I
See BRUCE LEE: THE MAN – THE MYTH
LIBERACE: BEHIND THE MUSIC
See LIBERACE: THE UNTOLD STORY
LIBIDO
See NIGHT PLEASURES
LICENSED TO TERMINATE
See NINJA OPERATION 3: LICENSED TO

TERMINATE

LIES OF THE TWINS
See LIES FROM THE
TWINS

LIFE IN THE PINK
See PETTICOAT AFFAIR

LIFE IS ROSY
See LA VIE EST BELLE

LIFE OF A NINJA, A
See LIFE OF NINJA, A

LIFE OF BRIAN, THE
See MONTY PYTHON'S
LIFE OF BRIAN

LIFESAVERS
See MIXED NUTS

LIGHTHOUSE
See STRANDED

L'ILE AU TRESOR
See TREASURE ISLAND

LIMBO
See REBEL ROUSERS, THE

LIMELIGHT
See CHARLIE CHAPLIN:
LIMELIGHT

LINE OF FIRE
See SWAP, THE

L'INFERMIERA
See I WILL IF YOU WILL

L'INNOCENTE
See INNOCENT, THE

**L'INTERNO DI UN
CONVENTO**
See BEHIND CONVENT
WALLS

LIONHEART
See A.W.O.L.

LIONMAN 2
See LIONMAN AND THE
WITCH QUEEN

LISA E IL DIAVOLO
See LISA AND THE DEVIL

LITTLE DEVILS
See WITCH ACADEMY

LITTLE HEROES
See FUZZ THE HERO

**LITTLE HEROES OF
SHAOLIN TEMPLE**
See LITTLE HERO OF
SHAOLIN TEMPLE, THE

**LITTLE MISS
INNOCENCE**
See TEENAGE
INNOCENCE

LITTLE MISS INNOCENT
See TEENAGE
INNOCENCE

**LITTLE NINJA DRAGON,
THE**
See MAGIC KID 2

LITTLE NINJAS
See THREE LITTLE NINJAS
AND THE LOST
TREASURE

**LITTLE RED RIDING
HOOD**
See RED RIDING HOOD

LITTLE SISTER
See MISTER SISTER

**LIVE A LITTLE, LOVE A
LITTLE**
See ELVIS PRESLEY: LIVE
A LITTLE, LOVE A LITTLE

**LIVE! FROM DEATH
ROW**
See LIVE FROM DEATH
ROW

LIVING DEAD, THE
See REVENGE OF THE
LIVING DEAD GIRLS, THE

LJUBAVNI SLUCAJ
See SWITCHBOARD
OPERATOR, THE

LO BALLO DA SOLA
See STEALING BEAUTY

LO SCEICCO BIANCO
See WHITE SHEIK, THE

LOADED WEAPON 1
See NATIONAL
LAMPOON'S LOADED
WEAPON 1

L'OCCHIO DEL MALE
See POSSESSED, THE

LOCO DE AMOR
See TWO MUCH

**L'ODEUR DE LA
PAPAYE VERTE**
See SCENT OF GREEN
PAPAYA, THE

L'OMBRE DU DOUTE
See SHADOW OF A
DOUBT, A

**LONELY HEARTS
KILLERS, THE**
See HONEYMOON
KILLERS, THE

LONELY WOMAN, THE
See VOYAGE TO ITALY

**LONG JOHN SILVER'S
RETURN TO TREASURE
ISLAND**
See TREASURE ISLAND

LONG ROAD HOME, THE
See BROTHERS' DESTINY

LONG XIONG HU DI
See ARMOUR OF GOD,
THE

LONGEST YARD, THE
See MEAN MACHINE, THE

LOOSE SCREWS
See SCREWBALLS 2:
LOOSE SCREWS

**LOOT... GIVE ME
MONEY, HONEY!**
See LOOT

LOOTERS
See TRESPASS

**LORELEY'S GRASP,
THE**
See LORELEI'S GRASP,
THE

**LOS NUEVOS EXTRA
TERRESTRES**
See EXTRA TERRESTRIAL
VISITORS

**LOS RITOS SEXUALES
DEL DIABLO**
See BLACK CANDLES

LOSER TAKES ALL
See MONEY TALKS

L'OSSESSA
See SEXORCIST

LOST ANGELS
See ROAD HOME, THE

**LOST HONOR OF
KATHRYN BECK, THE**
See ACT OF PASSION

LOST WOMEN
See MESA OF LOST
WOMEN, THE

**LOUIS L'AMOUR'S THE
SACKETTS**
See DAYBREAKERS, THE

L'OURS
See BEAR, THE

**LOVE & HUMAN
REMAINS**
See LOVE AND HUMAN
REMAINS

**LOVE AFFAIR: OR THE
CASE OF THE MISSING
SWITCHBOARD
OPERATOR**
See SWITCHBOARD
OPERATOR, THE

**LOVE AND DEATH IN
SAIGON**

See BETTER TOMORROW
3, A

LOVE AND LIES
See TRUE BETRAYAL

LOVE BAN, THE
See ANYONE FOR SEX

**LOVE CAN BUILD A
BRIDGE: THE JUDDS**
See NAOMI AND
WYNONA: LOVE CAN
BUILD A BRIDGE

LOVE DREAM
See LIEBESTRAUM

LOVE GAMES, THE
See DIRTY LOVE 2: THE
LOVE GAMES

LOVE HAPPY
See MARX BROTHERS:
LOVE HAPPY

LOVE IN LAS VEGAS
See ELVIS PRESLEY: VIVA
LAS VEGAS

LOVE MADNESS
See REEFER MADNESS

LOVE MATCH
See UNE PARTIE DE
PLAISIR

LOVE ME TENDER
See ELVIS PRESLEY: LOVE
ME TENDER

LOVE POTION #9
See LOVE POTION NO. 9

**LOVERS ON THE
BRIDGE**
See LES AMANTS DU
PONT NEUF

**LOVERS: A TRUE
STORY**
See LOVERS

**LOVES AND TIMES OF
SCARAMOUCHE, THE**
See SCARAMOUCHE

**LOVES OF A BLONDE,
THE**
See BLONDE IN LOVE, A

**LOVES OF A WALL
STREET WOMAN, THE**
See HIGH FINANCE
WOMAN

LOVING YOU
See ELVIS PRESLEY:
LOVING YOU

LUCI DEL VARIETA
See LIGHTS OF VARIETY

LULU
See PANDORA'S BOX

**LUNATICS: A LOVE
STORY**
See LUNATICS

LUNES DE FIEL
See BITTER MOON

LUNG MU TOU
See CHINESE BOXER, THE

**L'UOMO DALLE DUE
OMBRE**
See COLD SWEAT

L'UOMO DELLE STELLE
See STARMAKER, THE

LUSSURIA
See LUST

LUST
See ALCOVE, THE

LUST AT FIRST BITE
See DRACULA SUCKS

**LUST OCH FAGRING
STOR**
See LOVE LESSONS

MA NUIT CHEZ MAUD
See MY NIGHT WITH
MAUD

MABOROSHI NO HIKARI
See MABOROSI

MACABRO

See MACABRE

**MacARTHUR THE
REBEL GENERAL**
See MacARTHUR

MAD DOG TIME
See TRIGGER HAPPY

**MAD MAGAZINE
PRESENTS UP THE
ACADEMY**
See UP THE ACADEMY

MADAME BUTTERFLY
See M. BUTTERFLY

MADE IN L.A.
See L.A. CRIMEWAVE

**MADEMOISELLE
FRANCE**
See REUNION IN FRANCE

**MADEMOISELLE
STRIPTEASE**
See PLUCKING THE DAISY

**MADNESS OF GEORGE
III, THE**
See MADNESS OF KING
GEORGE, THE

MAGEE
See MAGEE AND THE
LADY

MAGICA AVENTURA
See MAGIC ADVENTURE,
THE

**MAGICAL MYSTERY
TOUR, THE**
See BEATLES: THE
MAGICAL MYSTERY
TOUR

**MAGICAL PRINCESS
GIGI, THE**
See GIGI AND THE
FOUNTAIN OF YOUTH

**MAGICAL WORLD OF
GIGI, THE**
See GIGI AND THE
FOUNTAIN OF YOUTH

**MAGNIFICENT SEVEN,
THE**
See SEVEN SAMURAI, THE

**MAGNIFICENT TWO,
THE**
See MORECOMBE AND
WISE: THE MAGNIFICENT
TWO

**MAIS NE NOUS
DELIVREZ PAS DU MAL**
See DON'T DELIVER US
FROM EVIL

MAKING IT
See LES VALSEUSES

MALIBU SPICE
See MALIBU NIGHTS

**MALPERTUIS: HISTOIRE
D'UNE MAISON
MAUDITE**
See MALPERTUIS

**MAMA, THERE'S A MAN
IN YOU BED**
See ROMUALD AND
JULIETTE

MAMBA
See FAIR GAME

**MAN CALLED RAINBO,
A**
See REBEL

**MAN ESCAPED, OR THE
WIND BLOWETH
WHERE IT LISTETH, A**
See MAN ESCAPED, A

MAN IN A COCKED HAT
See CARLTON-BROWNE
OF THE F.O.

MAN IN UNIFORM, A
See I LOVE A MAN IN
UNIFORM

MAN OF THE MOMENT

See NORMAN WISDOM:
MAN OF THE MOMENT
**MAN WHO BROKE 1,000
CHAINS, THE**
See UNCHAINED
**MAN WHO COULDN'T
GET ENOUGH, THE**
See CONFESSIONS OF A
SEX MANIAC
**MAN WHO LIVED AT
THE RITZ, THE**
See MAN WHO STAYED
AT THE RITZ, THE
MAN WITH A MILLION
See MILLION POUND
NOTE, THE
**MAN WITH THE
TWISTED LIP, THE**
See RETURN OF
SHERLOCK HOLMES, THE:
THE MAN WITH THE
TWISTED LIP
MANGIATI VIVI
See EATEN ALIVE
**MANGIATI VIVI DAI
CANNIBALI**
See EATEN ALIVE
MANGLED ALIVE
See FATAL EXPOSURE
MANHATTAN BABY
See POSSESSED, THE
**MANHATTAN PROJECT,
THE**
See DEADLY GAME
**MANHATTAN PROJECT,
THE: THE DEADLY
GAME**
See DEADLY GAME
**MANIACS ARE LOOSE,
THE**
See THRILL KILLERS, THE
MANIACS ON WHEELS
See ONCE A JOLLY
SWAGMAN
MANIKA
See TALE OF A VAMPIRE
MANKILLERS
See FASTER PUSSYCAT,
KILL... KILL
**MANNEQUIN 2: ON THE
MOVE**
See MANNEQUIN ON THE
MOVE
MANON OF THE SPRING
See MANON DES
SOURCES
**MANSION OF THE
DOOMED**
See MASSACRE MANSION
MANTIS FIST FIGHTER
See THUNDERING
MANTIS, THE
**MANY-SPLENDORED
THING, A**
See LOVE IS A MANY-
SPLENDORED THING
**MARCH OF THE
WOODEN SOLDIERS**
See LAUREL AND HARDY:
BABES IN TOYLAND
**MARIE HILLEY STORY,
THE**
See WIFE, MOTHER,
MURDERER
MARIHUANA
See MARIHUANA: THE
DEVIL'S WEED
**MARIHUANA: DEVIL'S
WEED WITH ROOTS IN
HELL**
See MARIHUANA: THE
DEVIL'S WEED
MARIJUANA

See ASSASSIN OF YOUTH
**MARIO PUZO'S THE
FORTUNATE PILGRIM**
See FORTUNATE PILGRIM,
THE
MARK OF THE WITCH
See MARK OF THE DEVIL
**MARLA HANSON
STORY, THE**
See FACE VALUE
**MARQUIS DE SADE'S
JUSTINE**
See CRUEL PASSION
MARRYING MAN, THE
See TOO HOT TO HANDLE
**MARSHAL
BRAVESTARR**
See BRAVESTARR: THE
LEGEND
**MARTIAL ARTS OF
SHAOLIN**
See SHAOLIN TEMPLE 3,
THE
**MARTIAL MONKS OF
SHAOLIN TEMPLE**
See MARTIAL MONKS OF
SHAOLIN
**MARTIAN CHRONICLES
PART 1, THE: THE
EXPEDITIONS**
See MARTIAN
CHRONICLES, THE
**MARTIAN CHRONICLES
PART 2, THE: THE
SETTLERS**
See MARTIAN
CHRONICLES, THE
**MARTIAN CHRONICLES
PART 3, THE: THE
MARTIANS**
See MARTIAN
CHRONICLES, THE
MARTIANS, THE
See MARTIAN
CHRONICLES, THE
**MARX BROTHERS GO
WEST, THE**
See MARX BROTHERS: GO
WEST
MARY FOREVER
See FOREVER MARY
**MARY SHELLEY'S
FRANKENSTEIN**
See FRANKENSTEIN
MASADA
See ANTAGONISTS, THE
MASCULIN/FEMININ
See MASCULINE/
FEMININE
MASK OF SATAN
See BLACK SUNDAY
MASKS OF DEATH, THE
See SHERLOCK HOLMES:
THE MASKS OF DEATH
MASONIC MYSTERIES
See INSPECTOR MORSE:
MASONIC MYSTERIES
**MASTER
BLACKMAILER, THE**
See ADVENTURES OF
SHERLOCK HOLMES, THE:
THE MASTER
BLACKMAILER
MASTER KILLER
See THIRTY-SIXTH
CHAMBER OF SHAOLIN,
THE
MASTER OF TERROR
See FOUR D MAN, THE
MASTERMIND, THE
See ITALIAN JOB, THE
**MASTERS OF THE
UNIVERSE: THE**

MOTION PICTURE
See MASTERS OF THE
UNIVERSE
MAT
See MOTHER
**MATRIARCH, THE:
MOTHER OF THE DEAD**
See MATRIARCH, THE
**MATRIMONIO
ALL'ITALIANA**
See MARRIAGE ITALIAN-
STYLE
**MATT MILLER: PARTY
DUDE**
See MISSING PARENTS
**MATT RIKER: MUTANT
HUNT**
See MUTANT HUNT
**MATTER OF
RESISTANCE**
See LA VIE DU CHATEAU
**MAUDITE: THE LEGEND
OF DOOM HOUSE**
See MALPERTUIS
MAUVAIS SANG
See NIGHT IS YOUNG, THE
**MAX HEADROOM
STORY, THE**
See MAX HEADROOM
FILM, THE
**MAX HEADROOM: THE
ORIGINAL STORY**
See MAX HEADROOM
FILM, THE
MAY FOOLS
See MILOU IN MAY
MAYBE BABY
See MAYBE BABY...
AGAIN?
MAYBE... MAYBE NOT
See MOST DESIRED MAN,
THE
McGUIRE GO HOME!
See HIGH BRIGHT SUN,
THE
McMARTIN TRIAL, THE
See INDICTMENT: THE
McMARTIN TRIAL
MEANING OF LIFE, THE
See MONTY PYTHON'S
MEANING OF LIFE
MEANS OF EVIL
See INSPECTOR
WEXFORD: MEANS OF
EVIL
MEAT
See SPARE PARTS
**MEATBALLS 3:
SUMMER JOB**
See MEATBALLS 3
MEATBALLS: PART 2
See MEATBALLS 2
MEET CAPTAIN KIDD
See ABBOTT AND
COSTELLO MEET
CAPTAIN KIDD
**MEET WHIPLASH
WILLIE**
See FORTUNE COOKIE,
THE
MEGA JUGS
See REAL MEN EAT
KEISHA
**MEKAGOJIRA NO
GYAKUSHU**
See GODZILLA:
MONSTERS FROM AN
UNKNOWN PLANET
MELODY OF YOUTH
See THEY SHALL HAVE
MUSIC
MEMOIRE TRAQUEE
See LAPSE OF MEMORY

MEN IN TIGHTS
See ROBIN HOOD: MEN IN
TIGHTS
MEN OF THE SEA
See MAN AT THE GATE,
THE
MENSONGE
See LIE, THE
**MERCENARIES OF
LOVE**
See GOOD EVENING
VIETNAM!
MERCENARY, THE
See DJANGO
**MERCHANTS OF
MENACE, THE**
See DEATH MERCHANTS,
THE
MERIDIAN
See PHANTOMS
MERLIN
See OCTOBER 32ND
**MERRY-GO-ROUND,
THE**
See LA RONDE
MERY PER SEMPERE
See FOREVER MARY
MESSENGER, THE
See MESSENGER OF
DEATH
**METAL SKIN PANIX
MADOX 01**
See MADOX 01 – METAL
SKIN PANIC
METEOR MONSTER
See TEENAGE MONSTER
MI FAMILIA
See MY FAMILY
MI VIDA LOCA
See MY CRAZY LIFE
MIAMI VICE: THE MOVIE
See MIAMI VICE
**MICHAEL JACKSON'S
MOONWALKER**
See MOONWALKER
MICROSCOPIA
See FANTASTIC VOYAGE
MIDDLE MAN
See IN THE LINE OF DUTY
2
MIDNIGHT MELODY
See MURDER IN THE
MUSIC HALL
**MIDNIGHT RUN FOR
YOUR LIFE**
See MIDNIGHT RUN 3:
MIDNIGHT RUN FOR
YOUR LIFE
MIDNIGHT RUNAROUND
See MIDNIGHT RUN 1:
MIDNIGHT RUNAROUND
MIDWAY
See BATTLE OF MIDWAY,
THE
**MIGHTY DUCKS ARE
THE CHAMPIONS, THE**
See MIGHTY DUCKS, THE
MILLION TO JUAN, A
See MILLION TO ONE, A
MILO AND OTIS
See ADVENTURES OF
MILO AND OTIS, THE
MILOU EN MAI
See MILOU IN MAY
MIND GAMES
See AGENCY
MINNIE
See TENNESSEE NIGHTS
**MIO IN THE LAND OF
FARAWAY**
See LAND OF FARAWAY,
THE
MIO, MIN MIO

See LAND OF FARAWAY
MIRACOLO A MILANO
See MIRACLE IN MILAN
**MIRROR, MIRROR 2:
RAVEN DANCE**
See MIRROR, THE
**MISADAVENTURES OF
MISTER WILT, THE**
See WILT
MISFIT BRIGADE, THE
See WHEELS OF TERROR
**MISS SHUMWAY JETTE
UN SORT**
See ROUGH MAGIC
MISTER ARKADIN
See CONFIDENTIAL
REPORT
**MISTER BUG GOES TO
TOWN**
See HOPPITY GOES TO
TOWN
MISTER COOL
See LOW DOWN DIRTY
SHAME, A
**MISTER HULOT'S
HOLIDAY**
See MONSIEUR HULOT'S
HOLIDAY
**MISTER JOLLY LIVES
NEXT DOOR**
See COMIC STRIP
PRESENTS: MISTER JOLLY
LIVES NEXT DOOR
MISTER SARDONICUS
See MR SARDONICUS
MISTER TEN PER CENT
See CHARLIE DRAKE:
MISTER TEN PER CENT
MISTER V
See PIMPERNEL SMITH
MISTY UNDERCOVER
See DICK TRACER
MITT LIV SOM HUND
See MY LIFE AS A DOG
**MO' MONEY: WHY
SETTLE FOR LESS?**
See MO' MONEY
MOBSTERS
See MOBSTERS: THE EVIL
EMPIRE
MODERN TIMES
See CHARLIE CHAPLIN:
MODERN TIMES
**MOHAMMED,
MESSENGER OF GOD**
See MESSAGE, THE
MOI UNIVERSITETI
See MY UNIVERSITIES
MOLL FLANDERS
See FORTUNES AND
MISFORTUNES OF MOLL
FLANDERS, THE
**MOMENT OF TRUTH:
MURDER OR MEMORY?**
See MURDER OR
MEMORY: THE TRUE
STORY OF A TEENAGER'S
DEADLY CONFESSION
**MOMENT OF TRUTH: TO
WALK AGAIN**
See TO WALK AGAIN
MOMMY DEAREST
See MOMMIE DEAREST
MON ONCLE
See MONSIEUR HULOT:
MON ONCLE
MON PERE, CE HEROS
See MY FATHER, MY
HERO
**MONASTERY OF SAINT
MICHAEL, THE**
See SISTERS OF SATAN
MONIKA

See SUMMER WITH
MONIKA
MONK, THE
See FINAL TEMPTATION
MONKEY BUSINESS
See MARX BROTHERS:
MONKEY BUSINESS
MONSIEUR VERDOUX
See CHARLIE CHAPLIN:
MONSIEUR VERDOUX
MONSTER CITY
See WICKED CITY
MONSTER FROM MARS
See ROBOT MONSTER
MONSTER MAKER
See MONSTER FROM THE
OCEAN FLOOR, THE
MONSTER SHOW, THE
See FREAKS
MONSTER SNOWMAN
See HALF HUMAN
MONSTER ZERO
See GODZILLA: INVASION
OF THE ASTRO-
MONSTERS
**MONSTERS ARE
LOOSE, THE**
See THRILL KILLERS, THE
**MONSTERS FROM AN
UNKNOWN PLANET**
See GODZILLA:
MONSTERS FROM AN
UNKNOWN PLANET
**MONSTERS FROM THE
MOON**
See ROBOT MONSTER
**MONTALVO ET
L'ENFANT**
See MONTALVO AND THE
CHILD
MONTANA
See F.T.W.
**MONTENEGRO: OR
PIGS AND PEARLS**
See MONTENEGRO
MOONLIGHTING
See MOONLIGHTING: THE
ORIGINAL TV MOVIE
MOONRISE
See MY GRANDAD'S A
VAMPIRE
MORE BAD NEWS
See COMIC STRIP
PRESENTS: MORE BAD
NEWS
MORGAN!
See MORGAN: A
SUITABLE CASE FOR
TREATMENT
MORT A VENISE
See DEATH IN VENICE
**MORTAL KOMBAT: THE
MOVIE**
See MORTAL KOMBAT
MORTE A VENEZIA
See DEATH IN VENICE
MOSCOW NIGHTS
See KATIA ISMAILOVA
MOSES
See MOSES THE
LAWGIVER
MOST WANTED GIRLS
See AMERICA'S MOST
WANTED GIRL
MOSURA TAI GOJIRA
See GODZILLA: GODZILLA
VERSUS MOTHRA
MOTHER'S DAY
See ALL SHE EVER
WANTED
**MOTHRA VERSUS
GODZILLA**
See GODZILLA: GODZILLA

VERSUS MOTHRA
**MOUSE IN THE
CORNER, THE**
See INSPECTOR
WEXFORD: THE MOUSE
IN THE CORNER
MOUSEY
See CAT AND MOUSE
MOVIE CRAZY
See HAROLD LLOYD:
MOVIE CRAZY
MOVIE MAKERS, THE
See HOLLYWOOD
DREAMING
MOVING GUILLOTINE
See TRAUMA
MOVING TARGET, THE
See HARPER
MR KLEIN
See MISTER KLEIN
MR MOM
See MR MUM
MRS HYDE
See SHE KILLED IN
ECSTASY
MS. DON JUAN
See IF DON JUAN WERE A
WOMAN
MUDHONEY
See MUD HONEY
MUI DU DU XANH
See SCENT OF GREEN
PAPAYA, THE
**MUJERES AL BORDE
DE UN ATAQUE DE
NERVIOS**
See WOMEN ON THE
VERGE OF A NERVOUS
BREAKDOWN
MURDER AHOY
See MISS MARPLE:
MURDER AHOY
**MURDER AT THE
GALLOP**
See MISS MARPLE:
MURDER AT THE GALLOP
**MURDER AT THE P.T.A.
LUNCHEON**
See MENU FOR MURDER
**MURDER BEING ONCE
DONE**
See INSPECTOR
WEXFORD: MURDER
BEING ONCE DONE
**MURDER BY
CONFESSION**
See ABSOLUTION
MURDER BY DEMAND
See MURDER ON
DEMAND
MURDER C.O.D.
See MURDER ON
DEMAND
**MURDER IN PARADISE
LAKE**
See PLEASURE IN
PARADISE
**MURDER IS
ANNOUNCED, A**
See MISS MARPLE: A
MURDER IS ANNOUNCED
MURDER MOST FOUL
See MISS MARPLE:
MURDER MOST FOUL
MURDER ON MONDAY
See HOME AT SEVEN
**MURDER ON THE RIO
GRANDE**
See HUNTED
MURDER SHE SAID
See MISS MARPLE:
MURDER SHE SAID
MURDER WITH

MIRRORS
See MISS MARPLE:
MURDER WITH MIRRORS
MURDER, INC.
See ENFORCER, THE
**MURDER, SMOKE AND
SHADOWS**
See COLUMBO: MURDER,
SMOKE AND SHADOWS
**MURMUR OF THE
HEART**
See DEAREST LOVE
**MUSGRAVE RITUAL,
THE**
See RETURN OF
SHERLOCK HOLMES, THE:
THE MUSGRAVE RITUAL
**MUSSOLINI: THE
DECLINE AND FALL OF
IL DUCE**
See MUSSOLINI AND I
MUTANT ACTION
See ACCION MUTANTE
**MY BIRD TO THE
RESCUE**
See MISTER BIRD TO THE
RESCUE
MY FAMILY
See EAST L.A.
MY FATHER, THE HERO
See MY FATHER, MY
HERO
**MY GRANDPA IS A
VAMPIRE**
See MY GRANDAD'S A
VAMPIRE
MY NAME IS IVAN
See IVAN'S CHILDHOOD
MY NIGHT AT MAUD'S
See MY NIGHT WITH
MAUD
MY UNCLE
See MONSIEUR HULOT:
MON ONCLE
**MY UNCLE MISTER
HULOT**
See MONSIEUR HULOT:
MON ONCLE
MY UNIVERSITY
See MY UNIVERSITIES
MY WONDERFUL LIFE
See MY FIRST FORTY
YEARS
**MYSTERIOUS AFFAIR
AT STYLES, THE**
See POIROT: THE
MYSTERIOUS AFFAIR AT
STYLES
**MYSTERY OF KASPAR
HAUSER, THE**
See ENIGMA OF KASPAR
HAUSER, THE
NACH DER FINSTERNIS
See AFTER DARKNESS
NAGISA NO SINDBAD
See LIKE GRAINS OF
SAND
NAKED DREAMS
See BLACK CANDLES
NAKED GUN, THE
See NAKED GUN, THE:
FROM THE FILES OF
POLICE SQUAD
**NAKED UNDER
LEATHER**
See GIRL ON A
MOTORCYCLE
NAKED VAMPIRE, THE
See NUDE VAMPIRE, THE
**NANCY CONN STORY,
THE**
See FIGHT FOR JUSTICE:
THE NANCY CONN

STORY
NANKAI NO DAIKETTO
See GODZILLA: EBIRAH –
HORROR OF THE DEEP
NANOOK
See KABLOONAK
NATE AND HAYES
See SAVAGE ISLANDS
**NATIONAL LAMPOON
GOES TO THE MOVIES**
See NATIONAL
LAMPOON'S MOVIE
MADNESS
**NATIONAL LAMPOON'S
SCUBA SCHOOL**
See NATIONAL
LAMPOON'S LAST
RESORT
NATTEVAGTEN
See NIGHTWATCH
NATTVARDSGASTERNA
See WINTER LIGHT
NATURE'S MISTAKES
See FREAKS
NAUGHTY KNIGHTS
See UP THE CHASTITY
BELT
NAVAL TREATY, THE
See ADVENTURES OF
SHERLOCK HOLMES, THE:
THE NAVAL TREATY
NAVIGATOR, THE
See BUSTER KEATON:
NAVIGATOR, THE
**NAVIGATOR, THE: A
MEDIEVAL ODYSSEY**
See NAVIGATOR, THE
NEAR MISFITS
See NEAR MRS
NEAR MISSES
See NEAR MRS
**NECROMANCER:
SATAN'S SERVANT**
See NECROMANCER
NECRONOMICON
See SUCCUBUS
**NECRONOMICON:
DREAMED SINS**
See SUCCUBUS
**NECRONOMICON:
GETRAUMTE SUNDEN**
See SUCCUBUS
**NEIL SIMON'S
BROADWAY BOUND**
See BROADWAY BOUND
**NEIL SIMON'S LOST IN
YONKERS**
See LOST IN YONKERS
**NEIL SIMON'S THE
SUNSHINE BOYS**
See SUNSHINE BOYS, THE
**NEOKONCHENNAYA
PYESSA DLYA
MEKHANICHESKOGO
PIANIN**
See UNFINISHED PIECE
FOR MECHANICAL
PIANO
**NEVER PICK UP A
STRANGER**
See BLOODRAGE
NEVER TO LOVE
See BILL OF
DIVORCEMENT, THE
**NEVERENDING STORY
2, THE: THE NEXT
CHAPTER**
See NEVERENDING STORY
2, THE
**NEVERENDING STORY
3, THE: ESCAPE FROM
FANTASIA**
See NEVERENDING STORY

3, THE
**NEVERENDING STORY
3, THE: RETURN TO
FANTASIA**
See NEVERENDING STORY
3, THE
NEVINOST BEZ ZASTITE
See INNOCENCE
UNPROTECTED
**NEW ADVENTURES OF
DON JUAN, THE**
See ADVENTURES OF DON
JUAN, THE
**NEW GAME OF DEATH,
THE**
See GAME OF DEATH 2
**NEW LEASE OF DEATH,
A**
See INSPECTOR
WEXFORD: A NEW LEASE
OF DEATH
NEWSIES
See NEWS BOYS, THE
**NI JU-SEIKI SHONEN
DOKUHON**
See CIRCUS BOYS
NICE DREAMS
See CHEECH AND
CHONG'S NICE DREAMS
NICO: ABOVE THE LAW
See NICO
**NIEBELUNGEN SAGA,
THE: PART 1**
See SIEGFRIED
**NIGHT AT THE OPERA,
A**
See MARX BROTHERS: A
NIGHT AT THE OPERA
NIGHT CRAWLERS, THE
See NAVY VERSUS THE
NIGHT MONSTERS, THE
**NIGHT HEAVEN FELL,
THE**
See LES BIJOUTIERS DU
CLAIR DE LUNE
**NIGHT IN
CASABLANCA, A**
See MARX BROTHERS: A
NIGHT IN CASABLANCA
NIGHT KILLER
See SILENT MADNESS
NIGHT LEGS
See FRIGHT
NIGHT OF ABANDON
See RED SHOE DIARIES 8:
NIGHT OF ABANDON
NIGHT OF ANUBIS
See NIGHT OF THE LIVING
DEAD
**NIGHT OF THE BLIND
DEAD**
See TOMBS OF THE BLIND
DEAD
**NIGHT OF THE DEATH
CULT**
See NIGHT OF THE
SEAGULLS
**NIGHT OF THE
DEMONS: ANGELA'S
REVENGE**
See NIGHT OF THE
DEMONS 2
**NIGHT OF THE FLESH
EATERS**
See NIGHT OF THE LIVING
DEAD
**NIGHT OF THE LIVING
DEAD: THE REMAKE**
See NIGHT OF THE LIVING
DEAD
**NIGHT OF THE
SHOOTING STARS**
See NIGHT OF SAN

LORENZO, THE
**NIGHT OF VARENNES,
THE**
See THAT NIGHT IN
VARENNES
NIGHT SLASHER
See NIGHT AFTER NIGHT
AFTER NIGHT
**NIGHT TRAIN TO
MURDER**
See MORECOMBE AND
WISE: NIGHT TRAIN TO
MURDER
NIGHT WARRIOR
See NIGHT OF THE
WARRIOR
NIGHT, THE
See LA NOTTE
NIGHTMARE
See KIDNAPPED
NIGHTMARE ISLAND
See SLAYER, THE
**NIGHTMARE OF
TERROR**
See TOURIST TRAP, THE
**NIGHTNARE IN THE
DAYLIGHT, A**
See NIGHTMARE IN
DAYLIGHT, A
NIGHTSCARE
See BEYOND BEDLAM
NIGHTWATCH
See NIGHT WATCH
**NINE SEVEN SIX: EVIL 2
– THE ASTRAL FACTOR**
See NINE SEVEN SIX: EVIL
2
NINJA 2
See NINJA 2: THE
REVENGE OF THE NINJA
NINJA CONNECTION
See NINJA IN THE
KILLING FIELDS
**NINJA KISS OF DEATH
KIDS**
See NINJA KIDS: KISS OF
DEATH
NINJA OPERATION 3
See NINJA OPERATION 3:
LICENSED TO
TERMINATE
NO CRYING HE MAKES
See INSPECTOR
WEXFORD: NO CRYING
HE MAKES
**NO ESCAPE, NO
RETURN**
See NO ESCAPE NO
RETURN
NO MERCY
See SIN COMPASION
NO MORE DYING THEN
See INSPECTOR
WEXFORD: NO MORE
DYING THEN
**NO PLACE LIKE
HOMICIDE**
See WHAT A CARVE UP!
NO PLACE TO HIDE
See REBEL
**NORMAN'S AWESOME
EXPERIENCE**
See SWITCH IN TIME, A
**NORWOOD BUILDER,
THE**
See ADVENTURES OF
SHERLOCK HOLMES, THE:
THE NORWOOD BUILDER
NOSFERATU A VENZIA
See VAMPIRE IN VENICE
NOSFERATU IN VENICE
See VAMPIRE IN VENICE
NOSFERATU THE

VAMPIRE
See NOSFERATU THE
VAMPYRE
**NOSFERATU, A
SYMPHONY OF TERROR**
See NOSFERATU
**NOSFERATU, EINE
SYMPHONIE DES
GRAUENS**
See NOSFERATU
**NOSFERATU, THE
VAMPIRE**
See NOSFERATU
**NOSFERATU: PHANTOM
DER NACHT**
See NOSFERATU THE
VAMPYRE
NOSTALGHIA
See NOSTALGIA
**NOT AGAINST THE
FLESH**
See VAMPYR
NOT IN MY FAMILY
See SHATTERING THE
SILENCE
**NOT TOO SCARED TO
DIE**
See EAGLE SHADOW FIST
**NOTHING BUT
TROUBLE**
See LAUREL AND HARDY:
NOTHING BUT TROUBLE
**NOTORIOUS
GENTLEMAN, THE**
See RAKE'S PROGRESS,
THE
**N'OUBLIE PAS QUE TU
VAS MOURIR**
See DON'T FORGET
YOU'RE GOING TO DIE
NOVECENTO
See 1900: PARTS 1 AND 2
NOW IS THE TIME
See DEAREST LOVE
NOZ W WODZIE
See KNIFE IN THE WATER
NRNA GOUYNE
See COLOUR OF
POMEGRANATES, THE
NUCLEAR COUNTDOWN
See TWILIGHT'S LAST
GLEAMING
NUIT ET JOUR
See NIGHT AND DAY
NUKE 'EM HIGH
See CLASS OF NUKE 'EM
HIGH
**NUN AND THE DEVIL,
THE**
See SISTERS OF SATAN
**NUNS OF SAINT
ARCHANGEL, THE**
See SISTERS OF SATAN
**NUOVO CINEMA
PARADISO**
See CINEMA PARADISO
NURSE, THE
See I WILL IF YOU WILL
**NUTTY PROFESSOR,
THE**
See JERRY LEWIS: THE
NUTTY PROFESSOR
OBCHOD OD NA KORZE
See SHOP ON THE HIGH
STREET, THE
OBERST REDL
See COLONEL REDL
OBSESSED, THE
See SEXORCIST, THE
OBSESSION, THE
See SEXORCIST
OCI CIORNIE
See DARK EYES

OFFICIAL STORY, THE
See OFFICIAL VERSION, THE

OFFRET
See SACRIFICE, THE

OFFSPRING, THE
See FROM A WHISPER TO A SCREAM

OGGI A ME... DOMANI A TE
See TODAY IT'S ME, TOMORROW IT'S YOU!

OGRE, THE
See DEMONS 3

OH LUCKY MAN
See O LUCKY MAN!

OH, ALFIE
See ALFIE DARLING

OH, CAROL
See COOL IT CAROL!

OKTYABR
See OCTOBER

OLD AND NEW
See GENERAL LINE, THE

OLD WOMAN WHO WALKED INT THE SEA, THE
See OLD LADY WHO WALKED IN THE SEA, THE

OLIVER'S KING LEAR
See KING LEAR

OLIVIER'S HAMLET
See HAMLET

OMAR MUKHTAR
See LION OF THE DESERT

OMEGA FACTOR
See SILENT MADNESS

OMEN 2
See DAMIEN: OMEN 2

ON HIS OWN
See MY APPRENTICESHIP

ON THE BEAT
See NORMAN WISDOM: ON THE BEAT

ON THE ROAD AGAIN
See HONEYSUCKLE ROSE

ON THE STREETS OF L.A.
See FATHER AND SON

ON THE THIRD DAY ARRIVED THE CROW
See FISTFUL OF DEATH, A

ONCE UPON A FRIGHTMARE
See FRIGHTMARE

ONCE UPON A LOVE
See FANTASIES

ONCE UPON A TIME THERE WAS A COUNTRY
See UNDERGROUND

ONE GOOD TURN
See NORMAN WISDOM: ONE GOOD TURN

ONE MILLION POUND NOTE, THE
See MILLION POUND NOTE, THE

ONE RIOT, ONE RANGER
See WALKER, TEXAS RANGER: ONE RIOT, ONE RANGER

ONE WOMAN OR TWO
See WOMAN OR TWO, A

ONE WOMAN'S STORY
See PASSIONATE FRIENDS, THE

ONEAMISU NO TSUBASA
See WINGS OF HONNEAMISE, THE

ONLY THE FRENCH CAN
See FRENCH CAN-CAN

OPEN CITY
See ROME, OPEN CITY

OPERA
See TERROR AT THE OPERA

OPERATION CONDOR: ARMOUR OF GOD 2
See ARMOUR OF GOD 2

OPERATION MONSTERLAND
See GODZILLA: DESTROY ALL MONSTERS

OPERATION NAM
See COBRA MISSION

OPERATION PETTICOAT
See PETTICOAT AFFAIR

OPPOSITE SEX... AND HOW TO LIVE WITH THEM, THE
See OPPOSITE SEX, THE

ORFEU NEGRO
See BLACK ORPHEUS

ORGY OF THE VAMPIRES
See ORGY OF THE DEAD

ORPHEUS
See ORPHEE

OSSESSIONE
See OBSESSION

OSTRE SLEDOVANE VLAKY
See CLOSELY WATCHED TRAINS

OTAC NA SLUZBENOM PUTU
See WHEN FATHER WAS AWAY ON BUSINESS

OTTO E MEZZO
See FELLINI'S 8$^{1}/_{2}$

OU LE VENT SOUFFLE OU IL VENT
See MAN ESCAPED, A

OUR HOSPITALITY
See BUSTER KEATON: OUR HOSPITALITY

OUR RELATIONS
See LAUREL AND HARDY: OUR RELATIONS

OUT IN THE WORLD
See MY APPRENTICESHIP

OUT OF DARKNESS: TRIUMPH OF COURAGE
See LIGHT IN THE JUNGLE, THE

OUT OF ROSENHEIM
See BAGDAD CAFE

OUT OF THE ASHES
See HIROSHIMA

OUTING, THE
See LAMP, THE

OUTLAW
See OUTLAW OF GOR

OUTLAWS OF LOVE
See BONNIE AND CLYDE: OUTLAWS OF LOVE

OUTLAWS, THE
See OUTLAW BROTHERS, THE

OUTPOST, THE
See MIND RIPPER

OUTSIDER, THE
See GUINEA PIG, THE

OXEN
See OX, THE

P.T. RAIDERS
See SHIP THAT DIED OF SHAME, THE

PACE THAT KILLS, THE
See COCAINE FIENDS, THE

PACK UP YOUR

TROUBLES
See LAUREL AND HARDY: PACK UP YOUR TROUBLES

PAINT JOB, THE
See PAINTED HEART

PAISA
See PAISAN

PALE KINGS AND PRINCES
See SPENSER: PALE KINGS AND PRINCES

PANGA
See WITCHCRAFT

PANIC IN THE TRANS-SIBERIAN TRAIN
See HORROR EXPRESS

PANIC ON THE TRANS-SIBERIAN EXPRESS
See HORROR EXPRESS

PANICO EN EL TRANSSIBERIANO
See HORROR EXPRESS

PAODA SHUANG DENG
See RED FIRECRACKER, GREEN FIRECRACKER

PAR-DELA LES NUAGES
See BEYOND THE CLOUDS

PARADISE, HAWAIIAN STYLE
See ELVIS PRESLEY: PARADISE, HAWAIIAN STYLE

PARADOX: BACK TO THE FUTURE 2
See BACK TO THE FUTURE: PART 2

PARASITE MURDERS, THE
See SHIVERS

PARDON US
See LAUREL AND HARDY: PARDON US

PARIS MATCH
See FRENCH KISS

PARIS VU PAR
See SIX IN PARIS

PARISIENNE, A
See UNE PARISIENNE

PARTY AT KITTY AND STUDS
See ITALIAN STALLION, THE

PARTY INCORPORATED
See PARTY GIRLS

PASQUALINO SETTEBELLEZZE
See SEVEN BEAUTIES

PASQUALINO: SEVEN BEAUTIES
See SEVEN BEAUTIES

PASSION FOR INNOCENCE, A
See TILL MURDER DO US PART: PART 2

PASSIONATE PLEASURES
See PLEASURE, THE

PASSIONE D'AMORE
See PASSION OF LOVE

PATSY, THE
See JERRY LEWIS: THE PATSY

PATTON: LUST FOR GLORY
See PATTON

PATTY HEARST
See PATTY HEARST: HER OWN STORY

PAULINE A LA PLAGE
See PAULINE AT THE BEACH

PAVRA NELLA CITTA

DEI MORTI VIVENTI
See CITY OF THE LIVING DEAD

PEACEMAKER
See AMBASSADOR, THE

PELLE EROVRARE
See PELLE THE CONQUEROR

PEPI, LUCI, BOM AND OTHER GIRLS ON THE HEAP
See PEPI, LUCI, BOM

PEPI, LUCI, BOM Y OTRAS CHICAS DEL MONTON
See PEPI, LUCI, BOM

PER QUAICHE DOLLARI IN PIU
See FOR A FEW DOLLARS MORE

PER UN PUGNO DI DOLLARI
See FISTFUL OF DOLLARS, A

PERFECT TRIBUTE
See GETTYSBURG

PERICOLO IN AGGUATO
See MAN ON FIRE

PERIL EN LA DEMEURE
See DEATH IN A FRENCH GARDEN

PERILS FROM THE PLANET MONGO
See FLASH GORDON CONQUERS THE UNIVERSE

PERRY MASON: THE CASE OF THE AVENGING ACE
See PERRY MASON: THE CASE OF THE AVENGING ANGEL

PERSECUTION AND ASSASSINATION OF JEAN-PAUL MARAT AS PERFORMED BY THE INMATES OF THE ASYLUM OF CHARENTON UNDER THE DIRECTION OF THE MARQUIS DE SADE, THE
See MARAT/SADE

PETTICOAT PIRATES
See CHARLIE DRAKE: PETTICOAT PIRATES

PHANTOM CHARIOT, THE
See PHANTOM CARRIAGE, THE

PHANTOM OF LOVE, THE
See AI NO BOREI

PHASE IV
See PHASE 4

PHILLY
See PRIVATE LESSONS

PHOENIX 2772
See SPACE FIREBIRD

PHOTOS SCANDALE
See PHOTO SCANDAL

PIADONE L'AFRICANO
See FLATFOOT

PICNIC, THE
See ZOMBIE ISLAND MASSACRE

PIDA HUIVISTA KIINNI, TATJANA
See TAKE CARE OF YOUR SCARF, TATJANA

PIER, THE
See LA JETEE

PIN: A PLASTIC

NIGHTMARE
See PIN

PIPPI LONGSTOCKING: THE NEW ADVENTURES
See NEW ADVENTURES OF PIPPI LONGSTOCKING, THE

PIRANHA 2: THE SPAWNING
See PIRANHA 2: FLYING KILLERS

PIXOTE: LA LEY DEL MAS DEBIL
See PIXOTE

PLANET OF INCREDIBLE CREATURES
See FANTASTIC PLANET

PLANET OF THE DINO-KNIGHTS
See JOSH KIRBY: TIME WARRIOR! – PLANET OF THE DINO-KNIGHTS

PLAYTIME
See MONSIEUR HULOT: PLAYTIME

PLEASE! MISTER BALZAC
See PLUCKING THE DAISY

PLEASURE PARTY
See UNE PARTIE DE PLAISIR

PLUCKING THE DAISIES
See PLUCKING THE DAISY

PO DEZJU
See BEFORE THE RAIN

POCKETFUL OF RYE, A
See MISS MARPLE: A POCKETFUL OF RYE

PODMOSKOVNYE VECHERA
See KATIA ISMAILOVA

POE'S TALES OF TERROR
See TALES OF TERROR

POINT OF IMPACT
See IN TOO DEEP

POINT OF NO RETURN
See ASSASSIN, THE

POISONED BY LOVE: THE KERN COUNTY MURDERS
See MURDER SO SWEET

POKOLENIE
See GENERATION, A

POLICE ACADEMY 2: THEIR FIRST ASSIGNMENT
See POLICE ACADEMY 2

POLICE ACADEMY 3: BACK IN TRAINING
See POLICE ACADEMY 3

POLICE ACADEMY 4: CITIZEN'S ON PATROL
See POLICE ACADEMY 4

POLICE ACADEMY 5: ASSIGNMENT MIAMI BEACH
See POLICE ACADEMY 5

POLICE ACADEMY 6: CITY UNDER SIEGE
See POLICE ACADEMY 6

POLICE FORCE
See POLICE STORY

POLICE STORY 4: CRIME STORY
See CRIME STORY

POLTERGEIST 2: THE OTHER SIDE
See POLTERGEIST 2

POOR LITTLE RICH GIRL: THE BARBARA HUTTON STORY
See POOR LITTLE RICH GIRL: PARTS 1 AND 2

POPE MUST DIET, THE
See POPE MUST DIE, THE

POPE, THE
See POPE MUST DIE, THE

POPIOL Y DIAMENT
See ASHES AND DIAMONDS

PORKY'S 2: THE NEXT DAY
See PORKY'S 2

PORKY'S 3
See PORKY'S REVENGE

PORTRAIT, THE
See GIRL'S SCHOOL SCREAMERS

POSING
See I POSED FOR PLAYBOY

POSING: INSPIRED BY THREE REAL LIFE STORIES
See I POSED FOR PLAYBOY

POSITIVELY TRUE ADVENTURES OF THE ALLEGED TEXAS CHEERLEADER-MURDERING MOM, THE
See CHEERLEADER-MURDERING MOM

POTEMKIN
See BATTLESHIP POTEMKIN, THE

POTOMOK CHINGIS-KHANA
See STORM OVER ASIA

POULET AU VINAIGRE
See COP AU VIN

POUND PUPPIES AND THE LEGEND OF BIG PAW
See POUND PUPPIES: THE LEGEND OF BIG PAW

POUND PUPPIES: THE MOVIE
See POUND PUPPIES: THE LEGEND OF BIG PAW

POWER PLAY
See MAY THE BEST MAN WIN

POWER PLAY: THE JACKIE PRESSER STORY
See TEAMSTER BOSS: THE JACKIE PRESSER STORY

POWER RANGERS: THE MOVIE
See MIGHTY MORPHIN POWER RANGERS: THE MOVIE

POWERS OF EVIL
See TALES OF MYSTERY AND IMAGINATION

PRAY DEATH COMES SWIFTLY
See VICTIM, THE

PRECINCT 45: LOS ANGELES POLICE
See NEW CENTURIONS, THE

PRED DOZDOT
See BEFORE THE RAIN

PRENOM CARMEN
See FIRST NAME: CARMEN

PREPAREZ VOS MOUCHOIRS
See GET OUT YOUR HANDKERCHIEFS

PRESCRIPTION MURDER
See COLUMBO:

PRESCRIPTION – MURDER
See OTHER HELL, THE

PRESIDENT'S WIFE, THE
See ASSASSINATION

PRESS FOR TIME
See NORMAN WISDOM: PRESS FOR TIME

PRESUME DANGEREU
See BELIEVED VIOLENT

PRICE OF JUSTICE, THE
See KOJAK: THE PRICE OF JUSTICE

PRICE OF PEACE, THE
See BLACK FOX: THE PRICE OF PEACE

PRICE OF VENGEANCE, THE
See IN THE LINE OF DUTY: THE PRICE OF VENGEANCE

PRICE SHE PAID, THE
See PLAN OF ATTACK

PRIMA DELLA RIVOLUZIONE
See BEFORE THE REVOLUTION

PRIMAL SCREAM
See DISTANT SCREAM, A

PRIORY SCHOOL, THE
See RETURN OF SHERLOCK HOLMES, THE: THE PRIORY SCHOOL

PRISONERS OF THE SUN
See BLOOD OATH

PRIVATE LESSONS: ANOTHER STORY
See X-TRA PRIVATE LESSONS

PRIVATE WORLD OF DARKNESS, THE
See TWILIGHT ZONE, THE: THE EYE OF THE BEHOLDER

PRIZE PULITZER, THE
See ROXANNE: THE PRIZE PULITZER

PROBLEM OF THOR BRIDGE, THE
See CASEBOOK OF SHERLOCK HOLMES, THE: THE PROBLEM OF THOR BRIDGE

PROCURER, THE
See ACCATTONE

PROFESSION: REPORTER
See PASSENGER, THE

PROFESSIONAL GUN, A
See DJANGO

PROFESSIONE: REPORTER
See PASSENGER, THE

PROFONDO ROSSO
See DEEP RED

PROGRAMME, THE
See PROGRAM, THE

PROGRAMMED TO KILL
See RETALIATOR

PROJECT INTERCEPT
See OPERATION INTERCEPT

PROJECT: SHADOWCHASER
See PROJECT SHADOWCHASER

PROMISE OF RED LIPS
See DAUGHTERS OF DARKNESS

PROMISED LAND
See INSPECTOR MORSE:

PROMISED LAND

PROMOTER, THE
See CARD, THE

PROPHET OF EVIL: THE ERVIL LeBARON STORY
See PROPHET OF EVIL

PROTOTYPE X29A
See PROTOTYPE

PROVA D'ORCHESTRA
See ORCHESTRA REHEARSAL

PROZZIE
See OLIVIA

PSYCHO COP
See PSYCHOCOP

PSYCHO COP 2
See PSYCHOCOP RETURNS

PSYCHOTHERAPY
See DON'T GET ME STARTED

PUMPKINHEAD 2: BLOODWINGS
See REVENGE OF PUMPKINHEAD, THE

PUNCH AND JUDY MAN, THE
See HANCOCK: THE PUNCH AND JUDY MAN

PUNK, THE
See PUNK AND THE PRINCESS, THE

PURPLE DEATH FROM OUTER SPACE
See FLASH GORDON CONQUERS THE UNIVERSE

PURPLE HEART
See DECORATION DAY

PURSUIT OF THE GRAF SPEE
See BATTLE OF THE RIVER PLATE, THE

PUT ON BY CUNNING
See INSPECTOR WEXFORD: PUT ON BY CUNNING

PUTAIN DU ROI
See KING'S WHORE, THE

PUTTING HER ASS ON THE LINE
See PUTTIN' HER ASSETS ON THE LINE

PYRAMIDS OF MARS, THE
See DOCTOR WHO: THE PYRAMIDS OF MARS

QIAN NU YOUHAN
See CHINESE GHOST STORY, A

QIUJU DA GUANSI
See STORY OF QIU JU, THE

QIUYUE
See AUTUMN MOON

QU MO JING CHAN
See MAGIC COP

QUAI DES BRUMES
See PORT OF SHADOWS

QUALCOSA DI BLONDA
See AURORA

QUATERMASS
See QUATERMASS CONCLUSION, THE

QUATRE AVENTURES DE REINETTE ET MIRABELLE
See FOUR ADVENTURES OF REINETTE AND MIRABELLE

QUE HE HECHO YO PARA MERECER ESTO?
See WHAT HAVE I DONE TO DESERVE THIS?

QUEEN

See ROOTS: ALEX HALEY'S
QUEEN – THE ROOTS
SAGA CONTINUES
QUEEN OF MEAN, THE
See LEONA HELMSLEY:
THE QUEEN OF MEAN
**QUEEN OF THE
GORILLAS**
See BRIDE AND THE
BEAST, THE
**QUEEN OF THE
VAMPIRES**
See RAPE OF THE
VAMPIRE, THE
**QUEI TEMERARI SULLE
LORO PAZZE,
SCATENATE,
SCALCINATE CARRIOLE**
See MONTE CARLO OR
BUST!
**QUELLA VILLA
ACCANTO AL CIMITERO**
See HOUSE BY THE
CEMETERY, THE
QUERELLE DE BREST
See QUERELLE
**QUEST FOR PEACE,
THE**
See SUPERMAN 4: THE
QUEST FOR PEACE
QUESTION OF FAITH
See LEAP OF FAITH
QUIEN SABE?
See BULLET FOR THE
GENERAL, THE
**QUIET VICTORY: THE
CHARLIE WEDEMEYER
STORY**
See QUIET VICTORY
**QUOI DE NEUF,
PUSSYCAT?**
See WHAT'S NEW
PUSSYCAT?
R.O.B.O.T.
See CHOPPING MALL
**RACE AGAINST THE
DARK**
See NURSES ON THE LINE
**RAGE OF ANGELS: THE
STORY CONTINUES**
See RAGE OF ANGELS 2:
PARTS 1 AND 2
RAGGED ANGELS
See THEY SHALL HAVE
MUSIC
**RAINBOW
PROFESSIONALS**
See COBRA MISSION
RAPE OF MALAYA
See TOWN LIKE ALICE, A
**RAPE OF THE RED
TEMPLE**
See BURNING PARADISE
RAPPRESAGLIA
See MASSACRE IN ROME
RAT FINK AND BOBO
See RAT PFINK A BOO-
BOO
**RAT FINK AND BOO-
BOO**
See RAT PFINK A BOO-
BOO
RATS
See RATS: NIGHT OF
TERROR
**RATS: NOTTE DI
TERROR**
See RATS: NIGHT OF
TERROR
RAW MEAT
See DEATH LINE
RAWHEAD
See RAWHEAD REX

READY TO WEAR
See PRET-A-PORTER
REAZIONE A CATENA
See LAST HOUSE ON THE
LEFT PART 2, THE
REBEL NUN, THE
See FLAVIA THE HERETIC
REBEL WARRIORS
See REBEL ROUSERS, THE
REBEL, THE
See HANCOCK: THE
REBEL
RED LIPS, THE
See DAUGHTERS OF
DARKNESS
RED POMEGRANATE
See COLOUR OF
POMEGRANATES, THE
**RED TARGET: THE
PLOT TO OVERTHROW
THE USSR**
See CRISIS IN THE
KREMLIN: THE LAST
DAYS OF THE SOVIET
UNION
RED TIDE, THE
See BLOOD TIDE
RED WEDDING
See LES NOCES ROUGES
**RED-HEADED LEAGUE,
THE**
See ADVENTURES OF
SHERLOCK HOLMES, THE:
THE RED-HEADED
LEAGUE
REF, THE
See HOSTILE HOSTAGES
REGARD D'ULYSSE
See ULYSSES' GAZE
REIGN OF FEAR
See SUMMER CITY
**REILLY – THE ACE OF
SPIES**
See REILLY – ACE OF
SPIES: AN AFFAIR WITH A
MARRIED WOMAN
REISE DER HOFFNUNG
See JOURNEY OF HOPE
**RELENTLESS 4: ASHES
TO ASHES**
See RELENTLESS: THE
REDEEMER
**REMO WILLIAMS: THE
ADVENTURE BEGINS**
See REMO: UNARMED
AND DANGEROUS
**REMO WILLIAMS: THE
ADVENTURE
CONTINUES**
See REMO: UNARMED
AND DANGEROUS
RENEGADE
See RENEGADE
MURDERER'S ROW
RENEGADE GIRLS
See CAGED HEAT
**REPLIKATOR, THE:
CLONED TO KILL**
See REPLIKATOR
**REQUIEM POUR UN
VAMPIRE**
See REQUIEM FOR A
VAMPIRE
**RESIDENT PATIENT,
THE**
See ADVENTURES OF
SHERLOCK HOLMES, THE:
THE RESIDENT PATIENT
**REST IN PIECES, MRS
COLUMBO**
See REST IN PEACE, MRS
COLUMBO
REST OF DANIEL, THE

See FOREVER YOUNG
**RETURN OF
PSAMMEAD, THE**
See FIVE CHILDREN AND
IT
**RETURN OF THE
ALIEN'S DEADLY
SPAWN**
See DEADLY SPAWN, THE
**RETURN OF THE BLIND
DEAD**
See RETURN OF THE EVIL
DEAD, THE
**RETURN OF THE
DRAGON**
See WAY OF THE
DRAGON
**RETURN OF THE
INCREDIBLE HULK**
See INCREDIBLE HULK
RETURNS, THE
**RETURN OF THE
KILLER TOMATOES:
THE SEQUEL**
See RETURN OF THE
KILLER TOMATOES!
**RETURN OF THE LIVING
DEAD: PART 3**
See RETURN OF THE
LIVING DEAD 3
**RETURN TO MACON
COUNTY LINE**
See RETURN TO MACON
COUNTY
**RETURN TO SNOWY
RIVER**
See UNTAMED, THE
**RETURN TO SNOWY
RIVER: PART 2**
See UNTAMED, THE
REUNION
See REUNION IN FRANCE
REVENGE
See PAYBACK
REVENGE AT EL PASO
See ACE HIGH
REVENGE IN EL PASSO
See ACE HIGH
**REVENGE OF DOCTOR
DEATH, THE**
See MADHOUSE
REVENGE OF DRACULA
See DRACULA: PRINCE OF
DARKNESS
**REVENGE OF MILADY,
THE**
See FOUR MUSKETEERS,
THE
REVENGE OF THE DAD
See NIGHT OF THE
GHOULS
**REVENGE OF THE
KILLER TOMATOES**
See RETURN OF THE
KILLER TOMATOES!
**REVENGE OF THE
NERDS 2: NERDS IN
PARADISE**
See REVENGE OF THE
NERDS 2
**REVENGE OF THE
NINJA, THE**
See NINJA 2: THE
REVENGE OF THE NINJA
**REVENGE OF THE
VAMPIRE**
See BLACK SUNDAY
**REVENGE OF
YUKINOJO, THE**
See ACTOR'S REVENGE,
AN
RICH AND POWERFUL
See BALBOA

RICHES AND ROMANCE
See AMAZING
ADVENTURE, THE
RIGET
See KINGDOM, THE:
PARTS 1 TO 3
RIGHTING WRONGS
See ABOVE THE LAW
**RING OF FIRE 2: BLOOD
AND STEEL**
See RAGE: RING OF FIRE 2
RINGER, THE
See SUCCESS
**RIPOUX CONTRE
RIPOUX**
See LE COP 2
**RIPPER OF NOTRE
DAME, THE**
See DEMONIAC
RITES OF DRACULA
See SATANIC RITES OF
DRACULA, THE
**RIVER OF RAGE: THE
TAKING OF MAGGIE
KEENE**
See MURDER ON THE RIO
GRANDE
ROAD HOME, THE
See BROTHERS' DESTINY
ROAD HOUSE
See ROADHOUSE
ROAD KILLERS
See ROADFLOWER
**ROAD TO CORINTH,
THE**
See LA ROUTE DE
CORINTHE
ROAD TO VENGEANCE
See WALKER, TEXAS
RANGER: ROAD TO
VENGEANCE
ROAD WARRIOR, THE
See MAD MAX 2
**ROALD DAHL'S DANNY,
THE CHAMPION OF THE
WORLD**
See DANNY, THE
CHAMPION OF THE
WORLD
**ROALD DAHL'S
MATILDA**
See MATILDA
ROARING TIMBERS
See COME AND GET IT
**ROBERT A. HEINLEIN'S
THE PUPPET MASTERS**
See PUPPET MASTERS,
THE
**ROBIN COOK'S
FORMULA FOR DEATH**
See FORMULA FOR
DEATH
**ROBIN COOK'S
HARMFUL INTENT**
See HARMFUL INTENT
**ROBIN COOK'S
MORTAL FEAR**
See MORTAL FEAR
**ROBINSON
CRUSOELAND**
See LAUREL AND HARDY:
UTOPIA
ROBOT WAR
See GUNHED: THE
ULTIMATE BATTLE
**ROCCO E I SUOI
FRATELLI**
See ROCCO AND HIS
BROTHERS
**ROCK 'N' ROLL
SWINDLE, THE**
See GREAT ROCK 'N'
ROLL SWINDLE, THE

ROCKET TO THE MOON
See CAT WOMEN OF THE
MOON
**RODTOTTERNE OG
TYRANNOS**
See REDTOPS
**ROGER CORMAN'S
FRANKENSTEIN
UNBOUND**
See FRANKENSTEIN
UNBOUND
ROJIN Z
See ROUJIN Z
**ROLLER BLADE
WARRIORS**
See ROLLER BLADE
ROMA, CITTA APERTA
See ROME, OPEN CITY
ROMANCE AND RICHES
See AMAZING
ADVENTURE, THE
**ROMMEL: THE DESERT
FOX**
See DESERT FOX, THE
ROMUALD ET JULIETTE
See ROMUALD AND
JULIETTE
**ROOFTOP HOPPER,
THE**
See HALFAOUINE: CHILD
OF THE TERRACES
ROOM SERVICE
See MARX BROTHERS:
ROOM SERVICE
**ROOSTER COGBURN
AND THE LADY**
See ROOSTER COGBURN
ROOTS: THE GIFT
See ROOTS: KUNTA
KINTE'S GIFT
ROPE
See MUD HONEY
ROPE OF FLESH
See MUD HONEY
**ROSE AND THE SWORD,
THE**
See FLESH AND BLOOD
ROSE IN THE MUD, THE
See BAD SLEEP WELL, THE
**ROSEANNE AND TOM: A
HOLLYWOOD
MARRIAGE**
See ROSEANNE AND TOM:
BEHIND THE SCENES
**ROSWELL: THE UFO
COVER-UP**
See ROSWELL
ROTE LIPPEN
See SADISTEROTICA
ROUGH STUFF
See MR NANNY
ROUSTABOUT
See ELVIS PRESLEY:
ROUSTABOUT
ROXIE
See VIRGIN ON THE RUN
ROXY
See VIRGIN ON THE RUN
ROYAL WARRIORS
See NINJA OPERATION 7:
ROYAL WARRIORS
**RUDYARD KIPLING'S
JUNGLE BOOK**
See JUNGLE BOOK, THE
**RUDYARD KIPLING'S
THE JUNGLE BOOK**
See JUNGLE BOOK, THE
RUE SAINT-SULPICE
See FAVOUR, THE WATCH
AND THE VERY BIG FISH,
THE
**RULES OF THE GAME,
THE**

See LA REGLE DU JEU
RUMPO KID, THE
See CARRY ON COWBOY
RUNNING MATES
See DIRTY TRICKS
RUTLES, THE
See COMPLEAT RUTLES,
THE
**RUTLES, THE: ALL YOU
NEED IS CASH**
See COMPLEAT RUTLES,
THE
**RYMDINVASION I
LAPPLAND**
See INVASION OF THE
ANIMAL PEOPLE
**S.I.S. – EXTREME
JUSTICE**
See EXTREME JUSTICE
S.O.S. CONCORDE
See CONCORDE AFFAIR,
THE
**SABRE TOOTH TIGER,
THE**
See DEEP RED
SACKETTS, THE
See DAYBREAKERS, THE
**SADIST OF NOTRE
DAME, THE**
See DEMONIAC
SAFETY LAST
See HAROLD LLOYD:
SAFETY LAST
**SAGA OF THE ROAD,
THE**
See PATHER PANCHALI
SAIKAKU ICHIDAI ONA
See LIFE OF OHARU, THE
SAIMT EL QUSUR
See SILENCES OF THE
PALACE, THE
**SALEM'S LOT: THE
MOVIE**
See SALEM'S LOT
SAM'S SONG
See SWAP, THE
SAMURAI, THE
See LE SAMOURAI
**SAN TAI YON X
JUGATSU**
See BOILING POINT
SAND FAIRY, THE
See FIVE CHILDREN AND
IT
SANDLOT, THE
See SANDLOT KIDS, THE
SANDOKAN: THE MOVIE
See SANDOKAN: THE
AMAZING ADVENTURES
OF SANDOKAN
SANDS OF THE DESERT
See CHARLIE DRAKE:
SANDS OF THE DESERT
SANS PITIE
See SIN COMPASION
**SANTA CLAUS: THE
MOVIE**
See SANTA CLAUS
**SANZU NO KAWA NO
UBAGURAMA**
See SHOGUN ASSASSIN
SAPORE DI DONNA
See SCENT OF PASSION
SAPS AT SEA
See LAUREL AND HARDY:
SAPS AT SEA
SARDONICUS
See MR SARDONICUS
SASOM I EN SPEGEL
See THROUGH A GLASS
DARKLY
SATAN
See MARK OF THE DEVIL

SATAN'S CLAW
See BLOOD ON SATAN'S
CLAW
SATAN'S DOG
See PLAY DEAD
SATAN'S PRINCESS
See MALEDICTION
SATAN'S SKIN
See BLOOD ON SATAN'S
CLAW
**SAUVE QUI PEUT (LA
VIE)**
See SLOW MOTION
SAVAGE PLACE, A
See SPENSER: A SAVAGE
PLACE
SAYAT NOVA
See COLOUR OF
POMEGRANATES, THE
**SCANDAL IN BOHEMIA,
A**
See ADVENTURES OF
SHERLOCK HOLMES, THE:
A SCANDAL IN BOHEMIA
**SCANNER 3: THE
TAKEOVER**
See SCANNER FORCE
**SCANNERS 3: THE
TAKEOVER**
See SCANNER FORCE
**SCANNERS: THE
SHOWDOWN**
See SCANNER COP 2:
VOLKIN'S REVENGE
**SCARFACE: THE
SHAME OF THE NATION**
See SCARFACE
SCARLET CLAW, THE
See SHERLOCK HOLMES:
THE SCARLET CLAW
**SCENER UR ETT
AKTENSKAP**
See SCENES FROM A
MARRIAGE
**SCHONER GIGOLO,
ARMER GIGOLO**
See JUST A GIGOLO
**SCHOOLGIRL REPORT
NO. 1**
See BLUE FANTASIES
**SCHULMADCHEN:
REPORT 12 TEIL**
See BLUE FANTASIES
**SCREAM IN THE
STREETS, A**
See GIRLS IN THE STREET
SCREAM STREET
See GIRLS IN THE STREET
SCREAMER
See SCREAM AND
SCREAM AGAIN
**SEA OF TERROR;
TERROR SQUAD**
See HIJACKING OF THE
ACHILLE LAURO, THE
SEA WYF AND BISCUIT
See SEA WIFE
**SEARCH FOR SPOCK,
THE**
See STAR TREK 3: THE
SEARCH FOR SPOCK
**SEARCHING FOR
BOBBY FISCHER**
See INNOCENT MOVES
SECOND STAIN, THE
See RETURN OF
SHERLOCK HOLMES, THE:
THE SECOND STAIN
SECOND TIME AROUND
See INSPECTOR MORSE:
SECOND TIME AROUND
SECONDS TO LIVE
See VIVA KNIEVEL!

SECRET FOUR, THE
See KANSAS CITY
CONFIDENTIAL
**SECRET LIFE OF IAN
FLEMING, THE**
See SPYMAKER: THE
SECRET LIFE OF IAN
FLEMING
**SECRET OF BAY 5B,
THE**
See INSPECTOR MORSE:
THE SECRET OF BAY 5B
**SECRET SHAOLIN
KUNG FU, THE**
See INVINCIBLE SHAOLIN
KUNG FU
SECRET YEARNINGS
See GOOD LUCK, MISS
WYCKOFF
**SECRETS OF A
SENSUOUS NURSE**
See I WILL IF YOU WILL
**SECRETS OF A SEXY
GAME**
See CONFESSIONS FROM
THE DAVID GALAXY
AFFAIR
**SECRETS OF SUMMER
SCHOOL TEACHERS**
See SUMMER HEAT
**SECRETS OF SWEET
SIXTEEN: WHAT
SCHOOLGIRLS DON'T
TELL**
See WHAT SCHOOLGIRLS
DON'T TELL
SECRETS OF THE HART
See HART TO HART:
SECRETS OF THE HART
**SECRETS OF THE
SATIN BLUES**
See NAUGHTY BLUE
KNICKERS
SEDMIKRASKY
See DAISIES
SEDUCE AND DESTROY
See DOLL SQUAD, THE
**SEDUCE ME: THE
PAMELA PRINCIPLE 2**
See PAMELA PRINCIPLE 2,
THE **
SEDUCERS, THE
See SOME LIKE IT SEXY
**SEDUCTION OF A
PRIEST**
See FINAL TEMPTATION
SEED OF TERROR
See GRAVE OF THE
VAMPIRE
SEEDPEOPLE
See SEED PEOPLE
SENGOKU JIEITAI
See TIME SLIP
SENIOR TRIP
See NATIONAL
LAMPOON'S SENIOR TRIP
SENSUOUS NURSE, THE
See I WILL IF YOU WILL
**SENSUOUS TEENAGER,
THE**
See NIGHT PLEASURES
SERGEANT STEINER
See BREAKTHROUGH
**SERIOUS CRIMES
SQUAD**
See CRIME STORY
**SERVICE OF ALL THE
DEAD**
See INSPECTOR MORSE:
SERVICE OF ALL THE
DEAD
SETTLERS, THE
See MARTIAN

CHRONICLES, THE
SETTLING OF THE SUN, THE
See INSPECTOR MORSE: THE SETTLING OF THE SUN
SEVEN BAD MEN
See RAGE AT DAWN
SEVEN BROTHERS MEET DRACULA, THE
See LEGEND OF THE SEVEN GOLDEN VAMPIRES, THE
SEVEN DOORS OF DEATH
See BEYOND, THE
SEX AND THE MARRIED DETECTIVE
See COLUMBO: SEX AND THE MARRIED DETECTIVE
SEX AND THE VAMPIRE
See THRILL OF THE VAMPIRES, THE
SEX HELL
See DEVIL IN MISS JONES PART 3: A NEW BEGINNING
SEX IN THE SNOW
See SKI SLUTS GO SNOWBALLING
SEX ON SKIES
See SKI SLUTS GO SNOWBALLING
SEX RITES OF THE DEVIL
See BLACK CANDLES
SEX SURROGATE
See MY THERAPIST
SEX VAMPIRES
See REQUIEM FOR A VAMPIRE
SEXORCISME
See DEMONIAC
SEXORCISTS, THE
See SEXORCIST, THE
SEXUAL HEALER
See BAD GIRLS DOWN UNDER
SHADOW HUNTER
See SHADOWHUNTER
SHADOW MAN
See STREET OF SHADOWS
SHADOW OF DOUBT, A
See SHADOW OF A DOUBT, A
SHADOW OF OBSESSION
See STALKER: SHADOW OF OBSESSION
SHADOWCHASER
See PROJECT SHADOWCHASER
SHAKA ZULU
See SHAKA ZULU: PARTS 1, 2 AND 3
SHAKE HANDS FOREVER
See INSPECTOR WEXFORD: SHAKE HANDS FOREVER
SHAKEDOWN
See BLUE JEAN COP
SHAKEDOWN ON SUNSET STRIP
See SHAKEDOWN ON THE SUNSET STRIP
SHAME
See INTRUDER, THE
SHAME 2: THE SECRET
See DREAM LOVER
SHAME OF A NATION, THE

See SCARFACE
SHAMEFUL SECRETS
See GOING UNDERGROUND
SHAMING, THE
See GOOD LUCK, MISS WYCKOFF
SHAOLIN AGAINST LAMA
See SHAOLIN VERSUS LAMA
SHAOLIN DEVIL, SHAOLIN ANGEL
See SHAOLIN DEVIL AND SHAOLIN ANGEL
SHAOLIN DRUNK FIGHTER
See SHAOLIN DRUNKEN FIGHTER
SHAOLIN TEMPLE STRIKES BACK
See SHAOLIN TEMPLE 2
SHAOLIN: THE STORY OF SHAOLIN
See SHAOLIN VERSUS NINJA
SHATRANJ KE KHILARI
See CHESS PLAYERS, THE
SHATTERED SILENCE
See LOVE IS NEVER SILENT
SHE-HSING ZIAO SHOU
See SNAKE IN THE EAGLE'S SHADOW
SHE'LL BE SWEET
See MAGEE AND THE LADY
SHERLOCK HOLMES AND THE BASKERVILLE CURSE
See SHERLOCK HOLMES: THE BASKERVILLE CURSE
SHERLOCK HOLMES AND THE SECRET CODE
See SHERLOCK HOLMES: DRESSED TO KILL
SHERLOCK HOLMES AND THE SIGN OF FOUR
See SHERLOCK HOLMES: THE SIGN OF FOUR
SHERLOCK HOLMES AND THE VALLEY OF FEAR
See SHERLOCK HOLMES: THE VALLEY OF FEAR
SHERLOCK HOLMES' GROSSTER FALL
See STUDY IN TERROR, A
SHERLOCK HOLMES' THE SIGN OF THE FOUR
See SHERLOCK HOLMES: THE SIGN OF FOUR
SHERLOCK HOLMES: THE ELIGIBLE BACHELOR
See ADVENTURES OF SHERLOCK HOLMES, THE: THE ELIGIBLE BACHELOR
SHERLOCK HOLMES: THE LAST VAMPIRE
See ADVENTURES OF SHERLOCK HOLMES, THE: THE VAMPIRE OF LAMBERLEY
SHERLOCK HOLMES: THE MASTER BLACKMAILER
See ADVENTURES OF SHERLOCK HOLMES, THE: THE MASTER BLACKMAILER
SHERLOCK HOLMES:

THE VAMPIRE OF LAMBERLEY
See ADVENTURES OF SHERLOCK HOLMES, THE: THE VAMPIRE OF LAMBERLEY
SHERLOCK JR
See BUSTER KEATON: SHERLOCK JR
SHE'S SEVENTEEN AND ANXIOUS
See CONFESSIONS OF EMMANUELLE
SHICHININ NO SAMURAI
See SEVEN SAMURAI, THE
SHIP WAS LOADED, THE
See CARRY ON ADMIRAL
SHISENJO NO ARIA
See BEDROOM, THE
SHOCK (TRANSFER SUSPENSE HYPNOS)
See SHOCK
SHOOT THE PIANO PLAYER
See SHOOT THE PIANIST
SHOOT TO KILL
See HUE AND CRY
SHOOTFIGHTER: FIGHT TO THE DEATH
See SHOOTFIGHTER
SHOP ON MAIN STREET, THE
See SHOP ON THE HIGH STREET, THE
SHORT FUSE
See GOOD TO GO
SHORT RUN
See CAT CHASER
SHOSCOMBE OLD PLACE
See CASEBOOK OF SHERLOCK HOLMES, THE: SHOSCOMBE OLD PLACE
SHRIMP ON THE BARBIE
See BOYFRIEND FROM HELL
SHROUD FOR A NIGHTINGALE
See DALGLIESH: SHROUD FOR A NIGHTINGALE – PARTS 1 AND 2
SI DON JUAN ETAIT UNE FEMME
See IF DON JUAN WERE A WOMAN
SIBAK
See MIDNIGHT DANCERS
SIDNEY SHELDON'S A STRANGER IN THE MIRROR
See STRANGER IN THE MIRROR, A
SIDNEY SHELDON'S MEMORIES OF MIDNIGHT
See MEMORIES OF MIDNIGHT
SIDNEY SHELDON'S THE OTHER SIDE OF MIDNIGHT
See OTHER SIDE OF MIDNIGHT, THE
SIE TOTETE IN EKSTASE
See SHE KILLED IN ECSTASY
SIGN OF FOUR, THE
See SHERLOCK HOLMES: THE SIGN OF FOUR
SIGN OF FOUR, THE
See ADVENTURES OF

SHERLOCK HOMES, THE: THE SIGN OF FOUR
SIGN OF THE FOUR, THE
See SHERLOCK HOLMES: THE SIGN OF FOUR
SIGNIFICANT OTHER
See WHEN A MAN LOVES A WOMAN
SILENT KILLER FROM ETERNITY
See RETURN OF THE TIGER
SILENT MOUSE
See CHRISTMAS MOUSE, THE
SILENT NIGHT, DEADLY NIGHT 4: THE INITIATION
See BUGS
SILENT NIGHT, HOLY NIGHT
See CHRISTMAS MOUSE, THE
SILENT WITNESS
See BLOOD BROTHERS
SILENT WORLD OF NICHOLAS QUINN, THE
See INSPECTOR MORSE: THE SILENT WORLD OF NICHOLAS QUINN
SILVER BLAZE
See RETURN OF SHERLOCK HOLMES, THE: SILVER BLAZE
SILVER FOX
See SAVIOUR OF THE SOUL
SILVER SENSATIONS
See SILVER SENSATION
SIMON DEL DESIERTO
See SIMON OF THE DESERT
SIMON WIESENTHAL STORY, THE
See MURDERERS AMONG US: THE SIMON WIESENTHAL STORY
SIN, THE
See GOOD LUCK, MISS WYCKOFF
SINS OF THE FATHERS, THE
See INSPECTOR MORSE: THE SINS OF THE FATHERS
SISSY AND EGBERT
See AT THE PORNIES
SITTIN' PRETTY
See SITTING PRETTY
SIX NAPOLEONS, THE
See RETURN OF SHERLOCK HOLMES, THE: THE SIX NAPOLEONS
SIX THREE THREE SQUADRON
See 633 SQUADRON
SIXTH OF JUNE, THE
See D-DAY THE SIXTH OF JUNE
SIZZLE BEACH
See MALIBU HOT SUMMER
SIZZLE BEACH, USA
See MALIBU HOT SUMMER
SKATEBOARD KID, THE: A MAGICAL MOMENT
See SKATEBOARD KID 2, THE
SKI SLUTS GO SNOWBALLIN'
See SKI SLUTS GO

SNOWBALLING
**SKINHEADS: THE
SECOND COMING OF
HATE**
See SKINHEADS
SLADE IN FLAME
See FLAME
SLAGSKAMPEN
See INSIDE MAN, THE
SLAMMER
See WHO'S THAT GIRL
SLAUGHTER HOSPITAL
See COLD BLOODED
BEAST
SLAVES
See BLACKSNAKE
SLEEP STALKER
See SLEEPSTALKER: THE
SANDMAN'S LAST RITES
**SLEEPING BEAUTY
AROUSED**
See SLEEPING BEAUTY
SLEEPING LIFE, A
See INSPECTOR
WEXFORD: A SLEEPING
LIFE
**SLIP SLIDE
ADVENTURES**
See WATER BABIES, THE
SLUGS, THE MOVIE
See SLUGS
SMART ALEC
See HOLLYWOOD
DREAMING
SMOKE SCREEN
See PALAIS ROYALE
**SMOKEY AND THE
BANDIT RIDE AGAIN**
See SMOKEY AND THE
BANDIT 2
SMOKEY IS THE BANDIT
See SMOKEY AND THE
BANDIT: PART 3
SMULTRONSTALLET
See WILD STRAWBERRIES
**SNAKE EATER 2: THE
DRUG BUSTER**
See SNAKE EATER'S
REVENGE
SNAKE EYES
See DANGEROUS GAME
**SNAKE IN EAGLE
SHADOW**
See SNAKE IN THE
EAGLE'S SHADOW
**SNAKE IN EAGLE'S
SHADOW**
See SNAKE IN THE
EAGLE'S SHADOW
SNOWBALLIN'
See SKI SLUTS GO
SNOWBALLING
SOFT BALLS
See SQUEEZE PLAY
SOFT SKIN, THE
See SILKEN SKIN
**SOGNI EROTICI DI
CLEOPATRA**
See EROTIC DREAMS OF
CLEOPATRA, THE
SOIL
See EARTH
SOLEIL TROMPEUR
See BURNT BY THE SUN
**SOLITARY CYCLIST,
THE**
See ADVENTURES OF
SHERLOCK HOLMES, THE:
THE SOLITARY CYCLIST
SOME GIRLS
See SISTERS
**SOME LIE AND SOME
DIE**

See INSPECTOR
WEXFORD: SOME LIE AND
SOME DIE
**SOMERSET
MAUGHAM'S QUARTET**
See QUARTET
SOMETHING BLONDE
See AURORA
**SOMETHING LIKE THE
TRUTH**
See OFFENCE, THE
**SOMETIMES A GREAT
NOTION**
See NEVER GIVE AN INCH
**SOMEWHERE IN
FRANCE**
See FOREMAN WENT TO
FRANCE, THE
**SOMMAREN MED
MONIKA**
See SUMMER WITH
MONIKA
SOMMARLEK
See SUMMER INTERLUDE
**SOMMARNATTENS
LEENDE**
See SMILES OF A SUMMER
NIGHT
SON OF GODZILLA
See GODZILLA: SON OF
GODZILLA
SON-IN-LAW, THE
See SON IN LAW
**SONG OF THE ROAD,
THE**
See PATHER PANCHALI
SONG OF TWO HUMANS
See SUNRISE
**SONO OTOKO, KYOBO
NI TSUKI**
See VIOLENT COP
SONS OF THE DESERT
See LAUREL AND HARDY:
SONS OF THE DESERT
**SORORITY BABES IN
THE SLIME-A-RAMA**
See IMP, THE
**SOTTO IL VESTITO
NIENTE**
See NOTHING
UNDERNEATH
SOUP TO NUTS
See WAITRESS!
SOUTH BEACH
See NIGHT CALLER
SOUTH CENTRAL, L.A.
See SOUTH CENTRAL
SOUTH HILL RAPIST
See SINS OF THE MOTHER
**SOUTHWEST TO
SONORA**
See APPALOOSA, THE
SPACE CAMP
See SPACECAMP
SPACE FIREBIRD 2772
See SPACE FIREBIRD
**SPACE INVASION FROM
LAPPLAND**
See INVASION OF THE
ANIMAL PEOPLE
**SPACE INVASION OF
LAPLAND**
See INVASION OF THE
ANIMAL PEOPLE
**SPACE SOLDIERS
CONQUER THE
UNIVERSE**
See FLASH GORDON
CONQUERS THE
UNIVERSE
**SPACEHUNTER:
ADVENTURES IN THE
FORBIDDEN ZONE**

See SPACEHUNTER
SPANISH ROSE
See IN TOO DEEP
SPAWNING, THE
See PIRANHA 2: FLYING
KILLERS
**SPEAKER OF
MANDARIN, THE**
See INSPECTOR
WEXFORD: THE SPEAKER
OF MANDARIN
SPECKLED BAND, THE
See ADVENTURES OF
SHERLOCK HOLMES, THE:
THE SPECKLED BAND
**SPECTER OF FREEDOM,
THE**
See PHANTOM OF
LIBERTY, THE
SPECTRES
See SPECTERS
SPEEDWAY
See ELVIS PRESLEY:
SPEEDWAY
SPEEDY
See HAROLD LLOYD:
SPEEDY
SPETTRI
See SPECTERS
SPIDER WOMAN
See SHERLOCK HOLMES
AND THE SPIDERWOMAN
SPIDER-MAN
See SPIDERMAN: THE
MOVIE
**SPIDERMAN AND THE
DRAGON'S CHALLENGE**
See SPIDERMAN: THE
DRAGON'S CHALLENGE
SPIKLENCI SLASTI
See CONSPIRATORS OF
PLEASURE
SPINOUT
See ELVIS PRESLEY:
SPINOUT
SPIRIT CHASER
See GO KIDS, THE
**SPIRIT OF THE DEAD,
THE**
See ASPHYX, THE
SPIRITS OF BRUCE LEE
See SPIRITS OF BRUCE LI
SPIRITS OF THE DEAD
See TALES OF MYSTERY
AND IMAGINATION
SPITFIRE
See FIRST OF THE FEW,
THE
**SPITFIRE: THE FIRST
OF THE FEW**
See FIRST OF THE FEW,
THE
SPIVS
See I VITELLONI
SPLATTER
See NIGHT OF THE ALIEN
SPYLARKS
See MORECOMBE AND
WISE: THE INTELLIGENCE
MEN
SQUARE PEG, THE
See NORMAN WISDOM:
THE SQUARE PEG
**SQUINTO: A
WARRIOR'S TALE**
See LAST GREAT
WARRIOR, THE
SREDNI VASHTAR
See CEMENT GARDEN,
THE
SS REPRESSAILLES
See MASSACRE IN ROME
STACHKA

See STRIKE
STAGEFRIGHT
See STAGE FRIGHT
STAIRWAY TO HEAVEN
See MATTER OF LIFE AND
DEATH, A
STAKEOUT 2
See ANOTHER STAKEOUT
STAND EASY
See DOWN AMONG THE
Z-MEN
STANNO TUTTI BENE
See EVERYBODY'S FINE
STAR SEX
See CONFESSIONS FROM
THE DAVID GALAXY
AFFAIR
**STAR TREK: THE
MOTION PICTURE**
See STAR TREK
STARFIRE
See SOLAR CRISIS
**STARLIGHT
SLAUGHTERS**
See DEATH TRAP
STAROYE I NOVOYE
See GENERAL LINE, THE
STAY AWAY, JOE
See ELVIS PRESLEY: STAY
AWAY, JOE
**STEPFATHER 2: MAKE
ROOM FOR DADDY**
See STEPFATHER 2
**STEPFATHER 3:
FATHER'S DAY**
See STEPFATHER 3
**STEPHEN KING'S
GRAVEYARD SHIFT**
See GRAVEYARD SHIFT
STEPHEN KING'S IT
See IT
**STEPHEN KING'S
NEEDFUL THINGS**
See NEEDFUL THINGS
**STEPHEN KING'S
SILVER BULLET**
See SILVER BULLET
**STEPHEN KING'S
SLEEPWALKERS**
See SLEEPWALKERS
**STEPHEN KING'S
SOMETIMES THEY
COME BACK**
See SOMETIMES THEY
COME BACK
**STEPHEN KING'S THE
DARK HALF**
See DARK HALF, THE
**STEPHEN KING'S THE
GOLDEN YEARS**
See GOLDEN YEARS, THE
**STEPHEN KING'S THE
LANGOLIERS**
See LANGOLIERS, THE
**STEPHEN KING'S THE
STAND**
See STAND, THE
**STEPHEN KING'S THE
TOMMYKNOCKERS**
See TOMMYKNOCKERS,
THE
STEPSISTER
See SINDERELLA: PART 2
STEPSISTERS
See SINDERELLA: PART 2
**STEPTOE AND SON:
THE FEATURE**
See STEPTOE AND SON
STILL SMOKIN'
See CHEECH AND
CHONG: STILL SMOKIN'
STITCH IN TIME, A
See NORMAN WISDOM: A

STITCH IN TIME
STO DNEI DO PRIKAZA
See ONE HUNDRED DAYS
BEFORE THE COMMAND
**STONING IN FULHAM
COUNTY, A**
See STONING, A
**STORIA DE UNA
CAPINERA**
See SPARROW
**STORIA DI UNA
MONACA DI CLAUSURA**
See STORY OF A
CLOISTERED NUN
**STORY OF A CITIZEN
ABOVE SUSPICION**
See INVESTIGATION OF A
CITIZEN ABOVE
SUSPICION
**STORY OF A LOVE
AFFAIR**
See CHRONICLE OF A
LOVE
STORY OF DAVID, THE
See KING DAVID
**STORY OF ROBIN
HOOD, THE**
See STORY OF ROBIN
HOOD AND HIS MERRIE
MEN, THE
STRAIGHT JACKET
See DARK SANITY
**STRANGE ADVENTURE
OF DAVID GREY, THE**
See VAMPYR
STRANGE BEHAVIOUR
See SMALL TOWN
MASSACRE
**STRANGE EXORCISM
OF LYNN HART, THE**
See PIGS
STRANGE JOURNEY
See FANTASTIC VOYAGE
**STRANGER AMONG US,
A**
See CLOSE TO EDEN
**STRANGER IN
BETWEEN, THE**
See HUNTED
STRANGER, THE
See INTRUDER, THE
STRANGERS
See STRANGERS: THE
STORY OF A MOTHER
AND DAUGHTER
STRANGERS
See VOYAGE TO ITALY
**STREET FIGHTER: THE
BATTLE FOR
SHADALOO**
See STREET FIGHTER
**STREET FIGHTER: THE
MOVIE**
See STREET FIGHTER
**STREET FIGHTER: THE
ULTIMATE BATTLE**
See STREET FIGHTER
STREET FORCE
See NIGHTMARE COUNTY
STREET WAR
See IN THE LINE OF DUTY:
STREET WARS
**STRICTLY
CONFIDENTIAL**
See BROADWAY BILL
STRIKE A POSE
See LETHAL OBSESSION
STRIKE, THE
See COMIC STRIP
PRESENTS: THE STRIKE
STROKE OF MIDNIGHT
See PHANTOM
CARRIAGE, THE

STRONGER THAN FEAR
See EDGE OF DOOM
SUBSPECIES 2
See BLOODSTONE:
SUBSPECIES 2
**SUCH A GORGEOUS
KID LIKE ME**
See UNE BELLE FILLE
COMME MOI
SUDDEN TERROR
See EYEWITNESS
SUICIDE RUN
See TOO LATE THE HERO
**SUITABLE CASE FOR
TREATMENT, A**
See MORGAN: A
SUITABLE CASE FOR
TREATMENT
SUMMER
See GREEN RAY, THE
SUMMER CAMP
See PIG'S TALE, A
**SUMMER CAMP
MASSACRE**
See BUTTERFLY
REVOLUTION, THE
**SUMMER CAMP
NIGHTMARE**
See BUTTERFLY
REVOLUTION, THE
SUMMER HOLIDAY
See CLIFF RICHARD:
SUMMER HOLIDAY
SUMMER MANOEUVRES
See LES GRANDES
MANOEUVRES
**SUMMER OF
INNOCENCE**
See BIG WEDNESDAY
SUMMER SCHOOL
See SCREWBALLS 2:
LOOSE SCREWS
**SUMMER SCHOOL
TEACHERS**
See SUMMER HEAT
SUMMERPLAY
See SUMMER INTERLUDE
SUNA NO ONNA
See WOMAN OF THE
DUNES
SUNDOWN
See SUNDOWN: THE
VAMPIRE IN RETREAT
**SUNNYBOY UND
SUGARBABY**
See SUNNYBOY AND
SUGARBABY
SUNSET HEAT
See MIDNIGHT HEAT
SUOR OMICIDI
See KILLER NUN
SUPER MARIO BROS
See SUPER MARIO
BROTHERS
SUPERMAN: THE MOVIE
See SUPERMAN
**SUPERNATURAL BEAST
CITY**
See WICKED CITY
SURE FIRE
See TIGER CAGE
SURF NAZIS
See SURF NAZIS MUST DIE
SURF WARRIORS
See SURF NINJAS
SURGEON, THE
See EXQUISITE
TENDERNESS
SUSPENSE
See SHOCK
SUSPIRIA 2
See DEEP RED
SVART LUCIA

See PREMONITION, THE
SWEET 16
See SWEET SIXTEEN
SWEET LIFE, A
See LA DOLCE VITA
SWINGSHIFT
See SWING SHIFT
SWISS MISS
See LAUREL AND HARDY:
SWISS MISS
SYMPHONY OF LOVE
See ECSTASY
T & A ACADEMY 2
See GIMME AN "F"
T.A.P.S.
See TAPS
TACONES LEJANOS
See HIGH HEELS
**TAGEBUCH EINER
VELORENEN**
See DIARY OF A LOST
GIRL
TAGGET
See DRAGONFIRE
TAILSPIN
See CODED HOSTILE
**TAILSPIN: BEHIND THE
KOREAN AIRLINER
TRAGEDY**
See CODED HOSTILE
TAKE ME HIGH
See CLIFF RICHARD: TAKE
ME HIGH
**TAKING CARE OF
BUSINESS**
See FILOFAX
**TAKING OF FLIGHT 847,
THE: THE ULI
DERICKSON STORY**
See TAKING OF FLIGHT
847, THE
TALE OF WINTER, A
See WINTER'S TALE, A
**TALES FROM THE
CRYPT 1**
See VAULT OF HORROR 1
**TALES FROM THE
CRYPT 2**
See VAULT OF HORROR 2
**TALES FROM THE
CRYPT 2**
See TALES FROM THE
CRYPT: VOL. 2
**TALES FROM THE
CRYPT 2**
See VAULT OF HORROR
**TALES FROM THE
CRYPT 3**
See VAULT OF HORROR 3
**TALES FROM THE
CRYPT 3**
See TALES FROM THE
CRYPT: VOL. 3
**TALES FROM THE
CRYPT 4**
See VAULT OF HORROR 4
**TALES FROM THE
CRYPT PRESENTS
DEMON KNIGHT**
See TALES FROM THE
CRYPT: DEMON KNIGHT
TALES OF MYSTERY
See TALES OF MYSTERY
AND IMAGINATION
**TALES THAT TEAR
YOUR HEART OUT**
See ZOMBIE HOLOCAUST
**TALK DIRTY TO ME
PART 8**
See TALK DIRTY TO ME
PLEASE!
**TALK DIRTY TO ME:
PART 3**

See TRIALS OF TRACI, THE
TALLINN PIMEDUSES
See DARKNESS IN
TALLINN
TALLINN PIMEYS
See DARKNESS IN
TALLINN
TASTE FOR DEATH, A
See DALGLIESH: A TASTE
FOR DEATH
TASTE OF SIN, A
See OLIVIA
TAXI TO THE JOHN
See TAXI TO THE TOILET
TAXI TO THE LOO
See TAXI TO THE TOILET
TAXI ZUM KLO
See TAXI TO THE TOILET
TCHIN TCHIN
See TOUCH OF
ADULTERY, A
TEEN KANYA
See TWO DAUGHTERS
TEEN WOLF 1
See TEEN WOLF
**TEENAGE
FRANKENSTEIN**
See I WAS A TEENAGE
FRANKENSTEIN
TEENAGE INNOCENTS
See TEENAGE
INNOCENCE
**TEENAGE PSYCHO
MEETS BLOODY MARY,
THE**
See INCREDIBLY
STRANGE CREATURES
WHO STOPPED LIVING
AND BECAME MIXED-UP
ZOMBIES, THE
**TEENAGE THEATER:
THE VIOLENT YEARS**
See VIOLENT YEARS, THE
TELE
See TIME SLIP
TELL YOUR CHILDREN
See REEFER MADNESS
**TEMPO DI
CHARLESTON:
CHICAGO 1929**
See THEY PAID WITH
BULLETS
TEMPO DI UCCIDERE
See TIME TO KILL, A
**TEN COMMANDMENTS:
PARTS 1 TO 5**
See DEKALOG: THE TEN
COMMANDMENTS –
PARTS 1 TO 5
**TEN COMMANDMENTS:
PARTS 6 TO 10**
See DEKALOG: THE TEN
COMMANDMENTS –
PARTS 6 TO 10
**TEN DAYS THAT SHOOK
THE WORLD**
See OCTOBER
TENDRES COUSINES
See COUSINS IN LOVE
**TENNESSEE WILLIAMS'
A STREETCAR NAMED
DESIRE**
See STREETCAR NAMED
DESIRE, A
TEOREMA
See THEOREM
TERROR BY NIGHT
See SHERLOCK HOLMES:
TERROR BY NIGHT
**TERROR CHAMBER OF
DR FAUSTUS, THE**
See EYES WITHOUT A
FACE

TERROR FACTOR, THE
See SCARED TO DEATH
TERROR IN THE MIDNIGHT SUN
See INVASION OF THE ANIMAL PEOPLE
TERROR OF DR CHANEY, THE
See MASSACRE MANSION
TERROR OF DRACULA
See NOSFERATU
TERROR OF GODZILLA, THE
See GODZILLA: MONSTERS FROM AN UNKNOWN PLANET
TERROR OF MECHAGODZILLA
See GODZILLA: MONSTERS FROM AN UNKNOWN PLANET
TERROR OF THE MUMMY
See MUMMY, THE
TERROR OF THE VAMPIRES, THE
See THRILL OF THE VAMPIRES, THE
TERROR SQUAD
See HIJACKING OF THE ACHILLE LAURO, THE
TERROR STRIKES, THE
See WAR OF THE COLOSSAL BEAST
TERRORISTS, THE
See RANSOM
TERRORVISION
See TERROR VISION
TEXAS, ADIO
See TEXAS ADIOS
THANK YOU, LIFE
See MERCI LA VIE
THAT LADY IS A TRAMP
See THAT GIRL IS A TRAMP
THAT RIVIERA TOUCH
See MORECOMBE AND WISE: THAT RIVIERA TOUCH
THAT'S THE WAY IT IS
See ELVIS PRESLEY: THAT'S THE WAY IT IS
THERE'S A ZULU ON MY STOEP
See YANKEE ZULU
THESE ARE THE DAMNED
See DAMNED, THE
THEY
See THEY WATCH
THEY CAME FROM WITHIN
See SHIVERS
THEY LOVED LIFE
See KANAL
THIEF
See VIOLENT STREETS
THIEVES OF FORTUNE
See MAY THE BEST MAN WIN
THING IN THE ATTIC, THE
See GHOUL, THE
THING, THE
See THING FROM ANOTHER WORLD, THE
THINGS ARE TOUGH ALL OVER
See CHEECH AND CHONG: THINGS ARE TOUGH ALL OVER
THINGS CHANGE 1: MY FIRST TIME

See THINGS CHANGE
THIRST OF BARON BLOOD, THE
See BARON BLOOD
THIRST: THE TASTE FOR BLOOD
See THIRST
THIRTY-SIX CRAZY FIST, THE
See THIRTY-SIX FIST, THE
THIS GIRL IS A TRAMP
See THAT GIRL IS A TRAMP
THIS MAN MUST DIE
See QUE LA BETE MEURE
THIS VAMPIRE SUCKS
See DRACULA SUCKS
THOMAS CROWN AND COMPANY
See THOMAS CROWN AFFAIR, THE
THOSE DARING YOUNG MEN IN THEIR JAUNTY JALOPIES
See MONTE CARLO OR BUST
THOSE WERE THE HAPPY DAYS
See STAR!
THOSE WERE THE HAPPY TIMES
See STAR!
THOU SHALT NOT KILL
See SHORT FILM ABOUT KILLING, A
THREE BAD MEN IN THE HIDDEN FORTRESS
See HIDDEN FORTRESS, THE
THREE COLORS: BLUE
See THREE COLOURS: BLUE
THREE COLORS: WHITE
See THREE COLOURS: WHITE
THREE MUSKETEERS, THE: THE QUEEN'S DIAMONDS
See THREE MUSKETEERS, THE
THREE NINJAS
See THREE NINJA KIDS
THREE OF US, THE
See NOI TRE
THREE RASCALS IN THE HIDDEN FORTRESS
See HIDDEN FORTRESS, THE
THROUGH THE SHADOWLANDS
See SHADOWLANDS
THROWAWAY WIVES
See LOVE AND BETRAYAL
THUNDER
See THUNDER WARRIOR
THUNDER 2
See THUNDER WARRIOR 2
THUNDER 3
See THUNDER WARRIOR 3
THUNDER IN PARADISE 3: SEALED WITH A KISMET
See THUNDER IN PARADISE 3
THUNDERBIRD SIX
See THUNDERBIRDS SIX: THE MOVIE
THUNDERBOLT ANGELS
See NINJA OPERATION 4: THUNDERBOLT ANGELS
THY SOUL SHALL BEAR WITNESS

See PHANTOM CARRIAGE, THE
TIANGUO NIEZI
See DAY THE SUN TURNED COLD, THE
TICKLE ME
See ELVIS PRESLEY: TICKLE ME
TIDAL WAVE
See PORTRAIT OF JENNIE
TIGER CAGE 2
See TIGER CAGE
TIGER, THE
See TIGER WARSAW
TIGHT LITTLE ISLAND
See WHISKY GALORE
TIM BURTON'S THE NIGHTMARE BEFORE CHRISTMAS
See NIGHTMARE BEFORE CHRISTMAS, THE
TIME KILLER, THE
See NIGHT STRANGLER, THE
TIME TO KILL: IN THE LINE OF DUTY 3
See IN THE LINE OF DUTY 3: TIME TO KILL
TIME TO LIVE AND A TIME TO DIE, A
See FIRE WITHIN, THE
TIME WARS
See TIME SLIP
TINA: WHAT'S LOVE GOT TO DO WITH IT
See WHAT'S LOVE GOT TO DO WITH IT
TINTIN ET LE LAC AUX REQUINS
See TINTIN: THE LAKE OF SHARKS
TIREZ SUR LE PIANISTE
See SHOOT THE PIANIST
TO FORGET PALERMO
See PALERMO CONNECTION, THE
TO HAVE AND TO HOLD
See WHEN A MAN LOVES A WOMAN
TO KILL A PRIEST: A TRUE STORY
See TO KILL A PRIEST
TO LOVE A VAMPIRE
See LUST FOR A VAMPIRE
TO THE VICTOR
See OWD BOB
TO VLEMMA TOU ODYSSEA
See ULYSSES' GAZE
TOBE HOOPER'S NIGHT TERRORS
See NIGHT TERRORS
TOKYO MONOGATARI
See TOKYO STORY
TOKYO NAGAREMONO
See TOKYO DRIFTER
TOM CLANCY'S OP CENTER
See OP CENTER
TOMCAT: DANGEROUS DESIRE
See DANGEROUS DESIRE
TOO BEAUTIFUL FOR YOU
See TROP BELLE POUR TOI!
TORMENTED, THE
See SEXORCIST, THE
TORMENTORS, THE
See SEXORCIST, THE
TORPEDO RAIDERS
See FOREVER ENGLAND
TORTURE CHAMBER OF

BARON BLOOD, THE
See BARON BLOOD
TORY WITHOUT A CAUSE
See TORRID WITHOUT A CAUSE
TOTO LE HEROS
See TOTO THE HERO
TOUCH OF HELL, A
See SERIOUS CHARGE
TOUCHE PAS A LA FEMME BLANCHE!
See DON'T TOUCH THE WHITE WOMAN!
TOUGH GUY
See KUNG FU THE HEAD CRUSHER
TOUS LES MATINS DU MONDE
See ALL THE MORNINGS OF THE WORLD
TOWARDS THE DICTATORSHIP OF THE PROLETARIAT
See STRIKE
TOWER OF DEATH
See GAME OF DEATH 2
TRAFFIC
See MONSIEUR HULOT: TRAFFIC
TRAFIC
See MONSIEUR HULOT: TRAFFIC
TRAGEDIJA SLUZBENICE P.T.T.
See SWITCHBOARD OPERATOR, THE
TRAGEDY OF A SWITCHBOARD OPERATOR, THE
See SWITCHBOARD OPERATOR, THE
TRAIN OF TERROR
See TERROR TRAIN
TRAIN, THE
See DEATH TRAIN
TRANCERS 2: THE RETURN OF JACK DETH
See TRANCERS 2
TRANCERS 3: DETH LIVES
See TRANCERS 3
TRANSVESTITE, THE
See GLEN OR GLENDA?
TRAPPED ON TOY WORLD
See JOSH KIRBY: TIME WARRIOR! – TRAPPED ON TOY WORLD
TRAVELLING MAN
See TRAVELING MAN
TRE PASSI NEL DELIRIO
See TALES OF MYSTERY AND IMAGINATION
TREASURE OF ALHEUS WINTERBORN, THE
See CLUE ACCORDING TO SHERLOCK HOLMES, THE
TREASURE OF SIERRA MADRE, THE
See TREASURE OF THE SIERRA MADRE, THE
TREE OF LIBERTY, THE
See HOWARDS OF VIGINIA, THE
TRINITY: TRACKING FOR TROUBLE
See FLATFOOT
TRIP TO ITALY, A
See VOYAGE TO ITALY
TROIS COULEURS: BLANC
See THREE COLOURS:

WHITE
TROIS COULEURS: BLUE
See THREE COLOURS: BLUE
TROIS COULEURS: ROUGE
See THREE COLOURS: RED
TROPICAL HEAT
See TROPICAL NIGHTS
TROUBLE IN STORE
See NORMAN WISDOM: TROUBLE IN STORE
TROUBLE WITH GIRLS (AND HOW TO GET INTO IT), THE
See ELVIS PRESLEY: THE TROUBLE WITH GIRLS
TROUBLE WITH GIRLS, THE
See ELVIS PRESLEY: THE TROUBLE WITH GIRLS
TRUE STORY OF CAMILLE, THE
See LADY OF THE CARMELIAS, THE
TRUE STORY OF THE MENENDEZ MURDERS, THE
See HONOR THY FATHER AND MOTHER: THE MENENDEZ KILLINGS
TRUMAN CAPOTE'S THE GLASS HOUSE
See GLASS HOUSE, THE
TRUST ME: MURDER IS A DYING ART
See TRUST ME
TSAREUBIITSA
See ASSASSIN OF THE TSAR
TSUBAKI SANJURO
See SANJURO
TSUI CHUAN
See DRUNKEN MASTER
TSVET GRANATA
See COLOUR OF POMEGRANATES, THE
TUCKER: THE MAN AND HIS DREAM
See TUCKER
TULITIKKUTEHTAAN TYTTO
See MATCH FACTORY GIRL, THE
TUNE IN TOMORROW
See AUNT JULIA AND THE SCRIPTWRITER
TUREG
See DESERT WARRIOR
TURKEY SHOOT
See ESCAPE 2000
TURKISH DELIGHT
See SENSUALIST, THE
TURKS FRUIT
See SENSUALIST, THE
TURNING TO STONE
See CONCRETE HELL
TWELFTH NIGHT
See NOSFERATU
TWELVE TASKS OF ASTERIX, THE
See ASTERIX AND THE TWELVE TASKS
TWILIGHT MURDERS, THE
See IN THE LINE OF DUTY 3: TIME TO KILL
TWILIGHT OF THE DEAD
See CITY OF THE LIVING DEAD
TWILIGHT OF THE GODS

See INSPECTOR MORSE: TWILIGHT OF THE GODS
TWILIGHT ZONE, THE
See TWILIGHT ZONE: THE MOVIE
TWINS OF DRACULA
See TWINS OF EVIL
TWISTED SOULS
See SPOOKIES
TWITCH OF THE DEATH NERVE
See LAST HOUSE ON THE LEFT PART 2, THE
TWO ENGLISH GIRLS
See ANNE AND MURIEL
TWO FACES OF EVIL
See CONFESSIONS: TWO FACES OF EVIL
TWO FISTS VERSUS SEVEN SAMURAI
See FISTS OF VENGEANCE, THE
TWO IF BY SEA
See STOLEN HEARTS
TYSON
See TYSON: THE TRUE STORY
TYSTNADEN
See SILENCE, THE
U.F.O.
See U.F.O. – THE MOVIE
U.S. CATMAN 2: BOXER BLOW
See BOXER BLOW
UCCELLACCI E UCCELLINI
See HAWKS AND SPARROWS
UGLY, DIRTY AND MEAN
See UGLY, DIRTY AND BAD
ULTIMAS IMAGENES DEL NAUFRAGIO
See LAST IMAGES OF THE SHIPWRECK
ULTIMO TANGO A PARIGI
See LAST TANGO IN PARIS
UN AMOUR DE SWANN
See SWANN IN LOVE
UN ARMOUR DE SWANN
See SWANN IN LOVE
UN BRUIT QUI REND FOU
See BLUE VILLA, THE
UN COEUR EN HIVER
See HEART IN WINTER, A
UN COEUR QUI BAT
See YOUR BEATING HEART
UN CONDAMNE A MORT S'EST ECHAPPE
See MAN ESCAPED, A
UN DIMANCHE A LA CAMPAGNE
See SUNDAY IN THE COUNTRY
UN HEROS TRES DISCRET
See SELF MADE HERO, A
UN HOMME AMOUREUX
See MAN IN LOVE, A
UN HOMME ET UNE FEMME
See MAN AND A WOMAN, A
UN HOMME ET UNE FEMME: VINGT ANS DEJA
See MAN AND A WOMAN, A: TWENTY YEARS LATER

UN MONDE SANS PITIE
See WORLD WITHOUT PITY, A
UN NOS OLA LEUAD
See ONE FULL MOON
UNA MAGNUM SPECIAL PER TONY SAITTA
See BLAZING MAGNUM
UNCAGED
See ANGEL IN RED
UNCLE HARRY
See STRANGE AFFAIR OF UNCLE HARRY, THE
UNCLE OF THE BRIDE
See LAST OF THE SUMMER WINE: UNCLE OF THE BRIDE
UNCONVENTIONAL LINDA
See HOLIDAY
UND DER HIMMEL STEHT STILL
See INNOCENT, THE
UNDER SIEGE 2: DARK TERRITORY
See UNDER SIEGE 2
UNDER THE OLIVE TREES
See THROUGH THE OLIVE TREES
UNDERCOVER DOG
See SHERLOCK: UNDERCOVER DOG
UNDICI GIORNI, UNDICI NOTTI
See ELEVEN DAYS ELEVEN NIGHTS
UNE FEMME FIDELE
See WHEN A WOMAN IS IN LOVE
UNE FEMME FRANCAISE
See FRENCH WOMAN, A
UNE FEMME OU DEUX
See WOMAN OR TWO, A
UNE FLAMME DANS MON COEUR
See FLAME IN MY HEART, A
UNE HISTOIRE IMMORTELLE
See IMMORTAL STORY, THE
UNE HISTOIRE INVENTEE
See IMAGINARY TALE, AN
UNFAITHFUL WIFE
See UNFAITHFUL WIVES
UNFINISHED PIECE FOR PLAYER PIANO, AN
See UNFINISHED PIECE FOR MECHANICAL PIANO
UNIVERSITY OF LIFE
See MY UNIVERSITIES
UNKINDNESS OF RAVENS, AN
See INSPECTOR WEXFORD: AN UNKINDNESS OF RAVENS
UNNAMABLE 2, THE: THE STATEMENT OF RANDOLPH CARTER
See UNNAMABLE RETURNS, THE
UNTIL DEATH
See CHANGELING 2: THE REVENGE
UNVANQUISHED, THE
See APARAJITO
UNWANTED WOMAN, AN
See INSPECTOR

WEXFORD: AN UNWANTED WOMAN
UP FRANKENSTEIN
See ANDY WARHOL'S FRANKENSTEIN
UP IN SMOKE
See CHEECH AND CHONG'S UP IN SMOKE
UP IN THE WORLD
See NORMAN WISDOM: UP IN THE WORLD
UP THE WORLD
See GOODBYE CRUEL WORLD
UPWORLD
See UP WORLD
UTOMLENNYE SOLNTSEM
See BURNT BY THE SUN
UTOML'ONY SONTSEM
See BURNT BY THE SUN
UTOPIA
See LAUREL AND HARDY: UTOPIA
V LYUDAKH
See MY APPRENTICESHIP
VACAS
See COWS
VACATION
See NATIONAL LAMPOON'S VACATION
VALDEZ HORSES, THE
See VALDEZ: THE HALF-BREED
VALE ABRAAO
See ABRAHAM VALLEY
VALERIE A TYDEN DIVU
See VALERIE AND HER WEEK OF WONDERS
VALKENVANIA
See NOTHING BUT TROUBLE
VALLEY OF FEAR, THE
See SHERLOCK HOLMES: THE VALLEY OF FEAR
VAMOS A MATAR, COMPANEROS!
See COMPANEROS
VAMPIRE
See TALE OF A VAMPIRE
VAMPIRE OF LAMBERLEY, THE
See ADVENTURES OF SHERLOCK HOLMES, THE: THE VAMPIRE OF LAMBERLEY
VAMPIRE THRILLS
See THRILL OF THE VAMPIRES, THE
VAMPIRE WOMEN, THE
See RAPE OF THE VAMPIRE, THE
VAMPIRE, THE
See VAMPYR
VAMPIRE'S THRILL
See THRILL OF THE VAMPIRES, THE
VAMPYR OU L'ETRANGE AVENTURE DE DAVID GREY
See VAMPYR
VAMPYR, DER TRAUM DER DAVID GREY
See VAMPYR
VANISHING BODY, THE
See BLACK CAT, THE
VANISHING SON: THE KLANSMAN
See VANISHING SON
VARIETY LIGHTS
See LIGHTS OF VARIETY
VEILED ONE, THE
See INSPECTOR

WEXFORD: THE VEILED ONE
VENDETTA: SECRETS OF THE MAFIA
See VENDETTA
VENGEANCE
See BRAIN, THE
VENGEANCE
See DANGEROUS ORPHANS
VENGEANCE OF THE DEMON
See VENGEANCE, THE DEMON
VENUS IM PELZ
See VENUS IN FURS
VERBRECHEN AM SEELENLEBEN EINES MENSCHENS
See KASPAR HAUSER
VERFUHRUNG: DIE GRAUSAME FRAU
See SEDUCTION: THE CRUEL WOMAN
VERITES ET MENSONGES
See F FOR FAKE
VIAGGIO IN ITALIA
See VOYAGE TO ITALY
VICTIMISED
See CALENDAR GIRL MURDERS, THE
VICTIMS: IN THE MIND OF A SERIAL KILLER
See WHEN THE BOUGH BREAKS
VIERGES ET VAMPIRES
See REQUIEM FOR A VAMPIRE
VIETNAM WAR STORY 2
See WAR STORY 2
VIETNAM WAR STORY: THE LAST DAYS
See VIETNAM WAR STORY
VILLAGE FAIR, THE
See JOUR DE FETE
VILLAIN, THE
See CACTUS JACK
VIOLATED
See SISTERHOOD, THE
VIOLATORS, THE
See ACT OF VENGEANCE
VIOLENT BREED, THE
See KEOMA
VIPER THREE
See TWILIGHT'S LAST GLEAMING
VIRGIN VAMPIRES
See TWINS OF EVIL
VIRGINS AND VAMPIRES
See REQUIEM FOR A VAMPIRE
VIRTUOUS TRAMPS, THE
See LAUREL AND HARDY: BOGUS BANDITS
VIRUS
See FORMULA FOR DEATH
VISION QUEST
See CRAZY FOR YOU
VISKINGAR OCH ROP
See CRIES AND WHISPERS
VIVA LAS VEGAS
See ELVIS PRESLEY: VIVA LAS VEGAS
VIVEMENT DIMANCHE
See FINALLY SUNDAY!
VOCI DEL PROFONDO
See VOICES FROM BEYOND
VOICES FROM THE

DEEP
See VOICES FROM BEYOND
VOINA I MIR
See WAR AND PEACE
VOSKHOZHDENIE
See ASCENT, THE
VOYAGE HOME, THE
See STAR TREK 4: THE VOYAGE HOME
W.R. – MISTERIJE ORGANIZMA
See W.R. – MYSTERIES OF THE ORGANISM
WAGA JINSEI SAIAKU NO TOKI
See MOST TERRIBLE TIME IN MY LIFE, THE
WAGNER: THE COMPLETE EPIC
See WAGNER
WAITING FOR MORNING
See PLACES IN THE HEART
WALK AGAIN
See TO WALK AGAIN
WALL STREET WOMAN
See HIGH FINANCE WOMAN
WALL, THE
See PINK FLOYD: THE WALL
WANDERING KID, THE
See UROTSUKIDOJI: LEGEND OF THE OVERFIEND
WANTED: DEAD OR ALIVE
See WANTED DEAD OR ALIVE
WAR IS OVER, THE
See LA GUERRE EST FINIE
WAR OF THE MONSTERS
See GODZILLA: WAR OF THE MONSTERS
WARHOL'S FRANKENSTEIN
See ANDY WARHOL'S FRANKENSTEIN
WARUI YATSU HODO YOKO NEMURU
See BAD SLEEP WELL, THE
WASP
See W.A.S.P. – WHITE ANGLO-SAXON PROSTITUTE
WASP WOMAN
See FORBIDDEN BEAUTY
WATCH ME WHEN I KILL
See CAT'S VICTIM, THE
WATCHMAKER OF LYON, THE
See WATCHMAKER OF SAINT-PAUL, THE
WATCHMAKER, THE
See WATCHMAKER OF SAINT-PAUL, THE
WATUSI
See QUEST FOR KING SOLOMON'S MINES, THE
WAXWORK 2: LOST IN TIME
See LOST IN TIME
WAY OF CHALLENGE
See NINJA OPERATION 2: WAY OF CHALLENGE
WAY OUT WEST
See LAUREL AND HARDY: WAY OUT WEST
WAY THROUGH THE WOODS, THE
See INSPECTOR MORSE:

THE WAY THROUGH THE WOODS
WAY, THE
See YOL
WEB
See MISSION MANILA
WEDDING BELLS
See ROYAL WEDDING
WEDDING IN BLOOD
See LES NOCES ROUGES
WEE GEORDIE
See GEORDIE
WEEKEND PASS
See RED SHOE DIARIES 5: WEEKEND PASS
WEIRD TALES
See KWAIDAN
WELCOME TO THE SUICIDE CLUB
See SUICIDE CLUB, THE
WELL-MADE MARRIAGE, THE
See LE BEAU MARIAGE
WEREWOLF AT MIDNIGHT
See WEREWOLF OF WASHINGTON, THE
WES CRAVEN PRESENTS: MIND RIPPER
See MIND RIPPER
WES CRAVEN'S THE MIND RIPPER
See MIND RIPPER
WHAT A CHILD SAW
See BLOOD BROTHERS
WHAT'S GOOD FOR THE GOOSE
See NORMAN WISDOM: GIRL TROUBLE
WHAT'S IN IT FOR HARRY?
See TARGET: HARRY
WHAT'S UP NO. 2
See WHAT'S UP SUPERDOC?
WHEN KNIGHTHOOD WAS IN FLOWER
See SWORD AND THE ROSE, THE
WHEN THE SCREAMING STOPS
See LORELEI'S GRASP, THE
WHERE IS THE FRIEND'S HOUSE?
See WHERE IS MY FRIEND'S HOUSE?
WHITE
See THREE COLOURS: WHITE
WHITE WHITE BOY, A
See MIRROR
WHO KILLED HARRY FIELD?
See INSPECTOR MORSE: WHO KILLED HARRY FIELD?
WHOSE CHILD IS THIS? – THE WAR FOR BABY JESSICA
See WAR FOR BABY JESSICA, THE
WHOSE DAUGHTER IS SHE?
See SEMI-PRECIOUS
WHY HAS BODHI-DHARMA LEFT FOR THE EAST?
See WHY DID BOHDI-DHARMA LEAVE FOR THE EAST?
WHY THE WHALES

CAME
See WHEN THE WHALES CAME
WIFE, MOTHER, MURDERER: THE MARIE HILLEY STORY
See WIFE, MOTHER, MURDERER
WILD CHILD
See WILD CHILD: BEVERLY HILLS 38-24-36
WILD FLOWER
See FIORILE
WILD IN THE COUNTRY
See ELVIS PRESLEY: WILD IN THE COUNTRY
WILD ORCHID: RED SHOE DIARIES 2
See RED SHOE DIARIES 2: DOUBLE DARE
WILD ORCHID: THE RED SHOE DIARY
See RED SHOE DIARIES
WILD REEDS, THE
See LES ROSEAUX SAUVAGES
WILDCATS
See DOLL SQUAD, THE
WILL O' THE WISP
See FIRE WITHIN, THE
WILLIAM SHAKESPEARE'S ROMEO AND JULIET
See ROMEO AND JULIET
WIND NAMED AMNESIA, A
See WIND OF AMNESIA, THE
WINGS OF FREEDOM
See W.B. BLUE AND THE BEAN
WINNER TAKES ALL
See MacSHAYNE: WINNER TAKES ALL
WINTER GAMES
See SKI SLUTS GO SNOWBALLING
WISE GUYS: IN THE LINE OF DUTY
See IN THE LINE OF DUTY: WISE GUYS
WISTERIA LODGE
See RETURN OF SHERLOCK HOLMES, THE: WISTERIA LODGE
WITCHBOARD 2: THE DEVIL'S DOORWAY
See WITCHBOARD: THE RETURN
WITCHCRAFT THROUGH THE AGES
See HAXAN
WITCHES, THE
See HAXAN
WITH MURDER IN MIND
See WITH SAVAGE INTENT
WITHIN A CLOISTER
See BEHIND CONVENT WALLS
WITHOUT CONSENT
See TRAPPED AND DECEIVED
WITHOUT MERCY
See MAN ON FIRE
WITHOUT WARNING
See WARNING, THE
WOLF TO THE SLAUGHTER
See INSPECTOR WEXFORD: WOLF TO THE SLAUGHTER
WOLVERCOTE

TONGUE, THE
See INSPECTOR MORSE:
THE WOLVERCOTE
TONGUE
WOMAN ALONE, A
See SABOTAGE
WOMAN HUNT, THE
See ESCAPE
WOMAN IN GREEN, THE
See SHERLOCK HOLMES
AND THE WOMAN IN
GREEN
WOMAN IN THE DUNES
See WOMAN OF THE
DUNES
WOMAN IS A WOMAN, A
See UNE FEMME EST UNE
FEMME
**WOMAN SCORNED, A:
THE BETTY BRODERICK
STORY**
See TILL MURDER US DO
PART
WOMAN WITH A PAST
See DARK SECRETS (A
TRUE STORY)
**WOMEN'S
PENITENTIARY 1**
See BIG DOLL HOUSE, THE
**WOMEN'S
PENITENTIARY 2**
See BIG BIRD CAGE, THE
**WOMEN'S
PENITENTIARY 3**
See WOMEN IN CAGES
WONDERFUL LIFE
See CLIFF RICHARD:
WONDERFUL LIFE
**WONDERFUL TO BE
YOUNG**
See CLIFF RICHARD: THE
YOUNG ONES
**WONDERWORKS:
BROTHER FUTURE**
See BROTHER FUTURE
**WONDERWORKS:
SWEET 15**
See SWEET 15
WOODEN SOLDIERS
See LAUREL AND HARDY:
BABES IN TOYLAND
WORDS BY HEART
See WONDERWORKS:
WORDS BY HEART
WORLD SONG, THE
See LE CHANT DU
MONDE
**WORSE YOU ARE THE
BETTER YOU SLEEP,
THE**
See BAD SLEEP WELL, THE
WRATH OF KHAN, THE
See STAR TREK 2: THE
WRATH OF KHAN

WRITE TO KILL
See WITH DEADLY
INTENT
WRONG IS RIGHT
See MAN WITH THE
DEADLY LENS, THE
**WRONG KIND OF GIRL,
THE**
See BUS STOP
WRONG MOVEMENT
See WRONG MOVE, THE
**X: THE MAN WITH X-
RAY EYES, THE**
See MAN WITH X-RAY
EYES, THE
XICH LO
See CYCLO
XIYAN
See WEDDING BANQUET,
THE
**XTRO 2: THE SECOND
ENCOUNTER**
See XTRO 2
**XTRO: A NEW
DIMENSION IN FEAR**
See XTRO
**YABU NO NAKA
KURONEKO**
See KURONEKO
**YAO A YAO YAO DAO
WAIPO QIAO**
See SHANGHAI TRIAD
YEAR 2889
See IN THE YEAR 2889
**YEAR OF THE KING
BOXER**
See YEAR OF THE
KICKBOXER, THE
**YELLOW FACED TIGER,
THE**
See SLAUGHTER IN SAN
FRANCISCO
**YELLOW MAN AND THE
GIRL, THE**
See BROKEN BLOSSOMS
YELLOW SUBMARINE
See BEATLES: YELLOW
SUBMARINE
YINGXIONG BENSE
See BETTER TOMORROW,
A
YINSHI NAN NU
See EAT DRINK MAN
WOMAN
YO, LA PEOR DE TODAS
See I, THE WORST OF ALL
YOB, THE
See COMIC STRIP
PRESENTS: THE YOB
YOIDORE TENSHI
See DRUNKEN ANGEL
YOJU TOSHI
See WICKED CITY
YOU CAN'T SLEEP

HERE
See I WAS A MALE WAR
BRIDE
YOUIJI TOSHI
See WICKED CITY
**YOUNG AND THE
PASSIONATE, THE**
See I VITELLONI
**YOUNG AND THE
RESTLESS, THE**
See SHAKE, RATTLE AND
ROCK!
YOUNG DRACULA
See ANDY WARHOL'S
DRACULA
YOUNG HERO
See BAREFOOT KID, THE
YOUNG HERO
See DEADLY STRIKE, THE
YOUNG MAN OF MUSIC
See YOUNG MAN WITH A
HORN
YOUNG NURSE, THE
See SECRETS OF YOUNG
NURSES
YOUNG ONES, THE
See CLIFF RICHARD: THE
YOUNG ONES
YOUNG SCARFACE
See BRIGHTON ROCK
**YOUNG SHERLOCK
HOLMES**
See YOUNG SHERLOCK
HOLMES AND THE
PYRAMID OF FEAR
**YOUNG WARRIORS,
THE**
See EAGLE WARRIORS
YOUNGEST SPY, THE
See IVAN'S CHILDHOOD
**YOU'RE IN THE ARMY
NOW**
See O.H.M.S.
**YOU'VE GOT TO HAVE
HEART**
See VIRGIN WIFE, THE
YUKINOJO HENGE
See ACTOR'S REVENGE,
AN
Z MEN
See ATTACK FORCE Z
Z.P.G.
See ZERO POPULATION
GROWTH
**ZAMRI, UMRI,
VOSKRESNI!**
See DON'T MOVE, DIE
AND RISE AGAIN!
ZATERYANI V SIBIRIY
See LOST IN SIBERIA
ZAZIE
See ZAZIE DANS LE
METRO
ZAZIE IN THE SUBWAY

See ZAZIE DANS LE
METRO
**ZAZIE IN THE
UNDERGROUND**
See ZAZIE DANS LE
METRO
ZEBRAHEAD
See COLOUR OF LOVE,
THE
ZEMLYA
See EARTH
**ZENDEGI VA DIGAR
HICH**
See AND LIFE GOES ON
ZENTROPA
See EUROPA
ZERKALO
See MIRROR
ZERO FOR CONDUCT
See ZERO DE CONDUITE
**ZIR-E DARAKHTAN-E
ZEYTON**
See THROUGH THE OLIVE
TREES
ZOLTAN
See ZOLTAN, HOUND OF
HELL
**ZOLTAN, HOUND OF
DRACULA**
See ZOLTAN, HOUND OF
HELL
ZOMBI 2
See ZOMBIE FLESHEATERS
ZOMBIE
See ZOMBIE FLESHEATERS
**ZOMBIE 4: VIRGIN
AMONG THE LIVING
DEAD**
See VIRGIN AMONG THE
LIVING DEAD
ZOMBIE FLESH-EATERS
See ZOMBIE FLESHEATERS
ZOMBIE GRAVEYARD
See CHILDREN
SHOULDN'T PLAY WITH
DEAD THINGS
ZOMBIE, THE
See PLAGUE OF THE
ZOMBIES, THE
**ZOMBIE: THE DEAD
AMONG US**
See ZOMBIE FLESHEATERS
ZOMBIES 2
See ZOMBIE FLESHEATERS
ZOMBIES, THE
See PLAGUE OF THE
ZOMBIES, THE
ZONHENG SIHAI
See ONCE A THIEF
**ZOO: A ZED AND TWO
NOUGHTS**
See ZED AND TWO
NOUGHTS, A

DISTRIBUTORS LISTING

Video distributors are alphabetically under the abbreviations used in the main body of the book. Addresses are included where known; however, some of the abbreviations refer to labels in which case no address is given, but the company handling the material is noted. In addition, companies often act as distributors for each other, and this is noted. If doubts exist as to the current status of a company/label as a distributor of material, the entry is marked "inactive" and no address is given. In such cases, additional information is provided if relevant.

20TH
20th Century Fox Home Entertainment
Status: Active
Twentieth Century House, 31–32 Soho Square, London W1V 6AP (Tel: 0171 753 8686
Fax: 0171 437 1625). See FOXVID or TECH for distribution.
20VIS
20–20 Vision Video UK
Status: Active
Sony Pictures Europe House, 25 Golden Square, London W1R 6LU (Tel: 0171 533 1200). See SONOP for distribution.
4-FRONT
4-Front Video
Status: Active
PO Box 1425, Chancellors House, 72 Chancellors Road, London W6 9QB (Tel: 0181 910 5000
Fax: 0181 910 5892).
ABBEY
Abbey Home Entertainment
Status: Active
Warwick House, 106 Harrow Road, London W2 1XD (Tel: 0171 262 1012
Fax: 0171 262 6020). See POLYREC for distribution.
ACAD
Academy Video
Status: Active
Connoisseur Video Ltd., 10a Stephen Mews, London W1P 0AX (Tel: 0171 957 8957/8958
Fax: 0171 957 8968). See RTM or DISC for distribution.
ADVID
A.D. Vision
Status: Active
19 High Street, Bangor, Gwynedd, North Wales LL57 1NP (Tel: 01248 353593
Fax: 01248 370046). See DISC for distribution.
AFILMS
Arrow Film Distributors Ltd.
Status: Active
18 Watford Road, Radlett, Herts. WD7 8LE (Tel: 01923 858306
Fax: 01923 859673). See RTM or DISC for distribution.
ALLIED
Allied Home Entertainment
Status: Active
4th Floor, Avon House, 360 Oxford Street, London W1N 9HA (Tel: 0171 491 2262
Fax: 0171 491 2282). See RTM or DISC for distribution.
ANIME
Anime Projects
Status: Active
See RTM or DISC for distribution.
ARENA
Arena Home Entertainment

Status: Active
155d Holland Park Avenue, London W11 4UX (Tel: 0171 610 4101
Fax: 0171 371 2022). See SPEAR, SONOP or RTM for distribution.
ARROW
Arrow Film Distributors Ltd.
Status: Active
18 Watford Road, Radlett, Herts. WD7 8LE (Tel: 01923 858306
Fax: 01923 859673). See RTM or DISC for distribution.
ARTIF
Artificial Eye Film Company Ltd.
Status: Active
13 Soho Square, London W1V 5FB (Tel: 0171 437 2552
Fax: 0171 437 2992). See 20TH or FOXVID for distribution.
ARTPRO
Arthouse Productions Ltd.
Status: Active
39/41 North Road, London N7 9DP (Tel: 0171 700 3388
Fax: 0171 609 2249). See RTM or DISC for distribution.
AUDIO
Audiovisual Enterprises Ltd.
Status: Active
The Studio, 13a Oaklands Farm, Goatsmoor Lane, Stock, Ingatestone, Essex CM4 9RS (Tel: 01277 630505
Fax: 01277 632040).
AVID
Avid Video
Status: Active
Unit 2, Boeing Way, Int. Trading Estate, Brent Road, Southall, Middlesex UB2 5LB (Tel: 0181 893 5767
Fax: 0181 893 5955). See BMGREC or TOTAL for distribution.
BANO
Bano Communications
Status: Active
See FUNNY or SGSVID for distribution.
BBC
BBC Worldwide Publishing
Status: Active
BBC Centre, Woodlands, 80 Wood Lane, London W12 0TT (Tel: 0181 576 2000
Fax: 0181 749 0538). See TECH for distribution.
BEST
Best Film And Video UK
Status: Inactive
Not applicable
BLACK
Black Diamond Films Ltd.
Status: Active
13 New Row, Covent Garden, London WC2N 4LF (Tel: 0171 497 3320
Fax: 0171 497 3069).
BLUE
Blue Chip Video

Status: Active
See QUANT or TOTAL for distribution.
BMGREC
BMG Distribution
Status: Active
Lyng Lane, West Bromwich B70 7ST (Tel: 0121 500 5678
Fax: 0121 553 6880).
BMGVID
BMG Video
Status: Active
Bedford House, 69/79 Fulham High Street, London SW6 3JW (Tel: 0171 384 7500
Fax: 0171 371 9298). See BMGREC for distribution.
BRAVE
Braveworld Ltd.
Status: Inactive
Not applicable
BRITHOM
British Home Entertainment
Status: Active
5 Broadwater Road, Walton-On-Thames, Surrey KT12 5DB (Tel: 01932 228832
Fax: 01932 247759).
BUENA
Buena Vista Home Entertainment
Status: Active
3 Queen Caroline Street, Hammersmith, London W6 9PE (Tel: 0181 222 1000
Fax: 0181 222 2795). See TECH for distribution.
CALECO
Caleco Direct
Status: Inactive
Not applicable
CAMERON
Cameron Williams
Status: Active
Kingsgate House, 536 Kings Road, London SW10 0TE (Tel: 0171 331 3920
Fax: 0171 331 3929). See BMGREC for distribution.
CAPIT
Capital Home Video
Status: Active
See GUILD for distribution.
CARL
Carlton Home Entertainment
Status: Active
The Waterfront, Elstree Road, Elstree, Herts. WD6 3BS (Tel: 0181 207 6207
Fax: 0181 207 0722 Sales: 0181 810 5061). See TECH for distribution.
CASPIC
Castle Pictures
Status: Inactive
Not applicable
CBS
CBS/Fox Video Ltd.
Status: Inactive
Not applicable

CENTV
Central Television Enterprises
Status: Active
Now known as CTE.

CH4
Channel Four Video
Status: Active
124 Horseferry Road, London SW1P
2TX (Tel: 0171 396 4444
Fax: 0171 306 8350). See RTM or DISC
for distribution.

CHRYS
Chrysalis Home Video
Status: Active
c/o The Hit Label, The Chrysalis
Building, Bramley Road, London W10
6SP (Tel: 0171 221 2213
Fax: 0171 221 6455). See CARL or
TECH for distribution.

CIC
CIC Video
Status: Active
4th Floor, Glenthorne House,
Hammersmith Grove, London W6
0ND (Tel: 0181 563 3500
Fax: 0181 563 3501). See SONOP,
TOTAL or GOLD for distribution.

CLEAR
Clear Vision Video
Status: Active
36 Queens Way, Ponders End,
Middlesex EN3 4FA (Tel: 0181 805 1354
Fax: 0181 805 9987). See DISC for
distribution.

COLUM
Columbia Tri-Star Home Video
Status: Active
Sony Pictures Europe House, 25
Golden Square, London W1R 6LU
(Tel: 0171 533 1200). See SONOP for
distribution.

CONNO
Connoisseur Video
Status: Active
10a Stephen Mews, London W1P 0AX
(Tel: 0171 957 8957/8958
Fax: 0171 957 8968). See RTM or DISC
for distribution.

CREA
Creation Entertainment Ltd.
Status: Active
Godalming Business Centre, Woolsack
Way, Godalming, Surrey GU7 1XW
(Tel: 01483 427366
Fax: 01483 419205). See DISC for
distribution.

CREMED
Creative Media Marketing
Status: Active
84/86 Grays Inn Road, London WC1X
8AA (Tel: 0171 405 6165
Fax: 0171 405 6168).

CRYSTAL
Crystal Sky Communications
Status: Active
See HIFLI or SONOP for distribution.

CTE
CTE Video
Status: Active
CTE (Carlton) Ltd., 35/38 Portman
Square, London W1H 9FH
(Tel: 0171 224 3339
Fax: 0171 486 1707). See CARL or
TECH for distribution.

CURB
Curb Esquire Films
Status: Active
See HIFLI or SONOP for distribution.

CURZON
Curzon Video
Status: Active
Mayfair Entertainment (UK) Ltd.,
13 Soho Square, London W1V 5FB

(Tel: 0171 437 2552
Fax: 0171 437 2992). See 20TH for
distribution.

DANTE
Dante Video
Status: Active
3 Dewhurst House, Winnett Street,
Soho, London W1V 7HS
(Tel: 0171 287 6869
Fax: 0171 287 6882). See TOTAL for
distribution.

DDVID
DD Video
Status: Active
5 Churchill Court, 58 Station Road,
North Harrow, Middlesex HA2 7SA
(Tel: 0181 863 8819
Fax: 0181 863 0463 Sales: 0181 863
8819).

DISC
Disc Distribution Ltd.
Status: Active
Unit 12, Brunswick Park Industrial
Estate, London N11 1HX
(Tel: 0181 362 8111 Sales: 0181 362
8122).

DTK
Dangerous To Know
Status: Active
17a Newman Street, London W1P 3HP
(Tel: 0171 255 1955
Fax: 0171 636 5717). See RTM or DISC
for distribution.

DUKE
Duke Marketing
Status: Active
Milbourne House, 13 St Georges Street,
Douglas, Isle of Man IM99 1DD
(Tel: 01624 623634
Fax: 01624 629745).

DV8
DV8
Status: Active
See VISION

E2WEST
East2West Films
Status: Active
See CREMED

EAST
Eastern Heroes Ltd.
Status: Active
96 Shaftesbury Avenue, London W1
(Tel: 0171-734 4554
Fax: 0171 734 4786). See DISC for
distribution.

ECU
ECU
Status: Active
See CREMED

EIV
Entertainment In Video
Status: Active
27 Soho Square, London W1V 5FL
(Tel: 0171 439 1979
Fax: 0171 734 2483). See SONOP for
distribution.

ELPIC
Electric Pictures
Status: Active
15 Percy Street, London W1P 9FD
(Tel: 0171 636 1231
Fax: 0171 636 1675). See POLYREC for
distribution.

ELV
Electric Video Ltd.
Status: Active
30 Castellain Road, London W9 1EZ
(Tel: 0171 286 0700
Fax: 0171 286 0707). See DISC for
distribution.

EMIMUS
EMI Music Services (UK)
Status: Active

Hermes Close, Tachbrook Park,
Leamington Spa, Warks. CV34 6RP
(Tel: 01926 888888
Fax: 0181 479 5992).

EMPIRE
Empire Entertainment
Status: Active
30 Lock Road, Marlow, Bucks. SL7 1SN
(Tel: 01494-881033
Fax: 01494 881033). See TOTAL for
distribution.

ENCORE
Encore Entertainment
Status: Active
Cattespoole Mill, Stoney Lane,
Tadrebigge, Nr. Bromsgrove, Worcs.
B60 1LZ (Tel: 0121 447 8223
Fax: 0121 447 8255). See SPEAR or
TOTAL for distribution.

ENTREE
Entree Pictures
Status: Active
See HIFLI or SONOP for distribution.

ENTUK
Entertainment (UK) Ltd.
Status: Inactive
Not applicable

ETL
Entertainment Today Ltd.
Status: Active
See 4-FRONT

EUREKA
Eureka Video
Status: Active
375 Harrow Road, London W9 3NA
(Tel: 0181 960 4890
Fax: 0181 969 7211). See GOLD for
distribution.

FABFIL
Fabulous Films Ltd.
Status: Active
Sandford House, 10 Maynard Close,
London SW6 2DB (Tel: 0171 384 2600
Fax: 0171 384 2700). See SPEAR or
SONOP for distribution.

FALCON
Falcon A.M. Entertainment
Status: Active
Production House, PO Box 834,
Wootton Bassett, Swindon, Wiltshire
SN4 8RU (Tel: 01793 849304
Fax: 01793 849309). See TOTAL for
distribution.

FEATFIL
Feature Film Company
Status: Active
See RTM or DISC for distribution.

FIFTH
Fifth Avenue Films
Status: Active
14 South Avenue, Hullbridge,
Hockley, Essex SS5 6HA
(Tel: 01702 232396
Fax: 01702 230944). See DISC for
distribution.

FILM4
Film Four Distributors Ltd.
Status: Active
Castle House, 75 Wells Street, London
W1 (Tel: 0171 436 9944
Fax: 0171 436 9955). See RTM or DISC
for distribution.

FIONA
Fiona Cooper Audio Visual Ltd.
Status: Inactive – no longer trading in
the UK.
Not applicable

FIRC
First Class Films
Status: Active
98 St Pancras Way, London NW1 9NF
(Sales: 01923 816511). See RTM or DISC
for distribution.

FIRST
First Independent Video
Status: Active
99 Baker Street, London W1M 1FB
(Tel: 0171 317 2500
Fax: 0171 317 2503 Sales: 0171 528
7768). See SONOP for distribution.

FMARK
Filmark Entertainment Group
Status: Active
Suite 401, 302 Regent Street, London
W15 5AL (Tel: 0171 580 4242). See
major wholesalers such as GOLD or
TRENT for distribution.

FOCUS
Focus Entertainment Ltd.
Status: Active
Swan Centre, 9 Fishers Lane, London
W4 1RX (Tel: 0181 995 1331
Fax: 0181 994 2193). See DISC for
distribution.

FOXVID
Fox Guild Home Entertainment
Status: Active
Twentieth Century House, 31/32 Soho
Square, London W1V 6AP
(Tel: 0171 753 0015
Fax: 0171 437 1625).

FULL
Full Moon Entertainment
Status: Inactive
Not applicable

FUNNY
Funny Dream Ltd.
Status: Active
PO Box 27, Heywood, Lancs. W12 7RQ
(Tel: 01706 628917
Fax: 01706 628917). See major
wholesalers such as GOLD or TRENT
for distribution.

FUTMED
Future Media Productions
Status: Active
PO Box 510, Shenley Church Road,
Milton Keynes, Bucks. MK5 6LD
(Tel: 01908 503449
Fax: 01908 503449). See TOTAL for
distribution.

FUTPRO
Future Promises
Status: Active
54 Wood Lane, London W12 7RQ
(Tel: 0181 749 9887
Fax: 0181 749 2686).

FUTUR
Futuristic Entertainment Ltd.
Status: Inactive
Not applicable

GAME
Game Entertainment Group Ltd.
Status: Active
38 Wychwood Court, Cotswold
Business Village, London Road,
Moreton-In-Marsh, Gloucs. GL56 0JQ
(Tel: 01608 652475
Fax: 01608 652476). See SPEAR or
SONOP for distribution.

GENESIS
Genesis Home Video Ltd.
Status: Active
Unit 4, Warren Yard, Warren Farm
Office Village, Wolverton Mill, Milton
Keynes MK12 5NW (Tel: 01908 225100
Fax: 01908 225125).

GOLD
S. Gold & Sons Ltd.
Status: Active – wholesaler
69 Flempton Road, London E10 7NL
(Tel: 0181 539 3600
Fax: 0181 539 2176 Sales: 0181 558
7133).

GORDON
Gordon Duncan

Status: Active
Record and Tape Distribution, Market
Place, Inveruie, Aberdeen AB51 3PU
(Tel: 01467 621517
Fax: 01467 625536)

GREEN
Green Umbrella Productions
Status: Active
The Old Forge, Ockham Lane,
Ockham, Surrey GU23 6PH
(Tel: 01483 223022
Fax: 01483 223099). See QUANT or
TOTAL for distribution.

GROS
Grosvenor React UK
Status: Active
Production House, PO Box 834,
Wootton Bassett, Swindon, Wiltshire
SN4 8RU (Tel: 01793 849304
Fax: 01793 849309). See TOTAL for
distribution.

GUER
Guerilla Films
Status: Active
Suite 229, Linen Hall, 162/168 Regent
Street, London W1R 5TB
(Tel: 0171 437 0457
Fax: 0181 758 9364). See PINN for
distribution.

GUILD
Guild Pathe Cinema Ltd.
Status: Active
4th Floor, Kent House, 14/17 Market
Place, Great Titchfield Street, London
W1N 8AR (Tel: 0171 323 5151
Fax: 0171 631 3568). See FOXVID or
20TH for distribution.

HAR
Harmony Video
Status: Active
31 Fernshaw Road, London SW10 0TN
(Tel: 0171 352 9717). See QUANT or
TOTAL for distribution.

HEND
Hendring Ltd.
Status: Inactive
Not applicable

HIFLI
Hi-Fliers Distribution Ltd.
Status: Active
16 Heathfield Terrace, London W4 4JE
(Tel: 0181 742 2023
Fax: 0181 742 1475). See SONOP for
distribution.

HOLPIC
Hollywood Pictures Home Video
Status: Active
(Tel: 0181 222 1000
Fax: 0181 222 2795). See TECH for
distribution.

ICAPRO
ICA Projects Ltd.
Status: Active
See MANGA or SONOP for
distribution.

IKON
Ikon Records
Status: Active
3 Meadow Avenue, Hale, Altrincham,
Cheshire WA15 8JS (Tel: 0161 903 9929
Fax: 0161 926 9520). See RTM or DISC
for distribution.

IMAG
Imagine Home Entertainments
Status: Active
71 Brushfield Street, London E1 6AA
(Tel: 0171 377 9088
Fax: 0171 247 3554). See RTM or DISC
for distribution.

IMC
IMC Video Ltd.
Status: Active
Godalming Business Centre, Woolsack

Way, Godalming, Surrey GU7 1XW
(Tel: 01483 427366
Fax: 01483 419205). See DISC for
distribution.

IMPENT
Imperial Entertainment (UK) Ltd.
Status: Active
GEC Estate, East Lane, Wembley,
Middlesex HA9 7FF
(Tel: 0181 904 0921
Fax: 0181 908 6785).

ISLPIC
Island World Communications
Status: Inactive
Not applicable

ITC
ITC Home Video (UK)
Status: Active
See POLYREC for distribution.

JEZ
Jezebel Films Ltd.
Status: Active
BCM Box 923, London WC1N 3XX. See
RTM or DISC for distribution.

JOKER
Joker
Status: Inactive
Not applicable

KISEKI
Kiseki Films
Status: Active
Units 5/6, Parkside, Ravenscourt Park,
London W6 0UU (Tel: 0181 741 2203
Fax: 0181 741 2204). See PARADOX or
TOTAL for distribution.

KISS
Kiss Video Productions
Status: Active
19 Kitchener Road, Thornton Heath,
Surrey CR7 8QN
(Tel: 0181 771 4121). See SCRN or DISC
for distribution.

KRL
KRL
Status: Active
9 Watt Road, Hillingdon Industrial
Estate, Hillingdon, Glasgow G52 4RY
(Tel: 0141 882 9060
Fax: 0141 883 3686 Sales: 0141 882
9986).

LARK
Lark Productions Ltd.
Status: Active
PO Box 2933, London N20 9LR
(Tel: 0181 446 2653
Fax: 0181 445 9594).

LEIS
Leisureview Ltd.
Status: Active
5 Churchill Court, 58 Station Road,
North Harrow, Middlesex HA2 7SA
(Tel: 0181 863 8819
Fax: 0181 863 0463). See DDVID for
distribution.

LIFE
Lifetime Vision Ltd.
Status: Active
247/257 Euston Road, London NW1
2HY (Tel: 0171 387 9808
Fax: 0171 387 9106). See IMC, TOTAL
or VISVID for distribution.

LIQUID
Liquid Gold
Status: Active
370 Eastwood Road, Rayleigh, Essex
SS6 7LW (Tel: 01702 420084
Fax: 01702 420084)
See SCRN or DISC for distribution.

LOAD
Load Incorporated
Status: Active
See QUANT or TOTAL for
distribution.

LUMI
Lumiere Pictures
Status: Inactive
Not applicable

LWV
LWV Video
Status: Active
See QUANT for distribution.

MADE
Made In Hong Kong
Status: Active
231 Portobello Road, London W11 1LT
(Tel: 0171 792 9791
Fax: 0171 792 9871). See RTM or DISC
for distribution.

MAGNUM
Magnum Music Group
Status: Active
Magnum House, High Street, Lane
End, Bucks. HP14 3JG
(Tel: 01494 882858
Fax: 01494 882631).

MAINPIC
Mainline Pictures
Status: Active
37 Museum Street, London WC1A 1LP
(Tel: 0171 242 5523
Fax: 0171 430 0170). See RTM or DISC
for distribution.

MANAGE
Management Sales Services
Status: Active
PO Box 341, Northampton, Northants.
NN4 0BA (Tel: 01604 705659
Fax: 01604 705679). See TOTAL for
distribution.

MANFOR
Man For Man
Status: Active
See GOLD for distribution.

MANGA
Manga Entertainment Ltd.
Status: Active
40 St Peters Road, London W6 9BD
(Tel: 0181 748 9000
Fax: 0181 748 0841). See SONOP for
distribution.

MANHAT
Manhattan Video
Status: Active
Unit 20a, Canada House, Blackburn
Road, London NW6 1RZ
(Tel: 0171 625 7113
Fax: 0171 624 3258). See GOLD for
distribution.

MARQ
Marquee Pictures
Status: Active
Unit 4, Warren Yard, Warren Farm
Office Village, Wolverton Mill, Milton
Keynes MK12 5NW
(Tel: 01908 225100
Fax: 01908 225125).
See TOTAL for distribution.

MAXSCAN
Max Scandinavia
Status: Active
See FALCON

MED
Medusa Communications and
Marketing Ltd.
Status: Active
51 Bancroft, Hitchin, Herts. SG5 1LL
(Tel: 01462 421818
Fax: 01462 420393). See DISC or
SONOP for distribution.

MEGAVID
Megatron Video
Status: Active
Suite 8511, 16/18 Circus Road, London
NW8 6PG (Tel: 0836 500566
Fax: 01272 328915). See TOTAL for
distribution.

MGM
MGM Home Entertainment (Europe)
Ltd.
Status: Active
Ground Floor, South Wing, Glenthorne
House, Hammersmith Grove, London
W6 0ND (Tel: 0181 563 8383
Fax: 0181 563 2896). See WHV or TECH
for distribution.

MIA
Missing In Action
Status: Active
70 Baker Street, London W1M 1DJ
(Tel: 0171 935 9225
Fax: 0171 935 9565). See DISC for
distribution.

MIST
Mistique Productions
Status: Active
351 Holloway Road, London N7 0RN
(Tel: 0171 619 0301
Fax: 0171 700 5717). See FALCON or
TOTAL for distribution (both retail
only).

MOPIC
Moonstone Pictures
Status: Active
See FUNNY, SGSVID or QUANT
(selected titles only) for distribution.

MOSAIC
Mosaic Movies
Status: Active
19/24 Manasty Road, Orton Southgate,
Peterborough, Cambs. PE2 6UP
(Tel: 01733 363010
Fax: 01733 363011). See 20VIS, COLUM
or SONOP for distribution.

NEWAGE
New Age Entertainment Ltd.
Status: Inactive
Not applicable

NORSTAR
Norstar Entertainment
Status: Active
See HIFLI or SONOP for distribution.

NORVID
Northern and Shell Video
Status: Active
Northern and Shell Tower, 4 Selsdon
Way, City Harbour, Isle of Dogs,
London E14 9GL (Tel: 0171 987 5090
Fax: 0171 987 0633/2160). See DISC for
distribution.

NWV
New World Video
Status: Active
See HIFLI or SONOP for distribution.

OCEAN
Ocean Pictures
Status: Active
St Peters Road, Northney, Hayling
Island, Hants. (Tel: 01705 462845
Fax: 01705 462845). See FIRST or
SONOP for distribution.

ODY
Odyssey Video Ltd.
Status: Active
15 Dufours Place, London W1V 1FE
(Tel: 0171 437 8251
Fax: 0171 734 6941). See SONOP for
distribution.

ONE
One On One Video
Status: Active
5 Percy Street, London W1P 9FB
(Tel: 0171 580 6751
Fax: 0171 580 6759).

OPTIK
Optik
Status: Active
See HIFLI or SONOP for distribution.

ORBIT
Orbit Media Ltd.

Status: Active
7/11 Kensington High Street, London
W8 5NP (Tel: 0171 221 5548
Fax: 0171 727 0515). See DISC for
distribution.

OURVID
Our Video Company
Status: Active
88 Berkeley Court, Baker Street,
London NW1 5ND (Tel: 0171 486 4054
Fax: 0181 455 6579). See SCRN or DISC
for distribution.

OVER
Overseas Film Group
Status: Active
See HIFLI or SONOP for distribution.

PAL
Palace Video Ltd.
Status: Inactive
Not applicable

PARADE
Parade Video
Status: Active
See SCRN or DISC for distribution.

PARADOX
Paradox Films (London) Ltd.
Status: Active
Units 5/6, Parkside, Ravencourt Park,
London W6 0UU (Tel: 0181 741 2203
Fax: 0181 741 2204). See TOTAL for
distribution.

PASSION
Passion Video
Status: Active
See IMC or DISC for distribution.

PATHE
Pathe Video Ltd.
Status: Inactive
Not applicable

PHASE
Phase One Film And Video Ltd.
Status: Active
37 Dundalk Road, London SE4 2JJ
(Tel: 0171 732 8794
Fax: 0171 732 8794). See RTM or DISC
for distribution.

PICMUS
Picture Music International
Status: Active
EMI House, 43 Brook Green, London
W6 7EF (Tel: 0171 605 5000
Fax: 0171 605 5050). See DISC or
EMIMUS (music tapes only) for
distribution.

PINN
Pinnacle Records
Status: Active
Electron House, Cray Avenue, St Mary
Cray, Orpington, Kent BR5 3PN
(Tel: 01689 870622
Fax: 01689 78269 Sales: 01689 873144).

PION
Pioneer L.D.C.E. Ltd.
Status: Active
Pioneer House, Hollybush Hill, Stoke
Poges, Slough, Bucks. SL2 4QP
(Tel: 01753 789635
Fax: 01753 789647). See RTM or DISC
for distribution.

PLAT
Platinum Films
Status: Active
370 Eastwood Road, Rayleigh, Essex
SS6 7LW (Tel: 01702-420084
Fax: 01702 420084). See SCRN or DISC
for distribution.

POLFIL
Polygram Filmed Entertainment
Status: Active
Chancellors House, 72 Chancellors
Road, London W6 9QB
(Tel: 0181 910 500
Fax: 0181 910 5890).

POLY
Polygram Video Ltd.
Status: Active
PO Box 1425, Chancellors House,
London W6 9QB (Tel: 0181 910 500
Fax: 0181 910 5890 Sales: 0990 310310).
See POLYREC for distribution.

POLYREC
Polygram Record Operations Ltd.
Status: Active
Clyde Works, Grove Road, Chadwell
Heath, Romford, Essex RM6 4QR
(Tel: 0181 910 1500
Fax: 0181 910 1675 Sales: 0990 310310).

POPRO
Popular Progress
Status: Active
Maingrip Ltd., PO Box 3987, London
SE3 8UE (Tel: 0181 293 3465
Fax: 0181 293 2465). See RTM or DISC
for distribution.

PRIDE
Pride Video Productions Ltd.
Status: Active
Units 5/6, Parkside, Ravenscourt Park,
London W6 0UU (Tel: 0181 741 2203
Fax: 0181 741 2204). See PARADOX or
TOTAL for distribution.

PRISCO
Prism Leisure Corporation
Status: Active
Unit 1, Baird Road, Enfield, Middlesex
EN1 1SQ (Tel: 804 8100
Fax: 0181 805 8001).

PRISM
Prism Pictures
Status: Active
See HIFLI or SONOP for distribution.

PROMARK
Promark Entertainment Group
Status: Active
See HIFLI or SONOP for distribution.

PROWL
Prowler Press
Status: Active
3 Broadbent Close, 20 Highgate High
Street, London N6 5JG
(Tel: 0181 340 7667
Fax: 0181 347 7667).

PURG
Purgatory Films
Status: Active
BCM Box 9235, London WC1N 3XX.
See DANTE or TOTAL for distribution.

QUANT
Quantum Leap Group Ltd.
Status: Active
Quantum House, 26 Darin Court,
Crownhill, Milton Keynes, Bucks. MK8
0AD (Tel: 01908 561133
Fax: 01908 564414 Sales: 01908 564418).
See TOTAL for distribution.

QUAVID
Quadrant Video
Status: Active
37a High Street, Carshalton, Surrey
SM5 3BB (Tel: 0181 669 1114
Fax: 0181 669 8831). See DISC for
distribution.

RAVEN
Raven Home Video
Status: Active
PO Box 27, Heywood, Lancs. OL10
4GZ (Tel: 01706 628917
Fax: 01706 628917). See QUANT or
TOTAL for distribution.

RCA
RCA/Columbia Pictures Video (UK)
Status: Inactive
Not applicable

REDEEM
Redeem
Status: Active

1 Broadway Court, Shifnal, Shropshire
TF11 8AZ (Tel: 01952 462747
Fax: 01952 462747). See TOTAL for
distribution.

REDEM
Redemption Films Ltd.
Status: Active
BCM Box 9235, London WC1N 3XX.
See RTM or DISC for distribution.

REDEYE
Red Eye Entertainment
Status: Active
Sandford House, 10 Maynard Close,
London SW6 2DB (Tel: 0171 384 2600
Fax: 0171 384 2700). See DISC for
distribution.

REEL
Reeltime Pictures
Status: Active
26 Crown Road, Virginia Water, Surrey
GU25 4HT (Tel: 01344 844601
Fax: 01344 843770).

REFLEC
Reflective Film Distribution
Status: Active
69 New Oxford Street, London WC1A
1DG (Tel: 0171 528 7767
Fax: 0171 528 7771). See FIRST or
SONOP for distribution.

RELREC
Rel Records Ltd.
Status: Active
40 Sciennes, Edinburgh EH9 1NH
(Tel: 0131 668 3366
Fax: 0131 662 4463). See GORDON or
TOTAL for distribution.

RETRO
Retro Video
Status: Active
Metro Tartan House, 79 Wardour
Street, London W1V 3TH
(Tel: 0171 494 1400
Fax: 0171 439 1922). See 20TH for
distribution.

RICHARD
Richard Wells Productions
Status: Active
See FALCON or TOTAL for
distribution.

RIO
Rio Pictures
Status: Inactive
Not applicable

RNK
Rank Home Video
Status: Inactive
Not applicable

RTM
RTM Sales and Marketing
Status: Active
98 St Pancras Way, London NW1 9NF
(Sales: 01923 816511). See DISC for
distribution.

RYSHER
Rysher Entertainment
Status: Active
See HIFLI or SONOP for distribution.

S4C
S4C
Status: Active
Parc Ty Glas, Llanishen, Cardiff CF4
5DU (Tel: 01222 747444
Fax: 01222 741474).

SAIN
FIDEO SAIN
Status: Active
Canolfan Sain, LLandwrog,
Caernarfon, Gwynedd LL54 5TG
(Tel: 01286 831111
Fax: 01286 831497).

SCEDGE
Screen Edge
Status: Active

28/30 The Square, Lytham St Annes,
Lancs. FY8 1RF (Tel: 01253 712453
Fax: 01253 712362). See RTM or DISC
for distribution.

SCOT
Scotdisc
Status: Active
BGS Productions Ltd., Newtown
Street, Kilsyth, Glasgow G65 0JX
(Tel: 01236 821081
Fax: 01236 825683). See PRISCO for
distribution.

SCREAM
Scream Time Video
Status: Active
c/o Encore Entertainment Ltd.

SCRN
Screen Multimedia Ltd.
Status: Active
Control House, 9 Station Road, Radlett,
Herts. WD7 8ED (Tel: 01923 469043
Fax: 01923 469044). See DISC for
distribution (retail only).

SECOND
Second Sight Films Ltd.
Status: Active
3rd Floor, 21/23 Crosby Row, London
SE1 3YD (Tel: 0171 378 9739
Fax: 0171 357 6879). See RTM or DISC
for distribution.

SGSVID
SGS Home Video
Status: Active
69 Flempton Road, London E10 7NL
(Tel: 0181 539 3600
Fax: 0181 539 2176 Sales: 0181 558
7133). See GOLD for distribution.

SONOP
Sony Music Operations
Status: Active
Rabans Lane, Aylesbury, Bucks. HP19
3BX (Tel: 01296 26151
Fax: 01296 395551 Sales: 01296 395151).

SONY
Sony Video Software
Status: Inactive
Not applicable

SPEAR
Spearhead Sales and Marketing
Status: Active
The Lodge, 18/21 Church Gate,
Thatcham, Berkshire Rg19 3 PN
(Tel: 01635 866488
Fax: 01635 866499). See SONOP for
distribution.

STABL
Stablecane Home Video
Status: Inactive
Not applicable

START
Start Audio-Visual
Status: Active
Unit 20a, Canada House, Blackburn
Road, London NW6 1RZ
(Tel: 0171 625 7113
Fax: 0171 624 3258). See DISC for
distribution.

STUDIO
Studio K7
Status: Active
See RTM or DISC for distribution.

STYL
Stylus Video
Status: Inactive
Not applicable

SUPVID
Superfly Video
Status: Active
2 Ovaltine Cottages, Bedmond Road,
Abbots Langley, Herts. WD5 0QD (Tel:
01923 267432
Fax: 01923 268360). See RTM or DISC
for distribution.

TART
Tartan Video
Status: Active
Metro Tartan House, 79 Wardour Street, London W1V 3TH
(Tel: 0171 494 1400
Fax: 1071 439 1922). See 20TH for distribution.

TCX
Tele Cine X
Status: Inactive
Not applicable

TECH
Technicolor Distribution Services
Status: Active
Unit 1, Perivale Industrial Park, Greenford, Middlesex UB6 7RU
(Tel: 0181 810 5030
Fax: 0181 810 5761 Sales: 0181 810 5061).

TELVID
Telstar Video Entertainment
Status: Active
The Studio, 5 King Edward Mews, Byfeld Gardens, London SW13 9HP
(Tel: 0181 846 9946
Fax: 0181 741 5584). See BMGREC for distribution.

TERRIF
Terrific Stuff Ltd.
Status: Active
23 Cambridge Road, East Twickenham, Middlesex TW 2HN (Tel: 0181 891 1872
Fax: 0181 891 1872).

TERRY
Terry Blood Distribution
Status: Inactive
Not applicable

THAMES
Thames Video Ltd.
Status: Active
See DISC for distribution.

THIRD
Third Millenium Distribution Ltd.
Status: Active
The Studio, Church Farm Office Village, 2 High Street, Eaton Bray, Bedfordshire LU6 2DL
(Tel: 01525 221296
Fax: 01525 220512).

TOTAL
Total Home Entertainment
Status: Active
National Distribution Centre, Rosevale Business Park, Newcastle-Under-Lyme, Staffs. ST5 7QT
(Tel: 01782 566566
Fax: 01782 565400 Sales: 01782 566511)

TOTREC
Total Record Company
Status: Active
Unit 7, Pepys Court, 84/96 The Chase, London SW4 0NF (Tel: 0171 978 2300
Fax: 0171 498 6420). See BMGREC for distribution.

TOUCH
Touchstone Home Video
Status: Active
(Tel: 0181 222 1000
Fax: 0181 222 2795). See TECH for distribution.

TRANSAT
Trans-Atlantic Entertainment
Status: Active
See HIFLI or SONOP for distribution.

TRENT
Trent Video (Wholesale)
Status: Active
PO Box 178, Chesterfield, Derbyshire S40 3YW (Tel: 01246-567467
Fax: 01246 567467).

TRIM
Trimark Pictures
Status: Active
See HIFLI or SONOP for distribution.

TRING
Tring International PLC
Status: Active
Triangle Business Park, Wendover Road, Aylesbury, Bucks. HP22 5BL
(Tel: 01296 615511
Fax: 01296 614250).

TROMA
Troma Video
Status: Active
4th Floor, Avon House, 360 Oxford Street, London W1N 9HA
(Tel: 0171 491 2262
Fax: 0171 491 2282). See RTM or DISC for distribution.

UNIQUE
Unique Films
Status: Active
39/41 North Road, London N& 6DP
(Tel: 0171 700 0068
Fax: 0171 609 2249). See RTM or DISC for distribution.

VCC
Video Collection International Ltd.
Status: Active
Royalty House, 72/74 Dean Street, London W1V 5HB (Tel: 0171 470 6666
Fax: 0171 470 0006/0007 Sales: 0181 362 8117). See DISC for distribution.

VES
Vestron Video (UK) Ltd.
Status: Inactive
Not applicable

VGM
Video Gems
Status: Inactive
Not applicable

VIDRI
Video Rights Ltd.
Status: Inactive
Not applicable

VIPCO
Video Instant Pictures Co.
Status: Active
15 St Margaret's Court, St Margaret's Road, Edgware, Middlesex HA8 9UT
(Tel: 0181 905 4535
Fax: 0181 905 4898). See SGSVID or GOLD for distribution.

VIR
Virgin Video
Status: Inactive
Not applicable

VISCOM
Visionary Communications Ltd.
Status: Active
28/30 The Square, Lytham St Annes, Lancs. FY8 1RF (Tel: 01253 712453
Fax: 01253 712362). See RTM or DISC for distribution.

VISCORP
Visual Corporation Ltd.
Status: Active
Suite 204, Premier House, 77 Oxford Street, London W1R 1Rb
(Tel: 0171 439 1188
Fax: 0171 439 0307). See BMGREC for distribution.

VISION
Vision
Status: Active
29 D'Arblay Street, London W1V 3FH
(Tel: 0171 434 3452
Fax: 0171 434 3518). See DISC for distribution.

VISVID
Vision Video Ltd.

Status: Active
PO Box 1425, Chancellors House, 72 Chancellors Road, London W6 9QB
(Tel: 0181 910 5000
Fax: 0181 910 5900 Sales: 0181 910 1799). See POLYREC for distribution.

VIVID
Vivid UK
Status: Active
See SCRN, DISC or FALCON (selected titles only) for distribution.

WARMUS
Warner Music Vision
Status: Active
35/38 Portman Square, London W1H 0EU (Tel: 0171 467 2566
Fax: 0171 467 2564 Sales: 0181 998 5929). See WARUK for distribution.

WARUK
Warner Music UK Ltd.
Status: Active
Alperton Lane, Wembley, Middlesex HA0 1FJ (Tel: 0181 998 5929
Fax: 0181 998 6535).

WARVIS
Warner Vision International
Status: Active
35/38 Portman Square, London W1H 0EU (Tel: 0171 467 2566
Fax: 0171 467 2564 Sales: 0181 998 5929). See WARUK for distribution.

WDV
Walt Disney Home Video
Status: Active
(Tel: 0181 222 1000
Fax: 0181 222 2795). See TECH for distribution.

WESCON
Western Connection
Status: Active
Film and Video Distribution, Suite 18, 37 Westbourne Terrace, London W2 3UR (Tel: 0171 262 8707
Fax: 0171 224 8997). See RTM or DISC for distribution.

WEST
West End Video
Status: Active
See FUNNY or SGSVID for distribution.

WHITPRO
Whitewing Productions
Status: Active
Empire Entertainment, PO Box 872, Marlow, Bucks. SL7 1SN
(Tel: 01494 881033
Fax: 01494 881033). See TOTAL for distribution.

WHV
Warner Home Video
Status: Active
135 Wardour Street, London W1V 4AP
(Tel: 0171 494 3441
Fax: 0171 494 3297). See TECH for distribution (budget titles only).

WILLPRO
Williams Productions
Status: Active
PO Box 179, Deal, Kent CT14 7GF
(Tel: 01304 382024
Fax: 01304 361568). See RTM or DISC for distribution.

YORK
Yorkshire Tyne Tees Enterprises
Status: Active – selected titles distributed by SGSVID, GOLD, TRENT and SPEAR.
The TV Centre, Leeds, Yorkshire LS3 1JS (Tel: 0113 243 8283
Fax: 0113 234 3125).